AMERICAN
MEN AND WOMEN
OF SCIENCE

AMERICAN MEN AND WOMEN OF SCIENCE

13TH EDITION

Edited by the Jaques Cattell Press

Volume 4 L-O

R. R. BOWKER COMPANY
A Xerox Publishing Company
New York & London, 1976

LIBRARY
LOS ANGELES COUNTY MUSEUM OF NATURAL HISTORY

Copyright © 1976 by Xerox Corporation
Published by R. R. Bowker Co. (a Xerox Publishing Company)
1180 Avenue of the Americas, New York, N.Y. 10036
International Standard Book Number: 0-8352-0869-9
International Standard Serial Number: 0065-9347
Library of Congress Catalog Card Number: 6-7326
All rights reserved, including those to reproduce this book or parts
thereof in any form, not excepting putting on addressing equipment,
for use by anyone other than the original publisher.
Printed and bound in the United States of America.

The publishers do not assume and hereby disclaim any liability to
any party for any loss or damage caused by errors or omissions in
American Men and Women of Science, whether such errors or
omissions result from negligence, accident or any other cause.

Contents

ADVISORY COMMITTEE — vi

PREFACE — vii

ABBREVIATIONS — ix

BIOGRAPHIES — 2467

Advisory Committee

Dr. Dael L. Wolfle, Chairman
Graduate School of Public Affairs
University of Washington

Dr. Randolph W. Bromery
Chancellor,
University of Massachusetts

Dr. Janet W. Brown
Program Head,
Office of Opportunities
in Science
American Association for the
Advancement of Science

Dr. Robert W. Cairns
Executive Director,
American Chemical Society

Dr. S. D. Cornell
Assistant to the President
National Academy of Sciences

Dr. Ruth M. Davis
Director,
Institute for Computer Science
and Technology
National Bureau of Standards

Dr. Carl D. Douglass
Deputy Director,
Division of Research Grants
National Institutes of Health

Dr. Richard G. Folsom
President Emeritus,
Rensselaer Polytechnic Institute

Dr. Robert E. Henze
Director,
Membership Division
American Chemical Society

Dr. Eugene L. Hess
Executive Director,
Federation of American Societies
for Experimental Biology

Dr. William C. Kelly
Executive Director,
Commission on Human Resources
National Research Council

Dr. Kenneth B. Raper
Department of Bacteriology
University of Wisconsin

Dr. A. L. Schawlow
Department of Physics
Stanford University

Dr. John F. Sherman
Vice President,
Association of American Medical
Colleges

Dr. Matthias Stelly
Executive Vice President,
American Society of Agronomy

Dr. John R. Whinnery
Department of Electrical Engineering
and Computer Sciences
University of California, Berkeley

Preface

The observance of an anniversary prompts reflection on the past, and it is natural to note here that the astounding growth in the dimension and stature of this 200 year old American nation has been influenced immeasurably by the achievements of her scientists. 1976 also marks the 70th anniversary of AMERICAN MEN & WOMEN OF SCIENCE as a chronicle of the lives and professional activities of those men and women most instrumental in affecting the shape and quality of science in America. The explosion in scientific activity over the past seven decades is clearly evident when one compares the 1906 edition, a single volume containing 4,000 entries, with the present edition, six volumes profiling nearly 110,000 men and women of importance in their fields.

This edition of AMERICAN MEN & WOMEN OF SCIENCE is a landmark in biographic achievement. The information contained in all seven volumes has been gathered, edited and compiled by the Jaques Cattell Press in the space of ten months. This is a radical, and beneficial, departure from the production of the 12th edition, which took three years to publish in its entirety. The acceleration was made possible by the use of a computerized printing method and more efficient production procedures.

The editors have not sacrificed quality in the interest of speed, however. The criteria were stringently applied in the selection of new entrants, all nominated by former biographees. The criteria follow.

> Achievement, by reason of experience and training, of a stature in scientific work equivalent to that associated with the doctoral degree, coupled with presently continued activity in such work;
>
> or
>
> Research activity of high quality in science as evidenced by publication in reputable scientific journals; or, for those whose work cannot be published because of governmental or industrial security, research activity of high quality in science as evidenced by the judgment of the individual's peers;
>
> or
>
> Attainment of a position of substantial responsibility requiring scientific training and experience to the extent described for (1) and (2).

All data forms submitted by biographees were carefully edited and proofread. Those whose information was appearing for the first time received a proof before publication. Information on scientists who did not supply current material is included only if verification of present professional activity was established by our researchers. A former biographee whose current status could not be verified is given a reference to the 12th edition if the probability exists of a continued activity in science. References are also given to scientists who have died since publication of the last edition. Omitted are the names of those previously listed as retired and those who have entered fields or activities not covered by the scope of the directory. More than 20,000 nominations for inclusion were received and over 12,000 of those nominated returned information and were selected for listing. This is a significant increase in both nominations and selections over any previous edition and gives fuller representation to the developing scientist.

Geographic and discipline indexes make up the seventh volume of the set. The discipline index has been rearranged and is organized with major subject headings, providing easier access to this information.

Certain disciplines previously included in the directory are not represented in this edition. Engineering and economics will appear in separate directories with expanded criteria, to enable even broader coverage of these areas. Others, including sociology, psychology and political science were omitted because they are fully covered by membership directories in each field.

Great appreciation is expressed to the AMERICAN MEN & WOMEN OF SCIENCE Advisory Committee for their guidance in the planning of the 13th edition. Their efforts have contributed to an unusually good response to our requests for information and nominations, which has enhanced the value of the publication. Also to be thanked are the many scientific societies that provided membership lists for the use of our researchers or that published announcements in their bulletins and journals.

The staff of the Jaques Cattell Press deserves the highest accolade for their sustained interest, devotion and good will through the many hours of learning and implementing the new procedures necessary to the successful completion of this book. The overwhelming workload was shared by temporary employees who performed with the greatest diligence and responsibility. The job could not have been completed without their fine help. Everyone involved with this project gave outstanding service, but the contributions of Alice Smith, Pauline Stump, Joyce Howell and Ila Martin cannot go without mention. Special acknowledgement is given to Fred Scott, former general manager of the Jaques Cattell Press, who was responsible for initiating the formation of the advisory committee and for overseeing the planning and early stages of work on the 13th edition.

Comments and suggestions are invited and should be addressed to The Editors, American Men & Women of Science, Jaques Cattell Press, P.O. Box 25001, Tempe, Arizona, 85282.

Renee Lautenbach, Supervising Editor
Anne Rhodes, Administrative Managing Editor

JAQUES CATTELL PRESS

Desmond Reaney, Manager Book Editorial

R.R. BOWKER COMPANY

October, 1976

Abbreviations

AAAS—American Association for the Advancement of Science
abnorm—abnormal
abstr—abstract(s)
acad—academic, academy
acct—account, accountant, accounting
acoust—acoustic(s), acoustical
ACTH—adrenocorticotrophic hormone
actg—acting
activ—activities, activity
addn—addition(s), additional
Add—Address
adj—adjunct, adjutant
adjust—adjustment
Adm—Admiral
admin—administration, administrative
adminr—administrator(s)
admis—admission(s)
adv—adviser(s), advisory
advan—advance(d), advancement
advert—advertisement, advertising
AEC—Atomic Energy Commission
aerodyn—aerodynamic(s)
aeronaut—aeronautic(s), aeronautical
aerophys—aerophysical, aerophysics
aesthet—aesthetic(s)
AFB—Air Force Base
affil—affiliate(s), affiliation
agr—agricultural, agriculture
agron—agronomic, agronomical, agronomy
agrost—agrostologic, agrostological, agrostology
agt—agent
AID—Agency for International Development
Ala—Alabama
allergol—allergological, allergology
alt—alternate
Alta—Alberta
Am—America, American
AMA—American Medical Association
anal—analysis, analytic, analytical
analog—analogue
anat—anatomic, anatomical, anatomy
anesthesiol—anesthesiology
angiol—angiology
Ann—Annal(s)
ann—annual
anthrop—anthropological, anthropology
anthropom—anthropometric, anthropometrical, anthropometry
antiq—antiquary, antiquities, antiquity
antiqn—antiquarian

apicult—apicultural, apiculture
APO—Army Post Office
app—appoint, appointed
appl—applied
appln—application
approx—approximate(ly)
Apr—April
apt—apartment(s)
aquacult—aquaculture
arbit—arbitration
arch—archives
archaeol—archaeological, archaeology
archit—architectural, architecture
Arg—Argentina, Argentine
Ariz—Arizona
Ark—Arkansas
artil—artillery
asn—association
assoc(s)—associate(s), associated
asst(s)—assistant(s), assistantship(s)
Assyriol—Assyriology
astrodyn—astrodynamics
astron—astronomical, astronomy
astronaut—astronautical, astronautics
astronr—astronomer
astrophys—astrophysical, astrophysics
attend—attendant, attending
atty—attorney
audiol—audiology
Aug—August
auth—author
AV—audiovisual
Ave—Avenue
avicult—avicultural, aviculture

b—born
bact—bacterial, bacteriologic, bacteriological, bacteriology
BC—British Columbia
bd—board
behav—behavior(al)
Belg—Belgian, Belgium
bibl—biblical
bibliog—bibliographic, bibliographical, bibliography
bibliogr—bibliographer
biochem—biochemical, biochemistry
biog—biographical, biography
biol—biological, biology
biomed—biomedical, biomedicine
biomet—biometric(s), biometrical, biometry
biophys—biophysical, biophysics

bk(s)—book(s)
bldg—building
Blvd—Boulevard
Bor—Borough
bot—botanical, botany
br—branch(es)
Brig—Brigadier
Brit—Britain, British
Bro(s)—Brother(s)
bryol—bryology
Bull—Bulletin
bur—bureau
bus—business
BWI—British West Indies

c—children
Calif—California
Can—Canada, Canadian
cand—candidate
Capt—Captain
cardiol—cardiology
cardiovasc—cardiovascular
cartog—cartographic, cartographical, cartography
cartogr—cartographer
Cath—Catholic
CEngr—Corps of Engineers
cent—central
Cent Am—Central America
cert—certificate(s), certification, certified
chap—chapter
chem—chemical(s), chemistry
chemother—chemotherapy
chmn—chairman
citricult—citriculture
class—classical
climat—climatological, climatology
clin(s)—clinic(s), clinical
cmndg—commanding
Co—Companies, Company
coauth—coauthor
co-dir—co-director
co-ed—co-editor
coeduc—coeducation, coeducational
col(s)—college(s), collegiate, colonel
collab—collaboration, collaborative
collabr—collaborator
Colo—Colorado
com—commerce, commercial
Comdr—Commander
commun—communicable, communication(s)
comn(s)—commission(s), commissioned

ABBREVIATIONS

comnr—commissioner
comp—comparative
compos—composition
comput—computation, computer(s), computing
comt(s)—committee(s)
conchol—conchology
conf—conference
cong—congress, congressional
Conn—Connecticut
conserv—conservation, conservatory
consol—consolidated, consolidation
const—constitution, constitutional
construct—construction, constructive
consult(s)—consult, consultant(s), consultantship(s), consultation, consulting
contemp—contemporary
contrib—contribute, contributing, contribution(s)
contribr—contributor
conv—convention
coop—cooperating, cooperation, cooperative
coord—coordinate(d), coordinating, coordination
coordr—coordinator
corp—corporate, corporation(s)
corresp—correspondence, correspondent, corresponding
coun—council, counsel, counseling
counr—councilor, counselor
criminol—criminological, criminology
cryog—cryogenic(s)
crystallog—crystallographic, crystallographical, crystallography
crystallogr—crystallographer
Ct—Court
Ctr—Center
cult—cultural, culture
cur—curator
curric—curriculum
cybernet—cybernetic(s)
cytol—cytological, cytology
CZ—Canal Zone
Czech—Czechoslovakia

DC—District of Columbia
Dec—December
Del—Delaware
deleg—delegate, delegation
delinq—delinquency, delinquent
dem—democrat(s), democratic
demog—demographic, demography
demogr—demographer
demonstr—demonstrator
dendrol—dendrologic, dendrological, dendrology
dent—dental, dentistry
dep—deputy
dept—department(al)
dermat—dermatologic, dermatological, dermatology
develop—developed, developing, development, developmental
diag—diagnosis, diagnostic
dialectol—dialectological, dialectology
dict—dictionaries, dictionary
Dig—Digest
dipl—diploma, diplomate
dir(s)—director(s), directories, directory
dis—disease(s), disorders
Diss Abstr—Dissertation Abstracts
dist—district
distrib—distributed, distribution, distributive
distribr—distributor(s)
div—division, divisional, divorced

DNA—deoxyribonucleic acid
doc—document(s), documentary, documentation
Dom—Dominion
Dr—Drive

e—east
ecol—ecological, ecology
econ(s)—economic(s), economical, economy
economet—econometric(s)
ECT—electroconvulsive or electroshock therapy
ed—edition(s), editor(s), editorial
ed bd—editorial board
educ—education, educational
educr—educator(s)
EEG—electroencephalogram, electroencephalographic, electroencephalography
Egyptol—Egyptology
EKG—electrocardiogram
elec—electric, electrical, electricity
electrochem—electrochemical, electrochemistry
electrophys—electrophysical, electrophysics
elem—elementary
embryol—embryologic, embryological, embryology
emer—emeriti, emeritus
employ—employment
encour—encouragement
encycl—encyclopedia
endocrinol—endocrinologic, endocrinology
eng—engineering
Eng—England, English
engr(s)—engineer(s)
enol—enology
Ens—Ensign
entom—entomological, entomology
environ—environment(s), environmental
enzym—enzymology
epidemiol—epidemiologic, epidemiological, epidemiology
equip—equipment
ESEA—Elementary & Secondary Education Act
espec—especially
estab—established, establishment(s)
ethnog—ethnographic, ethnographical, ethnography
ethnogr—ethnographer
ethnol—ethnologic, ethnological, ethnology
Europ—European
eval—evaluation
evangel—evangelical
eve—evening
exam—examination(s), examining
examr—examiner
except—exceptional
exec(s)—executive(s)
exeg—exegeses, exegesis, exegetic, exegetical
exhib(s)—exhibition(s), exhibit(s)
exp—experiment, experimental
exped(s)—expedition(s)
explor—exploration(s), exploratory
expos—exposition
exten—extension

fac—faculty
facil—facilities, facility
Feb—February
fed—federal
fedn—federation
fel(s)—fellow(s), fellowship(s)
fermentol—fermentology

fertil—fertility, fertilization
Fla—Florida
floricult—floricultural, floriculture
found—foundation
FPO—Fleet Post Office
Fr—French
Ft—Fort

Ga—Georgia
gastroenterol—gastroenterological, gastroenterology
gen—general
geneal—genealogical, genealogy
geod—geodesy, geodetic
geog—geographic, geographical, geography
geogr—geographer
geol—geologic, geological, geology
geom—geometric, geometrical, geometry
geomorphol—geomorphologic, geomorphology
geophys—geophysical, geophysics
Ger—German, Germanic, Germany
geriat—geriatric(s)
geront—gerontological, gerontology
glaciol—glaciology
gov—governing, governor(s)
govt—government, governmental
grad—graduate(d)
Gt Brit—Great Britain
guid—guidance
gym—gymnasium
gynec—gynecologic, gynecological, gynecology

handbk(s)—handbook(s)
helminth—helminthology
hemat—hematologic, hematological, hematology
herpet—herpetologic, herpetological, herpetology
Hisp—Hispanic, Hispania
hist—historic, historical, history
histol—histological, histology
HM—Her Majesty
hochsch—hochschule
homeop—homeopathic, homeopathy
hon(s)—honor(s), honorable, honorary
hort—horticultural, horticulture
hosp(s)—hospital(s), hospitalization
hq—headquarters
HumRRO—Human Resources Research Office
husb—husbandry
Hwy—Highway
hydraul—hydraulic(s)
hydrodyn—hydrodynamic(s)
hydrol—hydrologic, hydrological, hydrology
hyg—hygiene, hygienic(s)
hypn—hypnosis

ichthyol—ichthyological, ichthyology
Ill—Illinois
illum—illuminating, illumination
illus—illustrate, illustrated, illustration
illusr—illustrator
immunol—immunologic, immunological, immunology
Imp—Imperial
improv—improvement
Inc—Incorporated
in-chg—in charge
incl—include(s), including
Ind—Indiana
indust(s)—industrial, industries, industry
inf—infantry
info—information
inorg—inorganic

ABBREVIATIONS

ins—insurance
inst(s)—institute(s), institution(s)
instnl—institutional(ized)
instr(s)—instruct, instruction, instructor(s)
instrnl—instructional
int—international
intel—intelligence
introd—introduction
invert—invertebrate
invest(s)—investigation(s)
investr—investigator
irrig—irrigation
Ital—Italian

J—Journal
Jan—January
Jct—Junction
jour—journal, journalism
jr—junior
jurisp—jurisprudence
juv—juvenile

Kans—Kansas
Ky—Kentucky

La—Louisiana
lab(s)—laboratories, laboratory
lang—language(s)
laryngol—laryngological, laryngology
lect—lecture(s)
lectr—lecturer(s)
legis—legislation, legislative, legislature
lett—letter(s)
lib—liberal
libr—libraries, library
librn—librarian
lic—license(d)
limnol—limnological, limnology
ling—linguistic(s), linguistical
lit—literary, literature
lithol—lithologic, lithological, lithology
Lt—Lieutenant
Ltd—Limited

m—married
mach—machine(s), machinery
mag—magazine(s)
maj—major
malacol—malacology
mammal—mammalogy
Man—Manitoba
Mar—March
Mariol—Mariology
Mass—Massachusetts
mat—material(s)
mat med—materia medica
math—mathematic(s), mathematical
Md—Maryland
mech—mechanic(s), mechanical
med—medical, medicinal, medicine
Mediter—Mediterranean
Mem—Memorial
mem—member(s), membership(s)
ment—mental(ly)
metab—metabolic, metabolism
metall—metallurgic, metallurgical, metallurgy
metallog—metallographic, metallography
metallogr—metallographer
metaphys—metaphysical, metaphysics
meteorol—meteorological, meteorology
metrol—metrological, metrology
metrop—metropolitan
Mex—Mexican, Mexico
mfg—manufacturing

mfr(s)—manufacture(s), manufacturer(s)
mgr—manager
mgt—management
Mich—Michigan
microbiol—microbiological, microbiology
micros—microscopic, microscopical, microscopy
mid—middle
mil—military
mineral—mineralogical, mineralogy
Minn—Minnesota
Miss—Mississippi
mkt—market, marketing
Mo—Missouri
mod—modern
monogr—monograph
Mont—Montana
morphol—morphological, morphology
Mt—Mount
mult—multiple
munic—municipal, municipalities
mus—museum(s)
musicol—musicological, musicology
mycol—mycologic, mycology

n—north
NASA—National Aeronautics & Space Administration
nat—national, naturalized
NATO—North Atlantic Treaty Organization
navig—navigation(al)
NB—New Brunswick
NC—North Carolina
NDak—North Dakota
NDEA—National Defense Education Act
Nebr—Nebraska
nematol—nematological, nematology
nerv—nervous
Neth—Netherlands
neurol—neurological, neurology
neuropath—neuropathological, neuropathology
neuropsychiat—neuropsychiatric, neuropsychiatry
neurosurg—neurosurgical, neurosurgery
Nev—Nevada
New Eng—New England
New York—New York City
Nfld—Newfoundland
NH—New Hampshire
NIH—National Institutes of Health
NIMH—National Institute of Mental Health
NJ—New Jersey
NMex—New Mexico
nonres—nonresident
norm—normal
Norweg—Norwegian
Nov—November
NS—Nova Scotia
NSF—National Science Foundation
NSW—New South Wales
numis—numismatic(s)
nutrit—nutrition, nutritional
NY—New York State
NZ—New Zealand

observ—observatories, observatory
obstet—obstetric(s), obstetrical
occas—occasional(ly)
occup—occupation, occupational
oceanog—oceanographic, oceanographical, oceanography
oceanogr—oceanographer
Oct—October
odontol—odontology

OEEC—Organization for European Economic Cooperation
off—office, official
Okla—Oklahoma
olericult—olericulture
oncol—oncologic, oncology
Ont—Ontario
oper(s)—operation(s), operational, operative
ophthal—ophthalmologic, ophthalmological, ophthalmology
optom—optometric, optometrical, optometry
ord—ordnance
Ore—Oregon
org—organic
orgn—organization(s), organizational
orient—oriental
ornith—ornithological, ornithology
orthod—orthodontia, orthodontic(s)
orthop—orthopedic(s)
osteop—osteopathic, osteopathy
otol—otological, otology
otolaryngol—otolaryngological, otolaryngology
otorhinol—otorhinologic, otorhinology

Pa—Pennsylvania
Pac—Pacific
paleobot—paleobotanical, paleobotany
paleont—paleontological, paleontology
Pan-Am—Pan-American
parasitol—parasitology
partic—participant, participating
path—pathologic, pathological, pathology
pedag—pedagogic(s), pedagogical, pedagogy
pediat—pediatric(s)
PEI—Prince Edward Islands
penol—penological, penology
periodont—periodontal, periodontic(s), periodontology
petrog—petrographic, petrographical, petrography
petrogr—petrographer
petrol—petroleum, petrologic, petrological, petrology
pharm—pharmacy
pharmaceut—pharmaceutic(s), pharmaceutical(s)
pharmacog—pharmacognosy
pharmacol—pharmacologic, pharmacological, pharmacology
phenomenol—phenomenologic(al), phenomenology
philol—philological, philology
philos—philosophic, philosophical, philosophy
photog—photographic, photography
photogeog—photogeographic, photogeography
photogr—photographer(s)
photogram—photogrammetric, photogramme-photogrammetry
photom—photometric, photometrical, photometry
phycol—phycology
phys—physical
physiog—physiographic, physiographical, physiography
physiol—physiological, physiology
Pkwy—Parkway
Pl—Place
polit—political, politics
polytech—polytechnic(al)
pomol—pomological, pomology
pontif—pontifical
pop—population
Port—Portugal, Portuguese
postgrad—postgraduate

ABBREVIATIONS

PQ—Province of Quebec
PR—Puerto Rico
pract—practice
practr—practitioner
prehist—prehistoric, prehistory
prep—preparation, preparative, preparatory
pres—president
Presby—Presbyterian
preserv—preservation
prev—prevention, preventive
prin—principal
prob(s)—problem(s)
proc—proceedings
proctol—proctologic, proctological, proctology
prod—product(s), production, productive
prof—professional, professor, professorial
Prof Exp—Professional Experience
prog(s)—program(s), programmed, programming
proj—project(s), projection(al), projective
prom—promotion
protozool—protozoology
prov—province, provincial
psychiat—psychiatric, psychiatry
psychoanal—psychoanalysis, psychoanalytic, psychoanalytical
psychol—psychological, psychology
psychomet—psychometric(s)
psychopath—psychopathologic, psychopathology
psychophys—psychophysical, psychophysics
psychophysiol—psychophysiological, psychophysiology
psychosom—psychosomatic(s)
psychother—psychotherapeutic(s), psychotherapy
Pt—Point
pub—public
publ—publication(s), publish(ed), publisher, publishing
pvt—private

Qm—Quartermaster
Qm Gen—Quartermaster General
qual—qualitative, quality
quant—quantitative
quart—quarterly

radiol—radiological, radiology
RAF—Royal Air Force
RAFVR—Royal Air Force Volunteer Reserve
RAMC—Royal Army Medical Corps
RAMCR—Royal Army Medical Corps Reserve
RAOC—Royal Army Ordnance Corps
RASC—Royal Army Service Corps
RASCR—Royal Army Service Corps Reserve
RCAF—Royal Canadian Air Force
RCAFR—Royal Canadian Air Force Reserve
RCAFVR—Royal Canadian Air Force Volunteer Reserve
RCAMC—Royal Canadian Army Medical Corps
RCAMCR—Royal Canadian Army Medical Corps Reserve
RCASC—Royal Canadian Army Service Corps
RCASCR—Royal Canadian Army Service Corps Reserve
RCEME—Royal Canadian Electrical & Mechanical Engineers
RCN—Royal Canadian Navy
RCNR—Royal Canadian Naval Reserve
RCNVR—Royal Canadian Naval Volunteer Reserve
Rd—Road

RD—Rural Delivery
rec—record(s), recording
redevelop—redevelopment
ref—reference(s)
refrig—refrigeration
regist—register(ed), registration
registr—registrar
regt—regiment(al)
rehab—rehabilitation
rel(s)—relation(s), relative
relig—religion, religious
REME—Royal Electrical & Mechanical Engineers
rep—represent, representative
repub—republic
req—requirements
res—research, reserve
rev—review, revised, revision
RFD—Rural Free Delivery
rhet—rhetoric, rhetorical
RI—Rhode Island
Rm—Room
RM—Royal Marines
RN—Royal Navy
RNA—ribonucleic acid
RNR—Royal Naval Reserve
RNVR—Royal Naval Volunteer Reserve
roentgenol—roentgenologic, roentgenological, roentgenology
RR—Railroad, Rural Route
rte—route
Russ—Russian
rwy—railway

s—south
SAfrica—South Africa
SAm—South America, South American
sanit—sanitary, sanitation
Sask—Saskatchewan
SC—South Carolina
Scand—Scandinavia(n)
sch(s)—school(s)
scholar—scholarship
sci—science(s), scientific
SDak—South Dakota
SEATO—Southeast Asia Treaty Organization
sec—secondary
sect—section
secy—secretary
seismog—seismograph, seismographic, seismography
seismogr—seismographer
seismol—seismological, seismology
sem—seminar, seminary
sen—senator, senatorial
Sept—September
ser—serial, series
serol—serologic, serological, serology
serv—service(s), serving
silvicult—silvicultural, silviculture
soc(s)—societies, society
soc sci—social science
sociol—sociologic, sociological, sociology
Span—Spanish
spec—special
specif—specification(s)
spectrog—spectrograph, spectrographic, spectrography
spectrogr—spectrographer
spectrophotom—spectrophotometer, spectrophotometric, spectrophotometry
spectros—spectroscopic, spectroscopy
speleol—speleological, speleology
Sq—Square

sr—senior
St—Saint, Street(s)
sta(s)—station(s)
stand—standard(s), standardization
statist—statistical, statistics
Ste—Sainte
steril—sterility
stomatol—stomatology
stratig—stratigraphic, stratigraphy
stratigr—stratigrapher
struct—structural, structure(s)
stud—student(ship)
subcomt—subcommittee
subj—subject
subsid—subsidiary
substa—substation
super—superior
suppl—supplement(s), supplemental, supplementary
supt—superintendent
supv—supervising, supervision
supvr—supervisor
supvry—supervisory
surg—surgery, surgical
surv—survey, surveying
survr—surveyor
Swed—Swedish
Switz—Switzerland
symp—symposia, symposium(s)
syphil—syphilology
syst(s)—system(s), systematic(s), systematical

taxon—taxonomic, taxonomy
tech—technical, technique(s)
technol—technologic(al), technology
tel—telegraph(y), telephone
temp—temporary
Tenn—Tennessee
Terr—Terrace
Tex—Texas
textbk(s)—textbook(s)
text ed—text edition
theol—theological, theology
theoret—theoretic(al)
ther—therapy
therapeut—therapeutic(s)
thermodyn—thermodynamic(s)
topog—topographic, topographical, topography
topogr—topographer
toxicol—toxicologic, toxicological, toxicology
trans—transactions
transl—translated, translation(s)
translr—translator(s)
transp—transport, transportation
treas—treasurer, treasury
treat—treatment
trop—tropical
tuberc—tuberculosis
TV—television
Twp—Township

UAR—United Arab Republic
UK—United Kingdom
UN—United Nations
undergrad—undergraduate
unemploy—unemployment
UNESCO—United Nations Educational Scientific & Cultural Organization
UNICEF—United Nations International Childrens Fund
univ(s)—universities, university
UNRRA—United Nations Relief & Rehabilitation Administration

ABBREVIATIONS

UNRWA—United Nations Relief & Works Agency
urol—urologic, urological, urology
US—United States
USA—US Army
USAAF—US Army Air Force
USAAFR—US Army Air Force Reserve
USAF—US Air Force
USAFR—US Air Force Reserve
USAR—US Army Reserve
USCG—US Coast Guard
USCGR—US Coast Guard Reserve
USDA—US Department of Agriculture
USMC—US Marine Corps
USMCR—US Marine Corps Reserve
USN—US Navy
USNAF—US Naval Air Force
USNAFR—US Naval Air Force Reserve
USNR—US Naval Reserve
USPHS—US Public Health Service
USPHSR—US Public Health Service Reserve
USSR—Union of Soviet Socialist Republics
USWMC—US Women's Marine Corps
USWMCR—US Women's Marine Corps Reserve

Va—Virginia
var—various
veg—vegetable(s), vegetation
vent—ventilating, ventilation
vert—vertebrate
vet—veteran(s), veterinarian, veterinary
VI—Virgin Islands
vinicult—viniculture
virol—virological, virology
vis—visiting
voc—vocational
vocab—vocabulary
vol(s)—voluntary, volunteer(s), volume(s)
vpres—vice president
vs—versus
Vt—Vermont

w—west
WAC—Women's Army Corps
Wash—Washington
WAVES—Women Accepted for Voluntary Emergency Service
WHO—World Health Organization
WI—West Indies
wid—widow, widowed, widower
Wis—Wisconsin
WRCNS—Women's Royal Canadian Naval Service
WRNS—Women's Royal Naval Service
WVa—West Virginia
Wyo—Wyoming

yearbk(s)—yearbook(s)
YMCA—Young Men's Christian Association
YMHA—Young Men's Hebrew Association
yr(s)—year(s)
YWCA—Young Women's Christian Association
YWHA—Young Women's Hebrew Association

zool—zoological, zoology

AMERICAN MEN AND WOMEN OF SCIENCE

L

LAANE, JAAN, b Paide, Estonia, June 20, 42; US citizen; m 66; c 2. PHYSICAL CHEMISTRY, SPECTROCHEMISTRY. Educ: Univ Ill, BS, Urbana, 64; Mass Inst Technol, PhD(chem), 67. Prof Exp: Asst prof chem, Tufts Univ, 67-68; asst prof, 68-72, ASSOC PROF CHEM, TEX A&M UNIV, 72- Mem: Am Chem Soc; Am Phys Soc; Soc Appl Spectros; Coblentz Soc. Res: Far-infrared spectroscopy of small ring compounds; potential energy functions; organometallic syntheses; infrared and raman spectroscopy; force constant calculations. Mailing Add: Dept of Chem Tex A&M Univ College Station TX 77843

LAATSCH, RICHARD G, b Fairmont, Minn, July 14, 31; m 66; c 2. MATHEMATICAL ANALYSIS. Educ: Cent Mo State Col, BS, 53; Univ Mo, MA, 57; Okla State Univ, PhD(math), 62. Prof Exp: Instr math, Univ Tulsa, 57-60; asst prof, Okla State Univ, 62; from asst prof to assoc prof, 62-70, PROF MATH & ASST CHMN DEPT, MIAMI UNIV, 70- Mem: Am Math Soc; Math Asn Am. Res: Subadditive functions; topological vector spaces and cones of functions. Mailing Add: Dept of Math Miami Univ Oxford OH 45056

LABACH, WILLIAM ANDERSON, mathematics topology, see 12th edition

LABAHN, RAYMOND WILLIS, b Orange, Calif, Jan 18, 36; m 59; c 2. IONOSPHERIC PHYSICS. Educ: Univ Calif, Riverside, BA, 60, MA, 63, PhD(physics), 65. Prof Exp: Gen physicist, US Naval Ord Lab, Calif, 59-67; from asst prof to assoc prof physics, La State Univ, Baton Rouge, 67-72; PHYSICIST, NAVAL ELEC LAB CTR, SAN DIEGO, 72. Mem: Am Phys Soc; Brit Inst Physics. Mailing Add: NELC-La Posta Observ Rt 1 Box 591 Campo CA 92006

LABANA, SANTOKH SINGH, b Maritanda, India, Nov 15, 36; m 64. ORGANIC CHEMISTRY, POLYMER SCIENCE. Educ: Univ Panjab, India, BSc, 57, MSc, 59; Cornell Univ, PhD(org chem), 63. Prof Exp: Lectr chem, G H G Col, Sadhar, India, 58-60; res chemist, Univ Calif, Berkeley, 63-64; scientist, Xerox Corp, 64-67; prin scientist assoc, 67-70, staff scientist, 70-72, MGR POLYMER SCI DEPT, FORD MOTOR CO, 72- Mem: Am Chem Soc. Res: Synthetic organic chemistry; polymer syntheses; mechanism of organic reactions; physical and thermal properties of polymers with special reference to network polymers; radiation induced polymerizations, coating and composites. Mailing Add: Eng & Res Staff Ford Motor Co PO Box 2053 Dearborn MI 48121

LABANAUSKAS, CHARLES KAZYS, b Upyna, Lithuania, Jan 3, 28; nat US. PLANT PHYSIOLOGY. Educ: Hohenheim Agr Univ, dipl, 47; Univ Ill, MS, 53, PhD, 54. Prof Exp: Asst horticulturist, 55-68, HORTICULTURIST, CITRUS RES CTR, UNIV CALIF, RIVERSIDE, 68-, PROF HORT SCI, COL NATURAL & AGR SCI, 68- Concurrent Pos: Lectr, Univ Calif, Riverside, 65-68. Mem: Am Soc Hort Sci. Res: Mineral metabolism in plants. Mailing Add: Dept of Plant Sci Col of Natural & Agr Sci Univ of Calif Riverside CA 92502

LABAR, MARTIN, b Radisson, Wis, May 15, 38. POPULATION BIOLOGY. Educ: Wis State Univ, Superior, BA, 58; Univ Wis, MS, 63, PhD(genetics, zool), 65. Prof Exp: Assoc prof, 64-66, PROF SCI, CENT WESLEYAN COL, 66-, CHMN DIV SCI, 64- Mem: Am Sci Affiliation; Genetics Soc Am; Ecol Soc Am; Soc Study Evolution. Res: Use of computer simulation in teaching population biology. Mailing Add: Div of Sci Cent Wesleyan Col Central SC 29630

LABARGE, ROBERT GORDON, b Buffalo, NY, July 11, 40. APPLIED CHEMISTRY, INDUSTRIAL CHEMISTRY. Educ: Univ Rochester, BS, 62; Carnegie-Mellon Univ, PhD(chem), 66. Prof Exp: Res chemist, Consumer Prod Dept, 66-73, dir acad educ, 73-75, RES SPECIALIST, DESIGNED PROD DEPT, DOW CHEM CO, 75- Concurrent Pos: Adj prof chem, Cent Mich Univ, 74. Mem: Am Chem Soc. Res: New product exploration and development. Mailing Add: Designed Prod Dept Dow Chem Co Midland MI 48640

LABARRE, ANTHONY E, JR, b New Orleans, La, July 18, 22; m 43; c 2. MATHEMATICS. Educ: Tulane Univ, BE, 43, MS, 47; Univ Okla, PhD(math), 57. Prof Exp: Instr math, Tulane Univ, 46-48; asst prof, Univ Idaho, 48-50; instr, Univ Okla, 50-54 & Univ Wyo, 54-56; from asst prof to assoc prof, Univ Idaho, 56-61; chmn dept, 61-66, PROF MATH, CALIF STATE UNIV, FRESNO, 61- Concurrent Pos: NSF & Math Asn Am lectr, High Sch, 60; mem panel high sch teachers awards, NSF, 60. Mem: Am Math Soc; Math Asn Am. Res: Functional analysis, especially Hilbert spaces. Mailing Add: Dept of Math Calif State Univ Fresno CA 93710

LA BARRE, WESTON, b Uniontown, Pa, Dec 13, 11; m 39; c 3. ANTHROPOLOGY. Educ: Princeton Univ, AB, 33; Yale Univ, PhD(anthrop), 37. Prof Exp: Soc Sci Res Coun res intern, Menninger Clin, Kans, 38-39; instr anthrop, NJ Col Women, Rutgers Univ, 39-43; from asst prof to assoc prof, 46-58, CLIN PROF ANTHROP, MED SCH, DUKE UNIV, 55-, PROF UNIV, 58-, JAMES B DUKE PROF, 70- Concurrent Pos: Yale Univ Sterling fel Aymara Indians of Lake Titicaca Plateau, 37-38; Guggenheim fel, 46; vis lectr, Univ NC, Chapel Hill, 51-52, clin prof, Med Sch, 55-; consult, Group Advan Psychiat, 59-71; NSF sr fel, Europe, 62-63; ed-in-chief, Landmarks in Anthrop, 65-; NSF-Viking Fund grant, 66-67. Honors & Awards: Roheim Mem Award, 58. Mem: Fel Am Anthrop Asn. Res: Psychoanalytically-oriented cultural anthropology and human biology; Americanist studies, especially Amerindian psychotropic drugs; primitive religion, art and music; structural linguistics; Eurasiatic-American prehistory; cognitive anthropology; nonverbal communication; kinesics; ethnobotany. Mailing Add: Mt Sinai Rd Rt 1 Box 591 Durham NC 27705

LABARTHE, DARWIN RAYMOND, b Berkeley, Calif, Aug 5, 39. EPIDEMIOLOGY. Educ: Princeton Univ, AB, 61; Columbia Univ, MD, 65; Univ Calif, Berkeley, MPH, 67, PhD(epidemiol), 75. Prof Exp: Epidemiologist, Comn Corps, Heart Dis & Stroke Control Prog, San Francisco, 67-69; dep chief & sr epidemiologist, Epidemiol Field & Training Sta, USPHS Heart Dis & Stroke Control Prog, San Francisco, 69-70; from assoc res epidemiologist to assoc prof epidemiol, Sch Pub Health, Univ Tex Health Sci Ctr, Houston, 70-73; CONSULT EPIDEMIOL, DEPT MED STATIST & EPIDEMIOL, MAYO CLIN & MAYO FOUND, 74- Concurrent Pos: Dep dir, Coord Ctr, Hypertension Detection & Followup Prog, Nat Heart & Lung Inst, 71-73; consult, Task Force Automated Blood Pressure Devices, Nat Heart & Lung Inst, 73-74; consult, coord ctr, Hypertension Detection & Followup Prog, Nat Heart & Lung Inst, 74-; co-investr & co-dir, Study Incidence & Natural Hist Genital Tract Anomalies & Cancer in Offspring exposed in Utero to synthetic Estrogens, Nat Cancer Inst, 74-; chmn & fac mem, Am Heart Asn first US Seminar in Cardiovas Epidemiol, 75-76. Mem: Fel Am Heart Asn; fel Am Col Prev Med; Soc Epidemiol Res (pres, 72-73); Am Pub Health Asn; Int Soc Cardiol. Res: Epidemiology and prevention, especially of cardiovascular and other chronic conditions; drugs; intra-individual variability of blood pressure and other personal characteristics. Mailing Add: Dept of Med Statist & Epidemiol Mayo Clin Rochester MN 55901

LABATE, SAMUEL, b Easton, Pa, Dec 19, 18; m 49; c 2. ACOUSTICS. Educ: Lafayette Col, AB, 40; Mass Inst Technol, MS, 48. Prof Exp: Asst instr math, Univ Pa, 40-41; engr, E I du Pont de Nemours & Co, 41-42; consult engr, Bolt & Beranek, 48-49, consult engr, 49-53, exec vpres, 53-69, PRES & DIR, BOLT BERANEK & NEWMAN, INC, 69- Mem: Fel Acoust Soc Am; Acad Appl Sci. Res: Engineering; applied architectural and physical acoustics. Mailing Add: Bolt Beranek & Newman Inc 50 Moulton St Cambridge MA 02138

LABAW, GLENN DOUGLAS, food bacteriology, see 12th edition

LABAW, LOUIS WARNE, b Hopewell, NJ, Jan 27, 18; m 46; 2. PHYSICS. Educ: Albright Col, BS, 38; Brown Univ, MS, 40, PhD(physics), 42. Prof Exp: From instr to asst prof physics, Brown Univ, 42-45; PHYSICIST, NAT INST ARTHRITIS, METAB & DIGESTIVE DIS, 47- Mem: Fel Acoust Soc Am; Am Phys Soc. Res: Acoustics; supersonics; biophysics. Mailing Add: Nat Inst of Arthritis Metab & Digestive Dis Bethesda MD 20014

L'ABBE, MAURICE, b Ottawa, Ont, May 20, 20; m 51; c 4. MATHEMATICS. Educ: Univ Montreal, BA, 42, LSc, 45; Princeton Univ, MA, 47, PhD(math), 51. Prof Exp: From asst prof to assoc prof math, 48-56, chmn dept math, 57-68, vdean fac sci, 64-68, PROF MATH, UNIV MONTREAL, 56-, VRECTOR RES, 68- Concurrent Pos: Can Govt res fel, Univ Paris, 52-53. Mem: Am Math Soc; Math Asn Am; Asn Symbolic Logic; Can Math Cong (pres, 67-69); Math Soc France. Res: Mathematical logic. Mailing Add: VRectorship for Res Univ of Montreal Box 6128 Montreal PQ Can

LABBE, ROBERT FERDINAND, b Portland, Ore, Nov 12, 22; m 55; c 3. BIOCHEMISTRY. Educ: Univ Portland, BS, 47; Ore State Col, MS, 49, PhD(biochem), 51. Prof Exp: AEC fel med sci, Col Physicians & Surgeons, Columbia Univ, 51-53; res instr, Med Sch, Univ Ore, 53-55, res asst prof, 55-57; res asst prof pediat & lectr biochem, 57-60, res assoc pediat, 60-68, prof pediat, 68-74, PROF LAB MED, MED SCH, UNIV WASH, 74- Concurrent Pos: Vis asst prof, Inst Enzyme Res, Univ Wis, 56-57; vis researcher, Commonwealth Sci & Indust Res Orgn, Australia, 65; NIH spec fel, 65 & career develop award, 66-70. Mem: Am Chem Soc; Am Soc Biol Chemists; Soc Exp Biol & Med; Am Inst Chemists; Am Asn Clin Chemists. Res: Heme biosynthesis; iron metabolism; related metabolic diseases and nutrition. Mailing Add: Dept of Lab Med Univ of Wash Seattle WA 98195

LABBEE, MARCEL D, b Holyoke, Mass, Sept 6, 22; m; c 6. AGRICULTURAL CHEMISTRY, FOOD TECHNOLOGY. Educ: Univ Mass, BS, 49, MS, 50, PhD(food technol), 52. Prof Exp: Mem staff qual control, Old Deerfield Pickling Co, 51-52; MEM STAFF TECH SALES, ROHM AND HAAS CO, 52- Res: Peroxidase and its relation to off-flavors in food; enzymes; freezer jars; pickles. Mailing Add: Rohm and Haas Co 511 Cooper Pkwy Bldg W Pennsauken NJ 08109

LABBY, DANIEL HARVEY, b Portland, Ore, Sept 1, 14; m 40; c 3. MEDICINE, PSYCHIATRY. Educ: Reed Col, BA, 35; Univ Ore, MD, 39. Prof Exp: Intern, Johns Hopkins Hosp, 39-40; asst resident, New York Hosp, 43-44, chief resident & instr med, 44-45; asst, Rockefeller Inst & asst physician, Hosp, 45-47; asst clin prof med, 47-51, from assoc prof to prof, 51-71, chief div diabetes & metab dis, 51-70, PROF MED & PSYCHIAT, MED SCH, UNIV ORE, 71- Concurrent Pos: Fel med, Med Col, Cornell Univ & New York Hosp, 40-41; Brower traveling fel, 53; Commonwealth traveling fel, Univ Strasbourg, 60-61; trainee, Med Sch, Univ Pa, 71; instr, Med Col, Cornell Univ, 44-45; vis prof, Med Col Va, 58, Univ Strasbourg, 60-61 & Med Sch, Univ Colo, 64; assoc, Tavistock Inst & Clin, London, Eng, 72. Mem: AAAS; Am Psychiat Asn; Am Soc Clin Invest; AMA; fel Am Col Physicians. Mailing Add: 5931 SW Hamilton Portland OR 97221

LABELLA, FRANK SEBASTIAN, b Middletown, Conn, Sept 23, 31; m 52; c 3. PHARMACOLOGY. Educ: Wesleyan Univ, BA, 52, MA, 54; Emory Univ, PhD(basic health sci), 57. Prof Exp: Asst biol, Wesleyan Univ, 52-54; asst physiol, Emory Univ, 54-55, asst histol, 55-57, instr physiol, 57-58; lectr pharmacol, 58-60, from asst prof to assoc prof, 60-67, PROF PHARMACOL & THERAPEUT, FAC MED, UNIV MAN, 67- Concurrent Pos: Am Heart Asn fel, Emory Univ, 57-58; Can Rheumatism & Arthritis Soc res fel, Univ Man, 58-61; estab investr, Am Heart Asn, 61-66, mem coun arteriosclerosis; med res assoc, Med Res Coun Can, 66- Honors & Awards: John J Abel Award, Am Soc Pharmacol & Exp Therapeut, 67; E W R Steacie Prize in Natural Sci, Can, 69. Mem: Geront Soc; Am Soc Pharmacol & Exp Therapeut; Am Soc Cell Biol; Am Aging Asn; Int Soc Biochem Pharmacol. Res: Cellular pharmacology and biochemistry; connective tissue; aging; endocrine pharmacology; neuroendocrinology. Mailing Add: Dept of Pharmacol & Therapeut Univ of Man Fac of Med Winnipeg MB Can

LABELLE, ROBERT LAWRENCE, b Watertown, NY, June 24, 24; m50; c 4. FOOD SCIENCE. Educ: Cornell Univ, BChemE, 50. Prof Exp: Res assoc food technol, 50-53, chem engr, 53-56, from asst prof to assoc prof, 56-68, PROF FOOD SCI, NY STATE AGR EXP STA, CORNELL UNIV, 68- Mem: Inst Food Technologists. Res: Storage and handling, processing, and nutritional value of fruits, vegetables, and pulses; food engineering and technology. Mailing Add: NY State Agr Exp Sta Cornell Univ Geneva NY 14456

LABEN, ROBERT COCHRANE, b Darien Center, NY, Nov 16, 20; m 46; c 4. GENETICS. Educ: Cornell Univ, BS, 42; Okla Agr & Mech Col, MS, 46; Univ Mo, PhD(animal breeding), 50. Prof Exp: Asst animal husb, Okla Agr & Mech Col, 46-47; asst dairy husb, Univ Mo, 47-49, asst instr, 49-50; instr animal husb & jr animal husbandman, 50-52, asst prof & asst animal husbandman, 52-58, assoc prof & assoc animal husbandman, 58-64, prof, animal husbandman & dir comput ctr, 64-69, PROF ANIMAL SCI & GENETICIST, EXP STA, UNIV CALIF, DAVIS, 69- Mem: Am Soc Animal Sci; Biomet Soc; Am Dairy Sci Asn; Am Genetic Asn. Res: Breeding and genetics of farm livestock. Mailing Add: Dept of Animal Sci Univ of Calif Davis CA 95616

LABER, LARRY JACKSON, b Lincoln, Vt, July 9, 37; m 63. PLANT PHYSIOLOGY. Educ: Univ Vt, BS, 59, MS, 61; Univ Chicago, PhD(bot), 67. Prof Exp: Asst prof, Pa State Univ, 65-66; NSF trainee, Univ Ga, 67-69; NIH trainee, Brandeis Univ, 69-70; ASST PROF BOT, UNIV MAINE, ORONO, 70- Mem: AAAS; Am Soc Plant Physiol; Japanese Soc Plant Physiol. Res: Choroplast development; photophosphorylation; carbon dioxide fixation. Mailing Add: Dept of Bot & Plant Path Univ of Maine Orono ME 04473

LA BERGE, DONALD EMMANUEL, biochemistry, cereal chemistry, see 12th edition

LABERGE, GENE L, b Ladysmith, Wis, Mar 15, 32; m 62; c 2. GEOLOGY. Educ: Univ Wis, BS, 58, MS, 59, PhD(geol), 63. Prof Exp: Sponsored res officer, Commonwealth Sci & Indust Res Orgn, Melbourne, Australia, 63-64; Nat Res Coun Can fel, Geol Surv Can, 64-65; from asst prof to assoc prof, 65-74, PROF GEOL, UNIV WIS-OSHKOSH, 74- Concurrent Pos: Mem staff, Wis Geol & Natural Hist Surv, 72- Mem: AAAS; Geol Soc Am; Soc Econ Geologists. Res: Origin of Precambrian iron formations; Precambrian geology and mineral deposits of Wisconsin. Mailing Add: Dept of Geol Univ Wis-Oshkosh Oshkosh WI 54901

LABERGE, WALLACE E, b Grafton, NDak, Feb 7, 27; m 58; c 3. ENTOMOLOGY. Educ: Univ NDak, BSc, 49, MS, 51; Univ Kans, PhD(entom), 55. Prof Exp: Asst cur, Snow Entom Mus & instr entom, Univ Kans, 54-55, asst prof, 55-56; asst prof zool, Iowa State Univ, 56-59; assoc prof entom, Univ Nebr, 59-65; assoc taxonomist, 65-67, TAXONOMIST, ILL NATURAL HIST SURV, 67-; PROF ENTOM, UNIV ILL, URBANA, 70- Mem: Entom Soc Am; Soc Study Evolution; Soc Syst Zool; Am Entom Soc. Res: Systematics of Hymenoptera, Apoidea and Braconidae. Mailing Add: Faunistics Div Ill Natural Hist Surv Urbana IL 61803

LABES, MORTIMER MILTON, b Newton, Mass, Sept 9, 29; m 53; c 6. CHEMICAL PHYSICS. Educ: Harvard Univ, AB, 50; Mass Inst Technol, PhD, 54. Prof Exp: Asst, Mass Inst Technol, 51-54, res chemist, Sprague Elec Co, 54-57; sr res chemist, Franklin Inst, 57-59, sr staff chemist, 59-60, lab mgr, 60-61, tech dir chem div, 61-66; prof chem, Drexel Inst, 66-70; PROF CHEM, TEMPLE UNIV, 70- Mem: Am Chem Soc; Am Phys Soc; Sigma Xi; The Chem Soc. Res: Chemistry and physics of organic solid state; molecular complexes; liquid crystals; electronic properties of polymers. Mailing Add: Dept of Chem Temple Univ Philadelphia PA 19122

LABHSETWAR, ANANT PANDURANG, b Darwha, India, Sept 29, 36; m 63; c 2. REPRODUCTIVE ENDOCRINOLOGY. Educ: Univ Jabalpur, BS, 58; Univ Wis-Madison, MS, 61, PhD(reprod physiol), 63. Prof Exp: Fel sch med, Case Western Reserve Univ, 63-64; res assoc sch med, Univ Louisville, 64-65; res instr sch med, Washington Univ, 65-68; tech officer, Pharmaceut Div, ICI Ltd, Eng, 68-71; staff scientist, Worcester Found Exp Biol, 71-74; ADJ ASSOC PROF BIOL, KANS STATE UNIV, 74- Mem: Soc Study Reproduction; Brit Soc Study Fertil; Brit Soc Endocrinol. Res: Gonadotrophins; control of ovulation; implantation of blastocysts; role of monoamines in ovulation; contraceptives; prostaglandins. Mailing Add: Div of Biol Kans State Univ Manhattan KS 66506

LABIANCA, DOMINICK A, b Brooklyn, NY, Feb 4, 43. ORGANIC CHEMISTRY, POLYMER CHEMISTRY. Educ: Polytech Inst Brooklyn, BS, 65; Univ Mich, PhD(chem), 69. Prof Exp: NSF fel org photochem, Calif Inst Technol, 69-70; res chemist, Res & Develop, Bound Brook Tech Ctr, Union Carbide Corp, 70-72; ASST PROF, NEW SCH LIB ARTS, BROOKLYN COL, CITY UNIV NEW YORK, 72- Mem: Sigma Xi; Am Chem Soc; Nat Sci Teachers Asn. Res: Interdisciplinary teaching; drugs; environment. Mailing Add: 948 Cloud Ave Franklin Square NY 11010

LABIANCA, FRANK MICHAEL, b Brooklyn, NY, Aug 17, 39; m 70; c 1. UNDERWATER ACOUSTICS. Educ: Polytech Inst Brooklyn, BEE, 61, MS, 63, PhD(elec eng), 67. Prof Exp: Instr elec eng, Polytech Inst Brooklyn, 61-67; MEM TECH STAFF UNDERWATER ACOUST, BELL LABS INC, 67- Mem: Inst Elec & Electronic Engrs; Acoust Soc Am. Res: Propagation in surface ducts and underwater channel, scattering of sound from the ocean surface, radiation from cavitating propellers and the origins of ambient noise. Mailing Add: Bell Labs Inc Whippany NJ 07981

LABINE, PATRICIA ANNE, population biology, see 12th edition

LABISKY, RONALD FRANK, b Aberdeen, SDak, Jan 16, 34; m 58; c 2. WILDLIFE BIOLOGY. Educ: SDak State Univ, BS, 55; Univ Wis, MS, 56, PhD(wildlife ecol-zool), 68. Prof Exp: Field asst game bird res, 56-57, from asst proj leader to proj leader, 57-59, from asst wildlife specialist to assoc wildlife specialist, 59-72, WILDLIFE SPECIALIST, ILL STATE NATURAL HIST SURV, 72- Mem: Wildlife Soc; Am Ornith Union; Wilson Ornith Soc; Am Soc Mammal. Res: Ecology and physiology of gallinaceous game birds, doves and waterfowl; population ecology, social biology and spatial distribution of pheasants; ecological, ethological, physiological and nutritive factors influencing distribution and abundance of pheasants. Mailing Add: Sect of Wildlife Res Ill Natural Hist Surv Urbana IL 61801

LA BONTE, ANTON EDWARD, b Minneapolis, Minn, May 6, 35; m 59. COMPUTER SCIENCE. Educ: Univ Minn, Minneapolis, BS, 57, MSEE, 60, PhD(elec eng), 66. Prof Exp: Instr elec eng, Univ Minn, Minneapolis, 59-62, res fel micromagnetics, 62-63, instr elec eng, 63-65; sr scientist, 66-69, mgr systs anal, Aerospace, Navigation & Space Systs, 69-75, SR CONSULT, DIGITAL IMAGE SYSTS, CONTROL DATA CORP, 75- Mem: Inst Elec & Electronics Eng; Am Phys Soc. Res: Application of computers to image processing and astronomy, especially automation of large-scale stellar proper motion survey. Mailing Add: 11 River Terrace Court Apt 104 Minneapolis MN 55414

LABOURIAU, LUIZ F G, plant physiology, plant ecology, see 12th edition

LABOWS, JOHN NORBERT, JR, b Wilkes-Barre, Pa, June 27, 41; m 64; c 1. ORGANIC CHEMISTRY. Educ: Lafayette Col, BS, 63; Cornell Univ, PhD(org chem), 67. Prof Exp: Asst prof org chem, 67-70, ASSOC PROF CHEM, WILKES COL, 70- Concurrent Pos: Nat Cancer Inst fel, Fels Res Inst, Temple Univ, 70-71. Mem: Am Chem Soc; Sigma Xi. Res: Cycloaddition reactions; thermal and photochemical rearrangements. Mailing Add: Dept of Chem Wilkes Col Wilkes-Barre PA 18703

LA BRECQUE, GERMAIN C, b Drummondville, Can, July 9, 22; nat US; m 52; c 2. MEDICAL ENTOMOLOGY. Educ: Tufts Univ, BS, 49; Univ Tenn, MS, 50; Ohio State Univ, PhD, 57. Prof Exp: MED ENTOMOLOGIST, AGR RES SERV, USDA, 51- Mem: Entom Soc Am; Sigma Xi. Res: Control of disease carrying insects. Mailing Add: 213 NW 29th St Gainesville FL 32607

LABREE, THEODORE ROBERT, b Lafayette, Ind, June 25, 31; m 51; c 2. BACTERIOLOGY, FOOD TECHNOLOGY. Educ: Purdue Univ, BS, 58, MS, 60. Prof Exp: Res asst food technol, Purdue Univ, 58-59; assoc bacteriologist, Mead Johnson & Co, 59-62, scientist bact, 62-65, mgr med admin, 65-73; TECH MGR, REGULATORY SERV, RIVIANA FOODS, INC, 73- Mem: Am Soc Microbiol; Inst Food Technologists. Res: Spore destruction of food spoilage organisms; new methods development; microbiology; government regulations, processing, packaging, labeling; good manufacturing practices regulations; low acid; sanitation; food plant; warehouse evaluation. Mailing Add: Riviana Foods Inc PO Box 2636 Houston TX 77001

LABRIE, DAVID ANDRE, b Baltimore, Md, Mar 23, 37; m 62; c 2. MICROBIOLOGY, BIOCHEMICAL GENETICS. Educ: Bethany Col, WVa, AB, 61; NMex Highlands Univ, MS, 65; NC State Univ, PhD(microbiol), 68. Prof Exp: NIH fel, Univ Tex, Austin, 68-70; asst prof, 70-74, ASSOC PROF BIOL, W TEX STATE UNIV, 74- Mem: AAAS. Res: Isozymic protein biochemistry; origin and evolution of cell organelles. Mailing Add: Dept of Biol WTex State Univ Canyon TX 79015

LABRIE, FERNAND, b June 28, 37; Can citizen; m 63; c 4. ENDOCRINOLOGY, BIOCHEMISTRY. Educ: Laval Univ, BA, 57, MD, 62, PhD(endocrinol), 67; FRCP(C), 73. Prof Exp: From asst prof to assoc prof physiol, 66-73, HEAD LAB MOLECULAR ENDOCRINOL, HOSP CTR, LAVAL UNIV, 69-, PROF PHYSIOL, 73- Concurrent Pos: Med Res Coun Can fels, Laval Univ, 63-66, Univ Cambridge, 66-67, Univ Sussex, 67-68 & centennial fel, Lab Molecular Biol, Univ Cambridge, 68-69; Med Res Coun Can scholar, Laval Univ, 69-; Med Res Coun Can assoc, Laval Univ, 73-; dir molecular endocrinol, Med Res Coun Group, 73- Mem: Brit Biochem Soc; Am Physiol Soc; Endocrine Soc; Can Physiol Soc; Can Biochem Soc. Res: Mechanism of action of hypothalamic regulatory hormones in the anterior pituitary gland, mammalian messenger RNA; hormone dependent breast cancer; reproductive physiology and biochemistry. Mailing Add: Lab of Molecular Endocrinol Laval Univ Hosp Ctr Quebec PQ Can

LABROSSE, ELWOOD HENRY, b Mason, Mich, Oct 1, 21; m 47; c 3. BIOCHEMISTRY. Educ: Northwestern Univ, BS, 45, MS, 48, MD, 49; Univ Tex, PhD(biochem), 56. Prof Exp: Res scientist biochem, Univ Tex, 55-57; chief unit schizophrenia, NIMH, 57-63; RES ASSOC PROF SURG & ASST PROF BIOCHEM, SCH MED, UNIV MD, BALTIMORE CITY, 64- Concurrent Pos: Eleanor Roosevelt Int Cancer fel, Inst Gustave Roussy, Villejuif, France, 63-64 & Nat Cancer Inst spec fel, 74-75; vis scientist, Inst Gustave Roussy, 73-74; asst res, Nat Inst Health & Med Res, Paris, 74-75. Mem: AAAS; Am Chem Soc; AMA; NY Acad Sci. Res: Metabolism of catechol amines and aromatic amino aciochemistry of schizophrenia; cancer, neuroblastoma. Mailing Add: 9033 Sidehill Rd Elliott City MD 21043

LABUTE, JOHN PAUL, b Tecumseh, Ont, Feb 26, 38; m 61; c 3. MATHEMATICS. Educ: Univ Windsor, BSc, 60; Harvard Univ, MA, 61, PhD(math), 65. Prof Exp: Nat Res Coun Can res fel, Col France, 65-67; asst prof, 67-70, ASSOC PROF MATH, McGILL UNIV, 70- Mem: Can Math Cong; Am Math Soc. Res: Algebra and number theory. Mailing Add: Dept of Math McGill Univ PO Box 6070 Montreal PQ Can

LABUZA, THEODORE PETER, b Perth Amboy, NJ, Nov 10, 40; m 63; c 1. FOOD SCIENCE, PHYSICAL CHEMISTRY. Educ: Mass Inst Technol, SB, 62, PhD(food sci), 65. Prof Exp: From instr to assoc prof food eng, Mass Inst Technol, 65-71; assoc prof, 71-72, PROF FOODTECHNOL, UNIV MINN, ST PAUL, 72- Concurrent Pos: Food processing consult; pres, Physicians Nutrit Serv, Ohio. Honors & Awards: Samuel Cate Precott Res Award, Inst Food Technologists, 72. Mem: Inst Food Technologists; Am Inst Chem Eng; Am Chem Soc; Soc Nutrit Educ. Res: Physical factors involved in autoxidation of food lipids and prediction of food storage life; stability of intermediate moisture foods; nutrient degradation in processing; kinetics of microbial death. Mailing Add: Dept of Food Sci & Nutrit Univ of Minn St Paul MN 55108

LACAILLADE, CHARLES WILLIAM, b Lawrence, Mass, Nov 30, 04. MEDICAL ENTOMOLOGY. Educ: Conn Agr Col, BS, 29; Harvard Univ, MS, 30, PhD(entom), 33. Prof Exp: Instr zool, Harvard Univ, 31; fel, Rockefeller Inst, 31-34; from asst prof to prof, 35-71, chmn dept, 61-65, EMER PROF BIOL, ST JOHN'S UNIV, NY, 71- Mem: AAAS; NY Acad Sci. Res: Virus diseases of insects; reactions caused by insect venoms; transmission of equine encelphalomyelitis by mosquitoes. Mailing Add: Star Field Maple Ave Atkinson NH 03811

LACASCE, ELROY OSBORNE, JR, b Fryeburg, Maine, Jan 17, 23. ACOUSTICS. Educ: Bowdoin Col, AB, 43; Harvard Univ, AM, 51; Brown Univ, PhD(physics), 55. Prof Exp: Instr physics, Bowdoin Col, 43 & 47-49, instr math, 51; physicist, Naval Res Lab, 44; foreign serv officer, US Dept State, 45-46; teacher, High Sch, 46-47; asst, Brown Univ, 51-54; from instr to assoc prof physics, 54-69, PROF PHYSICS, BOWDOIN COL, 69- Concurrent Pos: Res assoc, Yale Univ, 60-61; NSF fac fel, 60-61; vis investr, Woods Hole Oceanog Inst, 68-69, guest investr, 75-76. Mem: Acoust Soc Am; Am Asn Physics Teachers. Res: Ultrasonics and underwater sound. Mailing Add: Dept of Physics Bowdoin Col Brunswick ME 04011

LACASSE, NORMAN L, b Berlin, NH, Nov 6, 33; m 59; c 5. PLANT PATHOLOGY. Educ: Univ NH, BS, 61, MS, 63; Pa State Univ, PhD(plant path), 66. Prof Exp: Asst prof, 66-73, ASSOC PROF PLANT PATH, PA STATE UNIV, 73-, RES ASSOC, CTR AIR ENVIRON, 66- Mem: Am Phytopath Soc; Air Pollution Control Asn. Res: Physiology of infectious and noninfectious disease. Mailing Add: 210A Buckhout Lab Pa State Univ University Park PA 16802

LACATE, DOUGLAS STEWART, forestry, see 12th edition

LACEFIELD, WILLIAM BRYANT, organic chemistry, see 12th edition

LACELLE, PAUL (LOUIS), b Syracuse, NY, July 4, 29; m 53; c 4. HEMATOLOGY, BIOPHYSICS. Educ: Houghton Col, BA, 51; Univ Rochester, MD, 59. Prof Exp: Intern & resident, Strong Mem Hosp, Univ Rochester, 59-62; from sr instr to asst prof, 67-71, PROF MED, RADIATION BIOL & BIOPHYS, SCH MED, UNIV ROCHESTER, 71- Concurrent Pos: USAEC res fel, 63-65; USPHS spec fel, Univ Saarland, 65-66; Buswell fel, Sch Med, Univ Rochester, 66-67; NIH res grant, 70-; consult, US Vet Admin Hosp, Batavia, NY, 68- Mem: Am Soc Hemat; Am Fedn Clin Res; Int Soc Hemat. Res: Hemolytic anemias; membrane biophysical properties of erythrocytes. Mailing Add: 260 Crittenden Blvd Rochester NY 14620

LACEY, BEATRICE CATES, b New York, NY, July 22, 19; m 38; c 2. PSYCHOPHYSIOLOGY. Educ: Cornell Univ, AB, 40; Antioch Col, MA, 58. Prof Exp: Res asst psychophysiol, 53-55, sr res asst, 55-58, res assoc psychophysiol-neurophysiol, 58-66, sr investr, 66-72, SR SCIENTIST PSYCHOPHYSIOL-NEUROPHYSIOL, FELS RES INST, ANTIOCH COL, 72- Concurrent Pos: Instr psychol, Antioch Col, 56-63, asst prof, 63-68, assoc prof, 68-73, prof, 73-; co-prin investr, USPHS grant, 60-; assoc ed, Psychophysiol, 75- Mem: Soc Psychophysiol Res; Soc Neurosci. Res: Psychophysiology of the autonomic nervous system. Mailing Add: Sect Behav Physiol 800 Livermore Samuel S Fels Res Inst Yellow Springs OH 45387

LACEY, HOWARD ELTON, b Leakey, Tex, Feb 9, 37; m 58; c 4. MATHEMATICS. Educ: Abilene Christian Col, BA, 59, MA, 61; NMex State Univ, PhD(math), 63. Prof Exp: Asst prof math, Abilene Christian Col, 63-64 & Univ Tex, Austin, 64-67; res assoc, NASA Manned Spacecraft Ctr, 67-68; assoc prof, 68-73, PROF MATH, UNIV TEX, AUSTIN, 73-, MEM GRAD FAC, 68-, VCHMN DEPT, 75- Concurrent Pos: Res assoc, Inst Math, Polish Acad Sci, Warsaw, 72-73. Mem: Am Math Soc; Math Asn Am. Res: Functional analysis; classical Banach spaces. Mailing Add: Dept of Math Univ of Tex Austin TX 78712

LACEY, JOHN IRVING, b Chicago, Ill, Apr 11, 15; m 38; c 2. PSYCHOPHYSIOLOGY, NEUROPHYSIOLOGY. Educ: Cornell Univ, BA, 37, PhD(psychol), 41. Prof Exp: Instr psychol, Queens Col, NY, 41-42; res assoc, 46-73, assoc prof, 48-56, SR SCIENTIST PSYCHOPHYSIOL, FELS RES INST, ANTIOCH COL, 73-, PROF COL, 56-, CHMN DEPT, 48- Concurrent Pos: Res assoc, Psychol Corp, 39-42; lectr, Ohio State Univ, 50 & sch med, Univ Louisville, 55; fel, Commonwealth Fund, 57-59; mem ment health, behav sci & exp psychol study sects, USPHS, 56-60, res career develop comt, 64-65; mem adv panel life sci facilities, NSF, 60-; mem clin prog-proj rev comt, NIMH, 66-71, chmn, 70-71. Honors & Awards: Award, Soc Psychophysiol Res, 70. Mem: AAAS; Am Psychosom Soc; Am Psychol Asn; Am Acad Neurol; Soc Psychophysiol Res (pres, 61). Res: Psychophysiology of the autonomic nervous system and psychomatic medicine; brain physiology and behavior. Mailing Add: Sect on Behav Physiol Fels Res Inst Yellow Springs OH 45387

LACEY, RICHARD FREDERICK, b Vallejo, Calif, May 29, 31; m 71. MAGNETISM. Educ: Mass Inst Technol, SB, 52, PhD(physics), 59. Prof Exp: Sr engr, Sylvania Lighting Prod Co, 59-62; sr scientist, Am Sci & Eng Co, 62-63; sr physicist, Varian Assocs, 63-67; physicist, 67-69, STAFF SCIENTIST, HEWLETT-PACKARD LABS, 69- Mem: Am Phys Soc. Res: Atomic structure; radio-frequency spectroscopy; physical and quantum electronics. Mailing Add: Hewlett-Packard Labs 3500 Deer Creek Rd Palo Alto CA 94304

LACH, JOHN LOUIS, b Blairmore, Alta, Feb 10, 27; nat US; m 53; c 4. PHYSICAL PHARMACY. Educ: Univ Alta, BSc, 50; Univ Wis, MS, 52, PhD(pharm), 54. Prof Exp: Instr pharm, Univ Wis, 54; from asst prof to assoc prof, 54-62, PROF PHARM, COL PHARM, UNIV IOWA, 62-, ASSOC DEAN, 72- Honors & Awards: Acad Pharmaceut Sci Res Achievement Award Pharmaceut, Am Pharmaceut Asn Found, 75. Mem: Am Chem Soc; Am Pharmaceut Asn. Res: Application of physical-chemical principles to pharmaceutical systems involving stability studies, complex formation, formulation and analytical techniques. Mailing Add: Col of Pharm Univ of Iowa Iowa City IA 52240

LACH, JOSEPH T, b Chicago, Ill, May 12, 34; m 65. PHYSICS. Educ: Univ Chicago, AB, 53, MS, 56; Univ Calif, Berkeley, PhD(physics), 63. Prof Exp: Res assoc physics, Yale Univ, 63-65, asst prof, 66-69; chmn physics depts, 74-75, PHYSICIST, FERMI NAT ACCELERATOR LAB, 69- Mem: Am Phys Soc. Res: Elementary particle physics; physics electronic data processing. Mailing Add: Physics Dept PO Box 500 Fermi Nat Accelerator Lab Batavia IL 60510

LACHANCE, DENIS, b Quebec, Que, Feb 2, 39; m 64; c 3. FOREST PATHOLOGY. Educ: Laval Univ, BSc, 62; Univ Wis-Madison, PhD(phytopath), 66. Prof Exp: RES SCIENTIST FOREST PATH, LAURENTIAN FOREST RES CTR, CAN FORESTRY SERV, 66- Mem: Can Phytopath Soc (secy-treas, 71-73); Can Inst Forestry; Int Soc Plant Path. Res: Decay of conifers; root diseases. Mailing Add: Laurentian Forest Res Ctr Can Forestry Serv PO Box 3800 Quebec PQ Can

LACHANCE, JEAN PAUL, b Berthier, Que, Feb 4, 23; m 55; c 4. BIOLOGY. Educ: Levis Univ, BA, 45; Laval Univ, BSc, 49, PhD(biol), 53. Prof Exp: Dir med tech training lab, Laval Univ, 52-54, res asst inst physiol, Fac Med, 54-55; res asst biol, Univ Montreal, 55-56; Nuffield fel, Med Res Coun Radiopath Res Univ, Hammersmith Hosp, London, Eng, 56-57; Nat Res Coun Can fel, Univ Munich, 57-58; asst prof biochem, 58-61, asst prof biol, 61-65, ASSOC PROF BIOCHEM, UNIV MONTREAL, 65- Mem: Can Biochem Soc; Can Physiol Soc. Res: Fat metabolism in vivo and fatty acid biosynthesis; carboxylating enzymes. Mailing Add: Dept of Biochem Univ Montreal 2900 Blvd Mt Royal Montreal PQ Can

LA CHANCE, LEO EMERY, b Brunswick, Maine, Mar, 1, 31; m 55; c 3. GENETICS. Educ: Univ Maine, AB, 53; NC State Col, MS, 55, PhD(genetics), 58. Prof Exp: Res assoc biol, Brookhaven Nat Lab, 58-60; insect geneticist, 60-63, proj leader, Insect Genetics & Radiation Biol Sect, Metab & Radiation Res Lab, Entom Res Div, Agr Res Serv, USDA, 63-69; sci officer & head, Insect Eradication & Pest Control Sect, Joint Food & Agr Orgn-Int Atomic Energy Agency, Austria, 69-71; PROJ LEADER, INSECT GENETICS & RADIATION BIOL SECT, METAB & RADIATION RES LAB, ENTOM RES DIV, AGR RES SERV, USDA, 71- Mem: AAAS; Genetics Soc Am; Radiation Res Soc; Entom Soc Am. Res: Insect genetics and radiation biology; genetic effect of chemical mutagens and chemical sterilization of insects; insect cytology and cytogenetic effects of radiation and chemicals; factors influencing chromosome aberrations and dominant lethal mutations induced by radiation and chemicals; insect reproduction. Mailing Add: Metab & Radiation Res Lab Agr Res Serv State Univ Sta Fargo ND 58102

LACHANCE, PAUL ALBERT, b St Johnsbury, Vt, June 5, 33; m 55; c 4. NUTRITION, FOOD SCIENCE. Educ: St Michael's Col, Vt, BSc, 55; Univ Ottawa, PhD(biol, nutrit), 60. Prof Exp: Res biologist, Aerospace Med Res Lab, Wright-Patterson AFB, Ohio, 60-63; coord flight food & nutrit, NASA Manned Spacecraft Ctr, 63-67; assoc prof nutrit, dir sch feeding effectiveness res proj, 69-72, PROF NUTRIT & FOOD SCI, RUTGERS UNIV, 72- Concurrent Pos: Lectr, Univ Dayton, 63. Mem: AAAS; Inst Food Technol; Am Inst Nutrit; NY Acad Sci; Am Dietetic Asn. Res: Aerospace food and nutrition; metabolic role of vitamin A; nutritional aspects of food processing; amino acid fortification and micronutrient nutrification; school food service. Mailing Add: 34 Taylor Rd RD 4 Princeton NJ 08540

LACHANCE, RENE ONESIME, b Quebec, Que, Mar 22, 09; m 38; c 4. PLANT PATHOLOGY. Educ: Laval Univ, BA, 32, BSA, 35; McGill Univ, MSc, 39, PhD(plant path), 40. Prof Exp: Plant pathologist sci serv lab, Can Dept Agr, 40-59, head plant path sect, Res Sta, 60-62; prof plant protection, Laval Univ, 62-71, dean, Fac Agr & Food Sci, 67-71, prof plant path, 71-76; RETIRED. Mem: Can Phytopath Soc; French-Can Asn Adv Sci. Res: Boron deficiency diseases; diseases of forage crops and of potatoes. Mailing Add: Fac of Agr & Food Sci Laval Univ Quebec PQ Can

LACHAPELLE, EDWARD RANDLE, b Tacoma, Wash, May 31, 26; m 50; c 1. METEOROLOGY. Educ: Col Puget Sound, BS, 49. Hon Degrees: ScD, Col Puget Sound, 67. Prof Exp: Sr scientist glaciol, 57-68, assoc prof geophys, 68-73, PROF GEOPHYS & ATMOSPHERIC SCI, UNIV WASH, 73- Concurrent Pos: Res assoc, Univ Colo, 73-75; vpres, Int Comn Snow & Ice. Mem: Am Geophys Union; Glaciol Soc (vpres). Res: Glaciology; snow physics; avalanche hazard forecasting and control. Mailing Add: Dept of Atmospheric Sci Univ of Wash Seattle WA 98105

LACHAPELLE, RENE CHARLES, b Joliette, Que, Jan 28, 30; US citizen; m 59; c 3. MEDICAL MICROBIOLOGY. Educ: Seminaire de Joliette, BA, 50; Univ Montreal, BSc, 53; Syracuse Univ, MS, 57, PhD(microbiol), 62. Prof Exp: Lab admin dir clin path, Syracuse Mem Hosp, 62-66; assoc prof biol, Univ Dayton, 66-74; CHAIRPERSON, DEPT MEDICAL MICROBIOLOGY, UNIV VT, 74- Mem: AAAS; Am Soc Med Tech; Am Soc Microbiol; Can Soc Microbiol. Res: Morphogenesis and serological properties of Candida albicans; monomine oxidase and serotonin in germfree animals; skin bacteria in long-term space flights; educational aspects of medical technology. Mailing Add: 302 Rowell Bldg Univ of Vt Burlington VT 05401

LACHAPELLE, BENOIT VINCENT, b Que, Jan 8, 30; m 53; c 3. APPLIED MATHEMATICS. Educ: Univ Montreal, BA, 51, BSc, 53. Prof Exp: Asst prof math, Univ Montreal, 56-64; APPL MATH CONSULT, SMA INC, 64- Mem: Am Math Soc; Math Asn Am; Can Math Cong. Res: Complex analysis; numbers theory. Mailing Add: SMA Inc 700 W Lagauchetiere Montreal PQ Can

LACHAT, LAWRENCE LYSLE, organic chemistry, see 12th edition

LACHENBRUCH, ARTHUR HEROLD, b New Rochelle, NY, Dec 7, 25; m 50; c 3. GEOPHYSICS. Educ: Johns Hopkins Univ, BA, 50; Harvard Univ, MA, 52, PhD(geophys), 58. Prof Exp: GEOPHYSICIST, US GEOL SURV, 51- Concurrent Pos: Vis prof, Dartmouth Col, 63. Honors & Awards: Kirk Bryan Award, Geol Soc Am, 63. Mem: Nat Acad Sci; fel AAAS; Am Geophys Union; fel Geol Soc Am; fel Arctic Inst NAm. Res: Solid earth geophysics. Mailing Add: Earthquake Res & Crustal Studies US Geol Surv 345 Middlefield Rd Menlo Park CA 94025

LACHENBRUCH, PETER ANTHONY, b Los Angeles, Calif, Feb 5, 37; m 62. BIOSTATISTICS. Educ: Univ Calif, Los Angeles, BA, 58, PhD(biostatist), 65; Lehigh Univ, MS, 61. Prof Exp: Asst math, Lehigh Univ, 58-59; programmer, Douglas Aircraft Co, 59-60; sr opers res analyst, Syst Develop Corp, 60-61; res scientist, Am Inst Res, 61-62; USPHS fel biostatist, Univ Calif, Los Angeles, 62-65; asst prof, 65-70, assoc prof, 70-75, PROF BIOSTATIST, UNIV NC, CHAPEL HILL, 75- Honors & Awards: Mortimer Spiegelman Gold Medal Award, Am Pub Health Asn, 71. Mem: AAAS; Am Statist Asn; Biomet Soc; Am Pub Health Asn; Royal Statist Soc. Res: Discriminant analysis; Monte Carlo methods; population dynamics; computer analysis of data. Mailing Add: Dept of Biostatist Univ of NC Chapel Hill NC 27515

LACHER, JOHN ROBERT, b Montrose, Colo, May 26, 11; m 31; c 2. PHYSICAL CHEMISTRY. Educ: Univ Colo, AB, 33; Harvard Univ, PhD(chem), 36. Prof Exp: Sheldon traveling fel, Cambridge Univ, 36-37; res asst, Harvard Univ, 37-38; instr chem, Brown Univ, 38-41; chemist, Mallinckrodt Chem Works, Mo, 41-44 & E I du Pont de Nemours & Co, 44-45; from asst prof to prof, 45-74, EMER PROF CHEM, UNIV COLO, BOULDER, 74- Mem: Am Chem Soc. Res: Photochemistry; thermochemistry; chemical kinetics. Mailing Add: Dept of Chem Univ of Colo Boulder CO 80304

LACHMAN, LEON, b Bronx, NY, Jan 29, 29; m 51; c 1. PHARMACY. Educ: Columbia Univ, BSc, 51, MSc, 53; Univ Wis, PhD, 56. Prof Exp: Asst dir pharm, Res & Develop Div, Ciba Pharmaceut Co, NJ, 56-68, dir, 68-71; VPRES DEVELOP & CONTROL, ENDO LABS INC, 71- Concurrent Pos: Vis scientist, Am Asn Cols Pharm; mem bd trustees, Col Pharm, Columbia Univ, 74- Honors & Awards: Indust Pharmaceut Technol Award, Acad Pharmaceut Sci, 70. Mem: Fel Acad Pharmaceut Sci. Res: Process and equipment design; research and development of pharmaceutical dosage forms; analytical research; quality control practices. Mailing Add: Endo Labs Inc 1000 Stewart Ave Garden City NY 11530

LACHMAN, WILLIAM HENRY, JR, b Ft Washington, Pa, Apr 14, 12; m 41; c 2.

PLANT BREEDING. Educ: Pa State Col, BS, 34, MS, 36. Prof Exp: From instr to prof veg crops, 36-72, PROF PLANT SCI, UNIV MASS, AMHERST, 72- Honors & Awards: Vaughan Mem Award, 47; All-Am Selections Award, 55. Mem: Am Soc Hort Sci. Res: Breeding in vegetable crops. Mailing Add: Dept of Plant & Soil Sci Univ of Mass Amherst MA 01002

LACHMANN, ALFRED, b Ger, Apr 14, 15; nat US; m 46. PHYSICAL CHEMISTRY. Educ: Univ Geneva, ChE, 40, PhD(phys chem), 42. Prof Exp: Chemist, Fedn Migros Coops, Switz, 42-46; asst specialist, Dairy Indust Div, Univ Calif, 47-51; sr chemist, Stein, Hall & Co, Inc, 51-54; div head food prod develop, Am Sugar & Refining Co, 54-58; consult chemist, 59-64; asst tech dir, SuCrest Corp, 64-68; consult chemist & sr assoc, S M Cantor Assocs, 69-72; proj dir, UN Indust Develop Orgn, 72-75; TECHNOLOGIST, ENVIRON RES SERV, USDA, WASHINGTON, DC, 75- Mem: Am Chem Soc; Am Dairy Sci Asn; Am Inst Chem; Inst Food Technol; Soc Indust Chem. Res: Carbohydrate technology, proteins, fats, food ingredients, additives and their effect on nutrition and quality of foods; evaluation of food systems in developing and developed countries; domestic and foreign food legislation. Mailing Add: Washington DC

LACHNER, ERNEST ALBERT, b New Castle, Pa, Apr 3, 15; m 39; c 4. SYSTEMATIC ICHTHYOLOGY. Educ: Pa State Teachers Col, Slippery Rock, BS, 37; Cornell Univ, PhD(ichthyol), 46. Prof Exp: Teacher high sch, Pa, 37-39; asst zool, Cornell Univ, 40-42; assoc prof fishery biol, Pa State Col, 47-49; assoc cur, 49-65, cur in charge, 65-66, SUPVR & CUR, DIV FISHES, NAT MUS NATURAL HIST, 66- Concurrent Pos: Guggenheim fel, 56 & 59. Mem: AAAS; Am Soc Ichthyol & Herpet (vpres, 54, pres-elect, 66, pres, 67); Am Fisheries Soc; Am Soc Limnol & Oceanog; Biol Soc Washington (vpres, 66, pres, 67). Res: Systematics and morphology of marine and fresh water fishes; ecology; life history of fishes. Mailing Add: Div of Fishes Nat Mus of Natural Hist Washington DC 20560

LACK, LEON, b New York, NY, Jan 7, 22; m 48; c 3. BIOCHEMISTRY. Educ: Brooklyn Col, AB, 43; Mich State Univ, MS, 48; Columbia Univ, PhD(biochem), 53. Prof Exp: Fel, Duke Univ, 53-55; from instr to asst prof pharmacol, Sch Med, Johns Hopkins Univ, 55-64; from asst prof to assoc prof, 65-71, PROF PHARMACOL, MED CTR, DUKE UNIV, 71-, CHIEF LAB MACROMOLECULAR PHARMACOL, 65- Mem: Am Soc Biol Chemists. Res: Metabolism of aromatic substances; intestinal active transport. Mailing Add: Lab of Molecular Pharmacol Duke Univ Med Ctr Durham NC 27706

LACKEY, HOMER BAIRD, b Freewater, Ore, Nov 23, 20; m 42; c 3. APPLIED CHEMISTRY. Educ: Ore State Univ, BS, 47, MS, 48. Prof Exp: Asst, Ore State Univ, 47-48; res chemist, Cent Res Dept, 48-55, supvr prod res, Chem Prod Div, 55-68, MGR PROD RES, CHEM PROD DIV, CROWN ZELLERBACH CORP, 68- Mem: Am Chem Soc; Am Soc Test & Mat. Res: Forest byproduct utilization. Mailing Add: Chem Prod Div Crown Zellerbach Corp Camas WA 98607

LACKEY, JAMES ALDEN, b Glens Falls, NY, Nov 25, 38; m 61; c 2. MAMMALOGY. Educ: Cornell Univ, BS, 61; Calif State Univ, San Diego, MA, 67; Univ Mich, PhD(zool), 73. Prof Exp: ASST PROF ZOOL, NY STATE UNIV COL, OSWEGO, 73- Mem: Ecol Soc Am; Am Soc Mammalogists; Soc Study of Evolution; Sigma Xi; AAAS. Res: Reproduction, growth, development and population ecology of mammals. Mailing Add: Dept of Zool State Univ Col Oswego NY 13126

LACKEY, LAURENCE, US citizen. GEOMORPHOLOGY, QUATERNARY GEOLOGY. Educ: Principia Col, BS, 69; Univ Mich, PhD(geol), 74. Prof Exp: Asst prof geol, Mich State Univ, 74-75; ASST PROF GEOL, MEMPHIS STATE UNIV, 75- Mem: Geol Soc Am; Am Quaternary Asn; Nat Asn Geol Teachers. Res: Slope processes and quaternary landform development. Mailing Add: Dept of Geol Memphis State Univ Memphis TN 38111

LACKEY, ROBERT T, b Kamloops, BC, May 18, 44; m 67; c 2. FISHERIES MANAGEMENT. Educ: Humboldt State Univ, BS, 67; Univ Maine, Orono, MS, 68; Colo State Univ, PhD(fisheries), 71. Prof Exp: Asst prof, 71-73, ASSOC PROF FISHERIES, VA POLYTECH INST & STATE UNIV, 73- Concurrent Pos: Fels, Off Econ Opportunity & Celanese Corp, Va Polytech Inst & State Univ, 71-, Off Water Resources Res, 72-; consult, Brandermill Corp, 74- & US Army Corps of Eng, 75- Mem: Am Fisheries Soc; Am Soc Limnol & Oceanog; Fisheries Soc Brit Isles; Wildlife Soc. Res: Fisheries science, including structure and management of aquatic renewable natural resources; systems analysis. Mailing Add: Dept of Fisheries & Wildlife Sci Va Polytech Inst & State Univ Blacksburg VA 24061

LACKMAN, DAVID BUELL, b Plymouth, Conn, Dec 29, 11; m 47; c 4. BACTERIOLOGY, PUBLIC HEALTH. Educ: Univ Conn, BS, 33; Univ Pa, PhD(med bact), 37; Am Bd Med Microbiol, dipl pub health & med lab immunol. Prof Exp: Asst instr bact, Med Sch, Univ Pa, 33-37; instr, 37-39, assoc, 39-41; asst bacteriologist, USPHS, 41-46, from sr scientist to sci dir, Rocky Mountain Lab, Nat Inst Allergy & Infectious Dis, Mont, 46-67, dir microbiol lab, 67-71; ADMINR LABS DIV, MONT STATE DEPT HEALTH & ENVIRON SCI, 71- Concurrent Pos: Vis lectr, Mont State Univ, 52- Mem: Am Soc Microbiol; Asn Mil Surg US; fel Am Pub Health Asn; Am Asn Immunol; fel Am Acad Microbiol. Res: Serology of rickettsial and viral diseases; antigenic structure of bacteria and rickettsiae. Mailing Add: Labs Div Mont State Dept Health & Environ Sci Helena MT 59601

LACKNER, HENRIETTE, b Vienna, Austria, Feb 27, 22; US citizen; m 49; c 3. HEMATOLOGY. Educ: Univ Leeds, MB & ChB, 45, MD, 48. Prof Exp: Jr lectr med, Univ Cape Town, 55-62; res assoc, 63-65, instr med, 65-67, from asst prof clin med to asst prof med, 67-75, ASSOC PROF CLIN MED, SCH MED, NY UNIV, 75- Concurrent Pos: Res asst, Groote Schuur Hosp, Cape Town, SAfrica, 55-62, asst physician, Arthritis Clin, 55-56, physician-in-chg & asst physician med outpatient clin, 56-62; res scientist, Am Nat Red Cross, 63-; clin asst vis physician, Bellevue Hosp, New York, 65-75, assoc vis physician, 75-; asst, Univ Hosp, 66-75, assoc med, 75- Mem: Soc Study Blood; Am Soc Hemat; Med Soc Ny. Res: Blood coagulation disorders and pathological fibrinolysis. Mailing Add: Dept of Med NY Univ Med Ctr New York NY 10016

LACKO, ANDRAS GYORGY, b Budapest, Hungary, Nov 10, 36; Can citizen; m 64; c 3. BIOCHEMISTRY, MICROBIOLOGY. Educ: Univ BC, BSA, 61, MSc, 63; Univ Wash, PhD(biochem), 68. Prof Exp: Res asst biochem, Univ Wash, 63-68; asst mem, Albert Einstein Med Ctr, 69-71; mem staff, 71-72, asst prof med, Med Sch, Temple Univ, 72-75; ASSOC PROF BIOCHEM, NORTH TEX STATE UNIV, 75- Concurrent Pos: NIH fel, Albert Einstein Col Med, 68-69. Mem: Can Biochem Soc. Res: Structure and function of enzymes and other proteins. Mailing Add: Dept Biochem NTex State Univ Denton TX 76203

LACKS, SANFORD, b New York, NY, Jan 28, 34; m 59; c 3. GENETICS, BIOCHEMISTRY. Educ: Union Univ, NY, BS, 55; Rockefeller Inst, PhD, 60. Prof Exp: Instr biol, Harvard Univ, 60-61; from asst to assoc geneticist, 61-67, GENETICIST, BROOKHAVEN NAT LAB, 67- Mem: Am Soc Microbiol; Am Soc Biol Chem; Genetics Soc Am. Res: Microbial genetics; mechanism of bacterial transformation; function of enzymes acting on nucleic acids. Mailing Add: Biol Dept Brookhaven Nat Lab Upton NY 11973

LA COMBE, EDWARD MARTIN, chemistry, see 12th edition

LACOSTE, RENE JOHN, b New York, NY, Feb 19, 27. ANALYTICAL CHEMISTRY, PESTICIDE CHEMISTRY. Educ: Rensselaer Polytech Inst, BS, 50; Univ Chicago, MS, 53. Prof Exp: Chemist, Am Dent Asn, 50-53; chemist, 53-68, sr chemist, 68-69, INT REGISTR AGR & SANIT CHEM, ROHM AND HAAS CO, 69- Mem: AAAS; Am Chem Soc; Am Inst Chem; NY Acad Sci. Res: Physical and chemical methods of analysis; electrochemical analysis; separations of organic mixtures. Mailing Add: Int Div Independence Mall W Rohm and Haas Co Philadelphia PA 19105

LACOSTE, ROGER G, organic chemistry, see 12th edition

LACOUNT, ROBERT BRUCE, b Martinsburg, WVa, Sept 16, 35; m 64; c 2. ORGANIC CHEMISTRY. Educ: Shepherd Col, BS, 57; Univ Pittsburgh, MLitt, 62, PhD(org chem), 65. Prof Exp: Res assoc fundamental org chem, Mellon Inst, 58-65; from asst prof chem to prof & chmn dept chem, 65-71, PROF & CHMN DEPT CHEM & PHYSICS, WAYNESBURG COL, 71- Concurrent Pos: Res grant, Petrol Res Fund, 65-67; res chemist, US Bur Mines, 70-75 & Energy Res & Develop Admin, 75- Mem: Am Chem Soc. Res: Synthetic organic chemistry; organic sulfur chemistry; production of low-sulfur fuels from coal. Mailing Add: Waynesburg Col Waynesburg PA 15370

LACROIX, GUY, b Que, Apr 10, 30; m 60; c 2. MARINE ECOLOGY. Educ: Laval Univ, BA, 52, LPh, 53; Univ Montreal, BSc, 57, MSc, 59, DSc, 68. Prof Exp: Zooplanktonologist, Grande Riviere Marine Biol Sta, Que, 58-68; asst prof, 68-71, assoc prof, 71-74, PROF MARINE ECOL, DEPT BIOL, LAVAL UNIV, 74- Concurrent Pos: Secy ed bd, Can Naturalist, 68-; exec secy, Interuniv Group Oceanog Res, Que, 70-; mem bd, Laval Univ, 74- & Sci Comt Oceanic Res, Can Nat Comt, 74- Mem: AAAS; Am Soc Limnol & Oceanog; Plankton Soc Japan. Res: Zooplankton; invertebrate zoology; primary production; marine invertebrates. Mailing Add: Dept of Biol Laval Univ Ste-Foy Quebec PQ Can

LACROIX, JOSEPH DONALD, b Windsor, Ont, Apr 7, 25; nat US; m 51; c 3. BOTANY. Educ: Univ Western Ont, BA, 47; Univ Detroit, MS, 50; Purdue Univ, PhD(bot), 53. Prof Exp: Res asst mycol, Parke, Davis & Co, 51; from instr to assoc prof, 53-74, PROF BIOL, UNIV DETROIT, 74- Concurrent Pos: Kellogg fel, Univ Mich, 71-72. Mem: Bot Soc Am; Am Inst Biol Sci. Res: Scanning electron microscopy and electron probe analysis of silicification patterns in plant species; effects of gravity on plant tissue; radiosensitivity of higher plants. Mailing Add: Dept of Biol Univ of Detroit Detroit MI 48221

LACROIX LUCIEN JOSEPH, b St Louis, Sask, May 14, 29; m 52; c 5. PLANT PHYSIOLOGY. Educ: Univ Sask, BSA, 57, MSc, 58; Iowa State Univ, PhD(plant physiol), 61. Prof Exp: Technician field husb, Univ Sask, 54-57; res assoc plant sci, 61-63, from asst to assoc prof, 63-72, PROF PLANT SCI, UNIV MAN, 72- Mem: Am Soc Plant Physiol; Can Soc Plant Physiol. Res: Biochemical and cytological studies of embryo growth and development and seed dormancy; mineral nutrition of plants; winter hardiness. Mailing Add: Dept of Plant Sci Univ of Man Winnipeg MB Can

LACROIX, NORBERT HECTOR JOSEPH, b Sarsfield, Ont, Oct 26, 40; m 65; c 3. MATHEMATICS. Educ: Univ Ottawa, BSc, 62; Univ Notre Dame, PhD(math), 66. Prof Exp: Instr math, Univ Notre Dame, 62-66; asst prof, 66-70, ASSOC PROF MATH & CHMN DEPT, LAVAL UNIV, 70- Mem: Can Math Cong; Am Math Soc. Res: Classical groups and number theory. Mailing Add: Dept of Math Laval Univ Quebec PQ Can

LACY, ANN MATTHEWS, b Boston, Mass, May 29, 32. MICROBIAL GENETICS. Educ: Wellesley Col, BA, 53; Yale Univ, MS, 56, PhD(microbiol), 59. Prof Exp: Asst dept genetics, Carnegie Inst, 53-54; instr genetics, 59-61, asst prof, 61-67, assoc prof, 67-73, chmn dept biol sci, 69-72, PROF BIOL SCI, GOUCHER COL, 73- Concurrent Pos: Res fel, Glasgow Univ, 68-69. Mem: AAAS; Genetics Soc Am; Bot Soc Am; Am Inst Biol Sci; Sigma Xi. Res: Gene structure and function, and gene regulation in Neurospora crassa. Mailing Add: Dept of Biol Sci Goucher Col Towson MD 21204

LACY, JULIA CAROLINE, b Detroit, Mich, July 10, 46. ENVIRONMENTAL BIOLOGY. Educ: Univ Mich, BA, 68, MS, 72. Prof Exp: Chief ecologist environ biol, Stearns Roger, Inc, 72-74; SR SCIENTIST BIOL ENVIRON BIOL, RADIAN CORP, 74- Mem: Ecol Soc Am; Am Inst Biol Sci; World Future Soc; Sigma Xi. Res: Forecasting regional scale impacts on biota from western energy development; allelopathic interactions in central Texas old-field succession; trace element pathways in the biosphere. Mailing Add: Radian Corp PO Box 9948 Austin TX 78766

LACY, MELVYN LEROY, b Henry, Nebr, Oct 24, 31; m 54; c 2. PLANT PATHOLOGY, SOIL MICROBIOLOGY. Educ: Univ Wyo, BS, 59, MS, 61; Ore State Univ, PhD(plant path), 64. Prof Exp: Asst prof, 65-71, ASSOC PROF PLANT PATH, MICH STATE UNIV, 71- Mem: Am Phytopath Soc. Res: Soil-borne fungus diseases, survival of pathogens, inoculum potential and rhizosphere effects; soil fumigants and pesticides for disease control; epidemiology and disease management. Mailing Add: Dept of Bot & Plant Path Mich State Univ East Lansing MI 48823

LACY, PATRICIA, b New York, NY, Apr 21, 33. MICROBIOLOGY, GENETICS. Educ: Marymount Col, Calif, BA, 56; Univ Calif, Los Angeles, MA, 60; St Louis Univ, PhD(microbiol), 66. Prof Exp: Teacher Latin & math, Acad Sacred Heart of Mary, NY, 52-55, Sacred Heart of Mary High Sch, 55-56 & Corvallis High Sch, 56-58; from instr to asst prof biol, Marymount Col, Calif, 60-66; asst prof, Loyola Univ, Los Angeles, 66-75; MEM FAC STAFF, DEPT BIOL SCI, FORDHAM UNIV, 75- Mem: AAAS; Am Soc Microbiol; Am Soc Zoologists; Am Inst Biol Sci; Nat Asn Biol Teachers. Res: Animal virology concerned with the genetic relatedness among tumorigenic and non-tumorigenic human adenoviruses. Mailing Add: Dept Biol Sci Fordham Univ Rosehill Bronx NY 10458

LACY, PAUL ESTON, b Trinway, Ohio, Feb 7, 24; m 45; c 2. PATHOLOGY. Educ: Ohio State Univ, BA, 45, MSc & MD, 48; Univ Minn, PhD(path), 55. Prof Exp: Asst instr anat, Ohio State Univ, 44-48; intern, White Cross Hosp, Columbus, Ohio, 48-49; Nat Cancer Inst fel, Med Sch, Washington Univ, 55-56; from instr to assoc prof, 56-61, asst dean, 59-61, MALLINCKRODT PROF PATH & CHMN DEPT, MED SCH, WASHINGTON UNIV, 61- Concurrent Pos: Mem path B study sect, NIH, 61-66, chmn, 66-67; Banting mem lectr, Brit Diabetic Asn, 63; Eliott Proctor Joslin mem lectr, 66; mem adv comt res personnel, Am Cancer Soc, 66-70; mem basic sci adv comt, Nat Cystic Fibrosis Res Found, 67-69; Ninth Richard M Jaffe lectr, 69; Banting mem lectr & Rollin Turner Woodyatt mem lectr, 70; assoc ed, Diabetes, 73-; mem nat comn diabetes, NIH, 74-75 & mem nat adv environ health sci coun, 74-77. Honors & Awards: Mayo Found Achievement Award, 64; Banting Award, 70. Mem: Am Asn Anat; Am Soc Exp Path; Am Asn Path & Bact; Am Diabetes Asn; assoc mem Royal Soc Med. Res: Endocrine pathology; experimental diabetes. Mailing Add: Dept of Path Washington Univ Med Sch St Louis MO 63110

LACY, ROBERT M, b Coalgate, Okla, Aug 27, 13; m 38; c 2. SURFACE CHEMISTRY. Educ: Austin Col, AB, 36, AM, 37; Univ Okla, PhD(org chem), 43. Prof Exp: Lab asst chem, Austin Col, 36-37; lab instr, Univ Okla, 37-42; engr works lab, Gen Elec Co, 42-44, head chem sect, 44-46, tech asst to mgr works lab, 46-48, mgr labs, 48-53; tech dir, Mich Chrome & Chem Co, 53-58; mem staff, Adv Eng & Major Appliance Lab, Gen Elec Co, Ky, 59-65; tech mgr, Parker Rust Proof Div, Hooker Chem Corp, Mich, 65-71; VPRES RES & DEVELOP, THE PARKER CO, OXY METAL INDUSTRIES CORP, 71- Mem: AAAS; Am Chem Soc; Am Soc Test & Mat; Soc Mfg Eng; NY Acad Sci. Res: Organic and chemical conversion coatings; plating; chemical engineering; waste control. Mailing Add: The Parker Co 32100 Stephenson Hwy Madison Heights MI 48071

LACY, WILLIAM WHITE, b Atlanta, Ga, Sept 18, 23; m 53; c 5. MEDICINE. Educ: Davidson Col, BS, 47; Harvard Med Sch, MD, 51. Prof Exp: Intern med, Sch Med, Johns Hopkins Univ, 51-52; asst resident, Duke Univ Hosp, 52-53 & Vanderbilt Univ Hosp, 53-54; from instr to asst prof, 54-73, ASSOC PROF MED, SCH MED, VANDERBILT UNIV, 73- Concurrent res: Am Heart Asn res fel, 57-61, estab investr, 61- Res: Metabolism in patients with cardiovascular and renal disease. Mailing Add: Dept of Med Vanderbilt Univ Sch of Med Nashville TN 37203

LAD, ROBERT AUGUSTIN, b Chicago, Ill, May 8, 19; m 44; c 9. CHEMISTRY. Educ: Univ Chicago, SB, 39, SM, 41, PhD(inorg chem), 46. Prof Exp: Asst, Nat Defense Res Comt, Univ Chicago, 42-46; aeronaut res scientist, Nat Adv Comt Aeronaut, 46-59, CHIEF MAT SCI BR, LEWIS RES CTR, NASA, 59- Concurrent Pos: Mem solid state sci panel, Nat Acad Sci-Nat Res Coun, 63- Mem: AAAS; Am Phys Soc; fel Am Inst Chem. Res: Physics and chemistry of surfaces; radiation chemistry; solid state physics. Mailing Add: Lewis Res Ctr NASA 21000 Brookpark Rd Cleveland OH 44135

LADA, ARNOLD, b New York, NY, May 26, 26; m 47; c 3. BIOCHEMISTRY, ORGANIC CHEMISTRY. Educ: Brooklyn Col, BS, 47; Georgetown Univ, MS, 51, PhD, 53. Prof Exp: Chemist, Glyco Prod Co, 47-48; biochemist, NIH, 49-50; res chemist, Food & Drug Admin, 50-54; biochemist, Toni Co Div, Gillette Co, 54-57; chief tech servs, Onyx Oil & Chem Co, 57-60, chief tech servs, Onyx Chem Corp, 60-65, GEN MGR, ONYX CHEM CO DIV, MILLMASTER ONYX CORP, 65-, PRES, 74- Mem: AAAS; Am Chem Soc; Soc Cosmetic Chem. Res: Surface active agents; industrial applications; antimicrobial agents. Mailing Add: Onyx Chem Corp Div Millmaster Onyx Corp 190 Warren Jersey City NJ 07302

LADANYI, BRANKA MARIA, b Zagreb, Yugoslavia, Sept 7, 47; Can citizen; m 74. THEORETICAL CHEMISTRY. Educ: McGill Univ, BSc, 69; Yale Univ, MPhil, 71, PhD(chem), 73. Prof Exp: Vis asst prof chem, Univ Ill, Urbana, 74; RES ASSOC CHEM, YALE UNIV, 74- Mem: Am Chem Soc; Am Phys Soc. Res: Statistical mechanics of fluids; structure of molecular liquids; propagation and scattering of light in fluids; statistical mechanics of polymer solutions. Mailing Add: Dept of Chem Yale Univ New Haven CT 06520

LADAS, GERASIMOS E, b Cefalonia, Greece, Apr 25, 37; m; c 2. MATHEMATICAL ANALYSIS. Educ: Nat Univ Athens, BS, 61; MS, NY Univ, 66, PhD(math), 68. Prof Exp: Fel, NY Univ, 64-68; asst prof math, Fairfield Univ, 68-69; asst prof, 69-72, assoc prof, 72-75, PROF MATH, UNIV RI, 75-, CHMN DEPT, 72- Mem: Am Math Soc; Greek Math Soc; Am Asn Univ Prof. Res: Ordinary, functional and abstract differential equations. Mailing Add: Dept of Math Univ of RI Kingston RI 02881

LADD, ANTHONY THORNTON, b New York, NY, Aug 25, 20; m 56; c 2. INTERNAL MEDICINE. Educ: Cornell Univ, MD, 45. Prof Exp: Asst path, Med Col, Vanderbilt Univ, 45; from asst to instr, Med Col, Cornell Univ, 47-49; from intern to asst resident med, Presby Hosp, New York, 49-52; asst resident, Mary Imogene Bassett Hosp, Cooperstown, NY, 51; from instr to assoc prof med, State Univ NY Upstate Med Ctr, 52-70; PROF MED, MED UNIV SC, 70-; CHIEF SERV & CHIEF STAFF, COUNTY HOSP, 70- Concurrent Pos: Chief hemat sect, Vet Admin Hosp, Syracuse, NY, 59-60, chief med serv, 60-70. Res: Experimental atherosclerosis; unsaturated fats; hematology; genetically controlled hemoglobin abnormalities. Mailing Add: County Hosp 326 Calhoun St Charleston SC 29401

LADD, HARRY STEPHEN, b St Louis, Mo, Jan 1, 99; m 34; c 2. GEOLOGY. Educ: Washington Univ, St Louis, AB, 22; Univ Iowa, MS, 24, PhD(geol), 25. Prof Exp: Asst geol, Univ Iowa, 22-25; asst prof geol, Univ Va, 26-29; paleontologist, Venezuela Gulf Oil Co, 29-31; pvt res, US Nat Mus, 31-33; res assoc geol, Univ Rochester, 33-35; assoc geologist, Nat Park Serv, 35-38, geologist, 38-40; assoc geologist, US Geol Surv, 40-41, geologist, 41-43, sr geologist, 44-45, prin geologist, Washington, DC, 45-69; RES ASSOC, US NAT MUS, SMITHSONIAN INST, 69- Concurrent Pos: Bishop Mus fel, Yale Univ, 25 & 28; mem staff, Oper Crossroads. Honors & Awards: Distinguished Serv Award, Dept Interior, 65. Mem: AAAS (vpres, 65); Paleont Soc (pres, 54); fel Geol Soc Am (vpres, 55); Am Asn Petrol Geol. Res: Tertiary paleontology and stratigraphy; geology of Pacific Islands; origin of coral reefs. Mailing Add: US Nat Mus Smithsonian Inst Washington DC 20560

LADD, JOHN HERBERT, b Kewanee, Ill, Sept 6, 18; m 39; c 3. PHYSICAL CHEMISTRY, PHYSICS. Educ: Univ Ill, BS, 40, MS, 42, PhD(phys chem), 47. Prof Exp: Asst phys chem, Univ Ill, 40-42; sr chemist, 47-48, develop engr, 48, sr develop engr, 48-53, tech assoc, 53-57, sr develop proj engr, 57-69, RES ASSOC, EASTMAN KODAK CO, 69- Concurrent Pos: Instr, Univ Rochester, 49-51. Mem: Inst Elec & Electronics Eng; Soc Photog Sci & Eng; Am Phys Soc. Res: Electron emission from metals; color photographic printers; color films in television; television test charts and standards; analog computers; digital equipment instrumentation, optical testing; digital hardward; photosensors; laser applications. Mailing Add: Physics Div Res Lab Bldg 81 Eastman Kodak Co Kodak Park Rochester NY 14650

LADD, JOHN ROBERT, organic chemistry, see 12th edition

LADD, THYRIL LEONE, JR, b Albany, NY, Oct 10, 31; m 56; c 3. ENTOMOLOGY. Educ: State Univ NY, Albany, AB, 56, MA, 58; Cornell Univ, PhD(entom), 63. Prof Exp: Res entomologist, 62-69, RES LEADER, JAPANESE BEETLE RES LAB, AGR RES SERV, USDA, 69- Mem: AAAS; Entom Soc Am; Am Entom Soc; Sigma Xi. Res: Effects of radiation and chemicals on insect reproduction; integrated insect control; insect responses to attractants and repellents; effects of insect feeding on plant yields. Mailing Add: Japanese Beetle Res Lab USDA Ohio Agr Res & Develop Ctr Wooster OH 44691

LADD, WILLIAM ALEXANDER, b Brantford, Ont, May 8, 17; nat US; m 40. ELECTRONICS, MICROSCOPY. Educ: Univ Toronto, BA, 39, MA, 40. Prof Exp: Asst, Nat Res Coun Can, 39-40; res physicist, Columbian Carbon Co, 41-55; pres, 55-63, VPRES & DIR RES, LADD RES INDUSTS, INC, 63- Mem: Am Phys Soc; Am Chem Soc; Electron Micros Soc Am. Res: Design and construction of electron microscope accessories. Mailing Add: Ladd Res Industs Inc PO Box 901 Burlington VT 05401

LADELL, JOSHUA, solid state physics, see 12th edition

LADEN, HYMAN NATHANIEL, b Philadelphia, Pa, July 26, 15; m 39; c 3. MATHEMATICS, ELECTRONICS. Educ: Univ Pa, BA, 36, PhD(math), 41; Univ Md, MA, 38; US Naval Postgrad Sch, MS, 49. Prof Exp: Asst instr math, Univ Md, 36-38; instr, Drexel Inst, 38-39; asst instr, Univ Pa, 40-41; instr, Univ Wis, 41-42; chief comput systs develop, staff of vpres finance, Chesapeake & Ohio Rwy Co, 54-57, chief new systs develop, 57-63; dir data systs, 64-65, dir res serv, 65-68, ASST VPRES RES, CHESAPEAKE & OHIO RWY CO-BALTIMORE & OHIO RR CO, 68- Concurrent Pos: Asst, comt bibliog orthogonal polynomials, div phys sci, Nat Res Coun, 36-40; mem bd dirs, Nat Comput Anal, 61-63, Belmont Nat Corp, 69- & Del Capital Group, 70-; lectr, ord eng mgt training prog, US Army, Cornell Univ, Am Univ & Case Inst Technol; consult, mgt systs, opers res, electronic data processing & transportation. Mem: AAAS; Am Math Soc; Math Asn Am; Soc Indust & Appl Math; Opers Res Soc Am. Res: Administration of transportation research and development, systems planning and engineering; economics of design, procurement, scheduling, use and maintenance; advanced management information and control systems; operations research and analytical management decisions; computer simulation models of transportation operations. Mailing Add: PO Box 67 Brooklandville MD 21022

LADEN, KARL, b Brooklyn, NY, Aug 10, 32; m 56; c 5. BIOCHEMISTRY. Educ: Univ Akton, BS, 54; Northwestern Univ, PhD(chem), 57. Prof Exp: Asst, Northwestern Univ, 54-57; res chemist, William Wrigley Jr Co, 57-58; res biochemist, Toni Co Div, 59-60, res supvr, 60-62, mgr biol res, 62-64, asst lab dir, Gillette Med Res Inst, 64-67, mgr biomed sci dept, Gillette Res Inst, 67-68, vpres biomed sci, 68-71, PRES, GILLETTE RES INST, GILLETTE CO, 71- Concurrent Pos: Consult, Indust Bio-Test Labs, Inc, 55-57; lectr, Northwestern Univ, 58-64 & Am Univ, 64-65. Mem: Am Chem Soc; Soc Invest Dermat; Soc Cosmetic Chem (ed, J, 67-71). Res: Biochemistry and physiology of skin and hair. Mailing Add: Gillette Res Inst 1413 Res Blvd Rockville MD 20850

LADENBURG, KURT, chemistry, see 12th edition

LADENDORF, AGNES J, b Moorhead, Minn, May 1, 07; m 52; c 3. MATHEMATICS. Educ: NDak State Univ, BS, 32; Northwestern Univ, MA, 39. Prof Exp: Teacher pub schs, Minn, 27-28, supvr, 28-30, prin & instr, 32-38; assoc prof & head dept, 40-73, EMER PROF MATH, MINOT STATE COL, 73- Mem: Math Asn Am; NY Acad Sci; Nat Coun Teachers Math. Res: Mathematical analysis; calculus. Mailing Add: 1101 Valley View Dr Minot ND 58701

LADENHEIM, HARRY, b Vienna, Austria, Oct 17, 32; US citizen; m 55; c 2. PHYSICAL ORGANIC CHEMISTRY. Educ: City Col New York, BS, 54; Polytech Inst Brooklyn, PhD(org polymer chem), 58. Prof Exp: Fel chem, Ill Inst Technol, 58-59; res chemist, Esso Res & Eng Co, 59-63; res chemist, 63-70, SR RES CHEMIST, HOUDRY PROCESS & CHEM CO, MARCUS HOOK, 70- Mem: Am Chem Soc. Res: Mechanisms in physical organic chemistry; exploratory and process study in petroleum technology; use of polymers as enzyme models; catalysis; catalytic chemistry. Mailing Add: 2276 N 51 St Philadelphia PA 19131

LADERMAN, JACK, b New York, NY, Jan 6, 14; m 47; c 1. MATHEMATICS. Educ: City Col New York, BS, 34; Columbia Univ, MA, 35, PhD(math statist), 53. Prof Exp: Statistician, Fed Commun Comn, 37; analyst, Div Unemployment Ins, State Dept Labor, NY, 38; mathematician, Comput Lab, Nat Bur Standards, 39-40; statistician, Chem Warfare Serv, 40-46; mathematician, Comput Lab, Nat Bur Standards, 46-49; res scientist, Columbia Univ, 49-51; mathematician, Off Naval Res, Washington, DC, 51-54, mathematician & dep chief scientist, NY Area Off, 54-74; RETIRED. Concurrent Pos: Teacher, War Emergency Training Prog, NY, 42-43; adj assoc prof, NY Univ, 57-64; tech dir, Serv Bur Corp, 62-63. Honors & Awards: Superior Civilian Serv Award, Dept of Navy, 74. Mem: Fel AAAS; Am Math Soc; Biomet Soc; Am Statist Asn; Inst Math Statist. Res: Mathematical computation; statistics; mathematical economics; probability; algebra; operations research. Mailing Add: 2630 Kingsbridge Terr Bronx NY 10463

LADINSKY, HERBERT, b New York, NY, July 22, 35; m 67. PHARMACOLOGY. Educ: City Col New York, BS, 58; State Univ NY, PhD(pharmacol), 66. Prof Exp: Fel pharmacol, Royal Caroline Inst, Stockholm, 66-67 & Mario Negri Inst, 67-69; NIH fel, 66-68; LAB CHIEF, MARIO NEGRI INST, 69- Mem: Ital Pharmacol Soc. Res: Physiology and pharmacology of the cholinergic system. Mailing Add: Inst for Res in Pharmacol Mario Negri Inst Via Eritrea 62 Milan Italy

LADINSKY, JUDITH L, b Los Angeles, Calif, June 16, 38; m 61; c 2. CYTOLOGY, ENDOCRINOLOGY. Educ: Univ Mich, BS, 61; Univ Wis-Madison, MS, 64, PhD(reprod physiol), 68. Prof Exp: Med technologist, Clin Labs, St Mary's Hosp, Mich, 55-56; res asst, Dept Neuropath, Univ Mich, 56-58, Dept Anat, 58-60 & Dept Surg, 60-61; proj assoc, Dept Gynec-Obstet, 61-68, from instr to asst prof prev med, 68-74, ASSOC PROF PREV MED, SCH MED, UNIV WIS-MADISON, 75- Mem: Am Soc Cell Biol; NY Acad Sci; Tissue Cult Asn; Am Pub Health Asn; Asn Teachers Prev Med. Res: Cell kinetics of normal and neoplastic tissues; automated methods of cancer detection; endocrinology of tumors; community medicine; neonatology; health care delivery. Mailing Add: Dept of Prev Med Univ of Wis Med Sch Madison WI 53706

LADISCH, ROLF KARL, b Meissen, Ger, Jan 29, 14; US citizen; m 45; c 3. BIOCHEMISTRY, ELECTROCHEMISTRY. Educ: Johns Hopkins Univ, MA, 40; Dresden Tech Univ, PhD(chem), 41. Prof Exp: Plant mgr & head dept prod control, Inst Testing Mat, Wanderer Works, Ger, 41-45; asst res, Inst Indust Res, Fuerth, Ger, 46-47; chief spec activ lab, Pioneering Div, US Qm Corps, 48-55; consult chemist, Res & Develop, Lansdowne, Pa, 55-66, CONSULT CHEMIST, RES & DEVELOP, MALVERN, 66- Concurrent Pos: Res assoc, Cancer Res Unit, Immaculata Col, Pa, 61-69. Mem: Fel AAAS; Am Chem Soc; fel Am Inst Chem; NY Acad Sci. Res: Insect-infested food; cancer; synthesized tetritoxin; medical electro-analytical instruments; methods of production of oxidation catalysts and microfibers. Mailing Add: 19 Laurel Circle Malvern PA 19355

LADMAN, AARON JULIUS, b Jamaica, NY, July 3, 25; m 48; c 2. ANATOMY. Educ: NY Univ, AB, 47; Ind Univ, PhD(anat), 52. Prof Exp: Res fel anat, Harvard Med Sch, 52-55, assoc anat, 55-61; assoc prof, Med Units, Univ Tenn, 61-64; PROF ANAT & CHMN DEPT, UNIV NMEX, 64- Concurrent Pos: USPHS career develop award, 62-64; mem res career award comt, Nat Inst Gen Med Sci, 67-71; managing ed, Anat Record, 68- Mem: Am Asn Anat; Am Soc Zoologists; Endocrine Soc; Histochem Soc; Soc Exp Biol & Med. Res: Cytochemistry; electron microscopy;

LADMAN

endocrinology; experimental cytology; retina. Mailing Add: Dept of Anat Univ of NMex Sch of Med Albuquerque NM 87131

LADNER, JANE ELLEN CRAWFORD, b San Jose, Calif, Mar 14, 44; m 72; c 1. PHYSICAL CHEMISTRY, MOLECULAR BIOLOGY. Educ: Univ Calif, Santa Barbara, AB, 66; Calif Inst Technol, PhD(chem), 71. Prof Exp: Res assoc chem, Nucleic Acid Res Inst, Int Chem & Nuclear Corp, 71; vis staff mem, 71-73, STAFF MEM CHEM, MED RES COUN LAB MOLECULAR BIOL, CAMBRIDGE UNIV, 73- Concurrent Pos: Wellcome Trust fel, 71-72. Res: Application of physical chemistry techniques, principally x-ray diffraction and nuclear magnetic resonance, to study the structure of biologically important molecules. Mailing Add: 410 Canon Dr Santa Barbara CA 93105

LADNER, SIDNEY JULES, b Houston, Tex, Mar 12, 36; m 59; c 2. PHYSICAL CHEMISTRY. Educ: Univ Houston, BS, 59, PhD(phys chem), 65. Prof Exp: Fel chem, Univ NMex, 65-66; chemist, Shell Develop Co, Tex, 66-67; asst prof, 67-69, ASSOC PROF CHEM, HOUSTON BAPTIST COL, 69- Mem: Am Chem Soc. Res: Molecular spectroscopy; decay processes and the decay kinetics of molecules in excited electronic energy states; chemical education. Mailing Add: Dept of Chem Houston Baptist Col 7502 Fondren Rd Houston TX 77036

LADO, FRED, b La Coruna, Spain, June 5, 38; US citizen; m 60; c 3. PHYSICS. Educ: Univ Fla, BS, 60, PhD(physics), 64. Prof Exp: Fel physics, Univ Fla, 64-65; staff mem, Los Alamos Sci Lab, 65-68; asst prof, 68-74, ASSOC PROF PHYSICS, NC STATE UNIV, 74- Concurrent Pos: Fulbright sr lectr, Spain, 71-72. Mem: Am Phys Soc. Res: Statistical mechanics; equilibrium and non-equilibrium theory of liquids; many-body problem. Mailing Add: Dept of Physics NC State Univ Raleigh NC 27607

LADSON, THOMAS ALVIN, b Hyattsville, Md, Sept 29, 17; m 48; c 2. VETERINARY MEDICINE. Educ: Univ Pa, VMD, 39. Prof Exp: Vet, 39-60; field rep, Livestock Sanit Serv, Md State Bd Agr, 60-62, dir, 62-73, head vet sci dept, Univ Md, College Park, 62-73; ACTG DIR, DIV ANIMAL INDUST, MD DEPT AGR, 73- Mem: Am Vet Med Asn; US Animal Health Asn. Res: Large animal medicine; regulatory veterinary medicine. Mailing Add: Div of Animal Indust Md Dept Agr Parole Plaza Off Bldg Annapolis MD 21401

LA DU, BERT NICHOLS, JR, b Lansing, Mich, Nov 13, 20; m 47; c 4. BIOCHEMICAL PHARMACOLOGY. Educ: Mich State Col, BS, 43; Univ Mich, MD, 45; Univ Calif, PhD(biochem), 52. Prof Exp: Intern, Rochester Gen Hosp, NY, 45-46; asst biochem, Mich State Col, 46-47 & Univ Calif, 47-50; from sr asst surgeon to med dir, NIH, 50-63; prof pharmacol & chmn dept, Med Sch, NY Univ, 63-74; PROF PHARMACOL & CHMN DEPT, MED SCH, UNIV MICH, ANN ARBOR, 74- Concurrent Pos: Res assoc, Goldwater Mem Hosp Res Serv, NY Univ, 50-54, instr, Bellevue Med Ctr, 51-54. Mem: Am Soc Biol Chemists; Am Chem Soc; Am Soc Pharmacol & Exp Therapeut; Am Soc Human Genetics; NY Acad Sci. Res: Drug metabolism; metabolism of tyrosine; inborn errors of metabolism; pharmacogenetics. Mailing Add: Dept of Pharmacol Univ of Mich Med Sch Ann Arbor MI 48104

LADUE, JOHN SAMUEL, b Minot, NDak, Sept 6, 11; m 37. INTERNAL MEDICINE. Educ: Univ Minn, BA, 32, MS, 40, PhD(med), 41; Harvard Univ, MD, 36. Prof Exp: Intern, Long Island Col Hosp, 36-37; resident, Minneapolis Gen Hosp, 37-40; instr med & physiol, Med Sch, Univ Minn, 40-41; from instr to asst prof med, La State Univ, 41-45; clin asst med, 45-46, from instr to asst prof clin med, 48-57, ASSOC PROF CLIN MED, MED COL, CORNELL UNIV, 57-; ASSOC CLINICIAN, SLOANKETTERING INST CANCER RES, 56- Concurrent Pos: Adj staff physician, Minneapolis Gen Hosp, 40-41; asst physician, Univ Minn Hosp, 40-41; asst vis physician, Charity Hosp, New Orleans, La, 41-43, vis physician & dir lung sta, 43-45; physician, Outpatient Dept, New York Hosp, 46-47; clin asst, Mem Hosp, New York, 45-46, asst attend physician, 47-48, assoc attend physician, 48-; asst attend physician, New York Hosp, 47-68, assoc attend physician, 68-; courtesy attend physician, Doctors Hosp, 50-, chief med serv; chief med serv, Med Arts Ctr Hosp, 51-71; dir heart sta, Mem Ctr, 51-68, consult, Dept Med, Strang Clin, 54-68, head sect cardiol, 59-; assoc, Sloan-Kettering Inst Cancer Res, 53-56; consult thoracic & cardiovasc surg, 70. Mem: Soc Exp Biol & Med; fel Am Col Physicians; Am Fedn Clin Res (secy, 47-49, vpres, 49, pres, 51); NY Acad Med; NY Acad Sci. Res: Mechanism of compensation by digitalis of the failing human heart; treatment of massive hemorrhage due to peptic ulcer; serum glutamic ovaloacetic transaminase activity, physiological and pathological significance; fibrinolytic enzymes in coronary atherosclerosis and myocardial infarction; precordial electrocardiogram during exercise. Mailing Add: 34 E 67th St New York NY 10021

LADWIG, HAROLD ALLEN, b Manilla, Iowa, May 11, 22; m 46; c 2. NEUROLOGY. Educ: Univ Iowa, MD, 47, BA, 52; Am Bd Psychiat & Neurol, dipl. Prof Exp: Clin instr neurol, Univ Minn, 50-53; from instr to asst prof, 54-66, ASSOC PROF NEUROL & PSYCHIAT, SCH MED, CREIGHTON UNIV, 66- Concurrent Pos: Dir electroencephalog lab, Creighton Mem St Joseph's Hosp, 54-, asst dir rehab ctr, 54-58, assoc dir, 58-64; attend physician, Vet Admin Hosp, 54-59, consult physician, 59-; mem med staff, Nebr Children's Ther Ctr, 56-; dir electroencephalog lab, Children's Mem Hosp, 63- & Archbishop Bergan Mercy Hosp, 64- Mem: Am Electroencephalog Soc; AMA; Am Col Physicians; Am Cong Rehab Med; Am Acad Neurol. Res: Diagnostic neurology and rehabilitation of neurological patients; care of the aged; electroencephalography. Mailing Add: 1743 S 85th Ave Omaha NE 68124

LADY, JAMES HAROLD, analytical chemistry, physical chemistry, see 12th edition

LAEMMLE, GEORGE JOSEPH, b New York, NY, Aug 30, 14. ORGANIC CHEMISTRY. Educ: Fordham Univ, AB, 36, MS, 38, PhD(biochem), 41. Prof Exp: Res biochemist, Winthrop Chem Co, NY, 41-43; plant chemist, Wheatena Corp, 43-46; develop chemist, Burroughs Wellcome & Co, 46-48; res chemist, 48-50, res supvr, 50-56, asst dir, 56-60, tech dir, 60-61, mgr tech serv, 61-66, MGR OPER, JEFFERSON CHEM CO, 66- Mem: Am Inst Chem Eng; Am Chem Soc. Res: Petrochemicals. Mailing Add: Jefferson Chem Co PO Box 847 Port Neches TX 77651

LAEMMLE, JOSEPH THOMAS, b Louisville, Ky, Feb 7, 41; m 65; c 2. ORGANOMETTALIC CHEMISTRY. Educ: Bellarmine Col, BA, 64; Ga Inst Technol, MS, 68, PhD(org chem), 71. Prof Exp: From asst res chemist to asst, Ga Inst Technol, 67-73; ASST PROF CHEM, KENNESAW JR COL, 73- Mem: Am Chem Soc. Res: Determination of the structure of organometallic compounds; descriptions of organometallic reaction mechanisms and sterechemistry of additions with Ketones. Mailing Add: Natural Sci Div Kennesaw Jr Col Marietta GA 30061

LAEMMLEN, FRANKLIN, b Reedley, Calif, Mar 8, 38; m 61; c 2. PLANT PATHOLOGY. Educ: Univ Calif, Davis, BS, 60, PhD(plant path), 70; Purdue Univ, West Lafayette, MS, 67. Prof Exp: Res asst plant path, Univ Calif, Davis, 66-70; asst prof, Univ Hawaii, 70-72; ASST PROF PLANT PATH, MICH STATE UNIV, 72- Mem: Am Phytopath Soc. Res: Extension plant pathology; ornamental plant diseases; plant disease diagnostic laboratory. Mailing Add: Dept of Bot & Plant Path Mich State Univ East Lansing MI 48824

LAESSIG, RONALD HAROLD, b Marshfield, Wis, Apr 4, 40; m 66; c 1. CLINICAL CHEMISTRY, PUBLIC HEALTH. Educ: Wis State Univ-Stevens Point, BS, 62; Univ Wis-Madison, PhD(anal chem), 65. Prof Exp: CHIEF CHEM, STATE LAB HYG, MED CTR, UNIV WIS-MADISON, 66-, ASST DIR, LAB, 70-, ASSOC PROF PREV MED, UNIV, 71- Concurrent Pos: Mem, Inst Bd Anal Chem, 71-; chmn diag prod comt, Food & Drug Admin, 72- Honors & Awards: DIFCO Award, Am Pub Health Asn, 74. Mem: Am Asn Clin Chem; Am Chem Soc; Am Pub Health Asn. Res: Automation; multiphasic screening; computerization of laboratory operation. Mailing Add: Dept of Prev Med Univ of Wis Med Ctr Madison WI 53706

LAESSLE, ALBERT MIDDLETON, biology, see 12th edition

LAETSCH, THEODORE WILLIS, b St Louis, Mo, Jan 7, 40; m 61. MATHEMATICAL ANALYSIS. Educ: Washington Univ, St Louis, BS, 61; Mass Inst Technol, SM, 62; Calif Inst Technol, PhD(appl math), 68. Prof Exp: Asst prof physics, Col of Idaho, 62-65; asst prof math, Ill State Univ, 68-70; ASSOC PROF MATH, UNIV ARIZ, 71- Concurrent Pos: Nat Acad Sci-Nat Res Coun resident res associateship, Wright-Patterson AFB, Ohio, 70-71. Mem: Am Math Soc; Soc Indust & Appl Math. Res: Functional analysis in partially ordered spaces; boundary value problems for ordinary and partial differential equations. Mailing Add: Dept of Math Univ of Ariz Tucson AZ 85721

LAETSCH, WATSON MCMILLAN, b Bellingham, Wash, Jan 19, 33; m 58; c 2. BOTANY. Educ: Wabash Col, AB, 55; Stanford Univ, PhD(biol), 61. Prof Exp: Asst prof biol, State Univ NY, Stony Brook, 61-63; asst prof bot, 63-68, assoc prof, 68-71, assoc dir, Lawrence Hall Sci, 69-72, Dir Univ Bot Garden, 69-73; PROF BOT, UNIV CALIF, BERKELEY, 71-, DIR LAWRENCE HALL SCI, 72- Concurrent Pos: NSF sr fel, Univ Col, London, 68-69; mem bd dir, Asn Sci & Technol Ctrs, 73- Mem: AAAS; Bot Soc Am; Am Soc Plant Physiol; Soc Develop Biol; Soc Exp Biol & Med. Res: Plant development; structure and function of the photosynthetic apparatus; science education. Mailing Add: Dept of Bot Univ of Calif Berkeley CA 94720

LAEVASTU, TAIVO, b Vihula, Estonia, Feb 26, 23; m 49; c 3. OCEANOGRAPHY, METEOROLOGY. Educ: Goshenburg & Lund, Fil Kand, 51; Univ Wash, Seattle, MS, 54; Univ Helsinki, PhD(oceanog), 61. Prof Exp: Fisheries officer, Swedish Migratory Fish Comt, 51-53; res assoc oceanog, Univ Wash, Seattle, 54-55; fisheries oceanogr, Food & Agr Orgn, UN, 55-62; assoc prof oceanog, Univ Hawaii, 62-64; res oceanogr, US Fleet Numerical Weather Facil, 64-71, CHIEF OCEANOG DIV, ENVIRON PREDICTION RES FACIL, NAVAL POSTGRAD SCH, 71- Concurrent Pos: UNESCO lectr, Bombay, India, 59; mem panel disposal radioactive waste into sea & fresh water, Int Atomic Energy Agency, 59-62; mem working group fisheries prob comn maritime meteorol, World Meteorol Orgn, 60-62; NSF & US Navy res grants, 62-64; mem, World Meteor Orgn/UNESCO Panel on Oceanic Water Balance, 72- & Sea Use Coun, Sci-Tech Bd, 71- Honors & Awards: Mil Oceanog Award, 69; US Naval Weather Serv Spec Award, 70. Mem: Am Geophys Union; Am Soc Limnol & Oceanog; Am Meteorol Soc; Geochem Soc. Res: Fisheries hydrography and oceanography; marine chemistry; sea-air interactions; oceanographic forecasting; numerical modeling in oceanography and meteorology. Mailing Add: Dept Oceanog Naval Postgrad Sch Environ Prediction Res Facil Monterey CA 93940

LA FARGE, TIMOTHY, b New York, NY, Mar 14, 30; m 66; c 1. FOREST GENETICS. Educ: Univ Maine, BS, 64; Yale Univ, MF, 65; Mich State Univ, PhD(forestry), 71. Prof Exp: Assoc silviculturist, 65-68, ASSOC PLANT GENETICIST, FOREST GENETICS, FOREST SERV, SOUTHEASTERN FOREST EXP STA, 68- Mem: Soc Am Foresters; Am Inst Biol Sci; AAAS. Res: Recombination of resistance to Cronartium fusiforme of Pinus echinata with rapid growth rate of Pinus taeda through repeated hybridization; breeding of Pinus taeda and Pinus elliottii. Mailing Add: PO Box 5106 Macon GA 31208

LA FEHR, THOMAS ROBERT, b Los Angeles, Calif, Feb 6, 34; m 57; c 5. GEOPHYSICS. Educ: Univ Calif, Berkeley, AB, 58; Colo Sch Mines, MSc, 62; Stanford Univ, PhD(geophys), 64. Prof Exp: Geophysicist, US Geol Surv, 62-64; geophysicist, Geophys Assocs, Int, 64-66; vpres tech develop, GAI-GMX Inc, 66-67, dir tech develop, GAI-GMX Div, EG&G Inc, 67-69; assoc prof, 69-75, ADJ PROF GEOPHYS, COLO SCH MINES, 75-; PRES, EDCON, 75- Concurrent Pos: Lectr, Stanford Univ, 64; consult, GAI-GMS Div, EG&G, Inc, 69-70 & Explor Data Consult, 70- Mem: Soc Explor Geophys (ed, 72-73); Am Asn Petrol Geol; Am Geophys Union. Res: Gravity and magnetic exploration; potential field theory. Mailing Add: RR 5 Golden CO 80401

LAFERRIERE, ARTHUR L, b Willimantic, Conn, Dec 3, 33; m 55; c 3. ORGANIC CHEMISTRY, INORGANIC CHEMISTRY. Educ: Brown Univ, BS, 55; Rutgers Univ, MS, 58; Univ RI, PhD(chem), 60. Prof Exp: Res chemist, Minerals & Chem Corp, 56-58 & Am Cyanamid Co, 60-62; PROF CHEM, RI COL, 62- Mem: Am Chem Soc. Res: Inorganic solution chemistry; organic redox mechanisms. Mailing Add: Dept of Chem RI Col Providence RI 02908

LAFEVER, HOWARD N, b Hagerstown, Ind, May 13, 38; m 58; c 2. PLANT BREEDING, GENETICS. Educ: Purdue Univ, BS, 59, MS, 61, PhD(plant breeding & genetics), 63. Prof Exp: Instr bot, Wis State Univ, LaCrosse, 63; asst prof genetics, Purdue Univ, 63; res geneticist, Boll Weevil Res Lab, USDA, 63-65; asst prof, 65-71, PROF GENETICS & PLANT BREEDING, OHIO AGR RES & DEVELOP CTR, 71- Concurrent Pos: Prof, Ohio State Univ. Mem: Am Soc Agron; Crop Sci Soc Am. Res: Breeding new wheat varieties for distribution and production in midwest; genetic studies of wheat. Mailing Add: Dept of Agron Ohio Agr Res & Develop Ctr Wooster OH 44691

LAFF, ROBERT ALLAN, b North Platte, Nebr, Aug 20, 30; m 54; c 3. SOLID STATE PHYSICS. Educ: Calif Inst Technol, BS, 52; Univ Ill, MS, 54; Purdue Univ, PhD(physics), 60. Prof Exp: RES STAFF MEM, THOMAS J WATSON RES CTR, IBM CORP, 60- Mem: Am Phys Soc; Inst Elec & Electronics Engrs. Res: Transport and optical properties of semiconductors; physics and technology of semiconductor devices; acoustic physics. Mailing Add: Thomas J Watson Res Ctr IBM Corp Box 218 Yorktown Heights NY 10598

LAFFER, NORMAN CALLENDER, b Meadville, Pa, Nov 26, 07; m 43; c 4. BACTERIOLOGY. Educ: Allegheny Col, BS, 29; Univ Maine, MS, 32; Univ Ill, PhD(bact), 37. Prof Exp: Asst bact, Univ Maine, 29-30, instr, 30-31, asst, 31-32; asst, Univ Ill, 33-35; from instr to asst prof, Univ Ariz, 35-41; mem staff, Lederle Labs, Am Cyanamid Co, 41-42; assoc prof, 46-65, asst dean, Col Arts & Sci, 63-66, assoc dean, 66-74, PROF MICROBIOL, UNIV MD, COLLEGE PARK, 65- Mem: Am Soc Microbiol; fel Am Acad Microbiol. Res: Pathogenic fungi; yeasts. Mailing Add: Dept of Microbiol Univ of Md College Park MD 20740

LAFFER, WALTER B, II, mathematics, see 12th edition

LAFFERTY, JAMES FRANCIS, b Pampa, Tex, Dec 23, 27; m 56; c 3. BIOMEDICAL ENGINEERING, MECHANICAL ENGINEERING. Educ: Univ Ky, BS, 55; Univ

Southern Calif, MS, 57; Univ Mich, MS, 66, PhD(nuclear eng), 67. Prof Exp: Asst, Wenner-Gren Res Lab, Univ Ky, 54-55; mem tech staff, Hughes Aircraft Co, 55-57; asst prof nuclear eng, 57-62, assoc prof mech eng, 62-73, actg dir, Wenner-Gren Res Lab, 67-73, PROF MECH ENG & DIR, WENNERGREN RES LAB, UNIV KY, 73- Concurrent Pos: NSF fac fels, Univ Mich, 61-62 & 65-66. Mem: AAAS; Am Soc Mech Eng. Res: Biomechanics of the skeletal and cardiovascular systems; response of biosystems to impact, vibration, acceleration and noise; development and application of biological probes. Mailing Add: Wenner-Gren Res Lab Univ of Ky Lexington KY 40506

LAFFERTY, JOHN J, organic chemistry, medicinal chemistry, see 12th edition

LAFFERTY, ROBERT HERVEY, JR, b Charlotte, NC, May 10, 16; m 43; c 2. ANALYTICAL CHEMISTRY, INORGANIC CHEMISTRY. Educ: Davidson Col, BS, 37; Cornell Univ, PhD(anal chem), 41. Prof Exp: Instr chem, Lehigh Univ, 41-43; res chemist, Kellex Corp, NJ, 43-45 & Union Carbide Nuclear Co, 45-62; res chemist, Oak Ridge Nat Lab, 62-72; RETIRED. Mem: Am Chem Soc. Res: Chemistry of uranium, fluorides and fluorocarbons; thermal decomposition of barium carbonate; radioisotopes. Mailing Add: 437 East Dr Oak Ridge TN 37830

LAFFERTY, WALTER J, b Wilmington, Del, Feb 10, 34; m 58; c 6. PHYSICAL CHEMISTRY. Educ: Univ Del, BS, 56; Mass Inst Technol, PhD(phys chem), 61. Prof Exp: Res assoc, Johns Hopkins Univ, 61-62; CHEMIST, NAT BUR STANDARDS, 62- Concurrent Pos: Leverhulme vis fel, Univ Reading, 70-71. Mem: AAAS; Am Phys Soc. Res: Infrared and microwave spectroscopy. Mailing Add: Infrared Spectros Sect Nat Bur of Standards Washington DC 20234

LAFFIN, ROBERT JAMES, b New Haven, Conn, Apr 16, 27; m 51; c 4. MICROBIOLOGY. Educ: Yale Univ, BS, 49, PhD(microbiol), 55. Prof Exp: Instr microbiol, Womans Col, Univ NC, 53-55; from instr to assoc prof, Creighton Univ, 55-62; instr obstet & gynec, med sch, Tufts Univ, 62-64; asst prof, 64-67, ASSOC PROF MICROBIOL, ALBANY MED COL, 67- Concurrent Pos: Immunologist, St Margaret's Hosp, 62-64. Res: Study of Host immune responses to tumor-specific transplantation antigens. Mailing Add: Dept of Microbiol Albany Med Col Albany NY 12208

LAFFITTE, HERBERT BONELL, b Nashwauk, Minn, Oct 6, 13; m 49; c 3. DENTISTRY, PERIODONTOLOGY. Educ: Univ Minn, DDS, 44; Baylor Univ, BSD, 56. Prof Exp: Assoc prof periodont & actg chmn dept, Dent Sch, Univ Calif, San Francisco, 67-69; PROF PERIODONT & CHMN DEPT, DENT SCH, UNIV ORE, 69- Concurrent Pos: Consult, Vet Admin Hosps, San Francisco, Calif, 67-69, Vancouver, BC & Portland, Ore, 69- & Madigan Gen Hosp, Tacoma, Wash. Mem: Int Asn Dent Res. Res: Clinical, anticalculus agents; periodontal dressings; bone grafts. Mailing Add: 6 Pimlico Terr Lake Oswego OR 97034

LAFFOON, JEAN LUTHER, entomology, deceased

LAFLEUR, KERMIT STILLMAN, b Waterville, Maine, Feb 14, 15; m 39; c 1. TEXTILE CHEMISTRY, SOIL SCIENCE. Educ: Colby Col, BA, 37; Clemson Univ, MS, 64, PhD, 66. Prof Exp: Asst chemist, Wyandotte Worsted Co, Maine, 37-40, chief chemist, 40-46; chief chemist, Deering Milliken Maine Mills, 47-52; res chemist, Excelsior Mills, 52-56, tech supt, 56-58; group leader wool res, Deering Milliken Res Corp, 59-62, consult chemist, 62-66, sr scientist, 66-67; assoc prof, 67-75, PROF SOIL CHEM, CLEMSON UNIV, 75- Mem: AAAS; Am Soc Agron; Soil Sci Soc Am. Res: Wool chemistry; soil chemistry. Mailing Add: 206 Hunter Ave Clemson SC 29631

LAFLEUR, LOUIS DWYNN, b Elton, La, Dec 28, 40; m 64; c 3. SOLID STATE PHYSICS. Educ: Univ Southwestern La, BS, 62; Univ Houston, PhD(physics), 69. Prof Exp: Res scientist assoc, Defense Res Lab, Univ Tex, Austin, 64-65; aerospace technologist, Manned Spacecraft Ctr, NASA, 65-66; asst prof physics, Drury Col, 69-70; asst prof, 70-74, ASSOC PROF PHYSICS, UNIV SOUTHWESTERN LA, 74- Mem: Am Phys Soc; Am Asn Physics Teachers. Res: Mössbauer spectroscopy; electroacoustics. Mailing Add: Dept of Physics Univ of Southwestern La Lafayette LA 70501

LAFLEUR, ROBERT GEORGE, b Albany, NY, Mar 31, 29; m 50; c 3. GEOMORPHOLOGY. Educ: Univ Rochester, AB, 50; Rensselaer Polytech Inst, MS, 53, PhD, 61. Prof Exp: Instr geol, 52-55, from asst prof to assoc prof stratig & paleont, 55-74, ASSOC PROF GEOMORPHOL, GLACIAL GEOL & WATER RESOURCES, RENSSELAER POLYTECH INST, 74- Concurrent Pos: Consult, NY State Educ Dept & US Geol Surv. Mem: Fel Geol Soc Am; Nat Asn Geol Teachers; Am Geophys Union; Arctic Inst NAm; Soc Econ Paleontologists & Mineralogists. Res: Sedimentology; glacial geology; photogeology. Mailing Add: Dept of Geol Rensselaer Polytech Inst Troy NY 12181

LAFON, EARL EDWARD, b Oklahoma City, Okla, Apr 24, 40; m 61. SOLID STATE PHYSICS. Educ: Univ Okla, BS, 62, MS, 64, PhD(physics), 67. Prof Exp: Adj asst prof physics & res assoc, Res Inst, Univ Okla, 67, NSF fel, 67-68; asst prof physics, 68-71, ASSOC PROF PHYSICS, OKLA STATE UNIV, 71- Mem: Am Phys Soc; Am Math Soc. Res: Electronic structure and band theory. Mailing Add: Dept of Physics Okla State Univ Stillwater OK 74074

LAFON, GUY MICHEL, b Bordeaux, France, June 5, 43. GEOCHEMISTRY. Educ: Paris Sch Mines, Civil Ing Mines, 64; Univ Alta, MSc, 65; Northwestern Univ, Ill, PhD(geol), 69. Prof Exp: Res Found fel, State Univ NY Binghamton, 69-70, asst prof geol, 70-72; ASST PROF GEOL, JOHNS HOPKINS UNIV, 72- Mem: Sigma Xi; AAAS; Geochem Soc; Soc Econ Paleont & Mineral. Res: Geochemistry of natural water systems; thermodynamic properties of brines; equilibrium models. Mailing Add: Dept of Earth & Planetary Sci Johns Hopkins Univ Baltimore MD 21218

LAFOND, ANDRE, b Montreal, Que, July 1, 20; m 46; c 3. FORESTRY. Educ: Jean-de-Brebeuf Col, BA, 42; Laval Univ, BA, 45, BASc, 46; Univ Wis, PhD, 51. Prof Exp: Forester, Que Forest Serv, 46-51; dir res, Res Found, 55, PROF FOREST ECOL & PHYSIOL, LAVAL UNIV, 51-, VDEAN FAC FORESTRY, 66- Concurrent Pos: Consult, Que Northsore Paper Co. Mem: Can Soc Soil Sci; Can Soc Plant Physiol; French-Can Asn Advan Sci; Can Inst Forestry; Int Soc Soil Sci. Res: Forest ecology, particularly soil vegetation relationships; forest physiology, particularly mineral nutrition of trees and fertilization; forest management, particularly site classification. Mailing Add: Dept of Forestry Laval Univ Quebec PQ Can

LA FOND, EUGENE CECIL, b Bridgport, Wash, Dec 4, 09; m 35; c 2. OCEANOGRAPHY. Educ: San Diego State Col, AB, 32; Andhra Univ, India, DSc, 56. Prof Exp: Asst, Scripps Inst, Univ Calif, 33-40, oceanogr, 40-47; prof oceanog, Andhra Univ, India, 52-53 & 55-56; specialist oceanog, US State Dept, 56-57; sr scientist, Atomic Submarine US Ship Skate, North Pole, 58; marine biologist, Scirpps Inst & Int Coop Admin, 60-61; chief scientist, US Prog Biol, Int Indian Ocean Exped, Woods Hole Oceanog Inst, 62-63; dep dir off oceanog & dep secy, Int Oceanog Comn, UNESCO, Paris, France, 63-64; supvry res oceanogr, Navy Electronics Lab, 64-68, SR SCIENTIST & CONSULT OCEANOG, NAVAL UNDERSEA RES & DEVELOP CTR, 68- Mem: AAAS; Am Soc Limnol & Oceanog (vpres, 54-55); Am Geophys Union; Sigma Xi (vpres, 65); Int Asn Phys Sci Ocean (secy, 70-). Res: Physical oceanography. Mailing Add: Naval Undersea Res & Develop Ctr San Diego CA 92132

LAFONTAINE, JEAN-GABRIEL, b Sherbrooke, Que, Aug 4, 28; m 52; c 3. CELL BIOLOGY, ELECTRON MICROSCOPY. Educ: Laval Univ, Lic es Sci, 50; Univ Wis, MS, 52, PhD(zool), 54. Prof Exp: Res asst cytol, Sloan Kettering Inst, 54-56, Rockefeller Inst, 56-58 & Montreal Cancer Inst, 58-60; asst prof path, Med Sch, 60-64, assoc prof biol, 64-68, PROF BIOL, SCI FAC, LAVAL UNIV, 68- Concurrent Pos: Damon Runyon fel, 54-56. Mem: Am Soc Cell Biol; Can Soc Cell Biol. Res: Cytochemistry and ultrastructure of the cell nucleus. Mailing Add: Dept of Biol Fac of Sci Laval Univ Quebec PQ Can

LAFORNARA, JOSEPH PHILIP, b Buffalo, NY, Dec 5, 42; m 67; c 2. CHEMISTRY, ENVIRONMENTAL SCIENCES. Educ: Canisius Col, BS, 64; Univ Fla, PhD(inorg chem), 70. Prof Exp: Res chemist, Edison Water Qual Lab, Fed Water Qual Admin, Dept Interior, 70-71; res chemist, Nat Environ Res Ctr, 71-75, RES CHEMIST, OIL & HAZARDOUS MAT SPILLS BR, INDUST ENVIRON RES LAB, US ENVIRON PROTECTION AGENCY, 75- Concurrent Pos: Tech adv, Hazardous Mat Adv Comt, Nat Res Coun-Nat Acad Sci, 71-; mem, Task Force for Nitrosamine Control & Task Force for Kepone Control, US Environ Protection Agency, 75- Mem: Am Chem Soc; Water Pollution Control Fedn. Res: Application of chemical technology to control of spills of hazardous materials; chemical analysis of inorganic and organic water and air pollutants; ultimate disposal of chemical wastes. Mailing Add: Indust Environ Res Lab US Environ Protection Agency Edison NJ 08817

LAFOUNTAIN, LESTER JAMES, JR, b Marinette, Wis, Sept 27, 42; m 64. GEOLOGY. Educ: Univ Wis, BS & MS, 64; Univ Colo, PhD(geol), 73. Prof Exp: Field geologist, US Steel Corp, 64, party chief geol, 65; geologist, Texaco Inc, 66; res assoc rock mech, Dept Geol, Univ NC, 71-74; ASST PROJ GEOLOGIST, E D'APPOLONIA CONSULT ENGRS INC, 75- Mem: Geol Soc Am; Am Geophys Union; AAAS; Sigma Xi. Res: The mechanisms and physical aspects of rock dilation, stick slip and earthquake precursors; the tectonics of the mid-continent and its relationship to seismicity. Mailing Add: E D'Appolonia Consult Engrs Inc 10 Duff Rd Pittsburgh PA 15235

LAFRAMBOISE, JAMES GERALD, b Windsor, Ont, July 26, 38; m 62; c 2. PLASMA PHYSICS. Educ: Univ Windsor, BSc, 57; Univ Toronto, BASc, 59, MA, 60, PhD(aerospace studies), 62. Prof Exp: Asst prof math, Univ Windsor, 65-67; asst prof physics, 67-71, ASSOC PROF PHYSICS, UNIV & CTR RES EXP SPACE SCI, YORK UNIV, 71- Mem: Can Asn Physicists. Res: Electrodes in plasmas; transport problems in fluid flows. Mailing Add: Physics Dept Univ 4700 Keele St Toronto ON Can

LAFRAMBOISE, MARC ALEXANDER, b Winsor, Ont, May 18, 15; m 49; c 2. MATHEMATICS. Educ: Univ Ottawa, BA, 42; Univ Mich, MA, 46, MSc, 49. Prof Exp: Prin & teacher pub & separate schs, Ont, 34-42; asst prof math, Assumption Col, 42-50; asst prof, 53-67, ASSOC PROF MATH, UNIV 67- Concurrent Pos: Dean eve div, Assumption Col, 46-49. Mem: Nat Coun Teachers Math. Res: Mathematics education. Mailing Add: Dept of Math Univ of Detroit Detroit MI 48221

LAFRANCE, CHARLES ROBERT, biology, see 12th edition

LAFUZE, HENRY HARVEY, b Liberty, Ind, Mar 3, 08; m 36; c 2. BOTANY. Univ, AB, 29; Univ Iowa, MS, 30, PhD(plant physiol), 36. Prof Exp: Asst bot, Univ Iowa, 29-35, res assoc plant physiol, 36-37; asst prof biol, Southwestern State Teachers Col, Okla, 37-39; chmn dept, 47-69, PROF BIOL, EASTERN KY UNIV, 47- Mem: Nat Asn Biol Teachers; Nat Sci Teachers Asn. Res: Nutrition of fungi; mineral nutrition of plants; pollination in milkweed; taxonomy of Kentucky trees; science education. Mailing Add: Dept of Biol Eastern Ky Univ Richmond KY 40475

LAGACE, ANDRE JACQUES JOSEPH, veterinary pathology, see 12th edition

LAGAKOS, STEPHEN WILLIAM, b Philadelphia, Pa, June 18, 46; m 68. BIOSTATISTICS. Educ: Carnegie-Mellon Univ, BS, 68; George Washington Univ, MPhil & PhD(math & statist), 72. Prof Exp: Math statistician, Naval Ord Sta, 68-70; STATISTICIAN BIOSTATIST, STATIST LAB, STATE UNIV NY BUFFALO, 72-, ASST PROF STATIST SCI, 73- Concurrent Pos: Coord statistician, Working Party Ther Lung Cancer, 72-; protocol statistician, Eastern Coop Oncol Group, 72- Mem: Biomet Soc; Royal Statist Soc; Inst Math Statist; Int Asn Study Lung Cancer; Am Statist Asn. Res: The planning, design and analysis of clinical trials with particular emphasis on survival-type data. Mailing Add: Statist Sci Div State Univ NY Buffalo Amherst NY 14226

LAGALLY, MAX GUNTER, b Darmstadt, Ger, May 23, 42; US citizen; m 69. MATERIALS SCIENCE, SURFACE PHYSICS. Educ: Pa State Univ, BS, 63; Univ Wis, MS, 65, PhD(physics), 68. Prof Exp: Vis fel physics, Fritz Haber Inst, Max Planck Soc, 68-69; instr physics & res assoc surface physics, 70-71, asst prof mat sci, 71-74, ASSOC PROF MAT SCI, UNIV WIS-MADISON, 74- Concurrent Pos: Sloan Found fel, 73-77. Mem: Am Chem Soc; Am Vacuum Soc. Res: Surface crystallography by low-energy electron diffraction; chemisorption; electronic properties of surfaces and interfaces. Mailing Add: Dept of Metall & Mineral Eng Univ of Wis-Madison Madison WI 53706

LAGALLY, PAUL, b Munich, Ger, Aug 4, 11; US citizen; m 39; c 3. POLYMER CHEMISTRY. Educ: Dresden Tech Univ, BS, 35, MS, 37, PhD(chem), 39. Prof Exp: Chemist, Kaiser Wilhelm Inst, Berlin, 39-40; group leader org chem, Rohm & Haas, Darmstadt, 40-45; dir org res, Paper Mills, Aschaffenburg, 46-52; dir org res, Linden Labs, Inc, 53-62, tech dir org chem, 62-66; RES SUPVR, NAVAL SHIP RES & DEVELOP CTR, 66- Mem: AAAS; Am Chem Soc; NY Acad Sci. Res: Silicon organics; fluoro elastomers; surface chemistry of cellulose, glass and metals; polymerizations. Mailing Add: 1911 Dulany Place Annapolis MD 21401

LAGALLY, RALPH WERNER, b Darmstadt, Ger, July 23, 40; US citizen. ORGANIC CHEMISTRY, PHARMACEUTICAL CHEMISTRY. Educ: Pa State Univ, BS, 61; Purdue Univ, PhD(org chem), 67. Prof Exp: Res chemist, Gulf Res & Develop Co, 67-71; NIH fel, Pa State Univ, 71-74; supvr org chem, Omni Res Inc, 74-75; ASST PROF, INTERAM UNIV, 75- Mem: Am Chem Soc. Res: Synthesis of corticosteroids, steroid intermediates, isoquinoline alkaloids. Mailing Add: PO Box 1156 San German PR 00753

LA GANGA, THOMAS S, b Caldwell, NJ, July 23, 27; m 75; c 4. ENDOCRINOLOGY, CLINICAL CHEMISTRY. Educ: Drew Univ, AB, 51; Rutgers Univ, MS, 66, PhD(animal sci), 67. Prof Exp: Clin chemist, Princeton Hosp, NJ, 67-

LA GANGA

69; ASST DIR DEPT ENDOCRINOL, BIO-SCI LABS, 69- Mem: AAAS. Res: Improved methodology in chemical and biological hormone assays; endocrine physiology of the mammary gland; experimental hypertension; automated analytical systems; protein metabolism in tumor hosts; anti-estrogenic and anti-androgenic compounds. Mailing Add: Bio-Sci Labs 7600 Tyrone Ave Van Nuys CA 91405

LAGE, GARY LEE, b Hinsdale, Ill, Nov 11, 41; m 64; c 2. PHARMACOLOGY. Educ: Drake Univ, BS, 63; Univ Iowa, MS, 65, PhD(pharmacol), 67. Prof Exp: From asst prof to assoc prof pharmacol, Sch Pharm, Univ Kans, 67-73; ASSOC PROF PHARM, UNIV WIS-MADISON, 73- Concurrent Pos: USPHS res career develop award, 75-80. Mem: AAAS; Am Soc Pharmacol & Exp Therapeut; Am Asn Col Pharm. Res: Use of radioactive isotopes for study of drug distribution and/or metabolism, especially cardiac glycosides. Mailing Add: Univ of Wis Sch of Pharm Madison WI 53706

LAGEMANN, ROBERT THEODORE, b Marion, Ohio, Aug 31, 12; m 38; c 1. PHYSICS. Educ: Baldwin-Wallace Col, AB, 34; Vanderbilt Univ, MS, 35; Ohio State Univ, PhD(physics), 40. Hon Degrees: DSc, Baldwin-Wallace Col, 62. Prof Exp: Asst physics, Ohio State Univ, 36-39; instr, Marshall Col, 40-41; from asst prof to prof, Emory Univ, 41-51; chmn dept physics & astron, 51-65, dean grad sch, 65-73, GARLAND PROF PHYSICS, VANDERBILT UNIV, 51- Concurrent Pos: Physicist, Manhattan Proj, Columbia Univ, 42-43; counr, Oak Ridge Inst Nuclear Studies, 49-51 & 53-61, mem bd dirs, 61-67; Ford Found fel, Harvard Univ, 52-53. Mem: Am Phys Soc; Am Asn Physics Teachers. Res: Magneto-optics; infrared spectroscopy; ultrasonics; separation of isotopes by diffusion. Mailing Add: Vanderbilt Univ Grad Sch Nashville TN 37203

LAGERGREN, CARL ROBERT, b St Paul, Minn, Nov 21, 22; m 47; c 3. PHYSICS. Educ: State Col, Wash, BS, 44, MS, 49; Univ Minn, PhD(physics), 55. Prof Exp: Sr physicist, Hanford Atomic Prod Oper, Gen Elec Co, 55-65; mgr mass spectrometry, 65-68, RES ASSOC, RADIOL SCI DEPT, PAC NORTHWEST LABS, BATTELLE MEM INST, 68- Mem: Am Phys Soc. Res: Mass spectrometry; electron impact phenomena; isotopic abundances; surface ionization; ion optics. Mailing Add: 2110 Howell Ave Richland WA 99352

LAGERSTEDT, HARRY BERT, b Glen Ridge, NJ, Aug 2, 25; m 52; c 5. PLANT PHYSIOLOGY, HORTICULTURE. Educ: Ore State Univ, BS, 54, MS, 57; Tex A&M Univ, PhD(plant physiol), 65. Prof Exp: Instr, 57-60, asst prof, 60-67, ASSOC PROF HORT, ORE STATE UNIV, 67-; RES HORTICULTURIST, AGR RES SERV, 67- Mem: Am Soc Plant Physiol; Am Soc Hort Sci. Res: Plant growth regulators; nut crops. Mailing Add: Dept of Hort Ore State Univ Corvallis OR 97331

LAGERSTROM, PACO (AXEL), b Oskarshamm, Sweden, Feb 24, 14; nat US. APPLIED MATHEMATICS. Educ: Stockholm Univ, MA, 35, PhD(philos), 39; Univ Münster, 37-38; Princeton Univ, PhD(math), 42. Prof Exp: Instr math, Princeton, 41-44; res aerodynamicist, Bell Aircraft Corp, 44-45 & Douglas Aircraft Co, Inc, 45-46; from asst prof to prof aeronaut, 47-67, PROF APPL MATH, CALIF INST TECHNOL, 67- Concurrent Pos: Consult, Douglas Aircraft Co, Inc, 46-66; vis prof, Univ Paris, 60-61; Guggenheim grant & Fulbright lectr, 60-61; consult, TRW Inc, 66-68. Mem: Am Math Soc; Soc Indust & Appl Math. Res: Perturbation methods; applications of group theory; fluid dynamics. Mailing Add: Calif Inst of Technol 101-50 Pasadena CA 91125

LAGERWERFF, JOHN VINCENT, soil science, see 12th edition

LAGLER, KARL FRANK, b Rochester, NY, Nov 15, 12; m 41; c 3. FISHERIES, ZOOLOGY. Educ: Univ Rochester, AB, 34; Cornell Univ, MS, 36; Univ Mich, PhD(zool), 40. Prof Exp: Asst bot, Univ Rochester, 34-35; investr fish mgt, 37-39, from instr to asst prof zool, 39-50, res assoc, Lab Vert Biol, 45-49, from assoc prof to prof fisheries & zool, 50-72, chmn dept fisheries, 50-65, PROF NATURAL RESOURCES & ZOOL, UNIV MICH, ANN ARBOR, 72- Concurrent Pos: Leader, Mich Coop Fish Res Unit, Am Wildlife Inst, 37-40; res assoc, Inst Fish Res, Dept Conserv, Mich, 39- & Cranbrook Inst Sci, 40-41, 44-48 & 54-; mem expeds, Upper Great Lakes, 45-49, Western Europe, 57, Alaska, 58 & SE Asia, 64-65 & 70; Billington lectr, 47; tech consult, Asn Fish Tackle Mfrs, 48-50; mem coun, Great Lakes Res Inst, 48-54 & 56-60; Guggenheim fel, 57-58; adv, US Opers Mission Kasetsart-Hawaii Univ Contract, Thailand, 64-65; consult, UN Food & Agr Orgn, UN Develop Prog & WHO, Africa & SAm, 65-; coordr African Lake projs, Food & Agr Orgn, UN, 66-67; dir, Mekong Basinwide Fishery Studies, Laos, Thailand, Cambodia & Vietnam, 74- Honors & Awards: Gold Medal & Dipl, Fr Acad Agr, 61. Mem: Am Soc Ichthyologists & Herpetologists; Am Soc Limnol & Oceanog; Am Fisheries Soc; fel Acad Zool; fel Am Inst Fishery Res Biol (pres, 64-65). Res: Fish predation; fishery biology; ecology; distribution and taxonomy of Great Lakes fishes; natural history of cold-blooded vertebrates. Mailing Add: Sch of Nat Resources Univ of Mich Ann Arbor MI 48104

LAGNESE, JOHN EDWARD, b Pittsburgh, Pa, Mar 7, 37; m 60; c 3. MATHEMATICS. Educ: Univ Dayton, BS, 59; Univ Md, College Park, MA, 61, PhD(math), 63. Prof Exp: Nat Acad Sci-Nat Res Coun fel, Nat Bur Standards, 63-64; from asst prof to assoc prof, 64-72, PROF MATH & CHMN DEPT, GEORGETOWN UNIV, 72- Concurrent Pos: Consult, AID, 70. Mem: Am Math Soc; Math Asn Am; Sigma Xi. Res: Partial differential equations; operator theory. Mailing Add: Dept of Math Georgetown Univ Washington DC 20057

LAGO, BARBARA (DRAKE), b Wheeling, WVa, Dec 18, 30; m 68. GENETICS. Educ: Univ Calif, BA, 52; Stanford Univ, PhD(biol), 59. Prof Exp: Res assoc biol, Stanford, 59-61; asst prof, San Jose State Col, 61-62; res assoc, Univ Calif, Berkeley, 62-63, asst prof, Univ Calif, Davis, 64-65; res microbiologist, 65-72, RES FEL, MERCK & CO, INC, 72- Mem: Genetics Soc Am; Am Soc Microbiol. Res: Genetics of microorganisms; genetic control of metabolism and development. Mailing Add: Merck & Co Inc Rahway NJ 07065

LAGO, RUDOLPH MICHAEL, physics, mathematics, see 12th edition

LAGOMARSINO, RAYMOND J, b Union City, NJ, Dec 20, 30. RADIOCHEMISTRY. Educ: Fairleigh Dickinson Univ, BS, 52; Seton Hall Univ, MS, 62. Prof Exp: Res asst, Seton Hall Univ, 54-58; res asst, Isotopes, Inc, 58-60, from res assoc to sr res assoc, 60-65, appln scientist, 65-66; mgr radiochem, Ledoux & Co, 66-72; RADIOCHEMIST, HEALTH & SAFETY LAB, US ENERGY RES & DEVELOP ADMIN, 72- Mem: AAAS; Am Chem Soc; Am Nuclear Soc. Res: Collection and analyses of radionuclides produced by nuclear weapons testing; neutron activation analysis; analysis of nuclear reactor corrosion products; synthesis of labeled organic compounds; determination of halocarbons and trace gases in the lower stratosphere by gas chromatography. Mailing Add: US Energy Res & Develop Admin Health & Safety Lab 376 Hudson St New York NY 10014

LAGOW, RICHARD JAMES, b Albuquerque, NMex, Aug 16, 45; m 67; c 1. CHEMISTRY. Educ: Rice Univ, BA, 67, PhD(chem), 69. Prof Exp: Instr, Rice Univ, 68-70; ASST PROF CHEM, MASS INST TECHNOL, 70- Concurrent Pos: NSF fel. Res: Inorganic and fluorine chemistry; high temperature and plasma synthesis; inorganic polymers. Mailing Add: Dept of Chem Mass Inst of Technol Cambridge MA 02139

LAGOWSKI, JEANNE MUND, b St Louis, Mo, Nov 17, 29; m 54. ORGANIC CHEMISTRY. Educ: Bradley Univ, BS, 51, MS, 52; Univ Mich, PhD(org chem), 57. Prof Exp: Instr anal chem, Bradley Univ, 51-52; res chemist, Mich State Univ, 56-57; res fel phys org chem, Cambridge Univ, 57-59; assoc res scientist biochem genetics, 59-63, res scientist, 63-73, lectr zool, 73-74, ASSOC PROF ZOOL, UNIV TEX, AUSTIN, 74-, ASST DEAN, DIV GEN & COMP STUDIES, 72- Concurrent Pos: Res career develop award, NIH, 64-69. Mem: Am Chem Soc; Int Soc Heterocyclic Chemists. Res: Chemistry of nitrogen heterocycles; biochemical and genetic aspects of metabolism. Mailing Add: Dept of Zool Univ of Tex Austin TX 78712

LAGOWSKI, JOSEPH JOHN, b Chicago, Ill, June 8, 30; m 54. INORGANIC CHEMISTRY. Educ: Univ Ill, BS, 52; Univ Mich, MS, 54; Mich State Univ, PhD(inorg chem), 57; Cambridge Univ, PhD(inorg chem), 59. Prof Exp: From asst prof to assoc prof, 59-67, PROF CHEM, UNIV TEX, AUSTIN, 67- Mem: Am Chem Soc; The Chem Soc. Res: Liquid ammonia solutions; organometallic compounds; borazines and derivatives; electrochemistry; development of computer-based teaching methods; non-aqueous solution chemistry; metal atom reactions. Mailing Add: Dept of Chem Univ of Tex Austin TX 78712

LAGRANGE, WILLIAM SOMERS, b Ames, Iowa, Apr 23, 31; m 54; c 2. FOOD MICROBIOLOGY, DAIRY BACTERIOLOGY. Educ: Iowa State Univ, BS, 53, PhD(dairy bact), 59. Prof Exp: Exten technologist dairy mfg, Univ Ky, 59-62; EXTEN FOOD TECHNOLOGIST, IOWA STATE UNIV, 62- Mem: AAAS; Int Asn Milk, Food & Environ Sanit; Inst Food Technol; Am Soc Microbiol; Am Dairy Sci Asn. Res: Dairy manufacturing quality control; dairy and foods microbiology; foods processing and control. Mailing Add: Dept of Food Technol Iowa State Univ Ames IA 50010

LAGUAITE, JEANNETTE KATHERINE, b New Orleans, La, June 20, 12. SPEECH PATHOLOGY. Educ: Tulane Univ, BA, 37, MA, 40; La State Univ, PhD(speech), 52. Prof Exp: Teacher pub schs, La, 30-47, speech therapist, 47-52; PROF SPEECH PATH & AUDIOL & DIR, SPEECH & HEARING CTR, SCH MED, TULANE UNIV, 52- Concurrent Pos: Speech pathologist, La Eval Ctr for Except Children, 56; consult, Cleft Palate Team, State Dept Health Planning Coun, 71- Honors & Awards: John Robinson Award, La Speech & Hearing Asn, 70. Mem: Fel Am Speech & Hearing Asn; Am Soc Clin Hypnosis; Int Soc Clin & Exp Hypnosis; Int Asn Logopedics & Phonatrics; Am Cong Rehab Med. Res: Audiology; voice problems and esophageal speech; hereditary deafness. Mailing Add: Dept of Otolaryngol Tulane Univ Sch of Med New Orleans LA 70112

LAGUEUX, ROBERT, b Levis, Que, May 23, 16; m 47; c 7. ICHTHYOLOGY, LIMNOLOGY. Educ: Laval Univ, BA, 40, BS, 45; Univ Montreal, MSc, 50. Prof Exp: Biologist biol bur, Fish & Game Dept, Que, 45-47, biologist & dir, Tadoussac Salmon Hatchery, 47-60; asst dir, Wildlife Serv, Que, 60-62; prof ichthyol, limnol & fishery biol, Fac Sci, 62-72, SECY SCH GRAD STUDIES, LAVAL UNIV, 72- Mem: Fisheries Soc; French-Can Asn Adv Sci; Can Soc Wildlife & Fishery Biol. Res: Fishery science and conservation. Mailing Add: Sch of Grad Studies Laval Univ Quebec PQ Can

LAGUNOFF, DAVID, b New York, NY, Mar 14, 32; m 58; c 3. PATHOLOGY. Educ: Univ Chicago, MD, 57. Prof Exp: Asst microbiol, Univ Miami, 51-53; intern, San Francisco Hosp, Calif, 57-58; from instr to assoc prof, 60-69, PROF PATH, UNIV WASH, 69- Concurrent Pos: Nat Heart Inst fel path, Univ Wash, 58-59, USPHS trainee, 59-60; Nat Heart Inst spec fel physiol, Carlsberg Lab, Denmark, 62-64; Nat Cancer Inst spec fel path, Sir William Dunn Sch Exp Path, Oxford Univ, 69-70. Mem: Histochem Soc; Am Soc Cell Biol; Am Soc Exp Path; Reticuloendothelial Soc. Res: Mast cell structure and function; phagocytosis; pinocytosis; immune response; mucopolysaccharide storage diseases. Mailing Add: Dept of Path Univ of Wash Sch of Med Seattle WA 98195

LAHA, RADHA GOVINDA, b Calcutta, India, Oct 1, 30; US citizen. ANALYTICAL MATHEMATICS, PURE MATHEMATICS. Educ: Univ Calcutta, BSc, 49, MSc, 51, PhD(math), 57. Prof Exp: Mem staff math, Res & Training Sch, Indian Statist Inst, Calcutta, 52-57, lectr, 57; Smith-Mundt-Fulbright fel, Cath Univ Am, 57-58, res asst prof, 58-60; reader, Div Theoret Res & Training, Indian Statist Inst, 60-61; vis res fel, Inst Statist, Univ Paris & Swiss Fed Inst Technol, 61-62; from asst prof to prof, Cath Univ Am, 62-72; PROF MATH, BOWLING GREEN STATE UNIV, 72- Concurrent Pos: Vis res fel, Mass Inst Technol, 68-69. Mem: Fel Inst Math Statist; Int Statist Inst; Am Math Soc. Res: Analytical and abstract probability; harmonic analysis and representation theory of groups; application of probability and analysis to number theory. Mailing Add: Dept of Math Bowling Green State Univ Bowling Green OH 43403

LAHAM, QUENTIN NADIME, b Oshkosh, Wis, Feb 18, 27; m 50; c 3. HISTOLOGY, EMBRYOLOGY. Educ: Ripon Col, BA, 49; Marquette Univ, MS, 51; Univ Ottawa, PhD, 59. Prof Exp: Lectr gen biol, 51-54, from asst prof to assoc prof histol & embryol, 54-58, PROF HISTOL & EMBRYOL, UNIV OTTAWA, 68-, CHMN DEPT BIOL, 69- Concurrent Pos: Nuffield fel, 60-61. Mem: Teratology Soc; Soc Develop Biol; Can Soc Zoologists; Can Soc Cell Biol. Res: Ontogeny of enzyme systems during embryogenesis; influence of heavy metals on developing embryos. Mailing Add: Dept of Biol Univ of Ottawa Laurier Ave E Ottawa ON Can

LAHAM, SOUHEIL, b Port-au-Prince, Haiti, Apr 17, 26; m 55; c 2. TOXICOLOGY, CANCER. Educ: Univ Haiti, BSc, 47; Univ Paris, PhD(toxicol), 54, dipl indust hyg, 55; Graz Univ, dipl microchem, 54. Prof Exp: Res assoc, Univ Paris, 54-56; guest scientist, 56-58, head biochem sect, Environ Health Directorate, 58-62, chief environ toxicol prog, Occup Health Div, 62-71, sr res scientist & consult, 62-74, HEAD INHALATION TOXICOL UNIT, ENVIRON HEALTH DIRECTORATE, CAN DEPT HEALTH & WELFARE, 74- Concurrent Pos: Guest scientist, Nat Res Coun Can, 56-58; prof, Univ Ottawa, 71- & Univ Quebec, 73-; consult, Nat Inst Prev Occup Accidents & Dis, France & Coal Tar Mfrs Union, France. Mem: Am Indust Hyg Asn; Pharmacol Soc Can; Environ Mutagen Soc; Soc Toxicol; Pan-Am Med Asn. Res: Toxicity and comparative metabolism of toxic and carcinogenic substances; biochemical aspects of environmental and occupational cancer; chemical carcinogenesis; structure-biological activity of toxic and carcinogenic chemicals; pulmonary clearance and inhalation toxicity of gases and vapors; telemetering techniques in cardiopulmonary physiology. Mailing Add: 249 Latchford Rd Ottawa ON Can

LAHAYE, PHILIP ARTHUR, b Leonville, La, July 29, 38; m 66. BOTANY. Educ: Univ Southwestern La, BS, 64; La State Univ, MS, 66, PhD(bot), 68. Prof Exp: Fel plant nutrit, Univ Calif, Davis, 68-70; ASST PROF BIOL, STATE UNIV NY COL POTSDAM, 70- Mem: Am Soc Plant Physiol; Bot Soc Am; Am Inst Biol Sci. Res:

Plant nutrition; diurnal rhythms. Mailing Add: Dept of Biol State Univ of NY Col Potsdam NY 13676

LAHEY, JAMES FREDERICK, b Two Rivers, Wis, Mar 7, 21; m 43; c 2. PHYSICAL GEOGRAPHY, METEOROLOGY. Educ: Univ Wis, PhB, 43, MS, 49, PhD, 58. Prof Exp: Asst prof geog, Univ Ga, 50-53; res assoc meteorol, Univ Wis, 53-59; asst prof geog & meteorol, Univ Ohio, 59-61; from assoc prof to prof, Northern Ill Univ, 61-67; assoc prof, Univ Ill, Urbana, 67-71; PROF GEOG & METEOROL, ORE STATE UNIV, 71- Concurrent Pos: Vis assoc prof, Univ Calif, Berkeley, 65-66. Mem: Am Meteorol Soc; Asn Am Geog; Am Soc Photogram. Res: Long range weather forecasting; upper level windfield, cloud and smog climatology; ocean tidal and weather interrelationships. Mailing Add: Dept of Geog Ore State Univ Corvallis OR 97330

LAHEY, M EUGENE, b Ft Worth, Tex, Dec 28, 17; m 42; c 6. PEDIATRICS. Educ: Univ Tex, BA, 39; St Louis Univ, MD, 43. Prof Exp: Nat Res Coun fel med sci, Univ Utah, 49-51, asst prof pediat, 51-52; from asst prof to assoc prof, Univ Cincinnati, 52-58; PROF PEDIAT & HEAD DEPT, UNIV UTAH, 58- Concurrent Pos: Mem med adv bd, Leukemia Soc, 58- & hemat training grant comt, Nat Inst Arthritis & Metab Dis, 59-63; mem, Scope Panel, US Pharmacopeia, 60-; mem residency rev comt, AMA, 61-65, pres, 65-; res dir, Children's Hosp, East Bay, 64-65; vis prof, Children's Hosp, Honolulu. Mem: Am Soc Hemat; Am Pediat Soc; Soc Pediat Res. Res: Pediatric hematology. Mailing Add: Dept of Pediat Univ of Utah Med Ctr Salt Lake City UT 84132

LAHIRI, SUKHAMAY, b Calcutta, India, Apr 1, 33; m 65. PHYSIOLOGY. Educ: Univ Calcutta, BSc, 51, MSc, 53, DPhil(physiol), 56; Oxford Univ, DPhil(physiol), 59. Prof Exp: Govt of WBengal scholar, Oxford Univ, 56-59; asst prof physiol, Presidency Col, Univ Calcutta, 59-65, hon lectr, Univ, 60-65; vis fel & asst prof, State Univ NY Downstate Med Ctr, 65-67; sr res assoc, Cardiovasc Inst, Michael Reese Hosp & Med Ctr, Chicago, Ill, 67-69; assoc prof environ physiol, 69-73, ASSOC PROF PHYSIOL & MED, UNIV PA, 73- Concurrent Pos: Mem comt aviation med, Indian Coun Med Res, 62- Honors & Awards: Premchand-Roychand Gold Medal, Univ Calcutta, 62. Mem: NY Acad Sci; Am Physiol Soc. Res: High altitude physiology; regulation and adaptation; gas exchange; chemoreceptors. Mailing Add: Dept of Physiol Univ of Pa Philadelphia PA 19104

LAHITA, ROBERT GEORGE, medicine, microbiology, see 12th edition

LAHR, CHARLES DWIGHT, b Philadelphia, Pa, Feb 6, 45; m 69; c 1. MATHEMATICAL ANALYSIS. Educ: Temple Univ, BA, 66; Syracuse Univ, MA, 68, PhD(math), 71. Prof Exp: Mathematician, Bell Labs, 71-73; vis asst prof math, Savannah State Col, 73-74 & Amherst Col, 74-75; ASST PROF MATH, DARTMOUTH COL, 75- Mem: Am Math Soc; Sigma Xi. Res: Commutative Banach algebras, particularly convolution algebras. Mailing Add: Dept of Math Dartmouth Col Hanover NH 03755

LAHR, JOHN CLARK, b Indianapolis, Ind, Nov 11, 44; m 66; c 2. SEISMOLOGY. Educ: Rensselaer Polytech Inst, BS, 66; Columbia Univ, MS, 71, PhD(seismol), 75. Prof Exp: GEOPHYSICIST SEISMOL, US GEOL SURV, 71- Mem: Seismol Soc Am; Am Geophys Union; AAAS. Res: Seismicity and tectonics of Alaska, especially as related to hazards assessment. Mailing Add: US Geol Surv 345 Middlefield Rd Menlo Park CA 94025

LAHR, PAUL HILLMON, physical chemistry, see 12th edition

LAHR, ROY (JEREMY), b Battle Creek, Mich, Nov 13, 29. RESEARCH ADMINISTRATION. Educ: Univ Mich, BSEE, 52; Ohio State Univ, MEE, 57. Prof Exp: Mgr advan develop off equip, Int Bus Mach Corp, 57-66; MGR SPEC PROJS OFF EQUIP DEVELOP, XEROX CORP, 66- Mem: Sigma Xi; Inst Elec & Electronics Engrs. Res: Invention, research and development into technologies and devices for office and related environments, both for stand-alone and system/network use, including their human factors integration. Mailing Add: Xerox Corp High Ridge Park Stamford CT 06904

LAI, DAVID CHUEN YAN, b Canton, China, Sept 16, 37; Can citizen; m 68; c 2. URBAN GEOGRAPHY, GEOGRAPHY OF CHINA. Educ: Univ Hong Kong, BA, 60, MA, 64; Univ London, PhD(indust geog), 67. Prof Exp: Tutor geog, Univ Hong Kong, 60-64, lectr, 67-68; asst prof, 68-72, ASSOC PROF GEOG, UNIV VICTORIA, 73- Mem: Asn Am Geogr; Can Asn Geogr; Can Soc Asian Studies; Can Asn Univ Teachers. Res: Urbanization in developing countries; industrial development in China; overseas Chinese communities; socio-economic structures and functions of Chinatowns in Canada. Mailing Add: Dept of Geog Univ of Victoria Victoria BC Can

LAI, DAVID YING FAT, b Honolulu, Hawaii, June 30, 31. PHYSICAL CHEMISTRY, SOLID STATE KINETICS. Educ: Univ Hawaii, BA, 53, MS, 55. Prof Exp: Res asst phys metall, Forrestal Res Ctr, Princeton, 58-59; RES CHEMIST, LAWRENCE LIVERMORE LAB, UNIV CALIF, 59- Mem: AAAS. Res: Mössbauer effects in alloys; diffusion in iron-alloys; aerosol and particle characterization. Mailing Add: Lawrence Livermore Lab Univ of Calif PO Box 808 Livermore CA 94550

LAI, FRANCIS MING-HUNG, b Tainan, Taiwan, Feb 8, 39; m 70; c 1. PHYSIOLOGY, BIOCHEMISTRY. Educ: Tunghai Univ, BS, 63; NC State Univ, MSc, 67; Univ NC, Chapel Hill, PhD(physiol), 71. Prof Exp: NIH fel, Univ Pittsburgh, 71-72; ASST PROF PHYSIOL, MEHARRY MED COL, 72- Concurrent Pos: Instr biol, NC Cent Univ, 68-70; Am Heart Inst grant-in-aid, Meharry Med Col, 74-77. Res: Intracellular compartmental redox states of free nicotinamide-adenine dinucleotide in isolated rat heart under hypoxia, effects of physical training; effects of sodium cyanate on cardiac function and myocardial metabolism; control of pH and salts concentration in peritoneal dialysis. Mailing Add: Dept of Physiol Meharry Med Col Nashville TN 37208

LAI, KWAN WU, b Nanking, China, Nov 18, 33; US citizen; m 59; c 4. HIGH ENERGY PHYSICS. Educ: Univ Ill, BS, 57; Univ Mich, PhD(physics), 62. Prof Exp: Res asst physics, Univ Mich, 58-62, res assoc, 62; res assoc, 62-64, from asst physicist to physicist, 64-74, SR PHYSICIST, BROOKHAVEN NAT LAB, 74- Concurrent Pos: Vis physicist, Lawrence Radiation Lab, Univ Calif, Berkeley, 59-61; adj assoc prof, City Col New York, 67-69; vis prof, Fla State Univ, 69; vis physicist, Calif Inst Technol, 73-74. Mem: Fel Am Phys Soc. Res: High energy particle physics. Mailing Add: Physics Dept Brookhaven Nat Lab Upton NY 11973

LAI, PETER CHENGLIANG, b Lunnan, China, Oct 27, 25; US citizen; m 50; c 5. APPLIED MATHEMATICS, CHEMICAL ENGINEERING. Educ: Chung Cheng Univ, China, BS, 49; Vanderbilt Univ, MS, 63, PhD(chem eng), 66. Prof Exp: Res assoc air pollution, Nat Air Pollution Ctr, 66-67; PROF APPL MATH, TENN STATE UNIV, 67- Concurrent Pos: NSF fel air pollution, Vanderbilt Univ, 67-68; res supvr, Nashville Metrop Govt, 69-70. Mem: Am Chem Soc; Am Inst Chem Engrs; Math Asn Am. Res: Water pollution control; solid waste management; analysis of chemical process. Mailing Add: Dept of Math Tenn State Univ Nashville TN 37203

LAI, PING-YUEN, b Chai-yi, Taiwan, Apr 4, 30; m 59; c 2. PLANT PATHOLOGY, ENTOMOLOGY. Educ: Chung Hsing Univ, Taiwan, BS, 57; Wash State Univ, PhD(plant path), 67. Prof Exp: Sr specialist & sect chief plant protection, Taiwan Prov Food Bur, 59-63; res asst plant path, Wash State Univ, 63-66, sr sci aide, 66-67; USDA grant res assoc, Iowa State Univ, 67-68; chief plant pathologist, Res Dept, 68-73, SR PLANT PATHOLOGIST, STANDARD FRUIT CO, 73- Mem: Am Phytopath Soc; Am Inst Biol Sci. Res: Epidemiology and control of wheat, soybean and banana diseases; aerial application of fungicides. Mailing Add: Honduras Div Standard Fruit Co PO Box 50830 New Orleans LA 70150

LAI, TZE LEUNG, b Hong Kong, June 28, 45; m 75. MATHEMATICS, STATISTICS. Educ: Univ Hong Kong, BA, 67; Columbia Univ, MA, 70, PhD(statist), 71. Prof Exp: Asst prof, 71-74, ASSOC PROF STATIST, COLUMBIA UNIV, 74- Concurrent Pos: Vis assoc prof math, Univ Ill, Urbana-Champaign, 75-76. Mem: Am Statist Asn; Inst Math Statist; Sigma Xi. Res: Sequential methods in statistics; statistical quality control and clinical trials; time series analysis; limit theorems in probability; renewal theory and random walks; martingales and potential theory. Mailing Add: Dept of Math Statist Columbia Univ New York NY 10027

LAI, YUAN-ZONG, b Taiwan, Repub of China, Mar 11, 41; m 68; c 2. WOOD CHEMISTRY. Educ: Nat Taiwan Univ, BS, 63; Univ Wash, MS, 66 & 67, PhD(wood chem), 68. Prof Exp: From asst to res assoc wood chem, Col Forest Resources, Univ Wash, 64-70; sr res assoc wood chem, Univ Mont, 70-75; ASST PROF WOOD CHEM, DEPT FORESTRY, MICH TECHNOL UNIV, 75- Mem: Tech Asn Pulp & Paper Indust; Am Chem Soc; Sigma Xi. Res: Lignin, cellulose and extractive chemistry; thermal properties of wood components. Mailing Add: Dept of Forestry Mich Technol Univ Houghton MI 49931

LAIBLE, CHARLES A, b Freeport, Ill, Jan 15, 36; m 59; c 4. GENETICS. Educ: Univ Ill, BS, 58, MS, 59; Univ Minn, PhD(genetics), 64. Prof Exp: Asst, Univ Ill, 58-59 & Univ Minn, 59-64; corn breeder, Funk Bros Seed Co, 64, mgr hybrid wheat dept, 64-67, MGR RES DATA & QUANT GENETICS, FUNK SEEDS INT, 67- Mem: Crop Sci Soc Am; Am Soc Agron; Soil Sci Soc Am; Am Genetics Asn. Res: Components of genetic variance for ear number; inheritance of ear number in four Zea Mays genotypes; genetic variance and selective value of ear number in corn; computer programming. Mailing Add: Funk Seeds Int Bloomington IL 61701

LAIBLE, JON MORSE, b Bloomington, Ill, July 25, 37; m 59; c 4. ALGEBRA. Educ: Univ Ill, Urbana, BS, 59, PhD(math), 67; Univ Minn, Minneapolis, MA, 61. Prof Exp: Asst prof math, Western Ill Univ, 61-64; asst prof, 64-70, ASSOC PROF MATH, EASTERN ILL UNIV, 70- Mem: Am Math Soc; Math Asn Am; Sigma Xi. Mailing Add: Dept of Math Eastern Ill Univ Charleston IL 61920

LAIBLE, ROY C, b Boston, Mass, June 16, 24. POLYMER PHYSICS. Educ: Northeastern Univ, BS, 45; Boston Univ, MA, 48; Mass Inst Technol, PhD, 70. Prof Exp: Res assoc polymerization, Univ RI, 50-52; org chemist, Cent Intel Agency, 52-53; org chemist, 53-58, phys sci adminr, 58-63, physics res scientist, 63-70, CHIEF TEXTILE RES SECT, US ARMY NATICK LABS, 70- Concurrent Pos: Secy of Army res & study fel viscoelastic properties polymers, Sweden & Scotland, 62-63. Res: Allyl polymerization; viscoelastic properties of fibrous and non-fibrous polymers; ballistic properties of polymers. Mailing Add: US Army Natick Res & Develop Command Kansas St Natick MA 01760

LAIBOWITZ, ROBERT (BENJAMIN), b Yonkers, NY, Mar 24, 37; m 58; c 2. APPLIED PHYSICS. Educ: Columbia Col, BA, 59; Columbia Univ, BS, 60, MS, 63; Cornell Univ, PhD(appl physics), 67. Prof Exp: RES STAFF MEM, IBM RES CTR, 60-63 & 66- Mem: Am Phys Soc. Res: Electrical and optical properties of materials; superconductivity. Mailing Add: IBM Res Ctr PO Box 218 Yorktown Heights NY 10598

LAIBSON, PETER R, b New York, NY, Dec 11, 33; m 63; c 1. OPHTHALMOLOGY. Educ: Univ Vt, BA, 55; State Univ NY Downstate Med Ctr, MD, 59; Am Bd Ophthal, dipl, 65. Prof Exp: NIH fel corneal dis, Retina Found & Mass Eye & Ear Infirmary, 64-65; ASSOC PROF OPHTHAL, SCH MED, TEMPLE UNIV, 66-; PROF OPHTHAL, SCH MED, THOMAS JEFFERSON UNIV, 73- Concurrent Pos: Attend surgeon & dir cornea serv, Wills Eye Hosp; consult lectr, US Naval Hosp, Philadelphia, 68-; mem ophthal staff, Lankenau Hosp, Philadelphia. Mem: Asn Res Ophthal; Am Acad Ophthal & Otolaryngol; AMA. Res: Corneal diseases and surgery of the cornea, particularly viral external diseases, herpes simplex and adenoviruses. Mailing Add: Wills Eye Hosp 1601 Spring Garden St Philadelphia PA 19130

LAIDERMAN, DONALD D, b Minneapolis, Minn, Feb 26, 26; m 48; c 3. CHEMISTRY. Educ: Univ Minn, BA, 48. Prof Exp: Chemist, Sherwin-Williams Co, 48-53; chemist, Toni Div, 53-60, mgr prod develop, 60-68, dir res, 68-71, VPRES RES & DEVELOP, TOILETRIES DIV, GILLETTE CO, 71- Mem: Fel Am Inst Chemists; Am Chem Soc; Soc Cosmetic Chem. Res: Cosmetic and paint chemistry; corrosion; detergents. Mailing Add: Gillette Co Toiletries Div Gillette Park Boston MA 02106

LAIDIG, KERMIT MCCLELLAN, b Mowersville, Pa, Jan 11, 22; m 49; c 2. GEOGRAPHY. Educ: Pa State Teachers Col, Shippensburg, BS, 43; Univ Nebr, MA, 47, PhD, 55. Prof Exp: Instr geog, Nebr State Teachers Col, Kearney, 47-48; vis instr, Univ Okla, 50; cartographer & team chief, Sci Land Utilization Prog, PR, 50-51; from asst prof to prof geog, Ill State Univ, 55-68; PROF GEOG & EARTH SCI, SHIPPENSBURG STATE COL, 68- Concurrent Pos: Field reader, US Off Educ, 65- Mem: Am Geog Soc; Nat Coun Geog Educ. Res: Economic geography; teaching of geography; tropical subsistence agriculture. Mailing Add: Dept of Geog Shippensburg State Col Shippensburg PA 17257

LAIDLAW, GEORGE MICHAEL, organic chemistry, see 12th edition

LAIDLAW, HARRY HYDE, JR, b Houston, Tex, Apr 12, 07; m 46; c 1. APICULTURE. Educ: La State Univ, BS, 33, MS, 34; Univ Wis, PhD(entom, genetics), 39. Prof Exp: From minor sci helper to agent, USDA, 29-34 & 35-39; asst zool & entom, La State Univ, 33-34, asst exp sta, 34-35; prof biol sci, Oakland City Col, 39-41; apiarist, State Dept Agr & Indust, Ala, 41-42; entomologist hqs, 1st Army, New York, 46-47; asst prof entom & asst apiculturist, 47-53, assoc prof entom & assoc apiculturist, 53-59, prof entom & apiculturist, 59-74, prof genetics, 71-74, assoc dean col agr, 60-64, EMER PROF ENTOM, EXP STA, UNIV CALIF, DAVIS, 74- Concurrent Pos: Wis Alumni Res Found asst, Univ Wis, 37-39; NIH grant, Univ Calif, Davis, 63-66, NSF grant, 66-73. Mem: Fel AAAS; Genetics Soc Am; Am Soc Zool; Entom Soc Am; Am Soc Nat. Res: Genetics, breeding and anatomy of the honeybee; factors influencing the development of queen bees; artificial insemination of queen bees. Mailing Add: 761 Sycamore Lane Davis CA 95616

LAIDLAW, JOHN COLEMAN, b Toronto, Ont, Feb 28, 21; m 57; c 2. ENDOCRINOLOGY. Educ: Univ Toronto, BA, 42, MD, 44, MA, 47; Univ London, PhD(biochem), 50; FRCP(C), 55, FRS(C), 75. Prof Exp: Jr intern, Toronto Gen Hosp, 44; demonstr biochem, Univ Toronto, 46-47; lectr, Univ London, 47-50; sr intern med, Toronto Gen Hosp, 50-51; res fel, Harvard Med Sch, 51-53, instr, 53-54; assoc, 54-56, from asst prof to prof med, Univ Toronto, 56-75, dir inst med sci, 67-75; PROF MED & CHMN DEPT, McMASTER UNIV, 75- Concurrent Pos: Asst, Peter Bent Brigham Hosp, Boston, 51-53, jr assoc, 53-54; physician, Toronto Gen Hosp, 54-59, sr physician, 59- Mem: Endocrine Soc; Am Soc Clin Invest; Am Fedn Clin Res; Can Soc Clin Invest (pres, 62); Can Physiol Soc. Mailing Add: Dept of Med McMaster Univ Med Ctr Hamilton ON Can

LAIDLAW, WILLIAM GEORGE, b Wingham, Ont, Mar 13, 36; m 61; c 2. THEORETICAL CHEMISTRY. Educ: Univ Western Ont, BSc, 59; Calif Inst Technol, MSc, 61; Univ Alta, PhD(theoret chem), 63. Prof Exp: NATO fel, Math Inst, Univ Oxford, 64-65; from asst prof to assoc prof chem, 65-73, PROF CHEM, UNIV CALGARY, 73- Concurrent Pos: Vis prof, Univ Waterloo & Free Univ Brussels, 71-72. Mem: The Chem Soc. Res: Hydrodynamics; light-scattering from chemically relaxing fluids, liquid crystals and systems near instabilities; molecular orbital calculations and instabilities. Mailing Add: Dept of Chem Univ of Calgary Calgary AB Can

LAIDLER, KEITH JAMES, b Liverpool, Eng, Jan 3, 16; m 43; c 3. PHYSICAL CHEMISTRY. Educ: Oxford Univ, BA, 37, MA, 55, DSc, 56; Princeton Univ, PhD(phys chem), 40. Prof Exp: Res chemist, Nat Res Coun Can, 40-42; sci officer, Can Armaments Res & Develop Estab, 42-44, chief sci officer & supt phys & math wing, 44-46; from asst prof to assoc prof chem, Cath Univ Am, 46-55; chmn dept, 61-66, vdean fac pure & appl sci, 62-66, PROF CHEM, UNIV OTTAWA, 55- Concurrent Pos: Commonwealth vis prof, Sussex Univ, 66-67. Honors & Awards: Chem Inst Can Medal, 71, Chem Educ Award, 74; Mfg Chemists Asn Chem Teaching Award, 75. Mem: The Chem Soc; fel Royal Soc Can; fel Chem Inst Can. Res: Chemical kinetics of gas reactions; surface, solution and enzyme reactions; photochemistry Mailing Add: Dept of Chem Univ of Ottawa Ottawa ON Can

LAI-FOOK, JOAN ELSA I-LING, b Port of Spain, Trinidad, Aug 3, 37. ZOOLOGY. Educ: Univ Col WI, BSc, 61; Western Reserve Univ, PhD(biol), 66. Prof Exp: Asst prof, 66-74, ASSOC PROF ZOOL, UNIV TORONTO, 74- Mem: Soc Develop Biol. Res: Fine structure of insect development and physiology. Mailing Add: Dept of Zool Univ of Toronto Toronto ON Can

LAIKEN, NORA DAWN, b Chicago, Ill, June 28, 46; m 67. MEDICAL PHYSIOLOGY, MEDICAL EDUCATION. Educ: Univ Chicago, BS, 67; Rockefeller Univ, PhD(life sci), 70. Prof Exp: USPHS fel & res assoc phys biochem, Inst Molecular Biol, Univ Ore, 70-71; USPHS fel & res assoc phys biochem, 71-72, sci curric adv physics, biol & chem, Adaptive Learning Prog, 72-73, DIR TUTORIAL PROG, SCH MED, UNIV CALIF, SAN DIEGO, 73-, LECTR MED, 74- Concurrent Pos: Mem test adv Asn Am Med Col, 74- Res: Development of innovative instructional materials and methods in the basic medical sciences, particularly in medical physiology and pharmacology. Mailing Add: Sch of Med M-002 Univ of Calif at San Diego LaJolla CA 92093

LAINE, ROGER ALLAN, b Cloquet, Minn, Jan 28, 41. BIOCHEMISTRY. Educ: Univ Minn, Minneapolis, BA, 64; Rice Univ, Houston, PhD(biochem), 70. Prof Exp: Fel biochem, Mich State Univ, ELansing, 70-72; fel pathobiol, Univ Wash, 72-74; ASST PROF BIOCHEM, COL MED, UNIV KY, 75- Mem: Am Chem Soc; Soc Complex Carbohydrates; Am Soc Mass Spectrometry. Res: Biochemistry of cell membrane components; gas-liquid-chromatography and mass spectrometry in carbohydrate analysis. Mailing Add: Dept of Biochem Col of Med Univ of Ky Lexington KY 40506

LAING, CHARLES CORBETT, b Brooklyn, NY, Dec 24, 25; m 59; c 3. ECOLOGY. Educ: Univ Chicago, PhB, 50, PhD, 54. Prof Exp: Instr bot, Univ Tenn, 54-56; asst prof, Univ Wyo, 56-59 & Univ Nebr, Lincoln, 59-66; mem fac biol, 66-68, ASSOC PROF BIOL, OHIO NORTHERN UNIV, 68- Mem: Ecol Soc Am; Brit Ecol Soc. Res: Ecology of sand dunes; population ecology of dune grasses; grasslands. Mailing Add: Dept of Biol Ohio Northern Univ Ada OH 45810

LAING, JOHN E, b Ottawa, Ont, Oct 17, 39; m 64; c 2. ENTOMOLOGY, ECOLOGY. Educ: Carleton Univ, BSc, 63, MSc, 64; Univ Calif, Berkeley, PhD(entom), 68. Prof Exp: Asst res entomologist & lectr, Div Biol Control, Univ Calif, Berkeley, 68-73; ASST PROF ENVIRON BIOL, UNIV GUELPH, 73- Mem: Entom Soc Can; Entom Soc Am; Ecol Soc Am; Int Asn Ecol; Int Orgn Biol Control. Res: Ecology of tetranychid mites; populations dynamics of arthropods; ecology and control of orchard pests. Mailing Add: Dept of Environ Biol Univ of Guelph Guelph ON Can

LAING, PATRICK GOWANS, b Barnes, Eng, Nov 8, 23; US citizen; m 56; c 4. ORTHOPEDIC SURGERY. Educ: Univ Southampton, MB & BS, 40; FRCS, 48; Royal Col Surgeons Can, cert orthop specialist, 54; Am Bd Orthop Surg, dipl, 60. Prof Exp: House surgeon, Kings Col Hosp, London, Eng, 45-46; registr orthop surg, Royal Hampshire County Hosp, Winchester, 46-47, gen & orthop surg, Queen Mary's Hosp, Sidcup, 48, orthop surg, Lewisham Hosp, London, 48-50 & Pembury Hosp, Kent, 50-52; sr registr, Bradford Hosp, Yorkshire, 52-54; chief resident surg, Vet Hosp, St John, NB, 54-55; assoc prof orthop surg, 56-63, CLIN PROF ORTHOP SURG, UNIV PITTSBURGH, 63-; CHIEF SERV, VET ADMIN HOSP, 56- Concurrent Pos: Fel cerebral palsy, Univ Pittsburgh, 55-56. Mem: Orthop Res Soc; Am Orthop Asn; Am Soc Testing & Mat; NY Acad Sci; Brit Orthop Asn. Res: Blood supply and the dynamics of circulation in bones and joints; metallurgy and engineering in orthopedics; radioisotopes in clinical orthopedics. Mailing Add: Aiken Med Bldg 532 S Aiken Ave Pittsburgh PA 15232

LAING, RONALD ALBERT, b Seattle, Wash, Dec 9, 33. BIOPHYSICS. Educ: Reed Col, BA, 56; Rice Univ, MA, 58, PhD(low temperature physics), 60. Prof Exp: Asst prof physics, Tulane Univ, 60-68; sr scientist, Space Sci Inc, 68-70; vis scientist, Univ Tokyo, 69-70; ASSOC PROF OPHTHAL, MED SCH, BOSTON UNIV, 70- Concurrent Pos: NSF sci fac fel, Harvard Univ, 65-66; NIH fel, Mass Inst Technol, 66-67; vis lectr, Univ Mass, Boston, 67-68; consult, Space Sci Inc, 67-68. Mem: Am Phys Soc; Am Asn Physics Teachers; Inst Elec & Electronics Eng. Res: Ophthalmic biophysics; bioengineering. Mailing Add: 25 Maple St Lexington MA 02173

LAIPIS, PHILIP JAMES, b Charleston, SC, Apr 20, 44; m 70. MOLECULAR BIOLOGY, VIROLOGY. Educ: Calif Inst Technol, BS, 66; Stanford Univ, PhD(genetics), 72. Prof Exp: Nat Cancer Inst fel, Princeton Univ, 72-74; ASST PROF BIOCHEM, UNIV FLA, 74- Mem: AAAS; Am Soc Microbiol; Sigma Xi. Res: Genetics and mechanism of mammalian cell DNA replication and repair; mechanism of DNA virus replication and transformation. Mailing Add: Box J-245 J H M Health Ctr Univ of Fla Gainesville FL 32610

LAIR, WILBERT MITCHELL, chemical engineering, polymer chemistry, see 12th edition

LAIRD, CHARLES DAVID, b Portland, Ore, May 12, 39; m 61; c 2. CELL BIOLOGY. Educ: Univ Ore, BA, 61; Stanford Univ, PhD(genetics), 66. Prof Exp: NIH fel genetics, Univ Wash, 67-68; asst prof zool, Univ Tex, Austin, 68-71; ASSOC PROF ZOOL, UNIV WASH, 71-, ADJ ASSOC PROF GENETICS, 73- Res: Evolution; gene action; chromosome structure. Mailing Add: Dept of Zool Univ of Wash Seattle WA 98195

LAIRD, CHRISTOPHER ELI, b Anniston, Ala, Nov 29, 42; m 66; c 1. NUCLEAR PHYSICS. Educ: Univ Ala, BS, 63, MS, 66, PhD(physics), 70. Prof Exp: ASSOC PROF PHYSICS, EASTERN KY UNIV, 67- Mem: Am Phys Soc. Res: Theoretical and experimental nuclear physics with primary emphasis in beta decay. Mailing Add: Dept of Physics PO Box 833 Eastern Ky Univ Richmond KY 40475

LAIRD, MARSHALL, b Wellington, NZ, Jan 26, 23; m 49; c 2. PARASITOLOGY, MEDICAL ENTOMOLOGY. Educ: Univ NZ, BSc, 45, MSc, 47, PhD(zool), 49, DSc(zool), 54. Prof Exp: Entomologist, Royal NZ Air Force, 45-48, 50-54; lectr parasitol, Univ Malaya, 54-57; from asst prof to assoc prof, Macdonald Col, McGill Univ, 57-61; chief environ biol, WHO, Geneva, 61-67; prof biol & head dept, 67-72, RES PROF BIOL & DIR RES UNIT VECTOR PATH, MEM UNIV NFLD, 72- Concurrent Pos: Mem sci adv panel insecticides, WHO, 53-61 & 67-, mem sci adv panel & sci & tech comt, Onchocerciasis Control Prog, 74-, head, WHO Collab Ctr for Biol Control, Identification, Ecol & Safety Non-Target Organisms, 75-; Franklin lectr, Auburn Univ, 75. Honors & Awards: Hamilton Prize, Royal Soc NZ, 51. Mem: Fel Royal Soc Trop Med & Hyg; Soc Protozoologists; Soc Invert Path; Can Soc Zoologists; Am Soc Parasitologists. Res: Protozoology; blood parasitology; biting fly larval ecology; insect control relating to aviation. Mailing Add: Res Unit on Vector Path Mem Univ of Nfld St John's NF Can

LAIRD, REGGIE JAMES, b Bassfield, Miss, Feb 11, 20; m 57; c 3. SOIL SCIENCE. Educ: Miss State Col, BS, 40; Univ Wis, MS, 42; Univ Calif, PhD(soils), 52. Prof Exp: Asst agronomist, Exp Sta, Miss State Col, 46-49; from asst soil scientist to assoc soil scientist, 52-55, SOIL SCIENTIST, ROCKEFELLER FOUND, 56- Mem: AAAS; Soil Sci Soc Am; Am Soc Agron; Int Soc Soil Sci. Res: Soil fertility and soil-plant-water relationships. Mailing Add: Londres 40 Mexico DF Mexico

LAIRD, WILLIAM MATTHEW, applied mathematics, see 12th edition

LAIRD, WILSON MORROW, b Erie, Pa, Mar 4, 15; m 38; c 4. GEOLOGY. Educ: Muskingum Col, BA, 36; Univ NC, MA, 36; Univ Cincinnati, PhD(geol), 42. Hon Degrees: DSc, Muskingum Col, 64. Prof Exp: Asst geol, Univ NC, 36-38 & Univ Cincinnati, 38-40; from asst prof to prof, 40-69, EMER PROF GEOL, UNIV NDAK, 69-; DIR COMT EXPLOR, AM PETROL INST, 71- Concurrent Pos: State geologist, NDak, 41-69, state geologist emer, 69-; geologist, US Geol Surv, 44-48; consult geologist, 47-48; dir off oil & gas, Dept Interior, Washington, DC, 69-71. Honors & Awards: Am Asn Petrol Geol Pres Award, 48. Mem: Fel Geol Soc Am; Am Asn Petrol Geol; hon mem Asn Am State Geol (vpres, 48, pres, 50). Res: Stratigraphy of the upper Devonian and lower Mississippian of southwestern Pennsylvania and the northern Rockies; paleontology of brachiopods; physiography and glacial geology; oil conservation; petroleum and ground water geology; geomorphology. Mailing Add: Am Petrol Inst 2101 L St NW Washington DC 20037

LAISHES, BRIAN ANTHONY, b Ottawa, Ont, Apr 18, 47; m 69; c 2. CANCER. Educ: Carleton Univ, BS, 69; Univ BC, MS, 71, PhD(genetics), 74. Prof Exp: RES FEL, NAT CANCER INST CAN, 74- Mem: Tissue Cult Asn. Res: Sequential analysis of chemical carcinogenesis using mammalian cell culture techniques; biochemical interpretations of biological behavior of normal, premalignant and malignant animals cells in vitro and relationships to carcinogenesis in vivo. Mailing Add: Dept of Path Med Sci Bldg Univ of Toronto Toronto ON Can

LAITINEN, HERBERT AUGUST, b Ottertail Co, Minn, Jan 27, 15; m 40; c 3. ANALYTICAL CHEMISTRY. Educ: Univ Minn, BCh, 36, PhD(phys chem), 40. Prof Exp: Asst phys chem, Univ Minn, 36-39; from instr to prof chem, 40-74, head div anal chem, 53-67, EMER PROF CHEM, UNIV ILL, URBANA, 74-; GRAD RES PROF, UNIV FLA, 74- Concurrent Pos: Guggenheim fel, 53, 62; ed, Anal Chem, 66-; Nat Acad Sci exchange visitor, Yugoslavia, 69. Honors & Awards: Fisher Award, 61; Synthetic Org Chem Mfrs Asn Award Environ Chem, 75. Mem: AAAS; Electrochem Soc; Am Chem Soc. Res: Electrochemistry; polarography; amperometric titrations; diffusion; polarization of microelectrodes; fused salts; environmental science. Mailing Add: Dept of Chem Univ of Fla Gainesville FL 32611

LAITY, JOHN LAWRENCE, b Helena, Mont, Feb 23, 42; m 64; c 2. INDUSTRIAL CHEMISTRY, PETROLEUM CHEMISTRY. Educ: Stanford Univ, BS, 64; Univ Wash, PhD(chem), 68. Prof Exp: CHEMIST & SUPVR, SHELL OIL CO, 68- Mem: Am Chem Soc. Res: Photochemical smog; automotive and engine research; combustion; gasoline and oil additives; compositions of fuels and solvents; exhaust emissions; catalysts; atmospheric reactions; air and water pollution; polymer chemistry. Mailing Add: Westhollow Res Lab Shell Develop Co Houston TX 77042

LAITY, RICHARD WARREN, b Mt Kisco, NY, Sept 16, 28; m 51; c 5. ELECTROCHEMISTRY, PHYSICAL CHEMISTRY. Educ: Haverford Col, AB, 50, MS, 51; Iowa State Univ, PhD(phys chem), 55. Prof Exp: From instr to asst prof chem, Princeton Univ, 55-65; PROF CHEM, RUTGERS UNIV, NEW BRUNSWICK, 65- Concurrent Pos: Consult, Monsanto Res Corp, 59-69 & Standard Oil Co (Ohio), 60-68; AEC res contract, 60-71; Frontiers in Chem lectr, Cleveland, Ohio, 62; consult ed, Prentice-Hall, Inc, 62-68; chmn, Gordon Res Conf Molten Salts, 67-69; NSF res grant, 72. Mem: Am Chem Soc; Electrochem Soc; The Chem Soc. Res: Properties of molten salts; electrochemistry; irreversible thermodynamics; transport properties of extremely concentrated aqueous electrolytes. Mailing Add: Sch of Chem Rutgers Univ New Brunswick NJ 08903

LAJINESS, THOMAS ANTHONY, organic chemistry, see 12th edition

LAJOIE, JEAN, b Montreal, Que, Aug 6, 34; m 62; c 1. GEOLOGY. Educ: Univ Montreal, BSc, 58; McGill Univ, PhD(stratig, sedimentation), 62. Prof Exp: PROF SEDIMENTOLOGY, UNIV MONTREAL, 62- Concurrent Pos: Nat Res Coun Can grant, 64. Mem: Am Asn Petrol Geol; fel Geol Asn Can; Soc Econ Paleont & Mineral; Geol Soc France; Int Asn Sedimentol. Res: Cambro-Ordovician paleogeography in the Northern Appalachians. Mailing Add: Fac of Sci Dept of Geol Univ of Montreal Box 6128 Montreal PQ Can

LAJTAI, EMERY ZOLTAN, b Hungary, Oct 28, 34; Can citizen; m 59. GEOLOGY. Educ: Univ Toronto, BASc, 50, MASc, 61, PhD(Pleistocene geol), 66. Prof Exp: Soils engr, Subway Construct Br, Toronto Transit Comn, 61-63; eng geologist, H G Acres & Co Ltd, Ont, 63-65; vis lectr eng geol, 65-67, asst prof, 67-70, ASSOC PROF ENG GEOL & ROCK MECH, UNIV NB, 70- Concurrent Pos: Nat Res Coun Can res grants, 66-68, 71-74 & 74-77; Govt Can, Geol Surv grants, 67-68 & 71-72. Mem: Can Geotech Soc; Can Rock Mech Group. Res: Brittle fracture of rocks under compressive loading with application in structural and engineering geology. Mailing Add: Dept of Geol Univ of NB Fredericton NB Can

LAJTHA, ABEL, b Budapest, Hungary, Sept 22, 22; nat US; m 53; c 2. BIOCHEMISTRY. Educ: Eötvös Lorand Univ, Budapest, PhD(chem), 45. Prof Exp: Asst prof biochem, Eötvös Lorand Univ, 45-47; asst prof, Inst Muscle Res, Mass, 49-50; sr res scientist, NY State Psychiat Inst, 50-57, assoc res scientist, 57-62; prin res scientist, 62-66, DIR, NY STATE RES INST NEUROCHEM & DRUG ADDICTION, 66-; ADJ PROF EXP PSYCHIAT, SCH MED, NY UNIV, 71- Concurrent Pos: Fel, Zool Sta, Italy, 47-48; res fel, Royal Inst Gt Brit, 48-49; asst prof, Col Physicians & Surgeons, Columbia Univ, 56-69. Mem: Int Brain Res Orgn; Am Soc Biol Chemists; Am Acad Neurol; Am Col Neuropsychopharmacol; Int Soc Neurochem. Res: Neurochemistry; amino acid and protein metabolism of the brain and the brain barrier system. Mailing Add: NY Inst for Neurochem & Drug Add Ward's Island New York NY 10035

LAKE, DONALD BIXBY, organic chemistry, see 12th edition

LAKE, JAMES ALBERT, b Nebr; m 67. BIOLOGY. Educ: Univ Colo, BA, 63; Univ Wis, Madison, PhD(physics), 67. Prof Exp: Fel physics, Univ Wis, 67; NIH fel molecular biol, Mass Inst Technol, 67-68 & Children's Cancer Res Found, Mass, 68-70; res fel, Harvard Univ, 69-70; asst prof cell biol, Rockefeller Univ, 70-72; asst prof, 72-73, ASSOC PROF CELL BIOL, MED SCH, NY UNIV, 73- Honors & Awards: Burton Award, Electron Micros Soc Am, 75. Mem: AAAS; Biophys Soc; Electron Micros Soc Am; Cell Biol Soc; Am Soc X-ray Crystallog. Res: Molecular structure of biological molecules; ribosome function and structure. Mailing Add: Dept of Cell Biol NY Univ Med Sch New York NY 10016

LAKE, KENNETH EUGENE, mathematics, see 12th edition

LAKE, LORRAINE FRANCES, b St Louis, Mo, Feb 12, 18. ANATOMY, REHABILITATION MEDICINE. Educ: Wash Univ, BS, 50, MA, 54, PhD(anat), 62. Prof Exp: Instr phys ther, 49-54, instr anat & phys ther, 54-58, dir phys ther, 59-60, asst dir, 60-67, ASSOC DIR, IRENE WALTER JOHNSON INST REHAB, WASH UNIV, 67-, ASST PROF ANAT & PHYS THER, SCH MED, 58- Concurrent Pos: Assoc dir phys ther curric & chg clin training, Sch Med, Wash Univ, 58-63; Woodcock Mem lectr, Univ Calif, 59; consult, Surgeon Gen, US Air Force, 65-67 & birth defects treatment ctr, Nat Found, 65-69. Mem: AAAS; Am Phys Ther Asn; Am Pub Health Asn; Int Soc Rehab Disabled. Res: Normal and abnormal neuromuscular function; electromyocardiographic investigations of normal human movement; human teratology. Mailing Add: Irene Walter Johnson Inst of Rehab Wash Univ Sch of Med St Louis MO 63110

LAKE, ROBERT D, b Lansing, Mich, Sept 7, 30; m 64; c 2. POLYMER CHEMISTRY. Educ: Mich State Univ, BS, 52; Ind Univ, PhD(org chem), 56. Prof Exp: Am Petrol Inst fel, Northwestern Univ, 56-57; fel chem res, Mellon Inst, 57-60; scientist, 60-66, group mgr explor res, 66-72, SR SCIENTIST, RES DEPT, KOPPERS CO, INC, 72- Mem: Am Chem Soc. Res: Synthesis and properties of vinyl and condensation polymers; preparation and properties of unsaturated polyester resins; smoke and flammability behavior of polymers; development of thermoset polyester molding compounds. Mailing Add: 169 Kelvington Dr Monroeville PA 15146

LAKE, ROBERT EDGAR, b Clanton, Ala, July 31, 07; m 41. PHYSICS. Educ: Univ Ala, BS, 32, MS, 34; Pa State Col, PhD(physics), 40. Prof Exp: Lab asst physics, Univ Ala, 32, instr, 33-36; asst, Pa State Col, 36-40; instr math, NC State Col, 40-41, instr mech eng, 41-42, instr math, 42-43, asst prof elec eng & physics, 43-47, assoc prof physics, 47-54, PROF PHYSICS, PRATT INST, 54- Mem: Am Soc Eng Educ; Am Asn Physics Teachers; Inst Elec & Electronics Eng. Res: Diffusion of gases through metals. Mailing Add: Dept of Physics Pratt Inst Brooklyn NY 11205

LAKE, ROBERT SAMUEL, b Wilkinsburg, Pa, Jan 23, 43. ANIMAL VIROLOGY, CELL BIOLOGY. Educ: Franklin & Marshall Col, BA, 64; Univ Del, MS, 67; Pa State Univ, PhD(microbiol), 70. Prof Exp: Res asst path, Dept Animal Sci, Univ Del, 64-67; staff fel virol, Lab Biol Viruses, Nat Inst Allergy & Infectious Dis, 70-73; group leader virol, Biosci Res, Dow Corning Corp, 73-75; SR RES ASSOC PATH, CHILDRENS HOSP AKRON, 75- Mem: Tissue Cult Asn; Am Soc Microbiol. Res: Characterization of in vitro mammalian bioassay systems for detection and quantitation of chemical carcinogens and mutagens. Mailing Add: Dept of Path Childrens Hosp of Akron Akron OH 44308

LAKE, ROBIN BENJAMIN, b Warren, Ohio, Sept 8, 38; m 63. BIOMETRICS, BIOMEDICAL ENGINEERING. Educ: Rensselaer Polytech Inst, BEE, 60; Case Western Univ, PhD(biomed eng), 69; Harvard Univ, AM, 64. Prof Exp: Researcher cell biol & motor neuron degeneration, Harvard Univ, 62-65; researcher systs theory, Systs Res Ctr & Cybernet Systs Group, 65-69, res assoc biomed eng, 69, sr instr, Schs Med & Eng, 69-70, sr instr biomet, Sch Med, 70-72, ASST PROF BIOMET, SCH MED, CASE WESTERN RESERVE UNIV, 72-, DIR BIOMET COMPUT LAB, 70- DIR COMPUT APPLN TRAINING PROG, 73- Concurrent Pos: Consult comput mfr, 70- Mem: AAAS; Am Soc Info Sci; Inst Elec & Electronics Engrs; Asn Comput Mach; Biomed Eng Soc. Res: Analysis and modelling of physiological systems; application of computers to clinical medicine; community health care delivery; mathematical modelling of complex systems. Mailing Add: Dept of Biomet Wearn Bldg Univ Hosps Cleveland OH 44106

LAKEIN, RICHARD BRUCE, b Baltimore, Md, Mar 5, 41; m 64; c 2. MATHEMATICS. Educ: Yale Univ, BA, 62; Univ Md, MA, 64, PhD(math), 67. Prof Exp: Lectr math, Univ Md, 67-68; asst prof math, State Univ NY Buffalo, 68-74 & Erie Community Col, 74-75; MATHEMATICIAN, NAT SECURITY AGENCY, 75- Mem: Am Math Soc; Asn Comput Mach. Res: Number theory; computational mathematics. Mailing Add: 8711 Bunnell Dr Potomac MD 20854

LAKEY, WILLIAM HALL, b Medicine Hat, Alta, Nov 12, 27; m 57; c 4. GENITO-URINARY SURGERY. Educ: Univ Alta, BSc, 49, MD, 53; FRCPS(C), 60. Prof Exp: PROF SURG, FAC MED, UNIV ALTA, 60-, DIR DIV UROL, UNIV HOSP, EDMONTON. Honors & Awards: Surg Res Medal, Royal Col Physicians & Surgeons Can, 56. Mem: Fel Am Col Surg; Am Urol Asn; Can Urol Asn; Can Acad Genito-Urinary Surg. Res: Renal transplantation; renal hypertension and use of diagnostic tests; kidney preservation. Mailing Add: Univ Hosp Univ of Alta Edmonton AB Can

LAKI, KOLOMAN, physical biochemistry, see 12th edition

LAKIN, WILBUR, b Plainfield, NJ, Nov 8, 22; m 66. THEORETICAL PHYSICS. Educ: Union Col, NY, AB, 43; Carnegie Inst Technol, PhD(physics), 54. Prof Exp: Res scientist, Res Lab, Gen Elec Co, 54-55; Courant Inst Math Sci, NY Univ, 56-65; Davidson Lab, Stevens Inst Technol, 65-67; PROF PHYSICS, WICHITA STATE UNIV, 67- Mem: Am Phys Soc. Res: Shock waves. Mailing Add: Dept of Physics Wichita State Univ Wichita KS 67208

LAKRITZ, JULIAN, b Antwerp, Belg, Feb 13, 30; US citizen; c 2. ORGANIC CHEMISTRY. Educ: NY Univ, BA, 52; Univ Mich, MS, 54, PhD(org chem), 60. Prof Exp: Res chemist, Esso Res & Eng Co, 58-68; DIR RES & DEVELOP, AM PERMAC INC, GARDEN CITY, NY, 68- Concurrent Pos: Tech dir, Anscott-Signal Chem Co, 75- Mem: Am Chem Soc; Am Asn Textile Chem & Colorists. Res: Chemistry and technology for solvent processing of textiles. Mailing Add: 2 Livingston Ave Edison NJ 08817

LAKS, HILLEL, b Pretoria, SAfrica, Oct 19, 42; m 65; c 3. THORACIC SURGERY. Educ: Univ Witwatersrand, MBBCh, 65. Prof Exp: Intern med & surg, Johannesburg Gen Hosp, 66; resident path, Hadassah Med Ctr, 67-68, resident surg, 68-69; resident surg, Peter Bent Brigham Hosp; Harvard Med Sch, 69-73, chief resident thoracic surg, 73; chief resident pediat cardiac surg, Children's Hosp, Med Ctr, Harvard Med Sch; ASST PROF SURG & THORACIC SURG, SCH MED, ST LOUIS UNIV, 74- Mem: Am Fedn Clin Res; Asn Acad Surgeons; AMA; Am Soc Artificial Int Organs. Res: Hemodilution; use of left atrial to aortic assist devices and left ventricular to aortic assist devices; long term hemodynamic effect of palliative and corrective pediatric and cardiac procedures; methods for long term preservations of the heart followed by transplantation. Mailing Add: Dept of Surg Sch Med St Louis Univ 1325 SGrand Blvd St Louis MO 63104

LAKSHMAN, AKARASI BHOJARAJ, b Bangalore, India, Apr 3, 32; m 63; c 1. ENDOCRINOLOGY, ANATOMY. Educ: Univ Mysore, BSc, 55, MSc, 56, PhD(endocrinol), 63. Prof Exp: Res fel endocrinol, Worcester Found Exp Biol, Mass, 61-62 & Rockefeller Inst, 62-63; res assoc, Univ Toronto, 63-67; MEM FAC BIOL, LAURENTIAN UNIV, 67- Mem: Endocrine Soc; Am Asn Anat; assoc Sex Info & Educ Coun US. Res: Induction and suppression of ovulation with natural and synthetic hormones; cell types in the pituitary gland; mammalian fertilization; Fallopian tubal physiology. Mailing Add: Dept of Biol Laurentian Univ Sudbury ON Can

LAKSHMANAN, FLORENCE LAZICKI, b New York, NY, Nov 20, 28; m 52; c 5. BIOCHEMISTRY. Educ: Col Mt St Vincent, BS, 50; Univ Md, PhD(biochem), 58. Prof Exp: RES CHEMIST, HUMAN NUTRIT RES DIV, AGR RES SERV, USDA, BELTSVILLE, 58- Concurrent Pos: Res assoc, Dept Nutrit & Foods, Mass Inst Technol, 71-72. Mem: Am Chem Soc. Res: Electrophoretic and ultracentrifugal studies of proteins in blood and urine; nutrition and longevity; human amino acid requirements; protein-carbohydrate interrelationships; malnutrition. Mailing Add: 4810 Blackfoot Rd College Park MD 20740

LAKSHMANAN, NANJA KADA, b India, Oct 23, 36; m 68; c 1. ECOLOGY. Educ: Madras Univ, BS, 56; Bombay Univ, MS, 58; NDak State Univ, PhD(bot), 70. Prof Exp: Res scholar & laectr malaria, Malaria Inst of India, 58-60; res scholar & Lectr ecol, Annamalai Univ, India, 60-63; instr biol, Brooklyn Col, 63-64; asst prof biol & chmn sci dept, Talladega Col, 64-67; asst prof biol, Bemidji State Univ, 67-68; from teaching asst to asst grad dean, NDak State Univ, 68-71; trainee, Environ Protection Dept, Montgomery County, Md, 71-73; PROF BIOL & DIR INST ENVIRON STUDIES, LeMOYNE-OWEN COL, 73- Concurrent Pos: Vis scientist, NSF, 64-67. Mem: Ecol Soc Am; Int Soc Trop Ecol; Am Inst Biol Sci; Am Asn Univ Prof; Human Ecol Soc. Res: Concentration on urban ecology; the effect of pollution on the environment. Mailing Add: Dir Inst of Environ Studies LeMoyne-Owen Col Memphis TN 38126

LAKSHMANAN, P R, b Jamshedpur, Bihar, India, Apr 28, 39. ORGANIC POLYMER CHEMISTRY. Educ: Univ Calcutta, BS, 58; Univ Bombay, BS, 61; NDak State Univ, MS, 65, PhD(polymers & coating), 66. Prof Exp: RES CHEMIST PLASTICS, SR RES CHEMIST COATINGS & ADHESIVES, GULF RES & DEVELOP CO, 66- Mem: Am Chem Soc; Oil & Color Chemists Asn. Res: Relationship between structure and performance of adhesives and coatings; mechanism of adhesion. Mailing Add: Gulf Oil Chem Co Houston Lab PO Box 79070 Houston TX 77079

LAKSHMIKANTHAM, VANGIPURAM, b Hyderabad, India, Aug 8, 26; m 42; c 3. MATHEMATICS. Educ: Osmania Univ, India, PhD(math), 59. Prof Exp: Res assoc math, Univ Calif, Los Angeles, 60-61; vis mem, Math Res Ctr, Univ Wis-Madison, 61-62; vis mem, Res Inst Advan Study, 62-63; assoc prof, Univ Alta, 63-64; prof & chmn dept, Marathwada Univ, India, 64-66; prof & chmn dept, Univ RI, 66-73; PROF MATH & CHMN DEPT, UNIV TEX, ARLINGTON, 73- Mem: Am Math Soc; Indian Math Soc; Indian Nat Acad Sci; Math Soc Can; Soc Indust & Appl Math. Res: Differential inequalities; theory and applications, including stability theory by Liapunov's second method. Mailing Add: Dept of Math Univ of Tex Arlington TX 76010

LAKSHMINARAYANA, J S S, b Penumantra, India, Sept 22, 31; m 60; c 2. PHYCOLOGY, WATER POLLUTION. Educ: Andhra Univ, India, BSc, 52; Banaras Hindu Univ, MSc, 54, PhD(bot), 60. Prof Exp: Lectr bot, Banaras Hindu Univ, 55; scientist & head biol, Cent Pub Health Eng Res Inst, India, 59-70; ASSOC PROF BIOL, UNIV MONCTON, 70- Concurrent Pos: Fr Govt fel, ASTEF, Paris, 65-66; lectr, Visvesvaraya Regional Col Eng, India, 66-69; fel, Mem Univ Nfld, 70. Mem: AAAS; Marine Biol Asn UK; Ecol Soc Am; Can Soc Microbiol; Int Phycol Soc. Res: Algology, limnology and oceanography in relation to pollution; biological treatment of waste waters; primary productivity in relation to fishery development. Mailing Add: Dept of Biol Univ of Moncton Moncton NB Can

LAKSHMINARAYANAIAH, NALLANNA, b Bangalore, India, July 26, 19; m 44; c 2. PHYSICAL CHEMISTRY. Educ: Univ Mysore, BSc, 40; Benares Hindu Univ, MSc, 45; Lowell Tech Inst, MS, 50; Univ London, PhD(phys chem), 56, dipl, Imp Col, 56. Hon Degrees: DSc, Univ London, 70; MA, Univ Pa, 71. Prof Exp: Lectr chem, Univ Mysore, 40-50, asst prof, 50-58; reader phys chem, Univ Madras, 59-62; asst prof pharmacol, 62-68, ASSOC PROF PHARMACOL, UNIV PA, 68- Res: Material transport across membranes; application of finishes to textile fabrics. Mailing Add: Dept of Pharmacol Univ of Pa Philadelphia PA 19104

LAKSHMINARAYANAN, KRISHNAIYER, b Bikshandarkoil, India, July 5, 24; m 60; c 2. BIOCHEMISTRY, INDUSTRIAL MICROBIOLOGY. Educ: Univ Madras, BSc, 45, MSc, 50, PhD(biochem), 55. Prof Exp: Jr chemist, King Inst Prev Med, India, 45-47, biochemist, Stanley Hosp, Madras, 50-51; asst prof microbiol, Birla Col, Pilani, 52; Imp Chem Industs fel, Nat Inst Sci India, 55-56; Nat Res Coun Can fel, Univ Manitoba, 56-58; sr fel, Coun Sci & Indust Res, Govt India, 59-60; plant biochemist, Cent Bot Lab, Allahabad, 60-61; res scientist indust microbiol, John Labatt Ltd, 61-62, sr res scientist, 62-63, proj leader, 63-67, sr indust enzymologist, Dawe's Fermentation Prod, Inc, 67-69, dir fermentation develop, 69-71, res & develop, 71; MGR PROCESS DEVELOP, SEARLE BIOCHEMICS, DIV G D SEARLE & CO, 71- Concurrent Pos: Hon lectr, Univ Western Ont, 64- Mem: Fel Chem Inst Can; fel Royal Inst Chemists. Res: Microbial enzymology; plant biochemistry; toxicology; immunology; chromatography; microtechniques; industrial fermentations; enzyme

LAKSHMINARAYANAN

production; immobilization. Mailing Add: Searle Biochemics 2634 S Clearbrook Dr Arlington Heights IL 60005

LAKSO, ALAN NEIL, b Auburn, Calif, Jan 3, 48. POMOLOGY, PLANT PHYSIOLOGY. Educ: Univ Calif, Davis, BS, 70, PhD(plant physiol), 73. Prof Exp: ASST PROF POMOL, NY STATE AGR EXP STA, 73- Mem: Am Soc Plant Physiologists; Am Soc Hort Sci; Am Soc Enologists; Sigma Xi. Res: Pruning and training in apple pertaining to tree and orchard productivity and mechanical harvest; plant-environment interaction and fruiting. Mailing Add: Dept of Pomol & Viticult NY State Agr Exp Sta Geneva NY 14456

LAL, DEVENDRA, b Banaras, India, Feb 14, 29; m 55. NUCLEAR PHYSICS, GEOCHEMISTRY. Educ: Banaras Hindu Univ, BSc, 47, MSc, 49; Univ Bombay, PhD(physics), 58. Prof Exp: Res asst, Tata Inst Fundamental Res, India, 50-53, fel, 57-60, prof, 63-67; PROF OCEANOG, SCRIPPS INST OCEANOG, UNIV CALIF, SAN DIEGO, 67-; DIR, PHYS RES LAB, AHMEDABAD, INDIA, 72- Concurrent Pos: Mem, Indian Nat Comt Oceanic Res, 60-; mem, Comn Atmospheric Chem & Radioactivity, 63-; Am Chem Soc Petrol Res Fund grant, 67-71; mem exec comt, Int Asn Phys Oceanog, 68-; sr prof, Tata Inst Fundamental Res, India, 69-72; chmn, Sci Comt on Oceanic Res Working Group, 73-; chmn, Earth Sci Res Comt, Coun Sci & Indust Res, Govt Of India, 74-; mem, River Inputs to Ocean Systs, UNESCO. Honors & Awards: Krishnan Medal Award, 65; Bhatnagar Award, Coun Sci & Indust Res, Govt of India, 67; Padma Shri, Govt of India, 71. Mem: Foreign assoc Nat Acad Sci; fel Indian Acad Sci. Res: Cosmic rays; astrophysics; meteoritics; oceanography; meteorology; hydrology. Mailing Add: GRD A-020 Scripps Inst of Oceanog La Jolla CA 92093

LAL, HARBANS, b Haripur, India, Jan 8, 31; m 64; c 3. PHARMACOLOGY, PSYCHOLOGY. Educ: Punjab Univ, India, BS, 52; Univ Kans, MS, 58; Univ Chicago, PhD(pharmacol), 62. Prof Exp: Pharmacologist, IIT Res Inst, 61-65; assoc prof pharmacol & toxicol, Univ Kans, 65-67; assoc prof pharmacol, 67-70, PROF PHARMACOL & TOXICOL, UNIV RI, 70-, PROF PSYCHOL, 71- Concurrent Pos: Res assoc, RI Inst Ment Health, 70- Mem: Soc Toxicol; Soc Biol Psychiat; Am Soc Pharmacol & Exp Therapeut; Am Pharmaceut Asn; fel Am Col Clin Pharmacol. Res: Psychopharmacology; neurochemistry; drug-addiction; environmental pharmacology; toxicology; aging; behavioral controls. Mailing Add: 38 Helme Rd Kingston RI 02881

LAL, JOGINDER, b Amritsar, India, July 2, 23; nat US; m 51; c 2. POLYMER CHEMISTRY. Educ: Punjab Univ, India, BSc, 44, MSc, 46; Polytech Inst Brooklyn, MS, 49, PhD, 51. Prof Exp: Prof chem, Jain Col, Ambala, India, 45-47 & Hindu Col Amritsar, 51-52; head polymer res, H D Justi & Son, Inc, Pa, 52-56; res scientist, Res Div, 56-67, sect head, 67-75, MGR BASIC RES IN POLYMERS, GOODYEAR TIRE & RUBBER CO, 75- Concurrent Pos: Mem adv bd, J Polymer Sci. Mem: Am Chem Soc. Res: Mechanism of polymerization; new polymerization; block copolymers; catalysts; relationship between structure and properties of polymers; vulcanization; monomer synthesis. Mailing Add: Res Div Goodyear Tire & Rubber Co Akron OH 44316

LAL, MOHAN, b Dharmkot, Punjab, May 8, 32; Can citizen; m 64. MATHEMATICS. Educ: D M Col, Punjab, India, BA, 52; Aligarh Muslim Univ, MSc, 55; Univ BC, PhD(nuclear physics), 62. Prof Exp: Lectr physics, D A V Col, Punjab, India, 55-57; res asst, Univ BC, 57-61; res assoc, Univ Alta, 62-63; asst prof math & physics, Mt Allison Univ, 63-64; from asst prof to assoc prof math, 64-75, PROF MATH, MEM UNIV NFLD, 75- Concurrent Pos: Comput specialist, Fed & Prov Land Inventory Studies, Dept Mines & Natural Resources, Can, 67. Mem: Can Math Cong. Res: Numerical analysis; applied mathematics and elementary number theory. Mailing Add: Dept of Math Mem Univ St John's NF Can

LAL, RAVINDRA BEHARI, b Agra, India, Oct 5, 35; m 62; c 1. PHYSICS. Educ: Agra Univ, BSc, 55, MSc, 58, PhD(physics), 63. Prof Exp: Lectr physics, REI Col, Agra Univ, 58-59 & Delhi Polytech, 63-64; Nat Acad Sci-Nat Res Coun resident res assoc, Marshall Space Flight Ctr, NASA, 64-67; asst prof, Indian Inst Technol, Delhi, 68-70; sr res assoc, Univ Ala, Huntsville, 71-73; asst prof physics, Paine Col, 73-75; ASSOC PROF PHYSICS, ALA A&M UNIV, 75- Mem: Am Phys Soc; Sigma Xi; Am Asn Crystal Growth. Res: solid state physics; crystal growth and characterization of materials; magnetic and electrical properties of II-VI and III-V compounds; manufacturing in space. Mailing Add: Dept of Physics & Math Ala A&M Univ Normal AL 35762

LAL, SAMARTHJI, b London, Eng, Mar 23, 38; Can citizen; m 74; c 1. NEUROPSYCHIATRY. Educ: Univ London, MB, BS, 62; McGill Univ, dipl psychiat, 67; FRCP(C), 70. Prof Exp: Med Res Coun Can res fel psychiat, 67-71; chief consultation serv, 71-75, DIR CLIN & BASIC RES PSYCHIAT, MONTREAL GEN HOSP, 75-, ASSOC PSYCHIATRIST, 74-; ASST PROF PSYCHIAT, McGILL UNIV, 73- Concurrent Pos: Consult psychiatrist, Queen Mary Vet Hosp, 71-; staff psychiatrist, Montreal Gen Hosp, 71-, consult, Psychiat Consultation Serv, 75- Mem: Brit Med Asn; Can Psychiat Asn; fel Am Psychiat Asn; Indian Psychiat Asn; Soc Biol Psychiat. Res: Monoaminergic mechanisms in anterior pituitary secretion and in neurological and psychiatric disorders; drug-induced sterotyped behavior in the rat. Mailing Add: Montreal Gen Hosp Dept Psychiat 1650 Cedar Ave Montreal PQ Can

LALA, PEEYUSH KANTI, b Chittagong, Bangladesh, Nov 1, 34; m 62; c 2. CANCER, CELL BIOLOGY. Educ: Univ Calcutta, MB, BS, 57, PhD(med biophysics), 62. Prof Exp: Demonstr path, Calcutta Med Col, 59-60; demonstr path & hemat, NRS Med Col, 61-62; res assoc biol & med res, Argonne Nat Lab, 63-64; res scientist, Radiobiol Lab, Univ Calif, San Francisco, 64-66; res assoc biol & health physics, Chalk River Nuclear Labs, Atomic Energy Can Ltd, 67-68; asst prof, 68-72, ASSOC PROF ANAT, McGILL UNIV, 72- Concurrent Pos: Fulbright travel scholar, 62; res dir, Med Res Coun Can grant, 68 & Nat Cancer Inst Can grant, 69; USPHS grant, 75. Mem: Am Soc Cell Biol; Reticuloendothelial Soc; Am Asn Anatomists; Int Soc Exp Hemat; Can Soc Cell Biol. Res: Studies on cell population kinetics during normal hematopoiesis and human leukemias; host-tumor cell interactions in vivo; control mechanisms in tumor growth; proliferation kinetics of cancer cells. Mailing Add: Dept of Anat McGill Univ Montreal PQ Can

LALANCETTE, JEAN-MARC, b Drummondville, Que, Apr 21, 34; m 59; c 3. INORGANIC CHEMISTRY, ENVIRONMENTAL CHEMISTRY. Educ: Univ Montreal, BSc, 57, MSc, 58, PhD(chem), 61. Prof Exp: From asst prof to assoc prof, 60-69, PROF CHEM, UNIV SHERBROOKE, 69- Mem: Chem Inst Can. Res: Organometallic chemistry; chemistry of graphite intercalates, both catalytic and synthetic properties; photochemical reactions; use of natural materials for protection of environment; peat moss. Mailing Add: Dept of Chem Univ of Sherbrooke Sherbrooke PQ Can

LALANCETTE, ROGER A, b Springfield, Mass, July 30, 39; m 67. ANALYTICAL CHEMISTRY, CRYSTALLOGRAPHY. Educ: Am Int Col, BA, 61; Fordham Univ, PhD(anal chem), 67. Prof Exp: Res fel, Brookhaven Nat Lab, 66-67; res chemist photopolymerization, Photo Prod Dept, E I du Pont de Nemours & Co, Inc, NJ, 67-69; ASST PROF ANAL CHEM, RUTGERS UNIV, NEWARK, 69- Res: Preparation and structural studies—magnetic susceptibility, x-ray powder and single crystal analysis, thermal stability of chelates of rare-earth and transition metals. Mailing Add: Dept of Chem Rutgers Univ Newark NJ 07102

LALAS, DEMETRIUS P, b Athens, Greece, Sept 28, 42; m 67; c 2. DYNAMIC METEOROLOGY. Educ: Hamilton Col, AB, 62; Cornell Univ, MAeroE, 65, PhD(Aerospace), 68. Prof Exp: From asst prof dept eng mech to asst prof dept mech eng, Wayne State Univ, 68-73; vis fel, Coop Inst Res Environ Eng, Univ Colo, 73-74; ASSOC PROF, DEPT MECH ENG, WAYNE STATE UNIV, 73- Concurrent Pos: Consult, Coop Inst Res Environ Eng, Univ Colo, 74-75. Mem: Am Meteorol Soc; Am Geophys Union; Greek Meteorol Soc. Res: Dynamics of micro and mesoscale wave dynamics, their excitation, stability and properties; physics and dynamics of two phase flows in the atmosphere and the laboratory. Mailing Add: Dept of Mech Eng Wayne State Univ Detroit MI 48202

LALEZARI, PARVIZ, b Hamadan, Iran, Aug 17, 31; m 58; c 2. MEDICINE, PHYSIOLOGY. Educ: Univ Teheran, MD, 54. Prof Exp: Asst prof, 67-70, ASSOC PROF MED, ALBERT EINSTEIN COL MED, 70-; DIR IMMUNOHEMAT & BLOOD BANK, MONTEFIORE HOSP & MED CTR, 60- Concurrent Pos: City New York Res Coun res grant, 60-64; NIH res grant, 65- Mem: Am Soc Hemat; Am Soc Clin Invest; Am Asn Immunol. Res: Leukocyte immunology; red cell immunology and autoimmune diseases. Mailing Add: Montefiore Hosp & Med Ctr 111 E 210th St Bronx NY 10467

LALIBERTE, LAURENT HECTOR, b Ottawa, Ont, Can, Nov 7, 43; m 66; c 2. ELECTROCHEMISTRY, CORROSION. Educ: Univ Ottawa, BSc, 66, PhD(chem), 69. Prof Exp: Fel chem, Univ Ottawa, 69-71; SCIENTIST CORROSION, PULP & PAPER RES INST CAN, 71- Mem: Electrochem Soc; Nat Asn Corrosion Engrs; Tech Section Can Pulp & Paper Asn; Tech Asn Pulp & Paper Indust. Res: Corrosion of materials used in pulp and paper industry process equipment. Mailing Add: Pulp & Paper Res Inst of Can 570 St John's Rd Pointe Claire PQ Can

LALIBERTE, REAL, biochemistry, chemistry, see 12th edition

LALICH, JOSEPH JOHN, b Slunj, Yugoslavia, Nov 23, 09; nat US; m 41. PATHOLOGY. Educ: Univ Wis, BS, 33, MS, 36, MD, 37. Prof Exp: Fel exp med, Univ Kans, 38-42; from instr to assoc prof, 46-56, PROF PATH, MED SCH, UNIV WIS-MADISON, 56- Mem: AAAS; fel Soc Exp Biol & Med. Res: Hemorrhagic and traumatic shock; hemostasis; coagulation; hemoglobinuric nephrosis; experimental lathyrism; myocardial necrosis after allylamine ingestion; monocrotaline induced cor pulmonale. Mailing Add: Dept of Path Univ of Wis Med Sch Madison WI 53705

LALL, AMRIT, b Lahore, Pakistan, Nov 10, 27; m 59; c 2. ECONOMIC GEOGRAPHY. Educ: Panjab Univ, BA, 48, MA, 50; Ind Univ, Bloomington, PhD(geog), 58. Prof Exp: Asst prof geog, State Col, Panjab, 51-58; asst prof, Panjab Univ, 58-62; assoc prof, Univ Delhi, 62-66; asst prof, Univ Wis-Oshkosh, 66-67; ASSOC PROF GEOG, UNIV WINDSOR, 67- Mem: Asn Am Geog; Can Asn Geog; Can Asn Asian Studies. Res: Urbanization process in South Asia; central-place functions in the Western Himalayas and rural transformation as a result of economic development. Mailing Add: Dept of Geog Univ of Windsor Windsor ON Can

LALL, SANTOSH PRAKASH, b Motihari, Bihar, India, Sept 8, 44; Can citizen; m 74; c 1. NUTRITIONAL BIOCHEMISTRY. Educ: Allahabad Univ, BSc, 64; Univ Guelph, MSc, 68, PhD(nutrit), 73. Prof Exp: Res asst animal nutrit, Allahabad Agr Inst, 64-65; RES SCIENTIST FISH NUTRIT, HALIFAX LAB, 74- Concurrent Pos: Res asst, Nutrit Dept, Univ Guelph, 68, fel, 73. Mem: Nutrit Soc Can; Nutrit Today Soc. Res: Nutrient requirements of salmonids in fresh water and sea water; nutritional evaluation of fats and oils; cardiopathological effects of lipid in rat and chickens. Mailing Add: Halifax Lab 1707 Lower Water St PO Box 429 Halifax NS Can

LALLI, ANTHONY, b Akron, Ohio, June 17, 30; m 55; c 2. ROENTGENOLOGY. Educ: Hiram Col, BA, 51; Univ Chicago, MD, 54. Prof Exp: Intern med, Toronto Gen Hosp & Hosp Sick Children, 54-55; gen pract, 55-57; resident radiol, Univ Chicago, 57-60; pvt pract, 60-63; from instr to asst prof, 63-68, ASSOC PROF RADIOL, UNIV MICH, ANN ARBOR, 68-; HEAD CLIN RADIOL, CLEVELAND CLIN, 69- Concurrent Pos: Asn Univ Radiologists travel grant, Karolinska Inst, Sweden, 65-66. Res: Uroradiology. Mailing Add: Cleveland Clin Dept of Radiol 9500 Euclid Ave Cleveland OH 44106

LALLI, CAROL MARIE, b Toledo, Ohio, Dec 5, 38. MARINE BIOLOGY. Educ: Bowling Green State Univ, BS & BEd, 60, MA, 62; Univ Wash, PhD(zool), 67. Prof Exp: Lectr zool, 68-69, asst prof marine sci, 69-73, ASSOC PROF MARINE SCI, McGILL UNIV, 73- Mem: Am Soc Zool; Can Soc Zool; Marine Biol Asn U K. Res: Ecological studies of planktonic molluscs; functional morphology of gymnosomatous and thecosomatous pteropods. Mailing Add: Marine Sci Ctr McGill Univ Montreal PQ Can

LALLY, VINCENT EDWARD, b Brookline, Mass, Oct 13, 22; m 53; c 3. METEOROLOGY, ELECTRONICS. Educ: Univ Chicago, BS, 44; Mass Inst Technol, BS, 48, MS, 49. Prof Exp: Engr, Bendix-Friez, Md, 49-51; chief meteorol instrument sect, Air Force Cambridge Res Labs, 51-58; res mgr, Tele-Dynamics Div, Am Bosch Arma Corp, 58-61; PROG HEAD, NAT CTR ATMOSPHERIC RES, 61-, COMT ON SPACE RES, 65- Mem: AAAS; fel Am Meteorol Soc; Sigma Xi. Res: Meteorological instruments and measurement systems. Mailing Add: 4330 Comanche Dr Boulder CO 80303

LALONDE, ROBERT THOMAS, b Bemidji, Minn, May 7, 31; m 57; c 7. ORGANIC CHEMISTRY. Educ: St John's Univ, Minn, BA, 53; Univ Colo, PhD, 57. Prof Exp: Sr res engr chem, Jet Propulsion Lab, Calif Inst Technol, 57-58; res assoc, Univ Ill, 58-59; from asst prof to assoc prof chem, 59-68, PROF CHEM, STATE UNIV NY COL ENVIRON SCI & FORESTRY, 68- Concurrent Pos: NIH fel, 65-66. Mem: Am Chem Soc; Phytochem Soc NAm. Res: Chemistry of natural products; stereochemistry; chemistry of alkaloids, terpenoids and steroids. Mailing Add: Dept of Chem State Univ of NY Col of Environ Sci & Forestry Syracuse NY 13210

LALOS, GEORGE THEODORE, b Harrisburg, Pa, June 5, 25; m 57; c 3. PHYSICS. Educ: Rensselaer Polytech Inst, BAeE, 46; Cath Univ Am, MS, 51. Prof Exp: Physicist, Nat Bur Standards, 47-53; PHYSICIST, US NAVAL ORD LAB, 53- Mem: Am Phys Soc; Soc Appl Spectros. Res: Spectroscopy of flames; high gas temperature measurement; pressure effects on spectral lines; hot dense gas emission spectra; chemical and high energy lasers. Mailing Add: Advan Chem Div US Naval Ord Lab 323-309 White Oak MD 20910

LAM, CHAN FUN, b Kwantung, China, Oct 23, 43; m 70; c 2. BIOMEDICAL ENGINEERING. Educ: Calif Polytech State Univ, BS, 65; Clemson Univ, MS, 67, PhD(elec & comp eng), 70. Prof Exp: Res asst, Grad Inst Technol, Univ Ark, 65-66; res asst comp anal, Clemson Univ, 66-70; dir opers & chief prog, 71-72, asst prof biomed eng, 70-75, ASSOC PROF, MED UNIV SC, 75- Mem: Inst Elec & Electronics Engrs; Pattern Recognition Soc; Soc Comput Simulation. Mailing Add: Dept of Biomet Med Univ of SC Charleston SC 29401

LAM, FUK LUEN, b Hong Kong, Nov 7, 37; US citizen; m 68; c 2. CHEMISTRY. Educ: Univ SC, PhD(org chem), 66. Prof Exp: Fel chem, Mass Inst Technol, 66-67; Brandeis Univ, 68-69; res assoc, 70-75, ASSOC CHEM ONCOGENESIS, SLOAN-KETTERING INST, 75- Mem: Am Chem Soc. Res: Photochemistry and chemical reactions. Mailing Add: Sloan-Kettering Inst 145 Boston Post Rd Rye NY 10543

LAM, GOW THUE, b China, Dec 28, 23; nat US; m 58; c 1. MICROBIOLOGY. Educ: Hope Col, BA, 50; Philadelphia Col Pharm, MS, 52; Univ Pa, PhD(microbiol), 57. Prof Exp: Asst instr bact, Philadelphia Col Pharm, 50-52; res fel microbiol, Univ Pa, 53-59; res assoc, Jefferson Med Col, 59-67; assoc clin bact, William Pepper Lab, Dept Path, Sch Med, Univ Pa, 67-75; DIR RES & DEVELOP, SERODIAGNOSTIC INC, 75- Mem: Am Soc Microbiol. Res: Immunology; biochemistry and immunology of Mycoplasma, Neisseria gonorrheae and other organisms; mechanisms of pathogenicity of the staphylococci and host-parasite relationships; mechanisms of drug-resistance, especially metabolic and enzymatic changes; developing rapid diagnostic tests for gonorrhea and mycoplasma. Mailing Add: 4001 Manayunk Ave Philadelphia PA 19128

LAM, HARRY CHI-SING, b Hong Kong, Nov 10, 36. THEORETICAL HIGH ENERGY PHYSICS. Educ: McGill Univ, BSc, 58; Mass Inst Technol, PhD(physics), 63. Prof Exp: Res assoc physics, Univ Md, 63-65; from asst prof to assoc prof, 65-75, PROF PHYSICS, McGILL UNIV, 75- Concurrent Pos: Asst ed, Can J Physics, 73- Mem: Am Phys Soc; Can Asn Physicists. Res: Quantum field theory. Mailing Add: Dept of Physics McGill Univ Montreal PQ Can

LAM, KWOK-WAI, b Kowloon, Hong Kong, Sept 21, 35; m 61; c 2. BIOCHEMISTRY. Educ: ETex Baptist Col, BS, 57; Univ Pittsburgh, PhD(biochem), 63. Prof Exp: Nat Inst Child Health & Human Develop fel enzymol & geront, 63-65, assoc, 65-66; assoc enzymol, Retina Found, Boston, 66-73; RES ASSOC PROF BIOCHEM, ALBANY MED COL, 73- Concurrent Pos: NIH career develop award, 67; asst prof biochem, Sch Med, Boston Univ, 70- Mem: AAAS; Am Chem Soc; Asn Res Vision & Ophthal; Fedn Am Socs Exp Biol; Nat Registry Clin Chem. Res: Mechanism of oxidative phosphorylation; clinical enzymology. Mailing Add: Dept of Biochem Albany Med Col Albany NY 12208

LAM, PING-FUN, b Hong Kong, Dec 16, 38. MATHEMATICS. Educ: Wash State Univ, BA, 62; Yale Univ, MA, 65, PhD(math), 67. Prof Exp: From instr to asst prof math, Wesleyan Univ, 66-69; asst prof, 69-71, ASSOC PROF MATH, UNIV MO-COLUMBIA, 71- Concurrent Pos: Asst, Sch Math, Inst Advan Study, 71-72. Mem: Am Math Soc. Res: Topological dynamics; Morse theory. Mailing Add: Dept of Math Univ of Mo Columbia MO 65201

LAM, ROBERT LEE, b Honolulu, Hawaii, Jan 4, 17; m 54, 63. NEUROLOGY. Educ: Univ Mich, AB, 41, MD, 44; Am Bd Psychiat & Neurol, dipl. Prof Exp: Instr neurol, Sch Med, Univ Colo, 47-48; asst neuropath, Neuropsychiat Inst, Med Sch, Univ Mich, 48-49; fel neurol, Sch Med, Washington Univ, 49-50, from instr to asst prof, 50-65, asst neurologist, Barnes & Affil Hosps & Univ Clin, 50-65; CHIEF ELECTROENCEPHALOG, DEACONESS HOSP, 60-, LUTHERAN HOSP, 62- & ST ELIZABETH HOSP & WARREN MURRAY CHILDREN'S CTR, 65- Concurrent Pos: Asst neurologist, St Louis Children's Hosp; consult, St Louis City Hosp, 49-, Homer G Phillips Hosp, 49- & St Louis Soc Crippled Children; vis neurologist, Malcolm Bliss Hosp, 50; assoc neuropsychiatrist, Mo Baptist Hosp, 55- & Deaconess Hosp, 59-; dir, St Louis Electroencephalog Lab; pres, Neuropsychiat Consults Ltd; consult neurol, psychiat & electroencephalog. Mem: AMA; Am Psychiat Asn; Am Acad Neurol; Am Col Physicians. Mailing Add: Suite 902 Clayton Inn Ctr 7777 Bonhomme Ave Clayton MO 63105

LAM, RONALD KA-WEI, b Hong Kong, Mar 3, 41; US citizen. OCEANOGRAPHY. Educ: Mass Inst Technol, SB, 63; Stanford Univ, MS, 65; Scripps Inst Oceanog, PhD(oceanog), 71. Prof Exp: Oceanogr, Scripps Inst Oceanog, 71-73; RES ASSOC OCEANOG, UNIV WASH, 73- Mem: Sigma Xi; The Coastal Soc. Res: Fjord dynamics; models of marine primary and secondary production. Mailing Add: Dept of Oceanog Univ Wash Seattle WA 98195

LAM, SHEUNG TSING, b Hong Kong, Dec 11, 34; Can citizen. NUCLEAR PHYSICS. Educ: Univ Hong Kong, BSc, 59; Univ Ottawa, MSc, 62; Univ Alta, PhD(physics), 67. Prof Exp: Demonstr physics, Univ Hong Kong, 59-60; Can Nat Coun fel & res assoc nuclear physics, Univ Toronto, 67-70; asst prof nuclear physics, Univ Va, 70-74; MEM FAC STAFF, NUCLEAR RES CTR, UNIV ALTA, 74- Concurrent Pos: Attached staff mem, Chalk River Nuclear Labs, Atomic Energy Can Ltd, 67-70; Frederick Gardner Cottrell Res Corp grant, 71-72. Mem: Am Phys Soc. Res: Nuclear structure studies using electrostatic accelerators. Mailing Add: Nuclear Res Ctr Univ of Alta Edmonton AB Can

LAM, SHUE-LOCK, b Hainan, China, Nov 19, 14; m 44; c 5. CYTOGENETICS, PLANT PHYSIOLOGY. Educ: Lingnan Univ, BS, 44; Okla State Univ, MS, 55; Univ Ill, PhD(hort), 58. Prof Exp: Technician & dir plant prod div, Lignan Univ, 33-46; dir tech dept, Garden & Forest Admin, 46-49; specialist, Joint Comn Rural Reconstruct, Chinese & Am Govts, 49-50; instr biol, Chung Chi Col, Hong Kong, 50-53; from instr to asst prof, 58-68, ASSOC PROF HORT, PURDUE UNIV, WEST LAFAYETTE, 68- Mem: AAAS; Am Soc Hort Sci; Bot Soc Am; Am Soc Plant Physiol; Am Genetic Asn. Res: Cytogenetics of horticulture crops. Mailing Add: Breeding Lab Dept of Hort Purdue Univ Agr Exp Sta West Lafayette IN 47906

LAM, TSIT-YUEN, b Hong Kong, Feb 6, 42; m 70. ALGEBRA. Educ: Hong Kong Univ, BA, 63; Columbia Univ, PhD(math), 67. Prof Exp: Fel math, Univ Ill, Urbana, 67; instr, Univ Chicago, 67-68; lectr, 68-69, asst prof, 69-72, ASSOC PROF MATH, UNIV CALIF, BERKELEY, 72-, VCHMN DEPT, 75- Concurrent Pos: Alfred P Sloan Found fel, 72-74. Mem: Am Math Soc. Res: Finite groups and group representation theory; quadratic forms. Mailing Add: Dept of Math Univ of Calif Berkeley CA 94720

LAM, VINH-TE, b Saigon, SVietnam, Dec 12, 39; Can citizen. PHYSICAL CHEMISTRY. Educ: Univ Montreal, BSc, 62, PhD(phys chem), 67. Prof Exp: Prof org chem, Col St Laurent, 66-67; fel, Nat Res Coun Can, 67-69; lectr phys chem & Nat Res Coun Can grant, Univ Sherbrooke, 69-72; PROF CHEM, COL BOIS-DE-BOULOGNE, 72- Mem: Am Chem Soc; Chem Inst Can. Res: Thermodynamics; thermochemistry; static and dynamic microcalorimetry; critical phenomena; surface and polymer chemistry; molecular interactions; structure of liquids and solutions. Mailing Add: 6728 Chateaubriand Montreal PQ Can

LAMANNA, CARL, b Brooklyn, NY, Dec 1, 16; m 42; c 2. MICROBIOLOGY, TOXICOLOGY. Educ: Cornell Univ, BS, 36, MS, 37, PhD(bact) 39. Prof Exp: Asst bact, Cornell Univ, 36-39; instr bact & pub health, State Col Wash, 40-41; instr bact, Ore State Col, 41-42; instr bact, Sch Med, La State Univ, 42-44, consult biologist, Fed Security Agency, 44; bacteriologist, Chem Corps, US War Dept, 45-48; from asst prof to assoc prof, Sch Hyg & Pub Health, Johns Hopkins Univ, 48-57; sci dir, Naval Biol Lab, Univ Calif, 57-61; dep & adv life sci div, Off Chief for Res & Develop, Dept Army, 61-74; ASSOC DIR PHARMACEUT RES & TESTING, FOOD & DRUG ADMIN, DEPT HEALTH, EDUC & WELFARE, 75- Concurrent Pos: Vis prof, Inst Hyg, Philippines, 54-55. Mem: AAAS; Am Soc Microbiol; Soc Exp Biol & Med; Am Acad Microbiol; NY Acad Sci. Res: Botulism; nature of bacterial neurotoxin; spore-forming bacteria. Mailing Add: 3812 37th St N Arlington VA 22207

LAMANNA, JOSEPH CHARLES, b Bronxville, NY, July 12, 49; m 71; c 1. NEUROSCIENCES. Educ: Georgetown Univ, BS, 71; Duke Univ, PhD(physiol), 75. Prof Exp: NIH FEL & RES ASSOC PHYSIOL, DUKE UNIV MED CTR, 75- Mem: Soc Neurosci. Res: Determining the role of oxygen and oxidative energy metabolism in the function of the central nervous system in mammals, utilizing optical monitoring techniques. Mailing Add: Dept of Physiol & Pharmacol Duke Univ Med Ctr Durham NC 27710

LAMAR, CARLOS, JR, medicine, biochemistry, see 12th edition

LAMAR, DONALD LEE, b Glendale, Calif, May 6, 30; m 53; c 2. GEOLOGY, GEOPHYSICS. Educ: Calif Inst Technol, BS, 52; Univ Calif, Los Angeles, MA, 59, PhD(geol), 61. Prof Exp: Phys scientist, Rand Corp, 60-64; VPRES, LAMAR-MERIFIELD, 64- Mem: Am Asn Petrol Geologists; Am Geophys Union; Geol Soc Am; Meteoritical Soc. Res: Internal structure of terrestrial bodies; electromagnetic effects of fireballs; effects of tides early in earth-moon history; structural geology of Southern California. Mailing Add: Lamar-Merifield Suite 27 1318 Second St Santa Monica CA 90401

LAMAR, EDWARD STONESTREET, physics, see 12th edition

LA MAR, GERD NEUSTADTER, b Brasov, Romania, Dec 21, 37; US citizen; m 64; c 2. STRUCTURAL CHEMISTRY. Educ: Lehigh Univ, BS, 60; Princeton Univ, PhD(chem), 64. Prof Exp: NSF fel, 64-66; NATO fel, 66-67; res chemist, Shell Develop Co, 67-70; from asst prof to assoc prof, 71-74, PROF CHEM, UNIV CALIF, DAVIS, 74- Concurrent Pos: Fel, Alfred P Sloan Found, 72, John Simon Guggenheim Mem Found, 75. Mem: Am Chem Soc. Res: The use of magnetic resonance spectroscopy as a tool for elucidating structure-function relationships in metallo-enzymes and their model complexes. Mailing Add: Dept of Chem Univ of Calif Davis CA 95616

LAMAR, JULE K, b Birmingham, Ala, Dec 5, 09; m 39; c 2. TOXICOLOGY. Educ: Birmingham-Southern Col, BS, 31; Univ Chicago, PhD(zool), 38. Prof Exp: Asst biol, Birmingham-Southern Col, 31-33, instr, 33-34; asst zool, Univ Chicago, 34-38; Nat Comt Maternal Health fel, Carnegie Lab Embryol, 38-40; instr obstet & gynec, Med Sch, Univ Tex, 40-43, from asst prof to assoc prof, 43-64; physiologist, Div Toxicol Eval, Bur Sci, 64-66, pharmacologist, Bur Med, 66-67, supvry pharmacologist, 67-70, PHARMACOLOGIST, BUR DRUGS, US FOOD & DRUG ADMIN, 70- Mem: Assoc Am Soc Zool; assoc Soc Exp Biol & Med; Soc Toxicol. Res: Sperm migration and purification; hormone assays; pregnancy testing; cancer and hormones; drug actions and toxicology; adverse effects on reproduction; mutagenicity testing; carcinogenicity testing. Mailing Add: US Food & Drug Admin 5600 Fishers Lane Rockville MD 20852

LAMARCA, MICHAEL JAMES, b Jamestown, NY, June 4, 31; m 54; c 3. DEVELOPMENTAL BIOLOGY. Educ: State Univ NY Albany, AB, 53; Cornell Univ, PhD(zool), 61. Prof Exp: Instr zool, Rutgers Univ, 61-63, asst prof, 63-65; asst prof biol, 65-67, chmn dept, 70, ASSOC PROF BIOL, LAWRENCE UNIV, 67- Concurrent Pos: NSF res grant, 63-65; resident dir, Assoc Cols Midwest Argonne Semester Prog, Argonne Nat Lab, 68-69; NSF sci fac fel biol sci, Purdue Univ, 71-72. Mem: AAAS; Am Soc Zoologists. Res: RNA synthesis in echinoderm and amphibian development. Mailing Add: Dept of Biol Lawrence Univ Appleton WI 54911

LAMARCHE, GUY, b Quebec City, Que, Dec 24, 28; m 53; c 2. NEUROPHYSIOLOGY. Educ: Univ Montreal, BA, 47; Laval Univ, BM, 49, MD, 52. Prof Exp: From asst prof to assoc prof physiol, Laval Univ, 58-67; chmn div basic med sci, 67-71, PROF PHYSIOL & CHMN DEPT, SCH MED, UNIV SHERBROOKE, 67- Concurrent Pos: Lederle med fac award, 58-61. Mem: AAAS; Am Col Physicians; French-Can Asn Advan Sci; Am Asn Anatomists; Asn French Speaking Physiologists. Res: Reticular formation and sensory projections to the brain stem by unit activity evoked by various modalities of physiological stimulation; experimental epilepsy in the brain stem and other subcortical structures of the cat. Mailing Add: Dept of Physiol Univ of Sherbrooke Sch of Med Sherbrooke PQ Can

LAMARCHE, J L GILLES, b Montreal, Que, May 31, 27. PHYSICS. Educ: Univ Montreal, BSc, 50; Univ BC, MA, 53, PhD(physics), 57. Prof Exp: From asst prof to assoc prof, 57-70, PROF PHYSICS, UNIV OTTAWA, 70- Mem: Can Asn Physicists. Res: Low temperature physics; adiabatic demagnetization; metal physics. Mailing Add: Dept of Physics Univ of Ottawa Ottawa ON Can

LAMARCHE, PAUL H, b Boston, Mass, Sept 5, 29; m 52; c 5. GENETICS, PEDIATRICS. Educ: Boston Col, BS, 56; Boston Univ, MD, 60; Mass Inst Technol, ScM, 74. Prof Exp: Res assoc path & dir genetics lab, RI Hosp, 63-74, med dir, Birth Defects Ctr, 65-74, med dir child develop ctr, 66-74, assoc physician-in-chief pediat, 69-74; PROF GENETICS, UNIV MAINE, ORONO, 74-; CHIEF PEDIAT & GENETICS, EASTERN MAINE MED CTR, BANGOR, 74- Concurrent Pos: Asst pediatrician, Providence Lying-In Hosp, 63-74, consult, 66-74; prin investr Nat Cancer Inst grant, 64-69; prog consult & site visitor, Nat Found, 66-74; consult, Child Study Ctr, Brown Univ, 67-74. Mem: AAAS; Genetics Soc Am; Tissue Cult Asn. Res: Genetics and cytogenetics of teratogenesis and oncogenesis; electron microscopy of fine structure of somatic cellular phenotypes normal and abnormal in the human. Mailing Add: 489 State St Bangor ME 04401

LAMARCHE, VALMORE CHARLES, JR, b Hurley, Wis, Aug 27, 37; m 57; c 4. GEOLOGY. Educ: Univ Calif, Berkeley, BA, 60; Harvard Univ, MA, 62, PhD(geol), 64. Prof Exp: Geologist, US Geol Surv, 62-67, hydrologist, 67; res assoc, 67-69, assoc prof dendrochronol, 69-74, PROF DENDROCHRONOL, UNIV ARIZ, 74- Mem: AAAS; Geol Soc Am; Ecol Soc Am; Int Asn Quaternary Res. Res: Geomorphology; hydrology; applications of tree-ring studies to geological problems; paleoclimatology; dendrochronology. Mailing Add: Lab for Tree-Ring Res Univ of Ariz Tucson AZ 85721

LAMAZE, GEORGE PAUL, b Algiers, Algeria, Jan 15, 45; US citizen; m 65; c 2. EXPERIMENTAL NUCLEAR PHYSICS. Educ: Fla State Univ, BA, 65; Duke Univ, PhD(physics), 72. Prof Exp: PHYSICIST NEUTRON STAND, NAT BUR STAND, 72- Mem: Am Phys Soc. Res: Measurement of standard neutron reaction cross sections; neutron resonance scattering; neutron capture cross sections. Mailing Add: Nat Bur of Stand Bldg 245 B119 Washington DC 20234

LAMB, ALBERT R, JR, b New York, NY, Dec 3, 13; m 42; c 4. MEDICINE. Educ: Yale Univ, BA, 35; Columbia Univ, MD, 40. Prof Exp: Instr & asst med, 46-54, asst prof clin med, 54-61, ASSOC PROF CLIN MED, PRESBY MED CTR, COLUMBIA UNIV, 61- Concurrent Pos: Dir student health serv, Col Physicians & Surgeons, Columbia Univ, 46-74; consult, Englewood Hosp, NJ, 55- Mem: AMA. Res: Internal medicine. Mailing Add: Columbia-Presby Med Ctr 161 Ft Washington Ave New York NY 10032

LAMB, DENNIS, b Chicago, Ill, Feb 3, 41. CLOUD PHYSICS. Educ: Kalamazoo Col, BA, 63; Univ Wash, PhD(atmospheric sci), 70. Prof Exp: Gen physicist data assessment, Naval Weapons Ctr, China Lake, Calif, 63-65; NATO res assoc meteorol, Univ Frankfurt, 71-72; ASST PROF ATMOSPHERIC PHYSICS, LAB ATMOSPHERIC PHYSICS, DESERT RES INST, UNIV NEV, RENO, 72- Mem: Am Meteorol Soc. Res: Nucleation and growth of solids from the liquid and vapor phases; formation and modification of cloud nuclei by gas phase reactions. Mailing Add: Lab of Atmospheric Physics Desert Res Inst Univ of Nev Reno NV 89507

LAMB, DONALD JOSEPH, b Pittsburgh, Pa, Oct 29, 31; m 56; c 2. PHARMACY. Educ: Ohio State Univ, BSc, 54, MSc, 55, PhD(pharm), 60. Prof Exp: Res assoc pharmaceut res & develop, 60-65, res head, 65-70, RES MGR PHARMACEUT RES & DEVELOP, UPJOHN CO, 70- Mem: Am Pharmaceut Asn; Am Acad Pharmaceut Sci; Am Chem Soc. Res: Design and evaluation of drug dosage forms, including design and evaluation of drugs to fit specific dosage forms. Mailing Add: Upjohn Co 7171 Portage Rd Kalamazoo MI 49001

LAMB, DONALD QUINCY, JR, b Manhattan, Kans, June 30, 45. THEORETICAL PHYSICS. Educ: Rice Univ, BA, 67; Univ Liverpool, MSc, 69; Univ Rochester, PhD(physics), 74. Prof Exp: Res asst prof, 73-75, ASST PROF PHYSICS, UNIV ILL, 75- Mem: Am Phys Soc; Am Astron Soc; fel Royal Astron Soc; Brit Inst Physics; Europ Phys Soc. Res: Evolution and structure of white dwarfs and neutron stars; physics of compact x-ray sources, novae and pulsars; properties of matter at high densities. Mailing Add: Dept of Physics Univ of Ill Urbana IL 61801

LAMB, FRANK BRUCE, b Cotopaxi, Colo, July 27, 13; m 41; c 1. FORESTRY. Educ: Univ Mich, BSF, 40, MF, 41, PhD(trop forestry), 54. Prof Exp: Engr, US Dept Eng, BWI, 41-42; field technician, Rubber Develop Corp, Brazil, 43-45; consult, PR Indust Develop Co, 46-47; forester, Inst Develop Prod, Guatemala, 50 & US Plywood Corp, Panama, 51-53; forester, Inst Colonization, Colombia, 53-54; consult, Industria Forestal, Colombia, 54, Guatemalan Forest Serv, 55, US Plywood Corp, Panama, 56, W R Grace & Co, 57, Int Coop Admin, Panama, 57 & Container Corp Am, Colombia, 58; res forester, Trop Forest Res Ctr, PR, US Forest Serv, 59-60; forester, Raw Mat Procurement, US Plywood Corp, 60-69, tech dir natural resources, Champion Int Corp, 69-75; RETIRED. Honors & Awards: Dipl de Honor, Ministry Agr, Guatemala. Mem: Soc Am Foresters; NY Acad Sci; Int Soc Trop Foresters. Res: Tropical forest surveys and management; study of mahogany; global ecology and international relations in forestry. Mailing Add: 3418 W. 37th St Topeka KS 66614

LAMB, FRANK WYMAN, b Akiak, Alaska, Aug 17, 18. PHYSICAL CHEMISTRY. Educ: Univ Wash, BS, 39; PhD(phys chem), 43. Prof Exp: Carnegie Inst fel, Northwestern Univ, 46; opers res, US Navy Proj, Div Indust Coop, Mass Inst Technol, 47-56; CONSULT SCIENTIST, LOCKHEED AIRCRAFT CORP, 56- Mem: Am Chem Soc; Opers Res Soc Am. Res: Compressibility and specific heat of aqueous solutions; operations research. Mailing Add: 2085 Emerson Palo Alto CA 94301

LAMB, FREDERICK KEITHLEY, b Manhattan, Kans, June 30, 45; m 71. THEORETICAL PHYSICS. Educ: Calif Inst Technol, BS, 67; Oxford Univ, PhD(theoret physics), 70. Prof Exp: Instr & res assoc, 70-72, asst prof, 72-75, ASSOC PROF PHYSICS, UNIV ILL, URBANA, 75- Concurrent Pos: Fel physics, Magdalen Col, Oxford Univ, 70-72; assoc, Ctr Advan Study, Univ Ill, Urbana, 73-74; res fel, Alfred P Sloan Found, 74- Mem: Am Phys Soc; Am Astron Soc; fel Royal Astron Soc. Res: Pulsars; compact x-ray sources; supernovae aad supernova remnants; magnetic white dwarfs. Mailing Add: Dept of Physics Univ of Ill Urbana IL 61801

LAMB, GEORGE ALEXANDER, b Glens Falls, NY, Sept 25, 34; m 56; c 3. PEDIATRICS, INFECTIOUS DISEASES. Educ: Swarthmore Col, BS, 55; State Univ NY Upstate Med Ctr, MD, 59. Prof Exp: Intern pediat, State Univ NY Upstate Med Ctr, 59-60, resident, 60-62, from asst prof to assoc prof, 64-72; ASSOC PROF PREV & SOCIAL MED, HARVARD MED SCH, 72- Concurrent Pos: Fel infectious dis, 64- Res: Infectious diseases of children, especially the epidemiology of respiratory illnesses; community child health. Mailing Add: Dept of Prev & Social Med Harvard Med Sch Boston MA 02115

LAMB, GEORGE LAWRENCE, JR, b Norwood, Mass, Apr 28, 31; m 59; c 4. PHYSICS. Educ: Boston Col, BS, 53, MS, 54; Mass Inst Technol, PhD(physics), 58. Prof Exp: Staff mem, Los Alamos Sci Lab, 58-63; physicist, United Aircraft Res Labs, Conn, 63-76; FAC MEM, DEPT MATH & OPTICAL SCI CTR, UNIV ARIZ, 76- Mem: Am Phys Soc. Res: Wave propagation. Mailing Add: Dept of Math & Optical Sci Ctr Univ of Ariz Tucson AZ 85721

LAMB, GEORGE MARION, b Little Rock, Ark, Dec 23, 28; m 53; c 2. MICROPALEONTOLOGY, STRATIGRAPHY. Educ: Emory Univ, BA, 50, MS, 54; Univ Colo, Boulder, PhD(geol), 64. Prof Exp: Geologist, Standard Oil Calif, Inc, 55-61; PROF GEOL & CHMN DEPT, UNIV SOUTH ALA, 64- Mem: Am Asn Petrol Geol; Geol Soc Am. Res: Ecology and paleoecology of Foraminifera; biostratigraphic relationships; groundwater and environmental geology. Mailing Add: Dept of Geol Univ of SAla Mobile AL 36608

LAMB, JAMES FRANCIS, b Denton, Tex, Oct 3, 37; m 58; c 2. NUCLEAR MEDICINE, NUCLEAR CHEMISTRY. Educ: NTex State Univ, BS, 60, MS, 61; Univ Calif, Berkeley, PhD(chem), 69. Prof Exp: Res chemist, Lawrence Radiation Lab, Univ Calif, 69-70; prin radiochemist, 70-74, DEP DIR RES & DEVELOP, MEDI-PHYSICS, INC, 74- Mem: Am Chem Soc; Soc Nuclear Med; AAAS. Res: Nuclear chemistry in nuclear medical and radiopharmaceutical applications. Mailing Add: Medi-Physics Inc PO Box 8684 Emeryville CA 94608

LAMB, JAMES L, b Los Angeles, Calif, Jan 17, 25; m 45; c 2. MICROPALEONTOLOGY. Educ: Univ Southern Calif, BS, 53. Prof Exp: Paleontologist, Richfield Oil Corp, 53-57 & Creole Petrol Corp, 57-64; PALEONTOLOGIST, EXXON PROD RES CO, 64- Mem: Soc Econ Paleont & Mineral; Am Asn Petrol Geologists; Venezuelan Asn Geol, Mining & Petrol. Res: Historical geology and paleontology; geologic distribution of planktonic foraminifera; tertiary microfossils; Pleistocene epoch. Mailing Add: Exxon Prod Res Co PO Box 2189 Houston TX 77001

LAMB, LAWRENCE EDWARD, b Fredonia, Kans, Oct 13, 26. MEDICINE. Educ: Univ Kans, MD, 49; Am Bd Internal Med, dipl, 58. Prof Exp: Intern, Med Ctr, Univ Kans, 49-50, resident internal med, 50-51; chief, Cardiovasc & Renal Sect, Sheppard AFB Hosp, 51-53; asst, Emory Univ, 54; dir cardiol, Dept Internal Med, Sch Aviation Med, Randolph AFB, Tex, 55-57, chief dept, 57-58, prof internal med & chief dept, Sch Aviation Med, Brooks AFB Aerospace Med Ctr, 58-61, dir consult serv, 59-61, chief, Aerospace Med Sci Div, US Air Force Sch Aerospace Med, 61-66; prof med, Col Med, Baylor Univ, 66-71; SYNDICATED MED COLUMNIST, NEWSPAPER ENTERPRISE ASN, 71- Concurrent Pos: Teaching fel, Emory Univ, 53; Am Heart Asn res fel, Geneva, Switz, 54-55; mem conf electrocardiog probs in aviation, Royal Can Air Force, 56; lectr, St Thomas, Manila, 57, Med Sch, Stanford Univ, 58 & Life Sci Div, Nat Acad Sci, 58; fel coun epidemiol, Am Heart Asn; consult, Mercury Proj, NASA, 60, consult to dir life sci, 65-; consult, President's Coun Phys Fitness & Sports, 62- Honors & Awards: Tuttle Award, Civil Aviation Med Asn, 59; Distinguished Civilian Serv Award, Dept Defense, 62; Meritorious Civilian Serv Award, Dept Air Force, 66. Mem: Fel Aerospace Med Asn; fel Am Col Chest Physicians; fel Am Col Cardiol; fel Am Col Physicians; fel Am Soc Clin Pharmacol & Chemother. Res: Myocardial infarction; cardiology; stresses and effects of influence of space flight on cardiovascular system; results of vectorcardiograms and electrocardiograms. Mailing Add: 135 Downing Dr San Antonio TX 78209

LAMB, MINA MARIE WOLF, b Sagerton, Tex, Aug 14, 10; m 41; c 1. NUTRITION. Educ: Tex Tech Col, BA, 32, MS, 37; Columbia Univ, PhD(nutrit, chem), 42. Prof Exp: Teacher, elem & high sch, 33-35; teacher & res worker food & nutrit, 35-37; from lab asst to prof, 40-69, Margaret W Weeks distinguished prof, 69-75, head dept food & nutrit, 55-69, lectr & adv foreign students, 60-71, EMER PROF FOOD & NUTRIT, TEX TECH UNIV, 75- Honors & Awards: Piper Award, 65. Mem: AAAS; Am Dietetic Asn; Am Home Econ Asn; Am Pub Health Asn; Am Inst Nutrit. Res: Basal metabolism of college girls and children of various ages older than two years; needs of children and adults; dietary studies of children, college girls and families; animal feeding work with albino rats determining growth and reproduction responses to various diets and foods; studies of motivation and behavioral modification. Mailing Add: 6002 W 34th St Lubbock TX 79407

LAMB, NEVEN P, b New York, NY, May 18, 32; m 57; c 2. BIOLOGICAL ANTHROPOLOGY. Educ: Pa State Univ, BA, 54; Univ Ariz, PhD, 69. Prof Exp: Res asst morphogenetics, Jackson Mem Lab, Maine, 60-61; from asst prof to assoc prof anthrop, Portland State Univ, 65-73; ASSOC PROF ANTHROP, TEX TECH UNIV, 73- Concurrent Pos: Vis prof, Univ Ariz, 71-72. Mem: AAAS; fel Am Anthrop Asn; Am Asn Phys Anthrop; Soc Study Evolution; Am Eugenics Soc. Res: Human evolution; population biology; mating patterns and genetic systems; anthropometry of North American Indians; socio-cultural and biological aspects of mate selection among Papago Indians. Mailing Add: Dept of Anthrop Tex Tech Univ Lubbock TX 79409

LAMB, RICHARD C, b Lexington, Ky, Sept 8, 33; m 59; c 4. COSMIC RAY PHYSICS, ELEMENTARY PARTICLE PHYSICS. Educ: Mass Inst Technol, BS, 55; Univ Ky, PhD(physics), 63. Prof Exp: Asst scientist, Argonne Nat Lab, 63-67; assoc prof physics, 67-72, PROF PHYSICS, IOWA STATE UNIV, 72- Concurrent Pos: Vis scientist, NASA-Goddard Space Flight Ctr, 75-76. Mem: Am Phys Soc. Res: Very high energy gamma ray astronomy. Mailing Add: Dept of Physics Iowa State Univ Ames IA 50010

LAMB, ROBERT CARDON, b Logan, Utah, Jan 8, 33; m 53; c 5. DAIRY SCIENCE. Educ: Utah State Univ, BS, 56; Mich State Univ, MS, 59, PhD(dairy cattle breeding), 62. Prof Exp: Instr dairy sci, Mich State Univ, 58-60; asst prof, Utah State Univ, 61-64; res dairy husbandman, 64-72, RES LEADER, ANIMAL HUSB RES DIV, AGR RES SERV, USDA, 72- Mem: Am Dairy Sci Asn. Res: Use of incomplete records in dairy cattle selection; genetics by nutrition interactions; inheritance of abnormalities in livestock; feed utilization efficiency in dairy cattle; dairy herd management; exercise for dairy cows. Mailing Add: Dept of Dairy Sci Utah State Univ Logan UT 84322

LAMB, ROBERT CHARLES, b Union Co, SC, Sept 28, 28; m 50; c 2. ORGANIC CHEMISTRY. Educ: Presby Col, SC, BS, 48; Univ Ga, MS, 55; Univ SC, PhD(chem), 58. Prof Exp: Instr chem, Presby Col, SC, 50-51; asst prof, Univ Ga, 58-66; assoc prof & chmn dept, Augusta Col, 66; PROF CHEM & CHMN DEPT, E CAROLINA UNIV, 66- Mem: Am Chem Soc. Res: Organic peroxides; free radicals in solution; chemical kinetics. Mailing Add: Dept of Chem ECarolina Univ PO Box 2787 Greenville NC 27834

LAMB, ROBERT CONSAY, b Saskatoon, Sask, May 11, 19; nat US; m 49; c 3. POMOLOGY. Educ: Univ Sask, BSA, 41; Univ Minn, MS, 47, PhD(hort), 54. Prof Exp: Asst hort, Univ Minn, 46-48; asst prof pomol, 48-55, ASSOC PROF POMOL, NY STATE COL AGR & LIFE SCI, CORNELL UNIV, 55- Concurrent Pos: Orgn Europ Econ Coop sr vis fel sci, John Innes Inst, Eng, 62. Mem: Am Soc Hort Sci; Int Soc Hort Sci; Can Soc Hort Sci. Res: Breeding peaches, apricots, pears and apples. Mailing Add: NY State Agr Exp Sta Geneva NY 14456

LAMB, ROBERT EDWARD, b Sharon, Pa, July 12, 45; m 73. ANALYTICAL CHEMISTRY. Educ: St Louis Univ, AB, 69, BS, 70; Univ Ill, MS, 74, PhD(anal chem), 75. Prof Exp: Lectr anal chem, Sch Chem Sci, Univ Ill, 75; ASST PROF CHEM, SOUTHERN METHODIST UNIV, 75- Mem: Am Chem Soc. Res: Pulse polarography and stripping analysis; ion-selective electrodes; analysis of trace metal complexes; environmental applications of analytical techniques. Mailing Add: Dept of Chem Southern Methodist Univ Dallas TX 75275

LAMB, ROBERT W, b Jasonville, Ind, Feb 22, 29; m 53; c 3. ORGANIC CHEMISTRY, PHYSICAL CHEMISTRY. Educ: Ariz State Univ, BS, 50; NMex Highlands Univ, MA, 51; Univ Colo, PhD(phys org chem), 61. Prof Exp: US Air Force, 53-, res chemist, US Army Chem Ctr, 53-54 & US Air Force Mat Lab, 54-55, instr chem, US Air Force Acad, 56-57, asst prof, 57-62, assoc prof, 63-65, res assoc, Frank J Seiler Res Lab, 62-63, proj officer chem, Europ Off Aerospace Res, Brussels, Belg, 65-68, assoc prof chem, US Air Force Acad, 68-70, PROF CHEM & HEAD DEPT, US AIR FORCE ACAD, 70- Mem: AAAS; Am Chem Soc. Res: Organic fluorine; absorption spectroscopy; chemical oceanography. Mailing Add: Dept of Chem US Air Force Academy CO 80840

LAMB, SANDRA INA, b New York, NY, Apr 20, 31; m 50; c 4. ORGANIC CHEMISTRY, PHARMACOLOGY. Educ: Univ Calif, Los Angeles, BS, 54, PhD(phys org chem), 59. Prof Exp: Instr chem, Santa Monica City Col, fall 59; asst prof, San Fernando Valley State Col, 60-61; instr, Exten Div, Univ Calif, 61-69; asst prof, 69-71, ASSOC PROF CHEM, MT ST MARY'S COL, CALIF, 71-, CHMN DEPT PHYS SCI & MATH, 69- Concurrent Pos: Asst res pharmacologist, Med Sch, Univ Calif, 66-, lectr, 70. Mem: AAAS; Am Chem Soc. Res: Analytical applications of gas chromatography in chemistry and medicine with special interest in analysis of acetylcholine and various cholinergic agents; mechanism of action of muscarinic agents; synthesis of small ring compounds. Mailing Add: Dept of Phys Sci & Math Mt St Mary's Col 12001 Chalon Rd Los Angeles CA 90049

LAMB, WALTER ROBERT, b Weiser, Idaho, Sept 26, 22; m 46; c 1. PHYSICS. Educ: Univ Calif, AB, 48. Prof Exp: Physicist, US Naval Radiol Defense Lab, 48-59; solid state physicist, Res & Develop Dept, Rheem Semiconductor Corp, 59-63, Fairchild Semiconductor, 63-64 & Union Carbide Corp, 64-65; mgr advan processing, Stewart-Warner Microcircuits, 65-68; physicist, Fairchild Semiconductor Corp, 68-71; PHYSICIST, RAYTHEON CO, 71- Mem: AAAS; Am Phys Soc. Res: Solid state, nuclear, atomic, optical, luminescent, thermodynamic and gravitational phenomena. Mailing Add: 148 Jacinto Way Sunnyvale CA 94086

LAMB, WILLIS EUGENE, JR, b Los Angeles, Calif, July 12, 13; m. QUANTUM MECHANICS, ATOMIC PHYSICS. Educ: Univ Calif, BS, 34, PhD(physics), 38; Oxford Univ, MA, 56; Yale Univ, MA, 61. Hon Degrees: ScD, Univ Pa, 54; LHD, Yeshiva Univ, 65; ScD, Gustavus Adolphus Col, 75. Prof Exp: Asst physics, Univ Calif, 34-35, 36-37; instr, Columbia Univ, 38-43, assoc, 43-45, from asst prof to prof physics, 45-52; prof, Stanford Univ, 51-56; Wykeham prof & fel, New Col, Oxford Univ, 56-62; Henry Ford II prof, Yale Univ, 62-72, Josiah Willard Gibbs prof, 72-74; PROF PHYSICS & OPTICAL SCI, UNIV ARIZ, 74- Concurrent Pos: Mem staff, Radiation Lab, Columbia Univ, 43-52; Loeb lectr, Harvard Univ, 53-54; Guggenheim fel, 60-61; consult, Philips Labs, Inc, NASA, Bell Tel Labs & Perkin-Elmer Corp. Honors & Awards: Rumford Medal, Am Acad Arts & Sci, 53; Nobel Prize Physics, 55; Award, Res Corp, 55. Mem: Nat Acad Sci; fel Am Phys Soc; fel NY Acad Sci; hon fel Brit Inst Physics; fel Optical Soc Am. Res: Theoretical physics; atomic and nuclear structure; microwave spectroscopy; fine structure of hydrogen and helium; magnetron oscillators; statistical mechanics; masers and lasers. Mailing Add: Dept of Physics Univ of Ariz Tucson AZ 85721

LAMBA, RAM SARUP, b Calcutta, India. INORGANIC CHEMISTRY, ORGANIC CHEMISTRY. Educ: Delhi Univ, India, BSc, 62, MSc, 64; ETex State Univ, DEd(inorg chem, educ), 73. Prof Exp: Res asst, Indian Inst Petrol, Dehradun, India, 64-65; chemist & supt dyeing & finishing, Beaunit Corp of NC, Humacao, PR, 68-69; instr chem, 69-70, asst prof & chmn dept, 70-71, ASSOC PROF CHEM, MATH & PHYSICS, INTER AM UNIV PR, 73-, CHMN DEPT NATURAL SCI, 73- Mem: Am Chem Soc; The Chem Soc. Res: To develop innovative methods in the teaching of college chemistry and to integrate with biological sciences; synthesis and study of chromium (III), complexes. Mailing Add: Dept of Chem Math & Physics Inter Am Univ PO Box 1293 Hato Rey PR 00919

LAMBA, SURENDAR SINGH, b India, Mar 3, 36; m 67; c 1. PHARMACY, PHARMACOGNOSY. Educ: Agra Univ, BSc, 54; Univ Rajasthan, BPharm, 57; Panjab Univ, India, MPharm, 60; Univ Nebr, MS, 63; Univ Colo, PhD(pharmacog), 66. Prof Exp: Assoc prof, 66-67, PROF PHARMACOG, FLA A&M UNIV, 67- Honors & Awards: Lederle Fac Award, 75. Mem: Am Pharmaceut Asn; Acad Pharmaceut Sci; NY Acad Sci; Am Soc Pharmacog. Res: Tissue culture studies; effects of growth retardants on growth and alkaloid biosynthesis; phytochemical investigation of some members of Papaveraceae; microbial transformation of Nucleosides. Mailing Add: Dept of Pharm-Pharmacog Fla A&M Univ Tallahassee FL 32307

LAMBDIN, MORRIS ARTHUR, b Painter, Va, May 5, 21; m 45; c 4. PEDIATRICS. Educ: Randolph-Macon Col, BS, 42; Univ Va, MD, 46; Am Bd Pediat, dipl, 54. Prof Exp: Intern med, Univ Va, 46-47, from asst resident to resident pediat, 50-52; pediatrician, Peninsula Gen Hosp, Salisbury, Md, 52-56; asst prof pediat, Univ Va, 56-62; CHIEF PEDIAT, MAINE COAST MEM HOSP, 62-; INSTR PEDIAT, HARVARD MED SCH, 68- Concurrent Pos: Fel hemat, Univ Va, 49-50. Mem: AMA; Am Acad Pediat. Res: Clinical pediatric hematology. Mailing Add: Maine Coast Mem Hosp Ellsworth ME 04605

LAMBDIN, PARIS LEE, b St Charles, Va, Oct 13, 41; m 64; c 2. ENTOMOLOGY. Educ: Lincoln Mem Univ, BA, 64; Va Polytech Inst & State Univ, MS, 72, PhD(entom), 74. Prof Exp: Teacher biol, Bassett High Sch, 64-66; ASST PROF ENTOM, DEPT AGR BIOL, UNIV TENN, 74- Mem: Entom Soc Am. Res: Systematics of species in the superfamily Coccoidea; biological control of vegetable insect pests. Mailing Add: Dept of Agr Biol Univ of Tenn Knoxville TN 37916

LAMBDIN, ROBERT WILLIAM, physical chemistry, see 12th edition

LAMBE, DWIGHT WILSON, JR, b Okeechobee, Fla, May 17, 30; m 64. MEDICAL MICROBIOLOGY. Educ: Fla State Univ, BS, 53, MS, 57; Wayne State Univ, PhD(med microbiol), 66. Prof Exp: Microbiologist bact, Ctr Dis Control, 60-68; ASSOC PROF PATH & LAB MED, EMORY UNIV, 68- Concurrent Pos: Res assoc, Orebro County Hosp, Sweden, 70. Mem: Am Soc Microbiol; NY Acad Sci; Sigma Xi. Res: Human immunological response to anaerobic infection; identification of anaerobic bacteria by the immunofluorescent technique; standardized antibiotic susceptibility testing of anaerobic bacteria; serological studies of anaerobes by the agglutation test. Mailing Add: Dept of Path Emory Univ Atlanta GA 30322

LAMBE, EDWARD DIXON, b Prince Rupert, BC, July 25, 24; m 50; c 4. PHYSICS. Educ: Univ BC, BASc, 48, MASc, 49; Princeton Univ, PhD(physics), 59. Prof Exp: Asst prof physics, Washington Univ, 56-61; assoc prof, 61-65, asst vchancellor, 66-70, PROF PHYSICS, STATE UNIV NY STONY BROOK, 65-, DIR INSTRUCTIONAL RESOURCES CTR, 67- Concurrent Pos: Exec secy, Comn Col Physics, 62-64, secy, 64- Mem: Am Phys Soc; Am Asn Physics Teachers. Res: Electron and nuclear magnetic resonance; beta and gamma ray polarization. Mailing Add: Dept of Physics State Univ of NY Stony Brook NY 11790

LAMBE, JOHN JOSEPH, physics, see 12th edition

LAMBE, ROBERT CARL, b Minneapolis, Minn, Nov 25, 27; m 50; c 2. PLANT PATHOLOGY. Educ: Univ Southern Calif, AB, 52; Univ Calif, MS, 55; Ore State Col, PhD(plant path), 60. Prof Exp: Jr plant pathologist, Ore State Col, 58-60; plant pathologist, Area Exten, Tex A&M Univ, 60-63; exten plant pathologist, Iowa State Univ, 63-67; ASSOC PROF PLANT PATH, VA POLYTECH INST & STATE UNIV, 67- Res: Fungicides and extension plant pathology. Mailing Add: Dept of Plant Path Va Polytech Inst & State Univ Blacksburg VA 24060

LAMBEK, JOACHIM, b Leipzig, Ger, Dec 5, 22; nat Can; m 48; c 3. MATHEMATICS. Educ: McGill Univ, BSc, 46, MSc, 47, PhD, 51. Prof Exp: Assoc prof math, 54-63, PROF MATH, McGILL UNIV, 63- Concurrent Pos: Mem, Inst Advan Study, 59-60. Mem: Am Math Soc; Math Asn Am; Can Math Cong. Res: Algebra. Mailing Add: Dept of Math McGill Univ Montreal PQ Can

LAMBERD, WILLIAM GORDON, b Wales, Oct 29, 21; Can citizen; m 53; c 3. PSYCHIATRY. Educ: Univ Liverpool, MB, ChB, 52; Univ Man, MSc, 57; CRCP(C), 57. Prof Exp: Clin dir psychiat, Hosp Ment Dis, Selkirk, Man, 58-60; chief psychiat, Can Dept Vet Affairs Hosp, Winnipeg, 60-69; clin dir dept psychiat, Health Sci Centre, 69-75, PROF PSYCHIAT, UNIV MAN, 69-, COORDR UNDERGRAD EDUC, DEPT PSYCHIAT, HEALTH SCI CENTRE, 75- Concurrent Pos: Am Psychiat Asn spec fel, Mayo Found, Univ Minn, 57-58; consult, Can Pensions Bd, 68- Royal Can Air Force, 70- & Dept Civil Aviation Med, 72-; mem rev bd, Govt Man, 69- Res: Psychotherapy, group therapy. Mailing Add: Dept of Psychiat Univ of Man Health Sci Centre Winnipeg MB Can

LAMBERG, STANLEY LAWRENCE, b Brooklyn, NY, Oct 2, 33; m 63; c 2. BIOCHEMISTRY, HISTOLOGY. Educ: Brooklyn Col, BS, 55; Oberlin Col, MA, 57; Tufts Univ, MS, 62; NY Univ, PhD(biol), 68. Prof Exp: Teaching asst biol, Oberlin Col, 55-57; chief technician biochem, Sch Med, Cornell Univ, 57-58; lectr biol, City Col New York, 66-67; asst prof biol, Conolly Col, Long Island Univ, 67-70; from asst prof to assoc prof, 70-75, PROF MED LAB TECHNOL, STATE UNIV NY AGR & TECH COL FARMINGDALE, 75- Concurrent Pos: Asst res scientist, Guggenheim Inst Dent Res, NY Univ, 68-69; adj asst prof, 70-73, adj assoc prof, Conolly Col, Long Island Univ, 73- Honors & Awards: Founder's Day Award, NY Univ, 69. Mem: AAAS; NY Acad Sci; Sigma Xi. Res: Mitochondrial phosphorylation reactions during embryonic development; effect of ultraviolet irradiation and various inhibitors and uncoupling reagents on mitochondrial phosphorylation reactions. Mailing Add: Dept of Med Lab Technol State Univ NY Agr & Tech Col Farmingdale NY 11735

LAMBERG-KARLOVSKY, CLIFFORD CHARLES, b Prague, Czech, Oct 2, 37; US citizen; m 59; c 2. ANTHROPOLOGY, ARCHAEOLOGY. Educ: Dartmouth Col, AB, 59; Univ Pa, AM, 64, PhD(anthrop), 65. Prof Exp: Asst prof anthrop, Franklin & Marshall Col, 64-65; asst prof, 65-69, PROF ANTHROP & CUR NEAR EASTERN ARCHAEOL, HARVARD UNIV, 69- Concurrent Pos: Wenner-Gren fel, Univ Pa, 64-65; NSF grants, Harvard Univ, 65-71; dir archaeol exped to Tepe Yahya, Southeastern Iran, 67-; assoc archaeol, Columbia Univ Sem, 68-; dir res, Am Sch Prehist Res; trustee, Am Inst Persian Studies & Am Sch Orient Res. Mem: Fel Am Anthrop Asn; fel Am Orient Soc. Mailing Add: Dept of Anthrop Harvard Univ Cambridge MA 02138

LAMBERT, BERND, b Frankfurt, Ger, Dec 28, 32; US citizen. ANTHROPOLOGY. Educ: Univ Calif, Berkeley, AB, 54, PhD(anthrop), 63. Prof Exp: Actg instr anthrop, Univ Calif, Berkeley, 62-63; Mellon fel, Univ Pittsburgh, 63-64; asst prof anthrop, 64-71, ASSOC PROF ANTHROP, CORNELL UNIV, 71- Mem: Am Anthrop Asn; Royal Anthrop Inst Gt Brit & Ireland; Int African Inst; Ethnol Soc; Polynesian Soc. Res: Ethnology of Oceania, especially Micronesia; comparative social organization, especially kinship studies. Mailing Add: Dept of Anthrop Cornell Univ Ithaca NY 14850

LAMBERT, CHARLES CALVIN, b Rockford, Ill, Apr 10, 35; m 65; c 2. DEVELOPMENTAL BIOLOGY, REPRODUCTIVE BIOLOGY. Educ: San Diego State Col, BA, 64, MS, 66; Univ Wash, PhD(zool), 70. Prof Exp: NIH traineeship, Univ Wash, 70; asst prof zool, 70-74, ASSOC PROF ZOOL, CALIF STATE UNIV, FULLERTON, 74- Mem: Am Soc Zoologists; Soc Develop Biol; Sigma Xi; AAAS. Res: Development and metamorphosis of marine invertebrates; reproductive physiology of tunicates. Mailing Add: Dept of Biol Calif State Univ Fullerton CA 92634

LAMBERT, EDWARD CARY, pediatrics, cardiology, deceased

LAMBERT, EDWARD HOWARD, b Minneapolis, Minn, Aug 30, 15; m 40, 75. MEDICAL PHYSIOLOGY. Educ: Univ Ill, BS, 36, MS, 38, MD, 39, PhD(physiol), 44. Prof Exp: Instr med technol, Herzl Jr Col, 41-42; assoc med, Off Sci Res & Develop, Col Med, Univ Ill, 42-43; res asst, 43-45, from instr to prof physiol, 45-58, prof physiol, Mayo Grad Sch Med, Univ Minn, 58-73, PROF PHYSIOL & NEUROL, MAYO MED SCH, 73- Concurrent Pos: Consult, Mayo Clin, 45-; mem pub adv group, NIH. Honors & Awards: Presidential Cert Merit, 47; Tuttle Award, Aerospace Med Asn, 52. Mem: Soc Exp Biol & Med; Soc Neurosci; Am Physiol Soc; Aerospace Med Asn; Am Asn Electromyography & Electrodiag (pres, 58). Res: Neurophysiology; neuromuscular disorders in man; electromyography. Mailing Add: Mayo Clin Rochester MN 55901

LAMBERT, FRANCIS LINCOLN, b Staunton, Va, Oct 8, 23; m 67. PHYSIOLOGY. Educ: George Washington Univ, BS, 49, MS, 51; Harvard Univ, PhD(biol), 58. Prof Exp: Instr zool, George Washington Univ, 49-52; asst prof biol, 55-61, ASSOC PROF PHYSIOL & BIOPHYS, UNION COL, NY, 61- Concurrent Pos: Jacques Loeb assoc marine biol, Rockefeller Inst, 60-61. Mem: AAAS; Am Soc Zool. Res: Invertebrate physiology; cellular neurophysiology. Mailing Add: Dept of Biol Sci Union Col Schenectady NY 12308

LAMBERT, FRANK (EDWIN), JR, b Denison, Tex, Jan 2, 14; m 42; c 1. SCIENCE EDUCATION. Educ: NTex State Col, BS, 36, MS, 38. Prof Exp: Asst chem, NTex State Col, 36-38; teacher, High Schs, Tex, 38-41; head sci dept, Victoria Jr Col, 41-46; instr chem, Ga Inst Technol, 46-50; instr, Biltmore Col, 50-52; prof, Jr Col Augusta, 52-63; PROF CHEM, RICHMOND ACAD, 63- Mem: Fel Am Chem Soc; fel Am Inst Chemists; fel NY Acad Sci; fel Am Asn Univ Prof. Res: Chemical apparatus; radiator anti-rusts; controlled fermentations; cockroach insecticides; sewage disposal effectiveness tests; science education curricula. Mailing Add: Richmond Academy Augusta GA 30904

LAMBERT, FRANK LEWIS, b Minneapolis, Minn, July 10, 18; m 43. ORGANIC CHEMISTRY. Educ: Harvard Univ, BA, 39; Univ Chicago, PhD(org chem), 42. Prof Exp: Res & develop chemist, Edwal Labs, Ill, 42-43, develop chemist, 43-44, head develop dept, 46-47; instr chem, Univ Calif, Los Angeles, 47-48; from asst prof to assoc prof, 48-56, PROF CHEM, OCCIDENTAL COL, 56- Concurrent Pos: NSF fac fel, 57-58, 70-71. Mem: Am Chem Soc. Res: Polarography of organic halogen compounds; halogenation of organic compounds. Mailing Add: Dept of Chem Occidental Col Los Angeles CA 90041

LAMBERT, GEORGE, b Etobicoke, Ont, Oct 8, 23; US citizen; m 48; c 3. VETERINARY MICROBIOLOGY. Educ: Univ Guelph, DVM, 47; Iowa State Univ, MS, 66. Prof Exp: Instr vet path, Ont Vet Col, Univ Guelph, 47-48; asst prof, WVa Univ, 48-50; coop agt, Univ Wis & USDA, 50-53; asst state vet epidemiol, Va Dept Agr, 53-57; res vet bact, Nat Animal Dis Lab, 57-65, res virol, 65-67; asst dir biol dept, Diamond Labs, Inc, 67-70; chief virol res lab, 70-75, ASST DIR, NAT ANIMAL DIS CTR, 75- Concurrent Pos: Tech adv, Agr Res Serv, USDA, 75- Mem: Am Vet Med Asn; Am Soc Microbiol; US Animal Health Asn; Conf Res Workers Animal Dis. Res: Administration of animal disease research. Mailing Add: Nat Animal Dis Ctr Box 70 Ames IA 50010

LAMBERT, GLENN FREDERICK, b Columbus, Ohio, Nov 21, 18; m 45; c 2. BIOCHEMISTRY. Educ: DePauw Univ, AB, 40, Univ Ill, PhD(biochem), 44. Prof Exp: Spec res asst, Univ Ill, 45-46; res chemist, 46-60, sr res pharmacologist, 61-73, RES CHEMIST, ABBOTT LABS, 73- Concurrent Pos: Mem coun arteriosclerosis & coun thrombosis, Am Heart Asn. Mem: Am Chem Soc. Res: Biochemistry and nutrition of amino acids; fat emulsions for intravenous therapy; atherosclerosis; thrombolytic drugs. Mailing Add: Anal Res Dept Abbott Labs North Chicago IL 60064

LAMBERT, HELEN HAYNES, b Baton Rouge, La, July 25, 39; m 59; c 2.

ENDOCRINOLOGY. Educ: Wellesley Col, BA, 61; Univ NH, MS, 63, PhD(zool), 69. Prof Exp: Instr zool, Univ NH, 67-68; asst prof biol, Simmons Col, 69-70; ASST PROF BIOL, NORTHEASTERN UNIV, 70- Mem: AAAS; Sigma Xi; Am Inst Biol Sci; Am Soc Zool. Res: Effect of lighting on reproduction; environmental factors affecting sexual behavior; neuroendocrine mechanisms of ovulation; sex determination and development of sex differences. Mailing Add: Dept of Biol Northeastern Univ Boston MA 02115

LAMBERT, HOWARD W, b Oakland, Calif, Aug 2, 37; m 57; c 3. TOPOLOGY. Educ: Univ Calif, Berkeley, BA, 60; Iowa State Univ, MS, 61; Univ Utah, PhD(math), 66. Prof Exp: Asst prof, 66-71, ASSOC PROF MATH, UNIV IOWA, 71- Mem: Am Math Soc. Res: Upper semi-continuous decompositions of topological spaces, 3-manifolds. Mailing Add: Dept of Math Univ of Iowa Iowa City IA 52240

LAMBERT, JACK LEEPER, b Pittsburg, Kans, Mar 2, 18; m 43; c 4. ANALYTICAL CHEMISTRY, INORGANIC CHEMISTRY. Educ: Kans State Teachers Col, Pittsburg, AB & MS, 47; Okla State Univ, PhD(chem), 50. Prof Exp: Instr chem, Kans State Teachers Col, Pittsburg, 47-48; asst, Okla State Univ, 48-50; from instr to assoc prof, 50-65, PROF CHEM, KANS STATE UNIV, 65- Concurrent Pos: Assoc prog dir, NSF, Washington, DC, 65-66. Res: Methods research in analytical chemistry; reagents for trace analysis in air, water and blood; insoluble, demand-type disinfectants for water. Mailing Add: Dept of Chem Kans State Univ Manhattan KS 66506

LAMBERT, JAMES LEBEAU, b Sanford, Fla, Feb 11, 34. ORGANIC CHEMISTRY, BIOCHEMISTRY. Educ: Spring Hill Col, BS, 59; Johns Hopkins Univ, PhD(chem), 63. Prof Exp: ASST PROF CHEM, SPRING HILL COL, 68- Mem: AAAS; Am Chem Soc. Res: Mechanisms of organic reactions; carbanions. Mailing Add: Dept of Chem Spring Hill Col Mobile AL 36608

LAMBERT, JAMES MORRISON, b Chicago, Ill, Feb 18, 28; m 53; c 3. NUCLEAR PHYSICS. Educ: Johns Hopkins Univ, BA, 55, PhD(physics), 61. Prof Exp: Instr physics, Johns Hopkins Univ, 60-61; asst prof, Univ Mich, 61-63; from asst prof to assoc prof, 64-74, PROF PHYSICS, GEORGETOWN UNIV, 74- Concurrent Pos: Res consult, Naval Res Lab, 66- Mem: Am Phys Soc; AAAS. Res: Experimental low energy nuclear physics; nuclear reaction studies using particle accelerators. Mailing Add: Dept of Physics Georgetown Univ Washington DC 20007

LAMBERT, JEAN WILLIAM, b Ewing, Nebr, June 10, 14; m 43; c 2. AGRONOMY. Educ: Univ Nebr, BS, 40; Ohio State Univ, MS, 42, PhD(agron), 45. Prof Exp: Instr agron, Ohio State Univ, 43-45; from asst prof to assoc prof agron & plant genetics, 46-59, PROF AGRON & PLANT GENETICS, UNIV MINN, ST PAUL, 59- Concurrent Pos: Consult, Am Soybean Asn, 63; res consult, Chilean Agr Prog, Rockefeller Found, 64; partic, vis scientist prog, Am Soc Agron; tech ed, Agron J, 71-73. Mem: Fel Am Soc Agron; Am Soybean Asn. Res: Bromegrass cultural research; varietal improvement in barley and soybeans; barley and soybean genetics. Mailing Add: Dept of Agron & Plant Genet Univ of Minn St Paul MN 55108

LAMBERT, JOHN CARLYLE, statistics, mathematics, see 12th edition

LAMBERT, JOSEPH B, b Ft Sheridan, Ill, July 4, 40; m 67; c 2. ORGANIC CHEMISTRY. Educ: Yale Univ, BS, 62; Calif Inst Technol, PhD(nuclear magnetic resonance spectros), 65. Prof Exp: From asst prof to assoc prof, 65-74, PROF CHEM, NORTHWESTERN UNIV, 74- Concurrent Pos: Alfred P Sloan Found fel, 68-70; Guggenheim fel, 73. Honors & Awards: Eastman Kodak Award, 65. Mem: AAAS; Sigma Xi; Am Chem Soc; The Chem Soc; fel Brit Interplanetary Soc. Res: Nuclear magnetic resonance spectroscopy, conformational analysis, organic reaction mechanisms, applications of analytical chemistry to archaeology. Mailing Add: Dept of Chem Northwestern Univ Evanston IL 60201

LAMBERT, JOSEPH MICHAEL, b Philadelphia, Pa, Nov 19, 42. MATHEMATICAL ANALYSIS. Educ: Drexel Univ, BS, 65; Cornell Univ, MA, 67; Purdue Univ, PhD(math), 70. Prof Exp: ASST PROF MATH, PA STATE UNIV, READING, 70- Mem: Am Math Soc. Res: Functional analysis with specialization in approximation theory and the geometry of the unit ball in Banach spaces. Mailing Add: PA State Univ Reading PA 19608

LAMBERT, JOSEPH PARKER, b Bronte, Tex, Oct 6, 21; m 45; c 4. DENTISTRY. Educ: Baylor Univ, DDS, 52. Prof Exp: Instr, 52-56, PROF PROSTHETICS & CHMN DEPT, COL DENT, BAYLOR UNIV, 56- Concurrent Pos: Consult, Vet Admin Hosp, Dallas. Mem: Am Dent Asn. Mailing Add: Dept of Prosthetics Baylor Univ Col of Dent Dallas TX 75226

LAMBERT, LLOYD MILTON, JR, b Olympia, Wash, May 10, 29; m 52; c 3. SOLID STATE PHYSICS. Educ: US Naval Acad, BS, 52; Univ Calif, MS, 58, MA, 63, PhD(physics), 64. Prof Exp: Res engr, Sperry Gyroscope Co, 57; assoc elec eng, Univ Calif, 58; res engr & mgr phys electronics dept, Aeronutronic Div, Ford Motor Co, 58-63; staff scientist, Aerospace Corp, 63-65; assoc prof elec eng, 65-71, PROF ELEC ENG, UNIV VT, 71- Concurrent Pos: Res fel, Norges Teknisk-Naturvitenskaplige Forskningsrad, 71-72. Mem: Am Phys Soc. Res: Solid state devices; magnetic storage devices; semiconductors; low temperature research; thin film devices; optical properties of solids; transport properties of semiconductors. Mailing Add: Dept of Elec Eng Univ of Vt Burlington VT 05401

LAMBERT, MARJORIE FERGUSON, b Colorado Springs, Colo, June 13, 10; m 50. ANTHROPOLOGY, ARCHAEOLOGY. Educ: Colo Col, BA, 30; Univ NMex, MA, 31. Prof Exp: Instr archaeol & museol, Univ NMex, 30-37; cur archaeol, Mus NMex & Sch Am Res, 38-59; cur gen anthrop, Lab Anthrop & Mus NMex, 60-69; EMER CUR GEN ANTHROP & RES ASSOC, MUS NMEX, 69-; ASSOC RES PROF ANTHROP, PALEO-INDIAN INST & DEPT ANTHROP, EASTERN NMEX UNIV, 69- Concurrent Pos: Res assoc Southwestern prehist & ethnol, Sch Am Res, NMex, 32-, hon fel Southwestern anthrop, 69- & mem exec & prog comts, Bd Mgrs, 72-; field supvr var archaeol exped, Univ NMex, Mus NMex & Sch Am Res, 34-37; cur, Palace of Gov & Anthrop Exhib, Mus NMex & Lab Anthrop, 55-60. Mem: Soc Am Archaeol. Res: Southwestern and Mesoamerican prehistory; Southwestern ethnology; Spanish colonial and contact period Southwestern history; Southwestern Indian art. Mailing Add: PO Box 578 Santa Fe NM 87501

LAMBERT, MARY PULLIAM, b Birmingham, Ala, Apr 27, 44; m 67; c 2. BIOCHEMISTRY. Educ: Birmingham-Southern Col, BS, 66; Northwestern Univ, PhD(biochem), 71. Prof Exp: INSTR BIOCHEM, NORTHWESTERN UNIV, ILL, 70- Mem: Am Soc Microbiol. Res: Control mechanisms; bacterial enzyme function; cell wall biosynthesis; enzyme kinetics. Mailing Add: Dept of Chem Northwestern Univ Evanston IL 60201

LAMBERT, MAURICE C, b Roosevelt, Utah, Apr 14, 18; m 42; c 5. PHYSICAL CHEMISTRY. Educ: Brigham Young Univ, BS, 39, MA, 41. Prof Exp: Assoc chemist, Indust Lab, Mare Island Naval Shipyard, 41-46; chemist, Hanford Atomic Prod Oper, 48-64; sr chemist, Gen Elec Co, 64; sr res scientist, Battelle Northwest Labs, 65-70; SR RES SCIENTIST, WESTINGHOUSE HANFORD CO, 70- Mem: Am Chem Soc; Soc Appl Spectros. Res: X-ray spectrometry, absorptiometry and diffraction; atomic absorption and flame emission spectrometry; separations of trace elements; gas-solid reactions; properties of inorganic oxides; fused salt studies; surface analysis by electron spectroscopy. Mailing Add: 1617 Hains Ave Richland WA 99352

LAMBERT, MAURICE REED, b Fillmore, Utah, July 13, 20; m 50; c 4. ANIMAL NUTRITION. Educ: Utah State Univ, BS, 48, MS, 49; Iowa State Univ, PhD(animal nutrit), 53. Prof Exp: From instr to asst prof animal nutrit, Iowa State Univ, 52-57; mgr animal nutrit res, Western Condensing Co, 57-65, mgr animal nutrit res, Foremost-McKesson Inc, 65-71, MGR NUTRIT, FOREMOST FOODS CO, 71- Res: Fat and carbohydrate metabolism of the calf, pig and chicken; lipid deficiency of the calf. Mailing Add: Foremost Foods Co R&D Ctr 6363 Clark Ave Dublin CA 94566

LAMBERT, PAUL DUDLEY, b Beaver Falls, Pa, Jan 27, 29; m 53; c 4. PUBLIC HEALTH ADMINISTRATION. Educ: Geneva Col, BS, 50; Univ Pa, VMD, 54; Johns Hopkins Univ, MPH, 60; Colo State Univ, PhD(physiol), 68. Prof Exp: Vet practitioner, 54-59; RES SCIENTIST & SCI ADMINR, USPHS, 59-, GEOG MED BR, NAT INST ALLERGY & INFECTIOUS DIS, 72- Concurrent Pos: Mem grad fac, Collab Radiol Health Lab, Colo State Univ, 64-71; officer, Off Res & Monitoring, Twinbrook Res Lab, Environ Protection Agency, 71-72. Res: Administration of mycobacterial immunology research. Mailing Add: Nat Inst Allergy & Infectious Dis Bldg 31 Room 1B62 Bethesda MD 20014

LAMBERT, PAUL WAYNE, b Ft Worth, Tex, Oct 27, 37; m 59; c 1. GEOMORPHOLOGY, QUATERNARY GEOLOGY. Educ: Tex Tech Col, BA, 59; Univ NMex, MS, 61, PhD(geol), 68. Prof Exp: Geologist, Texaco Inc, NMex, 61-62; asst prof geol, Cent Mo State Col, 65-68; assoc prof, WTex State Univ, 68-70; geologist, Dept Prehist, Nat Inst Anthrop & Hist, Mex, 72-73; GEOLOGIST, US GEOL SURV, 73- Concurrent Pos: Res grants, Geol Soc Am & Sigma Xi, 68-69 & NSF, 69-70. Mem: Geol Soc Am; Am Quaternary Asn; Soc Am Archaeol. Mailing Add: US Geol Surv Fed Ctr Denver CO 80225

LAMBERT, REGINALD MAX, b Delta, Ohio, Feb 25, 26; m 52; c 3. BACTERIOLOGY, IMMUNOLOGY. Educ: Butler Univ, BA, 50; Univ Buffalo, MA, 52, PhD(bact, immunol), 55. Prof Exp: Asst bact & immunol, Sch Med, State Univ NY Buffalo, 51-55, instr, 55-57, assoc, 57-59, asst prof, 59-64; asst prof path, Col Med, Univ Fla, 64-67; ASSOC PROF MICROBIOL, SCH MED, STATE UNIV NY BUFFALO, 67-, ASSOC DIR, BLOOD GROUP RES UNIT, 55-64 & 67- Concurrent Pos: Consult, E J Meyer Mem Hosp, Buffalo, 58, 60-63 & 67- & Buffalo Gen Hosp, 63-64; dir blood bank, Shands Teaching Hosp, Univ Fla, 64-67; dir, Buffalo Regional Red Cross Bldg Prog, 73- Mem: AAAS; Am Soc Microbiol; Int Soc Blood Transfusion; Int Soc Hemat. Res: Blood groups; immunohematology; transfusion genetics. Mailing Add: Dept of Microbiol Sch of Med State Univ of NY Buffalo NY 14214

LAMBERT, RICHARD BOWLES, JR, b Clinton, Mass, Apr 20, 39; m 64. PHYSICAL OCEANOGRAPHY. Educ: Lehigh Univ, AB, 61; Brown Univ, ScM, 64, PhD(physics), 66. Prof Exp: Fulbright fel, aerodyn, Munich Tech, 66-67; from asst prof to assoc prof oceanog, Univ RI, 67-75; PROG DIR PHYS OCEANOG, NSF, 75- Mem: AAAS; Am Phys Soc; Am Geophys Union. Res: Hydrodynamic stability; oceanic turbulence; diffusion energy transfer; air-sea interaction. Mailing Add: NSF Oceanog Sect 1800 G St NW Washington DC 20550

LAMBERT, RICHARD ST JOHN, b Trowbridge, Eng, Nov 11, 28; m 52; c 6. PETROLOGY, GEOCHEMISTRY. Educ: Univ Cambridge, BA, 52, PhD(petrol), 55, MA, 56; Oxford Univ, MA, 56. Prof Exp: Asst lectr geol, Univ Leeds, 55-56; lectr, Oxford Univ, 56-70; PROF GEOL, UNIV ALTA, 70-, CHMN DEPT, 74- Concurrent Pos: Vis prof, Univ Alta, 63-64; fels, Iffley Col, Oxford Univ, 65-66 & Wolfson Col, 66-70. Mem: Geol Soc London; Geochem Soc; Mineral Soc Gt Brit & Ireland. Res: Mineralogy, petrology, geochemistry and isotope geology of medium to high grade metamorphic rocks; theory of metamorphic processes; stable isotope studies; geological time-scale. Mailing Add: Dept of Geol Univ of Alta Edmonton AB Can

LAMBERT, ROBERT HENRY, b Bayshore, NY, Nov 3, 30; div; c 2. ATOMIC PHYSICS. Educ: St Lawrence Univ, BS, 52; Harvard Univ, MS, 54, PhD(physics), 63. Prof Exp: Instr physics, Univ NH, 55-57; asst, Harvard Univ, 57-60; from asst prof to assoc prof, 61-68, PROF PHYSICS, UNIV NH, 68- Concurrent Pos: Cent Univ res grants, Univ NH, 62-63, 65-66; NSF grant, 65-67, 67-71. Mem: Am Phys Soc. Res: Measurement of hyperfine structure using optical pumping. Mailing Add: Dept of Physics De Meritt Hall Univ of NH Durham NH 03824

LAMBERT, ROBERT J, b Dubuque, Iowa, Dec 23, 21; m 42; c 2. MATHEMATICS. Educ: Drake Univ, BA, 43; Iowa State Univ, MS, 48, PhD(math), 51. Prof Exp: Instr math, Drake Univ, 43-44; instr, Iowa State Univ, 46-51; mathematician, Nat Security Agency, 51-53; from asst prof to assoc prof math, Univ, 53-64, PROF MATH & COMPUT SCI & SR MATHEMATICIAN, AMES LAB, IOWA STATE UNIV, 64- Concurrent Pos: Consult, Nat Security Agency & Collins Radio Co. Mem: Am Math Soc; Math Asn Am. Res: Matrix theory; finite fields; numerical analysis; partial differential equations. Mailing Add: 3301 Ross Rd Ames IA 50010

LAMBERT, ROBERT JOHN, b Faribault, Minn, Mar 14, 27. PLANT GENETICS. Educ: Univ Minn, BS, 52, MS, 58; Univ Ill, PhD(plant breeding, genetics), 63. Prof Exp: Res asst plant breeding & genetics, Univ Minn, 56-58; from res asst to res assoc, 58-64, from instr to asst prof, 64-70, ASSOC PROF PLANT BREEDING & GENETICS, UNIV ILL, URBANA, 70- Concurrent Pos: Supvr world collection of maize mutants, Maize Genetics Coop. Mem: AAAS; Crop Sci Soc Am; Genetics Soc Am; Am Asn Cereal Chemists. Res: Investigations of plant geometry of maize in relation to yield; selection and development of modified protein maize strains. Mailing Add: S-116 Turner Hall Dept of Agron Univ of Ill Urbana IL 61822

LAMBERT, ROGER GAYLE, b Minneapolis, Minn, Jan 22, 30; m 56; c 3. PLANT PHYSIOLOGY. Educ: Univ Minn, BS, 53, MS, 57, PhD(plant physiol), 61. Prof Exp: Instr plant physiol, Univ Minn, 57-61; asst prof plant physiol & path, 61-64, from actg head to head dept biol, 63-66, assoc prof plant physiol, 64-68, PROF PLANT PHYSIOL, UNIV LOUISVILLE, 68- Concurrent Pos: Fel bot & plant path, Potato Virus Lab, Colo State Univ, 70-71. Mem: AAAS; Am Soc Plant Physiol. Res: Plant competition and trophic structure of ecosystems. Mailing Add: Dept of Biol Univ of Louisville Louisville KY 40208

LAMBERT, ROGERS FRANKLIN, b Kamas, Utah, July 12, 29; m 51; c 4. ORGANIC CHEMISTRY. Educ: Brigham Young Univ, BS, 53; Purdue Univ, PhD(org chem), 58. Prof Exp: Chemist, US Bur Mines, 53; res chemist, Ethyl Corp, 58-61; res supvr, Thiokol Chem Corp, 61-65; PROF CHEM, RADFORD COL, 65- Mem: Am Chem Soc. Res: Polymers; chemical reductions; reactions of heterocyclics. Mailing Add: Dept of Chem Radford Col Box 591 Radford VA 24141

LAMBERT, RONALD, b Brooklyn, NY, Mar 16, 39. PHOTOGRAPHIC CHEMISTRY. Educ: Columbia Univ, BS, 62; Univ Ill, MS, 64, PhD(flavin analog), 67. Prof Exp: Asst org chem, Univ Ill, 62-67; scientist, 67-70, SR SCIENTIST, POLAROID CORP, 70- Mem: Am Chem Soc. Mailing Add: Polaroid Corp Res Lab Osborn St Cambridge MA 02176

LAMBERT, ROYCE LEONE, b Coatesville, Ind, Nov 3, 33; m 53; c 3. SOILS, AGRONOMY. Educ: Purdue Univ, Lafayette, BS, 64, MS, 66, PhD(soil physics), 70. Prof Exp: ASSOC PROF SOILS, CALIF POLYTECH STATE UNIV, SAN LUIS OBISPO, 69- Concurrent Pos: Soil conservationist, Nat Park Serv, 71. Mem: Am Soc Agron; Soil Sci Soc Am; Soil Conserv Soc Am. Res: Soil management. Mailing Add: Dept of Soil Sci Calif Polytech State Univ San Luis Obispo CA 93401

LAMBERT, SHELDON MARVIN, b Cleveland, Ohio, Sept 27, 30; m 53; c 2. PHYSICAL CHEMISTRY. Educ: Ohio State Univ, BSc, 53, PhD(chem), 57. Prof Exp: Fel chem, Ohio State Univ, 58; chemist, Shell Develop Co, 58-66, supvr phys & anal chem, 66-70, mgr info servs, Shell Chem Co, 70-71, mem staff chem econ, 71-74, MGR ENERGY ECON & FORECASTING, SHELL OIL CO, 74- Concurrent Pos: Consult, Ore State Univ, 65- Mem: Am Chem Soc. Res: Chemistry of complex ions in solution; polyphosphate solution equilibria; instrumental methods of analysis; liquid-liquid chromatography; correlations of biological activity and chemical structure; movement and sorption of chemicals in soil. Mailing Add: 1415 Castle Rock Rd Houston TX 77090

LAMBERT, WILLIAM M, JR, b Wausau, Wis, Apr 6, 36. MATHEMATICS. Educ: Univ Wis, BA, 58; Univ Calif, Los Angeles, MA, 59, PhD(math), 65. Prof Exp: Teaching asst math, Univ Wis, 57-58; teaching asst, Univ Calif, Los Angeles, 59-60, res asst, 60-63; from asst prof to assoc prof, Loyola Univ, Calif, 63-69; assoc prof, Univ Detroit, 69-74; MEM FAC, DEPT MATH, UNIV COSTA RICA, 74- Mem: Am Math Soc; Math Asn Am; Asn Symbolic Logic. Res: Effective processes of general algebraic structures; metamathematics of algebra. Mailing Add: Dept of Math Univ of Costa Rica San Jose Costa Rica

LAMBERTI, JOSEPH W, b Toronto, Ont, Dec 20, 29; m 55; c 6. PSYCHIATRY. Educ: Univ Ottawa, MD, 54; Royal Col Physicians & Surgeons Can, cert psychiat, 61. Prof Exp: Asst psychiatrist, Winnipeg Psychiat Inst, 60-61, sr psychiatrist, 61-63; asst prof, 63-69, ASSOC PROF PSYCHIAT, MED CTR, UNIV MO-COLUMBIA, 69- Concurrent Pos: Dir consult serv & lectr, Med Ctr, Univ Mo, 63-67, consult, Peace Corps, 64- & Univ Press, 65- Mem: Am Psychiat Asn; cor mem Can Psychiat Asn. Res: Treatment of common sexual disorders; study of antisocial behavior; study of affective disorders. Mailing Add: Dept of Psychiat Univ of Mo Med Ctr Columbia MO 65201

LAMBERTI, VINCENT, organic chemistry, see 12th edition

LAMBERTS, AUSTIN E, b East Saugatuck, Mich, Nov 30, 14; div; c 4. MARINE ZOOLOGY, NEUROSURGERY. Educ: Calvin Col, AB, 36; Univ Mich, Ann Arbor, MD, 41, MS, 50; Am Bd Neurosurg, dipl, 52; Univ Hawaii, PhD(marine zool), 73. Prof Exp: Resident & instr neurosurg, Univ Mich, 45-50; pvt pract neurosurg, St Mary's Hosp, Grand Rapids, 50-68; teaching asst marine zool, Univ Hawaii, 69-73; INDEPENDENT RES, REEF ECOL, 73- Concurrent Pos: Consult neurosurg, St Mary's Hosp, Grand Rapids, 50-76. Honors & Awards: Res grant, Nat Geog Soc, 74. Mem: Paleont Res Inst; Am Asn Neurosurgeons; Cong Neurosurgeons; Am Med Asn. Res: Study of natural life cycles of reef corals and unexplained coral kills; coral growth using the dye alizarin; effects of pesticides on coral growth; collecting and identification of modern Pacific reef corals. Mailing Add: 1520 Leffingwell NE Grand Rapids MI 49505

LAMBERTS, BURTON LEE, b Fremont, Mich, Oct 24, 19; m 60; c 2. BIOCHEMISTRY. Educ: Calvin Col, BS, 49; Mich State Univ, PhD(chem), 58. Prof Exp: Chemist, Northern Regional Res Lab, Ill, 51-54; asst chem, Mich State Univ, 55-58, instr, 58-60; CHIEF BIOCHEMIST, DENT RES FACIL, NAVAL DENT RES INST, 60- Mem: AAAS; Am Chem Soc; Int Asn Dent Res. Res: Dental caries; salivary gland secretions; products of oral microorganisms. Mailing Add: Biochem Div Naval Dent Res Inst USN Base Great Lakes IL 60088

LAMBERTS, ROBERT L, b Fremont, Mich, Sept 8, 26; m 51; c 6. PHYSICS. Educ: Calvin Col, AB, 49; Univ Mich, MS, 51; Univ Rochester, PhD(optics), 69. Prof Exp: RES ASSOC, KODAK RES LABS, EASTMAN KODAK CO, 51- Mem: Fel Optical Soc Am; Soc Photog Sci & Eng. Res: Image structure of optical systems and photographic emulsions; physical optics. Mailing Add: Eastman Kodak Co Bldg 59 Kodak Park Rochester NY 14650

LAMBERTSEN, CHRISTIAN JAMES, b Westfield, NJ, May 15, 17; m 44; c 4. PHARMACOLOGY. Educ: Rutgers Univ, BS, 39; Univ Pa, MD, 43. Prof Exp: Intern, Hosp Univ Pa, 43; from instr to assoc prof, 46-53, PROF PHARMACOL, SCH MED, UNIV PA, 53-, PROF EXP THERAPEUT, 62-, ASSOC MED, UNIV HOSP, 48-, DIR INST ENVIRON MED, 70- Concurrent Pos: Markle scholar, 48-53; vis res assoc prof, Univ Col, London, 51-52; consult, US Army Chem Ctr, 55-59; consult & lectr, Off Surgeon Gen, US Navy, 57-60; consult neuropharmacol, Del State Hosp, 57-61; consult, Sci Adv Bd, US Air Force, 59-61. Mem adv panel med sci, Off Secy Defense; mem comt undersea warfare & comt naval med res, Nat Res Coun, 53- & panel underwater swimmers, 53-56; mem panel shipboard & submarine med, Off Secy Defense Res Develop Bd, 50-53; basic sci secy, Nat Bd Med Exam, 54-, mem pharmacol comt, 54-55; chmn man in space comt, Space Sci Bd, Nat Acad Sci, 60-62, consult, 62-; mem, US Oceanogr Adv Bd, 70-; mem comt undersea physiol & med, Nat Res Coun, 72- & comt hyperbaric oxygenation; chmn comt manned undersea activity, Off Secy Navy. Honors & Awards: Meritorious Pub Serv Citation, US Navy, 70; Distinguished Pub Serv Medal, US Govt, 72. Mem: Fel Am Soc Clin Pharmacol & Therapeut; Am Physiol Soc (pres, 54-55); Am Soc Clin Invest; Am Soc Pharmacol & Exp Therapeut. Res: Respiratory physiology and pharmacology; aerospace and diving medicine; breathing apparatus for underwater swimmers. Mailing Add: Dept of Pharmacol Univ of Pa Philadelphia PA 19104

LAMBETH, DAVID N, b Carthage, Mo, Mar 18, 47; m 69; c 1. MAGNETISM. Educ: Univ Mo-Columbia, BS, 69; Mass Inst Technol, PhD(physics), 73. Prof Exp: SR RES PHYSICIST, EASTMAN KODAK CO, 73- Mem: Inst Elec & Electronic Engrs; Magnetics Soc. Res: Magnetism and magneto-optics of and thin film materials. Mailing Add: Eastman Kodak Bldg 81 Kodak Park Rochester NY 14650

LAMBETH, VICTOR NEAL, b Sarcoxie, Mo, July 5, 20; m 46; c 2. VEGETABLE CROPS. Educ: Univ Mo, BS, 42, MA, 48, PhD(hort), 50. Prof Exp: Asst, 39-42, from instr to assoc prof, 46-59, PROF HORT, UNIV MO-COLUMBIA, 59- Mem: AAAS; Am Soc Plant Physiol; Am Soc Hort Sci; Sigma Xi. Res: Soil fertility and plant nutrition; raw product quality; vegetable breeding; water relationships; post harvest physiology. Mailing Add: 1-43 New Agr Bldg Univ of Mo Columbia MO 65201

LAMBIE, MARGARET B MCCLEMENTS, b Ayr, Scotland; US citizen; c 4. COSMIC RAY PHYSICS, POWER ENGINEERING. Educ: Univ Glasgow, BSc, 53; Univ London, DIC, 64, PhD(cosmic ray physics), 65. Prof Exp: Asst prof natural philos, Univ Glasgow, 53-55; res asst, Imp Col, London, 55-58; prof math, Univ Malaya, 63-64; consult & owner, Margaret B Lambie & Pathmakers, 65-71; physicist, 71-73, COMPUT SPECIALIST POWER SYST ENG, BONNEVILLE POWER ADMIN, 73- Mem: Brit Inst Physics; Inst Elec & Electronic Eng. Res: The application of computer techniques to the management of power systems; personal communication and the position accorded women in the community. Mailing Add: Bonneville Power Admin PO Box 3621 Portland OR 97208

LAMBOOY, JOHN PETER, b Kalamazoo, Mich, Dec 6, 14; m 42; c 4. BIOCHEMISTRY. Educ: Kalamazoo Col, AB, 37, MS, 38; Univ Ill, MA, 39; Univ Rochester, PhD(physiol chem), 42. Prof Exp: From instr to assoc prof physiol, Univ Rochester, 46-63; prof chem pharmacol & sect head biochem pharmacol, Eppley Inst Cancer Res, Col Med, Univ Nebr, 63-68, prof biochem, 64-69; assoc dean grad sch, Baltimore Campuses, 69-71, prof biol chem, Sch Med, 69-74, dean grad studies & res, 71-74, PROF BIOCHEM & CHMN DEPT, SCH DENT, UNIV MD, BALTIMORE, 74- Mem: AAAS; Am Chem Soc; Am Soc Biol Chem; Am Physiol Soc; Am Inst Nutrit. Res: Synthesis and biological acitivity of vitamin analogs, amino acid analogs, anesthetics, sympathomimetic amines, bacteriostatic agents, carcinolytic agents and carcinogenic agents. Mailing Add: Dept of Biochem Univ of Md Sch of Dent Baltimore MD 21201

LAMBORG, MARVIN, b Philadelphia, Pa, Aug 13, 27; m 56; c 1. BIOCHEMISTRY. Educ: Univ Rochester, BS, 51; Johns Hopkins Univ, PhD(biol), 58. Prof Exp: Jr instr biol, McCoy Col, Johns Hopkins Univ, 55-56; res fel biochem, Mass Gen Hosp, 58-65; investr, 65-69, chief cell specialization br, 69-72, MGR ENHANCEMENT PLANT PRODUCTIVITY MISSION, C F KETTERING RES LAB, 72- Concurrent Pos: Res fel, Harvard Univ, 60-65. Mem: Am Chem Soc; Am Soc Biol Chemists; AAAS; Am Soc Cell Biologists. Res: Cell biology of nitrogen fixation in legumes and microorganisms. Mailing Add: C F Kettering Res Labs Yellow Springs OH 45387

LAMBORN, BJORN N A, b Stockholm, Sweden, Apr 2, 37. PLASMA PHYSICS. Educ: Univ Calif, Berkeley, AB, 58, MA, 60; Univ Fla, PhD(physics, math), 62. Prof Exp: Instr physics, Univ Miami, 63; res physicist, Inst Plasmaphysik, GmbH, Munich, Ger, 63-65; from asst prof to assoc prof physics, 65-75, chmn dept, 70-73, PROF PHYSICS, FLA ATLANTIC UNIV, 75- Mem: Am Phys Soc. Res: Theoretical plasma physics; wave interaction in relativistic plasmas; nonadiabatic particle motion; diffusion; nonlinear wave coupling. Mailing Add: Dept of Physics Fla Atlantic Univ Boca Raton FL 33432

LAMBORN, CALVIN RAY, b Laketown, Utah, Dec 28, 33; m 62; c 4. PLANT VIROLOGY, PLANT BREEDING. Educ: Univ Utah, BS, 62; Utah State Univ, MS, 64, PhD(plant virol), 69. Prof Exp: Res asst plant virol, Utah State Univ, 62-68; asst res dir pea & bean res lab, 68-73, DIR RES, PEA & BEAN RES LAB, GALLATIN VALLEY SEED CO, 73- Mem: Am Phytopath Soc. Res: Biological assay of tobacco mosaic virus; breeding pea and bean varieties and research related to the seed industry. Mailing Add: Gallatin Valley Seed Co PO Box 167 Twin Falls ID 83301

LAMBOU, VICTOR WILLIAM, aquatic biology, see 12th edition

LAMBRECHT, RICHARD MERLE, b Salem, Ore, Apr 8, 43; m 64; c 2. PHYSICAL CHEMISTRY, NUCLEAR MEDICINE. Educ: Ore State Univ, BS, 65; Univ Nebr, PhD(phys chem), 69. Prof Exp: Res assoc chem, 69-70, assoc, 70-74, CHEMIST, BROOKHAVEN NAT LAB, 74- Concurrent Pos: Consult, Capintec, Inc, 75- Mem: Soc Nuclear Med; Am Chem Soc. Res: Chemical effects of nuclear transformations; application of scintillation phenomena and radioactive isotopes in nuclear medicine and radiopharmaceutical research; positron annihilation and positronium chemistry. Mailing Add: Dept of Chem Brookhaven Nat Lab Upton NY 11973

LAMBREMONT, EDWARD NELSON, b New Orleans, La, July 29, 28; m 51; c 4. ENTOMOLOGY, NUCLEAR SCIENCE. Educ: Tulane Univ, BS, 49, MS, 51; Ohio State Univ, PhD(entom), 58. Prof Exp: Asst zool, Tulane Univ, 48-51; asst entom, Ohio State Univ, 54-56; entomologist, Insect Physiol, Entom Res Div, Agr Res Serv, USDA, La, 58-66; assoc prof nuclear sci, 66-74, PROF NUCLEAR SCI & DIR NUCLEAR SCI CTR, LA STATE UNIV, BATON ROUGE, 74- Mem: AAAS; Entom Soc Am; Am Oil Chem Soc. Res: Physiology and biochemistry of insects, especially lipid metabolism, synthesis and utilization of fatty acids and other lipids; radiotracer and nuclear science methodology as applied to biological problems; insect radiation biology and physiology of tumorous tissues. Mailing Add: Nuclear Sci Ctr La State Univ Baton Rouge LA 70803

LAMBROPOULOS, MELISSA MARGARET, b Tampa, Fla, Nov 2, 43; m 67. ATOMIC PHYSICS. Educ: Univ Mich, BS, 65; Univ Chicago, MS, 68, PhD(chem physics), 72. Prof Exp: Res assoc multiphotoionization, Joint Inst Lab Astrophys, 72-75; ASST PROF PHYSICS, UNIV CALIF, SANTA BARBARA, 75- Concurrent Pos: Vis scholar molecular physics, Quantum Inst, Univ Calif, Santa Barbara, 75- Mem: Am Phys Soc. Res: Multiphoton processes; tunable dye lasers; molecular spectroscopy. Mailing Add: Dept of Physics Univ of Calif Santa Barbara CA 93106

LAMBROPOULOS, PETER POULOS, b Tripolis, Greece, Oct 5, 35; US citizen; m 67. THEORETICAL PHYSICS. Educ: Athens Tech Univ, dipl, 58; Univ Mich, MSE, 62, MS, 63, PhD(nuclear sci), 65. Prof Exp: Engr, Orgn Telecommun, Greece, 59-60; sr physicist, Bendix Res Lab, Mich, 65-67; asst physicist, Argonne Nat Lab, Ill, 67-72; vis fel, Joint Inst Lab Astrophys, 72-73; from asst to assoc prof, Tex A&M Univ, 73-75; ASSOC PROF PHYSICS, UNIV SOUTHERN CALIF, 75- Mem: AAAS; Am Phys Soc; Fedn Am Scientists. Res: Atomic physics; interaction of radiation with matter; quantum optics; strong electromagnetic fields. Mailing Add: Dept of Physics Univ of Southern Calif Los Angeles CA 90007

LAMBSON, ROGER O, b Provo, Utah, Feb 5, 39; m 59; c 3. ANATOMY. Educ: Univ Mont, BA, 61; Tulane Univ, PhD(anat), 65. Prof Exp: From instr to asst prof, 65-71, assoc dean student affairs & dir admis, 71-75, ASSOC PROF ANAT, COL MED, UNIV KY, 71-, ASSOC DEAN ACAD AFFAIRS, 75- Mem: AAAS; Electron Micros Soc Am; Soc Study Reproduction; Am Asn Anatomists. Res: Electron microscopic visualization of placental transport; ultrastructure of lung. Mailing Add: Dept of Anat Univ of Ky Col of Med Lexington KY 40506

LAMDEN, MERTON PHILIP, b Boston, Mass, Sept 7, 19; m 42; c 2. BIOCHEMISTRY. Educ: Univ Mass, BS, 41; Mass Inst Technol, PhD(food technol), 47. Prof Exp: Asst, Mass Inst Technol, 41-42, mem res staff, Food Technol Labs, 43-44; from asst prof to assoc prof, 47-72, PROF BIOCHEM, COL MED, UNIV VT, 72- Concurrent Pos: Commonwealth Fund fel & NSF-Orgn Europ Econ Coop sr vis fel, Univ Col, Univ London, 61-62; vis res biochem, Dept Food Sci & Technol, Univ Calif, Davis, 75. Mem: AAAS; Am Chem Soc; Am Inst Nutrit; Brit Biochem Soc. Res: Vitamin content and retention in foods; nutritional status of humans; biochemical studies on ascorbic and oxalic acids; role of ascorbic acid in metabolism. Mailing Add: Dept of Biochem Univ of Vt Col of Med Burlington VT 05401

LAMDIN, EZRA, b Cleveland, Ohio, Nov 25, 23; m; c 3. MEDICINE, PHYSIOLOGY. Educ: Harvard Univ, AB, 47, MD, 51; Am Bd Internal Med, dipl, 58. Prof Exp: Intern med, Harvard Med Serv, Boston City Hosp, 51-52, asst resident, 52-53; USPHS fel, Sch Med, Yale Univ, 53-55; asst, Sch Med, Boston Univ, 55-56, instr, 56-58; clin instr, Sch Med, Tufts Univ, 58-60; res fel biochem, Brandeis Univ, 61-62; asst prof med & physiol, Col Med, Univ Cincinnati, 62-67; assoc prof med, Sch Med, Univ Pittsburgh, 67-69; asst med dir, Am Heart Asn, 69-73, dir div sci affairs, 73-75; ASST MED DIR, AYERST LABS, 75- Concurrent Pos: Teaching fel med, Harvard Med Sch, 52-53; resident, Vet Admin Hosp, Boston, 55-56, staff physician, 56-60, clin investr, 58-60; clin & res fel med, Mass Gen Hosp, Boston, 58-59; res collabr, Med Res Dept, Brookhaven Nat Lab, 63-69; chief med serv, Vet Admin Hosp, Pittsburgh, 67-69; adj assoc prof physiol, Mt Sinai Sch Med, 69- Mem: AAAS; Am Fedn Clin Res; Am Soc Cell Biol; Am Heart Asn; NY Acad Sci. Res: Renal physiology and disease; biochemical aspects of membrane phenomena and active transport; intermediary metabolism; diabetes and obesity; mechanism of hormone action; clinical pharmacology. Mailing Add: Ayerst Labs 685 Third Ave New York NY 10017

LAME, EDWIN LEVER, b Evanston, Ill, Feb 23, 04; m 40; c 2. RADIOLOGY. Educ: Mass Inst Technol, BS, 26; Univ Pa, MD, 33; Am Bd Internal Med, dipl, 45; Am Bd Radiol, dipl, 45. Prof Exp: Asst med, Sch Med, Univ Pa, 36-42, asst radiol, 42-47; dir radiol, Presby Hosp, 47-66; CHIEF RADIOL, VET ADMIN HOSP, COATESVILLE, 66- Concurrent Pos: Fel internal med & chest dis, physician & asst to dir, Dept Res Respiratory Dis, Germantown Hosp, 36-42; asst physician, Pa Hosp, 37-43; asst pediatrist & chief chest clin, Children's Hosp, 38-45; fel radiol, Hosp Univ Pa, 42-45; chief radiol, Jeanes Hosp, 45-48, consult, 48-60; assoc prof clin radiol, Sch Med & asst prof radiol, Grad Sch Med, Univ Pa, 57-66. Mem: Radiol Soc NAm; Am Roentgen Ray Soc; AMA; fel Am Col Physicians; fel Am Col Radiol. Res: Pulmonary radiologic interpretation; pelvic and vertebral osteomyelitis arising from the urinary tract; cholecystitis; clinical and radiologic criteria; gastrointestinal barium; protection and dose reduction in diagnostic radiology; radiologic signs of preclinical heart failure. Mailing Add: Vet Admin Hosp Coatesville PA 19320

LAMEY, HOWARD ARTHUR, b Bloomington, Ind, Dec 20, 29; m 56; c 5. PLANT PATHOLOGY. Educ: Ohio Wesleyan Univ, BA, 51; Univ Wis, PhD(plant path), 54. Prof Exp: Asst plant path, Univ Wis, 51-53, proj assoc, 54-55 & 57-58; from plant pathologist to sr res plant pathologist, Crops Res Div, Agr Res Serv, USDA, Cuba, 58-60, La, 60-69; plant pathologist, Int Inst Trop Agr, Nigeria, 69-71; PROJ MGR, FOOD & AGR ORGN, UN, KOREA, 71- Concurrent Pos: From plant pathologist to proj mgr, Food & Agr Orgn, UN, Thailand, 66-68. Mem: Fel AAAS; Am Phytopath Soc; Indian Phytopath Soc; Mycol Soc Am; NY Acad Sci. Res: Plant virology; insect transmission of plant viruses; fungus diseases of plants; seed treatment fungicides; genetics and physiology of plant pathogenic fungi; research administration; rice diseases; varietal resistance. Mailing Add: UN Develop Programme CPO Box 143 Seoul Korea

LAMKIN, WILLIAM MEREDITH, biochemistry, see 12th edition

LAMM, MICHAEL EMANUEL, b New York, NY, May 19, 34; m 61; c 2. IMMUNOLOGY, PATHOLOGY. Educ: Univ Rochester, MD, 59; Western Reserve Univ, MS, 62; Am Bd Path, dipl, 65. Prof Exp: From intern to resident path, Univ Hosps, Cleveland, Ohio, 59-62; res assoc chem, NIH, 62-64; from asst prof to assoc prof, 64-73, PROF PATH, SCH MED, NY UNIV, 73- Mem: NY Acad Sci; Am Soc Exp Path; Soc Exp Biol & Med; Am Asn Immunol; Am Asn Path & Bact. Mailing Add: Dept of Path NY Univ Sch of Med New York NY 10016

LAMME, ARY JOHANNES, III, historical geography, cultural geography, see 12th edition

LAMMI, JOE OSCAR, b Daly City, Calif, Jan 24, 13; m 40; c 2. FORESTRY, INTERNATIONAL ECONOMICS. Educ: Ore State Univ, BS, 34, MS, 37; Univ Calif, Berkeley, PhD(agr econ), 54. Prof Exp: Forest adminr, Forest Serv, USDA, 34-43, farm forester, 47-50, forest economist, 50-55; forest economist, UN Europ Hq, 55-61; PROF FOREST ECON & REMOTE SENSING, NC STATE UNIV, 61- Concurrent Pos: Forest owner & operator, 41-; vis prof, Duke Univ, 65-; career consult, Peace Corps, 65-; consult, USAID, 66-; vis prof, Fulbright-Hays Prog, Univ Tampere, Finland, 70- Mem: Corresp mem Soc Forest Finland; Am Soc Photogram; Soc Am Foresters. Res: Remote sensing of the environment; forest resource policy and international development. Mailing Add: Dept of Forestry NC State Univ Raleigh NC 27607

LAMOLA, ANGELO ANTHONY, b Newark, NJ, Aug 12, 40; m 63; c 2. PHOTOBIOLOGY, PHOTOCHEMISTRY. Educ: Mass Inst Technol, BS, 61; Calif Inst Technol, PhD(chem), 65. Prof Exp: Asst prof chem, Univ Notre Dame, 64-66; MEM RES STAFF, BELL LABS, 66- Concurrent Pos: Mem photobiol comt, Nat Res Coun-Nat Acad Sci; ed, Molecular Photochem; adj prof, Dept Dermat, Columbia Col Physicians & Surgeons, 75- Mem: AAAS; Am Chem Soc; Am Soc Photobiol (pres, 76-77); Biophys Soc. Res: Molecular photobiology; molecular biophysics; medical applications of fluorescence. Mailing Add: Bell Labs Murray Hill NJ 07974

LAMON, EDDIE WILLIAM, b Yuba City, Calif, Aug 30, 39; c 2. CANCER, IMMUNOLOGY. Educ: Univ NAla, BS, 61; Med Col Ala, MD, 69; Karolinska Inst, Sweden, DSc, 74. Prof Exp: Asst biologist, Southern Res Inst, 64-65; from intern surg to resident gen surg, Univ Ala Sch Med, 69-71; asst prof, 74-75, ASSOC PROF SURG & MICROBIOL, UNIV ALA, BIRMINGHAM, 75- Concurrent Pos: Guest investr, Karolinska Inst, Sweden, 71-74; mem, Comt Cancer Immunodiagnosis, Nat Cancer Inst, Div Cancer Biol & Diag, 76-77; Nat Cancer Inst res career develop award, 75. Mem: Am Asn Immunologists. Res: Studies of the immune response to virus induced tumors with emphasis on the interactions of antibodies and lymphocytes in tumor cells destruction. Mailing Add: Dept of Surg Univ Sta Univ of Ala Birmingham AL 35294

LAMONDE, ANDRE M, b St Lambert, Que, Oct 5, 36; m 66. PHARMACY. Educ: Univ Montreal, BPharm, 61; Purdue Univ, MSc, 63, PhD(indust pharm), 65. Prof Exp: Asst prof pharm, Univ Montreal, 65-68; regulatory affairs coordr, Med Div, 68-70, qual control mgr, 70-72, QUAL CONTROL DIR, SYNTEX LTD, 72- Res: Basic pharmaceutics and pharmaceutical analysis. Mailing Add: Syntex Ltd Qual Control Dept 8255 Mountain Sights Ave Montreal PQ Can

LAMONICA, CARL J, b Colver, Pa, Aug 17, 20; m 47; c 4. COMPUTER SCIENCE. Educ: Univ Pittsburgh, BS, 48, MEA, 50. Prof Exp: Instr physics & math, Richland Schs, Pa, 48-53; physicist, US Naval Weapons Lab, 53-56; physicist, US Naval Ord Lab, 56-57; head res br, Missile Safety Staff, US Naval Weapons Lab, 57-63, supv physicist & missile launcher br head, 63-68, head fire control prog & exterior ballistic div, 68-71, head aeroballistic & weapons control divs, 71-72, DIR COMPUT FACILITIES, US NAVAL SURFACE WEAPONS CTR, 72- Concurrent Pos: Lectr, Univ Va, 54; mem grad physics panel, US Naval Weapons Lab, 59-63, comt improv naval nuclear weapons, 61, ord & coord subcomts develop Eagle & Phoenix weapon systs, 62. Mem: Am Inst Aeronaut & Astronaut; fel Am Inst Chem. Res: Liquid propellants; infrared systems; instrumentation; weapon control; general ordnance research; integrated electronics in computer design and computer architecture. Mailing Add: Comput Facilities US Naval Surface Weapons Ctr Dahlgren VA 22448

LAMONT, PATRICK JOHN COLL, b Dublin, Ireland, Aug 29, 36. ALGEBRA, NUMBER THEORY. Educ: Glasgow Univ, BSc, 58, PhD(math), 62. Prof Exp: Asst lectr math, Royal Col Sci & Technol, Scotland, 61-62; Dept Sci & Indust Res traveling fel, State Univ Utrecht & Univs Göttingen & Munich, 62-64; lectr pure math, Univ Birmingham, 64-70; ASSOC PROF MATH, ST MARY'S COL, INC, 70- Concurrent Pos: Asst prof, Univ Notre Dame, 66-68; instr, Deep Springs Col, California, Nev, 74-76. Mem: Am Math Soc; London Math Soc; Math Asn Am. Res: Arithmetic theory of nonassociative algebras. Mailing Add: 1012 Riverside Dr South Bend IN 46616

LAMONT, ROBERT ELLIS, b Columbus, Ohio, May 31, 24; m 50; c 2. GEOGRAPHY, MILITARY ENGINEERING. Educ: Harvard Univ, AB, 50; Boston Univ, MA, 54. Prof Exp: Mem, US Weather Bur Arctic Exped, 48; cartogr, Inst Geog Explor, Harvard Univ, 48-50; geogr, US Army Map Serv, 50-51; topog opers officer, Eng Intel Group, US Army, 51-52; managing ed, Univ Prints, Inc, 52-54; cartogr, US Army Eng Res & Develop Labs, 54-55, geogr, Hq, US Dept Army, 55-56, equip & mat analyst, Chem Corps, 56-58, cartogr, US Army Res & Develop Labs, 58-61; gen engr, US Army Mobility Equip Res & Develop Ctr, 61-67, gen engr, Hq, US Dept Army, 67-70, intel specialist, 70-73, geogr, US Army Concepts Anal Agency, 73-74; DISASTER PREPAREDNESS ENGR, NAVAL FACILITIES ENG COMMAND, 74- Concurrent Pos: Dir pub facilities, 352nd Civil Affairs Command, US Army Res, 74- Honors & Awards: Soc Am Mil Engrs Award of Merit, 65; Spec Serv Award, US Army Mobility Command Res & Develop Labs, 65. Mem: Asn Am Geogr; Soc Am Mil Engrs. Res: Regional studies; color science; cartography; photogrammetry; engineering management; graphics; civil-military operations; civil defense; disaster preparedness; damage assessment; location theory. Mailing Add: 4601 N 23rd St Arlington VA 22207

LAMOREAUX, PHILIP ELMER, b Chardon, Ohio, May 12, 20; m 43; c 3. GEOLOGY, HYDROLOGY. Educ: Denison Univ, BA, 43, DSc; Univ Ala, MA, 49. Prof Exp: Jr geologist, US Geol Surv, 43-45, asst geologist, 45-47, dist geologist, 47-48, div hydrologist, 58-59, chief groundwater br, 59-61; STATE GEOLOGIST & OIL & GAS SUPVR, GEOL SURV ALA, 61-; PROF GEOL, UNIV ALA, TUSCALOOSA, 66- Concurrent Pos: Lectr, Univ Ala, Tuscaloosa, 48-59, assoc prof, 61-66; consult, Egypt, 53, 59, 61, 63-64, 65, Thailand, 54, 61, Philippines, 61, Surinam, 63, Mauritania, Africa, Senegal & Colombia, 64. Mem: Int Union Geod & Geophys; Geol Soc Am; Soc Econ Geol; Am Asn Petrol Geologists; Am Inst Mining, Metall & Petrol Eng. Res: Groundwater geology; fluoride in groundwater; stratigraphy of Gulf Coastal Plain. Mailing Add: Dept of Geol & Geog Univ of Ala in Tuscaloosa University AL 35486

LAMOREE, MAURICE DEAN, mathematics, see 12th edition

LAMOREUX, WELFORD FORREST, genetics, deceased

LA MORI, PHILLIP NOEL, b Anaheim, Calif, Dec 9, 33; c 2. GEOPHYSICS. Educ: Univ Calif, Los Angeles, BS, 56, MA, 62; Northwestern Univ, PhD(mat sci), 67. Prof Exp: Res asst geol, Univ Calif, Los Angeles, 56-60; res chemist, Gen Chem Res Lab, Allied Chem Corp, 60-62; high pressure scientist, Northwestern Univ, 62-67; sr geophysicist, Battelle Mem Inst, 67-75; PROJ MGR GEOTHERMAL ENERGY, ELEC POWER RES INST, 75- Mem: AAAS; Am Geophys Union; Geothermal Resources Coun (vpres, 75-). Res: Solid earth geophysics; equation of state of geologic materials; effect of pressure and temperature on materials; high pressure calibration standards; geothermal energy, program planning and project management in research and development of utilization of geothermal energy; resource assessment. Mailing Add: 3412 Hillview Ave PO Box 10412 Palo Alto CA 94303

LAMOTTE, CAROLE CHOATE, b Washington, DC, May 15, 47; m 70. NEUROANATOMY, NEUROPHYSIOLOGY. Educ: Univ Okla, BS, 67; Georgetown Univ, MS, 69; Johns Hopkins Univ, PhD(physiol), 72. Prof Exp: NIH fel anat, Sch Med, Johns Hopkins Univ, 72-74, Inst res grant & instr anat, Sch Med, 74-75, ASST PROF ANAT, SCH MED, JOHNS HOPKINS UNIV, 75- Mem: Soc Neurosci. Res: Physiology and anatomy of pain and temperature sensation. Mailing Add: Dept of Anat Johns Hopkins Univ Sch of Med Baltimore MD 21205

LAMOTTE, CLIFFORD ELTON, b Alpine, Tex, June 24, 30; m 55; c 2. PLANT PHYSIOLOGY. Educ: Tex A&M Univ, BS, 53; Univ Wis, PhD(bot), 60. Prof Exp: Res assoc biol, Princeton Univ, 60-61; from instr to asst prof, Boston Univ, 61-66; ASSOC PROF BOT & PLANT PATH, IOWA STATE UNIV, 66- Concurrent Pos: NSF res grant, 64-66. Mem: AAAS; Am Soc Plant Physiol. Res: Hormonal regulation of vascular tissue formation and biochemistry of leaf abscission in plants; plant morphogenesis using tissue culture methods. Mailing Add: Dept of Bot & Plant Path Iowa State Univ Ames IA 50010

LAMOTTE, LOUIS COSSITT, JR, b Clinton, SC, Jan 21, 28; m 48; c 5. MICROBIOLOGY, EPIDEMIOLOGY. Educ: Duke Univ, AB, 48; Univ NC, MSPH, 51; Johns Hopkins Univ, ScD(virol, entom), 58. Prof Exp: Bacteriologist, State Bd Health, NC, 48-51; virologist, Chem Corps, US Dept Army, 51-58; chief virus invests unit, Dis Ecol Sect, Tech Br, Nat Commun Dis Ctr, 58-65, asst chief sect, Colo, 65-66, chief community studies, Pesticides Prog, Ga, 66-69, dep chief, 69-70, CHIEF MICROBIOL BR, LAB DIV, CTR DIS CONTROL, 70- Concurrent Pos: Mem grad fac, Colo State Univ, 59-66; adj prof, Ga State Univ, 71- Mem: AAAS; Sigma Xi; Am Soc Trop Med & Hyg; Am Pub Health Asn; Am Soc Microbiol. Res: Epidemiology of arthropod-borne viruses; virology, bacteriology, parasitology and epidemiology of infectious diseases. Mailing Add: Microbiol Br Lab Div Ctr Dis Control 1600 Clifton Rd Atlanta GA 30333

LAMOTTE, ROBERT HILL, b Washington, DC, Nov 4, 40; m 70. NEUROSCIENCES. Educ: Trinity Col, BS, 63; Kans State Univ, PhD(psychol), 68. Prof Exp: Fel neurophysiol, 68-70, instr, 70-73, ASST PROF NEUROPHYSIOL, SCH MED & ASST PROF PSYCHOL, JOHNS HOPKINS UNIV, 73- Mem: Soc Neurosci; AAAS. Res: Neurophysiology and psychophysics of somesthesis. Mailing Add: Dept of Physiol Johns Hopkins Univ Med Sch Baltimore MD 21205

LAMOUREUX, CHARLES HARRINGTON, b West Greenwich, RI, Sept 14, 33; m 54; c 2. BOTANY. Educ: Univ RI, BS, 53; Univ Hawaii, MS, 55; Univ Calif, Davis, PhD(bot), 61. Prof Exp: Asst bot, Univ Hawaii, 53-55; jr plant pathologist, Calif State Dept Agr, 55; asst bot, Univ Calif, Davis, 55-59; from asst prof to assoc prof, 59-71, PROF BOT, UNIV HAWAII, 71- Concurrent Pos: Hon assoc, Bernice P Bishop Mus, 62-; guest scientist, Nat Biol Inst Indonesia, 72-73. Mem: AAAS; Sigma Xi; Int Asn Plant Taxon. Res: Plant morphology; phenology; island biology. Mailing Add: 3190 Maile Way Honolulu HI 96822

LAMOUREUX, GERALD LEE, b Bottineau, NDak, Apr 13, 39; m 69. BIOCHEMISTRY. Educ: Minot State Col, BS, 61; NDak State Univ, PhD(chem), 66. Prof Exp: RES CHEMIST, METAB & RADIATION RES LAB, AGR RES SERV, USDA, 66- Concurrent Pos: Adj prof, NDak State Univ, 74- Mem: Am Chem Soc; Sigma Xi. Res: Elucidation of metabolic pathways utilized by plants and animals in the metabolism of herbicides, insecticides and other exenobiotics; glutathione-S-transferase mediated reactions; isolation and identification of natural products. Mailing Add: USDA Metab & Radiation Res Lab State Univ Sta Fargo ND 58102

LAMOUREUX, GILLES, b Marieville, Que, Mar 2, 34; c 5. MEDICINE, IMMUNOLOGY. Educ: St Mary's Col, BA, 56; Univ Montreal, MD, 61, MSc, 63, PhD(immunol), 67. Prof Exp: HEAD IMMUNODIAG LAB, INST MICROBIOL & HYG MONTREAL, 67-; ASST PROF IMMUNOL, UNIV MONTREAL, 68- Concurrent Pos: Fel, Walter & Eliza Hall Inst Med Res, Melbourne, Australia; Multiple Sclerosis Soc Can grant. Mem: Can Soc Immunol; Fr Soc Immunol; Transplantation Soc. Res: Immune studies in multiple sclerosis and experimental allergic encephalomyelitis; mechanism of action of lymphocytes in specific cell-mediated immunity; stimulation and immunosuppression; practical aspect in autoimmune diseases, transplantation and cancer. Mailing Add: Inst of Microbiol & Hyg PO Box 100 Laval-del Rapides Laval PQ Can

LAMP, HERBERT F, b Davenport, Iowa, Aug 6, 19; m 47; c 6. PLANT ECOLOGY. Educ: Chicago Teacher Col, BEd, 41; Univ Chicago, SM, 47, PhD(bot), 51. Prof Exp: Instr bot, Fla State Univ, 47-50; teacher biol, Ill Teachers Col Chicago-South, 50-59, from assoc prof to prof, 59-64, chmn dept natural sci, 56-64; prof biol, 64-66, PROF BIOL SCI & CHMN DEPT, NORTHEASTERN ILL UNIV, 66- Concurrent Pos: Res assoc, Univ Chicago, 52-56. Mem: AAAS; Bot Soc Am; Ecol Soc Am; Am Inst Biol Sci; Nat Asn Biol Teachers. Res: Physiological ecology of range grasses; Bromus inermis Leyss and prairie ecology. Mailing Add: Dept of Biol Northeastern Ill Univ Chicago IL 60625

LAMPE, FREDERICK WALTER, b Chicago, Ill, Jan 5, 27; m 49; c 5. PHYSICAL CHEMISTRY. Educ: Mich State Col, BS, 50; Columbia Univ, AM, 51, PhD(chem), 53. Prof Exp: Asst chem, Columbia Univ, 50-53; res chemist, Humble Oil & Refining Co, 53-56, sr res chemist, 56-60, res specialist, 60; assoc prof chem, 60-65, PROF CHEM, PA STATE UNIV, 65- Concurrent Pos: Consult, Socony Mobil Oil Co, 61-69, Sci Res Instruments Corp, 67- & W H Johnston Labs, Inc, 60-67; NSF sr fel & guest prof, Univ Freiburg, 66-67; Alexander von Humboldt Found US sr scientist award, 73-74. Mem: AAAS; Am Chem Soc; fel Am Phys Soc; The Chem Soc. Res: Photochemistry; radiation chemistry; reactions of free radicals and of gaseous ions; mass spectrometry. Mailing Add: 542 Ridge Ave State College PA 16801

LAMPE, ISADORE, b London, Eng, Nov 16, 06; nat US; m 43; c 2. RADIOLOGY. Educ: Western Reserve Univ, AB, 27, MD, 31; Univ Mich, PhD(roentgenol), 38. Prof Exp: PROF RADIOL & CHG RADIATION THERAPY, MED SCH, UNIV MICH, ANN ARBOR, 39- Mem: Radiol Soc NAm; Am Roentgen Ray Soc; Am Radium Soc; AMA. Res: Comparative evaluation of high-voltage x-ray, cesium-137 and cobalt-60 radiations. Mailing Add: Univ Hosp Radiation Ther Ctr 1405 E Ann St Ann Arbor MI 48104

LAMPE, KENNETH FRANCIS, b Dubuque, Iowa, Dec 3, 28; m 49; c 1. PHARMACOLOGY. Educ: Univ Iowa, BA, 49, MS, 51, PhD(pharm), 53. Prof Exp: Instr chem, Mont State Col, 49-50; instr pharm, Univ Iowa, 53; from asst prof to assoc prof pharmacol, 54-67, PROF PHARMACOL & ANESTHESIOL, SCH MED, UNIV MIAMI, 67- Concurrent Pos: Fel, Yale Univ, 53-54. Mem: Am Chem Soc. Res: Chemical constitution and biological activity; neuropharmacology. Mailing Add: Dept of Pharmacol Univ of Miami Sch of Med Miami FL 33152

LAMPEN, J OLIVER, b Holland, Mich, Feb 26, 18; m 44; c 3. MICROBIOLOGY. Educ: Hope Col, AB, 39; Univ Wis, MS, 41, PhD(biochem), 43. Hon Degrees: LHD, Hope Col, 74. Prof Exp: Biochemist, Am Cyanamid Co, 43-46; res assoc, Med Sch, Washington Univ, 46-47, instr biochem, 47-48, asst prof biol chem, 48-49; assoc prof microbiol, Sch Med, Western Reserve Univ, 49-53; dir div biochem res, Squibb Inst Med Res, Olin Mathieson Chem Corp, 53-58; DIR WAKSMAN INST MICROBIOL, RUTGERS UNIV, 58- Honors & Awards: Lilly Award, 52. Mem: AAAS; Am Soc Microbiol; Am Soc Biol Chemists; Am Acad Microbiol; Brit Biochem Soc. Res: Secretion of enzymes by microorganisms: site of exoenzyme formation, mechanism of secretion, control of synthesis; antibiotics and cell membrane. Mailing Add: Waksman Inst of Microbiol Rutgers Univ New Brunswick NJ 08903

LAMPERT, BERNARD B, organic chemistry, see 12th edition

LAMPERTI, ALBERT A, b Bronx, NY, Oct 24, 47; m 72; c 1. NEUROENDOCRINOLOGY. Educ: Manhattan Col, BS, 69; Univ Cincinnati, PhD(anat), 73. Prof Exp: ASST PROF ANAT, COL MED, UNIV CINCINNATI, 73- Mem: AAAS; Am Asn Anatomists; Soc Study Reprod; Sigma Xi. Res: Reproductive neuroendocrinology; heavy metals. Mailing Add: Col of Med Dept of Anat Univ of Cincinnati Cincinnati OH 45267

LAMPERTI, JOHN WILLIAMS, b Montclair, NJ, Dec 20, 32; m 57; c 4. MATHEMATICS. Educ: Haverford Col, BS, 53; Calif Inst Technol, PhD(math), 57. Prof Exp: From instr to asst prof math, Stanford Univ, 57-61; vis asst prof, Dartmouth Col, 61-62; res assoc, Rockefeller Inst, 62-63; assoc prof, 63-68, PROF MATH, DARTMOUTH COL, 68- Concurrent Pos: Sci exchange visitor, USSR, 70; vis prof, Aarhus Univ, 72-73. Mem: Fel Inst Math Statist. Res: Probability theory, particularly properties of stochastic processes, especially limit theorems for Markov processes. Mailing Add: Dept of Math Dartmouth Col Hanover NH 03755

LAMPI, EUGENE ELIAS, physics, see 12th edition

LAMPI, RAUNO ANDREW, b Gardner, Mass, Aug 12, 29; m 51; c 4. FOOD TECHNOLOGY. Educ: Univ Mass, BS, 51, MS, 55, PhD(food technol), 57. Prof Exp: Res instr food technol, Univ Mass, 53-57; tech dir food processing, New Eng Appl Prod Co, 59-62; mgr process develop, Food Technol Sect, Cent Eng Labs, FMC Corp, 62-66; packaging technologist, Container Div, 66-67, res phys scientist, Packaging Div, 67-69, CHIEF SYSTS DEVELOP BR, PACKAGING DIV, NATICK LABS, US ARMY, 69- Concurrent Pos: Asst mgr indust instrumentation, Food Div, Foxboro Co, 57-59. Honors & Awards: Rohland Isker Award, Res & Develop Assocs, 69; Sci Director's Silver Key for Eng, Natick Labs, 69, Gold Key for Eng, 73 & Meritorious Civilian Serv Decoration, 73. Res: Development of continuous applesauce and juice processes; stability characteristics of freeze dried foods; thermal processing of foods. Mailing Add: Packaging Div FEL US Army Natick Labs Natick MA 01760

LAMPIDIS, THEODORE JAMES, b New York, NY, June 10, 43; m 69. MICROBIOLOGY, CELL BIOLOGY. Educ: Brooklyn Col, BS, 65; NY Univ, MS, 69; Univ Miami, PhD(microbiol), 74. Prof Exp: Med res technician med microbiol, Sch Med, NY Univ, 65-67; NIH fel, 69-74; NAT INST AGING RES FEL PHYSIOL, SCH PUB HEALTH, HARVARD UNIV, 74- Mem: Am Soc Microbiol. Res: DNA repair in post-mitotic cells utilizing beating heart cells in vitro as a model for effects of insults (irradiations) on function; aging human diploid cells in plateau phase. Mailing Add: Lab of Radiobiol Sch of Pub Health Harvard Univ Boston MA 02115

LAMPING, JOHN ANDREW, zoology, see 12th edition

LAMPKIN, RICHARD HENRY, JR, general science, see 12th edition

LAMPKIN-ASAM, JULIA MCCAIN, b Tuscaloosa, Ala, Feb 27, 31; m 71. ONCOLOGY, PHYSIOLOGY. Educ: Univ Ala, BS, 52; George Washington Univ, MS, 54, PhD(oncol, physiol), 58. Prof Exp: Med asst, Med Div, Civil Aeronaut Admin, Washington, DC, 55-56; res instr endocrinol cancer res, Univ Miami, 58-62; vis scientist, Univ Okla, 62-63; dir cancer res, Appl Res Lab, Inc, Fla, 63-65, Miami Serpentarium Cancer Res Lab, 65-66 & Lampkin-Hibbard Cancer Inst, Inc, Miami, 65-71; DIR CANCER RES, LAMPKIN-ASAM CANCER INST, INC, 71- Concurrent Pos: Cancer Chemother Nat Serv Ctr contract, Univ Miami, 58-62, Am Cancer Soc instnl grant, 61-62; USPHS spec travel grant, Univ Wis, 61; Am Cancer Soc instnl grant, Univ Okla, 62-63; Am Cancer Soc grant, Appl Res Lab, 63-65, Miami Serpentarium & Lampkin-Hibbard Cancer Inst, Inc, 65-66; teacher exten div, Univ Tampa, Homestead AFB, 71. Mem: Am Asn Cancer Res. Res: Developer of p1798 model lymphoma systems in mice; chemotherapy; carcinogenesis; biochemistry; endocrinology; pharmacology; immunology; genetics; radioactive tracers; methods of obtaining anti-lymphoma antibodies. Mailing Add: Lampkin-Asam Cancer Inst Inc PO Box 2115 Tuscaloosa AL 35401

LAMPKY, JAMES ROBERT, b Battle Creek, Mich, June 19, 27; m 50, 71; c 6. MICROBIOLOGY. Educ: Eastern Mich Univ, BS, 59; Univ Mo, MA, 61, PhD(microbiol), 66. Prof Exp: From instr to asst prof bact, Wis State Univ, 63-66; asst prof, 66-70, ASSOC PROF BACT, CENT MICH UNIV, 70- Mem: Am Soc Microbiol; Soc Indust Microbiologists; Mycol Soc Am. Res: Cellulolytic fruiting myxobacteria of the genus Polyangium with emphasis on morphology and ultrastructure. Mailing Add: Dept of Biol Cent Mich Univ Mt Pleasant MI 48858

LAMPMAN, GARY MARSHALL, b South Gate, Calif, Oct 8, 37; m 71; c 1. ORGANIC CHEMISTRY. Educ: Univ Calif, Los Angeles, BS, 59; Univ Wash, PhD(chem), 64. Prof Exp: From asst prof to assoc prof, 64-73, PROF CHEM, WESTERN WASH STATE COL, 73- Mem: Am Chem Soc. Res: Conformational analysis in small ring compounds; synthesis and reactions of strained compounds. Mailing Add: Dept of Chem Western Wash State Col Bellingham WA 98225

LAMPORT, DEREK THOMAS ANTHONY, b Brighton, Eng, Dec 1, 33; m 63; c 5. BIOCHEMISTRY. Educ: Univ Cambridge, BA, 58, PhD(biochem), 63. Prof Exp: Staff scientist, Res Inst Advan Studies, Martin County, Md, 61-64; from asst prof to assoc prof biochem, 64-74, PROF BIOCHEM, AEC PLANT RES LAB, MICH STATE UNIV, 74- Mem: Am Chem Soc. Res: Plant cell wall proteins. Mailing Add: AEC Plant Res Lab Mich State Univ East Lansing MI 48823

LAMPORT, HAROLD, physiology, deceased

LAMPORT, JAMES EVERETT, b Seligman, Ariz, Nov 8, 21; m 45; c 4. COSMIC RAY PHYSICS. Educ: Ill Inst Technol, BS, 50. Prof Exp: Asst technologist, Sinclair Res Labs, Inc, 50-51; res physicist, Armour Res Found, Ill Inst Technol, 51-57; asst dir syst develop, Labs Appl Sci, 57-64, TECH SERV MGR, LAB ASTROPHYS & SPACE RES, UNIV CHICAGO, 64- Honors & Awards: NASA Pub Serv Award, 75. Res: Spectroscopy; infrared detection systems; cosmic ray measurement systems for space vehicles; cosmic ray instrumentation for space flight. Mailing Add: 700 Bruce Lane Glenwood IL 60425

LAMPRECH, EARL DUWAIN, b Stoddard, Wis, Feb 25, 31; m 56; c 3. FOOD SCIENCE. Educ: Univ Wis, BS, 58, MS, 59, PhD(bact), 62. Prof Exp: Scientist, 62-65, sr scientist, 65-68, tech mgr prod develop, 68-69, tech dir refrig foods res & develop, 69-72, MGR PROD DEVELOP REFRIG FOODS, PILLSBURY CO, 72- Mem: Am Asn Cereal Chemists; Inst Food Technologists. Res: Food and dairy microbiology; lactic acid bacteria; chemical analysis and physical testing of food products and ingredients; refrigerated dough technology. Mailing Add: Pillsbury Co 311 Second St SE Minneapolis MN 55414

LAMPREY, HEADLEE, b Snohomish, Wash, Oct 30, 08; m; c 4. CHEMISTRY. Educ: Univ Wash, BS & MS, 31; Univ Mich, PhD(chem), 35. Prof Exp: Chemist, Linde Air Prod Co, Union Carbide Corp, 35-50, chemist, Nat Carbon Co, 50-57, mgr chem res, Union Carbide Metals Co, 57-63; dir res packaging div, Olin Mathieson Chem Corp, Conn, 63-65; ASSOC PROF PHYS CHEM, STATE UNIV NY COL CERAMICS, ALFRED UNIV, 65- Mem: Am Chem Soc. Res: Inorganic chemistry. Mailing Add: State Univ of NY Col of Ceramics Alfred Univ Alfred NY 14802

LAMPSON, BUTLER WRIGHT, b Washington, DC, Dec 23, 43; m 67; c 1. COMPUTER SCIENCE. Educ: Harvard Univ, AB, 64; Univ Calif, Berkeley, PhD(comput sci), 67. Prof Exp: From asst prof to assoc prof comput sci, Univ Calif, Berkeley, 67-71; RES FEL COMPUT SCI, XEROX PALO ALTO RES CTR, 71- Concurrent Pos: Dir syst develop, Berkeley Comput Corp, 69-71. Mem: Asn Comput Mach. Res: Programming languages and operating systems. Mailing Add: Xerox Palo Alto Res Ctr 3333 Coyote Hill Rd Palo Alto CA 94304

LAMPSON, GEORGE PETER, b Colman, SDak, June 12, 19; m 48; c 3. BIOCHEMISTRY. Educ: SDak State Univ, BS, 42; Univ Wis, MS, 50. Prof Exp: Res assoc biochem, Ortho Pharmaceut Corp, 50-59; SR RES BIOCHEMIST, DEPT VIRUS & CELL BIOL, MERCK INST THERAPEUT RES, 59- Mem: NY Acad Sci; Am Chem Soc; Am Soc Biol Chemists. Res: Purification and characterization of chicken embryo interferon as a low molecular protein; synthetic and natural double-stranded RNA as inducers of interferon and host resistance; virus chemistry. Mailing Add: Dept of Virus & Cell Biol Merck Inst for Therapeut Res West Point PA 19486

LAMPTON, MICHAEL LOGAN, b Williamsport, Pa, Mar 1, 41. X-RAY ASTRONOMY. Educ: Calif Inst Technol, BS, 62; Univ Calif, Berkeley, PhD(physics), 67. Prof Exp: ASST RES PHYSICIST, SPACE SCI LAB, UNIV CALIF, BERKELEY, 67- Concurrent Pos: NSF fel, Univ Calif, Berkeley, 68-69. Mem: Am Geophys Union; Am Astron Soc. Res: Ultraviolet astronomy. Mailing Add: Space Sci Lab Univ of Calif Berkeley CA 94720

LAMPTON, ROBERT KOERBEL, b Zanesville, Ohio, Jan 30, 11; m 43. BOTANY, TAXONOMY. Educ: Univ Toledo, BA, 33; Univ Mich, MA, 35, PhD(bot), 53. Prof Exp: Instr biol, Univ Toledo, 35-38; high sch teacher, Ohio, 38-42, 46-48; instr bot, Univ Mich, 48-52; prof biol, Pa State Teachers Col, Millersville, 52-55; asst prof biol, Upsala Col, 55-59; asst prof, Fairleigh Dickinson Univ, 59-62; head dept, 62-66, PROF BIOL, W GA COL, 62-, CUR HERBARIUM, 74- Concurrent Pos: Biologist, S C Johnson & Sons, Inc, 51; vis prof, Western Med Col, 53; tissue culturist, Wyeth Labs, Inc, Marietta, Pa, 55- Mem: Bot Soc Am; Am Bryol & Lichenological Soc; Am

Fern Soc; Int Soc Plant Morphol; Torrey Bot Club. Res: Morphogenesis and growth; tissue culture; cryptogams; taxonomy of bryophytes and higher plants. Mailing Add: Dept of Biol W Ga Col Carrollton GA 30117

LAMSON, BALDWIN GAYLORD, b Berkeley, Calif, May 20, 16; m 42; c 4. PATHOLOGY. Educ: Univ Calif, AB, 38; Univ Rochester, MD, 44. Prof Exp: Asst path, Sch Med & Dent, Univ Rochester, 44-45; asst, Sch Med, Emory Univ, 45-46; clin instr exp oncol, Sch Med, Univ Calif, 48-49; from asst prof to assoc prof, 51-61, assoc dean, Sch Med, 65-66, PROF PATH, SCH MED, UNIV CALIF, LOS ANGELES, 62-, DIR UNIV HOSPS & CLINS, 66- Concurrent Pos: AEC & Nat Res Coun fel, Med Sch, Univ Calif, 48-49; fel, Sch Med & Dent, Univ Rochester, 49. Res: Automation in hospital management. Mailing Add: Director's Off Univ of Calif Hosp & Clins Los Angeles CA 90024

LAMUTH, HENRY LEWIS, physics, see 12th edition

LAMY, FRANCOIS, b Lyon, France, Jan 1, 22; nat; m 49; c 3. BIOCHEMISTRY. Educ: Univ Paris, Lic es Sc, 48; Amherst Col, MA, 50; Mass Inst Technol, PhD(physiol), 55. Prof Exp: Instr biol, Mass Inst Technol, 53-55; from instr to assoc prof anat, Univ Pittsburgh, 55-68; PROF BIOCHEM, SCH MED, UNIV SHERBROOKE, 68-, CHMN DEPT, 71- Mem: AAAS; Biophys Soc; Can Biochem Soc; Brit Biochem Soc. Res: Protein chemistry; enzymology. Mailing Add: Dept of Biochem Univ of Sherbrooke Sch of Med Sherbrooke PQ Can

LAMY, PETER PAUL, b Breslau, Ger, Dec 14, 25; US citizen; m 51; c 3. BIOPHARMACEUTICS, CLINICAL PHARMACY. Educ: Philadelphia Col Pharm & Sci, BSc, 56, MSc, 58, PhD(biopharmaceut), 64. Prof Exp: Instr pharm, Philadelphia Col Pharm & Sci, 56-63; assoc prof, Sch Pharm, 67-72, PROF PHARM, SCH PHARM & PROF, DEPT SOCIAL & PREV MED, SCH MED, UNIV MD, BALTIMORE, 72-, DIR INSTNL PHARM PROGS, 68- Concurrent Pos: Asst to dir pharm, Jefferson Med Col, 59-62; instr, Woman's Hosp, Philadelphia, 60-62; lectr, Sch Nursing, Cath Univ Am; consult, USPHS Hosp, 66-72; Am Asn Cols Pharm vis scientist, 68-72. Mem: Fel AAAS; fel Am Col Clin Pharmacol; Acad Pharmaceut Sci; Am Soc Hosp Pharmacists; Am Pharmaceut Asn. Res: Drug transport mechanisms in vivo and in vitro; drug interactions; drug equivalencies and efficiencies. Mailing Add: Univ of Md Sch of Pharm 636 W Lombard St Baltimore MD 21201

LAN, SHIH-JUNG, b Kwangtung, China, Sept 15, 38; m 67; c 2. DRUG METABOLISM. Educ: Univ Tunghai, BS, 60; Okla State Univ, MS, 64; Univ Minn, PhD(biochem), 68. Prof Exp: Fel physiol chem, Univ Wis, 67-69; res investr drug metab, 69-73, sr res investr, 73-74, RES GROUP LEADER, SQUIBB INST MED RES, 75- Res: Am Chem Soc; Am Soc Pharmacol & Exp Therapeut. Res: Microsomal drug metabolism enzyme systems; mechanism of drug action. Mailing Add: Drug Metab Dept Squibb Inst for Med Res New Brunswick NJ 08903

LANA, EDWARD PETER, b Duluth, Minn, Oct 17, 14; m 42; c 3. HORTICULTURE. Educ: Univ Minn, BS, 42, MS, 43, PhD, 48. Prof Exp: Canning crops res, Fairmont Canning Co, Minn, 43-47; asst prof hort, Iowa State Univ, 47-56; PROF HORT & CHMN DEPT, NDAK STATE UNIV, 56- Mem: Am Soc Hort Sci. Res: Crop breeding; cultural studies. Mailing Add: Dept of Hort NDak State Univ Fargo ND 58102

LANCASTER, DOUGLAS, b Fargo, NDak, Mar 30, 29; m 62; c 1. ORNITHOLOGY. Educ: Carleton Col, BA, 50; La State Univ, PhD(zool), 60. Prof Exp: Asst prof zool, Northwestern State Col, La, 60-62; Frank M Chapman res fel ornith, Am Mus Natural Hist, 62-64; asst dir lab ornith, 64-73, DIR LAB ORNITH, CORNELL UNIV, 73- Concurrent Pos: NSF grant, 66-68. Mem: Cooper Ornith Soc; Am Ornith Union; Am Soc Zool; Soc Study Evolution; Animal Behav Soc. Res: Tropical new world ornithology, especially general biology of family Tinamidae; comparative behavior and ecology of the family Ardeidae. Mailing Add: Lab of Ornith Cornell Univ Ithaca NY 14850

LANCASTER, GEORGE MAURICE, b Penrith, Eng, July 18, 34; m 64; c 2. MATHEMATICS. Educ: Univ Liverpool, BSc, 56; Univ Sask, PhD(math), 67. Prof Exp: Res analyst, Weapons Res Div, A V Roe & Co Ltd, Woodford, Eng, 56-58; Northeres analyst, Northern Elec Co Ltd, Montreal, 58-60; lectr math, Royal Roads Mil Col, 60-64; asst prof, Univ Sask, 67-70; ASSOC PROF MATH, ROYAL ROADS MIL COL, 70- Concurrent Pos: Nat Res Coun Can grant, Univ Sask, 68-7O; spec lectr, Univ Victoria, 71. Mem: Am Math Soc; Can Math Cong. Res: Differential geometry; imbedding of Riemannian manifolds. Mailing Add: Dept of Math Royal Roads Mil Col Victoria BC Can

LANCASTER, JAMES D, b Randolph, Miss, June 11, 19; m 42; c 3. AGRONOMY. Educ: Miss State Col, BS, 47, MS, 48; Univ Wis, PhD, 54. Prof Exp: Asst agronomist, Exp Sta. & asst prof agron, 51-57, AGRONOMIST, EXP STA & PROF AGRON, MISS STATE UNIV, 57- Mem: Am Soc Agron; Soil Sci Soc Am; AAAS. Res: Soil fertility and testing; fertilizer evaluation; crop fertilization. Mailing Add: Dept of Agron Miss State Univ Mississippi State MS 39762

LANCASTER, JANE, b Quebec City, PQ, Jan 19, 31; US citizen. TRANSPORTATION GEOGRAPHY. Educ: Columbia Univ, BA, 52, MA, 54, PhD(geog), 62. Prof Exp: Pub instr geog & geol, Am Mus Natural Hist, 52-53; teaching asst geol, Barnard Col, Columbia Univ, 54-58, from lectr to instr geog & geol, 59-64; instr grad fac, Univ, 60-64; instr, Brooklyn Col, 64-65; from asst prof to assoc prof geog, State Univ NY Binghamton, 65-69; ASSOC PROF GEOG, BOSTON UNIV, 69- Concurrent Pos: Commodity-indust analyst, US Bur Mines, Dept Interior, 55-; consult, Link Group, Gen Precision, Inc, NY, 67-; vis lectr, NSF Inst Geog Appln Remote Sensing, Univ Mich, 68-; consult, Comt Geog Appln Remote Sensing, Asn Am Geogr, 69-; vis prof geog, Univ Mass, Amherst, 71-; Fulbright res scholar award systs anal air transp, NZ, 72; transp consult, McGraw-Hill Bk Co, 72- Honors & Awards: Andrew McNally Award, Am Cong Surv & Mapping, 69. Mem: Am Cong Surv & Mapping; Am Geog Soc; Am Soc Photogram; Asn Can Studies in US; Asn Am Geogr. Res: Systems analysis of air transport in North America and abroad, including airport planning and development problems; third level carriers, and New Zealand's problems in particular; geographical applications of remote sensing. Mailing Add: Dept of Geog Boston Univ Boston MA 02215

LANCASTER, JESSIE LEONARD, JR, b Horatio, Ark, Jan 26, 23; m 46; c 4. ENTOMOLOGY. Educ: Univ Ark, BSA, 47; Cornell Univ, PhD(econ entom), 51. Prof Exp: Asst, Cornell Univ, 47-51; from asst prof to assoc prof entom, 51-60; PROF ENTOM, UNIV ARK, FAYETTEVILLE, 60- Concurrent Pos: NIH spec res fel, Rocky Mountain Lab, 63-64. Mem: Entom Soc Am. Res: Medical veterinary entomology and mosquito control. Mailing Add: Rte 9 Old Wire Rd N Fayetteville AR 72701

LANCASTER, JOHN, b Bolton, Miss, Aug 30, 37; m 64. MICROBIAL GENETICS. Educ: Miss State Univ, BS, 59, MS, 61; Univ Tex, PhD(microbiol), 64. Prof Exp: NIH trainee, Univ Tex, 63-64; asst prof microbiol, 64-68, ASSOC PROF MICROBIOL, UNIV OKLA, 68- Concurrent Pos: NIH grant, 66. Mem: AAAS; Am Soc Microbiol. Res: Mechanism of conjugation in Escherichia coli. Mailing Add: Dept of Bot & Microbiol Univ of Okla Norman OK 73069

LANCASTER, JOHN EDGAR, b Newell, WVa, Mar 18, 21; m 56; c 3. PHYSICAL CHEMISTRY. Educ: Duquesne Univ, BS, 42; Univ Minn, PhD(chem), 51. Prof Exp: Chemist, Chem Warfare Serv, 42-44; PHYS CHEMIST & INFRARED SPECTROSCOPIST, AM CYANAMID CO, 52- Mem: Am Chem Soc; Am Phys Soc. Res: Infrared spectroscopy; molecular vibrations; magnetic resonance. Mailing Add: Am Cyanamid Co 1937 W Main St Stamford CT 06904

LANCASTER, MALCOLM, b Amarillo, Tex, July 28, 31; m 59; c 4. AEROSPACE MEDICINE, CARDIOLOGY. Educ: Univ Tex Southwestern Med Sch, MD, 56; Univ Colo, Denver, MS, 60. Prof Exp: Chief med serv, 48th Tactical Hosp, Royal Air Force, Lakenheath, Eng, 60-63; chief cardiopulmonary serv & chmn dept med, US Air Force Hosp, Wright-Patterson AFB, 65-66; chief internal med br, 66-73, CHIEF CLIN SCI DIV, SCH AEROSPACE MED, BROOKS AFB, TEX, 72- Concurrent Pos: Prof clin med, Med Sch, Univ Tex, San Antonio, 72- Honors & Awards: Casimir Funk Award, Asn Mil Surgeons US, 71; USAF Res & Develop Award, 71; John Jeffries Award, Am Inst Aeronaut & Astronaut, 74; Arnold D Tuttle Award, Aerospace Med Asn, 75. Mem: AMA; fel Am Col Cardiol; fel Am Col Physicians; Am Heart Asn; fel Am Col Prev Med. Res: Medical aspects of aerospace operations; cardiovascular disease epidemiology; computers and electrocardiography. Mailing Add: US Air Force Sch Aerospace Med Brooks AFB TX 78235

LANCASTER, OTIS EWING, b Pleasant Hill, Mo, Jan 28, 09; m 42; c 5. AERONAUTICAL ENGINEERING, MATHEMATICS. Educ: Cent Mo State Teachers Col, 29; Univ Mo, MA, 34; Harvard Univ, PhD(math), 37; Calif Inst Technol, AeroE, 45. Prof Exp: Teacher, Oak Grove High Sch, Mo, 29-30 & Independence Jr High Sch, Mo, 30-33; instr & tutor math, Harvard Univ, 36-37; from instr to asst prof, Univ Md, 37-42; head appl math br, Res Div, Bur Aeronaut, Navy Dept, Washington, DC, 46-54, asst dir, 54; mem planning staff, Internal Revenue Serv, 54-55; dir statist & econ staff, Bur Finance, Post Off Dept, 55-57; GEORGE WESTINGHOUSE PROF ENG EDUC, PA STATE UNIV, UNIVERSITY PARK, 57-, ASSOC DEAN, 67- Concurrent Pos: Consult, NASA-Univ rels. Mem: Am Soc Eng Educ (vpres, 65-67); Math Asn Am; assoc fel Am Inst Aeronaut & Astronaut; Inst Math Statist; Am Soc Mech Engrs. Res: Engineering education; aircraft propulsion; statistics; improvement of teaching. Mailing Add: 268 Ellen Ave State College PA 16801

LANCE, JOHN FRANKLIN, b Vaughn, NMex, May 21, 16; m 42; c 2. GEOLOGY. Educ: Col Mines & Metal, Univ Tex, BA, 37; Calif Inst Technol, MS, 46, PhD(paleont, petrog), 49. Prof Exp: Asst prof geol, Whittier Col, 48-50; from asst prof to prof paleont, Univ Ariz, 50-67; staff assoc, Div Inst Prog, 63-65 & 67-70, exec asst div environ sci, 71-75, PROG DIR GEOL, NSF, 75- Concurrent Pos: Geologist, US Geol Surv, 52-63. Mem: Geol Soc Am; Paleont Soc; Soc Vert Paleontologists; Soc Study Evolution. Res: Pliocene and Pleistocene mammalian fossils. Mailing Add: Div of Earth Sci NSF Washington DC 20550

LANCEFIELD, REBECCA CRAIGHILL, b Ft Wadsworth, NY, Jan 5, 95; m 18; c 1. BACTERIOLOGY. Educ: Wellesley Col, AB, 16; Columbia Univ, AM, 18, PhD(immunol, bact), 25. Prof Exp: Tech asst, Rockefeller Inst, 18-19; tech asst, Sta Exp Evolution, Dept Genetics, Carnegie Inst, 19-21; instr bact, Univ Ore, 21-22; tech asst, 22-25, asst, 25-29, assoc, 29-42, assoc mem, 42-58, mem & prof microbiol, 58-65, EMER PROF MICROBIOL, ROCKEFELLER UNIV, 65- Concurrent Pos: Gehrmann lectr, Col Med, Univ Ill, 64; mem streptococcal dis comn, US Armed Forces Epidemiol Bd. Honors & Awards: T Duckett Jones Mem Award, Helen Hay Whitney Found, 60; Res Achievement Award, Am Heart Asn, 64. Mem: Nat Acad Sci; AAAS; hon mem Am Soc Microbiol (vpres, 42, pres, 43); Am Asn Immunologists (pres, 61-62); fel NY Acad Med. Res: Immunochemical studies of streptococci; chemical composition and antigenic structure of hemolytic streptococci; relationship of streptococcal components to biological properties of the organism; classification of streptococci into serological groups and types. Mailing Add: Rockefeller Univ New York NY 10021

LANCHANTIN, GERARD FRANCIS, b Detroit, Mich, Mar 27, 29; m 55; c 5. BIOCHEMISTRY. Educ: Seton Hall Univ, BS, 50; Univ Wyo, MS, 51; Univ Southern Calif, PhD(biochem), 54. Prof Exp: Res assoc, 54-55, from asst prof to assoc prof, 57-69, ADJ PROF BIOCHEM, SCH MED, UNIV SOUTHERN CALIF, 69-; DIR BIOCHEM, ST JOSEPH MED CTR, 74- Concurrent Pos: Consult, Los Angeles County Gen Hosp, 57-; chief biochem dept, Cedars-Sinai Med Ctr, 57-74. Mem: Fel AAAS; Am Asn Clin Chem; Am Chem Soc; Am Soc Hemat; Am Soc Biol Chemists. Res: Clinical biochemistry; blood coagulation. Mailing Add: Dept of Biochem Div of Labs St Joseph Med Ctr Burbank CA 91505

LANCIANI, CARMINE ANDREW, b Leominster, Mass, May 16, 41; m 64; c 2. ECOLOGY. Educ: Cornell Univ, BS, 63, PhD, 68. Prof Exp: Interim asst prof zool, 68-70, asst prof zool & biol sci, 70-73, ASSOC PROF ZOOL, UNIV FLA, 73- Mem: Ecol Soc Am; Entom Soc Am; Am Soc Limnol & Oceanog; Soc Study Evolution. Res: Population ecology of aquatic organisms, particularly life cycles, growth, reproduction and competition of parasitic water mites; effect of parasitism on host ecology. Mailing Add: Dept of Zool Univ of Fla Gainesville FL 32611

LAND, ANTHONY HAMILTON, b Lexington, Ky, Jan 15, 14; m 41; c 2. ORGANIC CHEMISTRY. Educ: Univ Ky, BS, 36; Univ Ill, PhD(org chem), 41. Prof Exp: Res chemist, Nat Defense Res Com, Ill, 41-42, Sharp & Dohme, Inc, Pa, 42-53, Sharp & Dohme Div, Merck & Co, Inc, 53-58 & Sci Info Dept, Merck Sharp & Dohme Res Labs, 58-65, RES CHEMIST, DEPT MED CHEM, MERCK SHARP & DOHME RES LABS, 65- Mem: Am Chem Soc. Res: Synthetic organic chemistry; bisulfite compounds of acetaldehyde and reactions of 3-phenylbenzisosulfonazole; derivatives of sulfathiazole. Mailing Add: Mounted Rte Box 168 Telford PA 18969

LAND, CHARLES EVEN, b San Francisco, Calif, July 13, 37; m 60; c 1. STATISTICS. Educ: Univ Ore, BA, 59; Univ Chicago, MA, 64, PhD(statist), 68. Prof Exp: Res assoc statist, Atomic Bomb Casualty Comn, Nat Acad Sci, 66-68; asst prof, Ore State Univ, 68-73; res assoc statist, Atomic Bomb Casualty Comn & Radiation Effects Res Found, 73-75; EXPERT MATH STATISTICIAN, BIOMETRY BR, NAT CANCER INST, 75- Mem: AAAS; Inst Math Statist; Am Statist Asn; Biomet Soc. Res: Mathematical statistics; inference problems associated with transformations of data; biometry. Mailing Add: Biometry Br Nat Cancer Inst Bethesda MD 20014

LAND, DAVID JOHN, b Boston, Mass, Feb 15, 39. THEORETICAL PHYSICS. Educ: Boston Col, BS, 59; Brown Univ, PhD(physics), 66. Prof Exp: Res asst physics, Brown Univ, 60-66; Nat Acad Sci res assoc, 66-68, RES PHYSICIST, US NAVAL ORD LAB, 68- Mem: Am Phys Soc. Res: Atomic physics; effects of meteorology of optical systems. Mailing Add: Nuclear Physics Br Naval Surface Weapons Ctr Silver Spring MD 20910

LAND, EDWIN HERBERT, b Bridgeport, Conn, May 7, 09; m; c 2. PHYSICS. Hon Degrees: ScD, Tufts Col, 47, Polytech Inst Brooklyn, 52, Colby Col, 55, Harvard Univ, 57, Northeastern Univ, 59, Yale Univ, 66, Columbia Univ, 67, Loyola Univ, 70 & NY Univ, 73; LLD, Bates Col, 53, Wash Univ, 66 & Univ Mass, 67; LHD, Williams Col, 68. Prof Exp: FOUNDER, CHMN BD & DIR RES, POLAROID CORP, 37- Concurrent Pos: Vis prof, Mass Inst Technol, 56-; mem, President's Sci Adv Comt, 57-59, consult-at-large, 60-73; mem, Nat Comn Technol, Automation & Econ Progress, 64-66; mem, Carnegie Comn Educ TV, 66-67; William James lectr psychol, Harvard Univ, 66-67, Morris Loeb lectr physics, 74; mem bd trustees, Ford Found, 67- Honors & Awards: Hood Medal, Royal Photog Soc Gt Brit, 35, Progress Medal, 57; Cresson Medal, Franklin Inst, 37, Potts Medal, 56 & Vermilye Medal, 74; John Scott Medal & Award, Philadelphia City Trusts, 38; Rumford Medal, Am Acad Arts & Sci, 45; Holley Medal, Am Soc Mech Engrs, 48; Duddell Medal, Phys Soc Gt Brit, 49; Progress Medal, Soc Photog Scientists & Engrs, 55; Progress Medal, Photog Soc Am, 60; Golden Soc Medal, Photog Soc Vienna, 61; Proctor Award, Sci Res Soc Am, 63; Presidential Medal of Freedom, 63; Indust Res Inst Medal, 65; Albert A Michelson Award, 66; Diesel Medal in Gold, Ger Asn Inventors, 66; Frederick Ives Medal, Optical Soc Am, 67; Nat Medal of Sci, 67; Kulturpreis, Photog Soc Ger, 67; Photog Sci & Eng J Award, 71; Founders Medal, Nat Acad Eng, 72; Perkin Medal, 74. Mem: Nat Acad Sci; Nat Acad Eng; fel Am Acad Arts & Sci (pres, 51-53); Am Philos Soc; hon mem Optical Soc Am. Res: Light polarization; synthetic polarizers; three-dimensional presentation; one-step photography and associated photochemical mechanisms; vision and color vision. Mailing Add: Polaroid Corp 730 Main St Cambridge MA 02138

LAND, GEOFFREY ALLISON, b Jeannette, Pa, July 9, 42; m 66; c 2. MEDICAL MYCOLOGY, MEDICAL MICROBIOLOGY. Educ: Univ Tex, Arlington, BSc, 68; Tex Christian Univ, MSc, 70; Tulane Univ, PhD(med mycol), 73. Prof Exp: Inhalation therapist, Baylor Univ Med Ctr, Dallas, 66-68; teaching asst microbiol, Tex Christian Univ, 68-70; res assoc, Lab Cell Surface Chem, Duke Univ Med Ctr, 73-74; CHMN & DIR MED MYCOL, WADLEY INST MOLECULAR MED, 74- Concurrent Pos: Vis asst prof, NC Cent Univ, 73-74; instr & lectr, Tex Soc Clin Microbiologists, 74- & NTex Soc Med Technologists, 75-; vis scientist, Dept Virol, Cent Pub Health Lab State Serum Inst, Helsinki, 75; adj prof biol & chem, NTex State Univ, 75-; assoc ed, J Oncol & Hematol, 75- Mem: Am Soc Microbiol; Med Mycol Soc Am. Res: The molecular basis and early diagnosis of fungal infections in the compromised host; production of the antiviral interferon and its possible chemotherapeutic role for treating viral diseases. Mailing Add: Dept of Mycol Wadley Inst of Molecular Med Dallas TX 75235

LAND, JAMES EDWARD, b Filbert, SC, Jan 5, 15; m 43. CHEMISTRY. Educ: Clemson Col, BS, 35; Tulane Univ, MS, 38; Univ NC, PhD, 49. Prof Exp: From assoc prof to prof, 38-75, EMER PROF CHEM, AUBURN UNIV, 75- Mem: Am Chem Soc; The Chem Soc. Res: Electrochemistry of germanium; coordination complex compounds of niobium; spectrophotometric studies of niobium compounds; alkoxides of niobium; measurement of activity coefficients by electromotive force measurements or solubility measurements. Mailing Add: Dept of Chem Auburn Univ Auburn IL 36830

LAND, LYNTON S, b Baltimore, Md, Dec 30, 40. GEOLOGY, GEOCHEMISTRY. Educ: Johns Hopkins Univ, AB, 62, MA, 63; Lehigh Univ, PhD(geol), 66. Prof Exp: Res fel geol, Calif Inst Technol, 66-67; ASST PROF GEOL, UNIV TEX, AUSTIN, 67- Mem: Soc Econ Paleontologists & Mineralogists; Int Asn Sedimentol. Res: Sedimentology; carbonate sedimentation; diagenesis; sedimentary geochemistry; stable isotope geochemistry. Mailing Add: Dept of Geol Univ of Tex Austin TX 78712

LAND, PETER L, b Leasburg, Mo, Nov 20, 29; m 66; c 3. SOLID STATE PHYSICS, CERAMICS. Educ: Univ Mo, BS, 58, MS, 60, PhD(physics), 64. Prof Exp: Res scientist, Metall & Ceramics Lab, Aerospace Res Labs, 74-75, RES SCIENTIST, METALL & CERAMICS BR, AIR FORCE MAT LABS, WRIGHT-PATTERSON AFB, 75- Mem: Am Phys Soc. Res: New or improved high temperature materials; optical properties of solids. Mailing Add: 1565 Woods Dr Dayton OH 45432

LAND, ROBERT H, b Portland, Maine, Sept 17, 24. NUCLEAR PHYSICS. Educ: Univ Maine, BS, 49; Mass Inst Technol, PhD(nuclear physics), 57. Prof Exp: Asst physicist, 56-60, ASSOC PHYSICIST, ARGONNE NAT LAB, 60- Mem: Am Phys Soc; Am Nuclear Soc; Asn Comput Mach. Res: Photoproduction of charged pi-mesons from deuterium; reactor neutron diffusion theory; molecular physics and solid state physics calculations using digital computers. Mailing Add: Argonne Nat Lab Chem Eng Div 9700 S Cass Ave Argonne IL 60439

LAND, WILLIAM EVERETT, b Baltimore, Md, Aug 23, 08; m 42. PHYSICAL CHEMISTRY. Educ: Johns Hopkins Univ, BS, 28, PhD(phys chem), 33. Prof Exp: Chemist, Devoe & Raynolds Co, 28-29; instr chem, Emory Univ, 33-37; asst dir res, Glidden Co, 37-42; dep head high explosives sect, Bur Ord, 42-51, head high explosives res & develop, 51-60, div engr, Mines & Explosives Div, Bur Naval Weapons, 60-65, asst dir, Mine Warfare Proj Off, 65-66, div engr, Mine Warfare Div, Naval Ord Systs Command, 66-68, CONSULT NAVAL ORD SYSTS COMMAND, BUR NAVAL WEAPONS, NAVY DEPT, 68- Concurrent Pos: Mem ammunition & high explosives panel, Res & Develop Bd, 48-51, chmn, 51-53; US leader explosives panel, Tripartite Tech Coop Prog, 61-65. Mem: Am Chem Soc. Res: Adsorption; catalysis; titanium pigments; microscopy; chemical engineering; ordnance engineering. Mailing Add: 9200 Beech Hill Dr Bethesda MD 20034

LANDAHL, HERBERT DANIEL, b Fancheng, China, Apr 23, 13; US citizen; m 40; c 3. MATHEMATICAL BIOLOGY. Educ: St Olaf Col, AB, 34; Univ Chicago, SM, 36, PhD(math biophys), 41. Prof Exp: Asst, Psychomet Lab, Univ Chicago, 37-38, asst math biophys, Dept Physiol, 39-42, res assoc, 42-45, asst prof, 45-48, from assoc prof to prof math biol, 48-58, prof biophys, 64-68, secy comt math biol, 48-64, actg chmn, 64-68; PROF BIOPHYS & BIOMATH, UNIV CALIF, SAN FRANCISCO, 68- Concurrent Pos: Res career award, NIH, 62-68. Mem: Biomet Soc; Biophys Soc; Soc Math Biol (vpres, 72-). Res: Mathematical biophysics of cell division, nerve excitation and central nervous system; removal of aerosols and vapors by the human respiratory tract; biological effects of radiation; population interaction; biological periodicities. Mailing Add: Dept of Biochem & Biophys Univ of Calif San Francisco CA 94122

LANDAU, BARBARA RUTH, b Pierre, SDak, Apr 28, 23. PHYSIOLOGY. Educ: Univ Wis, BS, 45, MS, 49, PhD, 56. Prof Exp: Instr phys educ, Rockford Col, 45-47; instr physiol, Mt Holyoke Col, 49-51; instr, St Louis Univ, 56-59; from instr to asst prof, Univ Wis, 59-62; asst prof zool, Univ Idaho, 62-64; asst prof physiol, biophys & biol struct, 64-72, ASSOC PROF PHYSIOL, BIOPHYS & BIOL STRUCT, UNIV WASH, 72- Mem: AAAS; Am Physiol Soc. Res: Neural aspects of temperature regulation, hibernation, cell activity at reduced temperature. Mailing Add: Dept of Physiol & Biophys Univ of Wash Seattle WA 98195

LANDAU, BERNARD ROBERT, b Newark, NJ, June 24, 26; m 56; c 3. MEDICINE, BIOCHEMISTRY. Educ: Mass Inst Technol, SB, 47; Harvard Univ, MA, 49, PhD(chem), 50; Harvard Med Sch, MD, 54. Prof Exp: Med house officer, Peter Bent Brigham Hosp, Boston, Mass, 54-55, sr res physician, 58-59; asst prof biochem & from asst prof to assoc prof med, Case Western Reserve Univ, 59-67; dir dept biochem, Merck Inst Therapeut Res, 67-69; PROF MED, CASE WESTERN RESERVE UNIV, 69-, PROF PHARMACOL, 70- Concurrent Pos: Clin assoc, Nat Cancer Inst, 55-57; USPHS res fel biochem, Harvard Med Sch, 57-58, tutor, 57-59; estab investr, Am Heart Asn, 59-64. Mem: Am Fedn Clin Res; Soc Biol Chem; Am Physiol Soc; Am Soc Clin Invest; Am Diabetes Asn. Res: Carbohydrate metabolism; endocrinology; diabetes mellitus. Mailing Add: Dept of Med Case Western Reserve Univ Cleveland OH 44106

LANDAU, BURTON JOSEPH, b Boston, Mass, May 6, 33; m 57; c 2. MICROBIOLOGY, VIROLOGY. Educ: Boston Univ, AB, 54; Univ NH, MS, 57; Univ Mich, PhD(microbiol), 67. Prof Exp: Sr res virologist, Merck Inst Therapeut Res, Merck, Inc, 64-65; res assoc microbiol, Univ Mich, 65-67; sr instr, 67-69, asst prof, 69-74, ASSOC PROF MICROBIOL, HAHNEMANN MED COL, 74- Mem: Am Soc Microbiol; Soc Gen Microbiol; Tissue Culture Asn; AAAS. Res: Use of differentiating cell cultures to grow viruses with restricted host ranges; virus-host cell interactions leading to virus induced transformation. Mailing Add: Dept of Microbiol & Immunol Hahnemann Med Col Philadelphia PA 19102

LANDAU, DAVID PAUL, b St Louis, Mo, June 22, 41; m 66; c 2. MAGNETISM, STATISTICAL MECHANICS. Educ: Princeton Univ, BA, 63; Yale Univ, MS, 65, PhD(physics), 67. Prof Exp: Asst res physics, Nat Ctr Sci Res, Grenoble, France, 67-68; lectr eng & appl sci, Yale Univ, 68-69; asst prof, 69-73, ASSOC PROF PHYSICS, UNIV GA, 73- Concurrent Pos: Guest scientist, KFA Jülich, WGer, 74; Alexander von Humboldt fel, Univ Saarland, 75. Mem: Am Phys Soc; Sigma Xi. Res: Critical phenomena associated with phase transitions; properties of magnetic solids. Mailing Add: Dept of Physics Univ of Ga Athens GA 30602

LANDAU, EDWARD FREDERICK, b New York, NY, Jan 25, 16; m 43; c 2. ORGANIC CHEMISTRY, TEXTILE CHEMISTRY. Educ: City Col New York, BS, 38; Polytech Inst Brooklyn, MS, 42, PhD(org chem), 45. Prof Exp: Clin chemist, Hosp Holy Family, NY, 38-41; develop & res chemist, Nopco Chem Co, NJ, 41-44; res chemist, Celanese Corp Am, 45-49; lab dir, United Merchants Labs, Inc, 49-57; prod appln mgr, Cent Res Lab, 58-60, mgr mkt eval, Plastics Div, 60-68, mgr res strategy, 68-70, mgr res planning & admin, 70-72, DIR RES PLANNING & ADMIN, ALLIED CHEM CORP, 72- Concurrent Pos: Instr, Polytech Inst Brooklyn, 42-52. Mem: AAAS; Am Chem Soc; Chem Mkt Res Asn; Com Develop Asn; Indust Res Inst. Res: Synthetic resins, plastics and fibers; market research and evaluations; research planning, evaluation and administration. Mailing Add: 67 S Munn Ave East Orange NJ 07018

LANDAU, EMANUEL, b New York, NY, Nov 28, 19; m 48; c 2. EPIDEMIOLOGY, BIOSTATISTICS. Educ: City Col New York, BA, 39; Am Univ, PhD(econ), 66. Prof Exp: Bus economist, Econ Date Anal Br, Off Price Admin, 41-42 & 46-47; chief, Family Statist Sect, Bur Census, 48-56; mem staff, Calif State Dept Pub Health, 57-59; chief, Biomet Sect, Div Air Pollution, USPHS, 59-62; head, Lab & Clin Trials Sect, Nat Cancer Inst, 62-64; statist adv, Nat Air Pollution Control Admin, 65-69; epidemiologist, Adminr Res & Develop, Environ Health Serv, 69-71; epidemiologist, Adminr Res & Monitoring, Environ Protection Agency, 71; chief, Epidemiol Studies Br, Bur Radiol Health, Food & Drug Admin, USPHS, 71-75; PROJ DIR, AM PUB HEALTH ASN, 75- Concurrent Pos: Mem, Career Serv Bd Math & Statist, Dept Health, Educ & Welfare, 65-69; mem, Comt Long-term Training Outside Serv, USPHS, 66-68; adv air qual criteria, WHO, Switz, 67; mem, Study Lung Cancer Among Uranium Miners, USPHS, 67; adv air qual criteria, Karolinska Inst, Sweden, 68; Nat Air Control Admin tech liaison rep, Adv Comt Toxicol, Nat Acad Sci, 68-69; adv, Dept Transp, 72-74; assoc ed, J Air Pollution Control Asn, 72 & J Clin Data & Anal, 74-; consult, Bur Radiol Health, Food & Drug Admin, 75- Honors & Awards: Superior Serv Award, Dept Health, Educ & Welfare, 63. Mem: Fel Am Pub Health Asn; fel Royal Soc Health; Air Pollution Control Asn; Am Statist Asn; Soc Occup & Environ Health. Res: Problems of environmental health; public health statistics; chronic disease epidemiology. Mailing Add: Am Pub Health Asn 1015 18th St NW Washington DC 20036

LANDAU, HENRY JACOB, b Lwow, Poland, Feb 11, 31; nat US; m 60; c 2. MATHEMATICS. Educ: Harvard Univ, AB, 53, AM, 55, PhD(math), 57. Prof Exp: MEM TECH STAFF, BELL TEL LABS, 57- Concurrent Pos. NSF fel, Inst Advan Study, 59-60, 67. Mem: Am Math Soc. Res: Theory of functions of a complex variable. Mailing Add: Bell Tel Labs Murray Hill NJ 07971

LANDAU, JOSEPH VICTOR, b New York, NY, Jan 9, 28; m 50; c 3. MOLECULAR BIOLOGY. Educ: City Col New York, BS, 47; NY Univ, MSc, 49, PhD, 53. Prof Exp: USPHS asst, NY Univ, 49-51; USPHS res asst, Naples Zool Sta, Italy, 52; Runyon Cancer Res fel, NY Univ, 52-55; instr physiol, Russell Sage Col, 56-57; res assoc oncol, Albany Med Col, Union, 57-66; PROF BIOL & HEAD ACCELERATED BIOMED PROG, RENSSELAER POLYTECH INST, 67-, CHMN DEPT BIOL, 72- Concurrent Pos: Chief biol sect, Basic Sci Res Lab, Vet Admin Hosp, Albany, 57-66; adj assoc prof, Rensselaer Polytech Inst, 64-67. Mem: Am Soc Cell Biol; Undersea Med Soc; Am Inst Biol Sci; Am Soc Microbiol. Res: Protein and nucleic acid synthesis; barobiology. Mailing Add: Dept of Biol Rensselaer Polytech Inst Troy NY 12181

LANDAU, JOSEPH WHITE, b Buffalo, NY, May 23, 30; m 64; c 5. MEDICINE, DERMATOLOGY. Educ: Cornell Univ, BA, 51, MD, 55; Am Bd Pediat, dipl, 62; Am Bd Dermat, dipl, 65. Prof Exp: Intern, Gen Hosp, Buffalo, 55-56; resident pediat, Children's Hosp, Buffalo, 56, Children's Hosp, Boston, Mass, 59-60 & Med Ctr, Univ Calif, Los Angeles, 60-61; asst res dermatologist, 64, from asst prof to assoc prof med & dermat, 64-74, ASSOC CLIN PROF MED DERMAT, MED CTR, UNIV CALIF, LOS ANGELES, 74-, ATTEND PHYSICIAN, STUDENT HEALTH SERV, 65- Concurrent Pos: USPHS fel hemat, Children's Hosp, Los Angeles, 61-62 & fel mycol, Med Ctr, Univ Calif, Los Angeles, 62-63; attend physician, Wadsworth Vet Admin Hosp, 66- Mem: Am Acad Dermat; Soc Invest Dermat. Res: Host-parasite relationships in mycology; genodermatoses. Mailing Add: Suite 106 Med Arts Bldg 2200 Santa Monica Blvd Santa Monica CA 90404

LANDAU, JUDAH, b Bronx, NY, Oct 23, 42; US & Israeli citizen; m 64; c 3. LOW TEMPERATURE PHYSICS. Educ: Mass Inst Technol, SB, 64; Ohio State Univ, PhD(physics), 69. Prof Exp: Vis assoc prof physics, Ohio State Univ, 75-76; SR LECTR PHYSICS, ISRAEL INST TECHNOL, 69- Mem: Am Phys Soc; Israel Phys Soc; Am Asn Physics Teachers. Res: Experiments on liquid and solid helium at ultra low temperatures. Mailing Add: Dept of Physics Israel Inst of Technol Haifa Israel

LANDAU, RICHARD LOUIS, b St Louis, Mo, Aug 8, 16; m 43; c 3. ENDOCRINOLOGY. Educ: Wash Univ, BS & MD, 40. Prof Exp: From asst prof to assoc prof, 48-59, PROF MED, UNIV CHICAGO, 59- Concurrent Pos: Ed, Perspectives in Biol & Med. Mem: Am Soc Clin Invest; Endocrine Soc; AMA. Res: Hormonal regulation of growth processes; reproductive endocrinology; metabolic influence of progesterone; effect of steroid hormones on electrolyte metabolism. Mailing Add: Dept of Med Univ of Chicago Sch of Med Chicago IL 60637

LANDAU, RONALD WOLF, b New York, NY, Aug 26, 32; m 63; c 2. PLASMA PHYSICS. Educ: Columbia Univ, BA, 53, MA, 57; Stevens Inst Technol, PhD(physics), 63. Prof Exp: Asst, Stevens Inst Technol, 58-62; res physicist, Lawrence Radiation Lab, Univ Calif, 63-65 & Physics Int Inc, 65-67; asst prof, San Francisco State Univ, 67-74; MEM STAFF, DEPT PHYSICS, QUEENS COL, 74- Concurrent Pos: Lab instr, Hunter Col, 58-61. Res: Theoretical research in support of a program to find a practical means of thermonuclear fusion; stability criteria of the negative mass instability. Mailing Add: Dept of Physics Queens Col Flushing NY 11367

LANDAU, WILLIAM, b Jersey City, NJ, July 3, 27; m 63; c 2. MICROBIOLOGY. Educ: Univ Conn, BA, 49; Yale Univ, MS, 51; Univ Pa, PhD(pub health, prev med), 58. Prof Exp: Res assoc biochem, Roswell Park Mem Inst, 58-61; assoc dir microbiol dept, Presby-St Luke's Hosp, 61-74; asst prof, 62-74, ASSOC PROF BACT, RUSH MED CTR, 74-; ASSOC SCIENTIST, PRESBY-ST LUKE'S HOSP, CHICAGO, 74- Mem: Am Soc Microbiol; NY Acad Sci. Res: Clinical bacteriology. Mailing Add: Microbiol Dept Rush Med Ctr 1753 W Congress Pkwy Chicago IL 60612

LANDAU, WILLIAM M, b St Louis, Mo, Oct 10, 24; m 47; c 4. NEUROLOGY, NEUROPHYSIOLOGY. Educ: Washington Univ, MD, 47. Prof Exp: From instr to assoc prof, 52-63, PROF NEUROL, SCH MED, WASHINGTON UNIV, 63-, HEAD DEPT, 70- Concurrent Pos: Sr asst surgeon & neurophysiologist, NIMH & Nat Inst Neurol Dis & Blindness, 52-54; vis prof, Univ Munich, 63; pres, Am Bd Psychiat & Neurol, 75. Mem: Am Physiol Soc; Am Electroencephalog Soc; Am Neurol Asn; Am Acad Neurol. Res: Sensory and motor systems. Mailing Add: 660 S Euclid Ave St Louis MO 63110

LANDAUER, JOSEPH K, b Highland Park, Ill, June 9, 27; m 51; c 2. PHYSICS. Educ: Univ Chicago, MS, 51, PhD(physics), 54. Prof Exp: Physicist, Snow, Ice & Permafrost Res Estab, US Army Corps Engrs, Ill, 53-58; PHYSICIST, LAWRENCE LIVERMORE LAB, UNIV CALIF, 58- Concurrent Pos: Asst to dep asst secy defense, Dept Defense, 75-76. Mem: Am Phys Soc. Res: Rheology of snow and ice; hydrodynamics; neutronics. Mailing Add: 241 La Questa Dr Danville CA 94526

LANDAUER, ROLF WILLIAM, b Stuttgart, Ger, Feb 4, 27; nat US; m 50; c 3. SOLID STATE PHYSICS, COMPUTER SCIENCE. Educ: Harvard Univ, SB, 45, AM, 47, PhD(physics), 50. Prof Exp: Physicist, Lewis Lab, Nat Adv Comt Aeronaut, 50-52; physicist, 52-61, dir phys sci, 61-66, asst dir res, 66-69, IBM FEL, THOMAS J WATSON RES CTR, IBM CORP, 69- Mem: Fel Am Phys Soc; fel Inst Elec & Electronics Eng. Mailing Add: Thomas J Watson Res Ctr IBM Corp PO Box 218 Yorktown Heights NY 10598

LANDAUER, WALTER, b Mannheim, Ger, July 15, 96; nat US; m 64. EXPERIMENTAL EMBRYOLOGY. Educ: Univ Heidelberg, PhD(zool), 21. Prof Exp: Asst, Zool Inst, Univ Heidelberg, 22-24; asst geneticist, Storrs Agr Exp Sta, Univ Conn, 24-28; prof genetics, 28-64; res assoc zool & comp anat, 64-66, HON RES ASSOC, UNIV COL, UNIV LONDON, 66-; EMER PROF GENETICS, UNIV CONN, 64- Concurrent Pos: Guest investr, Nuffield Inst Comp Med, Eng, 64-67. Honors & Awards: Borden Award, 54. Mem: AAAS; Am Soc Nat; Genetics Soc Am; Am Soc Zool; Soc Develop Biol. Res: Developmental genetics, especially in poultry; skeletal abnormalities; role of neuromuscular transmission in chick morphogenesis. Mailing Add: Wolfson House 4 Stephenson Way London England

LANDAW, STEPHEN ARTHUR, b Paterson, NJ, June 20, 36; m 62; c 2. INTERNAL MEDICINE, HEMATOLOGY. Educ: Univ Wis-Madison, BS, 57; George Washington Univ, MD, 59; Univ Calif, Berkeley, PhD(med physics), 69; Am Bd Internal Med, dipl, 72, Am Bd Nuclear Med, dipl, 72. Prof Exp: Intern, Mt Sinai Hosp, NY, 59-60, asst resident internal med, 60-61; NIH fel med & hemat, Med Col Va, 62-63; Nat Heart Inst fel med physics, Donner Lab, Univ Calif, Berkeley, 63-70; asst physician, Donner Lab, Univ Calif, Berkeley, 70-73, lectr med physics, Univ, 70-72; ASSOC PROF MED & RADIOL, STATE UNIV NY UPSTATE MED CTR, 73- Concurrent Pos: Attend staff physician, Alameda County Hosp, Oakland, Calif, 69-73, chief isotope lab, 71-73; Nat Heart & Lung Inst career develop award, 70-73; assoc chief staff res, Vet Admin Hosp, Syracuse, NY, 73-; mem attend staff med, Vet Admin Hosp, Univ Hosp & Crouse-Irving Mem Hosp, 73- Mem: Fel Am Col Physicians; Am Soc Hemat; Am Fedn Clin Res; Soc Nuclear Med; Soc Exp Biol & Med. Res: Endogenous carbon monoxide production; quantitative red blood cell kinetics. Mailing Add: Vet Admin Hosp Irving Ave & University Pl Syracuse NY 13210

LANDAY, MARSHALL EDWIN, b Pittsburgh, Pa, Oct 7, 27; m 58; c 3. MICROBIOLOGY. Educ: Univ Pittsburgh, BS, 49; WVa Univ, MS, 61; Duke Univ, PhD(microbiol), 65. Prof Exp: Sect leader microbiol, US Army, Ft Detrick, Md, 65-67; asst prof mycol, Med Sch, George Washington Univ, 67-70; microbiologist, Clin Lab, Jewish Hosp, 70-72; MICROBIOLOGIST, ST ELIZABETH'S HOSP, 72- Concurrent Pos: Res Corp fel, Med Sch, George Washington Univ, 68-69. Mem: Am Soc Microbiol; Med Mycol Soc of the Americas; Int Soc Human & Animal Mycol. Res: Immunology, serology and morphology of medical mycology; bacterial vaccines. Mailing Add: Bact Lab St Elizabeth's Hosp Covington KY 41014

LANDBORG, RICHARD JOHN, b Manchester, Iowa, May 13, 33; m 55; c 4. CHEMISTRY, SCIENCE EDUCATION. Educ: Luther Col, Iowa, BA, 55; Univ Iowa, MS, 57, PhD(chem), 59. Prof Exp: Part-time instr chem, Cornell Col, 57-58; asst prof, 59-63, chmn dept, 65-67, ASSOC PROF CHEM, AUGUSTANA COL, SDAK, 63- Concurrent Pos: Fulbright exchange prof, Univ Santa Maria Antigua, Panama, 67. Mem: AAAS; Am Chem Soc. Res: Chemistry of diazomethane particularly the addition cyclization reactions with activated olefinic systems. Mailing Add: Dept of Chem Augustana Col Sioux Falls SD 57102

LANDE, ALEXANDER, b Hilversum, Netherlands, Jan 5, 36; US citizen. THEORETICAL NUCLEAR PHYSICS. Educ: Cornell Univ, BA, 57; Mass Inst Technol, PhD(theoret physics), 64. Prof Exp: Instr, Palmer Phys Lab, Princeton Univ, 63-66; NSF fel, Niels Bohr Inst, 66-68, asst prof, 68-70; vis assoc prof, Nordic Inst Theoret Atomic Physics, 70-72; ASSOC PROF PHYSICS, INST THEORET PHYSICS, STATE UNIV GRONINGEN, 72- Mem: Am Phys Soc. Res: Theoretical nuclear structure. Mailing Add: State Univ Groningen Hoogbouw WSN PO Box 800 Groningen Netherlands

LANDE, KENNETH, b Vienna, Austria, June 5, 32; nat US. ASTROPHYSICS, ELEMENTARY PARTICLE PHYSICS. Educ: Columbia Univ, AB, 53, AM, 55, PhD(physics), 58. Prof Exp: Asst physics, Columbia Univ, 54-57; from instr to assoc prof, 59-74, PROF PHYSICS, UNIV PA, 74- Mem: Am Phys Soc. Res: Meson, nuclear and neutrino physics. Mailing Add: Dept of Physics Univ of Pa Philadelphia PA 19174

LANDE, SAUL, b Philadelphia, Pa, Aug 7, 30; m 54; c 4. BIOCHEMISTRY, ORGANIC CHEMISTRY. Educ: Ursinus Col, BS, 48; Univ Pittsburgh, PhD(biochem), 60. Prof Exp: Sr res chemist, Squibb Res Inst, 61-63; ASSOC PROF BIOCHEM IN MED, SCH MED, YALE UNIV, 63- Mem: Am Chem Soc. Res: Chemistry of biologically active peptides. Mailing Add: Dept of Med Yale Univ Sch of Med New Haven CT 06511

LANDE, SHELDON SIDNEY, b Chicago, Ill, July 16, 41; m 64; c 1. ENVIRONMENTAL CHEMISTRY. Educ: Ill Inst Technol, BS, 62; Mich State Univ, PhD(chem), 66. Prof Exp: Multiple fel petrol, Mellon Inst, 68-70; res chemist, Gulf Res & Develop Co, 70-71; res assoc water chem, Grad Sch Pub Health, Univ Pittsburgh, 71-72; pesticide info adminr, Allegheny County Health Dept, Pa, 72-75; RES ASSOC, SYRACUSE UNIV RES CORP, 75- ASSOC CHEM, SYRACUSE UNIV RES CORP, 75- Concurrent Pos: Consult environ health, Grad Sch Pub Health, Univ Pittsburgh, 72-, adj asst prof, 73- Mem: AAAS; Am Chem Soc. Res: Movement of organic substances in soil and water; analysis of organic chemicals in the environment; treatment of waste-water and landfill leachate. Mailing Add: Syracuse Univ Res Corp Merrill Ln Syracuse NY 13210

LANDEL, AURORA MAMAUAG, b Manila, Philippines, Feb 12, 26; m 53; c 6. BIOCHEMISTRY. Educ: Univ Philippines, BS, 49; Univ Wis-Madison, MS, 51, PhD(biochem), 55. Prof Exp: Instr org chem, Univ Philippines, 49-50; instr org chem & math, Far Eastern Univ, 49-50; res fel chem, 68-71, SR RES FEL BIOMED ENG, CALIF INST TECHNOL, 73- Mem: Am Chem Soc; Sigma Xi. Res: Preparation of fluorescent tracers in studies of capillary permeability; lipoprotein and lipoprotein lipase studies by means of fluorchromes. Mailing Add: Div of Appl Sci & Eng Calif Inst of Technol Pasadena CA 91109

LANDEL, ROBERT FRANKLIN, b Pendleton, NY, Oct 10, 25; m 53; c 6. PHYSICAL CHEMISTRY, RHEOLOGY. Educ: Univ Buffalo, BA, 50, MA, 51; Univ Wis, PhD(phys chem), 54. Prof Exp: Res assoc, Univ Wis, 54-55; sr res engr, 55-59, chief solid propellant chem sect, 59-61, chief polymer res sect, 61-65, MGR PROPULSION & MAT SECT, JET PROPULSION LAB, CALIF INST TECHNOL, 75- Concurrent Pos: Sr res fel, Calif Inst Technol, 65-69; sr Fulbright fel, Italy, 71-72; sr fel, Ctr Res Macromolecules, France, 72. Mem: Am Phys Soc; Am Chem Soc; Soc Rheol. Res: Mechanical properties of high polymers. Mailing Add: Jet Propulsion Lab 4800 Oak Grove Dr Pasadena CA 91103

LANDEN, ERNEST WILHELM, physics, see 12th edition

LANDER, JAMES FRENCH, b Bristol, Va, Aug 24, 31; m 60; c 3. GEOPHYSICS, SEISMOLOGY. Educ: Pa State Univ, BS, 58; Am Univ, MS, 62, MA, 68. Prof Exp: Geophysicist, US Coast & Geod Surv, Nat Oceanic & Atmospheric Admin, 58-62, chief seismol invests sect, 62-63, chief seismol invests br, Environ Res Labs, 63-73, chief, Nat Earthquake Info Ctr, 66-73; DEP DIR, NAT GEOPHYS & SOLAR-TERRESTRIAL DATA CTR, 73- Concurrent Pos: With Exec Off of President, Off Emergency Preparedness, 70-71; dir, World Data Ctr-A Solid Earth Geophys, 73-; mem, Earthquake Eng Res Inst. Mem: AAAS; Seismol Soc Am; Am Geophys Union; Sigma Xi. Res: Seismicity, earthquake intensity, earthquake engineering, strong motion seismology, volcanology, marine geology and geophysics, geomagnitism, geothermics, geodynamics, disaster studies, natural hazard risks, digital data bases, seismic reflections, computer graphics. Mailing Add: NOAA/EDS Nat Geophys & Solar Terrestrial Data Ctr Boulder CO 80302

LANDER, JOHN JOSEPH, physical chemistry, see 12th edition

LANDER, PATRICIA SLADE, b Washington, DC, May 28, 41. CULTURAL ANTHROPOLOGY, SOCIAL ANTHROPOLOGY. Educ: Columbia Univ, BS, 65, PhD(anthrop), 71. Prof Exp: From instr to asst prof anthrop, York Col, NY, 70-72; ASST PROF ANTHROP, BROOKLYN COL, 72- Mem: AAAS; Am Anthrop Asn; Am Ethnol Soc; Soc Advan Scand Studies. Res: Scandinavian ethnography; comparative social structure; sex roles; stratification. Mailing Add: Dept of Anthrop Brooklyn Col Brooklyn NY 11210

LANDER, RICHARD LEON, b Oakland, Calif, Apr 23, 28; m 55; c 3. PHYSICS. Educ: Univ Calif, Berkeley, BA, 50, PhD(physics), 58; Ohio State Univ, MA, 51. Prof Exp: Staff physicist, Lawrence Radiation Lab, Univ Calif, 58-60; res specialist nuclear physics, Boeing Co, 60-61; assoc res physicist, Univ Calif, San Diego, 61-66; assoc prof physics, 66-70, assoc dean res, Grad Div, 70-73, PROF PHYSICS, UNIV CALIF, DAVIS, 70- Concurrent Pos: Vis scientist, Europ Orgn Nuclear Res, Switz, 66-67. Mem: AAAS; Am Phys Soc. Res: Experimental elementary particle physics. Mailing Add: Dept of Physics Univ of Calif Davis CA 95616

LANDERL, HAROLD PAUL, b Pittsburgh, Pa, Apr 26, 22; m 44; c 3. ORGANIC CHEMISTRY. Educ: Carnegie Inst Technol, BS, 43, MS, 47, DSc(chem), 48. Prof Exp: Asst, Nat Defense Res Comt, Calif Inst Technol, 44-46; res chemist, Jackson Lab, 48-54, supvr res & develop, Tech Lab, 54-60, head textile dye appln div, 60-70, asst dir dyes & chem tech lab, 70-72, tech mgr dyes & chem, Belg, 72-76, TECH MGR DYES, ORCHEM, TECH LAB, E I DU PONT DE NEMOURS & CO, INC, 76- Mem: Am Chem Soc; Am Asn Textile Chemists & Colorists. Res: Application of dyes to fibers. Mailing Add: 1503 Fresno Rd Wilmington DE 19803

LANDERS, AUBREY WILFRED, b Manchester, NH, Sept 2, 06; m 32; c 3. MATHEMATICS. Educ: Acadia Univ, BA, 26; Brown Univ, MS, 29; Univ Chicago, PhD(math), 39. Prof Exp: Asst math, Brown Univ, 26-29; instr, Hunter Col, 29-30; from instr to prof, 30-74, dir off grants & res, 67-70, EMER PROF MATH, BROOKLYN COL, 74- Mem: AAAS; Am Math Soc; Math Asn Am; NY Acad Sci. Res: Calculus of variations. Mailing Add: Dept of Math Brooklyn Col Brooklyn NY 11210

LANDERS, EARL JAMES, b Greybull, Wyo, Dec 17, 21; m 51; c 2. INVERTEBRATE ZOOLOGY. Educ: Univ Wyo, AB, 50, MS, 52; NY Univ, PhD(zool), 58. Prof Exp: Instr zool, Univ Wyo, 55-56; from asst prof to assoc prof biol sci, Tex Western Col, 56-60, actg chmn dept, 59-60; assoc prof zool, 60-70, PROF ZOOL, ARIZ STATE UNIV, 70- Mem: AAAS; Am Soc Parasitol. Res: Parasitic protozoa life cycles; transmembrane electrolyte transport. Mailing Add: Dept of Zool Ariz State Univ Tempe AZ 85281

LANDERS, HOLBROOK, b Washington, DC, Aug 1, 17; m 46; c 5. METEOROLOGY. Educ: Syracuse Univ, BA, 48; Univ Calif, Los Angeles, MA, 50. Prof Exp: Asst meteorologist, Univ Calif, Los Angeles, 49-51; from instr to asst prof, Fla State Univ, 51-67; SC State climatologist, 67-73; forecaster, Nat Meteorol Ctr, Suitland, Md, 73-74; LEAD FORECASTER, NAT WEATHER SERV OFF, JACKSON, MISS, 74- Mem: Am Meteorol Soc. Res: Synoptic meteorology of low and middle latitudes; climatology; satellite and applied meteorology; biometeorology; weather forecasting. Mailing Add: Nat Weather Serv Forecast Off 204 Brenmar St Brandon MS 39042

LANDERS, JAMES WALTER, b Norfolk, Nebr, Oct 19, 27; m 52; c 3. PATHOLOGY. Educ: Univ Nebr, MD, 53. Prof Exp: ASSOC PROF PATH, SCH MED, WAYNE STATE UNIV, 62- Concurrent Pos: Assoc pathologist, William Beaumont Hosp, Royal Oak, Mich, 64-68 & St John Hosp, Detroit, 68- Mem: Am Soc Clin Path; Col Am Path; Int Acad Path. Res: Neuropathology. Mailing Add: 1507 Sunningdale Grosse Pointe MI 48236

LANDERS, JOHN HERBERT, JR, b Stockton, Mo, Jan 24, 21; m 43; c 3. ANIMAL NUTRITION. Educ: Univ Mo, BS, 42, MS, 50; Kans State Univ, PhD(animal nutrit), 66. Prof Exp: County agent agr, Univ Mo, 45-49, instr animal sci, 49-50; EXTEN ANIMAL SCIENTIST, ORE STATE UNIV, 50- Mem: Am Soc Animal Sci. Res: Counseling and advising livestock growers in more efficient production of meat and fiber. Mailing Add: 212 Withycombe Hall Ore State Univ Corvallis OR 97331

LANDERS, KENNETH EARL, b Leighton, Ala, Aug 31, 33; m 52; c 3. PLANT PHYSIOLOGY. Educ: Florence State Col, BS, 60; Auburn Univ, MS, 63, PhD(bot, zool), 66. Prof Exp: Inspector, Reynolds Metals Co, Ala, 52-54, 56-60; res asst bot & plant path, Auburn Univ, 60-61, instr, 62-63; assoc prof biol, 66-71, PROF BIOL, JACKSONVILLE STATE UNIV, 71-, HEAD DEPT, 73- Mem: Ecol Soc Am; Am Phytopath Soc. Res: Ecology. Mailing Add: Dept of Biol Jacksonville State Univ Jacksonville AL 36265

LANDERS, LESLIE CHARLES, physical chemistry, see 12th edition

LANDERS, MARY KENNY, b Fall River, Mass, Feb 5, 05; m 32; c 3. MATHEMATICS. Educ: Brown Univ, AB, 26, AM, 27; Univ Chicago, PhD(math), 39. Prof Exp: From instr to assoc prof, 27-63, PROF MATH, HUNTER COL, 64- Mem: AAAS; Am Math Soc; Math Asn Am. Res: Calculus of variations. Mailing Add: Dept of Math Hunter Col New York NY 10021

LANDERS, ROGER Q, JR, b Menard, Tex, July 23, 32; m 54; c 2. PLANT ECOLOGY, RANGE MANAGEMENT. Educ: Tex A&M Univ, BS, 54, MS, 55; Univ Calif, Berkeley, PhD(bot), 62. Prof Exp: From asst prof to assoc prof, 62-71, PROF PLANT ECOL, IOWA STATE UNIV, 71- Mem: Ecol Soc Am; Am Inst Biol Sci; Soc Range Mgt. Res: Grasslands; ecosystems analysis. Mailing Add: Dept of Bot & Plant Path Iowa State Univ Ames IA 50011

LANDERYOU, VICTOR ALLEN, organic chemistry, see 12th edition

LANDES, DOELAS RANDY, b Hope, Ark, Oct 10, 43; m 65; c 3. NUTRITION. Educ: Univ Ark, BSA, 65, MS, 66; Mich State Univ, PhD(food sci), 69. Prof Exp: ASST PROF NUTRIT, UNIV GA, 69- Mem: Inst Food Technologists; Soc Exp Biol Med; Nutrit Today Soc. Res: Trace mineral absorption and utilization and lipid composition of tissue as affected by dietary carbonhydrate and/or lipid. Mailing Add: Food Sci Dept Ga Sta Experiment GA 30212

LANDES, KENNETH KNIGHT, b Seattle, Wash, May 10, 99; m 24; c 3. GEOLOGY. Educ: Univ Wash, BS, 21; Harvard Univ, MS, 23, PhD(geol), 25. Prof Exp: Asst, Harvard Univ, 23-24; instr, Wellesley Col, 24-26; from asst prof to prof geol, Univ Kans, 26-41, chmn dept, 39-41; prof, 41-68, chmn dept, 41-51, EMER PROF GEOL, UNIV MICH, ANN ARBOR, 68-; GEOL CONSULT, 68- Concurrent Pos: From jr geologist to sr geologist, US Geol Surv, 21-52; asst state geologist to state geologist, State Geol Surv, Kans, 26-41; ed, J Asn Petrol Geologists, 51-53; consult, Oak Ridge Nat Lab, 72 & 75-76; mem task force, Natural Gas Div, Fed Power Comn, 73 & 75-76. Honors & Awards: Hardinge Award, Am Inst Mining, Metall & Petrol Engrs, 73. Mem: AAAS (vpres, 51); fel Geol Soc Am (vpres, 46); fel Mineral Soc Am (vpres, 38, pres, 45); hon mem Am Asn Petrol Geologists; Am Inst Mining, Metall & Petrol Engrs. Res: Petroleum geology; Michigan Basin subsurface geology; industrial geology. Mailing Add: 1200 Earhart Rd Suite 344 Ann Arbor MI 48105

LANDES, RUTH, b New York, NY, Oct 8, 08; m 29. ANTHROPOLOGY. Educ: NY Univ, BS, 28; New York Sch Social Work, MSW, 29; Columbia Univ, PhD(anthrop), 35. Prof Exp: Social worker, Child Care, NY, 29-31; instr anthrop, Brooklyn Col, 37; instr, Fisk Univ, 37-38; researcher, Carnegie Corp, NY, 39; res dir to coordr, Inter-Am Affairs, DC, 41; rep Negro & Mex-Am affairs, President Roosevelt's Comt Fair Employ Pract, 41-45; researcher Mex-Am youth gangs & families, Los Angeles Metrop Welfare, Coun, 46-47; study dir, Sci Res Dept, Am Jewish Comt, 48-51; Fulbright sr res scholar to UK, Dept Anthrop, Univ Edinburgh, 51-52; lectr, New Sch Social Res, 53-55; vis prof anthrop, Univ Southern Calif, 57-58; dir geriat prog, Los Angeles City Health Dept, 58-59; vis prof anthrop & dir anthrop & educ prog, Claremont Grad Sch, 59-62; exten lectr & consult, Univ Calif, Los Angeles & Berkeley, 62 & Los Angeles State Col, 63; vis prof, Tulane Univ, 64; PROF ANTHROP, McMASTER UNIV, 65- Concurrent Pos: Columbia Univ-Soc Sci Res Coun field studies, Ojibwa Reservations, Ont, 32-33 & Minn, 34-35 & 35-36, Santee Dakota Reservation, Minn, 35, Potawatomi Reservation, Kans, 35-36 & Negro communities, Brazil, 38-39; consult, Contemp Cult Proj, Columbia Univ, 49-51; consult to var Calif & nat agencies, 57-; consult, IBM Corp, 63-66; McMaster Univ grant, Brazil, 66; Can Coun grant res bilingualism, 68-72, Can Coun grant res, S Africa, 74; mem staff, Los Angeles Metrop Welfare Coun & Rosenberg Found Claremont Col Proj. Mem: Am Anthrop Asn. Res: Bilingualism; minority groups; personality and culture; education; status of women; religion; language and culture in officially plurilingual nations. Mailing Add: Dept of Anthrop McMaster Univ Hamilton ON Can

LANDESBERG, JOSEPH MARVIN, b New York, NY, Apr 21, 39; m 64; c 2. ORGANIC CHEMISTRY. Educ: Rutgers Univ, BS, 60; Harvard Univ, MA, 62, PhD(chem), 65. Prof Exp: NIH res fel, Columbia Univ, 64-66; asst prof, Univ, 66-70, ASSOC PROF CHEM, GRAD SCH ARTS & SCI, ADELPHI UNIV, 70- Mem: AAAS; Am Chem Soc; The Chem Soc; Am Asn Univ Prof. Res: Heterocyclic chemistry; synthetic applications of organometallic compounds; synthesis of strained, small-membered rings. Mailing Add: Dept of Chem Adelphi Univ Garden City NY 11530

LANDESMAN, EDWARD MILTON, b Brooklyn, NY, Mar 19, 38. MATHEMATICS. Educ: Univ Calif, Los Angeles, BA, 60, MA, 61, PhD(math), 65. Prof Exp: Asst prof in residence math, Univ Calif, Los Angeles, 65-66; asst prof, Univ Calif, Santa Cruz, 66-68; asst prof, Univ Calif, Los Angeles, 68-69; asst prof, 69-71, ASSOC PROF MATH, CROWN COL, UNIV CALIF, SANTA CRUZ, 71- Concurrent Pos: Air Force Off Sci Res grant, Univ Calif, Santa Cruz, 70-71. Mem: AAAS; Am Math Soc; Math Asn Am. Res: Partial differential equations; combinatorial theory; calculus. Mailing Add: Dept Math Crown Col Univ of Calif Santa Cruz CA 95060

LANDESMAN, HERBERT, b Newark, NJ, Apr 22, 27; m 53; c 2. INORGANIC CHEMISTRY. Educ: Harvard Univ, BS, 48; Purdue Univ, PhD(chem), 51. Prof Exp: Res chemist, Naval Ord Test Sta, 51-52, Olin Mathieson Chem Corp, 52-59 & Nat Eng Sci Co, 59-66; chem consult, West Precipitation Group, Joy Mfg Co, 66-68; vpres, Environ Resources, Inc, 68-69; PROF CHEM, LOS ANGELES SOUTHWEST COL, 69- Mem: Air Pollution Control Asn; Am Chem Soc. Res: Organosilicon chemistry; chemistry of boron hydrides; fire extinguishants; fluorocarbons; hazards analysis; air and water pollution. Mailing Add: Dept of Chem Los Angeles Southwest Col Los Angeles CA 90047

LANDESMAN, RICHARD, b Brooklyn, NY, Jan 30, 40; m 63; c 1. DEVELOPMENTAL BIOLOGY. Educ: NY Univ, BA, 61, MS, 63; Univ BC, PhD(zool), 66. Prof Exp: NIH fel biol, Mass Inst Technol, 66-69; ASST PROF ZOOL, UNIV VT, 69- Mem: Am Soc Zoologists; Soc Develop Biol. Mailing Add: Dept of Zool Univ of Vt Burlington VT 05401

LANDESMAN, ROBERT, b New York, NY, May 6, 16; m 47; c 3. MEDICINE. Educ: Columbia Univ, AB, 36; Cornell Univ, MD, 39; Am Bd Obstet & Gynec, dipl. Prof Exp: Res, Mt Sinai Hosp, New York, 39-40, intern, 40-41, asst resident surg, 41-42, asst resident obsobstet & gynec, 46-47, resident, 47-49; from instr to assoc prof obstet & gynec, 50-71, PROF CLIN OBSTET & GYNEC, MED COL, CORNELL UNIV, 71- Concurrent Pos: Fel, New York Hosp, 49-50; from asst attend physician to assoc attend physician, New York Hosp-Cornell Med Ctr, 50-64, attend physician, 64-; pvt pract, 56-; dir obstet & gynec, Jewish Mem Hosp, 63- Mem: Am Col Obstet & Gynec; Microcirc Soc; NY Acad Med. Res: Gonadotropins; premature labor; obstetrics and gynecology. Mailing Add: 449 E 68th St New York NY 10021

LANDGRAF, JOHN LESLIE, anthropology, academic administration, see 12th edition

LANDGRAF, WILLIAM CHARLES, b Elizabeth, NJ, Jan 10, 28; m 53; c 3. PHARMACEUTICAL CHEMISTRY. Educ: Seton Hall Univ, BS, 50; Stanford Univ, PhD(chem), 59; Univ Santa Clara, MBA, 75. Prof Exp: Sr scientist chem, Lockheed Res Labs, Lockheed Missile Systs Div, 58-61; proj leader & lab supt, Ampex Corp, 61-63; mgr & res scientist, Varian Assocs, Calif, 63-70; MGR, SYNTEX LABS, 70- Mem: Am Chem Soc; Am Pharmaceut Asn. Res: Physical, biophysical and organic chemistry; kinetics; computer assisted experimentation; analytical chemistry; physical organic chemistry. Mailing Add: 762 Stone Lane Palo Alto CA 94303

LANDGREBE, ALBERT R, b New Rochelle, NY, Mar 4, 33; m 58; c 2. CHEMISTRY. Educ: Fordham Univ, BS, 57; Univ Md, PhD(chem), 64. Prof Exp: Inorg chemist, USDA, 60-63; radiochemist, Nat Bur Standards, Md, 63-68; chemist & chmn comt sci & tech symposia, AEC, 68-75; BR CHIEF CHEM STORAGE, ENERGY RES & DEVELOP ADMIN, 75- Mem: AAAS; Soc Nuclear Med; Sigma Xi; Am Chem Soc. Res: Use of radioisotopes in analytical and inorganic chemistry; radio chromatographic methods; substoichiometric radioisotopic dilution analysis; removal of radioisotopes from milk; activation analysis; trace and micro analysis. Mailing Add: 3201 Dunnington Rd Beltsville MD 20705

LANDGREBE, JOHN A, b San Francisco, Calif, May 6, 37; m 61; c 1. ORGANIC CHEMISTRY. Educ: Univ Calif, Berkeley, BS, 59; Univ Ill, Urbana, PhD(org chem), 62. Prof Exp: From asst prof to assoc prof, 62-71, assoc chmn, 67-70, PROF CHEM, UNIV KANS, 71-, CHMN DEPT, 70- Mem: Am Chem Soc; The Chem Soc. Res: Organic reaction mechanisms; small ring compounds; carbene intermediates; electrophilic aliphatic substitution; organometallic compounds; neighboring group effects. Mailing Add: Dept of Chem Univ of Kans Lawrence KS 66045

LANDI, VINCENT RUSSELL, physical chemistry, see 12th edition

LANDIN, JOSEPH, b New York, NY, Jan 25, 13; m 34. MATHEMATICS. Educ: Brooklyn Col, BS, 37; NY Univ, MS, 41; Univ Notre Dame, PhD(math), 46. Prof Exp: Asst math, Univ Notre Dame, 41-43, instr, 43-46; from instr to prof, Univ Ill, Urbana, 46-64; prof & head dept, Univ Ill, Chicago Circle, 64-74; PROF MATH & CHMN DEPT, OHIO STATE UNIV, 74- Mem: Am Math Soc; Math Asn Am. Res: Algebra. Mailing Add: Dept of Math Ohio State Univ Columbus OH 43210

LANDING, BENJAMIN HARRISON, b Buffalo, NY, Sept 11, 20; m 49; c 4. PATHOLOGY. Educ: Harvard Univ, AB, 42; Harvard Med Sch, MD, 45. Prof Exp: Intern, Children's Hosp, Boston, 45-46; res pathologist, Children's Med Ctr, Boston, 48-50, asst pathologist, 50-52, assoc pathologist, 52-53; from asst to assoc path, Harvard Med Sch, 48-53; from asst prof to assoc prof path & pediat, Col Med, Univ Cincinnati, 53-61; prof path & pediat, 61-76, WINZER PROF PATH & PEDIAT, UNIV SOUTHERN CALIF, 76-; PATHOLOGISTINCHIEF & DIR LABS, CHILDREN'S HOSP, LOS ANGELES, 61- Concurrent Pos: Res pathologist, Free Hosp for Women & Boston Lying-In-Hosp, 49; dir pathologist, Children's Hosp & Res Found, Cincinnati, 53-61. Mem: Histochem Soc; Endocrine Soc; Am Asn Path & Bact (asst secy, 53-57); Int Acad Path. Res: Histochemistry of endocrine and metabolic diseases; pediatric pathology. Mailing Add: Children's Hosp 4650 Sunset Blvd Los Angeles CA 90027

LANDING, JAMES EDWARD, b Buffalo, NY, Jan 7, 28; m 70; c 6. CULTURAL GEOGRAPHY, HISTORICAL GEOGRAPHY. Educ: Manchester Col, BS, 52; Pa State Univ, MS, 63, PhD(geog), 67. Prof Exp: Assoc lectr geog, Ind Univ, South Bend, 64-68; ASSOC PROF GEOG, UNIV ILL, CHICAGO CIRLCE, 68- Concurrent Pos: US Off Educ fel, Ind, 66-69; A M Todd Found grant, 67-68; eval consult, US Off Educ, 67-70; assoc lectr geog, Ind Univ, Northwest, 68-69; Nat Coun Geog Educ fel, Ind, 68-69; Circle Ctr Res Bd fel, Univ Ill, Chicago Circle, 72-73. Mem: Asn Am Geogr; Nat Coun Geog Educ; Int Geog Union. Res: Spatial organization of small group minorities, especially those in an alien culture, with emphasis on linguistic, sectarian, ethnic and racial differences. Mailing Add: Dept of Geog 2102 BSB Univ of Ill Chicago Circle PO Box 4348 Chicago IL 60680

LANDIS, ABRAHAM L, b New York, NY, May 25, 28; m 57; c 2. CHEMISTRY. Educ: City Col New York, BS, 51; Univ Kans, PhD(chem), 55. Prof Exp: Asst chem, Univ Kans, 51-55; aeronaut res scientist, Nat Adv Comt Aeronaut, 55-56; sr res chemist, Atomics Int Div, NAm Aviation, Inc, 56-61; SR STAFF CHEMIST, HUGHES AIRCRAFT CO, CULVER CITY, 61- Mem: Am Chem Soc; Sigma Xi; Am Inst Chem. Res: High temperature polymers; polymer chemistry; vacuum technology; organic synthesis; organometallic polymers; aerospace materials. Mailing Add: 10935 Canby Ave Northridge CA 91324

LANDIS, CHARLES WALTER, b Logansport, Ind, Sept 16, 20; m 53; c 2. PSYCHIATRY. Educ: DePauw Univ, AB, 42; Ind Univ, MD, 51. Prof Exp: Instr psychiat, Sch Med & dir, Riley Child Guid Clin, Ind Univ, 56-58; med dir, Milwaukee County Hosp Ment Dis, 58; from asst prof to assoc prof psychiat, Med Col Wis, 64-71; MED DIR & CHIEF STAFF, ST MARY'S HILL HOSP, 71- Concurrent Pos: Consult, Ind State Hosps, 56-58; dir ment health, Milwaukee County Insts & Depts, 58-71; consult, NIMH, 64-66; clin prof psychiat & social welfare, Univ Wis, 65-69; deleg, White House Conf Aging, 71. Mem: AMA; Am Psychiat Asn; Am Col Psychiat. Res: Public and private mental health services. Mailing Add: St Mary's Hill Hosp 1445 S 32nd St Milwaukee WI 53215

LANDIS, EDWARD EVERETT, b Marion, Kans, June 15, 07; m 34; c 2. PSYCHIATRY. Educ: NCent Col, Ill, BA, 28; Northwestern Univ, MD, 34; Am Bd Psychiat & Neurol, dipl, 46. Prof Exp: Nat Comt Ment Hyg fel child psychiat,

Louisville Ment Hyg Clin, 37-38; instr, 38-39, from asst prof to prof psychiat, 41-75, actg chmn dept, 73, vchmn dept, 74-75, PROF EMER PSYCHIAT, SCH MED, UNIV LOUISVILLE, 75- Concurrent Pos: Rockefeller fel neurol & res, Johnson Found, Univ Pa, 39-40; founder & dir dept electroencephalog, Sch Med, Univ Louisville & Dept Psychiat, Louisville & Jefferson County Children's Home, 40-47, mem gov bd, 51-; clin dir psychiat, Louisville Gen Hosp, 42-45, med dir outpatient clin, 45-49; asst dir psychiat, Louisville Ment Hyg Clin, 42-51 & Norton Mem Infirmary, 45-; med dir psychiat, Norton Psychiat Clin, 49-75; psychiat consult, Louisville Vet Admin, Kosair Crippled Children's, Jewish & Our Lady of Peace Hosps. Mem: Fel Am Col Physician; fel Am Psychiat Asn; Am Acad Child Psychiat; AMA. Res: Grantham type pre-frontal lobotomy; D-Lysergic acid diethylamide; vitamin B-complex in alcohol addiction. Mailing Add: Rte 2 Box 77 Anchorage KY 40223

LANDIS, EUGENE MARKLEY, b New Hope, Pa, Apr 4, 01; m 34; c 1. PHYSIOLOGY, MEDICINE. Educ: Univ Pa, BS, 22, MS, 24, MD, 26, PhD, 27. Hon Degrees: MS, Yale Univ, AM, Harvard Univ, 43. Prof Exp: Asst zool, Univ Pa, 20-21, asst instr, 21-22; fel, Nat Res Coun, 26-27; intern, Hosp Univ Pa, 27-29; fel, Guggenheim Mem Found, 29-31; assoc med, Univ Pa, 31-35, res assoc, 32-39, asst prof, 35-39; prof internal med & head dept, Univ Va, 39-43; George Higginson prof physiol, 43-67, chmn div med sci, Fac Arts & Sci, 49-52, EMER PROF PHYSIOL, HARVARD MED SCH, 67- Concurrent Pos: Secy comt aviation med, Nat Res Coun, 40-46, chmn subcomt acceleration, 41-46; mem sci adv coun, Life Ins Med Res Fund, 45-51, chmn, 51; mem panel physiol, Res & Develop Bd, 48-51, chmn, 51; spec lectr, Univ London, 52; consult, AEC, 52; Graves lectr, Ind Univ, 53; consult, Vet Admin, 56-59; ed, Circulation Res, Am Heart Asn, 62-66; adj prof, Lehigh Univ, 67-72. Honors & Awards: Phillips Medal, Am Col Physicians, 36; Gold Heart Award, Am Heart Asn, 66. Mem: Nat Acad Sci; Am Soc Clin Invest (secy, 38-42, pres, 42); Am Physiol Soc (pres, 52); hon mem Harvey Soc; fel Am Col Physicians. Res: Physiology of circulation, particularly blood capillaries; hypertension; kidney disease; edema. Mailing Add: 1547 Silver Creek Dr Hellertown PA 18055

LANDIS, PHILLIP SHERWOOD, b York, Pa, July 29, 22; m 44; c 2. ORGANIC CHEMISTRY. Educ: Franklin & Marshall Col, BS, 43; Univ Ky, MS, 47; Northwestern Univ, PhD(chem), 58. Prof Exp: Chemist, Cities Serv Refining Corp, 43-45; res chemist, Mobil Oil Corp, 47-63, res assoc, 63-66, sr res assoc, 66-69, MGR PROD RES GROUP, MOBIL RES & DEVELOP CORP, 69- Mem: Am Chem Soc. Res: Mechanisms and kinetics of organic reactions; pyrolysis of organic compounds; organo-sulfur compounds; petrochemicals; radical reactions. Mailing Add: Mobil Res & Develop Corp Paulsboro NJ 08066

LANDIS, STORY CLELAND, b New York, NY, May 14, 45; m 69. NEUROBIOLOGY. Educ: Wellesley Col, BA, 67; Harvard Univ, MA, 70, PhD(biol), 73. Prof Exp: NIH fel neuropath, 73-75, RES FEL NEUROBIOL, HARVARD MED SCH, 75- Mem: Am Asn Anatomists; Soc Neurosci; Am Soc Cell Biol. Res: Developmental neurobiology; cell biology. Mailing Add: Dept of Neurobiol Harvard Med Sch Boston MA 02115

LANDIS, VINCENT J, b Minneapolis, Minn, Oct 27, 28; m 50; c 6. INORGANIC CHEMISTRY. Educ: Wash State Univ, BS, 50; Univ Minn, PhD(inorg chem), 57. Prof Exp: From instr to assoc prof, 54-65, PROF CHEM, SAN DIEGO STATE UNIV, 65- Concurrent Pos: Richland fac fel, Univ Wash, 64-65. Mem: Am Chem Soc. Res: Metal coordination compounds; radiochemistry. Mailing Add: Dept of Chem San Diego State Univ San Diego CA 92115

LANDMAN, DONALD ALAN, b New York, NY, Apr 23, 38; m 70; c 2. ATOMIC PHYSICS, SOLAR PHYSICS. Educ: Columbia Univ, AB, 59, MA, 61, PhD(physics), 65. Prof Exp: Asst prof physics, NY Univ, Bronx, 65-69; res scientist, Cornell Aeronaut Lab, Buffalo, 70 & Advan Res Instrument Systs Inc, 71; ASSOC ASTRONOMER SOLAR PHYSICS, INST ASTRON, UNIV HAWAII, 72- Mem: Am Phys Soc. Res: Application of theoretical and experimental atomic physics to astrophysical problems, especially in the area of solar physics research; development of modern astrophysical instrumentation. Mailing Add: Inst for Astron Univ of Hawaii 2680 Woodlawn Dr Honolulu HI 96822

LANDMAN, OTTO ERNEST, b Mannheim, Ger, Feb 15, 25; nat US; m 48; c 3. MICROBIAL GENETICS. Educ: Queens Col, BS, 47; Yale Univ, MS, 48, PhD(microbiol), 51. Prof Exp: USPHS fel, Calif Inst Technol, 51-52; res assoc bact, Univ Ill, 53-56; chief microbial genetics br, US Army Biol Labs, Ft Detrick, 56-61, sr investr, 61-63; assoc prof biol, 63-66, PROF BIOL, GEORGETOWN UNIV, 66- Concurrent Pos: NIH spec fel, Ctr Molecular Genetics, Nat Ctr Sci Res, Gif-Sur-Yvette, France, 68-69; vis investr, Nat Inst Med Res, Mill Hill, London, 75-76. Mem: AAAS; Am Soc Microbiol; Genetics Soc Am; Brit Soc Gen Microbiol. Res: Protoplasts and L forms of bacteria; cell division; wall biosynthesis and transformation in bacteria; phage infection of protoplasts. Mailing Add: Dept of Biol Georgetown Univ Washington DC 20007

LANDMAN, RUTH HALLO, b Kassel, Ger, Sept 8, 26; nat US; m 48; c 3. ANTHROPOLOGY. Educ: Vassar Col, BA, 47; Yale Univ, MA, 51, PhD, 54. Prof Exp: Field worker anthrop, Columbia Univ, 47-48; researcher, Ctr Alcohol Studies, Yale Univ, 50-51; research, Univ Ill, 54-56; lectr, Am Univ, 60-64; lectr, Howard Univ, 64-65; from asst prof to assoc prof, 65-74, chmn dept, 70-72, PROF ANTHROP, AM UNIV, 74- Mem: Fel Am Anthrop Asn; Am Ethnol Soc. Res: Cultural anthropology; acculturation; language and culture. Mailing Add: Dept of Anthrop Am Univ Washington DC 20016

LANDMANN, WENDELL AUGUST, b Waterloo, Ill, Dec 29, 19; m 44; c 2. BIOCHEMISTRY. Educ: Univ Ill, BS, 41; Purdue Univ, MS, 44, PhD(biochem), 51. Prof Exp: Chemist, US Naval Res Lab, 43-46; asst chem, Purdue Univ, 46-51; res chemist, Armour & Co, 51-55; assoc biochem, Argonne Nat Lab, 55-57; chief div anal & phys chem, Am Meat Inst Found, Ill, 57-64; prof animal sci, biochem & nutrit, 64-70, King Ranch Chair, 64-72, PROF BIOCHEM & BIOPHYS & HEAD DEPT, TEX A&M UNIV, 70- Mem: AAAS; Am Chem Soc; Inst Food Technol; Am Meat Sci Asn; Am Inst Nutrit. Res: Lysosomal enzymes; human nutrition; protein chemistry; chemistry of animal tissues. Mailing Add: Dept of Biochem & Biophys Tex A&M Univ College Station TX 77843

LANDMESSER, CHARLES MONROE, b Newark, NJ, June 24, 17; m 45; c 4. ANESTHESIOLOGY. Educ: Cornell Univ, BA, 39, MD, 42. Prof Exp: Instr anesthesiol, 49-52, assoc pharm, 49-52, from asst prof to assoc prof anesthesiol, 52-56, chmn dept, 56-68, PROF ANESTHESIOL, ALBANY MED COL, 56- Concurrent Pos: Asst attend anesthesiol, Albany Med Ctr Hosp, 49-53 & 53-56, anesthesiologist-in-chg, 56-68, attend anesthesiologist, 68-; attend physician, Vet Admin Hosp, Albany, 51-56, consult, 56-; asst attend, Brady Maternity Hosp, Albany, 53-61, attend, 61-; consult, Physicians Hosp, Plattsburgh, 57-63. Mem: Int Anesthesia Res Soc; AMA; Am Soc Anesthesiologists. Res: Respiratory physiology; muscle relaxants; narcotic antagonists. Mailing Add: Albany Med Ctr Albany NY 12208

LANDO, BARBARA ANN, b Elizabeth, NJ, Dec 7, 40; m 65. ALGEBRA. Educ: Georgian Court Col, BA, 62; Rutgers Univ, New Brunswick, MS, 64, PhD(math), 69. Prof Exp: Instr math, Douglass Col, Rutgers Univ, New Brunswick, 69; asst prof, 69-73, ASSOC PROF MATH, UNIV ALASKA, FAIRBANKS, 73- Mem: Am Math Soc; Math Asn Am. Res: Differential algebra. Mailing Add: Dept of Math Univ of Alaska Fairbanks AK 99701

LANDO, DAVID J, physical chemistry, see 12th edition

LANDO, JEROME B, b Brooklyn, NY, May 23, 32; m 62; c 2. POLYMER SCIENCE. Educ: Cornell Univ, BA, 53; Polytech Inst Brooklyn, PhD(chem), 63. Prof Exp: Fel, Polytech Inst Brooklyn, 63; res chemist, Camille Dreyfus Lab, Res Triangle Inst, 63-65; asst prof polymer sci & eng, 65-68, assoc prof macromolecular sci, 68-74, PROF MACROMOLECULAR SCI, CASE WESTERN RESERVE UNIV, 74- Concurrent Pos: Humboldt Found Sr Am Scientist Award, 74; vis prof, Univ Mainz, 74. Mem: Am Chem Soc; Am Crystallog Asn; Am Phys Soc. Res: Polymer physical chemistry; solid state reactions, especially polymerization reactions and polymer crystal structure. Mailing Add: Dept of Macromolecular Sci Case Western Reserve Univ Cleveland OH 44106

LANDOLL, LEO MICHAEL, b Cleveland, Ohio, Oct 11, 50; m 71; c 1. POLYMER CHEMISTRY. Educ: Kent State Univ, BA, 70; Univ Akron, PhD(polymer sci), 75. Prof Exp: RES CHEMIST POLYMER SYNTHESIS, HERCULES INC, 74- Mem: Am Chem Soc. Res: Polymer synthesis related to thermoplastic and thermoset systems, fibers, property correlations. Mailing Add: Hercules Res Ctr Hercules Inc Wilmington DE 19808

LANDOLT, ARLO UDELL, b Highland, Ill, Sept 29, 35; m; c 5. ASTRONOMY. Educ: Miami Univ, BA, 55; Ind Univ, MA, 60, PhD(astron), 62. Prof Exp: Scientist aurora & airglow, US Int Geophys Year Comt, 56-58; from asst prof to assoc prof physics & astron, 62-68, PROF PHYSICS & ASTRON, LA STATE UNIV, BATON ROUGE, 68-, DIR OBSERV, 70- Concurrent Pos: Mem first wintering-over party, Int Geophys Year Amundson-Scott S Pole Sta, Antarctica, 57; Grad Res Coun res grants, La State Univ, Baton Rouge, 64-74; res grants grants, NSF, 64-76, Res Corp, 65 & NASA, 65-67; prog dir, NSF, Washington, DC, 75-76. Mem: AAAS; Int Astron Union; Am Astron Soc; Royal Astron Soc; Am Polar Soc. Res: Photographic and photoelectric studies of galactic star clusters, variable stars, and eclipsing binaries; standard photometric systems. Mailing Add: Dept of Physics & Astron La State Univ Baton Rouge LA 70803

LANDOLT, MARSHA LAMERLE, b Houston, Tex, Jan 19, 48. FISH PATHOLOGY, WILDLIFE PATHOLOGY. Educ: Baylor Univ, BS, 69; Univ Okla, MS, 70; George Washington Univ, PhD(path), 76. Prof Exp: ASST PROF FISHERIES, COL FISHERIES, UNIV WASH, 75- Concurrent Pos: Histopathologist, Eastern Fish Dis Lab, US Dept Interior, Leetown, WVa, 70-74; path clerk, Dept Animal Health, Nat Zool Park, Smithsonian Inst, Washington, DC, 74-75. Mem: Wildlife Dis Asn; Nat Shellfish Asn. Res: Relationship of elevated hepatic microsomal enzyme levels, high tumor incidence and other pathological changes to the presence of trace organic stressors in flatfish along the Washington coast. Mailing Add: Col of Fisheries Univ of Wash Seattle WA 98195

LANDOLT, PAUL ALBERT, b Shubert, Nebr, July 10, 12; m 35; c 1. PHYSIOLOGY. Educ: Nebr State Teachers Col, Peru, BA, 33; Univ Nebr, MS, 51, PhD(zool, physiol), 60. Prof Exp: Teacher, High Schs, Nebr, 36-42; field dir mil welfare, Am Red Cross, Mariannas Islands, 42-46; instr biol sci, Scottsbluff Jr Col, Nebr, 46-53; from instr to assoc prof physiol, 53-74, PROF PHYSIOL, UNIV NEBR, LINCOLN, 74- Mem: AAAS; Am Soc Cell Biol; Am Soc Zool; Tissue Cult Asn. Res: Vertebrate physiology; tissue culture; problems related to effects of air pollutants on lung tissue. Mailing Add: Sch of Life Sci Univ of Nebr Lincoln NE 68588

LANDOLT, ROBERT GEORGE, b Houston, Tex, Apr 4, 39; m 62; c 3. ORGANIC CHEMISTRY. Educ: Austin Col, BA, 61; Univ Tex, PhD(org chem), 65. Prof Exp: Res assoc org chem, Univ Ill, 65-67; asst prof, 67-71, chmn dept, 71-74, ASSOC PROF ORG CHEM, MUSKINGUM COL, 71- Concurrent Pos: Resident consult, Columbus Labs, Battelle Mem Inst, 74-75. Mem: AAAS; Am Asn Univ Prof; Am Chem Soc. Res: Abnormal claisen rearrangement and reactions in aprotic polar solvents; nitroso aromatic compounds; oxidation of coal and coal model compounds. Mailing Add: Dept of Chem Muskingum Col New Concord OH 43762

LANDOLT, ROBERT RAYMOND, b Sherman, Tex, May 11, 37; m 70; c 3. BIONUCLEONICS. Educ: Austin Col, BA, 59; Univ Kans, MS, 61; Purdue Univ, PhD(bionucleonics), 68. Prof Exp: Reactor health physicist, Phillips Petrol Co, Idaho, 61-64; from instr to asst prof bionucleonics, 64-72, ASSOC PROF BIONUCLEONICS, PURDUE UNIV, WEST LAFAYETTE, 72- Mem: Health Physics Soc. Res: Development of autoradiography for quantification of histochemically demonstrated enzyme inhibition; measurement of neutron doses using threshold detectors. Mailing Add: Dept of Bionucleonics Purdue Univ West Lafayette IN 47906

LANDON, DONALD OMAR, b Champaign, Ill, Apr 19, 26; m 46. PHYSICS, CHEMISTRY. Prof Exp: Asst chemist, Armour Res Found, Ill Inst Technol, 49-54; group leader instrument anal, Sperry Gyroscope Co, 54-58; dir res, 58-69, VPRES RES, SPEX INDUSTS, INC, 69- Concurrent Pos: Consult, Wall St Authority, 55- Mem: AAAS; Soc Appl Spectros; Optical Soc Am. Res: Instrumentation for optical spectroscopy, particularly vacuum ultraviolet spectrometers and Raman instrumentation. Mailing Add: 6 Robin Rd Warren NJ 07060

LANDON, ERWIN JACOB, b Cleveland, Ohio, Jan 22, 25; m 65. BIOCHEMISTRY. Educ: Univ Chicago, BS, 45, MD, 48; Univ Calif, PhD(biochem), 53. Prof Exp: Intern, Harper Hosp, Detroit, Mich, 48-49; asst prof, 59-67, ASSOC PROF PHARMACOL, SCH MED, VANDERBILT UNIV, 67- Concurrent Pos: Sr res fel pharmacol, Sch Med, Yale Univ, 57-59. Mem: Am Chem Soc; Am Soc Pharmacol & Exp Therapeut. Res: Biochemistry of renal transport. Mailing Add: Dept of Pharmacol Vanderbilt Univ Nashville TN 37203

LANDON, HARRY HILL, JR, nuclear physics, see 12th edition

LANDON, JOHN CAMPBELL, b Hornell, NY, Jan 3, 37; m 58; c 4. VIROLOGY, CANCER. Educ: Alfred Univ, AB, 59; George Washington Univ, MS, 62, PhD(biol), 67. Prof Exp: Biologist, Nat Cancer Inst, 60-65; head virol, Litton Bionetics, Inc, 65-68, dir dept virol & cell biol, 68-71, dir spec prog develop, 71-72, dir sci, Frederick Cancer Res Ctr, 72-75; PRES, MASON RES INST, 75- Mem: AAAS; Tissue Cult Asn; Am Soc Cell Biol; NY Acad Sci; Am Soc Microbiol. Res: Viral oncology; tissue culture; general human and simian virology; cell biology; environmental biology. Mailing Add: 8213 Raymond Ln Potomac MD 20854

LANDON, ROBERT E, b Chicago, Ill, June 1, 05; m 35. GEOLOGY. Educ: Univ Chicago, SB, 26, PhD(geol), 29. Prof Exp: Instr, YMCA Col, 28-29; geologist,

Anaconda Copper Mining Co, 29-30; asst geologist, US Geol Surv, 30-31; instr geol, Colo Col, 31-33; consult mining geologist, 33-39; sr mining securities analyst, US Securities & Exchange Comn, 40-45; geologist, Mobil Oil Corp, NY, 45-70; GEOL CONSULT, 70- Mem: Fel Geol Soc Am; Am Asn Petrol Geol. Mailing Add: 3460 S Race St Englewood CO 80110

LANDOR, JOHN HENRY, b Canton, Ohio, Sept 30, 27; m 53; c 6. SURGERY. Educ: Univ Chicago, PhB, 48, MD, 53. TPXInstr surg, Sch Med, Univ Chicago, 58; from instr to prof, Sch Med, Univ Mo-Columbia, 59-69; prof, Col Med, Univ Fla, 69-72; PROF & CHIEF GEN SURG, COL MED & DENT NJ, RUTGERS MED SCH, 72-; CHIEF DEPT SURG, RARITAN VALLEY HOSP, GREEN BROOK, 73- Concurrent Pos: Commonwealth Found fel, Royal Postgrad Med Sch, London, 66-67. Mem: Am Col Surgeons; Soc Univ Surgeons; Am Gastroenterol Asn; Soc Surg Alimentary Tract; Int Soc Surgeons. Res: Physiology of the stomach. Mailing Add: Dept of Surg Col Med & Dent NJ Rutgers Med Sch Piscataway NJ 08854

LANDORF, ROBERT, physical chemistry, see 12th edition

LANDOVITZ, LEON FRED, b Brooklyn, NY, May 24, 32; m 59; c 2. SOLID STATE PHYSICS. Educ: Columbia Univ, AB, 53, PhD(physics), 58. Prof Exp: Mem physics, Inst Advan Study, 57-58; res assoc, Brookhaven Nat Lab, 58-60; from asst prof to assoc prof, Belfer Grad Sch Sci, 60-67, PROF PHYSICS, YESHIVA UNIV, 67- Mem: Am Phys Soc. Res: High energy physics; quantum field theory; elementary particles; astrophysics. Mailing Add: Dept of Physics Yeshiva Univ New York NY 10033

LANDOWNE, DAVID, b Chicago, Ill, Dec 26, 42; m 66; c 2. PHYSIOLOGY, BIOPHYSICS. Educ: Mass Inst Technol, BS, 63; Harvard Univ, PhD(physiol), 68. Prof Exp: Res assoc pharmacol, Sch Med, Yale Univ, 68-70, 71-72; asst prof physiol, 72-75, ASSOC PROF PHYSIOL, SCH MED, UNIV MIAMI, 75- Concurrent Pos: Grass Found fel, 70; NSF fel, Univ London, 70-71. Mem: Biophys Soc; Soc Gen Physiol. Res: Excitable membranes; ion movements and optical methods. Mailing Add: Dept of Physiol & Biophys Univ Miami Sch Med PO Box 875 Miami FL 33152

LANDOWNE, MILTON, b New York, NY, Nov 19, 12; m 41; c 5. INTERNAL MEDICINE. Educ: City Col New York, BS, 32; Harvard Univ, MD, 36. Prof Exp: Intern, Mt Sinai Hosp, New York, 36-39; Libman fel, Michael Reese Hosp, Chicago, 39-41; instr, Univ Chicago, 41-46, asst prof med, 46-48; chief cardiovasc res unit, Vet Admin Hosp, 48-49; assoc chief geront sect, Nat Heart Inst, 49-57; med dir, Levindale Hebrew Home & Infirmary, Baltimore, Md, 57-65; DIR MED LAB, US ARMY RES INST ENVIRON MED, 65- Concurrent Pos: Asst prof, Johns Hopkins Univ, 55-65; head div cardiol & chronic dis, Sinai Hosp, Baltimore, 58-65; asst clin prof, Harvard Univ, 65-74. Mem: AAAS; Am Soc Clin Invest; Am Physiol Soc; Soc Exp Biol & Med; Am Heart Asn. Res: Disorders of the circulation; biology aging; clinical medicine; physiology of blood and circulation; metabolic and renal diseases; environmental medicine. Mailing Add: 67 Woodchester Dr Weston MA 02193

LANDOWNE, ROBERT ALLEN, organic chemistry, analytical chemistry, see 12th edition

LANDRETH, HOBART F, physiology, see 12th edition

LANDRUM, BILLY FRANK, b Atlanta, Ga, June 7, 20; m 48; c 3. ORGANIC CHEMISTRY, POLYMER CHEMISTRY. Educ: Emory Univ, AB, 47, MS, 49, PhD(chem), 50. Prof Exp: Res chemist polymer chem, M W Kellogg Co, 50-53, res supvr pilot plant, 53-57; head polymer sect, Minn Mining & Mfg Co, 57-62; mgr advan projs, FMC Corp, NJ, 62-66; staff scientist, Whittaker Corp, 66-67; staff scientist, Com Develop Dept, 67-74, MGR MKT DEVELOP, PLASTICS & ADDITIVES DIV, CIBA-GEIGY CORP, 74- Mem: Am Chem Soc; Sigma Xi; Soc Advan Mat & Processing Eng. Res: Organo-metallic reactions; polymers; fluorocarbons; urethanes; coal and coke; activated carbon. Mailing Add: 65 Baylor Ave Hillsdale NJ 07642

LANDRUM, BOBBY L, b Taylor, Tex, Jan 18, 32; m 51; c 4. PHYSICS, MATHEMATICS. Educ: Tex A&M Univ, BS, 53, MS, 54, PhD(physics), 59. Prof Exp: Chief new detection tech sect, Wright-Patterson AFB, Ohio, 60-61; from res scientist to dir appl res, Northrop Corp, 61-69, chief engr, Electro-Optical Dept, Systs Labs, 69-72, dir advan sensors & displays, 72-74; MEM STAFF, MARTIN MARIETTA CORP, ORLANDO, 74- Mem: Am Phys Soc. Res: Applied research for military and scientific purposes; technical program management of electrooptical sensors and instruments. Mailing Add: Martin Marietta Corp MS 276 PO Box 5837 Orlando FL 32805

LANDRUM, RALPH AVERY, JR, b Memphis, Tenn, Oct 2, 26; m 49; c 3. GEOPHYSICS. Educ: Rice Inst, BS, 49; Univ Tulsa, MS, 64. Prof Exp: Asst seismic observer, Amerada Petrol Corp, 49-51; res seismic observer, Stanolind Oil & Gas Co, 51-56; res engr, Pan Am Petrol Corp, 56-63, staff res engr, 63-67, staff res scientist, 67-71; res assoc, Res Ctr, Amoco Prod Co, 71-74; SR RES GEOPHYSICIST, WESTERN GEOPHYS CO AM, 74- Mem: Seismol Soc Am; Am Soc Explor Geophys; Inst Elec & Electronic Eng; Europ Asn Explor Geophys. Res: Exploration geophysics; design of seismic instrumentation; mathematics of seismic data processing. Mailing Add: PO Box 2469 Houston TX 77001

LANDRY, EDWARD F, b Cambridge, Mass, Jan 9, 47; m 71. MOLECULAR GENETICS. Educ: State Col Boston, BA, 68; Univ NH, PhD(microbiol), 75. Prof Exp: FEL MOLECULAR GENETICS, SCH MED & DENT, UNIV ROCHESTER, 75- Mem: Am Soc Microbiol. Res: Study of control mechanisms in the translation of bacteriophage T4; studies center around the ribosome and its factors and proteins. Mailing Add: Dept of Radiation Biol & Biophys Univ of Rochester Sch Med & Dent Rochester NY 14642

LANDRY, FERNAND, b Levis, PQ, Can, Jan 13, 30; m 55; c 4. EXERCISE PHYSIOLOGY. Educ: Univ Ottawa, BSc, 54; Univ Ill, Urbana, MS, 55, PhD(phys educ exercise physiol), 68. Prof Exp: From teacher-researcher, phys educ to dept head, Univ Ottawa, 55-68; DEPT HEAD PHYS ACTIV SCI, LAVAL UNIV, 68- Honors & Awards: Medal, Que Govt, 74. Mem: Can Asn Sport Sci (pres, 71-72); Int Coun Sport & Phys Educ (vpres, NAm, 72-); Corp Int Cong Phys Activ Sci (pres, 74-). Res: Short-term and chronic effects of physical activity and sports; use of physical activity in the prevention and/or rehabilitation of the generated diseases. Mailing Add: Dept of Phys Activ Sci Laval Univ Quebec PQ Can

LANDRY, MARTHA MOSELEY, b Greensboro, Ala, Aug 31, 31; div; c 1. MICROBIOLOGY, GENETICS. Educ: Univ Ala, BS, 59; Tex Woman's Univ, MS, 65, PhD(radiation biol), 68. Prof Exp: Asst prof biol sci, Miss State Col Women, 68-71; PROF BIOL SCI, POLK COMMUNITY COL, 71- Concurrent Pos: Consult, Columbus Hosp, Miss, 68-70. Mem: Genetics Soc Am; Am Soc Microbiol; Wildlife Dis Asn; Am Inst Biol Sci; NY Acad Sci. Res: Gas ecology of bacteria, especially fluorine gases and equine leptospirosis. Mailing Add: 620 Seventh St NE Winter Haven FL 33880

LANDRY, PRESTON MYLES, b New Orleans, La, Apr 14, 30; m 53; c 4. OPERATIONS RESEARCH. Educ: Southeastern La Col, BS, 52; Univ Miss, MA, 57. Prof Exp: Assoc mathematician, Vitro Corp Am, 53-54, mathematician, 56-58; supvry mathematician, Air Proving Ground Ctr, 58-59, mathematician, 59-65, chief anal br, Math Serv Lab, 65-66, chief systs div, Directorate Sci Staff, Dept Chief Staff Eval, Ninth Aerospace Defense Div, Ent AFB, 66-68, asst dir, Directorate Sci Appln, 14th Aerospace Force, 68-73, OPERS ANALYST, AEROSPACE DEFENSE COMMAND, US AIR FORCE, 73- Concurrent Pos: Mem, Electronic Trajectory Measurements Group. Res: Data reduction and analysis related to weapons system evaluation; electromagnetic propagation; Space-track system; characteristics and projections of earth satellite population. Mailing Add: 1123 N Meade Ave Colorado Springs CO 80909

LANDRY, RICHARD GEORGES, b Manchester, NH, Nov 7, 42; m 66; c 3. APPLIED STATISTICS. Educ: Oblate Col, BA, 64; Boston Col, MEd, 67, PhD(res & statist), 70. Prof Exp: Asst prof, 69-73, ASSOC PROF MEASUREMENT & STATIST, CTR TEACHING & LEARNING, UNIV N DAK, 70- Concurrent Pos: Eval auditor, numerous ESEA Title III Projs, 70-; eval consult, Grand Rapids Sch Dist, Minn, 73-; res coordr, Nat Inst Educ Proj, Univ N Dak, 73- Mem: Am Asn Univ Prof; Am Educ Res Asn; Am Statist Asn; Nat Coun Measurement Educ. Res: Applied educational statistics; educational measurement and evaluation in affective domain; foreign language learning and creativity. Mailing Add: Ctr Teaching & Learning Univ of N Dak Grand Forks ND 58201

LANDRY, STUART OMER, JR, b New Orleans, La, Sept 30, 24; m 50; c 2. ZOOLOGY. Educ: Harvard Univ, BS, 49; Univ Calif, PhD(zool), 54. Prof Exp: Curatorial asst, Mus Vert Zool, Calif, 50-52; asst zool, Univ Calif, 52-53, assoc, 53-54; from instr to asst prof anat, Univ Mo, 54-59; assoc prof biol & chmn dept, La State Univ, 59-63; actg dean grad sch, 66-68, PROF BIOL, STATE UNIV NY BINGHAMTON, 63- Mem: AAAS; Soc Syst Zool; Am Soc Mammal; Am Asn Anat; Am Soc Zoologists. Res: Comparative anatomy and classification of mammals; functional anatomy of mammals. Mailing Add: Dept of Biol State Univ NY Binghamton NY 13901

LANDS, WILLIAM EDWARD MITCHELL, b Chillicothe, Mo, July 22, 30; c 4. BIOCHEMISTRY. Educ: Univ Mich, Ann Arbor, BS, 51; Univ Ill, PhD(biol chem), 54. Prof Exp: NSF fel, Calif Inst Technol, 54-55; from instr to assoc prof, 55-67, PROF BIOL CHEM, UNIV MICH, ANN ARBOR, 67- Concurrent Pos: Chmn subcomt biochem nomenclature, Nat Acad Sci, 62-64; assoc ed, Lipids, 65- & Can J Biochem, 72-; Danforth Assoc, 66-; ed, Biochimica Et Biophysica Acta, 71- Honors & Awards: Gold Medal Bond Award, Am Oil Chemists Soc, 65; Glycerine Res Award, 69. Mem: AAAS; Am Chem Soc; Am Soc Biol Chemists; Am Oil Chemists Soc. Res: Metabolism of glycerides and long-chain aliphatic acids and aldehydes; formation of membranes and regulation of membrane function; prostaglandin biochemistry and control of its biosynthesis. Mailing Add: Dept of Biol Chem Univ of Mich Ann Arbor MI 48109

LANDSBERG, HELMUT ERICH, b Frankfurt-am-Main, Ger, Feb 9, 06; nat US; m. CLIMATOLOGY. Educ: Univ Frankfurt, PhD(geophys, meteorol), 30. Prof Exp: Asst seismol & climatol, Inst Meteorol & Geophys, Univ Frankfurt, 30-31; supvr res seismol & meteorol, Taunus Observ, Ger, 31-34; from instr to asst prof geophys, Pa State Col, 34-41; assoc prof meteorol, Univ Chicago, 41-43, res assoc indust climatol, 45-46; chief sect indust climatol, US Weather Bur, Wash, 46; dep exec dir comt geophys sci, Res & Develop Bd, 46-48, exec dir comt geophys & geog, 48-51; dir geophys res directorate, Air Force Cambridge Res Ctr, 51-54; dir off climat, US Weather Bur, 54-65, dir environ data serv, Environ Sci Serv Admin, 65-66; RES PROF, INST FLUID DYNAMICS & APPL MATH, UNIV MD, COLLEGE PARK, 67-, ACTG DIR, 74- Concurrent Pos: Vis prof, Inst Fluid Dynamics & Appl Math, Univ Md, College Park, 64-67; pres comn spec applns meteorol & climatol, World Meteorol Orgn, 69-; mem, Nat Adv Comt Oceans & Atmosphere, 75-77. Honors & Awards: Award Bioclimat, Am Meteorol Soc, 64, Brooks Award, 72. Mem: Nat Acad Eng; AAAS; Am Meteorol Soc (vpres, 63-64); Meteoritical Soc; Am Geophys Univ (vpres, 66-68, pres, 68-70). Res: Microclimatology; bioclimatology; aerosol; meteorology; seismology. Mailing Add: Inst Fluid Dynamics & Appl Math Univ of Md College Park MD 20742

LANDSBERGER, FRANK ROBBERT, b Amsterdam, Netherlands, Aug 10, 43; US citizen; m 65. PHYSICAL BIOCHEMISTRY, VIROLOGY. Educ: Cornell Univ, BA, 64; Brown Univ, PhD(physics), 70. Prof Exp: Res asst physics, Brown Univ, 64-69; res fel biochem, Div Endocrinol, Sloan-Kettering Inst Cancer Res, 69-71; asst prof chem, Ind Univ, Bloomington, 71-74; ASST PROF BIOCHEM & ANDREW W MELLON FOUND FEL, ROCKEFELLER UNIV, 74- Mem: AAAS; Biophys Soc; Am Phys Soc; Am Soc Microbiol; Am Chem Soc. Res: Use of physical biochemical methods in studying the structure and function of biological and model membranes with emphasis on those of enveloped viruses. Mailing Add: Rockefeller Univ New York NY 10021

LANDSHOFF, ROLF, b Berlin, Ger, Nov 30, 11; nat US; m 41; c 4. MATHEMATICAL PHYSICS. Educ: Berlin Tech Inst, DrIng, 36; Univ Minn, PhD(theoret physics), 38. Prof Exp: Asst physics, Univ Minn, 36-40; prof, Col St Thomas, 40-44; scientist, Los Alamos Sci Lab, 44-56; SR MEM, LOCKHEED PALO ALTO RES LAB, 56- Concurrent Pos: Vis lectr, Weizmann Inst, 63-64. Mem: Fel Am Phys Soc. Res: Atomic physics; hydrodynamics; statistical mechanics. Mailing Add: Lockheed Palo Alto Res Lab 52-11 3251 Hanover St Bldg 202 Palo Alto CA 94304

LANDSKROENER, PETER ARMSTRONG, b Woodbury, NJ, Dec 3, 28; m 46; c 6. PHYSICAL CHEMISTRY. Educ: Univ Del, BS, 50, MS, 51; Cath Univ Am, PhD(chem), 54. Prof Exp: Asst, Cath Univ Am, 51-54; res chemist, Photo Prod Dept, E I du Pont de Nemours & Co, 54-59, tech serv rep, Sales Div, 59-61; sr res specialist, Ansco Div, Gen Aniline & Film Corp, 61-62, dir commercial develop dept, 63-65; assoc dir develop serv dept, Champion Papers Inc, 65-66; vpres res & develop, Anken Chem & Film Corp, 66-69; GEN MGR, CBS SOUNDCRAFT, DANBURY, 69- Mem: Am Chem Soc; Soc Photog Sci & Eng. Res: New product development, planning, mergers and acquisitions. Mailing Add: 118 Seney Dr Bernardsville NJ 07924

LANDSMAN, DOUGLAS ANDERSON, b Dundee, Scotland, May 31, 29; m 57; c 4. PHYSICAL CHEMISTRY. Educ: Univ St Andrews, BSc, 49, Hons, 50, PhD(thermodynamics), 57. Prof Exp: Sr sci officer, Chem Div, Atomic Weapon Res Estab, UK Atomic Energy Authority, Eng, 53-57; Nat Res Coun Can fel, 57-58; Harwell sr fel, Chem Div, Atomic Weapon Res Estab, UK Atomic Energy Authority, Eng, 58-60, prin sci officer, 60-65; res supvr, Mat Eng Res Lab, Pratt & Whitney Aircraft Div, Middletown, 65-72, SR PROJ ENGR, FUEL CELL FACILITY, POWER SYST DIV, UNITED TECHNOL CORP, UNITED AIRCRAFT CORP,

72- Mem: Fel Royal Inst Chem. Res: Thermodynamics of ionization in aqueous solutions; chemistry of the hydrogen isotopes; isotope separation; gas chromatography; chemonuclear reactors; energy conversion; fuel cells. Mailing Add: 19 Farmstead Ln West Hartford CT 06117

LANDSTREET, JOHN DARLINGTON, b Philadelphia, Pa, Mar 13, 40; m 72; c 2. ASTROPHYSICS. Educ: Reed Col, BA, 62; Columbia Univ, MA, 63, PhD(physics), 66. Prof Exp: Instr physics, Mt Holyoke Col, 65-66, asst prof, 66-67; res assoc astron, Columbia Univ, 67-70, asst prof, 70; asst prof, 70-72, ASSOC PROF ASTRON, UNIV WESTERN ONT, 72- Mem: Am Astron Soc; Can Astron Soc; Royal Astron Soc; Int Astron Union. Res: Observation of circular and linear polarization in stars and extra-galactic objects, especially white dwarfs; observation of stellar magnetism. Mailing Add: Dept of Astron Univ of Western Ont London ON Can

LANDT, JAMES FREDERICK, b Anniston, Ala, Jan 4, 26; m 50; c 2. BIOLOGY. Educ: Howard Col, AB, 50; Emory Univ, MS, 54, PhD(biol), 61. Prof Exp: Parasitologist, US Army Med Lab, Ga, 54; instr biol, 54-57, asst prof, 60-63, ASSOC PROF BIOL, OXFORD COL, EMORY UNIV, 63-, CHMN DIV SCI & MATH, 67- Concurrent Pos: Assoc prof, Univ Ala, 63-65. Mem: AAAS; Am Soc Parasitol; Am Soc Zoologists. Res: Host-parasite relationships; biochemistry of parasites. Mailing Add: Dept of Biol Oxford Col Emory Univ Oxford GA 30267

LANDUCCI, LAWRENCE L, b St Paul, Minn, May 20, 39; m 69; c 2. ORGANIC CHEMISTRY, WOOD CHEMISTRY. Educ: Univ Minn, BS, 62, PhD(org chem), 67. Prof Exp: RES CHEMIST, US FOREST PROD LAB, 67- Mem: Am Chem Soc. Res: Lignin and lignin model compound chemistry; influence of transition metals on lignin oxidation pathways. Mailing Add: US Forest Prod Lab N Walnut St Madison WI 53705

LANDWEBER, PETER STEVEN, b Washington, DC, Aug 17, 40; m 64; c 2. MATHEMATICS. Educ: Univ Iowa, BA, 60; Harvard Univ, MA, 61, PhD(math), 65. Prof Exp: Asst prof math, Univ Va, 65-68; asst prof, Yale Univ, 68-70; assoc prof, 70-74, PROF MATH, RUTGERS UNIV, NEW BRUNSWICK, 74- Concurrent Pos: Mem sch math, Inst Advan Study, 67-68; NATO fel, Univ Cambridge, 74-75. Mem: Am Math Soc. Res: Cobordism theory of differential manifolds. Mailing Add: Dept of Math Rutgers Univ New Brunswick NJ 08903

LANDY, ARTHUR H, b Philadelphia, Pa, Mar 17, 39; m 65; c 2. MOLECULAR BIOLOGY, BIOCHEMICAL GENETICS. Educ: Amherst Col, BA, 61; Univ Ill, PhD(microbiol & biochem), 66. Prof Exp: Res fel biochem genetics, Med Res Coun Lab Molecular Biol, Cambridge, Eng, 66-68; asst prof, 68-75, ASSOC PROF MED SCI, BROWN UNIV, 75- Concurrent Pos: NATO fel, 66-67; fel, Am Cancer Soc, 68, fac res assoc, 75-80. Mem: Am Soc Microbiol. Res: Gene structure and regulation in prokaryotes and eukaryotes; organization of eukaryote genes; mechanisms of site-specific recombination. Mailing Add: Div of Biol & Med Sci Box G Brown Univ Providence RI 02912

LANDY, DAVID, b Savannah, Ga, June 4, 17; m 49; c 3. ANTHROPOLOGY. Educ: Univ NC, BA, 49, MA, 50; Harvard Univ, PhD(anthrop), 56. Prof Exp: Field dir anthrop res, Family Life Proj, Social Sci Res Ctr, Univ PR, 51-53; lectr & res assoc, Sch Social Work, Boston Univ, 54-55; co-prin investr, Voc Rehab Admin Grant, Mass Ment Health Ctr, Boston, 56-60; from assoc prof to prof anthrop, Grad Sch Pub Health, Univ Pittsburgh, 60-70, chmn dept, 63-70; chmn dept, 70-75, PROF ANTHROP, COL II, UNIV MASS, BOSTON, 70- Concurrent Pos: Instr anthrop, Col Social Sci, Univ PR, 51-52; res assoc, Harvard Med Sch, 56-60; Voc Rehab Admin grant, Grad Sch Pub Health, Univ Pittsburgh, 61-65; assoc ed, Enthology: An Internation Jour Cult & Social Anthrop, 61-70; mem adv coun, Int Cong Social Psychiat, Eng, 64-69; mem adv comt, Grad Teaching Internship Prog, Univ Pittsburgh, 65-70 & Learning Res & Develop Ctr, 66-70; mem res adv comt, Psychol Eval Proj, Vet Admin Hosps, New York, 67-68. Mem: Fel AAAS; fel Am Anthrop Asn; Am Ethnol Soc; fel Am Soc Ethnohist; fel Soc Appl Anthrop. Res: Socialization as cultural transmission; medical systems as cultural systems; conformity and deviance in cultural systems. Mailing Add: Dept of Anthrop Univ of Mass Col II Boston MA 02125

LANDY, MAURICE, immunology, see 12th edition

LANDY, RICHARD ALLEN, b Clearfield, Pa, Sept 23, 31; m 59; c 2. MINERALOGY. Educ: Mass Inst Technol, AB, 53; Pa State Univ, MS, 55, PhD(mineral), 61. Prof Exp: Supvr engr, Bonded Abrasives Div, Carborundum Co, 60-62; group leader, Basic Inc, 62-63, mgr qual assurance, 63-65; asst prof geol, Allegheny Col, 65-67; mineralogist, 67-70, DIR RES, NAM REFRACTORIES CO, 70- Mem: Am Crystallog Asn; Am Soc Qual Control; Soc Appl Spectros. Res: Quality control; statistical applications to sampling problems in geology and ceramic industry; quantitative analysis of inorganic chemistry systems by means of x-ray diffraction analysis and other instrumental techniques. Mailing Add: 105 Pauline Dr W Clearfield PA 16830

LANE, ALEXANDER Z, b Detroit, Mich, July 22, 29; m 56. BIOCHEMISTRY, MEDICINE. Educ: Univ Detroit, BS, 50; Wayne State Univ, PhD(biochem), 54, MD, 58. Prof Exp: Intern med, Bon Secours Hosp, Grosse Pointe, Mich, 58-59; clin investr, Parke Davis & Co, Mich, 60-62, dir clin pharmacol, 62-66; dir med res, 66-70, VPRES & MED DIR, BRISTOL LABS, 70- Concurrent Pos: Nat Cancer Inst fel occup med, Wayne State Univ, 59-60; lectr, Univ Mich, 65-66. Mem: Fel Am Soc Clin Pharmacol & Therapeut; Am Chem Soc; Am Soc Microbiol. Res: Correlation of animal and clinical pharmacological data; experimental design of clinical studies; analysis, performance and interpretation of clinical chemical tests; antibiotics; cancer chemotherapy; narcotic antagonists. Mailing Add: Bristol Labs Box 657 Syracuse NY 13201

LANE, ALFRED GLEN, b Stoutland, Mo, Aug 21, 32; m 57; c 2. ANIMAL NUTRITION. Educ: Univ Mo, BS, 59, MS, 60, PhD(animal nutrit), 65. Prof Exp: Instr voc agr, Parkersburg Community Sch, Iowa, 60-63; asst dairy husb, Univ Mo-Columbia, 63-65, asst prof, 65-70; MGR DAIRY RES, ALLIED MILLS, INC, 70- Mem: Am Dairy Sci Asn; Am Soc Animal Sci. Res: Ruminant nutrition; physiology. Mailing Add: Allied Mills Inc PO Box 459 Libertyville IL 60048

LANE, ARDELLE CATHERINE, b Port Angeles, Wash, Mar 8, 22; m 67. PHYSIOLOGY. Educ: Seattle Pac Col, BS, 44; Northwestern Univ, MS, 47; Univ Ill, PhD(physiol), 54. Prof Exp: From instr to assoc prof physiol, 52-64, PROF PHYSIOL, DENT SCH, NORTHWESTERN UNIV, 64- Mem: Am Physiol Soc. Res: Controlling mechanisms of gastric secretion. Mailing Add: Dept of Physiol Northwestern Univ Dent Sch Chicago IL 60611

LANE, BENNIE RAY, b Deming, NMex, July 2, 35; m 56; c 4. MATHEMATICS. Educ: Colo State Col, BA, 56, MA, 57; George Peabody Col, PhD(math), 62. Prof Exp: Asst prof math, Univ Chattanooga, 59-61; instr appl math, Vanderbilt Univ, 61-62; asst prof math, Colo State Col, 62-63; from asst prof to assoc prof, George Peabody Col, 63-66; PROF MATH & CHMN DEPT, EASTERN KY UNIV, 66- Mem: Am Math Asn. Res: Mathematics education; teaching mathematics by television; abstract algebra; programmed instruction. Mailing Add: Dept of Math Eastern Ky Univ Richmond KY 40475

LANE, BERNARD OWEN, b Greensboro, NC, Oct 5, 25; m 51. INVERTEBRATE PALEONTOLOGY. Educ: Univ NC, BS, 50; Brown Univ, MSc, 55; Univ Southern Calif, PhD(geol), 60. Prof Exp: Lectr geol, Univ Nev, 59-60 & 61-62; from asst prof to assoc prof, 62-72, PROF GEOL, CALIF STATE POLYTECH UNIV, POMONA, 72- Concurrent Pos: Consult, Earth Sci Curriculum Proj, 65-66. Mem: Paleont Soc; Nat Asn Geol Teachers. Res: Paleontology and stratigraphy of the early Pennsylvanian in the cordilleran region of North America. Mailing Add: Dept of Physics & Earth Sci Calif State Polytech Univ Pomona CA 91768

LANE, BERNARD PAUL, b Brooklyn, NY, June 27, 38; m 62; c 3. PATHOLOGY. Educ: Brown Univ, AB, 59; NY Univ, MD, 63. Prof Exp: NIH trainee exp path, Sch Med, NY Univ, 65-66, from asst prof to assoc prof path, 66-71; ASSOC PROF PATH, HEALTH SCI CTR, STATE UNIV NY STONY BROOK, 71- Concurrent Pos: Attend pathologist, Bellevue & NY Univ Hosps, 69-71; attend pathologist, Vet Admin Hosp, Northport, NY, 71-; vis scientist, Armed Forces Inst Path, 71; chief cell injury labs, Armed Forces Inst Path. Mem: Am Soc Cell Biol; Am Soc Exp Path; Am Asn Path & Bact; Am Soc Clin Path. Res: Experimental pathology; electron microscopy; cellular injury; chemical carcinogenesis. Mailing Add: Dept of Path State Univ of NY Health Sci Ctr Stony Brook NY 11790

LANE, BYRON GEORGE, b Toronto, Ont, May 16, 33; m 61. BIOCHEMISTRY. Educ: Univ Toronto, BA, 56, PhD(biochem), 59. Prof Exp: Jr res asst biochem, Med Ctr, Univ Calif, San Francisco, 59-60; res assoc, Rockefeller Inst, 60-61; asst prof, Univ Alta, 61-63, assoc prof, 64-68; PROF BIOCHEM, UNIV TORONTO, 68- Mem: Am Soc Biol Chemists. Res: Chemistry of ribonucleates, especially the biochemical involvements of trace components. Mailing Add: Dept of Biochem Univ of Toronto Toronto ON Can

LANE, CARL LEATON, b Raleigh, NC, Feb 11, 28; m 52; c 1. FOREST SOILS. Educ: NC State Univ, BS, 52, MS, 61; Purdue Univ, PhD(forest soil microbiol), 68. Prof Exp: Forest mgr, State Hosp Butner, NC, 52-59; asst prof forestry, 60-70, ASSOC PROF FORESTRY, CLEMSON UNIV, 70. Mem: Soc Am Foresters. Res: Forest soils microbiology; forest soil tree disease relationships. Mailing Add: Dept of Forestry Clemson Univ Clemson SC 29631

LANE, CARLTON ANDREWS, mathematics, physics, see 12th edition

LANE, CECIL TAVERNER, physics, see 12th edition

LANE, CHARLES A, b Wichita, Kans, Nov 18, 32. ORGANIC CHEMISTRY. Educ: Univ Okla, BS, 54; Yale Univ, MS, 59; Univ Calif, PhD(chem), 63. Prof Exp: Org chemist, Lederle Labs, Am Cyanamid Co, 56-58; asst prof org chem, Univ Nigeria, 61-63; asst prof, Univ Calif, 63-64; asst prof, 64-73, ASSOC PROF ORG CHEM, UNIV TENN, KNOXVILLE, 73- Mem: AAAS; Am Chem Soc. Res: Theoretical organic chemistry. Mailing Add: Dept of Chem Univ of Tenn Knoxville TN 37916

LANE, CHARLES EDWARD, b Riverton, Wyo, Dec 17, 09; m 31; c 2. ZOOLOGY. Educ: Univ Wis, AB, 31, MA, 33, PhD(physiol, zool), 35. Prof Exp: Asst zool, Univ Wis, 31-36; from asst prof to assoc prof, Univ Wichita, 36-42; marine biologist, Borden Co, 45-49; from assoc prof to prof marine sci, 49-74, EMER PROF MARINE SCI, UNIV MIAMI, 74- Mem: AAAS; Am Physiol Soc; Soc Exp Biol & Med; Int Soc Toxicol. Res: General marine biology. Mailing Add: 2005 Mooringline Dr Vero Beach FL 32960

LANE, CHARLES FRANKLIN, b Knoxville, Tenn, Dec 10, 19; m 51; c 3. PHYSICAL GEOGRAPHY. Educ: Univ Tenn, AB, 44, MS, 45; Northwestern Univ, PhD(phys geog), 51. Prof Exp: Asst prof geog & geol, Univ Ga, 48-50; assoc prof, 50-53, PROF GEOG & GEOL, LONGWOOD COL, 53-, CHMN HIST & SOCIAL SCI DEPT, 63- Mem: Asn Am Geogr; Nat Coun Geog Educ. Res: Geomorphology. Mailing Add: Dept of Hist & Social Sci Longwood Col Farmville VA 23901

LANE, CLINTON FISHER, organic chemistry, see 12th edition

LANE, CONSTANCE A, b Rockport, Maine, Jan 7, 24. POLYMER CHEMISTRY, PLASTICS CHEMISTRY. Educ: Bates Col, BS, 46. Prof Exp: Jr chemist, 46-51, intermediate chemist, 51-61, sr chemist, 61-66, GROUP LEADER PLASTICS RES, ROHM AND HAAS CO, 66- Mem: Am Chem Soc. Res: Synthesis and application of polymeric intermediates for plastics. Mailing Add: Rohm and Haas PO Box 219 Bristol PA 19007

LANE, DONALD WILSON, b Fayetteville, Tenn, June 23, 34; m 60; c 3. PETROLEUM GEOLOGY. Educ: Dartmouth Col, BA, 56; Univ Ill, MS, 58; Rice Univ, PhD(geol), 61. Prof Exp: Geologist, Tenneco Oil Co, 61-70; regional geologist, Royal Resources Corp, 70; staff geologist, Geol Surv Wyo, 70-73; mgr exp geol & Rocky Mountain area, Mich Wis Pipe Line Co, 73-76; CONSULT GEOLOGIST, 76- Mem: Am Asn Petrol Geol; Soc Econ Paleont & Mineral; Geol Soc Am. Res: Lower Paleozoic stratigraphy and hydrocarbon potential of the northeastern United States; Wyoming stratigraphy and stratigraphic resources; Lower Cretaceous stratigraphy of northwestern Colorado. Mailing Add: 12214 Mossycup Dr Houston TX 77024

LANE, EDWIN DAVID, b Vancouver, BC, May 9, 34; m 58; c 2. AQUATIC ECOLOGY, FISHERIES. Educ: Univ BC, BSc, 59, MSc, 62, PhD(zool), 66. Prof Exp: Staff mem, Fisheries Invest Off, NZ Marine Dept, 62-63; res scientist, Res Br, Ont Dept Land & Forest, 66-68; assoc prof zool, Calif State Univ, Long Beach, 68-74, res grant, 68-69 & 71-74; PROG BIOLOGIST, ENVIRON PROTECTION SERV, CAN DEPT ENVIRON, 74- Concurrent Pos: Fisheries expert, Food & Agr Orgn, 67- Mem: Am Soc Ichthyologists & Herpetologists; Can Soc Zoologists; Royal Soc NZ; NZ Limnol Soc; Can Soc Environ Biologists. Res: Ecology of fishes, especially in streams, estuaries and coastal bay systems; population dynamics; environmental impact of development, especially in the North. Mailing Add: Environ Protection Serv 10025 Jasper Ave Edmonton AB Can

LANE, ERIC TRENT, b Baton Rouge, La, Aug 30, 38; m 60; c 2. THEORETICAL PHYSICS. Educ: La State Univ, BS, 60; Rice Univ, MA, 63, PhD(physics), 67. Prof Exp: Vis lectr physics, La State Univ, New Orleans, 63-65; ASST PROF PHYSICS, UNIV TENN, CHATTANOOGA, 67- Mem: Am Phys Soc; Am Asn Physics Teachers; Soc Gen Systs Res. Res: Mathematical physics; general systems research applied to improvement of teaching and human relationships. Mailing Add: Tenn Chattanooga TN 37403

LANE, ERNEST PAUL, b Greene Co, Tenn, Nov 14, 33; m 61; c 2. TOPOLOGY. Educ: Berea Col, BA, 55; Univ Tenn, MA, 57; Purdue Univ, PhD(math), 65. Prof Exp: Programmer, Army Ballistic Missile Agency, Ala, 57-58; instr math, Berea Col,

58-60; asst prof, Va Polytech Inst, 65-70; assoc prof, 70-75, PROF MATH, APPALACHIAN STATE UNIV, 75- Mem: Math Asn Am; Am Math Soc. Res: Abstract spaces; metrization; real-valued functions on abstract spaces. Mailing Add: Dept of Math Appalachian State Univ Boone NC 28607

LANE, FORREST EUGENE, b Enola, Ark, June 24, 34; m 54; c 5. PLANT PHYSIOLOGY, PLANT BIOCHEMISTRY. Educ: Univ Ark, BA, 56, MEd, 59, MS, 63; Univ Okla, PhD(plant physiol), 65. Prof Exp: Teacher, Hall High Sch, Ark, 57-58; instr biol, Univ Ark, 58-63; asst prof biol, Kans State Col Pittsburg, 65-67; mem fac, Dept Bot & Bact, 67-69, ASSOC PROF BOT & BACT, UNIV ARK, FAYETTEVILLE, 69- Mem: Am Soc Plant Physiol; Scand Soc Plant Physiol; Bot Soc Am; Phytochem Soc NAm. Res: Dormancy in plant structures such as seeds, fruits, tubers and buds; relationship between dormancy and plant phenolics; enzymes associated with hormone control and plant growth. Mailing Add: Dept of Bot & Bact Univ of Ark Fayetteville AR 72701

LANE, GARY (THOMAS), b Center, Ky, Nov 8, 41; m 63; c 3. ANIMAL NUTRITION, BIOCHEMISTRY. Educ: Berea Col, BS, 63; Purdue Univ, West Lafayette, MS, 65, PhD(animal nutrit), 68. Prof Exp: Res asst animal nutrit, Purdue Univ, West Lafayette, 63-67; asst prof, 67-73, ASSOC PROF ANIMAL NUTRIT, TEX A&M UNIV, 73- Mem: Am Dairy Sci Asn; Am Soc Animal Sci. Res: Ration additives for ruminants; ration and its relation to milk composition and yield; mechanisms of milk synthesis; chemical preservation of high-moisture grain. Mailing Add: Dept of Animal Sci Tex A&M Univ College Station TX 77843

LANE, GEORGE ASHEL, b Norman, Okla, May 9, 30; m 52. PHYSICAL CHEMISTRY, APPLIED CHEMISTRY. Educ: Grinnell Col, AB, 52; Northwestern Univ, PhD(phys chem), 55. Prof Exp: Asst chem, Grinnell Col, 51-52; asst, Northwestern Univ, 52-55; spec projs chemist, 55-56, staff asst, 56-58, chemist, 58-63, proj leader, 63-66, sr res chemist, 66-69, res specialist, 69-73, SR RES SPECIALIST, DOW CHEM USA, 73- Mem: AAAS; Am Chem Soc; Int Solar Energy Soc; Sigma Xi. Res: Solar energy; energy storage; oxygen isotope effects; calorimetry; auto crash protection; fog modification; rocket propellant testing and evaluation; pyrotechnics; aerosol dissemination. Mailing Add: 3802 Wintergreen Dr Midland MI 48640

LANE, GEORGE H, b Milford, NH, Feb 19, 24; m 48; c 2. PHYSICS. Educ: Amherst Col, BA, 47; Yale Univ, MS, 49; Univ Conn, PhD(physics), 61. Prof Exp: Asst instr physics, Univ Conn, Hartford Br, 49-51; instr, Franklin & Marshall Col, 54-57, asst prof, 57-60; from asst prof to assoc prof, 60-66, PROF PHYSICS, NORWICH UNIV, 66-, HEAD DEPT, 62-, DIR GRAD STUDIES, 68- Mem: AAAS; Am Phys Soc; Am Asn Physics Teachers; Nat Sci Teachers Asn. Res: Atomic collisions; mass spectrometry. Mailing Add: Dept of Physics Norwich Univ Northfield VT 05663

LANE, HAROLD HOOKER, astronomy, see 12th edition

LANE, HAROLD RICHARD, b Danville, Ill, Mar 7, 42; m 68. PALEONTOLOGY, STRATIGRAPHY. Educ: Univ Ill, Urbana, BS, 64; Univ Iowa, MS, 66, PhD(geol), 69. Prof Exp: SR RES SCIENTIST, RES CTR, AMOCO PROD CO, 68- Mem: Brit Palaeont Asn; Int Palaeont Asn; Soc Econ Paleont & Mineral. Res: The evolution, biostratigraphy and systematic paleontology of the microfossils, conodonts, especially in Devonian through Middle Pennsylvanian strata of North America. Mailing Add: Amoco Prod Co Res Ctr PO Box 591 Tulsa OK 74102

LANE, HARRY CLEBURNE, b Hugo, Okla, Jan 25, 22; m 48; c 1. PLANT PHYSIOLOGY. Educ: Tex A&M Univ, BS, 50, MS, 51; Iowa State Univ, PhD(plant physiol), 53. Prof Exp: Assoc prof plant physiol, Exp Sta, Tex A&M Univ, 53-61; res plant physiologist, Pioneering Physiol Lab, Md, 61-65, PLANT PHYSIOLOGIST, BOLL WEEVEL RES LAB, USDA, 65- Concurrent Pos: USDA fel, Pioneering Lab Plant Physiol, Md, 61-62. Mem: Am Soc Plant Physiol; Crop Sci Soc Am. Res: Growth and development of cotton and cotton fiber; effect of light on maturity in crops; physiology of phytochrome; leaf proteins of cotton. Mailing Add: Boll Weevil Res Lab USDA Agr Res Serv Box 53677 Mississippi State MS 39762

LANE, IRWIN WILLIAM, biochemistry, see 12th edition

LANE, JAMES DALE, b Las Cruces, NMex, Aug 28, 37; m 58. VERTEBRATE ZOOLOGY. Educ: NMex State Univ, BS, 59, MS, 62; Univ Ariz, PhD(zool), 65. Prof Exp: Asst biol, NMex State Univ, 59-62; asst zool, Univ Ariz, 62-65, asst geochronology, 65; asst prof biol, 65-70, PROF ZOOL, McNEESE STATE UNIV, 70- Mem: AAAS; Soc Syst Zool; Am Soc Mammal; Soc Vert Paleont; Am Inst Biol Sci. Res: Ecology and systematics of various mammalians taxons, especially rodents. Mailing Add: Dept of Biol McNeese State Univ Lake Charles LA 70601

LANE, JOHN EDWARD, b Washington, DC, Nov 22, 17; m 46; c 3. PUBLIC HEALTH ADMINISTRATION. Educ: Univ Md, BS, 48; Ohio State Univ, MS, 52, PhD(entom), 63. Prof Exp: Entomologist, Commun Dis Ctr, USPHS, 48-55, med entom adv, Int Coop Admin, Ethiopia, 55-59, Jamaica, 59, malaria eradication specialist, Indonesia, 59-61, exec secy, Toxicol Study Sect, Div Res Grants, NIH, 63-66, exec secy, US-Japan Coop Med Sci Prog, Off Int Res, 66-69; res adminr, Am Cancer Soc, 69-72; HEALTH SCI ADMINR, NAT CANCER INST, 72- Mem: Am Soc Trop Med & Hyg. Res: Malaria eradication; vector borne disease control; research grants administration. Mailing Add: Nat Cancer Inst Bethesda MD 20014

LANE, KEITH ALDRICH, b Gridley, Kans, Nov 11, 21; m 45; c 2. ANALYTICAL CHEMISTRY. Educ: Oglethorpe Univ, AB, 42, MA, 43. Prof Exp: Chemist, Mutual Chem Co Am, 43-51, group leader anal res, 51-58; chemist, Solvay Process Div, 58-64, group leader anal res, 64-70, ENVIRON CHEMIST, INDUST CHEMS DIV, ALLIED CHEM CORP CORP, 70- Mem: Am Chem Soc. Res: All phases of chromium chemistry; organic and inorganic analytical method development; environmental studies and pollution control. Mailing Add: 122 Royal Rd Liverpool NY 13088

LANE, LEONARD JAMES, b Tucson, Ariz, Apr 25, 45; m 64; c 2. HYDROLOGY. Educ: Univ Ariz, BS, 70, MS, 72; Colo State Univ, PhD(civil eng), 75. Prof Exp: HYDROLOGIST, USDA, 70- Concurrent Pos: Fac affil civil eng, Colo State Univ, 73-74. Mem: Am Geophys Union; Am Soc Civil Engrs; Am Water Resources Asn; Brit Geomorphol Res Group. Res: Hydrology of semiarid regions; runoff and sediment simulation models incorporating geomorphic features and geomorphic thresholds; prediction of runoff and sediment production. Mailing Add: US Dept of Agr 442 E Seventh St Tucson AZ 85705

LANE, LESLIE C, JR, organic chemistry, see 12th edition

LANE, LESLIE CARL, b Stamford, Conn, Apr 5, 42; m 66; c 1. PLANT VIROLOGY. Educ: Univ Wis, BS, 65, PhD(biochem), 71. Prof Exp: Fel virol, John Innes Inst, Norwich, Eng, 71-73; fel virol, 73-75, ASST PROF PLANT PATH, UNIV NEBR-LINCOLN, 75- Mem: Am Phytopath Soc. Res: Structure and replication of plant viruses; virus directed protein and nucleic acid synthesis; virus-host interactions. Mailing Add: Dept of Plant Path Univ of Nebr Lincoln NE 68583

LANE, MALCOLM DANIEL, b Chicago, Ill, Aug 10, 30; m 51; c 2. BIOCHEMISTRY. Educ: Iowa State Univ, BS, 51, MS, 53; Univ Ill, PhD, 56. Prof Exp: Res asst, Iowa State Univ, 51-53; assoc prof biochem & nutrit, Va Polytech Inst, 56-62, prof, 62-64; assoc prof biochem, Sch Med, NY Univ, 64-69, prof, 69-70; PROF PHYSIOL CHEM, SCH MED, JOHNS HOPKINS UNIV, 70- Concurrent Pos: Sr fel, Max Planck Inst Cell Chem, 62-63; mem biochem study sect, NIH, 70- Honors & Awards: Mead Johnson Award, Am Inst Nutrit, 66. Mem: AAAS; Am Chem Soc; Am Soc Biol Chem; Am Inst Nutrit; Harvey Soc. Res: Enzymology regulation of enzyme activity; enzymatic carboxylation; fatty acid and carbohydrate metabolism; cholesterol biosynthesis. Mailing Add: Dept of Physiol Chem Johns Hopkins Univ Sch of Med Baltimore MD 21205

LANE, MAX HERBERT, b Graceville, Fla, Mar 17, 29; m 48; c 2. APPLIED MATHEMATICS. Educ: Troy State Col, BS, 50; Auburn Univ, MS, 56. Prof Exp: Mathematician, Ballistics Directorate, Air Proving Ground Ctr, 58-64, chief ballistics tech br, Air Force Systs Command, 64-65, CHIEF, OFF ASTRODYNAMIC APPL, HQ 14TH AEROSPACE FORCE, AEROSPACE DEFENSE COMMAND, US AIR FORCE, 66- Res: Development of general perturbation artificial satellite orbit theories; atmospheric density representation studies. Mailing Add: 2214 Shalimar Dr Colorado Springs CO 80915

LANE, MONTAGUE, b New York, NY, Aug 28, 29; m 57. CLINICAL PHARMACOLOGY, THERAPEUTICS. Educ: NY Univ, BA, 47; Chicago Med Sch, MB, 52, MD, 53; Georgetown Univ, MS 57. Prof Exp: Res assoc radiobiol, Cancer Res Lab, City New York, 47-48; res assoc oncol, Chicago Med Sch, 50-52; intern med, Jewish Hosp Brooklyn, NY, 52-53, asst res, 53-54; clin assoc pharmacol, Nat Cancer Inst, 54-56; asst resident med, USPHS, Rochester, 56-57; investr clin pharmacol, Nat Cancer Inst, 57-60; from asst prof to assoc prof, 60-67, PROF PHARMACOL & MED, BAYLOR COL MED, 67-, HEAD DIV CLIN ONCOL, 69- Concurrent Pos: Instr, Sch Med, George Washington Univ, 57-60; consult, Vet Admin Hosp, 63-; mem pharmacol & therapeut study sect, Nat Cancer Inst, 66-69, consult chemother study sect, mem nat new agents & spec Krebiozen rev comts & ad hoc pharmacol adv comt, Colo-Rectal Task Force, mem cancer clin invests rev comt, 73-; mem subcomt med oncol, Am Bd Internal Med. Mem: Am Soc Hemat; Am Soc Pharmacol & Exp Therapeut; Soc Exp Biol & Med; Am Asn Cancer Res; Am Soc Clin Pharmacol & Therapeut (pres, 71-72). Res: Cancer chemotherapy alkylating agents; antimetabolites; riboflavin deficiency; drug screening; ferro-kinetics; internal medicine. Mailing Add: Baylor Col of Med 1200 Moursund Houston TX 77025

LANE, NANCY JANE, b Halifax, NS, Nov 23, 36; m 69; c 1. CYTOCHEMISTRY. Educ: Dalhousie Univ, BSc, 58, MSc, 60; Oxford Univ, DPhil(cytol), 63; Cambridge Univ, PhD, 68. Prof Exp: Res asst prof path, Albert Einstein Col Med, 64-65; res staff biologist, Yale Univ, 65-68; HEAD ELECTRON MICROS, AGR RES COUN RES UNIT, CAMBRIDGE UNIV, 68- Concurrent Pos: Res fel, Girton Col, Cambridge Univ, 68-70, fel & lectr, 70-, grad tutor, 75- Mem: AAAS; Am Soc Cell Biol; Histochem Soc; Soc Exp Biol & Med; fel Royal Micros Soc. Res: Cytochemical and ultrastructural cytology; experimental and cytological studies of neurosecretory systems in invertebrates; enzyme cytochemistry and freeze fracture studies of invertebrate cells; accessibility of nervous systems to trace molecules. Mailing Add: Agr Res Coun Res Unit Cambridge Univ Cambridge England

LANE, NEAL F, b Oklahoma City, Okla, Aug 22, 38; m 60; c 2. ATOMIC PHYSICS. Educ: Univ Okla, BSc, 60, MS, 62, PhD(physics), 64. Prof Exp: NSF res fel physics, Queen's Univ, Belfast, 64-65; vis fel, Joint Inst Lab Astrophys, Univ Colo, 65-66; from asst prof to assoc prof, 66-72, PROF PHYSICS, RICE UNIV, 72- Concurrent Pos: Alfred P Sloan res fel, 67-73; vis fel, Joint Inst Lab Astrophys, Univ Colo, 75-76. Mem: Am Phys Soc. Res: Theoretical studies of electron-atom and electron molecule elastic and inelastic collision processes; theoretical investigation of atom-atom collision processes. Mailing Add: Dept of Physics Rice Univ Houston TX 77001

LANE, NORMAN GARY, b French Lick, Ind, Feb 19, 30; m 58; c 3. PALEONTOLOGY. Educ: Oberlin Col, AB, 52; Univ Kans, MS, 54, PhD(geol), 58. Prof Exp: From asst prof to prof geol, Univ Calif, Los Angeles, 58-73; PROF PALEONT, IND UNIV, BLOOMINGTON, 73- Concurrent Pos: Fulbright scholar, Univ Tasmania, 55-56; Fulbright prof, Trinity Col, Dublin, 71-72; res assoc paleont, Smithsonian Inst, 71- Mem: Paleont Soc; Soc Econ Paleontologists & Mineralogists; Soc Vert Paleont. Res: Functional morphology and community relations of fossil crinoids. Mailing Add: Dept of Geol Ind Univ Bloomington IN 47401

LANE, PATRICIA ANN, ecology, limnology, see 12th edition

LANE, RAYMOND OSCAR, b Asbury Park, NJ, Sept 25, 24; m 49; c 3. NUCLEAR PHYSICS. Educ: Iowa State Univ, PhD(physics), 53. Prof Exp: Res asst, Inst Atomic Res, Iowa State Col, 49-53; assoc physicist, Argonne Nat Lab, 53-66; prof physics, 66-74, DISTINGUISHED PROF PHYSICS, OHIO UNIV, 74- Mem: Am Phys Soc. Res: Penetration of electrons in matter; beta ray spectroscopy; neutron scattering; neutron polarization; nuclear structure. Mailing Add: Dept of Physics Ohio Univ Athens OH 45701

LANE, RICHARD DALE, b Kansas City, Mo, Sept 13, 12; m 41. FOREST MANAGEMENT. Educ: Iowa State Col, BS, 41, MS, 42. Prof Exp: Field asst forestry, Cent States Forest Exp Sta, US Forest Serv, USDA, 41; asst, Iowa State Col, 41-42; jr forester, Cent States Forest Exp Sta, US Forest Serv, USDA, 42-43, forester, NCent Region, 43-44 & Cent States Forest Exp Sta, 44-47, chg Carbondale Res Ctr, 47-56, chief div forest mgt & fire res, Northeastern Forest Exp Sta, 56-59, asst dir div forest mgt res, 59-60, dir, Cent States Forest Exp Sta, 60-66, dir, Northeastern Forest Exp Sta, 66-70, dir, Far Eastern Regional Res Off, Int Progs Div, Agr Res Serv, USDA, 70-74; RETIRED. Concurrent Pos: Adj prof, Southern Ill Univ, 51-56; forest res & admin consult, Turkish Forest Serv, AID, 65. Mem: AAAS; Soil Conserv Soc Am; Soc Am Foresters; Forest Prod Res Soc. Res: Forest research administration. Mailing Add: PO Box 3086 Marathon Shores FL 33052

LANE, ROBERT K, b Brandon, Man, Feb 7, 37; m; c 1. ENVIRONMENTAL MANAGEMENT. Educ: Brandon Col, BSc, 57; Ore State Univ, MS, 62, PhD(phys oceanog), 65. Prof Exp: Meteorol officer, Meteorol Br, Dept Transport, Can, 57-58; asst scientist, Pac Oceanog Group, Fisheries Res Bd Can, 59-61; instr oceanog, Ore State Univ, 63-65; res scientist, Marine Sci Br, Dept Energy, Mines & Resources, Can, 65-67; head phys limnol sect, Great Lakes Div, Inland Waters Br, 67-72, head lake resources subdiv, Lakes Div, 72-73, chief sci opers div, Can Ctr Inland Waters, 73-75, POLICY & PROG DEVELOP OFFICER, WESTERN & NORTHERN REGION, ENVIRON MGT SERV, DEPT ENVIRON, CAN, 75- Concurrent Pos: Mem subcomt hydrol, Nat Res Coun, 68- Honors & Awards: Can Centennial Medal, 68. Mem: Am Soc Limnol & Oceanog; Am Geophys Union; Royal Meteorol Soc; Int Asn Gt Lakes Res (pres, 74-75); Am Meteorol Soc. Res: Physical oceanography and limnology, especially heat and radiation exchange; remote sensing. Mailing Add: Dept Environ Environ Mgt Serv 10025 Jasper Ave Edmonton AB Can

LANE, ROBERT SIDNEY, b Worcester, Mass, Mar 7, 44; m 68. MEDICAL ENTOMOLOGY. Educ: Univ Calif, Berkeley, BA, 66, PhD(entom), 74; San Francisco State Col, MA, 69. Prof Exp: ASST PUB HEALTH BIOLOGIST, VECTOR CONTROL SECT, STATE DEPT HEALTH, CALIF, 74- Mem: Entom Soc Am; Wildlife Dis Asn; Wildlife Soc. Res: Biology, control, ecology and taxonomy of tabanidae and other hematophagous Diptera; biology and taxonomy of immature insects; medical and veterinary entomology; aquatic entomology. Mailing Add: Vector Control Sect State Dept of Health Berkeley CA 94704

LANE, RONALD PATON, b Jackson, Ga, Oct 11, 37; m 61; c 2. HORTICULTURE, PLANT BREEDING. Educ: Univ Ga, BSA, 63, MS, 65; Mich State Univ, PhD(hort), 68. Prof Exp: ASST HORTICULTURIST, GA STA, COL AGR, UNIV GA, 68- Mem: Am Soc Hort Sci. Res: Grape and bramble breeding. Mailing Add: Dept of Hort Ga Exp Sta Experiment GA 30212

LANE, STUART MICHAEL, organic chemistry, physical chemistry, see 12th edition

LANE, WALLACE, b Chicago, Ill, Aug 31, 11; m 38; c 2. PREVENTIVE MEDICINE. Educ: Univ Kans, AB, 33, MA, 35, MD, 39; Johns Hopkins Univ, MPH, 51. Prof Exp: Staff physician student health serv, Univ Kans, 46-48, clin assoc prof med & microbiol, 52-56; dir Bi-County Health Dept, Kans, 48-52; dir, Div Adult Health, Seattle-King County Health Dept, 56-58; chief div, 56-68, DIR, WASH STATE DEPT PUB HEALTH, 68- Concurrent Pos: Dir, Health Dept, Kansas City, Kans, 52-56; lectr, Sch Med, Univ Kans, 55-56; clin assoc prof prev med, Univ Wash, 57- Mem: AAAS; Am Pub Health Asn. Res: Role of behavioral science in medicine and public health. Mailing Add: Wash State Dept of Pub Health Olympia WA 98501

LANE, WILLIAM DAVID, b Vernon, BC, July 27, 47; m 71. PLANT BREEDING. Educ: Univ BC, BSc, 69, MSc, 71, PhD(plant physiol), 74. Prof Exp: RES SCIENTIST PLANT BREEDING, AGR CAN, 74- Mem: Can Soc Plant Physiol; Int Asn Plant Tissue Cult; Am Soc Hort Sci; Can Soc Hort Sci. Res: Breeding improved cultivars of apple and cherry using hybridization, induced mutations and plant tissue culture including haploidy and in vitro differentiation and development. Mailing Add: Agr Can Res Sta Summerland BC Can

LANE, WILLIAM JAMES, b Zanesville, Ohio, Dec 5, 25; m 50; c 3. ANALYTICAL CHEMISTRY. Educ: Denison Univ, BA, 47; Miami Univ, MS, 53; Iowa State Univ, PhD(chem), 57. Prof Exp: Asst chemist, AEC, Mound Lab, Monsanto Chem Co, 48-51; res asst anal chem, Ames Lab, Iowa State Univ, 52-57; anal chemist, Columbia-Southern Chem Corp, Pittsburgh Plate Glass Co, 57-60, anal group supvr, 60-64, res anal chemist, Chem Div, 64-69; res coordr anal chem, Universal Oil Prod Co, 69-74, GROUP LEADER ANAL CHEM RES & DEVELOP, UOP, INC, 74- Mem: Am Chem Soc. Res: General wet chemical analysis; polarography; chromatography; spectrophotometry. Mailing Add: 444 W Norman Ct Des Plaines IL 60016

LANEGRAN, DAVID ANDREW, b St Paul, Minn, Nov 27, 41; m 64; c 4. URBAN GEOGRAPHY. Educ: Macalester Col, BA, 63; Univ Minn, MA, 66, PhD(geog), 70. Prof Exp: From teaching asst to teaching assoc geog, Univ Minn, 64-68, instr geog, Exten Div, 68-69; instr, 69-70, ASST PROF GEOG, MACALESTER COL, 70-, CHMN DEPT, 75- Concurrent Pos: Planner, Coop Res & Planning, 67-; instr geog, Univ Wis-River Falls, 68; Minn State coordr, Nat Coun Geog Educ, 69-; Endowment fac res, Macalester Col, 70-73; dir, Living Hist Mus St Paul, 74- Mem: Asn Am Geog; Am Geog Soc. Res: Internal structure of American cities with emphasis on residential perception of urban landscapes; qualitative geography; neighborhoods, spatial expansion of cities, commercial strips, urbanization. Mailing Add: Dept of Geog Macalester Col St Paul MN 55101

LANESE, JOHN GERALD, analytical chemistry, see 12th edition

LANFORD, CAROLINE SHERMAN, biochemistry, see 12th edition

LANFORD, OSCAR E, III, b New York, NY, Jan 6, 40; m 61; c 1. MATHEMATICAL PHYSICS. Educ: Wesleyan Univ, BA, 60; Princeton Univ, MA, 62, PhD(physics), 66. Prof Exp: Instr math, Princeton Univ, 65-66; asst prof, Univ Calif, Berkeley, 66-67; vis prof physics, Inst Advan Sci Studies, 67-68; asst prof math, 68-70, ASSOC PROF MATH, UNIV CALIF, BERKELEY, 70- Concurrent Pos: Alfred Sloan Found res fel, 69-71; mem, Inst Advan Studies, 70; exchange prof, Univ Aix Marseille, 71. Res: Mathematical physics, especially statistical mechanics and quantum field theory. Mailing Add: Dept of Math Univ of Calif Berkeley CA 94720

LANG, ANDREW GEORGE, b Dayton, Ohio, Nov 18, 08; m 38; c 4. PAPER CHEMISTRY. Educ: Miami Univ, AB, 32; Cornell Univ, MS, 33; Univ NC, PhD(bot), 36. Prof Exp: Asst, NC State Col, 33-36; agent, Bur Plant Indust, USDA, NC, 36-37; assoc botanist, Chesapeake Biol Lab, Md, 38-39; chief chemist, Biochem Labs, Ohio, 39-41; from sr res engr to mgr cent lab, Champion Papers Inc, 41-62; assoc prof, 62-70, PROF PAPER TECHNOL, MIAMI UNIV, 70- Mem: AAAS; Inter-Soc Color Coun; Tech Asn Pulp & Paper Indust; Optical Soc Am; Am Inst Chem. Res: Paper technology. Mailing Add: Dept of Paper Technol Miami Univ Oxford OH 45056

LANG, ANTON, b Petersburg, Russia, Jan 18, 13; nat US; m 46; c 3. DEVELOPMENTAL PLANT BIOLOGY. Educ: Univ Berlin, Dr Nat Sci, 39. Prof Exp: Sci asst plant physiol, Max-Planck Inst Biol, Ger, 39-49; res assoc genetics, McGill Univ, 49; vis prof genetics & agron, Agr Exp Sta, Agr & Mech Col Tex, 50; from res fel to sr res fel plant physiol, Calif Inst Technol, 50-52; from asst prof to assoc prof bot, Univ Calif, Los Angeles, 52-59; prof biol, Calif Inst Technol, 59-65; PROF BOT & PLANT PATH, MICH STATE UNIV & DIR MSU/ERDA PLANT RES LAB, 65- Concurrent Pos: Lady Davis Found Can fel, 49; NSF sr fel, Max-Planck Inst Biol, Ger & Hebrew Univ, Israel, 58-59; consult develop biol prog, NSF, 60-64, consult, adv comt biol & med sci, 68-70; partic sci exchange prog, Nat Acad Sci-Acad Sci Sci, USSR, 63, 68 & 75; trustee, Argonne Univs Asn, 65-69; chmn comt effects on herbicides in Vietnam, Nat Acad Sci-Nat Res Coun, 72-74. Mem: Nat Acad Sci; fel AAAS; Bot Soc Am; Am Soc Plant Physiol (vpres, 63, pres, 71-); Soc Develop Biol (pres, 69). Res: Physiology of cell differentiation and organ formation in plants; physiology of flowering; plant hormone physiology; control of plant development by environmental factors. Mailing Add: MSU/ERDA Plant Res Lab Mich State Univ East Lansing MI 48824

LANG, ARTHUR (HAMILTON), b Peachland, BC, July 3, 05. ECONOMIC GEOLOGY. Educ: Univ BC, BA, 27, MA, 28; Princeton Univ, PhD(econ geol), 30. Prof Exp: Geologist, 30-52, chief radioactive resources div, 52-54, chief mineral deposits div, 55-59, chief spec projs, 59-70, CONSULT, GEOL SURV CAN, 70- Mem: Royal Soc Can; Geol Soc Am; Geol Asn Can; Can Inst Mining & Metall. Res: Geological writing for laymen; geology of mineral deposits. Mailing Add: Geol Surv of Can Ottawa ON Can

LANG, BRUCE Z, b St Joseph, Mo, May 31, 37; m 59; c 1. PARASITOLOGY, IMMUNOLOGY. Educ: Chico State Col, BS, 60; Univ NC, Chapel Hill, MSPH, 61, PhD(parasitol), 66. Prof Exp: Vis asst prof zool, Univ Okla, 66-67; from asst prof to assoc prof biol, 67-74, PROF BIOL, EASTERN WASH STATE COL, 74- Concurrent Pos: NIH fel zool, Univ Okla, 66-67; NSF grants, 69-72. Mem: AAAS; Am Soc Parasitol; Am Soc Zoologists; Am Soc Trop Med & Hyg; Am Inst Biol Sci. Res: Host-parasite relationships; ecology of parasitism; ecology of fresh-water gastropod molluscs. Mailing Add: Dept of Biol Eastern Washington State Col Cheney WA 99004

LANG, CALVIN ALLEN, b Portland, Ore, June 13, 25; m 49; c 4. GERONTOLOGY. Educ: Princeton Univ, AB, 47; Johns Hopkins Univ, ScD(biochem), 54. Prof Exp: Res collabr, Brookhaven Nat Labs, 49-51; asst scientist insect biochem, Conn Agr Exp Sta, 54-56; res assoc, Sch Hyg & Pub Health, Johns Hopkins Univ, 56-59; from asst prof to assoc prof, 59-72, PROF BIOCHEM, SCH MED, UNIV LOUISVILLE, 72-, DIR BIOMED AGING RES PROG, 71- Concurrent Pos: Fel, Sch Hyg & Pub Health & McCollum-Pratt Inst, Johns Hopkins Univ, 56-59; NIH fel, 57-59 & res career develop award, 67-72. Mem: Am Soc Biol Chem; Am Inst Nutrit; Soc Exp Biol & Med; Soc Develop Biol; fel Geront Soc (vpres, 71). Res: Biochemistry of growth and aging; insect and nutritional biochemistry. Mailing Add: Dept of Biochem PO Box 1055 Univ of Louisville Sch of Med Louisville KY 40201

LANG, CHARLES E, physical chemistry, radiochemistry, see 12th edition

LANG, CONRAD MARVIN, b Chicago, Ill, July 1, 39; m 61; c 3. PHYSICAL CHEMISTRY. Educ: Elmhurst Col, BS, 61; Univ Wis-Madison, MS, 64; Univ Wyo, PhD(chem), 70. Prof Exp: Teaching asst chem, Univ Wis-Madison, 61-63, res asst, 63-64; from instr to asst prof, 64-70, ASSOC PROF CHEM, UNIV WIS-STEVENS POINT, 70- Concurrent Pos: Consult, Crowns, Merklin, Midthun & Hill, Attorneys at Law, 70-; NSF res grant, Univ Wis-Stevens Point, 71- Mem: Am Chem Soc. Res: Application of electron spin resonance to molecular structure and biological molecules; semiempirical quantum chemical calculations on systems of biological interest; physiochemical aspects of vision; macromolecular structure of binary fluids. Mailing Add: Dept of Chem Univ of Wis Stevens Point WI 54481

LANG, DAVID (VERN), b Wilmar, Minn, July 11, 43; m 68; c 1. SEMICONDUCTORS, PHYSICS. Educ: Corncordia Col, Moorhead, Minn, BA, 65; Univ Wis-Madison, PhD(physics), 69. Prof Exp: Res assoc physics, Univ Ill, Urbana, 69-70, res asst prof, 70-72; MEM TECH STAFF, BELL LABS, 72- Mem: Am Phys Soc. Res: Capacitance spectroscopy; defects in III-V semiconductors; recombination enhanced solid state defect reactions; radiation damage in semiconductors. Mailing Add: Bell Labs Murray Hill NJ 07974

LANG, DENNIS ROBERT, b Albany, NY, May 7, 44. CELL PHYSIOLOGY. Educ: Syracuse Univ, BS, 66, PhD(microbiol), 71. Prof Exp: Fel biochem, Cornell Univ, 71-73; ASST PROF MICROBIOL, SCH MED, UNIV CINCINNATI, 73- Mem: Am Soc Microbiol; Sigma Xi; Tissue Cult Asn. Res: Role of the magnesium stimulated adenosine triphosphatase in membrane bioenergetics of bacteria; regulation of metabolism in normal and transformed mammalian cells. Mailing Add: Dept of Microbiol Sch of Med Univ of Cincinnati Cincinnati OH 45267

LANG, DIMITRIJ ADOLF, b Berlin, Ger, Aug 30, 26; m 59; c 2. BIOPHYSICS, MOLECULAR BIOLOGY. Educ: Univ Frankfurt, MS, 53, PhD(biophys), 59. Prof Exp: Res asst biophys, Max Planck Inst Biophys, 53-58 & Hyg Inst, Univ Frankfurt, 58-65; from asst prof to assoc prof biol, 65-72, PROF BIOL, UNIV TEX, DALLAS, 72- Concurrent Pos: NIH res career develop awards, 67-71 & 72-76. Mem: Ger Soc Electron Micros; Electron Micros Soc Am; Biophys Soc. Res: Electronics; physics of ionizing radiations; high-output x-ray machines; standard dosimetry of x-rays; electron microscopy of bacteria, viruses and nucleic acids; physical chemistry of nucleic acids. Mailing Add: Univ Tex Dallas Molec Biol Prog PO Box 688 Richardson TX 75080

LANG, EDGAR REED, b Sioux Falls, SDak, Dec 11, 13; m 40; c 3. POLYMER CHEMISTRY. Educ: SDak Sch Mines & Technol, BS, 36; Ohio State Univ, MS & PhD(phys chem), 40. Prof Exp: Asst, Ohio State Univ, 36-40; res chemist, Rohm and Haas Co, 41-44, lab head, 44-70, sr res chemist, 70-76; RETIRED. Mem: Am Chem Soc. Res: Electrophoresis; ultraviolet microscopy; ultracentrifugal sedimentation of colloid systems; plastics; polymerization kinetics; polymer applications. Mailing Add: 2216 Wharton Rd Glenside PA 19038

LANG, ENID ASHER, b Los Angeles, Calif, Aug 28, 44; m 69. PSYCHIATRY. Educ: Radcliffe Col, AB, 66; Univ Southern Calif, MD, 70; Columbia Univ, MS, 74. Prof Exp: Med intern, Beth Israel Hosp, New York, 71-72; resident psychiat, Columbia Psychiat Inst, 72-74; res fel, Columbia Univ Health Serv, 74-75; FAC MEM PSYCHIAT, SCH MED, NY UNIV, 75- Concurrent Pos: Dir, Group Psychiat & Training Psychiat Residents, Bellevue Hosp, Sch Med, NY Univ, 75-, dir, Socialization Prog Psychiat Outpatients, 75- Mem: Am Psychiat Asn; NY Acad Sci. Res: A longitudinal comparative study of treatment-outcome of discharged psychiatric outpatients who receive group therapy with medication versus individual therapy with medication only. Mailing Add: 1520 York Ave Apt 29D New York NY 10028

LANG, ERICH KARL, b Vienna, Austria, Dec 7, 29; US citizen; m 56; c 2. RADIOLOGY. Educ: Columbia Univ, MS, 51; Univ Vienna, MD, 53. Prof Exp: Assoc radiol, Johns Hopkins Hosp & Univ, 56-59, assoc radiologist, 59-61; radiologist, Methodist Hosp, Indianapolis, Ind, 61-67; PROF RADIOL & CHMN DEPT, SCH MED, LA STATE UNIV, SHREVEPORT, 67- Mem: AMA; Radiol Soc NAm; Soc Nuclear Med; Am Col Radiol; Am Col Chest Physicians. Res: Diagnostic, vascular roentgenographic examinations; diagnostic roentgenographic evaluation of tumors and tumor diagnosis. Mailing Add: Dept of Radiol La State Univ Sch of Med Shreveport LA 71101

LANG, FRANCES SPARKS, physical chemistry, see 12th edition

LANG, FRANK ALEXANDER, b Olympia, Wash, May 14, 37; m 59; c 2. SYSTEMATIC BOTANY. Educ: Ore State Univ, BS, 59; Univ Wash, MS, 61; Univ BC, PhD(bot), 65. Prof Exp: Asst prof biol, Whitman Col, 65-66; asst prof, 66-71, ASSOC PROF BIOL, SOUTHERN ORE COL, 71- Mem: Am Soc Plant Taxon; Int Asn Plant Taxon. Res: Biosystematics and cytotaxonomy of vascular plants; flora of the Siskiyou Mountains. Mailing Add: Dept of Biol Southern Ore Col Ashland OR 97520

LANG, FRANK THEODORE, b New York, NY, Jan 25, 38; m 63; c 3. PHYSICAL CHEMISTRY. Educ: St Francis Col, NY, BS, 59; Rensselaer Polytech Inst, PhD(phys chem), 64. Prof Exp: Res assoc photochem, Univ Sheffield, 64-65; res assoc-instr energy transfer, Univ NC, Chapel Hill, 65-67; chmn dept chem, 70-73, ASSOC PROF PHYS CHEM, FAIRLEIGH DICKINSON UNIV, FLORHAM-MADISON CAMPUS, 67- Mem: Am Chem Soc. Res: Charge transfer complexes; flash photolysis; low temperature photochemistry; energy transfer processes; luminescene studies. Mailing Add: Dept of Chem Fairleigh Dickson Univ Madison NJ 07940

LANG, FREDERICK, b Albany, NY, Feb 1, 44; m 69; c 1. NEUROBIOLOGY, INVERTEBRATE ZOOLOGY. Educ: Antioch Col, BA, 66; Univ Ill, Urbana, MS & PhD(physiol, biophys), 70. Prof Exp: Muscular Dystrophy Asn Can fel zool, Univ Toronto, 70-71, Nat Insts Neurol Dis & Stroke fel, 71-72; ASST PROF BIOL, BOSTON UNIV, 72-, INSTR NEUROBIOL, MARINE BIOL LAB, 75- Concurrent Pos: Spec fel, Grass Found, 73. Mem: AAAS; Am Soc Zool; Soc Neurosci. Res: Developmental neurobiology of invertebrate neuromuscular systems, central nervous systems and hearts. Mailing Add: Boston Univ Marine Prog Marine Biol Lab Woods Hole MA 02543

LANG, GEORGE E, JR, b Chicago, Ill, June 29, 42; m 68; c 1. TOPOLOGY. Educ: Loyola Univ Chicago, BS, 64; Univ Dayton, MS, 66; Purdue Univ, PhD(math), 70. Prof Exp: Teaching asst, Univ Dayton, 64-66 & Purdue Univ, 67-69; ASSOC PROF MATH, FAIRFIELD UNIV, 70- Mem: Am Math Soc; Math Asn Am; Am Asn Univ Profs. Res: Homotopy theory; subgroups of homotopy groups; direct limits of CW complexes with an eye to group theoretic applications. Mailing Add: Dept of Math Fairfield Univ Fairfield CT 06430

LANG, GERALD EDWARD, b Chicago, Ill, Mar 1, 45; m 73. PLANT ECOLOGY. Educ: Western Ill Univ, BS, 67; Univ Wyo, MS, 69; Rutgers Univ, PhD(bot), 73. Prof Exp: RES INSTR TERRESTRIAL ECOL, DARTMOUTH COL, 73- Concurrent Pos: NSF res grant, 74-76. Mem: AAAS; Brit Ecol Soc; Ecol Soc Am; Sigma Xi. Res: Forest floor dynamics and log decomposition along an elevational gradient in New Hampshire; mechanisms underlying canopy influences on throughfall chemistry in subalpine fir forests of New Hampshire. Mailing Add: Dept of Biol Sci Dartmouth Col Hanover NH 03755

LANG, GERHARD PAUL, b Omaha, Nebr, Feb 20, 17; m 44; c 4. INORGANIC CHEMISTRY. Educ: Valparaiso Univ, BA, 43; Wash Univ, MA, 55. Prof Exp: Chemist, Visking Corp, 44-45; chemist, Uranium Div, Mallinckrodt Chem Works, 46-64; res chemist, Emerson Elec Mfg Co, 64-66; SR ENGR, McDONNELL DOUGLAS CORP, 66- Mem: Am Chem Soc. Res: Uranium chemistry; liquid-liquid extraction; radiochemistry; aerospace chemistry. Mailing Add: McDonnell Douglas Corp PO Box 516 St Louis MO 63166

LANG, GOTTFRIED OTTO, b Oberammergau, Bavaria, Mar 24, 19; US citizen; m 48; c 7. ANTHROPOLOGY. Educ: Brown Univ, AB, 44; Univ Chicago, MA, 48; Cornell Univ, PhD(anthrop), 53. Prof Exp: Instr anthrop, Univ Utah, 48-51; teaching asst, Cornell Univ, 52-53; Soc Sci Res Coun fel study Ute tribe in northern Utah, 53-54; from asst prof to prof anthrop, Cath Univ Am, 54-66; PROF ANTHROP & CO-CHMN AFRICAN & MIDEAST STUDIES PROG, UNIV COLO, BOULDER, 66-, DIR PROG RES CULT CHANGE, INST BEHAV SCI, 66- Concurrent Pos: Consult, Ute Indian Tribe, 50-54, Planning Div, Indian Bur, US Dept Interior, 53-55, USPHS, 57-59 & Med Sch, Howard Univ, 57-61; Fulbright prof anthrop, Sociol Inst, Univ Munich, 61-62; assoc ed, Anthrop Quart; NSF grant social & cult change in Sukumaland, Tanzania, Cath Univ Am, 61-66, NIH grant Utes & Basukuma, 63-64, NIH grant witchcraft, 64-65, Wenner-Gren Found grant, 66; US Off Educ grant Plains & Navaho area, Univ Colo, 67-69, UN Develop Prog-Food & Agr Orgn sr fel livestock develop in Tanzania, 69-71, Univ Colo grant pop dynamics in Asmat, W Irian, 72. Mem: AAAS; African Studies Asn; Am Anthrop Asn; Am Ethnol Soc; Am Sociol Asn. Res: Problems in socio-cultural change as they pertain to decisions in socio-economic development involving planning social organization that promote innovation in agriculture, technology, education and population planning. Mailing Add: Inst of Behav Sci Univ of Colo Boulder CO 80302

LANG, JAMES FREDERICK, b Dayton, Ohio, Mar 19, 31; m 58; c 2. DRUG METABOLISM. Educ: Univ Cincinnati, BS, 58, MS, 70. Prof Exp: Res asst toxicol & drug metab, Christ Hosp Inst Med Res, Subsid Elizabeth Gamble Deaconess Home Asn, 58-63; from res asst biochem to biochemist, 63-72, SECT HEAD DRUG METAB, MERRELL-NAT LABS, DIV RICHARDSONMERRELL, 72- Mem: Am Chem Soc; AAAS. Res: Isolation and identification of drug metabolites, pharmacokinetics; development and application of analytical methods for trace analysis of drug residues in biological media. Mailing Add: Merrell-Nat Labs 110 E Amity Rd Cincinnati OH 45215

LANG, JOSEPH EDWARD, b Covington, Ky, Aug 10, 42. THEORETICAL PHYSICS. Educ: Thomas More Col, AB, 64; Univ Ill, MS, 65, PhD(physics), 70. Prof Exp: NSF fel, Lawrence Radiation Lab, Univ Calif, Berkeley, 70-71; ASST PROF PHYSICS, THOMAS MORE COL, 71- Mem: Am Phys Soc; Am Asn Physics Teachers, Sigma Xi. Res: Elementary particle physics; magnetic anisotropy; computers in education. Mailing Add: Dept of Physics Thomas More Col Ft Mitchell KY 41017

LANG, JOSEPH HERMAN, b New York, NY, Oct 31, 23. BIOCHEMISTRY. Educ: Univ Calif, Los Angeles, BS, 45; Fla State Univ, PhD(biochem), 57. Prof Exp: Res assoc biochem, Grad Sch Pub Health, Univ Pittsburgh, 57-63, asst res prof radiol, Sch Med, 63-68; assoc res biochemist & radiologist, 68-74, RES BIOCHEMIST & RADIOLOGIST, SCH MED, UNIV CALIF, SAN DIEGO, 74- Mem: Am Chem Soc. Res: Physical chemistry of proteins; protein binding; pharmacology of radiopaque compounds. Mailing Add: Dept of Radiol Univ of Calif San Diego La Jolla CA 92037

LANG, KENNETH LYLE, b Cuba City, Wis, Apr 12, 36; m 61; c 1. FRESHWATER ECOLOGY. Educ: Iowa State Col, BS, 59; Univ Iowa, MS, 66, PhD(zool), 70. Prof Exp: ASSOC PROF BIOL, HUMBOLDT STATE UNIV, 70- Mem: Am Soc Zool; Am Soc Limnol & Oceanog. Res: Freshwater zooplankton populations; dispersion patterns and species diversity of benthic and planktonic Cladoceran assemblages. Mailing Add: Dept of Biol Humboldt State Univ Arcata CA 95521

LANG, LAWRENCE GEORGE, b Pittsburgh, Pa, Mar 25, 31; m 53; c 2. SOLID STATE PHYSICS. Educ: Carnegie Inst Technol, BS, 52, MS, 53, PhD(physics), 57. Prof Exp: Res physicist, Carnegie Inst Technol, 56-57; eng specialist, Philco Corp, 57-58; res physicist, Carnegie Inst Technol, 58-60, asst prof physics, 60-64; res physicist, Atomic Energy Res Estab, Eng, 63-65; assoc prof physics, Carnegie Inst Technol, 64-66; PROF PHYSICS, PA STATE UNIV, 73- Concurrent Pos: Nat Acad Sci-Nat Res Coun fel, 63-64. Mem: Am Phys Soc. Res: Angular correlation of radiation from positron annihilation in solids; Mössbauer effect in compounds; paramagnetic and diamagnetic salts; biological macromolecules. Mailing Add: Dept of Physics Davey Bldg Pa State Univ University Park PA 16802

LANG, NORMA JEAN, b Memphis, Tenn, July 25, 31. PHYCOLOGY. Educ: Ohio State Univ, BS, 52, MA, 58; Ind Univ, PhD(bot), 62. Prof Exp: NIH res fel algae, Univ Tex, 62-63; from asst prof to assoc prof bot, 63-74, PROF BOT, UNIV CALIF, DAVIS, 74- Concurrent Pos: NSF res grant, 65-67; Guggenheim fel, Westfield Col, London, 69-70. Mem: Phycol Soc Am (treas, 71-73, vpres, 74, pres, 75); Brit Phycol Soc; Electron Micros Soc Am; Int Phycol Soc. Res: Electron microscopic studies of cellular structure in unicellular green algae, colonial green algae and blue-green algae, especially differentiation of heterocysts. Mailing Add: Dept of Bot Univ of Calif Davis CA 95616

LANG, NORTON DAVID, b Chicago, Ill, July 5, 40; m 69. THEORETICAL SOLID STATE PHYSICS, SURFACE PHYSICS. Educ: Harvard Univ, AB, 62, AM, 65, PhD(physics), 68. Prof Exp: Asst res physicist, Univ Calif, San Diego, 67-69; STAFF MEM, IBM CORP, 69- Mem: Am Phys Soc. Res: Solid state physics; theory of the electron gas. Mailing Add: IBM Res Ctr Yorktown Heights NY 10598

LANG, PHILIP CHARLES, b Jamestown, NY, Nov 16, 34; m 55; c 3. ORGANIC CHEMISTRY. Educ: Allegheny Col, BS, 57; Ohio Univ, MS, 59; Rensselaer Polytech Inst, PhD(org chem), 66. Prof Exp: Res chemist, Diamond Alkali Co, 59-62; assoc res chemist, Sterling Winthrop Res Inst, 62-67; res chemist, GAF Corp, 67-73; RES CHEMIST, TOMS RIVER CHEM CORP, 73- Mem: Am Chem Soc. Res: Synthetic medicinal chemistry; heterocyclic and acteylene compounds; aromatics and synthetic dyes. Mailing Add: Toms River Chem Co Box 71 Toms River NJ 08753

LANG, RAYMOND W, b Syracuse, NY, Aug 1, 30; m 53; c 5. BACTERIOLOGY, IMMUNOLOGY. Educ: LeMoyne Col, NY, BS, 52; Mich State Univ, MS, 57, PhD(microbiol), 59. Prof Exp: Asst prof microbiol, St John's Univ, NY, 59-62; fel bact & immunol, Sch Med, State Univ NY Buffalo, 62-63, from instr to asst prof bact & immunol, 63-68; assoc prof, 68-72, PROF MED MICROBIOL, COL MED, OHIO STATE UNIV, 72- Concurrent Pos: Consult urol res sect, Millard Fillmore Hosp, 64-66; consult training prog, Nat Inst Dent Res, 72-75. Mem: AAAS; Am Soc Microbiol. Res: Immunochemistry of tissue antigens; transplantation, autoimmunity and aging. Mailing Add: Dept of Med Microbiol Ohio State Univ Col of Med Columbus OH 43210

LANG, ROBERT EUGENE, b Greenford, Ohio, Oct 18, 19; m 44; c 2. PHYSICAL CHEMISTRY. Educ: Youngstown Univ, BS, 42; NY Univ, MS, 50, PhD(phys chem), 53. Prof Exp: Process operator, Trinitrotoluene Mfr, Trojan Powder Co, 42-43; chemist, E Frederics, Inc, 46-50; microanalyst, Dept Chem, NY Univ, 52-53; proj leader, Instrumental Anal Sect, Gen Foods Corp, 53-58; mgr anal sect, Am Mach & Foundry Co, 58-67; asst dir chem div, Food & Drug Res Lab Inc, 67-69; DIR & CHIEF CHEMIST, NEW YORK CUSTOMS LAB, 69- Mem: Am Chem Soc; Sigma Xi; Am Soc Testing & Mat. Res: Instrumental analytical chemistry; spectroscopy; chromatography; electrochemistry; microanalysis. Mailing Add: US Customs Lab 6 World Trade Ctr New York NY 10048

LANG, ROBERT LEE, b Ely, Mo, Apr 8, 13; m 38; c 4. AGRONOMY. Educ: Univ Wyo, BS, 36, MS, 41; Univ Nebr, PhD, 55. Prof Exp: From instr to assoc prof agron, 36-55, head plant sci div, 61-70, PROF RANGE MGT, UNIV WYO, 55- Mem: Soc Range Mgt. Res: Range and pasture management and improvement. Mailing Add: Univ of Wyo Col of Agr Univ Sta Box 3354 Laramie WY 82070

LANG, ROBERT PHILLIP, b Chicago, Ill, June 15, 32. PHYSICAL CHEMISTRY. Educ: Univ Ill, BS, 55; Univ Chicago, MS, 60, PhD(phys chem), 61. Prof Exp: From instr to assoc prof 62-74, PROF CHEM, QUINCY COL, 74-, CHMN DEPT, 71- Concurrent Pos: NSF res grant, 65-68 & 70-72. Mem: AAAS; Am Chem Soc; Am Asn Physics Teachers. Res: Thermodynamic and electronic spectral characteristics of molecular complexes of iodine; amino acid sequences in enzymes. Mailing Add: Dept of Chem Quincy Col Quincy IL 62301

LANG, RUDOLF F, polymer chemistry, plastics, see 12th edition

LANG, SERGE, mathematics, see 12th edition

LANG, SOLOMON MAX, b Jersey City, NJ, Nov 21, 25; m 49; c 3. HYDROLOGY. Educ: Newark Col Eng, BS, 49. Prof Exp: Hydraul engr, 49-67, supv hydrologist & chief reports sect, Water Resources Div, 67-69, hydrologist & prin asst, Off Asst Chief Hydrologist for Sci Publ & Data Mgt, Washington, DC, 69-73, DEP ASST CHIEF HYDROLOGIST FOR SCI PUBL & DATA MGT, WATER RESOURCES DIV, US GEOL SURV, 73- Concurrent Pos: Mem comt X3L8 data representations, Am Nat Standards Inst, 73-; dep mem, US Bd Geog Names, 75- Mem: Am Water Works Asn; Am Geophys Union. Res: Management of publications program; design, implementation and management of water resources data system. Mailing Add: Water Resources Div US Geol Surv 440 Nat Ctr Reston VA 22092

LANG, STANLEY ALBERT, JR, b Cleveland, Ohio, Mar 30, 44. ORGANIC CHEMISTRY. Educ: John Carroll Univ, BS, 66; Brown Univ, PhD(org chem), 70. Prof Exp: Res fel, Ohio State Univ, 70, Nat Cancer Inst fel, 71; res chemist, Lederle Labs, 72-74, GROUP LEADER, INFO DIS THERAPY SECT, LEDERLE LABS DIV, AM CYANAMID INC, 74- Mem: AAAS; Am Chem Soc; The Chem Soc. Res: Synthetic organic chemistry; medicinal drugs; antituberculous; antiparasitics. Mailing Add: Lederle Labs Div Am Cyanamid Inc Bldg 65A Pearl River NY 10965

LANG, WILLIAM HARRY, b Etna, Pa, Mar 29, 18; m 46; c 2. PETROLEUM CHEMISTRY, FUEL SCIENCE. Educ: Grove City Col, BS, 40. Prof Exp: SR RES CHEMIST, CENT RES DIV, MOBIL RES & DEVELOP CORP, 40- Mem: Am Chem Soc; Catalysis Soc. Res: Development of new and alternate fuels and energy sources to use as a petroleum substitute. Mailing Add: Mobil Res & Develop Corp Box 1025 Princeton NJ 08540

LANG, WILLIAM WARNER, b Boston, Mass, Aug 9, 26; m 54; c 1. ACOUSTICS. Educ: Iowa State Univ, BS, 46, PhD(physics), 58; Mass Inst Technol, SM, 49. Prof Exp: Acoustical engr, Bolt, Beranek & Newman, Inc, 49-51; instr physics, US Naval Post-Grad Sch, 51-55; spec engr, E I du Pont de Nemours & Co, 55-57; adv physicist, 58-64, sr physicist & mgr acoust lab, 64-75, PROG MGR ACOUSTICS TECHNOL, IBM CORP, 75- Concurrent Pos: Mem eval panel, Mech Div, Nat Bur Standards, 74-75, chmn, 75-76. Honors & Awards: Achievement Award, Inst Elec & Electronics Engr Group Audio & Electroacoustics, 72. Mem: AAAS; Inst Noise Control Eng (exec vpres, 71-); Acoust Soc Am; Inst Elec & Electronics Engr; Audio Eng Soc. Res: Physical acoustics; effects and control of noise; theory and design of acoustical materials. Mailing Add: 29 Hornbeck Ridge Poughkeepsie NY 12603

LANGAGER, BRUCE ALLEN, b Willmar, Minn, Jan 17, 42; m 64; c 2. POLYMER CHEMISTRY. Educ: Augsburg Col, BA, 64; Univ Minn, Minneapolis, PhD(org chem), 68. Prof Exp: Sr res chemist, 68-74, RES SPECIALIST, 3M CO, 74- Mem: Sigma Xi. Res: Life sciences. Mailing Add: 3M Co Cent Res Labs PO Box 33221 St Paul MN 55133

LANGAN, THOMAS AUGUSTINE, b Providence, RI, July 25, 30; m 60; c 2. BIOCHEMISTRY. Educ: Fordham Univ, BS, 52; Johns Hopkins Univ, PhD(biochem), 59. Prof Exp: Mem res staff, Med Nobel Inst, Stockholm, Sweden, 59-60; guest investr biochem, Rockefeller Inst, 60-61; mem res staff, Wenner-Gren Inst, Stockholm, Sweden, 61-62; res assoc biochem, Rockefeller Inst, 62-65; staff scientist, C F Kettering Res Lab, 65-67, investr, 67-70, sr investr, 70-71; ASSOC PROF PHARMACOL, MED SCH, UNIV COLO, DENVER, 71- Concurrent Pos: NSF fel, 59-62; from asst prof to assoc prof, Antioch Col, 67-71. Mem: Am Soc Biol Chemists; Am Soc Cell Biol. Res: Metabolism and function of histones and nuclear phosphoproteins; control of histone phosphorylation by cyclic adenosine monophosphate; regulation of nucleic acid synthesis in eukaryotes. Mailing Add: Dept of Pharmacol Univ of Colo Med Ctr Denver CO 80220

LANGAN, WILLIAM BERNARD, b Wayne Co, Pa, Oct 31, 13; m 57; c 2. PHYSIOLOGY. Educ: Univ Scranton, BS, 36; Columbia Univ, MA, 37 & 44; Fordham Univ, PhD(zool), 42. Prof Exp: Asst sci, Teachers Col, Columbia Univ, 36-37; asst biol, Fordham Univ, 40-42; instr physiol, New York Med Col, 42-47, assoc, 47-50, asst prof prof, 50-60, asst prof pharmacol, 58-60; sect head physiol, Food & Drug Res Labs, Inc, 60-61; NIH spec fel, State Univ NY Downstate Med Ctr, 61-62, asst prof, 62-63; ASSOC PROF BIOL, VILLANOVA UNIV, 63- Concurrent Pos: Lectr, Hunter Col, 60-62; consult, Food & Drug Res Labs, Inc, 61-63; training prog partic, Int Lab Genetics & Biophys, Italy, 65. Mem: Fel AAAS; Harvey Soc; Endocrine Soc; Am Soc Zool; NY Acad Sci. Res: Steroids and cardiac electrophysiology; steroids and ovulation in sub-mammalian species; mechanism of hormone action at the cellular level. Mailing Add: Dept of Biol Villanova Univ Villanova PA 19085

LANGBEIN, WALTER, b Newark, NJ, Oct 17, 07; m 39; c 2. HYDROLOGY. Educ: Cooper Union, BS, 31. Prof Exp: Construct engr, Rosoff Construct Co, NY, 31-35; hydraul engr, US Geol Surv, NY, 35-36, hydrologist, 36-75. Concurrent Pos: Ed, Water Resources Res, 65- Honors & Awards: Distinguished Serv Award, US Dept Interior, 58. Mem: Nat Acad Sci; Am Meteorol Soc; Am Soc Civil Eng; Am Geophys Union; Int Asn Hydrol (vpres, 63-67). Res: Rivers; lakes; climatology. Mailing Add: 4452 N 38th St Arlington VA 22207

LANGDALE, GEORGE WILFRED, b Walterboro, SC, Sept 14, 30; m 55. SOIL FERTILITY, SOIL CHEMISTRY. Educ: Clemson Univ, BS, 57, MS, 61; Univ Ga, PhD(soil sci), 69. Prof Exp: Soil scientist, SC Exp Sta, 57-63, res soil scientist, Southeast Tidewater Exp Sta, Ga, 66-68, res soil scientist, Rio Grande Soil & Water Res Ctr, Tex, 68-71, RES SOIL SCIENTIST, SOUTHERN PIEDMONT CONSERV RES CTR, AGR RES SERV, USDA, 71- Honors & Awards: Cert of Merit, Agr Res Serv, USDA, 73. Mem: Am Soc Agron; Soil Sci Soc Am; Soil Conserv Soc Am. Res: Soil-water-plant interactions, with respect to optimizing food and fiber production; environmental pollution, herbicides, salanity and drainage and plant nutrition. Mailing Add: Southern Piedmont Conserv Res Ctr USDA PO Box 555 Watkinsville GA 30677

LANGDELL, ROBERT DANA, b Pomona, Calif, Mar 14, 24; m 48; c 2. PATHOLOGY. Educ: George Washington Univ, MD, 48. Prof Exp: Intern, Henry Ford Hosp, Detroit, Mich, 48-49; fel path, Sch Med, Univ NC, Chapel Hill, 49-51, from instr to assoc prof, 51-64, PROF PATH, SCH MED, UNIV NC, CHAPEL HILL, 64- Concurrent Pos: USPHS sr res fel, 56-62 & career develop award, 62-67; mem hemat study sect, USPHS, 67-70; ed-in-chief, Transfusion, 72; pres, Am Asn Blood Banks, 73-74; mem panel rev blood & blood derivatives, Food & Drug Admin, 75. Mem: Am Asn Path & Bact; AMA; Am Soc Clin Path; Am Soc Exp Path; Col Am Path. Res: Hematologic pathology; physiology of blood coagulation; hemorrhagic disorders. Mailing Add: Dept of Path Univ of NC Sch of Med Chapel Hill NC 27514

LANGDON, ALLAN BRUCE, b Edmonton, Alta, Dec 14, 41; m 66; c 2. PLASMA PHYSICS. Educ: Univ Man, BSc, 63; Princeton Univ, PhD(astrophys), 69. Prof Exp: Actg asst prof elec eng, Univ Calif, Berkeley, 67-69; STAFF PHYSICIST, LAWRENCE LIVERMORE LAB, 70- Concurrent Pos: Lectr elec eng, Univ Calif, Berkeley, 69-73. Mem: Am Phys Soc; Can Asn Physicists; Asn Comput Mach. Res: Plasma theory; computational physics; computer simulation of plasmas. Mailing Add: L-545 Lawrence Livermore Lab Box 808 Livermore CA 94550

LANGDON, EDWARD ALLEN, b Los Angeles, Calif, Feb 9, 22. MEDICINE, RADIOLOGY. Educ: Western Reserve Univ, BS, 42; Univ Mich, MD, 45. Prof Exp: From asst prof to assoc prof, 59-70, asst dean, 65-70, PROF RADIOL, CHIEF RADIOTHER DIV & ASST DEAN STUDENT AFFAIRS, SCH MED, UNIV CALIF, LOS ANGELES, 70- Mem: AMA; Am Col Radiol; Radiol Soc NAm; Soc Nuclear Med; Asn Univ Radiol. Res: Radiation therapy. Mailing Add: Dept of Radiol Radio Ther Div Univ of Calif Med Ctr Los Angeles CA 90024

LANGDON, GEORGE L J, b Plymouth, Pa, Jan 3, 21; m 59; c 2. GEOGRAPHY OF LATIN AMERICA, PHYSICAL GEOGRAPHY. Educ: Pa State Univ, BS, 42, MS, 47; Clark Univ, PhD(geog), 51. Prof Exp: Prof geog, Mansfield State Col, 47-56; PROF GEOG, WEST CHESTER STATE COL, 56- Concurrent Pos: Lectr geog, Exten, Pa State Univ, Univ Del & State Univ NY Buffalo. Mem: Asn Am Geogr; Nat Coun Geog Educ. Res: Production of educational filmstrips. Mailing Add: Dept of Geog West Chester State Col West Chester PA 19380

LANGDON, HERBERT LINCOLN, b Malone, NY, July 7, 35; m 72. GROSS ANATOMY, DEVELOPMENTAL ANATOMY. Educ: St Lawrence Univ, BS, 57; Univ Mo, MA, 63; Univ Miami, PhD(biol struct), 72. Prof Exp: Asst prof biol, Miami-Dade Community Col, 65-68; ASST PROF ANAT, SCH DENT MED, UNIV PITTSBURGH, 72- Concurrent Pos: Coordr lab sci, Cleft Palate Ctr, Univ Pittsburgh, 76. Mem: AAAS; Sigma Xi. Res: Normal and abnormal morphology and development of human tongue and velopharyngeal mechanism. Mailing Add: Univ of Pittsburgh 620 Salk Hall Pittsburgh PA 15261

LANGDON, KENNETH R, b Cache, Okla, Aug 20, 28; m 61; c 3. PLANT TAXONOMY, NEMATOLOGY. Educ: Okla State Univ, BS, 58, MS, 60; Univ Fla, PhD(plant path, nematol, bot), 63. Prof Exp: NEMATOLOGIST & BOTANIST, DIV PLANT INDUST, FLA DEPT AGR & CONSUMER SERV, 63- Mem: Soc Nematol; Int Asn Plant Taxon; Soc Europ Nematologists. Res: Nematode and plant systematics; nematode control. Mailing Add: Fla Dept Agr & Consumer Serv Box 1269 Div Plant Indust Gainesville FL 32602

LANGDON, ROBERT GODWIN, b Dallas, Tex, Jan 18, 23; m 45; c 4. BIOCHEMISTRY. Educ: Univ Chicago, MD, 45, PhD(biochem), 53. Prof Exp: From instr to prof physiol chem, Sch Med, Johns Hopkins Univ, 53-67; prof biochem & chmn dept, Col Med, Univ Fla, 67-69; PROF BIOCHEM, UNIV VA, 69- Concurrent Pos: USPHS fel, Univ Chicago, 51-53; Lederle award, 54-57. Mem: Am Soc Biol Chemists; Am Chem Soc. Res: Lipid metabolism; enzymology; mechanism of hormone action. Mailing Add: Dept of Biochem Univ of Va Charlottesville VA 22901

LANGE, ALLEN SALVATORE, aeronautical engineering, see 12th edition

LANGE, BENJAMIN OTTO, celestial mechanics, see 12th edition

LANGE, CHARLES FORD, JR, b Chicago, Ill, Feb 16, 29; m 53; c 3. BIOCHEMISTRY, IMMUNOLOGY. Educ: Roosevelt Univ, BS, 51, MS, 53; Univ Ill, PhD(biochem), 59. Prof Exp: Res assoc biochem, Univ Ill, 60-61; res assoc, Hektoen Inst Med Res, Cook County Hosp, 61-63, head phys chem, 63-69; ASSOC PROF MICROBIOL, STRITCH SCH MED, LOYOLA UNIV, 70- Mem: Am Chem Soc; Brit Biochem Soc; Transplantation Soc; Am Soc Microbiol; Am Asn Immunol. Res: Urinary glycoproteins; immunochemistry of streptococcal related glomerulonephritis; streptococcal M-proteins; transplantation antigens. Mailing Add: Dept of Microbiol Stritch Sch of Med Loyola Univ Maywood IL 60153

LANGE, CHARLES GENE, b Chattanooga, Tenn, Mar 30, 42; m 63; c 3. APPLIED MATHEMATICS. Educ: Tri-State Col, BS, 63; Case Inst Technol, MS, 65; Mass Inst Technol, PhD(appl math), 68. Prof Exp: Asst prof, 68-75, ASSOC PROF MATH, UNIV CALIF, LOS ANGELES, 75- Mem: Soc Indust & Appl Math; Am Math Soc. Res: Nonlinear random processes; singular perturbation techniques; nonlinear stability theory; clasticity; wave propagation. Mailing Add: Dept of Math Univ of Calif Los Angeles CA 90024

LANGE, CHARLES HENRY, b Janesville, Wis, July 3, 17; m 41; c 3. CULTURAL ANTHROPOLOGY, APPLIED ANTHROPOLOGY. Educ: Univ NMex, BA, 40, MA, 42, PhD(anthrop), 51. Prof Exp: From instr to asst prof anthrop, Univ Tex, Austin, 47-55; from asst prof to prof, Southern Ill Univ, Carbondale, 55-71, cur, 55-58, chmn dept anthrop, 66-71; chmn dept, 71-74, PROF ANTHROP, NORTHERN ILL UNIV, 71- Concurrent Pos: NATO fel, Univ Tübingen 60-61. Mem: Fel AAAS; fel Am Anthrop Asn; Soc Am Archaeol; Am Ethnol Soc. Res: Culture history of Southwestern pueblos; European, especially German, peasantry and culture history; educational anthropology; acculturation. Mailing Add: Dept of Anthrop Northern Ill Univ De Kalb IL 60115

LANGE, CHRISTOPHER STEPHEN, b Chicago, Ill, Feb 11, 40; m 64; c 1. RADIOBIOLOGY, BIOPHYSICS. Educ: Mass Inst Technol, BS, 61; Oxford Univ, DPhil(radiation biol), 68. Prof Exp: Res asst radiation biol, Churchill Hosp, Headington, Eng, 61-62; from res officer to sr res officer, Paterson Labs, Christie Hosp & Holt Radium Inst, Univ Manchester, 62-69; ASST PROF RADIOL, RADIATION BIOL & BIOPHYS, SCH MED & DENT, UNIV ROCHESTER, 69- Concurrent Pos: NIH cancer res career develop award, 72-; vis prof chem, Univ Calif, San Diego, 75-76. Mem: Brit Asn Radiation Res; Radiation Res Soc; Biophys Soc. Res: Induction and repair of radiation injury at the molecular, cellular and organismal levels; DNA structure in the chromosomes of eucaryotes; the cellular basis of aging; metabolic requirements for cell cycle progression; mechanisms of differentiation control in the organism. Mailing Add: Dept of Radiation Biol & Biophys Univ of Rochester Med Ctr Rochester NY 14642

LANGE, GAIL LAURA, b Chicago, Ill, June 28, 46. ALGEBRA. Educ: Univ Md, BS, 67; Univ NH, MS, 69, PhD(math), 75. Prof Exp: Instr, 72-75, ASST PROF MATH, UNIV MAINE-FARMINGTON, 75- Mem: Am Math Soc; Math Asn Am. Res: Investigation of which finite p-groups can be the Frattini subgroup of finite p-groups. Mailing Add: Dept of Math Univ of Maine Farmington ME 04938

LANGE, GORDON DAVID, b Douglas, Ariz, Jan 15, 36; m 58; c 3. NEUROPHYSIOLOGY, THEORETICAL BIOLOGY. Educ: Calif Inst Technol, BS, 58; Rockefeller Univ, PhD(biophys), 65. Prof Exp: Res assoc biophys, Rockefeller Univ, 65-66; asst res neuroscientist, 66-68, asst prof, 68-74, ASSOC PROF NEUROSCI, UNIV CALIF, SAN DIEGO, 74- Mem: Soc Neurosci; NY Acad Sci. Res: Neurophysiology of vision; studies of the dynamics of interactions among nerve cells; mathematical and computer models of interactions of organisms. Mailing Add: Scripps Inst Oceanog A-012 Univ of Calif San Diego La Jolla CA 92093

LANGE, GORDON LLOYD, b Edmonton, Alta, Mar 1, 37; m 64; c 2. ORGANIC CHEMISTRY. Educ: Univ Alta, BSc, 59; Univ Calif, Berkeley, PhD(org chem), 63. Prof Exp: Res chemist, Procter & Gamble Co, 62-65; lectr org chem, Univ Western Ont, 65-67; asst prof chem, 67-73, ASSOC PROF CHEM, UNIV GUELPH, 73- Mem: Am Chem Soc; Chem Inst Can. Res: Organic photochemistry; synthesis of natural products. Mailing Add: Dept of Chem Univ of Guelph Guelph ON Can

LANGE, IAN M, b New York, NY, Nov 11, 40. GEOLOGY, GEOCHEMISTRY. Educ: Dartmouth Col, BA, 62, MA, 64; Univ Wash, PhD(geol), 68. Prof Exp: Asst prof geol, Fresno State Col, 68-73; ASSOC PROF GEOL, UNIV MONT, 73- Mem: Geochem Soc; Geol Soc Am. Res: Isotope geology, economic geology. Mailing Add: Dept of Geol Univ of Mont Missoula MT 59801

LANGE, JAMES NEIL, JR, b Bridgeport, Conn, May 4, 38; m 58; c 1. PHYSICS. Educ: Pa State Univ, PhD(physics), 64. Prof Exp: From asst prof to assoc prof, 65-71, PROF PHYSICS, OKLA STATE UNIV, 71- Concurrent Pos: Sr vis fel, Univ Nottingham, 76. Mem: Am Phys Soc; Acoust Soc Am. Res: Low temperature solid state physics; physical acoustic and superconductivity; acoustics. Mailing Add: Dept of Physics Okla State Univ Stillwater OK 74074

LANGE, KLAUS ROBERT, b Berlin, Ger, Jan 15, 30; US citizen; m 51; c 2. PHYSICAL CHEMISTRY, SURFACE CHEMISTRY. Educ: Univ Pa, AB, 52; Univ Del, MS, 54, PhD(phys chem), 56. Prof Exp: Res chemist, Atlantic Refining Co, 55-59; sr res chemist, Philadelphia Quartz Co, 59-67, res assoc, 67-69; lab mgr, Betz Lab Inc, 69-74; LAB MGR, QUAKER CHEM CORP, 75- Concurrent Pos: Mem bd dir, Chem Data Systs. Mem: AAAS; Am Chem Soc. Res: Physical adsorption; heterogeneous catalysis; surface chemistry of silica and related solids; silicate solutions, fundamental properties; colloidal suspensions; detergency; pollution control; polymer applications; lignin and paper chemistry. Mailing Add: Quaker Chem Co Conshohocken PA 19428

LANGE, KURT, b Berlin, Ger, Oct 31, 06; nat US; m 36; c 2. INTERNAL MEDICINE. Educ: Univ Berlin, MD, 30. Prof Exp: Instr med, Univ Berlin, 31-33; from instr to assoc prof, 40-64, PROF MED & PEDIAT, NEW YORK MED COL, 64-, DIR MED & PEDIAT RENAL SERV, 55- Concurrent Pos: Vis physician, Metrop & Bird S Coler Hosps, 54-; assoc attend physician, Flower & Fifth Ave Hosps, 54-64, attend physician, 64-; consult physician, Chenango Mem Hosp, Norwich, NY, 66- & Horton Mem Hosp, Middletown; consult artificial kidney-uremia prog, NIH; consult, Nat Med Libr; vchmn, Int Comt Nomenclature & Nosology of Renal Dis. Honors & Awards: Bronze Medal, AMA, 46; Hektoen Gold Medal, 66. Mem: Fel AMA; fel Am Col Physicians; fel Am Col Cardiol; sr mem Am Fedn Clin Res; fel NY Acad Sci. Res: Renal and vascular diseases; immunology. Mailing Add: Dept of Med & Pediat New York Med Col 1 E 105th St New York NY 10029

LANGE, LEO JEROME, b New Rockford, NDak, Aug 29, 28; m 55; c 4. MATHEMATICS. Educ: Regis Col, Colo, BS, 52; Univ Colo, MA, 56, PhD(math), 60. Prof Exp: Instr math, Univ Colo, 52-56, asst, 58-60; mathematician, Boulder Labs, Nat Bur Standards, 56-60; asst prof, 60-65, ASSOC PROF MATH, UNIV MO-COLUMBIA, 65- Mem: Math Asn Am; Am Math Soc. Res: Continued fractions. Mailing Add: Dept of Math Univ of Mo Columbia MO 65201

LANGE, LESTER HENRY, b Concordia, Mo, Jan 2, 24; m 47, 62; c 5. MATHEMATICS. Educ: Valparaiso Univ, AB, 48; Stanford Univ, MS, 50; Univ Notre Dame, PhD, 60. Prof Exp: Instr math, Valparaiso Univ, 50-53, asst prof, 54-56; instr, Univ Notre Dame, 56-57 & 59-60; from asst prof to prof, 60-70, actg head dept, 61-62, chmn dept, 62-70, DEAN SCH SCI, SAN JOSE STATE UNIV, 70- Honors & Awards: L R Ford Sr Award, Math Asn Am, 72. Mem: Am Math Soc; Math Asn

Am; London Math Soc. Res: Complex variable; topology. Mailing Add: Sch of Sci San Jose State Univ San Jose CA 95192

LANGE, RAMON LINUS, internal medicine, cardiology, see 12th edition

LANGE, RAYMOND JOSEPH, organic chemistry, see 12th edition

LANGE, ROBERT CARL, b Stoneham, Mass, Aug 26, 35; m 59; c 2. NUCLEAR MEDICINE, RADIOLOGICAL PHYSICS. Educ: Northeastern Univ, BS, 57; Mass Inst Technol, PhD(chem), 62. Prof Exp: Group leader physics, Monsanto Res Corp, 62-69; ASST PROF RADIOL PHYSICS, SCH MED, YALE UNIV, 69-, TECH DIR NUCLEAR MED, YALENEW HAVEN HOSP, 69- Concurrent Pos: Consult, Stamford Hosp, 71- & R J Schulz Assocs, 73- Mem: AAAS; Am Chem Soc; Am Phys Soc; Soc Nuclear Med. Res: New radioisotopes for nuclear medicine; radiopharmaceutals; computer applications to medicine. Mailing Add: Dept Radiol Sect Nuclear Med Yale Univ Sch of Med New Haven CT 06520

LANGE, ROBERT DALE, b Redwood Falls, Minn, Jan 24, 20; m 44; c 2. HEMATOLOGY. Educ: Macalester Col, AB, 41; Washington Univ, MD, 44. Prof Exp: Dir, St Louis Regional Blood Ctr, 48-51; instr med, Washington Univ, 51-53; instr clin med, Univ Minn, 53-54; asst prof med, Washington Univ, 56-61; chief physician, Vet Admin Hosp, 61-62; assoc prof med, Med Col, Univ Ga, 62-65; RES PROF, MEM RES CTR, UNIV TENN, KNOXVILLE, 65- Concurrent Pos: Consult, Milledgeville State Hosp & Vet Admin Hosp, Augusta, Ga; hematologist, Atomic Bomb Casualty Comn, 51-53. Mem: Am Soc Hemat; Am Fedn Clin Res; fel Am Col Physicians; fel Int Soc Hemat. Res: Internal medicine. Mailing Add: Univ of Tenn Mem Res Ctr 1924 Alcoa Hwy Knoxville TN 37920

LANGE, ROBERT ECHLIN, JR, b Janesville, Wis; m 70. WILDLIFE DISEASES. Educ: Colo State Univ, BS, 68, 70, MS, 73, DVM, 74. Prof Exp: WILDLIFE VET WILDLIFE DIS, N MEX DEPT OF GAME & FISH, 74- Concurrent Pos: Wildlife dis consult, Colo Wild Animal Dis Ctr, 72-74; adj prof biol dept, NMex Highlands Univ, 75- Mem: Wild Animal Dis Asn; Am Vet Med Asn; Wildlife Soc. Res: Investigation into the game management implications of wildlife diseases concentrating at this time on elk and Rocky Mountain bighorn sheep. Mailing Add: NMex Dept of Game & Fish State Capitol Santa Fe NM 87501

LANGE, ROBERT WALTER, b Brooklyn, NY, Dec 19, 16; m 41; c 1. FOREST MANAGEMENT. Educ: Colo State Univ, BS, 39, MF, 55. Prof Exp: Mem staff, Int Paper Co, 39-41, dist forester, 46-51; forester, Colo State Forestry Dept, 53-54; asst prof forestry, Pa State Univ, 55-62; asst prof, 62-69, ASSOC PROF FORESTRY, UNIV MONT, 69- Mem: Soc Am Foresters. Res: Dendrology and forest mensuration. Mailing Add: Sch of Forestry Univ of Mont Missoula MT 59801

LANGE, RONALD FREDERICK, organic chemistry, see 12th edition

LANGE, WILLIAM HARRY, JR, b San Francisco, Calif, Sept 2, 12; m 37, 69; c 3. ENTOMOLOGY. Educ: Univ Calif, BS, 33, MS, 34, PhD(entom), 41. Prof Exp: Sci aide forest entom, Bur Entom & Plant Quarantine, USDA, 34-35; asst, Univ, 35-36, assoc, 36-43, jr entomologist, Exp Sta, 43-47, asst entomologist, 47-55, assoc prof entom & assoc entomologist, 55-57, ENTOMOLOGIST, EXP STA & PROF ENTOM, UNIV CALIF, DAVIS, 57- Concurrent Pos: Field assoc, Nat Res Coun, 47-48. Mem: Entom Soc Am; Am Entom Soc; NY Acad Sci; Am Inst Biol Sci; Am Soc Naturalists. Res: Biosystematics of Lepidoptera and Aphidoidea; integrated control of vegetable and field crop insects; mites and mollusks. Mailing Add: Dept of Entom Univ of Calif Davis CA 95616

LANGE, WILLIAM JAMES, b Sandusky, Ohio, Jan 20, 30; m 51; c 4. SURFACE PHYSICS. Educ: Oberlin Col, AB, 51; Mass Inst Technol, PhD(physics), 56. Prof Exp: Asst phys electronics, Mass Inst Technol, 51-56; physicist, 56-64, MGR VACUUM PHYSICS, RES LABS, WESTINGHOUSE ELEC CORP, 64- Mem: Am Phys Soc; Am Soc Mass Spectrometry; Am Vacuum Soc. Res: Ultrahigh vacuum; interaction of gases with surfaces. Mailing Add: Westinghouse Res Lab Beulah Rd Pittsburgh PA 15235

LANGE, WILLY, b Berlin, Ger, Oct 31, 00; nat US; m 25. MICROBIAL ECOLOGY. Educ: Univ Berlin, PhD(chem), 23. Prof Exp: Asst to Wilhelm Traube, Berlin, 23-25, Chem Inst, 25-30, pvt-docent, 30-37; head res dept, Henkel & Co, Ger, 35-39; res assoc chem, Univ Cincinnati, 39, assoc prof appl sci, 40-42; res chemist, Procter & Gamble Co, 40-45, head oil res sect, 45-55, assoc dir res div, 56-65; res prof, 65-71, EMER RES PROF BASIC SCI, UNIV CINCINNATI, 71- Mem: AAAS; Am Chem Soc; Sigma Xi. Res: Microbiology; eutrophication of lakes; blue green algae. Mailing Add: Tanner's Coun Lab Univ of Cincinnati Cincinnati OH 45221

LANGE, WINTHROP EVERETT, b Appleton, Wis, Sept 22, 25; m 48; c 2. PHARMACY. Educ: Univ Wis, BS, 52, MS, 53, PhD(pharm, chem), 55. Prof Exp: Asst pharm, Univ Wis, 52-54; asst prof, SDak State Col, 55-58; from asst prof to assoc prof, Mass Col Pharm, 58-66, prof & chmn dept, 66-68; dir labs, 68-74, VPRES, INT DIR TECH SERV, PURDUE FREDERICK CO, 74- Concurrent Pos: Adj prof, Col Pharmaceut Sci, Columbia Univ, 70-76. Mem: Am Chem Soc; Am Pharmaceut Asn; Soc Cosmetic Chem; Acad Pharmaceut Sci; Int Fedn Socs Cosmetic Chem (pres, 76-77). Res: Synthesis of metal chelates as pro-drugs; pharmaceutical analysis. Mailing Add: Purdue Frederick Co 50 Washington St Norwalk CT 06856

LANGE, YVONNE, b Durban, SAfrica, Apr 5, 41; m 66. BIOPHYSICS. Educ: Univ London, BSc, 62; Oxford Univ, DPhil(theoret physics), 65. Prof Exp: Res assoc physics, Brandeis Univ, 65-66; instr, Simmons Col, 66-67; NIH res fel, 67-70, instr, 70-72, LECTR BIOPHYS, HARVARD MED SCH, 72- Mem: Biophys Soc. Res: Membrane biophysics. Mailing Add: Biophys Lab Harvard Med Sch Boston MA 02115

LANGEBARTEL, RAY GARTNER, b Quincy, Ill, Apr 27, 21; m 45; c 4. MATHEMATICS, ASTRONOMY. Educ: Univ Ill, AB, 42, AM, 43, PhD(math), 48. Prof Exp: Asst math, 46-48, from instr to assoc prof, 48-70, PROF MATH, UNIV ILL, URBANA, 70- Concurrent Pos: Vis res assoc, Stockholm Observ, 50-51. Res: Function theory; stellar dynamics. Mailing Add: Dept of Math Univ of Ill Urbana IL 61803

LANGEBERG, JOHN CARL, physical chemistry, see 12th edition

LANGEL, ROBERT ALLAN, b Pittsburgh, Pa, May 25, 37; m 59; c 3. GEOPHYSICS. Educ: Wheaton Col, AB, 59; Univ Md, College Park, MS, 71, PhD(physics), 73. Prof Exp: Physicist, Electronics Optics Br, US Naval Res Lab, 59-62; physicist, Commun Br, 62-63, Fields & Plasmas Br, 63-74, GEOPHYSICIST GEOMAGNETISM, GEOPHYS BR, GODDARD SPACE FLIGHT CTR, NASA, 74- Mem: Am Geophys Union. Res: Utilization of surface and near-earth satellite magnetic field measurements to study lithospheric magnetic anomalies, upper mantle conductivity and core-mantle processes; derivation of geomagnetic field models. Mailing Add: Code 922 Goddard Space Flight Ctr Greenbelt MD 20771

LANGELAND, KAARE, b Saltdal, Norway, Nov 3, 16. DENTAL MATERIALS, EXPERIMENTAL PATHOLOGY. Educ: Vet Col Norway, grad, 38; Norweg State Dent Sch, DDS, 42; Univ Oslo, PhD, 57. Prof Exp: Pvt pract, 42; asst to Dr Ferdinand Ström, Dent Diag Dept, Gaustad Hosp, 42-52; res assoc, Norweg Inst Dent Res, 52-63; assoc prof oral histol & chmn dept, univ & proj dir, Minn Mining & Mfg, State Univ NY Buffalo, 63, prof oral biol, 64-69; prof gen dent, 69-71, PROF ENDODONT & CHMN DEPT, SCH DENT MED, UNIV CONN, 71- Concurrent Pos: USPHS grants, Res Found, State Univ NY Buffalo, 63; 3M grant, 63-69; USPHS grant, 63-68 & 69-70; Serco grant, 66-67; Univ Conn Res Found grants, 70-72; Off Naval Res Contract, 71-77; teacher, Norweg State Dent Sch, 48-49; ed, Scand Dent J, 57-63; vis lectr, Boston Univ; vchmn comn dent res, Int Dent Fedn, chmn working group biol testing dent mat; mem coun standardization dent mats & devices, Am Mat Standardization Inst; Norweg state rep, Inter-Nordic Comt Planning Nordic Bur Standards Dept Mat, 61-62; mem working group dent terminology, Int Orgn Standardization, 65. Honors & Awards: Badge of Honor in Silver & Prize, Norweg Dent Asn, 59. Mem: Norweg Dent Asn; hon mem Dent Asn SAfrica; hon mem Dent Asn South Rhodesia; hon mem SAfrican Prosthodont Soc; cor mem Finnish Dent Soc. Res: Experimental pathology regarding biomaterials; evaluation of the biologic properties of methods; devices, and materials used in dentistry before they are released for general use. Mailing Add: Sch of Dent Med Univ of Conn Health Ctr Farmington CT 06032

LANGELAND, WILLIAM ENBERG, b Massillon, Ohio, July 5, 23; m 74. ORGANIC CHEMISTRY. Educ: Hobart Col, BA, 44; Rochester Univ, PhD(chem), 50. Prof Exp: Res assoc, NY State Agr Exp Sta, 46; asst, Rochester Univ, 47-49; asst prof chem, St Lawrence Univ, 50-52; Cottrell res grant, 52; res chemist, Pa Salt Mfg Co, 52-54, develop chemist, 54-56, sr develop rep, 56-58; proj coordr, 58-62, DIR PROJ COORD DIV, WYETH LABS, INC, 63- Mem: AAAS; Am Chem Soc; Am Inst Chemists; NY Acad Sci. Res: Heterocyclic nitrogen compounds; pseudoaromaticity in polycyclic materials; vapor phase reactions of amines; correlation of biological activity with chemical structure; research administration. Mailing Add: Wyeth Labs Inc King of Prussia Rd Radnor PA 19807

LANGENAUER, HAVIVA DOLGIN, b New York, NY, Apr 3, 33; m 54; c 4. BOTANY. Educ: Brooklyn Col, BA, 53; Jewish Theol Sem Am, BRE, 58; Univ Mass, Amherst, MA, 67, PhD(bot), 72. Prof Exp: Blakeslee res fel, Smith Col, 71-73; vis scientist plant genetics, Weizmann Inst & bot, Tel Aviv Univ, Israel, 73-74; CABOT RES FEL, HARVARD UNIV, 74- Mem: AAAS; Bot Soc Am. Res: Plant meristems; morphogenesis; developmental anatomy. Mailing Add: Cabot Found Harvard Univ Petersham MA 01366

LANGENBERG, DONALD NEWTON, b Devils Lake, NDak, Mar 17, 32; m 53; c 4. SOLID STATE PHYSICS. Educ: Iowa State Univ, BS, 53; Univ Calif, Los Angeles, MS, 55, Berkeley, PhD(physics), 59. Hon Degrees: MA, Univ Pa, 71. Prof Exp: Actg instr physics, Univ Calif, Berkeley, 58-59; NSF fel, 59-60; from asst prof to assoc prof, 60-67, PROF PHYSICS, UNIV PA, 67-, VPROVOST GRAD STUDIES & RES, 74- Concurrent Pos: Sloan Found fel, 62-64; Guggenheim Found fel, 66-67; assoc lectr, Advan Normal Sch, Univ Paris, 66-67; univ adv, Tex Instruments, Inc, 68-70; distinguished vis scientist, Mich State Univ, 69; mem Nat Acad Sci-Nat Acad Eng-Nat Res Coun Panel Adv to Cryogenics Div, Nat Bur Standards, 69-70, chmn, 70-75; mem comn I, Int Union Radio Sci, 69-; vis prof, Calif Inst Technol, 71; dir lab res struct matter, Univ Pa, 72-74; guest researcher, Cent Inst Low Temperature Study, Bayer Acad Sci & Tech Univ Munich, 74; mem, NSF adv comt res, 74- Honors & Awards: John Price Wetherill Medal, Franklin Inst, 75. Mem: AAAS; fel Am Phys Soc. Res: Cyclotron resonance and Fermi surface studies in metals and semiconductors; tunneling and Josephson effects in superconductors; precision measurement and fundamental physical constants; low temperature physics; nonequilibrium phenomena in superconductors. Mailing Add: Dept of Physics Univ of Pa Philadelphia PA 19104

LANGENBERG, WILLEM G, b Djombang, Indonesia, Apr 16, 28; US citizen; m 55; c 3. PLANT PATHOLOGY, PLANT VIROLOGY. Educ: Calif State Col Long Beach, BS, 63; Univ Calif, Berkeley, PhD(plant path), 67. Prof Exp: RES PLANT PATHOLOGIST, AGR RES SERV, USDA, 67-; ASSOC PROF PLANT PATH, UNIV NEBR-LINCOLN, 67- Mem: Am Phytopath Soc. Res: Study of plant-virus-vector relationships with labeled antibodies or viruses; light and electron microscopy radioautography. Mailing Add: 304 P I Bldg East Campus Univ of Nebr Lincoln NE 68503

LANGENDORF, RICHARD, b Prague, Czech, July 11, 08; US citizen; m 44; c 3. CARDIOLOGY. Educ: Ger Univ Prague, MD, 32. Prof Exp: RES ASSOC CARDIOL, CARDIOVASC INST, MICHAEL REESE HOSP, 39- Concurrent Pos: Mem sci coun, Am Heart Asn, 48; Malcolm Rogers lectr, Univ Wis, 59; mem, Nat Conf Cardiovasc Dis, Washington, DC, 64; prof lectr, Univ Chicago, 66-73, clin prof med, 73- Mem: Soc Exp Biol & Med; Am Col Physicians; NY Acad Med. Res: Electrocardiography, especially interpretation of complex arrhythmias. Mailing Add: 111 N Wabash Ave Chicago IL 60602

LANGENHEIM, JEAN HARMON, b Homer, La, Sept 5, 25. PLANT ECOLOGY, PALEOBOTANY. Educ: Univ Tulsa, BS, 46; Univ Minn, MS, 49, PhD(bot, geol), 53. Prof Exp: Lectr, Coe Col, 51-52; investr, Nat Geol Serv, Colombia, 53; res assoc, Univ Calif, Berkeley, 54-59; teaching assoc, Univ Ill, 59-62; Asn Univ Women fel, Harvard Univ, 62-63; res assoc, Bot Mus, 63-66; from asst prof to assoc prof biol, 66-73, PROF BIOL, UNIV CALIF, SANTA CRUZ, 73- Concurrent Pos: Mem teaching staff & bd trustees, Rocky Mountain Biol Lab, 54-65; lectr, Mills Col, 55-56; from instr to asst prof, San Francisco Col Women, 56-59; scholar, Radcliffe Inst Independent Study, 63-64. Mem: AAAS; Bot Soc Am; Ecol Soc Am; Soc Study Evolution; Am Soc Plant Physiol. Res: Paleoecological studies of amber; evolutionary and physio-ecological studies of tropical resin-producing plants; concepts of ecology. Mailing Add: Div of Natural Sci Univ of Calif Santa Cruz CA 95060

LANGENHEIM, RALPH LOUIS, JR, b Cincinnati, Ohio, May 26, 22; m 46, 62, 70; c 2. PALEONTOLOGY, STRATIGRAPHY. Educ: Univ Tulsa, BS, 43; Univ Colo, MS, 47; Univ Minn, PhD(geol), 51. Prof Exp: Asst prof geol, Coe Col, 50-52; asst prof paleont, Univ Calif, 52-59; from asst prof to assoc prof, 59-67, PROF GEOL, UNIV ILL, URBANA, 67- Concurrent Pos: Foreign expert, Nat Inst Geol, Columbia Univ, 53; adv, Geol Surv Can, 57; foreign assoc, Geol Surv Iran, 73. Mem: AAAS; Geol Soc Am; Paleont Soc (secy, 62-70); Am Asn Petrol Geol; Soc Econ Paleont & Mineral. Res: Invertebrate paleontology and stratigraphy; Paleozoic of western and central North America; Tertiary of southern Mexico; Permian and Carboniferous of Iran. Mailing Add: Dept of Geol Univ of Ill Urbana IL 61801

LANGENHOP, CARL ERIC, b Richmond, Va, Dec 20, 22; m 46. MATHEMATICS. Educ: Univ LuLouisville, BA, 43; Iowa State Col, MS, 45, PhD(math), 48. Prof Exp: Instr math, Iowa State Col, 46-48; res asst, Princeton Univ, 48-49; asst prof, Iowa

LÅNGENHOP

State Univ, 49-51 & 52-53, from assoc prof to prof, 53-60; dir res, Mathematica Div, Mkt Res Corp Am, 60-61; prof math, Southern Ill Univ, 61-67 & Univ Ky, 67-70; PROF MATH, SOUTHERN ILL UNIV, 70- Concurrent Pos: Res assoc, Princeton Univ, 51-52. Mem: Am Math Soc; Math Asn Am; Soc Indust & Appl Math. Res: Matrix theory; control theory. Mailing Add: Dept of Math Southern Ill Univ Carbondale IL 62901

LANGER, ALOIS, molecular physics, chemistry, see 12th edition

LANGER, ARTHUR M, b New York, NY, Feb 18, 36; m 62; c 2. MINERALOGY, ENVIRONMENTAL SCIENCES. Educ: Hunter Col, BA, 56; Columbia Univ, MA, 62, PhD(mineral), 65. Prof Exp: Res assoc mineral, Columbia, 62-65; res assoc environ sci, Mt Sinai Hosp, 65-66, asst prof mineral, 66-68, ASSOC PROF MINERAL, MT SINAI SCH MED, 68-, ASSOC PROF COMMUNITY MED, 74- Concurrent Pos: Adj assoc prof, Queens Col, NY, 68-70. Mem: AAAS; Geol Soc Am; Electron Probe Anal Soc Am; Geochem Soc; Mineral Soc Am. Res: Metamorphic and igneous petrology; clay mineralogy; secondary mineralization; instrumentation; microparticulate identification, analysis and interaction in the human environment. Mailing Add: Environ Sci Lab Mt Sinai Sch of Med New York NY 10029

LANGER, ARTHUR WALTER, JR, b Manchester, NH, May 17, 23; m 45; c 2. ORGANIC CHEMISTRY. Educ: Univ NH, BS, 44, MS, 47; Ohio State Univ, PhD(chem), 51. Prof Exp: Instr, Univ NH, 47; res assoc, Ohio State Univ, 50-51; res chemist, Standard Oil Develop Co, 51-56, asst sect head, Esso Labs, 56-57, sect head, Esso Res & Eng Co, 57-59, res assoc, 59-64, sr res assoc, 64-67, assoc sci adv, 67-69, SCI ADV, EXXON RES & ENG CO, 69- Mem: Am Chem Soc. Res: Petrochemicals; polymerization; petroleum processing; catalysis. Mailing Add: Oakwood Rd Watchung NJ 07060

LANGER, BERNHARDT WILHELM, JR, biochemistry, nutrition, see 12th edition

LANGER, DIETRICH WILHELM JOSEF, b Berlin, Ger, Aug 13, 30; US citizen; m 61; c 4. SOLID STATE PHYSICS. Educ: Tech Univ Berlin, MS, 57, PhD(physics), 60. Prof Exp: From res physicist, to sr res physicist, 57-65, GROUP LEADER, ELECTRONIC PROPERTIES, SEMICONDUCTORS GROUP, AEROSPACE RES LABS, WRIGHT-PATTERSON AIR FORCE BASE, 65- Concurrent Pos: Eve lectr, Univ Dayton, 59-60; grant, Ecole Normale Superieure, Paris, 64-65; adv, Max Planck Inst Solid State Study, 72-73; fel Indust Col Armed Forces, 75-76. Mem: Fel Am Phys Soc. Res: Optical, electronic and electrooptical properties of semiconductors; high pressure, optical and electron spectroscopy; materials research; research and ddevelopment administration. Mailing Add: 1310 Rice Rd Yellow Springs OH 45387

LANGER, GLENN A, b Nyack, NY, May 5, 28; m 54; c 1. MEDICINE. Educ: Colgate Univ, BA, 50; Columbia Univ, MD, 54. Prof Exp: Intern, Mass Gen Hosp, 54-55; asst res, Columbia-Presby Med Ctr, 57-58; sr resident, Mass Gen Hosp, 59-60; clin instr, Los Angeles County Cardiovasc Res Lab & Med Ctr, Univ Calif, Los Angeles, 60-62; asst prof med, Columbia Univ, 63-66; assoc prof, 66-69, PROF MED & PHYSIOL, MED CTR, UNIV CALIF, LOS ANGELES, 69-, ASSOC DIR CARDIOVASC RES LAB, 66-, VCHMN DEPT PHYSIOL, 67- Concurrent Pos: Vchmn exec comt, Basic Sci Coun, Am Heart Asn, 74-76. Mem: Am Physiol Soc; Am Soc Clin Invest; Am Asn Physicians. Res: Myocardial physiology and metabolism. Mailing Add: Cardiovasc Res Lab Univ of Calif Med Ctr Los Angeles CA 90024

LANGER, HORST GÜNTER, b Breslau, Ger, Dec 29, 27; US citizen; m 55; c 2. INORGANIC CHEMISTRY. Educ: Brunswick Tech Univ, Dipl, 54, Dr rer nat(chem), 56. Prof Exp: Asst inorg anal chem, Brunswick Tech Univ, 51-56; res assoc inorg chem, Univ Ind, 56-58; res chemist, 58-64, sr res chemist, 64-68, ASSOC SCIENTIST, DOW CHEM USA, 68- Mem: Am Chem Soc; Am Soc Mass Spectrometry; Int Confedn Thermal Anal; NAm Thermal Anal Soc; Ger Chem Soc. Res: Analytical, dental and organometallic chemistry; mass spectrometry; thermal analysis. Mailing Add: 28 Joyce Rd Wayland MA 01778

LANGER, JAMES STEPHEN, b Pittsburgh, Pa, Sept 21, 34; m 58; c 3. STATISTICAL MECHANICS. Educ: Carnegie Inst Technol, BS, 55; Univ Birmingham, PhD, 58. Prof Exp: Instr, 58-64, assoc prof, 64-67, assoc dean, Mellon Inst Sci 71-74, PROF PHYSICS, CARNEGIE-MELLON UNIV, 67- Concurrent Pos: Vis assoc prof, Cornell Univ, 66-67; Guggenheim fel, Harvard Univ, 74-75. Res: Theoretical solid state physics; kinetics of phase transformations. Mailing Add: Dept of Physics Carnegie-Mellon Univ Pittsburgh PA 15213

LANGER, LAWRENCE MARVIN, b New York, NY, Dec 22, 13; m 36. NUCLEAR PHYSICS. Educ: NY Univ, BS, 34, MS, 35, PhD(physics), 38. Prof Exp: Asst physics, NY Univ, 34-38; from instr to assoc prof, 38-52, actg chmn dept, 61-62 & 65-66, chmn, 66-73, PROF PHYSICS, IND UNIV, BLOOMINGTON, 52- Concurrent Pos: Res assoc, Radiation Lab, Mass Inst Technol, 41-42; alternate group leader, Los Alamos Atomic Bomb Proj, 43-45; sci consult, US War Dept, 45; expert consult, AEC, 48-74; US Energy Res & Develop Admin, 74-; consult & observer, Greenhouse Atomic Bomb Tests, Marshall Islands, 51; adv consult, Nat Res Coun, 57-60; dir, Off Naval Res & NSF res nuclear spectros, 63-; adv consult, Nuclear Data Proj, Nat Acad Sci-Nat Res Coun; Mem: AAAS; fel Am Phys Soc. Res: Nuclear physics; artificial and natural radioactivity; beta ray spectra; neutron scattering; D-D reaction; beta and gamma coincidence measurements; Cockroft-Walton accelerator; Geiger Counter; cyclotron; counting equipment; microwave radar; underwater sound; ballistics; nuclear spectroscopy and nuclear weapons; double beta decay; mass of the neutrino; shapes of the allowed and forbidden beta spectra. Mailing Add: Dept of Physics Ind Univ Bloomington IN 47401

LANGER, SIDNEY, b New York, NY, Dec 15, 25; m 51; c 1. PHYSICAL CHEMISTRY. Educ: NY Univ, AB, 49; Ill Inst Technol, PhD(chem), 55. Prof Exp: Chemist, Oak Ridge Nat Lab, 54-60; mem res staff, Gen Atomic Div, Gen Dynamics Corp, 60-69, group leader, Nuclear Fuels Group, Res & Develop Div, Gulf Gen Atomic, 69-71, GROUP LEADER FUELS & MAT DEVELOP, GAS-COOLED FAST REACTOR PROJECT, GEN ATOMIC CO, 71-, MEM SR RES STAFF, 69- Mem: Am Chem Soc; Am Nuclear Soc. Res: Nuclear reactor chemistry; physical chemistry and thermodynamics of high temperature systems; phase equilibria; fission product behavior in fuels; fission product release; fuel processing and reprocessing. Mailing Add: Gen Atomic Co PO Box 81608 San Diego CA 92138

LANGER, STANLEY HAROLD, chemistry, see 12th edition

LANGERAK, ESLEY OREN, organic chemistry, see 12th edition

LANGERBECK, MARY THERESE, b Chicago, Ill. ASTRONOMY. Educ: Northwestern Univ, AB, 24; Univ Mich, MA, 30 & 46; Georgetown Univ, PhD(astron), 48. Prof Exp: Instr physics, Clarke Col, 30-36 & Mundelein Col, 36-44, prof physics & astron, 48-71; PROF MATH & PHYSICS, LIVINGSTONE COL, 71- Concurrent Pos: Off Naval Res contract, Mt Wilson & Mt Palomar Observs, 52-54; Res Corp grant, 53. Mem: Assoc Am Phys Soc; Am Astron Soc; assoc Am Asn Physics Teachers. Res: X-ray diffraction; statistical astronomy; interstellar absorption. Mailing Add: 324 W McCubbins St Salisbury NC 28144

LANGERMAN, NEAL RICHARD, b Philadelphia, Pa, Mar 11, 43; m 65; c 2. PHYSICAL BIOCHEMISTRY. Educ: Franklin & Marshall Col, AB, 65; Northwestern Univ, PhD(chem), 69. Prof Exp: NIH fel chem, Yale Univ, 69-70; asst prof biochem, Sch Med, Tufts Univ, 70-75; ASST PROF CHEM, UTAH STATE UNIV, 75- Res: Thermodynamic studies of protein quaternary structure and protein reactions, especially bacterial luciferase and thymidylate synthetase; microcalorimetry and ultracentrifugation. Mailing Add: Dept of Chem Utah State Univ Logan UT 84322

LANGEVIN, RAYMOND J FRANCOIS, b Beauport, Que, Mar 1, 19; m 46; c 5. FORESTRY. Educ: Laval Univ, Forest Engr, 44; Yale Univ, MFor, 46. Prof Exp: With Prov Que Govt, 44-46; prof, Duchesnay Forestry Sch, 46-54; PROF LOGGING & LUMBERING, LAVAL UNIV, 54- Concurrent Pos: Tech adv, Coun Eastern Forest Prod Asns, 52-; land surv, 54-; mem Laval Univ Res Found, 65-; consult, Soc Engr Mont, 71- Res: Development of lumber grading rules; environmental factors in logging. Mailing Add: Fac of Forestry & Geodesy Laval Univ Quebec PQ Can

LANGEVIN, ROBERT ARTHUR, b Toledo, Ohio, Nov 15, 18; m 42; c 2. MATHEMATICS, PHYSICS. Educ: Ohio State Univ, BA, 40. Prof Exp: Asst, Univ Va, 40-41; engr, Am Tel & Tel Co, 45-46; sr engr, Brush Develop Co, 46-50; dir develop, Clevite Transistor Corp, 53-54; opers analyst, Tech Opers, Inc, 54, mgr, Washington Res Off, 54-57, dir proj omega, 57-59, dir comput appl & res, 59-62, corporate fel, 62-66, vpres res syst sci div, 66-68; assoc tech dir, 68-70, vpres fed systs div, 70-71, vpres res & develop div, 71, VPRES & DEP TECH DIR, AUERBACH CORP, 71- Concurrent Pos: Vis lectr, Harvard Univ, 64-65, res fel, 65-66. Mem: Asn Comput Mach. Res: Computer applications; digital simulation; operations research. Mailing Add: Auerbach Corp 1501 Wilson Blvd Arlington VA 22209

LANGFITT, THOMAS WILLIAM, b Clarksburg, WVa, Apr 20, 27; m 53; c 3. NEUROSURGERY. Educ: Princeton Univ, AB, 49; Johns Hopkins Univ, MD, 53. Prof Exp: Intern gen surg, Johns Hopkins Univ Hosp, 53-54; asst resident, Vanderbilt Univ Hosp, 54-55; from asst resident to chief resident neurosurg, Johns Hopkins Univ Hosp, 57-61; assoc, 61-63, from asst prof to assoc prof, 63-68, CHARLES FRAZIER PROF NEUROSURG, MED SCH, UNIV PA, 68-, HEAD DEPT, UNIV HOSP, 68-, VPRES HEALTH AFFAIRS, UNIV, 74- Concurrent Pos: Res fel, Johns Hopkins Univ Hosp, 57-60; contractor, US Army Chem Corp, 57-; head sect neurosurg, Pa Hosp, 61-68. Mem: AMA; Cong Neurol Surg; Am Col Surgeons; Am Asn Neurol Surg; Asn Res Nerv & Ment Dis. Res: Pathophysiology and metabolism in acute brain injuries. Mailing Add: Div of Neurosurg Univ of Pa Med Sch Philadelphia PA 19104

LANGFORD, ARTHUR NICOL, b Ingersoll, Ont, July 30, 10; m 38; c 3. BOTANY. Educ: Queen's Univ, Ont, BA, 31; Univ Toronto, MA, 33, PhD(plant path, genetics), 36. Prof Exp: Asst bot, Univ Toronto, 33-36, bact, 37; lectr, 38-45, from asst prof to assoc prof, 46-51, head dept biol, 51-71, chmn dept, 71-75, PROF BIOL, BISHOP'S UNIV, 51- Mem: AAAS; Am Bot Soc; Ecol Soc Am; Am Ornith Union; Can Bot Soc. Res: Plant ecology of forests and peat bogs. Mailing Add: Dept of Biol Sci Bishop's Univ Lennoxville PQ Can

LANGFORD, COOPER HAROLD, III, b Ann Arbor, Mich, Oct 14, 34; m 59. PHYSICAL INORGANIC CHEMISTRY. Educ: Harvard Univ, AB, 56; Northwestern Univ, PhD(chem), 59. Prof Exp: NSF fel, Univ Col, London, 59-60; instr chem, Amherst Col, 60-61, from asst prof to assoc prof, 61-67; assoc prof, 67-70, PROF CHEM, CARLETON UNIV, 70- Concurrent Pos: Vis asst prof, Columbia Univ, 64; Alfred P Sloan Found fel, 68-70; consult, Inland Waters Directorate, Can, 73; mem, Chem Grants Comt, Nat Res Coun Can, 75-78. Mem: AAAS; Am Chem Soc; The Chem Soc; fel Chem Inst Can. Res: Inorganic reaction mechanisms; inorganic photochemistry; kinetics in analysis; solution physical chemistry. Mailing Add: Dept of Chem Carleton Univ Ottawa ON Can

LANGFORD, EDGAR VERDEN, b Parry Sound, Ont, Can, Apr 27, 21; m 44; c 4. VETERINARY BACTERIOLOGY. Educ: Mem, Royal Col Vet Surgeons, Univ Toronto, DVM, VS, 49, DVPH, 50. Prof Exp: Vet surgeon diag & res, Can Dept Agr, 50-56; animal pathologist, BC Dept of Agr, 56-67; RES SCI DIAG, CAN DEPT AGR, 67- Mem: Am Soc Microbiol; Can Soc Microbiologists; Can Vet Med Asn; Int Soc Mycoplasmologists; Brit Vet Asn. Res: Mycoplasma relationship to disease in the ruminant; incidence of infection in the systems of the ruminant; serological response of the host; development of diagnostic serological methods; pathogenicity of Mycoplasma. Mailing Add: Can Dept of Agr Box 640 Lethbridge AB Can

LANGFORD, ERIC SIDDON, b New York, NY, May 23, 38; m 59. MATHEMATICS. Educ: Mass Inst Technol, SB, 59; Rutgers Univ, MS, 60, PhD(math), 63. Prof Exp: Res specialist, Autonetics Div, NAm Rockwell, 63-64; asst prof math, Naval Postgrad Sch, 64-69; ASSOC PROF MATH, UNIV MAINE, ORONO, 69- Concurrent Pos: Vis assoc, Daniel H Wagner, Assocs, 66; collab ed, Am Math Monthly, 69-71, assoc ed, 71-75; vis assoc prof math, Calif Inst Technol, 72-73. Honors & Awards: L R Ford Award, Math Asn Am, 71. Mem: Math Asn Am; Am Math Soc. Res: Integration theory via linear functionals; linear spaces of operators; Riesz spaces. Mailing Add: Dept of Math Univ of Maine Orono ME 04473

LANGFORD, ERNEST ROBERT, b Hamilton, Ont, June 13, 34. PREVENTIVE MEDICINE, PUBLIC HEALTH. Educ: Queen's Univ, Ont, MD, CM, 58; Univ Toronto, DPH, 60. Prof Exp: Med officer health & dir, Northwestern Health Unit, Ont, 60-62; ASSOC PROF PUB HEALTH, SCH HYG, UNIV TORONTO, 62- Mem: Am Pub Health Asn; Can Med Asn; Can Pub Health Asn. Res: Health education. Mailing Add: Sch of Hyg Univ of Toronto Toronto ON Can

LANGFORD, FLORENCE, b Celina, Tex, Sept 20, 12. NUTRITION. Educ: Tex Woman's Univ, BS, 32, MA, 38; Iowa State Univ, PhD(nutrit), 60. Prof Exp: High sch teacher, Tex, 32-34; teacher home econ, Kilgore Jr Col, 35-37; instr nutrit & home mgt, Univ Tenn, 38-41; from asst prof to assoc prof, 42-61, actg dean, 70-72, PROF FOOD & NUTRIT, TEX WOMAN'S UNIV, 62- Mem: Am Chem Soc; Am Dietetic Asn; Am Home Econ Asn; Am Pub Health Asn; Int Fedn Home Econ. Res: Food and nutrition; human nutrition; energy metabolism. Mailing Add: Box 23835 Tex Woman's Univ Denton TX 76204

LANGFORD, FRED F, b Toronto, Ont, Dec 19, 29; m 53; c 3. ECONOMIC GEOLOGY. Educ: Univ Toronto, BA, 53; Queen's Univ, Ont, MA, 55; Princeton Univ, PhD(geol), 60. Prof Exp: Geologist, Imp Oil Co, 53-54; geologist, Ont Dept Mines, 54-57; assoc prof geol, Univ Kans, 58-62; assoc prof, 62-71, ASSOC PROF GEOL SCI, UNIV SASK, 71- Mem: Nat Asn Geol Teachers; Geol Asn Can. Res: Igneous and metamorphic petrology; Precambrian geology; exploration geology. Mailing Add: Dept of Geol Sci Univ of Sask Saskatoon SK Can

LANGFORD, GEORGE SHEALY, b Blythewood, SC, Mar 9, 01; m 26; c 2.

ENTOMOLOGY. Educ: Clemson Col, BS, 21; Univ Md, MS, 24; Ohio State Univ, PhD(entom), 29. Prof Exp: Teacher pub schs, NC, 21-22; asst entom, Univ Md, 22-24; dep state entomologist, Colo, 24-27; asst zool, Ohio State Univ, 27-28 & Ohio Exp Sta, 28-29; from assoc prof to prof entom, Univ Md, College Park, 29-71; actg dir, State Bd Agr Progs, 71-72; ENTOM CONSULT, 72- Concurrent Pos: Ed, Entoma, 44-54; pres, Eastern Plant Bd, 55; state entomologist, 56-71. Mem: AAAS; Entom Soc Am; Am Mosquito Control Asn. Res: Economic and applied entomology. Mailing Add: 4606 Hartwick Rd College Park MD 20740

LANGFORD, HERBERT GAINES, b Columbia, SC, Apr 22, 22; m 50; c 4. INTERNAL MEDICINE, PHYSIOLOGY. Educ: Univ SC, BS, 42; Med Col Va, MD, 45. Prof Exp: Jr res asst, Med Col Va, 48-49, res asst neurophysiol, 49-50, assoc med & res assoc neurol sci, 52-55; asst res, Osler Med Serv, Johns Hopkins Hosp, 50-52; from asst prof to assoc prof med, 55-64, from asst prof to assoc prof physiol, 61-71, chief endocrine & hypertension div, 58-64, actg chmn dept med, 63-65, PROF MED, MED SCH, UNIV MISS, 64-, PROF PHYSIOL, 71- Concurrent Pos: Chmn steering comt, Hypertension Detection & Follow-up Prog, Nat Heart & Lung Inst. Mem: Am Fedn Clin Res; Am Soc Clin Invest; Am Physiol Soc; Soc Exp Biol & Med. Res: Endocrine control of blood pressure; pre-eclampsia; epidemiology and hypertension. Mailing Add: Dept of Med Univ of Miss Sch of Med Jackson MS 39216

LANGFORD, PAUL BROOKS, b Lockesburg, Ark, Aug 11, 30; m 59; c 1. PHYSICAL ORGANIC CHEMISTRY. Educ: Okla State Univ, BS, 52, MS, 54; Ga Inst Technol, PhD(chem), 62. Prof Exp: Instr chem, Ga Inst Technol, 56-62; from asst prof to assoc prof, 62-70, PROF CHEM, DAVID LIPSCOMB COL, 70- Mem: Am Chem Soc. Res: Rates and mechanisms of reactions of organic halogen compounds; charge transfer complex compounds. Mailing Add: Dept of Chem David Lipscomb Col Nashville TN 37203

LANGFORD, RAYMOND ROBERT, b Warkworth, Ont, Sept 22, 08; m 36; c 2. LIMNOLOGY. Educ: Univ Sask, BSc, 32; Univ Toronto, PhD(plankton), 36. Prof Exp: From lectr to asst prof zool, 39-53, assoc prof, 53-60, PROF ZOOL, UNIV TORONTO, 60-, ASSOC CHMN DEPT, 64- Concurrent Pos: Consult, Ont Dept Lands & Forests, 50-60. Mem: Am Soc Limnol & Oceanog. Res: Lake productivity; general limnology; plankton. Mailing Add: Ramsay Wright Lab Univ of Toronto Dept of Zool Toronto ON Can

LANGFORD, ROBERT BRUCE, b San Francisco, Calif, Mar 7, 19; m 57. ORGANIC CHEMISTRY. Educ: Univ Calif, Los Angeles, BS, 48; Univ Southern Calif, MS, 63, PhD, 72. Prof Exp: Chemist petrol anal, Southern Pac Co, 49-54; res chemist pesticides, Stauffer Chem Co, 54-58; lab mgr org synthesis, Cyclo Chem Corp, 58-61; teacher chem, Los Angeles City Schs, 61-64; from instr to assoc prof, 64-73, chmn dept, 68-74, PROF CHEM, EAST LOS ANGELES COL, 73- Mem: AAAS; Am Chem Soc. Res: Organic synthesis; sulfur compounds; photochemistry. Mailing Add: 4037 Farmouth Dr Los Angeles CA 90027

LANGFORD, RUSSEL HAL, b North Platte, Nebr, Nov 14, 25; m 46; c 4. HYDROLOGY, ENVIRONMENTAL CHEMISTRY. Educ: Univ Nebr, BSc, 49. Prof Exp: Hydrologist, US Geol Surv, 49-66, asst chief, Off Water Data Coord, 66-68, CHIEF, OFF WATER DATA COORD, US GEOL SURV, 68- Mem: Am Chem Soc; Am Water Works Asn; Am Geophys Union; Water Pollution Control Fedn; AAAS. Res: Water chemistry; geochemistry. Mailing Add: Off of Water Data Coord US Geol Surv Reston VA 22092

LANGFORD, WALTER ROBERT, b Opp, Ala, Dec 20, 18; m 47; c 1. AGRONOMY. Educ: Ala Polytech Univ, BS, 41; Univ Mo, PhD(field crops), 50. Prof Exp: Asst agronomist, Univ Mo, 49-51; asst, Univ Fla, 51-53; assoc, Ala Polytech Univ, 53-58; AGRONOMIST & COORDR REGIONAL PROJ NEW PLANTS, STATE AGR EXP STA, AGR RES SERV, USDA, 58- Concurrent Pos: Mem, Nat Plant Germplasm Comt, 73. Mem: Am Soc Agron. Res: Adaptation and management of forage and pasture crops. Mailing Add: Agr Res Serv USDA Experiment GA 30212

LANGFORD, WILLIAM SIDDON, b Lakewood, NJ, Aug 15, 06; m 28; c 2. CHILD PSYCHIATRY. Educ: Harvard Univ, AB, 29; Columbia Univ, MD, 31; Am Bd Psychiat & Neurol, dipl. Prof Exp: Clin prof, 39-53, prof, 53-72, EMER PROF PSYCHIAT, COL PHYSICIANS & SURGEONS, COLUMBIA UNIV, 72- Concurrent Pos: Dir pediat psychiat clin, Babies Hosp, 35-72; attend pediatrician, Presby Hosp, 54-72, consult pediat servs, 72-; attend psychiatrist, NY State Psychiat Inst, 55-; dir child psychiat, Columbia Presby Med Ctr, 55-72. Mem: Am Pediat Soc; fel Am Psychiat Asn; hon assoc fel Am Acad Pediat; fel Am Orthopsychiat Asn (treas, 52-57, pres, 60-61); Am Acad Child Psychiat (pres, 57-59). Res: Clinical child psychiatry. Mailing Add: 120 Cabrini Blvd New York NY 10033

LANGHAAR, HENRY LOUIS, b Bristol, Conn, Oct 14, 09; m 37; c 3. APPLIED MECHANICS. Educ: Lehigh Univ, BS, 31, MS, 33, PhD(math), 40. Prof Exp: Test engr, Ingersoll Rand Co, NJ, 33-36; seismographer, Carter Oil Co, 36-37; instr math, Purdue Univ, 40-41; struct engr, Consol-Vultee Aircraft Corp, 41-47; assoc prof theoret & appl mech, 47-49, PROF THEORET & APPL MECH, UNIV ILL, URBANA, 49- Mem: AAAS; fel Am Soc Mech Eng. Res: Stress analysis; elasticity and buckling theories; theory of plates and shells; dimensional analysis and model theory. Mailing Add: Dept of Theoret & Appl Mech Univ of Ill Urbana IL 61801

LANGHAM, DERALD G, b Polk, Iowa, May 27, 13; m 39; c 4. GENETICS. Educ: Iowa State Univ, BS, 36; Cornell Univ, PhD(genetics), 39; US Int Univ, PhD(leadership & human behav), 70. Prof Exp: Secy, Maize Genetics Coop, Cornell Univ, 36-39; head, Dept Genetics, Agr Exp Sta, Caracas, Venezuela, 39-49; tech adv agr, Am Int Asn, 50-51; PRES, GENETICA VENEZOLANA, S A, 51-, SESAMUM FOUND, 54-, GENESA, 64-; PROF, US INT UNIV, 74- Concurrent Pos: From asst prof to prof genetics, Univ Caracas, Venezuela, 39-46; lectr econ bot, Yale Univ, 61-62; consult genetics, Marine Biol Labs, US Dept Interior, 63-64; res assoc plant breeding, Univ Calif, Riverside, 65-66; lectr, US Int Univ, 71-74. Honors & Awards: Spec award for develop human resources & econ crops, Venezuelan Govt, 61, Order Merit, First Class, 72. Mem: AAAS; Genetics Soc Am; NY Acad Sci. Res: Genesa, a thirteen dimensional process involving walk-into crystals of simple geometrical forms similar to those that living energies spiral through in natural growth patterns. Mailing Add: Genesa Fallbrook CA 92028

LANGHAM, ROBERT FRED, b Grand Ledge, Mich, Jan 31, 12; m 37; c 4. PATHOLOGY. Educ: Calvin Col, AB, 35; Mich State Univ, MS, 37, DVM, 42, PhD, 50. Prof Exp: From instr to assoc prof, 38-51, PROF VET PATH, MICH STATE UNIV, 51-, ACTG CHMN DEPT PATH, COL VET MED, 74- Mem: Am Vet Med Asn; Am Col Vet Path; Conf Res Workers Animal Dis; Int Acad Path. Res: Leptospirosis; neoplasms and joint disease in animals. Mailing Add: Col of Vet Med Mich State Univ East Lansing MI 48823

LANGHANS, ROBERT W, b Flushing, NY, Dec 29, 29; m 52; c 3. FLORICULTURE. Educ: Rutgers Univ, BS, 52; Cornell Univ, MS, 54, PhD(floricult), 56. Prof Exp: From asst prof to assoc prof, 56-68, PROF FLORICULT, CORNELL UNIV, 68- Honors & Awards: Blauvelt Award, 55; Kenneth Post Award, Am Soc Hort Sci, 65. Mem: Am Soc Hort Sci; Am Soc Plant Physiol; Int Soc Hort Sci. Res: Effects of photoperiod and temperature on growth and flowering. Mailing Add: Dept of Hort Cornell Univ Col of Agr Ithaca NY 14850

LANGHOFF, PETER WOLFGANG, b New York, NY, Jan 19, 37; m 62; c 3. THEORETICAL CHEMISTRY. Educ: Univ Hofstra, BS, 58; State Univ NY Buffalo, PhD(physics), 65. Prof Exp: Physicist, Cornell Aeronaut Labs, Inc, Cornell Univ, 62-65; fel, Harvard Univ, 67-69; asst prof, 69-72, ASSOC PROF CHEM, IND UNIV, BLOOMINGTON, 72- Mem: Am Phys Soc; Am Chem Soc; AAAS; Optical Soc Am. Res: Atomic and molecular physics; interaction of radiation and matter; atomic and molecular structure. Mailing Add: Dept of Chem Ind Univ Bloomington IN 47401

LANGHUS, WILLARD L, dairy industry, see 12th edition

LANGILLE, ALAN RALPH, b Amherst, NS, Apr 2, 38; m 67; c 1. CROP PHYSIOLOGY. Educ: McGill Univ, BS, 60; Univ Vt, MS, 62; Pa State Univ, PhD(agron), 67. Prof Exp: Asst prof agron, 67-73, ASSOC PROF AGRON & BOT, UNIV MAINE, ORONO, 73- Concurrent Pos: Maine State Hwy Comn grant, Univ Maine, Orono, 71-73. Mem: Am Soc Agron; Am Soc Hort Sci; Am Inst Biol Sci; Potato Asn Am. Res: Hormonal control of tuber initiation and subsequent growth in the potato; growth regulator physiology; salt tolerance in conifers. Mailing Add: Dept of Plant & Soil Sci Univ of Maine Orono ME 04473

LANGKAMMERER, CARL MARTIN, b Schwer, Ill, June 5, 05; m 29; c 2. ORGANIC CHEMISTRY. Educ: Univ Minn, BS, 29, PhD(org chem), 34. Prof Exp: Asst, Univ Minn, 30-34; prof chem, Concordia Col, 34-36; res chemist, E I du Pont de Nemours & Co, 36-71; RETIRED. Res: Action of sulfur dioxide and its salts on organic compounds; piria reaction; synthetic resin coatings; organic sulfur compounds; organic compounds of titanium and zirconium; fluorine compounds. Mailing Add: 2514 Deepwood Dr Foulkwoods Wilmington DE 19810

LANGLAND, OLAF ELMER, b Madrid, Iowa, May 30, 25; m 56; c 2. RADIOLOGY, DENTISTRY. Educ: Univ Iowa, DDS, 52, MS, 61. Prof Exp: From instr to assoc prof oral diag & radiol, Col Dent, Univ Iowa, 59-69, head dept, 64-69; prof oral diag-med-radiol & head dept, Sch Dent, La State Univ, New Orleans, 69-74; PROF DIAG & ROENTGENOL, SCH DENT, UNIV TEX HEALTH SCI CTR, SAN ANTONIO, 74- Concurrent Pos: USPHS grant, Col Dent, Univ Iowa, 64-66; consult, Wilford Hall US Air Force Hosp, Lackland, Tex, 68-71; mem subcomt proposed dent x-ray mach, Am Nat Standards Inst, 68-74; staff dentist, Charity Hosp, New Orleans, 69- Mem: Fel Am Col Dent; Am Acad Dent Radiol; Am Acad Oral Med; Int Asn Dent Res. Res: Panoramic radiography; educational research in dentistry; clinical research in oral manifestations of systemic disease. Mailing Add: Dept Diag & Roentgenol Sch Dent Univ of Tex Health Sci Ctr San Antonio TX 78284

LANGLANDS, ROBERT P, b New Westminster, BC, Oct 6, 36; m 56; c 4. MATHEMATICS. Educ: Univ BC, BA, 57, MA, 58; Yale Univ, PhD(math), 60. Prof Exp: Instr math, Princeton Univ, 60-61, lectr, 61-62, from asst prof to assoc prof, 62-67; prof math, Yale Univ, 67-72; PROF MATH, INST ADVAN STUDY, 72- Concurrent Pos: Mem, Inst Advan Study, 62-63; Miller fel, Univ Calif, Berkeley, 64-65; Sloan fel, 64-66. Res: Group representations; automorphic forms. Mailing Add: Inst for Advan Study Princeton NJ 08540

LANGLEBEN, MANUEL PHILLIP, b Poland, Apr 9, 24; nat US; m 48; c 3. GLACIOLOGY, MICROMETEOROLOGY. Educ: McGill Univ, BSc, 49, MSc, 50, PhD(physics), 53. Prof Exp: Res assoc atmospheric physics, 53-57, lectr, 57-59, from asst prof to assoc prof, 59-69, PROF PHYSICS, McGILL UNIV, 69-, RES ASSOC, 57- Mem: Can Asn Physicists; Royal Meteorol Soc; Glaciol Soc; Am Geophys Union. Res: Physics of ice; sea ice; ice drift. Mailing Add: Dept of Physics McGill Univ Montreal PQ Can

LANGLER, JAMES EDWARD, b Oakland, Calif, Oct 27, 36; m 63; c 2. FOOD SCIENCE. Educ: Univ Calif, Berkeley, BS, 59; Ore State Univ, MS, 64, PhD(food sci), 66. Prof Exp: Lab technician biochem, Univ Calif, Berkeley, 59-61; res asst food sci, Ore State Univ, 61-66, asst prof, Seafoods Lab, 66-69; DIR FOOD RES, WOODSTOCK RES CTR, MORTON SALT DIV, MORTON NORWICH PROD, INC, 69- Mem: Am Chem Soc; Inst Food Technologists; Am Oil Chemists Soc. Res: Chemistry and biochemistry of marine biological systems as they are related to food; lipid deterioration; flavor chemistry; new product development; flavor and chemistry of new food products. Mailing Add: Morton Norwich Prod Inc Woodstock Res Ctr Woodstock IL 60098

LANGLEY, G R, b Sydney, NS, Oct 6, 31; m; c 3. INTERNAL MEDICINE, HEMATOLOGY. Educ: Allison Univ, BA, 52; Dalhousie Univ, MD, 57; FRCP(C), 61. Prof Exp: Asst resident internal med, Victoria Gen Hosp, Toronto, 57-60; J Arthur Haatz fel hemat, Univ Melbourne, 60-61; Med Res Coun Can fel, Sch Med & Dent, Univ Rochester, 61-62 & Dalhousie Univ, 63; lectr internal med, 63-64, from asst prof to assoc prof med, 64-68, PROF MED, DALHOUSIE UNIV, 68-, HEAD DEPT, 74- Concurrent Pos: John & Mary R Markle scholar med, 63-68; consult, Armed Forces Hosp, Halifax, 64-; head dept med, Camp Hill Hosp, 69-74; head dept med, Victoria Gen Hosp, 74- Mem: Fel Am Col Physicians; Can Soc Clin Invest; Can Med Asn. Res; Hematological oncology. Mailing Add: Dept of Med Victoria Gen Hosp Halifax NS Can

LANGLEY, KENNETH HALL, b Ft Collins, Colo, Sept 1, 35; m 59; c 2. BIOPHYSICS. Educ: Mass Inst Technol, SB, 58; Univ Calif, Berkeley, PhD(physics), 66. Prof Exp: Actg asst prof & res assoc physics, Univ Calif, Berkeley, 66; asst prof, 66-73, ASSOC PROF PHYSICS, UNIV MASS, AMHERST, 73- Mem: Am Phys Soc; AAAS. Res: Experimental solid state physics; dynamic nuclear orientation; light scattering from critical point fluids macromolecules; biological systems. Mailing Add: Dept of Physics & Astron Univ of Mass Amherst MA 01002

LANGLEY, LEROY LESTER, physiology, see 12th edition

LANGLEY, NEAL ROGER, b Sumas, Wash, July 27, 39; m 61; c 3. POLYMER CHEMISTRY. Educ: Univ Wash, BS, 61; Univ Wis, PhD(chem), 68. Prof Exp: Sr res scientist, Pac Northwest Lab, Battelle Mem Inst, 67-68; PROJ CHEMIST, DOW CORNING CORP, 68- Mem: Am Chem Soc. Res: Viscoelastic properties of cross-linked polymers; mechanical behavior of silicone polymers; noise and vibration control. Mailing Add: Res Dept Dow Corning Corp Midland MI 48640

LANGLEY, ROBERT ARCHIE, b Athens, Ga, Oct 21, 37; m 59; c 3. SOLID STATE PHYSICS. Educ: Ga Inst Technol, BS, 59, MS, 60, PhD(physics), 63. Prof Exp: Asst physics, Ga Inst Technol, 59-63; physicist, Air Force Cambridge Res Labs, 63-65 & Oak Ridge Nat Lab, 66-68; STAFF MEM, SANDIA CORP, 68- Mem: Am Phys Soc; Am Nuclear Soc; Am Vacuum Soc; Health Physics Soc. Res: Atomic collisions: ion implantation; hydrogen and helium migration in metals; surface physics; ion backscattering; nuclear microanalysis. Mailing Add: Div 5111 Sandia Corp Albuquerque NM 87115

LANGLEY, ROBERT CHARLES, b NJ, Apr 11, 25; m 54; c 1. ORGANIC CHEMISTRY. Educ: St Peters Col, BS, 49. Prof Exp: Chemist, E I du Pont de Nemours & Co, 50-54; res dir, Hanovia Div, 54-62, SECT HEAD, RES & DEVELOP DIV, ENGELHARD INDUSTS, INC, 62- Mem: Am Chem Soc; Am Ceramic Soc. Res: Organic compounds of metals; thin films; gas purification. Mailing Add: 214 Old Forge Rd Millington NJ 07946

LANGLEY, THEODORE DUKE, neurophysiology, see 12th edition

LANGLOIS, BRUCE EDWARD, b Berlin, NH, Sept 16, 37; m 60; c 2. FOOD MICROBIOLOGY. Educ: Univ NH, BS, 59; Purdue Univ, PhD(dairy microbiol), 62. Prof Exp: Asst prof dairy, Purdue Univ, 62-64; asst prof dairy sci, 64-67, ASSOC PROF DAIRY SCI, UNIV KY, 67- Mem: Am Soc Microbiol; Am Meat Sci Asn; Int Asn Milk, Food & Environ Sanit; Nat Environ Health Asn; Am Dairy Sci Asn. Res: Effects of pesticides on physical and biological properties of foods; microflora of dairy and meat products. Mailing Add: Dept of Animal Sci Univ of Ky Lexington KY 40506

LANGLOIS, GORDON ELLERBY, b Burley, Idaho, Aug 30, 18; m 44; c 3. PHYSICAL CHEMISTRY. Educ: Northwestern Univ, BS, 42; Univ Calif, PhD(phys chem), 52. Prof Exp: Sr res chemist, Calif Res Corp, 43-69, sr res chemist, Chevron Res Co, 69-73, MGR, SYNTHETIC FUELS DIV, CHEVRON RES CO, 73- Mem: Am Chem Soc. Res: Catalytic reactions of hydrocarbons, as polymerization, alkylation and isomerization. Mailing Add: 3996 S Peardale Dr Lafayette CA 94549

LANGLOIS, WILLIAM EDWIN, b Providence, RI, Oct 23, 33; m 54; c 2. HYDRODYNAMICS. Educ: Univ Notre Dame, ScB, 53; Brown Univ, PhD(appl math), 57. Prof Exp: Res engr, Polychem Dept, E I du Pont de Nemours & Co, 56-59; STAFF MEM, SAN JOSE RES LAB, IBM CORP, 59- Concurrent Pos: Lectr, Univ Del, 56-58 & Univ Santa Clara, 61-; vis prof, Univ Notre Dame, 70-71. Mem: Soc Rheol; Soc Natural Philos. Res: Global circulation; theory of viscous and viscoelastic fluid flow; air pollution transport. Mailing Add: San Jose Res Lab IBM Corp San Jose CA 95114

LANGLYKKE, ASGER FUNDER, b Pleasant Prairie, Wis, July 17, 09; m 39; c 4. MICROBIOLOGY. Educ: Univ Wis, BS, 31, MS, 34, PhD(biochem), 36. Hon Degrees: ScD, Trinity Col, 65. Prof Exp: Foreman, Proctor & Gamble Co, Ill, 31-32; asst, Univ Wis, 32-36, Dow fel, 36-37; res chemist, Hiram Walker & Sons, Inc, 37-40; supt butyl alcohol plant, Cent Lafayette, PR, 40-43; div head, Northern Regional Res Lab, USDA, Ill, 43 & 45-47; chief pilot plant div & chief tech officer, Chem Warfare Serv, Md & Ind, 43-45; dir, Microbiol Develop, E R Squibb & Sons, 47-49, dir res & develop labs, 49-64, vpres, 64-68; exec dir, Am Acad Microbiol, Am Soc Microbiol, 68-74, PROJ MGR, FREDERICK CANCER RES CTR, 74- Concurrent Pos: Consult, Res & Develop Bd, US Army, 45-53 & Chem Corp, 53-66; off dir, Defense Res & Eng, 60-63; mem, Comt Agr Sci, USDA, 64-68; Adv Bd Res & Grad Educ, Rutgers Univ & Sci Adv Comt, Rutgers Inst Microbiol, 64-68 & Appl Chem Div Comt, Int Union Pure & Appl Chem, 73-; adj prof, Rutgers Univ, 68-74. Mem: Fel AAAS; Am Chem Soc; Am Soc Microbiol; fel Am Acad Microbiol; fel Am Inst Chem. Res: Microbial production of antitumor agents; antibiotics; steroids; pharmaceuticals. Mailing Add: Frederick Cancer Res Ctr PO Box B Frederick MD 21701

LANGMAN, JAN, b Bodegraven, Neth, Oct 21, 23; m 55; c 2. MEDICINE. Educ: Univ Amsterdam, MD, 49, PhD, 50. Prof Exp: Demonstr anat & embryol, Univ Groningen, 45; demonstr, Univ Amsterdam, 46; demonstr exp histol, Univ Leiden, 47; asst prof anat & embryol, Univ Amsterdam, 48-51; from asst prof to assoc prof, Free Univ Amsterdam, 51-54, head dept, 54; assoc prof anat, McGill Univ, 57-59, prof anat & embryol, 59-64; PROF ANAT & CHMN DEPT, UNIV VA, 64- Concurrent Pos: Vis prof, Univ Recife, 59; mem adv coun, Nat Inst Dent Res, 67-71, spec asst to the dir, 72-74; mem study sect, NIH, 68-72. Mem: Am Asn Anatomists (vpres, 72-74); Int Inst Embryol; Can Asn Anatomists; Teratology Soc (pres, 70-71). Res: Gross anatomy; embryology; causation congenital malformations; tissue culture; immunoembryology. Mailing Add: Dept of Anat Univ of Va Charlottesville VA 22903

LANGMUIR, ALEXANDER DUNCAN, b Santa Monica, Calif, Sept 12, 10; m 40; c 5. EPIDEMIOLOGY. Educ: Harvard Univ, AB, 31; Cornell Univ, MD, 35; Johns Hopkins Univ, MPH, 40. Hon Degrees: ScD, Emory Univ, 70. Prof Exp: Intern med, Boston City Hosp, 35-37; consult, Pneumonia Control, NY State Health Dept, Albany, 37-40, asst dist health officer, 40-41; dep comnr health, Westchester County, 41-42; epidemiologist, US Army Epidemiol Bd, 42-46; assoc prof, Johns Hopkins Univ, 46-49; dir epidemiol prog, Ctr Dis Control, USPHS, 49-70; VIS PROF EPIDEMIOL, HARVARD MED SCH, 70- Concurrent Pos: Consult, Secy Defense, 47-62. Honors & Awards: Distinguished Serv Award, US Dept Health, Educ & Welfare, 58; Bronfman Award, Am Pub Health Asn, 65. Mem: Am Epidemiol Soc (pres, 60); Am Soc Trop Med & Hyg; fel Am Pub Health Asn; Int Asn Epidemiol; hon mem Royal Soc Med. Res: Epidemiology of poliomyelitis; acute respiratory diseases; pneumonia; influenza; airborne infection; biological warfare defense; training epidemiologists. Mailing Add: 25 Shattuck St Boston MA 02115

LANGMUIR, DAVID BULKELEY, b Los Angeles, Calif, Dec 14, 08; m 42; c 3. PHYSICS. Educ: Yale Univ, BS, 31; Mass Inst Technol, ScD(physics), 35. Prof Exp: Res physicist, Radio Corp Am Mfg Co, NJ, 35-41; liaison officer, Off Sci Res & Develop, Washington, DC & London, 41-45; secy, Guided Missiles Comt, Joint Chiefs of Staff, Washington, DC, 45-46; dir planning div, Res & Develop Bd, 46-48; exec officer, Prog Coun, AEC, 48-50, liaison officer, Chalk River, 50-54; mem tech staff, Ramo-Wooldridge Corp, 54-56, dir res lab, Ramo-Wooldridge Div, Thompson Ramo Wooldridge Corp, 56-60 & TRW Space Technol Labs, 60-65, dir, Phys Res Ctr, TRW Systs, Calif, 65-70, dir res planning, 70-73; RES CONSULT, 73- Concurrent Pos: Prof lectr, George Washington Univ, 48-50; mem sci adv group, Off Aerospace Res, US Air Force, 63-70, chmn, 69-70. Mem: Fel Am Phys Soc; fel Inst Elec & Electronics Engrs; Am Inst Aeronaut & Astronaut. Res: Thermionics; television light valves; high frequency tubes; diffusion in metals; ionic propulsion. Mailing Add: 350 21st St Santa Monica CA 90402

LANGMUIR, DONALD, b Nashua, NH, Apr 5, 34; m 66; c 2. GEOCHEMISTRY. Educ: Harvard Univ, AB, 56, MA, 61, PhD(geol sci), 65. Prof Exp: Geochemist, Water Resources Div, US Geol Surv, 64-66; lectr water resources, Rutgers Univ, 66-67; asst prof geochem, 67-71, ASSOC PROF GEOCHEM, PA STATE UNIV, UNIVERSITY PARK, 71-, STAFF GEOCHEMIST, MINERAL CONSERV SECT, 67- Mem: AAAS; Am Chem Soc; Am Water Works Asn; Geochem Soc; Mineral Soc Am. Res: Chemical variability of natural waters and its relation to thermodynamic properties of minerals and dissolved and particulate species in solution; hydrogeology; water pollution. Mailing Add: Dept of Geosci Pa State Univ University Park PA 16802

LANGMUIR, MARGARET ELIZABETH LANG, b Chicago, Ill, Nov 11, 35; m 62; c 2. PHYSICAL CHEMISTRY, PHOTOCHEMISTRY. Educ: Culver-Stockton Col, BA, 56; Purdue Univ, PhD(chem), 63. Prof Exp: Instr anal & inorg chem, Wellesley Col, 60-63; phys chemist, Pioneering Res Div, US Army Natrick Labs, 63-69; CONSULT, 69- Mem: AAAS; Am Chem Soc. Res: Organic photochemistry; acidity functions; flash photolysis; fast reaction mechanisms; excited state proton transfer; fluorescence, phosphorescence and charge transfer spectra; isomerization. Mailing Add: 9 Bent Brook Rd Sudbury MA 01776

LANGMUIR, ROBERT VOSE, b White Plains, NY, Dec 20, 12; m 39; c 2. PHYSICS. Educ: Harvard Univ, AB, 35; Calif Inst Technol, PhD(physics), 43. Prof Exp: Physicist, Consol Eng Corp, Calif, 39-42 & Gen Elec Co, 42-48; sr res fel, 48-50, asst prof physics, 50-52, assoc prof elec eng, 52-57, PROF ELEC ENG, CALIF INST TECHNOL, 57- Concurrent Pos: Consult, TRW, Inc, 57- Mem: Am Phys Soc; Inst Elec & Electronics Eng. Res: Mass spectroscopy; synchrotrons; secondary emission cathode in magnetrons; starting of synchrotrons. Mailing Add: Dept of Elec Eng Calif Inst of Technol Pasadena CA 91125

LANGNER, PAUL HARRY, JR, b Folsom, Pa, June 16, 10; m 33; c 2. CARDIOLOGY. Educ: Univ Pa, BS, 31, MD, 34. Prof Exp: Asst to vpres med affairs & asst instr med, Sch Med, Univ Pa, 36-39; from asst med dir to med dir, Provident Mutual Life Ins Co, 54-73, chief med res, 44-73. Concurrent Pos: Instr, Sch Med, Univ Pa, 48-65, adj asst prof, 65, vis asst prof, 71; vchmn med sect, Am Life Conv, 60-61, chmn, 61-62; deleg nat assembly, Am Heart Asn, 63-64, fel coun clin cardiol, 63-, regional rep, 64- Honors & Awards: Achievement Award, AMA. Mem: Fel Am Col Physicians; fel Am Col Cardiol; Am Heart Asn; AMA; Asn Life Ins Med Dirs Am (secy-treas bd, 55-59). Res: Diagnosis by original new method of electrocardiography using wide frequency band, expanded time scale and high gain factor, which detects heart disease not revealed by conventional methods. Mailing Add: 1208 Edmonds Ave Drexel Hill PA 19026

LANGNER, RALPH ROLLAND, b Alta, Iowa, June 24, 25; m 48; c 5. ENVIRONMENTAL CHEMISTRY. Educ: Univ Iowa, BA, 50, MS, 52, PhD(biochem), 53; Am Bd Indust Hyg, cert, 75. Prof Exp: Asst, Univ Iowa, 51-53; instr biochem, Univ Tex, 53-56; indust hygienist, 56-60, res specialist, 60-72, mgr indust hyg, 72-75, DIR INDUST HYG LAB, DOW CHEM CO, 75- Mem: AAAS; Am Chem Soc; NY Acad Sci; Am Indust Hyg Asn; Soc Occup & Environ Health. Res: Environmental health and ecology. Mailing Add: Dow Chem Co 607 Bldg Midland MI 48640

LANGNER, RONALD O, b Chicago, Ill, May 10, 40; m 63; c 2. BIOCHEMICAL PHARMACOLOGY. Educ: Blackburn Col, BA, 62; Univ RI, MS, 66, PhD(pharmacol), 69. Prof Exp: ASSOC PROF PHARMACOL, UNIV CONN, 69- Res: Metabolism of collagen and its relationship to experimental atherosclerosis. Mailing Add: Dept of Pharmacol Univ of Conn Storrs CT 06268

LANGNER, THOMAS S, b New York, NY, Jan 1, 24; m 53; c 5. PSYCHIATRY, EPIDEMIOLOGY. Educ: Harvard Col, AB, 48; Columbia Univ, PhD(sociol), 54. Prof Exp: Res assoc pub health, Cornell Univ, 52-53, res assoc psychiat, 53-56, asst prof sociol, 56-63, res assoc anthrop & sociol, 58-63; asst prof psychiat & sociol, Sch Med, NY Univ, 63-66, res assoc & prof psychiat, 66-69; PROF EPIDEMIOL, SCH PUB HEALTH, COLUMBIA UNIV, 69- Concurrent Pos: Health Res Coun New York career scientist award, 63-72; NIMH res scientist award, 72-77; mem epidemiol studies rev comt, NIMH, 72-75. Mem: Fel Am Sociol Asn; fel Am Pub Health Asn. Res: Psychopathology of children; family and its effect on child behavior; value systems; class and ethnic differences in behavior; environmental stress and effect on strain in individual cross-cultural studies. Mailing Add: Div of Epidemiol Columbia Univ Sch of Pub Health New York NY 10032

LANGNESS, LEWIS L, b Spokane, Wash, Oct 29, 29; m 65; c 1. ANTHROPOLOGY. Educ: Univ Idaho, BS, 56; Univ Wash, MA, 59, PhD(anthrop), 64. Prof Exp: Instr anthrop, Univ Wash, 62-63, res instr psychiat, Sch Med, 63-64, instr psychiat & anthrop, 64-65, asst prof, 65-66; asst prof anthrop, Northwestern Univ, 66-68; assoc prof bus & anthrop, Univ Wash, 68-73; ASSOC PROF IN RESIDENCE, ANTHROP, UNIV CALIF, LOS ANGELES, 73- Concurrent Pos: NIMH res grant, 65-66. Mem: Fel Am Anthrop Asn; Am Ethnol Soc; fel Am Acad Polit & Soc Sci; Royal Anthrop Inst Gt Brit & Ireland. Res: Psychological and social anthropology; Oceania; northwest coast Indians. Mailing Add: Dept of Psychiat Univ of Calif Los Angeles CA 90024

LANGONE, JOHN JOSEPH, b Cambridge, Mass, Aug 20, 44; m 67; c 2. BIO-ORGANIC CHEMISTRY. Educ: Boston Col, BS, 66; Boston Univ, PhD(org chem), 72. Prof Exp: Fel org chem, Boston Univ, 71-72; sr res assoc biochem, Brandeis Univ, 72-75; STAFF FEL IMMUNOCHEM, NAT CANCER INST, 75- Mem: Am Chem Soc. Res: Immunopharmacology and immunochemistry of biologically active compounds; cancer immunochemistry. Mailing Add: Dept of Health Educ & Welfare NIH Bldg 34 Rm 2B13 Bethesda MD 20014

LANGRETH, DAVID CHAPMAN, b Greenwich, Conn, May 22, 37; m 66; c 1. PHYSICS. Educ: Yale Univ, BS, 59; Univ Ill, MS, 61, PhD(physics), 64. Prof Exp: Res assoc physics, Univ Chicago, 64-65 & Cornell Univ, 65-67; asst prof, 67-69, ASSOC PROF PHYSICS, RUTGERS UNIV, NEW BRUNSWICK, 69-, ASSOC CHMN DEPT, 70- Mem: AAAS; Am Phys Soc. Res: Theoretical solid state physics, specializing in the many-body problem. Mailing Add: Dept of Physics Rutgers Univ New Brunswick NJ 08903

LANGRIDGE, ROBERT, b Essex, Eng, Oct 26, 33; m 60; c 2. MOLECULAR BIOLOGY. Educ: Univ London, BSc, 54, PhD(crystallog), 57. Prof Exp: Res fel biophys, Yale Univ, 57-59; res assoc biophys, Mass Inst Technol, 59-61 & Children's Hosp, Med Ctr, Boston Univ, 61-66; res assoc, Harvard Univ, 63-64, lectr, 64-66; prof biophys & info sci, Univ Chicago, 66-68; PROF BIOCHEM, PRINCETON UNIV, 68- Mem: AAAS; Biophys Soc; Am Crystallog Asn. Res: X-ray diffraction and physical-chemical studies of the structures of biological macromolecules, particularly nucleic acids, nucleoproteins, viruses and ribosomes; applications of high speed digital computers. Mailing Add: Dept of Biochem Princeton Univ Princeton NJ 08540

LANGRIDGE, WILLIAM HENRY RUSSELL, b New York, NY, Jan 30, 38; m 60; c 1. VIROLOGY, DEVELOPMENTAL BIOLOGY. Educ: Univ Ill, Urbana, BS, 62, MS, 64; Univ Mass, PhD(biochem), 74. Prof Exp: NIH FEL VIROL, BOYCE THOMPSON INST, YONKERS, NY, 74- Mem: AAAS; Soc Invert Path; Am Soc Microbiol. Res: Metabolism of vertebrate and insect poxviruses; the mechanism of infection and the structure of the virus genome and virus protein. Mailing Add: Rural Delivery Tivoli NY 12583

LANGSDORF, ALEXANDER, JR, b St Louis, Mo, May 30, 12; m 41; c 2. NUCLEAR PHYSICS. Educ: Washington Univ, AB, 32; Mass Inst Technol, PhD(physics), 37. Prof Exp: Nat Res Found fel, Univ Calif, 38; instr physics, Washington Univ, 39-43; physicist, 43-45, SR PHYSICIST, ARGONNE NAT LAB, 45- Concurrent Pos: Asst physicist, Mallinckrodt Inst Radiol, 39-43; vis scientist, Atomic Energy Res Estab 59-60. Mem: Fel Am Phys Soc. Res: Neutron cross-section and polarization measurements; cyclotron, electrostatic and dynamitron construction, design and operation; theory of design of experiments; corona studies; diffusion cloud chamber. Mailing Add: R R 1 Box 228 Meachem Rd Schaumberg IL 60172

LANGSDORF, WILLIAM PHILIP, b Cambridge, Ohio, Apr 6, 19; m 41; c 1. PHYSICAL ORGANIC CHEMISTRY. Educ: Ohio State Univ, BSc, 41; Mass Inst Technol, PhD(chem), 49. Prof Exp: Res chemist, 49-62, RES ASSOC, INDUST & BIOCHEM DEPT, E I DU PONT DE NEMOURS & CO, 62- Mem: Am Chem Soc; Sigma Xi. Res: Organic reactions; mechanism of organic reactions; kinetics. Mailing Add: 2407 Rambler Rd Graylyn Crest Wilmington DE 19810

LANGSETH, MARCUS G, JR, b Lebanon, Tenn, Nov 24, 32; m 63. GEOPHYSICS, OCEANOGRAPHY. Educ: Waynesburg Col, BS, 54; Columbia Univ, PhD(geol), 64. Prof Exp: SR RES ASSOC GEOPHYS, LAMONT-DOHERTY GEOL OBSERV, COLUMBIA UNIV, 58- Mem: AAAS; Am Geophys Union; Geol Soc Am. Res: Terrestrial heat flow; lunar heat flow; oceanographic instrumentation; submarine geology. Mailing Add: Lamont-Doherty Geol Observ Palisades NY 10964

LANGSJOEN, ARNE NELS, b Dalton, Minn, Apr 6, 19; m 43; c 3. ORGANIC CHEMISTRY. Educ: Gustavus Adolphus Col, BA, 42; Univ Iowa, MS, 43, PhD, 49. Prof Exp: From asst prof to assoc prof, 48-56, chmn dept, 56-66, PROF CHEM, GUSTAVUS ADOLPHUS COL, 56- Concurrent Pos: NSF res fel, Uppsala & Royal Inst Technol, Sweden, 58-59. Mem: Am Chem Soc. Res: Biological chemistry. Mailing Add: Dept of Chem Gustavus Adolphus Col St Peter MN 56082

LANGSJOEN, PER HARALD, b Fergus Falls, Minn, Aug 9, 21; m 45; c 5. MEDICINE, CARDIOLOGY. Educ: Gustavus Adolphus Col, AB, 43; Univ Minn, Minneapolis, MB, 50, MD, 51. Prof Exp: Intern & residency internal med, Letterman Army Hosp, San Francisco, Calif, 50-54; staff physician, Coco Solo Hosp, Cristobal, CZ, 54-56; resident cardiol, Fitzsimmons Army Hosp, Denver, Colo, 56-58; chief cardiovasc serv, Wm Beaumont Gen Hosp, El Paso, Tex, 58-60; CHIEF CARDIOVASC SERV, SCOTT & WHITE CLIN, 60- Concurrent Pos: Fel coun clin cardiol, Am Heart Asn, 65-; consult, US Army Hosp, Ft Hood, Tex, 61- & Vet Admin Hosp, Temple, 61- Mem: Fel Am Col Physicians; fel Am Col Cardiol. Res: Rheologic aspects of the circulatory system in health and in disease states. Mailing Add: Scott & White Clin Temple TX 76501

LANGSLEY, DONALD GENE, b Topeka, Kans, Oct 5, 25; m 55; c 3. PSYCHIATRY, PSYCHOANALYSIS. Educ: State Univ NY Albany, AB, 49; Univ Rochester, MD, 53. Prof Exp: Intern, USPHS Hosp, San Francisco, 53-54; resident psychiat, Langley Porter Clin, Sch Med, Univ Calif, San Francisco, 54-59, NIMH career teacher award psychiat, 59-61; from asst to assoc prof psychiat, Sch Med, Univ Colo, 61-68; PROF PSYCHIAT & CHMN DEPT, SCH MED, UNIV CALIF, DAVIS, 68- Concurrent Pos: Dir inpatient serv, Colo Psychiat Hosp, Denver, 61-68; dir ment health serv, Sacramento Med Ctr, Univ Calif, Davis, 68-73; mem psychiat training comt, NIMH, 71-75, mem psychiat test comt, Nat Bd Med Exam, 72- & mem comt community ment health, Am Psychoanal Asn, 73-; chief staff, Sacramento Med Ctr, 74-75; chmn dept defense select comt psychiat care eval, NIMH, 75-; dir, Am Bd Psychiat & Neurol, 75- Mem: Am Psychiat Asn; Am Psychoanal Asn; AMA. Res: Family crisis therapy; evaluation of therapy, medical education and psychiatric education. Mailing Add: Sacramento Med Ctr Univ of Calif at Davis Sacramento CA

LANGSTON, CLARENCE WALTER, b Gainesville, Tex, July 4, 24; m 48; c 2. MICROBIOLOGY. Educ: Southern Methodist Univ, BS, 49; North Texas State Univ, MA, 51; Univ Wis, PhD(bact), 55. Prof Exp: Bacteriologist, Kraft Foods Co, 50-51; asst, Univ Wis, 51-54; bacteriologist, Dairy Cattle Res Br, Agr Res Serv, USDA, 54-62; chief virus & rickettsial div, Directorate Biol Opers, Pine Bluff Arsenal, US Army, Ark, 62-66; head bact sect, Midwest Res Inst, 66-71; PRES, LANGSTON LABS, INC, KS & PR, 71- Mem: Am Acad Microb; Am Soc Microbiol; Soc Indust Microbiol. Res: Dairy and food bacteriology; microbiology and chemistry of fermentations; physiology of bacteria; taxonomy and nomenclature of bacteria; research development and production of viruses and rickettsiae. Mailing Add: 4921 W 96th St Overland Park KS 66207

LANGSTON, HIRAM THOMAS, b Rio de Janeiro, Brazil, Jan 12, 12; US citizen; m 41; c 3. THORACIC SURGERY. Educ: Univ Louisville, AB, 30, MD, 34; Univ Mich, MS, 41; Am Bd Surg, dipl, 42; Am Bd Thoracic Surg, dipl, 48. Prof Exp: Intern, Garfield Mem Hosp, Washington, DC, 34-35, resident path, 35-37; from asst resident to resident surg, Univ Mich Hosp, 37-39, resident thoracic surg, 39-40, instr, Univ, 40-41; assoc surg, Med Sch, Northwestern Univ, Ill, 41-42, asst prof, 46-48; assoc prof, Col Med, Wayne State Univ, 48-52; ASSOC PROF SURG, UNIV ILL COL MED, 52- Concurrent Pos: Chief surgeon, Chicago State Tuberc Sanitarium, State Dept Pub Health, 52-71; mem staff, Grant & St Joseph's Hosps, 52- Mem: Am Thoracic Soc; Am Asn Thoracic Surg (secy, 56-61, vpres, 68-69, pres, 69-70); fel Am Col Surg; Soc Thoracic Surgeons; Am Surg Asn. Res: Surgery for diseases of the chest; tuberculosis and cancer of the lung. Mailing Add: 2913 NCommonwealth Ave Chicago IL 60657

LANGSTON, JAMES HORACE, b Garrison, Tex, Oct 8, 17. ORGANIC CHEMISTRY. Educ: Stephen F Austin State Col, BA, 37; Univ NC, MA, 39, PhD(org chem), 41. Prof Exp: Asst chem, Univ NC, 37-40; res chemist, Columbia Chem Div, Pittsburgh Plate Glass Co, 41-46; from assoc prof to prof textile chem & dyeing, Clemson Col, 46-58; chmn div natural sci & math, 68-72, PROF CHEM & HEAD DEPT, SAMFORD UNIV, 58- Concurrent Pos: Fulbright lectr, Cent Univ & Nat Polytech Sch, Ecuador, 59-60, Fulbright lectr & res consult, Nat Univ Honduras, 67-68. Mem: Am Chem Soc; fel Am Inst Chemists. Res: Drugs; polymers; plastics; plasticizers; catalysis; fibers; textile finishing materials; sulfone formation. Mailing Add: Dept of Chem Samford Univ Birmingham AL 35209

LANGSTON, JIMMY B, physiology, see 12th edition

LANGSTON, WANN, JR, b Oklahoma City, Okla, July 10, 21; m 46; c 2. VERTEBRATE PALEONTOLOGY. Educ: Univ Okla, BS, 43, MS, 47; Univ Calif, PhD(paleont), 51. Prof Exp: Instr geol, Tex Tech Col, 46-48; preparator, Mus Paleont, Univ Calif, 49-54, lectr, 51-52; vert paleontologist, Nat Mus Can, 54-62; RES SCIENTIST, TEX MEM MUS, 62-, DIR VERTEBRATE PALEONT LAB, 69-; PROF DEPT GEOL SCI, UNIV TEX, AUSTIN, 75- Concurrent Pos: Res assoc, Cleveland Mus Natural Hist, 74- Mem: Geol Soc Am; Soc Vert Paleont (vpres, 73-74, pres, 74-75); Am Soc Icthyol & Herpet; Am Asn Petrol Geologists. Res: Fossil amphibians and reptiles. Mailing Add: 4001 Rockledge Dr Austin TX 78731

LANGSTROTH, GEORGE FORBES OTTY, b Montreal, Que, July 13, 36; m 60; c 2. PHYSICS. Educ: Univ Alta, BSc, 57; Dalhousie Univ, MSc, 59; Univ London, PhD(physics), 62. Prof Exp: Res assoc physics, 62-63, from asst prof to assoc prof, 63-70, asst dean grad studies, 67-70, PROF PHYSICS & DEAN GRAD STUDIES, DALHOUSIE UNIV, 70- Mem: Can Asn Physicists. Res: Microwave breakdown in gases; ions in afterglows; positron annihilation; optical properties of metals. Mailing Add: Dept of Physics Dalhousie Univ Halifax NS Can

LANGWAY, CHESTER CHARLES, JR, b Worcester, Mass, Aug 15, 29; m 59; c 4. GEOLOGY. Educ: Boston Univ, AB, 55, MA, 56; Univ Mich, PhD(geol, glaciol), 65. Prof Exp: Res geologist, US Army Snow, Ice & Permafrost Res Estab, 56-59; res assoc properties of snow & ice, Res Inst, Univ Mich, 59-61; RES GLACIOLOGIST, US ARMY COLD REGIONS RES & ENG LAB, 61-, CHIEF SNOW & ICE BR, 66- Concurrent Pos: Mem panel glaciol, Comt Polar Res, Nat Acad Sci, 69-75, secy, Int Comt Ice Core Studies; chmn dept geol sci, State Univ NY Buffalo, 74- Mem: AAAS; fel Geol Soc Am; Am Geophys Union; Am Polar Soc; fel Arctic Inst NAm. Res: Basic and applied research related to the properties of snow and ice, including field and laboratory techniques of analyzing shallow and deep ice cores for stratigraphy and age dating; isotopic and ionic constituents, terrestrial and extraterrestrial inclusions. Mailing Add: US Army Cold Regions Res & Eng Lab Hanover NH 03755

LANGWIG, JOHN EDWARD, b Albany, NY, Mar 5, 24; m 46; c 1. FOREST PRODUCTS. Educ: Univ Mich, Ann Arbor, BS, 48; State Univ NY Col Forestry, Syracuse Univ, MS, 68, PhD(wood sci), 71. Prof Exp: Instr wood prod eng, State Univ NY Col Forestry, Syracuse Univ, 69-70; from asst prof to assoc prof, 71-75, PROF WOOD SCI & CHMN DEPT FORESTRY, OKLA STATE UNIV, 75- Mem: Soc Wood Sci & Technol; Forest Prod Res Soc; Soc Am Foresters. Res: Neutron activation analysis of trace elements in wood and effects on physical properties; physical properties of wood-polymer composites. Mailing Add: Dept of Forestry Okla State Univ Stillwater OK 74074

LANGWORTHY, HAROLD FREDERICK, b White Plains, NY, Aug 1, 40; m 65. MATHEMATICS, OPTICS. Educ: Rensselaer Polytech Inst, BS, 62; Univ Minn, PhD(math), 70. Prof Exp: SR PHYSICIST, RES LABS, EASTMAN KODAK CO, 67- Mem: Optical Soc Am. Res: Mathematical optics; rheology. Mailing Add: 157 Courtly Circle Rochester NY 14615

LANGWORTHY, THOMAS ALLAN, b Oak Park, Ill, Aug 7, 43; m 65; c 2. MICROBIAL PHYSIOLOGY. Educ: Grinnell Col, AB, 65; Univ Kans, PhD(microbiol), 71. Prof Exp: Res assoc, 71-72, ASST PROF MICROBIOL, UNIV SD, 72- Mem: Am Soc Microbiol; AAAS. Res: Structure and function of the membranes and cell surfaces of extremely thermoacidophilic microorganisms and mycoplasmas. Mailing Add: Dept of Microbiol Sch of Med Univ of SD Vermillion SD 57069

LANGWORTHY, WILLIAM CLAYTON, b Watertown, NY, Sept 3, 36; m 58; c 2. ORGANIC CHEMISTRY, ENVIRONMENTAL CHEMISTRY. Educ: Tufts Univ, BSChem, 58; Univ Calif, Berkeley, PhD(org chem), 62. Prof Exp: NIH fel chem, Mass Inst Technol, 61-62; asst prof, Alaska Methodist Univ, 62-65; from asst prof to prof, Calif State Col, Fullerton, 65-73; PROF CHEM & HEAD DEPT, CALIF POLYTECH STATE UNIV, SAN LUIS OBISPO, 73- Concurrent Pos: Assoc dean, Sch Letters, Arts & Sci, Calif State Col, Fullerton, 70-73, dir environ studies prog, 70-72. Mem: AAAS; Am Chem Soc; The Chem Soc. Res: Physical organic chemistry; organic reactions in liquid ammonia; environmental chemistry, especially analysis and effects of trace pollutants. Mailing Add: Dept of Chem Calif State Polytech Univ San Luis Obispo CA 93407

LANHAM, BETTY BAILEY, b Statesville, NC, Aug 12, 22. ETHNOLOGY. Educ: Univ Va, BS, 44, MA, 47; Syracuse Univ, PhD(anthrop), 62. Prof Exp: Instr sociol & US Govt, River Falls State Teachers Col, 48-49; instr sociol, Univ Md, 49-50; asst prof, Randolph-Macon Women's Col, 54-55; asst prof sociol & anthrop, Oswego Teacher's Col, 56-58, Hamilton Col, 61-62, anthrop, Ind Univ, 62-65 & Western Mich Univ, 65-67; assoc prof behav sci, AlabnAlbany Med Col, 67-70; ASSOC PROF ANTHROP, INDIANA UNIV, PA, 70- Concurrent Pos: Vis lectr anthrop & social psychol, Univ Guyana, 69-70. Mem: Fel Am Anthrop Asn; fel Soc Appl Anthrop; Asn Asian Studies. Res: Adult-child relations; inhibitions, freedom, and mechanisms of interpersonal control; culturally prescribed forms of association. Mailing Add: Dept of Sociol & Anthrop Indiana Univ of Pa Indiana PA 15701

LANHAM, URLESS NORTON, b Grainfield, Kans, Oct 17, 18; m 45; c 3. Univ Colo, BA, 40; Univ Calif, Berkeley, PhD(entom), 48. Prof Exp: Asst zool, Univ Calif, Los Angeles, 40-41, biol, Scripps Inst, 41-42, entom, 47-48; from instr to asst prof zool, Univ Mich, 48-56, res asst & consult, NSF Proj Insect Ecol, 56-61; vis cur entom, 61-62, assoc cur, 62-73, CUR ENTOM & PROF NATURAL HIST, UNIV COLO MUS, BOULDER, 73- Concurrent Pos: Consult, Biol Sci Curriculum Study, Univ Colo, Boulder, 63-66, lectr, Inst Develop Biol, 66-67, asst prof biophys, 68-71; adv ed, Columbia Univ Press, 64-72; assoc prof, Div Natural Sci, Monteith Col, Wayne State Univ, 59-61; consult, Smithsonian Inst, 67. Res: Faunistics and systematics of Colorado apoidea, especially of the genus Andrena. Mailing Add: Dept of Natural Hist Univ of Colo Mus Boulder CO 80302

LANIER, GERALD NORMAN, b Alamosa, Colo, Dec 9, 37; m 62; c 2. FOREST ENTOMOLOGY. Educ: Univ Calif, Berkeley, BSc, 60, MSc, 65, PhD(entom), 67. Prof Exp: Forester, US Forest Serv, 60-64; res scientist, Can Dept Forestry, 67-70; asst prof, 70-74, ASSOC PROF ENTOM, STATE UNIV NY COL ENVIRON SCI & FORESTRY, 74- Mem: AAAS; Entom Soc Am; Entom Soc Can. Res: Pheromones in dark beetles; biosystematics; karyology; management of forest insects, ecological impact of forest insects. Mailing Add: Dept of Forest Entom Syracuse NY 13210

LANIER, RANDOLPH D, physical chemistry, inorganic chemistry, see 12th edition

LANIER, ROBERT GEORGE, b Chicago, Ill, Oct 27, 40; m 69; c 2. NUCLEAR PHYSICS. Educ: Lewis Col, BS, 62; Fla State Univ, PhD(nuclear chem), 68. Prof Exp: US AEC fel, Chem Div, Oak Ridge Nat Lab, 67-69; fel, 69-71, STAFF MEM, RADIOCHEM DIV, LAWRENCE LIVERMORE LAB, UNIV CALIF, 71- Mem: AAAS; Am Phys Soc. Res: Experimental low energy nuclear structure studies; charged-particle reaction spectroscopy, neutron capture-gamma ray spectroscopy and decay scheme spectroscopy. Mailing Add: Radiochem Div Livermore Lab Univ of Calif Livermore CA 94550

LANIER, WARREN WOOD, JR, analytical chemistry, see 12th edition

LANIER, WAYNE BANKS, genetics, see 12th edition

LANIGAN, M REGINA, b Buffalo, NY, Oct 4, 18. BACTERIOLOGY, CELL PHYSIOLOGY. Educ: Col St Mary, Ohio, BS, 53; St Bonaventure Univ, PhD(biol chem), 59. Prof Exp: Teacher, Parochial Schs, Ohio & NY, 41-55; from instr to assoc prof biol, Rosary Hill Col, 58-70, dir res, 61-70, chmn dept, 62-70; DIR RES, SMM RES LAB, 70- Concurrent Pos: Grants, Res Corp, 60, Smith Kline & French Found, 61, NSF, 62-65, United Health Found, 63, AEC, 64 & Nat Inst Allergy & Infectious Dis, 64-66; res partic, Oak Ridge Nat Lab, 62 & 63, USPHS res fel, 63-64. Mem: AAAS; Am Soc Microbiol; Am Soc Plant Physiologists; Nat Asn Biol Teachers; Nat

Sci Teachers Asn. Res: Various aspects of metabolism of sulfate-reducing bacteria in the genus Desulfovibrio, and on biochemistry of sporulation in anaerobic bacteria of genus Clostridium. Mailing Add: SMM Sci Ctr 472 Emslie St Buffalo NY 14212

LANING, STEPHEN HENRY, b Albany, NY, Oct 18, 18; m 46; c 2. ANALYTICAL CHEMISTRY. Educ: Union Col, NY, BS, 41; Rutgers Univ, PhD(phys chem), 47. Prof Exp: Asst, Rutgers Univ, 41-44, instr, 45-46; res chemist, Chem Div, Pittsburg Plate Glass Co, 47-61, supvr res & tech serv, 61-67, SUPVR RES & TECH SERV, PPG INDUSTS, INC, 67- Honors & Awards: Hon Award, Pittsburgh Plate Glass Co, 62 Mem: Am Chem Soc. Res: Development of methods for x-ray analysis of glass, silica and titania pigments, minerals, cements and other types of materials. Mailing Add: PPG Industs Inc PO Box 31 Barberton OH 44203

LANK, ROBERT BYRON, b Kansas City, Mo, July 1, 19; m 46; c 2. VETERINARY MEDICINE. Educ: Kans State Col, BS, 40, DVM, 42; Iowa State Col, MS, 58. Prof Exp: Assoc prof, 48-54, actg head dept, 67. PROF VET SCI, LA STATE UNIV, 54-, ASSOC DEAN SCH VET MED & HEAD DEPT VET SCI, 68- Mem: Am Vet Med Asn. Res: Reproductive disorders and diseases of cattle. Mailing Add: Dept of Vet Sci La State Univ Baton Rouge LA 70803

LANKA, WAYNE ALLEN, organic chemistry, see 12th edition

LANKFORD, CHARLES ELY, b DeValls Bluff, Ark, Mar 22, 12; m 42; c 2. BACTERIOLOGY. Educ: Univ Tex, BA, 35, PhD(bact), 48; Univ Wis, MA, 43. Prof Exp: Instr bact, Univ Tex, 35-37; bacteriologist, State Dept Health Labs, Tex, 39-40; instr bact, Med Br, 40-45, asst prof med bact, 45-48, assoc prof, 48-49, assoc prof gen bact & pub health bact, 49-55, PROF PUB HEALTH BACT & MICROBIOL, UNIV TEX, AUSTIN, 55-, DIR LAB MED BACT, 52- Mem: AAAS; Soc Exp Biol & Med; Am Soc Microbiol; Am Acad Microbiol; NY Acad Sci. Res: Host-parasite biology; cell division control mechanisms; nutritional studies of bacteria; drug resistance of microorganisms; bacterial variation; biology and physiology of Vibrio cholera and Brucella. Mailing Add: Dept of Microbiol Univ of Tex Austin TX 78712

LANKFORD, PHILIP MARLIN, b St Louis, Mo, Dec 29, 45; m 67; c 2. URBAN GEOGRAPHY, ECONOMIC GEOGRAPHY. Educ: Univ Chicago, AB, 67, PhD(geog), 71. Prof Exp: Asst prof geog, Univ Calif, Los Angeles, 70-75; RES PLANNER, ASN BAY AREA GOVTS, CALIF, 75- Mem: Regional Sci Asn; Asn Am Geogrs; Urban Regional Info Systs Asn. Res: Spatial processes of urban-regional economic development; location of public health facilities; computer cartography. Mailing Add: Hotel Claremont Berkeley CA 94705

LANKFORD, ROBERT RENNINGER, geology, see 12th edition

LANKFORD, WILLIAM FLEET, b Charlottesville, Va, Jan 9, 38. NUCLEAR PHYSICS. Educ: Univ Va, BA, 60; Univ SC, MS, 64, PhD(physics), 69. Prof Exp: Instr physics, Univ NC, Greensboro, 62-63; fel, Col William & Mary, 69; asst prof, 59-72, ASSOC PROF PHYSICS, GEORGE MASON UNIV, 72- Concurrent Pos: NIH instnl res grant, George Mason Col, 70-71. Mem: Am Inst Physics. Res: Intermediate energy nuclear structure research. Mailing Add: Dept of Physics George Mason Univ Fairfax VA 22030

LANKS, KARL WILLIAM, b Philadelphia, Pa, Nov 1, 42. PATHOLOGY, MOLECULAR BIOLOGY. Educ: Pa State Univ, BS, 63; Temple Univ, MD, 67; Columbia Univ, PhD(path), 71. Prof Exp: Intern, Columbia-Presby Hosp, 67-68; instr path, Columbia Univ, 71-72; ASST PROF PATH, STATE UNIV NY DOWNSTATE MED CTR, 74- Concurrent Pos: NIH res fel, Dept Chem, Harvard Univ, 71-72. Mem: Am Chem Soc. Res: Structure and metabolism of messenger ribonucleic acids. Mailing Add: 211 Central Park West New York NY 10024

LANMAN, JONATHAN T, b Columbus, Ohio, Dec 3, 17; m; c 3. PEDIATRICS. Educ: Yale Univ, MD, 43. Prof Exp: Intern, Johns Hopkins Hosp, 43-44; asst resident, Sydenham Hosp, 44; intern, Bellevue Hosp, 46-47; asst resident, Univ Hosp, Univ Calif, San Francisco, 47-48; from instr to assoc prof pediat, Sch Med, NY Univ, 49-60; prof & chmn dept, State Univ NY Downstate Med Ctr, 60-72; PROF PEDIAT, SCH MED, NY UNIV, 72-; ASSOC DIR, POP COUN, 72- Mem: Am Pediat Soc; Soc Pediat Res; Endocrine Soc; Am Acad Pediat; NY Acad Med. Res: Endocrinology. Mailing Add: Pop Coun Rockefeller Univ York Ave & 66th St New York NY 10021

LANMAN, ROBERT CHARLES, b Bemidji, Minn, Oct 2, 30; m 57; c 4. BIOCHEMICAL PHARMACOLOGY. Educ: Univ Minn, BS, 56, PhD(pharmacol), 67. Prof Exp: Teaching asst, Col Pharm, Univ Minn, 58-59; pharmacologist, Sect Biochem Drug Action, Lab Chem Pharmacol, Nat Heart Inst, 62-66; ASSOC PROF PHARMACOL & MED, UNIV MO-KANSAS CITY, 66- Mem: AAAS; Am Asn Cols Pharm; Am Soc Pharmacol & Exp Therapeut. Res: Passage of drugs across body membranes; mechanism of active transport, drug absorption, distribution, metabolism and excretion. Mailing Add: 8202 W 72nd St Overland Park KS 66204

LANN, JOSEPH SIDNEY, b Washington, DC, Sept 16, 17; m 45; c 3. ORGANIC CHEMISTRY. Educ: Univ Md, BS, 37, PhD(org chem), 41. Prof Exp: Res chemist, Jackson Lab, 46-54, dir, Freon Prod Lab, 54-64, asst mgr new prod & mkt develop, Freon Prod Div, 64-65, asst dist mgr, 65-68, MGR DEVELOP PROD, FREON PROD DIV, E I DU PONT DE NEMOURS & CO, INC, 68- Mem: Am Chem Soc. Res: Surface active agents and neoprene; organic compounds; fluorinated hydrocarbons. Mailing Add: 608 Haverhill Rd Wilmington DE 19803

LANNER, RONALD MARTIN, b Brooklyn, NY, Nov 12, 30; m 57; c 2. FOREST GENETICS, TREE PHYSIOLOGY. Educ: Syracuse Univ, BS, 52, MF, 58; Univ Minn, Minneapolis, PhD(forestry), 68. Prof Exp: Res forester, Pac SouthewstSouthwest Forest & Range Exp Sta, US Forest Serv, 58-64; ASSOC PROF FOREST GENETICS & DENDROL, UTAH STATE UNIV, 67- Concurrent Pos: Consult, Forestry Proj, Food & Agr Orgn, UN, Taiwan, 69; res fel, Univ Fla, 73-74. Mem: AAAS; Soc Am Foresters; Am Inst Biol Sci. Res: Morphogenesis and growth of woody p)ants; evolution and natural hybridization of western conifers. Mailing Add: Dept of Forest Sci Utah State Univ Logan UT 84321

LANNERT, KENT PHILIP, b Belleville, Ill, Nov 29, 44; m 70; c 1. INDUSTRIAL ORGANIC CHEMISTRY. Educ: Southern Ill Univ, Carbondale, BA, 66; Vanderbilt Univ, PhD(chem), 69. Prof Exp: Res chemist, 69-72, RES SPECIALIST, MONSANTO CO, 72- Mem: Am Chem Soc. Res: Organic synthesis; chelation; synthesis of chelants and other detergent related chemicals. Mailing Add: Monsanto Co 800 N Lindbergh Blvd St Louis MO 63166

LANNI, YVONNE THERY, b Torino, Italy, Jan 18, 21; nat US; m 49. MICROBIOLOGY. Educ: Univ Brazil, MD, 46. Prof Exp: Am Red Cross fel, US & Can, 48-49; res affil bact, Duke Univ, 49-50 & Inst Oswaldo Cruz, Brazil, 50-51; res asst, Univ Ill, 51-53, res assoc, 53-54; instr bact & immunol, Sch Med, Emory Univ, 54-60, from asst prof to assoc prof microbiol, 60-67; assoc prof, Southwest Ctr Advan Studies, 67-69 & biol, Univ Tex, Dallas, 69-70; Guggenheim fel, France, 70-71, Am Cancer Soc grant, 71-72; master in residence, Gustave Roussy Inst, France, 72-73. Mem: Am Soc Biol Chem. Res: Bacteriophage; molecular biology. Mailing Add: 5 av Princesse Alice Monte Carlo Monaco

LANNING, EDWARD P, b Northville, Mich, Sept 21, 30; m 66; c 3. ANTHROPOLOGY, ARCHAEOLOGY. Educ: Univ Calif, BA, 53, PhD(anthrop), 60. Prof Exp: Lectr archaeol, San Marcos Univ, Lima, 58; instr anthrop, Sacramento State Col, 59-60; sr mus anthropologist, Robert H Lowie Mus Anthrop, Univ Calif, 60-61; from asst prof to prof anthrop, Columbia Univ, 63-73; PROF ANTHROP, STATE UNIV NY STONY BROOK, 73- Concurrent Pos: Fulbright fel, San Marcos Univ, Lima, 61-63; asst ed, Soc Am Archaeol, 63-65; NSF grants, Ecuador, 64-65, Peru, 64-66 & Chile, 66-73; vis assoc prof, Yale Univ, 67-68; vis prof, State Univ NY Stony Brook, 72-73. Mem: AAAS; Soc Am Archaeol; Inst Andean Studies (vpres, 60-61). Res: Andean archaeology; archaeology of western North America; development of ancient civilizations; archaeological theory; theory of culture change; human ecology. Mailing Add: Dept of Anthrop State Univ of NY Stony Brook NY 11790

LANNING, FRANCIS CHOWING, b Denver, Colo, Jan 5, 08; m 34; c 2. CHEMISTRY. Educ: Univ Denver, BS, 30, MS, 31; Univ Minn, PhD(phys chem), 36. Prof Exp: Asst, Univ Denver, 30-31; anal chemist, Minn State Hwy Dept, 36-42; from instr to asst prof, 42-60, ASSOC PROF CHEM, KANS STATE UNIV, 60- Mem: Am Chem Soc; Am Soc Plant Physiol; Am Soc Hort Sci. Res: Organosilicon compounds; silicon and other minerals in plants; chemical composition of limestones. Mailing Add: Dept of Chem Kans State Univ Manhattan KS 66506

LANNING, WILLIAM CLARENCE, b Boicourt, Kans, Dec 9, 13; m 36; c 3. PHYSICAL CHEMISTRY. Educ: Sterling Col, AB, 34; Univ Kans, AM, 36, PhD(chem), 38. Prof Exp: Asst instr chem, Univ Kans, 35-38; res chemist & sect mgr, Naval Res Lab, 38-45; res chemist, Phillips Petrol Co, 46-73, sect mgr, 57-73; RES CHEMIST, BARTLESVILLE ENERGY RES CTR, US ENERGY RES & DEVELOP ADMIN, 74- Mem: Am Chem Soc. Res: Non-aqueous solutions; inorganic preparations; catalytic hydrocarbon and petroleum processes; fundamentals of crude oil production; refining of synthetic crude oils. Mailing Add: 1609 Dewey Ave Bartlesville OK 74003

L'ANNUNZIATA, MICHAEL FRANK, b Springfield, Mass, Oct 14, 43. AGRICULTURAL CHEMISTRY, SOIL BIOCHEMISTRY. Educ: St Edward's Univ, BS, 65; Univ Ariz, MS, 67, PhD(agr chem, soils), 70. Prof Exp: Herbicide res chemist, Amchem Prod, Inc, 71-72; res assoc soils & natural fertilizers, Univ Ariz, 72-73; prof & res investr, Free Univ Chapingo, Mex, 73-75; RES INVESTR, NAT INST NUCLEAR ENERGY, MEX, 75- Concurrent Pos: Consult, Sanidad Vegetal, Caborca, Mex, 72- Mem: Am Chem Soc; Soil Sci Soc Am; Am Soc Agron; Int Soc Soil Sci. Res: Soil and plant relationships of strontium-90 fallout; plant and soil biochemistry of inositol phosphate stereoisomers; plant and soil microbial metabolism of herbicides; radioisotope techniques in agriculture and biology. Mailing Add: Nat Inst Nuclear Energy 1079 S Insurgentes Mexico DF Mexico

LANNUTTI, JOSEPH EDWARD, b Cedar Hollow, Pa, May 4, 26; m 54; c 3. PARTICLE PHYSICS. Educ: Pa State Univ, BS, 50; Univ Pa, MS, 53; Univ Calif, PhD(physics), 57. Prof Exp: Admin asst, Pa RR, 43-47; asst, Univ Pa, 51-52; physicist, NAm Aviation Inc, 52-53; asst, Univ Calif, 54-57; from asst prof to assoc prof, 57-64, PROF PHYSICS, FLA STATE UNIV, 64- Concurrent Pos: Physicist, Lawrence Radiation Lab, Univ Calif, 59-60; consult, Oak Ridge Nat Lab. Mem: Am Phys Soc. Res: Physics of elementary particles. Mailing Add: Dept of Physics Fla State Univ Tallahasee FL 32306

LANOU, ROBERT EUGENE, JR, b Burlington, Vt, Feb 13, 28; m 60; c 4. PHYSICS. Educ: Worcester Polytech Univ, BS, 52; Yale Univ, PhD(physics), 57. Prof Exp: Physicist, Lawrence Radiation Lab, Univ Calif, 57-59; mem fac, 59-66, PROF PHYSICS, BROWN UNIV, 66- Mem: AAAS; fel Am Phys Soc. Res: Elementary particle physics. Mailing Add: Dept of Physics Brown Univ Providence RI 02912

LA NOUE, KATHRYN F, b Camden, NJ, Dec 21, 34; m 58; c 4. BIOCHEMISTRY, CELL PHYSIOLOGY. Educ: Bryn Mawr Col, AB, 56; Yale Univ, PhD(biochem), 60. Prof Exp: Res chemist, US Army Surg Res Unit, 61-67; NIH fel, Johnson Res Found, Sch Med, Univ Pa, 68-70; res assoc, Johnson Res Found, Sch Med, Univ Pa, 70, asst prof, 71-74; ASSOC PROF PHYSIOL, MILTON S HERSHEY MED CTR, PA STATE UNIV, 74- Concurrent Pos: Dr W D Stroud estab investr, Am Heart Asn, 71- Mem: AAAS; Am Chem Soc; Am Soc Biol Sci; Biophys Soc. Res: Interaction of riboflavin with various aromatic compounds, mechanism of interaction of riboflavin with protein; metabolic effects of endotoxic shock; control of mitochondrial metabolism; membrane transport mechanisms. Mailing Add: Dept of Physiol Milton S Hershey Med Ctr Hershey PA 17033

LANOUX, SIGRED BOYD, b New Orleans, La, Nov 1, 31; m 54; c 2. INORGANIC CHEMISTRY. Educ: Southwestern La Univ, BS, 57; Tulane Univ, PhD(inorg chem), 62. Prof Exp: Res assoc, Univ Ill, Urbana, 61-62; res chemist, Textile Fibers Dept, E I du Pont de Nemours & Co, 62-66; from asst prof to assoc prof, 66-74, PROF CHEM, UNIV SOUTHWESTERN LA, 74-, HEAD DEPT CHEM, 72- Mem: Am Chem Soc. Res: Phosphonitrile chemistry. Mailing Add: 104 Ridgewood Lafayette LA 70501

LANPHEAR, FREDERICK ORVILLE, horticulture, see 12th edition

LANPHERE, MARVIN ADLER, b Spokane, Wash, Sept 29, 33; m 61; c 3. GEOLOGY, GEOCHEMISTRY. Educ: Mont Sch Mines, BS, 55; Calif Inst Technol, MS, 56, PhD(geol), 62. Prof Exp: Geologist, US Geol Surv, Calif, 62-67, dep asst chief geologist, Washington, DC, 67-69, RES GEOLOGIST, US GEOL SURV, CALIF, 69- Concurrent Pos: Vis prof, Stanford Univ, 72; vis fel, Australian Nat Univ, 75-76. Mem: Geol Soc Am; Am Geophys Union. Res: Geochronology, application of techniques of radioactive age determination of rocks and minerals to geological problems; isotope tracer studies of geological processes. Mailing Add: US Geol Surv 345 Middlefield Rd Menlo Park CA 94025

LANPHIER, EDWARD HOWELL, b Madison, Wis, May 29, 22. MEDICINE, PHYSIOLOGY. Educ: Univ Wis, BS, 46; Univ Ill, MS & MD, 49. Prof Exp: Am Col Physicians res fel physiol, Grad Sch Med, Univ Pa, 50-51; asst med officer & physiologist, Exp Diving Unit, US Navy, 52-58, diving med officer, Eniwetok Proving Ground, 58, med officer, Under-water Demolition Team, Norfolk, Va, 58-59; asst prof, 59-64, ASSOC PROF PHYSIOL, SCH MED, STATE UNIV NY BUFFALO, 64- Concurrent Pos: Leave of absence for theol study, 73-76. Mem: Am Physiol Soc; Aerospace Med Asn; Am Col Sports Med; Undersea Med Soc. Res: Respiratory physiology; submarine and diving medicine; physiological problems of exposure to increased pressure; hyperbaric medicine. Mailing Add: of Physiol State Univ of NY Buffalo NY 14214

LANSBURY, JOHN, b Cheddar, Eng, Apr 7, 97; nat US; m 38; c 3. INTERNAL MEDICINE. Educ: Queen's Univ, Ont, MD & CM, 26; Univ Minn, MS, 33. Prof

Exp: Asst, Philadelphia Inst Med Res, 33-35; PROF CLIN MED, SCH MED, TEMPLE UNIV, 35-; DIR MED SERV, PHILADELPHIA STATE HOSP, 67- Mem: Am Rheumatism Soc. Res: Methods of evaluation of drugs in rheumatoid arthritis; isolated studies in pathogenesis. Mailing Add: Philadelphia State Hosp Roosevelt Blvd & Southampton Rd Philadelphia PA 19114

LANSBURY, PETER THOMAS, organic chemistry, see 12th edition

LANSFORD, EDWIN MYERS, JR, b Houston, Tex, June 26, 23; m 50; c 3. BIOCHEMISTRY. Educ: Univ Calif, Los Angeles, BA, 46; Univ Tex, MA, 51, PhD(biochem), 51. Prof Exp: Fel, Univ Ill, 51-52; fel, Univ Tex, 52-53, instr chem, 56-57, res scientist, Clayton Found Biochem Inst, 53-67; PROF BIOCHEM, SOUTHWESTERN UNIV, TEX, 67- Mem: Am Chem Soc. Res: Microbial intermediary metabolism; amino acid activating enzymes; metabolic effects of alcohol; single carbon unit metabolism and its control. Mailing Add: Dept of Biol Southwestern Univ Georgetown TX 78626

LANSING, ALBERT INGRAM, b Woodbine, NJ, Mar 13, 15; m 41; c 5. ANATOMY. Educ: Univ Pa, AB, 37; Ind Univ, PhD(zool), 41. Prof Exp: Asst anat, Sch Med, Washington Univ, 46-47, from asst prof to assoc prof, 47-54; prof & chmn dept, Emory Univ, 54-56; PROF ANAT & CHMN DEPT, SCH MED, UNIV PITTSBURGH, 56- Concurrent Pos: Fel anat, Sch Med, Washington Univ, 41-45; trustee, Marine Biol Lab; mem adv coun aging, Vet Admin, 58-60; Nat Adv Child Health & Human Develop Coun, USPHS, 63-64; asst ed, J Geront. Mem: Geront Soc (pres, 58). Res: General physiology of aging; cytochemical procedures; physiological aspects of senescence; microincineration and electron microscopy. Mailing Add: Dept of Anat & Cell Biol Univ of Pittsburgh Sch of Med Pittsburgh PA 15213

LANSING, ALLAN M, b St Catherines, Ont, Sept 12, 29; m 51; c 3. CARDIOVASCULAR SURGERY. Educ: Univ Western Ont, MD, 53, PhD(physiol), 57; FRCS(C), 59. Prof Exp: Nat Res Coun Can scholar, Univ Western Ont, 55-57; asst prof surg & physiol, Fac Med, Univ Western Ont, 61-63; assoc prof, 63-69, chief sect cardiovasc surg, 69-74, PROF SURG, SCH MED, UNIV LOUISVILLE, 69- Concurrent Pos: Markle scholar med sci, 61- Mem: Fel Am Col Surg; fel Am Col Cardiol; Soc Univ Surg. Res: Cardiovascular physiology and shock; pulmonary atelectasis; renal transplantation; open heart surgery. Mailing Add: 718 Medical Towers Louisville KY 40202

LANSINGER, JOHN MARCUS, b July 20, 32; US citizen; m 53; c 2. GEOPHYSICS, ENVIRONMENTAL MANAGEMENT. Educ: Lewis & Clark Col, BS, 54; Univ Alaska, MS, 56. Prof Exp: Instr geophys, Univ Alaska, 56-57; sr engr, Philco Corp, 57-59; staff assoc, Boeing Sci Res Labs, 59-69, VPRES, NORTHWEST ENVIRON TECHNOL LABS, INC, 69- Mem: Am Geophys Union; Inst Elec & Electronics Engrs; Air Pollution Control Asn. Res: Environmental sciences; ionospheric research; atmospheric propagation at optical wavelengths. Mailing Add: 9301 26th Pl NW Seattle WA 98107

LANSKI, CHARLES PHILIP, b Chicago, Ill, Oct 19, 43. MATHEMATICS. Educ: Univ Chicago, SB, 65, SM, 66, PhD(math), 69. Prof Exp: Asst prof, 69-74, ASSOC PROF MATH, UNIV SOUTHERN CALIF, 74- Concurrent Pos: NSF grant, 71-73. Mem: Am Math Soc. Res: Noncommutative ring theory. Mailing Add: Dept of Math Univ of Southern Calif Los Angeles CA 90007

LANSON, HERMAN JAY, b Utica, NY, Feb 22, 13; m 35; c 3. ORGANIC CHEMISTRY. Educ: Syracuse Univ, BS, 34, MS, 36; Polytech Inst Brooklyn, PhD(org chem), 45. Prof Exp: Org res chemist, Nuodex Prod Co, Inc, NJ, 36-40; res chemist, HD Roosen Co, NY, 40-43; chief chemist, Crown Oil Co Prod Corp, 43-45; supvr, Vehicle Res & Prod, Grand Rapids Varnish Corp, 45-50; supvr, resin & plasticizer develop, Chem Mat Dept, Gen Elec Co, 50-57; vpres & tech dir, US Vehicle & Chem Co, 57-61; pres & res dir, Lanson Chem Corp, 61-70, PRES & RES DIR, WASHBURN LANSON CORP, 70- Concurrent Pos: Lectr, Washington Univ, Roosevelt Univ, Chicago, Univ Houston & Univ Mo. Mem: Am Oil Chem Soc; Soc Plastics Engrs; fel Am Inst Chemists. Res: Synthetic resins; drying oils; protective coatings; electrical insulation materials; development of water-soluble polymers, synthetic latexes, synthetic resins for electrical insulation. Mailing Add: 12020 Gardengate Dr St Louis MO 63141

LANTERMAN, ELMA, b Elkhart, Ill, Jan 25, 17. ANALYTICAL CHEMISTRY. Educ: Univ Ill, BS, 40; Univ Ind, MA, 48, PhD(anal chem), 51. Prof Exp: Chem technician, Mayo Clin, 41-42; chemist, Devoe & Raynolds Co, 42-43 & Metal & Thermit Corp, 43-46; asst, Univ Ind, 46-51; asst prof physics, NC State Col, 51-53; supvr, Metall Groups, Indianapolis Naval Ord Plant, 53-55; sr res chemist, Am Can Co, 55-59; sr res chemist, 59-61, scientist, 61-63, mgr anal chem, 63-74, STAFF SCIENTIST, RES CTR, BORG-WARNER CORP, 74- Concurrent Pos: Comnr, Environ Comn, Des Plains, Ill, 74-79. Mem: Am Chem Soc; Electron Micros Soc Am; Am Crystallog Asn; Soc Appl Spectros (treas, 70-74). Res: X-ray diffraction and spectroscopy; electron microscopy; technology forecasting. Mailing Add: Res Ctr Borg-Warner Corp Des Plains IL 60018

LANTERMAN, HAROLD H, b Berwick, Pa, Nov 23, 05; m 42. CHEMISTRY. Educ: Pa State Teachers Col, BS, 31; NY Univ, MA, 47; Pa State Univ, EdD, 54. Prof Exp: Teacher high sch, Pa, 31-46; prof, 46-73, EMER PROF CHEM, BLOOMSBURG STATE COL, 73- Mem: Am Chem Soc. Res: Analytical chemistry. Mailing Add: Dept of Phys Sci Bloomsburg State Col Bloomsburg PA 17815

LANTERMAN, WILLIAM STANLEY, JR, physics, meteorology, see 12th edition

LANTHIER, ANDRE, b Montreal, Que, Apr 13, 28; m 56; c 3. MEDICINE, BIOCHEMISTRY. Educ: Univ Montreal, BA, 47, MD, 53; FRCPS(C), 58. Prof Exp: Resident med, Notre Dame Hosp, Montreal, Que, 52-55; asst, Peter Bent Brigham Hosp, Boston, 55-57; clin instr, 57-58, from asst prof to assoc prof, 59-68, PROF MED & CHMN DEPT, UNIV MONTREAL, 68- Concurrent Pos: Res fel, Harvard Med Sch, 55-57; asst physician, Notre-Dame Hosp, 57-64, sr physician, 64-, dir endocrinol lab, 58- Mem: Endocrine Soc; Am Fedn Clin Res; NY Acad Sci; fel Am Col Physicians; Can Soc Clin Invest. Res: Ovarian steroid biosynthesis; biosynthesis and physiological effects of 18-hydroxylated corticosteroids. Mailing Add: Notre Dame Hosp 1560 Sherbrooke St E Montreal PQ Can

LANTIS, DAVID W, b Fort Thomas, Ky, Apr 13, 17; m 54. GEOGRAPHY. Educ: Adams State Col, BA, 39; Univ Cincinnati, MA, 48; Ohio State Univ, PhD(geog), 50. Prof Exp: Asst prof geog, Univ Southern Calif, 49-53, Los Angeles City Col, 53-55 & Compton Col, 55-57; from asst prof to assoc prof, 57-64, PROF GEOG, CALIF STATE UNIV, CHICO, 64- Honors & Awards: Award for Innovative Teaching Col Level, Nat Coun Geog Educ, 71. Mem: Asn Am Geogrs; Am Geog Soc; Royal Geog Soc. Res: Regional geography; California. Mailing Add: Dept of Geog Calif State Univ Chico CA 95926

LANTIS, MARGARET LYDIA, b Dayton, Ohio, Sept 1, 06. ANTHROPOLOGY. Educ: Univ Minn, BA, 30; Univ Calif, PhD(anthrop), 39. Prof Exp: Res assoc anthrop, Univ Calif, 39-41; instr, Univ Minn, 41-42; Soc Sci Res Coun fel, Univ Chicago, 42-43; instr sociol & anthrop, Reed Col, 43-44; community analyst, War Relocation Authority, 44-45; soc sci analyst, Bur Agr Econ, USDA, 45-46; fel, Arctic Inst NAm, 46; social econ analyst, Bur Census, US Dept Com, 47; res fel, Viking Fund, NY, 47-48; social anthropologist study adult develop, Harvard Univ, 48-52; anthropologist, USPHS, 54-63, vis prof anthrop, McGill Univ, 63-64; vis prof, Univ Calif, Berkeley, 64-65; prof anthrop, Univ Ky, 65-74; RETIRED. Concurrent Pos: Lectr, Sch Social Work, Boston Univ, 50; vis prof, Univ Calif, 50, Air Univ, 52-53, Univ Alaska, 55, Univ Minn, 58 & Univ Wash, 62. Mem: Fel Am Anthrop Asn; Arctic Inst NAm; Am Ethnol Soc (pres, 63-64); Soc Appl Anthrop (pres, 73-74). Res: Arctic ethnography and geography; United States culture studies; applied anthropology. Mailing Add: Dept of Anthrop Univ of Ky Lexington KY 40506

LANTZ, HAROLD J, b Fullerton, Pa, Sept 9, 13; m 43; c 1. DENTISTRY. Educ: Philadelphia Col Pharm, BSc, 39; Temple Univ, DDS, 50, MEd, 58; Am Bd Oral Med, dipl. Prof Exp: From instr to assoc prof, 50-62, clin coordr, 54-65, PROF PROSTHETIC DENT & CHMN DEPT, SCH DENT, TEMPLE UNIV, 62- Concurrent Pos: Consult, Vet Admin Hosp, Philadelphia, 62- & Dover AFB, Del, 64-; adv, Eaton Labs, Norwich Pharmacal Co, NY, 65-; consult, Dent Dept, Eastern Pa Psychiat Inst, 71-; mem med staff, Northern Div, Albert Einstein Med Ctr, 72-; mem exam comt, NE Regional Bd Dent Examr, 72-; assoc staff prosthodont, Shriners Hosp for Crippled Children, 73-74; mem bd, Adv Coun, Zion Br, Philadelphia Ctr Older People, 74- Mem: Fel Am Col Dent; master Acad Gen Dent; Am Dent Asn; Am Prosthodontic Soc; Am Acad Oral Med. Res: Prosthodontics. Mailing Add: Dept of Prosthetic Dent Temple Univ Sch of Dent Philadelphia PA 19140

LANYI, JANOS K, b Budapest, Hungary, June 5, 37; US citizen; m 62; c 3. BIOCHEMISTRY. Educ: Stanford Univ, BS, 59; Harvard Univ, MA, 61, PhD(biochem), 63. Prof Exp: NIH fel genetics, sch med, Stanford Univ, 63-65; Nat Acad Sci-Nat Res Coun res assoc biochem, 65-66, RES SCIENTIST, PLANETARY BIOL DIV, AMES RES CTR, NASA, 66- Mem: Am Soc Microbiol; Biophys Soc; Am Soc Biol Chem; AAAS. Res: Physical chemistry of proteins; structure and function of enzymes and membranes in halophilic microorganisms; electron transport system and ion carriers; energetics and mechanism of amino acid transport. Mailing Add: Ames Res Ctr 239-10 Moffett Field CA 94035

LANYON, HUBERT PETER DAVID, b Halesowen, Eng, June 25, 36; m 62; c 3. SOLID STATE PHYSICS. Educ: Cambridge Univ, BA, 58, MA, 62; Leicester Univ, PhD(physics), 61. Prof Exp: Res demonstr physics, Leicester Univ, 58-61; res assoc elec eng, Univ Ill, Urbana, 61-63; mem tech staff, RCA Labs, 63-66; assoc prof elec eng, Carnegie Inst Technol, 66-67; ASSOC PROF ELEC ENG, WORCESTER POLYTECH INST, 67- Mem: Am Phys Soc. Res: Photoconductivity; surface physics. Mailing Add: Dept of Elec Eng Worcester Polytech Inst Worcester MA 01609

LANYON, WESLEY EDWIN, b Norwalk, Conn, June 10, 26; m 51; c 2. ORNITHOLOGY. Educ: Cornell Univ, AB, 50; Univ Wis, MS, 51, PhD(zool), 55/. Prof Exp: Instr zool, Univ Ariz, 55-56; asst prof, Miami Univ, 56-57; asst cur, 57-63, assoc cur, 63-67, CUR, AM MUS NATURAL HIST, 67- Concurrent Pos: Res dir, Kalbfleisch Field Res Sta, Am Mus Natural Hist, Huntington, 58-; adj prof, City Univ New York, 68- Mem: AAAS; Cooper Ornith Soc; Wilson Ornith Soc; Ecol Soc Am; Soc Study Evolution. Res: Application of comparative behavior and ecology of closely related populations of birds to avian systematics, especially vocalizations in avian taxonomy. Mailing Add: Dept of Ornith Am Mus of Natural Hist New York NY 10024

LANZA, GIOVANNI, b Trieste, Italy, Aug 5, 26; m 50; c 4. PHYSICS. Educ: Univ Trieste, PhD, 50. Prof Exp: Asst prof quantum theory, Univ Trieste, 50-52; prof, Univ Cagliari, Univ Sardinia & Univ Padua, 52-54; vis physicist nuclear physics, Mass Inst Technol, 54-55; res fel, Harvard Univ, 55-58; assoc prof, 58-60, PROF PHYSICS, NORTHEASTERN UNIV, 60- Concurrent Pos: Fulbright & Smith Mundt scholars, 54; consult, Lab Electronics, Inc, 58- & Saunders Assocs, 63- Mem: Ital Phys Soc. Res: Magnetohydrodynamics; plasma physics; energy conversion techniques; nuclear physics; physics of upper atmosphere. Mailing Add: Dept of Physics Northeastern Univ Boston MA 02115

LANZA, GUY ROBERT, b Englewood, NJ, Jan 27, 39; m 68; c 2. AQUATIC ECOLOGY. Educ: Fairleigh Dickinson Univ, BS, 61; Univ Ky, MS, 69; Va Polytech Inst & State Univ, PhD(zool), 72. Prof Exp: Res biologist, Merck Inst Therapeut Res, 63-69; aquatic ecologist, Smithsonian Inst, 71-73 & NY Univ, 73-75; AQUATIC ECOLOGIST, UNIV TEX, DALLAS, 75-, ASSOC PROF, 75- Concurrent Pos: Consult ecologist, Int Ctr Med Res & Training, Malaysia, 72-73; asst dir, Aquatic ecol prog, NY Univ Med Ctr, 73-75; subcomt partic, Nat Comn Water Qual, 75. Mem: AAAS; Water Pollution Control Fedn. Res: Structure and function of aquatic ecosystems; pollution ecology and the environmental physiology and energetics of aquatic organisms. Mailing Add: Environ Sci Prog Univ of Tex Dallas Box 688 Richardson TX 75080

LANZA, RICHARD CHARLES, b New York, NY, Apr 28, 39; m 63. PHYSICS. Educ: Princeton Univ, AB, 59; Univ Pa, MS, 61, PhD(physics), 66. Prof Exp: Res assoc physics, 66-68, asst prof, 68-74, MEM RES STAFF, MASS INST TECHNOL, 74- Concurrent Pos: Assoc radiol, Harvard Med Sch & Peter Bent Brigham Hosp, 74-; res fel, Mass Gen Hosp, 75- Mem: AAAS; Am Phys Soc. Res: Experimental particle physics, nuclear and electronic instrumentation, medical instrumentation, especially in radiology and nuclear medicine. Mailing Add: Lab for Nuclear Sci Mass Inst of Technol Cambridge MA 02139

LANZA, VINCENT LEONARD, organic chemistry, deceased

LANZAFAME, FRANK MICHAEL, physical chemistry, nuclear chemistry, see 12th edition

LANZANO, PAOLO, b Cairo, Egypt, Nov 29, 23; nat US; m 57. APPLIED MATHEMATICS, SPACE PHYSICS. Educ: Univ Rome, BS, 42, PhD(math), 45. Prof Exp: Asst prof math, Univ Rome, 46-49 & St Louis Univ, 50-56; design specialist, Douglas Aircraft Co, Calif, 56-58; mem tech staff, Space Tech Labs, Calif, 58-60; res scientist, Nortronics Div, Northrop Corp, 60-61; prin scientist, Space & Info Systs Div, NAm Aviation, 61-71; assoc prof math, Nicholls State Univ, 71-72; HEAD MATH RES CTR, NAVAL RES LAB, 72- Concurrent Pos: Fel, Inst Advan Studies, Univ Rome, 46-48. Mem: Am Math Soc; Am Astronaut Soc; Soc Indust & Appl Math; Am Inst Aeronaut & Astronaut. Res: Celestial mechanics; theory of relativity; Reimannian geometry; space physics. Mailing Add: Naval Res Lab Washington DC 20375

LANZEROTTI, LOUIS JOHN, b Carlinville, Ill, Apr 16, 38; m 65; c 2. GEOPHYSICS, SPACE PHYSICS. Educ: Univ Ill, BS, 60; Harvard Univ, AM, 63, PhD(physics), 65. Prof Exp: Fel, 65-67, MEM TECH STAFF, BELL LABS, 67- Mem: AAAS; Am Phys Soc; Am Geophys Union; Inst Elec & Electronics Engrs; Soc

Terrestial Magnegtism & Elec Japan. Res: Particles and fields in the magnetosphere; solar cosmic ray composition and propagation; ionosphere-magnetosphere coupling; planetary magnetospheres. Mailing Add: Bell Labs Murray Hill NJ 07974

LANZEROTTI, MARY YVONNE DEWOLF, b Phoenix, Ariz, Nov 7, 38; m 65; c 2. PHYSICAL CHEMISTRY. Educ: Univ Calif, Berkeley, BS, 60; Harvard Univ, PhD(phys chem), 65. Prof Exp: Chemist, US Naval Ord Test Sta, 60; asst, Harvard Univ, 60-64; res chemist, Mithras Inc, Mass, 64-65; RES CHEMIST, EXPLOSIVES DIV, PICATINNY ARSENAL, 65- Mem: AAAS; Am Chem Soc; Am Phys Soc; Am Geophys Union; Sigma Xi. Res: Gas dynamic lasers; chemical lasers; explosive product lasers; energy conversion; thermal initiation of explosives; friction; chemical lasers; gas dynamic lasers; explosive product lasers. Mailing Add: Explosives Div Bldg 407 Picatinny Arsenal Dover NJ 07801

LANZI, LAWRENCE HERMAN, b Chicago, Ill, Apr 8, 21; m 47; c 2. PHYSICS, MEDICAL PHYSICS. Educ: Northwestern Univ, BS, 43; Univ Ill, MS, 47, PhD(physics), 51; Am Bd Health Physics, dipl; Am Bd Radiol, dipl. Prof Exp: Asst, Dearborn Observ, Ill, 41-42; asst, Northwestern Univ, 42-43; asst, Univ Chicago, 44; jr scientist, Los Alamos Sci Lab, 44-45; asst, Univ Ill, 46-50; assoc physicist, Argonne Nat Lab, 51; sr physicist, 51-55, from asst prof to assoc prof, Dept Radiol, Sch Med, 55-68, PROF MED PHYSICS, DIV BIOL SCI, PRITZKER SCH MED & FRANKLIN McLEAN MEM RES INST, UNIV CHICAGO, 68- Concurrent Pos: With Argonne Cancer Res Hosp, 51-; first officer, Int Atomic Energy Comn, Vienna, Austria, 67-68; bd mem, Am Int Sch, Vienna, 67-68; mem Nat Coun Radiation Protection, 67 & US Nat Comt Med Physics, 70-; chmn Radiation Protection Adv Coun, State of Ill, 71-; mem tech adv panel, Los Alamos Meson Physics Facil, Los Alamos, NMex & steering comt, radiol physics ctr, Univ Tex M D Anderson Hosp & Tumor Inst; consult, Int Atomic Energy Agency, Vienna; Hines Vet Admin Hosp, Ill & NIH. Mem: AAAS; Am Phys Soc; Radiation Res Soc; Am Asn Physicists in Med (past pres); Radiol Soc NAm. Res: Accelerators; radiation as applied to medicine; radiation physics; high energy x-rays and electrons; isotopes in medicine; nuclear reactors; health physics. Mailing Add: Franklin McLean Mem Res Inst Univ of Chicago Chicago IL 60637

LANZILOTTA, RAYMOND PHILIP, b Camden, NJ, Sept 26, 42; m 65; c 1. MICROBIOLOGY. Educ: Rutgers Univ, BA, 64, MS, 67, PhD(biochem, microbiol), 68. Prof Exp: Instr microbiol, Sch Med, Temple Univ, 68-70; MICROBIAL CHEMIST, SYNTEX RES DIV, PALO ALTO, 70- Mem: AAAS; Am Soc Microbiol; Am Chem Soc; Soc Indust Microbiol. Res: Microbial transformations of organic substrates; fermentation technology. Mailing Add: 856 Ferngrove Dr San Jose CA 95129

LANZKOWSKY, PHILIP, b Cape Town, SAfrica, Mar 17, 32; m 55; c 5. PEDIATRICS, HEMATOLOGY. Educ: Univ Cape Town, MB & ChB, 54, MD, 59; Royal Col Physicians & Surgeons, dipl child health, 60; Am Bd Pediat, dipl, 66, cert pediat hematol-oncol, 75; FRCP(E), 73. Prof Exp: From intern to sr intern, Groote Schuur Hosp, Univ Cape Town, 55-56; gen pract, 56-57; from registr to sr registr, Red Cross War Mem Children's Hosp, 57-60, consult pediatrician & pediat hematologist, 63-65; asst prof pediat, New York Hosp-Cornell Med Ctr, 65-67, assoc prof, 67-70; PROF PEDIAT, STATE UNIV NY STONY BROOK, 70- Concurrent Pos: Dr C L Herman res grants, 58 & 64; Cecil John Adams mem traveling fel & Hill-Pattison-Struthers bursary, 60; Benger Labs traveling grant, 61; registr pediat unit, St Mary's Hosp Med Sch, Univ London, 61; clin & res fel pediat hemat, Duke Univ, 61-62; res fel, Col Med, Univ Utah, 62-63; lectr, Univ Cape Town, 63-65; dir pediat hemat, New York Hosp-Cornell Med Ctr, 65-70; pediatrician-in-chief, chmn pediat & chief pediat hemat, Long Island Jewish-Hillside Med Ctr, 70-; pediatrician-in-chief, Queens Hosp Ctr, 70-; mem pediat adv comt NY City Dept Health, 70-73. Honors & Awards: Joseph Arenow Prize, 59. Mem: Am Soc Hemat; Am Acad Pediat; Am Soc Clin Oncol; Am Asn Cancer Res; Am Pediat Soc. Res: Nutritional anemias in children, especially iron, folate and protein deficiency; pediatric oncology. Mailing Add: Dept of Pediat Long Island Jewish-Hillside Med Ctr New Hyde Park NY 11040

LANZL, GEORGE FRANK, b Chicago, Ill, Aug 9, 18; m 42; c 2. PHYSICAL CHEMISTRY. Educ: De Pauw Univ, BS, 40; Northwestern Univ, PhD(chem), 46. Prof Exp: Instr chem, Northwestern Univ, 40-44; mem staff, 44-51, res mgr, 51-57, dir dacron res lab, 57-60, DIR, PIONEERING RES DIV, TEXTILE FIBERS DEPT, E I DU PONT DE NEMOURS & CO, 60- Mem: Am Chem Soc; Am Phys Soc. Res: Physical chemistry of polymers; fiber physics and processing. Mailing Add: 409 Crest Rd Wilmington DE 19803

LANZONI, VINCENT, b Kingston, Mass, Feb 23, 28; m 60; c 3. PHARMACOLOGY, CLINICAL MEDICINE. Educ: Tufts Univ, PhD(pharmacol), 53; Boston Univ, MD, 60. Prof Exp: Instr pharmacol, Sch Med, Tufts Univ, 53-54; from intern to resident, Boston City Hosp, 60-63; asst prof pharmacol & instr med, 63-66, assoc prof pharmacol & med, 66-73, assoc dean sch med, 69-75, prof pharmacol, Sch Med, Boston Univ, 73-75; DEAN & PROF MED, NJ MED SCH, 75- Concurrent Pos: Res fel, NIH, 53-54; fel med, Boston City Hosp. Res: Cardiovascular pharmacology. Mailing Add: 100 Bergen St Newark NJ 07103

LA PALME, DONALD WILLIAM, physical chemistry, see 12th edition

LAPATNICK, LEONARD NOAH, analytical chemistry, inorganic chemistry, see 12th edition

LAPEYRE, GERALD J, b Riverton, Wyo, Jan 3, 34; m 60; c 3. SOLID STATE PHYSICS. Educ: Univ Notre Dame, BS, 56; Univ Mo, MA, 58, PhD(physics), 62. Prof Exp: Asst prof, 62-69, ASSOC PROF PHYSICS, MONT STATE UNIV, 69- Mem: Am Phys Soc; Am Asn Physics Teachers. Res: Solid state physics with emphasis on optical properties and electronic structure. Mailing Add: Dept of Physics Mont State Univ Bozeman MT 59715

LAPHAM, DAVID MORTIMER, mineralogy, economic geology, see 12th edition

LAPHAM, LOWELL WINSHIP, b New Hampton, Iowa, Mar 20, 22; m 45; c 4. NEUROPATHOLOGY. Educ: Oberlin Col, BA, 43; Harvard Med Sch, MD, 48. Prof Exp: Sr instr path, Western Reserve Univ, 55-57, from asst prof to assoc prof, 57-64; assoc prof path, 64-69, PROF NEUROPATH, MED CTR, UNIV ROCHESTER, 69- Concurrent Pos: Nat Multiple Sclerosis Soc fel cytochem, Western Reserve Univ, 56-58. Mem: AAAS; Am Asn Neuropath; Am Asn Path & Bact; Am Acad Neurol; Tissue Cult Asn. Res: Studies of developmental diseases of nervous system; brain tumors; nature and function of glia. Mailing Add: Dept of Path Univ of Rochester Med Ctr Rochester NY 14642

LAPHAM, ROGER FULMER, b Utica, NY, June 8, 07. MEDICINE. Educ: Syracuse Univ, AB, 29, MD, 32. Prof Exp: MED DIR, GEN FOODS CORP, 45-; PRES & CLIN DIR, HUDSON LABS, INC, 55- Concurrent Pos: Dir, World Rehab Fund. Mem: AMA; Am Heart Asn; Indust Med Asn; NY Acad Med. Res: Internal and industrial medicine. Mailing Add: Hudson Labs Inc 77 Seventh Ave New York NY 10011

LAPIDES, JACK, b Rochester, NY, Nov 27, 14; m 48. UROLOGY. Educ: Univ Mich, BA, 36, MA, 38, MD, 41. Prof Exp: USPHS fel, Nat Cancer Inst, 48-50; from instr to assoc prof, 50-64, PROF SURG, MED SCH, UNIV MICH, ANN ARBOR, 64-, HEAD SECT UROL, 68- Concurrent Pos: Rockefeller res assoc, 36-38; assoc physician surg, Wayne County Gen Hosp, 50-, chief sect urol, 58- & Vet Admin Hosp, Ann Arbor, 54-; mem comt genito-urinary systs, Nat Acad Sci-Nat Res Coun, 66-70. Mem: AAAS; AMA; Am Urol Asn; fel Am Col Surg; Am Asn Genito-Urinary Surg. Res: Renal, ureteral and bladder physiology; fluid and electrolyte balance; urinary incontinence; urinary infection. Mailing Add: 1405 EAnn St Ann Arbor MI 48104

LAPIDUS, ARNOLD, b Brooklyn, NY, Nov 6, 33; m 52. MATHEMATICS. Educ: Brooklyn Col, BS, 56; NY Univ, MS, 60, PhD(math), 67. Prof Exp: Asst math, Courant Inst, NY Univ, 58-60, asst res scientist, AEC comput facility, 61-65, assoc res scientist, 61-68; math analyst, Comput Applns Inc-NASA, 68-69, sci prog mgr, 69-71; ASST PROF MATH, FAIRLEIGH DICKINSON UNIV, 71- Mem: Soc Indust & Appl Math; Math Asn Am; Am Math Soc. Res: Partial and ordinary differential equations; Monte Carlo methods; scientific programming; artificial intelligence; tedious algebra by computer; fluid dynamics by computer; shock calculations; numerical methods and analysis. Mailing Add: Dept of Math Fairleigh Dickinson Univ Teaneck NJ 07666

LAPIDUS, HERBERT, b New York, NY, Aug 10, 31; m 52; c 2. PHARMACY, PHARMACOLOGY. Educ: Columbia Univ, BS, 53, MS, 55; Rutgers Univ, PhD(pharm), 67. Prof Exp: Proj leader, Julius Schmid Co, 57-60; proj leader pharm, Bristol-Myers Co, 60-63, group leader, 63-67, dept head, 67-70; TECH DIR, COMBE INC, 70- Mem: Am Pharmaceut Asn; Sigma Xi; Soc Cosmetic Chemists; NY Acad Sci. Res: Development of pharmaceutical dosage forms, especially sustained release medication, biopharmaceutics and percutaneous absorption. Mailing Add: COMBE Inc 240 Westchester Ave White Plains NY 10604

LAPIDUS, IVAN RICHARD, b Brooklyn, NY, Apr 7, 35; m 59; c 2. MOLECULAR BIOPHYSICS, THEORETICAL PHYSICS. Educ: Univ Chicago, AB, 55, BS, 56, MS, 57; Columbia Univ, PhD(physics), 63. Prof Exp: Lectr physics, City Col, New York, 59-60; asst prof, 63-68, ASSOC PROF PHYSICS, STEVENS INST TECHNOL, 68- Concurrent Pos: Res grants, Stevens Inst Technol, 65, 68 & 71; NASA fac fel, 66; vis assoc prof, Inst Cancer Res, Columbia Univ, 69; vis fel biochem sci, Princeton Univ, 70-71. Mem: AAAS; Am Phys Soc; Am Asn Physics Teachers; Am Inst Biol Sci; Am Soc Microbiol. Res: Chemotaxis in microorganisms; cell cycle in bacteria; secondary structure of RNA. Mailing Add: Dept of Physics Stevens Inst of Technol Hoboken NJ 07030

LA PIDUS, JULES BENJAMIN, b Chicago, Ill, May 1, 31; m 54; c 2. MEDICINAL CHEMISTRY. Educ: Univ Ill, BS, 54; Univ Wis, MS, 57, PhD(pharmaceut chem), 58. Prof Exp: From asst prof to assoc prof, 58-67, assoc dean res, Grad Sch, 72-74, PROF MED CHEM, OHIO STATE UNIV, 67-, V PROVOST RES & DEAN GRAD SCH, 74- Concurrent Pos: Consult, Pharmacol & Toxicol Training Grants Comt, Nat Inst Gen Med Sci, NIH, 65-67, Prog Comt, 71-75. Mem: Am Chem Soc; AAAS. Res: Structure-action relationships; autonomic pharmacology. Mailing Add: Grad Sch 164 W 19th Ave Ohio State Univ Columbus OH 43210

LAPIDUS, LEO, b Boston, Mass, Feb 13, 10; m 40; c 3. MATHEMATICS. Educ: Harvard Col, AB, 31; Boston Univ, AM, 37; Mich State Univ, PhD(math), 56. Prof Exp: Instr math, Univ Maine, 46-48; instr, Mich State Univ, 48-56; sr res engr, Convair Div, Gen Dynamics Corp, 56-59, asst head, Digital Comput Lab, 59-60, head math anal & programming, 60-62; prof, 62-75, EMER PROF MATH, LEWIS & CLARK COL, 75- Concurrent Pos: Lectr, Univ Calif, Los Angeles, 57- Mem: Am Math Soc; Math Asn Am. Res: Abstract distance geometry; Boolean and Brouwerian algebras and geometries; lattice theory; theory of rings; numerical analysis and high speed digital computers. Mailing Add: 880 Bickner St Lake Oswego OR 97034

LAPIDUS, MILTON, b New York, NY, May 8, 22; m 58; c 2. BIOCHEMISTRY. Educ: Univ Wis, BS, 48, MS, 53, PhD(biochem), 56. Prof Exp: Res chemist, Abbott Labs, 48-51; sr fel, Eastern Regional Res Lab, USDA, 56-59; SR BIOCHEMIST, WYETH LABS, INC, 59- Mem: AAAS; Am Chem Soc. Res: Isolation of vitamin B12b; microbiological transformation of steroids; chromatographic purification of viruses; prostaglandin biosynthesis and isolation; synthesis of penicillins and sweeteners. Mailing Add: 412 Yorkshire Way Rosemont PA 19010

LAPIETRA, RICHARD ANDREW, b New York, NY, July 20, 32. PHYSICAL CHEMISTRY, THERMODYNAMICS. Educ: Marist Col, BA, 54; Cath Univ Am, PhD(chem), 61. Prof Exp: Teacher high sch, NY, 54-56; instr chem, Cath Univ Am, 60-61; asst prof, 64-68, ASSOC PROF CHEM, MARIST COL, 68-, ACAD DEAN, 69- Res: Chemistry of transition metal complexes; thermochemistry; electrochemistry. Mailing Add: Marist Col Poughkeepsie NY 12601

LAPIN, DAVID MARVIN, b New York, Apr 12, 39; m 67; c 2. BIOLOGY. Educ: NY Univ, BA, 60, MS, 63, PhD(biol), 68. Prof Exp: From instr to asst prof biol, 66-72, ASSOC PROF BIOL SCI, FAIRLEIGH DICKINSON UNIV, 72- Concurrent Pos: Grants in aid, Fairleigh Dickinson Univ, 68-72. Mem: AAAS. Res: Kinetics of hematopoiesis; humoral regulation of hematopoiesis. Mailing Add: Dept of Biol Fairleigh Dickinson Univ Teaneck NJ 07666

LAPIN, EVELYN P, b Montreal, Que, Aug 29, 33; wid; c 3. NEUROCHEMISTRY, ENZYMOLOGY. Educ: McGill Univ, BSc, 54, PhD(biochem), 57. Prof Exp: Am Cancer Soc fel, Albert Einstein Col Med, 57-59; instr math & chem, Herzliah Acad, 62-65; lectr biochem, McGill Univ, 65-66; NIH fel, Mt Sinai Sch Med, 70-73; INSTR NEUROCHEM, MT SINAI SCH MED, 74- Concurrent Pos: Lectr, Queen's Col, NY, 73-; vis asst prof, Stern Col Women, Yeshiva Univ, NY. Mem: Brit Biochem Soc. Res: Subcellular compartmentalization of respiratory activity and energy metabolism; protein chemistry of nervous system: their localization, separation and identification and function. Mailing Add: Dept of Neurol Mt Sinai Sch of Med New York NY 10029

LAPKIN, MILTON, organic chemistry, see 12th edition

LAPLANTE, ALBERT AUREL, JR, b Williamstown, Mass, Dec 21, 22; m 45; c 1. ECONOMIC ENTOMOLOGY. Educ: Mass State Col, BS, 44; Cornell Univ, PhD(econ entom), 49. Prof Exp: Asst, NY Agr Exp Sta, Geneva, 44-48; asst exten entomologist, Pa State Col, 48-49; exten entomologist, Cornell Univ, 49-56, entomologist, Govt Guam, 57-59, agriculturist & chief plant indust, 59-64; prof agr, Col Guam, 64-65; EXTEN ENTOMOLOGIST, UNIV HAWAII, 65- Mem: Entom Soc Am. Res: Biology and control of peach tree borer and oriental fruit moth; biological and integrated insect control of common vegetable crops under tropical conditions. Mailing Add: Dept of Entom Col of Trop Agr Univ of Hawaii Honolulu HI 96822

LAPLAZA, MIGUEL LUIS, b Zaragoza, Spain, Mar 20, 38; m 69; c 4. MATHEMATICS. Educ: Univ Barcelona, MD, 60; Univ Madrid, PhD(math), 65. Prof Exp: Instr math, Univ Barcelona, 60-61; from asst prof to assoc prof, Univ Madrid, 60-66; ASSOC PROF MATH, UNIV PR, MAYAGUEZ, 67- Mem: Am Math Soc; Math Asn Am. Res: Category theory. Mailing Add: 14-M-7-B Terrace Mayaguez PR 00708

LA POINTE, JOSEPH L, b Harvey, Ill, Sept 7, 34; m 54, 66; c 3. ZOOLOGY. Educ: Portland State Col, BA, 60; Univ Calif, Berkeley, PhD(zool), 66. Prof Exp: Assoc instr zool, Univ Calif, Berkeley, 64-66; Nat Inst Child Health & Human Develop fel, 66-68; asst prof, 68-71, ASSOC PROF BIOL, N MEX STATE UNIV, 71- Concurrent Pos: NIH res grant, 72-73. Mem: Am Soc Ichthyol & Herpet; Am Soc Zool; Brit Soc Endocrinol; Europ Soc Comp Endocrinol. Res: Effect of perietal eye on circadian rhythms in lizards; thermoregulation in antusiid lizards; ultrastructure of reptilian pituitary; physiology of neurohypophysial hormones in lower vertebrates; fat mobilization in lizards. Mailing Add: Dept of Biol NMex State Univ Las Cruces NM 88001

LAPOINTE, LEONARD LYELL, b Iron Mountain, Mich, June 28, 39; m 63; c 2. SPEECH PATHOLOGY. Educ: Mich State Univ, BA, 61; Univ Colo, MA, 66, PhD(speech path), 69. Prof Exp: Dir speech path commun dis, Bd Educ, Menasha, Wis, 61-64; speech pathologist, Gen Rose Mem Hosp, 66; asst prof phonetics, Univ Colo, Denver, 68-69; COORDR INSTR AUDIOL & SPEECH PATH, VET ADMIN HOSP, GAINESVILLE & ASST PROF COMMUN DIS, UNIV FLA, 69- Concurrent Pos: Fel neurogenic commun dis, Vet Admin Hosp, Denver, 68-69; res investr speech sci, Vet Admin Hosp, Gainesville, 71-; res fac neuroling, Ctr Neurol-Behav Ling Res, Univ Fla, 74- Honors & Awards: Award, Sci Exhib, XV World Cong Logopedics & Phoniatrics, 71. Mem: Acad Aphasia; Am Speech & Hearing Asn; Int Asn Logopedics & Phoniatrics; Int Neuropsychol Soc; Am Cleft Palate Asn. Res: Development of measurement strategies of human oral sensation-perception; oral physiology and neurolinguistics; diagnosis and treatment strategies in aphasia and related neurogenic communication impairments. Mailing Add: Audiol & Speech Path Serv Vet Admin Hosp Gainesville FL 32602

LAPOINTE, SERGE MICHEL, space physics, see 12th edition

LAPONSKY, ALFRED BAER, b Cleveland, Ohio, Nov 24, 21; m 57; c 2. PHYSICAL ELECTRONICS. Educ: Lehigh Univ, BS, 43, MS, 47, PhD(physics), 51. Prof Exp: Instr physics, Lehigh Univ, 47-51; physicist electron physics, Res Lab, Gen Elec Co, 51-63; assoc prof elec eng, Univ Minn, Minneapolis, 63-66; MEM STAFF, ELEC TUBE DIV, WESTINGHOUSE ELEC CORP, 66- Mem: Am Phys Soc. Res: Electron physics; electro-optics; image sensing and display techniques. Mailing Add: Westinghouse Elec Corp Electronic Tube Div Horseheads NY 14845

LA PORE, RICHARD FRANCIS, organic chemistry, see 12th edition

LAPORTE, LEO FREDERIC, b Englewood, NJ, July 30, 33; m 56; c 2. GEOLOGY. Educ: Columbia Univ, AB, 56, PhD(geol), 60. Prof Exp: From instr to prof geol, Brown Univ, 59-71; PROF EARTH SCI, UNIV CALIF, SANTA CRUZ, 71-, DEAN NATURAL SCI, 75- Concurrent Pos: Vis prof, Yale Univ, 64- Mem: AAAS; Am Asn Petrol Geologists; Geol Soc Am; Soc Econ Paleont & Mineral; Paleont Soc. Res: Paleoecology and environmental stratigraphy; history and evolution of life. Mailing Add: Div of Natural Sci Univ of Calif Santa Cruz CA 95064

LAPOSA, JOSEPH DAVID, b St Louis, Mo, July 21, 38; m 68; c 2. PHYSICAL CHEMISTRY. Educ: St Louis Univ, BS, 60; Univ Chicago, MS, 62; Loyola Univ, Ill, PhD(chem), 65. Prof Exp: NIH fel chem, Cornell Univ, 65-67; asst prof, 67-73, ASSOC PROF CHEM, McMASTER UNIV, 73- Mem: Am Phys Soc. Res: Molecular luminescence. Mailing Add: 17 Ravina Crescent Ancaster ON Can

LAPP, RALPH EUGENE, physics, see 12th edition

LAPP, THOMAS WILLIAM, b Joliet, Ill, Oct 6, 37; m 61. ENVIRONMENTAL CHEMISTRY, INDUSTRIAL CHEMISTRY. Educ: Coe Col, BA, 59; Kans State Univ, MS, 61, PhD(inorg chem), 63. Prof Exp: Fel radiation chem, NAm Aviation Sci Ctr, 63-64; asst prof nuclear chem, Univ WVa, 64-66; assoc chemist, Midwest Res Inst, Mo, 66-69; mem staff, Univ Mo-Kansas City, 70-74; ASSOC CHEMIST, MIDWEST RES INST, MO, 74- Res: Production and utilization of industrial chemicals suspected of possessing toxic properties; mechanisms whereby such materials are introduced to the environment, quantities introduced, and subsequent reactions in the environment. Mailing Add: Midwest Res Inst 425 Volker Blvd Kansas City MO 64110

LAPPANO-COLLETTA, ELEANOR RITA, cytology, histochemistry, see 12th edition

LAPPAS, LEWIS CHRISTOPHER, b Lynn, Mass, May 14, 21; m 49; c 3. PHARMACEUTICAL CHEMISTRY. Educ: Mass Col Pharm, BS, 43, MS, 48; Purdue Univ, PhD(pharmaceut chem), 51. Prof Exp: RES SCIENTIST, ELI LILLY & CO, 51- Mem: Am Chem Soc; Am Pharmaceut Asn. Res: Drug encapsulation processes; basic gelatin research as applied to capsular forms; study pf filmogens as drug release mechanisms; stabilization of drugs and drug forms; pharmaceutical aspects of drug absorption; investigation of antimicrobials in drug and cosmetic formulations. Mailing Add: 5331 Hawthorne Dr Indianapolis IN 46226

LAPPIN, GERALD R, b Caro, Mich, Apr 14, 19; m 45; c 2. CHEMISTRY. Educ: Alma Col, BS, 41; Northwestern Univ, PhD(org chem), 46. Prof Exp: Asst, Northwestern Univ, 41-43, interim instr chem, 44, asst, 44-46; asst prof, Antioch Col, 46-49 & Univ Ariz, 49-51; sr res chemist, 51-69, RES ASSOC, TENN EASTMAN CO DIV, EASTMAN KODAK CO, 69- Concurrent Pos: Consult, Vernay Labs, Ohio, 46-49. Mem: Am Chem Soc. Res: Additives for foods; plastics and petroleum products; chemistry of polyesters; technology forecasting as applied to research and development planning. Mailing Add: Tenn Eastman Co Kingsport TN 37664

LAPPING, MARK BARRY, b New York, NY, June 17, 46; m 67; c 2. RESOURCE MANAGEMENT, FORESTRY. Educ: State Univ NY Col New Paltz, BS, 67; Columbia Univ, BS, 69; Mass Inst Technol, cert urban & regional econ, 70; Emory Univ, PhD(urban studies), 72. Prof Exp: Asst prof planning, State Univ NY Col Oswego, 70-72 & Grad Sch Planning, Va Polytech Inst & State Univ, 72-73; assoc prof, Inst Man & Environ, State Univ NY Col Plattsburgh, 73-75; ASSOC PROF PLANNING & FORESTRY, UNIV MO-COLUMBIA, 75- Concurrent Pos: Planning consult, State Comn Tug Hill, NY, 73-75, Lake Champlain-Lake George Regional Planning Bd, 73-75, State Planning Off Vt, 74-75 & State Planning Off Mo, 75-; res assoc, Ctr Northern Studies, Vt, 73-; Environ Protection Agency grant, 74-75. Mem: Inst Ecol; Am Inst Planners; Am Soc Planning Officials; Soil Conserv Soc Am. Res: Developing tools and methods for comprehensive land use planning; working to integrate systems and methods for forest land use planning and regional resource managment. Mailing Add: Forestry Fisheries & Wildlife Univ of Mo 1-25 Agr Bldg Columbia MO 65201

LAPPORTE, SEYMOUR JEROME, b Chicago, Ill, Mar 26, 30; m 64; c 2. ORGANIC CHEMISTRY, PETROLEUM CHEMISTRY. Educ: Univ Chicago, MS, 53; Univ Calif, Los Angeles, PhD(chem), 57. Prof Exp: Asst, Univ Calif, Los Angeles, 53-56; sr res assoc, Chevron Res Co, 56-74, MGR PIONEERING DIV, CHEVRON RES CO, 74- Concurrent Pos: Teacher, Exten, Univ Calif, 60-; vis scholar, Stanford Univ, 68-69. Mem: Am Chem Soc; The Chem Soc. Res: Organic reaction mechanisms; organometallics; oxidation; transition metal chemistry; homogenous catalysis; ultraviolet stabilization. Mailing Add: 107 Ardith Dr Orinda CA 94563

LAPRADE, JESSE COBB, b Bainbridge, Ga, Apr 1, 41; m 69. PLANT PATHOLOGY. Educ: Va Polytech Inst & State Univ, BS, 66; NC State Univ, MS, 68; Univ Fla, PhD(plant path), 73. Prof Exp: Exten agent ornamental hort, Univ Fla, 68-70; ASST PROF PLANT PATH, CLEMSON UNIV, 73- Concurrent Pos: Landscape maintenance consult, Palms Nursery, Naples, Fla, 69-70. Res: Tobacco disease control; aflatoxin controlled research; epidemiology and control. Mailing Add: PeeDee Exp Sta PO Box 271 Florence SC 29501

LA PRADE, KERBY EUGENE, b Grand Saline, Tex, Feb 12, 28; div; c 1. GEOLOGY. Educ: Tex Tech Univ, BS, 51, MS, 54, PhD(geol), 69. Prof Exp: Geophysicist, Texaco, Inc, 53-55; geologist, Phillips Petrol Co, 55-63; from instr to teaching asst gen geol, Tex Tech Univ, 63-67; PROF EARTH SCI, E TEX STATE UNIV, 67- Concurrent Pos: Fac res grant geol, E Tex State Univ, 69-70; Am Geol Inst intern, Spain, 71. Honors & Awards: Antarctic Serv Medal, US Govt, 69. Mem: Am Asn Petrol Geologists; Geol Soc Am. Res: Structure, stratigraphy and sedimentation of Antarctica and the southwestern United States. Mailing Add: Dept of Earth Sci ETex State Univ Commerce TX 75428

LA PRADE, MARIE DOUGLAS, b Jerome, Ariz, Aug 21, 42; m 63. STRUCTURAL CHEMISTRY, INORGANIC CHEMISTRY. Educ: Univ Mich, BSc, 64; Univ Vt, MS, 67; Mass Inst Technol, PhD(chem), 70. Prof Exp: Res assoc chem, Mass Inst Technol, 69-73; ASST PROF CHEM, NASSON COL, 73- Concurrent Pos: Vis res assoc chem, Univ NH, 76- Mem: Am Crystallog Asn; Am Asn Univ Prof. Res: Single crystal x-ray diffraction applied to inorganic and organometallic molecules. Mailing Add: Dept of Chem Nasson Col Springvale ME 04083

LAPRADE, MARY HODGE, b Oakland, Calif, Feb 6, 29; m 58; c 2. ZOOLOGY. Educ: Wilson Col, AB, 51; Radcliffe Col, AM, 52, PhD(biol), 58. Prof Exp: Instr biol, Simmons Col, 52-55; instr zool, 58-60 & 64-65, LECTR BIOL SCI, SMITH COL, 65-DIR, CLARK SCI CTR, 73- Concurrent Pos: Instr, NSF In-Serv Inst High Sch Biol Teachers, 65-66. Mem: Sigma Xi. Res: Growth and regeneration, particularly in crustaceans; fine structure of endocrine organs in crustaceans. Mailing Add: Dept of Biol Sci Smith Col Northampton MA 01060

LAPSLEY, ALWYN COWLES, b Albemarle Co, Va, Mar 12, 20. PHYSICS. Educ: Univ Va, BEE, 41, MS, 44, PhD(physics), 47. Prof Exp: Instr physics, Univ Va, 41-43, asst, Manhattan Proj & Navy Fire Control, 43-46; sr physicist, Photo Prod Dept, E I du Pont de Nemours & Co, 47-51, res engr, Atomic Energy Div, 51-60; SR SCIENTIST, REACTOR FACIL, SCH ENG & APPL SCI, UNIV VA, 60-, LECTR NUCLEAR ENG, 60- Mem: Am Phys Soc; Am Nuclear Soc. Res: Mechanics; physical optics; Kerr effect with high frequency fields; nuclear physics; ion chambers; reactor kinetics; isotope separation. Mailing Add: Reactor Facil Univ of Va Sch of Eng & Appl Sci Charlottesville VA 22901

LAPUCK, JACK LESTER, b Jamaica Plain, Mass, Aug 28, 24; m 48; c 3. FOOD CHEMISTRY, BACTERIOLOGY. Educ: Northeastern Univ, BS, 46; Univ Mass, MS, 49; Calvin Coolidge Col, DSc(sci educ), 60. Prof Exp: Food sanitarian, Montgomery County Health Dept, Md, 50-51; food chemist, Food & Drug Res Labs, NY, 51; chemist, Waltham Labs, Inc, 51-55, lab dir & vpres, 55-66; OWNER & MGR, LAPUCK LABS, 66- Concurrent Pos: Instr, Univ Exten, Mass Dept Educ, 55-; instr, Boston State Col. Mem: AAAS; Am Chem Soc; Am Soc Microbiol; Inst Food Technologists; Am Pub Health Asn. Res: Food technology; microbiology; analytical chemistry. Mailing Add: Lapuck Labs 520 Main St Waltham MA 02154

LAQUER, HENRY L, b Frankfurt-am-Main, Ger, Nov 28, 19; nat US; m 47; c 4. CRYOGENICS. Educ: Temple Univ, AB, 43; Princeton Univ, MA, 45, PhD(phys chem), 47. Prof Exp: Res chemist, Ladox Labs, Pa, 46; MEM STAFF, LOS ALAMOS SCI LAB, UNIV CALIF, 47- Concurrent Pos: Adj prof, Los Alamos Residence Ctr, Univ NMex, 70-73. Mem: Am Phys Soc. Res: High magnetic fields; applied superconductivity; dielectric studies; elastic properties of metals; cryogenics. Mailing Add: Los Alamos NM

LAQUEUR, GERT LUDWIG, b Strasbourg, France, June 11, 12; nat US; m 42; c 3. PATHOLOGY. Educ: Univ Freiburg, MD, 37. Prof Exp: Asst anat, Univ Zurich, 37; asst gynec & obstet, Stanford Univ, 38, instr, 39-42, instr path, 42-46, asst prof, 46-50; pathologist, 50-54, chief path, Atomic Bomb Casualty Comn, Hiroshima, Japan, 54-57, pathologist, 57-60, asst chief, 60, CHIEF LAB EXP PATH, NAT INST ARTHRITIS, METAB & DIGESTIVE DIS, 61- Mem: Fel AAAS; Am Soc Exp Path; Soc Exp Biol & Med; Am Asn Cancer Res; Am Asn Path & Bact. Res: Action of androgens; hormone analyses in testicular tumors; pathology of the endocrine system; general and experimental pathology; geographic pathology; chemical carcinogenesis. Mailing Add: Lab of Exp Path Nat Inst of Arthritis Metab & Digestive Dis Bethesda MD 20014

LARA-BRAUD, CAROLYN WEATHERSBEE, b Waco, Tex, Jan 4, 40; m 70. BIOCHEMISTRYRY. Educ: Univ Tex, Austin, BA, 62, PhD(chem), 69. Prof Exp: Res assoc biochem, Clayton Found Biochem Inst & lectr home econ, Nutrit Div, Univ Tex, Austin, 71-73; asst res scientist biochem 73-75, ASST PROF HOME ECON, UNIV IOWA, 75- Mem: AAAS; Am Chem Soc. Res: Intermediary metabolism; regulation of inducible enzyme systems. Mailing Add: Dept of Home Econ Univ of Iowa Iowa City IA 52242

LARACH, SIMON, b Brooklyn, NY, Apr 21, 22; m 48; c 2. PHYSICAL INORGANIC CHEMISTRY, SOLID STATE CHEMISTRY. Educ: City Col New York, BS, 43; Princeton Univ, MA, 51, PhD(chem), 54. Prof Exp: Res chemist, Third Res Div, Goldwater Hosp, Col Med, NY Univ, 43 & 46; res chemist luminescence & solid state, David Sarnoff Res Ctr, Radio Corp Am, 46-59, head photoelectronic, magnetic & dielec res, 59-61, assoc lab dir, 61-67, overseas fel, 69, FEL, DAVID SARNOFF RES CTR, RCA CORP, 67- Concurrent Pos: Vis fel, Princeton Univ, 67-68; Indust Res Inst-Am Chem Soc liaison lectr, 68; vis prof, Hebrew Univ Jerusalem, 69-70, adj prof, 71-73; UN consult, UNESCO Div Tech Educ & Res, 71; prin lectr, NATO Adv Study Inst, Norway, 72; vis prof, Swiss Fed Inst Technol, 72; div ed electronics, J Electrochem Soc; vis prof, Hahnemann Med Col & Hosp, 74- Honors & Awards: Citation, Off Sci Res & Develop, 46; Sarnoff Gold Medal Team Award Sci, 66. Mem: Am Chem Soc; fel Am Phys Soc; fel Am Inst Chem; Electrochem Soc. Res: Synthesis

LARACH

and properties of electronically-active solids. Mailing Add: 139 Sycamore Rd Princeton NJ 08540

LARAGH, JOHN HENRY, b Yonkers, NY, Nov 18, 24; m 74; c 2. PHYSIOLOGY, MEDICINE. Educ: Cornell Univ, MD, 48. Prof Exp: Intern med, Presby Hosp, New York, 48-49, asst resident, 49-50, asst, 50-55, instr, 55-57, assoc, 57-59, from asst prof to prof clin med, Col Physicians & Surgeons, Columbia Univ, 67-75; HILDA ALTSCHUL MASTER PROF MED, MED COL, CORNELL UNIV, 75-, DIR CARDIOVASC CTR, NY HOSP-CORNELL MED CTR, 75- Concurrent Pos: Nat Heart Inst trainee, 50-51; asst physician, Presby Hosp, New York, 50-54, from asst attend physician to assoc attend physician, 54-69, attend physician, 69-75, vchmn in chg med affairs, Bd Trustees, 74; NY Heart Asn res fel, 51-52; mem med adv bd, Coun High Blood Pressure Res, Am Heart Asn, 61, chmn, 68-72; consult cardiovasc study sect, USPHS, 64-68 & heart prog proj A, 67-72; dir, Hypertension Ctr & Nephrol Div, Columbia-Presby Med Ctr, 71-75; mem policy adv bd, Hypertension Detection & Follow-Up Prog, Nat Heart & Lung Inst, 71-, mem bd sci coun, 74-; mem adv bd, Am Soc Contemporary Med & Surg, 74. Honors & Awards: Stouffer Prize Med Res, 69. Mem: Am Soc Clin Invest; fel Am Col Physicians; Am Soc Nephrol; assoc Harvey Soc; Asn Am Physicians. Res: Cardiovascular and renal diseases; endocrinology. Mailing Add: New York Hosp-Cornell Med Ctr 525 E68th St New York NY 10021

LARAMORE, GEORGE ERNEST, b Ottawa, Ill, Nov 5, 43; m 64; c 1. PHYSICS, MEDICINE. Educ: Purdue Univ, BS, 65; Univ Ill, Urbana, MS, 66, PhD(physics), 69; Univ Miami, MD, 76. Prof Exp: NSF fel, Univ Ill, Urbana, 69-70, res assoc physics, 70-71; res physicist, Sandia Labs, 71-75. Mem: Am Phys Soc; Am Vacuum Soc; AMA. Res: Theory of low-energy electron diffraction; interaction of fast electrons with solids. Mailing Add: 3830 SW 61st Ave Miami FL 33155

LARAN, ROY JOSEPH, physical chemistry, see 12th edition

LARCHER, HEINRICH, b Telfs, Austria, Dec 1, 25; nat US; m 64; c 2. PURE MATHEMATICS. Educ: Univ Innsbruck, Dr phil(math), 50. Prof Exp: Teacher pub sch, Eng, 50-51; master math, Seaford Col, 51-53; prof, Inst Montana, Switz, 54-56; from instr to asst prof, Mich State Univ, 56-64; MEM STAFF, UNIV MD, MUNICH, 64- Mem: Am Math Soc. Res: Orthogonal functions in two variables; theory of functions of a complex variable; theory of numbers.

LARD, EDWIN WEBSTER, b Ala, July 17, 21; m 45; c 4. ANALYTICAL CHEMISTRY. Educ: Ark State Col, BS, 49; Memphis State Univ, MA, 61. Prof Exp: Chemist, Ethyl Corp, 49-52 & Chemstrand Corp, 52-54; sr chemist, Nitrogen Prod Div, Tenn, 54-62, RES SUPVR, RES DIV, W R GRACE & CO, CLARKSVILLE, MD, 62- Mem: Am Chem Soc. Res: Trace gas analysis with infrared; separation and determination of argo, oxygen and nitrogen by chromatography; trace analysis of acetylene, methane, carbon monoxide and carbon dioxide; synthesis of aryl dimethyl sulfonium chloride compounds; chemical warfare agents; unsaturates in auto emissions. Mailing Add: 12703 Beaverdale Lane Bowie MD 20715

LARDNER, PETER JAMES, ecology, zoology, see 12th edition

LARDNER, ROBIN WILLMOTT, b Leicester, Eng, Feb 9, 38; m 58; c 4. APPLIED MATHEMATICS, SOLID MECHANICS. Educ: Cambridge Univ, BA, 59, PhD(appl math), 63. Prof Exp: Res assoc physics, Columbia Univ, 61-63; NATO fel appl math & theoret physics, Peterhouse Col, Cambridge Univ, 63-65; lectr math & physics, Univ East Anglia, 65-67; assoc prof, 67-70, chmn dept, 71-73, PROF MATH, SIMON FRASER UNIV, 70- Res: Solid mechanics, particularly dislocation theory and fracture; nonlinear vibrations of continuous media; asymptotic solutions of nonlinear partial differential equations. Mailing Add: Dept of Math Simon Fraser Univ Burnaby BC Can

LARDNER, THOMAS JOSEPH, b New York, NY, July 19, 38; m 64; c 3. ENGINEERING MECHANICS, BIOMECHANICS. Educ: Polytech Inst Brooklyn, BAeroE, 58, MS, 59, PhD(appl mech), 61. Prof Exp: Res assoc appl mech, Polytech Inst Brooklyn, 59-61; res engr, Jet Propulsion Lab, Calif Inst Technol, 62-63; instr math, Mass Inst Technol, 63-67, asst prof appl math, 67-70, assoc prof mech eng, 70-73; PROF THEORET & APPL MECH, UNIV ILL, URBANA, 73- Concurrent Pos: Fulbright lectr, Univ Nepal, 65-66; consult, Adelphi Res Ctr, Polytech Inst Brooklyn & Jet Propulsion Lab, Calif Inst Technol. Mem: Am Soc Mech Eng; Soc Indust & Appl Math; Am Soc Eng Educ. Res: Applied mathematics and mechanics in heat transfer, thermoelasticity, spacecraft temperature control and shell theory; applied solid mechanics. Mailing Add: Dept of Theoret & Appl Mech Talbot Lab Univ of Ill Urbana IL 61801

LARDY, HENRY ARNOLD, b Roslyn, SDak, Aug 19, 17; m 43; c 4. BIOCHEMISTRY. Educ: SDak State Col, BS, 39; Univ Wis, MS, 41, PhD(biochem), 43. Prof Exp: Fel, Nat Res Coun, Banting Inst, Univ Toronto, 44-45; from asst prof to prof, 45-66, VILAS PROF BIOL SCI, UNIV WIS-MADISON, 66-, CHMN, RES DEPT, ENZYME INST, 50- Honors & Awards: Neuberg Medal, 56; Lewis Award, Am Chem Soc, 49. Concurrent Pos: Harvey lect, 65. Mem: Nat Acad Sci; Am Chem Soc; Am Soc Biol Chem (pres, 64); Am Acad Arts & Sci. Res: Enzymes; intermediary metabolism; hormones. Mailing Add: Enzyme Inst Univ of Wis-Madison Madison WI 53706

LARDY, LAWRENCE JAMES, b Sentinel Butte, NDak, Aug 23, 34; m 56; c 2. NUMERICAL ANALYSIS. Educ: NDak State Col, Dickinson, BS, 57; Univ NDak, MS, 59; Univ Minn, PhD(math), 64. Prof Exp: Instr math, Univ NDak, 59-60 & Univ Minn, 62-64; from asst prof to assoc prof, 64-74, PROF MATH, SYRACUSE UNIV, 74- Concurrent Pos: Res fel, Yale Univ, 67-68; vis assoc prof, Univ Md, 73-74. Mem: Soc Indust & Appl Math; Math Asn Am; Am Math Soc. Res: Functional analysis. Mailing Add: Dept of Math Syracuse Univ Syracuse NY 13210

LARGE, ALFRED MCKEE, b Listowel, Ont, Mar 7, 12; nat US; m 43; c 1. SURGERY. Educ: Univ Toronto, BA, 33, MD, 36. Prof Exp: Instr surg, Sch Med, Washington Univ, 43-44; asst prof, 46-62, ASSOC PROF CLIN SURG, COL MED, WAYNE STATE UNIV, 62- Concurrent Pos: Attend surgeon, Grace Hosp, Detroit & St John Hosp; consult surgeon, Vet Admin, USPHS Hosp & Bon Secours Hosp, Detroit; pvt pract. Mem: AMA. Res: Clinical surgery. Mailing Add: 19515 Mack Grosse Pointe Woods MI 48236

LARGE, ROBERT F, b Kansas City, Kans, Apr 1, 36; m 60; c 2. ANALYTICAL CHEMISTRY, PHOTOGRAPHIC SCIENCE. Educ: Cent Mo State Col, BA & BS, 58; Mich State Univ, PhD(anal chem), 63. Prof Exp: Sr res chemist, 63-68, lab head, 68-72, ASST DIV HEAD, RES LABS, EASTMAN KODAK CO, 72- Mem: Am Chem Soc. Res: Electrochemistry. Mailing Add: Res Labs Eastman Kodak Co Rochester NY 14650

LARGENT, DAVID LEE, b San Francisco, Calif, Oct 30, 37; m 70. MYCOLOGY. Educ: San Francisco State Col, BA, 60, MA, 63; Univ Wash, PhD(bot), 68. Prof Exp: Instr bot, Foothills Jr Col, 63; instr bot & biol, Phoenix Jr Col, 63-64; asst prof bot, 68-74, ASSOC PROF BOT, HUMBOLDT STATE COL, 74- Mem: Am Soc Plant Taxon; Mycol Soc Am; Am Bryol & Lichenological Soc. Res: Taxonomy and ecology of the Rhodophylloid fungi on the Pacific coastal states of America; cryptogamic botany. Mailing Add: Dept of Biol Humboldt State Col Arcata CA 95521

LARGENT, MAX DALE, b Winchester, Va, Feb 28, 23; m 54; c 1. DENTISTRY. Educ: Med Col Va, DDS, 50. Prof Exp: Instr pedodontics, Med Col Va, 52-56, from asst prof to prof, 56-72; ASST DEAN, BAYLOR COL DENT, 72- Concurrent Pos: Chmn dept pedodontics, Med Col Va, 69-72, dir postgrad pedodontics, 57-69. Mem: Am Dent Asn; Am Asn Dent Schs; fel Am Col Dent. Mailing Add: Baylor Col of Dent 800 Hall St Dallas TX 75226

LARGHI, OSCAR PEDRO, biological chemistry, virology, see 12th edition

LARGMAN, THEODORE, b Philadelphia, Pa, Nov 16, 23; m 59; c 4. ORGANIC CHEMISTRY. Educ: Temple Univ, AB, 48; Ind Univ, PhD(org chem), 52. Prof Exp: Sr res chemist, Nitrogen Div, 62-66, scientist, Cent Res Labs, 66-68, RES GROUP LEADER, CORP CHEM RES LAB, ALLIED CHEM CORP, 68- Mem: Am Chem Soc. Res: Organic synthesis; fine chemicals; ozone reactions; design of micro-pilot plants; agricultural pesticides; flame retardant chemicals and polymers. Mailing Add: 7 Upper Field Rd Morristown NJ 07960

LARGUIER, EVERETT HENRY, b New Orleans, La, Jan 26, 10. MATHEMATICS. Educ: St Louis Univ, AB, 34, MS, 36; Univ Mich, PhD(math), 47. Hon Degrees: PhL, St Louis Univ, 37, STL, 42. Prof Exp: Instr math, Spring Hill Col, 37-38; asst, St Louis Univ, 38-42; chmn dept, 47-71, trustee, 52-59, PROF MATH, SPRING HILL COL, 47- Mem: Am Math Soc; Math Asn Am; Math Soc France. Res: Probability; foundations of mathematics; topology. Mailing Add: Dept of Math Spring Hill Col Mobile AL 36608

LARIMER, JAMES LYNN, b Washington Co, Tenn, Jan 7, 32. NEUROBIOLOGY. Educ: ETenn State Univ, BS, 53; Univ Va, MA, 54; Duke Univ, PhD, 59. Prof Exp: From asst prof to assoc prof, 59-68, actg chmn dept, 73-74, PROF ZOOL, UNIV TEX, AUSTIN, 68- Concurrent Pos: Guggenheim fel, Stanford Univ, 67-68; mem, Physiol Study Sect, NIH, 72-76. Mem: Soc Neurosci; Am Physiol Soc; Sigma Xi; Am Soc Zoologists. Res: Comparative physiology; behavior and neurophysiology of invertebrates. Mailing Add: Dept of Zool Univ of Tex Austin TX 78712

LARIMER, JOHN WILLIAM, b Pittsburgh, Pa, Sept 4, 39; m 65; c 1. GEOCHEMISTRY. Educ: Lehigh Univ, BA, 62, MS, 63, PhD(geol), 66. Prof Exp: NASA-AEC res assoc geochem, Enrico Fermi Inst, Univ Chicago, 66-69; asst prof, 69-74, ASSOC PROF GEOL, ARIZ STATE UNIV, 74- Mem: AAAS; Geochem Soc; Am Geophys Union. Res: Cosmochemistry; mineralogy and composition of meteorites. Mailing Add: Dept of Geol Ariz State Univ Tempe AZ 85281

LARIMORE, ANN EVANS, b St Louis, Mo, Dec 5, 31; m 58. GEOGRAPHY. Educ: Wellesley Col, BA, 52; Univ Chicago, MA, 55, PhD, 58. Prof Exp: Instr geog, Univ Chicago, 57-58, res assoc, 58-59; asst prof, Newark Col, Rutgers Univ, 61-64; lectr, 66-70, assoc prof, 70-72, PROF GEOG, RESIDENTIAL COL, UNIV MICH, ANN ARBOR, 72-, ASSOC DIR INSTR, 75- Mem: Asn Am Geogrs; Am Anthrop Asn; Nat Coun Geog Educ. Res: Cultural geography; cultural influences upon changing geographic patterns in non-Western areas; urban settlement patterns. Mailing Add: Residential Col E Quad Univ of Mich 212 Tyler Ann Arbor MI 48104

LARIMORE, RICHARD WELDON, b Rogers, Ark, Feb 10, 23; m 47; c 3. FISH BIOLOGY. Educ: Univ Ark, BS, 46; Univ Ill, MS, 47; Univ Mich, PhD(zool), 50. Prof Exp: Asst aquatic biol, 46-54, assoc, 54-58, AQUATIC BIOLOGIST, ILL STATE NATURAL HIST SURV, 58-, PROF ENVIRON ENG, UNIV ILL, URBANA, 69-, ZOOL, 70- Concurrent Pos: Fishery expert, Food & Agr Orgn, 63-64 & 72-73. Honors & Awards: Fisheries Pub Award, Wildlife Soc, 57; Am Fisheries Soc Award, 60. Mem: Ecol Soc Am; Am Soc Limnol & Oceanog; Am Soc Ichthyol & Herpet; Am Fisheries Soc; Am Inst Fishery Res Biol. Res: Ecology of stream and reservoir fishes; dynamics of cooling lakes. Mailing Add: Natural Hist Surv Univ of Ill Urbana IL 61801

LARIS, PHILIP CHARLES, b Perth Amboy, NJ, Sept 5, 31; m 56; c 4. PHYSIOLOGY. Educ: Rutgers Univ, BS, 52; Princeton Univ, MA, 54, PhD(physiol), 56. Prof Exp: Instr biol, Univ Calif, 56-58, from asst prof to assoc prof, 58-65; assoc prof, Franklin & Marshall Col, 65-66; ASSOC PROF BIOL, UNIV CALIF, SANTA BARBARA, 66- Mem: AAAS; Soc Gen Physiol. Res: Cell permeability and cell membrane structure; ion and sugar transport; shape change in membranes. Mailing Add: Dept of Biol Sci Univ of Calif Santa Barbara CA 93106

LARISEY, MARY MAXINE, b Terre Haute, Ind, July 1, 09. TAXONOMY, PLANT ANATOMY. Educ: Washington Univ, AB, 32, MS, 34, PhD(bot), 39. Prof Exp: Asst bot, Henry Shaw Sch Bot, Washington Univ, 34-36, instr, 36-38; instr, Wellesley Col, 39-40; from asst prof to prof biol, Judson Col, 40-47; from asst prof to assoc prof, 47-72, PROF BIOL, COL PHARM, MED UNIV SC, 72- Concurrent Pos: Chmn div sci, Judson Col, 43-47. Mem: Am Soc Plant Taxon; Bot Soc Am; Torrey Bot Club. Res: Botany; cytogenetics; taxonomy of flowering plants; Baptista; revision of North American species of Thermopsis. Mailing Add: 2 Franklin St Charleston SC 29401

LA RIVERS, IRA JOHN, b San Francisco, Calif, May 1, 15; m 51; c 1. INSECT TAXONOMY. Educ: Univ Nev, BS, 37; Univ Calif, PhD(entom), 48. Prof Exp: Asst, NC State Col, Raleigh, 37-38; entomologist, Univ Nev, 39-42; agr inspector, State Dept Agr, Calif, 42; asst, Univ Calif, 48; from asst prof to assoc prof, 48-61, chmn dept & dir mus biol, 53-63, PROF BIOL, UNIV NEV, RENO, 61- Concurrent Pos: Field agent, USDA, 36, 37, 39 & 42; mem, Univ Calif Trop Biogeog Exped, Mex, 47; field biologist, Nev Fish & Game Comn, 49; ecologist, Pac Sci Bd, Nat Sci Coun, Marshall Islands, 50, fire control, US Forest Serv, 52, 54 & 55; fisheries biologist, US Fish & Wildlife Serv, 53; NSF grant, 68-70; consult, Desert Res Inst, Winston Encycl, Handbook Biol, Parents Mag Encycl, Reinhold Encycl Biol, City San Francisco & Nat Lexicographic Bd. Mem: Am Entom Soc; Paleont Soc; Ecol Soc Am; Soc Vert Paleont; Soc Syst Zool. Res: Biological control Mormon cricket; taxonomy Tenebrionidae and Naucoridae; endemism in western Great Basin fishes; comparative chromatography and electrophoretic patterns in Nevada fishes; revision genus Ambrysus; algae of Nevada; amphibians and reptiles of Nevada; general zoology. Mailing Add: Biol Soc of Nev PO Box 167 Verdi NV 89439

LARK, CARL GORDON, b Lafayette, Ind, Dec 13, 30; m 51; c 3. MICROBIOLOGY, VIROLOGY. Educ: Univ Chicago, PhB, 49; NY Univ, PhD(microbiol), 53. Prof Exp: Am Cancer Soc res fel, Statenserum Inst, Denmark, 53-55; Nat Found res fel, Biophys Lab, Univ Geneva, 55-56; instr microbiol, Sch Med, St Louis Univ, 56-57, sr instr, 57-58, from asst prof to assoc prof, 58-63; prof, Kans State Univ, 63-70; CHMN DEPT BIOL, UNIV UTAH, 70- Concurrent Pos: Nat Inst Gen Med Sci career develop award, 63-70; mem, Genetics Panel, NSF, 66-69; consult, Eli Lilly & Co, 68-74. Mem: Biophys Soc; Am Soc Cell Biol; Am Soc Biol Chem; Am Soc Microbiol;

Am Cancer Soc. Res: Cell growth and division; DNA replication and segregation in bacteria and eucaryotes. Mailing Add: Dept of Biol Univ Utah Salt Lake City UT 84112

LARK, CYNTHIA ANN, b Shawnee, Okla, Dec 31, 28; m 51; c 3. MICROBIOLOGY. Educ: Mt Holyoke Col, BA, 50; St Louis Univ, PhD, 62. Prof Exp: Lab technician, Carnegie Inst, 50-51 & Sloan-Kettering Inst Cancer Res, 51-53; res asst, Univ Geneva, 55-56 & St Louis Univ, 59-62; NIH fel, Washington Univ, 62-63; asst prof microbiol, Kans State Univ, 63-70; ASSOC RES PROF BIOCHEM, UNIV UTAH, 70- Mailing Add: 1190 Gilmer Dr Salt Lake City UT 84105

LARK, JOHN CALLOW, organic chemistry, polymer chemistry, see 12th edition

LARK, NEIL LAVERN, b Baker, Ore, Sept 10, 34; m 58; c 2. NUCLEAR PHYSICS. Educ: Chico State Col, AB, 55; Cornell Univ, PhD(phys chem), 60. Prof Exp: Res asst, Los Alamos Sci Lab, 55-56; res asst, Brookhaven Nat Lab, 57, jr res assoc, 58-59, res assoc, 60-61; NATO fel, Inst Nuclear Res, Amsterdam, Netherlands, 61-62, res assoc, 62; from asst prof to prof natural sci, Raymond Col, 62-75, PROF PHYSICS, UNIV PAC, 75- Concurrent Pos: Res asst, Univ NMex, 54; consult & res collabr, Los Alamos Sci Lab, 69; Ford Found fel & Fulbright grantee, Niels Bohr Inst, Copenhagen, Denmark, 67-68; NSF lectr, Tex A&M Univ, 70; res collabr, Brookhaven Nat Lab, 71-72, vis physicist, 75. Mem: Am Phys Soc; Am Asn Physics Teachers. Res: Nuclear spectroscopy; teaching of physics and chemistry. Mailing Add: Dept of Physics Univ of the Pacific Stockton CA 95211

LARKE, ROBERT PETER BRYCE, b Blairmore, Alta, Nov 14, 36; m 60; c 3. VIROLOGY. Educ: Queen's Univ, Ont, MD & CM, 60; Univ Toronto, DClSc, 66. Prof Exp: Intern, St Michael's Hosp, Toronto, Ont, 60-61; resident pediat, Hosp for Sick Children, Toronto, Ont, 61-62, res fel virol, Res Inst, 62-66; instr prev med, Sch Med, Case Western Reserve Univ, 66-68, sr instr, 68; asst prof pediat, McMaster Univ, 69-72, assoc prof pediat & path, 72-75; assoc prof pediat, 75-76, PROF PEDIAT, UNIV ALTA, 76-, CHIEF, DIV DIAG VIROL, PROVINCIAL LAB PUB HEALTH, 75- Concurrent Pos: Fel Microbiol, Sch Hyg, Univ Toronto, 63-66; Med Res Coun Can res fel, 64-66; asst prof path, McMaster Univ, 71-72; dir virol lab, St Joseph's Hosp, Hamilton, 71-75. Honors & Awards: Parkin Prize, Royal Col Physicians, Edinburgh, 66. Mem: Am Soc Microbiol; Am Fedn Clin Res; Infectious Dis Soc Am; Soc Pediat Res; Can Paediat Soc. Res: Clinical virology and infectious diseases of children and adults; mechanisms of host resistance to viral infections; interaction between viruses and blood platelets. Mailing Add: Dept of Pediat Univ of Alta Edmonton AB Can

LARKIN, DAVID, b London, Eng, Oct 6, 41. ELECTROCHEMISTRY. Educ: Loughborough Univ Technol, BTech, 65, PhD(electrochem), 68; Royal Inst Chem, ARIC, 70. Prof Exp: Asst prof chem, Fla Technol Univ, 72-73; ASST PROF CHEM, TOWSON STATE COL, 73- Mem: Am Chem Soc; The Chem Soc. Res: Study of electro kinetics at solid metal electrodes. Mailing Add: Dept of Chem Towson State Col Baltimore MD 21204

LARKIN, DONALD RICHARD, organic chemistry, see 12th edition

LARKIN, EDWARD P, b Watertown, Mass, Sept 27, 20; m 49; c 3. FOOD MICROBIOLOGY, VIROLOGY. Educ: Mass State Col, BS, 46; Univ Mass, MS, 48, PhD(bact), 54. Hon Degrees: Hon BS, Mass State Col, 43. Prof Exp: Instr bact, Univ Mass, 48-54, asst prof, 54-61; adminr, & assoc mem virol, Inst Med Res, NJ, 61-68; asst prof viral oncol, Sch Vet Med, Univ Pa, 68-70; CHIEF VIROL BR, BUR FOODS, FOOD & DRUG ADMIN, 70- Concurrent Pos: Vis assoc, Univ Pa, 61-65; Leukemia Soc scholar, 65-70. Mem: Am Soc Microbiol; Sigma Xi; Int Asn Comp Res on Leukemia & Related Dis (secy, 69-71). Res: Food virology; physical characterization of viruses; kinetic studies of physical and chemical inactivation of viruses. Mailing Add: Virol Br Div of Microbiol Bur of Foods FDA Cincinnati OH 45226

LARKIN, FRANCES ANN, b Ohio, Jan 16, 30. NUTRITION. Educ: Ohio State Univ, BS, 52; Univ Minn, Minneapolis, MS, 58; Cornell Univ, PhD(educ), 67. Prof Exp: Clin dietitian, St Luke's Hosp, Cleveland, Ohio, 53-57; assoc prof nutrition, Long Beach State Col, 58-61; adv, Col Home Econ, Dacca, Bangladesh, 61-64; ASSOC PROF, SCH PUB HEALTH, UNIV MICH, ANN ARBOR, 68- Mem: Am Dietetic Asn; Am Pub Health Asn. Res: Food consumption studies; nutrition education techniques; social correlates of food consumption. Mailing Add: Dept of Nutrit Sch of Pub Health Univ of Mich Ann Arbor MI 48104

LARKIN, JAMES RICHARD, b Boston, Mass, June 22, 25; m 46; c 6. MATHEMATICS. Educ: Tulane Univ, AB, 45; Univ Kans, AM, 49, PhD(math), 52. Prof Exp: Asst instr math, Univ Kans, 48-49, asst, 49-51; mathematician, Bur Ord, Navy Dept, 52-55, mathematician & opers analyst, Opers Eval Group, 55-63, dep dir off prog appraisal, Off Secy Navy, 63-67; sr staff engr, 67-71, MGR WASH OPERS, TRW SYSTS, INC, 71- Honors & Awards: Navy Dept Distinguished Civilian Serv Award, 67. Mem: Am Math Soc. Res: Series expansions; determinants; weapons systems analysis; operations analysis. Mailing Add: 7725 Timon Dr McLean VA 22101

LARKIN, JEANNE HOLDEN, b Chicago, Ill, Jan 19, 31; m 57; c 1. DEVELOPMENTAL BIOLOGY. Educ: Univ Ill, BEd, 52, MS, 53; Univ Mich, PhD(zool), 58. Prof Exp: Instr zool, Univ Mich, 57-58 & Duke Univ, 58-61; asst prof, 66-70, ASSOC PROF ZOOL, WESTERN ILL UNIV, 71- Mem: Soc Develop Biol; Am Soc Zoologists; Am Inst Biol Sci. Res: Homotransplantation, especially differentiation of first and second set of testicular grafts and antigen-antibody reactions; sex differentiation in fish; gut associated lymphod tissue. Mailing Add: 913 Jamie Lane Macomb IL 61455

LARKIN, JOHN MICHAEL, b York, Nebr, Aug 11, 37; m 65; c 3. ORGANIC CHEMISTRY. Educ: Univ Nebr, Lincoln, BS, 59; Univ Colo, Boulder, MS, 63, PhD(org chem), 65. Prof Exp: Chemist, Qual Water Br, US Geol Surv, Nebr, 59-60; from chemist to sr chemist, 65-70, res chemist, 70-74, SR RES CHEMIST, TEXACO INC, BEACON, 74- Mem: Am Chem Soc. Res: Organic synthesis; structure elucidation; reaction mechanisms; organic nitrogen compounds; free radical rearrangements; steroidal heterocycles; air pollution; lubricant additive synthesis. Mailing Add: 7 Sidney Lane Wappingers Falls NY 12590

LARKIN, JOHN MONTAGUE, b Philadelphia, Pa, Apr 7, 36; m 62; c 2. MICROBIAL PHYSIOLOGY, MICROBIAL ECOLOGY. Educ: Ariz State Univ, BS, 61, MS, 63; Wash State Univ, PhD(microbiol), 67. Prof Exp: Asst microbiol, Ariz State Univ, 61-63 & Wash State Univ, 63-67; asst prof, 67-71, ASSOC PROF MICROBIOL, LA STATE UNIV, BATON ROUGE, 71- Concurrent Pos: Chmn int bacillus comt, Int Asn Microbiol Socs, 72- Mem: AAAS; Am Soc Microbiol; Soc Appl Microbiol; Can Soc Microbiol. Res: Physiology of psychophilic bacteria; bacterial taxonomy; biology of gliding bacteria. Mailing Add: Dept of Microbiol La State Univ Baton Rouge LA 70803

LARKIN, LYNN HAYDOCK, b Highland Co, Ohio, Jan 29, 34; m 65; c 2. ANATOMY, REPRODUCTIVE BIOLOGY. Educ: Otterbein Col, BS, 56; Univ Colo, PhD(anat), 67. Prof Exp: Instr anat, Sch Med, Univ Colo, 66-67, res assoc molecular, cellular & develop biol, 67-68; asst prof anat, 68-73, ASSOC PROF PATH, DIV ANAT SCI, COL MED, UNIV FLA, 73- Res: Role of relaxin in pregnancy and parturition. Mailing Add: Div of Anat Sci Univ of Fla Col of Med Gainesville FL 32601

LARKIN, PETER ANTHONY, b Auckland, NZ, Dec 11, 24; m 48; c 5. FISHERIES. Educ: Univ Sask, BA, 45, MA, 46; Oxford Univ, DPhil, 48. Prof Exp: Chief fisheries biologist, Game Comn, BC, 48-55; dir inst fisheries & prof zool, Univ BC, 55-63; dir biol sta, Fisheries Res Bd Can, BC, 63-66; dir inst fisheries, 66-69, prof zool, 66-75, actg head dept, 69-70, prof, Inst Animal Resource Ecol, 69-75, head dept zool, 72-75, DEAN FAC GRAD STUDIES, UNIV BC, 75- Concurrent Pos: Nuffield Found fel, 61-62; mem, Sci Coun Can, 71-; mem Can nat comt, Spec Comt on Probs of Environ, 71-; mem, Fisheries Res Bd Can, 72-; mem adv comt sci criteria environ qual, Nat Res Coun Can, 75- Honors & Awards: Can Centennial Medal, 67. Mem: Am Fisheries Soc; Can Nature Fedn; Can Soc Environ Biologists; Can Soc Zoologists (pres, 72). Res: Population studies in fisheries biology. Mailing Add: Inst Animal Resource Ecol Univ of BC Vancouver BC Can

LARKIN, ROBERT HAYDEN, b New York, NY, Mar 26, 46; m 67; c 2. ANALYTICAL CHEMISTRY. Educ: Providence Col, BS, 68; Univ Mass, PhD(phys chem), 72. Prof Exp: Res assoc chem, Mass Inst Technol, 72-73; RES CHEMIST, ROHM & HAAS CO, 73- Mem: Am Chem Soc. Res: Development of analytical methods, particularly separation methods, for the analysis of a wide variety of industrial products and/or processes. Mailing Add: Rohm & Haas Co Bristol Res Labs Bristol PA 19007

LARKIN, WILLIAM J, physics, mathematics, see 12th edition

LARKINS, THOMAS HASSELL, JR, b Dickson, Tenn, Mar 1, 39; m 60; c 3. INORGANIC CHEMISTRY. Educ: Austin Peay State Col, BS, 60; Vanderbilt Univ, MA, 62, PhD(chem), 63. Prof Exp: SR RES CHEMIST, TENN EASTMAN CO, 63- Mem: Am Chem Soc. Res: Coordination chemistry; catalysis. Mailing Add: 4408 Beechcliff Dr Kingsport TN 37664

LARKS, SAUL DAVID, b London, Eng, June 16, 10; nat US; m 31; c 2. NEONATOLOGY. Educ: Univ Ill, BS, 43; Northwestern Univ, MS, 51; Univ Calif, Los Angeles, PhD(biophys), 56. Prof Exp: Jr res biophysicist, Univ Calif, Los Angeles, 55-56, asst res biophysicist, 56-58, asst prof biophys, 58-61; prof elec eng, Marquette Univ, 61-66; prof vet physiol & pharmacol, Univ Mo-Columbia, 66-72; SCI RESEARCHER & WRITER, 72- Concurrent Pos: USPHS res grants, 55-; consult, Packard-Bell Corp, 58, Magnavox Co, 61-65 & Gen Elec Co, 62-65. Mem: AAAS; Biophys Soc; Inst Elec & Electronics Engrs; NY Acad Sci; Soc Study Reproduction. Res: Bioelectric studies in human reproduction; living systems; biotelemetry; biomathematics; computer-based prediction of birth condition, malformation; electrical activity of uterus; hormonal basis fetal rotation; perinatal physiology; developmental cardiology; oviduct function; reproductive physiology. Mailing Add: 1112 Ninth St Santa Monica CA 90403

LARMIE, WALTER ESMOND, b Smithfield, RI, Sept 6, 20; m 43; c 3. FLORICULTURE. Educ: Univ RI, BS, 49, MS, 54. Prof Exp: From asst prof to assoc prof, 49-72, PROF HORT & CHMN DEPT PLANT & SOIL SCI, UNIV RI, 72- Mem: Am Soc Hort Sci. Res: Storage of flowers and plants; weed control; growth regulators; propagation and culture of poinsettias. Mailing Add: Red Wing Park West Kingston RI 02892

LARMORE, LAWRENCE LOUIS, b Washington, DC, Nov 23, 41; m 64; c 2. MATHEMATICS. Educ: Tulane Univ, BS, 61; Northwestern Univ, PhD(math), 65. Prof Exp: Asst prof math, Univ Ill, Chicago Circle, 65-68 & Occidental Col, 68-70; ASSOC PROF MATH, CALIF STATE COL, DOMINGUEZ HILLS, 70- Mem: Am Math Soc. Res: Algebraic topology; obstruction theory; classification of liftings, embeddings and immersions; twisted extraordinary cohomology. Mailing Add: Dept of Math Calif State Col at Dominguez Hills Dominguez Hills CA 90246

LARMORE, LEWIS, physics, see 12th edition

LARNER, JOSEPH, b Brest-Litovsk, Poland, Jan 9, 21; nat US; m 47; c 3. BIOCHEMISTRY, PHARMACOLOGY. Educ: Univ Mich, BA, 42; Columbia Univ, MD, 45; Univ Ill, MS, 49; Wash Univ, PhD(biochem), 51. Prof Exp: Instr biochem, Wash Univ, 51-53; asst prof, Noyes Lab Chem, Univ Ill, 53-57; from assoc prof to prof pharmacol, Sch Med, Western Reserve Univ, 57-65; Hill prof metab enzym, Col Med Sci, Univ Minn, Minneapolis, 65-69; PROF PHARMACOL & CHMN DEPT, SCH MED, UNIV VA, 69- Concurrent Pos: Travel award, Int Cong Biochem, 55; NIH res career award, 63-64; Commonwealth Fund fel, Lab Molecular Biol, Cambridge Univ, 63-64; mem metab study sect, NIH, 62-66, mem training comt, Nat Inst Arthritis & Metab Dis, 66; mem subcomt enzymes, Nat Res Coun-Nat Acad Sci, 64-; mem rev bd, Am Cancer Soc, 70-74. Honors & Awards: Alumni Prof Pharmacol, Univ Va, 73. Mem: Am Chem Soc; Am Soc Biol Chemists; fel Royal Col Med; Am Soc Pharmacol & Exp Therapeut. Res: Enzymatic aspects of intermediary carbohydrate metabolism, genetic and hormonal control. Mailing Add: Dept of Pharmacol Univ of Va Sch of Med Charlottesville VA 22903

LARNER, KENNETH LEE, b Chicago, Ill, Nov 1, 38; m 61; c 2. EXPLORATION GEOPHYSICS. Educ: Colo Sch Mines, GpE, 60; Mass Inst Technol, PhD(geophys), 70. Prof Exp: Res scientist image enhancement, EG&G, Inc, 67-69; sr res geophysicist, 70-74, MGR RES & DEVELOP EXPLOR GEOPHYS, WESTERN GEOPHYS CO, 75- Concurrent Pos: Lectr geophysics, Univ Houston, 73-74. Mem: Soc Explor Geophysicists; Seismol Soc Am; Europ Asn Explor Geophysicists; Sigma Xi. Res: Seismic signal enhancement and wave propagation; estimation of geophysical parameters from seismic measurements. Mailing Add: Western Geophys Co PO Box 2469 Houston TX 77001

LARNEY, VIOLET HACHMEISTER, b Chicago, Ill, May 19, 20; m 50. MATHEMATICS. Educ: Ill State Univ, BEd, 41; Univ Ill, AM, 42; Univ Wis, PhD(math), 50. Prof Exp: Teacher high sch, Ill, 42-44; asst math, Univ Wis, 44-48, instr, Exten Div, 48-50; asst prof, Kans State Univ, 50-52; assoc prof, 52-59, PROF MATH, STATE UNIV NY, ALBANY, 59- Mem: Math Asn Am; Am Math Soc. Res: Abstract algebra, groups and matrices. Mailing Add: Dept of Math State Univ of NY Albany NY 12222

LARNTZ, KINLEY, b Coshocton, Ohio, Oct 2, 45; m 65; c 2. APPLIED STATISTICS, MATHEMATICAL STATISTICS. Educ: Dartmouth Col, AB, 67; Univ Chicago, PhD(statist), 71. Prof Exp: ASST PROF APPL STATIST, UNIV MINN, ST PAUL, 71- Mem: Am Statist Asn; Inst Math Statist. Res: Analysis of qualitative data; comparison of small sample distributions for chi-square goodness-of-fit statistics;

model building for small group problems; applied statistical methods. Mailing Add: Dept of Appl Statist Univ of Minn St Paul MN 55108

LAROCCA, ANTHONY JOSEPH, b New Orleans, La, May 15, 23; m 54; c 6. PHYSICS, ELECTRICAL ENGINEERING. Educ: Tulane Univ, BS in EE, 49; Univ Mich, MS, 52. Prof Exp: Res asst physics, Tulane Univ, 49-51; res asst, Inst Sci & Technol, Univ Mich, Ann Arbor, 56-58, res assoc, 58-60, from assoc res physicist to res physicist, 60-73; RES PHYSICIST, ENVIRON RES INST MICH, 73- Concurrent Pos: Adj prof, Univ Mich, 73- Mem: AAAS; Optical Soc Am. Res: Infrared and optical technology; radiation phenomena; propagation and attenuation; techniques of measurement of radiation; design and use of electrooptical devices; standards of radiation; study of techniques of remote sensing of environment. Mailing Add: Environ Res Inst of Mich Box 618 Ann Arbor MI 48107

LA ROCCA, JOSEPH PAUL, b La Junta, Colo, July 5, 20; m 47; c 4. MEDICINAL CHEMISTRY. Educ: Univ Colo, BS, 42; Univ NC, MS, 44; Univ Md, PhD(pharmaceut chem), 48. Prof Exp: Res chemist, Naval Res Lab, 47-49; assoc prof, 49-50, PROF PHARM, UNIV GA, 50-, HEAD DEPT MEDICINAL CHEM, 70- Mem: Am Chem Soc; Am Pharmaceut Asn. Res: Synthetic sedative-hypnotics; anticonvulsant compounds; chemotherapy of cancer. Mailing Add: 115 Fortson Circle Athens GA 30601

LA ROCHE, GILLES (JOSEPH REGENT), biochemistry, comparative physiology, see 12th edition

LA ROCHELLE, JOHN HART, b Longmeadow, Mass, Aug 17, 24; m 48; c 4. PHYSICAL CHEMISTRY. Educ: Univ Mass, BS, 48; Northeastern Univ, MS, 50; Univ Mich, PhD(chem), 55. Prof Exp: Res chemist, Shell Chem Co, 55-60, sr res chemist, 60-68, res supvr, 68-72, res supvr, Shell Develop Co, 72-74, CHEMIST, CHG GEISMAR PLANT, SHELL CHEM CO, 74- Mem: AAAS; Am Inst Chemists. Res: Dielectric polarization of gases. Mailing Add: Shell Chem Co Box 500 Geismar LA 70734

LAROCHELLE, RONALD WILLIAM, organic chemistry, see 12th edition

LAROCK, RICHARD CRAIG, b Berkeley, Calif, Nov 16, 44. ORGANIC CHEMISTRY. Educ: Univ Calif, Davis, BS, 67; Purdue Univ, PhD(chem), 72. Prof Exp: NSF fel chem, Harvard Univ, 71-72; instr, 72-74, ASST PROF CHEM, IOWA STATE UNIV, 74- Concurrent Pos: Du Pont Young fac scholar, Du Pont Chem Co, 75-76. Mem: Am Chem Soc; The Chem Soc; Sigma Xi. Res: Organic synthesis; new synthetic methods; organometallic chemistry. Mailing Add: Dept of Chem Iowa State Univ Ames IA 50011

LAROCQUE, JOSEPH ALFRED AURELE, b Ottawa, Ont, Apr 26, 09; m 40. GEOLOGY. Educ: Univ Ottawa, Can, BA, 45; Univ Mich, MSc, 47, PhD(geol), 48. Prof Exp: Asst, Mus Zool, Univ Mich, 45-48; from asst prof to assoc prof, 48-54, PROF GEOL, OHIO STATE UNIV, 54- Mem: AAAS; Geol Soc Am; Paleont Soc; Am Asn Petrol Geologists. Mem: Fossil Molusca; paleoecology; history of geology. Mailing Add: Dept of Geol Ohio State Univ Columbus OH 43210

LAROS, GERALD SNYDER, II, b Los Angeles, Calif, July 19, 30; m 58; c 3. ORTHOPEDIC SURGERY. Educ: Northwestern Univ, BS, 52, MD, 55; Univ Iowa, MS, 70. Prof Exp: Orthop resident, Vet Admin Hosp, Hines, Ill, 56-60 & Shriners Hosp, Honolulu, Hawaii, 60-61; pvt pract, 61-68; NIH fel, Univ Iowa, 68-70, asst prof orthop surg, 70-71; assoc prof, Univ Ark, 71-73; PROF SURG & CHMN SECT ORTHOP SURG, UNIV CHICAGO, 73- Mem: Am Acad Orthop Surg; Orthop Res Soc; Asn Bone & Joint Surg. Res: Electromicroscopy of bone tendon and ligaments. Mailing Add: Sect of Orthop Univ of Chicago Hosps & Clins Chicago IL 60637

LAROSA, RICHARD THOMAS, pharmacology, physiology, see 12th edition

LAROSE, ROGER, b Montreal, Que, July 28, 10; m 36, 61; c 1. PHARMACY. Educ: Univ Montreal, BSP, 32, Lic sci, 34. Prof Exp: Lectr pharm, 34-50, assoc prof, 50-60, dean fac pharm, 60-65, mem bd gov, 65-67, spec counr to rector, 68-69, VICE RECTOR ADMIN, UNIV MONTREAL, 69- Concurrent Pos: Adminr, Mt Royal Chem Ltd, 58-72; mem exec admin coun, Pharm Exam Bd Can, 64; mem, Sci Coun Can, 66-71; pres, Ciba Co Ltd, 68-71 & Ciba-Geigy Can Ltd, 71-74; mem, Nat Comn Pharmaceut Serv in Can & WHO Coun Experts Pharm; dir, Inst Diag & Clin Res Montreal; hon mem, Can Conf. Mem: Pharm Soc Gt Brit; Fr Acad Pharmaceut Sci. Res: Teaching and administration. Mailing Add: Off of Vice Rector of Admin Univ of Montreal PO Box 6128 Montreal PQ Can

LAROW, EDWARD J, b Albany, NY, Dec 22, 37; m 63; c 4. AQUATIC ECOLOGY, INVERTEBRATE ZOOLOGY. Educ: Siena Col, BS, 60; Kans State Univ, MS, 65; Rutgers Univ, PhD(zool), 68. Prof Exp: From asst prof to assoc prof, 68-74, PROF BIOL, SIENA COL, NY, 74-, CHMN DEPT, 71- Concurrent Pos: Nat Res Coun Int Biol Prog grant, 70-74; vis lectr, State Univ NY Albany, 70- Mem: Ecol Soc Am; Am Soc Limnol & Oceanog; Am Inst Biol Sci; Int Soc Limnol. Res: Biological rhythms and their role in the vertical migration of zooplankton; remineralization and respiration in zooplankton; secondary production of zooplankton. Mailing Add: Dept of Biol Siena Col Loudonville NY 12211

LARR, ALFRED LOUIS, b Terre Haute, Ind, Oct 30, 14; m 46; c 2. SPEECH PATHOLOGY. Educ: Ind State Teachers Col, AB, 39, MA, 44; Syracuse Univ, PhD(speech correction), 55. Prof Exp: Asst prof speech, Whittier Col, 48-50; asst, Southwest Mo State Col, 50-53; psychologist, Calif Sch Deaf, 53-56; asst prof speech, Univ Calif, Los Angeles, 56-60; assoc prof, 60-70, PROF SPEECH, CALIF STATE UNIV, LONG BEACH, 70- Concurrent Pos: Consult, US Med Ctr, Mo, 52-53; lectr, Univ Redlands, 55; consult, Los Angeles Speech & Hearing Ctr, 58. Mem: Am Psychol Asn; Am Speech & Hearing Asn. Res: Perceptual and conceptual abilities of the deaf; speech intelligibility; non-organic hearing loss; binaural hearing; hearing evaluation; speech reading in closed circuit television; tongue thrust therapy. Mailing Add: Dept of Speech Calif State Univ Long Beach CA 90804

LARRABEE, ALLAN ROGER, b Flushing, NY, Feb 24, 35; m 60; c 2. BIOCHEMISTRY. Educ: Bucknell Univ, BS, 57; Mass Inst Technol, PhD, 62. Prof Exp: Staff fel biosynthesis of fatty acids, NIH, 64-66; asst prof chem, Univ Ore, 66-72; ASSOC PROF CHEM, MEMPHIS STATE UNIV, 72- Concurrent Pos: NIH res grant, 67-70; NSF res grant, 70-75. Mem: Am Soc Biol Chemists. Res: Role of vitamin B-12, folic acid and pantothenate as coenzymes; biosynthesis of fatty acis; multienzyme complexes; protein turnover. Mailing Add: Dept of Chem Memphis State Univ Memphis TN 38152

LARRABEE, CLIFFORD EVERETT, b Springfield, Mass, June 17, 22; m 45; c 3. ORGANIC CHEMISTRY. Educ: Bates Col, BS, 44; Univ Rochester, PhD(org chem), 49. Prof Exp: Res chemist, Gen Aniline & Film Corp, 49-52; RES CHEMIST, PFIZER, INC, 52- Mem: Am Chem Soc. Res: Synthetic organic chemistry. Mailing Add: 362 Boston Post Rd East Lyme CT 06333

LARRABEE, GRAYDON B, b Sherbrooke, Que, Oct 2, 32; US citizen; m 54; c 2. RADIOCHEMISTRY. Educ: Bishop's Univ, Can, BSc, 53; McMaster Univ, MSc, 54. Prof Exp: Radiochemist, Atomic Energy Can Ltd, 54-56; analyst, Can Westinghouse, 56-59; HEAD RES BR, MAT CHARACTERIZATION LAB, TEX INSTRUMENTS, INC, 59- Mem: Soc Appl Spectros; Electrochem Soc. Res: Materials characterization; solid state chemistry. Mailing Add: Mat Characterization Lab M/S 147 Tex Instruments Inc PO Box 5936 Dallas TX 75222

LARRABEE, MARTIN GLOVER, b Boston, Mass, Jan 25, 10; m 32, 44; c 2. NEUROCHEMISTRY, NEUROPHYSIOLOGY. Educ: Harvard Univ, AB, 32; Univ Pa, PhD(biophys), 37. Hon Degrees: MD, Univ Lausanne, 74. Prof Exp: Asst, Univ Pa, 34-35, fel med physics, 37-40; asst prof physiol, Med Col, Cornell Univ, 40-41; fel, Johnson Found, Univ Pa, 41-42, assoc, 42-43, from asst prof to assoc prof biophys, 43-48; assoc prof, 49-63, PROF BIOPHYS, JOHNS HOPKINS UNIV, 63- Mem: Nat Acad Sci; Am Physiol Soc; Am Soc Neurochem; Int Soc Neurochem; Soc Neurosci (treas). Res: Metabolism in relation to physiological function and embryological development in sympathetic ganglia. Mailing Add: Dept of Biophys Johns Hopkins Univ Baltimore MD 21218

LARRABEE, RICHARD BRIAN, b Sacramento, Calif, Apr 29, 40; m 62; c 2. ORGANIC CHEMISTRY. Educ: Univ Santa Clara, BS, 61; Univ Chicago, MS & PhD(chem), 67. Prof Exp: NIH fel, Ohio State Univ, 66-67; STAFF MEM CHEM, IBM RES DIV, 67- Mem: AAAS; Am Chem Soc; The Chem Soc. Res: Investigation of symmetry forces underlying elementary chemical processes. Mailing Add: IBM Res Div 5600 Cottle Rd San Jose CA 95193

LARRISON, MILLARD SAMUEL, b Oak Hill, WVa, May 27, 05; m 31; c 4. ORGANIC CHEMISTRY. Educ: WVa Univ, AB, 28; Purdue Univ, PhD(org chem), 42. Prof Exp: Chemist, Evans Lead Co, WVa, 28-29; supv chemist, du Pont Ammonia Corp, 29; res chemist & supvr, Reilly Tar & Chem Corp, Ind, 31-38; res chemist, US Gypsum Co, Ill, 41-42; res supvr, Barrett Div, Allied Chem & Dye Corp, Pa, 42-44; head new proj dept, Ringwood Chem Corp, Ill, 44-46; assoc prof mech eng, Univ Toledo, 46-47; prod mgr, Charles Bruning Co, Inc, 47-56; vpres, Warren Develop Corp, 55-62; chief chemist & vpres, Weston Chem Corp, 63-69; PRES, LARRISON CHEM CO, 69- Mem: AAAS; Am Chem Soc. Res: Diazo compounds; tar products; vitamin syntheses; general organic synthesis; organo-phosphorus compounds; light sensitive products. Mailing Add: 1168 Charles Ave Morgantown WV 26505

LARRIVEE, JULES ALPHONSE, astronomy, mathematics, deceased

LARRY, JOHN ROBERT, b Mt Clare, WVa, Nov 13, 39; m 63; c 2. PHYSICAL CHEMISTRY. Educ: WVa Univ, BS, 61; Ohio State Univ, PhD(chem), 66. Prof Exp: Res chemist, 66-74, SR RES CHEMIST, PHOTO PROD DEPT, ELECTRONIC PROD DIV, E I DU PONT DE NEMOURS & CO, INC, NIAGARA FALLS, 74- Mem: Am Chem Soc. Res: Charge transfer and molecular complexes; ultracentrifugation; emulsion polymerization; colloid chemistry; rheology; solid state conductors. Mailing Add: 396 Dansworth Rd Youngstown NY 14174

LARSEN, ARNOLD LEWIS, b Audubon, Iowa, Sept 7, 27; m 50; c 2. BOTANY, AGRONOMY. Educ: Univ Iowa, BA, 50; Iowa State Univ, MS, 61, PhD(econ bot), 63. Prof Exp: Farmer, Audubon County, Iowa, 53-57; from assoc to res asst bot, Seed Lab, Iowa State Univ, 47-63; botanist, Field Crops & Animal Prod Seed Res Lab, Agr Res Serv, USDA, 63-70; RES ASSOC PROF SEED TECHNOL, COLO STATE UNIV, 70-, DIR, COLO SEED LAB, 70- Mem: Am Soc Agron; Crop Sci Soc Am; Asn Off Seed Analysts; Coun Agr Sci & Technol. Res: Developmental procedures for determining quality of seeds. Mailing Add: Colo Seed Lab Colo State Univ Ft Collins CO 80521

LARSEN, AUBREY ARNOLD, b Rockford, Ill, Sept 27, 19; m 43; c 5. MEDICINAL CHEMISTRY. Educ: Antioch Col, BS, 43; Mich State Col, MS, 44; Cornell Univ, PhD(org chem), 46. Prof Exp: Mem staff, Sterling-Winthrop Res Inst, 46-60; asst dir org chem, Mead Johnson & Co, Ind, 60-63, dir chem res, 63-67, vpres phys sci, 67-70; vpres & sci dir, Bristol-Myers Int, 70-75; V PRES RES & DEVELOP, MEAD JOHNSON & CO, 75- Mem: AAAS; NY Acad Sci; Am Chem Soc. Res: Pharmaceutical and nutritional research and development. Mailing Add: Mead Johnson & Co 2404 Pennsylvania St Evansville IN 47721

LARSEN, AUSTIN ELLIS, b Provo, Utah, Nov 1, 23; m 45; c 4. VETERINARY MEDICINE, MICROBIOLOGY. Educ: Wash State Univ, BS, 48, DVM, 49; Univ Utah, MS, 56, PhD(microbiol), 69. Prof Exp: Vet pvt pract, Utah, 49-68; clin instr microbiol, 61-68, ASST PROF MICROBIOL & DIR VIVARIUM, COL MED, UNIV UTAH, 68- Concurrent Pos: Vet & dir res lab, Fur Breeders Agr Coop Lab, 52-68, consult vet, 68-; consult, Schering Corp, 63-; non-med med asst res, Vet Admin, 69-; mem, Adv Comt Fur Farmers Res Inst; mem, Health Task Force of Utah. Res: Slow virus research. Mailing Add: Dept of Microbiol Univ of Utah Salt Lake City UT 84112

LARSEN, BODIL ASTRID, b Copenhagen, Denmark, Oct 17, 38; Norweg citizen. IMMUNOLOGY. Educ: Univ Bergen, Cand Real, 65, Dr Phil(immunol), 72. Prof Exp: Res fel immunol, Norweg Res Coun Sci & Humanities, 66-70; prof immunol, Univ Bergen, 70-73; fel, 73-74, ASST PROF IMMUNOL, MEM UNIV NFLD, 74- Res: Cell-cooperation in the immune response; human immunogenetics. Mailing Add: Fac of Med Mem Univ of Nfld St John's NF Can

LARSEN, CARL DONALD, biochemistry, deceased

LARSEN, CHARLES MCLOUD, b Staten Island, NY, Dec 6, 24; m 48; c 4. MATHEMATICS. Educ: Cornell Univ, AB, 45, AM, 50; Stanford Univ, PhD(educ), 60. Prof Exp: From instr to assoc prof, 54-67, PROF MATH, SAN JOSE STATE COL, 67- Mem: Math Asn Am; Soc Indust & Appl Math. Res: Teaching of elementary calculus; methods of problem solving in mathematics; mathematics education. Mailing Add: Dept of Math San Jose State Col San Jose CA 95114

LARSEN, DAVID M, b Hawthorne, NJ, Mar 8, 36; m 58; c 2. THEORETICAL PHYSICS. Educ: Mass Inst Technol, SB, 57, PhD(physics), 62. Prof Exp: Nat Res Coun-Nat Bur Standards fel physics, Nat Bur Standards, Washington, DC, 62-64; STAFF PHYSICIST, LINCOLN LAB, MASS INST TECHNOL, 64- Mem: Fel Am Phys Soc. Res: Impurities in semiconductors; polaron theory. Mailing Add: 6 Fessenden Way Lexington MA 02173

LARSEN, DAVID W, b Chicago, Ill, Feb 21, 36; m 63. PHYSICAL CHEMISTRY. Educ: Dana Col, BA, 58; Northwestern Univ, PhD(phys chem), 63. Prof Exp: Res assoc nuclear magnetic resonance spectros, Washington Univ, 63-64; asst prof, 64-66, ASSOC PROF CHEM, UNIV MO-ST LOUIS, 66- Mem: Am Chem Soc. Res: Nuclear magnetic resonance spectroscopy; exchange reactions of Lewis acids and bases in non-aqueous media; ionic interactions in aqueous media. Mailing Add: Dept of Chem Univ of Mo-St Louis St Louis MO 63121

LARSEN, DON HYRUM, b Provo, Utah, Sept 22, 17; m 41; c 3. INDUSTRIAL MICROBIOL, MEDICAL MICROBIOLOGY. Educ: Brigham Young Univ, BS, 40; Univ Nebr, MA, 42; Univ Utah, PhD, 50. Prof Exp: Bacteriologist, Commercial Solvents Co, Inc, 43-46; instr bact, Univ Utah, 47-50; asst prof, Univ Nebr, 50-52; from asst prof to assoc prof, 52-60, chmn dept, 55-60 & 65-72, PROF BACT, BRIGHAM YOUNG UNIV, 60- Concurrent Pos: Res assoc, Naval Biol Lab, Calif, 60-61; consult, Vitamins Inc, 74- Mem: AAAS; Am Soc Microbiol; Soc Indust Microbiol; Soc Appl Microbiol. Res: Microbial physiology; pathogenic microbiology; ergosterol production by yeasts; single cell protein from waste materials. Mailing Add: 893 Widtsoe Bldg Brigham Young Univ Provo UT 84602

LARSEN, EARL GEORGE, b Milltown, Wis, Oct 11, 21. BIOCHEMISTRY. Educ: Univ Wis, BS, 43, MS, 44; Wayne State Univ, PhD(biochem), 53. Prof Exp: Res chemist biochem, Armour Res Labs, Ill Inst Technol, 44-46; res chemist, Carnation Res Labs, Wis, 46-48; chief biochemist, Vet Admin Hosp, Dearborn, 48-; res assoc, Col Med, Wayne State Univ, 53-55; ASSOC PROF BIOCHEM, SCH MED, UNIV OKLA, 55- Mem: AAAS. Res: Chemistry and biosynthesis of porphyrins and heme-enzymes; physical biochemistry; chemical origins and evolution of life; chemical basis and mechanisms of the mind. Mailing Add: Dept of Biochem Univ of Okla Sch of Med Oklahoma City OK 73103

LARSEN, EDWIN MERRITT, b Milwaukee, Wis, July 12, 15; m 46; c 3. INORGANIC CHEMISTRY. Educ: Univ Wis, BS, 37; Ohio State Univ, PhD(chem), 42. Prof Exp: Chemist, Rohm and Haas, Pa, 37-38; asst chem, Ohio State Univ, 38-42; instr, Univ Wis, 42-43; group leader, Manhattan Dist Proj, Monsanto Chem Co, Ohio, 43-46; from asst prof to assoc prof, 46-58, PROF CHEM, UNIV WIS-MADISON, 58- Concurrent Pos: Vis prof, Univ Fla, 58; Fulbright lectr, Inst Inorg Chem, Vienna Tech Inst, 66-67. Mem: AAAS; Am Chem Soc. Res: Chemistry of the transitional elements; reduced states; solid state chemical and physical properties. Mailing Add: Dept of Chem Univ of Wis Madison WI 53706

LARSEN, ELEANOR MARIE, b Olivet, Mich, Oct 29, 00. MEDICAL PHYSIOLOGY. Educ: Oberlin Col, AB, 24; Univ Wis, MS, 42, PhD(physiol), 48. Prof Exp: From asst to asst prof, 40-70, EMER ASST PROF PHYSIOL, UNIV WIS-MADISON, 70- Mem: Fel AAAS; Am Physiol Soc. Mailing Add: Dept of Physiol Univ of Wis Med Sch Madison WI 53706

LARSEN, ELMER CONRAD, b Owen, Wis, Oct 23, 12; m 39, 60; c 5. PHYSICAL CHEMISTRY. Educ: St Olaf Col, BA, 34; Mont Sch Mines, MS, 36; Stuttgart Tech Univ, Dr tech sci(phys chem), 38; Univ Wis, PhD(inorg chem), 39. Prof Exp: Instr chem, Mont Sch Mines, 34-36; res chemist & mem tech staff, Bell Tel Labs, Inc, NY, 39-45; asst dir res, J T Baker Chem Co, NJ, 45-49, vpres, tech dir & mem bd dirs, 54-59; chief engr, Tungsten & Chem Div, Sylvania Elec Prod, Inc, 49-54; dir planning & develop, Coatings & Resins Div, Pittsburgh Plate Glass Co, 59-66; VPRES & GEN MGR, COATINGS & RESINS DIV, PPG INDUSTS, INC, 66- Concurrent Pos: Vpres, Salt Lake Tungsten Co, 53-54. Mem: Am Chem Soc; Soc Chem Indust; NY Acad Sci. Res: Electrochemistry; colloids; catalysis; metallurgy of tungsten and molybdenum; plastics. Mailing Add: Coatings & Resins Div PPG Industs Inc 1 Gateway Center Pittsburgh PA 15222

LARSEN, ERIC RUSSELL, b Port Angeles, Wash, July 7, 28; m 51; c 4. ORGANIC CHEMISTRY. Educ: Univ Wash, BS, 50; Univ Colo, PhD(org chem), 54. Prof Exp: Res chemist, Chem Eng Lab, 56-59, proj leader, 59-62, sr res chemist, 62-64, group leader, Halogens Res Lab, 64-68, assoc res scientist, 68-74, RES SCIENTIST, HALOGENS RES LAB, DOW CHEM CO, 74- Mem: AAAS; Am Chem Soc. Res: Organic fluorine and bromine chemistry, especially anesthesiology; discovery of methoxyflurane; fire research, especially flame retardancy in plastics. Mailing Add: Halogens Res Lab Dow Chem Co Midland MI 48640

LARSEN, FENTON E, b Preston, Idaho, Mar 22, 34; m 54; c 4. HORTICULTURE. Educ: Utah State Univ, BS, 56; Mich State Univ, PhD(hort), 59. Prof Exp: From asst horticulturist to assoc horticulturist, 59-73, PROF & HORTICULTURIST, WASH STATE UNIV, 73- Mem: Am Soc Hort Sci; Am Pomol Soc; Int Plant Propagators Soc. Res: Pomology; propagation; rootstocks; leaf abscission of nursery stock. Mailing Add: Dept of Hort Wash State Univ Pullman WA 99163

LARSEN, FREDERICK DUANE, b St Johnsbury, Vt, Mar 20, 30; m 52; c 4. GEOMORPHOLOGY. Educ: Middlebury Col, BA, 52; Boston Univ, MA, 60; Univ Mass, PhD(geol), 72. Prof Exp: From instr to asst prof, 57-64, ASSOC PROF GEOL, NORWICH UNIV, 64- Concurrent Pos: US Geol Surv, 68- Mem: AAAS; Geol Soc Am; Glaciol Soc. Res: Glacial geology; deglaciation of Muir Inlet, Alaska; glacial geology of central Vermont and Mount Tom Quadrangle, Massachusetts. Mailing Add: Dept of Geol Norwich Univ Northfield VT 05663

LARSEN, HARRY STITES, b Pittsburgh, Pa, Aug 12, 27; m 56; c 3. FORESTRY. Educ: Rutgers Univ, BS, 50; Mich State Univ, MS, 53; Duke Univ, PhD, 63. Prof Exp: Forester, Southern Timber Mgt Serv, 53-56; asst prof, 59-71, ASSOC PROF SILVICULT, DEPT FORESTRY, AUBURN UNIV, 71- Mem: Ecol Soc Am; Soc Am Foresters. Res: Tree physiology and silvics. Mailing Add: Dept of Forestry Auburn Univ Auburn AL 36830

LARSEN, HOWARD JAMES, b Duluth, Minn, Jan 21, 25; m 46; c 2. DAIRY NUTRITION. Educ: Univ Wis, BS, 50; Iowa State Col, MS, 52, PhD(dairy husb), 53. Prof Exp: Assoc & instr dairy husb, Iowa State Col, 54-55; from asst prof to assoc prof, 55-57, PROF DAIRY SCI, MARSHFIELD EXP STA, UNIV WIS, 75- Mem: Fel AAAS; Am Inst Biol Sci; Am Soc Animal Sci; Am Dairy Sci Asn. Res: Forage utilization by dairy cattle; forage and concentrate preservation and utilization by dairy cattle; environmental studies with ruminants. Mailing Add: Univ of Wis Marshfield Exp Sta Marshfield WI 54449

LARSEN, JAMES ARTHUR, b Rhinelander, Wis, Mar 14, 21; m 68; c 3. PLANT ECOLOGY. Educ: Univ Wis-Madison, BS, 46, MS, 56, PhD(ecol), 68. Prof Exp: SCI ED, UNIV WIS-MADISON, 56- Mem: Ecol Soc Am; Bot Soc Am; Arctic Inst NAm; Nat Asn Sci Writers. Res: Arctic ecology and bioclimatology; ecologico-economic studies of northern lands; arctic botany; problems of scientific communication and of public information on science. Mailing Add: 1215 WARF Bldg Univ of Wis-Madison Madison WI 53706

LARSEN, JAMES BOUTON, b Detroit, Mich, July 28, 41; m 64; c 2. COMPARATIVE PHYSIOLOGY, BIOCHEMISTRY. Educ: Kalamazoo Col, BA, 63; Univ Miami, MS, 66, PhD(marine biol), 68. Prof Exp: Fel biochem, Colo State Univ, 67-68; asst prof biol, Hamline Univ, 68-73; ASST PROF BIOL, UNIV SOUTHERN MISS, 73- Mem: AAAS; Sigma Xi. Res: Physiology, toxicology and biochemistry of coelenterate toxins; teleost endocrine systems; osmoregulation in aquatic animals. Mailing Add: Dept of Biol Univ Southern Miss Hattiesburg MS 39401

LARSEN, JAMES WILBURNE, inorganic chemistry, analytical chemistry, see 12th edition

LARSEN, JOHN ELBERT, b Watseka, Ill, Feb 7, 19; m 47; c 2. HORTICULTURE. Educ: Purdue Univ, BS, 42, MS, 46, PhD(hort), 57. Prof Exp: Agronomist, Stokeley-Van Camp, Inc, 46-51; asst hort, Purdue Univ, 51-55; horticulturist, J W Davis Co, 57-61; HORTICULTURIST, AGR EXTEN SERV, TEX A&M UNIV, 61-, ASSOC PROF HORT, 70- Mem: Am Soc Agron; Soil Sci Soc Am; Am Soc Hort Sci. Res: Field and glasshouse vegetables. Mailing Add: Agr Exten Serv Tex A&M Univ College Station TX 77843

LARSEN, JOHN HERBERT, JR, b Tacoma, Wash, July 20, 29; m 51; VERTEBRATE ZOOLOGY. Educ: Univ Wash, BA, 55, MS, 58, PhD(zool), 63. Prof Exp: Instr embryol, Univ Wash, 60; instr biol, Univ Puget Sound, 60-61; asst cur & instr zool, Univ Wash, 61-62, NIH res fel, 63-64, USPHS sr fel electron micros, 64-65; from asst prof to assoc prof, 65-75, PROF ZOOL, WASH STATE UNIV, 75- Mem: AAAS; Am Soc Zoologists; Soc Study Evolution; Soc Syst Zool; Am Soc Ichthyologists & Herpetologists. Res: Evolution and functional morphology of feeding systems in amphibians; implications of neoteny to urodele evolution; mechanisms of ovulation in lower vertebrates. Mailing Add: Dept of Zool Wash State Univ Pullman WA 99163

LARSEN, JOHN W, b Hartford, Conn, Oct 30, 40; m 62; c 2. ORGANIC CHEMISTRY, PHYSICAL CHEMISTRY. Educ: Tufts Univ, BS, 62; Purdue Univ, PhD(chem), 67. Prof Exp: Res fel chem, Univ Pittsburgh, 66-68; asst prof, 68-71, ASSOC PROF CHEM, UNIV TENN, KNOXVILLE, 71- Mem: AAAS; Am Chem Soc; The Chem Soc. Res: Thermodynamics of organic intermediates; chemistry in strong acid solutions; organic ion pairs; solvent-solute interactions; thermodynamics of micellar systems; coal chemistry. Mailing Add: Dept of Chem Univ of Tenn Knoxville TN 37916

LARSEN, JOSEPH REUBEN, b Ogden, Utah, May 21, 27; m 48; c 3. INSECT PHYSIOLOGY. Educ: Univ Utah, BS, 50, MS, 52; Johns Hopkins Univ, ScD(med entom), 58. Prof Exp: Asst biol, Univ Utah, 50-52; entomologist, US Army Chem Corps, 52-54; res assoc zool, Univ Pa, 58-62; asst prof, Univ Wyo, 62-63; from assoc prof to prof & head dept entom, 63-75, DIR SCH LIFE SCI, UNIV ILL, URBANA, 75- Mem: Fel AAAS; Entom Soc Am; Am Soc Zoologists; Electron Micros Soc Am; Micros Soc Am. Res: Endocrine relationships in insects; insect histology; microanatomy and physiology of insect sensory receptors and insect nervous system. Mailing Add: Sch of Life Sci Univ of Ill Urbana IL 61801

LARSEN, KARL DAVIS, b Bangor, Maine, Feb 3, 08; m 31; c 1. PHYSICS, ACOUSTICS. Educ: Univ Maine, BA, 29, MA, 30; Pa State Univ, PhD(physics), 34. Prof Exp: From asst to asst prof physics, Univ Maine, 29-46; asst, Pa State Univ, 30-34; res engr, Stevens Inst Technol, 46-47; prof & chmn dept, Lafayette Col, 47-58; staff scientist, Davidson Mem Lab, Stevens Inst Technol, 58-62; prof physics & dean fac, Onondaga Community Col, 62-65, actg pres, 65-66; prof physics & chmn dept, 66-74, actg dean col arts & sci, 67-69, asst dean grad studies, 69-73, EMER PROF PHYSICS & COORDR OCEANOG PROGS, UNIV BRIDGEPORT, 74- Concurrent Pos: Sci consult to legal firms & ins co, 70- Mem: Am Phys Soc; Am Soc Eng Educ; Am Asn Physics Teachers. Res: Electron emission from coated surfaces; Raman spectrum of heavy water vapor; microwave spectroscopy; experimental studies of sound transmission in liquids containing entrained gas bubbles; hobby study of transmission of vibrations by violin and cello bridges. Mailing Add: Dept of Physics Univ of Bridgeport Bridgeport CT 06602

LARSEN, KENNETH MARTIN, b Ogden, Utah, June 26, 27; m 55; c 9. APPLIED MATHEMATICS. Educ: Univ Utah, BA, 50; Brigham Young Univ, MA, 56; Univ Calif, Los Angeles, PbD(math), 64. Prof Exp: Teaching asst math, Brigham Young Univ, 54-55 & Univ Calif, Los Angeles, 56-60; from asst prof to assoc prof, 60-75, PROF MATH, BRIGHAM YOUNG UNIV, 75- Mem: Sigma Xi; Math Asn Am; Soc Indust & Appl Math. Res: Potential theory which is connected with harmonic functions; ordinary and partial differential equations, particularly existence and uniqueness properties; plasma confinement and stability. Mailing Add: 292 TMCB Dept of Math Brigham Young Univ Provo UT 84602

LARSEN, LAWRENCE HAROLD, b Staten Island, NY, July 22, 39; m 69; c 1. OCEANOGRAPHY, HYDRODYNAMICS. Educ: Stevens Inst Technol, BS, 61; Johns Hopkins Univ, PhD(hydrodyn), 65. Prof Exp: NSF fel meteorol, Univ Oslo, 65-66; vis res asst prof, Johns Hopkins Univ, 66-67; res asst prof, 67-71, RES ASSOC PROF OCEANOG, UNIV WASH, 71- Concurrent Pos: Prog dir phys & chem oceanog, NSF, 72-73. Mem: Am Meteorol Soc; Am Geophys Union. Res: Physical oceanography; wave motion; estuaries; sediment dynamics. Mailing Add: Rm 22A Dept of Oceanog Univ of Wash Seattle WA 98105

LARSEN, LELAND MALVERN, b Blair, Nebr, Aug 20, 15; m 40; c 3. ALGEBRA. Educ: Dana Col, BA, 41; Univ Nebr, MA, 48, PhD, 67. Prof Exp: Pub sch teacher, Nebr, 36-41, supt schs, 41-43; instr, Univ Nebr, 43-44; assoc prof, 48-67, PROF MATH & HEAD DEPT, KEARNEY STATE COL, 67- Mem: Math Asn Am. Res: Theory of fields; various algorithms. Mailing Add: Dept of Math Kearney State Col Kearney NE 68847

LARSEN, LEONARD HILLS, geology, see 12th edition

LARSEN, MARLIN LEE, b Grand Island, Nebr, Nov 22, 42; m 64; c 3. CLINICAL CHEMISTRY. Educ: Kearney State Col, BS, 64; Wash State Univ, PhD(chem), 68. Prof Exp: DIR RES & DEVELOP, UNITED MED LABS, INC, 68- Mem: Am Asn Clin Chemists; Asn Advan Med Instrumentation; Am Chem Soc. Res: Automation and methodology research of medical laboratory procedures. Mailing Add: United Med Labs Inc Res & Develop Box 3932 Portland OR 97208

LARSEN, MAX DEAN, b Pratt, Kans, Jan 23, 41; m 62. ALGEBRA. Educ: Kans State Teachers Col, BA, 61; Univ Kans, MA, 63, PhD(math), 66. Prof Exp: From asst prof to assoc prof, 66-73, PROF MATH, UNIV NEBR-LINCOLN, 73-, INTERIM DEAN COL ARTS & SCI, 74- Concurrent Pos: NSF res grant, 68-70. Mem: Am Math Soc; Math Asn Am; Nat Coun Teachers Math. Res: Extension of integral domain concepts to general commutative rings, particularly valuation theory and Prüfer rings; module theory over commutative rings. Mailing Add: Dept of Math Univ of Nebr Lincoln NE 68508

LARSEN, PETER FOSTER, b Mt Kisco, NY. BIOLOGICAL OCEANOGRAPHY. Educ: Univ Conn, BA, 67, MS, 69; Col William & Mary, PhD(marine sci), 74. Prof Exp: Lectr chem, Norwalk Community Col, 68-70; res asst marine biol, Va Inst Marine Sci, 70-72, asst marine sceintist ecol, 72-73; state oceanogr, Maine Dept Marine Resources, 73-76; RES SCIENTIST, BIGELOW LAB OCEAN SCI, 76- Concurrent Pos: Pres, Coastal Sci, 73-; consult, Res Inst Gulf Maine, 73- & Bigelow Lab Ocean Sci, 75-76. Mem: Am Soc Limnol & Oceanog; Ecol Soc Am; Estuarine & Brackish-Water Sci Asn; Atlantic Estuarine Res Soc. Res: The documentation of benthic community structure and function in the estuarine and marine environments of the Gulf of Maine region. Mailing Add: Bigelow Lab Ocean Sci West Boothbay Harbor ME 04575

LARSEN, PHILIP O, b Audubon Co, Iowa, Dec 1, 40; m 61; c 3. PLANT PATHOLOGY. Educ: Iowa State Univ, BS, 63; Univ Ariz, MS, 67, PhD(plant path), 69. Prof Exp: Asst prof plant path, 69-73, ASST PROF BOT, OHIO STATE UNIV, 73- Mem: Am Phytopath Soc. Res: Physiology of plant disease. Mailing Add: Dept of Bot Ohio State Univ Columbus OH 43210

LARSEN, ROBERT GORDON, organic chemistry, deceased

LARSEN, ROBERT PAUL, b Vineyard, Utah, Dec 1, 26; m 48; c 4. HORTICULTURE. Educ: Utah State Univ, BS, 50; Kans State Univ, MS, 51; Mich State Univ, PhD(hort), 55. Prof Exp: From asst prof to prof hort, Mich State Univ, 55-68; SUPT TREE FRUIT RES CTR, WASH STATE UNIV, 68- Mem: Fel Am Soc Hort Sci (pres, 75); Int Soc Hort Sci. Res: Physiology, nutrition and management of tree fruit crops. Mailing Add: Tree Fruit Res Ctr Wash State Univ Wenatchee WA 98801

LARSEN, ROBERT PETER, b Kalamazoo, Mich, Jan 1, 21; m 48; c 2. PHYSICAL CHEMISTRY. Educ: Kalamazoo Col, AB, 42; Brown Univ, PhD(chem), 48. Prof Exp: Instr phys chem, Brown Univ, 44-46; instr anal chem, Amherst Col, 47-48; asst prof phys chem, Ohio Wesleyan Univ, 48-51; ASSOC CHEMIST & ANAL GROUP LEADER, CHEM ENG DIV, ARGONNE NAT LAB, 51- Mem: Am Chem Soc. Res: Inorganic analytical chemistry in reactor fuel process development; uranium and plutonium analysis and the measurement of fission yields; determinations of plutonium and other transuranic elements in biological materials. Mailing Add: Radiol & Environ Res Div Argonne Nat Lab Argonne IL 60439

LARSEN, RONALD JOHN, b Chicago, Ill, Jan 1, 37; m 62. MATHEMATICAL ANALYSIS. Educ: Mich State Univ, BS, 57, MS, 59; Stanford Univ, PhD(math), 64. Prof Exp: Instr math, Yale Univ, 63-65; asst prof, Cowell Col, Univ Calif, Santa Cruz, 65-70; ASSOC PROF MATH, WESLEYAN UNIV, 70- Concurrent Pos: Fulbright-Hays advan res grant, Univ Oslo, 68-69, vis asst, 73-74; Fulbright-Hays travel award to Norway, 73-74; vis assoc prof, State Univ NY Binghamton, 75-76. Mem: Am Math Soc; Math Asn Am; Norweg Math Soc. Mailing Add: Dept of Math Wesleyan Univ Middletown CT 06457

LARSEN, RUSSELL D, b Muskegon, Mich, June 6, 36; m 58; c 2. CHEMICAL PHYSICS, STATISTICS. Educ: Kalamazoo Col, BA, 57; Kent State Univ, PhD(chem), 64. Prof Exp: Teaching asst chem, Univ Cincinnati, 58-60; asst instr, Kent State Univ, 64; res assoc, Princeton Univ, 64-65; Robert A Welch fel, Rice Univ, 65-66; asst prof, Ill Inst Technol, 66-72; ASST PROF CHEM, TEX A&M UNIV, 72- Mem: Am Chem Soc; Am Phys Soc; Am Statist Asn. Res: Chemical and biomedical signal processing; spectral analysis; Walsh functions; spline representations. Mailing Add: Dept of Chem Tex A&M Univ College Station TX 77843

LARSEN, SIGURD YVES, b Brussels, Belg, Aug 14, 33; US citizen. THEORETICAL PHYSICS. Educ: Columbia Univ, AB, 54, MA, 56, PhD(physics), 62. Prof Exp: Asst physics, Columbia Univ, 54-57 & 60-62; consult, Nat Bur Standards, 62; Nat Acad Sci-Nat Res Coun assoc, 62-63, physicist, Washington, DC, 63-68; assoc prof, 68-75, PROF PHYSICS & CHMN DEPT, TEMPLE UNIV, 75- Concurrent Pos: Consult, Los Alamos Sci Lab, 64, Lawrence Radiation Lab, 67-72 & Nat Bur Standards, 68-71; mem panel quantum fluids, Int Union Pure & Appl Chem, 66-; vis prof, Inst Mexicano del Petroleo, 71-72. Mem: Am Phys Soc; Ital Phys Soc; Sigma Xi. Res: Statistical physics; quantum theory; numerical analysis. Mailing Add: Dept of Physics Temple Univ Philadelphia PA 19122

LARSEN, TED LEROY, b Jerome, Idaho, Mar 18, 35; m 57; c 2. PHYSICS, SEMICONDUCTORS. Educ: Univ Calif, Berkeley, BS, 61, MS, 62; Stanford Univ, PhD(mat sci), 70. Prof Exp: Mem tech staff, 62-65, res & develop group leader, 68-71, RES & DEVELOP SECT MGR SEMICONDUCTORS, HEWLETT-PACKARD ASSOCS, HEWLETT-PACKARD CO, 71- Mem: Electrochem Soc. Res: III-V compound materials and devices for optoelectronic and microwave applications, including single crystal and epitaxial growth, crystalline defects, impurity diffusion and minority carrier recombination processes. Mailing Add: Hewlett-Packard Co 640 Page Mill Rd Palo Alto CA 94304

LARSEN, VICTOR ROBINSON, JR, b San Francisco, Calif, Dec 3, 15; m 45; c 1. BOTANY. Educ: Columbia Univ, BS & MA, 39, PhD, 53. Prof Exp: Instr high sch, NC, 38-39; instr biol & photog, Adelphi Col, 39-42; instr bot & econ bot, Barnard Col, Columbia Univ, 46-53; from asst prof to assoc prof, 53-68, PROF BIOL, ADELPHI UNIV, 68- Mem: Bot Soc Am; Torrey Bot Club. Res: Epidemiology; malariology; aviation medicine; cytology of latex plants. Mailing Add: Dept of Biol Adelphi Univ Garden City TX 11530

LARSEN, WAYNE AMMON, b Waynesboro, Va, Apr 2, 39; m 64; c 3. STATISTICS. Educ: Brigham Young Univ, BS, 61; Va Polytech Inst, PhD(statist), 68. Prof Exp: Statistician, Hercules Powder Co, Utah, 61-63; mem tech staff data anal, Bell Tel Labs, NJ, 67-69, supvr data anal, 69-75; SR SCIENTIST, EYRING RES INST, 75- Mem: Am Statist Asn. Res: Methodology of data analysis. Mailing Add: 186 W Oakridge Dr Farmington UT 84025

LARSEN, WESLEY P, b Twin Falls, Idaho, July 29, 16; m 42; c 4. ZOOLOGY. Educ: Univ Utah, BS, 40, MS, 52, PhD(zool), 60. Prof Exp: Instr biol, Carbon Col, Univ Utah, 48-58, dir sci, 58-60; from assoc prof to prof, Col Southern Utah, Utah State Univ, 60-65; assoc prog dir pre-col sci educ, NSF, 66-67, eval panelist, 65; dean, Sch Sci, 67-71, PROF BIOL, SOUTHERN UTAH STATE COL, 67-, GRANTS OFFICER, 71- Concurrent Pos: NSF fac fel sci, 58-59; prin investr, AEC res contract, 61-65; dir resource personnel workshop, NSF, 72-73. Mem: AAAS; Am Soc Zoologists; Entom Soc Am; Am Nature Study Soc. Res: Insect tissue and organ culture; radiation biology; cryobiology. Mailing Add: Sch of Sci Southern Utah State Col Cedar City UT 84720

LARSH, HOWARD WILLIAM, b East St Louis, Ill, May 29, 14; m 38; c 1. MEDICAL MYCOLOGY. Educ: McKendree Col, BA, 36; Univ Ill, MS, 38, PhD, 41. Prof Exp: Asst bot, Univ Ill, 38-41; instr, Dept Bot & Bact, Univ Okla, 41-42; plant pathologist, Bur Plant Indust, Soils & Agr Eng, USDA, 43-45; assoc prof bact, med, mycol & plant path, 45-48, chmn dept plant sci, 45-62, PROF PLANT SCI, UNIV OKLA, 48-, RES PROF, 62-, CHMN DEPT BOT & MICROBIOL, 66- Concurrent Pos: Spec consult, McKnight State Tuberc Hosp & Nat Communicable Disease Ctr, USPHS, 50-; med mycologist consult & co-dir lab, Mo State Chest Hosp, Mt Vernon, 55-; consult, Manned Spacecraft Ctr, NASA; res reviewer immunol & infectious diseases comt, Vet Admin Hosp, Oklahoma City, 72; cert bioanal lab dir, Am Bd Bioanal, 74-; ed rev bd, J Clin Microbiol, 74-; ed-in-chief, Sabouraudia, 75- Mem: Fel AAAS; fel Am Pub Health Asn; fel Am Acad Microbiol; Bot Soc Am; Mycol Soc Am. Res: Medical mycosis; systemic mycoses; histoplasmosis; cryptococcosis. Mailing Add: Dept of Bot & Microbiol Univ of Okla Norman OK 73069

LARSH, JOHN E, JR, b East St Louis, Ill, Oct 3, 17; m 39; c 3. MEDICAL PARASITOLOGY. Educ: Univ Ill, AB, 39, MS, 40; Johns Hopkins Univ, ScD(parasitol), 43; Am Bd Med Microbiol, dipl, 63. Prof Exp: From instr to assoc prof parasitol, Sch Pub Health, 43-47, asst dean acad progs, 64-70, actg dean sch pharm, 65-66, PROF PARASITOL & HEAD DEPT, SCH PUB HEALTH & SCH MED, UNIV NC, CHAPEL HILL, 47-, ASSOC DEAN ACAD PROGS, SCH PUB HEALTH, 71- Concurrent Pos: Assoc, Sch Med, Duke Univ, 43, assoc prof, 57-58, prof parasitol, 59-; consult, Nat Univ Eng, Peru, 56, Dept Health, Educ & Welfare, USPHS, Bur State Serv, 64-66 & Bur Health Manpower, 66-68; mem trop med & parasitol study sect, Div Res Grants, NIH, 69-73, chmn, 71-73; pres, Int Comn Trichinellosis, 72-76. Mem: Fel AAAS; Am Soc Trop Med & Hyg (secy-treas, 53-56, vpres, 63-64, pres, 65-66); Am Soc Parasitologists; Am Pub Health Asn; Am Acad Microbiol. Res: Immunity to Trichinella; host-parasite relations. Mailing Add: Dept of Parasitol Univ of NC Sch of Pub Health Chapel Hill NC 27514

LARSON, ALBERT JO, b Chicago, Ill, Nov 2, 34; m 69. CULTURAL GEOGRAPHY, HISTORICAL GEOGRAPHY. Educ: Univ Miami, BA, 56; Univ NC, MA, 58; Univ Nebr, PhD(geog), 69. Prof Exp: Asst prof geog, Univ Nebr, Omaha, 61-64; asst prof, Carroll Col, 64-66; ASST PROF GEOG, UNIV ILL, CHICAGO CIRCLE, 66- Concurrent Pos: Consult, Olson Travel Orgn, 68- & Encycl Britannica Educ Corp, 70- Mem: Asn Am Geog; Am Geog Soc; Nat Coun Geog Educ. Res: Cultural and historical geography of settlement and the Great Plains. Mailing Add: Dept of Geog Univ Ill Chicago Cir Chicago IL 60680

LARSON, ALLAN BENNETT, b Chicago, Ill, Feb 9, 43; m 71; c 1. INDUSTRIAL PHARMACY, COSMETIC CHEMISTRY. Educ: Drake Univ, BS, 66; Univ Wis, MS, 69; Purdue Univ, PhD(phys & indust pharm), 72. Prof Exp: Sr res pharmacist, Dorsey Labs Div, Sandoz-Wander Inc, 72-76; SR PHARMACEUT SCIENTIST, VICK DIVS RES & DEVELOP, RICHARDSON-MERRELL INC, 76- Mem: Am Pharmaceut Asn; Acad Pharmaceut Sci; Am Soc Cosmetic Chemists. Res: Pharmaceutical and skin care product development and stability, preformulation, uniformity of mixing, scale-up technology, process improvements. Mailing Add: 16 Harstrom Pl Rowayton CT 06853

LARSON, ALLEN CARROL, physical chemistry, see 12th edition

LARSON, BRUCE LINDER, b Minneapolis, Minn, June 24, 27; m 54; c 3. BIOCHEMISTRY. Educ: Univ Minn, BS, 48, PhD(biochem), 51. Prof Exp: Asst, Univ Minn, 48-51; from instr to assoc prof, 51-66, PROF BIOL CHEM, UNIV ILL, URBANA, 66- Concurrent Pos: Fulbright lectr, Arg, 65. Honors & Awards: Am Chem Soc Award, 66. Mem: AAAS; Am Chem Soc; Am Soc Biol Chemists; Am Soc Cell Biologists; Am Dairy Sci Asn. Res: Metabolism of the mammary gland; lactation. Mailing Add: Biochem Div Dept of Dairy Sci Univ of Ill Urbana IL 61801

LARSON, CARL LEONARD, b Butte, Mont, Jan 28, 09; m 28; c 2. BACTERIOLOGY. Educ: Mont State Col, BS, 32; Univ Minn, BM, 38, MD, 39. Hon Degrees: DSc, Mont State Univ, 54. Prof Exp: Res assoc snowshoe hare pop, Bur Biol Surv, USDA, 32-38; asst prof med bact, George Washington Univ, 39; med dir, Lab Infectious Dis, NIH, 39-50, dir, Rocky Mountain Lab, USPHS, 50-65; PROF MICROBIOL & DIR STELLA DUNCAN MEM INST, UNIV MONT, 65- Mem: Soc Exp Biol & Med; Am Asn Immunologists; Am Pub Health Asn; Am Acad Microbiol; NY Acad Sci. Res: Immunology; tuberculosis. Mailing Add: Stella Duncan Mem Inst Univ of Mont Missoula MT 59801

LARSON, CARROLL BERNARD, b Council Bluffs, Iowa, Sept 10, 09; m 35; c 4. ORTHOPEDIC SURGERY. Educ: Univ Iowa, BS, 31, MD, 33. Prof Exp: Instr orthop surg, Harvard Med Sch, 39-50; head dept, 50-73, PROF ORTHOP SURG, COL MED, UNIV IOWA, 50-, DIR REHAB, 55- Concurrent Pos: Traveling fel, Gt Brit, 48; jr asst, Mass Gen Hosp, 39-45, sr asst, 45-50. Mem: Orthop Res Soc; Am Acad Orthop Surg (pres, 66-). Res: Bone metabolism; development of arthroplastics. Mailing Add: Rehab Unit Univ of Iowa Col of Med Iowa City IA 52240

LARSON, CHARLES CONRAD, b Pettibone, NDak, Nov 17, 14; m 58; c 2. FORESTRY ADMINISTRATION. Educ: Univ Minn, BS, 40; Univ Vt, MS, 43; Inst Pub Admin, NY, cert, 50; State Univ NY Col Forestry, Syracuse Univ, PhD(forestry admin), 52. Prof Exp: Summer res asst, Lake States Forest Exp Sta, 40-41; asst forest supvr, Crossett Lumber Co, Ark, 44; res & state exten forester, Univ Vt, 44-47; res assoc, Inst Pub Admin, NY, 48-49; res assoc forestry, 50-58, assoc prof, 58-61, PROF WORLD FORESTRY, STATE UNIV NY COL ENVIRON SCI & FORESTRY, 61-, DIR INT FORESTRY, 65-, DEAN SCH ENVIRON & RESOURCE MGT, 71- Concurrent Pos: Ford Found oversease res fel, 55; consult, Syracuse Res Inst, 57-59; vis prof & proj leader, AID-State Univ NY Res Found Assistance Contract, Col Forestry, Univ Philippines, 59-62; mem bd dirs, Orgn Trop Studies, 71-; chmn educ comn, Int Union Socs Foresters, 71- Mem: Soc Am Foresters. Res: Forestry administration, policy and economics; tropical vegetation; world forestry development with emphasis on the tropics. Mailing Add: Sch Environ & Resource Mgt State Univ NY Col Environ Sci & Forestry Syracuse NY 13210

LARSON, CLARENCE EDWARD, b Cloquet, Minn, Sept 20, 09; m 34, 57; c 3. PHYSICAL CHEMISTRY. Educ: Univ Minn, BS, 32; Univ Calif, PhD(chem), 36. Prof Exp: Instr chem, Univ Calif, 32-36; res assoc, Mt Zion Res Found, Calif, 36-37; assoc prof chem, Col of the Pac, 37-39, prof & head dept, 39-40; chief anal sect, Radiation Lab, Univ Calif, 40-43; head tech staff, Electromagnetic Plant, Carbide & Carbon Chem Co Div, Union Carbide Corp, 43-46, dir res & develop, 46-49, supt, 49-50, dir, Oak Ridge Nat Lab, 50-55, vpres chg res, Nat Carbon Co Div, 55-59, assoc mgr res admin, 59-61, vpres, Union Carbide Nuclear Co Div, 61-64; ENERGY CONSULT, SYSTS CONTROL, INC, 64-; COMNR, ENERGY RES & DEVELOP ADMIN, WASHINGTON, DC, 69- Concurrent Pos: Am Nuclear Soc fel, 62; mem bd dirs, Oak Ridge Inst Nuclear Studies, 63-65; pres, Nuclear Div, Union Carbide Co, Tenn, 65-69; deleg, UN Conf Peaceful Uses of Atomic Energy. Mem: Nat Acad Eng; AAAS; Am Chem Soc; fel Am Inst Chemists; Am Nuclear Soc. Res: Inorganic chemistry of biological systems; separation methods for isotopes; radiochemistry; analytical methods; uranium chemistry; colloids; chemical separation methods. Mailing Add: 6514 Bradley Blvd Bethesda MD 20034

LARSON, DANIEL A, b Jersey City, NJ, July 7, 23; m 51; c 3. ATOMIC PHYSICS. Educ: St Peter's Col, NJ, BS, 44; Stevens Inst Technol, MS, 56. Prof Exp: Trainee physics, Schenectady Res Lab, Gen Elec Co, 47-48, physicist, Nela Park Lamp Div, 48-52; from assoc res engr to sr res engr, 52-65, ENGR FEL RES, LAMP DIV, WESTINGHOUSE ELEC CORP, 65- Mem: AAAS; Illum Eng Soc; Optical Soc Am. Res: Low pressure discharges in metal vapor; fluorescent lamp discharges; high pressure discharges in mercury metal and metal iodides in high pressure discharges; optical pumps for lasers; high pressure sodium lamp discharges. Mailing Add: Westinghouse Lamp Div Dept 82012.4 One Westinghouse Plaza Bloomfield NJ 07003

LARSON, DANIEL LEWIS, b Sioux City, Iowa, Nov 28, 21; m 55; c 4. MEDICINE. Educ: Columbia Univ, MD, 46; Am Bd Internal Med, dipl. Prof Exp: Intern med, Columbia-Presby Hosp, 46-47, resident, 47-50; instr, 50-52, assoc, 52-53 & 55-57, ASST PROF MED, COL PHYSICIANS & SURGEONS, COLUMBIA UNIV, 57-; DIR MED, ST BARNABAS HOSP, 65- Concurrent Pos: Markle scholar, 50-53 &

55-57. Mem: AAAS; Am Soc Clin Invest; Soc Exp Biol & Med; AMA; Am Asn Immunol. Res: Immunochemistry; cardiology; rheumatic fever, systemic lupus erythematosus; connective tissue disease. Mailing Add: St Barnabas Hosp Third Ave Bronx NY 10457

LARSON, DAVID L, b Moline, Ill, June 16, 42; m 66. VIROLOGY, BIOCHEMISTRY. Educ: Univ Northern Colo, BA, 64; St Louis Univ, PhD(biochem), 70. Prof Exp: NIH fels viral immunol, Scripps Clin & Res Found, Calif, 69-71; SR RES INVESTR VIROL, JOHN L SMITH MEM CANCER RES, PFIZER, INC, 71- Mem: AAAS; NY Acad Sci; Am Soc Microbiol. Res: Viral immunology and biochemistry; breast cancer; leukemia; lymphoma viruses; mechanism of action of estrogens. Mailing Add: John L Smith Mem for Cancer Res Pfizer Inc Maywood NJ 07607

LARSON, DENIS WAYNE, b Prince Albert, Sask, Dec 2, 38; m 62; c 4. PHYSICAL CHEMISTRY. Educ: Univ Sask, BA, 61, MA, 62; Univ Toronto, PhD(inorg chem), 65. Prof Exp: Mem res staff, Can Forces Inst Aviation Med, 65-68; asst prof, 68-71, ASSOC PROF CHEM, UNIV REGINA, 71- Mem: Chem Inst Can. Res: Kinetics; thermodynamics; mixed solvents; glycols; conductivity; viscosity studies. Mailing Add: Dept of Chem Univ of Regina Regina SK Can

LARSON, DONALD, high pressure physics, see 12th edition

LARSON, DONALD ALFRED, b Chicago, Ill, Sept 15, 30; m 53; c 3. BOTANY. Educ: Wheaton Col, Ill, BS, 53; Univ Ill, MS, 55, PhD(bot), 59. Prof Exp: Assoc prof bot, Univ Tex, Austin, 59-69, prof & dir health prof, 69-73; PROF BIOL & ASSOC VPRES HEALTH SCI, STATE UNIV NY BUFFALO, 73- Mem: Bot Soc Am; Am Soc Cell Biologists. Res: Electron microscopy; cytology; palynology; taxonomic uses; paleobotany. Mailing Add: 104 Farber Hall State Univ NY Buffalo NY 14214

LARSON, DONALD CLAYTON, b Wadena, Minn, Jan 29, 34; m 60; c 3. SOLID STATE PHYSICS. Educ: Univ Wash, BS, 56; Harvard Univ, SM, 57, PhD(appl physics), 62. Prof Exp: Asst prof physics, Univ Va, 62-67; ASSOC PROF PHYSICS, DREXEL UNIV, 67- Mem: Am Phys Soc. Res: Electrical properties of metallic, organic and amorphous semiconducting films; biomechanics; solar energy. Mailing Add: Dept of Physics Drexel Univ Philadelphia PA 19104

LARSON, DONALD W, b Aroca, Minn, Aug 7, 40; m 62; c 3. AGRICULTURAL ECONOMICS. Educ: SDak State Univ, BS, 62; Mich State Univ, MS, 64, PhD(agr econ), 68. Prof Exp: Asst prof agr econ, Mich State Univ, 68-70; ASST PROF AGR ECON, OHIO STATE UNIV, 70- Concurrent Pos: Mkt consult, 73, Loan consult, US AID, 74. Mem: Am Agr Econ Asn; Int Asn Agr Economists; Brazilian Agr Econ Asn. Res: Marketing and rural transportation systems in developing countries. Mailing Add: 1905 Tewksbury Rd Columbus OH 43221

LARSON, DUANE L, b Stoughton, Wis, June 19, 28; m 58;c 3. PLASTIC SURGERY, SURGERY. Educ: Univ Wis, BS, 51, MD, 54; Am Bd Surg, dipl, 62; Am Bd Plastic Surg, dipl, 65. Prof Exp: Intern, Med Col Va, 54-55; gen surg resident, Univ Wis, 55-57 & 59-61; plastic surg resident, 61-64, asst prof, 64-68, assoc prof, 69-73, PROF PLASTIC SURG, UNIV TEX MED BR GALVESTON, 73-; CHIEF SURGEON, SHRINERS BURNS INST, 65- Mem: AAAS; Am Col Surg; Am Fedn Clin Res; Plastic Surg Res Coun; Am Soc Head & Neck Surg. Res: Lymphatics; burns. Mailing Add: Shriners Burns Inst Galveston TX 77550

LARSON, EDGAR WILLIAM, b Argusville, NDak, Feb 3, 26; m 49; c 3. MICROBIOLOGY. Educ: Iowa State Univ, BS, 49. Prof Exp: Chief lab sect, Test Chamber Br, US Army Biol Labs, 49-54 & Test Chamber Br, Tech Eval Div, 54-55 & 57-60, asst dir tech serv, 60-64, chief tech eval div, 64-71, chief aerobiol & eval div, 71-72, AEROBIOL SCIENTIST, AEROBIOL DIV, US ARMY MED RES INST INFECTIOUS DISEASES, 72- Honors & Awards: Meritorious Civilian Serv Award, US Army. Mem: AAAS; Am Soc Microbiol; Biomet Soc; Sigma Xi. Res: Quantitative aerobiology; experimental design and analysis. Mailing Add: Aerobiol Div US Army Med Res Inst of Infectious Diseases Frederick MD 21701

LARSON, EDWARD RICHARD, b New York, NY, May 24, 20; m 43; c 3. GEOLOGY. Educ: Columbia Univ, BA, 42, MA, 47, PhD(geol), 51. Prof Exp: Jr geologist, US Geol Surv, 43-44; asst prof geol, Mo Sch Mines, 57-59; asst prof, 49-53, chmn dept, 52-66, assoc prof geol & geog, 53-59, PROF GEOL, UNIV NEV, RENO, 59- Mem: Geol Soc Am; Am Asn Petrol Geologists. Res: Structure; stratigraphy. Mailing Add: Dept of Geol & Geog Univ of Nev Reno NV 89507

LARSON, EDWIN E, b Los Angeles, Calif, Jan 5, 31. GEOPHYSICS, GEOLOGY. Educ: Univ Calif, Los Angeles, BA, 54, MA, 58; Univ Colo, PhD(geol), 65. Prof Exp: Explor geologist, Humble Oil & Refining Co, 57-60; NSF fel, 65-66; from asst prof to assoc prof, 66-75, PROF GEOL SCI, UNIV COLO, BOULDER, 75- Mem: AAAS; Geol Soc Am; Am Geophys Union. Res: Investigation of rock magnetic properties; paleomagnetism and its application to the solution of geological problems; lunar magnetism. Mailing Add: Dept of Geol Univ of Colo Boulder CO 80302

LARSON, EVERETT GERALD, b Logan, Utah, Nov 12, 35; m 63; c 4. SOLID STATE PHYSICS, MOLECULAR PHYSICS. Educ: Mass Inst Technol, SB, 57, SM, 59, PhD(electron correlation), 64. Prof Exp: Asst, Mass Inst Technol & mem staff, Lincoln Labs, 57-64; asst prof physics, Brigham Young Univ, 64-68; vis asst prof chem, Univ Ga, 68-69; assoc prof, 69-75, PROF PHYSICS, BRIGHAM YOUNG UNIV, 75- Mem: Am Phys Soc. Res: Atomic molecular and solid state theory; correlation effects; cooperative phenomena; theory of irreversible processes. Mailing Add: Dept of Physics & Astron Brigham Young Univ Provo UT 84601

LARSON, FRANK CLARK, b Columbus, Nebr, Jan 17, 20; m 48; c 3. MEDICINE. Educ: Nebr State Teachers Col, AB, 41; Univ Nebr, MD, 44; Am Bd Internal Med, dipl. Prof Exp: Instr, 50-51, asst clin prof, 51-56, from asst prof to assoc prof, 56-63, PROF MED & PATH, UNIV WIS-MADISON, 63-, DIR CLIN LABS, HOSP, 58- Concurrent Pos: Asst chief med & tuberc serv, Vet Admin Hosp, Madison, 51-56, chief invest med serv, 52-56. Mem: Endocrine Soc; Am Col Physicians. Res: Thyroid metabolism. Mailing Add: Dept of Med Univ of Wis Med Sch Madison WI 53706

LARSON, FREDERIC ROGER, b Los Angeles, Calif, Mar 26, 42; m 74; c 1. FOREST MANAGEMENT, SYSTEMS ANALYSIS. Educ: Northern Ariz Univ, BSF, 66, MS, 68; Colo State Univ, PhD(forestry), 75. Prof Exp: Forester, Nez Perce Nat Forest, 66-67, RES FORESTER, ROCKY MOUNTAIN FOREST & RANGE EXP STA, US FOREST SERV, 67- Concurrent Pos: Prof, Northern Ariz Univ Forestry Sch, 71-72. Mem: Soc Am Foresters. Res: Quantifying and simulating growth and management of southwestern coniferous forests. Mailing Add: Forest Sci Lab Northern Ariz Univ Flagstaff AZ 86001

LARSON, GARY EUGENE, b Jersey Shore, Pa, Aug 10, 36; m 60; c 1. BOTANY, BIOPHYSICS. Educ: State Univ NY Albany, BS, 58, MS, 60; Rutgers Univ, PhD(bot), 64. Prof Exp: Asst bot, Rutgers Univ, 61-62, instr biol, Douglass Col, 62-64; asst prof, 64-66, ASSOC PROF BOT, BETHANY COL, W VA, 66-, CHMN DEPT, 68- Concurrent Pos: WVa Heart Asn grant, 67-; teacher & dir, Col Educ Prog, WVa Penitentiary, 68-; Peace Corps sci curric adv, Gambian Govt, 70-71; dir, Nat Defense Educ Act Title I Proj, 72-73. Mem: Fel AAAS; Am Soc Plant Physiol; Am Inst Biol Sci; Torrey Bot Club. Res: Effect of work on the heart rate; effect of exercise on muscle bioelectric potentials; bioelectric potentials surrounding the roots of plants. Mailing Add: Dept of Biol Bethany Col Bethany WV 26032

LARSON, GERALD LOUIS, chemistry, see 12th edition

LARSON, GERALD WILLIS, b Halstad, Minn, Nov 21, 23; m 60. ORGANIC CHEMISTRY. Educ: Univ Minn, BChem, 49; Carnegie Inst Technol, PhD(chem), 53. Prof Exp: Chemist, E I du Pont de Nemours & Co, 53-55; CHEMIST, MINN MINING & MFG CO, ST PAUL, 55- Res: Polymers; photochemistry. Mailing Add: 10550 Quarry Ave Stillwater MN 55082

LARSON, GUSTAV OLOF, b Cedar City, Utah, Dec 24, 26; m 55; c 4. ORGANIC CHEMISTRY. Educ: Univ Utah, BS, 48, MA, 51; Wash State Univ, PhD, 59. Prof Exp: Asst prof chem, Utah State Univ, 57-62; assoc prof & Lead dept, Westminster Col, Utah, 62-65; NSF fac fel, Univ Colo, 65-66; assoc prof, 66-71, PROF CHEM, FERRIS STATE COL, 71- Mem: AAAS; Am Chem Soc. Res: Organic reaction mechanisms; stereochemistry; teaching aids, including patents on paper steromodels and pK-pH calculator. Mailing Add: Dept of Phys Sci Ferris State Col Big Rapids MI 49307

LARSON, HAROLD JOSEPH, b Eagle Grove, Iowa, Nov 16, 34; m 62; c 4. MATHEMATICAL STATISTICS. Educ: Iowa State Univ, BS, 56, MS, 57, PhD(math, statist), 60. Prof Exp: Instr statist, Iowa State Univ, 59-60; math statistician, Stanford Res Inst, 60-62; PROF OPERS ANAL, NAVAL POSTGRAD SCH, 62- Concurrent Pos: Consult, Autonetics Div, NAm Aviation, Inc, 63-64, Data Dynamics, Inc, 65-67 & Field Res Corp, 65-69; Fulbright prof, Univ Sao Paulo, 70-71. Mem: Am Statist Asn. Res: Probability theory; general statistical methods. Mailing Add: Dept of Opers Anal Naval Postgrad Sch Monterey CA 93940

LARSON, HAROLD OLAF, b Port Wing, Wis, May 27, 21. ORGANIC CHEMISTRY. Educ: Univ Wis, BS, 43; Purdue Univ, MS, 47; Harvard Univ, PhD, 50. Prof Exp: Navigator, Pan Am Airways, 44-45; chemist, Hercules Powder Co, 50-54; res fel chem, Harvard Univ, 54-55; asst prof, Univ WVa, 55-57; res fel, Purdue Univ, 57-58; from asst prof to assoc prof, 58-75, PROF CHEM, UNIV HAWAII, 75- Mem: Am Chem Soc; The Chem Soc. Mailing Add: Dept of Chem Univ of Hawaii Honolulu HI 96822

LARSON, HAROLD PHILLIP, b Hartford, Conn, July 13, 38; m 60; c 2. ASTRONOMY. Educ: Bates Col, BS, 60; Purdue Univ, MS, 63, PhD(physics), 67. Prof Exp: Res assoc physics, Purdue Univ, 67-68; fel, Aime-Cotton Lab, Nat Ctr Sci Res, France, 68-69; asst prof, 69-71, ASSOC PROF ASTRON, LUNAR & PLANETARY LAB, UNIV ARIZ, 71- Mem: Am Astron Soc. Res: Infrared astronomy of planetary atmospheres and surfaces. Mailing Add: Lunar & Planetary Lab Univ of Ariz Tucson AZ 85721

LARSON, HAROLD VINCENT, b Portland, Ore, Dec 15, 24; m 48; c 5. HEALTH PHYSICS. Educ: Ore State Col, BS, 50, MS, 57. Prof Exp: Physicist health physics, Gen Elec Co, 52-65; sr develop engr, Radiation Protection Dept, 65-68, mgr radiation protection dept, 68-70, MGR RADIATION STANDARDS & ENG SECT, BATTELLE MEM INST, 71- Mem: Am Phys Soc; Health Physics Soc; Radiation Res Soc; Am Asn Physicists in Med; Am Nuclear Soc. Res: Atomic and nuclear physics as applied to radiological and health physics. Mailing Add: 904 Cottonwood Ave Richland WA 99352

LARSON, INGEMAR W, b Clarissa, Minn, Dec 4, 28; m 62. ZOOLOGY, PARASITOLOGY. Educ: Concordia Col, Moorhead, Minn, AB, 51; Kans State Univ, MS, 57, PhD(parasitol), 63. Prof Exp: Instr biol, Concordia Col, 51-52; asst zool, Kans State Univ, 55-57; asst parasitol, Biol Sta, Univ Mich, 57; instr zool, Kans State Univ, 57-62, asst zool, 62-63; res assoc, Ore State Univ, 63-66; asst prof, 66-69, ASSOC PROF BIOL, AUGUSTANA COL, ILL, 69- Mem: Am Soc Parasitologists; Am Micros Soc; Sigma Xi. Res: Parasitic protozoa; floodplain insects. Mailing Add: Dept of Biol Augustana Col Rock Island IL 61201

LARSON, JEROME VALJEAN, b Norfolk, Nebr, Mar 8, 33; m 55; c 3. GEOPHYSICS, ELECTRICAL ENGINEERING. Educ: Univ Md, BSEE, 60, MSEE, 63, PhD(elec eng), 68. Prof Exp: Instr elec eng, 60-68, ASST PROF ELEC ENG & PHYSICS, UNIV MD, COLLEGE PARK, 68- Concurrent Pos: Consult, Amecom Div, Litton Systs, Inc, 60- Mem: Inst Elec & Electronics Engrs; Am Geophys Union. Res: Search for free mode oscillators of earth and moon, possibly caused by gravity waves. Mailing Add: Dept of Physics Univ of Md College Park MD 20742

LARSON, JERRY KING, b Willmar, Minn, May 15, 41; m 64; c 2. CLINICAL PHARMACOLOGY. Educ: Macalester Col, BA, 63; Mass Inst Technol, PhD(org chem), 67. Prof Exp: Res chemist, Chas Pfizer & Co, Inc, 67-70; ASST TO DIR CLIN RES, PFIZER INC, 70- Mem: Am Chem Soc; Am Heart Asn. Res: Clinical trials with new potential drug candidates, particularly the administration and monitoring of such trials. Mailing Add: Pfizer Inc Groton CT 06340

LARSON, JOHN GRANT, physical chemistry, see 12th edition

LARSON, JOHN ROBERT, organic chemistry, see 12th edition

LARSON, JOSEPH SBANLEY, b Stoneham, Mass, June 23, 33; m 58; c 2. WILDLIFE BIOLOGY. Educ: Univ Mass, BS, 56, MA, 58; Va Polytech Inst, PhD(zool), 66. Prof Exp: Exec secy, Wildlife Conserv Inc, Mass, 58-59; state ornithologist & asst to dir, Mass Div Fisheries & Fame, 59-60; head conserv educ div, Nat div, Natural Resources Inst, Univ Md, 60-62, res asst prof wildlife, 65-67; adj asst prof wildlife biol & asst unit leader, Mass Coop Wildlife Res Unit, 67-69, ASSOC PROF FORESTRY & WILDLIFE MGT, UNIV MASS, AMHERST, 69- Mem: AAAS; Wildlife Soc; Ecol Soc Am; Animal Behav Soc; Am Soc Mammal. Res: Wetland ecology and management; beaver behavior. Mailing Add: Dept of Forestry & Wildlife Mgt Univ of Mass Amherst MA 01002

LARSON, KENNETH ALLEN, b Havre, Mont, July 6, 35; m 61; c 4. IMMUNOLOGY, VETERINARY MEDICINE. Educ: Wash State Univ, DVM, 61, MS, 65, PhD(immunol), 66. Prof Exp: Vet pvt pract, Mont, 61-63; asst prof vet med, 66-69, ASSOC PROF VET MED & MICROBIOL, COLO STATE UNIV, 69- Mem: AAAS; Am Soc Microbiol; Am Vet Med Asn. Res: Immunology of tumors in animals. Mailing Add: Vet Hosp Colo State Univ Ft Collins CO 80521

LARSON, KENNETH CURTIS, b Madison, Wis, July 7, 40; m 64; c 2. PAPER CHEMISTRY, WOOD CHEMISTRY. Educ: Calif Inst Technol, BS, 62; Lawrence

Univ, MS, 64, PhD(wood chem), 67. Prof Exp: Sci specialist res & develop, 67-74, SCI ASSOC, SCOTT PAPER CO, PHILADELPHIA, 74- Mem: Tech Asn Pulp & Paper Indust; Am Chem Soc. Res: Catalysis; colloid and surface chemistry of wood pulp fibers. Mailing Add: Scott Paper Co Scott Plaza Philadelphia PA 19113

LARSON, LARRY LEE, b Horton, Kans, Nov 18, 39; m 63; c 2. REPRODUCTIVE PHYSIOLOGY. Educ: Kans State Univ, BS, 62, MS, 65, PhD(animal breeding), 68. Prof Exp: NIH fel, Cornell Univ, 68-70, asst prof reproductive physiol, 70-72; ASST PROF REPRODUCTIVE PHYSIOL, UNIV NEBR-LINCOLN, 72- Mem: Soc Study Reproduction; Brit Soc Study Fertil; Am Soc Animal Sci; Am Dairy Sci Asn. Res: Uterine metabolism, especially estrus control and determination and conception failures; management factors to improve reproductive performance. Mailing Add: Dept of Animal Sci Univ of Nebr Lincoln NE 68503

LARSON, LAURENCE ARTHUR, b Cleveland, Ohio, Mar 17, 30; m 56; c 2. PLANT PHYSIOLOGY. Educ: Ohio Univ, BS, 56; Univ Tenn, MS, 59; Purdue Univ, PhD(amino acid metab), 63. Prof Exp: Phys oceanogr, US Navy Hydrographic Off, 56-57; instr bot, Univ Tenn, 59; from asst prof to assoc prof, 63-75, PROF BOT, OHIO UNIV, 75- Mem: Am Soc Plant Physiologists. Res: Germination physiology. Mailing Add: Dept of Bot Ohio Univ Col Arts & Sci Athens OH 45701

LARSON, LAWRENCE T, b Waukegan, Ill, Dec 3, 30; m 57; c 3. ECONOMIC GEOLOGY, MINERALOGY. Educ: Univ Ill, BS, 57; Univ Wis, MS, 59, PhD(geol), 62. Prof Exp: From asst prof to prof geol, Univ Tenn, Knoxville, 61-75; PROF GEOL & CHMN DEPT, MACKAY SCH MINES, UNIV NEV, RENO, 75- Concurrent Pos: Consult, Oak Ridge Nat Lab, 63-69 & mining firms, 68-; partner, Appl Explor Concepts, 72- Mem: Geol Soc Am; Mineral Soc Am; Can Inst Mining & Metall; Am Inst Mining, Metall & Petrol Engrs; Soc Econ Geologists. Res: Manganese mineralogy and ore deposits; ore microscopy; optical properties of metals; geologic thermometry; ceramic and high alumina clay deposits. Mailing Add: Dept of Geol Univ of Nev Mackay Sch of Mines Reno NV 89507

LARSON, LEE EDWARD, b Bristol, Conn, Dec 2, 37; m 63; c 1. PHYSICS. Educ: Univ NH, PhD(ionospheric physics), 67. Prof Exp: Instr physics, Allegheny Col, 61-63; asst prof, 66-70, ASSOC PROF PHYSICS, DENISON UNIV, 70- Mem: Am Asn Physics Teachers; Am Geophys Union. Res: Ionospheric physics, experimental study using sounding rocket techniques. Mailing Add: Dept of Physics Denison Univ Granville OH 43023

LARSON, LESLIE L, b Preston, Idaho, Oct 18, 10; m 40; c 4. ORGANIC CHEMISTRY. Educ: Univ Idaho, BS, 34, MS, 36; Lawrence Col, PhD(pulp & paper technol), 40. Prof Exp: Res chemist, Kimberly-Clark Corp, 39-42; assoc chemist, Chem Warfare Serv, US Dept Army, Edgewood Arsenal, 42-45; tech dir, Detroit, Sulphite Pulp & Paper Co, 45-53 & Potlatch Forests, Inc, 53-65; tech dir, Pomona Div, Northwest Paper Co, Calif, 65-66; tech staff mem, Paper Div, 66-68, GROUP LEADER, ANAL DEPT, RES & NEW BUS DEVELOP, WEYERHAEUSER CO, 68- Mem: Tech Asn Pulp & Paper Indust. Res: Pulp bleaching; machine coating of paper; paper specialties. Mailing Add: Anal Dept Res & New Bus Dev Weyerhaeuser Co PO Box 601 Fitchburg MA 01421

LARSON, LESTER LEROY, b Amherst Junction, Wis, Feb 12, 23; m 51; c 3. VETERINARY MEDICINE. Educ: Univ Minn, BS, 50, DVM & MS, 53, PhD(vet med), 57; Am Col Theriogenologists, dipl, 70. Prof Exp: Vet, 53-54; from instr to asst prof vet med, Univ Minn, 54-58; assoc vet, 58-73, VET & HEAD VET DEPT, AM BREEDERS SERV, INC, 74- Concurrent Pos: Chmn comt exam, Am Col Theriogenologists, 73. Mem: Am Vet Med Asn. Res: Neuroanatomical and neurophysiological aspects of reproductive process-male; surgical technique for anesthesia of penis-bull; control of veneral diseases of cattle through artificial insemination; actuarial studies of bull under conditions of an artificial insemination center. Mailing Add: Am Breeders Serv Inc DeForest WI 53532

LARSON, LESTER MIKKEL, b Devils Lake, NDak, Aug 2, 18; m 43; c 3. CHEMISTRY. Educ: Lawrence Col, BA, 40; Univ Wis, MS, 50, PhD(biochem), 51. Prof Exp: Chemist, Abbott Labs, 41-48; instr biochem, Univ Wis, 51-52; biochemist, E I du Pont de Nemours & Co, 52-64; assoc prof, 65-69, PROF CHEM, DEL STATE COL, 69- Concurrent Pos: Pres, Larson Corp. Mem: Am Chem Soc; Am Inst Chem. Res: Arsenicals; antibiotics; polymers; separations. Mailing Add: 4619 Bailey Dr Limestone Acres Wilmington DE 19808

LARSON, LEWIS HENRY, JR, b St Paul, Minn, Jan 24, 27. ANTHROPOLOGY. Educ: Univ Minn, AB, 49, MA, 51, PhD(anthrop), 69. Prof Exp: Asst prof anthrop, Univ Ark, 51-52; archaeologist, Ga Hist Comn, 54-59; lectr anthrop, Ga Inst Technol, 60-61; from asst prof to assoc prof, Ga State Col, 61-67; assoc prof, Eastern Ky Univ, 70-71; assoc prof, 71-75, PROF ANTHROP, W GA COL, 71- Concurrent Pos: Mem, Ga Rev Bd, Nat Regist Hist Places, 71-; archaeologist, State of Ga, 72- Mem: Am Anthrop Asn; Soc Am Archaeol; Soc Hist Archaeol; Int African Inst. Res: Southeastern United States archaeology, ethnohistory and cultural ecology. Mailing Add: Dept of Anthrop WGa Col Carrollton GA 30117

LARSON, MERLYN MILFRED, b Story City, Iowa, Sept 11, 28; m 54; c 6. FOREST PHYSIOLOGY. Educ: Colo State Univ, BS, 54; Univ Wash, MF, 58, PhD, 62. Prof Exp: Jr forester, US Forest Serv, 55, res forester, Rocky Mountain Forest & Range Exp Sta, 55-64, forest physiologist, 64-66; assoc prof silvicult, 66-70, PROF FORESTRY, OHIO AGR RES & DEVELOP CTR, 70-; PROF NATURAL RESOURCES, OHIO STATE UNIV, 70- Concurrent Pos: Asst, Univ Wash, 58-59. Mem: Soc Am Foresters. Res: Forest regeneration; tree physiology. Mailing Add: Dept of Forestry Ohio Agr Res & Develop Ctr Wooster OH 44691

LARSON, MICHAEL OBERLIN, physics, see 12th edition

LARSON, MYRA JOAN, b Dayton, Ohio, June 29, 42. GENETICS. Educ: Univ Nebr, Omaha, BA, 64; Univ Ill, Urbana-Champaign, PhD(zool), 71; Case Western Reserve Univ, 74- Prof Exp: Asst prof biol, Kenyon Col, 71-74. Res: Analyses of variation in odontometric traits of rodents. Mailing Add: Sch of Dent Case Western Reserve Univ Cleveland OH 44106

LARSON, NANCY MARIE, b Dickinson, ND, Sept 30, 46; m 68. THEORETICAL PHYSICS. Educ: Mich State Univ, BS, 67, MS, 69, PhD(theoret physics), 72. Prof Exp: COMPUT SCI SPECIALIST PHYSICS, OAK RIDGE NAT LAB, NUCLEAR DIV, UNION CARBIDE CORP, 72- Mem: Am Phys Soc; Pattern Recognition Soc. Res: Computer modeling of environmental transport; nuclear structure theory; pattern recognition applied to chemical and biological data. Mailing Add: Comput Sci Div Oak Ridge Nat Lab Oak Ridge TN 37830

LARSON, NOAL P, b Castle Dale, Utah, Sept 17, 12; m 34; c 3. ECONOMIC ENTOMOLOGY. Educ: Ore State Col, BS, 34, MS, 36; Iowa State Col, PhD(insect toxicol), 41. Prof Exp: Entomologist, E Clemems Horst Co, Ore, 34; asst entom, Ore State Col, 34-36 & Iowa State Col, 37; agt grasshopper control, Bur Entom & Plant Quarantine, USDA, 38; asst prof entom & zool & asst entom, Exp Sta, SDak State Col, 38-43; entomologist, Rohm and Haas Co, 43-54; OWNER & MGR, STATE CHEM CO, 54- Mem: Entom Soc Am. Res: Insecticide research and development. Mailing Add: 4215 N 16th St Phoenix AZ 85016

LARSON, NORA LEONA, b Fillmore Co, Minn, Sept 19, 01. BACTERIOLOGY. Educ: St Olaf Col, BA, 23; Univ Minn, MS, 33, PhD(bact), 47. Prof Exp: Tech asst clin bact, Mayo Clin, 24-36; bacteriologist, Lahey Clin, Boston, 36-39; asst res bact, Mayo Clin, 39-41; bacteriologist, Lahey Clin, Boston, 41-43; asst bact, Univ Minn, 44-46; bacteriologist, Takamine Labs, Inc, 46-49; res fel, Hormel Inst, Univ Minn, 50-56, asst prof, 56-60; assoc prof, 60-72, EMER ASSOC PROF BIOL, ST OLAF COL, 72- Mem: AAAS; Am Soc Microbiol; Am Inst Biol Sci; NY Acad Sci. Res: Lipids of Enterobacter cloacae; effect of chlortetracycline on intestinal microorganisms of young swine. Mailing Add: 1110 W First St Northfield MN 55057

LARSON, OMER R, b Roseau, Minn, Dec 1, 31; m 60; c 4. PARASITOLOGY. Educ: Univ NDak, BA, 54; Univ Minn, MS, 60, P PhD(fish parasites), 63. Prof Exp: Instr biol, Minot State Col, 63-64; asst prof, 64-68, assoc dean arts & sci, 69-70 & 75-76, ASSOC PROF BIOL, UNIV NDAK, 68- Mem: Am Micros Soc; Am Soc Parasitologists; Wildlife Disease Asn. Res: Helminth life cycles; parasites and diseases of fish; biogeography of fleas. Mailing Add: Dept of Biol Univ of NDak Grand Forks ND 58201

LARSON, PAUL STANLEY, b Cannon Falls, Minn, June 9, 07; m 36; c 1. PHARMACOLOGY. Educ: Univ Calif, AB, 30, PhD(physiol), 34. Prof Exp: Instr physiol, Sch Med, Georgetown Univ, 34-39; assoc physiol & pharmacol, Med Col Va, 40; lectr pharmacol, Col Med, Wayne Univ, 40-41; res assoc, 41-46, from assoc prof to prof, 46-63, chmn dept pharmacol, 55-72, HAAG PROF RES PHARMACOL, MED COL VA, 63-, CHMN DEPT, 55- Mem: Am Physiol Soc; Am Soc Pharmacol & Exp Therapeut; Am Chem Soc; Soc Toxicol (pres, 63-64); NY Acad Sci. Res: Protein and purine metabolism; water balance; potassium metabolism; anti-spasmodics; biological actions of nicotine; chemical nature of irritants; toxicology. Mailing Add: Dept of Pharmacol Med Col of Va Richmond VA 23298

LARSON, PHILIP RODNEY, b North Branch, Minn, Nov 26, 23; m 48; c 2. FOREST PHYSIOLOGY. Educ: Univ Minn, BS, 49, MS, 52; Yale Univ, PhD(forestry), 57. Prof Exp: Res forester, Fla, 52-54, plant physiologist, Lake States Forest Exp Sta, 56-62, LEADER PHYSIOL WOOD FORMATION, PIONEERING RES UNIT, N CENT FOREST EXP STA, US FOREST SERV, 62- Honors & Awards: Distinguished Serv Award, USDA, 75; Barrington Moore Biol Res Award, Soc Am Foresters, 75. Mem: Am Soc Plant Physiologists; Soc Am Foresters; fel Int Acad Wood Sci; Bot Soc Am; Tech Asn Pulp & Paper Indust. Res: Wood formation; vascular anatomy; physiology of growth and development. Mailing Add: Pioneering Res Unit NCent Forest Exp Sta Star Rte 2 Rhinelander WI 54501

LARSON, RACHEL HARRIS (MRS JOHN WATSON HENRY), b Wake Forest, NC, Aug 1, 13; m 56, 71; c 3. BIOCHEMISTRY, DENTAL RESEARCH. Educ: Appalachian State Univ, BS, 40; Georgetown Univ, MS, 49, PhD(biochem), 58. Prof Exp: Teacher, High Sch, Ga, 40-41 & NC, 41-42; chemist, Lab Indust Hyg Res, NIH, 42-48, RES CHEMIST, NAT INST DENT RES, 48-, CHIEF, PREV METHODS DEVELOP SECT, NAT CARIES PROG, 72- Concurrent Pos: Vis scientist, Royal Dent Col, Denmark, 70-71. Mem: AAAS; fel Am Col Dent; Am Inst Nutrit; Int Asn Dent Res. Res: In vivo studies of caries inhibitory agents including fluorine, enzymes and phosphates; further refinement of the rat model for caries studies. Mailing Add: Nat Caries Prog Nat Inst of Dent Res Bethesda MD 20014

LARSON, RAYMOND GEORGE, b Grand Forks, NDak, June 1, 09; m 41; c 1. CHEMISTRY. Educ: Univ ND, BS, 31, MS, 34; Purdue Univ, PhD(phys chem), 38. Prof Exp: Asst prof chem, Manchester Col, 37-38; instr, Washington & Jefferson Col, 38-39; instr, Valparaiso Univ, 39-41; head anal div, Kingsbury Ord Plant, Ind, 41-43; assoc res chemist, Metall Lab, Univ Chicago, 43-45; from asst prof to assoc prof, 45-52, head dept, 52-55, PROF CHEM, VALPARAISO UNIV, 52- Mem: AAAS; Am Chem Soc. Res: Physical chemistry of solutions; radiation corrosion; process separation of fission products; heavier metals. Mailing Add: Dept of Chem Valparaiso Univ Valparaiso IN 46383

LARSON, RICHARD ALLEN, b Minot, NDak, July 9, 41; m 63; c 2. BIO-ORGANIC CHEMISTRY. Educ: Univ Minn, Minneapolis, BA, 63; Univ Ill, Urbana, PhD(org chem), 68. Prof Exp: USPHS fels, Univ Liverpool, 68-69 & Cambridge Univ, 69-70; res assoc bot, Univ Tex, Austin, 70-72; ASST CUR, STROUND WATER RES CTR, ACAD NATURAL SCI, 72- Concurrent Pos: Spec lectr, Univ Pa, 73-74. Mem: Phytochem Soc NAm; Brit Freshwater Biol Asn; Am Chem Soc; The Chem Soc. Res: Natural products; phytochemistry; aquatic ecology. Mailing Add: Stroud Water Res Ctr RD 1 Box 512 Avondale PA 19311

LARSON, RICHARD BONDO, b Toronto, Ont, Jan 15, 41. ASTROPHYSICS. Educ: Univ Toronto, BSc, 62, MA, 63; Calif Inst Technol, PhD(astron), 68. Prof Exp: Asst prof, 68-75, PROF ASTRON, YALE UNIV, 75- Mem: Am Astron Soc; Royal Astron Soc Can; Royal Astron Soc. Res: Theoretical studies of the gravitational collapse of gas clouds, especially formation of stars and galaxies; stellar dynamics, structure and evolution of galaxies. Mailing Add: Yale Univ Observ Box 2023 Yale Sta New Haven CT 06520

LARSON, RICHARD GUSTAVUS, b Pittsburgh, Pa, May 16, 40; m 70; c 3. MATHEMATICS. Educ: Univ Pa, AB, 61; Univ Chicago, MS, 62, PhD(math), 65. Prof Exp: Instr math, Mass Inst Technol, 65-67; asst prof, 67-70, ASSOC PROF MATH, UNIV ILL, CHICAGO CIRCLE, 70- Mem: Am Math Soc; Math Asn Am; Soc Indust & Appl Math. Res: Algebraic and arithmetic structure of Hopf algebras; applications of Hopf algebras to algebra; computational problems of algebra. Mailing Add: Dept of Math Univ of Ill at Chicago Circle Chicago IL 60680

LARSON, ROBERT ELOF, b Spokane, Wash, Oct 9, 32; m 57; c 3. PHARMACOLOGY, TOXICOLOGY. Educ: Wash State Univ, BS, & BPharm, 57, MS, 62; Univ Iowa, PhD(pharmacol), 64. Prof Exp: Pharmacist, Manito Pharm, Wash, 57-59; staff fel toxicol, Nat Cancer Inst, 64-65; asst prof, 65-69, ASSOC PROF PHARMACOL & TOXICOL, SCH PHARM, ORE STATE UNIV, 69-, CHMN DEPT, 70- Mem: Soc Toxicol; Am Soc Pharmacol & Exp Therapeut. Res: Hepatotoxicity and nephrotoxicity of halogenated hydrocarbons; toxicity of nitrosoureas. Mailing Add: Dept of Pharmacol & Toxicol Ore State Univ Sch of Pharm Corvallis OR 97331

LARSON, ROLAND EDWIN, b Ft Lewis, Wash, Oct 31, 41; m 60; c 2. MATHEMATICS. Educ: Lewis & Clark Col, BS, 66; Univ Colo, Boulder, MA, 68, PhD(math), 70. Prof Exp: ASSOC PROF MATH, BEHREND COL, PA STATE UNIV, 70- Mem: Am Math Soc; Math Asn Am. Res: General topology; lattice to topologies; minimal and maximal topological spaces. Mailing Add: Dept of Math Behrend Col Pa State Univ Erie PA 16510

LARSON, ROY AXEL, b Cloquet, Minn, Feb 5, 31; m 53; c 4. HORTICULTURE. Educ: Univ Minn, BS, 53, MS, 57; Cornell Univ, PhD(floricult), 61. Prof Exp: Assoc prof, 61-69, PROF HORT, NC STATE UNIV, 69- Mem: Am Soc Hort Sci. Res: Floriculture, particularly investigations on effects of environment and regulators on flowering and plant growth. Mailing Add: Dept of Hort Sci NC State Univ Raleigh NC 27607

LARSON, ROY FRED, b Chicago, Ill, May 29, 14; m 41. CARBOHYDRATE CHEMISTRY. Educ: Univ Ill, BS, 37. Prof Exp: Res chemist, 37-75, SR RES CHEMIST, A E STALEY MFG CO, 75- Mem: Am Chem Soc; Am Asn Cereal Chemists. Res: Research and process development of corn syrup and dextrose products from starch. Mailing Add: A E Staley Mfg Co Decatur IL 62523

LARSON, RUBY ILA, b Hatfield, Sask, Can, May 30, 14. CYTOGENETICS. Educ: Univ Sask, BS, 42 & 43, MA, 45; Univ Mo, PhD(genetics), 52. Prof Exp: Cytogeneticist, Cereal Div, Dom Exp Sta, Sask, 45-48; cytogeneticist, Sci Serv Lab, 48-59, CYTOGENETICIST, RES STA, CAN DEPT AGR, 59- Res: Wheat cytogenetics. Mailing Add: Res Sta Can Dept of Agr Lethbridge AB Can

LARSON, RUSSELL EDWARD, b Minneapolis, Minn, Jan 2, 17; m 39; c 3. AGRICULTURAL EDUCATION, ACADEMIC ADMINISTRATION. Educ: Univ Minn, BS, 39, MS, 40, PhD(genetics, plant breeding), 42. Prof Exp: Asst hort, Univ Minn, 39-41; asst res prof agron & asst agronomist, Exp Sta, RI State Col, 41-44; asst prof veg gardening, 44-45, assoc prof plant breeding, 45-47, head dept hort, 52-61, dir agr & home econ exten, 61-63, dean & dir, Col Agr, 63-74, interim provost, 72, PROF HORT, PA STATE UNIV, UNIVERSITY PARK, 47- Concurrent Pos: Sci aide, Mex Agr Prog, Rockefeller Found, 60; chmn comn educ agr & natural resources, Nat Acad Sci-Nat Res Coun, 66- Honors & Awards: Vaughan Award, Am Soc Hort Sci, 48. Mem: AAAS; Am Soc Hort Sci; Genetics Soc Am; Am Genetics Asn. Res: Administration of agricultural research development and education. Mailing Add: 201 Agr Admin Bldg Pa State Univ University Park PA 16802

LARSON, RUSSELL L, b Bridgewater, SDak, Dec 9, 28; m 64. BIOCHEMISTRY. Educ: SDak State Univ, BS, 57, MS, 59; Univ Ill, PhD(biochem), 62. Prof Exp: Fel microbiol, Ore State Univ, 62-64; RES CHEMIST, USDA, UNIV MO-COLUMBIA, 65- Mem: Am Chem Soc; Phytochem Soc NAm. Res: Metabolic processes in microorganisms; alteration in metabolic processes in maize as a result of alteration in the genetic systems in maize. Mailing Add: 304 Curtis Hall USDA Univ of Mo Columbia MO 65201

LARSON, SAGER DARYL, b Ritzville, Wash, Mar 7, 29; m 55; c 2. PHYSICAL CHEMISTRY. Educ: State Col Wash, BS, 51; Univ Ill, MS, 53, PhD(chem), 55. Prof Exp: Asst prof chem, Evangel Col, 55-59, assoc prof, 59-63; from asst prof to assoc prof, 63-71, PROF CHEM, DRURY COL, 71- Res: Phase studies of the two component carbon dioxide-water system involving the carbon dioxide hydrate. Mailing Add: Route 8 5810 Ellison Springfield MO 65804

LARSON, SANFORD J, b Chicago, Ill, Apr 9, 29; m 57; c 3. NEUROANATOMY, NEUROSURGERY. Educ: Wheaton Col, Ill, BA, 50; Northwestern Univ, MD, 54, PhD(anat), 62. Prof Exp: Resident neurosurg, Northwestern Univ, 55-57 & 59-61; USPHS res fel, 61-62; dir neurosurg educ, Cook County Hosp, Chicago, 62-63; assoc prof, 63-68, PROF NEUROSURG, MED COL WIS, 68-, CHMN DEPT, 63- Concurrent Pos: Chief neurosurg, Vet Admin Hosp, Wood, Wis, Milwaukee County Gen Hosp; consult, Milwaukee Lutheran, Columbia & Milwaukee Children's Hosps, Wis & Shriners Hosps Crippled Children, Chicago, Ill. Mem: Biophys Soc; Soc Univ Surg; Am Asn Neurol Surg; Am Col Surg; Soc Neurol Surgeons. Res: Neurophysiology; neurological surgery. Mailing Add: Dept of Neurosurg Med Col of Wis Milwaukee WI 53226

LARSON, THOMAS E, b Waupaca, Wis, Apr 13, 26; m 49; c 3. PHYSICAL CHEMISTRY, ORGANIC CHEMISTRY. Educ: Lewis & Clark Col, BA, 50; Johns Hopkins Univ, MA, 51, PhD(phys chem), 56. Prof Exp: Jr instr chem, Johns Hopkins Univ, 50-53, res asst, Inst Coop Res, 53-56; staff mem, 56-75, ALT GROUP LEADER WX-2, LOS ALAMOS SCI LAB, 75- Mem: Am Chem Soc. Res: Kinetics of exchange reactions in boron hydrides; sensitivity of explosives to various stimuli; radioactive materials. Mailing Add: Los Alamos Sci Lab Box 1663 Los Alamos NM 87544

LARSON, THURSTON E, b Chicago, Ill, Mar 3, 10; m 38; c 2. CHEMISTRY. Educ: Univ Ill, BS, 32, PhD(chem), 37. Prof Exp: From asst chemist to chemist, 32-47, head chem sect, 47-56, HEAD & ASST CHIEF CHEM SECT, ILL STATE WATER SURV, 56-; PROF SANIT ENG, UNIV ILL, URBANA, 61- Mem: Fel AAAS; Water Pollution Control Fedn; Am Chem Soc; hon mem Am Water Works Asn; Nat Asn Corrosion Engrs. Res: Corrosion; water treatment; water quality; analytical corrosion; environmental pollution. Mailing Add: Chem Sect Ill State Water Surv Box 232 Urbana IL 61801

LARSON, TYRONE RAY, US citizen. MICROWAVE PHYSICS. Educ: St Olaf Col, BA, 66; Cornell Univ, MS & PhD(physics), 71. Prof Exp: Res electronic engr, Naval Res Lab, Washington, DC, 71-74; SR MEM TECH STAFF MICROWAVE PHYSICS, BALL BROS RES CORP, 75- Mem: Am Phys Soc; Inst Elec & Electronics Engrs. Res: Principles and applications of microwave radiometry for earth remote sensing. Mailing Add: Ball Bros Res Corp PO Box 1062 Boulder CO 80302

LARSON, VAUGHN LEROY, b Mondovi, Wis, Feb 21, 31; m 61; c 1. VETERINARY MEDICINE. Educ: Univ Minn, BS, 58, DVM, 60, PhD(vet med), 68. Prof Exp: Veterinarian, 60-61; fel bovine leukemia, Univ Minn, 61-68; med scientist animal res, Brookhaven Nat Lab, 68; from asst prof to assoc prof, 68-74, PROF VET MED, UNIV MINN, ST PAUL, 74- Mem: Am Vet Med Asn; Am Asn Equine Practitioners; Int Leukemia Res Asn; Sigma Xi. Res: Transmission and pathogenesis studies on bovine leukemia; clinical, pathological and therapeutic studies on chronic obstructive pulmonary disease in the equine. Mailing Add: Vet Teaching Hosp Univ of Minn St Paul MN 55108

LARSON, VERNON C, b Stambough, Mich, Apr 8, 23; m 46; c 3. AGRICULTURE, ACADEMIC ADMINISTRATION. Educ: Mich State Univ, BS, 47, MS, 50, EdD(educ admin), 54. Prof Exp: From asst prof to assoc prof dairy & asst to dean, Mich State Univ, 47-59; prof agr & asst dean, Am Univ, Beirut, 59-62; prof dairy agr & dir int agr prog, Kans State Univ, 62-65; prof agr sci & dean agr, Ahmadu Bello Univ, Nigeria & chief party, AID & Kans State Univ Team in Nigeria, 66-68; dir int agr progs, Kans State Univ, 68-70; chief party, AID & Kans State Univ Team in India, 70-72; DIR INT AGR PROGS, KANS STATE UNIV, 72- Concurrent Pos: AID consult, Jordan, 60, Cyprus & Sudan, 61 & Colombia, 69. Mem: AAAS. Res: Agricultural administration. Mailing Add: Int Agr Progs Waters Hall 14 Kans State Univ Manhattan KS 66506

LARSON, VIVIAN M, b Erie, NDak, Oct 3, 31. VIROLOGY. Educ: NDak State Col, BS, 53; Univ Mich, MPH, 58, PhD(virol), 63. Prof Exp: Bacteriologist, Detroit Dept Health Labs, Mich, 53-59; from res fel to sr res fel virol, 63-71, DIR CANCER RES LABS, MERCK SHARP & DOHME RES LABS, 71- Mem: Am Soc Microbiol. Res: Cell biology; immunology; tumor viruses; viral biochemistry; herpes virus and leukemia virus vaccines. Mailing Add: Merck Sharp & Dohme Res Labs West Point PA 19486

LARSON, WILBUR JOHN, b Rockford, Ill, Nov 19, 21; m 45; c 3. ANALYTICAL CHEMISTRY. Educ: Augustana Col, Ill, BA, 46; Univ Wis, MS, 48, PhD, 51. Prof Exp: Chemist, Mallinckrodt Chem Works, 51-57, head anal develop lab, 57-63, asst dir qual control, 64-72; RES ASSOC, MALLINCKRODT INC, 72- Mem: Am Chem Soc; Sigma Xi. Res: Trace analysis; quality control; reagent chemicals. Mailing Add: Mallinckrodt Inc 3600 N Second St St Louis MO 63147

LARSON, WILBUR S, b Downing, Wis, Jan 28, 23; m 53; c 1. INORGANIC CHEMISTRY, ANALYTICAL CHEMISTRY. Educ: Wis State Univ, River Falls, BS, 45; Univ Wyo, MS, 58, PhD(inorg chem), 64. Prof Exp: Teacher high schs, Wis, 44-55; asst gen chem, Univ Wyo, 55-63; ASSOC PROF CHEM, UNIV WIS-OSHKOSH, 63- Mem: Am Chem Soc. Res: Colorimetric sulfide, sulfite and thiosulfate analysis; stability of antimony addition compounds; equilibrium exchange mechanisms of metallic sulfides; mechanism studies of antimony pentachloride. Mailing Add: Dept of Chem Univ of Wis-Oshkosh Oshkosh WI 54901

LARSON, WILLIAM DAY, b St Paul, Minn, Oct 3, 08. PHYSICAL CHEMISTRY, ANALYTICAL CHEMISTRY. Educ: Univ Minn, BA, 29, PhD(phys chem), 36; Mich State Col, MS, 31. Prof Exp: From instr to prof chem, Col St Thomas, 32-76; RETIRED. Concurrent Pos: NSF sci fac fel, 57-58; vis prof, Col Women, San Diego, 63-64. Mem: Am Chem Soc; Am Inst Chemists. Res: Solubility of salts in electrolyte solutions; electromotive force studies. Mailing Add: 2151 Grand Ave St Paul MN 55105

LARSON, WILLIAM EARL, b Creston, Nebr, Aug 7, 21; m 47; c 4. SOIL SCIENCE. Educ: Univ Nebr, BS, 44, MS, 46; Iowa State Univ, PhD, 49. Prof Exp: Asst prof agron, Iowa State Univ, 49-50; soil scientist, USDA, Mont State Col, 51-54; from assoc prof to prof soils, Iowa State Univ, 54-67; PROF SOIL SCI, UNIV MINN, ST PAUL, 67- Concurrent Pos: Soil scientist, USDA, Iowa State Univ, 54-67; vis prof, Univ Ill, 60 & Univ Minn, 63; Fulbright scholar, Australia, 65-66. Honors & Awards: Soil Sci Award, Am Soc Agron. Mem: AAAS; fel Am Soc Agron; Soil Sci Soc Am; Int Soc Soil Sci. Res: Soil structure and mechanics; water infiltration; nutrient interrelations in plants; crop response to soil moisture levels and soil temperature; tillage requirements of crops; utilization of sewage wastes on land. Mailing Add: 201 Soil Sci Bldg Univ of Minn St Paul MN 55101

LARSSEN, PER ALFRED, physical chemistry, see 12th edition

LARSSON, BJORN ERIC, organic chemistry, see 12th edition

LARSSON, ROBERT DUSTIN, b Everett, Mass, July 6, 18; m 53; c 5. MATHEMATICS. Educ: Univ Maine, AB, 41; Syracuse Univ, AM, 58. Prof Exp: From instr to prof math, Clarkson Tech Univ, 46-63; dean instr, Mohawk Valley Community Col, 63-70, actg pres, 66 & 68; PRES, SCHENECTADY COUNTY COMMUNITY COL, 70- Mem: Am Soc Eng Educ; Math Asn Am. Res: Higher education administration; mathematics undergraduate programs. Mailing Add: Schenectady County Community Col Washington Ave Schenectady NY 12305

LARTER, EDWARD NATHAN, b Can, Feb 13, 23; m 45; c 3. GENTICS, PLANT BREEDING. Educ: Univ Alta, BSc, 51, MSc, 52; State Col Wash, PhD(genetics, plant breeding), 54. Prof Exp: Assoc prof genetics & plant breeding, Univ Sask, 54-69; ROSNER RES CHAIR PROF PLANT SCI & DIR TRITICALE RES PROG, UNIV MAN, 69- Mem: Genetics Soc Can; Can Soc Agron; Agr Inst Can. Res: Plant breeding and cytogenetics of barley and related species. Mailing Add: Dept of Plant Sci Univ of Man Winnipeg MB Can

LARTIGUE, DONALD JOSEPH, b Baton Rouge, La, Sept 7, 34; m 62; c 2. BIOCHEMISTRY, ENZYMOLOGY. Educ: La State Univ, BS, 57, MS, 59, PhD(biochem), 65. Prof Exp: Analyst, US Food & Drug Admin, 59-60; marine biologist, Marine Lab, Univ Miami, 60; biochemist, USPHS Hosp, Carville, La, 60-62; asst biochem, La State Univ, 62-63, assoc, 63-65; biochemist, R J Reynolds Tobacco Co, NC, 65-70; sr clin biochemist, J T Baker Chem Co, 70-71; sr res biochemist, Corning Glass Works, 71-72; MEM STAFF, DEPT PATH, OCHSNER'S CLIN, 72- Concurrent Pos: Vis prof, Wake Forest Univ, 68. Mem: AAAS; Am Chem Soc; Am Asn Clin Chemists. Res: Clinical and industrial enzymology; immobilized enzymes. Mailing Add: Dept of Path Ochsner's Clin 1514 Jefferson Hwy New Orleans LA 70121

LARUE, JAMES ARTHUR, b Rivesville, WVa, Apr 24, 23; m 47; c 2. MATHEMATICS. Educ: WVa Univ, AB, 48, MS, 49; Univ Pittsburgh, PhD(math), 61. Prof Exp: Asst prof math, Morris Harvey Col, 49-54; asst prof, 54-65, PROF MATH & CHMN DEPT, FAIRMONT STATE COL, 65- Res: Mathematical analysis; divergent series. Mailing Add: 1114 S Park Dr Fairmont WV 26554

LA RUE, JERROLD A, b San Bernardino, Calif, June 22, 23; m 46; c 2. METEOROLOGY. Educ: Univ Calif, Los Angeles, BA, 48. Prof Exp: Gen meteorologist, US Weather Bur, 51-55, forecast meteorologist, 55-57, meteorologist proj analyst, 57-60, quant precipation meteorologist, 60-64, sect supvr meteorol, 64-68, br chief, 68-69, METEOROLOGIST CHG WASHINGTON, DC WEATHER SERV OFF, NAT WEATHER SERV, 69- Mem: Am Meteorol Soc; Am Geophys Union. Res: Objective methods adapted to operational meteoerology. Mailing Add: World Weather Bldg Marlow Heights MD 20233

LARUE, THOMAS A, b Winnipeg, Man, May 1, 35; m 58; c 4. BIOCHEMISTRY. Educ: Univ Man, BSc, 56, MSc, 58; Univ Iowa, PhD(biochem), 62. Prof Exp: From asst res officer to assoc res officer, 62-72, SR RES OFFICER, PRAIRIE REGIONAL LAB, NAT RES COUN CAN, 72- Mem: Am Soc Plant Physiologists; Can Soc Microbiol. Res: Analytical biochemistry; plant tissue culture; nitrogen fixation. Mailing Add: Prairie Regional Lab Nat Res Coun Can Saskatoon SK Can

LARUFFA, ANTHONY LOUIS, b Oyster Bay, NY, Oct 19, 33. CULTURAL ANTHROPOLOGY. Educ: City Col New York, BA, 55; Columbia Univ, MA, 59, PhD(anthrop), 66. Prof Exp: From lectr to asst prof, 61-71, ASSOC PROF ANTHROP, LEHMAN COL, 71- Concurrent Pos: Ed, Libr Anthrop, Gordon & Breach, 69-, co-ed, Ethnic Studies: Int Jour, 72- Mem: Fel Am Anthrop Asn; Am Ethnol Soc; fel NY Acad Sci. Res: Social stratification; ethnicity; religion; Latin America, especially the Caribbean; Latin American migrants overseas. Mailing Add: Dept of Anthrop Herbert H Lehman Col Bronx NY 10468

LASAGNA, LOUIS (CESARE), b New York, NY, Feb 22, 23; m 46; c 7. PHARMACOLOGY. Educ: Rutgers Univ, BS, 43; Columbia Univ, MD, 47. Prof Exp: Asst & instr pharmacol, Sch Med, Johns Hopkins Univ, 50-52, from asst prof to

LASAGNA

assoc prof med & from asst prof to assoc prof pharmacol & exp therapeut, 54-70; PROF PHARMACOL, TOXICOL & MED, UNIV ROCHESTER, 70- Concurrent Pos: Vis physician, Columbia Res Serv, Goldwater Mem Hosp, 51, 53 & 54; clin & res fel, Mass Gen Hosp, 52-54; lectr, Sch Med, Boston Univ, 52-54; res assoc, Harvard Univ, 53-54. Mem: Inst Med; Asn Am Physicians; Am Soc Pharmacol & Exp Therapeut; Am Soc Clin Invest; Am Fedn Clin Res. Res: Hypnotics; analgesics; psychological responses to drugs; placebos; clinical trials; prescribing patterns. Mailing Add: Dept of Pharmacol Univ of Rochester Sch of Med Rochester NY 14642

LASALA, EDWARD FRANCIS, b Lynn, Mass, June 15, 28; m 53; c 5. PHARMACEUTICAL CHEMISTRY, ORGANIC CHEMISTRY. Educ: Mass Col Pharm, BS, 53, MS, 55, PhD(pharmaceut chem), 58. Prof Exp: From instr to asst prof, 58-67, ASSOC PROF CHEM, MASS COL PHARM, 67- Mem: Am Chem Soc; Am Pharmaceut Asn. Res: Synthesis and biological studies of medicinal agents, chiefly analgesics and antiradiation agents. Mailing Add: Dept of Chem Mass Col of Pharm Boston MA 02115

LA SALLE, GERALD, b Ile du Calumet, Que, Dec 2, 15; m 41; c 9. MEDICAL ADMINISTRATION. Educ: Univ Montreal, BA, 35; Laval Univ, MD, 40; Univ Toronto, DHA, 51. Prof Exp: Asst dir, Royal Victoria Hosp, 50-52; dir, Univ Montreal Hosp, 52-56, dir & prof hosp admin, Inst Sup d'Admin Hosp, 56-60; exec dir, Col Physicians & Surgeons, Ont, 60-64; dean & dir med ctr, 64-68, VPRES IN CHG MED AFFAIRS, UNIV SHERBROOKE, 68- Concurrent Pos: Trustee, Can Hosp Asn, 52-54 & Can Asn Hosp Accreditation, 52-54; exec dir, Montreal Hosp Coun, 52-58 & Que Hosp Asn, 56-60; dir, Les Artisans' Coopvie; mem, State Comn Hosps, 54-58, Royal Comn Hosp Prob, 59-60 & Med Coun Can, 62-64; vpres, Unity Bank Can; dir, Sandoz Ltd, Exec Fund Can, Exec Int Investment Ltd, Pony Sporting Goods, Ft Steele Minerals & Zesta Food Corp. Mem: Can Med Asn. Res: Social aspects of medical care. Mailing Add: Sherbrooke Med Ctr Sherbrooke PQ Can

LASALLE, JOSEPH PIERRE, b State College, Pa, May 28, 16; m 42; c 2. SYSTEMS THEORY. Educ: La State Univ, BS, 37; Calif Inst Technol, PhD(math), 41. Prof Exp: Instr math, Univ Tex, 41-42 & Radar Sch, Mass Inst Technol, 42-43; mathematician, Design Div, Frankford Arsenal, 43-44 & Stromberg-Carlson Co, NY, 44-46; from asst prof to prof math, Univ Notre Dame, 46-58; assoc dir math ctr, Res Inst Advan Study, 58-64; chmn div appl math, 68-75, PROF APPL MATH & DIR CTR DYNAMICAL SYSTS, BROWN UNIV, 64- Concurrent Pos: Vis res assoc, Princeton Univ, 43-44, asst prof & res assoc, 47-49; res assoc, Cornell Univ, 44-46; sci adv, US Naval Forces, Ger, 50-52; Guggenheim Found fel, 75-76. Mem: Fel AAAS; Soc Indust & Appl Math (pres, 63); Am Math Soc; Math Asn Am. Res: Differential equations; theory of stability; control theory. Mailing Add: Div of Appl Math Brown Univ Providence RI 02912

LASALLE, MARJORIE, b Oregon City, Ore, July 1, 13. IMMUNOLOGY, IMMUNOHEMATOLOGY. Educ: Ore State Univ, BS, 48; Univ Ore, MS, 54; Stanford Univ, PhD(med microbiol), 60. Prof Exp: USPHS trainee, Univ Tex M D Anderson Hosp & Tumor Inst Houston, 60-62; asst scientist immunohemat, Ore Regional Primate Res Ctr, 63-71, assoc scientist, 71-75; RETIRED. Mem: AAAS. Res: Erythrocytes and leukocyte alloantigens in nonhuman primates; induction of tolerance to erythrocytes; experimental erythroblastosis; genetics of blood groups; experimental transfer reactions. Mailing Add: Ore Regional Primate Res Ctr 505 NW 185th Ave Beaverton OR 97005

LASALLE, PIERRE, b Que, July 17, 31; m. GEOMORPHOLOGY. Educ: Univ Montreal, BA, 52, BSc, 59; McGill Univ, MSc, 62; State Univ Leiden, PhD(geol), 66. Prof Exp: GEOLOGIST, QUE DEPT NATURAL RESOURCES, 66- Mem: AAAS; Geol Soc Am; Geol Asn Can; Geol Soc France. Mailing Add: Que Dept of Natural Resources 1620 Blvd de l'Entente Quebec PQ Can

LASATER, HERBERT ALAN, b Paris, Tenn, Sept 11, 31; m 59; c 2. MATHEMATICAL STATISTICS, APPLIED STATISTICS. Educ: Univ Tenn, BS, 57, MS, 62; Rutgers Univ, PhD(statist), 69. Prof Exp: Instr statist, Univ Tenn, Knoxville, 57-58; assoc statistician, Nuclear Div, Union Carbide Corp, 58-62; asst prof, 62-65 & 68-71, ASSOC PROF STATIST, UNIV TENN, KNOXVILLE, 71- Concurrent Pos: Lectr statist, Univ Tenn, 59-62; ed, J Qual Technol, 71-74. Mem: Am Statist Asn; fel Am Soc Qual Control; Sigma Xi. Res: Statistical quality control techniques. Mailing Add: Dept of Statist Univ of Tenn Knoxville TN 37916

LASCA, NORMAN P, JR, b Detroit, Mich, Oct 20, 34; m 65. QUATERNARY GEOLOGY, GEOMORPHOLOGY. Educ: Brown Univ, AB, 57; Univ Mich, MS, 61, PhD(geol), 65. Prof Exp: Res asst geol, Univ Mich, 60-61, teaching fel, 61-65; NATO res fel, Inst Geol, Univ Oslo, 65-66; asst prof, 66-71, ASSOC PROF GEOL, UNIV WIS-MILWAUKEE, 71-, ASST TO CHANCELLOR, 75- Concurrent Pos: Assoc scientis- scientist, Ctr Great Lakes Studies, Univ Wis-Milwaukee, 73-; fel acad admin, Am Coun Educ, 74-75. Mem: Geol Soc Am; Am Asn Quaternary Res; Int Water Resources Asn; Glaciol Soc; Swedish Soc Anthrop & Geog. Res: Glacial geology and geomorphology of polar regions; river and lake ice formation and processes; Glacial-Pleistocene geology in Wisconsin. Mailing Add: Dept of Geol Sci Univ of Wis Milwaukee WI 53201

LASCELLES, JUNE, b Sydney, Australia, Jan 23, 24. MICROBIOLOGY, BIOCHEMISTRY. Educ: Univ Sydney, BSc, 44, MSc, 47; Oxford Univ, DPhil(microbial biochem), 52. Prof Exp: Mem external sci staff, Med Res Coun, Eng, 53-60; lectr microbial biochem, Oxford Univ, 60-65; PROF BACT, UNIV CALIF, LOS ANGELES, 65- Concurrent Pos: Rockefeller fel, 56-57; consult panel microbial chem, NIH, 73-77. Mem: Am Soc Microbiol; Am Soc Biol Chemists; Soc Develop Biol; Brit Biochem Soc; Brit Soc Gen Microbiol. Res: Biochemistry of microorganisms; tetrapyrrole synthesis and regulation; bacterial photosynthesis. Mailing Add: Dept of Bact Univ of Calif Los Angeles CA 90024

LASCO, RALPH HARRELL, organic chemistry, see 12th edition

LASDAY, ALBERT HENRY, b Pittsburgh, Pa, Apr 8, 20; m 42; c 2. ENVIRONMENTAL MANAGEMENT. Educ: Univ Pittsburgh, BS, 41; Harvard Univ, MA, 47; Carnegie Inst Technol, DSc(physics), 51. Prof Exp: Engr, Carnegie Inst Technol, 47-50; physicist, Preston Labs, 51-55; dir res, Houze Glass Corp, 55-58; pres, A H Lasday Co, 58-63; proj mgr, Micarta Div, Westinghouse Elec Corp, SC, 63-65; group leader mat develop, Texaco Exp, Inc, 65-68; supvr res, Richmond Res Labs, 68-71, COORDR, ENVIRON PROTECTION DEPT, TEXACO INC, 71- Concurrent Pos: Consult, Res & Develop Bd, Off Secy Defense, 51; chmn phys transport oil task force, Am Petrol Inst, 71-, vchmn fate & effects of oil comt, 74- Mem: Am Phys Soc; Sigma Xi; Mfg Chemists Asn. Res: Air and water conservation; water pollution control for petroleum and petrochemical waste effluents. Mailing Add: 37 King Dr Poughkeepsie NY 12603

LASEK, RAYMOND J, b Chicago, Ill, Nov 25, 40; m 64; c 3. NEUROBIOLOGY. Educ: Utica Col, BA, 61; State Univ NY, PhD(anat), 67. Prof Exp: NIH res fel neuropath, McLean Hosp, Harvard Med Sch, 66-68 & neurobiol, Univ Calif, San Diego, 68-69; asst prof, 69-73, ASSOC PROF ANAT, CASE WESTERN RESERVE UNIV, 74- Res: Axoplasmic transport; regulation of growth and differentiation in neurons; neuroanatomy; developmental biology. Mailing Add: Med Ctr Case Western Reserve Univ Cleveland OH 44106

LASETER, JOHN LUTHER, b Houston, Tex, Sept 9, 37; m 69; c 2. BIOCHEMISTRY. Educ: Univ Houston, BS, 59, MS, 62, PhD(biochem), 68. Prof Exp: Instr biol, Univ Houston, 59-60; asst prof, Sacred Heart Col, Houston, 61-64; dir chem res, Teledyne, Inc, 64-65; instr biol, Univ Houston, 65-66, NASA fel biophys, 68-69; asst prof, 69-70, ASSOC PROF BIOL SCI & CHMN DEPT, LA STATE UNIV, NEW ORLEANS, 70- Concurrent Pos: Pres, J L Laseter & Assocs, Inc, 66-; Res Corp & New Orleans Cancer Soc grants, La State Univ, New Orleans, 70-71, NASA res grant, 70-72. Mem: AAAS; Am Chem Soc; Am Soc Mass Spectrometry; NY Acad Sci; Am Oil Chem Soc. Res: Lipid biochemistry; analytical biochemistry and mass spectrometry; control mechanisms in steroid metabolism. Mailing Add: Dept of Biol Sci La State Univ New Orleans LA 70122

LA SEUR, NOEL EDWIN, b Stanhope, Iowa, June 25, 22; m 44; c 3. METEOROLOGY. Educ: Univ Chicago, SB, 47, SM, 49, PhD(meteorol), 53. Prof Exp: From asst to instr, Univ Chicago, 48-52; asst prof, 53-56, assoc prof, 56-58, PROF METEOROL, FLA STATE UNIV, 58- Mem: AAAS; Am Meteorol Soc; Am Geophys Union; Royal Meteorol Soc. Res: Synoptic meteorology of temperature and tropical latitudes. Mailing Add: Dept of Meteorol Fla State Univ Tallahassee FL 32306

LASFARGUES, ETIENNE YVES, b Milhars, France, May 5, 16; nat US; m 50; c 2. MICROBIOLOGY, ONCOLOGY. Educ: Univ Paris, BS, 35, DVM, 41. Prof Exp: Roux Found res fel microbiol, Pasteur Inst, France, 42-44, asst virol, 44-47, head lab virol, 50-55; Am Cancer Soc res fel cytol, Inst Cancer Res, Pa, 47-50; assoc microbiol, Col Physicians & Surgeons, Columbia Univ, 55-59, asst prof, 59-66; ASSOC MEM, DEPT CYTOL BIOPHYS, INST MED RES, 66- Honors & Awards: Jensen Prize, Fr Acad Med, 46; Silver & Bronze Medal, Pasteur Inst, 69. Mem: AAAS; Am Soc Cell Biologists; Tissue Cult Asn; Int Soc Cell Biologists; Am Asn Cancer Res. Res: Viral oncology; fine structure electron microscopy; molecular biochemistry. Mailing Add: Dept of Cytol Biophys Inst for Med Res Copewood St Camden NJ 08103

LASH, ABRAHAM FAE, b Chicago, Ill, Nov 25, 98; m 25; c 2. OBSTETRICS & GYNECOLOGY. Educ: Univ Chicago, BS, 19; Rush Med Col, MD, 21; Univ Ill, MS, 25, PhD(obstet & gynec), 29. Prof Exp: Instr gynec, Col Med, Univ Ill, 23-26, assoc obstet & gynec, 26-29, from asst prof to assoc prof, 29-53, clin prof, 53-65; prof, 65-73, EMER PROF OBSTET & GYNEC, MED SCH, NORTHWESTERN UNIV, CHICAGO, 73- Concurrent Pos: Dir div obstet & gynec, Cook County Hosp, 66-73. Mem: Am Col Surgeons; Int Col Surg. Res: Pathology; bacteriology; immunology. Mailing Add: 4250 N Marine Chicago IL 60613

LASH, EDWARD DAVID, clinical chemistry, see 12th edition

LASH, JAMES (JAY) W, b Chicago, Ill, Oct 24, 29; m 55; c 1. DEVELOPMENTAL BIOLOGY, ANATOMY. Educ: Univ Chicago, PhD(zool), 55. Hon Degrees: MA, Univ Pa, 71. Prof Exp: Nat Cancer Inst, Univ Pa, 55-57, instr, 57-59, assoc, 59-61, from asst prof to assoc prof, 61-69, PROF ANAT, SCH MED, UNIV PA, 69- Concurrent Pos: Helen Hay Whitney Found fel, 58-61; investr, Helen Hay Whitney Found, 61-66; NIH career develop award, 66-70; mem maternal & child health res comt, Nat Inst Child Health & Human Develop, 71-75; mem sci adv comt, Ctr Oral Health Res, Univ Pa, 73-; mem nat adv comt, Primate Res Ctr, Davis, Calif, 73- Mem: Soc Cell Biol; Soc Develop Biol. Res: Developmental biology: tissue interactions during chondrogenesis; Ascidian embryology. Mailing Add: Dept of Anat Univ of Pa Sch of Med Philadelphia PA 19174

LASHBROOK, ROBERT VERN, chemistry, see 12th edition

LASHEEN, ALY M, b Cairo, Egypt, Dec 27, 19; nat US; m 54; c 3. PLANT PHYSIOLOGY, HORTICULTURE. Educ: Cairo Univ, BS, 42; Univ Calif, Los Angeles, 49; Agr & Mech Col Tex, PhD(plant physiol, hort), 54. Prof Exp: Asst hort, Agr & Mech Col Tex, 50-53, res assoc plant physiol, 54-55; jr plant pathologist, Wash State Univ, 55-57, asst prof hort, 57-61; assoc prof plant physiol, AID Contract-Univ Ky, Indonesia, 61-65, prof, 65-67, ASSOC PROF HORT, UNIV KY, 67- Concurrent Pos: Hort adv, Univ Ky team, Northeast Agr Ctr, Tha Phra-Khon Kaen, Thailand, 70- Mem: Am Soc Hort Sci; Am Soc Plant Physiol. Res: Chemical analysis of macro and micro elements in plants; biochemical analysis of sugars and amino, organic and nucleic acids in plants; dormancy in seeds; effects of additives on plants; cold hardiness in plants; nature of dwarfing in apples. Mailing Add: Dept of Hort Univ of Ky Lexington KY 40506

LASHEN, EDWARD S, b New York, NY, Aug 11, 34; m 57; c 3. MICROBIOLOGY. Educ: Brooklyn Col, BS, 56; Rutgers Univ, MS, 62, PhD(microbiol), 65. Prof Exp: Res asst blood res, Jewish Chronic Disease Hosp Brooklyn, NY, 56-57; res asst hemat & pharmacol, Wallace Labs Div, Carter Prod, Inc, 57-61; sr microbiologist, 64-75, HEAD PROJ LEADER BIOCIDES, RES LABS, ROHM AND HAAS CO, 75- Mem: Am Soc Testing & Mat; Am Soc Microbiol. Res: Microbial transformation of thiourea and substituted thioureas; biodegradation of surfactants by sewage sludge and river water microflora; broad spectrum anti-microbial agents for application as industrial biocides. Mailing Add: Rohm and Haas Res Labs Spring House PA 19477

LASHER, GORDON (JEWETT), b Denver, Colo, Feb 1, 26; m 53; c 3. ASTROPHYSICS. Educ: Rensselaer Polytech Inst, BS, 49; Cornell Univ, PhD(theoret physics), 54. Prof Exp: Staff physicist, Lawrence Radiation Lab, Univ Calif, 53-55; assoc physicist, 55-56, PROJ PHYSICIST, RES LAB, IBM CORP, 56- Concurrent Pos: Vis prof, Cornell Univ, 69-70. Mem: Am Phys Soc; Am Astron Soc. Res: Applied mathematics with computer applications; solid state physics; superconductivity; theory of liquid crystals; general relativity; theory of supernovae. Mailing Add: IBM-Watson Res Ctr Yorktown Heights NY 10598

LASHER, SIM, b Chicago, Ill, May 16, 16. MATHEMATICS. Educ: Cent YMCA Col, Chicago, BS, 39; Univ Chicago, MS, 47, PhD(math), 67. Prof Exp: Teacher high sch, Ill, 50-59; instr, 59-67, asst prof, 69-71, ASSOC PROF MATH, UNIV ILL, CHICAGO CIRCLE, 71- Concurrent Pos: NSF sci fac fel, Univ Chicago, 64-65. Mem: Am Math Soc; Math Asn Am. Res: Generalized differentiation and integration in real analysis. Mailing Add: Dept of Math Univ of Ill at Chicago Circle Chicago IL 60680

LASHINSKY, HERBERT, b New York, NY, Dec 14, 21; div; c 1. PHYSICS. Educ: City Col New York, BSc, 50; Columbia Univ, PhD(physics), 61. Prof Exp: Engr, Western Elec Co, 40-43; res asst physics, Radiation Lab, Columbia Univ, 52-60; res assoc, Plasma Physics Lab, Princeton Univ, 61-64; from res asst prof to res assoc prof, 64-69, RES PROF PHYSICS, INST FLUID DYNAMICS & APPL MATH, UNIV MD, COLLEGE PARK, 69- Concurrent Pos: Instr, City Col New York, 54-55; translr, Rev Plasma Physics, Vols I-V, 66-70; ed, Soviet Physics-Tech Physics. Mem:

Am Phys Soc; Inst Elec & Electronics Engrs. Res: Plasma physics; microwave electronics; nonlinear distributed systems; translation. Mailing Add: Inst of Fluid Dynamics & Appl Math Univ of Md College Park MD 20742

LASHLEY, GERALD ERNEST, b Johnstown, Pa, Sept 26, 35; m 55; c 3. NUMERICAL ANALYSES. Educ: Eastern Nazarene Col, BS, 57; Boston Univ, AM, 61; Brown Univ, EdD(math educ), 69. Prof Exp: High sch teacher, Mass, 58-63; assoc prof math, Eastern Nazarene Col, 64-72, chmn dept, 70-72; CHMN DIV NATURAL SCI, MT VERNON NAZARENE COL, 72- Concurrent Pos: Dir & instr, NSF In-Serv Inst Secondary Teachers, 64-72. Mem: Math Asn Am. Mailing Add: Mt Vernon Nazarene Col Mt Vernon OH 43050

LASHOF, JOYCE COHEN, b Philadelphia, Pa, Mar 27, 26; m 50; c 3. MEDICINE. Educ: Duke Univ, AB, 46. Prof Exp: Intern, Bronx Hosp, New York, 50-51, asst resident med, 51-52; asst resident med, Montefiore Hosp, 52-53; Nat Found Infantile Paralysis fel, Yale Univ, 53-54; asst med & physician, Student Health Serv, Univ Chicago, 54-56, from instr to asst prof, Sch Med, 56-60; dir sect prev med & clin labs, Rush-Presby-St Luke's Med Ctr, 61-66, dir sect community med, 66-72, chmn dept prev med, 72-73; DIR DEPT PUB HEALTH, STATE OF ILL, 73-; PROF PREV MED, RUSH MED COL, 71- Concurrent Pos: Staff physician, Union Health Serv, Inc, 60-61; asst attend physician, Presby-St Luke's Hosp, 60-61, assoc attend physician, 61-; clin asst prof med, Sch Med, Univ Ill, 61-64, assoc prof prev med, 64-71. Res: Internal medicine; medical care. Mailing Add: Dept of Pub Health 160 NLa Salle Chicago IL 60601

LASHOF, RICHARD KENNETH, b Philadelphia, Pa, Nov 9, 22; m 50; c 3. MATHEMATICS. Educ: Univ Pa, BS, 43; Columbia Univ, PhD(math), 54. Prof Exp: From instr to assoc prof, 54-64, chmn dept, 67-70, PROF MATH, UNIV CHICAGO, 64- Concurrent Pos: NSF fel, 60-61; mem-at-large, Nat Res Coun. Mem: Am Math Soc. Res: Algebraic topology; differential geometry. Mailing Add: Dept of Math Univ of Chicago Chicago IL 60637

LASHOF, THEODORE WILLIAM, b Philadelphia, Pa, June 27, 18; m 50; c 3. PHYSICS. Educ: Univ Pa, BS, 39, PhD(physics), 42. Prof Exp: Mem staff, Radiation Lab, Mass Inst Technol, 42-45; from instr to asst prof physics, Reed Col, 45-49; assoc prof, Mich Col Mining & Technol, 49-50; physicist, Mass Sect, 50-54, paper sect, 54-62, appl polymer standards sect, 62-64 & eval criteria sect, 64-67, actg chief paper standards sect, 67-70, chief performance criteria sect, 70-74, PROG MGR LAB PERFORMANCE, NAT BUR STANDARDS, WASHINGTON, DC, 74- Honors & Awards: Testing Div Medal, Tech Asn Pulp & Paper Indust, 72; Award of Merit, Am Soc Testing & Mat, 75. Mem: Fel Am Soc Testing & Mat; fel Tech Asn Pulp & Paper Indust. Res: Physical properties of paper; interlaboratory stnadardization; consumer product performance and safety; testing laboratory evaluation. Mailing Add: 10125 Ashburton Lane Bethesda MD 20034

LASHOMB, JAMES HAROLD, b Potsdam, NY, Oct 25, 42; m 69; c 1. ENTOMOLOGY. Educ: Cornell Univ, BS, 70; Univ Md, College Park, MS, 73, PhD(entom), 75. Prof Exp: RES ASSOC ENTOM, MISS STATE UNIV, 75- Mem: AAAS; Entom Soc Am; Ecol Soc Am. Res: Sampling of insect populations and antural enemies; parasitic insect distribution within plants and parasitic insect biology. Mailing Add: Miss State Univ Drawer EM Mississippi State MS 39762

LASKA, EUGENE, b New York, NY, Mar 17, 38; m 59; c 3. MATHEMATICS, STATISTICS. Educ: City Col New York, BS, 59; NY Univ, MS, 61, PhD(math), 63. Prof Exp: Asst res scientist, Comput Lab, Res Div, NY Univ, 59-61, res assoc math, Courant Inst Math Sci, 61-62; systs engr, Int Bus Mach Corp, NJ, 62-63; DIR INFO SCI DIV, RES CTR, ROCKLAND STATE HOSP, ORANGEBURG, 63- Concurrent Pos: NIMH grant, 67-72; consult comput in psychiat, USSR, Israel, Peru & Indonesia, 68. Mem: Fel AAAS; Inst Math Statist; Am Statist Asn; Soc Indust & Appl Math; Biomet Soc. Res: Mathematical statistics, including estimation theory; applied statistics, including psychopharmacology. Mailing Add: 34 Dante St Larchmont NY 10538

LASKAR, AMULYA LAL, b Dacca, India, June 11, 31; m 62; c 2. PHYSICS. Educ: SN Col, India, BS, 50; Univ Calcutta, MS, 52; Inst Elec Engrs, Gt Brit, grad, 58; Univ Ill, Urbana, PhD(physics), 60. Prof Exp: Lectr physics, Indian Inst Technol, Kharagpur, 53-60, asst prof, 60-65; sr fel & sr res assoc, Univ NC, Chapel Hill, 65-68; ASSOC PROF PHYSICS, CLEMSON UNIV, 68-, RES CORP GRANT, 71-, ARMY RES OFF-DURHAM GRANT, 72- Concurrent Pos: Vis res scientist, Picatinny Arsenal, NJ, 69 & Argonne Nat Lab, Ill, 70. Mem: Am Phys Soc; Indian Phys Soc. Res: Solid state physics, especially transport phenomena; defect structures and their role on the physical properties in various solids, especially ionic and semiconductors. Mailing Add: Dept of Physics Clemson Univ Clemson SC 29631

LASKAR, RENU CHAKRAVARTI, b Bhagalpur, India, Aug 8, 30; m 62; c 2. MATHEMATICS. Educ: Univ Bihar, BS, 54; Univ Bhagalpur, MS, 57; Univ Ill, Urbana, PhD(group theory), 62. Prof Exp: Lectr math, Ranchi Women's Col, India, 57-59; lectr, Indian Inst Technol, Kharagpur, 62-65; fel, Univ NC, Chapel Hill, 65-68; ASSOC PROF MATH, CLEMSON UNIV, 68- Mem: Am Math Soc; Math Asn Am. Res: Group theory; combinatorial mathematics, especially graph theory. Mailing Add: Dept of Math Clemson Univ Clemson SC 29631

LASKER, BARRY MICHAEL, b Hartford, Conn, Aug 12, 39; m 70; c 1. ASTROPHYSICS. Educ: Yale Univ, BS, 61; Princeton Univ, MA, 63, PhD(astrophys sci), 64. Prof Exp: NSF fel, Mt Wilson & Palomar Observ, 65-67; asst prof astron, Univ Mich, 67-69; STAFF ASTRONR, CERRO TOLOLO INTERAM OBSERV, CHILE, 69- Mem: Am Astron Soc. Res: Magnetohydrodynamics; extragalactic astronomy and cosmology; ultrashort period stellar oscillations; astronomical instrumentation. Mailing Add: Kitt Peak Nat Observ 950 N Cherry Ave Tucson AZ 85717

LASKER, GABRIEL (WARD), b Huntington, Eng, Apr 29, 12; US citizen; m 49; c 3. HUMAN ANATOMY, BIOLOGICAL ANTHROPOLOGY. Educ: Univ Mich, AB, 34; Harvard Univ, MA, 41, PhD(phys anthrop), 45. Prof Exp: From instr to assoc prof, 46-64, PROF ANAT, SCH MED, WAYNE STATE UNIV, 64- Concurrent Pos: Viking Fund grant, Paracho, Michoacan, Mex, 48; ed, Human Biol, 53-; mem staff, Dept Anthrop, Univ Wis, 54-55; Fulbright fel, Peru, 57-58. Mem: Fel AAAS (vpres, 68); Am Asn Phys Anthrop (secy-treas, 46-50, vpres, 60-62, pres, 63-65); Am Asn Anat; fel Am Anthrop Asn. Res: Human genetics; physical anthropology; physical characteristics of Chinese, Mexicans and Peruvians; demographic aspects of human biology. Mailing Add: Dept of Anat Wayne State Univ Sch of Med Detroit MI 48201

LASKER, GEORGE ERIC, b Prague, Czech; Can citizen. INFORMATION SCIENCES, PSYCHOLOGY. Educ: Prague Tech Univ, EC, 57; Charles Univ, Prague, DP, 61. Prof Exp: Asst prof math, Univ Sask, 65-66; assoc prof comput sci, Univ Man, 66-68; PROF COMPUT SCI, UNIV WINDSOR, 68- Concurrent Pos: Nat Res Coun Can grant, Univ Man, 66-68; ed bd, Int J Gen Systs, 74-; vis scholar, Dept Comput & Commun Sci, Univ Mich, Ann Arbor, 74- Mem: Can Comput Sci Asn; World Orgn Gen Systs & Cybernet; NY Acad Sci. Res: Diagnostic methodology; pattern recognition techniques; threshold logic; automata; systems interaction; simulation models; behavioral prediction; computer controlled conditioning; mathematical psychology; forecasting methodology. Mailing Add: Sch of Comput Sci Univ Windsor Windsor ON Can

LASKER, REUBEN, b Brooklyn, NY, Dec 1, 29; m 52; c 2. COMPARATIVE PHYSIOLOGY. Educ: Univ Miami, BS, 50, MS, 52; Stanford Univ, PhD(biol), 56. Prof Exp: Rockefeller Found res fel marine biol, Scripps Inst, Univ Calif, 56-57; instr biol, Compton Dist Jr Col, 57-58; physiologist, Bur Commercial Fisheries, US Fish & Wildlife Serv, 58-70, asst dir, Fishery-Oceanog Ctr, 68-70, PHYSIOLOGIST, NAT MARINE FISHERIES SERV, 70- Concurrent Pos: Assoc prof in residence, Univ Calif, San Diego, 67-74, adj prof marine biol, Scripps Inst Oceanog, 74-; res fel natural hist, Aberdeen Univ, 66-67; ed, Fishery Bull, 70-75. Honors & Awards: Gold Medal Award, US Dept Com, 74. Mem: Fel AAAS; Am Soc Limnol & Oceanog; Soc Gen Physiol; Am Soc Zoologists; Marine Biol Asn UK. Res: Nutrition, biochemistry and general physiology of marine organisms. Mailing Add: Southwest Fisheries Ctr Nat Marine Fisheries Serv La Jolla CA 92037

LASKER, SIGMUND E, b New York, NY, Sept 5, 23; m 65. PHYSICAL CHEMISTRY. Educ: Brooklyn Col, BS, 49; NY Univ, MS, 51; Stevens Inst Technol, PhD(phys chem), 65. Prof Exp: Res assoc surg, NY Med Col, 57-60, asst instr biochem, 58-60; res assoc phys chem, Stevens Inst Technol, 60-66; asst prof biophys, 66-72, ASST PROF PHARMACOL, NY MED COL, 72- Mem: AAAS; Am Inst Chemists; Am Chem Soc; Am Asn Clin Chem; Biophys Soc. Res: Solution properties of biological macromolecules; polyelectrolytes; experimental thermal injuries; anticoagulants. Mailing Add: Dept of Pharmacol NY Med Col New York NY 10029

LASKIN, ALLEN I, b Brooklyn, NY, Dec 7, 28; m 54, 73; c 2. MICRIBIOLOGY. Educ: City Col New York, BS, 50; Univ Tex, MA, 52, PhD, 56. Prof Exp: Res scientist, Univ Tex, 53-55; sr res microbiologist, Squibb Inst Med Res, 55-64, res supvr microbiol, 64-67, asst dir microbiol, 67-69; res assoc, 69-74, HEAD BIOSCI RES, EXXON RES & ENG CO, 71-, SR RES ASSOC, 74- Concurrent Pos: Found microbiol lect, Am Soc Microbiol, 72; Selman A Waksman Hon Lect, Theobald Smith Soc, 74. Mem: Soc Indust Microbiol (secy, 73-76); Am Soc Microbiol; Am Chem Soc; Am Acad Microbiol; NY Acad Sci. Res: Cell-free protein synthesis; mode of action of antibiotics; mechanisms of bacterial resistance to antibiotics; microbial transformations of steroids; petroleum microbiology; microbial enzymes. Mailing Add: Exxon Res & Eng Co PO Box 45 Linden NJ 07036

LASKIN, DANIEL M, b New York, NY, Sept 3, 24; m 45; c 3. ORAL SURGERY, MAXILLOFACIAL SURGERY. Educ: Ind Univ, BS & DDS, 47; Univ Ill, MS, 51; Am Bd Oral Surg, dipl. Prof Exp: Intern oral surg, Jersey City Med Ctr, 47-48; clin asst, 49-50, res asst, 50-51, from instr to assoc prof, 51-60, PROF ORAL SURG, COL DENT, UNIV ILL MED CTR, 60-, HEAD DEPT ORAL & MAXILLOFACIAL SURG, 73-, DIR TEMPOROMANDIBULAR JOINT RES CTR, 61- Concurrent Pos: Resident, Cook County Hosp, 50-51, chmn dept oral surg, 67-; attend oral surg, Hosps, 52-; clin prof, Col Med, Univ Ill, Chicago, 58-, head dept dent, Hosps, 70-; ed, Am Soc Oral Surg Forum & J Oral Surg. Mem: Am Soc Oral Surg; fel Am Col Dent; fel Int Col Dent; Am Soc Exp Path; Am Dent Asn. Res: Temporomandibular joint; metabolism of bone and cartilage; sutural growth; calcification and resorption of bone. Mailing Add: Dept Oral & Maxillofacial Surg Univ of Ill at the Med Ctr Chicago IL 60680

LASKOWSKI, LEONARD FRANCIS, JR, b Milwaukee, Wis, Nov 16, 19; m 46; c 3. MEDICAL MICROBIOLOGY, CLINICAL MICROBIOLOGY. Educ: Marquette Univ, BS, 41, MS, 48; St Louis Univ, PhD(med bact), 51; Am Bd Med Microbiol, dipl. Prof Exp: Instr bact, 46-48 & 51-53, sr instr, 53-54, asst prof microbiol, 54-57, from asst prof to assoc prof path, 57-69, PROF PATH, SCH MED, ST LOUIS UNIV, 69- Concurrent Pos: China Med Bd fel, Latin Am, 57; fel trop med, La State Univ, 57; consult, St Mary's Group Hosp, 57-; attend, Vet Admin Hosp, health & tech training coordr, Latin Am Peace Corps Projs, 62-66; dir clin microbiol sect, St Louis Univ Hosps, 65; consult clin microbiol, John Cochran Vet Hosp, 66-, St Elizabeth's Hosp, 68- & St Louis County Hosp, 69-; referee mycol, USPHS Commun Dis Ctr Prog for Testing Clin Diag Labs, 68-; consult, Jefferson Barracks, Vet Hosp, 72-, Mogul Diag Div, Mogul Corp, 72- & St Francis Hosp, 74- Mem: AAAS; Am Soc Microbiol; NY Acad Sci; fel Am Acad Microbiol; Med Mycol Soc Americas. Res: Mechanism of intracellular parasitism; mechanism of action of therapeutic compounds. Mailing Add: Dept of Path St Louis Univ Sch of Med St Louis MO 63103

LASKOWSKI, MICHAEL, b Smolensk, Poland, Mar 18, 05; nat US; m 27; c 1. BIOCHEMISTRY. Educ: Univ Warsaw, PhD(biochem), 29. Prof Exp: Sr asst animal physiol, Col Agr, Warsaw, 29-35, privatdocent, 35, actg head dept, 37-39; assoc prof biochem, New Sch Social Res, 41; fel, Univ Minn, 42; assoc prof, Sch Med, Univ Ark, 43-45; from assoc prof to prof, Sch Med, Marquette Univ, 45-60, Am Cancer Soc res prof, 60-65; res prof, 66-74, EMER RES PROF, STATE UNIV NY BUFFALO, 74-; AM CANCER SOC RES PROF & HEAD LAB ENZYM, ROSWELL PARK MEM INST, 66- Concurrent Pos: Polish Nat Cult Fund fel, Lister Inst, London, 35-36; Rockefeller Found fel, Univ Basel, 36-37. Honors & Awards: Co-recipient E K Frey Award, 72; Schoellkopf Medal, 73. Mem: Fel AAAS; Am Chem Soc; Am Soc Biol Chem; Soc Exp Biol & Med; for mem Polish Acad Sci. Res: Calcium and phosphorus metabolism; enzymes, mainly proteolytic and nucleolytic. Mailing Add: Roswell Park Mem Inst 666 Elm St Buffalo NY 14263

LASKOWSKI, MICHAEL JR, b Warsaw, Poland, Mar 13, 30; nat US; m; c 2. BIOCHEMISTRY. Educ: Lawrence Col, BS, 50; Cornell Univ, PhD(phys chem), 54. Prof Exp: Asst chem, Cornell Univ, 50-52, USPHS fel, 54-56, instr, 56-57; from asst prof to assoc prof, 57-65, PROF CHEM, PURDUE UNIV, 65- Concurrent Pos: Chmn, Gordon Res Conf Physics & Phys Chem Biopolymers, 66; mem, Biophys & Biophys Chem Study Sect, NIH, 67-71. Honors & Awards: McCoy Award, Purdue Univ, 75. Mem: AAAS; Am Chem Soc; Am Soc Biol Chemists; Polish Acad Arts & Sci in Am. Res: Protein chemistry; role of individual amino acid residues in proteinase inhibitor-proteinase interaction; evolution. Mailing Add: Dept of Chem Purdue Univ Lafayette IN 47907

LASKOWSKI, MICHAEL BERNARD, b Chicago, Ill, Apr 20, 43; m 65; c 4. NEUROPHYSIOLOGY, NEUROANATOMY. Educ: Loyola Univ, BS, 66; Univ Okla, PhD(physiol), 70. Prof Exp: Muscular Dystrophy Asn fel, Northwestern Univ, 70-71; res assoc, 71-72, instr, 72-74, ASST PROF PHARMACOL, VANDERBILT UNIV, 74- Honors & Awards: Career Develop Award, Andrew Mellon Found, 74. Mem: Am Physiol Soc. Res: Electrophysiology; electron microscopy. Mailing Add: Dept of Pharmacol Vanderbilt Univ Sch of Med Nashville TN 37203

LASKY, JACK SAMUEL, b New York, NY, Mar 14, 30; m 52; c 2. ORGANIC CHEMISTRY, POLYMER CHEMISTRY. Educ: City Col New York, BS, 51; Univ Md, PhD(chem), 55. Prof Exp: Res chemist, US Rubber Co, 55-69; dir polymer res, 69-70, VPRES RES, OKONITE CO, 70- Mem: Am Chem Soc; Inst Elec & Electronics Engrs; Power Eng Soc. Res: Synthetic polymers; rubber; plastics;

stereospecific polymerization; heterogeneous catalysis; kinetics of reactions; electrical insulating and covering materials. Mailing Add: 29 Newman Ave Verona NY 07044

LASLETT, LAWRENCE JACKSON, b Boston, Mass, Jan 12, 13; m 39; c 3. PHYSICS. Educ: Calif Inst Technol, BS, 33; Univ Calif, PhD(physics), 37. Prof Exp: Rockefeller Found fel, Inst Theoret Physics, Copenhagen Univ, 37-38, Oersted fel, 38; instr physics, Univ Mich, 39 & Ind Univ, 39-42; mem staff, Radiation Lab, Mass Inst Technol, 41-45; from asst prof to prof physics, Iowa State Univ, 46-63; STAFF MEM THEORET PHYSICS DIV, LAWRENCE BERKELEY LAB, UNIV CALIF, 63-, LECTR PHYSICS, UNIV, 65- Concurrent Pos: Head nuclear physics br & consult, Off Naval Res, DC, 52-53, sci liaison officer, London Br, 60-61; vis res prof, Univ Ill, 55-56 & Univ Wis, 56-57; head high energy physics br, US AEC, 61-63. Mem: AAAS; fel Am Phys Soc. Res: Cyclotron design and construction; beta-ray spectra; electron and gamma-ray spectra; photonuclear research; design particle accelerators. Mailing Add: Theoret Physics Div Univ of Calif Lawrence Berkeley Lab Berkeley CA 94720

LASLEY, BETTY JEAN, b Winston-Salem, NC, July 10, 27. MICROBIOLOGY. Educ: Drew Univ, AB, 49; Rutgers Univ, MS, 53; NY Univ, PhD, 68. Prof Exp: Sr med technologist, Path Lab, NJ State Hosp, Greystone Park, 49-51; assoc investr chem microbiol, Warner Chilcott Res Labs Div, Warner-Lambert Co, 53-56; asst physiol, Med Col, NY Univ, 56; asst leukemia div, Sloan-Kettering Inst Cancer Res, 56-57; asst abstractor, Am Cyanamid Co, 57-58; asst metab & endocrinol, Med Ctr, 58-70, res assoc, 70-74, INSTR, MED CTR, NY HOSP, CORNELL UNIV, 74- Mem: NY Acad Sci. Res: Metabolism; immunobiology and endocrinology. Mailing Add: Payne Whitney Cornell Med Ctr NY Hosp 525 E 68th St New York NY 10021

LASLEY, BILL LEE, b Ottumwa, Iowa, June 4, 41. REPRODUCTIVE PHYSIOLOGY. Educ: Calif State Univ, Chico, BA, 63; Univ Calif, Davis, PhD(physiol), 72. Prof Exp: Teacher math & sci, Roseville Union High Sch, 64-67; fel reproductive biol, Rockefeller Found, 72-75; RES ENDOCRINOLOGIST, SAN DIEGO ZOOL GARDEN, 75- Res: Reproductive endocrinology with emphasis on the regulatory mechanisms of the hypothalamic-pituitary-gonadal axis in human and nonhuman primates. Mailing Add: San Diego Zool Res Dept PO Box 551 San Diego CA 92112

LASLEY, EARL LOREN, genetics, see 12th edition

LASLEY, JOHN FOSTER, b Liberal, Mo, Jan 26, 13; m 40; c 1. ANIMAL BREEDING. Educ: Univ Mo, BSA, 38, AM, 40, PhD(physiol reprod), 43. Prof Exp: Asst animal husb, Univ Mo, 38-43; agr exten agent & supvr agr & stock raising, US Indian Serv, 43-49, PROF ANIMAL HUSB, UNIV MO-COLUMBIA, 49- Concurrent Pos: Mem, Gov Sci Adv Comts, Mo. Mem: AAAS; Am Soc Animal Sci; Am Genetic Asn. Res: Physiology of spermatozoa; breeding problems of range cattle; staining method for differentiation of live and dead spermatozoa; improvement of beef cattle through selection and crossing. Mailing Add: 2207 Bushnell Dr Columbia MO 65201

LASS, HARRY, b Toronto, Ont, Dec 19, 14; m 49; c 2. MATHEMATICS. Educ: Univ Calif, Los Angeles, BA, 38, MA, 40; Calif Inst Technol, PhD, 48. Prof Exp: Instr math, Univ Calif, 49-50; asst prof, Univ Ill, 50-52; res mathematician, Naval Ord Lab, 52-54; res mathematician, Res Lab, Motorola, Inc, 54-55; RES SPECIALIST, JET PROPULSION LAB, CALIF INST TECHNOL, 55- Concurrent Pos: Assoc prof, Univ Southern Calif, 56-58; prof systs anal, West Coast Univ, 64-73. Mem: Math Asn Am; Am Math Soc. Res: Applied mathematics; relativity theory. Mailing Add: 2025 Meadowbrook Rd Altadena CA 91001

LASS, NORMAN JAY, b New York, NY, Sept 20, 43; m 67; c 2. SPEECH PATHOLOGY, AUDIOLOGY. Educ: Brooklyn Col, BA, 65; Purdue Univ, MS, 66, PhD(speech & hearing sci), 68. Prof Exp: Res fel, Bur Child Res Labs, Univ Kans Med Ctr, Kansas City, 68-69; asst prof, 69-73, ASSOC PROF SPEECH & HEARING SCI, W VA UNIV, 73-, DIR SPEECH & HEARING SCI LAB, 69-, COORDR SPEECH PATH-AUDIOL PROG, 74- Concurrent Pos: Abstractor, Cleft Palate J, Am Cleft Palate Asn, 72-; abstractor, dsh Abstracts, Am Speech & Hearing Asn, 73-, ed consult, J Speech & Hearing Res & J Speech & Hearing Disorders, 74- & Perceptual & Motor Skills, 75-; consult speech path, Audiol & Speech Path Serv, Vet Admin Ctr, Martinsburg, WVa, 75- Mem: Am Speech & Hearing Asn; Acoust Soc Am; Am Asn Phonetic Sci; Int Soc Phonetic Sci; Am Cleft Palate Asn. Res: Determination of the acoustic and perceptual cues involved in speaker race, sex, height, weight, age and photograph identification by listeners; sex differences in speech rate production and perception; usefulness of time-expanded speech. Mailing Add: Speech & Hearing Sci Lab WVa Univ Morgantown WV 26506

LASSEN, LAURENCE E, b Milwaukee, Wis, Dec 16, 32; m 59; c 2. FORESTRY. Educ: Iowa State Univ, BS, 54, MS, 58; Univ Mich, PhD(forestry), 67. Prof Exp: Res forest prod technologist, Forest Prod Lab, Wis, 58-67, proj leader, 67-71; staff asst res admin, 71-72, chief forest prod technol res, Washington, DC, 72-74, DEP DIR, SOUTHERN FOREST EXP STA, US FOREST SERV, 74- Mem: Soc Am Foresters. Res: Forestry research. Mailing Add: Southern Forest Exp Sta US Forest Serv 701 Loyola Ave New Orleans LA 70113

LASSER, ELLIOTT CHARLES, b Buffalo, NY, Nov 30, 22; m 44; c 4. MEDICINE, RADIOLOGY. Educ: Harvard Univ, BS, 43; Univ Buffalo, MD, 46; Univ Minn, MS, 53. Prof Exp: Instr radiol, Grad Sch Med, Univ Minn, 52-53; assoc, Sch Med, Univ Buffalo, 53-54, asst prof, 54-56; prof & chmn dept, Sch Med, Univ Pittsburgh, 56-58; PROF RADIOL & CHMN DEPT, SCH MED, UNIV CALIF, SAN DIEGO, 68- Concurrent Pos: Consult, Vet Admin Hosp, Pittsburgh, Pa, 57-68. Mem: AAAS; Radiol Soc NAm; Soc Nuclear Med; AMA; Am Col Radiol. Res: Radiology of the vascular system. Mailing Add: Univ Hosp of San Diego County 225 W Dickinson St San Diego CA 92103

LASSER, MARVIN ELLIOTT, b Brooklyn, NY, Feb 9, 26; m 54; c 2. SOLID STATE PHYSICS. Educ: Brooklyn Col, AB, 49; Syracuse Univ, MS, 51, PhD(physics), 54. Prof Exp: Res assoc physics, Syracuse Univ, 49-54; proj scientist, Philco Corp, 54-56, sect head, 56-57, mgr res sect, 57-63, asst dir physics lab, 63-64, dir appl res lab, 64-66; CHIEF SCIENTIST, US DEPT ARMY, 66-, EXEC DIR, ARMY SCI ADV PANEL, 70-, DIR ARMY RES, 74- Concurrent Pos: Mem comt, Int Conf Semiconductors, NY, 58; mem, Int Conf Photoconductivity, 61; consult, US Army Electronics Command & adv comt optical commun & tracking, instrumentation & data processing, NASA, 66. Mem: Fel Am Phys Soc. Res: Photoconductivity; optical and surface properties; metallurgy and measurements of electrical characteristics; lasers; electrooptics; semiconductor devices. Mailing Add: Off Dep Chf Staff Res Dev & Acq Rm 3E 364 The Pentagon Washington DC 20310

LASSER, TOBIAS, plant taxonomy, see 12th edition

LASSETER, KENNETH CARLYLE, b Jacksonville, Fla, Aug 12, 42; m 63; c 2. CLINICAL PHARMACOLOGY, INTERNAL MEDICINE. Educ: Stetson Univ, BS, 63; Univ Fla, MD, 67. Prof Exp: From intern to resident med, Univ Ky Hosp, 67-69, USPHS fel cardiol, 69-70; fel clin pharmacol, 70-71, asst prof, 71-74, ASSOC PROF PHARMACOL & MED, UNIV MIAMI, 74- Concurrent Pos: Attend physician, Jackson Mem Hosp, 71- Honors & Awards: Res Award, Interstate Postgrad Med Asn, 74. Mem: Am Soc Pharmacol & Exp Therapeut; Am Soc Clin Pharmacol & Therapeut; Am Col Physicians; Am Col Clin Pharmacol; Sigma Xi. Res: Cardiovascular pharmacology. Mailing Add: Dept of Pharmacol Univ of Miami Sch of Med PO Box 520875 Biscayne Annex Miami FL 33152

LASSETTER, JOHN STUART, b Jackson, Miss, Jan 7, 44; m 65; c 2. BOTANY. Educ: Miss Col, BS, 66; Univ Miss, MS, 68; Iowa State Univ, PhD(plant taxon), 72. Prof Exp: ASST PROF BOT, EASTERN KY UNIV, 73- Concurrent Pos: Res assoc, Surface Mine Pollution Abatement & Land Use Impact Invest proj, 74-75. Mem: Am Soc Plant Taxonomists; Int Asn Plant Taxon; Sigma Xi. Res: Biosystematics of North American Vicia, local Kentucky flora. Mailing Add: Dept of Biol Sci Eastern Ky Univ Richmond KY 40475

LASSETTRE, EDWIN NICHOLS, b Monroe Co, Ga, Oct 26, 11; m 33, 51; c 2. CHEMICAL PHYSICS. Educ: Mont State Col, BSc, 33; Calif Inst Technol, PhD(chem), 38. Prof Exp: Instr chem, Ohio State Univ, 37-40, asst prof, 40-54, group leader & res scientist, Manhattan Proj, SAM Labs, Columbia Univ, 44 & Union Carbide & Carbon Chem Corp, 45; assoc prof chem, Ohio State Univ, 46-49, prof, 50-62; staff fel, Mellon Inst Sci, 62-74, dir ctr spec studies, 71-73, prof, 67-74, UNIV PROF CHEM PHYSICS, CARNEGIE-MELLON UNIV, 74- Concurrent Pos: Consult, Oak Ridge Gaseous Diffusion Plant, Union Carbide Nuclear Co. Mem: Am Chem Soc; fel Am Phys Soc. Res: Elastic and inelastic electron scattering by molecular gases; electron impact spectroscopy; theoretical chemistry. Mailing Add: Mellon Inst Sci 4400 Fifth Ave Carnegie-Mellon Univ Pittsburgh PA 15213

LASSILA, KENNETH EINO, b Hancock, Mich, Apr 27, 34; m 57; c 3. HIGH ENERGY PHYSICS, THEORETICAL NUCLEAR PHYSICS. Educ: Univ Wyo, BS, 56; Yale Univ, MS, 59, PhD(theoret physics), 61. Prof Exp: Res assoc theoret physics, Case Inst Technol, 61-63; asst prof physics, Iowa State Univ, 63-64; sr res assoc, Res Inst Theoret Physics, Univ Helsinki, 64-66; from asst prof to assoc prof physics, 66-69, PROF PHYSICS, IOWA STATE UNIV, 69- Concurrent Pos: Fulbright res fel, Res Inst Theoret Physics, Univ Helsinki, 65; prof, Nordic Inst Theoret Atomic Physics, 72; Fulbright lectr, Univ Oulu, Finland & Univ Oslo, Norway, 73. Mem: AAAS; fel Am Phys Soc. Res: Nucleon-nucleon interaction; radiation theory; elementary particle interactions. Mailing Add: Dept of Physics Iowa State Univ Ames IA 50010

LASSITER, CHARLES ALBERT, b Murray, Ky, Feb 20, 27; m 46; c 2. DAIRY SCIENCE. Educ: Univ Ky, BS, 49, MS, 50; Mich State Univ, PhD, 52. Prof Exp: From asst prof to assoc prof dairy sci, Univ Ky, 52-55; assoc prof, 56-59, PROF DAIRY SCI & CHMN DEPT, MICH STATE UNIV, 59- Mem: Am Soc Animal Sci; Am Dairy Sci Asn. Res: Dairy cattle nutrition, especially pasture and forage studies dealing with more applied aspects; nutritional problems of young dairy calves. Mailing Add: Dept of Dairy Sci Mich State Univ East Lansing MI 48823

LASSITER, JAMES WILLIAM, b Covington, Ga, May 28, 20; m 47; c 2. ANIMAL NUTRITION. Educ: Univ Ga, BSA, 41, MSA, 52; Univ Ill, PhD, 55. Prof Exp: Asst, Univ Ill, 52-54; asst prof animal husb, 54-61, assoc prof animal sci, 61-70, PROF ANIMAL SCI, UNIV GA, 70- Concurrent Pos: Gold Kist study & res grant, Cambridge & Aberdeen Univs, 72; mem manganese panel, Comt Biol Effects of Atmospheric Pollutants, Nat Acad Sci. Mem: Am Inst Nutrit; Am Soc Animal Sci; Am Dairy Sci Asn; Soc Environ Chem & Health; Coun Agr Sci & Technol. Res: Protein requirements of swine; effects of antibiotics in swine; factors influencing body composition; interrelationships of nutrients; ruminant nutrition; mineral nutrition of animals; metabolism of mineral elements by animals. Mailing Add: Dept of Animal & Dairy Sci Univ of Ga Athens GA 30602

LASSITER, WILLIAM EDMUND, b Wilmington, NC, July 21, 27; m 56; c 4. PHYSIOLOGY, NEPHROLOGY. Educ: Harvard Univ, AB, 50, MD, 54. Prof Exp: Intern & asst resident med, Mass Gen Hosp, Boston, 54-56; sr asst resident, NC Mem Hosp, 56-57; Donner res fel, Mass Gen Hosp, Boston, 57-58; Life Ins Med Res Fund fel, Univ NC, Chapel Hill, 58-60; from instr to assoc prof, 60-70, PROF MED, SCH MED, UNIV NC, CHAPEL HILL, 70- Concurrent Pos: Estab investr, Am Heart Asn, 62-67, mem coun clin cardiol & coun kidney in cardiovasc dis, 71-; vis investr, Physiol Inst, Berlin, 63-64; Markle scholar, 63-68; Nat Inst Arthritis & Metab Dis career develop award, 67-72; mem cardiovasc & pulmonary res A study sect, NIH, 69-73; sect ed, Renal & Electrolyte Physiol, Am J Physiol & J Appl Physiol, 74- Mem: AAAS; Am Soc Clin Invest; Am Physiol Soc; fel Am Col Physicians; Am Soc Nephrol. Res: Micropuncture studies of mammalian kidney function; transport of cations, organicanions, and urea in mammalian kidney. Mailing Add: Dept of Med Univ of NC Sch of Med Chapel Hill NC 27514

LASSLO, ANDREW, b Mukacevo, Czech, Aug 24, 22; nat US; m 55; c 1. MEDICINAL CHEMISTRY. Educ: Univ Ill, MSc, 48, PhD(pharmaceut chem), 52, MSLS, 61. Prof Exp: Asst chem, Univ Ill, 47-51; res chemist, Org Chem Div, Monsanto Chem Co, 52-54; asst prof med chem, Emory Univ, 54-60; PROF MED CHEM & CHMN DEPT, COL PHARM, MED UNITS, UNIV TENN, MEMPHIS, 60- Concurrent Pos: Dir postgrad training prog for sci librn, Nat Libr Med, 66-72. Mem: Fel AAAS; fel Am Inst Chemists; sr mem Am Chem Soc; Am Soc Pharmacol & Exp Therapeut; Am Pharmaceut Asn. Res: Synthesis and study of compounds with pharmacodynamic potentialities; exploration of relationships between the molecular constitution of synthetic entities and their biochemical response; processing and coordination of scientific and technical information. Mailing Add: Dept Med Chem Col Pharm Univ Tenn Ctr Health Sci Memphis TN 38163

LAST, JEROLD ALAN, b New York, NY, June 5, 40; m 68; c 2. BIOCHEMISTRY. Educ: Univ Wis, BS, 59, MS, 61; Ohio State Univ, PhD(biochem), 65. Prof Exp: Biochemist, Corn Prod Co, Ill, 61-62; SR RES SCIENTIST, SQUIBB INST MED RES, NJ, 67- Concurrent Pos: Am Cancer Soc fel biochem, Med Sch, NY Univ, 66-67. Mem: Am Chem Soc; Brit Biochem Soc. Res: Antibiotics; biosynthesis and mechanism of action; protein biosynthesis; nucleic acids; fermatation biochemistry. Mailing Add: Squibb Inst of Med Res Georgus Rd New Brunswick NJ 08903

LAST, JOHN MURRAY, b Tailem Bend, Australia, Sept 22, 26; m 57; c 3. EPIDEMIOLOGY. Educ: Univ Adelaide, MB & BS, 49, MD, 68; Univ Sydney, DPH, 60. Prof Exp: Australian Postgrad Med Found fel, Social Med Res Unit, Med Res Coun UK, 61-62; lectr pub health, Univ Sydney, 62-63; asst prof epidemiol, Univ Vt, 63-64; sr lectr social med, Univ Edinburgh, 65-69; PROF EPIDEMIOL & COMMUNITY MED & CHMN DEPT, UNIV OTTAWA, 70- Concurrent Pos: Mem, Nat Health Grant Rev Comt, 70- & Epidemiol Study Sect, NIH, 72-; consult med educ, WHO, 72-73 & 74. Mem: Int Asn Epidemiol; fel Am Pub Health Asn; fel Royal Australasian Col Physicians; Brit Asn Study Med Educ. Res: Medical care and education; cancer epidemiology; reproductive behavior Mailing Add: 495 Island Park Dr Ottawa ON Can

LASTER, HOWARD JOSEPH, b Jersey City, NJ, Mar 13, 30; m 52; c 4. COSMIC RAY PHYSICS. Educ: Harvard Univ, AB, 51; Cornell Univ, PhD(theoret physics), 57. Prof Exp: Asst physics, Cornell Univ, 51-56; from asst prof to assoc prof, 56-65, chmn dept physics & astron, 65-75, PROF PHYSICS, UNIV MD, COLLEGE PARK, 65- Concurrent Pos: Instr, Wells Col, 53-54; mem adv comn manpower, Am Inst Physics; mem bd dirs, Atlantic Res Corp, 72-; chmn, Gov Sci Adv Coun, Md, 73-75, exec comt, 75-; vis prog assoc, Off Int Progs, NSF, 75-76. Mem: AAAS; Am Phys Soc; Am Asn Physics Teachers; Am Astron Soc; Royal Astron Soc. Res: Cosmic ray theory; astrophysics; space theory; origins and propagation of cosmic rays; interplanetary modulation of cosmic rays. Mailing Add: Dept of Physics & Astron Univ Md College Park MD 20742

LASTER, LEONARD, b New York, NY, Aug 25, 28; m 56; c 3. MEDICINE, SCIENCE POLICY. Educ: Harvard Univ, AB, 49, MD, 50; Am Bd Internal Med, dipl, 61; Am Bd Gastroenterol, dipl, 66. Prof Exp: From intern to resident med, Mass Gen Hosp, 50-53; vis investr purine metab, Pub Health Res Inst, New York, Inc, 53-54; sr clin investr, Nat Inst Arthritis & Metab Dis, 54-58; chief sect gastroenterol, Metab Dis Br, Nat Inst Arthritis & Metab Dis, 59-69; staff mem, Off Sci & Technol, Exec Off of President, 69-71, asst dir human resources, 71-74; VPRES ACAD & CLIN AFFAIRS & DEAN COL MED, STATE UNIV NY DOWNSTATE MED CTR, 74- Concurrent Pos: Res fel gastroenterol, Mass Mem Hosps, 58-59; clin instr, Sch Med, George Washington Univ, 55-58, prof lectr, 66- Mem: AAAS; Am Fedn Clin Res; Am Gastroenterol Asn; Am Col Physicians; Am Soc Clin Invest. Res: Biochemical aspects of human disease; inborn errors of metabolism; disturbances of the gastrointestinal tract. Mailing Add: State Univ NY Downstate Med Ctr Brooklyn NY 11203

LASTER, MARION LOGAN, b Gravelly, Ark, Nov 19, 31; m 58; c 2. ENTOMOLOGY. Educ: Univ Ark, Fayetteville, BS, 58, MS, 59; Miss State Univ, PhD(entom), 66. Prof Exp: ENTOMOLOGIST, DELTA BR, MISS AGR & FORESTRY EXP STA, 59- Mem: Entom Soc Am. Res: Biology and control of insect pests of agricultural crops with emphasis on Heliothis species. Mailing Add: Delta Br Miss Agr & Forestry Exp Sta Stoneville MS 38776

LASTER, STANLEY JERRAL, b Fittstown, Okla, June 28, 37; m 60; c 3. GEOPHYSICS. Educ: Univ Tulsa, BS, 59; Southern Methodist Univ, MS, 62; Mass Inst Technol, PhD(geophys), 70. Prof Exp: Sr res geophysicist, Tex Instruments Inc, 59-74; ASSOC PROF GEOPHYS, UNIV TULSA, 74- Mem: Soc Explor Geophys; Seismol Soc Am. Res: Theoretical seismology; signal processing; numerical analysis. Mailing Add: Dept of Earth Sci Univ of Tulsa Tulsa OK 74104

LASTER, WILLIAM RUSSELL, JR, b Ala, Oct 20, 26; m 61; c 1. VETERINARY MEDICINE. Educ: Auburn Univ, DVM, 51. Prof Exp: Virologist, 51-56, head cancer screening sect, 56-66, HEAD CANCER SCREENING DIV, SOUTHERN RES INST, 66- Mem: Am Asn Cancer Res. Res: Cancer chemotherapy. Mailing Add: Cancer Scrn Div Southern Res Inst 2000 Ninth Ave S Birmingham AL 35205

LASTHUYSEN, WILLEM, b Netherlands, July 18, 10; nat US; m 59. ORGANIC CHEMISTRY. Educ: Univ Bonn, MS, 32. Prof Exp: Chemist, Indonesia, 32-34; chemist & tech salesman, Netherlands, 34-40; res chemist, NV Chem Works, 40-45; res dir, Great Atlantic & Pac Tea Co, NY, 47-52; chief chemist, Dodge & Olcott, Inc, 52-55; res dir, Colgate-Palmolive Co, NJ, 55-57; tech dir, Rhodia, Inc, 57-64; vpres & tech dir, Firmenich, Inc, NY, 64-66; PRES, FLAVOR RES LABS, 66- Concurrent Pos: Lectr, Columbia & Rutgers Univs, 50-; indust consult. Mem: Am Chem Soc; Inst Food Technologists. Res: Essential oils; aromatic chemicals; pharmaceuticals; food, beverage and feed problems; flavors and nutrition. Mailing Add: 318 Ocean Blvd Atlantic Highlands NJ 07716

LASURE, LINDA LEE, b Bartlesville, Okla, Nov 23, 46. MICROBIAL GENETICS. Educ: St Cloud State Col, BS, 68; Syracuse Univ & State Univ NY, PhD(genetics), 73. Prof Exp: Res fel, New York Bot Garden, 72-74; RES SCIENTIST MICROBIOL, MILES LABS INC, 74- Mem: Sigma Xi; AAAS; Genetics Soc Am; Am Soc Microbiol; Mycol Soc Am. Res: Studies of mutagenesis, inheritance, and physiological control of sexual reproduction in primitive fungi, fungal spore germination, and selection of strains of fungi with improved yields of enzymes. Mailing Add: Dept of Microbiol & Enzymol Marschall Div Miles Labs Inc Elkhart IN 46514

LASWELL, TROY JAMES, b Ottawa, Ky, Nov 12, 20; m 43; c 1. GEOLOGY. Educ: Berea Col, AB, 42; Oberlin Col, AM, 48; Univ Mo, PhD(geol), 53. Prof Exp: Instr geol, Univ Mo, 48-53; from asst prof to assoc prof, Washington & Lee Univ, 53-57; from assoc prof to prof, La Polytech Inst, 57-62; PROF GEOL & GEOG & HEAD DEPT, MISS STATE UNIV, 62- Mem: Geol Soc Am; Am Asn Petrol Geologists; Am Asn Geol Teachers. Res: Stratigraphy; sedimentation. Mailing Add: PO Box 824 Mississippi State MS 39762

LASZEWSKI, RONALD M, b Chicago, Ill, June 22, 47. EXPERIMENTAL NUCLEAR PHYSICS. Educ: Univ Ill, Urbana, BS, 69, MS, 72, PhD(physics), 75. Prof Exp: FEL PHYSICS, ARGONNE NAT LAB, 75- Mem: Am Phys Soc. Res: Photonuclear physics. Mailing Add: Argonne Nat Lab Argonne IL 60403

LASZLO, CHARLES A, b Budapest, Hungary, July 8, 35; Can citizen; m 64. BIOMEDICAL ENGINEERING. Educ: McGill Univ, BE, 61, ME, 66, PhD(biomed eng), 68. Prof Exp: Design engr, Northern Elec Co, Que, 61; RCA Victor Co, Ltd, 61-62; biomed engr, Res Labs, Otolaryngol Inst, Royal Victoria Hosp, 62-70, assoc scientist, 70-74; ASSOC DIR, DIV HEALTH SYSTS, UNIV BC, 74- Concurrent Pos: Assoc prof eng & med, McGill Univ, 68-74. Mem: Inst Elec & Electronics Eng; Acoustical Soc Am; Can Med & Biol Eng Soc. Res: Assessment of technology in the health care system; system and computer applications in physiology, biology and medicine; experimental and theoretical investigation of electrophysiological phenomena in the auditory system. Mailing Add: Fourth Floor IRC Bldg Univ of BC Vancouver BC Can

LASZLO, JOHN, b Cologne, Ger, May 28, 31; US citizen; m 62; c 3. MEDICINE, BIOCHEMISTRY. Educ: Columbia Univ, AB, 52; Harvard Med Sch, MD, 55; Am Bd Internal Med, dipl, 63. Prof Exp: Intern, Univ Chicago Clins, 55-56; clin assoc clin chemother, Nat Cancer Inst, 56-58; USPHS fel cytochem, Nat Cancer Inst, 58; resident med, 59-60, from asst prof to assoc prof, 62-70, PROF MED, SCH MED, DUKE UNIV, 70- Concurrent Pos: Chief hemat sect, Vet Admin Hosp, Durham, 63-68, chief med serv, 68-73; mem pharmacol study sect, Nat Cancer Inst, 70-74. Mem: Am Soc Hemat; Am Asn Cancer Res; fel Am Col Physicians; Asn Am Physicians. Res: Nucleic acids; leukemia. Mailing Add: Dept of Med Duke Univ Med Ctr Durham NC 27706

LASZLO, PIERRE, b Algiers, Algeria, Aug 15, 38; m 62; c 2. ORGANIC CHEMISTRY. Educ: Univ Paris, Lic es sci, 60, dipl, 61, DEtat, 65. Prof Exp: Instr chem, Univ Paris, 61-62; res assoc with Prof P von R Schleyer, Princeton Univ, 62-63; instr chem, Univ Paris, 63-66; asst prof, Princeton Univ, 66-70; PROF CHEM, UNIV LIEGE, 70- Mem: Am Chem Soc; Chem Soc France; Am Inst Chemists; NY Acad Sci; The Chem Soc. Res: Applications of nuclear magnetic resonance in organic chemistry. Mailing Add: Inst de Chimie Sart-Tilman par 4000 Liege Belgium

LASZLO, TIBOR S, b Oraviczafalu, Hungary, Apr 25, 12; US citizen. INDUSTRIAL CHEMISTRY. Educ: Royal Hungarian Univ Technol & Econ Sci, DSc(chem eng), 35. Prof Exp: Res & develop engr oil industs, Hungary, 35-48; asst dir inorg high temperature chem, Fordham Univ, 48-51, process engr res dept, Calif Tex Oil Co, 51-53; dir high temperature lab, Fordham Univ, 53-57; proj specialist high temperature ceramics, Wright Aeronaut Div, Curtiss Wright Corp, 57-58; prin staff scientist, High Temperature Inorg Lab, Res & Adv Develop Div, Avco Corp, 58-65, sr consult scientist, 65-66; PRIN SCIENTIST, RES CTR, PHILIP MORRIS, INC, 66- Concurrent Pos: Consult specialist, 53-57; NSF grant, 55-57; foreign ed, La Revue des Hautes Temperatures et des Refractaires, 65- Mem: Solar Energy Soc; Am Soc Testing & Mat. Res: Petroleum and vegetable oil processing; high temperature generation; solar furnaces; high refractory compounds; solar radiation simulation; temperature control coatings; thermal radiation measurements; cellulose-water bond. Mailing Add: Res Ctr Philip Morris Inc PO Box 26583 Richmond VA 23261

LATA, GENE FREDERICK, b New York, NY, May 17, 22; m 51; c 5. BIOCHEMISTRY. Educ: City Col New York, BS, 42; Univ Ill, MS, 48, PhD(biochem), 50. Prof Exp: Jr chemist, Gen Foods Corp, 42, chemist, 46; asst chem, Univ Ill, 47, asst biochem, 48-50; from instr to asst prof, 50-62, fel med, 65-66, ASSOC PROF BIOCHEM, UNIV IOWA, 62- Concurrent Pos: Vis lectr, Huntington Labs, Harvard Med Sch, 65-66; consult, Am Dent Asn. Mem: Fel AAAS; Am Chem Soc; Am Soc Biol Chemists; NY Acad Sci. Res: Peroxisomal enzymes and their control; hormonal control of enzyme actions; steroid interactions and transport. Mailing Add: Dept of Biochem Univ of Iowa Iowa City IA 52242

LATADY, WILLIAM ROBERTSON, b Birmingham, Ala, Feb 14, 18; m 49; c 5. OPTICS, MECHANICS. Prof Exp: Mem staff radar, Radiation Lab, Mass Inst Technol, 41-46; mem, Am Antarctic Res Exped, 47-48; dir, Juneau Iceland Res Proj, 48; vpres motion picture syst, Cinerama, Inc, 52-55, Cinerama, Int, 56-59; PRES, LATADY INSTRUMENTS, INC, 60- Honors & Awards: Latady Mt mountain range, named by US Govt; Latady Island, named by Brit Govt. Mem: AAAS; Am Soc Photogram; Soc Motion Picture & TV Engrs; Am Inst Physics. Res: Design, development and manufacture of optical-mechanical instruments, especially in photographic field; medical instruments. Mailing Add: 220 Prospect St Hingham MA 02043

LATCH, DANA MAY, b New York, NY, Aug 29, 43; c 1. ALGEBRA, TOPOLOGY. Educ: Harpur Col, BA, 65; Queens Col, NY, MA, 67; City Univ New York, PhD(math), 71. Prof Exp: Teaching asst math, Queens Col, NY, 65-66, lectr, 66-67; asst prof, Douglass Col, Rutgers Univ, 71-74 & Lawrence Univ, 74-76; ASST PROF MATH, NC STATE UNIV, 76- Concurrent Pos: NSF res grant, Rutgers Univ, 72-73; Carnegie Found fac develop award, Lawrence Univ, 75. Mem: Am Math Soc; Asn Women Math. Res: Algebraic topology of small categories; applications of ring theoretic methods to topological questions; applications of categorical methods to homotopy theory and the theory of localizations. Mailing Add: Dept of Math NC State Univ Raleigh NC 27607

LATEEF, ABDUL BARI, b Lyallpur, WPakistan, Apr 4, 39; m 70. FORENSIC SCIENCE, CHEMISTRY. Educ: Punjab Univ, Pakistan, BS, 59, MS, 61; Univ Newcastle, PhD(chem), 66. Prof Exp: Nat Res Coun Can fel, Univ Calgary, 66-69; from instr to asst prof chem, 69-71, ASST PROF FORENSIC SCI, YOUNGSTOWN STATE UNIV, 71- Mem: Am Acad Forensic Sci; Am Chem Soc; Forensic Soc, London; Am Soc Testing & Mat. Res: Spectroscopic and chromatographic analytical techniques in forensic science; role of forensic science in criminal justice system. Mailing Add: Dept of Criminal Justice Youngstown State Univ Youngstown OH 44503

LATHAM, ARCHIE J, b Blackfoot, Idaho, June 26, 26; m 63; c 3. PLANT PATHOLOGY, MYCOLOGY. Educ: Idaho State Col, BS, 56; Univ Idaho, MS, 59; Univ Ill, PhD(plant path), 61. Prof Exp: Ranger, Yellowstone Park, Wyo, 56; res asst plant path, Univ Idaho, 56-58 & Univ Ill, 58-61; res biologist, Gulf Res & Develop Co, Kans, 61-67; ASST PROF BOT & PLANT PATH, AUBURN UNIV, 67- Mem: Am Phytopath Soc; Mycol Soc Am. Res: Fruit and nut diseases. Mailing Add: Dept of Bot & Microbiol Auburn Univ Auburn AL 36830

LATHAM, DAVID WINSLOW, b Boston, Mass, Mar 19, 40; m 60; c 5. ASTRONOMY. Educ: Mass Inst Technol, BS, 61; Harvard Univ, MA, 65, PhD(astron), 70. Prof Exp: ASTRONOMER, SMITHSONIAN ASTROPHYS OBSERV, 65-; LECTR ASTRON, HARVARD UNIV, 71- Mem: Am Astron Soc; Soc Photog Scientist & Engrs; Royal Astron Soc. Res: Stellar spectroscopy; stellar chemical abundances and nucleosynthesis; detection of light at low levels. Mailing Add: Ctr for Astrophys 60 Garden St Cambridge MA 02138

LATHAM, DEWITT ROBERT, b Chugwater, Wyo, Oct 26, 28; m 55; c 2. PETROLEUM CHEMISTRY. Educ: Univ Wyo, BS, 50. Prof Exp: Anal chemist, Rocky Mountain Arsenal, US Army Chem Corps, Colo, 52-54; chemist, Laramie Energy Res Ctr, US Bur Mines, 54-60, res chemist, 60-64, PROJ LEADER LARAMIE ENERGY RES CTR, ENERGY RES & DEVELOP ADMIN, 64- Mem: Am Chem Soc; Sigma Xi. Res: Nitrogen and oxygen compounds in petroleum; development of methods of analysis for petroleum; separation and characterization of fossil fuel energy sources. Mailing Add: Laramie Energy Res Ctr ERDA PO Box 3395 Laramie WY 82070

LATHAM, DON JAY, b Lewiston, Idaho, Dec 21, 38; m 60; c 3. ATMOSPHERIC SCIENCES. Educ: Pomona Col, BA, 60; NMex Inst Mining & Technol, MS, 64, PhD, 68. Prof Exp: Res asst, NMex Inst Mining & Technol, 61-67; sr res scientist, 67-68, ASST PROF ATMOSPHERIC SCI, ROSENSTIEL SCH MARINE & ATMOSPHERIC SCI, UNIV MIAMI, 68- Concurrent Pos: NSF grants, Univ Miami, 69-72. Mem: Am Meteorol Soc; Am Geophys Union. Res: Atmospheric electricity; radar meteorology. Mailing Add: 207 Comput Ctr Div Atmos Sci Univ of Miami Coral Gables FL 33124

LATHAM, GARY V, b Conneautville, Pa, Sept 2, 35; m 71; c 1. GEOPHYSICS. Educ: Pa State Univ, BS, 58; Columbia Univ, PhD(geophys), 65. Prof Exp: Sr res scientist, Columbia Univ, 66-71; PROF GEOPHYS, UNIV TEX, 72-, ASSOC DIR GEOPHYS LAB, MARINE SCI INST, 75- Concurrent Pos: Prin investr, Apollo Lunar Seismic Exp, 68-; trustee, Palisades Geophys Inst, 71-; consult, NASA, 72- Honors & Awards: Medal for Except Sci Achievement, NASA, 71 & Expert, 73. Mem: AAAS; Soc Explor Geophysicists; Seismol Soc Am; Am Geophys Union; Am Soc Oceanog. Res: Seismic experiments on the moon and Mars, Apollo and Viking; seismic and land deformation studies in Central America; ocean bottom seismic stations. Mailing Add: 700 The Strand Galveston TX 77550

LATHAM, JAMES PARKER, b Collingdale, Pa, June 2, 18; m 45. GEOGRAPHY. Educ: Univ Pa, BS, 49, MS, 50, MA, 51, PhD(econ), 59. Prof Exp: Lectr geog &

indust & res assoc geog, Univ Pa, 51-59; assoc prof geog, Bowling Green State Univ, 59-64; chmn dept, 64-70, PROF GEOG, FLA ATLANTIC UNIV, 64-, DIR REMOTE SENSING & ENVIRON EVAL LAB, 70- Concurrent Pos: Consult remote sensing, Champlain Technol, Inc; mem Nat Res Coun Comt, Remote Sensing Progs, 73- Mem: Asn Am Geog; fel Am Geog Soc; Am Soc Photogram; Inst Elec & Electronic Eng; Nat Coun Geog Educ. Res: Electronic instrumentation of geographic research; remote sensing; economic land use; military strategic area analysis. Mailing Add: Dept of Geog Fla Atlantic Univ Boca Raton FL 33432

LATHAM, MICHAEL CHARLES, b Kilosa, Tanzania, May 6, 28; m 74; c 2. NUTRITION, TROPICAL PUBLIC HEALTH. Educ: Univ Dublin, BA, 49, MB, BCh & BAO, 52; Univ London, DTM&H, 58; Harvard Univ, MPH, 65. Prof Exp: House surgeon, High Wycombe Hosp, Buckinghamshire, Eng, 52-53; rotating physician, Methodist Hosp, Los Angeles, 53-54; sr house officer, NMiddlesex Hosp, London, 54-55; med officer, Tanzania Ministry of Health, 55-64, dir nutrit unit, 62-64; res assoc & asst prof nutrit, Harvard Univ, 64-68; PROF INT NUTRIT, GRAD SCH NUTRIT, CORNELL UNIV, 68- Concurrent Pos: Vis exchange fel, Methodist Hosp, Los Angeles, 53-54; contrib ed, Nutrit Rev, 68-; chmn panel, White House Conf Food, Nutrit & Health & vchmn panel, Follow-Up Conf, 69-71; consult, WHO, Manila, 70 & UN Food & Agr Orgn, 64 & Zambia, 70; mem, Nat Acad Sci-Nat Res Coun Int Nutrit Comt, 70- UNICEF consult, Thailand & Malaysia, 73; fel, Fac Community Med, Univ London, 73; mem exec comt pest control, Nat Acad Sci, 73-; mem expert adv panel nutrit, WHO, 74- Honors & Awards: Officer of the Order of the Brit Empire, 65. Mem: Am Inst Nutrit; Brit Nutrit Soc; Am Soc Clin Nutrit; fel Royal Soc Trop Med & Hyg; fel Am Pub Health Asn. Res: International nutrition problems; nutrition and health of low income populations; xerophthalmia; nutrition and intellectual development; protein-calorie malnutrition of children; evaluation of applied nutrition programs; lactose intolerance; energy expenditure. Mailing Add: Div of Nutrit Sci Cornell Univ Ithaca NY 14853

LATHAM, ROGER ALAN, b Parsons, WVa, July 6, 39; m 66; c 2. POLYMER CHEMISTRY. Educ: Harvard Univ, AB, 61; Mass Inst Technol, PhD(org chem), 66. Prof Exp: SR RES CHEMIST, E I DU PONT DE NEMOURS & CO, INC, 66- Res: Development of engineering plastics. Mailing Add: E I du Pont de Nemours & Co PO Box 1217 Parkersburg WV 26101

LATHAM, ROSS, JR, b Chicago, Ill, Dec 18, 32; m 61; c 2. INORGANIC CHEMISTRY. Educ: Principia Col, BS, 55; Univ Ill, MS, 57, PhD(inorg chem), 61. Prof Exp: Instr chem, Lafayette Col, 59-60; chemist, Esso Res & Eng Co, NJ, 61-66; assoc prof, 66-72, PROF CHEM, ADRIAN COL, 72- Res: Titanium alkoxide-halide chemistry. Mailing Add: 434 Westwood Dr W Adrian MI 49221

LATHEM, WILLOUGHBY, b Atlanta, Ga, Oct 9, 23; m 51; c 4. MEDICINE. Educ: Emory Univ, BS, 44, MD, 46; Am Bd Internal Med, dipl, 54. Prof Exp: Asst med, Col Physicians & Surgeons, Columbia Univ, 52-53; asst clin prof, Yale Univ, 53-56; from asst prof to assoc prof, Sch Med, Univ Pittsburgh, 56-64; sci rep, Off Int Res, NIH, Eng, 62-66; dep dir biomed sci, 66-72, ASSOC DIR HEALTH SCI, ROCKEFELLER FOUND, 72-, REGIONAL OFFICER, ASIA, BANGKOK & THAILAND, 75- Concurrent Pos: Hon res assoc, Univ Col, Univ London, 62-65. Mem: Am Soc Clin Invest; Harvey Soc; Am Fedn Clin Res. Res: International health. Mailing Add: Rockefeller Found 111 W50th St New York NY 10020

LATHRAP, DONALD WARD, b Oakland, Calif, July 4, 27; m 58; c 1. ANTHROPOLOGY. Educ: Univ Calif, AB, 50; Harvard Univ, PhD, 62. Prof Exp: Instr anthrop, Univ Ill, Urbana, 59-62, from asst prof to assocprof, 62-70, PROF ANTHROP & ASSOC, CTR ADVAN STUDY, UNIV ILL, URBANA, 70- Concurrent Pos: NSF grant, 64-65. Mem: Am Anthrop Asn; Soc Am Archaeol. Res: Rise of agricultural societies in the New World; basis of New World civilizations; agricultural societies of Tropical Forest South America; ceramic technology and underlying cognitive systems. Mailing Add: Dept of Anthrop Univ of Ill Urbana IL 61801

LATHROP, ARTHUR LAVERN, b Kittitas, Wash, Nov 21, 18; m 46. PHYSICS. Educ: Wash State Univ, BS, 43, Univ Ill, MS, 46; Rice Univ, PhD(physics), 52. Prof Exp: Instr physics, Wash State Univ, 43-44; physicist, Ames Aeronaut Lab, Moffett Field, Calif, 44-45; instr physics, Univ Tulsa, 47-49; res engr, Boeing Airplane Co, Wash, 52-53; res asst physics, Inst Paper Chem, Lawrence, 53-58, res assoc, 58-65; from asst prof to assoc prof, 65-71, PROF PHYSICS, WESTERN ILL UNIV, 71- Concurrent Pos: Res Corp grant. Mem: Am Asn Physics Teachers. Res: Superconductivity; physical properties of paper; radiative transfer theory. Mailing Add: 700 Auburn Dr Macomb IL 61455

LATHROP, EARL WESLEY, b Oakley, Kans, Mar 1, 24; m 49; c 1. SYSTEMATIC BOTANY. Educ: Walla Walla Col, BA, 50, MA, 52; Univ Kans, PhD(bot), 57. Prof Exp: Instr biol, Can Union Col, 52-54; assoc prof bot, La Sierra Col, 57-64; ASSOC PROF BIOL, LOMA LINDA UNIV, 64- Mem: Bot Soc Am; Ecol Soc Am. Res: Floristics. Mailing Add: 11740 Valiant St Riverside CA 92505

LATHROP, KATHERINE AUSTIN, b Lawton, Okla, June 16, 15; m 38; c 5. NUCLEAR MEDICINE. Educ: Okla State Univ, BS, 36 & 39, MS, 39. Prof Exp: Asst, Univ Wyo, 42-44; jr chemist, Biomed Div, Manhattan Proj Metall Lab, 45-47; assoc biochemist, Argonne Nat Lab, 47-54, RES ASSOC, FRANKLIN McLEAN MEM RES INST, 54-, ASSOC PROF RADIOL, UNIV CHICAGO, 67- Concurrent Pos: Mem comt nuclear med, Am Nat Standards Inst; adv comt radiopharmaceut, US Pharmacopeia; from instr to asst prof, Univ Chicago, 54-67; mem med internal radiation dose comt, Soc Nuclear Med, 67- Mem: AAAS; Radiation Res Soc; Tissue Cult Asn; Soc Nuclear Med; Sigma Xi. Res: Development of radionuclides for diagnostic and therapeutic purposes, including their production, purification and incorporation into various chemical forms; biological behavior in laboratory animals and humans. Mailing Add: Univ Chicago & F M Mem Res Hosp 950 E 59th St Box 420 Chicago IL 60637

LATHROP, KAYE DON, b Bryan, Ohio, Oct 8, 32; m 57; c 2. REACTOR PHYSICS. Educ: US Mil Acad, BS, 55; Calif Inst Technol, MS, 59, PhD(mech eng, physics), 62. Prof Exp: Staff mem reactor math, Los Alamos Sci Lab, 62-67; staff mem & group leader reactor physics methods develop, nuclear anal & reactor physics dept, Gen Atomic Div, Gen Dynamics Corp, 67-68; T-1 alt group leader, 68-72, T-1 GROUP LEADER, LOS ALAMOS SCI LAB, 72-, T-DIV ASST DIV LEADER, 73- Concurrent Pos: Vis prof, Univ NMex, 64-65, adj prof, 66-67; mem, US Energy Res & Develop Admin Adv Comt for Reactor Physics, 73- Mem: Fel Am Nuclear Soc; Am Phys Soc. Res: Analytic and numerical solutions of equations of neutron and photon transport. Mailing Add: Los Alamos Sci Lab PO Box 1663 Los Alamos NM 87545

LATHROUM, LEO BADEN, pharmaceutical chemistry, see 12th edition

LATHWELL, DOUGLAS J, b Mich, Mar 28, 22; m 48; c 3. SOIL SCIENCE. Educ: Mich State Univ, BS, 47; Ohio State Univ, PhD(soil sci), 50. Prof Exp: From asst prof to assoc prof, 50-61, PROF SOIL SCI, CORNELL UNIV, 61- Concurrent Pos: Fulbright res scholar, Neth, 64. Mem: Soil Sci Soc Am; Am Soc Agron; Am Soc Plant Physiol; Int Soc Soil Sci. Res: Soil fertility; plant nutrition. Mailing Add: 130 Northview Rd Ithaca NY 14850

LATIES, ALAN M, b Beverly, Mass, Feb 8, 31; m; c 2. OPHTHALMOLOGY. Educ: Harvard Col, AB, 54; Baylor Univ, MD, 59; Am Bd Ophthal, dipl, 65. Prof Exp: Intern, Mt Sinai Hosp, New York, 59-60; resident ophthal, Hosp Univ Pa, 60-63, instr, 63-64, assoc, 64-66, asst prof, 66-70, GIVEN PROF OPHTHAL, MED SCH, UNIV PA, 70- Concurrent Pos: NIH trainee, 61-63, spec fel, 63-64; assoc ophthal, Children's Hosp Philadelphia, 63-; asst attend physician, Philadelphia Gen Hosp, 63-; attend ophthal, Vet Admin Hosp, 63-; mem vision res & training comt, Nat Eye Inst. Honors & Awards: Res to Prevent Blindness Professorship Award, 64; Friedenwald Award for Res in Ophthal, 72. Mem: AMA; Asn Res Vision & Ophthal; Histochem Soc; Am Asn Anat; Am Acad Ophthal & Otolaryngol. Res: Histochemistry; visual pathways. Mailing Add: Scheie Eye Inst Presby-Univ of Pa Med Ctr Philadelphia PA 19104

LATIES, GEORGE GLUSHANOK, b Sevastopol, Russia, Jan 17, 20; US citizen; m 47. PLANT PHYSIOLOGY. Educ: Cornell Univ, BS, 41; Univ Minn, MS, 42; Univ Calif, PhD(plant physiol), 47. Prof Exp: Asst, Div Plant Nutrit, Univ Calif, 42-43, sr asst, 43-47; res fel biol, Calif Inst Technol, 47-50, sr res fel, 50-52 & 55-58; asst prof bot, Univ Mich, 52-55; plant physiologist, Exp Sta & assoc prof hort sci, Univ, 59-63, PROF PLANT PHYSIOL, DEPT BIOL, UNIV CALIF, LOS ANGELES, 63- Concurrent Pos: Rockefeller Found fel, Sheffield & Cambridge Univs, 49-50; res botanist, Univ Mich, 52-55; mem physiol chem study sect, USPHS, 63-65; Guggenheim fel, Commonwealth Sci & Indust Res Orgn, Australia, 66-67. Mem: AAAS; Am Soc Plant Physiol (vpres, 64-65); Bot Soc Am; Scand Soc Plant Physiol. Res: Respiratory regulatory mechanisms and respiratory pathways in plant tissues; permeability and salt transport; biochemical aspects of growth and development. Mailing Add: Dept of Biol Kinsey Hall Univ of Calif Los Angeles CA 90024

LATIES, VICTOR GREGORY, b Racine, Wis, Feb 2, 26; m 56; c 3. PSYCHOPHARMACOLOGY. Educ: Tufts Univ, AB, 49; Univ Rochester, PhD(psychol), 54. Prof Exp: Res assoc, Univ Rochester, 53-54; teaching intern, Brown Univ, 54-55; from instr to asst prof, Sch Med, Johns Hopkins Univ, 55-65; assoc prof radiation biol, biophys, pharmacol & psychol, 65-71, PROF RADIATION BIOL, BIOPHYS, PHARMACOL & PSYCHOL, UNIV ROCHESTER, 71- Concurrent Pos: Ed, J Exp Anal Behav, 73-77; mem, Nat Acad Sci-Nat Res Coun Panel on Carbon Monoxide, 73- Mem: Am Psychol Asn; Am Soc Pharmacol & Exp Therapeut; Behav Pharmacol Soc; Soc Neurosci; Psychonomic Soc. Res: Behavioral pharmacology; experimental analysis of behavior. Mailing Add: Dept of Radiation Biol & Biophys Univ of Rochester Sch Med & Dent Rochester NY 14642

LATIMER, CLINTON NARATH, b New York, NY, Aug 30, 24; m 56; c 3. NEUROPHYSIOLOGY. Educ: Columbia Univ, AB, 48; Syracuse Univ, PhD(physiol), 56. Prof Exp: GROUP LEADER NEUROPHARMACOL, LEDERLE LABS DIV, AM CYANAMID CO, 58- Concurrent Pos: Nat Inst Neurol Dis & Blindness res fel neurophysiol, Univ Wash, 56-58. Mem: AAAS; Am Soc Pharmacol & Exp Therapeut; Soc Neurosci; Am Physiol Soc. Res: Neuropharmacology; function of the central nervous system as delineated by extra and intracellular recordings of neuronal activity under influence of drugs and in control states; psychopharmacology; computer applications. Mailing Add: Lederle Labs Div Am Cyanamid Co Pearl River NY 10965

LATIMER, DONALD ANDREW, b Timaru, NZ, Jan 26, 21; Can citizen; m 44; c 3. PHYSICAL ORGANIC CHEMISTRY, BIOCHEMISTRY. Educ: Mt Allison Univ, BSc, 53; Univ Western Ont, PhD(chem), 59. Prof Exp: Chemist, Ruakura Animal Res Sta, NZ, 38-41; res asst, NS Res Found Can, 50-53; res asst, Can Dept Agr, 54-59; sr res scientist, J Labatt Ltd, 59-67; V PRES RES, J E SIEBEL SONS' CO, 67- Concurrent Pos: Demonstr, Mt Allison Univ, 52-53; mem teaching staff, J E Siebel Inst Technol. Mem: Inst Food Technologists; Soc Indust Microbiol; Am Soc Brewing Chem; Master Brewers Asn; Am Chem Soc. Res: Analytical chemistry and instrumentation; pesticides; biochemistry of brewing; cereal grains; industrial microbiology and enzymology. Mailing Add: J E Siebel Sons' Co 4055 W Peterson Ave Chicago IL 60646

LATIMER, GEORGE WEBSTER, JR, analytical chemistry, see 12th edition

LATIMER, HOWARD LEROY, b Seattle, Wash, July 18, 29; m 57; c 4. PLANT GENETICS, PLANT ECOLOGY. Educ: Wash State Univ, BS, 51, MS, 55; Claremont Cols, PhD(bot), 59. Prof Exp: From asst prof to assoc prof, 58-68, PROF BIOL, CALIF STATE UNIV, FRESNO, 68- Mem: Soc Study Evolution; Ecol Soc Am. Res: Gene ecology of plant populations and the impact of man on natural ecosystems. Mailing Add: Dept of Biol Calif State Univ Fresno CA 93740

LATIMER, PAUL HENRY, b New Orleans, La, Nov 25, 25; m 52; c 3. BIOPHYSICS. Educ: Univ Ill, MS, 50, PhD(biophys), 56. Prof Exp: Instr physics, Col William & Mary, 50-51; asst bot, Univ Ill, 53-56; res fel plant biol, Carnegie Inst Technol, 56-57; asst prof physics, Vanderbilt Univ, 57-62; assoc prof, 62-71, PROF PHYSICS, AUBURN UNIV, 71- Concurrent Pos: Investr, Howard Hughes Med Inst, 57-62. Mem: Biophys Soc; Am Phys Soc; Optical Soc Am. Res: Light scattering; biological optics; fluorescence; photosynthesis. Mailing Add: Dept of Physics Auburn Univ Auburn AL 36830

LATIMER, STEVE B, b Oklahoma City, Okla, Nov 4, 27; m 52; c 3. BIOCHEMISTRY, ORGANIC CHEMISTRY. Educ: Tuskegee Inst, BS, 53, MS, 55; NC State Univ, PhD(biochem), 67; Fed Exec Inst, cert, 73. Prof Exp: Asst prof chem, Shaw Univ, 55-62; res asst biochem, NC State Univ, 62-66; prof chem, Langston Univ, 66-75, chmn dept phys sci, 70-72, dir coop res, 72-75, res coordr, Coop State Res Serv, 71-75; NIH MARC fac fel, 75-77, MEM MARC FAC, OKLA MED RES FOUND, 75- Mem: Am Chem Soc; Nat Inst Sci; Am Inst Biol Scientists; fel Am Inst Chemists. Res: Isolation, purification and characterization of short-chain fatty acid enzymes; medical research; biomembranes; carcinogenesis. Mailing Add: Box 780 Langston OK 73050

LATORELLA, A HENRY, b Winthrop, Mass, Mar 12, 40; m 64; c 2. GENETICS, ALGOLOGY. Educ: Boston Col, BS, 61, MS, 64; Univ Maine, Orono, PhD(zool, genetics), 71. Prof Exp: Asst prof biol, Salem State Col, 66-68; fel, 71-72, ASST PROF BIOL, STATE UNIV NY COL GENESEO, 70-, RES FOUND GRANT, 72- Mem: AAAS; Genetics Soc Am. Res: Algal genetics and physiology; genetics and biochemistry of salinity adaptation by phytoflagellates. Mailing Add: Dept of Biol State Univ NY Col Geneseo Geneseo NY 14454

LATORRE, DONALD RUTLEDGE, b Charleston, SC, May 4, 38; m 60; c 2. MATHEMATICS. Educ: Wofford Col, BS, 60; Univ Tenn, MA, 62, PhD(math), 64. Prof Exp: Asst prof math, Univ Tenn, 67; ASST PROF MATH, CLEMSON UNIV,

67- Mem: Am Math Soc; Math Asn Am. Res: Abstract algebra, especially semigroups. Mailing Add: Dept of Math Sci Clemson Univ Clemson SC 29631

LATOUR, ALBERT ANDRE, organic chemistry, textiles, see 12th edition

LATOUR, ROGER, b Gravelbourg, Sask, Mar 20, 30; m 58; c 2. PHYSIOLOGY. Educ: Univ Montreal, BPharm, 54; Univ Paris, DPharm, 57. Prof Exp: Asst prof physiol, Fac Med, Univ Montreal, 57-67; clin coordr, Med Div, 67-70, DIR SCI AFFAIRS, SYNTEX LTD, 70- Mem: NY Acad Sci; Pharmacol Soc Can; Can Soc Chemother; Can Asn Res Toxicol; Int Pharmaceut Fedn. Res: Neurophysiology; clinical research in contraceptive field and with anti-inflammatory agents. Mailing Add: Syntex Ltd 8255 Mountain Sights Ave Montreal PQ Can

LATOURETTE, HAROLD KENNETH, b Seattle, Wash, Apr 10, 24; m 44; c 5. ORGANIC CHEMISTRY. Educ: Whitman Col, AB, 47; Univ Wash, PhD(chem), 51. Prof Exp: Res assoc, Univ Wash, 48-51; res chemist, Westvaco Chem Div, FMC Corp, WVa, 51-54, dir pioneering res, Westvaco Chlor-Alkali Div, 54-56, supvr org res, Cent Res Labs, NJ, 56-57, mgr org res & develop, Becco Chem Div, NY, 57-58, mgr org chem res, Inorg Chem Div, 58-62, Europ tech dir, Chem Div & vpres, FMC Chem, SA, Switz, 62-65, mgr planning & eval, Cent Res Dept, 65-72, MGR EVAL, CHEM GROUP, FMC CORP, 72- Mem: AAAS; Sigma Xi; Am Chem Soc. Res: Organic reaction mechanisms; aromatic substitution; peroxides; epoxides; isocyanates; phosphorus and sulfur organics; halogenation; industrial processes; polymerization. Mailing Add: FMC Corp PO Box 8 Princeton NJ 08540

LATOURETTE, HOWARD BENNETT, b Detroit, Mich, Aug 26, 18; m 42; c 4. RADIOLOGY. Educ: Oberlin Col, AB, 40; Univ Mich, MD, 43. Prof Exp: From instr to assoc prof radiol, Univ Mich, 49-59; PROF RADIOL, COL MED, UNIV IOWA, 59- Concurrent Pos: Mem staff radiation ther sect, Univ Hosps, Univ Iowa. Mem: AAAS; Radiol Soc NAm. Res: Clinical use of radiation in treatment of cancer and resulting survival studies; tumor registry organization and function alteration of radiosensitivity of tumors. Mailing Add: C142 Radiation Ther Sect Univ of Iowa Hosps Iowa City IA 52240

LA TOURRETTE, JAMES THOMAS, b Miami, Ariz, Dec 26, 31; m 55; c 4. PHYSICS. Educ: Calif Inst Technol, BS, 53; Harvard Univ, MA, 54, PhD(physics), 58. Prof Exp: Res fel physics, Harvard Univ, 57-58, lectr, 58-59; NSF fel, Univ Bonn, 59-60; physicist, Gen Elec Res Lab, 60-62; sr supvry scientist, TRG, Inc Div, Control Data Corp, 62-66, sect head gas laser, 66-67; PROF ELECTROPHYS, POLYTECH INST NEW YORK, 67- Mem: AAAS; Am Phys Soc; Inst Elec & Electronics Engrs. Res: Gas laser research and applications; saturated resonance spectroscopy and laser frequency stabilization; solar power; production and distribution of synthetic fuels; energy conservation techniques. Mailing Add: Polytech Inst of New York Route 110 Farmingdale NY 11735

LATSCHAR, CARL ERNEST, b Newton, Kans, May 24, 19; m 41; c 5. PHYSICAL CHEMISTRY. Educ: Kans State Univ, BS, 41, MS, 47; Univ Wis, PhD(phys chem), 50. Prof Exp: Res chemist, 50-62, SR RES CHEMIST, E I DU PONT DE NEMOURS & CO, INC, 62- Mem: Am Chem Soc. Res: Textile fiber process and product development. Mailing Add: 2905 Ginger Rd Kinston NC 28501

LATSHAW, DAVID RODNEY, b Allentown, Pa, Nov 4, 39; m 71. ANALYTICAL CHEMISTRY. Educ: Muhlenberg Col, BS, 61; Lehigh Univ, MS, 63, PhD(chem), 66. Prof Exp: Res asst, Lehigh Univ, 63-66; res chemist, 66-69, GROUP LEADER, AIR PROD & CHEM, INC, 69- Mem: Am Chem Soc; Sigma Xi. Res: Gas chromatography; infrared spectroscopy; atomic absorption. Mailing Add: Air Prod & Chem Inc PO Box 538 Allentown PA 18105

LATTA, BRUCE MCKEE, b Oil City, Pa, Dec 31, 40; m 68; c 1. PHYSICAL ORGANIC CHEMISTRY. Educ: Muskingum Col, BS, 62; Univ Calif, Irvine, PhD(phys org chem), 67. Prof Exp: Res chemist, Film Dept, E I du Pont de Nemours & Co, Inc, 67-70; res assoc textile finishing, 70-74, MGR EXPLORATORY RES, J P STEVENS & CO, INC, GARFIELD, 74- Mem: AAAS; Am Chem Soc; Sigma Xi; Am Inst Chemists. Res: Molecular structure-property relationships; morphological modifications of polymer films and fibers and effects on properties; chemical modification of textile fiber surfaces; textile finishing and dyeing. Mailing Add: 75 Seneca Trail Wayne NJ 07470

LATTA, GORDON, b Vancouver, BC, Mar 8, 23; m 45; c 3. MATHEMATICS. Educ: Calif Inst Technol, PhD(math), 51. Prof Exp: Lectr math, Univ BC, 47-48; asst, Calif Inst Technol, 48-51; Fine instr math & hydrodyn, Princeton Univ, 51-52; asst prof, Univ BC, 52-53; from asst prof to prof math, Stanford Univ, 53-67; PROF MATH, UNIV VA, 67- Res: Singular perturbation problems for differential equations; linear integral equations and Wiener-Hopf techniques. Mailing Add: Dept of Appl Math & Comput Sci Univ of Va Charlottesville VA 22901

LATTA, HARRISON, b Los Angeles, Calif, Apr 5, 18; m 41; c 4. PATHOLOGY, BIOPHYSICS. Educ: Univ Calif, Los Angeles, AB, 40; Johns Hopkins Univ, MD, 43. Prof Exp: Intern, Church Home & Hosp, Md, 44; from asst resident to resident path, Johns Hopkins Hosp, 44-46, instr, Johns Hopkins Univ, 45-46; res assoc biol, Mass Inst Technol, 49-51; asst prof path, Sch Med, Case Western Reserve Univ, 51-54; assoc prof, 54-60, PROF PATH, SCH MED, UNIV CALIF, LOS ANGELES, 60- Concurrent Pos: Res fel, Children's Hosp, Boston & Harvard Med Sch, 48-49. Mem: AAAS; Electron Micros Soc Am; Am Soc Cell Biol; Am Soc Exp Path; Am Asn Path & Bact. Res: Foreign protein behavior and hypersensitivity; mechanisms of hemolysis and of cytologic injury; tissue culture and electron microscopy; ultrastructure and diseases of the kidney. Mailing Add: Dept of Path Ctr for Health Sci Univ of Calif Sch of Med Los Angeles CA 90024

LATTA, JOHN NEAL, b Ottumwa, Iowa, Apr 11, 44; m 66; c 1. OPTICS, ELECTRICAL ENGINEERING. Educ: Brigham Young Univ, BES, 66; Univ Kans, MS, 69, PhD(elec eng), 71. Prof Exp: Mem tech staff holography, RCA Labs, NJ, 67; res asst optics, Ctr Res Eng Sci, Univ Kans, 67-68; mem tech staff holography, Bell Tel Labs, NJ, 69; res engr, Radar & Optics Lab, Univ Mich, Ann Arbor, 69-73; SR RES ENGR, ENVIRON RES INST MICH, 73- Mem: Inst Elec & Electronics Engrs; Optical Soc Am; Brit Comput Soc; Soc Info Display; Sigma Xi. Res: Holography; applications of computers to engineering analysis and design; optical design; digital image processing; visual displays; computer graphics. Mailing Add: Environ Res Lab of Mich PO Box 618 Ann Arbor MI 48107

LATTA, WILLIAM CARL, b Niagara Falls, NY, May 18, 25; m 50; c 5. FISH BIOLOGY. Educ: Cornell Univ, BS, 50; Univ Okla, MS, 52; Univ Mich, PhD(fishery biol), 57. Prof Exp: Aquatic biologist, State Conserv Dept, NY, 50; FISHERY BIOLOGIST, INST FISHERIES RES, STATE DEPT NATURAL RESOURCES, MICH, 55-, IN-CHG INST FISHERIES RES, 66- Concurrent Pos: Adj prof fisheries & wildlife, Sch Natural Resources, Univ Mich, Ann Arbor, 73-74. Mem: Am Fisheries Soc; Am Soc Ichthyol & Herpet; Wildlife Soc; Ecol Soc Am; Am Inst Fishery Res Biologists. Res: Fish population dynamics; management of freshwater fisheries. Mailing Add: Inst for Fisheries Res Univ Mus Annex Ann Arbor MI 48104

LATTER, ALBERT L, b Kokomo, Ind, Oct 17, 20; m 49; c 2. THEORETICAL PHYSICS. Educ: Univ Calif, Los Angeles, BA, 41, PhD(physics), 51. Prof Exp: Staff physicist radiation lab, Univ Calif, 41-46; staff physicist, Rand Corp, 52-60, head physics dept, 60-71; PRES, R&D ASSOCS, 71- Concurrent Pos: Mem Air Force Sci Adv Bd, 56-72, chmn nuclear panel, 63-67; mem US sci deleg to test ban negotiations, Geneva, 59; mem adv group, Ballistic Systs Div, 61-67; mem sci adv group on effects, Defense Nuclear Agency, 64-; mem adv group, Space & Missile Systs Orgn, 66-; mem Defense Sci Bd, 67-70, chmn penetration panel, 66-68, mem vulnerability task force, 66-, chmn strategic task force, 69-72, mem net assessment task force, 71-72; mem sci adv group, Joint Strategic Target Planning Staff, 68-72. Res: Nuclear physics; quantum mechanics; nuclear weapons design and effects; strategic weapons systems. Mailing Add: R&D Assocs PO Box 9695 Marina del Rey CA 90291

LATTER, RICHARD, b Chicago, Ill, Feb 29, 23; m 62; c 4. THEORETICAL PHYSICS. Educ: Calif Inst Technol, BS, 42, PhD(theoret physics), 49. Prof Exp: Physicist, Rand Corp, 49-56, chief physics div, 56-60, mem res coun, 60-71; VPRES, R&D ASSOCS, 71- Concurrent Pos: Lectr, Calif Inst Technol, 55-57; adv, AEC, 58-71 & Dept Defense, 58- Mem: Am Phys Soc; Am Math Soc. Res: Solid state, atomic, and nuclear physics; hydrodynamics. Mailing Add: 1815 N Fort Meyer Dr Suite 1100 Arlington VA 22209

LATTERELL, FRANCES MEEHAN, b Kansas City, Mo, Dec 21, 20; m 51. PLANT PATHOLOGY. Educ: Univ Kansas City, BA, 42; Iowa State Col, MS, 47, PhD, 50. Prof Exp: Seed analyst, Iowa State Univ, 42-44; res assoc cereal path, State Agr Exp Sta, Iowa, 45-49; res plant pathologist, US Dept Army, 49-71; PLANT PATHOLOGIST, PLANT DIS RES LAB, AGR RES SERV, USDA, 71- Mem: Am Phytopath Soc; Am Inst Biol Sci; Mycol Soc Am. Res: Cereal crop pathology; rice blast; Helminthosporium diseases. Mailing Add: Plant Dis Res Lab ARS-USDA Box 1209 Frederick MD 21701

LATTERELL, JOSEPH J, b St Cloud, Minn, Nov 2, 32; m 64; c 4. ANALYTICAL CHEMISTRY. Educ: St John's Univ, Minn, BA, 59; Purdue Univ, MS, 62; Univ Colo, PhD(anal chem), 64. Prof Exp: Instr anal chem, Univ Wis, 64; asst prof, John Carroll Univ, 64-67; asst prof, 67-71, ASSOC PROF ANAL CHEM, UNIV MINN, MORRIS, 71- Mem: Am Chem Soc. Res: Separation of amines by ligand exchange; chemical analyses of the bottom deposits of Minnesota lakes; removal of dissolved phosphate from lake water by bottom deposits; land application of sewage sludge. Mailing Add: Dept of Chem Univ of Minn Morris MN 56267

LATTERELL, RICHARD L, b Paynesville, Minn, Mar 14, 28; m 51. GENETICS. Educ: Univ Minn, Duluth, BA, 50; Pa State Univ, MS, 55; Cornell Univ, PhD(genetics), 58. Prof Exp: Nat Cancer Inst fel genetics, Brookhaven Nat Lab, 58-60; geneticist, Div Radiation & Organisms, Smithsonian Inst, 60-63; geneticist, Union Carbide Res Inst, 63-68; ASSOC PROF BIOL SHEPHERD COL, 68- Mem: Genetics Soc Am; Bot Soc Am. Res: Radiation genetics of maize; space biology; stress tolerance of higher plants; plant cytogenetics; ecology; evolution. Mailing Add: Dept of Biol Shepherd Col Shepherdstown WV 25443

LATTES, RAFFAELE, b Italy, May 22, 10; nat US; m 36; c 2. MEDICINE. Educ: Univ Turin, MD, 33; Columbia Univ, MedSciD, 46. Prof Exp: Asst gen surg, Univ Turin, 34-38; instr path, Woman's Med Col Pa, 41-43; instr surg & surg path, 43-46, asst prof path, Postgrad Hosp, 46-48, asst prof, 48-51, PROF SURG PATH, COL PHYSICIANS & SURGEONS, COLUMBIA UNIV, 51- Concurrent Pos: Consult, US First Army, Roswell Park Mem Inst, Knickerbocker Hosp, NY, Roosevelt Hosp, St Lukes Hosp, Hosp Joint Dis & Vet Admin Hosp. Mem: Harvey Soc; Am Asn Path & Bact; Am Asn Cancer Res; Col Am Path; Asn Am Med Cols. Res: Surgical pathology. Mailing Add: Col of Physicians & Surgeons Columbia Univ New York NY 10032

LATTIMER, JOHN KINGSLEY, b Mt Clemens, Mich, Oct 14, 14; m 49; c 3. UROLOGY. Educ: Columbia Col, BA, 35; Columbia Univ, MD, 38, ScD(med), 43; Am Bd Urol, dipl, 47. Prof Exp: Asst urol, 40-46, instr, 48-50, assoc, 50-52, asst prof clin urol, 52-55, PROF UROL & CHMN DEPT, COL PHYSICIANS & SURGEONS, COLUMBIA UNIV, 55-, INSTR, SCH NURSING, COLUMBIA-PRESBY HOSP, 46- Concurrent Pos: Urologist, Presby Hosp & Babies Hosp, 40-; attend urologist & sr consult urol in charge res unit genito-urinary tuberc, Vet Admin Hosp, Kingsbridge, 46-; dir, Squier Urol Clin, Presby Hosp, New York, 55-, trustee hosp, 74-; dir urol serv, Delafield Hosp, 55-; presidential appointee expert adv panel human reprod, WHO, 68-; mem, Coun Med TV Inc, 68-; rep, Study of Surg Serv US, 71; vis prof, Kansas City Gen Hosp & Med Ctr, 72; mem coord comt, Int Cancer Res, 72-73; urol liaison rep, Nat Res Coun. Honors & Awards: Mather Smith Prize, 43; Award of Hon, Am Acad Tuberc Physicians; Belfield Medal, Chicago Urol Asn, 69; William P Burpeau Award, Acad Med NJ, 71; Gold Medal Award, Columbia Med Sch Alumni, 71; Award, Buffalo Urol Soc, 72. Mem: Am Col Surg; Int Soc Urol (pres, 73-76); Soc Pediat Urol (pres); Soc Univ Urol (pres, 68-69); Am Urol Asn (pres, 75-76). Res: Treatment of kidney tuberculosis and kidney function; treatment of urological disorders in children; cancer of prostate gland. Mailing Add: Col of Physicians & Surgeons Columbia Univ New York NY 10032

LATTIN, DANNY LEE, b Smith Center, Kans, Jan 9, 42; m 61; c 2. MEDICINAL CHEMISTRY. Educ: Univ Kans, BS, 65; Univ Minn, Minneapolis, PhD(med chem), 70. Prof Exp: Asst prof med chem, 70-75, ASSOC PROF MED CHEM, COL PHARM, MED SCI, UNIV ARK, LITTLE ROCK, 75- Mem: Am Chem Soc. Res: Synthesis of new organic medicinal agents; study of stereochemical properties of drug receptors. Mailing Add: Dept of Med Chem Col of Pharm Univ of Ark Med Sci Little Rock AR 72201

LATTIN, JOHN D, b Chicago, Ill, July 27, 27; m 53; c 3. ENTOMOLOGY. Educ: Iowa State Univ, BS, 50; Univ Kans, MA, 51; Univ Calif, Berkeley, PhD(entom), 64. Prof Exp: Aquatic entomologist, Dept Limnol, Acad Natural Hist, Philadelphia, 51; jr vector control specialist, Bur Vector Control, Calif Dept Pub Health, 54-55; asst entomologist, Agr Exp Sta, 55-61, from instr to assoc prof, 55-68, PROF ENTOM, ORE STATE UNIV, 68-, ASST DEAN SCI, 67- Concurrent Pos: Cur entom, Ore State Mus, 61-66, dir sci honors prog, 63-65, dir univ hons prog, 66-67; NSF fac fel, Univ Wageningen, 65-66. Honors & Awards: Lyod Carter Award, Ore State Univ, 61. Mem: Entom Soc Am; Soc Syst Zool. Res: Systematics of the Pentatomoidea and Leptopododoidea; origin, distribution and phylogeny of the Heteroptera; evolution and zoogeography of the insecta; aquatic entomology; scientific education; education of talented students; impact of science on society. Mailing Add: Dept of Entom Ore State Univ Corvallis OR 97331

LATTMAN, EATON EDWARD, b Chicago, Ill, May 15, 40; m 66. MOLECULAR BIOPHYSICS. Educ: Harvard Col, BA, 62; Johns Hopkins Univ, 69. Prof Exp: NIH fel biophysics, Johns Hopkins Univ, 69-70, res scientist, 70-73; fel, Max Planck Inst

Biochem, 74; NIH FEL BIOPHYS, BRANDEIS UNIV, 74- Mem: Am Phys Soc; Am Crystallog Asn. Res: Determination of the three-dimensional structure of biological macromolecules by x-ray diffraction and/or electron microscopy; globins, muscle proteins and viruses. Mailing Add: Rosenstiel Ctr Brandeis Univ Waltham MA 02154

LATTMAN, LAURENCE HAROLD, b New York, NY, Nov 30, 23; m 46; c 2. GEOLOGY. Educ: City Col New York, BChE, 48; Univ Cincinnati, MS, 51, PhD(geol), 53. Prof Exp: Instr geol, Univ Cincinnati, 51 & Univ Mich, 52-53; photogeologist, Gulf Oil Corp, 53-56, asst head photogeol sect, 56-57; from asst prof to prof geomorphol, Pa State Univ, 57-70; prof geol & head dept, Univ Cincinnati, 70-75; DEAN COL MINES & MINERAL INDUST, UNIV UTAH, 75- Concurrent Pos: Mem, Nat Res Coun, 59-62; Fulbright lectr, Moscow State Univ, 75. Mem: Fel Geol Soc Am; Am Soc Photogram; Am Asn Petrol Geologists. Res: Remote sensing of environment; geomorphology; fracture analysis on aerial photographs. Mailing Add: 209 Mineral Sci Univ of Utah Salt Lake City UT 84112

LATTUADA, CHARLES P, b Danville, Ill, May 8, 33; m 55; c 4. BACTERIOLOGY. Educ: Ind State Univ, BS, 56; Univ Wis, MS, 58, PhD(bact), 63. Prof Exp: Instr bact, Univ Wis, 58-59; consult microbiol, Res Prod Corp, Wis, 59-61; bacteriologist, Wis Alumni Res Found, 61-64; consult, A R Schmidt Co, Inc, 62-64, dir res, 64-69; dir germ-free res, ARS/Sprague-Dawley Div, 69-74, gen mgr microbiol lab, 68-74, VPRES, GIBCO DIAG DIV, MOGUL CORP, 74- Concurrent Pos: Vpres & mgr, A R S Serum Co, Inc, 62-68; mem prog adv comt & lectr, Madison Area Tech Col. Mem: Am Soc Microbiol; Animal Care Panel; Asn Gnotobiotics; NY Acad Sci. Res: Dairy and food industries; botulinum food poisoning; derivation of new strains of germfree animals and their maintenance and associated technology; intestinal flora of laboratory animals; bacterial culture media. Mailing Add: 1512 Simpson St Madison WI 53713

LATZ, ARJE, b Lithuania, July 1, 27; US citizen; m 60; c 1. PSYCHOPHARMACOLOGY. Educ: Univ Maine, BA, 58; Boston Univ, MA, 60, PhD(psychol), 63. Prof Exp: Res assoc, 63-67, asst prof, 67-71, ASSOC PROF PSYCHOPHARMACOL, SCH MED, BOSTON UNIV, 71- Concurrent Pos: NIH fel, 63-65; Boston Med Found fel, 65-67; consult ed, Psychopharmacologia, 69- Mem: Am Psychol Asn. Res: Effects of amphetamine on behavior differentially maintained by excitatory or inhibitory processes; the effects of morphine and morphine antagonists on the cortical EEG of free moving rats. Mailing Add: Dept of Psychiat Boston Univ Sch of Med Boston MA 02118

LATZ, HOWARD W, b Rochester, NY, Jan 9, 33; m 53; c 3. ANALYTICAL CHEMISTRY. Educ: Rochester Inst Technol, BS, 59; Univ Fla, MS, 61, PhD(chem), 63. Prof Exp: Res specialist, Union Carbide Corp, 63-66; assoc prof, 66-74, PROF ANAL CHEM, OHIO UNIV, 74- Mem: Am Chem Soc. Res: Luminescence methods of analysis; electrophoresis of organic ions; analytical applications of dye lasers. Mailing Add: Dept of Chem Ohio Univ Athens OH 45701

LAU, BRIAN RICHARD, b Waukegan, Ill, July 2, 44; m 68. MATHEMATICS. Educ: Northwestern Univ, Ill, BS, 65, MS, 67, PhD(math), 70. Prof Exp: ASST PROF MATH, UNIV UTAH, 71- Mem: Am Math Soc; Soc Indust & Appl Math; Math Asn Am. Res: Linear water waves and Wiener-Hapf equations. Mailing Add: Dept of Math Univ of Utah Salt Lake City UT 84112

LAU, FRANCIS YOU KING, b Honolulu, Hawaii, Jan 5, 24; m 48; c 4. MEDICINE. Educ: Loma Linda Univ, MD, 47. Prof Exp: From clin instr to clin asst prof med, Loma Linda Univ, 54-59; asst prof, Sch Med, Univ Calif, San Francisco, 59-60; assoc prof, 64-70, PROF MED & RADIOL, SCH MED, UNIV SOUTHERN CALIF, 70-; PROF MED, LOMA LINDA UNIV, 72- Concurrent Pos: Assoc prof med, Loma Linda Univ, 60-72; mem attend staff & chief adult cardiovasc catheterization lab, Los Angeles County-Univ South Calif Med Ctr, 65-, chief cardiol, 70-; consult, Vet Admin & Glendale Hosps; fel coun clin cardiol, Am Heart Asn. Mem: Fel Am Col Cardiol; fel Am Col Physicians; AMA; Am Fedn Clin Res. Res: Cardiology; cardiac catheterization, arrhythmias and pacemakers; artificial heart-lung preparations. Mailing Add: 1200 NState St Los Angeles CA 90033

LAU, KENNETH KWOK-KWAN, b Hong Kong, Aug 22, 45; m 71; c 1. MATHEMATICS. Educ: Chinese Univ Hong Kong, BSc, 68; Univ Miami, PhD(math), 73. Prof Exp: ASST PROF MATH, CENTRAL COL, 73- Mem: Am Math Soc; Sigma Xi; Math Asn Am. Res: Topological dynamics; number theory. Mailing Add: Dept of Math Central Col Pella IA 50219

LAU, KENNETH W, b Lamoure, NDak, Nov 27, 41; m 69; c 1. CHEMISTRY. Educ: Univ NDak, BS, 63, PhD(chem), 69; Univ Sask, Regina, MSc, 68. Prof Exp: Chemist, Rock Island Arsenal Lab, US Army, 63-64; res chemist, 69-73, DEVELOP SUPVR, FILM DEPT, E I DU PONT DE NEMOURS & CO, INC, 73- Mem: Am Chem Soc. Res: Ylid chemistry; new packaging films and packaging methods. Mailing Add: Film Dept E I du Pont de Nemours & Co Inc Richmond VA 23261

LAU, NORMAN EUGENE, b Harvey, Ill, July 8, 30; wid; c 1. ENVIRONMENTAL MANAGEMENT. Educ: Colo State Univ, BS, 53, MS, 55; Rutgers Univ, PhD(entom), 58. Prof Exp: Entomologist, NJ Dept Agr, 58-60, Ratner Pest Control, 61-63 & Hooker Chem Corp, 63-65; ASSOC PROF ENTOM, VA POLYTECH INST & STATE UNIV & EXTEN COORDR CHEM, DRUGS & PESTICIDES, COOP EXTEN SERV, 65- Mem: Entom Soc Am; Sigma Xi. Res: Applied science and administration in fields of entomology, plant pathology and botany. Mailing Add: 17 Laurel Dr Blacksburg VA 24060

LAU, PHILIP T S, b Kuala Lumpur, Malaysia, Feb 13, 35; US citizen; m 59; c 3. ORGANIC CHEMISTRY. Educ: Alfred Univ, BS, 58; Syracuse Univ, PhD(org chem), 62. Prof Exp: Fel & res assoc chem, Univ Calif, Berkeley, 62-63; sr res chemist, 63-68, RES ASSOC, COLOR PHOTOG DIV, RES LABS, EASTMAN KODAK CO, 68- Mem: Am Chem Soc. Res: New synthetic routes to novel heterocyclic compounds; photographic developers, couplers, stabilizers and dyes. Mailing Add: Color Photog Div Eastman Kodak Res Labs Rochester NY 14650

LAU, RICHARD LEWIS, b Kansas City, Mo, Sept 30, 41. Educ: Yale Univ, BA, 63, PhD(math), 67. Prof Exp: Fel math, Calif Inst Technol, 67-69; MATHEMATICIAN, OFF NAVAL RES, US DEPT NAVY, 69- Mem: Am Math Soc; Soc Indust & Appl Math. Res: Partial differential equations and numerical analysis. Mailing Add: Off of Naval Res Dept of Navy 1030 E Green St Pasadena CA 91106

LAU, SHIU CHI, analytical chemistry, see 12th edition

LAUBACH, GERALD DAVID, b Bethlehem, Pa, Jan 21, 26; m 53; c 3. ORGANIC CHEMISTRY. Educ: Univ Pa, AB, 47; Mass Inst Technol, PhD(chem), 50. Prof Exp: Res chemist, Chas Pfizer & Co, Inc, 50-54, res supvr, 54-59, mgr med prod sect, 59-61, dir med chem res, 61-63, group dir med prod res, 63-64, vpres med prod res & develop, 64-68, vpres & dir, 68-69, pres, Pfizer Pharmaceut, 69-71, exec vpres, 71-72, PRES, PFIZER INC, 72- Mem: AAAS; Am Chem Soc; NY Acad Sci. Res: Medicinal chemistry; steroids. Mailing Add: Pfizer Inc 235 E 42nd St New York NY 10017

LAUBENGAYER, ALBERT WASHINGTON, b Saline Co, Kans, Feb 22, 99; m 30; c 2. INORGANIC CHEMISTRY. Educ: Cornell Univ, BChem, 21, PhD(inorg chem), 26. Prof Exp: Asst, Cornell Univ, 20-21; instr chem, Ore State Col, 21-23; asst, Cornell Univ, 23-26; Heckscher fel, 27, lectr inorg chem, 27-28, from asst prof to prof, 28-67, EMER PROF INORG CHEM, CORNELL UNIV, 67- Concurrent Pos: Consult, Standard Oil Co (Ohio), 46 & Olin Mathieson Chem Corp, 52. Mem: Am Chem Soc. Res: Chemistry of boron, aluminum, gallium, indium, silicon, germanium, fluorine, molecular addition compounds and donor-acceptor bonding; hydrous oxide systems; solid-solid reactions; organo-metallic compounds; inorganic polymers. Mailing Add: Dept of Chem Cornell Univ Ithaca NY 14850

LAUBER, JEAN KAUTZ, b Seattle, Wash, Aug 30, 26; m 56; c 3. ZOOLOGY, PHYSIOLOGY. Educ: Whitman Col, BA, 48; Washington Univ, MA, 51; Univ NMex, PhD(zool), 59. Prof Exp: Sr lab technician electron micros, Sch Med, Univ Wash, 51-53, res assoc, 53-56; instr zool, Univ Idaho, 58-60, res fel, 60-62; asst prof, Wash State Univ, 62-65; assoc prof, 65-70, ASSOC PROF ZOOL & ASSOC CHMN DEPT, UNIV ALTA, 70-, HON ASSOC PROF OPHTHAL, 75- Mem: AAAS; Am Soc Zool; Am Physiol Soc; Soc Exp Biol & Med; Electron Micros Soc Am. Res: Endocrine physiology; photobiology; ophthalmology; electron microscopy; embryology. Mailing Add: Dept of Zool Univ of Alta Edmonton AB Can

LAUBSCHER, ANER NEARHOOD, b Mediapolis, Iowa, Mar 29, 16; m 41; c 2. ORGANIC CHEMISTRY. Educ: Iowa Wesleyan Col, BS, 36; Ind Univ, PhD(org chem), 42. Prof Exp: Chemist, Am Can Co, 42-48; chemist, 48-60, mgr chem metall, Res Lab, 60-75, SR RES CONSULT, TIN MILL PROD, RES LAB, US STEEL CORP, 75- Mem: Am Chem Soc; Inst Food Technol. Res: Physical chemistry of surfaces in relation to enamel coating performance; corrosion of steel products. Mailing Add: Res Lab US Steel Corp Monroeville PA 15146

LAU-CAM, CESAR A, b Lima, Peru, Nov 24, 40; m 67; c 2. PHARMACOGNOSY, PHYTOCHEMISTRY. Educ: San Marcos Univ, Lima, BS, 63; Univ RI, MS, 66, PhD(pharmacog), 69. Prof Exp: Instr pharm bot, San Marcos Univ, Lima, 62-63; lab instr pharmacog, Univ RI, 67; asst prof phytochem, 69-73, ASSOC PROF PHARMACEUT SCI, COL PHARM & ALLIED HEALTH PROFESSIONS, ST JOHN'S UNIV, NY, 73- Mem: AAAS; Am Soc Pharmacog; Am Pharmaceut Asn; Phytochem Soc NAm; Phytochem Soc London. Res: Alkaloids; terpenes; phytosterols; phenolic compounds; chemotaxonomy; analytical methods applied to natural products; phytochemical screenings for pharmacologically active compounds. Mailing Add: Dept Pharmaceut Sci St John's Univ Col Pharm & Allied Health Prof Jamaica NY 11439

LAUCIUS, JOSEPH FRANCIS, organic chemistry, see 12th edition

LAUCK, ALBERT JOHN, organic chemistry, see 12th edition

LAUCK, DAVID R, b Alton, Ill, June 6, 30; m 53; c 2. ENTOMOLOGY. Educ: Univ Ill, BS, 55, MS, 58, PhD(entom), 61. Prof Exp: Cur invert entom, Chicago Acad Sci, 59-61; from asst prof to assoc prof, 61-70, chmn div biol sci, 66-72, PROF ENTOM, HUMBOLDT STATE UNIV, 70- Mem: Entom Soc Am; Entom Soc Can. Res: Aquatic and forest entomology. Mailing Add: Div of Biol Sci Humboldt State Univ Arcata CA 95521

LAUDANI, HAMILTON, b Sicily, Italy, Aug 18, 15; US citizen; m 48. ECONOMIC ENTOMOLOGY. Educ: Univ Mass, BS, 41, MS, 42. Prof Exp: Plant pest suppressor, Mass Dept Agr, 40; termite control inspector, Muirhead & Halway, Inc, 41; field aide, Div Forest Forest Insect, Bur Entom & Plant Quarantine, USDA, NJ, 42; consult entomologist, United Chem & Exterminating Co, 46-47; entomologist, Div Insect Affecting Man & Animals, Bur Entom & Plant Quarantine, USDA, 47-53, leader, Savannah Stored-Prod Insects Lab, Agr Res Serv, 54-58, asst br chief, 58-65, dir, Stored-Prod Insects Res & Develop Lab, 65-68, dir, Europ Regional Res Off, 68-74; RETIRED. Mem: AAAS; Entom Soc Am; Am Registry Prof Entomologists. Res: Medical entomology; forest insects; stored-product insects. Mailing Add: 531 Bowspirit Lane Longboat Key Sarasota FL 33577

LAUDE, HORTON MEYER, b Beaumont, Tex, Feb 25, 15; m 46; c 2. AGRONOMY, BOTANY. Educ: Kans State Univ, BS, 37; Univ Chicago, PhD(bot), 41. Prof Exp: Agent, USDA, Univ Chicago, 40-41; asst prof agron & asst agronomist, 46-52, assoc prof & assoc agronomist, 52-58, PROF AGRON & AGRONOMIST, EXP STA, UNIV CALIF, DAVIS, 58- Mem: Fel Am Soc Agron; Am Soc Plant Physiol; Soc Range Mgt; Crop Sci Soc Am. Res: Physiology and ecology of dry range plants, irrigated forages and cereals; plant growth substances; seed production; resistance to environmental stresses of heat, cold and drought. Mailing Add: Dept of Agron & Range Sci Univ of Calif Davis CA 95616

LAUDENSLAGER, JAMES BISHOP, b Harrisburg, Pa, June 8, 45. CHEMICAL PHYSICS. Educ: Temple Univ, AB, 67; Univ Calif, Santa Barbara, PhD(phys chem), 71. Prof Exp: Res assoc space sci, 71-73, SR RES SCI CHEM PHYSICS, JET PROPULSION LAB, 73- Mem: Am Phys Soc. Res: Fundamental properties of charge transfer and penning ionization reactions in the gas phase for use in laser development and mass spectrometry. Mailing Add: Jet Propulsion Lab 4800 Oak Grove Dr Pasadena CA 91103

LAUDENSLAGER, MARK LEROY, b Charlotte, NC, May 13, 47; m 69. NEUROPSYCHOLOGY. Educ: Univ NC, AB, 69; Univ Calif, Santa Barbara, PhD(psychol), 75. Prof Exp: Teaching asst introductory psychol, Univ Calif, Santa Barbara, 70-71, res asst neuropsychol, 71-72, teaching asst physiol psychol, 73, lectr, 74; NIMH FEL, SCRIPPS INST OCEANOG, 75- Mem: AAAS. Res: Hypothalamic factors influencing autonomic and behavioral thermoregulatory responses in mammals, reptiles and birds. Mailing Add: Scripps Inst of Oceanog Physiol Res Lab Mail Code A-004 La Jolla CA 92093

LAUDERBACK, SANFORD KEITH, b Dayton, Ohio, Feb 1, 42; m 74; c 2. PHYSICAL ORGANIC CHEMISTRY, TEXTILE CHEMISTRY. Educ: Otterbein Col, BS, 64; Wayne State Univ, MS, 70, PhD(chem), 74. Prof Exp: Chemist anal metall, Lincoln Elec Co, 64-67; instr chem, Hiram Col, 67-68; RES CHEMIST PROD DEVELOP, E I DU PONT DE NEMOURS & CO, 74- Mem: Am Chem Soc. Res: Nuclear magnetic resonance and studies of conformationally mobile molecular systems; chiroptical phenomena; pseudoasymmetry; sulfur-nitrogen chemistry. Mailing Add: Textile Fiber Dept Christina Lab E I du Pont de Nemours & Co Wilmington DE 19898

LAUDERDALE, JAMES W, JR, b Washington, DC, Dec 21, 37; m 62; c 3. REPRODUCTIVE PHYSIOLOGY, ENDOCRINOLOGY. Educ: Auburn Univ, BS, 62; Univ Wis, MS, 64, PhD(reproductive physiol, endocrinol), 68. Prof Exp: SCIENTIST, UPJOHN CO, 67- Mem: Am Soc Animal Sci; Soc Study Reproduction.

Res: Reproductive and endocrinological function of large and small animals. Mailing Add: Upjohn Co 7000 Portage Rd Kalamazoo MI 49001

LAUDISE, ROBERT ALFRED, b Amsterdam, NY, Sept 2, 30; m 57; c 5. SOLID STATE CHEMISTRY. Educ: Union Univ, NY, BS, 52; Mass Inst Technol, PhD(inorg chem), 56. Prof Exp: Mem tech staff, 56-60, head crystal chem res dept, 60-73, ASST DIR MAT RES, BELL LABS, 73- Concurrent Pos: Consult, President's Sci Adv Comt, 60-70; panel mem mat adv bd, Nat Acad Sci, 65-70; vis comt, Nat Bur Standards, 68-75; ed, J of Crystal Growth, 74- Honors & Awards: Sawyer Award, Conf Frequency Control, 74. Mem: Am Asn Crystal Growth (pres, 68-75); fel Mineral Soc Am; Am Ceramic Soc; Am Chem Soc. Res: Materials research; crystal growth; hydrothermal chemistry; quartz; ferroelectrics; non-linear optical materials; magnetic materials. Mailing Add: 65 Lenape Lane Berkeley Heights NJ 07922

LAUDON, LOWELL ROBERT, b Redwood Falls, Feb 5, 05; m 30; c 3. GEOLOGY. Educ: Univ Iowa, BA, 28, MA, 29, PhD(paleont), 30. Prof Exp: Asst prof geol, Univ Tulsa, 30-32, asst registr, 32-34, dean men, 34-36, from assoc prof to prof, 36-39; prof, Univ Kans, 39-40, chmn dept, 40-48; PROF GEOL, UNIV WIS-MADISON, 48- Mem: Fel Geol Soc Am; Paleont Soc; Am Asn Petrol Geologists. Res: Mississippian stratigraphy; paleontology of Crinoidea; paleotectonic history. Mailing Add: Dept of Geol Univ of Wis-Madison Madison WI 53706

LAUDON, THOMAS S, b Sac City, Iowa, June 14, 32; m 56; c 5. GEOLOGY. Educ: Univ Wis, BSc, 55, MSc, 57, PhD(geol), 63. Prof Exp: Chmn dept geol, 69-72, PROF GEOL, UNIV WIS-OSHKOSH, 63- Concurrent Pos: NSF res grants, 62-67 & 70-71. Mem: Sigma Xi; Geol Soc Am; Am Asn Petrol Geologists. Res: Geologic exploration of Antarctica; gravity, tectonics and sedimentation in modern and ancient mobile belts. Mailing Add: Dept of Geol Univ of Wis Oshkosh WI 54901

LAUENROTH, WILLIAM KARL, b Carthage, Mo, July 31, 45; m. PLANT ECOLOGY. Educ: Humboldt State Col, BS, 68; NDak State Univ, MS, 70; Colo State Univ, PhD(plant ecol), 73. Prof Exp: FEL & RES ASSOC PLANT ECOL, NAT RESOURCE ECOL LAB, COLO STATE UNIV, 75- Mem: Ecol Soc Am; Soc Range Mgt. Res: Primary production and water relations of native plant communities, particularly temperate grasslands. Mailing Add: Nat Resource Ecol Lab Colo State Univ Ft Collins CO 80521

LAUER, DOLOR JOHN, b St Paul, Minn, July 8, 12; m 40; c 7. INDUSTRIAL MEDICINE, TOXICOLOGY. Educ: Univ Minn, BS & MB, 37, MD, 38; Univ Pittsburgh, IMD, 50; Am Bd Prev Med, dipl, 51. Prof Exp: Kemper fel, Univ Pittsburgh, 46-47; from instr to asst prof indust med, Col Med, Univ Cincinnati, 48-51; asst prof occup med, Kettering Lab, 48-51; asst prof indust med, Sch Med, Univ Pittsburgh, 51-55, clin assoc prof, 55-60; MED DIR, INT TEL & TEL CORP, 60- Concurrent Pos: Med dir, Jones & Laughlin Steel Corp, 51-60; clin assoc prof occup med, Postgrad Med Sch, NY Univ, 60-; trustee, Nat Health Coun; mem Permanent Comn & Int Asn Occup Health; del, Int Conf Health & Health Educ; mem, Am Bd Prev Med Inc. Mem: AMA; Indust Med Asn (pres, 59-60); Am Acad Occup Med; fel Am Pub Health Asn; Am Col Prev Med (pres, 62). Res: Toxicology of selenium related to occupational uses; observations on epidemiology and various drugs in the treatment of lead intoxication. Mailing Add: Med Dept Int Tel & Tel Corp 320 Park Ave New York NY 10022

LAUER, EDWARD WILLARD, b Petersburg, Mich, Nov 27, 02; m 26; c 1. ANATOMY. Educ: Adrian Col, BS, 23; Univ Mich, AM, 42, PhD(neuroanat), 44. Prof Exp: Teacher pub schs, Mich, 24-25, prin, 25-29, supt, 29-31 & 33-41; instr neuroanat & head lab, 43-50, from asst prof to prof, 50-74, EMER PROF ANAT, MED SCH, UNIV MICH, ANN ARBOR, 74- Mem: AAAS; Am Asn Anat; Soc Neurosci; Am Acad Neurol. Res: Comparative neuroanatomy of ungulates; mammalian nuclear pattern and fiber connections and function of the amygdala; nuclear patterns in the spinal cord of mammals. Mailing Add: Dept of Anat Univ of Mich Ann Arbor MI 48104

LAUER, EUGENE JOHN, b Red Bluff, Calif, Apr 11, 20; m 48; c 5. PLASMA PHYSICS. Educ: Univ Calif, BS, 42, PhD(physics), 51. Prof Exp: Asst physics, 46-51, PHYSICIST, LAWRENCE LIVERMORE LAB, UNIV CALIF, 51- Mem: Am Phys Soc. Res: Discharge through gases; particle accelerator design; nuclear and plasma physics; controlled thermonuclear energy; relativistic beams. Mailing Add: 2221 Martin Ave Pleasanton CA 94566

LAUER, FLORIAN ISIDORE, b Richmond, Minn, Sept 13, 28; m 55; c 3. HORTICULTURE. Educ: Univ Minn, BS, 51, PhD, 57. Prof Exp: From asst prof to assoc prof potato breeding, 59-66, PROF POTATO BREEDING, UNIV MINN, ST PAUL, 66- Mem: Am Soc Hort Sci; Potato Asn Am. Res: Potato breeding and genetics. Mailing Add: Dept of Hort Inst of Agr Univ of Minn St Paul MN 55101

LAUER, GEORGE, b Vienna, Austria, Feb 18, 36; US citizen; m 57; c 1. PHYSICAL CHEMISTRY. Educ: Univ Calif, Los Angeles, BS, 61; Calif Inst Technol, PhD(chem), 67. Prof Exp: Staff assoc, Sci Ctr, NAm Rockwell Corp, 62-66, mem tech staff, 66-71, group leader, Measurement Sci, 71-72, DIR AIR QUAL MONITORING RES, ROCKWELL INT SCI CTR, 72- Mem: AAAS; Am Chem Soc. Res: Development of chronocoulometric techniques in electrochemistry; study of electroactive adsorbed species at electrodes; development of computer controlled instrumentation; instrumental methods in air pollution monitoring. Mailing Add: 23119 Gainford Woodland Hills CA 91364

LAUER, GERALD J, b Montgomery City, Mo, Oct 18, 34; m 63. AQUATIC ECOLOGY, LIMNOLOGY. Educ: Quincy Col, BS, 56; Univ Wash, MS, 59, PhD(zool), 63. Prof Exp: Lab aide limnol, Quincy Col, 54-56; asst, Univ Wash, 56-59; teacher high sch, Mo, 59-60; staff biologist, USPHS, 60-62, prin biologist, Southeast Water Lab, 63-66, chief training br, 66; leader, Fisheries Coop Unit & assoc prof fisheries & limnol res, Ohio State Univ, 66-67; assoc cur limnol dept, Acad Natural Sci, Philadelphia, 67-69; asst dir, Lab Environ Studies, Med Ctr, NY Univ, 69-75; SR SCIENTIST & VPRES, ECOL ANALYSTS, INC, 75- Concurrent Pos: Adj assoc prof biol, NY Univ, 70- Mem: AAAS; Am Soc Limnol & Oceanog; Am Fisheries Soc; Am Littoral Soc; Ecol Soc Am. Res: Population dynamics and community diversity of aquatic organisms; effects of pollutants and other environmental stresses on aquatic life. Mailing Add: Ecol Analysts Inc R D 2 Goshen Turnpike Middletown NY 10940

LAUER, JAMES LOTHAR, b Vienna, Austria, Aug 2, 20; nat US; m 55. PHYSICAL CHEMISTRY. Educ: Temple Univ, AB, 42, MA, 44; Univ Pa, PhD(physics), 48. Prof Exp: Asst instr org chem, Temple Univ, 42-44; phys chemist, 44-45, res physicist, 47-54, sr physicist, 54-58, res assoc, 58-62, RES SCIENTIST, SUN OIL CO, 62- Concurrent Pos: Lectr, Univ Del, 52-56; asst prof, Univ Pa, 52-54; fel aerospace eng, Univ Calif, San Diego, 64-65; prin investr, Air Force Off Sci Res & NASA-Lewis Res Ctr, 73- Mem: Am Chem Soc; Am Phys Soc; Soc Appl Spectros. Res: Fourier spectroscopy; Raman and infrared spectroscopy; refraction and dispersion; theory of molecular structure; mathematical physics; x-ray spectra of polymers; combustion; shock waves in gases; adsorption and desorption; applications of molecular, mainly infrared emission, spectroscopy to problems of lubrication. Mailing Add: 226 Harrogate Rd Philadelphia PA 19151

LAUER, ROBERT B, b York, Pa, Mar 9, 42; m 60; c 3. MATERIALS SCIENCE, SOLID STATE SCIENCE. Educ: Franklin & Marshall Col, AB, 64; Univ Del, MS, 66, PhD(physics), 70. Prof Exp: Sr physicist, Itek Corp, 69-74; MEM TECH STAFF, GTE LABS, 74- Mem: Am Phys Soc. Res: Luminescence and photoconductivity of semiconductors, semi-insulators and insulators; defect structure of solids; materials preparation; liquid phase epitaxial growth; injection lasers and LED's. Mailing Add: GTE Labs 40 Sylvan Rd Waltham MA 02154

LAUER, RUDOLPH FRANK, b Vienna, Austria, Aug 28, 48; US citizen. PHARMACEUTICAL CHEMISTRY. Educ: Polytech Inst Brooklyn, BS, 70; Mass Inst Technol, PhD(org chem), 74. Prof Exp: SR SCIENTIST PHARMACEUT CHEM, HOFFMANN-LaROCHE INC, 74- Mem: Am Chem Soc. Res: The synthesis of new and potentially useful pharmaceutical agents. Mailing Add: Hoffmann-LaRoche Inc Nutley NJ 07110

LAUERMAN, LLOYD HERMAN, JR, b Everett, Wash, Feb 5, 33; m 55; c 3. VETERINARY MICROBIOLOGY, VETERINARY IMMUNOLOGY. Educ: Wash State Univ, BA, 56, DVM, 58; Univ Wis, MS, 59, PhD, 68; Am Col Vet Microbiol, dipl, 74. Prof Exp: Res asst vet med, Univ Wis, 58-60, proj asst, 60-63; res dir vet biol, Biol Specialties Corp, 63-68; Colo State Univ-AID prof microbiol, Univ Nairobi, 68-72; ASSOC PROF, COLO STATE UNIV, 72-, HEAD BACT SECT, DIAG LAB, DEPT PATH, 73- Concurrent Pos: Vis assoc prof microbiol, Sch Dent, Univ Colo, Denver, 74-75. Mem: Am Soc Microbiologists; Am Asn Vet Lab Diagnosticians; Am Vet Med Asn; Conf Res Workers Animal Dis; Wildlife Dis Asn. Res: Diagnosis and prevention of infectious diseases of animals; research and development of laboratory diagnostic techniques and animal vaccines. Mailing Add: Diag Lab Colo State Univ Ft Collins CO 80523

LAUFELD, SVEN, b Helsingborg, Sweden, Apr 25, 39. INVERTEBRATE PALEONTOLOGY. Educ: Univ Lund, Fil Kand, 62, Fil Lic(paleont), 66, Fil Dr(paleont), 74. Prof Exp: Geologist petrol explor, Axel Johnson Group of Co-Nynas Petrol, Stockholm, 66-67; sr asst paleont, Univ Lund, 67-73, docent hist geol & paleont, 74; fel geol & mineral, Ohio State Univ, 74-75; ASST PROF GEOL, UNIV & CUR PALEONT & GEOL, MUS, UNIV ALASKA, FAIRBANKS, 75- Concurrent Pos: Ed & managing ed, Trans Geol Soc Sweden, 66-67; geol consult, LN-Co, Consult Architects & Bldg Engrs, Hoganas, Sweden, 66-74; mem Silurian-Devonian boundary comt, Int Union Geol Sci, 67-72, mem subcomn for Silurian system, 72-, voting mem, Ordovician-Silurian Boundary Working Group, 74-; mem, Subcomn on Chitinozoa, Int Comn Paleozoic Microflora, 71-, chmn, 73-; adv & contribr, BBB Encycl, Sweden, 72-74. Mem: Paleont Soc; WGer Paleont Soc; Brit Palaeont Asn; Geol Soc Norway; Geol Soc Sweden. Res: Paleozoic micropaleontology, especially Chitinozoa; the Silurian system worldwide; lower and middle Paleozoic system boundaries; Silurian reefs of North America and Europe. Mailing Add: Dept of Geol Univ of Alaska Fairbanks AK 99701

LAUFER, ALLAN HENRY, b New York, NY, Mar 27, 36; m 59; c 2. PHOTOCHEMISTRY, CHEMICAL KINETICS. Educ: NY Univ, BA, 56; Lehigh Univ, MS, 58, PhD(phys chem), 62. Prof Exp: Res chemist, Gulf Res & Develop Co, 62-64; RES CHEMIST, NAT BUR STANDARDS, 64- Mem: AAAS; Am Chem Soc. Res: Vacuum ultraviolet photochemistry; gas phase radical reactions and kinetics; properties of materials in the vacuum ultraviolet region; chemistry of excited states; fluorescence. Mailing Add: Div 316.03 Nat Bur Standards Washington DC 20234

LAUFER, ARTHUR RUSSELL, b New York, NY, June 21, 17; m 47; c 3. PHYSICS, SCIENCE ADMINISTRATION. Educ: Yale Univ, MS, 47; NY Univ, PhD(physics), 49. Prof Exp: Instr physics, Yale Univ, 40-44, Mich State Univ, 44-46 & NY Univ, 47-49; from asst prof to assoc prof, Univ Mo, 49-53; phys sci coordr, Off Naval Res, 53-57, head sci dept, 57-59, chief scientist, 59-66, DEP DIR & CHIEF SCIENTIST, OFF NAVAL RES, 66- Mem: Am Phys Soc; Acoust Soc Am. Res: Ultrasonics; cavitation; nuclear physics; infrared. Mailing Add: 1030 E Green St Pasadena CA 91106

LAUFER, DANIEL A, b Affula, Israel, May 30, 38; US citizen; m 68. ORGANIC CHEMISTRY. Educ: Mass Inst Technol, BS, 59; Brandeis Univ, PhD(org chem), 64. Prof Exp: Res assoc org chem, Columbia Univ, 63-64; res fel biol chem, Harvard Med Sch, 64-66; asst prof, 66-72, ASSOC PROF CHEM, UNIV MASS, BOSTON, 72- Concurrent Pos: NIH trainee, 65-66, res grant, 67-70. Mem: Am Chem Soc. Res: Mechanism of organic reactions; peptide synthesis; protein conformation; enzymic activity. Mailing Add: Dept of Chem Univ of Mass Boston MA 02116

LAUFER, HANS, b Ger, Oct 18, 29; nat US; m 53; c 3. DEVELOPMENTAL BIOLOGY. Educ: City Col New York, BS, 52; Brooklyn Col, MA, 54; Cornell Univ, PhD(zool), 58. Prof Exp: Asst, Cornell Univ, 55-57; Nat Res Coun fel embryol, Carnegie Inst, 57-59; asst prof biol, Johns Hopkins Univ, 59-65; assoc prof biol, 65-72, PROF BIOL, UNIV CONN, 72- Concurrent Pos: Vis scholar, Case Western Reserve Univ, 62; Lalor fel, Marine Biol Lab, Woods Hole, Mass, 62-63, staff embryol course, 67-72, mem corp marine biol lab; assoc ed, J of Exp Zool, 69; mem nat bd on grad educ, Conf Bd Assoc Res Couns, 71-75; NATO sr fel, Karolinska Inst, 72; NIH spec fel, 73; NATO fel rev panel, NSF, 74; partic, Nat Acad Sci-Czech Acad Sci Exchange Prog, 74. Honors & Awards: Rosenstiel vis scholar, Brandeis Univ, 73; hon prof, Charles Univ, Prague, 74. Mem: Am Soc Zool; Soc Develop Biol; Am Soc Cell Biol; fel NY Acad Sci; Europ Soc Comp Endocrinol. Res: Developmental physiology and biochemistry; molecular interactions in development; proteins and enzymes in ontogeny, regeneration and metamorphosis; chromosomal puffing in Diptera; gene action as related to development. Mailing Add: Biol Sci Group U-42 Univ of Conn Storrs CT 06268

LAUFER, MAURICE WALTER, b Brooklyn, NY, May 2, 14; m 41; c 4. PSYCHIATRY. Educ: Univ Wis, BA, 33; Long Island Col Med, MD, 39; Am Bd Pediat, dipl, 44; Am Bd Psychiat & Neurol, dipl, 57, dipl child psychiat, 59. Prof Exp: Exec dir & physician-in-chief, Emma Pendleton Bradley Hosp, Riverside, RI, 48; MEM INST LIFE SCI, BROWN UNIV, 58-, PROF PSYCHIAT, 74- Concurrent Pos: Consult, Butler, Miriam, Roger Williams Gen & St Joseph's Hosps; clin prof psychiat, Brown Univ, 72-74. Mem: Fel AAAS; Am Psychiat Asn; fel Am Orthopsychiat Asn; fel Am Acad Child Psychiat; fel Am Acad Pediat. Res: Child psychiatry; fusion of organic and psychological components; electroencephalography. Mailing Add: 1011 Vet Mem Pkwy Riverside RI 02915

LAUFER, ROBERT J, b Pittsburgh, Pa, May 10, 32; m 67; c 1. ORGANIC CHEMISTRY. Educ: Carnegie Inst Technol, BS, 53, MS, 56, PhD(org chem), 58. Prof Exp: Proj supvr res div, Consol Coal Co, 58-68; proj leader, Int Flavors & Fragrances Inc, 68-69, assoc dir fragrance res, 69-72, dir, 72-75; DIR AROMATIC TECHNOL, NORDA, INC, 75- Mem: AAAS; Am Chem Soc; fel Am Inst Chemists. Res: Research management; aroma chemicals; terpenoids; fragrance applications research. Mailing Add: Norda Inc 140 Rte 10 East Hanover NJ 07936

LAUFERSWEILER, JOSEPH DANIEL, b Columbus, Ohio, Aug 13, 30; m 59; c 2. ECOLOGY, BOTANY. Educ: Univ Notre Dame, BS, 52; Ohio State Univ, MSc, 54, PhD(ecol), 60. Prof Exp: Instr bot & ecol, Ohio State Univ, 59-61; asst prof biol, Drake Univ, 61-63; ASST PROF BIOL, UNIV DAYTON, 63- Mem: AAAS; Ecol Soc Am; Am Inst Biol Sci. Res: Reproduction of plant communities; distribution of original vegetation and its influence on man. Mailing Add: Dept of Biol Univ of Dayton Dayton OH 45469

LAUFF, GEORGE HOWARD, b Milan, Mich, Mar 23, 27. LIMNOLOGY, ZOOLOGY. Educ: Mich State Univ, BS, 49, MS, 51; Cornell Univ, PhD(limnol, zool), 53. Prof Exp: Fisheries res technician, Mich State Dept Conserv, 50; asst phycol, Point Barrow, Alaska, 51; biol asst, Cornell Univ, 51-52, zool, 52-53; instr, Univ Mich, 53-57, from asst prof to assoc prof, 57-62; res coordr, Sapelo Island Res Found, Ga, 62-64; PROF ZOOL, FISHERIES & WILDLIFE & DIR, W K KELLOGG BIOL STA, MICH STATE UNIV, 64- Concurrent Pos: Res assoc, Great Lakes Res Inst, 54-59, Oak Ridge Inst Nuclear Studies & Oak Ridge Nat Lab, 60; assoc prof & dir marine inst, Univ Ga, 60-62. Mem: AAAS; Am Fisheries Soc; Am Soc Limnol & Oceanog (treas, 58-61, secy, 58-64, 67-70, vpres, 71-72, pres, 72-73); Ecol Soc Am; Int Asn Theoret & Appl Limnol. Res: Estuarine ecology. Mailing Add: W K Kellogg Biol Sta Mich State Univ Hickory Corners MI 49060

LAUFFENBURGER, JAMES C, b Buffalo, NY, Aug 23, 38; m 61; c 5. SOLID STATE PHYSICS. Educ: Canisius Col, BS, 60; Univ Notre Dame, PhD(solid state physics), 65. Prof Exp: Asst prof, 64-72, ASSOC PROF PHYSICS, CANISIUS COL, 72- Mem: Am Asn Physics Teachers. Res: Thermionic emission. Mailing Add: 1337 N French Rd North Tonawanda NY 14120

LAUFFER, DONALD EUGENE, b Lebanon, Pa, July 29, 40; m 64; c 3. PHYSICS. Educ: Ohio State Univ, BS, 64, PhD(physics), 68. Prof Exp: SR RES PHYSICIST, PHILLIPS PETROL CO, 68- Mem: Am Phys Soc; Am Asn Physics Teachers. Res: Magnetic resonance, especially applications to oil production research. Mailing Add: Phillips Petrol Co PRC RB-4 Bartlesville OK 74003

LAUFFER, MAX AUGUSTUS, JR, b Middletown, Pa, Sept 2, 14; m 36, 64; c 2. BIOPHYSICS. Educ: Pa State Univ, BS, 33, MS, 34; Univ Minn, PhD(biochem), 37. Prof Exp: Asst, Univ Minn, 35-36, instr biochem, 36-37; fel plant path, Rockefeller Inst, 37-38, asst, 38-41, assoc, 41-44; assoc res prof, 44-47, res prof physics & physiol chem, 47-49, chmn dept physics, 47-48, prof biophys, 49-63, head dept, 49-56 & chmn dept, 63-67, assoc dean res in natural sci, 53-54, dean div natural sci, 56-63, chmn dept biophys & microbiol, 71-76, ANDREW MELLON PROF BIOPHYS, UNIV PITTSBURGH, 63- Concurrent Pos: Spec lectr, Stanford Univ, 41; prin investr, Comt Med Res, 44-46; Priestley lectr, Pa State Univ, 46; Gehrmann lectr, Univ Ill, 51; vis prof, Theodor Kocher Inst, Bern Univ, 52, Max Planck Inst Virus Res, Tübingen, 65-66 & Univ Philippines, 67; mem Nat Res Coun comn macromolecules, 47-53, mem panel virol & immunol, sect biol, comt growth, 53-56; mem sci adv comt, Boyce Thompson Inst Plant Res Inc, 53-; co-ed, Adv Virus Res, 53-; mem prog proj comt, Nat Inst Gen Med Sci, 61-63, chmn, 62-63, mem adv coun, 63-67; mem sci adv bd, Delta Regional Primate Res Ctr, Tulane Univ, 64-67; ed, Biophys J, 69-73. Honors & Awards: Award, Eli Lilly & Co, 45; Pittsburgh Award, Am Chem Soc, 58; Outstanding Achievement Award, Univ Minn, 64. Mem: Am Chem Soc; Am Soc Biol Chemists; Biophys Soc (pres elect, 60, pres, 61); Soc Exp Biol & Med; Fedn Am Sci. Res: Electrokinetics; ultracentrifugation; viscometry; biophysics of viruses; kinetics of virus disintegration; size and shape of macromolecules; polymerization of virus protein; hydration of proteins; entropy-driven processes in biology. Mailing Add: 327 Clapp Hall Univ of Pittsburgh Pittsburgh PA 15260

LAUFMAN, HAROLD, b Milwaukee, Wis, Jan 6, 12; m 40; c 2. SURGERY. Educ: Univ Chicago, BS, 32, Rush Med Col, MD, 37; Northwestern Univ, MS, 46, PhD(surg), 48; Am Bd Surg, dipl. Prof Exp: From clin asst to prof surg, Med Sch, Northwestern Univ, 40-65; PROF SURG, ALBERT EINSTEIN COL MED, 65- DIR INST SURG STUDIES, MONTEFIORE HOSP, 65- Concurrent Pos: Assoc attend surgeon, Cook County Hosp, 46-48; attend, Hines Vet Admin Hosp, 48-50; adj attend, Michael Reese Hosp, Chicago, 48-54; attend, Passavant Mem & Vet Admin Res Hosps, 54-65; James IV traveling prof, Israel, 62. Mem: Am Surg Asn; Soc Vascular Surg; fel Am Col Surg; Am Med Writers Asn (pres, 69); Asn Advan Med Instrumentation (pres, 74-75). Res: Surgical physiology, especially mesenteric and peripheral vascular diseases; surgical design and facilities engineering. Mailing Add: Inst for Surg Studies Montefiore Hosp Bronx NY 10467

LAUG, GEORGE MILTON, b Ossining, NY, Feb 23, 23; m 48; c 1. BIOLOGY. Educ: Syracuse Univ, BS, 47, MS, 48, PhD(sci educ), 60. Prof Exp: Instr sci educ & genetics, Syracuse Univ, 48-49; asst prof sci, 49-60, PROF BIOL, STATE UNIV NY COL BUFFALO, 60- Mem: AAAS; Nat Asn Biol Teachers. Res: Environmental biology; genetics; successional replacement of elms which have died in natural communities as a result of the Dutch elm disease. Mailing Add: 517 Windermere Blvd Buffalo NY 14226

LAUGHLAND, DONALD HARTNEY, biochemistry, see 12th edition

LAUGHLIN, ALEXANDER WILLIAM, b Hot Springs, Ark, Nov 9, 36; m 69; c 3. GEOCHEMISTRY, ECONOMIC GEOLOGY. Educ: Mich Technol Univ, BSc, 58; Univ Ariz, MSc, 60, PhD(geol), 69. Prof Exp: Res assoc isotope geochem, Univ Ariz, 66-69; res assoc geol, Univ NMex, 69-70; from asst prof to assoc prof, Kent State Univ, 70-74; STAFF MEM, LOS ALAMOS SCI LAB, 74- Concurrent Pos: Adj prof, Univ NMex, 75. Mem: Geochem Soc; Am Geophys Union; Mineral Asn Can; Int Asn Geochem & Cosmochem; Geol Soc Am. Res: Geochronology; trace element geochemistry; origin of ultramafic inclusions and basalts; planetary evolution; geothermal energy extraction from dry hot rock, petrology of pre-Cambrian rocks. Mailing Add: Q-22 Geothermal Energy Group Los Alamos Sci Lab Univ of Calif Los Alamos NM 87545

LAUGHLIN, ALICE, b Malone, NY, Feb 19, 18. BIOCHEMISTRY, ANALYTICAL CHEMISTRY. Educ: St Joseph Col, Conn, BS, 49; Univ Vt, MS, 54; Columbia Univ, EdD(col sci teaching), 65. Prof Exp: Lab technician, Clin Lab, Staten Island Hosp, 49-50; teaching asst biochem, Univ Vt, 50-52; asst biochemist, Vt State Agr Exp Sta, 52-56; res chemist, Nat Biscuit Co, 56-57; res asst hemat & chemother, Columbia-Presby Med Ctr, Columbia Univ, 57-61; sci instr, Sch Nursing, St Michael Hosp, 61-62; from asst prof sci to assoc prof chem, 62-74, chmn dept, 69-70, PROF CHEM, JERSEY CITY STATE COL, 74- Concurrent Pos: Resource mem long range planning bd, Sch Nursing, St Francis Hosp, 63-; consult, NSF meeting on chem curric in jr cols, Rutgers Univ, 69. Mem: Am Chem Soc; Am Asn Higher Educ; fel Am Inst Chemists; NY Acad Sci; Am Microchem Soc. Res: Human nutrition; science education. Mailing Add: Dept of Chem Jersey City State Col Jersey City NJ 07305

LAUGHLIN, CHARLES WILLIAM, b Iowa City, Iowa, Dec 9, 39; m 66; c 1. PLANT NEMATOLOGY. Educ: Iowa State Univ, BS, 63; Univ Md, MS, 66; Va Polytech Inst & State Univ, PhD(plant path), 68. Prof Exp: Asst prof nematol, Univ Fla, 68-70; asst prof, Mich State Univ, 70-73, assoc prof nematol & asst dir acad & student affairs, Col Agr & Natural Resources, 73-75, actg dir, Inst Agr Technol, 75; CONSULT, MINISTRY EDUC & CULT, POSTGRAD AGR EDUC, DEPT UNIV AFFAIRS, 75- Mem: Soc Nematol; Am Phytopath Soc; Orgn Trop Am Nematol. Res: Ecology and pathogenicity of phytoparasitic nematodes; nematode control; nematode-nematode interactions; nematode-fungal interactions. Mailing Add: Ministry of Educ & Cult Dept Univ Affairs Brasilia DF Brazil

LAUGHLIN, ETHELREDA R, b Cleveland, Ohio, Nov 13, 22; m 51; c 1. BIOCHEMISTRY, SCIENCE EDUCATION. Educ: Case Western Reserve Univ, AB, 42, MS, 44, PhD(sci educ), 62. Prof Exp: Instr chem, anat & physiol, St John Col, 49-51; teacher high sch, Ohio, 53-62; assoc prof sci educ & org chem, Ferris State Col, 62-63; instr chem, 63-65, HEAD DEPT SCI, CUYAHOGA COMMUNITY COL, WESTERN CAMPUS, 65-, PROF CHEM, 68- Concurrent Pos: Vis prof biochem, Case Western Reserve Univ, 70-71. Mem: Nat Educ Asn; Am Chem Soc; Audubon Soc; Sigma Xi; Sierra Club. Mailing Add: 6486 State Rd 12 Parma OH 44134

LAUGHLIN, HAROLD EMERSON, b Tulsa, Okla, Feb 11, 32; m 52; c 3. ANIMAL ECOLOGY, HERPETOLOGY. Educ: Univ Tulsa, BSc, 55; Univ Tex, MA, 58, PhD(zool), 66. Prof Exp: Asst prof biol, Southeastern State Col, 61-63; DIR, HEARD NATURAL SCI MUS & WILDLIFE SANCTUARY, 66- Mem: Am Soc Ichthyol & Herpet. Res: Interspecific lizard ecology; vertebrate population dynamics; vertebrate dietary studies. Mailing Add: Heard Mus Rte 7 McKinney TX 75069

LAUGHLIN, JAMES STANLEY, b Guilford, Mo, Sept 23, 36; m 60; c 2. STATISTICS, OPERATIONS RESEARCH. Educ: Northwest Mo State Univ, BS, 58; Univ Northern Colo, MA, 61; Univ Denver, PhD(higher educ, math), 68. Prof Exp: Teacher chem & math, Grand Community Schs, Boxholm, Iowa, 58-59; teacher math, Denver Pub Schs, Colo, 59-67, res asst, 67-68; assoc prof math, Kans State Teachers Col, 68-75, dir instnl studies, 70-75; DIR INSTNL RES, IDAHO STATE UNIV, 75- Concurrent Pos: Kans State Dept Educ grant, 69-70. Mem: Asn Instnl Res; Am Educ Res Asn. Res: Higher education research in college management. Mailing Add: Idaho State Univ Pocatello ID 83201

LAUGHLIN, JOHN SETH, b Canton, Mo, Jan 26, 18; m 43; c 3. MEDICAL PHYSICS. Educ: Willamette Univ, AB, 40; Haverford Col, MS, 42; Univ Ill, PhD(physics), 47. Hon Degrees: DSc, Willamette Univ, 68. Prof Exp: Asst, Haverford Col, 40-42; asst, Univ Ill, 42-44, res assoc, Off Sci Res & Develop, 44-45, asst, 45-46; asst prof spec res, 46-48, asst prof radiol, Col Med, 48-51, assoc prof, 51-52; assoc prof biophys, 52-55, PROF BIOPHYS, SLOAN-KETTERING DIV, MED COL, CORNELL UNIV, 55-; CHIEF DIV BIOPHYS, SLOAN-KETTERING INST CANCER RES, 52-, VPRES, 66- Concurrent Pos: Attend physicist & chmn dept med physics, Mem Hosp, 52- Mem: Radiol Soc NAm; Radiation Res Soc (pres, 70-71); Soc Nuclear Med; Am Asn Physicists in Med (pres, 64-65); NY Acad Sci. Res: Neutron-proton and neutron-deutron interaction; high pressure cloud chamber design; interaction of high energy electrons with nuclei; application of betatron to medical therapy; radiation dosimetry; high energy gamma ray scanning; development of digital and computer controlled scanning; isotope metabolic studies. Mailing Add: Div of Biophys 444 E 68th St Mem Sloan-Kettering Cancer Ctr New York NY 10021

LAUGHLIN, KENNETH CLIFFORD, b Council Bluffs, Iowa, Dec 5, 05; m 34; c 2. TEXTILES. Educ: Huron Col, BS, 26; Pa State Univ, MS, 32, PhD(org chem), 33. Prof Exp: Res chemist, Esso Labs, Standard Oil Co (La), 33-43; res technician res dept, Hercules Powder Co, 43, mgr pilot plant div, Exp Sta, 43-55; mgr process develop, Fibers Co Div, Celanese Corp Am, 55-61, mgr fibers res, Fibers Div, Allied Chem Corp, 61-65, asst to dir res & develop, 65-66; sr lectr consumer sci, Univ Calif, Davis, 66-68; prof textiles, 68-75, RES PROF TEXTILES, WINTHROP COL, 75- Mem: AAAS; Am Chem Soc; Am Inst Chem Engrs; Fiber Soc; Am Asn Textile Chem & Colorists. Res: Consumer and laboratory evaluation of textile products; carpets and children's sleepwear. Mailing Add: Sch of Home Econ Winthrop Col Rock Hill SC 29733

LAUGHLIN, ROBERT DAVID, b Altoona, Pa, Mar 9, 24; m 50; c 3. PHYSICS. Educ: Pa State Univ, BS, 48, MS, 49. Prof Exp: Staff physicist & proj dir, HRB-Singer Inc, 54-67, staff physicist & sect mgr optics & appl physics, 67-70; EXEC DIR, PA SCI & ENG FOUND & DIR OFF SCI & TECHNOL, DEPT COM, 70- Concurrent Pos: Consult, 58- Res: Semiconductor physics; electron devices; optical design; infrared photometry physics of thin films; electrooptical devices; research management and program development. Mailing Add: Off of Sci & Technol 400 S Off Bldg Dept of Com Harrisburg PA 17120

LAUGHLIN, ROBERT GENE, b Sullivan, Ind, Aug 9, 30; m 58; c 2. PHYSICAL ORGANIC CHEMISTRY. Educ: Purdue Univ, BS, 51; Cornell Univ, PhD(org chem), 55. Prof Exp: Fel org chem, Hickrill Res Labs, NY, 55-56; res chemist, 56-68, SECT HEAD, MIAMI VALLEY LABS, PROCTER & GAMBLE CO, 68- Mem: AAAS; Am Chem Soc. Res: Organic synthesis, especially of surfactant-like molecules and correlation of molecular structure with solubility and other physical properties; synthesis of aliphatic organophosphorus compounds; chemistry of positive-halogen compounds; synthesis of organophosphorus and organosulfur compounds. Mailing Add: Procter & Gamble Co Miami Valley Labs Cincinnati OH 45239

LAUGHLIN, ROBERT MOODY, b Princeton, NJ, May 29, 34; m 60; c 2. ANTHROPOLOGY. Educ: Princeton Univ, AB, 56; Harvard Univ, MA, 59, PhD(anthrop), 63. Prof Exp: ASSOC CUR ETHNOL MESOAM, DEPT ANTHROP, SMITHSONIAN INST, 62- Concurrent Pos: Smithsonian res award, 66-68. Mem: Am Anthrop Asn. Res: Oral literature and language, as well as ethnography, of Tzotzil-speaking peoples of Chiapas, Mexico. Mailing Add: Dept of Anthrop Smithsonian Inst Washington DC 20560

LAUGHLIN, WILLIAM SCEVA, b Canton, Mo, Aug 26, 19; m 44; c 2. PHYSICAL ANTHROPOLOGY. Educ: Willamette Univ, BA, 41; Haverford Col, MA, 42; Harvard Univ, AM, 48, PhD(anthrop), 49. Hon Degrees: DSc, Willamette Univ, 68. Prof Exp: Asst anthrop, Harvard Univ, 47-48; asst prof, Univ Ore, 49-53, assoc prof, 53-55; from assoc prof to prof, Univ Wis, 55-69, chmn dept, 60-62; prof bio-behav sci & anthrop, Univ Conn, 69-75; MEM STAFF, STANFORD RES INST, 75- Concurrent Pos: Ed, J Am Asn Phys Anthrop, 58-63; mem anthrop study sect, Nat Res Coun, 59-62; fel, Ctr Advan Study Behav Sci, Stanford Univ, 64-65. Mem: AAAS; fel Am Soc Human Genetics; Soc Am Archaeol; Am Asn Phys Anthrop; Am Anthrop Asn. Res: Population genetics; blood group genetics and skeletal history of human isolates, Indians and Eskimo-Aleut stock; skeletal analysis of Eskimos and Indians; biological distance between groups. Mailing Add: Stanford Res Inst Menlo Park CA 94025

LAUGHLIN, WINSTON MEANS, b Fountain, Minn, May 2, 17; m 47; c 4. SOIL SCIENCE. Educ: Univ Minn, BS, 41; Mich State Univ, MS, 47, PhD(soil sci), 49. Prof Exp: Soil surveyor, Univ Minn, 40-41; asst, Mich State Univ, 41-42; SOIL SCIENTIST, ALASKA AGR EXP STA, 49- Mem: AAAS; Am Soc Agron; Soil Sci Soc Am; Am Sci Affil; Int Soc Soil Sci. Res: Soil fertility, chemistry and classification with emphasis on Arctic conditions. Mailing Add: Inst of Agr Sci Box AE Palmer AK 99645

LAUGHNAN, JOHN RAPHAEL, b Spring Green, Wis, Sept 27, 19; m 42; c 3. GENETICS. Educ: Univ Wis, BS, 42; Univ Mo, PhD(genetics), 46. Prof Exp: Asst prof biol, Princeton Univ, 47-48; asst prof bot, Univ Ill, 48-51, assoc prof, 51-54; prof field crops, Univ Mo, 54-55; prof bot & chmn dept, 55-59, head dept bot, 63-65, PROF BOT & PLANT GENETICS, UNIV ILL, URBANA, 59- Concurrent Pos: Gosney fel, Calif Inst Technol, 48; Guggenheim fel, 60-61. Mem: Genetics Soc Am; Bot Soc Am; Am Soc Naturalists. Res: Functional genetics; fine structure and recombination in maize and Drosophila; intrachromosomal recombination mechanisms; development of sweet corn carrying sh2 gene. Mailing Add: Dept of Bot Univ of Ill Urbana IL 61801

LAUGHON, ROBERT BUSH, b Greensboro, NC, Apr 20, 34. GEOLOGY. Educ: Colo Col, BA, 60; Univ Colo, Boulder, MS, 63; Univ Ariz, PhD(geol), 70. Prof Exp: Geologist, US Geol Surv, 62-63 & Anaconda Co, 64-65; instr, Univ Ariz, 66-67; GEOLOGIST, MANNED SPACECRAFT CTR, NASA, 67- Concurrent Pos: Lectr, Moody Col, Tex A&M Univ, 74. Mem: Geol Soc Am; Mineral Soc Am. Res: Mineralogy; structural crystallography. Mailing Add: Curator's Off Mail Code TL Johnson Space Ctr NASA Houston TX 77058

LAUGHTON, PAUL MACDONELL, b Toronto, Ont, Sept 8, 23; m 46; c 4. ORGANIC CHEMISTRY. Educ: Univ Toronto, BA, 45; Dalhousie Univ, MSc, 47; Univ Wis, PhD(chem), 50. Prof Exp: Res assoc, Univ Wis, 50; Nat Res Coun Can fel, Dalhousie Univ, 50-51; from asst prof to assoc prof chem, 51-65, PROF CHEM, CARLETON UNIV, 65- Concurrent Pos: Vis prof, Stanford Univ, 62 & Kings Col, London, 72-73; Am Chem Soc-Petrol Res Fund fel, Univ Calif, Berkeley, 62-63. Mem: Am Chem Soc; fel Chem Inst Can; The Chem Soc; Sigma Xi. Res: Mechanism studies; isotope effects. Mailing Add: 928 Muskoka Ave Ottawa ON Can

LAUKONIS, JOSEPH VAINYS, b Mich, Apr 1, 25; m 54. PHYSICS. Educ: Univ Detroit, BS, 51; Univ Cincinnati, PhD, 57. Prof Exp: SR RES PHYSICIST, GEN MOTORS CORP, 57- Res: Iron whiskers; metal surfaces; high temperature oxidation; electron microscopy; ultrahigh vacuums; oxidation-reduction catalysts. Mailing Add: 32405 Northampton Warren MI 48093

LAUNCHBAUGH, JOHN L, JR, b Sheridan Co, Kans, Nov 8, 22; m 46; c 2. PLANT ECOLOGY. Educ: Ft Hays Kans State Col, AB, 47, MS, 48; Agr & Mech Col Tex, PhD(range mgt), 52. Prof Exp: Instr bot, Ft Hays Kans State Col, 48-49; asst range mgt, Dept Range & Forestry, Agr & Mech Col Tex, 49-52; asst res specialist, Sch Forestry, Univ Calif, 52-55; assoc prof, 55-67, PROF RANGE MGT, FT HAYS BR, AGR EXP STA, KANS STATE UNIV, 67- Mem: Soc Range Mgt. Res: Grassland succession; grazing management; range reseeding. Mailing Add: Ft Hays Br Exp Sta Kans State Univ Hays KS 67601

LAUNER, PHILIP JULES, b Philadelphia, Pa, Nov 20, 22; m 47; c 3. ANALYTICAL CHEMISTRY. Educ: Drew Univ, AB, 43; Columbia Univ, MA, 47. Prof Exp: Teaching asst quant anal, Columbia Univ, 47-48; proj chemist petrol chem, Res Dept, Standard Oil Co (Ind), 48-55; specialist spectros, Silicone Prod Dept, Gen Elec Co, 55-72; PRES, LAB FOR MATS INC, 73- Concurrent Pos: Consult, Lazare Kaplan & Sons Inc, 75- Mem: Am Chem Soc; Coblentz Soc; Soc Appl Spectros. Res: Infrared spectroscopy; silicone technology; analysis of urinary tract calculi; identification of industrial materials. Mailing Add: Lab For Mats Inc PO Box 14 Burnt Hills NY 12027

LAUPUS, WILLIAM E, b Seymour, Ind, May 25, 21; m 48; c 4. MEDICINE. Educ: Yale Univ, BS, 43, MD, 45. Prof Exp: Instr pediat, Med Col, Cornell Univ, 50-52; from asst prof to prof, Med Col Ga, 59-63; PROF PEDIAT & CHMN DEPT, MED COL VA, 63- Mem: Fel Am Acad Pediat; Am Fedn Clin Res; AMA; Am Pediat Soc. Res: Pediatric medicine, especially problems of newborn and premature infants; cardiovascular physiology. Mailing Add: Dept of Pediat Med Col of Va Richmond VA 23219

LAURANCE, NEAL L, b Winsted, Minn, Aug 19, 32; m 53; c 4. COMPUTER SCIENCE. Educ: Marquette Univ, BS, 54, MS, 55; Univ Ill, PhD(physics), 60. Prof Exp: Res scientist, Sci Lab, 60-73, MGR CONTROL SYSTS DEPT, ENG & RES STAFF, FORD MOTOR CO, 73- Mem: AAAS; Am Phys Soc; Asn Comput Mach; Inst Elec & Electronics Eng. Res: Computer languages; time-sharing systems; simulation systems; computer design; real-time systems; micro processor systems. Mailing Add: Eng & Res Staff Ford Motor Co Dearborn MI 48121

LAURENCE, GEORGE CRAIG, physics, see 12th edition

LAURENCE, KENNETH ALLEN, b Cleveland, Ohio, Nov 4, 28; m 49. MICROBIOLOGY. Educ: Marietta Col, AB, 51; Univ Iowa, MS, 53, PhD, 56. Prof Exp: NIH fel immunol, Univ Iowa, 56-57; instr microbiol & immunol, Med Col, Cornell Univ, 57-59, asst prof, 59-60; asst med dir, 60-68, ASSOC DIR BIOMED DIV, POP COUN, INC, 68- Concurrent Pos: Ford Found consult physiol reprod, Egyptian Univs Prog, 66-67 & 70-; ed bd, J Fertil & Sterility, 71-; proj specialist, Ford Found, Cairo, Egypt, 73-74. Mem: AAAS; Am Soc Microbiol; Soc Study Reprod; Int Soc Res Reprod; Am Fertil Soc. Res: Physiology of reproduction; immunology; parasitology; medical bacteriology; immunologic studies of the reproductive processes. Mailing Add: Biomed Div York Ave & 66th St Pop Coun Inc Rockefeller Univ New York NY 10021

LAURENCE, MARIA (MAHER), b Lenox, Pa, July 16, 97. BOTANY. Educ: Marywood Col, BS, 26; Villanova Col, MS, 40; Fordham Univ, PhD(biol), 46. Prof Exp: Teacher chem & physics, Laurel Hill Acad, 32-38; from asst prof to assoc prof biol sci, 38-56, PROF BIOL SCI, MARYWOOD COL, 56- Mem: AAAS; Nat Asn Biol Teachers; Phycol Soc Am; Bot Soc Am. Res: Effect of environmental factors on growth of algae. Mailing Add: Dept of Sci Marywood Col Scranton PA 18509

LAURENCE, RICHARD, pharmacology, physiology, see 12th edition

LAURENCOT, HENRY JULES, JR, b Brooklyn, NY, Dec 14, 29; m 61; c 4. DRUG METABOLISM. Educ: St Peter's Col, NJ, BS, 51; Fordham Univ, MS, 55, PhD(biol), 65. Prof Exp: Asst plant physiologist, Boyce Thompson Inst Plant Res, Inc, 57-66; SR BIOCHEMIST, HOFFMANN-LA ROCHE, INC, 66- Mem: Am Soc Pharmacol & Exp Therapeut; NY Acad Sci. Res: In vivo and in vitro metabolic studies of radioactive experimental drugs. Mailing Add: Hoffmann-La Roche Inc Kingsland St Nutley NJ 07110

LAURENE, ANDERS HENNING, physical chemistry, analytical chemistry, see 12th edition

LAURENSON, RAE DUNCAN, b Lerwick, Scotland, Sept 10, 22; m 53; c 3. ANATOMY, MEDICAL EDUCATION. Educ: Univ Aberdeen, MB & Chb, 53, MD, 63. Prof Exp: Lectr anat, Univ Aberdeen, 54-57; asst prof, Queen's Univ, Ont, 57-63; from assoc prof to prof anat, Univ Alta, 68-74; COORDR MED INSTRNL RESOURCES, FAC MED, UNIV CALGARY, 74- Concurrent Pos: Dir health sci AV educ, Health Sci Ctr, Univ Alta, 69-74. Res: Communications; health sciences AV education. Mailing Add: Off Med Instrnl Resources Univ of Calgary Fac of Med Calgary AB Can

LAURENT, ANDRE GILBERT (LOUIS), b Paris, France, July 13, 21; nat US; m 45, 70; c 4. MATHEMATICAL STATISTICS. Educ: Univ Paris, BA & BSc, 38, statistician economist dipl, 46; Univ Lyon, MSc, 42; Superior Nat Sch Mines, Saint-Etienne, PhD(math eng), 44. Prof Exp: Adminr & head prices sect, Nat Inst Statist & Econ, France, 46-54; from instr to asst prof statist, Mich State Univ, 54-57; assoc prof math statist, 57-63, PROF MATH STATIST, WAYNE STATE UNIV, 63- Concurrent Pos: Adj prof inst statist, Univ Pa, 52-54; Rockefeller Award fel comt statist, Univ Chicago, 53-54; prof fac law, Royal Univ Khmere, Cambodia, 63-65; statist adv, Cambodian Govt for Asia Found, 63-65. Mem: Opers Res Soc Am; fel Am Statist Asn; Inst Math Statist; Soc Indust & Appl Math; Am Acad Relig. Res: Mathematical statistics; foundations of statistics; reliability and failure theory; bombing and firing theory; statistical multivariate analysis. Mailing Add: 10230 Dartmouth Oak Park MI 48237

LAURENT, ROGER, b Geneva, Switz, June 23, 38; m 62; c 2. GEOLOGY. Educ: Univ Geneva, Lic es sci, 62, Ing Geol, 64, Dr es Sci(geol & mineral sci), 67. Prof Exp: Asst mineral, Univ Geneva, 63-65, res asst geochronology for Sci Res Nat Corp of Switz, 65-67; asst prof mineral & petrog, Middlebury Col, 67-71; ADJ PROF PETROL, LAVAL UNIV, 71- Mem: Mineral Soc Am; Swiss Soc Mineral & Petrog; Swiss Geol Soc; Geol Soc Am. Res: Study of metamorphic rocks and granites in the field and in the labs by petrographic and x-ray analyses methods; geochronometric determinations and geochemical studies. Mailing Add: Dept of Geol Laval Univ Quebec PQ Can

LAURENZI, BERNARD JOHN, b Philadelphia, Pa, Dec 23, 38. CHEMICAL PHYSICS. Educ: St Joseph's Col, Pa, BS, 60; Univ Pa, PhD(chem), 65. Prof Exp: Res chemist, Rohm & Haas Chem Co, 60-61; NSF fel, Pa State Univ, 65-66; asst prof chem, Univ Tenn, Knoxville, 66-68 & Bryn Mawr Col, 68-69; asst prof, 69-71, ASSOC PROF CHEM, STATE UNIV NY ALBANY, 71- Mem: Am Chem Soc; Am Phys Soc. Res: Quantum chemistry; use of Green's functions in atomic and molecular calculations; properties of isoelectronic molecules. Mailing Add: Dept of Chem State Univ of NY Albany NY 12203

LAURENZI, GUSTAVE, b Orange, NJ, July 19, 26; m; c 2. MEDICINE. Educ: NY Univ, BA, 49; Georgetown Univ, MD, 53; Am Bd Internal Med, dipl; Am Bd Pulmonary Dis, dipl. Prof Exp: Intern path, Mallory Inst Path, Boston City Hosp, 53-54; intern med, Yale Med Serv, Grace-New Haven Hosp, Conn, 54-55; asst resident, Columbia Med Serv, Bellevue Hosp, NY, 55-56; chief resident physician chest serv, Bellevue Med Ctr, 58-59; asst prof med & dir div respiratory dis, NJ Col Med & Dent, 60-63, assoc prof med, 63-68; ASSOC PROF MED, SCH MED, TUFTS UNIV, 70- Concurrent Pos: Res fel, Cardiopulmonary Lab, Columbia-Presby Med Ctr, 56-57; USPHS res fel, 57-58; Nat Found training fel, Am Trudeau Soc, 58-59; Channing res fel bact & immunol, Mallory Inst Path, Harvard Univ & res fel, Am Thoracic Soc, 59-60; Am Thoracic Soc Edward L Trudeau fel, 62-64; consult, Harvard Med Serv, Boston City Hosp, 59-60; chief med, St Vincent Hosp, Worcester, Mass, 68-; dir respiratory care serv, Newton-Wellesley Hosp. Mem: Fel Am Col Physicians; Am Thoracic Soc; Am Fedn Clin Res. Res: Chest disease; chronic bronchitis and pulmonary emphysema. Mailing Add: Newton-Wellesley Hosp 2000 Washington St Newton Lower Falls MA 01778

LAURIA, ANTHONY THOMAS, applied mathematics, see 12th edition

LAURIE, JOHN SEWALL, b Gloucester, Mass, May 30, 25. EXPERIMENTAL BIOLOGY. Educ: Ore State Univ, BS, 50; Johns Hopkins Univ, ScD, 56. Prof Exp: Res fel parasite physiol, Inst Parasitol, McGill Univ, 56-57; instr zool, Tulane Univ, 57-59; asst prof exp biol, Univ Utah, 59-62; mem staff water pollution study, USPHS, 62-63; ASSOC PROF BIOL, E CAROLINA UNIV, 63- Mem: Am Soc Parasitol; Wildlife Dis Asn; Am Soc Zoologists; Sigma Xi. Res: Physiology of parasites; physiology, ultrastructure and ecology of helminth parasites. Mailing Add: Dept of Biol E Carolina Univ Greenville NC 27834

LAURIE, VICTOR WILLIAM, b Columbia, SC, June 1, 35; m 65; c 2. PHYSICAL CHEMISTRY. Educ: Univ SC, BS, 54; Harvard Univ, AM, 56, PhD(chem), 58. Prof Exp: Fel, Nat Res Coun-Nat Bur Standards, 57-59; NSF fel, Univ Calif, 59-60; asst prof chem, Stanford Univ, 60-66; assoc prof, 66-71, PROF CHEM, PRINCETON UNIV, 71- Concurrent Pos: Alfred P Sloan fel, 63-67; John S Guggenheim fel, 70. Mem: AAAS; Am Chem Soc; Am Phys Soc. Res: Molecular spectroscopy and structure. Mailing Add: Dept of Chem Princeton Univ Princeton NJ 08540

LAURILA, SIMO HEIKKI, b Tampere, Finland, Jan 15, 20; m 46; c 5. GEODESY. Educ: Inst Technol, Finland, BSc, 46, MSc, 48, PhD(electronic surv, photogram), 53, LLSc, 56. Prof Exp: Surveyor, Finnish Govt, 46-48; res assoc photogram, Hq Staff, Finnish Defense Forces, 48-52; geodesist, Finnish Geod Inst, 53; res assoc electronic surv, Res Found, Ohio State Univ, 53-56, assoc prof geod, 56-59, assoc prof & res assoc, 60-62, prof photogram, 62-66; chmn dept geosci, 69-73, PROF GEOD SCI, UNIV HAWAII, 66-, HEAD GEOD DIV, HAWAII INST GEOPHYS, 66- Concurrent Pos: Sr sci ed, Finnish Broadcasting Corp, 60; Hays-Fulbright sr award, 73. Honors & Awards: Found Advan Technol Award, Finland, 60; Kaarina & W A Heiskanen Award, Ohio State Univ, 65. Mem: Am Soc Photogram; Am Geophys Union. Res: Applications of electronics in geodesy photogrammetry and navigation; laser lunar ranging, laser faults measurements. Mailing Add: Dept of Geol & Geophys Univ of Hawaii Honolulu HI 96822

LAURIN, ANDRE FREDERIC, b Ste Anne de Bellevue, Que, Jan 18, 29; m; c 3. PETROLOGY, STRUCTURAL GEOLOGY. Educ: Univ Montreal, BSc, 51; McGill Univ, MSc, 54; Laval Univ, DSc(petrol), 57. Prof Exp: Geologist, Que Dept Natural Resources, 56-65, regional geologist, Grenville, 65-69, dir mineral deposits serv, 69-71, DIR GEOL SERV, QUE DEPT NATURAL RESOURCES, 71- Concurrent Pos: Lectr tour numerous univs, France & Ger, 68. Mem: Fel Geol Soc Am; Geol Asn Can; Can Inst Mining & Metall; NY Acad Sci. Res: Mapping and supervision of the mapping of the Grenville geological province, Quebec on a regional scale. Mailing Add: 2011 Chapdelaine St Quebec PQ Can

LAURIN, PUSHPAMALA, b Bangalore City, India; US citizen; m 64; c 2. ELECTROMAGNETISM. Educ: Gujarat Univ, India, BSc, 56; Karnatak Univ, India, MSc, 58; Univ Mich, Ann Arbor, MSE, 62, PhD(physics), 67. Prof Exp: Jr sci asst, Nat Sugar Inst, Kanpur, India, 58-59; sales engr, Toshniwal Bros, Bombay, 59-60; res asst meteorol, Univ Mich, Ann Arbor, 61-62, asst res physicist, Radiation Lab, 62-67; asst prof math, Eastern Mich Univ, 67-68; res scientist physics, McDonnell-Douglas Corp, Mo, 68-69; lectr elec eng & physics, Southern Ill Univ, 70-71; INSTR ELECTRONICS & PHYSICS, HARPER COL, 71- Mem: Inst Elec & Electronics

Engrs. Res: Electromagnetic wave interaction with different media and environmental effects of radiation. Mailing Add: 903 Greenfield Lane Mt Prospect IL 60056

LAURITSEN, THOMAS, physics, deceased

LAURMANN, JOHN ALFRED, b Cambridge, Eng, Aug 8, 26; nat US; m 57; c 2. FLUID DYNAMICS. Educ: Cambridge Univ, BA, 47, MA, 51; Cranfield Univ, MSc, 51; Univ Calif, PhD(eng sci), 58. Prof Exp: Asst res engr, Univ Calif, 53-58; head aerodyn res, Space Technol Labs, Inc, 58-59; aeronaut res scientist, NASA, 59-60; staff scientist, Lockheed Missile & Space Co, 60-63; tech staff mem, Gen Res Corp, 63-69; staff mem, Inst Defense Anal, 69-70; sr staff mem, Nat Acad Sci, Washington, DC, 70-74; CONSULT SCIENTIST, 75- Concurrent Pos: Lectr, Univ Calif, 55-58, San Jose State Col, 60 & Calif Exten Div, 60-63. Mem: Am Phys Soc; Am Geophys Union; Am Meteorol Soc; Sci Policy Found. Res: Rarefied gas dynamics; systems analysis; climatology; urban affairs. Mailing Add: 3372 Martin Rd Carmel CA 93921

LAURO, GABRIEL JOSEPH, b New York, NY, Oct 26, 30; m 58; c 2. FOOD SCIENCE, MICROBIOLOGY. Educ: NY Univ, AB, 52; Univ Idaho, MS, 55; Rutgers Univ, PhD(food sci), 60. Prof Exp: Microbiologist, Sixth Army Med Lab, Calif, 55-57; food scientist prod develop, Lever Bros & Co, 60-63, food microbiologist, 63-65, asst new prod coord, 65-67; mgr new prod & Pkg coord, Thomas J Lipton, Inc, 67-69; dir res & develop, Roman Prod Corp, NJ, 69-71, vpres res & develop, 71-72; ASSOC RES DIR PROD DEVELOP, HUNT-WESSON FOODS INC, 72- Mem: Am Soc Microbiol; Inst Food Technologists. Res: Microbiology of foods and food processes; new product development and marketing; treponema pallidum immobilization test. Mailing Add: Hunt-Wesson Foods 1645 W Valentia Dr Fullerton CA 92634

LAURS, ROBERT MICHAEL, b Oregon City, Ore, Jan 27, 39; m 63; c 2. OCEANOGRAPHY, FISHERIES. Educ: Ore State Univ, BS, 61, MS, 63, PhD(oceanog), 67. Prof Exp: OCEANOGR, SOUTHWEST FISHERIES CTR, LA JOLLA LAB, NAT MARINE FISHERIES SERV, 67- Mem: AAAS; Marine Biol Asn UK; Am Soc Limnol & Oceanog; Am Inst Fishery Res Biologists. Res: Fishery forecasting; environmental conditions affecting the distribution and abundance of tunas; albacore tuna ecology; vertical distribution and migration of micronektonic organisms. Mailing Add: La Jolla Lab PO Box 271 Nat Marine Fisheries Serv La Jolla CA 92037

LAURSEN, PAUL HERBERT, b Ord, Nebr, Mar 28, 29; m 59; c 2. ORGANIC CHEMISTRY. Educ: Dana Col, BA, 54; Ore State Univ, PhD(org chem), 61. Prof Exp: From asst prof to assoc prof chem, 59-64, PROF CHEM, NEBR WESLEYAN UNIV, 64- Concurrent Pos: NSF sci fac fel, Univ Calif, Los Angeles, 67-68. Mem: AAAS; Am Chem Soc. Res: Synthesis of nitrogen heterocycles; identification of natural products. Mailing Add: Dept of Chem Nebr Wesleyan Univ Lincoln NE 68504

LAURSEN, RICHARD ALLAN, b Normal, Ill, May 1, 38; m 71. BIO-ORGANIC CHEMISTRY. Educ: Univ Calif, Berkeley, BS, 61; Univ Ill, PhD(chem), 64. Prof Exp: NIH fel, Harvard Univ, 64-66; asst prof chem, 66-72, ASSOC PROF CHEM, BOSTON UNIV, 72- Concurrent Pos: NIH res career develop award, 69-74; guest scientist, Max Planck Inst Molecular Genetics, 71; Alfred P Sloan fel, 72-74; mem sci adv comt on clin invest, Am Cancer Soc, 75- Mem: AAAS; Am Chem Soc; Am Inst Chem; Am Soc Biol Chemists; NY Acad Sci. Res: Development of new methods for protein sequence analysis; sequence studies on human plasmin, elongation factor tu, molluscan myoglobins; mechanisms of enzyme catalysis. Mailing Add: Dept of Chem Boston Univ Boston MA 02215

LAUSCH, ROBERT NAGLE, b Chambersburg, Pa, Feb 22, 38; m 68; c 2. IMMUNOBIOLOGY. Educ: Muhlenberg Col, BS, 60; Pa State Univ, MS, 62; Univ Fla, PhD(microbiol), 66. Prof Exp: Fel virol, Baylor Col Med, 66-69; asst prof, 69-75, ASSOC PROF MICROBIOL, COL MED, PA STATE UNIV, 75- Mem: Am Soc Microbiol; Am Asn Cancer Res; Am Asn Immunologists. Res: Tumor immunology; study of the host immune response to membrane antigens found on the surface of virus transformed cells and how tumor cells escape destruction. Mailing Add: Milton S Hershey Med Ctr Col Med Pa State Univ Hershey PA 17033

LAUSH, GEORGE, b Barrackville, WVa, Sept 17, 21; m 56. MATHEMATICS. Educ: Univ Pittsburgh, BS, 43; Cornell Univ, PhD, 49. Prof Exp: Asst chem, Univ Pittsburgh, 43-44; res assoc, Manhattan Dist, Univ Rochester, 44-46; asst math, Cornell Univ, 46-49; from asst prof to assoc prof, 49-62, PROF MATH, UNIV PITTSBURGH, 62- Mem: Am Math Soc; Math Asn Am. Res: Infinite series; real functions; functional analysis. Mailing Add: Dept of Math Univ of Pittsburgh Pittsburgh PA 15213

LAUSON, HENRY DUMKE, b New Holstein, Wis, Aug 20, 12; m 36. PHYSIOLOGY. Educ: Univ Wis, BS, 36, PhD(physiol), 39, MD, 40. Prof Exp: Asst med, Univ Wis, 36-39; intern, Univ Kans Hosps, 40-41; asst resident med, Henry Ford Hosp, 41-42; Off Sci Res & Develop fel physiol & med, Col Med, NY Univ, 42-43, instr physiol, 43-46; assoc & assoc physician, Rockefeller Inst, 46-50; assoc prof physiol in pediat, Med Col, Cornell Univ, 50-55; PROF PHYSIOL & CHMN DEPT, ALBERT EINSTEIN COL MED, 55- Concurrent Pos: Assoc prof physiol, Med Col, Cornell Univ, 51-55; consult prog-proj comt, Nat Inst Arthritis & Metab Dis, 61-65. Mem: Am Physiol Soc; Am Soc Exp Biol & Med; Am Soc Clin Invest; Harvey Soc (secy, 52-55); Asn Am Med Cols. Res: Pituitary-ovary interrelations; blood pressure in human right heart; renal physiology; nephrotic syndrome; metabolism of antidiuretic hormone. Mailing Add: Dept of Physiol Albert Einstein Col of Med Bronx NY 10461

LAUTENBERGER, WILLIAM J, b Flushing, NY, Mar 11, 43; m 67; c 2. APPLIED STATISTICS, PHYSICAL CHEMISTRY. Educ: Muhlenberg Col, BS, 64; Univ Pa, PhD(phys chem), 67. Prof Exp: Res chemist, Dye Div, Org Chem & Res & Develop Dept, Jackson Labs, NJ, 67-71, res chemist, Org Chem Dept, Exp Sta, Wilmington, 71-74, STATIST PROG CONSULT, DU PONT FABRICS & FINISHES DEPT, APPL TECHNOL DIV, E I DU PONT DE NEMOURS & CO, WILMINGTON, 74- Mem: Am Chem Soc; Sigma Xi. Res: Reaction mechanisms; photochemistry; heterogenous catalysis; mechanisms of dyeing; emulsion science; solid-solid adsorption; consulting in design and analysis of experiments. Mailing Add: Fabrics & Finishes Dept Du Pont Co Wilmington DE 19898

LAUTENSCHLAEGER, FRIEDRICH KARL, b Gefell, Ger, June 27, 34; m 60; c 2. ORGANIC CHEMISTRY. Educ: Univ Heidelberg, BA, 56; Univ Toronto, MA, 60. Prof Exp: Res asst org chem, Univ Toronto, 57-59; res chemist, NAm Res Centre, Dunlop Co, Ltd, 61-72, GROUP LEADER, DUNLOP RES CENTRE, 72- Mem: Chem Inst Can; Am Chem Soc. Res: Stereochemistry of organic compounds; synthesis of small ring compounds; organic sulfur chemistry; reactive intermediates; vulcanization chemistry and physics. Mailing Add: Dunlop Res Centre Sheridan Park ON Can

LAUTENSCHLAGER, EDWARD WALTER, b Amsterdam, NY, Mar 1, 27; m 48; c 5. PARASITOLOGY, HELMINTHOLOGY. Educ: Franklin & Marshall Col, BS, 50; Univ Va, MS, 56, PhD(biol), 63. Prof Exp: Instr biol, Franklin & Marshall Col, 50-51; from instr to asst prof, Univ Va, 57-65, registr, 59-65; PROF BIOL & DEAN, ROANOKE COL, 65- Mem: Am Soc Parasitol; Am Soc Trop Med & Hyg. Res: Trematode infection and tumor development in host; trematode histochemistry and ultrastructure; larval schistosome and snail host relationship. Mailing Add: Dept of Biol Roanoke Col Salem VA 24153

LAUTENSCHLAGER, EUGENE PAUL, b Chicago, Ill, Apr 5, 37; m 61; c 1. BIOMATERIALS. Educ: Ill Inst Technol, BS, 58; Northwestern Univ, MS, 60, PhD(mat sci), 66. Prof Exp: Res metallurgist, Allis-Chalmers Mfg Co, 60-62; from asst prof to assoc prof, 66-74, PROF BIOL MAT, NORTHWESTERN UNIV, 74- Concurrent Pos: NIH career develop award, 71-75; consult, Bioeng Comt, Am Acad Orthop Surgeons, 73-76. Mem: Am Soc Testing & Mat; Am Inst Mining, Metall & Petrol Eng; Int Asn Dent Res. Res: Biological and dental materials; medical implant materials; kinetics of cementing media for implant stabilization; x-radiopacity; computer assisted instruction. Mailing Add: Dept of Biol Mat Northwestern Univ 311 E Chicago Chicago IL 60611

LAUTENSCHLAGER, HERMAN KENNETH, b Ohio, June 5, 18; m 42; c 3. GEOLOGY. Educ: Miami Univ, Ohio, AB, 42; Ohio State Univ, PhD(geol), 52. Prof Exp: Asst dept geol, Miami Univ, Ohio, 39-42; geologist, Ohio Fuel Gas Co, 45-48; asst dept geol, Ohio State Univ, 49-51; geologist, Standard Oil Co, Calif, 52-58; PROF GEOL, BAKERSFIELD COL, 58-, CHMN DEPT PHYS SCI, 67- Concurrent Pos: Vis prof, Ohio State Univ, 61-71. Mem: Geol Soc Am; Soc Econ Paleontologists & Mineralogists; Am Soc Testing & Mat; Am Asn Petrol Geologists; Nat Asn Geol Teachers. Res: Geology of Pavant Range of Utah, central coast ranges and San Joaquin Valley of California and Colorado plateaus. Mailing Add: 3012 Pomona St Bakersfield CA 93305

LAUTER, FELIX H, b Somerville, NJ, Feb 12, 19; m; c 1. PARASITOLOGY. Educ: Southwestern Col, Kans, BA, 50; La State Univ, MS, 52, PhD(parasitol), 59. Prof Exp: Asst prof Biol, Birmingham-Southern Col, 55-57; assoc prof & chmn dept, Ill Col, 58-61; ASSOC PROF PARASITOL & HISTOL, UNIV SC, 61- Mem: Am Soc Zoologists; Am Soc Parasitol. Res: Hemoflagellates and intestinal flagellates from anuran hosts; electron microscope studies in parasitology. Mailing Add: Dept of Biol Sci Univ of SC Columbia SC 29208

LAUTERBACH, GEORGE ERVIN, b Bushnell, Ill, June 13, 27; m 49; c 5. PHYSICAL CHEMISTRY. Educ: Monmouth Col, BS, 49; Bradley Univ, MS, 53; Purdue Univ, PhD(biochem), 58. Prof Exp: Chemist, Starch & Dextrose Div, Northern Regional Res Lab, 49-53; res org chemist, Paper Lab, Kimberly Clark Corp, 57-65 & Pioneering & Advan Develop Lab, 65; ASSOC PROF CHEM & RES ASSOC, DIV NATURAL MAT & SYSTS, INST PAPER CHEM, 65- Mem: Am Chem Soc; Tech Asn Pulp & Paper Indust; Am Asn Cereal Chemists. Res: Polysaccharide chemistry; enzymic and chemical modification of starch; high temperature starch cooking; hemicelluloses; starch in paper coatings; top sizes and internal sizing of paper products. Mailing Add: Inst of Paper Chem Box 1048 Appleton WI 54911

LAUTERBUR, PAUL CHRISTIAN, b Sidney, Ohio, May 6, 29; m 58; c 2. PHYSICAL CHEMISTRY. Educ: Case Inst, BS, 51; Univ Pittsburgh, PhD(chem), 62. Prof Exp: Res asst, Mellon Inst, 51-52, res assoc, 52-53, jr fel, 53, fel, 55-63; assoc prof chem, 63-69, PROF CHEM, STATE UNIV NY STONY BROOK, 69- Mem: AAAS; Am Chem Soc; Am Phys Soc. Res: Nuclear magnetic resonance studies of structure and properties of molecules, crystals and biological systems; imaging by magnetic resonance zeugmatography, including biological and medical applications. Mailing Add: Dept of Chem State Univ NY Stony Brook NY 11794

LAUTMAN, DON ANTHONY, astronomy, see 12th edition

LAUTSCH, ELIZABETH VIRGINIA, b Prairie Grove, Man, May 19, 19; m 44; c 2. PATHOLOGY. Educ: Univ Man, BA, 40; Laval Univ, Md, 44; McGill Univ, MSc, 51, PhD(path), 53. Prof Exp: Demonstr path, Laval Univ, 48-50; demonstr, McGill Univ, 50-52, lectr, 53-54; assoc, Woman's Med Col Pa, 55-57; from asst prof to prof path, Sch Med, Temple Univ, 57-73, dir cardiovasc teaching, Health Sci Ctr, 69-73; PROF PATH, COL MED & DENT NJ, 73- Res: Experimental atherosclerosis; histogenesis of atherosclerosis in man. Mailing Add: Dept of Path Col of Med & Dent of NJ Piscataway NJ 08854

LAUTT, WILFRED WAYNE, b Lethbridge, Alta, Can, June 29, 46; m 68; c 2. PHYSIOLOGY, PHARMACOLOGY. Educ: Univ Alta, BSc, 68; Univ Man, MSc, 70, PhD(pharmacol), 72. Prof Exp: Med Res Coun Can fel toxicol, Univ Montreal, 72-74; ASST PROF PHYSIOL & RES SCHOLAR CAN HEPATIC FOUND, UNIV SASK, 74- Mem: Can Physiol Soc. Res: Peripheral vascular physiology and hepatic physiology, pharmacology and toxicology; vascular and metabolic consequences of autonomic nerve activity in the liver; local control of intestinal and hepatic blood flow. Mailing Add: Dept of Physiol Col of Med Univ of Sask Saskatoon SK Can

LAUTZ, WILLIAM, b New York, NY, Nov 17, 17; m 49. PLANT PATHOLOGY, AIR POLLUTION. Educ: Yale Univ, BS, 40; Cornell Univ, MS, 50. Prof Exp: Plant pathologist, USDA, 51-57, nematologist, Fla, 57-60, plant pest control inspector, 60-63, PLANT PATHOLOGIST, USDA, 63- Mem: AAAS; Am Phytopath Soc; Air Pollution Control Asn. Res: Plant pathology including breeding for disease resistance; plant nematology including chemical control of nematodes. Mailing Add: Forest Serv USDA St & Pvt Forestry 6816 Market St Upper Darby PA 19082

LAUVER, MILTON RENICK, b Springfield, Ohio, Sept 14, 20; m 48; c 4. PLASMA PHYSICS. Educ: Wittenberg Col, AB, 42; Western Reserve Univ, MS, 44, PhD(phys chem), 48. Prof Exp: Res chemist, Westvaco Div, Food Mach & Chem Corp, 48-53; res chemist chromium chem div, Diamond Alkali Co, 53-58; RES CHEMIST, LEWIS RES CTR, NASA, 58- Mem: AAAS; Am Chem Soc. Res: Adsorbent bleaching; electrode-position of copper; detergency chromium chemicals; gas dynamics; nuclear fusion plasma studies. Mailing Add: 28385 Holly Dr North Olmsted OH 44070

LAUVER, RICHARD WILLIAM, b Monmouth, Ill, Mar 15, 43; m 72. PHYSICAL CHEMISTRY. Educ: Knox Col, AB, 65; Univ Ill, Urbana, PhD(chem), 70. Prof Exp: Nat Res Coun res assoc, 72-74, RES CHEMIST, LEWIS RES CTR, NASA, 74- Mem: Am Chem Soc. Res: Physical and chemical characterization of polymer materials. Mailing Add: 20090 Carolyn Ave Rocky River OH 44116

LAUX, DAVID CHARLES, b Sarver, Pa, Jan 1 45; m 70; c 1. IMMUNOLOGY, ONCOLOGY. Educ: Washington & Jefferson Col, BA, 66; Miami Univ, MS, 68; Univ Ariz, PhD(microbiol), 71. Prof Exp: Fel immunol, Dept Microbiol, Sch Med, Pa State Univ, 71-73; ASST PROF IMMUNOL, DEPT MICROBIOL, UNIV RI, 73- Concurrent Pos: Res grant, Nat Cancer Inst, 74. Mem: Am Asn Cancer Res; Am Soc Microbiol. Res: Investigation of factors responsible for tumor mediated suppression of cellular immune reactivity. Mailing Add: Dept of Microbiol 318 Morrill Hall Univ of RI Kingston RI 02881

LAUZAU, WILBUR R, b Niagara Falls, NY, Dec 11, 28; m 54; c 3. INORGANIC CHEMISTRY, PHYSICAL CHEMISTRY. Educ: Niagara Univ, BS, 50; Univ Toledo, MS, 62, MBA, 65. Prof Exp: Develop engr, 51-54 & 56-57, develop group leader, 58-67, asst to dir prod eng, 67-69, mgr chem prod eng, 69-71, SR DEVELOP ENGR, CARBON PROD DIV, UNION CARBIDE CORP, CLEVELAND, 71- Concurrent Pos: Phi Kappa Phi gen scholar & Beta Gamma Sigma scholar bus studies, 65. Mem: Fel Am Inst Chemists. Res: Industrial carbon and graphite products, including arc carbons, pipe, carbon and graphite fibers, spectroscopic and battery electrodes, welding carbons, activated carbon and metallurgical graphite. Mailing Add: 8308 Fair Rd Strongsville OH 44136

LAUZIER, LOUIS MARCEL, b Quebec, Que, Feb 16, 17; m 45; c 3. OCEANOGRAPHY. Educ: Laval Univ, BA, 37, BSc, 41, MSc, 42, DSc(physics, chem), 46. Prof Exp: Oceanogr, Sta Biol St Laurent, Laval Univ, 42-44, lectr, Laval Univ, 45; oceanogr, Que & Atlantic Herring Invest Comt, Dept Fisheries, 45-49; oceanogr, Biol Sta, 49-70, PROG COORDR, FISHERIES RES BD CAN, 70- Concurrent Pos: Secy, Can Comt Oceanog, 70- Mem: Marine Technol Soc; NY Acad Sci; Royal Soc Can. Res: Circulation; continental shelf; fisheries oceanography. Mailing Add: 59 Meadowbrook Dr Ottawa ON Can

LAVAGNINO, EDWARD RALPH, b Fall River, Mass, Apr 3, 30; m 58; c 4. ORGANIC CHEMISTRY. Educ: Southeastern Mass Technol Inst, BS, 52; Univ Mass, MS, 55. Prof Exp: Assoc sr org chemist, 54-68, sr org chemist, 68-75, RES SCIENTIST, CHEM RES DIV, ELI LILLY & CO, 75- Mem: Am Chem Soc. Res: Preparative organic chemistry; hydrogenation and high pressure reactions. Mailing Add: Chem Res Div Eli Lilly & Co 307 E McCarty St Indianapolis IN 46206

LAVAIL, JENNIFER HART, b Evansville, Ind, Apr 2, 43; m 70. NEUROANATOMY, NEUROEMBRYOLOGY. Educ: Trinity Col, DC, BA, 65; Univ Wis, PhD(anat), 70. Prof Exp: Instr, 73-74, ASST PROF NEUROPATH, HARVARD MED SCH, 74- Concurrent Pos: Vis fel anat, Sch Med, Washington Univ, 68-69, res fel, 69-70; Nat Inst Neurol Dis & Stroke res fel neuropath, Harvard Med Sch, 70-73, spec fel, 73-76. Res: Development of the central nervous system; retrograde axonal transport. Mailing Add: Dept of Neuropath Harvard Med Sch Boston MA 02115

LAVAIL, MATTHEW MAURICE, b Abilene, Tex, Jan 7, 43; m 70. NEUROSCIENCES, CELL BIOLOGY. Educ: NTex State Univ, BA, 65; Univ Tex Med Br, PhD(anat), 69. Prof Exp: Res fel, Harvard Med Sch, 69-73, asst prof neuropath, 73-76; ASSOC PROF ANAT, MED SCH, UNIV CALIF, SAN FRANCISCO, 76- Concurrent Pos: NatEye Inst fel neuropath, Harvard Med Sch, 70-73; res assoc neurosci, Children's Hosp Med Ctr, 73-76; Nat Eye Inst res career develop award, 74- Honors & Awards: Res Award, Sigma Xi, 70. Mem: Asn Res Vision & Ophthal; Am Asn Anat; Soc Neurosci. Res: Retrograde axonal transport; neuroembryology; retinal development; inherited retinal degeneration; photoreceptor-pigment epithelial cell interactions. Mailing Add: Children's Hosp Med Ctr 300 Longwood Ave Boston MA 02115

LAVAL, WILLIAM NORRIS, b Seattle, Wash, Jan 27, 22; m 63; c 2. GEOLOGY. Educ: Univ Wash, BS, 43, MS, 48, PhD(geol), 56. Prof Exp: Field asst, US Geol Surv, 42, geologist, 43-45 & 48-49; geologist, Corps Eng, US Army, 49-51; resident geologist, Yale Dam, Ebasco Serv, Inc, Wash, 51-53; asst prof geol, Colo State Univ, 56-60; assoc prof geol & geol eng, SDak Sch Mines & Technol, 60-62; PROF GEOL & EARTH SCI & CHMN DIV NATURAL SCI, LEWIS-CLARK STATE COL, 63- Concurrent Pos: Consult, 54-56 & 62-63. Mem: Fel Geol Soc Am. Res: Stratigraphy, structure and petrology of Columbia Plateau; environmental geology; earth sciences teaching. Mailing Add: Lewis-Clark State Col Lewiston ID 83501

LAVALLE, PLACIDO DOMINICK, b New York, NY, May 13, 37; m 58; c 3. GEOMORPHOLOGY. Educ: Columbia Univ, BA, 59; Univ Southern Ill, MA, 61; Univ Iowa, PhD(geog), 65. Prof Exp: Res asst geog, Univ Iowa, 62-63; asst prof, Univ Calif, Los Angeles, 64-67 & Univ Ill, Urbana, 67-69; ASSOC PROF GEOG, UNIV WINDSOR, 69- Mem: AAAS; Asn Am Geogrs; Am Geophys Union; Nat Speleol Soc. Res: Soil geography; quantitative analysis of karst geomorphology in Kentucky and Puerto Rico; spatial patterns of soil toxin distribution in Lebec, California; application of remote sensors in environmental research. Mailing Add: Dept of Geog Univ of Windsor Windsor ON Can

LAVALLEE, ANDRE, forest pathology, see 12th edition

LAVALLEE, LORRAINE DORIS, b Holyoke, Mass, May 31, 31. MATHEMATICS. Educ: Mt Holyoke Col, AB, 53; Univ Mass, MA, 55; Univ Mich, PhD(math), 62. Concurrent Pos: From instr to asst prof math, 59-70, assoc head dept math & statist, 71-72, ASSOC PROF MATH, UNIV MASS, AMHERST, 70- Mem: Am Math Soc; Math Asn Am. Res: General topology. Mailing Add: Dept of Math Univ of Mass Amherst MA 01002

LAVALLEE, MARC, b Montreal, Que, Sept 12, 33. BIOPHYSICS, BIOMEDICAL ENGINEERING. Educ: Univ Montreal, BA, 52, MD, 58, BSc & MSc, 60; Univ Southern Calif, PhD(biophys), 62. Prof Exp: Asst physiol, Univ Montreal, 58-59, instr pharmacol, 59-60, asst prof biophys, 62-66; assoc prof, 66-70, head dept, 66-69, head div basic sci, 69-72, assoc dean res, 68-72, VDEAN EDUC, FAC MED, UNIV SHERBROOKE, 73-, PROF BIOPHYS, 69- Concurrent Pos: Can Ins Officers Asn fel, 59; Los Angeles County Heart Asn fel, Univ Southern Calif, 60-61, assoc prof, 61-63; invited prof, Univ Heidelberg, 63; Ger Res Asn fel, 63; assoc comt biophys, Nat Res Coun Can, 64-66; Can Coun Interdisciplinary Sci fel, 64-66; consult, Corning Glass Works, 66-70; assoc, Med Res Coun Can, 66-71, mem comts physiol, pharmacol & med eng, 69-, chmn comt biomed eng, 71-73; mem working party med comput, Dept Nat Health & Welfare Can, 70-72; mem bd dir, Univ Clin, Univ Sherbrooke, 71-72; pres & chmn bd, Center d'Informatique de la Sante pour l'Estrie, 71-; mem bd gov, Prof Corp Physicians Que. Mem: NY Acad Sci; Biophys Soc; Can Pharmacol Soc; Int Union Pure & Appl Physics. Res: Computer science; cellular biophysics; microelectrode techniques; electrobiology of excitable membranes; cations sensitive glass microelectrodes. Mailing Add: Dept of Biophys Univ of Sherbrooke Fac of Med Sherbrooke PQ Can

LA VALLEE, WILLIAM ALFRED, b Portland, Maine, May 25, 41; m 64; c 2. PULP & PAPER TECHNOLOGY. Educ: Bates Col, BS, 63; Boston Univ, PhD(phys chem), 70. Prof Exp: CHEMIST & ASST LICENSING MGR, S D WARREN RES LAB, SCOTT PAPER CO, WESTBROOK, 68- Mem: Tech Asn Pulp & Paper Indust. Res: Paper coatings; graphic arts; colloid chemistry. Mailing Add: 8 Birch Lane Cumberland Foreside ME 04110

LAVANISH, JEROME MICHAEL, b Cleveland, Ohio, Mar 10, 40; m 65. ORGANIC CHEMISTRY. Educ: Case Inst Technol, BS, 62; Yale Univ, MS, 63, PhD(chem), 66. Prof Exp: SR RES CHEMIST, PPG INDUSTS, 66- Mem: Am Chem Soc. Res: Synthetic organic chemistry; synthesis and action of herbicides and plant growth regulators. Mailing Add: Chem Div PPG Indust PO Box 31 Barberton OH 44203

LAVAPPA, KANTHARAJAPURA S, b India, Mar 8, 38; m 68; c 3. BIOLOGY. Educ: Univ Mysore, BS, 59, MS, 60; Boston Univ, PhD(biol), 68. Prof Exp: Lectr zool, Univ Mysore, 60-62; jr sci officer, Ministry of Sci Res, Govt India, 63-64; res asst, Children's Cancer Res Found, Boston, Mass, 65-68, fel, Lab Cytogenetics, 68-70; res assoc path, Children's Hosp Med Ctr, Boston, 70-72; asst prof pharmacol, Med Sch, Case Western Reserve Univ, 72-74; RES ASSOC, AM TYPE CULT COLLECTION, 74- Concurrent Pos: Nat Found-March of Dimes Basil O'Connor starter res grant, 73-74; ed reviewer, Lab Animal Sci, 73 & Can J Genetics & Cytol, 74; consult, Stanford Res Inst, 74. Mem: Genetics Soc Am; Tissue Cult Asn. Res: In vivo and in vitro human and mammalian cytogenetics; environmental mutagenesis, teratogenesis and carcinogenesis. Mailing Add: Am Type Cult Collection 12301 Parklawn Dr Rockville MD 20852

LAVATELLI, LEO SILVIO, b Mackinaw Isaland, Mich, Aug 15, 17; m 41; c 2. PHYSICS. Educ: Calif Inst Technol, BS, 39; Harvard Univ, PhD(physics), 50. Prof Exp: Instr physics, math & chem, Deep Springs Jr Col, Calif, 39-41; instr physics, Princeton Univ, 41, asst, 42-43; jr mem staff, Los Alamos Sci Lab, Univ Calif, 43-46; asst, Harvard Univ, 46-50; from asst prof to assoc prof, 50-59, sr staff mem, Control Systs Lab, 51-56, PROF PHYSICS, UNIV ILL, URBANA, 59- Concurrent Pos: Guggenheim fel, 57-58; consult, Phys Sci Study Comt, Nat Acad Sci, 57-59; consult, Sci Teaching Ctr, Mass Inst Technol & Harvard Project Physics, 66 & MacMillan Sci Series, 70. Mailing Add: Dept of Physics Univ of Ill Urbana IL 61801

LAVEGLIA, JAMES GARY, b Cleveland, Ohio, May 8, 47; m 68; c 3. ENTOMOLOGY. Educ: Bowling Green State Univ, BS, 69, MA, 70; Iowa State Univ, PhD(entom), 75. Prof Exp: RES ASSOC ENTOM, IOWA STATE UNIV, 72- Mem: Entom Soc Am; Sigma Xi. Res: The interaction of insecticides with soil, including their effect on soil microorganisms, as well as their breakdown in the soil. Mailing Add: Dept of Entom Insectary Bldg Iowa State Univ Ames IA 50011

LAVELLE, FAITH WILSON, b St Johnsbury, Vt, Mar 14, 21; m 47; c 1. HISTOLOGY, NEUROEMBRYOLOGY. Educ: Mt Holyoke Col, BA, 43, MA, 45; Johns Hopkins Univ, PhD(biol), 49. Prof Exp: Lab instr zool, Mt Holyoke Col, 43-45; admin asst zool, Univ Pa, 48-51, instr anat, Med Sch, 51-52; lectr, Univ Ill Col Med, 52-53, instr, 53-55, res assoc anat, 55-70; ASST PROF ANAT, STRITCH SCH MED, LOYOLA UNIV CHICAGO, 70- Concurrent Pos: USPHS res grant, Univ Ill Col Med, 53-70. Mem: Am Asn Anat. Res: Experimental alteration of development of nerve cells; proteins in neural development. Mailing Add: Dept of Anat Loyola Univ Med Ctr Maywood IL 60153

LAVELLE, GEORGE ARTHUR, b Fargo, NDak, Nov 29, 21; m 47; c 1. NEUROANATOMY. Educ: Univ Wash, BS, 46; Johns Hopkins Univ, MA, 48; Univ Pa, PhD(anat), 51. Prof Exp: Asst zool, Univ Wash, 44-46; jr instr biol, Johns Hopkins Univ, 46-48; asst instr anat, Sch Med, Univ Pa, 48-51; from instr to assoc prof, 52-65, PROF ANAT, UNIV ILL COL MED, 65- Concurrent Pos: USPHS fel, Univ Pa, 51-52; USPHS-NIH res grant, Univ Ill, 53-70; vis prof dept anat & brain res inst & Guggenheim fel, Univ Calif, Los Angeles, 68-69. Mem: AAAS; Am Asn Anat; Biol Stain Comn; Soc Develop Biol; Am Soc Cell Biol. Res: Neurocytology; cytological development of nerve cells; experimental alteration of development of nerve cells. Mailing Add: Dept of Anat Univ of Ill Col of Med Chicago IL 60612

LAVELLE, GEORGE CARTWRIGHT, b Minneapolis, Minn, Dec 1, 37; m 66; c 4. VIROLOGY. Educ: St Johns Univ, Minn, BA, 59; Univ Notre Dame, PhD(microbiol), 67. Prof Exp: Res assoc, Sch Hyg & Pub Health, Johns Hopkins Univ, 67-69; sr staff fel NIH, 69-73; RES SCIENTIST VIROL, OAK RIDGE NAT LAB, 73- Concurrent Pos: Mem fac, Univ Tenn-Oak Ridge Grad Sch Biomed Sci, 75- Mem: Am Soc Exp Path; Am Soc Microbiol. Res: Teratogenesis by Parvoviruses and Parvovirus DNA replication; proviral DNA of murine RNA tumor viruses; isolation and integration into cellular DNA. Mailing Add: Biol Div Oak Ridge Nat Lab PO Box Y Oak Ridge TN 37830

LAVELLE, JAMES W, b Lamar, Colo, Aug 11, 30; m 50; c 4. LIMNOLOGY. Educ: Abilene Christian Col, BS, 53; Univ Tex, MA, 55, PhD(stream biol), 68. Prof Exp: Instr biol, Abilene Christian Col, 55-56; PROF BIOL & HEAD DEPT, SOUTHERN COLO STATE COL, 56- Res: Primary and secondary productivity of the upper Arkansas River; stream biology and limnology. Mailing Add: Dept of Biol Southern Colo State Col Pueblo CO 81005

LAVELLE, JOHN WILLIAM, b Sacramento, Calif, Apr 26, 43; m 71. GEOLOGICAL OCEANOGRAPHY. Educ: Univ Calif, Berkeley, BA, 65; Univ Calif, San Diego, MS, 68, PhD(physics), 71. Prof Exp: Marine geophysicist, 72-73, GEOL OCEANOG, NAT OCEANOG & ATMOSPHERIC ADMIN, ENVIRON RES LABS, 73- Mem: Am Geophys Union; Am Phys Soc. Res: Experimental and theoretical studies of sediment movement and bed forms on the continental shelf. Mailing Add: Atlantic Oceanog & Meteorol Lab 15 Rickenbacker Causeway Miami FL 33149

LAVENDA, NATHAN, b New York, NY, Dec 10, 18; m 43; c 7. PHYSIOLOGY, ENDOCRINOLOGY. Educ: City Col New York, BS, 42; NY Univ, MS, 47, PhD(biol), 52. Prof Exp: Aquatic biologist, US Fish & Wildlife Serv, 44-47; biologist, Venereal Dis Res Lab, USPHS, 47-50; lectr physiol, Howard Univ, 52-56; asst dir pathophysiol, Jewish Mem Hosp, 56-60; assoc prof biol & chmn dept, NAdams State Col, 61-67; asst prof, Univ Wis-Oshkosh, 67-69; ASSOC PROF PHYSIOL & DIR RES, ILL COL PODIATRIC MED, 69- Concurrent Pos: Fel path, Albert Einstein Col Med, 57-59; prin investr cancer res proj, AEC, 53-56; consult, NY Acad Med. Mem: AAAS; Am Soc Zool; NY Acad Sci. Res: Physiological changes induced by sex hormones; study of cytological changes during oestrus; demonstration of viruses in body fluids associated with normal and malignant tissues. Mailing Add: Dept of Physiol Ill Col of Podiat Med Chicago IL 60610

LAVENDER, ARDIS RAY, b Bedford, Ind, July 2, 27; m 48; c 3. INTERNAL MEDICINE, NEPHROLOGY. Educ: Ind Univ, MD, 53. Prof Exp: Intern, Sch Med, Univ Chicago, 53-54, asst resident med, 54-56, from instr to assoc prof, 57-68, prof med, Loyola Univ Chicago, 68-75; DIR, MOSES TAYLOR KIDNEY & HYPERTENSION INST, 76- Concurrent Pos: Mem renal sect & coun circulation, Am Heart Asn; Am Heart Asn fel, Univ Chicago, 56-57; USPHS career develop award, 65-68; chief nephrol, Hines Vet Admin Hosp, 68-75. Mem: AAAS; Am Fedn Clin Res; NY Acad Sci; Int Soc Nephrol. Res: Renal physiology; biomedical engineering. Mailing Add: Moses Taylor Kidney & Hyper Inst 737 Quincy Ave Scranton PA 18510

LAVENDER, DENIS PETER, b Seattle, Wash, Oct 13, 26; m 57. PLANT PHYSIOLOGY. Educ: Univ Wash, BS, 49; Ore State Univ, MSc, 58, PhD, 62. Prof Exp: Res asst, Ore State Bd Forestry, 50-57, in chg forest physiol, Forest Res Ctr, 57-63; assoc prof, Ore Forest Res Lab, 63-70, PROF FOREST PHYSIOL, SCH FORESTRY, ORE STATE UNIV, 70- Res: Development of hardy coniferous seedlings; nutrition of second growth Douglas fir stands; reduction of the juvenile period of conifers; dormancy in Douglas fir seedlings and conifers; mineral nutrition and precocious flowering in conifers. Mailing Add: Sch of Forestry Ore State Univ Corvallis OR 97331

LAVENDER, DEWITT EARL, b Jackson Co, Ga, Nov 9, 38; m 58; c 3. MATHEMATICS, STATISTICS. Educ: Univ Ga, BS, 62, MA, 63, PhD(math, statist), 66. Prof Exp: Asst prof math, 66-68, ASSOC PROF MATH, GA SOUTHERN COL, 68-, HEAD DEPT, 70- Mem: Am Math Soc; Math Asn Am; Inst Math Statist. Res: Mathematical statistics. Mailing Add: Dept of Math Ga Southern Col Statesboro GA 30458

LAVENDER, JOHN FRANCIS, b Nov 16, 29; US citizen; m 69; c 3. VIROLOGY, MICROBIOLOGY. Educ: Drake Univ, BA, 51; Univ Ill, Champaign, MS, 53; Univ Calif, Los Angeles, PhD(infectious dis), 62. Prof Exp: NIMH fel virol, Univ Calif, Los Angeles, 62-63; sr virologist, 64-72, RES VIROLOGIST, ELI LILLY & CO, 72- Mem: NY Acad Sci. Res: Psychological stress and viral disease resistance; drugs and the entry of viruses across the blood brain barrier; development of parainfluenza, rabies, canine distemper and measles vaccines; Herpes Simplex vaccines types 1 and 2. Mailing Add: Dept MC 932 Virus Chemother Eli Lilly & Co Indianapolis IN 46206

LAVER, MURRAY LANE, b Warkworth, Ont, Mar 7, 32; m 63; c 2. ORGANIC CHEMISTRY. Educ: Ont Agr Col, BScA, 55; Ohio State Univ, PhD(org chem), 59. Prof Exp: Res chemist food sci, Westreco Co, 59-63; res chemist wood sci, Rayonier Can, Inc, 63; res scientist, Weyerhaeuser Co, 64-66, prof specialist res div, 66-68; res instr chem, Univ Wash, 68-69; ASSOC PROF FOREST PRODS CHEM, ORE STATE UNIV, 69- Mem: Am Chem Soc. Res: Pulp and paper, carbohydrate, food and wood chemistry. Mailing Add: Sch of Forestry Ore State Univ Corvallis OR 97331

LAVER, MYRON B, b Bucharest, Romania, Aug 17, 26; US citizen; m 54; c 3. ANESTHESIOLOGY. Educ: Earlham Col, AB, 48; Univ Basel, MD, 56. Prof Exp: Asst prof, 65-70, PROF ANESTHESIA, HARVARD MED SCH, 70-; ANESTHETIST, MASS GEN HOSP, 68- Concurrent Pos: NIH grant, Anesthesia Ctr, Harvard Med Sch, 68; consult, Vet Admin Respiratory Syst Res Eval Comt, 69. Mem: Asn Univ Anesthetists; Am Physiol Soc; Am Soc Anesthesiol. Res: Respiratory physiology in heart disease, during and following surgery; clinical management of patients undergoing open-heart surgery. Mailing Add: Dept of Anesthesia Mass Gen Hosp Boston MA 02114

LAVERELL, WILLIAM DAVID, b Philadelphia, Pa, July 1, 41; m 68; c 1. MATHEMATICS. Educ: Ursinus Col, BS, 63; Lehigh Univ, MS, 65, PhD(math), 69. Prof Exp: Instr math, Temple Univ, 66-70; ASST PROF MATH, IND UNIV-PURDUE UNIV, INDIANAPOLIS, 70- Mem: Am Math Soc; Math Asn Am. Res: Relating properties of a topological space to properties of its space of continuous real valued functions. Mailing Add: Dept of Math Ind Univ-Purdue Univ Indianapolis IN 46205

LAVERGNE, EDGAR ALBERT, chemical engineering, see 12th edition

LAVERTY, JOHN JOSEPH, b Chicago, Ill, May 27, 38; m 67; c 2. ORGANIC CHEMISTRY. Educ: Eastern Ill Univ, BS, 64; Univ Ariz, MS, 66. Prof Exp: Res chemist, 66-71, ASSOC SR RES CHEMIST, POLYMER DEPT, GEN MOTORS RES LABS, 71- Mem: Am Chem Soc. Res: Monomer and polymer synthesis; structure and property relationship of block and graft copolymers; polyurethanes. Mailing Add: Gen Motors Res Labs Polymers Dept 12 Mile & Mound Rd Warren MI 48075

LAVERTY, S G, b Hertfordshire, Eng, Dec 19, 22; m 48; c 2. PSYCHIATRY. Educ: Univ Edinburgh, BSc, 44, MBChB, 46; Univ London, DPM, 56. Prof Exp: Lectr psychiat, Univ Edinburgh, 56-58; asst prof, Queen's Univ, Ont, 59-60, assoc prof, 60-67, chmn comt human experimentation, 67, PROF PSYCHIAT, QUEEN'S UNIV, ONT, 67- Concurrent Pos: Dir, Addiction Studies Unit, Queen's Univ, Ont, 67-; scientist grade 4, Alcohol & Drug Addiction Res Found, Ont, 69- Mem: Can Psychiat Asn. Res: Effects of drugs on behavior, sleep and social interaction; evaluation studies of treatment methods in addictions; role of muscle activity in relation to addiction. Mailing Add: Fac of Med Queen's Univ Kingston ON Can

LAVERY, BERNARD JAMES, b New York, NY, Sept 7, 40; m 62; c 3. PHYSICAL CHEMISTRY. Educ: Mt St Mary's Col, BS, 62; Pa State Univ, PhD, 67. Prof Exp: RES CHEMIST, E I DU PONT DE NEMOURS & CO, INC, 66- Res: Nuclear magnetic resonance; liquid crystals; polymer solution rheology; textile fiber end-use research and product development. Mailing Add: Textile Fibers Dept E I du Pont de Nemours & Co Inc Wilmington DE 19898

LAVIA, ANTHONY L, b Brooklyn, NY, Aug 10, 18; m 48; c 3. PHARMACEUTICAL CHEMISTRY. Educ: St John's Univ, BS, 41. Prof Exp: Explosives chemist, Plum Brook Ord Works, Ohio, 41-42; plant supvr, Int Vitamin Corp, NY, 45-47; pharmacist, Hancock Pharm, NY, 48-49; sr res chemist, E R Squibb & Sons Div, Olin Mathieson Chem Corp, 49-63, res supvr proprietary prod develop labs, 63-68; mgr tech servs, Beech-Nut Inc, 68-69; mgr pharmaceut liaison, 69-71, mgr prod develop fluid dosage form, 71-72, HEAD PHARMACEUT FORMULATION DESIGN SECT, E R SQUIBB & SONS, 72- Mem: Acad Pharmaceut Sci; Soc Cosmetic Chem; Inst Food Technol; Am Pharmaceut Asn. Res: Development and research of pharmaceutical dosage forms specializing in the flavoring aspects and aerosol forms. Mailing Add: 66 Farms Rd Circle East Brunswick NJ 08816

LA VIA, MARIANO FRANCIS, b Rome, Italy, Jan 29, 26; nat US; m 59; c 7. PATHOLOGY. Educ: Univ Messina, MD, 49. Prof Exp: Asst gen path, Univ Palermo, 50-52; asst path, Univ Chicago, 52-57, instr anat, 57-60; from asst prof to assoc prof, Sch Med, Univ Colo, Denver, 60-68; prof, Bowman Gray Sch Med, 68-71; PROF PATH, SCH MED, EMORY UNIV, 71- Mem: Sigma Xi; Am Asn Immunol; Am Soc Exp Path; Am Soc Cell Biol; Soc Exp Biol & Med. Res: Cellular and biochemical mechanism of antibody production. Mailing Add: Dept of Path Emory Univ Sch of Med Atlanta GA 30322

LAVIER, EUGENE CLARK, b Ogdensburg, NY, Dec 26, 15; m 41; c 3. PHYSICS. Educ: US Air Force Inst Technol, BSc, 50; Johns Hopkins Univ, PhD(physics), 53. Prof Exp: Chief radiation br, Weapons Effects Div, Armed Forces Spec Weapons Proj, 53-57, dep comdr opers, Air Force Off Sci Res, 57-59, dep dir res vehicles & instrumentation, Air Res & Develop Command, 59-60, dir anal, 60-61; mgr res div, Nat Co, 61-65; mgr optical signal processing br, Perkin-Elmer Corp, 65-66; asst dir res & develop, Page Commun Engrs, 66-70; prin scientist, Northrop Corp Labs, 70-71; DIR VULNERABILITY, DEFENSE NUCLEAR AGENCY, 71- Mem: Am Phys Soc; Inst Elec & Electronics Eng. Res: Physical sciences. Mailing Add: 5800 Plainview Rd Bethesda MD 20034

LAVIETES, BEVERLY BLATT, b Pittsburgh, Pa, Mar 17, 44; m 66; c 1. DEVELOPMENTAL BIOLOGY, ONCOLOGY. Educ: Vassar Col, AB, 65; Case Western Reserve Univ, PhD(biol), 69. Prof Exp: ASST PROF PATH, SCH MED, NY UNIV, 71- Concurrent Pos: Am Cancer Soc fel, NY Univ, 69-70, Nat Cancer Inst fel, 70-71, res grant, 72-77. Mem: AAAS; Am Soc Zool; NY Acad Sci; Fedn Am Sci; Soc Develop Biol. Res: Control of cellular differentiation and phenotypic expression; alteration of cellular control mechanisms in neoplasia. Mailing Add: Dept of Path NY Univ Med Ctr New York NY 10016

LAVIGNE, ANDRE ANDRE, b Manchester, NH, Sept 6, 32; m 53; c 3. ORGANIC CHEMISTRY. Educ: St Anselm's Col, BA, 58; Univ NH, MS, 60; Lowell Tech Inst, PhD(org chem), 64. Prof Exp: Head chemist, Sylvania Elec Co, 60-61; asst prof, 63-70, ASSOC PROF ORG CHEM, ST ANSELM'S COL, 70-, DIR, INST RES & SERVS, 71- Concurrent Pos: Nat Defense Ed Act lectr adv high sch srs, 66; consult water pollution study, Merrimack River, Pub Serv Co, NH, 67-70; surv analyst effectiveness educ progs relevant to small town planning bd, 67-; dir, Water Resources Bd, State NH, 75- Mem: Am Chem Soc; NY Acad Sci. Res: Interaction of organocadmium reagents on various substituted phthalides; action of halogermanes on cyclic ethers. Mailing Add: Dept of Chem St Anselm's Col Manchester NH 03102

LAVIGNE, JOE BRYSON, organic chemistry, see 12th edition

LAVIGNE, ROBERT JAMES, b Herkimer, NY, May 30, 30; m 52; c 2. ENTOMOLOGY. Educ: Am Int Col, BA, 52; Univ Mass, MS, 58, PhD(entom), 61. Prof Exp: Res instr entom, Univ Mass, 56-59; res asst prof, 59-65, assoc prof, 65-71, PROF ENTOM, UNIV WYO, 71- Concurrent Pos: Ed, Crop Care in Wyo. Mem: Entom Soc Am; Animal Behav Soc; Soc Syst Zool. Res: Insect taxonomy, especially Diptera; insect behavior, especially robber flies, Asilidae, and horse flies, Tabanidae. Mailing Add: Box 3354 University Sta Laramie WY 82070

LAVIK, PAUL SOPHUS, b Camrose, Alta, Feb 11, 15; US citizen; m 41; c 4. BIOCHEMISTRY. Educ: St Olaf Col, AB, 37; Univ Wis, MS, 41, PhD(biol chem), 43; Western Reserve Univ, MD, 59. Prof Exp: Instr biochem, Sch Med, La State Univ, 43; from instr to asst prof, Baylor Col Med, 43-47; asst prof biochem, 47-52, asst prof, 52-70, ASSOC PROF RADIOL, SCH MED, CASE WESTERN RESERVE UNIV, 70- Mem: Am Soc Biol Chemists; Radiation Res Soc; Am Asn Cancer Res. Res: Nucleic acid metabolism; radiation biochemistry; radiation therapy. Mailing Add: Dept of Radiol Case Western Reserve Univ Cleveland OH 44106

LAVILLA, ROBERT E, b New York, NY, May 8, 26. PHYSICAL CHEMISTRY. Educ: Bethany Col, BS, 53; Cornell Univ, PhD(phys chem), 60. Prof Exp: RES CHEMIST, NAT BUR STANDARDS, 60- Concurrent Pos: Nat Res Coun-Nat Bur Standards fel, 60-61. Mem: Am Phys Soc. Res: Electron diffraction of solids and gases; optical properties of materials; x-ray absorption and emission. Mailing Add: Nat Bur of Standards Washington DC 20234

LAVIN, EDWARD, b Springfield, Mass, Jan 6, 16; m 48; c 3. POLYMER CHEMISTRY. Educ: Univ Mass, BS, 36; Tufts Col, MS, 37. Prof Exp: Res chemist, Shawinigan Resins Corp, 38-53, group leader, 53-56, sect leader, 56-60, dir appln res, 60-63, res mgr, 63-65; MGR RES, MONSANTO CO, 65- Mem: AAAS; Am Chem Soc; Inst Elec & Electronics Eng; Soc Aerospace Mat & Process Eng. Res: Vinyl high polymers; acetal resins. Mailing Add: Monsanto Co Springfield MA 01051

LAVIN, FRED, b Chicago, Ill, May 23, 14; m 54; c 2. RANGE SCIENCE. Educ: Utah State Agr Col, BS, 37; Univ Chicago, PhD(bot), 53. Prof Exp: Jr tech foreman, Soil Conserv Serv, USDA, 34-38, asst ranger, US Forest Serv, 41-42, range conservationist, Rocky Mt Forest & Range Exp Sta, 46-53, res range scientist, Agr Res Serv, 54-67, RANGE SCIENTIST, AGR RES SERV, ROCKY MT FOREST & RANGE EXP STA, USDA, 67- Mem: Soc Range Mgt; Am Soc Agron; Soil Sci Soc Am; Crop Sci Soc Am. Res: Range ecology and revegetation. Mailing Add: Rocky Mt Forest & Range Exp Sta Flagstaff AZ 86001

LAVIN, GEORGE ISRAEL, b Clifton Forge, Va, Sept 14, 03. CHEMISTRY. Educ: Univ Va, BS, 25; Johns Hopkins Univ, PhD, 28. Prof Exp: Res assoc chem, Princeton Univ, 29-34; fel, Rockefeller Inst, 34-35, asst, 35-50; res chemist, Emerson Drug Co, 50-51; chemist, Aberdeen Proving Ground, 51-75; RETIRED. Mem: AAAS; Am Chem Soc; Harvey Soc. Res: Atomic hydrogen; dissociated water vapor; catalysis in relation to atoms and radicals; ammonia discharge tube; spectrochemistry of enzymes and proteins; ultraviolet microscopy; chemistry of the upper atmosphere; animal bio-structure simulation in regard to the mechanism of task performance. Mailing Add: 1562 NE 191st St Apt 310 North Miami Beach FL 33179

LAVIN, PETER MASLAND, b Philadelphia, Pa, Apr 16, 35; m 58; c 2. GEOPHYSICS. Educ: Princeton Univ, BSE, 57; Pa State Univ, PhD(geophys), 62. Prof Exp: From instr to asst prof, 60-71, ASSOC PROF GEOPHYS, PA STATE UNIV, 71- Mem: Soc Explor Geophys; Am Geophys Union. Res: Exploration geophysics with emphasis on gravity and magnetic interpretation; time-series analysis; applications of paleomagnetism. Mailing Add: Dept Geosci 202 Mineral Sci Bldg Pa State Univ University Park PA 16802

LAVIN, PHILIP TODD, b Rochester, NY, Nov 21, 46. BIOSTATISTICS. Educ: Univ Rochester, AB, 68; Brown Univ, PhD(appl math), 72. Prof Exp: Res asst prof appl math, Brown Univ, 72-74; RES ASST PROF BIOSTATIST, STATE UNIV NY BUFFALO, 74- Concurrent Pos: Coord statistician, Gastrointestinal Tumor Study Group, 74-; protocol statistician, Eastern Coop Oncol Group, 74-; biostatistician, US deleg Japan, Nat Cancer Inst Sci Exchange Comt Gastric Oncol, 75. Mem: Inst Math Statist; Am Statist Asn. Res: Biometry; clinical trials; pattern analysis; experimental design; statistical computing; time series; stochastic modeling; econometrics. Mailing Add: Statist Lab State Univ of NY at Buffalo Amherst NY 14226

LAVINE, JAMES PHILIP, b Syracuse, NY, Dec 3, 44; m 71; c 1. THEORETICAL PHYSICS. Educ: Mass Inst Technol, BS, 66; Univ Md, PhD(physics), 71. Prof Exp: Res assoc physics, Univ Liege, 71-73; res asst prof, Laval Univ, 73-74; RES ASSOC PHYSICS, UNIV ROCHESTER, 74- Mem: Am Phys Soc; Belgian Phys Soc; Europ Phys Soc. Res: Relativistic effects in low-energy nucleon-nucleon scattering; pion-nucleon amplitudes and radiative scattering; properties of two-body relativistic wave-equations; inverse scattering problems; numerical analysis. Mailing Add: Dept of Physics & Astron Univ of Rochester Rochester NY 14627

LAVINE, LEROY S, b Jersey City, NJ, Oct 28, 18; m 46; c 2. ORTHOPEDIC SURGERY. Educ: NY Univ, AB, 40, MD, 43; Am Bd Orthop Surg, dipl, 55. Prof Exp: Res instr orthop surg, Col Med, Ind Univ, 51-52; from instr to assoc prof, 52-65, PROF ORTHOP SURG & HEAD DEPT, COL MED, STATE UNIV NY DOWNSTATE MED CTR, 65- Concurrent Pos: Consult, Am Mus Natural Hist, 55- & Brooklyn Vet Admin Hosp, 65-; attend orthop surgeon, Long Island Jewish Hosp, 57-; vis surgeon & dir, Kings County Hosp, 65-; adj prof biol, Grad Sch, NY Univ, 66-; clin prof, Med Sch, State Univ NY Stony Brook, 72- Mem: Fel Am Col Surg; Am Asn Phys Anthrop; fel Am Acad Orthop Surg; fel NY Acad Sci; fel NY Acad Med. Res: Bone growth and metabolism and mechanisms of calcification; clinical research in bone healing and nerve root compression syndromes; physical properties of and piezoelectricity in bone; electrical enhancement of bone growth. Mailing Add: 1300 Union Turnpike New Hyde Park NY 11040

LAVINE, RICHARD BENGT, b Philadelphia, Pa, June 27, 38; m 65; c 1. MATHEMATICAL ANALYSIS. Educ: Princeton Univ, AB, 61; Mass Inst Technol, PhD(math), 65. Prof Exp: Instr math, Aarhus Univ, 65-66; asst prof, Cornell Univ, 66-71; vis prof, Inst Theoret Physics, Univ Geneva, 71; mem staff, Inst Advan Study, 71-72; ASSOC PROF MATH, UNIV ROCHESTER, 72- Mem: Am Math Soc. Res: Mathematics of quantum mechanics; functional analysis. Mailing Add: 360 Rockingham St Rochester NY 14620

LAVINE, ROBERT ALAN, b Chicago, Ill, Feb 18, 41. NEUROPHYSIOLOGY, PSYCHOLOGY. Educ: Univ Chicago, BS, 62, PhD(physiol), 69. Prof Exp: Instr, 69-70, ASST PROF PHYSIOL, SCH MED, GEORGE WASHINGTON UNIV, 70- Concurrent Pos: NIH fel, George Washington Univ, 69-70. Res: Neuron discharge patterns and averaged evoked potentials in audition; human psychophysiology using computer analysis, averaged evoked potentials; evolution and modification of behavior. Mailing Add: Dept of Physiol George Washington Univ Med Ctr Washington DC 20005

LAVINE, THEODORE FREDERICK, b Cambridge, Mass, Aug 25, 06; m 31; c 2. ORGANIC CHEMISTRY. Educ: Clarkson Tech Inst, BS, 27; Univ Pa, PhD(org chem), 34. Prof Exp: Chemist, Tex Co, 27-29; chemist, Res Inst, Lankenau Hosp, 29-58; chemist, 58-72, sr mem, 72-75, EMER SR MEM, DIV BIOCHEM, INST CANCER RES, 75- Concurrent Pos: Consult, Am Oncol Hosp, 53- Mem: Am Chem Soc; Am Soc Biol Chemists. Res: Amino acids; proteins; chemistry of sulfur compounds of biological interest. Mailing Add: Inst for Cancer Res 7701 Burholme Rd Fox Chase Philadelphia PA 19111

LAVIOLETTE, FRANCIS A, b Bellingham, Wash, Nov 22, 19; m 46; c 2. PLANT PATHOLOGY. Educ: Purdue Univ, BS, 53, MS, 56. Prof Exp: Res asst plant path, 55-56, from instr to asst prof, 56-71, ASSOC PROF PLANT PATH, PURDUE UNIV, 71- Concurrent Pos: Consult, Ind Crop Improv Asn, 67-71. Mem: Am Phytopath Soc; Am Inst Biol Sci. Res: Host relationships and methods of control of diseases of soybeans and diseases of forage crops. Mailing Add: Dept of Bot & Plant Path Lilly Hall Life Sci Purdue Univ West Lafayette IN 47906

LAVKULICH, L M, b Coaldale, Alta, Apr 28, 39; m 62; c 2. SOIL SCIENCE. Educ: Univ Alta, BSc, 61, MSc, 63; Cornell Univ, PhD(soil sci), 67. Prof Exp: From asst prof to assoc prof, 66-75, PROF SOIL SCI, UNIV BC, 75- Mem: Am Soc Agron; Soil Sci Soc Am; Can Soc Soil Sci; Int Soc Soil Sci. Res: Soil classification, genesis and mineralogy, including stability relationships in common soil minerals; soil-landscape-plant inter-relationships and environmental classification. Mailing Add: Dept of Soil Sci Univ of BC Vancouver BC Can

LAVOIE, MARCEL ELPHEGE, b Manchester, NH, July 16, 17; m 42; c 2. ZOOLOGY. Educ: St Anselm's Col, BA, 40; Univ NH, MS, 52; Syracuse Univ, PhD(zool), 56. Prof Exp: Instr chem, St Anselm's Col, 46-47; instr biol, Univ NH, 50-52; lectr zool, Syracuse Univ, 52-55; asst prof, 55-61, ASSOC PROF ZOOL, UNIV NH, 61- Res: Mammalian anatomy and physiology. Mailing Add: Dept of Zool Univ of NH Durham NH 03824

LAVOIE, RONALD LEONARD, b Manchester, NH, Apr 21, 33; m 59; c 2. METEOROLOGY. Educ: Univ NH, BA, 54; Fla State Univ, MS, 56; Pa State Univ, PhD, 68. Prof Exp: Chief observer, Mt Wash Observ, 57-59; asst prof meteorol, Univ Hawaii, 59-68; assoc prof, Pa State Univ, 68-72; assoc dir meteorol prog, NSF, 72-73; DIR ENVIRON MODIFICATION OFF, NAT OCEANIC & ATMOSPHERIC ADMIN, 73- Concurrent Pos: NSF sci fac fel, 63-64. Mem: AAAS; Am Meteorol Soc; Am Geophys Union. Res: Cloud physics and weather modification; numerical modeling on the mesoscale; tropical meteorology. Mailing Add: Nat Oceanic & Atmospheric Admin 6010 Exec Blvd Rockville MD 20852

LAVOIE, VICTORIN, b Petite Riviere, Que, Apr 8, 28; m 58; c 4. PLANT ECOLOGY. Educ: Laval Univ, BS, 53; Univ Montreal, MSc, 57; Univ Madrid, DSc(natural sci), 59. Prof Exp: Res officer ecol, Que Dept Agr, 60-62, chief admin div, 62-65; PROF PLANT ECOL, LAVAL UNIV, 65-, DEAN FAC AGR & FOOD SCI, 71- Res: Classification of plants in flood plain in New Jersey; ecological study of pine stands in the Sierra de Guadarrama in Spain; blueberries; sugar maple stands. Mailing Add: Fac of Agr & Food Sci Laval Univ Ste-Foy PQ Can

LAVROFF, VIACHESLAV V, b Biisk, Russia, Sept 22, 10; nat US. MATHEMATICS. Educ: Ga Inst Technol, BS, 33; Emory Univ, MA, 38. Prof Exp: Prof math & comptroller, 30-72, vpres financial affairs, 72-75, EMER PROF MATH & EMER V PRES FINANCIAL AFFAIRS, GA STATE UNIV, 75- Mailing Add: 4250 Carmain Dr NE Atlanta GA 30342

LAVY, TERRY LEE, b Greenville, Ohio, Feb 9, 36; m 55; c 3. WEED SCIENCE, SOIL CHEMISTRY. Educ: Ohio State Univ, BS, 58, MS, 59; Purdue Univ, PhD(plant nutrit), 62. Prof Exp: Lab supvr soil chem classification, Ohio State Univ, 58-59; from asst prof to assoc prof, 62-75, PROF AGRON, UNIV NEBR-LINCOLN, 75- Mem: Weed Sci Soc Am; Am Soc Agron. Res: Factors affecting the mobility and degradation of pesticides in the soil profile. Mailing Add: 318 Keim Hall Univ of Nebr Lincoln NE 68503

LAW, ALVIN GEORGE, b Hill City, Kans, July 26, 15; m 40; c 3. AGRONOMY. Educ: Kans State Col, BS, 38, MS, 40. Prof Exp: From instr to prof farm crops, Wash State Univ, 41-69; mkt specialist, US Agency Int Develop, Miss State Univ Contract, India, 69-71; PROF AGRON, WASH STATE UNIV, 71- Mem: Fel AAAS; fel Am Soc Agron; Soil Conserv Soc Am; Soc Range Mgt. Res: Pasture management; breeding of forage crops; effect of clipping on the vegetative development of some perennial grasses; forage seed. Mailing Add: Dept of Agron & Soils Wash State Univ Pullman WA 99163

LAW, AMY STAUBER, b Philadelphia, Pa, June 26, 38; m 65. CLINICAL BIOCHEMISTRY. Educ: Mt Holyoke Col, AB, 59; Univ Del, MS, 63, PhD(biochem), 69. Prof Exp: Res chemist, AviSun Corp, 61-65; chief lab sect, Meat Inspection Div, Del State Bd Agr, 68-69; CLIN BIOCHEMIST, WILMINGTON MED CTR, 69- Mem: AAAS; Am Chem Soc; Am Asn Clin Chemists; Asn Women Sci; NY Acad Sci. Res: Fetal and neonatal medicine; development of clinical methods with special application to fetal and neonatal patients; endocrine assays for clinical application; clinical applications of protein biochemistry; hemoglobinopathies. Mailing Add: Res Lab Mem Div Wilmington Med Ctr PO Box 1548 Wilmington DE 19899

LAW, CECIL E, b Vancouver, BC, Nov 27, 22; m 45; c 6. OPERATIONS RESEARCH. Educ: Univ BC, BA, 50. Prof Exp: Head animal field exp sect, Suffield Exp Sta, Defence Res Bd, 51, head arctic oper res sect, Defence Res Northern Lab, 51-54, head weapons effects & field trials sect, Can Army Oper Res Estab, 55-58, head oper gaming & tactics sect, 58-60; supvr opers res, Can Industs Ltd, 60-61, opers res mgr, 61-62; sr opers res analyst, Can Nat Rwy, 62-64, coordr opers anal, 64-66; PROF OPERS RES, SCH BUS, QUEEN'S UNIV, ONT, 66-, PROF COMPUT SCI, 69-, EXEC DIR, CAN INST GUIDED GROUND TRANSPORT, 71- Concurrent Pos: Lectr, Exten Dept, McGill Univ, 64-66. Mem: Opers Res Soc Am; Wildlife Soc; Am Inst Indust Eng; Can Soc Wildlife & Fishery Biol; Can Opers Res Soc (vpres, 66, pres, 67). Res: Wildlife ecology and population dynamics, particularly Arctic; operations research, especially military and civil operational gaming and simulation; theoretical and applied critical path analysis and program evaluation and review technique; transportation research. Mailing Add: Can Inst Guided Ground Transport Queen's Univ Kingston ON Can

LAW, DAVID BARCLAY, b Menominee, Mich, Nov 28, 14; m 44; c 2. DENTISTRY. Educ: Northwestern Univ, BS & DDS, 38, MS, 41. Prof Exp: From instr to assoc prof pedodont, Dent Sch, Northwestern Univ, 38-47; from asst prof to assoc prof pedodont, 47-64, exec off, 49-73, PROF PEDODONT, DENT SCH, UNIV WASH, 64- Concurrent Pos: Consult, US Army, 50; USPHS, 55; mem exam bd, Am Bd Pedodont, 55. Mem: Am Soc Dent for Children; Int Asn Dent Res; Am Acad Pedodont (pres, 56). Res: Clinical and histological studies in pulp therapy in the young primary and permanent tooth; polygraph studies of emotional reactions of children undergoing dental stress. Mailing Add: Dept of Pedodont Univ of Wash Dent Sch Seattle WA 98105

LAW, DAVID H, b Milwaukee, Wis, July 24, 27; m 49; c 5. INTERNAL MEDICINE, GASTROENTEROLOGY. Educ: Cornell Univ, AB, 50, MD, 54. Prof Exp: Intern med, NY Hosp, 54-55, asst resident, 55-57, asst physician to outpatients, 57-58, physician to out-patients & dir personnel health serv, 58-60; med dir out-patient dept & chief div gastroenterol, Vanderbilt Univ Hosp, 60-69; PROF MED, SCH MED, UNIV NMEX, 69-; CHIEF MED SERV, ALBUQUERQUE VET ADMIN HOSP, 69- Concurrent Pos: NIH fel, Nat Cancer Inst, 57-58; spec consult interdept comt nutrit for nat defense, NIH, 62-63; attend physician, Thayer Vet Admin Hosp, 62-69. Mem: Fel Am Col Physicians; Am Fedn Clin Res; fel Am Pub Health Asn; Am Gastroenterol Asn; Pan-Am Med Asn. Res: Inflammatory bowel disease; malabsorption; gastric secretion; medical care; nutrition; out-patient clinics; delivery of health care. Mailing Add: Dept of Med Univ of NMex Sch of Med Albuquerque NM 87106

LAW, GEORGE ROBERT JOHN, b Vermilion, Alta, June 4, 28; US citizen; m 50; c 2. POULTRY GENETICS. Educ: Univ BC, BSA, 50; Wash State Univ, MS, 57; Univ Calif, Davis, PhD(genetics), 61. Prof Exp: Immuno-geneticist, Hy-Line Int, Pioneer Hi-Bred Int Inc, 61-72; ASSOC PROF ANIMAL SCI, COLO STATE UNIV, 73- Mem: AAAS; Genetics Asn Am; Poultry Sci Asn; World Poultry Sci Asn. Res: Immuno-genetic studies of turkeys and chickens including blood type variation, serum protein, egg white protein and isozyme polymorphism; teaching, research and application of studies to breeding of poultry. Mailing Add: Dept of Animal Sci Colo State Univ Ft Collins CO 80521

LAW, HAROLD BELL, b Douds, Iowa, Sept 7, 11, M 42; c 3. PHYSICS. Educ: Kent State Univ, BS, 34; Ohio State Univ, MS, 36, PhD(physics), 41. Prof Exp: Asst physics dept, Ohio State Univ, 38-41; physicist, RCA Labs, 41-62, DIR MAT & DISPLAY DEVICE LAB, PICTURE TUBE DIV, RCA CORP, 62- Honors & Awards: TV Broadcasters Asn Medal, 46; Zworkin Prize, 55; consumer electronics Group Award, Inst Elec & Electronics Eng, 66; Francis Rice Darne Mem Award, Soc Info Display, 75; Lamme Medal, Inst Elec & Electronics Engrs, 75. Mem: Am Phys Soc; fel Inst Elec & Electronics Eng; fel Soc Info Display. Res: Color television kinescopes. Mailing Add: Mat & Display Device Lab David Sarnoff Res Ctr RCA Corp Princeton NJ 08540

LAW, JAMES PIERCE, JR, b Atlanta, Tex, Aug 4, 16; m 41; c 1. WATER POLLUTION. Educ: Univ Tex, BS, 39; Southern Methodist Univ, MS, 61; Tex A&M Univ, PhD(soil physics), 65. Prof Exp: Chemist, Convair Div, Gen Dynamics Corp, Tex, 56-61; res asst, Agr Exp Sta, Tex A&M Univ, 61-65; soil scientist, 65-67, res soil scientist, 67-73, SUPVRY SOIL SCIENTIST, ROBERT S KERR ENVIRON RES LAB, OFF RES & DEVELOP, ENVIRON PROTECTION AGENCY, 73- Mem: Am Soc Agron; Soil Sci Soc Am; Int Soc Soil Sci; Am Soc Agr Eng. Res: Water pollution problems of irrigation return flows; soil and plant systems for treating wastewaters such as cannery, sewage effluents and animal wastes. Mailing Add: Robert S Kerr Environ Res Lab PO Box 1198 Ada OK 74820

LAW, JIMMY, b Seremban, Malaysia, Sept 23, 42; m 69. THEORETICAL PHYSICS. Educ: Univ London, BSc, 63, PhD(physics), 68. Prof Exp: Teaching fel physics, McMaster Univ, 66-69; asst prof, 69-74, ASSOC PROF PHYSICS, UNIV GUELPH, 74- Mem: Brit Inst Physics; Can Asn Physicists. Res: Theoretical calculations in nuclear and hypernuclear physics; inner shell vacancy creation mechanisms in atomic physics. Mailing Add: Dept of Physics Univ of Guelph Guelph ON Can

LAW, JOHN HAROLD, b Cleveland, Ohio, Feb 27, 31; m 56. BIOCHEMISTRY. Educ: Case Inst Technol, BS, 53; Univ Ill, PhD(chem), 57. Prof Exp: Res fel, Harvard Univ, 57-58; instr chem, Northwestern Univ, 58-59; from instr to asst prof, Harvard Univ, 59-65; PROF BIOCHEM, UNIV CHICAGO, 65-, PROF CHEM, 67- Mem: Am Chem Soc; Am Soc Biol Chemists; Am Soc Microbiol. Res: Insect biochemistry; lipid metabolism; enzymology. Mailing Add: Dept of Biochem Univ of Chicago Chicago IL 60637

LAW, LLOYD WILLIAM, b Ford City, Pa, Oct 28, 10; m 42; c 2. ONCOLOGY. Educ: Univ Ill, BS, 31; Harvard Univ, AM, 35, PhD(biol), 37. Prof Exp: Instr high sch, Ill, 31-33; asst, Harvard Univ, 36-37; res assoc, Stanford Univ, 37-38; Finney-Howell med res fel physiol genetics, Jackson Mem Lab, 38-41, Commonwealth Fund fel cancer res, 41-42; sci dir, Jackson Mem Lab, 46-47; sr geneticist, 47-54, SCIENTIST DIR, NAT CANCER INST, 54-, CHIEF LAB CELL BIOL & MEM SCI DIRECTORATE, 71- Concurrent Pos: Harvard Univ Parker fel, Stanford Univ, 37-38; trustee, Jackson Mem Lab, 47-; mem study sect cancer chemother, NIH, 56-59, drug eval panel, Nat Serv Ctr, 56-59, sci adv bd, Roswell Park Mem Inst, 57-, adv bd, Children's Cancer Found, expert adv panel cancer, WHO, 60- & adv sci bd, Hektoen Inst Chicago, 64-; mem, US Nat Comt, Int Union Against Cancer, 68-72 & 72- Honors & Awards: A F Rosenthal Award, AAAS, 58; G H A Clowes Award, Am Asn Cancer Res, 65; Meritorious Serv Award, USPHS, 65, Distinguished Serv Award, 69; Alexander Pascoli Prize, Univ Perugia, 69; G B Mider Lect Award, 70. Mem: Am Soc Exp Path; Am Asn Cancer Res (pres, 68-69); Transplantation Soc; Soc Exp Biol & Med; Soc Exp Leukemia. Res: Genetics; factors affecting development of leukemia and breast tumors; immunogenetics of the mouse; tumor immunology; chemotherapy of neoplasms. Mailing Add: Lab of Cell Biol Nat Cancer Inst Bethesda MD 20014

LAW, MARGARET ELIZABETH, b Birmingham, Eng, May 6, 34; m 57. EXPERIMENTAL HIGH ENERGY PHYSICS. Educ: Univ Birmingham, Eng, BSc, 55, PhD(high energy physics), 58. Prof Exp: Nat Res Coun fel nuclear physics, McMaster Univ, 58-60; res fel, 61-67, res assoc, 67-71, SR RES ASSOC HIGH ENERGY PHYSICS, HARVARD UNIV, 71-, LECTR, 72- Mem: Am Phys Soc; Brit Inst Physics. Res: Experimental research in strong interactions; polarization effects of hadron scattering at energies greater than 100 giga electron volts. Mailing Add: Lyman Lab of Physics Harvard Univ Cambridge MA 02138

LAW, O THOMAS, physiological psychology, see 12th edition

LAW, PAUL ARTHUR, b Lowell, Mass, Sept 19, 34; m 60; c 4. PHOTOGRAPHIC CHEMISTRY. Educ: Lowell Tech Inst, BS, 56; Mich State Univ, PhD(org chem), 62. Prof Exp: Res chemist, Dow Corning Corp, 56-58; fel polypeptide synthesis, Fla State Univ, 62-63; sr res chemist, 63-69, TECH ASSOC, EASTMAN KODAK CO, 69- Res: Color photographic chemistry. Mailing Add: 56 Briar Hill Rochester NY 14636

LAW, RONALD DEE, b Brigham City, Utah, July 14, 29; m 51; c 3. ANALYTICAL CHEMISTRY. Educ: Brigham Young Univ, BS, 54. Prof Exp: Chemist, Riker Labs Div, Rexall Drug Co, Calif, 55-57; engr, Douglas Circraft Co, Inc, 57; chemist, Res Ctr, Richfield Oil Co, 57-58; from chemist to sr chemist, 58-67, ASSOC SCIENTIST, WASATCH DIV, THIOKOL CHEM CORP, 67- Mem: Am Chem Soc; fel Am Inst Chem. Res: Characterization and analysis of polymers and curing agents used in solid propellants; separation and identification of low molecular weight impurities using gel permeation chromatography, liquid chromatography, infrared spectrophotometry, and chemical analysis; functional group analysis. Mailing Add: 408 North Second E Brigham City UT 84302

LAW, WILLIAM BROUGH, b Elko, Nev, Oct 11, 32; m 56; c 3. PLASMA PHYSICS. Educ: Univ Nev, BSc, 54; Ohio State Univ, PhD(nuclear physics), 60. Prof Exp: Physicist, Armour Res Found, 60; staff mem, Sandia Lab, 60-65; asst prof, 65-68, ASSOC PROF PHYSICS, COLO SCH MINES, 68- Res: Gamma ray spectroscopy; accelerator physics. Mailing Add: Dept of Physics Colo Sch of Mines Golden CO 80401

LAWANI, SAMUEL ADETUNJI, b Ipetu, Nigeria, Mar 7, 39. PHYSICAL CHEMISTRY. Educ: Morehouse Col, BS, 64; Howard Univ, PhD(phys chem), 72. Prof Exp: ASST PROF CHEM, STATE UNIV NY BUFFALO, 70- Mem: Am Chem Soc. Res: Kinetics and mechanisms of fast reactions; kinetics and mechanisms of adsorption on solid surfaces. Mailing Add: Dept of Chem State Univ Col 1300 Elmwood Ave Buffalo NY 14222

LAWFORD, GEORGE ROSS, b Toronto, Ont, Feb 27, 41; m 66. BIOCHEMISTRY, CELL BIOLOGY. Educ: Univ Toronto, BSc, 63, PhD(biochem), 66. Prof Exp: Can Med Res Coun fel, 66-68; asst prof biochem, McMaster Univ, 68-73; MEM STAFF, WESTON RES CTR, 73- Mem: Can Biochem Soc; Brit Biochem Soc. Res: Functional significance of interactions between subcellular components; regulation of protein biosynthesis and the adenyl cyclase system. Mailing Add: Weston Res Ctr 3456 Yonge St Toronto ON Can

LAWHEAD, JAMES STOUT, b Sedalia, Mo, Aug 9, 19; m 42; c 2. ORGANIC CHEMISTRY. Educ: Park Col, AB, 40; Univ Nebr, AM, 42. Prof Exp: Org chemist, Merck & Co, 42-46 & George A Breon & Co, 46-47; org chemist, 47-57, head fine chem mfg, 57-61, mgr chem mfg, 61-65, dir mfg opers, 65-69, PRES, SEARLE & CO, 69- Concurrent Pos: Mem res staff, Nat Defense Res Coun, 41-42. Mem: Am Chem Soc. Res: Organic syntheses; pharmaceuticals; bile acids and other steroids; production administration. Mailing Add: Searle & Co GPO Box 3826 San Juan PR 00936

LAWHEAD, ROBERT BLAYNEY, physics, acoustics, see 12th edition

LAWING, WILLIAM DENNIS, b Charlotte, NC, Mar 29, 35; m 57; c 3. STATISTICS. Educ: NC State Col, BS, 57, MS, 59; Iowa State Univ, PhD(statist), 65. Prof Exp: Statistician, Res Triangle Inst, 65-69; ASSOC PROF INDUST & EXP STATIST, UNIV RI, 69- Concurrent Pos: Adj prof, Duke Univ, 66-67; vis lectr, Iowa State Univ, 67-68; adj assoc prof, NC State Univ, 68-69. Mem: Inst Math Statist; Am Statist Asn; Math Asn Am. Res: Industrial applications of statistics; quality control; operations research; sequential analysis; decision theory; survey sampling. Mailing Add: Dept Comput Sci & Ext Statist Univ of RI Kingston RI 02881

LAWLER, CHARLES WESLEY, b San Antonio, Tex, Nov 3, 24; m 52; c 3. CHEMISTRY. Educ: Tex Col Arts & Indust, BS & MS, 55. Prof Exp: From asst phys chemist to assoc phys chemist, Southwest Res Inst, 55-59, from res engr to sr res engr, 59-65; mgr, Fuels & Lubricants Div, Alcor, Inc, 65-69, Vpres, 69-71; owner, Lubri-Tech Lab, 71-73; owner, Spectro-Chem Labs, 73-74, PRES, SPECTRO-CHEM LABS, INC, 74- Mem: Am Chem Soc; Sigma Xi; Am Soc Lubrication Eng; Soc Automotive Eng. Res: Fuels and lubricants research, testing and evaluations. Mailing Add: Spectro-Chem Labs, Inc PO Box 17035 San Antonio TX 78217

LAWLER, GEORGE HERBERT, b Kingston, Ont, June 13, 23; m 50; c 5. FISHERIES. Educ: Queen's Univ Ont, BA, 46; Univ Western Ont, MSc, 48; Univ Toronto, PhD, 59. Prof Exp: Demonstr, Zool Lab, Queen's Univ Ont, 44-46 & Univ Western Ont, 46-48; assoc div ichthyol, Royal Ont Mus Zool, 48-50; asst scientist, Fisheries Res Bd, Man, 50-57, assoc scientist, Ont, 57-61, sr scientist, Man, 60-72, from asst dir to dir, Freshwater Inst, 72-74, GEN DIR, WESTERN REGION, DEPT ENVIRON, FISHERIES & MARINE SERV, 74- Concurrent Pos: Hon prof, Dept Zool, Univ Man, 68- Mem: Am Fisheries Soc; Am Inst Fishery Res Biol; Int Asn Great Lakes Res; Int Asn Theoret & Appl Limnol. Res: Population dynamics; parasitology and taxonomy of fishes. Mailing Add: Freshwater Inst 501 Univ Crescent Winnipeg MN Can

LAWLER, HELEN CLAIRE, biochemistry, deceased

LAWLER, JOHN C, organic chemistry, see 12th edition

LAWLER, RONALD GEORGE, b Centralia, Wash, May 19, 38; m 61; c 2. PHYSICAL ORGANIC CHEMISTRY. Educ: Calif Inst Technol, BS, 60; Univ Calif, Berkeley, PhD(chem), 64. Prof Exp: Res assoc chem, Columbia Univ, 63-65; from asst prof to assoc prof, 65-73, PROF CHEM, BROWN UNIV, 73- Concurrent Pos: NSF fel, 63-64; Alfred P Sloan res fel, 70-71. Mem: Am Chem Soc. Res: Theoretical organic chemistry; electron and nuclear magnetic resonance; chemistry of free radicals and radical ions. Mailing Add: Dept of Chem Brown Univ Providence RI 02912

LAWLESS, EDWARD WILLIAM b Jacksonville, Ill, Apr 9, 31; m 59; c 6. SCIENCE POLICY, PHYSICAL INORGANIC CHEMISTRY. Educ: Ill Col, AB, 53; Univ Mo, PhD(phys chem), 60. Prof Exp: Assoc chemist, 59-64, sr chemist, 64-66, prin chemist, 66-73, HEAD TECHNOL ASSESSMENT PROGS, MIDWEST RES INST, 73- Mem: AAAS; Am Chem Soc; World Future Soc; Int Soc Technol Assessment; Sigma Xi. Res: Technology forecast, risk assessment and societal effects analysis; environmental chemistry and pollution control; chemistry of pesticides, fluorine, metal hydrides; correlations of chemical structures with properties; spectroscopic analysis; vacuum techniques; kinetics. Mailing Add: Phys Sci Div Midwest Res Inst 425 Volker Blvd Kansas City MO 64110

LAWLESS, GREGORY BENEDICT, b Covington, Va, Jan 5, 40; m 66; c 1. PHARMACEUTICAL CHEMISTRY. Educ: Fordham Univ, BS, 62; St Johns Univ NY, MS, 65; Temple Univ, PhD(pharmaceut chem), 69. Prof Exp: Tech rep, 69-71, regional sales mgr, 71-73, nat sales mgr, 73-74, MKT MGR, INSTRUMENT PROD, SCI & PROCESS DIV, E I DU PONT DE NEMOURS & CO, INC, 74- Mem: Am Chem Soc. Res: The development of analytical and process instrumentation and their application to current measurement problems. Mailing Add: Instrum Prod Sci & Process Div E I du Pont de Nemours & Co Inc Wilmington DE 19898

LAWLESS, JAMES GEORGE, b Brooklyn, NY, Aug 18, 42; m 66; c 1. ANALYTICAL CHEMISTRY. Educ: Lafayette Col, BS, 64; Purdue Univ, MS, 66; Kans State Univ, PhD(chem), 69. Prof Exp: RES SCIENTIST MASS SPECTROMETRY, AMES RES CTR, NASA, 69- Concurrent Pos: Co-investr returned lunar samples, NASA, 70- Mem: Am Soc Mass Spectrometry; Geochem Soc; Meteoritic Soc. Res: Mass spectrometry of organic compounds; analysis of lunar samples and meteorites for carbon compounds. Mailing Add: Ames Res Ctr NASA Moffet Field CA 94035

LAWLESS, KENNETH ROBERT, b Key West, Fla, Aug 21, 22; m 52; c 4. MATERIALS SCIENCE. Educ: Lynchburg Col, BS, 46; Univ Va, PhD(chem), 51. Prof Exp: Fulbright fel, Univ Norway, 51-52; res assoc chem, 52-60, from asst prof to assoc prof, 60-68, PROF CHEM, UNIV VA, 68- Mem: Electron Micros Soc Am; Am Crystallog Asn; Int Asn Dent Res; Inst Mining, Metall & Petrol Engrs; Microbeam Anal Soc. Res: Chemistry and physics of solids and surfaces; x-ray diffraction; electron diffraction and electron microscopy. Mailing Add: Thornton Hall Univ of Va Charlottesville VA 22901

LAWLESS, PHILIP AUSTIN, b Tulsa, Okla, June 7, 43; m 72; c 2. ENGINEERING PHYSICS. Educ: Rice Univ, BA, 65; Duke Univ, PhD(physics), 74. Prof Exp: PHYSICIST, RES TRIANGLE INST, 74- Mailing Add: PO Box 12194 Res Triangle Park NC 27709

LAWLESS, ROBERT DALE, b Tulsa, Okla, Oct 4, 37; m 63; c 2. ANTHROPOLOGY. Educ: Northwestern Univ, BSJ, 59; Univ Philippines, Ma, 68; New Sch Social Res, PhD(anthrop), 75. Prof Exp: Lectr anthrop, Univ Philippines, 67-68; acquisitions ed, Prentice-Hall Inc, 68-73; res fel, Social Sci Res Coun, 73-75; RES ASSOC ANTHROP, NEW SCH SOCIAL RES, 75- Mem: Am Anthrop Asn; Asn Asian Studies. Res: Investigating theoretical formulations, concepts and models of interrelationships among population; organization, environment and technology among hunters-gatherers and nascent agriculturalists. Mailing Add: Dept of Anthrop New Sch for Social Res New York NY 10003

LAWLESS, WILLIAM N, b Denver, Colo, Sept 15, 36; m 57; c 3. SOLID STATE PHYSICS. Educ: Colo Sch Mines, EMet, 59; Rensselaer Polytech Inst, PhD(physics), 64. Prof Exp: Fel solid state physics, Swiss Fed Inst Technol, 64-66; sr res physicist, 66-68, RES ASSOC PHYSICS, RES & DEVELOP LABS, CORNING GLASS WORKS, 69- Concurrent Pos: Guest worker, Cryogenics Div, Nat Bur Stand, Boulder, 73-75. Mem: AAAS; Am Inst Physics; Cryogenic Soc Am. Res: Ferroelectricity; doped alkali halides; glass-ceramic technology. Mailing Add: Res & Develop Labs Corning Glass Works Corning NY 14830

LAWLIS, JOHN FRANK, JR, b Indianapolis, Ind, Oct 23, 23; m 46; c 3. MICROBIOLOGY. Educ: Burler Univ, BS, 49, MS, 51; Ind Univ, PhD(microbiol), 58. Prof Exp: Assoc parasitologist, Eli Lilly & Co, 46-52; cir cancer res, Pitman-Moore Co, Ind, 58-59; head, Dept Biol Control, Lederle Labs, Am Cyanamid Co, 60-61; mgr qual control biol prod, Merck Sharp & Dohme, 61-70; VPRES OPERS, MERRELL-NAT LABS, DIV RICHARDSON-MERRELL, INC, 70- Mem: Am Soc Microbiol; Pharmaceut Mfrs Asn; NY Acad Sci. Res: Role of virus and nucleic acid in tumorogenesis; virology as applied to vaccines and general medical microbiology. Mailing Add: Merrell-Nat Labs Div of Richardson-Merrell Inc Swiftwater PA 18370

LAWLOR, ANNA CATHERINE, b Waterbury, Conn, May 20, 98. VERTEBRATE ZOOLOGY. Educ: Col St Elizabeth, BS, 31; Columbia Univ, AM, 34; Cath Univ Am, PhD, 37. Prof Exp: Instr, 33-35, asst prof, 35-37, PROF BIOL & CHMN DEPT, COL ST ELIZABETH, 37- Concurrent Pos: NSF grant, 59-60. Mem: AAAS; Am Soc Human Genetics; Am Soc Microbiol; Am Pub Health Asn; Nat Asn Biol Teachers. Res: Sensory structures of the proboscis and tongue of the ground mole, Scalopus aquaticus; cancer chemotherapy. Mailing Add: Dept of Biol Col of St Elizabeth Convent NJ 07961

LAWRASON, F DOUGLAS, b St Paul, Minn, July 30, 19; m 44; c 3. INTERNAL MEDICINE. Educ: Univ Minn, BA, 41, MA & MD, 44. Prof Exp: Instr anat, Med Sch, Univ Minn, 41-43; from intern to resident, Sch Med, Yale Univ, 44-49, from instr to asst prof med, 49-50; prof assoc, Nat Res Coun, 50-53; asst prof & asst dean, Sch Med, Univ NC, 53-55; provost med affairs & dean med ctr, Univ Ark, 55-60; exec dir med res, Merck Sharp & Dohme Res Labs, Pa, 61-66, vpres acad med affairs, 66-69; prof internal med, Univ Tex Health Sci Ctr Dallas, 69-73, assoc dean acad affairs, 69-72, dean, 72-73; SR VPRES SCI AFFAIRS, SCHERING-PLOUGH CORP, 73- Concurrent Pos: James Hudson Brown res fel, Yale Univ, 48-49; mem hemat study sect, NIH, 51-53; comt blood & related probs, Nat Acad Sci, 53-57; inst grant comt, Am Cancer Soc, 55-60; training grant comt, Nat Res Coun, 59-64. Mem: AAAS; Am Fedn Clin Res. Res: Cancer and leukemia in inbred strains of mice; hematology; medical education and administration. Mailing Add: Sci Affairs Schering-Plough Corp Bloomfield NJ 07003

LAWRENCE, ADDISON LEE, b Cape Girardeau, Mo, Dec 19, 35; m 59; c 3. CELL PHYSIOLOGY, COMPARATIVE PHYSIOLOGY. Educ: Southwest Mo State Col, BS, 56; Univ Mo, MA, 58, PhD(physiol), 62. Prof Exp: Asst prof physiol, Westminster Col Mo, 61-62; Nat Insts Health fel, Stanford Univ, 62-64; assoc prof, 64-74, PROF PHYSIOL, UNIV HOUSTON, 74-, ASSOC DIR OFF RES, 75- Concurrent Pos: Adj res biologist, Gulf Coast Fisheries Ctr, Nat Marine Fisheries Serv, 75- Mem: AAAS; World Maricult Soc; Am Soc Zoologists. Res: Digestion and absorption primarily in marine invertebrates. Mailing Add: Dept of Biol Univ of Houston Houston TX 77004

LAWRENCE, AUBREY WILFORD, b Cedar City, Utah, Jan 18, 16; m 43; c 5. CHEMISTRY. Educ: Utah State Univ, BS, 39, MS, 48; Okla State Univ, PhD(chem), 54. Prof Exp: Teacher pub schs, Utah, 39-41; asst chem, Utah State Univ, 41-42, from instr to asst prof, 47-50; res chemist & precision analyst, Basic Magnesium, Inc, Nev, 42-44; teacher high sch, 44-45; asst chem, Utah State Univ, 46-47; PROF CHEM & HEAD DEPT, WESTERN STATE COL COLO, 53-, CHMN DIV NATURAL SCI & MATH, 62- Concurrent Pos: Res assoc, Med Div, Oak Ridge Inst Nuclear Studies, 59. Mem: Am Chem Soc. Res: Amide nitrogen metabolism; ammonia metabolism and toxicity. Mailing Add: Dept of Chem Western State Col of Colo Gunnison CO 81230

LAWRENCE, BARBARA, b Boston, Mass, July 30, 09; m 38; c 2. ZOOLOGY. Educ: Vassar Col, BA, 31. Prof Exp: Mus asst, 31-37, asst cur, 38-41, assoc & actg cur, 42-52, CUR MUS COMP ZOOL, HARVARD UNIV, 52- Mem: Am Soc Mammal; Am

Soc Zoologists. Res: Mammalian systematics, particularly Canidae; zooarcheology of early domestic sites. Mailing Add: Mammal Dept Mus of Comp Zool Harvard Univ Cambridge MA 02138

LAWRENCE, CARL ADAM, microbiology, see 12th edition

LAWRENCE, CARTERET, ophthalmology, see 12th edition

LAWRENCE, CHARLES HILLMAN, sanitary engineering, environmental health, see 12th edition

LAWRENCE, CHARLOTTE WOLF, biochemistry, see 12th edition

LAWRENCE, CHRISTOPHER WILLIAM, b London, Eng, Oct 2, 34; m 61; c 3. GENETICS, RADIOBIOLOGY. Educ: Univ Wales, BSc, 56; Univ Birmingham, PhD(genetics), 59. Prof Exp: Sci officer radiation biol, Wantage Labs, UK Atomic Energy Authority, 59-61, sr sci officer, 61-70; ASSOC PROF RADIATION BIOL, UNIV ROCHESTER, 70- Concurrent Pos: Vis asst prof radiation biol, Univ Rochester, 69. Mem: AAAS; Genetics Soc Am; Brit Genetical Soc; Brit Asn Radiation Res; Europ Radiation Res Soc. Res: Radiation molecular genetics of Saccharomyces cerevisiae. Mailing Add: Dept of Radiation Biol & Biophys Univ of Rochester Med Ctr Rochester NY 14642

LAWRENCE, DAVID A, b Paterson, NJ, Jan 9, 45; m 67. IMMUNOLOGY. Educ: Rutgers Univ, BA, 66; Boston Col, MS, 68, PhD(biol), 71. Prof Exp: USPHS fel, Scripps Clin & Res Found, 71-74; ASST PROF MICROBIOL & IMMUNOL, ALBANY MED COL, 74- Mem: AAAS; Am Soc Microbiol; Am Asn Immunologists. Res: Cellular and subcellular events resulting from antigen activation and regulation of immune response; tumor immunology. Mailing Add: Dept of Microbiol & Immunol Albany Med Col Albany NY 12208

LAWRENCE, DAVID REED, b Woodbury, NJ, Oct 11, 39; m 66. GEOLOGY, INVERTEBRATE PALEONTOLOGY. Educ: Johns Hopkins Univ, AB, 61; Princeton Univ, PhD(geol), 66. Prof Exp: Asst geol, Princeton Univ, 63-64, prof assoc, 66; asst prof, 66-69, ASSOC PROF GEOL & MARINE SCI, UNIV SC, 69- Concurrent Pos: NSF sci fac fel, Univ Tübingen, Ger, 71-72. Mem: Int Paleont Union; Geol Soc Am; Paleont Soc. Res: Evolutionary, ecologic and biogeographic aspects of fossil invertebrates; taphonomy; historiography of the earth sciences. Mailing Add: Dept of Geol Univ of SC Columbia SC 29208

LAWRENCE, DONALD BUERMANN, b Portland, Ore, Mar 8, 11; m 35. ETHNOBOTANY, PLANT ECOLOGY. Educ: Johns Hopkins Univ, PhD(plant physiol), 36. Prof Exp: Researcher Sigma Xi grant, 36-37; from instr to assoc prof, 37-50, PROF BOT, UNIV MINN, MINNEAPOLIS, 50- Concurrent Pos: Mem, Johns Hopkins Univ exped, Jamaica, 32, Am Geog Soc exped, Alaska, 41, 49, 50, 52, 55 & SChile, 59; spec consult, US Air Force, 48; dir terrestrial ecosyst proj, Hill Found, 57-61; Fulbright res fel, NZ, 64-65; US del Int Sea Ice Conf, Reykjavik, Nat Acad Sci Nat Res Coun, 71. Mem: AAAS; Ecol Soc; Am Geog Soc; Arctic Inst NAm; Brit Glaciol Soc. Res: Vegetation development; physiographic ecology; ecological life histories of plants; causes of climatic change; glaciology; Asian plant names and place names in the New World; history of plant dispersal by man. Mailing Add: Dept of Bot Univ of Minn St Paul MN 55108

LAWRENCE, DONALD GILBERT, b Kingston, Ont, Jan 18, 32; m 56; c 3. NEUROLOGY. Educ: Bishop's Univ, BSc, 53; McGill Univ, MDCM, 57; Royal Col Physicians & Surgeons, FRCP(C), 74. Prof Exp: Res fel neuroanat, Western Reserve Univ, 65-66; Nat Multiple Sclerosis Soc fel neurophysiol, Univ Lab Physiol, Oxford Univ, 66-68; from asst prof to assoc prof neuroanat, Erasmus Univ, 68-72; ASSOC PROF NEUROL & NEUROANAT, McGILL UNIV, 72-; ASST PHYSICIAN, MONTREAL GEN HOSP, 72- Honors & Awards: Osler Medal, Am Asn Hist Med, 58. Mem: AAAS; Am Asn Hist Med; Am Asn Anat; Soc Neurosci; Am Acad Neurol. Res: Anatomical, behavioral and clinical investigations of motor pathways in the central nervous system. Mailing Add: Lab Neuroanat Montreal Neuro Inst 3801 University St Montreal PQ Can

LAWRENCE, EDMOND FRANCIS, b Bessemer, Ala, Oct 9, 18; m 43; c 1. EXPLORATION GEOLOGY. Educ: Univ Ala, BS, 49; Univ Calif, Los Angeles, MS, 67; Univ Calif, Riverside, PhD, 69. Prof Exp: Metallurgist, Tenn Coal, Iron & RR Co, 42-46; res geologist, Anaconda Co, 49-51; consult geologist, Idaho Md Mining Co, Calif, 52-55; econ geologist, Univ Nev, 55-61; CONSULT MINING GEOLOGIST, 61- Mem: Geol Soc Am; Soc Econ Geologists; Mineral Soc Am; Geochem Soc; Sigma Xi. Res: Causes for localization of ore deposits, including geochemistry of ore fluids and exploration geochemistry. Mailing Add: PO Box 8044 University Sta Reno NV 89507

LAWRENCE, FRANCIS JOSEPH, b Glen Arm, Md, May 12, 25; m 51; c 4. PLANT BREEDING. Educ: Univ Md, BS, 51, MS, 58, PhD(hort, bot), 65. Prof Exp: Asst hort, Univ Md, 53-62, from instr to asst prof, 62-65; RES HORTICULTURIST, CORVALLIS RES STA, USDA, 65- Mem: Am Soc Hort Sci; Am Pomol Soc. Res: Breeding of Fragaria and Rubus. Mailing Add: Dept of Hort USDA Ore State Univ Corvallis OR 97331

LAWRENCE, FRANKLIN ISAAC LATIMER, b Brooklyn, NY, July 14, 05; m 34; c 1. ORGANIC CHEMISTRY. Educ: NY Univ, BS, 28, MS, 30. Prof Exp: Jr engr, Atlantic Ref Co, Pa, 30-34, tech sales supvr, 34-42; develop engr, Permutit Co, NY, 42-44; res dir, Kendall Ref Div, Witco Chem Corp, 44-68, vpres res & develop, 68-70, CONSULT, 70- Honors & Awards: Cert Appreciation, Am Petrol Indust. Mem: Am Chem Soc; Soc Automotive Eng; fel Am Inst Chem. Res: Petroleum chemistry; ion exchange. Mailing Add: 600 McClellan Circle Elizabethton TN 37643

LAWRENCE, FRED PARKER, b Lebanon, Tenn, Mar 13, 11; m 34; c 2. HORTICULTURE, ENTOMOLOGY. Educ: Univ Fla, BS, 34, MAg, 53. Prof Exp: County supvr, Fla Rural Rehab Corp, 34-36; dist supvr, Farm Security Admin, 36-40, asst state dir, 40-42; prof hort & citriculturist, US Coop Exten Serv, 46-74, EMER PROF, FRUIT CROPS DEPT, UNIV FLA, 74- Mem: Int Soc Citricult; Am Soc Hort Sci. Res: Citrus culture, expecially nutrition, insect and disease control. Mailing Add: 2805 SW 1st Ave Gainesville FL 32607

LAWRENCE, GEORGE EDWIN, b Berkeley, Calif, Mar 25, 20; m 43; c 4. ZOOLOGY. Educ: Univ Calif, AB, 46, MA, 49, PhD, 60. Prof Exp: Asst zool, Univ Calif, 46-47; from instr to assoc prof, 47-67, PROF LIFE SCI, BAKERSFIELD STATE COL, 67-, HEAD DEPT, 70- Mem: Wilderness Soc; Cooper Ornith Soc; Ecol Soc Am; Nat Biol Teachers Asn. Res: Natural history of mammals and birds; ecology of chaparral association; temperature tolerance of small mammals. Mailing Add: Dept of Life Sci Bakersfield Col Bakersfield CA 93305

LAWRENCE, GEORGE HILL MATHEWSON, b East Greenwich, RI, June 19, 10; m 34; c 2. SYSTEMATIC BOTANY. Educ: RI State Col, BS, 32, MS, 33; Cornell Univ, PhD(syst bot), 39. Hon Degrees: DSc, Univ RI, 52. Prof Exp: Asst, Exp Sta, RI State Col, 32-33; tech supvr emergency conserv work, USDA, 33-34; asst bot, Cornell Univ, 36-38, res instr, Bailey Hortorium, 39-40, from asst prof to prof bot, 40-60; dir, Hunt Bot Libr, Carnegie-Mellon Univ, 60-70, res assoc bot hist, 71-75. Concurrent Pos: Secy, Am Hort Coun, 48-54; dir, Bailey Hortorium, Cornell Univ, 51-60; Guggenheim fel, 55; dir, Fairchild Trop Garden, 58-67, pres, 67-71; mem exec comt, Int Comn Nomenclature Cultivated Plants, 57-66 & Int Orchid Comn Classification, Nomenclature & Registr, 57-60; mem Int Comn Nomenclature & Registr Hort Plants, 60-66; trustee, R H Montgomery Found, 65-73. Mem: Bot Soc Am; Am Soc Plant Taxon (pres, 65); Soc Econ Botanists (pres, 70-71); Am Hort Soc (secy, 70-71); Am Inst Biol Sci. Res: Botanical and horticultural bibliography. Mailing Add: PO Box 177 East Greenwich RI 02818

LAWRENCE, GEORGE MELVIN, b Salt Lake City, Utah, Mar 26, 37; m 59; c 3. PHYSICS. Educ: Univ Utah, BS, 59; Calif Inst Technol, PhD(physics), 63. Prof Exp: Res assoc astrophys sci, Princeton Univ, 63-65, staff physicist, 65-67; res scientist, McDonnell Douglas Advan Res Lab, Calif, 67-70; vis fel, Joint Inst Lab Astrophys-Lab Atmospheric & Space Physics, 70-71, res assoc, 71-74, FEL, LAB ATMOSPHERIC & SPACE PHYSICS, UNIV COLO, BOULDER, 74- Concurrent Pos: Vis assoc, Calif Inst Technol, 68-70. Mem: Fel Am Phys Soc; Am Phys Union. Res: Transition probabilities; cross sections; physical chemistry. Mailing Add: Lab Atmospheric & Space Phys Univ of Colo Boulder CO 80302

LAWRENCE, HENRY SHERWOOD, b New York, NY, Sept 22, 16; m 43; c 3. MEDICINE. Educ: NY Univ, AB, 38, MD, 48; Am Bd Internal Med, dipl. Prof Exp: Intern, 3rd Med Div, Bellevue Hosp, New York, 43-44, from asst resident to chief resident, 46-48; asst, 47-49, from instr to assoc prof, 49-61, PROF MED, MED SCH, NY UNIV, 61-, HEAD INFECTIOUS DIS & IMMUNOL UNIT, 59-, CO-DIR, NY UNIV-BELLEVUE MED SERV, 64-, DIR, NY UNIV CANCER INST, 74- Concurrent Pos: Wyckoff fel, NY Univ, 48-49; dir student health serv, NY Univ-Bellevue Med Serv, 49-57; Commonwealth Fund fel, Univ Col, Univ London, 59; USPHS career develop award, 60-65; assoc mem streptococcal comt, Armed Forces Epidemiol Bd; mem comt cutaneous syst & comt tissue transplantation, Div Med Sci, Nat Res Coun-Nat Acad Sci, chmn comt tissue transplantation, 68-71, mem, Nat Res Coun, 70; consult, Allergy & Immunol Study Sect, USPHS, 60-65, chmn, 63-65; consult & chmn allergy & infectious dis panel, Health Res Coun City of New York, infectious dis prog comt res serv, Vet Admin & res comts, Arthritis Found, Am Cancer Soc & Am Thoracic Soc. Honors & Awards: Von Pirquet Gold Medal Award, Annual Forum Allergy, 72; Am Col Physicians Award, 73; Am Acad Allergy Achievement Award, 74; NY Acad Med Sci Medal, 74; Bristol Award, Infectious Dis Soc Am, 74; Hon Fel Royal Col Phys & Surg, Glasgow, 75; Chapin Medal, 75; Lila Gruber Award Cancer Res, 75. Mem: Nat Acad Sci; Am Soc Clin Invest; Soc Exp Biol & Med; Harvey Soc (secy, 57-60; Asn Am Physicians. Res: Infection and immunity; delayed bacterial hypersensitivity; homograft reactions; tissue transplantation. Mailing Add: Infectious Dis & Immunol Div NY Univ Med Ctr New York NY 10016

LAWRENCE, IRVIN E, JR, b Raleigh, NC, Apr 18, 26. EMBRYOLOGY, HISTOLOGY. Educ: Univ NC, AB, 50; Univ Wyo, MS, 55; Univ Kans, PhD(anat), 63. Prof Exp: Teacher high sch, NC, 51-54; instr biol, Louisburg Col, 55-57; asst prof zool, Univ Wyo, 60-64; assoc prof biol, 64-70, ASSOC PROF ANAT, EAST CAROLINA UNIV, 70- Concurrent Pos: Univ Res fel, Univ Wyo, 63-64; USPHS res grant, 64-65; NIH res grant, 75- Mem: Am Soc Zool; Soc Develop Biol; Int Soc Stereol; Pan-Am Asn Anat; NY Acad Sci. Res: Biogenic amines in development; experimental analysis of morphogenetic patterns and histochemical and hormonal correlations in the chicken comb; comparative histology of native mammals in Wyoming; ovarian nerves and reproductive function. Mailing Add: Dept of Anat ECarolina Univ Sch of Med Greenville NC 27834

LAWRENCE, JAMES D, JR, atmospheric physics, see 12th edition

LAWRENCE, JAMES FRANKLIN, b Okemah, Okla, Aug 20, 50. MATHEMATICS. Educ: Okla State Univ, BS, 72; Univ Wash, PhD(math), 75. Prof Exp: INSTR MATH, UNIV TEX, AUSTIN, 75- Mem: Math Asn Am; Am Math Soc. Res: Field of combinatorics; study of oriented matroids. Mailing Add: Dept of Math Univ of Tex Austin TX 78712

LAWRENCE, JAMES LESTER, b New York, NY, Oct 22, 41. PARASITOLOGY, INVERTEBRATE ZOOLOGY. Educ: Tex Christian Univ, BA, 63, MA, 65; Univ Maine, PhD(parasitol), 68. Prof Exp: Assoc dean col, 70-74, DEAN SPEC PROGS, HARTWICK COL, 74-, ASST PROF BIOL, 68-, ASST VPRES EDUC AFFAIRS, 70- Mem: AAAS; Am Soc Parasitol; Am Micros Soc; Wildlife Dis Asn. Res: Ecology of host-parasite relationships; biology of the caryophyllaceidae; parasites of the cypriniformes. Mailing Add: Dept of Biol Hartwick Col Oneonta NY 13820

LAWRENCE, JAMES NEVILLE PEED, b Norfolk, Va, May 29, 29; m 48; c 1. HEALTH PHYSICS, PHYSICS. Educ: Johns Hopkins Univ, BA, 50; Vanderbilt Univ, MA, 58, PhD(physics), 68. Prof Exp: Res asst health physics, 51-54, mem staff, 54-68, ASSOC GROUP LEADER, LOS ALAMOS SCI LAB, UNIV CALIF, 68- Mem: Health Physics Soc. Res: Theoretical treatment of nuclear fission, especially liquid drop applications; health physics, especially dosimetry, internal exposure calculations and radio-nuclide identification. Mailing Add: 206 El Conejo Los Alamos NM 87544

LAWRENCE, JAMES VANTINE, b Middletown, Ohio, July 2, 18; m 42; c 3. BACTERIOLOGY. Educ: Univ Ill, BS, 43; Ohio State Univ, MS, 48, PhD(bact), 50. Prof Exp: From asst prof to assoc prof, 50-70, PROF BACT, OHIO UNIV, 70- Concurrent Pos: Vis prof, WVa Univ, 51 & 52; clin lab dir, Athens State Hosp, 53, 54 & 56. Mem: Am Soc Microbiol; Am Pub Health Asn. Res: General and pathogenic bacteriology and parasitology. Mailing Add: Dept of Zool & Microbiol Ohio Univ Athens OH 45701

LAWRENCE, JEAN MCVAY, b Independence, Kans, July 15, 14; m 43; c 3. COMPARATIVE PHYSIOLOGY, ZOOLOGY. Educ: Yankton Col, BA, 36; Wellesley Col, MA, 37; Northwestern Univ, PhD(zool), 42. Prof Exp: Physiologist, Bauer & Black Div, Kendall Co, 42-43; from instr to asst prof biol, 59-65, ASSOC PROF BIOL, WESTERN MICH UNIV, 65- Res: Invertebrate neurosecretion. Mailing Add: Dept of Biol Western Mich Univ Kalamazoo MI 49001

LAWRENCE, JOHN FRANCIS, b Oakland, Calif, Dec 27, 34. ENTOMOLOGY. Educ: Univ Calif, Berkeley, BA, 57, PhD(entom), 65. Prof Exp: Asst cur insects, 64-71, COORDR ENTOM COLLECTIONS, MUS COMP ZOOL, HARVARD UNIV, 73- Mem: AAAS; Soc Study Evolution; Soc Syst Zoologists; Entom Soc Am. Res: Ecology of fungus feeding insects; classification and evolution of Coleoptera; systematic theory; biosystematics of Ciidae and other beetle groups. Mailing Add: Mus of Comp Zool Harvard Univ Cambridge MA 02138

LAWRENCE, JOHN HUNDALE, b Canton, SDak, Jan 7, 1904. MEDICINE. Educ: Univ SDak, AB, 26; Harvard Univ, MD, 30. Hon Degrees: DSc, Univ SDak, 42 &

Cath Univ Am, 59; Dr, Univ Bordeaux, France, 58. Prof Exp: Intern, Peter Bent Brigham Hosp, Boston, 30-31; trainee, Strong Mem Hosp, Univ Rochester, 31-32; trainee, New Haven Hosp, Yale Univ, 32-34, assoc physician hosp & instr med univ, 34-37; res assoc, 37, EMER PROF MED PHYSICS, UNIV CALIF, BERKELEY, 50-, EMER DIR, DONNER LAB, 72- Concurrent Pos: Regent, Univ Calif, Berkeley; William H Welch lectr; Ludwig Kast Mem lectr, 49; Stephen Walter Ranson Mem lectr; Von Hevesy Mem lectr; Richardson lectr, Harvard Univ, 63; Pasteur Inst lectr, 63; AEC cons & vis prof AEE of India, Univ Bombay, 61; vis prof med, Am Univ, Beirut, Lebanon, 63 & Ohio State Univ, 65; mem sci & tech adv comt, NASA, 65-67, led 3-man med mission to USSR, 74. Honors & Awards: Decorated Silver Cross, Royal Order Phoenix, Greece; Caldwell Medal, US; Davidson Medal, Eng; Cert Appreciation, War & Navy Depts, 47; Medal, Univ Bordeaux, 58; Pasteur Medal, Pasteur Inst, Paris, 63; Nuclear Pioneer Award, Soc Nuclear Med, 70; First Marshall Brucer Medal & Katharine Berkan Judd Award, Sloan-Kettering Mem Hosp, New York, 75. Mem: Fel Am Nuclear Soc; Am Soc Clin Invest; hon mem Soc Nuclear Med (pres, 66); Am Asn Neurol Surgeons; Endocrine Soc. Res: Aerospace medicine; hematology; metabolism; physiology; medical physics; radioisotopes and nuclear radiations in research, diagnosis and therapy. Mailing Add: Donner Lab Univ of Calif Berkeley CA 94720

LAWRENCE, JOHN M, b Cape Girardeau, Mo, Oct 11, 37. PHYSIOLOGY. Educ: Southeast Mo State Col, BS, 58; Univ Mo, AM, 60; Stanford Univ, PhD(biol), 66. Prof Exp: Instr physiol, Stanford Univ, 64-65; asst prof, 65-71, ASSOC PROF PHYSIOL, UNIV S FLA, 71- Concurrent Pos: Fel, Marine Biol Lab, Hebrew Univ Israel, 69-70. Mem: AAAS; Am Soc Zool; Marine Biol Asn UK. Res: Nutritional and reproductive physiology of marine invertebrates. Mailing Add: Dept of Biol Univ of SFla Tampa FL 33620

LAWRENCE, JOHN MCCUNE, b Carmichaels, Pa, Feb 17, 16; m 38; c 3. BIOCHEMISTRY. Educ: Carnegie Inst Technol, BS, 37, MS, 39; Univ Pittsburgh, PhD(biochem), 43. Prof Exp: Res asst, Mellon Inst, 39-41; Nutrit Found fel, Dept Dairy Indust, Cornell Univ, 43-45, from instr to asst prof biochem, 45-48; assoc chemist, 48-58, AGR CHEMIST, WASH STATE UNIV, 58- Mem: AAAS; Am Chem Soc; Am Soc Plant Physiol; Am Ornith Union. Res: Enzymes; proteins; amino acids; biochemistry of seed germination and nutritional quality of seeds. Mailing Add: Dept of Agr Chem Wash State Univ Pullman WA 99163

LAWRENCE, JOHN MEDLOCK, b Cedar Bluff, Ala, Sept 25, 19; m 47; c 3. FISHERIES. Educ: Ala Polytech Inst, BS, 41, MS, 43; Iowa State Univ, PhD, 56. Prof Exp: Asst fish culturist, 46-57, assoc fish culturist, 57-63, PROF FISHERIES, AUBURN UNIV, 63- Mem: Am Fisheries Soc; Weed Sci Soc Am. Res: Methods for control of aquatic weeds and their effects upon fish production. Mailing Add: Dept Fisheries & Allied Aquacult Auburn Univ Auburn AL 36830

LAWRENCE, LOUISE DE KIRILINE, b Sweden, Jan 30, 94; Can citizen; m 18, 39. ORNITHOLOGY. Hon Degrees: LittD, Laurentian Univ, 71. Prof Exp: Ornithologist; RESEARCHER & WRITER. Mem: Am Ornith Union; Wilson Ornith Soc. Res: Breeding history of the Canada Jay, Warblers, the Red-eyed Vireo and Woodpeckers; body weight of the Black-capped Chickadee. Mailing Add: Pimisi Bay R R 1 Rutherglen ON Can

LAWRENCE, MERLE, b Remsen, NY, Dec 26, 15; m 42; c 3. PHYSIOLOGY. Educ: Princeton Univ, AB, 38, MA, 40, PhD(psychol), 41. Prof Exp: Nat Res Coun fel, Johns Hopkins Univ, 41; from asst prof to assoc prof psychol, Princeton Univ, 46-52; assoc prof physiol acoust, 52-57, PROF OTOLARYNGOL & PSYCHOL, MED SCH, UNIV MICH, ANN ARBOR, 57-, PROF PHYSIOL, 59-, RES ASSOC, INST INDUST HEALTH, 52-, DIR KRESGE HEARING RES INST, 61- Concurrent Pos: Consult, Surgeon Gen Off, 53- & Secy Defense, 55-58; mem commun disorders res training comt, Nat Inst Neurol Dis & Stroke, 61-65 & commun sci study sect, NIH, 65-70, mem communicative disorders rev comt, Nat Inst Neurol & Communicative Disorders & Stroke. Mem: Am Physiol Soc; fel Acoust Soc Am; Am Laryngol, Rhinol & Otol Soc; Am Otol Soc; Col Oto-Rhino-Laryngol Amicitiae Sacrum. Res: Physiology of hearing. Mailing Add: Kresge Hearing Res Inst Univ of Mich Med Sch Ann Arbor MI 48104

LAWRENCE, MONTAGUE SCHIELE, b Laurel, Miss, Apr 22, 23. MEDICINE, SURGERY. Educ: Alcorn Agr & Mech Col, BS, 43; Meharry Med Col, MD, 46; Am Bd Surg, dipl; Am Bd Thoracic Surg, dipl. Prof Exp: From intern to resident, Homer Phillips Hosp, 46-51, supvr surg, 53-54; instr, 54-55, assoc, 55-56, res asst prof, 56-57, from asst prof to assoc prof thoracic & cardiovasc surg, 57-66, PROF SURG, COL MED, UNIV IOWA, 66- Mem: Nat Tuberc Asn; Nat Med Asn; Am Heart Asn; fel Am Col Surg; fel Am Col Chest Physicians. Res: Congenital and acquired cardiac disease; arterial peripheral vascular disease. Mailing Add: 1030 Fifth Ave SE Cedar Rapids IA 52403

LAWRENCE, PAUL J, b Hazleton, Pa, Dec 18, 40; m 63; c 1. BIOCHEMISTRY. Educ: King's Col, Pa, BS, 62; Univ Wis, Madison, MS, 64, PhD(biochem), 67. Prof Exp: ASST PROF BIOCHEM, COL MED, UNIV UTAH, 68- Concurrent Pos: Fel biochem, Univ Wis, 67-68; NIH fel, 67-69. Mem: AAAS; Am Chem Soc. Res: Mechanisms of drug action. Mailing Add: Rm 1C 218 Univ of Utah Col of Med Salt Lake City UT 84112

LAWRENCE, PHILIP LINWOOD, b New Bedford, Mass, Mar 27, 23; m 48; c 2. GEOPHYSICS. Educ: Colo Sch Mines, GeolE, 49, Southern Methodist Univ, MS, 60. Prof Exp: Res physicist, Magnolia Petrol Co, Tex, 49-62; staff adv, Mobil Petrol Co, NY, 63-65; unit supvr geophys serv ctr, Mobil Oil Co, Tex, 66-68, CORP GEOPHYSICIST, MOBIL OIL CORP, 69- Concurrent Pos: Lectr elec eng, Southern Methodist Univ, 60-62. Mem: Sigma Xi. Res: Seismic, magnetic, gravity data gathering, processing and interpretation techniques and development for mineral exploration. Mailing Add: Mobil Oil Corp 221 Nassau St Princeton NJ 08540

LAWRENCE, RALPH WALTER, chemistry, see 12th edition

LAWRENCE, RICHARD MANLEY, b Indianapolis, Ind, June 14, 30; m 51; c 2. STRUCTURAL CHEMISTRY. Educ: Earlham Col, BS, 52; Ball State Univ, MA, 58; Univ Ark, PhD(phys chem), 65. Prof Exp: Asst prof sci, 58-65, assoc prof chem, 65-70, PROF CHEM, BALL STATE UNIV, 70- Mem: Soc Appl Spectros; Am Chem Soc. Res: Liquid state; structure of electrolyte solutions. Mailing Add: Dept of Chem Ball State Univ Muncie IN 47306

LAWRENCE, ROBERT D, b Ithaca, NY, May 24, 43; m 66. GEOLOGY. Educ: Earlham Col, BA, 65; Stanford Univ, PhD(geol), 68. Prof Exp: Asst prof geol, Earlham Col, 68-70; ASST PROF GEOL, ORE STATE UNIV, 70- Mem: AAAS; Geol Soc Am; Nat Asn Geol Teachers; Am Geophys Union. Res: Major faults of western North America; deformation features of minerals; structural petrology; tectonic history of Pacific Northwest. Mailing Add: Dept of Geol Ore State Univ Corvallis OR 97331

LAWRENCE, ROBERT G, b Wilmington, NY, Feb 14, 21; m 46; c 2. ZOOLOGY. Educ: Eastern Nazarene Col, AB, 44; Boston Univ, MA, 46; Okla State Univ, PhD, 64. Prof Exp: Teacher, Henry Ford's Boys Sch, Mass, 45-46; prof biol & head dept, Bethany Nazarene Col, 47-68; assoc dean, 68-71, PROF BIOL SCI, MID-AM NAZARENE COL, 68-, DIR INSTNL RES, 71- Concurrent Pos: Chmn div natural sci, Bethany Nazarene Col, 49-68. Mem: AAAS; Am Ornith Union; Nat Audubon Soc; Nat Asn Biol Teachers. Res: Ornithology; relation of weather factors to migration of water fowl. Mailing Add: Dept Biol Mid-Am Nazarene Col Box 1776 Olathe KS 66061

LAWRENCE, ROBERT HOWARD, JR, b Memphis, Tenn, July 16, 42; m 63; c 2. PLANT PHYSIOLOGY, BIOCHEMISTRY. Educ: Vanderbilt Univ, BA, 65; Univ Ga, MS, 67, PhD, 71. Prof Exp: Res asst, Sch Forest Resources, Univ Ga, 66-71; res assoc & mem fac biochem, Okla State Univ, 71-75; RES SCIENTIST, CORP RES LAB, UNION CARBIDE CORP, 75- Concurrent Pos: Researcher with Dr George R Waller, NSF grant, Okla State Univ, 71-73; Nat Inst Ment Health res grant, 73-74; USDA res grant, 73-74. Mem: Am Chem Soc; Tissue Cult Asn; AAAS; Am Soc Plant Physiol; Phytochem Soc NAm. Res: Biochemistry and physiology of plant growth and development; plant cell culture in agricultural research; biochemical applications of mass spectrometry; plant secondary metabolites. Mailing Add: Corp Res Lab Union Carbide Corp Tarrytown Tech Ctr Tarrytown NY 10591

LAWRENCE, ROBERT MARSHALL, b Kennecott, Alaska, June 28, 23; m 50; c 9. ANESTHESIOLOGY. Educ: Univ Rochester, MD, 49. Prof Exp: Resident surg, Strong Mem Hosp, 49-54, resident anesthesiol, 54-56; from instr to assoc prof surg & med, 56-68, PROF ANESTHESIOL, SCH MED, UNIV ROCHESTER, 68- Concurrent Pos: Consult, Vet Admin Hosps, Batavia & Canandaigua, 56- Mem: Am Soc Anesthesiol; Am Asn Respiratory Ther. Res: Oxygen toxicity; respiratory therapy. Mailing Add: Dept of Anesthesiol Univ of Rochester Sch of Med Rochester NY 14642

LAWRENCE, THOMAS, b Colonsay, Sask, July 19, 27; m 51; c 2. PLANT BREEDING. Educ: Univ Sask, BSc, 50, MSc, 52; Univ Alta, PhD(plant breeding, genetics), 55. Prof Exp: Asst, Univ Sask, 50-52 & Univ Alta, 52-54; RES SCIENTIST, RES STA, CAN DEPT AGR, 54- Mem: Can Soc Agron; Agr Inst Can. Res: Grass breeding, genetics and seed production. Mailing Add: Res Sta Swift Current Sask Can

LAWRENCE, VINNEDGE MOORE, b Bangor, Maine, Feb 19, 40; m 66; c 1. ENTOMOLOGY, ECOLOGY. Educ: Miami Univ Ohio, BS, 62, MA, 64; Purdue Univ, PhD(entom), 68. Prof Exp: Instr biol, Xavier Univ Ohio, 64-65; asst prof, 68-72, ASSOC PROF BIOL, WASHINGTON & JEFFERSON COL, 72- Concurrent Pos: Mem citizen's adv coun, Pa Dept Environ Resources, 71- Mem: AAAS; Am Inst Biol Sci; Ecol Soc Am. Res: Population dynamics of Odonata naiads in farm-pond ecosystems; distribution of Odonata in Pennsylvania. Mailing Add: Dept of Biol Washington & Jefferson Col Washington PA 15301

LAWRENCE, WALTER, JR, b Chicago, Ill, May 31, 25; m 47; c 4. SURGERY. Educ: Univ Chicago, PhB, 45, SB, 46, MD, 48. Prof Exp: Intern surg, Johns Hopkins Hosp, 48-49, asst resident & asst, Sch Med, Johns Hopkins Univ, 49-50, Halsted fel, Johns Hopkins Hosp, 50; resident, Mem Ctr Cancer & Allied Dis, 51-52 & 54-56; res fel exp surg, Mem Ctr Cancer & Allied Sci, 56; instr surg, Med Col, Cornell Univ, 57-58, asst prof, 58-63, clin assoc prof, 63-66; PROF SURG & CHMN DIV SURG ONCOL, MED COL VA, 66-, AM CANCER SOC PROF CLIN ONCOL, 72-, DIR CANCER CTR, 74- Concurrent Pos: Asst mem, Sloan-Kettering Inst Cancer Res, 57-60, assoc mem & assoc chief div exp surg, 60-66; clin asst attend surgeon, Mem Hosp, 57-59, asst attend surgeon, 59-62, assoc vis surgeon, 62-; asst vis surgeon, James Ewing Hosp, 57-62, assoc vis surgeon, 62-; mem surg staff, NY Hosp, 57- Honors & Awards: Sloan Award Cancer Res, 64; Horsley Award, 73. Mem: Halsted Soc (pres, 75); fel Am Col Surg; fel NY Acad Sci; fel NY Acad Med; fel Royal Soc Med. Res: Surgery, particularly cancer and cancer research. Mailing Add: Box 11 Dept of Surg Med Col of Va Richmond VA 23298

LAWRENCE, WALTER EDWARD, b Albany, NY, May 22, 42; m 69. Educ: Carnegie Inst Technol, BS, 64; Cornell Univ, PhD(physics), 70. Prof Exp: Res assoc physics, Stanford Univ, 69-71; ASST PROF PHYSICS, DARTMOUTH COL, 71- Mem: Am Phys Soc. Res: Solid state theory, principally superconductivity; transport theory of metals. Mailing Add: Dept of Physics Dartmouth Col Hanover NH 03755

LAWRENCE, WILLARD EARL, b Chassell, Mich, Apr 8, 17; m 43; c 5. STATISTICS. Educ: Marquette Univ, BS, 51, MS, 53; Univ Wis, MS, 62, PhD(statist), 64. Prof Exp: From instr to assoc prof, 53-69, asst chmn dept, 58-63, PROF MATH, MARQUETTE UNIV, 69-, CHMN DEPT MATH & STATIST, 73- Concurrent Pos: Consult, NSF, 67-69. Mem: Math Asn Am; Am Statist Asn. Res: Experimental design; response surface designs which minimize variance and bias errors; designs for mixtures. Mailing Add: Dept of Math & Statist Marquette Univ 540 N 15th St Milwaukee WI 53233

LAWRENCE, WILLIAM CHASE, b Cambridge, Mass, July 10, 34; m 55; c 3. VIROLOGY, MOLECULAR BIOLOGY. Educ: Univ Mass, BS, 55; Univ Pa, VMD, 59, PhD(microbiol), 66. Prof Exp: Asst prof, 65-72, ASSOC PROF MIBROBIOL, UNIV PA, 72- Concurrent Pos: USPHS res grant, 67-74. Mem: AAAS; Am Vet Med Asn; Am Soc Microbiol; NY Acad Sci. Res: Biochemistry of viral infection. Mailing Add: Dept of Pathobiol Sch Vet Med Univ of Pa 3800 Spruce St Philadelphia PA 19104

LAWRENCE, WILLIAM HOBART, b Washington, DC, Nov 9, 21; m 47; c 5. FORESTRY, WILDLIFE MANAGEMENT. Educ: Univ Mich, BSF, 47, MF, 48, PhD(wildlife mgt), 54. Prof Exp: Res assoc, Univ Mich, 54-56; forest wildlife biologist, 56-66, group leader forest regeneration res, 66-71, mgr, 72-74, MGR, FOREST ENVIRON SCI RES, FORESTRY RES CTR, WEYERHAEUSER CO, 74- Mem: Wildlife Soc; Am Soc Mammal. Res: Forest wildlife management; ecological control of wildlife damage; chemical repellents; physiology of repellent reactions; environmental protection; water quality. Mailing Add: Forest Res Ctr Weyerhaeuser Co Centralia WA 98531

LAWRENCE, WILLIAM HOMER, b Magnet Cove, Ark, Mar 20, 28; div 73; c 4. TOXICOLOGY, PHARMACOLOGY. Educ: Col of Ozarks, BS, 50; Univ Md, MS, 52, PhD(pharmacol), 55. Prof Exp: Instr materia medica, Sch Nursing, Univ Md, 51, asst pharmacol, Sch Pharm, 51-54; from asst prof to assoc prof pharmacol & physiol, Col Pharm, Univ Houston, 56-66; assoc prof, 66-73, asst dir mat sci toxicol labs, 67-75, PROF TOXICOL, UNIV TENN CTR FOR HEALTH SCI, 73-, ASSOC DIR MAT SCI TOXICOL LABS, 75-, HEAD ANIMAL TOXICOL SECT, 67- Concurrent Pos: Consult, Drug-Plastic Res & Toxicol Labs, Univ Tex, 63-68, lectr, Med Br, 64-67; consult, Vet Admin Hosp, Houston, 64-66 & Memphis, 67- Mem: AAAS; Soc Toxicol; Am Pharmaceut Asn. Res: Toxicity of biomaterials, especially dental materials, blood bags, intravenous administration tubings, extracorporeal devices, implantable devices and carcinogenic studies of plastics. Mailing Add: Mat Sci Toxicol Lab Univ of Tenn Ctr for Health Memphis TN 38163

LAWRENCE, WILLIAM MASON, b Brooktondale, NY, Oct 2, 18; m 42; c 2. FISHERIES, NATURAL RESOURCES. Educ: Cornell Univ, BS, 38, PhD(fishery biol), 41. Prof Exp: Biometrician bur game, NY State Conserv Dept, 41-42, sr aquatic biologist, 46-52, chief bur fish, 52-55, dir fish & game, 55-58, asst comnr, 58-64, dep comnr, NY State Dept Environ Conserv, 64-74; CONSULT, 74- Concurrent Pos: US comnr & chmn, Great Lakes Fishery Comn, 72-74; chmn, Atlantic States Marine Fisheries Comn, 71-73; comnr, Great Lakes River Basin Comn; alt comnr, Del River Basin Comn. Mem: Am Fisheries Soc (pres, 59); Wildlife Soc; NY Acad Sci; Int Asn Game, Fish & Conserv Comnrs (pres, 69). Res: Fish and wildlife conservation; water resources. Mailing Add: 40 Albin Rd Delmar NY 12054

LAWRENCE (ROCHE), MARY ANNA, b Newton, Mass, June 11, 97. BIOLOGY. Educ: Cath Univ Am, PhD(biol), 33. Prof Exp: Instr biol, 27-30, prof biol & chmn div natural sci, 33-68, EMER PROF BIOL & ARCHIVIST, REGIS COL MASS, 68- Concurrent Pos: Atomic Energy grant, 60. Mem: Soc Am Arch. Res: Histology; good teaching of biology and improvement of biology. Mailing Add: Regis Col 235 Wellesley St Weston MA 02193

LAWRY, JAMES VORIS, b San Francisco, Calif, May 8, 40; m 65; c 2. NEUROPHYSIOLOGY, NEUROANATOMY. Educ: Stanford Univ, AB, 62, PhD(biol), 66. Prof Exp: ASST PROF ANAT, SCH MED, UNIV CALIF, SAN FRANCISCO, 68- Concurrent Pos: NIH fel, Gatty Marine Lab, Scotland, 65-68. Mem: AAAS; Am Asn Anat; Am Soc Zool. Res: Behavioral correlates of structure and function of invertebrate nervous systems; pacemaker systems; function of neuropil; bioluminescence and olfaction of myctophid fishes. Mailing Add: Dept of Anat Univ of Calif Sch of Med San Francisco CA 94143

LAWS, E HAROLD, b Milan, Ind, Sept 10, 15; m 39; c 2. MEDICINE. Educ: Ind Univ, BS, 38, MD, 40. Prof Exp: CLIN PROF MED, SCH MED, UNIV WASH, 46-, ASSOC DEAN, 68-; MED DIR, HARBORVIEW HOSP, 68- Mem: AMA; assoc Am Col Physicians; fel Am Col Chest Physicians. Res: Internal medicine. Mailing Add: 325 Ninth Ave Seattle WA 98104

LAWS, EDWARD ALLEN, b Columbus, Ohio, Feb 4, 45. OCEANOGRAPHY. Educ: Harvard Univ, BA, 67, PhD(chem physics), 71. Prof Exp: Instr oceanog, Fla State Univ, 71-74; ASST PROF OCEANOG, UNIV HAWAII, 74- Mem: AAAS; Am Soc Limnol & Oceanog; Ecol Soc Am; Phycol Soc Am. Res: Metabolism of carbon and nitrogen by marine phytoplankton; importance of conditioning in regulating growth characteristics and metabolism; response of phytoplankton communities to nutrient enrichments. Mailing Add: Dept of Oceanog Univ of Hawaii Honolulu HI 96822

LAWS, EDWARD RAYMOND, JR, b New York, NY, Apr 29, 38; m 62; c 3. TOXICOLOGY. Educ: Princeton Univ, AB, 59; Johns Hopkins Univ, MD, 63. Prof Exp: Intern, Johns Hopkins Hosp, 63-64; asst chief toxicol, Commun Dis Ctr, USPHS, Ga, 64-66; ASST PROF NEUROL SURG & NEUROL SURGEON, JOHNS HOPKINS HOSP, 66- Concurrent Pos: USPHS res grants, 60-62; Henry Strong Denison fel, 62-63; fel surg, Johns Hopkins Univ, 63-64. Res: Toxicology of economic poisons; neurosurgery. Mailing Add: Div of Neurosurg Johns Hopkins Hosp Baltimore MD 21205

LAWS, KENNETH LEE, b Pasadena, Calif, May 30, 35; m 65; c 2. SOLID STATE PHYSICS, METEOROLOGY. Educ: Calif Inst Technol, BS, 56; Univ Pa, MS, 59; Bryn Mawr Col, PhD(physics), 62. Prof Exp: Instr physics, Hobart & William Smith Cols, 58-59; asst prof, 62-66, asst dean, 71-72, ASSOC PROF PHYSICS, DICKINSON COL, 66-, ASSOC DEAN, 72- Mem: Am Asn Physics Teachers; Am Meteorol Soc; Sigma Xi. Res: Solid state physics; ionic diffusion; electrostriction; photoconductivity. Mailing Add: Dickinson Col Carlisle PA 17013

LAWS, LEONARD STEWART, b Pocasset, Okla, Dec 29, 17; m 43; c 4. MATHEMATICS. Educ: Willamette Univ, AB, 39; Stanford Univ, MA, 41; Mich State Univ, EdD, 53. Prof Exp: Asst math, Stanford Univ, 39-41; asst math & mech, Univ Minn, 41-42, from instr to asst prof, 42-52; dean-regstr, 53-55, chmn natural sci div, 55-74, PROF MATH, SOUTHWESTERN COL, KANS, 55- Concurrent Pos: Asst, Mich State Univ, 47 & 53; NSF fel, Stanford Univ, 61-62; mgt consult, 64- Mem: Am Soc Qual Control; Am Statist Asn. Res: Industrial reliability; design of experiments. Mailing Add: Dept of Math Southwestern Col 100 Col St Winfield KS 67156

LAWS, PRISCILLA WATSON, b New York, NY, Jan 18, 40; m 65. NUCLEAR PHYSICS. Educ: Reed Col, BA, 61; Bryn Mawr Col, MA, 63, PhD(physics), 66. Prof Exp: Asst physics, Bryn Mawr Col, 61-63; sr tech aide, Bell Labs, 62; asst prof, 65-70, ASSOC PROF PHYSICS, DICKINSON COL, 70- Concurrent Pos: Mem med radiation adv comt, Bur Radiation Health, Food & Drug Admin, Dept Health, Educ & Welfare, 74-78. Mem: Am Asn Physics Teachers. Res: Nuclear beta decay; environmental radiation; effects of medical x-rays, energy and environment. Mailing Add: Dept of Physics Dickinson Col Carlisle PA 17013

LAWS, WILFORD DERBY, JR, b Colonia Diaz, Mex, Nov 19, 11; US citizen; m 47; c 2. SOIL FERTILITY. Educ: Brigham Young Univ, BS, 39; Utah State Agr Col, MS, 41; Ohio State Univ, PhD(soil chem), 44. Prof Exp: Asst, Utah State Agr Col, 39-41; res assoc, Res Found, Ohio State Univ, 43-47; sr soil scientist, Tex Res Found, 47-56, prin soil scientist, 56-58, chmn dept soil sci, 47-58, chmn dept soil sci & chem, 58-60; chmn dept agron, 65-66, PROF AGRON, BRIGHAM YOUNG UNIV, 60- Concurrent Pos: Dir agr res, Tex Res Found, 67-68. Mem: Am Soc Agron; Soil Sci Soc Am; Int Soc Soil Sci. Res: Farming systems for Texas Blacklands; slow release nitrogen fertilizers for stony soils; corn planting density and silage quality; fall versus spring phosphate on irrigated alfalfa; soil chemistry. Mailing Add: 519 East 600 S Orem UT 84057

LAWSON, ANDREW WERNER, b San Francisco, Calif, Mar 3, 17; m 45; c 3. PHYSICS. Educ: Columbia Univ, AB, 36, PhD(physics), 40. Prof Exp: Asst physics, Columbia Univ, 36-39; from instr to asst prof, Univ Pa, 40-44; mem staff, Radiation Lab, Mass Inst Technol, 44-46; asst prof, Inst Study Metals & Dept Physics, Univ Chicago, 46-47, from assoc prof to prof, 47-61, from actg chmn to chmn dept, 48-56, assoc dir, Inst, 52-56; chmn dept physics, 61-64 & 67-70, PROF PHYSICS, UNIV CALIF, RIVERSIDE, 61- Concurrent Pos: Consult, E I du Pont de Nemours & Co, Inc, 47-67; chmn appl sci prog, Univ Calif, Riverside, 71-74. Mem: AAAS; fel Am Phys Soc; Am Geophys Union. Res: Solid state physics. Mailing Add: Dept of Physics Univ of Calif Riverside CA 92502

LAWSON, BENJAMIN F, b Montgomery, Ala, May 29, 31; m 56; c 3. ORAL MEDICINE, PERIODONTOLOGY. Educ: Auburn Univ, BS, 53; Emory Univ, DDS, 61; Ind Univ, MSD, 63, cert periodont, 68. Prof Exp: Asst prof oral diag & chmn dept, Sch Dent, Emory Univ, 63-66; asst prof, Sch Dent, Univ Ala, 66-68; from assoc prof to prof oral med & periodont, Sch Dent, 68-72, chmn dept, 68-70, chmn dept, Col Dent Med, 70-72, DEAN COL ALLIED HEALTH SCI, MED UNIV SC, 72- Concurrent Pos: NIH teacher training grant, 61-63; Eli Lilly & Co grant, 62-66; gen pract, 65-66; consult, Ft Jackson & Ft Benning Vet Admin Hosp Coastal Ctr, 68-; Warner-Lambert grant, 70- Mem: Am Acad Oral Path; Am Dent Asn; Am Acad Oral Med; Inst Asn Dent Res; Am Acad Periodont. Res: Histology-histopathology of dental pulp; ceramic and titanium implants. Mailing Add: Col of Allied Health Sci Med Univ of SC Charleston SC 29401

LAWSON, BOB LEROY, b Pryor, Okla, Jan 13, 35; m 53; c 3. PHYSICS. Educ: Oklahoma City Univ, BA, 56; Tex Christian Univ, MA, 62. Prof Exp: Nuclear physicist, Gen Dynamics/Ft Worth, 56-60, sr nuclear physicist, 60-62; res physicist, 62-69, SECT MGR, PHILLIPS PETROL CO, 69- Res: Use of shielding methods, especially Monte Carlo, in calculating neutron and gamma penetration in earth formations; nuclear and acoustical well logging; geophysics. Mailing Add: 1325 Harris Dr Bartlesville OK 74003

LAWSON, CHESTER ALVIN, b Salamanca, NY, Apr 22, 08; m 34; c 2. BIOLOGY, PSYCHOLOGY. Educ: Thiel Col, BS, 30; Univ Mich, MS, 31, PhD(zool), 34. Hon Degrees: LLD, Thiel Col, 59. Prof Exp: Asst, Univ Mich, 31-33; instr zool, Mich State Col, 34-35; from asst prof to assoc prof biol, Wittenberg Col, 35-42; from instr to asst prof zool, Mich State Univ, 42-44; prof biol sci & head dept, 44-52, prof natural sci, 52-63, res prof, 63-65; dir biol, Sci Curric Improv Study, 65-75, EMER DIR BIOL, SCI CURRIC IMPROV STUDY, UNIV CALIF, BERKELEY, 75- Concurrent Pos: Vis res biologist, Sci Curric Improv Study, Univ Calif, Berkeley, 65-67; consult, Sci Teaching Ctr, Philippines, 65-66. Mem: AAAS; Am Soc Zoologists; Am Genetics Soc. Res: Chromosomes of aphids; microscopic and experimental embryology of aphids; differential development in regeneration of chicory roots; evolution of ideas; learning. Mailing Add: 1624 Golden Rain Rd 4 Walnut Creek CA 94595

LAWSON, DANIEL DAVID, b Tucson, Ariz, Jan 13, 29; m 57; c 2. ORGANIC CHEMISTRY, POLYMER CHEMISTRY. Educ: Univ Southern Calif, BS, 57, MS, 59. Prof Exp: Biomed res fel, Charles Cook Hastings Found, 59-61; res polymer scientist, 61-71, SR SCIENTIST & MEM TECH STAFF, POLYMER RES SECT, JET PROPULSION LAB, CALIF INST TECHNOL, 71- Mem: AAAS; Am Chem Soc; The Chem Soc; fel Am Acad Forensic Sci; Brit Soc Chem Indust. Res: Physical organic chemistry of polymers; synthesis of new biomaterials; use of thermoluminescence as applied to criminalistics. Mailing Add: 5542 Halifax Rd Arcadia CA 91006

LAWSON, DAVID EDWARD, b Moncton, NB, Sept 17, 39; m 67. SEDIMENTOLOGY. Educ: Univ NB, BSc, 60, MSc, 62; Univ Reading, PhD(geol), 71. Prof Exp: Res geologist, Sedimentology Res Lab, Univ Reading, 62-66; lectr, 66-68, ASST PROF SEDIMENTOLOGY, UNIV WATERLOO, 68- Mem: Int Asn Sedimentol; Soc Econ Paleontologists & Mineralogists; fel Geol Soc London; fel Geol Asn Can. Res: Environmental fluvial sedimentology; nearshore sedimentation; primary sedimentary structures; volcanic sediments; sedimentary geochemistry; Torridonian sediments of northwest Scotland. Mailing Add: Dept of Earth Sci Univ of Waterloo Waterloo ON Can

LAWSON, DAVID FRANCIS, b Chicago, Ill, June 24, 45; m 67; c 2. ORGANIC CHEMISTRY. Educ: Lewis Univ, BA, 67; Iowa State Univ, PhD(org chem), 71. Prof Exp: Instr chem, Iowa State Univ, 68-69; res scientist, 70-75; SR RES SCIENTIST, CENT RES LABS, FIRESTONE TIRE & RUBBER CO, 75- Mem: Am Chem Soc; Combustion Inst; Sigma Xi. Res: Organic polymer chemistry; polymer stabilization; combustion and flammability; additives for flame and smoke retardation; synthetic and physical organic chemistry. Mailing Add: Cent Res Labs Firestone Tire & Rubber Co Akron OH 44317

LAWSON, DEWEY TULL, b Kinston, NC, Feb 6, 44; m 66; c 2. LOW TEMPERATURE PHYSICS. Educ: Harvard Univ, BA, 66; Duke Univ, PhD(physics), 72. Prof Exp: Res assoc physics, Lab Atomic & Solid State Physics, Cornell Univ, 72-74; ASST PROF PHYSICS, DUKE UNIV, 74- Mem: Am Phys Soc; Am Asn Physics Teachers. Res: Low temperature physics, especially liquid and solid helium. Mailing Add: Dept of Physics Duke Univ Durham NC 27706

LAWSON, ELMER JOHN, chemistry, see 12th edition

LAWSON, FAY A, cell biology, genetics, see 12th edition

LAWSON, FRED AVERY, b Washington Co, Ark, Oct 25, 19; m 62; c 4. ENTOMOLOGY. Educ: Univ Ark, BSc, 43; Ohio State Univ, MS, 47, PhD(entom), 49. Prof Exp: Asst prof entom, Univ Tenn, 49-52; from asst prof to assoc prof, Kans State Univ, 52-60; assoc prof, Colo State Univ, 61-63; assoc prof, 63-70, PROF ENTOM, UNIV WYO, 70- Mem: Entom Soc Am; Electron Micros Soc Am; Coleopterists Soc. Res: Insects; electron microscopy; morphology, physiology, biology and taxonomy of Coleoptera. Mailing Add: Dept of Entom & Plant Sci Univ of Wyo Box 3354 Univ Sta Laramie WY 82070

LAWSON, HERBERT BLAINE, JR, b Norristown, Pa, Jan 4, 42; m 64; c 2. MATHEMATICS. Educ: Brown Univ, ScB & AB, 64; Stanford Univ, PhD(math), 69. Prof Exp: Lectr math, Univ Calif, Berkeley, 68-70; vis prof, Inst Pure & Appl Math, Rio de Janeiro, Brazil, 70-71; vis asst prof, State Univ NY Stony Brook, 71; assoc prof, 71-75, PROF MATH, UNIV CALIF, BERKELEY, 75- Concurrent Pos: Sloan Found fel, Univ Calif, Berkeley, 70-; mem, Inst Advan Study, Princeton, 72-73. Honors & Awards: Steele Prize, Am Math Soc, 75. Mem: Am Math Soc. Res: Minimal surfaces; minimal varieties in Riemannian manifolds; Riemannian geometry; foliations; several complex variables. Mailing Add: Dept of Math Univ of Calif Berkeley CA 94720

LAWSON, JAMES EVERETT, b Derby, Va, Jan 8, 33; m 60; c 2. INVERTEBRATE ZOOLOGY. Educ: ETenn State Univ, BS, 58, MA, 59; Va Polytech Inst, PhD(zool), 67. Prof Exp: Instr biol, ETenn State Univ, 59-61; instr, Va Polytech Inst, 61-62; ASSOC PROF BIOL, E TENN STATE UNIV, 64- Res: Ecology and systematics of pseudoscorpions. Mailing Add: Dept of Biol ETenn State Univ Johnson City TN 37601

LAWSON, JAMES LLEWELLYN, b Pasumalai, SIndia, Dec 17, 15; US citizen; wid; c 3. PHYSICS. Educ: Univ Kans, AB, 35, AM, 36; Univ Mich, PhD(physics), 39. Prof Exp: Res physicist, Univ Mich, 39-40; mem staff, Radiation Lab, Mass Inst Technol, 40-45; physicist, Res & Develop Lab, 45-65, mgr electron physics res, 52-65, mgr, Info Lab, Res & Develop Ctr, 65-70, res & develop mgr, Info Sci & Eng, 70-74, MGR RES & DEVELOP PLANNING, GEN ELEC CORP RES & DEVELOP CTR, GEN ELEC CO, 74- Concurrent Pos: Consult, Sci Adv Bd, 53-59. Mem: AAAS; Am Phys Soc; Inst Elec & Electronics Engrs. Res: Nuclear physics and electronics, especially electronic components and systems, high energy radiation, particle acceleration, thermonuclear physics and information systems. Mailing Add: Gen Elec Res & Devel Ctr PO Box 8 Schenectady NY 12301

LAWSON, JIMMIE BROWN, b Checotah, Okla, June 29, 34; m 54; c 2. PETROLEUM CHEMISTRY. Educ: Southeastern State Col, BS, 56; Tex Christian Univ, MA, 60; Rice Univ, PhD(chem), 64. Prof Exp: Chemist, Gen Dynamics Corp, 56-60; res chemist, Shell Develop Co, 64-72, sr chem engr, Shell Oil Co, 72-74, SR PETROL ENGR, SHELL DEVELOP CO, 74- Mem: Am Chem Soc; Soc Petrol Eng. Res: Properties of semiconductors; x-ray crystallography; corrosion and metal coatings; petroleum production; development of chemical systems for the tertiary recovery of petroleum; surface chemistry. Mailing Add: 4626 Briarbend Houston TX 77001

LAWSON, JIMMIE DON, b Waukegan, Ill, Dec 6, 42; m 64; c 1. MATHEMATICS. Educ: Harding Col, BS, 64; Univ Tenn, PhD(math), 67. Prof Exp: Asst prof math, Univ Tenn, 67-68; asst prof, 68-70, ASSOC PROF MATH, LA STATE UNIV, BATON ROUGE, 70- Mem: Am Math Soc. Res: Topological algebra; algebraic topology and semigroups; topology. Mailing Add: Dept of Math La State Univ Baton Rouge LA 70803

LAWSON, JOHN DOUGLAS, b Meaford, Ont, Sept 2, 37; m 60; c 3. MATHEMATICS. Educ: Univ Toronto, BASc, 59; Univ Waterloo, MSc, 60, PhD(appl math), 65. Prof Exp: Teaching fel math, 59-60, lectr, 60-64, from asst prof to assoc prof, 64-73, assoc dean math, 68-71, PROF MATH, UNIV WATERLOO, 73-, CHMN DEPT COMPUT SCI, 74- Concurrent Pos: Asst scientist, Med Div, Oak Ridge Inst Nuclear Studies, 64-65; vis lectr, Univ Dundee, 71-72. Mem: Soc Indust & Appl Math; Asn Comput Math; fel Brit Inst Math & Applns. Res: Numerical solution of ordinary differential equations. Mailing Add: Dept of Comput Sci Univ of Waterloo Waterloo ON Can

LAWSON, JOHN EDWARD, b Detroit, Mich, Feb 25, 31; m 57; c 2. MEDICINAL CHEMISTRY. Educ: Wayne State Univ, BS, 53; Vanderbilt Univ, MA, 56, PhD(org chem), 57. Prof Exp: Fel, Mass Inst Technol, 57-58; SR INVESTR, MEAD JOHNSON & CO, 58- Mem: Am Chem Soc. Res: Natural products; organic synthesis. Mailing Add: 6711 Hogue Rd Evansville IN 47712

LAWSON, JUAN (OTTO), b Bluefield, WVa, Apr 18, 39; m 63; c 2. PHYSICS. Educ: Va State Col, BS, 60; Howard Univ, MS, 62, PhD(physics), 66. Prof Exp: From asst prof to assoc prof, 67-74, asst dean grad sch, 70-71, PROF PHYSICS, UNIV TEX, EL PASO, 74- Mem: Am Phys Soc; Am Asn Physics Teachers; Sigma Xi. Res: Mathematical physics; solid state theory. Mailing Add: Dept of Physics Univ of Tex El Paso TX 79968

LAWSON, JULIAN KEITH, JR, b Washington, DC, Nov 10, 17; m 42; c 4. ORGANIC CHEMISTRY. Educ: Univ Md, BS, 38; Univ Minn, PhD, 42. Prof Exp: Res chemist, E I du Pont de Nemours & Co, NJ, 42-43, Ill, 43 & Am Viscose Corp, Pa, 43-51; RES CHEMIST, GROUP LEADER & SCI FEL, MONSANTO TRIANGLE PARK DEVELOP CTR, INC, 51- Mem: Am Chem Soc. Res: Synthetic organic chemistry; polymers; synthetic textiles and finishes; pressure equipment and reactions; composite structures; reverse osmosis; flameproofing treatments. Mailing Add: Monsanto Triangle Pk Dev Ctr Box 12274 Research Triangle Park NC 27709

LAWSON, KATHERYN EMANUEL, b Shreveport, La, Sept 15, 26; div; c 2. MICROSCOPY. Educ: Dillard Univ, AB, 45; Tuskegee Inst, MS, 47; Univ NMex, PhD(chem), 57. Prof Exp: Asst prof chem, Bishop Col, A&T Col, Savannah State Col, Talladega Col & Grambling Col, 47-51; from asst prof to assoc prof chem, Cent State Col, 51-54; res asst chem, Univ NMex, 54-57; staff mem biochem, Vet Admin Hosp, Albuquerque, 57-58; STAFF MEM MAT SCI, SANDIA LABS, 58- Res: Quantitative television microscopy; image analysis including automatic and interactive pattern recognition; quantification of microstructural data for correlation with physical properties; stereology. Mailing Add: 1738 Lafayette Dr NE Albuquerque NM 87106

LAWSON, KENNETH DARE, b Clinchport, Va, May 12, 34; m 56; c 1. ELECTRON MICROSCOPY. Educ: East Tenn State Univ, BA, 59; Univ Fla, MS, 61, PhD(chem), 63. Prof Exp: SECT HEAD, MIAMI VALLEY LABS, PROCTER & GAMBLE CO, 63- Mem: AAAS; Am Chem Soc. Res: Nuclear magnetic resonance spectroscopy; structure of mesomorphic phases; biophysics. Mailing Add: Miami Valley Labs Procter & Gamble Co Cincinnati OH 45247

LAWSON, KENT DELANCE, b Binghamton, NY, Feb 17, 21; m 42; c 3. THEORETICAL PHYSICS, EXPERIMENTAL PHYSICS. Educ: Cornell Univ, BA, 43; Rensselaer Polytech Inst, MS, 51, PhD(physics), 56. Prof Exp: Instr physics, Rensselaer Polytech Inst, 46-52; prof, Bennington Col, 53-66; prof physics, 65-75, DISTINGUISHED TEACHING PROF, STATE UNIV NY COL ONEONTA, 75- Concurrent Pos: Res & educ prof consult, 56-70; dir eduction prog, 70- Mem: AAAS; Am Phys Soc; Am Asn Physics Teachers; Philos Sci Asn; Hist Sci Soc. Res: Theoretical and experimental studies of eduction theory, eduction procedures and their applications of particles, materials and fields, and of the mechanisms of irreversible phenomena. Mailing Add: Dept of Physics State Univ of NY Col Oneonta NY 13820

LAWSON, LOUIS RUSSELL, JR, b Columbia, SC, Sept 27, 21; m 48; c 3. ENERGY SYSTEMS & TECHNOLOGY, PAPER CHEMISTRY. Educ: Univ of the South, BS, 42. Prof Exp: Proj dir paper res, Westvaco Corp, SC, 46-57, res & develop mgr, NY, 57-59, regional mgr, Va, 59-72; chief exec officer, Med Serv, Am Biophys Corp, Va, 72-74; DIR, VA ENERGY OFF, 74- Concurrent Pos: Energy adv, Commonwealth of Va, 74-; energy consult, State of SC & State of Ky, 75; mem, Energy Comts, Nat Gov Conf, 75- & Natural Gas Comt, Fed Power Comn, 75- Mem: AAAS. Res: Energy resources, technologies systems, uses, conservation, supplies; alternate systems and resources; petroleum, natural gas, electricity, solar, coal, geothermal, nuclear technologies for energy generation or conversion. Mailing Add: Va Energy Off 823 E Main St Richmond VA 23219

LAWSON, MERLIN PAUL, b Jamestown, NY, Jan 12, 41; m 64; c 3. CLIMATOLOGY. Educ: State Univ NY Buffalo, BA, 63; Clark Univ, MA, 66, PhD(geog climat), 73. Prof Exp: Instr climat, Northeastern Univ, 67-68; asst prof, 68-74, ASSOC PROF CLIMAT, UNIV NEBR-LINCOLN, 74- Mem: Sigma Xi; Am Meteorol Soc; Asn Am Geogrs. Res: Historical climate of the Great American Desert; severe droughts since 1700 in the western United States; descriptive climatic change; dendroclimatology. Mailing Add: 350 Avery Hall Univ of Nebr Lincoln NE 68588

LAWSON, MILDRED WIKER, b New London, Conn, Nov 10, 22; m 63. MATHEMATICS, COMPUTER SCIENCES. Educ: Univ Md, BS, 47, MA, 49. Prof Exp: Asst math, Univ Md, 47-49, instr, 49-50; cartog compilation aide, Corps Engrs, Army Map Serv, 50-51, mathematician, 51-55, proj leader math & comput prog, 55-57, asst chief prog, 57-58; assoc mathematician, 58-65, SR MATHEMATICIAN, APPL PHYSICS LAB, JOHNS HOPKINS UNIV, LAUREL, 65- Mem: Asn Comput Mach. Res: Analysis and programming of computer solutions of problems arising in scientific projects; computer language training.

LAWSON, NELSON ERNEST, organic chemistry, see 12th edition

LAWSON, NORMAN C, b Glasgow, Scotland, Nov 3, 29; Can citizen; m 60; c 3. PLANT BREEDING. Educ: Glasgow Univ, BSc, 53; Univ Reading, dipl agr, 54; McGill Univ, MSc, 58, PhD, 61. Prof Exp: Res off, Exp Farm, Can Dept Agr, BC, 61-65, res scientist, Res Sta, Sask, 65-67; ASST PROF AGRON, McGILL UNIV, 67- Mem: Am Soc Agron; Crop Sci Soc Am; Genetics Soc Can; Agr Inst Can; Can Soc Agron. Res: Genetics and breeding of forage and oil crop species. Mailing Add: Dept of Agron McGill Univ Montreal PQ Can

LAWSON, RALPH WILLARD, b Albert Lea, Minn, June 4, 21; m 45; c 2. GEOLOGY. Educ: Iowa State Univ, BS, 47, MS, 49; Univ Wis, PhD(geol), 53. Prof Exp: Instr geol, Iowa State Univ, 49; geologist, Socony-Vacuum Oil Co, Inc, Venezuela, 54-55, Mobil Oil Co, Colombia, 55-58, Mobil Explor, Inc, 58-63, Mobil Northern & Southeastern Europe, 63-65, sr staff geologist, Mobil Mediterranean & Africa, Inc, 65-69, sr staff geologist, Mobil Explor Nigeria, Inc, 69-75, EXPLORATION TECH MGR, MOBIL OIL LIBYA LTD, 75- Mem: AAAS; Am Asn Petrol Geologists; fel Brit Geol Soc; fel Brit Inst Petrol. Res: Petrogenesis; physical and chemical properties of carbonate rocks. Mailing Add: Box 404 Tripoli Libya

LAWSON, ROBERT BARRETT, b Oakland, Calif, Aug 24, 11; m 39; c 2. PEDIATRICS. Educ: Harvard Univ, BA, 32, MD, 36; Am Bd Pediat, dipl, 41. Prof Exp: From asst prof to prof pediat & dir dept, Bowman Gray Sch Med, 40-54; prof & chmn dept, Sch Med, Univ Miami, 54-62, actg dean, Sch Med, 61-62; prof pediat, Sch Med, Northwestern Univ, 62-71, chmn dept, 62-70, vpres health sci, 70-71; CHIEF STAFF, VARIETY CHILDREN'S HOSP, 71- Concurrent Pos: Pro Sch Pub Health, Univ NC & assoc, Sch Med, Duke Univ, 40-42; pediat consult, State Bd Health, NC, 40-42; Nat Res Coun fel, Univ Calif & Yale Univ, 45; chief staff, Children's Mem Hosp, Chicago, Ill, 62-71; clin prof pediat, Sch Med, Univ Miami, 71- Mem: Am Soc Pediat Res; Am Pediat Soc; AMA; Am Acad Pediat. Res: Infectious disease. Mailing Add: Variety Children's Hosp 6125 S W 31st St Miami FL 33155

LAWSON, ROBERT DAVIS, b Sydney, Australia, July 14, 26; nat US; m 50; c 3. THEORETICAL PHYSICS. Educ: Univ BC, BASc, 48, MASc, 49; Stanford Univ, PhD, 53. Prof Exp: Asst, Univ BC, 47-48; asst, Stanford Univ, 49-53; jr res physicist, Univ Calif, 53-57; res physicist, Enrico Fermi Inst Nuclear Studies, Ill, 57-59; assoc physicist, 59-66, SR PHYSICIST, ARGONNE NAT LAB, 66- Concurrent Pos: Vis physicist, UK Atomic Energy Authority, Harwell, Eng, 62-63; Weizmann sr fel, Weizmann Inst Sci, Israel, 67-68. Mem: Am Phys Soc. Res: Nuclear physics. Mailing Add: Physics Div Argonne Nat Lab Bldg 203 Argonne IL 60439

LAWSON, VERNA REBECCA, b Crossville, Tenn, Apr 7, 43. PLANT PHYSIOLOGY. Educ: Tenn Technol Univ, BS, 66, MS, 69; George Washington Univ, PhD(biol sci), 73. Prof Exp: ASSOC PROF BIOL, ALCORN STATE UNIV, 73- Mem: Sigma Xi; Am Soc Plant Physiol; Bot Soc Am; Am Inst Biol Sci. Res: Control of plant cell morphogenesis by growth hormones and the phytochrome system. Mailing Add: PO Box 98 Alcorn State Univ Lorman MS 39096

LAWSON, WILLIAM BURROWS, b Detroit, Mich, June 8, 29; m 63; c 1. BIOCHEMISTRY. Educ: Wayne State Univ, BS, 51; Univ Md, PhD(chem), 56. Prof Exp: Res assoc, Mass Inst Technol, 55-56 & 57-58, Nat Cancer Inst fel, 56-57; sr asst scientist, Nat Insts Health, Md, 58-60; Nat Cancer Inst fel, Max Planck Inst Biochem, 60-61; sr res scientist, 61-66, ASSOC RES SCIENTIST, DIV LABS & RES, NY STATE DEPT HEALTH, 66- Mem: AAAS; Am Chem Soc; Am Soc Biol Chem; NY Acad Sci. Res: Amino acids; synthesis and degradation of peptides; chemical modification of enzymes. Mailing Add: Div of Labs & Res NY State Health Dept Albany NY 12201

LAWTON, ALEXANDER R, III, b Nov 8, 38. MEDICINE. Educ: Yale Univ, BA, 60; Vanderbilt Univ, MD, 64. Prof Exp: Intern, Vanderbilt Univ Hosp, 64-65, asst resident, 65-66; clin assoc, Lab Clin Invest, Nat Inst Allergy & Infectious Dis, 66-68, clin investr, 68-69; asst prof, 71-73, ASSOC PROF PEDIAT & MICROBIOL, SCH MED, UNIV ALA, BIRMINGHAM, 73- Concurrent Pos: NIH spec fel immunol, Sch Med, Univ Ala, Birmingham, 69-71. Res: Pediatrics; microbiology; immunology. Mailing Add: Dept of Pediat Univ of Ala Sch of Med Birmingham AL 35233

LAWTON, ALFRED HENRY, b Carson, Iowa, July 26, 16; m 40; c 3. GERIATRICS. Educ: Simpson Col, AB, 37; Northwestern Univ, MS, 39, BM, 40, MD, 41, PhD(physiol), 43. Hon Degrees: ScD, Simpson Col, 58. Prof Exp: Intern, Passavant Mem Hosp, Chicago, 40-41; resident, Henry Ford Hosp, Detroit, 41-42; asst prof med, physiol & pharm, Sch Med, Univ Ark, 46-47; dean & prof physiol & pharm, Univ NDak, 47-48; chief res div, US Vet Admin, Washington, DC, 48-51; med res adv, US Dept Air Force, 51-55; asst dir prof serv res & educ & chief intermediate serv, US Vet Admin Ctr, Bay Pines, 55-62; dir study ctr, Nat Inst Child Health & Human Develop, 62-66; from asst to assoc dean acad affairs, Univ SFla, 66-70, actg vpres acad affairs, 70-73; exec dir tech adv comn aging res, US Dept Health, Educ & Welfare, 73-74; DIR GERIAT RES, EDUC & CLIN CTR & ASSOC CHIEF STAFF RES & DEVELOP, VET ADMIN CTR, 75- Concurrent Pos: Asst clin prof, Sch Med, George Washington Univ, 48-55; liaison mem coun arthritis & metab dis, USPHS, 50-55; mem exec coun, Nat Res Coun, US Armed Forces Vision Comt & Nat Coun Aging; actg dean, Col Med, Univ SFla, 68-70. Mem: Am Physiol Soc; Am Soc Pharmacol & Exp Therapeut; Am Geriat Soc; Geront Soc; Am Chem Soc. Res: Aging; chronic diseases. Mailing Add: Vet Admin Ctr Bay Pines FL 33504

LAWTON, EMIL ABRAHAM, inorganic chemistry, see 12th edition

LAWTON, ERNEST LINWOOD, II, organic chemistry, polymer chemistry, see 12th edition

LAWTON, GERALD WARREN, b Coloma, Wis, Oct 22, 06; m 36; c 1. CHEMISTRY. Educ: Marquette Univ, BS, 31, MS, 33; Univ Wis, PhD(chem), 53. Prof Exp: Asst chem, Marquette Univ, 31-33, from instr to asst prof, 38-45; petrol chemist, Cities Serv Oil Co, Wis, 33-36; res chemist, Kolmar Labs, 36-38; from sr chemist to pub health engr, State Bd Health, Wis, 45-50; from instr to assoc prof civil eng, 50-61, from assoc prof to prof prev med, 61-74, EMER PROF PREV MED, UNIV WIS-MADISON, 74- Mem: Am Chem Soc; Am Water Works Asn; Water Pollution Control Fedn; Am Inst Chemists. Res: Fertilization of lakes; aerobic sludge digestion; water, sewage and sanitation; water pollution; fluoride, detergents; nitrates and sodium in water; sanitary chemistry. Mailing Add: Dept of Prev Med Univ of Wis Madison WI 53706

LAWTON, IRENE ELIZABETH, b New York, NY, June 9, 40. PHYSIOLOGY, ANATOMY. Educ: Hunter Col, AB, 62; Univ Ill Col Med, PhD(physiol), 66. Prof Exp: From instr to assoc prof physiol, Stritch Sch Med, Loyola Univ Chicago, 71-73. Concurrent Pos: NIH fel anat, Brain Res Inst, Univ Calif, Los Angeles, 66-67, res grant, Stritch Sch Med, Loyola Univ Chicago, 67-73, NSF res grant, 68-73; traveling fels, Endocrine Soc, 68 & NSF, 69; ad hoc consult, myocardial infarction prog, NIH & prog-proj comt, Nat Heart & Lung Inst, 70-73. Mem: AAAS; Endocrine Soc; Am

Physiol Soc; Soc Neurosci; Soc Study Reproduction. Res: Reproductive neuroendocrinology; regulation of gonadotrophin secretion by the hypothalamus, limbic system and ovarian and adrenal steroids; development and maintenance of reproductive cyclicity; circadian rhythms. Mailing Add: Northwestern Univ Med Sch 303 E Chicago Ave Chicago IL 60611

LAWTON, KIRKPATRICK, b St Paul, Minn, Sept 6, 17; m 42; c 3. AGRONOMY. Educ: Univ Minn, BS, 39; Mich State Col, PhD(soil sci), 45. Prof Exp: Jr tech asst, Soil Conserv Serv, USDA, Minn, 38; asst soils, Mich State Col, 40-43; res assoc agron, Iowa State Col, 43-46; from asst prof to assoc prof soils, 46-54, coordr agr prog, 64-66, dir inst int agr, 64-69, PROF SOILS, MICH STATE UNIV, 54-, ASST DEAN INT PROGS, 73- Concurrent Pos: Tenn Valley Authority lectr; consult, Int Atomic Energy Agency & USAID; dean agr, Mich State Univ-AID-Univ Nigeria, 61-64; prog coordr Indonesia proj, Midwest Univs Consortium Int Activ-USAID, 70-73& Mich State Univ-Brazil-Univ Ceara, 75- Mem: Soil Sci Soc Am; fel Am Soc Agron; Soil Conserv Soc Am. Res: Soil testing; minor element availability to plants; field fertilizer trials; relation between potassium supply in Michigan soils and the growth of alfalfa and field beans; radioactive phosphorous fertilizer; tropical soil; institutional development in developing nations. Mailing Add: 202 Int Ctr Mich State Univ East Lansing MI 48824

LAWTON, RICHARD G, b Berkeley, Calif, Aug 29, 34; m 58; c 5. SYNTHETIC ORGANIC CHEMISTRY, BIO-ORGANIC CHEMISTRY. Educ: Univ Calif, Berkeley, BS, 56; Univ Wis, PhD(chem), 62. Prof Exp: Asst isolation & identification, Merck Sharp & Dohme Res Labs, 56-57; asst org chem, Univ Wis, 59-62; from instr to assoc prof, 62-70, PROF ORG CHEM, UNIV MICH, ANN ARBOR, 70- Concurrent Pos: Vis prof, Univ Wis, 70-71; consult, Colgate-Palmolive Res Labs, 70- Mem: AAAS; Am Chem Soc. Res: Synthetic organic chemistry including peptide chemistry, alkaloids, terpenes and polycyclic aromatic hydrocarbons. Mailing Add: Dept of Chem Univ of Mich Ann Arbor MI 48104

LAWTON, RICHARD L, b Council Bluffs, Iowa, June 18, 18; m 43; c 7. SURGERY. Educ: Univ Omaha, AB, 39; Univ Nebr, BS & MD, 43. Prof Exp: Intern, St Mary's & St Louis Hosps, Mo, 43-44; resident, US Vet Hosp, Omaha, Nebr, 46-50, asst chief surg, 50-51; pvt pract, 52-53; staff physician, 53-58, CHIEF CANCER CHEMOTHER, VET ADMIN HOSP, 58-, ASST CHIEF SURG, 63-, DIR RENAL DIALYSIS, 64- Concurrent Pos: Instr, Creighton Univ, 50-51; from clin asst prof to clin assoc prof, Univ Iowa, 54-64, assoc prof, 64-68, prof surg, 68-; res grant angiol, Univ Iowa, 64-65. Mem: Am Asn Cancer Res; Am Soc Artificial Internal Organs; fel Am Col Surg; Soc Exp Biol & Med; Am Soc Clin Oncol. Res: Renal dialysis; cancer chemotherapy; transplant and preservation. Mailing Add: Dept of Surg Univ Hosp Iowa City IA 52240

LAWTON, RICHARD WOODRUFF, b New York, NY, June 22, 20; m 46; c 2. PHYSIOLOGY. Educ: Dartmouth Col, AB, 42; Cornell Univ, MD, 44. Prof Exp: Fel physiol sci, Med Sch, Dartmouth Col, 46-48; from instr to asst prof physiol, Med Col, Cornell Univ, 48-54; assoc prof, Sch Med, Univ Pa, 54-58; mgr bioastronaut sect, Missile & Space Div, Pa, 58-67, mgr bioastronaut sect, Res & Eng Space Systs Orgn, 67-69, Life Systs, 69-70, med res dir, Med Develop Oper, Chem & Med Div, 70-71, mgr ventures develop, Med Ventures Oper, Med Syst Bus Div, 72-74, CONSULT MED SYSTS, CORP RES & DEVELOP, GEN ELEC CO, 74- Concurrent Pos: Head physiol sect, Aviation Med Lab, Naval Air Develop Ctr, Pa, 54-58; adj assoc prof, Sch Med, Univ Pa, 58-70. Mem: Am Physiol Soc. Res: Elasticity of body tissues; aerospace physiology; cardiovascular physiology. Mailing Add: 1340 Stanley Lane Schenectady NY 12309

LAWTON, STEPHEN LATHAM, b Milwaukee, Wis, Nov 12, 39; m 67; c 2. STRUCTURAL CHEMISTRY. Educ: Univ Wis, BS, 63; Iowa State Univ Sci & Technol, MS, 66. Prof Exp: Res chemist, Socony Mobil Oil Co, Inc, 66-69, sr res chemist, 69-75, ASSOC CHEM, MOBIL RES & DEVELOP CORP, 75- Mem: Am Chem Soc; Am Crystallog Asn. Res: X-ray crystallography; crystal and molecular structures of inorganic and organometallic compounds and zeolites; computer applications in chemistry. Mailing Add: Mobil Res & Develop Corp Res Dept Paulsboro NJ 08066

LAWTON, WALLACE CLAYTON, b Can, July 20, 23; m 43; c 1. DAIRY BACTERIOLOGY. Educ: Univ Sask, BSA, 50, MSc, 51; Iowa State Univ, PhD(dairy bact), 54. Prof Exp: Dir lab, Qual Control Comt, 54-60; tech serv, Twin City Milk Producers Asn, 60-70, mgr opers, Northern Div, Mid-Am Dairymen, Inc, 70-75; GEN MGR, A&L LABS, INC, 75- Concurrent Pos: Lectr, Univ Minn, 58-60. Mem: Am Dairy Sci Asn; Int Asn Milk, Food & Environ Sanit (past pres). Res: Laboratory control of dairy products; laboratory methods and procedures. Mailing Add: 1001 Glenwood Minneapolis MN 55405

LAWTON, WILLIAM HARVEY, b Indianapolis, Ind, Nov 1, 37. APPLIED STATISTICS. Educ: Univ Calif, Berkeley, AB, 59, MA, 62, PhD(statist), 65. Prof Exp: Teaching asst statist, Univ Calif, Berkeley, 60-62, teaching assoc, 63-65; math analyst, 62-63, consult math & statist, 65-75, SUPVR APPL MATH SECT, EASTMAN KODAK CO, 75- Concurrent Pos: Lectr, Univ Rochester, 65-70; ed, Technometrics, 74- Honors & Awards: Shewell Award, Am Soc Qual Control, 71; Wixcuxon Award, Am Statist Asn, 71 & 75. Mem: Fel Am Statist Asn; Sigma Xi. Res: Probability; new tools for mathematical model building for problems in the physical sciences. Mailing Add: Eastman Kodak Co MSDD Bldg 56 Kodak Park Div Rochester NY 14650

LAWVER, DONALD ALLEN, b Dayton, Ohio, Dec 8, 39; m 64; c 3. MATHEMATICS. Educ: Ohio State Univ, BSc, 61; Univ Ariz, MSc, 63; Univ Wis, PhD(math), 67. Prof Exp: Instr math, Univ Wis, 67; asst prof, Univ Ariz, 67-74; MEM STAFF, DATA PROCESSING DIV, IBM CORP, 74- Mem: Am Math Soc; Math Asn Am. Res: Modern Algebra. Mailing Add: Data Processing Div IBM Corp 4502 N Central Ave Phoenix AZ 85012

LAWVERE, FRANCIS WILLIAM, b Muncie, Ind, Feb 9, 37; m 66; c 5. ALGEBRA. Educ: Ind Univ, BA, 60; Columbia Univ, MA & PhD(math), 63. Prof Exp: Syst analyst, Litton Industs, Inc, 62-63; asst prof math, Reed Col, 63-64; NATO fel, Swiss Fed Inst Technol, 64-65, res assoc, 65-66; asst prof, Univ Chicago, 66-67; assoc prof, Grad Ctr, City Univ New York, 67-68; Sloan fel, Swiss Fed Inst Technol, 68-70; res prof, Dalhousie Univ, 70-71; vis prof, Aarhus Univ, 71-72 & Nat Res Inst Italy, 72-74; PROF MATH, STATE UNIV NY BUFFALO, 74- Mem: Am Math Soc. Res: Foundations of category theory; categorical foundations of mathematics; algebraic theories and equational doctrines; axiomatic theory of topoi; closed categories and metric spaces. Mailing Add: Dept of Math State Univ NY 4246 Ridge Lea Rd Buffalo NY 14226

LAWWILL, STANLEY JOSEPH, b London, Ohio, May 23, 16; m 40; c 5. MATHEMATICS. Educ: Univ Cincinnati, AB, 37, MA, 39, PhD(math), 41. Prof Exp: Instr math, Northwestern Univ, 41-44; mathematician appl math group, Columbia Univ, 44; gunnery analyst, 2nd & 20th air forces, US Air Force, 44-46, opers analyst, HQ Strategic Air Command, 46-48, dep chief opers anal off, 48-50, chief atomic capabilities div, 50-54, dep chief scientist, 54-58; tech dir sci anal off, Melpar, Inc Div, Westinghouse Air Brake Co, 58; PRES, ANAL SERVS, INC, 58- Mem: Opers Res Soc Am. Res: Orthogonal functions; overconvergence of approximations in terms of rational harmonic functions; operations analysis; weapon systems evaluation. Mailing Add: 6532 Copa Ct Falls Church VA 22044

LAWYER, TIFFANY, JR, b Albany, NY, Jan 3, 15; m 42; c 4. NEUROLOGY. Educ: Albany Med Col, MD, 42. Prof Exp: Instr neurol & psychiat, Albany Med Col, 48-49; assoc neurol, Columbia Univ, 49-52; assoc prof, Georgetown Univ, 52-54; from assoc prof to prof clin neurol, Columbia Univ, 54-64; PROF NEUROL, ALBERT EINSTEIN COL MED, 64- Concurrent Pos: Chief sect neurol, Vet Admin Cent Off, DC, 52-53; chief div neurol, Montefiore Hosp & Med Ctr, 54-69, consult neurol & dep dir, 69- Mem: AMA; Am Neurol Asn; Asn Res Nerv & Ment Dis; Am Psychiat Asn; Am Acad Neurol. Res: Convulsive disorders; multiple and amyotrophic lateral sclerosis. Mailing Add: Montefiore Hosp & Med Ctr Bronx NY 10467

LAX, ANNELI, b Kattowitz, Ger, Feb 23, 22; nat US; m 48; c 2. MATHEMATICS. Educ: Adelphi Col, BS, 42; NY Univ, MA, 45, PhD, 55. Prof Exp: Asst aeronaut, 43-44, asst math, Inst Mech Sci, 45-55, asst res scientist, 55-65, assoc prof, 65-71, PROF MATH, WASH SQ COL, NY UNIV, 72- Concurrent Pos: Instr, Wash Sq Col, 45-55. Mem: Am Math Soc; Math Asn Am. Res: Mathematical analysis and exposition; partial differential equations. Mailing Add: Dept of Math Wash Sq Col NY Univ New York NY 10003

LAX, BENJAMIN, b Miskolz, Hungary, Dec 29, 15; m 42; c 2. SOLID STATE PHYSICS, PLASMA PHYSICS. Educ: Cooper Union, BME, 41; Mass Inst Technol, PhD(physics), 49. Prof Exp: Mech engr, US Eng Off, 41-42; consult, Sylvania Elec Prod, Inc, 46 & Air Force Cambridge Res Ctr, 46-51; mem staff solid state group, Lincoln Lab, 51-53, head ferrites group, 53-55, solid state group, 55-57, assoc head commun div, 57-58, head solid state div, 58-60, DIR, FRANCIS BITTER MAGNET LAB, MASS INST TECHNOL, 60-, PROF PHYSICS, 65- Concurrent Pos: Assoc dir, Lincoln Lab, 64-65; chmn, Comn Quantum Electronics, Int Union Pure & Appl Physics, 75- Honors & Awards: Outstanding Achievement Award, US Air Force Off Aerospace Res, 70; Buckley Prize, Am Phys Soc, 60. Mem: Nat Acad Sci; fel Am Phys Soc; fel Am Acad Arts & Sci. Res: Radar; microwave ferrites; semiconductors; cyclotron resonance; magnetospectroscopy; high magnetic fields; fusion; x-rays. Mailing Add: Francis Bitter Nat Mag Lab MIT Bldg NW14-3220 Cambridge MA 02139

LAX, EDWARD, b Toronto, Ont, Aug 29, 31; US citizen; m 60; c 1. CRYOGENICS. Educ: Univ Calif, Los Angeles, AB, 52, MA, 59, PhD(physics), 60. Prof Exp: Sr physicist, Ultrasonic Systs, Inc, 60-61; mem tech staff, Aerospace Corp, 61-67; MEM TECH STAFF, AUTONETICS DIV, ROCKWELL INT CORP, 67- Mem: Am Phys Soc; Inst Elec & Electronics Engrs. Res: Electron-phonon effects in metals at low temperatures; acoustical-optical effects; cryogenic heat transfer and thermodynamics; hypersonics and delay lines; infra-red systems. Mailing Add: 5637 Wilhelmina Ave Woodland Hills CA 91364

LAX, LOUIS CARL, b Toronto, Ont, Apr 25, 30; m 58. PHYSIOLOGY, MEDICINE. Educ: Univ Toronto, BA, 52, MA, 53, MD, 57, PhD(physiol), 66. Prof Exp: Asst, Banting & Best Dept Med Res, Univ Toronto, 52-57; med assoc, Dept Med, Brookhaven Nat Lab, 61-62, asst scientist, 62-64; asst prof surg & physiol, Sch Med, Univ Calif, Los Angeles, 64-69; ASSOC PROF SURG, ABRAHAM LINCOLN SCH MED, UNIV ILL MED CTR, 70-; VPRES, TELEMED CORP, 73- Concurrent Pos: Fel, Sunnybrook Vet Hosp, Toronto, 57-61; res collabr dept med, Brookhaven Nat Lab, 64-71; consult, Rand Corp, Calif, 65-66 & Epoxylite Corp, 68-69; sr clin investr med dept, Hosp Prod Div, Abbott Labs, Ill, 69-71; med dir, Telemed Corp, 71-73. Mem: AAAS; Biophys Soc; NY Acad Sci; Can Physiol Soc; Inst Elec & Electronics Eng. Res: Design of clinical protocols, analysis of experimental data; study of compartmental kinetics in living systems; influence of hormones and postsurgical trauma on fluid and electrolyte distribution in the body; mathematical simulation of living systems; computer processed electrocardiography; biomedical applications of computers. Mailing Add: Telemed Corp 2345 W Pembroke Hoffman Estates IL 60172

LAX, MELVIN, b New York, NY, Mar 8, 22; m 49; c 4. THEORETICAL SOLID STATE PHYSICS, QUANTUM OPTICS. Educ: NY Univ, BA, 42; Mass Inst Technol, SM, 43, PhD(physics), 47. Prof Exp: Res physicist, Underwater Sound Lab, Mass Inst Technol, 42-45, res assoc physics, 47; from asst prof to prof, Syracuse Univ, 47-55; mem tech staff, Bell Tel Labs, 55-71, head theoret physics dept, 62-64; DISTINGUISHED PROF PHYSICS, CITY COL NEW YORK, 71- Concurrent Pos: Consult, Crystal Br, US Naval Res Lab, 51-55 & US Army Missile Command, 71-; lectr, Oxford Univ, 61-62; mem basic res adv comt, Nat Acad Sci, 66-69. Mem: AAAS; fel Am Phys Soc. Res: Meson creation and absorption; multiple scattering; phase transitions; optical and electrical properties of solids; impurity bands; classical and quantum relaxation and noise; group theory in solids; quantum communication theory; nonlinear optical properties in deforming solids; optics, lasers, continuum mechanics. Mailing Add: Dept of Physics City Col of New York New York NY 10031

LAX, PETER DAVID, b Hungary, May 1, 26; nat US; m 48; c 2. MATHEMATICS. Educ: NY Univ, AB, 47, PhD, 49. Prof Exp: From asst prof to assoc prof math, 49-58, asst to dir math ctr, 59-63, PROF MATH, NY UNIV, 58-, DIR AEC COMPUT & APPL MATH CTR, COURANT INST MATH SCI, 63-, HEAD ALL-UNIV MATH DEPT, 72- Concurrent Pos: Mem staff, Los Alamos Sci Lab, 50; consult, Radiation Lab, Univ Calif. Mem: Nat Acad Sci; Am Math Soc. Res: Theory of partial differential equations; functional analysis; fluid dynamics. Mailing Add: Courant Inst of Math Sci NY Univ 251 Mercer St New York NY 10003

LAY, DAVID CLARK, b Los Angeles, Calif, Mar 1, 41; m 70. MATHEMATICS. Educ: Aurora Col, BA, 62; Univ Calif, Los Angeles, MA, 65, PhD(math), 66. Prof Exp: Teaching asst math, Univ Calif, Los Angeles, 63-64, asst prof, 66; asst prof, 66-70, ASSOC PROF MATH, UNIV MD, COLLEGE PARK, 70- Concurrent Pos: NSF res grants, Univ Md, College Park, 69-72; res grant, Neth Orgn Advan Pure Res, 73. Mem: AAAS; Am Math Soc; Math Asn Am. Res: Functional analysis; spectral theory of linear operators; operator-valued analytic functions. Mailing Add: Dept of Math Univ of Md College Park MD 20742

LAY, DOUGLAS M, b Jackson, Miss, July 3, 36; m 61; c 2. ANATOMY, ZOOLOGY. Educ: Millsaps Col, BS, 58; La State Univ, MS, 61; Univ Chicago, PhD(anat), 68. Prof Exp: Instr anat, Univ Chicago, 68-69; asst prof zool & cur mammals, Univ Mich, 69-73; ASST PROF ANAT, UNIV NC, CHAPEL HILL, 73- Mem: Am Soc Mammal; Am Soc Zool; Soc Study Evol; Soc Syst Zool; Soc Vert Palaeont. Res: The adaptive significance of specializations of mammals for life in deserts, particularly the structure and function of the ear in desert rodents; origin, evolution, functional anatomy, biology and systematics of rodents. Mailing Add: Div of Health Sci Univ of NC Dept of Anat Chapel Hill NC 27514

LAY, JOHN CHARLES, b Ponca City, Okla, Mar 6, 48; m 68; c 2. VETERINARY MEDICINE. Educ: Univ Mo, Columbia, BS, 71, DVM, 75. Prof Exp: ASSOC SCIENTIST VET MED, INHALATION TOXICOL RES INST, LOVELACE FOUND FOR MED EDUC & RES, 75- Mem: Am Vet Med Asn. Res: Long term effects of inhaled radionuclides. Mailing Add: Inhalation Toxicol Res Ints PO Box 5890 Albuquerque NM 87115

LAY, STEVEN R, b Los Angeles, Calif, Nov 28, 44; m 71. MATHEMATICS. Educ: Aurora Col, BA, 66; Univ Calif, Los Angeles, MA, 68, PhD(math), 71. Prof Exp: ASSOC PROF MATH, AURORA COL, 71- Mem: Am Math Soc; London Math Soc; Math Asn Am. Res: Combinatorial geometry and convexity; the separation of convex sets. Mailing Add: Dept of Math Aurora Col Aurora IL 60507

LAYBOURNE, PAUL C, b Akron, Ohio, Dec 21, 19; m 45; c 3. PSYCHIATRY. Educ: NY Med Col, MD, 44; Ind Univ, MS, 49. Prof Exp: Assoc, 49-51, from asst prof to assoc prof, 51-65, PROF CHILD PSYCHIAT, MED CTR, UNIV KANS, 65-, ASSOC PROF PEDIAT, 54- Concurrent Pos: Consult, Spofford Home, Kansas City, Mo, 50-; dir, Atchison County Guid Clin, 51-59; supvr extern prog, State Hosps, 53-55; dir child guid clin, Mercy Hosp, 57-66. Mem: AMA; fel Am Psychiat Asn. Res: Child psychiatry. Mailing Add: Dept of Psychiat Univ of Kans Med Ctr Kansas City KS 66103

LAYCOCK, ARLEIGH HOWARD, b Strathmore, Alta, May 10, 24; m 50; c 3. GEOGRAPHY, HYDROLOGY. Educ: Univ Toronto, BA, 49; Univ Minn, PhD, 57. Prof Exp: Lectr geog, Univ Minn, 52; hydrologist, Eastern Rockies Forest Conserv Bd, 52-55; hydrologist, Prairie Provs Water Bd, 55; lectr geog, 55-57, from asst prof to assoc prof, 57-67, PROF GEOG, UNIV ALTA, 67- Concurrent Pos: Contract, Fed Govt, Alta, 55-69; pvt consult, 55-; chmn water resources res ctr comt & mem nat adv comt water resources res & Can Coun grants comt, Univ Alta, 67-69; Can mem int hydrol decade comt, Int Geog Union. Mem: Am Geog Soc; Am Soc Photogram; Asn Am Geogr; Can Asn Geogr; Am Water Resources Asn (pres, 71). Res: Air photograph interpretation. Mailing Add: Dept of Geog Univ of Alta Edmonton AB Can

LAYCOCK, DAVID GERALD, bAurora, Ill, Apr 14, 43; m 66; c 2. MOLECULAR GENETICS. Educ: St Procopius Col, BS, 65; Univ Hawaii, MS, 67, PhD(genetics), 69. Prof Exp: Res assoc molecular biol, Northwestern Univ, 69-70; staff fel molecular hemat, Nat Heart & Lung Inst, 70-72; ASST PROF BIOCHEM, COL MED, UNIV S ALA, 72- Concurrent Pos: Ala Heart Asn res grant in aid, 73; NIH res grants, 74 & 76. Mem: Am Chem Soc. Res: Control of globin gene expression assessed by in vitro transcription and translation. Mailing Add: Dept of Biochem Univ of S Ala Col of Med Mobile AL 36688

LAYCOCK, WILLIAM ANTHONY, b Ft Collins, Colo, Mar 17, 30; m 55; c 2. PLANT ECOLOGY. Educ: Univ Wyo, BS, 52, MS, 53; Rutgers Univ, PhD(bot), 58. Prof Exp: Range technician, State Game & Fish Comn, Wyo, 55; asst, Rutgers Univ, 55-58; range scientist, Intermt Forest & Range Exp Sta, 58-74, ASST DIR, ROCKY MT FOREST RANGE EXP STA, US FOREST SERV & FAC AFFIL, COLO STATE UNIV, 74- Concurrent Pos: Collabr range sci, Utah State Univ, 64-74; NZ Nat Res Adv Coun-NZ Forest Serv sr res fel, 69-70; coordr site dir western coniferous biome, US Int Biol Prog, 71-72. Mem: Ecol Soc Am; Soc Range Mgt; Australian Ecol Soc; NZ Ecol Soc. Res: Ecology and management of high elevation rangelands; autecology of range species; natural areas. Mailing Add: Rocky Mt Forest & Range Exp Sta 240 W Prospect St Ft Collins CO 80521

LAYDE, DURWARD CHARLES, b La Crosse, Wis, Dec 29, 12; m 43; c 5. INORGANIC CHEMISTRY. Educ: St Norbert Col, AB, 33; Univ Wis, PhD(gen chem), 40. Prof Exp: Instr math, St Norbert Col, 33-34; asst chem, Univ Wis, 36-39, instr exten, 39-43 & army specialized training prog, 43-44, asst prof exten, 44-51, assoc prof, 51-62, PROF CHEM, UNIV WIS-MILWAUKEE, 62- Mem: AAAS; Am Chem Soc. Mailing Add: 4968 N Idlewild Ave Milwaukee WI 53217

LAYER, ROBERT WESLEY, b Brooklyn, NY, Aug 11, 28; m 55; c 3. ORGANIC CHEMISTRY. Educ: NY Univ, AB, 50; Univ Cincinnati, PhD, 55. Prof Exp: Control chemist, Naugatuck Chem Co, 50-52; sr tech man, 55-57, sr res chemist, 57-71, RES ASSOC, B F GOODRICH CO, 71- Mem: Am Chem Soc. Res: Rubber chemicals; reactions of ozone; chemistry of p-phenylenediamines and of anils. Mailing Add: B F Goodrich Co Res Ctr Brecksville OH 44141

LAYLOFF, THOMAS, b Granite City, Ill, Jan 29, 37; m 61; c 1. ANALYTICAL CHEMISTRY. Educ: Wash Univ, AB, 58, MA, 61; Univ Kans, PhD(anal chem), 64. Prof Exp: From asst prof to assoc prof chem, 64-71, PROF CHEM, ST LOUIS UNIV, 71- Concurrent Pos: Sci adv nat ctr drug anal, Food & Drug Admin, 67- Mem: Am Chem Soc. Res: Organic electrode reactions and follow up processes; solution behavior of reactive organic intermediates; biopotentiometry; electron transport; trace metals in biological systems. Mailing Add: Dept of Chem St Louis Univ St Louis MO 63103

LAYMAN, WILBUR A, b Blair, Nebr, Jan 9, 29; m 53; c 5. ANALYTICAL CHEMISTRY, PHYSICAL CHEMISTRY. Educ: Dana Col, BS, 53; Univ Nebr, MS, 58; Mont State Univ, PhD(anal chem), 63. Prof Exp: Chemist, Harris Labs, 54-58; instr chem, Hastings Col, 58-60 & Dana Col, 62-63; asst prof, S Dak State Univ, 63-66; assoc prof, Adams State Col, 66-67; PROF CHEM & CHMN PHYS SCI DEPT, EASTERN MONT COL, 67- Mem: Am Chem Soc. Res: Stability constants of metal complexes; polarized infrared spectroscopy of thin crystal films. Mailing Add: Dept of Phys Sci Eastern Mont Col Billings MT 59101

LAYMAN, WILLIAM ARTHUR, b West New York, NJ, Feb 8, 29; c 1. PSYCHIATRY. Educ: St Peter's Col, NJ, BS, 51; Georgetown Univ, MD, 55; Am Bd Psychiat & Neurol, dipl, 62. Prof Exp: Intern, Hackensack Hosp, NJ, 55-56; jr resident psychiat, Vet Admin Hosp, Lyons, 56-57; sr resident, Fairfield State Hosp, Newtown, Conn, 57-59; from instr to assoc prof, 59-74, CLIN PROF PSYCHIAT, COL MED & DENT NJ, 74- Concurrent Pos: Clin fel psychiat, Sch Med, Yale Univ, 58-59. Mem: AAAS; Am Asn Univ Prof; AMA; Am Psychiat Asn. Res: Nonverbal communication; psycotherapeutic technique. Mailing Add: Dept of Psychiat Col of Med & Dent of NJ Newark NJ 07103

LAYNE, DONALD SAINTEVAL, b Lime Ridge, Que, Apr 5, 31; m 59; c 3. BIOCHEMISTRY. Educ: McGill Univ, BSc, 53, MSc, 55, PhD, 57. Prof Exp: Fel biochem, Univ Edinburgh, 57-58; res assoc psychiat, Queen's Univ Ont, 58-59; from scientist to sr scientist, Worcester Found Exp Biol, 59-66; head physiol & endocrinol sect, Food & Drug Directorate, Can, 66-68; PROF BIOCHEM, UNIV OTTAWA, 68- Mem: AAAS; Am Physiol Soc; Endocrine Soc; Am Soc Biol Chemists; Royal Soc Can. Res: Biochemistry of estrogenic hormones. Mailing Add: Dept of Biochem Univ of Ottawa Ottawa ON Can

LAYNE, ENNIS C, b Lynchburg, Va, Apr 8, 27; m 49; c 3. BIOCHEMISTRY, PHYSCIAL CHEMISTRY. Educ: George Washington Univ, BS, 49, MS, 53, PhD(biochem, phys chem), 55. Prof Exp: Asst, Res Found, Children's Hosp, Washington, DC, 48-54; asst prof pediat res & asst prof biochem, Sch Med, Univ Md, 56-69; PROF PHARMACOL, UNIV SOUTHERN CALIF, 69- Concurrent Pos: Nat Found Infantile Paralysis fel phys chem, McCollum-Pratt Inst, Johns Hopkins Univ, 55-56. Mem: Am Chem Soc. Res: Factors associated with the regulation and maintenance of metabolism on the cellular level and biochemistry of sleep. Mailing Add: Dept of Pharmacol Univ of Southern Calif Los Angeles CA 90033

LAYNE, JAMES NATHANIEL, b Chicago, Ill, May 16, 26; m 50; c 5. VERTEBRATE BIOLOGY. Educ: Cornell Univ, BA, 50, PhD(zool), 54. Prof Exp: Asst vert zool, Cornell Univ, 50-54; asst prof zool, Southern Ill Univ, 54-55; from asst prof to assoc prof biol, Univ Fla, 55-63; assoc prof zool, Cornell Univ, 63-67; DIR ARCHBOLD BIOL STA & ARCHBOLD CUR DEPT MAMMAL & MEM ADV BD, ARCHBOLD EXPEDS, AM MUS NATURAL HIST, 67- Concurrent Pos: From asst cur to assoc cur biol sci, Fla State Mus, 55-63, res assoc, 63-; adj prof, Univ S Fla, 68- Mem: AAAS; Am Soc Zoologists; Am Soc Mammal (vpres, 65-70, pres, 70-72); Wildlife Soc; Ecol Soc Am. Res: Mammalian ecology; systematics; behavior and morphology; general vertebrate zoology. Mailing Add: Archbold Biol Sta Rte 2 Box 180 Lake Placid FL 33852

LAYNE, RICHARD C, b St Vincent, WI, Dec 14, 36; Can citizen; m 63; c 2. PLANT BREEDING. Educ: McGill Univ, BSc, 59; Univ Wis, MS, 60, PhD(plant path, agron), 63. Prof Exp: RES SCIENTIST, RES STA, AGR CAN, 63- Concurrent Pos: Chmn, NAm working group on winterhardiness woody perennials, Int Soc Hort Sci. Honors & Awards: Shepard Award, Am Pomol Soc, 67. Mem: Can Soc Hort Sci (vpres & pres elect, 75-76); Am Soc Hort Sci; Am Pomol Soc; Am Phytopath Soc; Int Soc Hort Sci. Res: Fruit breeding of scions and rootstocks of peach, nectarine and apricot; breeding for cold hardiness and disease resistance; environmental and genetic factors affecting cold acclimation and deacclimation. Mailing Add: Can Agr Res Sta Harrow ON Can

LAYNG, EDWIN TOWER, b Greenville, Pa, Jan 20, 09; m 46; c 2. CHEMISTRY. Educ: Allegheny Col, ScB, 30; NY Univ, PhD(chem), 33. Prof Exp: Asst, NY Univ, 30-33; res chemist, M W Kellogg Co, NJ, 34-42, assoc dir res, 43; dir res, Hydrocarbon Res, Inc, 43-44, asst to the pres, 44-46, vpres, 47-63, exec vpres, 64-72, pres, 72-74; vpres, Dynalectron Corp, 64-74; CONSULTANT, 74- Mem: Am Chem Soc; Am Inst Chem Eng; Am Petrol Inst. Res: Reaction kinetics; catalysis in petroleum processes; synthetic fuels; coal gasification; coal liquefaction. Mailing Add: 8 Surrey Rd Summit NJ 07901

LAYTON, EDWIN THOMAS, JR, b Sept 13, 28; m 52; c 1. HISTORY OF TECHNOLOGY. Educ: Univ Calif, Los Angeles, BA, 50, MA, 53, PhD(hist), 55. Prof Exp: Instr hist, Univ Wis, 56-57; instr hist, Ohio State Univ, 57-60; asst prof, Purdue Univ, 60-65; assoc prof hist sci & technol, Case Western Reserve Univ, 65-75; PROF HIST SCI & TECHNOL, UNIV MINN, 75- Concurrent Pos: Comt mem partic, Fourteenth Int Cong His Sci, Tokyo, 74; consult, Nat Mus Hist & Technol, Smithsonian Inst, 74; Comt mem partic, Int Comn Sci Policy Studies Meetings, Paris, 74-75 & Critical Issues Hist of Technol, NSF, Soc Hist Technol, 75-; consult, James Hill Reference Library, 75- Mem: Soc Hist Technol; Hist Sci Soc. Res: Interaction of science and technology in nineteenth century America. Mailing Add: Dept of Mech Eng Univ of Minn Minneapolis MN 55455

LAYTON, HERBERT WALLACE, b Morristown, NJ, Apr 30, 37; m 58; c 3. BACTERIOLOGY. Educ: Rutgers Univ, AB, 59, MS, 61, PhD(bact), 63. Prof Exp: NIH fel microbiol, Med Sch, Northwestern Univ, 63-65; RES BACTERIOLOGIST, AM CYANAMID CO, 65- Mem: AAAS; Am Soc Microbiol. Res: Immunological and nutritional studies on tuberculosis; pathogenesis and treatment of animal diseases caused by mycoplasmas, spirochetes, Salmonellae or Escherichia coli. Mailing Add: Am Cyanamid Co PO Box 400 Princeton NJ 08540

LAYTON, JACK MALCOLM, b Ossian, Iowa, Sept 27, 17; m 43; c 2. PATHOLOGY. Educ: Luther Col, AB, 39, DSc, 74; Univ Iowa, MD, 43; Am Bd Path, dipl, 50. Prof Exp: Intern, Univ Iowa Hosps, 43, asst path, 46-47, instr, 47-49, assoc, 49-50, from asst prof to prof, 50-67; PROF PATH & HEAD DEPT, COL MED, UNIV ARIZ, 67- Concurrent Pos: Actg dean col med & actg dir med ctr, Univ Ariz, 71-73; trustee, Am Bd Path, 74- Mem: Am Soc Clin Path (pres, 73); Col Am Path; Am Asn Path & Bact; Am Soc Exp Path; Int Acad Path (pres, 75-76). Res: Virology; host-parasite relationships in viral and rickettsial diseases; comparative pathology of inflammation; ultramicroscopic pathologic anatomy of infectious diseases; biological activities of teratomas; influenza and psittacosis-lymphogranuloma groups of viruses. Mailing Add: Dept of Path Univ of Ariz Col of Med Tucson AZ 85724

LAYTON, LAURENCE LAIRD, m 41; c 4. BIOCHEMISTRY, ALLERGY. Educ: WVa Inst Technol, BSc, 37; Univ WVa, MSc, 39; Pa State Univ, PhD(biochem), 42. Prof Exp: Asst vitamin res, Pa State Col, 41-42; res phys chemist, Distillation Prod, Inc, 42-43; asst prof inorg chem, Univ Md, 43-46; asst prof biochem, Sch Hyg & Pub Health, Johns Hopkins Univ, 46-51; chief biochem br, Chem Warfare Div, Chem Corps, Dugway Proving Ground, 51-52, chief chemist & chief, Chem Warfare Div, 52-54; assoc dir res & develop, US Navy Powder Factory, Md, 54-57; prin chemist, 57-60, HEAD ALLERGENS INVESTS, PHARMACOL LAB, WESTERN REGIONAL RES LAB, AGR RES SERV, USDA, 60- Concurrent Pos: Chem consult, Woods Hole Marine Biol Lab, 47, 50 & 51; lectr, Univ Calif, Berkeley. Honors & Awards: Superior Serv Award, US Navy, 56; Superior Serv Award, USDA, 63. Mem: Am Chem Soc; Am Acad Allergy; Int Col Allergy; Int Primatol Soc. Res: Cancerous growth; mucopolysaccharide metabolism; pharmacology; passive transfer of human allergies to monkeys; industrial and public health immunology; allergy. Mailing Add: Western Regional Res Ctr 800 Buchanan St Berkeley CA 94710

LAYTON, LIONEL H, b Layton, Utah, Mar 3, 24; m 54; c 2. PHYSICAL CHEMISTRY. Educ: Brigham Young Univ, BS, 49; Rutgers Univ, PhD(chem), 53. Prof Exp: Chemist, Standard Oil Co Ind, 52-54; instr chem, Colo Sch Mines, 54-56; prof, Ricks Col, 56-58; SUPVR, THIOKOL CHEM CORP, 58- Concurrent Pos: Adj prof, Univ Utah. Mem: AAAS. Res: Mechanical properties of elastomeric materials; solution properties of high polymeric soaps; solvent extraction; combustion; chemistry of material aging. Mailing Add: 448 S Seventh W Brigham City UT 84302

LAYTON, RICHARD GARY, b Salt Lake City, Utah, Dec 24, 35; m 63; c 2. PHYSICS. Educ: Univ Utah, BA, 60, MA, 62; Utah State Univ, PhD(physics), 65. Prof Exp: Asst physics, Univ Utah, 60-62; asst res physicist electro-dynamics Labs, Utah State Univ, 63-64; from asst prof to assoc prof physics, State Univ NY Col Fredonia, 65-69; ASSOC PROF PHYSICS, NORTHERN ARIZ UNIV, 69- Mem: Am Asn Physics Teachers; Am Meteorol Soc; Royal Meteorol Soc; Am Geophys Union. Res: Applications of physics to atmospheric problems; ice nucleation and cloud seeding; atmospheric optics; planetary atmospheres. Mailing Add: Dept of Physics Box 5763 Northern Ariz Univ Flagstaff AZ 86001

LAYTON, ROBERT L, b Salt Lake City, Utah, May 31, 25; m 51; c 7. GEOGRAPHY. Educ: Univ Utah, BS, 51, MS, 52; Syracuse Univ, PhD, 62. Prof Exp: Asst geog,

Syracuse Univ, 52-53, asst instr, 53-54; assoc prof, 54-67, chmn dept, 60-65, PROF GEOG & CHMN DEPT, BRIGHAM YOUNG UNIV, 67- Concurrent Pos: Adv, AID, Guatemala, 65-66. Mem: Nat Coun Geog Educ; Asn Am Geogrs; Conf Latin Am Geogrs. Res: Economic and urban geography. Mailing Add: Dept Geog 167 Heber Grant Bldg Brigham Young Univ Provo UT 84601

LAYTON, ROGER, b Idaho Falls, Idaho, May 7, 42; m 61; c 5. ORGANIC CHEMISTRY. Educ: Idaho State Univ, BS, 64; Ohio State Univ, PhD(org mechanisms), 67. Prof Exp: NIH fel, Univ Utah, 67-68; res chemist, Res & Develop Div, Kraftco Corp, 68-71 & Searle Biochemics, Div G D Searle & Co, 71-74, LECTR, HARPER COL, 74- Mem: Am Chem Soc. Res: Synthetic aspects of vitamin E; formaldehyde reactions; hydrogenation; fatty chemicals; food additives; photochemistry. Mailing Add: 400 Castle Wood Lane Buffalo Grove IL 60090

LAYTON, THOMAS NUTTER, b Rochester, NY, Nov 13, 42. ANTHROPOLOGY. Educ: Univ Calif, Davis, BA, 65, MA, 66; Harvard Univ, PhD(anthrop), 71. Prof Exp: Instr anthrop, Harvard Univ, 70-71; asst prof, La State Univ, 73; dir, Nev State Mus, 74; coordr, Nev Archaeol Surv, 74; ASST PROF ANTHROP, CALIF STATE COL, DOMINGUEZ HILLS, 75- Concurrent Pos: Fel Near Eastern archaeol, Harvard Univ, 71-72; vpres, Western Regional Conf Am Asn Mus, 74-75. Mem: Soc Am Archaeol; Am Anthrop Asn. Res: North American archaeology; cultural ecology of desert lands. Mailing Add: Dept of Anthrop Calif State Col Dominguez Hills CA 90747

LAYTON, THOMAS WILLIAM, b Kaysville, Utah, Feb 24, 27; m 47; c 1. PHYSICS. Educ: Calif Inst Technol, BS, 51, PhD(physics, math), 57. Prof Exp: Res engr, Jet Propulsion Lab, Calif Inst Technol, 53-55; mem tech staff, Inertial Guid Dept, Thompson-Ramo-Wooldridge, Inc, 55-59, mgr, 59-64, SR STAFF ENGR, ELECTRONICS DIV, TRW, INC, 64- Mem: Am Phys Soc; Am Inst Aeronaut & Astronaut. Res: Cosmic rays; navigation and guidance systems for ballistic missiles and space flight vehicles. Mailing Add: 4836 W Elmdale Dr Rolling Hills CA 90274

LAYTON, WILLIAM ISAAC, b Cameron, Mo, Sept 26, 13; m 41; c 2. GEOMETRY. Educ: Univ SC, BS, 34, MS, 35; George Peabody Col, PhD(math), 48. Prof Exp: Asst math, Univ SC, 34-35; head dept high sch, SC, 35-36; asst, Univ Chicago, 36-37; instr high sch, Ga, 37-39, head dept, 39-40; instr, Amarillo Jr Col, 41-42, head dept, 42-46; prof math & head dept, Austin Peay State Col, 46-48; assoc prof, Auburn Univ, 48-49; dean instruction, Frostburg State Teachers Col, Md, 49-50; coordr data processing ctr, 63-68, Distinguished Alumni Asn prof, 75, PROF MATH & HEAD DEPT, STEPHEN F AUSTIN STATE UNIV, 50- Concurrent Pos: Instr eng, sci & mgt war training, Tex Tech Col, 42-45; NSF lectr, Tex Acad Sci, 60. Mem: Math Asn Am; Am Math Soc. Res: Higher geometry; teaching of mathematics. Mailing Add: Stephen F Austin Sta Box 3040 Nacogdoches TX 75961

LAYTON, WILLIAM MALLOY, JR, b Mansfield, Ohio, Apr 28, 21; m 43; c 3. PATHOLOGY, MARINE BIOLOGY. Educ: Harvard Univ, BS, 43; Western Reserve Univ, MD, 46; Am Bd Path, dipl, 54. Prof Exp: Dir labs, Stamford Hosp, Conn, 53-56; pathologist dept exp path, Lederle Labs Div, Am Cyanamid Co, 56-65, assoc dir toxicol res, 65-66; prof path, 70-71, ASSOC PROF ANAT, DARTMOUTH MED SCH, 71- Concurrent Pos: Res assoc marine lab, Duke Univ, 63. Mem: AAAS; AMA; Am Soc Zool; NY Acad Sci. Res: Experimental pathology and teratology; morphology of primitive mollusca, especially monoplacophora. Mailing Add: Dept of Anat Dartmouth Med Sch Hanover NH 03755

LAYZER, ARTHUR JAMES, b Cleveland, Ohio, Aug 21, 27; m 64; c 1. PHYSICS. Educ: Columbia Univ, PhD(physics), 60. Prof Exp: Res scientist, Courant Inst, NY Univ, 60-63; asst prof, 64-67, ASSOC PROF PHYSICS, STEVENS INST TECHNOL, 68- Concurrent Pos: Vis res scientist, Brookhaven Nat Lab, 66; resident visitor comput music, Acoust Div, Bell Labs, 67- Mem: AAAS; Am Phys Soc. Res: Computer music; theory of quantum many-particle systems; mathematical physics; quantum electrodynamics. Mailing Add: Stevens Inst of Technol Castle Point Sta Hoboken NJ 07030

LAYZER, DAVID, b Ohio, Dec 31, 25; m 49, 59; c 6. ASTRONOMY. Educ: Harvard Univ, AB, 47, PhD(theoret astrophys), 50. Prof Exp: Nat Res Coun res fel, 50-51; lectr astron, Univ Calif, Berkeley, 51-52; res assoc physics, Princeton Univ, 52-53; res assoc, 53-55, res fel & lectr, 55-60, PROF ASTRON, HARVARD UNIV, 60- Concurrent Pos: Consult, Geophys Corp Am, 59-65. Honors & Awards: Bok Prize, 60. Mem: Am Acad Arts & Sci; Am Astron Soc; Int Astron Union; Royal Astron Soc. Res: Cosmology and cosmogony; theoretical astrophysics and atomic physics; ionospheric physics. Mailing Add: Harvard Col Observ 60 Garden St Cambridge MA 02138

LAZAR, JAMES TARLTON, JR, b Whiteville, NC, Oct 24, 22; m 47. DAIRY SCIENCE. Educ: Clemson Univ, BS, 43; Cornell Univ, MS, 49; NC State Univ, PhD(dairy sci), 55. Prof Exp: Supt off prod testing, Clemson Univ, 46-48, asst prof dairy sci, 49-51; teaching asst, Cornell Univ, 48-49; res asst, NC State Univ, 51-53; assoc prof, 53-65, PROF DAIRY SCI, CLEMSON UNIV, 65- Mem: Am Dairy Sci Asn. Res: Dairy management; labor efficiencies and economies of dairy plant operation; pricing plans for purchasing milk, processing for flavor control; influence of vacuum pasteurization upon the freezing point value, total solids and concentration of milk; use of vacuum pasteurizer to increase solids-not-fat in low fat milk. Mailing Add: Dept of Dairy Sci Clemson Univ Clemson SC 29631

LAZAR, JOSEPH, organic chemistry, see 12th edition

LAZAR, NORMAN HENRY, b Brooklyn, NY, June 21, 29. PHYSICS. Educ: City Col New York, BS, 49; Ind Univ, MS, 51, PhD(physics), 53. Prof Exp: PHYSICIST, OAK RIDGE NAT LAB, UNION CARBIDE NUCLEAR CO, 53- Mem: Fel Am Phys Soc. Res: Controlled thermonuclear reactions; beta and gamma ray spectroscopy; plasma physics. Mailing Add: Oak Ridge Nat Lab PO Box Y Oak Ridge TN 37831

LAZAR, ROBERT, chemistry, economics, see 12th edition

LAZARETH, OTTO WILLIAM, JR, b Brooklyn, NY, Sept 16, 38. SOLID STATE PHYSICS, REACTOR PHYSICS. Educ: Wagner Col, BS, 61; Queens Col, MA, 68; City Univ NY, PhD(physics), 73. Prof Exp: Res asst physics, Queens Col, 65-67; physics assoc, 67-73, ASST PHYSICIST, BROOKHAVEN NAT LAB, 74- Mem: Am Phys Soc; Am Nuclear Soc. Res: Radiation damage and effects in solids; point defects in solids; crystallites and crystal surfaces. Mailing Add: Bldg 129 Brookhaven Nat Lab Upton NY 11973

LAZARIDIS, CHRISTINA NICHOLSON, b New York, NY, Jan 12, 42; m 66; c 2. ORGANIC CHEMISTRY, PHOTOGRAPHIC CHEMISTRY. Educ: Mt Holyoke Col, BA, 62; Columbia Univ, MA, 63, PhD(chem), 66. Prof Exp: Res chemist, Colgate-Palmolive Co, 66-68; RES CHEMIST E I DU PONT DE NEMOURS & CO INC, 68- Mem: Am Chem Soc. Res: Photosensitive systems, including conventional silver halide as well as novel photopolymeric materials. Mailing Add: Photo Prod Dept Du Pont Co Exp Sta Wilmington DE 19898

LAZARO, ERIC JOSEPH, b Muttra, India, Dec 28, 21; m 49; c 3. SURGERY. Educ: Univ Madras, MBBS, 46; Georgetown Univ, MS, 55; Am Bd Surg, dipl, 55; FRCS(C), 56. Prof Exp: Instr surg, Georgetown Univ, 54-57; assoc prof, All India Inst Med Sci, 58-60, prof thoracic surg, 60-61, prof surg, 61-62; assoc prof, 62, PROF SURG & ASSOC DEAN STUD AFFAIRS, COL MED & DENT NJ, 62- Concurrent Pos: Rockefeller Found fels, 57 & 62; Colombo Plan fel, 61. Mem: AAAS; fel Am Col Surg; Soc Surg Alimentary Tract; AMA. Res: Pathogenesis of pancreatitis; effects of placental extracts on blood cells; blood vessel preservation. Mailing Add: Dept of Surg Col of Med & Dent NJ Newark NJ 07103

LAZAROFF, NORMAN, b Brooklyn, NY, Nov 24, 27; m 58; c 2. MICROBIOLOGY. Educ: Syracuse Univ, AB, 50, MS, 52; Yale Univ, PhD(microbiol), 60. Prof Exp: Asst enzymol, Res Found, State Univ NY, 55; bacteriologist, Schwarz Labs, Inc, 55-56; proj leader, Evans Res & Develop Corp, 56-57, consult, 57-59; fel microbiol, Brandeis Univ, 60-61; microbiologist, BC Res Coun, 61-62; asst prof biol sci, Univ Southern Calif, 62-64; sr res scientist, Res Corp, Syracuse, 64-66; ASSOC PROF BIOL, STATE UNIV NY BINGHAMTON, 66- Mem: Am Soc Microbiol; Am Inst Biol Sci; Phycol Soc Am; Brit Soc Gen Microbiol. Res: Microbial ecology and physiology of growth and development; algal physiology. Mailing Add: Dept of Biol Sci State Univ NY Binghamton NY 13901

LAZAROW, ARNOLD, anatomy, cell biology, deceased

LAZARUK, WILLIAM, b Willingdon, Alta, Jan 15, 12; m 54. HORTICULTURE, BIOLOGY. Educ: Univ Alta, BSc, 47, BEd, 48; SDak State Univ, MS, 50; Rutgers Univ, PhD(hort), 61. Prof Exp: Asst hort, SDak State Univ, 48-50; instr adult educ, Univ NDak, 50-52 & Exten Serv, Univ Ill, 52-53; instr animal sci, Berea Col, 53-56; asst plant nutrit, Rutgers Univ, 56-60, asst exp biol, Douglass Col, 59-60; assoc prof biol, Trenton State Col, 60-61; from assoc prof to prof, High Point Col, 61-68; ASSOC PROF BIOL, FAIRFIELD UNIV, 68- Concurrent Pos: Partic, NSF Genetics Conf, Colo State Univ, 65 & Yale Univ, 69; NSF grant, 64; Piedmont Univ Ctr res grants, 64-67; Fairfield Univ grants, 69- Mem: Am Soc Plant Physiol. Res: Plant growth functions as influenced by environment; plant nutrition; microstudy of lower plants and their relationship and usefulness to man in regards to food supply and water contamination; gypsy moth control by use of systemic insecticides. Mailing Add: Dept of Biol Fairfield Univ Fairfield CT 06430

LAZARUS, ALLAN KENNETH, b Bangor, Maine, May 20, 31; m 57; c 3. ORGANIC CHEMISTRY. Educ: NY Univ, BA, 52, MS, 55, PhD(org chem), 57. Prof Exp: Chemist, Cities Serv Res & Develop Co, 57-59, inorg chem div, FMC Corp, 59-65 & Exxon Res & Eng Co, 65-66; group leader synthetic lubricants, Intermediates Div, Tenneco Chem, Inc, 66-71; ASST PROF CHEM, TRENTON STATE COL, 72- Mem: Am Chem Soc. Res: Stereochemistry; organic synthesis; product and process development; fuels; automatic transmission fluids; synthetic lubricants. Mailing Add: Dept of Chem Trenton State Col Trenton NJ 08625

LAZARUS, DAVID, b Buffalo, NY, Sept 8, 21; m 43; c 4. SOLID STATE PHYSICS. Educ: Univ Chicago, PhD(physics), 49. Prof Exp: Instr electronics, Univ Chicago, 42-43; res assoc, Radio Res Lab, Harvard Univ, 43-45; asst physics, Univ Chicago, 46-49, instr, 49; from instr to assoc prof, 49-59, PROF PHYSICS, UNIV ILL, URBANA, 59- Concurrent Pos: Guggenheim fel, 68-69; consult ed, Addison-Wesley Publ Co, 60- Mem: Fel Am Phys Soc; Am Asn Physics Teachers; AAAS. Res: Defect and electronic properties of solids; high pressure physics. Mailing Add: Dept of Physics Univ of Ill Urbana IL 61801

LAZARUS, GERALD SYLVAN, b New York, NY, Feb 16, 39; m 61; c 4. DERMATOLOGY. Educ: Colby Col, BS, 59; George Washington Univ, MD, 63. Prof Exp: Med intern, Univ Mich Med Ctr, 63-64, med resident, 64-65; clin assoc, Med Neurol Br, Nat Inst Neurol Dis & Blindness, 65-67; prin investr, Lab Histol & Path, Nat Inst Dent Res, 67-68; clin & res assoc, Dept Dermat, Harvard Med Sch, 68-70, chief resident dermat, 69-70; vis scientist, Strangeways Labs, Univ Cambridge, Eng, 70-72; assoc prof med & co-dir dermat training prog, Albert Einstein Col Med, 72-75; PROF MED & CHMN DIV DERMAT, DUKE UNIV MED CTR, 75- Concurrent Pos: Carl Herzog fel, 70-72; res fel, Arthritis Found, 70-72, sr investr, 72-77; consult dermat, Addenbrookes Hosp, Cambridge, Eng, 70-72; vis fel, Clare Hall, Cambridge; head sect dermat, Dept Med, Montefiore Hosp, 72-75. Mem: Am Rheumatism Asn; Soc Invest Dermat; Am Fedn Clin Res; fel Am Col Physicians; Royal Soc Med. Res: Study of the role of lysosomal proteinases in catabolic processes in skin and evaluation of the mechanisms by which these proteinases can instigate an inflammatory response. Mailing Add: Dept of Med Div of Dermat Duke Univ Med Ctr Box 2987 Durham NC 27710

LAZARUS, MARC SAMUEL, b Brooklyn, NY, Sept 9, 46. PHYSICAL INORGANIC CHEMISTRY. Educ: City Univ New York, BS, 68; Princeton Univ, MA, 71, PhD(chem), 74. Prof Exp: Res assoc chem, Lawrence Berkeley Lab, Univ Calif, 73-74; ASST PROF CHEM, HERBERT H LEHMAN COL, CITY UNIV NEW YORK, 74- Concurrent Pos: Res collabr, Brookhaven Nat Lab, 74- Mem: Am Chem Soc; Am Phys Soc; AAAS. Res: Applications of x-ray photoelectron spectroscopy to the study of transition metal compounds and alloys. Mailing Add: Dept of Chem Herber H Lehman Col Bronx NY 10468

LAZARUS, ROGER BEN, b New York, NY, June 3, 25; m 46; c 5. THEORETICAL PHYSICS. Educ: Harvard Univ, AB, 47, MA, 48, PhD(physics), 51. Prof Exp: Mem staff, 51-58, group leader, 58-68, DIV LEADER, LOS ALAMOS SCI LAB, 68- Mem: NY Acad Sci; Am Phys Soc; Asn Comput Mach. Res: Simultaneous partial differential equations; computing machines. Mailing Add: Los Alamos Sci Lab PO Box 1663 Los Alamos NM 87545

LAZARUS, SAM, organic chemistry, see 12th edition

LAZARUS, SYDNEY SIMON, b Glasgow, Scotland, Apr 28, 19; nat US; m 45; c 1. PATHOLOGY. Educ: City Col NY, 38; Chicago Med Sch, MD, 43; Queen's Univ, Ont, MSc, 56; Am Bd Path, dipl, 57. Prof Exp: Asst resident path, Coney Island Hosp, Brooklyn, NY, 51-52; asst to dir labs, 53-54, asst dir dept labs, 57-68, ASSOC DIR DEPT LABS, KINGSBROOK JEWISH MED CTR, 68-, CHIEF ANAT PATH & ASST DIR ISAAC ALBERT RES INST, 57- Concurrent Pos: Fel clin path, Kingsbrook Jewish Med Ctr, 56-57; vis asst prof path, Albert Einstein Col Med, 57-60; clin assoc prof, State Univ NY Downstate Med Ctr, 69-73, clin prof, 73- Mem: Am Soc Exp Path; Am Asn Path & Bact; Am Diabetes Asn; Col Am Path; Histochem Soc. Res: Human and experimental diabetes mellitus; pancreatic morphology; histochemistry; experimental morphology; pancreatic islets; tissue culture cells; electron microscopy. Mailing Add: Isaac Albert Res Inst Kingsbrook Jewish Med Ctr Brooklyn NY 11203

LAZAY, PAUL DUANE, b Philadelphia, Pa, June 2, 39; m 67. SOLID STATE

LAZAY

PHYSICS. Educ: Trinity Col, Conn, BS, 61; Mass Inst Technol, PhD(physics), 68. Prof Exp: MEM TECH STAFF, BELL LABS, 69- Mem: Am Phys Soc; Optical Soc Am; Am Physicists Asn. Res: Inelastic light scattering studies in solids; optical fiber research and optical communications. Mailing Add: Bell Labs 1D-150 Murray Hill NJ 07974

LAZDA, VELTA ABULS, b Riga, Latvia, Dec 16, 39; US citizen; m 62. IMMUNOLOGY, MOLECULAR BIOLOGY. Educ: Purdue Univ, BS, 62; Northwestern Univ, PhD(microbiol), 67. Prof Exp: RES ASSOC, IMMUNOL DIV, RES INST, AM DENT ASN HEALTH FOUND, 69- Concurrent Pos: Am Cancer Soc fel ribosome struct, Northwestern Univ, 67-69; USPHS career develop award. Mem: AAAS; Am Asn Immunologists; Am Soc Microbiol. Mailing Add: Immunol Div Res Inst Am Dent Asn 211 E Chicago Ave Chicago IL 60611

LAZEAR, EDWARD JESSE, immunology, see 12th edition

LAZEN, ALVIN GORDON, b Baltimore, Md, May 28, 35; m 60; c 3. BIOCHEMISTRY, RESEARCH ADMINISTRATION. Educ: Univ Md, BS, 57; Ohio State Univ, PhD(microbiol), 63. Prof Exp: Opers res analyst, Opers Res Group, US Dept Defense, 58-60; microbiologist, Ft Detrick, 63; NIH fel, Weizmann Inst, 63-64; Arthritis Found fel, Sch Med, Johns Hopkins Univ, 64-67; asst prof & asst adminr, Johns Hopkins Univ, 67-70, from asst dean to dean res progs, Sch Med, 70-75; ASSOC EXEC DIR, ASSEMBLY LIFE SCI, NAT ACAD SCI-NAT RES COUN, 75- Mem: AAAS; Am Soc Microbiol; Fedn Am Scientists. Res: Science policy-planning and decision making. Mailing Add: Nat Acad Sci 2101 Constitution Ave Washington DC 20418

LAZERTE, JAMES DONALD, chemistry, see 12th edition

LAZIER, CATHERINE BEATRICE, b Galt, Ont, June 14, 39; m 61; c 2. BIOCHEMISTRY. Educ: Univ Toronto, BA, 61; Univ BC, MSc, 63; Dalhousie Univ, PhD(biochem), 68. Prof Exp: Med Res Coun Can fel, Univ Southampton, 68-71; ASST PROF BIOCHEM, DALHOUSIE UNIV, 71- Res: Mechanism of action of estrogenic hormones; biosynthesis of egg yolk proteins. Mailing Add: Dept of Biochem Dalhousie Univ Halifax NS Can

LAZO-WASEM, EDGAR ARTHUR, b Guatemala City, Guatemala, Jan 18, 26; nat US; m 50; c 3. PHARMACEUTICS. Educ: Carroll Col, Mont, BA, 49; Mont State Univ, MA, 51; Purdue Univ, PhD(endocrinol, pharmacol), 54. Prof Exp: Chief physiologist, Wilson Labs, 54-58, dir control, 58-63; dir sci serv, Strong Cobb Arner Inc, 63-64, vpres & sci dir, 65-67; vpres & tech dir, Wampole Labs Div, 67-74, HEAD LAB, CO RES LABS & VPRES & TECH DIR, DENVER CHEM MFG CO, 74- Mem: AAAS; Endocrine Soc; Soc Exp Biol & Med. Res: Enzyme analysis; pharmaceutical development; immunological and chemical clinical diagnostic reagents. Mailing Add: Co Res Labs Denver Chem Mfg Co 35 Commerce Rd Stamford CT 06904

LAZZARA, RALPH, b Tampa, Fla, Aug 14, 34; m 59; c 3. MEDICINE. Educ: Univ Chicago, BA, 55; Tulane Univ, MD, 59. Prof Exp: Instr med, Tulane Univ, 60-67; asst prof, 71-72, ASSOC PROF MED, SCH MED, UNIV MIAMI, 72-; CHIEF SECT CARDIOL, VET ADMIN HOSP, 74- Concurrent Pos: Asst, Charity Hosp, New Orleans, 60-64; resident, Med Sch, Tulane Univ & Charity Hosp, 63-64; fel, Col Physicians & Surgeons, Columbia Univ, 64-65; staff mem & dir cardiovasc res lab, Ochsner Clin & Ochsner Found Hosp, New Orleans, 65-67; staff cardiologist & chief sect electrophysiol, Mt Sinai Hosp, Miami Beach, 70-72; dir coronary care unit, Vet Admin Hosp, Miami, 72-74. Mem: Am Physiol Soc. Res: Cardiac electrophysiology. Mailing Add: Dept of Med Univ of Miami Sch of Med Miami FL 33152

LAZZARI, EUGENE PAUL, b Archbald, Pa, Mar 12, 31. BIOCHEMISTRY. Educ: Scranton Univ, BS, 53; Williams Col, MA, 55; Iowa State Univ, PhD(biochem), 61. Prof Exp: Res assoc biol, Brookhaven Nat Lab, 61-62; res assoc biochem, City of Hope Med Ctr, 62-63; from asst prof to assoc prof, 63-75, PROF BIOL CHEM, UNIV TEX DENT BR, HOUSTON, 75- Mem: AAAS; Am Chem Soc; Int Asn Dent Res; Sigma Xi. Res: Isolation and sequence determination of human pancreatic enzymes, human dentin and enamel proteins; amino acid chemistry and chemical modification of enzymes. Mailing Add: Dept of Physiol-Biochem Univ of Tex Dent Br PO Box 20068 Houston TX 77025

LEA, ARDEN OTTERBEIN, b Cleveland, Ohio, Oct 19, 26; m 52; c 3. ENTOMOLOGY. Educ: Univ Rochester, BA, 48; Ohio State Univ, MSc, 50, PhD(entom), 57. Prof Exp: Res assoc insecticide testing, Ohio State Univ, 50-51, insect nutrit, 56-58; USPHS med entomologist, Onchocerciasis Proj, Pan Am Sanit Bur, Guatemala, 51-53; chief physiol sect, Entom Res Ctr, State Bd Health, Fla, 58-69; assoc prof entom, 69-74, PROF ENTOM, UNIV GA, 74- Concurrent Pos: USPHS spec res fel, Denmark, 60-61; mem trop med & parasitol study sect, NIH, 74-78; mem sci adv panel onchocerciasis, WHO. Mem: AAAS; Entom Soc Am; Am Mosquito Control Asn. Res: Endocrine physiology of Diptera; physiology and behavior of mosquitoes. Mailing Add: Dept of Entom Univ of Ga Athens GA 30602

LEA, DAVID CHESTER, b Silver Creek, Nebr, Jan 11, 17; m 46; c 4. PAPER CHEMISTRY. Educ: Mont State Univ, BA, 49; Inst Paper Chem, Lawrence Col, MS, 51, PhD, 53. Prof Exp: Tech dir, Potlatch Forests, Inc, Idaho, 53-56; dir res & develop, Forest Prod Div, 56-60, dir res & develop, Ecusta Paper Div, 60-69, dir tech & eng dept, 69-75, DIR LONG RANGE PROJ, FINE PAPER & FILM GROUP, OLIN CORP, 75- Mem: Am Chem Soc; Tech Asn Pulp & Paper Indust. Res: Paper and paperboard, cellulose chemistry; surface coatings. Mailing Add: Olin Corp Fine Paper & Film Grp PO Box 200 Pisgah Forest NC 28768

LEA, JAMES DIGHTON, b Monticello, Ill, Apr 9, 33; m 53; c 2. SYSTEMS SCIENCE. Educ: Tex Western Col, BA, 57; Univ Tex, MA, 60, PhD(physics), 63. Prof Exp: Sr res scientist physics, Esso Prod Res Co, Stand Oil NJ, 63-69, sr prof systs analyst, Humble Oil & Refining Co, 69-73, explor systs adv, Exxon Co USA, 73-75, RES ASSOC SYSTS, EXXON PROD RES CO, EXXON CORP, 75- Mem: Am Math Soc; Am Phys Soc; Soc Explor Geophysicists; Am Asn Petrol Geologists. Res: Application of computer science to geological problems. Mailing Add: Exxon Prod Res Co PO Box 2189 Houston TX 77001

LEA, JAMES WESLEY, JR, b Lebanon, Tenn, Mar 17, 41; m 66; c 1. TOPOLOGY, ALGEBRA. Educ: Tenn Polytech Inst, BS, 63, MS, 65; La State Univ, PhD(math), 71. Prof Exp: Instr math, Univ Tenn, Martin, 65-66; instr, Tenn Technol Univ, 66-67; asst prof, 71-74, ASSOC PROF MATH, MIDDLE TENN STATE UNIV, 74- Mem: Am Math Soc; Math Asn Am; Sigma Xi. Res: Lattice theory. Mailing Add: Middle Tenn State Univ Murfreesboro TN 37132

LEA, MALCOLM SINCLAIR, b Evanston, Ill, Dec 1, 31; m 58; c 2. BIOLOGY. Educ: Northwestern Univ, Ill, BA, 57, MS, 59, PhD(biol), 64. Prof Exp: NIH fel, Australian Nat Univ, 64-65; ASST PROF BIOL, PORTLAND STATE UNIV, 65- Mem: AAAS; Am Soc Zoologists. Res: Environmental toxicology; cellular physiology. Mailing Add: Dept of Biol Portland State Univ PO Box 751 Portland OR 97207

LEA, MICHAEL ANTHONY, b Leeds, Eng, Dec 26, 39; m 61; c 2. BIOCHEMISTRY. Educ: Univ Birmingham, BSc, 61, PhD(biochem), 64. Prof Exp: Res assoc pharmacol, Sch Med, Ind Univ, 64-66, instr, 66-67; asst prof, 67-71, ASSOC PROF BIOCHEM, COL MED & DENT NJ, 71- Mem: AAAS; Am Chem Soc; Brit Biochem Soc; Am Asn Cancer Res; Am Soc Biol Chemists. Res: Metabolism of the nuclear proteins and the control of tissue growth rate. Mailing Add: Dept of Biochem Col of Med & Dent of NJ Newark NJ 07103

LEA, ROBERT MARTIN, b New York, NY, Nov 4, 31; m 53; c 1. PHYSICS. Educ: Union Col, NY, BS, 53; Yale Univ, PhD(physics), 57. Prof Exp: PROF PHYSICS, CITY COL NEW YORK, 57-, CHMN DEPT, 70- Concurrent Pos: Vis physicist, Brookhaven Nat Lab, 59-70; prin investr, NSF res grants, City Col New York, 59-70, dir, NSF dept develop grant, 70- Mem: Am Phys Soc; Am Asn Physics Teachers. Res: High energy experimental physics. Mailing Add: Dept of Physics City Col of New York New York NY 10031

LEA, SUSAN MAUREEN, b Cardiff, Wales, UK, July 10, 48; m 74. ASTROPHYSICS. Educ: Cambridge Univ, BA, 69, MA, 73; Univ Calif, Berkeley, PhD(astron), 74. Prof Exp: RES ASSOC ASTROPHYS, AMES RES CTR, NASA, 74- Mem: Am Astron Soc; Royal Astron Soc. Res: High energy astrophysics, especially x-ray and radio astronomy; numerical hydrodynamics; compact galactic x-ray sources, clusters of galaxies, intergalactic matter and cosmology. Mailing Add: NASA Ames Res Ctr MS 245-3 Moffett Field CA 94035

LEA, WILLIAM LIEF, b Chippewa Falls, Wis, May 19, 09; m 36. RADIOLOGICAL HEALTH. Educ: Univ Wis, BS, 33, PhD(sanit chem), 40. Prof Exp: Anal chemist, State Lab Hyg, Wis, 34-40; chemist, State Bd Health, 40-48; asst prof civil eng, Univ Wis, 48-50; DIR INDUST HYG DIV, STATE BD HEALTH, WIS, 50- Concurrent Pos: Asst prof prev med, Univ Wis, 63- Mem: Am Chem Soc; fel Am Inst Chem. Res: Biochemical oxidation; analytical methods. Mailing Add: 5222 Hammersley Rd Madison WI 53711

LEABO, DICK ALBERT, b Walcott, Iowa, Oct 30, 21; m 55; c 1. APPLIED STATISTICS, ECONOMIC STATISTICS. Educ: Univ Iowa, BS, 49, MA, 50, PhD(statist, econ), 53. Prof Exp: Res asst econ, Bur Bus & Econ Res, Univ Iowa, 48-49, res assoc, 49-53, asst prof econ & asst dir, 53-56; asst prof econ & asst dir, Bur Econ & Bus Res, Mich State Univ, 56-57; from asst prof to assoc prof statist, 57-63, assoc dean grad sch bus, 62-65, PROF STATIST, UNIV MICH, ANN ARBOR, 63-, DIR PhD PROG, 65- Concurrent Pos: Consult, Brookings Inst, 57-59; exchange prof, Rotterdam Sch Econ, 65. Mem: Am Statist Asn; Am Econ Asn; Am Finance Asn. Res: Regional economic research and the application of regression and correlation techniques. Mailing Add: Grad Sch of Bus Admin Univ of Mich Ann Arbor MI 48104

LEACH, BERTON JOE, b Tuscola, Ill, Mar 30, 32; m 55; c 2. BIOLOGY, SCIENCE ADMINISTRATION. Educ: Washington Univ, AB, 57; Univ Mo, MA, 60, PhD(zool), 63. Prof Exp: Asst zool, Univ Mo, 58-60, instr, 60-62, USPHS res fel, 62-63; asst prof, George Washington Univ, 63-66; asst prog dir, Undergrad Student Prog, NSF, 66-67, asst prog dir col sci improv prog, 67-68; prof biol, Cent Methodist Col, 68-70, F H Dearing prof, 70-74, chmn dept biol & geol, 68-74; EXEC SECY, CARDIOVASC & PULMONARY STUDY SECT, DIV RES GRANTS, NIH, 74- Concurrent Pos: NSF res grant, 63-65; assoc prof, George Washington Univ, 66; USPHS res evaluator, 66-67; vis scholar, Harvard Univ, 69; med technol educ adv, Jewish Hosp, St Louis, 72. Mem: AAAS; NY Acad Sci; Nat Asn Biol Teachers. Mailing Add: 12707 Weiss St Rockville MD 20853

LEACH, BYRON ELWOOD, b Kosciusko, Miss, Apr 5, 15; m 43; c 2. BIOCHEMISTRY. Educ: Miss State Univ, BS, 38; Va Polytech Inst, MS, 39; Univ Ill, PhD(biochem), 44. Prof Exp: Asst biochem, Med Col Va, 39-41; res chemist, Upjohn Co, 44-53; assoc prof biochem, psychiat & neurol, Sch Med, Tulane Univ, 53-63; res biochemist, Eastern Utilization Div, USDA, Washington, DC, 63-65; PROF BIOCHEM, UNIV TENN CTR HEALTH SCI, MEMPHIS & CLIN CHEMIST, DEPT PATH, BAPTIST MEM HOSP, 65- Mem: AAAS; Soc Neurol; Am Chem Soc; NY Acad Sci. Res: Metabolism of amino acids; antibiotics; metabolism of certain amino acids and their derivatives; biological psychiatry; virology; kidney metabolism. Mailing Add: Dept of Path Baptist Mem Hosp Memphis TN 38146

LEACH, CAROLYN SUE, b Leesville, La, Aug 25, 40; m 69; c 1. ENDOCRINOLOGY, PHYSIOLOGY. Educ: Northwestern State Col, La, BS, 62; Baylor Univ, MS, 66, PhD(physiol), 68. Prof Exp: Med technologist, Univ Tex M D Anderson Hosp & Tumor Inst, Houston, 62-64, spec med technologist, 64-68; HEAD ENDOCRINE & BIOCHEM LABS, L B JOHNSON SPACE CTR, NASA, 68- Concurrent Pos: Consult, Proj Sea Lab, US Navy, 68-69, Proj Tektite I, US Navy-NASA-Gen Elec-Dept of Interior, 69 & Proj Tektite II, NASA-Dept of Interior, 70; Nat Res Coun-Nat Acad Sci res assoc, Manned Spacecraft Ctr, NASA, 68-70; adj instr physiol, Baylor Col Med, 68-70, adj asst prof, 70-; assoc investr, Inst Environ Med, Sch Med, Univ Pa, 70; mem res staff, Univ Tex Marine Biomed Inst. Mem: Am Inst Biol Sci; Endocrine Soc; Am Physiol Soc; Am Soc Med Technol; Am Soc Clin Path. Res: The study of the physiological adaptation of man to changing environments, particularly the endocrine mechanisms involved in adaptation; aerospace medicine. Mailing Add: Endocrine Lab DB7 Biomed Res Div NASA L B Johnson Space Ctr Houston TX 77058

LEACH, CHARLES MORLEY, b Sacramento, Calif, Oct 28, 24; m 49; c 3. PLANT PATHOLOGY. Educ: Queen's Univ, Ireland, BS, 49, BAgr, 50; Ore State Univ, PhD(plant path), 56. Prof Exp: Instr bot, 51-57, from asst plant pathologist to assoc plant pathologist, 57-66, PROF PLANT PATH, ORE STATE UNIV, 66- Concurrent Pos: NSF fel, Univ Bristol, 62-63; NZ sr sci fel, 73-74. Mem: Am Phytopath Soc; Mycol Soc Am; Brit Mycol Soc; Int Seed Testing Asn. Res: Biology of plant pathogenic fungi, especially reproduction and spore discharge; seed-borne diseases of agricultural crops. Mailing Add: Dept of Bot & Plant Path Ore State Univ Corvallis OR 97331

LEACH, CHARLES WILLARD, b Fairbury, Nebr, June 13, 16; m 46; c 6. FORESTRY. Educ: Univ Mo, BS, 41; Yale Univ, MF, 47. Prof Exp: Asst forestry, Univ Mo, 42-43, instr, 45-46; asst prof, Auburn Univ, 47-50; SR RES ENGR FOREST PROD, APPL RES LAB, US STEEL CORP, 52- Mem: Soc Am Foresters; Forest Prod Res Soc. Res: Wood utilization and preservation; biological evaluation of water quality. Mailing Add: US Steel Corp Res 125 Jamison Lane Monroeville PA 15146

LEACH, DON EDWARD, mathematics, see 12th edition

LEACH, EDDIE DILLON, b Clovis, NMex, Aug 12, 36; m 59; c 4. MICROBIOLOGY, TOXICOLOGY. Educ: Baylor Univ, BA, 60, MA, 62; Tex A&M Univ, PhD(zool), 65. Prof Exp: Asst prof biol, Am Univ, 65-70; ASSOC PROF BIOL & CHMN AREA SCI LEARNING, MILLIGAN COL, 70- Concurrent Pos: Res assoc, R Schattner Found Med Res, Washington, DC. Mem: AAAS. Res: Effects of x-irradiation on mice testes and the possibility of repairing the damage by administering grafts of unirradiated testicular tissue into the irradiated testes; pursuit of cold sterilizing solution; pine oil disinfectants and toxicity of same. Mailing Add: Milligan College TN 37682

LEACH, ERNEST BRONSON, b Huchow, China, Dec 21, 24; US citizen; div. MATHEMATICS. Educ: Case Inst Technol, BS, 49; Mass Inst Technol, PhD(math), 53. Prof Exp: From instr to asst prof, 53-59, ASSOC PROF MATH, CASE WESTERN RESERVE UNIV, 59- Concurrent Pos: Partic, Indo-Am Prog, Indian Inst Technol, Kanpur, 63-64; mem staff, Northwestern Univ Proj, Univ Khartoum, 66-67. Res: Algebraic topology; functional analysis. Mailing Add: Dept of Math Case Western Reserve Univ Cleveland OH 44106

LEACH, FRANKLIN ROLLIN, b Gorman, Tex, Apr 2, 33; m 56, 70; c 5. BIOCHEMISTRY. Educ: Hardin-Simmons Col, BA, 53; Univ Tex, PhD(chem), 57. Prof Exp: Res scientist I, Biochem Inst, 53-56; Nat Acad Sci fel med sci, Univ Calif, 57-59; res assoc, 59-60, from asst prof to assoc prof biochem, 60-68, PROF BIOCHEM, OKLA STATE UNIV, 68- Concurrent Pos: Soc Am Bacteriologists pres fel, Univ Ill, 60; NIH res career develop award, 62-72; res fel, Calif Inst Technol, 65-66. Mem: AAAS; Am Chem Soc; Am Soc Microbiol; Tissue Cult Asn; Am Asn Cancer Res. Res: Biochemical genetics of microbial and cell culture cells; transport mechanisms; enzymology. Mailing Add: Dept of Biochem Okla State Univ Stillwater OK 74074

LEACH, JAMES MOORE, b Littleton, NC, Nov 29, 24; m 53; c 3. ORGANIC CHEMISTRY. Educ: High Point Col, BS, 45. Prof Exp: Res asst, Morton Chem Co, 45-46, res chemist, 46-49, chief chemist, Morton-Withers Chem Co, 49-58; res chemist, Greensboro Plant, Pfizer Inc, 58-60, mgr appln res, 60-62, mgr prod develop, 62-71; RES DIR, PIEDMONT CHEM INDUSTS, INC, 71- Concurrent Pos: Consult, George C. Brown Co, 53-55. Mem: Am Chem Soc; Am Asn Textile Chemists & Colorists. Res: Antistatic agents for plastics and textiles; vinyl plasticizers and plastics; organic sulphonates; organic esters; textile chemistry. Mailing Add: 603 Florham Dr High Point NC 27260

LEACH, JOE TRAVIS, analytical chemistry, see 12th edition

LEACH, JOHN KLINE, b Buffalo, NY, July 11, 22; m 45; c 5. CARDIOLOGY, PHYSIOLOGY. Educ: Baldwin-Wallace Col, BS, 43; Albany Med Col, MD, 47; Am Bd Internal Med, dipl, 68; Am Bd Cardiovasc Dis, dipl, 69. Prof Exp: Instr med, Albany Med Col, 55-62, asst prof physiol, 62-63; assoc chief staff res, Vet Admin Ctr, Wadsworth, Kans, 63-64; asst prof, 66-70, ASSOC PROF MED, MED SCH, UNIV NMEX, 70- Concurrent Pos: Clin asst & asst attend, Albany Hosp, 55-62; NIH res fel physiol, Albany Med Col, 61-63, res grant cardiol, Univ NMex, 69-72; lectr, Med Ctr, Univ Kans, 63-64; chief cardiol sect, Vet Admin Hosp, Albuquerque, 66-72, assoc chief res staff, 69-73; attend med, Bernalillo County Med Ctr, Albuquerque, NMex, 66-; consult med, Bataan Hosp, Albuquerque, 69-; consult med & cardiol, St Joseph Hosp, Albuquerque, 69-; consult cardiol, Presby Hosp, Albuquerque, 69-; vis assoc prof, Med Ctr, Univ Calif, Los Angeles, 71-72. Mem: Fel Am Col Cardiol; Am Fedn Clin Res; Am Heart Asn; fel Am Col Physicians; Am Physiol Soc. Res: Cardiovascular research; cardiac muscle mechanics and hemodynamics. Mailing Add: Cardiol Sect Vet Admin Hosp Albuquerque NM 87108

LEACH, LARRY LAMONT, b Kansas City, Mo, Mar 23, 38; m 67; c 1. ANTHROPOLOGY, ARCHAEOLOGY. Educ: Univ Colo, BA, 63, MA, 65, PhD(anthrop), 70. Prof Exp: Teaching assoc anthrop, Univ Colo, 65-66, field archaeologist, 67-68; assoc prof anthrop, 68-74, PROF ANTHROP, SAN DIEGO STATE UNIV, 74- Concurrent Pos: Dir archaeol field sch, Colo Col, 70. Mem: AAAS; Am Anthrop Asn; Soc Am Archaeol. Res: Archaeology of Great Basin, Northwestern Plains, Fremont culture and North and West Africa. Mailing Add: Dept of Anthrop San Diego State Univ San Diego CA 92115

LEACH, LEONARD JOSEPH, b Rochester, NY, Aug 3, 24; m 53; c 3. TOXICOLOGY, ENVIRONMENTAL HEALTH. Educ: Brigham Young Univ, BS, 49. Prof Exp: Phys chemist, Army Chem Ctr, Md, 51-52; assoc indust hyg, Atomic Energy Proj, 52-55, unit chief toxicol, 55-57, instr indust hyg, 57-65, ASST PROF RADIATION BIOL & BIOPHYS, SCH MED & DENT, UNIV ROCHESTER, 65- Concurrent Pos: Speaker, Gordon Res Conf Toxicol & Safety Eval, NH, 60. Mem: Am Soc Toxicol; Am Acad Indust Hyg; Pan-Am Med Asn; Am Indust Hyg Asn; NY Acad Sci. Res: Inhalation toxicity of airborne agents related to air pollution and all aspects of environmental health. Mailing Add: Dept Radiation Biol & Biophys Sch Med Dent Univ of Rochester Rochester NY 14642

LEACH, LYSLE DOUGLAS, b Stuart, Nebr, Nov 10, 00; m 28; c 3. PLANT PATHOLOGY. Educ: Kans State Univ, BS, 23; Iowa State Univ, MS, 26, PhD(plant path), 30. Prof Exp: Agt, USDA, 26-28; assoc plant path, 28-31, instr & jr plant pathologist, 31-34, asst prof & asst plant pathologist, 34-40, assoc prof & assoc plant pathologist, 40-47, prof & plant pathologist, 47-68, dean students, 52-58, chmn dept, 60-68, EMER PROF PLANT PATH, UNIV CALIF, DAVIS, 68- Concurrent Pos: Consult, Denmark, 51; Ireland, 51, 56, 70; Uruguay, 54; Chile, 62, 65 & 67; acad mem, Univ Concepcion, 62; instr, OECD/Grad Sch, Zaragoza, Spain, 75. Mem: AAAS; fel Am Phytopath Soc; Am Soc Sugar Beet Technol. Res: Seedling diseases of field and vegetable crops and their control by fungicidal treatments; fungus diseases of sugar beets. Mailing Add: 234 Rice Lane Davis CA 95616

LEACH, ROBERT ELLIS, b Sanford, Maine, Nov 25, 31; m 55; c 6. MEDICINE, ORTHOPEDIC SURGERY. Educ: Princeton Univ, BA, 53; Columbia Univ, MD, 57. Prof Exp: Chmn orthop, Lahey Clin Found, 67-70; PROF ORTHOP SURG, MED SCH, BOSTON UNIV, 70- Concurrent Pos: Dir orthop serv, Boston City Hosp, 70-72; lectr, Tufts Univ, 70-; Am, Brit & Can traveling fel, 71. Mem: Orthop Res Soc; Am Acad Orthop Surgeons. Res: Joint transplantation. Mailing Add: Dept of Orthop Surg Boston Univ Hosp Boston MA 02118

LEACH, ROLAND MELVILLE, JR, b Framingham, Mass, Aug 27, 32; m 54; c 3. ANIMAL NUTRITION. Educ: Univ Maine, BS, 54; Purdue Univ, MS, 56; Cornell Univ, PhD(nutrit), 60. Prof Exp: Asst prof animal nutrit, Cornell Univ, 60-68, chemist, Plant Soil & Nutrit Lab, USDA, 59-68; assoc prof poultry sci, 68-73, PROF POULTRY SCI, PA STATE UNIV, UNIVERSITY PARK, 73- Mem: Poultry Sci Asn; Am Inst Nutrit. Res: Mineral nutrition of animals; role of trace elements in bone formation. Mailing Add: 205 Animal Industs Bldg Pa State Univ University Park PA 16802

LEACH, WILLIAM MATTHEW, b Pine Mountain, Ky, June 26, 33; m 60; c 3. CELL BIOLOGY, RADIOBIOLOGY. Educ: Berea Col, BA, 56; Univ Tenn, MS, 62, PhD(zool), 65. Prof Exp: USAEC res assoc zool & entom, Inst Radiation Biol, Univ Tenn, 64-66; res biologist, Radiation Bio-effects Prog, Nat Ctr Radiol Health, 66-67, chief radiation cytol lab, Div Biol Effects, Bur Radiol Health, 67-71, CHIEF EXP STUDIES BR, BUR RADIOL HEALTH, FOOD & DRUG ADMIN, USPHS, 71- Mem: AAAS; Am Inst Biol Sci; Am Soc Cell Biologists. Res: Cell responses to radiation in relation to the cell cycle; cell synthetic activities during the cell cycle; behavior of particulates and molecules in cells. Mailing Add: USPHS Bur of Radiol Health 5600 Fishers Lane Rockville MD 20852

LEACHMAN, ROBERT BRIGGS, b Lakewood, Ohio, June 11, 21; m 45; c 3. NUCLEAR PHYSICS. Educ: Case Inst Technol, BS, 42; Iowa State Univ, PhD(physics), 50. Prof Exp: Staff mem, Radiation Lab, Mass Inst Technol, 43-46; staff mem, Los Alamos Sci Lab, Univ Calif, 50-67, leader cyclotron group, 57-67; dir nuclear sci labs, Kans State Univ, 67-72, head dept physics, 67-71; asst dep dir, Defense Nuclear Agency, US Dept Defense, 72-74; SPEC ASST, NUCLEAR REGULATORY COMN, 74- Concurrent Pos: Guggenheim fel, Nobel Inst, Sweden, 55-56; Fulbright fel, Inst Theoret Physics, Denmark, 62-63. Mem: Fel Am Phys Soc; fel Am Nuclear Soc; AAAS. Mem: Nuclear fission; nuclear materials safeguards. Mailing Add: Nuclear Regulatory Comn Washington DC 20555

LEACOCK, ELEANOR BURKE, b Weehawken, NJ, July 2, 22; m 41; c 4. APPLIED ANTHROPOLOGY. Educ: Barnard Col, BA, 44; Columbia Univ, MA, 46, PhD(anthrop), 52. Prof Exp: Res asst social psychol, Dept Psychiat, Cornell Univ Med Col, 52-55; lectr anthrop, Queens Col, 55-56; lectr, City Col New York, 56-57; sr res assoc educ, Bank St Col Educ, 58-65; assoc prof anthrop, Polytech Inst Brooklyn, 63-67, prof, 67-72; PROF ANTHROP & CHAIRWOMAN DEPT, CITY COL NEW YORK, 72- Concurrent Pos: Spec consult, Behav Studies Sect, US Dept Health, Educ & Welfare, 57-58; Albert M Greenfield Ctr Human Rels, Univ Pa-Am Jewish Comt-Anti-Defamation League of B'nai B'rith-New World Found-Soc Psychol Study Social Issues grants, NJ, 58-60; Carnegie Found grant, Polytech Inst Brooklyn, 66-68; Rabinowitz Found-Basic Res in Educ Prog grant, Zambia, EAfrica, 70-71. Mem: AAAS; Am Anthrop Asn; Soc Appl Anthrop; Am Indian Ethnohist Conf (secy/treas, 61-65). Res: American Indians; education, anthropology and contemporary social problems. Mailing Add: Dept of Anthrop City Col of New York New York NY 10031

LEACOCK, ROBERT A, b Detroit, Mich, Oct 3, 35; m 61; c 3. THEORETICAL PHYSICS. Educ: Univ Mich, BS, 57, MS, 60, PhD(physics), 63. Prof Exp: Instr physics, Univ Mich, 63-64; Am-Swiss Found Sci Exchange fel theoret physics, Europ Orgn Nuclear Res, 64-65; assoc physics, 65-67, asst prof, 67-71, ASSOC PROF PHYSICS, IOWA STATE UNIV & PHYSICIST, AMES LAB, 71- Mem: Am Phys Soc. Res: Theoretical high energy physics. Mailing Add: Dept of Physics Iowa State Univ Ames IA 50011

LEACOCK, ROBERT JAY, b New York, NY, Mar 1, 39; m 64; c 1. ASTRONOMY. Educ: Univ Fla, BS, 60, MS, 62, PhD(astron), 71. Prof Exp: Instr physics, Pensacola Jr Col, 62-63; res asst astron, 63-71, ASST PROF PHYS SCI & ASTRON, UNIV FLA, 71- Mem: AAAS; Am Astron Soc. Res: Nonthermal radio observations of the major planets; optical variations of extragalactic radio sources. Mailing Add: 358 Winston W Little Hall Univ of Fla Dept of Astron Gainesville FL 32611

LEACOCK, SETH, b Angleton, Tex, Aug 10, 24; m 49. ANTHROPOLOGY. Educ: Univ Calif, BA, 52, PhD(anthrop), 58. Prof Exp: Instr anthrop, Univ Calif, 58-59; asst prof, Univ Chicago, 59-64; ASSOC PROF ANTHROP, UNIV CONN, 64- Concurrent Pos: NSF res grants, Brazil, 62-63. Mem: Am Anthrop Asn; Am Ethnol Soc. Res: South American ethnology; religion; Afro-Brazilian religions. Mailing Add: Dept of Anthrop Univ of Conn Storrs CT 06268

LEADBETTER, EDWARD RENTON, b Barnesboro, Pa, Jan 26, 34; m 56; c 4. MICROBIOLOGY. Educ: Franklin & Marshall Col, BS, 55; Univ Tex, PhD(bact), 59. Hon Degrees: MA, Amherst Col, 70. Prof Exp: Instr, 59-61, from asst prof to assoc prof, 61-70, chmn dept, 67-71, PROF BIOL, AMHERST COL, 70- Concurrent Pos: NSF fel, Hopkins Marine Sta, Pacific Grove, Calif, 62-63; NIH spec fel, Univ Mass, 66-67; vis prof, Hampshire Col, 71; instr, Marine Biol Lab, Woods Hole, 71-72, mem corp, 71-; NATO sr fel, Univ Seville, 72. Mem: AAAS; Am Soc Microbiol; Soc Develop Biol; Brit Soc Gen Microbiol. Res: Microbial ecology, physiology, and biochemistry; amine metabolism; photosynthesis; myxobacteria; oral microbiology; hydrocarbon oxidation; ultrastructure. Mailing Add: Dept of Biol Amherst Col Amherst MA 01002

LEADER, GORDON ROBERT, b Milwaukee, Wis, Jan 27, 16; m 46; c 4. PHYSICAL CHEMISTRY. Educ: Univ Wis, BS, 37; Univ Minn, PhD(phys chem), 40. Prof Exp: Res chemist, Monsanto Chem Co, Mo, 40-42, Nat Defense Res Comt, Northwestern Univ, 42-43 & Manhattan Dist Proj, Univ Chicago, 43-47; asst prof chem, Univ Ky, 47-51; res chemist, Mallinckrodt Chem Works, Mo, 51-53 & Olin Mathieson Chem Corp, 53-58; sr res chemist, Thiokol Chem Corp, 58-64; SR RES CHEMIST, PENNWALT CORP, 64- Mem: Am Chem Soc. Res: Chemical process development; Raman and nuclear magnetic resonance spectroscopy; radiochemistry; conductance and dielectric constants of organic solutions. Mailing Add: 1661 Weedon Rd Wayne PA 19087

LEADER, ROBERT WARDELL, b Tacoma, Wash, Jan 16, 19; m 40, 69; c 3. COMPARATIVE PATHOLOGY. Educ: Wash State Univ, BS & DVM, 52, MS, 55. Hon Degrees: DMedSci, Univ Toledo, 76. Prof Exp: Instr vet path, Wash State Univ, 52-55; USPHS fel, Univ Calif, 55-56; asst prof vet path, Wash State Univ, 56-60; assoc prof, Rockefeller Univ, 65-71; prof animal path, Univ Conn, 71-75; PROF PATH & CHMN DEPT, MICH STATE UNIV, 75- Concurrent Pos: Mem path training comt, NIH, 66-71, mem virol study sect, 71-; mem bd div, Mark Morris Found, 71- Mem: Am Vet Med Asn; Am Soc Exp Path; Am Col Vet Path; NY Acad Sci; Int Acad Path. Res: Studies of model diseases in animals with objective of elucidating pathogenetic mechanisms of similar diseases in man; chronic degenerative and connective tissue diseases. Mailing Add: Dept of Path Mich State Univ East Lansing MI 48824

LEADER, SOLOMON, b Spring Lake, NJ, Nov 14, 25. MATHEMATICS. Educ: Rutgers Univ, BS, 49; Princeton Univ, MA, 51, PhD(math), 52. Prof Exp: From instr to assoc prof, 52-61, PROF MATH, RUTGERS UNIV, NEW BRUNSWICK, 61- Mem: Am Math Soc; Math Asn Am. Res: Functional analysis; general topology. Mailing Add: Dept of Math Rutgers Univ New Brunswick NJ 08903

LEADERS, FLOYD EDWIN, JR, b Denison, Iowa, Dec 11, 31; m 53; c 1. PHARMACOLOGY. Educ: Drake Univ, BS, 55; Univ Iowa, MS, 60, PhD(pharmacol), 62. Prof Exp: From instr to asst prof pharmacol, Med Ctr, Univ Kans, 62-67; head pharmacol res, Alcon Labs, Tex, 67-72; dir res serv, Plough, Inc, Tenn, 72-73; DIR RES & DEVELOP LABS, PHARMACEUT DIV, PENNWALT CORP, 73- Concurrent Pos: NIH res grant, 62-; consult, Midwest Res Inst, 65-; adj asst prof, Univ Tex Southwestern Med Sch; adj prof, Tex Christian Univ. Mem: AAAS; Am Soc Pharmacol & Exp Therapeut; Soc Exp Biol & Med; Asn Res Vision & Ophthal. Res: Pharmaceutical drug development, both ethical and proprietary; drug-vehicle systems; physiology and pharmacology of the eye, cardiovascular system and autonomic nervous systems. Mailing Add: Res & Develop Labs Pennwalt Corp Pharmaceut Div 755 Jefferson Rd Rochester NY 14623

LEADERS, WILLIAM M, b Holyoke, Mass, Mar 18, 16; m; c 3. INORGANIC CHEMISTRY, CHEMICAL ENGINEERING. Educ: Ohio Univ, BS, 37; Mass Inst Technol, PhD(chem), 40. Prof Exp: Res chemist, Swift & Co, 41-46; chief chemist, Anderson Clayton & Co, 46-47; res chemist, Union Carbide Corp, Tenn, 48-51; tech dir, Mallinckrodt Chem Works, 51-59 & Spencer Chem Co, 59-62; sr res group leader solvent extraction, Kerr-McGee Corp, Okla, 62-71; VPRES, URANIUM RECOVERY CORP, 71- Concurrent Pos: Mem, Atomic Indust Forum. Mem: Am Chem Soc; Am Nuclear Soc. Res: Uranium purification and solvent extraction technology; isotope separation. Mailing Add: 4021 Carlisle Rd Lakelake FL 33803

LEADLEY, JOHN DAVID, b Batavia, NY, Feb 14, 27; m 53; c 3. MATHEMATICS. Educ: St Lawrence Univ, BS, 50; Univ Wash, MS, 56; Univ Ore, PhD(math), 67. Prof Exp: PROF MATH, REED COL, 56- Mem: Am Math Soc; Math Asn Am. Res: Structure in categories and applications in group theory; linear spaces; implications for undergraduate education. Mailing Add: Dept of Math Reed Col Portland OR 97202

LEADON, ROLAND EUGENE, solid state physics, see 12th edition

LEAF, ALBERT LAZARUS, b Seattle, Wash, May 16, 28; m 52; c 3. FOREST SOILS. Educ: Univ Wash, BSF, 50, MF, 52; Univ Wis, PhD(soils), 57. Prof Exp: Jr res forester, Univ Wash, 51-52; asst soils, Univ Wis, 52-57; from instr to assoc prof silvicult, 57-65, chmn forest resources coun, Sch Environ & Resources Mgt, 72-73, PROF SILVICULT, STATE UNIV NY COL ENVIRON SCI & FORESTRY, 65- Concurrent Pos: Forestry vchmn, Nat Joint Comt on Fertilizer Appln, 58-60; forestry chmn, Am Coun Fertilizer Appln, 60-62; Japan Soc Prom Sci vis prof, Tokyo Univ Agr & Technol, 72-73. Mem: Fel AAAS; Soc Am Foresters; Soil Sci Soc Am; Am Soc Agron; NY Acad Sci. Res: Tree nutrition; soil and foliar diagnosis of nutrient deficiencies; forest fertilization and soil chemistry; nursery soil management. Mailing Add: Sch of Environ & Resource Mgt SUNY Col Environ Sci & Forestry Syracuse NY 13210

LEAF, ALEXANDER, b Yokohama, Japan, Apr 10, 20; nat US; m 43; c 3. INTERNAL MEDICINE. Educ: Univ Wash, BS, 40; Univ Mich, MD, 43. Hon Degrees: AM, Harvard Univ, 61. Prof Exp: Instr med, Univ Mich, 47-49; assoc, 53-56, from asst prof to assoc prof, 56-65, JACKSON PROF CLIN MED, HARVARD MED SCH, 66-; CHIEF MED SERV, MASS GEN HOSP, 66- Concurrent Pos: From asst physician to assoc physician, Mass Gen Hosp, 53-62, physician, 62-; John Simon Guggenheim Mem Found fel, Balliol Col, Oxford Univ, 71-72. Mem: Nat Acad Sci; Am Soc Clin Invest; Am Physiol Soc; Asn Am Physicians; Am Acad Arts & Sci. Res: Ion transport and membrane physiology; kidney physiology. Mailing Add: Mass Gen Hosp Boston MA 02114

LEAF, BORIS, b Yokohama, Japan, Mar 4, 19; nat US; m 47; c 3. STATISTICAL MECHANICS. Educ: Univ Wash, BS, 39; Univ Ill, PhD(phys chem), 42. Prof Exp: Spec asst chem, Univ Ill, 42-43, instr phys chem, 43-44; assoc chemist, Metall Lab, Univ Chicago, 44-45; Jewett fel, Yale Univ, 45-46; assoc prof physics, Kans State Univ, 46-54, prof, 54-65; PROF PHYSICS & CHMN DEPT, STATE UNIV NY COL CORTLAND, 65- Concurrent Pos: Spec asst, Nat Defense Res Comt, Univ Ill, 45; res fel, Brussels, 58-60; prof, State Univ NY Binghamton, 67-71; fac fel & grants-in-aid, Res Found, State Univ NY, 70-72, scholar exchange prof, 74-; vis prof, Cornell Univ, 73-74. Mem: Fel AAAS; fel Am Phys Soc; Am Asn Physics Teachers; NY Acad Sci. Res: Thermodynamic theory; transport processes. Mailing Add: Dept of Physics State Univ of NY Col Cortland NY 13045

LEAF, CLYDE WILLIAM, b Evansville, Ind, Jan 12, 10; m 35; c 1. ORGANIC CHEMISTRY. Educ: Evansville Col, AB, 34; Columbia Univ, PhD(chem), 41. Prof Exp: Org chemist, Armour Res Found, 41-43; res chemist, Sharples Chems, Inc, 43; group leader, 43-50; staff mem, BASF Wyandotte Corp, 50-51, sect head, 51-57, supvr, 57-58, staff consult, 58-75; RETIRED. Mem: Am Chem Soc. Res: Adsorption; analytical methods; organic chemistry. Mailing Add: 1909 S Trenton Dr Trenton MI 48183

LEAGUS-CAPACI, BERNICE, virology, bacteriology, see 12th edition

LEAHY, EDWARD PRIOR, b New York, NY, Apr 1, 23; m 69; c 2. GEOGRAPHY. Educ: Univ Va, BME, 48; Columbia Univ, MBA, 58; Univ Fla, PhD(geog), 67. Prof Exp: Asst prof geog, Univ Miami, 68-69; ASSOC PROF GEOG, E CAROLINA UNIV, 69-, COORDR LATIN AM STUDIES, 70- Concurrent Pos: ECarolina Univ Res Coun fel, Chile & Arg, 71-72. Honors & Awards: ECarolina Univ Res Coun Award, Amazon, 75. Mem: Asn Am Geographers; Am Geog Soc. Res: Latin America. Mailing Add: Dept of Geog East Carolina Univ Greenville NC 27834

LEAHY, MARY GERALD, b San Francisco, Calif, Oct 11, 17. INSECT PHYSIOLOGY, ACAROLOGY. Educ: Univ Southern Calif, BA, 45; Cath Univ Am, MA, 47; Univ Notre Dame, PhD(biol), 62. Prof Exp: From asst prof to assoc prof, 47-70, chmn dept, 62-65, PROF BIOL, MT ST MARY'S COL, CALIF, 70- Concurrent Pos: NSF res grants, 62-64, 65-67 & 70-71; fel trop pub health, Harvard Univ, 66; WHO grant, Israel Inst Biol Res, Ness Ziona & Hebrew Univ Jerusalem, 68-69; sr res scientist, Nairobi, Kenya, 73-74; collab scientist, EAfrican Vet Res Orgn, Kenya, 73-74; prin investr, NIH grant, 74-77; vis scientist, Ga Southern Col, 74-75. Mem: AAAS; Entom Soc Am; Am Inst Biol Sci. Res: Mosquitoes and ticks; pheromones and reproductive physiology; interspecific insemination sterility and genetic control. Mailing Add: Dept of Biol Mt St Mary's Col Los Angeles CA 90049

LEAHY, RICHARD GORDON, b Buffalo, NY, Mar 6, 29; m 53; c 3. GEOCHEMISTRY. Educ: Yale Univ, BS, 52; Harvard Univ, AM, 54, PhD(geol), 57. Prof Exp: Asst geochem, Yale Univ, 52-53; asst to dir & res assoc geol, Woods Hole Oceanog Inst, 56-60; dir labs, Div Eng & Appl Physics, 60-68, asst to pres civic & govt rels, 70-71, ASSOC DEAN FAC ARTS & SCI, HARVARD UNIV, 68- Concurrent Pos: Mem US tech panel geochem, Int Geophys Year, 57-58. Mem: AAAS; Am Geophys Union; Am Soc Limnol & Oceanog. Res: Geochemistry of heavy isotopes in sea water and marine sediments; chemical processes of submarine weathering; variation of carbon dioxide in the atmosphere and its relation to air mass properties. Mailing Add: Univ Hall 20 Harvard Univ Cambridge MA 02138

LEAHY, SIDNEY MARCUS, b Denver, Colo, Aug 19, 30; m 59; c 3. ORGANIC CHEMISTRY. Educ: Reed Col, BA, 52; Univ Wash, PhD, 56. Prof Exp: Chemist, Gen Elec Co, 52; sr res chemist, 56-60, group supvr, 60-62, dept mgr, 62-68, tech dir traffic control prod div, 68-73, mgr traffic control mat dept, 73-75, GEN MGR TRAFFIC CONTROL MAT DIV, 3M CO, 75- Mem: Inst Traffic Engrs; Am Chem Soc. Res: Organic reaction mechanisms; photochemistry; condensation polymers; lubrication; abrasion. Mailing Add: Traffic Control Mat Div 3M Co - 3M Ctr St Paul MN 55101

LEAK, JOHN CLAY, JR, b Washington, DC, Aug 31, 28; m 64; c 3. ORGANIC CHEMISTRY. Educ: Univ Vt, BS, 49; Univ Ill, PhD(chem), 54. Prof Exp: Asst, Univ Ill, 51-54, res assoc animal nutrit, 55-56; Fulbright scholar, Ger, 54-55; chemist, Isotopes Specialties Co, 56-59, dir carbon-14 dept, 59-60; res dir, Cyclo Chem Corp, 60-61; tech dir, ChemTrac Corp, 61-62, vpres, 62-65, oper mgr, Baird-Atomic, Inc, Mass, 62-65; mgr chem dept, Tracerlab, 65-67, sr staff chemist, 67-69; sr staff chemist, 69-75, MGR PROD OPERS, LIFE SCI GROUP, ICN PHARMACEUT, INC, 75- Mem: Am Chem Soc. Res: Mechanism of organic reactions; metabolic fate of labeled hydroxy-proline in rats; synthesis of labeled compounds; applications for stable and radioactive isotopes. Mailing Add: ICN Pharmaceut Inc 2727 Campus Dr Irvine CA 92664

LEAK, LEE VIRN, b Chesterfield, SC, July 22, 32; m 64; c 2. CELL BIOLOGY, ELECTRON MICROSCOPY. Educ: SC State Col, BS, 54; Mich State Univ, MS, 59, PhD(cell biol), 62. Prof Exp: Asst prof biol sci, Mich State Univ, 62; res fel electron micros, Mass Gen Hosp & Harvard Med Sch, 62-64; asst surg, Mass Gen Hosp, 64-65; asst biol, Harvard Med Sch, 65-68, from instr to asst prof anat, 67-71; PROF ANAT & CHMN DEPT, COL MED, HOWARD UNIV, 71- Concurrent Pos: USPHS res grant, 66-; consult, Shriners Burns Res Inst, 67-; mem, Anat Sci Training Comt; Am Heart Asn res grant, 67-70; mem, Div Biol & Agr, Nat Res Coun, 72-75; ed staff, Anat Bd; mem, Nat Bd Med Examr, 73; mem, Marine Biol Lab Corp, 73- Mem: Am Asn Anat; Am Soc Cell Biol; Am Soc Zool; Genetics Soc Am; Sigma Xi; NY Acad Sci. Res: Biology of the lymphatic vascular system and its role during the inflammatory response; comparative ultrastructural and cytochemical study of vertebrate heart tissue; location of DNA in primitive cell types and the effect of irradiation on its replication. Mailing Add: Dept of Anat Howard Univ Col of Med Washington DC 20001

LEAK, ROBERT J, physical chemistry, see 12th edition

LEAKE, CHARLES ROBERT, mathematics, philosophy of science, see 12th edition

LEAKE, CHAUNCEY D, b Elizabeth, NJ, Sept 5, 96; m 21; c 2. PHARMACOLOGY, HISTORY OF SCIENCE. Educ: Princeton Univ, LittB, 17; Univ Wis, MS, 20, PhD(physiol, pharmacol), 23. Hon Degrees: LHD, Kenyon Col, 59; DSc, Med Col Pa; LLD, Univ Calif; ScD, Philadelphia Col Pharm & Sci & Ohio State Univ, 75. Prof Exp: Asst prof physiol & pharmacol, Univ Wis, 23-27; prof pharmacol & hist med & librn, Univ Calif, 28-42; exec vpres, Univ Tex Med Br, Galveston, 42-55; prof pharmacol & lectr hist med, 56-62, EMER PROF PHARMACOL, OHIO STATE UNIV, 62-; SR LECTR PHARMACOL & HIST HEALTH PROFESSIONS, UNIV CALIF, SAN FRANCISCO, 63- Concurrent Pos: Pres, Family Rels Ctr, 35-40; pres bd, St Luke's Hosp, San Francisco, 37-39; chmn hon consult, Nat Libr Med, 46-47; Miller lectr, Univ Wis, 49, Waters lectr, 70; Clendenning lectr, Univ Kans, 52; Trent lectr, Duke Univ, 55; prof med jurisp, Hastings Col Law, Univ Calif, San Francisco, 63-65. Mem: AAAS (pres, 61); Hist Sci Soc (pres, 36-38); Am Physiol Soc; Am Soc Pharmacol & Exp Therapeut (pres, 59); Am Asn Hist Med (pres, 61). Res: Central nervous system activity; narcotics; anesthesia; chemotherapy; geriatrics; mechanisms of drug action; principles of functional activity; history and philosophy of health professions and science. Mailing Add: Univ of Calif San Francisco CA 94143

LEAKE, LOWELL, JR, b Denver, Colo, May 25, 28; m 59; c 2. MATHEMATICS. Educ: Tufts Col, AB, 50; Univ Wis, MS, 56, PhD(math, educ), 62. Prof Exp: Traffic chief, Northwestern Bell Tel Co, 50-54; high sch teacher, Ill, 56-58; from instr to assoc prof math, 60-74, PROF MATH, UNIV CINCINNATI, 74- Mem: Math Asn Am; Nat Coun Teachers Math; Asn Teachers Math (Eng). Res: Training of secondary and elementary mathematics teachers at undergraduate and graduate levels; learning of mathematics. Mailing Add: Dept of Math Univ of Cincinnati Cincinnati OH 45221

LEAKE, PRESTON HILDEBRAND, b Proffit, Va, Aug 8, 29; m 54; c 2. ORGANIC CHEMISTRY. Educ: Univ Va, BS, 50; Duke Univ, MA, 53, PhD(chem), 54. Prof Exp: Res supvr org chem, Nitrogen Div, Allied Chem Corp, 54-60; asst res dir, Albemarle Paper Mfg Co, 60-65; asst to managing dir, Res & Develop Dept, 65-68, asst managing dir, 68-70, ASST DIR RES & DEVELOP DEPT, AM TOBACCO CO, 70- Concurrent Pos: Adj prof, Richmond Prof Inst, 63-64. Mem: Am Chem Soc; Am Inst Chemists; Tech Asn Pulp & Paper Indust. Res: Polycyclic aromatic chemistry: Psychorr synthesis; amino acids and cyanuric acid derivatives; polyethylene; sizing; silica fume; specialty and filter papers; tobacco. Mailing Add: Am Tobacco Co R&D Dept PO Box 899 Hopewell VA 23860

LEAKE, WILLIAM WALTER, b Johnstown, Pa, Apr 24, 26; m 57; c 2. ORGANIC CHEMISTRY. Educ: Duquesne Univ, BS, 51; Duke Univ, MA, 53; Univ Pittsburgh, PhD(chem), 58. Prof Exp: Res chemist, Monsanto Chem Co, 57-60; asst prof, 61-66, ASSOC PROF CHEM, WASHINGTON & JEFFERSON COL, 66-, ASSOC DEAN ACAD AFFAIRS, 68- Mem: Am Chem Soc. Res: Instrumental methods of analysis. Mailing Add: Dept of Chem Washington & Jefferson Col Washington PA 15301

LEAL, JOSEPH ROGERS, b New Bedford, Mass, Sept 14, 18; m 44; c 4. ORGANIC CHEMISTRY. Educ: Univ Mass, BS, 49; Ind Univ, PhD(chem), 53. Prof Exp: Res asst, Corn Prod Refining Co, 40-42; asst chemist, Revere Copper & Brass Co, 42-43 & 45-46; res chemist, Am Cyanamid Co, 52-57, tech rep govt res liaison, Washington, DC, 57-63, mgr contract rels, 63-67; SR STAFF ASSOC, CELANESE RES CO, 67- Mem: AAAS; Am Chem Soc; NY Acad Sci. Res: High temperature resistant aromatic and heterocyclic polymers; nonflammable fibers; high strength, high modulus reinforcement materials. Mailing Add: 10 S Crescent Maplewood NJ 07040

LEAMAN, WILBUR KAUFFMAN, b Neffsville, Pa, Feb 17, 21; m 42; c 1. PETROLEUM CHEMISTRY. Educ: Franklin & Marshall Col, BS, 47. Prof Exp: Res chemist, Res Labs, Mobil Oil Corp, 47-51, sr res chemist, 51-67, staff asst to mgr process res & develop sect, 67-75, ADV BUDGETS & ADMIN, PROCESS RES & TECH SERV DIV, MOBIL RES & DEVELOP CORP, 75- Mem: Am Chem Soc; Int Cong Catalysis. Res: Catalysis. Mailing Add: Process Res & Tech Serv Div Mobil Res & Develop Corp Paulsboro NJ 08066

LEAMER, ROSS WILSON, b Ellendale, NDak, Apr 23, 15; m 44; c 1. SOIL SCIENCE. Educ: NDak Agr Col, BS, 37; NC State Col, MS, 39; Ohio State Univ, PhD(soils), 42. Prof Exp: Tech agr rep, US Rubber Co, NY, 44-45; soil physicist irrig exp sta, Bur Plant Indust, Soils & Agr Eng, 45-53, soil scientist, Western Soil & Water Mgt Res Br, Soil & Water Conserv Res Div, Agr Res Serv, 53-65, soil scientist, Southern Plains Br, 65-73, SOIL SCIENTIST, SOUTHERN REGION SUBTROP TEX AREA, AGR RES SERV, USDA, 73- Mem: Soil Sci Soc Am; Am Soc Agron; Am Soc Photogram; AAAS. Res: Remote sensing; automated identification of soil and surface cover from aircraft and satellite signals. Mailing Add: USDA Agr Res Serv PO Box 267 Weslaco TX 78596

LEAMY, LARRY JACKSON, b Alton, Ill, Nov 15, 40; m 65; c 1. QUANTITATIVE GENETICS. Educ: Eastern Ill Univ, BS, 62; Univ Ill, Urbana, MS, 65, PhD(zool), 67. Prof Exp: Asst prof, 67-71, ASSOC PROF BIOL, CALIF STATE UNIV, LONG BEACH, 71- Concurrent Pos: Calif State Univ Found new fac grant, Calif State Univ, Long Beach, 67-68, fac grant-in-aid, 69-70. Mem: Genetics Soc Am; Am Genetic Asn; Soc Study Evolution; Behavior Genetics Soc; Soc Syst Zool. Res: Quantitative genetics of mice. Mailing Add: Dept of Biol Calif State Univ Long Beach CA 90840

LEAN, ERIC GUNG-HWA, b Fukien, China, Jan 1, 38; m 65; c 1. ACOUSTICS, OPTICS. Educ: Cheng Kung Univ, Taiwan, BS, 59; Univ Wash, MS, 63; Stanford Univ, PhD(elec eng), 67. Prof Exp: Res asst elec eng, Univ Wash, 61-63; res asst microwave acoust, Hanson Lab Physics, Stanford Univ, 63-67, res assoc microwave acoust & laser, 67; mem res staff nonlinear optics, T J Watson Res Ctr, 67-69, MGR ACOUST & OPTICAL PHYSICS, T J WATSON RES CTR, IBM CORP, 69- Mem: Inst Elec & Electronic Engrs; Sigma Xi; Optical Soc Am. Res: Microwave acoustic waves in solids; nonlinear optics; optical signal processing devices; surface wave devices; laser applications; integrated optics; fiber optics. Mailing Add: IBM Corp Watson Res Ctr PO Box 218 Yorktown Heights NY 10598

LEANDER, JOHN DAVID, b Mt Vernon, Wash, Apr 8, 44; m 65; c 2. PSYCHOPHARMACOLOGY. Educ: Pac Lutheran Univ, BA, 66; Western Wash State Col, MA, 67; Univ Fla, PhD(psychol), 71. Prof Exp: Fel neurobiol prog, 71-73, instr pharmacol, 73-74, ASST PROF PHARMACOL, UNIV NC, CHAPEL HILL, 74- Concurrent Pos: Mem bd ed, J Exp Anal of Behavior, 74- Mem: AAAS; Am Psychol Asn; Behavioral Pharmacol Soc. Res: Behavioral pharmacology; effects of drugs on behavior and the interaction of drugs with ongoing behavior. Mailing Add: Dept of Pharmacology Univ of NC Chapel Hill NC 27514

LEANING, WILLIAM HENRY DICKENS, b Whakatane, NZ, Feb 24, 34; m 56; c 4. RESEARCH ADMINISTRATION. Educ: Univ Sydney, BVSc, 56. Prof Exp: Vet gen pract, Nth Canterbury Vet Club, NZ & Putaruru Vet Club, NZ, 57-62; vet tech dir appl res parasitol, Merck Sharp & Dohme NZ Ltd, 62-69; from dir mkt develop large animal prod, 69-72, sr dir clin res animal sci res, 72-74, EXEC DIR ANIMAL SCI RES DEVELOP RES & ADMIN, MERCK SHARP & DOHME RES LABS, 75- Mem: NZ Vet Asn; NZ Soc Animal Prod; World Asn Adv Vet Parasitol; Am Vet Med Asn; Indust Vet Asn. Res: Concepts of applied preventive medicine on whole herd flock basis throughout productive life of animal/bird; primary areas helminthology, coccidiosis; swine dysentery, production improvers; innovation in formulation and treatment application. Mailing Add: Animal Sci Res Merck & Co Inc Rahway NJ 07065

LEAP, WILLIAM L, b Philadelphia, Pa, Nov 28, 46. ANTHROPOLOGY. Educ: Fla State Univ, BA, 67; Southern Methodist Univ, PhD(anthrop), 70. Prof Exp: Instr ling, Southern Methodist Univ, 68-70; instr social sci, El Centro Col, Tex, 69-70; asst prof anthrop, 70-75, ASSOC PROF ANTHROP, AM UNIV, 75- Concurrent Pos: Consult, Rio Puerco-San Juan Anasazi Origins Proj, 71-; mem bd dirs, SE Mass Univ Consortium for Portuguese, Am Studies, 72-; coordr, Indian Educ Clearinghouse, Ctr Appl Ling, 74- Res: Development of language policy and language maintenance programs; Southwestern Indian languages; Portuguese immigrant communities in United States, Canada and Great Britain; Southwest Indian English. Mailing Add: Dept of Anthrop Am Univ Washington DC 20016

LEAPHART, CHARLES DONALD, b Sheridan, Wyo, Sept 19, 22; m 44; c 3. FOREST PATHOLOGY. Educ: Univ Mont, BS, 48; Yale Univ, MF, 49, PhD(forest path), 54. Prof Exp: Forest pathologist, Bur Plant Indust, Soil Eng, USDA, Mont, 50-54, plant pathologist, US Forest Serv, Wash, 54-60, plant pathologist & proj leader, Ohio, 60-62, asst chief forest dis res, Utah, 62-65, PROJ LEADER, FORESTRY SCI LAB, US FOREST SERV, IDAHO, 65- Mem: AAAS; Soc Am Foresters; Mycol Soc Am; Am Phytopath Soc. Res: Forest tree diseases; diseases of western white pine and associated tree species. Mailing Add: US Forest Serv INT Sta Forestry Sci Lab 1221 S Main St Moscow ID 83843

LEAR, BERT, b Logan, Utah, June 10, 17; m 50; c 1. PLANT PATHOLOGY. Educ: Utah State Univ, BS, 41; Cornell Univ, PhD(plant path), 47. Prof Exp: Agt, Exp Sta, USDA, 41-43; Dow Chem Co fel & res assoc, Cornell Univ, 47-48, asst prof plant path, 48-52; nematologist, NMex Col, 52-53; from asst nematologist to nematologist, 53-74, PROF NEMATOL, UNIV CALIF, DAVIS, 63- Mem: Am Phytopath Soc; Soc Nematol; Orgn Trop Am Nematologists. Res: Soil treatment for control of nematodes; fate of chemicals in soils and plants when applied for control of nematodes; role of plant parasitic nematodes in diseases of plants. Mailing Add: Dept of Plant Path Univ of Calif Davis CA 95616

LEAR, CLEMENT S C, b Christchurch, NZ, Oct 24, 29; m 67. ORTHODONTICS. Educ: Univ NZ, BDS, 53; Harvard Univ, DMD, 63. Prof Exp: Pvt pract, 54-58; res assoc orthod, Harvard Univ, 62-64, assoc, 64-67; PROF ORTHOD & HEAD DEPT, FAC DENT, UNIV BC, 67- Concurrent Pos: Clin fel dent med, Forsyth Dent Ctr, Sch Dent Med, Harvard Univ, 58-59, res fel orthod, 59-62; mem univ comt, Med Res Coun Can. Mem: AAAS; Am Dent Asn; Int Asn Dent Res; Int Soc Cranio-Facial Biol; Can Asn Orthod. Res: Oro-facial muscle physiology and mechanisms of tooth support, particularly as they relate to dental arch form. Mailing Add: Dept of Orthod Fac of Dent Univ of BC Vancouver BC Can

LEAR, JAMES BERNARD, analytical chemistry, see 12th edition

LEARN, ARTHUR JAY, b Lewistown, Mont, Mar 25, 33; m 59; c 2. SOLID STATE PHYSICS. Educ: Reed Col, BA, 54; Mass Inst Technol, PhD(physics), 58. Prof Exp: Mem tech staff, TRW Systs, Calif, 58-67; mem staff, Electronics Res Ctr, NASA, 67-70; SR MEM RES STAFF, FAIRCHILD CAMERA & INSTRUMENT CORP, 70- Mem: Am Phys Soc; Electrochem Soc. Res: X-ray diffraction; electron microscopy; properties of thin films; thin film superconductor and semiconductor devices; metallization. Mailing Add: Fairchild Camera & Instrum Corp 4001 Miranda Ave Palo Alto CA 94304

LEARNED, ROBERT EUGENE, b Glendale, Calif, July 3, 28; m 56; c 2. ECONOMIC GEOLOGY, GEOCHEMISTRY. Educ: Occidental Col, AB, 55; Univ Calif, Los Angeles, MA, 62; Univ Calif, Riverside, PhD(geol), 66. Prof Exp: Geologist, Aerogeophys Co, Calif, 55-56; asst prof geol, Chapman Col, 65-67; GEOLOGIST, US GEOL SURV, 67- Mem: Geol Soc Am; Geochem Soc; Asn Explor Geochemists; Soc Econ Geologists. Res: Geology and geochemistry of ore deposits; geochemical exploration methods. Mailing Add: 614 Wyoming St Golden CO 80401

LEARY, JOHN DENNIS, b New Bedford, Mass, Dec 6, 34; m 57; c 4. BIOCHEMISTRY. Educ: Mass Col Pharm, BS, 56, MS, 58; Univ Conn, PhD(pharmacog), 64. Prof Exp: Asst pharmacog, Univ Conn, 58-59; asst prof, Ore State Univ, 59-61; from asst prof to assoc prof phytochem, St John's Univ, NY, 63-68; assoc prof pharmacog & bot, 68-74, ASSOC PROF BIOCHEM, MASS COL PHARM, 74- Mem: Am Pharmaceut Asn; Am Soc Pharmacog. Res: Phytochemical studies, primarily Solanaceae: chemotaxonomy and biogenesis; microbial transformation of organic compounds. Mailing Add: 179 Longwood Ave Boston MA 02115

LEARY, JOHN SYLVESTER, b Savannah, Ga, Feb 8, 21; m 50; c 4. PHARMACOLOGY, TOXICOLOGY. Educ: The Citadel, BS, 47. Prof Exp: Res asst pharmacol, Med Col SC, 47-49; pharmacologist, Pesticide Regulation Div, USDA, 49-55, asst to chief pharmacologist, 55-57, chg lab, 57-61, chief staff officer, Washington, DC, 61-67; supvr toxicol, Agr Chem Div, Shell Chem Co, Tex, 67-72, toxicologist, Polymers Div, 72-74; CONSULT TOXICOL, 74- Concurrent Pos: Consult, WHO, Venezuela, 63. Honors & Awards: Fed Govt Outstanding Serv Award, 60. Mem: Am Asn Pesticide Control Off; Soc Toxicol; Am Soc Vet Physiol & Pharmacol. Res: Development of standard laboratory procedures with regard to pharmacology and toxicology of pesticides and the evaluation of the hazards of pesticide chemicals through the use of these standards; evaluations relating to both human and animal health in terms of environmental exposure to pesticide chemicals. Mailing Add: 2002 Sea Cove Ct Nassau Bay Dr Houston TX 77058

LEARY, JOHN VINCENT, b Buffalo, NY, Dec 27, 37; m 60; c 3. GENETICS. Educ: State Univ NY Buffalo, BS, 59, MA, 64; Mich State Univ, PhD(bot, plant path), 69. Prof Exp: NIH fel genetics, Cornell Univ, 69-70; ASST PROF PLANT PATH, UNIV CALIF, RIVERSIDE, 70- Mem: AAAS; Am Phytopath Soc; Genetics Soc Am. Res: Molecular basis of fungal morphogenesis; molecular basis of incompatibility; genetics of fungi. Mailing Add: Dept of Plant Path Univ of Calif Riverside CA 92502

LEARY, JOSEPH ALOYSIUS, b New York, NY, Nov 22, 19; m 43; c 2. PHYSICAL CHEMISTRY, CHEMICAL ENGINEERING. Educ: Newark Col Eng, BS, 43; Univ NMex, PhD(phys chem), 56. Prof Exp: Prod supvr, Am Cyanamid Co, 43-44; res sect leader, Los Alamos Sci Lab, Univ Calif, 46-47, alt res group leader, 47-74; MEM STAFF, ENERGY RES & DEVELOP ADMIN, WASHINGTON, DC, 74- Mem: Am Chem Soc; fel Am Inst Chemists; Am Nuclear Soc. Res: Materials of interest to the nuclear field, particularly transuranium elements and special isotopes. Mailing Add: US Energy Res & Develop Admin Washington DC 20545

LEARY, RALPH JOHN, b Elizabeth, NJ, Nov 3, 29; m 52; c 6. ORGANIC CHEMISTRY. Educ: Seton Hall Univ, BS, 51; Univ Ill, PhD(chem), 57. Prof Exp: Jr chemist, Merck & Co, Inc, 51-54; group leader, Esso Res & Eng Co, 57-67, group leader & res assoc, 67-75, LAB HEAD, EXXON CHEM CO USA, 75- Mem: Am Chem Soc. Res: Gas chromatography and automation of laboratory instruments. Mailing Add: Chem Plant Lab Exxon Chem Co USA PO Box 222 Linden NJ 07036

LEARY, ROBERT EUGENE, organic chemistry, see 12th edition

LEARY, ROLFE ALBERT, b Waterloo, Iowa, Mar 5, 38; m 67; c 2. FOREST MENSURATION. Educ: Iowa State Col, BS, 59; Purdue Univ, MS, 61, PhD(forest mgt), 68. Prof Exp: Vol forester, US Peace Corps, St Lucia, WIndies, 61-63; instr forest mensuration, Southern Ill Univ, 64-65; mensurationist, 68-72, PRIN MENSURATIONIST FORESTRY, US FOREST SERV, NCENT FOREST EXP STA, 72- Mem: AAAS; Ecol Soc Am; Sigma Xi; Coun Unified Res & Educ. Res: Development and testing methods of synthesis and analysis applied to forestry problems; evaluation of the periodic coordinate system for assembling results of research on interactions into a total system map. Mailing Add: NCent Forest Exp Sta Folwell Ave St Paul MN 55108

LEARY, THOMAS RICHARD, biochemistry, see 12th edition

LEASE, ELMER JOHN, b Madison, Wis, Oct 18, 07; m 31; c 3. BIOCHEMISTRY. Educ: Univ Wis, BS, 31, MS, 33, PhD(biochem), 35. Prof Exp: Asst agr chem, Univ Wis, 32-35, fel, 35-38; assoc chemist, Exp Sta, Clemson Univ, 38-66, prof nutrit, 43-66; res assoc, Sch Pharm, Univ SC, 66-70, res assoc, Malnutrit & Parasite Proj, 70-73; RETIRED. Concurrent Pos: Mem cereal comt & food & nutrit bd, Nat Res Coun, 42. Res: Organic chemical synthesis; vitamin assays; analyses for biochemical constituents; isolation and identification of nutritional factors; blood analyses; technical problems connected with enrichment of cereals; public health nutrition; eradication of intestinal parasites. Mailing Add: 1220 Jennings Ct Columbia SC 29204

LEASURE, ELDEN EMANUEL, b Solomon, Kans. PATHOLOGY, PHYSIOLOGY. Educ: Kans State Col, DVM, 23, MS, 30. Prof Exp: Asst prof path, 26-35, prof physiol, 35-64, head dept, 44-64, dean col vet med, 48-64, EMER PROF PATH, PARASITOL & PUB HEALTH & EMER DEAN COL VET MED, KANS STATE UNIV, 64- Mem: AAAS; Asn Am Cols Vet Med (pres, 62-64); Asn Deans Am Cols Vet Med (pres, 51-62); Am Vet Med Asn (pres, 60). Res: Feline infectious enteritis; anaplasmosis; equine encephalomyelitis; bovine allergic dermatitis. Mailing Add: Col of Vet Med Kans State Univ Manhattan KS 66502

LEATH, KENNETH T, b Providence, RI, Apr 29, 31; m 55; c 4. PLANT PATHOLOGY. Educ: Univ RI, BS, 59; Univ Minn, MS & PhD(phytopath), 66. Prof Exp: Res technician cereal rusts, Coop Rust Lab, Minn, 59-66, PLANT PATHOLOGIST, REGIONAL PASTURE RES LAB, USDA, 66- Concurrent Pos: Adj assoc prof, Dept Plant Path, Pa State Univ, 66- Mem: Am Phytopath Soc; Am Soc Agron; Am Forage & Grassland Coun. Res: Clover and alfalfa diseases; host-parasite interaction. Mailing Add: Regional Pasture Res Lab Northeastern Region USDA University Park PA 16802

LEATH, PAUL LARRY, b Moberly, Mo, Jan 9, 41; m 62; c 2. SOLID STATE PHYSICS. Educ: Univ Mo-Columbia, BS, 61, MS, 63, PhD(physics), 66. Prof Exp: Res assoc theoret physics, Oxford Univ, 66-67; asst prof, 67-71, assoc chmn dept, 73-75, ASSOC PROF PHYSICS, RUTGERS UNIV, NEW BRUNSWICK, 71- Mem: AAAS; Am Phys Soc. Res: Theoretical solid state physics; inelastic neutron scattering; vibrational and electronic properties of alloys; anharmonic crystals; percolation processes. Mailing Add: Dept of Physics Rutgers Univ New Brunswick NJ 08903

LEATHEM, JAMES HAIN, b Lebanon, Pa, Apr 10, 11; m 46. REPRODUCTIVE ENDOCRINOLOGY. Educ: Lebanon Valley Col, BS, 32; Princeton Univ, MA, 36, PhD(biol), 37. Hon Degrees: DSc, Lebanon Valley Col, 65. Prof Exp: Procter fel biol, Princeton Univ, 37-38; res assoc anat, Med Sch, Columbia, 38-41; from instr to assoc prof zool, Univ, 41-48, from asst dir to assoc dir, Bur Biol Res, 59-64, PROF ZOOL, RUTGERS UNIV, NEW BRUNSWICK, 48-, DIR, BUR BIOL RES, 65- Concurrent Pos: NSF sr fac fel instr, Middlesex Gen Hosp, 42-45 & Columbia Univ, 43-45; consult histologist, Ciba Pharmaceut Prod, Inc, 44-; vis assoc prof, Med Sch, Univ Tenn, 47, vis prof, 51; vis prof, Princeton Univ, 58 & Med Sch, Wayne State Univ, 57; adv coun, Nat Inst Child Health & Human Develop, 72-; mem, President's Biomed Res Panel Nutrit Interdisciplinary Cluster, 75-; mem, Fulbright Comn; adv coun, Inst Lab Animal Resources; mem div biol & agr, Nat Res Coun; trustee, Lebanon Valley Col & Seeing Eye, Inc. Honors & Awards: Distinguished Res Award, Rutgers Res Coun, 64, Lindback Award for Res, 70. Mem: AAAS; Am Fertil Soc; Geront Soc; Endocrine Soc; Soc Study Reproduction (pres, 70-71). Res: Endocrinology; control of reproduction; polycystic ovary disease; placental growth; neonatal maturation; protein nutrition and endocrines. Mailing Add: Bur of Biol Res Rutgers Univ New Brunswick NJ 08903

LEATHEM, WILLIAM DOLARS, b Chicago, Ill, Jan 6, 31; m 52; c 3. ZOOLOGY. Educ: Univ Wis, BS, 61, MS, 63, PhD(zool), 65. Prof Exp: Asst zool, Univ Wis, 62; asst prof biol, Wis State Univ, Whitewater, 65-66; asst prof, Univ Wis, Waukesha Ctr, 66-69, NSF grants, 67-69; res assoc, Norwich Pharmacal Co, 69-74; ASST DIR CLIN NUTRIT, EATON LABS, 74- Mem: AAAS; Am Soc Parasitol; Soc Protozool; Am

Soc Trop Med & Hyg. Res: General parasitology and protozoology. Mailing Add: Norwich Pharmacal Eaton Labs PO Box 191 Norwich NY 13815

LEATHEN, WILLIAM WARRICK, b Pittsburgh, Pa, July 7, 12; m; c 3. MICROBIOLOGY. Educ: Univ Pittsburgh, BS, 36, MS, 37. Prof Exp: Bacteriologist, Morris Knowles, Inc, Pa, 36-42; chief bacteriologist, Hektoen Inst, Cook County Hosp, Chicago, 46; fel bact, Mellon Inst, 46-52, head microbiol & micros sect, 52-63, sr fel, 63-69; supvr microbiol res, Gulf Res & Develop Co, 69-73; MICROBIOLOGIST, GULF OIL CORP, 73- Concurrent Pos: Bacteriologist, West Penn Hosp, Pittsburgh, 37-42. Mem: Fel AAAS; Am Chem Soc; Soc Indust Microbiol; Am Soc Microbiol; Inst Food Technologists. Res: Variation of hemolytic streptococci; bacteriological oxidation of ferrous iron; effect of industrial wastes on bacteriological flora of streams; microbiological deterioration of materials; petroleum microbiology; single cell protein research; microbial pesticides; geomicrobiology. Mailing Add: Gulf Oil Corp PO Box 3240 Pittsburgh PA 15230

LEATHERMAN, ANNA D, b Centre Square, Pa, Jan 17, 09. PLANT ECOLOGY. Educ: Goshen Col, BS, 38, AB, 39; Cornell Univ, MA, 47; Univ Tenn, PhD, 55. Prof Exp: Teacher, Rural Sch, Va, 39-44 & Pub Sch, 44-47; from instr to prof biol, Upland Col, 47-64; asst, Univ Tenn, 52-53; prof, Bethel Col, 64-66; PROF BIOL, SPRING ARBOR COL, 66- Mem: Ecol Soc Am. Res: Ecological life history of Lonicera japonica Thunb. Mailing Add: 7991 Kevin Dr Parma MI 49269

LEATHERS, CHESTER RAY, b Claremont, Ill, May 15, 29; m 53; c 4. MYCOLOGY. Educ: Eastern Ill Univ, BS, 50; Univ Mich, MS, 51, PhD(bot), 55. Prof Exp: Res mycologist, Biol Warfare Labs, US Army, Md, 55-57; asst prof, 57-61, ASSOC PROF BOT, ARIZ STATE UNIV, 61- Mem: Fel AAAS; Mycol Soc Am; Am Phytopath Soc; Am Inst Biol Sci. Res: Mycology and plant pathology, particularly fleshy fungi, cereal and vegetable diseases; allergenic fungi; medical mycology. Mailing Add: Dept of Bot & Microbiol Ariz State Univ Tempe AZ 85281

LEATHERS, JOEL MONROE, b Guy's Store, Tex, Jan 10, 20; m 45; c 3. CHEMISTRY. Educ: Sam Houston State Teachers Col, BS, 41. Hon Degrees: DEng, Mich Technol Univ, 72. Prof Exp: Chemist, 42-45, proj leader, Org Lab, 45-48, asst supt vinyl & vinylidene prod, 48-50, asst dir org lab, 50-53, dir org pilot plant lab, 54-61, dir res & develop, 61-66, gen mgr, Tex Div, 66-68, DIR US AREA OPERS, DOW CHEM CO, 68-, VPRES, 70-, DIR, 71-, EXEC VPRES, DOW CHEM USA, 71- Mem: Am Chem Soc; Am Inst Chem Engrs. Res: Chlorination of saturated hydrocarbons. Mailing Add: 2020 Dow Ctr Abbott Rd Midland MI 48640

LEATHERWOOD, JAMES M, b Waynesville, NC, Mar 22, 30; m 56; c 1. ANIMAL NUTRITION. Educ: Berea Col, BS, 52; NC State Univ, MS, 57, PhD(animal sci), 61. Prof Exp: From instr to assoc prof, 57-69, PROF ANIMAL SCI, NC STATE UNIV, 69- Concurrent Pos: Fel biochem, Duke Univ, 60-61. Mem: Am Soc Microbiol; Am Soc Animal Sci; Am Inst Nutrit. Res: Carbohydrate metabolism; enzymatic cellulose degradation; bacterial cell walls and microbiology of the rumen; animal efficiency and energetics. Mailing Add: Dept of Animal Sci NC State Univ Raleigh NC 27607

LEATON, JOHN ROGER, b Kansas City, Mo, Nov 6, 40; m 62; c 3. ANALYTICAL CHEMISTRY, PHARMACEUTICAL CHEMISTRY. Educ: Baker Univ, BS, 62. Prof Exp: Chemist agr chem, Chemagro Corp, Bayer, WGer, 62-64; chemist petrol chem, Mobil Oil Corp, 65-66; SR RES CHEMIST PHARM CHEM, BAYVET CORP, BAYER, W GER, 66- Mem: Am Chem Soc; Appl Spectros Soc. Res: Analytical method development of animal health pharmaceutical products oriented to instrumental analysis such as gas-chromatography, liquid-liquid chromatography and ultra-violet visible spectroscopy. Mailing Add: Bayvet Corp 12707 W 63rd St Shawnee KS 66216

LEAVELL, BYRD STUART, b Washington, DC, Dec 29, 10; m 39; c 3. MEDICINE. Educ: Va Mil Inst, BS, 31; Univ Va, MD, 35. Prof Exp: From instr to assoc prof, 38-54, asst dean, 58-61, chmn dept internal med, 66-68, PROF MED, SCH MED, UNIV VA, 54- Mem: Am Soc Hemat; Int Soc Hemat; AMA; Am Clin & Climat Asn; fel Am Col Physicians. Res: Hematology. Mailing Add: Dept of Med Univ of Va Hosp Charlottesville VA 22903

LEAVELL, ULLIN WHITNEY, JR, b Wu Chow, China, Nov 10, 22; US citizen; m 53; c 2. DERMATOLOGY, PATHOLOGY. Educ: Vanderbilt Univ, BS, 42; Duke Univ, MD, 45. Prof Exp: Internship, Duke Univ, 45-46; chief dermat resident, Western Reserve Univ Hosp, 48-49 & 50-51, dermat preceptor, Univ, 49-50; PROF MED & CHMN DIV DERMAT, MED CTR, UNIV KY, 61- Concurrent Pos: NIH grant, Med Ctr, Univ Ky, 64-70. Honors & Awards: Bronze Award, Am Acad Dermat, 70. Mem: Am Acad Dermat; Am Asn Prof Dermat; AMA; NY Acad Sci; Soc Invest Dermat. Res: Viruses of the skin and cutaneous carcinogens. Mailing Add: 807 S Limestone St Lexington KY 40508

LEAVENS, PETER BACKUS, b Summit, NJ, June 20, 39; m 58; c 1. MINERALOGY. Educ: Yale Univ, BA, 61; Harvard Univ, MA, 64, PhD, 66. Prof Exp: Res assoc, Dept Mineral Sci, Smithsonian Inst, 65-67; asst prof, 67-70, ASSOC PROF GEOL, UNIV DEL, 70- Concurrent Pos: Res assoc, Smithsonian Inst, 67- Mem: AAAS; Mineral Soc Am. Res: Descriptive mineralogy; description of new mineral species; conditions of mineral occurrence and stability; mineralogy and geochemistry of pegmatites; phosphate mineralogy; carbonate metamorphism. Mailing Add: Dept of Geol Univ of Del Newark DE 19711

LEAVER, FREDERICK WILSON, b Aspen, Colo, 17; m 45; c 5. BIOCHEMISTRY. Educ: Western Reserve Univ, PhD(biochem), 51. Prof Exp: NIH fel, 51-52; from instr to asst prof biochem, Univ Pa, 52-60; res biochemist, Vet Admin Hosp, Denver, 60-68; hon asst prof pharmacol, 60-68, RES ASSOC PHARMACOL, MED CTR, UNIV COLO, DENVER, 68- Mem: AAAS; Am Chem Soc; NY Acad Sci. Res: Reproduction sterility and fertility. Mailing Add: Dept of Pharmacol Univ of Colo Med Ctr Denver CO 80220

LEAVITT, BENJAMIN BURTON, b Brookline, Mass, Apr 5, 06; m; c 2. MARINE BIOLOGY. Educ: Dartmouth Col, AB, 29; Harvard Univ, PhD(biol), 37. Prof Exp: Instr zool, Dartmouth Col, 29-31; master biol chem & gen sci, Berkshire Sch, 34-37; biologist, State Div Fish & Game, Mass, 37-38; head dept sci, physics, chem & biol, Asheville Sch, 38-41 & Trinity Sch, 41-42; asst prof zool, 46-54, ASSOC PROF ZOOL, UNIV FLA, 54- Concurrent Pos: Fel, Oceanog Inst, Woods Hole, 33-35, res staff mem, 53- Mem: AAAS; Wilson Ornith Soc; Am Soc Mammal; Wildlife Soc; Am Ornith Union. Res: Quantitative vertical distribution of bathypelagic macrozooplankton in deep water. Mailing Add: Dept of Zool Univ of Fla Gainesville FL 32603

LEAVITT, CHRISTOPHER PRATT, b Boston, Mass, Nov 20, 27; m 59; c 5. PHYSICS. Educ: Mass Inst Technol, BS, 48, PhD(physics), 52. Prof Exp: Res assoc physics, Brookhaven Nat Lab, NY, 52-54, assoc physicist, 54-56; from asst prof to assoc prof physics, 56-65, actg chmn dept physics & astron, 58-60, PROF PHYSICS, UNIV NMEX, 65- Concurrent Pos: Consult, Res Directorate, Phyiscs Div, Kirtland AFB, 56-; directorate res & develop, Air Force Missile Develop Ctr, Holloman AFB, 56-60; mem particles & fields subcomt, NASA, 65-67; mem tech adv panel, Los Alamos Meson Physics Facility, 69-71. Mem: Am Phys Soc. Res: Nuclear and high energy physics; cosmic rays; space physics. Mailing Add: Dept of Physics & Astron Univ of NMex Albuquerque NM 87106

LEAVITT, FREDERICK CARLTON, b Albany, NY, Nov 14, 29; m 51; c 2. POLYMER CHEMISTRY. Educ: Rensselaer Polytech Inst, BChE, 51; State Univ NY, PhD, 56. Prof Exp: Res chemist, Hercules Powder Co, 51-53; res chemist, 56-63, mgr res placement, 63-64, LAB DIR, DOW CHEM USA, 64- Mem: Am Chem Soc; NY Acad Sci. Res: Organometallic chemistry; organic chemistry of polymers. Mailing Add: 4405 Arbor Dr Midland MI 48640

LEAVITT, JOHN ADAMS, b Lewis, Colo, Dec 8, 32; m 55; c 5. PHYSICS. Educ: Univ Colo, BA, 54, Harvard Univ, MA, 56, PhD, 60. Prof Exp: From asst prof to assoc prof, 60-71, PROF PHYSICS, UNIV ARIZ, 71- Res: Atomic physics. Mailing Add: Dept of Physics Univ of Ariz Tucson AZ 85721

LEAVITT, JULIAN JACOB, b Boston, Mass, Sept 4, 18; m 43; c 3. CHEMISTRY. Educ: Harvard Univ, AB, 39, AM, 40, PhD(org chem), 42. Prof Exp: Res chemist, Nat Defense Res Comt, Harvard Univ, 42 & Univ Pa, 42-44; res chemist, Calco Chem Div, 44-54, Res Div, 54-58, Org Chem Div, 58-64, mgr explor res, 64-69, tech dir decision making systs dept, 69-70, asst to mgr com develop, 70-74, MGR LICENSING & TECHNOL, AM CYANAMID CO, 74- Concurrent Pos: Ed, Sect 40, Chem Abstr Serv, 61-; mem, Adv Bd Mil Personnel Supplies & Comt Textile Dyeing & Finishing, Nat Acad Sci-Nat Res Coun, 63-68. Mem: AAAS; Com Develop Asn; Am Chem Soc; Am Asn Textile Chemists & Colorists; Licensing Exec Soc. Res: Dyes; applied photochemistry. Mailing Add: Org Chem Div American Cyanamid Co Bound Brook NJ 08805

LEAVITT, LEWIS A, b Epping, NDak, May 13, 19; m 46; c 2. BIOLOGY, MEDICINE. Educ: Loyola Univ, Calif, BS, 41; St Louis Univ, MD, 45; Am Bd Phys Med & Rehab, dipl, 53. Prof Exp: Resident phys med & rehab serv, Vet Admin Hosps, New Orleans, 48-51, asst chief, Houston, 51-52, chief, 52-59; PROF PHYS MED & REHAB & CHMN DEPT PHYS MED, BAYLOR COL MED, 59-; PROF, TEX WOMAN'S UNIV, 63-; PROF BIOMED ENG, TEX A&M UNIV, 72- Concurrent Pos: Consult, Tex Inst Res & Rehab, 59, mem active med staff, 59-; dir dept phys med, Methodist Hosp; mem consult staff, St Luke's Episcopal & Tex Children's Hosps, chief phys med & rehab, 69-; chief phys med & rehab serv & sr attend, Ben Taub Gen & Jefferson David Hosps. Mem: Fel AMA; Am Asn Electromyog & Electrodiag; Am Asn Rehab Ther; Am Cong Rehab Med (pres, 65-66); Am Acad Phys Med & Rehab. Res: Kinesiology; physical medicine; rehabilitation; electrodiagnosis; prostheses; ortheses. Mailing Add: Dept of Phys Med Baylor Col of Med Houston TX 77025

LEAVITT, MILO DAVID, JR, b Beloit, Wis, June 24, 15. MEDICINE, SCIENCE ADMINISTRATION. Educ: Univ Wis, BA, 38; Univ Pa, MD, 40; Univ Minn, MSc, 48; Harvard Univ, MPH, 59. Prof Exp: Mayo Clin fel, Univ Minn, 45-49; asst chief perinatal res br, Nat Inst Neurol Dis & Blindness, 59-62, head spec int progs sect, Off Int Res, 62-66, dep asst secy sci & pop, off secy, Dept Health, Educ & Welfare, 66-67, dir off prog planning, 67-68, DIR FOGARTY INT CTR ADVAN STUDIES HEALTH SCI, 68- Concurrent Pos: Del, World Health Assembly, Geneva, 72, 73 & 74. Mem: Am Diabetes Asn; Asn Am Med Cols; Indust Med Asn; Am Heart Asn; Am Pub Health Asn. Res: Government administration; internal medicine; cardiology. Mailing Add: Fogarty Int Ctr Bldg 31 Rm B2C02 NIH Bethesda MD 20014

LEAVITT, RICHARD IRWIN, b Brooklyn, NY, Mar 22, 30; m 53; c 4. MICROBIOLOGY. Educ: Brooklyn Col, BA, 54; Univ Pa, PhD(microbiol), 58. Prof Exp: Res fel, Harvard Med Sch, 58-60; res assoc, Biol Lab, L I Biol Asn, 60-62; SR RES BIOCHEMIST, CENT RES DIV LAB, MOBIL OIL CORP, PRINCETON, NJ, 62- Concurrent Pos: Res assoc, Ethyl Corp Cent Res Div, 74-75. Mem: Am Soc Microbiol; Am Chem Soc. Res: Bacterial physiology; intermediary metabolism; biosynthesis of amino acids; single-cell protein; enzyme oxidation of hydrocarbons; industrial fermentations. Mailing Add: 306 Arborlea Ave Morrisville PA 19067

LEAVITT, WENDELL W, b North Conway, NH, Jan 15, 38; m 59; c 4. ENDOCRINOLOGY, REPRODUCTIVE PHYSIOLOGY. Educ: Dartmouth Col, AB, 59; Univ NH, MS, 61, PhD(zool), 63. Prof Exp: Res analyst zool, Agr Exp Sta, Univ NH, 60-63, res assoc endocrinologist & instr zool, 63-64; asst prof, 64-67, from asst prof to assoc prof physiol & dir grad studies, Col Med, 67-75, PROF PHYSIOL, COL MED, UNIV CINCINNATI, 75- Concurrent Pos: Spec fel endocrinol & reprod physiol, Univ Wis, 67-68; grants, NSF, Pop Coun & NIH; vis scientist, Ctr Pop Res & Studies in Reprod Biol, Vanderbilt Univ, 72-73; consult, Merrell Nat Labs, Cincinnati, Oh. Mem: Sigma Xi; Am Soc Zool; Endocrine Soc; Soc Study Reprod; Am Physiol Soc. Res: Mechanism of pituitary function in relation to gonadotrophin secretion; control of female reproductive cycle; estrogens and pituitary function; aging and the reproductive system; steroid hormone receptor systems. Mailing Add: Dept of Physiol Univ of Cincinnati Col of Med Cincinnati OH 45267

LEAVITT, WILLIAM GRENFELL, b Omaha, Nebr, Mar 19, 16; m 41; c 3. ALGEBRA. Educ: Univ Nebr, AB, 37, MA, 38; Univ Wis, PhD(math), 47. Prof Exp: Actg instr math, Univ Wis, 46; from instr to assoc prof, 47-56, chmn dept, 54-64, PROF MATH, UNIV NEBR-LINCOLN, 56- Concurrent Pos: NSF fel, 59-60; Univ Nebr Res Coun vis fel, Leeds, Eng, 73. Mem: Am Math Soc; Math Asn Am. Res: Ring theory; theory of modules; theory of radicals. Mailing Add: Dept of Math Univ of Nebr Lincoln NE 68508

LEAVY, THOMAS A, b Turkey City, Pa, Sept 19, 34; m 58; c 1. EARTH SCIENCE. Educ: Slippery Rock State Col, BS, 60; Pa State Univ, MS, 62; Univ Pittsburgh, PhD, 71. Prof Exp: Asst prof geog, Glassboro State Col, 63-64; assoc prof, 64-71, PROF GEOG & EARTH SCI, CALIF STATE COL, PA, 71- Mem: Asn Am Geogr; Am Meteorol Soc. Res: Weather and climate. Mailing Add: Dept of Geog & Earth Sci Calif State Col California PA 15419

LEBAR, FRANK M, anthropology, see 12th edition

LEBARON, FRANCIS NEWTON, b Framingham, Mass, July 26, 22; m 53; c 1. BIOCHEMISTRY. Educ: Mass Inst Technol, BS, 44; Boston Univ, MA, 48; Harvard Univ, PhD(biochem), 51. Prof Exp: Asst biochemist, McLean Hosp, Mass, 52-53 & 54-57, assoc biochemist, 57-64; assoc prof, 64-69, PROF BIOCHEM, SCH MED, UNIV NMEX, 69-, CHMN DEPT, 71- Concurrent Pos: USPHS fel, McLean Hosp, Waverley, Mass, 51-52 & Maudsley Hosp, 53-54; res assoc, Harvard Med Sch, 56-59, assoc, 59-64, tutor, Harvard Univ, 57-64; vis scholar, Mass Inst Technol, 74-75. Mem: AAAS; Am Chem Soc; Am Soc Biol Chem; Brit Biochem Soc. Res: Biochemistry of the nervous system, especially the chemistry of proteins and lipids and their nervous complexes as they occur in mammalian nervous tissues. Mailing Add: 1713 Morningside Dr NE Albuquerque NM 87110

LEBARON, HOMER MCKAY, b Barnwell, Alta, May 13, 26; US citizen; m 52; c 6. PESTICIDE CHEMISTRY. Educ: Utah State Univ, BS, 56, MS, 58; Cornell Univ, PhD(chem). Prof Exp: Plant physiologist, Va Truck Exp Sta, 60-63; mem field res & develop, Geigy Chem Corp, 64; group leader herbicide res, 64-75, SR STAFF SCIENTIST, CIBA-GEIGY CORP, 75- Mem: Weed Sci Soc Am; Am Soc Agron; Am Soc Plant Physiologists; Am Chem Soc. Res: Field, greenhouse and laboratory studies on a wide variety of basic and applied research on control of weeds. Mailing Add: Agr Div Ciba-Geigy Corp PO Box 11422 Greensboro NC 27409

LEBARON, MARSHALL JOHN, b Spokane, Wash, Sept 22, 20; m 48; c 2. FIELD CROPS. Educ: Univ Idaho, BS, 47, MS, 50. Prof Exp: Asst agronomist, 47-50, SUPT TWIN FALLS BR EXP STA, UNIV IDAHO, 50-, ASSOC & EXTEN AGRONOMIST, 65-, RES PROF AGRON, 70-, SUPT KIMBERLY RES & EXTEN CTR & EXTEN PROF, 74- Mem: Sigma Xi. Res: Bean production, including variety, fertility, population and damage assessment studies; breeding for disease and plant improvement in dry edible beans. Mailing Add: Twin Falls Br Exp Sta Rte 1 Kimberly ID 83341

LEBARON, ROBERT (FRANCIS), b Binghamton, NY, Oct 31, 91; m 26. CHEMISTRY. Educ: Union Col, NY, BS, 15; Princeton Univ, MS, 17; Nat Sch Mines, Paris, grad, 19. Hon Degrees: DSc, Union Col, NY & Thiel Col, 54. Prof Exp: Asst chem, Princeton Univ, 14-16; res chemist, Arthur D Little, Inc, Mass, 19-21, res sales, 21-23, coordr sci & tech activities, 23-25, asst to pres, 24-25; tech vpres, Petrol Chem Corp & vpres, Nat Distillers Prod Corp, NY, 26-32; sales dir, Va Smelting Co, 37-41, dir res & develop, 45-48, dir, 48-49; dep to Secy Defense in atomic energy, 49-54; mem vis comt, Brookhaven Nat Lab, 55-57; managing dir, LeBaron Assocs, 56-72, PRES, LeBARON FOUND, 65- Concurrent Pos: Mem planning comt, Va Smelting Co, 38-41, exec comt, 48-49; consult, War Dept, 42-44; tech consult, Oronite Chem Co, 45-49; mem chem res comt, Calif Res Corp, 45-49; mem adv bd dirs, Southern Res Inst, Ala, 47-49; chmn mil liaison comt, AEC, 49-54, consult to chmn comn, 54-58; chmn comt atomic energy, Res & Develop Bd, 49-54; indust consult, 54. Mem: Am Chem Soc. Res: Chemicals from petroleum; atomic energy. Mailing Add: Suite I-842 Sheraton Park Hotel Washington DC 20008

LEBEAU, JACK BERTRAM, b Vulcan, Alta, Nov 29, 15; m 41; c 3. PLANT PATHOLOGY. Educ: Univ Alta, BSA, 50; Univ Wis, MSc, 51, PhD, 53. Prof Exp: Plant pathologist, Sci Serv, 53-60, HEAD PLANT PATH SECT, AGR RES STA, CAN DEPT AGR, 60- Mem: Can Phytopath Soc. Res: Forage crop diseases. Mailing Add: Dept of Plant Path Agr Res Sta Can Dept of Agr Lethbridge AB Can

LE BEAU, LEON JOSEPH, b Kankakee, Ill, Dec 30, 19; m 48; c 5. MICROBIOLOGY, IMMUNOLOGY. Educ: Univ Ill, PhD(bact), 52. Prof Exp: From instr bact to assoc prof microbiol, Univ Ill, Col Med, 52-63; vis prof, Fac Med, Univ Chiengmai, 63-66; ASSOC DIR LABS, UNIV HOSP & PROF PATH & MICROBIOL, UNIV ILL COL MED, 66- Mem: AAAS; Biol Photog Asn (vpres & pres elect, 75). Res: Medical bacteriology; epidemiology. Mailing Add: Dept of Microbiol Univ of Ill Col of Med Chicago IL 60612

LEBEL, JACK LUCIEN, b Montreal, Que, Sept 16, 33; US citizen; m 60; c 2. RADIOLOGY, RADIATION BIOLOGY. Educ: Univ Montreal, DVM, 58; Colo State Univ, MS, 66, PhD(radiation biol), 67; Am Col Vet Radiol, dipl, 68. Prof Exp: Asst prof vet med, Univ Montreal, 61-64, assoc prof radiol, 67-68; assoc prof radiol & radiation biol, 68-73, asst dean curric col, 70-72, PROF RADIOL & RADIATION BIOL, COL VET MED & BIOMED SCI, COLO STATE UNIV, 73- Concurrent Pos: Consult, Orthop Found for Animals, 69- Mem: Am Vet Med Asn; Am Vet Radiol Soc. Res: Bone pathology; biological effects of plutonium contamination; nuclear medicine. Mailing Add: Vet Hosp Colo State Univ Ft Collins CO 80521

LEBEL, JEAN EUGENE, b Can, Mar 21, 22. MATHEMATICS. Educ: McGill Univ, BSc, 44; Univ Toronto, MA, 50, PhD(appl math), 58. Prof Exp: Theoret physicist, Newmont Explor, Ltd, Ariz, 52-54; lectr, McGill Univ, 55-57; from asst prof to assoc prof math, Georgetown Univ, 58-65; ASSOC PROF MATH, UNIV TORONTO, 65- Mem: Am Math Soc; Can Math Cong. Res: Analysis; applied mathematics. Mailing Add: Dept of Math Univ of Toronto Toronto ON Can

LEBEL, NORMAN ALBERT, b Augusta, Maine, Mar 22, 31; m 52; c 3. ORGANIC CHEMISTRY. Educ: Bowdoin Col, AB, 52; Mass Inst Technol, PhD, 57. Prof Exp: Chemist, Merck & Co, Inc, 52-54; from asst prof to assoc prof, 57-64, PROF CHEM, WAYNE STATE UNIV, 64-, CHMN DEPT, 71- Concurrent Pos: Sloan Found fel, 61-65. Mem: AAAS; Am Chem Soc; The Chem Soc. Res: Stereochemistry and mechanism of elimination reactions; chemistry of nitrones and isoxazolidines; additions to olefins; bridged polycyclic molecules. Mailing Add: 277 Chem Bldg Wayne State Univ Detroit MI 48202

LEBEL, RONALD GUY, b Edmundston, NB, Feb 10, 32; m 64; c 1. PHYSICAL CHEMISTRY, CHEMICAL ENGINEERING. Educ: NS Tech Col, BEng, 55; Mass Inst Technol, SM, 56; McGill Univ, PhD(phys chem), 62. Prof Exp: Res engr, Fraser Pulp & Paper Co Ltd, 58-59; demonstr, McGill Univ, 59-62; sr develop engr, Anglo Paper Prod Ltd, 63-65; res & develop proj coordr pulp & paper, 65-69, TECH DIR, PAPETERIE REED LTE, 69- Mem: Sr mem Can Pulp & Paper Asn. Res: Planning, organization and coordination of research and development of projects related to the manufacture of newsprint and other paper products. Mailing Add: Papeterie Reed Ltd 10 Bldes Capucins, Quebec PQ Can

LEBEN, CURT (CHARLES), b Chicago, Ill, July 7, 17; m 44; c 2. PLANT PATHOLOGY. Educ: Ohio Univ, BS, 40; Univ Wis, PhD(plant path), 46. Prof Exp: Asst bot, Ohio Univ, 39-40; asst, Univ Wis, 42, asst plant path, 42-46, res assoc, 46-49, asst prof, 49-55; plant pathologist, Eli Lilly & Co, Ind, 55-57, head agr res labs, 57-59; prof bot & plant path & assoc chmn dept, 59-67, actg chmn, 67-68, PROF PLANT PATH, AGR RES & DEVELOP CTR, OHIO STATE UNIV, 67- Mem: AAAS; Am Phytopath Soc. Res: Antibiotics and antibiosis in relation to plant diseases; microbiology; epiphytic microorganisms; bacterial diseases; forest tree diseases. Mailing Add: Dept of Plant Path Ohio Agr Res & Develop Ctr Wooster OH 44691

LEBENSOHN, ZIGMOND MEYER, b Kenosha, Wis, Sept 8, 10; m 40; c 4. PSYCHIATRY. Educ: Northwestern Univ, BS, 30, MB, 33, MD, 34; Am Bd Psychiat & Neurol, dipl, 41. Prof Exp: Instr neurol, Med Sch, George Washington Univ, 36-41; PROF CLIN PSYCHIAT, MED SCH & MEM STAFF, UNIV HOSP, GEORGETOWN UNIV, 41-; CHIEF DEPT PSYCHIAT, SIBLEY MEM HOSP, 57- Concurrent Pos: Med officer, St Elizabeth's Hosp, 35-39; mem staff, Doctors Hosp, 41-74; consult, Vet Admin, 46-49, US Naval Hosp, Bethesda, Md, 52-, US Info Agency, 58-61, Walter Reed Army Hosp, Washington, DC, 58-66 & NIMH, 68-; consult med adv panel, Fed Aviation Agency, 61-66. Mem: Fel Am Psychiat Asn; fel AMA; Am Psychopath Asn; Asn Res Nerv & Ment Dis. Res: Legal aspects of psychiatry; psychiatric hospital design; trans-cultural psychiatry. Mailing Add: 2015 R St NW Washington DC 20009

LEBERFELD, DORIS TREPEL, speech pathology, deceased

LEBERMAN, PAUL R, b NY, Mar 1, 04; m 32. UROLOGY. Educ: NY Univ, BS, 25; Univ Pa, MS, 27, MD, 31; Am Bd Urol, dipl, 42. Prof Exp: Assoc prof clin urol & asst prof urol, Grad Sch Med, 51-66, PROF CLIN UROL, UNIV PA, 66-, CHIEF UROL SURG, OUTPATIENT DEPT, HOSP, 51- Concurrent Pos: Chief urol, Philadelphia Gen Hosp, 59-; consult, US Naval Hosp, Philadelphia, 60- Mem: Fel Am Col Surg; fel Royal Soc Med; fel Int Soc Urol; fel Am Acad Pediat; fel Pan-Pac Surg Asn. Res: Urological surgery. Mailing Add: Univ Hosp 3400 Spruce St Philadelphia PA 19104

LEBERMANN, KENNETH WAYNE, b Davenport, Iowa; m 62; c 2. FOOD SCIENCES. Educ: Univ Ill, Urbana, BS, 59, PhD(food sci), 64; Univ Calif, Davis, MS, 61. Prof Exp: Scientist, CPC Int, 64-66; group leader cereals, 66-67, sr group leader bakery prod, 67-69, sect mgr bakery prod, 69-70, sect mgr pet foods, 70-75, MGR PET FOOD RES, QUAKER OATS CO, 75- Mem: Inst Food Technologists; Sigma Xi. Res: Industrial research related to new product development, pet food research. Mailing Add: Quaker Oats Co 617 W Main St Barrington IL 60010

LEBHERZ, HERBERT G, b San Francisco, Calif, July 27, 41; m 71. BIOCHEMISTRY. Educ: San Francisco State Univ, BA, 64, MA, 66; Univ Wash, PhD(biochem), 70. Prof Exp: From res assoc to sr res assoc cell biol, Swiss Fed Inst Technol, 71-75; cancer res sci biochem, Roswell Park Mem Inst, 75-76; ASSOC PROF BIOCHEM, SAN DIEGO STATE UNIV, 76- Res: Elucidation of the mechanisms involved in the regulation of protein synthesis and protein degradation in developing adult and ageing organisms. Mailing Add: Dept of Chem San Diego State Univ San Diego CA 92115

LEBLANC, ADRIAN DAVID, b Salem, Mass, May 21, 40; m 66; c 2. RADIOLOGICAL PHYSICS, NUCLEAR MEDICINE. Educ: Univ Mass, BA, 62; Iowa State Univ, MS, 66; Univ Kans, PhD(radiation biophys), 72; Am Bd Health Physics, cert; Am Bd Radiol, cert in nuclear med physics. Prof Exp: HEALTH PHYSICIST, HOSP & RADIOL PHYSICIST, DEPT NUCLEAR MED, METHODIST HOSP, 66-; ASST PROF MED, BAYLOR COL MED, 72- Concurrent Pos: Radiation physicist, Vet Admin Hosp, 67- Mem: Health Physics Soc; Soc Nuclear Med; Am Asn Physicists in Med; Am Col Radiol; Sigma Xi. Res: Coronary blood flow; neutron activation analysis; x-ray fluorescence; health physics. Mailing Add: 7810 Braeburn Valley Dr Houston TX 77036

LEBLANC, ARTHUR EDGAR, b Moncton, NB, Sept 29, 23; US citizen; m 52; c 2. GEOLOGY, PALYNOLOGY. Educ: Univ Mass, Amherst, BS, 52, MS, 54. Prof Exp: Paleontologist, Shell Oil Co, Tex, 54-62 & Calif, 62-65; sr res scientist, Res Ctr, Pan Am Petrol Corp, Okla, 65-69; PROJ PALYNOLOGIST, GULF OIL RES & DEVELOP CO, 69- Mem: AAAS; Am Asn Petrol Geologists; Paleont Soc; Soc Econ Paleontologists & Mineralogists; Am Asn Stratig Palynologists. Res: Stratigraphic Paleozoic palynology; Mesozoic and Cenozoic stratigraphic palynology. Mailing Add: Gulf Oil Res & Develop Co HTSC PO Box 36506 Houston TX 77036

LEBLANC, FABIUS, b Nashua, NH, Jan 27, 18; Can citizen. BRYOLOGY, ECOLOGY. Educ: Univ Montreal, BA, 46, cert, 50, PhD(ecol), 60; Iowa State Univ, MSc, 57. Prof Exp: Instr biol, Norm Sch, Granby, Que, 50-57; from asst prof to assoc prof, 60-72, PROF BIOL, UNIV OTTAWA, 72- Mem: Am Bryol & Lichenological Soc; Ecol Soc Am; Air Pollution Control Asn; Int Asn Plant Taxon; Can Bot Asn. Res: Ecology of mosses and lichens; air pollution and epiphytes. Mailing Add: Dept of Biol Univ of Ottawa Ottawa ON Can

LE BLANC, FLOYD JOSEPH, b Medelia, Minn, May 30, 97; m 31. PHARMACY. Educ: SDak State Col, BS, 24, MS, 27; Purdue Univ, PhD(pharm), 38. Prof Exp: From asst to prof pharmaceut chem, 22-40, from actg dean to dean pharm, 40-67, prof pharmaceut chem, 67-70, EMER PROF PHARMACEUT CHEM, S DAK STATE UNIV, 70- Mem: Am Pharmaceut Asn. Res: Toxic principles of red squill; chenopodium as a worm eradicator. Mailing Add: 624 12th Ave Brookings SD 57006

LEBLANC, FRANCIS ERNEST, b North Sydney, NS, June 10, 35; m 61; c 3. NEUROPHYSIOLOGY, NEUROSURGERY. Educ: St Francis Xavier Univ, BSc, 55; Univ Ottawa, MD, 59; Univ Montreal, MSc, 62, PhD(neurophysiol), 64; FRCS(C), 68. Prof Exp: Intern, Montreal Gen Hosp, Que, 59-60, jr asst resident surg, Queen Mary Vet Hosp, Montreal, 60-61; res asst, Neurol Sci Lab, Univ Montreal, 61-62, lectr physiol, Univ, 61-64; demonstr neurol, McGill Univ, 64-67, lectr neurosurg, 67-68, asst prof, 68-70; assoc prof, 71-74, ASSOC PROF SURG, UNIV CALGARY, 74-; CHIEF, DIV NEUROSURG, FOOTHILLS HOSP, CALGARY, 74- Concurrent Pos: Med Res Coun fel, Univ Montreal, 61-64, res fel, 62-64; clin fel & chief resident, Montreal Neurol Inst, 66-67, res scholar, 67-70, asst neurosurgeon, 68; consult, Queen Mary Vet Hosp, Montreal, 67; vis neurosurgeon, Royal Victoria Hosp, Montreal, 68; consult neurosurgeon, Foothills Hosp, 71-74. Mem: Can Neurosurg Soc; Cong Neurol Surg; Asn Acad Surg; fel Am Col Surg. Res: Cerebrovascular physiology; epilepsy; movement disorders. Mailing Add: Div of Neurosurg Foothills Hosp Calgary AB Can

LEBLANC, GABRIEL, b Montreal, Que, June 24, 27; m 68. GEOPHYSICS, SEISMOLOGY. Educ: Univ Montreal, BA, 52; L'Immaculee-Conception, Montreal, LPh, 53; Boston Col, MSc, 58, SThL, 60; Pa State Univ, PhD(geophys), 66. Prof Exp: Res asst seismol, Pa State Univ, 63-66; assoc prof geophys, Laval Univ, 66-71; RES SCIENTIST, SEISMOL DIV, EARTH PHYSICS BR, DEPT ENERGY, MINES & RESOURCES, 71- Concurrent Pos: Res dir geophys observ, Jean-de-Brebeuf Col, 66-68. Mem: Am Geophys Union; Seismol Soc Am; Soc Explor Geophys. Res: Solid earth; crustal studies; spectral analysis of seismic waves; local seismicity. Mailing Add: Seismol Div Earth Physics Br Dept of Energy Mines & Resources Ottawa ON Can

LEBLANC, JACQUES ARTHUR, b Quebec, Que, Aug 23, 21; nat US; m 51; c 3. PHYSIOLOGY. Educ: Laval Univ, BA, 43, BSc, 47, PhD(physiol), 51. Prof Exp: Physiologist human physiol, Defence Res Bd, Can Dept Nat Defence, 49-56; PROF HUMAN PHYSIOL, FAC MED, LAVAL UNIV, 58- Mem: AAAS; Am Physiol Soc; Soc Exp Biol & Med; Fr-Can Asn Advan Sci. Res: Amines in stress conditions; tranquilizers; mast cells; basic and applied work in environmental physiology. Mailing Add: Dept of Physiol Fac of Med Laval Univ Quebec PQ Can

LEBLANC, JERALD THOMAS, b Baton Rouge, La, Mar 10, 43; m 68. PHOTOGRAPHIC SCIENCE, ORGANIC CHEMISTRY. Educ: Birmingham-Southern Col, BS, 65; Fla State Univ, PhD(org chem), 70. Prof Exp: SR CHEMIST, RES LABS, EASTMAN KODAK CO, 70- Mem: Am Chem Soc; Soc Photog Scientists & Engr. Res: Silver halide chemistry and physics; organic dye synthesis. Mailing Add: Eastman Kodak Co Res Labs 343 State St Rochester NY 14650

LE BLANC, JOHN ROGER, b Montreal, Can, Feb 7, 27; US citizen; m 48; c 7. ORGANIC POLYMER CHEMISTRY. Educ: Boston Col, BS, 50; Ohio State Univ, MS, 60. Prof Exp: Supvr anal chem, 51-52, supvr prod, 52-53, res chemist cent res, 53-59 & plastics res, 61-63, GROUP LEADER PLASTICS RES, MONSANTO CO, INDIAN ORCHARD, 63- Mem: Am Chem Soc. Res: Exploratory research in organic-aromatic chemistry; kinetics; photochemistry; free radical polymerization; condensation polymers; phenolic chemistry. Mailing Add: 34 Decorie Dr Wilbraham MA 01095

LEBLANC, LEONARD JOSEPH, b Moncton, NB, Nov 6, 37; m 59; c 3. SOLID STATE PHYSICS. Educ: St Joseph's Univ, NB, BSc, 59; Univ Notre Dame, PhD(physics), 64. Prof Exp: From asst prof to assoc prof physics, 64-71, vdean fac sci, 69-74, PROF PHYSICS, UNIV MONCTON, 71-, DEAN FAC SCI, 75- Mem: Optical Soc Am; Am Asn Physics Teachers; Can Asn Physicists. Res: Optical and photoelectric properties of metals in the vacuum ultraviolet. Mailing Add: Dept of Physics & Math Univ de Moncton Moncton NB Can

LE BLANC, MARCEL A R, b Gravelbourg, Sask, Mar 25, 29; m 62; c 3. PHYSICS. Educ: Univ Ottawa, BA, 49; Univ Sask, BA, 52, MA, 54; Univ BC, PhD(physics), 58. Prof Exp: Res assoc physics, Stanford Univ, 59-63; asst prof elec eng, Univ Southern Calif, 63-68; PROF PHYSICS, UNIV OTTAWA, 68- Concurrent Pos: Asst prof, San Jose State Col, 61-62; consult, Varian Assocs, 61-62 & Spectromagnetic Industs, 62-63; mem tech staff & consult, Aerospace Corp, 63-68. Mem: Am Phys Soc; Can Asn Physicists. Res: Photonuclear cross-sections; nuclear polarization; low temperature physics; superconductivity. Mailing Add: Dept of Physics Univ of Ottawa Ottawa ON Can

LEBLANC, NORMAN FRANCIS, b Boston, Mass, June 28, 26; m 52; c 4. ANALYTICAL CHEMISTRY, RESEARCH ADMINISTRATION. Educ: Tufts Univ, BS, 47; Mass Inst Technol, PhD(anal chem), 50. Prof Exp: Res chemist, 50-56, res chemist solid rocket res & develop, 56-68, dir develop, Polymers Dept, 68-73, DIR HOME FURNISHINGS DIV, HERCULES, INC, 73- Mem: Am Chem Soc. Mailing Add: Hercules Inc 910 Market St Wilmington DE 19899

LEBLANC, OLIVER HARRIS, JR, b Beaumont, Tex, Nov 14, 31; m 56; c 1. PHYSICAL CHEMISTRY. Educ: Rice Univ, BA, 53; Univ Calif, PhD, 57. Prof Exp: PHYS CHEMIST, RES & DEVELOP CTR, GEN ELEC CO, 57- Mem: Am Phys Soc; Am Chem Soc. Res: Organic semiconductors; dielectrics; membrane biophysics; biomedical instrumentation. Mailing Add: 1173 Phoenix Ave Schenectady NY 12308

LEBLANC, ROBERT BRUCE, b Alexandria, La, Jan 28, 25; m 68; c 7. INORGANIC CHEMISTRY. Educ: Loyola Univ, BS, 47; Tulane Univ, MS, 49, PhD(chem), 50. Prof Exp: Asst prof chem, Tex A&M Univ, 50-52; res specialist, Org Res Dept, Dow Chem Co, 52-63, sr textile specialist, 63-67; textile chem develop mgr, ADM Chem Div, Ashland Oil Co, 67-68; res mgr, Nat Cotton Coun, 68-70; PRES, LEBLANC RES CORP, 70- Concurrent Pos: Mem, Info Coun Fabric Flammability. Mem: Am Chem Soc; Am Asn Textile Chemists & Colorists; Am Soc Testing & Mat; Nat Fire Protection Asn. Res: Textile chemistry and phosphorus chemistry; flammability and fire retardance of textiles and plastics. Mailing Add: LeBlanc Res Corp 5454 Post Rd East Greenwich RI 02818

LEBLANC, ROBERT GEORGE, b Nashua, NH, Oct 28, 30; m 59; c 3. GEOGRAPHY. Educ: Univ NH, BA, 59; Univ Minn, MA, 62, PhD(geog), 68. Prof Exp: Res asst geophys, Air Force Cambridge Res Inst, 57-58; from instr to asst prof geog, 63-73, ASSOC PROF GEOG, UNIV NH, 73- Mem: Asn Am Geogr; Am Geog Soc; Can Asn Geogr. Res: Historical geography of New England. Mailing Add: Dept of Geog Univ of NH Durham NH 03824

LEBLANC, RUFUS JOSEPH, b Erath, La, Oct 12, 17; m 40; c 4. GEOLOGY. Educ: La State Univ, BS, 39, MS, 41. Prof Exp: Asst geologist, La State Univ, 41-43; geologist, Miss River Comn, US War Dept, 44-46, chief geol sect, 47-48; sr res geologist, 48-52, mgr dept geol res, 53-56, sr geologist, Tech Servs, 56-60, staff geologist, 60-63, SR STAFF GEOLOGIST, EXPLOR & PROD RES DIV, SHELL OIL CO, 63- Concurrent Pos: Mem comt fundamental res occurrence and recovery petrol, Am Petrol Inst, 52-57. Mem: Fel Geol Soc Am; Soc Econ Paleont & Mineral; Am Asn Petrol Geol. Res: Fundamental research in stratigraphy; structural geology and geological chemistry; exploration for oil and gas; quaternary geology. Mailing Add: Shell Develop Co 3737 Bellaire Blvd Box 481 Houston TX 77001

LEBLEU, RONALD EUGENE, b Lexington, Ky, Feb 26, 37; m 60; c 1. ORGANIC CHEMISTRY. Educ: Transylvania Col, BA, 57; Duke Univ, PhD(org chem), 61. Prof Exp: Res chemist, Org Chem Dept, 60-63, col rels rep, Employee Rels Dept, 63-67, res asst, 67-68, proj leader, 68-70, prod mgr, Indust Finishes Dept, 70-71, MGR CONSULT SERV GROUP, CENT RES & DEVELOP DEPT, E I DU PONT DE NEMOURS & CO, INC, 71- Mem: Am Chem Soc. Res: Nitrogen-containing heterocycles arising from aromatic cyclodehydration, synthesis and characterization of fluoropolymers; doctoral recruitment and college relations. Mailing Add: Cent Res & Develop Dept E I du Pont de Nemours & Co Inc Wilmington DE 19898

LEBLOND, CHARLES PHILIPPE, b Lille, France, Feb 5, 10; m 36; c 4. ANATOMY. Educ: Univ Nancy, Lic es S, 32; Univ Paris, MD, 34; Univ Montreal, PhD(iodine metab), 45; Univ Sorbonne, DSc, 45. Prof Exp: Asst histol, Med Sch Paris, 34-35; Rockefeller fel, Sch Med, Yale Univ, 36-37; asst, Lab de Synthese Atomique, Paris, 38-40; lectr histol & embryol, 41-43, from asst prof to assoc prof, 43-48, chmn dept, 57-75, PROF ANAT, McGILL UNIV, 48- Concurrent Pos: Assoc ed, J Histochem. Mem: Am Asn Anat; Am Asn Cancer Res; fel Royal Soc; fel Royal Soc Can. Res: Histological localization of vitamin C; uptake of iodine by thyroid; tracing of radio elements and labelled precursors of nucleic acids; proteins and glycoproteins by means of radioautography. Mailing Add: Dept of Anat McGill Univ PO Box 6070 Montreal PQ Can

LEBLOND, PAUL HENRI, b Quebec, Que, Dec 30, 38; m 63; c 3. PHYSICAL OCEANOGRAPHY. Educ: Laval Univ, BA, 57; McGill Univ, BSc, 61; Univ BC, PhD(physics), 64. Prof Exp: Nat Res Coun Can fel, Inst Meereskunde, Kiel, Ger, 64-65; from asst prof to assoc prof physics, Dept Physics & Inst Oceanog, 65-75, PROF PHYSICS & OCEANOG, INST OCEANOG, UNIV BC, 75- Concurrent Pos: Vis assoc prof, Simon Fraser Univ, 70; vis scientist, Inst Oceanology, USSR Acad Sci, Moscow, 73-74; assoc ed, Atmosphere, 75- Mem: Can Meteorol Soc. Res: Surface, internal, planetary waves; ocean currents; tides; estuarine circulation. Mailing Add: Inst of Oceanog Univ of BC Vancouver BC Can

LEBO, GEORGE ROBERT, b Chadron, Nebr, Sept 27, 37; m 58; c 2. PHYSICS, RADIO ASTRONOMY. Educ: Wheaton Col, BS, 59; Univ Ill, MS, 60; Univ Fla, PhD(physics), 64. Prof Exp: Res assoc radio astron, 64-65, ASST PROF ASTRON, UNIV FLA, 65- Mem: Am Astron Soc; Am Geophys Union; Am Phys Soc. Res: Study of decametric radiation from the planets, particularly Jupiter. Mailing Add: Dept of Physics & Astron Univ of Fla Gainesville FL 32601

LEBOFSKY, LARRY ALLEN, b Brooklyn, NY, Aug 31, 47; m 73. PLANETARY SCIENCES. Educ: Calif Inst Technol, BS, 69; Mass Inst Technol, PhD(earth & planetary sci), 74. Prof Exp: RES ASSOC PLANETARY SCI, JET PROPULSION LAB, 75- Mem: Am Astron Soc; Am Geophys Union. Res: Remote sensing of the visual, near infrared and thermal infrared spectra of asteroids and satellites for the study of composition; related studies of laboratory frost reflection spectra. Mailing Add: Jet Propulsion Lab MS 183-501 4800 Oak Grove Dr Pasadena CA 91103

LEBOUTON, ALBERT V, b La Salle, Ill, July 10, 37; m 59; c 3. MICROSCOPIC ANATOMY, CELL BIOLOGY. Educ: San Diego State Col, BS, 60; Univ Calif, Los Angeles, PhD(anat), 66. Prof Exp: Asst prof anat, Univ Calif, Los Angeles, 66-72, actg head dept, 72-74, ASSOC PROF ANAT, UNIV ARIZ, 72- Mem: AAAS; Am Inst Biol Sci; Fedn Am Sci; Am Soc Cell Biol; Am Asn Anat. Res: Radioautography and radiobiochemistry of protein metabolism and growth in the liver. Mailing Add: Dept of Anat Univ of Ariz Col of Med Tucson AZ 85724

LEBOVITZ, NORMAN RONALD, b New York, NY, Sept 27, 35; m 71. APPLIED MATHEMATICS, ASTROPHYSICS. Educ: Univ Calif, Los Angeles, AB, 56; Univ Chicago, MS, 57, PhD(physics), 61. Prof Exp: Moore instr math, Mass Inst Technol, 61-63; from asst prof to assoc prof, 63-69, PROF MATH, UNIV CHICAGO, 69- Concurrent Pos: Sloan Found fel, 67. Mem: Am Astron Soc; Am Math Soc; Soc Indust & Appl Math. Res: Stability theory; rotating fluid masses; singular perturbation theory. Mailing Add: Dept of Math Univ of Chicago Chicago IL 60637

LEBOVITZ, ROBERT MARK, b Scranton, Pa, May 6, 37. NEUROPHYSIOLOGY, BIOENGINEERING. Educ: Calif Inst Technol, BS, 59, MS, 60; Univ Calif, Los Angeles, PhD(neurophysiol), 67. Prof Exp: Mem tech staff, Hughes Aircraft Co, 59-62; resident consult, Dept Math, Rand Corp, 67; NSF fel, Sch Med, NY Univ, 67-69; res assoc neural models, Ctr Theoret Biol, State Univ NY Buffalo, 69; staff scientist auditory & visual info processing, Recognition Equip, Inc, 69-70; ASSOC PROF NEUROPHYSIOL, UNIV TEX HEALTH SCI CTR DALLAS, 70- Concurrent Pos: Consult, Dept Path & Eng Sci, Rand Corp, 69, Recognition Equip, Inc, 69, Neurosysts, Inc, 71, Dallas Epilepsy Asn, 71 & Energy Conversion Devices, 71; chief technical officer, Centra-Guard Inc & Neighborhood Coop Patrol, 69-71; NSF fel, Univ Tex Regents, 70-72; adj prof, Inst Technol, Southern Methodist Univ, 71-74; NIH grants, 71-76; Food & Drug Admin grant, 74-77. Mem: Am Physiol Soc; NY Acad Sci; Biophys Soc; Inst Elec & Electronics Eng; Soc Neurosci. Res: Neurophysiology of epilepsy and behavior; modification and control of behavior via drugs and implanted or extraneous brain stimulating arrays; microwave interactions with the nervous system; electronic medicine. Mailing Add: Dept of Physiol Univ of Tex Health Sci Ctr Dallas TX 75235

LEBOW, ARNOLD, b Detroit, Mich, Mar 7, 32; m 65. MATHEMATICS. Educ: Univ Mich, PhD(math), 62. Prof Exp: Asst prof math, NY Univ, 62-65 & Univ Calif, Irvine, 65-68; ASSOC PROF MATH, BELFER GRAD SCH SCI, YESHIVA UNIV, 68- Mem: Am Math Soc. Res: Functional analysis; operator theory. Mailing Add: Dept of Math Belfer Grad Sch of Sci Yeshiva Univ New York NY 10033

LEBOWITZ, ELLIOT, b Monticello, NY, June 19, 41; m 63; c 2. NUCLEAR MEDICINE, NUCLEAR CHEMISTRY. Educ: Columbia Univ, BA, 61, PhD(nuclear chem), 67. Prof Exp: Res assoc chem kinetics & asst chemist, Dept Chem, Brookhaven Nat Lab, NY, 67-69, assoc chemist, Dept Appl Sci, 69-75; ASST TO VPRES, NEW ENG NUCLEAR CORP, 75- Res: Nuclear technology. Mailing Add: New England Nuclear Corp North Billerica MA 01862

LEBOWITZ, JACOB, b Brooklyn, NY, Oct 20, 35. PHYSICAL CHEMISTRY, BIOCHEMISTRY. Educ: Brooklyn Col, BS, 57; Purdue Univ, PhD(phys chem), 62. Prof Exp: Res fel biophys chem, Calif Inst Technol, 62-66; from asst prof to assoc prof biochem, Syracuse Univ, 66-74; ASSOC PROF MICROBIOL, MED CTR, UNIV ALA, BIRMINGHAM, 74- Mem: AAAS; Am Chem Soc; Am Soc Microbiol; Biophys Soc. Res: Physical chemistry and molecular biology of nucleic acids and proteins in relation to their biological structure-function relationships, with particular emphasis on circular DNA from viral systems. Mailing Add: Dept of Microbiol Univ of Ala Med Ctr Birmingham AL 35294

LEBOWITZ, JACOB MORDECAI, b New York, NY, Mar 21, 36; m 65; c 2. NUCLEAR PHYSICS. Educ: Yeshiva Univ, BA, 57; Columbia Univ, MA, 60, PhD(physics), 65. Prof Exp: From instr to asst prof, 59-70, ASSOC PROF PHYSICS, BROOKLYN COL, 70- Mem: Am Phys Soc. Res: Nuclear forces; fission. Mailing Add: Dept of Physics Brooklyn Col Brooklyn NY 11210

LEBOWITZ, JOEL LOUIS, b Taceva, Czech, May 10, 30; nat US; m 53. PHYSICS. Educ: Brooklyn Col, BS, 52; Syracuse Univ, MS, 55, PhD(physics), 56. Prof Exp: NSF res fel, Yale Univ, 56-57; asst prof physics, Stevens Inst Technol, 57-59; from asst prof to assoc prof, 57-65, PROF PHYSICS, GRAD SCH SCI, YESHIVA UNIV, 65- Concurrent Pos: Vis prof, Sch Med, Cornell Univ. Mem: AAAS; NY Acad Sci; Am Phys Soc. Res: Statistical mechanics of equilibrium and nonequilibrium processes; theory of liquids; biomathematics and mathematical economics. Mailing Add: Dept of Physics Belfer Grad Sch of Sci Yeshiva Univ New York NY 10033

LEBOWITZ, MICHAEL DAVID, b New York, NY, Dec 21, 39; m 60; c 3. EPIDEMIOLOGY, PULMONARY DISEASES. Educ: Univ Calif, Berkeley, AB, 61, MA, 65; Univ Wash, PhC, 69, PhD(epidemiol), 71. Prof Exp: Pub health statistician, Alameda County Health Dept, 62-63; biostatistician, Calif Dept Pub Health, 67; res assoc environ health, Univ Wash, 67-71; asst prof internal med, 71-75, ASSOC PROF INTERNAL MED, COL MED, UNIV ARIZ, 75-, ASST DIR SPECIALIZED CTR RES, 71-, ASST DIR DIV RESPIRATORY SCI, 74-, ASST DIR TOXICOL PROG COMT, 74- Concurrent Pos: Partic, NSF-Japan Soc Promotion Sci Coop Sci Group in Air Pollution, 69-70; ad hoc reviewer, Nat Air Pollution Control Admin, 69-71; consult, NIH & Nat Heart & Lung Inst, 72-; mem dist adv coun, Pima County Air Pollution Control, 75-77; mem environ coun, Univ Ariz, 75-; mem environ health effects comt, Ariz State Dept Health, 75-; mem epidemiol study sect, NIH, 76-79. Mem: Int Epidemiol Asn; Asn Teachers Prev Med; Am Thoracic Soc; Biomet Soc; Soc Epidemiol Res. Res: Pulmonary and chronic disease epidemiology; etiology and natural history of pulmonary diseases and other chronic diseases; air pollution health effects research. Mailing Add: Ariz Med Ctr Sect Pulmonary Dis Univ Ariz Col of Med Tucson AZ 85724

LEBOY, PHOEBE STARFIELD, b Brooklyn, NY, July 29, 36; m 57. BIOCHEMISTRY. Educ: Swarthmore Col, AB, 57; Bryn Mawr Col, PhD(biochem), 62. Hon Degrees: MA, Univ Pa, 71. Prof Exp: Res assoc biochem, Bryn Mawr Col, 61-63; res assoc, Sch Med, 63-66, from instr to asst prof, 65-70, ASSOC PROF BIOCHEM, SCH DENT MED, UNIV PA, 70- Concurrent Pos: NATO fel, Weizmann Inst Sci, 66-67; USPHS res grant, Univ Pa, 68-, NIH res career develop award, 71- Mem: Am Soc Biol Chem; Am Soc Microbiol; Am Soc Cell Biol. Res: Control of protein and nucleic acid metabolism; RNA methylation; bacterial genetics. Mailing Add: Dept of Biochem Sch Dent Med Univ of Pa Philadelphia PA 19174

LEBRA, TAKIE SUGIYAMA, b Ito, Japan, Feb 6, 30; US citizen; m 63. ANTHROPOLOGY. Educ: Gakushuin Univ, Tokyo, BA, 54; Univ Pittsburgh, MA, 60, PhD(sociol), 67. Prof Exp: Lectr anthrop, 68-71, RESEARCHER, SOCIAL SCI RES INST, UNIV HAWAII, HONOLULU, 69-, ASSOC PROF ANTHROP, 71- Mem: Fel Am Anthrop Asn; Japanese Soc Ethnol; Am Sociol Asn. Res: Social system theory; social movements; religion; ethnopsychiatry; Japanese culture and behavior; Japanese-American acculturation. Mailing Add: Dept of Anthrop Univ of Hawaii Honolulu HI 96822

LEBRA, WILLIAM PHILIP, b St Paul, Minn, Sept 2, 22; m 63. ANTHROPOLOGY. Educ: Univ Minn, BA, 48, MA, 49; Harvard Univ, PhD(anthrop), 58. Prof Exp: From instr to asst prof anthrop, Univ Pittsburgh, 57-61; vis scholar, East-West Ctr, 61-62, assoc prof anthrop & Asian studies, 62-64, dir, Social Sci Res Inst, 62-70, PROF ANTHROP, UNIV HAWAII, HONOLULU, 64-, PROG DIR CULT & MENT HEALTH, SOCIAL SCI RES INST, 70- Concurrent Pos: NIMH grant, Okinawa, 59-62 & Hawaii & Asia, 67-73. Mem: AAAS; fel Am Anthrop Asn; fel Royal Anthrop Inst; Japanese Soc Ethnol; Asiatic Soc Japan. Res: Culture and mental health; religion and behavior; Japan and Ryukyu Islands, East Asia. Mailing Add: Dept of Anthrop Univ of Hawaii Honolulu HI 96822

LEBRASSEUR, ROBIN JOHN, biological oceanography, see 12th edition

LEBRIE, STEPHEN JOSEPH, b Long Island City, NY, Jan 23, 31; m 56; c 2. MEDICAL PHYSIOLOGY. Educ: Long Island Univ, BS, 53; Princeton Univ, MA, 55, PhD(biol), 56. Prof Exp: Assoc surg res, Sch Med, Univ Pa, 56-57; from instr to assoc prof, Sch Med, Tulane Univ, 57-66; assoc prof, 66-70, PROF PHYSIOL, OHIO STATE UNIV, 70- Honors & Awards: Lederle Med Fac Award, 59-62. Mem: Am Physiol Soc; Am Soc Nephrology; Int Soc Nephrology; Soc Exp Biol & Med. Res: Renal lymphatics; comparative renal; renalendocrines. Mailing Add: Dept of Physiol Ohio State Univ Columbus OH 43210

LEBSOCK, KENNETH L, b Brush, Colo, Oct 19, 21; m 43; c 2. AGRONOMY. Educ: Mont State Col, BS, 49; NDak Agr Col, MS, 51; Iowa State Col, PhD(plant breeding), 53. Prof Exp: AGR ADMINR, AGR RES SERV, USDA, 53- Mem: Fel Am Soc Agron. Mailing Add: 3045 Shorewood Lane St Paul MN 55113

LE CAM, LUCIEN MARIE, b Croze, France, Nov 18, 24; m 52; c 3. MATHEMATICAL STATISTICS. Educ: Univ Paris, Lic, 45; Univ Calif, PhD, 52. Prof Exp: Statistician, Elec of France, 45-50; instr math & asst, Statist Lab, 50-52, instr & jr res statistician, 52-53, from asst prof to prof statist, 53-73, PROF STATIST & MATH, UNIV CALIF, BERKELEY, 73- Concurrent Pos: Dir, Ctr Math Res, Univ Montreal, 72. Mem: Am Math Soc; Inst Math Statist (pres, 72-73); Int Statist Inst. Res: General statistics. Mailing Add: Dept of Statist Univ of Calif Berkeley CA 94720

LECAR, HAROLD, b Brooklyn, NY, Oct 18, 35; m 58; c 2. BIOPHYSICS. Educ: Columbia Univ, AB, 57, PhD(physics), 63. Prof Exp: BIOPHYSICIST, LAB BIOPHYS, NAT INST NEUROL, COMMUNICABLE DISEASES & STROKE, 63- Concurrent Pos: Fel commoner, Churchill Col, Cambridge Univ, 75-76. Mem: AAAS; Am Phys Soc; Biophys Soc; Sigma Xi. Res: Masers; biophysics of excitable membranes. Mailing Add: Rm 2A29 Bldg 36 Nat Inst Neurol Communicable Diseases & Stroke Bethesda MD 20014

LECAR, MYRON, b Brooklyn, NY, Apr 10, 30. ASTROPHYSICS. Educ: Mass Inst Technol, BS, 51; Case Inst Technol, MS, 53; Yale Univ, PhD(astron), 63. Prof Exp: Lectr astrophys, Yale Univ Observ, 62-65; LECTR ASTROPHYS, COL OBSERV, HARVARD UNIV, 65-; ASTRONR, SMITHSONIAN ASTROPHYS OBSERV, 65- Concurrent Pos: Astronr, Inst Space Studies, NASA, 62-65. Mem: Am Astron Soc; Fedn Am Scientists; Royal Astron Soc; Int Astron Union. Res: Dynamics of the solar system; stellar dynamics and galactic structure; cosmology. Mailing Add: Ctr Astrophys Harvard Col Observ & Smithsonian Astrophys Observ Cambridge MA 02138

LECATO, GEORGE LEONARD, III, b Nasawadox, Va, Aug 28, 44; m 66. ENTOMOLOGY. Educ: Lynchburg Col, BS, 66; Va Polytech Inst & State Univ, MS, 68, PhD(entom), 70. Prof Exp: RES ENTOMOLOGIST, USDA, 71- Mem: Entom Soc Am. Res: Develops, directs and conducts studies of the population ecology(population dynamics) of stored-product insects that infest raw and processed commodities such as bulk grain, peanuts and packaged, processed foods. Mailing Add: Stored-Prod Insects Res & Develop Agr Res Serv USDA PO Box 5125 Savannah GA 31403

LECCE, JAMES GIACOMO, b Williamsport, Pa, Jan 11, 26; m 50. MICROBIOLOGY. Educ: Dartmouth Col, BA, 49; Pa State Univ, MS, 51; Univ Pa, PhD(microbiol), 53. Prof Exp: Instr prev med, Sch Vet Med, Univ Pa, 53-55; from asst prof to assoc prof microbiol, 55-63, PROF MICROBIOL, NC STATE UNIV, 63- Mem: Am Soc Microbiol. Res: Pleuropneumonia-like organisms; large viruses; plasma proteins. Mailing Add: Dept of Animal Sci NC State Univ Raleigh NC 27607

LECH, JOHN JAMES, b Passaic, NJ, June 21, 40. PHARMACOLOGY. Educ: Rutgers Univ, Newark, BS, 62; Marquette Univ, PhD(pharmacol), 67. Prof Exp: From instr to asst prof, 67-74, ASSOC PROF PHARMACOL, MED COL WIS, 74- Concurrent Pos: Am Heart Asn grant, Med Col Wis, 72-75, Sea grant, 71-75. Mem: AAAS; Soc Toxicol; Am Fisheries Soc; Am Soc Pharmacol & Exp Therapeut. Res: Cardiac triglyceride metabolism; metabolism of foreign compounds by fish. Mailing Add: Dept of Pharmacol Med Col of Wis Milwaukee WI 53233

LECHER, DAVID WAYNE, b Dover, NJ, May 5, 41; m 63; c 4. METEOROLOGY. Educ: Rutgers Univ, BS, 63; Univ Nebr, MS, 65; Cornell Univ, PhD(meteorol), 71. Prof Exp: Grad asst agr climat, Univ Nebr, 63-65; grad asst meteorol, Cornell Univ, 65-68; asst prof, 68-72, ASSOC PROF METEOROL, PHYSICS DEPT, TRENTON STATE COL, 72- Mem: AAAS; Am Meteorol Soc; Sigma Xi. Res: Precipitation composition and acidity; investigation into the causes of acid rain and its potential impacts upon soil, forest and fresh water ecosystems. Mailing Add: Dept of Physics Trenton State Col Trenton NJ 08625

LECHEVALIER, HUBERT ARTHUR, b Tours, France, May 12, 26; nat US; m 50; c 2. MICROBIOLOGY. Educ: Laval Univ, MS, 48; Rutgers Univ, PhD(microbiol), 51. Prof Exp: Asst prof, 51-56, assoc prof, Inst Microbiol, 56-66, PROF MICROBIOL, INST MICROBIOL, RUTGERS UNIV, NEW BRUNSWICK, 66- Concurrent Pos: Exchange scientist, Acad Sci USSR, 58-59; USPHS spec fel, Pasteur Inst, Paris, 61-62. Mem: Mycol Soc Am; Am Soc Microbiol; Can Soc Microbiol; assoc foreign mem Fr Soc Microbiol. Res: Morphology, classification and products of actinomycetes, including antibiotics; history of microbiology; ecology of actinomycetes. Mailing Add: 28 Juniper Lane Piscataway NJ 08854

LECHEVALIER, MARY P, b Cleveland, Ohio, Jan 27, 28; m 50; c 2. MICROBIOLOGY. Educ: Mt Holyoke Col, BA, 49; Rutgers Univ, MS, 51. Prof Exp: Res microbiologist, E R Squibb & Sons, 60-61; res assoc, 62-75, ASST RES PROF MICROBIOL, RUTGERS UNIV, NEW BRUNSWICK, 75- Mem: AAAS; Brit Soc Gen Microbiol; Am Soc Microbiol. Res: Antiviral substances; microbial transformation of steroids; classification and natural products of actinomycetes. Mailing Add: Waksman Inst Microbiol Rutgers Univ New Brunswick NJ 08903

LECHNER, JAMES ALBERT, b Danville, Pa, Aug 6, 33; m 56; c 3. MATHEMATICAL STATISTICS, APPLIED STATISTICS. Educ: Carnegie Inst Technol, BS, 54; Princeton Univ, PhD(math statist), 59. Prof Exp: Instr math, Princeton Univ, 57-58; sr mathematician, Res Labs, Westinghouse Elec Corp, Md, 60-63, adv mathematician, Aerospace Div, 63-67; mem tech staff, Res Anal Corp, Va, 67-71; MATH STATISTICIAN, STATIST ENG LAB, NAT BUR STANDARDS, 71- Mem: Inst Math Statist; Am Statist Asn. Res: Probability; theory of reliability; systems analysis; stochastic processes. Mailing Add: Statist Eng Lab Nat Bur of Standards Washington DC 20234

LECHOWICH, RICHARD V, b Chicago, Ill, June 23, 33; m 57; c 5. FOOD SCIENCE, FOOD MICROBIOLOGY. Educ: Univ Chicago, AB, 52, MS, 55; Univ Ill, PhD(food sci), 58. Prof Exp: Microbiologist, Am Meat Inst Found, Ill, 52-55; res asst food sci, Univ Ill, Urbana, 55-58; res microbiologist, Continental Can Co, Inc, Ill, 58-63; asst prof food sci, Mich State Univ, 63-66, from assoc prof to prof, 66-71; PROF FOOD SCI & TECHNOL & HEAD DEPT, VA POLYTECH INST & STATE UNIV, 71- Mem: AAAS; Int Asn Milk, Food & Environ Sanitarians; Inst Food Technologists; Am Soc Microbiologists; Brit Soc Appl Bact. Res: Mechanisms of bacterial spore formation and germination; chemical composition and thermal resistance phenomena of bacterial spores; food poisoning microorganisms, especially Clostridium botulinum; thermal and radiation resistances of microorganisms. Mailing Add: Dept of Food Sci & Technol Va Polytech Inst & State Univ Blacksburg VA 24061

LECHTENBERG, VICTOR LOUIS, b Butte, Nebr, Apr 14, 45; m 67; c 3. AGRONOMY. Educ: Univ Nebr, BS, 67; Purdue Univ, PhD(agron), 71. Prof Exp: From instr to asst prof, 69-75, ASSOC PROF AGRON, PURDUE UNIV, WEST LAFAYETTE, 75- Mem: Am Soc Agron; Crop Sci Soc Am; Am Soc Animal Sci; Am Forage & Grassland Coun; Sigma Xi. Res: Factors that affect forage crop quality and utilization; environmental physiology of forage crops; genetic improvement of crop quality. Mailing Add: Dept of Agron Purdue Univ West Lafayette IN 47906

LECHTMAN, MAX DRESSLER, b Providence, RI, Apr 24, 35; m 62; c 3. MICROBIOLOGY. Educ: Univ RI, AB, 57; Univ Mass, MS, 59; Univ Southern Calif, PhD(microbiol), 68. Prof Exp: Microbiologist, Douglas Aircraft Co, Calif, 61-62 & Res & Develop Div, Magna Chem Co, 62-64; instr microbiol, Univ Southern Calif, 64; microbiol consult, Garrett Corp, 64-65, microbiologist, AiRes Mfg Co Div, 65-67; mem tech staff aerospace microbiol, Autonetics Div, NAm, Rockwell Corp, Anaheim, 67-71; INDUST CONSULT, CALIF LAB INDUSTS, 71- Concurrent Pos: Instr, Calif Community Col Syst, 71- Mem: Am Soc Microbiol; Soc Indust Microbiol; Am Asn Contamination Control; NY Acad Sci. Res: Microbial cytology, cytochemistry and physiology; bioluminescent bacteria for detection of toxic chemicals. Mailing Add: 8641 Delray Circle Westminster CA 92683

LECK, CHARLES FREDERICK, b Princeton, NJ, June 20, 44. ORNITHOLOGY, ECOLOGY. Educ: Muhlenberg Col, BS, 66; Cornell Univ, PhD(vert zool), 70. Prof Exp: Vis res assoc, Smithsonian Trop Res Inst, 68-69; ASSOC PROF ECOL, RUTGERS UNIV, NEW BRUNSWICK, 70-, DIR ECOL PROG, 74- Mem: Wilson Ornith Soc; Cooper Ornith Soc; Am Ornith Union; Ecol Soc Am; Asn Trop Biologists. Res: Avian behavior and feeding ecology; tropical biology. Mailing Add: Dept of Zool Rutgers Univ New Brunswick NJ 08903

LECKIE, DONALD STEWART, b Chicago, Ill, June 3, 20; m 52; c 1. OPERATIONS RESEARCH, STATISTICS. Educ: Purdue Univ, BS, 48; Case Inst Technol, MS, 55, PhD(opers res), 60. Prof Exp: Qual control engr, 48-57, planning engr, Res & Planning Div, 57-65, HEAD DIV MATH, RES CTR, REPUB STEEL CORP, 65- Concurrent Pos: Lectr, Case Western Reserve Univ, 62- Mem: Sr mem Am Soc Qual Control; Am Iron & Steel Inst; Inst Math Statist; assoc Opers Res Soc Am; Inst Mgt Sci. Res: Application of mathematical and statistical methods to the solution of management problems; application of computers to process control and to problems in research and development; systems engineering. Mailing Add: Repub Steel Res Ctr 6801 Brecksville Rd Independence OH 44131

LECKRONE, DAVID STANLEY, b Salem, Ill, Nov 30, 42; m 64; c 2. ASTROPHYSICS. Educ: Purdue Univ, BS, 64; Univ Calif, Los Angeles, MA, 66, PhD(astron), 69. Prof Exp: ASTROPHYSICIST, LAB OPTICAL ASTRON, GODDARD SPACE FLIGHT CTR, 69- Mem: Am Astron Soc; Sigma Xi. Res: Ultraviolet stellar spectroscopy and photometry from space vehicles; magnetic and chemically peculiar stars; stellar atomspheres; abundances of the elements in astronomical objects; instrumentation for space astronomy. Mailing Add: Code 671 NASA-Goddard Space Flight Ctr Greenbelt MD 20771

LECLAIR, HUGH GRENVILLE, physical chemistry, see 12th edition

LECLAIRE, CLAIRE DEAN, b Huron, SDak, Aug 26, 10; m 41. ORGANIC CHEMISTRY. Educ: SDak State Sch Mines, BS, 33; Univ Minn, PhD(org chem), 39. Prof Exp: Chemist, Coal Res Lab, Carnegie Inst Technol, 38-41 & Rohm and Haas Co, Pa, 41-42; rubber fel, Mellon Inst, 42-47; res chemist, Firestone Tire & Rubber Co, 47-49; asst dir rubber res proj, Univ Minn, 49-51; sect leader appl res, Adhesives & Coatings Div, Minn Mining & Mfg Co, 51-58; sr res chemist, Int Latex Corp, 58-69 & Standards Brands Chem Industs, Inc, 69-76; RES CHEMIST, REICHHOLD POLYMERS INC, 76- Mem: AAAS; Am Chem Soc. Res: Organic synthesis; structure; polymerization; properties and utilization of high polymers; bituminous coals; elastomers; resins; adhesives. Mailing Add: Reichhold Polymers Inc Dover DE 19906

LECLAIRE, ROGER, b Montreal, Que, Jan 15, 12. ASTRONOMY, PHYSICS. Educ: Col Ste-Marie, Montreal, BA, 36; Sault-au-Recollet, BEd, 37; Univ Montreal, Lic math, 40; Immaculate Conception Sem, lic phil & theol, 44; Georgetown Univ, PhD, 49. Prof Exp: Astronr, Vatican Observ, Italy, 50-52; prof physics, Col Sacred Heart, Ont, 52-55 & Ethiopia, 55-57; chmn, Inst Astron, 67-74, PROF PHYSICS & HEAD DEPT, LAURENTIAN UNIV, 57-, DIR, INST ASTRON, 74- Mem: Am Phys Soc; Am Astron Soc; Inst Elec & Electronics Engr; Can Asn Physicists. Res: Milky Way absorption problems; celestial mechanics; ionospheric absorption problems. Mailing Add: Dept of Physics Laurentian Univ Sudbury ON Can

LECLERCQ, GEORGE MORROW, organic chemistry, see 12th edition

LECONTE, JOSEPH NISBET, organic chemistry, see 12th edition

L'ECUYER, JACQUES, b St-Jean, Que, Mar 6, 37; m 59; c 3. NUCLEAR PHYSICS. Educ: Col St Jean, BA, 56; Univ Montreal, BSc, 59, MSc, 61, PhD(physics), 66. Prof Exp: Lectr, Univ Montreal, 61-63 & Univ Sherbrooke, 63-64; asst prof physics, Laval Univ, 64-67; Nat Res Coun Can fel, Oxford Univ, 67-69; from asst prof to assoc prof, 69-73, PROF PHYSICS, UNIV MONTREAL, 73- Mem: Am Phys Soc; Can Asn Physicists; Fr-Can Asn Advan Sci. Res: Experimental nuclear physics. Mailing Add: Nuclear Physics Lab Univ of Montreal CP 6128 Montreal PQ Can

L'ECUYER, PHILIBERT, organic chemistry, see 12th edition

LEDBETTER, HARVEY DON, b Pierson, Ill, June 26, 26; m 47; c 3. ORGANIC CHEMISTRY. Educ: Univ Ariz, BS, 49, MS, 50; Univ Tenn, PhD(chem), 54. Prof Exp: Asst gen chem, Univ Tenn, 50-52; res group leader, 53-67, res supvr, 67-71, TECH MGR, DESIGNED PROD DEPT, DOW CHEM USA, 71- Mem: AAAS; Am Chem Soc; Soc Plastics Engrs. Res: Polymer nucleation and stabilization; vacuum processes; preparation and physics of composite structures. Mailing Add: 1006 Sterling Dr Midland MI 48640

LEDBETTER, MYRON C, b Ardmore, Okla, June 25, 23. CELL BIOLOGY, BOTANY. Educ: Okla State Univ, BS, 48; Univ Calif, MA, 51; Columbia Univ, PhD, 58. Prof Exp: Fel plant physiol, Boyce Thompson Inst, 53-57, asst plant anatomist, 57-60; guest investr & fel, Rockefeller Inst, 60, res assoc, 61; res assoc, Harvard Univ, 61-65; cell biologist, 65-74, SR CELL BIOLOGIST, BROOKHAVEN NAT LAB, 74- Concurrent Pos: Guest assoc, Brookhaven Nat Lab, 58-60. Mem: AAAS; Am Soc Cell Biol; Bot Soc Am; Electron Micros Soc Am; Am Inst Biol Sci. Res: Morphology of physiologically dwarfed tree seedlings; feeding damage to plant tissues by lygus bugs; histopathology of ozone on plants; distribution of fluorine in plants; plant fine structure, microtubules in plants. Mailing Add: Biol Dept Brookhaven Nat Lab Upton NY 11973

LEDDY, JAMES JEROME, b Detroit, Mich, July 28, 29; m 51; c 6. INDUSTRIAL CHEMISTRY. Educ: Univ Detroit, BS, 51; Univ Wis, PhD(chem), 55. Prof Exp: Chemist inorg res, 55-59, proj leader, 59-60, proj leader, Electrochem & Inorg Res Lab, 60-63, sr res chemist, 63-67, assoc res scientist, 67-71, RES SCIENTIST, ELECTROCHEM & INORG RES LAB, DOW CHEM USA, 71- Concurrent Pos: Asst prof, Assumption Univ, 59-60. Mem: AAAS; Am Chem Soc; Electrochem Soc. Res: Chemistry of less familiar elements, especially titanium, zirconium and hafnium; coordination compounds; unfamiliar oxidation states; amalgam chemistry; electrochemistry; inorganic polymers; industrial inorganic and electrochemistry; chlor-alkali. Mailing Add: Dow Chem USA Midland MI 48640

LEDDY, JOHN PLUNKETT, b New York, NY, Sept 10, 31; m 56; c 3. IMMUNOLOGY. Educ: Fordham Univ, BA, 52; Columbia Univ, MD, 56. Prof Exp: From intern to resident internal med, Boston City Hosp, Harvard Med Serv, 56-59; USPHS trainee hemat, Med Ctr, Univ Rochester, 59-60; sr instr, 62-64, from asst prof to assoc prof med, 64-73, PROF MED & MICROBIOL, MED CTR, UNIV ROCHESTER, 73-, DIR CLIN IMMUNOL UNIT, 70- Concurrent Pos: Res med officer immunochem, Walter Reed Army Inst Res, 60-62; Nat Found fel, 62-64; NIH res grant, 65-; sr investr, Arthritis Found, 65-70; dir, USPHS Training Grant, 70- Mem: Am Soc Clin Invest; Am Asn Immunologists. Res: Biology of complement system in man; erythrocyte autoantibodies in human diseases. Mailing Add: Dept of Med Univ of Rochester Med Ctr Rochester NY 14642

LEDEBOER, FREDERICK BERNHARD, agronomy, see 12th edition

LEDEEN, ROBERT, b Denver, Colo, Aug 19, 28. BIOCHEMISTRY, NEUROBIOLOGY. Educ: Univ Calif, Berkeley, BS, 49; Ore State Univ, PhD(org chem), 53. Prof Exp: Fel, Univ Chicago, 53-54; res chemist, Mt Sinai Hosp, New York, 56-59; res chemist, 59-61, from asst prof to assoc prof biochem in neurol, 62-75, PROF BIOCHEM IN NEUROL, ALBERT EINSTEIN COL MED, 75- Mem: AAAS; Am Chem Soc; Am Soc Biol Chem; Am Soc Neurochem; Int Soc Neurochem. Res: Chemistry of the nervous systems; gangliosides and other lipids; myelin lipids; methods in structure determination and stereochemistry. Mailing Add: Dept of Neurol Albert Einstein Col of Med Bronx NY 10461

LEDER, IRWIN GORDON, b New York, NY, June 16, 20; m 45; c 3. BIOCHEMISTRY. Educ: Brooklyn Col, AB, 42; NY Univ, MS, 47; Duke Univ, PhD(biochem), 51. Prof Exp: Res assoc, Duke Univ, 51; USPHS fel, NY Univ, 51-52 & Yale Univ, 52-53; asst, Pub Health Res Inst, Inc, NY, 53-54; BIOCHEMIST, NAT INST ARTHRITIS, METAB & DIGESTIVE DIS, BETHESDA, 54- Res: Niacin metabolism; pyridine nucleotide synthesis; intermediary carbohydrate metabolism; enzymology. Mailing Add: 4004 Wexford Dr Kensington MD 20795

LEDER, LEWIS BEEBE, b Brooklyn, NY, Oct 20, 28; m 48; c 1. EXPERIMENTAL PHYSICS. Educ: Univ Idaho, BS, 43. Prof Exp: Instr physics, Williams Col, 43-44; physicist gas flow, Inst Gas Technol, 44-45; physicist, Dielectric Studies, Indust Condenser Corp, 45-46; physicist electron heat controls, Wheelco Instruments Co, 46-48; physicist nuclear physics, Inst Nuclear Studies, 48-51; physicist electron physics, Nat Bur Standards, 51-62; physicist, Appl Res Lab, Philco Corp, 62-65; physicist electron scattering & superconductivity, Ford Sci Lab, 65-69; RES SCIENTIST, XEROX CORP, 69- Mem: Am Vacuum Soc; Am Phys Soc. Res: Electron scattering; superconductivity; thin films preparation and properties. Mailing Add: Xerox Corp Bldg 103 Xerox Square Rochester NY 14644

LEDERBERG, ESTHER MIRIAM, b New York, NY, Dec 18, 22; m 46. GENETICS, MICROBIOLOGY. Educ: Hunter Col, AB, 42; Stanford Univ, MA, 46; Univ Wis, PhD(genetics), 50. Prof Exp: Proj assoc genetics, Univ Wis, 50-59; res geneticist, Med Sch, 59-68, res assoc, 68-71, sr scientist, 71-74, ADJ PROF MED MICROBIOL, STANFORD UNIV, 74- Concurrent Pos: Fulbright fel, Australia, 57; Am Cancer Soc Sr Dernham fel, 68-70. Honors & Awards: Co-recipient, Pasteur Award, Soc Ill Bact, 56. Mem: Fel AAAS; Genetics Soc Am; Brit Soc Gen Microbiol; Am Soc Microbiol. Res: Genetics of microorganisms; lysogenicity; bacterial recombination and transformation; DNA repair; phase variation of Flagellar antigens in Salmonella; R plasmids. Mailing Add: Dept of Med Microbiol Stanford Univ Stanford CA 84305

LEDERBERG, JOSHUA, b NJ, May 23, 25; m 68. GENETICS. Educ: Columbia Univ, BA, 44; Yale Univ, PhD(microbiol), 47. Hon Degrees: ScD, Yale Univ, 60 & Columbia Univ, 67; MD, Univ Turin, 69; ScD, Univ Wis, 67 & Albert Einstein Col Med, 70. Prof Exp: From asst prof to prof genetics, Univ Wis, 47-58, prof med & genetics & chmn dept, 58-59; PROF GENETICS & CHMN DEPT, MED SCH, STANFORD UNIV, 59- Concurrent Pos: Consult, Ctr Advan Study Behav Sci. Honors & Awards: Nobel Prize, 58. Mem: Nat Acad Sci. Res: Genetics and evolution; science policy; computer science. Mailing Add: Dept of Genetics Stanford Univ Sch of Med Stanford CA 94305

LEDERBERG, SEYMOUR, b New York, NY, Oct 30, 28; m 59; c 2. MOLECULAR BIOLOGY, GENETICS. Educ: Cornell Univ, BA, 41; Univ Ill, PhD(bact), 55. Prof Exp: Am Cancer Soc fel, Univ Calif, 55-57, vis asst prof bact, 57-58; from asst prof to assoc prof, 58-66, PROF BIOL, BROWN UNIV, 66-, LECTR, LAW SCH CTR FOR LAW & HEALTH SCI, 73- Concurrent Pos: USPHS fel, Inst Biol Phys Chem, Paris, 65-66; consult, Gen Med & Sci Adv Coun, Cystic Ribrosis Found, 72-, Comt Genetic Screening Cystic Fibrosis, Nat Acad Sci-Nat Res Coun, 74-75 & Nat Inst Arthritis, Metab & Digestive Diseases, 75- Mem: Am Soc Microbiol; Genetics Soc Am; Am Soc Human Genetics; Soc Study Social Biol; Environ Mutagen Soc. Res: Human, microbial and viral genetics; cystic fibrosis; human chromosome pathology; normal and abnormal chromosomal, microtubule and lysosomal functions. Mailing Add: Biomed Sci Box G Brown Univ Providence RI 02912

LEDERER, LUDWIG GEORGE, b Chicago, Ill, Jan 19, 11; m 32; c 5. PHYSIOLOGY. Educ: Univ Ill, BA, 31; Northwestern Univ, MS, 32, MD, 38, PhD(physiol), 39; Am Bd Prev Med, dipl & cert aerospace med. Prof Exp: Assoc res physiol, Med Sch, Northwestern Univ, 31-39, instr, 39-41; med dir, Capital Airlines, 40-60; MED DIR, AM AIRLINES, INC, 60- Concurrent Pos: Asst prof, Med Sch & Hosp, George Washington Univ, 52-60; chmn med adv panel to adminr, Fed Aviation Agency, 64-66. Mem: Airline Med Dirs Asn (past pres); Aerospace Med Asn (past pres). Res: Physiology of heavy metals; aviation cardiology. Mailing Add: Am Airlines Inc 633 Third Ave New York NY 10017

LEDERER, SEYMOUR JERRALD, organic chemistry, see 12th edition

LEDERIS, KARL, b Lithuania, Aug 1, 20; m 52; c 2. PHARMACOLOGY, ENDOCRINOLOGY. Educ: Bristol Univ, BSc, 58, PhD(pharmacol), 61, DSc(endocrinol), 68. Prof Exp: Jr fel pharmacol, Bristol Univ, 61-63, lectr, 63-66, sr lectr, 66-68, reader, 68-69; PROF PHARMACOL & THERAPEUT, MED SCH, UNIV CALGARY, 69- Concurrent Pos: Wellcome Trust & Ger Res Asn fel, Univ Kiel, 61-62; NSF fel, Univ Calif, Berkeley, 67-68; mem grants comt endocrinol & univs consult comt, Med Res Coun Can, 70, assoc, Coun, 70- Mem: Int Brain Res Orgn; Endocrine Soc; Brit Soc Endocrinol; Brit Pharmacol Soc; Can Physiol Soc. Res: Hypothalamo-neurohypophysial system, mechanisms of hormone storage and secretion; caudal neurosecretory system of teleosts, chemistry and pharmacology of urotensin peptides. Mailing Add: Div of Pharmacol Univ of Calgary Med Sch Calgary AB Can

LEDERKREMER, ROSA MUCHNIK, organic chemistry, see 12th edition

LEDERLE, HENRY FREDERICK, organic chemistry, see 12th edition

LEDERMAN, DAVID MORDECHAI, b Bogota, Colombia, May 26, 44; m 67; c 2. BIOMEDICAL ENGINEERING, BIOPHYSICS. Educ: Univ Los Andes, BEng, 66; Cornell Univ, BSc, 66, MEng, 67, PhD(aerospace eng), 73. Prof Exp: From prof appl math, Fac Arts & Sci, to dir div biomed eng, Fac Eng, Univ Los Andes, 72-73; SR SCIENTIST MED RES COMT, AVCO-EVERETT RES LAB INC, AVCO CORP, 73- Mem: Am Phys Soc; Am Soc Artificial Internal Organs. Res: Physico-chemical phenomena during genesis and growth of thrombi on artificial surfaces; cardiovascular fluiddynamics and development of cardiac assist devices; biomechanics of brain suspension mechanisms; applications of lasers in medicine. Mailing Add: 20 Cornell Rd Marblehead MA 01945

LEDERMAN, LEON MAX, b NY, July 15, 22; m 45; c 3. NUCLEAR PHYSICS. Educ: City Col New York, BS, 43; Columbia Univ, AM, 50, PhD(physics), 51. Prof Exp: Assoc, 51-52, from asst prof to assoc prof, 52-58, PROF PHYSICS, COLUMBIA UNIV, 58-, DIR, NEVIS LABS, 68- Concurrent Pos: Guggenheim fel, 58-59; NSF sr fel, 67. Honors & Awards: Nat Medal of Sci, 65. Mem: Nat Acad Sci; Am Phys Soc. Res: Properties and interactions of elementary particles. Mailing Add: Dept of Physics Columbia Univ New York NY 10027

LEDFORD, RICHARD ALLISON, b Charlotte, NC, June 30, 31; m 57; c 5. FOOD MICROBIOLOGY. Educ: NC State Univ, BS, 54, MS, 56; Cornell Univ, PhD(food sci), 62. Prof Exp: Dir, NY State Food Lab, NY State Dept Agr & Mkts, 61-64; asst prof, 64-71, ASSOC PROF FOOD SCI, CORNELL UNIV, 71-, CHMN DEPT, 72-, ASSOC HEAD INST FOOD SCI, 75- Mem: Inst Food Technologist; Am Dairy Sci Asn. Res: Microbiological aspects of food science, especially food fermentations and analytical methods. Mailing Add: Dept of Food Sci Cornell Univ Ithaca NY 14850

LEDIAEV, JOHN P, b Goorgan, Iran, Sept 8, 40; US citizen; m 65; c 1. MATHEMATICS. Educ: Occidental Col, BA, 63; Univ Calif, Riverside, MA, 65, PhD(noether lattices), 67. Prof Exp: Asst prof, 67-71, ASSOC PROF MATH, UNIV IOWA, 71- Mem: Am Math Soc. Res: Structure; representation and embedding of Noether lattices; primary decomposition in multiplicative lattices; semi-prime operations in Noether lattices. Mailing Add: Dept of Math Univ of Iowa Iowa City IA 52240

LEDIG, F THOMAS, b Dover, NJ, Aug 13, 38; m 56; c 3. GENETICS. Educ: Rutgers Univ, BS, 62; NC State Univ, MS, 65, PhD(genetics), 67. Prof Exp: Lectr, 66-67, asst prof, 67-72, ASSOC PROF FOREST GENETICS, YALE UNIV, 72- Mem: Scand Soc Plant Physiologists; Soc Am Foresters; Bot Soc Am. Res: Genecology, including population genetics, hybridization, and evolution; physiological mechanisms of adaptation in wild populations; physiological analysis of genetic variation in productivity; photosynthetic physiology; quantitative genetics and breeding. Mailing Add: Greeley Mem Lab Yale Univ New Haven CT 06511

LEDIN, GEORGE, JR, b Seekirchen, Austria, Jan 28, 46; US citizen; m 68; c 1. COMPUTER SCIENCE, STATISTICS. Educ: Univ Calif, Berkeley, BS, 67. Prof Exp: Statistician & mathematician, 65-70, lectr math & comput sci, 68-73, SR RES ASSOC STATIST & MATH BIOSCI, INST CHEM BIOL, UNIV SAN FRANCISCO, 70-, ASST PROF COMPUT SCI, 73- Concurrent Pos: Consult statist & comput sci, 66-; US rep, Int Fedn Info Processing, 72-74. Mem: AAAS; Am Statist Asn; Asn Comput Mach; Math Asn Am; NY Acad Sci. Res: Algorithmic languages; programming methodology; heuristic programming; pattern recognition; robotics; mathematical models for biosciences; number theory; combinatorics; graph theory; game theory; information theory; algebra. Mailing Add: Dept of Comput Sci Univ of San Francisco San Francisco CA 94117

LEDINGHAM, GEORGE FILSON, b Moose Jaw, Sask, Jan 31, 11; m 42; c 1. SYSTEMATIC BOTANY. Educ: Univ Sask, BSc, 34, MSc, 36; Univ Wis, PhD(genetics), 39. Prof Exp: Instr biol, Moose Jaw Jr Col, 39-42; from instr to assoc prof, 45-67, PROF BIOL, UNIV REGINA, 67- Mem: AAAS; Am Soc Plant Taxon; Bot Soc Am; Genetics Soc Am; Can Bot Asn. Res: Cytotaxonomy of Astragalus and Oxytropis. Mailing Add: Dept of Biol Univ of Regina Regina SK Can

LEDINKO, NADA, b Girard, Ohio, Dec 16, 25. VIROLOGY. Educ: Ohio State Univ, BS, 46; Pa State Col, MS, 49; Yale Univ, PhD(microbiol), 52. Prof Exp: Res asst virol, Yale Univ, 52-53; Nat Found Infantile Paralysis fel, Walter & Eliza Hall Inst, Australia, 53-55; virologist, Pub Health Res Inst, 56-62; USPHS fel, Carnegie Inst Genetics Res Unit & Salk Inst Biol Studies, 63-65; assoc investr & NIH res career develop awardee, Putnam Mem Hosp Inst Med Res, Bennington, Vt, 65-71; PROF BIOL, UNIV AKRON, 71- Mem: AAAS; Am Asn Path & Bact; Tissue Cult Asn; Am Soc Microbiol; Am Asn Cancer Res. Res: Genetical and biochemical aspects of viral growth. Mailing Add: Dept of Biol Univ of Akron Akron OH 44304

LEDLEY, BRIAN G, b Brisbane, Australia, Sept 29, 38; m 58; c 1. MAGNETOSPHERIC PHYSICS. Educ: Univ Queensland, BSc, 50; Univ Birmingham, PhD(physics), 55. Prof Exp: Asst lectr physics, Univ Birmingham, 55-57;

res assoc nuclear physics, Fermi Inst, Univ Chicago, 57-60; res fel cosmic rays, Univ Sydney, 60-62; Nat Acad Sci-NASA res fel space sci, 62-63; sect head, 63-70, PHYSICIST, NASA, 70- Mem: Am Geophys Union. Res: Geomagnetic field measurements using satellites; magnetometer instrumentation. Mailing Add: Code 625 Goddard Space Flight Ctr Greenbelt MD 20771

LEDLEY, ROBERT STEVEN, b New York, NY, June 28, 26; m 49; c 2. BIOPHYSICS, MATHEMATICS. Educ: NY Univ, DDS, 48; Columbia Univ, MA, 49. Prof Exp: Res physicist, Radiation Lab, Columbia Univ, 48-50, instr physics, 49-50; vis scientist, Nat Bur Standards, 51-52, physicist, External Control Group, Electronic Comput Lab, 53-54; opers res analyst, Opers Res Off, Strategic Div, Johns Hopkins Univ, 54-56; assoc prof elec eng, Sch Eng, George Washington Univ, 56-60; instr pediat, Sch Med, Johns Hopkins Univ, 60-63; prof elec eng, Sch Eng & Appl Sci, George Washington Univ, 68-70; PROF PHYSIOL & BIOPHYS, MED CTR, GEORGETOWN UNIV, 70-, RADIOL, 74-, DIR MED COMPUT & BIOPHYS DIV, 75- Concurrent Pos: Consult mathematician, Data Processing Systs Div, Nat Bur Standards, 57-60; mem staff, Nat Acad Sci-Nat Res Coun, 57-61; pres & res dir, Nat Biomed Res Found, 60-; pres, Digital Info Sci Corp, 70-75. Mem: Soc Math Biophys; Am Statist Asn; Biophys Soc; NY Acad Sci; Pattern Recognition Soc. Res: Mathematical methods for application of symbolic logic; data processing techniques for computers; digital computer engineering; mathematical biophysics applied to physiology and other fields; use of computers in biomedical research and medical data processing; development of whole body computerized tomography. Mailing Add: 1002 La Grande Rd Silver Spring MD 20903

LEDLIE, DAVID B, b Jersey City, NJ, June 4, 40. ORGANIC CHEMISTRY. Educ: Middlebury Col, AB, 62; Mass Inst Technol, PhD(org chem), 66. Prof Exp: ASST PROF CHEM, MIDDLEBURY COL, 67- Concurrent Pos: Petrol Res grants, 67-69 & 72-74; Res Corp grant, 70-71. Mem: Am Chem Soc. Res: Stereochemistry of sodium-liquid ammonia reductions; isonitriles; electrocyclic ring openings; carbenes. Mailing Add: Dept of Chem Middlebury Col Middlebury VT 05753

LEDNEY, GEORGE DAVID, b Sharon, Pa, June 25, 37. BIOLOGY. Educ: Youngstown Univ, BS, 60; Univ Notre Dame, PhD(biol), 65. Prof Exp: Asst prof radiation biol, Med Units, Univ Tenn, Memphis, 65-67, assoc prof, 70-73; HEAD DIV IMMUNOL, ARMED FORCES RADIOBIOL RES INST, 73- Concurrent Pos: Nat Cancer Inst fel, 65-67; Am Cancer Soc grants, 68 & 70. Mem: Radiation Res Soc; Transplantation Soc; Asn Gnotobiotics; Inst Soc Exp Hemat. Res: Radiation biology; transplantation immunology. Mailing Add: Div of Immunol Armed Forces Radiobiol Res Inst Bethesda MD 20014

LEDNICER, DANIEL, b Antwerp, Belg, Oct 15, 29; nat US; m 56; c 2. ORGANIC CHEMISTRY. Educ: Antioch Col, BS, 52; Ohio State Univ, PhD(chem), 55. Prof Exp: Sr chemist, G D Searle & Co, 55-56; res assoc, Duke Univ, 56-58; Esso Res & Develop Co fel, Univ Ill, 58-59; chemist, 59-73, SR SCIENTIST, UPJOHN CO, 73- Mem: Am Chem Soc. Res: Stereochemistry; medicinal chemistry; hypotensives; analgesics. Mailing Add: Cardiovasc Diseases Res Upjohn Co PO Box 831 Kalamazoo MI 49002

LEDOUX, ROBERT LOUIS, b Marieville, Que, Apr 19, 33; m 61; c 2. MINERALOGY. Educ: Univ Montreal, BSc, 56; Laval Univ, MScA, 60; Purdue Univ, PhD(mineral), 64. Prof Exp: From assoc prof to prof mineral, 64-74, PROF PETROL & MINERAL, LAVAL UNIV, 74- Res: Infrared studies of layered silicates. Mailing Add: Dept of Geol Laval Univ Quebec PQ Can

LEDSOME, JOHN R, b Bebington, Eng, June 18, 32; m 57; c 3. PHYSIOLOGY, MEDICINE. Educ: Univ Edinburgh, MB, ChB, 55, MD, 62. Prof Exp: Lectr physiol, Univ Leeds, 59-68; PROF PHYSIOL, UNIV BC, 68- Concurrent Pos: USPHS int fel, 64-65, res grant, 66-68; Med Res Coun Eng res grant, 65-68; Med Res Coun Can res grant, 68-75; BC Heart Found res grant, 72-75. Mem: Can Physiol Soc; Brit Physiol Soc. Res: Control of the cardiovascular system; function of left atrial receptors. Mailing Add: Dept of Physiol Univ of BC Vancouver BC Can

LEDUC, ELIZABETH, b Rockland, Maine, Nov 19, 21. CELL BIOLOGY. Educ: Univ Vt, BS, 43; Wellesley Col, MA, 45; Brown Univ, PhD(biol), 48. Prof Exp: Res assoc biol, Brown Univ, 48-49; instr & assoc anat, Harvard Med Sch, 49-53; from asst prof to assoc prof, 53-64, PROF BIOL, BROWN UNIV, 64-, DEAN DIV BIOL & MED, 73- Concurrent Pos: Mem adv coun, Nat Inst Gen Med Sci, 72- Mem: AAAS; Am Soc Cell Biol; Soc Francaise de Microscopie Electronique; Histochem Soc; Am Soc Exp Path. Res: Histophysiology and pathology of the liver; cellular mechanism in antibody production; ultrastructural and cytochemical effects of cancer chemotherapeutic compounds on normal and neoplastic cells. Mailing Add: Div of Biol & Med Sci Brown Univ Providence RI 02912

LEDUC, GERARD, b Verdun, Que, Sept 7, 34; m 59; c 3. FISHERIES, BIOCHEMISTRY. Educ: Univ Montreal, BSc, 58, MSc, 60; Ore State Univ, PhD(fisheries), 66. Prof Exp: Biologist, Que Wildlife Serv, 63-66; asst prof biol sci, 66-72, chmn dept, 69-72, ASSOC PROF BIOL SCI, SIR GEORGE WILLIAMS UNIV, 72- Mem: Am Fisheries Soc. Res: Fisheries problems in water pollution, mainly the long-term effects of sublethal concentrations of toxicant; development of a new fish toxicant for lake reclamation. Mailing Add: Dept of Biol Sci Sir George Williams Univ Montreal PQ Can

LEDUC, HENRIETTA MARIE, physical chemistry, see 12th edition

LE DUC, J ADRIEN MAHER, b Can, July 2, 24; nat US; m 49; c 2. INORGANIC CHEMISTRY, PHYSICAL CHEMISTRY. Educ: Sir George William Univ, BSc, 47; Polytech Inst Brooklyn, MS, 51; Univ Dijon, DSc(phys inorg chem), 53. Prof Exp: Anal paper chemist, Howard Smith Paper Co, Ltd, 43-44; anal paper chemist dyes & colors, L B Holliday & Co, 45; res chemist zirconium alloys, St Lawrence Alloys & Metals Co, Ltd, 46; consult chemist, Milton Hersey Co, Ltd, 47; res consult chemist, Wyssmont Co, 47-49; sr inorg chemist, J T Baker Chem Co, 49-50; sr electroplating chemist, MacDermid, Inc, 53-54; head electrolytic sect, Diamond Alkali Co, 54-56, group leader res, Electrochem Div, 56-59; supvr explor & process res, M W Kellogg Co, 59-62, dir electrochem res, 62, mgr electrochem res & develop, 63-70; CONSULT CHEM & ELECTROCHEM TECHNOL, LICENSING & PROCESS TECHNOL EXCHANGE, 70- Honors & Awards: Order of Merit Medal & Res & Invention Award, Paris, 68, Gold Medal Award Res & Invention, 75. Mem: Sr mem Am Chem Soc; Electrochem Soc; NY Acad Sci; Soc Indust Chem; Sigma Xi. Res: Inorganic and organic chemicals; electrochemistry; electrometallurgy; metal finishing molten salts; process licensing; technical exchange; negotiation and contract agreements; research and development management; metals, including rare-earths. Mailing Add: 189 Parsonage Hill Rd Short Hills NJ 07078

LEDWITZ-RIGBY, FLORENCE INA, b New York, NY, Feb 14, 46; m 68. REPRODUCTIVE ENDOCRINOLOGY. Educ: City Col NY, BS, 66; Case-Western Reserve Univ, MS, 68; Univ Wis-Madison, PhD(endocrinol & reprod physiol), 72. Prof Exp: Res fel reprod physiol, Sch Med, Dept Physiol, Univ Pittsburgh, 72-74, vis asst prof physiol, Dept Biol, 74-75; ASST PROF PHYSIOL, DEPT BIOL SCI, NORTHERN ILL UNIV, 75- Mem: Soc Study Reprod. Res: Endocrine and physiological control mechanisms of ovarian cell function and differentiation. Mailing Add: Dept of Biol Sci Northern Ill Univ DeKalb IL 60115

LEE, ADDISON EARL, b Maydelle, Tex, June 18, 14; m 37; c 2. BOTANY, SCIENCE EDUCATION. Educ: Stephen F Austin State Teachers Col, BS, 34; Agr & Mech Col Tex, MS, 37; Univ Tex, PhD(bot), 49. Prof Exp: Lab instr, Stephen F Austin State Teachers Col, 33-36 & Agr & Mech Col Tex, 36-37; from instr to assoc prof bot, 46-59, morphologist & asst dir plant res inst, 49-53, PROF SCI EDUC & BIOL & DIR SCI EDUC CTR, UNIV TEX, AUSTIN, 59- Concurrent Pos: High sch teacher, Tex, 34-36; vis prof, Univ Va, 57-58. Honors & Awards: Robert H Carleton Award, Nat Sci Teachers Asn, 75. Mem: Nat Sci Teachers Asn (pres, 67); Nat Asn Biol Teachers (pres, 73); Nat Asn Res Sci Teaching. Res: Experimental plant morphology. Mailing Add: Sci Educ Ctr Univ of Tex Austin TX 78712

LEE, AGNES C, b July 8, 45; US citizen. BIOCHEMISTRY. Educ: Tamkang Col, Taiwan, BS, 66; Bryn Mawr Col, MA, 68, PhD(chem), 70. Prof Exp: RES BIOCHEM, UNIV IOWA, 70- Mem: Am Phys Soc. Res: TAT enzyme purification and reaction. Mailing Add: Internal Med Univ of Iowa Iowa City IA 52240

LEE, ALFRED TZE-HAU, b Hong Kong, July 22, 39; US citizen; m 70; c 2. ANALYTICAL CHEMISTRY, ORGANIC CHEMISTRY. Educ: Univ Calif, Berkeley, BS, 63; Univ Calif, Los Angeles, PhD(chem), 68. Prof Exp: Instr chem, East Los Angeles Col, 67-68; PROF CHEM, CITY COL OF SAN FRANCISCO, 68- Mem: Am Chem Soc. Res: Oxidation-state diagrams; pulse polarography; analytical methods in general, vacuum-ultraviolet spectra of olefins. Mailing Add: Dept of Chem City Col of San Francisco San Francisco CA 94112

LEE, ANTHONY, b Canton, China, Dec 8, 41; US citizen; m 69; c 2. PLASMA PHYSICS. Educ: Drexel Univ, BS, 66; Stevens Inst Technol, MS, 69, PhD(physics), 71. Prof Exp: Fel plasma physics, Univ Sask, 71-73; adj asst prof, 73-75, VIS ASST PROF PHYSICS, UNIV S FLA, 75- Mem: Am Phys Soc; Inst Elec & Electronic Engrs. Res: Nonlinear plasma wave theory; linear and nonlinear low frequency waves in plasmas with density, potential and temperature gradients; catalytic turbulent heating of plasmas as supplementary tokamac heating. Mailing Add: Dept of Physics Univ of S Fla Tampa FL 33620

LEE, ARTHUR CLAIR, b Abilene, Kans, Aug 3, 23; m 51; c 5. VETERINARY MEDICINE, RADIOLOGICAL HEALTH. Educ: Colo State Univ, DVM, 52, MS, 63, PhD(radiation biol), 70. Prof Exp: Practitioner vet med, 52-60; Morris Found fel, 60-62; vet radiologist, AEC Proj, 63-66, VET SECT LEADER, COLLAB RADIOL HEALTH LAB, USPHS, COLO STATE UNIV, FOOTHILLS CAMPUS, 64- Mem: AAAS; Am Vet Med Asn; Radiation Res Soc Am. Res: Late effects of ovarian irradiation; radiation effects on canine growth and development; ocular lesions as a result of age at exposure. Mailing Add: Collab Radiol Health Lab Colo State Univ Foothills Campus Ft Collins CO 80523

LEE, BENJAMIN KWOK-CHU, pharmaceutical chemistry, see 12th edition

LEE, BENJAMIN W, b Seoul, Korea, Jan 1, 35; US citizen; m 61; c 2. THEORETICAL PHYSICS. Educ: Miami Univ, BS, 56; Univ Pittsburgh, MS, 58; Univ Pa, PhD(physics), 60. Prof Exp: From res assoc to prof physics, Univ Pa, 60-66; prof physics & mem, Inst Theoret Physics, State Univ NY Stony Brook, 66-73; HEAD THEORET PHYSICS DEPT, FERMI NAT ACCELERATOR LAB, 73-; PROF PHYSICS, ENRICO FERMI INST & DEPT PHYSICS, UNIV CHICAGO, 74- Concurrent Pos: Mem, Inst Advan Study, 61-62 & 64-66; Sloan Found res fel, Pa, 63-66; Guggenheim fel, 68-69; vis prof, Fac Sci d'Orsay, Paris & Inst Advan Sci Study, France, 68-69. Mem: Am Phys Soc. Res: Elementary particle physics; scattering theory; quantum field theory; symmetry groups; high energy physics. Mailing Add: Fermi Nat Accelerator Lab Batavia IL 60510

LEE, BYUNGKOOK, b Korea, Feb 7, 41; m 64. X-RAY CRYSTALLOGRAPHY, BIOLOGICAL STRUCTURE. Educ: Seoul Nat Univ, BS, 61; Cornell Univ, PhD(phys chem), 67. Prof Exp: USPHS res fel, Yale Univ, 69-70; asst prof, 70-74, ASSOC PROF CHEM, UNIV KANS, 74- Mem: Am Chem Soc; Am Crystallog Asn. Res: Structural studies of large and small molecules through x-ray crystallographic techniques. Mailing Add: Dept of Chem Univ of Kans Lawrence KS 66045

LEE, CHANG LING, b Chiang King, China, Nov 4, 16; nat US; m 45; c 3. IMMUNOLOGY, PATHOLOGY. Educ: West China Union Univ, MD, 43; Univ Liverpool, DTM&H, 50; Am Bd Path, dipl, 62, cert blood banking, 73. Prof Exp: Instr internal med, West China Union Univ, 49; res assoc & instr microbiol, Univ Chicago, 53-57; head div immunohemat, Mt Sinai Hosp Med Ctr, 57-69; DIR BLOOD CTR, MT SINAI HOSP MED CTR, 69-; PROF MED & PATH, RUSH MED COL, 75- Concurrent Pos: Res asst prof, Chicago Med Sch, 58-68, prof path, 68-74; sci dir, Mid-Am Regional Red Cross Blood Prog, 75- Honors & Awards: Morris H Parker Award, 65 & 68. Mem: Am Soc Exp Path; Am Asn Immunol; Am Soc Clin Path; AMA. Res: Immunochematology; immunopathology. Mailing Add: Mt Sinai Blood Ctr 2746 W 15th St Chicago IL 60608

LEE, CHANG YONG, b Korea, Apr 12, 35; m 68; c 1. FOOD SCIENCE. Educ: Cent Univ, Korea, BS, 58, MS, 61; Utah State Univ, PhD(food sci & technol), 67. Prof Exp: Chemist, Nat Indust Res Inst, Korea, 59-62; asst food sci, Cornell Univ, 63-64 & Utah State Univ, 64-67; NIH res fel food sci & technol, 67-69, asst prof, 69-74, ASSOC PROF FOOD SCI, EXP STA, NY STATE COL AGR & LIFE SCI, CORNELL UNIV, 74- Concurrent Pos: Res assoc, Cent Univ, Korea, 61-62; int travel grant awards, Inst Food Technologists, 74 & Am Inst Nutrit, 75. Mem: Am Chem Soc; Inst Food Technologists; Sigma Xi. Res: Biochemical studies of nitrogen compounds in food of plant origin; nitrate, nitrite glutamine, pyrrolidon carboxylic acid and carbohydrate chemistry of foods, flavor chemistry and enology; nutritional quality of processing vegetables. Mailing Add: NY State Col Agr & Life Sci Agr Exp Sta Cornell Univ Geneva NY 14456

LEE, CHARLES ALBERT, b Lindsay, Calif, Feb 7, 08; m 31; c 3. ECONOMIC GEOLOGY. Educ: Univ Idaho, BS, 34; Univ Ariz, MS, 35. Prof Exp: Mine foreman, Sombrero Butte Mining Co, Ariz, 34-35; field asst geol, US Geol Surv, Idaho, 35; engr-geologist, Soil Conserv Serv, USDA, Ariz, 36; engr & geologist, Idaho-Md Gold Mines Corp, Calif, 36-37; geologist-engr, Duranzo Gold Mines, Mex, 38 & Am Smelting & Refining Co, Ariz, 38-40; asst prof geol, Southern Br, Univ Idaho, 40-42; chief engr-geologist, Peru Mining Co & NMex Consol Mining Co, NMex, 42-44; geologist, Co Minera Asarco, Am Smelting & Refining Co, Mex, 44-45; prof goel & head dept, Idaho State Col, 45-51; prof geol & phys sci, 53-73, EMER PROF GEOL GEOG, SANTA MONICA COL, 73- Concurrent Pos: Geologist, Ymir Consol Mines Co, Idaho, 41-42; lectr, Exten Serv, Idaho State Col; mining geologist, Union Pac RR Co, 52-; lectr, Univ Calif Exten, 55- Mem: Geol Soc Am; Soc Econ Geol; Nat Asn Geol Teachers; Am Inst Mining, Metall & Petrol Eng. Res: Phosphate deposits of

Idaho; ore deposits, particularly as applied to discovery of new deposits. Mailing Add: Santa Monica Col 1815 Perl St Santa Monica CA 90405

LEE, CHARLES ALEXANDER, b New York, NY, Aug 28, 22; m 53; c 1. PHYSICS. Educ: Rensselaer Polytech Inst, BEE, 44; Columbia Univ, PhD(physics), 54. Prof Exp: Res assoc molecular beam spectros, Columbia Univ, 52-53; mem tech staff semiconductors, Bell Tel Labs, 53-67; PROF ELEC ENG, CORNELL UNIV, 67- Mem: Am Phys Soc; Inst Elec & Electronics Engr. Res: Solid state device physics; molecular beam spectroscopy. Mailing Add: Sch of Elec Eng Cornell Univ Phillips Hall Ithaca NY 14850

LEE, CHARLES BRUCE, environmental sciences, see 12th edition

LEE, CHARLES RICHARD, b Tarrytown, NY, Dec 3, 42; m 65; c 3. SOIL CHEMISTRY, PLANT NUTRITION. Educ: Univ Tampa, BS, 64; Clemson Univ, MS, 65, PhD(agron), 68. Prof Exp: Res scientist, Can Dept Agr, 68-73; RES SOIL SCIENTIST, US ARMY CORPS OF ENGRS, ENVIRON EFFECTS LAB, WATERWAYS EXP STA, 73- Mem: Am Chem Soc; Am Soc Agron; Can Soc Soil Sci. Res: Land treatment of wastewater; heavy metal uptake by marsh plants; soil fertility; minor elements; influence of soil acidity on potato production; plant growth in high zinc, aluminum or manganese media. Mailing Add: Corps Lab US Army Environ Effects Engr Waterways Exp Sta PO Box 631 Vicksburg MS 39180

LEE, CHARLES WAH, organic chemistry, see 12th edition

LEE, CHARLOTTE ELIZABETH OUTLAND, b Boligee, Ala, July 13, 30; m 59; c 4. BIOCHEMISTRY. Educ: Knoxville Col, BS, 53; Tuskegee Inst, MS, 55; Univ Kans, PhD(biochem), 59. Prof Exp: Asst prof nutrit, Univ Kans, 59-62; assoc prof chem, Ala Agr & Mech Col, 63-64, prof, 64-68; asst prof, Nassau Community Col, 70-71; FEL, MED SCH, ST LOUIS UNIV, 71- Concurrent Pos: NIH res grant, 64-67; NASA res grant, 65-68. Mem: Am Chem Soc; NY Acad Sci. Res: Lipid chemistry and metabolism; biochemistry of behavior. Mailing Add: 6806 Waterman Ave University City MO 63130

LEE, CHEN HUI, b Taipei, Taiwan, Dec 2, 29; m 62; c 2. FORESTRY. Educ: Nat Taiwan Univ, BS, 53; Mich State Univ, MS, 60, PhD(forestry), 66. Prof Exp: Asst forestry, Nat Taiwan Univ, 54-59, instr, 59-62; res asst, Mich State Univ, 62-66; asst prof, 66-70, ASSOC PROF FORESTRY, UNIV WIS-STEVENS POINT, 70- Mem: Soc Am Foresters; Chinese Soc Forestry; Japanese Forestry Soc. Res: Forest genetics and tree improvement, especially tree physiology, pine leaf anatomy and wood quality. Mailing Add: Col of Natural Resources Univ of Wis Stevens Point WI 54481

LEE, CHENG-CHUN, b Youngchow, China, May 24, 22; nat US; m 58; c 2. PHARMACOLOGY, TOXICOLOGY. Educ: Nat Cent Univ, China, BS, 45, MS, 48; Mich State Univ, MS, 50, PhD(physiol), 52. Prof Exp: Asst vet med, Nat Cent Univ, China, 45-48; asst physiol, Mich State Univ, 49-51; from pharmacologist to sr pharmacologist, Eli Lilly & Co, 52-62; from sr pharmacologist to prin pharmacologist, 62-67, HEAD PHARMACOL & TOXICOL, MIDWEST RES INST, 67- Concurrent Pos: Lectr, Univ Mo-Kansas City, 65-66 & Med Ctr, Univ Kans, 66- Mem: Am Physiol Soc; Am Soc Pharmacol & Exp Therapeut; Soc Toxicol; Soc Exp Biol & Med; NY Acad Sci. Res: Teratology; carcinogenesis; drug metabolism and disposition; antimalarials; antineoplastics; pharmaceuticals; chemicals; drug development; mechanism of drug action and toxicity; liver and renal functions. Mailing Add: Midwest Res Inst 425 Volker Blvd Kansas City MO 64110

LEE, CHEUK MAN, b China, Feb 22, 29. ORGANIC CHEMISTRY, PHARMACEUTICAL CHEMISTRY. Educ: Univ Hong Kong, BSc, 54, MSc, 57; Univ Mich, PhD(pharmaceut chem), 60. Prof Exp: SR RES CHEMIST, ABBOTT LABS, 60- Mem: Am Chem Soc. Res: Synthesis of organic compounds of neuropharmacological activities; heterocyclic chemistry; neurotropic drugs. Mailing Add: Abbott Labs 1400 Sheridan Rd North Chicago IL 60064

LEE, CHIA-MING, b Yun-Nan, China, Nov 16, 45; m 73. ATOMIC PHYSICS. Educ: Nat Taiwan Univ, BS, 68; Univ Chicago, MS, 73, PhD(physics), 74. Prof Exp: Res assoc theoret atomic physics, Univ Chicago, 74; RES ASSOC QUANTUM ELECTRODYNAMICS ATOMIC PHYSICS, UNIV PITTSBURGH, 74- Mem: Am Phys Soc. Res: Collision theory and its application to atomic and molecular structures, specifically photoexcitation and photoionization processes; bremsstrahlung and radiative electron capture processes in neutral ionized atoms. Mailing Add: Dept of Physics Univ of Pittsburgh Pittsburgh PA 15260

LEE, CHI-HANG, b Vinh Long, SVietnam, Jan 1, 39; nat US; m 64; c 2. NATURAL PRODUCTS CHEMISTRY. Educ: Southern Ill Univ, Carbondale, BA, 60; Rutgers Univ, New Brunswick, PhD(natural prod chem), 66. Prof Exp: Res asst, Rutgers Univ, 61-62, res assoc, 66; sr chemist, 67-71, RES SPECIALIST, CORP RES DEPT, GEN FOODS CORP, 72- Concurrent Pos: Vis prof, King's Col, 73- Mem: Am Chem Soc; Am Sci Affiliation. Res: Amino sugars; polyene macrolide antibiotics; nitrogen heterocycles; structural elucidation of natural products. Mailing Add: Corp Res Dept Gen Foods Corp White Plains NY 10625

LEE, CHIN-CHIU, b Hunan, China, Aug 10, 34; m 64; c 1. MICROBIOLOGY, ELECTRON MICROSCOPY. Educ: Taiwan Norm Univ, BSc, 55; Loyola Univ, MS, 64; La State Univ, PhD(parasitol), 68. Prof Exp: Biol teacher, Taiwan Prov Agr Sch, 55-56 & High Sch, Taiwan, 56-58; asst instr biol, Taiwan Norm Univ, 58-59 & Nanyang Univ, 59-61; res technologist biochem, La State Univ, 63, res assoc parasitol, 64; ASSOC PROF BIOL, KING'S COL, PA, 68- Mem: Am Soc Parasitol; Electron Micros Soc Am; AAAS. Res: Medical parasitology; studies on the physiological and ultrastructural aspects of parasites, particularly of parasitic nematodes. Mailing Add: Dept of Biol King's Col Wilkes-Barre PA 18702

LEE, CHING TSUNG, b Taiwan, July 1, 37; m 67; c 2. QUANTUM OPTICS, MATHEMATICAL PHYSICS. Educ: Nat Taiwan Univ, BS, 62; Rice Univ, MA, 65, PhD(physics), 67. Prof Exp: Welch Found fel, Tex A&M Univ, 67-68; NASA fel, Rice Univ, 68-69; assoc prof, 69-73, PROF PHYSICS & MATH, ALA A&M UNIV, 73- Mem: Am Phys Soc; Optical Soc Am. Res: X-ray laser; superradiance. Mailing Add: Dept of Physics Ala A&M Univ Normal AL 35762

LEE, CHING-TSE, b Sinchu, Taiwan, China, May 21, 40; m 69; c 2. ANIMAL BEHAVIOR. Educ: Nat Taiwan Univ, BS, 63; Bowling Green State Univ, MA, 67, PhD(psychol), 69. Prof Exp: Fac assoc psychol, Univ Tex-Austin, 69-71; ASST PROF PSYCHOL, BROOKLYN COL, CITY UNIV NEW YORK, 71- Concurrent Pos: Fac res award, City Univ New York, 71, 73, 74; fel, Dept Health Educ & Welfare, 74. Mem: Animal Behav Soc; Am Psychol Asn; AAAS; Behav Genetics Asn. Res: Investigation of animal communication processes through olfaction and hormonal determinants of the production of olfactory signals; effects of neonatal hormones on behavioral differentiation; mathematical models applied to animal behavior. Mailing Add: Dept of Psychol Brooklyn Col City Univ of New York Brooklyn NY 11210

LEE, CHIUNG PUH, b China; US citizen. Educ: Univ Minn, PhD(physiol), 53. Prof Exp: Res mem physiol, Nat Acad Sci, 41-43; res assoc, Med Sch, Univ Minn, 57-70; res assoc biochem, Med Sch, Univ Pittsburgh, 56-57; ASSOC PROF PHYSIOL, MED SCH, UNIV MINN, 72- Mem: Am Physiol Soc; Sigma Xi; AAAS; Soc Exp Biol Med. Res: Mechanism of blood cell destruction, controlling antibody formation, aging and nutrition; cardiac muscle physiology; energy metabolism, ion fluxes and excitation-contraction coupling. Mailing Add: Dept of Physiol Univ Minn 424 Millard Hall Minneapolis MN 55455

LEE, CHOI CHUCK, b Vancouver, BC, Apr 27, 24; m 53; c 4. ORGANIC CHEMISTRY. Educ: Univ Sask, BE, 47, MSc, 49; Mass Inst Technol, ScD(chem), 52. Prof Exp: Res off chem, Prairie Regional Lab, Nat Res Coun, 52-53; spec lectr chem, Univ Sask, 53-54; asst prof, Univ BC, 54-55; from asst prof to assoc prof, 55-63, PROF CHEM, UNIV SASK, 63- Mem: Am Chem Soc; Chem Inst Can. Res: Mechanisms of organic reactions; applications of radioactive isotopes as tracers; cereal and agricultural chemistry. Mailing Add: Dept of Chem Univ of Sask Saskatoon SK Can

LEE, CHONG SUNG, b Seoul, Korea, Sept 4, 39; nat US; m 72; c 2. BIOLOGICAL CHEMISTRY. Educ: Seoul Nat Univ, BS, 64; Calif Inst Technol, PhD(chem), 70. Prof Exp: Grad res asst biophys, Calif Inst Technol, 65-69; fel biochem, Harvard Med Sch, 69-72; ASST PROF MOLECULAR GENETICS, UNIV TEX, AUSTIN, 72- Concurrent Pos: Jane Coffin Childs Mem Fund for Med Res fel, 70-72. Mem: Genetics Soc Am; Am Soc Cell Biol; NY Acad Sci; Korean Chem Soc. Res: Structure and function of DNA and chromosomes in Drosophila. Mailing Add: Dept of Zool Univ of Tex Austin TX 78712

LEE, CHUAN-PU, b Tsing-Tao, China, Sept 24, 31. BIOCHEMISTRY, PHYSICAL CHEMISTRY. Educ: Nat Taiwan Univ, BS, 54; Ore State Univ, PhD(biochem), 61. Prof Exp: Instr chem, Nat Taiwan Univ, 54-56; res assoc biochem, Ore State Univ, 60-61; Johnson Found res fel, Univ Pa, 61-63; Jane Coffin Childs Mem Fund Med Res fel physiol chem, Wenner-Gren Inst, Stockholm, 63-65, docent, 65-66; mem staff, Johnson Found, Univ Pa, 66-75, from assoc prof to prof biochem, 70-75; PROF BIOCHEM, SCH MED, WAYNE STATE UNIV, 75- Concurrent Pos: USPHS career develop award, 68-73; ed, Biochimica et Biophysica Acta & Biochimica et Biophysica Acta Review on Bioenergetics, 73- Honors & Awards: Silver Medal, Chinese Chem Soc, 55; Merck Index Award, 60. Mem: AAAS; Chinese Chem Soc; Am Soc Biol Chem; Biophys Soc; NY Acad Sci. Res: Reaction mechanisms of electron and energy transfer in oxidative phosphorylation. Mailing Add: Dept of Biochem Sch of Med Wayne State Univ Detroit MI 48201

LEE, CHUNG, b Shanghai, China, Sept 18, 36; US citizen; m 66; c 2. REPRODUCTIVE ENDOCRINOLOGY, NUTRITION. Educ: Nat Taiwan Univ, BS, 59; WVa Univ, MS, 66, PhD(nutrit & endocrinol), 69. Prof Exp: USPHS fel, Albany Med Col, 69-71; assoc obstet & gynec, 71-74, ASST PROF UROL, MED SCH, NORTHWESTERN UNIV, CHICAGO, 74-; ASSOC STAFF MEM RES, EVANSTON HOSP, 74- Concurrent Pos: Am Cancer Soc grant, Med Sch, Northwestern Univ, Chicago, 73-74, USPHS grant, 74-75. Mem: Endocrine Soc; Soc Study Reprod; Sigma Xi. Res: Hormonal regulation of breast cancer; mechanism of estrogen action; regulation of ovarian function. Mailing Add: Dept of Urol Northwestern Univ Med Sch Chicago IL 60611

LEE, CHUNG N, b Sinuiju, Korea, Nov 7, 31. MATHEMATICS. Educ: Seoul Nat Univ, BA, 54; Univ Va, MA, 57, PhD, 59. Prof Exp: From instr to asst prof, 60-68, ASSOC PROF MATH, UNIV MICH, ANN ARBOR, 68- Mem: Am Math Soc. Res: Algebraic topology; transformation groups; topology of manifolds. Mailing Add: 1372 Pine Valley Ct Ann Arbor MI 48104

LEE, CLARENCE EDGAR, b San Jose, Calif, Aug 18, 31; c 3. MATHEMATICAL PHYSICS. Educ: Univ Calif, Berkeley, BA, 53; Cornell Univ, MA, 62; Univ Colo, Boulder, PhD(physics), 73. Prof Exp: STAFF MEM, LOS ALAMOS SCI LAB, 53-, TEAM LEADER, 71- Mem: Am Physics Soc; Am Nuclear Soc. Res: Theoretical, nuclear, plasma, solid state, chemical reactor, radiation and applied physics; hydrodynamics, chemical metallurgy and kinetics, heat transfer; computer sciences and numerical analysis; fuel technology. Mailing Add: Los Alamos Sci Lab PO Box 1663 Los Alamos NM 87545

LEE, CLARENCE MATTHEWS, zoology, see 12th edition

LEE, DAISY SI, b Peiping, China, July 21, 34; US citizen; m 66; c 2. PEDIATRICS, ALLERGY. Educ: Okla Baptist Univ, BA, 56; Bowman Gray Sch Med, MD, 61; Am Bd Pediat, dipl. Prof Exp: Intern med, Georgetown Div, Washington Gen Hosp, 61-62; resident pediat, St Luke's Hosp Ctr, New York, 62-64, NY Heart Asn fel med, 64-65; fel, Inst Nutrit Sci, Columbia Univ & Dept Med, St Lukes Hosp Ctr, 65-66; fel pediat, Sch Med, Stanford Univ, 66-69; pediatrician, Ctr Develop & Learning Disorders, 69-72, ASST PROF PEDIAT, MED CTR, UNIV ALA, BIRMINGHAM, 69-, DIR PEDIAT ALLERGY PROG, 72- Mem: NY Acad Sci; Sigma Xi. Res: Quantitation of milk antibodies in atopic children. Mailing Add: Children's Hosp 1601 Sixth Ave S Birmingham AL 35233

LEE, DANIEL DIXON, JR, b Dillon, SC, Sept 27, 35; m 58; c 4. ANIMAL NUTRITION. Educ: Clemson Univ, BS, 57, MS, 64; NC State Univ, Raleigh, PhD(biochem & nutrit), 70. Prof Exp: Res supvr biochem, NC State Univ, 67-70; asst prof, 70-73, ASSOC PROF MINERAL METAB, DEPT ANIMAL INDUST, SOUTHERN ILL UNIV, 73- Mem: Am Soc Animal Sci. Res: Trace mineral metabolism; nonprotein nitrogen utilization; wintering of cattle on crop residues and feeding of recycled animal wastes to ruminants. Mailing Add: Dept of Animal Indust Southern Ill Univ Carbondale IL 62901

LEE, DAVID ALLAN, applied mathematics, see 12th edition

LEE, DAVID MALLIN, b Brooklyn, NY, Jan 18, 44; m 66; c 3. EXPERIMENTAL NUCLEAR PHYSICS. Educ: Manhattan Col, BS, 66; Univ Va, PhD(physics), 71. Prof Exp: Res assoc physics, Univ Va, 71-74; STAFF MEM, LOS ALAMOS SCI LAB, 74- Mem: Am Phys Soc; Sigma Xi. Res: Medium energy nuclear physics; position sensitive detectors; P-P polarization; beam line instrumentation. Mailing Add: MP-7 Los Alamos Sci Lab Los Alamos NM 87544

LEE, DAVID MORRIS, b Rye, NY, Jan 20, 31; m 60; c 2. PHYSICS. Educ: Harvard Univ, AB, 52; Univ Conn, MS, 55; Yale Univ, PhD(physics), 59. Prof Exp: From instr to assoc prof, 59-68, PROF PHYSICS, CORNELL UNIV, 68- Concurrent Pos: Guggenheim fel, 66-67; guest assoc physicist, Brookhaven Nat Lab, 66-67. Mem: Am Phys Soc; NY Acad Sci. Res: Low temperature physics. Mailing Add: Lab Atomic & Solid State Physics Cornell Univ Dept of Physics Ithaca NY 14850

LEE, DAVID RAYMOND, b Salt Lake City, Utah, Dec 31, 37; m 62; c 2. CULTURAL GEOGRAPHY, GEOGRAPHY OF AFRICA. Educ: Chico State Col, BA, 60; Ohio State Univ, MA, 62; Univ Calif, Los Angeles, PhD(geog), 67. Prof Exp:

Lectr geog, Univ Khartoum, 64-66; asst prof, Univ Calif, Davis, 66-71; ASSOC PROF GEOG, FLA ATLANTIC UNIV, 71- Mem: Asn Am Geogr; Am Geog Soc; African Studies Asn. Res: Cultural landscapes of Africa. Mailing Add: Dept of Geog Fla Atlantic Univ Boca Raton FL 33432

LEE, DEAN RALPH, b Jackson, Tenn, July 31, 13; m 40. ORGANIC CHEMISTRY. Educ: Tex A&M Univ, MS, 39. Prof Exp: Instr, 41-43, chmn annex dept, 46-48, asst prof, 48-65, ASSOC PROF CHEM & DIR LAB, TEX A&M UNIV, 65- Concurrent Pos: Asst chief chemist, Crown Cent Petrol Corp, 43-46; consult, Int Consult, Inc. Mem: AAAS; Am Chem Soc. Res: Natural petroleum dye; gasoline stabilizing inhibitors; secondary recovery methods of petroleum crudes; production of furfural from cottonseed hulls; coordination in amines of the hexachloroplumbates; petroleum chemistry; tetra-iodo cadmates of amines. Mailing Add: RD 3 Box 309 Bryan TX 77803

LEE, DO-JAE, b Namwon, Korea, Jan 24, 28; US citizen; m 57; c 2. PHYSICAL ORGANIC CHEMISTRY. Educ: Long Beach State Col, BS, 60; San Diego State Col, MS, 64; Univ Calif, San Diego, PhD(chem), 67. Prof Exp: RES CHEMIST, TOMS RIVER CHEM CORP, 68- Mem: Am Chem Soc. Res: Development of new dyestuff and economic process for plant production. Mailing Add: 34 Oakside Dr Toms River NJ 08753

LEE, DONALD E, b Seoul, Korea, Dec 10, 22; US citizen; m 46; c 2. SCIENCE EDUCATION. Educ: Emmanuel Missionary Col, BA, 44; Ohio State Univ, MA, 51; NY Univ, PhD(sci educ, physics), 59. Prof Exp: Asst physics, Ill Inst Technol, 44; instr, Emmanuel Missionary Col, 46-47; instr sci & math, Mt Vernon Acad, Ohio, 47-53; instr, Greater NY Acad, 53-55; instr sci educ, NY Univ, 55-59; from asst prof to assoc prof physics & sci educ, 59-65, ASSOC PROF SCI EDUC & REGISTR, LOMA LINDA UNIV, LA SIERRA CAMPUS, 65-, ASSOC PROF PHYSICS, 73- Mem: Nat Asn Res Sci Teaching. Res: Mathematics education. Mailing Add: Loma Linda Univ La Sierra Campus Riverside CA 92505

LEE, DONALD EDWARD, b Detroit, Mich, Jan 31, 21; m 71. MINERALOGY. Educ: Carleton Col, AB, 43; Univ Minn, MS, 47; Stanford Univ, PhD(mineral, ore deposits), 54. Prof Exp: Sci consult, Nat Resources Sect, Supreme Comdr Allied Powers, Japan, 47-51; instr phys sci, Stanford Univ, 53-56, asst prof, 56-57; prof geol, Univ Bahia, 57-58; GEOLOGIST, EXP GEOCHEM & MINERAL BR, US GEOL SURV, 59- Mem: Geol Soc Am; Mineral Soc Am. Res: Manganese minerals of Japan; glaucophane schists of California; intrusive rocks of eastern great basin; applied mineralogy and geochemistry; granitoid rocks of the basin-range area. Mailing Add: 12979 W Ohio Ave Lakewood CO 80228

LEE, DONALD GARRY, b Midale, Sask, June 21, 35; m 59; c 3. PHYSICAL CHEMISTRY, ORGANIC CHEMISTRY. Educ: Univ Sask, BA, 58, MA, 60; Univ BC, PhD(chem), 63. Prof Exp: Instr chem, Camrose Lutheran Col, 62-65; res assoc, Harvard Univ, 65-66; assoc prof, Pac Lutheran Univ, 66-67; PROF CHEM, UNIV REGINA, 67- Mem: Chem Inst Can; Am Chem Soc. Res: Oxidation mechanisms; protonation studies. Mailing Add: Dept of Chem Univ of Regina Regina SK Can

LEE, DONALD GIFFORD, b Bedford, Pa, Dec 25, 14; m 38; c 4. COMPARATIVE ANATOMY. Educ: Univ Pa, VMD, 36. Prof Exp: From instr to assoc prof, 37-51, assoc dean sch vet med, 62-71, PROF ANAT, UNIV PA, 51- Mem: Am Vet Med Asn; Am Asn Anat; Am Asn Vet Anat (pres, 58-59). Res: Histology. Mailing Add: Dept of Animal Biol Univ Pa Sch Vet Med 3800 Spruce St Philadelphia PA 19174

LEE, DONALD JACK, b Goldendale, Wash, Jan 28, 32; m 58; c 3. NUTRITIONAL BIOCHEMISTRY. Educ: Wash State Univ, BS, 58, MS, 60; Univ Ill, PhD(nutrit, biochem), 65. Prof Exp: Assoc prof food sci & technol, Food Protection Sect, Ore State Univ, 65-75; ASST DIR, AGR RES CTR, WASH STATE UNIV, 75- Concurrent Pos: USPHS res grant, 66-75. Mem: AAAS; Am Inst Nutrit. Res: Nutritional biochemistry, especially lipid metabolism; toxicity and carcinogenicity of natural compounds. Mailing Add: Agr Res Ctr Wash State Univ Pullman WA 99163

LEE, DOROTHY DEMETRACOPOULOU, anthropology, see 12th edition

LEE, DOUGLAS HARRY KEDGWIN, b Bristol, Eng, Feb 22, 05; nat US; m 52. ENVIRONMENTAL SCIENCES. Educ: Univ Queensland, MSc, 27; Univ Sydney, MB & BS, 29, dipl trop med, 33, MD, 40; FRACP, 40; Am Bd Indust Hyg, dipl. Prof Exp: Med officer, Commonwealth Dept Health, Australia, 30-33; prof physiol, King Edward VII Col Med, Singapore, 35-36; prof physiol, Univ Queensland, 36-48, dean fac med, 38-42; prof physiol climat & lectr environ med, Johns Hopkins Univ, 48-55; chief res br, Off Qm Gen, 55-58; assoc sci dir res, Qm Res & Eng Command, 58-60; chief occup health res & training facility, USPHS, 60-66, assoc dir, Nat Inst Environ Health Sci, 66-73; CONSULT, 74- Concurrent Pos: Consult, US Qm Corps, 47-55 & Food & Agr Orgn, UN, 47-60; Cutter lectr, Sch Pub Health, Harvard Univ, 50; adj prof, NC State Univ, 68-74. Mem: AAAS; Am Physiol Soc; fel NY Acad Sci. Res: Climatic physiology, effects of climate on man and animals and application to clothing, housing and tropical development. Mailing Add: Deer Hill Rd Star Route St Thomas VI 00801

LEE, EDWARD KYUNG CHAI, b Seoul, Korea, Jan 24, 35; m 65. PHYSICAL CHEMISTRY. Educ: Kans Wesleyan Univ, BA, 59; Univ Kans, PhD(chem), 63. Prof Exp: Res assoc chem, Univ Kans, 63-65; from asst prof to assoc prof, 65-71, PROF CHEM, UNIV CALIF, IRVINE, 71- Mem: AAAS; Am Chem Soc; Am Phys Soc. Res: Photochemistry; atomic and molecular energy transfer; reaction kinetics; spectroscopy; lasers. Mailing Add: Dept of Chem Univ of Calif Irvine CA 92664

LEE, EDWARD LOUIS, physical chemistry, see 12th edition

LEE, EMERSON HOWARD, b Okmulgee, Okla, Feb 23, 21; m 48; c 4. PHYSICAL CHEMISTRY. Educ: Univ Tex, BS, 52, PhD(chem), 55. Prof Exp: Chemist, Darco Div, Atlas Powder Co, 46-50; res engr, Develop & Res Dept, Continental Oil Co, 54-56; res chemist, 56-59, res specialist, 59-60, group leader, 60-65, scientist, 65-75, SCI FEL, MONSANTO CO, 75- Mem: Am Chem Soc; Sigma Xi. Res: Surface chemistry and catalysis. Mailing Add: Monsanto Co 800 N Lindbergh Blvd St Louis MO 63166

LEE, FANG AN, b Shanghai, China, Oct 9, 31; US citizen; m 60. HYDRODYNAMICS, PHYSICAL OCEANOGRAPHY. Educ: Nat Taiwan Univ, BS, 54; Va Polytech Inst & State Univ, MS, 56; Univ Wash, PhD(aeronaut, astronaut), 66. Prof Exp: Res engr, Boeing Co, 58-63; res assoc, 66-70, SR RES ASSOC HYDRODYN, UNIV WASH, 71- Mem: AAAS; Am Geophys Union. Res: Stability of stratified shear flows, wind-generated waves. Mailing Add: Dept of Oceanog Univ of Wash Seattle WA 98105

LEE, FLOYD DENMAN, b Hays, Kans, Apr 27, 38; m 65. NUCLEAR PHYSICS. Educ: Univ Kans, BS, 60, PhD(physics), 66. Prof Exp: Instr physics, Univ Kans, 65-66; Nat Acad Sci-Nat Res Coun assoc, 66-68; ASST PROF PHYSICS, MONT STATE UNIV, 68- Mem: Am Asn Physics Teachers; Am Phys Soc. Res: Low-energy nuclear research with Van-de-Graaf accelerators; nuclear structure. Mailing Add: Dept of Physics Mont State Univ Bozeman MT 59715

LEE, FRED GUEY HONG, organic chemistry, see 12th edition

LEE, FREDERICK STRUBE, b Baltimore, Md, Dec 26, 27; m 52; c 3. PHYSICAL CHEMISTRY. Educ: Johns Hopkins Univ, AB, 50; Brown Univ, PhD(chem), 58. Prof Exp: Res chemist, Agr Div, W R Grace & Co, 58-60 & Anal & Phys Div, 60-61; prof chem, Baltimore Jr Col, 61-71; DIR GEN STUDIES, COMMUNITY COL BALTIMORE, 71- Mem: Am Chem Soc. Res: X-ray crystallography; inorganic synthesis; physical inorganic chemistry. Mailing Add: Rt 2 Box 199 Owings Mills MD 21117

LEE, GARTH LORAINE, b Hinkley, Utah, Sept 25, 20; m 43; c 9. PHYSICAL CHEMISTRY. Educ: Univ Utah, BA, 44, MA, 47; Univ Toronto, PhD(phys chem), 49. Prof Exp: Asst chem, Univ Utah, 45-46; asst, Univ Toronto, 46-48, spec lectr, 48-49; instr, Univ Colo, 49-51, asst prof, 52-54; assoc prof, 54-60, head dept chem, 69-73, PROF CHEM, UTAH STATE UNIV, 61-, HEAD DEPT CHEM & BIOCHEM, 73- Mem: Am Chem Soc. Mailing Add: Dept of Chem & Biochem Utah State Univ Logan UT 84321

LEE, GARY ALBERT, b Scottsbluff, Nebr, May 18, 41; m 62; c 2. WEED SCIENCE. Educ: Univ Wyo, BS, 64, MS, 65, PhD(agron), 71. Prof Exp: Instr weed sci, Univ Wyo, 65-71, from asst prof to assoc prof, 71-75; PROF WEED SCI, UNIV IDAHO, 75- Concurrent Pos: US Borax Concurrent Pos: Consult, US Borax Res Corp, 75- Mem: Weed Sci Soc Am; Am Soc Sugarbeet Technologists; Soc Range Mgt. Res: Mechanisms of herbicide selectivity in agronomic crops and perennial weed control; population dynamics of weeds in agronomic crops and rangeland; influence of herbicides on the metabolism of weed species. Mailing Add: Dept of Plant & Soil Sci Univ of Idaho Col of Agr Moscow ID 83843

LEE, GEORGE FRED, b Delano, Calif, July 27, 33; m 53; c 1. WATER CHEMISTRY, ENVIRONMENTAL QUALITY. Educ: San Jose State Col, BA, 55; Univ NC, MSPH, 57; Harvard Univ, PhD(water chem-environ qual), 60. Prof Exp: Res assoc water chem, Harvard Univ, 59-60; asst prof, Univ Pittsburgh, 60-61; from asst prof to prof, Univ Wis-Madison, 61-73; prof, Tex A&M Univ, 73-74; PROF WATER CHEM, UNIV TEX, DALLAS, 74- Mem: AAAS; Am Chem Soc; Am Fisheries Soc; Am Soc Microbiol; Ecol Soc Am. Res: Chemistry of natural waters; water supply and pollution control. Mailing Add: Ctr for Environ Studies Univ of Tex Dallas PO Box 688 Richardson TX 75080

LEE, GEORGE H, II, b Ithaca, NY, Feb 26, 39; m 64; c 2. PHYSICAL CHEMISTRY, ANALYTICAL CHEMISTRY. Educ: Rensselaer Polytech Inst, 61, PhD(phys chem), 65. Prof Exp: Res assoc, Cornell Univ, 65-67; res chemist, Res Ctr, Hercules Inc, Del, 67-71; Sr Res Chemist, Dept Phys & Biol Sci, Southwest Res Inst, 71-73; ASSOC FOUND SCIENTIST, SOUTHWEST FOUND RES & EDUC, 73- Mem: Am Chem Soc. Res: Determination of structure and properties of certain boron-nitrogen ring compounds and use of electron diffraction and mass spectrometric instrumentation in structure analyses; quantitative microchemical analysis of trace elements and other environmental pollutants. Mailing Add: 11107 Whispering Wind San Antonio TX 78230

LEE, GERHARD BJARNE, b Deerfield, Wis, Nov 27, 17; m 48; c 4. SOILS. Educ: Univ Wis, BS, 48, MS, 49, PhD(soils), 55. Prof Exp: Asst agronomist, Soil Surv, SDak State Col, 49-51; from instr to asst prof 51-60, ASSOC PROF SOILS, UNIV WIS-MADISON, 60- Concurrent Pos: Soil scientist, USDA, 51-59. Mem: Fel AAAS; Soil Sci Soc Am; Am Soc Agron. Res: Soil morphology; genesis of podzolic profiles; relationship of soils to their parent materials; classification and correlation of soils; methodlogy of soil survey; characteristics of lake sediments as related to adsorption of pesticides; use of soil maps in zoning. Mailing Add: 204 Soils Bldg Univ of Wis Col of Agr Madison WI 58706

LEE, GLENN RICHARD, b Ogden, Utah, May 18, 32; m 69; c 2. INTERNAL MEDICINE, HEMATOLOGY. Educ: Univ Utah, BS, 53, MD, 56. Prof Exp: Intern med, Boston City Hosp, 56-57, asst resident, 57-58; clin fel hemat, 60-61, res fel, 61-63, from instr to assoc prof, 63-73, assoc dean acad affairs, 73-76, PROF MED, COL MED, UNIV UTAH, 73- Mem: Am Fedn Clin Res; Am Soc Hemat; Am Col Physicians; Am Soc Clin Invest. Res: Clinical and experimentally induced abnormalities in heme biosynthesis; physiologic consequences of copper deficiency; iron metabolism. Mailing Add: Dept of Med Univ of Utah Col of Med Salt Lake City UT 84112

LEE, HAROLD HON-KWONG, b China, Jan 31, 34; m 66; c 1. DEVELOPMENTAL BIOLOGY, VIROLOGY. Educ: Okla Baptist, AB, 56; Univ Tenn, MS, 58, PhD(embryol), 65. Prof Exp: USPHS fel, Carnegie Inst, 65-67; DIR, NSF INSTR SCI EQUIP GRANT, UNIV TOLEDO, 67-, PROF BIOL, 75- Concurrent Pos: Am Cancer Soc grant. Honors & Awards: Labor Found Res Award. Mem: Soc Develop Biologists; Tissue Culture Asn; Soc Reprod Biol. Res: Cell interactions and biochemical differentiation; development of endocrine system in testis. Mailing Add: Dept of Biol Univ of Toledo Toledo OH 43606

LEE, HAYNES A, b Johnson City, Tenn, Oct 14, 32; m 59; c 3. LASERS, GLASS TECHNOLOGY. Educ: Emory & Henry Col, BS, 54; State Univ NY Col Ceramics, Alfred Univ, MS, 61. Prof Exp: Glass technologist, Thatcher Glass Mfg Co, NY, 61-63; glass technologist, 63-66, glass scientist, 66-68, CHIEF LASER SCIENTIST, OWENS-ILL, INC, TOLEDO, 68- Mem: Am Ceramic Soc; Sigma Xi. Res: Electronic pheonmena in glasses, particularly laser phenomena. Mailing Add: 5946 Gillingham Dr Sylvania OH 43560

LEE, HENRY C, b China, Nov 22, 38; US citizen; m 63; c 2. FORENSIC SCIENCE, BIOCHEMISTRY. Educ: John Jay Col NY, BS, 72; NY Univ, MS, 74, PhD(biochem), 75. Prof Exp: ASST PROF FORENSIC SCI, UNIV NEW HAVEN, 75- Concurrent Pos: Consult, Conn State Police Forensic Lab & NY Univ, 75. Res: Protein biosynthesis; blood individualization and forensic science. Mailing Add: Univ of New Haven 300 Orange Ave West Haven CT 06516

LEE, HENRY FOSTER, b Sargentville, Maine, Sept 13, 13; m 39; c 1. PEDIATRICS. Educ: Syracuse Univ, BA, 36, MD, 39. Prof Exp: Intern, Bryn Mawr Hosp, 39-40; intern, Children's Hosp, Philadelphia, 40-41, from asst chief resident to chief resident, 41-43, asst & actg dir out-patient dept, 43-45; assoc, 48-54, asst prof, 54-69, ASSOC PROF PEDIAT, SCH MED, UNIV PA, 69- Concurrent Pos: Chief pediatrician, Pa Hosp, 48-58; assoc pediatrician, Children's Hosp, Philadelphia, 48-62, sr assoc gen pediat, 62-; chief pediatrician, Chestnut Hill Hosp, 58- Mem: Fel Am Acad Pediat. Res: Health insurance. Mailing Add: 8236 Germantown Ave Philadelphia PA 19118

LEE, HENRY LAWRENCE, JR, polymer chemistry, see 12th edition

LEE, HERBERT CARL, b Madison, Wis, Jan 18, 11; m 42; c 4. SURGERY. Educ:

Univ Wis, BA, 33, MD, 35; Am Bd Surg, dipl, 42. Prof Exp: From asst prof to assoc prof, 41-50, clin prof, 50-64, PROF SURG, MED COL VA, 64- Mem: Am Col Surgeons. Res: Clinical and experimental surgery. Mailing Add: Dept of Surg Med Col of Va Richmond VA 23219

LEE, HOONG-CHIEN, b Hong Kong, Aug 12, 41; m 65; c 2. PHYSICS. Educ: Nat Taiwan Univ, BSc, 63; McGill Univ, MSc, 67, PhD(physics), 69. Prof Exp: Tech collabr physics, Brookhaven Nat Lab, 67-68; RES OFFICER THEORET PHYSICS, CHALK RIVER NUCLEAR LABS, ATOMIC ENERGY CAN, LTD, 68- Mem: Am Phys Soc; Can Asn Physicists. Res: Theoretical nuclear physics; structure of nuclei. Mailing Add: Physics Div Chalk River Nuclear Labs Chalk River ON Can

LEE, HOWARD AUGUSTUS, b New Brunswick, NJ, Oct 27, 35; m; c 3. BIOCHEMISTRY. Educ: Denison Univ, BS, 55; Univ Ky, PhD(biochem), 70. Prof Exp: Res asst chem, Ohio State Univ, 57-60; res assoc biochem, Univ Mich, 60-63; sr res chemist, Anheuser-Busch, Inc, 68-73; dir labs, Reliable Chem Co, 73-75; PRES, LEE SCI, INC, ST LOUIS, 75- Mem: Am Chem Soc; AAAS; Sigma Xi. Res: Industrial use and preparation of enzymes, mainly for the food industry; mechanism of enzyme action; coenzyme B12 dependent enzymes; immobilization of enzymes; preparation of clinical diagnostic enzymes and procedures. Mailing Add: 9856 Madison Ave St Louis MO 63119

LEE, HSIN-YI, b Hsin-chu, Taiwan. DEVELOPMENTAL BIOLOGY. Educ: Nat Taiwan Univ, BS, 59; Oberlin Col, MA, 64; Univ Minn, Minneapolis, PhD(zool), 67. Prof Exp: Res assoc tissue cult, Cardiovasc Inst, Michael Reese Hosp & Med Ctr, 68; asst prof biol, 68-73, res coun grant, 69-76, ASSOC PROF ZOOL, RUTGERS UNIV, CAMDEN, 73- Mem: AAAS; Am Soc Zoologists; Sigma Xi. Res: Cell differentiation and biochemistry of chick development. Mailing Add: Dept of Biol Rutgers Univ Camden NJ 08102

LEE, HUA-TSUN, b Nanking, China, May 11, 37; m 63; c 4. MATHEMATICS. Educ: Tunghai Univ, Taiwan, BS, 59; Univ Pittsburgh, PhD(math), 71. Prof Exp: Asst physics, Tunghai Univ, Taiwan, 61; asst physics, Univ Pittsburgh, 61-64, asst math, 65-67 & 68-69, instr biostatist, Grad Sch Pub Health, 67-68; asst prof, 69-74, ASSOC PROF MATH, POINT PARK COL, 74- Mem: Math Asn Am. Res: Summability methods of infinite series. Mailing Add: Dept of Math Point Park Col Pittsburgh PA 15219

LEE, HULBERT AUSTIN, b Chelsea, Que, June 17, 23; m 47; c 5. GEOLOGY. Educ: Queen's Univ, Ont, BSc, 49; Univ Chicago, PhD(geol), 53. Prof Exp: Geologist, Geol Surv Can, 50-69; CONSULT GEOLOGIST & PRES, LEE GEO-INDICATORS LTD, 69- Concurrent Pos: Vis lectr, Univ NB, 64-65. Mem: Fel Geol Soc Am. Res: Correlation of quaternary events around Hudson Bay, the Tyrrell Sea and Keewatin ice divide; quaternary studies in New Brunswick; esker and till methods of mineral exploration; kimberlite petrology. Mailing Add: Lee Geo-Indicators Ltd 94 Alexander St Box 68 Stittsville ON Can

LEE, HYUNG JOON, solid state physics, astrophysics, see 12th edition

LEE, HYUNG MO, b Tanchon, Korea, Sept 27, 26; US citizen; m 59; c 2. MEDICINE, SURGERY. Educ: Keijo Imp Univ, BS, 45; Seoul Nat Univ, MD, 49. Prof Exp: Res fel surg, Med Col Va, 59-61; from instr to assoc prof, 63-70, PROF SURG, MED COL VA, 70- Mem: Am Col Surg; Am Soc Nephrology; Transplantation Soc; NY Acad Sci. Res: Renal homotransplantation. Mailing Add: Dept of Surg Med Col of Va Richmond VA 23219

LEE, HYUN-JAE, biochemistry, see 12th edition

LEE, ILBOK, biostatistics, see 12th edition

LEE, JAMES A, b Troy, NY, July 11, 25; m 46; c 1. HUMAN ECOLOGY, PUBLIC HEALTH. Educ: Union Col, BS, 49; Cornell Univ, MS, 51; George Washington Univ, MPh, 69, PhD, 70. Prof Exp: Res biologist, State Dept Fish & Game, NH, 51-53, sr res biologist, 53-56; tech asst dir, State Conserv Dept, Minn, 56-61, from dep comnr, State Dept Resources Devleop, 61-63; scientist adminr, USPHS, Washington, DC, 63-66, asst environ health to asst secy health & sci affairs, Dept Health, Educ & Welfare, 67-69, dir human ecol, 69-70; DIR ENVIRON & HEALTH, THE WORLD BANK, 70- Concurrent Pos: Adj assoc prof, Sch Med, George Washington Univ, 70-; vis prof, Sch Med, Cornell Univ, 76-; chmn Great Lakes panel & vchmn coastal zone comn, US Nat Coun Marine Sci; vchmn, Int Biol Prog, Nat Res Coun. Mem: Am Pub Health Asn; Ecol Soc Am; Am Soc Trop Med & Hyg; Asn Mil Surg US; Am Acad Health Admin. Res: Environmental, public health, and socio-cultural aspects of international economic development; human ecology with emphasis on multi-environmental causation of diseases; social anthropology and medical sociology; natural resources planning and management. Mailing Add: The World Bank Washington DC 20433

LEE, JAMES B, b Ware, Mass, June 30, 30; m 64; c 2. ENDOCRINOLOGY, METABOLISM. Educ: Col Holy Cross, AB, 51; Jefferson Med Col, MD, 56. Prof Exp: Intern med, St Vincent Hosp, Worcester, Mass, 56-57; resident, Pa Hosp, Philadelphia, 57-58; resident, Georgetown Univ Hosp, 58-59; USPH res fel renal metab, Peter Bent Brigham Hosp, Boston, 59-62; dir res metab & endocrinol, St Vincent Hosp, Worcester, 62-68; assoc prof med & chief sect exp med, Sch Med, St Louis Univ, 68-71; PROF MED, STATE UNIV NY BUFFALO, 71- Concurrent Pos: Mass Heart Asn res grant, 62-; USPHS res grant, 62-, develop training grant, 64-; asst prof, Georgetown Univ Hosp, 63-68. Mem: Am Physiol Soc; Am Soc Clin Invest; Am Fedn Clin Res; Endocrine Soc; Am Heart Asn. Res: Hypertension; in vitro metabolism of kidney cortex and medulla related to sodium excretion; isolation and identification of medullin, a peripheral visodilator, from renal medulla. Mailing Add: 100 High St Buffalo NY 14203

LEE, JAMES K, b Wuchang, China, Feb 4, 17; m 39; c 4. PHYSICAL CHEMISTRY, ANALYTICAL CHEMISTRY. Educ: Cent China Univ, BS, 37; Cath MS, 49, PHD(phys chem), 52. Prof Exp: Instr chem, Yale Col, 37-39; res chemist, China Indust Co, 40-44, chief chemist, 44-47; res chemist, Houdry Process Corp, Pa, 51-57; phys chemist, Monsanto Chem Co, Mo, 57-58; chmn radiation safety, 64-70, SR ENGR, TUBE DIV, RCA CORP, 58- Concurrent Pos: Prof phys chem, Taylor Univ, 59-72. Mem: AAAS; Am Chem Soc; Electrochem Soc; Sigma Xi. Res: Thermodynamics; reaction kinetics; surface chemistry; corrosion; radioactive tracer techniques; infrared spectroscopy; analytical methods; environmental science. Mailing Add: 3322 Huntington Rd Marion IN 46952

LEE, JAMES STEWART, b Washington, DC, Sept 18, 19; m 43; c 3. RESOURCE GEOGRAPHY, CARTOGRAPHY. Educ: Univ Ill, BA, 40; Indust Col Armed Forces, dipl, 63. Prof Exp: Cartog aide, New York Off, US War Dept, 41-44, cartog specialist, Off Chief of Staff, War Dept, 44-45, cartog specialist, Off Strategic Servs, 45; intel res specialist, Asst Chief of Staff for Intel, US Dept Army, 45-50; air intel specialist, Air Targets Div, Hq US Air Force, 50-62; supvry intel opers specialist & chief, Targets Systs Div, Targets Off, Defense Intel Agency, US Dept Defense, 63-74; RETIRED. Concurrent Pos: Lectr, US Dept Defense Schs, Cols & Overseas Units, 47-; head US deleg, Int Intel Prod Conf, London, 55; hon mem fac, US Army Intel Sch, Md, 66. Mem: Fel Am Geog Soc; Asn Am Geogr. Res: Military requirements. Mailing Add: 6813 Felix McLean VA 22101

LEE, JAMES WILLIAM, b North Vancouver, BC, Jan 16, 24; m 52; c 3. GEOLOGY. Educ: Univ BC, BASc, 47, MASc, 49; Stanford Univ, PhD(geol), 51. Prof Exp: Jr engr, Pac Lime Co, BC, 47; jr geologist, Kelowna Explor Co, 48-50; geologist, Kaiser Aluminum & Chem Corp, 51-58, Kaiser Bauxite Co, 58-63, mine engr, Jamaica, WI, 63-66; property mgt supt, Alumina Partners, 66-68; gen mgr, Farms Jamaica, Ltd, 68-73; LANDS MGR, ALPART FARMS, JAMAICA, LTD, 73- Concurrent Pos: Mem, Jamaica Sci Res Coun, 60-72. Mem: Geol Soc Am; Soc Am Archaeol; Soc Econ Geol; Archaeol Soc Jamaica (pres, 70-). Res: Petrography; petrology; Caribbean archaeology. Mailing Add: Alpart Farms Jamaica Ltd PO Box 206 Mandeville Jamaica

LEE, JEAN CHOR-YIN, b Canton, China, Aug 26, 41; m 67; c 2. BIOCHEMISTRY. Educ: Chung Chi Col, Hong Kong, dipl, 62; Univ Nebr, Lincoln, PhD(chem), 67. Prof Exp: Res instr biochem, Col Med, Univ Nebr, Omaha, 67-70; FEL CHEM & RES ASSOC, UNIV NEBR, LINCOLN, 70- Mem: Am Chem Soc. Res: Biomembranes, structure and transport. Mailing Add: Dept of Chem Univ of Nebr Lincoln NE 68504

LEE, JEAN T, b Milwaukee, Wis, Jan 31, 22; m 46; c 2. GEOPHYSICS, METEOROLOGY. Educ: Univ Chicago, BS, 44, MS, 62. Prof Exp: Instr math, Univ Southern Miss, 46-47; forecaster, US Weather Bur, 47-50, ed specialist, 50-54, forecaster, 54-59, RES METEOROLOGIST, NAT SEVERE STORMS LAB, ENVIRON SCI SERV ADMIN, NAT OCEANIC & ATMOSPHERIC ADMIN, 59- Honors & Awards: US Weather Bur Suggestion Award, 51; US Dept Com Meritorious Serv Medal, 58. Mem: Am Meteorol Soc; AAAS; Am Meteorol Soc. Res: Planning, development and research in severe local storms including atmospheric turbulence in the vicinity of thunderstorms, its effect on aircraft and its relationship to radar reflectivity characteristics. Mailing Add: Nat Severe Storm Lab Nat Ocean & Atmospheric Admin 1313 Halley Ave Norman OK 73070

LEE, JEN-SHIH, b Kwangtung, China, Aug 22, 40; nat US; m 66; c 3. BIOMEDICAL ENGINEERING. Educ: Nat Taiwan Univ, BS, 61; Calif Inst Technol, MS, 63, PhD(aeronaut, math), 66. Prof Exp: Asst res engr, Univ Calif, San Diesgo, 66-69; asst prof, 69-74, ASSOC PROF BIOMED ENG, UNIV VA, 74- Concurrent Pos: San Diego County Heart Asn advan res fel, 66-69; USPHS res grant, Univ Va, 71; Nat Heart & Lung Inst career develop award, 75- Mem: Am Physiol Soc; Biomed Eng Soc; Am Soc Mech Eng; Am Optical Soc; Microcirc Soc. Res: Hemodynamics; pulmonary mechanics and edema; noninvasive method for detecting atherosclerosis; indicator dilution technique as applied to microcirculation and transcapillary exchange in microvessels. Mailing Add: Div of Biomed Eng Univ of Va Charlottesville VA 22901

LEE, JOHN ALEXANDER HUGH, b Isle of Wight, Eng, Oct 10, 25; m 49; c 3. EPIDEMIOLOGY. Educ: Univ Edinburgh, BSc, 47, MB, ChB, 49, MD, 55; Univ London, DPH, 52. Prof Exp: Fel epidemiol, London Sch Hyg & Trop Med, 52-55; mem sci staff, Social Med Res Unit, Med Res Coun, London Hosp, 55-66; PROF EPIDEMIOL, UNIV WASH, 66- Mem: Am Epidemiol Soc; Brit Soc Social Med; Brit Med Asn; Int Epidemiol Asn. Res: Epidemiology of neoplastic diesease; operational research in health care programs. Mailing Add: Dept of Epidemiol & Int Health Univ of Wash Seattle WA 98195

LEE, JOHN CHUNG, b Shanghai, China, Mar 2, 37; US citizen; m 63; c 2. BIOCHEMISTRY. Educ: Taylor Univ, AB, 61; Purdue Univ, West Lafayette, MSc, 64, PhD(molecular biol), 67. Prof Exp: Res assoc biochem, Mass Inst Technol, 67-69; asst prof, 69-72, ASSOC PROF BIOCHEM, UNIV TEX MED SCH SAN ANTONIO, 72- Concurrent Pos: USPHS res grant, Univ Tex Med Sch San Antonio, 71- Mem: AAAS; Am Chem Soc; Am Soc Biol Chem. Res: Structure and function of nucleic acids; differentiation. Mailing Add: Dept of Biochem Univ of Tex Med Sch San Antonio TX 78284

LEE, JOHN DENIS, b Trinidad, WI, Apr 22, 29; m 58; c 3. METEOROLOGY. Educ: Fla State Univ, BS, 70, MS, 71, PhD(meteorol), 73. Prof Exp: Fel, Nat Ctr Atmospheric Res, 73-74; ASST PROF METEOROL, PA STATE UNIV, 74- Mem: Am Meteorol Soc. Res: Numerical modeling of cooling tower plumes and urban pollution; time series analysis. Mailing Add: Dept of Meteorol Pa State Univ University Park PA 16802

LEE, JOHN JOSEPH, b Philadelphia, Pa, Feb 23, 33; m 56; c 2. MARINE MICROBIOLOGY, PROTOZOOLOGY. Educ: Queens Col, NY, BS, 55; Univ Mass, MA, 57; NY Univ, PhD(biol), 60. Prof Exp: Asst prof, NY Univ, 61-66; from asst prof to assoc prof, 66-72, PROF BIOL, CITY COL NEW YORK, 72- Concurrent Pos: Res fel & dir, Living Foraminifera Lab, Am Mus Natural Hist, 60-68, res assoc, 70-; dir, Marine Microbiol Ecol Lab, Inst Oceanog, City Univ New York, 68-; res assoc, Lamont-Doherty Geol Observ, 70- Mem: Fel AAAS; Soc Protozool; Am Soc Parasitol; Am Soc Microbiol; Am Micros Soc. Res: Morphology, cytology, fine structure, life history, ecology, cultivation and nutrition of Foraminifera, intestinal flagellates, slime molds, dinoflagellates, diatoms and other marine organisms. Mailing Add: Dept of Biol City Col of New York New York NY 10031

LEE, JOHN WILLIAM, b Sydney, Australia, Apr 7, 35; m 60; c 2. CHEMICAL PHYSICS, BIOPHYSICS. Educ: Univ New South Wales, BSc, 56, PhD(phys chem), 60. Prof Exp: Res assoc biochem, McCollum-Pratt Inst, Johns Hopkins Univ, 61-63; staff scientist, New Eng Inst Med Res, 63-69; ASSOC PROF BIOCHEM, UNIV GA, 69- Mem: Am Chem Soc; Am Phys Soc. Res: Positron annihilation in matter; radiation chemistry; energy exchange processes in chemical and biological systems; bioluminescence; chemiluminescence; radiation physics. Mailing Add: Dept of Biochem Univ of Ga Athens GA 30601

LEE, JONATHAN K P, b Kiangsu, China, July 13, 37; m 67; c 2. NUCLEAR PHYSICS. Educ: McGill Univ, BEng, 60, MSc, 62, PhD(nuclear physics), 65. Prof Exp: Nat Res Coun Can overseas fel, 65-66; res asst, Univ Toronto, 66-68; ASST PROF NUCLEAR PHYSICS, McGILL UNIV, 68- Mem: Can Asn Physicists. Res: Nuclear structure studies. Mailing Add: Dept of Physics McGill Univ Montreal PQ Can

LEE, JONG SUN, b Suwon, Korea, July 10, 32; m 58; c 3. MICROBIOLOGY. Educ: Univ Calif, Berkeley, BA, 58; Ore State Univ, MS, 62, PhD(microbiol), 63. Prof Exp: Asst prof, 63-69, ASSOC PROF FOOD MICROBIOL, ORE STATE UNIV, 69- Mem: AAAS; Am Soc Microbiol; Inst Food Technol; Brit Soc Appl Bact. Res: Microbiology of seafoods. Mailing Add: Dept of Food Sci & Technol Ore State Univ Corvallis OR 97331

LEE, JORDAN GREY, JR, b Lafayette, La, Feb 28, 14; m 42; c 4. CHEMISTRY.

Educ: La State Univ, BS, 34, MS, 35; Univ Mo, PhD(agr chem), 40. Prof Exp: Asst agr chem, Univ Mo, 38-40; from instr to asst prof chem, Okla Agr & Mech Col, 40-43; from asst prof to assoc prof, 43-57, head dept biochem, 62-67, PROF BIOCHEM, LA STATE UNIV, BATON ROUGE, 57- Mem: AAAS; Am Chem Soc; Am Inst Nutrit; Am Pub Health Asn. Res: Nutriton; general biochemistry. Mailing Add: Dept of Biochem Coates Labs La State Univ Baton Rouge LA 70803

LEE, JOSEPH CHING-YUEN, b Hutow, Anchi, China, Feb 25, 22; m 54; c 1. ANATOMY. Educ: Lingnan Univ, BSc, 47; Univ Sask, MSc, 58, PhD(exp neuropath), 61, MD, 62. Prof Exp: Asst anat, Med Col, Lingnan Univ, 47-51, lectr, 51-54; demonstr, Med Fac, Univ Hong Kong, 55-57; res asst exp neuropath, Univ Sask, 57-58, spec lectr anat, 58-62, asst prof, 62-63; from asst prof neurosurg to assoc prof anat, 63-68, PROF ANAT & ASSOC RES PROF NEUROSURG, STATE UNIV NY BUFFALO, 68- Concurrent Pos: Lederle med fac award, Univ Sask & Buffalo Gen Hosp, NY, 62-65; Am Cancer Soc & Nat Inst Neurol Dis & Stroke res grant, State Univ NY Buffalo, 65-73. Mem: AAAS; Am Asn Anat; Am Soc Cell Biol; Electron Micros Soc Am; Am Asn Neurol Surg. Res: Experimental neuropathology; various types of cerebral edema, including those associated with intracranial tumors studied by combination of electron microscopy, autoradiography and cytochemistry; permeability of the cerebral blood vessels under normal and pathological conditions; pathogenesis of experimental brain tumors. Mailing Add: Dept of Anat Sci State Univ of NY Buffalo NY 14214

LEE, JOSEPH CHUEN KWUN, b Chungking, China, Oct 6, 38; m. PATHOLOGY, CELL BIOLOGY. Educ: Univ Hong Kong, MB, BS, 64; Univ Rochester, PhD(path), 70; FRCP(C), 71. Prof Exp: Intern surg, Queen Elizabeth Hosp, Hong Kong, 64-65; intern med, Grantham Hosp, Hong Kong, 65; rotating intern, St Francis Hosp, New York, 66; resident path, New York Hosp-Cornell Med Ctr, 66-67, Toronto Gen Hosp, Univ Toronto, 70-71 & Princess Margaret Hosp, Ont Cancer Inst Can, 71-72; ASST PROF PATH, MED CTR, UNIV ROCHESTER, 72- Mem: AAAS; Am Soc Cell Biol; Am Soc Exp Path; Am Asn Path & Bact. Res: Experimental pathology of iron metabolism; control of cell replication in normal cells and in neoplastic cells. Mailing Add: Dept of Path Univ of Rochester Med Ctr Rochester NY 14642

LEE, JOSEPH ROSS, mathematics, see 12th edition

LEE, JOSHUA ALEXANDER, b Rocky Ford, Ga, Oct 30, 24; m 56; c 2. GENETICS. Educ: San Diego State Col, AB, 50; Univ Calif, PhD(genetics), 58. Prof Exp: Technician, Univ Calif, 51-53, asst, 54-56; GENETICIST, AGR EXP STA, NC STATE UNIV, 58-, PROF CROP SCI, UNIV, 71- Mem: Crop Sci Soc Am; Soc Study Evolution. Res: Genetical problems pertaining to the improvement of domesticated cotton species. Mailing Add: 5104 New Castle Rd Raleigh NC 27606

LEE, JUI SHUAN, b Hopei, China, Aug 14, 13; US citizen. PHYSIOLOGY. Educ: Tsinghua Univ, Peking, BS, 36; Univ Minn, Minneapolis, PhD(physiol), 53. Prof Exp: Res fel, Biol lab, Sci Soc China, 36-38; res fel physiol, 39-41, from lectr physiol to assoc prof, Col Med, Nat Cent Univ, Chengtu, 42-47; res fel physiol, Univ Minn, Minneapolis, 53-56; res assoc biochem nutrit, Grad Sch Pub Health, Univ Pittsburgh, 56-57; from res assoc to asst prof physiol, 57-68, ASSOC PROF PHYSIOL, UNIV MINN, MINNEAPOLIS, 69- Mem: Am Physiol Soc; Microcirculatory Soc; Sigma Xi; AAAS. Res: Mechanism of water absorption and secretion by the mammalian small intestine; microvascular pressure and lymph pressure of the intestinal villi are determined during water transport under various pathophysiological conditions. Mailing Add: Dept of Physiol Univ of Minn Minneapolis MN 55455

LEE, KAH-HOCK, b Jan 28, 41; US citizen; m 67; c 3. INORGANIC CHEMISTRY. Educ: Nanyang Univ, BS, 64; Georgetown Univ, PhD(inorg chem), 70. Prof Exp: ENVIRON CHEMIST, DC DEPT ENVIRON SERV, 69- Concurrent Pos: Res fel, Dept Chem, Grad Sch, Georgetown Univ, 71-73; vis asst prof, Sch Eng, Dept Civil Eng, Howard Univ, 73-74. Mem: Am Chem Soc. Res: Environmental toxic trace metals; effect of x-ray film developer on radiation protection; heteropoly inorganic anions exchange mechanisms. Mailing Add: 6201 S Crain Hwy Mitchellville MD 20216

LEE, KAI NIEN, b New York, NY, Oct 19, 45; m 71. ENVIRONMENTAL MANAGEMENT, SCIENCE POLICY. Educ: Columbia Univ, AB, 66; Princeton Univ, PhD(physics), 71. Prof Exp: Soc Sci Res Coun res training fel, 71-72; asst res social scientist, Inst Governmental Studies, Univ Calif, Berkeley, 72-73; res asst prof, Prog Social Mgt of Technol & Dept Polit Sci, 73-75, ASST PROF INST ENVIRON STUDIES & DEPT POLIT SCI, UNIV WASH, 75- Concurrent Pos: Mem reading comt, Kent Fel Prog, Danforth Found, 72-; mem policy adv bd, Environ Impact Assessment Proj, Inst Ecol, 74-75; mem adv coun environ educ, HEW, 76-; mem int adv bd, Policy Sci, 74-; consult nuclear waste, Human Affairs Res Ctrs, Seattle & Pac Northwest Labs, Battelle Mem Inst, Richland, 74-; consult power plant siting, US Nuclear Regulatory Comn, 75. Mem: Fel Soc Religion in Higher Educ; AAAS; Fedn Am Scientists. Res: Energy and environmental policy and politics, especially nuclear energy-power plant siting, regional power development and nuclear waste management; influence of technological change on American political and economic life. Mailing Add: Inst for Environ Studies FM-12 Univ of Wash Seattle WA 98195

LEE, KAI-LIN, b Nanking, China, Sept 16, 35; m 64. BIOCHEMISTRY, ENDOCRINOLOGY. Educ: Nat Taiwan Univ, BS, 60; Tulane Univ, PhD(biochem), 66. Prof Exp: Teaching asst bot, Nat Taiwan Univ, 61-62; res asst biochem, Tulane Univ, 62-66, fel biochem endocrinol, 66-67, instr, 67-68; Hoffmann-La Roche fel, Biochem, Biol Div, 68-69, res assoc, 69-70, BIOCHEMIST, BIOL DIV, OAK RIDGE NAT LAB, 70- Mem: AAAS; Endocrine Soc; Am Chem Soc; Am Soc Biol Chemists. Res: Hormonal regulation of metabolic processes. Mailing Add: Biol Div Oak Ridge Nat Lab PO Box Y Oak Ridge TN 37830

LEE, KEENAN, b Huntington, NY, Nov 20, 36; m 66; c 2. GEOLOGY. Educ: La State Univ, Baton Rouge, BS, 60, MS, 63; Stanford Univ, PhD(geol), 69. Prof Exp: Geophys trainee, Cuban Stanalind Oil Co, Pan Am Petrol Corp, 57-58; geologist, Mobil Oil Libya, Ltd, 63-66; asst prof, 69-74, ASSOC PROF GEOL, COLO SCH MINES, 74- Mem: AAAS; Geol Soc Am; Am Soc Photogram. Res: Hydrogeology; remote sensing. Mailing Add: Dept of Geol Colo Sch of Mines Golden CO 80401

LEE, KENNETH, b San Francisco, Calif, July 3, 37; m 59; c 2. SOLID STATE PHYSICS. Educ: Univ Calif, Berkeley, AB, 59, PhD(physics), 63. Prof Exp: Res physicist, Varian Assocs, 63-68; RES STAFF MEM, RES LAB, IBM, CORP, 68- Concurrent Pos: Consult, Lawrence Radiation Lab, Univ Calif, 63-68; comt mem, Magnetism & Magnetic Mat Conf, 64-65, prog comt 71 & adv comt, 71-74 & 74-77, prog co-chmn, 75. Mem: Fel Am Phys Soc. Res: Antiferromagnetism; crystal defects and motion of nuclei in solids; electron and nuclear magnetic resonance; magnetism in thin films; amorphous magnetism. Mailing Add: Res Lab IBM Corp Monterey & Cottle Rds San Jose CA 95193

LEE, KEUM HWI, theoretical physics, see 12th edition

LEE, KIUCK, b Hamhung, Korea, Jan 15, 22; m; c 5. NUCLEAR PHYSICS. Educ: Seoul Nat Univ, BS, 47, MS, 49; Fla State Univ, PhD(physics), 55. Prof Exp: Res assoc physics, Fla State Univ, 55-56 & Argonne Nat Lab, 56-57; from asst prof to assoc prof, 57-68, PROF PHYSICS, MARQUETTE UNIV, 68- Concurrent Pos: Res assoc, Argonne Nat Lab, 60. Mem: Am Phys Soc. Res: Nuclear structure studies on deformed nucleus, fission and the superheavy nucleus. Mailing Add: 516 W Apple Tree Rd Glendale WI 53217

LEE, KOTIK KAI, b Chungking, China, May 30, 41; US citizen; m 67; c 1. MATHEMATICAL PHYSICS. Educ: Chung-Yuan Col, BSc, 64; Univ Ottawa, MSc, 67; Syracuse Univ, PhD(physics), 72. Prof Exp: Res asst physics, Syracuse Univ, 68-72, instr, 72-73; asst prof, Rio Grande Col, 73-74; asst physics, 65-67, VIS PROF MATH, UNIV OTTAWA, 74- Mem: Am Math Soc; Am Phys Soc; Can Asn Physicists; Soc Indust & Appl Math. Res: Global structures of spacetimes, singularities in general relativity, gravitational collapse, quantization of the gravitational field, cosmology, mathematical foundations of quantum field theory and statistical mechanics, and astrophysics. Mailing Add: Dept of Math Univ of Ottawa Ottawa ON Can

LEE, KUO-HSIUNG, b Taiwan, Jan 4, 40; m 68; c 1. MEDICINAL CHEMISTRY, NATURAL PRODUCTS CHEMISTRY. Educ: Kaohsiung Med Col, Taiwan, BS, 61; Kyoto Univ, MS, 65; Univ Minn, Minneapolis, PhD(med chem), 68. Prof Exp: Scholar chem, Univ Calif, Los Angeles, 68-70; asst prof, 70-74, ASSOC PROF MED CHEM, SCH PHARM, UNIV NC, CHAPEL HILL, 74- Concurrent Pos: USPHS res grant, Univ NC, 71- Mem: Am Chem Soc; Am Soc Pharmacog; Am Pharmaceut Asn; Acad Pharmaceut Sci; The Chem Soc. Res: Isolation, structure determination, synthesis and structure-activity relationships of plant antitumor agents. Mailing Add: Sch of Pharm Univ of NC Chapel Hill NC 27514

LEE, KWANG, b Seoul, Korea, Jan 18, 42; m 70; c 2. POULTRY NUTRITION. Educ: Seoul Nat Univ, BS, 64; Southern Ill Univ, MS, 69; Mich State Univ, PhD(poultry sci & nutrit), 73. Prof Exp: ASSOC PROF POULTRY SCI & NUTRIT, UNIV ARK, PINE BLUFF, 73- Mem: Poultry Sci Asn; World Poultry Sci Asn. Res: Nutritional and environmental factors affecting liver fat accumulation and liver hemorrhages associated with fatty liver problems in laying hens; production efficiency of laying hens as influenced by Marek's disease vaccination and debeaking. Mailing Add: Dept of Agr Univ of Ark Pine Bluff AR 71601

LEE, KWANG SOO, b Seoul, Korea, Feb 1, 18; m 40; c 1. PHARMACOLOGY. Educ: Keijo Imp Univ, Korea, MD, 42, PhD, 45; Johns Hopkins Univ, 56. Prof Exp: Asst prof pharmacol, Seoul Nat Univ, 48-49; instr, Jefferson Med Col, 49-50, assoc, 50-51, from asst prof to assoc prof, 51-56; PROF PHARMACOL, STATE UNIV NY DOWNSTATE MED CTR, 62- Mem: Am Soc Pharmacol & Exp Therapeut. Res: Cardiac metabolism; mechanisms of drug actions. Mailing Add: Dept of Pharmacol State Univ NY Downstate Med Ctr Brooklyn NY 11203

LEE, KWANG WOO, b Gimchun, Korea, Apr 20, 37; m 67; c 2. WATER CHEMISTRY, SOIL CHEMISTRY. Educ: Seoul Nat Univ, BS, 60; Univ Minn, St Paul, MS, 64, PhD(soil chem), 68. Prof Exp: Res assoc soil phys & chem, Purdue Univ, Lafayette, 67-68 & biochem, 68-70; res assoc water chem, Univ Wis-Madison, 70-75; RES WATER CHEMIST, CRANBROOK INST SCI, 75- Mem: Am Chem Soc; Am Soc Agron; Am Soc Limnol & Oceanog; Water Pollution Control Fedn. Res: Studies on the behavior of nutrients and hazardous substances in the Great Lakes. Mailing Add: EPA Lab 9311 Groh Rd Cranbrook Inst Sci Grosse Ile MI 48138

LEE, KWANG-YUAN, b Shantung, China, Oct 8, 19; nat US; m 52; c 2. PETROLOGY, SEDIMENTOLOGY. Educ: Nat Southwest Assoc Univs, China, BSc, 42; Ohio State Univ, MSc, 50, PhD, 53. Prof Exp: Geologist, Nat Geol Surv China, 43-48; instr geol, Univ Ark, 53-54; geologist, Ohio State Geol Surv, 54-55 & SDak State Geol Surv, 55-59; geologist, 59-73, PROJ CHIEF CULPEPER BASIN VA & MD, BR EASTERN ENVIRON GEOL, US GEOL SURV, 73- Mem: AAAS; Geol Soc Am; Am Asn Petrol Geol. Res: Economic Geology. Mailing Add: Nat Ctr US Geol Surv Mail Stop 926 Reston VA 22092

LEE, KWAN-HUA, b China, Sept 2, 12; m 42; c 2. PHARMACEUTICAL CHEMISTRY. Educ: Nanking Univ, BS, 34; Univ Calif, MS, 49, PhD(pharmaceut chem), 51. Prof Exp: Asst demonstr & asst instr, Peiping Union Med Col, China, 34-41; from instr to asst prof, Nat Kweigang Med Col, 42-44; vis prof, Nat Shanghai Med Col, 44-45; prof biochem & head dept, Nat Defence Med Ctr, 45-48; asst & asst res biochemist, 48-53, from asst prof to assoc prof pharmaceut chem, 53-69, PROF PHARMACEUT CHEM, SCH PHARM, UNIV CALIF, SAN FRANCISCO, 69-, PROF PHARM, 74- Mem: AAAS; Am Chem Soc; Sigma Xi; Am Soc Pharmacol & Exp Therapeut; NY Acad Sci; Chinese Soc Physiol Sci. Res: Immunological and enzyme chemistry; mode of drug actions. Mailing Add: 1336 Fourth Ave San Francisco CA 94122

LEE, KYU MYUNG, b Kimpo, Korea, May 27, 21; m 45; c 2. VIROLOGY. Educ: Seoul Nat Univ, MD, 46; Cornell Univ, PhD(virol), 52. Prof Exp: Instr microbiol, Col Vet Med, Seoul Nat Univ, 47-49; res assoc virol, NY State Vet Col, Cornell Univ, 54-56; from asst prof to assoc prof microbiol, Sch Pub Health & Col Med, Seoul Nat Univ, 60-63; assoc prof virol, 63-74, PROF VIROL, NY STATE VET COL, CORNELL UNIV, 74- Concurrent Pos: Consult, US Opers Mission, Korea, 56-60; NIH spec res fel, Nat Cancer Inst, 70. Mem: AAAS; Am Soc Microbiol; Tissue Cult Asn; fel Am Acad Microbiol. Res: Animal virology with emphais in feline respiratory and leukemia viruses. Mailing Add: Dept of Microbiol Cornell Univ NY State Vet Col Ithaca NY 14850

LEE, KYU TAI, organic chemistry, see 12th edition

LEE, KYU TAIK, b Taegu, Korea, Sept 17, 21; m 44; c 2. MEDICINE, PATHOLOGY. Educ: Severence Union Med Col, MD, 43; Wash Univ, PhD(path), 56. Prof Exp: Physician in chief, Presby Gen Hosp, Taegu, Korea, 50-53; prof med & chmn dept, Kyung-Pook Nat Univ, 56-60; assoc prof, 60-66, PROF PATH, ALBANY MED COL, 66-, DIR, GEOG PATH DIV, 60-, DIR SECT MOLECULAR BIOL & PATH, 68- Concurrent Pos: NIH res grants, 56-; dir, Saratoga Conf Molecular Biol & Path, 68; managing ed, Exp & Molecular Path, 69; mem comt comp path, Nat Res Coun, 69-71, mem comt path, 70-71; mem nutrit study sect, NIH, 69-73. Mem: Am Heart Asn; Am Soc Exp Path; Int Acad Path; Am Soc Cell Biol; Int Soc Cardiol; Am Inst Nutrit. Res: Ulstrastructural and biochemical aspects of atherosclerosis and thrombosis; geographic aspects of atherosclerosis. Mailing Add: Dept of Path Albany Med Col Albany NY 12208

LEE, KYU YAWP, b Pyongnam, Korea, Apr 8, 25; US citizen; m 53; c 3. BIOCHEMISTRY. Educ: Mich State Univ, BA, 53, BS, 54, MS, 57, PhD(sci), 59. Prof Exp: Res assoc oncol, Inst Med Res, Chicago Med Sch, 60-65; res assoc biochem, Med Res Coun Lab, London, Eng, 65; asst prof oncol, Chicago Med Sch, 66-68; asst prof biochem, 68-70, ASSOC PROF BIOCHEM & MEM GRAD FAC, MED SCH, UNIV NEBR AT OMAHA, 71- Concurrent Pos: Am Cancer Soc grant,

Chicago Med Sch, 62-63; consult, Packard Instrument Co, 68; vis scientist, Courtauld Inst Biochem, London, Eng, 68 & Inst Med Sci, Tokyo Univ, 71. Mem: AAAS; Am Oil Chem Soc; Brit Biochem Soc; NY Acad Sci. Res: Interdisciplinary experimental cell biology and biochemistry related to chemical insult and cellular alteration leading to permanent injury such as cancer. Mailing Add: Eppley Inst Univ of Nebr Med Sch Omaha NE 68105

LEE, LEAVIE EDGAR, JR, b Norfolk, Va, June 23, 30; m 64; c 2. PATHOLOGY, MEDICAL ADMINISTRATION. Educ: Univ Va, BA, 52, MD, 56. Prof Exp: Intern med, Columbia-Presby Med Ctr, 56-57; res assoc, Nat Inst Arthritis & Metab Dis, 57-58, exec secy res training br, Div Gen Med Sci, NIH, 58-60; from instr to asst prof path, Yale Univ, 63-66; asst chief clin progs, Res Grants Br, Nat Inst Gen Med Sci, 66-68, assoc chief, 68-70; ASST PROF PATH & ASSOC DEAN ADMIN, CASE WESTERN RESERVE UNIV SCH MED, 70-, ASST PROF COMMUNITY HEALTH, 74- Concurrent Pos: NIH trainee, Sch Med, Yale Univ, 60-63; consult & collabr, Dept Animal Dis, Univ Conn, 63-65; asst dir, Jane Coffin Childs Mem Fund Med Res, 63-66. Res: Support of biomedical research and the development of medical scientists; comparative pathology; oncology; fish tumors; ultrastructural aspects of disease; analysis and planning of health education, research and care delivery programs. Mailing Add: Sch of Med Case Western Reserve Univ Cleveland OH 44106

LEE, LEROY WILLIAM, b Omaha, Nebr, July 29, 14; m 37; c 4. UROLOGY, SURGERY. Educ: Univ Nebr, BSc, 36, MSc, 38, MD, 39; Am Bd Urol, dipl, 47. Prof Exp: Intern, Philadelphia Gen Hosp, 39-41; resident urol, Hosp Univ Pa, 41-43; resident surg, Philadelphia Gen Hosp, 43-44; chmn dept, 51-70, PROF UROL, COL MED, UNIV NEBR MED CTR, OMAHA, 51-, VCHMN DEPT, 70- Concurrent Pos: Consult, Strategic Air Command, 48-65 & Vet Admin Hosp, 48-; chief med staff, Clarkson Hosp, 60; urologist, Lutheran Hosp, Immanuel Hosp & Clarkson Hosp. Mem: AMA; Am Urol Asn; Soc Univ Urol. Res: New and improved techniques in urological surgery and treatment. Mailing Add: 800 Doctors Bldg Omaha NE 68131

LEE, LESTER TSUNG-CHENG, b Tsing-tao, China, Dec 16, 34; US citizen; m 59; c 2. POLYMER CHEMISTRY, ORGANIC CHEMISTRY. Educ: Nat Taiwan Univ, BS, 56; State Univ NY Syracuse, MS, 60; State Univ NY Buffalo, PhD(org chem), 68. Prof Exp: Chemist monogram indust, Spaulding Fiber Co, 60-68; res chemist, Dept Polymer Sci, Cent Res Lab, Chem Res Ctr, 68-72, sr res chemist, 72-75, COORDR, ALLIED CHEM CORP, 75- Mem: Am Chem Soc. Res: Synthesis and characterization of new polymers, polymer structure-property relationship, polymer applications; fiber, industrial laminates, flame-retardant, high temperature performance, membrane preparation and technology. Mailing Add: Dept of Polymer Sci Cent Res Lab Allied Chem Corp Morristown NJ 07960

LEE, LIENG-HUANG, b Fukien, China, Nov 6, 24; m 49; c 4. ORGANIC CHEMISTRY. Educ: Amoy Univ, BSc, 47; Case Inst Technol, MSc, 54, PhD(chem), 55. Prof Exp: Jr chemist, Nantou Sugar Factory, China, 47-48; asst res chemist, Chia-yee Solvent Works, 48-51 & Rain Stimulation Res Inst, 51-52; res assoc, Case Inst Technol, 55-56; lectr, Tunghai Univ, 56-57; vis prof, Taiwan Prov Norm Univ, 57-58; res org chemist, Dow Chem Co, 58-63, sr res chemist, 63-68; SR SCIENTIST, XEROX CORP, 68- Concurrent Pos: Consult, Union Res Inst, Taiwan, China, 57-58. Mem: Am Chem Soc; Sigma Xi; Am Phys Soc; Fel Am Inst Chemists. Res: Polymer friction and wear; adhesion; electrophotography. Mailing Add: 796 John Glenn Blvd Webster NY 14580

LEE, LIH-SYNG, b China, Oct 28, 45; m 74; c 1. BIOCHEMISTRY, BIOPHYSICS. Educ: Nat Taiwan Univ, BS, 68; Yale Univ, MPh, 71, MS, 72, PhD(chem), 74. Prof Exp: RES SCIENTIST BIOCHEM, ROSWELL PARK MEM INST, 74- Concurrent Pos: Fel, Roswell Park Mem Inst, 74-75. Mem: AAAS; Am Chem Soc. Res: Network of DNA synthesis; enzyme purification and characterization; cancer chemotherapy and drug combination; stochastic theory of cell proliferation; biophysical models of chemical oscillations; metabolic regulations. Mailing Add: Dept of Exp Therapeut Roswell Park Mem Inst Buffalo NY 14263

LEE, LINWOOD LAWRENCE, JR, b Trenton, NJ, Aug 5, 28; m 57; c 1. NUCLEAR PHYSICS. Educ: Princeton Univ, AB, 50; Yale Univ, MS, 51, PhD(physics), 55. Prof Exp: Asst physicist, Argonne Nat Lab, 54-59, assoc physicist, 60-65; vis asst prof physics, Univ Minn, 59-60; PROF PHYSICS, STATE UNIV NY STONY BROOK, 65-, DIR NUCLEAR STRUCTURE LAB, 65- Mem: Am Phys Soc. Res: Experimental studies of nuclear structure and spectroscopy; nucleon transfer reactions; gamma ray transitions in nuclei. Mailing Add: Dept of Physics State Univ of NY Stony Brook NY 11790

LEE, LONG CHI, b Kaohsiung, Taiwan, Oct 19, 40; m 67; c 2. EXPERIMENTAL PHYSICS. Educ: Taiwan Normal Univ, BS, 64; Univ Southern Calif, MA, 67, PhD(physics), 71. Prof Exp: From res asst to res assoc, 67-72, RES STAFF PHYSICIST, UNIV SOUTHERN CALIF, 72- Mem: Am Phys Soc; Sigma Xi. Res: Study on the photoionization and photodissociation processes of small molecules by atomic emission lines and synchrotron radiation in the vacuum ultraviolet region; developing carbon dioxide laser line modulator. Mailing Add: Dept of Physics Univ of Southern Calif Los Angeles CA 90007

LEE, LYNDON EDMUND, JR, b Islip, NY, Aug 11, 12; m 43; c 3. SURGERY, PHARMACOLOGY. Educ: Duke Univ, BS, 37, MD, 38. Prof Exp: Asst cardiol, Univ Va, 38; resident obstet & gynec, Duke Univ Hosp, 40; intern surg, Med Col Va, 41; instr, Med Sch, Univ Mich, 42-43, instr pharmacol & surg, 42-47; instr post-grad educ comt, State Med Asn, Tenn, 47-49; dir cancer control, PR Dept Health, 49-54; instr pharmacol & surg, Med Sch, Univ Mich, 54-57; coordr res, 57-69, dir surg servs, 65-69, asst chief med dir res & educ, 69-71, ASST CHIEF MED DIR PROF SERV, US VET ADMIN, 71- Concurrent Pos: Nat Res Coun fel, 38-40 & 41-42; assoc physician, Blue Ridge Sanatorium, Charlottesville, Va, 38; physician, Am Hosp in Brit, Oxford, Eng, 41 & Pondville State Hosp Cancer, Walpole, Mass; assoc surg, Mass Gen Hosp, Boston; prof, Sch Med, Univ PR, 52-54; dir surg, Wayne County Gen Hosp, 54-57; mem med sci div, Nat Acad Sci, 69- & White House Fed Coun Sci & Technol, 69- Consult, Smithsonian Inst, 58-, nat adv cancer coun, NIH, 58- & training grants rev bd & cancer chemother nat serv ctr, Nat Cancer Inst, 58-; mem nat adv coun child health & human develop, NIH. Mem: AAAS; Am Pub Health Asn; Pub Health Cancer Asn Am; fel Am Col Surgeons; fel Int Soc Surg. Res: Analgesics and sedatives; neoplasms. Mailing Add: US Vet Admin Washington DC 20420

LEE, MARTIN JEROME, b Bayonne, NJ, May 24, 43; m 67; c 1. BIOCHEMISTRY. Educ: Rutgers Univ, New Brunswick, BA, 65, MS, 68, PhD(biochem), 69. Prof Exp: Nat Inst Gen Med Sci fel, Univ Wis-Madison, 69-70, res assoc biochem, Enzyme Inst, 70-71; sr res assoc, Pharmacia Fine Chem, Inc, 71-73; SR SCIENTIST, TECHNICON INSTRUMENTS CORP, 73- Mem: Biophys Soc; Am Chem Soc; Sigma Xi; NY Acad Sci. Res: Bioenergetics; chromatography; separational techniques and instrumentation; automated cytochemistry, immunology, clinical chemistry and enzymology. Mailing Add: Technicon Corp 511 Benedict Ave Tarrytown NY 10591

LEE, MARTIN J G, b Kings Lynn, Eng, Mar 16, 42. SOLID STATE PHYSICS. Educ: Cambridge Univ, BA, 63, MA & PhD(physics), 67. Prof Exp: Instr, 67-69, ASST PROF SOLID STATE PHYSICS, JAMES FRANCK INST & DEPT PHYSICS, UNIV CHICAGO, 69- Mem: Am Asn Physics Teachers; Brit Inst Physics. Res: Experimental and theoretical study of fermi surfaces and electronic structure of metals. Mailing Add: James Franck Inst Univ Chicago 5640 S Ellis Ave Chicago IL 60637

LEE, MARY ANN, mathematics, see 12th edition

LEE, MATHEW HUNG MUN, b Hawaii, July 28, 31; m 58; c 3. PHYSICAL MEDICINE & REHABILITATION. Educ: Johns Hopkins Univ, AB, 53; Univ Md, MD, 56; Univ Calif, MPH, 62; Am Bd Phys Med & Rehab, dipl, 66. Prof Exp: Resident, Inst Phys Med & Rehab, Med Ctr, NY Univ, 62-64, NY State Health Dept assignee, Rehab Serv, 64-65, from asst prof to assoc prof rehab med, 65-73, dir educ & training, Dept Rehab Med, 66-68, assoc dir, 68, DIR DEPT REHAB MED, GOLDWATER MEM HOSP, 68-, PROF REHAB MED, SCH MED, NY UNIV, 73- Concurrent Pos: Assoc vis physician, Goldwater Mem Hosp, 65-68, vis physician, 68-, chief electrodiag unit, 66-, vpres med bd, 69-70, pres, 71-; asst clin prof, Col Dent, NY Univ, 66-69, clin asst prof, 69-70, clin assoc prof, 70-; consult, Daughters of Israel Hosp, New York, 65-72, Bur Adult Hyg, 65- & Human Resources Ctr, 66-; asst attending physician, Hosp, NY Univ, 68-; World Rehab Fund consult, Gordon Seagrave & Maryknoll Hosps, Korea, 69; attend physician, Bellevue Hosp Ctr, 71-; consult, US Dept Interior. Mem: AAAS; fel Am Acad Phys Med & Rehab; fel Am Col Physicians; Pan-Am Med Asn; fel Am Pub Health Asn. Mailing Add: 80-39 188th St Jamaica Estates NY 11423

LEE, MATTHEW CHIEH-YEN, mathematics, see 12th edition

LEE, MELVIN, b New York, NY, Jan 5, 26; m 49; c 4. NUTRITION, BIOCHEMISTRY. Educ: Univ Calif, Los Angeles, BA, 47; Univ Calif, Berkeley, MA, 52, PhD(nutrit), 58. Prof Exp: From instr to asst prof prev med, Sch Med, Univ Calif, San Francisco, 58-67, asst prof biochem, 63-67, lectr dent, 61-67; prof nutrit & dir, Sch Home Econ, 67-74, PROF HOME ECON, SCH HOME ECON, UNIV BC, 74- Concurrent Pos: USPHS res fel, 66. Mem: AAAS; Am Inst Nutrit. Res: Relation of diet to metabolic patterns; factors influencing growth. Mailing Add: Sch of Home Econ Univ of BC Vancouver BC Can

LEE, MERLIN RAYMOND, b Hoytsville, Utah, Mar 21, 28; m 49; c 2. EVOLUTIONARY BIOLOGY. Educ: Univ Utah, BS, 52, MS, 54, PhD, 60. Prof Exp: Asst zool, biol & genetics, Univ Utah, 52-56, instr biol, 58-59; field collector, Mus Natural Hist, Univ Kans, 60-61; res assoc, Mus Natural Hist, 61-63, asst prof zool, 63-67, ASSOC PROF ZOOL, UNIV ILL, URBANA, 67- Concurrent Pos: Ed, J Mammal, 64-66. Mem: Am Soc Mammal. Res: Mammalogy, karyology, chromosome evolution and mammalian systematics. Mailing Add: Dept of Zool Univ of Ill Urbana IL 61801

LEE, MICHAEL JOHN, b Exeter, Eng, Oct 6, 32; m 66; c 2. BIOCHEMISTRY. Educ: Oxford Univ, BA, 54, MA, 60, DPhil, 64. Prof Exp: INSTR BIOCHEM, HARVARD MED SCH, 69- Concurrent Pos: Fel biochem, Univ Calif, Los Angeles, 65-67; fel, Univ Pittsburgh, 67-69; asst biochemist, Mass Gen Hosp, 69- Res: Control mechanisms in RNA and protein synthesis; regulation of growth and cell division. Mailing Add: Dept of Path Mass Gen Hosp Boston MA 02114

LEE, MILFORD RAY, physics, see 12th edition

LEE, MIN-SHIU, b Taipei, Taiwan, June 30, 40; US citizen; m 66; c 2. PHYSICAL CHEMISTRY, POLYMER CHEMISTRY. Educ: Nat Taiwan Univ, BS, 62; NMex Highlands Univ, MS, 66; Case Western Reserve Univ, PhD(macromolecular sci), 69. Prof Exp: Res fel chem, Chinese Ord Res Inst, 62-64; teaching asst, NMex Highlands Univ, 64-65, res fel hot atom chem, Inst Sci Res, 65-66; res asst polymer res, Case Western Reserve Univ, 66-69; res chemist, FMC Corp, 69-73, sr res chemist, 73-76; SR RES SCIENTIST, JELCO LABS, DIV, JOHNSON & JOHNSON, 76- Mem: Sigma Xi; Am Chem Soc; Am Asn Univ Prof. Res: Medical products improvement and development; collagen research and development; chemical process improvement; new products development; textile and non-woven applications; polymer characterization; polymer modification and application; process/quality control; biomaterial development; colloidal macromolecular phenomena. Mailing Add: 4 Knollwood Dr East Windsor NJ 08520

LEE, MONHE HOWARD, b Pusan, Korea, May 21, 37; US citizen; m 67; c 1. THEORETICAL PHYSICS. Educ: Univ Pa, BS, 59, PhD(physics), 67. Prof Exp: Fel physics, Theoret Physics Inst, Univ Alta, 67-69; res assoc, Dept Physics & Mat Sci Ctr, Mass Inst Technol, 69-71; NIH res grant & investr biomat, Health Sci & Technol, Mass Inst Technol & Harvard Univ, 71-73; ASST PROF PHYSICS, UNIV GA, 73- Concurrent Pos: Guest lectr, Inst Theoret Physics, Univ Leuven, Belgium, 76; res grant, NATO, 76. Mem: Am Phys Soc; Biophys Soc. Res: Many-body theory; statistical mechanics of phase transitions; biophysics of membrane transport. Mailing Add: Dept of Physics Univ of Ga Athens GA 30602

LEE, MYUNG WOO, physical chemistry, see 12th edition

LEE, NANCY ZEE-NEE MA, b Shanghai, China, Oct 28, 40; US citizen; m 65; c 2. BIOCHEMISTRY. Educ: Southwestern Univ, BS, 63; Univ Tex, PhD(chem), 67. Prof Exp: Fel biochem, Northwestern Univ, 68; RES BIOCHEM PHARMACOL, UNIV CALIF, SAN FRANCISCO, 69- Concurrent Pos: Res biochem, Bay Area Heart Res, 70-72. Mem: Sigma Xi; Soc Neurosci. Res: Biochemical mechanism for narcotic addiction. Mailing Add: Dept of Pharmacol Univ of Calif San Francisco CA 94143

LEE, NEVILLE KA-SHEK, b Hong Kong, Jan 2, 47. PHYSICS. Educ: Univ Calif, Los Angeles, BS, 67; Mass Inst Technol, PhD(physics), 73. Prof Exp: RES STAFF PHYSICS, FRANCIS BITTER NAT MAGNET LAB, MASS INST TECHNOL, 73- Mem: Sigma Xi. Res: Quantum optics; working on noncollinear nonlinear light mixing to generate infrared coherent sources. Mailing Add: NW14-4117 Mass Inst Technol Cambridge MA 02139

LEE, NORMAN DAVID, b Bridgeport, Conn, May 25, 20; m 41; c 3. BIOCHEMISTRY. Educ: Univ Calif, Los Angeles, AB, 46; Univ Calif, Berkeley, PhD(biochem), 50. Prof Exp: Instr med, Med Sch, Univ Wash, 49-53; asst chief radioisotope serv, Vet Admin Hosp, Memphis, 53-59; chief radioisotope div, Biosci Labs, 59-65, asst to dir, 65-67, dir dept endocrinol, 67-68, asst dir br labs, 68-70, ASST DIR AFFIL LABS, BIO-SCI ENTERPRISES, 70- Mem: Am Soc Biol Chem; Soc Nuclear Med; Am Asn Clin Chem; Am Thyroid Asn. Res: Thyroid methodology; laboratory management. Mailing Add: Bio-Sci Enterprises 7600 Tyrone Ave Van Nuys CA 91405

LEE, NORMAN K, b Frankfort, Ind, Feb 3, 34; m 56; c 2. APPLIED MATHEMATICS. Educ: Hanover Col, BA, 56; Vanderbilt Univ, MA, 58; Purdue

Univ, PhD(bionucleonics), 69. Prof Exp: Asst prof, 58-66, ASSOC PROF MATH, BALL STATE UNIV, 69- Concurrent Pos: NSF fel, 63-64, partic, Acad Year Inst, 62-63; sr instr, Somerset Community Col, 68-69. Mem: AAAS; Sigma Xi. Res: Mathematical models. Mailing Add: Dept of Math Sci Ball State Univ Muncie IN 47306

LEE, NORMAN KUNHAN, b China, July 11, 17; m 50; c 2. PATHOLOGY. Educ: Nat Chung Cheng Med Col, China, MD, 47. Prof Exp: From instr to asst prof path, Sch Med, Univ Okla, 59-62, dir blood bank & sch med technol, 60-61; asst prof path, Univ Kans, 62-63; CHIEF LAB SERV, VET ADMIN HOSP, 63- Mailing Add: Vet Admin Hosp Lab Serv Perry Hill Rd Montgomery AL 36109

LEE, PANG-KAI, b Kiangsu, China, Nov 3, 22; m 49; c 3. PHYSICAL CHEMISTRY. Educ: Nat Kwangsi Univ, BS, 45; Oberlin Col, MA, 62; Pa State Univ, PhD(chem), 67. Prof Exp: Teaching asst chem, Nat Kwangsi Univ, 45-48; chemist, Taiwan Indust & Mining Corp, 48-53; factory supt, Soong-San Chem Works, 53-58; lectr chem, Nanyang Univ, Singapore, 58-60; SR SCIENTIST, WESTINGHOUSE RES LABS, 67- Mem: Am Chem Soc. Res: High temperature and surface chemistry ; ASTROPHYSICS. Educ: Univ Ill, BS, 63, MS, 65, PhD, 68. Prof Exp: Asst prof, 68-73, ASSOC PROF ASTRON, LA STATE UNIV, 73- Mem: Royal Astron Soc; Am Astron Soc. Res: Spectrophotometry of stellar and nonstellar objects; stellar atmospheres and chemical abundances in stars. Mailing Add: Dept of Physics & Astron La State Univ Baton Rouge LA 70803

LEE, PAUL D, b Ina, Ill, Feb 15, 40; m 61. ASTRONOMY, ASTROPHYSICS. Educ: Univ Ill, BS, 63, MS, 65, PhD, 68. Prof Exp: Asst prof, 68-73, ASSOC PROF ASTRON, LA STATE UNIV, BATON ROUGE, 73- Mem: Royal Astron Soc; Am Astron Soc. Res: Spectrophotometry of stellar and nonstellar objects; stellar atmospheres and chemical abundances in stars. Mailing Add: Dept of Physics & Astron La State Univ Baton Rouge LA 70803

LEE, PETER E, b Trinidad, WI, Oct 18, 30; m 60; c 2. PLANT VIROLOGY. Educ: Univ Man, BSc, 58; Univ Wis, MSc, 59, PhD(entom), 61. Prof Exp: Jr res entomologist, Univ Calif, Berkeley, 61; res officer virus-vector studies, Can Dept Agr, 61-65; asst prof, 65-67, ASSOC PROF BIOL, CARLETON UNIV, 67- Mem: Electron Micros Soc Am. Res: Characterization of leafhopper-transmitted viruses and insect viruses; virus purification and electron microscopy. Mailing Add: Dept of Biol Carleton Univ Ottawa ON Can

LEE, PETER VAN ARSDALE, b San Francisco, Calif, Mar 31, 23; m 51; c 4. PHARMACOLOGY, MEDICINE. Educ: Stanford Univ, AB, 44, MD, 47. Prof Exp: Intern, San Francisco Hosp, 46-47; asst resident path, Stanford Univ Hosps, 49-50, resident med, 50-51; clin asst, Col Med, State Univ NY, 51-52; instr pharmacol, Sch Med, Stanford Univ, 52-54, asst prof, 54-55; asst prof & asst dean, 55-58, assoc prof pharmacol, 58-67, assoc prof med, 60-67, assoc dean, 58-60, admis officer, 60-65, PROF MED & PHARMACOL, SCH MED, UNIV SOUTHERN CALIF, 67- Concurrent Pos: Resident, King's County Hosp, NY, 51-52; consult, Commonwealth Fund, 58-59; vis fel, Brit Asn Study Med Educ, 71-72. Mem: AAAS; Am Fedn Clin Res; Asn Am Med Cols; Brit Asn Study Med Educ; fel Royal Soc Med. Res: Medical education; clinical pharmacology. Mailing Add: Univ of Southern Calif Sch of Med Los Angeles CA 90033

LEE, PHILIP CALVIN, b Roanoke, Va, Dec 6, 33. MYCOLOGY. Educ: Roanoke Col, BS, 60; Univ Richmond, MA, 62; Va Polytech Inst, PhD(mycol), 66. Prof Exp: Vet bacteriologist, Regional Regulatory Lab, Va State Dept Agr, 61-62; instr biol, genetics & bot, Va Polytech Inst, 63-65; asst prof, 65-69, ASSOC PROF BIOL, ROANOKE COL, 69- Res: Effects of environmental factors on the sexual and asexual sporulation of phycomycetous fungi. Mailing Add: Dept of Biol Roanoke Col Salem VA 24153

LEE, PHILIP RANDOLPH, b San Francisco, Calif, Apr 17, 24; m 53; c 4. INTERNAL MEDICINE. Educ: Stanford Univ, AB, 45, MD, 48; Univ Minn, MS, 56. Prof Exp: Asst prof clin phys med & rehab, Sch Med, NY Univ, 55-56; clin instr med, Sch Med, Stanford Univ, 56-59, asst clin prof, 59-67; chancellor, 69-72, PROF SOCIAL MED, MED CTR, UNIV CALIF, SAN FRANCISCO, 69- Concurrent Pos: Mem dept internal med, Palo Alto Med Clin, Calif, 56-65; consult, Bur Pub Health Serv, USPHS, 58-63; dir health serv, Off Tech Coop & Res, AID, 63-65; dep asst secy health & sci affairs, Dept Health, Educ & Welfare, 65, asst secy, 65-69. Mem: Inst of Med of Nat Acad Sci; AAAS; AMA; Am Pub Health Asn; Am Fedn Clin Res. Res: Arthritis and rheumatism, especially Rubella arthritis; cardiovascular rehabilitation; academic medical administration; health policy. Mailing Add: Dept of Social Med Univ Calif Med Ctr San Francisco CA 94143

LEE, PUI KUM, b Peking, China, June 22, 16; US citizen; m 41; c 4. OPTICS. Educ: Lingnam Univ, BS, 40; Columbia Univ, MS, 49. Prof Exp: Asst chem, Nat Kwangsi Univ, China, 40-41; res chemist, Inst Indust Res, 41-42; mgr, China Chem Corp, 42-46; res assoc chem, Columbia Univ, 49-56; sr chemist reprography, 56-63, res specialist imaging, 63-67, SR RES SPECIALIST, CENT RES LABS, 3M CO, 67- Mem: Am Chem Soc; Soc Photog Sci & Eng; imaging optics. Mailing Add: Cent Res Labs 3M Co PO Box 33221 St Paul MN 55133

LEE, RALPH EDWARD, b Gilliam, Mo, July 1, 21; m 42; c 5. COMPUTER SCIENCE. Educ: Mo Valley Col, BS, 42; Univ Mo, MS, 49; Ind Univ, MA, 53. Prof Exp: From instr to assoc prof, Sch Mines, 46-59, PROF MATH, UNIV MO-ROLLA, 59-, DIR COMPUT CTR, 60- Concurrent Pos: NSF fel, Nat Bur Standards, 59. Mem: Math Asn Am; Asn Comput Mach; Soc Indust & Appl Math; Asn Educ Data Systs. Res: Numerical analysis; matrix computations. Mailing Add: Comput Ctr Univ of Mo-Rolla Rolla MO 65401

LEE, RALPH HEWITT, b Lexington, Ky, Apr 10, 38; m 59; c 4. INORGANIC CHEMISTRY, ACADEMIC ADMINISTRATION. Educ: Morehouse Col, BS, 57; Univ Kans, PhD(inorg chem), 64. Prof Exp: From assoc prof to prof chem, Ala Agr & Mech Col, 64-69, head dept, 65-69; acad dean, Morehouse Col, 69; dir SEEK prog, Queens Col, City Univ New York, 69-71; PRES, FOREST PARK COMMUNITY COL, 71- Concurrent Pos: NSF India Prog, 67. Mem: AAAS; Am Chem Soc; Sigma Xi. Res: Structure of nitrogen-donor transition metal complex compounds; education of secondary school science teachers. Mailing Add: 5600 Oakland Ave St Louis MO 63110

LEE, RAYMOND CURTIS, b Dallas, Tex, Oct 22, 29; m 53; c 2. PHYSICAL CHEMISTRY. Educ: Rice Univ, BA, 50, MA, 53; Univ Tenn, PhD(chem), 61. Prof Exp: Tech engr, Gen Elec Co, 56-61; asst prof chem, San Jose State Col, 61-63; res specialist, Lockheed Missiles & Space Co, 63-66; mem tech staff, Aerospace Corp, 66-70; SCIENTIST, SCI APPLNS, INC, 70- Mem: Am Chem Soc; Am Nuclear Soc; Am Inst Aeronaut & Astronaut; Am Sci Affil. Res: Reactor engineering, in-pile experiment design; testing and evaluation for aircraft nuclear propulsion reactor; nuclear safety evaluation for nuclear rocket project; systems analysis; computer systems modeling and simulation; advanced development engineering. Mailing Add: 5275 Soledad Rancho Ct San Diego CA 92109

LEE, RICHARD, b Washington, Pa, Feb 7, 26; m 56; c 3. HYDROLOGY, MICROCLIMATOLOGY. Educ: WVa Univ, BS, 59; Colo State Univ, PhD(forest hydrol), 62. Prof Exp: NATO fel meteorol, Univ Munich, 62-63; res assoc microclimatology, Univ Colo, 63; res hydrologist, US Forest Serv, 63-64; asst prof forest hydrology, Univ Conn, 64-66 & Pa State Univ, 66-68; assoc prof, 68-73, PROF FOREST HYDROLOGY, WVA UNIV, 73- Mem: Am Meteorol Soc; Am Geophys Union. Res: Applied meteorology and hydrology; plant-water relations; forest microclimatology; disturbed lands. Mailing Add: Dept of Forest Hydrol WVa Univ Morgantown WV 26506

LEE, RICHARD FAYAO, b Shanghai, China, July 13, 41; US citizen; m 70; c 2. ENVIRONMENTAL CHEMISTRY, BIOLOGICAL OCEANOGRAPHY. Educ: San Diego State Col, BA, 64, MA, 66; Univ Calif, San Diego, PhD(marine biol), 70. Prof Exp: Res assoc biochem, Pa State Univ, 71-72 & Scripps Inst Oceanog, 72-73; ASST PROF OCEANOG, SKIDAWAY INST OCEANOG, 74- Concurrent Pos: Lectr oceanog, San Diego State Univ, 71-73; mem adv comt, Marine Resources Res Group Biol Accumulators, UN Food & Agr Orgn, 74-; consult, Exxon Corp, 75. Mem: AAAS; Am Chem Soc; Am Soc Limnol & Oceanog; Am Oil Chemists Soc; Sigma Xi. Res: Fate of petroleum hydrocarbons in the marine food web; role of lipids in the ecology of marine zooplankton. Mailing Add: Skidaway Inst of Oceanog PO Box 13687 Savannah GA 31406

LEE, RICHARD J, b Minot, NDak, July 23, 44; m 60. SOLID STATE PHYSICS. Educ: Univ NDak, BSEd, 66; Colo State Univ, PhD(physics), 70. Prof Exp: Instr physics, Lake Regional Jr Col, 66; res asst, Colo State Univ, 68-70; asst prof, Purdue Univ, Ft Wayne, 70-74; MEM STAFF, US STEEL CORP, 74- Mem: Am Phys Soc; Am Asn Physics Teachers. Res: Theory of quantum solids; phase transition; light scattering. Mailing Add: US Steel Corp 125 Jamison Lane Monroeville PA 15146

LEE, RICHARD JUI-FU, b China, May 26, 19; nat US; m 40; c 3. ORGANIC CHEMISTRY. Educ: Loyola Univ, Ill, BS, 42; Ohio State Univ, PhD(chem), 54. Prof Exp: Sr res chemist, Sherwin-Williams Co, 46-49; fel, Ohio State Univ, 54-55; res chemist, Armour & Co, 55-59; mgr polymer sect, Am Marietta Co, 59-61; SR RES CHEMIST, AMOCO CHEM CORP, STANDARD OIL CO, IND, 61- Mem: Am Chem Soc; Sigma Xi. Res: Free radical chemistry in organic mechanisms and synthesis; transition metal chemistry; catalysis. Mailing Add: 1513 62nd St Downers Grove IL 60515

LEE, RICHARD K C, b Honolulu, Hawaii, Oct 2, 09; m 52; c 3. PUBLIC HEALTH. Educ: Tulane Univ, MD, 33; Yale Univ, DrPH, 38. Hon Degrees: DSc, Tulane Univ, 73. Prof Exp: Instr anat, Sch Med, Tulane Univ, 33-35; intern, Hotel Dieu Hosp, New Orleans, 35-36; dep comnr health, Territory Hawaii, 36-43, dir pub health, 43-53; pres, Hawaii Bd Health, 53-60; dir health, Hawaii Dept Health, 60-62; dir pub health & med activities, 62-65, dean sch pub health, 65-69, prof, 62-69, EMER PROF PUB HEALTH & EMER DEAN SCH PUB HEALTH, UNIV HAWAII, 69-, EXEC DIR, RES CORP OF UNIV HAWAII, 70- Concurrent Pos: Lectr, Univ Hawaii, 37-55; WHO fel, 52; mem US deleg, Western Pac Regional Comt Meetings, WHO, 52-65, chief US rep, Manila, 66 & Taiwan, 67, mem US deleg, Assembly Meetings, 57-61; Dept State specialist, Int Educ Exchange Prog, Far East, 55; mem, Western Interstate Comn Higher Educ, Western Ment Health Coun, 60-; coordr-consult health & med prog, East-West Ctr Inst Tech Interchange, 62-; mem task force pub health & med educ, World Affairs of New York & Food Found, 65-; consult, WHO, Manila, 67, Alexandria, UAR, 69 & Am Pub Health Asn, Korea, 69; chief div environ health & occup med & med coord dept commun med, Straub Clin, 69-71; consult, Water Qual Mgt Prog, City & County of Honolulu, 69-71 & Southern Calif Coastal Water Res Proj, 69-; actg dir, Cancer Ctr Hawaii, 71-73; mem, Nat Adv Coun, Nat Inst Aging, NIH, 75-; chmn adv bd, Health Manpower Planning Proj, Hawaii, 75- Honors & Awards: Pfizer Award of Merit, US Civil Defense Coun, 60; Samuel J Crumbine Award, Kans State Univ Interfraternity Coun, 63. Mem: AAAS; AMA; Am Pub Health Asn (vpres, 62-63). Res: Public health administration; international health activities in the Pacific and Asian areas of the world. Mailing Add: Res Corp of the Univ of Hawaii 402 Varsity Blvd 1110 Univ Ave Honolulu HI 96814

LEE, RICHARD NORMAN, b Waukegan, Ill, Nov 3, 39; m 64; c 1. ATMOSPHERIC CHEMISTRY. Educ: Park Col, BA, 61; Univ Kans, PhD(chem), 68. Prof Exp: Asst prof chem, St Norbert Col, 66-72; PRES, INTERN ATMOSPHERIC ANAL & RES SCIENTIST, BATTELLE PAC NORTHWEST LABS, 72- Mem: Am Chem Soc. Res: Reaction kinetics; inorganic synthesis; environmental chemistry; chemical analysis; atmospheric pollutants and tracers. Mailing Add: Atmospheric Sci Dept Battelle Pac Northwest Labs Richland WA 99352

LEE, ROBERT E, JR, b Albany, NY, Sept 21, 36; m 60; c 2. PHYSICAL CHEMISTRY, AIR POLLUTION. Educ: Siena Col, BS, 58; George Washington Univ, MEA, 64; Univ Cincinnati, MS, 67, PhD(phys chem), 69. Prof Exp: Chemist, US Army Biol Labs, Md, 58-62; scientist, Melpar, Inc, Va, 62-64; RES CHEMIST, US ENVIRON PROTECTION AGENCY, 64- Mem: AAAS; Air Pollution Control Asn; Am Chem Soc. Res: Aerosol and biological chemistry; diffusion and ion interactions; surface and colloid chemistry; technical administration; fuels and source sample analysis; air pollution chemistry; pesticides and toxic substances. Mailing Add: US EPA Nat Environ Res Ctr Research Triangle Park NC 27711

LEE, ROBERT JAMES, b San Jose, Ill, Jan 4, 16; m 38; c 3. INDUSTRIAL ORGANIC CHEMISTRY. Educ: St Olaf Col, AB, 37; Ind Univ, AM, 38, PhD(chem), 41. Prof Exp: Asst, Ind Univ, 37-41; res chemist, Pan Am Ref Corp, 41-42, res group leader, 42-45, sect head explor chem res, 45-52 & petrol res sect, 52-56; sect head, Chem Sect, Am Oil Co, 56-60, SECT LEADER NEW CHEM RES, AMOCO CHEM CORP, 60- Concurrent Pos: With Oak Ridge Inst Nuclear Sci, 52. Mem: Am Chem Soc. Res: Research and development on production of purified terephthalic acid and dimenthyl terephthalate, and purified isopathalic acid used for manufacture of polyester fibers and film. Mailing Add: Res Dept Amoco Chem Corp PO Box 400 Naperville IL 60540

LEE, ROBERT JEROME, b New York, NY, Nov 2, 14; m 41; c 2. VETERINARY MEDICINE. Educ: NY Univ, BS, 35; Kans State Univ, DVM, 39. Prof Exp: Vet med off, Fed Meat Grading Br, USDA, 49-58, head, Regulatory Sect, Fed Poultry Inspection, 58-61, Poultry Prod Sect, 61-64, training off, 64-68; CHIEF, MEAT & POULTRY INSPECTION SECT, MD DEPT AGR, 68- Concurrent Pos: Mem, Expert Comt Food Hyg Comt, Codex Alimentarius, WHO-UN Food & Agr Org, 65-68, alt deleg, Poultry Comt, 66-68; mem, Food Hyg Comt, US Animal Health Asn, 70- Honors & Awards: Spec Merit Award, USDA, 68, Super Serv Award, 70. Mem: Nat Asn State Meat & Food Inspection Dirs (pres, 70-73); Am Asn Food Hyg Vets (secy, 73-); Nat Asn Fed Vets (secy-treas, 60-61); Am Vet Med Asn. Mailing Add: Md Dept of Agr Rm 0104B Symons Hall College Park MD 20742

LEE, ROBERT JOHN, b Worcester, Mass, July 2, 29; m 60; c 3. MEDICAL

PHYSIOLOGY. Educ: Adelphi Univ, AB, 58; Yeshiva Univ, MS, 59; State Univ NY, PhD(physiol), 67. Prof Exp: Teacher, NY Schs, 58-59; teaching asst surg, State Univ NY, 59-63; sr scientist, Geigy Res Labs, 66-70; RES GROUP LEADER, SQUIBB INST MED RES, 70- Mem: AAAS; Am Soc Pharmacol & Exp Therapeut; Am Col Cardiol; Am Heart Asn. Res: Mechanical aspects of cardiovascular physiology; ventricular function. Mailing Add: Dept of Pharmacol Squibb Inst for Med Res New Brunswick NJ 08903

LEE, ROBERT K S, b Lanai City, Hawaii, Mar 16, 31; m 54; c 2. MARINE BOTANY. Educ: Univ Hawaii, BA, 59, MSc, 62; Univ BC, PhD(bot), 65. Prof Exp: Asst prof biol, Mem Univ, 64-66; CUR ALGAE, NAT MUS NATURAL SCI, 66- Mem: AAAS; Bot Soc Am; Phycol Soc Am; Can Bot Soc; Int Phycol Soc. Res: Systematics of melobesioid algae; systematics and ecology of Arctic algae. Mailing Add: Bot Div Nat Mus of Natural Sci Metcalfe & McLeod Sts Ottawa ON Can

LEE, ROBERT W, b Cedar Rapids, Iowa, Feb 28, 31; m 51; c 2. EXPERIMENTAL PHYSICS Educ: Mich State Univ, BS, 53, MS, 55. Prof Exp: SR RES PHYSICIST, RES LABS, GEN MOTORS CORP, 55- Res: Permanent magnets; electro-optics; gas diffusion in solids. Mailing Add: Physics Dept Tech Ctr Gen Motors Res Warren MI 48090

LEE, RONALD NORMAN, b Springfield, Mo, Oct 21, 35; m 59. SURFACE PHYSICS. Educ: Univ Ill, BS, 58, MS, 60; Brown Univ, PhD(physics), 65. Prof Exp: Res assoc physics, Coord Sci Lab, Univ Ill, 60 & Brown Univ, 64-65; fel phys chem, Battelle Mem Inst, Ohio, 65-68; physicist, US Naval Ord Lab, 68-74, PHYSICIST, SOLID STATE DIV, NAVAL SURFACE WEAPONS CTR, 74- Mem: Am Phys Soc; Am Vacuum Soc. Res: Surface science; semiconductor thin film phenomena. Mailing Add: Solid State Br Naval Surf Weapons Ctr White Oak Silver Spring MD 20910

LEE, RONALD S, b Ames, Iowa, Dec 29, 38; m 60; c 2. SOLID STATE PHYSICS. Educ: Luther Col, Iowa, BA, 61; Iowa State Univ, PhD(physics), 67. Prof Exp: Asst prof, 67-74, ASSOC PROF PHYSICS, KANS STATE UNIV, 74- Mem: Am Phys Soc; Am Asn Univ Prof. Res: Electrical properties of thin films; radiation effects in solids. Mailing Add: Dept of Physics Kans State Univ Manhattan KS 66502

LEE, RONNIE, b China, Nov 6, 42. TOPOLOGY. Educ: Chinese Univ Hong Kong, BS, 65; Univ Mich, PhD(math), 68. Prof Exp: Mem, Inst Advan Studies, 68-70; asst prof, 70-73, ASSOC PROF MATH, YALE UNIV, 73- Concurrent Pos: Sloan Found fel, 73. Mem: Am Math Soc. Res: Differential topology. Mailing Add: Dept of Math Yale Univ New Haven CT 06520

LEE, ROY, b St Catharines, Ont, Mar 7, 24; m 49; c 3. METEOROLOGY. Educ: Univ Toronto, BA, 46, MA, 48 & 49. Prof Exp: Forecaster, Meteorol Serv Can, 49-52, res meteorologist, 52-57, supvr field training, 57-70, supvr sci develop & eval, 62-69; supt, Weather & Ice Servs, 70-75, DIR, ADMIN BR, ATMOSPHERIC ENVIRON SERV, 75- Concurrent Pos: Spec lectr physics, Univ Toronto, 63-70; Can rep, Comn Maritime Meteorol, World Meteorol Orgn, 70- Honors & Awards: Co-recipient Darton Prize, Royal Meteorol Soc, 54, Darton Prize, 55; co-recipient Pres Prize, 57. Mem: Am Meteorol Soc; fel Royal Meteorol Soc (secy, Can Br, 55-57); Can Meteorol Soc. Res: Development of weather services; weather prediction; general circulation of the atmosphere; forecasting system design. Mailing Add: Atmospheric Environ Serv 4905 Dufferin St Downsview ON Can

LEE, RUPERT ARCHIBALD, b Guyana, Jan 21, 32; m 56; c 6. RADIATION CHEMISTRY. Educ: Univ West Indies, BSc, 54; Univ London, MSc, 59, PhD(chem), 67. Prof Exp: Asst prof chem, Univ PR, Mayaguez, 65-67; res assoc, 65-67, scientist, 67-70, HEAD RADIATION CHEM PROG, PR NUCLEAR CTR, 70-; ASSOC PROF CHEM, UNIV PR, MAYAGUEZ, 67- Concurrent Pos: AEC res grant, PR Nuclear Ctr, 70- Mem: Am Chem Soc. Mailing Add: 46 Urb El Regreo Mayagüez PR 00708

LEE, SAMUEL HUNT, JR, physical organic chemistry, deceased

LEE, SHAW-GUANG LIN, b Miao-Li, Taiwan, Oct 9, 44; US citizen; m 68; c 3. BIOCHEMISTRY. Educ: Nat Taiwan Univ, BS, 67; Northwestern Univ, PhD(biochem), 72. Prof Exp: Fel biochem, Northwestern Univ, 73-75; RES BIOCHEMIST MOLECULAR VIROL, ABBOTT LABS, 75- Mem: Am Chem Soc. Res: Cyclic AMP and protein kinase; RNA tumor virus replication; drug metabolism and disposition. Mailing Add: Molecular Virol Lab Abbott Labs North Chicago IL 60064

LEE, SHERIDAN HSIO-TAO, helminthology, see 12th edition

LEE, SHUI LUNG, b Canton, China, Sept 15, 38; m 69; c 1. ORGANIC CHEMISTRY. Educ: Univ Western Australia, BS, 65, PhD(org chem), 69. Prof Exp: Fel org chem, McMaster Univ, 69-71, Queens Univ, Can, 71-74; sr chemist, Aldrich Chem Co, 74-75; SR RES CHEMIST, PIGMENTS DIV, CHEMETRON CORP, 75- Mem: Am Chem Soc. Res: New organic pigments and intermediates. Mailing Add: Chemetron Corp Pigments Div 491 Columbia Ave Holland MI 49423

LEE, SHUNG-YAN LUKE, b China, Sept 10, 38; US citizen; m 62; c 2. SURFACE CHEMISTRY. Educ: Univ Wis, BS, 59; Ohio State Univ, MS, 62, PhD(phys org chem), 66. Prof Exp: RES CHEMIST PHOTOG SYST, E I DU PONT DE NEMOURS & CO, INC, 66- Mem: Am Chem Soc. Res: Study of surface characteristics with electron spectroscopy for chemical analysis and other techniques for improving adhesion, wetting, sensitization and stabilization of photoimaging systems. Mailing Add: Exp Sta Bldg 352 E I du Pont de Nemours & Co Inc Wilmington DE 19898

LEE, SI DUK, b Ham Hung, Korea, Jan 2, 32; US citizen; m 57; c 3. ENVIRONMENTAL SCIENCES, BIOLOGICAL CHEMISTRY. Educ: Seoul Nat Univ, BS, 55; Univ Md, MS, 59, PhD(biochem), 62. Prof Exp: Res assoc biochem, Med Ctr, Duke Univ, 61-62, NIH fel, 62-63, Am Heart Asn adv res fel, 63-64; res chemist, USPHS, 64-65, supvry res chemist, 65-67, chief biochem unit, 67-69, chief biochem sect, Nat Air Pollution Control Admin, 69-71; DEP BR CHIEF, BIOL EFFECTS BR, NAT ENVIRON RES CTR, ENVIRON PROTECTION AGENCY, 71- Concurrent Pos: Adj asst prof, Dept Biol, Col Med, Univ Cincinnati, 67- & Dept Environ Health, 75- Mem: AAAS; Am Chem Soc; Am Oil Chemists Soc. Res: Effects of air pollutants on metabolism; lipid metabolism; effects of pollutants on aging; effects of sulfur dioxide on subcellular metabolism. Mailing Add: Nat Environ Res Ctr Environ Protection Agency Cincinnati OH 45268

LEE, SIU-LAM, b Macao, China, Oct 3, 41; m 68. INSECT ECOLOGY, EVOLUTION. Educ: Chung Chi Col, Chinese Univ, Hong Kong, BSc, 62; Oberlin Col, AM, 63; Cornell Univ, PhD(entom), 67. Prof Exp: ASST PROF BIOL, UNIV LOWELL, S CAMPUS, 67- Concurrent Pos: NSF res award, 69-72. Mem: Am Entom Soc; Animal Behav Soc; Bee Res Asn. Res: Learning ability of fruit fly; behavioral ecology of the leaf-cutter bee. Mailing Add: Dept of Biol Univ of Lowell S Campus Lowell MA 01854

LEE, SOOK, b Pukchong, Korea, July 13, 29; m 60; c 2. SOLID STATE PHYSICS. Educ: Korea Univ, BSc, 57; Brown Univ, PhD(physics), 63. Prof Exp: Res assoc physics, Brown Univ, 63-64; from asst prof to assoc prof, 64-70, PROF PHYSICS, ST LOUIS UNIV, 70- Concurrent Pos: Consult, McDonnell Douglas Res Labs, McDonnell Douglas Corp, 66- Mem: Am Phys Soc. Res: Electron spin and nuclear magnetic resonances in various types of solids. Mailing Add: Dept of Physics St Louis Univ St Louis MO 63103

LEE, SPENCER HON-SUN, b Hong Kong, Dec 18, 36; m 64; c 1. MEDICAL BIOLOGY, ANIMAL VIROLOGY. Educ: McGill Univ, BSc, 61; Dalhousie Univ, MSc, 63, PhD(virol), 65. Prof Exp: Fed Prov res grant virol, 65-66, lectr, 66-68, asst prof microbiol, 68-73, ASSOC PROF MICROBIOL, FAC MED, DALHOUSIE UNIV, 73- Concurrent Pos: Med Res Coun Can grant, 67-78; Wellcome Res Found fel, Dept Med Microbiol, Univ Liverpool, 69-70. Mem: Can Soc Microbiol; Am Soc Microbiol; Can Soc Cell Biol. Res: Viral pathogenesis; the role of interferon. Mailing Add: Dept of Microbiol Fac Med Dalhousie Univ Halifax NS Can

LEE, STANLEY L, b Newburgh, NY, Aug 27, 19; m 47; c 3. INTERNAL MEDICINE, HEMATOLOGY. Educ: Columbia Univ, AB, 39; Harvard Univ, MD, 43. Prof Exp: Intern, Mt Sinai Hosp, New York, 43-44, resident med, 46-48, Georg Escherich fel path, 48-49, asst, 49-53, asst attend hematologist, 53-59; assoc prof, 59-68, PROF MED, STATE UNIV NY DOWNSTATE MED CTR, 68-; DIR MED, JEWISH HOSP & MED CTR BROOKLYN, 71- Concurrent Pos: Dir hemat, Maimonides Med Ctr, Brooklyn, 59-71; treas, Int Cong Hemat, NY, 65-68. Mem: Am Soc Hemat; Am Rheumatism Asn; Soc Human Genetics; Am Fedn Clin Res; fel Am Col Physicians. Res: Systematic lupus erythematosus; leukemia. Mailing Add: Jewish Hosp & Med Ctr Brooklyn NY 11238

LEE, SUE YING, b Schenectady, NY, Jan 11, 40. VERTEBRATE MORPHOLOGY. Educ: State Univ NY Albany, BS, 61, MS, 63; Univ Ill, Urbana, PhD(zool), 68. Prof Exp: Instr vert morphol & human anat, Univ Ill, Chicago, 67-69; asst prof, 69-72, ASSOC PROF VERT MORPHOL & HUMAN ANAT, HUMBOLDT STATE UNIV, 73- Mem: AAAS; Am Soc Zool; Am Inst Biol Sci. Res: Reproductive biology. Mailing Add: Dept of Biol Humboldt State Univ Arcata CA 95521

LEE, SUN, b Seoul, Korea, June 2, 20; US citizen; m 45; c 6. SURGERY. Educ: Seoul Nat Univ, MD, 45. Prof Exp: From instr to asst prof, Univ Pittsburgh, 57-64; assoc prof surg, 68-74, PROF EXP SURG, UNIV CALIF, SAN DIEGO, 74-; ASSOC, SCRIPPS CLIN & RES FOUND, 64- Concurrent Pos: Surg fel, Univ Pittsburgh, 55-57. Res: Development of organ transplant in the rat to study transplantation immunology and associated physiology; techniques of heart-lung, liver, spleen, pancreas, testicle, kidney and stomach transplantation. Mailing Add: 6462 Cardeno Dr La Jolla CA 92037

LEE, SUNG GUE, b Hwachon, Korea, Mar 1, 37; m 65; c 2. BIOCHEMISTRY. Educ: Southern Ill Univ, BS, 62; Iowa State Univ, MS, 64, PhD(biochem), 68. Prof Exp: Fel biochem, Charles F Kettering Res Lab, 68-71; res assoc, 71-74, ASST PROF BIOCHEM, ROCKEFELLER UNIV, 74- Res: Mechanism of nonribosomal synthesis of peptides through enzymatic activation of amino acids and their condensation on protein templates; mechanism of cell transformation-dependent increase of sugar uptake. Mailing Add: Dept Biochem Rockefeller Univ 66th St & York Ave New York NY 10021

LEE, SUNG KI, b Seoul, Korea, Feb 17, 33; US citizen; m 60; c 3. ORGANIC CHEMISTRY, POLYMER CHEMISTRY. Educ: Baylor Univ, BS, 58, MS, 60; Carnegie Inst Technol, PhD(org chem), 64. Prof Exp: Jr fel polymer chem, Mellon Inst, 61-63; proj chemist, Carnegie Inst Technol, 63-64; res chemist, Cent Res, Hooker Chem Corp, 64-65, sr res chemist, 65-66, res supvr, 66-68, res sect mgr, 68-71; dir res & develop, 3 B Develop Inc, 71-72; vpres, Andco Res & Develop, 72-74; PRES, FRONTIER CHEM CORP, 74- Mem: Am Chem Soc; Am Mgt Asn; Electrochem Soc; Asn Consult Chemists & Chem Engrs. Res: Nitrogen heterocyclics; Fischer indole synthesis; reactions of polymer electroplating; electrosynthesis; electro-organic synthesis; environmental control and pollution control systems; energy and resource recovery systems; detoxification of hazardous materials. Mailing Add: 4626 Royal Ave Niagara Falls NY 14302

LEE, SUNG MOOK, b Seoul, Korea, Mar 2, 33; m 58; c 3. THEORETICAL MECHANICS, SOLID STATE PHYSICS. Educ: Chosun Christian Univ, BSc, 55; Ohio State Univ, MSc, 59, PhD(crystal dynamics), 65. Prof Exp: Teacher, Hansung Boy's High Sch, Korea, 54-55; asst prof physics, Denison Univ, 61-65; from asst prof to assoc prof, 65-72, PROF PHYSICS, MICH TECHNOL UNIV, 72- Concurrent Pos: NATO sr fel sci, 74. Mem: Am Phys Soc; Am Asn Physics Teachers; Am Acad Mechanics. Res: Vibrational analysis of periodic systems, crystal lattices, and molecules; wave propagation in solids; mechanical properties of solids; mechanics. Mailing Add: Dept of Physics Mich Technol Univ Houghton MI 49931

LEE, SUNGHEE K, nutrition, see 12th edition

LEE, SYLVAN BURTON, b Homen, Wis, Jan 5, 16; m 43; c 1. MICROBIOLOGY, BIOCHEMISTRY. Educ: Univ Wis, BS, 38, MS, 39, PhD(biochem bact), 43. Prof Exp: Asst, Univ Wis, 36-39 & 41-42; res biochemist, Com Solvents Corp, 39-41, dir microbiol res & develop, 50-51; head microbiol res, Gen Mills, Inc, 42-44; supt fermentation, Merck & Co, 46-49, asst dir microbiol res & develop, 49-50; gen supt mfg, New Brunswick Plant, E R Squibb & Sons, 51-52, plant mgr, 52-54, asst to pres, E R Squibb & Sons, Olin Mathieson Chem Corp, 54-58; vpres & dir opers, Hales & Hunter Co, 58-62, vpres & gen mgr, Northern Div, 62-65; dir opers fine chem, Gen Mills, Inc, 65-70, VPRES BIOCHEM, GEN MILLS CHEM, INC, 70- Mem: AAAS; Am Chem Soc; Am Soc Microbiol; Am Mgt Asn; NY Acad Sci. Res: Antibiotic and vitamin research and production; biochemistry in relation to the pharmaceutical, chemical, food, feed and agricultural industries; food and feed technology; general management; biochemicals. Mailing Add: Gen Mills Chem Inc 4620 W 77th St Minneapolis MN 55435

LEE, TALMAGE HOYLE, b Belwood, NC, July 22, 12; m 39; c 1. MATHEMATICS. Educ: Wake Forest Col, BA, 33; Univ NC, MA, 36, PhD(math), 53. Prof Exp: Teacher, High Sch, NC, 33-35; asst math, Univ NC, 36; instr, Brown Univ, 36-37; asst, Univ Wis, 37-40; assoc prof, Univ SC, 40-41; asst, Univ Wis, 41-44, instr, 44-45; assoc prof, La State Univ, 45-46; adj prof, 46-53, ASSOC PROF MATH, UNIV SC, 53- Concurrent Pos: Dir math, US Armed Forces Inst, Univ Wis, 44-45. Mem: Am Math Soc; Math Asn Am. Res: Matrix theory. Mailing Add: Dept of Math Univ of SC Columbia SC 29208

LEE, TEE-PING, b Taipei, Taiwan, Aug 3, 39; m 67; c 2. BIOCHEMISTRY, PHARMACOLOGY. Educ: Nat Taiwan Univ, BS, 62; Cornell Univ, MS, 66; Univ NC, PhD(biochem), 70. Prof Exp: Assoc in res, Dent Sch, Harvard Univ, 66-67;

ASST PROF MED, UNIV WIS-MADISON, 72- Concurrent Pos: NIH fel, Yale Univ, 70-72. Mailing Add: Allergy Lab Dept of Med Univ of Wis Madison WI 53706

LEE, TEH HSUN, b Shaoshin, China, Mar 25, 17; m 45; c 1. BIOCHEMISTRY. Educ: Chekiang Univ, BS, 38; Univ Mich, PhD, 54. Prof Exp: Res assoc, Sch Med, Univ Ore, 54-55; res assoc, Sch Med, Yale Univ, 55-60, asst prof exp med, 60-62; sr biochemist, Merck, Sharp & Dohme, 62-64; asst dir, Vet Admin Human Protein Hormone Bank, 64-66, CHIEF PROTEIN HORMONE RES LAB, VET ADMIN HOSP, 66-; VIS ASSOC PROF BIOCHEM, ALBERT EINSTEIN COL MED, 67- Concurrent Pos: Assoc prof, Sch Med, Univ Colo, Denver, 64-66. Mem: Am Chem Soc; Am Soc Biol Chem; Endocrine Soc. Res: Pituitary hormones. Mailing Add: Vet Admin Hosp 130 W Kingsbridge Rd Bronx NY 10468

LEE, TEH-HSUANG, b Shanghai, China, Aug 15, 36; m 61; c 2. SOLID STATE PHYSICS. Educ: Nat Taiwan Univ, BS, 58; Purdue Univ, West Lafayette, PhD(physics), 67. Prof Exp: SR PHYSICIST, RES LABS, EASTMAN KODAK CO, 67- Mem: Am Phys Soc. Res: Optical properties of solids; semiconductors; magnetic semiconductors. Mailing Add: Res Labs Eastman Kodak Co 1669 Lake Ave Rochester NY 14650

LEE, THOMAS ALAN, b Chippewa Falls, Wis, Sept 6, 39; m 68. ASTRONOMY. Educ: Wis State Univ, Eau Claire, BS, 61; Univ Ariz, PhD(astron), 67. Prof Exp: Asst prof astron, Lunar & Planetary Lab, Univ Ariz, 67-68; Nat Res Coun resident res assoc, Goddard Inst Space Studies, NASA, 68-69; res assoc astron, Lunar & Planetary Lab, Univ Ariz, 69, asst prof, 69-74; vis asst prof, Univ Minn, Minneapolis, 74; SR PROJ DATA COORDR, MIDREX CORP, NC, 75- Mem: AAAS; Am Astron Soc. Res: Interstellar extinction; infrared astronomy. Mailing Add: Midrex Corp One NCNB Plaza Bldg Charlotte NC 28280

LEE, THOMAS HENRY, b China, May 11, 23; nat US; m 48; c 3. PHYSICS. Educ: Nat Chiao-Tung Univ, China, BS; Union Col, MS, 50; Rensselaer Polytech Inst, PhD(elec eng, physics), 54. Prof Exp: Eng analyst, Pa, 54-55, sr res physicist, 55-59, mgr eng res, 59-67, mgr lab -pers, 67-71, mgr group tech resources oper, 71-74, MGR STRATEGIC PLANNING OPER, GEN ELEC CO, 74- Concurrent Pos: Adj prof, Rensselaer Polytech Inst, 54-55; lectr, Univ Pa. Honors & Awards: Received 30 US patents. Mem: Nat Acad Eng; Am Phys Soc; Inst Elec & Electronics Engrs; Power Eng Soc (pres). Res: Electron physics; gaseous discharges; magnetohydrodynamics; ultra high vacuum technology; plasma physics. Mailing Add: 1402 Melville Ave Fairfield CT 06430

LEE, THOMAS SEYMOUR, analytical chemistry, see 12th edition

LEE, THOMAS W, b New Britain, Conn, Sept 12, 37; m 63; c 2. COMPARATIVE PHYSIOLOGY. Educ: Bates Col, BS, 59; Duke Univ, MA, 61; Rice Univ, PhD(biol), 64. Prof Exp: Lectr biol, Rice Univ, 64-65; from asst prof to assoc prof, 65-72, actg chmn dept, 69-74, PROF BIOL, CENT CONN STATE COL, 72-, ASST TO CHMN, 74- Mem: AAAS; Am Soc Zool; Ecol Soc Am. Res: Embryonic development of amphibians; nitrogen metabolism in invertebrate animals. Mailing Add: Dept of Biol Cent Conn State Col New Britain CT 06050

LEE, TIEN-CHANG, b Nantou, Taiwan, July 1, 43; m 69; c 2. GEOPHYSICS. Educ: Nat Taiwan Univ, BS, 65; Univ Southern Calif, PhD(geophys), 74. Prof Exp: Fel marine geophys, Woods Hole Oceanog Inst, 73-74; ASST PROF GEOPHYS, UNIV CALIF, RIVERSIDE, 74- Mem: Am Geophys Union; Geol Soc Am. Res: Geothermal study using heat flow measurement, magnetotelluric survey, and microearthquake monitoring. Mailing Add: Dept of Earth Sci Univ of Calif Riverside CA 92502

LEE, TONG-NYONG, b July 22, 27; US citizen; m 59; c 3. PLASMA PHYSICS, ATOMIC PHYSICS. Educ: Seoul Nat Univ, BS, 50; Univ London, PhD(physics), 59. Prof Exp: Asst prof physics, Seoul Nat Univ, 60-63; assoc prof appl physics, Cath Univ Am, 64-70; RES PHYSICIST PLASMA PHYSICS & OPTICAL SCI, NAVAL RES LAB, 70- Mem: Am Phys Soc. Res: Short wavelength laser generation; plasma physics and spectroscopy of high temperature; high density plasma and solar flare study. Mailing Add: Naval Res Lab Washington DC 20375

LEE, TSUNG DAO, b China, Nov 25, 26; m 50; c 2. THEORETICAL PHYSICS. Educ: Univ Chicago, PhD(physics), 50. Hon Degrees: DSc, Princeton Univ, 58. Prof Exp: Res assoc astrophys, Univ Chicago, 50; res assoc physics, Univ Calif, 50-51; mem, Inst Advan Study, 51-53; from asst prof to prof, Columbia Univ, 53-60; prof, Inst Advan Study, 60-63; ENRICO FERMI PROF PHYSICS, COLUMBIA UNIV, 63- Honors & Awards: Nobel Prize in Physics, 57; Einstein Award Sci, 57. Mem: Nat Acad Sci. Res: Field theory; statistical mechanics; hydrodynamics; astrophysics. Mailing Add: Dept of Physics Columbia Univ New York NY 10027

LEE, TSUNG TING b Anhwei, China, Mar 21, 23; m 50; c 3. PLANT PHYSIOLOGY. Educ: Nat Cent Univ, China, BS, 47; Univ Wis, MS, 59, PhD(plant physiol), 62. Prof Exp: Res asst plant physiol, Taiwan Sugar Exp Sta, China, 47-52, asst plant physiologist, 52-57; plant physiologist, Res Sta, Ont, 62-68, PLANT PHYSIOLOGIST, RES INST, CAN DEPT AGR, 68- Mem: Am Soc Plant Physiol; Tissue Cult Asn; Scan Soc Plant Physiol; Can Soc Plant Physiol. Res: Plant growth regulators; auxin metabolism; tissue and cell culture; pesticide-plant interaction. Mailing Add: Res Inst Agr Can Univ Western Ont Campus London ON Can

LEE, TSUNG-SHUNG HARRY, b Taipei, Taiwan, June 7, 43; m 68; c 1. NUCLEAR PHYSICS. Educ: Taiwan Norm Univ, BS, 65; Nat Tsing-Hua Univ, MS, 67; Univ Pittsburgh, PhD(physics), 73. Prof Exp: Res assoc physics, Bartol Res Found, 73-75; RES ASSOC PHYSICS, ARGONNE NAT LAB, 75- Mem: Sigma Xi. Res: Intermediate-energy nuclear physics and low-energy nucleon-nucleon interaction. Mailing Add: Physics Div Argonne Nat Lab Argonne IL 60439

LEE, TZOONG-CHYH, b Taiwan, Jan 2, 36; m 62; c 3. ORGANIC CHEMISTRY, BIO-ORGANIC CHEMISTRY. Educ: Yamagata Univ, Japan, BSc, 63; Tohoku Univ, Japan, MSc, 65; Australian Nat Univ, PhD(med chem), 68. Prof Exp: USPHS fel, 68-71, Damon Runyon res fel, 69-70, res assoc, 71-75, ASSOC ORG CHEM, SLOAN-KETTERING INST CANCER RES, 75- Mem: Am Chem Soc. Res: Nitrogen heterocyclic chemistry; organic synthesis; structure-activity relationship. Mailing Add: Sloan-Kettering Inst Cancer Res 145 Boston Post Rd Rye NY 10580

LEE, VERNON HAROLD, b Flandreau, SDak, Oct 27, 31; m 62. MEDICAL ENTOMOLOGY. Educ: SDak State Col, BS, 53, MS, 57; Univ Wis, PhD(med entom), 62. Prof Exp: Res fel med entom, Univ Wis, 61-62; med entomologist, Rockefeller Found, 62; hon vis asst prof, Univ del Valle, Colombia, 62-66; hon vis asst prof, Univ Ibadan, 67-73; WITH US NAVAL MED RES UNIT, ADDIS ABABA, ETHIOPIA, 73- Mem: Entom Soc Am; Mosquito Control Asn; Am Soc Trop Med & Hyg; Wildlife Dis Asn. Res: Biological and ecological study of haematophagous diptera, in relationship to their role as vectors of Arboviruses. Mailing Add: US Naval Med Res Unit 5 Ethiopia APO New York NY 09319

LEE, VIRGINIA ANN, b Grand Rapids, Mich, Oct 30, 22. BIOCHEMISTRY. Educ: Univ Ill, BS, 44; Univ Colo, MS, 46, PhD(biochem), 52. Prof Exp: Instr biochem, Sch Med, Univ Colo, Denver, 55-59, asst prof, 59-67; ASST PROF FOOD SCI & NUTRIT, COLO STATE UNIV, 67- Concurrent Pos: Res biochemist, Child Res Coun, 44- Mem: AAAS. Res: Biochemistry of human growth. Mailing Add: Dept of Food Sci & Nutrit Colo State Univ Ft Collins CO 80521

LEE, WARREN FORD, b Harriston, Ont, Aug 25, 41; m 66; c 3. AGRICULTURAL ECONOMICS. Educ: Univ Toronto, BSA, 63; Univ Ill, MS, 67; Mich State Univ, PhD(agr econ), 70. Prof Exp: Credit adv, Farm Credit Corp, 63-65; ASSOC PROF AGR FINANCE, OHIO STATE UNIV, 70- Concurrent Pos: Economist, Econ Br, Agr Can, 75-76. Mem: Am Agr Econ Asn. Res: Agricultural credit and finance; farm firm growth; rural capital markets; bank structure and performance; financial institutions. Mailing Add: Dept of Agr Econ Ohio State Univ 2120 Fyffe Rd Columbus OH 43210

LEE, WARREN G, b Abington, Pa, June 10, 24; m 51; c 3. PHYSICAL ORGANIC CHEMISTRY. Educ: Roosevelt Univ, BS, 50; Ill Inst Technol, PhD(chem), 61. Prof Exp: Sr res scientist, Res & Develop Lab, Gen Am Transportation Co, 51-63; mgr chem res & instrumentation, Liquid Carbonic Div, Gen Dynamic Corp, 63-67; asst to dir res & com develop, 67-68 & planning & facilities eng, 68-70, adminr corp pollution control & tech adv, 70-73, MKT & TECH SPECIALIST, LIQUID CARBONIC CORP, 73- Mem: AAAS; Am Chem Soc; Am Inst Chemists. Res: Radiochemistry; electroless methods of metal plating; corrosion protection; catalytic processes; chemical kinetics and isotope studies; organic synthesis; gas purification processes and applied instrumentation for process control; environmental medicine, health and safety; specialty gases and chemical products. Mailing Add: Indust & Med Div Liquid Carb Corp 135 S LaSalle St Chicago IL 60603

LEE, WEI HWA, b Shanghai, China, July 7, 32; US citizen; m 59; c 2. MICROBIOLOGY, FOOD TECHNOLOGY. Educ: Cornell Univ, BS, 54; Univ Ill, MS & PhD(food technol), 61. Prof Exp: RES MICROBIOLOGIST, FOOD & DRUG ADMIN, 71- Mem: Am Soc Microbiol. Res: Basic and applied research in microbiology. Mailing Add: Food & Drug Admin BF214 200 C St SW Washington DC 20204

LEE, WEI-LI S, b Kiangsi, China, Feb 14, 45; m 70. BIOCHEMISTRY. Educ: Tunghai Univ, BS, 66; State Univ NY Buffalo, MA, 69, PhD(biol), 72. Prof Exp: Asst biol sci, State Univ NY Buffalo, 66-72; res assoc immunochem, Col Physicians & Surgeons, Columbia Univ, 72-75; INSTR DERMATOL BIOCHEM, DOWNSTATE MED CTR, BROOKLYN, STAtE UNIV NY, 75- Mem: Sigma Xi. Res: Control and regulation of basic cell metabolism at enzymatic level; purification and structural studies of yeast cell surface antigens and their immunological functions. Mailing Add: Dept of Med Dermatol Div PO Box 46 State Univ of NY Brooklyn NY 11203

LEE, WEI-MING, b Kiangsu, China, June 11, 36; m 62. PHYSICAL CHEMISTRY, POLYMER CHEMISTRY. Educ: Nat Taiwan Univ, BS, 57; Southern Ill Univ, MA, 61; Univ Ill, PhD(phys chem), 64. Prof Exp: Teacher chem & math, Univ High Sch, Taiwan Norm Univ, 59; res assoc theoret chem kinetics, Univ Calif, Santa Barbara, 64-65; sr res engr styrene molding polymers res & develop, Dow Chem USA, 65-74, STAFF MEM OLEFIN PLASTICS RES & DEVELOP, DOW CHEM CO, 74- Mem: Am Chem Soc; Soc Rheol. Res: Mechanical properties of polymeric materials and plastic foams; computer simulation of chemical and physical processes. Mailing Add: Olefin Plastics Res & Develop Dow Chem Co Bldg 1702 Midland MI 48640

LEE, WELTON LINCOLN, b San Francisco, Calif, Feb 12, 33. MARINE ECOLOGY, PHYSIOLOGICAL ECOLOGY. Educ: Univ Calif, Berkeley, AB, 60; Stanford Univ, PhD(physiol ecol), 65. Prof Exp: Asst prof marine & physiol ecol, Hopkins Marine Sta, Stanford Univ, 66-73; CHMN, CUR & FEL INVERT ZOOL, CALIF ACAD SCI, 73- Concurrent Pos: Consult sci adv comt, Monterey Peninsula Water Qual Control Agency, 73-; chmn, NSF Adv Comt Systs Resources Invert Zool, 74-75; chmn subcomt stand, Asn Systs Collections-Coun Stand Systs Collections, 74-75. Mem: Am Soc Zoologists. Res: Color and color change in marine invertebrates and its ecological significance; effects of various pollutants on marine benthic communities. Mailing Add: Dept of Invert Zool Calif Acad of Sci San Francisco CA 94118

LEE, WILLIAM GLEN, b Knoxville, Tenn, Nov 15, 45; m 67. FOOD SCIENCE. Educ: Univ Tenn, Knoxville, BS, 68; Miss State Univ, MS, 71, PhD(food sci), 73. Prof Exp: Food technologist, US Army Natick Res Labs, 73-74; sr food technologist, 74-76, GROUP LEADER FLOUR & BAKERY, PEAVEY CO, 76- Mem: Sigma Xi. Res: Cereal grains and bakery technology; general food technology. Mailing Add: Peavey Tech Ctr 11 Peavey Rd Chaska MN 55318

LEE, WILLIAM HALL, JR, b Charleston, SC, Apr 28, 29; m 52; c 3. THORACIC SURGERY, CARDIOVASCULAR SURGERY. Educ: Col Charleston, BS, 50; Med Col SC, MD, 54. Prof Exp: Intern, Roper Hosp, Charleston, SC, 54-55; resident surg, Med Ctr Hosps, 55-59; teaching fel, Med Col SC, 59-60; fel thoracic surg, Med Ctr, Univ Calif, Los Angeles, 60-61; asst prof surg, Sch Med, Univ Tenn, 61-63; asst prof, 63-64, PROF THORACIC SURG & CHMN DIV, MED UNIV SC, 64- Concurrent Pos: Consult, Kennedy Vet Admin Hosp, Memphis, 61-63; Markle Found scholar med sci, 62-67; chief consult, Vet Admin Hosp, Charleston, SC, 66-; assoc fac, Columbia Univ, 67-; pres, Dirs Thoracic Surg Residency Progs, 74-75. Mem: Am Col Surg; Soc Univ Surg; Am Soc Thoracic Surg; Am Asn Thoracic Surg; Soc Vascular Surg. Res: Biophysics and physical chemistry of blood-prosthesis interaction; physiology of blood flow; pathophysiology of lung function; disorders of clotting mechanisms. Mailing Add: Div of Thoracis Surg Med Univ of SC Charleston SC 29401

LEE, WILLIAM HUNG KAN, b Kwangsi, China, Oct 6, 40; m 66; c 2. GEOPHYSICS. Educ: Univ Alta, BSc, 62; Univ Calif, Los Angeles, PhD(planetary & space physics), 67. Prof Exp: Asst res geophysicist, Univ Calif, Los Angeles, 67; RES GEOPHYSICIST, US GEOL SURV, 67- Concurrent Pos: Secy, Heat-Flow Comt, Int Union Geod & Geophys, 63-65; ed, Geophys Monogr 8, Am Geophys Union, 65; guest lectr, Stanford Univ, 69. Mem: AAAS; Am Geophys Union; Seismol Soc Am; Soc Explor Geophys. Res: Terrestrial heat-flow; thermal evolution of the planets; computer modeling of geologic processes; earthquake seismology. Mailing Add: US Geol Surv Menlo Park CA 94025

LEE, WILLIAM ORVID, b Brigham City, Utah, July 2, 27; m 48; c 4. FIELD CROPS. Educ: Utah State Univ, BS, 50, MS, 54; Ore State Univ, PhD(farm crops), 65. Prof Exp: Soil scientist, Bur Reclamation, US Dept Interior, 50-51; agronomist, Utah, 51-54 & Wyo, 54-56, RES AGRONOMIST, AGR RES SERV, USDA, 56- Mem: Weed Sci Soc Am. Res: Crop science; control of weeds in forage and turf seed crops; legumes and grasses. Mailing Add: Agron Crop Sci Dept Agr Res Serv Ore State Univ Corvallis OR 97331

LEE, WILLIAM RICHARD, b Waukegan, Ill, Aug 2, 31; m 53; c 5. ANALYTICAL CHEMISTRY. Educ: Wayne State Univ, BS, 56. Prof Exp: Technician, 53-54, res technician, 54-56, chemist, 56-59, res chemist, 59-62, ASSOC SR RES CHEMIST,

GEN MOTORS RES LAB, 62- Mem: Am Chem Soc; Am Soc Testing & Mat. Res: Analytical chemistry-classical methods, instrumental; gases in metals; spectroscopy-emission, x-ray fluorescence; low angle light scattering; vapor pressure osmometry. Mailing Add: Chem Dept Gen Motors Res Labs 12 Mile & Mound Rds Warren MI 48090

LEE, WILLIAM ROSCOE, b Little Rock, Ark, Feb 14, 30; m 53; c 3. GENETICS. Educ: Univ Ark, BSA, 53; Univ Wis, MS, 53, PhD(genetics, entom), 56. Prof Exp: Asst, Univ Wis, 52-56; from asst prof to assoc prof entom, Univ NH, 56-63; asst prof zool, Univ Tex, Austin, 63-67; assoc prof, 67-73, PROF ZOOL & PHYSIOL, LA STATE UNIV, BATON ROUGE, 73- Concurrent Pos: Res exec for H J Muller, Ind Univ, 62-63. Mem: Genetics Soc Am; Radiation Res Soc; Environ Mutagen Soc. Res: Radiation genetics; mutagenesis. Mailing Add: Dept of Zool & Physiol La State Univ Baton Rouge LA 70803

LEE, WILLIAM THAD, b Gentry, Ark, Aug 4, 23; m 62; c 2. APPLIED MATHEMATICS. Educ: US Merchant Marine Acad, BS, 47; Univ Tex, MA, 55. Prof Exp: Marine engr, Socony-Vacuum Oil Corp, 47-49; test engr, Westinghouse Elec Corp, 52-53; lubrication engr, Douglas Aircraft Co, Inc, 53-54; teacher math & physics, Okla Mil Acad, 54-56; nuclear engr, Martin Co, 56-58 & ACF Industs, Inc, 58-59; eng mathematician, Marquardt Corp, 59; sr mathematician, Comput Div, Bendix Corp, 59-61; comput specialist, Litton Syst, Inc, 61-62; programming specialist, Packard Bell Comput, 62-64; tech asst to hybrid simulation br chief, 64-68, HEAD, COMPUT APPLN SECT, GUID & CONTROL DIV, JOHNSON SPACE CTR, NASA, 68- Concurrent Pos: Mem, Simulation Coun. Mem: Soc Indust & Appl Math; Am Math Asn; Asn Comput Mach. Res: Use of electronic computers to solve engineering and scientific problems. Mailing Add: 2204 Webster St League City TX 77573

LEE, WILLIAM WEI, b San Francisco, Calif, May 17, 23; m 47; c 3. ORGANIC CHEMISTRY. Educ: Univ Calif, BS, 47; PhD(org chem), 52. Prof Exp: Jr chemist org anal chem, Shell Develop Co, 47-48; asst org chem, Univ Minn, 48-51; org res chemist, Cent Res Dept, Monsanto Chem Co, 52-54; assoc chemist org chem, 54-56, SR ORG CHEMIST, STANFORD RES INST, 56- Mem: Am Chem Soc; Sigma Xi. Res: Allylic and acetylenic compounds; nucleosides, amino acids and alkylating agents; enzyme chemistry; active halogen compounds; folic acid antagonists; heterocyclic chemistry; chemotherapy, particularly cancer chemotherapy. Mailing Add: Bioorg Dept Stanford Res Inst 333 Ravenswood Ave Menlo Park CA 94025

LEE, WONYONG, b Korea, Dec 29, 30; m 61; c 1. HIGH ENERGY PHYSICS. Educ: Calif Inst Technol, BS, 57; Univ Calif, Berkeley, PhD(physics), 61. Prof Exp: Res assoc physics, Lawrence Radiation Lab, Univ Calif, 61-62; res assoc, 62-64, from asst prof to assoc prof, 64-72, PROF PHYSICS, COLUMBIA UNIV, 72- Concurrent Pos: Sloan Found fel, 65-67. Mem: Am Phys Soc. Res: High energy experimental physics. Mailing Add: Dept of Physics Columbia Univ New York NY 10027

LEE, WYLIE IN-WEI, b Taiwan, Aug 18, 41; c 2. POLYMER PHYSICS. Educ: Taiwan Normal Univ, BS, 63; Univ Mass, MS, 67, PhD(physics), 71. Prof Exp: Res assoc physics, Manchester Univ, 70-72; res & teaching assoc chem phys, 72-75, NIH FEL & RES ASSOC BIOENG, UNIV WASH, 75- Mem: Am Phys Soc. Res: The dynamics of macromolecules in solution or network state by laser light scattering; applications of the intensity fluctuation spectroscopy in biomedical studies such as gamete transport and cutaneous blood flow. Mailing Add: Ctr for Bioeng Univ of Wash RK-15 Seattle WA 98195

LEE, YA PIN, b Taipei, Formosa, Oct 23, 24; m 50; c 3. BIOCHEMISTRY. Educ: Nat Taiwan Univ, MD, 49; Kyushu Univ, DrMedSci, 56. Prof Exp: Proj assoc, Inst Enzyme Res, Univ Wis, 55-57, asst prof, 60-63; res asst prof, Sch Med, Washington Univ, 58-59; HILL RES PROF BIOCHEM, SCH MED, UNIV NDAK, 63- Concurrent Pos: Res fel, Univ Wis, 60-63. Mem: Am Soc Biol Chem. Mailing Add: Dept of Biochem Univ of NDak Grand Forks ND 58202

LEE, YAT-SHIR, b Kwangtung, China. INORGANIC CHEMISTRY. Educ: Nat Taiwan Univ, BS, 58; Kent State Univ, MS, 67, PhD(chem), 71. Prof Exp: MEM TECH STAFF LIQUID CRYSTAL RES, HUGHES AIRCRAFT CO, 72- Concurrent Pos: Fel, Harvard Univ, 72. Mem: Am Chem Soc. Res: Structure, hydrodynamics and electro-optical effects of liquid crystals. Mailing Add: Hughes Aircraft Co 500 Superior Ave Newport Beach CA 92663

LEE, YIEN-HWEI, b Taiwan, China, Oct 20, 37; m 68; c 3. PHARMACOLOGY, MEDICINE. Educ: Nat Taiwan Univ, MD, 63; Univ Calif, Los Angeles, PhD(pharmacol), 68. Prof Exp: Res asst pharmacol, Univ Calif, Los Angeles, 64-66, res pharmacologist, 66-68, res assoc, Neuropsychiat Inst, 66-68; sr investr pharmacol, Searle Res Lab, G D Searle & Co, 68-72; SECT HEAD, ABBOTT LABS, 72- Mem: AAAS; Am Soc Pharmacol & Exp Therapeut; NY Acad Sci; Sigma Xi; Am Chem Soc. Res: Gastrointestinal pharmacology and physiology; neuropsychophysiology and pharmacology. Mailing Add: Abbott Labs North Chicago IL 60064

LEE, YING KAO, b Shanghai, China, Dec 14, 32; US citizen; m 61; c 3. POLYMER CHEMISTRY. Educ: Tai Tung Univ, BSc, 52; Univ Cincinnati, PhD(chem), 61. Prof Exp: Res chemist, Tex-US Chem Co, 60-63, proj leader, 65; res chemist, 65-68, staff chemist, 68-70, RES ASSOC, MARSHALL RES & DEVELOP LAB, E I DU PONT DE NEMOURS & CO, INC, 70- Mem: Am Chem Soc. Res: Polymers or polymeric systems used in coating field. Mailing Add: Marshall Res & Develop Lab E I du Pont de Nemours & Co Inc Philadelphia PA 19146

LEE, YOON CHAI, b Seoul, Korea, Dec 12, 29; US citizen; m 60; c 2. CHEMISTRY. Educ: RI Sch Design, BS, 57; Lowell Technol Inst, MS, 58; Carnegie Inst Technol, PhD(org chem), 62. Prof Exp: SR SPECIALIST POLYMER RES, MONSANTO POLYMERS & PETROCHEM CO, 61- Mem: Am Chem Soc. Res: Macromolecular chemistry; synthesis of polymeric materials; flammability of organic polymers. Mailing Add: Monsanto Polymers & Petrochem Co 730 Worcester St Indian Orchard MA 01151

LEE, YOUNG CHANG, b Chonju, Korea, Mar 17, 30; m 54; c 5. CELL BIOLOGY, RADIATION BIOLOGY. Educ: Seoul Nat Univ, BS, 53; Univ Calif, Los Angeles, MA, 66, PhD(zool), 69. Prof Exp: Instr gen chem, Korean Air Force Acad, 53-57; lectr biol, Seoul Nat Univ, 58-60; asst prof, Kyung-Hee Univ, Korea, 60-63; ASST RES BIOLOGIST, DIV RADIOBIOL, LAB NUCLEAR MED & RADIATION BIOL, UNIV CALIF, LOS ANGELES, 69- Mem: Am Soc Cell Biol; Am Soc Zool; Soc Protozool. Res: DNA repair; nucleic acid metabolism and protein synthesis. Mailing Add: Lab Nuclear Med & Radiation Biol Univ of Calif 900 Veteran Ave Los Angeles CA 90024

LEE, YOUNG JACK, b Seoul, Korea, Feb 25, 42; m 67; c 2. STATISTICS. Educ: Seoul Nat Univ, BSE, 64; Ohio State Univ, MS, 72, PhD(statist), 74. Prof Exp: Instr electronics eng, Korean Air Force Acad, 67-69; ASST PROF STATIST, UNIV MD, COLLEGE PARK, 74- Mem: Am Statist Asn; Inst Math Statist. Res: Nonparametric/robust design of experiment and statistical analysis in hypothesis testing, ranking and selection and estimation; applications of statistics to social science and life science. Mailing Add: Dept of Math Univ of Md College Park MD 20742

LEE, YOUNG-HOON, b Korea, Sept 18, 35; m 65; c 1. SOLID STATE PHYSICS. Educ: Dong-Guk Univ, BS, 61, MS, 63; State Univ NY Albany, PhD(physics), 72. Prof Exp: Sr asst physics, Dong-Guk Univ, 63-66; FEL PHYSICS, STATE UNIV NY ALBANY, 73- Mem: Am Phys Soc; Korean Phys Soc. Res: Defects in solids and electron spin resonance. Mailing Add: Dept of Physics State Univ NY Albany NY 12222

LEE, YOUNG-JIN, b Seoul, Korea, Nov 22, 46; m 72. CHEMISTRY, ORGANIC CHEMISTRY. Educ: Millikin Univ, BA, 68; Univ Rochester, MS, 70; State Univ NY Albany, PhD(chem), 74. Prof Exp: Fel chem, Cornell Univ, 74-75; CHEMIST, UNION CARBIDE CORP, 75- Mem: Am Chem Soc. Res: Syntheses and process development of agricultural chemicals; new pesticides. Mailing Add: Union Carbide Corp Tech Ctr South Charleston WV 25303

LEE, YUAN CHUAN, b Taiwan, China, Mar 30, 32; m 58; c 1. BIOCHEMISTRY. Educ: Nat Taiwan Univ, BS, 55, MS, 57; Univ Iowa, PhD(biochem), 62. Prof Exp: Res assoc biochem, Univ Iowa, 62 & Univ Calif, Berkeley, 62-65; from asst prof to assoc prof, 65-74, PROF BIOL, JOHNS HOPKINS UNIV, 74- Mem: Am Chem Soc; Am Soc Biol Chem; Brit Biochem Soc. Res: Complex carbohydrates. Mailing Add: Dept of Biol Johns Hopkins Univ Baltimore MD 21218

LEE, YUAN TSEH, b Hsinchu, Taiwan, Nov 29, 36; m 63; c 3. CHEMISTRY. Educ: Nat Taiwan Univ, BS, 59; Nat Tsing Hua Univ, Taiwan, MS, 61; Univ Calif, Berkeley, PhD(chem), 65. Prof Exp: From asst prof to prof chem, Univ Chicago, 68-74; PROF CHEM, UNIV CALIF, BERKELEY, 74- Concurrent Pos: Alfred P Sloan res fel, Univ Chicago, 69-71; consult, Chem Div, Argonne Nat Lab, 69-72; assoc ed, J Chem Phys, 75-78. Mem: Am Chem Soc; Am Phys Soc; fel Am Acad Arts Sci. Res: Chemical kinetics and molecular interaction. Mailing Add: Dept Chem Lawrence Berkeley Lab Univ of Calif Berkeley CA 94720

LEE, YUEN SAN, b Taipei, Taiwan, Oct 13, 39; m 67; c 2. FOODS, BIOCHEMISTRY. Educ: Nat Taiwan Univ, BS, 62; Utah State Univ, MS, 65; Univ Md, College Park, PhD(food sci, biochem), 68. Prof Exp: CHEMIST, HEALTH SERV ADMIN, DC GOVT, 68- Concurrent Pos: lectr food sanit, Fed City Col, DC, 75- Mem: Inst Food Technol; Am Chem Soc; Am Dietetic Asn. Res: Method development in the determination of pesticides in meat, milk and water; quality control of detecting adulteration in meat and meat products for consumer protection; heavy metals in foods. Mailing Add: 1 C Westway Greenbelt MD 20770

LEE, YUK, urban geography, see 12th edition

LEE, YUNG-CHANG, plasma physics, nuclear physics, see 12th edition

LEE, YUNG-KEUN, b Seoul, Korea, Sept 26, 29; m 58; c 4. NUCLEAR PHYSICS. Educ: Johns Hopkins Univ, BA, 56; Univ Chicago, MS, 57; Columbia Univ, PhD(physics), 61. Prof Exp: Res scientist, Columbia Univ, 61-64; from asst prof to assoc prof, 64-71, PROF PHYSICS, JOHNS HOPKINS UNIV, 71- Mem: Am Phys Soc. Res: Nuclear beta decay; nuclear reactions; Mössbauer effects. Mailing Add: Dept of Physics Johns Hopkins Univ Baltimore MD 21218

LEE, YU-SUN, b Taipei, Taiwan, Feb 6, 41; m 69. ORGANIC CHEMISTRY. Educ: Chung Yuan Christian Col Sci & Eng, Taiwan, BS, 65; La State Univ, Baton Rouge, PhD(org chem), 74. Prof Exp: Instr chem, Fu Jen Univ, 66-68; RES SPECIALIST ORG CHEM, CENT RES LABS, MEAD CORP, 74- Mem: Am Chem Soc; Sigma Xi. Res: Encapsulation and coating on carbonless copy paper system; kinetics and mechanisms of free radical halogenations of hydrocarbons. Mailing Add: Cent Res Labs Mead Corp Eighth & Hickory Sts Chillicothe OH 45601

LEECH, GEOFFREY BOSDIN, b Montreal, Que, Aug 28, 18; m 46; c 1. GEOLOGY. Educ: Univ BC, BASc, 42; Queen's Univ, Ont, MSc, 43; Princeton Univ, PhD(petrol, econ geol), 49. Prof Exp: Field asst, Geol Surv Can, 40-41 & 42; geologist, Int Nickel Co Can, Ltd, Ont, 43-46; chief party, BC Dept Mines, 47-48; geologist, 49-72, HEAD, ECON GEOL SUBDIV, DEPT ENERGY, MINES & RESOURCES, GEOL SURV CAN, 73- Mem: Fel Geol Soc Am; Soc Econ Geol; fel Royal Soc Can; Can Inst Mining & Metall; fel Geol Asn Can. Res: Regional metallogeny; mineral resource evaluation; problems of resource adequacy. Mailing Add: Geol Surv Can Ottawa ON Can

LEECH, JOHN G, b Norwood, Pa, May 28, 23; m 44; c 3. PULP CHEMISTRY, PAPER CHEMISTRY. Educ: Pa State Univ, BS, 47; Inst Paper Chem, Lawrence Col, MS, 49, PhD(chem), 52. Prof Exp: Proj leader, WVa Pulp & Paper Co, 51-55, res lab dir, 55-62; asst dir res, 62-64, dir proj planning, 64-70, DIR RES DEVELOP ADMIN, UNION CAMP CORP, 70- Mem: Am Chem Soc; Am Inst Chem Eng; Tech Asn Pulp & Paper Indust. Res: Pulp and paper; packaging; silvichemicals. Mailing Add: 305 Prospect Ave Princeton NJ 08540

LEECH, JOHN WATSON, b London, Eng, July 4, 17; m 43; c 2. SOLID STATE PHYSICS, THEORETICAL PHYSICS. Educ: Univ London, BSc, 38, PhD(physics), 47. Prof Exp: Lectr physics, Queen Mary Col, London, 45-60, sr lectr, 60-68; PROF PHYSICS & CHMN DEPT, UNIV WATERLOO, 68- Concurrent Pos: Assoc res officer, Nat Res Coun Can, 57-58; vis prof, Univ Edinburgh, Scotland, 74-75. Mem: Brit Inst Physics. Res: Lattice dynamics. Mailing Add: Dept of Physics Univ of Waterloo Waterloo ON Can

LEECH, WILLIAM DALE, b Kingsville, Mo, Oct 3, 94. CHEMISTRY. Educ: Union Col, Nebr, AB, 19; Univ Southern Calif, AM, 26; Calif Inst Technol, PhD(chem), 42. Prof Exp: Instr chem, Pasadena Jr Col, 26-32, from instr to mem staff, 38-47, Pasadena City Col, 47-52; mem staff, Australasian Food Res Lab, Australia, 32-38; prof & chmn dept, La Sierra Col, 52-60; dir natural prod prog, 60-75, chmn dept chem, Grad Sch, 65-75, EMER PROF CHEM, LOMA LINDA UNIV, 75- Res: Bio-organic chemistry; vitamins; plant-growth hormones; chlorophyll; neoplasma. Mailing Add: Dept of Chem Loma Linda Univ Grad Sch Loma Linda CA 92354

LEECHMAN, DOUGLAS, b London, Eng, Dec 20, 90; m 28. ANTHROPOLOGY, LINGUISTICS. Educ: Univ Ottawa, BSc, 37, MA, 39, PhD(anthrop), 40. Prof Exp: Anthropologist, Nat Mus Can, 24-55; dir, Glenbow Found, 55-57; CONSULT ANTHROPOLOGIST, 57- Concurrent Pos: Res assoc, Dept Univ Victoria, BC, 67; outside consult, Oxford Eng Dictionary, Ling. Res: Anthropology of Canada; migration routes; Dorset Eskimo culture. Mailing Add: 2078 68. Mem: Am Anthrop Asn; Soc Am Archaeol; Royal Soc Can. St Victoria BC Can

LEED, RUSSELL ERNEST, b Denver, Pa, Dec 24, 15. PHYSICAL CHEMISTRY. Educ: Franklin & Marshall Col, BS, 37; Univ Md, MS, 40, PhD(phys chem), 41. Prof Exp: From instr to asst prof chem, Va Mil Inst, 41-48; assoc prof, Va Polytech Inst,

48-51; chemist, Prod Div, AEC, 51-71, dep dir, 58-73, ASST/ASST MGR DEVELOP & PLANNING, ENERGY RES & DEVELOP ADMIN, OAK RIDGE OPERS, 73- Mem: Am Chem Soc. Res: Electrochemistry; war gases; physical properties of alloys; solubility of mercurous iodate; thermal conductivity; phase diagrams; process development; weapons and separation of uranium isotopes. Mailing Add: 125 Nesper Rd Oak Ridge TN 37830

LEEDER, JOSEPH GORDON, b Oil City, Pa, July 4, 16; m 39; c 2. DAIRY CHEMISTRY. Educ: Ohio State Univ, BS, 38; Univ Vt, MS, 40; Pa State Univ, PhD, 44. Prof Exp: Dir labs & Supvr zone qual control, Nat Dairy Prod Co, Ohio, 44-46, chief res chemist, Ramsey Labs, 46-48; from assoc prof to prof dairy indust, Dept Animal Sci, 48-65, RES PROF DAIRY INDUST, DEPT FOOD SCI, RUTGERS UNIV, 65- Mem: Am Dairy Sci Asn; Inst Food Technologists. Res: Immobilized enzymes for milk and whey utilization; dairy chemistry and manufacturing; chemistry of butter oil used in deep-fat frying. Mailing Add: Dept of Food Sci Rutgers Univ New Brunswick NJ 08903

LEEDOM, JOHN MILTON, b Peoria, Ill, Oct 18, 33; m 56; c 2. INTERNAL MEDICINE. Educ: Univ Ill, BA, 55, BS, 56, MD, 58; Am Bd Internal Med, dipl, 67. Prof Exp: Resident med, Univ Ill Res & Educ Hosps, 59-60 & 61-62; asst prof, 62-68, ASSOC PROF MED, SCH MED, UNIV SOUTHERN CALIF, 68- Concurrent Pos: Res fel, Univ Ill Res & Educ Hosps, 60-61; officer res proj, Epidemic Intel Serv, USPHS, Infectious Dis Lab, Univ Southern Calif, 62-64, consult health facil construction div, Health Serv & Ment Health Admin. Mem: Am Fedn Clin Res; Am Soc Microbiol; Infectious Dis Soc Am. Res: Infectious disease, particularly viral and bacterial diseases of the central nervous system. Mailing Add: Hastings Found Infect Dis Lab Univ of Southern Calif Med Sch Los Angeles CA 90033

LEEDS, ANTHONY, b New York, NY, Jan 26, 25; m 48, 67; c 5. ANTHROPOLOGY. Educ: Columbia Univ, BA, 49, PhD(anthrop), 57. Prof Exp: Lectr anthrop, Columbia Univ, 53-54; instr sci, Baldwin Sch, 54-56; instr anthrop & sociol, Hofstra Univ, 56-59; asst prof, City Univ New York, 59-61, chief, Prog Urban Develop, Dept Social Affairs, Pan Am, 61-63; from assoc prof to prof anthrop, Univ Tex, 63-72; PROF ANTHROP, BOSTON UNIV, 72- Concurrent Pos: Lectr anthrop & sociol, City Univ New York, 56-59; lectr, Columbia Univ, 59-60; Am Philos Soc grant hist Yaruro Indians, 60; lectr, Cath Univ Am, 61-62; prof lectr, Am Univ, 62-63; Wenner-Gren Found grant systs anal complex socs, 65; Ford Found-Soc Sci Res Coun foreign area studies develop grant, Rio de Janeiro, 65-66; chief eval team & tech consult, Brazil Community Develop Proj, AID, 66; mem joint US-Brazilian comn develop indust res, Nat Acad Sci-Nat Res Coun, 66-68; vis prof anthrop, Univ Rio de Janeiro, 69; Fulbright grants, Lima, Peru, 69-70 & Gt Brit, 72-73; prof Latin Am studies, Univ London & Oxford Univ, 72-73. Mem: AAAS (secy sect anthrop, 65-73); Am Ethnol Soc; Soc Hist Technol; Royal Anthrop Inst; Soc Gen Systs Res. Res: Homeostatic and dynamic analysis of societal systems; ecology; cultural evolution; complex social systems; specialization, resources, power and institution; migration and urbanization as complex systems; human biology and divisional labor. Mailing Add: Dept of Anthrop Boston Univ Boston MA 02215

LEEDS, MORTON W, b Brooklyn, NY, Dec 18, 16; m 45; c 1. ORGANIC CHEMISTRY. Educ: Polytech Inst Brooklyn, BS, 38, MS, 39, PhD(org chem), 44. Prof Exp: Chemist, Bio-Med Res Lab, NY, 35-39; sr chemist, Res Lab, Interchem Corp, 39-45, head develop dept amino acids, Biochem Div, NJ, 45-48; sr chemist, E I du Pont de Nemours & Co, Inc, 48-50; asst chief chemist, Schwarz Labs, NY, 50-52; head appln & develop, Res Labs, Air Reduction Co, 52-56, supvr org chem develop & res, 56-60, from asst dir to assoc dir chem res, 60-71; MGR CLIN DEVELOP, MED RES DIV, CIBA PHARMACEUT CO, 71- Concurrent Pos: Permanent adj prof, Kean Col, 71- Mem: Am Chem Soc; fel Am Inst Chem; NY Acad Sci. Res: Organic synthesis and development; acetylene and pharmaceutical chemistry; petrochemicals. Mailing Add: 100 Chestnut Hill Dr Murray Hill NJ 07974

LEEDY, DANIEL LONEY, b North Liberty, Ohio, Feb 17, 12; m 45; c 2. ZOOLOGY, ECOLOGY. Educ: Miami Univ, AB, 34, BSc, 35; Ohio State Univ, MSc, 38, PhD(wildlife mgt), 40. Prof Exp: Lab asst geol & zool, Miami Univ, 34-35; asst leader, Ohio Wildlife Res Unit, Ohio State Univ, 40-42; leader, Ohio Unit, US Fish & Wildlife Serv, 45-48, coordr, Coop Wildlife Res Unit Prog, 49-57, chief br wildlife res, 57-63; chief div res & educ, Bur Outdoor Recreation, 63-65, WATER RESOURCES RES SCIENTIST, OFF WATER RESOURCES RES, US DEPT INTERIOR, 65- Honors & Awards: Am Motors Conserv Award, 58. Mem: Fel AAAS; assoc Am Soc Mammalogists; Wildlife Soc (pres, 53, exec secy, 54-57); Am Fisheries Soc; Ecol Soc Am. Res: Wildlife ecology; outdoor recreation; socioeconomics of fish and wildlife and recreation; natural resources training and employment; wildlife-land use relationships; water resources; biology; ecologic impacts of water development. Mailing Add: 10707 Lockridge Dr Silver Spring MD 20901

LEE-FRANZINI, JULIET, b Paris, France, May 18, 33; US citizen; m 64; c 1. EXPERIMENTAL PHYSICS. Educ: Hunter Col, BA, 53; Columbia Univ, MA, 57, PhD(physics), 60. Prof Exp: Res assoc physics, Columbia Univ, 60-62; res fel astrophys, Nat Acad Sci-Nat Res Coun, 62-63; from asst prof to assoc prof elem particle physics, 63-74, PROF PHYSICS, STATE UNIV NY STONY BROOK, 74- Concurrent Pos: State Univ NY Res Found grant, 63-66; vis assoc physicist, Brookhaven Nat Lab, 64- Res: Elementary particle physics; weak interactions. Mailing Add: Dept of Physics State Univ of NY Stony Brook NY 11790

LEEHEY, PATRICK, b Waterloo, Iowa, Oct 27, 21; m 44; c 6. APPLIED MECHANICS. Educ: US Naval Acad, BSc, 42; Brown Univ, PhD(appl math), 50. Prof Exp: Proj officer, US Off Naval Res, DC, 51-53, prog officer, David Taylor Model Basin, 53-56, design supt, Puget Sound Naval Shipyard, Wash, 56-58, head ship silencing br, Bur Ships, DC, 58-63, head acoustics & vibration lab, David Taylor Model Basin, 63-64; assoc prof mech eng & naval archit, 64-67, prof naval archit, 67-71, PROF APPL MECH, MASS INST TECHNOL, 71- Honors & Awards: Gold Medal, Am Soc Naval Eng, 62. Mem: Am Math Soc; fel Acoust Soc Am; Am Soc Naval Eng. Res: Hydrodynamics; hydrofoil craft development, unsteady airfoil theory; supercavitating flow theory; acoustics; ship silencing, underwater acoustics, boundary layer noise; mathematics; hyperbolic partial differential equations; singular integral equations; boundary layer stability. Mailing Add: Dept of Ocean Eng Rm 5-222 Mass Inst Technol Cambridge MA 02139

LEE-HUANG, SYLVIA, b Shanghai, China, July 14, 30; US citizen; m 57; c 3. BIOCHEMISTRY, MOLECULAR BIOLOGY. Educ: Nat Taiwan Univ, BS, 52; Univ Idaho, MS, 57; Univ Pittsburgh, PhD(biophys), 61. Prof Exp: NIH fel microbiol, Sch Med, Univ Pittsburgh, 61-62; res assoc chem physics, Sloan-Kettering Inst, 62-64; instr biochem, Med Col, Cornell Univ, 64-66; res scientist, 66-67, from instr to asst prof, 67-70, res assoc prof, 70-71, ASSOC PROF BIOCHEM, SCH MED, NY UNIV, 71- Mem: AAAS; Am Soc Biol Chemists; Biophys Soc; Harvey Soc; NY Acad Sci. Res: Molecular mechanism of transmission and expression of genetic information; control and mechanism of differentiation and development. Mailing Add: Dept of Biochem NY Univ Sch of Med New York NY 10016

LEEKLEY, ROBERT MITCHELL, b Montrose, SDak, July 18, 11; m 41; c 3. ORGANIC CHEMISTRY. Educ: Dakota Wesleyan Univ, BS, 33; Univ SDak, MA, 34; Univ Minn, PhD(chem), 38. Prof Exp: Asst, State Chem Lab, SDak, 33-34 & Univ Minn, 34-37; res chemist, E I du Pont de Nemours & Co, 38-45; chem consult, Springdale Labs, Time Inc, 45-50, assoc res dir, 51-62; res assoc, 63-68, SR RES ASSOC, INST PAPER CHEM, 68- Mem: Am Chem Soc; Tech Asn Pulp & Paper Indust. Res: Light sensitive plastics and coatings; papers for graphic reproduction processes; paper coatings. Mailing Add: Inst of Paper Chem Appleton WI 54911

LEELA, SRINIVASA (G), b Mysore, India. MATHEMATICS. Educ: Osmania Univ, India, BSc, 55, MSc, 57; Marathwada Univ, India, PhD(math), 65. Prof Exp: Lectr math, Women's Col, Kurnool, India, 59-65; instr, Calgary Univ, 65-66; asst prof, Univ RI, 66-68; assoc prof, 68-73, PROF MATH, STATE UNIV NY COL GENESEO, 73- Mem: Am Math Soc; Math Asn Am. Res: Qualitative analysis in differential equations; stability theory. Mailing Add: Dept of Math State Univ of NY Col Geneseo NY 14454

LEELING, JERRY L, b Ottumwa, Iowa, July 11, 36; m 57; c 3. BIOCHEMISTRY. Educ: Parsons Col, BS, 59; Univ Iowa, MS, 62, PhD(biochem), 64. Prof Exp: Res biochemist, 64-70, SR RES SCIENTIST, TOXICOL DEPT, MILES LABS, INC, 70-, SECT HEAD, 73- Mem: Am Chem Soc. Res: Disposition and metabolism of drugs; biochemical pharmacology of drugs. Mailing Add: Toxicol Dept Miles Labs Inc Elkhart IN 46514

LEEMAN, SUSAN EPSTEIN, b Chicago, Ill, May 9, 30; m 57; c 1. PHYSIOLOGY. Educ: Goucher Col, BA, 51; Radcliffe Col, MA, 54, PhD, 58. Prof Exp: Instr physiol, Harvard Med Sch, 58-59; fel neurochem, Brandeis Univ, 59-62, sr res assoc biochem, 62-68, adj asst prof, 66-68, asst res prof, 68-71; asst prof, 72-73, ASSOC PROF PHYSIOL, LAB HUMAN REPROD & REPROD BIOL, HARVARD MED SCH, 73- Concurrent Pos: USPHS career develop award, 62- Mem: Am Physiol Soc; Endocrine Soc. Res: Neuroendocrinology. Mailing Add: Lab Human Reprod & Reprod Biol Havard Med Sch Boston MA 02115

LEEMING, DAVID JOHN, b Victoria, BC, June 8, 39; m 66; c 1. MATHEMATICS. Educ: Univ BC, BSc, 61; Univ Ore, MA, 63; Univ Alta, PhD(math), 69. Prof Exp: Instr, 63-66, ASST PROF MATH, UNIV VICTORIA, BC, 69- Mem: Math Asn Am; Can Math Cong; Soc Indust & Appl Math. Res: Approximation theory; error bounds for interpolation schemes. Mailing Add: Dept of Math Univ of Victoria Victoria BC Can

LEENHEER, JERRY ALYN, soil chemistry, microbiology, see 12th edition

LEEPER, GEORGE FREDERICK, b Detroit, Mich, Sept 27, 43; m 66; c 3. BOTANY. Educ: Univ Iowa, BA, 65, MS, 66; Wash Univ, St Louis, PhD(bot), 69. Prof Exp: BOTANIST, FRUIT & VEG LAB, RICHARD B RUSSELL AGR RES CTR, AGR RES SERV, USDA, 69- Concurrent Pos: Fel, Wash Univ, St Louis, 69. Res: Cell wall and polysaccharide ultrastructure; fruit histochemistry; post-harvest fruit handling; new crops utilization. Mailing Add: Hort Crops Lab R B Russell Agr Res Ctr PO Box 5677 Athens GA 30604

LEEPER, JOHN ROBERT, b Hackensack, NJ, July 12, 47; m 73; c 1. ENTOMOLOGY. Educ: Carthage Col, BA, 69; Univ Hawaii, MS, 71, PhD(entom), 75. Prof Exp: RES ASSOC ENTOM, TREE FRUIT RES CTR, WASH STATE UNIV, 75- Mem: Entom Soc Am. Res: Psyllid ecology and biological control; coccinellid systematics; tree fruit entomology. Mailing Add: Tree Fruit Res Ctr 1100 N Western Ave Wenatchee WA 98801

LEEPER, LEMUEL CLEVELAND, biochemistry, see 12th edition

LEEPER, ROBERT DWIGHT, b Lewiston, Idaho, Nov 9, 24; m 52; c 4. NUCLEAR MEDICINE, ENDOCRINOLOGY. Educ: Univ Idaho, BS, 40; Columbia Univ, MD, 53. Prof Exp: Intern, Brooklyn Hosp, 53-54, from resident to sr resident, 54-57; fel clin invest, 57-60, assoc, 60-66, ASSOC MEM, SLOAN-KETTERING INST CANCER RES, 66- Concurrent Pos: NY Heart Asn fel, 58-60; asst attend physician, Mem Hosp & James Ewing Hosp, 58-66, assoc attend physician, 66-; Am Cancer Soc scholar, 61-66; chief endocrine clin, Mem Hosp, 65-; asst attend physician, New York Hosp, 66-; asst prof, Med Col, Cornell Univ, 63-74, clin assoc prof med, 74- Mem: AAAS; Endocrine Soc; Am Thyroid Asn; NY Acad Sci. Res: Thyroid physiology; effects of exogenous enzymes in man; use of radionuclides in diagnosis and therapy in cancer. Mailing Add: Sloan-Kettering Inst for Cancer Res 444 E 68th St New York NY 10021

LEEPER, ROBERT WALZ, b Waterloo, Iowa, Apr 28, 15; m 42; c 3. ORGANIC CHEMISTRY. Educ: Univ Iowa, BA, 36, MS, 38; Iowa State Univ, PhD(org chem), 42. Prof Exp: Nat Defense Res Coun fel, Iowa State Univ, 42-43; org chemist, Gelatin Prod Corp, Mich, 43-45; dir res, Boyle-Midway Div, Am Home Prod Co, NJ, 45-48; org chemist, Pineapple Res Inst, Hawaii, 48-66, head chem dept, 61-66; chem consult, 66-69, GROUP LEADER, AGR DIV, AMCHEM PROD, INC, 69- Concurrent Pos: Vis prof, Mich State Univ, 55; vis scientist, NZ, 62. Mem: AAAS; Am Chem Soc. Res: Organometallics; synthesis of potential herbicides and plant growth regulators. Mailing Add: Amchem Prods Inc Agr Div Ambler PA 19002

LEER, JOHN ADDISON, JR, b Carlisle, Pa, Dec 8, 22; m 46; c 2. MEDICINE, PEDIATRICS. Educ: Temple Univ, MD, 46, MSc, 52; Am Bd Pediat, dipl, 52. Prof Exp: Pvt pract, Pa, 52-63; physician monitor, 63-65, assoc dir, 65-68, dir clin invest, 68-73, SR FELLOW MED RES, SCHERING CORP, 73- Concurrent Pos: Clin asst prof pediat, Col Med & Dent NJ, 67-72. Mem: Fel Am Acad Pediat; Am Acad Dermat; Am Soc Clin Pharmacol & Therapeut; Soc Pediat Dermat; Am Soc Photobiol. Res: Clinical investigation of new drugs. Mailing Add: Schering Corp 60 Orange St Bloomfield NJ 07003

LEERBURGER, BENEDICT ALAN, JR, b New York, NY, Jan 2, 32; m 58; c 2. SCIENCE WRITING, JOURNALISM. Educ: Colby Col, BA, 54. Prof Exp: Asst ed, Prod Eng Mag, 54-59; sci ed, Grolier, Inc, 59-61; ed, Cowles Ed Corp, 61-68; proj dir, CCM Info Sci, Inc, 68-70; vpres & ed dir, Nat Micro-Publ Corp, 70-72; dir publ, NY Times, 72-74; PUBL, KRAUS-THOMSON ORG, LTD, 74- Concurrent Pos: Consult, Sci Digest, 59-61, Cross, Hinshaw & Lindberg, Inc, 65-67, Storrington Printing & Pub Co, Inc, 67-70 & NSF Deep Freeze Prog, Antarctica, 67. Mem: Nat Asn Sci Writers; Am Hist Asn; Nat Sci Teachers Asn. Res: Physical science; Antarctica; American scientific history. Mailing Add: 338 Heathcote Rd Scarsdale NY 10583

LEERS, WOLF-DIETRICH, b Halle, Ger, Aug 9, 27; Can citizen; m 56; c 4. MEDICAL MICROBIOLOGY. Educ: Univ Würzburg, MD, 55; Univ Toronto, dipl bact, 63, PhD(microbiol), 67; Am Bd Med Microbiol, cert pub health, 71; FRCP(C). Prof Exp: Intern, Univ Clin, Hamburg-Eppendorf, Ger, 55-56; resident otolaryngol, City Hosp, Verden, Ger, 56-58; intern, Victoria Hosp, London, Ont, 58-59; asst resident med, pediat & surg, St Jospeh's Hosp, 59-60; resident clin microbiol, Victoria

Hosp, 60-61; resident path & microbiol, Lab Serv Br, Ont Dept Pub Health, 61-62; Fitzgerald Mem fel, 63-65, ASST PROF MED MICROBIOL, FAC MED & ASST PROF MICROBIOL, SCH HYG, UNIV TORONTO, 66-; CHIEF MICROBIOLOGIST, WELLESLEY HOSP, 67- Concurrent Pos: Consult, Ontario Cancer Inst & Princess Margaret Hosp, 67. Mem: Can Asn Med Microbiol; NY Acad Sci; Am Soc Microbiol; Can Pub Health Asn; Can Soc Microbiol. Res: Reoviruses; hepatitis B; urinary tract infections; opportunistic fungal and bacterial infections. Mailing Add: Dept of Microbiol Wellesley Hosp Toronto ON Can

LEE-RUFF, EDWARD, b Shanghai, China, Jan 4, 44; Can citizen; m 69. CHEMISTRY. Educ: McGill Univ, BSc, 64, PhD(org chem), 67. Prof Exp: Nat Res Coun fel, Columbia Univ, 67-69; asst prof, 69-74, ASSOC PROF CHEM, YORK UNIV, 74- Mem: Am Chem Soc; Chem Inst Can. Res: Organic photochemistry; reactions of strained molecules; gas-phase organic ion-molecule reactions. Mailing Add: Dept of Chem York Univ Downsview ON Can

LEES, HELEN, b Detroit, Mich, June 1, 25. BIOCHEMISTRY. Educ: Wayne State Univ, BS, 47, PhD(physiol chem), 56. Prof Exp: Fel biochem, Edsel B Ford Inst Med Res, 56-58; biochemist, Lafayette Clin, State Dept Health, Mich, 58-62; asst prof biochem, Med Sch, Northwestern Univ, 62-68; ASSOC PROF MED TECHNOL, STATE UNIV NY BUFFALO, 68- Mem: AAAS; Am Chem Soc; Am Soc Med Technol; Am Soc Clin Chem. Res: Intermediary metabolism; clinical chemistry. Mailing Add: Dept of Med Technol State Univ of NY Buffalo NY 14214

LEES, HOWARD, b Leeds, Eng, May 13, 17; m 39; c 3. BIOCHEMISTRY. Educ: Univ Liverpool, BSc, 38; Univ London, PhD(biochem), 44. Prof Exp: Res scientist, Lever Bros & Unilever, Ltd, Eng, 38-42; sci off, Rothamsted Exp Sta, 42-47; sr lectr, Imp Col Trop Agr, Trinidad, BWI, 47-49; lectr & sr lectr biochem, Aberdeen Univ, 48-58; assoc prof, Ont Agr Col, Can, 58-60; assoc prof, 60-61, head dept, 61-71, chmn div biol sci, 64-69, PROF MICROBIOL, UNIV MAN, 61- Mem: Am Soc Microbiol; Can Soc Microbiol; fel Royal Soc Edinburgh; Brit Soc Gen Microbiol. Res: Biochemistry of autotrophic bacteria. Mailing Add: Dept of Microbiol Univ of Man Winnipeg MB Can

LEES, JOSEPH KOLB, b Philadelphia, Pa, Apr 21, 38; m 63; c 4. SOLID STATE PHYSICS. Educ: Lafayette Col, BS, 60; Carnegie-Mellon Univ, MS, 62, PhD(physics), 66. Prof Exp: Res physicist, Res & Develop Div, 66-69 & Plastics Prod Div, 69-71, sr res physicist, 71-74, SR RES SUPVR, PLASTICS DEPT, E I DU PONT DE NEMOURS & CO, INC, 74- Mem: Am Phys Soc. Res: Mössbauer spectroscopy applied to solid state determinations; physical properties of composite structures; research management in nylon processes and products; chemical and physical property engineering. Mailing Add: Plastics Dept PO Box 1217 E I du Pont de Nemours & Co Inc Parkersburg WV 26101

LEES, MARJORIE BERMAN, b New York, NY, Mar 17, 23; m 46; c 3. NEUROCHEMISTRY. Educ: Hunter Col, BA, 43; Univ Chicago, MS, 45; Harvard Univ, PhD(med sci), 51. Prof Exp: Asst, Univ Chicago, 43-45; asst, Col Physicians & Surgeons, Columbia Univ, 45-46; Am Cancer Soc res fel, 51-53; asst res lab, McLean Hosp, 53-55, asst biochemist, 55-58, assoc biochemist, 58-62; sr res assoc pharmacol, Dartmouth Med Sch, 62-66; ASSOC BIOCHEMIST, McLEAN HOSP, 66-; SR RES ASSOC, HARVARD MED SCH, 75- Concurrent Pos: Instr neuropath, Harvard Med Sch, 55-59, res assoc, 59-62 & 66-71, sr res assoc, 71-75. Mem: Am Soc Biol Chem; Am Soc Neurochem; Soc Neurosci; Am Soc Neuropath; Int Soc Neurochem. Res: Chemistry of the nervous system; lipid chemistry; brain proteins. Mailing Add: Biol Res Lab McLean Hosp Belmont MA 02178

LEES, MARTIN H, b London, Eng, May 11, 29; m 59; c 3. PEDIATRICS, CARDIOLOGY. Educ: Univ London, MB, BS, 55. Prof Exp: Assoc prof, 62-71, PROF PEDIAT, MED SCH, UNIV ORE, 71- Res: Pediatric cardiology; newborn and infant cardiopulmonary physiology and pathophysiology. Mailing Add: Dept of Pediat Univ of Ore Med Sch Portland OR 97201

LEES, ROBERT S, b New York, NY, July 16, 34; m 60; c 4. MEDICINE, BIOCHEMISTRY. Educ: Harvard Univ, AB, 55, MD, 59; Am Bd Internal Med, dipl. Prof Exp: Intern surg, Mass Gen Hosp, Boston, 59-60, asst resident, 61-62; hon asst registr cardiol, Nat Heart Hosp, Eng, 62-63; staff assoc & attend physician med, Nat Heart Inst, 63-66; asst prof med & attend physician, Rockefeller Univ, 66-69; dir clin res ctr, 69-74, assoc prof, 69-71, PROF CARDIOVASC DIS & DIR ARTERIOSCLEROSIS CTR, MASS INST TECHNOL, 71- Concurrent Pos: USPHS res fel med, 60-61; Dalton scholar, Harvard Med Sch, 60; USPHS fel, 62-63; fel coun arteriosclerosis, Am Heart Asn, 65-; lectr, Harvard Med Sch; assoc in med, Peter Bent Brigham Hosp; asst in med, Mass Gen Hosp. Mem: Am Heart Asn; Am Fedn Clin Res; Am Soc Pharmacol & Exp Therapeut. Res: Cardiology, especially ischemic heart disease; lipid and lipoprotein metabolism. Mailing Add: Mass Gen Hosp Fruit St Boston MA 02142

LEES, RONALD MILNE, b Sutton, Eng, Oct 28, 39; Can citizen; m 62; c 2. MOLECULAR SPECTROSCOPY. Educ: Univ BC, BSc, 61, MSc, 65; Bristol Univ, PhD(physics), 67. Prof Exp: Nat Res Coun Can fel, Nat Res Coun, Ottawa, 66-68; ASSOC PROF PHYSICS, UNIV NB, FREDERICTON, 68- Concurrent Pos: Nat Res Coun Can grant, Univ NB, Fredericton, 68- vis assoc prof, Physics Dept, Univ BC, Vancouver, 74-75. Mem: Can Asn Physicists; Am Asn Physics Teachers. Mailing Add: Dept of Physics Univ of NB Fredericton NB Can

LEES, THOMAS MASSON, b New York, NY, June 16, 17; m 43; c 2. BIOCHEMISTRY. Educ: Long Island Univ, BS, 39; Iowa State Univ, MS, 42, PhD(biophys chem), 44. Prof Exp: Res chemist, Am Distilling Co, 44-46; ANAL RES CHEMIST, PFIZER, INC, 46- Mem: Am Chem Soc. Res: Fermentative production of glycerol; antibiotics development, production and identification; analysis of medicinal compounds. Mailing Add: 35 Woodridge Circle Gales Ferry CT 06335

LEES, WAYNE LOWRY, b Washington, DC, July 18, 14; m 39; c 2. EXPERIMENTAL PHYSICS. Educ: Swarthmore Col, BA, 37; Harvard Univ, MA, 40, PhD(physics), 49. Prof Exp: Asst, Bartol Res Found, Pa, 39-40; physicist, Geophys Lab, Wash, DC, 42-44, Nat Bur Standards, 44-46, Tracerlab, Inc, 49-50, Metall Proj, Mass Inst Technol, 50-54, Nuclear Metals, Inc, 54-58, Instrumentation Lab, Mass Inst Technol, 58-65 & Electronics Res Ctr, NASA, Cambridge, 65-70; assoc prof math, Wash Tech Inst, 71-72; proj engr, Design Automation, Inc, Lexington, Mass, 72-73; STAFF MEM, LAB PHYS SCI, P R MALLORY & CO, INC, BURLINGTON, MASS, 74- Mem: AAAS; Am Phys Soc; Fedn Am Sci; Am Geophys Union; Inst Elec & Electronics Engrs. Res: Quasistatic electrical systems and dielectric properties; physics of high pressures and metals; electrode phenomena; electron and ion transport; engineering physics. Mailing Add: 29 Tower Rd Lexington MA 85003

LEES, WILLIAM MORRIS, b Chicago, Ill, Aug 27, 14; m 41; c 4. THORACIC SURGERY, CARDIOVASCULAR SURGERY. Educ: Univ Chicago, BS, 35, MD, 39. Prof Exp: Instr physiol, Univ Chicago, 32-37; instr surg, Loyola Univ Chicago, 40-41; instr thoracic surg, Med Sch, Univ Mich, 46-49; from instr to assoc prof, 50-67, clin prof surg, 67-68, PROF SURG & HEAD SECT CARDIOPULMONARY SURG, STRITCH SCH MED, LOYOLA UNIV CHICAGO, 68- Concurrent Pos: Chief surg, Munic Tuberc Sanitarium, Chicago, 50-; consult, Fifth US Army, 64-; lectr, Sch Med, Univ Chicago, 66-; mem speakers bur & del, AMA, 70- Mem: AMA; Am Col Surgeons; Am Col Chest Physicians; Am Asn Thoracic Surg; Soc Thoracic Surg. Res: Thoracic and cardiovascular aspects and the pathologic physiology of pulmonary diseases; chronic bronchitis; pulmonary emphysema. Mailing Add: 6518 N Nokomis Ave Lincolnwood IL 60646

LEESE, BERNARD M, b Keyser, WVa, Jan 17, 25; m 50; c 3. PLANT PHYSIOLOGY. Educ: George Washington Univ, BS, 51. Prof Exp: Res botanist, 51-55, plant explorer, 56-59, plant variety specialist, 59-62, head fed seed lab, 62-70, CHIEF PLANT EXAM, PLANT VARIETY PROTECTION OFF, USDA, 70- Concurrent Pos: Mem plant nomenclature comt, USDA, 61-; consult, Am Soc Hort Sci, 65-; plant explorer for world sorghum germ plasm collection in Ethiopia, 67; observer, Int Union Protection New Varieties Plants, 72- Mem: Asn Off Seed Analysts (chmn, 70); Int Seed Testing Asn; NY Acad Sci. Res: Identification of plant and seed material by morphological characteristics; identification of plant germ plasm; plant variety nomenclature and plant variety identification. Mailing Add: Plant Var Protect Off Grain Div Agr & Mkt Serv Nat Agr Libr USDA Beltsville MD 20705

LEESE, ERIC LESLIE, b London, Eng, Feb 16, 12; Can citizen; m 41; c 2. MATHEMATICS. Educ: Cambridge Univ, BA, 32, MA, 36. Prof Exp: Statist analyst, London Passanger Transp Bd, Eng, 35-37; res scientist, Brit Admiralty, Whitehall, London, 37-51; RES SCIENTIST, CAN DEFENCE RES BD, OTTAWA, ONT, 51-; DIR MATH & STATIST, CAN FORCES HQ, OTTAWA, 65- Concurrent Pos: Sr oper res scientist, Royal Can Air Force Hq, Ottawa, 54-57, dir oper res, 60-65; dep dir opers anal, NAm Air Defense Command Hq, Colorado Springs, Colo, 57-60; lectr, Ottawa Univ, 65- Mem: Opers Res Soc Am; Can Oper Res Soc. Res: Operational research in military matters; currently used techniques in operational research. Mailing Add: Can Defence Res Bd 101 Colonel By Dr Ottawa ON Can

LEESE, JOHN ALBERT, meteorology, see 12th edition

LEESON, BRUCE FRANK, b Florence, Ont, Aug 20, 43; m 66; c 1. ECOLOGY, ENVIRONMENTAL MANAGEMENT. Educ: Univ Guelph, BSc, 67, MSc, 69; Mont State Univ, PhD(natural resources), 72. Prof Exp: Researcher soil sci, Ont Dept Agr & Food, 66-69; RES ECOLOGIST, PARKS CAN, CAN GOVT NAT & HIST PARKS BR, 72- Mem: Can Nature Fedn; Soil Conserv Soc Am; Am Forestry Asn. Res: Scientific management of wilderness environments in Canada's Rocky Mountain National Parks where the objective is to facilitate recreation without disturbing wildland ecosystems. Mailing Add: Parks Can 134-11th Ave SE Calgary AB Can

LEESON, CHARLES ROLAND, b Halifax, Eng, Jan 26, 26; m 54; c 5. ANATOMY. Educ: Cambridge Univ, BA, 47, MB, BChir, 50, MA, 50, MD, 59, PhD(anat), 71. Prof Exp: Lectr anat, Univ Col SWales, 55-58; assoc prof, Dalhousie Univ, 58-61; assoc prof anat & histol, Queen's Univ, Ont, 61-63; prof anat, Univ Iowa, 63-66; PROF ANAT & CHMN DEPT, UNIV MO-COLUMBIA, 66- Concurrent Pos: Vis prof anat, London Hosp Med Col, Eng, 73-74. Mem: Anat Soc Gt Brit & Ireland; Am Asn Anat; Electron Micros Soc Am. Res: Post natal development, particularly in marsupials and with reference to certain organ systems. Mailing Add: Dept of Anat Univ of Mo Sch of Med Columbia MO 65201

LEESON, LEWIS JOSEPH, b Paterson, NJ, Apr 26, 27; m 53; c 3. PHARMACY. Educ: Rutgers Univ, BS, 50, MS, 54; Univ Mich, PhD(pharmaceut chem), 57. Prof Exp: Intern pharm, Mack Drug Co, 50-51; pharmacist, Silver Rod Drugs, 51-52; asst, Rutgers Univ, 52-54 & Univ Mich, 55-56; res chemist, Lederle Labs, Am Cyanamid Co, 57-67; proj leader pharmaceut, Union Carbide Res Inst, 67-69; asst dir pharmaceut develop, Geigy Chem Corp, 69-71, asst dir, 71-73, DIR PHARM RES & DEVELOP, CIBA-GEIGY PHARMACEUT CO, 73- Concurrent Pos: Relief pharmacist, Frieds Pharm, 52-54. Mem: Am Chem Soc; Am Pharmaceut Asn; fel Acad Pharmaceut Sci; Sigma Xi. Res: Pharmaceutical product development; application of physical chemical techniques for developing various pharmaceutical dosage forms. Mailing Add: Pharm Res Ciba-Geigy Pharm Co Morris Ave Summit NJ 07901

LEESON, THOMAS SYDNEY, b Halifax, UK, Jan 26, 26; m 52; c 3. ANATOMY. Educ: Cambridge Univ, BA, 46, MA, 49, MD & BCh, 50, MD, 59, PhD, 71. Prof Exp: Asst lectr anat, Univ Wales, 55-57; from asst prof to assoc prof, Univ Toronto, 57-63; PROF ANAT & HEAD DEPT, UNIV ALTA, 63- Mem: Am Asn Anat; Can Fedn Biol Soc; Anat Soc Gt Brit & Ireland. Res: Electron microscopy, histology and embryology. Mailing Add: Dept of Anat Univ of Alta Edmonton AB Can

LEESTMA, DAVID JOSEPH, physics, operations research, see 12th edition

LEET, DUANE GARY, b Duncan, Okla, May 29, 44; m 66; c 1. BIONICS, INFORMATION SCIENCE. Educ: Mich State Univ, BS, 66, MS, 68, PhD(systs sci), 71. Prof Exp: Res assoc environ mgt, Mich State Univ, 71-72; asst prof comput sci, Univ Evansville, 72-74; ASSOC RES ANALYST BIONICS, UNIV DAYTON RES INST, 74- Res: Investigation of information processing principles of mammalian auditory system; computer simulation of guinea pig peripheral auditory systems; behavioral characteristics of neuromines and their networks; scientific computer programming. Mailing Add: Univ of Dayton Res Inst 300 College Park Dayton OH 45469

LEET, HENRY PETER, b El Dorado, Kans, Dec 4, 32; m 63; c 2. OPTICS. Educ: Univ Kans, BS, 55. Prof Exp: Physicist, 55-61, head, Detection Br, 61-70 & Optics Technol Br, 70-74, HEAD, OPTICS & SENSORS BR, MICHAELSON LABS, NAVAL WEAPONS CTR, 74- Mem: AAAS; Sigma Xi; Optical Soc Am. Res: Infrared detection; spectro-radiometric analysis of all tvpes of thermal emitters; visual and infrared image converters and trackers; infrared and electro-optical active and passive sensors. Mailing Add: 627 Kevin Ct Ridgecrest CA 93555

LEETCH, JAMES FREDERICK, b Butler, Pa, Sept 27, 29; m 58; c 2. MATHEMATICS. Educ: Grove City Col, BS, 51; Ohio State Univ, MA, 57, PhD(math), 61. Prof Exp: From asst prof to assoc prof, 61-71, PROF MATH, BOWLING GREEN STATE UNIV, 71- Mem: Math Asn Am. Res: Mathematical analysis; integration; set functions. Mailing Add: 120 S College Dr Bowling Green OH 43402

LEETE, EDWARD, b Leeds, Eng, Apr 18, 28; nat US; m 54; c 2. ORGANIC CHEMISTRY. Educ: Univ Leeds, BSc, 48, PhD(chem), 50, DSc, 65. Prof Exp: Goldsmith fel, Nat Res Coun Can, 50-52, res fel, 52-54; from instr to asst prof org chem, Univ Calif, Los Angeles, 54-58; from asst prof to assoc prof, 58-63, PROF ORG CHEM, UNIV MINN, MINNEAPOLIS, 63- Concurrent Pos: Mem med chem study sect, NIH, 61-65; Alfred P Sloan fel, 61, 63 & 64; Guggenheim mem fel, 65-66; consult, Philip Morris Res Ctr, Richmond, Va, 74- Mem: Am Chem Soc; Am Soc

Pharmacog; The Chem Soc; Brit Soc Chem Indust; Royal Inst Chem. Res: Biosynthesis of natural substances, especially alkaloids; synthesis of heterocyclic compounds; isolation of enzymes from plants; use of radioactive isotopes. Mailing Add: Dept of Chem Univ of Minn Minneapolis MN 55455

LEEVY, CARROLL M, b Columbia, SC, Oct 13, 20; m 56; c 2. MEDICINE, NUTRITION. Educ: Fisk Univ, AB, 41; Univ Mich, MD, 44. Prof Exp: Intern med, Jersey City Med Ctr, 44-45, resident, 45-48, dir clin invest & outpatient dept, 48-58; res assoc, Harvard Univ, 58-59; assoc prof, 59-62, actg chmn dept med, 66-68, PROF MED, COL MED NJ, 62-, DIR DIV HEPATIC METAB & NUTRIT, 59- Concurrent Pos: USPHS spec res fel, 58-59; consult, US Naval Hosp, St Albans, 48; consult & mem med adv comt, Vet Admin Hosp, East Orange, NJ, 64; physician-in-chief, Martland Hosp, 66-68; mem dean's comt, East Orange Vet Admin Hosp, 66-68, chief med, 66-71; mem clin cancer training comt, NIH, 69-73; consult, Food & Drug Admin, 70- Honors & Awards: Mod Med Award, 72. Mem: AAAS; Soc Exp Biol & Med; Nat Med Asn; fel AMA; fel Am Col Physicians. Res: Pathogenesis of cirrhosis of alcoholics; factors which control hepatic desoxyribonucleic acid synthesis and regeneration; mechanism of portal hypertension and malutilization of vitamins and proteins. Mailing Add: Col of Med 100 Bergen St Newark NJ 07103

LEFAR, MORTON SAUL, b New York, NY, Apr 11, 37; m 61; c 2. ORGANIC CHEMISTRY. Educ: Brooklyn Col, BS, 58, MA, 62; Rutgers Univ, PhD(chem), 65. Prof Exp: Res chemist, Inst Environ Med, Med Sch, NY Univ, 58-60 & Photoprod Dept, E I du Pont de Nemours & Co, 65-66; sect head org chem div nutrit, Food & Drug Admin, Washington, DC, 66-68; sr scientist, Warner-Lambert Res Inst, 68-69; mgr anal chem, Rhodia, Inc, 69-74; DIR QUAL CONTROL, HOECHST-ROUSSEL PHARMACEUT, INC, 75- Mem: Am Pharmaceut Asn; Am Chem Soc. Res: Natural products chemistry; analytical chemistry; photochemistry; environmental health; analytical methods development on new pharmaceuticals, identification and proof of structure. Mailing Add: Hoechst-Roussel Pharmaceut Inc Rt 202-206 N Somerville NJ 08876

LEFCOE, NEVILLE, b Montreal, Que, July 19, 25; m 54; c 4. PHYSIOLOGY. Educ: McGill Univ, BSc, 46; Vanderbilt Univ, MD, 50; FRCP(C), 56. Prof Exp: From instr to assoc prof, 57-72, PROF MED, UNIV WESTERN ONT, 72- Mem: Am Fedn Clin Res; Can Soc Clin Invest. Res: Pulmonary physiology, chiefly exercise physiology, cellular mechanisms in bronchial mouth muscle and the domestic microenvironment. Mailing Add: Dept of Med Victoria Hosp London ON Can

LEFEBRE, VERNON GLEN, b Hammond, Ind, July 2, 36; m; c 4. THERMODYNAMICS. Educ: Purdue Univ, BS, 58; Univ Utah, PhD(phys chem), 63. Prof Exp: Asst prof, 64-68, ASSOC PROF PHYSICS, NMEX INST MINING & TECHNOL, 68- Res: Mixing theory in aquifers. Mailing Add: Dept of Physics NMex Inst of Mining & Technol Socorro NM 87801

LE FEBVRE, EUGENE ALLEN, b St Paul, Minn, Oct 18, 29; m 54. ZOOLOGY. Educ: Univ Minn, BS, 52, AP, 53, MS, 58, PhD(zool), 62. Prof Exp: Teaching asst ornith & zool, Univ Minn, Minneapolis, 53-59, res fel ornith, Mus Natural Hist, 60-61, res assoc, 61-66; asst prof, 66-72, ASSOC PROF ZOOL, SOUTHERN ILL UNIV, 72- Concurrent Pos: Res assoc, NIH grant, 60-65, coprin investr, 63-65; NSF res grant, Midway Island, 69-73. Mem: AAAS; Am Ornith Union; Cooper Ornith Soc; Ecol Soc Am; Am Inst Biol Sci. Res: Physiological ecology of birds and mammals, especially bioenergetics of flight and migratory behavior. Mailing Add: Dept of Zool Southern Ill Univ Carbondale IL 62901

LEFEBVRE, RENE, b Verdun, Que, Apr 5, 23; m 51; c 2. MEDICINE. Educ: Col Montreal, BA, 44; Univ Montreal, MD, 50; Univ Pa, DSc(med), 54. Prof Exp: From asst prof to assoc prof, 60-70, PROF PATH, FAC MED, UNIV MONTREAL, 70-; PATHOLOGIST, HOTEL DIEU HOSP, 54- Mem: Can Asn Path (pres-elect, 65-66, pres, 66-). Res: Pathology of kidney. Mailing Add: Dept of Path Hotel Dieu Hosp Montreal PQ Can

LEFEBVRE, RICHARD HAROLD, b Detroit, Mich, Dec 11, 33; m 59; c 3. GEOLOGY. Educ: Univ Mich, BS, 57; Univ Kans, MS, 61; Northwestern Univ, PhD(geol), 66. Prof Exp: Asst prof geol, Univ Ga, 65-67; from asst prof to assoc prof, 67-73, chmn dept, 70-75, PROF GEOL, GRAND VALLEY STATE COLS, 75- Mem: Geol Soc Am; Nat Asn Geol Teachers; Int Asn Volcanology. Res: Flood basalts of the northwestern United States; remote sensing of Holocene basaltic lava flows, especially Craters of the Moon National Monument, Idaho. Mailing Add: Dept of Geol Grand Valley State Cols Allendale MI 49401

LEFEBVRE, YVON, b Montreal, Que, June 20, 31; m 62; c 2. ORGANIC CHEMISTRY, INFORMATION SCIENCE. Educ: Col Stanislas, BA, 50; Univ Mont, BSc, 53, MSc, 55, PhD(chem), 57. Prof Exp: Res chemist, Ayerst Labs, 58-75, group leader, 68-75, DIR INFO DEPT, AYERST RES LABS, 75- Res: Steroids chemistry; progestational agents; estrogens; oxidation of furan derivatives in steroid and non-steroid series. Mailing Add: Ayerst Res Labs PO Box 6115 Montreal PQ Can

LEFELHOCZ, JOHN F, inorganic chemistry, see 12th edition

LEFER, ALLAN MARK, b New York, NY, Feb 1, 36; m 59; c 4. CARDIOVASCULAR PHYSIOLOGY. Educ: Adelphi Univ, BA, 57; Western Reserve Univ, MA, 59; Univ Ill, PhD(physiol), 62. Prof Exp: Instr physiol, Case Western Reserve Univ, 62-64; from asst prof to prof, Sch Med, Univ Va, 64-72; PROF PHYSIOL & CHMN DEPT, JEFFERSON MED COL, THOMAS JEFFERSON UNIV, 74- Concurrent Pos: USPHS fel, 62-64; estab investr, Am Heart Asn, 67-72; vis prof & USPHS sr fel, Hadassah Med Sch, Hebrew Univ, Israel, 71-72; mem comt pub affairs, Fedn Am Socs Exp Biol; ed, Circulatory Shock; consult, Task Group on Shock, NIH; mem, Int Study Group Res Cardiac Metab & Pancreatic Study Group; mem coun basic sci, Am Heart Asn. Mem: Am Physiol Soc; Cardiac Muscle Soc; Soc Exp Biol & Med; Am Soc Pharmacol & Exp Therapeut; Reticuloendothelial Soc. Res: Cardiovascular effects of adrenal hormones; humoral regulation of myocardial contractility; corticosteroid pharmacology; experimental myocardial infarction; metabolic alterations in shock; pathogenesis of circulatory shock. Mailing Add: Jefferson Med Col Thomas Jefferson Univ Philadelphia PA 19107

LE FEVER, HERMON MICHAEL, b Pomona, Calif, Dec 22, 37; m 57; c 2. GENETICS, ELECTRON MICROSCOPY. Educ: Okla State Univ, BS, 60, MS, 62; Univ Tex, PhD(zool), 66. Prof Exp: Res scientist asst genetics, Univ Tex, 63-65; asst prof biol, 65-70, ASSOC PROF BIOL, EMPORIA KANS STATE COL, 70- Mem: Genetics Soc Am. Res: Genetics of Drosophila; pseudoallelism; chromosome ultrastructure; electron microscopy of chromosomes. Mailing Add: Dept of Biol Emporia Kans State Col Emporia KS 66801

LEFEVER, ROBERT ALLEN, b York, Pa, May 29, 27; m 46; c 3. SOLID STATE CHEMISTRY, MATERIALS SCIENCE. Educ: Juniata Col, BS, 50; Mass Inst Technol, PhD(inorg chem), 53. Prof Exp: Res chemist, Linde Co, 53-56; sr scientist, Va Inst Sci Res, 56-58; mem tech staff, Hughes Res Labs, 58-59, head chem physics group, 59-61; staff mem, Gen Tel & Electronics Labs, Inc, 61-63; supvr, Mat Res Div, Sandia Labs, NMex, 63-74; DIR MAT PREPARATION, SCH ENG, UNIV SOUTHERN CALIF, 74- Concurrent Pos: Consult, Spectrotherm Corp, 74- Mem: Am Phys Soc; Am Chem Soc; Am Ceramic Soc; fel Am Inst Chemists; NY Acad Sci. Res: Single crystal growth; growth mechanisms and characterization; sintering processes and mechanisms; ferrites; garnets; metal and rare earth oxides; semiconductors; phosphors; thermoelectrics. Mailing Add:

LE FEVRE, CECIL W, agronomy, plant physiology, see 12th edition

LEFEVRE, GEORGE, JR, b Columbia, Mo, Sept 13, 17; m 43; c 3. GENETICS. Educ: Univ Mo, AB, 37, AM, 39, PhD(genetics), 49. Prof Exp: Asst zool, Columbia Univ, 41-42; res biologist, Oak Ridge Nat Lab, 46-47; instr zool, Univ Mo, 47-48; from asst prof to assoc prof biol, Univ Utah, 49-56; prog dir genetic biol, NSF, 56-59; dir biol labs, Harvard Univ, 59-65; PROF BIOL & CHMN DEPT, CALIF STATE UNIV, NORTHRIDGE, 65- Concurrent Pos: Consult, NSF, 59-62 & NIH, 62-66. Mem: Genetics Soc Am (treas, 72-75). Res: Radiation genetics of Drosophila melanogaster; mutation of individual loci; cytogenetics. Mailing Add: Dept of Biol Calif State Univ Northridge CA 91324

LEFEVRE, HARLAN W, b Great Falls, Mont, May 19, 29; m 51; c 8. NUCLEAR PHYSICS. Educ: Reed Col, BA, 51; Univ Idaho, MS, 57; Univ Wis, PhD(physics), 61. Prof Exp: Physicist, Hanford Atomic Prod Oper, Gen Elec Co, Wash, 51-58; assoc prof, 61-71, PROF PHYSICS, UNIV ORE, 71- Concurrent Pos: Consult, Lawrence Radiation Lab, Univ Calif, 62- Mem: Am Phys Soc. Res: Experimental nuclear physics; nuclear reactions; fast neutron spectrometry. Mailing Add: Dept of Physics Univ of Ore Eugene OR 97403

LEFEVRE, MARIAN E WILLIS, b Washington, DC, Jan 21, 23; m 48; c 3. PHYSIOLOGY. Educ: Iowa State Univ, BS, 44; Univ Pa, MS, 47; Univ Louisville, PhD(physiol), 69. Prof Exp: Assoc, 68-73, ASST PROF PHYSIOL, MT SIANI SCH MED, 73-; ASSOC SCI, BROOKHAVEN NAT LAB, 75- Concurrent Pos: Res collabr, Brookhaven Nat Lab, 68-75. Mem: Am Physiol Soc; NY Acad Sci; Biophys Soc; Am Soc Cell Biol; Soc Exp Biol & Med. Res: Structure and function of multicellular membranes; ion transport and metabolism; fluid transport and accumulation. Mailing Add: Med Dept Brookhaven Nat Lab Upton NY 11973

LE FEVRE, PAUL GREEN, b Baltimore, Md, Dec 27, 19; m 48; c 4. CELL PHYSIOLOGY. Educ: Johns Hopkins Univ, AB, 40; Univ Pa, PhD(physiol, zool), 45. Prof Exp: Asst zool, Univ Pa, 43-45; from instr to asst prof physiol, Col Med, Univ Vt, 45-49, assoc prof physiol & biophys, 49-52; asst to chief med br, AEC, 52-55; scientist, Med Res Ctr, Brookhaven Nat Lab, 55-60; prof pharmacol, Sch Med, Univ Louisville, 60-68; PROF PHYSIOL, HEALTH SCI CTR, STATE UNIV NY STONY BROOK, 68- Concurrent Pos: Mem corp, Marine Biol Lab, Woods Hole, Mass. Mem: AAAS; Soc Gen Physiol; Am Physiol Soc; Biophys Soc; Am Soc Cell Biologists. Res: Mechanisms, kinetics and model systems for cell membrane mediated transport; phospholipid-carbohydrate complexing. Mailing Add: Dept of Physiol & Biophys State Univ NY Health Sci Ctr Stony Brook NY 11794

LE FEVRE, PAUL HENRY, organic chemistry, see 12th edition

LEFF, HARVEY SHERWIN, b Chicago, Ill, July 24, 37; m 58; c 6. THERMAL PHYSICS. Educ: Ill Inst Technol, BS, 59; Northwestern Univ, MS, 60; Univ Iowa, PhD(physics), 63. Prof Exp: Res assoc physics, Case Inst Technol, 63-64, from asst prof to assoc prof, 64-71; ASSOC PROF PHYSICS & CHMN DEPT PHYS SCI, CHICAGO STATE UNIV, 71- Mem: Am Asn Physics Teachers; Am Phys Soc. Res: Statistical mechanics and thermodynamics of classical and quantum fluids and Heisenberg-Ising magnets; connections between entropy, forces and disorder; entropy production and efficiency for cyclic processes. Mailing Add: Dept of Phys Sci Chicago State Univ 95th at King Dr Chicago IL 60628

LEFF, JUDITH, b Vienna, Austria, July 6, 35; US citizen; m 61; c 3. PLANT PHYSIOLOGY. Educ: Sorbonne, Lic natural sci, 58, PhD(photobiol), 61. Prof Exp: Jr researcher photobiol seed germination, Nat Ctr Sci Res, Paris, 60-61, res assoc photobiol, 61; fel biol, Brandeis Univ, 62-63; fel pharmacol, Sch Med, Tufts Univ, 65, res assoc, 66-67; res assoc plant morphogenesis, Manhattan Col, 67-71, NY Univ, 71-72 & Hebrew Univ, Jerusalem, 72-73; NIH SPEC RES FEL, ALBERT EINSTEIN COL MED, 74- Mem: Sigma Xi. Res: Photobiology; chloroplast development; molecular biology; microbiology; nucleic acids as tools on solving physiological or developmental questions; microbiology of crown gall; replication of mitochondrial DNA in yeast. Mailing Add: 5829 Liebig Ave Bronx NY 10471

LEFFEK, KENNETH THOMAS, b Nottingham, Eng, Oct 15, 34; m 58; c 2. PHYSICAL ORGANIC CHEMISTRY. Educ: Univ London, BSc, 56, PhD(chem), 59. Prof Exp: Nat Res Coun Can fel, 59-61; from asst prof to assoc prof, 61-72, PROF CHEM & DEAN GRAD STUDIES, DALHOUSIE UNIV, 72- Concurrent Pos: Leverhulme vis fel, Univ Kent, Canterbury, 67-68. Mem: Fel Chem Inst Can; The Chem Soc. Res: Kinetics and mechanisms of organic reactions; primary and secondary kinetic deuterium isotope effects. Mailing Add: Dept of Chem Dalhousie Univ Halifax NS Can

LEFFEL, CLAUDE SPENCER, JR, b Pearisburg, Va, Dec 21, 21; m 55; c 2. PHYSICS. Educ: St John's Col, Md, BA, 43; Johns Hopkins Univ, PhD(physics), 60. Prof Exp: Tutor math & physics, St John's Col, Md, 46-50; SR PHYSICIST, APPL PHYSICS LAB, JOHNS HOPKINS UNIV, 60- Mem: AAAS; Am Phys Soc. Res: Plasma physics; nuclear physics; applied physics; atmospheric physics. Mailing Add: Appl Physics Lab Johns Hopkins Univ Laurel MD 20810

LEFFEL, EMORY CHILDRESS, b Pearisburg, Va, July 31, 23; m 56; c 2. ANIMAL SCIENCE. Educ: Univ Md, BS, 43, MS, 47, PhD(animal nutrit), 53. Prof Exp: Asst animal & dairy husb, Univ Md, 46-50; dairy husbandman, USDA, 50-51; instr dairy husb, 51-52, from asst prof to assoc prof animal husb, 53-68, PROF ANIMAL SCI, COL AGR, UNIV MD, COLLEGE PARK, 68- Mem: AAAS; Am Soc Animal Sci; Am Dairy Sci Asn. Res: Ruminant nutrition and physiology; protein and energy intake and effect of diet on rumen function in cattle and sheep; etiology of bloat in ruminants; bioenergetics. Mailing Add: Dept of Animal Sci Univ of Md College Park MD 20742

LEFFEL, ROBERT CECIL, b Woodbine, Md, Apr 26, 25; m 59; c 2. AGRONOMY, PLANT BREEDING. Educ: Univ Md, BS, 48; Iowa State Univ, MS, 50, PhD, 52. Prof Exp: Res agronomist, Agr Res Serv, 52-57; assoc prof agron, Univ Md, 57-62; res leader, 62-72, chief plant nutrit lab, 72-75, RES AGRONOMIST, AGR RES SERV, 62-, CHIEF CELL CULT & NITROGEN FIXATION LAB, 75- Mem: AAAS; Am Soc Agron. Res: Soybean, forage crop and clover genetics; breeding and production. Mailing Add: Cell Cult Nitrogen Fixation Lab Plant Physiol Inst USDA Beltsville MD 20705

LEFFERT, CHARLES BENJAMIN, b Logansport, Ind, May 22, 22; m 45. ENERGY CONVERSION, CHEMICAL PHYSICS. Educ: Purdue Univ, BS, 43; Univ Pittsburgh, MS, 57; Wayne State Univ, PhD(chem eng), 74. Prof Exp: Chem engr, Res Dept, Union Oil Co Calif, 43-49; chem engr, Pittsburgh Consol Coal Co, 49-52; asst physics, Univ Pittsburgh, 52-56; sr res physicist, Res Labs, Gen Motors Corp, 57-70; res asst, Res Inst Eng Sci, 70-74, ASSOC PROF CHEM ENG, WAYNE STATE UNIV, 74-, DIR COL ENG ENERGY CTR, 74- Mem: Am Phys Soc; Am Inst Chem Engrs. Res: Chemical engineering. Mailing Add: Wayne State Col Eng Energy Ctr 5050 Anthony Wayne Dr Detroit MI 48202

LEFFERT, HYAM LERNER, b New York, NY, May 11, 44. CELL BIOLOGY. Educ: Univ Rochester, BA, 65; Brandeis Univ, MA, 67; Albert Einstein Col Med, MD, 71. Prof Exp: Fel, 71-72, res assoc, 72-73, ASST RES PROF CELL BIOL, SALK INST BIOL STUDIES, 73- Concurrent Pos: Res grant, Nat Cancer Inst, NSF & Diabetes Asn Southern Calif, 74; consult cell biol, Dept Nutrit Path, Mass Inst Technol, 76, Nat Heart & Lung Inst, 73- & Dept Med, Vet Admin Hosp, Dallas, 74- Mem: Int Study Group Carcinoembryonic Proteins. Res: Mechanism of liver regeneration and differentiation in mammals. Mailing Add: Cell Biol Lab Salk Inst Biol Studies La Jolla CA 92037

LEFFINGWELL, JOHN C, b Evanston, Ill, Feb 16, 38; m 60; c 3. ORGANIC CHEMISTRY. Educ: Rollins Col, BS, 60; Emory Univ, MS, 62, PhD(org chem), 63. Prof Exp: Res assoc org chem, Columbia Univ, 63-64; res chemist, Org Chem Div, Glidden Co, Fla, 64-65 & R J Reynolds Tobacco Co, 65-70; sect head res dept, R J Reynolds Industs Inc, 70-73, HEAD FLAVOR DEVELOP, R J REYNOLDS TOBACCO CO, 73- Concurrent Pos: NIH fel, 63-64. Honors & Awards: Philip Morris Award for Distinguished Achievement in Tobacco Sci, Philip Morris Inc & Tobacco Sci, 74. Mem: Am Chem Soc; The Chem Soc; NY Acad Sci; Sigma Xi. Res: Natural products; terpenes and terpenoids; new synthetic reactions; flavor chemistry; olfaction; natural products; organic synthesis. Mailing Add: 3730 Northriding Rd Winston-Salem NC 27104

LEFFINGWELL, LOIS MENDLE, virology, see 12th edition

LEFFINGWELL, THOMAS PEGG, JR, b San Marcos, Tex, June 27, 26; m 60; c 1. CELL BIOLOGY. Educ: Southwest Tex State Col, BS, 48, MA, 49; Univ Tex, Austin, PhD(biol sci), 70. Prof Exp: Asst dept head & instr aerospace physiol & radiobiol, US Air Force Sch Aerospace Med, Gunter AFB, 51-53; res assoc physiol, Univ Tex, 53-55; radiobiologist, Nat Hq, Fed Civil Defense Admin, 55-56; res physiologist & group leader, Radiobiol Lab, Univ Tex, Austin & US Air Force Sch Aerospace Med, 56-65; RES SCIENTIST, CELL RES INST, UNIV TEX, AUSTIN, 65- Mem: AAAS; Electron Micros Soc Am; Am Soc Cell Biologists. Res: Biosynthesis and physicochemical characteristics of extracellular protein-polysaccharides; matrix materials, antigenic and other recognition factors on the cell surface; specific receptor sites for biological molecules. Mailing Add: Cell Res Inst Biol Labs Bldg 12-G Univ of Tex Austin TX 78712

LEFFLER, AMOS J, b New York, NY, Sept 9, 24; m 49; c 3. INORGANIC CHEMISTRY, PHYSICAL CHEMISTRY. Educ: Brooklyn Col, BS, 49; Univ Chicago, PhD(inorg chem), 53. Prof Exp: Chemist, Callery Chem Co, 52-55; res chemist, Stauffer Chem Co, 55-60; res assoc chem, Arthur D Little, Inc, 60-65; assoc prof, 65-75, PROF CHEM, VILLANOVA UNIV, 75- Concurrent Pos: Am Inst Chemists Award, Brooklyn Col, 49; sr res assoc, Nat Acad Sci-Nat Res Coun, 73. Mem: Am Chem Soc; The Chem Soc. Res: Inorganic and physical chemistry, especially boron and metallorganic chemistry, catalysis and metal oxides and fluorine chemistry. Mailing Add: Dept of Chem Villanova Univ Villanova PA 19085

LEFFLER, CHARLES WILLIAM, b Cleveland, Ohio, May 21, 47; m 68; c 1. PHYSIOLOGY, PHYSIOLOGICAL ECOLOGY. Educ: Univ Miami, BS, 69; Univ Fla, MS, 71, PhD(zool), 74. Prof Exp: Teaching asst physiol, Univ Fla, 69-73, coun fel, 73-74, fel, 74-76; ASST PROF PHYSIOL & BIOPHYS, UNIV LOUISVILLE, 76- Res: Pulmonary hemodynamics and the role of prostaglandins in the pulmonary vasculature; function of prostaglandins in the pulmonary vasculature in the perinatal period; significance of prostaglandins in shock hypotension. Mailing Add: Dept of Physiol & Biophys Health Sci Ctr Univ Louisville Louisville KY 40201

LEFFLER, ESTHER BARBARA, b Clearfield, Pa, Feb 1, 25. PHYSICAL CHEMISTRY. Educ: Pa State Univ, BS, 45; Univ Va, PhD(chem), 50. Prof Exp: Asst chemother, Stanford Res Labs, Am Cyanamid Co, 45-46; instr chem, Randolph-Macon Woman's Col, 49-53; from asst prof to prof chem, Sweet Briar Col, 53-66, chmn dept, 56-59 & 60-66; from asst prof to assoc prof, 67-75, actg chmn dept, 73-74, PROF CHEM, CALIF STATE POLYTECH UNIV, 75- Concurrent Pos: Res assoc & vis lectr, Stanford Univ, 66-67; res assoc, Oxford Univ, 74-75; resident dir, Calif State Univ Int Prog in UK, 74-75. Mem: Am Chem Soc; Sigma Xi. Res: Dissociation constants and reaction rates of bio-inorganic compounds; biochemical kinetics. Mailing Add: Dept of Chem Calif State Polytech Univ Pomona CA 91768

LEFFLER, JOHN EDWARD, b Brookline, Mass, Dec 27, 20; m 52. CHEMISTRY. Educ: Harvard Univ, BS, 42, PhD(org chem), 48. Prof Exp: Res assoc chem, Harvard Univ, 42-44 & Univ Chicago, 44-45; res assoc rocket fuels, US Navy Proj, Mass Inst Technol, 45-46; du Pont fel, Cornell Univ, 48-49; mem fac, Brown Univ, 49-50; from asst prof to assoc prof, 50-59, PROF CHEM, FLA STATE UNIV, 59- Mem: Am Chem Soc; The Chem Soc. Res: Reaction rate theory; polar and radical reactions; reactive intermediates of organic chemistry; peroxides; reactions of adsorbed organic compounds. Mailing Add: Dept of Chem Fla State Univ Tallahassee FL 32306

LEFFLER, MARLIN TEMPLETON, b College Corner, Ind, Feb 28, 11; m 33; c 3. ORGANIC CHEMISTRY. Educ: Miami Univ, AB, 32; Univ Ill, MA, 33, PhD(org chem), 36. Prof Exp: Asst chemist, Univ Ill, 33-35; res chemist, Abbott Labs, Ill, 36-46, head org dept, 46-49, asst dir res, 49-50, assoc dir, 50-57, dir chem & agr res, 57-59, dir res liaison, 59-71; CONSULT, 71- Mem: Am Chem Soc. Res: Local anesthetics; antiseptics; chemotherapy; optical activity of organic deuterium compounds; organic medicinal chemistry; biochemistry; overseas technological advances. Mailing Add: 510 E Prospect Lake Bluff IL 60044

LEFFLER, RICHARD GORDON, physics, see 12th edition

LEFKOVITCH, LEONARD PHILIP, b London, Eng, Feb 27, 29; m 56, 67; c 3. ENTOMOLOGY, STATISTICS. Educ: Univ London, BSc, 54. Prof Exp: Prin sci officer entom, Pest Infestation Lab, UK Agr Res Coun, 54-67; res scientist, 67-72, DIR, STATIST RES SERV, CAN DEPT AGR, 72- Concurrent Pos: Consult ecologist, Fed Govt Nigeria, 62; vis prof, Univ Man, 65-66. Mem: Biomet Soc; Zool Soc London; Royal Entom Soc London. Res: Taxonomy, biology and ecology of stored products beetles; mathematical and experimental population dynamics; numerical taxonomy. Mailing Add: Statist Res Serv Can Dept Agr Carling Bldg Ottawa ON Can

LEFKOWITZ, ISSAI, b New York, NY, Mar 13, 26; m 52; c 2. SOLID STATE PHYSICS. Educ: Brooklyn Col, BA, 59; Cambridge Univ, PhD(physics), 64. Prof Exp: Chief engr, Haines Industs, Inc, NJ, 53-54; physicist, Gulton Industs, Inc, 54-60; physicist res physics, Cavendish Lab, Cambridge Univ, 60-64; physicist, Pitman-Dunn Lab, Frankford Arsenal, Philadelphia, 64-72; pres, Princeton Mat Sci Inc, 72-75; PROG MGR, ARMY RES OFF, 75- Concurrent Pos: Guest scientist, Brookhaven Nat Lab, 54-60; fel neutron physics, Europ Atomic Energy Comn, Italy, 62-63; adj prof, Hunter Col; co-ed, Int J Ferroelec. Mem: Am Phys Soc; assoc Brit Inst Physics & Phys Soc. Res: Ferroelectricity; solid state; lattice dynamics; biophysics; administration with computer assistance. Mailing Add: Dept of Physics & Astron Hunter Col PO Box 387 New York NY 10021

LEFKOWITZ, LEWIS BENJAMIN, JR, b Dallas, Tex, Dec 18, 30; m 61; c 3. MEDICINE. Educ: Denison Univ, BA, 51; Univ Tex Southwest Med Sch Dallas, MD, 56. Prof Exp: USPHS res fel med, Univ Tex Southwest Med Sch Dallas, 59-60, instr, 60-61; USPHS res fel, Univ Ill, 61-62, USPHS trainee infectious dis, 62-65; asst prof, 65-70, ASSOC PROF PREV MED, SCH MED, VANDERBILT UNIV, 70-, ASST PROF MED, 71- Concurrent Pos: Asst clin prof internal med, Meharry Med Col, 66-, assoc clin prof family & community med, 70-; consult, US Army Hosp, Ft Campbell, Ky, 69- Mem: AAAS; Am Col Chest Physicians; NY Acad Sci; Am Col Prev Med; Am Pub Health Asn. Res: Epidemiology and pathogenesis of infectious diseases; health care delivery. Mailing Add: Dept of Prev Med Vanderbilt Univ Sch of Med Nashville TN 37232

LEFKOWITZ, ROBERT JOSEPH, b New York, NY, Apr 15, 43; m 63; c 4. MOLECULAR PHARMACOLOGY, MEDICAL SCIENCE. Educ: Columbia Univ, BA, 62, MD, 66. Prof Exp: From intern to jr asst resident, Columbia Presbyterian Med Ctr, NY, 66-68; clin & res assoc, Nat Inst Arthritis & Metab Dis, 68-70; sr asst resident, Mass Gen Hosp, Harvard Univ, 70-71, fel cardiol, 71-73; ASSOC PROF MED & ASST PROF BIOCHEM, MED CTR, DUKE UNIV, 73- Concurrent Pos: Estab investr, Am Heart Asn, 73. Mem: Am Soc Clin Invest; Am Soc Biol Chem; Am Fedn Clin Res; Am Heart Asn. Res: Molecular pharmacology of drug and hormone receptors. Mailing Add: Med Ctr Duke Univ PO Box 3325 Durham NC 27710

LEFKOWITZ, RUTH SAMSON, b Cincinnati, Ohio, Oct 7, 10; m 40; c 2. MATHEMATICS. Educ: Hunter Col, BA, 30; Columbia Univ, MA, 60, EdD(math educ), 66. Prof Exp: Sec Sch teacher math, New York City Bd Educ, 38-59; from asst prof to assoc prof, Bronx Community Col, 60-67; assoc prof, 67-70, chmn dept, 73-75, PROF MATH, JOHN JAY COL CRIMINAL JUSTICE, 70- Mem: AAAS; Math Asn Am. Res: Mathematics for open admissions students. Mailing Add: 900 W 190 St New York NY 10040

LEFKOWITZ, STANLEY A, b Philadelphia, Pa, Aug 5, 43. PHYSICAL INORGANIC CHEMISTRY. Educ: Temple Univ, AB, 65; Princeton Univ, PhD(chem), 70. Prof Exp: Asst to vchancellor Urban Affairs, City Univ New York, 70-73; asst dir instruct develop, Queens Col, 73-75; EXEC ASST TO CHMN BD, MOCATTA METALS CORP, 75- Concurrent Pos: Environ consult, NY State Temp Comn Powers Local Govt, 72-73; consult, Prof Exam Serv, 74-75 & Guana Island Hotel Corp, 76- Mem: Am Phys Soc; Fedn Am Scientists. Res: Alternative techniques for the extraction, refining and analysis of precious metals; the development and design of a solar-wind energy installation on Guana Island in the British Virgin Islands. Mailing Add: Mocatta Metals Corp 25 Broad St New York NY 10004

LEFKOWITZ, STANLEY S, b New York, NY, Nov 26, 33; m 57; c 2. MICROBIOLOGY, VIROLOGY. Educ: Univ Miami, BS, 55, MS, 57; Univ Md, PhD(plant path), 61; Am Bd Med Microbiol, dipl, 74. Prof Exp: Fel viral oncol, Variety Childrens Res Found, 61-64, res assoc, 64-65; from asst prof to assoc prof virol, Med Col Ga, 65-69; ASSOC PROF VIROL, SCH MED, TEX TECH UNIV, 72-, ASSOC DEAN GRAD STUDIES, 75- Mem: AAAS; fel Am Acad Microbiol; Am Soc Microbiol; Tissue Cult Asn; Soc Exp Biol & Med; Reticuloendothelial Soc (treas); NY Acad Sci; Am Asn Immunol. Res: Viral oncogenesis including its physical and biological implications; cancer immunology; effects of highly abused drugs on immunity. Mailing Add: 3801 67th St Lubbock TX 79413

LEFLORE, WILLIAM B, b Mobile, Ala, Feb 22, 32; m 65. PARASITOLOGY. Educ: St Augustine's Col, BS, 50; Univ Atlanta, MSc, 52; Univ Southern Calif, MS, 61, PhD(biol), 65. Prof Exp: Instr biol, Bennett Col, NC, 52-57; assoc prof, 64-68, PROF BIOL, SPELMAN COL, 68- Concurrent Pos: Consult, Off Educ, USPHS, 68, 71-72; vis prof, Col St Teresa, 68-69. Mem: Am Micros Soc; Am Soc Zoologists; Am Soc Parasitologists; Am Physiol Soc. Res: Serology of marine trematodes; life cycle of Cloacitrema michiganensis; histochemical demonstration of cercarial morphology; serology of Cysticercus fasciolaris; histochemistry of whole mounts of Cysticercus fasciolaris. Mailing Add: Dept of Biol Spelman Col Atlanta GA 30314

LEFOR, WILLIAM MATHEW, b Blue Earth, Minn, Oct 25, 37; m 61; c 2. IMMUNOLOGY. Educ: Mankato State Col, BA, 63; Ind Univ, Indianapolis, MD, 65, PhD(immunol), 68. Prof Exp: CHIEF IMMUNOCHEMIST, VET ADMIN HOSP, 68- Concurrent Pos: Asst prof med, Sch Med, Ind Univ, Indianapolis, 70-73. Mem: AAAS. Res: Quantitative and qualitative examination of humoral antibody response; immunological parameters in organ transplantation and rejection, as well as diseases of an autoimmune nature. Mailing Add: Vet Admin Hosp 1481 W Tenth St Indianapolis IN 46202

LEFRANCOIS, PHILIP ANDREW, chemistry, see 12th edition

LEFSCHETZ, SOLOMON, mathematics, deceased

LEFTAULT, CHARLES JOSEPH, JR, b Douglas, Ariz, Dec 4, 32; m 64; c 2. ANALYTICAL CHEMISTRY, MECHANICAL ENGINEERING. Educ: Univ Ariz, BA, 54, BS, 55; Purdue Univ, MS, 59. Prof Exp: Res chemist, 59-65, develop engr, Closure Develop Div, 65-68, mgr div, 68-75, MGR BLDG PROD SECT, CORP PROD DIV, ALCOA RES LABS, ALUMINUM CO AM, 75- Mem: Am Chem Soc. Res: Product-interior and exterior residential construction products such as siding, windows, gutters, shutters, bathroom and kitchen applications; process-plastic injection molding, blow-molding and profile extrusion including thermoplastic and thermoset technology. Mailing Add: Alcoa Res Labs PO Box 772 New Kensington PA 15068

LEFTIN, HARRY PAUL, b Beverly, Mass, Oct 23, 26; m 54; c 3. PHYSICAL ORGANIC CHEMISTRY, INDUSTRIAL CHEMISTRY. Educ: Boston Univ, AB, 50, PhD(chem), 55. Prof Exp: Res fel, Mellon Inst, 54-59; res chemist, 59-60, supvr chem res, 60-67, sr res assoc, Res & Develop Lab, 67-73, MGR RES, M W KELLOGG CO, 73- Concurrent Pos: Instr, Fairleigh Dickinson Univ, 60-75; ed, Catalysis Rev. Mem: AAAS; Am Chem Soc; Catalysis Soc NAm; Sigma Xi. Res: Heterogeneous catalysis; chemisorption; electronic infrared and nuclear magnetic resonance spectra of molecules in the adsorbed state; petroleum and petrochemical process development; gas phase kinetics. Mailing Add: Pullman Kellogg Res & Develop Ctr PO Box 79513 Houston TX 77079

LEFTON, PHYLLIS, b Neptune, NJ, Feb 10, 49. MATHEMATICS, NUMBER

THEORY. Educ: Columbia Univ, BA, 71, MA, 72, MPhil & PhD(math), 75; Jewish Theol Sem, BHL, 75. Prof Exp: Teaching asst calculus, Columbia Univ, 70-73; INSTR MATH, BELFER GRAD SCH & STERN COL, YESHIVA UNIV, 75- Mem: Am Math Soc; Math Asn Am; Asn Women in Math. Res: Algebraic number theory; analytic number theory; group representation theory; theory of polynomials and field theory. Mailing Add: Belfer Grad Sch of Sci Yeshiva Univ 2495 Amsterdam Ave New York NY 10033

LEGAL, CASIMER CLAUDIUS, JR, b Farrell, Pa, Feb 3, 15; m 38; c 4. INORGANIC CHEMISTRY. Educ: Thiel Col, BS, 37. Prof Exp: Anal chemist soap & raw mat, Lever Bros Co, 37-42; res chemist & supvr res lab & chem eng dept, Davison Chem Corp, 42-47, supvr res eng, 47-56; supvr agr chem res, 56-65, mgr fertilizer res, 65-74, MGR INDUST RES, W R GRACE & CO, 74- Mem: Am Chem Soc; Am Inst Chem Engrs. Res: Fertilizer; superphosphate; wet process phosphoric acid; silica gel; land reclamation. Mailing Add: 5331 Landing Rd Elkridge MD 21227

LEGALLAIS, DONALD RICHMOND, plant physiology, see 12th edition

LEGAN, SANDRA JEAN, b Cleveland, Ohio, Sept 10, 46. REPRODUCTIVE PHYSIOLOGY, NEUROENDOCRINOLOGY. Educ: Univ Mich, BS, 67, MS, 70, PhD(physiol), 74. Prof Exp: Lab asst, Geigy Co, Basel, Switz, 67-68; instr physiol, Univ Mich, 71-72, teaching fel, 72-73; NIH fel physiol, Emory Univ, 74-75; NIH FEL, REPRODUCTIVE ENDOCRINOL PROG, UNIV MICH, ANN ARBOR, 75- Mem: Assoc Am Physiol Soc; assoc Soc Study Reproduction; AAAS. Res: Neuroendocrine control of gonadotrophin secretion, specifically how modulations in steroid concentrations, environmental stimuli and neural input are transduced into endocrine events in the hypothalamo-hypophyseal axis. Mailing Add: Reproductive Endocrinol Prog Dept of Path Univ of Mich Ann Arbor MI 48109

LEGARE, RICHARD J, b Central Falls, RI, Dec 27, 34; m 57; c 1. POLYMER CHEMISTRY, CHEMICAL KINETICS. Educ: Providence Col, BS, 56; Univ Minn, MS, 60, PhD(phys chem), 62. Prof Exp: Sr res chemist, Allegany Ballistics Lab, Md, 62-71, STAFF SCIENTIST, BACCHUS WORKS, HERCULES INC, 71- Mem: Am Chem Soc. Res: High speed kinetics; biophysical and polymer chemistry; rocket propellants; high temperature resins; composite materials. Mailing Add: 3581 E Millstream Lane Salt Lake City UT 84109

LEGAT, WILHELM HANS, b Graz, Austria, Oct 31, 14; US citizen; m 61; c 1. SOLID STATE PHYSICS. Educ: Graz Univ, PhD(physics), 38. Prof Exp: From asst prof to assoc prof physics, Mining & Metall Col, Austria, 47-58; physicist, Allen Bradley Co, Wis, 58-59; sr engr, Mountain View Oper, 60-62, prin engr, 62-71, prin res scientist, Res Div, Waltham, 71-72, PRIN ENGR, MISSILE SYST DIV, RAYTHEON CO, 72- Concurrent Pos: Guest prof, Graz Tech Univ, 56-58; sr engr, Rheem Co, 59-62. Mem: Electrochem Soc; Austrian Phys Soc. Res: Heat conduction; x-ray diffraction; semiconductor materials; vapor deposition; device development of semiconductors. Mailing Add: Missile Syst Div Raytheon Co Hartwell Rd Bedford MA 01730

LEGATE, CARL EUGENE, geochemistry, see 12th edition

LEGATES, JAMES EDWARD, b Milford, Del, Aug 1, 22; m 44; c 4. ANIMAL GENETICS. Educ: Univ Del, BS, 43; Iowa State Col, MS, 47, PhD, 49. Prof Exp: Asst, Iowa State Col, 48-49; from asst prof to prof animal indust, 49-56, actg head dairy husb sect, 55-58, WILLIAM NEAL REYNOLDS PROF ANIMAL INDUST, NC STATE UNIV, 56-, HEAD ANIMAL BREEDING SECT, 58-, DEAN SCH AGR & LIFE SCI, 71- Concurrent Pos: Consult agr prog, Rockefeller Found, Colombia, 59; consult, Exp Sta Div, USDA, 59-65; vis prof, Nat Inst Animal Sci, Copenhagen, 63 & State Agr Univ, Wageningen, 71. Honors & Awards: Borden Award, Am Dairy Sci Asn, 67. Mem: AAAS; Biomet Soc; Am Soc Animal Sci; Am Dairy Sci Asn. Res: Selection in dairy cattle; genetics of mastitis resistance; quantitative inheritance in mice. Mailing Add: Sch Agr & Life Sci NC State Univ PO Box 5126 Raleigh NC 27607

LEGATOR, MARVIN (SEYMOUR), b Chicago, Ill, June 27, 26; m 50; c 2. BIOCHEMISTRY, BACTERIOLOGY. Educ: Univ Ill, BS & MS, 48, PhD(bact), 51. Prof Exp: Chief microbiologist, Agr Res Lab, Shell Develop Co, 52-63; head tissue cult & genetics sect, Food & Drug Admin, 63-66, chief cell biol br, 66-69, head gen toxicol br, 69-72; PROF GENETICS, BROWN UNIV, 72- Concurrent Pos: Prof, Med Sch, George Washington Univ, 69-72. Mem: AAAS; Tissue Cult Asn; NY Acad Sci; Am Soc Cell Biol. Res: Mutagenicity studies; microbial genetics; tissue culture methodology; cell physiology. Mailing Add: Div Biol & Med Sci Roger Williams Hosp Brown Univ Providence RI 02912

LEGAULT, ALBERT, b Hull, Que, June 7, 19; m 57. SYSTEMATIC BOTANY, PHYTOGEOGRAPHY. Educ: Univ Montreal, BA, 48, BPed, 53, BSc, 55, MSc, 58; Yale Univ, MSc, 59. Prof Exp: Teacher biol, Montreal-St Louis Col, Montreal, 50-57; researcher palynol, Serv Biogeog, Prov of Que, 61-62; from asst prof to assoc prof, 62-69, PROF BOT, UNIV SHERBROOKE, 69- Mem: Can Bot Asn; Int Asn Plant Taxon; Fr-Can Asn Advan Sci. Res: Floristics of southeastern, arctic and subarctic Quebec. Mailing Add: Dept of Biol Univ of Sherbrooke Sherbrooke PQ Can

LEGENDY, CHARLES RUDOLF, b Budapest, Hungary, Nov 2, 36; US citizen. THEORETICAL PHYSICS. Educ: Princeton Univ, BSE, 59; Cornell Univ, PhD(theoret physics), 64. Prof Exp: Theoret physicist, Res Labs, United Aircraft Corp, 64-67; sr physicist, Westinghouse Cambridge Lab, 67-68, fel physicist, Westinghouse Aerospace Div, 68-71; vis fel, Cybernet Lab, Nat Res Coun, Italy, 71-73; vis lectr, Univ Tübingen, 73-75; RES ASSOC, MAX PLANCK INST BIOPHYS CHEM, GÖTTINGEN, 75- Res: Solid state plasmas; boundary value problems; organization of the brain; electrophysiology of cat visual cortex. Mailing Add: 83-45 Broadway Elmhurst NY 11373

LEGEROS, RACQUEL ZAPANTA, chemistry, see 12th edition

LEGERTON, CLARENCE W, JR, b Charleston, SC, July 8, 22; m 58; c 3. GASTROENTEROLOGY. Educ: Davidson Col, BS, 43; Med Col SC, MD, 46. Prof Exp: Instr med, Duke Univ Hosp, 51-53; instr, 56-58, asst, 58-60, from asst prof to assoc prof, 61-70, PROF MED & CHIEF DIV GASTROENTEROL, MED UNIV SC, 70- Concurrent Pos: Consult, Vet Admin Hosp, Charleston, 66- Mem: Am Gastroenterol Asn; Am Fedn Clin Res; Am Col Physicians. Res: Mechanism of pain in peptic ulcer; effects of anticholinergic drugs on gastric motility and secretion. Mailing Add: Med Univ of SC Charleston SC 29401

LEGG, DAVID ALAN, b Elwood, Ind, Sept 7, 47. MATHEMATICAL ANALYSIS. Educ: Purdue Univ, BS, 69, MS, 70, PhD(math), 73. Prof Exp: ASST PROF MATH, IND UNIV-PURDUE UNIV, FT WAYNE, 74- Mem: Am Math Soc; Math Asn Am. Res: Approximation theory in the space of operators on Hilbert space. Mailing Add: Dept of Math Ind Univ-Purdue Univ Ft Wayne IN 46805

LEGG, JAMES C, b Kokomo, Ind, Sept 17, 36; m 73; c 3. NUCLEAR PHYSICS. Educ: Ind Univ, BS, 58; Princeton Univ, MA, 60, PhD(physics), 62. Prof Exp: Instr physics, Princeton Univ, 61-62; res assoc, Rice Univ, 62-63, asst prof, 63-67; assoc prof, 67-73, PROF PHYSICS, KANS STATE UNIV, 73-, DIR NUCLEAR SCI LAB, 72- Mem: Am Phys Soc. Res: Nuclear spectroscopy as studied by nuclear reactions. Mailing Add: Dept of Physics Kans State Univ Manhattan KS 66502

LEGG, JOHN IVAN, b New York, NY, Oct 15, 37; m 62; c 2. BIOINORGANIC CHEMISTRY. Educ: Oberlin Col, BA, 60; Univ Mich, MS, 63, PhD(inorg chem), 65. Prof Exp: Res assoc inorg chem, Univ Pittsburgh, 65-66; from asst prof to assoc prof, 66-75, PROF INORG CHEM & ASSOC IN BIOCHEM, WASH STATE UNIV, 75- Concurrent Pos: NIH spec fel, Harvard Med Sch, 72-73. Mem: Am Chem Soc. Res: Use of metalions to probe structure-function relationships in metalloenzymes; development of models for metalion binding sites in proteins. Mailing Add: Dept of Chem Wash State Univ Pullman WA 99163

LEGG, JOHN WALLIS, b Minter City, Miss, Sept 20, 36; m 56; c 3. PHYSICAL CHEMISTRY. Educ: Miss Col, BS, 58; Univ Fla, MS, 60, PhD, 64. Prof Exp: Chemist, Shell Oil Co, Tex, 58; asst, Univ Fla, 58-60; asst prof, Miss Col, 60-62; asst, Univ Fla, 62-64, instr, 63; assoc prof, 64-71, PROF CHEM, MISS COL, 71- Mem: Am Chem Soc. Res: Adsorption at solid surfaces and heterogeneous catalysis, specifically reactions over thorium oxide catalysts, primarily of the alcohols; dielectric properties of freon hydrates. Mailing Add: Dept of Chem Miss Col Clinton MS 39056

LEGG, JOSEPH OGDEN, b Tex, Oct 16, 20; m 44. SOIL SCIENCE. Educ: Univ Ark, BS, 50, MS, 51; Univ Md, PhD(soil fertil), 57. Prof Exp: SOIL SCIENTIST, USDA, 51- Concurrent Pos: USDA exchange scientist to USSR, 63-64. Mem: Fel AAAS; Soil Sci Soc Am; Am Soc Agron; Int Soc Soil Sci; Coun Agr Sci & Technol. Res: Soil nitrogen and organic matter. Mailing Add: Biol Waste Mgt & Soil Nitrogen Lab BARC-W Agr Res Serv Beltsville MD 20705

LEGG, KENNETH DEARDORFF, b Ogdensburg, NY, Feb 19, 43; m 67. ANALYTICAL CHEMISTRY. Educ: Union Col, BS, 64; Mass Inst Technol, PhD(chem), 69. Prof Exp: Vis prof chem, Univ Southern Calif, 74-75; asst prof, 69-74, ASSOC PROF CHEM, CALIF STATE UNIV, LONG BEACH, 75- Mem: Am Chem Soc. Res: Study of fast photophysical processes using laser excitation; electrogenerated chemiluminescence. Mailing Add: Dept of Chem Calif State Univ Long Beach CA 90840

LEGG, PAUL DAVIS, genetics, statistics, see 12th edition

LEGG, THOMAS HARRY, b Kamloops, BC, May 4, 29; m 57; c 2. PHYSICS, RADIO ASTRONOMY. Educ: Univ BC, BASc, 53; McGill Univ, MSc, 56, PhD(physics), 60. Prof Exp: Radar engr, Can Aviation Electronics Ltd, 53-54; sci officer radio physics, Defense Res Bd Can, 56-57; SR RES OFFICER, HERZBERG INST ASTROPHYS, NAT RES COUN CAN, 60- Mem: AAAS; Am Astron Soc; Can Asn Physicists; Royal Astron Soc Can; Inst Elec & Electronics Eng. Res: Microwave diffraction; electronic circuitry; radio interferometry. Mailing Add: 84 Delong Dr Ottawa ON Can

LEGGE, NORMAN REGINALD, b Edmonton, Alta, Apr 20, 19; nat US; m 42; c 5. POLYMER CHEMISTRY. Educ: Univ Alta, BSc, 42, MSc, 43; McGill Univ, PhD(phys chem), 45. Prof Exp: Jr res chemist, Nat Res Coun Can, 45; sect leader, Polymer Corp, 45-51; dir res & develop, Ky Synthetic Rubber Corp, 51-55; res supvr, Shell Develop Co, 55-58, from asst dept head to dept head, 58-61, res dir, Synthetic Rubber Div, 61-65, mgr res & develop, 65-69, mgr res & develop, Polymers Div, 69-73, spec assignment technol forecasting, 73-74, MGR NEW BUS, SHELL CHEM CO, 74- Mem: AAAS; Am Chem Soc; NY Acad Sci; Soc Chem Indust. Res: Industrial research administration; plastics; synthetic rubber; molecular and physical properties of high polymers; technological forecasting; organizational development. Mailing Add: Shell Chem Co PO Box 2463 Houston TX 77001

LEGGE, THOMAS NELSON, b Erie, Pa, Sept 23, 36; m 60; c 2. ZOOLOGY, LIMNOLOGY. Educ: Edinboro State Col, BS, 59; Miami Univ, MAT, 62; Univ Vt, PhD(zool), 69. Prof Exp: Instr biol, Northwestern Mich Col, 62-64; from asst prof to assoc prof, 67-70, PROF BIOL, EDINBORO STATE COL, 70- Mem: Am Soc Limnol & Oceanog; Int Asn Gt Lakes Res; Int Asn Theoret & Appl Limnol. Res: Physical and biological limnology, especially the distribution and ecology of calanoid copepods. Mailing Add: Dept of Biol Edinboro State Col Edinboro PA 16412

LEGGETT, ANNE MARIE, b Columbus, Ohio, May 28, 47. MATHEMATICAL LOGIC. Educ: Ohio State Univ, BA, 69; Yale Univ, PhD(math), 73. Prof Exp: Instr math, Mass Inst Technol, 73-75; ASST PROF MATH, UNIV TEX, AUSTIN, 75- Mem: Asn Symbolic Logic; Am Math Soc. Res: Recursion theory on admissible ordinals, especially the lattice of recursively enumerable sets and its relationships with the upper semi-lattice of degrees. Mailing Add: Dept of Math RLM 8-100 Univ of Tex Austin TX 78712

LEGGETT, GLEN EUGENE, b Brigham City, Utah, Mar 13, 22; m 43; c 2. SOIL SCIENCE. Educ: Utah State Univ, BS, 50, MS, 51; Wash State Univ, PhD(agron), 58. Prof Exp: Jr soil scientist, Wash State Univ, 52-57; SOIL SCIENTIST, SNAKE RIVER CONSERV RES CTR, AGR RES SERV, USDA, 57- Mem: Am Soc Agron; Soil Sci Soc Am; Soil Conserv Soc Am. Res: Chemistry of soil. Mailing Add: Agr Res Serv USDA Rte 1 Box 186 Kimberly ID 83341

LEGGETT, JAMES EVERETT, b West Union, WVa, Oct 20, 26; m 48; c 3. PLANT PHYSIOLOGY. Educ: Glenville State Col, AB, 49; Univ Md, College Park, MS, 54, PhD(bot), 65. Prof Exp: Plant physiologist, USDA, 53-70; assoc prof, 70-72, PROF AGRON, UNIV KY, 72- Mem: AAAS; Am Soc Plant Physiologists; Soil Sci Soc Am; Am Soc Agron; Crop Sci Soc Am. Res: Ion transport; application of ion transport mechanism to growth of plants and relationships to other environmental factors. Mailing Add: Dept of Agron Agr Sci Ctr Univ of Ky Lexington KY 40506

LEGGETT, JOSEPH EDWIN, b Warren, Ark, Sept 9, 37; m 60; c 2. ENTOMOLOGY. Educ: Univ Ark, BS, 60; Univ Ariz, MS, 68; Miss State Univ, PhD(entom), 75. Prof Exp: RES ENTOMOLOGIST, USDA AGR RES SERV, 68- Honors & Awards: Cert Merit, Agr Res Serv, 73. Mem: Entom Soc Am. Res: The physiological and behavioral aspects of boll weevil migration, and the relative importance of color and pheromone on boll weevil traps. Mailing Add: USDA Agr Res Serv PO Box 271 Florence SC 29501

LEGLER, DONALD WAYNE, b Minneapolis, Minn, Oct 2, 31; m 57; c 3. IMMUNOLOGY, PHYSIOLOGY. Educ: Univ Minn, BS, 54, DDS, 56; Univ Ala, PhD(physiol), 66. Prof Exp: From instr pedodont to assoc prof oral biol, 63-71, asst dean sch dent, 71-74, PROF ORAL BIOL & CHMN DEPT, SCH DENT, UNIV ALA, BIRMINGHAM, 71-, ASSOC DEAN ADMIN AFFAIRS, 74- Concurrent Pos: NIH trainee, 62-66; Swedish Med Res Coun fel, 67-68. Mem: Am Dent Asn; fel Am Col Dent; Am Soc Microbiol. Res: Comparative immunology; germ free research;

preventive dentistry. Mailing Add: Dept of Oral Biol Univ of Ala Sch of Dent Birmingham AL 35294

LEGLER, JOHN MARSHALL, b Minneapolis, Minn, Sept 9, 30; m 52; c 4. ZOOLOGY. Educ: Gustavus Adolphus Col, AB, 53; Univ Kans, PhD(zool), 59. Prof Exp: Asst human anat, Gustavus Adolphus Col, 52-53; asst zool, Univ Kans, 53-57, asst cur herpet, Mus Natural Hist, 55-59, asst instr zool, Univ, 58; from asst prof to assoc prof, 59-69, PROF ZOOL, UNIV UTAH, 69-, CUR HERPET, 59- Mem: Soc Study Evolution; Brit Herpet Soc. Res: Systematics and ecology amphibians and reptiles; turtles of Mexico and Central America; ecology of emyid and kinosternid turtles; chelonian morphology; histology of reptilian integumentary organs. Mailing Add: Dept of Biol Univ of Utah Salt Lake City UT 84112

LEGNER, E FRED, b Chicago, Ill, Oct 17, 32; m 60; c 1. ENTOMOLOGY, ECOLOGY. Educ: Univ Ill, Urbana, BS, 54; Utah State Univ, MS, 58; Univ Wis, PhD(entom), 61. Prof Exp: Asst entom, Univ Wis, 61-62; asst entomologist, 62-68, ASSOC ENTOMOLOGIST, UNIV CALIF, RIVERSIDE, 68-, ASSOC PROF ENTOM, 70-, PROF BIOL CONTROL, 73- Concurrent Pos: Consult, Africa, Australasia, SAm, Mid-E, Micronesia, WI & Europe, 62, 63 & 65-75; USPHS grants, 64-70 & NSF, 72-74. Mem: Entom Soc Am; Entom Soc Can; Am Mosquito Control Asn. Res: Population dynamics of arthropods and their biological control. Mailing Add: Div of Biol Control Univ of Calif Riverside CA 92502

LEGOFF, EUGENE, b Passaic, NJ, Aug 18, 34; m 60; c 2. ORGANIC CHEMISTRY. Educ: Rutgers Univ, BS, 56; Cornell Univ, PhD(org chem), 59. Prof Exp: Fel, Harvard Univ, 59-60; fel org chem, Mellon Inst, 60-65; ASSOC PROF ORG CHEM, MICH STATE UNIV, 65- Mem: Am Chem Soc. Res: Synthesis of pseudoaromatics, non-benzenoid aromatics, organic semi-conductors and organic superconductors, new synthetic methods. Mailing Add: Dept of Chem Mich State Univ East Lansing MI 48824

LEGOLVAN, PAUL CELESTIN, b Iron Mountain, Mich, Nov 18, 15; m 41; c 1. MEDICINE, PATHOLOGY. Educ: Univ Mich, MD, 40. Prof Exp: Intern, Walter Reed Gen Hosp, Washington, DC, 40-41; intern, US Army Med Field Serv Sch, Pa, 41-43, exec officer field units, Madison Barracks, NY, 43-44, command officer & med officer, Belg, Europe, Eng, France, Ger, 44-47, Ft McPherson, 47-48; residency path, Letterman Gen Hosp, San Francisco, 48-51, asst chmn path lab serv, Brooke Gen Hosp, Ft Sam Houston, 51-56, res, US Naval Med, 56-58, asst chmn dept & asst to dir, consult, Armed Forces Inst, Off Surg Gen, 58-65; chief lab serv, Walter Reed Hosp, 65-66; MEM STAFF, PATH LAB, VET ADMIN, 66- Concurrent Pos: Asst prof path, Baylor Univ, 53-56; consult, Off Surgeon Gen US Army, 62-66, chief sci br, 63-65. Mem: AMA; Am Soc Clin Path; fel Col Am Path; Am Pub Health Asn. Res: General and schistosomiasis associated carcinoma of the bladder; pathology administration, education and research. Mailing Add: Path & Allied Serv Vet Admin Cent Off 113 Washington DC 20420

LEGORE, RICHARD STEPHEN, b Vineland, NJ, Jan 4, 43; m 64; c 2. POLLUTION BIOLOGY, INVERTEBRATE PATHOLOGY. Educ: Univ Corpus Christi, BA, 68; Univ Wash, MS, 70, PhD(pollution biol), 74. Prof Exp: Trainee molluscan larval cult, Bur Com Fisheries, Milford, Conn, 69; res asst, Water Pollution Info Ctr, Univ Wash, 71-72, instr, Col Fisheries, 73-74; TECH DIR ENVIRON CONSULT, PARAMETRIX INC, 74- Concurrent Pos: Consult to var corps, fed govt, cities & comns, 71- Mem: Nat Shellfisheries Asn; Soc Invertebrate Path; Fedn Am Scientists. Res: Effects of pollutants on aquatic organisms and interpretation of pertinent data into legitimate and appropriate protective regulatory policy; comparative analysis of wound repair and regeneration in leeches versus other annelids. Mailing Add: 4242 NE 107th St Seattle WA 98125

LEGRAND, DONALD GEORGE, b Springfield, Mass, Apr 3, 30; m 51; c 4. PHYSICAL CHEMISTRY. Educ: Boston Univ, BA, 52; Univ Mass, PhD(chem), 59. Prof Exp: Res chemist, Mallinckrodt Chem Works, 52; asst prof chem, Univ Mass, 58-59; RES CHEMIST, RES LABS, GEN ELEC CO, 59- Mem: Am Chem Soc; Am Phys Soc; Soc Rheol. Res: Polymer physics; surface physics; rheo-optics. Mailing Add: Gen Elec Res Lab PO Box 1088 Schenectady NY 12301

LEGRAND, FRANK EDWARD, b Mayfield, Okla, Dec 18, 26; m 49; c 5. GENETICS, ECOLOGY. Educ: Okla State Univ, BS, 59; NDak State Univ, PhD(plant breeding), 63. Prof Exp: EXTEN AGRONOMIST, OKLA STATE UNIV, 63-, PROF AGRON, 74- Mem: Am Soc Agron. Res: Genetic studies of wheat in relation to the inheritance of several quantitative and qualitative characters. Mailing Add: Dept of Agron Okla State Univ Stillwater OK 74074

LEGRAND, HARRY E, b Concord, NC, May 19, 17; m 45; c 2. HYDROGEOLOGY. Educ: Univ NC, BS, 38. Prof Exp: Geol aide, US Geol Surv, 38-40, geologist, Ground Water Br, 46-49, dist geologist, 49-56, consult geologist, 56-59, res geologist, 59-60, chief radiohydrol sect, 60-62, res geologist, 62-74; CONSULT HYDROLOGIST, 74- Mem: AAAS; Geol Soc Am; Soc Econ Geologists; Am Geophys Union; Am Water Works Asn. Res: Contamination and geochemistry of ground water; ground water geology; ground water in igneous and metamorphic rocks; pollution and ground waste disposal. Mailing Add: 331 Yadkin Dr Raleigh NC 27609

LEGROW, GARY EDWARD, b Toronto, Ont, Mar 9, 38; m 63: c 2. CHEMISTRY. Educ: Univ Toronto, BA, 60, MA, 62, PhD(organosilicon chem), 64. Prof Exp: Res assoc metalloorganosiloxanes, Dept Chem, Univ Sussex, 64-65; res chemist, 65-68, group leader organo-functional silicon chem, 68-70, group leader resins & chem res, 70-73, SR GROUP LEADER RESINS RES, DOW CORNING CORP, 73- Concurrent Pos: Lectr, Mich State Univ, 66-68. Mem: Am Chem Soc; Sigma Xi; The Chem Soc. Res: Synthesis, kinetics and molecular rearrangements of organo-functional silanes; influence of the proximity of silicon on the reactivity of organic functions; silicon resin process. Mailing Add: 3612 Westbrier Terr Midland MI 48640

LEGTERS, LLEWELLYN J, b Clymer, NY, May 23, 32; m 56; c 2. PREVENTIVE MEDICINE, TROPICAL PUBLIC HEALTH. Educ: Univ Buffalo, BA & MD, 56; Harvard Univ, MPH, 61. Prof Exp: Rotating intern, Akron Gen Hosp, Ohio, 56-57; US Army, 57-, surgeon, 82nd Airborne Div, 57-58, 504th Infantry, Ger, 58-59, prev med officer, 8th Infantry Div, Ger, 59-60 & John F Kennedy Ctr Mil Assistance, 63-66, chief, Walter Reed Amy Inst Res Field Epidemiol Surv Team, Vietnam, 66-68, prev med officer, 68-70, chief volar eval group, Training Ctr, Infantry & Ft Ord, 71-72, chief ambulatory health serv, Silas B Hays Army Hosp, Ft Ord, 72-74, with US Army War Col, Carlisle Barracks, Pa, 74-75, CHIEF HEALTH & ENVIRON DIV, OFF SURGEON GEN, US ARMY, WASHINGTON, DC, 75- Concurrent Pos: La State Univ fel trop med & parasitol, Cent Am, 63. Mem: AAAS; fel Am Col Prev Med; Am Soc Trop Med & Hyg; Am Pub Health Asn; Soc Int Develop. Res: Epidemiology of infectious diseases, especially malaria and plague; behavioral sciences and organizational behavior, especially behavior modification in army training. Mailing Add: 4341 Ashford Ln Fairfax VA 22030

LEGVOLD, SAM, b Huxley, Iowa, Jan 8, 14; m 41; c 3. PHYSICS. Educ: Luther Col, AB, 35; Iowa State Univ, MS, 36, PhD(physics), 46. Hon Degrees: DSc, Luther Col, 75. Prof Exp: Asst prof physics, Luther Col, 37-39; assoc prof, 46-56, PROF PHYSICS, AMES LAB, IOWA STATE UNIV, 56- Honors & Awards: Meritorious Civilian Serv Award, US Navy, 46. Mem: AAAS, fel Am Phys Soc: Sigma Xi. Res: Cryogenics; magnetism; electricity; solid state physics. Mailing Add: Ames Lab Iowa State Univ Ames IA 50010

LEHMAN, ALFRED BAKER, b Cleveland, Ohio, Mar 21, 31. MATHEMATICS. Educ: Ohio Univ, BS, 50; Univ Fla, PhD(math), 54. Prof Exp: Instr math, Tulane Univ, 54; mem staff, Acoust Lab & Res Lab Electronics, Mass Inst Technol, 55-57; asst prof math, Case Inst Technol, 57-61; vis mem, Math Res Ctr, Univ Wis, 61-63; res assoc, Rensselaer Polytech Inst, 63; res mathematician, Walter Reed Army Inst Res, 64-67; PROF MATH & COMP SCI, UNIV TORONTO, 67- Concurrent Pos: Vis prof, Univ Toronto, 65-67. Mem: Am Math Soc; Math Asn Am; Soc Indust & Appl Math; Asn Comput Mach; Can Math Cong. Res: Combinatorics. Mailing Add: Dept of Math Univ of Toronto Toronto ON Can

LEHMAN, DAVID HERSHEY, b Lancaster, Pa. GEOLOGY. Educ: Franklin & Marshall Col, AB, 68; Univ Tex, Austin, PhD(geol), 74. Prof Exp: RES GEOLOGIST, EXXON PROD RES CO, 74- Mem: Geol Soc Am. Res: Expulsion of hydrocarbons from their source rocks and fluid flow in sedimentary basins. Mailing Add: Exxon Prod Res Co Houston TX 77001

LEHMAN, DENNIS DALE, b Youngstown, Ohio, July 14, 45; m 69; c 2. INORGANIC CHEMISTRY, ORGANOMETALLIC CHEMISTRY. Educ: Ohio State Univ, BSc, 67; Northwestern Univ, MS, 68, PhD(chem), 73. Prof Exp: ASST PROF CHEM, LOOP COL, 68- Concurrent Pos: Vis scholar, Dept Chem, Northwestern Univ, 73- Mem: Am Chem Soc; Chem Soc London; AAAS. Res: The use of organometallic complexes as catalyst for a variety of inorganic reactions; the reaction of transition metal complexes with small molecules and structural studies on the resulting products. Mailing Add: Loop Col 64 East Lake St Chicago IL 60601

LEHMAN, DONALD RICHARD, b York, Pa, Dec 13, 40; m 62. THEORETICAL NUCLEAR PHYSICS. Educ: Rutgers Univ, BA, 62; Air Force Inst Technol, MS, 64; George Washington Univ, PhD(physics), 70. Prof Exp: Proj scientist nuclear physics, Air Force Off Sci Res, 64-68; instr physics, George Washington Univ, 69-70; Nat Acad Sci-Nat Res Coun res assoc nuclear physics, Nat Bur Standards, 70-72; ASST PROF PHYSICS, GEORGE WASHINGTON UNIV, 72- Concurrent Pos: Vis staff mem, Los Alamos Sci Lab, 74- Mem: Am Phys Soc. Res: Nuclear few-body problem; photonuclear physics; intermediate energy physics; hypernuclei; scattering theory. Mailing Add: Dept of Physics George Washington Univ Washington DC 20052

LEHMAN, DUANE STANLEY, b Berne, Ind, Jan 18, 32; m 55; c 3. INORGANIC CHEMISTRY. Educ: Wheaton Col, Ill, BS, 54; Ind Univ, PhD(chem), 59. Prof Exp: Res chemist, 58-65, proj leader, Chem Dept Res Lab, 65-67, group leader, Chem Eng Lab, 67-71, LAB DIR, CHEM ENG LAB, DOW CHEM CO, 71- Mem: Am Chem Soc; Sigma Xi. Res: Coordination chemistry; brine chemistry; inorganic process research; basic refractories; new product development. Mailing Add: 704 Linwood Dr Midland MI 48640

LEHMAN, ERNEST DALE, b Woodward, Okla, Mar 2, 42; m 65; c 2. BIOCHEMISTRY. Educ: Northwestern State Col, Okla, BS, 65; Okla State Univ, PhD(biochem), 71. Prof Exp: Res assoc biochem, Okla State Univ, 71-72; NIH fel, Case Western Reserve Univ, 72-74; SR RES BIOCHEM, DEPT VIRUS & CELL BIOL RES, MERCK, SHARP & DOHME RES LABS, 74- Mem: AAAS; Am Chem Soc. Res: Biochemistry and function of glycoproteins; isolation and identification of bacterial and viral antigens. Mailing Add: Dept of Virus & Cel Biol Res Merck Sharp & Dohme Res Labs West Point PA 19486

LEHMAN, EUGENE H, b New York, NY, Jan 26, 13; m 61; c 4. MATHEMATICAL STATISTICS. Educ: Yale Univ, BA, 33; Columbia Univ, MA, 37; NC State Univ, PhD(math statist), 61. Prof Exp: Res assoc math, Univ Alaska, 49-51; asst prof, Univ Fla, 55-57 & Univ San Diego, 57-58; consult statistician, Los Angeles Area, 61-64; consult biostatistician, Cedars of Lebanon Hosp, Hollywood, 64-66; assoc prof math, Northern Mich Univ, 66-69; prof math, Mo Southern Col, 69-70; PROF STATIST, UNIV QUE, TROIS-RIVIERES, 70- Concurrent Pos: Corresp abstractor, Math Rev, 60-; referee, La Rev Can de Statist. Mem: Am Math Soc; Am Statist Asn; Fr-Can Asn Advan Sci; Can Asn Statist Sci. Res: Goodness of fit; theory of codage. Mailing Add: Dept of Math Univ of Que PO Box 500 Trois-Rivieres PQ Can

LEHMAN, GRACE CHURCH, b Mt Holly, NJ, June 10, 41. ZOOLOGY, ENDOCRINOLOGY. Educ: Drew Univ, AB, 63; Ind Univ, Bloomington, PhD(zool), 67. Prof Exp: USPHS fel, 67 & 68-70, univ fel, 70-71, res assoc, 71-74, RES INVESTR ZOOL, UNIV MICH, ANN ARBOR, 74- Mem: Am Soc Zool. Res: Endocrine interactions; influence of thyroid activity on reproduction; comparative and developmental endocrinology; reproductive biology of the amphibia. Mailing Add: Dept of Zool Univ of Mich Ann Arbor MI 48104

LEHMAN, GUY WALTER, b Walkerton, Ind, Sept 21, 23. THEORETICAL PHYSICS. Educ: Purdue Univ, BSEE, 48, MS, 50, PhD(physics), 54. Prof Exp: Jr engr electronics, Eastman Kodak Co, 48; asst physicist, Cornell Aeronaut Lab, 51; asst physics, Purdue Univ, 51-54; res specialist, Res Dept, Atomics Int Div, NAm Aviation, Inc, 54-62, group leader theoret physics, Sci Ctr, 63-67, mem tech staff, 67-70; PROF PHYSICS, UNIV KY, 70- Mem: Fel Am Phys Soc. Res: Solid state; mathematical physics; electronic structure; statistical mechanics; electromagnetic theory; lattice dynamics. Mailing Add: Dept of Physics Univ of Ky Lexington KY 40506

LEHMAN, HARVEY EUGENE, b Yuhsien, China; US citizen; m 58. DEVELOPMENTAL BIOLOGY, EMBRYOLOGY. Educ: Maryville Col, Tenn, BA, 41; Univ NC, MA, 44; Stanford Univ, PhD(embryol), 48. Prof Exp: From asst prof to assoc prof, 48-59, chmn dept, 62-67, PROF ZOOL, UNIV NC, 59- Concurrent Pos: Fel, Univ Berne, 52-53; chg exp embryol course, Bermuda Biol Sta, 60-; vis prof zool, Univ Vienna, 75. Mem: AAAS; Soc Develop Biol; Am Soc Zoologists; Am Micros Soc; Am Soc Cell Biol. Res: Rhabdocoele parasitology; amphibian pigmentation; nuclear transplantation in Triton; hybridization in Echinoderms; tissue culture; cell migration and differentiation of the neural crest; cytochemistry of embryonic differentiation. Mailing Add: Dept of Zool Univ of NC Chapel Hill NC 27514

LEHMAN, ISRAEL ROBERT, b Tauroggen, Lithuania, Oct 5, 24; US citizen; m 59; c 3. BIOCHEMISTRY. Educ: Johns Hopkins Univ, AB, 48, PhD(biochem), 54. Prof Exp: Am Cancer Soc fel, 55-57; instr microbiol, Wash Univ, 57-59; from asst prof to assoc prof, 59-66, PROF BIOCHEM, SCH MED, STANFORD UNIV, 66-, CHMN DEPT, 74- Mem: Am Soc Biol Chem. Res: Nucleic acid metabolism; biochemistry of virus infection. Mailing Add: Dept of Biochem Stanford Univ Sch of Med Stanford CA 94305

LEHMAN, JOE JUNIOR, b Versailles, Mo, July 1, 21; m 43; c 4. ORGANIC

CHEMISTRY. Educ: Bethel Col, Kans, AB, 43; Wash State Univ, MS, 47, PhD, 49. Prof Exp: From instr to assoc prof, 49-64, PROF CHEM, COLO STATE UNIV, 63- Concurrent Pos: Res fel, Midwest Res Inst, 58-59; vis prof, US Naval Acad, 61-62. Mem: AAAS; Am Chem Soc; Am Soc Microbiol. Res: Organic synthesis; modification of compounds by microorganisms; steric acceleration of hydrolytic reactions. Mailing Add: Dept of Chem Colo State Univ Ft Collins CO 80521

LEHMAN, JOHN MICHAEL, b Abington, Pa, June 19, 42. EXPERIMENTAL PATHOLOGY, VIROLOGY. Educ: Philadelphia Col Pharm & Sci, BS, 64; Univ Pa, PhD(path), 70. Prof Exp: NIH fel, Wistar Inst Anat & Biol, 70; NIH fel, 70-71, instr, 71-72, ASST PROF PATH, MED SCH, UNIV COLO MED CTR, DENVER, 72- Concurrent Pos: Vis staff mem, Los Alamos Sci Lab, 72- Mem: Am Soc Microbiol; Tissue Cult Asn; Am Asn Cancer Res; Am Asn Exp Path; Am Soc Cell Biol. Res: Tumor biology and virus transformation with oncogenic DNA viruses. Mailing Add: Dept of Path Univ of Colo Med Ctr Denver CO 80220

LEHMAN, JOHN WILLIAM, forestry, see 12th edition

LEHMAN, LILLIAN MARGOT YOUNGS, b Charleston, WVa, Jan 3, 28; m 58. EMBRYOLOGY. Educ: Catawba Univ, BA, 49; Univ NC, MA, 52, PhD, 54. Prof Exp: USPHS fel, Stanford Univ, 54-55; res asst embryol, Univ NC, 55-57; instr zool, Univ Vt, 57-58; from vis asst prof to vis assoc prof, 58-74, REGISTR & DIR INSTNL RES, UNIV NC, CHAPEL HILL, 75- Mem: AAAS. Res: Differentiation of pigment cells in amphibians. Mailing Add: South Bldg Univ of NC Chapel Hill NC 27514

LEHMAN, RICHARD LAWRENCE, b Portland, Ore, Nov 7, 29; m 63; c 4. BIOPHYSICS. Educ: Univ Ore, BS, 51, MA, 53; Univ Calif, Berkeley, PhD(biophys), 63. Prof Exp: High sch instr math & sci, Calif, 53-56; physicist, Lawrence Radiation Lab, Univ Calif, 57-64, asst prof in residence biophys & nuclear med, med ctr, Univ Calif, Los Angeles, 64-68; res physicist, lab nuclear sci, Mass Inst Technol, 68-71; DEPT DIR OFF ECOL, NAT OCEANIC & ATMOSPHERIC ADMIN, 71- Concurrent Pos: Vis scientist, Am Inst Biol Sci, 61-; vis scientist, Swiss Fed Inst Technol, 63 & Cambridge Univ, 66. Mem: AAAS; Radiation Res Soc; Biophys Soc; Am Geophys Union; Health Physics Soc. Res: Neutron spectrometry and activation analysis; radiological physics; molecular electronic structure; environmental impact of new technologies; energy-environment conflicts. Mailing Add: Rm 5813 Off of Ecol US Dept Com Nat Oceanic & Atmospheric Admin Washington DC 20230

LEHMAN, ROBERT ALONZO, pharmacology, see 12th edition

LEHMAN, ROBERT C, b Harrisonburg, Va, Mar 22, 29; m 54; c 4. BIOPHYSICS. Educ: Eastern Mennonite Col, AB, 50; Pa State Univ, MEd, 59, DEd(biophys), 62. Prof Exp: PROF PHYS SCI, EASTERN MENNONITE COL, 62- Concurrent Pos: Summer res assoc, Pa State Univ, 62-66. Mem: Am Asn Physics Teachers; Am Sci Affil. Res: Effects of ionizing radiation on macromolecules. Mailing Add: 1520 College Ave Harrisonburg VA 22801

LEHMAN, ROBERT HAROLD, b Duncannon, Pa, Nov 15, 29; m 52; c 1. PHYSIOLOGY, ECOLOGY. Educ: Bloomsburg State Col, BS, 60; Univ Okla, MNS, 65, PhD(physiol, ecol), 70. Prof Exp: Asst prof biol, 66-70, assoc prof bot, 70-74, ASSOC PROF BIOL, LONGWOOD COL, 74- Mem: Brit Ecol Soc. Res: Allelopathic effects of caffeoylquinic acids and scopolin on vegetational patterning. Mailing Add: 1205 Fifth Ave Farmville VA 23901

LEHMAN, ROBERT NATHAN, b Lancaster, Pa, Oct 22, 11; m 38; c 1. MEDICINE, SURGERY. Educ: Franklin & Marshall Col, BS, 33; Temple Univ, MD, 37. Prof Exp: Chief ophthal, Murphy Gen Hosp, Waltham, Mass, 51-54; CHIEF OPHTHAL, VET ADMIN HOSP, 54-; PROF OPHTHAL, UNIV PITTSBURGH, 54- Mem: AMA; Asn Res Vision & Ophthal; Am Col Surg; Am Acad Ophthal & Otolaryngol. Res: Surgical treatment of glaucoma using gelfilm setons; anatomical variations in iris as a cause of failure in surgery for glaucoma. Mailing Add: Vet Admin Hosp Pittsburgh PA 15240

LEHMAN, ROGER H, b Neosho, Wis, Apr 24, 21; m 57; c 3. MEDICINE, OTOLARYNGOLOGY. Educ: Univ Wis, BA, 42, MD, 44. Prof Exp: Resident otolaryngol, Vet Admin Ctr, Wood, Wis, 48-51; resident ophthal, Milwaukee County Gen Hosp, 51-52; chief otolaryngol, Vet Admin Ctr, Wood, 54-74; PROF OTOLARYNGOL & CHMN DEPT, MED COL WIS, 66- Concurrent Pos: Chief otolaryngol, Milwaukee County Hosp, Wis, 60-74 & Milwaukee Children's Hosp, 71-74. Mem: AMA; Am Col Surg; Am Laryngol, Rhinol & Otol Soc; Soc Univ Otolaryngol; Am Acad Ophthal & Otolaryngol. Mailing Add: Vet Admin Ctr Wood WI 53193

LEHMAN, RUSSELL SHERMAN, b Ames, Iowa, Jan 25, 30; m 52; c 6. MATHEMATICS. Educ: Stanford Univ, BS, 51, MS, 52, PhD, 54. Prof Exp: Prob analyst, Comput Lab, Ballistic Res Labs, Aberdeen Proving Ground, 55-56; Fulbright res grant, Univ Göttingen, 56-57; from asst prof to assoc prof, 58-66, PROF MATH, UNIV CALIF, BERKELEY, 66- Concurrent Pos: Consult, Rand Corp, 54-65. Mem: Am Math Soc; Math Asn Am; Asn Symbolic Logic. Res: Partial differential equations; number theory; dynamic programming; computing; mathematical logic. Mailing Add: Dept of Math Univ of Calif Berkeley CA 94720

LEHMAN, WILLIAM FRANCIS, b Montgomery, Minn, Apr 25, 26; m 50; c 2. PLANT GENETICS, AGRONOMY. Educ: Wartburg Col, BS, 50; Univ Minn, Minneapolis, MS & PhD(plant genetics), 56. Prof Exp: Grade sch teacher, Iowa, 50-51; asst plant genetics, Univ Minn, 53-56; from jr agronomist to assoc agronomist, 56-74, AGRONOMIST, UNIV CALIF, DAVIS, 74- Mem: Am Soc Agron; Am Genetic Asn; Genetics Soc Can; Crop Sci Soc Am. Res: Genetics, breeding, agronomy, and disease and insect resistance primarily on alfalfa but also on rice, wheat, and sunflower. Mailing Add: Dept of Agron & Range Sci Univ of Calif 1004 E Holton Rd El Centro CA 92243

LEHMANN, DONALD LEWIS, parasitology, see 12th edition

LEHMANN, ELROY PAUL, b Tigerton, Wis, June 22, 28; m 51; c 2. RESOURCE MANAGEMENT, PETROLEUM GEOLOGY. Educ: Univ Wis, BS, 50, MS, 51, PhD(geol), 55. Prof Exp: Asst prof geol, Wesleyan Univ, 52-59, actg chmn, 55-57; paleontologist, Mobil Oil Can, Libya, 59-60, sr paleontologist, 60-61, geol lab supvr, 61-63, staff geologist, Mobil Oil Libya Ltd, 63-65, sr staff geologist, Mobil Latin Am Inc, 65-67, sr res geologist, Mobil Res & Develop Corp, 67-69, chief geoscientist, Int Div, Mobil Oil Corp, 69-72, sr staff explorationist, 72-74, EXPLOR MGR, MOBIL OIL LIBYA LTD, 74- Concurrent Pos: Mem educ comt, Am Geol Inst, 55-57; Fulbright lectr, Karachi, 58-59. Mem: AAAS; fel Geol Soc Am; Am Asn Petrol Geologists; Soc Econ Paleontologists & Mineralogists; Libya Petrol Explor Soc (treas, 64, pres, 65). Res: Geoscience aspects of energy resource identification and evaluation. Mailing Add: Mobil Oil Libya Ltd PO Box 690 Tripoli Libya

LEHMANN, ERICH LEO, b Strasbourg, France, Nov 20, 17; nat US; m 39; c 3. MATHEMATICAL STATISTICS. Educ: Univ Calif, MA, 42, PhD(math statist), 46. Prof Exp: From asst to assoc prof math, 42-54, chmn dept statist, 73-76, PROF STATIST, UNIV CALIF, BERKELEY, 54- Concurrent Pos: Vis assoc prof, Columbia Univ, 50 & Stanford Univ, 51; vis lectr, Princeton Univ, 51. Mem: Am Math Soc; Am Statist Asn; Inst Math Statist; Int Statist Inst; Am Acad Arts & Sci. Res: Theories of testing hypotheses and of estimation; nonparametric statistics. Mailing Add: Dept of Statist Univ of Calif Berkeley CA 94720

LEHMANN, HEINZ EDGAR, b Berlin, Ger, July 17, 11; nat Can; m 40; c 1. PSYCHIATRY. Educ: Univ Berlin, MD. Prof Exp: From asst prof to assoc prof, 52-65, chmn dept, 70-74, PROF PSYCHIAT, McGILL UNIV, 65- Concurrent Pos: Dir med educ & res, Douglas Hosp, 48-; vis prof, Univ Cincinnati, 58- Honors & Awards: Lasker Award, Am Pub Health Asn, 57. Mem: Fel Am Psychiat Asn; Can Psychiat Asn; Can Ment Health Asn; fel Am Col Neuropharmacol; fel Int Col Neuropsychopharmacol. Res: Diagnosis and therapy of psychotic conditions; effects of drugs on mental processes. Mailing Add: 6603 Lasalle Blvd Montreal PQ Can

LEHMANN, HERMANN PETER, b London, Eng, June 24, 37; m 61. CLINICAL BIOCHEMISTRY. Educ: Univ Durham, BSc, 59, PhD(phys chem), 64. Prof Exp: Weizmann fel, Weizmann Inst Sci, 65-66; Volkswagen Found fel, Max Planck Inst, Mülheim Ruhr, WGer, 66-67; res assoc, Radiation Lab, Univ Notre Dame, 67-69; NIH sr fel biochem, Univ Wash, 69-71; ASST PROF PATH, LA STATE UNIV MED CTR, NEW ORLEANS, 71- Concurrent Pos: Vis scientist, Charity Hosp La, New Orleans, 71-; consult, Vet Admin Hosp, New Orleans, 73- Mem: Am Chem Soc; The Chem Soc; Am Asn Clin Chem; Brit Asn Clin Biochem. Res: Clinical chemistry; physical biochemistry. Mailing Add: Dept of Path La State Univ Med Ctr New Orleans LA 70112

LEHMANN, JUSTUS FRANZ, b Koenigsberg, Ger, Feb 27, 21; nat US; m 43; c 3. PHYSICAL MEDICINE. Educ: Univ Frankfurt, MD, 45. Prof Exp: Asst physician internal med, Univ Frankfurt, 45-46; res asst, Max Planck Inst Biophys, 46-48; asst physician internal med, Univ Frankfurt, 48-51; asst prof med, Mayo Clinic, 51-55; asst prof & assoc dir dept, Ohio State Univ, 55-57; PROF PHYS MED & CHMN DEPT PHYS MED & REHAB, UNIV WASH, 57- Concurrent Pos: Fel phys med, Mayo Clin, 51-55. Mem: Biophys Soc; AMA; Am Asn Electromyog & Electrodiag; Am Acad Phys Med & Rehab; Am Cong Rehab Med. Res: Biophysics of physical agents used in medicine; rehabilitation. Mailing Add: Univ Hosp Univ of Wash Seattle WA 98105

LEHMANN, WALTER JULIUS, physical chemistry, spectroscopy, see 12th edition

LEHMANN, WILLI, physics, see 12th edition

LEHMANN, WILLIAM FREDRICK, b Bingen, Wash, Oct 20, 32; m 58; c 3. FOREST PRODUCTS. Educ: Wash State Univ, BS, 58; NC State Univ, MS, 61; Colo State Univ, PhD(wood sci), 70. Prof Exp: Asst prof forest prod, Forest Res Lab, Ore State Univ, 60-68; RES FOREST PROD TECHNOLOGIST, FOREST PROD LAB, US FOREST SERV, 70- Concurrent Pos: Mem comt wood-based panels, Food & Agr Orgn UN, 74- Honors & Awards: Wood Award, Vance Publ Co, 71. Mem: Forest Prod Res Soc; Soc Wood Sci & Technol (exec secy, 75-); Am Soc Testing & Mat; Int Union Forestry Res Orgn. Res: Research and development in the areas of wood adhesives and adhesively-bonded wood products, particularly wood-base composite panels such as particleboards, hardboards and fiberboards. Mailing Add: Forest Prod Lab PO Box 5130 Madison WI 53705

LEHMANN, WILLIAM LEONARDO, physics, see 12th edition

LEHMANN, WILMA HELEN, b Chicago, Ill, Nov 14, 29. VERTEBRATE MORPHOLOGY. Educ: Mundelein Col, BA, 51; Northwestern Univ, MS, 54; Univ Ill, PhD(zool), 61. Prof Exp: Res asst allergy, Med Sch, Northwestern Univ, 54-56; asst prof zool, Pa State Univ, 61-64; asst prof natural sci, Mich State Univ, 64-67; from asst prof to assoc prof, 67-73, PROF BIOL, NORTHEASTERN ILL UNIV, 73- Concurrent Pos: Indexer, Evolution, 60-66; NSF instnl res grant, 63-64; vis res assoc, Argonne Nat Lab, 69-70. Mem: Fel AAAS; Soc Study Evolution; Am Soc Zool. Res: Comparative vertebrate anatomy; adaptive radiation of primates and rodents; functional mammalian anatomy; functional morphology, gross and microscopic, of bone. Mailing Add: Dept of Biol Northeastern Ill Univ Chicago IL 60625

LEHMER, DERRICK HENRY, b Berkeley, Calif, Feb 23, 05; m 28; c 2. MATHEMATICS. Educ: Univ Calif, AB, 27; Brown Univ, ScM, 29, PhD(math), 30. Prof Exp: Asst, Brown Univ, 28; Nat Res Coun fel, Calif Inst Technol, 30-32; res worker, Inst Advan Study, 33-34; from instr to asst prof math, Lehigh Univ, 34-38 & 39-40; Guggenheim fel, Cambridge Univ, 38-39; from asst prof to prof, 40-72, EMER PROF MATH, UNIV CALIF, BERKELEY, 72- Concurrent Pos: Nat Res Coun fel, Stanford Univ, 30-32; mathematician, Aberdeen Proving Ground, Md, 45-46; dir, Inst Numerical Analysis, 51-53; adv panel, NSF, 52- Mem: AAAS; Soc Indust & Appl Math; Am Math Soc (vpres, 53); Math Asn Am; Asn Comput Mach (vpres, 54-57). Res: Theory of numbers; computing devices; mathematical tables and other aids to computation. Mailing Add: 1180 Miller Ave Berkeley CA 94708

LEHMICKE, DAVID JOHN, b Stillwater, Minn, May 11, 13; m 44; c 6. TEXTILE PHYSICS. Educ: Calif Inst Technol, BS, 35; Univ Minn, PhD(anal chem), 46. Prof Exp: From asst to instr phys chem, Univ Minn, 41-44; res chemist, Textile Fibers Dept, E I du Pont de Nemours & Co, Del, 45-55; mgr eval lab, Firestone Plastics Co, Pa, 55-60; tech mgr, Firestone Synthetic Fibers Co, 60-62; group leader, 62-66, RES ASSOC, FIRESTONE CENT RES LABS, 67- Mem: Am Chem Soc. Res: Instrumented analysis; polarography; vinyl and polyethylene plastics; polyamide and polyester synthetic fibers. Mailing Add: 1830 Massillon Rd Akron OH 44312

LEHMKUHL, DENNIS MERLE, b Pierre, SDak, Aug 22, 42; m 65. ENTOMOLOGY, ECOLOGY. Educ: Univ Mont, BA, 64, MS, 66; Ore State Univ, PhD(entom), 69. Prof Exp: ASST PROF BIOL, UNIV SASK, 69- Mem: AAAS; Entom Soc Can; Entom Soc Am; Ecol Soc Am; Can Soc Zool. Res: Taxonomy and biology of Ephemeroptera; ecology of rivers; arctic and northern aquatic insects, especially ecological adaptations and limiting factors and the resulting zoogeographical implications. Mailing Add: Dept of Biol Univ of Sask Saskatoon SK Can

LEHMKUHL, L DON, physiology, see 12th edition

LEHNE, RICHARD KARL, b Newark, NJ, Nov 18, 20; m 45; c 4. ORGANIC CHEMISTRY. Educ: Muhlenberg Col, BS, 41; Yale Univ, PhD(org chem), 49. Prof Exp: Process develop chemist, Gen Aniline & Film Co, 49-52; from assoc dir to dir res & develop, Wildroot Co, 52-59; mgr hair prod, Colgate-Palmolive Co, 59-63; dir res & develop, Mennen Co, 63-66; dir consumer prod res & develop, Cyanamid Int, 66-71; DIR REGULATORY AFFAIRS, CHURCH & DWIGHT CO, INC, 71- Mem: AAAS; Am Chem Soc; Soc Cosmetic Chemists (secy, 60-63). Res: Emulsion technology and viscosity versus stability; effects of phenolic additives. Mailing Add: Church & Dwight Co Inc 2 Pennsylvania Plaza New York NY 10001

LEHNER, GUYDO R, b Chicago, Ill, Apr 14, 28. TOPOLOGY. Educ: Loyola Univ, Ill, BS, 51; Univ Wis, MS, 53, PhD, 58. Prof Exp: Instr math, Univ Wis-Milwaukee, 57-58; from instr to assoc prof, 58-68, PROF MATH, UNIV MD, COLLEGE PARK, 68- Mem: Am Math Soc; Math Asn Am. Res: Abstract spaces; continua; point set topology. Mailing Add: Dept of Math Univ of Md College Park MD 20742

LEHNER, JOSEPH, b New York, NY, Oct 29, 12; m 38; c 1. MATHEMATICAL ANALYSIS, NUMBER THEORY. Educ: NY Univ, BS, 38; Univ Pa, PhD(math), 41. Prof Exp: Instr math, Cornell Univ, 41-43; mathematician, Kellex Corp, NY, 43-46 & Hydrocarbon Res, Inc, 46-49; assoc prof math, Univ Pa, 49-53; mem staff, Los Alamos Sci Lab, NMex, 53-58; prof math, Mich State Univ, 57-63 & Univ Md, College Park, 63-72; MELLON PROF MATH, UNIV PITTSBURGH, 72- Concurrent Pos: Consult, Sandia Corp, NMex & Nat Bur Standards, Washington, DC. Mem: Am Math Soc; Math Asn Am; London Math Soc. Res: Analytic theory of numbers; theory of automorphic functions; neutron transport. Mailing Add: Dept of Math Univ of Pittsburgh Pittsburgh PA 15260

LEHNER, PHILIP NELSON, b NH, July 5, 40; m 67; c 1. ANIMAL BEHAVIOR, ECOLOGY. Educ: Syracuse Univ, BS, 62; Cornell Univ, MS, 64; Utah State Univ, PhD(wildlife), 69. Prof Exp: Biologist, Bur Sport Fisheries & Wildlife, 62; res asst, Cornell Univ, 62-64 & Smithsonian Inst, 64-65; biologist, USPHS, 65; asst prof, 69-74, ASSOC PROF ANIMAL BEHAV, COLO STATE UNIV, 74- Concurrent Pos: NIH grant, Colo State Univ, 69-; consult, Stearns-Roger Corp, 70- Mem: AAAS; Animal Behav Soc; Soc Exp Anal Behav; Wildlife Soc; Am Inst Biol Sci. Res: Animal behavior, its description, analysis and the effects of environmental variables. Mailing Add: Dept of Zool Colo State Univ Ft Collins CO 80521

LEHNERT, JAMES PATRICK, b Mecosta, Mich, Mar 29, 36. ZOOLOGY. Educ: Univ Mich, BS, 58, MA, 61; Univ Ill, PhD(parasitol), 67. Prof Exp: Asst prof, 67-72, ASSOC PROF BIOL, FERRIS STATE COL, 72- Res: Immunology of Ascaris suum in small rodents. Mailing Add: 221 Sci Ferris State Col Big Rapids MI 49307

LEHNERT, SHIRLEY MARGARET, b London, Eng, June 2, 34; m 61; c 2. RADIOBIOLOGY. Educ: Univ Nottingham, BSc, 55; Univ London, MSc, 58, PhD(biophys), 61. Prof Exp: Fel, Univ Rochester, 61-63; res biophysicist, Montreal Gen Hosp, 63-65; sci serv officer, Defence Bd Can, 65-67; res assoc phys biol, Sloan-Kettering Inst Cancer Res, 68-71; asst prof radiol, Radiol Res Lab, Col Physicians & Surgeons, Columbia Univ, 71-74; ASST PROF THERAPEUT RADIOL, McGILL UNIV, 74- Mem: AAAS; Radiation Res Soc. Res: Biological and biochemical effects of ionizing radiation. Mailing Add: Dept of Therapeut Radiol Royal Victoria Hosp 597 Pine Ave Montreal PQ Can

LEHNHOFF, HENRY JOHN, JR, b Lincoln, Nebr, Sept 13, 11; m 39; c 2. INTERNAL MEDICINE. Educ: Univ Nebr, AB, 33; Northwestern Univ, MD, 38; Am Bd Internal Med, dipl, 44. Prof Exp: PROF INTERNAL MED, COL MED, UNIV NEBR, OMAHA, 53- Concurrent Pos: Med dir, Northwestern Bell Tel Co & Woodman of the World Life Ins Co, 58. Mem: Am Soc Internal Med; Indust Med Asn; AMA; fel Am Col Physicians; Am Fedn Clin Res. Res: Nephrosis; infectious hepatitis; liver disease; arthritis; cardiac decompensation. Mailing Add: 530 Doctors Bldg Omaha NE 68131

LEHNINGER, ALBERT LESTER, b Bridgeport, Conn, Feb 17, 17; m 42; c 2. BIOCHEMISTRY. Educ: Wesleyan Univ, BA, 39; Univ Wis, MS, 40, PhD(biochem), 42. Hon Degrees: DSc, Wesleyan Univ, 54, Univ Notre Dame, 68, Acadia Univ, 72 & Mem Univ Nfld, 73; MD, Univ Padua, 66. Prof Exp: From instr to asst prof physiol chem, Univ Wis, 42-45; from asst prof to assoc prof biochem, Univ Chicago, 45-52; DELAMAR PROF PHYSIOL CHEM & DIR DEPT, SCH MED, JOHNS HOPKINS UNIV, 52- Concurrent Pos: Exchange prof, Univ Frankfurt, 51; Guggenheim fel & Fulbright scholar, Cambridge Univ, 51-52; Fulbright fel, Weizmann Inst & Univs Rome, Padua & Göttingen, 64; assoc mem, Neurosci Res Prog. Honors & Awards: Paul-Lewis Award, Am Chem Soc, 48, Remsen Award, 69; Distinguished Serv Award, Univ Chicago, 65; La Madonnina Award Sci, City Milan, 76. Mem: Inst of Med of Nat Acad Sci; Am Acad Arts & Sci; Am Philos Soc (vpres, 75-); Am Chem Soc; Am Soc Biol Chem (pres, 72-73). Res: Biological oxidations and phosphorylations; biochemistry of mitochondria; bioenergetics. Mailing Add: Dept of Physiol Chem Johns Hopkins Univ Sch Med Baltimore MD 21205

LEHOCZKY, JOHN PAUL, b Columbus, Ohio, June 29, 43; m 66; c 1. STATISTICS. Educ: Oberlin Col, BA, 65; Stanford Univ, MS, 67, PhD(statist), 69. Prof Exp: Asst prof, 69-73, ASSOC PROF STATIST, CARNEGIE-MELLON UNIV, 73- Mem: Inst Math Statist; Am Statist Asn. Res: Applied probability theory; stochastic processes and their application to computer, communication, and repair systems; diffusion approximations. Mailing Add: Dept of Statist Carnegie-Mellon Univ Schenley Park Pittsburgh PA 15213

LEHOTAY, JUDITH MONA, b Hungary, Apr 20, 26; US citizen; m 51; c 2. FORENSIC MEDICINE, PATHOLOGY. Educ: St Angela Inst, Budapest, BA, 46; Univ Budapest, MD, 56. Prof Exp: Resident anat path, Children's Hosp Buffalo, 62-63; resident, Sisters Hosp, Buffalo, 63-65; assoc pathologist, E J Meyer Mem Hosp, Buffalo, 66-69; assoc med examr forensic path, Chief Med Examr Off, New York, 69-70; actg chief med examr, 70-71, CHIEF MED EXAMR FORENSIC PATH, ERIE COUNTY LABS, 72- Concurrent Pos: Buswell fel, State Univ NY Buffalo, 65-66, clin assoc prof path, Med Sch, 69-; consult, E J Meyer Mem Hosp, 70-, Buffalo Columbus Hosp, 71- & Emergency Hosp & Sheridan Park Hosp, 72- Mem: AMA; fel Am Col Path; fel Am Acad Forensic Sci. Res: Anatomic pathology and cytology. Mailing Add: Div of Forensic Path Erie County Labs Buffalo NY 14215

LEHOUX, JEAN-GUY, b St Severin, Que, Jan 9, 39; m 63; c 3. BIOCHEMISTRY, ENDOCRINOLOGY. Educ: Univ Montreal, BSc, 63, MSc, 67, PhD(biochem), 69. Prof Exp: Chief chemist, Cyanamid Can Ltd, 63-65; res asst biochem, Univ Montreal, 65-69, lectr med, 69-71; asst prof obstet & gynec, 71-75, ASSOC PROF OBSTET & GYNEC, UNIV SHERBROOKE, 75-, HEAD CLIN ENDOCRINOL LAB, 74- Concurrent Pos: Biochemist, Hosp Maisonneuve, 69-70; Med Res Coun Can fels, Fac Med, Univ Montreal, 69-70 & Dept Zool, Univ Sheffield, 70-71; Med Res Coun Que & Med Res Coun Can grant, Univ Sherbrooke, 71-74; Med Res Coun Can scholar, 74. Mem: Brit Soc Endocrinol. Res: Studies on steroid hydroxylation with a special interest to aldosterone regulation. Mailing Add: Fac of Med Univ of Sherbrooke Sherbrooke PQ Can

LEHR, DAVID, b Sadagura, Austria, Mar 22, 10; US citizen; div; c 2. PHARMACOLOGY, MEDICINE. Educ: Univ Vienna, BA, 29, MD, 35. Prof Exp: Asst pharmacol, Univ Vienna, 34-48; instr, Univ Lund, 38-39; pharmacologist & res assoc, Path Dept, Newark Beth Israel Hosp, NJ, 39-42; from instr to asst prof pharmacol & med, 41-49, assoc prof pharmacol, 49-54, chmn dept, 54-56, prof physiol & pharmacol & chmn dept, 56-64, chmn res comt, Metrop Med Ctr, 54-72, ASSOC PROF MED, NEW YORK MED COL, 49-, PROF PHARMACOL & CHMN DEPT, 64- Concurrent Pos: Asst vis physician, Metrop Hosp, Welfare Island, NY, 42-54, vis physician, 54-; asst attend physician, Flower & Fifth Ave Hosps, 44-49, assoc attend physician, 49-; vis physician, Bird S Coler Hosp, 54-; Claud Bernard prof, Inst Exp Med & Surg, Univ Montreal, 61; mem rev comt, Health Res Coun New York, 61-65; vchmn panel neurol & psychiat dis, 61-; chmn ad hoc comt use of new therapeut agents & procedures in human beings, Assoc Med Schs, NY, 67-; mem coun arteriosclerosis, Am Heart Asn. Honors & Awards: Cert Merit, Asn Mil Surg US. Mem: Fel AAAS; fel Am Col Physicians; fel Am Col Cardiol; fel AMA; fel Am Soc Clin Pharmacol & Therapeut. Res: Cardiology; hypertension; experimental arteriosclerosis; toxicity of sulfonamides; sulfonamide mixtures; chemotherapy; experimental cardiovascular necrosis; parathyroid hormone interrelations; tissue electrolytes. Mailing Add: Dept of Pharmacol NY Med Col Valhalla NY 10595

LEHR, HANNS H, b Sadagora, Austria, Jan 1, 08; nat US; m 34; c 3. PHARMACEUTICAL CHEMISTRY. Educ: Univ Vienna, PhD(chem), 31, MPharm, 32. Prof Exp: Asst org chem, Univ Vienna, 30-32; managing dir pharmaceut lab, Salvatorapotheke, Vienna, 33-38; asst pharmacol, Paris, 38-40 & French Pub Health Serv, 40; asst biochem, Univ Aix Marseille, 40-42; asst org chem, Univ Basle, Switz, 43-46; sr res chemist, 46-66, asst to vpres chem res, 66-68, asst dir, 68-73, CONSULT, HOFFMANN-LA ROCHE, INC, 73- Mem: AAAS; Am Chem Soc; fel Am Inst Chemists. Res: Organic chemistry; biochemistry; chemotherapy; antibiotics; synthetic drugs; research administration. Mailing Add: 10 Tuers Pl Upper Montclair NJ 07043

LEHR, JAY H, b Teaneck, NJ, Sept 11, 36; m 57; c 2. HYDROLOGY, GROUNDWATER GEOLOGY. Educ: Princeton Univ, BSE, 57; Univ Ariz, PhD(hydrol), 62. Prof Exp: Hydrol field asst, Groundwater Br, US Geol Surv, NY, 55-56; res assoc hydrol, Univ Ariz, 59-62, from instr to asst prof, 62-64; asst prof, Ohio State Univ, 64-67; EXEC DIR, NAT WATER WELL ASN, 67- Concurrent Pos: Ed, Ground Water, 66-; ed-in-chief, Water Well J, 72- Res: Groundwater model studies utilizing consolidated porous medias; groundwater pollution; groundwater and surfacewater law; water well construction techniques. Mailing Add: Suite 135 Nat Water Well Asn 500 W Wilson Bridge Rd Worthington OH 43085

LEHR, MARVIN HAROLD, b Brooklyn, NY, Mar 17, 33; m 56; c 4. POLYMER PHYSICS, POLYMER CHEMISTRY. Educ: Reed Col, BA, 54; Yale Univ, MS, 55, PhD(kinetics), 59. Prof Exp: Res chemist, B F Goodrich Res Ctr, 59-61, sr res chemist, 61-66, res assoc, 66-73, SR RES ASSOC CORP RES, B F GOODRICH RES & DEVELOP CTR, 73- Res: Viscoelastic-fracture behavior; polymer morphology; relation of polymer structure to properties. Mailing Add: B F Goodrich Res & Develop Ctr Brecksville OH 44141

LEHR, RAYMOND BRUCE, b Philadelphia, Pa, Oct 2, 33; m 60. ANTHROPOLOGY, ARCHAEOLOGY. Educ: Bucknell Univ, AB, 55; Univ of the Americas, MA, 60. Prof Exp: Instr lang & social sci, 61-63, asst prof social sci, 65-66, chmn dept sociol-anthrop, 65-70, ASSOC PROF SOCIOL, ELIZABETHTOWN COL, 66- Concurrent Pos: Instr, Exten Serv, Elizabethtown Col York Mem Hosp, Pa, 66-67; instr, Exten Serv, Col Univ Ctr, Harrisburg, 66-68 & vis lectr, Indust Rels Inst, Col, 69-73. Mem: AAAS; Am Anthrop Asn; Soc Am Archaeologists. Res: Ethnohistory of prehispanic northern Yucatan; native chronicles of Mesoamerica. Mailing Add: Dept of Sociol-Anthrop Elizabethtown Col Elizabethtown PA 17022

LEHR, ROLAND E, b Quincy, Ill, Nov 7, 42. ORGANIC CHEMISTRY. Educ: Princeton Univ, AB, 64; Harvard Univ, AM, 66, PhD(chem), 69. Prof Exp: ASST PROF CHEM, UNIV OKLA, 68- Concurrent Pos: Res grants, NASA, 68-69 & Am Chem Soc, 68-70. Mem: Am Chem Soc; The Chem Soc; Sigma Xi. Res: Stereochemistry of pentadienyl radical cyclizations; development of assay procedures for psychoactive tertiary amine drugs in biological fluids. Mailing Add: Dept of Chem Univ of Okla Norman OK 73069

LEHRER, GERARD MICHAEL, b Vienna, Austria, May 29, 27; US citizen; m 60; c 2. NEUROLOGY. Educ: City Col New York, BS, 50; NY Univ, MD, 54. Prof Exp: Res asst neurol, Col Physicians & Surgeons, Columbia Univ, 53-54; intern, Mt Sinai Hosp, NY, 54-55, asst resident neurologist, 55-57, resident, 58; asst neurol, Sch Med, Wash Univ, 58-60; from asst attend neurologist to assoc attend neurologist, Mt Sinai Hosp, NY, 60-68; assoc prof, 66-67, PROF NEUROL, MT SINAI SCH MED, 67-, DIR DIV NEUROCHEM, 66-; ATTEND NEUROLOGIST, MT SINAI HOSP, NY, 68- Concurrent Pos: NIH trainee, Mt Sinai Hosp, NY, 56-58; NIH spec trainee neurochem & res fel pharmacol, Sch Med, Wash Univ, 58-61; consult, Preclin Psychopharmacol Res Rev Comt, NIMH, 65-69; res collabr, Brookhaven Nat Lab, NY, 67-69; consult neurologist, Vet Admin Hosp, Bronx, NY, 67- Mem: Fel Am Acad Neurol; Am Asn Neuropath; Int Soc Neurochem; Soc Neurosci; Am Neurol Asn. Res: Brain maturation and metabolism; molecular mechanisms of central nervous system differentiation and disease, especially demyelination. Mailing Add: Dept of Neurol Mt Sinai Sch of Med New York NY 10029

LEHRER, HAROLD Z, b New York, NY, Aug 22, 27. RADIOLOGY. Educ: Columbia Univ, AB, 47; NY Univ, MD, 53; Am Bd Radiol, dipl, 60. Prof Exp: Am Cancer Soc fel, 58; asst adj radiologist, Beth Israel Hosp & Med Ctr, New York, 59-63; instr neuroradiol, NY Univ-Bellevue Med Ctr, 63-65; from asst to assoc prof, Sch Med, Tulane Univ, 65-67; ASSOC PROF NEURORADIOL, NY MED COL, 68-; DIR DEPT RADIOL, BIRD S COLER HOSP, NEW YORK, 73- Concurrent Pos: NIH spec fel neuroradiol, NY Univ-Bellevue Med Ctr, 63-65. Mem: AAAS; Radiol Soc NAm; Am Roentgen Ray Soc; Am Soc Neuroradiol. Res: Neuroradiology, especially analysis of clinical data mathematically. Mailing Add: Dept of Radiol NY Med Col New York NY 10029

LEHRER, HARRIS IRVING, b Boston, Mass, May 28, 39. BIOCHEMISTRY, IMMUNOCHEMISTRY. Educ: Brandeis Univ, BA, 60, PhD(biochem), 65. Prof Exp: Sr biochemist, Monsanto Corp, 68-69; SR SCIENTIST IMMUNOCHEM, ORTHO DIAG, 69- Concurrent Pos: NIH fels, Marine Biol Lab, Woods Hole, 65, Univ Palermo, 65-67 & Brandeis Univ, 67-68. Mem: AAAS; Am Chem Soc. Res: Development of new immunochemical techniques and their application to diagnostic testing. Mailing Add: Ortho Diag Inc Div of Immunol Rte 202 Raritan NJ 08869

LEHRER, PAUL LINDNER, b Chicago, Ill, Feb 9, 28; m 53; c 4. PHYSICAL GEOGRAPHY. Educ: Univ Cincinnati, BS, 49; Ohio State Univ, MA, 51; Univ Nebr, PhD(geog), 62. Prof Exp: Instr geog, Ohio Univ, 56-59; from instr to asst prof, Univ Wis-Milwaukee, 60-66; assoc prof, 66-69, PROF GEOG, UNIV NORTHERN COLO, 69- Concurrent Pos: NSF sci fac fel, Univ Witwatersrand, 64-65. Mem: Asn Am Geogr; Sigma Xi. Res: Soils and regional geography of Subsaharan Africa. Mailing Add: Dept of Geog Univ of Northern Colo Greeley CO 80639

LEHRER, SHERWIN SAMUEL, b New York, NY, Apr 2, 34; m 60; c 2. BIOCHEMISTRY. Educ: Univ Pittsburgh, BS, 56; Univ Calif, Berkeley, PhD(chem), 61. Prof Exp: Staff scientist thin magnetic films, Lincoln Lab, Mass Inst Technol, 61-62; fel biochem, Brandeis Univ, 63-66; res assoc, Retina Found, Mass, 66-70; SR

STAFF SCIENTIST BIOCHEM, BOSTON BIOMED RES INST, 70- Concurrent Pos: USPHS res grant, Retina Found, Mass & Boston Biomed Res Inst, 67-; assoc, Harvard Med Sch, 68- Mem: AAAS; NY Acad Sci; Am Soc Biol Chem; Am Chem Soc; Biophys Soc. Res: Application of fluorescence techniques to protein conformation and interactions; muscle protein interactions. Mailing Add: Dept of Muscle Res Boston Biomed Res Inst Boston MA 02114

LEHRER, WILLIAM PETER, JR, b Brooklyn, NY, Feb 6, 16; m 45; c 1. ANIMAL SCIENCE. Educ: Pa State Univ, BS, 41; Univ Idaho, MS, 46 & 54; Wash State Univ, PhD(animal nutrit, chem), 51; Blackstone Sch of Law, LLB, 72, JD, 74; Pepperdine Univ, MBA, 75. Prof Exp: Salesman, Swift & Co, WVa, 41-42; mgr livestock farm, NY, 44-45; from asst prof & asst animal husbandman to assoc prof animal husb & assoc animal husbandman, Univ Idaho, 46-60, prof, 60; dir nutrit, Albers Milling Co, 60-62, dir nutrit & res, 62-74, DIR NUTRIT & RES, ALBERS MILLING CO & JOHN W ESHELMAN & SONS, CARNATION CO, 74- Concurrent Pos: Mem, Comt Animal Nutrit & Comt Dog Nutrit, Nat Acad Sci-Nat Res Coun & Tech Comt, Western Livestock Range Livestock Nutrit, USDA; mem, Res Adv Coun, US Brewers Asn, 69- & Adv Coun to pres & deans, Calif State Polytech Univ, Pomona, 69- Honors & Awards: WAP Award, Agr-Bus Award, 64. Mem: Fel AAAS; Am Inst Nutrit; Animal Nutrit Res Coun; Sigma Xi; fel Am Soc Animal Sci. Res: Animal production and nutrition; reproduction and growth of beef cattle and dairy cattle, dogs, horses, sheep and swine. Mailing Add: Carnation Co Carnation Bldg 5045 Wilshire Blvd Los Angeles CA 90036

LEHRMAN, GEORGE PHILIP, b New York, NY, Nov 28, 26; m 48; c 3. PHARMACY ADMINISTRATION. Educ: Univ Conn, BS, 50, PhD, 55; Purdue Univ, MS, 52. Prof Exp: Asst pharm, Purdue Univ, 50-52, instr chem, 52-53; mkt analyst, Mead Johnson & Co, 55-57; res chemist, Am Cyanamid Co, 57-59; head develop, Cent Pharmacal Co, Ind, 59-61; pharmaceut develop mgr, Baxter Labs, 61-62; dir labs, Conal Pharmaceut, 62-64; vpres, Owen Labs, 64-67; assoc prof pharm, Univ Okla, 67-75; ASST DEAN COL PHARM, UNIV NMEX, 75- Mem: Am Chem Soc; Am Pharmaceut Asn; Soc Cosmetic Chemists. Res: Product development; economics. Mailing Add: 8431 Palo Duro NE Albuquerque NM 87111

LEHRMAN, LEO, b New York, NY, July 3, 00; m 30; c 1. ORGANIC CHEMISTRY. Educ: City Col New York, BS, 21; Columbia Univ, MA, 23, PhD(chem), 26. Prof Exp: Instr, 21-24, from asst prof to prof, 34-74, EMER PROF CHEM, CITY COL NEW YORK, 74- Concurrent Pos: Instr, NY Univ, 21 & Wash Sq Col, 25-29; instr, Exten Dept, Columbia Univ, 23-25. Mem: Am Chem Soc. Res: Fatty acids associated with starches; methods of analysis; organic reagents in qualitative analysis; coprecipitation studies; inhibiting action of sodium silicate on corrosion of metal by water; synthetic production of stibnite for use as a primer in small arms ammunition. Mailing Add: Dept of Chem City Col of New York New York NY 10028

LEI, KAI YUI, b Macau, July 30, 44; m 66; c 1. NUTRITION. Educ: Univ London, BS, 68; Univ Guelph, MS, 70; Mich State Univ, PhD(human nutrit), 73. Prof Exp: Res asst nutrit, Mich State Univ, 70-73; res assoc hemat, Wayne State Univ, 74-75; ASST PROF NUTRIT, MISS STATE UNIV, 75- Mem: Sigma Xi. Res: Oral contraceptives and nutrient interactions; trace mineral metabolism; carbohydrate and lipid metabolism. Mailing Add: PO Drawer HE Miss State Univ Starkville MS 39762

LEIBACH, FREDRICK HARTMUT, b Kitzingen, Ger, Sept 21, 30; US citizen; m 61; c 3. BIOCHEMISTRY, ENDOCRINOLOGY. Educ: Southwest Mo State Col, BS, 59; Emory Univ, PhD(biochem), 64. Prof Exp: Nat Acad Sci-Nat Res Coun res assoc, Ames Res Ctr, NASA, 64-67; ASSOC PROF BIOCHEM, MED COL GA, 67- Mem: AAAS; Am Chem Soc. Res: Enzymes in protein and amino acid metabolism; peptidases, transpeptidases and esterases; protein turnover; hormone effects on proteolytic enzymes. Mailing Add: Dept of Biochem Med Col of Ga Augusta GA 30902

LEIBACHER, JOHN WILLIAM, b Chicago, Ill, May 28, 41. ASTROPHYSICS, SOLAR PHYSICS. Educ: Harvard Univ, AB, 63, PhD(astron), 70. Prof Exp: Res assoc astrophys, Joint Inst Lab Astrophys, Nat Bur Stand, Univ Colo, 71-72, Nat Ctr Sci Res, France, 72-74; CONSULT ASTROPHYS, LOCKHEED PALO ALTO RES LAB, 75- Res: Radiative hydrodynamics of stellar atmospheres, observed motions and heating; x-ray observations. Mailing Add: Space Astron Group Dept 52-12 Lockheed Palo Alto Res Lab Palo Alto CA 94304

LEIBBRANDT, VERNON DEAN, b McCook, Nebr, Oct 31, 44; m 67; c 1. ANIMAL NUTRITION. Educ: Univ Nebr, BS, 66; Iowa State Univ, PhD(animal nutrit), 72. Prof Exp: ASST PROF ANIMAL NUTRIT, UNIV FLA, 75- Concurrent Pos: Fel, Res Div, Cleveland Clin Found, 72-75. Mem: Am Soc Animal Sci; AAAS. Res: Husbandry and nutritional aspects of swine production. Mailing Add: Agr Res Ctr Univ of Fla Rte 3 Box 383 Marianna FL 32446

LEIBEL, BERNARD S, b Toronto, Ont, Dec 14, 14; m 43; c 3. PHYSIOLOGY. Educ: Univ Toronto, BA, 36, MA, 38, BSc & MD, 38. Prof Exp: Res assoc, Banting & Best Dept Res, Univ Toronto, 36-38; Lester Mem fel blood circulation, Royal Col Surgeons Eng, 39-41; dir med res & clin invests, Mt Sinai Hosp, Toronto, 54-56; ASSOC PROF CLIN INVESTS, BANTING & BEST DEPT MED RES, FAC MED, UNIV TORONTO, 61- Concurrent Pos: Consult, Sunnybrook Hosp, Toronto, 44-; mem orgn coun & co-chmn sci prog comt, Int Diabetes Fedn, 61-64; consult, Mt Sinai Hosp, 65- Mem: Am Diabetes Asn; NY Acad Sci; Can Diabetic Asn. Res: Clinical physiology; anatomy of the human heart; diabetes and its effect on newborn children; oral agents in the treatment of diabetes; computer therapy in diabetes. Mailing Add: 291 Forest Hill Rd Toronto ON Can

LEIBFRIED, RAYMOND THOMAS, organic chemistry, see 12th edition

LEIBHARDT, EDWARD, b New Rome, Wis, Oct 13, 19; m 61; c 2. ASTRONOMY, SPECTROSCOPY. Educ: Northwestern Univ, BA, 54, PhD(astron), 59. Prof Exp: PRES, DIFFRACTION PROD, INC, 51- Mem: Optical Soc Am; Soc Appl Spectros. Res: Spectroscopy and photometry in astronomy; diffraction grating manufacture. Mailing Add: 9416 Bull Valley Rd Woodstock IL 60098

LEIBLER, RICHARD ARTHUR, b Chicago, Ill, Mar 18, 14; m 54; c 2. MATHEMATICS. Educ: Northwestern Univ, BS, 35, AM, 36; Univ Ill, PhD(math), 39. Prof Exp: Asst, Univ Ill, 37-40; instr math, Purdue Univ, 40-42 & 46; mem, Inst Advan Study & vis lectr, Princeton Univ, 46-48; mathematician, US Dept Navy, 48-51, Sandia Corp, 51-53 & Nat Security Agency, 53-58; dep dir, 58-63, DIR, COMMUN RES DIV, INST DEFENSE ANAL, 63- Concurrent Pos: Vis lectr, Princeton Univ, 59- Mem: Am Math Soc. Res: Probability; statistics; computing machines; managing of computing centers; statistical testing of hypotheses. Mailing Add: Inst for Defense Anal Thanet Rd Princeton NJ 08540

LEIBMAN, KENNETH CHARLES, b New York, NY, Aug 7, 23; m 46; c 2. BIOCHEMICAL PHARMACOLOGY. Educ: Polytech Inst Brooklyn, BS, 43; Ohio State Univ, MSc, 48; NY Univ, PhD(biochem), 53. Prof Exp: Org res chemist, Nat Lead Co, 43-44; asst chem, Ohio State Univ, 46-48; instr, Univ Louisville, 48-49; fel oncol, Univ Wis, 53-54, proj assoc, 54-55; specialist tracer techniques, US Tech Coop Mission, India, 55-56; from instr to assoc prof, 56-68, PROF PHARMACOL, COL MED, UNIV FLA, 68- Concurrent Pos: Ed, Drug Metab & Disposition, 72- Mem: AAAS; Am Soc Pharmacol & Exp Therapeut. Res: Drug metabolism; enzymology; toxicology. Mailing Add: Dept of Pharmacol & Therapeut Univ of Fla Col of Med Gainesville FL 32610

LEIBMAN, LAWRENCE FRED, b Bronx, NY, Sept 10, 47. ORGANIC CHEMISTRY. Educ: City Univ New York, PhD(chem), 76. Prof Exp: FEL ORG BIOCHEM, COLUMBIA UNIV, 75- Mem: Am Chem Soc; Sigma Xi. Res: Organic reaction mechanisms.

LEIBO, STANLEY PAUL, b Pawtucket, RI, Apr 8, 37; m 61; c 2. CRYOBIOLOGY, EMBRYOLOGY. Educ: Brown Univ, AB, 59; Univ Vt, MS, 61; Princeton Univ, MA, 62, PhD(biol), 63. Prof Exp: Res assoc, 63-64, USPHS res fel, 64-65, STAFF BIOLOGIST, BIOL DIV, OAK RIDGE NAT LAB, 65- Concurrent Pos: Vis scientist health sci & technol, Mass Inst Technol, 74- Mem: AAAS; Soc Cryobiol; Soc Study Reproduction; Biophys Soc. Res: Biology of bacteriophage; cryobiology of bacteriophage, proteins and algae; cryobiology and physiology of mammalian embryos, erythrocytes and tissue-culture cells. Mailing Add: Biol Div Oak Ridge Nat Lab Oak Ridge TN 37830

LEIBOVIC, K NICHOLAS, b Plunge, Lithuania, June 14, 21; m 43; c 3. NEUROSCIENCES, BIOPHYSICS. Prof Exp: Mathematician, Dulwich Col, Eng, 46-53 & Courtaulds, 53-56; proj leader indust math, Brit Oxygen Res & Develop Co, 56-60; sr mathematician, Westinghouse Res Labs, 60-63; prin mathematician, Cornell Aeronaut Lab, 63-64; assoc prof biophys, 64-74, asst dir ctr theoret biol, 67-68, PROF BIOPHYS, STATE UNIV NY BUFFALO, 74- Concurrent Pos: Lectr, Norwood Col, Eng, 52-53; mem math adv coun, Battersea Col Technol, 59-60; vis prof, Univ Calif, Berkeley, 69 & Hadassah Med Sch, Hebrew Univ, 71. Mem: AAAS; Biophys Soc; NY Acad Sci. Res: Information processing in the nervous system; mathematical models in biology; models of the nervous system; industrial mathematics; applications to chemical, mechanical and electrical engineering; computers and operations research. Mailing Add: Dept of Biophys State Univ of NY Buffalo NY 14226

LEIBOVITZ, ALBERT, b New York, NY, Nov 30, 15; m 48. MEDICAL MICROBIOLOGY, ONCOLOGY. Educ: Univ Conn, BS, 38; Univ Minn, MS, 40. Prof Exp: Sr microbiologist, Southbury Training Sch, Conn, 40-42; chief serv, 6th Army Med Lab, 60-70; DIR MICROBIOL SECT, SCOTT & WHITE CLIN, 70- Concurrent Pos: Consult, Naval Biol Lab, Calif, 64-70 & Vet Admin Hosp, Temple, Tex, 75- Mem: AAAS; Am Soc Microbiologists; Tissue Cult Asn; Asn Mil Surgeons US; NY Acad Sci. Res: Pathogenic and diagnostic bacteriology and virology; tissue culture media for enhancement of both cell and virus growth; new medium enabling cell growth in free gas exchange with atmosphere; transport medium for diagnostic virology; establish permanent cell lines from human solid tumors. Mailing Add: Microbiol Sect Scott & White Clin Temple TX 76501

LEIBOVITZ, LOUIS, b Philadelphia, Pa, May 29, 21; m 52; c 3. VETERINARY MEDICINE. Educ: Pa State Col, BS, 46; Univ Pa, VMD, 50. Prof Exp: Pvt vet pract, 50-56; prof avian dis & dir regional poultry diag lab, Del Valley Col, 56-63; AVIAN PATHOLOGIST, NY STATE VET COL, CORNELL UNIV, 63- Mem: Am Vet Med Asn; Am Asn Avian Path; Am Soc Parasitol; Soc Protozool; Am Soc Microbiol; Wildlife Dis Asn. Res: Diseases of birds, their history, etiology, pathogenesis, immunology prevention, treatment and cure. Mailing Add: Dept of Avian Dis NY State Vet Col Cornell Univ Ithaca NY 14850

LEIBOWITZ, GERALD MARTIN, b New York, NY, Feb 17, 36; m 63; c 4. MATHEMATICAL ANALYSIS. Educ: City Col New York, BS, 57; Mass Inst Technol, SM, 59, PhD(math), 63. Prof Exp: Instr math, Mass Inst Technol, 63; from instr to asst prof, Northwestern Univ, 63-68; assoc dir, CUPM, Calif, 68-69; ASSOC PROF MATH, UNIV CONN, 69- Mem: Am Math Soc; Math Asn Am. Res: Functional analysis; Banach algebras. Mailing Add: Dept of Math Univ of Conn Storrs CT 06268

LEIBOWITZ, JACK RICHARD, b Bridgeport, Conn, July 21, 29; m 54; c 2. SOLID STATE PHYSICS. Educ: NY Univ, BA, 51, MS, 55; Brown Univ, PhD(physics), 62. Prof Exp: Physicist, Signal Corps Eng Labs, 51-54 & Electronics Corp Am, 55-56; res physicist, Lincoln Labs, Mass Inst Technol, 56-61 & Westinghouse Res Lab, 61-64; asst prof physics, Univ Md, College Park, 64-69; PROF PHYSICS, CATH UNIV AM, 69- Mem: Am Phys Soc; Sigma Xi. Res: Superconductivity; ultrasonic interactions in solids; intermediate and mixed states; Fermi surfaces; electron-phonon interaction; excitation spectra of inhomogeneous superconductors. Mailing Add: Dept of Physics Cath Univ of Am Washington DC 20064

LEIBOWITZ, LEONARD, b New York, NY, Feb 5, 31. PHYSICAL CHEMISTRY. Educ: NY Univ, AB, 51, MS, 54, PhD(chem), 56. Prof Exp: Chemist, Pigments Dept, E I du Pont de Nemours & Co, 56-58; asst chemist, 58-61, ASSOC CHEMIST, CHEM ENG DIV, ARGONNE NAT LAB, 61- Mem: Am Chem Soc. Res: Kinetics of surface processes; catalysis; metal oxidation and ignition processes; high temperature chemistry. Mailing Add: Chem Eng Div Argonne Nat Lab Argonne IL 60439

LEIBOWITZ, LILA, b US. BEHAVIORAL ANTHROPOLOGY. Educ: Brooklyn Col, BA, 52; Columbia Univ, MA, 58, PhD(anthrop), 71. Prof Exp: ASSOC PROF SOCIOL & ANTHROP, NORTHEASTERN UNIV, 72- Mem: Am Anthrop Asn. Res: Bio-cultural evolution. Mailing Add: Dept of Sociol & Anthrop Northeastern Univ Boston MA 02115

LEIBOWITZ, MARTIN ALBERT, b New York, NY, Oct 15, 35. APPLIED MATHEMATICS, OPERATIONS RESEARCH. Educ: Columbia Univ, BA, 56; Harvard Univ, MA, 57, PhD(appl math), 61. Prof Exp: Staff scientist, Int Bus Mach Res Ctr, 60-63; mem tech staff, Bellcomm, Inc, Washington, DC, 63-66; assoc prof eng, 66-73, ASSOC PROF APPL MATH & STATISTICS, STATE UNIV NY STONY BROOK, 73- Mem: Opers Res Soc Am; Asn Comput Mach. Res: Random processes with application to control theory; guidance and communications; scientific programming. Mailing Add: Col of Eng State Univ of NY Stony Brook NY 11790

LEIBOWITZ, SARAH FRYER, b White Plains, NY, May 23, 41; m 66; c 1. PSYCHOPHARMACOLOGY. Educ: NY Univ, BA, 64, PhD(physiol psychol), 68. Prof Exp: USPHS fel & guest investr, 68-70, ASST PROF PHYSIOL PSYCHOL, ROCKEFELLER UNIV, 70- Honors & Awards: First Prize, Div Psychopharmacol, Am Psychol Asn, 69. Mem: AAAS; Am Psychol Asn; NY Acad Sci; Am Soc Pharmacol & Exp Therapeut; Soc Neurosci. Res: Study of neurochemical mechanisms in the brain which regulate behavioral and physiological responses. Mailing Add: Rockefeller Univ New York NY 10021

LEIBSON, FANNIE FREIREICH, biochemistry, see 12th edition

LEIBU, HENRY J, b Schlesiengrube, Ger, Apr 22, 17; m; c 2. CHEMISTRY. Educ: Swiss Fed Inst Technol, ChemE, 42, ScD(chem), 45. Prof Exp: Instr & res assoc indust chem, Swiss Fed Inst Technol, 45-49; chemist res plant develop, Polychems Dept, 49-59, CHEMIST, ELASTOMER CHEM DEPT, E I DU PONT DE NEMOURS & CO, 59- Mem: Am Chem Soc. Res: High pressure synthesis; carbon monoxide and polymer chemistry; elastomers; urethanes. Mailing Add: 4905 Threadneedle Rd Sedgely Farms Wilmington DE 19807

LEIBY, CLARE C, JR, b Ashland, Ohio, May 4, 24; m 52; c 5. PHYSICS. Educ: Mass Inst Technol, SB, 54; Univ Ill, MS, 58. Prof Exp: Physicist, Air Force Cambridge Res Labs, 54-56; res assoc, Univ Ill, 60-61 & Sperry Rand Res Ctr, 61-64; RES ASSOC, AIR FORCE CAMBRIDGE RES LABS, 64- Concurrent Pos: Consult, Leghorn Labs, 67- Mem: Sigma Xi. Res: Experimental and theoretical research in plasma physics in area of gas lasers; gravitational theory. Mailing Add: Optical Physics Lab Air Force Cambridge Res Labs Hanscom AFB MA 01731

LEIBY, GEORGE MARTIN, b Reading, Pa, Apr 20, 02; m 30; c 2. EPIDEMIOLOGY. Educ: Univ NC, BS, 29; Vanderbilt Univ, MD, 31; Harvard Univ, MPH, 35; Johns Hopkins Univ, DrPH, 38; Am Bd Internal Med, dipl, 47; Am Bd Prev Med & Pub Health, dipl, 49. Prof Exp: Intern, Roper Hosp, SC, 31-32; resident, Vanderbilt Univ Hosp, 32-33, State Bd Health, NC, 36-40, Dist Health Dept, Washington, DC, 40-41 & State Bd Health, La, 41-44; chief physician, US Vet Admin, 46-59; spec asst to surgeon-gen, US Air Force, 59-62; chief pub health adv, AID, Brazil, 62-64; DIR DIV LABS, STATE BD HEALTH, NC, 65-; HEALTH DIR, STANLY COUNTY HEALTH DEPT, 66- Concurrent Pos: Assoc prof, Sch Med, Johns Hopkins Univ, 47; clin prof, Med Sch, Univ Calif, Los Angeles, 47-54; consult, WHO, SE Asia & Africa, 49 & 58; clin assoc prof med, Univ NC, Chapel Hill, 65-70, clin assoc prof family med, 70- Mem: Fel Am Col Physicians; fel Am Col Prev Med; NY Acad Med; NY Acad Sci. Mailing Add: 907 Honeysuckle Ln Albermarle NC 28001

LEIBY, PAUL D, parasitology, helminthology, see 12th edition

LEIBY, ROBERT WILLIAM, b Allentown, Pa, Apr 2, 49; m 74. ORGANIC CHEMISTRY. Educ: Albright Col, BS, 71; Lehigh Univ, MS, 73, PhD(org chem), 75. Prof Exp: RES ASSOC ORG CHEM, DARTMOUTH COL, 75- Mem: Am Chem Soc; Sigma Xi. Res: Organic mass spectroscopy; heterocyclic synthesis; synthesis of pharmaceutical agents particularly central nervous system agents and antineoplastic agents. Mailing Add: Dept of Chem Dartmouth Col Hanover NH 03755

LEICESTER, HENRY MARSHALL, b San Francisco, Calif, Dec 22, 06; m 41; c 3. BIOCHEMISTRY. Educ: Stanford Univ, AB, 27, MA, 28, PhD, 30. Prof Exp: Instr chem, Oberlin Col, 30-31; chemist, Carnegie Inst, 31-32; chemist, Midgeley Found, Ohio State Univ, 34-38; asst prof biochem, Col Physicians & Surgeons of San Francisco, 38-46, assoc prof, 46-48; PROF BIOCHEM, DENT SCH, UNIV OF THE PAC, 48- Honors & Awards: Dexter Award, Div Hist Chem, Am Chem Soc, 62. Mem: Am Chem Soc; Int Asn Dent Res; Hist Sci Soc. Res: Trace elements in teeth; history of chemistry. Mailing Add: 2155 Webster St San Francisco CA 94115

LEICHNETZ, GEORGE ROBERT, b Buffalo, NY, Oct 15, 42; m 67; c 3. NEUROANATOMY. Educ: Wheaton Col, Ill, BS, 64; Ohio State Univ, MS, 66, PhD(anat), 70. Prof Exp: Instr anat, Ohio State Univ, 69-70; ASST PROF ANAT, MED COL VA, VA COMMONWEALTH UNIV, 70- Concurrent Pos: A D Williams res grant, Med Col Va, Va Commonwealth Univ, 71-72. Mem: Am Asn Anatomists; Soc Neurosci. Res: Comparative neuroanatomy of primates. Mailing Add: Dept of Anat Med Col of Va Va Commonwealth Univ Richmond VA 23298

LEICHTER, JOSEPH, b Feb 4, 32; US citizen. NUTRITION, FOOD SCIENCE. Educ: Cracow Col, Poland, BS, 56; Univ Calif, Berkeley, MS, 66, PhD(nutrit), 69. Prof Exp: Chemist, Pharmaceut Plant, Cracow, Poland, 56-57; chemist, Ministry Com & Indust, Haifa, Israel, 57-60; chemist, Anresco Lab, Calif, 60-65; ASST PROF NUTRIT, UNIV BC, 69- Mem: Nutrit Soc Can; Am Inst Nutrit. Res: Folic acid metabolism; effect of protein-calorie malnutrition on carbohydrate digestion and absorption; effect of dietary lactose on intestinal lactase activity. Mailing Add: Human Nutrit Div Univ of BC Sch Home Econ Vancouver BC Can

LEICHTLING, BENJAMIN H, biochemistry, see 12th edition

LEIDECKER, HENNING WILLIAM, JR, b Birmingham, Ala, Sept 9, 41; m 66; c 1. PHYSICS. Educ: Cath Univ Am, BA, 63, PhD(theoret physics), 69. Prof Exp: Asst prof, 68-74, ASSOC PROF PHYSICS, AM UNIV, 74- Concurrent Pos: Consult, Cath Univ Am, 68-69. Res: Statistical mechanics of fluids and polymers. Mailing Add: Dept of Physics Am Univ Washington DC 20016

LEIDER, HERMAN R, b Detroit, Mich, Jan 14, 29; m 60; c 2. SOLID STATE CHEMISTRY. Educ: Wayne State Univ, BS, 51, PhD(chem), 54. Prof Exp: Asst, Wayne State Univ, 51-52, res assoc, 52-54; aeronaut res scientist, Solid State Physics Br, Chem Mat Sec, Nat Adv Comt Aeronaut, 54-56; CHEMIST, LAWRENCE LIVERMORE LAB, UNIV CALIF, 56- Mem: Am Phys Soc. Res: Alkali halides; color center; luminescence; radiation effects. Mailing Add: Dept of Chem Lawrence Livermore Lab Livermore CA 94550

LEIDERMAN, P HERBERT, b Chicago, Ill, Jan 30, 24; m 47; c 4. PSYCHIATRY. Educ: Calif Inst Technol, MS, 49; Univ Chicago, MA, 49; Harvard Med Sch, MD, 53. Prof Exp: Asst psychol, Univ Chicago, 48-49; intern med, Beth Israel Hosp, 53-54; resident neurol, Boston City Hosp, 54-56; resident psychiat, Mass Gen Hosp, 56-57; res fel, Mass Ment Health Ctr, 57-58; assoc, Harvard Med Sch, 58-63; assoc prof, 63-68, PROF PSYCHIAT, MED SCH, STANFORD UNIV, 68- Concurrent Pos: Consult, USPHS. Mem: AAAS; Am Psychosom Soc; Am Psychiat Asn; Soc Res Child Develop. Res: Child development; psychology; transcultural psychiatry; psychophysiology. Mailing Add: Dept of Psychiat Stanford Univ Med Sch Palo Alto CA 94306

LEIDHEISER, HENRY, JR, b Union City, NJ, Apr 18, 20; m 44; c 2. PHYSICAL CHEMISTRY. Educ: Univ Va, BS, 41, MS, 43, PhD(phys chem), 46. Prof Exp: Res worker, Nat Adv Comt Aeronaut Proj, Univ Va, 43-45, res assoc, Cobb Chem Lab, 46-49; proj dir, Va Inst Sci Res, 49-52, mgr lab, 52-58, dir res, Lab, 58-60, dir res, Inst, 60-68; PROF CHEM & DIR CTR SURFACE & COATINGS RES, LEHIGH UNIV, 68- Concurrent Pos: Mem adv comt, Oak Ridge Nat Lab; chmn, Gordon Conf Corrosion, 64; NATO sr scientist fel, Cambridge Univ, 69; consult, Marshall Space Flight Ctr, NASA, 71- Honors & Awards: Award, Oak Ridge Inst Nuclear Studies, 48; J Shelton Horsley Res Prize, Va Acad Sci, 49; Young Auth Prize, Electrochem Soc. Mem: Am Chem Soc; Electrochem Soc; The Chem Soc; Am Soc Metals. Res: Corrosion; surface science; electrodeposition; Mössbauer spectroscopy; polymer coatings; long-term food storage; paint adherence. Mailing Add: Ctr for Surface & Coatings Res Lehigh Univ Bethlehem PA 18015

LEIDY, ROSS BENNETT, b Newark, Ohio, June 1, 39; m 71; c 2. PESTICIDE CHEMISTRY. Educ: Tex A&M Univ, BS, 63, MS, 66; Auburn Univ, PhD(biochem), 72. Prof Exp: Res asst radiation biol, Radiation Biol Lab, Tex A&M Univ, 65-66; instr & lab supvr, Biol Br, Microbiol Lab, US Army Chem Ctr & Sch, Ft McClellan, Ala, 66-68; supvr pesticide chem, Lab Div, NC Dept Human Resources, 73-74; CHEMIST PESTICIDE CHEM, PESTICIDE RESIDUE RES LAB, NC STATE UNIV, 74- Concurrent Pos: NIH res assoc, Dept Animal Sci, NC State Univ, 72-73. Mem: Sigma Xi. Res: Methodology and analyses of pesticide residues on plant and animal products relating to the laboratory's research projects. Mailing Add: Pesticide Residue Res Lab PO Box 5215 NC State Univ Raleigh NC 27607

LEIES, GERARD M, b Chicago, Ill, Aug 19, 18. NUCLEAR PHYSICS. Educ: Loyola Univ, Ill, BS, 40; Univ Calif, MA, 53; Georgetown Univ, PhD(physics), 62. Prof Exp: TECH DIR, AIR FORCE TECH APPLNS CTR, 57- Mem: Am Phys Soc. Res: Nuclear weapon test detection. Mailing Add: Air Force Tech Applns Ctr/TD Patrick AFB FL 32925

LEIF, ROBERT CARY, b New York, NY, Feb 27, 38; m 63; c 2. IMMUNOHEMATOLOGY, BIOMEDICAL ENGINEERING. Educ: Univ Chicago, BS, 59; Calif Inst Technol, PhD(chem), 64. Prof Exp: Fel, Univ Calif, Los Angeles, 64-66; res assoc microbiol, Sch Med, Univ Southern Calif, 66-67; asst prof chem & biochem, Fla State Univ, 67-71; assoc scientist, 71-72, SR SCIENTIST, PAPANICOLAOU CANCER RES INST, 72- Concurrent Pos: Consult, Int Equip Corp, 65-73, Xerox Corp, 66-67, Damon Eng, 69-73, Solid State Radiation, Calif, 67-73, Coulter Electronics, 75- & Photometrics, 75-; res scientist, Dept Microbiol, Univ Miami, 71-72, adj asst prof microbiol, 73- & adj asst prof biomed eng, 74- Mem: AAAS; Reticuloendothelial Soc; Biomed Eng Soc; Am Chem Soc; Bop Biophys Soc. Res: Cellular differentiation; cytology automation; cytophysical and histochemical techniques to separate, purify and analyze heterogeneous cell populations; identifying biological activities with cell morphology. Mailing Add: Papanicolaou Cancer Res Inst 1155 NW 14th St PO Box 6188 Miami FL 33136

LEIFER, CALVIN, b New York, NY, Mar 4, 29; m 63; c 2. EXPERIMENTAL PATHOLOGY, ELECTRON MICROSCOPY. Educ: NY Univ, BA, 50, DDS, 54; State Univ NY Buffalo, PhD(exp path), 71. Prof Exp: Pvt pract, 58-65; USPHS fels, State Univ NY Buffalo, 65-70; ASSOC PROF PATH, SCH DENT, TEMPLE UNIV, 70- Mem: AAAS; Int Asn Dent Res; Am Acad Oral Path. Res: Ultrastructural and biochemical alterations of the rat parotid gland following single and multiple doses of x-irradiation; ultrastructural and histochemical features of tumors of the salivary glands in humans. Mailing Add: Dept of Path Temple Univ Sch of Dent Philadelphia PA 19140

LEIFER, HERBERT NORMAN, b New York, NY, Jan 30, 25; m 48; c 3. SOLID STATE PHYSICS. Educ: Univ Calif, Los Angeles, BA, 48, PhD(physics), 52. Prof Exp: Asst physics, Univ Calif, Los Angeles, 48-51, res engr, 51-52; res assoc, Res Lab, Gen Elec Corp, 52-55; staff scientist, Lockheed Missiles & Space Co, 55-59, mgr solid state electronics dept, 61-62; mgr basic physics, Fairchild Semiconductor Corp, 62-65; staff scientist, Electro optical Lab, Autonetics Div, NAm Aviation, Inc, Calif, 65-67 & High Energy Laser Lab, TRW Systs Group, 68-72; SR STAFF SCIENTIST, RAND CORP, 72- Concurrent Pos: Mem staff, Physics Lab, Ecole Normale Superieure, Paris, 60. Mem: Am Phys Soc. Res: Semiconductors; thermoelectric effects; electron-acoustic interactions; electrooptic effects; lasers. Mailing Add: Rand Corp 1700 Main St Santa Monica CA 90406

LEIFER, LESLIE, b New York, NY, Apr 13, 29; m 57; c 1. PHYSICAL CHEMISTRY. Educ: City Col New York, BS, 50; Univ Kans, PhD(phys chem), 59. Prof Exp: Res assoc nuclear & inorg chem, Mass Inst Technol, 56-59; asst prof chem, Clark Univ, 59-60; res assoc & staff mem, Lab Nuclear Sci, Mass Inst Technol, 61-63; assoc prof, Boston Col, 63-66; PROF CHEM, MICH TECHNOL UNIV, 66- Mem: AAAS; Am Chem Soc. Res: Solution physical chemistry; Mössbauer spectroscopy; quantum chemistry. Mailing Add: Dept of Chem & Chem Eng Mich Technol Univ Houghton MI 49931

LEIFF, MORRIS, b Ottawa, Ont, Feb 25, 15; nat US; m 38; c 3. ORGANIC CHEMISTRY. Educ: Queen's Univ, Ont, BA, 34, MA, 35; McGill Univ, PhD(wood & cellulose chem), 38. Prof Exp: Res chemist, Elmendorf Corp, Ill, 38-40, dir res, 40-47; tech dir, Smith & Kanzler Corp, 47-50; vpres & tech dir, S K Insulrock Corp, 50-56, tech dir, Insulrock Div, Flintkote Co, 56-59; vpres & tech dir, Smith & Kanzler Corp, 59-65; assoc prof sci, Jersey City State Col, 65-66; prof chem & chmn dept sci, Middlesex Co Col, 66-68; PROF CHEM & PHYSICS & CHMN DIV MATH, PHYS SCI & INDUST TECHNOL, COUNTY COL OF MORRIS, 68- Concurrent Pos: Lectr, Ill Inst Technol, 43-45, Sch Art Inst, Ill, 45-47 & Sch Design, 46-47. Mem: AAAS; Am Chem Soc; Am Soc Testing & Mat. Res: Wood chemistry; development of building materials, especially those related to wood; sprayed fiber fireproofing. Mailing Add: 100 Stonehill Rd Apt A-12 Springfield NJ 07081

LEIFIELD, ROBERT FRANCIS, b St Louis, Mo, Jan 29, 28; m 52; c 7. INORGANIC CHEMISTRY. Educ: St Louis Univ, BS, 52, MS, 59. Prof Exp: Chemist, Great Lakes Carbon Co, 52-56; chemist, 56-58, supvr metall & ceramics, 58-61, group leader process develop, 61-62, res chemist, 62-66, assoc mgr res, 66-69, res & develop mgr, Calsicat Div, 69-70, TECH MGR, CALSICAT DIV, MALLINCKRODT CHEM WORKS, 70- Mem: Catalysis Soc; Am Chem Soc. Res: Column and thin layer chromatography; analytical reagents; product and process research and development; uranium metallurgical chemistry; heterogeneous catalysis. Mailing Add: Mallinckrodt Chem Works 1707 Gaskell Ave Erie PA 16503

LEIFSON, OLAF S, solid state physics, see 12th edition

LEIGA, ALGIRD GEORGE, b New York, NY, Mar 25, 33; m 55; c 4. PHYSICAL CHEMISTRY. Educ: NY Univ, BA, 55, MS, 60, PhD(phys chem), 63. Prof Exp: Res scientist, Dept Chem, NY Univ, 62-64; sr scientist, Mat Sci Lab, NY, 64-73, MGR MAT DEVELOP, XERORADIOGRAPHY, XEROX CORP, PASADENA, 73- Mem: Am Chem Soc; Optical Soc Am; Soc Photog Scientists & Engrs. Res: Vacuum ultraviolet photochemistry and spectroscopy; decomposition reactions of solids; materials development for electrophotography. Mailing Add: 289 San Luis Pl Claremont CA 91711

LEIGH, EGBERT G, JR, b Richmond, Va, July 27, 40; m 68. ECOLOGY, POPULATION GENETICS. Educ: Princeton Univ, AB, 62; Yale Univ, PhD(biol), 66. Prof Exp: Actg instr biol, Stanford Univ, 66; asst prof, Princeton Univ, 66-72; BIOLOGIST, SMITHSONIAN TROP RES INST, 69- Res: Ecological aspects of population genetics; patterns of evolution in communities and in individual species; evolutionary biology; physiognomy and trophic organization of tropical rain forests. Mailing Add: Smithsonian Trop Res Inst Box 2072 Balboa CZ

LEIGH, ROGER, b Warrington, Eng, Aug 29, 40; m 62; c 2. GEOGRAPHY. Educ: Univ London, BSc, 63, PhD(urban geog), 69; Univ BC, MA, 65. Prof Exp: Asst prof urban geog, Univ BC, 69-73; SR LECTR GEOG, MIDDLESEX POLYTECHNIC, 73- Concurrent Pos: Nat Adv Comt Geog res grant, 69. Mem: Regional Studies Asn;

Asn Pac Coast Geogr; Can Asn Geogr; Asn Am Geogr. Res: Industrial location studies. Mailing Add: Dept of Geog Middlesex Polytechnic London England

LEIGH, THOMAS FRANCIS, b Loma Linda, Calif, Mar 7, 23; m 54; c 2. ENTOMOLOGY. Educ: Univ Calif, BS, 49, PhD(entom), 56. Prof Exp: Asst entom, Univ Calif, 52-54; asst prof, Univ Ark, 54-58; from asst entomologist to assoc entomologist, 58-68, ENTOMOLOGIST, UNIV CALIF, 68- Concurrent Pos: NIH, NSF-Int Biol Prog & Rockefeller Found grants. Mem: AAAS; Entom Soc Am; Ecol Soc Am; Mex Soc Entom. Res: Biology, ecology and control of cotton insects; insect resistance in crop plants; plant nutrition and insect abundance. Mailing Add: 242 Pine St Shafter CA 93263

LEIGH, WALTER HENRY, b Lake City, Mich, Apr 16, 12; m 38; c 2. PARASITOLOGY. Educ: Greenville Col, AB, 34; Univ Ill, PhD(zool), 38. Prof Exp: Asst zool, Univ Ill, 35-38; teacher biol, Wright Jr Col, 38-43, teacher zool & anat, 46-49; prof zool, 49-70, chmn dept, 58-70, PROF BIOL, UNIV MIAMI, 70- Mem: Am Soc Parasitologists; Soc Syst Zool. Res: Helminthology; trematode life histories. Mailing Add: Dept of Biol Univ of Miami Coral Gables FL 33124

LEIGHT, WALTER GILBERT, b New York, NY, Nov 19, 22; m 48; c 2. OPERATIONS RESEARCH, METEOROLOGY. Educ: City Col New York, BS, 42. Prof Exp: High sch instr, NY, 42; res meteorologist, US Weather Bur, 46-53; from sci analyst to dir opers anal div, Opers Eval Group & sr sci analyst, Systs Eval Group, Ctr Naval Anal, 53-70; prog mgr decision systs, Tech Anal Div, 71-74, CHIEF OFF CONSUMER PROD SAFETY, NAT BUR STANDARDS, 74- Mem: Fel AAAS; Opers Res Soc Am; Am Meteorol Soc. Res: Decision systems; criminal justice; search and rescue; nuclear safeguards; military systems; extended forecasting; consumer product safety. Mailing Add: 9416 Bulls Run Pkwy Bethesda MD 20034

LEIGHTON, ALEXANDER HAMILTON, b Philadelphia, Pa, July 17, 08. PSYCHIATRY, CULTURAL ANTHROPOLOGY. Educ: Princeton Univ, BA, 32; Cambridge Univ, MA, 34; Johns Hopkins Univ, MD, 36. Hon Degrees: AM, Harvard Univ, 66; SD, Acadia Univ, 74. Prof Exp: Social Sci Res Coun fel field work among Navajos & Eskimos, Columbia Univ, 39-40; Guggenheim fel, 46-47; dir, Southwest Proj, Cornell Univ, 48-53, dir, Prog Social Psychiat, 55-66, prof sociol & anthrop, Col Arts & Sci & Sch Indust & Labor Rels, 56-66, prof social psychiat, Med Col, 56-66; prof social psychiat & Head dept behav sci, 66-75, EMER PROF SOCIAL PSYCHIAT, HARVARD SCH PUB HEALTH, 75-; PROF PSYCHIAT & EPIDEMIOL, DALHOUSIE UNIV, 75- Concurrent Pos: Consult, Bur Indian Affairs, US Dept Interior, 48-50; mem bd dirs, Social Sci Res Coun, 48-58, chmn comt psychiat & social sci, 50-58; dir, Stirling County Proj, 48-; consult, Surgeon Gen Adv Comt Indian Affairs, 56-59; tech adv, Milbank Mem Fund, 56-63; fel, Ctr Advan Study Behav Sci, 57-58; mem expert adv panel ment health, WHO, 57-; Thomas W Salmon Mem Lectr, NY Acad Med, 58; mem sub-panel behav sci, President's Sci Adv Comt, 61-62; consult, Peace Corps, 61-63; reflective fel, Carnegie Corp NY, 62-63; vis lectr, Cath Univ Louvain, 71; mem comt effects of herbicides in Vietnam, Nat Acad Sci, 71-73. Honors & Awards: Human Rels Award, AAAS, 46. Mem: Fel AAAS; fel Am Psychiat Asn; Asn Am Indian Affairs; fel Am Anthrop Asn; Am Psychopath Asn. Res: Social and cultural change; social psychiatry; psychiatric epidemiology. Mailing Add: Dept of Psychiat Dalhousie Univ Halifax NS Can

LEIGHTON, ALVAH THEODORE, JR, b Portland, Maine, Apr 17, 29; m 53; c 4. GENETICS, PHYSIOLOGY. Educ: Univ Maine, BS, 51; Univ Mass, MS, 53; Univ Minn, PhD(poultry genetics, physiol), 60. Prof Exp: Asst poultry genetics, Univ Mass, 51-52 & Univ Minn, 55-59; assoc prof, 59-71, PROF POULTRY SCI, VA POLYTECH INST & STATE UNIV, 71- Mem: AAAS; Am Inst Biol Sci; World Poultry Sci Asn; Am Genetic Asn; Poultry Sci Asn. Res: Reproductive physiology and management of turkey populations. Mailing Add: Dept of Poultry Sci Va Polytech Inst & State Univ Blacksburg VA 24061

LEIGHTON, DOROTHEA CROSS, b Lunenburg, Mass, Sept 2, 08; m 37; c 2. MEDICINE, PSYCHIATRY. Educ: Bryn Mawr Col, AB, 30; Johns Hopkins Univ, MD, 36. Prof Exp: Chem technician, Univ Hosp, Johns Hopkins Univ, 30-32, house officer psychiat, 37-39; intern med, Baltimore City Hosps, 36-37; Social Sci Res Coun res fel, 39-40; spec physician, US Indian Serv, 41-45; soc sci analyst, US Off of War Info, 45; Guggenheim fel, 46-47; prof home econ, Cornell Univ, 49-52, res assoc sociol & anthrop, 52-65; assoc prof, 65-66, PROF MENT HEALTH, SCH PUB HEALTH, UNIV NC CHAPEL HILL, 66-, CHMN DEPT, 72- Concurrent Pos: From asst prof to assoc prof psychiat, Med Col, Cornell Univ, 54-65. Mem: AAAS; Am Sociol Asn; fel Am Psychiat Asn; Am Anthrop Asn; Am Asn Social Psychiat (vpres, 71-72). Res: Relationship between socio-cultural environment and psychiatric symptoms; child development. Mailing Add: Dept of Ment Health Univ of NC Sch of Pub Health Chapel Hill NC 27514

LEIGHTON, FREEMAN BEACH, b Champaign, Ill, Dec 19, 24; m 46; c 4. GEOLOGY. Educ: Univ Va, BS, 46; Calif Inst Technol, MS, 48, PhD(geol), 50. Prof Exp: From asst prof to assoc prof, 50-60, PROF GEOL, WHITTIER COL, 60- Concurrent Pos: Dir undergrad res prog, NSF, 59-64; pres, F Beach Leighton & Assocs, Inc. Mem: AAAS; Geol Soc Am; Am Asn Petrol Geologists; Am Asn Prof Geologists; Nat Asn Geol Teachers. Res: Active faulting; environmental planning; landslides; hillside development; geomorphology; engineering geology. Mailing Add: Dept of Geol Sci Whittier Col Whittier CA 90806

LEIGHTON, HENRY GEORGE, b London, Eng, May 2, 40; Can citizen; m 62; c 3. METEOROLOGY. Educ: McGill Univ, BS, 61, MS, 64; Univ Alta, PhD(nuclear physics), 68. Prof Exp: Res assoc nuclear physics, R J van de Graaff Lab, Holland, 68-70; vis asst prof, Univ Ky, 70-71; res assoc, 71-72, ASST PROF METEOROL, McGILL UNIV, 72- Mem: Can Meteorol Soc; Am Meteorol Soc. Res: Development of precipitation; weather modification, particularly hail suppression; interaction of solar radiation with atmospheric aerosols. Mailing Add: Dept of Meteorol McGill Univ Montreal PQ Can

LEIGHTON, JOSEPH, b New York, NY, Dec 13, 21; m 46; c 2. PATHOLOGY, ONCOLOGY. Educ: Columbia Univ, AB, 42; Long Island Univ, MD, 46. Prof Exp: From assoc prof to prof path, Sch Med, Univ Pittsburgh, 46-71; PROF PATH & CHMN DEPT, MED COL PA, 71- Concurrent Pos: Intern, Mt Sinai Hosp, New York, 46-47; resident path anat, Mass Gen Hosp, Boston, 48-49; resident clin path, USPHS Hosp, Baltimore, 50, exp pathologist, Path Lab, Nat Cancer Inst, 51-56, consult, Nat Serv Ctr, USPHS, 59-; mem coun, Gordon Res Confs, 63-66, chmn, Gordon Res Conf Cancer, 63. Mem: Am Soc Exp Path; Soc Develop Biol; Am Asn Path & Bact; Tissue Cult Asn; Am Asn Cancer Res. Res: Development of matrix methods and meniscus gradient methods for tissue culture; pathogenesis of tumor invasion and metastasis. Mailing Add: Dept of Path Med Col of Pa Philadelphia PA 19129

LEIGHTON, MORRIS WELLMAN, b Champaign, Ill, June 17, 26; m 47; c 3. EXPLORATION GEOLOGY. Educ: Univ Ill, BS, 47; Univ Chicago, MS, 48, PhD(geol), 51. Prof Exp: Res geologist, Jersey Prod Res Co, 51-58, geol sect head, 58-61, geologist-in-chg Europ study group, Esso Mediter, 61-63, sr res geologist, Jersey Prod Res Co, 63-64, geol adv, Esso Explor Inc, 64-68, asst explor mgr, Esso Standard Oil Ltd, Australia, 69-70, explor mgr, Esso Australia Ltd, 70-72, div mgr, Esso Prod Res Co, 72-74, CHIEF GEOLOGIST, ESSO INTERAMERICA, 74- Mem: Am Asn Petrol Geologists; fel Geol Soc Am; Soc Econ Paleontologists & Mineralogists. Res: Petroleum geology; basin studies; sedimentary and igneous petrology; basin and play assessment; estimating hydrocarbon potential. Mailing Add: Esso InterAmerica 396 Alhambra Circle Miami FL 33134

LEIGHTON, ROBERT BENJAMIN, b Detroit, Mich, Sept 10, 19; m 43; c 2. ASTROPHYSICS. Educ: Calif Inst Technol, BS, 41, MS, 44, PhD(physics), 47. Prof Exp: Mem res staff, 43-45, res fel, 47-49, from asst prof to assoc prof, 49-59, PROF PHYSICS, CALIF INST TECHNOL, 59- Mem: Nat Acad Sci; AAAS; Am Phys Soc; Am Astron Soc; Am Acad Arts & Sci. Res: Millimeter-wave; submillimeter; infrared-astronomy. Mailing Add: Calif Inst of Technol Pasadena CA 91109

LEIGHTON, ROBERT LYMAN, b Boston, Mass, Feb 27, 17; m 45; c 1. VETERINARY SURGERY. Educ: Univ Pa, VMD, 41; Am Col Vet Surg, dipl, 68. Prof Exp: Intern, Angell Mem Animal Hosp, 41-42; mem staff, Rowley Mem Animal Hosp, 46-56; head surg serv, Animal Med Ctr, 56-65; from asst prof to assoc prof, 65-68, PROF CLIN SCI, SCH VET MED, UNIV CALIF, DAVIS, 68- Honors & Awards: Merit Award, Am Animal Hosp Asn, 74. Mem: Am Vet Med Asn; NY Acad Sci. Res: Small animal surgery, orthopedics, surgical instrumentation and surgical prostheses. Mailing Add: Dept of Surg Univ of Calif Sch of Vet Med Davis CA 95616

LEIGHTON, RUDOLPH ELMO, b Welch, Okla, July 8, 10; m 33; c 2. DAIRY SCIENCE. Educ: Okla State Univ, BS, 32, MS, 43; Agr & Mech Col Tex, PhD, 56. Prof Exp: Pub sch teacher, Okla, 33-36; dairy herdsman, Okla State Univ, 36-39; supt, Dairy Exp Sta, Bur Dairy Indust, USDA, 39-47; from asst prof to assoc prof, 47-70, PROF DAIRY SCI, TEX A&M UNIV, 70- Concurrent Pos: Tex A&M Univ chief adv, Pakistan, 61-63. Mem: Am Dairy Sci Asn; Am Soc Animal Sci. Res: Dairy cattle nutrition and management including calf nutrition, hot-weather feeds for dairy cows and forage utilization. Mailing Add: Dept of Animal Sci Tex A&M Univ College Station TX 77843

LEIGHTON, WALTER (WOODS), b Toledo, Ohio, Sept 6, 07; m 37; c 2. MATHEMATICS. Educ: Northwestern Univ, BA, 31, MA, 32; Harvard Univ, AM, 33, PhD(math), 35. Prof Exp: Instr & tutor math, Harvard Univ, 33-36; instr, Univ Rochester, 36-37; lectr, Rice Inst, 37-43; dir appl math group, Northwestern Univ, 44-45; prof & chmn dept, Washington Univ, 46-54 & Carnegie Inst Technol, 54-59; Elias Loomis prof, Western Reserve Univ, 59-67; chmn math sci, 68-72, DEFOE PROF MATH, UNIV MO-COLUMBIA, 67- Concurrent Pos: Secy bd trustees, Rice Inst, 41-43; res mathematician, Appl Math Group, Off Sci Res & Develop, Columbia Univ, 43-44; res assoc, Brown Univ, 45-46; chief math div, Off Sci Res, US Air Force, 53-54; consult, 54-62. Mem: Am Math Soc; Math Asn Am; Soc Indust & Appl Math. Res: Stability theory for ordinary nonlinear differential equations; calculus of variations and associated problems of oscillation of solutions of self-adjoint differential equations. Mailing Add: Math Sci Bldg Univ of Mo Columbia MO 65201

LEIGHTY, EDITH GARDNER, b Zanesville, Ohio, Nov 4, 28; div; c 1. BIOCHEMISTRY. Educ: Marshall Univ, BS, 62; Ohio State Univ, MS, 65, PhD(physiol chem), 67. Prof Exp: Technician, Cabell Huntington Hosp, 56-63; res chemist, Holland-Suco Color Co, 62-63; asst physiol chem, Ohio State Univ, 63-65; PRIN BIOCHEMIST, BATTELLE-COLUMBUS LABS, 67- Mem: Am Chem Soc; NY Acad Sci; Sigma Xi. Res: Metabolism and distribution of marihuana and other drugs; identification of drug metabolites; development of analytical methods for drug detection; effect of magnetic fields on red blood cells, white blood cells, and malignant tumors. Mailing Add: Path Pharmacol Toxicol & Animal Resources Sect Battelle-Columbus Columbus OH 43201

LEIMANIS, EUGENE, b Koceni, Latvia, Apr 10, 05; m 42; c 6. APPLIED MATHEMATICS. Educ: Univ Latvia, Mag Math, 29; Univ Hamburg, Dr rer nat(math), 47. Prof Exp: Asst math, Univ Latvia, 29-35, privat-docent, 35-37, docent, 37-44; docent, Univ Greifswald, 44-45; assoc prof, Baltic Univ, Ger, 46-48; from asst prof to prof, 49-74, EMER PROF MATH, UNIV BC, 74- Concurrent Pos: Mem nat comt, Int Union Theoret & Appl Mech. Mem: Am Math Soc; Math Asn Am; London Math Soc; Asn Advan Baltic Studies; Am Astron Soc. Res: Dynamical systems; differential equations; non-linear and celestial mechanics. Mailing Add: Dept of Math Univ of BC Vancouver BC Can

LEIMGRUBER, WILLY, b Zurich, Switz, Sept 4, 30; m 61. ORGANIC CHEMISTRY. Educ: Swiss Fed Inst Technol, PhD(org chem), 59. Prof Exp: Sr chemist, 61-64, group leader, 65-67, sect chief, 67-69, dir, 69-73, DIR CHEM RES & ASST VPRES, HOFFMANN-LA ROCHE INC, 73- Mem: Swiss Chem Soc; Am Chem Soc. Res: Synthesis and structure elucidation. Mailing Add: 166 Highland Ave Montclair NJ 07042

LEIN, ALLEN, b New York, NY, Apr 15, 13; m 41; c 2. PHYSIOLOGY. Educ: Univ Calif, Los Angeles, BA, 35, MA, 38, PhD(zool), 41. Prof Exp: Assoc zool, Univ Calif, Los Angeles, 37, asst, 37-40; instr surg res, Ohio State Univ, 41-42, res assoc, Sch Med, 42-43; asst prof physiol, Vanderbilt Univ, 46-47; asst prof, Med Sch, Northwestern Univ, 47-52, Abbott assoc prof, 52-61, prof, 61-68, dir student affairs, 60-63, asst dean, 64-68, asst dean grad sch, 66-68; prof med & assoc dean sch med, 68-73, PROF REPROD MED, SCH MED, UNIV CALIF, SAN DIEGO, 73-, ASSOC DEAN GRAD STUDIES HEALTH SCI, 74- Concurrent Pos: Vis prof, Calif Inst Technol, 54-55; Guggenheim fel, Col France, 58-59; consult, Vet Admin Res Hosp, Chicago, 64-68; mem adv comt to Nat Res Coun, Nat Inst Gen Med Sci. Mem: AAAS; Am Physiol Soc; Soc Exp Biol & Med; Endocrine Soc; Am Inst Biol Sci. Res: Endocrinology. Mailing Add: Off of Assoc Dean Univ of Calif Sch of Med La Jolla CA 90237

LEIN, JOSEPH, b New York, NY, Oct 28, 19; m 47; c 3. MICROBIOLOGY. Educ: City Col New York, BS, 41; Princeton Univ, PhD(biol), 47. Prof Exp: Biologist, Ciba Pharmaceut Prod, NJ, 41-44; from instr to assoc prof zool, Syracuse Univ, 46-52; sr scientist, Bristol Labs, 52, dir microbiol res, 52-65, asst dir labs, 58-65, dir res int, 62-70; pres, 70-74, CHMN, PANLABS, INC, 74- Concurrent Pos: Merck fel, Calif Inst Technol, 47, Gosney fel, 50; Lalor fel, Marine Biol Lab, 48. Mem: AAAS; Am Chem Soc; Am Soc Microbiol; fel Am Acad Microbiol. Res: Enzymology; biochemical genetics; antibiotics. Mailing Add: Panlabs Inc PO Box 81 Fayetteville NY 13066

LEINBACH, HAROLD, b Ft Collins, Colo, Jan 7, 29; m 53; c 2. PHYSICS. Educ: SDak State Univ, BS, 49; Calif Inst Technol, MS, 50; Univ Alaska, PhD(geophys), 62. Prof Exp: Geophysicist, Geophys Inst, Alaska, 50-53 & 56-62; asst prof physics, Univ Iowa, 62-66; PHYSICIST, SPACE ENVIRON LAB, NAT OCEANOG & ATMOSPHERIC ADMIN, 66- Mem: Am Astron Soc; Am Geophys Union; Am Asn Physics Teachers. Res: High latitude ionospheric absorption of cosmic radio noise; solar cosmic rays and their interaction with the ionosphere; solar physics. Mailing Add: Space Environ Lab Nat Oceanic & Atmospheric Admin Boulder CO 80302

LEINBACH, THOMAS RAYMOND, b Reading, Pa, Nov 28, 41. GEOGRAPHY. Educ: Pa State Univ, BS, 63, MS, 66, PhD(geog), 71. Prof Exp: Instr geog, Pa State Col, Shippensburg, 66-68; asst prof, 71-74, ASSOC PROF GEOG, UNIV VT, 75- Concurrent Pos: Southeast Asian res fel, Ford Found, 74-75. Mem: Asn Am Geogrs; Asn Asian Studies. Res: Transport and regional development. Mailing Add: Dept of Geog Univ of Vt Burlington VT 05401

LEINEWEBER, JAMES PETER, b Cicero, Ill, Nov 30, 22; m 48. PHYSICAL CHEMISTRY. Educ: Ohio Univ, BS, 43; NY Univ, MS, 47; Polytech Inst Brooklyn, PhD, 55. Prof Exp: Res chemist, Am Cyanamid Co, 45-56; sr res chemist, 56-57, chief basic chem res, 67-69, mgr corp res & develop, Res & Eng Ctr, 69-72, mgr appl res, 72-73, MGR APPL RES, RES & DEVELOP CTR, JOHNS-MANVILLE CORP, 73- Mem: Am Chem Soc; Am Ceramic Soc. Res: Crystal growth; chemical kinetics; silicate chemistry. Mailing Add: Res & Develop Ctr Johns-Manville Corp Denver CO 80217

LEINFELDER, PLACIDUS JOSEPH, b La Crosse, Wis, Aug 26, 05; m 29, 60; c 3. OPHTHALMOLOGY. Educ: Univ Wis, BA, 26, MD, 29. Prof Exp: Intern, Gen Hosp, Univ Wis, 29-30; intern, Univ Iowa, 30-34; from asst prof to prof, 35-73, EMER PROF OPHTHAL, COL MED, UNIV IOWA, 73- Concurrent Pos: Prof, Peking Union Med Col, 40; ophthalmologist, Iowa Braille & Sight Seeing Sch, 46; med consult, Ophthal Res Found, 54; consult, Atomic Bomb Casualty Comn, 61-, mem adv comt, 63-67; chmn, Comt Classification Visual Function Rehab Codes Inc, New York, 65- Mem: Nat Soc Prev Blindness; Am Ophthal Soc; fel AMA; Asn Res Vision & Ophthal; Am Acad Ophthal & Otolaryngol. Res: Medical and clinical ophthalmology; neuro-ophthalmology; cellular metabolism. Mailing Add: Dept of Ophthal Univ of Iowa Hosp Iowa City IA 52240

LEINHARDT, THEODORE EDWARD, b Gretna, La, Sept 5, 21; m 52; c 3. SOLID STATE PHYSICS. Educ: La Polytech Inst, BS, 50; La State Univ, MS, 52, PhD(physics), 56. Prof Exp: Assoc prof physics, La Polytech Inst, 55-56; res engr, Sperry Rand Res & Develop Lab, NY, 56-57; PROF PHYSICS, VA POLYTECH INST & STATE UNIV, 58- Mem: Am Phys Soc. Res: Low temperature experiments; behavior of superconducting compounds and alloys near critical temperatures and critical magnetic fields. Mailing Add: Dept of Physics Va Polytech Inst & State Univ Blacksburg VA 24061

LEININGER, CHARLES W, b Los Angeles, Calif, Aug 14, 13; m 42; c 5. MATHEMATICS. Educ: Univ Ariz, BS, 36, MS, 37; Univ Tex, PhD(math), 63. Prof Exp: Var eng & acct pos, 37-42; pvt pract cert pub acct, 45-55; from asst prof to assoc prof math, Univ Tex, Arlington, 56-66; assoc prof, Univ Dallas, 66-68; PROF MATH, STATE UNIV NY COL CORTLAND, 68- Mem: Asn Comput Mach; Am Math Soc; Asn Symbolic Logic. Res: Summability; ordered systems; mathematical logic; foundations of mathematics. Mailing Add: Dept of Math State Univ of NY Cortland NY 13045

LEININGER, HAROLD VERNON, b Baton Rouge, La, June 18, 25; m 50; c 1. MICROBIOLOGY. Educ: La State Univ, BS, 48, MS, 51. Prof Exp: Lab technician, Dairy Improv Ctr, La State Univ, 51; food & drug inspector, 51, from bacteriologist to dir biol warfare proj, 52-63, res microbiologist, 63-71, DIR MINNEAPOLIS CTR MICROBIOL ANAL, FOOD & DRUG ADMIN, 71- Concurrent Pos: Proj officer, Test Site, AEC, Nev, 57. Mem: Asn Off Anal Chemists; Am Soc Microbiologists; Inst Food Technologists; Int Asn Milk, Food & Environ Sanit. Res: Microbiological research in food toxicity, decompsition, natural flora, and sanitation. Mailing Add: Food & Drug Admin 240 Hennepin Ave Minneapolis MN 55401

LEININGER, LESTER NORMAN, plant breeding, see 12th edition

LEININGER, MADELEINE MONICA, b Sutton, Nebr, July 13, 25. ANTHROPOLOGY. Educ: Mt St Scholastica Col, BS, 50; Cath Univ Am, MSN, 53; Univ Wash, PhD(anthrop), 66. Hon Degrees: LHD, Scholastic Col, Kans & Benedictine Col, Kans, 75. Prof Exp: Staff nurse, St Anthony's Hosp, Denver, 48 & Col Health Nursing, Mt St Scholastica Col, 48-50; supvr psychiat unit & dir educ progs, Creighton Univ, 50-54; assoc prof psychiat nursing & dir grad progs, Univ Cincinnati, 54-60; res assoc anthrop, Univ Wash, 65-66; prof nursing & anthrop, Univ Colo, Boulder, 66-69; lectr prof anthrop, prof nursing & dean sch nursing, Univ Wash, 69-74; DEAN & PROF NURSING & ADJ PROF ANTHROP, UNIV UTAH, 74- Concurrent Pos: Distinguished vis prof, Sch Nursing, Univ Hawaii, 72; consult psychiat & anthrop nursing, Kans Neurol Inst, Topeka, Boulder Pub Health Dept, Colo, Dept Nursing, Med Ctr, Univ Kans, Vet Admin Hosp, Denver & Ft Lysons, Colo & Sioux Falls, SDak, Sch Nursing, Univ Iowa, Ft Logan Ment Health Ctr, Denver, Intercult Health Prog, Alaska Psychiat Inst, Anchorage, Univ Hawaii & others; curric consult, Undergrad & Grad Curricula Nursing Progs, several schs nursing & int health progs. Mem: NY Acad Sci; fel Am Anthrop Asn; Soc Int Develop; World Fedn Ment Health; fel Soc Appl Anthrop. Res: Ethnological health-illness systems; culture and personality; culture and mental health; culture change; social structure and personality; ethnoadministration; initiated transcultural nursing subfield; transcultural health; nursing education; child care; interdisciplinary health care. Mailing Add: 2789 Thunderbird Dr Salt Lake City UT 84109

LEININGER, PAUL MILLER, b Pa, Oct 29, 11; m 37; c 3. PHYSICAL CHEMISTRY. Educ: Univ Pa, BS, 32, MS, 36, PhD(phys chem), 39. Prof Exp: Asst instr chem, Univ Pa, 34-38; chemist, E I du Pont de Nemours & Co, 39-49; asst prof chem, Lafayette Col, 49-54; assoc prof, 54-58, PROF CHEM, ALBRIGHT COL, 58-, CHMN DEPT, 69- Concurrent Pos: Lectr, Sch Nurses, Reading Hosp, Pa, 54-61; mem teaching staff exp prog teacher educ, Temple Univ, 55-60; consult, George W Bollman Co. Honors & Awards: Lindback Award for Distinguished Teaching, Albright Col, 67. Mem: AAAS; Am Chem Soc; Am Soc Metals; Am Leather Chemists Asn. Res: Case hardening and heat treatment of metals in molten salts; new uses and processes for sodium cyanide and related cyanogen compounds; reaction kinetics in solutions; chemical education; electrolytic dissociation and technology of felting. Mailing Add: 1726 Hampden Blvd Reading PA 19604

LEININGER, ROBERT IRVIN, b Cleveland, Ohio, May 11, 19; m 42; c 4. BIOMEDICAL ENGINEERING, POLYMER CHEMISTRY. Educ: Fenn Col, BChE, 40; Western Reserve, MS, 41, PhD, 43. Prof Exp: Instr chem eng, Fenn Col, 40-43; res chemist, Monsanto Chem Co, 43-48; prin chemist, Battelle Mem Inst, 48-51, asst chief rubber & plastics div, 51-60, chief polymer res sect, 60-65, mgr, 65-69; tech adv, Korean Inst Sci & Technol, Seoul, 69-70; proj dir biomat, Dept Biol, Environ & Chem, 70-73, MEM RES COUN, BATTELLE-COLUMBUS LABS, 74- Concurrent Pos: Mem adv comt, Div Technol Devices, Nat Heart & Lung Inst, 72-73; mem adv comt, Nuclear Powered Artificial Heart Prog, Energy Res & Develop Admin, 75- Honors & Awards: IR-100 Indust Res Award, 74. Mem: AAAS; Am Soc Artificial Internal Organs; Am Heart Asn; Am Chem Soc; NY Acad Sci. Res: Biomaterials; biomedical engineering; polymer chemistry. Mailing Add: Battelle-Columbus Labs 505 King Ave Columbus OH 43201

LEININGER, VERNON EDWARD, radiation physics, see 12th edition

LEINONEN, ELLEN A, b Oct 15, 12; US citizen. ANATOMY, PHYSIOLOGY. Educ: Univ Mich, BS, 56, MS, 62; Ohio State Univ, PhD(anat), 67. Prof Exp: ASST PROF DENT, SCH DENT, UNIV MICH, ANN ARBOR, 65-, ASST PROF ANAT, MED SCH, 71- Mem: Am Asn Anat; NY Acad Sci; fel Am Inst Chemists; AAAS. Res: DNA and RNA metabolism in the primitive cells of acute lymphocytic leukemia; L-asparaginase activity in the cells of acute lymphocytic leukemia. Mailing Add: 3093 Lexington Dr Ann Arbor MI 48105

LEINWEBER, CHARLES LEE, b Bandera, Tex, Oct 29, 22; m 47; c 4. PLANT PHYSIOLOGY. Educ: Agr & Mech Col Tex, BS, 52, MS, 53, PhD, 56. Prof Exp: Asst, Agr & Mech Col Tex, 52-53; asst prof plant physiol, Okla State Univ, 55-59, assoc prof, 59-60; head dept, 60-74, PROF RANGE SCI, TEX A&M UNIV, 60-, DIR ENVIRON QUAL PROG, 72- Concurrent Pos: With Agr Res Serv, USDA, 55- Mem: Am Soc Animal Sci; Soc Range Mgt; Am Inst Biol Sci; Am Soc Agron. Res: Physiology of range plants; range animal nutrition; ecology and biochemistry of range plants. Mailing Add: Environ Qual Prog FE Box 119 Tex A&M Univ College Station TX 77843

LEINWEBER, FRANZ JOSEF, b Berlin, Ger, Jan 18, 31; US citizen; m 60; c 2. DRUG METABOLISM. Educ: Univ Tübingen, Dr rer nat(biol), 56. Prof Exp: Fel biochem, Tex A&M Univ, 57-60 & Johns Hopkins Univ, 60-63; res assoc, Univ Tenn, 63-65; res scientist, McNeil Labs, Inc, 65-69; SR SCIENTIST, WARNER-LAMBERT RES INST, 69- Mem: Am Soc Pharmacol & Exp Therapeut. Res: Photoperiodism and biological clocks; enzymology, intermediary metabolism and metabolic regulation of sulfur amino acid biosynthesis in bacteria and molds; drug metabolism and separation methods. Mailing Add: Dept of Drug Metab Warner-Lambert Res Inst Morris Plains NJ 07950

LEIPNIK, ROY BERGH, b Los Angeles, Calif, May 6, 24; m 44; c 3. MATHEMATICAL ANALYSIS, MATHEMATICAL PHYSICS. Educ: Univ Chicago, SB, 45, SM, 48; Univ Calif, PhD(math), 50. Prof Exp: Asst math statist & econ, Univ Chicago, 45-46; from asst to assoc math, Univ Calif, 46-48; fel, Sch Math, Inst Advan Study, 48-50; asst prof, Univ Wash, 50-57; sr res scientist, Naval Weapons Ctr, Calif, 57-75; PROF MATH & DIR CYBERNETICS INST, UNIV CALIF, SANTA BARBARA, 75- Concurrent Pos: Fulbright res prof, Univ Adelaide, 55, 63 & 68; lectr, Univ Calif, Los Angeles, 59-; prof, Univ Fla, 61-62, 64-65 & 70; consult, Decisional Controls Assocs & Commun Res Labs. Mem: Am Math Soc; Math Asn Am; Inst Math Statist; Inst Elec & Electronics Engrs; Soc Indust & Appl Math. Res: Operator analysis; mathematical physics; control systems; stochastic processes; information theory; plasma physics; solid state physics; microprocessor design; transportation theory; allocation and assignment algorithms; recursive algorithms; differential equations. Mailing Add: Univ of Calif at Santa Barbara Goleta CA 93017

LEIPOLD, HORST WILHELM, b Muehlhausen, Ger, Mar 17, 34; m 60; c 2. ANATOMIC PATHOLOGY. Educ: Univ Giessen, Tierarzt, 60, DVM, 63; Kans State Univ, MS, 67, PhD(path), 68. Prof Exp: Res asst develop path, Univ Giessen, 63-64; asst prof vet path, Univ Sask, 68-71; from asst prof to assoc prof & res pathologist, 71-75, PROF VET PATH, KANS STATE UNIV, 75- Concurrent Pos: Can Med Res Coun grant, 69-70. Mem: Soc Study Reproduction; Am Dairy Sci Asn; Am Genetic Asn; Conf Res Workers Animal Diseases. Res: Nature, cause and importance of congenital defects in domesticated animals. Mailing Add: Dept of Path Kans State Univ Manhattan KS 66502

LEIPPER, DALE F, b Salem, Ohio, Sept 8, 14; m 42; c 4. OCEANOGRAPHY. Educ: Wittenberg Col, BS, 37; Ohio State Univ, MA, 39; Univ Calif, PhD(oceanog), 50. Hon Degrees: DSc, Wittenberg Col, 68. Prof Exp: Weight & balance engr, Consol Aircraft, Calif, 40; sch teacher, Calif, 40-41; oceanogr, Scripps Inst, Univ Calif, 46-49; from assoc prof to prof oceanog & actg head dept, Tex A&M Univ, 49-68; CHMN DEPT OCEANOG, NAVAL POSTGRAD SCH, 68- Concurrent Pos: Head dept, Tex A&M Univ, 50-64, assoc exec dir, Res Found, 53-54, trustee, Univ Corp Atmospheric Res, 59-65; dir, World Data Ctr Oceanog, 57-60; consult, Comt Sci & Astronaut, US House Rep, 60; mem joint panel sea-air interaction, Nat Acad Sci, 60-62. Mem: Am Meteorol Soc; Am Soc Limnol & Oceanog (pres, 58); Am Soc Oceanog (pres, 67); Am Geophys Union. Res: Coastal fog forecasting; analysis of sea temperature variations; use of the bathythermograph; interaction between ocean and atmosphere; physical oceanography; marine meteorology. Mailing Add: Dept of Oceanog Naval Postgrad Sch Code 58 Monterey CA 93940

LEIPUNER, LAWRENCE BERNARD, b Long Beach, NY, May 27, 28; m 48; c 3. HIGH ENERGY PHYSICS, ELEMENTARY PARTICLE PHYSICS. Educ: Univ Pittsburgh, BS, 50; Carnegie Inst Technol, MS, 54, PhD(physics), 62. Prof Exp: SR RES PHYSICIST, BROOKHAVEN NAT LAB, 55- Concurrent Pos: Vis prof, Yale Univ, 67-68; ed, J Comput & Physics, 75- Mem: Fel Am Phys Soc. Res: Lepton and quark experiments. Mailing Add: Brookhaven Nat Lab Upton NY 11973

LEIPZIGER, FREDRIC DOUGLAS, b New York, NY, Aug 26, 29; m 51; c 2. ANALYTICAL CHEMISTRY. Educ: Univ Conn, BA, 51; Univ Mass, MS, 53, PhD(chem), 56. Prof Exp: Res assoc chem, Gen Elec Co, 55-62; head anal chem dept, Sperry Rand Res Ctr, 62-66; MGR ANAL SERV, LEDGEMONT LAB, KENNECOTT COPPER CORP, 66- Mem: Am chem Soc; Soc Appl Spectros. Res: Electron microscopy; mass spectrometry; atomic absorption; automated analyses; process control. Mailing Add: Ledgemont Lab Lexington MA 02173

LEIR, CHARLES MARTIN, organic chemistry, see 12th edition

LEIS, DONALD GEORGE, b Jeannette, Pa, Aug 26, 19; m 45; c 3. ORGANIC CHEMISTRY. Educ: St Vincent Col, BS, 41; Univ Notre Dame, MS, 42, PhD(org chem), 45. Prof Exp: Chemist, Carbide & Carbon Chem Co, 46-55; group leader, 55-63, mgr mkt develop-cellular prod, 63-71, COMMERCIAL MKT MGR, CHEM DIV, UNION CARBIDE CORP, 71- Mem: AAAS; Nat Fire Protection Asn; Soc Plastics Engrs; Am Chem Soc; Soc Plastics Indust. Res: Urethane products; polyethers; polyglycols; alkylene oxides; alkylene oxide derivatives; ethylene oxide; propylene oxide; urethanes; surfactants; lubricants and coatings. Mailing Add: Union Carbide Chem Div 270 Park Ave 33rd Floor New York NY 10017

LEIS, PHILIP EDWARD, b Newburgh, NY, Oct 19, 31. CULTURAL ANTHROPOLOGY. Educ: Antioch Col, BA, 54; Northwestern Univ, PhD(anthrop), 62. Prof Exp: From instr to asst prof social & anthrop, Iowa State Univ, 60-62; from asst prof to assoc prof, 62-72, dir anthrop, 67-70, PROF ANTHROP, BROWN UNIV, 72-, CHMN DEPT, 70- Concurrent Pos: NSF fel, Cameroon, 65-66; NIMH spec res fel, Eng, 69-70. Mem: Am Anthrop Asn; Am Ethol Soc; African Studies Asn; Royal Anthrop Inst. Res: Interethnic relation; comparative religion; political organizations. Mailing Add: Dept of Anthrop Brown Univ Providence RI 02906

LEISE, JOSHUA MELVIN, b Baltimore, Md, Mar 12, 19; m 48; c 2. MEDICAL MICROBIOLOGY. Educ: Univ Md, BS, 40, MS, 43; Yale Univ, PhD(bact), 47. Prof Exp: Bacteriologist, Flavorex, Inc, 40-41; res bacteriologist, Univ Md, 42-45; lab asst, Yale Univ, 45-47; med bacteriologist, Walter Reed Army Med Ctr, 51-52; chief, Bio-

Detection Br, Ft Detrick, Md, 52-59; chief, Biophys-Biochem Br, Army Res Off, 59-60; prog dir for life sci facil, 60-62, head sci facil sect, 62-70, dep dir, Sci Develop Div, 70-71, SR STAFF ASSOC, OFF DEP ASST DIR MATH & PHYS SCI & ENG, NSF, DC, 71- Mem: AAAS; fel Am Acad Microbiol; Am Soc Microbiol; Am Inst Biol Sci. Res: Microbial virulence; methodology; research administration. Mailing Add: 7813 Winterberry Pl Bethesda MD 20034

LEISEMER, RONALD NEWELL, b Toledo, Ohio, Mar 23, 38; m 61; c 2. POLYMER CHEMISTRY. Educ: NCent Col, Ill, BA, 60; Wayne State Univ, MS, 63, PhD(chem), 65. Prof Exp: Res chemist, Plastics Dept, 65-69, tech rep, 69-72, mkt rep, 72-75, TECH SUPVR, E I DU PONT DE NEMOURS & CO, INC, 75- Mem: Am Chem Soc. Res: Polyolefins; films and coatings. Mailing Add: 14 Tenby Chase Dr Newark DE 19711

LEISENRING, KENNETH BAYLIS, b Waldron, Mich, Oct 29, 03. MATHEMATICS. Educ: Univ Mich, AB, 28, MA, 31, PhD(math), 46. Prof Exp: From instr to prof, 46-73, EMER PROF MATH, UNIV MICH, ANN ARBOR, 73- Mem: Assoc Am Math Soc; assoc Math Asn Am. Res: Geometry in the complex domain. Mailing Add: Dept of Math Univ of Mich Ann Arbor MI 48104

LEISERSON, LEE, b Toledo, Ohio, Mar 30, 16; m 43, 71; c 6. ORGANIC CHEMISTRY, PHYSICAL CHEMISTRY. Educ: Antioch Col, BS, 37; Univ NC, MA, 40, PhD(org chem), 41. Prof Exp: Res chemist, Eastman Kodak Co, NY, 41-45; fel, Va Smelting Co, NC, 45-47; chemist, Am Cyanamid Co, NJ, 47-51; chief org chemist, Liggett & Myers Tobacco Co, 51-55; chemist & res adminr, Air Force Off Sci Res, Washington, DC, 55-62; chief chem div, Off Saline Water, Washington, DC, 62-74; CONSULT, 74- Mem: AAAS; Am Chem Soc. Res: Synthetic organic research; surface active agents; organic synthesis; development of natural products; turpentine and tall oil separation processes; reactions in liquid sulfur dioxide; tobacco; water and aqueous solutions. Mailing Add: 9912 Holmhurst Rd Bethesda MD 20034

LEISMAN, GILBERT ARTHUR, b Washington, DC, May 12, 24; m 52. BOTANY. Educ: Univ Wis, BS, 49; Univ Minn, MS, 52, PhD, 55. Prof Exp: Asst plant physiol, Univ Wis, 49-50; asst bot, Univ Minn, 50-55; from asst prof to assoc prof, 55-64; PROF BIOL, EMPORIA KANSAS STATE COL, 64- Concurrent Pos: Mem world orgn paleobot, Int Union Biol Sci. Mem: AAAS; Bot Soc Am; Nat Asn Biol Teachers; Int Asn Plant Taxon. Res: Coal ball plants; morphology of pteridosperm leaves and fructifications; plant succession and soil development of mine dumps. Mailing Add: Dept of Biol Emporia Kans State Col Emporia KS 66801

LEISNER, ROBERT STANLEY, b Washington, DC, Aug 9, 27. BOTANY. Educ: Va Polytech Inst, BS, 53, MS, 55; PhD(bot), 58. Prof Exp: Asst to exec dir, Am Inst Biol Sci, 58-59, dir ed, 59-61, dir film serv, 61-63, co-ed-in-chief BioSci, 63-66, exec ed, 66-68, assoc ed, 68-71, asst dir publ div, 71-72; ASST EXEC ED, FEDN PROC, FEDN AM SOCIETIES EXP BIOL, 73- Mem: Am Inst Biol Sci; Bot Soc Am; Sigma Xi. Res: Plant anatomy and physiology; morphology; anatomy and phylogeny of the Empetraceae. Mailing Add: 4600 S Four Mile Run Dr Apt 1019 Arlington VA 22204

LEISS, JAMES ELROY, b Youngstown, Ohio, June 2, 24; m 45; c 4. PHYSICS. Educ: Case Inst Technol, BS, 49; Univ Ill, MS, 51, PhD(physics), 54. Prof Exp: Lab asst, Gen Elec Co, 48-49; asst physics, Univ Ill, 49-54; DIR, CTR RADIATION RES, NAT BUR STANDARDS, 54- Mem: Am Phys Soc. Res: Nuclear physics, especially photonuclear reactions and photomeson reactions; design of particle accelerators. Mailing Add: Ctr for Radiation Res Nat Bur of Standards Washington DC 20234

LEISTER, HARRY M, b Quakertown, Pa, Mar 3, 41; m 62; c 3. PHYSICAL CHEMISTRY. Educ: Pa State Univ, BS, 63; Drexel Univ, MS, 65; Temple Univ, PhD(phys chem), 70. Prof Exp: Res chemist, E I du Pont de Nemours & Co, 69-70; chemist, 71-73, GROUP LEADER, RORER-AMCHEM INC, 73- Res: Organic coatings; inorganic coatings. Mailing Add: Amchem Prod Inc Ambler PA 19002

LEISTER, NORMAN ANDREW, organic chemistry, see 12th edition

LEISURE, ROBERT GLENN, b Cromwell, Ky, Jan 29, 38; m 62. METAL PHYSICS. Educ: Western Ky Univ, BS, 60; Wash Univ, PhD(physics), 67. Prof Exp: Res scientist, Boeing Sci Res Lab, 67-70; asst prof, 70-73, ASSOC PROF PHYSICS, COLO STATE UNIV, 73- Mem: AAAS; Am Phys Soc. Res: Ultrasonics; magnetic resonance; propagation of ultrasound in metals; acoustic excitation of nuclear magnetic resonance. Mailing Add: Dept of Physics Colo State Univ Ft Collins CO 80523

LEITCH, LEONARD CHRISTIE, b Ottawa, Ont, Aug 22, 14; m 42; c 2. ORGANIC CHEMISTRY. Educ: Univ Ottawa, BSc, 35; Laval Univ, DSc, 49. Prof Exp: Res chemist, Mallinckrodt Chem Works, 37-46; res chemist, 46-70, SR RES CHEMIST, DIV CHEM, NAT RES COUN CAN, 70- Mem: Chem Inst Can; The Chem Soc; Chem Soc France. Res: Synthetic drugs and plant hormones; synthesis of organic compounds with stable isotopes; reaction mechanisms. Mailing Add: Div of Chem Nat Res Coun 100 Sussex Dr Ottawa ON Can

LEITCH, ROBERT EDGAR, JR, b Annapolis, Md, Nov 12, 40; m 63; c 1. ANALYTICAL CHEMISTRY. Educ: Washington Col, BS, 62; Rutgers Univ, PhD(anal chem), 67. Prof Exp: Res asst anal chem, Rutgers Univ, 62-67; RES CHEMIST, INDUST & BIOCHEM DEPT, E I DU PONT DE NEMOURS & CO, INC, WILMINGTON, 67- Res: Chromatographic separation of racemic mixtures; methods of separation and analysis of organic chemicals, especially agricultural chemicals and pesticide residues. Mailing Add: 230 Jupiter Dr North Star Newark DE 19711

LEITER, EDWARD HENRY, b Columbus, Ga, Apr 17, 42; m 64. CELL BIOLOGY. Educ: Princeton Univ, BS, 64; Emory Univ, MS, 66, PhD(biol), 68. Prof Exp: NIH trainee, Univ Tex, Austin, 68-71; asst prof biol, Brooklyn Col, 71-74; assoc staff scientist, 74-75, STAFF SCIENTIST, JACKSON LAB, 75- Concurrent Pos: Nat Inst Arthritis & Metab Dis res grant, 74. Mem: Tissue Cult Asn; Am Soc Cell Biol. Res: Function of normal and diabetic pancreatic endocrine cells in vitro; genetic, viral, and environmental parameters producing pancreatic pathologies in the mouse. Mailing Add: Jackson Lab Bar Harbor ME 04609

LEITER, ELLIOT, b Brooklyn, NY, May 24, 33; m 63; c 3. MEDICINE, UROLOGY. Educ: Columbia Univ, AB, 54; NY Univ, MD, 57. Prof Exp: Asst urol, Johns Hopkins Hosp, 58-59; asst urol, NY Univ, 60-63; instr, Columbia Univ, 66-67; asst prof, 66-69, ASSOC PROF UROL, MT SINA SCH MED, 69- Concurrent Pos: Fel, NY Univ, 60-61, USPHS trainee, 61-62, fel hypertensive renal group, 63; asst vis surgeon, Bellevue Hosp, New York, 63 & Greenpoint Hosp, Brooklyn, 64; asst attend urologist, Mt Sinai Hosp, New York, 63-69, assoc attend urologist, 69- Mem: Asn Acad Surg; Am Urol Asn; Am Col Surg; Soc Univ Urol; Soc Pediat Urol. Res: Renal disease; pediatric urology; kidney transplantation; hypertension. Mailing Add: 19 E 98th St New York NY 10029

LEITER, HOWARD ALLEN, b Mt Gilead, Ohio, Feb 16, 18; m 52; c 3. PHYSICS. Educ: Miami Univ, AB, 40; Univ Ill, AM, 42, PhD(physics), 49. Prof Exp: Asst physics, Univ Ill, 40-42; mem staff, Radiation Lab, Mass Inst Technol, 42-45; asst physics, Univ Ill, 45-48, res assoc, 48-49; physicist, Res Labs, Westinghouse Elec Corp, 49-58; sr engr, Labs, Int Tel & Tel Corp, 58-69; ASSOC PROF PHYSICS, TRI-STATE COL, 70- Concurrent Pos: Guest lectr, Off-campus Grad Prog, Purdue Univ, 59-60. Mem: Am Phys Soc. Res: Charged particle scattering; microwave components; interaction of electromagnetic radiations with matter; infrared detectors and systems; image tubes; cryogenic equipment; field emission microscopy; satellite instrumentation. Mailing Add: 2703 Capitol Ave Ft Wayne IN 46806

LEITER, JOSEPH, b New York, NY, May 14, 15; m 39; c 2. BIOCHEMISTRY. Educ: Brooklyn Col, BS, 34; Georgetown Univ, PhD(biochem), 49. Prof Exp: Jr chemist org & fibrous mat, Nat Bur Standards, 35-38; carcinogenesis, Nat Cancer Inst, 38-40, asst chemist, 40-42, assoc chemist chemotherapy, 46-47, chemist, 47-49, from sr chemist to sr scientist & chief biochem sect, Lab Chem Pharmacol, 49-55, scientist dir & asst chief lab activities, Cancer Chemother Nat Serv Ctr, 55-63, chief, Ctr, 63-65, ASSOC DIR LIBR OPERS, NAT LIBR MED, 65- Mem: AAAS; Soc Pharmacol & Exp Therapeut; Am Chem Soc; Am Asn Cancer Res. Res: Carcinogenesis, production of tumors with chemical agents, air dust; chemotherapy of cancer; drug metabolism, effect of chemical agents on enzymes in normal and malignant tissues. Mailing Add: Nat Libr of Med NIH Bethesda MD 20014

LEITER, LOUIS, b Falticeni, Rumania, Oct 24, 98; nat US; m 25; c 2. INTERNAL MEDICINE. Educ: Univ Chicago, SB, 19, SM, 20, PhD(path), 24; Rush Med Col, MD, 22. Prof Exp: Instr path, Univ Chicago, 23-24; asst res physician, Rockefeller Inst Hosp, 24-26; from asst prof to assoc prof med, Univ Chicago, 26-41, clin prof, 42; clin prof, Col Physicians & Surgeons, Columbia Univ, 42-63; prof, 63-71, EMER PROF MED, ALBERT EINSTEIN COL MED, 71-; chief med div, 42-66, CONSULT, MONTEFIORE HOSP, 67- Concurrent Pos: In charge sect renal & vascular dis, Dept Med, Billings Hosp, 26-41; attend physician, Michael Reese Hosp, Chicago, 41-42; assoc ed, Archives Internal Med, Am Med Asn, 51-61; vchmn med adv bd, Hebrew Univ-Hadassah Med Sch, 52-70. Mem: Am Soc Clin Invest; Asn Am Physicians; Am Heart Asn; NY Acad Med; Am Soc Nephrol. Res: Experimental and clinical studies in Bright's disease and congestive heart failure. Mailing Add: Montefiore Hosp & Med Ctr 111 E 210th St New York NY 10467

LEITH, CARLTON JAMES, b Madison, Wis, Sept 24, 19; m 41; c 2. GEOLOGY. Educ: Univ Wis, BA, 40, MA, 41; Univ Calif, PhD(geol), 47. Prof Exp: Asst geol, Univ Calif, 41-42; from jr mineral economist to asst mineral economist, Mineral Prod & Econ Div, US Bur Mines, 42-43; geologist, Standard Oil Co, Tex, 46; asst geol, Univ Calif, 46-47; from instr to asst prof geol, Univ Ind, 47-49; chief petrog unit, US Engrs Testing Lab, 49-51; geologist, Standard Oil Co, Calif, 51-60 & Holmes & Narver, Inc, 60-61; assoc prof, 61-65, PROF GEOL ENG, NC STATE UNIV, 65-, HEAD DEPT GEOSCI, 67- Mem: AAAS; Geol Soc Am; Soc Econ Paleont & Mineral; Am Asn Petrol Geol; Am Geophys Union. Res: Engineering geology; sedimentary petrology; areal geology; gravity and magnetics. Mailing Add: Dept of Geosci NC State Univ Box 5966 Raleigh NC 27607

LEITH, CECIL ELDON, JR, b Boston, Mass, Jan 31, 23; m 42; c 3. MATHEMATICAL PHYSICS. Educ: Univ Calif, AB, 43, PhD(math), 57. Prof Exp: Physicist, Lawrence Radiation Lab, Univ Calif, 46-68, SR SCIENTIST, NAT CTR ATMOSPHERIC RES, NAT SCI FOUND, 68- Concurrent Pos: US Comt Global Atmospheric Res Prog, Nat Acad Sci, 70-72 & 74-76, comt atmospheric sci, 71-73; mem Int Comn Dynamic Meteorol, 72-; mem Adv Bd, Off Earth Sci, Nat Res Coun, 75-76. Honors & Awards: Meisinger Award, Am Meteorol Soc, 67. Mem: AAAS; fel Am Phys Soc; fel Am Meteorol Soc. Res: Numerical simulation of the flow of the atmosphere and ocean; geophysical fluid dynamics; statistical hydrodynamics; turbulence and convection. Mailing Add: Nat Ctr for Atmospheric Res PO Box 3000 Boulder CO 80303

LEITH, DAVID W G S, b Glasgow, Scotland, Sept 5, 37; m 62; c 3. HIGH ENERGY PHYSICS. Educ: Univ Glasgow, BSc, 59, PhD(natural philos), 62. Prof Exp: Glasgow Univ res fel physics, Europ Orgn Nuclear Res, Geneva, Switz, 62-63, staff physicist, 63-66; assoc prof, 66-70, PROF PHYSICS, LINEAR ACCELERATOR CTR, STANFORD UNIV, 70- Mem: Am Phys Soc; Brit Inst Physics & Phys Soc. Res: Strong interaction physics with emphasis on scattering experiments and investigations of resonance properties, their classification and the associated phenomenological analysis. Mailing Add: Stanford Linear Accelerator Ctr PO Box 4349 Stanford CA 94305

LEITH, EMMETT NORMAN, b Detroit, Mich, Mar 12, 27; m 56; c 1. OPTICS. Educ: Wayne State Univ, BS, 49, MS, 52. Prof Exp: Lab instr physics, Wayne State Univ, 51; asst eng res inst, 52-56, res assoc, 56-59, assoc res engr, 59-65, assoc prof, 65-68, PROF ELEC ENG, INST SCI & TECH, UNIV MICH, ANN ARBOR, 68- Honors & Awards: Gordon Memorial Award, Soc Photo-Optical Instrumentation Eng, 65, Liebmann Award, Inst Elec & Electronics Eng, 67; Daedalion Award, 68; Stuart Ballantine Medal, Franklin Inst, 69. Mem: Fel Optical Soc Am; fel Inst Elec & Electronics Eng. Res: Wavefront reconstruction; electronic physics; electromagnetics; radar; resonant-cavity design; data processing; optical system design; coherent optics; interferometry; holography. Mailing Add: Dept Elec & Comput Eng Univ of Mich Ann Arbor MI 48107

LEITH, JOHN DOUGLAS, JR, b Grand Forks, NDak, Apr 20, 31; m 57; c 2. PATHOLOGY, CELL BIOLOGY. Educ: Lehigh Univ, BA, 52; Univ Pa, MD, 56; Univ Wis, PhD(cytol), 64; Am Bd Path, cer anat & clin path, 74, cert radioisotopic path, 75. Prof Exp: Intern, Med Ctr, Univ Calif, San Francisco, 56-57; asst zool, Univ Wis, 59-60, NSF fel, 60-63; Nat Cancer Inst spec fel, 63-64; asst prof anat & cell biol, Med Sch, Univ Pittsburgh, 64-67; from asst prof to assoc prof biol, Univ Wis-Oshkosh, 67-71; resident path, Peter Bent Brigham Hosp, Boston, 71-74; ASSOC PATHOLOGIST, BROCKTON HOSP, 75- Concurrent Pos: Am Cancer Soc Inst res grant, 65-66; Health Res Serv Found res grant, 66-67; NSF grant, 68-70; Univ Wis res grants, 68-71. Mem: Col Am Path; Am Soc Clin Path; Am Soc Cell Biol; assoc Radiation Res Soc. Res: Structure and function of chromosomes. Mailing Add: Dept of Path Brockton MA 02402

LEITH, JOHN THAYER, physiology, radiation biology, see 12th edition

LEITH, THOMAS HENRY, b Toronto, Ont, Sept 11, 27; m 53; c 3. PHILOSOPHY OF SCIENCE, GEOPHYSICS. Educ: Univ Toronto, BA, 49, MA, 50; Boston Univ, PhD(philos sci), 62. Prof Exp: Instr geol & physics, Gordon Col, 52-55, asst prof, 55-58, assoc prof, 58-61; asst prof philos, Univ RI, 62-64; ASSOC PROF GEN EDUC & NATURAL SCI, YORK UNIV, 64-, COORDR, 65- Concurrent Pos: Chmn div natural sci, Gordon Col, 55-61; consult, R Woike Assocs, 61-62; coordr honors prog, Univ RI, 63-64; assoc dean, Atkinson Col, & actg dir div natural sci, 66-69, dir, 69-71. Mem: AAAS; Am Sci Affil; Philos Sci Asn; Hist Sci Soc; Sigma Xi. Res: Tectonic theories and theory formation in geology and other physical sciences; techniques and principles of science education for the nonscientist in college; relationships of theorizing in science and in religion. Mailing Add: Atkinson Col York Univ Downsview ON Can

LEITHEISER, ROBERT H, organic chemistry, polymer chemistry, see 12th edition

LEITMAN, MARSHALL J, b Yonkers, NY, Jan 16, 41. APPLIED MATHEMATICS, CONTINUUM PHYSICS. Educ: Rensselaer Polytech Inst, BS, 62; Brown Univ, PhD(appl math), 65. Prof Exp: Res assoc appl math, Brown Univ, 65-66; asst prof, 66-71, ASSOC PROF MATH, CASE WESTERN RESERVE UNIV, 71- Concurrent Pos: Vis asst prof, Cath Univ Louvain, 70-71. Mem: Soc Natural Philos. Res: Mechanics; viscoelasticity. Mailing Add: Dept of Math Case Western Reserve Univ Cleveland OH 44106

LEITNAKER, JAMES MACLEAN, physical chemistry, see 12th edition

LEITNER, ALFRED, b Vienna, Austria, Nov 3, 21; nat US; m 48; c 3. MATHEMATICAL PHYSICS, SCIENCE EDUCATION. Educ: Univ Buffalo, BA, 44; Yale Univ, MS, 45, PhD(physics), 48. Prof Exp: Asst, Yale Univ, 43-47; res scientist, Courant Inst Math Sci, NY Univ, 47-51; from asst prof to prof physics, Mich State Univ, 51-67; PROF PHYSICS, RENSSELAER POLYTECH INST, 67- Concurrent Pos: Guggenheim fel & vis prof, Aachen Tech Univ, 58-59; res assoc, Harvard Univ, 65-66, consult proj physics, 66-69. Mem: Fel Am Phys Soc; Am Asn Physics Teachers; Am Asn Univ Prof. Res: Special functions; boundary value problems; waves, history of science; educational films. Mailing Add: Dept of Physics Rensselaer Polytechnic Inst Troy NY 12181

LEITNER, FELIX, b Oradea, Rumania, Oct 8, 21; nat US; m 62; c 2. MICROBIOLOGY, CHEMOTHERAPY. Educ: Univ Geneva, MA, 46, MS, 50, PhD(chem), 54. Prof Exp: Res fel pharmacol, Col Med, NY Univ, 54-56, from res asst to res assoc, 56-58; res assoc microbiol, Michael Reese Hosp, 58-65, asst dir biochem, 65-68; ASST DIR MICROBIOL RES, BRISTOL LABS, 68- Mem: AAAS; Am Soc Microbiol; NY Acad Sci. Res: Cellular regulatory mechanisms; antibiotics; beta-lactamase: biosynthesis and properties. Mailing Add: Microbiol Res Dept Bristol Labs Thompson Rd Syracuse NY 13201

LEITNER, NATHAN BARR, b Brooklyn, NY, Apr 14, 23; m 47; c 3. BIOCHEMISTRY. Educ: City Col New York, BS, 47; Polytech Inst Brooklyn, MS, 51. Prof Exp: Asst, City Col New York, 47-51; sr res fel, Polytech Inst Brooklyn, 51-52; res chemist, 52-59, chief chemist, 59-62, V PRES & TECH DIR, AM ADHESIVES PROD CO, 62- Concurrent Pos: Mem, Book Mfr Inst, Inc. Mem: AAAS; Am Chem Soc; Tech Asn Pulp & Paper Indust. Res: Adhesives. Mailing Add: Am Adhesives Prod Co 1855 E 63rd St Los Angeles CA 90006

LEITNER, PHILIP, b Peking, China, June 16, 36; US citizen; m 60; c 2. VERTEBRATE ZOOLOGY. Educ: St Mary's Col, Calif, BS, 58; Univ Calif, Los Angeles, MA, 60, PhD(zool), 61. Prof Exp: Jr res zoologist, Univ Calif, Los Angeles, 61-62; from instr to asst prof, 62-68, ASSOC PROF BIOL, ST MARY'S COL, CALIF, 68-, CHMN DEPT, 70- Concurrent Pos: NIH res grant, 63-65, NSF res grants, 65-70. Mem: AAAS; Am Soc Zoologists; Soc Study Evolution; Am Soc Mammalogists. Res: Environmental physiology of mammals, especially physiological responses to temperature and photoperiod. Mailing Add: Dept of Biol St Mary's Col Moraga CA 94575

LEITZ, FRED JOHN, JR, b Portland, Ore, Feb 2, 21; m 45; c 3. PHYSICAL CHEMISTRY. Educ: Reed Col, BA, 40; Univ Calif, PhD(phys chem), 43. Prof Exp: Instr chem, Univ Calif, 43-44; sr res chemist, Monsanto Chem Co, Ohio, 44-46; sr chemist, Oak Ridge Nat Lab, Tenn, 46-48; chemist, Radiochem & Reactor Metall Res, Hanford Works, Gen Elec Co, Wash, 48-56; nuclear engr, Atomic Power Develop Assocs, Mich, 56-58; develop proj engr, Atomic Power Equip Dept, Gen Elec Co, 58-64, mgr fast reactor core eng & test, Adv Prod Oper, 64-66, mgr steam reactor technol, 66-68; consult to dir, Battelle Northwest Lab, 69-7O; SR STAFF SCIENTIST, WESTINGHOUSE HANFORD CO, 70- Mem: Am Chem Soc; Am Nuclear Soc. Res: Heavy element and fission product chemistry; nuclear fuel cycle development; fast and steam cooled reactor design and technology. Mailing Add: Westinghouse Hanford Co PO Box 1970 Richland WA 99352

LEITZ, FREDERICK HENRY, b Hastings, Mich, Nov 20, 28; m 70. PHARMACOLOGY. Educ: Kalamazoo Col, BA, 52; Univ Calif, Los Angeles, PhD(chem), 62. Prof Exp: Res grant, Inst Org Chem, Royal Inst Technol, Sweden, 62-63; res scientist, Lamont Geol Observ, 63-65; staff fel, Lab Chem Pharmacol, Nat Heart Inst, 65-70; PRIN SCIENTIST, SCHERING CORP, 70- Mem: Am Chem Soc. Res: Role of biogenic amines in function of sympathetic nervous systems, including synthesis and metabolism; nature of drug action on nerve endings. Mailing Add: Schering Corp 86 Orange St Bloomfield NJ 07003

LEITZEL, JAMES ROBERT C, b Shenandoah, Pa, May 27, 36; m 65; c 2. MATHEMATICS. Educ: Pa State Univ, BA, 58, MA, 60; Ind Univ, PhD(math), 65. Prof Exp: Asst prof math, Bloomsburg State Col, 59-63; asst prof, 65-69, ASSOC PROF MATH, OHIO STATE UNIV, 69- Mem: Am Math Soc; Math Asn Am. Res: Algebra, especially class field theory and algebraic function fields. Mailing Add: Dept of Math Ohio State Univ Columbus OH 43210

LEITZEL, JOAN PHILLIPS, b Valparaiso, Ind, July 2, 36; m 65; c 2. MATHEMATICS. Educ: Hanover Col, AB, 58; Brown Univ, AM, 61; Univ Ind, PhD(algebra), 65. Prof Exp: Instr math, Oberlin Col, 61-62; asst prof, 65-70, ASSOC PROF MATH, OHIO STATE UNIV, 70- Concurrent Pos: Vchmn math dept, Ohio State Univ, 73- Mem: Am Math Soc; Math Asn Am. Res: Field theoretical proofs for cohomological results in class field theory. Mailing Add: Dept of Math Ohio State Univ Columbus OH 43210

LEITZMANN, CLAUS, b Dahlenburg, Ger, Feb 6, 33; US citizen; m 57; c 4. BIOCHEMISTRY, NUTRITION. Educ: Capital Univ, BS, 62; Univ Minn, MS, 64, PhD(biochem), 67. Prof Exp: Nat Inst Gen Med Sci res asst molecular biol inst, Univ Calif, Los Angeles, 67-69; vis prof biochem, Mahidol Univ, Thailand, 69-71; chief labs, Anemia & Malnutrit Res Ctr, Thailand, 71-74; ASSOC, INST NUTRIT, UNIV GIESSEN, 74- Concurrent Pos: Mem, Trop Inst, Univ Giessen, 74- Mem: AAAS; Am Soc Microbiol. Res: Malnutrition; interaction of nutrition and infection; adaptations to changes in food intake; hunger and satiety. Mailing Add: Inst for Nutrit Univ Giessen 6300 Giessen Germany

LEIVE, LORETTA, b New York, NY, Apr 12, 36; c 1. BIOCHEMISTRY, MICROBIOLOGY. Educ: Barnard Col, Columbia Univ, AB, 56; Harvard Univ, AM, 61, PhD, 63. Prof Exp: RES BIOLOGIST, LAB BIOCHEM PHARMACOL, NAT INST ARTHRITIS & METAB DIS, NIH, 63- Mem: Am Soc Biol Chemists; Am Soc Microbiol. Res: Bacterial permeability; membrane and cell wall structure and function. Mailing Add: Lab of Biochem Pharmacol Nat Inst Arthritis & Metab Dis Bethesda MD 20014

LEIVO, WILLIAM JOHN, b New Castle, Pa, Sept 11, 15; m 39; c 2. PHYSICS. Educ: Carnegie Inst Technol, BS, 39, MS, 45, DSc(physics), 48. Prof Exp: Supt bldg construct, Matthew Leivo & Sons, Inc, Pa, 33-35 & 39-42; from instr to asst prof physics, Carnegie Inst Technol, 42-55; PROF PHYSICS, OKLA STATE UNIV, 55- Mem: Fel Am Phys Soc; Am Asn Physics Teachers. Res: Color centers in crystals; radiation effects in solids; optics; solid state physics; semiconducting diamond; ESR studies of blood cell membranes. Mailing Add: Dept of Physics Okla State Univ Stillwater OK 74074

LEJA, STANISLAW, b Grodzisko, Poland, Jan 3, 12; nat US; m 39, 64; c 5. MATHEMATICS. Educ: Jan Kazimierz Univ, Poland, MA, 38; Cornell Univ, PhD(math), 58. Prof Exp: Instr math, Jan Kazimierz Univ, 38-40; teacher high schs, Palestine & Eng, 45-51; from asst to instr, Cornell Univ, 53-57; asst prof, 57-67, PROF MATH, WESTERN MICH UNIV, 67- Mem: Am Math Soc; Math Asn Am. Res: Real variable; Fourier analysis. Mailing Add: 931 Oakland Dr Kalamazoo MI 49001

LEJEUNE, ANDRE JOSEPH, b Can, Aug 24, 18; m 39; c 2. AGRONOMY, GENETICS. Educ: Univ Man, BS, 38, MS, 45. Prof Exp: From lectr to asst prof agron, 39-47; asst prof & agronomist, NDak Agr Col, 47-50; agronomist, Malting Barley Improv Asn, 50-56, dir, 56-69; COLLABR, DEPT AGRON & SOILS, WASH STATE UNIV, 72- Mem: AAAS; Am Soc Agron. Res: Crops breeding; barley breeding; especially malting and brewing quality. Mailing Add: Dept of Agron & Soils Wash State Univ Pullman WA 99163

LEKLEM, JAMES ERLING, b Rhinelander, Wis, Aug 1, 41; m 67; c 1. NUTRITION. Educ: Univ Wis, BS, 64, MS, 66, PhD(nutrit), 73. Prof Exp: Proj assoc clin oncol, Univ Wis, 66-71, res assoc, 73-75; ASST PROF NUTRIT, ORE STATE UNIV, 75- Mem: Sigma Xi. Res: Vitamin B6; metabolism of tryptophan; nutrient relationship to cancer etiology. Mailing Add: Dept of Foods & Nutrit Ore State Univ Corvallis OR 97331

LELACHEUR, ROBERT MURRAY, b Ottawa, Ont, Oct 12, 20; US citizen; m 46; c 4. PHYSICS. Educ: Mt Allison Univ, BSc, 42; Dalhousie Univ, MSc, 47; Univ Va, PhD(physics), 49. Prof Exp: Physicist, Nat Res Coun Can, 49-53; mem tech staff, Bell Labs, NJ, 53-58; asst supt eng, 58-62, dir mat & chem processes res & develop, NY, 62-66, MGR DEVELOP & MFG ENG, WESTERN ELEC CO, INC, READING, 66- Mem: Am Phys Soc; Inst Elec & Electronics Engrs. Res: Materials properties and processing; chemical processing; semiconductor device engineering. Mailing Add: Western Elec Co Inc PO Box 241 Reading PA 19603

LELAND, FRANCES ELBRIDGE, b Chicago, Ill, Apr 22, 32. PHYSICAL CHEMISTRY. Educ: Swarthmore Col, BA, 54; Northwestern Univ, PhD(phys chem), 59. Prof Exp: Instr chem, Brooklyn Col, 59-61; from asst prof to assoc prof, 62-73, PROF CHEM, MacMURRAY COL, 73- Mem: Am Chem Soc. Res: Molecular quantum mechanics. Mailing Add: Dept of Chem MacMurray Col Jacksonville IL 62650

LELAND, STANLEY EDWARD, JR, b Chicago, Ill, Aug 1, 26; m 50; c 3. PARASITOLOGY. Educ: Univ Ill, BS, 49, MS, 50; Mich State Univ, PhD(parasitol), 53. Prof Exp: Asst parasitol, Mich State Univ, 50-53; from assoc parasitologist to parasitologist, Univ Ky, 53-60, prof animal path, 60-63; assoc parasitologist, Univ Fla, 63-67; actg head dept infectious dis, 70-72, PROF PARASITOL, KANS STATE UNIV, 67-, ASSOC DIR AGR EXP STA, 75- Concurrent Pos: Coop agent animal dis & parasite res div, USDA, 53-59; consult, Eli Lilly Co, 62-66. Honors & Awards: Col Vet Med Res Award, Kans State Univ, 71. Mem: Am Soc Parasitol. Res: Electrophoresis; drug testing; pathology; physiology; biochemistry; in vitro cultivation; immunology as related to parasitology. Mailing Add: Agr Exp Sta Kans State Univ Manhattan KS 66506

LELAND, THOMAS MIKELL, b Greenwood, SC, June 23, 44; m 66; c 4. PHYSIOLOGY, ENDOCRINOLOGY. Educ: Presby Col, BS, 66; Clemson Univ, MS, 68, PhD(physiol), 70; Med Univ SC, MD, 75. Prof Exp: Physiologist, Toxicol Div, US Army Environ Hyg Agency, Edgewood Arsenal, 70-72; RESIDENT OPHTHAL, MED UNIV SC, 74- Concurrent Pos: Consult, US Army Environ Hyg Agency; NDEA fel med, Med Univ Sc, 72- Mem: Soc Study Reprod; Am Soc Animal Sci. Res: Female reproductive physiology; ovarian fine structure and endocrinology; prenatal toxicology; experimental teratology; toxicology hazard evaluation. Mailing Add: Dept of Ophthal Med Univ SC 80 Barre St Charleston SC 29401

LELAND, WALLACE THOMPSON, b Minn, Jan 21, 22; m 43; c 4. LASERS. Educ: Univ Minn, BEE, 43, PhD(physics), 50. Prof Exp: Head, Instrument Develop Dept, Carbide & Carbon Chem Corp, 46-47; MEM STAFF NUCLEAR RES, LOS ALAMOS SCI LAB, 50- Mem: Am Phys Soc. Res: Mass spectroscopy and nuclear reactions; high energy lasers. Mailing Add: Los Alamos Sci Lab Los Alamos NM 87545

LELEK, ANDREW STANISLAUS, b Radom, Poland, Nov 3, 34. TOPOLOGY. Educ: Univ Wroclaw, MS, 55, PhD(math), 59; Polish Acad Sci, Habilitated Dozent, 64. Prof Exp: Asst prof math, Inst Math, Polish Acad Sci, 59-64, assoc prof, 64-65; vis assoc prof, La State Univ, Baton Rouge, 65-66; assoc prof, Polish Acad Sci, 66-70; vis prof, Univ Houston, 70-71, prof math, 71-74; PROF MATH, WAYNE STATE UNIV, 74- Concurrent Pos: Ed, Houston J Math, 75- Honors & Awards: Mazurkiewicz Award, Polish Math Soc, 64. Mem: Am Math Soc. Res: Topological properties of metric spaces; dimension theory; theory of curves; theory of convexity. Mailing Add: Dept of Math Wayne State Univ Detroit MI 48202

LELES, BYRON K, physical chemistry, see 12th edition

LE LEVIER, ROBERT ERNEST, b Los Angeles, Calif, Nov 7, 23; m 45; c 3. THEORETICAL PHYSICS. Educ: Univ Calif, Los Angeles, PhD(physics), 51. Prof Exp: Mem staff, Lawrence Radiation Lab, Univ Calif, 51-57 & Rand Corp, 57-71; MEM STAFF R&D ASSOCS, 71- Res: Ionospheric physics; nuclear physics; geophysics. Mailing Add: R&D Associates PO Box 9695 Marina del Rey CA 90291

LELL, EBERHARD, b Linz, Austria, Oct 3, 27; m 62; c 3. PHOTOCHEMISTRY. Educ: Graz Tech Univ, dipl, 52; Tech Univ Stuttgart, Ger, PhD, 56. Prof Exp: Instr inorg chem, Stuttgart Tech Univ, 56-57; res chemist, Bausch & Lomb, Inc, NY, 57-69; MEM STAFF, ITEK CORP, 69- Mem: Am Ceramic Soc; Optical Soc Am; Austrian Chem Soc. Res: Molecular spectroscopy; radiation and glass chemistry. Mailing Add: Itek Corp Ten Maguire Rd Lexington MA 02173

LELLINGER, DAVID BRUCE, b Chicago, Ill, Jan 24, 37; m 63; c 2. TAXONOMIC BOTANY. Educ: Univ Ill, AB, 58; Univ Mich, MS, 60, PhD(bot), 65. Prof Exp: ASSOC CUR FERNS, US NAT HERBARIUM, SMITHSONIAN INST, 63- Concurrent Pos: Ed-in-chief, Am Fern Soc, 66-; Nat Geog Soc & Smithsonian Res Found explor & res grants, 71, 74; counr, Las Cruces Trop Bot Garden, Costa Rica, 75- Mem: Am Soc Plant Taxon; Int Asn Plant Taxon; Brit Pterid Soc. Res: Taxonomy of ferns and fern allies, especially those of the New World tropics. Mailing Add: US Nat Herbarium NHB 166 Smithsonian Inst Washington DC 20560

LELONG, MICHEL GEORGE, b Casablanca, Morocco, Mar 20, 32; US citizen; m 59; c 3. PLANT TAXONOMY. Educ: Univ Algiers, baccalaureat, 50; Northwestern State Col, La, BS, 59, MS, 60; Iowa State Univ, PhD(syst bot), 65. Prof Exp: ASSOC PROF BIOL, UNIV S ALA, 65- Mem: Am Soc Plant Taxon; Int Asn Plant Taxon. Res: Systematics of Panicum subgenus Dichanthelium; floristics of the Mobile Bay region. Mailing Add: Dept of Biol Sci Univ of SAla Mobile AL 36608

LE MAHIEU, RONALD, organic chemistry, see 12th edition

LEMAIRE, HENRY, b New York, NY, Jan 24, 21; m 53; c 2. ORGANIC CHEMISTRY. Educ: Mass Inst Technol, BS, 42; Calif Inst Technol, PhD(org chem), 50. Prof Exp: Fel org synthesis, Calif Inst Technol, 50; mem staff, Wyandotte Chem Corp, 50-52; fel, Univ Wis, 52-53; mem staff, Lever Bros Co, 53-60, sr res assoc, 60-67; asst prof, 67-69, ASSOC PROF CHEM, UNIV NEW HAVEN, 69- Mem: Am Chem Soc; Sigma Xi. Res: Organic synthesis. Mailing Add: Dept of Chem Univ of New Haven West Haven CT 06516

LEMAIRE, MINNIE ETHEL, b Taunton, Mass, July 15, 08. GEOGRAPHY. Educ: Wheaton Col, Mass, BA, 30; Clark Univ, MA, 32, PhD(geog), 35. Prof Exp: Instr geog, Wis State Teachers Col, LaCrosse, 35-43; chmn dept, Pa State Teachers Col, E Stroudsburg, 43-47; prof geog, 47-73, EMER PROF, MT HOLYOKE COL, 73- Mem: Nat Coun Geog Educ; Asn Am Geogr; Am Geog Soc; Soc Woman Geogr; Latin Am Geog Soc. Res: Geographical implications of tropical agriculture. Mailing Add: 190 Brattle St Holden MA 01520

LEMAIRE, NORMAND ARTHUR, physical chemistry, see 12th edition

LEMAISTRE, CHARLES AUBREY, b Lockhart, Ala, Feb 10, 24; m 52; c 4. INTERNAL MEDICINE, EPIDEMIOLOGY. Educ: Univ Ala, BA, 44; Cornell Univ, MD, 46. Prof Exp: From instr to asst prof internal med, Med Col, Cornell Univ, 51-54; assoc prof, Sch Med, Emory Univ, 54-59, prof prev med & chmn dept, 57-59; prof, Univ Tex Southwestern Med Sch, Dallas, 59-66, assoc dean, 65-66; vchancellor health affairs, Univ Tex, Austin, 66-68; from exec vchancellor to chancellor-elect, 68-70, CHANCELLOR, UNIV TEX SYST, 71- Concurrent Pos: Mem human ecol study sect, NIH, 62-65; mem, Surgeon Gen Adv Comt Smoking & Health, 63-64; mem, Gov Comt Eradication Tuberc, 64-65; mem comt res tobacco & health, AMA Educ & Res Found, 64-; mem, Nat Citizens Comn Int Coop, 65-; mem Surgeon Gen emergency health preparedness adv comt, Dept Health, Educ & Welfare, 67, consult, Div Physician Manpower, 67-68; mem, President's Comn White House Fel, 71; mem, Comn Non-Traditional Study, 71-; mem joint task force continuing competence pharm, Am Pharmaceut Asn-Am Asn Col Pharm, 73-74; mem bd comnr, Nat Comn Accrediting, 73-76; trustee, Biol Humanics Found, Dallas, 73-; mem, Nat Coun Educ Res, 73-75, consult, 75; chmn subcomt diversity & pluralism, Nat Coun Educ Res, 73-75; mem, United Negro Col Fund Develop Coun, 74-; mem, Nat Adv Coun, Inst Serv Educ, 74- Mem: Am Thoracic Soc (vpres, 64-65). Res: Chest diseases. Mailing Add: Univ of Tex Syst 601 Colorado St Austin TX 78701

LE MAISTRE, JOHN WESLEY, b Lockhart, Ala, Sept 27, 09; m 47; c 4. CHEMISTRY. Educ: Univ Mich, BSE, 30, MS, 31; Duke Univ, PhD(chem), 34. Prof Exp: Res chemist, Dow Chem Co, Mich, 31-32; group leader, Oxford Paper Co, Maine, 34-36; res dir, Swann & Co, Ala, 36-42; chief org chemist, Indust Res Inst, Chattanooga, 46-48; res dir, Chattanooga Med Co, 47-50; chem res dept, Atlas Chem Industs, 53-70, dir biomed res labs, ICI Am, Wilmington, Del, 70-75; RETIRED. Concurrent Pos: Lectr, Univ Chattanooga, 47-50. Mem: AAAS; Am Chem Soc. Res: Sugar derivatives; industrial chemicals; drug synthesis and metabolism. Mailing Add: 6 Glenrock Dr Claymont DE 19703

LEMAL, DAVID M, b Plainfield, NJ, Feb 20, 34; m 63; c 4. ORGANIC CHEMISTRY. Educ: Amherst Col, AB, 55; Harvard Univ, PhD, 59. Prof Exp: From instr to asst prof chem, Univ Wis, 58-65; assoc prof, 65-69, PROF CHEM, DARTMOUTH COL, 69- Concurrent Pos: A P Sloan Found res fel, 68-70; trustee, Gordon Res Conf, 73- Mem: AAAS; Am Chem Soc. Res: Unusual species, stable and short-lived, in organic chemistry; organic reaction mechanisms; organic photochemistry; organofluorine chemistry. Mailing Add: Dept of Chem Dartmouth Col Hanover NH 03755

LEMAN, ALLEN DUANE, b Peoria, Ill, Jan 15, 44. VETERINARY MEDICINE. Educ: Univ Ill, BS, 66, DVM, 68, PhD(physiol), 74. Prof Exp: Instr, Univ Ill, 69-75; ASSOC PROF VET MED, UNIV MINN, 75- Concurrent Pos: Dir, Nat Pork Producers Coun Res Comt, 72-; prog chmn, Int Pig Vet Cong, 73-; ed, Dis Swine, Iowa State Press, 74-; dir, Soc Study of Breeding Soundness & Nat Swine Improv Fedn, 75- Mem: Am Asn Swine Practitions (pres, 75); Am Vet Med Asn; Soc Study Reproduction; Am Soc Animal Sci; Soc Study Breeding Soundness. Res: Causes of swine infertility; causes of lameness in boars; maximum economic returns from pork production. Mailing Add: Col of Vet Med Univ of Minn St Paul MN 55108

LEMANSKI, LARRY FREDRICK, b Madison, Wis, June 5, 43; m 66; c 1. CELL BIOLOGY, DEVELOPMENTAL BIOLOGY. Educ: Univ Wis-Platteville, BS, 66; Ariz State Univ, MS, 68, PhD(zool), 71. Prof Exp: NIH fel cell biol, Univ Pa, 71-73, Muscular Dystrophy Asn fel, 73-75; ASST PROF ANAT, UNIV CALIF MED CTR, SAN FRANCISCO, 75- Concurrent Pos: Estab investr, Am Heart Asn-Univ Calif, 76-81. Mem: Am Soc Cell Biol; Biophys Soc; Sigma Xi; Am Asn Anatomists; Soc Develop Biol. Res: Vertebrate heart development with emphasis on factors controlling embryonic heart formation; modern techniques in cellular and developmental biology as well as electron microscopy are used in the research. Mailing Add: Dept of Anat Univ of Calif San Francisco CA 94143

LEMASTER, EDWIN WILLIAM, b Perryton, Tex, Apr 27, 40; m 64; c 1. SOLID STATE PHYSICS. Educ: West Tex State Univ, BS, 62; Tech Tech Univ, MS, 66; Univ Tex, PhD(physics), 70. Prof Exp: Asst prof physics, Gen Motors Inst, 64-66; ASST PROF PHYSICS, PAN AM UNIV, 70-, CHMN PHYS SCI DEPT, 73- Mem: Am Phys Soc; Am Asn Physics Teachers. Res: Metalammonia solution properties; amorphous semiconductors; remote sensing of vegetative canopies; mathematical modeling. Mailing Add: Dept of Phys Sci Pan Am Univ Edinburg TX 78539

LEMASURIER, WESLEY ERNEST, b Wash, DC, May 3, 34; m 63; c 3. GEOLOGY. Educ: Union Col, BS, 56; Univ Colo, MS, 62; Stanford Univ, PhD(geol), 65. Prof Exp: Asst prof geol, Cornell Univ, 64-68; ASSOC PROF GEOL, DIV NAT & PHYS SCI, UNIV COLO, DENVER, 68- Mem: AAAS; Geol Soc Am; Am Geophys Union; Int Asn Volcanol & Chem Earth's Interior. Res: Subglacial volcanism; petrology and tectonic relationships of volcanism in Antarctica. Mailing Add: Div Nat & Phys Sci Univ of Colo Denver CO 80202

LEMAY, CHARLOTTE ZIHLMAN, b Ft Worth, Tex, June 30, 19; m 44; c 3. SOLID STATE PHYSICS. Educ: Tex Christian Univ, AB, 40; Mt Holyoke Col, MA, 41; La State Univ, PhD(physics), 50. Prof Exp: Res physicist, Monsanto Chem Co, 43-44; instr physics, Mt Holyoke Col, 45-46; instr physics, La State Univ, 46-48, asst, 48-50; engr, Tex Instruments, Inc, 52-53, 55-57; res physicist, Stanford Res Inst, 53-54; engr, Westinghouse Elec Corp, 58-60 & Int Bus Mach Corp, 60-63; from asst prof sci to assoc prof, 63-69, PROF PHYSICS & CHMN DEPT, WESTERN CONN STATE COL, 69- Mem: Am Phys Soc; Am Asn Physics Teachers; Inst Elec & Electronics Eng; Sigma Xi. Res: Dielectric liquids; transistors. Mailing Add: Rte 2 Chestnut Ridge Rd Mt Kisco NY 10549

LEMAY, HAROLD E, JR, b Tacoma, Wash, May 28, 40; m 64; c 2. INORGANIC CHEMISTRY. Educ: Pac Lutheran, BS, 62; Univ Ill, MS, 64, PhD(inorg chem), 66. Prof Exp: Asst prof, 66-70, ASSOC PROF CHEM, UNIV NEV, RENO, 70-, VCHMN DEPT, 74- Mem: Am Chem Soc. Res: Preparation and characterization of coordination compounds; reactions of coordination compounds in the solid phase; kinetic studies of ligand exchange reactions. Mailing Add: Dept of Chem Univ of Nev Reno NV 89507

LEMAY, JEAN-PAUL, b St Hyacinthe, Que, Apr 7, 23; m 52; c 2. ANIMAL PHYSIOLOGY, ANIMAL BREEDING. Educ: Classical Col St Hyacinthe, BA, 45; Univ Montreal, BSA, 49; Univ Mass, MSc, 51. Prof Exp: Mem artificial insemination unit, Classical Col St Hyacinthe, 48-49; from instr to prof animal sci, Res Sta La Pocatiere, Que, 51-62; PROF ANIMAL SCI, LAVAL UNIV, 62- Mem: Am Soc Animal Sci; Can Soc Animal Prod. Res: Early weaning of sheep; histophysiology of sperm atogenesis and ovogenesis in sheep; sterility in dairy cattle. Mailing Add: Dept of Animal Sci Laval Univ Fac of Agr Ste Foy PQ Can

LEMAY, YVAN, b Montreal, Que, Mar 24, 31; m 55; c 2. BOTANY, ZOOLOGY. Educ: Univ Montreal, BA, 52, BSc, 55, MSc, 56. Prof Exp: Botanist, 56-67, electron microscopist, 67-74, SR MICROSCOPIST, CIP RES LTD, 74- Mem: Assoc Tech Asn Pulp & Paper Indust; assoc Brit Inst Wood Sci. Res: Tree physiology; wood anatomy; microscopy; pulp and paper products fiber analysis; wood identification. Mailing Add: CIP Res Ltd 179 Main St W Hawkesbury ON Can

LEMBACH, KENNETH JAMES, b Rochester, NY, June 16, 39; m 65; c 2. BIOCHEMISTRY. Educ: Mass Inst Technol, BS, 61; Univ Pa, PhD(biochem), 66. Prof Exp: USPHS fel, Mass Inst Technol, 66-68, res assoc biochem, 68-69; ASST PROF BIOCHEM, SCH MED, VANDERBILT UNIV, 69- Concurrent Pos: US Nat Cancer Inst res grants, 71-74 & 75-79. Mem: AAAS; NY Acad Sci; Am Soc Cell Biol; Sigma Xi. Res: Growth regulation in normal and malignant cells; metabolic control mechanisms. Mailing Add: Dept of Biochem Vanderbilt Univ Nashville TN 37232

LEMBECK, WILLIAM JACOBS, b Kansas City, Mo, Aug 29, 28; m 60; c 1. MICROBIOLOGY, ACADEMIC ADMINISTRATION. Educ: La State Univ, BS, 50, MS, 56, PhD(bact), 62. Prof Exp: Supvry bacteriologist, US Army Chem Corps, Pine Bluff Arsenal, Ark, 57-59; from asst prof to assoc prof biol, Ark Agr & Mech Col, 61-62; asst prof bact, McNeese State Col, 62-65, assoc prof microbiol, 65-66; asst prof bot, Baton Rouge, 66-68, assoc prof biol & head div sci, 68-75, PROF BIOL, LA STATE UNIV, EUNICE, 75- Mem: AAAS; Am Inst Biol Sci; Sigma Xi. Res: Effects of herbicides on normal soil microflora; biological catalysis of herbicides in soil; effects of herbicides on cellulose decomposition by Sporocytophaga myxococcoides. Mailing Add: Div of Sci La State Univ PO Box 1129 Eunice LA 70535

LEMBERG, HOWARD LEE, b Queens, NY, July 29, 49; m 70. CHEMICAL PHYSICS. Educ: Columbia Univ, BA, 69; Univ Chicago, PhD(chem physics), 73. Prof Exp: Res chem physics, Bell Tel Lab, 73-75; ASST PROF CHEM, UNIV NC, 75- Mem: Am Chem Soc; Am Phys Soc; AAAS. Res: Statistical mechanics of liquids; electronic structure of surfaces; fluctuations and instabilities in chemical systems. Mailing Add: Dept of Chem Univ of NC Chapel Hill NC 27514

LEMBERG, LOUIS, b Chicago, Ill, Dec 27, 16; m 39. PHYSIOLOGY. Educ: Univ Ill, BS, 38, MD, 40; Am Bd Internal Med, dipl, 50; Am Bd Cardiovasc Dis, dipl, 55. Prof Exp: Intern, Mt Sinai Hosp, Chicago, Ill, 40-41, res, 45-48; assoc prof med & physiol, Sch Med, Univ Fla, 56-69; PROF CLIN CARDIOL, SCH MED, UNIV MIAMI, 69- Concurrent Pos: Dir cardiol, Dade County Hosp, 55-57; attend specialist, Vet Admin Hosp, 55-64; chief staff, Nat Children's Cardiac Hosp, Miami, 55-66; attend cardiologist, Mercy & Cedars of Lebanon Hosp; chief div electrophysiol, 56-74, dir coronary care unit, Jackson Mem Hosp, 68-74; mem coun clin cardiol, Am Heart Asn. Honors & Awards: Savage Award, 60; Luis Guerrero Mem Award, Philippines. Mem: Hon mem Philippine Med Asn; fel Am Col Physicians; Am Col Chest Physicians; Am Col Cardiol; NY Acad Sci. Res: Cardiology. Mailing Add: Sect of Cardiol Dept of Med Univ of Miami Sch of Med Miami FL 33152

LEMBERG, SEYMOUR, organic chemistry, see 12th edition

LEMBERGER, AUGUST PAUL, b Milwaukee, Wis, Jan 25, 26; m 47; c 7. PHARMACEUTICS. Educ: Univ Wis, BS, 48, PhD(pharm), 52. Prof Exp: Sr chemist pharmaceut res, Merck & Co, Inc, 52-53; from instr to prof pharm, Univ Wis-Madison, 53-69, coordr exten serv, 65-69; PROF PHARM & DEAN COL, UNIV ILL MED CTR, 69- Concurrent Pos: Mem & secy, Wis Pharm Internship Comn, 65-69; consult, Dept Health, Educ & Welfare, 72- Honors & Awards: Kiekhofer Award, Univ Wis, 57; Distinguished Serv Award, Wis Pharm Asn, 69. Mem: Am Pharmaceut Asn; fel Acad Pharm Sci. Mailing Add: Col of Pharm PO Box 6998 Univ of Ill Med Ctr Chicago IL 60680

LEMBERGER, LOUIS, b Monticello, NY, May 8, 37; m 59; c 2. CLINICAL PHARMACOLOGY. Educ: Long Island Univ, BS, 60; Albert Einstein Col Med, PhD(pharmacol), 64, MD, 68. Prof Exp: Fel pharmacol, Albert Einstein Col Med, 64-68; med intern, Metropolitan Hosp Ctr-NY Med Col, 68-69; pharmacol & toxicol res assoc clin pharmacol, Lab Clin Sci, NIMH, 69-71; clin pharmacologist, 71-75, CHIEF CLIN PHARMACOL, LILLY LAB CLIN RES, LILLY RES LABS, 75- Concurrent Pos: Dir, Clin Pharmacol Training Prog, Sch Med, Ind Univ, 72-75, asst prof pharmacol & med, 72-73, assoc prof, 73-, assoc prof, Grad Fac, 75-; adj prof clin pharmacol, Ohio State Univ, 75- Mem: Am Soc Pharmacol Exp Ther; Am Soc Clin Pharmacol Ther; Am Col Neuro Psychopharm; Am Col Physicians. Res: Drug metabolism and drug-drug interactions; synthesis and metabolism of biogenic amines; biochemical mechanisms of drug action; psychopharmacology; marihuana & cannabinoids. Mailing Add: Lilly Lab for Clin Res Lilly Res Labs Indianapolis IN 46206

LEMBO, NICHOLAS J, b Boston, Mass, Aug 18, 29; m 56; c 4. ANALYTICAL CHEMISTRY. Educ: Boston Col, BS, 51; Teachers Col City Boston, EdM, 52; Northeastern Univ, MS, 62. Prof Exp: Instr pub schs, Mass, 52-56; instr phys sci, 56-59, asst prof chem, 59-65, ASSOC PROF CHEM, BOSTON STATE COL, 65- Concurrent Pos: Lectr, Lincoln Col, Northeastern Univ, 53-65. Res: Improvement of science teaching in the elementary schools. Mailing Add: Dept of Chem Boston State Col Boston MA 02115

LE MEHAUTE, BERNARD J, b Brieuc, France, Mar 29, 27; m 53; c 2. HYDRODYNAMICS. Educ: Univ Rennes, Baccalaureat, 47; Univ Toulouse, lic es sc, 51; Univ Grenoble, Dr es Sc(hydrodyn), 57. Prof Exp: Res engr, Neyrpic-Sogreah, France, 53-57; assoc prof hydrodyn, Polytech Sch, Montreal, 57-59; res prof, Queen's Univ, Ont, 59-61; mem tech staff, Nat Eng Sci Co, 61-62; mem sr staff, 62-64, assoc dir hydrodyn, 64-66; VPRES, TETRA TECH, INC, 66- Mem: Am Geophys Union; Am Soc Civil Eng; Marine Technol Soc; Int Asn Hydraul Res. Res: Hydrodynamics and hydraulic and coastal engineering, ranging from theoretical fluid mechanics applied to physical oceanography to the design of engineering structures for water power, harbors of offshore drilling. Mailing Add: Tetra Tech Inc 630 N Rosemead Blvd Pasadena CA 91107

LEMIER, EMANUEL H, b Breckenridge, Minn, Aug 2, 22; m 46; c 3. FISHERIES. Educ: Univ Wash, Seattle, BS, 48. Prof Exp: Res biologist, 48-51, aquatic biologist, 51-52, fishery biologist, 52-62, SR BIOLOGIST, RES DIV, WASH STATE DEPT FISHERIES, 62- Concurrent Pos: App to tech adv comt for biol effects, Secy of Interior; mem res staff, Pac Marine Fisheries Comn; aquatic plant & insect control comt & water supply & water pollution control comt, Pac Northwest River Basin Comn; tech adv comt for Klickitat River, Yakima Indian Nation; agr pesticide adv bd, State of Wash. Honors & Awards: Cert pesticides, US Pub Health Serv, 64; cert, Wash State Univ, 67; cert fisheries scientist, Am Fisheries Soc, 71. Mem: AAAS; Am Inst Fishery Res Biol; Am Soc Ichthyol & Herpet; Am Soc Limnol & Oceanog; Am Fisheries Soc. Res: Fishery research, management, planning, coordination and supervision; water quality, pollution control, pesticides and offshore seismic; Pacific salmon, fresh-water rearing, stream appraisal and improvements, and stock identification, distribution and contribution; scrapfish control; shellfish studies. Mailing Add: Washington State Dept Fisheries Rm 115 Gen Admin Bldg Olympia WA 98504

LEMIEUX, GUY, b Montreal, Que, Oct 31, 27; m 56; c 1. INTERNAL MEDICINE. Educ: Univ Montreal, BA, 48, MD, 53. Prof Exp: Intern med, Montreal Hosps, 52-53; resident, Hotel-Dieu Hosp, 53-54, fel, 54-56; fel, Sch Med, Tufts Univ, 56-59; from asst prof to assoc prof, 59-71, PROF MED, FAC MED, UNIV MONTREAL, 71-, ASSOC MEM DEPT EXP MED, 70- Concurrent Pos: Asst physician, Hotel-Dieu Hosp, 59-63, dir, Renal Clin, 59-, physician, 63-, chief, Nephrol-Metab Div & Nephrol Lab, 70-; Markle scholar med sci, 59-64. Mem: Am Fedn Clin Res; NY Acad Sci; Can Soc Clin Invest; Am Soc Nephrol; Can Soc Nephrol; Int Soc Nephrol. Res: Kidney and electrolyte metabolism. Mailing Add: Fac of Med Univ of Montreal Montreal PQ Can

LEMIEUX, JEAN-MARIE, b Bienville, Que, Nov 27, 11. SURGERY. Educ: Laval Univ, MD, 37. Prof Exp: Assoc prof, 45-64, PROF SURG, FAC MED, LAVAL UNIV, 64-, VDEAN FAC MED, 67-, CHIEF SURGEON, LAVAL HOSP, 66- Res: Thoracic and general surgery. Mailing Add: Fac of Med Laval Univ Quebec PQ Can

LEMIEUX, PAUL E, b Levis, Que, Mar 16, 24; m 54; c 2. PHYSICS. Educ: Laval Univ, BA, 46, BASc, 50. Prof Exp: Spectroscopist, Bur Mines, Que, 50-52; from spectroscopist to chief spectroscopist, Aluminum Labs, Ltd, 52-74, CHIEF SPECTROSCOPIST, ALUMINUM CO CAN, 74- Mem: Optical Soc Am; Soc Appl Spectros; Can Asn Appl Spectros (secy, 64-65, vpres, 65-66, pres, 66-67); Chem Inst Can. Res: Emission spectrochemical analysis of aluminum, its alloys and various associated products. Mailing Add: Aluminum Co of Can Tech Info Ctr PO Box 250 Arvida PQ Can

LEMIEUX, RAYMOND URGEL, b Lac la Biche, Alta, June 16, 20; m 48; c 6. ORGANIC CHEMISTRY. Educ: Univ Alta, BSc, 43; McGill Univ, PhD(org chem), 46. Hon Degrees: DSc, Univ NB, 67 & Laval Univ, 70. Prof Exp: Res assoc carbohydrate chem, Ohio State Univ, 46-47; asst prof org chem, Univ Sask, 47-49; res officer chem natural prod, Prairie Regional Lab, Nat Res Coun, 49-54; prof chem, chmn dept & vdean fac pure & appl sci, Ottawa Univ, Can, 54-61; PROF ORG CHEM, UNIV ALTA, 61- Concurrent Pos: Merck lectr, 56; Folkers lectr, 58; Karl Pfister lectr, 68; Purves lectr, 70; pres & dir res, Raylo Chem Ltd, Alta. Honors & Awards: Medal, Chem Inst Can, 64; C S Hudson Award, Am Chem Soc, 66; Medal of Serv, Order of Can, 68. Mem: AAAS; The Chem Soc; Am Chem Soc; fel Chem Inst Can; fel Royal Soc Can. Res: Stereochemistry; conformational analysis; carbohydrate chemistry; chemistry of natural products; especially antibiotics; mechanisms of organic replacement and oxidation reactions. Mailing Add: Dept of Chem Univ of Alta Edmonton AB Can

LEMIRE, RONALD JOHN, b Portland, Ore, Apr 20, 33; m 55; c 5. TERATOLOGY, PEDIATRICS. Educ: Univ Wash, MD, 62. Prof Exp: Intern, King County Hosp, Seattle, Wash, 62-63; NIH fel teratology & embryol, Univ Wash, 63-65, asst resident pediat, 65-67; chief resident, Children's Orthop Hosp & Med Ctr, 67-68; asst prof, 68-72, ASSOC PROF PEDIAT, UNIV WASH, 72- Mem: Soc Pediat Res; Teratology Soc. Res: Neuroembryology; neuroteratology. Mailing Add: Dept of Pediat Univ of Wash Seattle WA 98195

LEMISH, JOHN, b Rome, NY, July 4, 21; m 46; c 5. ECONOMIC GEOLOGY, GEOCHEMISTRY. Educ: Univ Mich, BS, 47, MS(geol), 48, PhD, 55. Prof Exp: Geologist, US Geol Surv, 48, 49-51; instr geol, Univ Mich, 53-55, from asst prof to assoc prof, 55-62, PROF GEOL, IOWA STATE UNIV, 62- Concurrent Pos: Mem comts, Hwy Res Bd, Nat Acad Sci-Nat Res Coun, 58-; chmn, State Mining Bd, 64-; trans comt, Am Geol Inst; mem adv comt, Iowa Coal Proj, 74-; mem tech adv comt, Nat Gas Surv, Fed Power Comn, 75- Mem: AAAS; fel Geol Soc Am; Geochem Soc; Am Asn Petrol Geol; Am Inst Mining, Metall & Petrol Eng. Res: Weathering studies of concrete; behavior of carbonate aggregates in Portland cement concrete; aggregate-cement reactions; physical and chemical phenomena related to ore deposition; structural geology; trace elements in Pennsylvania Shales. Mailing Add: Dept of Earth Sci Iowa State Univ Ames IA 50010

LEMKAU, PAUL VICTOR, b Springfield, Ill, July 1, 09; m 34; c 5. PSYCHIATRY, PUBLIC HEALTH. Educ: Baldwin-Wallace Col, AB, 31; Johns Hopkins Univ, MD, 35; Am Bd Psychiat & Neurol, dipl; Am Bd Prev Med, dipl. Hon Degrees: ScD, Baldwin-Wallace Col, 54; DPH, Dickinson Col, 58. Prof Exp: Instr psychiat, Sch Med, 37-40, from asst prof to prof pub health admin, 41-61, chmn dept ment hyg, 61-75, PROF MENT HYG, SCH HYG & PUB HEALTH, JOHNS HOPKINS UNIV, 61- Concurrent Pos: Asst resident, Johns Hopkins Hosp, 37-40; Commonwealth Fund fel, Phipps Psychiat Clin, Johns Hopkins Univ, 37-40; chief div ment hyg, State Dept Health, Md, 47-54; dir, Community Ment Health Bd, New York, 55-57; mem expert comt ment health, WHO, 63-; mem nat ment health adv coun, Dept Health, Educ & Welfare, 64-67; mem exec bd, World Fedn Ment Health, 66-70. Mem: Fel Am Psychiat Asn (vpres, 69-70); Am Pub Health Asn. Res: Administration of mental hygiene programs; epidemiology of mental hygiene problems. Mailing Add: Johns Hopkins Univ Sch of Hyg & Pub Health Baltimore MD 21205

LEMKE, CARLTON EDWARD, b Buffalo, NY, Oct 11, 20; m 55; c 2. MATHEMATICS. Educ: Univ Buffalo, BA, 49; Carnegie Inst Technol, MA, 51, PhD(math), 53. Prof Exp: Instr math, Carnegie Inst Technol, 52-54; res assoc anal, Knolls Atomic Power Lab, Gen Elec Co, NY, 54-55; engr Radio Corp Am, NJ, 55-56; from asst prof to prof math, 56-67, FORD FOUND PROF MATH, RENSSELAER POLYTECH INST, 67- Mem: Am Math Soc; Soc Indust & Appl Math; Opers Res Soc Am; Economet Soc; Math Asn Am. Res: Algebra; mathematical programming; probability and statistics; operations research. Mailing Add: Dept of Math Rensselaer Polytech Inst Troy NY 12180

LEMKE, PAUL ARENZ, b New Orleans, La, July 14, 37; m; c 1. GENETICS, MICROBIOLOGY. Educ: Tulane Univ, BS, 60; Univ Toronto, MA, 62; Harvard Univ, PhD(biol), 66. Prof Exp: Instr biol, Tulane Univ, 62-63; sr microbiologist, Eli Lilly & Co, Ind, 66-72; ASSOC PROF BIOL SCI, CARNEGIE-MELLON UNIV, 72- Concurrent Pos: Instr, Franklin Col, 67; sr fel, Carnegie-Mellon Inst Res, 72- Mem: Am Soc Microbiol; Genetics Soc Am; Mycol Soc Am; Bot Soc Am; Soc Indust Microbiol (treas, 76-78). Res: Genetics and viruses of fungi; biochemistry of double-stranded RNA; biosynthesis of antibiotics; cytoplasmic inheritance in fungi; fluorescent staining of fungal nuclei and chromosomes. Mailing Add: Dept of Biol Sci Carnegie-Mellon Univ Pittsburgh PA 15213

LEMKE, THOMAS FRANKLIN, b Tremont, Pa, July 28, 42; m 65. ORGANIC CHEMISTRY, CORROSION. Educ: Wake Forest Univ, BS, 64; Marshall Univ, MS, 66; Lehigh Univ, PhD(chem), 68. Prof Exp: Biochemist, Med Res Labs, Edgewood Arsenal, 68-70; asst prof chem, Marshall Univ, 70-72; TECH SERV SPECIALIST, HUNTINGTON ALLOYS, INC, 72- Mem: AAAS; Am Chem Soc; Nat Asn Corrosion Engrs; Sigma Xi; Am Nuclear Soc. Res: Synthesis of heterocyclic compounds of medicinal interest; corrosion of nickel base alloys. Mailing Add: 6233 Highland Dr Huntington WV 25705

LEMKE, THOMAS LEE, b Waukesha, Wis, June 1, 40; m 63; c 3. MEDICINAL CHEMISTRY. Educ: Univ Wis, BS, 62; Univ Kans, PhD(med chem), 66. Prof Exp: Res assoc org chem & patent liaison, Upjohn Co, Mich, 66-70; asst prof, 70-75, ASSOC PROF PHARM, UNIV HOUSTON, 75- Mem: Am Chem Soc; Am Pharmaceut Asn. Res: Heterocyclic chemistry; anticancer agents; Favorskii rearrangement; drugs for mental disease. Mailing Add: Col of Pharm Univ of Houston Houston TX 77004

LEMMER, KENNETH ELLERY, b Spooner, Wis, Nov 11, 06; m 35; c 2. SURGERY. Educ: Univ Wis, BS, 28, MD, 30. Prof Exp: Intern, Med Col Va, 30-31; resident surg, Wis Gen Hosp, 31-34; from instr to assoc prof, 34-54, PROF SURG, MED SCH, UNIV WIS-MADISON, 54- Concurrent Pos: Civilian consult, US Dept Army. Mem: Am Asn Surg Trauma; Am Col Surg; Aerospace Med Asn; Asn Mil Surg US; NY Acad Sci. Res: Clinical aspects of pancreatic disease. Mailing Add: Dept of Surg Univ of Wis Madison WI 53706

LEMMERMAN, KARL EDWARD, b Willoughby, Ohio, May 30, 23; m 46; c 3. PHYSICAL CHEMISTRY. Educ: Oberlin Col, AB, 47; Cornell Univ, PhD(chem), 51. Prof Exp: Asst gen chem, Cornell Univ, 47-50; RES CHEMIST, PROCTER & GAMBLE CO, 51- Mem: Am Chem Soc. Res: Kinetics of gas-phase photochemical reactions; complex inorganic electrolytes; surfactant solutions; colloids. Mailing Add: 1952 Compton Rd Cincinnati OH 45231

LEMMING, JOHN FREDERICK, b Dayton, Ohio, Oct 31, 43. NUCLEAR PHYSICS. Educ: Univ Dayton, BS, 66; Ohio Univ, MS, 68, PhD(physics), 72. Prof Exp: Fel physics, Ohio Univ, 72-74; SR PHYSICIST NUCLEAR SPECTROS, MONSANTO RES CORP, MOUND LAB, 74- Concurrent Pos: Nuclear infor res assoc, Nat Acad Sci, Nat Res Coun Comt Nuclear Sci, 72-74. Mem: Am Inst Physics. Res: Nuclear safeguards. Mailing Add: Mound Lab PO Box 32 Miamisburg OH 45342

LEMMON, DONALD H, b Sugar Grove, Pa, Oct 19, 35; m 56; c 1. SPECTROSCOPY. Educ: Univ Pittsburgh, PhD(chem), 66. Prof Exp: Fel, State Univ NY, Stony Brook, 66-67; SR ENGR, WESTINGHOUSE RES CTR, 67- Mem: Am Chem Soc. Res: Infrared, Raman and nuclear magnetic resonance spectroscopy; mass spectrometry. Mailing Add: 304 Sixth St Oakmont PA 15139

LEMMON, DWIGHT MOULTON, b Fair Oaks, Calif, Jan 23, 12; m 66. GEOLOGY. Educ: Stanford Univ, AB, 33, AM, 35, PhD(econ geol), 37. Prof Exp: Instr geol, Univ Nev, 38; instr econ geol, Stanford Univ, 39; from asst geologist to assoc geologist, 42-43, geologist, US Geol Surv, 43-73; RETIRED. Mem: AAAS; fel Geol Soc Am; fel Mineral Soc Am; fel Soc Econ Geologists; Am Inst Mining, Metall & Petrol Engrs. Res: Economic geology. Mailing Add: c/o US Geol Surv 80 Amador Ave Atherton CA 94025

LEMMON, RICHARD MILLINGTON, b Sacramento, Calif, Nov 24, 19; m 49; c 3. RADIATION CHEMISTRY. Educ: Stanford Univ, AB, 41; Calif Inst Technol, MS, 43; Univ Calif, PhD(chem), 49. Prof Exp: Res chemist, Calif Inst Technol, 43-45; fel, Med Sch, Univ Calif, 49-50; USPHS fel, Fed Inst Tech, Switz, 50-51; RES CHEMIST, LAWRENCE BERKELEY LAB, UNIV CALIF, 51- Concurrent Pos: Guggenheim fel, Helsinki, 65; assoc dir, Lab Chem Biodynamics, Univ Calif, 57- Mem: AAAS; Am Chem Soc. Res: Radiochemistry; hot-atom chemistry; radiation decomposition of organic compounds; chemical evolution. Mailing Add: Lawrence Berkeley Lab Univ of Calif Berkeley CA 94720

LEMOINE, ALBERT N, JR, b Nelson, Nebr, Apr 13, 18; m 40; c 3. OPHTHALMOLOGY. Educ: Univ Kans, AB, 39; Wash Univ, MD, 43; Am Bd Ophthal, dipl. Prof Exp: Teaching & res fel ophthal, Harvard Med Sch, 45-46; PROF OPHTHAL & CHMN DEPT, SCH MED, UNIV KANS MED CTR, KANSAS CITY, 50- Mem: AAAS; Asn Res Vision & Ophthal; fel Am Col Surg; Am Acad Ophthal & Otolaryngol. Res: Surgical and applied anatomy of the eye and orbit. Mailing Add: Dept of Ophthal Univ of Kans Med Ctr Kansas City KS 66103

LEMON, EDGAR ROTHWELL, b Buffalo, NY, Aug 22, 21; m 44; c 3. SOIL SCIENCE. Educ: Cornell Univ, BS, 43, MS, 49; Mich State Univ, PhD(soil physics), 54. Prof Exp: Prof agron, Tex A&M Univ, 51-56; PROF AGRON, CORNELL UNIV, 56-; SOIL SCIENTIST, AGR RES SERV, USDA, 51- Concurrent Pos: Guggenheim & Fulbright fel, Australia, 62-63; USSR-US exchange scientist, 69; Dept Sci & Indust Res fel, NZ, 70-71. Mem: Fel AAAS; fel Am Soc Agron; Soil Sci Soc Am; Am Meteorol Soc. Res: Applied physics, particularly physical processes in the micrometeorological field associated with agricultural crops. Mailing Add: Dept of Agron Cornell Univ Col Agr Ithaca NY 14850

LEMON, FRANK RAYMOND, b San Bernardino, Calif, Oct 16, 17; m 40; c 4. EPIDEMIOLOGY. Educ: Col Med Evangelists, MD, 50; Tulane Univ, MPH, 55; Am Bd Prev Med, dipl, 64. Prof Exp: From instr to assoc prof prev med, Sch Med, Loma Linda Univ, 55-68; assoc dean continuing educ, 68-72, PROF COMMUNITY MED, CLIN MED & FAMILY PRACT, COL MED, UNIV KY, 68-, ASSOC DEAN EXTRAMURAL AFFAIRS & ASST TO VPRES MED CTR, 72- Concurrent Pos: Consult, Inst Inter-Am Affairs, Mex, 53-54. Mem: Fel Am Pub Health Asn; fel Am Col Physicians; Asn Teachers Prev Med. Res: Epidemiology of cancer; atherosclerotic disease. Mailing Add: Dept Commun Med & Internal Med Univ of Ky Col of Med Lexington KY 40506

LEMON, HENRY MARTYN, b Chicago, Ill, Dec 23, 15; m 41; c 5. MEDICINE. Educ: Univ Chicago, BS, 38; Harvard Univ, MD, 40; Am Bd Internal Med, dipl, 50, cert med oncol, 73. Prof Exp: From intern to asst resident med, Univ Chicago Clins, 40-42; asst, Sch Med, Univ Chicago, 42-43; mem comn airborne infection, Army Epidemiol Bd, 44-45; from instr to assoc prof, Sch Med, Boston Univ, 46-61; dir, Eugene Eppley Cancer Inst, 61-68, PROF INTERNAL MED, COL MED, UNIV NEBR, OMAHA, 61-, DIR DIV CLIN ONCOL, 68- Concurrent Pos: Consult, Vet Admin, 50-; mem & consult rev comt clin cancer training, Regional Med Prog Serv, Dept Health, Educ & Welfare, 68-72. Honors & Awards: Cert of Merit, AMA, 57-62. Mem: Endocrine Soc; Am Asn Cancer Res; Am Asn Cancer Educ; Am Col Physicians; Am Soc Clin Oncol. Res: Metabolism; connective tissue; bone; tumor; endocrinology; estrogen metabolism parathormone; adrenal steroids; neoplastic disease prevention; chemotherapy detection cancer. Mailing Add: 10805 Poppleton Ave Omaha NE 68144

LEMON, JAMES THOMAS, b West Lorne, Ont, July 2, 29; m 58; c 3. URBAN GEOGRAPHY. Educ: Univ Western Ont, BA, 55; Yale Univ, BD, 58; Univ Wis, MS, 61, PhD(geog), 64. Prof Exp: Asst prof geog, Univ Calif, Los Angeles, 64-67, from asst prof to assoc prof, 67-74, PROF GEOG, UNIV TORONTO, 74- Mem: AAAS. Res: Early America—social, economic and cultural; historical urbanization, especially residential areas. Mailing Add: Dept of Geog Univ of Toronto Toronto ON Can

LEMON, PAUL CLIPFELL, plant ecology, see 12th edition

LEMON, ROY RICHARD HENRY, b Birmingham, Eng, July 13, 27; Can citizen; m 59. GEOLOGY. Educ: Univ Wales, BSc, 51; Univ Toronto, MA, 53, PhD, 55. Prof Exp: Geologist, Ghana Geol Surv, 56-57; asst cur invert paleont, Royal Ont Mus, 57-58, 59-61, assoc cur, 61-67; staff geologist, Texaco Oil Co, Trinidad, WI, 67-68; PROF GEOL, FLA ATLANTIC UNIV, 68- Concurrent Pos: Nat Res Coun res grant, 63-66; asst prof, Queen's Univ, Ont, 58-59; assoc prof, Univ Toronto, 62-67. Mem: Geol Soc Am; Am Asn Petrol Geol. Res: Pliocene and Pleistocene geology and faunas of the west coast of South America and the Caribbean; world wide Pleistocene sea level changes; origin of sedimentary phosphates. Mailing Add: Dept of Geol Fla Atlantic Univ Boca Raton FL 33432

LEMONDE, ANDRE, b Saint-Liboire, Que, May 30, 21; m 53; c 2. BIOCHEMISTRY, PHYSIOLOGY. Educ: Univ Montreal, BA, 42; Laval Univ, BS, 47, ScD(biol), 51. Prof Exp: Demonstr physiol, Laval Univ, 47-51; hon fel biochem & entom, Cornell Univ, 51-52; from asst prof to assoc prof, 52-66, PROF BIOCHEM, SCH MED, LAVAL UNIV, 66- Mem: AAAS; Am Physiol Soc; Can Biochem Soc; Nutrit Soc Can. Res: Comparative biochemistry and physiology. Mailing Add: Dept of Biochem Laval Univ Sch of Med Quebec PQ Can

LEMONDE, PAUL, b St Libiore, Que, Nov 4, 13; m 47; c 5. CANCER. Educ: St Hyacinthe Col, BA, 33; Univ Montreal, lic, 38, PhD(exp med), 54. Prof Exp: From asst to assoc prof biol, Univ Montreal, 38-54; RES PROF MED, INST MICROBIOL & HYG MONTREAL, 54- Concurrent Pos: Prof, Que Sch Vet Med, 47-49. Honors & Awards: Award, Prov Que, 57. Mem: AAAS; Soc Exp Biol & Med; Can Soc Microbiol; Am Asn Cancer Res. Res: Endocrine factors in infections; experimental tuberculosis; viruses, immunological and environmental factors in cancer. Mailing Add: Inst Microbiol & Hyg Montreal Laval-des-Rapides PQ Can

LEMONE, DAVID V, b Columbia, Mo, Apr 16, 32; m 55; c 2. INVERTEBRATE PALEONTOLOGY, PALEOBOTANY. Educ: NMex Inst Mining & Technol, BS, 55; Univ Ariz, MS, 59; Mich State Univ, PhD(geol), 64. Prof Exp: Geologist, Stanolind Oil & Gas Co, 55-56 & Tex Co, 58-59; assoc prof geol, Southern Miss Univ, 61-64; ASSOC PROF GEOL, UNIV TEX, EL PASO, 64- Mem: AAAS; Am Paleont Soc; Soc Econ Paleontologists & Mineralogists; Paleont Soc Japan. Res: Paleophycology; stratigraphic paleontology; systematic invertebrate paleontology and paleobotany; palynology; paleoecology; numerical taxonomy. Mailing Add: Dept of Geol Univ of Tex El Paso TX 79999

LEMONE, MARGARET ANNE, b Columbia, Mo, Feb 21, 46. METEOROLOGY. Educ: Univ Mo, AB, 67; Univ Wash, PhD(atmos sci), 72. Prof Exp: Fel, 72-73, SCIENTIST ATMOSPHERIC RES, ADVAN STUDY PROG, NAT CTR ATMOSPHERIC RES, 73- Mem: Am Meteor Soc; AAAS; Am Geophys Union; Sigma Xi. Res: Structure and dynamics of atmospheric planetary boundary layer; interaction of planetary boundary layer with moist convection; aircraft measurements of atmosphere. Mailing Add: Nat Ctr for Atmospheric Res PO Box 3000 Boulder CO 80303

LEMONICK, AARON, b Philadelphia, Pa, Feb 2, 23; m 50; c 2. HIGH ENERGY PHYSICS. Educ: Univ Pa, BA, 50; Princeton Univ, MA, 52, PhD, 54. Prof Exp: Instr physics, Princeton Univ, 53-54; asst prof, Haverford Col, 54-57, assoc prof & chmn dept, 57-61; assoc prof, 61-64, assoc dir, Princeton-Penn Accelerator, 61-67, assoc chmn dept, 67-69, dean grad sch, 69-73, PROF PHYSICS, PRINCETON UNIV, 64-, DEAN FAC, 73- Concurrent Pos: NSF sci fac fel, Univ Calif, Berkeley, 60-61. Mem: Fel Am Phys Soc, Am Asn Physics Teachers. Mailing Add: Rm 9 Nassau Hall Princeton Univ Princeton NJ 08540

LEMONS, HOYT, b Eldorado, Ill, Mar 11, 14; m 36, 66; c 3. PHYSICAL GEOGRAPHY. Educ: Southern Ill Univ, BEd, 36; Univ Nebr, MA, 38, PhD(geog & climatol), 41. Prof Exp: From instr to asst prof geog, Okla State Univ, 39-42; asst prof, Wash State Univ, 42-45; consult environ res, Off Qm Gen, US Army, 45-46, res dir environ protection br, 46-52; res coordr, Fed Civil Defense Admin, 52-54; asst for res, Off Surgeon Gen, USPHS, 54-57; br chief, US Army Res Off, 57-69, CHIEF SCIENTIST, EUROP RES OFF, US ARMY, 69- Concurrent Pos: Mem comt rev comt earth sci for Fulbright grants, Nat Res Coun, 50-53, mem comt phys & biogeog, 52-54; part-time prof lectr climatol, Univ Md, 52-65; chmn, US-Can Steering Comt Proj HARP, 65-66; mem comt, Army Res Coun, 65-67 & US Nat Comt Int Geog Union, 67-69. Mem: Asn Am Geog (treas, 52-55); fel Am Geog Soc; Am Meteorol Soc; Inst Brit Geog; Arctic Inst NAm. Res: Climatology; military and medical geography; geomorphology; environmental research. Mailing Add: 71 Gunnersbury Ave Ealing W5 London England

LEMONS, ROSS ALAN, b Los Alamos, NMex, Jan 14, 49; m 70. APPLIED PHYSICS. Educ: Univ Colo, BA, 70; Stanford Univ, MS, 72, PhD(appl physics), 75. Prof Exp: RES ASSOC APPL PHYSICS, HANSEN MICROWAVE LAB, STANFORD UNIV, 74- Res: Acoustic microscopy; development of new imaging systems and techniques; application of physical techniques to problems in biology. Mailing Add: Hansen Microwave Lab Stanford Univ Stanford CA 94305

LEMOS, ANTHONY M, b Arlington, Mass, Aug 31, 30; m 53; c 3. THEORETICAL PHYSICS, SOLID STATE PHYSICS. Educ: Boston Col, AB, 52; Univ Chicago, MS, 56; Ill Inst Technol, PhD(physics), 64. Prof Exp: Instr physics, Lake Forest Col, 58-63; assoc prof, 64-71, PROF PHYSICS, ADELPHI UNIV, 71- Concurrent Pos: Res collabr, Brookhaven Nat Lab, 65- Mem: Am Phys Soc. Res: Theoretical understanding of the phenomena surrounding the F-center; color centers in the alkali halides. Mailing Add: Dept of Physics Adelphi Univ Grad Sch Arts & Sci Garden City NY 11530

LEMP, JOHN FREDERICK, JR, b Alton, Ill, May 25, 28; m 53; c 3. MICROBIOLOGY, BIOENGINEERING. Educ: Univ Ill, BS, 51; Nat Registry Microbiologists, Regist. Prof Exp: Bacteriologist, Commercial Solvents Corp, Ind, 51, fermentation supt & microbiologist, Ill, 53-57; microbiologist, Pilot Plants Div, US Army Biol Labs, Ft Detrick, Md, 57-61, prin investr, Process Develop Div, 61-63; sr microbiologist & asst br chief, Biol Ctr, 63-71; proj mgr RNA tumor virus, Electro-Nucleonics Labs, Inc, Md, 71-72; DIR, VIRAL SCI LAB, 72- Mem: Am Soc Microbiol; Am Chem Soc; Sigma Xi. Res: Ultrasonic defoaming, purification microbial products, B-12, riboflavin, penicillin; bacitracin, alcohols, fungal amylase, continuous sterilization and culture, pH control, polarographic dissolved oxygen; mammalian tissue culture; virus propagation and purification; electrophoresis; polymer phase. Mailing Add: RR 1 Lovettsville VA 22080

LEMPER, ANTHONY LOUIS, b Buffalo, NY, June 3, 39; m 62; c 4. ORGANIC CHEMISTRY. Educ: Univ Buffalo, BA, 60; State Univ NY Buffalo, PhD(org chem), 66. Prof Exp: Res chemist, Res Div, Goodyear Tire & Rubber Co, 65-66, sr res chemist, 66; res chemist, Corp Res Ctr, 66-68, sr res chemist, 68-71, res group leader, 71-75, RES MGR, HOOKER CHEM CORP, 75- Mem: Am Chem Soc. Res: Engineering thermoplastics research and development; polyvinyl chloride chemistry and technology; organophosphorus chemistry; organofluorine chemistry. Mailing Add: Corp Res Ctr Hooker Chem Corp Box 8 MPO Niagara Falls NY 14302

LEMPERT, JOSEPH, b North Adams, Mass, July 3, 13; m 41; c 3. MAGNETOHYDRODYNAMICS. Educ: Mass Inst Technol, BS, 35; Stevens Inst Technol, MS, 42. Prof Exp: Engr, Lamp Div, 36-44, sect engr, 44-53, mgr eng sect, Electronic Tube Div, 53-56, adv develop, 56-58, sect mgr camera tubes, 58, res engr, 58-66, ADV ENGR, WESTINGHOUSE RES & DEVELOP CTR, 66- Mem: Am Phys Soc; fel Inst Elec & Electronics Eng. Res: Photoemission; photoconductivity; secondary emission; thin films; vacuum tube electronics; electronic imaging; x-ray image intensification; x-rays; storage techniques; electron beams; electron beam welding; thermionic emission. Mailing Add: Westinghouse Res & Develop Ctr Beulah Rd Pittsburgh PA 15235

LEMPICKI, ALEXANDER, b Warsaw, Poland, Jan 26, 22; nat US; m 52; c 2. PHYSICS. Educ: Imp Col, Univ London, MSc, 52, PhD, 60. Prof Exp: Res physicist, Electronic Tube Co, Ltd, Eng, 49-54; head quantum physics group, 65-72, MGR ELECTROOPTICS LAB, SPEC PROJ OFF, GEN TEL & ELECTRONICS LABS, INC, 73- Concurrent Pos: Mem adv subcomt electrophys, NASA, 69-71. Mem: Fel Am Phys Soc; Optical Soc Am. Res: Thermionic and secondary emission; solid state physics, particularly luminescence; electro luminescence; optical properties of solids; spectroscopy and molecular structure of organo metallic complexes; optical maser materials, particularly liquid luminescence. Mailing Add: GTE Labs Inc 40 Sylvan Rd Waltham MA 02154

LEMPKE, ROBERT EVERETT, b Dover, NH, Nov 27, 24; m 49; c 4. SURGERY. Educ: Yale Univ, MD, 48. Prof Exp: Intern surg, Johns Hopkins Hosp, Baltimore, 48-49; resident, Med Ctr, Ind Univ, Indianapolis, 49-51; med officer, Army Med Res Lab, Ft Knox, Ky, 51-53; resident, Med Ctr, Ind Univ, Indianapolis, 53-55; assoc chief of staff med res, Vet Admin Hosp, Indianapolis, 56-71; from instr to assoc prof, 56-65, PROF SURG, SCH MED, IND UNIV, INDIANAPOLIS, 65-; CHIEF OF SURG, VET ADMIN HOSP, INDIANAPOLIS, 59- Concurrent Pos: Vis prof surg, Jinnah Postgrad Med Ctr, Karachi, Pakistan, 64-65. Mem: Soc Surg Alimentary Tract; Am Asn Surg of Trauma; Am Col Surg. Res: Diseases of the alimentary tract. Mailing Add: 1481 W Tenth St Indianapolis IN 46202

LEMUNYAN, COBERT DUANE, b Hornell, NY, May 3, 27; m 46; c 4. ENVIRONMENTAL HEALTH. Educ: Cornell Univ, BS, 50; Univ Md, MA, 55, EdD(human develop), 61. Prof Exp: Naval Med Res Inst, 51-62; chief animal prod sect, NIH, 62-67, actg chief animal sci & tech br & head reprod physiol sect, Nat Environ Health Sci Ctr, 67-70, scientist adminr, Spec Progs Br, 70-74, CHIEF, SCI PROGS, NAT INST ENVIRON HEALTH SCI, 74-, ACTG ASSOC DIR, EXTRAMURAL PROG, 75- Concurrent Pos: Mem panel US Bd Civil Serv Comn, NIH, 54-67; adj prof animal sci & assoc mem grad fac, NC State Univ, 67-72. Honors & Awards: US Navy Meritorious Civilian Serv Award, 55. Mem: AAAS; Am Asn Lab Animal Sci; Am Psychol Asn; NY Acad Sci. Res: Ecology and behavior; reproductive physiology as influenced by photoperiodism, light sources and intensity, auditory and other external stimuli; environmental toxicology. Mailing Add: Nat Inst of Environ Health Sci Nat Inst Health PO Box 12233 Research Triangle Park NC 27709

LENANDER, CARL JOHN, applied physics, mathematical physics, see 12th edition

LENARD, ANDREW, b Balmazujvaros, Hungary, July 18, 27; US citizen; m 53; c 2. MATHEMATICAL PHYSICS. Educ: State Univ Iowa, BA, 49, PhD(physics), 53. Prof Exp: Res assoc physics, Columbia Univ, 55-57; res staff mem, Plasma Physics Lab, Princeton Univ, 57-65; PROF MATH PHYSICS, IND UNIV, BLOOMINGTON, 66- Mem: Am Phys Soc. Res: Kinetic theory; statistical mechanics; fundamental problems of quantum physics; mathematical problems related to physics. Mailing Add: Dept of Physics & Math Ind Univ Bloomington IN 47401

LENARD, JOHN, b Vienna, Austria, May 17, 37; US citizen; m 59; c 3. BIOCHEMISTRY. Educ: Cornell Univ, BA, 58, PhD(biochem), 64. Prof Exp: Res assoc biochem, Cornell Univ, 63-64; fel biol, Univ Calif, San Diego, 64-65, Am Heart Asn advan res fel, 65-67; asst prof biochem, Albert Einstein Col Med, 67-68; assoc, Sloan-Kettering Inst Cancer Res, 68-72; ASSOC PROF PHYSIOL, RUTGERS MED SCH, COL MED & DENT NJ, 73- Concurrent Pos: Am Heart Asn estab investr, 70-72. Mem: Am Phys Soc; Am Soc Biol Chemists. Res: Protein chemistry; structure of biological membranes; biological membranes and enveloped viruses; red cell membrane proteins. Mailing Add: Dept of Physiol Rutgers Med Sch Piscataway NJ 08854

LENCE, RICHARD HAYES, mathematics, physics, see 12th edition

LENDE, RICHARD ALLAN, neurosurgery, neurophysiology, deceased

LENER, WALTER, b New York, NY, Mar 20, 25. ENTOMOLOGY. Educ: NY Univ, BA, 48, MA, 50, PhD(biol), 57; Rutgers Univ, MS, 60. Prof Exp: Instr biol, State Univ NY Col Oneonta, 50-51, sci consult, New Paltz, 51-52, from instr to prof biol, Geneseo, 52-64, coordr biol sci, 62-64; PROF BIOL, NASSAU COMMUNITY COL, 64- Concurrent Pos: Res grant, Res Found, State Univ NY, 63-64, 65-67; fel trop med, Sch Med, La State Univ, 64; NSF res grant, 66-68; consult, Choice. Mem: AAAS; Ecol Soc Am; Animal Behav Soc; NY Acad Sci; Sigma Xi. Res: Investigating

LENER

the physiology, genetics and ethology of large milkweed bug, Oncopeltus fasciatus. Mailing Add: Dept of Biol Nassau Community Col Garden City NY 11530

LENEY, LAWRENCE, b New York, NY, Dec 14, 17; m 45; c 5. WOOD SCIENCE, MICROSCOPY. Educ: State Univ NY Col Forestry, Syracuse, BS, 42, MS, 48, PhD, 60. Prof Exp: Instr wood tech, State Univ NY Col Forestry, Syracuse, 46-52; asst prof, Univ Mo, 52-60; ASSOC PROF WOOD & FIBER SCI, COL FOREST RESOURCES, UNIV WASH, 60- Mem: AAAS; Forest Prod Res Soc; Tech Asn Pulp & Paper Indust; Int Asn Wood Anat; Soc Wood Sci & Technol. Res: Wood anatomy; microtechnique; machining wood; photomicrography of woody tissue; seasoning and preservation of wood; pulp and paper fiber analysis. Mailing Add: Col of Forest Resources Univ of Wash Seattle WA 98195

LENFANT, CLAUDE J M, b Paris, France, Oct 12, 28; US citizen; m 40; c 5. PHYSIOLOGY. Educ: Univ Rennes, BS, 48; Univ Paris, MD, 56. Prof Exp: Intern clin med, Univ Paris, 51-52; resident, Hosp Rothschild, Paris, 53; from res asst to dir res, Ctr Marie Lannelongue, France, 54-57; res fel, Univ Buffalo, 57-58; res fel, Columbia Univ, 58; asst prof physiol, Univ Lille, 59-60; from clin instr to clin asst prof, Univ Wash, 61-67, assoc prof, 68-71; assoc dir lung progs & actg assoc dir collab res & develop prog, 70-72, actg chief pulmonary res br, 72-74, DIR DIV LUNG DIS, NAT HEART & LUNG INST, 72-; PROF MED, PHYSIOL & BIOPHYS, UNIV WASH, 71- Concurrent Pos: Scholar, Univ Paris, 56; Fulbright fel, 56-58; assoc dir, Inst Respiratory Physiol & staff physician, Firland Sanitorium, Seattle, 61-68; mem physiol study sect, NIH, 69-70; consult, Univ Hosp, Seattle, 69- Mem: Am Physiol Soc; Am Soc Zool; Soc Exp Biol & Med; Undersea Med Soc; Am Soc Clin Invest. Res: Respiratory physiology, especially in gas exchange; comparative physiology related to the development and environmental adaptation of the respiratory system. Mailing Add: Westwood Bldg Rm 6A-18 Nat Heart & Lung Inst Bethesda MD 20016

LENG, EARL REECE, b Williamsfield, Ill, June 12, 21; m 44; c 3. PLANT GENETICS, PLANT BREEDING. Educ: Univ Ill, BS, 41, MS, 46, PhD(agron), 48. Prof Exp: Spec asst agron, 41-42, asst plant genetics, 46-48, from asst prof to assoc prof, 48-58, from asst dir to assoc dir int prog, 69-73, PROF AGRON, UNIV ILL, URBANA, 58-; CROP SPECIALIST, USAID, 75- Concurrent Pos: Fulbright sr res fel, Max Planck Inst, Ger, 61; consult, Fed Govt Yugoslavia & USAID, 60-61; res adv, USAID & Uttar Pradesh Agr Univ, India, 64-66; adv, USAID & Midwest Univs Consortium for Int Activities, Indonesia, 71; consult, Food & Agr Orgn, Thailand, 71; consult, Int Coffee Orgn, 72; consult, UNDP/FAO, Yugoslavia, 73, 75; consult, World Bank, 74. Mem: Crop Sci Soc Am; Am Soc Agron. Res: Comparative international agriculture; genetics and breeding of maize; breeding systems; evolution of maize and relatives; international soybean improvement; international agricultural development, emphasis on major cereal crops. Mailing Add: Off of Agr Tech Assistance Bur USAID Washington DC 20523

LENG, MARGUERITE LAMBERT, b Edmonton, Alta, Can, Sept 25, 26; m 55; c 3. AGRICULTURAL BIOCHEMISTRY, ANALYTICAL BIOCHEMISTRY. Educ: Univ Alta, BSc, 47; Univ Sask, MSc, 50; Purdue Univ, PhD(biochem), 56. Prof Exp: Ed asst chem & physics, Nat Res Coun Can, 47-48, anal chemist, 48-49; sr chemist, Allergy Res Lab, Univ Mich Hosp, Ann Arbor, 50-53; anal chemist, Agr Dept, 56-59, registr specialist, Ag-Org Dept, 66-73; SR REGIST SPECIALIST, GOVT REGIST-HEALTH & ENVIRON RES, DOW CHEMICAL USA, 73- Mem: Am Chem Soc. Res: Pesticides, their toxicology, metabolism, residues, analytical methods and realistic evaluation of hazard to the environment; chemistry and nature of steroid hormone action; meaningful communication of scientific information. Mailing Add: Dept of Health & Environ Res Dow Chemical Co PO Box 1706 Midland MI 48640

LENGEMANN, FREDERICK WILLIAM, b New York, NY, Apr 8, 25; m 50; c 2. PHYSICAL BIOLOGY. Educ: Cornell Univ, BS, 50, MNS, 51; Univ Wis, PhD(dairy husb), 54. Prof Exp: Res assoc radiation biol, Univ Tenn, 54-55, asst prof chem, 55-59; PROF PHYS BIOL, NY STATE VET COL, CORNELL UNIV, 59- Concurrent Pos: Mem comt 33, Nat Coun Radiation Protection & Measurements. Mem: Fel AAAS; Am Dairy Sci Asn; Am Inst Nutrit. Res: Environmental contamination; fission product and mineral metabolism; milk secretion; mineral absorption; bone calcification. Mailing Add: Dept Phys Biol NY State Vet Col Cornell Univ Ithaca NY 14853

LENGYEL, BELA ADALBERT, b Budapest, Hungary, Oct 5, 10; nat US; m 42, 62; c 2. PHYSICS. Educ: Pazmany Univ, Hungary, PhD(math), 35. Prof Exp: Asst math, Univ Tech Sci, Budapest, 31-34; res fel, Harvard Univ, 35-36, asst actuary, 36-37; asst statistician, Worcester State Hosp, Mass, 38-39; instr math, Rensselaer Polytech Inst, 39-42; instr physics, City Col New York, 42-43; asst prof, Univ Rochester, 43-46; physicist, Naval Res Lab, 46-50 & Off Naval Res, 50-52; mem tech staff, Res Labs, Hughes Aircraft Co, Calif, 52-63; chmn dept, 63-70, PROF PHYSICS, CALIF STATE UNIV, NORTHRIDGE, 63- Concurrent Pos: Vis prof, Univ Lund & Chalmers Univ Technol, Sweden, 71-72. Mem: Am Math Soc; Am Phys Soc; Inst Elec & Electronics Engrs. Res: Theory of operators; electromagnetic theory; quantum electronics. Mailing Add: Dept of Physics & Astron Calif State Univ Northridge CA 91324

LENGYEL, ISTVAN, b Kaposvar, Hungary, July 12, 31; US citizen; m 62; c 2. ORGANIC CHEMISTRY. Educ: Eötvös Lorand, Dipl, 55; Mass Inst Technol, PhD(org chem), 64. Prof Exp: Res chemist, G Richter Pharmaceut Co, 54-55; sci coworker geochem, Geophys Res Inst of Hungary, 55-56; lab chemist, Kundl Tirol Austria Pharmaceut Co, 57-58; res asst biochem, Sch Med, Johns Hopkins Univ, 58-59; res assoc org synthesis, Mass Inst Technol, 59-64; fel, Munich Tech Univ, 64-65; res assoc mass spectrometry, Mass Inst Technol, 65-67; from asst prof to assoc prof, 67-73, PROF CHEM, ST JOHN'S UNIV, NY, 73- Honors & Awards: Sr Awardee, Alexander Von Humboldt Found, Germany, 73-74. Mem: Fel The Chem Soc; Ger Chem Soc. Res: Organic mass spectrometry; amino acids; peptides; synthesis, reactions and spectroscopic characteristics of small heterocycles; skeletal rearrangements upon electron impact. Mailing Add: Dept of Chem St John's Univ Jamaica NY 11439

LENGYEL, PETER, b Budapest, Hungary, May 24, 29; US citizen; m 56; c 2. BIOCHEMISTRY. Educ: Budapest Tech Univ, Dipl, 51; NY Univ, PhD(biochem), 62. Prof Exp: Instr biochem, Sch Med, NY Univ, 62-63, asst prof, 63-65; assoc prof molecular biophys, 65-69, PROF MOLECULAR BIOPHYS & BIOCHEM, YALE UNIV, 69- Concurrent Pos: NIH spec fel, Pasteur Inst, Paris, 63-64. Mem: Am Chem Soc; Am Soc Biol Chem. Res: Protein biosynthesis; nucleic acid and protein metabolism of animal cells and viruses; interferon defense mechanism. Mailing Add: Dept of Biophys & Biochem Yale Univ Box 1937 Yale Sta New Haven CT 06520

LENHARD, JOSEPH ANDREW, b Detroit, Mich, June 18, 29; m 70; c 4. HEALTH PHYSICS, NUCLEAR PHYSICS. Educ: Vanderbilt Univ, BA, 53, MS, 57; Am Bd Health Physics, dipl, 60. Prof Exp: Health physicist radiation protection, 57-61, sr health physicist broad nuclear safety, US AEC, 61-67, dir safety & environ control div, 67-72, DIR RES, OAK RIDGE OPERS OFF, ENERGY RES & DEVELOP ADMIN, 72- Mem: Health Physics Soc. Res: Research administration; physical, life and engineering sciences. Mailing Add: 125 Newell Ln Oak Ridge TN 37830

LENHARDT, MARTIN LOUIS, b Elizabeth, NJ, Dec 14, 44; m 66; c 4. SPEECH PATHOLOGY, SPEECH & HEARING SCIENCES. Educ: Seton Hall Univ, BS, 66, MS, 68; Fla State Univ, PhD(audiol, speech sci), 70. Prof Exp: Nat Inst Neurol Dis & Stroke fel, Johns Hopkins Univ, 70-71; asst prof, 71-75, ASSOC PROF OTORHINOLARYNGOL, MED COL VA, VA COMMONWEALTH UNIV, 71- Mem: AAAS; Acoust Soc Am; Am Audiol Soc; Animal Behav Soc. Res: Psychological ahd physiological acoustics; speech communication; bioacoustics and linguistics. Mailing Add: Med Col of Va Box 157 MCV Sta Richmond VA 23298

LENHER, SAMUEL, b Madison, Wis, June 19, 05; m 29; c 3. CHEMISTRY. Educ: Univ Wis, AB, 24; Univ London, PhD(chem), 26. Hon Degrees: DSc, Univ Wis, 59; DSc, Univ Del, 61. Prof Exp: Int Ed Bd fel, Univ Berlin, 26-27; Nat Res Coun fel & res assoc, Univ Calif, 27-28; from res chemist chem dept to asst gen mgr, Chambers Works, E I du Pont de Nemours & Co, Inc, 29-55, vpres, 55-70, dir, 55-74; RETIRED. Concurrent Pos: Dir, Du Pont SA de CV, 56-61 & Cia Mexicana de Explosivos, SA, 57-61; consult, US Dept Health, Educ & Welfare, 57-58; trustee, res found, Univ Del, 54-75, pres, 55-66, trustee univ, 63-; Wis Alumni Res Found, 57-, Inst, 68-; Wistar Inst, 59-74; Johns Hopkins Univ, 59-; Marine Biol Lab, Woods Hole, 67-; dir, Block Eng Inc, 62- Mem: AAAS; Am Chem Soc; Soc Indust Chem; NY Acad Sci; Am Philos Soc. Res: Manufacturing of organic chemicals. Mailing Add: 1900 Woodlawn Ave Wilmington DE 19806

LENHERT, ANNE GERHARDT, b Lynchburg, Va, Apr 1, 36; m 67; c 1. ORGANIC CHEMISTRY. Educ: Hollins Col, BA, 58; Univ NMex, MS, 63, PhD(chem), 65. Prof Exp: Res fel, Univ NMex, 64-65; asst prof chem, Cent Mo State Col, 65-67; ASST PROF CHEM, KANS STATE UNIV, 67- Mem: Am Chem Soc; Int Heterocyclic Chem. Res: Synthesis of heterocyclic ring systems as potential purine and pteridine antagonists; anti-cancer agents and anti-radiation drugs. Mailing Add: Dept of Chem Kans State Univ Manhattan KS 66502

LENHERT, P GALEN, b Dayton, Ohio, July 31, 33; m 56; c 2. CRYSTALLOGRAPHY, BIOPHYSICS. Educ: Wittenburg Univ, AB, 55; Johns Hopkins Univ, PhD(biophys), 60. Prof Exp: USPHS res fel chem crystallog, Oxford Univ, 60-61; asst prof physics, Wittenburg Univ, 61-64; asst prof, 64-68, ASSOC PROF PHYSICS, VANDERBILT UNIV, 68- Concurrent Pos: USPHS grants, 62-63, 64-72. Mem: AAAS; Am Crystallog Asn; Biophys Soc. Res: Determination of molecular structures by x-ray crystallographic methods; biologically important molecules of medium to high molecular weight. Mailing Add: Dept of Physics Vanderbilt Univ Box 1807 Sta B Nashville TN 37235

LENHOFF, HOWARD MAER, b North Adams, Mass, Jan 27, 29; m 54; c 2. BIOCHEMISTRY, INVERTEBRATE ZOOLOGY. Educ: Coe Col, BA, 50; Johns Hopkins Univ, PhD(biol), 55. Hon Degrees: DSc, Coe Col, 76. Prof Exp: USPHS fel, Loomis Lab, Nat Cancer Inst, 54-56; actg chief, Biochem Sect, Armed Forces Insts Path, 56-57; assoc consult res, George Washington Univ, 57-58; fel, Dept Terrestrial Magnetism, Carnegie Inst Technol, 58; assoc prof biol, Univ Miami, Fla, 59-65, prof, 66-69, dir lab quant biol, 63-69; assoc dean sch biol sci, 69-71, dean grad div, 71-73, PROF DEVELOP & CELL BIOL, FAC RES FACIL, UNIV CALIF, IRVINE, 69- Concurrent Pos: Vis lectr, Howard Univ, 57-58; investr, Biochem Labs, Howard Hughes Med Inst, 58-63; USPHS career develop award, 65-69; vis scientist, Polymer Lab, Weizmann Inst Sci, Israel, 68-69; vis prof, Hebrew Univ Jerusalem, 70, 71; vis prof chem eng, Israel Inst Technol, 73-74. Mem: Am Soc Biol Chem; Am Soc Cell Biol; Am Chem Soc; Biophys Soc; Soc Develop Biol. Res: Invertebrate biology; chemoreception; symbiosis; cellular differentiation; immobilized enzymes. Mailing Add: Fac Res Facil Univ of Calif Irvine CA 92664

LENIART, DANIEL STANLEY, b Norwich, Conn, Jan 5, 43; m 71. PHYSICAL CHEMISTRY. Educ: The Citadel, BS, 64; Cornell Univ, PhD(phys chem), 69. Prof Exp: Fel, 69-70, applications engr, 70-75, MGR EPR RES & DEVELOP, VARIAN ASSOCS, 75- Mem: Am Phys Soc. Res: Study of relaxation phenomena using the techniques of electron spin resonance, electron nuclear double resonance and electron electron double resonance. Mailing Add: Varian Assocs 611 Hansen Way Palo Alto CA 94303

LENIE, CAMILLE ARMAND, inorganic chemistry, see 12th edition

LENK, CARL THEODORE, chemistry, see 12th edition

LENKER, EARLE SCOTT, geology, see 12th edition

LENKOSKI, L DOUGLAS, b Northampton, Mass, May 13, 25; m 52; c 4. MEDICINE, PSYCHIATRY. Educ: Harvard Univ, AB, 48; Western Reserve Univ, MD, 53. Prof Exp: Fel psychiat, Yale Univ, 55-56; teaching fel, 57-60, from instr to assoc prof, 60-69, PROF PSYCHIAT, SCH MED, CASE WESTERN RESERVE UNIV, 69-, CHMN DEPT, 70-; DIR DEPT, CLEVELAND METROP GEN HOSP & UNIV HOSPS CLEVELAND, 69- Concurrent Pos: Consult, DePaul Maternity & Infant Home, 58- & Cleveland Ctr on Alcoholism, 58-61; actg dir dept psychiat, Univ Hosps Cleveland, 62-66, assoc dir dept, 66-69; dir dept, 69-; consult, St Ann's Hosp & Cleveland Vet Admin Hosp, 65-; indust personnel, US Dept Defense, 66- & White Cliff Nursing Home, 66- Mem: AAAS; Am Col Psychiat; fel Am Psychiat Asn. Res: Psychiatric education; community mental health planning. Mailing Add: Dept of Psychiat Sch of Med Case Western Reserve Univ Cleveland OH 44106

LENN, NICHOLAS JOSEPH, b Chicago, Ill, Nov 26, 38; m 64; c 3. NEUROLOGY, ANATOMY. Educ: Univ Chicago, SB, 59, MS & MD, 64, PhD(anat), 67. Prof Exp: Res assoc neuroanat, NIH, 64-66; asst prof pediat & med, Univ Chicago, 70-74; ASST PROF NEUROL, UNIV CALIF, DAVIS, 74- Mem: AAAS; Am Asn Anat; Am Acad Neurol. Res: Synaptic organization of mammalian brain; cerebral lipidoses. Mailing Add: Dept of Neurol Univ of Calif Davis CA 95616

LENNARZ, WILLIAM JOSEPH, b New York, NY, Sept 28, 34; m 56; c 3. BIOCHEMISTRY. Educ: Pa State Univ, BS, 56; Univ Ill, PhD(chem), 59. Prof Exp: Teaching asst, Univ Ill, 56-57; from asst prof to assoc prof, 62-71, PROF PHYSIOL CHEM, SCH MED, JOHNS HOPKINS UNIV, 71- Concurrent Pos: Res fel, Harvard Univ, 59-62; NSF fel, 59-60, grant, 69-71; NIH fel, 60-62, grants, 62-75; Clayton scholar, 62-65; consult, NIH, 71-75. Honors & Awards: Lederle Med Fac Award, 66. Mem: Am Chem Soc; Am Soc Microbiol; Am Soc Biol Chemists. Res: Structure and function of cell membranes; biochemistry of fertilization. Mailing Add: Dept of Physiol Chem Johns Hopkins Univ Sch of Med Baltimore MD 21205

LENNEBERG, ERIC HEINZ, physiological psychology, biological anthropology, deceased

LENNERT, ANDREW E, b Jugoslavia, Jan 28, 21; US citizen; m 58; c 2. PHYSICS. Educ: Drury Col, BS, 48. Prof Exp: Engr nuclear propulsion, Sverdrup & Parcel,

Consult Engrs, 50-53; physicist, Olin Mathieson Chem Co, 53-55; engr, Aircraft Nuclear Propulsion Dept, Gen Elec Co, 55-56; design specialist & physicist, Nuclear Div, Martin Co, 56-60; MGR, ARO, INC, 60- Mem: Am Nuclear Soc; Am Phys Soc; Am Inst Aeronaut & Astronaut; NY Acad Sci; Nat Safety Coun. Res: Research and development of electrooptical instrumentation for fluid flow and other diagnostic applications; conceptual and detailed design of nuclear and nuclear-electric propulsion systems for aircraft and space applications. Mailing Add: ARO Inc Arnold Air Force Sta TN 37389

LENNETTE, EDWIN HERMAN, b Pittsburgh, Pa, Sept 11, 08; m 30; c 2. EPIDEMIOLOGY, EXPERIMENTAL PATHOLOGY. Educ: Univ Chicago, BS, 31, PhD(bact), 35; Rush Med Col, MD, 36. Prof Exp: Instr bact, Univ Chicago, 36-37, res assoc, 37-38; instr path, Wash Univ, 38-39; mem staff, Int Health Div, Rockefeller Found, 39-46; chief med-vet div, Chem Corps, US Dept Army, 46-47; CHIEF VIRAL & RICKETTSIAL DIS LAB, CALIF STATE DEPT HEALTH, 47-, CHIEF BIOMED LABS, 73- Concurrent Pos: Lectr, Sch Pub Health, Univ Calif, 47-, lectr, Univ, 48-58; consult physician, Highland-Gen Hosp, 48-; consult, Sixth Army Surgeon, 48-; assoc mem comn influenza, Armed Forces Epidemiol Bd, 48-51, mem, 51-73, mem comn rickettsial dis, 51-72, adv mem, 72-73; dep dir & chief lab serv prog, Calif State Dept Health, 72-73; mem, Armed Forces Epidemiol Bd, Off Surgeon Gen, 70-, pres, 73-; mem adv panel, Naval Biol Lab, 48-56; dir regional lab, Influenza Study Prog, WHO, 49-, mem expert adv panel virus dis, 51-, mem expert panel zoonoses, 52-62, mem WHO sci group on virus dis, 66 & 75-, mem sci adv comt to WHO team, EAfrican Virus Res Inst, 71-; mem viral & rickettsial registry, Am Type Cul Collection, 49-, mem coun, 60-63; coordr sect XII, sect res prog, 51-56; mem virus & rickettsial study sect, Commun Dis Ctr, USPHS, 51-53, chmn, 52-53, consult, 52-62, mem med lab serv adv comt, 68-72; consult, NIH, 51-, co-chmn microbiol & immunol study sect, Div Res Grants & Fels, 53-54, chmn, 55-56, chmn NIH panel virus dis, 65-70; mem adv comt poliomyelitis vaccine eval ctr, Nat Found Infantile Paralysis, 53-54; chmn bd sci counsr, Nat Inst Allergy & Infectious Dis, 57-61, mem training grant comt, 60-61, chmn, 62, mem panel respiratory & related viruses, 60-63, mem comt vaccine develop, 63-68, chmn, 67-68, mem & chmn subcomt rubella virus, 65-66, mem nat adv allergy & infectious dis coun, 63-66; consult physician, Peralta Hosp, 57-; mem microbiol panel study manpower needs in basic health sci, Fedn Am Socs Exp Biol, 60-61; chmn ad hoc comt rubella vaccine, Nat Inst Neurol Dis & Blindness, 63-64; mem microbiol panel, Wooldridge Comt, White House, 64; chmn panel virus dis, US-Japan Coop Med Sci Prog, Dept of State, 65-70, mem US del, 70-; mem sci adv comt, Hastings Found, 66-; consult, Univ Tex, MD Anderson Hosp & Tumor Inst, 66-; mem solid tumor-virus segment, Spec Virus-Leukemia Prog, Nat Cancer Inst, 66-72, consult Nat Cancer Inst, 67-73; mem, Am Biol Coun, 68-70, chmn, 69-70; mem bd trustees, Rush Med Col, 71-74; consult, Bur of Biologics, Fed Drug Admin & mem Bo B Panel on Viral & Ricket Vaccines, 73-; consult, US-Israel Binat Sci Fedn, 74-; mem sci adv bd, Leonard Wood Mem, 75-; mem cancer res safety symp planning comt, Frederick Cancer Res Ctr, 76-; mem viruses in water comt, Am Water Works Asn, 75-; mem margin of safety & extrapolation subcomt of safe drinking water comt, Nat Acad Sci/Nat Res Coun, 76- Mem: Fel Am Pub Health Asn; Am Soc Clin Path; Soc Gen Microbiol; Tissue Cult Asn (pres, 76-); fel Royal Soc Trop Med & Hyg. Res: Virology; clinical, epidemiologic and immunologic research on viral and rickettsial diseases, including poliomyelitis, enteroviruses, respiratory disease, Q fever and virus-cancer relationships. Mailing Add: Biomed Labs Calif State Dept of Health Berkeley CA 94704

LENNEY, JAMES FRANCIS, b St Louis, Mo, Oct 11, 18; m 42; c 2. PHARMACOLOGY. Educ: Wash Univ, BA, 39; Mass Inst Technol, PhD(gen physiol), 46. Prof Exp: Asst zool, Wash Univ, 41; asst, Mass Inst Technol, 42-43; res chemist, 44-45; res biochemist, Fleischmann Labs, 46-49, head, Enzyme Dept, 49-56; sect head, Union Starch & Ref Co, 56-62; PROF PHARMACOL, UNIV HAWAII, 63- Mem: AAAS; Am Chem Soc. Res: Biochemical pharmacology; enzyme and protein chemistry. Mailing Add: Dept of Pharmacol Univ of Hawaii Sch of Med Honolulu HI 96822

LENNON, EDWARD JOSEPH, b Chicago, Ill, Aug 2, 27; m 51; c 2. INTERNAL MEDICINE. Educ: Univ Ill, BA, 47, MA, 48; Northwestern Univ, MD, 52. Prof Exp: Intern med, Milwaukee County Hosp, Wis, 52-53, resident internal med, 55-58; from instr to assoc prof, 58-68, PROF MED & ASSOC DEAN, MED COL WIS, 68- Concurrent Pos: Fel, Dr Arnold Relman's Dept, Mass Mem Hosp, Boston, 60-61; dir clin res ctr, Milwaukee County Hosp, 61-68, chief renal serv, 63-70. Mem: AAAS; Am Soc Clin Invest; Am Physiol Soc; Am Fedn Clin Res; Am Diabetes Asn. Res: Endocrine disorders; renal disease; acid-base metabolism. Mailing Add: Milwaukee County Hosp 8700 W Wisconsin Ave Milwaukee WI 53226

LENNON, HARRY DEGENER, b Elizabeth, NJ, Mar 28, 28; m 57; c 3. PHYSIOLOGY, ENDOCRINOLOGY. Educ: Rutgers Univ, BS, 53, PhD, 58. Prof Exp: Asst, Rutgers, 54-58; res investr, Res Div, Armour & Co, 58-60; RES INVESTR, G D SEARLE & CO, 60- Mem: AAAS; Soc Exp Biol & Med. Res: Anabolic-androgenic effects of steroidal hormones; liver function testing; lactation and mammary gland studies. Mailing Add: Searle Labs Div of G D Searle & Co PO Box 5110 Chicago IL 60680

LENNON, ROBERT EARL, b Berlin, NH, July 11, 18; m 46; c 4. FISH BIOLOGY. Educ: Univ NH, BS, 41, MS, 48; Univ Mich, PhD(fisheries), 53. Prof Exp: Chief Appalachian sport fishery invests, 52-58, dir, Fish Control Lab, 59-73, DIR ATLANTIC SALMON INVESTS, US FISH & WILDLIFE SERV, 73- Concurrent Pos: Mem, res adv comt, Wis Conserv Comn, 61-73, chmn, 66-67; mem subcomt vert pests, Nat Acad Sci-Nat Res Coun, 64-70; res assoc, Univ Wis-La Crosse, 70-73; lectr zool, Univ Maine, Orono, 74- Mem: Am Fisheries Soc; NY Acad Sci; Am Inst Fishery Res Biologists. Res: Atlantic salmon biology; selective control of freshwater fishes and sea lamprey. Mailing Add: US Fish & Wildlife Serv Murray Hall Univ of Maine Orono ME 04473

LENNON, VANDA ALICE, b Sydney, Australia, Aug 1, 43; m 75. NEUROIMMUNOLOGY. Educ: Univ Sydney, MB, BS, 66; Univ Melbourne, PhD(immunol), 73. Prof Exp: Res asst nuclear med, Univ Sydney, 66; from jr intern to asst med resident, Montreal Gen Hosp, 66-68; fel immunol, Walter & Eliza Hall Inst Med Res, 68-72; res assoc, 72-73, ASST RES PROF, SALK INST BIOL STUDIES, 73-; ASSOC ADJ PROF, DEPT NEUROSCI, UNIV CALIF, SAN DIEGO, 76- Mem: Am Asn Immunol; Soc Neurosci; Tissue Culture Asn. Res: Autoimmunity to antigens of central and peripheral nervous systems and muscle; identification of neural cells using immunologic markers; immunologic studies of patients with neurological diseases of presumed autoimmune basis. Mailing Add: Lab Molecular Neuropath Salk Inst Biol Studies PO Box 1809 San Diego CA 92112

LENNOX, ARLENE JUDITH, b Cleveland, Ohio, Dec 3, 42. ELEMENTARY PARTICLE PHYSICS. Educ: Notre Dame Col, Ohio, BS, 63; Univ Notre Dame, MS, 73, PhD(physics), 74. Prof Exp: Teacher, Marymount High Sch, 63-64, Regina High Sch, 64-65 & Shrine High Sch, 65-69; RES ASSOC PHYSICS, FERMI NAT ACCELERATOR LAB, 74- Mem: Am Phys Soc; AAAS; Am Asn Physics Teachers. Res: Experiments to study backward peak in pi-p elastic scattering; experiments to measure pion form factor. Mailing Add: Fermi Lab PO Box 500 Batavia IL 60510

LENNOX, DONALD HAUGHTON, b Toronto, Ont, June 7, 24; m 47; c 2. HYDROLOGY. Educ: Univ Toronto, BA, 49; Univ Alta, MSc, 60. Prof Exp: Tech off, Occup Health Lab, Dept Nat Health & Welfare Can, 50-57; asst res off, Res Coun Alta, 57-61, head groundwater div, 61-68; maritime res sect, 68-70, head groundwater subdiv, 70-72; CHIEF HYDROL RES DIV, INLAND WATERS DIRECTORATE, ENVIRON CAN, 72- Mem: Soc Explor Geophys; Geol Soc Am; Geol Asn Can; Nat Water Well Asn. Res: Application of geophysical techniques to shallow groundwater exploration; investigation of analytical methods for the determination of aquifer and well characteristics. Mailing Add: Inland Waters Directorate Environ Can Ottawa ON Can

LENNOX, EDWIN SAMUEL, b Savannah, Ga, June 9, 20; m 47; c 5. BIOCHEMISTRY. Educ: Vanderbilt Univ, BS, 42; Cornell Univ, PhD(physics), 48. Prof Exp: Res physicist, Manhattan Proj, Univ Rochester, 43-44; res physicist, Los Alamos Sci Lab, NMex, 44-46; from instr to asst prof physics, Univ Mich, 48-53; res assoc microbiol & biochem, Univ Ill, 53-57, from instr to assoc prof biochem, 56-60; NSF fel, 60; assoc prof, Sch Med, NY Univ, 60-61; vis investr, Pasteur Inst, Paris, 61-63; RESIDENT FEL, SALK INST BIOL STUDIES, 63- Concurrent Pos: Fel, Nat Res Coun, 46-48 & Nat Found Infantile Paralysis, 53-56; Nat Found fel, Univ Ill, 52-56; consult adv panel molecular biol, NSF, 59-61 & 65; consult panel molecular biol, Nat Acad Sci, 66; adj prof biol, Univ Calif, San Diego, 67-, vis fel, All Souls Col, Oxford Univ, 69-70; vis scientist, Basel Inst Immunol, 74; vis scientist, Med Res Coun Lab Molecular Biol, Postgrad Med Sch, Cambridge Univ, 74- Mem: Am Chem Soc; Am Inst Biol Sci; NY Acad Sci; Biophys Soc; Am Soc Biol Chem. Res: Genetics and regulation of antibody formation; photo reactivation of bacteriophage; bacterial genetics and transduction; diphtheria toxicity in tissue culture; biosynthesis of immunoglobulins; tumor immunology. Mailing Add: Salk Inst for Biol Studies PO Box 1809 San Diego CA 92112

LE NOBLE, WILLIAM JACOBUS, b Rotterdam, Netherlands, July 19, 28; nat US; m 71. ORGANIC CHEMISTRY. Educ: Advan Tech Sch, Netherlands, BS, 49; Univ Chicago, PhD, 57. Prof Exp: Res chemist, Indust Lab, Rohm & Haas Co, 57; instr chem, Rosary Col, 58; NSF fel & res asst, Purdue Univ, 58-59; from asst prof to assoc prof, 59-69, PROF ORG CHEM, STATE UNIV NY STONY BROOK, 69- Mem: Am Chem Soc. Res: Chemical kinetics, mechanisms and equilibria in liquid systems under high pressure. Mailing Add: Dept of Chem State Univ of NY Stony Brook NY 11794

LENOIR, WILLIAM CANNON, JR, b Loudon, Tenn, Sept 22, 29; m 56; c 3. BOTANY. Educ: Maryville Col, BS, 51; Univ Ga, MS, 62, PhD(bot), 65. Prof Exp: Instr high sch, Tenn, 57-59; asst prof biol, 60-62, from asst prof to assoc prof bot, 62-73, PROF BOT, COLUMBUS COL, 73-, CHMN DIV SCI & MATH, 73- Mem: Am Inst Biol Sci; Bot Soc Am. Res: Role of light in morphogenesis; organogenesis in pine; morphogenesis of the leaf of Lygodium japonicum. Mailing Add: Dept of Biol Columbus Col Columbus GA 31907

LENON, HERBERT LEE, b Battle Creek, Mich, June 8, 39; m 62; c 3. FISH BIOLOGY. Educ: Albion Col, AB, 61; Wayne State Univ, MS, 64, PhD(fisheries), 68. Prof Exp: Asst prof, 67-74, ASSOC PROF FISHERIES BIOL & ICHTHYOL, CENT MICH UNIV, 74- Concurrent Pos: Mem Great Lakes Found. Mem: Am Fisheries Soc; Nat Audubon Soc; Int Asn Great Lakes Res; Cousteau Soc; Nat Wildlife Fedn. Res: Freshwater fish population dynamics, especially concerning Salmonid species and smallmouth bass; management evaluation. Mailing Add: Dept of Biol Cent Mich Univ Mt Pleasant MI 48858

LENOX, RONALD SHEAFFER, b Lancaster, Pa, Jan 25, 48; m 71; c 1. ORGANIC CHEMISTRY. Educ: Juniata Col, BS, 69; Univ Ill, PhD(org chem), 73. Prof Exp: ASST PROF CHEM, WABASH COL, 73- Mem: Am Chem Soc; Sigma Xi; Nat Speleol Soc. Res: Synthesis of natural products, particularly insect pheremones; new synthetic reactions. Mailing Add: Dept of Chem Wabash Col Crawfordsville IN 47933

LENSCHOW, DONALD HENRY, b LaCrosse, Wis, July 17, 38; m 64; c 2. METEOROLOGY. Educ: Univ Wis, BS, 60, MS, 62, PhD(meteo- rol), 66. Prof Exp: SCIENTIST, NAT CTR ATMOSPHERIC RES, 66- Mem: Am Meteorol Soc. Res: Atmospheric boundary layer, especially airplane measurements of turbulent energy fluxes. Mailing Add: 95 Pawnee Dr Boulder CO 80302

LENT, CHARLES M, comparative physiology, neurophysiology, see 12th edition

LENT, PETER C, b Englewood, NJ, Dec 13, 36; m 64; c 2. BEHAVIORAL BIOLOGY, MAMMALIAN ECOLOGY. Educ: Univ Alaska, BA, 60; Univ Alta, PhD(zool), 64. Prof Exp: Lectr zool, Univ Botswana, Lesotho & Swaziland, 64-66, African Wildlife Leadership Found grant, 65-66; asst prof zool, Mem Univ Newfoundland, 66-67; asst prof biol, Dowling Col, 67-68; asst prof biol, 68-69, actg unit leader, 71-72, ASSOC PROF WILDLIFE MGT, UNIV ALASKA, 69-, ASST UNIT LEADER, US FISH & WILDLIFE SERV, 68- Mem: Fel AAAS; Am Soc Mammal; Animal Behav Soc; Wildlife Soc. Res: Mammalian social behavior and ontogeny of behavior, particularly in ungulates and carnivores; ungulate ecology and management. Mailing Add: Alaska Coop Wildlife Res Unit Univ of Alaska Fairbanks AK 99701

LENTINI, EUGENE ANTHONY, b Boston, Mass, July 6, 29; m 51; c 4. PHYSIOLOGY. Educ: Boston Univ, AB, 51, MA, 55, PhD(myocardial metab), 58. Prof Exp: Instr physiol, Med Sch, Univ Ore, 58-64; asst prof, Med Col Va, 64-68; ASSOC PROF PHYSIOL, ALBANY COL PHARM, 68- Concurrent Pos: Nat Heart & Lung Inst fel, 68-72. Mem: AAAS; Am Physiol Soc. Res: Bioelectronics, electronic micrometer, chart viewer; biophysics determination of oxygen diffusion coefficient through heart muscle; biochemical interrelation between ventricular dynamics and oxidative metabolism; physiological myocardial contract; substrate utilization. Mailing Add: Dept of Biol Sci Albany Col of Pharm Albany NY 12208

LENTNEK, BARRY, b Brooklyn, NY, Dec 14, 38; m 60; c 1. ECONOMIC GEOGRAPHY, GEOGRAPHY OF LATIN AMERICA. Educ: Brooklyn Col, AB, 60, MA, 61; Johns Hopkins Univ, PhD(geog), 66. Prof Exp: Instr geog, Essex Community Col, 62-63; asst prof, Univ Ariz, 64-65; asst prof, Ohio State Univ, 65-70; ASSOC PROF GEOG, GEOG, STATE UNIV NY BUFFALO, 70- Concurrent Pos: Ohio State Univ Ctr Natural Resources fel, 66-68; NSF fel, Ohio State Univ & Mex, 68-70; consult, Inst Brasileiro de Geografia, Brazil, 71-73. Mem: Asn Am Geogr; Conf Latin Am Geog. Res: Spatial aspects of economic development; location theory; recreation geography. Mailing Add: Dept of Geog State Univ of NY at Buffalo Amherst NY 14226

LENTO, HARRY G, analytical chemistry, food chemistry, see 12th edition

LENTS, JAMES MARCELLUS, b Knoxville, Tenn, Oct 18, 43; m 64; c 1. PHYSICS.

LENTS

Educ: Univ Tenn, Knoxville, BS, 65, MS, 67, PhD(physics), 70. Prof Exp: Engr, Arnold Eng Develop Ctr, 67-70; fel, Space Inst, Univ Tenn, Knoxville, 70-71; MGR TECH DIV, CHATTANOOGA-HAMILTON COUNTY AIR POLLUTION CONTROL BUR, 71- Mem: Air Pollution Control Asn; Am Phys Soc; Sigma Xi; Int Solar Energy Soc. Res: Atmospheric dispersion of air pollutants and the direct utilization of solar energy. Mailing Add: 2938 Old Britain Circle Chattanooga TN 37421

LENTZ, CHARLES WESLEY, b Mt Pleasant, Mich, May 6, 24; m 47; c 6. CHEMISTRY. Educ: Mich State Univ, BS, 46. Prof Exp: Chemist, Mich Chem Corp, 46-52 & Columbia-Southern Div, Pittsburgh Plate Glass Co, 52-55; chemist, 55-61, supvr develop, 61-68, mgr develop, 68-70, mgr res, 70-75, MGR LIFE SCI RES & DEVELOP, DOW CORNING CORP, 75- Concurrent Pos: Mem comt MC-B5, Hwy Res Bd, Nat Acad Sci-Nat Res Coun, 67- Honors & Awards: Sigma Xi Award, 65. Mem: AAAS; Am Chem Soc; Sigma Xi. Res: Study of silica as a reinforcing agent for silicone rubber, silicate minerals and the silicate structure changes that occur in portland cement during hydration. Mailing Add: Dow Corning Corp Mail 20 Midland MI 48640

LENTZ, GARY LYNN, b North Hollywood, Calif, July 15, 43; m 65; c 2. ECONOMIC ENTOMOLOGY. Educ: Univ Mo-Columbia, AB, 65; Iowa State Univ, PhD(entom), 73. Prof Exp: Res assoc entom, Iowa State Univ, 68-72; asst prof, Univ Ariz, 72-74; ASST PROF ENTOM, AGR EXP STA, UNIV TENN, 74- Mem: Entom Soc Am; Sigma Xi. Res: Pest management of cotton and soybean insects. Mailing Add: WTenn Exp Sta 605 Airways Blvd Jackson TN 38301

LENTZ, KENNETH EUGENE, b Chicago, Ill, Dec 9, 23; m 47; c 3. BIOCHEMISTRY. Educ: Iowa State Col, BS, 47, MS, 49; Western Reserve Univ, PhD(biochem), 56. Prof Exp: Asst chem, Iowa State Col, 48-49; from asst to asst prof, 49-70, ASSOC PROF BIOCHEM, SCH MED, CASE WESTERN RESERVE UNIV, 70-; RES BIOCHEMIST, HYPERTENSION UNIT, VET ADMIN HOSP, 55- Mem: Am Chem Soc; NY Acad Sci. Res: Enzymes, proteins and peptides of the renin-angiotensin system; mechanism of action of biological active principles. Mailing Add: Hypertension Unit Vet Admin Hosp 10701 East Blvd Cleveland OH 44106

LENTZ, PATRICK EDMUND, b Elma, Iowa, Dec 15, 40; m 62; c 3. PHYSIOLOGY. Educ: Loras Col, BA, 62; Marquette Univ, MS, 64, PhD(physiol), 67. Prof Exp: Res fel biochem path, Brown Univ, 67-68; staff scientist, Hazleton Labs, Inc, 68; ASST PROF PHYSIOL, SCH MED, TULANE UNIV, 69- Mem: AAAS; assoc Am Physiol Soc; Reticuloendothelial Soc. Res: Role of the endoplasmic reticulum in synthesis and segregation of proteins for intracellular deposition; attachment of enzymes to membranes of endoplasmic reticulum; structure of cell membranes; mitochondrial protein synthesis. Mailing Add: Dept of Physiol Tulane Univ Sch of Med New Orleans LA 70112

LENTZ, PAUL LEWIS, b Indianapolis, Ind, May 26, 18; m 43; c 2. MYCOLOGY. Educ: Butler Univ, AB, 40; Univ Iowa, MS, 42, PhD(mycol), 53. Prof Exp: Asst bot lab, Butler Univ, 38-40, bact lab, 40; asst mycol, Univ Iowa, 40-42, 46-47; assoc mycologist, Plant Indust Sta, 47-56, mycologist, Plant Sci Res Div, 56-72, instr advan educ sci, Grad Sch-Found, 58-71, CHIEF, MYCOL LAB, AGR RES SERV, USDA, 72- Mem: Bot Soc Am; Mycol Soc Am; Int Soc Plant Taxon. Res: Basidiomycete taxonomy, anatomy, morphology and biology; Aphyllophorales; National Fungus Collections. Mailing Add: Mycol Lab Agr Res Ctr-West US Dept of Agr Beltsville MD 20705

LENZ, ALFRED C, b Olds, Alta, Jan 6, 29; m 54; c 2. GEOLOGY. Educ: Univ Alta, BSc, 54, MSc, 56; Princeton Univ, PhD(paleont), 59. Prof Exp: Paleontologist, Calif Standard Co, 59-64; asst prof, 64-67, ASSOC PROF PALEONT & STRATIG, UNIV WESTERN ONT, 67- Concurrent Pos: Lectr, Univ Alta, 60-61. Mem: Am Asn Petrol Geologists; Paleont Soc; Can Palaeont Asn. Res: Lower Paleozoic biostratigraphy; Devonian stratigraphy and paleontology; graptolite biostratigraphy; Upper Silurian and Lower Devonian brachiopods. Mailing Add: Dept of Geol Univ of Western Ont London ON Can

LENZ, DAVID E, physical chemistry, biochemistry, see 12th edition

LENZ, GEORGE H, b Irvington, NJ, Oct 9, 39; m 61; c 2. NUCLEAR PHYSICS. Educ: Rutgers Univ, AB, 61, MS, 63, PhD(physics), 67. Prof Exp: Asst prof physics, Univ Va, 67-71; ASSOC PROF PHYSICS & CHMN DEPT, SWEET BRIAR COL, 71- Mem: Am Phys Soc; Am Asn Physics Teachers. Res: Analogue states; compound nucleus and direct reactions; Coulomb energy systematics. Mailing Add: Dept of Physics Sweet Briar Col Sweet Briar VA 24595

LENZ, GEORGE RICHARD, b Chicago, Ill, Nov 22, 41; m 70; c 3. ORGANIC CHEMISTRY. Educ: Ill Inst Technol, BS, 63; Univ Chicago, MS, 65, PhD(chem), 67. Prof Exp: Nat Cancer Inst fel, Yale Univ, 67-69; res investr, 69-71, RES CHEMIST, G D SEARLE & CO, 71- Mem: AAAS; Am Chem Soc; The Chem Soc. Res: Photochemistry; medicinal chemistry. Mailing Add: Div of Chem Res G D Searle & Co PO Box 5110 Chicago IL 60680

LENZ, LEE WAYNE, b Bozeman, Mont, Oct 12, 15. BOTANY. Educ: Mont State Col, BS, 37; Univ La, MS, 39; Wash Univ, PhD(bot), 48. Prof Exp: Assoc prof, 51-58, cytologist & plant breeder, 48-59, actg dir, 59-60, PROF BOT, CLAREMONT GRAD SCH, 58-, DIR, RANCHO SANTA ANA BOT GARDEN, 60- Concurrent Pos: Rockefeller Found fel, Mexico. Res: Plant breeding and cytology of ornamental plants with special reference to iris and orchids; taxonomy of genus Iris. Mailing Add: Rancho Santa Ana Bot Garden 1500 N College Ave Claremont CA 91711

LENZ, PAUL HEINS, b Newark, NJ, Mar 29, 38; m 60; c 4. PHYSIOLOGY, ENDOCRINOLOGY. Educ: Franklin & Marshall Col, BS, 60; Rutgers Univ, MS, 64, PhD(endocrinol), 66. Prof Exp: Asst prof, 66-70, ASSOC PROF PHYSIOL, FAIRLEIGH DICKINSON UNIV, 70- Concurrent Pos: Univ res grants, 68-71, Eli Lilly grant, 70; Ciba grants, 70-71. Mem: Endocrine Soc; Am Oil Chemists' Soc; Am Asn Clin Chemists; Am Heart Asn. Res: Development of micro-chemical techniques; hormonal and biochemical control of lipid metabolism; platelet aggregation and its control. Mailing Add: Dept of Biol Sci Fairleigh Dickinson Univ Madison NJ 07940

LENZ, ROBERT WILLIAM, b New York, NY, Apr 28, 26; m 53; c 4. POLYMER CHEMISTRY. Educ: Lehigh Univ, BS, 49; Inst Textile Technol, MS, 51; State Univ NY, PhD(polymer chem), 56. Prof Exp: Res chemist, Chicopee Mfg Corp, 51-53; res chemist, Polymer Res Lab, Dow Chem Co, 55-61, Eastern Res Lab, 61-63; asst dir, Fabric Res Labs, Inc, 63-66; assoc prof, 66-69, PROF CHEM ENG, UNIV MASS, AMHERST, 69- Concurrent Pos: Vis prof, Univ Mainz, Germany, 72-73 & Royal Inst Technol, Stockholm, Sweden, 75. Mem: Am Chem Soc; Fiber Soc; The Chem Soc; Am Inst Chem Eng. Res: Monomer and polymer synthesis; kinetics and mechanism of polymerization; structure-property relations of polymers; reactions of polymers; new polymeric materials and applications. Mailing Add: Dept of Chem Eng Univ of Mass Amherst MA 01002

LEO, MICAH WEI-MING, b Hangchow, China, Apr 18, 26; m 59; c 2. CHEMISTRY. Educ: Nat Taiwan Univ, BS, 50; Univ RI, MS, 57; Rutgers Univ New Brunswick, PhD(soils, agr biochem), 60. Prof Exp: Chemist, Taiwan Fertilizer Co, 50-55; res officer soil physics, Can Dept Agr, 60-62; sr res scientist, Atomic Energy Comn fel radioactive fallout in precipitation & Dept Defense Stardust Proj fel, Isotopes Inc, NJ, 62-64; assoc prof chem, Fla Mem Col, 64-65 & Biola Col, 65-67; plant nutritionist & NSF fel exudation, Univ Calif, Los Angeles, 67-68; assoc prof, 68-69, PROF CHEM, BARRINGTON COL, 69- Concurrent Pos: Abstractor, Chem Abstr Serv, 67-; provost, Chung Yuan Col Sci & Eng, 74-75. Mem: Am Chem Soc; Am Nuclear Soc; Am Asn Physics Teachers. Res: Chemistry curriculum research in self-paced studies; environmental and food chemistry and related subjects; relationship between science and Bible. Mailing Add: Sci Div Barrington Col Barrington RI 02806

LEON, ARTHUR SOL, b Brooklyn, NY, Apr 26, 31; m 56; c 3. MEDICAL RESEARCH, CLINICAL PHARMACOLOGY. Educ: Univ Fla, BS, 52; Univ Wis-Madison, MS, 54, MD, 57. Prof Exp: Intern, Henry Ford Hosp, Detroit, 57-58; fel internal med, Lahey Clin, Boston, 58-60; fel cardiol, Sch Med, Univ Miami & Jackson Mem Hosp, 60-61; chief gen med & cardiol, 34th Gen Hosp, US Army, France, 61-64, cardiol consult, US Armed Forces, France, 61-64, res cardiologist, Dept Cardiorespiratory Dis, Walter Reed Army Inst Res, 64-67, mem med eval team, Gemini & Apollo Projs, 66-67; dir clin pharmacol, Roche Spec Treatment Unit, Newark Beth Israel Med Ctr, 67-73; ASSOC PROF CLIN PHARMACOL & CARDIOL & DIR LABS & APPL RES, LAB PHYSIOL HYG, SCH PUB HEALTH, UNIV MINN, MINNEAPOLIS, 73- Concurrent Pos: Res assoc, Dept Clin Pharmacol, Hoffmann-La Roche Inc, 67-73; from instr to assoc prof, Col Med & Dent NJ, 67-73; chief med serv, 322nd Gen Hosp, US Army Reserve, Newark, 67-73; sr investr multiple coronary risk factor intervention traits, Univ Minn, Minneapolis, 73-; chief cardiol, 5501st Army Hosp, Ft Snelling, Minn, 73-; attd physicisn, Univ Hosp, Minneapolis & St Paul Ramsay Hosp, 73-; med dir, Minn Sports-Health Rehab Ctr, 75- Mem: Am Col Cardiol; Am Col Chest Physicians; Am Physiol Asn; Am Soc Pharmacol & Exp Therapeut; Am Col Sports Med. Res: Prevention of coronary heart disease by risk factor modification; cardiac rehabilitation; exercise testing; effects of exercise conditioning; evaluation of new cardiovascular and lipid-lowering drugs. Mailing Add: Lab Phys Hyg Stadium Gate 27 Univ of Minn Sch of Pub Health Minneapolis MN 55455

LEON, BURKE, b Brooklyn, NY, Feb 26, 38; m 63; c 3. PHYSICAL CHEMISTRY. Educ: Syracuse Univ, BS, 60, MS, 62; Univ Pittsburgh, PhD(solid state chem), 69. Prof Exp: Mem tech staff, Bell Tel Labs, 70-74; mgr appl res photopolymers, 74-75, MGR RES & DEVELOP, DEA EQUIP DIV, HUNT CHEM CO, 75- Mem: Am Chem Soc. Res: Use of physics, chemistry and engineering as they apply to the integrated circuit industry, particularly processing, etching, photolithography automation and equipment design. Mailing Add: 945 W 23rd St Tempe AZ 85281

LEON, EDWARD, organic chemistry, see 12th edition

LEON, HENRY A, b San Francisco, Calif, Sept 25, 28; m 58; c 3. ENVIRONMENTAL PHYSIOLOGY, AEROSPACE BIOLOGY. Educ: Univ Calif, Berkeley, BS, 52, PhD(physiol), 60. Prof Exp: Nat Cancer Inst fel, Wenner-Gren Inst, Stockholm, Sweden, 60-61; Milton res fel path, Harvard Med Sch, 61-62; RES SCIENTIST AEROSPACE BIOL, AMES RES CTR, NASA, 62- Concurrent Pos: Mem staff, Mass Gen Hosp, Boston, 61-62. Mem: AAAS; Am Physiol Soc; Undersea Med Soc; Aerospace Med Asn. Res: Effect of space cabin environments on blood elements; stress and the control of liver protein synthesis; nutrition and stress. Mailing Add: Ames Res Ctr NASA Moffett Field Mountain View CA 94035

LEON, JORGE, botany, see 12th edition

LEON, KENNETH ALLEN, b New York, NY, Nov 19, 37; m 63; c 2. FISH BIOLOGY. Educ: Ohio State Univ, BS, 60; Col William & Mary, MS, 63; Univ Wash, PhD(fisheries mgt), 70. Prof Exp: Biol consult, Ichthyol Assocs, 70-71; RES BIOLOGIST, TUNISON LAB FISH NUTRIT, US BUR SPORT FISHERIES & WILDLIFE, 71- Concurrent Pos: Mem, Comt on Improv Salmonid Broodstocks, 71- Mem: Am Fisheries Soc. Res: Effects of physical and nutritional factors on the growth, stamina, reproduction and survival of trout and salmon. Mailing Add: Tunison Lab of Fish Nutrit US Bur Sport Fisheries & Wildlife Cortland NY 13045

LEON, MELVIN, b Brooklyn, NY, Sept 2, 36; m 63. THEORETICAL PHYSICS. Educ: Univ Md, BS, 57; Cornell Univ, PhD(physics), 61. Prof Exp: Imp Chem Industs res fel & NSF fel theoret physics, Univ Birmingham, 61-63; res physicist, Carnegie Inst Technol, 63-66; asst prof physics, Rensselaer Polytech Inst, 66-74; MEM STAFF, LOS ALAMOS SCI LAB, 74- Mem: Am Phys Soc. Res: Elementary particle theory. Mailing Add: Los Alamos Sci Lab Mp 3 Los Alamos NM 87544

LEON, MYRON A, b Troy, NY, July 13, 26. IMMUNOLOGY. Educ: Columbia Univ, BS, 50, PhD(biochem), 54. Prof Exp: Assoc surg res, 53-64, ASSOC HEAD PATH RES, ST LUKE'S HOSP, 64-74; PROF IMMUNOL, SCH MED, WAYNE STATE UNIV, 74- Concurrent Pos: Fel, Univ Lund, Sweden, 58. Mem: Am Asn Immunol. Res: Immunochemistry; mechanisms of natural resistance to infection; complement; myeloma proteins; lymphocyte stimulation. Mailing Add: Dept of Immunol Wayne State Univ Sch of Med Detroit MI 48201

LEON, ROBERT LEONARD, b Denver, Colo, Jan 18, 25; m 47; c 4. MEDICINE, PSYCHIATRY. Educ: Univ Colo, MD, 48. Prof Exp: Intern, Univ Hosp, Ann Arbor, Mich, 48-49; resident psychiat, Med Ctr, Univ Colo, 49-52; resident child psychiat, State Dept Health, Conn, 52-53; asst dir child psychiat, Greater Kansas City Ment Health Found, 53-54; prof psychiat, Southwest Med Sch, Univ Tex, 57-67; PROF PSYCHIAT & CHMN DEPT, UNIV TEX HEALTH SCI CTR SAN ANTONIO, 67- Concurrent Pos: Chief ment health serv, USPHS, Mo, 54-57; consult, Bur Indian Affairs, 62-67; consult regional off VI, NIMH, 57-, mem psychiat training rev comt, 70-74. Mem: Fel Am Psychiat Asn; Am Pub Health Asn; Am Orthopsychiat Asn; Am Acad Child Psychiat. Res: Social psychiatry. Mailing Add: Dept of Psychiat Univ of Tex Health Sci Ctr San Antonio TX 78284

LEON, SHALOM A, b Sofia, Bulgaria, Apr 7, 35; m 62; c 3. BIOCHEMISTRY, RADIOBIOLOGY. Educ: Hebrew Univ, Jerusalem, MSc, 60, PhD(pharmacol), 64. Prof Exp: Jr res asst pharmacol, Med Sch, Hebrew Univ, Jerusalem, 60-64; res assoc biochem, Ind Univ, 65-67; MEM BIOSCI STAFF, ALBERT EINSTEIN MED CTR, 68- Mem: AAAS; Radiation Res Soc; Am Chem Soc; NY Acad Sci. Res: Mechanism of antibiotic action; biosynthesis of nucleic acids and proteins; relationship between structure and biological activity of toxins from microorganisms; effect of radioprotective agents against ionizing radiation. Mailing Add: Radiation Res Lab Albert Einstein Med Ctr Philadelphia PA 19141

LEONARD, ALBERT CLARK, cryogenics, mechanical engineering, see 12th edition

LEONARD, ALVIN ROBERT, b New York, NY, Jan 9, 18; m 42; c 2. MEDICINE, PUBLIC HEALTH. Educ: Univ Southern Calif, MD, 42; Harvard Univ, MPH, 47.

Prof Exp: Health officer, Solano County, Calif, 47-48; pub health physician, State Dept Pub Health, Calif, 48-50, asst dir pub health, San Diego County, 50-57, dir pub health, Berkeley, 57-70; PROF COMMUNITY MED, COL MED, UNIV ARIZ, 70- Concurrent Pos: Clin prof, Sch Pub Health, Univ Calif, Berkeley, 57-70. Mem: Fel Am Pub Health Asn. Res: Application of public health and administration principles in operating programs; community and social forces in health and disease. Mailing Add: Dept of Community Med Univ of Ariz Col of Med Tucson AZ 85721

LEONARD, ARNOLD S, b Minneapolis, Minn, Oct 26, 30; m 50; c 4. SURGERY. Educ: Univ Minn, Minneapolis, BA, 52, BS, 53, MD, 55, PhD(surg path), 63. Prof Exp: Univ fel, 56-63, from asst prof to assoc prof, 63-73, PROF SURG, UNIV MINN, MINNEAPOLIS, 73- Mem: Am Soc Artificial Internal Organs; Am Soc Exp Path; Int Soc Hist Med; Soc Univ Surg; Am Pediat Surg Asn. Res: Gastrointestinal physiology; hypothalamic stimulation and study of gastric secretion; transplantation; extracorporeal organ perfusion; pediatric surgery; computer technology. Mailing Add: Dept of Surg Univ of Minn Minneapolis MN 55455

LEONARD, ARTHUR BYRON, b Manhattan, Kans, Apr 11, 04. ZOOLOGY. Educ: Cent Teachers Col, Okla, BS, 31; Univ Kans, AM, 33, PhD, 37. Prof Exp: From asst instr to assoc prof zool, 31-49, prof biol sci, 49-73, EMER PROF PHYSIOL & CELL BIOL, SYSTEMATICS & ECOL, UNIV KANS, 73- Concurrent Pos: Corp res fel, Mus Comp Zool, Harvard Univ, 48-49; Fulbright res scholar, Univ Bologna, 56-57; coordr develop sci prog, Univ Oriente, Venezuela, 65-69, hon prof biol. Res paleontologist, Kans State Geol Surv, 45-55, res geologist, Ill State Geol Surv, 55-65 & 69- Mem: Am Malacol Union. Res: Stratigraphic paleontology and ecology of late Cenozoic and modern Mollusca. Mailing Add: 562 Snow Hall Univ of Kans Lawrence KS 66045

LEONARD, BENJAMIN FRANKLIN, b Dobbs Ferry, NY, May 12, 21; m 50; c 2. GEOLOGY. Educ: Hamilton Col, BS, 42; Princeton Univ, MA, 47, PhD(geol), 51. Prof Exp: Geol field asst, Geol Surv Nfld, 42; from jr geologist to geologist, 43-62, GEOLOGIST-IN-CHARGE, ORE MICROS LAB, US GEOL SURV, 62- Concurrent Pos: Vis prof, Colo Sch Mines, 67-68; mem, Int Comn Ore Micros, 68-70; mem Paragenetic Comn, Int Asn Genesis of Ore Deposits, 74- Mem: Fel Mineral Soc Am; fel Geol Soc Am; Soc Econ Geol; Soc Geol Appl Mineral Deposits; Mineral Asn Can. Res: Ore deposits, especially gold, iron and tungsten; geology of central Idaho and northwest Adirondaks; ore minerals; rock-forming minerals, especially amphiboles and iron borates. Mailing Add: Cent Mineral Resources Br US Geol Survey Box 25046 Stop 912 Denver CO 80225

LEONARD, BOWEN RAYDO, JR, b Houston, Tex, Mar 7, 26; div; c 2. PHYSICS. Educ: Tex Western Col, BS, 47; Univ Wis, MS, 49, PhD(physics), 52. Prof Exp: Asst, Univ Wis, 47-51; physicist, Hanford Labs, Gen Elec Co, 52-53, sr scientist, 53-57, mgr exp physics res, 57-64; mgr exp physics res, Pac Northwest Lab, 65-67, STAFF SCIENTIST, BATTELLE-NORTHWEST, BATTELLE MEM INST, 67- Concurrent Pos: Mem, nuclear cross sect adv group, Atomic Energy Comn, 57-63, ad-hoc mem, 69-, mem cross sect eval working group, 66- Mem: Fel Am Phys Soc; Am Nuclear Soc; Sigma Xi. Res: Neutron cross section measurements; nuclear physics; x-ray scattering; slow neutron in-elastic scattering studies of solids and liquids. Mailing Add: 212 S Morain St Kennewick WA 99336

LEONARD, BYRON PETER, b Morgan City, La, Feb 26, 25; m 46. PHYSICS. Educ: Southwestern La Univ, BS, 43; Univ Tex, MA, 52, PhD, 53. Prof Exp: Proj engr, US Naval Ord Test Sta, 46-47; instr physics, Southwestern La Univ, 48-50 & Univ Tex, 50-53; chief nuclear res & develop, Gen Dynamics Corp, 53-59; sr staff engr, Space Technol Labs, 59-60; dir satellite-missile observation syst prog, 60-65, vpres & gen mgr man orbiting lab, Systs Eng Off, 65-68, VPRES & GEN MGR EL SEGUNDO TECH OPERS, AEROSPACE CORP, 68- Mem: Am Nuclear Soc; Am Inst Aeronaut & Astronaut. Res: Nuclear shielding; radiation effects to materials and operating components; radiation hazards of fission products released to the atmosphere; design of research reactors; design and use of satellite systems, particularly for surveillance applications. Mailing Add: Aerospace Corp 2350 E El Segundo Blvd El Segundo CA 90045

LEONARD, CHARLES ARTHUR, b Penns Grove, NJ, June 23, 21; m 55. PHARMACOLOGY. Educ: Philadelphia Col Pharm, BSc, 50, MSc, 51, DSc(pharmacol), 54. Prof Exp: Asst tech lab, E I du Pont de Nemours & Co, NJ, 39-42, 46-48; res asst, Philadelphia Col Pharm, 54-55; sr pharmacologist, Smith Kline & French Labs, Pa, 55-66; GROUP MGR GEN SCREENING, A H ROBINS RES LABS, 66- Mem: AAAS; Am Pharmaceut Asn; NY Acad Sci. Res: Pharmacology of antiemetic agents; appetite depressants; nasal decongestants and thrombolytics. Mailing Add: A H Robins Res Labs 1211 Sherwood Ave Richmond VA 23220

LEONARD, CHARLES BROWN, JR, b Woodbury, NJ, May 28, 34; m 55; c 2. BIOCHEMISTRY. Educ: Rutgers Univ, AB, 55; Univ Md, MS, 57, PhD(biochem), 63. Prof Exp: Asst, 55-58, from instr to assoc prof, 58-76, PROF BIOCHEM, DENT SCH, UNIV MD, BALTIMORE CITY, 76-, DIR OFF ADMIS, 75- Concurrent Pos: Consult, Dr H L Wollenweber, clin pathologist, 59-61. Mem: AAAS; Am Chem Soc; NY Acad Sci; Am Inst Chem. Res: Amino acid-incorporation into rat liver ribosomes; effect of divalent ions on structure of rat liver RNA; effect of o,p'-DDD on cellular metabolism; metabolic products of o,p'-DDD. Mailing Add: Dept of Biochem Univ of Md Dent Sch Baltimore MD 21201

LEONARD, CHARLES GRANT, b Detroit, Mich, Apr 7, 39; m 64; c 2. ASTRONOMY. Educ: Eastern Mich Univ, BS, 63; Wayne State Univ, MA, 66. Prof Exp: Instr physics, Wis State Univ-Whitewater, 66-68; ASST PROF ASTRON & PHYSICS, JACKSON COMMUNITY COL, 68- Mem: Am Asn Physics Teachers. Mailing Add: Dept of Physics Jackson Community Col Jackson MI 49201

LEONARD, CHESTER D, b New Sharon, Iowa, Aug 27, 07. SOILS. Educ: Colo Agr & Mech Col, BS, 31; Rutgers Univ, PhD(soils), 50. Prof Exp: From asst county agt to county agt, Colo Agr Mech Col, 34-35, asst exten agronomist, 35-37; state compliance supvr, Agr Adj Admin & Prod & Mkt Admin, 37-40 & 46-47, assoc horticulturist, Agr Res & Educ Ctr, 50-66, HORTICULTURIST, AGR RES & EDUC CTR, USDA, 66- Mem: Am Soc Hort Sci. Res: Mineral nutrition of plants. Mailing Add: Agr Res & Educ Ctr PO Box 1088 Lake Alfred FL 33850

LEONARD, CHRISTIANA MORISON, b Boston, Mass, Jan 22, 38; div; c 2. NEUROANATOMY, PSYCHOLOGY. Educ: Radcliffe Col, BA, 59; Mass Inst Technol, PhD(psychol), 67. Prof Exp: USPHS trainee, Rockefeller Univ, 67-70, res assoc, 70-71, asst prof neuropsychol, 71-74; asst prof anat, Mt Sinai Sch Med, 74-76; ASSOC PROF NEUROSCI, COL MED, UNIV FLA, 76- Mem: AAAS; Soc Neurosci; NY Acad Sci. Res: Neurological basis of behavior. Mailing Add: JHM Health Ctr Col of Med Univ of Fla Gainesville FL 32610

LEONARD, DAVID E, b Greenwich, Conn, Dec 28, 34; m 57; c 2. ENTOMOLOGY. Educ: Univ Conn, BS, 56, MS, 60, PhD(entom), 64. Prof Exp: Asst entomologist, Conn Agr Exp Sta, 64-69, assoc entomologist, 69-70; ASSOC PROF ENTOM, UNIV MAINE, ORONO, 70- Mem: Entom Soc Am; Entom Soc Can; Japanese Soc Pop Ecol. Res: Speciation, biosystematics, biology and ecology of insects. Mailing Add: 313 Deering Hall Dept of Entom Univ of Maine Orono ME 04473

LEONARD, EDWARD CHARLES, JR, b Burlington, NC, Aug 21, 27; m 52; c 1. POLYMER CHEMISTRY. Educ: Univ NC, BS, 47, PhD(chem), 51; Univ Chicago, MBA, 74. Prof Exp: Asst, Univ NC, 47-50; sr res chemist, Res Dept, Bakelite Co, 51-56, group leader, Union Carbide Plastic Co, 56-64; res mgr, Borden Chem Co, 64-67; mgr indust chem prod lab, Res & Develop Div, 67-74, DIR OPER KRAFTCO CORP, ENENCO, INC, 74-; TECH DIR, HUMKO SHEFFIELD CHEM CO, 73- Mem: Am Chem Soc. Res: Synthetic surface active agents; ionic polymerizations; graft polymers; fatty acids; homogeneous catalysis. Mailing Add: Box 398 White Sta Tower Memphis TN 38101

LEONARD, EDWARD H, b Berwick, Maine, Feb 21, 19; m 51; c 1. ANALYTICAL CHEMISTRY. Educ: Dartmouth Col, AB, 42; Tufts Univ, MA, 54; Univ NH, MS, 61. Prof Exp: Res & develop engr, Elec Res Lab, Simplex Wire & Cable Co, 42-46, Eng Dept, 46-51; head sci dept high sch, NJ, 51-60, sci coord, 60-64; ASSOC PROF PHYSICS & NAT SCI, WORCESTER STATE COL, 64- Mem: AAAS; Am Chem Soc; Am Asn Physics Teachers; Nat Sci Teachers Asn. Res: Design and development of apparatus and aids for the teaching of physical science. Mailing Add: Dept of Physics Worcester State Col Worcester MA 01602

LEONARD, EDWARD JOSEPH, b Boston, Mass, Mar 20, 26; m 56; c 3. MEDICINE. Educ: Harvard Med Sch, MD, 49. Prof Exp: Investr, Nat Heart Inst, 53-69; investr, 69-73, HEAD TUMOR ANTIGEN SECT, BIOL BR, NAT CANCER INST, 73- Concurrent Pos: From instr to assoc clin prof, George Washington Univ, 57-74. Mem: Am Fedn Clin Res; Cardiac Muscle Soc; Am Asn Immunol. Res: Tumor immunology. Mailing Add: Tumor Antigen Sect Biol Br Nat Cancer Inst Bethesda MD 20014

LEONARD, FRED, b Brooklyn, NY, Apr 20, 15; m 45; c 2. CHEMISTRY. Educ: Univ Ark, BS, 38; Polytech Inst Brooklyn, MS, 42, PhD(chem), 47. Prof Exp: Res chemist, Westvaco Chlorine Prod Corp, 40-42; res assoc, Polytech Inst Brooklyn, 42-46; res assoc, Princeton Univ, 46-48; chief plastics develop br, Prosthetics Res Lab, 48-66, sci dir med biomech res lab, US Dept Army, Walter Reed Army Med Ctr, Washington, DC, 66-74; ASSOC DEAN RES, GEORGE WASHINGTON UNIV SCH MED & HEALTH SCI, 74- Concurrent Pos: Mem comt prosthetics res & develop, Nat Acad Sci-Nat Res Coun, 50-; sci adv comt mat, US Dept Army, 58-; basic res biomat, Clemson Univ, 73. Honors & Awards: Superior Accomplishment Awards, US Army, 55-58, Meritorious Civilian Serv Award, 62 & Exceptional Civilian Serv Award, 66; Presidential Distinguished Fed Civilian Serv Award, 71. Mem: AAAS; Am Chem Soc; NY Acad Sci. Res: Polymer chemistry; biomechanics; biomaterials; calcification mechanisms. Mailing Add: 7713 Beech Tree Rd Bethesda MD 20034

LEONARD, FREDERIC ADAMS, b Bangor, Maine, Mar 14, 21; m 52. PHYSIOLOGY. Educ: Univ Maine, BS, 43, MS, 47; George Washington Univ, PhD, 55. Prof Exp: Res bacteriologist, US Army Biol Labs, Ft Detrick, 48-55, asst dir biol res, 56-61, chief, Med Bact Div, 61-64; staff assoc, Sci Facil Eval Group, 64-72, ASSOC PROG DIR NEUROBIOL, NSF, 72- Mem: Am Soc Microbiol; Am Acad Microbiol. Res: Infectious diseases; combined infections; immunology. Mailing Add: Behav & Neurol Sci Div Nat Sci Found Washington DC 20550

LEONARD, GUY WILLIAM, b Bloomington, Ind, Aug 3, 21; m 47; c 3. ANALYTICAL CHEMISTRY. Educ: Univ Ind, BS, 45, AM, 46; Mass Inst Technol, PhD(chem), 49. Prof Exp: From asst prof to assoc prof anal chem, Kans State Univ, 49-57; head, Gen Res Br, Chem Div, Res Dept, US Naval Ord Test Sta, 57-59, head Propellants Div, Propulsion Develop Dept, 59-60, consult, bur weapons, 60-61, from asst head to assoc head, 61-63, ASST TECH DIR PROPULSION & EXP DEVELOP & HEAD PROPULSION DEVELOP DEPT, NAVAL WEAPONS CTR, 63- Mem: Am Chem Soc. Res: Propellants; explosives; pyrotechnics. Mailing Add: Propulsion Develop Dept Naval Weapons Ctr China Lake CA 93555

LEONARD, HENRY SIGGINS, JR, b Needham, Mass, Oct 12, 30; m 54; c 1. MATHEMATICS. Educ: Mich State Univ, BS, 52; Harvard Univ, AM, 53, PhD(math), 58. Prof Exp: From asst prof to assoc prof math, Carnegie Inst Technol, 58-68; PROF MATH, NORTHERN ILL UNIV, 68- Concurrent Pos: Prin investr, NSF grants, 59-68; vis assoc prof, Univ Ill, Urbana, 67- Mem: Am Math Soc; Math Asn Am. Res: Theory of groups of finite order. Mailing Add: Dept of Math Northern Ill Univ DeKalb IL 60115

LEONARD, HERBERT ARTHUR, b Thorndike, Maine, July 1, 17; m 41; c 2. ANIMAL SCIENCE. Educ: Univ Maine, BS, 39; Cornell Univ, MS, 50. Prof Exp: County 4H Club agent, 39-41, asst county agent, 41-42, county agent, 42-50, exten dairyman, 50-51, assoc prof dairy sci, 51-63, PROF ANIMAL SCI, UNIV MAINE, ORONO, 63-, FARM MGR, 70- Mem: Am Dairy Sci Asn. Res: Dairy feeding and management of dairy cattle. Mailing Add: Dept of Animal Sci Univ of Maine 28 Rogers Hall Orono ME 04473

LEONARD, JACQUES WALTER, b Montreal, Que, Aug 7, 36; m 63; c 2. POLYMER CHEMISTRY, PHYSICAL CHEMISTRY. Educ: Univ Montreal, BSc, 60, MSc, 61, PhD(chem), 64. Prof Exp: Can Nat Res Coun fel, Univ Leeds, 64-66; from asst prof to assoc prof chem, 66-75, PROF CHEM, LAVAL UNIV, 75- Mem: Chem Inst Can. Res: Kinetics and thermodynamics of polymerizations in solution; effect of the medium on the equilibrium of reversible of homo- and copolymerizations; thermodynamics of polymer solutions and binary liquid mixtures. Mailing Add: Dept of Chem Laval Univ Quebec PQ Can

LEONARD, JAMES JOSEPH, b Schenectady, NY, June 17, 24; m 54; c 4. INTERNAL MEDICINE. Educ: Georgetown Univ, MD, 50. Prof Exp: From intern to jr asst resident med, Georgetown Univ Hosp, 50-52; asst resident med serv, Boston City Hosp, Mass, 52-53; resident, Pulmonary Dis Div, DC Gen Hosp, 54-55; instr, Sch Med, Georgetown Univ, 55-56; instr, Med Sch, Duke Univ, 56-57; asst prof & coordr cardiovasc teaching, Sch Med & dir sect cardiol, Univ Med Ctr, Georgetown Univ, 57-59; from asst prof to assoc prof med, Univ Tex Med Br, 59-62; dir cardiopulmonary lab, 61-62; assoc prof med & dir cardiac diag lab, Ohio State Univ, 62-63; assoc prof med & dir div cardiol, 63-67, actg chmn dept med, 70-71, PROF MED, SCH MED, UNIV PITTSBURGH, 67-, CHMN DEPT, 71- Concurrent Pos: Washington Heart Asn fel cardiol, Georgetown Univ Hosp, 53-54; Am Trudeau Soc fel, Pulmonary Dis Div, DC Gen Hosp, 54-55; NIH cardiac trainee, Duke Univ Hosp, 56-57; med officer, DC Gen Hosp, 55-56, chief cent heart sta, 57-59; attend cardiol, Mt Alto's Vet Hosp, DC, 57-59. Mem: Am Fedn Clin Res; Am Col Physicians; Am Heart Asn; Am Col Cardiol. Res: Cardiopulmonary physiology. Mailing Add: Dept of Med Univ of Pittsburgh Sch of Med Pittsburgh PA 15261

LEONARD, JOHN ALEX, b Swindon, Eng, Dec 13, 37; m 61; c 2. INDUSTRIAL CHEMISTRY. Educ: Univ London, BSc, 59, PhD(chem), 62. Prof Exp: Res chemist polymers, Shell Develop Co, Calif, 63-66; from sr scientist catalysis to bus planning,

Imperial Chem Indust, UK, 66-74; RES ADV, CAN INDUST LTD, 74- Concurrent Pos: Fel, Harvard Univ, 62-63. Mem: Am Chem Soc; Chem Can; The Chem Soc. Res: Catalytic, electrochemical and biological processes and research management. Mailing Add: Can Indust Ltd 630 Dorchester Blvd Montreal PQ Can

LEONARD, JOHN EDWARD, b Great Falls, Mont, Apr 18, 18; m 40; c 2. ORGANIC CHEMISTRY, PHYSICAL CHEMISTRY. Educ: Antioch Col, BS, 42; Ohio State Univ, PhD(chem), 49. Prof Exp: Res engr, Battelle Mem Inst, 42-46; res fel, Calif Inst Technol, 49-52; res scientist, Beckman Instruments, Inc, 52-56, chief proj engr, 56-62, sr scientist, 62-66, mgr appl res, Med Develop Activity, 66-69; chief scientist, Int Biophysics Corp, 69-71; CONSULT ELECTROCHEM SENSORS & INSTRUMENTATION, 71- Mem: Am Chem Soc. Res: Analytical instruments, particularly electrochemical, for chemical research and industrial use; biomedical engineering; medical instrumentation research. Mailing Add: PO Box 5036 Fullerton CA 92635

LEONARD, JOHN JOSEPH, b Philadelphia, Pa, Feb 12, 49; m 72; c 2. PHYSICAL ORGANIC CHEMISTRY. Educ: Drexel Univ, BS, 72, PhD(phys org chem), 72. Prof Exp: Res assoc chem, Univ Pa, 72-73; SR RES CHEMIST, ARCO CHEM CO, DIV ATLANTIC RICHFIELD CO, 73- Concurrent Pos: Adj prof math, Drexel Univ Evening Div, 73- Mem: Am Chem Soc; Int Catalysis Soc; AAAS. Res: Kinetics and mechanisms of organic reactions especially catalysis of organic oxidation reactions(heterogeneous and homogeneous catalysis); spectroscopy of organic molecules. Mailing Add: Arco Chem Co 500 S Ridgeway Ave Glenolden PA 19036

LEONARD, JOHN LANDER, b Jamaica, NY, Oct 20, 35; m 65; c 1. MATHEMATICS. Educ: Carnegie Inst Technol, BS, 57; Univ Calif, Santa Barbara, MA, 63, PhD(math), 66. Prof Exp: Opers analyst, Comput Dept, Gen Elec Co, 59-60, mem tech staff, Tech Mil Planning Oper, 60-61; asst math, Univ Calif, Santa Barbara, 61-63, 64-66; ASST PROF MATH, UNIV ARIZ, 66- Mem: Am Math Soc. Res: Graph theory, extremal problems, connectivity; real function theory; mathematical analysis. Mailing Add: Dept of Math Univ of Arizona Tucson AZ 85721

LEONARD, JOHN MORRISON, chemistry, see 12th edition

LEONARD, JON NORMAN, biomathematics, biophysics, see 12th edition

LEONARD, JOSEPH THOMAS, b Scranton, Pa, Aug 8, 32; m 58; c 4. FUEL SCIENCE. Educ: Univ Scranton, BS, 54; Pa State Univ, University Park, PhD(fuel technol), 59. Prof Exp: Res asst chem, Pa State Univ, University Park, 54-59; RES CHEMIST FUELS, NAVAL RES LAB, WASHINGTON, DC, 59- Mem: Am Chem Soc. Res: Electrostatic charging of hydrocarbon liquids and fuels; suppression of evaporation of hydrocarbons and smoke abatement techniques. Mailing Add: 6424 Rotunda Ct Springfield VA 22150

LEONARD, JUSTIN WILKINSON, b Moulton, Iowa, Oct 28, 09; m 36. ZOOLOGY. Educ: Grinnell Col, AB, 31; Univ Mich, AM, 32, PhD(zool, entom), 37. Prof Exp: Fisheries investr, State Dept Conserv, Mich, 34-35, asst aquatic biologist, 36-38, assoc aquatic biologist chg Hunt Creek Exp Sta, 39-42, assoc fisheries biologist, 46-48, fisheries biologist, 49-50, asst dept dir, 51-64; chmn dept wildlife & fisheries, 65-67, chmn dept resource planning & conserv, 67-70, PROF NATURAL RESOURCES, SCH NATURAL RESOURCES & PROF ZOOL & RES ASSOC, MUS ZOOL, UNIV MICH, ANN ARBOR, 64- Concurrent Pos: Spec consult, USPHS, 58-61; lectr, Mich State Univ, 58-64. Mem: Fel AAAS; hon mem Am Fisheries Soc; fel Entom Soc Am; Wildlife Soc (pres, 56); Am Soc Ichthyologists & Herpetologists. Res: Conservation of renewable natural resources; systematics and ecology of aquatic insects; fishery biology; impact of cultural influences on environmental attitudes. Mailing Add: Sch of Natural Resources Univ of Mich Ann Arbor MI 48104

LEONARD, MARGARET IVES, b Lebanon, Pa, June 5, 18; m 64. CHEMISTRY. Educ: Milwaukee-Downer Col, AB, 40; Univ Wis, MS, 43, PhD(biochem), 45. Prof Exp: Clin chemist, 40-41; biochemist, Am Can Co, Ill, 45-57, sr biochemist & toxicologist, 57-59, group leader biochem, 59-61; vpres, Wedge's Creek Res Farm, Inc, Wis, 61-66; ASSOC PROF CHEM, UNIV WIS CTR, MARSHFIELD-WOOD COUNTY, 66-; PRES, RED PINES, INC, 66- Concurrent Pos: Dipl, Am Bd Nutrit. Mem: Am Chem Soc; Inst Food Technol; Soc Toxicol; NY Acad Sci. Res: Microbiological assays; physiology; new drugs; food additives; animal nutrition; biochemistry, vitamin assay. Mailing Add: Red Pines Inc Willard WI 54493

LEONARD, MARTHA FRANCES, b New Brunswick, NJ, May 10, 16. PEDIATRICS. Educ: NJ Col Women, BSc, 36; Johns Hopkins Univ, MD, 40. Prof Exp: Intern, Baltimore City Hosp, 40-41; asst resident med, Vanderbilt Univ Hosp, 42-43; asst resident pediat, New York Hosp, 43-46; pvt pract, 46-60; from instr to asst prof pediat, 62-69, ASSOC PROF CLIN PEDIAT, CHILD STUDY CTR, YALE UNIV, 69- Concurrent Pos: Fel, Child Study Ctr, Yale Univ, 60-62. Mem: Am Acad Pediat; Am Orthopsychiat Asn. Res: Normal and deviant child development; effects of deprivation; failure to thrive; child abuse; developmental impact of conditions such as genetic, metabolic and endocrine disorders. Mailing Add: Child Study Ctr Yale Univ New Haven CT 06510

LEONARD, NELSON JORDAN, b Newark, NJ, Sept 1, 16; m 47; c 4. ORGANIC BIOCHEMISTRY. Educ: Lehigh Univ, BS, 37; Oxford Univ, BSc, 40; Columbia Univ, PhD(org chem), 42. Hon Degrees: DSc, Lehigh Univ, 63. Prof Exp: Fel & asst chem, Univ Ill, 42-43, instr, 43-44, assoc, 44-45; sci consult & spec investr, Field Intel Agency Tech, US Army Dept & US Dept Commerce, 45-46; assoc, 46-47, from asst prof to prof chem, 47-73, head div org chem, 54-63, PROF BIOCHEM, UNIV ILL URBANA, 73-, MEM CTR ADVAN STUDY, 68- Concurrent Pos: Mem, Comt Med Res, 44-46; ed, Org Synthesis, 51-58, ed-in-chief, 56; Am-Swiss Found lectr, 53, 70; Guggenheim Mem Found fel, 59, 67; mem prog comt basic phys sci, Alfred P Sloan Found, 61-66; Stieglitz lectr, 62; mem educ adv bd, John Simon Guggenheim Mem Found, 69-; Edgar Fahs Smith Mem lectr, Univ Pa, 75. Honors & Awards: Award, Am Chem Soc, 63, Edgar Fahs Smith Award, 75; Synethetic Org Chem Mfrs Award, 70. Mem: Nat Acad Sci; AAAS; Am Chem Soc; The Chem Soc; Swiss Chem Soc. Res: Structure, synthesis and biological activity of cytokinins; modification of nucleic acid bases; fluorescent probes of coenzyme and nucleic acid structures; intramolecular interactions. Mailing Add: Dept of Biochem Univ of Ill Urbana IL 61801

LEONARD, OLIVER ANDREW, plant physiology, deceased

LEONARD, RALPH AVERY, b Louisburg, NC, Mar 2, 37; m 58; c 3. SOIL CHEMISTRY. Educ: NC State Univ, BS, 59, PhD(soil chem), 66; Purdue Univ, MS, 62. Prof Exp: Instr soil sci, NC State Univ, 62-66; RES SOIL SCIENTIST, USDA, 66- Mem: Am Chem Soc; Soil Sci Soc Am; Am Soc Agron. Res: Physical chemistry of soils; fate of pesticides in soil and water; soil chemical aspects of waste disposal and utilization on the land. Mailing Add: USDA Southern Piedmont Conserv Res Ctr PO Box 555 Watkinsville GA 30677

LEONARD, RALPH BEAUMONT, biomaterials, see 12th edition

LEONARD, REID HAYWARD, b Littleton, NH, Aug 28, 18; m 46; c 3. CHEMISTRY. Educ: Univ Vt, BS, 40; Univ WVa, MS, 42; Univ Wis, PhD(biochem), 47. Prof Exp: Asst, Exp Sta, Univ WVa, 40-42; asst, Univ Wis & Forest Prod Lab, US Forest Serv, 43-45; res chemist, Salvo Chem Corp, Wis, 46-47; res chemist, Newport Industs, 47-56; CONSULT BIOCHEMIST, 56- Mem: AAAS; Am Chem Soc; fel Am Inst Chem. Res: Chemistry of wood; sugars from wood; lignin; levulinic acid; clinical biochemistry; kidney stones; blood lipids; gas chromatography. Mailing Add: 537 Brent Lane Pensacola FL 32503

LEONARD, ROBERT MEYER, b Picatello, Idaho, Oct 2, 22; m 46; c 3. PHARMACOGNOSY, PHARMACOLOGY. Educ: Idaho State Univ, BS, 44; Univ Minn, PhD(pharmacog), 53. Prof Exp: Pharmacist, retail pharms, 45; asst pharm, Univ Minn, 45-46, asst pharmacog, 46-47; pharmacist, retail pharms, 47-48; from asst prof to prof pharmacol & pharmacog, George Washington Univ, 51-64, asst dean Sch Pharm, 56-61, actg dean, 61-62, dean, 62-64; grants assoc, 64-65, HEALTH SCI ADMINR, DIV RES GRANTS, NIH, 65- Mem: AAAS; NY Acad Sci. Res: General pharmacology; biological origin of drugs; general metabolism. Mailing Add: 1009 Laredo Rd Silver Spring MD 20901

LEONARD, ROBERT STUART, b Berkeley, Calif, Jan 20, 30; m 56; c 2. GEOPHYSICS, AERONOMY. Educ: Univ Nev, BS, 52, MS, 53; Univ Alaska, PhD(geophys), 61. Prof Exp: Res asst auroral studies, Geophys Inst, Univ Alaska, 53-58, instr, 58-60; radio physicist, 61-62, sr ionospheric physicist, 62-69, prog mgr, 69-72, ASST DIR, STANFORD RES INST, 72- Mem: Int Union Radio Sci; Am Geophys Union; Am Phys Soc. Res: Chemical seeding in the ionosphere; transionospheric propagation; ionospheric disturbances. Mailing Add: Stanford Res Inst 333 Ravenswood Ave Menlo Park CA 94025

LEONARD, ROBERT THOMAS, b Providence, RI, Dec 18, 43; m 66; c 1. PLANT PHYSIOLOGY. Educ: Univ RI, BS, 65, MS, 67; Univ Ill, Urbana, PhD(biol), 71. Prof Exp: Fel plant physiol, Univ Ill & Purdue Univ, 71-73; ASST PROF PLANT PHYSIOL, UNIV CALIF, RIVERSIDE, 73- Mem: Am Soc Plant Physiologists; Am Inst Biol Sci; AAAS. Res: Physiology and biochemistry of ion transport in plants. Mailing Add: Dept of Plant Sci Univ of Calif Riverside CA 92502

LEONARD, SAMUEL LEESON, b Elizabeth, NJ, Nov 26, 05; m 34; c 2. ZOOLOGY. Educ: Rutgers Univ, BSc, 27; Univ Wis, MSc, 29, PhD(zool), 31. Prof Exp: Asst zool, Univ Wis, 27-31; fel, Nat Res Coun, Col Physicians & Surg, Columbia Univ, 31-33; asst prof biol, Union Univ, NY, 33-37; asst prof zool, Rutgers Univ, 37-41; from assoc prof to prof, 41-71, EMER PROF ZOOL, CORNELL UNIV, 71- Mem: AAAS; Am Soc Zool; Soc Exp Biol & Med; Endocrine Soc; Am Physiol Soc; Am Asn Anat. Res: Endocrinology; physiology of reproduction; mechanisms of hormone action. Mailing Add: Div of Biol Sci Emerson Hall Cornell Univ Ithaca NY 14850

LEONARD, STANLEY LEE, b Oakland, Calif, Aug 27, 26; m 48; c 4. PLASMA PHYSICS. Educ: Principia Col, BS, 47; Univ Calif, PhD(physics), 53. Prof Exp: Physicist, Radiation Lab, Univ Calif, 52-53; instr physics, Principia Col, 53-55, asst prof, 55-56; mem tech staff, Ramo-Wooldridge Corp, 56-59 & Space Technol Labs, Inc, 59-60; mem tech staff, 60-64, HEAD, PLASMA RADIATION DEPT, PLASMA RES LAB, AEROSPACE CORP, 64- Mem: Am Phys Soc. Res: Experimental plasma physics. Mailing Add: 5019 Rockvalley Rd Palos Verdes Peninsula CA 90274

LEONARD, THOMAS JOSEPH, b Watertown, Mass, July 27, 37; m 65; c 1. DEVELOPMENTAL BIOLOGY. Educ: Clark Univ, AB, 62; Ind Univ, PhD(microbiol), 67. Prof Exp: NIH fel, Harvard Univ, 67-68; assoc prof mycol, Univ Ky, 68-74; ASSOC PROF BOT & BACT, UNIV WIS-MADISON, 74- Mem: AAAS; Genetics Soc Am; Mycol Soc Am; Am Soc Microbiol. Res: Physiology and genetics of fungi as applied to development; genetics and physiological aspects of cell differentiation. Mailing Add: Dept of Bot Univ of Wis Madison WI 53705

LEONARD, WALTER RAYMOND, b Scott Co, Va, July 5, 23; m 51; c 2. ZOOLOGY, PHYSIOLOGY. Educ: Tusculum Col, BA, 46; Vanderbilt Univ, MS, 47, PhD(zool), 49. Prof Exp: Asst prof biol, 49-50, assoc prof & acting chmn dept, 50-53, REEVES PROF BIOL & CHMN DEPT, WOFFORD COL, 54- Res: Respiratory metabolism of Allomyces arbuscula; effects of activity on growth in hydra. Mailing Add: Dept of Biol Wofford Col Spartanburg SC 29301

LEONARD, WILLIAM J, JR, b Ravenna, Ohio, Apr 28, 36. PHYSICAL CHEMISTRY. Educ: Kent State Univ, BS, 58; Purdue Univ, PhD(chem), 63. Prof Exp: Fel phys polymer chem, Stanford Univ, 63-65; chemist, Shell Develop Co, 65-72; DIR POLYMER SCI, DYNAPOL, INC, 72- Mem: AAAS; Am Chem Soc; NY Acad Sci; Inst Food Technologists. Res: Protein conformation; polymer chain statistics; solution thermodynamics; optical rotatory dispersion; liquid crystals. Mailing Add: 1454 Page Mill Rd Palo Alto CA 94304

LEONARD, WILLIAM WILSON, b Portland, Maine, May 1, 34; m 61; c 2. MATHEMATICS. Educ: Univ Tampa, BS, 60; Univ SC, MS, 63, PhD(math), 65. Prof Exp: Asst prof math, Susquehanna Univ, 64-65; from asst prof to assoc prof, 65-74, PROF MATH, GA STATE UNIV, 74- Mem: Am Math Soc; Math Asn Am; Math Soc France. Res: Module theory; homological algebra. Mailing Add: Dept of Math Ga State Univ Atlanta GA 30303

LEONARDS, JACK RALPH, b Montreal, Que, Feb 25, 19; m 41; c 5. BIOCHEMISTRY. Educ: McGill Univ, BSc, 39; Va Polytech Inst, MS, 40 & 41; Western Reserve Univ, PhD(biochem), 43, MD, 57. Prof Exp: From instr to sr instr biochem, Sch Med, Case Western Reserve Univ, 43-46, from sr instr to asst prof clin biochem, 46-53; assoc res dir, 43-48, RES DIR, BEN VENUE LABS, INC, 48-; ASSOC PROF CLIN BIOCHEM, SCH MED, CASE WESTERN RESERVE UNIV, 53- Mem: AAAS; Am Soc Biol Chem. Res: Intestinal absorption of foods; treatment of renal insufficiency; mechanism of action of salicylates; chemistry of rayon; isolation of antibiotics; biochemistry of diabetes. Mailing Add: Dept of Biochem Sch of Med Case Western Reserve Univ Cleveland OH 44106

LEONE, CHARLES ABNER, b Camden, NJ, July 13, 18; m 41; c 3. IMMUNOLOGY, RADIATION BIOLOGY. Educ: Rutgers Univ, BS, 40, MS, 42, PhD, 49. Prof Exp: Asst zool, Rutgers Univ, 40-42, instr, 46-49; from asst prof to prof, Univ Kans, 49-68; PROF BIOL & DEAN GRAD SCH, BOWLING GREEN STATE UNIV, 68-, VPROVOST RES & GRAD STUDIES, 71- Concurrent Pos: Resident res assoc, Argonne Nat Lab, 55, consult, 55-60; adj prof, Med Col Ohio, Toledo, 69- Mem: Fel AAAS; Am Asn Immunologists. Res: Immunochemistry; radiation biophysics; comparative serology among arthropods, mollusks and mammals. Mailing Add: Off of Grad Sch Bowling Green State Univ Bowling Green OH 43403

LEONE, FRED CHARLES, b New York, NY, Aug 3, 22; m 45; c 7. STATISTICS. Educ: Manhattan Col, BA, 41; Georgetown Univ, MS, 43; Purdue Univ, PhD(math statist, educ), 49. Prof Exp: Instr, Georgetown Univ, 42-43; instr, Purdue Univ, 43-44 & 46-49; from instr to prof math, Case Western Reserve Univ, 49-66, dir statist lab, 51-65, actg chmn dept, 63-65; prof statist & indust eng, Univ Iowa, 66-75; EXEC

DIR & SECY-TREAS, AM STATIST ASN, 73- Concurrent Pos: Fulbright prof, Univ Sao Paulo, Brazil, 68-69; ed, Technometrics, 63-68; NAm ed, Statist Theory & Methods Abstracts, 69-73. Mem: Fel AAAS; fel Am Soc Qual Control; fel Am Statist Asn; Sigma Xi; Math Asn Am. Res: Experimental design and statistics applied to engineering; order statistics, especially in analysis of variance. Mailing Add: Am Statist Asn 806 15th St NW Washington DC 20005

LEONE, IDA ALBA, b Elizabeth, NJ, Apr 28, 22. POLLUTION BIOLOGY. Educ: Rutgers Univ, BS, 44, MS, 46. Prof Exp: Asst plant path, Col Agr, Rutgers Univ, New Brunswick, 46-50, res assoc, 50-58, asst res specialist, 58-70, assoc res prof, 70-76, PROF PLANT BIOL, COOK COL, RUTGERS UNIV, NEW BRUNSWICK, 76- Concurrent Pos: Consult, NY State Environ Protection Bur, 75-76. Mem: Sigma Xi; Am Phytopath Soc; Am Soc Plant Physiologists; Air Pollution Control Asn. Res: Effect of air pollution; nutritional, physiological and environmental factors on plant growth; plants as sources of air pollution; undergraduate and graduate courses in air pollution effects; effect of cooling-tower salt spray on crops; phytotoxicity of anaerobic landfill gases. Mailing Add: 876 Rayhon Terr Rahway NJ 07065

LEONE, JAMES A, b Braddock, Pa, Dec 11, 37; m 61; c 1. PHYSICAL CHEMISTRY, INSTRUMENTATION. Educ: Univ Cincinnati, BS, 61; Johns Hopkins Univ, MA, 63, PhD(phys chem), 65. Prof Exp: Res assoc, Univ Notre Dame, 65-67; from asst prof to assoc prof phys chem, 67-74, DIR MED TECHNOL, CANISIUS COL, 74- Concurrent Pos: Vis assoc prof, Va Polytech Inst & State Univ, 75-76. Mem: Am Chem Soc; Sigma Xi; Soc Appl Spectros. Res: Radiation chemistry; ESR; on-line minicomputers; minicomputer and microprocessor interfacing; minicomputers and microprocessors in instrumentation automation. Mailing Add: Dept of Med Technol Canisius Col Buffalo NY 14208

LEONE, MARK PAUL, b Waltham, Mass, June 26, 40. ANTHROPOLOGY, ARCHAEOLOGY. Educ: Tufts Univ, AB, 63; Univ Ariz, MA, 65, PhD(anthrop), 68. Prof Exp: Asst prof anthrop, Princeton Univ, 68-75; ASSOC PROF ANTHROP, UNIV MD, 75- Concurrent Pos: Nat Inst Ment Health grant, 71-72; fel, Independent Study & Res, Nat Endowment for the Humanities, 75-76. Mem: AAAS; Am Anthrop Asn; Soc Am Archaeol. Res: Southwestern archaeology and ecology; cultural evolution; development of Mormon religion; historical archaeology. Mailing Add: Dept of Anthrop Univ of Md College Park MD 20742

LEONE, RONALD EDMUND, b New York, NY, Aug 11, 42. ORGANIC CHEMISTRY. Educ: Northwestern Univ, BA, 64; Princeton Univ, MA, 67, PhD(org chem), 70. Prof Exp: Fel org chem, Yale Univ, 69-71; SR RES CHEMIST, EASTMAN KODAK CO, 71- Mem: Am Chem Soc; Sigma Xi. Res: Aspects of physical organic chemistry including organic reaction mechanisms, carbonium ions, nuclear magnetic resonance spectroscopy, kinetics, synthesis of heterocyclic compounds for photographic applications. Mailing Add: Eastman Kodak Co Res Labs 1669 Lake Ave Rochester NY 14650

LEONE, STEPHEN ROBERT, b New York, NY, May 19, 48. CHEMICAL PHYSICS. Educ: Northwestern Univ, BA, 70; Univ Calif, Berkeley, PhD(chem), 74. Prof Exp: ASST PROF CHEM, UNIV SOUTHERN CALIF, 74- Mem: Am Phys Soc; Am Chem Soc; Sigma Xi; Optical Soc Am. Res: Laser-excited chemical reactions and isotope separation; kinetics and spectroscopic investigations of excited states using specific laser excitation; energy transfer and dynamical processes of small gas phase molecules. Mailing Add: Dept of Chem Univ of Southern Calif Los Angeles CA 90007

LEONG, BASIL K J, b Canton, China, June 13, 33; US citizen; m 61. PHARMACOLOGY, TOXICOLOGY. Educ: Univ Toronto, BSc, 59, MA, 61; Univ Ottawa, PhD(physiol), 64. Prof Exp: Res chemist, Occup Health Div, Dept Nat Health & Welfare, Can, 61-64, res officer, 64-65; staff physiologist, Dept Pharmacol, Hazleton Labs Inc, 65, Sect supvr, Inhalation Div, 65-67; res toxicologist, Health & Environ Res Lab, Dow Chem Co, 67-75; RES ASSOC PROF ENVIRON MED & DEP DIR INHALATION TOXICOL LAB, NY UNIV MED CTR, 75- Mem: AAAS; Am Indust Hyg Asn. Res: Inhalation toxicology; pulmonary carcinogenesis. Mailing Add: NY Univ Med Ctr PO Box 817 Tuxedo NY 10987

LEONG, GEORGE FRANCIS, physiology, see 12th edition

LEONG, JO-ANN CHING, b Honolulu, Hawaii, May 15, 42; c 1. VIROLOGY. Educ: Univ Calif, Berkeley, BA, 64; Univ Calif, PhD(microbiol), 71. Prof Exp: Sr res asst virol, Dept Surg, Stanford Univ Sch Med, 65-67; from teaching assoc microbiol to res biochemist, Univ Calif, San Francisco, 71-73, res fel biochem, 73-75; ASST PROF MICROBIOL, ORE STATE UNIV, 75- Concurrent Pos: Dernham fel, Am Cancer Soc, Calif Div, 73-75; Honors & Awards: Giannini Found fel for med res, 73. Mem: Am Soc Microbiologists; AAAS. Res: Virus-cell interactions; tumor virology. Mailing Add: Dept of Microbiol Ore State Univ Nash Hall Corvallis OR 97331

LEONG, KAM CHOY, b Honolulu, Hawaii, Dec 17, 20; m 50; c 3. BIOCHEMISTRY, POULTRY NUTRITION. Educ: Wash State Univ, BS, 49, MS, 50; Univ Wis, PhD(biochem, poultry), 58. Prof Exp: Asst, Wash State Univ, 48-50; jr animal husbandman, Univ Hawaii, 51-54; asst, Univ Wis, 54-57; fel, Wash State Univ, 57-58; jr poultry scientist, 58-61; res chemist, Bur Commercial Fisheries, 61-65; NUTRITIONIST, ALBERS MILLING CO DIV, CARNATION CO, 65- Mem: Am Poultry Sci Asn; Am Inst Nutrit. Res: Amino acids; enzymes; vitamins; pro- tein; metabolizable energy. Mailing Add: Albers Milling Co Div Carnation Co 5045 Wilshire Blvd Los Angeles CA 90036

LEONHARD, FREDERICK WILHELM, b Rheinhausen, Ger, Oct 25, 14; US citizen; m 50; c 4. PHYSICS. Educ: Univ Tübingen, dipl, 49, Dr rer nat, 54. Prof Exp: Asst exp physics & electron micros, Univ Tübingen, 48-50; scientist, Deutsche Gold und Silber Scheideanstalt, 50-56; res team leader, US Army Signal Res & Develop Lab, NJ, 56-60; scientist lab head, McDonnell Aircraft Corp, Mo, 60-69; PROF ELEC ENG, UNIV MO-COLUMBIA, 69- Mem: Am Vacuum Soc. Res: Crystallography; metallography; physical electronics of semiconductors; dielectrics; metals; thin films and surfaces; growth of single crystal thin films; electronic devices; thin film and semiconductor microelectronic materials, techniques and circuitry. Mailing Add: Dept of Elec Eng Univ of Mo Columbia MO 65201

LEONHARDT, EARL A, b Council Bluffs, Iowa, Apr 18, 19; m 41; c 3. MATHEMATICS. Educ: Union Col, Nebr, BA, 50; Univ Nebr, ME, 52, PhD(sec educ, math), 62. Prof Exp: High sch instr, Nebr, 51-52; from instr to assoc prof, 52-62, PROF MATH, UNION COL, NEBR, 62- Mem: Math Asn Am. Mailing Add: Dept of Math Union Col Lincoln NE 68506

LEONORA, JOHN, b Milwaukee, Wis, Jan 30, 28; m 52; c 2. ENDOCRINOLOGY. Educ: Univ Wis, BS, 49, MS, 54, PhD(zool), 57. Prof Exp: Asst endocrinol, Univ Wis, 52-57; from instr to assoc prof, 59-69, PROF MED, SCH MED, LOMA LINDA UNIV, 69-, CO-CHMN DEPT PHYSIOL & PHARMACOL, 74- Concurrent Pos: NIH fel, Univ Wis, 57-59. Mem: AAAS; NY Acad Sci; Endocrine Soc; Sigma Xi. Res: Hypothalamic-parotid endocrine axis; relationship of dentinal fluid movement to dental caries. Mailing Add: Dept of Physiol Loma Linda Univ Sch of Med Loma Linda CA 92354

LEONS, MADELINE BARBARA, b New York, NY, Aug 17, 38; m 60; c 2. ANTHROPOLOGY, CULTURAL ANTHROPOLOGY. Educ: Brandeis Univ, BA, 59; Univ Calif, Los Angeles, MA, 62, PhD(anthrop), 67. Prof Exp: Asst anthrop, Univ Calif, Los Angeles, 60-62; instr, Univ Southern Calif, 65-66; asst prof, Univ Pittsburgh, 66-70; assoc prof, 70-75, PROF ANTHROP, TOWSON STATE COL, 75- Mem: Am Anthrop Asn. Res: Peasantry; Latin America, especially Andean countries; class stratification and ethnic pluralism; local level politics; land reform; economic change and modernization. Mailing Add: Dept of Sociol & Anthrop Towson State Col Baltimore MD 21204

LEOPOLD, ALDO CARL, b Albuquerque, NMex, Dec 18, 19; m 42; c 3. PLANT PHYSIOLOGY. Educ: Univ Wis, BA, 41; Harvard Univ, MA, 47, PhD(biol), 48. Prof Exp: Plant physiologist, Hawaiian Pineapple Co, Hawaii, 48-49; from asst prof to prof hort, Purdue Univ, 49-75; GRAD DEAN & ASST VPRES RES, UNIV NEBR, 75- Concurrent Pos: Carnegie vis prof, Univ Hawaii, 62; mem panel regulatory biol, NSF, 65; bd govs, Am Inst Biol Sci; sr policy analyst NSF, 74-75; mem bd agr & renewable resources, Nat Res Coun, 75- Mem: AAAS; Am Soc Hort Sci; Bot Soc Am; Am Soc Plant Physiol (vpres, 59, pres, 65); Scand Soc Plant Physiol. Res: Plant hormones; growth regulators; plant growth and development; seed viability. Mailing Add: Univ of Nebr 3835 Holdrege Lincoln NE 68503

LEOPOLD, ALDO STARKER, b Burlington, Iowa, Oct 22, 13; m 38; c 2. ZOOLOGY, FORESTRY. Educ: Univ Wis, BS, 36; Univ Calif, PhD(zool), 44. Prof Exp: Jr biologist, Soil Erosion Serv, USDA, Wis, 34-35; field biologist, State Conserv Comn, Mo, 39-44; dir field res, Conserv Sect, Pan Am Union, Mex, 44-46; from instr to prof zool, 46-68, assoc dir, Mus Vert Zool, 59-69, PROF ZOOL & FORESTRY, UNIV CALIF, BERKELEY, 68-, DIR, SAGEHEN CREEK FIELD STA & CONSERVATIONIST, 69- Concurrent Pos: Guggenheim fel, 48. Mem: Nat Acad Sci; Wildlife Soc (pres, 57-58); Cooper Ornith Soc (pres, 58-60); Wilson Ornith Soc; assoc Am Soc Mammalogists. Res: Wildlife ecology and management; field ornithology and mammalogy; comparative anatomy and behavior of wild and domestic turkeys; nature of heritable wildness in turkeys; ecology and management of deer, moose and caribou; wildlife of Mexico. Mailing Add: Sch of Forestry & Conserv Univ of Calif 145 Mulford Hall Berkeley CA 94720

LEOPOLD, BENGT, b Valbo, Sweden, Dec 23, 22; m 45; c 2. PAPER CHEMISTRY, PULP & PAPER TECHNOLOGY. Educ: Royal Inst Tech, Sweden, BChem Eng, 47, MS, 49, PhD(org chem), 52. Prof Exp: Sr res chemist, Columbia-Southern Chem Corp, Ohio, 52-53; mgr pioneering res div, Indust Cellulose Res Ltd, Can Inst Paper Co, 53-58; mgr basic res div, Mead Corp, Ohio, 58-60, assoc dir res, 60-61; PROF PULP & PAPER RES & DIR, EMPIRE STATE PAPER RES INST, STATE UNIV NY COL ENVIRON SCI & FORESTRY, SYRACUSE UNIV, 61-, CHMN DEPT PAPER SCI & ENG, 74- Concurrent Pos: Ed, J Tech Asn Pulp & Paper Indust, 66- Mem: AAAS; Am Chem Soc; fel Tech Asn Pulp & Paper Indust; Paper Indust Mgt Asn; Can Pulp & Paper Asn. Res: Fiber physics; structure of lignin; mechanical properties of wood fibers; cellulose-water interactions. Mailing Add: Dept of Paper Sci & Eng State Univ of NY Col Forestry Syracuse NY 13210

LEOPOLD, ESTELLA (BERGERE), b Madison, Wis, Jan 8, 27. BOTANY. Educ: Univ Wis, PhB, 48; Univ Calif, MS, 50; Yale Univ, PhD(bot), 55. Prof Exp: Asst physiol & embryol, Genetics Exp Sta, Smith Col, 51-52; asst biol, Yale Univ, 52-53, asst animal ecol, 54; BOTANIST, PALEONT & STRATIG BR, US GEOL SURV, 55- Concurrent Pos: Asst, Lab Tree Ring Res, Univ Ariz, 51 & US Forest Prod Lab, 52; NSF travel grants, Int Asn Quaternary Res, Spain, 57 & Poland, 61; adj prof, Univ Colo, Boulder, 67-; vis prof, Inst Environ Studies, Univ Wis-Madison, 71-72. Honors & Awards: Co-recipient, Conservationist of Year Award, Nat Wildlife Fedn, 69. Mem: Nat Acad Sci; Ecol Soc Am; Bot Soc Am. Res: Paleobotany, palynology and paleoecology; pollen and spore floras of late Cenozoic age in Wyoming, Colorado and Alaska; palynology research in late quaternary deposits of Connecticut; upper cretaceous pollen and spore floras of Alabama and Wyoming. Mailing Add: Br Paleont & Stratig US Geol Surv Denver CO 80225

LEOPOLD, IRVING HENRY, b Philadelphia, Pa, Apr 19, 15; m 37; c 2. OPHTHALMOLOGY. Educ: Univ Pa, MD, 38, DSc, 43; Am Bd Ophthal, dipl, 43. Prof Exp: Fel ophthal, Univ Pa, 40-46, from assoc to assoc prof, Grad Sch Med, 46-55, prof & chmn dept, 55-64; clin prof, Col Physicians & Surgeons, Columbia Univ, 64-66; prof ophthal & chmn dept, Mt Sinai Sch Med, 66-75, dir dept ophthal, Hosp, 64-75; PROF OPHTHAL SURG & CHIEF OPHTHAL, UNIV CALIF, IRVINE, 75- Concurrent Pos: Clin asst, Wills Eye Hosp, 46-47, from asst surgeon to sr asst surgeon, 47-52, attend surgeon, 52-64, dir res, 49-64; chief ophthalmologist, Episcopal Hosp, 48-50; chief dept ophthal, St Joseph's Hosp, 55-59, consult, 59-; from vchmn to chmn, Am Bd Ophthal, 71-73. Consult, Med Res Div, US Chem Warfare, 50-53; consult, Training Grants Div, Surg Gen, USPHS, 52-58, Spec Sensory Study Sect for Res Neurol Dis & Blindness, 54-58, & Comt Neurol Field Invests, 59- Mem, Field Invest Comt, NIH, 59-61, Neurol Prog Proj Comt, 61-63, Subcomt Vision & Its Dis, 66-67, Training Grant Comt, 66-68 & Subcomt Impaired Vision & Blindness, 67-69, chmn, Task Force on Ocular Pharmacol, 67-69, mem visual sci study sect, 68-70, mem vision res training comt, Nat Inst Neurol Dis & Blindness, 63-68, chmn, 67-68; chmn ophthal panel, US Pharmacopeia, 60-70, mem, 71-; mem grants adv comt, The Seeing Eye, Inc, 67-70; mem comt admissions, Nat Drug Formulary, 70-; mem adv panel, Food & Drug Admin, 71-; six name lecture awards, 62-74. Honors & Awards: Zentmayer Award, 45, 49; Award of Merit, Am Acad Ophthal & Otolaryngol, 55; Holmes Citation & Award, 57; Friedenwald Medal, Asn Res Vision & Ophthal, 60; Lucien Howe Medal, Am Ophthal Soc, 74. Mem: Am Ophthal Soc; Geront Soc; Am Diabetes Asn; Asn Res Vision & Ophthal; Am Acad Ophthal & Otolaryngol. Res: Ocular pharmacology, especially antibiotics and chemotherapeutic agents, autonomic drugs, enzymes and antienzymes; ocular physiology, especially intra-ocular pressure, ocular circulation, intraocular fluid formation; ocular disease, particularly glaucoma, uveitis, diabetic retinopathy, infections, neoplasm. Mailing Add: Dept of Ophthal Univ of Calif Irvine CA 92664

LEOPOLD, LUNA BERGERE, b Albuquerque, NMex, Oct 8, 15; m 40; c 2. GEOMORPHOLOGY. Educ: Univ Wis, BS, 36; Harvard Univ, PhD(geol), 50; Univ Calif, Los Angeles, MA, 45. Hon Degrees: DrGeog, Univ Ottawa, 70; DSc, Iowa Wesleyan Univ, 72. Prof Exp: From jr engr to assoc engr, Soil Conserv Serv, USDA, NMex, 36-40; assoc engr, US Eng Off, Los Angeles, 41-42; assoc engr, bur reclamation, US Dept Interior, Washington, DC, 46-47; head meteorologist, Pineapple Res Inst, 47-50; hydraul engr, US Geol Surv, 50-56, chief hydrologist, 56-66, sr res hydrologist, 66-72; PROF GEOL, UNIV CALIF BERKELEY, 72- Honors & Awards: Dept Interior Distinguished Serv Award, 58; Bryan Award, Geol Soc Am, 58; Royal Netherlands Geog Soc Veth Medal, 63; Liege Univ Medal, 66; Cullum Medal, Am Geog Soc, 68; Rockefeller Pub Serv Award, 71. Mem: Nat Acad Sci; Am Soc Civil Eng; Geol Soc Am (pres, 71); Am Philos Soc; Am Acad Arts & Sci. Res: Hydrology of arid regions; rainfall characteristics; river morphology, erosion and sedimentation. Mailing Add: Dept of Geol Univ of Calif Berkeley CA 94720

LEOPOLD, ROBERT L, b Philadelphia, Pa, Oct 5, 22; m 44; c 3. PSYCHIATRY. Educ: Harvard Univ, AB, 43; Univ Pa, MD, 46. Prof Exp: Intern neurol, Grad Hosp, 47-50, from instr to assoc prof psychiat, 50-68, clin psychiat, 68-69, PROF COMMUNITY PSYCHIAT & COMMUNITY MED, DIV GRAD MED, SCH MED, UNIV PA, 68-, CHMN DEPT COMMUNITY MED, 71-, DIR DIV COMMUNITY PSYCHIAT, 65- Concurrent Pos: Resident, Philadelphia Psychoanal Inst, 49-55; fel, Psychiat Inst Pa, 50-51; resident, Univ Hosp, Univ Pa, 51-52; psychiat consult, Am Friends Serv Comt, 56-; sr psychiat consult, Peace Corps, 61-67; dir, WPhiladelphia Community Ment Health Consortium, 67-72. Mem: Am Sci Affil; Am Psychoanal Asn; fel Am Psychiat Asn; AMA. Res: Community psychiatry; psychoanalysis. Mailing Add: Dept of Community Med Univ of Pa Philadelphia PA 19174

LEOPOLD, ROBERT SUMMERS, b Dayton, Ohio, June 21, 15; m 43; c 3. ORGANIC CHEMISTRY. Educ: Miss State Univ, BS, 37; Univ NC, MA, 40; Univ Fla, PhD(chem), 42. Prof Exp: Instr, Va Mil Inst, 42-43; asst prof chem, Ga Inst Tech, 46; assoc prof chem, Fla State Univ, 46-49; mem staff, US Naval Dent Sch, 49-52; head chemist, Naval Med Field Res Lab, 52-57, head personnel protection div, 59-62; ASSOC PROF CHEM, THE CITADEL, 63- Mem: Am Chem Soc. Res: Biochemistry. Mailing Add: Dept of Chem The Citadel Charleston SC 29409

LEOPOLD, ROGER ALLEN, b Redwood Falls, Minn, Mar 23, 37; m 58; c 2. ENTOMOLOGY, HISTOCHEMISTRY. Educ: Concordia Col, Minn, BA, 62; Mont State Univ, PhD(entom), 67. Prof Exp: Res asst stress physiol, Mont State Univ, 62-67; PROF ZOOL, N DAK STATE UNIV, 71-; RES ENTOMOLOGIST, METAB & RADIATION RES LAB, AGR RES SERV, USDA, 67- Mem: AAAS; Entom Soc Am; Am Soc Zool. Res: Insect cytochemistry and genetics; reproductive physiology. Mailing Add: Metabolism & Radiation Res Lab USDA Fargo ND 58102

LEOSCHKE, WILLIAM LEROY, b Lockport, NY, May 2, 27; m 56; c 2. BIOCHEMISTRY. Educ: Valparaiso Univ, BA, 50; Univ Wis, MS, 52, PhD(biochem), 54. Prof Exp: Proj assoc biochem, Univ Wis, 54-59; from asst prof to assoc prof, 59-69, PROF CHEM, VALPARAISO UNIV, 69- Concurrent Pos: Consult, Nat-Northwood, Inc, Ill, 55-; mem, Nat Res Coun Sub-comt Fur Animal Nutrit. Mem: Am Chem Soc. Res: Biochemistry and nutrition of mink; fundamental nutritional requirements of mink; mink diseases of nutritional origin; composition of blood and urine of mink. Mailing Add: Dept of Chem Valparaiso Univ Valparaiso IN 46383

LEOVY, CONWAY B, b Hermosa Beach, Calif, July 16, 33; m 58; c 4. METEOROLOGY. Educ: Univ Southern Calif, BA, 54; Mass Inst Technol, PhD(meteorol), 63. Prof Exp: Res asst meteorol, Mass Inst Technol, 58-63; meteorologist, Rand Corp, Calif, 63-69; assoc prof atmospheric sci, 69-74, PROF ATMOSPHERIC SCI & GEOPHYS & ADJ PROF ASTRON, UNIV WASH, 74- Concurrent Pos: Mem, Comt on Atmospheric Sci, Nat Acad Sci, 72-75 & comt on Lunar & Planetary Exploration, 74-; assoc ed, J Atmos Sci; consult, NASA. Honors & Awards: NASA Outstanding Sci Achievement Award, 1972. Mem: Am Meteorol Soc; Am Geophys Union; AAAS. Res: Dynamics, radiation and photochemistry of earth and planetary atmospheres. Mailing Add: Dept of Atmospheric Sci & Geophys Col of Arts & Sci Univ of Wash Seattle WA 98795

LEPAGE, GERALD ALVIN, b Medicine Hat, Alta, Can, Oct 9, 17; m 44; c 2. BIOCHEMISTRY. Educ: Univ Alta, BSc, 40, MSc, 41; Univ Wis, PhD(bact & biochem), 43. Prof Exp: Lectr biochem, Univ Alta, 40-41; prof assoc, Dept Oncol, Off Sci Res & Develop contract, Univ Wis, 43-44; biochemist & bacteriologist, Lab Hyg, Can Food & Drug Admin, 44-45; from asst prof to prof oncol, Univ Wis, 45-58; proj dir, Stanford Res Inst, 58-61, chmn dept biochem oncol, 61-69; prof pharmacol & chief pharmacol sect, Univ Tex M D Anderson Hosp & Tumor Inst, 69-72; DIR McEACHERN LAB, UNIV ALTA, 72- Mem: AAAS; Am Chem Soc; Am Asn Cancer Res; Am Soc Pharmacol & Exp Therapeut; NY Acad Sci. Res: Enzymes of cancer tissue; mechanisms of action of carcinostatic agents. Mailing Add: McEachern Lab Univ of Alta Edmonton AB Can

LEPAGE, MARIUS, b Rimouski, Que, July 26, 30; m 57; c 2. AGRICULTURAL CHEMISTRY, BIOLOGICAL CHEMISTRY. Educ: Laval Univ, BA, 53, BScA, 57; Pa State Univ, MSc, 59, PhD(agr & biol chem), 61. Prof Exp: Res chemist, Food Res Inst, Cent Exp Farm, Dept Agr Can, 61-68; assoc prof food sci, Fac Agr, 68-69, ASSOC PROF BIOCHEM, MED SCH, LAVAL UNIV, 69- Mem: Am Oil Chemists' Soc. Res: Biochemistry. Mailing Add: Dept of Biochem Laval Univ Med Sch Ste-Foy PQ Can

LEPARD, DAVID WILLIAM, b Newmarket, Ont, Nov 1, 37; m 61; c 2. MOLECULAR PHYSICS. Educ: Univ Toronto, BA, 59, MA, 60, PhD(physics), 64. Prof Exp: Asst prof physics, Mem Univ, 64-65; fel spectros sect, Div Pure Physics, Nat Res Coun Can, 65-67; asst prof, 67-72, ASSOC PROF PHYSICS, BROCK UNIV, 72- Concurrent Pos: Nat Res Coun Can res grants, 64-65, 67-; Ont Dept Univ Affairs res grants, 68-69. Mem: Can Asn Physicists; Am Phys Soc. Res: Theory; infrared and Raman spectra of polyatomics; electronic spectra of diatomics. Mailing Add: Dept of Physics Brock Univ St Catherines ON Can

LEPESCHKIN, EUGENE, b Kazan, Russia, Apr 15, 14; nat US; m 49; c 3. CARDIOLOGY. Educ: Univ Vienna, MD, 39. Prof Exp: Asst physiol, Univ Vienna, 39-40; asst, Balneolog Inst, Ger, 40-42; asst, I Med Clin, Vienna, 42-44; cardiologist, Hosp Team 1064, UNRRA, Munich, Ger, 45-47; from asst prof to prof exp med, 47-65, PROF MED, COL MED, UNIV VT, 65- Concurrent Pos: Nat Heart Inst res career award, 62-; res cardiologist, Life Ins Hosp, Bad Nauheim, Ger, 40-42; chief cardiographer, de Goesbriand Hosp, Burlington, 52-62, consult, 62-; consult, Middlebury Hosp, 52-68; estab investr, Am Heart Asn, 53-58, mem basic sci clin cardiol coun, Comt Standard Electrocardiograph Vectorcardiograph Leads, 53- Mem: Am Physiol Soc; Am Col Cardiol; NY Acad Sci. Res: Physiology and pathology of the heart and circulation, especially electrophysiology of the heart, arrhythmias, electrocardiography and phonocardiography. Mailing Add: Dept of Med Univ of Vt Col of Med Burlington VT 05401

LEPIE, ALBERT HELMUT, b Malapane, Ger, Aug 6, 23; US citizen; m 56; c 1. PHYSICAL CHEMISTRY. Educ: Aachen Tech Univ, MS, 58; Munich Tech Univ, PhD(chem), 61. Prof Exp: Res chemist, Ger Inst Res Aeronaut, 61-63 & US Naval Propellant Plant, Md, 63-64; RES CHEMIST, NAVAL WEAPONS CTR, 64- Mem: Am Chem Soc; fel Am Inst Chem; NY Acad Sci. Res: Performance calculations of propellants; hypergolic ignitions; mechanical behavior of polymers; dynamic testing methods. Mailing Add: PO Box 5272 China Lake CA 93555

LEPLAE, LUC A, b Hammemille, Belgium, Nov 27, 30; m 59; c 4. THEORETICAL SOLID STATE PHYSICS. Educ: Cath Univ Louvain, Lic en Theoret Physics, 55; Univ Md, PhD(physics), 62. Prof Exp: Res assoc physics, Inst Theoret Physics, Naples, Italy, 62-66; res assoc, 66-67, vis asst prof, 67-68, ASST PROF PHYSICS, UNIV WIS-MILWAUKEE, 68- Res: Application of the Boson method to superconductivity, superfluidity, magnetism and phase transitions; many body problem; solid state physics. Mailing Add: Dept of Physics Univ of Wis Milwaukee WI 53201

LEPLEY, ARTHUR RAY, b Peoria, Ill, Nov 1, 33; m 57; c 4. PHYSICAL ORGANIC CHEMISTRY. Educ: Bradley Univ, AB, 54; Univ Chicago, SM, 56, PhD(chem), 58. Prof Exp: Res assoc org chem, Univ Munich, 58-59 & Univ Chicago, 59-60; asst prof chem, State Univ NY Stony Brook, 60-65; assoc prof, 65-68, PROF CHEM, MARSHALL UNIV, 68- Concurrent Pos: NSF fel, 58-59; USPHS gen med fel, 60; vis prof, Univ Utah, 69-71; res chemist, Lab Chem Phys, Nat Inst Arthritis, Metabolism & Digestive Dis, Md, 75-76. Mem: AAAS; Am Chem Soc; The Chem Soc; Am Inst Chem; Sigma Xi. Res: Rearrangements of quaternary ammonium salts; benzyne additions to tertiary amines; direct alpha alkylation of tertiary amines; free radical intermediates; nuclear magnetic resonance emission spectroscopy; chemically induced dynamic nuclear polarization. Mailing Add: Dept Of Chem Marshall Univ Huntington WV 25701

LEPLEY, DERWARD, JR, b Viola, Wis, Jan 10, 24; m 49; c 5. SURGERY. Educ: Marquette Univ, MD, 49; Am Bd Surg, dipl, 57; Am Bd Thoracic Surg, dipl, 59. Prof Exp: Intern, Wis Gen Hosp, Madison, 49-50; gen surg resident, Wood Vet Admin Hosp, Milwaukee, 52-56, thoracic surg resident, 56-58; teaching asst & Nat Heart Inst res fel, Grad Sch, Univ Minn, Minneapolis, 58-59; instr surg, 56-58, clin instr surg, 59-60, from asst prof to assoc prof, 60-68, PROF SURG & CHMN DEPT THORACIC & CARDIOVASC SURG, MED COL WIS, 68- Concurrent Pos: Attend surgeon, Wood Vet Admin Hosp, Milwaukee, 59-; mem attend staff, Columbia Hosp, 59-; attend surgeon, Milwaukee County Hosp, 59-, chief cardiovasc surg, 62-; asst surg, St Luke's Hosp, 61-; mem consult staff, St Joseph's Hosp, 61-; mem active staff, Milwaukee Children's Hosp, 64- Mem: Am Heart Asn; Am Col Surg; Am Thoracic Soc; Am Asn Thoracic Surg; Am Col Cardiol. Res: Cardiovascular surgery; evaluation of artificial heart valves for use in prosthetic replacement; study of various perfusates in pump oxygenator to improve homeostasis. Mailing Add: Dept of Surg Med Col Wis Milwaukee WI 53233

LEPOFF, JACK H, b Portland, Maine, July 22, 23; m 47; c 2. PHYSICS. Educ: Univ NH, BS, 43; Columbia Univ, MA, 48. Prof Exp: Electronic scientist, Nat Bur Standards, 49-50, Naval Res Lab, 50-51, Nat Bur Standards, 51-53 & Naval Ord Lab, 53-54; sr staff mem, Motorola, Inc, 54-59; eng specialist, Sylvania Electronic Defense Lab, Gen Tel & Electronics Corp, 59-65; diode appln mgr, HPA Div, 65-73, DIODE APPLNS ENGR, HEWLETT PACKARD CO, 73- Mem: Sigma Xi; Inst Elec & Electronics Engrs. Res: Microwaves; semiconductors. Mailing Add: 595 Templeton Dr Sunnyvale CA 94087

LEPORE, JOSEPH VERNON, b Detroit, Mich, Oct 9, 22. PHYSICS. Educ: Allegheny Col, BS, 43; Harvard Univ, PhD(physics), 48. Prof Exp: Instr physics, Princeton Univ, 43-44; physicist, Tenn Eastman Corp, 44-46; AEC fel, Inst Advan Study, 48-50; asst prof physics, Ind Univ, 50-51; lectr, Univ, 54-65, PHYSICIST, LAWRENCE BERKELEY LAB, UNIV CALIF, 51- Concurrent Pos: Adv to test dir, Nev Test Site, 57. Mem: Am Phys Soc. Res: Nuclear physics; quantum field theory; scattering theory. Mailing Add: Lawrence Berkeley Lab Univ of Calif Berkeley CA 94720

LEPORE, PAUL D, biochemistry, see 12th edition

LEPOVETSKY, BARNEY CHARLES, b Ridgway, Pa, Jan 17, 26; m 50; c 2. SCIENCE ADMINISTRATION. Educ: Ohio State Univ, BSc, 49, MSc, 51, PhD(microbiol), 54; Ohio Northern Univ, JD, 63. Prof Exp: Prof microbiol, Ohio Northern Univ, 54-64; scientist adminstr, NIH, 64-69, dep assoc dir extramural progs, Nat Inst Dent Res, 69-71, chief, Off Collab Res, 71-75, CHIEF RES TRAINING GRANTS BR, NAT CANCER INST, 75- Mem: Int Acad Law & Sci. Res: Law. Mailing Add: 4816 Mori Dr Rockville MD 20853

LEPOW, IRWIN HOWARD, b New York, NY, Sept 2, 23; m 58; c 3. IMMUNOLOGY, EXPERIMENTAL MEDICINE. Educ: Pa State Univ, BS, 42; Case Western Reserve Univ, PhD(immunochem), 51, MD, 58. Prof Exp: Biol res & develop, Lederle Labs, 43, 46 & 48; sr instr biochem, Case Western Reserve Univ, 51-52, asst prof biochem & exp path, 52-59, assoc prof exp path, 59-65, from assoc prof to prof med, 65-67; prof path & head dept, 67-73, PROF MED & HEAD DEPT, HEALTH CTR, UNIV CONN, 73- Concurrent Pos: USPHS res career award, 63-67; mem allergy & immunol study sect, NIH, 62-65 & 69-73, chmn, 70-73; dir comn immunization, Armed Forces Epidemiol Bd, 63-67, adv mem, 67-; mem, Conn Comn Medico-legal Invests, 69-73; mem bd trustees, Trudeau Inst, NY; assoc ed, J Immunol, 71-73; elect mem coun, Am Asn Immunol. Mem: AAAS; Asn Am Physicians; Am Asn Immunol; NY Acad Sci; fel Am Col Physicians. Res: Complement; mechanisms of immunological injury. Mailing Add: Dept of Med Univ of Conn Health Ctr Farmington CT 06032

LEPOWSKY, JAMES IVAN, b New York, NY, July 5, 44; m 68. MATHEMATICS. Educ: Harvard Univ, AB, 65; Mass Inst Technol, PhD(math), 70. Prof Exp: Lectr & res assoc math, Brandeis Univ, 70-72; ASST PROF MATH, YALE UNIV, 72- Concurrent Pos: Mem sch of math, Inst Advan Study, 75-76. Mem: Am Math Soc; Math Asn Am. Res: Representations of semisimple lie algebras and lie groups. Mailing Add: Dept of Math Yale Univ New Haven CT 06520

LEPP, ALBERT, b Chicago, Ill, Feb 11, 14; m 51; c 4. BIOCHEMISTRY. Educ: Univ Ill, BS, 38; Univ Calif, Los Angeles, PhD(biochem), 46; Univ Wis, MS, 47. Prof Exp: Asst biochem, Univ Wis, 46-48; asst hort, Iowa State Col, 49-50; asst physiol chem, Med Ctr, Univ Calif, Los Angeles, 51-53, asst biochem, 53-55; biochemist, Med Sch, Univ Southern Calif, 56-57, vis instr biochem & nutrit, 57-60; biochemist, Radioisotope Serv, Vet Admin Hosp, Indianapolis, Ind, 60-69, asst chief thyroid res, DC, 69-71, CHIEF RADIOIMMUNO ASSAY RES LAB, 71- Concurrent Pos: Biochemist, Los Angeles County Hosp, 58-60; from instr to asst prof, Sch Med, Univ Ind, Indianapolis, 60-63. Mem: Am Thyroid Asn; Endocrine Soc; NY Acad Sci. Res: Immunoassays of hormones in biological fluids. Mailing Add: Radioimmuno Assay Res Lab Vet Admin Hosp Washington DC 20422

LEPP, CYRUS ANDREW, b Brooklyn, NY, Aug 11, 46; m 72; c 1. CLINICAL BIOCHEMISTRY, CLINICAL CHEMISTRY. Educ: Syracuse Univ, BS, 68, PhD(biochem), 74. Prof Exp: Lab technician clin chem, Nassau Hosp, 68-69; SR BIOCHEMIST CLIN CHEM & BIOCHEM, CORNING GLASS WORKS, 74- Mem: Am Chem Soc. Res: Electrophoretic separations of isoenzymes and hemoglobins; development of specific isoenzyme assay procedures. Mailing Add: Corning Glass Works Sullivan Park Res Ctr FR-6 Corning NY 14830

LEPP, HENRY, b Russia, Mar 4, 22; m 52; c 4. GEOLOGY. Educ: Univ Sask, BSc, 44; Univ Minn, PhD(geol), 54. Prof Exp: Geologist & mining engr, Consol Mining & Smelting Co, Ltd, 44-46; mining engr, N W Byrne, Consult Mining Engr, 46-48; geologist, Aluminum Labs, Ltd, 48-50 & Freeport Sulphur Co, 54; from asst prof to prof, Univ Minn, Duluth, 54-64; PROF GEOL & CHMN DEPT, MACALESTER COL, 64- Mem: AAAS; Geol Soc Am; Soc Econ Geol. Res: Mining geology; sedimentary iron formations, particularly the geochemistry of these formations and its possible relation to the evolutionary pattern of the atmosphere. Mailing Add: Dept of Geol Macalester Col St Paul MN 55105

LEPP, HERMAN I, biology, science education, see 12th edition

LEPPARD, GARY GRANT, b Medicine Hat, Can, Aug 6, 40; m 70; c 2. CELL BIOLOGY. Educ: Univ Sask, BA, 62, BA hons, 63, MA, 64; Yale Univ, MS, 66, MPhil, 67, PhD(biol), 68. Prof Exp: RES SCIENTIST BIOL, CAN DEPT ENVIRON, 71- Concurrent Pos: NATO Sci fel, Fac Med, Univ Paris, 68, Inst Pharmacol, Univ Milan, 69; Nat Res Coun Sci fel, Fac Sci, Univ Laval, 69-70, Biochem Lab, Nat Res Coun Can, 70-71; sci res exec mem, Prof Inst Pub Serv Can, 71-73; adj prof, Dept Biol, Univ Ottawa, 74-75. Mem: Sigma Xi; Can Fedn Biol Soc; Can Soc Cell Biol. Res: Physical relationships between living cells and materials in their external milieu. Mailing Add: Process Res Div Ctr Inland Waters Environ Can Burlington ON Can

LEPPER, MARK (HUMMER), b Washington, DC, June 12, 17; m 41; c 2. MEDICINE. Educ: George Washington Univ, AB, 38, MD, 41. Prof Exp: Intern, Sibley Mem Hosp, Washington, DC, 41-42; fel med, Sch Med, George Washington Univ, 42-43, clin instr med, 46-48; med supt, Munic Contagious Dis Hosp, Chicago & clin assoc prof prev med, Univ Chicago, 50-52; from assoc prof to prof, Med Ctr, Univ Ill, 52-70, head dept prev med, Col Med, 55-66; EXEC VPRES PROF & ACAD AFFAIRS, RUSH PRESBY ST LUKE'S MED CTR, 66-, PROF INTERNAL MED & DEAN, RUSH MED COL, 70- Concurrent Pos: Practicing physician, 46-50; consult, Walter Reed Gen Hosp, 48-50. Mem: Fel AMA; Am Pub Health Asn; Asn Am Med Cols; Am Fedn Clin Res; Am Col Physicians. Res: Infectious diseases. Mailing Add: Rush Med Col 1725 W Harrison Chicago IL 60612

LEPPER, ROBERT, JR, b Pawtucket, RI, Apr 13, 14; m 44; c 4. CYTOLOGY, GENETICS. Educ: Univ RI, BS, 36, MS, 38; Univ Conn, PhD(cytol), 54. Prof Exp: From instr to assoc prof, 48-57, head dept, 54-57, chmn dept, 57-71, PROF BOT, UNIV RI, 57-, DEAN COL ARTS & SCI, 71- Mem: AAAS; Bot Soc Am; Am Genetic Asn; Am Inst Biol Sci. Res: Centrosome in plants; plant breeding. Mailing Add: Col of Arts & Sci Univ of RI Kingston RI 02881

LEPPI, THEODORE JOHN, b Mountain Iron, Minn, May 30, 33; m 59; c 3. ANATOMY. Educ: Albion Col, BA, 59; Yale Univ, PhD(anat), 63. Prof Exp: From asst prof to assoc prof anat, Sch Med, Univ NMex, 66-71; PROF BIOMED ANAT & CHMN DEPT & ASSOC DEAN, SCH MED, UNIV MINN, DULUTH, 71- Concurrent Pos: Staff fel, Lab Exp Path, Nat Inst Arthritis & Metab Dis, 63-66; Lederle Med Fac Award, 68-71; guest lectr, Sch Med, Georgetown Univ, 63-64; asst prof lectr, Sch Med, George Washington Univ, 65-66. Mem: Am Asn Anat; Am Soc Cell Biol; Histochem Soc; Electron Microscopy Soc Am. Res: Effects of hormones on connective tissues; histo- chemistry and cytochemistry of epithelial mucins and connective tissue mucopolysaccharides. Mailing Add: Dept of Biomed Anat Univ of Minn Sch of Med Duluth MN 55812

LEPPIK, ELMAR EMIL, US citizen; m; c 3. PLANT PATHOLOGY. Educ: Univ Bern & Tech Univ, Zurich, MBot & ScD(natural sci), 28. Prof Exp: Prof appl biol, Hort Col & Tech Univ Munich, 45-50; prof gen biol, Augustana Col, 50-55 & Univ Minn, 55-57; plant pathologist, Iowa, 57-64, RES BOTANIST, AGR RES SERV, USDA, 64- Concurrent Pos: Rockefeller Found fel, 26-29; guest investr, Trop Res Inst, El Salvador, 53-54; prof, Iowa State Univ, 57-64; expert adv, Int Agr Inst Rome, League Nations, Geneva, 38-39. Mem: Fel AAAS; Bot Soc Am; Mycol Soc Am; Soc Study Evolution; Ger Soc Appl Entom. Res: Mycology; phylogeny of rust fungi; hologeny; floral ecology; behavior of pollinating insects; sensory behavior of bees; origin and evolution of cultivated plants. Mailing Add: Plant Genetics & Germplasm Inst Agr Res Ctr-West Bldg 001 Beltsville MD 20705

LEPSE, PAUL ARNOLD, b Seattle, Wash, Mar 18, 37; m 61; c 2. ORGANIC CHEMISTRY. Educ: Seattle Pac Col, BS, 58; Univ Wash, PhD(org chem), 62. Prof Exp: NSF fel, Univ Munich, 62; from asst prof to assoc prof, 63-72, PROF CHEM, SEATTLE PAC COL, 72- Mem: AAAS; Am Chem Soc. Res: Organic reaction mechanisms; carbene chemistry. Mailing Add: Dept of Chem Seattle Pac Col Seattle WA 98119

LEPSON, BENJAMIN, b New York, NY, Mar 4, 24; m 48; c 1. MATHEMATICS, STATISTICS. Educ: Yale Univ, BS, 43, MS, 44; Columbia Univ, PhD(math), 50. Prof Exp: Lab asst physics, Yale Univ, 43-44; math physicist, Naval Ord Lab, 44-46; lectr math, Columbia Univ, 46-48; mem, Inst Advan Study, 50-52; mathematician, Off Naval Res, 52-53; asst prof math, Cath Univ Am, 53-54; head res comput ctr, 54-61, numerical anal br, 61-65, math consult, nucleonics div, 65-66, nuclear physics div, 66-67, math & info sci div, 67-69, space sci div, 69-74, res mathematician, Math Res Ctr, 74-75, RES MATHEMATICIAN, APPL MATH STAFF, US NAVAL RES LAB, 75- Concurrent Pos: Instr, Univ Ill, 48; lectr, Univ Md, 52-53 & Am Univ, 52-53, 56-57; lectr & res assoc, Cath Univ Am, 54-64, adj prof, 64-73; vis prof math & statist, Univ Md, 74-75. Honors & Awards: Res Publ Award, US Naval Res Lab, 69. Mem: Am Math Soc; Math Asn Am; Inst Math Statist. Res: Complex function theory, entire and meromorphic functions; Dirichlet type series; real function theory; potential theory; numerical analysis; applied mathematics; probability; mathematical statistics; application of mathematics and statistics to physical sciences. Mailing Add: Appl Math Staff Code 7707 US Naval Res Lab Washington DC 20375

LE QUESNE, PHILIP WILLIAM, b Auckland, NZ, Jan 6, 39; m 65; c 1. ORGANIC CHEMISTRY. Educ: Univ Auckland, MSc, 61, PhD(chem), 64. Prof Exp: Res assoc org chem, Oxford Univ, 64-65; res assoc, Univ BC, 65-66, teaching fel, 66-67; asst prof, Univ Mich, Ann Arbor, 67-73; ASSOC PROF ORG CHEM, NORTHEASTERN UNIV, 73- Mem: Am Chem Soc; Phytochem Soc NAm; The Chem Soc; assoc NZ Inst Chem; Am Soc Pharmacog. Res: Natural product chemistry, especially steroids, alkaloids, terpenoids, fungal metabolites; comparative phytochemistry; physiologically active compounds. Mailing Add: Dept of Chem Northeastern Univ Boston MA 02115

LEQUIRE, VIRGIL SHIELDS, b Maryville, Tenn, June 15, 21; m 46; c 4. PATHOLOGY. Educ: Maryville Col, BA, 42; Vanderbilt Univ, MD, 46. Prof Exp: NIH res asst, 49-50, from asst prof to assoc prof anat, 50-66, actg chmn dept path, 71-73; PROF ANAT & EXP PATH, SCH MED, VANDERBILT UNIV, 66- Concurrent Pos: USPHS sr fel. Mem: AAAS; Am Asn Anat; Am Fedn Clin Res; Am Soc Exp Path; fel Royal Micros Soc. Res: Lipoprotein transport; fat embolization; membrane structure. Mailing Add: Dept of Path Vanderbilt Univ Sch of Med Nashville TN 37232

LERAAS, HAROLD J, b Barrett, Minn, Dec 30, 08; m 35; c 3. BIOLOGY, DENTISTRY. Educ: Luther Col, Iowa, BA, 30; Univ Mich, MS, 32, PhD(mammal), 35, DDS, 46. Prof Exp: Asst zoologist, Cranbrook Inst Sci, Mich, 34-35; prof biol, Pac Lutheran Col, 35-42; prof zool, Luther Col, Iowa, 42-43; private practice, 47-60; lectr biol, Pac Lutheran Univ, 47-60, prof, 60-75; RETIRED. Mem: Am Soc Mammal; Am Dent Asn. Res: Ecology and behavior of mammals and birds; bone formation in small mammals. Mailing Add: Dept of Biol Pac Lutheran Univ Tacoma WA 98447

LERBEKMO, JOHN FRANKLIN, b Alta, Can, Dec 8, 24; m 49; c 4. GEOLOGY. Educ: Univ BC, BASc, 49; Univ Calif, Berkeley, PhD(geol), 56. Prof Exp: Asst prof geol, 56-59, assoc prof sedimentary geol, 59-68, PROF SEDIMENTARY GEOL, UNIV ALTA, 68- Mem: Soc Econ Paleontologists & Mineralogists; Int Asn Sedimentol; Geol Asn Can. Res: Sedimentary petrology; detrital sediments. Mailing Add: Dept of Geol Univ of Alta Edmonton AB Can

LERCH, IRVING A, b Chicago, Ill, June 29, 38; m 63. MEDICAL PHYSICS. Educ: US Mil Acad, BS, 60; Univ Chicago, SM, 66, PhD(med physics), 69. Prof Exp: RES ASSOC MED PHYSICS, UNIV CHICAGO, 69-, NIH GRANT ANALOG COMPUT, 72- Concurrent Pos: Cancer Res Found jr auxiliary grant, 70-71. Mem: AAAS; Am Asn Physicists in Med; Radiation Res Soc. Res: Diagnostic radiological physics as applied to problems in image quality; cell kinetic and modelling studies as applied to problems in radiation therapy. Mailing Add: Dept of Radiol Univ of Chicago Chicago IL 60637

LERCHE, IAN, b Corbridge, Eng, Aug 24, 41; m 62; c 3. ASTROPHYSICS, PLASMA PHYSICS. Educ: Univ Manchester, BSc, 62, PhD(astron), 65. Prof Exp: Res assoc, 65-66, asst prof, 66-68, ASSOC PROF PHYSICS, UNIV CHICAGO, 68- Mem: Am Geophys Union; Am Astron Soc; fel Royal Astron Soc. Res: Geomagnetic field and plasma effects; instabilities and power spectra of relativistic plasmas; dynamical state of interstellar gas; radiation from cosmic ray air showers. Mailing Add: Enrico Fermi Inst & Dept Physics Univ of Chicago Chicago IL 60637

LERCHER, BRUCE L, b Milwaukee, Wis, June 7, 30; m 60; c 2. MATHEMATICAL LOGIC. Educ: Univ Wis, BS, 51, MS, 52; Pa State Univ, PhD(math), 63. Prof Exp: Instr math, Univ Rochester, 59-62; asst prof, 62-67, ASSOC PROF MATH, STATE UNIV NY BINGHAMTON, 67- Mem: Am Math Soc; Math Asn Am; Asn Symbolic Logic. Res: Combinatory logic. Mailing Add: Dept of Math State Univ of NY Binghamton NY 13901

LERIBAUX, HENRI ROMAIN, b Brussels, Belgium, Feb 7, 36; m 60; c 3. SOLID STATE PHYSICS, FLUID PHYSICS. Educ: Univ Louvain, BS, 57, Dipl Ing, 60, PhD(nuclear sci, eng), 63. Prof Exp: AEC fel, Inst Atomic Res, Iowa State Univ, 63-65; res assoc physics, Defense Res Bd Can, 65-68; asst prof, 67-71, ASSOC PROF NUCLEAR ENG & PHYSICS, TEX A&M UNIV, 71- Concurrent Pos: Res Coun grant, Tex A&M Univ, 67-69, Tex Water Qual Bd grant, 71-72; vis scientist, Battelle Advan Studies Ctr, Switz, 73-74. Mem: Am Phys Soc; Am Nuclear Soc. Res: Thermodynamic properties of liquid metals; transport phenomena in metals and liquids; neutron transport theory; statistical mechanics; safety of breeder and fusion reactors. Mailing Add: Dept of Nuclear Eng Tex A&M Univ College Station TX 77843

LE RICHE, WILLIAM HARDING, b Dewetsdorp, SAfrica, Mar 21, 16; Can citizen; m 43; c 5. MEDICINE. Educ: Univ Witwatersrand, BSc, 37, MB, BCh, 43, MD, 49; Harvard Univ, MPH, 50; FRCP(C), 72. Prof Exp: Med officer, Union Health Dept, 44-49, epidemiologist, 50-52; consult epidemiol, Dept Nat Health & Welfare, Ottawa, Can, 52-54; res med officer, Physicians Serv, Inc, 54-57; res assoc pub health, 57-59, prof, 59-62, prof epidemiol & biomet & Head Dept, Sch Hyg, 62-75, PROF EPIDEMIOL, DEPT PREV MED, FAC MED, UNIV TORONTO, 75- Concurrent Pos: Carnegie fel, Bur Educ & Social Res, SAfrica, 37-39; consult, Physicians Serv Inc, 57-66, Ont Med Serv Ins Plan, 66- & Res Inst, Hosp Sick Children, Toronto. Mem: Fel Am Pub Health Asn; Can Med Asn; fel Am Col Physicians; fel Royal Statist Soc. Res: Child growth, nutrition and infectious diseases; medical care studies; epidemiology; cardiovascular disease; education in public health and preventive medicine; hospital infections. Mailing Add: Fac of Med Fitzgerald Bldg Univ of Toronto 15 College St Toronto ON Can

LERKE, PETER A, b Paris, France, June 6, 28; US citizen; m 53; c 3. BACTERIOLOGY. Educ: Univ Calif, Berkeley, AB, 51, MA, 53, PhD(bact), 58. Prof Exp: Asst res microbiologist, 55-68, assoc res microbiologist, 68-75, DIR LAB RES IN CANNING INDUST, GEORGE WILLIAMS HOOPER FOUND, UNIV CALIF, SAN FRANCISCO, 75- Mem: AAAS; Am Soc Microbiol; Inst Food Technol; Royal Soc Health. Res: Food microbiology; bacterial spoilage of protein foods; antibiotics and food preservation; food-borne intoxications; nutrition; public health aspects of canned foods. Mailing Add: George Williams Hooper Found Univ of Calif 1950 Sixth St Berkeley CA 94710

LERMAN, ABRAHAM, b Harbin, China, Nov 14, 35; c 1. MARINE GEOCHEMISTRY, LIMNOLOGY. Educ: Hebrew Univ, Israel, MSc, 60; Harvard Univ, PhD(geol), 64. Prof Exp: Lectr geol, Johns Hopkins Univ, 64, asst prof, 64-65; asst prof, Univ Ill, Chicago, 65-68; sr scientist, Weizmann Inst, Israel, 66-69; res scientist chem limnol, Can Centre Inland Waters, Can Dept Environ, 69-71; assoc prof, 71-75, PROF GEOL SCI, NORTHWESTERN UNIV, 75- Concurrent Pos: Geochemist-oceanogr, Israel Geol Surv, 66-68. Mem: AAAS; Geochem Soc; Am Chem Soc; Geol Soc Am; Am Geophys Union. Res: Chemical and physical dynamics of sediments; fresh and ocean waters and saline brines; mathematical models of transport processes, material balance and water quality in natural water systems. Mailing Add: Dept of Geol Sci Northwestern Univ Evanston IL 60201

LERMAN, CHARLES LEW, b Elizabeth, NJ, Apr 23, 48. BIO-ORGANIC CHEMISTRY. Educ: Yale Univ, BS, 69; Harvard Univ, AM, 70, PhD(org chem), 74. Prof Exp: Asst prof chem, Juniata Col, 74-76; ASST PROF CHEM, HAVERFORD COL, 76- Mem: Am Chem Soc; AAAS. Res: Enzyme mechanisms and model systems; specificity of molecular interactions; mechanism of 5-aminolevulinic acid dehydratase; model for the active site of ribonuclease. Mailing Add: Dept of Chem Haverford Col Haverford PA 19041

LERMAN, LEONARD SOLOMON, b Pittsburgh, Pa, June 27, 25; m 74; c 3. MOLECULAR BIOLOGY. Educ: Carnegie Inst Technol, BS, 45; Calif Inst Technol, PhD(chem), 50. Prof Exp: Asst org chem & explosives, Explosives Res Lab, Carnegie Inst Technol, 45; asst chem, Calif Inst Technol, 45-49; Schenley fel, Univ Chicago, 49-51; instr pediat, Univ Colo, 51-52, asst prof, 52-53, asst prof biophys, 53-59, assoc prof, 59-63, prof, 63-65; PROF MOLECULAR BIOL, VANDERBILT UNIV, 65- Concurrent Pos: USPHS Res Career Award, 63-65, mem, Nat Sci Found Adv Panel, 65-68; NIH study sect, 69-73; Guggenheim fel, 71-72; mem, ed bd, Molecular & General Genetics, 71-76. Mem: Biophys Soc. Res: DNA microbial genetics; DNA structure, mutagenesis; DNA complexes, genetics. Mailing Add: Dept of Molecular Biol Vanderbilt Univ Box 1818 Sta B Nashville TN 37235

LERMAN, MANUEL, b New York, NY, Feb 5, 43. MATHEMATICAL LOGIC. Educ: City Col New York, BS, 64; Cornell Univ, PhD(math logic), 68. Prof Exp: Instr math, Mass Inst Technol, 68-70; asst prof, Yale Univ, 70-74; ASSOC PROF MATH, UNIV CONN, 74- Mem: Am Math Soc; Math Asn Am; Asn Symbolic Logic. Res: Recursive function theory. Mailing Add: Dept of Math Univ of Conn Storrs CT 06268

LERMAN, STEPHEN PAUL, b Philadelphia, Pa, Oct 3, 44. IMMUNOLOGY. Educ: Philadelphia Col Pharm & Sci, BS, 66; Hahnemann Med Col, MS, 70, PhD(microbiol), 73. Prof Exp: Fel, 72-74, asst res scientist, 74-75, RES ASST PROF PATH, SCH MED, NY UNIV, 75- Mem: Am Asn Immunologists; Am Soc Microbiol. Res: Tumor immunology; immunological tolerance; immunological deficiencies; suppressor cells; cell cooperation in the immune response. Mailing Add: NY Univ Med Ctr Dept of Path 550 First Ave New York NY 10016

LERNER, AARON BUNSEN, b Minneapolis, Minn, Sept 21, 20; m 45; c 4. DERMATOLOGY, BIOCHEMISTRY. Educ: Univ Minn, BA, 41, MS, 42, MB & PhD(physiol chem), 45, MD, 46. Prof Exp: Asst physiol chem, Univ Minn, 41-45; Am Cancer Soc fel, Sch Med, Western Reserve Univ, 48-49; asst prof dermat, Med Sch, Univ Mich, 49-52; assoc prof, Univ Ore, 52-55; assoc prof, 55-57, PROF, SCH MED, YALE UNIV, 57- Mem: Am Soc Biol Chem; Soc Invest Dermat; Am Acad Dermat. Res: Plasma proteins associated with disease; metabolism of phenylalanine and tyrosine; biochemistry of melanin pigmentation; mechanism of endocrine control of pigmentation; malignant melanomas; cryoglobulins; biochemistry of skin. Mailing Add: Dept of Dermat Yale Univ Sch of Med New Haven CT 06510

LERNER, ALBERT MARTIN, b St Louis, Mo, Sept 3, 29; m 57; c 4. INTERNAL MEDICINE. Educ: Wash Univ, BA, 50, MD, 54; Am Bd Internal Med, dipl, 61. Prof Exp: Intern, Barnes Hosp, St Louis, Mo, 54-55; lab investr, Nat Inst Allergy & Infectious Dis, 55-57; asst resident, Harvard Med Serv, Boston City Hosp, Mass, 57-58; sr asst resident, Barnes Hosp, Mo, 58-59; res assoc biol, Mass Inst Technol, 62-63; assoc prof med & assoc microbiol & path, Col Med, Wayne State Univ, 63-67; assoc med & path & dir bact lab, 64-69, CLIN CONSULT BACT LAB, DETROIT GEN HOSP, 69-; PROF MED, COL MED, WAYNE STATE UNIV, 67-, CHIEF, HUTZEL HOSP MED UNIT, 70- Concurrent Pos: Res fel med, Thorndike Mem Lab, Boston City Hosp & Harvard Med Sch, 59-62; fel, Med Found Greater Boston, Inc, 60-63; consult, Vet Admin Hosp, Allen Park, Mich, 63- Mem: Fel Am Col Physciians; Am Fedn Clin Res; Am Soc Clin Invest; NY Acad Sci; dipl mem Pan-Am Med Asn. Res: Infectious diseases. Mailing Add: Dept of Med Hutzel Hosp 432 E Hancock Detroit MI 48201

LERNER, DAVID EVAN, b Kansas City, Mo, Mar 21, 44. THEORETICAL PHYSICS. Educ: Haverford Col, BA, 64; Univ Pittsburgh, PhD(math), 72. Prof Exp: Instr math, Univ Pittsburgh, 72-73; res assoc physics, Syracuse Univ, Relativity Group, 73-75; ASST PROF MATH, UNIV KANS, 75- Mem: Am Phys Soc; Am Math Soc. Res: Application of Lie groups and differential geometry to general relativity. Mailing Add: Dept of Math Univ of Kans Lawrence KS 66045

LERNER, EDWARD CLARENCE, b Brooklyn, NY, Sept 10, 24. THEORETICAL PHYSICS. Educ: Mass Inst Technol, BS, 45, PhD(physics), 52. Prof Exp: Mem staff, Lincoln Lab, Mass Inst Technol, 52-58; assoc prof, 57-62, PROF PHYSICS, UNIV SC, 62- Mem: AAAS; Am Phys Soc. Res: Electrodynamics; field theory; classical and quantum dynamics. Mailing Add: Dept of Physics & Astron Univ of SC Columbia SC 29208

LERNER, EUGENIO, physics, see 12th edition

LERNER, HARRY, b New York, NY, Apr 24, 30; m 57; c 4. ELECTROCHEMISTRY. Educ: City Univ New York, BS, 51; Polytech Inst Brooklyn, MS, 53; Pa State Univ, PhD(chem), 64. Prof Exp: Phys chemist, Air Reduction Labs, 56-57; chemist, US Testing Co, 57-58 & Nopco Chem Co, 58-60; res assoc electrochem, Pa State Univ, 64-65; mem sci staff, Itek Corp, 65-76; SR ENGR, SOLAR ENERGY SYSTS, INC, 76- Concurrent Pos: Consult, US Marine Labs, 64-65. Mem: Am Chem Soc; Electrochem Soc. Res: Thermodynamics and kinetics of electrode reactions; kinetics and mechanisms of photographic development reactions; research and development on electrochemical and solar energy systems. Mailing Add: Solar Energy Systs Inc One Tralee Indust Park Newark DE 19711

LERNER, I MICHAEL, b Harbin, China, May 15, 10; nat US; m 37. GENETICS. Educ: Univ BC, BSA, 31, MSA, 32; Univ Calif, PhD(genetics), 36. Hon Degrees: DSc, Univ BC, 62; DSc, Univ Edinburgh, 73. Prof Exp: From instr to prof poultry husbandry, 36-58, prof genetics, 58-73, chmn dept, 58-63, RES GENETICIST, INST PERSONALITY ASSESSMENT & RES, UNIV CALIF, BERKELEY, 70-, EMER PROF GENETICS, 73- Concurrent Pos: Guggenheim fel, Univ Edinburgh, 48, Univ Pavia, 53, 57; Can Club lectr, Univ BC, 50; fel, Ctr Advan Study Behav Sci, 67-68; US rep, Permanent Int Comt, Genetics Cong, 58-63, secy, 53-58; foreign mem, Georgofili Agr Econ Acad, Florence, Italy. Honors & Awards: Prize, Poultry Sci Asn, 37; Belling Prize, 40; Borden Award & Medal, 51; Mendel Mem Medal, Czech Acad Sci, 65; Weldon Mem Prize & Medal, Oxford Univ, 69. Mem: Nat Acad Sci; AAAS; Am Acad Arts & Sci; Soc Study Evolution (ed, Evolution, 59-61, vpres, 62, pres, 64); Am Soc Nat (vpres, 57). Res: Evolution; animal improvement; behavior genetics. Mailing Add: Dept of Genetics Univ of Calif Berkeley CA 94720

LERNER, JOSEPH, b Wilkes-Barre, Pa, Jan 16, 42; m 63; c 2. BIOCHEMISTRY. Educ: Rutgers Univ, BS, 63, PhD(biochem), 67. Prof Exp: Sr res investr biochem, Eastern Utilization Res & Develop Div, USDA, 67-68; asst prof, 68-71, ASSOC PROF BIOCHEM, UNIV MAINE, ORONO, 71- Concurrent Pos: Coe Res Fund grant, 68-69; Hatch Fund grant, 69-71; NIH res grant, 73; res assoc, Dept Avian Sci, Univ Calif, Davis, 74. Mem: Am Chem Soc; Am Inst Nutrition; NY Acad Sci; Am Soc Zoologists; Soc Develop Biol. Res: Intestinal absorption of amino acids in chicken; metabolism of small intestine; seperation of nucleotide derivatives by column chromatography. Mailing Add: 236 Hitchner Hall Univ of Maine Orono ME 04473

LERNER, JULES, b Englewood, NJ, Oct 24, 41; m 69. GENETICS, CYTOLOGY. Educ: Bowdoin Col, BA, 63; Johns Hopkins Univ, PhD(biol), 67. Prof Exp: Asst prof, 67-71, ASSOC PROF BIOL, NORTHEASTERN ILL UNIV, 71- Mem: AAAS; Am Soc Zoologists. Res: Developmental biology. Mailing Add: Dept of Biol Northeastern Ill Univ Chicago IL 60625

LERNER, LAWRENCE ROBERT, b New York, NY, Mar 17, 43; m 64; c 2. INDUSTRIAL ORGANIC CHEMISTRY. Educ: City Col New York, BS, 64; Mich State Univ, PhD(org chem), 68. Prof Exp: Res chemist org pigments, E I du Pont de Nemours & Co, 68-72; group leader, 72-73, SUPVR ORG PIGMENTS, ALLIED CHEM CORP, 73- Concurrent Pos: Adj asst prof, County Col Morris, NJ, 73-74. Mem: Sigma Xi; Am Chem Soc; AAAS; Inter-Soc Color Coun. Res: Synthesis of colored organic pigments; study of the effects of structure on the photostability, color and physical properties of organic pigments. Mailing Add: Allied Chem Corp PO Box 14 Hawthorne NJ 07507

LERNER, LAWRENCE S, b New York, NY, Mar 10, 34; m 59. SOLID STATE PHYSICS, HISTORY OF SCIENCE. Educ: Univ Chicago, AB, 53, MS, 55, PhD(physics), 62. Prof Exp: Staff mem, Labs Appl Sci, Univ Chicago, 58-60; physicist, Hughes Res Labs, 62-65 & Hewlett-Packard Labs, Calif, 65-67; res scientist, Lockheed Palo Alto Res Lab, 67-69; assoc prof, 69-73, PROF PHYSICS & ASTRON, CALIF STATE UNIV, LONG BEACH, 73- Concurrent Pos: Assoc, Danforth Found, 75- Mem: AAAS; Am Phys Soc; Am Asn Physics Teachers; Hist Sci Soc; Renaissance Soc Am. Res: Fermi surfaces of metals and semimetals; preparation and properties of ternary compound semiconductors; semiconductor physics; influence of non-scientific philosophical movements on early scientific revolution. Mailing Add: Dept of Physics & Astron Calif State Univ Long Beach CA 90840

LERNER, LEON MAURICE, b Chicago, Ill, Feb 2, 38; m 59; c 3. BIOCHEMISTRY. Educ: Ill Inst Technol, BS, 59, MS, 61; Univ Ill, PhD(biochem), 64. Prof Exp: Res assoc biochem, Col Med, Univ Ill, 64-65; from instr to asst prof, 65-73, ASSOC PROF BIOCHEM, STATE UNIV NY DOWNSTATE MED CTR, 73- Mem: AAAS; Am Chem Soc. Res: Potential nucleic acid antimetabolites; nucleoside analogs; chemistry and biochemistry of carbohydrates. Mailing Add: 450 Clarkson Ave Brooklyn NY 11203

LERNER, LEONARD JOSEPH, b Roselle, NJ, Sept 26, 22. ENDOCRINOLOGY. Educ: Rutgers Univ, BS, 43, AB, 51, MS, 53, PhD(zool), 54. Prof Exp: Pharmacist, 46-51; asst, Bur Biol Res, Rutgers Univ, 53-54; endocrinologist, Wm S Merrell Co, 54-58; head endocrine res, Squibb Inst Med Res, 58-71; DIR ENDOCRINOL, GRUPPO LEPETIT SPA, 71- Concurrent Pos: Assoc mem, Bur Biol Res, Rutgers Univ; vis prof obstet & gynec, Hahnemann Med Sch. Mem: Endocrine Soc; Am Physiol Soc; NY Acad Sci; Am Fertil Soc; Soc Study Reprod. Res: Hormone antagonists; fertility control; reproduction; placenta; prostaglandins; ovulation; steroids; endocrine-tumor relationships; hormone treatment of newborn; endocrine biochemistry; central nervous system-endocrine system relationship; adrenal physiology; pregnancy physiology and biochemistry. Mailing Add: Gruppo Lepetit SpA Via Durando 38 20158 Milano Italy

LERNER, LOUIS LEONARD, b Chicago, Ill, Feb 25, 15; m 49; c 1. COSMETIC CHEMISTRY. Educ: Cent YMCA Col, BS, 42. Prof Exp: Chem asst, Universal Merchandise Co, 34, assoc chemist, 35-37; chief chemist, Russian Duchess Labs, 37; pres & dir res, LaLerne Labs, 37-40; dir res & prod, Consol Royal Chem Corp, 40-46; exec vpres & dir res & prod, Allied Home Prods Corp, 46-49; vpres & dir res, Kalech Res Labs, 49-50; vpres & dir res & new prod develop, Bymart, Inc, 50-52; sr scientist, Personal Care Div, Gillette Co, 52-74; consumer prod specialist, 74, PHYSICAL SCIENTIST, FED TRADE COMN, US GOVT, 75- Concurrent Pos: Dir, Brokers, Inc, 37-40; instr, Cent YMCA Col, 42-44; dir, Allied Home Prods Corp, Ill, 47-49; vpres, Phil Kalech Co, 49-50; dir, AD Prods Corp, 64-66; consult, Seaquist Valve Co, 74-; sect ed, Chem Bull, Am Chem Soc, 71- Mem: Fel AAAS; Am Chem Soc; Soc Cosmetic Chem; NY Acad Sci; Nat Asn Sci Writers. Res: Product development exploratory research; pharmaceuticals, proprietaries, cosmetics, detergents, emulsions, waving compositions, dyes and pigments; chemistry of polymers, proteins and enzymes; mechanical devices; surface chemistry; consumer products. Mailing Add: 900 N Lake Shore Dr Chicago IL 60611

LERNER, MARGUERITE RUSH, b Minneapolis, Minn, May 17, 24; m 45; c 4. DERMATOLOGY. Educ: Univ Minn, Minneapolis, BA, 45; Case Western Reserve Univ, Md, 50. Prof Exp: From intern to resident dermat, Univ Mich Hosp, Ann Arbor, 50-52; resident, Multnomah County Hosp, Univ Ore, 52-54; from clin instr to assoc clin prof dermat, 47-73, PROF CLIN DERMAT, SCH MED, YALE UNIV, 73- Mem: Am Acad Dermat; Soc Invest Dermat. Res: Psoriasis. Mailing Add: Dept of Dermat Yale Univ Sch of Med New Haven CT 06510

LERNER, MELVIN, b Milwaukee, Wis, June 8, 17; m 68; c 2. ANALYTICAL CHEMISTRY. Educ: City Col NY, BS, 37. Prof Exp: Res chemist, Univ Patents, Inc, 38-39; chemist, Hoover & Strong, Inc, 39-41; chemist, US Customs Lab, 41-50, from asst chief chemist to chief chemist, 50-67; DIR TECH SERV, BUR CUSTOMS, 67- Mem: AAAS; Am Chem Soc; Am Soc Test & Mat. Res: Analytical chemistry of marijuana; narcotics and dangerous drugs; sampling of bulk materials; instrumental analytical chemistry. Mailing Add: US Customs Serv Washington DC 20229

LERNER, MICHAEL PAUL, b Los Angeles, Calif, May 2, 41; m 65; c 2. VIROLOGY, CELL BIOLOGY. Educ: Univ Calif, Los Angeles, BA, 63; Kans State Univ, MS, 67; Northwestern Univ, PhD(microbiol), 70. Prof Exp: Nat Inst Neurol Dis & Stroke fel, Univ Calif, Los Angeles, 70-71, asst res biologist neurochem, Ctr Health Sci, 72-73; ASST PROF MICROBIOL, HEALTH SCI CTR, UNIV OKLA, 73- Concurrent Pos: Fel, NATO Advan Study Inst, Italy, 72. Mem: Am Soc Microbiol. Res: Mammalian cell biochemistry and development. Mailing Add: Dept of Microbiol & Immunol Univ Okla Health Sci Ctr Oklahoma City OK 73190

LERNER, MORRIS WOLFE, analytical chemistry, see 12th edition

LERNER, NARCINDA REYNOLDS, b Brooklyn, NY, Oct 10, 33; m 59. POLYMER CHEMISTRY. Educ: Hofstra Univ, BA, 56; Univ Chicago, MS, 59, PhD(chem), 62. Prof Exp: Mem tech staff, Hughes Res Labs Labs, Calif, 62-63; res scientist, Lockheed Palo Alto Res Lab, 66-70; RES SCIENTIST, AMES RES CTR, NASA, 70- Mem: AAAS; Am Phys Soc. Res: Paramagnetic resonance; electron nuclear double resonance; crystalline field theory; crystal preparation; polymer degradation. Mailing Add: Ames Res Ctr NASA MS 223-6 Moffett Field CA 94035

LERNER, RICHARD ALAN, b Chicago, Ill, Aug 26, 38; m 66; c 3. IMMUNOLOGY. Educ: Stanford Univ, MD, 64. Prof Exp: Intern med, Stanford Univ, 64-65; assoc cell biol, Wistar Inst, 68-70; assoc immunol, 70-71, assoc mem, 71-73, MEM IMMUNOL, SCRIPPS CLIN & RES FOUND, 74- Concurrent Pos: USPHS grant, Scripps Clin & Res Found, 65-68; consult, Nat Cancer Inst, 72- Mem: Am Soc Immunol; Am Soc Path; Biophys Soc; Am Soc Microbiol. Res: Molecular medicine; differentiation. Mailing Add: Scripps Clin 476 Prospect St La Jolla CA 92037

LERNER, RITA GUGGENHEIM, b New York, NY, May 7, 29; m 54; c 2. INFORMATION SCIENCE. Educ: Radcliffe Col, AB, 49; Columbia Univ, MA, 51, PhD(chem phys), 56. Prof Exp: Res assoc chem phys, Columbia Univ, 56-64; staff physicist info sci, Am Inst Physics, 65, dep dir, Info Anal & Retention Div, 66-67; dir labs, Dept Biol Sci, Columbia Univ, 68; mgr planning & develop, 69-73, MGR SPEC PROJS INFO SCI, AM INST PHYSICS, 73- Mem: Am Phys Soc; Am Chem Soc; AAAS; Am Soc Info Sci; NY Acad Sci. Res: Information systems and information retrieval; applications of new technology to publishing and distribution of information. Mailing Add: Am Inst of Physics 335 E 45th St New York NY 10017

LERNER, SIDNEY ISAAC, b Baltimore, Md, May 2, 32; m 56; c 4. OCCUPATIONAL MEDICINE. Educ: Univ Md, BS, 53, MD, 57; Ohio State Univ, MS, 60. Prof Exp: Staff occup med, Army Environ Hyg Agency, 60-62 & Corp Med Dept, Ethyl Corp, 62-74; MEM FAC ENVIRON HEALTH, COL OF MED, UNIV CINCINNATI, 62- Concurrent Pos: Dir personal health, Cincinnati Gen Hosp, 64-; consult occup med, 62- Mem: AMA; Am Acad Occup Med; Am Occup Med Asn;

Am Indust Hyg Asn; Am Conf Govt Indust Hygienists. Res: Clinical effects of inorganic lead and other heavy metals, clinical industrial toxicology; medical surveillance; clinical practice of occupational medicine. Mailing Add: Dept Environ Health Col of Med Univ of Cincinnati Cincinnati OH 45267

LEROI, GEORGE EDGAR, b London, Eng, June 23, 36; US citizen; m; c 3. CHEMICAL PHYSICS. Educ: Univ Wis, BA, 56; Harvard Univ, AM, 58, PhD(chem), 60. Prof Exp: Res assoc chem, Univ Calif, Berkeley, 60-62; lectr, Princeton Univ, 62-64, asst prof, 64-67; assoc prof, 67-72, PROF CHEM, MICH STATE UNIV, 72- Concurrent Pos: Guest Prof, Lab Phys Chem, Swiss Fed Inst Technol, 74-75. Honors & Awards: Coblentz Award, 72. Mem: Am Phys Soc. Res: Molecular spectroscopy and structure; vacuum ultraviolet, visible, infrared, far infrared, Raman; photoionization mass spectrometry laser photochemistry. Mailing Add: Dept of Chem Mich State Univ East Lansing MI 48824

LEROUX, EDGAR JOSEPH, b Ottawa, Ont, Jan 24, 22; m 44; c 3. INSECT ECOLOGY. Educ: Carleton Univ, Can, BA, 50; McGill Univ, MSc, 52, PhD(entom), 54. Hon Degrees: DSc, McGill Univ, 73. Prof Exp: Asst entomologist, Fruit Insect Invests, Sci Serv, 49-53, assoc entomologist, Orchard Insect Ecol, 53-60, entomologist, Res Lab, 60-62, sr entomologist, 62-66, res coordr entom, 66-67, asst dir gen, Insts, Res Br, 68-72, DIR GEN, PLANNING & COORD, RES BR, CAN DEPT AGR, 72- Concurrent Pos: Demonstr, Macdonald Col, McGill Univ, 50-51, asst, 53-54, lectr, 58-62, assoc prof, 62-65, hon prof, 70-71; Grace Griswold lectr, Cornell Univ, 71. Mem orchard protection comt, Info & Res Serv, Que Dept Agr, 59-63; mem adv comt entom probs, Defence Res Bd, Dept Nat Defence, 64-67; sci ed, Can J Plant Sci, 65-68; mem panel experts integrated pest control, Food & Agr Orgn, 66-71; dir, Biol Coun Can, 66-70, pres, 70-71; off cor entom, Commonwealth Inst Biol Control, 67-73; Can rep, Int Soc Hort Sci, 67-73; mem, World Hort Coun, 68-70; chmn panel insect ecol, Study of Basic Biol in Can, Biol Coun Can, Sci Coun Can, 69-70; mem, Can Govt Tech Apple Mission, Japan, Australia, NZ & SAfrica, 71; mem, Asn Sci, Eng & Technol Community Can Mgt Comt, Sci Coun Can Spec Study on Sci Soc, 72; mem negotiated grants comt & adv comt on biol, Nat Res Coun Can, 73- Mem: Entom Soc Can (pres, 69-70); Fr-Can Asn Advan Sci; Agr Inst Can; Can Soc Zoologists; Asn Sci Eng & Technol Community Can (hon treas, 70-72). Res: Insect ecology; integrated pest control; morphology; toxicology. Mailing Add: Rang St Antoine Perkins PQ Can

LEROUX, EDMUND FRANK, b Muskegon Heights, Mich, Mar 8, 25; m 49; c 3. HYDROLOGY. Educ: Mich State Univ, BS, 48. Prof Exp: Geologist, 49-60, chief manpower sect, 61-64, ASST DIST CHIEF HYDROL, US GEOL SURV, 64- Mem: Am Geophys Union; Int Asn Hydrogeologists; AAAS. Res: Ground water temperature; hydrology of glacial terrane in a semiarid climate. Mailing Add: Water Resources Div US Geol Surv PO Box 1412 Huron SD 57350

LEROY, DONALD JAMES, b Detroit, Mich, Mar 5, 13; m 40; c 4. CHEMICAL KINETICS, PHOTOCHEMISTRY. Educ: Univ Toronto, BA, 35, MA, 36, PhD, 39. Hon Degrees: LLD, Trent Univ, 71; DSc, Laurentian Univ, 73, McMaster Univ, 74. Prof Exp: Ont Res Found, 39; chem div, Nat Res Labs, 40-44; from asst prof to prof chem, Univ Toronto, 44-74, chmn dept, 60-71; sci vpres, 69-74, PRIN RES OFFICER, NAT RES COUN CAN, 75- Honors & Awards: Palladium Medal, Chem Inst Can; Centennial Medal, Govt of Can, 67. Mem: Am Chem Soc; Chem Inst Can; The Chem Soc; Am Phys Soc; fel Royal Soc Can. Res: Reactions of atoms and free radicals; photosensitization; energy transfer processes. Mailing Add: Nat Res Coun Can Ottawa ON Can

LEROY, GEORGE VEACH, b Wilkinsburg, Pa, Oct 28, 19; m 30; c 1. MEDICINE. Educ: Univ Pittsburgh, BS, 32; Univ Chicago, MD, 34. Prof Exp: Asst med, Univ Chicago, 35-37; from instr to asst prof, Med Sch, Northwestern Univ, 37-51; assoc dean div biol, Univ Chicago, 51-62, prof med, 55-66; med dir, Metrop Hosp, Detroit, 66-70; PROF MED, UNIV CHICAGO, 70- Concurrent Pos: Mem, Joint Comn Invest Effects of Atomic Bomb in Japan, US Army, 45-46; dir radioisotope unit, Vet Admin Hosp, Hines, Ill, 47-51; consult, AEC, 48-; dir biomed prog, Oper Greenhouse, 50-51; clin prof med, Sch Med, Wayne State Univ, 66-; mem life sci comt, Space Sci Bd, Nat Acad Sci; dir-mem, Nat Coun Radiation Protection; Searle res fel, Med Sch, Northwestern Univ, 37-46. Mem: Fel AMA; Asn Am Physicians; Radiation Res Soc; Am Pub Health Asn. Res: Internal medicine; radiation biology; medical administration; hematology. Mailing Add: Dept of Med Univ of Chicago Chicago IL 60637

LERSTEN, NELS R, b Chicago, Ill, Aug 6, 32; m 58; c 3. BOTANY. Educ: Univ Chicago, BS, 58, MS, 60; Univ Calif, Berkeley, PhD(bot), 63. Prof Exp: From asst prof to assoc prof, 63-70, PROF BOT, IOWA STATE UNIV, 70- Mem: Bot Soc Am; Sigma Xi. Res: Systematic and developmental anatomy of angiosperms; embryology of flowering plants. Mailing Add: Dept of Bot & Plant Path Iowa State Univ Ames IA 50011

LES, EDWIN PAUL, b Adams, Mass, Dec 28, 23; m 67; c 2. GENETICS. Educ: Northeastern Univ, BS, 52; Ohio State Univ, MS, PhD(genetics), 59. Prof Exp: Assoc staff scientist, Jackson Lab, 59-60; biologist, Biol Div, Oak Ridge Nat Lab, 60-62; staff scientist, 62-75, SR STAFF SCIENTIST, JACKSON LAB, 75- Mem: AAAS; Am Genetic Asn; Am Asn Lab Animal Sci; Am Inst Biol Sci. Res: Effect of environment on reproduction, growth and survival of inbred laboratory mice; mouse husbandry techniques and practices; laboratory animal health. Mailing Add: Jackson Lab Animal Health Serv Bar Harbor ME 04609

LE SENEY, CATHERINE COSGRAY, b Eaton Rapids, Mich, June 23, 21; m 41. PREVENTIVE MEDICINE, PUBLIC HEALTH. Educ: Univ Denver, BS, 55; Univ Colo, Denver, MD, 61; Univ Calif, Berkeley, MPH, 69; Am Bd Prev Med, dipl, 70. Prof Exp: Intern, St Luke's Hosp, Denver, Colo, 61-62, resident internal med & path, 62-65; clinician migrant health, Pinal County Health Dept, Ariz, 65-66, health officer, 66-67; resident prev med, Calif State Health Dept, 67-69; dist health officer, Maricopa County Health Dept, Ariz, 70-71; med planning specialist, Ariz Health Planning Auth, 71-74; commun dis control officer, Pima County Health Dept, Ariz, 74-75; CONSULT EPIDEMIOL FOR HEALTH & ENVIRON SCI, 75- Mem: Am Col Prev Med; Am Pub Health Asn. Res: Comprehensive health planning; human ecology; population dynamics; nutrition; use and conservation of natural resources. Mailing Add: 2534 E Cochise Rd Phoenix AZ 85028

LESENSKY, LEONARD, b New York, NY, Sept 22, 22; m 48; c 1. PHYSICS. Educ: Brooklyn Col, BA, 42; Columbia Univ, MA, 49, PhD(physics), 56. Prof Exp: Asst, Radiation Lab, Columbia Univ, 42-44, physics dept, 50-53; physicist, Microwave Tubes, Raytheon Co, 53-57, mgr mat res, Spencer Lab, 57-61; prin scientist mat, Avco Corp, 61-64; PRIN ENGR MAT, MICROWAVE & POWER TUBE DIV, RAYTHEON CO, 64- Mem: Am Phys Soc. Res: Physical, chemical and electronic properties of microwave tube materials; microwave tubes; low temperature physics. Mailing Add: Microwave & Power Tube Div Raytheon Co Waltham MA 02154

LESER, ERNST GEORGE, b Mineola, NY, May 3, 43; m 69. ORGANIC CHEMISTRY. Educ: Bucknell Univ, BS, 65; Fordham Univ, PhD(org chem), 70. Prof Exp: Res chemist, Jackson Lab, 69-74, PROD SUPVR, CHAMBERS WORKS, E I DU PONT DE NEMOURS & CO, INC, 74- Mem: Am Chem Soc. Res: Supervision of dyes and intermediates production. Mailing Add: Chambers Works E I du Pont de Nemours & Co Inc Wilmington DE 19898

LESER, RALPH ULRICH, b Bloomington, Ind, Dec 31, 05; m 66; c 2. INTERNAL MEDICINE. Educ: Ind Univ, AB, 27, MD, 30; Am Bd Internal Med, dipl, 44. Prof Exp: Intern, Philadelphia Gen Hosp, Pa, 30-32; fel internal med, Mayo Clin, 34-37; from instr to assoc med, 38-50, asst prof, 50-67, ASSOC PROF MED, SCH MED, IND UNIV, INDIANAPOLIS, 67- Concurrent Pos: Vis physician, Marion County Gen Hosp, 38-, chief, Diag Clin, 46-58; vis physician, Methodist, St Vincent's & Community Hosps. Mem: AMA; fel Am Col Physicians. Res: Cardiology; diseases of metabolism; gastroenterology. Mailing Add: 5434 Ashurst Indianapolis IN 46220

LESER, TADEUSZ, b Lwow, Poland, Dec 14, 08; nat US; m 46. MATHEMATICS. Educ: Warsaw Tech Univ, ME, 39; Univ London, PhD(eng), 44. Prof Exp: Res engr, Petrol Warfare Dept, Imp Col Res Sta, Eng, 42-46; res engr, Indust Res Co, 47; assoc prof math, Emory & Henry Col, 47-48; asst prof, Univ Ky, 48-55; RES MATHEMATICIAN, ABERDEEN PROVING GROUND, 55- Res: Ballistics and space flight; numerical analysis; communication theory; elasticity of plasticity; programming for computers. Mailing Add: Moores Mill Rd Bel-Air MD 21014

LESH, JANET ROUNTREE, b Chicago, Ill, Aug 14, 37. ASTROPHYSICS. Educ: Cornell Univ, AB, 58; Univ Chicago, PhD(astron, astrophys), 67. Prof Exp: Res assoc astron, Yerkes Observ, Univ Chicago, 67-68; sci officer astron, Leiden Observ, Univ Leiden, Netherlands, 68-70; astronr adjoint, Meudon Observ, Observ Paris, France, 70-71; vis fel astrophys, Joint Inst Lab Astrophys, Univ Colo, 71-72; LECTR ASTRON & RES ASTRONR, UNIV DENVER, 72-, DIR OBSERV OPERS, 74- Concurrent Pos: Eng lang ed & translr, Astron & Astrophys, 69-72; translr, D Reidel Co, Dordrecht, Netherlands, 69-72 & Joint Publ Res Serv, 73- Mem: Int Astron Union; Am Astron Soc; Royal Astron Soc; Netherlands Astron Soc. Res: Spectral classification of early type stars; short-period variable stars; local galactic structure; stellar associations and clusters; infrared astronomy. Mailing Add: Dept of Physics & Astron Univ of Denver Denver CO 80210

LESH, THOMAS ALLAN, b Chicago, Ill, Aug 6, 29. PHYSIOLOGY. Educ: Mich State Univ, BS, 51; Ind Univ, PhD(physiol), 68. Prof Exp: Assoc ed, Howard W Sams & Co, Inc, Ind, 55-63; USPHS cardiovasc trainee, Bowman Gray Sch Med, 68-70; asst prof physiol, Med Ctr, Univ Ark, Little Rock, 70-75; ASST PROF PHYSIOL & HEALTH SCI, BALL STATE UNIV, 75- Mem: Assoc Am Physiol Soc; Microcirc Soc. Res: Control of vascular conductance in skeletal muscle. Mailing Add: Dept of Physiol Ball State Univ Muncie IN 47306

LESHER, DEAN ALLEN, b Endicott, NY, Feb 18, 27; m 49; c 3. CLINICAL PHARMACOLOGY, NEPHROLOGY. Educ: Colgate Univ, AB, 48; Univ Wis, PhD(pharmacol), 56; Univ Buffalo, MD, 62. Prof Exp: Asst pharmacol, Univ Mich, 51-52; asst, Univ Wis, 52-55; instr, Univ Buffalo, 55-58, assoc, 58-62; from intern to resident, Henry Ford Hosp, Detroit, 62-63, assoc med, 64-72; assoc dir med res, Lederle Labs, 71-73; ASSOC DIR CLIN PHARMACOL, CIBA GEIGY CORP, 74- Mem: Am Soc Clin Invest; Int Soc Nephrol; Am Soc Nephrol; Am Soc Clin Pharmacol & Therapeut; fel Am Col Clin Pharm. Res: Renal transport of acids and bases; renal potassium transport; diuretics; renal disease; hemodialysis; peritoneal dialysis; renal transplantation; drug intoxication. Mailing Add: 8 Cedar Green Lane Berkeley Heights NJ 07922

LESHER, GEORGE YOHE, b Mt Erie, Ill, Feb 22, 26; m 57; c 2. PHARMACEUTICAL CHEMISTRY, SYNTHETIC ORGANIC CHEMISTRY. Educ: Univ Ill, BA, 50; Dartmouth Col, MS, 52; Rensselaer Polytech Inst, PhD(chem), 56. Prof Exp: Sr res assoc & sect head, 52-67, ASST DIR CHEM DIV, STERLING-WINTHROP RES INST, 67- Mem: Am Chem Soc. Res: Heterocyclic chemistry. Mailing Add: Sterling-Winthrop Res Inst Rensselaer NY 12144

LESHER, SAMUEL WALTER, b Spokane, Wash, June 9, 16; m 70; c 2. CELL BIOLOGY, RADIOBIOLOGY. Educ: Univ Ill, BS, 42; Wash Univ, PhD(zool), 50. Prof Exp: Biologist, Div Biol & Med Res, Argonne Nat Lab, Ill, 54-68; DIR, CELL & RADIOBIOL LABS & CANCER RES UNIT, ALLEGHENY GEN HOSP, PITTSBURGH, 68-, ASSOC DIR, CLIN RADIATION THER RES CTR, 73- Concurrent Pos: Fel, Inst Cancer Res, Royal Cancer Hosp, Sutton, Eng, 62. Mem: Radiation Res Soc; Am Asn Cancer Res; Am Soc Cell Biol. Res: Cell kinetics; effects of radiation and drugs on normal and malignant cells; combined modality research; cancer biology. Mailing Add: Allegheny Gen Hosp Cancer Res Unit 320 E North Ave Pittsburgh PA 15213

LESHIN, RICHARD b New York, NY, June 16, 24; m 50; c 3. ORGANIC CHEMISTRY. Educ: City Col New York, BS, 44; NY Univ, MS, 47, PhD(chem), 52. Prof Exp: Sr res chemist, 52-65, SECT HEAD, GOODYEAR TIRE & RUBBER CO, 65- Concurrent Pos: Instr, eve sch, Univ Akron, 54-56. Mem: Am Chem Soc. Res: Organic synthesis; rubber additives; vulcanization of rubber. Mailing Add: 1527 Shanabrook Dr Akron OH 44313

LESH-LAURIE, GEORGIA ELIZABETH, b Cleveland, Ohio, July 28, 38; m 69. DEVELOPMENTAL BIOLOGY. Educ: Marietta Col, BS, 60; Univ Wis, MS, 61; Case Western Reserve Univ, PhD(biol), 66. Prof Exp: Instr biol, Case Western Reserve Univ, 65-66; asst prof biol sci, State Univ NY Albany, 66-69; asst prof, 69-73, ASST DEAN, WESTERN RESERVE COL, CASE WESTERN RESERVE UNIV, 73-, ASSOC PROF BIOL, 74- Concurrent Pos: NY State Res Found res fel, 66-67; USPHS instnl grant, 70-71; Am Cancer Soc grant, 68-71; Res Corp grant, 71; Am Cancer Soc instnl grant, 73. Mem: AAAS; Am Soc Zool; Soc Develop Biol; NY Acad Sci. Res: Study of the neurosecretory control of cellular differentiation in cnidarian systems and the role of these differentiations in the establishment and maintenance of polarized form. Mailing Add: Dept of Biol Case Western Reserve Univ Cleveland OH 44106

LESHNER, ALAN IRVIN, b Lewisburg, Pa, Feb 11, 44; m 69. PHYSIOLOGICAL PSYCHOLOGY. Educ: Franklin & Marshall Col, AB, 65; Rutgers Univ, MS, 67, PhD(psychol), 69. Prof Exp: Asst prof, 69-73, ASSOC PROF PSYCHOL, BUCKNELL UNIV, 73- Concurrent Pos: NIMH res grant, 70-71, NSF res grant, 70-72, 75- Mem: AAAS; Int Soc Psychoneuroendocrinol; Animal Behav Soc; Am Soc Mammal. Res: Hormonal basis of social behavior of primates and rodents. Mailing Add: Dept of Psychology Bucknell Univ Lewisburg PA 17837

LESIEWICZ, JEANNE LEE, b Teaneck, NJ, Nov 9, 48; m 70. MOLECULAR BIOLOGY. Educ: Univ Del, BA, 70, PhD(biol sci), 75. Prof Exp: RES ASSOC BIOCHEM, STATE UNIV NY STONY BROOK, 75- Mem: Am Soc Microbiol. Res: Molecular biology and evolution of subcellular genetic systems, particularly the sequence and structure of viral RNA and the sites of synthesis of chloroplast enzymes. Mailing Add: Dept of Biochem State Univ of NY Stony Brook NY 11790

LESINS, KARLIS A, b Latvia, July 30, 06; m 39. CYTOGENETICS, PLANT BREEDING. Educ: Royal Agr Col, Sweden, Lic Agr, 50; Univ Alta, DSc, 59. Prof Exp: Headmaster, Sch Agr Bebrene, Latvia, 31-41; supt, Agr Res Sta, Osupe, Latvia, 41-44; agronomist, Swedish Seed Asn, 44-51; res fel, 51-54, res assoc forage crops, 54-60, assoc res prof, 60-65, prof, 66-71, EMER PROF GENETICS, UNIV ALTA, 71- Mem: Am Soc Agron; Am Genetic Asn; Can Agr Inst Res: Specialty Medicago. Mailing Add: 9727-65 Ave Edmonton AB Can

LESINSKI, JOHN SILVESTER, b Philadelphia, Pa, Mar 29, 13; m 40; c 4. OBSTETRICS & GYNECOLOGY. Educ: Jagellonian Univ, Poland, MD, 39; Johns Hopkins Univ, MPH, 67. Prof Exp: Dep dir, Nat Res Inst Mother & Child Health, Warsaw, 52-65; chmn obstet & gynec, Med Acad, Warsaw & Inst Postgrad Training Physicians, Warsaw, 58-65; ASSOC PROF MOTHER & CHILD HEALTH, OBSTET & GYNEC, JOHNS HOPKINS UNIV, 67- Concurrent Pos: Expert, Mother & Child Health, WHO, 60- Honors & Awards: Medal Exemplary Serv Health, Govt Poland, 62, Order Polonia Restituta, 50 & 63. Mem: Fel Am Acad Reproductive Med; fel Am Pub Health Asn; fel Am Col Obstetricians & Gynecologists; fel Royal Soc Health; assoc fel Am Col Prev Med. Res: High risk factors in reproductive failure; sequelae of induced abortion; study of reproductive performance of individuals born prematurely or with low birth weight. Mailing Add: 615 N Wolfe St Baltimore MD 21205

LESKO, STEPHEN ALBERT, b Cassandra, Pa, Dec 30, 31. BIOCHEMISTRY. Educ: Ind Univ, Pa, BS, 59; Univ Md, PhD(biochem), 65. Prof Exp: Instr, 65-68, res assoc, 68-73, ASST PROF BIOCHEM & BIOPHYS SCI, JOHNS HOPKINS UNIV, 73- Mem: AAAS; Am Chem Soc; Biophys Soc. Res: Chemical carcinogenesis; nucleic acid chemistry and biology. Mailing Add: Dept of Biochem & Biophys Sci Johns Hopkins Univ Baltimore MD 21205

LESKOWITZ, SIDNEY, b New York, NY, Nov 15, 22; m 48; c 3. IMMUNOLOGY. Educ: City Col New York, BS, 43; Columbia Univ, MA, 48, PhD(chem), 50. Prof Exp: Res asst microbiol, Col Physicians & Surgeons, Columbia Univ, 50-54; from res assoc to asst prof bact & immunol, Harvard Med Sch, 54-70; PROF PATH, MED SCH, TUFTS UNIV, 70- Concurrent Pos: Asst immunol, Mass Gen Hosp, 54-57, assoc, 57-70. Mem: AAAS; Am Asn Immunol; Soc Exp Biol & Med. Res: Delayed hypersensitivity; induction of immunologic tolerance; structure of antigens. Mailing Add: Dept of Path Tufts Med Sch Boston MA 02155

LESLEY, FRANK DAVID, b El Paso, Tex, Dec 20, 44; m 67; c 1. MATHEMATICS. Educ: Stanford Univ, BS, 66; Univ Calif, San Diego, MA, 68, PhD(math), 70. Prof Exp: Asst prof, 70-73, ASSOC PROF MATH, SAN DIEGO STATE UNIV, 73- Mem: Am Math Soc. Res: Boundary behavior of conformal mappings, including minimal surfaces and approximation theory. Mailing Add: Dept of Math San Diego State Univ San Diego CA 92182

LESLEY, STANLEY M, virology, biochemistry, see 12th edition

LESLIE, CHARLES MILLER, b Lake Village, Ark, Nov 8, 23; m 46; c 3. MEDICAL ANTHROPOLOGY. Educ: Univ Chicago, PhB, 49, MA, 50, PhD(anthrop), 59. Prof Exp: Instr anthrop, Southern Methodist Univ, 50-51; instr, Univ Minn, 54-56; from instr to assoc prof, Pomona Col, 56-65; vis prof, Univ Wash, 65; assoc prof, Case Western Reserve Univ, 66-67; chmn dept, Univ Col, NY Univ, 67-71, PROF ANTHROP, NY UNIV, 67- Concurrent Pos: NSF fel, Sch Oriental & African Studies, Univ London, 62-63; res assoc, Dept Anthrop, Univ Chicago, 74-75; NSF res grant, 74; chmn ed bd, Comparative Studies of Health Systems & Med Care, 72-; assoc ed, J Cross-Cult Med & Psychiat, 76- Mem: Fel AAAS; fel Am Anthrop Asn; fel Royal Anthrop Inst Gt Brit & Ireland; fel Asn Asian Studies; Fel Soc Med Anthrop. Res: World view and social change in India and Latin America; comparative study of medical systems. Mailing Add: 110 Bleeker St Apt 20A New York NY 10012

LESLIE, GERRIE ALLEN, b Red Deer, Alta, Nov 19, 41; m 65; c 2. IMMUNOLOGY, IMMUNOCHEMISTRY. Educ: Univ Alta, BSc, 62, MSc, 65; Univ Hawaii, PhD(microbiol), 68. Prof Exp: From asst prof to assoc prof microbiol, Sch Med, Tulane Univ, 70-74; ASSOC PROF MICROBIOL, MED SCH, UNIV ORE, 74- Concurrent Pos: USPHS fel, Col Med, Univ Fla, 68-70; adj assoc prof microbiol, Sch Med, Tulane Univ, 74-; res affil, Delta Regional Primate Res Ctr, Covington, La. Mem: Am Asn Immunol; Am Soc Microbiol; Am Soc Zoologists. Res: Phylogeny of immunoglobulin structure and function; regulation of the immune response; secretory immunologic system. Mailing Add: Dept of Microbiol & Immunol Univ of Ore Med Sch Portland OR 97201

LESLIE, JAMES b Belfast, Ireland, Apr 25, 34; m 64; c 3. BIOPHYSICAL CHEMISTRY. Educ: Queen's Univ, Belfast, BSc, 56, PhD(chem), 59. Prof Exp: Fel, Okla State Univ, 59-61, asst prof & res assoc, 61-62; asst prof chem, Wash Col, 62-63; asst prof, 63-66, ASSOC PROF MED CHEM, SCH PHARM, UNIV MD, BALTIMORE, 66- Concurrent Pos: NIH res grant, 64, vis, Dept Clin Physics & Bioeng, Western Regional Hosp Bd, Glasgow, Scotland, 71-72. Mem: Am Chem Soc; Am Asn Cols Pharm. Res: Kinetics of processes of biological interest. Mailing Add: Univ of Md Sch of Pharm 636 W Lombard St Baltimore MD 21201

LESLIE, JAMES D, b Toronto, Ont, July 6, 35; m 64; c 3. SOLID STATE PHYSICS. Educ: Univ Toronto, BASc, 57; Univ Ill, MS, 60, PhD(physics), 63. Prof Exp: Asst prof, 63-68, ASSOC PROF PHYSICS, UNIV WATERLOO, 68- Mem: Am Phys Soc; Can Asn Physicists. Res: Low temperature physics; far infrared spectroscopy; superconductivity; electron tunneling. Mailing Add: Dept of Physics Univ of Waterloo Waterloo ON Can

LESLIE, STEPHEN HOWARD, b New York, NY, Nov 6, 18; m 43; c 2. MEDICINE. Educ: NY Univ, BS, 38, MD, 42. Prof Exp: Clin asst, 46-49, clin instr med, 49-54, asst prof, 54-69, ASSOC PROF CLIN MED, SCH MED, NY UNIV, 69- Concurrent Pos: Fel, Sch Med, NY Univ, 44-46. Mem: Endocrine Soc; Am Diabetes Asn; Am Fedn Clin Res. Res: Metabolic diseases; endocrinology. Mailing Add: 120 E 34th St New York NY 10016

LESLIE, STEVEN WAYNE, b Franklin, Ind, Jan 23, 46; m 70; c 1. PHARMACOLOGY. Educ: Purdue Univ, BS, 69, MS, 72, PhD(pharmacol), 74. Prof Exp: ASST PROF PHARMACOL, UNIV TEX, AUSTIN, 74- Mem: Sigma Xi; AAAS. Res: Investigations concerning the role of cellular organelles in calcium-mediated termination mechanisms in secretory tissues and the effects of various drugs on these termination mechanisms. Mailing Add: Col of Pharm Univ of Tex Austin TX 78712

LESLIE, WALLACE DEAN, b Dacoma, Okla, Nov 9, 22; m 48; c 3. ANALYTICAL CHEMISTRY. Educ: Northwestern State Col, Okla, BS, 47; Okla State Univ, MS, 50. Prof Exp: Asst chem, Okla State Univ, 47-49; instr, Northwestern State Col, Okla, 49-51; anal chemist, Mfg Dept, 51-52, from assoc res chemist to sr res chemist, 52-62, RES GROUP LEADER, RES & DEVELOP DEPT, CONTINENTAL OIL CO, 62- Mem: Am Chem Soc. Res: Analytical research and development; petroleum and petroleum products; petrochemicals. Mailing Add: Res & Develop Dept Continental Oil Co Ponca City OK 74601

LESNAW, JUDITH ALICE, b Chicago, Ill, July 30, 40. VIROLOGY. Educ: Univ Ill, BS, 62, MS, 64, PhD(cell biol), 69. Prof Exp: Res assoc virol, Univ Ill, 69-74; ASST PROF VIROL, UNIV KY, 74- Mem: Am Soc Microbiol; AAAS. Res: Structure and function of viral proteins and RNA; replication of RNA viruses; defective interfering particles. Mailing Add: Sch of Biol Sci Univ of Ky Lexington KY 40506

LESNIAK, LINDA MARIE, b Gary, Ind, Aug 14, 48. MATHEMATICS. Educ: Western Mich Univ, BA, 70, MA, 71, PhD(math), 74. Prof Exp: Asst math, Western Mich Univ, 70-74; ASST PROF MATH, LA STATE UNIV, BATON ROUGE, 74- Mem: Am Math Soc; Math Asn Am. Res: Extremal problems in graph theory; generalized Ramsey numbers; degree sets for graphs and digraphs. Mailing Add: Dept of Math La State Univ Baton Rouge LA 70803

LESNINI, DAVID GUIDO, physical chemistry, see 12th edition

LESOINE, L GRANT, chemical engineering, polymer chemistry, see 12th edition

LESPERANCE, PIERRE J, b Montreal, Que, Aug 16, 34; m 60; c 3. INVERTEBRATE PALEONTOLOGY. Educ: Univ Montreal, BSc, 56; Univ Mich, MS, 57; McGill Univ, PhD(geol), 61. Prof Exp: Geologist, Dept Natural Resources, Que, 60-61; from asst prof to assoc prof, 61-71, PROF GEOL, UNIV MONTREAL, 71-, CHMN DEPT, 75- Mem: Geol Soc Am; Am Asn Petrol Geol; Soc Econ Paleont & Mineral; Paleont Soc; Brit Paleont Asn. Res: Low Paleozoic field mapping; paleontology and biostratigraphy of Upper Ordovician to Lower Devonian trilobites and brachiopods. Mailing Add: Dept of Geol Univ of Montreal Montreal PQ Can

LESSA, WILLIAM ARMAND, b Newark, NJ, Mar 3, 08. ANTHROPOLOGY. Educ: Harvard Univ, AB, 28; Univ Chicago, AM, 41, PhD(anthrop), 47. Prof Exp: Res assoc human constitution, Columbia-Presby Med Ctr, 29-30; res assoc anthrop, Univ Hawaii, 30-33; instr, Brooklyn Col, 41-42; from instr to prof, 47-70, EMER PROF ANTHROP, UNIV CALIF, LOS ANGELES, 70- Mem: Fel AAAS; fel Am Anthrop Asn (secy, 51-53); Am Asn Phys Anthrop; Fr Soc Oceanists; Soc Hist Discoveries. Res: Micronesian cultures; oceanic folklore; myth and ritual; ethnohistory. Mailing Add: Dept of Anthrop Univ of Calif Los Angeles CA 90024

LESSARD, JAMES LOUIS, b Eau Claire, Wis, Mar 9, 43; m 65; c 2. BIOCHEMISTRY. Educ: Marquette Univ, BS, 65, PhD(biochem), 70. Prof Exp: Fel, Riche Inst Molecular Biol, Nutley, NJ, 69-71; res scholar biochem, Children's Hosp Res Found, 71-72; ASST PROF RES PEDIAT, MED SCH, UNIV CINCINNATI, 72-, ASST PROF BIOL CHEM, 74- Concurrent Pos: Fel pharmacol-morphol, Pharmaceut Mfrs Asn Found, Cincinnati, Ohio, 72-74. Mem: Am Chem Soc; AAAS; Sigma Xi. Res: Regulation processes in development; cell motility. Mailing Add: Fetal Pharm Div Children's Hosp Res Found Elland Ave & Bethesda Cincinnati OH 45229

LESSARD, JEAN, b East-Broughton, Que, Apr 29, 36; m 60; c 2. ORGANIC CHEMISTRY. Educ: Laval Univ, BA, 56, BSc, 60, PhD(org chem), 65. Prof Exp: Nat Res Coun Can fel, Imp Col, Univ London, 65-67; asst res officer org chem, Nat Res Coun Can, 67-69; asst prof, 69-71, ASSOC PROF ORG CHEM, UNIV SHERBROOKE, 71- Mem: Chem Inst Can; The Chem Soc; Am Chem Soc. Res: Transition metals; electrochemistry and photochemistry used to study new methods of effecting organic reactions or new organic reactions, investigation of the mechanism, scope and synthetic utility of these reactions. Mailing Add: Dept of Chem Univ of Sherbrooke Sherbrooke PQ Can

LESSARD, MAURICE, b Quebec, Que, Oct 21, 07; m 37; c 4. CHEMISTRY, FOOD TECHNOLOGY. Educ: Que Sem, BA, 26; Laval Univ, MSc, 30. Prof Exp: Chemist, Aluminum Co Can, 30-32; pvt consult, 32-35; chief chemist, Dept Agr Que, 35-42; tech dir, Dept Fisheries, Prov Que, 42-46; gen mgr, St Lawrence Sea Prod Co, 46-61; asst dep minister, Dept Game & Fisheries, Prov Que, 61-63; asst dep minister, Dept Indust & Com, 63-64, assoc dep minister, 64-72; consult, 72-76; RETIRED. Mem: Am Chem Soc; Inst Food Technologists; fel Chem Inst Can. Mailing Add: 430 Lemesurier Quebec PQ Can

LESSE, HENRY, b Philadelphia, Pa, Feb 7, 26; m 50; c 1. PSYCHIATRY, NEUROPHYSIOLOGY. Educ: Jefferson Med Col, MD, 50; Am Bd Psychiat & Neurol, dipl, 57. Prof Exp: Asst physiol, Jefferson Med Col, 46-50; asst neurophysiol, Col Physicians & Surgeons, Columbia Univ, 48-49; staff psychiatrist, US Med Ctr for Fed Prisoners, 51-53; vis scientist, 58-59, asst prof, 59-62, ASSOC PROF PSYCHIAT, UNIV CALIF, LOS ANGELES, 62-, CHIEF OF RES, NEUROPSYCHIAT INST, 59- Concurrent Pos: USPHS fel psychiat, Tulane Univ, 53-54, NIMH career investr award, 54-59; vis psychiatrist, Charity Hosp, New Orleans, 53-58; vis scientist, Vet Admin Hosps, Long Beach, Calif, 58-59; consult, Vet Admin Ctr, Brentwood Hosp, 65-; res adv comt mem, State Dept Ment Hyg, 66-69; mem adv comt brain res inst, Univ Calif, Los Angeles; consult, Vet Admin Hosp, Sepulveda. Mem: AAAS; Am Psychiat Asn; NY Acad Sci; Soc Neurosci. Res: Electrophysiology and behavior. Mailing Add: Neuropsychiat Inst Univ of Calif Los Angeles CA 90024

LESSELL, SIMMONS, b Brooklyn, NY, May 25, 33; m 55; c 4. NEUROLOGY, OPHTHALMOLOGY. Educ: Amherst Col, BA, 54; Cornell Univ, MD, 58. Prof Exp: Intern med, Cornell Univ, 58-59; resident neurol, Univ Vt, 59-60; resident ophthal, Mass Eye & Ear Hosp, 63-66; assoc prof neurol, 67-70, PROF OPHTHAL, SCH MED, BOSTON UNIV, 70- Concurrent Pos: Physician, NIH, 59-60; lectr, Sch Med, Tufts Univ, 66-; vis surgeon & dir dept ophthal, Boston City Hosp; vis surgeon, Univ Hosp; consult ophthal, Vet Admin Hosp & Tufts-New England Med Ctr. Mem: Asn Res Vision & Ophthal. Res: Optic neuropathies, clinical and experimental; histochemistry and experimental pathology of the optic nerve. Mailing Add: Dept of Ophthal Boston Univ Sch of Med Boston MA 02118

LESSEPS, ROLAND JOSEPH, b New Orleans, La, Aug 13, 33. DEVELOPMENTAL BIOLOGY. Educ: Spring Hill Col, BS, 58; Johns Hopkins Univ, PhD(biol), 62. Prof Exp: Asst prof, 67-71, ASSOC PROF BIOL, LOYOLA UNIV , LA, 71- Concurrent Pos: Vis prof, Roman Cath Univ, Nijmegen, Mem: AAAS; Am Soc Zool; Electron Micros Soc Am; Soc Develop Biol. Res: Morphogenetic movements of embryonic cells; electron microscopy of the cell surface. Mailing Add: Dept of Biol Loyola Univ New Orleans LA 70118

LESSER, ALEXANDER, b New York, NY, Oct 4, 02; m 40; c 3. CULTURAL ANTHROPOLOGY. Educ: Columbia Univ, AB, 23, PhD(anthrop), 29. Prof Exp: Lectr anthrop, Columbia Univ, 34-49; instr, Brooklyn Col, 39-46; exec dir, Asn Am Indian Affairs, 47-55; vis assoc prof, Brandeis Univ, 56-59; chmn dept sociol & anthrop, 60-65, prof, 60-64, EMER PROF ANTHROP, HOFSTRA UNIV, 64- Concurrent Pos: Soc Sci Res Coun fel Am Indian, field res & Columbia Univ, 29-30, Comt Am Indian Lang res fel, Okla & Columbia Univ, 29 & 33-34, Columbia Univ Coun Res Soc Sci, 30-33, Am Coun Learned Soc fel, 31-32; lectr, Inst Race Rels, 35;

vis lectr, NY Univ, 35; lectr & seminar leader, YMHA, NY, 37-39; vis assoc prof, Northwestern Univ, 38; soc sci analyst, Off Coordr Inter-Am Affairs, 43-44; chief econ studies sect, Latin Am Div, Off Strategic Serv, 44-45; chief north & northwest coast br, Div Res Am Repubs, US Dept State, 45-47; ed, The Am Indian, 47-59; Bollingen Found fel, Plains Indian Res & Hofstra Univ, 60-62; speaker cult eval symp, Am Acad Arts & Sci, 61, Conf Anthrop & Educ, Am Anthrop Asn, 63, 66 & partic, Symp on War, 67; vis prof anthrop, Columbia Univ, 64-68; speaker, Nat Coun Social Studies, 65; vis prof anthrop, John Jay Col, City Univ New York, 68-69 & Grad Sch, New Sch Soc Res, 69; Am Philos Soc grant, Boas Archives & Sources, 70-72. Honors & Awards: Cert of Merit, Off Strategic Serv, 46. Mem: Fel Am Anthrop Asn; Am Ethnol Soc (secy-treas, 33-39, ed, 37-40, vpres, 39-41, dir, 42-44); Am Folklore Soc; Soc Appl Anthrop. Res: American Indian ethnology, religion and languages; anthropological history and methods; cultural evolution; kinship and social organization. Mailing Add: Dept of Anthrop Hofstra Univ Hempstead NY 11550

LESSER, ELLIOTT, b Brooklyn, NY, Apr 20, 29. MEDICAL PARASITOLOGY, PROTOZOOLOGY. Educ: Brooklyn Col, BS, 50; McGill Univ, MS, 51, PhD(parasitol), 53. Prof Exp: Res assoc parasitol, Chicago Acad Sci, 57; res parasitologist, Regional Animal Dis Lab, Agr Res Serv, USDA, 57-58, Animal Parasite Lab, 58-64; mem staff, Extramural Prog, Nat Inst Dent Res, 64-69; health scientist adminr, Nat Ctr Health Serv Res & Develop, 69-74; HEALTH SCIENTIST ADMINR, BUR HEALTH MANPOWER, USPHS, DEPT HEALTH, EDUC & WELFARE, 74- Mem: AAAS; Am Soc Parasitol; Soc Protozool; Sigma Xi; Am Inst Biol Sci. Res: Cultivation of parasitic protozoa; relationships between hormones and endoparasites; chemotherapy of parasitic diseases; life cycles and transmission of parasites; grants administration. Mailing Add: Rm 3300 Bur of Health Manpower USPHS HEW 26 Federal Plaza New York NY 10007

LESSER, GERSON THEODORE, b Brooklyn, NY, Mar 29, 21; m 46; c 3. INTERNAL MEDICINE, PHYSIOLOGY. Educ: NY Univ, AB, 41, MD, 44; Am Bd Internal Med, dipl, 53. Prof Exp: Res fel physiol, Sch Med, 50-51, from asst to clin instr med, 51-55, asst prof clin med, 55-60 & 61-67, asst attend physician, Univ Hosp, 58-67, ASSOC PROF CLIN MED, SCH MED, NY UNIV, 67-, ASSOC ATTEND PHYSICIAN, UNIV HOSP, 68- Concurrent Pos: Adj physician, Lenox Hill Hosp, 55-64, assoc physician, 64-; assoc attend Goldwater Mem Hosp, 57-60, attend, 60- Mem: AAAS. Res: Body composition, particularly with aging; uptake and transport of inert gases by mammals; metabolism with relationship to body composition. Mailing Add: NY Univ Res Serv Goldwater Mem Hosp New York NY 10017

LESSHAFFT, CHARLES THOMAS, JR, b Louisville, Ky, Oct 6, 18; m 39; c 2. PHARMACY. Educ: Univ Ky, BS, 41; Purdue Univ, MS, 53, PhD(pharm), 55. Prof Exp: From instr to assoc prof, 47-70, PROF PHARM, UNIV KY, 70- Res: Formulation and properties of ointment bases; pharmaceutical applications of suspending agents. Mailing Add: Univ of Ky Col of Pharm Lexington KY 40506

LESSIE, THOMAS GUY, b New York, NY, Dec 14, 36; m 62; c 3. MICROBIAL PHYSIOLOGY. Educ: Queens Col, NY, BS, 58; Harvard Univ, AM, 61, PhD(biol sci), 63. Prof Exp: Res asst microbiol, Haskins Labs, NY, 58-59; NIH fels biochem, Oxford Univ, 63-65 & biol sci, Purdue Univ, 65-67; res assoc microbiol, Univ Wash, 67-68; asst prof, 68-74, ASSOC PROF MICROBIOL, UNIV MASS, AMHERST, 74- Concurrent Pos: Grants, Nat Sci Found, 68-70; NIH grant, 68-71, Inst Arthritis & Metab Dis, 71-74, Inst Gen Med Sci, 74-77. Mem: AAAS; Am Soc Microbiol; Am Chem Soc; Brit Soc Gen Microbiol. Res: Regulation of photopigment synthesis in photosynthetic bacteria; enzyme regulatory mechanisms in Pseudomonas and Bacillus species. Mailing Add: Dept of Microbiol Univ of Mass Amherst MA 01002

LESSING, PETER, b Englewood, NJ, June 15, 38; m 65; c 2. ENVIRONMENTAL GEOLOGY. Educ: St Lawrence Univ, BS, 61; Dartmouth Col, MA, 63; Syracuse Univ, PhD(geol), 67. Prof Exp: Asst prof geol, St Lawrence Univ, 66-71; environ geologist, 71-73, HEAD ENVIRON GEOL SECT, W VA GEOL SURV, 73- Concurrent Pos: Mem, Environ Geol Comt, Am Inst Prof Geol, 75-; adj prof, WVa Univ, 73- Mem: Geol Soc Am; Mineral Asn Can; Am Inst Prof Geologists. Res: Geologic field mapping; environmental geology investigations; land use planning; landslide evaluation; geologic hazard studies. Mailing Add: WVa Geol Surv PO Box 879 Morgantown WV 26505

LESSLER, MILTON A, b New York, NY, May 18, 15; m 43; c 3. PHYSIOLOGY. Educ: Cornell Univ, BS, 37, MS, 38; NY Univ, PhD(biochem cell physiol), 50. Prof Exp: Technician cardiac res, NY State Health Dept, 40-42; from asst prof to assoc prof, 51-63, PROF PHYSIOL, COL MED, OHIO STATE UNIV, 63- Concurrent Pos: Nat Cancer Inst fel, NY Univ-Washington Square Col, 49-50; NSF fac fel, Univ Mich, Ann Arbor, 58-59; vis lectr, Am Physiol Soc, 62-66; consult, Yellow Springs Instrument Co, 65-; ed-in-chief, Ohio J Sci, 74- Mem: Fel AAAS; fel NY Acad Sci; Am Physiol Soc; Am Asn Cancer Res; Am Soc Cell Biol. Res: Cell physiology; effects of environmental pollutants on the hemopoietic system; cellular radiobiology; erythropoiesis and lead poisoning. Mailing Add: Dept of Physiol Ohio State Univ Col of Med Columbus OH 43210

LESSLER, RICHARD MARSHALL, b New York, NY, Sept 1, 30; m 68. NUCLEAR CHEMISTRY. Educ: Univ Calif, Los Angeles, AB, 52; Univ Calif, MS, 54, PhD(chem), 59. Prof Exp: Nuclear chemist, Lawrence Livermore Lab, Univ Calif, 58-73; radiol physicist, Nuclear Regulatory Comn, Wash, DC, 73-75; NUCLEAR CHEMIST, LAWRENCE LIVERMORE LAB, UNIV CALIF, 75- Mem: Am Phys Soc. Res: Reactor health physics; environmental physics; nuclear safeguards; radiation protection; energy applications; neutron induced activation; nuclear reactions; Monte Carlo calculations; nuclear effects; neutron physics. Mailing Add: L-20 Lawrence Livermore Lab Univ of Calif PO Box 808 Livermore CA 94550

LESSLIE, THOMAS ELLIS, organic chemistry, see 12th edition

LESSMAN, GARY M, b Hillsboro, Ill, July 15, 38. SOIL FERTILITY, PLANT NUTRITION. Educ: Southern Ill Univ, BS, 60, MS, 62; Mich State Univ, PhD(soil sci), 67. Prof Exp: Exten agronomist, Purdue Univ, 67-68; ASST PROG AGRON, UNIV TENN, KNOXVILLE, 69- Mem: Am Soc Agron; AAAS. Res: Micronutrient nutrition and waste disposal on soils. Mailing Add: Dept of Plant & Soil Sci Univ of Tenn Knoxville TN 37916

LESSMAN, KOERT J, b Bay City, Mich, Jan 11, 31; m 53; c 5. PLANT BREEDING, PLANT GENETICS. Educ: Mich State Univ, BS, 58; Iowa State Univ, MS, 60, PhD(plant breeding), 62. Prof Exp: Assoc prof, 62-70, PROF AGRON, PURDUE UNIV, WEST LAFAYETTE, 70- Mem: Am Soc Agron; Am Genetic Asn; Crop Sci Soc Am. Res: Breeding and improvement of grain sorghums and forage grasses; chromosome analysis of the sorghums; discovery and development of alternative crops for agriculture of a non-food nature. Mailing Add: Dept of Agron Purdue Univ West Lafayette IN 47906

LESSOFF, HOWARD, b Boston, Mass, Sept 23, 30; m 59; c 2. SOLID STATE SCIENCE. Educ: Northeastern Univ, BS, 53, MS, 57. Prof Exp: Staff engr, Radio Corp Am, Mass, 57-60, sr staff mem, 61-64; staff mem, Bell Tel Labs, 60-61; aerospace technologist, Electronic Res Ctr, NASA, 64-70; supvry physicist, 70-75, BR HEAD ELECTRONIC MAT TECHNOL, NAVAL RES LAB, 75- Concurrent Pos: Lectr, Lincoln Col, Northeastern Univ, 57-70; consult, Datacove Corp, NJ, 69- Mem: AAAS; Am Ceramic Soc; Sci Res Soc Am; Inst Elec & Electronics Eng. Res: Magnetic materials for computer; microwave and optic properties; dielectric materials; solid state physics. Mailing Add: Code 5220 Electronic Technol Div Naval Res Lab Washington DC 20375

LESSOR, ARTHUR EUGENE, JR, b Schenectady, NY, Apr 2, 25; m 55; c 2. ANALYTICAL CHEMISTRY, CRYSTALLOGRAPHY. Educ: Union Univ, NY, BS, 49; Indiana Univ, PhD(chem), 55. Prof Exp: Fel crystallog, Indiana Univ, 55; asst prof chem, Univ Cincinnati, 55-56; anal chemist, Gen Elec Co, 56-59; mgr crystallog lab, Fed Systs Div, Int Bus Mach Corp, 59-61, film electronics develop, Components Div, 61-65, interconnection & packaging develop systs develop div, Oswego, 65-66, mgr tech serv, Components Div, 66-70, mgr mfg, 71-73, mgr semiconductor mat prod, 73-75, MGR CHEM ENERGY RESOURCES, E FISHKILL FACILITY, IBM CORP, 75- Mem: AAAS; Am Chem Soc; Am Crystallog Asn. Res: Thin film electronic and solid state devices; solid state electronic materials; conservation of energy and process materials. Mailing Add: 7 Rock Garden Way Poughkeepsie NY 12603

LESSOR, EDITH DORA, b Chicago, Ill, Aug 5, 30; m 55; c 2. ANALYTICAL CHEMISTRY. Educ: Valparaiso Univ, BS, 52; Indiana Univ, Bloomington, PhD(anal chem), 55. Prof Exp: Instr chem, Ulster Community Col, 64-65; lectr, Harpur Col, State Univ New York, Binghamton, 64-66; asst prof, 67-70, ASSOC PROF CHEM, MT ST MARY COL, NY, 70-, CHMN DEPT, 68-, CHMN DIV NATURAL SCI & MATH, 73- Mem: Am Chem Soc; Sigma Xi. Res: Spectrophotometry of organic analytical reagents and analytical chemistry of water pollution control. Mailing Add: Dept of Chem Mt St Mary Col Newburgh NY 12550

LESTER, CHARLES TURNER, b Covington, Ga, Nov 10, 11; m 36; c 2. ORGANIC CHEMISTRY. Educ: Emory Univ, AB, 32, MA, 34; Pa State Univ, PhD(org chem), 41. Prof Exp: Teacher high sch, Ga, 34-35; instr chem, Emory Jr Col, 35-39; res chemist, Calco Div, Am Cyanamid Corp, NJ, 41-42; from asst prof to assoc prof, 42-50, chmn dept chem, 54-57, PROF CHEM, EMORY UNIV, 50-, DEAN GRAD SCH, 57-, VPRES GRAD STUDIES, 70- Concurrent Pos: Mem bd dirs, Oak Ridge Assoc Univs, 62-65; mem Ga Sci & Technol Comn, 63; mem bd dirs, Atlanta Speech Sch, 65-; mem exec comt, Coun Grad Schs US, 65-68; mem bd dirs, Southeastern Educ Lab, 66-68; bd trustees, Reinhardt Col, 66-; vchmn, Ocean Sci Ctr, Atlantic Comn, 67-; mem bd trustees, Huntingdon Col, 68; chief acad progs br, Bur Higher Educ, 69-70; chmn elect, Coun Grad Sch, 73, chmn, 74; coun mem, Oak Ridge Assoc Univs, 58-62, 70- Honors & Awards: Herty Medal, 65. Mem: Am Chem Soc; Am Inst Chem. Res: Sterically hindered ketones; indigosol dyes; biphenyl mercaptan; oxetanones; alkyl aryl ketones; anti-microbial compounds. Mailing Add: 281 Chelsea Circle Decatur GA 30030

LESTER, DAVID, b New Haven, Conn, Jan 22, 16; m 38; c 2. BIOCHEMISTRY, PHARMACOLOGY. Educ: Yale Univ, BS, 36, PhD(org chem), 40; Am Bd Clin Chem, dipl. Prof Exp: Res asst drug metab & toxicol, Lab Appl Physiol, Yale Univ, 40-45, res assoc, 45-58, res assoc, Lab Appl Biodyn, 58-62; PROF BIOCHEM, CTR ALCOHOL STUDIES, RUTGERS UNIV, NEW BRUNSWICK, 62- Concurrent Pos: Consult, Allied Chem Corp; assoc ed, Quart J Studies Alcoholism. Mem: Fel AAAS; Am Chem Soc; Am Soc Pharmacol & Exp Therapeut; Am Indust Hyg Asn; NY Acad Sci. Res: Analytical biochemistry; metabolism of drugs; toxicology of plastics, resins and industrial materials; behavioral pharmacology; alcoholism. Mailing Add: Ctr of Alcohol Studies Rutgers Univ New Brunswick NJ 08903

LESTER, DONALD THOMAS, b New London, Conn, Aug 26, 34; m 62; c 2. FORESTRY. Educ: Univ Maine, BS, 55; Yale Univ, MF, 57, PhD(forest genetics), 62. Prof Exp: Asst prof, 62-67, ASSOC PROF FORESTRY, UNIV WIS-MADISON, 67- Res: Genetic variation, hybridization, disease resistance breeding and morphogenesis in forest tree species. Mailing Add: 126 Russell Labs Univ of Wis Madison WI 53706

LESTER, GABRIEL, microbiology, deceased

LESTER, GEORGE RONALD, b War Eagle, WVa, Sept 6, 34; m 56; c 4. PHYSICAL CHEMISTRY, PETROLEUM CHEMISTRY. Educ: Berea Col, BA, 54; Univ Ky, MS, 56, PhD(chem), 58. Prof Exp: Chemist, Universal Oil Prod Co, 58-63, assoc res coordr, 63-74, MGR APPL CATALYSIS, UOP, INC, 74- Mem: AAAS; Am Chem Soc; The Chem Soc; Sigma Xi; Catalysis Soc. Res: Conductivity of nonaqueous solutions; adsorption of gases on solids; heterogeneous catalysis; petrochemical processes; exhaust gas conversion catalysis. Mailing Add: 318 Meacham Ave Park Ridge IL 60068

LESTER, HENRY ALLEN, b New York, NY, July 4, 45. NEUROBIOLOGY, BIOPHYSICS. Educ: Harvard Col, AB, 66; Rockefeller Univ, PhD(biophys), 71. Prof Exp: Res fel molecular neurobiol, Inst Pasteur, 71-73; ASST PROF BIOL, CALIF INST TECHNOL, 73- Mem: Soc Neurosci; Biophys Soc; Soc Gen Physiologists. Res: Mechanisms and development of synaptic transmission. Mailing Add: Div of Biol Calif Inst of Technol Pasadena CA 91125

LESTER, JOSEPH EUGENE, b Bay City, Tex, July 2, 42; m 59; c 2. PHYSICAL CHEMISTRY. Educ: Rice Univ, BA, 64; Univ Calif, Berkeley, PhD(chem), 68. Prof Exp: Asst prof chem, Northwestern Univ, Evanston, 67-74; MEM STAFF, GTE LABS, 74- Mem: Am Chem Soc; Am Phys Soc. Res: Kinetics and mechanisms of surface reactions on crystalline solids; photoelectron spectroscopy. Mailing Add: GTE Labs 40 Sylvan Rd Waltham MA 02154

LESTER, LARRY JAMES, b Bay City, Tex, July 15, 47; m 69, 75; c 1. POPULATION GENETICS. Educ: Univ Tex, Austin, BA, 69, PhD(pop genetics), 75. Prof Exp: ASST PROF BIOL, UNIV HOUSTON, 75- Concurrent Pos: Adj res scientist, Nat Marine Fisheries Serv, Nat Oceanic & Atmospheric Admin, 75- Mem: Soc Study Evolution; Genetic Soc Am. Res: Genetics of gametic selection; genetic population structure of marine invertebrates. Mailing Add: Dept of Biol Univ of Houston Houston TX 77004

LESTER, RICHARD GARRISON, b New York, NY, Oct 24, 25; m 49; c 2. RADIOLOGY. Educ: Princeton Univ, AB, 46; Columbia Univ, MD, 48. Prof Exp: From instr to assoc prof radiol, Univ Minn, 54-61; prof & chmn dept, Med Col Va, 61-65; PROF RADIOL & CHMN DEPT, DUKE UNIV, 65- Concurrent Pos: Mem comt acad radiol, Nat Acad Sci, 66; mem steering comt, Soc Chmn Acad Radiol Dept, 67; mem bd trustees, Am Bd Radiol; mem bd trustees, Meharry Med Col, 75- Mem: AMA; Am Roentgen Ray Soc; Am Col Radiol; Soc Pediat Radiol (secy-treas, 58-62); regent Am Col Chest Physicians. Res: Cardiovascular and pediatric radiology. Mailing Add: Dept of Radiol Duke Univ Durham NC 27710

LESTER, ROBERT LEONARD, b New Haven, Conn, Aug 21, 29; m 54; c 2. BIOCHEMISTRY. Educ: Yale Univ, BS, 51; Calif Inst Technol, PhD(biochem), 56. Prof Exp: Asst prof biochem, Univ Wis, 58-60; from asst prof to assoc prof, 60-68, PROF BIOCHEM, MED SCH, UNIV KY, 68-, CHMN DEPT, 74- Concurrent Pos: Res fel, Inst Enzyme Res, Univ Wis, 55-58; NIH res grants, 60-; vis res biologist, Univ Calif, San Diego, 69-70. Mem: AAAS; Am Soc Biol Chemists; Am Chem Soc; Am Soc Microbiol; Fedn Am Sci. Res: Electron transport; mitochondrial functions; lipid metabolism. Mailing Add: Univ of Ky Med Sch Lexington KY 40506

LESTER, ROGER, b Brooklyn, NY, Dec 26, 29; m 54; c 2. MEDICINE. Educ: Princeton Univ, AB, 50; Yale Univ, MD, 55. Prof Exp: From intern to resident med, Col Med, Univ Utah, 55-57, resident, 59-60; NIH fel, Thorndike Mem Lab, Harvard Univ, 60-62; asst prof, Sch Med, Univ Chicago, 62-65; from asst prof to assoc prof med, Sch Med, Boston Univ, 65-74; PROF GASTROENTEROL & CHMN DEPT, SCH MED, UNIV PITTSBURGH, 74- Concurrent Pos: NIH career develop award, 64-65, res grant, 65- Mem: Am Fedn Clin Res; Am Asn Study Liver Dis; Am Soc Clin Invest; Am Gastroenterol Asn; Int Asn Study Liver. Res: Study of metabolism of bilirubin and bilirubin derivatives; fetal hepatic and intestinal function; relation of glycolipids to membrane function. Mailing Add: Dept of Gastroenterol Univ of Pittsburgh Sch of Med Pittsburgh PA 15313

LESTER, THOMAS WILLIAM, JR, b Chicago, Ill, Oct 12, 15; m 41; c 2. MEDICINE. Educ: Univ Chicago, BS, 38, MD, 41. Prof Exp: Intern, US Marine Hosp, 41-42; from instr to assoc prof med, Sch Med, Univ Chicago, 46-55, clin assoc prof, 55-62; from clin assoc prof to prof med, Sch Med, Univ Colo, Denver, 62-72; PROF MED, SCH MED, UNIV CHICAGO, 73- Concurrent Pos: Consult, Naval Med Res Unit 4, 49-54 & Commun Dis Ctr, USPHS, Ga, 50-55; dir student health serv, Sch Med, Univ Chicago, 50-55; assoc mem comn acute respiratory dis, Armed Forces Epidemiol Bd, 52-55; chief of staff, Suburban Cook County Tuberc Hosp Sanitarium, 55-62; chief chest med, Nat Jewish Hosp, Denver, 62-72; consult, Fitzsimons Army Hosp, 63-72. Mem: Am Thoracic Soc; Am Pub Health Asn; Am Acad Microbiol; fel Am Col Chest Physicians. Res: Aerial disinfection; survival of pathogenic agents in environment; air hygiene; airborne infections; preventive medicine; pulmonary diseases; tuberculosis; atypical mycobacterial infections. Mailing Add: Dept of Med Univ of Chicago Sch of Med Chicago IL 60637

LESTER, WILLIAM ALEXANDER, JR, b Chicago, Ill, Apr 24, 37; m 59; c 2. PHYSICAL CHEMISTRY. Educ: Univ Chicago, BS, 58, MS, 59; Catholic Univ, PhD(chem), 64. Prof Exp: Proj asst physics, lab molecular struct & spectra, Univ Chicago, 57-59; asst chem, Wash Univ, 59-60 & Catholic Univ, 60-62; phys chemist, Phys Chem Div, Nat Bur Standards, 61-64; proj assoc, Theoret Chem Inst, Univ Wis-Madison, 64-65, asst dir theoret chem, 65-68; MEM PERMANENT PROF STAFF THEORET CHEM, IBM RES LAB, 68- Concurrent Pos: Lectr, Univ Wis-Madison, 66-68. Mem: Am Chem Soc; Am Phys Soc; Sigma Xi; NY Acad Sci. Res: Molecular quantum mechanics and molecular collision theory. Mailing Add: IBM Res Lab San Jose CA 95193

LESTER, WILLIAM LEWIS, b Webster City, Iowa, July 21, 32; m 64; c 3. MICROBIOLOGY. Educ: San Jose State Col, BA, 58; Univ Calif, Davis, PhD(microbiol), 68. Prof Exp: Lab technician pharmacol, Univ Calif, Davis, 62-66; supvr res & develop, Cutter Labs, 68-70; asst prof, 70-74, ASSOC PROF MICROBIOL, HUMBOLDT STATE UNIV, 74- Concurrent Pos: Nat Oceanic & Atmospheric Admin sea grant, Samoa & Calif, 70-73; bd dirs, Redwood Health Consortium, 73-75; univ rep, Conf Assist Undergrad Sci Educ, 75- Mem: AAAS; Am Soc Microbiol; Wildlife Soc; Am Soc Allied Health Prof. Res: Biodegradation of kraft pulp mill effluent; microbial ecology; marine bioassays utilizing echino embryo. Mailing Add: Dept of Biol Humboldt State Col Arcata CA 95521

LESTER, WILLIAM WRIGHT, b Lansing, Mich, May 14, 34; m 65; c 2. APPLIED PHYSICS. Educ: Mich State Univ, BS, 56, MS, 58, PhD(physics), 63. Prof Exp: Res fel, Acoustics Lab, Harvard Univ, 64-65; physicist, Fundamental Physics Dept, Corning Glass Works, 65-68; sr scientist, Tracor, Inc, 68-72; CHIEF RES & DEVELOP SCIENTIST, GILLETTE CO, 72- Mem: Acoust Soc Am; Inst Elec & Electronics Eng. Res: Ultrasonics; physical and engineering acoustics; propagation phenomena, transducers and applications for industrial uses; mechanical radiation; optics; shaving processes; instrumentation for applied physics. Mailing Add: 189 Benvenue St Wellesley MA 02181

LESTON GERD, b Germany, Sept 19, 24; nat US; m 50; c 2. ORGANIC CHEMISTRY. Educ: City Col New York, BS, 48; Purdue Univ, MS, 49, PhD(chem), 52. Prof Exp: Chemist, 52-54, sr chemist, 54-58, group mgr, 58-66, sr group mgr, 67-72, SR PROJ SCIENTIST, KOPPERS CO, INC, 72- Mem: Am Chem Soc; Sigma Xi. Res: Synthetic organic chemistry, particularly phenol chemistry, aromatic substitution; aromatic alkylation and dealkylation; hydrogenation; aromatic acylation; ultraviolet stabilizers; antioxidant synthesis and testing; homogenous and heterogenous catalysis; pesticides; drugs. Mailing Add: 1219 Raven Dr Pittsburgh PA 15243

LE STRANGE, RAYMOND J, analytical chemistry, mathematical statistics, see 12th edition

LESUER, WILLIAM MONROE, b Ingram, Pa, Oct 13, 20; m 44; c 4. PETROLEUM CHEMISTRY. Educ: Monmouth Col, BS, 42; Indiana Univ, PhD(org chem), 48. Prof Exp: Res chemist, 48-52, dir org res labs, 52-60, asst div head res & develop, 60-68, div head, 68-69, VPRES RES & DEVELOP, LUBRIZOL CORP, 69- Mem: Am Chem Soc. Res: Oil additive chemistry; phosphorus chemistry; sulfonation methods; dispersants. Mailing Add: Lubrizol Corp Euclid Sta Box 17100 Cleveland OH 44117

LESUK, ALEX, biochemistry, see 12th edition

LESURE, FRANK GARDNER, b Camden, SC, Jan 28, 27; m 63; c 2. GEOLOGY. Educ: Va Polytech Inst, BS, 51; Yale Univ, MS, 52, PhD(geol), 55. Prof Exp: GEOLOGIST, US GEOL SURV, 55- Mem: AAAS; Soc Econ Geol; Mineral Soc Am; Geol Soc Am; Geochem Soc. Res: Geology of Oriskany iron deposits in Virginia, uranium deposits in southeastern Utah, gold deposits in North Carolina and Georgia, mica pegmatites of southeastern United States; mineral resources of eastern wilderness areas; geochemistry. Mailing Add: US Geol Surv 954 Nat Ctr Reston VA 22092

LE SURF, JOSEPH ERIC, b London, Eng, July 21, 29; Can citizen; m 52; c 3. PHYSICAL CHEMISTRY. Educ: Univ London, BSc, 50 & 51. Prof Exp: Sci officer corrosion, Royal Naval Sci Serv, 51-57; sr sci officer, UK Atomic Energy Authority, 57-64; HEAD SYST MAT BR, ATOMIC ENERGY CAN LTD, 64- Mem: Nat Asn Corrosion Engrs. Res: Marine corrosion; corrosion and material selection for nuclear processing plants and nuclear power plants. Mailing Add: Chalk River Nuclear Lab Atomic Energy of Can Ltd Chalk River ON Can

LETANG, NICHOLAS JOSEPH, chemistry, see 12th edition

LETARTE, JACQUES, b Montreal, Que, Aug 19, 34; m 60; c 2. PEDIATRICS, BIOCHEMISTRY. Educ: Univ Montreal, BS, 57, MD, 62. Prof Exp: Resident med, Notre Dame Hosp, Montreal, 62; resident pediat, St Justine Hosp, 63-64; resident, Royal Postgrad Med Sch, London, 68; ASSOC PROF PEDIAT, UNIV MONTREAL, 69- Concurrent Pos: Mead-Johnson fel pediat, Univ Montreal, 63-64; res fel biochem, Children's Hosp, Zurich, 64-65; Queen Elizabeth II res fel, Can 64-68; res fel, Clin Biochem Inst, Geneva, 65-67; res fel metab, Royal Postgrad Med Sch, London, 68-69; Med Res Coun Can fel, 68-69, scholar, 69-74. Mem: AAAS; Royal Soc Med; Can Soc Clin Invest; Soc Pediat Res; Endocrine Soc. Res: Mode of action of insulin on adipose tissue; hyperammonemia in children; lipid and carbohydrate metabolism in children. Mailing Add: Dept Endocrinol St Justine Hosp 3175 Chemin St Catherine Montreal PQ Can

LETARTE-MUIRHEAD, MICHELLE, b Quebec, Can, Oct 12, 47. BIOCHEMISTRY. Educ: Laval Univ, BS, 68; Ottawa Univ, PhD(biochem), 72. Prof Exp: Fel immunochem, Med Res Coun Immunochem Unit, Oxford Univ, 72-75; ASST PROF IMMUNOCHEM, DEPT MED BIOPHYS, UNIV TORONTO, 75- Mem: Can Biochem Soc; Brit Biochem Soc. Res: Characterization and isolation of antigens from lymphocytes coded for by the immune response region of the major histocompatibility complex of the mouse. Mailing Add: Dept of Med Biophys Univ of Toronto 500 Sherbourne St Toronto ON Can

LETBETTER, WILLIAM DEAN, b Manhattan, Kans, Sept 21, 40; m 62; c 4. NEUROPHYSIOLOGY, NEUROANATOMY. Educ: Tex A&M Univ, BS, 63, MS, 64; Univ Tex, PhD(biophys), 69. Prof Exp: ASST PROF PHYS MED & ANAT, SCH MED, EMORY UNIV, 69- Concurrent Pos: USPHS fel, Univ Tex Southwestern Med Sch Dallas, 69; Nat Inst Neurol Dis & Stroke res grants, 71- Mem: AAAS; Am Asn Lab Animal Sci; Soc Neurosci. Res: Mammalian spinal cord neurophysiology and neuroanatomy; sensory-motor organization; muscle innervation. Mailing Add: Dept of Anat Emory Univ Sch of Med Atlanta GA 30322

LETCHER, JOHN HENRY, b Wilkes-Barre, Pa, July 18, 36; m 60; c 2. PHYSICS, COMPUTER SCIENCE. Educ: Univ Tulsa, BS, 57; Univ Mo, MS, 59, PhD(physics), 63. Prof Exp: Mem staff, Advan Electronics Techniques Div, McDonnell Corp, 63-64; mem staff, Cent Res Dept, Monsanto Co, 64-68; vpres systs & res, Data Res Corp, 68-70; PRES, SYNERGISTIC CONSULTS, INC, 70- Res: Computer software; hardware systems development; quantum physics and chemistry. Mailing Add: 7421 S Marion Ave Tulsa OK 74136

LETCHER, STEPHEN VAUGHAN, b Chicago, Ill, Dec 13, 35; m 59; c 2. PHYSICS. Educ: Trinity Col, BS, 57; Brown Univ, PhD(physics), 64. Prof Exp: From asst prof to assoc prof, 63-75, PROF PHYSICS, UNIV R I, 75- Mem: Am Phys Soc; Am Asn Physics Teachers; Acoust Soc Am. Res: Physical acoustics; physics of fluids. Mailing Add: Dept of Physics Univ of R I Kingston RI 02881

LETEY, JOHN, JR, b Carbondale, Colo, June 13, 33; m 55; c 3. BIOPHYSICS. Educ: Colo State Univ, BS, 55; Univ Ill, PhD, 59. Prof Exp: Asst agron, Univ Ill, 55-59; asst prof, Univ Calif, Los Angeles, 59-64; assoc prof, 64-68, chmn div environ sci, 68-75, PROF SOIL PHYSICS, UNIV CALIF, RIVERSIDE, 68-, CHMN DEPT SOIL SCI & AGR ENG, 75- Honors & Awards: Award, Soil Sci Soc Am, 70. Mem: Fel Am Soc Agron; Soil Sci Soc Am. Res: Soil aeration; soil-water-plant relationships; soil wettability, infiltration; environmental pollutants. Mailing Add: Dept of Soil Sci & Agr Eng Univ of Calif Riverside CA 92502

LETKEMAN, PETER, b Winkler, Man, Feb 12, 38; m 64; c 3. CHEMISTRY. Educ: Univ Man, BSc, 60, MSc, 61, PhD(chem), 69. Prof Exp: Teacher high sch, Man, Can, 61-63; lectr, 63-66, asst prof, 66-71, ASSOC PROF CHEM, BRANDON UNIV, 71-, DEPT HEAD, 72- Concurrent Pos: Mem sci curriculum coun, Dept Educ, Man, 68-; grant, Univ Calif, Riverside, 70; consult, Christie Sch Supplies, Man, 70-; mem bd of gov & senate, Brandon Univ, 73-; vpres, Western Man Sci Fair; judge-in-chief, Can Wide Sci Fair. Mem: Chem Inst Can. Res: The polarography and nuclear magnetic resonance of metal complexes in aqueous media; environmental research with regard to water and soil analysis. Mailing Add: Dept of Chem Brandon Univ Brandon Man Can

LETO, JOSEPH REDER, physical-inorganic chemistry, see 12th edition

LETOURNEAU, DUANE JOHN, b Stillwater, Minn, July 12, 26; m 47; c 3. PLANT BIOCHEMISTRY. Educ: Univ Minn, BS, 48, MS, 51, PhD(agr bot), 54. Prof Exp: Asst, Univ Minn, 48-53; asst prof agr chem & asst agr chemist, 53-58, assoc prof & assoc agr chemist, 58-63, prof agr biochem & agr biochemist, 63-73, actg head dept agr biochem & soils, 61-62, PROF BIOCHEM & BIOCHEMIST, UNIV IDAHO, 73- Concurrent Pos: Resident res assoc, USDA, 64-65. Mem: Fel AAAS; Am Soc Plant Physiol; Am Chem Soc; Am Phytopath Soc; Mycol Soc Am. Res: Biochemistry and physiology of plant pathogenic fungi. Mailing Add: Dept of Bacteriol & Biochem Univ of Idaho Moscow ID 83843

LETOURNEAU, ROBERT LOUIS, b Portland, Ore, Sept 12, 18; m 42; c 3. CHEMISTRY. Educ: Wheaton Col, BS, 39; Univ Akron, MS, 40; Univ Ill, PhD(anal chem), 42. Prof Exp: Res assoc proj, Nat Defense Res Comt, Illinois, 42-45; res chemist & group leader, Anal Lab, 45-52, lab supvr, 52-60, supv chemist, 60-66, SR RES ASSOC, CHEVRON RES CO, 66- Mem: Am Chem Soc. Res: Pollution chemistry; agricultural chemistry; hydrocarbon type analysis; distillation; systematic study by x-rays of elastomer addition agents; molecular spectroscopy. Mailing Add: Chevron Res Co 576 Standard Ave Richmond CA 94802

LETOURNEUX, JEAN, b Quebec, Que, Mar 23, 35; m 70. THEORETICAL NUCLEAR PHYSICS. Educ: Laval Univ, BSc, 59; Oxford Univ, DPhil(physics), 62. Prof Exp: Ciba fel, Inst Theoret Physics, Copenhagen, 62-64; res assoc physics, Univ Va, 64-65, asst prof, 65-66; from asst prof to assoc prof, 66-74, PROF PHYSICS, UNIV MONTREAL, 74- Mem: Am Phys Soc; Can Asn Physicists. Res: Nuclear theory. Mailing Add: Dept of Physics Univ of Montreal PO Box 6128 Montreal PQ Can

LETSINGER, ROBERT LEWIS, b Bloomfield, Ind, July 31, 21; m 43; c 3. ORGANIC CHEMISTRY. Educ: Mass Inst Technol, BS, 43, PhD(org chem), 45. Prof Exp: Asst, Mass Inst Technol, 43-45, res assoc, 45-46; res chemist, Tenn Eastman Corp, 46; from instr to assoc prof, 46-59, chmn dept, 72-75, PROF CHEM, NORTHWESTERN UNIV, 59- Concurrent Pos: Guggenheim fel, 56; mem, NIH Fel Rev Panel, 65-69; med chem study sect, NIH, 71-75; bd of eds, J Am Chem Soc, 69-71. Mem: Am Chem Soc; Am Soc Biol Chemists; AAAS. Res: Bioorganic chemistry; synthesis of polynucleotides and nucleotide analogs; photochemistry; organoboron and organoalkali metal compounds. Mailing Add: Dept of Chem, Molecular Biol & Biochem Northwestern Univ Evanston IL 60201

LETT, JOHN TERENCE, b London, Eng, Dec 23, 33; m 56; c 1. BIOPHYSICS, RADIATION BIOLOGY. Educ: Univ London, BSc, 56, PhD(phys org chem), 60. Prof Exp: Sr lectr, Inst Cancer Res, Univ London, 56-67; PROF RADIOL & RADIATION BIOL, GRAD SCH, COLO STATE UNIV, 68- Concurrent Pos: Res assoc, Univ Calif, 61; vis scientist, Oak Ridge Nat Lab, 64. Mem: Radiation Res Soc; Brit Biophys Soc; Biophys Soc; Brit Asn Radiation Res. Res: DNA structure of the chromosome; repair of radiation damage to cellular DNA; radiation and aging. Mailing Add: Dept of Radiol & Radiation Biol Colo State Univ Ft Collins CO 80521

LETTAU, HEINZ HELMUT, b Koenigsberg, Ger, Nov 4, 09; m 37; c 3. GEOPHYSICS, METEOROLOGY. Educ: Univ Leipzig, PhD(geophys, meteorol), 31, Dr Phil Habil(geophys, meteorol), 36. Prof Exp: Asst, Geod Inst Potsdam, Ger, 31-33 & Dept Geophys, Univ Leipzig, 33-38; chief geophys sect, Univ Koenigsberg, 38-43; dir & prof geophys, Geophys Inst, Tech Univ Graz, 43-45; chief res sect, Ger Weather Orgn, US Occupied Zone, 47-48; proj scientist, Geophys Res Div, Air Force Cambridge Res Ctr, 47-58; PROF METEOROL, UNIV WIS-MADISON, 58- Concurrent Pos: Lectr, Mass Inst Technol, 49-58. Mem: Sigma Xi; Am Meteorol Soc; Am Geophys Union. Res: Atmospheric dynamics and turbulence. Mailing Add: Dept of Meteorol Univ of Wis Madison WI 53706

LETTAU, KATHARINA, b Plauen, Ger, July 20, 10; US citizen; m 37; c 3. CLIMATOLOGY. Educ: Univ Leipzig, PhD(meteorol, math physics), 35. Prof Exp: Meteorologist, Ger Weather Serv, 35-37, 39-44; METEOROLOGIST, CTR CLIMATIC RES, UNIV WIS-MADISON, 62- Res: Bioclimatology; phenology. Mailing Add: Ctr for Climatic Res Univ of Wis Madison WI 53706

LETTERMAN, GORDON SPARKS, b St Louis, Mo, Aug 17, 14; m 47; c 1. SURGERY. Educ: Wash Univ, AB, 37, BS, 40, MD, 41; Am Bd Surg, dipl; Am Bd Plastic Surg, dipl. Prof Exp: Asst surg, Wash Univ, 43-48; instr, 49-53, from assoc to asst prof, 53-60, from assoc clin prof to clin prof, 60-64, PROF SURG, GEORGE WASHINGTON UNIV, 64- Concurrent Pos: Mem, Int Cong Plastic Surgeons, 55. Mem: Am Soc Plastic & Reconstruct Surg; Am Asn Plastic Surg; AMA; Asn Mil Surg US; fel Am Col Surgeons. Res: Plastic surgery. Mailing Add: 2700 Q St NW Washington DC 20007

LETTERMAN, HERBERT, b Brooklyn, NY, Oct 8, 36; m 57; c 4. ANALYTICAL CHEMISTRY. Educ: City Col New York, BS, 58; Brooklyn Col, MA, 62; Seton Hall Univ, MS, 67, PhD(anal chem), 73. Prof Exp: Anal chemist, Brooklyn Jewish Hosp, NY, 58-59; anal chemist, Ciba Pharmaceut Co, NJ, 59-63; group leader phys chem res & develop, 63-66, HEAD QUAL CONTROL, BRISTOL-MYERS PROD, 66- Mem: Sigma Xi; Am Chem Soc; Am Pharmaceut Asn; Acad Pharmaceut Sci; sr mem Am Soc Qual Control. Res: Quality control; analytical method development. Mailing Add: 44 Delaware Ave New Providence NJ 07974

LETTON, JAMES CAREY, b Lexington, Ky, June 9, 33; m 56; c 3. PHARMACEUTICAL CHEMISTRY. Educ: Ky State Col, BS, 55; Univ Ill, Chicago Circle, PhD(chem), 71. Prof Exp: Prod foreman, Julian Labs, 57-62, supt prod, Smith Kline & French Labs, 62-64, res & develop chemist, Julian Res Inst, 64-69; instr org chem, Triton Col, 68-70; assoc prof, 70-73, PROF ORG CHEM, KY STATE COL, 73-, CHMN DEPT CHEM, 71- Mem: Fel Am Inst Chemists; Am Chem Soc. Res: Medicinal chemistry, especially beta amino ketones and analgesic properties; morphine-like compounds; steroid synthesis. Mailing Add: Dept of Chem Ky State Univ Frankfort KY 40501

LETTVIN, JEROME Y, b Chicago, Ill, Feb 23, 20; m 47; c 3. NEUROPHYSIOLOGY. Educ: Univ Ill, BS, 42, MD, 43. Prof Exp: Intern neurol, Boston City Hosp, 43-44; physiologist, Dept of Psychol, Univ Rochester, 47-48; neuropsychiatrist & physiologist, Manteno State Hosp, 48-51; PROF COMMUN PHYSIOL, DEPTS BIOL, ELEC ENG & HUMANITIES, MASS INST TECHNOL, 66-, NEUROPHYSIOLOGIST, LAB ELECTRONICS, 51- Mem: Am Physiol Soc. Res: Experimental epistemology. Mailing Add: 8 Blanchard Cambridge MA 02139

LEU, RICHARD WILLIAM, b Argonia, Kans, Jan 5, 35; m 60; c 2. MICROBIOLOGY, IMMUNOLOGY. Educ: Northwestern State Col, Okla, BS, 60; Univ Okla, MS, 63, PhD(microbiol, immunol), 70. Prof Exp: USPHS res training fel pediat & path, Med Sch, Univ Minn, Minneapolis, 70-74; MEM STAFF, NOBLE FOUND, 74- Res: Cellular immunity; effector molecules associated with macrophage inhibition, proliferation and activation; role of cytophilic antibody in cellular immunity; localized immunity in the lung. Mailing Add: Noble Found Rte 1 Ardmore OK 73401

LEUBNER, GERHARD WALTER, b Walton, NY, Aug 31, 21; m 44; c 3. ORGANIC CHEMISTRY. Educ: Union Col, BS, 43; Univ Ill, PhD(chem), 49. Prof Exp: Chemist, Winthrop Chem Co, 43-45; asst, Univ Ill, 45-46; RES ASSOC, EASTMAN KODAK CO, 48- Mem: Am Chem Soc. Res: Patent information storage and retrieval systems. Mailing Add: Eastman Kodak Co Res Labs B 85 1669 Lake Ave Kodak Park Rochester NY 14650

LEUBNER, INGO HERWIG, b Prittlbach, Ger, Apr 9, 38; m 69. PHYSICAL CHEMISTRY. Educ: Munich Tech Univ, Dipl, 63, PhD(phys chem), 66. Prof Exp: Ger Res Asn res fel phys chem, Munich Tech Univ, 66-68; Welch Found fel & lectr photochem, Tex Christian Univ, 68-69; RES CHEMIST, RES LABS, EASTMAN KODAK CO, 69- Mem: Soc Photog Scientists & Engrs; Am Chem Soc. Res: Photochemistry of organic and inorganic compounds. Mailing Add: Res Labs Eastman Kodak Co 343 State St Rochester NY 14650

LEUCHTENBERGER, CECILE, b Leipzig, Ger, Mar 17, 06; nat US; m 33. CYTOLOGY. Educ: Columbia Univ, MA, 46, PhD(biol), 49. Prof Exp: Biologist, Cancer Res Lab, Mt Sinai Hosp, NY, 36-47; res assoc, Columbia Univ, 47-50; head, Cytochem Lab, Inst Path, Western Reserve Univ, 50-59; sr biologist & cytochemist, Children's Cancer Res Found, Inc, Boston, Mass, 59-63; PROF CYTOCHEM, UNIV LAUSANNE & HEAD DEPT, SWISS INST EXP CANCER RES, 63- Mem: Am Chem Soc; Am Soc Zool; Am Soc Exp Path; Am Asn Cancer Res. Res: Cancer research; cytochemistry; microspectrophotometry; biology. Mailing Add: Chemin du Triolet 3 1110 Morges Switzerland

LEUNG, ALBERT YUK-SING, b Hong Kong, May 24, 38; nat US; m 68; c 1. PHARMACOGNOSY, MICROBIOLOGY. Educ: Nat Taiwan Univ, BS, 61; Univ Mich, Ann Arbor, MS, 65, PhD(pharmacog), 67. Prof Exp: NIH res chemist, Med Ctr, Univ Calif, 67-69; res supvr microbial protein prod, Bohna Eng & Res, Inc, 69-71; tech dir chem & microbiol consult, Sci Res Info Serv, Inc, 71-74; DIR RES & DEVELOP, DR MADIS LABS, INC, 74- Mem: AAAS; Am Chem Soc; Am Soc Pharmacog; Am Pharmaceut Asn; Sigma Xi. Res: Production of food products from wastes by fermentation; isolation of active principles from plants and microorganisms; commercial production of Morchella species; biosynthesis of plant products. Mailing Add: South Hackensack NJ

LEUNG, BENJAMIN SHUET-KIN, b Hong Kong, June 30, 38; US citizen; m 64; c 3. ENDOCRINOLOGY, ONCOLOGY. Educ: Seattle Pac Col, BS, 63; Colo State Univ, PhD(biochem), 69. Prof Exp: Res asst steroid hormones, Pac Northwest Res Found, 63-66; asst prof, 71-74, ASSOC PROF SURG, MED SCH, UNIV ORE, 74-, DIR LAB, CANCER RES, CLIN RES CTR, 71- Concurrent Pos: NIH & Ford Found res fel reprod endocrinol, Med Sch, Vanderbilt Univ, 69-71; Med Res Found Ore grant, Med Sch, Univ Ore, 71-72, Am Cancer Soc Ore Div res grants, 72-74; Cammack Trust Fund grant, 74-75; NIH grant, 75-78. Mem: AAAS; Endocrine Soc; NY Acad Sci. Res: Mechanism of steroid hormones action related to male and female reproductive physiology in normal and neoplastic state. Mailing Add: Dept of Surg Univ of Ore Med Sch Portland OR 97201

LEUNG, CHRISTOPHER CHUNG-KIT, b Hong Kong, Jan 3, 39; m 70; c 1. EMBRYOLOGY, IMMUNOLOGY. Educ: Howard Univ, BSc, 64; Jefferson Med Col, PhD(anat, embryol), 69. Prof Exp: Res asst, Sch Med, Univ Rochester, 64-65; from instr to asst prof pediat, 69-74, INSTR ANAT, JEFFERSON MED COL, 69-, INSTR, COL ALLIED HEALTH SCI, THOMAS JEFFERSON UNIV, 70-, RES ASSOC PROF PEDIAT, 74- Concurrent Pos: NIH fel, Stein Res Ctr, Thomas Jefferson Univ, 69- Mem: Teratol Soc. Res: Teratology; immunopathology; cell biology. Mailing Add: Jefferson Med Col Thomas Jefferson Univ Philadelphia PA 19107

LEUNG, IRENE SHEUNG-YING, b Hong Kong, July 10, 34. MINERALOGY. Educ: Univ Hong Kong, BA, 57; Ohio State Univ, MA, 63; Univ Calif, Berkeley, PhD(geol), 69. Prof Exp: Res staff geologist, Yale Univ, 69-71; ASST PROF GEOL, LEHMAN COL, 71- Mem: Sigma Xi; Mineral Soc Am; Am Geophys Union; Geochem Soc; Asian Environ Soc. Res: X-ray investigation of mineral inclusions in natural diamonds; magmatic crystallization and sector-zoning in crystals; deformation structures and glide mechanisms in deformed minerals. Mailing Add: Dept of Geol & Geog Herbert H Lehman Col Bronx NY 10468

LEUNG, KAM-CHING, b Hong Kong, June 16, 35; m 63; c 1. ASTRONOMY, ASTROPHYSICS. Educ: Queen's Univ, Ont, BSc, 61; Univ Western Ont, MA, 63; Univ Pa, PhD(astron), 67. Prof Exp: Nat Acad Sci-Nat Res Coun res fel astron, Inst Space Studies, NASA, 68-70; asst prof physics, 70-72, ASSOC PROF PHYSICS & ASTRON, UNIV NEBR, LINCOLN, 72- Concurrent Pos: NSF res grant, Univ Nebr, Lincoln, 70-71 & 75, dir observ, 72-75. Mem: Fel AAAS; Int Astron Union; Am Astron Soc. Res: Stellar photometry and spectroscopy; intrinsic variable stars; binary stars. Mailing Add: Behlen Lab of Physics Univ of Nebr Lincoln NE 68508

LEUNG, KIN-VINH, b Aug 1, 20; Chinese citizen; m 52; c 4. NUMERICAL ANALYSIS. Educ: Spec Sch Pub Works, Paris, dipl eng, 41; Univ Paris, Lic es Sci, 41, Dr Ing, 46. Prof Exp: Lectr, Univ Hong Kong, 53-70; assoc prof comput sci, Univ Alta, 70-74; ASSOC PROF COMPUT SCI, CONCORDIA UNIV, 74- Concurrent Pos: Consult engr, E E Keen & Partners, 65-68; vis prof, Univ Alta, 68-69. Honors & Awards: Fr Asn Eng Medal, 46. Res: Queueing process; auditory system model; optical data processing; structure analysis; finite element method. Mailing Add: Dept of Comput Sci Concordia Univ Sir George Williams Campus Montreal PQ Can

LEUNG, PAK SANG, b Shanghai, China, June 8, 35; US citizen; m 65; c 2. COLLOID CHEMISTRY. Educ: Nat Taiwan Univ, BSc, 57; Columbia Univ, MA, 62, PhD(phys chem), 67. Prof Exp: Dyes lab asst, Imp Chem Industs, 57-59; demonstr chem, Hong Kong Baptist Col, 59-61; res scientist, Brookhaven Nat Lab-Columbia Univ, 66-67; RES SCIENTIST, UNION CARBIDE CORP, TARRYTOWN, 67- Concurrent Pos: Mem chem adv bd, Harriman Col, 74- Mem: Sigma Xi. Res: Colloidal chemistry; ultrafiltration; membrane technology. Mailing Add: 15 Woodland Rd Highland Mills NY 10930

LEUNG, PHILIP MIN BUN, b Canton, China, July 31, 34; m; c 3. NUTRITION, BIOCHEMISTRY. Educ: Chung Hsing Univ, Taiwan, BSc, 56; McGill Univ, MSc, 59; Mass Inst Technol, PhD(nutrit biochem), 65. Prof Exp: Res asst, McGill Univ, 56-59; res asst, Mass Inst Technol, 59-65; group leader biochem res, Med Sci Res Lab, Miles Labs Inc, Ind, 65-67; ASST RES NUTRITIONIST, SCH VET MED, UNIV CALIF, DAVIS, 67- Mem: AAAS; Am Inst Nutrit; Inst Food Technol; NY Acad Sci. Res: Nutrition and biochemistry of amino acid imbalance; nutritional regulation of protein metabolism; influence of nutrition, especially amino acid balance on food intake regulation. Mailing Add: Dept of Physiol Sci Sch of Vet Med Univ of Calif Davis CA 95616

LEUNG, SO WAH, b China, Nov 2, 18; nat US; m 57; c 1. DENTISTRY, PHYSIOLOGY. Educ: McGill Univ, DDS, 43, BSc, 45; Univ Rochester, PhD(physiol), 50; FRCDent(C). Prof Exp: Intern dent, Royal Victoria Hosp, Montreal, 43-44; from assoc prof to prof physiol, Sch Dent, Univ Pittsburgh, 50-61, head dept, 52-61, prof dent res & dir grad educ, 57-61; prof oral biol, Sch Dent & lectr physiol, Sch Med, Univ Calif, Los Angeles, 61-62; PROF ORAL BIOL & DEAN SCH DENT, UNIV BC, 62- Concurrent Pos: Mem comt dent, Nat Acad Sci-Nat Res Coun, 57-61; consult, Colgate-Palmolive Co, 58-62; mem dent study sect, NIH, 59-63; consult, Nat Bd Dent Exam, 60-66 & Lever Bros, 63-65; mem assoc comt dent res, Nat Res Coun Can, 63-68, exec comt, 65-68; mem, Nat Dent Exam Bd Can, 65-67, chmn exam comt, 67-71; chmn res comt, Asn Can Fac Dent, 68-70, pres, 70-72. Mem: Am Dent Asn; fel Am Col Dent; fel Int Col Dent; NY Acad Sci; Sigma Xi. Res: Salivary chemistry; oral calculus formation; physiology of salivary glands. Mailing Add: Fac of Dent Univ of BC Vancouver BC Can

LEUNG, WOOT-TSUEN WU, b China, Dec 14, 14; nat US; m 43; c 1. NUTRITION. Educ: Lingnan Univ, BA, 36; Pa State Univ, MS, 37, PhD, 40. Prof Exp: Tech asst mass nutrit study, Children's Hosp, Philadelphia, Pa, 41-43; nutritionist, Cmt Food Composition, Nat Res Coun, 43-44; interim cmn food & agr, UN, 44; commodity specialist, Off Food Prog & nutritionist, Cmt Int Food Value Probs, For Econ Admin, 45; nutrit analyst, Agr Res Serv, USDA, 45-64; nutrit analyst, Off Int Res, NIH, 64; chief food sci info, Nutrit Prog, Ctr Dis Control, Dept Health, Educ & Welfare, 65-73; nutrit consult, Pan Am Health Orgn-World Health Orgn, 74; RETIRED. Concurrent Pos: Assoc mem Chinese del, Food & Agr Orgn, UN Conf, Can, 45, del comt calorie conversion factors & food composition tables, 47; del, Int Assembly Women, NY, 46. Mem: Am Inst Nutrit; Inst Food Technol. Res: Tables of food composition for Armed Forces, Far East, Latin America, Africa and East Asia; nutritional value of foods grown in world-regions; conserving nutritive value. Mailing Add: 3114 Oliver St NW Washington DC 20015

LEUNG, YUEN PO, chemistry, see 12th edition

LEURGANS, PAUL JOHN, physics, see 12th edition

LEUSSING, DANIEL, JR, b Cincinnati, Ohio, Oct 8, 24; m 57; c 3. ANALYTICAL CHEMISTRY. Educ: Univ Cincinnati, BA, 45; Univ Ill, MS, 47; Univ Minn, PhD(chem), 53. Prof Exp: Instr anal chem, Univ Minn, 51-52; chemist, Am Cyanamid Co, 53; instr anal chem, Mass Inst Technol, 53-55; from instr to asst prof, Univ Wis, 55-60; chemist, Nat Bur Standards, 60-62; from asst prof to assoc prof, 62-70, PROF CHEM, OHIO STATE UNIV, 70- Mem: Am Chem Soc. Res: Physical chemistry of aqueous solutions; coordination chemistry; metal mercaptide complexes; Schiff base complexes; kinetics. Mailing Add: Dept of Chem Ohio State Univ Columbus OH 43210

LEUTERT, WERNER WALTER, b Ottenback, Switz, Nov 9, 22; nat US; m 48; c 5. APPLIED MATHEMATICS. Educ: Swiss Fed Inst Technol, DSc, 48. Prof Exp: Asst prof math, Univ Md, 48-51; mathematician, Comput Lab, Ballistic Res Labs, Aberdeen Proving Ground, Md, 51-53, chief, 53-55; mgr math & comput serv dept, Lockheed Missile Systs Div, 55-57; sr mathematician, Tidewater Oil Co, 57-59; mgr adv prog & appl math, Remington Rand Univac Div, Sperry Rand Corp, NY, 59-61, dir systs prog, 61-62; CONSULT, 62- Mem: Soc Indust & Appl Math; Math Asn Am; Asn Comput Mach; Inst Mgt Sci. Res: Large real time systems; applications of electronic computers; optimization of operations; automatic programming of computers. Mailing Add: Laurel Ledge Park Stamford CT 06903

LEUTGOEB, ROSALIA ALOISIA, b Vienna, Austria, Apr 2, 01; nat US; m 20; c 1. CHEMISTRY. Educ: Marquette Univ, BS, 35, MS, 36, PhD(electroeng chem), 38. Prof Exp: Instr chem & math, St Ambrose Col, 40-42; res chemist dried yeast, Red Star Yeast Co, 42-44; res chemist synthetic rubber, US Govt Labs, 44-45; asst prof chem, Mundelein Col, 45-49; res chemist, Froedtert Grain & Malting Co, Inc, 50-52; prof chem & chmn dept, 53-70, EMER PROF CHEM, NORTHLAND COL, 70- Mem: AAAS; Am Chem Soc. Res: Synthesis of glucuronic acid; oxidation mechanism. Mailing Add: Route 1 Box 162 Ashland WI 54806

LEUTNER, FREDERICK STANLEY, b Cleveland, Ohio, Sept 30, 13; m 39, 69; c 2. CLINICAL BIOCHEMISTRY. Educ: Harvard Univ, AB, 34; Western Reserve Univ, PhD(chem), 42. Prof Exp: Instr chem, Univ Cincinnati, 41-43; res chemist, Goodyear Tire & Rubber Co, 43-51; group leader fundamental res, Arco Co, 51-60, asst dir res, 60; lab mgr, Air Reduction Carbon & Carbide Co, 60-64; asst dir polymer & plastics res & develop, Coatings & Resins Group, Glidden Co, Ohio, 64-68, dir tech serv, Macco Chem Sales Div, 66-68; chem consult, 68-70; PRES & CO-DIR, KLEIN MED LAB, CLEVELAND, 70- Mem: Am Chem Soc. Res: Chemistry of high polymers; coatings; low temperature distillation of gases; high vacuum technique; hematology; serology; clinical chemistry; bacteriology; testing service to the medical profession. Mailing Add: 1796 Wilton Rd Cleveland Heights OH 44118

LEUTRITZ, JOHN, JR, b Saginaw, Mich, June 22, 03; m 28, 61. CHEMISTRY. Educ: Bowdoin Col, BS, 29; Columbia Univ, MA, 34, PhD(bot), 46. Prof Exp: Res engr, Bell Tel Labs, Inc, 29-66; chem engr, 65-72, TIMBER PROD SPECIALIST, RURAL ELECTRIFICATION ADMIN, USDA, 72- Mem: AAAS; Am Chem Soc; Soc Indust Microbiol; Am Wood Preservers Asn; NY Acad Sci. Res: Microbiological deterioration; wood preservation; climatic effects on materials; miscellaneous materials preservation; quality control. Mailing Add: 5405 Duke St Alexandria VA 22304

LEUTZE, WILLARD PARKER, b Burlington, Vt, Mar 2, 27; m 51; c 2. GEOLOGY. Educ: Syracuse Univ, BS, 51, MS, 55; Ohio State Univ, PhD(geol), 59. Prof Exp: Geologist, US Geol Surv, 54-55; asst geol, Ohio State Univ, 55-58; instr geol & soil sci, Earlham Col, 58-60; geologist, Texaco, Inc, 60-66; biostratigrapher, 66-71, SR GEOLOGIST, ATLANTIC RICHFIELD CO, 71- Mem: Geol Soc Am; assoc Soc Econ Paleont & Mineral; Soc Prof Well Log Analysts. Res: Paleontology, particularly foraminifera, arthropods and echinoderms; stratigraphy of Upper Silurian and of Gulf Coast; subsurface stratigraphy of south Louisiana. Mailing Add: Atlantic Richfield Co PO Box 51408 Lafayette LA 70501

LEV, MAURICE, b St Joseph, Mo, Nov 13, 08; m 47; c 2. PATHOLOGY. Educ: NY Univ, BS, 30; Creighton Univ, MD, 34; Northwestern Univ, MA, 66. Prof Exp: From instr to assoc prof path, Col Med, Univ Ill, 39-41; asst prof, Sch Med, Creighton Univ, 46-47; from assoc prof to prof, Sch Med, Univ Miami, 51-57; PROF PATH, MED SCH, NORTHWESTERN UNIV, CHICAGO, 57-; DIR CONGENITAL HEART DIS RES & TRAINING CTR, HEKTOEN INST MED RES, 57- Concurrent Pos: Pathologist, Chicago State Hosp, 40-42; pathologist & dir res labs, Mt Sinai Hosp, Miami Beach, 51-57; career investr & educr, Chicago Heart Asn, 66-; consult, Children's Mem Hosp, Chicago, 57-; prof lectr, Univ Chicago, 59-; lectr, Col Med, Univ Ill, 63-, Chicago Med Sch, Univ Health Sci, 70- & Stritch Sch Med, Loyola Univ, 71-; distinguished prof pediat, Rush Med Col, 74-, distinguished prof int med, 75- Mem: Am Soc Clin Path; Am Asn Path & Bact; AMA; Am Heart Asn; NY Acad Sci. Res: Cardiac pathology; pathology of congenital heart disease and of conduction system. Mailing Add: 629 S Wood St Chicago IL 60612

LEVAN, HOA, radiation physics, radiobiology, see 12th edition

LEVAN, MARIJO O'CONNOR, b Detroit, Mich, Oct 27, 36; m 59; c 3. MATHEMATICS. Educ: Spring Hill Col, BS, 59; Univ Ala, MA, 61; Univ Fla, PhD(math), 64. Prof Exp: From instr to asst prof math, Univ Fla, 62-67; asst prof, Southeast Mo State Col, 67-69; assoc prof, 69-74, PROF MATH, EASTERN KY UNIV, 74- Mem: Am Math Soc. Res: Number theory; partition functions and translated geometric progressions. Mailing Add: Dept of Math Eastern Ky Univ Richmond KY 40475

LEVAN, NORMAN E, b Cleveland, Ohio, Mar 17, 16; m 50. MEDICINE. Educ: Univ Southern Calif, AB, 36, MD, 40. Hon Degrees: MA, St John's Col, NMex, 74. Prof Exp: PROF DERMAT & CHMN DEPT, SCH MED, UNIV SOUTHERN CALIF, 61- Mailing Add: Dept of Med Sch of Med Univ of Southern Calif Los Angeles CA 90007

LEVAND, OSCAR, b Parnu, Estonia, Nov 3, 27; US citizen. ORGANIC CHEMISTRY. Educ: Miss State Col, BS, 54; Purdue Univ, MS, 58; Univ Hawaii, PhD(org chem), 63; Univ Minn, Mineapolis, MPH, 70. Prof Exp: Jr res chemist, Mead Johnson Co, Ind, 54-56; res chemist, Knoll Pharmaceut Co, NJ, 58-59; fel NIH, 62-63; res chemist, Dole Co, Hawaii, 63-68; consult, Air Pollution Control Prog, Govt of Guam, 70-74; ASST PROF CHEM, UNIV GUAM, 74- Mem: Am Chem Soc; Air Pollution Control Asn. Res: Air and water chemistry; environmental problems. Mailing Add: Dept of Chem Univ of Guam PO Box EK Agana GU 96910

LEVANDER, ORVILLE ARVID, b Waukegan, Ill, Apr 6, 40. BIOCHEMISTRY. Educ: Cornell Univ, BA, 61; Univ Wis-Madison, MS, 63, PhD(biochem), 65. Prof Exp: Res fel biochem, Col Physicians & Surgeons, Columbia Univ, 65-66; res assoc, Sch Public Health, Harvard Univ, 66-67; res chemist, Food & Drug Admin, 67-69, RES CHEMIST, NUTRIT INST, AGR RES CTR, 69- Concurrent Pos: Mem, Nat Res Coun Comt Biol Effects Environ Pollutants, 74- Mem: AAAS; Am Inst Nutrit; Am Chem Soc. Res: Toxicology and nutrition of selenium; pharmacology of heavy metals; trace mineral nutrition; vitamin E; drug metabolism; lead poisoning. Mailing Add: Nutrit Inst Agr Res Ctr Beltsville MD 20705

LEVANDOWSKI, DONALD WILLIAM, b Stockett, Mont, Dec 20, 27; m 55; c 2. GEOLOGY. Educ: Mont Col Mineral Sci & Technol, BS, 50; Univ Mich, MS, 52, PhD(mineral), 56. Hon Degrees: Geol Engr, Mont Col Mineral Sci & Technol, 68. Prof Exp: Res geologist, Calif Res Corp, Standard Oil Co, 55-64, staff asst to mgr explor res, Chevron Res Co, 64-65, geophysicist, Western Opers, Inc, 65-67; assoc prof, 67-75, PROF GEOSCI, PURDUE UNIV, 75-, ASSOC HEAD DEPT, 70- Mem: Fel Geol Soc Am; AAAS; Am Asn Petrol Geologists; fel Geol Asn Can; Am Soc Photogrammetry. Res: Mineral deposits; remote sensing; igneous and metamorphic petrology; geophysics. Mailing Add: Dept of Geosci Purdue Univ Lafayette IN 47907

LEVANDOWSKY, MICHAEL, b Knoxville, Tenn, Aug 15, 35. MARINE ECOLOGY, MATHEMATICAL BIOLOGY. Educ: Antioch Col, AB, 61; Columbia Univ, Ma, 65, PhD(biol), 70; NY Univ, MS, 73. Prof Exp: Instr biol, Bard Col, 67-69; instr, Bronx Community Col, 69-70; asst prof biol, Col, 70-71, RES ASSOC, HASKINS LABS, PACE UNIV, 70- Concurrent Pos: Nat Sci Found sci fac fel, Courant Inst Math Sci, NY Univ, 71-72; asst prof biol, York Col, NY, 73-74. Mem: AAAS; Ecol Soc Am; NY Acad Sci; Torrey Bot Club; Am Soc Limnol & Oceanog. Res: Mathematical models in ecology and evolution; microbial ecology; marine biology; gnotobiotic systems; chemoreception of Protista; human evolution. Mailing Add: Haskins Labs Pace Univ 41 Park Row New York NY 10038

LEVANONI, MENACHEM, b Jerusalem, Israel, Feb 20, 38; US citizen; m 62; c 3. ACOUSTICS, FLUID PHYSICS. Educ: Hebrew Univ Jerusalem, BSc, 64; Calif Inst Technol, PhD(physics), 70. Prof Exp: Fel nuclear physics, Calif Inst Technol, 70-71; RES STAFF MEM, IBM T J WATSON RES CTR, 71- Res: Applications of physical interactions in acoustics, optics and magnetism. Mailing Add: IBM T J Watson Res Ctr PO Box 214 Yorktown Heights NY 10598

LEVEAU, BARNEY FRANCIS, b Denver, Colo, Oct 2, 39; m 61; c 3. PHYSICAL MEDICINE. Educ: Univ Colo, BS, 61, MS, 66; Mayo Clin, RPT, 65; Pa State Univ, PhD(phys educ), 73. Prof Exp: Teacher math & sci, Colo Springs Sch Dist, 61-63; from asst prof to assoc prof phys educ, WChester State Col, 66-70; ASST PROF PHYS THER, SCH MED, UNIV NC, CHAPEL HILL, 72- Mem: Am Phys Ther Asn; Am Col Sports Med; Am Asn Health, Phys Educ & Recreation. Res: Biomechanics as it applies to physical therapy and physical education; sports medicine. Mailing Add: Div of Phys Ther Univ of NC Sch of Med Chapel Hill NC 27514

LEVEDAHL, BLAINE HESS, b Ogden, Utah, Aug 11, 19; m 40; c 2. BIOCHEMISTRY. Educ: Brigham Young Univ, BS, 41; Univ Utah, MA, 48, PhD(biochem), 49. Prof Exp: Physicist internal ballistics, Off Sci Res & Develop, 42-45; physicist, Naval Bur Ord, 45-46; from instr to assoc prof gen physiol, Univ Calif, Los Angeles, 50-67; PROF CHEM & CHMN DEPT, UNIV ALA, BIRMINGHAM, PROF BIOCHEM, PHYSIOL & BIOPHYS, MED CTR, 67- Concurrent Pos: AEC fel, Princeton Univ, 49-50. Mem: Biophys Soc; Am Soc Zool; Am Chem Soc. Res: Cell biology; cellular metabolism; structure of macromolecules. Mailing Add: Dept of Chem Univ of Ala Birmingham AL 35233

LEVEE, RICHARD DOUGLAS, b Los Angeles, Calif, Sept 17, 25; m 59; c 2. DATA PROCESSING. Educ: Univ Calif, AB, 48, MA & PhD(astron), 51. Prof Exp: Res asst, Princeton Univ, 51-52; asst prof astron, Univ Mo, 52-54; res astronomer, Leuschner Observ, Univ Calif, 54-55, sr physicist, Lawrence Radiation Lab, 55-63; dir theoret physics div, Physics Int Inc, 63-64; mgr, Europ Orgn Nuclear Res Support Group, Control Data Corp, Minn, 64-66, prod line mgr, 66-67, gen mgr 6000/7000 prod lines, 67-68, staff gen mgr mkt info systs, 68-70, gen mgr data serv planning, 70; EXEC VPRES, CULLER-HARRISON, INC, 70- Concurrent Pos: Fulbright fel, Weizmann Inst, Israel, 59-60; consult, Atomic Energy Comn. Res: Problems in stellar interiors; numerical solution of partial differential equations of radiation flow; hydrodynamics and neutron transport; numerical analysis. Mailing Add: Culler-Harrison Inc 150 Aero Camino Goleta CA 93017

LEVEEN, HARRY HENRY, b Woodhaven, NY, Aug 10, 14; m; c 2. SURGERY. Educ: Princeton Univ, BA, 36; NY Univ, MD, 40; Univ Chicago, MS, 47; Am Bd Surg, dipl. Prof Exp: Instr & res assoc, Univ Chicago, 45-47; instr surg, Col Med, NY Univ, 47-50; assoc prof physiol, Sch Med, Loyola Univ, Ill, 50-55; assoc prof surg, Chicago Med Sch, 55-56; assoc prof, 57-59, PROF SURG, COL MED, STATE UNIV NY DOWNSTATE MED CTR, 60- Concurrent Pos: Assoc prof, NMex Mil Inst, 52-55; chief surgeon, Vet Admin Hosp, 57- Mem: Soc Exp Biol & Med; Am Physiol Soc; Int Soc Surg; fel Am Col Surg; NY Acad Med. Res: Surgical physiology. Mailing Add: Surg Dept Vet Admin Hosp Brooklyn NY 11209

LEVEILLE, GILBERT ANTONIO, b Fall River, Mass, June 3, 34; m 56; c 3. NUTRITION, BIOCHEMISTRY. Educ: Univ Mass, BVA, 56; Rutgers Univ, MS, 58, PhD(nutrit), 60. Prof Exp: Biochemist, US Army Med Res & Nutrit Lab Colo, 60-66; assoc prof nutrit biochem, Univ Ill, Urbana, 66-69, prof, 69-71; PROF FOOD SCI & HUMAN NUTRIT & CHMN DEPT, MICH STATE UNIV, 71- Honors & Awards: Res Award, Poultry Sci Asn, 65; Mead Johnson Res Award, Am Inst Nutrit, 71. Mem: AAAS; Am Inst Nutrit; Am Soc Clin Nutrit; Am Chem Soc; Poultry Sci Asn. Res: Lipid metabolism; protein and amino acid nutrition and metabolism; atherosclerosis. Mailing Add: Dept of Food Sci & Human Nutrit Mich State Univ East Lansing MI 48823

LEVENBERG, MILTON IRWIN, b Chicago, Ill, Nov 5, 37; m 65. MASS SPECTROMETRY, COMPUTER SCIENCE. Educ: Ill Inst Technol, BS, 58; Calif Inst Technol, PhD(chem), 65. Prof Exp: SR CHEM PHYSICIST, ABBOTT LABS, 65- Mem: Am Chem Soc; Am Soc Mass Spectrometry. Res: Computer applications to instrumentation; instrumentation; electronics; nuclear magnetic resonance spectroscopy. Mailing Add: Chemical Physics Sect D-482 Abbott Labs North Chicago IL 60064

LEVENBOOK, LEO, b Kobe, Japan, Dec 29, 19; nat US; m 50; c 1. BIOCHEMISTRY. Educ: Univ London, BSc, 41; Cambridge Univ, PhD(biochem), 49. Prof Exp: Asst insect biochem, Cambridge Univ, 46-50; fel, Harvard Univ, 50-51; res assoc biochem genetics, Inst Cancer Res, Philadelphia, 51-54; asst prof biochem, Jefferson Med Col, 54-58; BIOCHEMIST, NAT INST ARTHRITIS, METAB & DIGESTIVE DIS, NIH, 58- Concurrent Pos: Lectr, Haverford Col, 53-54. Mem: Am Soc Biol Chemists; Brit Biochem Soc. Res: Insect physiology and biochemistry. Mailing Add: Nat Inst Arthritis Metab & Digestive Dis NIH Bethesda MD 20014

LEVENE, CYRIL, b Gateshead, Eng, May 27, 26; m 52; c 3. ANATOMY. Educ: Queen's Univ Belfast, MB, BCh & BAO, 48, MD, 60. Prof Exp: Demonstr anat, Queen's Univ Belfast, 51-52, asst lectr, 52-54; lectr human anat, Univ Col WIndies, 54-65, sr lectr anat, 65-67; assoc prof, Univ Western Ont, 67-69; assoc prof, 69-74, PROF ANAT, DIV MORPHOL SCI, FAC MED, UNIV CALGARY, 74- Concurrent Pos: WHO fel human genetics, 66. Mem: Am Asn Anat; Can Asn Anat; Anat Soc Gt Brit & Ireland. Res: Vascularization and growth of hyaline cartilage. Mailing Add: Div of Morphol Sci Univ of Calgary Calgary AB Can

LEVENE, HOWARD, b New York, NY, Jan 17, 14. MATHEMATICAL STATISTICS. Educ: NY Univ, BA, 41; Columbia Univ, PhD(math statist), 47. Prof Exp: Exten lectr zool & math statist, 47-48, from instr to assoc prof math statist & biomet, 48-70, PROF MATH STATIST & BIOMET, COLUMBIA UNIV, 70- Mem: AAAS; Am Math Soc; Soc Study Evolution; Biomet Soc; Soc Human Genetics. Res: Mathematical genetics; nonparametric tests; biometrics. Mailing Add: Dept of Math Statist Columbia Univ New York NY 10027

LEVENE, JOHN REUBEN, b Hull, Eng, Dec 7, 29; m 59; c 2. OPTOMETRY, PHYSIOLOGICAL OPTICS. Educ: City Univ, London, dipl ophthalmic optics, 54; Ind Univ, MS, 62; Oxford Univ, PhD(biol sci), 66. Prof Exp: Asst prof optom & physiol optics, Univ Houston, 62-63; lectr optom, City Univ, London, 65-67; from assoc prof to prof optom, Ind Univ, Bloomington, 67-75, chmn physiol optics prog & dir low vision clin, 70-75; DEAN FAC & PROF OPTOM, SOUTHERN COL OPTOM, MEMPHIS, TENN, 75- Honors & Awards: Obrig Labs Mem Award, 69. Mem: Am Acad Optom (vpres, Brit chap, 66); Brit Soc Hist Sci; Royal Micros Soc; Optom Hist Soc (vpres, 69); Am Asn Hist Med. Res: Pathological processes concerning vision; history of visual science. Mailing Add: Southern Col Optom 1245 Madison Ave Memphis TN 38104

LEVENE, LEON, organic chemistry, physical organic chemistry, see 12th edition

LEVENE, MARTIN BARRACK, b New York, NY, Mar 5, 22; m 46; c 3. RADIOTHERAPY. Educ: Mass Inst Technol, BS, 42; Boston Univ, MD, 50; Am Bd Radiol, 54. Prof Exp: Intern, Bellevue Hosp, 50-51; resident radiol, Mass Gen Hosp, Boston, 51-54, clin asst, 54-56; clin asst prof, Sch Med, Boston Univ, 56-68; instr, Sch Med, Tufts Univ, 54-68; asst clin prof radiol, 68-70, ASSOC PROF RADIATION THER, HARVARD UNIV, 71-, DEP DIR, JOINT CTR RADIATION THER, 68- Concurrent Pos: Resident phys med div, Oak Ridge Inst Nuclear Studies, 52-53; assoc vis radiologist, Beth Israel Hosp, Boston, 54-64, radiation therapist & head dept radiother, 64-; dir radiother, Lemuel Shattuck Hosp, Jamaica Plain, 55-68; consult, Sturdy Mem Hosp, Attleboro, 60; assoc staff radiol, Peter Bent Brigham Hosp, 63-; consult, USPHS Hosp, Brighton, Mass, 63-68 & Boston Vet Admin Hosp, 63-68; lectr, Harvard Univ, 65-66; mem cancer clin invest rev comt, Nat Cancer Inst, 68-71. Mem: Fel Am Col Radiol; Radiol Soc NAm; Am Radium Soc; Am Soc Therapeut Radiol. Res: Radiation therapy. Mailing Add: Joint Ctr for Radiation Ther Binney St Boston MA 02115

LEVENE, RALPH ZALMAN, b Winnipeg, Man , May 17, 27; nat US; m 54; c 1. OPHTHALMOLOGY. Educ: Univ Man, MD, 49; NY Univ, DSc(ophthal), 57; Am Bd Ophthal, dipl, 55. Prof Exp: Intern, Winnipeg Gen Hosp, Can, 49-50, resident ophthal, 51-55; from instr to assoc prof, Med Sch, NY Univ, 55-73; PROF & CHMN DEPT OPHTHAL, UNIV ALA, BIRMINGHAM, 73- Mem: AMA; Asn Res Vision & Ophthal; Am Acad Ophthal & Otolaryngol; NY Acad Med. Res: Clinical and basic science aspects of glaucoma. Mailing Add: Dept of Ophthal Univ of Ala Birmingham AL 35233

LEVENGOOD, CLAUDE ANDERSON, zoology, deceased

LEVENGOOD, WILLIAM CAMBURN, b Jackson, Mich, Mar 13, 25; m 43; c 1. BIOPHYSICS. Educ: Univ Toledo, BS, 57; Ball State Univ, MA, 61; Univ Mich, Ann Arbor, MS, 70. Prof Exp: Res physicist, Libbey Owens Ford Glass Co, Ohio, 44-57; physicist, Res Dept, Ball Res Corp, Ind, 57-61; res assoc solid state physics, Inst Sci & Technol, Univ Mich, Ann Arbor, 61-64, assoc res physicist, 64-69, res physicist, Sch Natural Resources, 69-71; DIR BIOPHYS RES, SENSORS, INC, 71- Mem: AAAS. Res: Biophysical experimentation involving induced genetically transferred alterations in embryogenesis; influence of subtle exogenous environmental factors on living organisms; defect structures in noncrystalline solids. Mailing Add: Biophys Res Sensors Inc 3908 Varsity Dr Ann Arbor MI 48104

LEVENSON, ALAN IRA, b Boston, Mass, July 25, 35; m 60; c 2. PSYCHIATRY. Educ: Harvard Univ, AB, 57, MD, 61, MPH, 65; Am Bd Psychiat & Neurol, Dipl, 67. Prof Exp: Intern, Univ Hosp, Ann Arbor, 61-62; resident in psychiat, Mass Ment Health Ctr, Boston, 62-65; staff psychiatrist, NIMH, 65-66, dir servs div, 67-69; PROF PSYCHIAT & HEAD DEPT, COL MED, UNIV ARIZ, 69- Concurrent Pos: Consult, Vet Admin, 69-; pres, Palo Verde Found Ment Health, 71- Mem: Fel Am Pub Health Asn; fel Am Psychiat Asn; fel Am Col Psychiat; Group Advan Psychiat. Res: Organization and delivery of mental health services. Mailing Add: Dept of Psychiatry Ariz Med Ctr Univ of Ariz Tucson AZ 85724

LEVENSON, GORDON EDWARD, b Philadelphia, Pa, Apr 2, 36. DENTAL RESEARCH. Educ: Univ Pa, BA, 57, PhD(embryol), 60, DDS, 64. Prof Exp: Asst instr anat-embryol, Univ Pa, 57-59, instr histol-embryol, 61-64, assoc, 64-66; asst prof, 68-72, ASSOC PROF HISTOL-EMBRYOL, SCH DENT MED, UNIV PA, 72- Concurrent Pos: Nat Cancer Inst spec fel, Strangeways Res Lab, Cambridge, Eng, 66-68. Mem: AAAS; Am Soc Zool; Am Dent Asn; Int Asn Dent Res. Res: Early embryonic vascular development; cell reaggregation and histogenesis in disaggregated cell systems; tissue and organ culture; role of ascorbic acid in chondrogenesis in vitro; histogenesis of dental tissues. Mailing Add: Univ of Pa Sch of Dent Med Philadelphia PA 19174

LEVENSON, HAROLD SAMUEL, b Allentown, Pa, July 12, 16; m 49; c 3. FOOD SCIENCE. Educ: Lehigh Univ, BSChE, 37, MS, 39, PhD(chem physics), 41. Prof Exp: Asst chem, Lehigh Univ, 37-41; res chemist, NJ, 41-46, chief chemist, Maxwell House Div, Calif, 46-51, res mgr, NJ, 51-64, DIR COFFEE RES, TECH CTR, GEN FOODS CORP, 65- Mem: AAAS; Am Chem Soc; Inst Food Technol. Res: Antioxidants; food spoilage; kinetics of saponification; hydrocaffeic acid and esters as antioxidant for edible materials; coffee technology. Mailing Add: Tech Ctr Gen Foods Corp White Plains NY 10625

LEVENSON, LEONARD L, b San Francisco, Calif, Sept 18, 28; m 57; c 3. PHYSICS. Educ: Univ Calif, Berkeley, AB, 52, MS, 55; Univ Paris, PhD(physics), 68. Prof Exp: Physicist, US Naval Ord Test Sta, 52; res engr, Univ Calif, Berkeley, 52-58, physicist, Lawrence Radiation Lab, 58-62; physicist, Nuclear Res Ctr, Saclay, France, 62-68; asst prof, 68-71, ASSOC PROF PHYSICS, UNIV MO-ROLLA, 71-, DIR GRAD CTR MAT RES, 75- Mem: AAAS; Am Phys Soc. Res: Condensation and evaporation of gases at solid surfaces; epitaxial growth of semiconductor and insulator solid surfaces. Mailing Add: Grad Ctr Mat Res Univ of Mo Rolla MO 65401

LEVENSON, MARC DAVID, b Philadelphia, Pa, May 28, 45; m 71. LASERS, QUANTUM ELECTRONICS. Educ: Mass Inst Technol, BS, 67; Stanford Univ, MS, 68, PhD(physics), 72. Prof Exp: Res fel non-linear optics, Gordon McKay Lab, Harvard Univ, 71-74; ASST PROF PHYSICS, UNIV SOUTHERN CALIF, 74- Concurrent Pos: Alfred P Sloan fel, 75-77. Honors & Awards: Adolph Lomb Award, Optical Soc Am, 76. Mem: Am Phys Soc; Inst Elec & Electronics Engrs; Optical Soc Am. Res: Development and application of new techniques of laser spectroscopy to problems in atomic, molecular and condensed matter physics. Mailing Add: Seaver Sci Ctr Univ of Southern Calif Los Angeles CA 90007

LEVENSON, MORRIS E, b New York, NY, Nov 13, 14; m 43. MATHEMATICS. Educ: NY Univ, PhD(math), 48. Prof Exp: Asst, Duke Univ, 37-38; instr math, NY Univ, 43-44; mathematician, David Taylor Model Basin, 44-46; instr math, Cooper Union, 46-49; from instr to assoc prof, 49-71, PROF MATH, BROOKLYN COL, 71- Mem: Am Math Soc; Math Asn Am. Res: Nonlinear vibrations. Mailing Add: 160 W 73rd St New York NY 10023

LEVENSON, STANLEY MELVIN, b Dorchester, Mass, May 25, 16; m 42; c 2. SURGERY. Educ: Harvard Univ, AB, 37, MD, 41; Am Bd Nutrit, dipl, 52; Am Bd Surg, dipl, 57. Prof Exp: Surg house officer, Beth Israel Hosp, Boston, Mass, 41-42; res assoc physiol, Sch Pub Health, Harvard Univ, 41-43; resident burn serv & res assoc surg, Boston City Hosp, 42-43; surg scientist, Med Nutrit Lab, Univ Chicago, 47-49; from asst resident to sr asst resident surg, Med Col Va, 50-52; chief dept surg metab & physiol, Walter Reed Army Inst Res, 56-61, from assoc dir to dir dept germfree res, 56-61, dir div basic surg res, 61; PROF SURG, ALBERT EINSTEIN COL MED, 61-, VIS SURGEON, COL HOSP, 66-, DEP DIR RES SURG, COL MED, 67- Concurrent Pos: Res fel med, Thorndike Mem Lab, Harvard Univ, 44-47; NIH res career award, 62-; chmn subcomt burns & radiation injury, Food & Nutrit Bd, Nat Res Coun, 49-50, comt on trauma, 56-; dir surg metab lab & clin assoc prof, Georgetown Univ, 59-61; vis surgeon, Bronx Munic Hosp Ctr, 61-; consult, Walter Reed Army Inst Res, 61-63; Am Surg Asn rep, Nat Res Coun-Nat Acad Sci, 71-75. Honors & Awards: Achievement Award, Army Res & Develop Bd, 61; Harvey Allen Distinguished Serv Medal, Am Burn Asn, 76. Mem: AAAS; AMA; Am Inst Nutrit; Am Soc Clin Nutrit; Am Col Surgeons. Res: Metabolic and clinical response to trauma; wound healing; germfree life. Mailing Add: Dept of Surg Albert Einstein Col of Med Bronx NY 10461

LEVENSTEIN, IRVING, b Fair Lawn, NJ, Aug 14, 12; m 37; c 2. ENDOCRINOLOGY, TOXICOLOGY. Educ: NY Univ, BA, 34, MSc, 36, PhD, 38. Prof Exp: Instr biol sci, NY Univ, 36-40; PRES & DIR, LEBERCO LABS, 42- Concurrent Pos: Res fel, Nat Comt Maternal Health, 38-42. Mem: Am Soc Zoologists; Am Asn Anatomists; Am Pharmaceut Asn; Soc Toxicol; Am Chem Soc. Res: Hormones; histology; anatomy. Mailing Add: Leberco Labs 123 Hawthorne St Roselle Park NJ 07204

LEVENTHAL, BRIGID GRAY, b London, Eng, Aug 31, 35; US citizen; m 62; c 4. PEDIATRICS, ONCOLOGY. Educ: Univ Calif, Los Angeles, BA, 55; Harvard Univ, MD, 60. Prof Exp: Sr investr leukemia serv, Med Br, 65-73, HEAD CHEMOIMMUNOTHER SECT, PEDIAT ONCOL BR, NAT CANCER INST, 73- Concurrent Pos: Fel pediat, Harvard Univ, 60-62; fel, Boston Univ, 62-63; fel med, Tufts Univ, 63-64; fel hemat, Nat Cancer Inst, 64. Mem: Am Fedn Clin Res; Am Soc Hemat; Am Asn Cancer Res; Am Soc Clin Oncol; Am Soc Clin Invest. Res: Pediatric hematology and oncology. Mailing Add: Pediat Oncol Br Nat Cancer Inst Bethesda MD 20014

LEVENTHAL, CARL M, b New York, NY, July 28, 33; m 62; c 4. NEUROLOGY, NEUROPATHOLOGY. Educ: Harvard Univ, AB, 54; Univ Rochester, MD, 59. Prof Exp: Intern med, Johns Hopkins Hosp, 59-60, asst res physician, 60-61; asst resident neurol, Mass Gen Hosp, 61-62, resident, 63-64; assoc neuropathologist, Nat Inst Neurol Dis & Blindness, 64-66, neurologist, Nat Cancer Inst, 66-68, asst to dep dir sci, NIH, 68-74, actg dep dir sci, 73-74; DEP DIR, BUR DRUGS, FOOD & DRUG ADMIN, 74- Concurrent Pos: Fel, Johns Hopkins Univ, 59-61; fel, Harvard Univ, 61-64, clin & res fel neuropath, 62-63; instr, Georgetown Univ, 64-66, asst prof, 67- Mem: Am Acad Neurol; Am Asn Neuropath. Res: Government science; research and medical administration; clinical neuropathology; brain tumors. Mailing Add: Bur of Drugs Food & Drug Admin Rockville MD 20852

LEVENTHAL, EDWIN ALFRED, b Brooklyn, NY, Jan 26, 34; m 56; c 2. SOLID STATE PHYSICS. Educ: Cornell Univ, BEng Phys, 56; Polytech Inst Brooklyn, MS, 59; NY Univ, PhD(physics), 63. Prof Exp: Sr physicist, Philips Labs Div, NAm Philips Co, 61-70; ed & publ, Med Instrument Reports, 70-74; DIR SYSTS PLANNING, GORDON A FRIESEN INT, 74- Mem: Am Phys Soc; Asn Advan Med Instrumentation; NY Acad Sci. Res: Materials handling and information processing in hospital management and design. Mailing Add: 12710 Saddlebrook Dr Silver Spring MD 20906

LEVENTHAL, JACOB J, b Brooklyn, NY, Dec 18, 37; m 62. ATOMIC PHYSICS, MOLECULAR PHYSICS. Educ: Wash Univ, BS, 60; Univ Fla, PhD(physics), 65. Prof Exp: Res assoc physics & chem, Brookhaven Nat Lab, 65-67, assoc chemist, 67-68; asst prof, 68-71, ASSOC PROF PHYSICS, UNIV MO-ST LOUIS, 71- Mem: Am Phys Soc. Res: Interactions of positive ions with neutral molecules; spectroscopic observations of excited state production in low energy atomic and molecular collision processes. Mailing Add: Dept of Physics Univ of Mo St Louis MO 63121

LEVENTHAL, LEON, b New York, NY, Jan 25, 22; m 52; c 4. RADIOCHEMISTRY, CHEMICAL ENGINEERING. Educ: Univ Calif, BS, 42; Va Polytech Inst, BS, 44; Univ Calif, Los Angeles, MS, 48. Prof Exp: Control chemist, Richfield Oil Corp, Calif, 42-43; res chemist, Metall Lab, Chicago, 44-45; jr chem engr, Oak Ridge Nat Lab, 45; chem engr, Atomic Bomb Lab, Los Alamos Sci Lab, 45-46; res chemist, Radiol Defense Lab, San Francisco Naval Shipyard, 47-49; res radiochemist, Tracerlab, Inc, 49-50, sr chemist, 50-57, dept head, 57-59, div mgr tech serv, 59-67, GEN MGR, LFE ENVIRON ANALYSIS LABS, DIV LFE CORP, 67-, VPRES, 75- Mem: Am Chem Soc; Am Health Phys Soc; Am Nuclear Soc; Am Mgt Asn; fel Am Inst Chem. Res: Nuclear, plutonium and semimicro chemistry; plutonium metallurgy; complex compounds of zinc with zinc 65; general radiochemistry of radiological defense and radioactive waste problems; fission products; environmental and fallout studies; particle analysis; mass spectrometry of plutonium and uranium; applications of radioisotopes to science and industry; transuranium nuclides in the environment. Mailing Add: LFE Environ Anal Labs 2030 Wright Ave Richmond CA 94804

LEVENTHAL, MARVIN, b New York, NY, Dec 4, 37; m 61. ASTROPHYSICS, ATOMIC PHYSICS. Educ: City Col New York, BS, 58; Brown Univ, PhD(physics), 64. Prof Exp: Res assoc physics, Yale Univ, 63-67, asst prof, 67-68; MEM TECH STAFF, BELL LABS, 68- Mem: Am Phys Soc; Am Astronom Soc. Res: Precision measurements of atomic physics quantities which have bearing on quantum electrodynamics; experimental and theoretical gamma ray astronomy. Mailing Add: Bell Labs Rm 1C-302 Murray Hill NJ 07974

LEVEQUE, PHILLIP EDWIN, b Fresno, Calif, Feb 22, 23; m 48; c 5. PHARMACOLOGY. Educ: Ore State Col, BS, 47, MS, 50; Univ Ore, MS, 52, PhD(pharmacol), 54. Prof Exp: Lab instr chem, Ore State Col, 46-49; instr pharmacol, Med Sch, Univ Ore, 50-54; asst prof pharmacol, Med Col Ga, 55-57; asst prof, Sch Med, Univ PR, 57-58; asst res prof biochem, Univ Pittsburgh, 58-59; asst prof pharmacol, Georgetown Univ, 59-61; asst prof, Univ SDak, 61-62; from asst prof to assoc prof, Col Med, Ohio State Univ, 62-68; vis sr lectr, Makerere Med Sch, Uganda, 68-69; vis sr lectr & actg chmn, Univ Col, Dar es Salaam, Tanzania, 69-70; ASST PROF PHARMACOL & DIR DRUG & POISON LAB, COL OSTEOP MED & SURG, 71- Concurrent Pos: From instr to asst prof physiol, Dent Sch, Univ Ore, 53-55. Mem: Am Soc Pharmacol & Exp Therapeut; Soc Exp Biol & Med. Res: Cardiac pharmacology; cellular physiology; antiseptics; fungicides; forensic toxicology. Mailing Add: Dept of Pharmacol Col of Osteop Med & Surg Des Moines IA 50312

LEVEQUE, THEODORE FRANCOIS, b Lewiston, Maine, June 30, 21; m 47; c 2. ANATOMY. Educ: Univ Denver, BA, 49, MS, 50; Univ Colo, PhD(anat), 54. Prof Exp: Instr histol, Dept Anat, McGill Univ, 54-55; from asst prof to prof anat, Sch

Med, Univ Md, 55-68; PROF ANAT & CHMN DEPT, FAC MED, UNIV SHERBROOKE, 68- Concurrent Pos: Secy fac med, Univ Sherbrooke, 71- Mem: AAAS; Am Asn Anat. Res: Endocrinology; neuroendocrinology; neurosecretion and connective tissue; wound healing. Mailing Add: Dept of Anat Fac of Med Univ of Sherbrooke Sherbrooke PQ Can

LEVEQUE, WILLIAM JUDSON, b Boulder, Colo, Aug 9, 23; m 49; c 1. MATHEMATICS. Educ: Univ Colo, BA, 44; Cornell Univ, MA, 45, PhD(math), 47. Prof Exp: Benjamin Pierce instr math, Harvard Univ, 47-49; from instr to prof, Univ Mich, Ann Arbor, 49-70, chmn dept, 67-70; PROF MATH, CLAREMONT GRAD SCH, 70- Concurrent Pos: Fulbright res scholar, 51-52; Sloan res fel, 57-60; exec ed, Math Rev, 65-66; chmn, Conf Bd Math Sci, 73-74. Mem: Am Math Soc; Math Asn Am. Res: Theory of numbers. Mailing Add: Dept of Math Claremont Grad Sch Claremont CA 91711

LEVER, ALFRED B P, b London, Eng, Feb 21, 36; m 63; c 3. INORGANIC CHEMISTRY. Educ: Univ London, BSc & ARCS, 57, dipl, Imp Col & PhD(chem), 60. Prof Exp: Hon res asst, Univ Col, London, 60-61, hon res assoc, 61-62; lectr chem, Inst Sci & Tech, Univ Manchester, 62-66; vis lectr, Ohio State Univ, 67; assoc prof, 67-72, PROF CHEM, YORK UNIV, 72- Concurrent Pos: Ed, Coord Chem Rev, 66-; prog chmn, XIVth Int Conf Coord Chem, Toronto, 72. Mem: Am Chem Soc; Chem Inst Can; The Chem Soc; AAAS; Can Inst Mining. Mailing Add: Dept of Chem York Univ Downsview ON Can

LEVER, CYRIL, JR, b Abington, Pa, June 5, 29; m 61; c 2. ORGANIC CHEMISTRY. Educ: Pa Mil Col, BS, 53. Prof Exp: Asst treas & asst dir res, 53-57, PRES & DIR RES, C LEVER CO, INC, 57- Mem: Soc Am Mil Eng. Res: Dyes and colors for paper. Mailing Add: C Lever Co Inc The Lever Bldg 736 Dunks Ferry Rd Cornwells Heights PA 19020

LEVER, REGINALD FRANK, b Birmingham, Eng, July 5, 30; m 57; c 1. MATERIALS SCIENCE. Educ: Oxford Univ, BA, 51, MA, 54. Prof Exp: Sci officer, Serv Electronics Res Lab, Baldock, Eng, 51-57; sr sci officer, UK Atomic Energy Agency Indust Group, Lancashire, 57-58; staff mem, Res Div, Philco Corp, 58-60; staff mem, Thomas J Watson Res Ctr, 60-70, STAFF MEM COMPONENTS DIV, IBM CORP, 70- Mem: Am Phys Soc. Res: Growth of crystals from vapor by chemical deposition; gaseous diffusion; semiconductors; surfaces; silicon device processing; material analysis by MeV ion backscattering. Mailing Add: IBM SPD Dept 171 BLDG 300-95 Route 52 Hopewell Junction NY 12533

LEVER, WALTER FREDERICK, b Erfurt, Ger, Dec 13, 09; US citizen; m 40; c 2. DERMATOLOGY. Educ: Univ Leipzig, MD, 34. Prof Exp: Asst dermat, Harvard Med Sch, 44-49, instr, 49-51, clin assoc, 51-55, asst clin prof, 55-59; clin prof, 59-61, PROF DERMAT & CHMN DEPT, SCH MED, TUFTS UNIV, 61-; CHMN DERMAT, NEW ENG MED CTR, 59- Concurrent Pos: Lectr, Harvard Med Sch, 59-; mem bd consult, Mass Gen Hosp, Boston, 59-; mem gen med study sect, NIH, 59-63; dir dermat, Boston City Hosp, 61-74. Mem: Am Acad Dermat; Am Dermat Asn; Soc Invest Dermat. Res: Histopathology of the skin; electron microscopy of bullous diseases and of appendage tumors of the skin; clinical, biochemical and immunological manifestations and treatment of pemphigus and pemphigoid. Mailing Add: Dept of Dermat Tufts Univ Sch of Med Boston MA 02111

LEVER, WILLIAM EDWIN, b Skewen, Wales, Dec 21, 35; US citizen; m 64; c 2. STATISTICS. Educ: Col Steubenville, BA, 58; Fla State Univ, MS, 63, PhD(statist), 68. Prof Exp: STATISTICIAN, NUCLEAR DIV, UNION CARBIDE CORP, 66- Mem: Am Statist Asn; Biometric Soc. Res: Risk analysis associated with the fuel cycle of nuclear and coal-fired power plants. Mailing Add: Math & Statist Res Dept Comput Sci Div Nuclear Div Union Carbide Corp PO Box Y Oak Ridge TN 37830

LEVERE, RICHARD DAVID, b Brooklyn, NY, Dec 13, 31; m 56; c 3. INTERNAL MEDICINE, HEMATOLOGY. Educ: State Univ NY, MD, 56. Prof Exp: From intern to asst resident med, Bellevue Hosp, 56-58; resident, Kings County Hosp, 60-61; instr med, State Univ NY, 62-63; res assoc biochem, Rockefeller Inst, 62-63, asst prof, 64-65; from asst prof to assoc prof, 65-73, PROF MED, STATE UNIV NY DOWNSTATE MED CTR, 73-, CHIEF HEMAT SECT, 70- Concurrent Pos: Fel hemat, State Univ NY, 61-62; NIH grant, 65-75; adj prof, Rockefeller Univ, 73- Mem: AAAS; Am Soc Clin Invest; Am Fedn Clin Res; Am Soc Hemat; Am Col Physicians. Res: Control mechanisms in heme and porphyrin synthesis; metabolism of normal and abnormal hemoglobins; diseases of porphyrin metabolism. Mailing Add: Dept of Med State Univ NY Downstate Med Ctr Brooklyn NY 11203

LEVERENZ, HUMBOLDT WALTER, b Chicago, Ill, July 11, 09; m 40; c 4. SOLID STATE SCIENCE. Educ: Stanford Univ, AB, 30. Prof Exp: Res chemico-physicist, Radio Corp Am, 31-54, dir phys & chem lab, 54-57, asst dir res, 57-59, dir, 59-61, assoc dir, RCA Labs, 61-66, staff vpres, Res & Bus Eval, 66-68, staff vpres & chmn educ aid comt, RCA Corp, 68-74; RETIRED. Concurrent Pos: With advan mgt prog, Bus Sch, Harvard Univ, 58; mem, Mat Adv Bd, Nat Acad Sci, 64-68; mem conf comt, Nat Conf Admin Res, 64-68. Honors & Awards: Brown Medal, Franklin Inst, 54. Mem: Nat Acad Eng; fel AAAS; Am Chem Soc; fel Am Phys Soc; fel Inst Elec & Electronics Eng. Res: Syntheses and applications of solids used in electronics; phosphors; secondary-emitters; photoconductors; semiconductors; nonmetallic magnetic materials; scotophors; crystals used in electronics. Mailing Add: 35 Westcott Road Princeton NJ 08540

LEVERETT, SIDNEY DUNCAN, JR, b Houston, Tex, Nov 27, 25; m 48; c 2. PHYSIOLOGY. Educ: Agr & Mech Col Tex, BS, 49; Ohio State Univ, MS, 55, PhD, 60. Prof Exp: Chief acceleration sect, Aeromed Lab, Wright Air Develop Ctr, 55-58; aviation physiologist, 60-63, CHIEF BIODYN BR, US AIR FORCE SCH AEROSPACE MED, 63- Concurrent Pos: Consult, Manned Space Ctr, NASA, Houston, Tex, 65- Honors & Awards: Achievement Award, US Air Force Sch Aerospace Med, 62; Eric Liljencrantz Award, 70; Meritorious Civilian Serv Award, Dept of the Air Force, 72; Award for Excellence, Life Sci & Biomed Eng Br, 75. Mem: Fel Aerospace Med Asn; Int Acad Aviation & Space Med. Res: Cardiovascular physiology, particularly hemodynamics; aviation medicine, particularly acceleration stress. Mailing Add: Sch of Aerospace Med Brooks AFB TX 78235

LEVERING, DALE FRANKLIN, JR, b Millstone, WVa, Sept 23, 43; m 63; c 2. BOTANY, PLANT ECOLOGY. Educ: Glenville State Col, BA, 65; Ohio State Univ, MSc, 67, PhD(bot), 72. Prof Exp: ASST PROF BIOL, NORTHEASTERN UNIV, 72- Mem: Sigma Xi. Res: Investigation of vegetational community dynamics of Boston Harbor Islands, physiological ecology of Polygonatum species and air pollutants effects upon crop plants. Mailing Add: Dept of Biol 403 Richards Hall Northeastern Univ Boston MA 02115

LEVERING, DEWEY ROBERT, organic chemistry, see 12th edition

LEVERTON, RUTH MANDEVILLE, b Minneapolis, Minn, Mar 23, 08. NUTRITION, FOOD. Educ: Univ Nebr, BS, 28; Univ Ariz, MS, 32; Univ Chicago, PhD(nutrit), 37. Hon Degrees: DSc, Univ Nebr, 61. Prof Exp: Teacher high sch, Nebr, 28-30; asst home econ, Exp Sta, Univ Ariz, 32-34; asst prof, Univ Nebr, 37-40; assoc specialist, Bur Home Econ, USDA, 40-41; assoc prof home econ, Univ Nebr, 41-49, prof, 49-54; prof home econ & asst dir agr exp sta, Okla Agr & Mech Col, 54-57; asst dir human nutrit res div, USDA, 57-58, assoc dir inst home econ, Agr Res Serv, 58-61, asst dept adminr, 61-71, sci adv, 71-74; RETIRED. Concurrent Pos: Fulbright prof, Univ Philippines, 49-50. Honors & Awards: Borden Award, 53. Mem: Am Dietetic Asn; Am Home Econ Asn; Am Pub Health Asn; Am Inst Nutrit. Res: Human metabolism and requirements of minerals; nutritive value of Nebraska food products; blood regeneration and prevention of anemia; nutritional status; iron metabolism; amino acid requirements of women. Mailing Add: Apt 240 3900 16th St NW Washington DC 20011

LEVESQUE, CHARLES LOUIS, b Manchester, NH, Feb 16, 13; m 38; c 3. ORGANIC CHEMISTRY. Educ: Dartmouth Col, AB, 34, AM, 36; Univ Ill, PhD(org chem), 39. Prof Exp: Instr anal chem, Dartnouth Col, 34-36; sr chemist, Resinous Prod & Chem Co, 39-41, group leader, 41-45, lab head, 45-48; res supvr, Rohm & Haas Co, 48-69, asst dir res, 69-71; PROF APPL SCI & DIR EVE SCH, URSINUS COL, 71- Mem: Am Chem Soc. Res: Structures of vinyl polymers; polyester resins and raw materials; new organic synthesis; surface active agents; pharmaceuticals. Mailing Add: 965 Dale Rd Meadowbrook PA 19046

LEVESQUE, RENE J A, b St-Alexis, Que, Oct 30, 26; m 56; c 3. NUCLEAR PHYSICS. Educ: Sir George Williams Col, BSc, 52; Northwestern Univ, PhD(physics), 57. Prof Exp: Res assoc physics, Univ Md, 57-59; from asst prof to assoc prof, 59-67, dir lab nuclear physics, 65-67, 68-69, dir dept, 68-73, vdean res fac arts & sci, 73-75, DEAN FAC ARTS & SCI, 75-, PROF PHYSICS, UNIV MONTREAL, 67- Concurrent Pos: Asst ed, Can J Physics, 73-75. Mem: Am Phys Soc; Can Asn Physicists (vpres elect, 74-75, vpres, 75-76). Res: Nuclear spectroscopy; nuclear reactions at low energy. Mailing Add: Dept of Physics Univ of Montreal PO Box 6128 Montreal PQ Can

LEVETIN-AVERY, ESTELLE, b Boston, Mass, Mar 24, 45; m 74. MYCOLOGY, BOTANY. Educ: State Col Boston, BS, 66; Univ RI, PhD(bot & mycol), 71. Prof Exp: Lab instr & teaching asst bot, Univ RI, 69-71, asst prof, Exten Div, 71-72, fel res assoc, Dept Plant Path, 71-72; asst prof physiol, Mt St Joseph Col, 72; ASST PROF BOT, UNIV TULSA, 72- Concurrent Pos: Res grant, Univ Tulsa, 74; consult, Joint Res Prog, Allergy Clin Tulsa, Inc, 75-76. Mem: Mycol Soc Am; Bot Soc Am; Am Inst Biol Sci; AAAS; Brit Mycol Soc. Res: Physiology and development of fungi; the distribution of fleshy fungi in Oklahoma; the distribution of air-borne fungi in Tulsa county. Mailing Add: Fac of Nat Sci Univ of Tulsa 600 S College Tulsa OK 74104

LEVEY, GERALD SAUL, b Jersey City, NJ, Jan 9, 37; m 61; c 2. INTERNAL MEDICINE, ENDOCRINOLOGY. Educ: Cornell Univ, AB, 57; NJ Col Med, MD, 61. Prof Exp: Intern med, Jersey City Med Ctr, 61-62, resident, 62-63; resident, Mass Gen Hosp, Boston, 65-66; clin assoc endocrinol, Nat Inst Arthritis & Metab Dis, 66-68; sr investr endocrinol, Nat Heart & Lung Inst, 69-70; assoc prof, 70-73, PROF MED, SCH MED, UNIV MIAMI, 73- Concurrent Pos: NIH fel biochem, Med Sch, Harvard Univ, 63-65; consult med, Vet Admin Hosp, Miami, Fla, 70-; investr, Howard Hughes Med Inst, 71- Mem: Am Soc Clin Invest; Am Col Physicians; Am Thyroid Asn; Am Fedn Clin Res; Soc Exp Biol & Med. Res: Mechanism of hormone action; cyclic adenosine monophosphate. Mailing Add: Dept of Med Sch of Med Univ of Miami PO Box 520875 Miami FL 33152

LEVEY, GERRIT, b Friesland, Wis, Jan 9, 24; m 52; c 3. PHYSICAL CHEMISTRY. Educ: Hope Col, BA, 46; Univ Wis, PhD(chem), 49. Prof Exp: Assoc prof, 49-58, PROF CHEM, BEREA COL, 58- Concurrent Pos: NSF sci fac fel, Mass Inst Technol, 57-58 & Univ Leeds, 65-66; fac res partic, Argonne Nat Lab, 73-74. Mem: Am Chem Soc; The Chem Soc. Res: Chemical reactions resulting from nuclear energy activation; use of radiotracers in reaction kinetics studies; radiation chemistry of inorganic peroxides. Mailing Add: Dept of Chem Berea Col Berea KY 40403

LEVEY, HAROLD ABRAM, b Boston, Mass, Aug 14, 24; m 59; c 2. ENDOCRINE PHYSIOLOGY. Educ: Harvard Univ, AB, 47; Univ Calif, Los Angeles, PhD(zool), 53. Prof Exp: Jr & asst res physiol chemist, Univ Calif, Los Angeles, 53-56; from instr to asst prof, 56-64, ASSOC PROF PHYSIOL, COL MED, STATE UNIV NY DOWNSTATE MED CTR, 64- Concurrent Pos: USPHS fel, 53-; China Med Bd vis prof physiol, Fac Med, Univ Singapore, 66-67. Mem: AAAS; Am Physiol Soc; Endocrine Soc; NY Acad Sci; Harvey Soc. Res: Pituitary chemistry and physiology; pituitary-thyroid interrelationships; factors influencing metabolism of endocrine organs; electrophysiology of thyroid. Mailing Add: Dept of Physiol State Univ NY Downstate Med Ctr Brooklyn NY 11203

LEVEY, SAMUEL, b Cape Town, SAfrica, July 11, 32; US citizen; m 65; c 3. HOSPITAL ADMINISTRATION, HEALTH ADMINISTRATION. Educ: Bowdoin Col, AB, 55; Columbia Univ, AM, 56; Univ Iowa, MA, 59, PhD(hosp & health admin), 61; Harvard Univ, SM, 63. Prof Exp: Admin assoc, Univ Iowa Hosps, 58-60; instr hosp & health admin, Univ Iowa Grad Prog, 60-61, asst prof, 61-62; dir div nursing homes & related facil, Mass Dept Pub Health, 63-67; asst dir med care planning, Harvard Med Sch, 67-68; asst comnr med div, Mass Dept Pub Welfare, 68-69; PROF & CHMN HEALTH CARE ADMIN, BARUCH COL, CITY UNIV NEW YORK, 69- Concurrent Pos: USPHS grant, Harvard Univ, 62-63; assoc prof admin med, Mt Sinai Sch Med, City Univ New York, 69-73, prof, 73-; prof consult to hosp & health care orgn, 73- Mem: Am Pub Health Asn; Am Hosp Asn; Soc Gen Systs Res; Am Pub Welfare Asn. Res: Health systems management; long term care administration. Mailing Add: Baruch Col City Univ of New York 17 Lexington Ave New York NY 10010

LEVI, BARBARA GOSS, b Washington, DC, May 5, 43; m 66; c 2. HIGH ENERGY PHYSICS. Educ: Carleton Col, BA, 65; Stanford Univ, MS, 67, PhD(physics), 71. Prof Exp: Asst ed, 69-70, CONTRIB ED, PHYSICS TODAY MAGAZINE, AM INST PHYSICS, 71- Concurrent Pos: Lectr physics, Fairleigh Dickinson Univ, 70-; mem task force energy, Am Asn Univ Women, 75-77. Mem: Am Asn Physics Teachers. Mailing Add: 16 Weather Vane Dr Convent Station NJ 07961

LEVI, DAVID WINTERTON, b Berryville, Va, Sept 2, 21; m 47; c 2. POLYMER CHEMISTRY. Educ: Randolph-Macon Col, BS, 43; Va Polytech Inst, MS, 51, PhD(chem), 54. Prof Exp: From instr to assoc prof chem, Va Polytech Inst, 46-59; SUPVRY CHEMIST, PICATINNY ARSENAL, DOVER, 59- Mem: Am Chem Soc. Res: Solution properties of high polymers; polymer-energetic compatibility; adhesives; thermal degradation of polymers. Mailing Add: 2 Oak Hill Dr Succasunna NJ 07876

LEVI, ENRICO, b Milano, Italy, May 20, 18; US citizen; m 41. ENERGY CONVERSION, PLASMA PHYSICS. Educ: Israel Inst Technol, BSc, 41, Ing, 42; Polytech Inst Brooklyn, MEE, 56, DEE, 58. Prof Exp: Foreman elec shop, Shipwrights & Engrs Ltd, Israel, 42-44; mech engr, Palestine Elec Co, 44-45; sect head elec eng, Mouchly Eng Co, 45-48; lectr, Israel Inst Technol, 48-55; fel, Microwave Res Inst, Polytech Inst Brooklyn, 56-57; sr scientist, Elec & Electronic

Res Found, Westbury, 57-58; assoc prof, 58-64, PROF ELECTROPHYS, POLYTECH INST NY, 64- Concurrent Pos: Consult, Lever Bros, Israel, 48-55, Hudson Paper Mill Co, 54-55, Am Mach & Foundry Co, 60-62, Westinghouse Elec Astronuclear Labs, Pa, 62-64, Gen Appl Sci Labs, 65, Van Karman Inst Fluid Dynamics, 66, Consol Edison Co, New York, 72 & Long Island Lighting Co, 73-74; mem, Israel Govt Comt, 50-51 & Elec Wire Standard Comt, Israel, 50-55. Mem: Inst Elec & Electronics Eng. Res: Electromechanical power conversions; magnetic amplifiers; automatic control; linear electric propulsion; variable speed drives, electric power. Mailing Add: Apt 620 110-20 71st Rd Forest Hill NY 11375

LEVI, HERBERT WALTER, b Frankfurt am Main, Ger, Jan 3, 21; nat US; m 49; c 1. ARACHNOLOGY, SYSTEMATICS. Educ: Univ Conn, BS, 46; Univ Wis, MS, 47, PhD(zool), 49. Hon Degrees: AM, Harvard Univ, 70. Prof Exp: From instr to assoc prof bot & zool, Exten Div, Univ Wis, 49-56; from asst cur to assoc cur, Mus, 55-66, mem fac educ, Univ, 64-66, lectr biol, 64-70, PROF BIOL, HARVARD UNIV, 70-, AGASSIZ PROF ZOOL, 72-, CUR ARACHNOL, MUS COMP ZOOL, 66- Concurrent Pos: Secy, Rocky Mountain Biol Lab, 59-65; vpres, Ctr Int Document Arachnol, 65-68; vis prof, Hebrew Univ Jerusalem, 75. Mem: AAAS; Am Arachnol Soc; Am Ecol Soc; Am Inst Biol Sci; Am Micros Soc. Res: Evolution; systematic zoology; spiders and other arachnids; animal transplantation; systematic studies of orb-weaving spiders in the family Araneidae. Mailing Add: Mus of Comp Zool Harvard Univ Cambridge MA 02138

LEVI, HOWARD, b New York, NY, Nov 9, 16; m 35, 63, c 3. MATHEMATICS. Educ: Columbia Univ, AB, 37, PhD(math), 42. Prof Exp: Lectr math, Columbia Univ, 39-41; instr math & physics, US Navy Pre-Flight Sch, Iowa, 41-42; res scientist, Sam Labs, Columbia Univ, 43-46, from instr to prof math, 42-61; consult, Orgn Econ Coop & Develop, Paris, 61-62; prof math, Hunter Col, 62-69; PROF MATH, LEHMAN COL, 69- Concurrent Pos: NSF sr sci fac fel, 60-61; vis prof, Univ Turin, 68-69. Mem: Am Math Soc; Math Asn Am. Res: Geometry; mathematics education. Mailing Add: Dept of Math Herbert H Lehman Col Bronx NY 10468

LEVI, IRVING, b Winnipeg, Man, Dec 15, 14; m 44; c 4. MEDICINAL CHEMISTRY. Educ: Univ Man, BSc, 38, MSc, 39; McGill Univ, PhD(chem), 42. Prof Exp: Carnegie Corp res fel, McGill Univ, 42-43, res assoc, 44-46, lectr, 46-47; sr res chemist, Charles E Frosst & Co, 48-68; PRES, ALMEDIC LTD, 68- Concurrent Pos: Civilian with Can Govt, 40-44. Mem: Am Chem Soc; fel Chem Inst Can. Res: Organic synthesis; carbohydrates; synthetic analgesics and sedatives; antibiotic and cancer chemotherapy; amino acids and derivatives; steroids and hormones; medicinal applications of natural products and derivatives. Mailing Add: Almedic Ltd 8255 Mountain Sights Montreal PQ Can

LEVI, LEO, physics, see 12th edition

LEVI, LEO, pharmaceutical chemistry, see 12th edition

LEVI, MICHAEL PHILLIP, b Leeds, Eng, Feb 5, 41; m 66; c 2. FOREST PRODUCTS. Educ: Univ Leeds, BS, 61, PhD(biophys), 64. Prof Exp: Fulbright travel scholar & res fel wood prod path, Sch Forestry, Yale Univ, 65; res fel, NC State Univ, 65-66; Sci Res Coun-NATO res fel, Univ Leeds, 66-67; sr biologist, Timber Res & Develop Lab, Hickson & Welch, Eng, 67-68, head res wood preservation, 68-71; ASSOC PROF FORESTRY, NC STATE UNIV, 71- Mem: Forest Prod Res Soc; The Chem Soc; Am Phytopath Soc. Res: Wood preservation; mode of action of fungicides; wood deterioration by fungi. Mailing Add: Sch of Forest Resources NC State Univ Raleigh NC 27607

LEVI, RALPH SIGMUND, b Chicago, Ill, June 15, 30; m 60; c 4. PHARMACEUTICAL CHEMISTRY, PHYSICAL PHARMACY. Educ: Univ Ill, BS, 51; Univ Fla, MS, 52, PhD(pharm), 55. Prof Exp: Res pharmacist, Div Indian Health, USPHS, 55-56 & NIH, 56-58; group supvr parenteral prod, 58-60, proj coordr res & develop div, 60-63, MGR PHARM DEVELOP SECT, WYETH LABS, AM HOME PROD CORP, 63- Mem: AAAS; Am Pharmaceut Asn; Acad Pharmaceut Sci. Res: Pharmaceutical dosage form design including chemical and physical properties related to biological availability, stability and preservation, specifically lyophilization, parenteral products, powder flow and compaction. Mailing Add: Pharm Develop Sect Wyeth Labs PO Box 8299 Philadelphia PA 19101

LEVI, ROBERTO, b Milano, Italy, Mar 2, 34; m 62; c 2. PHARMACOLOGY. Educ: Univ Florence, MD, 60. Prof Exp: Asst pharmacol, Univ Florence, 60-61; asst prof, 66-71, ASSOC PROF PHARMACOL, MED COL, CORNELL UNIV, 71- Concurrent Pos: Fulbright travel fel pharmacol & exp therapeut, Sch Med, Johns Hopkins Univ, 61-63; sr res fel electrophysiol, Univ Florence, 63-66; prin investr, USPHS grant, 67-68; co-investr, NIH grant, 67-69; prin investr, NY Heart Asn grant, 68-71, 71-73 & 74-76; Nat Inst Gen Med Sci grant, 74-77; fel, Polachek Found Med Res, 73-76. Honors & Awards: Alberico Benedicenti Prize, 64; J Murray Steele Prize, 70. Mem: Am Soc Pharmacol & Exp Therapeut; Harvey Soc. Res: Cardiovascular pharmacology; heart electrophysiology; neuropharmacology; immunopharmacology. Mailing Add: Dept of Pharmacol Cornell Univ Med Col New York NY 10021

LEVICH, CALMAN, b Iowa City, Iowa, May 26, 21; m 46; c 4. BIOPHYSICS. Educ: Morningside Col, BS, 49; Cath Univ Am, PhD(physics), 66. Prof Exp: Biophysicist, Naval Med Res Inst, 50-61; proj dir, Armed Forces Radiobiol Res Inst, 61-67; assoc prof physics, Cent Mich Univ, 67-68; chmn dept, Seton Hall Univ, 68-70; PROF PHYSICS, CENT MICH UNIV, 70- Mem: AAAS; Biophys Soc; Radiation Res Soc; Am Asn Physics Teachers. Res: Mechanical properties of muscle; radiation biophysics; reactor operator education. Mailing Add: 805 Douglas St Mt Pleasant MI 48858

LEVIÉ, HAROLD WALTER, b Augusta, Ga, Jan 17, 49. SURFACE PHYSICS. Educ: William Marsh Rice Univ, BA, 71, MS, 73, PhD(mat sci), 76. Prof Exp: Physicist, Phys Sci Lab, US Army Missile Command, 71; MAT SCIENTIST SURFACE TECHNOL, INORG MAT DIV, LAWRENCE LIVERMORE LAB, 75- Concurrent Pos: Instr corrosion eng, Nat Asn Corrosion Engrs, 75. Mem: Nat Asn Corrosion Engrs. Res: Analysis and characterization of solid surfaces; kinetics of surface reactions and interface formation. Mailing Add: L-503 Lawrence Livermore Lab PO Box 808 Livermore CA 94550

LEVIEN, ROGER ELI, b Brooklyn, NY, Apr 16, 35; m 60; c 2. SYSTEMS ANALYSIS, INFORMATION SCIENCE. Educ: Swarthmore Col, BS, 56; Harvard Univ, MS, 58, PhD(appl math), 62. Prof Exp: Engr, Rand Corp, 60-67, head syst sci dept, 67-71, mgr, Washington Domestic Progs, 71-74; proj leader, 74-75, DIR, INT INST APPL SYSTS ANAL, AUSTRIA, 75- Concurrent Pos: Adj prof, Univ Calif, Los Angeles, 70- Mem: Asn Comput Mach; Inst Elec & Electronics Engrs; Opers Res Soc Am. Res: Data retrieval; operations research; research and development management. Mailing Add: Int Inst for Appl Systs Res A-2361 Laxenburg Austria

LEVI-MONTALCINI, RITA, b Torino, Italy, Apr 22, 09; nat US. NEUROLOGY. Educ: Univ Turin, MD, 40. Prof Exp: Res assoc, 47-51, assoc prof, 51-58, PROF ZOOL, WASHINGTON UNIV, 58- Mem: AAAS; Soc Develop Biol; Am Soc Zoologists; Am Asn Anatomists; Tissue Cult Asn. Res: Experimental neurology; effect of a nerve growth factor isolated from the mouse salivary gland on the sympathetic nervous system and of an antiserum to the nerve growth factor; study of other specific growth factors. Mailing Add: Dept of Biol Washington Univ St Louis MO 63130

LEVIN, AARON R, b Johannesburg, SAfrica, Mar 19, 29; m 55; c 3. PEDIATRICS, CARDIOLOGY. Educ: Univ Witwatersrand, BSc, 48, MBBCh, 53, MD, 68; Royal Col Physicians & Surgeons, dipl child health, 60; FRCPS(E), 61. Prof Exp: Intern, Edenvale Hosp, SAfrica, 54-55; sr intern, Johannesburg Fever Hosp, 55; pediat intern, Coronation Hosp, 55-56; pediat registr, 56-60; pediat registr, Charing Cross Hosp, Eng, 61; gen pract, 62-63; instr pediat, Med Ctr, Duke Univ, 64-66; from asst prof to assoc prof, 66-74, PROF PEDIAT, MED CTR, CORNELL UNIV, 74- Concurrent Pos: NIH fel cardiol, Med Ctr, Duke Univ, 64-66; attend physician, Pediat Intensive Care Unit, New York Hosp-Cornell Med Ctr. Mem: Assoc fel Am Acad Pediat; Am Heart Asn; Soc Pediat Res. Res: Pediatric cardiology, specifically related to studies of pressure-flow dynamics in various forms of congenital heart disease; extra cardiac factors in congenital heart disease; right ventricular hypertrophy at cellular level. Mailing Add: Pediat Cardiopulmonary Lab New York Hosp-Cornell Med Ctr New York NY 10021

LEVIN, ALFRED P, microbial physiology, biochemical gen etics, see 12th edition

LEVIN, BARBARA CHERNOV, b Providence, RI, May 5, 39; m 61; c 2. MICROBIAL GENETICS, MOLECULAR BIOLOGY. Educ: Brown Univ, AB, 61; Georgetown Univ, PhD(microbial genetics), 73. Prof Exp: Lab teaching asst, Brown Univ, 59-61; res assoc endocrinol, Sch Med, Johns Hopkins Univ, 62-63; NIH fel, 73-75, STAFF FEL MOLECULAR BIOL, NIH, 75- Mem: Sigma Xi; Am Soc Microbiol. Res: Examination of effects of lipophilic acids on membrane functions in Bacillus Subtilis; membrane functions include respiration, adenosine triphosphate synthesis and active transport. Mailing Add: Lab Molecular Biol NINCDS NIH Bethesda MD 20014

LEVIN, BERTRAM, b Chicago, Ill, May 2, 20; m 50; c 3. RADIOLOGY. Educ: Cent YMCA Col, BS, 41; Univ Ill, MD, 44; Am Bd Radiol, dipl. Prof Exp: Asst dir dept radiol, Michael Reese Hosp & Med Ctr, 50-51; clin instr radiol, Univ Minn Hosps, 52-53; Nat Heart Inst trainee, Univ Minn, 53; from asst prof to assoc prof, Chicago Med Sch, 55-69; PROF RADIOL, UNIV CHICAGO, 69-; DIR DEPT DIAG RADIOL, MICHAEL REESE HOSP & MED CTR, 54- Mem: Radiol Soc NAm; Am Col Radiol. Res: Design of in vivo techniques of bone mineral determination. Mailing Add: Dept of Diag Radiol Michael Reese Hosp & Med Ctr Chicago IL 60616

LEVIN, DANIEL H, molecular biology, biochemistry, see 12th edition

LEVIN, EDWIN ROY, b Philadelphia, Pa, Nov 4, 27; m 51; c 3. SOLID STATE SCIENCE. Educ: Temple Univ, AB, 49, MA, 51, PhD(physics), 59. Prof Exp: Asst physics, Temple Univ, 49-51; physicist, Frankford Arsenal, US Army, 51-63; MEM TECH STAFF, RCA LABS, 63- Concurrent Pos: Secy Army res & study fel, Cavendish Lab, Cambridge Univ, 61-62; guide prof, World Univ, 73- Mem: AAAS; Am Phys Soc; Electron Micros Soc Am. Res: Solid state physics; theory of dielectrics; photoconductivity; quantum electronics; analysis of solid materials for electronics, including electron microscopy and related methodologies. Mailing Add: RCA Labs Princeton NJ 08540

LEVIN, ELINOR A, biochemistry, see 12th edition

LEVIN, EUGENE (MANUEL), b New York, NY, Aug 14, 34; m 60; c 3. PHYSICS. Educ: Univ Vt, BA, 56; Columbia Univ, MA, 59; NY Univ, PhD(physics), 67. Prof Exp: ASSOC PROF PHYSICS, YORK COL, NY, 67- Mem: Am Phys Soc; Am Asn Physics Teachers; Sigma Xi. Res: Excited states and fluorescence properties of organic molecules; applications of fluorescence techniques to charged particle dosimetry. Mailing Add: Dept of Physics York Univ Jamaica NY 11432

LEVIN, FRANK, mathematics, see 12th edition

LEVIN, FRANK S, b Bronx, NY, Apr 14, 33; m 55; c 2. NUCLEAR PHYSICS. Educ: Johns Hopkins Univ, AB, 55; Univ Md, PhD(physics), 61. Prof Exp: Res assoc physics, Rice Univ, 61-63 & Brookhaven Nat Lab, 63-65; temporary res assoc, Atomic Energy Res Estab, Eng, 65-67; ASSOC PROF PHYSICS, BROWN UNIV, 67- Concurrent Pos: Lectr, Latin Am Sch Physics, Mex, 68. Res: Nuclear reaction theory; scattering theory; atomic collision theory. Mailing Add: Dept of Physics Brown Univ Providence RI 02912

LEVIN, FRANKLYN KUSSEL, b Terre Haute, Ind, June 28, 22; m 46; c 3. EXPLORATION GEOPHYSICS. Educ: Purdue Univ, BS, 43; Univ Wis, PhD(physics), 49. Prof Exp: Physicist, Sam Labs, Columbia Univ, 43-44, Carbide & Carbon Chem Corp, 44-46; asst physics, Univ Wis, 46-47; physicist, Carter Oil Co, 49-53; asst dir, Hudson Labs, Columbia Univ, 53-54; physicist, Carter Oil Co, 54-58; physicist, Jersey Prod Res Co, Standard Oil Co (NJ), 58-59, res assoc, 59-63, sr res assoc, 63-64, res assoc, Esso Prod Res Co, 65-67, res scientist, 67-73, SR RES SCIENTIST, EXXON PROD RES CO, 73- Concurrent Pos: Lectr, Univ Tulsa, 58-63; ed, Geophys, 69-71. Mem: AAAS; Soc Explor Geophys; Seismol Soc Am; Am Geophys Union; Europ Asn Explor Geophys. Mailing Add: Exxon Prod Res Co PO Box 2189 Houston TX 77001

LEVIN, GERSON, b Philadelphia, Pa, Oct 27, 39; m 69. MATHEMATICS. Educ: Univ Pa, AB, 61; Univ Chicago, MS, 62, PhD(math), 65. Prof Exp: NSF fel, Univ Ore, 66, vis asst prof math, 66-67; asst prof, NY Univ, 67-74; ASST PROF MATH, BROOKLYN COL, 74- Res: Commutative rings and homological algebra. Mailing Add: Dept of Math Brooklyn Col Brooklyn NY 11210

LEVIN, GILBERT VICTOR, b Baltimore, Md, Apr 23, 24; m 53; c 3. ENVIRONMENTAL HEALTH, ENGINEERING. Educ: Johns Hopkins Univ, BE, 47, MS, 48, PhD(sanit eng), 63. Prof Exp: Jr asst sanit engr, State Dept Health, Md, 48-50; asst sanit engr, Dept Pub Health, Calif, 50-51; pub health engr, DC, 51-56; vpres, Resources Res, Inc, 56-63; dir spec res, Hazleton Labs, Inc, 63-65, dir life systs div, 65-67; PRES, BIOSPHERICS INC, 67- Concurrent Pos: Res asst biochem, Schs Med & Dent, Georgetown Univ, 52-61, clin asst prof, 53-60; biochemist, Dept Sanit Eng, DC, 62-63; consult, Dept Interior, 63-71; NASA planetary quarantine adv, 65-74; NASA experimenter, Mariner 9, Viking 1975 Missions to Mars. Honors & Awards: IR100 Indust Res Mag, 75. Mem: Am Soc Civil Eng; Am Pub Health Asn; Am Water Works Asn; NY Acad Sci; Am Inst Biol Sci. Res: Inventor PhoStrip process for wastewater phosphorus removal; life sciences; applied biology; water supply; waste disposal; sanitary biology; environmental sanitation; life detection techniques; public health and medical microbiology; instrumentation; space biology. Mailing Add: Biospherics Inc 4928 Wyaconda Rd Rockville MD 20852

LEVIN, HAROLD LEONARD, b St Louis, Mo, Mar 11, 29; m 54; c 3. GEOLOGY, PALEONTOLOGY. Educ: Univ Mo, AB, 51, MA, 52; Wash Univ, PhD(paleont), 56.

Prof Exp: Geologist, Standard Oil Co Calif, 56-61; asst prof, 61-64, ASSOC PROF PALEONT, WASH UNIV, 64-, CHMN DEPT EARTH & PLANETARY SCI, 73-, GEOLOGIST, 61- Concurrent Pos: Res grants, Wash Univ, 61-72; consult, Ecol Serv, Mo Bot Garden, 73- Mem: AAAS; Soc Econ Paleont & Mineral; Paleont Soc. Res: Foraminifera, Coccolithophoridae and related microfossils; biostratigraphy of microorganisms; geological education. Mailing Add: Dept Earth & Planetary Sci Wash Univ St Louis MO 63130

LEVIN, HERMAN WESTLEY, b Philadelphia, Pa, Dec 23, 29; m 53; c 3. BIOCHEMISTRY, ENZYMOLOGY. Educ: Temple Univ, AB, 51, AM, 55, PhD(biochem), 58. Prof Exp: Asst, Temple Univ, 51-54; jr res fel, Inst Cancer Res, 54-57, asst, 57-58; res biochemist, Lederle Labs, Am Cyanamid Co, 58-63; sr biochemist, Smith Kline & French Labs, 63-69; PRIN SCIENTIST, LEEDS & NORTHRUP CO, 69- Mem: AAAS; Am Chem Soc; NY Acad Sci. Res: Adopting biochemical and enzymatic reactions to electronic instrumentation for clinical, laboratory, industrial and pollution analysis; multianalysis systems based on biosensors. Mailing Add: Tech Ctr Corp Res Dept Leeds & Northrup Co North Wales PA 19454

LEVIN, IRA WILLIAM, b Washington, DC, Sept 20, 35; m 61; c 1. CHEMICAL PHYSICS. Educ: Univ Va, BS, 57; Brown Univ, PhD(chem), 61. Prof Exp: Res instr chem, Univ Wash, 61-62; guest worker, 63-65, staff fel, 65-66, res chem, Phys Biol Lab, 66-72, RES CHEMIST, LAB CHEM PHYSICS, NIH, 72- Concurrent Pos: Lectr, Georgetown Univ, 64-65, assoc mem grad fac chem, 74- Mem: Am Phys Soc. Res: Vibrational spectroscopy; absolute intensities; molecular dynamics and structure; spectra; spectroscopy of biomembranes. Mailing Add: Lab Chem Physics Nat Inst Health Bethesda MD 20014

LEVIN, IRVIN, b Baltimore, Md, Dec 18, 12; m 49; c 3. PHYSICAL CHEMISTRY. Educ: Johns Hopkins Univ, BS, 35; Univ Md, MS, 40, PhD(chem), 48. Prof Exp: Res chemist, Nat Dairy Prod Corp, Md, 35-42; phys chemist, Signal Corps, US Army, Camp Evans, NJ, 42-45; instr chem, Univ Md, 45-46, res assoc, 48-50; chief dept biophys instrumentation, Army Med Serv Grad Sch, Walter Reed Army Med Ctr, 50-55, DIR INSTRUMENTATION DIV, WALTER REED ARMY INST RES, 55- Concurrent Pos: Consult, Power Condenser & Electronics Corp, Washington, DC, 49-50. Mem: Am Chem Soc; Am Phys Soc. Res: Electrochemistry; photochemistry; design of laboratory apparatus; semiconductors; photovoltaic behavior of chemical substances. Mailing Add: 1404 Billman Ln Wheaton MD 20902

LEVIN, ISADOR, b Baltimore, Md, Dec 14, 09; m 39; c 2. PHYSICAL CHEMISTRY. Educ: Johns Hopkins Univ, PhD(phys chem), 33. Prof Exp: Inspector, US Food & Drug Admin, 36-37; chem engr, Mutual Chem Co, 37-46; plant engr, Baltimore Paint & Chem Corp, 46-75. Mem: Am Chem Soc; Am Inst Chem. Mailing Add: Apt 14 2905 Fallstaff Rd Baltimore MD 21209

LEVIN, JACK, b Newark, NJ, Oct 11, 32. INTERNAL MEDICINE, HEMATOLOGY. Educ: Yale Univ, BA, 53, MD, 57; Am Bd Internal Med, dipl, 65, recert, 74. Prof Exp: Chief resident & instr, Yale Univ, 64-65; from instr to asst prof, 65-70, ASSOC PROF MED, DIV HEMAT, JOHNS HOPKINS UNIV, 70- Concurrent Pos: Fel med, Sch Med, Johns Hopkins Univ, 62-64; Markle scholar acad med, 68-73; mem corp, Marine Biol Lab, 65-; physician chg hemat out-patient clin, Johns Hopkins Hosp, 67-71; consult, Vet Admin Hosp, Baltimore, Md, 68- Mem: Am Fedn Clin Res; Am Soc Hemat; fel Am Col Physicians; Am Soc Exp Path; Am Soc Clin Invest. Res: Blood coagulation, platelets; thrombopoiesis; endotoxin and endotoxemia; Shwartzman phenomenon; thrombocytosis; invertebrate coagulation; von Willebrand's disease. Mailing Add: Div of Hemat Johns Hopkins Hosp Baltimore MD 21205

LEVIN, JACOB JOSEPH, b New York, NY, Dec 21, 26; m 52; c 3. MATHEMATICAL ANALYSIS. Educ: City Col New York, BEE, 49; Mass Inst Technol, PhD, 53. Prof Exp: Instr math, Mass Inst Technol, 52-53; instr, Purdue Univ, 53-55; vis lectr, Mass Inst Technol, 55-56, staff mem, Lincoln Lab, 56-63; assoc prof, 63-66, PROF MATH, UNIV WIS-MADISON, 66- Concurrent Pos: NSF sr fel, Univ Calif, Los Angeles, 70-71. Mem: Am Math Soc; Soc Indust & Appl Math. Res: Differential equations; integral equations. Mailing Add: 1110 Frisch Rd Madison WI 53711

LEVIN, JEROME ALLEN, b Washington, DC, Aug 25, 39; m; c 2. BIOCHEMICAL PHARMACOLOGY. Educ: Philadelphia Col Pharm & Sci, BSc, 61; Univ Mich, PhD(pharmacol), 66. Prof Exp: Res assoc pharmacol, State Univ NY Downstate Med Ctr, 66-68; asst prof, 68-74, interim chmn, 73-75, ASSOC PROF PHARMACOL, MED COL OHIO, 74- Concurrent Pos: USPHS fel, State Univ NY Downstate Med Ctr, 66-68; Am Heart Asn res grant, Med Col Ohio, 69-75, USPHS res grant, 70-76. Mem: AAAS; Am Heart Asn; Am Soc Pharmacol & Exp Therapeut. Res: Inactivation of norepinephrine in vascular tissue; drug-receptor interactions. Mailing Add: Dept Pharmacol & Exp Therapeut Med Col of Ohio PO Box 6190 Toledo OH 43614

LEVIN, JOSEPH DAVID, b New York, NY, Feb 7, 18; m 47; c 2. INDUSTRIAL MICROBIOLOGY. Educ: Queens Col, NY, BS, 41. Prof Exp: Tech aide, E R Squibb & Sons, 47-50, res asst, 50-53, res asst supvr, Squibb Div, Olin Mathieson Chem Corp, 53-59, res scientist, 59-64, sr res scientist, 64-69, LAB SUPVR, INST MED RES, SQUIBB CORP, NEW BRUNSWICK, 68- Mem: AAAS; Am Soc Microbiol. Res: Analytical microbiology; test and develop microbiological assays of antibiotics including traces in mammalian tissues; test and development methods for pharmaceutical preservative efficacy; co-patentee, diagnostic aid for fungal infection. Mailing Add: 244 Benner St Highland Park NJ 08904

LEVIN, JUDITH GOLDSTEIN, b Brooklyn, NY, Nov 8, 34; m 57; c 2. BIOCHEMISTRY, VIROLOGY. Educ: Barnard Col, Columbia Univ, BA, 55; Harvard Univ, MA, 57; Columbia Univ, PhD(biochem), 62. Prof Exp: Sr scientist molecular biol viruses, Nat Cancer Inst, 69-73, SR SCIENTIST, LAB MOLECULAR GENETICS, NAT INST CHILD HEALTH & HUMAN DEVELOP, 73- Concurrent Pos: Nat Heart Inst res fel biochem genetics, 62-69; USPHS fel, 63-65; Am Heart Asn advan res fel, 66-68; consult lab path, Nat Cancer Inst, 69; estab investr, Am Heart Asn, 69-74. Mem: AAAS; Am Soc Biol Chem; Am Chem Soc; Am Soc Microbiol. Res: Mechanisms of protein biosynthesis, in particular, studies on the codon recognition step; biochemistry of animal viruses; regulation of transcription and translation in mammalian cells. Mailing Add: Nat Inst of Child Health & Human Develop Bethesda MD 20014

LEVIN, KATHRYN J, b Lawrence, Kans, Feb 25, 44; m 69. SOLID STATE PHYSICS. Educ: Univ Calif, Berkeley, BA, 66; Harvard Univ, PhD(physics), 70. Prof Exp: Res assoc physics, Univ Rochester, 70-72; asst res physicist, Univ Calif, Irvine, 72-75; ASST PROF PHYSICS, UNIV CHICAGO, 75- Mem: Am Phys Soc. Res: Phase transitions in disordered systems; superconductivity. Mailing Add: James Franck Inst Univ of Chicago Chicago IL 60637

LEVIN, LOUIS, b Milwaukee, Wis, May 9, 08; m 34; c 2. BIOCHEMISTRY. Educ: Kalamazoo Col, AB, 29; St Louis Univ, PhD(biol chem), 34. Prof Exp: Asst, Sch Med, St Louis Univ, 29-34; res assoc anat, Col Physicians & Surgeons, Columbia Univ, 34-45; asst dir hemat, Michael Reese Hosp, Chicago, 45-46; asst prof anat, Col Physicians & Surgeons, Columbia Univ, 46-48; head sci div, Off Naval Res, NY, 48-51, head biochem br, DC, 51-52; prog dir regulatory biol, NSF, 52-58, prog dir metab biol, 57-59, dep asst dir, Div Biol & Med Sci, 57-60, head off inst prog, 60; dean sci & assoc dean fac, Brandeis Univ, 60-64; head off prog develop & anal, NSF, 64-66, assoc dir instnl rels, 66-68, exec assoc dir, NSF, 68-70, asst dir instnl progs, 70-72; prof & spec consult eval, Tex Tech Univ, 72-74; CONSULT, 74- Concurrent Pos: Instr, City Col New York, 42-45; mem endocrinol sect, Growth Comt, Nat Res Coun, 52-53; mem nat adv coun health res facilities, NIH, 59-60; mem grad facilities panel, NSF, 60-63; mem adv comt instnl grants, Am Cancer Soc, 61-64 & 65-67, chmn, 66-67; mem bd trustees, Carver Res Found, Tuskegee Inst, 64-; mem, Nat Adv Gen Med Sci Coun, NIH, 74- Honors & Awards: Distinguished Serv Award, NSF, 72. Mem: Fel AAAS; Am Soc Biol Chem; Soc Exp Biol & Med; Endocrine Soc; fel NY Acad Sci. Res: Purification and isolation of estrogens; purification of gonadotrophins; bioassay technics; hormonal control of serum protein and of lipid metabolism; biochemistry and physiology of endocrines; science and academic administrations. Mailing Add: 1730 La Coronilla Dr Santa Barbara CA 93109

LEVIN, MARK DAVID, mathematics, see 12th edition

LEVIN, MARSHALL DAVID, b Brownwood, Tex, May 18, 22; m 45; c 3. ENTOMOLOGY, APICULTURE. Educ: Univ Conn, AB, 47; Univ Minn, MS, 49, PhD(entom, agr biochem), 56. Prof Exp: Asst entom, Conn Agr Exp Sta, 47; asst apicult, Univ Minn, 46-50; entomologist, 50-64, honey bee pollination invests leader, Honey Bee Res Lab, Ariz, 64-69, chief apicult res br, Plant Indust Sta, 69-72, SCIENTIST, NAT PROG STAFF, AGR RES SERV, USDA, 72- Mem: AAAS; Entom Soc Am; Bee Res Asn. Res: Biology and behavior of Bombus, Osmia, Nomia; biology, foraging behavior, management, pollinating activities of honey bees; honey bee nutrition; effects of insecticides on honey bees and wild bees. Mailing Add: Nat Prog Staff Agr Res Serv USDA Agr Res Ctr-West Beltsville MD 20705

LEVIN, MICHAEL HOWARD, b New York, NY, Sept 25, 36; m 59; c 1. ECOLOGY, RESOURCE MANAGEMENT. Educ: Univ Vt, BS, 58; Rutgers Univ, MS, 60, PhD(bot), 64. Prof Exp: Res assoc taxon, NY Bot Garden, 63-64; cur, Greene-Nieuwland Herbarium & asst prof biol, Univ Notre Dame, 64-66; asst prof bot & cur herbarium, Univ Man, 66-68; asst prof landscape archit & regional planning, Univ Pa, 68-73; PRES, ENVIRON RES ASSOCS, INC, 73-, DIR RES & CHMN BD, 74- Concurrent Pos: Coop investr, Delta Waterfowl Res Sta, Man, 65-66. Mem: Fel AAAS; Ecol Soc Am; Am Soc Plant Taxon; Int Asn Plant Taxon; Brit Ecol Soc. Res: Forensic ecology; ecology of altered communities and ecosystems; ecological management; application of gradient analysis to terrestrial communities; wetlands ecology. Mailing Add: Environ Res Assoc Inc Devon PA 19333

LEVIN, MORRIS A, b New York, NY, May 15, 34; m 57; c 2. MICROBIOLOGY. Educ: Univ Chicago, BS, 59; Univ RI, PhD(microbiol), 70. Prof Exp: Microbiologist aerobiol, Dept Defense, 57-66; microbiologist marine microbiol, Dept Health Educ & Welfare, 66-70; MICROBIOLOGIST HEALTH EFFECTS, ENVIRON PROTECTION AGENCY, 70- Concurrent Pos: Adj prof civil eng & microbiol, Univ RI, 75. Prof Exp: Res: Quantitating of microorganisms in the environment, dose-response relationships and epidemiological considerations correlating the public health effects of exposure to microbial populations under natural conditions. Mailing Add: Greenwood Dr Peacedale RI 02883

LEVIN, MORTON LOEB, b Russia, Aug 25, 03; div; c 2. PREVENTIVE MEDICINE, PUBLIC HEALTH. Educ: Univ Md, MD, 30; Johns Hopkins Univ, DrPH, 34. Prof Exp: Intern & asst resident, Mt Sinai Hosp, Baltimore, Md, 30-32; asst dispensary physician, Johns Hopkins Hosp, 32-33; comnr health, Ottawa County, Mich, 34-35; instr epidemiol, Sch Hyg & Pub Health, Johns Hopkins Univ, 35-36; assoc physician, Roswell Park Mem Inst, NY, 36-39; asst dir div cancer control, State Dept Health, NY, 39-46, dir, 46 46-47, asst comnr med serv, 47-60; chief dept epidemiol, Roswell Park Mem Inst, 60-67; VIS PROF EPIDEMIOL, SCH HYG & PUB HEALTH, JOHNS HOPKINS UNIV, 67- Concurrent Pos: Dir Comn Chronic Illness, 50-51. Mem: Am Epidemiol Soc; Am Pub Health Asn. Res: Epidemiology of malignant tumors. Mailing Add: Dept of Epidemiol Johns Hopkins Univ Baltimore MD 21205

LEVIN, MURRAY LAURENCE, b Boston, Mass, Nov 14, 35; m 61; c 2. INTERNAL MEDICINE, NEPHROLOGY. Educ: Harvard Col, AB, 57; Tufts Univ, MD, 61. Prof Exp: Intern med, Beth Israel Hosp, Boston, Mass, 61-62, resident, 62-64; assoc, 66-69, asst prof, 69-72, ASSOC PROF MED, MED SCH, NORTHWESTERN UNIV CHICAGO, 72- Concurrent Pos: Res fel renal dis, Univ Tex Southwestern Med Sch Dallas, 64-66; NIH res fel, 65-66; Chicago Heart Asn res grants, 66-70 & 73-75; Nat Inst Arthritis & Metab Dis res grant, 67-70; attend physician, Vet Admin Res Hosp, Chicago, Ill, 66-, chief renal sect; adj staff, Passavant Mem Hosp, 68-, assoc attend physician, 75- Mem: AAAS; Am Fedn Clin Res; Int Soc Nephrol; Am Soc Nephrol. Res: Salt and water metabolism; uremia; membrane transport; calcium and phosphorus metabolism. Mailing Add: Vet Admin Res Hosp Renal Sect 333 E Huron St Chicago IL 60611

LEVIN, NATHAN, b Baltimore, Md, July 30, 15; m; c 1. PHARMACEUTICAL CHEMISTRY. Educ: Univ Md, BS, 36, MS, 38, PhD(org pharmaceut chem), 41. Prof Exp: Asst instr bact, Sch Pharm, Univ Md, Baltimore City, 36-37, asst org chem, 37-41; fel synthetic org chem, Upjohn Co, Mich, 41-42, res chemist, 42-44; res chemist, Schering Corp, NJ, 44-45; res develop chemist, Burroughs Wellcome & Co, Inc, NY, 45-48; assoc prof pharmaceut chem, Col Pharm, Howard Univ, 48-59; pharmacist, Washington, DC, 59-68; HEAD PHARM CHEM LAB, MD STATE BUR LABS, 68- Concurrent Pos: Abstractor, Pharmaceut Abstr, Am Pharmaceut Asn, 39-48 & Chem Abstr, Am Chem Soc, 53-55. Mem: Am Pharmaceut Asn; Am Chem Soc. Res: Analysis of drugs and pharmaceutical preparations; identification of narcotics and/or dangerous drugs; quality control and methods of analysis of pharmaceutical products. Mailing Add: Md State Bur Labs PO Box 2355 Baltimore MD 21203

LEVIN, NORMAN LEWIS, b Hartford, Conn, Mar 31, 24; m 50; c 2. ZOOLOGY, PARASITOLOGY. Educ: Univ Conn, BS, 48, MS, 49; Univ Ill, PhD(zool, parasitol), 56. Prof Exp: Asst zool, Univ Ill, 53-56, instr, 56-57; asst prof biol, Westminster Col, Mo, 57-60; from instr to asst prof, 60-71, ASSOC PROF BIOL, BROOKLYN COL, 71- Mem: Fel AAAS; Am Soc Zool; Am Soc Parasitol; Am Soc Trop Med & Hyg; Am Micros Soc. Res: General taxonomy; morphology; life cycles; interrelationship of larval trematodes and marine snails. Mailing Add: Dept of Biol Brooklyn Col Brooklyn NY 11210

LEVIN, RICHARD ALEXANDER, b Hartford, Conn, Oct 26, 32; m 64; c 3. MICROBIAL PHYSIOLOGY, GENETICS. Educ: Harvard Univ, AB, 54; Univ Wash, MS, 56, MA, 63; Univ Iowa, PhD(microbiol), 68. Prof Exp: Lectr microbiol, Univ Wash, 60-62; ASST PROF BIOL, OBERLIN COL, 68- Mem: AAAS; Am Soc Microbiol. Res: Rhizobium genetics; lipid biosynthesis in autotrophic bacteria; bacterial degradation of sterols. Mailing Add: Dept of Biol Oberlin Col Oberlin OH 44074

LEVIN, ROBERT AARON, b New York, NY, July 25, 29; m 55; c 4. CLINICAL CHEMISTRY. Educ: St John's Univ, NY, BS, 51, MS, 55. Prof Exp: Res toxicologist, 55-58, sr res clin path & toxicol, 58-62, UNIT LEADER CLIN PATH, NORWICH PHARMACAL CO, 62- Mem: Am Chem Soc; Am Asn Clin Chem; fel Am Inst Chem; Am Soc Vet Clin Pathologists. Res: Automation and computerization of chemical technics; new drug toxicity testing; establishing effects on clinical pathology parameters; veterinary hematology. Mailing Add: Clin Path Unit Norwich Pharmacal Co Norwich NY 13815

LEVIN, ROBERT E, b Boston, Mass, Dec 1, 30; m; c 2. MICROBIOLOGY, FOOD SCIENCE. Educ: Los Angeles State Col, BS, 52; Univ Southern Calif, MS, 54; Univ Calif, Davis, PhD(microbiol), 63. Prof Exp: Asst prof microbiol, Ore State Univ, 63-64; asst prof, 64-72, ASSOC PROF FOOD SCI, UNIV MASS, AMHERST, 72- Concurrent Pos: NIH res grant, 65-68. Mem: Am Soc Microbiol; Inst Food Technologists; Soc Cryobiol. Res: Microbiological sulfate reduction; yeast cytology; psychophilic bacteria. Mailing Add: Dept of Food Sci Univ of Mass Amherst MA 01002

LEVIN, ROBERT HAROLD, b Chicago, Ill, Nov 1, 15; m 41; c 4. ORGANIC CHEMISTRY. Educ: Univ Ill, AB, 37; Univ Wis, PhD(org chem), 41. Prof Exp: Chem libr asst, Univ Ill, 34-36; asst chem, Univ Wis, 37-41; res chemist, Upjohn Co, Mich, 41-46, group leader chem res, 46-52, head dept chem, 52-58, asst dir res, 58-68; VPRES RES, RICHARDSON-MERRELL, INC, 68- Concurrent Pos: Mem subcomt steroid nomenclature, Nat Res Coun, 50-55; mem coun, Gordon Res Conf. Mem: AAAS; Am Chem Soc; NY Acad Sci. Res: Chemistry of steroids, especially the cortical hormones; biomedical research and new drug development long range planning for pharmaceutical research. Mailing Add: Merrell Nat Labs Richardson-Merrell Inc Cincinnati OH 45215

LEVIN, ROBERT MARTIN, b New York, NY, Apr 6, 45; m 67; c 2. PHARMACOLOGY. Educ: Albright Col, BS, 67; Univ Pa, MS, 69, PhD(pharmacol), 74. Prof Exp: FEL, DEPT PHARMACOL, MED COL PA, 74- Mem: Soc Neurosci. Res: The role of the cyclic nucleotide system in the response of the central nervous system to pharmacological agents. Mailing Add: Dept of Pharmacol Med Col of Pa Philadelphia PA 19129

LEVIN, ROBERT WARREN, b Parsons, Kans, Jan 6, 21; m 42; c 1. NUCLEAR CHEMISTRY. Educ: Harvard Univ, AB, 42. Prof Exp: Shift chemist, Hercules Powder Co, 42-44; process foreman, Union Carbide Chem Co Div, 44-48, head lab dept, Union Carbide Nuclear Co Div, 48-51, LAB DIR GASEOUS DIFFUSION, UNION CARBIDE NUCLEAR CO DIV, UNION CARBIDE CORP, 51- Mem: Am Chem Soc; Am Nuclear Soc. Res: Uranium enrichment technology; gaseous diffusion; uranium chemistry. Mailing Add: 3937 Pines Rd Paducah KY 42001

LEVIN, RONALD HAROLD, b San Francisco, Calif, Sept 26, 45; m 69; c 2. ORGANIC CHEMISTRY. Educ: Case Western Reserve Univ, BS, 67; Princeton Univ, PhD(chem), 70. Prof Exp: Fel chem, Univ Freiburg, 70-71 & Calif Inst Technol, 71-72; ASST PROF CHEM, HARVARD UNIV, 72- Mem: Am Chem Soc; Chem Soc London. Res: Reactive intermediates; thermal and photochemical transformations; applications of magnetic resonance. Mailing Add: Dept of Chem Harvard Univ 12 Oxford St Cambridge MA 02138

LEVIN, SAMUEL JOSEPH, b Detroit, Mich, Sept 19, 35; m 63; c 2. BIOCHEMISTRY. Educ: Wayne State Univ, BA, 58, PhD(chem), 61; Am Bd Clin Chem, dipl. Prof Exp: Res assoc chem, Col Med, Wayne State Univ, 55-61; scientist, Warner Lambert Pharmaceut Co, 61-62; from instr to asst prof biochem, Div Grad Studies, Med Col, Cornell Univ, 63-66; from asst prof to assoc prof biochem, Sch Dent, Univ Mo-Kansas City, 67-74, from asst prof to assoc prof, Sch Med, 68-74; ASSOC DIR CLIN PATH, MICHAEL REESE HOSP, 74-; MEM, MICHAEL REESE INST, 74- Concurrent Pos: Asst attend biochemist, Mem Hosp Cancer & Allied Dis, 62-66; assoc, Sloan-Kettering Inst, 63-66; chief biochemist, Dept Path, Kansas City Gen Hosp, 66-74. Mem: AAAS; Am Asn Clin Chem; Am Chem Soc. Res: Analysis and metabolism of steroids; enzymology and enzyme reaction mechanisms. Mailing Add: Dept of Path Michael Reese Hosp Chicago IL 60616

LEVIN, SEYMOUR R, b Chicago, Ill, Apr 27, 34; m 57; c 3. INTERNAL MEDICINE. Educ: Univ Ill, BS, 56, MD, 61; Am Bd Internal Med, dipl internal med, 70 & endocrinol, 73. Prof Exp: Intern, Cook Co Hosp, Chicago, 61-62; resident, Wadsworth Vet Admin Hosp, Los Angeles, 62-65; physician, US Army Hosp, Ft Carson, 65-67; res fel endocrinol, Univ Calif, San Francisco, 67-69, asst res physician, 69-73; DIR DIABETES CLIN & CHIEF METAB UNIT, WADSWORTH VET ADMIN HOSP, 73-; ASSOC PROF MED, UNIV CALIF, LOS ANGELES, 75- Concurrent Pos: Consult, Fresno Valley Med Ctr & Endocrine Div, Univ Calif, Los Angeles, 73- Honors & Awards: Grant, Am Diabetes Asn Southern Calif, 74 & 75. Mem: Fel Am Col Physicians; Am Fedn Clin Res; Am Diabetes Asn; Endocrine Soc. Res: Studies of secretion and mechanisms of secretion by the endocrine pancreas. Mailing Add: Wadsworth Vet Admin Hosp 691/ 111K Los Angeles CA 90073

LEVIN, SIDNEY, b Washington, DC, May 4, 31; div; c 3. PEDIATRICS, NEPHROLOGY. Educ: Univ Md, BS, 53; George Washington Univ, MS, 54; Baylor Univ, MD, 58. Prof Exp: From intern to sr resident pediat, Johns Hopkins Hosp, 58-61, chief resident, 62-63; pediatrician, Nat Heart Inst, 63-64; asst prof pediat, Johns Hopkins Univ & dir pediat renal & chronic dis progs, Johns Hopkins Hosp, 64-68; assoc prof pediat, 68-73, chmn dept, 70-73, PROF PEDIAT & CHMN JACKSONVILLE HOSPS EDUC PROG, COL MED, UNIV FLA, 73-; CHMN DEPT, DUVAL MED CTR, 68- Concurrent Pos: Med officer, DC Gen Hosp, 61-62; consult, Kimbrough Army Hosp, Ft George G Meade, Md & USPHS Hosp, Wyman Park, Baltimore, 65-68. Mem: Am Soc Nephrol. Res: Clinical renal disease in children; basic mechanisms of renal disease, clinical, pathologic and histologic correlation; dietary management and nutritional aspects of uremia. Mailing Add: Dept of Pediat Univ of Fla Gainesville FL 32603

LEVIN, SIDNEY SEAMORE, b Philadelphia, Pa, Mar 29, 29; m 62; c 2. PHYSIOLOGY, PHARMACOLOGY. Educ: Univ Pittsburgh, BS, 51, MS, 53, PhD(biol sci), 55. Hon Degrees: MA, Univ Pa, 71. Prof Exp: RES ASSOC, HARRISON DEPT SURG RES, SCH MED, UNIV PA, 58- Mem: AAAS; NY Acad Sci. Res: Adrenal output in shock; effect of hypertension on adrenal cortical steroids; cytochrome P-450. Mailing Add: 565 Dulles Bldg Hosp of the Univ of Pa Philadelphia PA 19104

LEVIN, SIMON ASHER, b Baltimore, Md, Apr 22, 41; m 64; c 2. MATHEMATICS, BIOLOGY. Educ: Johns Hopkins Univ, BA, 61; Univ Md, PhD(math), 64. Prof Exp: Asst math, Univ Md, 61-62; NSF fel biomath, Univ Calif, Berkeley, 64-65; asst prof math, 65-70, ASSOC PROF APPL MATH, CORNELL UNIV, 71-, ASSOC PROF ECOL & SYST & THEORET & APPL MECHANICS, 72-, CHMN SECT ECOL & SYST, 74- Concurrent Pos: Res assoc, Univ Md, College Park, 64; co-chmn board, Gordon Res Conf, 70, chmn theoret biol & biomath, 71; vis scholar, Univ Wash, 73-74; assoc ed, Ecol & Ecol Monographs, Ecol Soc Am, 73- & Theoret Pop Biol, 76-; managing ed, Lecture Notes Biomath, 73- & Siam J Appl Math, 75-; mem US comt, Israel Environ, 75- Mem: AAAS; Am Math Soc; Am Soc Naturalists; Ecol Soc Am; Soc Indust & Appl Math. Res: Applied mathematics; ecology; population biology; mathematical physics and biology; partial differential equations. Mailing Add: Sect of Ecol & Syst Cornell Univ Ithaca NY 14853

LEVIN, SIMON EUGENE, b Philadelphia, Pa, Nov 29, 20; m 48; c 2. BACTERIOLOGY. Educ: Philadelphia Col Pharm, BS, 41; Pa State Col, MS, 42, PhD, 49. Prof Exp: Bacteriologist, La Wall & Harrison Res Labs, 38-41; lab asst bact, Pa State Col, 41-42; bioassayist, La Wall & Harrison Res Labs, 42-43 & 46; asst bact, Pa State Col, 46-49; res assoc chemother, E R Squibb & Sons, 49-50; head div biol, La Wall & Harrison Res Labs, 50-56; pres, Huntingdon Farms, Inc, West Conshohocken, 57-74; DIR LIFE SCI DIV, AM STANDARDS TESTING BUR, 74- Concurrent Pos: Dir, Syndot Labs, 57-74; consult, Decker Corp, 59-70. Mem: AAAS; Am Soc Microbiol; Am Asn Lab Animal Sci; Am Chem Soc; NY Acad Sci. Res: Medical and industrial pharmacology and toxicology. Mailing Add: 222 Glendalough Rd Erdenheim PA 19118

LEVIN, VICTOR ALAN, b Milwaukee, Wis, Nov 22, 41; m 63; c 2. CANCER, NEUROLOGY. Educ: Univ Wis-Madison, BS, 63, MD, 66. Prof Exp: Intern med, St Louis City Hosp, Washington Univ, 66-67; staff assoc chem pharm, Nat Cancer Inst, 67-69; resident neurol, Mass Gen Hosp, 69-71, Nat Inst Neurol Dis & Stroke fel, 71-72; instr, 72-73, ASST PROF NEUROL, NEUROSURG & PHARM CHEM, SCH MED & PHARM, UNIV CALIF, SAN FRANCISCO, 73- Mem: Am Acad Neurol; Am Asn Cancer Res. Res: Pharmacology and pharmacokinetics of brain tumor chemotherapy and experimental brain tumor chemotherapy. Mailing Add: Brain Tumor Res Ctr HSW-783 Univ of Calif Sch of Med San Francisco CA 94143

LEVIN, WILLIAM COHN, b Waco, Tex, Mar 2, 17; m 41; c 2. INTERNAL MEDICINE. Educ: Univ Tex, BA, 38, MD, 41. Prof Exp: From instr to assoc prof internal med, 44-65, dir hemat res lab & blood bank, 46-74, PROF INTERNAL MED, UNIV TEX MED BR GALVESTON, 65-, PRES, 74- Concurrent Pos: Consult, USPHS Hosp, Galveston, 52, US Air Force, 54- & Wilford Hall, US Air Force Hosp, San Antonio; dir clin res ctr, John Sealy Hosp, 62-74; consult aerospace med div, Air Force Systs Command, Lackland AFB, 64-; vis prof, William Beaumont Gen Hosp, El Paso, 65-; mem cancer clin invest rev comt, Nat Cancer Inst. Mem: AAAS; Soc Exp Biol & Med; AMA; fel Am Col Physicians; fel Int Soc Hemat. Res: Hematology; immunology; oncology. Mailing Add: Off of Pres Univ of Tex Med Br Galveston TX 77550

LEVINE, AARON WILLIAM, b New York, NY, July 14, 43; m 64; c 3. ORGANIC CHEMISTRY. Educ: Yeshiva Univ, BA, 63; City Col New York, MA, 66; Seton Hall Univ, PhD(org chem), 70. Prof Exp: Teacher, High Schs, NY, 63-66; res chemist, M&T Chem, Inc, 66-69; teaching asst chem, Seton Hall Univ, 69; MEM TECH STAFF, DAVID SARNOFF RES CTR, RCA LABS, 69- Mem: Am Chem Soc. Res: Liquid crystals; kinetics of organic reactions in solution; polymer syntheses and reaction; organic electrochemistry. Mailing Add: RCA Labs David Sarnoff Res Ctr Princeton NJ 08540

LEVINE, ALVIN SAUL, b Hamlet, NC, Aug 29, 25; m 51; c 4. VIROLOGY. Educ: Wake Forest Col, BS, 48; Univ NC, MSPH, 50; Rutgers Univ, PhD(microbiol), 54. Prof Exp: Res asst biochem, Duke Univ, 50-51; instr bact & immunol, Harvard Med Sch, 56-58; from asst prof to assoc prof microbiol, 58-64, PROF MICROBIOL, SCH MED, IND UNIV, INDIANAPOLIS, 64-; PROF LIFE SCI & DIR TERRE HAUTE CTR MED EDUC, IND STATE UNIV, TERRE HAUTE, 71- Concurrent Pos: Res fel microbiol, Rutgers Univ, 51-54; teaching fel bact & immunol, Harvard Med Sch, 54-56; Fulbright vis prof, Univ West Indies, 67-68. Mem: Fel Am Soc Microbiol; Am Soc Exp Path; Am Asn Immunol; Am Asn Cancer Res; Tissue Cult Asn. Res: Infectious diseases; viral oncology; RNA viruses; biochemical, biophysical and immunological studies. Mailing Add: Terre Haute Ctr for Med Educ Ind State Univ Terre Haute IN 47809

LEVINE, ARNOLD DAVID, b Brooklyn, NY, Oct 24, 25; m 62. THEORETICAL PHYSICS. Educ: Columbia Univ, PhD(physics), 58. Prof Exp: Asst prof physics, WVa Univ, 57-60; asst prof, Wayne State Univ, 60-62; from asst prof to assoc prof, 62-71, PROF PHYSICS, WVA UNIV, 71- Concurrent Pos: Consult, Columbia Liquified Natural Gas Corp, 71-73; consult, Am Gas Asn, currently. Mem: Combustion Inst; Am Phys Soc; Am Asn Physics Teachers. Res: Meson physics; quantum field theory; non-equilibrium thermodynamics; fluid dynamics; combustion. Mailing Add: Dept of Physics WVa Univ Morgantown WV 26505

LEVINE, ARNOLD JAY, b Brooklyn, NY, July 30, 39; m 63; c 1. VIROLOGY. Educ: State Univ NY Binghamton, BA, 61; Univ Pa, PhD(microbiol), 66. Prof Exp: NIH res fel biophys, Calif Inst Technol, 66-68; ASSOC PROF BIOCHEM, PRINCETON UNIV, 68- Concurrent Pos: Ed, Virology. Honors & Awards: C & H Dryfus Teacher-Scholar Award, 73. Mem: Am Soc Microbiol; Fedn Am Soc Exp Biol. Res: Mechanism of replication and oncogenesis with tumor viruses. Mailing Add: Dept of Biochem Sci Princeton Univ Princeton NJ 08540

LEVINE, BARRY FRANKLIN, b Brooklyn, NY, Sept 5, 42; m 68. LASERS. Educ: Polytech Inst Brooklyn, BS, 63; Harvard Univ, PhD(physics), 69. Prof Exp: PHYSICIST, BELL LABS, 68- Mem: Am Phys Soc. Res: Experimental and theoretical nonlinear optics of crystals and liquids; relationship of optical harmonic generation to chemical and physical microscopic molecular properties. Mailing Add: Bell Labs Murray Hill NJ 07974

LEVINE, BERNARD BENJAMIN, b New York, NY, Nov 8, 28. IMMUNOLOGY, MEDICINE. Educ: City Col New York, BS, 50; NY Univ, MD, 54. Prof Exp: From asst prof to assoc prof, 62-70, PROF MED, MED CTR, NY UNIV, 70-, DIR ALLERGY, 62- Concurrent Pos: Res fel path, Med Ctr, NY Univ, 60-62. Mem: Am Asn Immunol; Soc Exp Biol & Med; Am Soc Clin Invest; Am Acad Allergy. Res: Immunopathology; hypersensitivity; antigenicity; immune response; allergy. Mailing Add: Dept of Med NY Univ Med Ctr New York NY 10016

LEVINE, CHARLES (ARTHUR), b Des Moines, Iowa, Dec 25, 22; m 48; c 2. PHYSICAL CHEMISTRY. Educ: Iowa State Col, BS, 47; Univ Calif, PhD(chem), 51. Prof Exp: Asst, Univ Calif, 48-49, asst, Radiation Lab, 49-51; res chemist, 51-65, ASSOC SCIENTIST, DOW CHEM CO, 65- Mem: AAAS; Electrochem Soc; Am Chem Soc; Am Phys Soc; Am Inst Chem Eng. Res: Nuclear chemistry; radiation chemistry; electrochemistry. Mailing Add: 5869 Pine Hollow Rd Concord CA 94521

LEVINE, DAVID MORRIS, b Boston, Mass, Dec 15, 37; m 65; c 2. PUBLIC HEALTH, MEDICAL RESEARCH. Educ: Brandeis Univ, AB, 59; Univ Vt, MD, 64; Johns Hopkins Univ, MPH, 69, SCD, 72; Nat Bd Med Examr, dipl, 65; Am Col Prev Med, dipl, 71; Pan Am Med Assoc, dipl. Prof Exp: Intern, Montefiore Hosp, Pittsburgh, 64-65; resident, Waltham Hosp, Mass, 65-66; US Army Med Corp, 66-68; resident prev med, 68-70, ASSOC PROF PUB HEALTH, MED EDUC & INTERNAL MED, JOHNS HOPKINS UNIV, 72-, DIR MANPOWER STUDIES,

CTR HEALTH SERV RES & DEVELOP, 72- Concurrent Pos: Fel pub health serv, Sch Hyg & Pub Health, Johns Hopkins Univ, 68-71; consult, Nat Ctr Health Serv Res, 72- & Am Asn Med Col, 73-; mem study sect, Nat Heart-Lung Inst, NIH, 75- Mem: AAAS; Am Pub Health Asn; Am Fedn Clin Res; Am Col Prev Med; Pan Am Med Asn. Res: Health care manpower-services, quality of care, productivity, planning health education strategies in managing chronic disease process and outcome of medical education. Mailing Add: Ctr Health Serv Res & Develop Johns Hopkins Med Inst Baltimore MD 21205

LEVINE, DONALD MARTIN, b Boston, Mass, Oct 17, 29. ZOOLOGY, PARASITOLOGY. Educ: Univ Vt, BA, 51; Univ RI, MS, 53; Univ Pa, PhD(zool), 58. Prof Exp: USPHS fel, 58-60; helminthologist, Liberian Inst, Am Found Trop Med, 60-62; assoc prof, 62-74, PROF BIOL SCI, WILLIAM PATERSON COL NJ, 74- Mem: Am Soc Trop Med & Hyg; Am Inst Biol Sci; Royal Soc Trop Med & Hyg. Res: Immunology and ecology of parasitic infections. Mailing Add: Dept of Biol Sci William Paterson Col of NJ Wayne NJ 07470

LEVINE, ELI MORRIS, b Russia, Dec 28, 10; nat US; m; c 3. CHEMISTRY. Educ: Yeshiva Univ, AB, 32; Columbia Univ, AM, 37; Polytech Inst Brooklyn, PhD(chem), 43. Prof Exp: From instr to assoc prof, 33-50, PROF CHEM & CHMN DEPT, YESHIVA UNIV, 50- Concurrent Pos: Res chemist, Unified Labs, 43-45; tech dir, Wimpole Chem Labs, 45-48; consult chemist, Hema Drug Co, 48- Mem: AAAS; Am Chem Soc. Res: Male and female sex hormones; correlation of the ionization constants of a series of aromatic benzenesulfonamides and their structure; strength of acids of some benzenesulfonamides. Mailing Add: Dept of Chem Yeshiva Univ New York NY 10033

LEVINE, ELLIOT MYRON, cell biology, see 12th edition

LEVINE, EUGENE, b Brooklyn, NY, Jan 11, 25; m 48; c 3. ANALYTICAL STATISTICS, OPERATIONS RESEARCH. Educ: City Col New York, BBA, 48; NY Univ, MPA, 50; Am Univ, PhD(pub admin), 60. Prof Exp: Statistician, New York City Dept Health, 47-50; CHIEF MANPOWER ANAL & RESOURCES BR, DIV NURSING, USPHS, 50- Honors & Awards: Super Serv Award, Dept Health, Educ & Welfare, 62. Mem: Am Pub Health Asn; Am Statist Asn. Res: Health manpower analysis; problems of health services organization and delivery; psychometric analysis into problems of job satisfaction; career choice and motivation; evaluation of health care programs. Mailing Add: 8135 Iverness Ridge Rd Potomac MD 20854

LEVINE, HAROLD, b New York, NY, Mar 24, 22; m 47. APPLIED MATHEMATICS. Educ: City Col New York, BS, 41; Cornell Univ, PhD(physics), 44. Prof Exp: Res fel physics, Harvard Univ, 45-54; assoc prof, 55-70, PROF MATH, STANFORD UNIV, 70- Concurrent Pos: Lectr, Harvard Univ, 52-54; consult, Lawrence Radiation Lab, Univ Calif. Mem: Am Phys Soc. Res: Boundary value problems of classical field theories, particularly acoustics, electrodynamics and hydrodynamics. Mailing Add: Dept of Math Stanford Univ Stanford CA 94305

LEVINE, HAROLD IRVING, b Lynn, Mass, Dec 14, 28; m 61. MATHEMATICS. Educ: Univ Chicago, PhD(math), 57. Prof Exp: Fulbright fel & Ger Res Asn grant, Univ Bonn, 57-59; instr math, Yale Univ, 59-60; from asst prof to assoc prof, 60-70, PROF MATH, BRANDEIS UNIV, 70- Mem: Am Math Soc. Res: Differential topology. Mailing Add: Dept of Math Brandeis Univ Waltham MA 02154

LEVINE, HARRY, b New York, NY, July 22, 11. INTERNAL MEDICINE. Educ: City Col New York, BS, 34; Univ Iowa, MS, 37, PhD(biochem), 38; Univ Tex, MD, 44; Am Bd Internal Med, dipl, 51. Prof Exp: Asst med, Univ Iowa, 36-38; res assoc path chem, Univ Tex, 38-41; biochemist & dir labs, Galveston State Psychiat Hosp, 41-42; instr pharmacol, Univ Tex Med Br Galveston, 42-44, from instr to asst prof med, 47-52; sect chief metab dis, Vet Admin Hosp, East Orange, NJ, 52-57; ASSOC PROF MED, UNIV TEX MED BR, GALVESTON, 57- Concurrent Pos: Attend physician, John Sealy Hosp; consult, USPHS Hosp, Galveston. Mem: AAAS; AMA; fel Am Col Physicians; NY Acad Sci; Brit Biochem Soc. Res: Metabolic diseases. Mailing Add: Dept of Internal Med Univ of Tex Med Br Galveston TX 77551

LEVINE, HARVEY ROBERT, b New York, NY, Sept 15, 31; m 56; c 2. PARASITOLOGY, ENTOMOLOGY. Educ: City Col New York, BS, 53; Univ Mass, MS, 55, PhD(entom), 58. Prof Exp: Instr entom, Univ Mass, 55; from asst prof to prof biol, Bemidji State Col, 58-68; asst dean acad affairs, Sch Sci, 71-72, PROF BIOL & CHMN DEPT, QUINNIPIAC COL, 68- Mem: Am Inst Biol Sci; Entom Soc Am. Res: Freshwater insects; medical entomology. Mailing Add: Dept of Biol Sci Quinnipiac Col Hamden CT 06518

LEVINE, HERBERT JEROME, b Boston, Mass, July 22, 28; m 58; c 1. CARDIOLOGY. Educ: Harvard Univ, AB, 50; Johns Hopkins Univ, MD, 54; Am Bd Internal Med, dipl, 63. Prof Exp: Intern med, Peter Bent Brigham Hosp, 54-55, sr resident, 58-59; resident, Mass Gen Hosp, 57-58; res fel, Harvard Med Sch, 59-61; sr instr, 61-63, from asst prof to assoc prof, 63-70, PROF MED, SCH MED, TUFTS UNIV, 70-; CHIEF CARDIOL SERV, NEW ENG MED CTR HOSPS, 66- Concurrent Pos: Res fel cardiol, Peter Bent Brigham Hosp, 56-61; consult, Vet Admin Hosp, Mass, 66-; lectr, US Naval Hosp, Mass, 67- Mem: Fel Am Fedn Clin Res; fel Asn Univ Cardiol; fel Am Col Cardiol; fel Am Soc Clin Invest; Asn Am Physicians. Res: Clinical cardiology; physiology of congestive heart failure; muscle mechanics and energetics in the intact heart. Mailing Add: New Eng Med Ctr Hosp 171 Harrison Ave Boston MA 02111

LEVINE, HERMAN SAUL, b Jeannette, Pa, Feb 11, 22; m 47; c 3. PHYSICAL CHEMISTRY, HIGH TEMPERATURE CHEMISTRY. Educ: Univ Pittsburgh, BS, 43; Univ Ill, PhD(phys chem), 48. Prof Exp: Res asst, Ill State Geol Surv, 44-46; staff mem, NY State Col Ceramics, Alfred Univ, 48-51 & USPHS, R A Taft Sanit Eng Ctr, 51-57; STAFF MEM, SANDIA LABS, 57- Mem: Am Chem Soc. Res: Radioactive waste disposal processes and heterogeneous catalysis in coal liquefaction. Mailing Add: Sandia Labs Div 5824 Albuquerque NM 87115

LEVINE, HILLEL BENJAMIN, b Montreal, Que, July 9, 23; nat US; m 49; c 2. BACTERIOLOGY. Educ: McGill Univ, BSc, 45; Univ Wis, MS, 46, PhD(bact, biochem), 49. Prof Exp: Chief biochem div, 53-58, chief mycol div, 58-70, RES BACTERIOLOGIST, NAVAL BIOMED RES LAB, UNIV CALIF, 50-, CHMN MED MICROBIOL DEPT, 70- Concurrent Pos: Consult, Sir Herbert Reddy Mem Hosp, Montreal, 49; secy comt on mycoses, Pan Am Health Orgn, 71. Honors & Awards: Lab Award, Am Pub Health Asn, 63. Mem: Am Soc Microbiol; NY Acad Sci; Int Soc Human & Animal Mycol. Res: Bacterial and fungal physiology and virulence; fungal vaccines; respiratory metabolism. Mailing Add: Univ Calif Naval Biosci Lab Naval Supply Ctr Oakland CA 94625

LEVINE, HOWARD ALLEN, b St Paul, Minn, Jan 15, 42. MATHEMATICS. Educ: Univ Minn, Duluth, BA, 64; Cornell Univ, MA, 67, PhD(math), 69. Prof Exp: Asst prof math, Univ Minn, Minneapolis, 69-73; asst prof, 73-75, ASSOC PROF MATH, UNIV RI, 75- Concurrent Pos: Vis scientist, Battelle Advan Studies Ctr, Switz, 71 & 72; Sci Res Coun Gt Brit grant, Univ Dundee, 72; NSF res grant, 74-76. Mem: Am Math Soc. Res: Partial differential equations; numerical analysis. Mailing Add: Dept of Math Univ of RI Kingston RI 02881

LEVINE, HOWARD BERNARD, b Brooklyn, NY, Apr 15, 28; m 67; c 1. PHYSICAL CHEMISTRY. Educ: Univ Ill, BS, 50; Univ Chicago, MS, 52, PhD(chem), 55. Prof Exp: Res fel chem, Inst Atomic Res, Iowa State Univ, 55-56; chemist, Lawrence Radiation Lab, Univ Calif, 56-62; mem tech staff, NAm Aviation Sci Ctr, 62-70; proj assoc, Theoret Chem Inst & Space Sci & Eng Ctr, Univ Wis, 70-71; prof chem eng, Va Polytech Inst & State Univ, 71-73; PROG MGR CHEM SYSTS, SYSTS, SCI & SOFTWARE, 73- Concurrent Pos: Consult, Tech Adv Bd Supersonic Transport, Dept Com; mem ad hoc comt ozone & environ studies bd, Nat Acad Sci. Mem: AAAS; Am Chem Soc; fel Am Phys Soc; Am Inst Aeronaut & Astronaut. Res: Thermodynamics; statistical mechanics; quantum mechanics; spectroscopy; atmospheric chemistry; molecular physics; applied mathematics; chemical kinetics. Mailing Add: Systs Sci & Software PO Box 1620 La Jolla CA 92038

LEVINE, IRA NOEL, b Brooklyn, NY, Sept 8, 37. PHYSICAL CHEMISTRY. Educ: Carnegie Inst Technol, BS, 58; Harvard Univ, AM, 59, PhD(chem), 63. Prof Exp: Res assoc chem, Univ Pa, 63-64; instr, 64-66, asst prof, 67-70, ASSOC PROF CHEM, BROOKLYN COL, 71- Concurrent Pos: Am Chem Soc Petrol Res Fund starter grant, 65-66. Mem: Am Chem Soc; Am Phys Soc. Res: Quantum chemistry. Mailing Add: Dept of Chem Brooklyn Col Brooklyn NY 11210

LEVINE, ISIDORE, b New York, NY, Dec 1, 22; m 51; c 4. INTERNAL MEDICINE. Educ: City Col New York, BS, 48; Univ Rochester, MD, 52. Prof Exp: Instr med & psychiat, Sch Med & Dent, Univ Rochester, 56-58, asst prof prev med & community health, 58-62, assoc prof med, health serv, prev med & community health, 62-67; ASSOC PROF PREV MED, ALBERT EINSTEIN COL MED, 67-; DEP DIR, MONTEFIORE HOSP & MED CTR, 67- Concurrent Pos: Commonwealth Fund fel med & psychiat, Sch Med & Dent, Univ Rochester, 56-58, asst prof med & assoc physician, 58-64; sr assoc physician, 64-67; consult, Genesee Hosp, Rochester, 58-; assoc med dir, Strong Mem Hosp, 65- Res: Educational programs in comprehensive patient care. Mailing Add: Montefiore Hosp & Med Ctr 111 E 210th St Bronx NY 10467

LEVINE, JACK, b Philadelphia, Pa, Dec 15, 07; m 38. MATHEMATICS. Educ: Univ Calif, Los Angeles, AB, 29; Princeton Univ, PhD(math), 34. Prof Exp: Asst math, Univ Calif, Los Angeles, 29-30; instr, Princeton Univ, 30-35; from instr to assoc prof, 35-47, PROF MATH, NC STATE UNIV, 47- Concurrent Pos: Res analyst, US Dept War, 42-43. Mem: Am Math Soc; Am Soc Eng Educ; Math Asn Am. Res: Differential geometry; tensor analysis; combinatorial analysis. Mailing Add: Dept of Math NC State Univ Raleigh NC 27607

LEVINE, JEFFREY, b Brooklyn, NY, Feb 7, 45; m 66. MATHEMATICS. Educ: State Univ NY Stony Brook, BS, 66; Rutgers Univ, New Brunswick, PhD(math), 70. Prof Exp: Asst prof math, Monmouth Col, NJ, 69-71; ASST PROF MATH, STATE UNIV NY COL GENESEO, 71- Mem: Am Math Soc; Math Asn Am. Res: Ring theory. Mailing Add: Dept of Math State Univ of NY Geneseo NY 14454

LEVINE, JEROME PAUL, b New York, NY, May 4, 37; m 58; c 3. TOPOLOGY. Educ: Mass Inst Technol, BS, 58; Princeton Univ, PhD(math), 62. Prof Exp: Instr math, Mass Inst Technol, 61-63; NSF fels, 63-64; from asst prof to assoc prof math, Univ Calif, Berkeley, 64-66; assoc prof, 66-69, PROF MATH, BRANDEIS UNIV, 69-, CHMN DEPT, 74- Concurrent Pos: Sloan Found fel, 66-68. Mem: Am Math Soc. Res: Differential topology; knot theory. Mailing Add: Dept of Math Brandeis Univ Waltham MA 02154

LEVINE, JOEL STEWART, b Brooklyn, NY, May 14, 42; m 68; c 1. PLANETARY ATMOSPHERES. Educ: Brooklyn Col, BS, 64; NY Univ, MS, 67; Univ Mich, MS, 73. Prof Exp: Res scientist atmospheric sci, Goddard Inst Space Studies, 64-70, SPACE SCIENTIST PLANETARY ATMOSPHERES, LANGLEY RES CTR, NASA, 70- Concurrent Pos: Instr physics & dir astron observ, Brooklyn Col, 64-70; res scientist atmospheric sci, Geophys Res Lab, NY Univ, 64-70; consult, Mars Aeronomy, NASA Viking Proj, 70-; space sci rep, NASA-Langley Basic Res Adv Comt, 74-; guest investr, Orbiting Astron Observ-Copernicus, 75- Honors & Awards: H J E Reid Award, NASA Langley Res Ctr, 74. Mem: Am Geophys Union; Am Meteorol Soc; Am Astron Soc; Sigma Xi; AAAS. Res: Origin, evolution, physics and chemistry of planetary atmospheres; ozone photochemistry; spacecraft ultraviolet measurements of planetary atmospheres. Mailing Add: Planetary Physics Br Environ & Spa Sci Div NASA Langley Res Ctr Hampton VA 23665

LEVINE, JON HOWARD, b Toronto, Ont, July 13, 41; m 64; c 3. ENDOCRINOLOGY. Educ: Univ Toronto, MD, 65, MSc, 69; Royal Col Physicians & Surgeons Can, FRCP(C), 71. Prof Exp: Instr, Vanderbilt Univ, 71-73; ASST PROF ENDOCRINOL, MED UNIV SC, 73- Concurrent Pos: Fel, Med Res Coun Can, 71-73. Mem: Am Fedn Clin Res; Can Soc Endocrin & Metab. Res: Hormonal regulation of polyamine biosynthesis; pituitary regulation of adrenal steroidogenesis. Mailing Add: Dept of Med Med Univ of SC Charleston SC 29401

LEVINE, JOSEPH, meteorology, see 12th edition

LEVINE, JULES IVAN, b Brooklyn, NY, Apr 17, 38; m 62; c 2. HEALTH SCIENCES, MEDICAL ADMINISTRATION. Educ: Univ Va, BEE, 60, PhD(biomed eng), 72; Johns Hopkins Univ, MS, 68. Prof Exp: Sr engr aerospace electronics, Westinghouse Elec Corp, 63-68; ASST PROF PEDIAT, UNIV VA, 72-, ASST DEAN ALLIED HEALTH, 74- Concurrent Pos: Consult, Region III Off, Philadelphia, Pa, Dept Health, Educ & Welfare, 74- Mem: Am Pub Health Asn; Am Soc Allied Health Prof. Res: Planning and evaluation of health resources and the health care delivery system. Mailing Add: Div of Health Serv Res Univ of Va Med Ctr Charlottesville VA 22901

LEVINE, LAURENCE, b New York, NY, July 10, 26; m 51; c 3. CELL BIOLOGY. Educ: NY Univ, BA, 49; Univ Wis, MA, 52, PhD(zool, biochem), 55. Prof Exp: Asst parasitol, Univ Wis, 50-55; from instr to assoc prof biol, 55-66, PROF BIOL, WAYNE STATE UNIV, 66-, COORDR FRESHMAN BIOL, 68- Honors & Awards: Fac Res Recognition Award, Wayne State Univ Fund, 60. Mem: AAAS; NY Acad Sci; Am Soc Cell Biol. Res: Cell contractility; chromosome motion; chromosome structure and function; mechanism of meiosis. Mailing Add: Dept of Biol Wayne State Univ Detroit MI 48202

LEVINE, LAWRENCE, b Hartford, Conn, July 18, 24. IMMUNOCHEMISTRY. Educ: Univ Conn, BA, 48; Univ Mich, MS, 50; Johns Hopkins Univ, DSc(microbiol), 53. Prof Exp: Instr microbiol, Johns Hopkins Univ, 53-54; res scientist, Div Labs & Res, State Dept Health, NY, 54-57; from asst prof to assoc prof, 57-70, PROF BIOCHEM, BRANDEIS UNIV, 70- Res: Blood proteins and their immunological properties. Mailing Add: Grad Dept of Biochem Brandeis Univ Waltham MA 02154

LEVINE, LAWRENCE ELLIOTT, b Chelsea, Mass, June 23, 41; m 65; c 5. APPLIED MATHEMATICS. Educ: Rensselaer Polytech Inst, BS, 63; Univ Md, PhD(appl math), 68. Prof Exp: ASSOC PROF MATH, STEVENS INST TECHNOL, 68- Mem: Am Math Soc; Soc Indust & Appl Math. Res: Fluid dynamics; partial differential equations; perturbation methods. Mailing Add: Dept of Math Stevens Inst of Technol Hoboken NJ 07030

LEVINE, LEO, b Boston, Mass, Jan 25, 11; m 46; c 2. IMMUNOLOGY, MICROBIOLOGY. Educ: Harvard Univ, BS, 33. Prof Exp: Chemist, Mass Div Occup Hyg, 43-46; from asst chemist to sr chemist, 46-56, chief bacteriologist, 56-59, CHIEF LAB, MASS PUB HEALTH BIOLOGIC LABS, 59-; MEM STAFF IMMUNOL SCH PUB HEALTH, HARVARD UNIV, 60- Concurrent Pos: Prin investr, Nat Inst Allergy & Infectious Dis res grant, 61-; prin prof asst, Comn Immunization, Armed Forces Epidemiol Bd Contract, 61-; dir br lab, WHO Reference Serum Bank, Sch Med, Yale Univ, 63-; mem admin comt, Mass Health Res Inst, 63-66. Mem: Am Asn Immunol; Am Soc Microbiol. Res: Adult immunization; development of soluble pertussis vaccine; serologic surveys of immunity; factors affecting the immune response; active-passive tetanus immunization using human immune globulin; sensitizing activities of B-pertussis. Mailing Add: Biologic Labs 375 South St Forest Hills Boston MA 02130

LEVINE, LEO MEYER, b Brooklyn, NY, May 26, 22; m 49; c 3. MATHEMATICS. Educ: City Col New York, BS, 42; NY Univ, PhD(math), 60. Prof Exp: Asst physicist, Signal Corps Labs, Eatontown, NJ, 42-43; sr physicist, Mat Lab, NY Naval Shipyard, 47-59; from asst res scientist to assoc res scientist, Courant Inst Math Sci, NY Univ, 59-63, from asst prof to assoc prof, 63-70; ASSOC PROF MATH, QUEENSBOROUGH COMMUNITY COL, 70- Concurrent Pos: Consult, Radio Corp Am, 61-62. Mem: Am Math Soc; Math Asn Am. Res: Applied mathematics; ordinary and partial differential equations; acoustics; electromagnetic theory. Mailing Add: Queensborough Community Col Bayside NY 11364

LEVINE, LEON, organic chemistry, polymer chemistry, see 12th edition

LEVINE, LEONARD, chemistry, see 12th edition

LEVINE, LEONARD, b Atlantic City, NJ, Jan 28, 29; m 52; c 2. NEUROPHYSIOLOGY. Educ: Rutgers Univ, BS, 50; Columbia Univ, PhD(physiol), 59. Prof Exp: Instr physiol, Columbia Univ, 57-60; from asst prof to assoc prof, Univ Va, 61-66; PROF PHYSIOL, PAC UNIV, 66- Concurrent Pos: USPHS fel physiol, Columbia Univ, 59-60; fel biophys, Univ Col, Univ London, 60-61; USPHS res grants, 62-65, 67-68 & 70-72. Mem: AAAS; Am Physiol Soc; Biophys Soc; Am Soc Zool; Asn Res Vision & Ophthal. Res: Electrophysiology and pharmacology of ocular tissues; trophic interrelations between nerve and muscle tissues. Mailing Add: Col of Optom Pac Univ Forest Grove OR 97116

LEVINE, LEONARD P, b Newark, NJ, July 24, 32; m 54; c 1. SPECTROSCOPY, SURFACE PHYSICS. Educ: Queens Col, NY, BS, 54; Syracuse Univ, MS, 56, PhD(physics), 60. Prof Exp: Engr, Sperry-Gyroscope Co, 59-60; sr scientist, Honeywell Res Ctr, 60-64, prin res scientist, 64-66; PROF ELEC ENG, UNIV WIS-MILWAUKEE, 66- Mem: Am Phys Soc; Am Vacuum Soc. Res: Vacuum surface interface; interaction of ions, photons and electrons with the surface in vacuum; high voltage breakdown problem; digital and analog computation and hardware. Mailing Add: Dept of Elec Eng Univ of Wis Milwaukee WI 53201

LEVINE, LESLIE S, plasma physics, see 12th edition

LEVINE, LOUIS, b New York, NY, May 14, 21. GENETICS, ANIMAL BEHAVIOR. Educ: City Col New York, BS, 42, MS, 47; Columbia Univ, MA, 49, PhD(zool), 55. Prof Exp: From instr to assoc prof, 55-67, PROF BIOL, CITY COL NEW YORK, 68- Concurrent Pos: NSF grants, 60-; AEC grant, 63- Mem: Fel AAAS; Animal Behav Soc; Am Genetic Asn; Genetics Soc Am; Am Soc Naturalists. Res: Genetics of animal behavior and population genetics. Mailing Add: Dept of Biol City Col of New York New York NY 10031

LEVINE, MAITA FAYE, b Cincinnati, Ohio, Oct 17, 30. MATHEMATICS. Educ: Univ Cincinnati, BA, 52, BE, 53, MAT, 66; Ohio State Univ, PhD(math educ), 70. Prof Exp: Teacher, High Sch, Ohio, 53-63; instr, 63-70, ASST PROF MATH, UNIV CINCINNATI, 70- Concurrent Pos: NSF res grant, 74. Mem: Math Asn Am; Am Educ Res Asn. Res: Relationship between mathematical competence and mathematical confidence; measurement of mathematical maturity; reasons why qualified women do not pursue mathematical careers. Mailing Add: 1106 Louis Dr Cincinnati OH 45237

LEVINE, MAX, organic chemistry, see 12th edition

LEVINE, MELVIN, biological chemistry, see 12th edition

LEVINE, MELVIN MORDECAI, b Richmond, Va, Nov 20, 25; m 50; c 3. ENGINEERING PHYSICS. Educ: Mass Inst Technol, BS, 46; Univ Va, PhD(physics), 55. Prof Exp: Instr physics, Pa State Univ, 46-48; reactor physicist, Babcock & Wilcox Co, 55-59; PHYSICIST, BROOKHAVEN NAT LAB, 59- Mem: Am Nuclear Soc. Res: Neutronic and thermo-hydraulic studies of nuclear reactor power plant safety; development of computational models for computer simulation of reactor transients. Mailing Add: Appl Sci Dept Brookhaven Nat Lab Upton NY 11973

LEVINE, MICHAEL S, b Brooklyn, NY, Sept 22, 44; m 66; c 1. NEUROSCIENCES. Educ: Queens Col, BA, 66; Univ Rochester, PhD(physiol psychol), 70. Prof Exp: Fel neurophysiol, Brain Res Inst, 70-72, ASST RES NEUROPHYSIOLOGIST, 72-, LECTR PSYCHOL, UNIV CALIF, LOS ANGELES, 75- Concurrent Pos: Consult neurophysiologist, Hereditary dis Found, 75. Mem: Soc Neurosci; Am Psychol Asn; Am Asn Anatomists. Res: Neurophysiology of basal ganglia in mature and developing animals; role of basal ganglia in regulation of behavior; development and prediction of learning ability in developing animals. Mailing Add: Ment Retardation Res Ctr Dept of Psychiat Univ of Calif Los Angeles CA 90024

LEVINE, MICHEAL JOSEPH, b Oak Park, Ill, Dec 1, 40; m 68. NUCLEAR PHYSICS. Educ: Yale Univ, BS, 62, MS, 64, PhD(physics), 68. Prof Exp: PHYSICIST NUCLEAR PHYSICS, BROOKHAVEN NAT LAB, 68- Concurrent Pos: Consult, High Voltage Eng Corp, 72-75; guest physicist, Max Planck Inst Nuclear Physics, Heidelberg, Ger, 75. Mem: Am Phys Soc. Res: Study of nuclear reactions induced by charged particles, principally heavy ions; development of magnetic spectrometers and associated focal plane detectors. Mailing Add: Brookhaven Nat Lab Bldg 901 A Upton NY 11973

LEVINE, MILTON ISRA, b Syracuse, NY, Aug 15, 02; m 36; c 2. PEDIATRICS. Educ: City Col New York, BS, 23; Cornell Univ, MD, 27. Prof Exp: Instr pediat, 33-44, asst prof clin pediat, 44-54, assoc prof, 54-66, clin prof pediat, 66-71, EMER PROF PEDIAT, MED COL, CORNELL UNIV, 71- Concurrent Pos: Dir poliomyelitis comn, Bur Labs, New York City Dept Health, 31, dir vaccine study, 31-42, consult pediatrician, 34-, chmn adv comn vaccine, 47-; pediatrician, Children's Tuberc Clin, Harlem Hosp, 32-42; asst attend pediatrician, New York Hosp, 32-52, assoc attend pediatrician, 52-57, attend pediatrician, 57-71, consult pediatrician, 71-; mem adv comt, Mid-Century White House Conf Children & Young, Washington, DC, 50 & 60. Mem: AAAS; Am Thoracic Soc; fel Am Pub Health Asn; fel Am Col Chest Physicians; fel Am Acad Pediat. Res: Tuberculosis and pulmonary conditions in children; Bacillus Calmette-Guerin vaccination against tuberculosis; psychosomatic disorders in children. Mailing Add: 1111 Park Ave New York NY 10028

LEVINE, MYRON, b Brooklyn, NY, July 28, 26; m 50; c 2. GENETICS, VIROLOGY. Educ: Brooklyn Col, BA, 47; Ind Univ, PhD(zool), 52. Prof Exp: Res assoc microbiol, Univ Ill, 54-56; asst to assoc biologist, Brookhaven Nat Lab, 56-61; assoc prof, 61-66, PROF HUMAN GENETICS, SCH MED, UNIV MICH, ANN ARBOR, 66- Concurrent Pos: Am Cancer Soc fel, Johns Hopkins Univ, 53-54; Commonwealth Fund fel, Univ Geneva, 66-67; ed, J Virol, 72-; vis scientist, Imp Cancer Res Fund, London, 73-74; chmn grad prog cell & molecular biol in health sci, Univ Mich, 74-; mem mammalian cell lines adv comt, Nat Inst Gen Med Sci, NIH, 75- Mem: Genetics Soc Am; Am Soc Microbiol; Am Inst Biol Sci. Res: Genetics and biochemistry of phage and animal viruses; lysogeny; tumor virology. Mailing Add: Dept of Human Genetics Univ of Mich Sch of Med Ann Arbor MI 48104

LEVINE, NATHAN B, chemistry, polymer science, see 12th edition

LEVINE, NORMAN DION, b Boston, Mass, Nov 30, 12; m 35. PARASITOLOGY, PROTOZOOLOGY. Educ: Iowa State Col, BS, 33; Univ Calif, PhD(zool), 37; Am Bd Med Microbiol, cert pub health & med lab parasitol. Prof Exp: Asst zool, Univ Calif, 33-37; asst animal parasitologist, 37-41, assoc animal pathologist, 41-42, from asst prof to assoc prof vet parasitol, 46-53, asst to dean col vet med, 47-57, sr mem, Ctr Zoonoses Res, 60-74, dir, Ctr Human Ecol, 68-74, PROF VET PARASITOL, COL VET MED, UNIV ILL, URBANA, 53-, PROF VET RES & ZOOL, 65- Concurrent Pos: Mem, Nat Res Coun, 56-62; mem bd gov, Am Bd Med Microbiol, 59-64; vis prof, Univ Hawaii, 62 & Santa Catalina Marine Biol Lab, 72; mem comt health sci achievement award prog, NIH, 65-66, mem trop med & parasitol study sect, 65-69, chmn, 66-69, mem animal resources adv comt, 71-75; ed, J Protozool, 65-71. Mem: AAAS; Am Soc Parasitol; hon mem Soc Protozool (secy, 52-58, vpres, 58-59, pres, 59-60, actg secy, 60-62); Micros Soc Am (pres, 69-70); fel Am Acad Microbiol. Res: Protozoan and roundworm parasites of domestic animals; malaria and other insect-borne diseases. Mailing Add: Col of Vet Med Univ of Ill Urbana IL 61801

LEVINE, OSCAR, b Brooklyn, NY, Feb 6, 23; m 48; c 2. PHYSICAL CHEMISTRY. Educ: City Col New York, BS, 43; Columbia Univ, AM, 48; Georgetown Univ, PhD(chem), 57. Prof Exp: With Nat Adv Comt Aeronaut, Ohio, 48-52; chemist, US Naval Res Lab, 52-58; CHEMIST, CHEM & MAT RES, GILLETTE SAFETY RAZOR CO, BOSTON, 58- Mem: AAAS; Am Chem Soc. Res: Chemistry and physics of solid and liquid surfaces and interfaces; lubrication; adhesion. Mailing Add: 43 Connolly St Randolph MA 02368

LEVINE, PAUL HERSH, b New York, NY, Sept 27, 35; m 63. THEORETICAL PHYSCIS, APPLIED PHYSICS. Educ: Mass Inst Technol, BS, 56; Calif Inst Technol, MS, 57, PhD(theoret physics), 63. Prof Exp: Sr scientist, Jet Propulsion Lab, Calif Inst Technol, 63-64; chief scientist, Astrophys Res Corp, 64-72; CHIEF SCIENTIST, MEGATEK CORP, 72- Concurrent Pos: Vis prof physics & biomed computation, Maharishi Europ Res Univ, 74- Mem: Am Phys Soc. Res: Ionospheric physics; over-the-horizon radar; quantum many-body problem; exploding wire phenomena; electron field emission; radiative transport; electromagnetic propagation; navigation and communication systems analysis; minicomputer applications; electroencephalography; psychobiology of consciousness. Mailing Add: Megatek Corp 1055 Shafter St San Diego CA 92106

LEVINE, PHILIP, b Kletsk, Russia, Aug 10, 00; nat US; m 38; c 4. IMMUNOLOGY, CANCER. Educ: City Col New York, BS, 19; Cornell Univ, MD, 23, MA, 26; FRCP, 73. Hon Degrees: DS, Mich State Univ, 67. Prof Exp: Asst, Rockefeller Inst, 25-28, assoc, 28-32; instr path & bact, Univ Wis, 32-35; bacteriologist & serologist, Newark Beth Israel Hosp, 35-44; dir, 44-66, EMER DIR IMMUNOHEMAT DIV, ORTHO RES FOUND, 66- Concurrent Pos: Seminar assoc, Columbia Univ, 53- Honors & Awards: Johnson Award; Lasker Award, Am Pub Health Asn; Award, Passano Found; Landsteiner Award; Harris Award; Kennedy Award; Burdick Award, Am Soc Clin Path; Gold Medal, Norwegian Soc Immunohemat, 75; Allan Award, Am Soc Human Genetics, 75. Mem: Nat Acad Sci; Am Soc Clin Path; fel Am Col Physicians; Am Soc Human Genetics (pres, 69); fel NY Acad Sci. Res: Blood groups, individual blood differences and their heredity; the Rh-Hr system and role in erythroblastosis; differences in Rh and ABO disease of newborn; blood group antigens in malignancy and use of their antibodies as cytotoxic agents in malignancies. Mailing Add: Ortho Res Found Rte 202 Raritan NJ 08869

LEVINE, PHILIP THEODORE, b Malden, Mass, Apr 20, 15; m 46; c 2. PROTEIN CHEMISTRY, ORAL BIOLOGY. Educ: Tufts Univ, BS, 38; Harvard Univ, DMD, 42. Prof Exp: From instr to asst prof oral path & periodont, Sch Dent Med, Tufts Univ, 48-53, from asst prof to assoc prof oral diag, 61-64; PROF ORAL BIOL, SCH DENT MED, UNIV CONN, 67- Concurrent Pos: Res fel orthop surg, Mass Gen Hosp, 64-67; consult, Vet Admin, 53-65 & Mass Gen Hosp, 61-63, consult dept orthop surg, 67- Mem: Am Dent Asn; Int Asn Dent Res; Am Inst Chem. Res: Enamel protein chemistry directed toward understanding the mechanism of hard tissue calcification, an understanding of the mechanism could possibly relate to an understanding of tissue ageing in general. Mailing Add: Univ of Conn Sch of Dent Med 263 Farmington Ave Farmington CT 06032

LEVINE, PHILLIP J, b Providence, RI, Jan 7, 34; m 55; c 2. PHARMACY. Educ: Univ RI, BS, 55; Univ Md, MS, 57, PhD(pharm), 63. Prof Exp: Instr pharm, Sch Pharm, Univ Md, 57-63; from asst prof to assoc prof, 63-70, PROF PHARM, COL PHARM, DRAKE UNIV, 70- Concurrent Pos: Consult, Dr Salsbury's Labs, Charles City, Iowa, 65-70; dir, Coop IV Additive Proj, 67-69; chmn, Mayor's Task Force on Drugs, Des Moines, Iowa, 69-70; consult, Gov, State of Iowa, 70-72. Mem: Am Pharmaceut Asn. Res: Development of topical anesthetic suspensions to test their applicability to long duration of anesthesia in dental patients; product development in area of suspension and formulations. Mailing Add: Col of Pharm Drake Univ Des Moines IA 50311

LEVINE, PINCUS PHILIP, b New York, NY, Aug 25, 07; m 33; c 2. PATHOLOGY, BACTERIOLOGY. Educ: City Col New York, 27; Cornell Univ, MS, 32, DVM, 32, PhD(parasitol), 37. Hon Degrees: DVM, Univ Munich, 70. Prof Exp: Asst, State Univ NY Vet Col, Cornell Univ, 28-32; wildlife res pathologist, State Conserv Dept, NY, 32-34; instr res poultry dis, 34-41, asst prof, 41-43, from assoc prof to prof poultry dis, 43-74, head dept avian dis, 61-66, EMER PROF AVIAN DIS, NY STATE VET COL, CORNELL UNIV, 74- Concurrent Pos: Ed, Cornell Vet, 42-47; Guggenheim fel, Inst Biol, Brazil, 47-48; res partic, Oak Ridge Inst Nuclear Studies, 54-55; consult,

US Army, 54-57; ed, Avian Dis, 57-61; consult, Rockefeller Found Mex Proj, 60; poultry virologist, AID, Israel Prog, 61-62; collabr, Fish & Wildlife Serv; consult, Food & Agr Orgn, Colombia & Mex, 69, Peru, 69-72 & Israel, 71 & 72; mem, Northeastern Conf Avian Dis. Honors & Awards: White Prize, 32. Mem: AAAS; Am Soc Zoologists; Am Soc Parasitologists; Am Asn Avian Pathologists (pres, 62-63); World Vet Poultry Asn (vpres, 62-69, pres, 69-). Res: Helminthology; Ascaridia; Capillaria; Davainea; biology; chemotherapy; protozoology; virology; bronchitis; Newcastle disease; mycoplasmosis. Mailing Add: NY State Vet Col Cornell Univ Ithaca NY 14850

LEVINE, RACHMIEL, b Poland, Aug 26, 10; nat US; m 43; c 2. ENDOCRINOLOGY. Educ: McGill Univ, BA, 32, MD, 36. Hon Degrees: MD, Univ Ulm, 69. Prof Exp: Asst dir dept metab & endocrine res, Michael Reese Hosp, 39-42, dir, 42-58, chmn dept med & dir med educ, 52-60; prof & chmn dept, NY Med Col, 60-70; DIR, CITY OF HOPE MED CTR, 70- Concurrent Pos: Williams fel, Michael Reese Hosp, 36-37, res fel, 37-39; Endocrine Soc Upjohn scholar, 57; Guggenheim Found fel, 71-72; consult, NSF, 56-59 & 70-; pres, Int Fedn Diabetes, 67-70; mem bd dirs, Found Fund Psychiat Res. Honors & Awards: Thompson Award, Am Geriat Soc, 71; Gairdner Found Award, 71. Mem: Am Physiol Soc; Soc Exp Biol & Med; Endocrine Soc; Am Diabetes Asn (pres, 64-65); fel Am Acad Arts & Sci. Res: Hormonal control of metabolism; mode of action of insulin; diabetes. Mailing Add: City of Hope Med Ctr Duarte CA 91010

LEVINE, RALPH MANUEL, b Galsgow, Scotland, Oct 30, 28; m 59; c 2. POLYMER CHEMISTRY. Educ: Univ Glasgow, BSc, 50; ARIC, 55; FRIC, 61. Prof Exp: Asst res mgr polymer res & develop, Vinyl Prod Ltd, 52-59; group leader latex & coatings res & develop, Marbon Chem Div, Borg Warner, Inc, 60-62; prod mgr, ABS resins & Plastics, Marbon Chem Div, Borg Warner UK Ltd, 62-64; mgr polymerisation lab, Nat Starch & Chem Corp, 65-70; tech dir coatings & resins, Dutch Boy Paints, Div NL Indust Inc, 70-74; tech dir coatings & resins, 74-76, CORP TECH DIR, FULLER-O'BRIEN DIV, O'BRIEN CORP, 76- Mem: Royal Inst Chem; The Chem Soc; Am Chem Soc; Nat Paint & Coatings Asn. Res: Development of resins and coatings for decorative and protective architectural finishes and for industrial applications. Mailing Add: 12603 Cheverly Ct Saratoga CA 95070

LEVINE, RANDOLPH HERBERT, b Denver, Colo, Nov 20, 46; m 70; c 2. ASTROPHYSICS. Educ: Univ Calif, Berkeley, AB, 68; Harvard Univ, AM, 69, PhD(physics), 72. Prof Exp: Vis scientist solar physics, High Altitude Observ, Nat Ctr Atmospheric Res, Boulder, Colo, 72-74; res fel astrophys, 74-75, RES ASSOC ASTROPHYS, CTR ASTROPHYS, HARVARD COL OBSERV, 75- Mem: Am Astron Soc; Am Phys Soc. Res: Physics and astrophysics of magnetic fields, especially pertaining to the sun. Mailing Add: Ctr for Astrophys Harvard Col Observ Cambridge MA 02138

LEVINE, RAPHAEL BERG, b Minneapolis, Minn, Dec 8, 20. BIOPHYSICS. Educ: Univ Minn, BA, 41, MA, 50, PhD(biophys), 51. Prof Exp: Degaussing physicist, Brooklyn Navy Yard, 41-43; instr physics, Univ Minn, 43-44, res assoc biophys, 51-54; res assoc neurophysiol, Neuropsychiat Inst, Univ Ill, 54-56; res biophysicist & asst prof psychiat, Columbus Psychiat Inst & Ohio State Univ, 57-58; from sr analyst opers res to staff scientist, Lockheed-Ga Co, 58-70; exec dir, Metrop Atlanta Coun for Health, 70-71; DIR HEALTH & SOCIAL SERV PLANNING DEPT, ATLANTA REGIONAL COMN, 72- Concurrent Pos: Dir comprehensive health planning, Community Coun Atlanta Area, 69-70. Mem: AAAS; Biophys Soc; Human Factors Soc; Am Asn Physics Teachers; Am Asn Comp Health Planning. Res: Health planning methodology; adaptive computer pattern recognition; biophysical reactions to stress; quantitative bioelectricity; space physiology. Mailing Add: Atlanta Regional Comn Suite 910 100 Peachtree St Atlanta GA 30303

LEVINE, RAPHAEL DAVID, b Egypt, Mar 29, 38; m 62; c 1. THEORETICAL CHEMISTRY. Educ: Hebrew Univ, Israel, MSc, 60; Univ Nottingham, PhD, 64; Oxford Univ, DPhil, 66. Prof Exp: Ramsay Mem fel, 64-66; vis asst prof math, theoret chem & chem, Univ Wis, 66-67; assoc prof chem, 68-70, PROF MATH & BATTELLE PROF CHEM, OHIO STATE UNIV, 70- Concurrent Pos: Sr lectr, Hebrew Univ, Israel, 68-, prof, 69-; Alfred P Sloan fel, 70; prof, Weizmann Inst, Israel, 71- Honors & Awards: Int Acad Quantum Molecular Sci Award, 68. Mem: Am Phys Soc; Am Chem Soc. Res: Quantum mechanics of molecular rate processes. Mailing Add: Dept of Chem Ohio State Univ Columbus OH 43210

LEVINE, RHEA JOY COTTLER, b Brooklyn, NY, Nov 26, 39; m 60; c 3. CELL BIOLOGY, CYTOCHEMISTRY. Educ: Smith Col, AB, 60; NY Univ, MS, 63, PhD(biol), 66. Prof Exp: Lab instr biol, Sch Com, Acct & Finance, Wash Sq Col, NY Univ, 63-64; res assoc neuropath, Sch Med, Univ Pa, 68-69; asst prof, 69-74, ASSOC PROF ANAT, MED COL PA, 74- Concurrent Pos: A H Robins Co fel biochem res, Manhattan State Hosp, Ward's Island, New York, 66; USPHS fel, Sch Med, Yale Univ, 66-68; Nat Heart & Lung Inst grant, Pa Muscle Inst, 73; Nat Inst Neurol Commun Dis & Stroke career develop award, 74-79; Nat Inst Gen Med Sci res grant, 75- Mem: AAAS; Histochem Soc; Am Asn Anat; NY Acad Sci; Biophys Soc; Am Soc Cell Biol. Res: Ultrastructure; muscle structure and function; comparative aspects of immunohistochemistry and cytochemistry. Mailing Add: Med Col of Pa Dept of Anat 3300 Henry Ave Philadelphia PA 19129

LEVINE, ROBERT, b Boston, Mass, July 30, 19; m 50; c 3. ORGANIC CHEMISTRY. Educ: Dartmouth Col, BA, 40, MA, 42; Duke Univ, PhD(org chem), 45. Prof Exp: Asst, Dartmouth Col, 40-42; asst, Duke Univ, 42-45; chemist, Mathieson Chem Corp, NY, 45-46; from instr to assoc prof, 46-59, PROF CHEM, UNIV PITTSBURGH, 59- Concurrent Pos: Consult, Monsanto Co, 52-62, Schering Corp, 59-63, Reilly Tar & Chem Corp, 64-66, FMC Corp, 65-67, Columbia Org Chem Co, 70-73, Pressure Chem Co, 70-75, Fike Chem Inc, 71- & Mallinckrodt Chem Works, 74- Mem: Am Chem Soc; Int Asn Heterocyclic Chem; NY Acad Sci; Israel Chem Soc. Res: Heterocyclic nitrogen chemistry, including pyridine, pyrazine, pyrimidine and triazine; synthesis of organic fluorine compounds; chemistry of organometallic compounds; synthesis of potential medicinals. Mailing Add: Dept of Chem Univ of Pittsburgh Pittsburgh PA 15260

LEVINE, ROBERT, b New York, NY, Nov 10, 26; m 54; c 2. PEDIATRIC CARDIOLOGY. Educ: City Col New York, BS, 48; Western Reserve Univ, MD, 54. Prof Exp: From intern to resident pediat, State Univ NY Upstate Med Ctr, 54-57; from instr to assoc prof, Col Physicians & Surgeons, Columbia Univ, 62-72; PROF PEDIAT & DIR PEDIAT CARDIOL, NJ MED SCH, COL MED & DENT NJ, 72- Concurrent Pos: NIH trainee pediat cardiol, Col physicians & Surgeons, Columbia Univ, 59-61 & NIH fel cardiorespiratory physiol, 61-62; NY City Health Res Coun career scientist award, 62-72; John Polachek Found fel, 68-69; prin investr, NIH Grad Training Prog Pediat Cardiol, Columbia Univ, 70-72, responsible investr, Nat Heart & Lung Inst-SCOR, Col Physicians & Surgeons, 71-72. Honors & Awards: Borden Award, Western Reserve Univ, 54. Mem: Am Acad Pediat; Am Pediat Soc; Am Physiol Soc; Am Heart Asn. Res: Cardiorespiratory physiology. Mailing Add: NJ Med Sch Col of Med & Dent of NJ Newark NJ 07103

LEVINE, ROBERT ALAN, b New York, NY, Mar 27, 32; m 68; c 2. ANTHROPOLOGY, PSYCHOLOGY. Educ: Univ Chicago, BA, 51, MA, 53; Harvard Univ, PhD(social anthrop), 58. Prof Exp: From instr to asst prof anthrop & polit sci, Northwestern Univ, 58-60; from asst prof to prof anthrop & human develop, Univ Chicago, 60-76, prof psychiat, 72-76; ROY E LARSEN PROF EDUC & HUMAN DEVELOP, HARVARD UNIV, 76- Concurrent Pos: Found Fund Res Psychiat fel, Inst Psychoanal, Chicago, 62-65; NIMH res scientist develop award, Univ Chicago, 62-72, res scientist award, 72-76; mem ment health small grant comt, NIMH, 66-68; fel, Ctr Advan Study Behav Sci, 71-72. Mem: Fel Am Anthrop Asn; Am Psychoanal Asn. Res: Cross-cultural study of personality development; culture and personality in Africa; socialization of the child; psychoanalytic research. Mailing Add: Lab of Human Develop Harvard Univ Cambridge MA 02138

LEVINE, ROBERT ALAN, b New York, NY, June 12, 32; m 56; c 3. MEDICINE, PHARMACOLOGY. Educ: Cornell Univ, AB, 54, MD, 58; Am Bd Gastroenterol, cert. Prof Exp: Intern med, NY Hosp-Cornell Med Ctr, 58-59, asst resident, 59-60; clin fel med, Liver Study Unit, Sch Med, Yale Univ, 61-62, res fel, 62-63; from asst chief to chief metab div, Army Med Res & Nutrit Lab, Fitzsimons Gen Hosp, 63-65; chief div gastroenterol, Brooklyn-Cumberland Med Ctr, 65-71, assoc prof med, 69-71; PROF MED, STATE UNIV NY UPSTATE MED CTR, 71-, CHIEF DIV GASTROENTEROL, STATE UNIV HOSP, 71- Concurrent Pos: Clin fel gastroenterol, NY Hosp-Cornell Med Ctr, 60-61. Mem: Am Soc Pharmacol & Exp Therapeut; Am Fedn Clin Res; Am Gastroenterol Asn. Res: Basic and clinical research in gastroenterology, metabolism and pharmacology; cyclic adenosine 3', 5'-monophosphate in vivo and in vitro; isolated perfused rat liver; chronic hepatitis; hormone regulation of gastrointestinal function. Mailing Add: State Univ Hosp 750 E Adams St Syracuse NY 13210

LEVINE, ROBERT JOHN, b New York, NY, Dec 29, 34; m 55; c 2. INTERNAL MEDICINE, PHARMACOLOGY. Educ: George Washington Univ, MD, 58; Am Bd Internal Med, dipl, 65. Prof Exp: Intern internal med, Peter Bent Brigham Hosp, Boston, Mass, 58-59, asst resident, 59-60; clin assoc clin pharmacol, Nat Heart Inst, 60-62; resident internal med, Vet Admin Hosp, West Haven, Conn, 62-63; investr clin pharmacol, Nat Heart Inst, 63-64; from instr to assoc prof internal med & pharmacol, 64-73, chief sect clin pharmacol, 66-74, dir physician's assoc prog, 73-75, PROF INTERNAL MED & LECTR PHARMACOL, SCH MED, YALE UNIV, 73- Concurrent Pos: Clin asst, Yale-New Haven Hosp, 64-65, asst attend physician, 65-68, attend physician, 68-; clin investr, Vet Admin Hosp, West Haven, Conn, 64-66, attend physician, 66-; mem myocardial infarction comt, Nat Heart & Lung Inst, 69-72; mem div med sci, Nat Res Coun, 71-75; ed, Clin Res, Am Fedn Clin Res, 71-; mem adv bd, Clin Resources Inc, 72-; consult, Nat Comn Protection Human Subj Biomed & Behav Res, 74- Mem: Am Soc Pharmacol; Am Soc Clin Pharmacol & Therapeut; Am Fedn Clin Res; fel Am Col Physicians; fel Am Col Cardiol. Res: Clinical pharmacology; metabolism of biologically active aromatic amines in man; ethics of human experimentation. Mailing Add: Dept of Internal Med Yale Univ Sch of Med New Haven CT 06510

LEVINE, ROBERT PAUL, b Brooklyn, NY, Dec 18, 26; m 69. GENETICS. Educ: Univ Calif, Los Angeles, AB, 49, PhD(genetics), 51. Hon Degrees: AM, Harvard Univ, 57. Prof Exp: Instr biol, Amherst Col, 51-53; from asst prof to assoc prof, 53-63, chmn dept, 67-70, PROF BIOL, HARVARD UNIV, 63- Concurrent Pos: NSF sr fel, 63-64. Mem: AAAS; Genetics Soc Am; Am Soc Naturalists; Am Soc Plant Physiologists; Am Soc Cell Biol. Res: Genetics of photosynthesis in unicellular algae; mechanism of photosynthetic electron transport; molecular structure of chloroplast membranes. Mailing Add: Biol Labs Harvard Univ Cambridge MA 02138

LEVINE, RUTH R, b New York, NY, m 53. PHARMACOLOGY. Educ: Hunter Col, BA, 38; Columbia Univ, MA, 39; Tufts Univ, PhD(pharmacol), 55. Prof Exp: From instr to asst prof pharmacol, Sch Med, Tufts Univ, 55-58; from asst prof to prof, 58-72, UNIV PROF PHARMACOL, SCH MED, BOSTON UNIV, 72-, CHMN DIV MED & DENT SCI, GRAD SCH, 64- Mem: Am Soc Pharmacol & Exp Therapeut (secy-treas elect, 74, secy-treas, 75); Biophys Soc; Acad Pharmaceut Sci; Am Chem Soc. Res: Mechanisms of transport of drugs across biological barriers, particularly the intestinal epithelium; biochemical, histological and physiological factors influencing intestinal absorption. Mailing Add: Div of Med & Dent Sci Boston Univ Sch of Med Boston MA 02118

LEVINE, SAMUEL, b Brooklyn, NY, Jan 21, 21; m 53; c 3. PHYSICAL CHEMISTRY. Educ: Brooklyn Col, BA, 46; Columbia Univ, MA, 52, PhD(chem), 55. Prof Exp: Electrochemist, Arc Anodying & Plating Co, 46-47; phys chemist thermodyn, Nat Bur Standards, 47-51, proj leader, Macromolecular Properties Unit, Northern Regional Res & Develop Div, 55-58; assoc prof chem, Western Ill Univ, 58-59; chemist, Dow Chem Co, 59-61; prof chem & dir sci, Delta Col, 61-64; PROF CHEM & DEAN, SAGINAW VALLEY STATE COL, 64- Mem: AAAS; Am Chem Soc. Res: Physical chemistry of polymers; thermodynamics; kinetics. Mailing Add: Saginaw Valley State Col 2250 Pierce Rd University Center MI 48710

LEVINE, SAMUEL GALE, b Malden, Mass, Nov 1, 28; m 53; c 4. ORGANIC CHEMISTRY. Educ: Tufts Univ, BS, 50; Harvard Univ, MA, 52, PhD(org chem), 54. Prof Exp: Res assoc, Forrestal Res Ctr, Princeton Univ, 53-54; res chemist, Walter Reed Army Inst Res, 54-56; res chemist, Eastern Regional Res Br, USDA, 56-60; sr chemist, Natural Prod Lab, Res Triangle Inst, 60-64; assoc prof, 64-68, PROF CHEM, NC STATE UNIV, 68- Concurrent Pos: Consult, Res Triangle Inst, 64-; Weizmann fel, Weizmann Inst Sci, 71-72. Mem: AAAS; Am Chem Soc. Res: New methods in organic synthesis; stereochemistry and conformational analysis; structure determination and synthesis of natural products. Mailing Add: Dept of Chem NC State Univ Raleigh NC 27607

LEVINE, SAMUEL HAROLD, b Hazlehurst, Ga, Nov 30, 25; m 55; c 3. NUCLEAR PHYSICS, REACTOR PHYSICS. Educ: Va Polytech Inst, BS, 47; Univ Ill, MS, 48; Univ Pittsburgh, PhD(physics), 54. Prof Exp: Instr physics, Va Polytech Inst, 49-50; sr scientist, Bettis Atomic Power Lab, Westinghouse Elec Corp, 54-55, supv scientist, 55-57, mgr, 57-59; physicist in charge, Gen Atomic Div, Gen Dynamics Corp, 59-61; group physicist, Rocketdyne Div, NAm Aviation, Inc, 61-62; lab head nuclear sci, Northrop Space Labs, 62-68; PROF NUCLEAR ENG & DIR NUCLEAR REACTOR FACILITY, PA STATE UNIV, UNIVERSITY PARK, 68- Concurrent Pos: Lectr, Univ Calif, Los Angeles, 64-68. Honors & Awards: Invention Award, NASA, 73. Mem: Am Phys Soc; Am Nuclear Soc; Am Soc Eng Educ. Res: Nuclear-electric and magnetic field anti-pollution devicedevices, measuring electric fields in space; neutron detection; design of space superconducting magnets; experimental reactor physics; nuclear reactor fuel management. Mailing Add: Breazeale Nuclear Reactor PA State Univ University Park PA 16802

LEVINE, SAMUEL J, bacteriology, see 12th edition

LEVINE, SAMUEL W, b Dallas, Tex, May 15, 16; m 44; c 1. PHYSICAL CHEMISTRY. Educ: Agr & Mech Col Tex, BS, 38, MS, 41; Mass Inst Technol, PhD(phys chem), 48. Prof Exp: Combustion engr, Lone Star Gas Co, Tex, 38-39; instr thermodyn, Agr & Mech Col Tex, 41-42; assoc chemist, Atlantic Refining Co, 48-51; dir develop labs, Fisher Sci Co, 51-53; assoc dir res & develop, Fairchild Camera & Instrument Corp, 53-55, dir res & eng, Graphic Equip Div, 55-59, dir res & eng, Defense Prod Div, 59-61, tech dir, Corp, NY, 61-70; vpres technol, Varadyne, Inc, Calif, 70-72; vpres corp develop, Mass, 72-73, V PRES TECHNOL, DATEL SYSTS, 73- Mem: Am Chem Soc; Optical Soc Am; Inst Elec & Electronics Engrs; NY Acad Sci; fel Am Inst Chemists. Res: X-ray spectroscopy; emission spectroscopy; petroleum reservoir characteristics; thermodynamic properties of hydrocarbons; radar systems research and development; instrumentation physics; radioactive tracers; photogrammetry instrumentation; corporate technical management; semiconductors; integrated circuits. Mailing Add: 11 Melby Lane East Hills NY 11576

LEVINE, SEYMOUR, b New York, NY, Mar 13, 25; m 45; c 2. PATHOLOGY, NEUROPATHOLOGY. Educ: NY Univ, BA, 46; Chicago Med Sch, MB, 47, MD, 48. Prof Exp: Pathologist, St Francis Hosp, Jersey City, NJ, 56-64; PROF PATH, NY MED COL, 64-, PATHOLOGIST & CHIEF LABS, BIRD S COLER HOSP, CTR CHRONIC DIS, 64- Mem: Am Asn Path & Bact; Soc Exp Biol & Med; Am Soc Exp Path; Am Asn Neuropath. Res: Demyelinating diseases; autoimmune disease. Mailing Add: Dept of Path NY Med Col Ctr for Chronic Dis Roosevelt Island NY 10017

LEVINE, SEYMOUR, b New York, NY, Jan 25, 25; m 49; c 3. PSYCHOPHYSIOLOGY. Educ: NY Univ, PhD(psychol), 52. Prof Exp: Res assoc, Queens Col, NY, 51-52; asst prof, Boston Univ, 52-53; lectr, Northwestern Univ, 54-56; asst prof psychiat, Med Sch, Ohio State Univ, 56-60; assoc prof, 62-69, PROF PSYCHOL, SCH MED, STANFORD UNIV, 69- Concurrent Pos: USPHS fel, 53-55; res assoc, Inst Psychosom & Psychiat, Michael Reese Hosp, 55-56; consult, Nat Cancer Inst, 57-; Found Fund Res Psychiat fel, Dept Neuroendocrinol, Inst Psychiat, Maudsley Hosp, London, 60; consult, Nat Inst Child Health & Human Develop, 66-67, mem neuropsychol res, 67-70, consult, Nat Comt Causes & Prev Violence, 68. Honors & Awards: Hoffheimer Res Award, 61. Mem: Am Psychol Asn; Endocrine Soc; Int Soc Develop Psychobiol. Res: Infantile experience development physiology and endocrinology. Mailing Add: Dept of Psychiat Stanford Univ Sch of Med Stanford CA 94305

LEVINE, SEYMOUR, b Chicago, Ill, Apr 30, 22; m 43, 66; c 2. VIROLOGY. Educ: Univ Chicago, BS, 43; Univ Ill, MS, 45, PhD(bact), 49; Am Bd Med Microbiol, cert pub health & med lab virol. Prof Exp: Asst bact, Med Sch, Univ Ill, 45-49; from instr to asst prof biophys, Univ Colo, 51-56; res biologist, Lederle Labs, Am Cyanamid Co, 56-65; sr res scientist, Upjohn Co, Mich, 65-71; ASSOC PROF MICROBIOL, SCH MED, WAYNE STATE UNIV, 71- Concurrent Pos: AEC fel, Univ Colo, 49-50; Nat Res Coun fel, Case Western Reserve Univ, 50-51. Mem: Am Soc Microbiol; Tissue Cult Asn; Am Acad Microbiol; Soc Exp Biol & Med. Res: Viral-host cell interactions; tissue culture; viral vaccines and chemoprophylaxis; viral interference and interferon. Mailing Add: Dept of Immunol & Microbiol Wayne State Univ Sch of Med Detroit MI 48201

LEVINE, SEYMOUR D, organic chemistry, see 12th edition

LEVINE, SOLOMON LEON, b Schenectady, NY, Jan 7, 40; m 60; c 3. ANALYTICAL CHEMISTRY. Educ: Rensselaer Polytech Inst, BS, 61; Univ RI, PhD(anal chem), 66. Prof Exp: Sr assoc engr, Components Div, 65-66, sr assoc chemist, Systs Develop Div, 66-68, staff chemist, 68-69, proj chemist, 69-72, develop chemist, 72-74, ADV CHEMIST, IBM CORP, 74- Mem: Am Chem Soc; Sigma Xi; Soc Appl Spectros. Res: Spectroscopy, absorption and emission; electroanalytical chemistry; environmental chemistry. Mailing Add: IBM Corp PO Box 390 Poughkeepsie NY 12601

LEVINE, STEPHEN ALAN, b Brooklyn, NY, Dec 24, 38; m 61; c 4. ORGANIC CHEMISTRY. Educ: City Col New York, BS, 61; Purdue Univ, PhD(org chem), 66. Prof Exp: Res chemist, Acme Shellac Prod Co, 61; chemist, 66-67, sr chemist, 67-73, RES CHEMIST, TEXACO RES CTR, 73- Mem: Fel Am Inst Chem; Am Chem Soc; NY Acad Sci. Res: Polymer chemistry; catalytic oxidation; organic synthesis through catalytic conversion; process research; lubricant additive synthesis. Mailing Add: Texaco Res Ctr PO Box 509 Beacon NY 12508

LEVINE, SUMNER NORTON, b Boston, Sept 5, 23; m 52; c 1. PHYSICAL CHEMISTRY. Educ: Brown Univ, BS, 46; Univ Wis, PhD(phys chem), 49. Prof Exp: Instr phys chem, Univ Chicago, 49-50; sr res fel, Columbia Univ, 50-54; dir res labs, US Vet Admin Hosp, East Orange, NJ, 54-56; mgr chem & physics lab, Gen Eng Labs, Am Mach & Foundry Co, 56-58; sr staff scientist, Surface Commun Div, Radio Corp Am, 58-60, head solid state devices & electronics, 60-61; chmn dept, 61-67, PROF MAT SCI, STATE UNIV NY STONY BROOK, 61- Concurrent Pos: Childs fel, Univ Chicago, 49; Runyan fel, Columbia Univ, 52; lectr, Atomic Indust Forum, 56, Albert Einstein Med Col, 57 & Grad Div, Univ Conn, 57-58; instr, Grad Div, Brooklyn Col, 60 & City Col New York, 60; vis prof & dir urban res, Grad Ctr, City Univ New York, 67-68; ed-in-chief, Advan in Biomed Eng & Med Physics, J Socio-Econ Planning Sci & J Biomed Mat Res; NSF guest lectr, Berlin Acad Sci. Mem: Am Chem Soc; Electrochem Soc; Sigma Xi; sr mem Inst Elec & Electronics Engrs; Inst Mgt Sci. Res: Biophysical investigation of reaction mechanisms and isotopes; semiconductor physics; solid state high frequency devices; thermoelectric materials and devices; energy conversion techniques. Mailing Add: Dept of Mat Sci State Univ of NY Stony Brook NY 11790

LEVINE, WALTER (GERALD), b Detroit, Mich, Dec 18, 30; m 55; c 3. PHARMACOLOGY. Educ: Wayne State Univ, BS, 52, MS, 54, PhD(physiol, pharmacol), 58. Prof Exp: Res assoc physiol & pharmacol, Wayne State Univ, 54-56 & 57-58, asst, 56-57; asst prof pharmacol, 61-67, ASSOC PROF PHARMACOL, ALBERT EINSTEIN COL MED, 67- Concurrent Pos: Fel pharmacol, Albert Einstein Col Med, 58-61; USPHS career develop award. Mem: Am Soc Pharmacol & Exp Therapeut; NY Acad Sci; Int Soc Biochem Pharmacol. Res: Biochemical pharmacology; drug metabolism and disposition. Mailing Add: Dept Pharmacol Albert Einstein Col of Med Yeshiva Univ Bronx NY 10461

LEVINGER, BERNARD WERNER, b Berlin, Ger, Sept 3, 28; nat US; m 54; c 3. MATHEMATICS. Educ: Lehigh Univ, BS, 48; Mass Inst Technol, MS, 50; NY Univ, PhD(math), 60. Prof Exp: Asst metallurgist, Armour Res Found, Ill Inst Technol, 51-52; res metallurgist, Tung-Sol Elec, Inc, 52-57; res engr, Labs, Gen Tel & Electronics Corp, 57-62; asst prof math, Case Western Reserve Univ, 62-68; ASSOC PROF MATH, COLO STATE UNIV, 68- Mem: Am Math Soc; Math Asn Am; Soc Indust & Appl Math. Res: Matrix theory; numerical analysis; group theory. Mailing Add: Dept of Math Colo State Univ Ft Collins CO 80521

LEVINGER, JOSEPH S, b New York, NY, Nov 14, 21; m 43; c 4. PHYSICS. Educ: Univ Chicago, BS, 41, MS, 44; Cornell Univ, PhD(physics), 48. Prof Exp: Jr physicist, Metall Lab, Univ Chicago, 42-44; physicist, Franklin Inst, 45-46; asst, Cornell Univ, 46-48, instr physics, 48-51; from asst prof to prof, La State Univ, 51-61; Avco vis prof, Cornell Univ, 61-64; PROF PHYSICS, RENSSELAER POLYTECH INST, 64- Concurrent Pos: Guggenheim fel, 57-58; Fulbright travel grant, 72-73; assoc prof, Univ Paris, 72-73. Mem: Am Phys Soc. Res: Theoretical physics. Mailing Add: Dept of Physics Rensselaer Polytech Inst Troy NY 12181

LEVINGS, CHARLES SANDFORD, III, b Madison, Wis, Dec 1, 30; m; c 4. GENETICS. Educ: Univ Ill, BS, 53, MS, 56, PhD(agron), 63. Prof Exp: Res instr, 62-64, from asst prof to assoc prof, 64-72, PROF GENETICS, NC STATE UNIV, 72- Mem: Am Soc Agron; Am Genetic Asn; Genetics Soc Am; Am Soc Plant Physiologists. Res: Autotetraploid genetics; maize biochemical genetics. Mailing Add: Dept of Genetics NC State Univ Raleigh NC 27607

LEVINGS, WILLIAM STEPHEN, b Denver, Colo, Dec 28, 96; m 32; c 2. GEOLOGY. Educ: Colo Sch Mines, EM, 20, MSc, 30, DSc(geol), 51. Prof Exp: Asst geologist, Mex Eagle Oil Co, 20-22; field engr & asst, Shell Oil Co, 22-23; draftsman, Standard Oil Col, 23-24, asst geologist, Standard Oil Co Calif, 24-25; geologist, Gulf Oil Corp, 25-27; instr geol, Colo Sch Mines, 27-28; torsion balance party chief, Gulf Oil Corp, 28-29; geologist, Sinclair Explor Co, 30-31; asst geol, Harvard Univ, 32-33; tech adv grade 13, Petrol Admin Bd, Washington, DC, 33-36; from instr to assoc prof geol, Colo Sch Mines, 36-55; assoc prof math, 55-57, prof geol, 57-75, EMER PROF GEOL, REGIS COL, 75- Honors & Awards: Bartlett Medal, Am Soc Photogram, 52. Mem: Am Soc Photogram; fel Geol Soc Am. Res: Geomorphology of the Raton Mesa region, New Mexico and Colorado; photogeology. Mailing Add: Dept of Geol Regis Col Denver CO 80221

LEVINS, PHILIP LEO, physical organic chemistry, see 12th edition

LEVINS, RICHARD, b New York, NY, June 1, 30; m 50; c 3. POPULATION BIOLOGY, MATHEMATICAL BIOLOGY. Educ: Cornell Univ, AB, 51; Columbia Univ, PhD(zool), 65. Prof Exp: Res assoc pop genetics, Univ Rochester, 60-61; assoc prof biol, Univ PR, 61-66; from assoc prof to prof math biol, Univ Chicago, 67-75; JOHN ROCK PROF POP SCI, SCH PUB HEALTH, HARVARD UNIV, 75- Concurrent Pos: NIH res grant, 63-66; consult genetics prog, Cuban Acad Sci, 64-65; NSF res grant, 64-66. Mem: Soc Study Evolution; Genetics Soc Am; Soc Gen Syst Res; Am Soc Naturalists. Res: Ecology and genetics; complex systems; agriculture. Mailing Add: Dept of Pop Sci Harvard Sch Pub Health Boston MA 02115

LEVINSKAS, GEORGE JOSEPH, b Tariffville, Conn, July 8, 24; m 46; c 3. TOXICOLOGY, PHARMACOLOGY. Educ: Wesleyan Univ, AB, 49; Univ Rochester, PhD(pharmacol), 53. Prof Exp: Res assoc biol sci, USAEC, Univ Rochester, 52-53; dept occup health, Grad Sch Pub Health, Univ Pittsburgh, 53-54, res assoc & lectr, 54-56, asst prof appl toxicol, 56-58; res pharmacologist, Cent Med Dept, Am Cyanamid Co, 58, chief indust toxicologist & dir environ health lab, 59-71; mgr prod eval, 71, MGR ENVIRON ASSESSMENT & TOXICOL, DEPT MED & ENVIRON HEALTH, MONSANTO CO, 72- Mem: AAAS; Am Chem Soc; Am Soc Pharmacol & Exp Therapeut; Soc Toxicol; NY Acad Sci. Res: Pharmacology and toxicology of boron compounds; organic phosphates; industrial chemicals; food additives; insecticides; chemistry of bone mineral. Mailing Add: Med Dept Monsanto Co 800 N Lindbergh Blvd St Louis MO 63166

LEVINSKY, NORMAN GEORGE, b Boston, Mass, Apr 27, 29; m 56; c 3. MEDICINE. Educ: Harvard Univ, AB, 50, MD, 54. Prof Exp: Intern & resident med, Beth Israel Hosp, Boston, Mass, 54-56; clin assoc, Nat Heart Inst, 56-58; NIH spec fel med, Boston Univ Hosp, 58-60; from instr to assoc prof, 60-68, Wesselhoeft prof, 68-72, WADE PROF MED & CHMN DIV, SCH MED, BOSTON UNIV, 72-, DIR EVANS MEM DEPT CLIN RES & PREV MED & PHYSICIAN-IN-CHIEF, UNIV HOSP, 72- Concurrent Pos: Asst dir, Univ Med Serv, Boston City Hosp, 61-68, dir, 68- Mem: Am Fedn Clin Res; Am Soc Clin Invest; Asn Am Physicians. Res: Renal physiology and medical research. Mailing Add: Boston Univ Med Ctr 80 E Concord St Boston MA 02118

LEVINSKY, WALTER JOHN, b Meadville, Pa, Sept 16, 20; m 48; c 4. MEDICINE. Educ: Allegheny Col, BS, 42; Temple Univ, MD, 45, MS, 52; Am Bd Internal Med, dipl, 54 & 74. Prof Exp: Intern med, Hamot Hosp, Erie, Pa, 45-46; resident path, Univ Hosp, 48-49, resident internal med, 49-52, instr, Sch Med, 52-54, assoc, 54-58, from asst prof to assoc prof internal med, 58-74, CLIN PROF MED, SCH MED, TEMPLE UNIV, 74- Concurrent Pos: Chief dept med, Northeastern Hosp, Philadelphia, Pa, 54-58. Honors & Awards: Christian Lindback Found Award, Temple Univ, 65. Mem: Sr mem am Fedn Clin Res; fel Am Col Physicians; fel Royal Soc Med. Res: Internal medicine; clinical research. Mailing Add: Dept of Internal Med Temple Univ Sch of Med Philadelphia PA 19140

LEVINSON, ALFRED ABRAHAM, b Staten Island, NY, Mar 31, 27. MINERALOGY. Educ: Univ Mich, BS & MS, 49, PhD(mineral), 52. Prof Exp: Asst, Univ Mich, 50-52, res assoc, 52-53; asst prof mineral, Ohio State Univ, 53-56; mineralogist, Dow Chem Co, 56-62; sr res geologist, Gulf Res & Develop Co, 62-67; PROF GEOL, UNIV CALGARY, 67- Concurrent Pos: Lectr, Univ Houston, 57-59; exec ed, Geochimica et Cosmochimica Acta, 67-70; ed, Proc Apollo 11 & Second Lunar Sci Conf. Mem: Fel Mineral Soc Am; Geochem Soc; fel Geol Soc Am; Mineral Asn Can. Res: General mineralogy and geochemistry with industrial application; economic geology; clay mineralogy; gemology; exploration geochemistry. Mailing Add: Dept of Geol Univ of Calgary Calgary AB Can

LEVINSON, ALFRED STANLEY, b Portland, Ore, Aug 27, 32; m 58; c 3. ORGANIC CHEMISTRY. Educ: Reed Col, BA, 54; Wesleyan Univ, MA, 57; Ind Univ, PhD(org chem), 63. Prof Exp: Res assoc chem, Ind Univ, 62-63; from asst prof to assoc prof, 63-73, PROF CHEM, PORTLAND STATE UNIV, 73- Mem: AAAS; Sigma Xi; Am Chem Soc; The Chem Soc. Res: Isolation and characterization of natural products; organic synthesis. Mailing Add: Dept of Chem Portland State Univ Box 751 Portland OR 97207

LEVINSON, CARL ANSELL, physics, see 12th edition

LEVINSON, CHARLES, b San Antonio, Tex, Dec 31, 36; m 67; c 2. CELL PHYSIOLOGY. Educ: Univ Tex, BA, 58; Trinity Univ, MA, 60; Rutgers Univ, PhD(physiol), 64. Prof Exp: Nat Cancer Inst fel, Med Col, Cornell Univ, 65-66; sr cancer res scientist, Roswell Park Mem Inst, 66-68; ASSOC PROF PHYSIOL, MED SCH, UNIV TEX, SAN ANTONIO, 68- Mem: Biophys Soc; Soc Gen Physiol; Am Physiol Soc. Res: Membrane phenomena; ion transport in tumor cells. Mailing Add: 942 Serenade San Antonio TX 78213

LEVINSON, GERALD STANLEY, physical chemistry, see 12th edition

LEVINSON, GILBERT E, b New York, NY, Jan 25, 28; m 50; c 2. MEDICINE, CARDIOLOGY. Educ: Yale Univ, AB, 48; Harvard Med Sch, MD, 53. Prof Exp: Intern med, Harvard Med Serv, Boston City Hosp, 53-54, asst resident, 54-55, chief resident, Thorndike Mem Ward, 58-59; from asst prof to prof med, Col Med & Dent NJ, NJ Med Sch, 68-76, assoc dean admin affairs, 70-73; PROF MED, MED SCH, UNIV MASS, 76-; CHIEF MED, ST VINCENT HOSP, WORCESTER, MASS, 76- Concurrent Pos: Teaching fel, Harvard Med Sch, 54-55; Nat Heart Inst res fel,

Thorndike Mem Lab, Boston City Hosp, 57-59; Nat Heart & Lung Inst res career develop award, 67-70; assoc dir, T J White Cardiopulmonary Inst, B S Pollak Hosp, Jersey City, NJ, 61-71; consult, USPHS Hosp, Staten Island, NY, 63-; estab investr, Union County Heart Asn, NJ, 61-66 & 70-75. Mem: Am Fedn Clin Res; fel Am Col Cardiol; Am Physiol Soc; Am Soc Clin Invest; fel Am Col Physicians. Res: Hemodynamics in valvular heart disease; indicator-dilution theory and methodology; cardiopulmonary blood volumes; relations between myocardial performance and metabolism. Mailing Add: Dept of Med Col of Med of NJ Newark NJ 07103

LEVINSON, HILLEL SALMON, b New York, NY, Feb 16, 18; m 43; c 1. MICROBIOLOGY. Educ: City Col New York, BS, 37, MS, 39; Univ Pa, PhD(med microbiol), 54. Prof Exp: With Nat Bur Standards, 42-43; bacteriologist, Biol Labs, Qm Res & Eng Ctr, 46-60, head bact group, Pioneering Res Lab, US Army Natick Lab, 60-74, RES MICROBIOLOGIST, FOOD SCI LAB, US ARMY NATICK DEVELOP CTR, 74- Concurrent Pos: Lectr microbiol, Sch Med, NY Univ, 66-; adj prof biol, Northeastern Univ, 75- Mem: Am Soc Microbiol; Am Acad Microbiol; NY Acad Sci; Soc Appl Bact. Res: Physiology and biochemistry of anaerobic spore formers; physiology and biochemistry of bacterial sporulation, spore germination and growth. Mailing Add: Food Sci Lab US Army Natick Develop Ctr Natick MA 01760

LEVINSON, LIONEL MONTY, b Johannesburg, SAfrica, Mar 12, 43. SOLID STATE PHYSICS. Educ: Univ Witwatersrand, BSc, 65, MSc, 66; Weizmann Inst Sci, PhD(solid state physics), 70. Prof Exp: PHYSICIST, GEN ELEC CORP RES & DEVELOP, 70- Mem: Am Phys Soc. Res: Electronic ceramics; varistors; Mössbauer effect. Mailing Add: Gen Elec Corp Res & Develop PO Box 8 Schenectady NY 12301

LEVINSON, NORMAN, mathematics, deceased

LEVINSON, SIDNEY BERNARD, b Russia, July 4, 11; nat US; m 36; c 2. CHEMISTRY, CHEMICAL ENGINEERING. Educ: City Col New York, BS(chem) & BS(eng), 32, Chem Engr, 33. Prof Exp: Consult, Protective Coating Lab, Joachim Res Labs, 33-36; consult, Indust Consult Labs, 36-42; mgr & tech dir, ADCO Chem Co, 42-48; mgr & tech dir, Garland Co, 48-52; vpres & tech dir, D H Litter Co, 52-73, PRES, DAVID LITTER LABS, INC, 74- Concurrent Pos: Dir, Artists Tech Res Inst, 65- Mem: Am Chem Soc; Am Soc Testing & Mat; Nat Asn Corrosion Engrs; Fedn Socs Coating Technol; Com Develop Asn. Res: Protective coatings; thermosetting and reinforced plastics, sealants and allied products; evaluation of raw materials; formulation; testing of finished products; preparation of specifications and manuals, personnel training, investigation of complaints and legal assistance. Mailing Add: David Litter Labs Inc 116 E 16th St New York NY 10003

LEVINSON, STEVEN R, b Brooklyn, NY, Oct 13, 47. ANALYTICAL CHEMISTRY, PHOTOGRAPHY. Educ: Rensselaer Polytech Inst, BS, 68, PhD(anal chem), 73. Prof Exp: SR RES CHEMIST, PHOTOG RES DIV, KODAK RES LABS, EASTMAN KODAK CO, 73- Concurrent Pos: Instr, Rochester Inst Technol, 76- Mem: Am Chem Soc; Sigma Xi; Soc Photog Scientists & Engrs. Res: Research and development of photographic materials. Mailing Add: Eastman Kodak Co Kodak Park-82 Rochester NY 14650

LEVINSON, STUART ALAN, b Detroit, Mich, Oct 29, 20; m 47; c 3. PALEONTOLOGY, GEOLOGY. Educ: Wayne State Univ, BS, 47; Washington Univ, AM, 49, PhD(geol), 51. Prof Exp: Asst geol, Washington Univ, 47-51; sr geologist, Humble Oil & Refining Co, 51-64; res supvr, 64-66, RES ASSOC, ESSO PROD RES CO, 66- Concurrent Pos: Instr, Washington Univ, 50-51. Mem: AAAS; Paleont Soc; Soc Econ Paleontologists & Mineralogists (vpres, 57); Geol Soc Am; Am Asn Petrol Geologists. Res: Invertebrate paleontology, micropaleontology, palynology, and zoology. Mailing Add: Esso Prod Res Co Box 2189 Houston TX 77001

LEVINSON, WARREN E, b Brooklyn, NY, Sept 28, 33; m 65. MICROBIOLOGY, CELL BIOLOGY. Educ: Cornell Univ, BS, 53; Univ Buffalo, MD, 57; Univ Calif, Berkeley, PhD(virol), 65. Prof Exp: ASSOC PROF MICROBIOL, MED CTR, UNIV CALIF, SAN FRANCISCO, 65- Concurrent Pos: Am Cancer Soc fel tumor viruses, Univ Col, Univ London, 65-67. Res: Tumor viruses. Mailing Add: Dept of Microbiol Univ of Calif Med Ctr San Francisco CA 94122

LEVINSTEIN, HENRY, b Themar, Ger, Dec 4, 19; nat US; m 62; c 3. PHYSICS. Educ: Univ Mich, BS, 42, MS, 43, PhD(physics), 47. Prof Exp: Lectr physics, Univ Mich, 46; from asst prof to assoc prof, 47-55, PROF PHYSICS, SYRACUSE UNIV, 55- Concurrent Pos: Consult, Westinghouse Elec Corp, 55-60; mem tech adv bd, Aerojet-Gen Corp, 61-66; univ adv, Tex Instruments, Inc, 61-; consult, Jet Propulsion Lab, Calif Inst Technol, 62-66, Int Bus Mach Corp & Gen Elec Co, 65-68. Mem: Am Phys Soc; Am Asn Physics Teachers; Optical Soc Am. Res: Photoconductivity; formation and structure of thin metallic films; infrared detectors. Mailing Add: Dept of Physics Syracuse Univ Syracuse NY 13210

LEVINTHAL, CYRUS, b Philadelphia, Pa, May 2, 22; m 44, 63; c 4. BIOPHYSICS. Educ: Swarthmore Col, BA, 43; Univ Calif, PhD(physics), 50. Prof Exp: From instr to assoc prof physics, Univ Mich, 50-57; prof biophys, Mass Inst Technol, 57-68; PROF BIOL & CHMN DEPT BIOL SCI, COLUMBIA UNIV, 68- Mem: Fel Am Acad Arts & Sci; Am Phys Soc; Am Soc Cell Biol; Genetics Soc Am; Biophys Soc. Res: Molecular biophysics and genetics. Mailing Add: Dept of Biol Sci Columbia Univ New York NY 10027

LEVINTHAL, ELLIOTT CHARLES, b Brooklyn, NY, Apr 13, 22; m 44; c 4. PHYSICS. Educ: Columbia Univ, BA, 42; Mass Inst Technol, MS, 43; Stanford Univ, PhD(physics), 49. Prof Exp: Proj engr, Sperry Gyroscope Co, NY, 43-46; res assoc nuclear physics, Stanford Univ, 46-48; res physicist, Varian Assocs, 49-50, res dir, 50-52; chief engr, Century Electronics & Instruments, Inc, 52-53; pres, Levinthal Electronic Prod, Inc, 53-61; assoc dean res affairs, Sch Med, 71-74, DIR, INSTRUMENTATION RES LAB & ADJ PROF GENETICS, STANFORD UNIV, 61- Mem: AAAS; Am Phys Soc; sr mem Inst Elec & Electronics Eng; Optical Soc Am; Biomed Eng Soc. Res: Measurements of nuclear moments; applications of computers to image processing and medical instrumentation; exobiology and planetary sciences. Mailing Add: Dept of Genetics Stanford Univ Sch Med Stanford CA 94305

LEVINTHAL, MARK, b Brooklyn, NY, Mar 3, 41; m 62; c 2. MICROBIAL GENETICS. Educ: Brooklyn Col, BS, 62; Brandeis Univ, PhD(biol), 66. Prof Exp: Fel genetics, Johns Hopkins Univ, 66-68; staff fel genetics lab molecular biol, Nat Inst Arthritis & Metab Dis, 68-72; ASSOC PROF BIOL, PURDUE UNIV, WEST LAFAYETTE, 72- Mem: Am Soc Microbiol; assoc Sigma Xi. Res: Regulation of enzyme synthesis of biosynthetic pathways and its relationship to general metabolic controls in bacteria. Mailing Add: Dept of Biol Sci Purdue Univ West Lafayette IN 47906

LEVINTON, JEFFREY SHELDON, b New York, NY, Mar 20, 46; m 67. ECOLOGY, PALEONTOLOGY. Educ: City Col New York, BS, 66; Yale Univ, MPhil, 69, PhD(paleoecol), 71. Prof Exp: From instr to asst prof, 70-74, ASSOC PROF PALEOECOL, STATE UNIV NY STONY BROOK, 74- Concurrent Pos: State Univ NY Stony Brook Res Found fel & grant-in-aid, 71; managing ed, Am Naturalist, 74- Mem: AAAS; Paleont Soc; Genetics Soc Am; Am Soc Naturalists; Soc Study Evolution. Res: Marine benthic ecology; paleoecology; isoenzymes of marine benthos; fossil population dynamics; benthic deposit feeder-detritus-bacteria interactions. Mailing Add: Dept of Ecol & Evolution State Univ NY Stony Brook NY 11790

LEVINTOW, LEÓN, b Philadelphia, Pa, Nov 10, 21; m 46; c 4. BIOCHEMISTRY, VIROLOGY. Educ: Haverford Col, AB, 43; Jefferson Med Col, MD, 46. Prof Exp: Intern, Jefferson Hosp, Philadelphia, Pa, 46-47; chief of lab, US Army Hepatitis Res Ctr, Ger, 47-49; biochemist, Nat Cancer Inst, 49-56, asst chief lab cell biol, Nat Inst Allergy & Infectious Dis, 56-61, asst chief lab biol viruses, 61-65; PROF MICROBIOL, SCH MED, UNIV CALIF, SAN FRANCISCO, 65- Concurrent Pos: Res fel, Biochem Res Lab, Mass Gen Hosp, Boston, 51-52. Mem: Am Soc Microbiol; Am Chem Soc; Am Soc Biol Chemists; Am Soc Cell Biol. Res: Biochemistry of viruses. Mailing Add: Dept of Microbiol Univ of Calif Sch of Med San Francisco CA 94122

LEVIS, WILLIAM WALTER, JR, b Chicago, Ill, May 14, 18; m 41; c 2. ORGANIC CHEMISTRY. Educ: Univ Fla, BS, 41. Prof Exp: Res chemist, Fla Chem Indust, 41-42; sr res chemist, Sharples Chem, Inc, 42-52; sr res chemist, 52-55, sect head, 55-56, RES SUPVR, BASF WYANDOTTE CORP, 56- Mem: Am Chem Soc. Res: Organic synthesis; catalysis; hydrogenation; amination; oxyalkyation. Mailing Add: 2233 17th St Wyandotte MI 48192

LEVI SETTI, RICCARDO, b Milan, Italy, July 11, 27; m 59; c 2. PHYSICS. Educ: Univ Pavia, Dr, 49. Prof Exp: Asst physics, Univ Pavia, 49-51 & Univ Milan, 52-56; res assoc, Inst, 56-57, from asst prof to assoc prof, Univ, 57-64, PROF PHYSICS, ENRICO FERMI INST, UNIV CHICAGO, 65- Concurrent Pos: Guggenheim fel, 63-64. Mem: Fel Am Phys Soc; Ital Phys Soc. Res: Elementary particles; high energy nuclear physics. Mailing Add: Enrico Fermi Inst Univ of Chicago 5630 Ellis Ave Chicago IL 60637

LEVISON, MATTHEW EDMUND, b New York, NY, May 18, 37; m 66; c 1. MEDICAL SCIENCE, HEALTH SCIENCES. Educ: Columbia Univ, BA, 58; State Univ NY, MD, 62. Prof Exp: Asst instr med, Downstate Med Ctr, State Univ NY, 65-67; asst physician, NY Hosp, 67-69; instr, Med Col, Cornell Univ, 68-69; clin instr, Downstate Med Ctr, State Univ, NY, 69-70; asst prof med & chief, 70-73, ASSOC PROF MED & CHIEF INFECTIOUS DIS DIV, MED COL PA, 73- Concurrent Pos: Attend physician & chief infectious dis unit, Queens Hosp Ctr, Long Island Jewish Med Ctr affil, 69-70; attend staff, Philadelphia Vet Admin Hosp, 70- Mem: Am Soc Microbiol; Am Fedn Clin Res; Infectious Dis Soc Am; fel Am Col Clin Pharmacol; fel Am Col Physicians. Res: Anaerobic bacteria, the pathogenesis of the renal concentrating defect in experimental pylonephritis and the pathogenesis of experimental endocarditis. Mailing Add: Med Col Pa 3300 Henry Ave Philadelphia PA 19129

LEVISON, STUART ALLEN, physical chemistry, biochemistry, see 12th edition

LEVIT, EDITHE J, b Wilkes-Barre, Pa, Nov 29, 26; m 52; c 2. MEDICINE. Educ: Bucknell Univ, BS, 46; Woman's Med Col Pa, MD, 51. Prof Exp: Intern med, Philadelphia Gen Hosp, 51-52, resident endocrinol, 52-53, clin instr, 53-57, dir med educ, 57-61; asst dir, 61-67, secy & assoc dir, 67-75, VPRES NAT BD MED EXAMR, 75- Mem: Fel Am Col Physicians; Asn Am Med Cols; AMA. Res: Evaluation and research in medical education. Mailing Add: Nat Bd of Med Examr 3930 Chestnut St Philadelphia PA 19104

LEVIT, LAWRENCE BRUCE, b Cleveland, Ohio, Sept 24, 42; m 67; c 1. PHYSICS. Educ: Case Western Reserve Univ, BS, 64, PhD(physics), 71. Prof Exp: Res assoc physics, Case Western Reserve Univ, 66-69; asst prof, La State Univ, Baton Rouge, 69-74; MEM STAFF, LECROY RES SYST CORP, 74- Concurrent Pos: Resident physicist, High Energy Physics Lab, Climax, Colo, 69-71. Mem: Am Phys Soc; Am Inst Physics. Res: Ultrahigh energy physics research using cosmic rays as a particle source. Mailing Add: Lecroy Res Syst Corp 126 North Rte 303 West Nyack NY 10994

LEVIT, ROBERT JULES, b San Francisco, Calif, Aug 17, 16; m 43; c 3. NUMBER THEORY. Educ: Calif Inst Technol, BS, 38, MS, 39; Univ Calif, PhD(math), 41. Prof Exp: Asst math, Univ Calif, 40-41; from asst prof to assoc prof, Univ Ga, 46-53; vis asst prof, Mass Inst Technol, 54-55; mem staff, Appl Sci Div, Int Bus Mach Corp, 55-57; from asst prof to prof math, 57-72, EMER PROF MATH, SAN FRANCISCO STATE UNIV, 72- Mem: Am Math Soc; Math Asn Am; Asn Symbolic Logic. Res: Foundations of mathematics; abstract algebra; digital computers. Mailing Add: 148 Miraloma Dr San Francisco CA 94127

LEVITAN, MAX, b Tverai, Lithuania, Mar 1, 21; nat US; m 47; c 3. GENETICS, ANATOMY. Educ: Univ Chicago, AB, 44; Univ Mich, MA, 46; Columbia Univ, PhD(zool), 51. Prof Exp: Statistician, USPHS, 44-45; asst zool, Columbia Univ, 46-49; assoc prof genetics, Va Polytech Inst, 49-55; from asst prof to assoc prof anat, Woman's Med Col Pa, 55-62, prof anat & med genetics, 62-66; prof biol & chmn dept, George Mason Col, Univ Va, 66-68; assoc prof anat, 68-70, PROF ANAT, MT SINAI SCH MED, 70- Concurrent Pos: Seminar assoc, Columbia Univ, 58-; spec lectr, Univ Pa, 62-63; actg chmn dept anat, Woman's Med Col Pa, 64-66; adj prof anat & genetics, George Washington Univ & Sch Med, Univ Va, 66-68. Mem: Am Soc Nat; Am Soc Human Genetics; Genetics Soc Am; Soc Study Social Biol; Soc Study Evolution. Res: Cytogenetics; population genetics of linked loci; chromosome breakage; cytoplasmic inheritance; medical genetics. Mailing Add: 1212 Fifth Ave New York NY 10029

LEVITAN, MICHAEL LEONARD, b Brooklyn, NY, Sept 12, 41; m 64; c 2. MATHEMATICS. Educ: Rensselaer Polytech Inst, BS, 62; Univ Minn, MS, 66, PhD(math), 67. Prof Exp: Asst prof math, Drexel Univ, 67-70; asst prof, 70-74, ASSOC PROF MATH, VILLANOVA UNIV, 74- Mem: Am Math Soc; Math Asn Am. Res: Probability theory; Markov processes; operations research; statistics. Mailing Add: Dept of Math Villanova Univ Villanova PA 19085

LEVITAN, RUVEN, b Kaunas, Lithuania, Mar 12, 27; US citizen; m 49; c 3. INTERNAL MEDICINE, GASTROENTEROLOGY. Educ: Hebrew Univ, Israel, MD, 53; Am Bd Internal Med, dipl & cert gastroenterol. Prof Exp: Resident, Mt Sinai Hosp, NY, 56-57; resident, Beth Israel Hosp, Boston, 58-59; dir gastroenterol res, New Eng Med Ctr Hosps, 64-68; assoc prof, 68-70, PROF MED, ABRAHAM LINCOLN SCH MED, UNIV ILL MED CTR, 70-; CHIEF GASTROENTEROL SECT, VET ADMIN WEST SIDE HOSP, 68- Concurrent Pos: Spec fel med neoplasia, Mem Ctr Cancer & Allied Dis, NY, 57-58; fel gastroenterol & res fel med, Mass Mem Hosps & Sch Med, Boston Univ, 59-61, sr res fel, Mass Mem Hosps, 61-62; from asst prof to assoc prof, Sch Med, Tufts Univ, 64-69; lectr, Sch Med, Boston Univ, 65-68; pres, Chicago Soc Gastroenterol. Mem: Fel Am Col Physicians; Am

Physiol Soc; Am Gastroenterol Asn; Am Asn Study Liver Dis; Am Soc Clin Invest. Res: Water electrolyte absorption from the intestine; hormonal influences on absorption; lymphomas, including involvement of liver and gastrointestinal tract. Mailing Add: Vet Admin West Side Hosp PO Box 8195 Chicago IL 60680

LEVITAN, STEPHEN ROY, organic chemistry, see 12th edition

LEVITAS, ALFRED DAVE, b New York, NY, Mar 27, 20; m 43; c 1. PHYSICS. Educ: Syracuse Univ, BA, 47, MS, 50, PhD(physics), 58. Prof Exp: Res engr solid state physics, Sylvania Elec Corp, 53-55 & Sprague Elec Corp, 55-56; physicist, Honeywell Res Ctr, 56-58; PROF PHYSICS, STATE UNIV NY ALBANY, 58- Concurrent Pos: Consult, Naval Res Lab, 60-63. Mem: Am Phys Soc. Res: Solid state and statistical physics; thermodynamics. Mailing Add: Dept of Physics State Univ of NY Albany NY 12203

LEVITON, ALAN EDWARD, b Brooklyn, NY, Jan 11, 30; m 52; c 2. SYSTEMATIC ZOOLOGY, ZOOGEOGRAPHY. Educ: Stanford Univ, AB, 49, AM, 53, PhD, 60. Prof Exp: From asst cur to assoc cur, 57-62, CUR HERPET & CHMN DEPT, CALIF ACAD SCI, 62- Concurrent Pos: Assoc cur div syst biol, Stanford Univ, 62-63, lectr, 62-70; lectr, San Francisco State Univ, 67- Mem: Fel AAAS (secy-treas, Pac Div, 75-); Soc Syst Zool; Soc Study Amphibians & Reptiles; Am Soc Ichthyol & Herpet; Am Inst Biol Sci. Res: Herpetology of Asia; Tertiary paleogeography; phylogeny and taxonomy of reptiles. Mailing Add: Dept of Herpet Calif Acad of Sci San Francisco CA 94118

LEVITSKY, JOHN M, b Youngstown, Ohio, Nov 2, 23; div; c 3. INTERNAL MEDICINE, PREVENTIVE MEDICINE. Educ: Univ Nebr, MD, 49; Am Bd Internal Med, dipl, 57. Prof Exp: NIH fel med genetics, Johns Hopkins Univ, 59-60; asst prof, 60-65, ASSOC PROF PREV MED, UNIV ILL COL MED, 65- Mem: AAAS; fel Am Col Physicians; Asn Teachers Prev Med; Am Soc Human Genetics. Res: Inheritance of various human disorders; host factors in the causation of illness. Mailing Add: 835 S Wolcott Chicago IL 60612

LEVITSKY, MICHAEL, analytical chemistry, see 12th edition

LEVITSKY, SIDNEY, b New York, NY, Mar 3, 36; m 67; c 3. CARDIOVASCULAR SURGERY, SURGERY. Educ: Albert Einstein Col Med, MD, 60; Bd Surg & Bd Thoracic Surg, dipl, 68. Prof Exp: Instr surg, Sch Med, Yale Univ, 64-66; chief surg, Third Surg Hosp, Vietnam, 66-67; thoracic surgeon, Valley Forge Army Hosp, 67-68; sr investr cardiac surg, Nat Heart Inst, NIH, 68-70; assoc prof surg, 70-75, PROF SURG & PHARMACOL, COL MED, UNIV ILL, 75-, CHIEF DIV CARDIOTHORACIC SURG, MED CTR, 74-, LECTR SURG, GRAD SCH, 70- Concurrent Pos: Estab investr, Am Heart Asn, 71; attend surgeon, Cook County Hosp, 73-; sr consult, West Side Vet Hosp, 75- Mem: Soc Univ Surgeons; Am Physiol Soc; Soc Thoracic Surgeons; Am Asn Thoracic Surg; Asn Acad Surg. Res: Thoracic surgery; non-invasive methods of monitoring myocardial contractility; intra-operative protection of myocardium; myocardial ischemia and metabolism. Mailing Add: Dept of Surg PO Box 6990 Col Med Univ of Ill Chicago IL 60680

LEVITT, ABEL, b Lithuania, Oct 20, 97; nat US; m. INTERNAL MEDICINE. Educ: McGill Univ, MD & CM, 21; Am Bd Internal Med, dipl. Prof Exp: Intern, Maine Gen Hosp, Portland, Ore, 21-22; resident physician, Buffalo Gen Hosp, NY, 22-23; chief resident physician, Buffalo City Hosp, 23-28, externe in charge med serv, 28-35; CHIEF VIS PHYSICIAN, MEYER MEM HOSP, 35-; ASSOC PROF THERAPEUT & CLIN PROF MED, SCH MED, STATE UNIV NY BUFFALO, 41- Concurrent Pos: Consult, US Selective Serv Syst, 42-46; vis physician, Millard Fillmore Hosp, 46-; assoc, Buffalo Gen Hosp. Mem: AAAS; AMA; Am Diabetes Asn; fel Am Col Physicians; fel Am Col Chest Physicians. Mailing Add: 118 Crosby Blvd Buffalo NY 14226

LEVITT, ARNOLD EDWIN, analytical chemistry, see 12th edition

LEVITT, BARRIE, b Brooklyn, NY, Aug 19, 35; m 68; c 1. PHARMACOLOGY, INTERNAL MEDICINE. Educ: State Univ NY Downstate Med Ctr, MD, 59. Prof Exp: Rotating intern, Mt Sinai Hosp, New York, 59-60, resident med, 60-63; fel pharmacol, State Univ NY Downstate Med Ctr, 63-64; fel, Med Col, Cornell Univ, 64-65, from instr to asst prof pharmacol, 65-69; asst prof med, 69-70, ASSOC PROF MED & PHARMACOL & DIR DIV CLIN PHARMACOL, NEW YORK MED COL, 70- Concurrent Pos: NY Heart Asn sr investr, Med Col, Cornell Univ, 66-69; consult, Bur Drugs, US Food & Drug Admin, 71- Mem: Am Soc Pharmacol & Exp Therapeut; Am Heart Asn. Res: Clinical and cardiovascular pharmacology; cardiology. Mailing Add: 1249 Fifth Ave New York NY 10029

LEVITT, DAVID GEORGE, b Minneapolis, Minn, May 9, 42; m 64; c 2. PHYSIOLOGY. Educ: Univ Minn, BS, 66, MD & PhD(physiol), 68. Prof Exp: ASSOC PROF PHYSIOL, UNIV MINN, MINNEAPOLIS, 68- Res: Theoretical transport processes across membranes and in capillary beds; intestinal absorption; microcirculation in skeletal muscle. Mailing Add: Dept of Physiol Univ of Minn Minneapolis MN 55455

LEVITT, GEORGE, b Newburg, NY, Feb 19, 25; m 50; c 4. ORGANIC CHEMISTRY. Educ: Duquesne Univ, BS, 50, MS, 52; Mich State Univ, PhD, 57. Prof Exp: Res chemist, Exp Sta, 56-63, res chemist, Stine Lab, 63-66, res chemist, Exp Sta, 66-68, SR RES CHEMIST, EXP STA, E I DU PONT DE NEMOURS & CO, INC, 68- Res: Organic syntheses; agricultural chemicals medicinals; pesticides. Mailing Add: 3218 Romilly Rd Wilmington DE 19810

LEVITT, ISRAEL MONROE, b Philadelphia, Pa, Dec 19, 08; m 37; c 2. ASTRONOMY. Educ: Drexel Inst Technol, BS, 32; Univ Pa, AM, 37, PhD(astron), 48. Hon Degrees: DSc, Temple Univ, 58, Drexel Inst Technol, 58, Philadelphia Col Pharm, 63. Prof Exp: Engr, Abrasive Co, 29-30; astronr, Franklin Inst, 33-39, asst dir, Fels Planetarium, 39-48, dir, 48-70, vpres, Inst, 70-72; EXEC DIR, SCI & TECHNOL COUN, MAYOR PHILADELPHIA, PA, 72- Concurrent Pos: Engr, Eclipse Exped, Franklin Inst, 32, asst assoc dir astron, photog & seismol, 38-48 & assoc dir astron & seismol, 49-70; astronr, Cook Observ, Univ Pa, 35-46; mem, Air Pollution Control Bd, Philadelphia, 64-, chmn, 66- Honors & Awards: Joseph Priestly Award, Spring Garden Inst, 63; Samuel S Fels Medal, 70. Mem: AAAS; fel Am Astron Soc; Am Inst Aeronaut & Astronaut; Nat Asn Sci Writers; Brit Astron Soc. Res: Lunar studies; scientific museum and planetarium operation. Mailing Add: 223 Melrose Circle Merion PA 19066

LEVITT, JACOB, b Montreal, Que, Sept 22, 11; nat US; m 42; c 2. PLANT PHYSIOLOGY. Educ: McGill Univ, BSc, 32, MSc, 33, PhD(bot), 35. Prof Exp: Asst plant physiol, Macdonald Col, McGill Univ, 34, Nat Res Coun Can bursary, 35-36; Royal Soc Can res fel, Univ Minn, 36-37; lectr bot, Macdonald Col, McGill Univ, 38-40; instr plant physiol, Univ Minn, 40-42, asst prof bot, 42-47; from assoc prof to prof, Univ Mo-Columbia, 47-73, chmn dept, 65-68; sr scientist, Volcani Ctr, Israel, 73-74; VIS PROF HORT SCI, UNIV MINN, ST PAUL, 74- Concurrent Pos: Guggenheim fel, 54-55; NSF sr fel, 61-62. Mem: AAAS; Am Soc Plant Physiol; Soc Cryobiol (pres, 71-72); Scand Soc Plant Physiol. Res: Frost hardiness and drought resistance of plants; cell physiology; water relations of plants; ion absorption. Mailing Add: Dept of Hort Sci Univ of Minn St Paul MN 55101

LEVITT, LEONARD SIDNEY, b Philadelphia, Pa, Feb 1, 26; m 52; c 1. INORGANIC CHEMISTRY, PHYSICAL ORGANIC CHEMISTRY. Educ: Univ Pa, BA, 46; Pa State Univ, BS, 47; Haverford Col, MS, 48; Temple Univ, PhD(phys org chem), 53. Prof Exp: Asst chem, Haverford Col, 47-48, instr, 52; assoc prof & head dept, Union Col, Ky, 53; instr, Stevens Inst Technol, 53-54, from asst prof to assoc prof, 54-61; prof, Seton Hall Univ, 61-65; PROF CHEM, UNIV TEX, EL PASO, 65- Concurrent Pos: Asst, Temple Univ, 53; vis lectr, Farleigh Dickinson Univ, 59-60; vis lectr, St John's Univ, NY, 60. Mem: Am Phys Soc; Am Chem Soc. Res: Kinetics and mechanism of persulfate oxidation; mechanism of organic oxidations; photoelectric theory of photosynthesis; equation of state of gases, liquids, and solids at extreme pressures; ionization potentials of organic molecules; base strengths of weak organic bases. Mailing Add: Dept of Chem Univ of Tex El Paso TX 79999

LEVITT, LEROY P, b Plymouth, Pa, Jan 8, 18; m 71; c 4. PSYCHIATRY. Educ: Pa State Univ, BS, 39; Chicago Med Sch, MD, 43; Inst Psychoanal, cert, 59. Prof Exp: Pvt pract, 49-66; prof psychiat & dean, Chicago Med Sch, 66-73; DIR DEPT MENT HEALTH, STATE OF ILL, 73- Concurrent Pos: Consult, Chicago Am Red Cross, 50-54, Asn Family Living, 50-54 & Nat Coun Aging, 52-; mem, Mayor's Comn Aging, 60- & Gov Comn Ment Health Planning Bd, 66- Honors & Awards: Chicagoan of Year Award in Med, 71; Gold Medal Sci Award, Phi Lambda Kappa, 74. Mem: Fel Am Psychiat Asn; fel Am Psychoanal Asn; fel Acad Psychoanal; Am Col Psychiat; Am Col Psychoanal (treas, 71-73). Res: Process of aging; medical education and administration and study of personality of medical students; psychoanalysis; geriatric psychiatry; mental health. Mailing Add: State Dept of Ment Health 160 N LaSalle St Chicago IL 60601

LEVITT, MARVIN FREDERICK, b New York, NY, Dec 9, 20; c 2. NEPHROLOGY. Educ: Cornell Univ, BA, 41; NY Univ, MD, 44. Prof Exp: Res asst med, Mt Sinai Sch Med, 50-53; asst attend physician, Mt Sinai Hosp, 53-60; CHIEF DIV NEPHROLOGY, DEPT MED, MT SINAI SCH MED, 60-, PROF MED, 68- Concurrent Pos: Mem cardio-vascular renal panel, Mayor's Res Coun, 69-72; mem sci adv bd, NY State Kidney Dis Inst, 69-72; emer mem, Nat Heart Inst Training Comt; chmn med adv bd, NY Kidney Dis Found. Mem: NY Acad Sci; Harvey Soc; Am Fedn Clin Res; fel Am Col Physicians; Asn Am Physicians. Mailing Add: 1176 Fifth Ave New York NY 10029

LEVITT, MELVIN, b Chicago, Ill, Mar 13, 25. NEUROBIOLOGY. Educ: Roosevelt Univ, BS, 49, MA, 53; Mich State Univ, PhD(psychol), 58. Prof Exp: Res asst neurol & psychiat, Med Sch, Northwestern Univ, 52-54; res assoc neurophysiol, Rockefeller Inst, 61; assoc anat, Sch Med, Univ Pa, 61-65, asst prof anat & mem, Inst Neurol Sci, 65-70; ASSOC PROF PHYSIOL, BOWMAN GRAY SCH MED, 70- Concurrent Pos: USPHS fel, Inst Neurol Sci, Sch Med, Univ Pa, 57-61. Mem: Am Physiol Soc; Am Asn Anat; Soc Neurosci; Int Asn Study Pain. Res: Sensory organization in central nervous system of vertebrates. Mailing Add: Bowman Gray Sch of Med Wake Forest Univ Winston-Salem NC 27103

LEVITT, MICHAEL D, b Chicago, Ill, May 10, 35; m 56; c 3. GASTROENTEROLOGY. Educ: Univ Minn, BS, 58, MD, 50. Prof Exp: Intern, Univ Minn Hosp, 60-61; resident, Boston Univ Hosp, 61-64; resident, Beth Israel Hosp, Boston, 64-65; fel gastroenterol, Boston City Hosp, 65-68; from asst to assoc prof, 68-74, PROF MED, MED SCH, UNIV MINN, 74- Concurrent Pos: Guest lectr, Gastroenterol Res Group, 72; counr, Am Fedn Clin Res, 74-76; consult med, Minneapolis Vet Admin Hosp, 74- Mem: Am Fedn Clin Res; Am Soc Clin Invest; Am Gastroenterol Soc. Res: Studies employing gas to investigate gastrointestinal physiology and studies of serum and urinary isoamylases. Mailing Add: Gastroenterol Sect Dept of Med Box 36 Univ of Minn Hosp Minneapolis MN 55455

LEVITT, NEIL HILLIARD, b Philadelphia, Pa, Aug 12, 41; m 64; c 2. VIROLOGY, IMMUNOLOGY. Educ: Temple Univ, AB, 63; Hahnemann Med Col, PhD(virol), 68. Prof Exp: RES VIROLOGIST, US ARMY MED RES INST, 69- Mem: Am Soc Microbiol; NY Acad Sci. Res: Antigens of specific arboviruses, their isolation, purification and use in the development of rapid diagnostic tests of arbovirus infections. Mailing Add: US Army Med Res Inst Ft Detrick MD 21701

LEVITT, NORMAN JAY, b New York, NY, Aug 27, 43. MATHEMATICS. Educ: Harvard Univ, AB, 63; Princeton Univ, AM, 65, PhD(math), 67. Prof Exp: Instr math, Courant Inst Math Sci, NY Univ, 67-69; asst prof, 69-71, ASSOC PROF MATH, RUTGERS UNIV, NEW BRUNSWICK, 72- Mem: Am Math Soc. Res: Relationships of various manifold-like categories and their corresponding bundle theories. Mailing Add: Dept of Math Rutgers Univ New Brunswick NJ 08903

LEVITT, SEYMOUR H, b Chicago, Ill, July 18, 28; m 52; c 3. RADIOTHERAPY. Educ: Univ Colo, BA, 50, MD, 54. Prof Exp: Instr radiation ther & radiol, Med Sch, Univ Mich, 61-62; asst radiotherapist, Sch Med & Dent, Univ Rochester, 62-63; assoc prof radiation ther & chief div, Sch Med, Univ Okla, 63-66; prof radiol & chmn div radiation ther, Med Col Va, 66-70; PROF THERAPEUT RADIOL & HEAD DEPT, UNIV MINN, MINNEAPOLIS, 70- Concurrent Pos: Consult radiother, Vet Admin Hosp, Minneapolis. Mem: Fel Am Col Radiol; Am Radium Soc; Soc Nuclear Med; Radiol Soc NAm; Am Soc Therapeut Radiol. Res: Experimental and clinical radiation therapy; radiation biology. Mailing Add: Dept of Therapeut Radiol Univ of Minn Hosps Minneapolis MN 55455

LEVITZ, HILBERT, b Lebanon, Pa, Nov 13, 31; m 66. MATHEMATICS. Educ: Univ NC, BA, 53; Pa State Univ, PhD(math), 65. Prof Exp: Instr math, Williams Col, 65; asst prof, NY Univ, 65-69; ASSOC PROF MATH, FLA STATE UNIV, 69- Mem: Am Math Soc; Asn Symbolic Logic. Res: Mathematical logic; concrete systems of ordinal notations. Mailing Add: Dept of Math Fla State Univ Tallahassee FL 32304

LEVITZ, KATHLEEN BULGER, mathematics, see 12th edition

LEVITZ, MORTIMER, b New York, NY, May 11, 21; m 47; c 2. BIOCHEMISTRY, ENDOCRINOLOGY. Educ: City Col New York, BS, 41; Columbia Univ, MA, 47, PhD(org chem), 51. Prof Exp: Res assoc steroid biochem, Col Physicians & Surgeons, Columbia Univ, 51-52; res assoc, 52-56, from asst prof to assoc prof, 56-67, PROF STEROID BIOCHEM, MED CTR, NY UNIV, 67- Concurrent Pos: NIH res career award, 62-72; consult, Endocrine Study Sect, NIH, 66-70 & 73-75. Mem: Am Chem Soc; Am Soc Biol Chemists; Endocrine Soc; Soc Gynec Invest. Res: Estrogen metabolism and mechanisms of action in pregnancy and cancer. Mailing Add: NY Univ Med Ctr 550 First Ave New York NY 10016

LEVKOV, JEROME STEPHEN, b New York, NY, June 12, 39; m 70. PHYSICAL CHEMISTRY. Educ: City Col New York, BS, 61; Univ Pa, PhD(phys chem), 67. Prof Exp: Swiss Copper Inst fel, Swiss Fed Inst Technol, 67-68; asst prof gen & phys chem, Drexel Univ, 68-69; ASST PROF GEN & PHYS CHEM, IONA COL, 70- Mem: Am Chem Soc. Res: Transport properties in electrolyte solutions; structure of solutions of electrolytes in solvents of low dielectric constant polymorphic transitions; forensic chemistry. Mailing Add: Dept of Chem Iona Col New Rochelle NY 10801

LE VON, ERNEST FRANKLIN, b Chicago, Ill, Dec 17, 31; m 55; c 3. ORGANIC CHEMISTRY. Educ: Univ Ill, BS, 54; Univ Mich, MS, 56, PhD(chem), 59. Prof Exp: RES SCIENTIST, G D SEARLE & CO, 58- Mem: AAAS; Am Chem Soc. Res: Synthesis and chemistry of organic compounds having therapeutic activity. Mailing Add: Searle Labs Searle Pkwy Skokie IL 60076

LEVOW, ROY BRUCE, b Richmond, Va, June 3, 43; m 62; c 1. MATHEMATICS, COMPUTER SCIENCE. Educ: Univ Pa, AB, 64, PhD(math), 69. Prof Exp: Sci programmer, Atlantic-Richfield Co, 64-65; asst prof math, Univ Hawaii, 69-70; asst prof, 70-75, ASSOC PROF MATH & CHMN DEPT, FLA ATLANTIC UNIV, 75- Mem: Am Math Soc; Math Asn Am. Res: Combinatorics and graph theory with emphasis on combinatorial optimization problems and the permanent function. Mailing Add: Dept of Math Fla Atlantic Univ Boca Raton FL 33432

LEVY, ALAN, b New York, July 25, 37; m 62; c 2. POLYMER CHEMISTRY. Educ: City Col New York, BS, 58; Purdue Univ, PhD(chem), 62. Prof Exp: Sr res chemist, Cent Res Lab, Allied Chem Corp, NJ, 62-66; sr res scientist org polymer chem, 66-70, PRIN SCIENTIST & GROUP LEADER, ETHICON INC, 70-, MGR POLYMER DEPT, 75- Mem: AAAS; Am Chem Soc; NY Acad Sci. Res: Biomedical materials; polymer and synthetic organic chemistry. Mailing Add: Ethicon Inc Rte 22 Somerville NJ 08876

LEVY, ALAN B, b San Francisco, Calif, Apr 12, 45; m 69; c 1. ORGANOMETALLIC CHEMISTRY. Educ: Univ Calif, Berkeley, BS, 67; Univ Colo, Boulder, PhD(chem), 71. Prof Exp: Fel chem, Purdue Univ, 71-74; ASST PROF CHEM, STATE UNIV NEW YORK STONY BROOK, 74- Mem: Am Chem Soc; Sigma Xi; AAAS. Res: The use of organoboranes and organocopper reagents for the development of new synthetic methods; the total synthesis of natural products. Mailing Add: Dept of Chem State Univ of New York Stony Brook NY 11794

LEVY, ALAN C, b Baltimore, Md, Feb 24, 30; m 56; c 2. PHYSIOLOGY, TOXICOLOGY. Educ: Univ Md, BS, 52; George Washington Univ, MS, 56; Georgetown Univ, PhD(physiol), 58. Prof Exp: Instr physiol, Sch Med, Howard Univ, 58-60; sect head, Dept Endocrinol, Wm S Merrell Co, 60-67; dir labs, Woodard Res Corp, 67-69; group chief, 69-74, SECT HEAD, DEPT TOXICOL, HOFFMANN-LA ROCHE INC, 74- Mem: AAAS; Endocrine Soc; Am Chem Soc; NY Acad Sci; Soc Toxicol. Res: Inflammation; anti-inflammation; adrenal cortex; neuroendocrinology; lipid metabolism; acute and chronic toxicology; teratology. Mailing Add: Dept of Toxicol Hoffmann-La Roche Inc Nutley NJ 07110

LEVY, ALLAN HENRY, b New York, NY, Nov 2, 29; m 61; c 1. COMPUTER SCIENCES, VIROLOGY. Educ: Columbia Univ, AB, 49; Harvard Med Sch, MD, 53. Prof Exp: From intern to asst resident, Harvard Med Serv, Boston City Hosp, 53-55; clin assoc, Nat Cancer Inst, 55-57; from instr to asst prof microbiol, Johns Hopkins Univ, 59-65; assoc prof virol & comput sci, Baylor Col Med, 65-71, prof comput sci, 71-73, prof virol & Epidemiol, 73-75; PROF CLIN SCI & PROF COMPUT SCI, COL MED, UNIV ILL, URBANA, 75- Concurrent Pos: Res fel, Sch Med, Johns Hopkins Univ, 57-59; USPHS res career develop award, 60-65; consult div hosp & med facil, Bur State Serv, USPHS, 65- Mem: Am Fedn Clin Res. Res: Hosp resistance to virus diseases; viral interference; medical models of virus cell interaction; general applications of digital computers to medicine and biology. Mailing Add: Med Comput Lab Sch Basic Med Sci Univ of Ill Urbana IL 61801

LEVY, ARTHUR, b New York, NY, Sept 29, 21; m 49; c 4. ATOMOSPHERIC CHEMISTRY, FUEL SCIENCE. Educ: Queen's Col, BS, 43; Univ Minn, MS, 48. Prof Exp: Chemist, Los Alamos Nat Lab, 44-46; aeronaut res scientist, Nat Adv Comt Aeronaut, 48-50; phys chemist, Brookhaven Nat Lab, 50-51; prin phys chemist, 51-59, asst chief, 56-69, fel, 69-71, sr fel, 71-73, SR RES LEADER, COLUMBUS LABS, BATTELLE MEM INST, 73- Mem: Am Chem Soc; Combustion Inst; Air Pollution Control Asn. Res: Kinetics of hydrogen and hydrocarbon oxidation; combustion chemistry; kinetics of radiation and ionic reactions; boron hydride chemistry; induced reactions; flame structure; air pollution kinetics; coal-oil combustion and environmental assessments. Mailing Add: Columbus Labs Battelle Mem Inst 505 King Ave Columbus OH 43201

LEVY, ARTHUR LOUIS, b Bridgeport, Conn, Aug 2, 17; m 43; c 1. ANALYTICAL CHEMISTRY, PHYSICAL CHEMISTRY. Educ: Univ Mo, AB, 38; Yale Univ, PhD(phys chem), 48. Prof Exp: From instr to asst prof phys chem, Rensselaer Polytech Inst, 48-54; chemist, Hodgkins Dis Res Lab, 54-58, CHIEF CHEMIST, ST VINCENT'S HOSP, 58- Concurrent Pos: Ford Found fel, 53-54; dir labs, New York Dept Health, 64- Mem: Fel AAAS; Am Asn Clin Chem; Am Chem Soc; NY Acad Sci; Asn Clin Sci. Res: Electrolyte solutions; immunochemistry of Hodgkins disease; human ribonuclease; standards and methodologies in clinical chemistry including automation and data processing. Mailing Add: St Vincent's Hosp Chem Lab 153 W 11th St New York NY 10011

LEVY, ARTHUR MAURICE, b New York, NY, Nov 20, 30; c 3. CARDIOLOGY. Educ: Harvard Univ, BA, 52; Cornell Univ, MD, 56; Am Bd Internal Med, dipl, 66. Prof Exp: Intern, Cornell Med Div, Bellevue Hosp, 56-57, resident med, 57-58; resident, 58-59, from instr to asst prof, 63-68, ASSOC PROF MED, COL MED, UNIV VT, 68-, ASSOC PROF PEDIAT, 69- Concurrent Pos: NIH fel cardiol, Col Med, Univ Vt, 59-60; Nat Heart Inst res fel, 59-60; trainee cardiol, Harvard Med Sch, Boston Children's Hosp, 62-63; teaching scholar, Am Heart Asn, 66-71; fel coun clin cardiol, Am Heart Asn, 69- Mem: Am Fedn Clin Res; fel Am Col Physicians; fel Am Col Cardiol. Res: Clinical electrophysiology. Mailing Add: Cardiopulmonary Lab Med Ctr Hosp of Vt Burlington VT 05401

LEVY, BARNET M, b Scranton, Pa, Jan 13, 17; m 40. HISTOPATHOLOGY. Educ: Univ Pa, AB, 38, DDS, 42; Med Col Va, MS, 44; Am Bd Oral Path, dipl. Prof Exp: Instr bact, path & clin dent, Med Col Va, 42-44; asst prof bact & path, Wash Univ, 44-47, assoc prof path. 47-49; prof dent & dir res & postgrad studies, Sch Dent & Oral Surg, Columbia Univ, 49-57; PROF PATH, UNIV TEX DENT BR HOUSTON, 57-, DIR, DENT SCI INST, 64- Concurrent Pos: Assoc attend dent surgeon, Presby Hosp, New York, 49-57; consult-instr, US Naval Hosp, St Albans, NY, 52-57; USPHS Hosp, Staten Island, 51-57; consult, Vet Admin Hosp, Bronx, 54-57, Houston, 57-, Univ Tex M D Anderson Hosp & Tumor Inst, 57-; mem Nat Res Coun, 52; chmn dent study sect, NIH, 57-62, training grants comt, 62-67; mem adv comt dent, Comt Int Exchange Persons; pres, Am Bd Oral Path, 65-66. Honors & Awards: Isaac Schour Mem Award, Int Asn Dent Res, 75. Mem: Am Soc Exp Path; Soc Exp Biol & Med; Am Asn Cancer Res; fel Am Acad Oral Path (pres, 69-70); Int Asn Dent Res (pres, 65-66). Res: Experimental pathology; inflammation; immunopathology and oncology. Mailing Add: 3736 Underwood Houston TX 77025

LEVY, BERNARD, pharmacology, deceased

LEVY, BERTRAM RAYMOND, mathematical physics, deceased

LEVY, BORIS, b New York, NY, Nov 24, 27; m 56; c 3. PHOTOGRAPHIC CHEMISTRY. Educ: NY Univ, BA, 48, MS, 50, PhD(phys chem), 55. Prof Exp: Res chemist, Sylvania Elec Co, 50-51; sr res chemist, Radio Corp Am, 55-56; sr scientist, Westinghouse Elec Corp, 56-60; sr res chemist, Socony Mobil Oil Co, 60-65; RES CHEMIST, POLAROID CORP, 65- Concurrent Pos: Assoc prof, Trenton Jr Col, 62-65; assoc ed, Photog Sci & Eng, 75. Honors & Awards: Honorable Mention, Soc Photog Scientists & Engr, 74. Mem: Am Chem Soc; Soc Photog Scientists & Engr. Res: Radiotracers; surface chemistry; electrokinetics; photoconductivity; photoelectron emission from semiconductors; spectral sensitization; energy and electron transfer reactions across phase boundaries; photographic emulsion preparation and characterization; preparation of novel image rector layers in diffusion transfer photography. Mailing Add: Polaroid Corp Res Div 750 Main St Cambridge MA 02139

LEVY, CHARLES KINGSLEY, b Boston, Mass, Dec 25, 24; m 58; c 3. RADIATION ECOLOGY. Educ: George Washington Univ, BSc, 48, MSc, 51; Univ NC, Chapel Hill, PhD(physiol), 56. Prof Exp: Instr physiol, Vassar Col, 56-58; staff scientist, Worcester Found Exp Biol, 58-62; ASSOC PROF RADIOL & BIOL, BOSTON UNIV, 62- Concurrent Pos: Res collabr, Brookhaven Nat Lab, 57-61; Am Physiol Soc fel, Boston Univ, 58; staff scientist, Worcester Found Exp Biol, 58-62; consult, Mass Gen Hosp, 62-; consult bioinstrumentation, NASA, 67-; Fulbright prof zool, Univ Nairobi, 69-70; proj dir avian radioecol nuclear reactor site, AEC, 73- Mem: Am Physiol Soc; Radiation Res Soc; Soc Gen Physiol. Res: Effect of high energy particulate radiation mammalian systems; dose-rate phenomena and responses of sensory and neural tissues to ionizing radiation; biological impact of reactor effluents on free ranging populations of wild birds. Mailing Add: Dept of Biol Boston Univ Boston MA 02215

LEVY, DANIEL, b New York, NY, Nov 27, 40; m 68. BIOCHEMISTRY, PROTEIN CHEMISTRY. Educ: City Col New York, BS, 61; Brandeis Univ, MS, 63, PhD(chem), 65. Prof Exp: Res biochemist, Univ Calif, Berkeley, 67-68; ASSOC PROF BIOCHEM, SCH MED, UNIV SOUTHERN CALIF, 74- Concurrent Pos: NIH fel biochem, Univ Calif, Berkeley, 65-67; Am Diabetes Asn res grant, 69-71; Am Cancer Soc res grant, 69-75; NIH res grant, 73-76. Mem: AAAS; Am Soc Biol Chem; Am Chem Soc. Res: Chemical modifications of proteins and membranes; nuclear magnetic resonance of macromolecules; mechanism of hormone action; membrane structure and function. Mailing Add: Dept of Biochem Sch of Med Univ of Southern Calif Los Angeles CA 90033

LEVY, DAVID ALFRED, b Washington, DC, Aug 27, 30; m 51; c 3. IMMUNOLOGY, ALLERGY. Educ: Univ Md, BS, 52, MD, 54; Am Bd Internal Med, cert, 62; Am Bd Allergy, cert, 74, cert immunol, 74. Prof Exp: From intern to chief resident med, Univ Hosp, Baltimore, Md, 54-59; physician, Pulmonary Dis Serv, Fitzsimons Gen Hosp, Denver, 59-61; staff physician, Chest Serv, Vet Admin Hosp, Baltimore, 61-62; USPHS fel, Sch Med, 62-66, asst prof radiol sci, 66-68, from assoc prof to prof radiol sci & epidemiol, 68-73, PROF BIOCHEM & BIOPHYS SCI & EPIDEMIOL, SCH HYG & PUB HEALTH, JOHNS HOPKINS UNIV, 73- Mem: AAAS; Am Asn Immunol; fel Am Acad Allergy; Am Fedn Clin Res; Soc Exp Biol & Med. Res: Mechanisms of allergic reactions; mechanisms of immunotherapy for allergic diseases; alpha-antitrypsin and pulmonary disease. Mailing Add: Dept Biochem Biophy Sci Sch Hyg & Pub Health Johns Hopkins Univ Baltimore MD 21205

LEVY, DAVID EDWARD, b Washington, DC, May 10, 41; m 67. NEUROLOGY. Educ: Harvard Univ, AB, 63; Harvard Med Sch, MD, 68; Am Bd Internal Med, dipl, 72; Am Bd Psychiat & Neurol, dipl, 75. Prof Exp: From intern to resident, New York Hosp, 68-72; fel & instr, 72-75, ASST PROF NEUROL, MED COL, CORNELL UNIV, 75- Concurrent Pos: Asst attend neurologist, New York Hosp, 75- Honors & Awards: Teacher-Scientist Award, Andrew W Mellon Found, 75. Mem: Am Col Physicians; Am Acad Neurol; Am Fedn Clin Res; AAAS. Res: Brain carbohydrate and energy metabolism in cerebral ischemia; prediction of outcome from stroke and coma. Mailing Add: Dept of Neurol New York Hosp Cornell Med Ctr New York NY 10021

LEVY, DONALD HARRIS, b Youngstown, Ohio, June 30, 39; m 61; c 1. CHEMICAL PHYSICS. Educ: Harvard Univ, BA, 61; Univ Calif, Berkeley, PhD(chem), 65. Prof Exp: Asst prof, 67-74, ASSOC PROF CHEM, UNIV CHICAGO, 74- Concurrent Pos: NIH fel, Cambridge Univ, 65-66, NATO fel, 66-67, Alfred P Sloan fel; Guggenheim fel, Univ Leiden, 75- Mem: Am Phys Soc. Res: Electron resonance spectroscopy of gas phase free radicals; microwave spectroscopy of free radicals and other short-lived species; optical spectroscopy in supersonic molecular beams; microwave-optical double resonance. Mailing Add: Dept of Chem Univ of Chicago Chicago IL 60637

LEVY, EDWARD ROBERT, b New York, NY, Oct 3, 27; m 51; c 4. ORGANIC CHEMISTRY. Educ: City Col New York, BS, 49; Univ Kans, PhD(org chem), 63. Prof Exp: Asst instr chem, Univ Kans, 49-50 & 51-53; res chemist, Glyco Prod, Inc, 53-57; process chemist, Chemagro Corp, 57-65, asst supvr, Process Develop Lab, 65-66, supvr, 66-68, asst mgr, 68-70, mgr, 70-73, PRIN CHEMIST, CHEMAGRO AGR DIV, MOBAY CHEM CORP, 73- Mem: AAAS; Am Chem Soc; Sigma Xi. Res: Organophosphorus insecticides; carbamates; chelating agents; synthesis and process development. Mailing Add: Chemagro PO Box 4913 Kansas City MO 64120

LEVY, EDWIN Z, b Highland Park, Ill, Nov 19, 29; m 51; c 3. PSYCHIATRY, PSYCHOANALYSIS. Educ: Northwestern Univ, BS, 51, MS, 53, MD, 54; Topeka Inst Psychoanal, grad, 70. Prof Exp: Intern internal med, Second Med Div, Bellevue Hosp, Cornell Univ, 54-55, resident pediat, Children's Med Div, 55-56; res psychiatrist, Aero-Med Lab, Wright Air Develop Ctr, Wright Patterson AFB, 57-59; res psychiat, Vet Admin Hosp, Topeka, Kans, 56-57 & 59-60; fel child psychiat, 60-62, staff psychiatrist, 62-70, NIMH career develop award psychiat, 63-67, dir res, Children's Div, 70-73, DIR THER, CHILDREN'S DIV, MENNINGER CLIN, 73-; MEM FAC, MENNINGER SCH PSYCHIAT, 64- Mem: AAAS; fel Am Psychoanal Asn; fel Am Psychiat Asn; fel Am Orthopsychiat Asn; fel Am Acad Child Psychiat. Res: Clinical psychiatry; skin resistance; human isolation; space psychiatry; residential treatment of emotionally disturbed children and adolescents and follow-up research. Mailing Add: Box 829 Menninger Clin Topeka KS 66601

LEVY, EZRA, b Bagdad, Iraq, Feb 2, 31; m 58; c 2. INORGANIC CHEMISTRY, PHYSICAL CHEMISTRY. Educ: Hebrew Univ, Israel, MSc, 58, PhD(chem), 63. Prof Exp: Instr chem, Hebrew Univ, Israel, 58-63; res scientist, NY Univ, 63-65; from instr to asst prof, Polytech Inst Brooklyn, 65-68; ASSOC PROF CHEM, YESHIVA UNIV, 68-, CHMN DEPT, 75- Concurrent Pos: Fel, NY Univ, 63-65, adj assoc prof, 68- Mem: Fel Am Inst Chemists. Res: Synthesis of inorganic compounds under very high pressures and temperatures; electrochemistry in fused salts; solid state. Mailing Add: Yeshiva Univ 500 W 185th St New York NY 10033

LEVY, GABOR BELA, b Budapest, Hungary, July 16, 13; nat US; m 38; c 2. CHEMISTRY. Educ: Karlsruhe Tech Univ, Dipl Ing, 38; Inst Divi Thomae, PhD(chem), 53. Prof Exp: Asst physics, NY Univ, 38-41; sr res chemist & sect head, Schenley Labs, Inc, 42-50, head anal & phys chem res, 50-55; head chem div, Consumers Union US, 55-57; asst to pres, 57-64, SR VPRES, PHOTOVOLT CORP, 64- Concurrent Pos: Adj prof, Polytech Inst NY. Mem: AAAS; Am Chem Soc; Sigma Xi. Res: Applied colloid chemistry; spectrophotometry; polarography; electron microscopy; swelling of casein; determination of antibiotics; physical methods in organic chemistry; enzymes; optical rotation. Mailing Add: 48 Roseville Rd Westport CT 06880

LEVY, GEORGE CHARLES b Brooklyn, NY, June 4, 44. PHYSICAL ORGANIC CHEMISTRY. Educ: Syracuse Univ, AB, 65; Univ Calif, Los Angeles, PhD(chem), 68. Prof Exp: Mem res staff, Gen Elec Res & Develop Ctr, 68-73; ASSOC PROF CHEM, FLA STATE UNIV, 73- Concurrent Pos: Alfred P Sloan fel, 75-77. Mem: Am Chem Soc; The Chem Soc. Res: NMR spectroscopy; carbon-13 NMR; organic and biological molecular structures; intermolecular interactions; molecular dynamics; hydrogen bonding and solution effects. Mailing Add: Dept of Chem Fla State Univ Tallahassee FL 32306

LEVY, GERALD FRANK, b Paterson, NJ, June 20, 38; m 60; c 4. ECOLOGY. Educ: Bowling Green State Univ, BS, 60, MA, 61; Univ Wis, PhD(bot), 66. Prof Exp: Tech asst, Univ Wis, 63-65; asst prof bot & zool, Univ Wis, Marinette Campus, 65-67; asst prof biol & ecol, 67-71, ASSOC PROF BIOL & ECOL, OLD DOM UNIV, 71- Concurrent Pos: Bot consult, Animal Ecol Proj, 68-69; vpres, Environ Consult Inc, 73- Mem: Am Forestry Soc; Ecol Soc Am. Res: Phytosociology; tick ecology research; small mammals; transpiration; terpine emission; ecology of red heart disease. Mailing Add: Dept of Biol Sci Old Dom Univ Norfolk VA 23508

LEVY, GERHARD, b Wollin, Ger, Feb 12, 28; nat US; m 58; c 3. PHARMACOLOGY. Educ: Univ Calif, BS, 55, PharmD, 57. Honors & Awards: DHonCausa, Univ Uppsala, 75. Prof Exp: Res pharmacist, Med Ctr, Univ Calif, 57-58; from asst prof to assoc prof pharm, 58-64, prof biopharmaceut, 64-72, actg chemn dept, 59-60, chmn, 66-70; DISTINGUISHED PROF PHARMACEUT, SCH PHARM, STATE UNIV NY BUFFALO, 72- Concurrent Pos: Vis prof, Hebrew Univ Jerusalem, 66-; consult, Bur Drugs, Food & Drug Admin, 71-73; mem comt probs drug safety, Nat Acad Sci-Nat Res Coun, 71-; vis prof, Univ Rochester, 72-73; grad prof, Victorian Col Pharm, Melbourne, Australia, 73- Honors & Awards: Richardson Pharm Award, 57; Found Achievement Award, Am Pharmaceut Asn, 69, Ebert Prize, 69. Mem: AAAS; Am Chem Soc; fel Am Pharmaceut Asn; Am Soc Pharmacol & Exp Therapeut; Am Soc Clin Pharmacol & Therapeut. Res: Biopharmaceutics; clinical pharmacology; pharmacokinetics. Mailing Add: Dept of Pharmaceut State Univ NY Sch Pharm Buffalo NY 14214

LEVY, HANS RICHARD, b Leipzig, Ger, Oct 22, 29; nat US; m 60; c 1. BIOCHEMISTRY. Educ: Rutgers Univ, BSc, 50; Univ Chicago, PhD(biochem), 56. Prof Exp: USPHS fel, Ben May Lab, Univ Chicago, 56-58 & Hammersmith Hosp, London, Eng, 58-59; from instr to asst prof biochem, Ben May Lab, Univ Chicago, 59-63; from asst prof to assoc prof, 63-71, PROF BIOCHEM, SYRACUSE UNIV, 71- Mem: AAAS; Am Soc Biol Chem; Am Chem Soc; Brit Biochem Soc. Res: Mechanism of action of enzymes; mechanism of control of enzymes. Mailing Add: Biol Res Labs Dept Biol Syracuse Univ Syracuse NY 13210

LEVY, HARRIS BENJAMIN, b Philadelphia, Pa, Nov 29, 28; m 62; c 2. NUCLEAR CHEMISTRY. Educ: Univ Pa, BS, 50; Univ Calif, PhD(chem), 53. Prof Exp: Asst chem, Univ Calif, Berkeley, 51-53; CHEMIST, LAWRENCE LIVERMORE LAB, 53- Mem: Am Phys Soc. Res: General radiochemical research; plowshare applications of nuclear explosives; data treatment of analysis. Mailing Add: Radiochem Div L233 Lawrence Livermore Lab PO Box 808 Livermore CA 94550

LEVY, HARRY, b Boston, Mass, Jan 9, 02; m 28. MATHEMATICS. Educ: Harvard Univ, AB, 20, AM, 23; Princeton Univ, PhD(math), 24. Prof Exp: Instr math, Harvard Univ, 20-22, Nat Res Coun, fel, 24-27; from asst prof to prof, 27-70, EMER PROF MATH, UNIV ILL, URBANA, 70- Mem: Am Math Soc; Math Asn Am. Res: Differential and Riemannian geometry. Mailing Add: 712 W Nevada St Urbana IL 61801

LEVY, HARVEY LOUIS, b Augusta, Ga, Oct 3, 35; m 61; c 2. MEDICINE, PEDIATRICS. Educ: Med Col Ga, MD, 60. Prof Exp: Intern pediat, Boston City Hosp, 60-61; asst resident path, Columbia-Presby Med Ctr, 61-62; asst resident pediat, Johns Hopkins Hosp, 64-65; chief resident, Boston City Hosp, 65-66; instr, 68-70, ASST PROF NEUROL, HARVARD MED SCH, 70- Concurrent Pos: NIH fel neurol, Harvard Med Sch, 66-68; consult, Walter E Fernald Sch Ment Retardation, 67-; lectr, Grad Sch Dent, Boston Univ, 68-; prin investr, Mass Dept Pub Health, 69-; mem amniocentesis registry, NIH, 70-; assoc, Ctr Human Genetics, Harvard Med Sch, 71- Mem: Fel Am Acad Pediat; Soc Pediat Res. Res: Inborn errors of metabolism; biochemical and genetic disorders. Mailing Add: Dept of Neurol Mass Gen Hosp Boston MA 02114

LEVY, HARVEY MERRILL, b Pittsburgh, Pa, May 12, 28; m 57; c 1. BIOCHEMISTRY, PHYSIOLOGY. Educ: Univ Calif, Los Angeles, BA, 50, PhD(biochem), 55. Prof Exp: Asst res biochemist, Army Med Res Lab, Ky, 54-56; assoc res biochemist, Brookhaven Nat Lab, 56-58; asst prof pharmacol, Sch Med, NY Univ, 58-60, from asst prof to prof physiol & biophys, 60-71; PROF PHYSIOL & BIOPHYS, STATE UNIV NY STONY BROOK, 71- Mem: Am Chem Soc; Am Soc Biol Chemists; Harvey Soc; Biophys Soc; Soc Gen Physiol. Res: Muscle biochemistry; enzymology; kinetics. Mailing Add: Dept of Physiol State Univ of NY Stony Brook NY 11790

LEVY, HENRI ARTHUR, b Oxnard, Calif, Sept 12, 13; m 47; c 2. STRUCTURAL CHEMISTRY. Educ: Calif Inst Technol, BS, 35, PhD(chem), 38. Prof Exp: Asst chem, Calif Inst Technol, 36-38, res fel, 38-43; res group leader, Clinton Labs, Tenn, 43-48; res group leader, 48-52, CHIEF RES SCIENTIST, OAK RIDGE NAT LAB, 52- Concurrent Pos: Lectr, Univ Tenn, 46-50; mem, US Nat Comt Crystallog, 62-68. Mem: Sigma Xi; Am Crystallog Asn (secy, 62-64, pres, 65). Res: Electron, x-ray and neutron diffraction; crystal and molecular structure; structure of liquids and solutions. Mailing Add: Chem Div Oak Ridge Nat Lab Oak Ridge TN 37830

LEVY, HILTON BERTRAM, b New York, NY, Sept 21, 16; m 42; c 2. VIROLOGY. Educ: City Col New York, BS, 35; Columbia Univ, MA, 36; Polytech Inst Brooklyn, PhD(biochem), 46. Prof Exp: Chief chemist, Gen Sci Labs, NY, 37-41; res biochemist, Mem Hosp Cancer & Allied Dis, 41-46; res biochemist, Overly Biochem Res Found, 46-52; HEAD SECT MOLECULAR VIROL, NAT INST ALLERGY & INFECTIOUS DIS, 52- Concurrent Pos: Prof, Med Sch, Howard Univ. Mem: AAAS; Soc Exp Biol & Med; Am Asn Immunol; Soc Gen Physiol; Am Soc Biol Chem. Res: Enzymes; cancer; nucleic acid metabolism; infectious diseases; virus reproduction; interferon action and induction. Mailing Add: Sect of Molecular Virol Nat Inst of Allergy & Infect Dis Bethesda MD 20014

LEVY, HIRAM, II, b Rochester, NY, Sept 2, 40; m 64; c 2. ATMOSPHERIC CHEMISTRY, ATOMIC PHYSICS. Educ: Iowa State Univ, BS, 61; Harvard Univ, MA, 65, PhD(chem), 66. Prof Exp: Res assoc chem, Mass Inst Technol, 66-68; res scientist, Smithsonian Astrophys Observ, 68-73; RES SCIENTIST, GEOPHYS FLUID DYNAMICS LAB, NAT OCEANIC & ATMOS ADMIN, 73- Res: Atmospheric and molecular physics; atmospheric chemistry; geochemical cycles of trace constituents. Mailing Add: Geophys Fluid Dynamics Lab PO Box 308 Princeton NJ 08540

LEVY, IRA SHELDON, physical chemistry, see 12th edition

LEVY, IRWIN, b St Louis, Mo, Mar 29, 07; m 40; c 3. NEUROLOGY, PSYCHIATRY. Educ: Cornell Univ, AB, 27; St Louis Univ, MD, 31; Columbia Univ, DSc(neurol), 35. Prof Exp: Asst neurol, Columbia Univ, 34-35; instr neurol & psychiat, St Louis Univ, 38-40; from instr to assoc prof neurol, 40-72, PROF NEUROL, SCH MED, WASH UNIV, 72-; ASSOC NEUROLOGIST, BARNES HOSP, 50-, ASST PSYCHIATRIST, 67- Concurrent Pos: Consult, US Air Force, 50-; dir dept neurol & psychiat, Jewish Hosp, 53-59. Mem: Fel Am Psychiat Asn; Am Neurol Asn; Asn Res Nerv & Ment Dis; fel Am Acad Neurol. Res: Cerebral vascular diseases; headaches; vertigo. Mailing Add: Dept of Neurol Sch of Med Wash Univ St Louis MO 63110

LEVY, JACK BENJAMIN, b Savannah, Ga, Jan 17, 41; m 63. ORGANIC CHEMISTRY. Educ: Duke Univ, AB, 62, NC State Univ, MS, 64, PhD(chem), 67. Prof Exp: From asst prof to assoc prof, 68-73, PROF CHEM, UNIV NC, WILMINGTON, 73-, CHMN DEPT, 75- Mem: AAAS; Am Chem Soc; Am Inst Chemists. Res: Synthesis of heterocyclic organophosphorus compounds; synthesis, spectral properties and biological testing of new phenoxaphosphine derivatives. Mailing Add: Dept of Chem Univ of NC Wilmington NC 28401

LEVY, JEROME FREDERICK, organic chemistry, see 12th edition

LEVY, JOSEPH, b New Haven, Conn, June 30, 13; m 41; c 3. ORGANIC CHEMISTRY. Educ: Yale Univ, BS, 35, PhD(org chem), 38. Prof Exp: Asst chem, Yale Univ, 35-38; res chemist, Polyxor Chem Co, 39-40; dir org res, Ernst Bischoff Co, 40-46; res chemist, Nopco Chem Co, 46-50; sr res assoc, Chem Div, Universal Oil Prod Co, 50-72, SR RES ASSOC, RES CTR, UOP INC, DES PLAINES, 72- Mem: Am Chem Soc. Res: Organic synthesis; fine organics; aromatics; process development; pharmaceuticals. Mailing Add: 3900 Dundee Rd Apt 106 Northbrook IL 60062

LEVY, JOSEPH BENJAMIN, b Manchester, Eng, Feb 23, 23; nat US; m 48; c 3. PHYSICAL CHEMISTRY, ORGANIC CHEMISTRY. Educ: Univ NH, BS, 43; Harvard Univ, MA, 45, PhD(chem), 48. Prof Exp: Mem sci staff, Columbia Univ, 47-49; res chemist, US Naval Ord Lab, 49-56 & Atlantic Res Corp, 56-65; PROF CHEM, GEORGE WASHINGTON UNIV, 65- Mem: Am Chem Soc. Res: Thermal decomposition of nitrate esters; reactions of free radicals; chemistry of rocket propellants; fluorine chemistry. Mailing Add: Dept of Chem George Washington Univ Washington DC 20006

LEVY, JOSEPH VICTOR, b Los Angeles, Calif, Apr 7, 28; m 54; c 2. PHYSIOLOGY, PHARMACOLOGY. Educ: Stanford Univ, BA, 50; Univ Calif, Los Angeles, MS, 56; Univ Wash, PhD(pharmacol), 59. Prof Exp: Asst physiol, Stanford Univ, 51-53, asst pharmacol, 54-56; asst, Univ Wash, 56-57; pharmacologist, Western Labs Resources Res, 60; sr res pharmacologist, Res Labs, Presby Med Ctr, 60-65; ASSOC PROF PHYSIOL & PHARMACOL, SCH MED SCI, PAC MED CTR, UNIV PAC, 69-, DIR LAB PHARMACOL & EXP THERAPEUT, INST MED SCI, PAC MED CTR, 61- Concurrent Pos: NIH res trainee, 58-59; Am Heart Asn res fel, 59-60; Nat Heart Inst, res career prog scientist, 65-70; mem drug interaction panel, Am Pharmaceut Asn; mem, Coun Basic Res, Am Heart Asn. Mem: Am Soc Pharmacol & Exp Therapeut; Biophys Soc; Am Soc Cell Biol; Cardiac Muscle Soc; Am Soc Clin Pharmacol & Therapeut. Res: Inflammation; hypertension; chemotherapy; drug structure-activity relationship; computerized drug information systems. Mailing Add: Pac Med Ctr Inst Med Sci PO Box 7999 San Francisco CA 94120

LEVY, JULIA GERWING, b Singapore, May 15, 34; nat Can; m 55, 69; c 3. MICROBIOLOGY. Educ: Univ BC, BA, 55; Univ London, PhD(bact), 58. Prof Exp: From instr to assoc prof, 58-74, PROF MICROBIOL, UNIV BC, 74- Mem: Am Asn Immunologists. Res: Characterization of antigenic determinants on natural antigens and the effect of these determinants on the cellular immune response. Mailing Add: Dept of Bact & Immunol Univ of BC Vancouver BC Can

LEVY, LAWRENCE, b Cleveland, Ohio, Oct 2, 33; m 61; c 2. ALGEBRA. Educ: Juilliard Sch Music, BS, 54, MS, 56; Univ Ill, MA, 58, PhD(math), 61. Prof Exp: Instr math, Univ Ill, 61; from asst prof to assoc prof, 61-71, PROF MATH, UNIV WIS-MADISON, 71- Mem: Am Math Soc. Res: Structure of associative rings and their modules. Mailing Add: Dept of Math Univ of Wis Madison WI 53706

LEVY, LEO, b New York, NY, July 11, 28; m 57; c 2. PSYCHOLOGY, PREVENTIVE MEDICINE. Educ: City Col New York, BS, 50, MA, 51; Univ Wash, PhD(psychol), 58; Harvard Univ, SMHyg, 64. Prof Exp: Instr psychol, Univ Mich, 58-60; adminr & chief psychologist, Pueblo Guid Ctr, Pueblo, Colo, 60-63; dir planning & eval, Ill Dept Ment Health, 64-69; asst prof psychiat, 65-69, assoc prof prev med, 69-75, PROF PREV MED, UNIV ILL MED CTR, 75- Concurrent Pos: NIMH fels, Univ Mich, 58-60 & Harvard Univ, 63-64; Fulbright Hays res grant, State Univ Leiden, 72-73; vis assoc prof psychiat, McMaster Univ, 69-71. Mem: AAAS; Am Psychol Asn; Am Pub Health Asn. Res: Promotion and maintenance of mental health; social planning; drug abuse; social ecology; population problems; problems of urban mental health. Mailing Add: Dept Prev Med & Community Health Univ of Ill Med Ctr PO Box 6998 Chicago IL 60680

LEVY, LEON BRUCE, b New York, NY, July 20, 37. INDUSTRIAL ORGANIC CHEMISTRY. Educ: NY Univ, BA, 58; Harvard Univ, AM, 59, PhD(chem), 62. Prof Exp: Res chemist, Clarkwood Res Lab, 62-65, sr res chemist, Tech Ctr, 65-71, RES ASSOC, TECH CTR, CELANESE CHEM CO, 72- Mem: Catalysis Soc. Res: Kinetics and mechanisms of homolytic organic reactions; vapor phase oxidations of hydrocarbons; heterogeneous catalysis. Mailing Add: Celanese Chem Co Box 9077 Corpus Christi TX 78408

LEVY, LEON SHOLOM, b Perth Amboy, NJ, June 28, 30; m 54; c 3. COMPUTER SCIENCE. Educ: Yeshiva Univ, BA, 52; Harvard Univ, SM, 55, ME, 57; Univ Pa, PhD(comput sci), 70. Prof Exp: Engr, RCA Corp, 55-58; sr staff engr, Hughes Aircraft Co, 58-63; mgr comput & displays sect, Aerospace Corp, 63-66; systs architect, IBM Corp, 66-67; asst prof, 70-74, ASSOC PROF STATIST & COMPUT SCI, UNIV DEL, 74- Concurrent Pos: Consult, Linguistics Proj, Univ Pa, 70- Mem: Asn Comput Mach; Inst Elec & Electronics Eng. Res: Relationship of machines and their languages; relationship of functional and structural aspects of computers. Mailing Add: 1021 Mt Pleasant Way Cherry Hill NJ 08034

LEVY, LOUIS, b Bound Brook, NJ, Sept 20, 27; m 51; c 5. PHARMACOLOGY. Educ: Univ Pa, AB, 47, MD, 52; Univ Calif, San Francisco, PhD(pharmacol), 65. Prof Exp: Intern, Philadelphia Gen Hosp, 52-53; med officer tuberc, Indian Health Serv, 53-56; resident med, USPHS Hosp, Staten Island, NY, 56-59; asst chief dept med, 59-62, CHIEF LEPROSY RES UNIT, USPHS HOSP, SAN FRANCISCO, 65-, NIH GRANTS LEPROSY & TUBERC PHARMACOL, 63- Concurrent Pos: Asst prof in residence, Univ Calif, San Francisco, 66-73, assoc prof in residence, Dept Pharmacol, 74-; mem, US Leprosy Panel, US-Japan Coop Med Sci Prog, 70-74. Mem: Am Soc Pharmacol & Exp Therapeut; Int Leprosy Asn; Am Soc Microbiol; Am Thoracic Soc; Soc Exp Biol & Med. Res: Pharmacology and immunology of leprosy. Mailing Add: Leprosy Res Unit USPHS Hosp San Francisco CA 94118

LEVY, LOUIS, b Brooklyn, NY, Feb 1, 23; m 54; c 5. PHARMACOLOGY. Educ: Univ Iowa, BS, 49, MS, 51, PhD(pharmacol), 54. Prof Exp: Asst chem, Syracuse Univ, 49-50; asst pediat, Univ Iowa, 50-52, instr pharmacol, 53-54; asst prof, Med Sch, Georgetown Univ, 54-55; asst prof, Col Med, Univ Cincinnati, 55-59; sr pharmacologist, Riker Labs, Calif, 59-71; ASSOC PROF PHARMACOL, SCH MED, UNIV CALIF, LOS ANGELES, 71- Concurrent Pos: Res collabr, Brookhaven Nat Lab, 53-55. Mem: AAAS; Am Soc Pharmacol & Exp Therapeut; NY Acad Sci. Res: Biochemical pharmacology. Mailing Add: Dept of Med Univ of Calif Sch of Med Los Angeles CA 90024

LEVY, LOUIS, II, b New Orleans, La, Nov 24, 19; m 45; c 4. CARDIOLOGY. Educ: La State Univ, BS, 40, MD, 43. Prof Exp: Asst dir heart sta, Charity Hosp, 47-59, from clin asst prof to clin assoc prof, 50-63, CLIN PROF MED, SCH MED, UNIV NEW ORLEANS, 63-, DIR HEART STA, CHARITY HOSP, 59- Concurrent Pos: Asst dir cardiovasc clin, La Heart Asn, 51- Mem: Soc Exp Biol & Med; fel Am Col Physicians; fel Am Col Chest Physicians; fel Am Col Cardiol; Am Fedn Clin Res. Res: Clinical cardiology and electrocardiography; cardiovascular diseases. Mailing Add: Dept of Med Univ of New Orleans Sch of Med New Orleans LA 70122

LEVY, LOUIS A, b New York, NY, Mar 6, 41; m 67; c 1. ORGANIC CHEMISTRY. Educ: City Col New York, BS, 61; Univ Colo, PhD(chem), 66. Prof Exp: Proj leader synthesis, Int Flavors & Fragrances, Inc, 65-66; scientist, Nat Air Pollution Control Admin, USPHS, 66-67, SCIENTIST, ENVIRON BIOL & CHEM BR, NAT INST ENVIRON HEALTH SCI, 67- Mem: Am Chem Soc; The Chem Soc. Res: Chemistry and synthesis of natural products; heterocyclic chemistry. Mailing Add: Nat Inst Environ Health Sci PO Box 12233 Research Triangle Park NC 27709

LEVY, LUIS WERNER, organic chemistry, biochemistry, see 12th edition

LEVY, MATTHEW NATHAN, b New York, NY, Dec 2, 22; m 46; c 3. PHYSIOLOGY, BIOMEDICAL ENGINEERING. Educ: Western Reserve Univ, BS, 43, MD, 45. Prof Exp: From instr to asst prof physiol, Western Reserve Univ, 49-53; from asst prof to assoc prof, Albany Med Col, 53-57; dir res, St Vincent Charity Hosp, Cleveland, Ohio, 57-67; assoc prof, 61-68, PROF PHYSIOL & BIOMED ENG, CASE WESTERN RESERVE UNIV, 68-; CHIEF DEPT INVESTIGATIVE MED, MT SINAI HOSP, 67- Concurrent Pos: Res fel, Western Reserve Univ, 48-49; assoc prof, Case Inst Technol, 63-67; assoc ed, Circulation Res, 70-74; sect ed, Am J Physiol, 75- Honors & Awards: Lederle Med Fac Award, 55-57. Mem: Am Physiol Soc; Soc Exp Biol & Med; Am Heart Asn. Res: Cardiovascular and renal physiology. Mailing Add: Dept of Investigative Med Mt Sinai Hosp Cleveland OH 44106

LEVY, MICHAEL GREEN, b Brooklyn, NY, May 8, 50; m 72; c 1. PARASITOLOGY. Educ: State Univ NY Col Buffalo, BA, 72; Rice Univ, PhD(biol), 75. Prof Exp: NIH trainee, Rice Univ, 75; RES FEL TROP PUB HEALTH, HARVARD SCH PUB HEALTH, 75- Mem: Am Soc Parasitologists. Res: Dynamics of host-parasite interface; membrane transport; vector ecology and control; biochemical basis of parasitism; schistosomiasis and Chagas' disease. Mailing Add: Dept of Trop Pub Health Harvard Sch Pub Health Boston MA 02115

LEVY, MICHAEL R, b Los Angeles, Calif, Aug 8, 35; m 62; c 2. CELLULAR BIOLOGY. Educ: Univ Calif, Los Angeles, BS, 57, MA, 59, PhD(zool), 63. Prof Exp: USPHS fel, 63-65, trainee, 65-66; teaching fel, Univ Mich, 66-67; from asst prof to assoc prof, 67-74, PROF BIOL, SOUTHERN ILL UNIV, EDWARDSVILLE, 74- Concurrent Pos: USPHS res grant, 69-72. Mem: AAAS; Am Soc Protozoologists; Am Inst Biol Sci; Am Soc Cell Biol. Res: Regulation of cellular metabolism; environmental effects on cell biochemistry and ultrastructure; physiological role of cell organelles. Mailing Add: Dept of Biol Southern Ill Univ Edwardsville IL 62025

LEVY, MILTON, b St Louis, Mo, July 13, 03; m 24; c 2. BIOCHEMISTRY. Educ: Wash Univ, BS, 25; St Louis Univ, PhD(biochem), 29. Prof Exp: Asst biochem, Sch Med, St Louis Univ, 25-29; from instr to prof biochem, Sch Med, 30-56, prof & chmn dept, Col Dent, 56-71, EMER PROF BIOCHEM, COL DENT & RES PROF, SCH MED, NY UNIV, 71- Concurrent Pos: Nat Res Coun fel med sci, Harvard Univ, 29-30; Guggenheim fel, Carlsberg Lab, Copenhagen, 56; investr, Off Sci Res & Develop, 42-48; consult, Vet Admin, NY, 50-; consult common fels, Nat Res Coun, 54-60, mem comt selections fels in med sci, 59-65; sci counr, Nat Inst Dent Res, 63-67; mem, Marine Biol Lab, Woods Hole, Mass. Mem: AAAS; Am Chem Soc; Am Soc Biol Chem; Soc Exp Biol & Med; NY Acad Sci. Res: Proteins; enzymes; chemical embryology; nerve growth factor; physical chemistry of aqueous solutions. Mailing Add: 39-95 48th St Long Island City NY 11104

LEVY, MOISES, b Panama, Apr 8, 30; US citizen; m 59. SOLID STATE PHYSICS. Educ: Calif Inst Technol, BS, 52, MS, 55; Univ Calif, Los Angeles, PhD(physics), 63. Prof Exp: Res chemist, Speciality Resins, Inc, 53-54; mem tech staff, Semiconductor Div, Hughes Aircraft Co, 56-58; asst prof solid state physics, Univ Pa, 64-65; asst prof ultrasonic invest solid state, Univ Calif, Los Angeles, 65-70; ASSOC PROF PHYSICS, UNIV WIS-MILWAUKEE, 71-, CHMN DEPT, 75- Concurrent Pos: Vis assoc prof, Univ Calif, Los Angeles, 71. Res: Experimental investigation of electron phonon interaction in superconductors and normal metals; spin phonon interaction in the rare earth metals. Mailing Add: Dept of Physics Univ of Wis Milwaukee WI 53201

LEVY, MORRIS, b Chicago, Ill, May 22, 44; m 74; c 1. EVOLUTIONARY BIOLOGY, BIOSYSTEMATICS. Educ: Univ Ill, Chicago Circle, 67; Yale Univ, MPh, 72, PhD(ecol, evolution), 73. Prof Exp: ASST PROF BIOL SCI, PURDUE UNIV, 73- Concurrent Pos: Dir, Kriebel Herbarium, Purdue Univ, 73-; NSF grant, Res Prog Biomed Sci, 75. Mem: Bot Soc Am; Soc Study Evolution; Soc Am Naturalists; AAAS. Res: Systematics and biochemical ecology of plants; novel chemistry and reticulate evolution in allopolyploids; genetic variation and population biology of permanent translocation heterozygotes. Mailing Add: Dept of Biol Sci Purdue Univ Lafayette IN 47907

LEVY, MORTIMER, b Rochester, NY, July 7, 24; m 50; c 2. RESEARCH ADMINISTRATION. Educ: Cornell Univ, BSEE, 49; Columbia Univ, MA, 51. Prof Exp: Physicist, Xerox Corp, 54-57, sect leader, 58-61; dir appl res, Mat Res Corp, 61-63; res scientist, 63-64, sr scientist, 64-67, mgr explor res, 67-73, mgr process sect, 73-75, SECT MGR, PROCESS ELEMENT SECT, XEROX CORP, 75- Mem: Soc Photog Sci & Eng. Res: Electrostatic photography. Mailing Add: Wilson Ctr Res & Technol Xerox Corp Webster NY 14580

LEVY, MORTON FRANK, b New York, NY, May 31, 25; m 55; c 3. ORGANIC CHEMISTRY. Educ: Queens Col, NY, BS, 50; Columbia Univ, MA, 51; Yale Univ, PhD(chem), 56. Prof Exp: Group leader org synthesis & anal develop, Argus Chem Co, 55-60; group leader org synthesis, Harchem Div, Wallace & Tiernan, NJ, 60-64; SR CHEMIST, MAT SCI COMPLEX, IBM CORP, SAN JOSE, 64- Concurrent Pos: Res assoc, Univ Calif, Berkeley, 75-76. Mem: AAAS; Am Chem Soc. Res: Utilization of new raw materials in organic synthesis; photosensitive materials; dibasic acids; new polymers; application of computers to chemistry. Mailing Add: 105 Stacia St Los Gatos CA 95030

LEVY, NELSON LOUIS, b Somerville, NJ, June 19, 41; m 74; c 3. IMMUNOLOGY, NEUROSCIENCES. Educ: Yale Univ, BA & BS, 63; Columbia Univ, MD, 67; Duke Univ, PhD(immunol), 73. Prof Exp: Intern surg, Univ Colo Med Ctr, 67-68; res assoc virol & immunol, NIH, 68-70; resident neurol, 71-72, asst prof immunol, 72-76, ASSOC PROF IMMUNOL, DUKE UNIV, 76- Concurrent Pos: Mem gastrointestinal cancer study group, Nat Cancer Inst, 74- Mem: Am Asn Cancer Res; Am Asn Immunologists. Res: Immunologic and non-immunologic defenses against human cancer; pathogenesis and etiology of multiple sclerosis. Mailing Add: Div of Immunol Box 3010 Duke Univ Med Ctr Durham NC 27710

LEVY, NORMAN B, b New York, NY, May 28, 31; m 58, 70; c 3. PSYCHIATRY, PSYCHOSOMATIC MEDICINE. Educ: NY Univ, BA, 52; State Univ NY Downstate Med Ctr, MD, 56. Prof Exp: Teaching fel med, Sch Med, Univ Pittsburgh, 57-58; dir med serv, US Air Force Hosp, Ashiya, Japan, 58-60; resident physician psychiat, Kings County Hosp Ctr, Brooklyn, NY, 60-63; from instr to asst prof med & psychiat, 63-73, ASSOC PROF PSYCHIAT, STATE UNIV NY DOWNSTATE MED CTR, 73-, DIR CONTINUING EDUC PSYCHIAT, 74-, PRESIDING OFFICER FAC, COL MED, 75- Concurrent Pos: NIMH career develop award, 66; ed consult, Psycosomatics, 72- & Kidney Int, 72-; consult psychiat educ, NIMH, 74-; examr psychiat, Am Bd Psychiat & Neurol, 74- Mem: Fel Am Col Physicians; fel Am Psychiat Asn; fel Int Col Psychosom Med; fel Am Asn Psychoanal Physicians; Asn Appl Psychoanal. Res: Effects of psychological stresses on kidney transplant rejections; psychological adaptation to hemodialysis; attitudes of students and physicians on informing patients of their fatal diagnosis; problems of spouses of psychoanalytical candidates. Mailing Add: State Univ NY Downstate Med Ctr Box 127 450 Clarkson Ave Brooklyn NY 11203

LEVY, PAUL, b New York, NY, May 25, 41; m 65; c 1. APPLIED MATHEMATICS. Educ: Rensselaer Polytech Inst, BS, 63, MS, 65, PhD(math), 68. Prof Exp: Asst prof math, NY Univ, 67-74; ASST PROF MATH, NY INST TECHNOL, 74- Mem: Am Math Soc; Soc Indust & Appl Math. Res: Investigation of problems in wave propagation and elasticity. Mailing Add: Dept of Math NY Inst of Technol Old Westbury NY 11568

LEVY, PAUL F, b New York, NY, Dec 9, 34; m 59; c 4. ANALYTICAL CHEMISTRY, INSTRUMENTATION. Educ: City Col New York, BS, 59; Columbia Univ, MA, 61, PhD(anal chem), 65. Prof Exp: Lectr chem, City Col New York, 59-65; proj engr, 65-68, sr proj engr, 68-69, supvr appln lab, 69-73, prod mgr thermal anal, 73-75, PROD MGR LIQUID CHROMATOG, INSTRUMENT PROD DIV, E I DU PONT DE NEMOURS & CO, 75- Mem: Am Chem Soc; Am Soc Testing & Mat. Res: Theory, applications, design and development of thermal analysis and other material characterization instrumentation; coulostatic impulse-chain and other forms of polarography; electrochemical instrumentation. Mailing Add: Instrument Prod Div E I du Pont de Nemours & Co Wilmington DE 19898

LEVY, PAUL SAMUEL, b New Haven, Conn, May 30, 36; m 75; c 2. BIOSTATISTICS, EPIDEMIOLOGY. Educ: Yale Univ, BS, 58; Columbia Univ, MA, 62; Johns Hopkins Univ, ScD(biostatist), 64. Prof Exp: Statistician, Ctr for Dis Control, Dept Health, Educ & Welfare, 60-62; res fel biomath, Med Sch, Harvard Univ, 64-65, from instr to asst prof biostatist, 66-70; math statistician, Nat Ctr Health Statist, Dept Health, Educ & Welfare, 70-72; assoc prof, 72-76, ASSOC PROF BIOMETRY, SCH PUB HEALTH, UNIV ILL MED CTR, 76- Concurrent Pos: Consult, Nat Ctr Health Statist, Dept Health, Educ & Welfare, 72- & Ill Col Podiatric Med, 75-; statist expert, US Food & Drug Admin, 75- Mem: AAAS; Am Statist Asn; Biometric Soc. Res: Epidemiology of accidents and trauma; analysis of data from the National Health Surveys; epidemiology of foot conditions; methods of obtaining estimates for local populations. Mailing Add: Sch of Pub Health PO Box 6998 Univ of Ill Med Ctr Chicago IL 60680

LEVY, PAUL WARREN, b Chicago, Ill, Mar 17, 21; m 44; c 4. SOLID STATE PHYSICS. Educ: Univ Chicago, BS, 43; Carnegie Inst Technol, PhD, 54. Prof Exp: Jr physicist, Metall Lab, Univ Chicago, 43-44; physicist beta-ray spectros, Oak Ridge Nat Lab, 44-48; assoc physicist, 52-58, PHYSICIST RADIATION DAMAGE INSULATORS, BROOKHAVEN NAT LAB, 58- Concurrent Pos: Indust consult to civilian & mil agencies, 55-; adj prof, Adelphi Univ, 66- Mem: Fel Am Phys Soc; Optical Soc Am. Res: Nuclear physics; luminescence of solids; optical and defect properties of solids; optical spectrophotometry; radiation effects in insulators, metals, explosives and propellants. Mailing Add: Dept of Physics Brookhaven Nat Lab Upton NY 11973

LEVY, PETER MICHAEL, b Frankfurt, Ger, Jan 10, 36; US citizen; m 65; c 2. SOLID STATE PHYSICS. Educ: City Col New York, BME, 58; Harvard Univ, MA, 60, PhD(appl physics), 63. Prof Exp: Res assoc physics, Lab Electrostatics & Physics of Metals, Grenoble, France, 63-64; res assoc, Univ Pa, 64-66; asst prof, Yale Univ, 66-70; assoc prof, 70-75, PROF PHYSICS, NY UNIV, 75- Concurrent Pos: Fel, CNRS, France, 63-64; Air Force Off Sci Res grant, Yale Univ & NY Univ, 67-72; NSF grant, NY Univ, 72-; Fulbright-Hays res scholar, France, 75-76; res exchange scientist, CNRS/NSF, 75-76. Mem: Am Phys Soc; NY Acad Sci. Res: Anisotropy of exchange interactions involving rare-earth ions; magnetic behavior of rare-earth salts at low temperatures; nature of the phase transitions in ferromagnets and antiferromagnets; theory of the magnetic behavior of insulators; magneto-elastic and thermodynamic behavior of the rare-earth pnictides. Mailing Add: Dept of Physics NY Univ New York NY 10003

LEVY, RENE HANANIA, b Casa Blanca, Morocco, Sept 30, 42; US citizen; m 64; c 2. PHARMACODYNAMICS. Educ: Univ Bordeaux, Baccalaureat, 60; Univ Paris, Pharm, 65; Univ Calif, San Francisco, PhD(pharm, pharmaceut chem), 70. Prof Exp: Intern, Hosps of Paris, Hopital Corentin Celton, 64-66; asst prof pharm, Col Pharm, 70-74, ASSOC PROF PHARMACEUT SCI, SCH PHARM & ASSOC PROF NEUROL SURG, SCH MED, UNIV WASH, 74- Mem: Am Pharmaceut Asn; Acad Pharmaceut Sci. Res: Pharmacokinetic evaluation of anticonvulsants prior to efficacy testing. Mailing Add: Sch of Pharm BG-20 Univ of Wash Seattle WA 98195

LEVY, RICHARD H, b London, Eng, Oct 25, 32; Can citizen; m 55; c 3. LASERS,

NUCLEAR ENGINEERING. Educ: Cambridge Univ, BA & MA, 54; Princeton Univ, MA & PhD(aeronaut eng), 59. Prof Exp: Mem res staff, Canadair, Ltd, 58-60; prin res scientist, Avco Everett Res Lab, 60-72, vpres, Avco Corp, 72-74; MGR, JNAI PLANNING, EXXON NUCLEAR CO, INC, 74- Concurrent Pos: Sr vpres & dir, Jersey Nuclear-Avco Isotopes, Inc, 72- Mem: Am Nuclear Soc; Am Phys Soc; Inst Elec & Electronics Engr. Res: Laser enrichment of uranium. Mailing Add: Exxon Nuclear Co Inc C-00777 777 106th NE Bellevue WA 98009

LEVY, ROBERT, b Montreal, Que, Apr 12, 38; m 64; c 1. BIOCHEMISTRY, CLINICAL CHEMISTRY. Educ: McGill Univ, BS, 59, PhD(biochem), 65. Prof Exp: NIH fel, Vet Admin Hosp/Univ Mo-Kansas City, 64-66; chief chemist, Vet Admin Hosp, Washington, DC, 66-67; asst prof neurobiol, Psychiat Inst, Univ Md, Baltimore City, 67-71; DIR LAB SERV, PATH DEPT, CHURCH HOME & HOSP, 71- Concurrent Pos: Asst prof, George Washington Univ, 66-67; guest lectr, Towson State Col, 69. Mem: AAAS; Am Asn Clin Chem. Res: Neurochemistry of membranes; neurotransmitters; neuroenzymology; clinical enzymology. Mailing Add: Dept of Path Church Home & Hosp Baltimore MD 21231

LEVY, ROBERT AARON, b El Paso, Tex, Nov 15, 26; m 56; c 4. SOLID STATE PHYSICS. Educ: Univ Tex, BS, 47, MA, 48; Univ Calif, MA, 50, PhD(physics), 55. Prof Exp: Physicist, US Naval Radiol Defense Lab, 50-53; asst physics, Univ Calif, 53-55; physicist, Tex Instruments, Inc, 55-57; proj engr, Motorola, Inc, 57-59; mem tech staff, Hughes Aircraft Co, 59-60; physicist, Nat Co, 61-62; assoc prof, Univ Cincinnati, 63-69. Concurrent Pos: Assoc fac mem, Ariz State Univ, 58-59 & Univ Southern Calif, 60-61; consult, US Naval Radiol Defense Lab, 59; fel, Israel AEC, 62-63. Mem: AAAS; fel Am Phys Soc; Am Asn Physics Teachers. Res: Magnetic resonance spectroscopy; quantum electronics. Mailing Add: PO Box 432 Cloudcroft NM 88317

LEVY, ROBERT I, b New York, NY, May 3, 37; m 58; c 4. MEDICINE, BIOCHEMISTRY. Educ: Cornell Univ, BA, 57; Yale Univ, MD, 61. Prof Exp: Intern med, Yale-New Haven Med Ctr, 61-62, resident, 62-63; clin asst med res, 63-65, chief resident med, 65-66, dep clin dir, 68-69, chief clin serv, Molecular Dis Br, 69-73, chief lipid metab br, 70-74, dir, Div Heart & Vascular Dis, 73-75, HEAD SECT LIPOPROTEINS, NAT HEART & LUNG INST, 66-, DIR, 75- Concurrent Pos: Mem coun arteriosclerosis, Am Heart Asn. Mem: Am Fedn Clin Res; Am Soc Clin Invest; fel Am Col Cardiol. Res: Lipid metabolism; lipid transport; atherosclerosis; hyperlipoproteinemia. Mailing Add: Lipid Metab Br Bldg 10-7N220 Nat Heart & Lung Inst Bethesda MD 20014

LEVY, ROBERT ISAAC, b New York, NY, June 1, 24. ANTHROPOLOGY, PSYCHIATRY. Educ: NY Univ, MD, 47. Prof Exp: Res grant psychiat, Bellevue Hosp, New York, 48-51; asst clin prof, Univ Calif, San Francisco, 56-64; sr fel social psychiat, East-West Ctr, Univ Hawaii, Honolulu, 64-66, assoc, Social Sci Res Inst, 66-69; PROF ANTHROP, UNIV CALIF, SAN DIEGO, 69- Mem: Fel Am Psychiat Asn; fel Am Anthrop Asn. Res: Psychological anthropology; interrelationships between sociocultural and psychological phenomena; geographic areal interests—Polynesia, Nepal. Mailing Add: Dept of Anthrop Univ of Calif at San Diego La Jolla CA 92037

LEVY, ROBERT SIGMUND, b Fresno, Calif, Nov 3, 21; m 52; c 2. BIOCHEMISTRY, NUTRITION. Educ: Univ Calif, Berkeley, AB, 48, AM, 52; Univ Southern Calif, PhD(biochem, nutrit), 57. Prof Exp: Asst zool, Univ Calif, Berkeley, 49-52; asst biochem & nutrit, Sch Med, Univ Southern Calif, 55-57; from asst prof to assoc prof biochem, 57-72, PROF BIOCHEM, SCH MED, UNIV LOUISVILLE, 72- Concurrent Pos: Assoc emergency med, Sch Med, Univ Louisville; fel coun arteriosclerosis, Am Heart Asn. Mem: Asn Multidiscipline Educ Health Sci; Sigma Xi; Am Chem Soc; Am Soc Biol Chem. Res: Structure, function and lipid-binding sites of apoproteins from plasma lipoproteins; complexing of polysaccharides with lipoproteins and implications of this association for atherogenesis. Mailing Add: Dept of Biochem Sch of Med Univ Louisville Health Sci Ctr Louisville KY 40201

LEVY, ROLAND ALBERT, b Cairo, Egypt, Nov 21, 44; US citizen; m 68; c 1. SOLID STATE PHYSICS. Educ: Queens Col, BA, 67; Columbia Univ, MS, 69, DEng Sc(physics), 73. Prof Exp: Fel, Carnegie-Mellon Univ, 73-75; ASST PROF PHYSICS, RENSSELAER POLYTECH INST, 75- Concurrent Pos: Consult, Rensselaer Res Corp, Burgess Anal Lab, 75- Mem: Am Phys Soc; Am Ceramic Soc; Am Inst Mining Metall & Petrol Engrs; Am Soc Metals; Sigma Xi. Res: Local moments and onset of ferromagnetism in dilute alloys; magnetic interactions in metallic spin glasses; Mössbauer spectroscopy of iron in silicate glasses; structural aspects of irradiation produced voids. Mailing Add: Dept of Physics Rensselaer Polytech Inst Troy NY 12181

LEVY, RONALD FRED, b St Louis, Mo, Dec 11, 44; m 66. TOPOLOGY. Educ: Wash Univ, AB, 66, AM, 70, PhD(math), 74. Prof Exp: Asst prof math, Goucher Col, 74-75; INSTR MATH, WASH UNIV, 75- Mem: Am Math Soc; Math Asn Am. Res: Compact Hausdorff spaces; almost-P-spaces; linearly ordered topological spaces. Mailing Add: Dept of Math Wash Univ St Louis MO 63130

LEVY, SAMUEL C, b Far Rockaway, NY, Jan 5, 37; m 58; c 2. ELECTROCHEMISTRY, ELECTROANALYTICAL CHEMISTRY. Educ: Hofstra Col, BA, 58; Iowa State Univ, PhD(inorg chem), 62. Prof Exp: STAFF MEM, SANDIA CORP, 62- Mem: Am Chem Soc; Electrochem Soc. Res: Chemical to electrical energy conversion; mechanism of electrical reactions. Mailing Add: Sandia Labs Div 2523 PO Box 5800 Albuquerque NM 87115

LEVY, SAMUEL WOLFE, b Montreal, Que, Feb 26, 22; m 67; c 2. CLINICAL CHEMISTRY. Educ: McGill Univ, BSc, 49, PhD(physiol), 54; Univ Sask, MSc, 51. Prof Exp: Multiple Sclerosis Soc Can res fel biochem, McGill-Montreal Gen Hosp Res Inst, 54-56; res assoc, Hotel-Dieu Hosp, Montreal, 56-61; DIR DEPT BIOCHEM, QUEEN MARY VET HOSP, MONTREAL, 61- Concurrent Pos: Dom-Prov Health grant, 56-61; Dept Vet Affairs grant, 64-71. Honors & Awards: Ames Award, Can Soc Clin Chem, 75. Mem: Can Biochem Soc; Can Soc Clin Chem (pres, 70-71); Am Asn Clin Chem; NY Acad Sci. Res: Lysosomal enzymes in blood in inflammation disease; effects of heparin in vivo on enzymes and lipid in blood. Mailing Add: Dept of Biochem Queen Mary Vet Hosp Montreal PQ Can

LEVY, SHELDON LEONARD, mathematics, physics, see 12th edition

LEVY, STANLEY SIDNEY, polymer chemistry, organic chemistry, see 12th edition

LEVY, SUSANNA AGNES, b Budapest, Hungary, Mar 20, 34; m 58; c 2. CLINICAL CHEMISTRY, TOXICOLOGY. Educ: Budapest Tech Univ, BSc, 57; Hebrew Univ, Jerusalem, MSc, 61; Polytech Inst Brooklyn, PhD(anal & phys chem), 68. Prof Exp: Asst chief biochem, Kingsbrook Jewish Med Ctr, 68-69; asst chief clin chem, St Vincent's Hosp, New York, 68-72; CHIEF CLIN CHEM, ST BARNABAS HOSP, BRONX, 72- Concurrent Pos: Consult, Biometric Labs, NY, 70-, Med Anal, 71, Brinkman Corp & Technicon Corp, 73- & Union Labs, 74-75. Mem: Am Asn Clin Chem; Am Chem Soc; Asn Clin Sci. Res: Analytical biochemistry; automation; development of new methods for clinical chemical use; detection and identification of drug and metabolites. Mailing Add: St Barnabas Hosp 183rd St & Third Ave Bronx NY 10011

LEW, BAAK WAI, b Vancouver, BC, Sept 15, 13; nat US; m 50; c 2. ORGANIC CHEMISTRY. Educ: Lingnan Univ, BSc, 36; Univ Mich, MSc, 38; Univ Minn, PhD(agr biochem), 41. Prof Exp: Res assoc, Ohio State Univ, 41-45; res chemist, Atlas Powder Co, 45-60, SR RES CHEMIST, ICI-US, 60- Mem: AAAS; Am Chem Soc; Am Oil Chemists Soc. Res: Carbohydrates; chromatography; resins; surfactants; organophosphorus compounds; furan chemistry; textile chemistry, fluroine chemistry, detergents. Mailing Add: 1507 The Mall Ardentown Wilmington DE 19810

LEW, CHEL WING, b San Antonio, Tex, Dec 9, 35; m 59; c 4. CHEMISTRY. Educ: Tex A&M Univ, BS, 60. Prof Exp: Technician chem, 60-61, asst res chemist, 61-65, res chemist, 65-73, SR RES CHEMIST, SOUTHWEST RES INST, 73- Mem: Sigma Xi. Res: Microencapsulation. Mailing Add: 9218 Old Homestead San Antonio TX 78230

LEW, HENRY Y, organic chemistry, see 12th edition

LEW, JOHN S, b New York, NY, Sept 9, 34; m 63, '75; c 2. APPLIED MATHEMATICS. Educ: Yale Univ, BS, 55; Princeton Univ, PhD(physics), 60. Prof Exp: C L E Moore instr math, Mass Inst Technol, 62-64; asst prof appl math, Brown Univ, 64-70; RES STAFF MEM MATH SCI, T J WATSON RES CTR, IBM CORP, 70- Mem: Math Asn Am; Soc Indust & Appl Math. Res: Asymptotic expansion of integral transforms; densities and distributions for algebraic combinations of random variables. Mailing Add: T J Watson Res Ctr IBM Corp Yorktown Heights NY 10598

LEWANDOWSKI, THADDEUS, b Brooklyn, NY, Aug 12, 16; m 45; c 2. BACTERIOLOGY. Educ: Long Island Univ, BS, 37; Univ Md, MS, 39; Ohio State Univ, PhD(bact), 44. Prof Exp: Lab technician, Int Health Div Labs, Rockefeller Found, NY, 39-41; asst bact, Ohio State Univ, 41-44; asst prof animal diseases, Univ Conn, 46-47; bacteriologist, Nat Drug Co, 47-48 & Pennwalt Chem Corp, 48-54; res assoc microbiol, Jefferson Med Col, 54-55; tech dir, Bonewitz Chem, Inc, 55-65; head microbiologist, Diamond Shamrock Co, 65-69; TECH DIR, CHEMIDYNE CORP, 69- Res: Hemolytic spectrococci; virus stability; pullorum control; vaccine and toxoid oil emulsions; food industry stabilitzers and cleaners. Mailing Add: 8228 Windham Dr Mentor OH 44060

LEWARS, ERROL GEORGE, Can citizen. ORGANIC CHEMISTRY. Educ: London Univ, BSc, 64; Univ Toronto, PhD(chem), 68. Prof Exp: Fel chem, Harvard Univ, 68-70, Univ Western Ont, 70-72; fel, 72-73, ASST PROF CHEM, TRENT UNIV, 73- Mem: Am Chem Soc. Res: Synthetic organic chemistry; compounds of theoretical interest. Mailing Add: Dept of Chem Trent Univ Peterbore ON Can

LEWBART, MARVIN LOUIS, b Philadelphia, Pa, May 28, 29; m 57; c 4. BIOCHEMISTRY. Educ: Philadelphia Col Pharm & Sci, BSc, 51, MSc, 53; Jefferson Med Col, MD, 57; Univ Minn, PhD(biochem), 61. Prof Exp: Intern pharm, Jefferson Med Col, 51-52, res assoc biochem, 53-57; intern med, Lankenau Hosp, 57-58; fel biochem, Mayo Found, Univ Minn, 58-61; USPHS spec res fel, Univ Basel, 61-62; res assoc, 62-67, asst prof, 67-70, ASSOC PROF MED, JEFFERSON MED COL, 70- Concurrent Pos: USPHS res fel, 59-61. Mem: Am Chem Soc. Res: Steroid chemistry and metabolism. Mailing Add: Dept of Med Rm 812 Jefferson Med Col Philadelphia PA 19107

LEWELLEN, ROBERT THOMAS, b Nyssa, Ore, Apr 27, 40; m 62. GENETICS, PLANT BREEDING. Educ: Ore State Univ, BS, 62; Mont State Univ, PhD(genetics), 67. Prof Exp: Asst agronomist, Mont State Univ, 65-66; RES GENETICIST, AGR RES SERV, USDA, 66- Mem: AAAS; Am Soc Agron; Crop Sci Soc Am; Am Phytopath Soc; Am Soc Sugar Beet Technol. Res: Genetics of virus resistance in sugar beet, Beta vulgaris, and development of resistant lines. Mailing Add: USDA Agr Res Sta PO Box 5098 Salinas CA 93901

LEWENZ, GEORGE F, b Berlin, Ger, Aug 29, 20; US citizen; m 55; c 4. ORGANIC CHEMISTRY. Educ: Western Reserve Univ, BS, 47, MS, 52. Prof Exp: Aeronaut res scientist, NASA, 48-53; chemist, Texaco, Inc, 53-58 & Esso Res & Eng Co, NJ, 59-68; res logician, 68-71, sr info chemist, 71-73, RES SPECIALIST, DOW CHEM CO, 73- Mem: Am Soc Info Sci; Am Chem Soc; fel Am Inst Chemists; Nat Microfilm Asn. Res: Organic synthesis; abstracting, indexing and information science. Mailing Add: Dow Chem Co 566 Bldg Midland MI 48640

LEWERT, ROBERT MURDOCH, b Scranton, Pa, Sept 30, 19; m 48; c 2. MEDICAL PARASITOLOGY, IMMUNOLOGY. Educ: Univ Mich, BS, 41; Lehigh Univ, MS, 43; Johns Hopkins Univ, ScD(parasitol), 49. Prof Exp: Asst instr biol, Lehigh Univ, 42-43; instr zool, Cols of Seneca, 43-44; instr parasitol, Dept Bact & Parasitol, 48-52, asst prof, 52-54, microbiol, 54-56, assoc prof, 57-61, PROF MICROBIOL, UNIV CHICAGO, 61- Concurrent Pos: Fulbright res fel, Philippines, 61; Guggenheim fel, 61; vis prof, Inst Hyg, Univ Philippines, 61 & 63-65; consult, Surg Gen, US Army, 59-75 & clin parasitol, Hines Vet Admin Hosp, 75-; mem, Comn parasitic diseases, Armed Forces Edpidemiol Bd, 59-66, parasitol study sect, NIH Trop Med, 65-69, Am Bd Microbiol, 65 & Training Grant Study Sect, Nat Inst Allergy & Infectious Dis, 69-73. Mem: AAAS; fel Am Acad Microbiol; Am Soc Parasitol; Am Soc Trop Med & Hyg; Royal Soc Trop Med & Hyg. Res: Host parasite relationships with emphasis on immunity, tolerance and immunopathology of schistosomiasis; histochemical and cytochemical studies of parasite effects on host; immunity and invasiveness of helminths; schistosomiasis. Mailing Add: Dept of Microbiol Univ of Chicago Chicago IL 60637

LEWIN, ANITA HANA, b Bucarest, Rumania, Oct 27, 35; m 56; c 2. PHYSICAL ORGANIC CHEMISTRY. Educ: Univ Calif, Los Angeles, BS, 59, PhD(phys org chem), 63. Prof Exp: Res asst prof chem, Univ Pittsburgh, 64-66; asst prof, 66-70, ASSOC PROF CHEM, POLYTECH INST BROOKLYN, 70- Mem: AAAS; Am Chem Soc; The Chem Soc. Res: Reaction mechanisms; catalysis by transition metals and their salts; organometallic reaction intermediates; conformational analysis; hindered rotation; diazonium ion decompositions. Mailing Add: Dept of Chem Polytech Inst of Brooklyn Brooklyn NY 11201

LEWIN, ELLEN, b Brooklyn, NY, Jan 19, 46. ANTHROPOLOGY. Educ: Univ Chicago, AB, 67; Stanford Univ, AM, 68, PhD(anthrop), 75. Prof Exp: Res anthropologist, 74-75, ASST RES ANTHROPOLOGIST & LECTR, UNIV CALIF, SAN FRANCISCO, 75- Mem: Fel Am Anthrop Asn; Soc Appl Anthrop; Soc Med Anthrop. Res: Women and sex roles; urban and medical anthropology; US culture; cultural and ethnic minorities; sexuality; feminist theory. Mailing Add: Dept of Social & Behav Sci Univ of Calif San Francisco CA 94143

LEWIN, ISAAC, b Lublin, Poland, Apr 7, 16; nat US; m 47; c 2. MEDICINE. Educ: Univ Ill, BS, 42; Univ Pa, MD, 50. Prof Exp: Bacteriologist, Queen's Gen Hosp, NY,

42-44; asst bact & enzymol, Princeton Univ, 44-46; intern, Mt Sinai Hosp, NY, 50-51; resident, Montefiore Hosp, Bronx, 51-52; resident med, Vet Admin Hosp, 52-54; Am Cancer soc fel, Chester Beatty Res Inst & Inst Cancer Res, Univ London, 54-56; res assoc, Montefiore Hosp, 56-62; from asst prof to assoc prof med, Stritch Sch Med, Loyola Univ Chicago, 62-69; cancer consult, Sch Med, Univ Ariz, 69-74; ONCOLOGIST, SIDNEY FARBER CANCER CTR, BOSTON, 74- Concurrent Pos: Nat Cancer Inst trainee, Montefiore Hosp, Bronx, 51-52; assoc mem, Hektoen Inst Med Res, 62-69; assoc chief metab sect, Hines Vet Admin Hosp, 62-69; chief oncol serv, Cook County Hosp, 66-69; prof dir, Cancer Control Prog, Springfield. Mem: Am Soc Microbiol; Asn Cancer Res; fel Am Geriat Soc; NY Acad Sci; Royal Soc Med. Res: Endocrine therapy of cancer; role of enzymes in cancerogenesis; metabolic effects of hormones. Mailing Add: 333 Merriweather Dr Longmeadow MA 01106

LEWIN, JOYCE CHISMORE, b Ilion, NY, Nov 13, 26; m 50. MICROBIOLOGY. Educ: Cornell Univ, BS, 48; Yale Univ, MS, 50, PhD(bot), 53. Prof Exp: Guest res worker biol, Lab, Nat Res Coun Can, 52-55; res assoc marine biol, Woods Hole Oceanog Inst, 56-60; asst res biologist, Scripps Inst, Univ Calif, San Diego, 60-65; from asst prof to assoc prof, 65-69, PROF OCEANOG, UNIV WASH, 69- Mem: Am Soc Limnol & Oceanog; Phycol Soc Am; Am Inst Biol Sci; Marine Biol Asn UK; Int Phycol Soc. Res: Physiology and culture of algae, especially diatoms; uptake and decomposition of silica in biological systems. Mailing Add: Dept of Oceanog Univ of Wash Seattle WA 98195

LEWIN, LAWRENCE M, b New York, NY, Mar 3, 32; m 58; c 2. BIOCHEMISTRY, MICROBIOLOGY. Educ: Mass Inst Technol, BS, 53; Cornell Univ, PhD(biochem), 59. Prof Exp: Res asst biochem, Cornell Univ, 53-54; res asst biochem, NY State Agr Exp Sta, 56-59; res assoc, Mass Inst Technol, 59-61; NIH fel, Weizmann Inst, 61-62; ASST PROF BIOCHEM, SCHS MED & DENT, GEORGETOWN UNIV, 63- Concurrent Pos: Mem staff biochem sect, Dept Oral Biol, Walter Reed Army Inst Res, 64- Mem: AAAS; Am Chem Soc. Res: Biosynthesis and metabolic functions of vitamins, particularly thiamine and inositol; structure and function of inositol lipids; food chemistry; microbial metabolism. Mailing Add: Dept of Biochem Georgetown Univ Sch of Med Washington DC 20007

LEWIN, RALPH ARNOLD, b London, Eng, Apr 30, 21; m 69. PHYCOLOGY. Educ: Cambridge Univ, BA, 42, MA, 46, ScD, 72; Yale Univ, MSc, 49, PhD, 50. Prof Exp: Spec lectr phycol, Yale Univ, 50-51, instr bot, 51-52; asst res off biol, Maritime Regional Lab, Nat Res Coun Can, 52-55; investr phycol, NIH grant, Marine Biol Lab, Woods Hole, 55-60; assoc prof marine biol, 59-67, PROF EXP PHYCOL, SCRIPPS INST OCEANOG, UNIV CALIF, 67- Concurrent Pos: Mem, Corp Marine Biol Lab, Woods Hole. Honors & Awards: Darbaker Prize, Bot Soc Am, 58. Mem: Soc Gen Microbiol; Marine Biol Asn UK; Brit Phycol Soc; Phycol Soc Am (pres, 70); Indian Phycol Soc. Res: Experimental phycology; microbiology; microbial genetics; marine biology. Mailing Add: Scripps Inst of Oceanog Univ of Calif La Jolla CA 92109

LEWIN, SEYMOUR Z, b New York, NY, Aug 16, 21; m 43; c 2. PHYSICAL CHEMISTRY, ANALYTICAL CHEMISTRY. Educ: City Col New York, BS; Univ Mich, MS, 42, PhD(chem), 50. Prof Exp: Lectr chem, Univ Mich, 47; from instr to assoc prof, 51-60, PROF CHEM, NY UNIV, 60- Concurrent Pos: Belg-Am Educ Found fel, 62; hon prof, Inst Quimico Sarria, Barcelona, Spain, 62; ed, Art & Archeol Tech Abstr, 66-69; consult, US Army Chem Corp. Honors & Awards: A Cressy Morrison Prize, NY Acad Sci, 56; Medal Hon, Govt Ethiopia, 74. Mem: AAAS; Am Chem Soc; Soc Appl Spectros; Am Inst Chemists; fel NY Acad Sci. Res: Crystal growth; spectroscopy; instrumentation; materials of art and archaeology; polymorphism; solid state chemistry; stone decay and preservation. Mailing Add: Dept of Chem New York Univ New York NY 10003

LEWIN, VICTOR, b San Francisco, Calif, Sept 8, 30; m 50; c 3. ECOLOGY. Educ: Univ Calif, AB, 53, PhD(zool), 58. Prof Exp: Asst zool, Univ Calif, 54-58; from asst prof to assoc prof, 58-70, PROF ZOOL, UNIV ALTA, 70- Concurrent Pos: Asst cur, Mus Vert Zool, Univ Calif, 55-56. Honors & Awards: Painton Award, Cooper Ornith Soc, 65. Mem: Wildlife Soc; Am Soc Mammalogists; Cooper Ornith Soc; Am Ornithologists Union; Can Soc Wildlife & Fishery Biol. Res: Wildlife ecology; ecology of game birds and mammals, particularly reproductive anatomy and physiology of gallinaceous birds; effects of chlorinated hydrocarbon residues on birds. Mailing Add: Dept of Zool Univ of Alta Edmonton AB Can

LEWIN, WALTER H G, b The Hague, Netherlands, Jan 29, 36; m 59; c 4. EXPERIMENTAL NUCLEAR PHYSICS. Educ: Univ Delft, Ir, 60, Dr(physics), 65. Prof Exp: Res assoc physics, Univ Delft, 59-66; fel space res, 66, from asst prof to assoc prof x-ray astron, 68-74, PROF PHYSICS, MASS INST TECHNOL, 74- Concurrent Pos: Teacher, Libanon Lyceum Rotterdam, 60-66. Mem: Am Astron Soc; Am Phys Soc. Res: Radioactive isotope applications; nuclear and atomic physics; x-ray astronomy; high-altitude ballooning; satellite observations, orbital solar observatory-7, small astronomical satellite-C, high energy astronomy observatory-A; astrophysics. Mailing Add: Ctr for Space Res Mass Inst of Technol Cambridge MA 02139

LEWINSON, VICTOR A, b New York, NY; m 57; c 2. OPERATIONS RESEARCH. Educ: Harvard Col, AB, 39; Columbia Univ, MA, 45, PhD(chem), 50. Prof Exp: Asst chem, Columbia Univ, 39-42; res scientist & sect leader, Manhattan Proj, 42-45; res fel chem, Calif Inst Technol, 50-51; fel, Mellon Inst, 51-54; analyst opers res, Nat Acad Sci, 54-61; MEM PROF STAFF, ARTHUR D LITTLE, INC, 61- Concurrent Pos: Mem, Transp Res Forum. Mem: Opers Res Soc Am; Am Inst Mgt Sci. Res: Freight transportation, especially maritime. Mailing Add: Arthur D Little Inc Acorn Park Cambridge MA 02140

LEWIS, AARON, b Calcutta, India, Oct 14, 45; US citizen; m. BIOPHYSICS. Educ: Univ Mo, BS, 66; Case Western Reserve Univ, PhD(phys chem), 70. Prof Exp: Instr phys chem, Case Western Reserve Univ, 70; NIH fel, 70-72, instr, 71-72, ASST PROF BIOPHYS, CORNELL UNIV, 72- Concurrent Pos: Sloan fel, 74-76. Mem: AAAS; Am Chem Soc; Biophys Soc; Asn Res Vision & Ophthal; Am Photobiol Soc. Res: Primary physical processes of importance in visual transduction. Mailing Add: Dept of Appl Physics Cornell Univ Ithaca NY 14850

LEWIS, ALAN ERVIN, b Milwaukee, Wis, Feb 1, 36; m 61; c 2. MEDICINE. Educ: Univ Wis-Madison, BS, 57; Marquette Univ, MD, 60. Prof Exp: Intern, Hosp Univ Pa, 60-61; resident, Med Ctr, Univ Mich, 61 & 63-65; from instr to sr instr internal med, 67-71, ASST PROF MED, HAHNEMANN MED COL & HOSP, 71- Concurrent Pos: Fel endocrinol, Sch Med, Tufts Univ, 65-67. Mem: Am Diabetes Asn; Am Fedn Clin Res; Am Col Physicians. Mailing Add: 191 Presidential Blvd Bala Cynwyd PA 19004

LEWIS, ALAN GRAHAM, b Pasadena, Calif, Mar 14, 34; m 57; c 2. BIOLOGICAL OCEANOGRAPHY, ZOOLOGY. Educ: Univ Miami, BSc, 56, MSc, 58; Univ Hawaii, PhD(zool), 61. Prof Exp: Asst prof zool, Univ NH, 61-64; asst prof, 64-68, ASSOC PROF OCEANOG & ZOOL, UNIV BC, 68- Mem: Soc Syst Zool. Res: Ecology of marine zooplankton. Mailing Add: Inst of Oceanog Univ of BC Vancouver BC Can

LEWIS, ALVIN EDWARD, b New York, NY, Nov 21, 16; m 43; c 2. PHYSIOLOGY. Educ: Univ Calif, Los Angeles, AB, 38; Stanford Univ, AM, 39, MD, 44. Prof Exp: Asst path, Stanford Univ, 47-48; clin instr & chief path sect, AEC Proj, Univ Calif, Los Angeles, 49-53; dir labs, Mt Zion Hosp, 53-66; prof path, Mich State Univ, 66-72; prof path & chmn dept, Med Sch, Univ SAla, 72-74; PROF PATH, UNIV CALIF, DAVIS, 74- Concurrent Pos: Am Cancer Soc fel, Stanford Univ, 48-49; vis physician, Los Angeles County Harbor Hosp, 49-53; attend physician, Wadsworth Gen Hosp, 50-53; asst clin prof, Med Ctr, Univ Calif, San Francisco, 59-66. Mem: AAAS; Am Physiol Soc; AMA; fel Col Am Path. Res: Hepatic function tests; plasma volume and distribution. Mailing Add: Dept of Path Univ of Calif Davis CA 95616

LEWIS, ANDREW MORRIS, JR, US citizen. VIROLOGY. Educ: Duke Univ, MD, 61. Prof Exp: RES VIROLOGIST, LAB VIRAL DIS, NAT INST ALLERGY & INFECTIOUS DIS, 63- Mailing Add: Bldg 7 Rm 313 Lab of Viral Dis Nat Inst of Allergy & Infect Dis Bethesda MD 20014

LEWIS, ANTHONY JAMES, b Drexel Hill, Pa, Jan 12, 41; m 65; c 4. PHYSICAL GEOGRAPHY. Educ: West Chester State Col, BS, 62; Ore State Univ, MS, 67; Univ Kans, PhD(phys geog), 71. Prof Exp: Res asst chem, Scott Paper Res Lab, Pa, 61; sci teacher, Sparrows Point Jr-Sr High Sch, Md, 62-63; res asst geog, Ore State Univ, 63-64, instr, 64-65, teaching asst, Univ Kans, 66-67, res asst remote sensing, 65-69; asst prof, 69-74, ASSOC PROF GEOG, LA STATE UNIV, BATON ROUGE, 74- Concurrent Pos: Invited lectr, Remote Sensing Workshop, Int Geog Cong, 72. Mem: Am Soc Photogram; Asn Am Geog; Geol Soc Am. Res: Applications of remote sensing to geographic research; river and coastal morphology. Mailing Add: Dept of Geog & Anthrop La State Univ Baton Rouge LA 70808

LEWIS, ARMAND FRANCIS, b Fairhaven, Mass, May 22, 32; m 58; c 2. PHYSICAL CHEMISTRY. Educ: Southeastern Mass Univ, BS, 53; Okla State Univ, MS, 55; Lehigh Univ, PhD(chem), 58. Prof Exp: Res asst rheol, Lehigh Univ, 58-59; res chemist, Cent Res Div, Am Cyanamid Co, 59-63, sr res chemist, Plastics & Resins Div, 63-64, group leader polymer physics & adhesion, 64-69, proj leader noise control mat, 70-71; sr res assoc, 71-73, SR MAT SCIENTIST, LORD CORP, 73- Honors & Awards: Union Carbide Award, Am Chem Soc, 63. Mem: Am Chem Soc; Soc Rheol (treas, 66-). Res: Polymer physics; rheology; surface chemistry and adhesion; dynamic mechnical properties of polymers; glass transition phenomena in polymeric systems; polymer to metal adhesion and fracture of adhesive joints; vibration and noise control materials; rubber chemicals; engineering composites. Mailing Add: Lord Corp 2000 W Grandview Blvd Erie PA 16512

LEWIS, ARNOLD D, b Philadelphia, Pa, May 6, 20; m 45; c 2. ANALYTICAL CHEMISTRY. Educ: Philadelphia Col Pharm, BS, 40; Polytech Inst Brooklyn, MS, 47. Prof Exp: Control chemist, Hance Bros & White, 40-41; pilot plant chemist, United Gas Improv Corp, 41-43; asst scientist to Dr E A H Friedheim, 43-44; jr scientist, G D Res Inst, 44-47; scientist, Dept Org Chem, 47-54, sr scientist, 54-63, sr res assoc, Chem Res Div, 63-64, DIR ANAL & PHYS CHEM, PROF PROD GROUP, WARNER-LAMBERT CO, 64- Mem: AAAS; Am Chem Soc; Am Microchem Soc; Soc Appl Spectros; NY Acad Sci. Res: Organic synthesis of heterocycles; infrared and ultraviolet absorption spectrophotometry; microanalysis; gas, paper, thin-layer and column chromatography; proton magnetic resonance spectroscopy. Mailing Add: Warner-Lambert Co 170 Tabor Rd Morris Plains NJ 07950

LEWIS, ARTHUR EDWARD, b Jamestown, NY, Jan 11, 29; m 53; c 3. EARTH SCIENCE. Educ: St Lawrence Univ, BS, 50; Calif Inst Technol, MS, 55, PhD(geol), 58. Prof Exp: Sr engr, Curtiss Wright Corp, 58-60; scientist, Hoffman Sci Ctr, 60-62; mem tech staff, Fairchild Semiconductor, Calif, 62-67; geologist, 67-69, group leader, Plowshare Prog, Peaceful Appln Nuclear Explosives, 69-73, PROJ LEADER, OIL SHALE, LAWRENCE LIVERMORE LAB, UNIV CALIF, 73- Concurrent Pos: Consult, Fed Energy Agency, 74. Honors & Awards: Peele Award, Am Inst Mining & Metall Engrs, 74. Mem: Geol Soc Am; Am Nuclear Soc. Res: In-site recovery of oil from oil shale; energy resource development. Mailing Add: 884 Santa Rita Ave Los Altos CA 94022

LEWIS, B KENNETH, b Plano, Iowa, Apr 23, 11; m 34; c 2. INORGANIC CHEMISTRY. Educ: Wichita State Univ, BS, 31, MS, 32; Univ Okla, PhD(chem), 59. Prof Exp: Teacher high sch, Kans, 37-41; prof chem, Northern Okla Jr Col, 41-45; instr, Univ Okla, 45; chemist, Continental Oil Co, 45-46; prof chem, Phillips Univ, 46-47 & Northwestern State Col, Okla, 47-49; prof, 49-60, sci educ, 70-76, EMER PROF SCI EDUC, PHILLIPS UNIV, 76- Concurrent Pos: Dean, Col Arts & Sci, Phillips Univ, 60-70. Mem: Am Chem Soc; Nat Sci Teachers Asn. Res: Nonaqueous solvation. Mailing Add: Dept of Sci Educ Phillips Univ Enid OK 73701

LEWIS, BARBARA-ANN GAMBOA, b Manila, Philippines; US citizen; m 66; c 3. ENVIRONMENTAL SCIENCES, SOIL SCIENCE. Educ: Philippine Women's Univ, BS, 53; Univ Calif, Berkeley, MS, 63, PhD(soil sci), 71. Prof Exp: Soil physicist, Bur Soils, Dept Agr, Philippines, 53-61, supvr soil technologist, 61-64; staff res assoc soils, Dept Forestry, Univ Calif, Berkeley, 71-72; appointee, 72-73, ASST BIOLOGIST, ARGONNE NAT LAB, 73- Mem: Soil Sci Soc Am; Am Soc Agron; Scientists Inst Pub Info; Sigma Xi. Res: Environmental effects of steam-electric power generation, including environmental effects of cooling alternatives. Mailing Add: 923 Asbury Ave Evanston Il 60202

LEWIS, BENJAMIN MARZLUFF, b Scranton, Pa, Oct 7, 25; m 56, 69; c 3. PHYSIOLOGY. Educ: Univ Pa, MD, 49. Prof Exp: Res fel med, Harvard Univ, 50-52; actg chmn dept, 70-71, from asst prof to assoc prof, 56-62, PROF MED, WAYNE STATE UNIV, 62- Concurrent Pos: Teaching fel med, Harvard Univ, 52-53; fel, Grad Sch Med, Univ Pa, 53-55; consult, Vet Admin Hosp, 56. Mem: Fel Am Col Physicians; Am Physiol Soc; Am Soc Clin Invest. Res: Pulmonary physiology, particularly gas diffusion and pulmonary circulation. Mailing Add: Dept of Med Wayne State Univ Sch of Med Detroit MI 48201

LEWIS, BERNARD, b London, Eng, Nov 1, 99; nat US; m 34; c 2. PHYSICAL CHEMISTRY. Educ: Mass Inst Technol, BS, 23; Harvard Univ, MA, 24; Cambridge Univ, PhD(phys chem), 26. Hon Degrees: ScD, Cambridge Univ, 53. Prof Exp: Demonstr phys chem, Cambridge Univ, 25-26; Nat Res Coun fel, Univ Berlin & Univ Minn, 26-29; phys chemist, US Bur Mines, 29-42, chief, Explosives & Phys Sci Div, 46-53; PRES, COMBUSTION & EXPLOSIVES RES, INC, 53- Concurrent Pos: Dir res powder & explosives, Ord Dept, US Army, 51-52; consult, US Army, US Navy, US Air Force & Nat Bur Standards; mem, Sci Adv Comt, Ord Corps, Aberdeen Proving Ground, Md, Combustion Comt, Nat Adv Comt Aeronaut, Fire Res Conf, Nat Acad Sci; pres, Comt High Temperature, Int Union Pure & Appl Chem; US ed, J Combustion & Flame; co-ed, Phys Measurements in Gas Dynamics & Combustion, 54 & Combustion Processes, 56. Honors & Awards: Gold Medal, Ital Thermotech Asn; Pittsburgh Award, Am Chem Soc, 74; Orleans Medal, City Orleans, France, 75; Lewis Gold Medal, Combustion Inst. Mem: Emer mem, Am Chem Soc; Am Phys Soc; fel Am Inst Aeronaut & Astronaut; Combustion Inst (pres, 54-66, hon pres, 66-); fel NY Acad Sci. Res: Chemical kinetics of gas reactions; thermodynamics of explosives;

flame propagation; ignition; explosion limits and hazards; combustion in jet propulsion; propellants; detonation; internal combustion engines; oxidation of hydrocarbons; fuels, interior ballistics; combustion and flame phenomena. Mailing Add: Combust & Explos Res Inc 1016 Oliver Bldg Pittsburgh PA 15222

LEWIS, BERTHA ANN, b Lewisville, Minn, Oct 21, 27. BIOCHEMISTRY. Educ: Univ Minn, BChem, 49, MS, 54, PhD(biochem), 57. Prof Exp: Res fel biochem, Univ Minn, St Paul, 57-65, res assoc, 65-67; assoc prof design & environ anal, 67-70, ASSOC PROF DIV NUTRIT SCI, CORNELL UNIV, 70-, ASSOC DEAN, COL HUMAN ECOL, 74- Mem: Am Chem Soc; Inst Food Technologists; Soc Complex Carbohydrates. Res: Carbohydrate chemistry and biochemistry; chemistry of glycoproteins; polysaccharide-protein interactions. Mailing Add: Col of Human Ecol Cornell Univ Ithaca NY 14853

LEWIS, BILLY M, b Atlanta, Ga, Dec 18, 27; m 47; c 1. METEOROLOGY. Educ: Univ Ga, BS, 51; NY Univ, MS, 54. Prof Exp: Meteorologist, Extended Forecast Div, Nat Meteorol Ctr, US Weather Bur, 47-65, Nat Hurricane Res Lab, 65-74, METEOROLOGIST, NAT HURRICANE & EXP METEOROL LAB, US WEATHER BUR, 75- Mem: Am Meteorol Soc. Res: Analysis of tropical meteorological systems with emphasis on their inner relationships and the application of remote sensing such as radar and satellite data.

LEWIS, BRIAN KREGLOW, b SAfrica, Sept 2, 32; US citizen; m 53; c 6. HUMAN PHYSIOLOGY. Educ: Ohio State Univ, BS, 54; Tufts Univ, PhD(physiol), 71. Prof Exp: Res assoc physiol, Sch Med, Tufts Univ, 71, May Inst Med Res, Jewish Hosp Cincinnati, 71-74 & Col Med, Univ Cincinnati, 74-75; ASST PROF HEALTH SCI, GRAND VALLEY STATE COL, 75- Concurrent Pos: Adj asst prof physiol, Col Med, Univ Cincinnati, 72-75. Mem: AAAS; Am Inst Biol Sci; Soc Study Reprod; Study Fertil; Sigma Xi. Res: Changes in endocrine function related to the pathogenesis of cardiovascular disease; role of the female reproductive tract in capacitation of sperm. Mailing Add: Sch of Health Sci Grand Valley State Col Allendale MI 49401

LEWIS, BURNADINE LANGSTON, b Texarkana, Tex, Nov 23, 21; m 42; c 1. FOOD SCIENCE, NUTRITION. Educ: Prairie View Agr & Mech Col, BS, 41; Colo State Univ, MS, 52; Kans State Univ, PhD(foods, nutrit), 55. Prof Exp: Teacher, High Sch, NC, 45 & Tex, 46-52; asst instr, Kans State Univ, 52-55; prof foods & nutrit & chmn div home econ, Tex Southern Univ, 57-68; prof food & nutrit, Dept Home Econ, Southwest Tex State Univ, 68-75; PROF FOODS & NUTRIT, TEX SOUTHERN UNIV, 75- Mem: AAAS; Am Dietetic Asn; Poultry Sci Asn; Inst Food Technologists. Res: Chemistry, zoology; food research including organoleptic and chemical testing; histological study of animal tissues. Mailing Add: Dept of Home Econ Tex Southern Univ Houston TX 77004

LEWIS, CAMERON DAVID, b Staunton, Va, June 1, 20; m 45; c 2. ORGANIC CHEMISTRY. Educ: Univ Buffalo, AB, 42; Univ Ill, AM, 45, PhD(org chem), 47. Prof Exp: Asst chemist, Ill Geol Surv, 42-46; asst chem, War Prod Bd Prog, Univ Ill, 46-47; chemist, 47-74, SR RES CHEMIST, E I DU PONT DE NEMOURS & CO, W VA, 74- Mem: Am Chem Soc. Res: Analytical test methods; polymer intermediates. Mailing Add: Potomac River Develop Lab E I du Pont de Nemours & Co Martinsburg WV 25401

LEWIS, CARMIE PERROTTA, b New Castle, Pa, June 9, 29; m 65. HISTOLOGY. Educ: Thiel Col, BS, 51; Univ NH, MS, 53; Univ Wis, PhD(anat, zool), 56. Prof Exp: Res asst, Univ Wis, 53-56; Am Asn Univ Women res fel, Cambridge Univ, 56-57; lectr embryol & histol, Fac Med, Queen's Univ, Ont, 57-58; instr anat, Sch Med, Yale Univ, 58-61; asst radiobiologist, Brookhaven Nat Lab, 61-64, res collabr, 64-67; assoc prof, 67-74, PROF BIOL, SUFFOLK COUNTY COMMUNITY COL, 74- Concurrent Pos: USPHS res fel, 61-64; asst prof, Queens Col, NY, 64-67. Mem: Am Asn Anat. Res: Radiobiology; endocrines of reproduction. Mailing Add: Dept of Biol Suffolk County Community Col 533 College Rd Selden NY 11784

LEWIS, CHARLES BERNARD, JR, b Worcester, Mass, Apr 9, 13; m 40; c 4. NATURAL HISTORY. Educ: Brown Univ, BA, 35. Prof Exp: CUR SCI MUS, INST OF JAMAICA, 39-, DIR INST, 50- Concurrent Pos: Mem, Wildlife Protection Comt, Nat Trust Comn, Sci Res Coun, UNESCO Comn & Pub Gardens Adv Comt, Jamaica; chmn, Jamaica Zoo Comt; Mem, Oxford Biol Exped, Cayman Islands, 38. Honors & Awards: Hon officer, Order Brit Empire, 57. Res: Biogeography of Caribbean area. Mailing Add: Inst of Jamaica 14-16 East St Kingston Jamaica

LEWIS, CHARLES E, b Kansas City, Mo, Dec 28, 28; m 63; c 4. MEDICINE, PREVENTIVE MEDICINE. Educ: Harvard Med Sch, MD, 53; Univ Cincinnati, MS, 57, ScD(prev med), 59. Prof Exp: House officer med, Univ Kans Hosps, 53-54; resident occup med, Eastman Kodak Co, 58-59, plant physician, Tex Div, 59-60; asst prof epidemiol, Col Med, Baylor Univ, 60-61; assoc prof med, Med Ctr, Univ Kans, 61-62, prof prev med, 62-69; prof social med, Harvard Med Sch, 69-70; PROF MED & PUB HEALTH, SCH MED & PUB HEALTH, UNIV CALIF, LOS ANGELES, 70- Concurrent Pos: Fel prev med, Kettering Lab, Univ Cincinnati, 56-58; USPHS trainee, 57-58; dir, Kans Regional Med Prog, 67-69. Honors & Awards: Ginsberg Prize, 54. Mem: Am Pub Health Asn; Asn Teachers Prev Med; Am Acad Occup Med. Res: Medical care and education. Mailing Add: Sch of Med & Pub Health Univ of Calif Los Angeles CA 90024

LEWIS, CHARLES EDWARD, b Hagerstwon, Md, Oct 18, 12; m 36; c 3. ORGANIC CHEMISTRY. Educ: Univ Md, BS, 34; Pa State Col, MS, 39, PhD(org chem), 41. Prof Exp: Chemist, Western Md Rwy Co, Md, 35-38; res chemist, Calco Chem Div, Am Cyanamid Co, NJ, 41-45, asst chief develop chemist, 45-48, chief develop chemist, 48-54, group leader dyes, Res Dept, 54-71; CONSULT, 71- Mem: Am Chem Soc. Res: Metallized azo-dyes; organic pigments; basic dyes, disperse dyes. Mailing Add: 404 William St Somerville NJ 08876

LEWIS, CHARLES J, b Park River, NDak, May 20, 27; m 50; c 2. ANIMAL SCIENCE. Educ: Utah State Univ, BS, 52; Iowa State Univ, MS, 54, PhD(animal sci), 56. Prof Exp: Dir nutrit, Kent Feeds, Inc, 56-58, vpres & nutritionist, 58-67, mem, Bd Dirs, 60-67; prof animal sci & head dept, SDak State Univ, 67-68; EXEC VPRES RES & DEVELOP, GRAIN PROCESSING CORP & KENT FEEDS, INC, 68- Mem: AAAS; Am Inst Biol Sci; Am Soc Animal Sci; Inst Food Technologists; Poultry Sci Asn Am. Res: Animal nutrition and research. Mailing Add: 7 Colony Dr Muscatine IA 52761

LEWIS, CHARLES JOSEPH, b New York, NY, Sept 18, 17; m 67; c 2. MATHEMATICS. Educ: Georgetown Univ, AB, 41, MS, 45; Brown Univ, PhD(math), 57. Prof Exp: Instr math, Georgetown Univ, 43-44 & St Peters Col, 45; instr, Fordham Univ, 54-56, asst prof, 56-65, chmn dept, 58-65; assoc prof, 65-66, PROF MATH, MONMOUTH COL, NJ, 66- Concurrent Pos: NSF fac fel math, Harvard Univ, 59-60; chmn dept math, Monmouth Col, NJ, 68-74. Mem: Am Math Soc; Math Asn Am. Res: Complex function theory; extremal problems; growth of entire functions; generalized potential theory; special functions and differential equations of mathematical physics. Mailing Add: 8 Timothy Lane Tinton Falls NJ 07724

LEWIS, CHARLES WILLIAM, b New York, NY, Oct 29, 20; m 42; c 1. PHYSICAL CHEMISTRY. Educ: City Col, New York, BS, 41; Polytech Inst Brooklyn, PhD(chem), 50. Prof Exp: Res chemist, Res Labs, Westinghouse Elec Corp, 49-58; assoc dir basic res, Int Resistance Co, 58-65; STAFF SCIENTIST, PPG INDUSTS, INC, 65- Mem: Am Chem Soc. Res: Polymer chemistry; solid and liquid dielectrics; kinetcs and mechanism of organic reactions; physics of thin films; mechanical behavior of polymers. Mailing Add: 2137 Beechwood Blvd Pittsburgh PA 15217

LEWIS, CLARENCE E, b Oneida, NY, July 19, 07; m, 38; c 2. HORTICULTURE. Educ: Cornell Univ, BS, 34; Hofstra Col, MS, 46. Prof Exp: Asst landscape architect hort, USDA, 35-37; assoc prof ornamental hort, State Univ NY, 37-57; assoc prof, 57-64, prof hort, 64-72, EMER PROF HORT, MICH STATE UNIV, 72-; WRITER & LECTR, 72- Concurrent Pos: Lectr, NY Univ, 43-54. Honors & Awards: Auth Citation, Int Arboricult Soc, 66, Award Merit, 71. Mem: Int Arboricult Soc (mem, Bd Govs, 68-71); Am Hort Soc (mem, Bd Govs, 70-73). Res: Experimentation with the hardiness of questionable shrubs, trees and vines. Mailing Add: 1520 Ridgewood Dr East Lansing MI 48823

LEWIS, CLAUDE IRENIUS, b Stanley, NC, Apr 21, 35; m 56; c 2. ANALYTICAL CHEMISTRY. Educ: Duke Univ, BS, 57; Va Poltech Inst, MS, 59, PhD(chem), 62. Prof Exp: Res chemist, Texaco, Inc, 61-62 & E I du Pont de Nemours & Co, 62-65; res chemist, 65-66, sr res chemist, 66-70, SUPVR ANAL DEVELOP, P LORILLARD CO, 70- Concurrent Pos: Instr, Guilford Col, 67-70. Mem: AAAS; Am Chem Soc; Am Inst Chem Eng. Res: Cigarette tobacco technology; tobacco smoke chemistry; polyester fiber technology; alkyl benzene synthesis; synthesis of polycyclic aromatic compounds. Mailing Add: 3001 Shadylawn Dr Greensboro NC 27408

LEWIS, CLIFFORD EUGENE, b Phoenix, Ariz, Dec 23, 29; m 57; c 2. RANGE MANAGEMENT, BOTANY. Educ: Abilene Christian Col, BS, 58; Utah State Univ, MS, 61. Prof Exp: Asst range scientist, Fla, 60-61, assoc range scientist, 61-66, Ga, 66-72, Fla, 72-73, RANGE SCIENTIST, SOUTHEASTERN FOREST EXP STA, US FOREST SERV, USDA, FLA, 73- Mem: Soc Range Mgt; Wildlife Soc. Res: Wildlife habitat and range management research in the pine-wiregrass vegetation type of the southeastern United States. Mailing Add: Southeastern Forest Exp Sta PO Box 900 Marianna FL 32446

LEWIS, CORNELIUS CRAWFORD, b Appomattox, Va, May 24, 21; m 49. AGRONOMY, SOIL SCIENCE. Educ: Va State Col, BS, 42; Mich State Univ, MS, 45; Univ Mass, PhD(agron), 48. Prof Exp: Prof agron, Ft Valley State Col, 47-48; head dept agr, WVa State Col, 48-49; prof agron, Univ Md, 49-50; anal chemist, New York Testing Lab, 50-51; soil specialist, USDA For Serv, Liberia, WAfrica, 51-53; head dept plant industs, Agr & Tech Col, NC, 54-56; head dept agr, Grambling Col, 56-63; PROF AGRON & NUCLEAR SCI, VA STATE COL, 63- Mem: AAAS; Am Soc Plant Physiol; Am Soc Agron; Soil Sci Soc Am; NY Acad Sci. Res: Field crops; soil fertility; plant nutrient relationship, particularly fertility levels and nutrient requirements for economic crops. Mailing Add: VA State Col PO Box 43 Petersburg VA 23806

LEWIS, DANIEL RALPH, b Camden, Ark, Oct 31, 44. MATHEMATICS. Educ: La State Univ, Baton Rouge, 66, MS, 68, PhD(math), 70. Prof Exp: Asst prof math, Va Polytech Inst & State Univ, 70-72; ASST PROF MATH, UNIV FLA, 72- Mem: Am Math Soc. Res: Functional analysis. Mailing Add: Dept of Math Univ of Fla Gainesville FL 32601

LEWIS, DANIEL WILLIAM, b Wilkes-Barre, Pa; m 44; c 5. ORGANIC CHEMISTRY. Educ: Univ Scranton, BA, 39; Bucknell Univ, MS, 41; Univ Chicago, PhD(org chem), 48. Prof Exp: Instr org microanal, Univ Chicago, 46-47; RES CHEMIST, RES LABS, WESTINGHOUSE ELEC CORP, 48- Mem: Am Chem Soc. Res: Organometallics; synthetic rubber; factors affecting the Grignard reaction. Mailing Add: Westinghouse Res Labs Pittsburgh PA 15235

LEWIS, DANNY HARVE, b Decatur, Ala, Apr 9, 48; m 68; c 1. POLYMER CHEMISTRY. Educ: Univ NAla, BS, 69; Univ Ala, PhD(chem), 73. Prof Exp: Res chemist textile fibers, E I du Pont de Nemours Co Inc, 73-75; SR CHEMIST POLYMER CHEM, SOUTHERN RES INST, 75- Mem: Am Chem Soc; Sigma Xi. Res: Control-release delivery systems; dental materials for extraoral and intraoral use; synthesis and characterization of new polymers; physical properties of polymers; polymers for fiber spinning and polymers as adhesives. Mailing Add: Southern Res Inst 2000 Ninth Ave S Birmingham AL 35205

LEWIS, DAVID, chemistry, deceased

LEWIS, DAVID HAROLD, b New York, NY, Dec 22, 25; m 47, 63; c 2. CARDIOLOGY. Educ: Columbia Univ, AB, 44, MD, 47. Prof Exp: Intern med, Bellevue Hosp, New York, 47-48; intern, Kings County Hosp, 48-49, resident, 49-50; from instr to asst prof physiol, Sch Med, Univ Pa, 50-57, assoc cardiol, Grad Sch Med, 55-63; GUEST INVESTR, FIRST SURG DEPT, UNIV GÖTEBORG, 63- Concurrent Pos: Chief hemodynamics sect, Div Cardiol, Philadelphia Gen Hosp, 55-63; estab investr, Am Heart Asn, 57-62. Mem: Am Physiol Soc; Am Fedn Clin Res; Am Heart Asn. Res: Cardiovascular physiology. Mailing Add: Univ of Göteborg Sch of Med Göteborg Sweden

LEWIS, DAVID JAMES, b Montreal, Que, May 28, 20; c 5. PSYCHIATRY, MEDICAL EDUCATION. Educ: McGill Univ, BA, 41; Univ Toronto, MD, 50; FRCP(C), 57. Prof Exp: Jr intern med, Toronto Gen Hosp, 50-51; sr intern, Sunnybrook Hosp, 51-52; resident psychiat, Johns Hopkins Hosp, 52-54; sr house officer, Bethlem, Royal & Maudsley Hosps, 54-56; asst psychiatrist, St Michael's Hosp, 56-65; asst prof psychiat, Univ Toronto, 64-65; assoc prof, McGill Univ, 65-71; PROF PSYCHIAT, UNIV CALGARY, 71- Concurrent Pos: R S McLaughlin Found res fel, Bethlem, Royal & Maudsley Hosps, 54-56; coordr postgrad educ psychiat, McGill Univ, 65-71, dir behav growth & develop course, 67-71; clin dir psychiat, Allan Mem Inst, 66-71; consult, Rehab Inst Montreal, 67, St Lawrence State Hosp, NY, 67, St Anne's Dept Vet Asmin Hosp, 68 & Banff Mineral Springs Hosp, 72; clin dir psychiat, Foothills Hosp, Calgary, 71- Mem: Fel Am Psychiat Asn; Royal Col Psychiat; Can Psychiat Asn; Royal Soc Med; sci assoc Acad Psychoanal. Res: Psychiatric and interdisciplinary teaching and research; rural psychiatric services. Mailing Add: Serendip RR 1 Calgary AB Can

LEWIS, DAVID KENNETH, b Poughkeepsie, NY, Feb 11, 43; m 64; c 3. PHYSICAL CHEMISTRY. Educ: Amherst Col, AB, 64; Cornell Univ, PhD(phys chem), 70. Prof Exp: ASST PROF CHEM, COLGATE UNIV, 69- Mem: Am Chem Soc; Sigma Xi. Res: Chemical kinetcis and energy transfer in gases at high temperatures; atmospheric chemistry and physics; innovative teaching methods. Mailing Add: Dept of Chem Colgate Univ Hamilton NY 13346

LEWIS, DAVID KENT, b Madison, Wis, June 11, 38; m 62; c 3. FOREST MANAGEMENT, FOREST ECONOMICS. Educ: Univ Minn, BS, 60; Yale Univ, MF, 66. Prof Exp: Forester, Ore, 63-65, SILVICULTURIST, FORESTRY RES CTR, WEYERHAEUSER CO, WASH, 67- Mem: Soc Am Foresters; Am Econ Asn. Res: Economics of producing timber crops. Mailing Add: Forestry Res Ctr Weyerhaeuser Co PO Box 420 Centralia WA 98531

LEWIS, DAVID THOMAS, b Downing, Mo, Sept 27, 35; m 68; c 1. AGRONOMY, SOIL MORPHOLOGY. Educ: Univ Maine, BS, 60, MS, 62; Univ Nebr, PhD(agron), 71. Prof Exp: Instr soil sci, Dept Agron, Univ Maine, 60-62; soil scientist, Soil Conserv Serv, USDA, 62-67; instr agron, 67-71, asst prof soil classification, 71-75, ASSOC PROF SOIL CLASSIFICATION, DEPT AGRON, UNIV NEBR, 75- Mem: Soil Sci Soc Am; Soil Conserv Soc Am; Sigma Xi. Res: Studies relating to the genesis and classification of soils and to the solution of problems that relate to proper correlation of survey mapping units. Mailing Add: Dept of Agron Rm 235 Keim Hall Univ of Nebr Lincoln NE 68503

LEWIS, DENNIS ALLEN, b Morristown, NJ, Dec 25, 42; m 70; c 2. ORGANIC CHEMISTRY. Educ: St Peters Col, BS, 64; Univ Conn, PhD(org chem), 72. Prof Exp: Instr & sr instr, Nuclear Weapons Employ Div, Ft Sill, Okla, 70-71; ASST PROF CHEM, ROSE-HULMAN INST TECHNOL, 72- Concurrent Pos: Instr, US Army Reserve Sch, Ft Benjamin Harrison, Ind, 73- Mem: Am Chem Soc. Res: Synthesis of small-ring compounds via photochemical reactions involving carbene and nitrene intermediates; investigation of chemiluminescent systems. Mailing Add: Dept of Chem Rose-Hulman Inst of Technol Terre Haute IN 47803

LEWIS, DENNIS OSBORNE, b Wales, UK, July 7, 39; m 69; c 1. ORGANIC CHEMISTRY. Educ: Univ Wales, BS, 60, PhD(chem), 64. Prof Exp: Chemist, Brit Drug Houses, Ltd, London, 63-65; staff scientist, Worcester Found Exp Biol, 65-66, Nat Heart Inst fel plant biochem, 66-67; res chemist, Toms River Chem Corp, 67-75; SUPT PROD, PHARMACEUT DIV, CIBA-GEIGY CORP, RI, 75- Mem: Am Chem Soc. Res: Medium membered ring chemistry; steroids; phytochemistry; anthraquinone synthesis. Mailing Add: 69 Brookside Dr East Greenwich RI 02818

LEWIS, DONALD EVERETT, b Paducah, Tex, July 3, 31; m 68; c 2. BIOCHEMISTRY. Educ: Abilene Christian Col, BS, 52; Fla State Univ, MS, 54, PhD(biochem), 57. Prof Exp: From assoc prof to prof chem, Queen's Col, NC, 57-66; assoc prof, 66-68, PROF CHEM, ABILENE CHRISTIAN COL, 68- Concurrent Pos: Vis assoc prof, Abilene Christian Col, 63-64. Mem: AAAS; Am Chem Soc. Res: Synthesis and biological assay of amino acid analogues. Mailing Add: 2541 Campus Courts Abilene TX 79601

LEWIS, DONALD HOWARD, b Stamford, Tex, May 31, 36; m 60; c 3. FISH PATHOLOGY, MICROBIOLOGY. Educ: Univ Tex, Austin, BA, 59; Southwest Tes State Univ, MA, 64; Tex A&M Univ, PhD(vet microbiol), 67. Prof Exp: Res assoc, 66-68, asst prof, 69-75, ASSOC PROF MICROBIOL, TEX A&M UNIV, 75- Concurrent Pos: Consutl, TerEco Corp, 75. Mem: AAAS; Am Soc Microbiol; Am Fisheries Soc; Soc Invert Path; World Maricult Soc. Res: Microbial diseases and immune mechanisms of aquatic animals; role of microflora upon host welfare; antibiotic resistance. Mailing Add: Dept of Vet Microbiol Col of Vet Microbiol Tex A&M Univ College Station TX 77843

LEWIS, DONALD JOHN, b Adrian, Minn, Jan 25, 26; m 53. MATHEMATICS. Educ: Col St Thomas, BS, 46; Univ Mich, MS, 49, PhD(math), 50. Prof Exp: Instr math, Ohio State Univ, 50-52; NSF fel, Inst Adv Study, 52-53; from asst prof to assoc prof, Univ Notre Dame, 53-61; assoc prof, 61-63, PROF MATH, UNIV MICH, ANN ARBOR, 63- Concurrent Pos: NSF sr fel, Manchester & Cambridge Univs, 59-61; sr vis fel, Cambridge Univ, 65, 69. Mem: Am Math Soc; London Math Soc. Res: Diophantine equations; finite fields; algebraic number theory. Mailing Add: Dept of Math Univ of Mich Ann Arbor MI 48104

LEWIS, DONALD RICHARD, b New Leipzig, NDak, May 18, 20; m 43; c 2. PHYSICAL CHEMISTRY. Educ: Univ Wis, BS, 42, MS, 47, PhD(chem), 48. Prof Exp: Ballistics supvr, Hercules Powder Co, 42-46; asst, Univ Wis, 46; chemist, 48-50, sr chemist, 50-63, res assoc, 63-67, GROUP LEADER INSTRUMENTATION, EXPLOR & PROD RES CTR, SHELL DEVELOP CO DIV, 67- Concurrent Pos: Shell exchange scientist, Amsterdam, 56-57. Mem: AAAS; Am Chem Soc; Am Phys Soc; Mineral Soc Am; Geochem Soc; Instrument Soc Am. Res: Interfacial chemistry; defect dominated properties of insulating solids; computer control of experiments and data acquisition. Mailing Add: Explor & Prod Res Ctr Shell Develop Co PO Box 481 Houston TX 77001

LEWIS, DONALD W, b Tiffin, Ohio, July 26, 36; m 58; c 2. ECONOMIC GEOGRAPHY. Educ: Bowling Green State Univ, BSc, 58; Ohio State Univ, MA, 63, PhD(geog), 66. Prof Exp: Res analyst cartog, Aeronaut Chart & Info Serv, US Air Force, Mo, 58-60; cartographer, Ohio Div Water, 60-61; asst geog, Ohio State Univ, 61-62, asst instr, 62-63, res assoc, Natural Resources Inst, 63-64; instr geog, Calif State Col Long Beach, 64-65, asst prof, 65-66; from asst prof to assoc prof, 66-73, PROF GEOG, UNIV TOLEDO, 73- Mem: Asn Am Geogrs; Int Asn Gt Lakes Res. Res: Economic geography, particularly urban geography; conservation and management of natural resources; regional economic development; geography of the United States and Canada; urban land resources. Mailing Add: Dept of Geog Univ of Toledo Col of Arts & Sci Toledo OH 43606

LEWIS, EDWARD B, b Wilkes-Barre, Pa, May 20, 18; m 46; c 3. BIOLOGY. Educ: Univ Minn, BA, 39; Calif Inst Technol, PhD(genetics), 42. Prof Exp: From instr genetics to assoc prof genetics, 46-56, prof biol, 56-66, THOMAS HUNT MORGAN PROF BIOL, CALIF INST TECHNOL, 66- Concurrent Pos: Rockefeller Found fel, Sch Bot, Cambridge Univ, 48-49; mem, Nat Adv Comt Radiation, 58-61. Mem: AAAS; Am Soc Nat; Genetics Soc Am. Res: Genetics; somatic effects of radiation. Mailing Add: Dept of Biol Calif Inst of Technol Pasadena CA 91109

LEWIS, EDWARD LYN, b Aberystwyth, UK, Oct 9, 30; m 59; c 3. OCEANOGRAPHY. Educ: Univ London, BSc, 51, MSc, 58, PhD(physics), 62. Prof Exp: Physicist, Mullard Res Labs, 52-56; res assoc microwave electronics, Univ BC, 59-62; RES SCIENTIST, MARINE SCI BR, DEPT ENVIRON, CAN, 62- Mem: Am Geophys Union; Glaciol Soc. Res: Arctic oceanography; ice physics; energy exchange ocean-atmosphere; arctic instrument development. Mailing Add: Frozen Sea Res Group Marine Sci Br Fed Bldg Victoria BC Can

LEWIS, EDWARD SHELDON, b Berkeley, Calif, May 7, 20; m 55; c 2. CHEMISTRY. Educ: Univ Calif, BS, 40; Harvard Univ, MA, 47, PhD(chem), 47. Prof Exp: Nat Res Coun fel, Univ Calif, Los Angeles, 47-48; from asst prof to assoc prof, 48-58, PROF CHEM, RICE UNIV, 58- Concurrent Pos: Vis prof, Univ Southampton, 57; chmn dept chem, Rice Univ, 65-67; Guggenheim fel, 67; vis prof, Phys Chem Lab, Oxford Univ, 67-68. Mem: Am Chem Soc; The Chem Soc. Res: Mechanism of reactions of organic compounds, especially diazonium salts and hydrogen isotope effects. Mailing Add: Dept of Chem Rice Univ Houston TX 77001

LEWIS, EDWIN AUGUSTUS STEVENS, b Baltimore, Md, Mar 4, 40; m 68; c 2. THERMAL PHYSICS. Educ: Princeton Univ, BA, 61; Makerere Univ, Uganda, dipl educ, 62; Univ Ill, Urbana, MS, 65, PhD(physics), 69. Prof Exp: Educ off physics & math, Tanganyika Govt, 62-63; asst prof physics, Trinity Col, Conn, 69-70; ASST PROF PHYSICS, UNION COL, NY, 70- Mem: Am Phys Soc; Am Asn Physics Teachers. Res: Liquid crystals; cooperative behavior. Mailing Add: Dept of Physics Union Col Schenectady NY 12308

LEWIS, EDWIN REYNOLDS, b Los Angeles, Calif, July 14, 34; m 60; c 2. BIOENGINEERING. Educ: Stanford Univ, AB, 56, MS, 57, PhD(elec eng), 62. Prof Exp: Mem res staff neural modeling, Lab Automata Res, Gen Precision, 61-67; PROF BIOENG, UNIV CALIF, BERKELEY, 67- Mem: Fel Inst Elec & Electronics Engrs; Biomed Eng Soc; Soc Neurosci; AAAS; Electron Micros Soc Am. Res: Applications of engineering analytical tools to problems in neurobiology, ecology; network models of dynamical biological systems; morphology and physiology of vestibular and auditory apparatus. Mailing Add: Dept of Elec Eng & Comput Sci Univ of Calif Berkeley CA 94720

LEWIS, ELMER JAMES, b Cleveland, Ohio, Mar 23, 12; m 41; c 1. ORGANIC CHEMISTRY, PHYSICS. Educ: Univ Buffalo, BA, 43. Prof Exp: Asst to plant engr, Lumen Bearing Co, NY, 28-30; lab asst, 30-32, physicist, 32-36, lab dept head, 36-59, tech mgr performance testing, 59-69, MGR PHYS TESTING, PRATT & LAMBERT, INC, 69- Concurrent Pos: Teacher, eve, Buffalo Mus Sci, 74. Mem: Am Chem Soc; Am Soc Testing & Mat; Sea Horse Inst; Nat Asn Corrosion Engrs. Res: Physical, instrumental and performance tests on organic coatings and chemical interpretation of such data; corrosion studies; weathering and its accelerators; color; viscosity. Mailing Add: Pratt & Lambert Inc 79 Tonawanda St Buffalo NY 14207

LEWIS, ERNEST EUGENE, b Colorado Springs, Colo, Apr 11, 16; m 38; c 3. PLASTICS CHEMISTRY. Educ: Colo Col, BA, 36; Columbia Univ, PhD(org chem), 40. Prof Exp: Asst chem Columbia Univ, 36-39; chemist, E I du Pont de Nemours & Co, Inc, 39-46, group leader, 46-50, supvr, 50-55, sr supvr, 55-58, tech supt, Polyethylene Film Plant, 59, plant supt, 59-60, lab dir, Sabine Film Res & Develop Lab, 60-64, res mgr, Spruance Film Res & Develop Lab, Va, 64-68, asst mgr, Patent Sect, Film Dept, 68-69, MGR, PATENT SECT, FILM DEPT, E I DU PONT DE NEMOURS & CO, INC, 69- Mem: Am Inst Chemists; Am Chem Soc; Sigma Xi. Res: Development and application of high polymers for plastics and films; polyolefin films; cellophane; polyester bottles and films; thermoplastics. Mailing Add: Film Dept E I du Pont de Nemours & Co Inc Wilmington DE 19898

LEWIS, EVERETT VERNON, b Malden, Mass, Aug 3, 07; m 31; c 1. STATISTICS. Educ: Mass Inst Technol, SB, 28, PhD(phys chem), 31. Prof Exp: Res chemist, E I du Pont de Nemours & Co, 31-43, mem staff, Manhattan Proj, 43-44, technologist, 45-46, from jr res assoc to res assoc, 46-50; spec instr math, Univ Del, 46-50, from asst prof to assoc prof, 50-64; assoc prof, 64-74, EMER PROF MATH, URSINUS COL, 74- Mem: AAAS; Am Statist Asn; Inst Math Statist; Math Asn Am; Soc Indust & Appl Math. Res: Applied statistics. Mailing Add: Dept of Math Ursinus Col Collegeville PA 19426

LEWIS, FLORENCE SCOTT, b West Point, Va, May 28, 25; m 47; c 2. BIOCHEMISTRY. Educ: St Augustine's Col, BS, 45; Univ Pa, PhD(molecular biol), 75. Prof Exp: Res asst cancer chemother, 51-72, NIH FEL BIOCHEM, UNIV PA, 75- Mem: AAAS; Am Chem Soc; NY Acad Sci. Res: The isolation, characterization and origin of mammalian mitochondrial messenger ribonucleic acid. Mailing Add: Lippincott Bldg Univ of Pa 2425 Locust St Philadelphia PA 19103

LEWIS, FLOYD JOHN, b Waseca, Minn, Nov 26, 16; m 45; c 3. SURGERY. Educ: Univ Minn, BS, 38, BM & MS, 41, MD, 42, PhD(surg), 50. Prof Exp: Asst anat, Med Sch, Univ Minn, 38-39, intern surg, 41-42, asst, 48-49, from instr to assoc prof, 49-56; assoc prof, 56-57, PROF SURG, SCH MED, NORTHWESTERN UNIV, CHICAGO, 57- Concurrent Pos: Chief thoracic surg, State Hosp, Anoka, Minn, 52; attend surgeon, Univ Minn Hosp, 50-56, Ancker Hosp, Minneapolis, 54-56, Passavant Hosp, Chicago, 56- & Vet Res Hosp, 56- Mem: Soc Clin Surg; Soc Vascular Surg; Soc Univ Surg; Am Surg Asn; Asn Thoracic Surg. Res: Computer applications to patient care. Mailing Add: Dept of Surg Med Sch Northwestern Univ Chicago IL 60611

LEWIS, FRANCIS HOTCHKISS, JR, b Milwaukee, Wis, Aug, 14, 37; m 59; c 2. PHYSICS, MECHANICAL ENGINEERING. Educ: Stevens Inst Technol, ME, 59; Stanford Univ, PhD(physics), 64; Univ San Francisco, JD, 74. Prof Exp: Res asst prof physics, Univ Wash, 64-66; PHYSICIST, LAWRENCE LIVERMORE LAB, UNIV CALIF, 66- Mem: Am Phys Soc. Res: Theoretical nuclear and particle physics. Mailing Add: Lawrence Livermore Lab Univ of Calif PO Box 808 Livermore CA 94550

LEWIS, FRANK, b New York, NY, Jan 7, 18; m 42; c 3. METEOROLOGY. Educ: Univ Vt, BS, 39. Prof Exp: Hydrometeorologist, Bur Reclamation, US Dept Interior, 46-52, meteorologist, Air Weather Serv, US Dept Air Force, 52-56, chief, Electronic Comput Br, 56-65, chief comput systs br, Nat Weather Serv, Nat Oceanic & Atmospheric Admin, 65-75; SR ANALYST, MGT & TECH SERV CO, 75- Concurrent Pos: Mem, Comt Probability & Statist, Am Meteorol Soc, 73-75. Honors & Awards: Dept Com Silver Medal, 70. Res: Development of design storms for design of flood spillways; evaluation of weather forecasts; development of physical-statistical forecasting techniques on electronic data processing systems. Mailing Add: Mgt & Tech Serv Co 5030 Herzel Pl Beltsville MD 20705

LEWIS, FRANK HARLAN, b Redlands, Calif, Jan 8, 19; m 45, 68; c 2. BOTANY. Educ: Univ Calif, Los Angeles, BA, 41, MA, 42, PhD(bot), 46. Prof Exp: Asst instr bot, Univ Calif, Los Angeles, 42-44, instr, 46-47; Nat Res Coun fel, John Innes Hort Inst, London, 47-48; from asst prof to assoc prof, 48-56, PROF BOT, UNIV CALIF, LOS ANGELES, 56-, DEAN, DIV LIFE SCI, 62- Concurrent Pos: Teaching fel, Calif Inst Technol, 43-44; Guggenheim fel, 54-55; consult, NSF, 58-69; chmn dept bot, l)niv Calif, Los Angeles, 59-62; vpres, Int Orgn Biosyst, 64-69, pres, 69- Mem: AAAS; Bot Soc Am; Am Soc Naturalists (pres, 71); Am Soc Plant Taxonomists (pres, 69); Soc Study Evolution (secy, 53-58, vpres, 59, pres, 61). Res: Mechanisms of evolution; systematics of flowering plants. Mailing Add: Dept of Biol Univ of Calif Los Angeles CA 90024

LEWIS, FRANK HERBERT, b Galivants Ferry, SC, Oct 28, 13; m 39; c 3. PLANT PATHOLOGY. Educ: Clemson Col, BS, 37; Cornell Univ, PhD(plant path), 43. Prof Exp: Asst plant path, Cornell Univ, 37-43; from asst prof to assoc prof, 43-52, PROF PLANT PATH & MEM FAC, GRAD SCH, 52-, SCIENTIST-IN-CHARGE, FRUIT RES LAB, 63- Mem: AAAS; Am Phytopath Soc; Am Soc Hort Sci; Am Soc Testing & Mat. Res: Diseases of apples, cherries and peaches; fungicides. Mailing Add: Fruit Res Lab Biglerville PA 17307

LEWIS, FRANKLIN BEACH, b Derby, Conn, Nov 10, 23; m 54; c 2. PATHOLOGY, MEDICAL ENTOMOLOGY. Educ: Union Univ, NY, BS, 48; Univ Conn, MS, 50, PhD(zool), 55. Prof Exp: Asst zool & entom, Univ Conn, 49-53, parasitol, 54-55;

entomologist plant pest, Forest Insect Lab, US Forest Serv, 55-69, PRIN INSECT PATHOLOGIST, NORTHEASTERN FOREST EXP STA, US FOREST SERV, 69-, PROJ LEADER & LAB CHIEF, 74- Mem: AAAS; Entom Soc Am; Sigma Xi; Soc Invert Path. Res: Insect pathology; epidemiology of insect disease; forest insect parasites; biological control of forest insects. Mailing Add: Hallsey Lane Woodbridge CT 06525

LEWIS, FREDERICK D, b Boston, Mass, Aug 12, 43; m 68. PHOTOCHEMISTRY. Educ: Amherst Col, BA, 65; Rochester Univ, PhD(chem), 68. Prof Exp: USPHS res fel chem, Columbia Univ, 68-69; asst prof, 69-74, ASSOC PROF, NORTHWESTERN UNIV, 74- Concurrent Pos: Fel, Dreyfus Found, 73-78 & Sloan Found, 75-77. Mem: Am Chem Soc. Res: Organic photochemistry; energy transfer mechanisms; free radicals; cycloaddition reactions; exciplexes. Mailing Add: Dept of Chem Northwestern Univ Evanston IL 60201

LEWIS, GEORGE CAMPBELL, JR, b Williamsburg, Ky, Mar 25, 19; m 45; c 6. OBSTETRICS & GYNECOLOGY, ONCOLOGY. Educ: Haverford Col, BS, 42; Univ Pa, MD, 44; Am Bd Obstet & Gynec, dipl, 53. Prof Exp: Intern med, Hosp Univ Pa, 44-45, resident obstet & gynec, 47-50, instr, Sch Med, Univ Pa, 50-53, instr radium ther, 51-63, res asst, 53-56, asst prof obstet & gynec, 56-63; prof obstet & gynec & chmn dept, Hahnemann Med Col & Hosp, 62-73, dir div gynec oncol, 71-73; PROF GYNEC ONCOL & DIR DIV, JEFFERSON MED COL, 73- Concurrent Pos: Am Cancer Soc fel gynec oncol, Hosp Univ Pa, 50-52; consult lectr, US Naval Hosp Philadelphia, 56-; consult, Lankenau Hosp & Philadelphia Gen Hosp, 62-, Am Oncol Hosp, 63- & Magee Mem Hosp Rehab Ctr, 68-; mem div gynec oncol, Am Bd Obstet & Gynec; chmn, Gynec Oncol Group, 75- Mem: Am Cancer Soc; Soc Gynec Oncol (pres, 69); Am Gynec Soc; Am Asn Obstet & Gynec; Am Col Obstet & Gynec. Res: Etiology, early diagnosis and evaluation of modes of therapy of gynecologic oncology. Mailing Add: 1025 Walnut St Philadelphia PA 19107

LEWIS, GEORGE EDWARD, b Lorain, Ohio, Oct 27, 08; m 37; c 3. STRATIGRAPHY, VETEBRATE PALEONTOLOGY. Educ: Yale Univ, PhB, 30, PhD, 37. Prof Exp: Instr geol, Yale Univ, 38-43, asst prof, 43-45; GEOLOGIST, US GEOL SURV, 44- Concurrent Pos: Cur, Peabody Mus, Yale Univ, 39-45. Mem: Fel Geol Soc Am; Soc Vert Paleont. Res: Continental stratigraphy; Cenozoic mammals, primates; strategic and terrain intelligence. Mailing Add: Geol Div US Geol Surv Bldg 25 Fed Ctr Denver CO 80225

LEWIS, GEORGE EDWIN, b Decatur, Ga, Jan 6, 33; m 56. ORGANIC CHEMISTRY. Educ: Emory Univ, AB, 52, MS, 53; Fla State Univ, PhD(chem), 58. Prof Exp: Res asst, Ga Inst Technol, 58-59; asst prof chem, La State Univ, 59-66; ASSOC PROF CHEM, JACKSONVILLE UNIV, 66- Mem: Am Chem Soc; The Chem Soc. Res: Mechanisms of organic reactions. Mailing Add: Div of Sci & Math Jacksonville Univ Jacksonville FL 32211

LEWIS, GEORGE KENNETH, b Garden City, Kans, Dec 5, 40; m 65; c 2. CHEMICAL PHYSICS. Educ: Univ Wis-Madison, BS, 63; Univ Ill-Urbana, MS, 66, PhD(chem eng), 68. Prof Exp: Res engr, Eastern Labs, 68-72, RES ENGR TEXTILE RES LAB, CHESTNUT RUN LABS, E I DU PONT DE NEMOURS & CO, INC, 72- Mem: AAAS; Am Phys Soc. Res: Effect of ultrahigh dynamic and static pressure on matter; synthesis of new forms of matter; properties of shock waves in composite solids. Mailing Add: Textile Res Lab Chestnut Run Labs E I du Pont de Nemours & Co Inc Wilmington DE 19898

LEWIS, GEORGE KNOWLTON, b Waltham, Mass, Jan 11, 23; m 42; c 3. GEOGRAPHY. Educ: Harvard Univ, AB, 44, MA, 47, PhD, 56. Prof Exp: From instr to assoc prof, 50-66, PROF GEOG, BOSTON UNIV, 66- Concurrent Pos: Lectr, Wellesley Col, 56 & 58, US Naval War Col, 58-59 & Harvard Univ, 60-67; consult, Itek Corp, 58-60. Honors & Awards: Order of Simon Bolivar, Govt Venezuela, 69. Mem: Asn Am Geogrs; Am Geog Soc. Res: Land use changes at rural-urban fringe; urban-regional analysis; New England. Mailing Add: Dept of Geog Boston Univ Boston MA 02215

LEWIS, GEORGE LEOUTSACOS, b Cotrona, Greece, Oct 23, 11; nat US; m 50; c 4. CHEMISTRY. Educ: Wesleyan Univ, BA, 34, MA, 35; Princeton Univ, PhD(chem), 39. Prof Exp: Fel, Am Philos Soc, Univ Pa, 40; RES ASSOC, EXP STA, E I DU PONT DE NEMOURS & CO, INC, 40- Mem: Fel AAAS; Am Chem Soc; Sigma Xi. Res: Inorganic fibers; inorganic polymers, white pigments; titaniferous ores. Mailing Add: E I du Pont de Nemours & Co Inc Henry Clay DE 19898

LEWIS, GEORGE MCCORMICK, b Los Angeles, Calif, Sept 14, 40; m 64; c 3. GEOMETRY. Educ: Stanford Univ, BA, 61; Univ Southern Calif, MA, 64, PhD(math), 70. Prof Exp: Asst prof, 67-72, ASSOC PROF MATH, CALIF POLYTECH STATE UNIV, SAN LUIS OBISPO, 72- Mem: AAAS; Am Math Soc; Math Asn Am; Soc Indust & Appl Math. Res: Synthetic differential geometry. Mailing Add: Dept of Math Calif State Polytech Univ San Luis Obispo CA 93407

LEWIS, GEORGE W, b Detroit, Mich, Dec 15, 34; m 62; c 3. INTERNAL MEDICINE, MICROBIOLOGY. Educ: Southwestern Univ, BS, 56; Johns Hopkins Univ, MD, 60. Prof Exp: Fel med, Johns Hopkins Univ, 62-65, instr med, 65-66; asst prof, Univ Iowa, 67-68; asst prof, 68-71, ASSOC PROF MED, UNIV CONN, 71- Mem: AAAS; Am Soc Microbiol; Am Fedn Clin Res. Res: Infectious diseases; medical education. Mailing Add: Univ of Conn-McCook Hosp 2 Holcomb St Hartford CT 06112

LEWIS, GLENN C, b Oakley, Idaho, July 13, 20; m 56; c 3. SOIL CHEMISTRY. Educ: Univ Idaho, BS, 46, MS, 49; Purdue Univ, PhD(soils), 62. Prof Exp: Anal agr chem, 47-52, from asst prof org chem to assoc prof soils, 52-67, PROF SOILS, UNIV IDAHO, 67- Mem: Soil Sci Soc Am; Int Soc Soil Sci. Res: Chemical and mineralogical studies on slick spot soils; water quality, including effects of irrigation water quality on soil characteristics; phosphorus reactions in calcareous soils; mineralogical studies on loess. Mailing Add: Dept of Plant & Soil Sci Univ of Idaho Moscow ID 83843

LEWIS, GLENN EDWIN, applied mathematics, see 12th edition

LEWIS, GORDON DEPEW, b Charlottesville, Va, July 22, 29; m 54. FOREST ECONOMICS, MARKETING. Educ: Va Polytech Inst, BS, 51; Duke Univ, MFor, 57; Mich State Univ, PhD(forest econ), 61. Prof Exp: Asst prof forest econ, Univ Mont, 59-62; proj leader, Southeastern Forest Exp Sta, US Forest Serv, 62-66, economist, Washington, DC, 66-67, br chief, 68-71, PROJ LEADER SPEC STUDIES, ROCKY MOUNTAIN FOREST EXP STA, US FOREST SERV, 71- Mem: Soc Am Foresters; Am Econ Asn. Res: Economic evaluations of alternative methods of exploiting natural resources for regional development consistent with the maintenance of the quality of rural and wildlife environments. Mailing Add: Rocky Mt Forest & Range Exp Sta Forest Serv USDA Ft Collins CO 80521

LEWIS, GWYNNE DAVID, b Hackensack, NJ, June 12, 28; m 60; c 1. PLANT PATHOLOGY. Educ: Rutgers Univ, BS, 51; Purdue Univ, MS, 53; Cornell Univ, PhD, 58. Prof Exp: Asst plant path, Purdue Univ, 51-53 & Cornell Univ, 53-58; from asst prof to assoc prof, 58-70, PROF PLANT PATH, RUTGERS UNIV, NEW BRUNSWICK, 70- Mem: Am Phytopath Soc; Soc Nematol. Res: Diseases of vegetable crops; plant nematology; control of plant and vegetable diseases. Mailing Add: Dept of Plant Biol Rutgers Univ New Brunswick NJ 08903

LEWIS, HARLAN LEE, inorganic chemistry, organometallic chemistry, see 12th edition

LEWIS, HAROLD RALPH, b Chicago, Ill, June 7, 31; m 61; c 2. PHYSICS. Educ: Univ Chicago, AB, 51, SB, 53; Univ Ill, MS, 55, PhD(physics), 58. Prof Exp: Res assoc physics, Univ Heidelberg, 58-60; instr, Princeton Univ, 60-63; mem staff, 63-75, ASSOC GROUP LEADER, LOS ALAMOS SCI LAB, 75- Concurrent Pos: Ger Acad Exchange Serv fel, Univ Heidelberg, 58-59. Mem: Am Phys Soc. Res: Plasma physics; nulcear spectroscopy; superconductivity. Mailing Add: Los Alamos Sci Lab PO Box 1663 Los Alamos NM 87544

LEWIS, HAROLD WALTER, b Keene, NH, May 7, 17; m 46; c 2. NUCLEAR PHYSICS. Educ: Middlebury Col, BS, 38; Univ Buffalo, AM, 40; Duke Univ, PhD(physics), 50. Prof Exp: Vis instr & res assoc, 46-49, from asst prof to assoc prof, 49-59, PROF PHYSICS, DUKE UNIV, 59-, V PROVOST, 63-, DEAN FAC, 69- Concurrent Pos: Dean arts & sci, Duke Univ, 63-69. Mem: Am Phys Soc; Am Asn Physics Teachers. Mailing Add: Dept of Physics Duke Univ Durham NC 27706

LEWIS, HAROLD WARREN, b New York, NY, Oct 1, 23; m 47. PHYSICS. Educ: NY Univ, AB, 43; Univ Calif, AM, 44, PhD(physics), 48. Prof Exp: Asst prof physics, Univ Calif, 48-53; mem tech staff, Bell Tel Labs, NJ, 51-56; from assoc prof to prof physics, Univ Wis, 56-64; PROF PHYSICS, UNIV CALIF, SANTA BARBARA, 64- Concurrent Pos: Mem staff, Inst Advan Study, 47-48 & 50-51; dir, Quantum Inst, Univ Calif, Santa Barbara, 69-73. Mem: Am Phys Soc. Res: Theoretical physics. Mailing Add: Dept of Physics Univ of Calif Santa Barbara CA 93106

LEWIS, HARRISON FLINT, ornithology, mammalogy, deceased

LEWIS, HARVYE FLEMING, b Hodge, La, Dec 24, 17. NUTRITION. Educ: La Polytech Inst, BS, 38; Univ Tenn, MS, 42; Iowa State Col, PhD(nutrit), 50. Prof Exp: Teacher high sch, La, 38-40; nutritionist, State Dept Pub Health, Tenn, 42; res assoc, Agr Exp Sta, La State Univ, 43-47, asst nutritionist, 50-52; assoc prof food & nutrit, Fla State Univ, 52-65; PROF FOOD & NUTRIT, SCH HOME ECON, LA STATE UNIV, 65- Mem: AAAS; Am Dietetic Asn; Am Home Econ Asn; Inst Food Technologists; Am Inst Nutrit. Res: Vitamin content of foods; nutritional requirements; food patterns and nutritional health of children. Mailing Add: Sch of Home Econ La State Univ Baton Rouge LA 70803

LEWIS, HENRY RAFALSKY, b Yonkers, NY, Nov 19, 25; m 57; c 3. PHYSICS. Educ: Harvard Univ, AB, 48, MA, 49, PhD(physics), 56. Prof Exp: Mem staff opers res, Opers Eval Group, Mass Inst Technol, 51-53, 56; group head quantum electronics, David Sarnoff Res Ctr, RCA Corp, 57-66, dir, electronic res lab, 66-70; VPRES RES & DEVELOP, ITEK CORP, 70- Mem: Am Phys Soc; Opers Res Soc Am; Inst Elec & Electronics Eng. Res: Paramagnetic resonance; quantum electronics; operations research; molecular beams. Mailing Add: Itek Corp Bldg 21 10 Maguire Rd Lexington MA 02173

LEWIS, HERBERT SAMUEL, b Jersey City, NJ, May 8, 34; m 57; c 3. ANTHROPOLOGY, ETHNOGRAPHY. Educ: Brandeis Univ, AB, 55; Columbia Univ, PhD(anthrop), 63. Prof Exp: Instr anthrop & polit sci, Northwestern Univ, 61-63; from asst prof to assoc prof, 63-73, PROF ANTHROP, UNIV WIS-MADISON, 73- Concurrent Pos: Soc Sci Res Coun travel grant, Int Conf, 63; NSF res grant, Ethiopia, 65-66; NIMH small grant, Israel, 70; NSF res grant, 75-77. Mem: Fel Am Athrop Asn; Am Ethnol Soc; fel African Studies Asn; fel Royal Anthrop Inst Gt Brit & Ireland. Res: Political anthropology; cultural and social change; ethnicity in the modern world; African culture history and political systems; Ethiopia, Israel and Wisconsin. Mailing Add: Dept of Anthrop Univ of Wis-Madison Madison WI 53706

LEWIS, HERMAN WILLIAM, b Chicago, Ill, July 10, 23; c 2. GENETICS, ZOOLOGY. Educ: Univ Ill, BS, 47, MS, 49; Univ Calif, PhD(genetics), 53. Prof Exp: USPHS res fel, Univ Calif, 52-54; asst prof biol, Mass Inst Technol, 54-61; prof life sci & chmn dept, Mich State Univ, 61-62; prog dir genetic biol, NSF, DC, 62-66, HEAD CELLULAR BIOL SECT, NSF, 66- Mem: AAAS; Biophys Soc; Genetics Soc Am; Am Soc Cell Biol. Res: Biochemical, physiological and molecular genetics; biophysics and cytology of genetic material; human cell biology. Mailing Add: Cellular Biol Sect Nat Sci Found Washington DC 20550

LEWIS, HOWARD PHELPS, b San Francisco, Calif, Feb 18, 02; m 27; c 2. INTERNAL MEDICINE. Educ: Ore State Col, BS, 24; Univ Ore, MD, 30; Am Bd Internal Med, dipl, 52. Prof Exp: Asst anat, 26-30, instr med, 29-30, instr clin med, 32-36, assoc, 36-38, asst prof, 38-46, assoc prof med, 46-47, prof & head dept, 47-71, EMER PROF MED, MED SCH, UNIV ORE, 71- Concurrent Pos: Mem, Nat Adv Heart Coun, Nat Heart Inst, 56-60 & Bd Dirs, Am Bd Family Pract, 70-; mem, Adv Bd Internal Med, 52, chmn, 59-61. Honors & Awards: Merit Award, Am Heart Asn, 60; Alfred Stengel Award, Am Col Physicians, 66. Mem: Asn Am Physicians; Am Heart Asn (ed, Modern Concepts of Cardiovasc Dis, 56-); Am Clin & Climat Asn (pres, 68); Am Col Physicians (3rd vpres, 51, pres, 59-60). Mailing Add: 2151 SW Laurel St Portland OR 97201

LEWIS, IRVING JAMES, b Boston, Mass, July 9, 18; m 41; c 3. COMMUNITY HEALTH. Educ: Harvard Univ, AB, 39; Univ Chicago, AM, 41. Prof Exp: With US Govt, 42 & 46-55, dep chief, Int Div, Bur Budget, 55-57, dept head, Intergovt Comn Europ Migration, Geneva, Switz, 57-59, dep chief, Int Div, Bur Budget, 59-65, chief, Health & Welfare Div, 65-67, dep asst dir, 67-68, dep adminr health serv & ment health admin, Dept Health, Educ & Welfare, 68-70; PROF COMMUNITY HEALTH, ALBERT EINSTEIN COL MED, 70- Honors & Awards: Except Serv Award, Bur Budget, 64; Career Serv Award, Nat Civil Serv League, 69. Mem: Inst of Med of Nat Acad Sci; Am Polit Sci Asn; Am Soc Pub Adminr; Am Pub Health Asn; Asn Am Med Cols. Mailing Add: Dept of Community Health Albert Einstein Col of Med New York NY 10641

LEWIS, IRWIN CHARLES, organic chemistry, see 12th edition

LEWIS, JACK A, b Brooklyn, NY, Apr 8, 39; m 68; c 2. PLANT PATHOLOGY, SOIL MICROBIOLOGY. Educ: Brooklyn Col, BS, 60; Rutgers Univ, PhD(microbiol), 65. Prof Exp: MICROBIOLOGIST, USDA, 65- Mem: Am Soc Microbiol; Am Phytopath Soc. Res: Biological control of soil-borne plant pathogenic fungi; microbial decomposition of natural materials in soil. Mailing Add: Soilborne Dis Lab Plant Prot Inst Agr Res Serv USDA Beltsville MD 20705

LEWIS, JACK SMITH, b Whitesburg, Ky, June 10, 26; m 48; c 3. MASS

SPECTROMETRY. Educ: Univ Ky, BS, 50. Prof Exp: Jr res physicist, Tenn Eastman Co, 51-52, sales corresp, Eastman Chem Prod, Inc, 52-53, res physicist, Tenn Eastman Co, 53-63, sr res physicist, 63-73, RES ASSOC, TENN EASTMAN CO, 73- Mem: Am Chem Soc; Am Soc Testing & Mat; Am Soc Mass Spectrometry. Res: Determination of structure of organic compounds; gas chromatography-mass spectrometry; high resolution mass spectrometry; storage and retrieval of gas chromatographic data; computer interpretation of mass spectral data. Mailing Add: Tenn Eastman Co Res Labs Kingsport TN 37662

LEWIS, JAMES BRYAN, b York, Pa, Dec 14, 45; m 67; c 2. MOLECULAR BIOLOGY. Educ: Univ Pa, BA, 67; Harvard Univ, MA, 68, PhD(chem), 71. Prof Exp: Damon Runyan-Walter Winchell Cancer Fund fel, Swiss Inst Exp Cancer Res, 71-73; fel, 73-74, staff investr, 74-75, SR STAFF INVESTR, COLD SPRING HARBOR LAB, 75- Res: Mapping the adenovirus genes for specific polypeptides by cell-free translation of messenger RNA fractionated by hybridization to fragments of adenovirus DNA. Mailing Add: Cold Spring Harbor Lab Box 100 Cold Spring Harbor NY 11724

LEWIS, JAMES CHESTER, b Kalamazoo, Mich, Jan 31, 36; m 57; c 3. WILDLIFE ECOLOGY. Educ: Univ Mich, BS, 57; Mich State Univ, MS, 63; Okla State Univ, PhD(wildlife ecol), 74. Prof Exp: Biologist aide, Mich Game Div, US Forest Serv, 57-59, dist biologist, Tenn Game Div, 59-60, res proj leader game mgt, 60-64, res supvr, 64-67; asst unit leader & asst prof, 67-75, ACTG LEADER, OKLA COOP WILDLIFE RES UNIT & ASSOC PROF LIFE SCI, SCH BIOPHYS SCI, OKLA STATE UNIV, 75- Mem: Wildlife Soc; Soc Range Mgt; Soc Am Foresters. Res: Endangered species research; deer and turkey management; ecology of wildlife rabies; mourning dove and sandhill crane behavior and ecology. Mailing Add: Okla Coop Wildlife Res Unit Okla State Univ Stillwater OK 74074

LEWIS, JAMES CLEMENT, b Lewisville, Minn, Aug 10, 15; m 39; c 3. BIOCHEMISTRY. Educ: Univ Minn, BCh, 36; Ore State Col, MS, 39, PhD(soils, agr chem), 40. Prof Exp: Analyst, Univ Minn, 36-37; asst chemist animal nutrit, Exp Sta, Ore State Col, 40-41; biochemist, Western Regional Res Lab, Bur Agr & Indust Chem, USDA, 41-53, BIOCHEMIST, WESTERN REGIONAL RES LAB, AGR RES SERV, USDA, 53- Mem: AAAS; Am Chem Soc; Am Soc Microbiol; Am Soc Biol Chem. Res: Microbial biochemistry; trace elements in plant and microbiology; bacterial spores. Mailing Add: 1 Harvard Circle Berkeley CA 94708

LEWIS, JAMES EDWARD, b Ashland, Ky, July 11, 27; m 52; c 3. PHYSICAL CHEMISTRY. Educ: Centre Col, AB, 50; Purdue Univ, MS, 54, PhD(chem), 56. Prof Exp: Sr chemist, E I du Pont de Nemours & Co, 55-56; gen mgr, Radiochem Inc, Ky, 56-57, pres, 57-65; sr res assoc, United Carbon Co Div, Ashland Oil & Refining Co, 65-66, exec asst to pres res & develop, 66-67, dir res & develop, 67-70, Ashland Oil Inc, 70-74, VPRES, ASHLAND CHEM CO, 74- Concurrent Pos: Mem, Adv Comt Nuclear Sci, Ky Atomic Energy Authority, 59-65. Mem: AAAS; Am Chem Soc; Am Phys Soc; Am Nuclear Soc. Res: Radiochemistry; thermodynamics and transport properties of liquids; instrumentation; high vacuum phenomena; carbon black and rubber chemistry. Mailing Add: Ashland Chem Co PO Box 2458 Columbus OH 43216

LEWIS, JAMES ELDON, b Pontiac, Mich, Jan 5, 38; m 63; c 2. ECONOMIC GEOGRAPHY, TRANSPORTATION GEOGRAPHY. Educ: Eastern Mich Univ, BA, 61; Univ Ga, MA, 63, PhD(geog), 66. Prof Exp: Asst prof geog, La State Univ, Baton Rouge, 66-69; asst prof environ sci, Univ Va, 69-70, community & family med, 70-73; vpres, Enviro-Med Inc, 73-74; SR PROF ASSOC, INST MED, NAT ACAD SCI, 74- Concurrent Pos: Consult, Gulf South Res Inst, La, 66-69, Va Off Comprehensive Health Planning, 69-70, Med Facility Comn, Va Gen Assembly, 70-72 & Va Hwy Res Coun, 70-; grants, US Dept Health, Educ & Welfare Acad Ctr Comprehensive Health Planning, Univ Va, 70-76, Va Hwy Safety Div Eval Emergency Med Serv, 72 & US Dept Health, Educ & Welfare, 72-73. Mem: Asn Am Geogrs. Res: Health planning; health services delivery; regionalization; management of academic health science centers. Mailing Add: Nat Acad of Sci Inst of Med Washington DC 20418

LEWIS, JAMES KELLEY, b Waco, Tex, Oct 24, 24; m 49; c 4. RANGE SCIENCE. Educ: Colo State Univ, BS, 48; Mont State Univ, MS, 51. Prof Exp: Asst prof, 50-58, ASSOC PROF ANIMAL SCI, SDAK STATE UNIV, 58- Mem: Soc Range Mgt; Am Soc Animal Sci; Ecol Soc Am; Wildlife Soc; Brit Grassland Soc. Res: Structure, function, measurement, manipulation, uses and systems analysis of range ecosystems; range animal nutrition and management; coupling of range and agronomic ecosystems. Mailing Add: Dept of Animal Sci SDak State Univ Brookings SD 57006

LEWIS, JAMES VERNON, b Neligh, Nebr, May 2, 15; m 46; c 3. MATHEMATICS. Educ: Univ Calif, AB, 37, MA, 39, PhD(math), 42. Prof Exp: Asst, Univ Calif, 39-42; jr physicist, US Navy, Calif, 42-43; mathematician, Radiation Lab, Univ Calif, 43-45; mathematician, Aberdeen Proving Ground, 45-53; ASSOC PROF MATH, UNIV N MEX, 53- Concurrent Pos: Asst prof, Univ Nev, 46-47. Mem: Am Math Soc; Math Asn Am; Sigma Xi. Res: Calculus of variations; urban planning; iterative methods for decision making in urban planning. Mailing Add: Dept of Math & Statist Univ of NMex Albuquerque NM 87131

LEWIS, JAMES W L, b Natchez, Miss, May 3, 38; m 61; c 3. MOLECULAR PHYSICS, ACOUSTICS. Educ: Univ Miss, BS, 60, MS, 64, PhD(physics), 66. Prof Exp: Physicist, US Naval Weapons Lab, 61-62 & Arnold Res Orgn, 66-68; UK Sci Res Coun fel physics, Queen's Univ, Belfast, 68-69; PHYSICIST, ARNOLD RES ORGN, 69-; ASSOC PROF PHYSICS, SPACE INST, UNIV TENN, 66- Mem: Am Phys Soc. Res: Vibrational relaxation processes in gases; molecular processes in hypersonic flow phenomena; molecular and atomic beam collision processes using high temperature shock tube source. Mailing Add: Arnold Res Orgn Arnold AFB TN 37389

LEWIS, JANE SANFORD, b Pasdena, Calif, Dec 26, 18; m 42; c 4. NUTRITION. Educ: Pomona Col, BA, 40; Cornell Univ, MS, 42; Univ Calif, Los Angeles, MPH, 66, DrPH(nutrit), 69. Prof Exp: Home economist, Wilson & Co, Ill, 42-43; anal chemist, Nat Defense Res Coun, Northwestern Univ, 43-45; anal chemist, Nat Defense Res Coun Proj, Calif Inst Technol, 45-46; technician, Nutrit Lab, Sch Pub Health, Univ Calif, Los Angeles, 65; nutritionist, Head Start Prog, Fedn Settlements & Recreation Ctr, Calif, 66; technician, Nutrit Lab, Sch Pub Health, Univ Calif, Los Angeles, 67; assoc prof, 68-72, PROF NUTRIT, CALIF STATE COL, LOS ANGELES, 72- Concurrent Pos: Chmn, Task Force, Calif Nutrit Coun, 70- Mem: Am Pub Health Asn; Am Dietetic Asn; Soc Nutrit Educ; Am Inst Nutrit; Am Home Econ Asn. Res: Nutritional status of children of varying backgrounds; effect of oral contraceptives and anticonvulsant drugs on nutritional status; food habits of various ethnic groups, anthropometric measurements of Oriental children. Mailing Add: Dept of Home Econ Calif State Univ Los Angeles CA 90032

LEWIS, JASPER PHELPS, b Danville, Va, Nov 8, 17; m 50. CHEMISTRY, BIOCHEMISTRY. Educ: Univ Va, BSChem, 46; Univ Louisville, MS, 58; Med Col Ga, PhD(biochem), 66. Prof Exp: Res chemist biochem, Sch Med, Univ Va, 46-50; clin chemist, Vet Admin Hosp, Bay Pines, Fla, 50-52, clin chemist, Louisville, Ky, 52-60 & res chemist, St Louis, Mo, 60-62; BASIC SCIENTIST ERYTHROPOIESIS RES, VET ADMIN HOSP, AUGUSTA, 62- Concurrent Pos: Assoc res prof, Med Col Ga, 67- Mem: AAAS; Am Chem Soc; Am Asn Clin Chemists. Res: Erythropoiesis regulatory factors. Mailing Add: Vet Admin Hosp Med Res Forest Hills Div Augusta GA 30904

LEWIS, JERRY PARKER, b Terre Haute, Ind, Sept 20, 31; m 56; c 4. MEDICINE. Educ: James Millikin Univ, 52; Univ Ill, BS, 53, MD, 56. Prof Exp: Asst prof med, Univ Ill, 64-67; assoc prof, 67-69, lectr clin path & vet med, 68-74, PROF MED, UNIV CALIF, DAVIS, 69-, CHIEF SECT HEMAT & ONCOL, 67- Concurrent Pos: NIH fel hemat, Presby-St Luke's Hosp, Chicago, 61-63, res fel, 63-65; actg chief clin hemat & chief spec hemat, Presby-St Luke's Hosp, Chicago, 65-67; consult, David Grant Hosp, Travis, AFB, Calif, 68- Mem: Am Soc Hemat; Transplantation Soc; fel Am Col Physicians; Soc Exp Hemat; Am Soc Human Genetics. Res: Bone marrow transplantation; preservation of marrow using cryobiologic techniques; erythropoietin. Mailing Add: Sch of Med Univ of Calif Davis CA 95616

LEWIS, JESSE C, b Vaughan, Miss, June 26, 29; m 59; c 1. MATHEMATICS, COMPUTER SCIENCES. Educ: Univ Ill, MS, 55, MA, 59; Syracuse Univ, PhD(math), 66. Prof Exp: Instr math, Southern Univ, 55-57 & Prairies View Agr & Mech Col, 57-58; asst prof, Jackson State Col, 59-61; res asst, Comput Ctr, Syracuse Univ, 63-66; DIR COMPUT CTR & CHMN DEPT COMPUT SCI, JACKSON STATE COL, 66-, PROF MATH & CHMN DIV NATURAL SCI, 67- Concurrent Pos: Consult, Comt Undergrad Prog Math, Jackson State Col. Mem: Math Asn Am; Am Math Soc; Asn Comput Mach; Asn Educ Data Systs. Res: Computer study of permanents of n-square (0,1)-matrices with k l's in each row and column. Mailing Add: Dept of Math Jackson State Col Jackson MS 39217

LEWIS, JESSICA HELEN, b Harpswell, Maine, Oct 26, 17; m 46; c 5. MEDICINE. Educ: Goucher Col, AB, 38; Johns Hopkins Univ, MD, 42. Prof Exp: Intern, Hosp Women, Baltimore, Md, 42-43; asst resident, Univ Calif Hosp, 43-44; res fel, Thorndike Mem Lab & Harvard Univ, 44-46; res assoc physiol, Univ NC, 48-55; res assoc med, 55-58, res assoc prof, 58-70, RES PROF MED, UNIV PITTSBURGH, 70- Concurrent Pos: USPHS res fel, Univ NC, 47-48; asst med, Boston City Hosp, 44-46; res assoc, Med Sch, Emory Univ, 46-47; assoc med, Med Sch, Duke Univ, 51-55; staff mem, Presby-Univ Hosp, 55-; dir res, Cent Blood Bank Pittsburgh, 69- Mem: Am Soc Hemat; World Fedn Hemophilia; Am Physiol Soc; Am Soc Clin Invest; Am Fedn Clin Res. Res: Blood coagulation; enzyme and protein chemistry. Mailing Add: Dept of Med Univ of Pittsburgh Sch of Med Pittsburgh PA 15261

LEWIS, JOHN ALBERT, b Glanworth, Ont, May 3, 12; m 39; c 2. MEDICINE. Educ: FRCP(C), 49. Prof Exp: Asst prof internal med, 45-64, from clin assoc prof to clin prof med, 64-68, PROF MED, UNIV WESTERN ONT, 68-; CHIEF MED, DEPT VET AFFAIRS, WESTMINSTER HOSP, 45- Concurrent Pos: Attend physician, Victoria Hosp, London, Ont, 39- & St Joseph's Hosp, 45- Mem: Am Heart Asn; fel Am Col Physicians; fel Am Col Chest Physicians; Can Med Asn; affil Royal Soc Med. Res: Cardiovascular renal disease. Mailing Add: Westminster Hosp Box 5701 Terminal A London ON Can

LEWIS, JOHN ALLEN, b Detroit, Mich, Jan 21, 23; m 48; c 3. APPLIED MATHEMATICS. Educ: Worcester Polytech Inst, BS, 44; Brown Univ, ScM, 48, PhD(appl math), 50. Prof Exp: Asst appl math, Brown Univ, 46-50; res physicist, Corning Glass Works, 50-51; MEM TECH STAFF MATH, BELL TEL LABS, 51- Mem: Am Math Soc; Soc Indust & Appl Math. Res: Applied mechanics; viscous flow, elasticity; heat transfer; piezoelectricity. Mailing Add: Bell Tel Labs Rm 2-C456 Murray Hill NJ 07971

LEWIS, JOHN BRADLEY, b Ottawa, Ont, Jan 12, 25; m 52; c 3. MARINE BIOLOGY. Educ: McGill Univ, BSc, 40, MSc, 50, PhD(zool), 54. Prof Exp: Asst marine biol, Inst Marine Sci, Univ Miami, 51-54; dir Bellairs Res Inst, McGill Univ, 54-71, assoc prof, 61-69, PROF MARINE SCI, MARINE SCI CTR, McGILL UNIV, 69-, DIR REDPATH MUS, 71- Mem: AAAS; Can Soc Zool; Asn Island Marine Labs Caribbean. Res: Tropical marine ecology and physiology; tropical marine organisms and coral reef ecology. Mailing Add: Marine Sci Ctr McGill Univ Montreal PQ Can

LEWIS, JOHN ERWIN, b Elmsford, NY, May 27, 25; m 56; c 2. MICROBIOLOGY. Educ: Wagner Col, BS, 50; Rutgers Univ, MS, 51; Nat Registry Microbiol, registered, 74. Prof Exp: Bacteriologist, Genesee Hosp, Univ Rochester, 51-55, instr, Sch Nursing, 52-55; SR BACTERIOLOGIST, HIGHLAND VIEW HOSP & DEPT MICROBIOL, SCH MED, CASE WESTERN RESERVE UNIV, 55-, RES ASST, DIV RADIATION BIOL, SCH MED, 57- Mem: Sigma Xi; Am Acad Microbiol. Res: Enzyme systems of anaerobic gram negative bacteria implicated in human disease processes. Mailing Add: Highland View Hosp 3901 Ireland Dr Cleveland OH 44128

LEWIS, JOHN GOVER, nuclear physics, see 12th edition

LEWIS, JOHN HUBBARD, b Jamestown, NY, Apr 13, 29; m 56; c 4. GEOLOGY. Educ: Allegheny Col, BS, 56; Univ Colo, PhD(geol), 65. Prof Exp: From instr to assoc prof, 58-74, PROF GEOL, COLO COL, 74-, CHMN DEPT, 70- Concurrent Pos: Lectr, Exten Div, Univ Colo, 58-66; dir, NSF Sec Sci Training Prog, Colo Col, 65-67; US Antarctic res partic, Tex Tech Col, 67-68. Mem: Geol Soc Am. Res: Sedimentary petrology; petrology and diagenesis of upper Cambrian rocks of Colorado; structural geology. Mailing Add: Dept of Geol Colo Col Colorado Springs CO 80903

LEWIS, JOHN L, JR, b San Antonio, Tex, June 5, 29; m 55; c 3. OBSTETRICS & GYNECOLOGY. Educ: Harvard Univ, BA, 52, MD, 57; Am Bd Obstet & Gynec, dipl, 67. Prof Exp: Clin assoc endocrinol br, Nat Cancer Inst, 59-61, sr investr surg br, 65-67; assoc prof, 68-71, PROF OBSTET & GYNEC, MED COL, CORNELL UNIV, 71-; CHIEF GYNEC SERV, MEM HOSP CANCER & ALLIED DIS & JAMES EWING HOSP, 68- Concurrent Pos: Sr investr clin ctr, NIH, 65-67; assoc attend gynecologist, Francis Delafield Hosp, 67; assoc attend obstetrician & gynecologist, Presby Hosp, NY, 67; assoc attend obstetrician & gynecologist, New York Lying-in Hosp, 68-71, attend obstetrician & gynecologist, 71-; attend surgeon, Mem Hosp Cancer & Allied Dis, 68-; assoc, Sloan-Kettering Inst Cancer Res, 68-73, mem, 73-; assoc prof, Col Physicians & Surgeons, Columbia Univ, 67, lectr, 68-; dir, Am Bd Obstet & Gynec, 70-, dir, Div Gynec Oncol, 70-; consult Am joint comt cancer staging & end result reporting. Mem: Soc Gynec Invest; AMA; Soc Gynec Oncol; Soc Pelvic Surg; James Ewing Soc. Res: Gynecologic cancer; hormonal, immunologic and therapeutic aspects of gestational trophoblastic neoplasms. Mailing Add: Mem Hosp for Cancer & Allied Dis 1275 York Ave New York NY 10021

LEWIS, JOHN MORGAN, b Joliet, Ill, June 5, 20; m 44; c 3. ANIMAL HUSBANDRY. Educ: Univ Ill, BS, 43. Prof Exp: Asst supt, 43-59, actg supt, 59-62, ASSOC PROF ANIMAL SCI, DIXON SPRINGS EXP STA, UNIV ILL, URBANA,

62- Mem: Am Soc Animal Sci. Res: Sheep breeding, feeding and management. Mailing Add: Dixon Springs Exp Sta Univ of Ill Simpson IL 62985

LEWIS, JOHN RAYMOND, b Philadelphia, Pa, July 25, 18; m 52; c 1. POLYMER CHEMISTRY. Educ: Franklin & Marshall Col, BS, 42. Prof Exp: Chemist, Naval Stores Div, Res Ctr, Hercules Inc, 42-44, shift supvr explosives dept, Sunflower Ord Works, Kans, 44-45, chemist, Naval Stores Div, 45-49, res chemist, 49-55, res supvr, 55-59, res mgr, Plastics & Elastomers Div, 59-64, res assoc, Cent Res Div, 64, mgr develop, Res Dept, 64-69, venture projs, New Enterprise Dept, 69-75, MGR PLANNING & ACQUISITIONS, NEW ENTERPRISE DEPT, HERCULES, INC, 75- Mem: Am Chem Soc; Financial Analysts Asn; Sigma Xi; Com Develop Asn. Res: Commercial development; polymers, energy and raw materials. Mailing Add: New Enterprise Dept Hercules Inc Wilmington DE 19899

LEWIS, JOHN REED, b Ottawa, Kans, Dec 27, 15; m 38; c 3. PHARMACOLOGY. Educ: Ottawa Univ, AB, 37; Mich State Col, MS, 40; Univ Mich, PhD, 49. Prof Exp: Asst chem exp sta, Mich State Col, 39-41; supvr biol control, Frederick Stearns & Co, 41-45; sr biologist, Sterling-Winthrop Res Inst, 47-53, assoc dir sect coord & integration, 53-60; asst secy coun drugs, 60-64, assoc dir dept drugs, 64-72, SR SCIENTIST, DEPT DRUGS, AMA, 72- Mem: Am Soc Pharmacol & Exp Therapeut; Soc Toxicol; NY Acad Sci; Drug Info Asn. Res: Vitamin assays; pharmacology of sympathominetics and analgesics; diuretics; anticholinesterases; coordination of research projects in development of new drugs; medical writing. Mailing Add: Am Med Asn 535 N Dearborn St Chicago IL 60610

LEWIS, JOHN SIMPSON, b Trenton, NJ, June 27, 41; m 64; c 3. GEOCHEMISTRY, METEORITICS. Educ: Princeton Univ, AB, 62; Dartmouth Col, MA, 64; Univ Calif, San Diego, PhD, 68. Prof Exp: Asst prof, 68-72, ASSOC PROF CHEM, EARTH & PLANETARY SCI, MASS INST TECHNOL, 72- Concurrent Pos: Mem, Working Group Outer Planet Probe Sci, NASA-Ames Res Ctr, 74-, NASA Phys Sci Comt, 75- & Space Sci Bd spec panels outer solar syst & explor Venus, Nat Acad Sci-Nat Res Coun; chmn, Uranus Sci Adv Comt, NASA-Jet Propulsion Lab, 74-75, mem, Sci Adv Group Outer Solar Syst; consult, Aerospace Div, Martin-Marietta Corp, 72 & Avco Systs Div, Avco Corp; Guggenheim lectr, Nat Air & Space Mus, Smithsonian Inst, 73; sci lectr, Div Planetary Sci, Am Astron Soc, 74. Mem: AAAS; Am Chem Soc; NY Acad Sci; Am Astron Soc. Res: Composition, structure and origin of planetary atmospheres; atmosphere-lithosphere interactions; application of thermodynamics to problems of composition and origin of meteorites. Mailing Add: Rm 54-1220 Mass Inst of Technol Cambridge MA 02139

LEWIS, KEITH HARROWER, bacteriology, see 12th edition

LEWIS, LAWRENCE A, b Pittsburgh, Pa, June 26, 39. GEOGRAPHY. Educ: Antioch Col, BA, 61; Northwestern Univ, MS, 63, PhD(geog), 65. Prof Exp: Asst prof geog, Univ Ind, 65-70; ASSOC PROF GEOG, CLARK UNIV, 70- Mem: Asn Am Geogrs; Geol Soc Am. Res: Slope processes in the tropics; erosion control in tropical agriculture; fluvial processes; geomorphology. Mailing Add: Grad Sch of Geog Clark Univ Worcester MA 01610

LEWIS, LAWRENCE GUY, mathematical analysis, see 12th edition

LEWIS, LENA ARMSTRONG, b Lancaster, Pa, July 12, 10. PHYSIOLOGY. Educ: Lindenwood Col, AB, 31; Ohio State Univ, MA, 38, PhD(physiol), 40. Hon Degrees: LLD, Lindenwood Col, 52. Prof Exp: Asst biochem, Sch Med, Johns Hopkins Univ, 31-32; bacteriologist & technologist, Gen Hosp, Lancaster, Pa, 32-36; asst physiol, Ohio State Univ, 36-41; mem res staff & supvr electrophoresis lab, 43-45, MEM STAFF & SUPVR ELECTROPHORESIS LAB, CLEVELAND CLIN FOUND, 45- Concurrent Pos: Spec fel endocrinol, Cleveland Clin Found, 41-43; adj prof, Cleveland State Univ, 71-74, clin prof chem, 74-; mem arteriosclerosis coun, Am Heart Asn. Honors & Awards: Award Outstanding Contrib Clin Chem in Field Lipids & Lipoproteins, Am Asn Clin Chem, 74. Mem: AAAS; Am Physiol Soc; Soc Exp Biol & Med. Res: Relation of adrenal to electrolyte metabolism; changes in plasma proteins in endocrine disease hypertension; factors regulating lipid and protein metabolism, especially their relation to atherosclerosis; electrophoresis in physiology. Mailing Add: Cleveland Clin Found 9500 Euclid Ave Cleveland OH 44106

LEWIS, LEONDA LAMONTE, b Ripley, NY, July 14, 17; m 43; c 2. CHEMISTRY. Educ: Greenville Col, AB, 42; Purdue Univ, PhD(org chem), 47. Prof Exp: Res chemist, Yerkes Lab, E I du Pont de Nemours & Co, Inc, 46-50, res supvr, 50-54, tech investr, Patent Sect, 54-55, spec rep, Sales Div, 55-67, TECH CONSULT, E I DU PONT DE NEMOURS & CO, INC, 67- Mem: Am Chem Soc. Res: Adhesives; laminations; reinforced plastics; high temperature insulation for wire and cable; organic synthesis; high polymers; polymers for films and coatings; preparation and polymerization of fluorinated styrenes. Mailing Add: Film Dept E I du Pont de Nemours & Co Inc Wilmington DE 19898

LEWIS, LEROY CRAWFORD, b Pocatello, Idaho, Mar 18, 40; m 62; c 2. PHYSICAL CHEMISTRY, INORGANIC CHEMISTRY. Educ: Col Idaho, BS, 62; Ore State Univ, PhD(phys chem), 68. Prof Exp: Sr res chemist, Idaho Nuclear Corp, 68-71; sr res chemist, 71-72, GROUP SUPVR, ALLIED CHEM CORP, 72- Mem: Am Chem Soc. Res: Nuclear fuel reprocessing chemistry; chemical waste handling chemistry; actinide chemistry; electrochemistry. Mailing Add: Allied Chem Corp CPP-637 Box 2204 Idaho Falls ID 83401

LEWIS, LESLIE ARTHUR, b Castries, WI, May 17, 40; m 68; c 1. GENETICS, MICROBIOLOGY. Educ: Univ Toronto, BSA, 63, MSA, 64; Columbia Univ, PhD(genetics), 68. Prof Exp: NIH fel, Mich State Univ, 68-69; lectr, 69-70, asst prof, 70-74, ASSOC PROF BIOL, YORK COL (NY), 74- Mem: Genetics Soc Am. Res: Non-reciprocal recombination in the fungus Sordaria. Mailing Add: Div of Natural Sci York Col 150-14 Jamaica Ave Jamaica NY 11432

LEWIS, LEWIS JAMES, b Washington, Pa, Sept 6, 16; m 52; c 4. VIROLOGY, BACTERIOLOGY. Educ: Waynesburg Col, BS, 39; Univ Pittsburgh, PhD(med bact), 43. Prof Exp: Res assoc, Sharpe & Dohme Div, Merck & Co, Inc, 43-44; asst dir biol res, Nat Drug Co, 44-46; res assoc, E R Squibb & Sons, 47-49; res assoc, Sch Med, Univ Pittsburgh, 49-57; dir virus res, Nat Drug Co, 57-59; asst dir biol div & res fel virol, Sterling-Winthrop Res Inst, 59-67; dir virus res, Abbott Labs, 67-72; DIR CELL CULT LAB, COL MED, UNIV IOWA, 72- Mem: AAAS; Am Soc Microbiol; Brit Biochem Soc. Res: Inactivated poliomyelitis vaccine; rubella viruses; physiological products from tissue culture; fibrinolytic agents; endothelial cells, platelets and lipids. Mailing Add: Cell Cult Lab Dept of Med Univ of Iowa Col of Med Iowa City IA 52240

LEWIS, LEYBURN F, b Raymondville, Mo, Dec 11, 11; m 46; c 3. MEDICAL ENTOMOLOGY. Educ: Iowa State Univ, BS, 49, MS, 59. Prof Exp: MED ENTOMOLOGIST, ENTOM RES DIV, AGR RES SERV, USDA, 49- Mem: AAAS; Entom Soc Am; Am Mosquito Control Asn. Res: Basic and applied research in medical entomology, primarily mosquitoes, other blood-sucking Diptera and cattle lice. Mailing Add: 5544 Air Terminal Dr Fresno CA 93727

LEWIS, LLOYD GEORGE, b Brocton, Ill, Dec 22, 17; m 41; c 5. NUCLEAR PHYSICS. Educ: Univ Chicago, BS, 39, PhD(physics), 46. Prof Exp: Assoc physicist, Armour Res Found, 40-42 & Univ Chicago, 43-45; instr physics, Princeton Univ, 46-50; sect head, Eng Res Dept, Standard Oil Co, Ind, 50-60; physicist, Electronic Assocs, Inc, 60-61; SR PHYSICIST, ARGONNE NAT LAB, 61- Mem: Am Phys Soc. Res: Cosmic ray air showers using the method of coincident bursts in two unshielded ionization chambers.

LEWIS, LON D, physiology, see 12th edition

LEWIS, LOWELL N, b Kingston, Pa, July 9, 31; m 53; c 3. PLANT PHYSIOLOGY. Educ: Pa State Univ, BS, 53; Mich State Univ, MS, 58, PhD(hort, biochem), 60. Prof Exp: Asst horticulturist, 60-65, assoc prof hort, 66-70, PROF PLANT PHYSIOL, UNIV CALIF, RIVERSIDE, 70-, ASSOC DEAN RES, COL NATURAL & AGR SCI, 71- Concurrent Pos: Guggenheim res fel, Mich State Univ-AEC Plant Res Lab, 67-68. Mem: Am Soc Hort Sci; Am Soc Plant Physiol; Japanese Soc Plant Physiol. Res: Hormonal regulation of plant cell development, especially senescence and abscission. Mailing Add: Dept of Plant Sci Univ of Calif Riverside CA 92502

LEWIS, LYNN LORAINE, b Terra Alta, WVa, Mar 2, 29; m 54; c 4. ANALYTICAL CHEMISTRY. Educ: WVa Wesleyan Col, BS, 50; Marshall Univ, MS, 52; Univ Tenn, PhD(chem), 55. Prof Exp: Engr semiconductors, Westinghouse Elec Corp, 55-56; sr res chemist, Res Lab, US Steel Corp, Pa, 56-66; ASST HEAD ANAL CHEM DEPT, GEN MOTORS RES LAB, 66- Mem: Am Chem Soc; Am Soc Testing & Mat; Soc Appl Spectros. Res: Behavior and determination of gases in metals; analysis of metals; instrumentation for chemical analysis. Mailing Add: Res Lab Gen Motors Corp Warren MI 48090

LEWIS, MARC SIMON, b Cleveland, Ohio, Oct 30, 26; m 48; c 3. BIOCHEMISTRY, BIOPHYSICS. Educ: Western Reserve Univ, BS, 46, MS, 47; Georgetown Univ, PhD(biochem), 55. Prof Exp: Guest scientist, Nat Inst Arthritis & Metab Dis, 52-55, biochemist, 57-58, biochemist, Nat Inst Dent Res, 58-62, head sect opthal chem, Nat Inst Neurol Dis & Blindness, 62-70, SR RES INVESTR, LAB VISION RES, NAT EYE INST, 70- Concurrent Pos: USPHS fel, 55-57; lectr, Found Advan Educ in Sci Grad Prog, NIH. Mem: Am Chem Soc; Asn Res Vision & Opthal; NY Acad Sci; Biophys Soc; Am Soc Biol Chem. Res: Physical biochemistry; physical chemistry of proteins; opthalmological biochemistry; biophysical techniques, especially ultracentrifugation. Mailing Add: Lab Vision Res Nat Eye Inst Bethesda MD 20014

LEWIS, MARGARET NAST, b Baltimore, Md, Aug 20, 11. PHYSICS. Educ: Goucher Col, AB, 31; Johns Hopkins Univ, PhD(physics), 37. Prof Exp: Asst physics, Vassar Col, 37-38; Am Asn Univ Women Berliner fel, Univ Calif, 38-39, fel, Crocker Radiation Lab, 39-40, Howell fel, 40-42; instr, Vassar Col, 42-43; instr physics, Univ Pa, 43-48, assoc physics res, 53-54; lectr, Boston Univ, 48-50; physicist, Nat Bur Stand, 50-52; asst prof res, Brown Univ, 54-58; assoc prof, Univ Mass, 58-61; res fel, 61-70, ASSOC PHYSICS, HARVARD COL OBSERV, 70- Concurrent Pos: Radioisotopes res, Mass Mem Hosp, 48-49 & Haverford Col, 52-54; consult, Cushing Vet Hosp, 49-50. Mem: Am Phys Soc. Res: Spectroscopy; atomic structure. Mailing Add: Harvard Col Observ Cambridge MA 02138

LEWIS, MARTIN GWENT, b Abertillery, Wales, Jan 15, 39; m 63; c 3. CANCER, PATHOLOGY. Educ: London Univ, MB, BS, 62, Dr Med, 68. Prof Exp: House physician & surgeon, Woolwich Mem Hosp, London, Eng, 62-63; demonstr & resident pathologist, Bristol Univ, Eng, 63-65; lectr path, Makerere Med Col, Kampala, Uganda, 65-68; mem staff cancer res, Chester Beatty Inst, London, 68-69; sr lectr path, Westminster Med Sch, London, 69-71; prof & chmn dept, Mem Univ Nfld, 71-73; DIR CANCER RES, McGILL UNIV & NAT CANCER INST CAN, 73- Mem: Europ Soc Dermat Res; Am Soc Cancer Res; AAAS; Can Oncol Soc; Int Acad Path. Res: Tumor immunology, with special emphasis on the role of humoral immunity and causes of failure in control of metatases in human cancer, particularly malignant melanoma; immunotherapy of cancer. Mailing Add: McGill Cancer Res Unit 3655 Drummond St Montreal PQ Can

LEWIS, MARVIN BURTON, b Chicago, Ill, Aug 28, 27; m 53; c 3. THEORETICAL PHYSICS. Educ: Northwestern Univ, PhD, 56. Prof Exp: Asst prof physics, Yale Univ, 55-60; assoc prof mech eng, 60-65, PROF MECH ENG, NORTHWESTERN UNIV, 65-, PROF ASTROPHYS, 68- Mem: Am Phys Soc. Res: Statistical mechanics. Mailing Add: Dept of Mech Eng Northwestern Univ Evanston IL 60201

LEWIS, MELISSA, physiology, physiological psychology, see 12th edition

LEWIS, MERLE LOVELL, biochemistry, see 12th edition

LEWIS, MILTON, b New York, NY, Dec 30, 21; m 43; c 3. PHYSICAL CHEMISTRY. Educ: Univ Wash, BS; Univ Calif, PhD(chem), 50. Prof Exp: Field serv consult, Off Sci Res Develop, 43-46; asst, Univ Calif, 46-48; chemist, Gen Elec Co, 48-51, chg pile coolant studies, 51-54, supvr, Nonmetallic Mat Develop, 54-56, sr engr prog, 56-62, mgr chem & metall, 62-67; mgr chem & metall, Douglas United Nuclear, Inc, 67-68, fuel & target technol, 68, asst chief, Mat Br, Donald W Douglas Labs, 68-70, mgr, Betacel Prog, 70-74; PRES, COLUMBIA ENGRS SERV, INC, WASH, 74- Concurrent Pos: Vis lectr, Univ Calif, Los Angeles, 60-61. Mem: Am Nuclear Soc. Res: Mechanism of irreversible reactions; analytical chemistry of fission products; corrosion in aqueous media; radiation effects on materials. Mailing Add: 2600 Harris Ave Richland WA 99352

LEWIS, MORRIS LEON, economic geography, see 12th edition

LEWIS, MORTON, b Oak Park, Ill, June 28, 36; m 63; c 3. ORGANIC CHEMISTRY. Educ: Purdue Univ, BS, 58; Univ Chicago, PhD(org chem), 62. Prof Exp: Res chemist, 62-74, HEAD, SPECIALTY CHEM RES DIV, SWIFT & CO, 74- Mem: Am Chem Soc; The Chem Soc. Res: Synthesis and reactions of steroid derivatives; products of fats and oils; surface active agents; quaternary ammonium salts and organophosphorous compounds; flame retardants; plastics and coatings; specialty chemicals. Mailing Add: Swift & Co Res & Develop Ctr 1919 Swift Dr Oakbrook IL 60521

LEWIS, NEIL JEFFREY, b New York, NY, Feb 10, 45. MEDICINAL CHEMISTRY, ORGANIC CHEMISTRY. Educ: City Col New York, BS, 66; Univ Kans, PhD(med chem), 72. Prof Exp: NIH res assoc, 72, ASST PROF MED CHEM, DIV MED CHEM, COL PHARM, OHIO STATE UNIV, 72- Mem: Am Chem Soc; Am Pharmaceut Asn. Res: Immunochemotherapy; aminosugars. Mailing Add: Div of Med Chem Ohio State Univ Col of Pharm Columbus OH 43210

LEWIS, OTIS GRIFFIN, b Bridgeport, Conn, Mar 6, 28; m 54; c 3. POLYMER CHEMISTRY. Educ: Mass Inst Technol, BS, 49; Univ Akron, PhD(polymer chem),

60. Prof Exp: Chemist, Sponge Rubber Prods Co, 49-54; proj engr protective clothing, US Army Chem Ctr, 54-56; group leader polymer plastics, Cent Res, 59-72, GROUP LEADER PHYS TESTING, DAVIS & GECK DEPT, AM CYANAMID CO, 72- Mem: Am Soc Testing & Mat; Asn Advan Med Instrumentation. Res: Polymer characterization; fiber structure characterization; development of new surgical sutures; development of new test methods for sutures. Mailing Add: Davis & Geck 1 Casper St Danbury CT 06810

LEWIS, PAUL EDWARD, b Beverly, WVa, Feb 10, 43; m 69; c 1. REPRODUCTIVE PHYSIOLOGY. Educ: Berea Col, BS, 66; WVa Univ, MS, 68, PhD(reprod physiol), 72. Prof Exp: Res asst reprod phys, WVa Univ, 66-68; with med res labs, Physiol Sec, Fitzsimmons Gen Hosp, 68-69; vet specialist, Ft Benning, 69-70; from res asst to res assoc reprod phys, WVa Univ, 70-73; ASST PROF REPROD PHYSIOL, LINCOLN UNIV, MO, 73- Mem: Am Soc Animal Sci. Res: Investigations concerning the physiological factors regulating corpus luteum function, behavioral estrus and ovulation in domestic animals. Mailing Add: Dept of Agr & Nat Resources Lincoln Univ Jefferson City MO 65101

LEWIS, PAUL EDWIN, b West Fork, Ark, Oct 6, 11; m 46; c 1; m 62; c 2. MATHEMATICS. Educ: Okla State Teachers Col, BS, 31; Okla Agr & Mech Col, MS, 36; Univ Ill, PhD(math), 40. Prof Exp: Teacher high sch, Ikla, 31-36; asst math, Univ Ill, 36-40; from instr to assoc prof, Okla Agr & Mech Col, 40-47; assoc prof, NC State Univ, 47-55; supvr digital comput lab, Convair Astronaut Div, Gen Dynamics Corp, 55-65; dir comput ctr, 65-67, PROF MATH, NC STATE UNIV, 65-, HEAD DEPT COMPUT SCI, 67- Mem: Soc Indust & Appl Math; assoc Am Math Soc; assoc Asn Comput Mach; assoc Math Asn Am. Res: Algebra; characters of Abelian groups. Mailing Add: Dept of Comput Sci NC State Univ Raleigh NC 27607

LEWIS, PAUL HERBERT, b New York, NY, Jan 19, 24; m 55; c 3. PHYSICAL CHEMISTRY. Educ: Columbia Univ, AB, 47, MA, 48; Iowa State Col, PhD(chem), 52. Prof Exp: Chemist paints, E I du Pont de Nemours & Co, Inc, 48; PETROL CHEMIST, TEXACO INC, 52- Mem: Am Chem Soc. Res: X-ray analysis; catalysts. Mailing Add: 9 Creek Bend Rd Poughkeepsie NY 12603

LEWIS, PAUL KERMITH, JR, b Monticello, Ark, Jan 24, 31; m 55; c 3. MEAT SCIENCE. Educ: Okla State Univ, BS, 53; Univ Wis, MS, 55, PhD, 58. Prof Exp: Res asst animal husb & biochem, Univ Wis, 53-57; from asst prof to assoc prof animal indust, 57-68, PROF ANIMAL SCI, UNIV ARK FAYETTEVILLE, 68- Mem: AAAS; Am Soc Animal Sci; Am Meat Sci Asn; Inst Food Technologists; Inst Briquetting & Agglomeration. Res: Pre-slaughter stress and storage life of beef and pork; sensory characteristics of hamburger cooked over various charcoal. Mailing Add: Dept of Animal Sci Univ of Ark Fayetteville AR 72701

LEWIS, PAUL WELDON, b Dallas, Tex, Jan 31, 43; m 65; c 2. MATHEMATICS. Educ: NTex State Univ, BA, 65, MS, 66; Univ Utah, PhD(math), 70. Prof Exp: Asst prof, 70-74, ASSOC PROF MATH, N TEX STATE UNIV, 74- Mem: Am Math Soc. Res: Vector measures; functional analysis; operators on function spaces. Mailing Add: Dept of Math N Tex State Univ Denton TX 76203

LEWIS, PEIRCE FEE, b Detroit, Mich, Oct 26, 27; m 52; c 1. CULTURAL GEOGRAPHY. Educ: Albion Col, BA, 50; Univ Mich, MA, 52, PhD(geog), 58. Prof Exp: Geographer, US Army Intel, Far East, 53-55; asst prof geog, 58-62, PROF GEOG, PA STATE UNIV, UNIVERSITY PARK, 62- Concurrent Pos: NSF fel, Northwestern Univ, 62; US Off Educ grant, Pa State Univ, 67-68; Asn Am Geog grant, diffusion of folk architecture in New York State, 68-69; vis lectr geog, Univ Mich, 58; dir, US Off Educ Exp Teacher Fel Prog, Pa State Univ, 67-68; vis scientist, NSF-Asn Am Geog, 69-72. Mem: Asn Am Geog; Am Geog Soc. Res: Origin and evolution of American landscapes. Mailing Add: Dept of Geog Pa State Univ 439 Deike University Park PA 16802

LEWIS, PETER ADRIAN WALTER, b Johannesburg, SAfrica, Oct 3, 32; US citizen; m 60; c 2. STATISTICS. Educ: Columbia Univ, BA, 54, BS, 55, MS, 57; Univ London, PhD(statist), 64. Prof Exp: Res staff mem statist, Int Bus Mach Res Labs, 55-71; PROF STATIST & OPERS RES, NAVAL POSTGRAD SCH, 71- Concurrent Pos: NIH spec fel, Imp Col, Univ London, 69-70. Mem: Inst Math Statist; Royal Statist Soc; Am Statist Asn. Res: Stochastic process; applications of statistics in computer applications. Mailing Add: Dept of Oper Res Naval Postgrad Sch Monterey CA 93940

LEWIS, PHILLIP ALBERT, b Indianapolis, Ind, Feb 11, 21; m 44; c 5. ANALYTICAL CHEMISTRY. Educ: Aurora Col, BS, 42; Okla State Univ, MS, 54, PhD(chem), 56. Prof Exp: Instr chem & math, Duluth Jr Col, 46; asst prof chem & head dept, Aurora Col, 46-47 & 49-51; asst, Ill Inst Technol, 47-49 & Okla State Univ, 51-54; asst prof, Univ Kansas City, 54-56; sr chemist, Midwest Res Inst, 56-59; prof chem & head dept, Iowa Wesleyan Col, 59-64; vis Fulbright lectr, Meerut Col, Agra, 64-65; prof chem, Hastings Col, 65-68; PROF CHEM & DEAN, WESTMINSTER COL, PA, 68- Concurrent Pos: Lectr, Univ Kansas City, 56-57, William Jewell Col, 57-58 & Nat Col Mo, 58-59. Mem: AAAS; Am Chem Soc; Sigma Xi. Res: Nonaqeous polarography; spectrophotometric analytical methods; electrochemical phenomena; spectrophotometry. Mailing Add: Dept of Chem Westminster Col New Wilmington PA 16142

LEWIS, PHILLIP H, b Chicago, Ill, July 31, 22; m 49; c 3. ANTHROPOLOGY, ETHNOLOGY. Educ: Sch of Art Inst Chicago, BFA, 47; Univ Chicago, MA, 53, PhD(anthrop), 66. Prof Exp: From asst cur to cur primitive art, 57-66, CUR PRIMITIVE ART & MELANESIAN ETHNOL, FIELD MUS NATURAL HIST, 66- Concurrent Pos: Instr primitive art, Sch of Art Inst Chicago, 59; instr art of nonliterate peoples, Northwestern Univ, 66 & 72; lectr anthrop, Univ Chicago, 66-71; study grants, Field Mus Natural Hist, various mus in Ger & Switz, 66, Wenner-Gren Found Anthrop Res, mus in Holland, WBerlin, Leipzig & Dresden, 68 & NSF field mus study, Honolulu, Australia & NZ, 69; fel ethnol field work, New Ireland & Melanesia, 70. Mem: Fel Am Anthrop Asn; fel Royal Anthrop Inst Gt Brit & Ireland. Res: Anthropological study of primitive art, especially in New Ireland, Territory of New Guinea and Melanesia; art in social context; change in art and society; styles of New Ireland art. Mailing Add: Dept of Anthrop Field Mus of Natural Hist Chicago IL 60605

LEWIS, RALPH KEPLER, b Mar 7, 12; m 49; c 3. ANTHROPOLOGY. Educ: Southwest Mo State col, BA, 34; Univ Southern Calif, MA, 39; Columbia Univ, PhD(anthrop), 67. Prof Exp: Dep asst chief transp, Europ Theater, US Army, 45-46; field assoc anthrop, Morden African Exped, Am Mus Natural Hist, 47-48; field dir, Mid East Res Proj, Columbia Univ, 49-51; prof foreign serv inst, US Dept State, 51-53, area specialist, Lang & Area Training Ctr, Beirut, Lebanon, 53-55, dir, 55-57, consult, Am Consulate Gen, Dhahran, Saudi Arabia, 57-60, chief NAfrican div, Bur Anal & Res, DC, 60-61, chief Near East, Bur Educ & Cultural Exchange, 61-64; assoc prof anthrop, 64-70, chmn dept, 68-70, PROF ANTHROP, GEORGE WASHINGTON UNIV, 70- Mem: Fel Am Anthrop Asn; Am Ethnol Soc. Res: Social and economic organizations of Near Eastern societies; man-land relationships and social organization in East Africa. Mailing Add: Dept of Anthrop George Washington Univ Washington DC 20006

LEWIS, RALPH WILLIAM, b Marion, Mich, May 21, 11; m 37; c 2. PLANT PATHOLOGY. Educ: Mich State Col, BS, 34, MS, 37, PhD(plant path), 45. Prof Exp: From instr to asst prof bot, 37-44, from asst prof to assoc prof biol, 44-50, PROF BIOL, MICH STATE UNIV, 50- Concurrent Pos: Fel, Calif Inst Technol, 47; NIH spec res fel, Instituto Superiore Sanita, Rome, 58-59. Mem: Fel AAAS; Bot Soc Am; Am Soc Naturalists; Nat Sci Teachers Asn; Nat Asn Biol Teachers. Res: Differentiation; nutrition; parasitism; study of patterns in the molecular environment of differentiational events in fungi; study of the structure of biological knowledge. Mailing Add: Dept of Natural Sci Mich State Univ East Lansing MI 48824

LEWIS, RAYMOND ALWYN, elementary particle physics, see 12th edition

LEWIS, RICHARD JOHN, b Chicago, Ill, Jan 20, 35; m 61; c 3. HEMATOLOGY, CLINICAL PHARMACOLOGY. Educ: Univ Notre Dame, BS, 56; Northwestern Univ, Chicago, MD, 60. Prof Exp: Rotating intern, Cook County Hosp, Chicago, 60-61; med intern & resident, Columbia Div, Bellevue Hosp, 61-63; med resident, Presby Hosp, New York, 63-64; chief nuclear med lab, USPHS Hosp, San Francisco, 66-68; clin instr, 68-70, ASST PROF MED, MED CTR, UNIV CALIF, SAN FRANCISCO, 70- Concurrent Pos: Fel hemat, Montefiore Hosp, 64-65; NIH res fel, Med Sch, Univ Wash, 65; clin investr hemat, Vet Admin Hosp, San Francisco, 70-74. Mem: Int Soc Hemat; Soc Nuclear Med; Am Fedn Clin Res; NY Acad Sci; Am Soc Hemat. Res: Metabolism of coumarin anticoagulant drugs. Mailing Add: 465 N Roxbury Dr Beverly Hills CA 90210

LEWIS, RICHARD MULLINGTON, horticulture, botany, see 12th edition

LEWIS, RICHARD NEWTON, b Berkeley, Calif, May 3, 16; m 43; c 4. POLYMER CHEMISTRY. Educ: Univ Calif, Berkeley, AB, 37; Calif Inst Technol, PhD(org chem), 43. Prof Exp: Res chemist, Gen Elec Co, 42-48; asst prof chem, Univ Del, 48-51; res assoc, Olin Mathieson Chem Corp, 52-65; SR RES ASSOC POLYMER RES, SWS SILICONES CORP, ADRIAN, 65- Mem: Am Chem Soc. Res: Polymer reactions; polycyclic structures; interfacial bonding. Mailing Add: 503 Seminole Dr Tecumseh MI 49286

LEWIS, RICHARD THOMAS, b East Cleveland, Ohio, Jan 9, 43. PHYSICAL CHEMISTRY. Educ: Case Western Reserve Univ, BS, 64; Univ Chicago, PhD(phys chem), 70. Prof Exp: Staff scientist chem, 70-74, GROUP LEADER HYDROCARBON PYROLYSIS, UNION CARBIDE CARBON PROD DIV, 75- Mem: Am Chem Soc. Res: Surface chemistry; chemistry of hydrocarbon pyrolysis and carbonization. Mailing Add: Union Carbide Corp 12900 Snow Rd Parma OH 44130

LEWIS, RICHARD WHEATLEY, JR, b Montclair, NJ, Apr 15, 25; m 53; c 4. ECONOMIC GEOLOGY. Educ: Bowdoin Col, AB, 49; Stanford Univ, MS, 51, PhD, 64. Prof Exp: Geologist, US Geol Surv, 51-72, geochemist & tech adv, 72-73; PRES, GEOQUIMICA, SERVICOS GEOLOGICOS E ANALITICOS, LTD, 73- Mem: AAAS; Geol Soc Am; Mineral Soc Am; Geochem Soc; Am Inst Mining, Metall & Petrol Eng. Res: Geochemical exploration of tropical terrain; statistical interpretation of geochemical data. Mailing Add: Rua Aguaraiba 86 Bonsucesso Rio de Janeiro Brazil

LEWIS, ROBERT ALLEN, b Dunkirk, NY, July 27, 43; m 69. PETROLEUM CHEMISTRY. Educ: Carnegie Inst Technol, BS, 65; Princeton Univ, MS, 67, PhD(chem), 69. Prof Exp: NIH fel, Mass Inst Technol, 69-70; CHEMIST, CHEVRON RES CO, STANDARD OIL CO CALIF, 70- Mem: Am Chem Soc; The Chem Soc. Res: Fuel additives; air pollution control. Mailing Add: Chevron Res Co Standard Ave Richmond CA 94806

LEWIS, ROBERT DONALD, b Wyalusing, Pa, Nov 4, 97; m 22; c 2. AGRONOMY. Educ: Pa State Col, BS, 19; Cornell Univ, PhD(plant breeding), 26. Prof Exp: Instr exp agron, Pa State Col, 19-24; asst plant breeding, Cornell Univ, 23-24, instr, 24-26, exten asst prof, 26-30; exten prof agron, Ohio State Univ, 30-39, from assoc prof to prof, 33-46; dir, 46-62, EMER DIR, TEX AGR EXP STA, TEX A&M UNIV, 62-, CONSULT, 62- Concurrent Pos: Agent, Div Cereal Crops & Dis, Bur Plant Indust, USDA, Md, 36-46; assoc, Ohio Exp Sta, 39-46; vchmn dept agron, Ohio State Univ, 39-40, chmn, 40-46; mem, Nat Rice Res & Mkt Adv Comn, 53-62 & Comt of Nine, 53-55; chmn, Southern Regional Exp Sta Dirs, USDA, 55-57; adv, Venezuela Ministry Agr, 56; Ford Found sr consult, Univ Aleppo, 62-65; consult agr res & educ, 65-67 & agr res, USAID, Korea, 67-70. Honors & Awards: Distinguished Serv Award, Grain Sorghum Producers Asn, 67; awards, Ministry Agr, Repub Korea, 68, Off Rural Develop, 70 & Seoul Nat Univ, 70. Mem: AAAS; fel Am Soc Agron; Genetics Soc Am; hon mem Int Crop Improv Asn. Res: Genetics of oats; corn breeding; seeding methods; forage crops; seed production and distribution; administration of agricultural research. Mailing Add: 102 Greenway Bryan TX 77801

LEWIS, ROBERT EARL, b Richmond, Ind, Dec 1, 29; m 52. ENTOMOLOGY, VERTEBRATE ZOOLOGY. Educ: Earlham Col, AB, 52; Univ Ill, MS, 56, PhD(entom), 59. Prof Exp: From asst prof to assoc prof zool, Am Univ, Beirut, 59-67; assoc prof entom, 67-71, PROF ENTOM, IOWA STATE UNIV, 71- Concurrent Pos: Consult, US Naval Med Res Unit, Egypt, 63-; grants, Off Naval Res, 66-71 & NIH, 64-67. Mem: Entom Soc Am; Am Entom Soc; Am Soc Mammal; Soc Syst Zool; Am Soc Parasitol. Res: Siphonaptera of the world, their host relationships and zoogeography. Mailing Add: 306 21st St Ames IA 50010

LEWIS, ROBERT FRANK, b Wis, Dec 13, 20; m 45; c 2. PUBLIC HEALTH. Educ: Univ Calif, BS, 50, MPH, 53; Univ Mich, PhD(public health statist), 58. Prof Exp: Asst sanitarian epidemiol res, Commun Dis Ctr, USPHS, Ga, 49-52; health analyst, San Joaquin Local Health Dist, Calif, 53-54; state health analyst, Calif, 54-56; res assoc cerebral palsy, Univ Mich, 58; from asst prof to prof biomet, Sch Med, Tulane Univ, 58-67, head div, 60-67; prof human med, Col Human Med, Mich State Univ, 67-75; PROF & DIR HEALTH MEASUREMENT PROGS, SCH OF PUB HEALTH, UNIV SC, 75- Concurrent Pos: Consult, Div Radiol Health, NIH, 66-70, div manpower intel, 71; chief, Mich Ctr Health Statist, 67-69; consult, Nat Ctr Health Servs Res, 75- Mem: Fel AAAS; fel Am Pub Health Asn; Am Statist Asn; Biomet Soc; Soc Epidemiol Res. Res: Public health and medicine, especially development of methodology in community health research. Mailing Add: 3128 Chinaberry Dr Columbia SC 29204

LEWIS, ROBERT GLENN, b Morehead City, NC, Nov 11, 37; m 60; c 3. ORGANIC CHEMISTRY. Educ: Univ NC, BS, 60; Univ Wis, PhD(org chem), 64. Prof Exp: Res chemist, Chemstrand Res Ctr, Inc, 64-69, res specialist, 69-71; SECT CHIEF, ENVIRON PROTECTION AGENCY, 71- Mem: Am Chem Soc. Res: Mechanistic photochemistry; reaction mechanisms; photodegradation of dyes and polymers; air pollution analysis; organic photochemistry; ultra violet-visible absorption and luminescence spectroscopy; organic analyses; pesticide chemistry and analysis; mass spectrometry. Mailing Add: Health Effects Res Lab US Environ Protection Agency Research Triangle Park NC 27711

LEWIS, ROBERT MILLER, b Flushing, NY, May 20, 37; m 58; c 2. IMMUNOLOGY. Educ: Wash State Univ, DVM, 61. Prof Exp: Intern, Angell Mem Animal Hosp, 61-62; res fel, Harvard Med Sch, 62-65; from instr to sr instr surg, Sch Med, Tufts Univ, 65-67, from asst prof to assoc prof, 67-75, dir lab animal sci, 69-75; PROF PATH & CHMN DEPT, NY STATE COL VET MED, CORNELL UNIV, 75- Concurrent Pos: Res assoc path, Angell Mem Animal Hosp, 62-63, assoc pathologist, 65-67, affil in med, 68-75; consult surg res, New Eng Med Ctr Hosp, 62-65, mem spec sci staff, 66-75, chief vet serv, 70-75; asst path, Harvard Med Sch, 65-68, clin asst, 68-76. Honors & Awards: Mary Mitchell Award Outstanding Res, 61. Mem: Am Vet Med Asn; Am Col Vet Pathologists; Int Acad Path; Am Asn Lab Animal Sci; Am Soc Vet Clin Pathologists. Res: Investigations on the etiology and pathogenesis of spontaneous immunologic diseases of animals which mimic human diseases. Mailing Add: Dept of Path Schurman Hall NY State Col Vet Med Cornell Univ Ithaca NY 14853

LEWIS, ROBERT MINTURN, b Hempstead, NY, Aug 23, 24; m 53; c 3. FISH BIOLOGY. Educ: Cornell Univ, BS, 54, MS, 56. Prof Exp: Res fishery biologist, Striped Bass Prog, Mid-Atlantic Coastal Fisheries Res Ctr, 56-63, Menhaden Prog, 63-70, RES FISHERY BIOLOGIST, ATLANTIC ESTUARINE FISHERIES CTR, NAT MARINE FISHERIES SERV, 63- Mem: Am Fisheries Soc; Am Inst Fisheries Res Biol. Res: Effects of environmental conditions on larval and juvenile marine fishes; population dynamics of marine fishes; electrophoretic studies of fish protein; analysis of large scale tagging; analysis and identification of off-shore and estuatine larval fish populations. Mailing Add: Atlantic Estuarine Fish Ctr Nat Marine Fisheries Serv Beaufort NC 28516

LEWIS, ROBERT RICHARDS, JR, b New Haven, Conn, Mar 7, 27; m 50; c 4. THEORETICAL PHYSICS. Educ: Univ Mich, BS, 50, MS, 51, PhD(physics), 54. Prof Exp: Asst prof physics, Univ Notre Dame, 54-58; from asst prof to assoc prof, 58-65, PROF PHYSICS, UNIV MICH, ANN ARBOR, 65- Concurrent Pos: Mem, Inst Advan Study, 56-58. Res: Quantum theory; angular correlation theory; parity nonconservation in atoms; partial coherence theory. Mailing Add: Dept of Physics Univ of Mich Ann Arbor MI 48104

LEWIS, ROBERT TABER, b New York, NY, May 12, 32; m 58; c 1. PHYSICS. Educ: Alfred Univ, BS, 54; Univ Calif, Berkeley, PhD(solid state physics), 64. Prof Exp: Res physicist, 63-72, SR RES PHYSICIST, CHEVRON RES CO, 72- Mem: Am Phys Soc. Res: Magnetism; heterogenous catalysts; x-ray photoelectron spectroscopy. Mailing Add: Chevron Res Co 576 Standard Ave Richmond CA 94802

LEWIS, ROBERT WILLIAM, chemistry, see 12th edition

LEWIS, ROGER ABBOTT, b Far Rockaway, NY, Jan 29, 12; m; c 3. PHARMACOLOGY. Educ: Johns Hopkins Univ, AB, 34, MD, 38. Prof Exp: Asst prof pediat, Sch Med, Johns Hopkins Univ, 48-51; WHO vis prof pharmacol, Med Col, Airlangga Univ, Indonesia, 52-63; PROF PHARMACOL & HEAD DEPT, GHANA MED SCH, 63- Concurrent Pos: Fel, Harvard Med Sch; Archbold, Ciba, Commonwealth & Am Cancer Soc fels, Sch Med, Johns Hopkins Univ. Mem: Royal Soc Med. Res: Chemotherapy; endocrinology; application of chemical principles to medical problems; clinical investigation. Mailing Add: Ghana Med Sch PO Box 4236 Accra Ghana

LEWIS, ROGER ALLEN, b Wellington, Kans, June 1, 41; m 62; c 3. BIOCHEMISTRY. Educ: Phillips Univ, BA, 63; Ore State Univ, PhD(biochem), 68. Prof Exp: Res assoc pyrimidine nucleotide metab, Stanford Univ, 68-69; asst prof, 69-75, ASSOC PROF BIOCHEM, UNIV NEV, RENO, 75- Mem: Am Chem Soc; Sigma Xi. Res: Purine deoxynucleotide biosynthesis and its control; toxicology of pesticides with respect to nucleotide metabolism and DNA and/or RNA synthesis. Mailing Add: Dept of Biochem Univ of Nev Reno NV 89507

LEWIS, ROGER WOLCOTT, b Mont, June 2, 20. BIOCHEMISTRY, MARINE BIOLOGY. Educ: Pomona Col, BA, 48; Univ Southern Calif, MS, 61; Univ Calif, San Diego, PhD(marine biol), 65. Prof Exp: Chemist, Chaney Chem Labs, 53-57 & Stanford Res Inst, 57-59; field biologist, Univ Southern Calif, 59-61; fel, Inst Marine Resources, 62-65; asst prof zoophysiol, Univ Alaska, 65-67; NIH fel, Div Food Chem, Dept Sci & Indust Res, NZ, 67-68; mem staff, Div Food Res, Commonwealth Sci & Indust Res Orgn, 69-71 & Appl Biochem Div, Dept Sci & Indust Res, NZ, 71-73; MEM STAFF, DEPT CLIN BIOCHEM, MED SCH, UNIV OTAGO, NZ, 73- Res: Lipids of marine organisms; lipids of mammaliam mucus secretions. Mailing Add: Dept of Clin Biochem Univ of Otago Med Sch Dunedin New Zealand

LEWIS, RONALD GENE, organic chemistry, see 12th edition

LEWIS, ROSCOE WARFIELD, b Beaumont, Tex, Dec 27, 20; m 42; c 1. ANIMAL NUTRITION, BIOCHEMISTRY. Educ: Prairie View State Col, BS, 39; Kans State Univ, MS, 52, PhD, 55. Prof Exp: Teacher high sch, 42-44 & Bowie County Training Sch, 46-51; asst, Kans State Univ, 52-55; prof nutrit & biochem, Tex Agr Exp Sta & Prairie View Exp Sta, Tex A&M Univ, 55-69 & Southwest Tex State Univ, 69-70; PROF NUTRIT & BIOCHEM, TEX A&M UNIV, 70- Honors & Awards: Fribourg Award, 55. Mem: AAAS; Poultry Sci Asn; Inst Food Technologists; Am Chem Soc; NY Acad Sci. Res: Chicken flavor; skin homografts in the chicken; tolerance of anticarcinogenic purine derivatives in the chick; nutrition of pre-adolescent females. Mailing Add: Dept of Biochem & Biophys Tex A&M Univ College Station TX 77843

LEWIS, ROY STEPHEN, b Oakland, Calif, Aug 10, 44; m 66; c 3. METEORITICS. Educ: Univ Calif, Berkeley, AB, 67, PhD(atmospheric & space sci), 73. Prof Exp: RES ASSOC METEORITICS, DEPT CHEM, UNIV CHICAGO, 73- Res: Isotopic composition and elemental abundances of noble gases in meteorites and other samples. Mailing Add: Enrico Fermi Inst Univ of Chicago 5630 Ellis Ave Chicago IL 60637

LEWIS, RUSSELL J, b Liberty Road, Ky, Jan 23, 29; m 54; c 2. SOIL CHEMISTRY. Educ: Univ Ky, BS, 56, MS, 57; NC State Col, PhD(soils), 61. Prof Exp: Int Atomic Energy Agency fel chem, Univ NC, 61-62; asst prof, 62-66, ASSOC PROF SOIL CHEM, UNIV TENN, KNOXVILLE, 66- Mem: Am Soc Agron; Soil Sci Soc Am; Clay Minerals Soc. Res: Surface chemistry of colloids; ion exchange, fixation and nutrient availability. Mailing Add: Dept of Plant & Soil Sci Univ of Tenn Knoxville TN 37916

LEWIS, SALLY, b Calcutta, India, July 21, 14; m 44; c 1. BOTANY, HORTICULTURE. Educ: Univ Calcutta, BSc, 35, MSc, 37; Univ Mo, PhD(plant path), 64. Prof Exp: Head dept bot & vprin col, Bethune Col, Univ Calcutta, 39-59; instr bot & res asst, Univ Mo, 59-60, asst hort, 60-63, asst prof, 64-66; ASSOC PROF BIOL, CLEVELAND STATE UNIV, 66- Concurrent Pos: Govt India overseas scholar bot, Eng, 44-47; mem Pub Serv Comn, Bengal, India, 57-59; NSF grant, 65-67. Mem: Int Fedn Univ Women; fel Royal Hort Soc. Res: Electromicroscopy; plant phytopathology, especially diseases related to horticultural plants; gross morphology; anatomy; taxonomy; tropical and medicinal plants. Mailing Add: 4082 Okalona South Euclid Cleveland OH 44121

LEWIS, SEYMOUR, b New York, NY, Mar 14, 19; m 57; c 2. MICROBIOLOGY, BOTANY. Educ: Univ Wis, BA, 49, MS, 50; Columbia Univ, MA, 55, PhD(biol sci), 68. Prof Exp: Bacteriologist, Med Sch, Cornell Univ, 52; res asst, Nat Agr Col, 52-53; bacteriologist, Columbia Univ, 53-55; sci teacher, Bd Educ, City of New York, 55-59; microbiologist, Dept Urol, Columbia Univ, 59-60, bacteriologist, Lamont Geol Observ, 60-64, res asst marine biol, 66-69; ASSOC PROF BIOL, BERGEN COMMUNITY COL, 69- Mem: Am Soc Microbiol; Sigma Xi. Mailing Add: Dept of Biol Bergen Community Col Paramus NJ 07652

LEWIS, SHELDON NOAH, b Chicago, Ill, July 1, 34; m 57; c 3. PHYSICAL CHEMISTRY, ORGANIC CHEMISTRY. Educ: Northwestern Univ, BA & MS, 56; Univ Calif, Los Angeles, PhD(phys & org chem), 59. Prof Exp: NSF fel, Univ Basel, 59-60; sr chemist, 60-61, group leader org chem, 61-63, lab head, 63-68, res supvr, 68-73, dir specialty chem res, 73-74, gen mgr, DCL Lab AG, Switz, 74-75, dir, European Labs, France, 75, CORP DIR RES POLYMERS, RESINS & MONOMERS WORLDWIDE, ROHM & HAAS CO, 75- Mem: Am Chem Soc. Res: Reaction mechanisms; organic synthesis; process development; agricultural chemicals; polymers and surface coatings; leather, paper, textile, cosmetic and petroleum chemicals; plastics and modifiers; ion exchange resins; adhesives; building products. Mailing Add: Rohm & Haas Co Res Labs Spring House PA 19477

LEWIS, SILAS DAVIS, b Gastonia, NC, June 26, 30; m 62; c 1. ORGANIC CHEMISTRY. Educ: Wake Forest Col, BS, 52; Ga Inst Technol, PhD(org chem), 59. Prof Exp: Sr chemist, Atlast Chem Industs, Inc, 59-63; assoc prof chem, Del Valley Col, 63-66; ASSOC PROF CHEM, AUGUSTA COL, 66- Mem: Am Chem soc. Res: Polyphenyls; Ullmann reaction; nitro and nitrato compounds and explosives. Mailing Add: Dept of Chem Augusta Col Augusta GA 30904

LEWIS, STANDLEY EUGENE, b Twin Falls, Idaho, Nov 15, 40; m 65; c 3. ENTOMOLOGY, PALEONTOLOGY. Educ: Univ Nebr, Omaha, BA, 62, MA, 64; Wash State Univ, PhD(entom), 68. Prof Exp: Teaching asst biol, Univ Nebr, Omaha, 63-64; teaching asst zool-entom, Wash State Univ, 64-68; asst prof biol, 68-74, Sigma Xi, Geol Soc Am & St Cloud instnl grants, 69-70, ASSOC PROF BIOL, ST CLOUD STATE COL, 74- Concurrent Pos: Res asst mosquito control, Adams County Abate Dist, 65- Mem: Entom Soc Am. Res: Paleobiology, specifically paleoentomology. Mailing Add: Dept of Biol St Cloud State Col St Cloud MN 56301

LEWIS, STEPHEN ALBERT, b Sodus, NY, Sept 9, 42; m 68; c 1. PLANT NEMATOLOGY. Educ: Pa State Univ, BS, 64; Rutgers Univ, MS, 69; Univ Ariz, PhD(plant path), 73. Prof Exp: Sales rep, Stand Oil Calif, 65-66; ASST PROF NEMATOL, DEPT PLANT PATH & PHYSIOL, CLEMSON UNIV, 73- Mem: Sigma Xi. Res: Host-parasite relations of the phytoparasitic nematodes, Hoplolaimus columbus and Criconemoides xenoplax on field crops and peach trees, respectively; nematode-mycorrhizae-rhizobium relationships; gnotobiotic culture of nematodes. Mailing Add: Dept of Plant Path & Physiol Clemson Univ Clemson SC 29631

LEWIS, STEPHEN ROBERT, b Mt Horeb, Wis, Aug 26, 20; m 48; c 2. PLASTIC SURGERY. Educ: Carroll Col, Wis, BA, 41; Marquette Univ, MD, 44. Prof Exp: Instr surg, 50-53, from asst prof to assoc prof plastic & maxillofacial surg, 53-61, asst dean med, 58-62, chief staff, 69-73, PROF SURG, UNIV TEX MED BR GALVESTON, 61-, DIR POSTGRAD EDUC, 56-, CHIEF PLASTIC SURG, 61- Concurrent Pos: Consult, St Mary's Infirmary, Galveston, 54-, Galveston County Mem Hosp, 54-, US Air Force, 57- & USPHS Hosp, 58-; mem, Am Bd Plastic Surg, 66-72, chmn, 71-72. Honors & Awards: First Sci Exhibit Award, Tex Med Asn, 63. Mem: Am Soc Plastic & Reconstruct Surg (past-pres, 56-); fel Am Col Surg; AMA; Am Asn Plastic Surg. Res: Tissue culture studies on human skin; burns; multiple studies in systemic and local problems; lymphatics in Lymphedema; congenital deformities of face, neck and hands. Mailing Add: Div Plastic & Maxillofacial Surg Univ of Tex Med Br Galveston TX 77550

LEWIS, TED EDWIN, textile chemistry, soil chemistry, see 12th edition

LEWIS, THOMAS BRINLEY, b Cleveland, Ohio, Nov 3, 38; m 65; c 3. POLYMER SCIENCE. Educ: John Carroll Univ, BS, 60, MS, 62; Mass Inst Technol, PhD(physics), 65. Prof Exp: Res assoc phys chem, Cornell Univ, 65-66; group leader, Cent Res Dept, 66-72, PROJ MGR, RUBBER CHEM DIV, MONSANTO CO, 72- Mem: Am Phys Soc; Am Chem Soc; Soc Rheol. Res: Rates for characteristic processes in polymer systems of biological interest using fast reaction chemical kinetic techniques; methods, including ultrasonic absorption relaxation and temperature jump; mechanical and physical properties of composite materials; dynamic properties of polymers and elastomers. Mailing Add: Rubber Chem Div Monsanto Co Akron OH 44313

LEWIS, THOMAS HOWARD, b Red Lodge, Mont, July 28, 19; m 44; c 5. PSYCHIATRY. Educ: Univ Wash, BS, 41; Duke Univ, MD, 46. Prof Exp: Chief dept psychiat & neurol, US Naval Hosp, Corona, Calif, 53-57, admin psychiatrist, US Naval Hosp, Nat Naval Med Ctr, Bethesda, Md, 57-60, dir residency training, 60-73, chief dept psychiat, 69-73; CLIN PROF PSYCHIAT, SCH MED, GEORGETOWN UNIV, 74- Mem: AAAS; fel Am Col Physicians; fel Am Psychiat Asn; AMA; NY Acad Sci. Res: Neurology; vertebrate zoology; anthropology. Mailing Add: Georgetown Univ Sch of Med 3800 Reservoir Rd NW Washington DC 20007

LEWIS, TRENT R, b Baltimore, Md, Feb 3, 32; m 65; c 2. OCCUPATIONAL HEALTH. Educ: Univ Md, BS, 54, MS, 57; Mich State Univ, PhD(nutrit), 61. Prof Exp: Res asst dairy sci, Univ Md, 55-57; res instr, Mich State Univ, 57-61; asst prof animal sci, Univ Maine, 61-63; res chemist, R A Taft Sanit Eng Ctr, USPHS, 63-68; chief chronic & explor toxicol, Nat Air Pollution Control Admin, 68-71; CHIEF EXPLOR TOXICOL SECT, NAT INST OCCUP SAFETY & HEALTH, 71- Concurrent Pos: Consutl, Hilltop Res Inc, 75- Mem: Am Dairy Sci Asn; Am Soc Microbiol; Soc Toxicol; Sigma Xi; Am Conf Govt Indust Hygienists. Res: Occupational toxicology; environmental toxicology; mammalian cardiopulmonary physiology; biochemical mechanisms; development or modification of standard toxicity testing regimens; mutagenic chemical agents via reproductive and teratogenic methodology; development of dose-response criteria. Mailing Add: Toxicol Br Nat Inst for Occup Safety & Health Cincinnati OH 45202

LEWIS, TREVOR JOHN, b Vancouver, BC, Jan 19, 40; m 64; c 2. GEOPHYSICS. Educ: Univ BC, BASc, 63, MSc, 64; Univ Western Ont, PhD(geophys), 75. Prof Exp: RES SCIENTIST GEOTHERMAL STUDIES, DEPT ENERGY MINES & RESOURCES, EARTH PHYSICS BR, 64- Mem: Geol Asn Can; Can Geophys Union; Can Geotech Soc. Mailing Add: 1 Observ Crescent Ottawa ON Can

LEWIS, URBAN JAMES, b Flagstaff, Ariz, Apr 28, 23; m 50; c 2. ENDOCRINOLOGY. Educ: San Diego State Col, BA, 48; Univ Wis, MS, 50,

PhD(biochem), 52. Prof Exp: NIH fel, Med Nobel Inst, Stockholm, 52-53; instr biochem & biochemist, Am Meat Found, Univ Chicago, 53-54; sr biochemist, Merck & Co, Inc, 54-61; MEM, SCRIPPS CLIN & RES FOUND, 61- Mem: Am Soc Biol Chem; Am Chem Soc; Endocrine Soc. Res: Proteolytic enzymes; pituitary hormones. Mailing Add: Scripps Clin & Res Found 475 Prospect St La Jolla CA 92037

LEWIS, VANCE DE SPAIN, b Los Angeles, Calif, June 26, 09; m 36; c 2. PHYSICS. Educ: Univ Calif, Berkeley, BA, 33, MA, 40; Univ Southern Calif, PhD(educ), 54. Prof Exp: From asst prof to prof, 46-72, EMER PROF PHYSICS, CALIF POLYTECH STATE UNIV, SAN LUIS OBISPO, 72- Concurrent Pos: Assoc dean, Sch Sci & Math, Calif Polytech State Univ, San Luis Obispo, 68-72. Mem: Am Phys Soc; NY Acad Sci. Res: Optics; statistical analysis. Mailing Add: 1386 Oceanaire Dr San Luis Obispo CA 93401

LEWIS, WALLACE JOE, b Smithdale, Miss, Oct 30, 42; m 65; c 2. ENTOMOLOGY. Educ: Miss State Univ, BS, 64, MS, 65, PhD(entom), 68. Prof Exp: ENTOMOLOGIST, SOUTHERN GRAIN INSECTS RES LAB, ENTOM RES DIV, AGR RES SERV, USDA, 67- Concurrent Pos: Asst prof, Univ Fla, 70- Mem: Entom Soc Am. Res: Ecological and physiological relationships between parasitic insects and their hosts; development of methods for the use of parasitic insects for control of insect pests. Mailing Add: So Grain Insect Lab Agr Res Serv USDA Coastal Plains Exp Sta Tifton GA 31794

LEWIS, WALTER HEPWORTH, b Carleton Place, Ont, June 26, 30; m 57; c 2. BOTANY. Educ: Univ BC, BA, 51, MA, 54; Univ Va, PhD(bot), 57. Prof Exp: Asst prof biol & dir herbarium, Stephen F Austin State Col, 57-61, assoc prof biol, 61-64; assoc prof, 64-69, PROF BOT, WASH UNIV, 69- Concurrent Pos: Guggenheim fel, 63-64; mem, Int Orgn Biosyst; dir herbarium, Mo Bot Garden, 64-72, sr botanist, 72- Honors & Awards: Horsley Res Award, Va Acad Sci, 57. Mem: Bot Soc Am; Am Soc Plant Taxon; Asn Trop Biol; Int Asn Plant Taxon. Res: Cytotaxonomy of Rosa, the Rubiaceae, palynotaxonomy of angiosperms and southern flora; medical plants; allergy. Mailing Add: Dept of Biol Wash Univ St Louis MO 63130

LEWIS, WARREN BURTON, b Pomona, Calif, Feb 24, 18; m 42; c 5. CHEMISTRY. Educ: Pomona Col, BA, 40; Univ Calif, Los Angeles, MA, 42; Mass Inst Technol, PhD(chem), 49. Prof Exp: Res assoc nat defense res comt proj, Northwestern Univ, 42-45; Nat Res Coun fel, Mass Inst Technol, 46-48; MEM STAFF, LOS ALAMOS SCI LAB, 48- Mem: Fel Am Inst Chem. Res: Magnetic properties of the transuranic elements; electron spin resonance; nuclear magnetic resonance; kinetics of gas phase reactions; chemical lasers. Mailing Add: Los Alamos Sci Lab Box 1663 Los Alamos NM 87544

LEWIS, WILFRID BENNETT, b Cumberland, Eng, June 24, 08; Can citizen. PHYSICS. Educ: Univ Cambridge, BA, 30, MA & PhD(physics), 34. Hon Degrees: DSc, Queen's Univ, Ont, 60, Univ Sask, 64, McMaster Univ, 65, Dartmouth Col, 67 & McGill Univ, 69; LLD, Dalhousie Univ, 60, Carleton Univ, 62 & Trent Univ, 68. Prof Exp: Demonstr, Univ Cambridge, 35-37, lectr, 37-39; chief supt, Telecommun Res Estab, Ministry Aircraft Prod, UK, 45-46; dir atomic energy div, Nat Res Coun Can, 46-52; vpres res & develop, 52-63, SR V PRES SCI, ATOMIC ENERGY CAN, LTD, 63- Concurrent Pos: Res fel, Gonville & Caius Cols, Univ Cambridge, 34-40, hon fel, 71; Can rep, Sci Adv Comt, Secy Gen, UN, 55-; mem sci adv comt to dir gen, Int Atomic Energy Agency, 57- Honors & Awards: Comdr, Order of Brit Empire, 46; Am Medal of Freedom with Silver Palms, 46; Outstanding Achievement Award, Pub Serv Can, 66; Companion of Order of Can, 67; Atoms for Peace Award, 67; Spec Gold Medal, Can Asn Physicists, 70. Mem: Am Phys Soc; fel Am Nuclear Soc (vpres, 60, pres, 61-62); fel Royal Soc; fel Royal Soc Can. Res: Radioactivity; radio; electronics; radar; radiation; nuclear reactor physics; nuclear physics; fluctuations; reactor economics and materials; fission gas behavior; economics of nuclear power; high power accelerators. Mailing Add: 13 Beach Ave Box 189 Deep River ON Can

LEWIS, WILLARD DEMING, b Augusta, Ga, Jan 6, 15; m 41; c 5. PHYSICS, ACADEMIC ADMINISTRATION. Educ: Harvard Univ, AB, 35, AM, 39, PhD(math physics), 41; Oxford Univ, BA, 38, MA, 45; Moravian Col, LHD, 66. Hon Degrees: LLD, Lafayette Col, 65, Rutgers Univ, 66 & Hahnemann Med Col, 66; DEng, Lehigh Univ, 74. Prof Exp: Mem tech staff, Bell Tel Labs, 41-51, dir switching res, 51-58, dir res, Commun Systs, 58-60, exec dir, 60-62; managing dir systs studies, Bellcomm, Inc, 62-64; PRES, LEHIGH UNIV, 64- Concurrent Pos: Mem, Vis Comt, Div Appl Sci, Harvard Univ, 56-66 & Coun, Harvard Found Advan Study & Res, 59-64, chmn, 61-62, mem, Harvard Grad Soc Advan Study & Res, 66-69, Defense Indust Adv Comt, 62-63 & Naval Res Adv Comt, 64-70, from vchmn to chmn, 65-69; dir, Pa Power & Light Co, 67-; mem, Defense Sci Bd, 67-69; chmn, State Bd Educ Commonwealth Pa, 68-73 & Comt Power Plant Siting, Nat Acad Eng, 70; mem coun, 72-; dir, Fairchild Industs, 73- Mem: Nat Acad Eng; Am Phys Soc; fel Inst Elec & Electronics Eng. Res: Magnetism; phonograph distortion; radar antennas; microwave filters and radio repeaters; telephone switching. Mailing Add: Off of the Pres Lehigh Univ Bethlehem PA 18015

LEWIS, WILLIAM C, b Chicago, Ill, July 22, 17; m 40; c 7. PSYCHIATRY. Educ: Univ Chicago, BS, 39, MD, 41. Prof Exp: Intern, St Lukes Hosp, Chicago, 41-42; resident psychiat, Winter Vet Admin Hosp, Topeka, Kans, 46-48, mem staff, 48-51; PROF PSYCHIAT, SCH MED, UNIV WIS-MADISON, 62- Concurrent Pos: Instr, Menninger Sch Psychiat, 48; consult, Lakeview Sanatorium, 51-55, Madison Vet Admin Hosp, 51-64, Mendota State Hosp, 57-58 & Wis Diag Ctr, 58-59; mem, Hoffheimer Prize Bd, Am Psychiat Asn, 64-65. Mem: AMA; Am Psychiat Asn; Am Psychoanal Asn; Am Psychosom Soc. Res: Psychoanalysis; psychosomatic medicine. Mailing Add: Dept of Psychiat Univ of Wis Madison WI 53706

LEWIS, WILLIAM JAMES, b Talahassee, Fla, Feb 11, 45; m 66; c 1. MATHEMATICS. Educ: La State Univ, Baton Rouge, BS, 66, PhD(math), 71. Prof Exp: Instr math, La State Univ, 71; ASST PROF MATH, UNIV NEBR-LINCOLN, 71- Mem: Am Math Soc. Res: Commutative algebra; valuation theory; ring theory. Mailing Add: Dept of Math Univ of Nebr Lincoln NE 68508

LEWIS, WILLIAM MADISON, b Faison, NC, Nov 26, 22; m 43; c 4. FISHERIES. Educ: NC State Col, BS, 43; Iowa State Col, MS, 48, PhD(zool), 49. Prof Exp: Sci bact aide, USDA, 42; asst prof, 49-60, PROF ZOOL, SOUTHERN ILL UNIV, 60-, DIR, COOP FISHERIES LAB, 49-, CHMN DEPT ZOOL, 72- Mem: Am Fisheries Soc. Res: Aquaculture and fish management. Mailing Add: Dept of Zool Southern Ill Univ Carbondale IL 62901

LEWIS, WILLIAM MASON, b Ithaca, NY, Aug 13, 29; m 57; c 5. WEED SCIENCE. Educ: Tex A&M Univ, BS, 52; Univ Minn, MS, 56, PhD(plant genetics), 57. Prof Exp: Asst agron & plant genetics, Univ Minn, 52-56; from instr to assoc prof, 56-69, PROF CROP SCI, NC STATE UNIV, 69- Mem: Am Soc Agron; Crop Sci Soc Am; Weed Sci Soc Am. Res: Agronomy; turf weed control. Mailing Add: Dept of Crop Sci NC State Univ Raleigh NC 27607

LEWIS, WILLIAM PERRY, b Swatow, China, Aug 12, 29; US citizen; m 51; c 2. MEDICAL MICROBIOLOGY. Educ: Univ Redlands, BS, 51; Univ Calif, Los Angeles, PhD(infectious dis), 62. Prof Exp: Asst res parasitologist & instr parasitol, Sch Pub Health, Univ Calif, Los Angeles, 62-69; ASST PROF PATH, SCH MED, UNIV SOUTHERN CALIF & SUPV MED MICROBIOLOGIST, MICROBIOL II, LOS ANGELES COUNTY-UNIV SOUTHERN CALIF MED CTR, 69- Mem: AAAS; Am Soc Trop Med & Hyg; Am Soc Microbiol; Am Soc Parasitol. Res: Immunology of parasitic diseases, especially toxoplasmosis, amebiasis and filariases; diagnostic bacteriology, parasitology and immunology. Mailing Add: Mycobact Lab Los Angeles County- Univ Southern Calif Med Ctr 1200 N State St Los Angeles CA 90033

LEWIS, WILLIAM WESTON, b Marblehead, Mass, Feb 2, 32; m 68. LABORATORY MEDICINE. Educ: Bates Col, BA, 59. Prof Exp: Res asst, Protein Found, 61-63, Tufts Med Sch, 63-66; Harvard Med Sch, 66-69; PRIN SCIENTIST, INSTRUMENTATION LAB, INC, 69- Mem: Am Asn Clin Chemists. Res: Development and implementation of state of the art methods for the measurement of clinically important materials in human blood and body fluids. Mailing Add: Instrumentation Lab Inc 113 Hartwell Ave Lexington MA 02178

LEWKE, ROBERT EDWARD, b Decatur, Ill, June 29, 44; m 68; c 1. VERTEBRATE ECOLOGY. Educ: Univ Ill, Urbana, BS, 66, MS, 68; Wash State Univ, PhD(zool), 75. Prof Exp: Teacher biol, Urbana High Sch, 68-70; ASST PROF ZOOL, UNIV WIS-EAU CLAIRE, 75- Mem: Nat Wildlife Fedn; Am Ornithologists Union; Wildlife Soc. Res: Niche-segregation studies of bird communities with emphasis on feeding behavior; constriction behavior of newborn snakes. Mailing Add: Dept of Biol Univ of Wis Eau Claire WI 54701

LEWONTIN, RICHARD CHARLES, b New York, NY, Mar 29, 29; m 47; c 4. GENETICS, POPULATION BIOLOGY. Educ: Harvard Univ, AB, 51; Columbia Univ, MA, 52, PhD(zool), 54. Prof Exp: Reader biomet, Columbia Univ, 53-54; asst prof genetics, NC State Col, 54-58; from asst prof to prof biol, Univ Rochester, 58-64; prof biol, Univ Chicago, 64-73; PROF BIOL, HARVARD UNIV, 73- Concurrent Pos: NSF fel, 54-55, sr fel, 61-62 & 71-72; lectr, Columbia Univ, 59, seminar assoc, 59-61; Fulbright fel, 61-62; co-ed, Am Naturalist, Am Soc Nat, 65. Mem: Nat Acad Sci; AAAS; fel Am Acad Arts & Sci; Genetics Soc Am; Soc Study Evolution (pres, 70). Res: Population genetics, ecology and evolution. Mailing Add: Mus of Comp Zool Harvard Univ Cambridge MA 02138

LEWTHWAITE, GORDON ROWLAND, b Oamaru, NZ, Aug 12, 25; m 53; c 2. HISTORICAL GEOGRAPHY, GEOGRAPHY OF THE PACIFIC. Educ: Univ Canterbury, BA & MA, 48; Auckland Teacher's Training Col, cert, 48; Univ Auckland, MA, 50; Univ Wis-Madison, PhD(geog), 56. Prof Exp: Lectr geog, Univ Auckland, 55-59; PROF GEOG, CALIF STATE UNIV, NORTHRIDGE, 59- Concurrent Pos: NSF grant, Pac Sci Cong, Honolulu, 61; San Fernando Valley State Col Found instnl grants, Pac, 63-64; Wenner-Gren Found fel hist geog Maori, NZ, 64-68; vis prof, Univ BC, 66-67 & Univ Newcastle, 73. Mem: AAAS; Am Sci Affil; Asn Am Geogr; Am Geog Soc; Polynesian Soc. Res: Historical geography of Polynesia; economic geography of dairying. Mailing Add: Dept of Geog Calif State Univ Northridge CA 91324

LEWY, HANS, b Breslau, Ger, Oct 20, 04; nat US; m. MATHEMATICS. Educ: Univ Göttingen, PhD, 26. Prof Exp: Privat-docent, Univ Göttingen, 27-33; assoc, Brown Univ, 33-35; lectr, 35-37, from asst prof to prof, 37-72, EMER PROF MATH, UNIV CALIF, BERKELEY, 72- Res: Calculus of variations; partial differential equations; hydrodynamics. Mailing Add: Dept of Math Univ of Calif Berkeley CA 94720

LEX, BARBARA WENDY, b Buffalo, NY, Mar 12, 41; m 70. CULTURAL ANTHROPOLOGY. Educ: Syracuse Univ, AB, 64, MA, 68, PhD(anthrop), 69. Prof Exp: Res assoc, Soc Sci Inst, Wash Univ, 66-68, lectr anthrop, Univ Col, 67-68; lectr behav studies, Utica Col, 68-69; asst prof anthrop, Lehigh Univ, 69-70; asst prof anthrop, Western Mich Univ, 70-75; ASSOC PSYCHIAT, ALCOHOL & DRUG ABUSE CTR, MCLEAN HOSP, HARVARD MED SCH, 75- Concurrent Pos: Fac res grant, Western Mich Univ, 71-73; vis res anthropologist, Langley Porter Neuropsychiat Inst, 73; NIH, Nat Inst Drug Abuse fel, 75- Honors & Awards: Phillips Fund res grant, Onondaga Indian Reservation, 66. Mem: AAAS; fel Am Anthrop Asn; Soc Appl Anthrop; Soc Med Anthrop. Res: Native North America; Iroquois; urban United States; white migration from south; cross-cultural study of altered states of consciousness; revitalization movements; neurobiology. Mailing Add: 26 Townsend Rd Belmont MA 02178

LEY, ALLYN BRYSON, b Springfield, Mass, Dec 5, 18; m 43; c 4. MEDICINE. Educ: Dartmouth Col, AB, 39; Columbia Univ, MD, 42; Am Bd Internal Med, dipl. Prof Exp: Asst med, Med Col, Cornell Univ, 47-49; instr, 51-52, asst dir, Sloan-Kettering Div, 54-55, cancer coord, 54-63, from asst prof to assoc prof, 54-63, PROF MED, MED COL, CORNELL UNIV, 63-, CLIN DIR UNIV HEALTH SERV, 71- Concurrent Pos: Asst, Boston City Hosp, 49-51; dir blood bank, Mem Hosp, 51-63, dir hemat labs, 55-63; asst attend physician, New York Hosp, 54-63, attend physician & dir ambulatory serv, 63-69; consult, Manhattan Vet Admin Hosp, 58-60 & Hosp Spec Surg, 58-71; assoc vis physician, Bellevue Hosp, 60-67. Mem: AAAS; Harvey Soc; Am Soc Hemat; Am Fedn Clin Res; Int Soc Blood Transfusion. Res: Immunohematology; erythrocyte biochemistry; medical education and care. Mailing Add: Univ Health Serv Cornell Univ Ithaca NY 14850

LEYDA, FRANK ARTHUR, chemistry, see 12th edition

LEYDA, JAMES PERKINS, b Youngstown, Ohio, Oct 2, 35; m 67; c 2. PHARMACY, PHARMACEUTICAL CHEMISTRY. Educ: Ohio Northern Univ, BS, 57; Ohio State Univ, MS, 59, PhD(pharm), 62. Prof Exp: Develop chemist, Lederle Labs Div, Am Cyanamid Co, 62-66, mgr prod develop, Int Med Res & Develop, 66-69; MGR NEW PROD DEVELOP, MERRELL INT DIV, RICHARDSON-MERRELL, INC, 69- Mem: AAAS; Am Pharmaceut Asn; NY Acad Sci; Acad Pharmaceut Sci. Res: Barbiturates, antibiotics; vitamins; complexation; stability and analysis of pharmaceutical products. Mailing Add: Merrell Int Div Richardson-Merrell Inc 10 Westport Rd Wilton CT 06897

LEYDEN, DONALD E, b Gadsden, Ala, June 26, 38; m 61; c 2. ANALYTICAL CHEMISTRY. Educ: Kent State Univ, BS, 60; Emory Univ, MS, 61, PhD, 64. Prof Exp: Res assoc, Univ NC, 64-65; asst prof, 65-71, ASSOC PROF CHEM, UNIV GA, 71- Mem: Am Chem Soc. Res: Nonaqueous titrations; ion-exchange; applications of nuclear magnetic resonance to the study of chemical systems of analytical importance. Mailing Add: Dept of Chem Univ of Ga Athens GA 30601

LEYDEN, ROBERT FULLERTON, b Glasgow, Scotland, Apr 26, 21; nat US; m; c 3. AGRONOMY, SOILS. Educ: NMed State Univ, BSc, 51; Kans State Univ, MSc, 53; Rutgers Univ, PhD(soil chem), 56. Prof Exp: From asst prof to assoc prof, 56-70, PROF AGR, CITRUS CTR, TEX A&I UNIV, 56- Res: Soil fertility and plant nutrition. Mailing Add: Citrus Ctr Tex A&I Univ Weslaco TX 78596

LEYENDECKER, PHILIP JORDON, b Albuquerque, NMex, July 8, 14; m 37; c 4. PHYTOPATHOLOGY. Educ: Univ NMex, BS, 38; Iowa State Univ, MS, 40, PhD(phytopath), 48. Prof Exp: Assoc prof biol, 41-42, assoc biologist, 45-50, assoc plant pathologist & head dept agr serv, 50-59, asst dean agr, 59-60, DEAN & DIR COL AGR & HOME ECON, NMEX STATE UNIV, 60- Res: Soil borne pathogens. Mailing Add: Col of Agr & Home Econ NMex State Univ Box 3AG Las Cruces NM 88003

LEYLAND, HARRY MOURS, b Newark, Ohio, July 8, 25; m 48. ORGANIC CHEMISTRY. Educ: Williams Col, AB, 46; Univ Cincinnati, MS, 48, PhD(chem), 53. Prof Exp: Assoc prof biol sci, Col Pharm, Univ Cincinnati, 48-50, instr chem, 51-52; asst dir res, Lloyd Bros, Inc, 53-55, dir prof serv & clin invest, 55-59; assoc dir res, Lakeside Labs, Inc, 59-61; actg dir clin pharmacol & dir toxicol lab, Miles Labs, Inc, 61-63; assoc dir clin res, 63-66, dir prod develop, 66-71, DIR CLIN THERAPEUT, WM S MERRELL CO DIV, RICHARDSON-MERRELL, INC, 71- Mem: AAAS; Am Chem Soc; Am Soc Clin Pharmacol & Therapeut; NY Acad Sci. Mailing Add: 8870 Paw Paw Lane Cincinnati OH 45236

LEYMASTER, GLEN RONALD, b Aurora, Nebr, Aug 7, 15; m 42; c 3. MEDICINE. Educ: Univ Nebr, AB, 38; Harvard Univ, MD, 42; Johns Hopkins Univ, MPH, 50; Am Bd Prev Med, dipl. Prof Exp: Intern, Boston City Hosp, 42-43, asst resident, 43, resident, 44; clin instr, Sch Med, Johns Hopkins Univ, 44-46; from instr to asst prof bact, Sch Hyg & Pub Health, 46-48; assoc prof pub health & prev med, Sch Med, Univ Utah, 48-50, prof prev med & head dept, 50-60; assoc secy, Coun Med Educ & Hosps, AMA, 60-63; pres, dean, prof prev med & assoc prof med, Med Col, Pa, 64-70; dir dept undergrad med Educ, AMA, 70-75; EXEC DIR, AM BD MED SPECS, 75- Concurrent Pos: Clin asst, Harvard Med Sch, 42-44; asst prof & dir univ health serv, Univ Utah, 50-60; med educ adv, US Dept State, Int Coop Admin, Thailand, 57-58. Mem: AMA. Res: Clinical and epidemiological character of influenza and of data regarding encephalitis; experimental immunity and epidemiology of mumps; industrial toxicology; epidemiology of gastroenteritis; medical education. Mailing Add: Am Bd of Med Specs 1603 Orrington Ave Evanston IL 60201

LEYON, ROBERT EDWARD, b Newton, Mass, July 28, 36; m 62; c 2. ANALYTICAL CHEMISTRY. Educ: Williams Col, BA, 58; Princeton Univ, MA, 60, PhD(chem), 62. Prof Exp: Instr chem, Princeton Univ, 61-62; from instr to asst prof, Swarthmore Col, 62-69; asst prof, 69-72, ASSOC PROF CHEM, DICKINSON COL, 72- Concurrent Pos: Res assoc, Univ NC, 67-68 & Colo State Univ, 75-76. Mem: AAAS; Am Chem Soc. Res: Environmental chemistry; trace metal analysis by atomic absorption. Mailing Add: Dept of Chem Dickinson Col Carlisle PA 17013

LEYSIEFFER, FREDERICK WALTER, b Milwaukee, Wis, Jan 30, 33; m 64; c 2. MATHEMATICS. Educ: Univ Wis-Madison, BA, 55, MA, 56; Univ Mich, PhD(math), 64. Prof Exp: Asst prof, 64-70, ASSOC PROF STATIST, FLA STATE UNIV, 70-, ASSOC HEAD DEPT, 69- Concurrent Pos: Vis lectr, Sheffield Univ, Sheffield, Eng, 73-74, Leverhulme Commonwealth-Am fel, 73. Mem: Am Math Soc; Math Asn Am; Inst Math Statist; AAAS. Res: Probability theory; stochastic processes; sampling theory. Mailing Add: Dept of Statist Fla State Univ Tallahassee FL 32306

LEYTON, MORLEY KAMLER, b Butler, Pa, Apr 2, 31; m 65; c 2. BIOSTATISTICS. Educ: City Col New York, BS, 52; Columbia Univ, MS, 62; Johns Hopkins Univ, PhD(biostatist), 66. Prof Exp: Asst prof biostatist, Univ Wash, 66-69; exchange prof, Inst Ecol, Polish Acad Sci, Warsaw, 69-70; ASST PROF BIOMET, HEALTH SCI CTR, TEMPLE UNIV, 70- Mem: Biomet Soc. Res: Stochastic models in biology and medicine; contagious distributions in plants and parasites. Mailing Add: Dept of Biomet Temple Univ Health Sci Ctr Philadelphia PA 19140

LEZAK, EDWARD HERMAN, mathematics, statistics, see 12th edition

LEZER, LEON ROBERT, b Graniteville, Vt, July 27, 17; m 45; c 2. MEDICINE. Educ: Univ Vt, BS, 39, Md, 42; Harvard Univ, MPH, 54. Prof Exp: Chief med officer, Presby Hosp, Philadelphia, Pa, 48-50; asst dir, Mass Gen Hosp, Boston, 50-54; assoc prof prev med, head dept & dir health studies, Col Med, Univ Vt, 54-65; dept supt med serv, Boston City Hosp, 65-66; dep comnr hosp serv, Dept Health & Hosps, City of Boston, 66-69; med coordr, Regional Health Coun Eastern Appalachia, 69-70; dir dept community med, South Nassau Communities Hosp, 70-73; ASSOC DIR, OFF NEW YORK CITY AFFAIRS, NY STATE DEPT HEALTH, 73- Concurrent Pos: Commonwealth Fund NY fel, 60; instr, Univ Pa, 49-50; clin prof, Sch Med, Boston Univ, 63-69; lectr, Univ Mass, 66-69; adj prof, Roth Grad Sch Bus Admin, C W Post Ctr, Long Island Univ, NY. Mem: Fel Am Pub Health Asn. Res: Community medicine in medical administration and education; gerontology. Mailing Add: 23 Seneca East Nassau Shores Massapequa NY 11758

LEZNOFF, ARTHUR, b Montreal, Que, Apr 7, 30; m 54; c 4. MEDICINE, IMMUNOLOGY. Educ: McGill Univ, BSc, 51, MD, CM, 55, MSc, 60; FRCPS(C), 60. Prof Exp: ASSOC PROF MED, UNIV TORONTO, 71- Concurrent Pos: Physician, Dept Med, St Michael's Hosp, Toronto, 71- Mem: Am Acad Allergy; Can Med Asn; Can Soc Immunol. Res: Immunology; clinical allergy; histoplasmosis. Mailing Add: 277 Victoria St Toronto ON Can.

LEZNOFF, CLIFFORD CLARK, b Montreal, Que, May 30, 40; m 63; c 2. ORGANIC CHEMISTRY. Educ: McGill Univ, BSc, 61, PhD(org chem), 65. Prof Exp: Fel org chem, Northwestern Univ, 64-65; Nat Res Coun Can overseas fel, Cambridge Univ, 65-67; asst prof, 67-71, ASSOC PROF ORG CHEM, YORK UNIV, 71- Mem: Am Chem Soc; Chem Inst Can; The Chem Soc. Res: Polymer supports in organic synthesis; studies in anion transport across mitochondria membranes; studies in the photochemistry of conjugated macrocyclic olefins. Mailing Add: Dept of Chem York Univ Downsview ON Can

LHAMON, WILLIAM TAYLOR, b Washington, DC, June 11, 15; m 42; c 5. PSYCHIATRY. Educ: Stanford Univ, AB, 36, MD, 40; Am Bd Psychiat & Neurol, dipl, 45. Prof Exp: Intern, San Francisco City & County Hosp, 39-40; asst resident neuropsychiat, Stanford Univ Hosp, 40-41; jr asst resident psychiat, Payne Whitney Clin, NY Hosp, 41-42; asst resident, 46-47, resident, Hosp & instr, Med Col, Cornell Univ, 47-48; assoc psychiat & instr med, Sch Med, Univ Pa & Hosp Univ Pa, 48-50, asst prof psychiat & assoc med, 50-51, assoc prof, 51-53, prof clin psychiat, 53-54; prof psychiat & chmn dept, Baylor Col Med; prof psychiat, Med Col Cornell Univ, 62-70, chmn dept, 62-74, Barklie McKee Henry prof psychiat, 70-74, psychiatrist-in-chief, NY Hosp, 62-74, med dir, NY Hosp-Cornell Med Ctr, Westchester Div, 74-75. Concurrent Pos: Consult, Vet Admin Hosp, Coatesville, Pa, 49-51, Wilmington, Del, 49-54 & Houston, Tex, 54-62; psychiatrist-in-chief, Jefferson Davis Hosp, Harris County, Tex, 54-62; dir, Houston State Psychiat Inst, 59-62. Mem: Am Psychopath Asn; fel Am Psychiat Asn. Res: Perception of time; psychopathology of schizophrenia. Mailing Add: Dept of Psychiat Westchester Div NY Hosp-Cornell Med Ctr White Plains NY 10605

LHERMITTE, ROGER M, b Pontchartrain, France, May 28, 20; US citizen; m 45; c 2. PHYSICAL METEOROLOGY. Educ: Univ Paris, MS, 51, DrSci(meteorol), 54. Prof Exp: Physicist meteorol res, Meteorol Nat, France, 46-60, Air Force Cambridge Res Lab, 60-63; head res br, Nat Oceanic & Atmospheric Admin, 63-70; PROF PHYS METEOROL, UNIV MIAMI, 70- Concurrent Pos: Mem, Active Microwave Workshop, NASA, 74-75; chmn, Nat Ctr Atmospheric Res Adv Panel, 74- Honors & Awards: Authorship Award, Nat Oceanic & Atmospheric Admin, 69, Spec Achievement Award, 70. Mem: Fel Am Meteorol Soc; Int Union Geodesy & Geophys; Int Union Radio Sci. Res: Use of Doppler radars for the observation and study of atmospheric motion; the observation of three dimensional motion fields inside these systems by use of three Doppler radars operated simultaneously. Mailing Add: Univ of Miami Coral Gables FL 33134

L'HEUREUX, JACQUES (JEAN), b Trois-Rivieres, Que, Dec 20, 39; m 62; c 3. COSMIC RAY PHYSICS, ASTROPHYSICS. Educ: Univ Montreal, BSc, 61; Univ Chicago, MSc, 62, PhD(physics), 66. Prof Exp: Res assoc physics, Univ Chicago, 66-69; ASST PROF PHYSICS, UNIV ARIZ, 69- Mem: Fel Am Phys Soc; Am Geophys Union. Res: Primary cosmic ray electrons; solar modulation of cosmic rays; primary heavy nuclei at high energies. Mailing Add: Dept of Physics Univ of Ariz Tucson AZ 85721

L'HEUREUX, MAURICE VICTOR, b Lewiston, Maine, May 23, 14; m 46; c 5. BIOCHEMISTRY. Educ: Col of the Holy Cross, BS, 36, MS, 37; Yale Univ, PhD(biochem), 44. Prof Exp: Control chemist, Stokely Bros-Van Camp, Inc, Ind, 40-41; assoc, 46-49, from asst prof to assoc prof, 49-59, PROF BIOCHEM, STRITCH SCH MED, LOYOLA UNIV CHICAGO, 59- Mem: Am Chem Soc; Am Soc Biol Chem. Res: Lipid metabolism; modifying and regulatory effects of parathyroid hormone, calcitonin and vitamin D upon calcium metabolism. Mailing Add: Dept of Biochem & Biophys Loyola Univ Stritch Sch of Med Maywood IL 60153

LHOTKA, JOHN FRANCIS, JR, b Butte, Mont, May 13, 21; m 51. HISTOCHEMISTRY, HUMAN ANATOMY. Educ: Univ Mont, BA, 42; Northwestern Univ, MS, 48, MB, 49, MD, 51, PhD(anat), 53. Prof Exp: Asst anat, Northwestern Univ, 47-50; mem house staff, Minneapolis Gen Hosp, Minn, 50-51; from asst prof to assoc prof, 51-69, PROF ANAT, SCH MED, UNIV OKLA, 69- Concurrent Pos: Mem, Biol Stain Comn. Mem: Fel Int Acad Path; fel Royal Soc Health; Am Asn Anatomists; Histochem Soc. Res: Localization of tissue polysaccharides in degenerative processes; argyrophilic reactions; embryological histochemistry; histochemical polysaccharide methods; heavy metal histochemical toxicology. Mailing Add: Univ of Okla Health Sci Ctr 801 NE 13th St PO Box 26901 Oklahoma City OK 73190

LI, CHIA-YU, b Shanghai, China, May 5, 41; m 69; c 2. ELECTROANALYTICAL CHEMISTRY. Educ: Taiwan Normal Univ, BS, 62; Univ Louisville, MS, 67; Wayne State Univ, PhD(anal chem), 72. Prof Exp: Res fel electrochem, Univ Ariz, 72-73; ASST PROF CHEM, ECAROLINA UNIV, 73- Mem: Am Chem Soc; Sigma Xi. Res: Electrochemistry of biological model compounds-phthalocyanines and porphyrins; spectroelectrochemistry and digital data acquisition techniques. Mailing Add: Dept of Chem ECarolina Univ Greenville NC 27834

LI, CHING CHUN, b Tientsin, China, Oct 27, 12; nat US; m 41; c 2. POPULATION GENETICS, BIOMETRICS. Educ: Nanking Univ, BS, 36; Cornell Univ, PhD(plant breeding), 40. Prof Exp: Plant breeder, Agr Exp Sta, Yenching Univ, 36-37; asst prof, Agr Col, Nat Kwangsi Univ, 42-43; prof genetics & biomet, Agr Col, Nanking Univ, 43-46; prof agron & head dept, Peking Univ, 46-50; from res fel to asst prof, 51-58, from assoc prof to prof, 58-75, UNIV PROF BIOSTATIST, GRAD SCH PUB HEALTH, UNIV PITTSBURGH, 75-, HEAD DEPT, 69- Mem: Am Soc Human Genetics (pres, 60); Biomet Soc; fel Am Statist Asn. Res: Human genetics; biometry; design of experiments. Mailing Add: Dept of Biostatist Univ of Pittsburgh Grad Sch Pub Health Pittsburgh PA 15261

LI, CHI-TANG, b Ningtu, Kiangsi, China, Oct 16, 34; m 62; c 4. PHYSICAL CHEMISTRY, CRYSTALLOGRAPHY. Educ: Nat Taiwan Univ, BS, 55; Univ Louisville, MS, 59; Mont State Univ, PhD(chem), 64. Prof Exp: Scientist, 64-67, SR SCIENTIST, ADV MAT RES SECT, OWENS-ILL TECH CTR, 67- Mem: Am Chem Soc; Am Crystallog Asn. Res: Crystal structure and chemistry; studies of glass ceramic materials, research on high temperature materials. Mailing Add: Glass & Ceramic Dept Corp Res Owens-Ill Tech Ctr 1700 NWestwood Toledo OH 43607

LI, CHOH HAO, b Canton, China, Apr 21, 13; nat US; m. BIOCHEMISTRY. Educ: Nanking Univ, BS, 33; Univ Calif, Berkeley, PhD(chem), 38. Hon Degrees: Dr, Cath Univ Chile, 62; LLD, Chinese Univ Hong Kong, 70; DSc, Univ Pac, Marquette Univ & St Peter's Col, 71. Prof Exp: Instr chem, Nanking Univ, 33-35; res asst, Univ Calif, Berkeley, 35-38, res assoc, 38-42, lectr chem morphol, 42-44, from asst prof to assoc prof exp biol, 44-49, PROF BIOCHEM & EXP ENDOCRINOL & DIR HORMONE RES LAB, UNIV CALIF, BERKELEY, 50- & UNIV CALIF, SAN FRANCISCO, 50- Concurrent Pos: Claude Bernard vis prof, Univ Mont, 47; mem sci adv bd, Sloan Kettering Inst Cancer Res, 52-55; vis scientist, Children's Cancer Res Found, Boston, 55, sci adv, 63-; China Found vis prof, Nat Taiwan Univ, 58, mem adv bd, Chem Res Ctr, 64-; sect ed, Chem Abstracts, 60-63; mem acad adv bd, Chinese Univ Hong Kong, 63-; Albert Lasker Award lectr, Salk Inst, 69; Pfizer lectr, Clin Res Inst Montreal, 71; co-chmn, Conf Glycoproteins with Hormonal Activity, NIH, 71; chmn adv bd, Inst Biochem, Acad Sinica, 71-; Nord lectr, Fordham Univ, 72. Honors & Awards: Ciba Award, 47; John Simon Guggenheim Mem Found Fel, 48; Francis Emory Prize, Am Acad Arts & Sci, 55; Repub China Minister Educ Gold Medal, 58; Chinese-Am Citizen Alliance Award, 61; Lasker Award, 62; Am Acad Achievement Golden Plate Award, 64; Chinese Soc Eng Achievement Award, 65; Univ Milan & City Milan Gold Medals, 67; Repub China Distinguished Achievement in Sci Award, 67; Univ Liege Medal, 68; Chinese Eng & Sci Asn Sci Achievement Award, 69; City of Hope Award, 70; AMA Sci Achievement Award, 70; Am Cancer Soc Nat Award, 71. Mem: Nat Acad Sci; fel AAAS; fel NY Acad Sci; fel Am Acad Arts & Sci; fel Am Inst Chem. Res: Protein chemistry; chemistry and biology of pituitary hormones; biochemistry of protein hormones. Mailing Add: Hormone Res Lab Univ of Calif San Francisco CA 94122

LI, CHOH-LUH, b Canton, China, Sept 19, 19; nat US; m 48; c 3. NEUROPHYSIOLOGY, NEUROSURGERY. Educ: Nat Med Col Shanghai, China, MD, 42; McGill Univ, MSc, 51, PhD, 54. Prof Exp: Asst resident surg, Nat Med Col Shanghai Hosp, China, 42 & 44-45, resident, 45-46; sr asst resident neurosurg, Montreal Inst, 47-51, resident, 48, res EEG, 51-52; ASSOC NEUROSURGEON, NAT INST NEUROL DIS & STROKE, 54- Concurrent Pos: Clin prof neurol surg, George Washington Univ. Mem: Int Brain Res Orgn; Am Physiol Soc; Res Soc Am Neurosurg; AAAS; Int Asn Pain Res. Res: Neurology. Mailing Add: Nat Inst of Neurol Dis & Stroke Bethesda MD 20014

LI, EDWARD HSIEN-CHI, b Taiwan, Aug 31, 35; m 67; c 2. PLANT PHYSIOLOGY, CELL BIOLOGY. Educ: Nat Taiwan Univ, BS, 59; Univ Kans, MA, 66; Univ Okla, PhD(bot), 69. Prof Exp: Lab instr gen bot & taxon, Nat Taiwan Univ, 61-64; teaching asst gen bot & physiol, Univ Okla, 66-69; ASSOC PROF

CELLULAR PHYSIOL, CENT METHODIST COL, MO, 69- Mem: AAAS; Am Soc Plant Physiol. Res: Utilization of tissue culture technique to study the biosynthesis of phenolic compounds and purine catabolism. Mailing Add: Dept of Biol Cent Methodist Col Fayette MO 65248

LI, HSIAO-JUN, organic chemistry, see 12th edition

LI, HSUEH MING, b Taiwan, Rep China, Oct 25, 39; US citizen; m 63; c 2. POLYMER CHEMISTRY. Educ: Tunghai Univ, Taiwan, BS, 62; Southern Methodist Univ, MS, 66; Polytech Inst Brooklyn, PhD(polymer chem), 71. Prof Exp: Fel x-ray diffraction, Polytech Inst Brooklyn, 70-72; res assoc polymer chem, Midland Macromolecular Inst, 72-73; RES CHEMIST POLYMER RES, ETHYL CORP, 73- Mem: Am Chem Soc. Res: Opacifying plastic pigment; polymeric flame retardants based on phosphazene-synthesis and evaluation; synthesis, characterization and mechanism of linear and cyclic phosphonitrilic chloride oligomers. Mailing Add: Ethyl Corp PO Box 341 Baton Rouge LA 70821

LI, HUI-LIN, b Soochow, China, July 15, 11; US citizen; m 46; c 2. BOTANY. Educ: Soochow Univ, BS, 30; Yenching Univ, MS, 32; Harvard Univ, PhD(biol), 42. Prof Exp: Instr biol, Soochow Univ, 32-40, prof, 46-47; Harrison res fel, Univ Pa, 42-46; prof bot & chmn dept, Nat Taiwan Univ, 47-50; Blandy Farm res fel, Univ Va, 50-51; US State Dept res fel, 51-52; res assoc bot, Morris Arboretum, 52-55, taxonomist, 55-74, dir, 72-75, from assoc prof to prof bot, 58-74, JOHN BARTRAM PROF BOT, UNIV PA, 74- Concurrent Pos: John Simon Guggenheim Found fel, 61-62; academician, Acad Sinica, Taiwan, 64. Mem: AAAS; Bot Soc Am; Soc Econ Bot; Am Soc Plant Taxon. Res: Plant taxonomy and geography; biosystematics; economic botany. Mailing Add: Dept of Biol Univ of Pa Philadelphia PA 19174

LI, HUNG CHIANG, b Kinhwa, China, Dec 10, 21; US citizen; m 57; c 2. STATISTICS, ANALYTICAL MATHEMATICS. Educ: Univ Chekiang, BS, 46; Mich State Univ, MS, 64; Purdue Univ, PhD(statist), 69. Prof Exp: Asst math, Taiwan Normal Univ, 47-50; instr, Nat Taiwan Univ, 50-52; asst prof, Taiwan Inst Technol, 52-55; assoc prof, Tunghai Univ, 55-62; PROF STATIST, UNIV SOUTHERN COLO, 69- Concurrent Pos: Consult, Univ Southern Colo, 69- Mem: Inst Math Statist; Am Math Soc; Math Asn Am; Sigma Xi. Res: Multivariate analysis, particularly interested in normal distributions and the asymptotic expansions for distributions of characteristic roots of normal populations. Mailing Add: Dept of Math Univ of Southern Colo Pueblo CO 81001

LI, JANE CHIAO, b Shanghai, China, May 1, 39; US citizen; m 63; c 2. APPLIED STATISTICS. Educ: Hunter Col, BS, 63; Rutgers Univ, MS, 65, PhD(statist), 71. Prof Exp: Res chemist org chem, Endo Labs, Long Island, NY, 61-63; STATIST CONSULT COMPUT SCI & STATIST, RUTGERS UNIV, 74- Mem: Sigma Xi; Am Statist Asn; Inst Math Statist. Res: Design and analysis of mixture experiments with process variables; evaluation of computer programs for efficient statistical analysis, data analysis in medical and bioscience research. Mailing Add: Comput Ctr Busch Campus Rutgers Univ New Brunswick NJ 08903

LI, JEANNE B, b New York, NY, Apr 15, 44. BIOCHEMISTRY, PHYSIOLOGY. Educ: Vassar Col, AB, 66; Harvard Univ, PhD(biochem), 71. Prof Exp: Res fel physiol, Harvard Med Sch, 71-73; ASST PROF PHYSIOL, HERSHEY MED CTR, PA STATE UNIV, 73- Concurrent Pos: Nat Cancer Inst & Muscular Dystrophy Asn fels, Harvard Med Sch, 71-73; Am Diabetes Asn grant, Hershey Med Ctr, Pa State Univ, 74-75. Mem: AAAS. Res: Regulation of protein synthesis and degradation in mammalian tissues. Mailing Add: Dept of Physiol Hershey Med Ctr Pa State Univ 500 Univ Dr Hershey PA 17033

LI, KELVIN K, b Kwantung, China, Mar 25, 34; stateless; m 65; c 4. PHYSICS. Educ: McGill Univ, BEng, 58; Mass Inst Technol, PhD(physics), 64. Prof Exp: PHYSICIST, BROOKHAVEN NAT LAB, 65- Mem: Am Phys Soc. Res: Elementary particle interactions. Mailing Add: Dept of Physics Brookhaven Nat Lab Upton NY 11973

LI, KOIBONG, b Pingtung, Taiwan, Jan 12, 14; US citizen; m 46; c 2. MICROBIOLOGY. Educ: Seoul Nat Univ, MD, 40; Tokyo Univ, PhD(microbiol), 47. Prof Exp: Res mem microbiol, Inst Infectious Dis, Univ Tokyo, 40-46; sr microbiologist, 406th Med Gen Lab , US Army, 46-58; res assoc microbiol, Sch Med, NY Univ, 58-61, fac mem, 61-64; chief bact lab, Fordham Hosp, New York, 64-66; asst prof & dir bact lab, Bird S Coler Hosp, Ctr Chronic Dis, NY Med Col, 67-69; ASST PROF, STATE UNIV NY DOWNSTATE MED CTR, 69- Mem: AAAS; Harvey Soc; Am Soc Microbiol. Res: Leptospirosis; streptococci; interaction of bacteriophage and Shigella dysenteriae; toxin production of Shigella dysenteriae; a novel type of resistant bacteria induced by gentamicin; anti-evolutionary effect of gentamicin on bacteria. Mailing Add: 74-11 44th Ave Elmhurst NY 11373

LI, KUANG-PANG, b Kwang-tung, China, Oct 11, 38; m; c 2. ANALYTICAL CHEMISTRY. Educ: Nat Taiwan Univ, BS, 61; Univ Ill, MS, 68, PhD(anal chem), 70. Prof Exp: Lectr chem, Kaohsiung Prov Inst Technol, Taiwan, 64-65; res assoc, Ariz State Univ, 70-72; res assoc, Univ Ill, 72-73; ASST PROF CHEM, UNVI FLA, 73- Mem: Am Chem Soc. Res: Metallic ion transport in biomembranes; membrane interactions of chemical carcinogens; theoretical and practical developments of chromatographic methods; analysis of polynuclear aromatics in auto exhaust particulates. Mailing Add: Dept of Chem Univ of Fla Gainesville FL 32611

LI, LI-HSIENG, b Peking, China, Dec 31, 33; m 60; c 2. BIOCHEMISTRY. Educ: Nat Taiwan Univ, BS, 55; Va Polytech Inst, MS, 62, PhD(biochem), 64. Prof Exp: Res assoc, Ind Univ, 64-65; SR RES SCIENTIST, UPJOHN CO, 65- Mem: Am Chem Soc; Am Asn Cancer Res. Res: Enzymology; mechanism of action of anticancer agent; virus and cancer; cell biology. Mailing Add: Upjohn Co Cancer Res 301 Henrietta St Kalamazoo MI 49001

LI, LING-FONG, b Fukien, China, Apr 17, 44. THEORETICAL HIGH ENERGY PHYSICS. Educ: Nat Taiwan Univ, BS, 65; Univ Pa, MS, 67, PhD(physics), 70. Prof Exp: Res assoc physics, Rockefeller Univ, 70-72; res assoc, Stanford Linear Accelerator Ctr, Stanford Univ, 72-74; ASST PROF PHYSICS, CARNEGIE-MELLON UNIV, 74- Mem: Am Phys Soc. Res: Unified theories of weak and electromagnetic interactions in relation to the fundamental structure of the elementary particles. Mailing Add: Dept of Physics Carnegie-Mellon Univ Pittsburgh PA 15213

LI, LU KU, b Honan, China, Apr 26, 36; m 61; c 1. BIOCHEMISTRY, PROTEIN CHEMISTRY. Educ: Nat Taiwan Univ, BS, 58; Princeton Univ, PhD(biol), 64. Prof Exp: Res asst biol, Princeton Univ, 63-64; res assoc chem, Cornell Univ, 64-66; instr opthal, 66-68, assoc, 68-69, ASST PROF OPTHAL, COLUMBIA UNIV, 69- Mem: Am Chem Soc. Res: Maturation of lens fiber cells and its relation to the subunits interactions of lens proteins; vision and opthalmology. Mailing Add: Dept of Opthal Columbia Univ New York NY 10032

LI, MIN CHIU, b Canton, China, Sept 21, 19; nat US; m 40; c 3. INTERNAL MEDICINE, ONCOLOGY. Educ: Mukden Med Col, China, MB & ChB, 44; Univ Southern Calif, MSc, 48. Prof Exp: Resident med, Presby Hosp, Chicago, Ill, 50-53; Runyon fel, Sloan-Kettering Inst Cancer Res, 53-55; clin asst, NIH, 55-57; assoc, Sloan-Kettering Inst Cancer Res, 57-63; DIR MED RES, NASSAU HOSP, 63- Concurrent Pos: Asst prof med, State Univ NY Downstate Med Ctr, 57-65. Mem: AAAS; Endocrine Soc; Am Asn Cancer Res; Am Fedn Clin Res. Res: Endocrine relationship of cancer and chemotherapy. Mailing Add: Dept of Med Res Nassau Hosp Mineola NY 11501

LI, MING CHIANG, b Ningpo, China, June 18, 35; US citizen; m 65; c 2. PHYSICS, MATHEMATICS. Educ: Peking Univ, BS, 58; Univ Md, PhD(physics, math), 65. Prof Exp: Lectr physics, Norm Col Inner Mongolia, China, 58-61; res asst, Univ Md, 64-65; fel & mem sci, Inst Advan Study, 65-67; asst prof, 68-72, ASSOC PROF PHYSICS, VA POLYTECH INST & STATE UNIV, 72- Concurrent Pos: NSF res grant, Va Polytech Inst & State Univ, 67-; Energy Res & Develop Admin res grant, 75. Mem: Am Phys Soc. Res: Theoretical high energy physics; quantum physics; mathematical physics. Mailing Add: Dept of Physics Va Polytech Inst & State Univ Blacksburg VA 24061

LI, MING FANG, b Chickiang, China, Oct 28, 22; m 56; c 2. ANIMAL NUTRITION, MICROBIOLOGY. Educ: St John's Univ, Shanghai, BS, 47; Univ Man, MSc, 57; Univ Alta, PhD(animal nutrit), 63. Prof Exp: Dairy specialist, Taiwan Animal Prod Co, Formosa, 49-54; assoc scientist, 63-71, RES SCIENTIST, HALIFAX LAB, FISHERIES & MARINE SCI SERV CAN, 71- Mem: Can Soc Microbiol; Tissue Cult Asn; Can Inst Food Sci & Technol. Res: Tissue culture; fish diseases; fatty acid metabolism in rumen microorganisms; keeping quality of pasteurized milk. Mailing Add: Halifax Lab Fisheries & Marine Sci Serv Environ Can PO Box 429 Halifax NS Can

LI, NORMAN CHUNG, b Foochow, China, Jan 13, 13; nat US; m 37; c 5. PHYSICAL CHEMISTRY. Educ: Kenyon Col, BS, 33; Univ Mich, MS, 34; Univ Wis, PhD(chem), 36. Prof Exp: Prof chem, Anhwei Univ, China, 36-38; lectr, Yenching Univ, 38-40; lectr, Cath Univ, China, 40-43, from assoc prof to prof, 43-46; from asst prof to assoc prof, St Louis Univ, 46-52; PROF CHEM, DUQUESNE UNIV, 52- Concurrent Pos: Consult, Argonne Nat Lab, 56-58; vis scientist, NIH, 62; tech asst expert, Int Atomic Energy Agency, 64; vis prof, Tsing Hua Univ, China, 64; adv chem res ctr, Nat Taiwan Univ, 66-; consult, Nat Res Coun of Repub China, 74-75; adv, Inst Chem, Nat Tsing Hua Univ, Repub China, 74-75. Mem: Fel AAAS; Am Chem Soc; Sigma Xi. Res: Nuclear magnetic resonance studies of hydrogen and metal binding; hydrogen bonding in coal and asphaltenes; research on metalloproteins. Mailing Add: Dept of Chem Duquesne Univ Pittsburgh PA 15219

LI, PAUL H, b China, May 4, 33; US citizen; m 63; c 2. HORTICULTURE, PLANT PHYSIOLOGY. Educ: Ore State Univ, PhD(hort & plant physiol), 63. Prof Exp: ASSOC PROF HORT & PLANT PHYSIOL, UNIV MINN, 63- Concurrent Pos: Vis prof, Int Potato Ctr, 73. Honors & Awards: Dow Chem Co Award, Am Soc Hort Sci, 65, Alex Laurie Award, 66. Mem: Am Soc Hort Sci; Am Soc Plant Physiologists; Soc Cryobiol; Potato Asn Am. Res: Research, teaching and advising in the area of plant cold hardiness, potato frost resistance and physiological aspects of potato proteins. Mailing Add: Dept of Hort Sci Univ of Minn St Paul MN 55101

LI, STEVEN SHOEI-LUNG, b Taiwan, China, Oct 20, 38; m 67. GENETICS, BIOCHEMISTRY. Educ: Univ Taiwan, BS, 61, MS, 63; Univ Mo, PhD(genetics), 68. Prof Exp: Res assoc, Univ Tex, Austin, 68-70; res assoc biol sci, Stanford Univ, 70-74; ASSOC PROF BIOL SCI, MT SINAI SCH MED, 74- Mem: AAAS; Genetics Soc Am. Res: Biochemical genetics; structure, function and evolution of hemoglobins, enzymes and hemagglutinins. Mailing Add: Dept of Microbiol Mt Sinai Sch of Med New York, NY 10029

LI, SU-CHEN, b Taipei, Taiwan, June 8, 35; US citizen; m 62; c 2. BIOCHEMISTRY. Educ: Nat Taiwan Univ, BS, 58; Univ Okla, PhD(biochem), 65. Prof Exp: Asst prof, 72-76, ASSOC PROF, DEPT BIOCHEM, SCH MED, TULANE UNIV, 76-, RES SCIENTIST, DELTA REGIONAL PRIMATE RES CTR, 63- Concurrent Pos: Career Develop Award, NIH, 75-80. Mem: Am Soc Biol Chemists; AAAS; Soc Complex Carbohydrates. Res: Biochemical studies of glycoconjugates and glycosidases. Mailing Add: Delta Primate Res Ctr Tulane Univ Three Rivers Rd Covington LA 70433

LI, TAO PING, b Szechwan, China, Nov 16, 20; m 48; c 3. ORGANIC CHEMISTRY. Educ: Nat Szechwan Univ, China, BS, 41; Univ Tex, MA, 59, PhD(org chem), 60. Prof Exp: Fel, Univ Tex, 60-61; sr proj chemist, Am Oil Co, Ind, 61-64; res specialist, 64-66, group leader, 66-69, SR GROUP LEADER, MONSANTO CO, 69- Mem: AAAS; Am Chem Soc; Catalysis Soc; NY Acad Sci. Res: Chemical kinetics, heterogeneous catalysis and chemistry of metal organic compounds. Mailing Add: Monsanto Co 800 N Lindbergh Blvd St Louis MO 63166

LI, TA-YUNG, b China, Sept 22, 45. METEOROLOGY. Educ: Chung Yuan Col, BS, 67; Drexel Univ, MS, 69, PhD(physics & atmospheric sci), 72. Prof Exp: Instr, 72-73, res fel, 73-75, VIS ASST PROF METEOROL, GRAD PROG, UNIV MD, 75- Mem: Am Meteorol Soc; Am Geophys Union; Sigma Xi. Res: Power plant plume modeling; atmospheric turbulence and diffusion in the planetary boundary layer; acidity of rainfall near a power plant; fuel switching and various associated environmental studies. Mailing Add: Grad Prog Meteorol Univ of Md College Park MD 20742

LI, TIEN-YIEN, b Hunan, China, June 28, 45; m 71; c 1. MATHEMATICS. Educ: Nat Tsing-Hua Univ, BS, 68; Univ Md, PhD(math), 74. Prof Exp: INSTR MATH, UNIV UTAH, 74- Res: Differential equations, dynamical systems and numerical analysis. Mailing Add: Dept of Math Univ of Utah Salt Lake City UT 84112

LI, TING KAI, b Nanking, China, Nov 13, 34; m 60; c 2. MEDICINE, BIOCHEMISTRY. Educ: Northwestern Univ, AB, 55; Harvard Med Sch, MD, 59; Mass Inst Technol, 60-61. Prof Exp: House officer, Peter Bent Brigham Hosp, 59-60, asst med, 60-63, jr assoc, 63-65; instr med, Harvard Med Sch, 65-67, assoc, 67-69; dep dir div biochem, Walter Reed Army Inst Res, 69-71; PROF MED & BIOCHEM, SCH MED, IND UNIV, INDIANAPOLIS, 71- Concurrent Pos: Helen Hay Whitney Found fel, 60-64; Med Found Boston fel, 64-68; Markle scholar acad med, 67-73; asst med, Harvard Med Ach, 60-63, res assoc biochem, 63-65; chief resident, Peter Bent Brigham Hosp, 65-66; guest scientist, Nobel Med Inst, Sweden, 68. Mem: Am Chem Soc; Am Soc Clin Invest; Endocrine Soc; Am Soc Biol Chem; Am Fedn Clin Res. Res: Enzymology; metabolism; chemical basis of biological specificity; alcohol metabolism. Mailing Add: Dept of Med Ind Univ Med Ctr Indianapolis IN 46202

LI, TSUNG HAN, pharmacology, anesthesiology, see 12th edition

LI, WEN-CH'ING WINNIE, b China, Dec 25, 48. NUMBER THEORY. Educ: Nat Taiwan Univ, BS, 70; Univ Calif, Berkeley, PhD(math), 74. Prof Exp: ASST PROF MATH, HARVARD UNIV, 74- Mem: Am Math Soc. Res: Modular forms and representation theory. Mailing Add: Dept of Math Harvard Univ Cambridge MA 02138

LI, WEN-HSIUNG, b Ping-Tung, Taiwan, Sept 22, 42; m 75. POPULATION GENETICS, APPLIED MATHEMATICS. Educ: Chung-Yuang Col Sci & Eng, Taiwan, BE, 65; Nat Cent Univ, Taiwan, MS, 68; Brown Univ, PhD(appl math), 72. Prof Exp: Proj assoc, Univ Wis-Madison, 72-73; ASST PROF POP GENETICS, UNIV TEX, HOUSTON, 73- Mem: Genetics Soc Am; AAAS; Am Soc Human Genetics. Res: Molecular evolution; biomathematics; human genetics. Mailing Add: Ctr Demog & Pop Genetics Univ of Tex Houston TX 77030

LI, WU-SHYONG, b Taipei, Taiwan, Aug 20, 43; m 75. PHYSICAL ORGANIC CHEMISTRY. Educ: Nat Taiwan Univ, BS, 66; Kent State Univ, MS, 69; Univ Minn, PhD(org chem), 73. Prof Exp: Fel, Ohio State Univ, 73-75; SR CHEMIST PLASTIC, ROHM & HAAS CO, 75- Mem: Am Chem Soc. Res: Organic reaction mechanism and kinetics. Mailing Add: Rohm & Haas Co Res Lab Box 219 Bristol PA 19007

LI, YU-TEH, b Hsin-Chu City, Formosa, Apr 1, 34; m 62; c 2. BIOCHEMISTRY. Educ: Nat Taiwan Univ, BS, 57, MS, 60; Univ Okla, PhD(biochem), 63. Prof Exp: From instr to asst prof biochem, Sch Med, Univ Okla, 63-66; CHIEF, DELTA REGIONAL PRIMATE RES CTR, 66-, PROF BIOCHEM, SCH MED, TULANE UNIV, 74- Concurrent Pos: Fel biochem, Sch Med, Univ Okla, 63-64; Nat Cancer Inst grant, 64-66; NSF grant, 68-; NIH grant, 71-; USPHS res career develop award, 71-76. Mem: Am Chem Soc; Am Soc Biol Chem. Res: Biochemical studies on glycoproteins and various glycosidases. Mailing Add: Dept of Biochem Sch of Med Tulane Univ New Orleans LA 70112

LIAN, HAROLD MAYNARD, b Fairdale, NDak, Dec 25, 22; m 53; c 4. GEOLOGY. Educ: Univ Calif, Los Angeles, PhD(geol), 53. Prof Exp: VPRES EXPLOR, INT DIV, UNION OIL CO, CALIF, 52- Concurrent Pos: Fulbright res scholar, Austria, 54-55. Mem: Geol Soc Am; Am Asn Petrol Geol. Res: Stratigraphy; petroleum geology. Mailing Add: Union Oil Co of Calif Box 7600 Los Angeles CA 90054

LIANG, CHARLES CHI, b Nanking, China, June 9, 34; m 61; c 3. PHYSICAL CHEMISTRY, ANALYTICAL CHEMISTRY. Educ: Nat Taiwan Univ, BS, 56; Baylor Univ, PhD(phys chem), 62. Prof Exp: Res chemist, Houdry Process & Chem Co, 62-63; from asst prof to assoc prof phys chem, WVa Inst Technol, 63-65; sr staff mem electrochem, 65-73, ASST TECH DIR BATTERIES, LAB PHYS SCI, P R MALLORY CO, INC, 73- Honors & Awards: IR 100 Award, Indust Res Mag, 71. Mem: Am Chem Soc; Electrochem Soc. Res: Mechanisms and kinetics of electrode processes; chemical thermodynamics; analytical techniques; solid state chemistry. Mailing Add: Lab for Phys Sci P R Mallory Co Inc Burlington MA 01803

LIANG, CHING YU, b Hupeh, China, Dec 16, 16; US citizen; m 49; c 1. PHYSICS. Educ: Nat Cent Univ, China, BS, 43; Univ Okla, MS, 50, PhD(physics), 52. Prof Exp: Instr physics, Nat Cent Univ, China, 43-48; fel, Univ Mich, 52-55; res assoc physics, Am Viscose Corp, Pa, 55-63; sr res scientist, Lear Siegler, Inc, Calif, 63-64; assoc prof, 64-68, PROF PHYSICS, CALIF STATE UNIV, 68- Concurrent Pos: Consult & intermittent physicist, Navy Electronic Lab, San Diego, Calif, 65- Mem: Am Phys Soc. Res: Infrared spectra and structures of polymers; semiconductivity of organic crystals and biological molecules; solid state physics. Mailing Add: Dept of Physics & Astron Calif State Univ Northridge CA 91324

LIANG, EDISON PARK-TAK, b Canton, China, July 22, 47; m 71. ASTROPHYSICS, COSMOLOGY. Educ: Univ Calif, Berkeley, BA, 67, PhD(physics), 71. Prof Exp: Res assoc relativity, Univ Tex, Austin, 71-73; res assoc & assoc instr astrophys & relativity, Univ Utah, 73-75; ASST PROF ASTROPHYS, MICH STATE UNIV, 75- Res: Origin and dynamics of primordial density fluctuations in early universe; astrophysics of compact objects (x-ray sources); gravitational collapse; galaxy formation problems; space-time singularities. Mailing Add: Dept of Astron & Astrophys Mich State Univ East Lansing MI 48824

LIANG, GEORGE H L, b Peiping, China, Oct 1, 34; m 63; c 1. PLANT GENETICS, PLANT BREEDING. Educ: Taiwan Prov Col Agr, BS, 56; Univ Wyo, MS, 61; Univ Wis, PhD(agron), 65. Prof Exp: Agronomist, Taiwan Prov Res Inst Agr, 58-59; asst prof, 64-69, ASSOC PROF PLANT GENETICS & CYTOGENETICIST, KANS STATE UNIV, 69- Mem: Am Soc Agron; Crop Sci Soc Am; Am Genetic Asn; Genetics Soc Am; Genetics Soc Can. Res: Quantitative genetics in plant species; cytogenetics and breeding aspects in cultivated crops; taxonomy of plant in relation to cytology and biochemistry. Mailing Add: Dept of Agron Kans State Univ Manhattan KS 66506

LIANG, JOSEPH JEN-YIN, b China; US citizen; m 65; c 2. MATHEMATICS. Educ: Nat Taiwan Univ, BA, 58; Univ Detroit, MA, 62; Ohio State Univ, PhD(math), 69. Prof Exp: Res fel math, Calif Inst Technol, 69-70; asst prof, 70-73, ASSOC PROF MATH, UNIV S FLA, 73- Res: Number theory; coding theory; algorithms. Mailing Add: Dept of Math Univ of SFla Tampa FL 33620

LIANG, KAI, b Hunan, China, Mar 23, 34; m 64. PHYSICAL CHEMISTRY. Educ: Nat Taiwan Univ, BS, 56; Univ Utah, PhD(phys chem), 64. Prof Exp: Res asst phys chem, Univ Utah, 60-64; sr res chemist, 64-70, MEM SR STAFF, COLOR PHOTOG DIV, RES LABS, EASTMAN KODAK CO, 70- Mem: Am Chem Soc. Res: Diffusion and transport processes; mathematical modelling of diffusion kinetics; color photography. Mailing Add: Res Labs Color Photog Div Eastman Kodak Co Kodak Park Rochester NY 14650

LIANG, KENG-SAN, b Tainan, Taiwan, Dec 17, 43; m 69; c 1. SOLID STATE SCIENCE. Educ: Nat Taiwan Univ, BS, 66; Stanford Univ, MS, 70, PhD(appl physics), 73. Prof Exp: Assoc scientist, 73-75, SCIENTIST SOLID STATE SCI, WEBSTER RES CTR, XEROX CORP, 75- Mem: Am Phys Soc; Am Vacuum Soc. Res: Electronic structure of solids; surface and interface physics and chemistry; electron spectroscopy; structure of amorphous materials. Mailing Add: Webster Res Ctr Xerox Corp Webster NY 14580

LIANG, SHOU CHU, b Foochow, China, May 14, 20; m 50; c 2. CHEMISTRY. Educ: Cent Univ, China, BS, 42; Princeton Univ, MA, 46, PhD(chem), 47. Prof Exp: Teacher high sch, China, 42; anal chemist, China Match Raw Mat Mfg Co, 42-44; asst, Princeton Univ, 45-47, Int Nickel Co fel, 47-48; res chemist, Merck & Co, NJ, 48-49; fel, Nat Res Coun Can, 49-51, asst res officer II, 51-53; group leader, Res Lab, Dom Tar & Chem Co, 53-56; res engr, Consol Mining & Smelting Co Can, Ltd, 56-64, head gen metall res, 64-70; GEN MGR, ELECTRONIC MAT DIV, COMINCO AM INC, 70- Mem: AAAS; Am Chem Soc; Electrochem Soc; NY Acad Sci; Chem Inst Can. Res: Flotation of ores; preparation of inorganic reagents; chemical method of analysis; surface catalysis; heterogeneity of catalyst surfaces for chemisorption; fast drying paint; metallurgy; semiconductors. Mailing Add: Electronic Mat Div Cominco Am 818 WRiverside Ave Spokane WA 99201

LIANG, TUNG, b Peking, China, June 7, 32; m 58; c 2. OPERATIONS RESEARCH, BIOENGINEERING. Educ: Nat Taiwan Univ, BS, 56; Mich State Univ, MS, 63; NC State Univ, PhD(biol eng), 67. Prof Exp: ASSOC PROF AGR ENG, UNIV HAWAII, 68- Mem: Opers Res Soc Am; Am Soc Agr Eng. Mem: Biological system modeling and optimization. Mailing Add: Dept of Agr Eng Univ of Hawaii 2560 Dole St Honolulu HI 96822

LIANIDES, SYLVIA PANAGOS, b Lynn, Mass, Sept 2, 31; m 56; c 3. PHYSIOLOGY, BIOCHEMISTRY. Educ: Tufts Univ, BS, 53, PhD(physiol), 59. Prof Exp: Res biologist, US Naval Radiol Defense Lab, 59-60; lectr biol, Col Notre Dame, Calif, 62-71; instr biol sci, De Anza Col, 71-73; INSTR BIOL, WEST VALLEY COL, 73- Mem: AAAS; Sigma Xi. Res: Hormonal and environmental influences upon mitochondrial oxidative phosphorylation; effects of environmental cold on lipid and carbohydrate metabolism; radiation physiology. Mailing Add: 19643 Montauk Dr Saratoga CA 95070

LIAO, CHIEN-WEI, organic chemistry, see 12th edition

LIAO, HSIANG PENG, b Nanping, China, May 4, 24; nat US; m 50; c 3. ORGANIC CHEMISTRY. Educ: Fukien Christian Univ, BS, 45; Northwestern Univ, PhD(chem), 52. Prof Exp: Res chemist, Standard Oil Co, Ind, 52-60; PROJ LEADER, FMC CORP, 60- Res: Stereochemistry; oxidation; hydrogenation. Mailing Add: Indust Chem Div FMC Corp PO Box 8 Princeton NJ 08540

LIAO, HSUEH-LIANG, b Silo, Taiwan, Jan 24, 41; US citizen; m; c 1. ANALYTICAL CHEMISTRY. Educ: Cheng Kong Univ, BS, 65; Drexel Univ, MS, 69; Georgetown Univ, PhD(chem), 72. Prof Exp: Sr res assoc chem, Northeastern Univ, 72-74; SR RES CHEMIST, NORWICH PHARMACAL CO, MORTON-NORWICH PROD INC, 74- Mem: Am Chem Soc; Am Inst Chem Eng. Res: Chromatographic methods of separation and quantitation; analytical and physical chemistry of drug compounds; solution thermodynamics. Mailing Add: Norwich Pharmacal Co PO Box 191 Norwich NY 13815

LIAO, JI-CHIA, b Chia-Yi, Taiwan, Feb 16, 36; m 63; c 3. ANESTHESIOLOGY, PHARMACOLOGY. Educ: Nat Taiwan Univ, MD, 61; Univ Minn, PhD(pharm), 71. Prof Exp: Resident surg, Nat Taiwan Univ Hosp, 62-64; resident anesthesiol, Univ Hosp, 65-67, instr, 70-71, ASST PROF ANESTHESIOL, UNIV MINN, MINNEAPOLIS, 71- Concurrent Pos: Consult, Vet Admin Hosp, Minn, 70-71. Res: Sympathetic effect of angiotensin; cardiovascular effect of ketamine; uptake of local anesthetics. Mailing Add: Dept of Anesthesiol Univ of Minn Hosp Box 158 Minneapolis MN 55455

LIAO, SHU-CHUNG, b Tainan, Taiwan, Oct 18, 39; m 71. PHYSICAL-ANALYTICAL CHEMISTRY. Educ: Nat Taiwan Univ, BSc, 63; Univ Western Ont, PhD(phys chem), 70. Prof Exp: Res assoc polymer chem, Univ Mich, Ann Arbor, 70-71, human nutrit prog, Sch Pub Health, 72-73; SR RES ASSOC, CLIMAX MOLYBDENUM CO, MICH, 73- Mem: Am Chem Soc. Res: Physical chemistry of molecular complexes and macromolecules; food chemistry and nutrition; instrumental analysis; analytical chemistry. Mailing Add: Climax Molybdenum Co of Mich Ann Arbor MI 48106

LIAO, SHUTSUNG, b Taiwan, Formosa, Jan 1, 31; m 60; c 4. BIOCHEMISTRY, ENDOCRINOLOGY. Educ: Nat Taiwan Univ, BSc, 54, MSc, 56; Univ Chicago, PhD(biochem), 61. Prof Exp: From asst prof to assoc prof, 64-71, PROF BIOCHEM, BEN MAY LAB CANCER RES, UNIV CHICAGO, 72- Concurrent Pos: NIH res grant, 63-; Am Cancer Soc res grant, 74-76. Mem: Am Soc Biol Chem; Endocrine Soc. Res: Mechanism of hormone action; control of intermediate reactions involved in the biosynthesis of proteins and nucleic acids; enzymology. Mailing Add: Ben May Lab for Cancer Res Univ of Chicago Chicago IL 60637

LIAO, SUNG JUI, b Changsha, China, Nov 15, 17; nat US; m 53; c 4. PHYSICAL MEDICINE. Educ: Hsiang Ya Med Col, China, MD, 42; London Sch Hyg & Trop Med, Univ London, DPH, 46, dipl bact, 47; Am Bd Phys Med & Rehab, dipl. Prof Exp: Asst prof prev med, Sch Med, Yale Univ, 50-54; assoc res prof bact, Col Med, Univ Utah, 49-50; dir phys med & rehab, Waterbury Hosp, 57-73; assoc clin prof, 67-73, LECTR REHAB MED, SCH MED, BOSTON UNIV, 73-; CLIN ASSOC PROF, INST REHAB MED, NY UNIV, 71- Concurrent Pos: Milbank Mem fel prev med, Sch Med, Yale Univ, 47-49; sr clin fel phys med, Mass Gen Hosp, Boston, 55-57; consult psychiatrist, Middlesex Mem Hosp, Middletown, Conn, 57-60; med dir, Waterbury Area Rehab Ctr, 57-62; dir phys med & rehab, St Mary's Hosp, 57-67 & Danbury Hosp, 57-69; med consult, Waterbury Dist, Conn State Div Voc Rehab, 63-72, chief admin med consult, State Div, 69-73; hon consult biomech, NY Univ Inst Rehab Med, 69-; chmn ad hoc comt acupuncture, Conn State Med Soc; secy, Am Acad Acupuncture, Inc; pres, Res Inst Acupuncture & Chinese Med. Mem: Fel Am Col Physicians; fel Am Pub Health Asn; Am Asn Immunol; sr mem Am Fedn Clin Res; fel Am Acad Phys Med & Rehab. Res: Excitability and conduction nerve and muscle; biomedical engineering; acupuncture. Mailing Add: Rte 188 & N Benson Rd Middlebury CT 06762

LIAO, TSUNG-KAI, b Chiayi, Taiwan, Aug 1, 23; m 63; c 3. ORGANIC CHEMISTRY, PHARMACEUTICAL CHEMISTRY. Educ: Nat Taiwan Univ, BS, 52; Wesleyan Univ, MA, 57; Univ Kans, PhD(chem), 60. Prof Exp: Asst chem, Nat Taiwan Univ, 53-55; fel, Wesleyan Univ, 55-57; asst, Univ Kans, 57-60; res assoc, Univ Mich, 60-61; assoc chemist, 61, SR CHEMIST, MIDWEST RES INST, 61- Concurrent Pos: Fel, Res Inst, Univ Mich, 60-61. Mem: Am Chem Soc; Sigma Xi. Res: Synthesis of biologically active organic compounds; chemistry of nitrogen heterocyclic compounds. Mailing Add: 1317 E 101st Terr Kansas City MO 64131

LIAUW, KOEI-LIANG, b Indonesia, May 4, 35; US citizen; m 61; c 1. ORGANIC CHEMISTRY. Educ: Nanyang Univ, Singapore, BSc, 60; Univ Calif, Berkeley, MS, 62, PhD(chem), 64. Prof Exp: Res chemist, Gen Chem Div, Allied Chem Corp, 64-66; sr res chemist, Mobil Chem Co, Div Mobil Oil Corp, 66-68; PROJ LEADER CENT RES, TECH CTR, WITCO CHEM CORP, 69- Mem: Am Chem Soc. Res: Textile treating agents; paper sizings; process development; synthetic organic chemistry; organotin chemistry; vapor phase catalysis. Mailing Add: Tech Ctr Witco Chem Corp 100 Bauer Dr Oakland CA 07436

LIBACKYJ, ANFIR, b Mosuriwci, Ukraine, Sept 8, 26; US citizen. PHYSICAL CHEMISTRY. Educ: Univ Liege, BSc, 54; Polytech Inst Brooklyn, PhD(phys chem), 65. Prof Exp: Asst phys chem, Univ Liege, 54-55; chemist, Am Viscose Corp, 56-58; res chemist, E I du Pont de Nemours & Co, Inc, 63-68; adj prof, NY Inst Technol, 68-70, asst prof, 70-73; M DIV CAND, UNION THEOL SEM, NEW YORK, 74- Mem: Am Chem Soc; NY Acad Sci; Am Acad Polit & Soc Sci; Acad Polit Sci; Am Acad Relig. Res: Physical chemistry of colloids and polymers. Mailing Add: 84-22 107th Ave Ozone Park NY 11417

LIBAN, ERIC, b Vienna, Austria, June 20, 21; nat US; m 54; c 3. APPLIED MATHEMATICS. Educ: NY Univ, BA, 48, MS, 49, PhD(math), 57. Prof Exp: Instr math, Long Island Univ, 49 & NY Univ, 49-50; asst, Ind Univ, 50-51; mathematician, Naval Res Lab, 51-52; assoc mathematician, Proj Cyclone, Reeves Instrument Corp, NY, 52; sr dynamics engr, Repub Aviation Corp, 52-55; staff mem analog comput & consult ctr, Dian Labs, Inc, 55-58; assoc prof eng sci, Pratt Inst, 58-61; res scientist,

LIBAN

Grumman Aircraft Eng Corp, 61-67; assoc prof, 67-71, PROF MATH, YORK COL, 71-, CHMN DEPT, 75- Concurrent Pos: Lectr, Univ Md, 50; consult, Avco Res & Develop Corp, Mass, 59-60 & Comput Systs, Inc, NJ, 59-61; adj lectr, Polytech Inst Brooklyn, 62-65; adj prof, Adelphi Univ, 66-67; adj assoc prof, Queens Col, 67-68. Mem: Am Math Soc; Asn Comput Mach. Res: Applications and methods of simulation on analog computers; logical design of computing systems; automata studies; theory of servo and feedback systems; information and communication theory; operations research. Mailing Add: 251-37 43rd Ave Little Neck NY 11363

LIBBER, LEONARD MITCHELL, b Boston, Mass, Jan 29, 19; m 46; c 3. PHYSIOLOGY. Educ: Cath Univ, BS, 47; Univ Pa, PhD(gen physiol), 52. Prof Exp: Physiologist, Air Crew Equip Lab, Philadelphia Naval Base, 51-59; head physiol br, 59-67, sci liaison officer, London, 67-69, DIR PHYSIOL PROG, US OFF NAVAL RES, DC, 69- Mem: Am Inst Biol Sci; Aerospace Med Asn; Undersea Med Soc. Res: Stress physiology. Mailing Add: Off of Naval Res 800 N Quincy St Arlington VA 22217

LIBBEY, LEONARD MORTON, b Boston, Mass, Apr 17, 30; m 71. FOOD SCIENCE. Educ: Univ Mass, BVA, 53; Univ Wis, MS, 54; Wash State Univ, PhD(food technol), 61. Prof Exp: Asst prof, 61-69, ASSOC PROF FOOD SCI & TECHNOL, ORE STATE UNIV, 69- Mem: AAAS; Inst Food Technologists; Am Dairy Sci Asn; Am Chem Soc; Am Oil Chemists Soc. Res: Food chemistry; chromatographic and spectrometric analysis, especially gas chromatography and mass spectrometry. Mailing Add: Dept of Food Sci & Technol Ore State Univ Corvallis OR 97331

LIBBEY, WILLIAM JERRY, b Grand Rapids, Minn, Mar 18, 42; m 64; c 1. ORGANIC CHEMISTRY. Educ: Carleton Col, BA, 64; Univ Wis, PhD(org chem), 69. Prof Exp: Res chemist, 68-72, SR RES CHEMIST, CONTINENTAL OIL CO, 72- Mem: Am Chem Soc. Res: Carbonium ion chemistry; thermal rearrangements; alkyl halide chemistry; Fischer-Tropsch chemistry; hydrocarbon pyrolysis. Mailing Add: Continental Oil Co Ponca City OK 74601

LIBBY, DOROTHY, b Honolulu, Hawaii, Nov 22, 23; div. ANTHROPOLOGY. Educ: Univ Calif, AB, 46, PhD(anthrop), 52. Prof Exp: Res analyst, US Govt, DC, 52-55; res assoc, Great Lakes-Ohio Valley Res Proj, Univ Ind, 57-64; asst prof anthrop, Vassar Col, 64-67; ASSOC PROF ANTHROP, CALIF STATE UNIV, LONG BEACH, 67- Concurrent Pos: Ed, Newsletter, Am Indian Ethnohist Conf, 61- Mem: Fel AAAS; fel Am Anthrop Asn; Soc Am Archaeol; Am Ethnol Soc; Am Indian Ethnohist Conf (pres, 60). Res: Northern Asian and northern North America cultures and cultural relationships; use of historical approaches to cultural materials and problems. Mailing Add: Dept of Anthrop Calif State Univ Long Beach CA 90804

LIBBY, JAMES WILLIAM, JR, b Swampscott, Mass, Feb 17, 14; m 39; c 3. CHEMISTRY. Educ: Mass Inst Technol, SB, 35, PhD(org chem), 38. Prof Exp: Res chemist, Jackson Lab, Del, 38-42, supvr neoprene plant, Ky, 42-45, chemist, Jackson Lab, Del, 45-50, group leader, 50-52, div head anal res, 52-53, new prod develop, 53-56, mgr develop conf, 56-57, mem staff develop dept, 57-67, MGR LICENSING, CENT RES & DEVELOP DEPT, E I DU PONT DE NEMOURS & CO, INC, 68- Mem: Am Chem Soc. Res: Dye chemistry; synthetic elastomers; organic pigments; azo dyes; methyl ethers of 3-aminophthal hydrazide; new product development; utilization of solar energy. Mailing Add: Cent Res & Develop Dept E I du Pont de Nemours & Co Inc Wilmington DE 19898

LIBBY, JOHN LESTER, b South Milwaukee, Wis, May 27, 32; m 56; c 2. ENTOMOLOGY. Educ: Univ Wis, BS, 54; Rutgers Univ, MS, 58, PhD(entom), 60; Am Registry Prof Entom, Cert. Prof Exp: Exten assoc entom, Rutgers Univ, 59-60, asst exten specialist, 60-62; from asst prof to assoc prof, 62-74, PROF ENTOM, UNIV WIS-MADISON, 74-, EXTEN SPECIALIST, 62- Concurrent Pos: From lectr to sr lectr, Univ Ife, Nigeria, 65-68. Mem: Entom Soc Am; Entom Soc Nigeria; Asn Trop Biol; Potato Asn Am; Sigma Xi. Res: Vegetable crops entomological research and extension. Mailing Add: Dept of Entom Univ of Wis-Madison Madison WI 53706

LIBBY, LEONA MARSHALL, b LaGrange, Ill, Aug 9, 19; m 43, 66; c 2. PHYSICS. Educ: Univ Chicago, BS, 38, PhD(chem), 43. Prof Exp: Res assoc, Metall Lab, Manhattan Dist, 42-44; consult physicist, E I du Pont de Nemours & Co, Wash, 44-46; fel, Inst Nuclear Studies, Chicago, 46-47, res assoc, 47-54, asst prof, 54-57; fel, Inst Advan Studies, NJ, 57-58; vis scientist, Brookhaven Nat Lab, NY, 58-60; from assoc prof to prof physics, NY Univ, 60-63; assoc prof, Univ Colo, Boulder, 63-72; ADJ PROF PHYSICS, UNIV CALIF, LOS ANGELES, 72- Concurrent Pos: Consult, Los Alamos Sci Lab, 51-; consult, Rand Corp, 57-, staff mem, 66-70; consult, TRW Space Systs Group, 60-; vis prof eng, Univ Calif, Los Angeles, 70-72; staff mem, R&D Assocs, 70- Mem: Fel Am Phys Soc; fel Royal Geog Soc. Res: High energy nuclear physics; nuclear reactions; fundamental particles; astrophysics; neutron physics. Mailing Add: Sch of Eng Univ of Calif Los Angeles CA 90024

LIBBY, PAUL ROBERT, b Torrington, Conn, Sept 2, 34; m 59; c 3. BIOCHEMISTRY. Educ: Yale Univ, BS, 56; Univ Chicago, PhD(biochem), 62. Prof Exp: Fel biochem, Univ Calif, Davis, 62-63; sr cancer res scientist, 63-72, ASSOC CANCER RES SCIENTIST, ROSWELL PARK MEM INST, 72- Mem: AAAS; Am Asn Cancer Res; Endocrine Soc; Am Chem Soc. Res: Biochemical mechanisms of hormone action; biochemical mechanisms of chemical carcinogenesis. Mailing Add: Roswell Park Mem Inst 666 Elm St Buffalo NY 14203

LIBBY, WILLARD FRANK, b Grand Valley, Colo, Dec 17, 08; m 40, 66; c 2. CHEMISTRY. Educ: Univ Calif, BS, 31, PhD(chem), 33. Hon Degrees: ScD, Wesleyan Univ, 55, Syracuse Univ, 57, Univ Dublin, 57, Carnegie Inst Technol, 59, Georgetown Univ, 62, Manhattan Col, 63, Univ Newcastle, 65, Gustavus Adolphus Col, 70. Prof Exp: Instr chem, Univ Calif, 33-38, from asst prof to assoc prof, 38-45; prof, Enrico Fermi Inst Nuclear Studies, Univ Chicago, 45-59; PROF CHEM, UNIV CALIF, LOS ANGELES, 59-, DIR INST GEOPHYS & PLANETARY PHYSICS, 62- Concurrent Pos: Guggenheim fel, Princeton Univ, 41; mem staff, SAM, labs, Columbia Univ, 41-45; consult, Esso Res & Eng Co, 46-; res assoc, Carnegie inst Geophys Lab, 54-59; mem comt, 54-59, comt sr reviewers, 45-52, gen adv comt, 50-54, 60-62, Plowshare Adv Comt, 59-72; mem comt of selection, Guggenheim Mem Found, 59-; pres & dir, Isotope Found, 60-; consult, Douglas Aircraft Co, 62-68, dir, 63-67; consult, Electro-Optical Systs, 63-66 & Univ Denver, 66-69; spec vis prof, Univ Colo, 67-70; mem air resources bd, State of Calif, 67-, Gov Earthquake Coun, 72-; dir, Sci Res Instruments Corp, 67-; mem acad adv bd, Rand Corp, 68-; vis comt, Jet Propulsion Lab, 68-; dir, Hedge Fund Am, 68-; dir, Summit Capital Fund, Inc, 68-; spec vis prof, La State Univ, 68-; mem, Task Force Air Pollution, 69-70; dir, Nuclear Systs, Inc, 69-; consult, Rand Corp, 69-; sci adv, State of Idaho, 69-72; mem US deleg, US-Japan Comt Sci Coop, 70-; dir, Res-Cottrell, 71-; adv, Idaho Nuclear Comn, 72- Honors & Awards: Res Corp award, 51; Chandler medal, Columbia Univ, 54; Remsen Mem lectr award, 55; Bicentennial lectr award, City Col New York, 56; Cresson medal, Franklin Inst, 57; Priestley Mem award, Dickinson Col, 59; Albert Einstein medal award, 59; Nobel Prize in chem, 60; Day medal, Geol Soc Am, 61; Lehman Award, NY Acad Sci, 71; Nuclear appln in chem award, 56 & Willard Gibbs medal, 58, Am Chem Soc; Gold medal, Am Inst Chem, 70. Mem: Nat Acad Sci; AAAS; Am Chem Soc; Am Inst Aeronaut & Astronaut; fel Am Inst Chem. Res: Physical, inorganic and nuclear chemistry; radiochemistry. Mailing Add: Dept of Chem Univ of Calif 405 Hilgard Ave Los Angeles CA 90024

LIBBY, WILLARD GURNEA, b Eugene, Ore, July 18, 29; m 65; c 3. GEOLOGY. Educ: Ore State Col, BS, 51; Northwestern Univ, MS, 59; Univ Wash, PhD(geol), 64. Prof Exp: Explor geologist, Stand Oil Co Calif, 56-59; asst prof geol, Univ SC, 63-65; lectr, Univ BC, 65-68; from asst prof to assoc prof, San Diego State Col, 69-71; PETROLOGIST, GEOL SURV WESTERN AUSTRALIA, 71- Mem: Geol Soc Am; Mineral Soc Am; Mineral Asn Can; Geol Soc Australia. Res: Igneous and metamorphic petrology. Mailing Add: Geol Surv of Western Australia Mineral House 66 Adelaide Terr Perth Australia

LIBBY, WILLIAM HARRIS, b Winona, Minn, Jan 25, 22; m 44; c 2. ORGANIC CHEMISTRY. Educ: Carleton Col, AB, 46; Univ Ill, PhD(chem), 49. Prof Exp: Res chemist, Gen Aniline & Film Corp, 49-53; res chemist, Minn Mining & Mfg Co, 53-66, lab mgr, 65-73, PROD MKT MGR, MICROFILM PROD DIV, 3M CO, 73- Mem: Am Chem Soc. Res: Hindered ketones; fluorescent dyes; cyclic monomers; photochemistry; imaging systems. Mailing Add: 3M Co 3M Ctr Bldg 220-9E St Paul MN 55101

LIBBY, WILLIAM JOHN, (JR), b Oak Park, Ill, Sept 10, 32; m 56; c 3. FORESTRY, GENETICS. Educ: Univ Mich, BS, 54; Univ Calif, Berkeley, MS, 59, PhD(genetics), 61. Prof Exp: NSF fel genetics, NC State Col, 61-62; asst prof forestry, 62-67, assoc prof forestry & genetics, 67-72, PROF FORESTRY & GENETICS, UNIV CALIF, BERKELEY, 72- Concurrent Pos: Pack lectr, Yale Univ, 67; L T Murray distinguished vis lectr forest resources, Univ Wash, 68; Fulbright res scholar, NZ Forest Res Inst, Univ Canterbury, Australian Forest Res Inst, 71. Mem: Soc Am Foresters; Genetics Soc Am. Res: Quantitative genetics of forest trees; gene conservation; vegetation propagation of conifers; maturation of woody plant meristens. Mailing Add: Mulford Hall Univ of Calif Dept of Forestry & Conserv Berkeley CA 94720

LIBELO, LOUIS FRANCIS, b Brooklyn, NY, Oct 12, 30; m 54; c 4. THEORETICAL PHYSICS. Educ: Brooklyn Col, BS, 53; Univ Md, MS, 56; Rensselaer Polytech Inst, PhD(physics), 64. Prof Exp: Engr physics, Md Electronics Co, 54; proj engr, Ahrendt Instrument Co, 55; physicist, Opers Res Off, Johns Hopkins Univ, 57-58; asst prof, 65-68, ADJ PROF PHYSICS, AM UNIV, 68-; RES PHYSICIST, US NAVAL SURFACE WEAPONS CTR, 64- Concurrent Pos: Consult physicist, Entron Inc, 55-58. Mem: Am Phys Soc. Res: Scattering theory for finite, non-spherical targets and by apertures; theory of cooperative phenomena in solids; theory of nonlinear phenomena in magnetic materials. Mailing Add: Physics Res Dept US Naval Surface Weapons Ctr Silver Spring MD 20910

LIBERA, RICHARD JOSEPH, b Thorndike, Mass, Aug 26, 29; m 54; c 2. MATHEMATICS. Educ: Am Int Col, BA, 56; Univ Mass, MA, 58; Rutgers Univ, PhD(math), 62. Prof Exp: Instr math, Rutgers Univ, 60-62; from asst prof to assoc prof, 62-73, PROF MATH, UNIV DEL, 73- Mem: Am Math Soc; Math Asn Am. Res: Geometric function theory. Mailing Add: Dept of Math Univ of Del Newark DE 19711

LIBERATORE, FREDERICK ANTHONY, b Framingham, Mass, Dec 11, 44; m 68. BIOCHEMISTRY. Educ: Mass State Col Framingham, BA, 70; Univ NH, PhD(biochem), 74. Prof Exp: FEL, OHIO STATE UNIV, 74- Mem: Am Chem Soc. Res: Enzyme immobilization to solid supports; use of enzymes to hydrolyze and sequence proteins from the carboxyl terminus, especially with immobilized carboxypeptidase yttrium. Mailing Add: Dept of Biochem Ohio State Univ 484 W12th Ave Columbus OH 43210

LIBERATORE, LAURENCE COLUMBUS, b Rochester, NY, Aug 22, 13; m 52; c 2. CHEMISTRY. Educ: Univ Rochester, BS, 34, MS, 36, PhD(phys chem), 41. Prof Exp: Lab asst, Univ Rochester, 39-40; asst prof chem, Canisius Col, 40-43; res scientist, Manhattan Proj, Columbia, 43-45; res engr, 45-52, develop engr, 52-55, gen foreman in charge glass prod eng & mfg, 55-64, factory mgr, Consumer Prod Div, 64-68, VPRES MFG, CONSUMER PROD DIV, TAYLOR INSTRUMENT CO, INC, 68- Concurrent Pos: Lectr, Univ Sch, Univ Rochester, 51-62; consult glass breakage, 57- Mem: Am Chem Soc; Am Ceramic Soc; Am Soc Testing & Mat. Res: Photochemistry; high vacuum techniques; electrochemistry; spectroscopy; glass stabilization as applied to thermometry; manufacturing techniques for liquid in glass thermometry; thermometer design. Mailing Add: Consumer Prods Div Taylor Instrument Co Inc Arden NC 28704

LIBERLES, ARNO, b Aschaffenburg, Ger, July 7, 34; US citizen; m 66; c 1. ORGANIC CHEMISTRY. Educ: Univ Mass, BS, 56; Yale Univ, MS, 59, PhD(chem), 60. Prof Exp: Fel chem, Col of France, 60-61; res chemist, W R Grace & Co, 61-62; from asst prof to assoc prof, 62-71, PROF CHEM, FAIRLEIGH DICKINSON UNIV, 71- Mem: Am Chem Soc; NY Acad Sci. Res: Photooxidation of organic compounds; theoretical organic chemistry. Mailing Add: Dept of Chem Fairleigh Dickinson Univ Teaneck NJ 07666

LIBERMAN, ARTHUR DAVID, b Newark, NJ, Oct 13, 40; m 68; c 2. HIGH ENERGY PHYSICS. Educ: Dartmouth Col, AB, 62; Harvard Univ, MA, 63, PhD(physics), 69. Prof Exp: Res assoc high energy physics, Linear Accelerator Lab, Univ Paris, 69-70; adj asst prof particle physics, Univ Calif, Los Angeles, 70-74; RES PHYSICIST, HIGH ENERGY PHYSICS LAB, STANFORD UNIV, 74- Mem: Am Phys Soc. Res: The study of gamma rays and entirely neutral final states in the annihilation interactions at electron-positron storage rings by utilizing a large solid angle, good energy resolution detector now under development. Mailing Add: High Energy Physics Lab Stanford Univ Stanford CA 94305

LIBERMAN, DANIEL FRANKLIN, b Boston, Mass, Apr 19, 45; m 67; c 2. CLINICAL MICROBIOLOGY, INDUSTRIAL MICROBIOLOGY. Educ: Suffolk Univ, BS, 67; Wash Univ, MA, 70; Univ Rochester, PhD(radiation biol & biophysics), 72. Prof Exp: Fel clin microbiol, Univ Rochester, Med Ctr, 72-75; RES MICROBIOLOGIST ANTIBIOTIC DEVELOP, LEDERLE LAB, AM CYANAMID CO, 75- Mem: Sigma Xi; Am Soc Microbiol. Res: Application radioimmunochemical techniques in clinical and industrial microbiology; characterization of novel antibiotics by biophysical techniques; development of automated methods in clinical microbiology. Mailing Add: Lederle Lab Am Cyanamid Co Pearl River NY 10965

LIBERMAN, DAVID ARTHUR, b Los Angeles, Calif, Nov 8, 26; m 61. THEORETICAL PHYSICS. Educ: Calif Inst Technol, BS, 49, PhD(physics), 55. Prof Exp: STAFF MEM, LOS ALAMOS SCI LAB, 55- Mem: Am Phys Soc. Res: Atomic and solid state physics. Mailing Add: Los Alamos Sci Lab Los Alamos NM 87544

LIBERMAN, IRVING, b New York, NY, June 24, 37; m 70. LASERS, OPTICAL

PHYSICS. Educ: City Col New York, BEE, 58; Northwestern Univ, MS, 60, PhD(elect eng), 65. Prof Exp: FEL SCIENTIST OPTICAL PHYSICS, WESTINGHOUSE RES LAB, 63-; INDUST STAFF MEM, LOS ALAMOS SCI LAB, 74- Mem: Sigma Xi; Am Phys Soc; Optical Soc Am. Res: High power carbon dioxide lasers as applied to laser fusion, particularly the design of all optical components and isolators and the determination of laser beam quality. Mailing Add: MS 548 Los Alamos Sci Lab Los Alamos NM 87545

LIBERMAN, ROBERT PAUL, b Newark, NJ, Aug 16, 37; m 61, 73; c 2. PSYCHIATRY, CLINICAL PSYCHOLOGY. Educ: Dartmouth Col, AB, 59; Dartmouth Med Sch, dipl med, 60; Univ Calif, MS, 61; Johns Hopkins Univ, MD, 63. Prof Exp: Intern internal med, Bronx Munic Hosp Ctr, Albert Einstein Col Med, 63-64; res scientist, NIMH, 64-70; asst prof psychiat, Sch Med, Univ Calif, Los Angeles, 70-72, ASSOC PROF PSYCHIAT, SCH MED, UNIV CALIF, LOS ANGELES, 72-, RES PSYCHIATRIST, CAMARILLO-NEUROPSYCHIAT INST RES PROG, 70- Concurrent Pos: NIMH res grant, 67-68, appl res grant, Camarillo-Neuropsychiat Res Prog, 72-75 & hosp improvement grant, Camarillo State Hosp, 72-75; consult, behav modification, NIMH, 70-; staff psychiatrist, Ventura Gen Hosp, 70-; consult training, Los Angeles County Ment Health Dept & Calif State Dept Ment Hyg Ctr Training in Community Psychiat, 72-; dep prog leader & staff psychiatrist, Oxnard Ment Health Ctr, 72-; Fogarty Sr Res Int fel, 75. Mem: Am Psychiat Asn; Asn Advan Behav Ther. Res: Experimental analysis of behavior in clinical psychiatry and psychology; interactions between drug effects and behavior modification; community mental health. Mailing Add: Camarillo-Neuropsy Inst Res Prog Camarillo State Hosp Box A Camarillo CA 93010

LIBERS, ROLAND, organic chemistry, see 12th edition

LIBERSON, WLADIMIR THEODORE, b Kieff, Russia, Aug 2, 04; m 29; c 2. NEUROPHYSIOLOGY. Educ: Univ Paris, MD, 36; Univ Montreal, PhD(physiol), 50. Prof Exp: Dir physiol res lab, Inst Living, Conn, 41-48; prof neurophysiol, Univ Conn, 51-57; prof therapeut & pharmacol & chief sect phys med, Stritch Sch Med, 59-73, prof pharmacol, Grad Sch, Loyola Univ Chicago, 70-73. Concurrent Pos: Consult, Hartford Hosp, Conn, 48-54, St Francis Hosp, 48-59, Newington Home Crippled Children, 48-59, Mt Sinai Hosp, 54-59, Mercy & Springfield Hosps, Mass, 54-59 & Providence Hosp, 58-59; dir EEG res lab, Vet Admin Hosp, Northampton, Mass, 53-59; chief phys med & rehab & dir neurophysiol lab, Vet Admin Hosp, Chicago, 59-73. Honors & Awards: Prix Bouchard Award, Biol Soc Paris, 38. Mem: Am Electroencephalog Soc (secy, 53-56, pres, 59); Asn Res Nerv & Ment Dis; Am Physiol Soc; Am Soc Med Psychiat. Res: Neuropharmacology; electroencephalography and electromyography; experimental psychology; psychiatry; physical medicine. Mailing Add: 3262 NE 166th St North Miami Beach FL 23160

LIBERTA, ANTHONY E, b La Salle, Ill, May 17, 33; m 60; c 2. MYCOLOGY. Educ: Knox Col, Ill, AB, 55; Univ Ill, Urbana, MS, 59, PhD(bot), 61. Prof Exp: Res mycologist, Ill State Natural Hist Surv, 61; assoc prof, 61-67, PROF MYCOL, ILL STATE UNIV, 67- Concurrent Pos: NSF grants, 63-71. Mem: Am Soc Plant Taxon; Int Asn Plant Taxon; Mycol Soc Am. Res: Biogeography and taxonomy of resupinate basidiomycetes. Mailing Add: Dept of Biol Ill State Univ Normal IL 61761

LIBERTI, ALFRED VINCENT, b New York, NY, Dec 6, 16; m 55; c 2. VERTEBRATE ANATOMY, VERTEBRATE PHYSIOLOGY. Educ: Niagara Univ, BS, 39; St John's Univ, NY, MS, 41; Fordham Univ, PhD(biol), 49. Prof Exp: Asst, 39-41, from actg chmn to chmn, Univ Col, 44-58, dept rep, 58-69, chmn dept, 69-74, PROF BIOL, ST JOHN'S UNIV, NY, 58- Concurrent Pos: Comt mem, Col Proficiency Exam Anat & Physiol, NY State Dept Educ, 73- Mem: AAAS; Am Soc Zool; Sigma Xi; Am Inst Biol Sci. Res: Anatomical sciences; vertebrate anatomy; physiology. Mailing Add: Dept of Biol Sci St John's Univ Grand Central & Utopia Pkwy Jamaica NY 11432

LIBERTI, FRANK NUNZIO, b Warsaw, NY, Nov 2, 39; m 66; c 1. POLYMER CHEMISTRY, PHYSICAL CHEMISTRY. Educ: Rensselaer Polytech Inst, BChE, 61, PhD(phys chem), 67. Prof Exp: Develop chemist, 67-69, specialist prod develop, 69-73, mgr qual assurance, 73-75, SPECIALIST ADVAN DEVELOP, PLASTICS DEPT, GEN ELEC CO, 75- Mem: Am Chem Soc; Soc Plastics Eng; Am Soc Qual Control. Res: Stabilization of polymers; flame retardant polymers; solid state of polymers; polymer crystallinity; thermal analysis of polymers. Mailing Add: Plastics Dept Gen Elec Co Lexan Lane Mt Vernon IN 47620

LIBERTI, PAUL A, b Lyndhurst, NJ, Mar 18, 36; m 61; c 4. IMMUNOLOGY, PHYSICAL BIOCHEMISTRY. Educ: Columbia Col, AB, 59; Loyola Univ, Ill, MS, 61; Stevens Inst Technol, PhD(phys chem), 66. Prof Exp: From instr to assocprof, 67-76, PROF BIOCHEM, JEFFERSON MED COL, 76- Concurrent Pos: Res fel phys chem, Stevens Inst TEchnol, 66; res fel biochem, NJ Col Med & Dent, 66-67; Nat Inst Allergy & Infectious Dis res career develop award, 73; lectr, Fairleigh Dickinson Univ, 64-67 & Temple Univ, 67-; adv ed, Immunochemistry, 76-81. Honors & Awards: Co-recipient Ottens Res Award, 69. Mem: Am Asn Immunol; Am Asn Biol Chem. Res: Physical chemistry of antigen-antibody interactions and complement proteins; proteins and polyelectrolytes; design of biophysical instruments. Mailing Add: Jefferson Med Col 1020 Locust St Philadelphia PA 19107

LIBERTY, BRUCE ARTHUR, b Toronto, Ont, Sept 29, 19; m 45; c 3. STRATIGRAPHY. Educ: Univ Toronto, BA, 47, MA, 49, PhD(geol), 53. Prof Exp: Instr geol, Univ Toronto, 47-51; sr res geologist, Int Petrol, Ltd, Peru, 51-52; tech officer, Geol Surv Can, 52-53, geologist, 53-66; assoc prof geol, Univ Guelph, 66-68; prof geol, 69-75, chmn dept geol sci, 73-75, PROF GEOL SCI, BROCK UNIV, 75- Concurrent Pos: Asst, Royal Ont Mus, 47-51. Mem: Geol Soc Am; Paleont Soc; Am Asn Petrol Geol; fel Geol Asn Can; Soc Econ Paleontologists & Mineralogists. Res: Stratigraphy; stratigraphic paleontology; paleozoic stratigraphy and biostratigraphy in Southern Ontario Canada. Mailing Add: Dept of Geol Sci Brock Univ St Catharines ON Can

LIBET, BENJAMIN, b Chicago, Ill, Apr 12, 16; m 39; c 4. PHYSIOLOGY. Educ: Univ Chicago, BS, 36, PhD(physiol), 39. Prof Exp: Asst physiol, Univ Chicago, 37-39; instr, Albany Med Col, 39-40; res assoc physiol & biochem, Inst Pa Hosp, 40-43; instr physiol, Sch Med, Univ Pa, 43-44; mat engr, Personal Equip Lab, US Air Force, Ohio, 44-45; from instr to asst prof physiol, Univ Chicago, 45-48; staff physiologist, Kabat-Kaiser Inst, 48-49; from asst prof to assoc prof, 49-62, PROF PHYSIOL, MED SCH, UNIV CALIF, SAN FRANCISCO, 62- Concurrent Pos: Consult, Mt Zion Neurol Inst, 56- Mem: AAAS; Am Physiol Soc; Soc Neurosci; Int Brain Res Orgn. Res: Neurophysiology; electrical and metabolic aspects of neural function; synaptic mechanisms; cerebral mechanisms in sensation. Mailing Add: Dept of Physiol S-762 Univ of Calif Med Ctr San Francisco CA 94122

LIBOFF, ABRAHAM R, b Paterson, NJ, Aug 27, 27; m 52; c 1. MEDICAL PHYSICS, BIOPHYSICS. Educ: Brooklyn Col, BS, 48; NY Univ, MS, 52, PhD(physics), 64. Prof Exp: Jr physicist, Naval Ord Lab, Md, 48-50; sr physicist, Metall Res Lab, Sylvania Elec Prod, Inc, 51-58; res asst cosmic ray lab, NY Univ, 59-64, assoc res scientist, 64-68, assoc dir environ radiation lab, 68-69, sr res scientist & proj coordr, Biophys Res Lab, 69-72; PROF PHYSICS & CHMN DEPT, OAKLAND UNIV, 72-; DIR MED PHYSICS PROG, 73- Concurrent Pos: Adj assoc prof physics, Hunter Col, NY, 68-72. Mem: Am Phys Soc; Biophys Soc; Am Geophys Union; Am Asn Physicists in Med. Res: Physics of collagenous tissues; biophysics of growth and development; electrically induced osteogenesis; standard measurement on sea-level cosmic ray ionization; environmental radiation; pyroelectric properties of bone. Mailing Add: 345 Berwyn Rd Birmingham MI 48009

LIBOFF, RICHARD L, b New York, NY, Dec 30, 31; m 54; c 2. THEORETICAL PHYSICS. Educ: Brooklyn Col, AB, 53; NY Univ, PhD(physics), 61. Prof Exp: Res asst appl math, Courant Inst Math Sci, NY Univ, 56-61, asst prof physics, NY Univ, 62-64; assoc prof, 64-69, PROF ELEC ENG & APPL PHYSICS, CORNELL UNIV, 69-, PRIN INVESTR, OFF NAVAL RES CONTRACT, 66- Concurrent Pos: Chief consult, NRA, Inc, 63-; Solvay fel, Univ Brussels, 71. Mem: AAAS; fel Am Phys Soc. Res: Nonequilibrium statistical mechanics; electrodynamics; quantum physics; plasma physics. Mailing Add: Dept of Appl Physics Clark Hall Cornell Univ Ithaca NY 14850

LIBONATI, JOSEPH PETER, b Philadelphia, Pa, Nov 16, 41; m 69; c 1. CLINICAL MICROBIOLOGY. Educ: St Joseph's Col, Pa, BS, 63; Duquesne Univ, MS, 65; Univ Md, Baltimore City, PhD(microbiol), 68. Prof Exp: Instr, 68-70, ASST PROF CLIN MICROBIOL, SCH MED, UNIV MD, BALTIMORE CITY, 70-, SPEC LECTR MICROBIOL, SCH DENT, 69- Mem: AAAS; Am Soc Microbiol; NY Acad Sci. Res: Enteric bacterial diseases; pathophysiology; immunologic response and vaccine development. Mailing Add: Div of Infectious Dis Univ of Md Sch of Med Baltimore MD 21201

LIBOWITZ, GEORGE GOTTHART, b Brooklyn, NY, June 18, 23; m 49; c 2. SOLID STATE CHEMISTRY. Educ: Brooklyn Col, BA, 45, MA, 50; Cornell Univ, PhD(phys chem), 54. Prof Exp: Chemist, Chromium Corp Am, 45-46, R Kann Chem Lab, 47-48 & Picatinny Arsenal, US Dept Army, 49; asst physics, Cornell Univ, 49-53; sr engr chem, Sylvania Elec Prod, Inc, 54; res assoc, Tufts Univ, 54-57; res supvr, Atomics Int Div, NAm Aviation, Inc, 57-61; sect head, Mat Sci Lab, Aerospace Corp, 61-63; staff scientist, Ledgemont Lab, Kennecott Copper Corp, 63-73; MGR SOLID STATE CHEM DEPT, MAT RES CTR, ALLIED CHEM CORP, 73- Mem: AAAS; Am Chem Soc; Am Phys Soc; Am Inst Chem; NY Acad Sci. Res: Solid state chemistry; metal hydrides; nonstoichiometric compounds; materials for energy conversion and storage. Mailing Add: Mat Res Ctr Allied Chem Corp Morristown NJ 07960

LICARI, JAMES JOHN, b Norwalk, Conn, July 22, 30; m 59. ORGANIC CHEMISTRY. Educ: Fordham Univ, BS, 52; Princeton Univ, PhD, 55. Prof Exp: Res chemist, Am Cyanamid Co, 55-57; res proj chemist, Am Potash & Chem Corp, 57-59; sr res engr, NAm Aviation, Inc, 59-61, supvr org chem, 61-67, group scientist, Res & Eng Div, NAm Rockwell Corp, 67-70, supvr chem lab, 70-72, SUPVR MICROCIRCUIT ENG LABS, ROCKWELL INT CORP, ANAHEIM, 72- Concurrent Pos: Asst prof, Fordham Univ, 55-56. Mem: Am Chem Soc. Res: Materials and processes for microelectronics. Mailing Add: 15711 Arbela Dr Whittier CA 90603

LICASTRO, PASQUALE HALLISON, geophysics, see 12th edition

LICH, ROBERT, JR, b Sutton, Nebr, Feb 8, 09; m 41; c 3. UROLOGY. Educ: Univ Calif, AB, 30; Long Island Col Med, MD, 36; Univ Louisville, MS, 41. Prof Exp: From instr to prof urol & chmn sect, Sch Med, Univ Louisville, 48-74; RETIRED. Concurrent Pos: Consult, Second Army Sta Hosp, Ft Knox, Ky, 46- & Vet Admin Hosp, 46- Honors & Awards: Ostermann Res Prize, 40. Mem: Am Urol Asn; Am Asn Genito-Urinary Surg; Clin Soc Genito-Urinary Surg; Int Soc Urol. Res: Renal function; congenital obstructive uropathies; urologic surgery. Mailing Add: 9402 US Hwy 42 Prospect KY 40059

LICHSTEIN, HERMAN CARLTON, b New York, NY, Jan 14, 18; m 42; c 2. MEDICAL MICROBIOLOGY. Educ: NY Univ, AB, 39; Univ Mich, MS, 40, ScD(bact), 43; Am Bd Microbiol, dipl. Prof Exp: Asst bact, Univ Mich, 40-43; instr, Univ Wis, 43-46; Nat Res Coun fel, Cornell Univ, 46-47; from assoc prof to prof, Univ Tenn, 47-50; from assoc prof to prof, Univ Minn, 50-61; PROF MICROBIOL & DIR DEPT, COL MED, UNIV CINCINNATI, 61-, DIR GRAD STUDIES, 62- Concurrent Pos: Lewis lectr; Novy lectr; fel, Grad Sch, Univ Cincinnati; consult, Carbide & Carbon Chem Corp, Oak Ridge Nat Lab, 48-54 & Vet Admin Hosp, 57-65; consult ed, Life Sci Series, Burgess Pub Co, 60-67; mem sci fac fel panel, NSF, 60-63, Microbiol Training Comt, Nat Inst Gen Med Sci & Adv Bd Methods Biochem Anal; mem microbiol fels rev comt, NIH; trustee, Asn Med Sch Microbiol chmn, 74-77. Mem: Fel AAAS; Am Soc Biol Chem; Am Soc Microbiol; Soc Exp Biol & Med; fel Am Acad Microbiol. Res: Microbial physiology and metabolism. Mailing Add: 2108 Med Sci Bldg Univ of Cincinnati Col of Med Cincinnati OH 45267

LICHT, ARTHUR LEWIS, b Hartford, Conn, Dec 18, 34; m 58. PHYSICS. Educ: Brown Univ, BSc, 57; Univ Md, PhD(physics), 63. Prof Exp: Physicist, Nat Bur Standards, 57-59; res physicist, NASA, 59-61 & US Naval Ord Lab, 61-70; MEM FAC, DEPT PHYSICS, UNIV ILL, 70- Concurrent Pos: Asst prof, Univ Md, 63-65; mem sch math, Inst Adv Study, 65-66. Mem: Am Phys Soc. Res: Space physics; quantum field theory. Mailing Add: Dept of Physics Univ of Ill Box 4348 Chicago IL 60680

LICHT, PAUL, b St Louis, Mo, Mar 12, 38; m 63. ZOOLOGY. Educ: Washington Univ (Mo), AB, 59; Univ Mich, MS, 61, PhD(zool), 64. Prof Exp: From asst prof to assoc prof, 64-73, PROF ZOOL, UNIV CALIF, BERKELEY, 73- Concurrent Pos: Lalor Found grant, 67-68; NSF grants, 64-72; consult ed, Col Div, McGraw-Hill Book Co, 68- Mem: Fel AAAS; Soc Study Reproduction; Ecol Soc Am; Am Soc Zool; Am Soc Ichthyol & Herpet. Res: Physiological adaptations to various aspects of the physical environment among the vertebrates; vertebrate endocrinology. Mailing Add: Dept of Zool Univ of Calif Berkeley CA 94720

LICHT, SIDNEY, b New York, NY, Apr 18, 07; m 37; c 2. PHYSICAL MEDICINE. Educ: City Col New York, BS, 27; NY Univ, MD, 31. Prof Exp: Lectr phys med, Columbia Univ, 35-42; lectr, Boston Univ, 47-53; asst clin prof, Yale Univ, 57-63, cur, Med Libr, 61-72; HON PROF MED LIBR SCI, UNIV MIAMI, 72- Concurrent Pos: Lectr, Tufts Univ, 47-53; Coulter lectr, 70; Zeiter lectr, 72; ed, Phys Med Libr. Honors & Awards: Gold Key, Am Cong Rehab Med, 69; Krusen Gold Medal, Am Acad Phys Med & Rehab, 74. Mem: AMA; Am Cong Rehab Med (pres, 68); Int Rehab Med Asn (secy, 70). Mailing Add: 4601 University Dr Coral Gables FL 33146

LICHTEN, WILLIAM LEWIS, b Philadelphia, Pa, Mar 5, 28; m 50; c 3. PHYSICS. Educ: Swarthmore Col, BA, 49; Univ Chicago, MS, 53, PhD(physics), 56. Prof Exp: NSF fel, 56-57; res physicist, Radiation Lab, Columbia Univ, 57-58; from asst prof to assoc prof physics, Univ Chicago, 58-64; dir undergrad studies, 69-71, PROF PHYSICS, YALE UNIV, 64- Mem: Fel Am Phys Soc. Res: Psychology of perception;

biophysics; chemical physics; atomic physics. Mailing Add: Dept of Physics Yale Univ New Haven CT 06520

LICHTENBAUM, STEPHEN, b Brooklyn, NY, Aug 24, 39; m 61; c 4. NUMBER THEORY. Educ: Harvard Univ, AB, 60, AM, 61, PhD(math), 64. Prof Exp: Lectr math, Princeton Univ, 64-67; from asst prof to assoc prof, 67-73, PROF MATH, CORNELL UNIV, 73- Concurrent Pos: Guggenheim fel, John Simon Guggenheim Mem Found, 73-74. Mem: Am Math Soc. Res: Algebraic number theory and algebraic geometry, particularly the study of the values of zeta and L-functions. Mailing Add: Dept of Math Cornell Univ Ithaca NY 14853

LICHTENBERG, DON BERNETT, b Passaic, NJ, July 2, 28; m 54; c 2. THEORETICAL PHYSICS. Educ: NY Univ, BA, 50; Univ Ill, MS, 51, PhD(physics), 55. Prof Exp: Res assoc physics, Ind Univ, 55-57; guest prof, Univ Hamburg, 57-58; from asst prof to assoc prof, Mich State Univ, 58-63; physicist, Linear Accelerator Ctr, Stanford Univ, 62-63; assoc prof, 63-66, PROF PHYSICS, INC UNIV, BLOOMINGTON, 66- Concurrent Pos: Vis prof, Tel-Aviv Univ, 67-68 & Imp Col, Univ London, 71. Mem: Fel Am Phys Soc. Res: Physics of the elementary particles. Mailing Add: Dept of Physics Ind Univ Bloomington IN 47401

LICHTENBERG, FRANZ VON, b Miskolc, Hungary, Nov 29, 19; nat US; m 49; c 6. PATHOLOGY. Educ: Nat Univ Mex, MD, 45; Am Bd Path, dipl, 51. Hon Degrees: Dr, Nat Univ Nicaragua, 59; MA, Harvard Univ, 68. Prof Exp: Pathologist, Hosp Exp Nutrit, Mex, 47; prof path, Nat Univ Mex, 48-52; from asst prof to assoc prof, Univ PR, 53-58; instr, 58-59, assoc, 59-62, from asst prof to assoc prof, 62-74, PROF PATH, HARVARD MED SCH, 74- Concurrent Pos: Fel, Mex Dept Health, 46; Kellogg Found, Am Col Physicians & Latin Am fels, 50-51; pathologist, Clin Hosp, Bahia, Brazil, 51-52, Gen Hosp, Mex, 52-53 & San Juan City Hosp, PR, 53-58; assoc pathologist, Peter Bent Brigham Hosp, 58-62, sr assoc, 62-68, pathologist, 68-; assoc mem comn parasitol, Armed Forces Epidemiol Bd, 59, mem, 64-71; consult, Div Parasitic Dis, WHO, 65; mem study sect trop med & parasitol, NIH, 68-73; mem parasitic dis panel, USJapan Coop Med Sci Prof, 71-74, chmn, 74- Mem: Am Soc Trop Med & Hyg; Am Asn Path & Bact; Fedn Am Socs Exp Biol. Res: Tropical and parasitic diseases; schistosomiasis; filariasis; liver pathology; immunopathology. Mailing Add: 721 Huntington Ave Boston MA 02115

LICHTENBERGER, HAROLD V, b Decatur, Ill, Apr 22, 20; m 43; c 5. NUCLEAR PHYSICS. Educ: Millikin Univ, AB, 42. Prof Exp: From mem staff to dir, Idaho Div, Metall Lab & Argonne Nat Lab, 42-56; vpres, Gen Nuclear Eng Corp, 56-61; asst div dir & mgr mfg, Nuclear Div, 61-69, dir nuclear prod mfg div, 69-74, VPRES MFG NUCLEAR POWER SYSTS, COMBUSTION ENG, INC, 74- Mem: Fel Am Nuclear Soc. Res: Design and use of nuclear reactors for production of heat and electrical power. Mailing Add: Nuclear Prod Mfg Div Combustion Eng Inc PO Box 500 Windsor CT 06095

LICHTENFELS, JAMES RALPH, b Robinson, Pa, Feb 14, 39; m 61; c 2. PARASITOLOGY, TAXONOMY. Educ: Ind Univ Pa, BS, 62; Univ Md, MS, 66, PhD(zool), 68. Prof Exp: ZOOLOGIST, ANIMAL PARASITOL INST, AGR RES SERV, USDA, 67-, CUR NAT PARASITE COLLECTION, AGR RES CTR, 71- Concurrent Pos: Instr, USDA Grad Sch, 71-; res assoc, Div Worms, Mus Natural Hist, Smithsonian Inst, Wash, DC, 72-; res affiliate, Div Parasitol, State Mus, Univ Nebr, Lincoln, 72-; mem coun resources, Asn Syst Collections, 75- Mem: Am Soc Parasitol; AAAS; Sigma Xi; Wildlife Dis Asn; Am Micros Soc. Res: Intra and interspecific variation in parasitic nematodes; effects of host on morphology of parasitic nematodes; identification, classification and description of parasitic nematodes. Mailing Add: 12311 Whitehall Dr Bowie MD 20715

LICHTENSTEIN, EMANUEL PAUL, b Selters, WGer, Feb 24, 15; nat US; m 51; c 2. ENTOMOLOGY. Educ: Hebrew Univ, Israel, MSc, 41, PhD(entom, biochem), 48. Prof Exp: Lectr biol, Sch Educ, Israel, 41-53; asst prof physiol & anat, Ill Wesleyan Univ, 53-54; proj assoc, 54-56, from asst prof to assoc prof, 56-65, chmn dept, 67-71, PROF ENTOM, UNIV WIS-MADISON, 65-, ASSOC DIR, CTR ENVIRON TOXICOL, 72- Mem: Entom Soc Am; Am Chem Soc; Soc Toxicol. Res: Pesticidal residues and their effect on the biological complex on our environment; factors affecting persistence and breakdown of pesticides in soils, crops and water; naturally occurring toxicants. Mailing Add: Dept of Entom Univ of Wis Madison WI 53706

LICHTENSTEIN, IVAN EDGAR, b Brooklyn, NY, Dec 17, 33; m 63; c 2. ANALYTICAL CHEMISTRY, INORGANIC CHEMISTRY. Educ: Columbia Univ, AB, 55; Univ Calif, Davis, PhD(chem), 60. Prof Exp: Res assoc chem, Univ Kans, 60-61; asst prof, Villanova Univ, 61-66; RES CHEMIST, CORNING GLASS WORKS, 66- Mem: Am Chem Soc; Am Inst Chem. Res: Classical and instrumental silicate analysis; coordination chemistry; especially structure of and bonding in transition metal complexes. Mailing Add: Chem Anal Res Dept Corning Glass Works Corning NY 14830

LICHTENSTEIN, LAWRENCE M, b Washington, DC, May 31, 34; m 56; c 3. MEDICINE, IMMUNOLOGY. Educ: Univ Chicago, BA, 54, MD, 60; Johns Hopkins Univ, PhD(immunol), 65. Prof Exp: Intern med, 60-61, fel microbiol, 61-65, resident, 65-66, asst prof, 66-70, ASSOC PROF MED, SCH MED, JOHNS HOPKINS UNIV, 70- Mem: Am Acad Allergy; Am Asn Immunol; Am Fedn Clin Res; Am Soc Clin Invest. Res: Mechanisms of reactions of immediate hypersensitivity and relationship to clinical problems. Mailing Add: Dept of Med Johns Hopkins Univ Sch of Med Baltimore MD 21218

LICHTENSTEIN, PEARL RUBENSTEIN, b Boston, Mass, June 13, 17; m 46; c 2. ASTROPHYSICS. Educ: Mass Inst Technol, SB, 38; Radcliffe Col, MA, 40, PhD(astrophys), 42. Prof Exp: Mem staff, Radiation Lab, Mass Inst Technol, 42-46; res assoc physics, Franklin Inst, Pa, 46-47; engr, Gen Eng Lab, Gen Elec Co, 48; res assoc astron, Rensselaer Polytech Inst, 57-69; asst prof, Schenectady County Community Col, 70-73; consult, NY State Educ Dept, 74; vis asst prof physics, Union Col, NY, 75-76. Mem: Am Astron Soc. Res: Stellar spectra; solar physics; radio astronomy; microwave propagation; solar-terrestrial relations; aurora. Mailing Add: 1200 Van Antwerp Rd Schenectady NY 12309

LICHTENSTEIN, ROLAND MAX, b Darmstadt, Ger, May 12, 14; nat US; m 46; c 2. PHYSICS. Educ: Darmstadt Tech Hochsch, Ger, dipl, 38; Mass Inst Technol, PhD(physics), 47. Prof Exp: Instr physics, Reed Col, 42-44 & Univ Pa, 46-47; develop engr, Gen Elec Co, 47-55; PROF PHYSICS, RENSSELAER POLYTECH INST, 55- Concurrent Pos: Consult, Gen Elec Co, 55- Mem: Am Phys Soc. Res: Mathematical physics; electro-magnetic theory and devices. Mailing Add: 1200 Van Antwerp Rd Schenectady NY 12309

LICHTENWALTER, GLEN, b Columbus, Kans, Oct 7, 31; m 58; c 3. ORGANIC CHEMISTRY. Educ: Univ Tex, BS, 54; Iowa State Col, PhD(org chem), 58. Prof Exp: Res chemist, Shell Chem Corp, Tex, 58-64, NJ, 64-67; asst res dir, Armour Indust Chem Co, 67-74, ASSOC DIR RES, ARMAK CO, 74- Mem: Am Chem Soc; Am Oil Chem Soc. Res: Organic syntheses; fats and oils. Mailing Add: 9S 758 Lorraine Hinsdale IL 60521

LICHTENWALTER, MYRL, organic chemistry, see 12th edition

LICHTER, EDWARD A, b Chicago, Ill, June 5, 28; m 52; c 2. PREVENTIVE MEDICINE, COMMUNITY HEALTH. Educ: Univ Chicago, PhB, 47; Roosevelt Univ, BS, 49; Univ Ill, MS, 51, MD, 55. Prof Exp: Asst physiol, Col Med, Univ Ill, 50-51, resident internal med, 58-61, instr med, 60-61; USPHS fel immunochem, Nat Inst Allergy & Infectious Dis, 61-63, mem staff, 63-66; assoc prof, 66-68, PROF PREV MED & HEAD DEPT, COL MED, UNIV ILL MED CTR, 68-, PROF HEALTH CARE SERV & HEAD DEPT, SCH PUB HEALTH, 72- Mem: AAAS. Res: Radiation effects on peripheral circulation; chronic pulmonary infections; clinical pharmacology and therapeutic evaluation of antibiotics; immunochemistry; immunogenetics of immunoglobulins and other serum proteins; structure and function of health care services. Mailing Add: Dept of Prev Med Col of Med Univ of Ill at the Med Ctr Chicago IL 60680

LICHTER, JAMES JOSEPH, b Algona, Iowa, Apr 29, 39. MOLECULAR PHYSICS, COMPUTER SCIENCE. Educ: Loras Col, BS, 61; Fordham Univ, MS, 63; Duke Univ, PhD(physics), 69. Prof Exp: Physicist, US Naval Ord Lab, Md, 60-65, res physicist, 65; res asst, Duke Univ, 65-69; ASSOC SCIENTIST, ITT FED ELEC CORP, 69- Mem: Am Phys Soc; Am Chem Soc. Res: Radiation damage to bases of DNA; molecular biophysics; electron paramagnetic resonance; nuclear magnetic resonance; quantum biochemistry; missile range instrumentation; radar; computer simulation models; lasers; electrooptical systems; atmospheric processes. Mailing Add: ITT Fed Elec Corp PO Box 1886 Vandenberg AFB CA 93437

LICHTER, ROBERT (LOUIS), b Cambridge, Mass, Oct 26, 41; m 66; c 1. ORGANIC CHEMISTRY. Educ: Harvard Univ, AB, 62; Univ Wis-Madison, PhD(chem), 67. Prof Exp: USPHS fel, Brunswick Tech Univ, 67-68; res fel chem, Calif Inst Technol, 68-70; asst prof, 70-74, ASSOC PROF CHEM, HUNTER COL, 75- Mem: Am Chem Soc; The Chem Soc; NY Acad Sci. Res: Organonitrogen chemistry; nuclear magnetic resonance spectroscopy; application of carbon and nitrogen nuclear magnetic resonance to organic chemistry. Mailing Add: Dept of Chem Hunter Col 695 Park Ave New York NY 10021

LICHTI-FEDEROVICH, SIGRID, ecology, see 12th edition

LICHTIN, J LEON, b Philadelphia, Pa, Mar 5, 24; m 50; c 2. PHARMACEUTICAL CHEMISTRY, COSMETIC CHEMISTRY. Educ: Philadelphia Col Pharm, BS, 44, MS, 47; Ohio State Univ, PhD(pharmaceut chem), 50. Prof Exp: Asst prof, Cincinnati Col Pharm, 50-51, assoc prof pharm, 51-55; from assoc prof to prof, 55-71, ANDREW JERGENS PROF PHARM, UNIV CINCINNATI, 72- Mem: AAAS; fel Soc Cosmetic Chem; Am Pharmaceut Asn. Res: Dermatologicals; formulation of pharmaceutical products; cosmetics. Mailing Add: Dept of Pharm Univ of Cincinnati Cincinnati OH 45221

LICHTIN, NORMAN NAHUM, b Newark, NJ, Aug 10, 22; m 47; c 3. PHYSICAL ORGANIC CHEMISTRY. Educ: Antioch Col, BS, 44; Purdue Univ, MS, 45; Harvard Univ, PhD(phys org chem), 48. Prof Exp: Teaching fel, Harvard Univ, 45-47; lectr, 47, from instr to prof, 48-73, UNIV PROF CHEM & CHMN DEPT, 73- Concurrent Pos: Vis chemist, Brookhaven Nat Lab, 57-58, res collab, 58-70; NSF sr fel, 62-63; guest scientist, Weizmann Inst, 62-63; vis prof, Hebrew Univ Jerusalem, 62-63 & 70-72. Mem: Fel AAAS; Am Chem Soc; fel Am Inst Chem; Radiation Res Soc; Int Solar Energy Soc. Res: Radiation chemistry; atomic nitrogen chemistry; photochemical conversion of solar energy. Mailing Add: 195 Morton St Newton Centre MA 02159

LICHTMAN, DAVID, b New York, NY, Feb 7, 27; m 48; c 3. SURFACE PHYSICS. Educ: City Col New York, BS, 49; Columbia Univ, MS, 50. Prof Exp: Physicist, Airborne Instruments Lab, 50-56; res engr, Sperry Gyroscope Co, 56-62; sr prin res scientist, Honeywell Res Ctr, 62-67; assoc prof, 67-70, PROF PHYSICS, UNIV WIS, MILWAUKEE, 70- Concurrent Pos: NATO sr sci fel, 71. Mem: AAAS; Am Phys Soc; Am Vacuum Soc. Res: Mass spectrometry; beam-surface interactions; thin films; metal-ceramic seals; dark trace tubes; gaseous discharge phenomena; high and ultra-high vacuum; surface physics; electron spectroscopy; photodesorption. Mailing Add: Dept of Physics Univ of Wis Milwaukee WI 53201

LICHTMAN, HERBERT CHARLES, b New York, NY, Sept 6, 21; m 46; c 3. INTERNAL MEDICINE, CLINICAL PATHOLOGY. Educ: Brooklyn Col, BA, 42; Long Island Col Med, MD, 45; Am Bd Internal Med, dipl, 53. Prof Exp: Intern, Long Island Col Serv, Kings County Hosp Ctr, 45-46; asst resident path, Montefiore Hosp, Bronx, NY, 48-49; asst resident med, Long Island Col Div, Kings County Hosp Ctr, 49-50; from instr to prof med, Col Med, State Univ NY Downstate Med Ctr, 51-70; PROF MED, BROWN UNIV, 70- Concurrent Pos: Res fel clin med, Long Island Col Med, 50; clin fel hemat, Col Med, Univ Utah, 50-51; clin asst vis physician, Kings County Hosp, 51-53, asst attend physician, 53-59, attend physician, 59-; chief hemat & blood bank, State Univ NY Hosp, 66-70; dir div clin path, Dept Lab Med & chief div lab med, Miriam Hosp, 70-74, physician-in-chief, 74- Mem: AAAS; Am Soc Hemat; Soc Exp Biol & Med; Am Fedn Clin Res; Harvey Soc. Res: Hematology; leukemia and malignant lymphona; heme synthesis. Mailing Add: Miriam Hosp Dept of Lab Med 164 Summit Ave Providence RI 02906

LICHTMAN, IRWIN A, b New York, NY, Nov 3, 20; m 48; c 1. PHYSICAL CHEMISTRY. Educ: City Col New York, BS, 43; NY Univ, MS, 48, PhD(phys chem), 53. Prof Exp: Instr chem, Seton Hall Col, 47-48; asst prof, Community Col, NY, 48-52; sr res chemist, Lever Bros Res Ctr, 52-55; group leader phosphates & detergents, Food Mach & Chem Co, 55-60; sr res chemist, Shell Chem Co, 60-64; MGR PHYS CHEM LAB, PROCESS CHEM DIV, DIAMOND SHAMROCK CHEM CO, MORRISTOWN, 64- Mem: Am Chem Soc; Am Inst Chem. Res: Surface and colloid chemistry; defoamers; insecticide decomposition mechanisms; reaction kinetics; mechanism of defoamer action, particularly role of hydrophobic particles. Mailing Add: 773 River Dell Rd Oradell NJ 07649

LICHTMAN, MARSHALL A, b New York, NY, June 23, 34; m 57; c 3. HEMATOLOGY, BIOPHYSICS. Educ: Cornell Univ, AB, 55; Univ Buffalo, MD, 60; Am Bd Internal Med, dipl, 67. Prof Exp: Resident internal med, Med Ctr, Univ Rochester, 60-63; res assoc epidemiol, Sch Pub Health, Univ NC, 63-65; instr med, Sch Med & chief resident, Med Ctr, 65-66, sr instr med, Sch Med, 66-67, asst prof, 68-71, assoc prof med, Radiation Biol & Biophys, 71-74, PROF MED, RADIATION BIOL & BIOPHYS, SCH MED, UNIV ROCHESTER, 74- CHIEF HEMATOL UNIT, 75- Concurrent Pos: USPHS res fel, Univ Rochester, 67-69; Leukemia Soc scholar, 69-74; from asst physician to sr assoc physician, Strong Mem Hosp, 65-71, sr physician, 74- Mem: Fel Am Col Physicians; Am Soc Hemat; Am Soc Clin Invest; Asn AmPhysicians; Am Physiol Soc. Res: Biochemical and biophysical studies of human erythrocytes and leukocytes. Mailing Add: Dept of Med Sch of Med Univ of Rochester Rochester NY 14642

LICHTON, IRA JAY, b Chicago, Ill, Sept 18, 28; m 49; c 2. PHYSIOLOGY. Educ: Univ Chicago, PhB, 47; Univ Ill, BS, 50, MS, 51, PhD(physiol), 54. Prof Exp: Res assoc obstet & gynec, Univ Chicago, 54-56; Am Heart Asn res fel cardiovasc physiol, Med Res Inst, Michael Reese Hosp, Chicago, Ill, 56-58; instr physiol, Stanford Univ, 58-62; assoc prof, 62-68, PROF NUTRIT, UNIV HAWAII, 68- Mem: AAAS; Am Physiol Soc; Soc Study Reprod; NY Acad Sci. Res: Reproduction, water and electrolyte metabolism in pregnancy; renal physiology; appetite control. Mailing Add: Dept Food & Nutrit Sci Univ of Hawaii 1825 Edmondson Rd Honolulu HI 96822

LICHTWARDT, ROBERT WILLIAM, b Rio de Janeiro, Brazil, Nov 27, 24; US citizen; m 51; c 2. MYCOLOGY. Educ: Oberlin Col, AB, 49; Univ Ill, MS, 51, PhD(bot), 54. Prof Exp: Fel, NSF, 54-55; res assoc bot, Iowa State Univ, 55-57, asst prof, 57; from asst prof to assoc prof, 57-65, chmn dept, 71-74, PROF BOT, UNIV KANS, 65- Concurrent Pos: NSF sr fel, 63-64; ed-in-chief, Mycologia, 65-70. Mem: AAAS; Bot Soc Am; Mycol Soc Am (pres, 71-72); Mycol Soc Japan. Res: Fungi association with arthropods, particularly those inhabiting their guts; deterioration fungi. Mailing Add: Dept of Bot Univ of Kans Lawrence KS 66045

LICHY, CHARLES THORNE, b Pittsburgh, Pa, Nov 4, 31; m 53; c 3. AGRICULTURAL CHEMISTRY. Educ: Pa State Univ, BS, 53; Rutgers Univ, PhD(plant nutrit), 56. Prof Exp: Res chemist, Monsanto Chem Co, 56-58; starch chemist, Anheuser-Busch, Inc, 58-60; dist agronomist, Calif Chem Co, 60-63; regional tech specialist, Dow Chem USA, 63-66, mgr US herbicide develop, 66-68, res contract specialist, 68-73, sr res specialist, Toxicol Res Lab, 73-75, AREA RES & DEVELOP SPECIALIST, DOW CHEM CO, 75- Mem: AAAS; Weed Sci Soc Am. Res: Agricultural chemicals; crop production; nutrition of plants and animals; agricultural chemical labeling; environmental sciences; ecology and pollution; environmental impact of chemicals; weed science; herbicides; plant physiology. Mailing Add: Dow Chem Co PO Box 350 Moorestown NJ 08057

LICK, DALE W, b Marlette, Mich, Jan 7, 38; m 56; c 4. PURE MATHEMATICS, APPLIED MATHEMATICS. Educ: Mich State Univ, BS, 58, MS, 59; Univ Calif, Riverside, PhD(math, partial differential equations), 65. Prof Exp: Instr math & chmn dept, Port Huron Jr Col, 59-60; asst to comptroller, Mich Bell Tel Co, 60-61; from instr to asst prof math, Univ Redlands, 61-63; asst prof, Univ Tenn, 65-67; asst res mathematician, Dept Appl Math, Brookhaven Nat Lab, 67-68; assoc prof math, Univ Tenn, 68-69; assoc prof & head dept, Drexel Univ, 69-72; vpres acad affairs, Russell Sage Col, 72-74; DEAN, SCH SCI & HEALTH PROFESSIONS & PROF MATH, OLD DOMINION UNIV, 74- Concurrent Pos: Consult, Union Carbide Corp, AEC, Oak Ridge Nat Lab, 66-67; adj assoc prof, Med Sch, Temple Univ. Mem: AAAS; Am Math Soc; Asn Comput Mach; Math Asn Am; Soc Indust & Appl Math. Res: Singular non-linear hyperbolic second order partial differential equations; non-linear Dirichlet problems; systems of non-linear boundary and initial value problems; partial differential equations and their numerical solution. Mailing Add: Sch Sci & Health Professions Old Dominion Univ Norfolk VA 23508

LICK, DON R, b Marlette, Mich, Sept 3, 34; m 61; c 2. MATHEMATICS. Educ: Mich State Univ, BS, 56, MS, 57, PhD(math), 61. Prof Exp: Asst prof math, Purdue Univ, 61-63 & NMex State Univ, 63-66; vis assoc prof, 65-66, assoc prof, 66-72, PROF MATH, WESTERN MICH UNIV, 72- Concurrent Pos: NSF res grant, 69-70; US Army Res Off Conf grant, 71-72; vis prof, Univ Calif, Irvine, 72-73 & Calif State Univ, Los Angeles, 72-73. Mem: Math Asn Am; Am Math Soc; London Math Soc; Am Asn Univ Prof. Res: Complex analysis; sets of convergence of series; representation of measurable functions by series; graph theory; connectivity; structural problems. Mailing Add: Dept of Math Western Mich Univ Kalamazoo MI 49001

LICK, WILBERT JAMES, b Cleveland, Ohio, June 12, 33; m 65; c 2. ENVIRONMENTAL SCIENCES, APPLIED MATHEMATICS. Educ: Rensselaer Polytech Inst, BA, 55, MA, 57, PhD(aeronaut eng), 58. Prof Exp: Res fel & lectr mech eng, Harvard Univ, 59-61, asst prof, 61-66; sr res fel aeronaut, Calif Inst Technol, 66; assoc prof eng, 66-70, PROF GEOPHYS & ENG, CASE WESTERN RESERVE UNIV, 70-, CHMN DEPT EARTH SCI, 73- Concurrent Pos: Guggenheim fel, 65. Mem: Am Geophys Union; Am Soc Mech Eng. Res: Applied mathematics. Mailing Add: 7610 Wilson Mills Rd Chesterland OH 44026

LICKO, VOJTECH, b Banska Stiavnica, Czech, Aug 30, 32; US citizen; m 59; c 1. MATHEMATICAL BIOLOGY. Educ: Czech Acad Sci, CSc(biophys), 63; Univ Chicago, PhD(math biol), 66. Prof Exp: Chief radioisotope lab, Inst Endocrinol, Slovak Acad Sci, Bratislava, 54-63; fel math biol, Univ Chicago, 63-66; scientist & assoc prof biophys, Inst Physics, Comenius Univ, Bratislava, 66-68; res fel biomath, Dept Biochem & Biophys, 68-71, res asst, 73-74, RES ASST BIOMATH, CARDIOVASC INST, UNIV CALIF, SAN FRANCISCO, 74- Mem: Biophys Soc; Soc Math Biol. Res: Mathematical modeling of biochemical and physiological processes; theory of secretory mechanisms; dynamics of glucose-insulin control in man; kinetics of transport of substances through epithelia. Mailing Add: Cardiovasc Res Inst Univ of Calif San Francisco CA 94143

LIDDELL, ROBERT WILLIAM, JR, b Pittsburgh, Pa, Sept 11, 13; m 40; c 3. ORGANIC CHEMISTRY, BIOCHEMISTRY. Educ: Univ Pittsburgh, BS, 34, PhD(chem), 40. Prof Exp: Chem engr, Swindell-Dressler Corp, 34-35; chemist, Hall Labs, 35-36; res chemist, Hagan Chem & Controls, Inc, 40-55, asst res mgr, 55-63; mgr prod eng, 63-70, MGR PILOT RES & DEVELOP, CALGON CORP, 70- Mem: Am Chem Soc. Res: Water treatment; phosphate chemicals. Mailing Add: Calgon Ctr Calgon Corp Box 1346 Pittsburgh PA 15230

LIDDICOAT, RICHARD THOMAS, JR, b Kearsage, Mich, Mar 2, 18; m 39. GEMOLOGY, MINERALOGY. Educ: Univ Mich, BS, 39, MS, 40; dipl, Gemol Inst Am, 41. Prof Exp: Asst mineral, Univ Mich, 37-40; instr, 40-41, dir ed, 41-42, 46-49, asst dir, 49-52, EXEC DIR, GEMOL INST AM, 52- Mem: AAAS; fel Geol Soc Am; fel Mineral Soc Am. Res: Gem identification and grading. Mailing Add: Gemological Inst of Am 11940 San Vicente Blvd Los Angeles CA 90049

LIDDICOET, THOMAS HERBERT, b Placerville, Calif, Nov 1, 27; m 56; c 5. PESTICIDE CHEMISTRY. Educ: Col of Pac, AB, 49; Univ Wash, PhD(chem), 53. Prof Exp: Instr chem, Univ Calif, 53-54 & San Francisco State Col, 54-55; from res chemist to sr res chemist, Chevron Res Co Div, 55-69, SUPVR PESTICIDE FORMULATIONS, ORTHO DIV, CHEVRON CHEM CO, STAND OIL CO CALIF, 69- Mem: Am Chem Soc; Am Oil Chem Soc. Res: Pesticide formulations; surface active agents. Mailing Add: Chevron Chem Co 940 Hensley St Richmond CA 94804

LIDDLE, CHARLES GEORGE, b Detroit, Mich, Mar 22, 36; m 60; c 4. VETERINARY MEDICINE, RADIATION BIOLOGY. Educ: Mich State Univ, BS, 58, DVM, 60; Univ Rochester, MS, 63. Prof Exp: Vet, Pvt Pract, Mich, 60-61; chief radioisotopes div, Fourth Army Med Lab, Vet Corps, US Army, Ft Sam Houston, 61-62, res vet, Walter Reed Army Inst Res, 63-65, chief small animal test, Dept Med Chem, 65, chief radioisotope sect, Army Med Res & Nutrit Lab, Denver, 65-69, lab vet, Navy Prev Med Unit, Viet Nam, 69; chief biophys unit, Twinbrook Res Lab, 70-73, RES VET, EXP BIOL DIV, HEALTH EFFECTS RES LAB, ENVIRON RES CTR, ENVIRON PROTECTION AGENCY, 73- Mem: Am Vet Med Asn. Res: The effects of microwaves on the immunologic competence of laboratory animals. Mailing Add: Rm H-210 Exp Biol Div Environ Res Ctr Environ Protection Agency Research Triangle Park NC 27711

LIDDLE, GRANT WINDER, b American Fork, Utah, June 27, 21; m 42, 71; c 5. ENDOCRINOLOGY. Educ: Univ Utah, BS, 43; Univ Calif, MD, 48. Prof Exp: Asst med & asst resident, Sch Med, Univ Calif, 49-51, instr, 53; from sr asst surgeon to surgeon sect clin endocrinol, Nat Heart Inst, 53-56; assoc prof med & chief endocrine serv, 56-61, PROF MED & DIR ENDOCRINOL, SCH MED, VANDERBILT UNIV, 61-, CHMN DEPT MED, 68- Concurrent Pos: USPHS res fel, Metab Unit, Univ Calif, 51-53; USPHS res career award, 62; mem endocrinol study sect, USPHS, 58-62, consult to Surg Gen, 58-66, mem diabetes & metab training grants comt, 62-66, chmn, 63-66, mem nat adv arthritis & metab dis coun, 67-71; mem coun, Asn Prof Med, 73-77. Honors, & Awards: Upjohn Award, Endocrine Soc, 62, Distinguished Leadership Award Endocrinol, 71; Sir Henry Hallet Dale Medal, Brit Soc Endocrinol, 73. Mem: Am Soc Clin Invest (secy-treas, 63-66, pres, 66-67); Endocrine Soc (pres, 73-74); Asn Am Physicians; Asm Profs Med; Int Soc Endocrinol. Res: Adrenal physiology; methods of assaying steroids; pituitary physiology; renal function as influenced by steroids; steroid hypertension; endocrinology of tumors. Mailing Add: 770 Norwood Dr Nashville TN 37204

LIDDLE, LARRY BROOK, b Prairie Creek, Ind, Dec 24, 35. MARINE PHYCOLOGY. Educ: Albion Col, BA, 57; Univ Chicago, MS, 62; Univ Calif, Santa Barbara, PhD(marine phycol), 67. Prof Exp: Instr biol, Univ Ill, Chicago, 60-64; asst prof, 68-71, ASSOC PROF BIOL, UNIV PR, RIO PIEDRAS, 71- Concurrent Pos: NIH res training grant, 73-74; res vis, Max Planck Inst Cell Biol, 74-75; Fulbright lectr, Govt of Uruguay, 76. Mem: AAAS; Am Phycol Soc; Brit Phycol Soc; Int Phycol Soc. Res: Development of marine benthic algae in the laboratory and the field, with special emphasis on reproductive cell development. Mailing Add: Dept of Biol Univ of PR Rio Piedras PR 00931

LIDE, DAVID REYNOLDS, JR, b Gainesville, Ga, May 25, 28; m 55; c 4. CHEMICAL PHYSICS. Educ: Carnegie Inst Technol, BS, 49; Harvard Univ, AM, 51, PhD(chem physics), 52. Prof Exp: Fulbright scholar & Ramsay mem fel, Oxford Univ, 52-53; res fel, Harvard Univ, 53-54; physicist, 54-63, chief infrared & microwave spectros sect, 63-68, CHIEF OFF STAND REF DATA, NAT BUR STANDARDS, 68- Concurrent Pos: Lectr, Univ Md, 56-66; NSF sr fel, Univ London, 59-60 & Univ Bologna, 67-68; US nat deleg, Comt Data Sci & Technol, Int Coun Sci Unions, 73-, assoc ed, CODATA Bulletin, 74-; ed, J Phys & Chem Ref Data. Honors & Awards: Silver Medal, US Dept Com, 65, Gold Medal, 68; Stratton Award, Nat Bur Standards, 68. Mem: AAAS; Am Chem Soc; fel Am Phys Soc. Res: Free radicals, high temperature, microwave and infrared spectroscopy; molecular structure; critical data evaluation; molecular lasers- Mailing Add: Off of Stand Ref Data Nat Bur of Standards Washington DC 20234

LIDE, ROBERT WILSON, b Hwanghsien, Shantung, China, June 27, 22; US citizen; m 55; c 3. NUCLEAR PHYSICS. Educ: Wake Forest Col, BS, 43; Univ Mich, MS, 50, PhD, 59. Prof Exp: Asst prof, 57-65, ASSOC PROF PHYSICS, UNIV TENN, KNOXVILLE, 65- Mem: Am Phys Soc. Res: Low-energy nuclear physics; gamma-gamma angular correlation; gamma-ray spectroscopy. Mailing Add: Dept of Physics Univ of Tenn Knoxville TN 37916

LIDER, LLOYD A, b Woodland, Calif, July 14, 21; m 46; c 2. GENETICS. Educ: Univ Calif, BS, 48, PhD, 52. Prof Exp: Asst viticult, 49-51, instr viticult & jr viticulturist, 51-54, asst prof, viticult univ & asst viticulturist, 54-60, assoc prof viticult & assoc viticulturist, 60-73, PROF VITICULT & VITICULTURIST, EXP STA, UNIV CALIF, DAVIS, 73- Res: Inheritance of phylloxera and nematode resistance in species of vitis; grape rootstock investigation in California vineyards. Mailing Add: Dept of Viticult Univ of Calif Davis CA 95616

LIDIAK, EDWARD GEORGE, b La Grange, Tex, Mar 14, 34. GEOLOGY. Educ: Rice Univ, BA, 56, MA, 60, PhD(geol), 63. Prof Exp: Res scientist, Univ Tex, 62-64; asst prof, 64-70, ASSOC PROF GEOL, UNIV PITTSBURGH, 70- Concurrent Pos: Geologist, US Geol Surv, Pa, 65- Mem: AAAS; Geochem Soc; Geol Soc Am. Res: Petrology of island are volcanic rocks; geology of buried Precambrian rocks of United States; phase qequilibria in mineral systems. Mailing Add: Dept of Earth & Planetary Sci Univ of Pittsburgh Pittsburgh PA 15213

LIDICKER, WILLIAM ZANDER, JR, b Evanston, Ill, Aug 19, 32; m 56. POPULATION BIOLOGY, MAMMALOGY. Educ: Cornell Univ, BS, 53; Univ Ill, MS, 54, PhD(zool), 57. Prof Exp: From instr to assoc prof, Univ, 57-69, vchmn dept zool, 66-67, from asst cur to assoc cur, Mus, 57-69, actg dir, 74-75, PROF ZOOL, UNIV CALIF, BERKELEY, 69-, ASSOC DIR, MUS VERTEBRATE ZOOL, 68-, CUR MAMMALS, 69- Concurrent Pos: NSF sr fel, 63-64; assoc res prof, Miller Inst Basic Res Sci, 67-68; hon res fel, Dept Animal Genetics, Univ Col London, 71-72; hon lectr, Dept Biol, Royal Free Hosp Sch Med, London, 71-72. Mem: Am Soc Mammal (2nd vpres, 74-76); Ecol Soc Am; Soc Study Evolution; Am Soc Naturalists; Am Soc Zoologists. Res: Evolution of mammals. Mailing Add: Mus of Vertebrate Zool Univ of Calif Berkeley CA 94720

LIDZ, THEODORE, b New York, NY, Apr 1, 10; m 39; c 3. PSYCHIATRY. Educ: Columbia Univ, AB, 31, MD, 36; Am Bd Psychiat & Neurol, dipl, Hon Degrees: MA, Yale Univ, 51. Prof Exp: From instr to assoc prof psychiat, Johns Hopkins Univ, 40-51; PROF PSYCHIAT, YALE UNIV, 51- Concurrent Pos: Examr, Am Bd Psychiat & Neurol, 46-51; psychiatrist-in-chief, Grace-New Haven Hosp, 51-61 & Yale Psychiat Inst, 51-61; chmn comt educ, Am Psychiat Asn, 52-55; mem study sect res grants, NIMH, 52-56, mem training grants comt, 59-63, career investr, 61-, mem ment health prog-proj comt, 63-67; consult, Off Surgeon Gen, 58-72; fel, Ctr Advan Study Behav Sci, 65-66; chmn dept psychiat, Sch Med, Yale Univ, 67-69; Honors & Awards: Frieda Fromm-Reichmann Award, Acad Psychoanal, 61; William C Menninger Award, Am Col Physicians, 72; Stanley R Dean Award, Am Col Psychiat, 73; Van Gieson Award, NY State Psychiat Inst, 73; Psychiatric Outpatients Ctrs Am Award, 75. Mem: Am Psychosom Soc (secy-treas, 52-56, pres, 57-58); fel Am Psychiat Asn; fel Am Col Psychoanal; fel Am Col Psychiat; Am Psychoanal Asn. Res: Schizophrenia; family. Mailing Add: Dept of Psychiat Sch of Med Yale Univ New Haven CT 06510

LIEB, ELLIOTT HERSHEL, b Boston, Mass, July 31, 32; c 2. MATHEMATICAL PHYSICS. Educ: Mass Inst Technol, BSc, 53; Univ Birmingham, PhD(physics), 56. Prof Exp: Fulbright fel physics, Kyoto Univ, 56-57; res assoc, Univ Ill, 57-58; lab nuclear studies, Cornell Univ, 58-60; staff physicist, Res Lab, IBM Corp, 60-63; assoc prof physics, Belfer Grad Sch Sci, Yeshiva Univ, 63-66; prof, Northeastern Univ, 66-68; prof math, Mass Inst Technol, 68-75; PROF MATH & PHYSICS, PRINCETON UNIV, 75- Concurrent Pos: Sr lectr, Univ Col Sierra Leone, 61-62; consult, IBM Corp, 63-65; vis mem staff, Los Alamos Sci Lab, 66-; guest prof, Inst Advan Sci Studies, 72-73. Honors & Awards: Boris Pregel Award, NY Acad Sci, 70. Mem: Fel

Am Phys Soc. Res: Field theory; solid state physics; statistical mechanics. Mailing Add: Dept of Physics-Jadwin Princeton Univ PO Box 708 Princeton NJ 08540

LIEB, MARGARET, b Bronxville, NY, Nov 28, 23. GENETICS. Educ: Smith Col, BA, 45; Ind Univ, MA, 46; Columbia Univ, PhD, 50. Prof Exp: Asst prof biol, Brandeis Univ, 55-60; vis assoc prof, 60-62, assoc prof, 62-67, PROF MICROBIOL, SCH MED, UNIV SOUTHERN CALIF, 67- Concurrent Pos: USPHS fel, Calif Inst Technol, 50-52, Nat Found Infantile Paralysis fel, 52-53; fel, Inst Pasteur, 53-54; French Govt fel, Inst Radium, 54-55; NIH res career award, 62-72; prog dir genetic biol, NSF, 72-73. Mem: Fel AAAS; Genetics Soc Am; Am Soc Microbiol. Res: Bacteriophage genetics; lysogeny. Mailing Add: Dept of Microbiol Sch Med Univ of Southern Calif Los Angeles CA 90033

LIEB, WILLIAM ROBERT, b Chicago, Ill, Aug 31, 40; m 62. BIOPHYSICS. Educ: Univ Ill, BS, 62, MS, 63, PhD(biophys), 67. Prof Exp: Air Force Off Sci Res-Nat Res Coun fel biochem, Univ Manchester, 67-68; mem staff, Dept Polymer Sci, Weizmann Inst Sci, 69; inst life sci, Hebrew Univ, Jerusalem, 69-70; STAFF SCIENTIST, MED RES COUN BIOPHYS UNIT, KING'S COL, LONDON, 70- Res: Mechanisms and regulation of transport across biological membranes. Mailing Add: Med Res Coun Biophysics Unit King's Col 26-29 Drury Lane London England

LIEBAN, RICHARD WARREN, b New York, NY, Jan 23, 21; m 45. CULTURAL ANTHROPOLOGY. Educ: Univ Mo, BJ, 43; Columbia Univ, MA, 51, PhD(anthrop), 56. Prof Exp: Info specialist, US Civil Admin Ryukyu Islands, 52-53; asst prof anthrop, Woman's Col, Univ NC, 55-59, assoc prof, 59-61; vis assoc prof, Univ NC, Chapel Hill, 61-62, assoc prof, 63-65; prog dir anthrop, NSF, 65-70; PROF ANTHROP & CHMN DEPT, UNIV HAWAII, 71- Concurrent Pos: Fulbright grant, Philippines, 58-59 & NSF grant, 62-63; mem adv panel, Conf Cross-Cultural Psychopharmacol, NIMH, 61; mem foreign currency adv coun, Smithsonian Inst, 65-67; mem fel review comt, NIMH, 74-75; mem bd trustees, Bishop Mus, 75-78. Mem: Fel AAAS; fel Am Anthrop Asn; Asn Asian Studies; Soc Appl Anthrop; Soc Med Anthrop. Res: Social and cultural aspects of health and medicine; social control; social and cultural change; religion and magic; the Philippines; Southeast Asia. Mailing Add: Dept of Anthrop Univ of Hawaii Honolulu HI 96822

LIEBE, DONALD CHARLES, b Cleveland, Ohio, Nov 16, 42; m 64; c 2. BIOPHYSICAL CHEMISTRY. Educ: Case Western Reserve Univ, BA, 66, MA, 68, PhD(phys chem), 70. Prof Exp: Res assoc chem, Yale Univ, 70-71, NIH res fel, 71-73, asst instr, 74; RES INVESTR, SEARLE LABS, G D SEARLE & CO, 74- Mem: Am Chem Soc; The Chem Soc; NY Acad Sci. Res: Physical chemistry of nucleic acids, conformations of RNA and DNA; protein-nucleic acid interactions; small molecule binding to nucleic acids; mechanism of animal virus replication; mechanism of growth factor action in cell culture. Mailing Add: Searle Labs G D Searle & Co Box 5110 Chicago IL 60680

LIEBE, RICHARD MILTON, b Norwalk, Conn, May 26, 32; m 55; c 3. GEOLOGY. Educ: Bates Col, BS, 54; Univ Houston, MS, 59; Univ Iowa, PhD(geol), 62. Prof Exp: Assoc prof geol, Col Wooster, 61-67; ASSOC PROF GEOL, STATE UNIV NY COL BROCKPORT, 67- Mem: Paleont Soc; Soc Econ Paleont & Mineral; Am Asn Geol Teachers; Geol Soc Am. Res: Stratigraphic paleontology of the Paleozoic era using conodonts; shallow water sedimentology and coral reef ecology. Mailing Add: Dept of the Earth Sci State Univ NY Col at Brockport Brockport NY 14420

LIEBELT, ANNABEL GLOCKLER, b Washington, DC, June 27, 26; m 54; c 4. ANATOMY, CANCER. Educ: Western Md Col, BA, 48; Univ Ill, MS, 55; Baylor Col Med, PhD(anat), 60. Prof Exp: Biologist, Path Sect, Nat Cancer Inst, 49-52; asst anat, Col Med, Univ Ill, 52-54; asst, Col Med, Baylor Univ, 54-58, from instr to assoc prof, 58-71; assoc prof cell & molecular biol, Med Col Ga, 71-74; PROF ANAT, NORTHEASTERN OHIO UNIVS COL MED, 74- Concurrent Pos: Dir, Kirschbaum Mem Lab, Col Med, Baylor Univ, 62-71. Mem: Am Asn Cancer Res; Am Asn Anat; NY Acad Sci; Am Soc Exp Path; Am Asn Lab Animal Sci. Res: Transplantation of normal and neoplastic tissue in inbred mice; genetic, hormonal, viral and environmental factors in the etiology and growth of spontaneous and induced tumors in inbred mice, mainly leukemia, breast cancer, hepatomas and endocrine glands. Mailing Add: Basic Med Sci Northeastern Ohio Univs Col Med Kent OH 44240

LIEBELT, ROBERT ARTHUR, b Chicago, Ill, Feb 3, 27; m 54; c 4. ANATOMY, EXPERIMENTAL PATHOLOGY. Educ: Loyola Univ, Ill, BS, 50; Wash State Univ, MS, 52; Baylor Univ, PhD(anat), 57, MD, 58. Prof Exp: Asst, Wash State Univ, 50-52; asst, Col Med, Baylor Univ, 54-57, from instr to prof anat & chmn dept, 57-71; prof cell & molecular biol & exp med & assoc dean curriculum, Med Col Ga, 71-72, provost, 72-74; PROF ANAT & DEAN, NORTHEASTERN OHIO UNIVS COL MED, 74- Concurrent Pos: Vis prof, Okayama Univ, 61. Mem: AAAS; Soc Exp Biol & Med; Am Asn Anat; Am Asn Cancer Res; NY Acad Sci. Res: Adipose tissue in obesity; relationship between nutrition and neoplasia; hypothalamus and appetite control; hypothalmic-pituitary relationships in experimental neoplasia; effects of pressure on food intake and body composition. Mailing Add: Northeastern Ohio Univs Col Med 275 Martinel Kent OH 44240

LIEBEN, JAN, b Trnovany, Czech, Mar 13, 15; US citizen; div; c 2. OCCUPATIONAL MEDICINE, PREVENTIVE MEDICINE. Educ: Univ Liverpool, MB, ChP, 43; Harvard Univ, MPH, 49; Am Bd Indust Hyg, dipl. Prof Exp: Asst health officer, New York City Dept Health, 49-50; epidemiologist, State Dept Health, Conn, 50-51, indust physician, 51-54; plant physician, Am Cyanamid Co, Pa, 54-55; dir occup health, State Dept Health, Pa, 55-69; med dir, Chem Group, FMC Corp, 69-75; PROF OCCUP MED, MED COL, THOMAS JEFFERSON UNIV, 75- Concurrent Pos: Vis prof, Jefferson Med Col, 61-; consult, Radioactivity Ctr, Mass Inst Technol, 62-; consult div radiobiol, Argonne Nat Lab; former chmn manganese panel comt biol effects air pollution, Nat Acad Sci. Mem: Fel Indust Med Asn. Res: Epidemiology of beryllium, pneumoconiosis, pesticides and other chemical exposures, including radium. Mailing Add: Dept of Prev Med 1025 Walnut St Med Col Thomas Jefferson Univ Philadelphia PA 19107

LIEBENAUER, PAUL (HENRY), b Cleveland, Ohio, Sept 21, 35; m 62; c 2. EXPERIMENTAL NUCLEAR PHYSICS. Educ: Case Western Reserve Univ, BS, 57, MS, 60, PhD(physics), 71. Prof Exp: Instr physics, Clarkson Col Technol, 60-62; asst prof, 68-70, ASSOC PROF PHYSICS, STATE UNIV NY COL OSWEGO, 70- Concurrent Pos: Consult, NASA, 71-; NSF grant, 72. Mem: Am Phys Soc; Am Asn Physics Teachers; Sigma Xi. Res: Low energy nuclear physics. Mailing Add: Dept of Physics State Univ of NY Col at Oswego Oswego NY 13126

LIEBENBERG, DONALD HENRY, b Madison, Wis, July 10, 32; m 57; c 2. PLASMA PHYSICS, LASERS. Educ: Univ Wis, BS, 54, MS, 56, PhD, 71. Prof Exp: Asst, Univ Wis, 54-61; STAFF MEM PHYSICS, LOS ALAMOS SCI LAB, 61- Concurrent Pos: Solar-terrestrial res prog dir, NSF, 67-68; app liaison to Geophys Res Bd, Nat Acad Sci & US Comt Solar Terrestrial Res; US coord, 1970 Solar Eclipse. Mem: AAAS; Am Astron Soc; Am Phys Soc; Am Geophys Union; Am Soc Testing & Mat. Res: Low temperature physics, especially superfluidity and helium films; solar physics; magneto optics; laser plasma interaction; high pressure physics. Mailing Add: Los Alamos Sci Lab PO Box 1663 Los Alamos NM 87545

LIEBER, CHARLES SAUL, b Antwerp, Belg, Feb 13, 31; US citizen; m 74; c 3. INTERNAL MEDICINE, NUTRITION. Educ: Univ Brussels, MD, 55. Prof Exp: Asst resident med, Univ Hosp Brugmann Brussels, Belg, 55-56; instr med, Harvard Med Sch, 61-62, assoc, 62-63; assoc prof, Med Col, Cornell Univ, 63-68; assoc prof, 68-69, PROF MED, MT SINAI SCH MED, 69-; CHIEF SECT LIVER DIS & NUTRIT, VET ADMIN HOSP, 68- Concurrent Pos: Belg Coun Sci Res fel internal med, Med Found Queen Elizabeth, 56-58; Belg-Am Found res fel med, Harvard Med Sch, 58-60; mem fat comt, Food & Nutrit Bd, Nat Acad Sci-Nat Res Coun, 61-67; dir liver dis & nutrit unit, Bellevue Hosp, 63-68; NIH res career develop award, 64-68. Honors & Awards: Laureate, Belg Govt, 56; Distinguished Achievement Award, Am Gastroenterol Asn, 73; McCollum Award, Am Soc Clin Nutrition, 73. Mem: Am Fedn Clin Res; Am Med Soc Alcoholism (pres, 75); Am Soc Clin Nutrit (pres, 75); Asn Am Physicians; fel Am Col Physicians. Res: Diseases of the liver; nutrition and intermediary metabolism, especially alcoholic cirrhosis, fatty liver, hyperlipemia, hyperuricemia, pathogenesis and treatment of hepatic coma and ascites, and pathophysiology of liver regeneration and drug abuse. Mailing Add: Vet Admin Hosp Sect of Liver Dis 130 W Kingsbridge Rd Bronx NY 10468

LIEBER, MICHAEL, b Brooklyn, NY, Dec 28, 36; m 64; c 2. THEORETICAL PHYSICS. Educ: Cornell Univ, AB, 57; Harvard Univ, AM, 58, PhD(physics), 67. Prof Exp: Sr scientist, Res & Advan Develop Div, Avco Corp, 63-66, chief sci probs, 66-67; assoc res scientist & adj asst prof physics, NY Univ, 67-70; asst prof, 70-75, ASSOC PROF PHYSICS, UNIV ARK, FAYETTEVILLE, 75-, PLANETARIUM LECTR, 72-, DIR, REACH KIT PROJ, 74- Mem: Am Phys Soc; Am Asn Physics Teachers. Res: Quantum scattering theory; few body problems; quantum electrodynamics and field theory; mathematical methods; atomic structure; cosmic rays; elementary particles; general relativity and cosmology. Mailing Add: Dept of Physics Univ of Ark Fayetteville AR 72701

LIEBERMAN, ARTHUR STUART, b Brooklyn, NY, Feb 24, 31; m 56; c 3. ORNAMENTAL HORTICULTURE. Educ: Cornell Univ, BS, 52, MS, 58. Prof Exp: Teacher high sch, NY, 52-53; asst prof, 58-64, EXTEN SPECIALIST ORNAMENTAL HORT, CORNELL UNIV, 56-, ASSOC PROF FLORICULT, 64- Mem: Am Soc Hort Sci; fel Am Soc Landscape Archit. Res: Community resource development, physical environmental quality; utilization of ecological knowledge in the planning process; rhododendron soils and nutrition; hardiness of landscape plant materials. Mailing Add: Dept of Hort Cornell Univ Ithaca NY 14850

LIEBERMAN, BURTON BARNET, b Boston, Mass, Sept 28, 38; m 63; c 2. MATHEMATICS. Educ: Harvard Univ, BA, 60; NY Univ, MS, 62, PhD(math), 67. Prof Exp: Asst prof, 65-69, ASSOC PROF MATH, POLYTECH INST NY, 69- Mem: Am Math Soc; Sigma Xi. Res: Ordinary differential equations; random differential equations. Mailing Add: Dept of Math Polytech Inst NY 333 Jay St Brooklyn NY 11201

LIEBERMAN, DANIEL, b Gunnison, Utah, Feb 21, 19; c 1. PSYCHIATRY. Educ: Univ Calif, AB, 42, MD, 46. Prof Exp: Chief hosp serv, Sonoma State Hosp, Calif, 49-54; supt & med dir, Mendocino State Hosp, 54-60; from chief dep dir to dir, State Dept Ment Health, 60-63; pvt pract, 63-64; comnr ment health, Del Dept Ment Health, 64-67; PROF PSYCHIAT & ACTG CHMN DEPT PSYCHIAT & HUMAN BEHAV, JEFFERSON MED COL, THOMAS JEFFERSON UNIV, 67-, DIR, JEFFERSON COMMUNITY MENT HEALTH-MENT RETARDATION CTR, 67- Concurrent Pos: Consult forensic psychiat, Calif Superior Courts, 54-63; consult ment hosp serv, Am Psychiat Asn, 60-62; consult forensic psychiat, US Fed Court, 61-; consult, Ment Health Res Inst, Palo Alto, Calif, 63-64; consult state ment progs, NIMH, 65-68, consult alcohol rev comt, 67-70, consult, Nat Coun Community Health Ctrs, 72-73; consult, Vet Admin Hosp, Coatesville, Pa, 67- Mem: Fel AAAS; fel Am Col Psychiat; fel Am Asn Ment Deficiency; Am Med Soc Alcoholism. Mailing Add: Jefferson Health Sci Ctr 130 S Ninth St Philadelphia PA 19107

LIEBERMAN, EDWARD MARVIN, b Lowell, Mass, Feb 10, 38; m 60; c 3. PHYSIOLOGY. Educ: Tufts Univ, BS, 59; Univ Mass, MA, 61; Univ Fla, PhD(physiol), 65. Prof Exp: Res assoc physiol, Col Med, Univ Fla, 66; asst prof, 68-72, ASSOC PROF PHYSIOL, BOWMAN GRAY SCH MED, 72- Concurrent Pos: Swed Med Res Coun fel, Col Med, Univ Uppsala, 66-68. Mem: Soc Neurosci; Biophys Soc; Am Heart Asn; Am Physiol Soc; NY Acad Sci. Res: Cellular nerve physiology; membrane ion and water transport and metabolism; ultraviolet radiation effects on membranes. Mailing Add: Dept of Physiol Bowman Gray Sch of Med Winston-Salem NC 27103

LIEBERMAN, EDWIN JAMES, b Milwaukee, Wis, Nov 21, 34; m 59; c 2. PSYCHIATRY, SOCIAL PSYCHOLOGY. Educ: Univ Calif, Berkeley, AB, 55; Univ Calif, San Francisco, MD, 58; Harvard Univ, MPH, 63; Am Bd Psychiat & Neurol, dipl, 66. Prof Exp: Psychiat fel, Mass Ment Health Ctr, Boston, 59-61; child psychiat fel, Putnam's Children Ctr, Boston, 61-62 ; psychiatrist & chief, Ctr Child & Family Ment Health, NIMH, 63-70; dir family ther, Hillcrest Children's Ctr, DC, 71-74, dir ment health proj, 72-75, DIR FAMILY PLANNING PROJ, AM PUB HEALTH ASN, 75- Concurrent Pos: Child psychiat fel, Hillcrest Children's Ctr, DC, 65-66; mem bd dirs, Sex Info & Educ Coun US, 66-69 & 73-76; clin asst prof psychiat, Sch Med, Howard Univ, 67-; mem bd dirs, Nat Coun Family Rels, 69-73; vis lectr maternal & child health, Harvard Sch Pub Health, 69-73. Mem: AAAS; fel Am Psychiat Asn; fel Am Pub Health Asn; Am Asn Marriage & Family Coun; Esperanto League NAm (pres, 72-75). Res: Mental health; preventive psychiatry; family planning; nonviolence; Esperantic studies; international language planning. Mailing Add: 6451 Barnaby St N W Washington DC 20015

LIEBERMAN, EMANUEL ROY, b New York, NY, Aug 1, 19; m 42; c 3. BIOENGINEERING. Educ: City Col New York, BS, 40; Univ Ky, BSEE, 46; Rutgers Univ, PhD(food sci), 71. Prof Exp: Sr chemist, Interchem Corp, 47-52; dir res, Spec Prod Div, Ethicon, Inc, 52-64; vpres res & develop, Devro Inc, Johnson & Johnson Co, 64-74; PRES, BIO-TECHNIQUES, INC, 74- Mem: AAAS; Am Chem Soc; Am Leather Chem Asn; NY Acad Sci. Res: Polymer and protein chemistry; packaging materials and concepts; diffusion of gases through polymers; bioengineering; immobilized enzymes. Mailing Add: 119 Branch Rd Bridgewater NJ 08807

LIEBERMAN, GERALD ALAN, b Minneapolis, Minn, Oct 19, 50. ECOLOGY. Educ: Univ Calif, Los Angeles, BA, 71; Princeton Univ, MA & PhD(biol), 74. Prof Exp: Ecologist, 74-76, PRIN INVESTR FIELD INVENTORY STUDIES, NATURE CONSERVANCY, 75- Mem: AAAS; Am Soc Mammalogists; Ecol Soc Am. Res: Development of ecological inventory techniques for natural areas, through the development of field survey techniques and other data gathering and management methodologies; this methodology is being developed for the protection of biological diversity through planning. Mailing Add: Nature Conservancy 325 W 15th St Minneapolis MN 55403

LIEBERMAN, GERALD J, b New York, NY, Dec 31, 25; m 50; c 4. OPERATIONS RESEARCH, STATISTICS. Educ: Cooper Union, BME, 48; Columbia Univ, AM, 49; Stanford Univ, PhD(statist), 53. Prof Exp: Math statistician, Nat Bur Standards, 49-50; from asst prof to prof statist & indust eng, 53-67, chmn dept opers res, 67-75, PROF STATIST & OPERS RES, STANFORD UNIV, 67-, ASSOC DEAN, SCH HUMANITIES & SCI, 75- Concurrent Pos: Consult, Stanford Res Inst; mem adv panel math sci, NSF, 68-73. Honors & Awards: Shewhart Medal, Am Soc Qual Control, 72. Mem: Opers Res Soc Am; fel Am Soc Qual Control; fel Am Statist Asn; fel Inst Math Statist; Int Statist Inst. Res: Industrial statistics; quality control; reliability. Mailing Add: Dept of Statist & Opers Res Stanford Univ Stanford CA 94305

LIEBERMAN, HERBERT A, b New York, NY, Aug 6, 20; m 51; c 2. PHARMACEUTICAL CHEMISTRY. Educ: Univ Ark, BS, 40; Columbia Univ, AM, 48, BS, 51, MS, 52; Purdue Univ, PhD(pharmaceut chem), 55. Prof Exp: Res fel biochem, Beth Israel Hosp, New York, 40-41; chemist, Pine Bluff Arsenal, Ark, 41-43; instr & assoc anal chem, Col Pharm, Columbia Univ, 46-52, res pharmacist, Res Inst, Wyeth Labs, 54-57; mgr pharmaceut prod develop, Isodine Pharmacal Co, 57-61; sr res assoc, 61-63, dir pharmaceut res & develop, 63-72, VPRES RES & DEVELOP, PERSONAL PRODS DIV, WARNER LAMBERT CO INC, 72- Mem: Am Chem Soc; Am Pharmaceut Asn; fel Acad Pharmaceut Sci. Res: Industrial pharmacy; pharmaceutical technology, particularly process and product development; analytical methods development for pharmaceutical products; biochemical research, particularly metabolic pathways. Mailing Add: Personal Prods Div Warner Lambert Co 170 Labor Rd Morris Plains NJ 07050

LIEBERMAN, HILLEL, b Philadelphia, Pa, Jan 24, 42; m 66; c 1. ORGANIC CHEMISTRY, MICROBIAL BIOCHEMISTRY. Educ: Temple Univ, BS, 63, MS, 65, PhD(med org chem), 70. Prof Exp: Var admin & sci pos, 65-73, ASST DIR RES, BETZ LABS, TREVOSE, 73- Mem: Am Soc Microbiol; Am Chem Soc; Sigma Xi; Tech Asn Pulp & Paper Indust. Res: Development of chemical agents of an antimicrobial and/or antipollution nature to be employed in industrial water systems; development of conceptual information to aid in application of the aforementioned. Mailing Add: 772 Cotlar Lane Warminster PA 18974

LIEBERMAN, IRVING, b Brooklyn, NY, Oct 25, 21; m 47; c 2. CELL BIOLOGY. Educ: Brooklyn Col, BA, 44; Univ Ky, MS, 48; Univ Calif, PhD(bact), 52. Prof Exp: Asst prof bact, Miami Univ, 48-49; from instr to asst prof microbiol, Sch Med, Washington Univ, 53-56; from asst prof to prof microbiol, 56-66, PROF ANAT & CELL BIOL, SCH MED, UNIV PITTSBURGH, 66- Concurrent Pos: Mem cell biol study sect, USPHS, 60-64, chmn, 71-73; fel engr, Westinghouse Res Ctr, 69- Mem: Am Soc Biol Chem. Res: Chemistry. Mailing Add: Dept of Anat & Cell Biol Univ of Pittsburgh Sch of Med Pittsburgh PA 15213

LIEBERMAN, JACK, b Chicago, Ill, Jan 4, 26; m 55; c 4. PULMONARY DISEASES, ENZYMOLOGY. Educ: Univ Calif, Los Angeles, AB, 49; Univ Southern Calif, MD, 54; Am Bd Internal Med, dipl, 62. Prof Exp: Intern, Harbor Gen Hosp, 54-55, resident internal med, 55-58; clin investr, Vet Admin Hosp, Long Beach, Calif, 60-63, sect chief internal med, 63-68; assoc clin prof med, Sch Med, Univ Calif, Los Angeles, 68-71; ASSOC CLIN PROF MED, SCH MED, UNIV CALIF, IRVINE, 71-; ASSOC DIR DEPT RESPIRATORY DIS, CITY OF HOPE MED CTR, 71- Concurrent Pos: Long Beach Heart Asn res fel, Harbor Gen Hosp, 58-60. Mem: AAAS; fel Am Col Physicians; fel Am Col Chest Physicians; Am Fedn Clin Res. Res: Cystic fibrosis; emphysema; antitrypsin deficiency; blood test for sarcoidosis. Mailing Add: City of Hope Med Ctr Duarte CA 91010

LIEBERMAN, JAMES, b New York, NY, June 2, 21; m 43; c 1. PUBLIC HEALTH. Educ: Middlesex Univ, DVM, 44; Univ Minn, MPH, 47. Prof Exp: Sr consult vet, UNRRA, 46; regional milk & food consult, USPHS, Kansas City, 48-50, asst to the chief, Milk & Food Br, DC, 50-51, from asst chief to chief spec proj br, Bur State Servs, 51-52, detailed, US Navy, 52, poultry hyg consult, 52-54, detailed, Vet Epidemiologist, Div Epidemiol & Commun Dis Control, NY State Health Dept, 54-55, vet training consult, Health Servs Training Sect, Commun Dis Ctr, 55, actg sect chief, 55-58, asst to the chief, Training Br, 58-59, asst chief training br & chief AV sect, 59-62, chief med AV Br (dir, Pub Health Serv AV Fac), 62-67, dir, Nat Med AV Ctr & assoc dir AV & telecommun, Nat Libr Med, 67-70, asst surgeon gen, USPHS, 68-70; vpres, med div, Videorecord Corp Am, 70-73; CONSULT HEALTH SCI EDUC & COMMUN, 73- Concurrent Pos: Secy, Conf Pub Health Vets, 53-57; consult, WHO, Geneva, 55; chmn, Fed Adv Coun Med Training Aids, 60; mem task force sci commun, Surgeon Gen Conf Health Commun, 62; secy AV Conf Med & Allied Sci, 62-70; pres, Metrop Atlanta Commun Coun, 65; chmn conf biomed commun, NY Acad Sci, 67; mem comt bio-technol, Ga Sci & Technol Comn, 69-70; mem prof adv coun, Nat Easter Seal Soc Crippled Children & Adults, 69-; vis prof, Hahnemann Med Col, 73-; mem bd gov, Conn Inst Health Manpower Resources, 74-; mem task force energy med servs, Conn Heart Asn, 75- Honors & Awards: Cert Commendation, UNRRA, 46; Letter of Commendation, Surgeon Gen, US Navy, 52; Citation, Nat League Nursing, 62; Meritorious Serv Medal, USPHS, 65; Citation, Fulton County Med Soc, Ga, 66 & Nat AV Asn, 67; Brensa Award, Theta Sigma Phi, 68; Myrtle Wreath Award, Hadassah, 69; Except Serv Award, Ga Easter Seal Soc, 70. Mem: Fel Am Pub Health Asn; Asn Mil Surg US; Am Vet Med Asn; NY Acad Sci; Asn Am Med Cols. Res: Biomedical communication and education; public health practice and epidemiology; training and administration; relationship of animal health to human welfare. Mailing Add: 12 Silver Brook Rd Westport CT 06880

LIEBERMAN, LEONARD, b Los Angeles, Calif, Sept 20, 25; m 47; c 3. ANTHROPOLOGY, SOCIOLOGY. Educ: Univ Calif, Berkeley, BA, 56, MA, 59; Mich State Univ, PhD(sociol, anthrop), 70. Prof Exp: Asst prof anthrop, Bakersfield Col, 59-62; assoc prof, 64-74, PROF SOCIOL & ANTHROP, CENT MICH UNIV, 74- Res: Social mobility; kinship; race. Mailing Add: Dept of Sociol & Anthrop Cent Mich Univ Mt Pleasant MI 48858

LIEBERMAN, LESLIE SUE, b Rockville Ctr, NY, June 23, 44. BIOLOGICAL ANTHROPOLOGY. Educ: Univ Colo, BA, 65; Univ Ariz, MA, 71; Univ Conn, PhD(biobehav sci), 75. Prof Exp: PROJ ASSOC BODY COMPOSITION, HUMAN PERFORMANCE RES LAB, PA STATE UNIV, UNIVERSITY PARK, 75- Concurrent Pos: Fel, Nat Inst Gen Med Sci Human Performance Res Lab & Dept Anthrop, Pa State Univ, 74-75. Mem: AAAS; Am Anthrop Asn; Am Asn Phys Anthropologists; Human Biol Coun; Soc Med Anthrop. Res: Study of body composition and the effects of nutritional behavior and diet on adaptation and microevolution in human populations. Mailing Add: Human Performance Res Lab Noll Lab Pa State Univ University Park PA 16802

LIEBERMAN, MELVYN, b Brooklyn, NY, Feb 4, 38; m 61; c 2. PHYSIOLOGY. Educ: Cornell Univ, BA, 59; State Univ NY, PhD(physiol), 65. Prof Exp: Instr biol, Queen's Col, NY, 60; asst physiol, State Univ NY Downstate Med Ctr, 60-64; res assoc, Div Biomed Eng, Sch Eng & Dept Physiol & Pharmacol, 67-68, asst prof, 68-73, ASSOC PROF PHYSIOL & PHARMACOL, MED CTR, DUKE UNIV, 73- Concurrent Pos: Nat Heart Inst fel, 64-65; Carnegie Inst fel, 65; Nat Heart Inst fel, Biophys Inst, Brazil, 65-67; Nat Heart Inst spec fel, Med Ctr, Duke Univ, 67-68; lectr, Queen's Col, NY, 63-64; vis investr, Jan Swammerdam Inst, Netherlands, 75; Soc Gen Physiol rep, Nat Res Coun, 71-; estab investr, Am Heart Asn, 71-76, chmn res rev comt, NC Heart Asn, 75-76; co-coordinator, USJapan Coop Sci Prog, 74; Porter Develop Prog, Am Physiol Soc, 74-77. Mem: Am Heart Asn; Am Physiol Soc; Biophys Soc; Cardiac Muscle Soc; Soc Gen Physiol (secy, 69-71). Res: Electrophysiology of the heart; developmental and comparative physiology of heart; structure-function relationships of heart cells in tissue culture; synthetic models of cardiac activity. Mailing Add: Dept of Physiol & Pharmacol Duke Univ Med Ctr Durham NC 27710

LIEBERMAN, MICHAEL MERRIL, b Chicago, Ill, June 10, 44; m 63; c 1. MICROBIOLOGY. Educ: Univ Chicago, BS, 66, PhD(microbiol), 69. Prof Exp: Nat Res Coun res assoc, Ames Res Ctr, NASA, Calif, 69-71; SR RES MICROBIOLOGIST, CUTTER LABS, INC, 71- Mem: Am Soc Microbiol. Res: Bacterial antigen and toxin purification; enzymology of halophilic bacteria; biochemical genetics and metabolic regulation. Mailing Add: Microbiol Res Div Cutter Labs Inc Fourth & Parker St Berkeley CA 94710

LIEBERMAN, MIRIAM, b Bordeaux, France; nat US. CANCER. Educ: Univ Calif, PhD, 56. Prof Exp: Damon Runyon Mem Fund fel radiol, 56-57, RES ASSOC, SCH MED, STANFORD UNIV, 57-59 & 62- Concurrent Pos: NSF fel, Inst Radium, France & Med Res Coun Labs, Carshalton, Eng, 60-61. Res: Experimental cancer research, virology. Mailing Add: Dept of Radiol Stanford Univ Sch of Med Stanford CA 94305

LIEBERMAN, MORRIS, b Brooklyn, NY, Sept 4, 19; m 47; c 1. PLANT PHYSIOLOGY. Educ: City Col New York, BS, 41; Rutgers Univ, MS, 47; Univ Md, PhD(plant physiol), 52. Prof Exp: Chemist, E R Squibb Co, 47-48; from asst plant physiologist to prin plant physiologist, 48-72, CHIEF, POST-HARVEST PLANT PHYSIOL LAB, PLANT INDUST STA, USDA, 72- Concurrent Pos: Res assoc, Univ Calif, 53-55 & low temperature res sta, Cambridge Univ, 61-62. Mem: AAAS; Am Soc Plant Physiol; Am Chem Soc; Am Soc Biol Chem; Brit Biochem Soc. Res: Post harvest physiology of fruits and vegetables; respiratory metabolism of plants; oxidation and phosphorylation by mitochondrial systems isolated from fruits and vegetables; oxidative enzymes; ethylene biosynthesis and physiology of ethylene action in plants; plant hormones; metabolic regulation. Mailing Add: Beltsville Agr Res Ctr West USDA Beltsville MD 20705

LIEBERMAN, MORTON LEONARD, b Chicago, Ill, Nov 22, 37; m 62; c 2. PHYSICAL CHEMISTRY. Educ: Ill Inst Technol, BS, 59, MS, 63, PhD(phys chem), 65. Prof Exp: Sr chemist, Res & Develop Labs, Corning Glass Works, 65-68; STAFF MEM TECH RES, SANDIA LABS, 68- Mem: Am Chem Soc; Inst Elec & Electronics Eng. Res: High-temperature chemistry, thermodynamics; phase transitions; thin films; optical properties; carbon research. Mailing Add: Sandia Labs Albuquerque NM 87115

LIEBERMAN, ROBERT, b Columbus, Ohio, Apr 9, 24; m 62; c 3. RADIOCHEMISTRY. Educ: Ohio State Univ, BA, 48, MSc, 52. Prof Exp: Chemist, Plastics Div, Battelle Mem Inst, 55-58, res scientist, Chem Physics Div, 58-64, sr chemist, 64-67, chief bioassay sect, Southeastern Radiol Health Lab, USPHS, 67-69, chief chem & biol, 69-71; chief phys sci br, Eastern Environ Radiation Lab, 71-74, CHIEF QUAL ASSURANCE SECT, EASTERN ENVIRON RADIATION FACIL, ENVIRON PROTECTION AGENCY, 74- Mem: Fel Am Inst Chemists; Health Physics Soc. Res: Polyurethane foams; fission gas release; neutron dosimetry; radiation effects on plastics; use of radiotracers on wear studies. Mailing Add: 3707 Laconia Lane Montgomery AL 36111

LIEBERMAN, SAMUEL VICTOR, b Philadelphia, Pa, Nov 3, 14; m 38; c 2. CHEMISTRY. Educ: Univ Pa, BS, 36, MS, 37, PhD(chem), 48. Prof Exp: Asst, Sharp & Dohme, Inc, 41-42; res chemist, Wyeth Inst Med Res, 45-47, sr res chemist, 48-55; scientist in charge, Phys Anal Dept, Prod Div, Bristol-Myers Co, 55-57, dir develop & phys sci, 57-60; CONSULT PHARMACEUT PROD, 61- Mem: AAAS; Sigma Xi; Am Statist Asn. Res: Research, development and testing of pharmaceutical products. Mailing Add: 33 Grand St Clark NJ 07066

LIEBERMAN, SEYMOUR, b New York, NY, Dec 1, 16; m 44; c 1. BIOCHEMISTRY. Educ: Brooklyn Col, AB, 36; Univ Ill, MS, 37; Stanford Univ, PhD(chem), 41. Prof Exp: Chemist, Schering Corp, 38-39; Rockefeller Found asst, Stanford Univ, 39-41; spec res assoc, Harvard Univ, 41-45; assoc, Sloan-Kettering Inst, 45-50; from asst prof to assoc prof, 50-62, PROF BIOCHEM, COL PHYSICIANS & SURGEONS, COLUMBIA UNIV, 62- Concurrent Pos: Mem panel steroids, Comt on Growth, Nat Res Coun, 46-50 & panel endocrinol, 55-56; traveling fel from Mem Hosp, New York to Basel, Switz, 46-47; mem endocrinol study sect, NIH, 58-63, mem, Insts, 59-65, chmn, 63-65, chmn gen clin res ctrs comn, 67-70; mem med adv comt, Pop Coun, 61-; assoc ed, J Clin Endocrinol & Metab, 63-67. Honors & Awards: Ciba Award, Endocrine Soc, 52 & Koch Award, 70. Mem: AAAS; Am Chem Soc; Am Soc Biol Chem; Endocrine Soc (vpres, 67, pres, 74). Res: Steroid chemistry and biochemistry; metabolism of hormones; steroid hormone-protein conjugates. Mailing Add: Col of Physicians & Surgeons Columbia Univ New York NY 10032

LIEBERMANN, LEONARD NORMAN, b Ironwood, Mich, May 14, 15; m 41; c 3. PHYSICS. Educ: Univ Chicago, BS, 37, MS, 38, PhD(physics), 40. Prof Exp: Instr physics, Wash Univ, 40-41; instr, Univ Kans, 41-43, asst prof, 43-44; prin physicist bur ships, Woods Hole Oceanog Inst, Mass, 44-46; res assoc, marine phys lab, 46-48, assoc prof geophys, 48-54, PROF PHYSICS, UNIV CALIF, SAN DIEGO, 54- Concurrent Pos: Guggenheim Found fel, 52-53; dir, Proj Sorrento, 59. Mem: Fel Am Phys Soc; fel Acoust Soc Am. Res: Ultrasonics; underwater sound; hydrodynamics; properties of liquids; electromagnetic propagation; solid state. Mailing Add: Dept of Physics Univ of Calif La Jolla CA 92037

LIEBERMANN, ROBERT C, b Ellwood City, Pa, Feb 6, 42; m 64; c 2. GEOPHYSICS. Educ: Calif Inst Technol, BS, 64; Columbia Univ, PhD(geophys), 69. Prof Exp: Res scientist, Lamont-Doherty Geol Observ, 69-70; res fel geophys, Calif Inst Technol, 70; MEM FAC, AUSTRALIAN NAT UNIV, 70- Concurrent Pos: Assoc ed, J Geophys Res, 73-76. Mem: Fel Royal Astron Soc; Am Geophys Union; Seismol Soc Am. Res: Relative excitation of seismic waves by earthquakes and underground explosions; elastic properties of minerals and rocks as a function of pressure and temperature; composition and mineralogy of earth's mantle. Mailing Add: Res Sch Earth Sci Australian Nat Univ Box 4 Canberra Australia

LIEBERSTEIN, HERBERT MELVIN, mathematics, see 12th edition

LIEBERT, WOLFGANG, b Woelfersheim, Ger, Jan 8, 36; m 68. ALGEBRA. Educ: Univ Frankfurt, PhD(math), 66. Prof Exp: Asst prof, 66-70, ASSOC PROF MATH, N MEX STATE UNIV, 70- Concurrent Pos: NSF grant, 69-71. Mem: Am Math Soc.

Res: Endomorphism rings of abelian groups. Mailing Add: Dept of Math NMex State Univ Las Cruces NM 88001

LIEBES, SIDNEY, JR, b San Francisco, Calif, Dec 13, 29; m 58; c 2. PHYSICS. Educ: Princeton Univ, BSE, 52; Stanford Univ, PhD(physics), 58. Prof Exp: Instr physics, Princeton Univ, 57-61, asst prof, 61-64; RES ASSOC-PHYSICIST, DEPT GENETICS, MED CTR, STANFORD UNIV, 64- Mem: Am Phys Soc; Am Asn Physics Teachers. Res: Experimental atomic and electron physics; gravitation experiments; mass spectrometry; physical microanalysis; techniques applied to biomedical research; computer imagery processing; Martian Lander imagery. Mailing Add: Dept of Genetics Stanford Med Ctr Stanford CA 94304

LIEBESKIND, HERBERT, b New York, NY, Nov 24, 21; m 43; c 2. PHYSICAL CHEMISTRY. Educ: NY Univ, BS, 41. Prof Exp: Asst instr chem, NY Univ, 43-45; from instr to assoc prof, 45-66, asst dean, 68-72, PROF CHEM, COOPER UNION, 66-, DIR ADMISSIONS & REGISTR, 70-, DEAN ADMISSIONS & RECORDS, 72- Concurrent Pos: Vis lectr, Stevens Inst Technol, 52-54; vis assoc prof, Yeshiva Univ, 61-62. Mem: AAAS; Am Chem Soc; Am Soc Eng Educ; NY Acad Sci. Mailing Add: Off of Admissions & Registr Cooper Union New York NY 10003

LIEBHAFSKY, HERMAN ALFRED, b Zwittau, Austria-Hungary, Nov 18, 05; US citizen; m 35; c 2. CHEMISTRY. Educ: Agr & Mech Col, Tex, BS, 26; Univ Nebr, MS, 27; Univ Calif, PhD(chem), 29. Prof Exp: Instr phys & inorg chem, Univ Calif, 29-34; inorg & anal chemist, Consult & Res Assoc, Gen Elec Co, 34-51, mgr phys chem res, 51-65, electrochem br, 65-67; prof, 67-74, EMER PROF CHEM, TEX A&M UNIV, 74- Honors & Awards: Fisher award, Am Chem Soc, 61. Mem: AAAS; Am Chem Soc; Electrochem Soc; Soc Appl Spectros. Res: Photoelectric spectrophotometry and x-ray methods in chemical analysis; kinetics of reactions in solution; chemistry of the mercury boiler and of amalgrams; rocket propellants; catalyses in homogeneous system; hydrolysis and hydration equilibria of halogens; statistics in analytical chemistry; analytical methods of silicones; corrosion; constitution of hydroxyanthraquinone lakes; fuel cells and batteries. Mailing Add: Dept of Chem Tex A&M Univ College Station TX 77843

LIEBHARDT, WILLIAM C, b Duluth, Minn, Feb 16, 36; m 61; c 4. SOILS, PLANT PHYSIOLOGY. Educ: Univ Wis-Madison, BS, 58, MS, 64, PhD(soils), 66. Prof Exp: Agronomist, Stand Fruit Co, 66-68; sr agronomist, Allied Chem Corp, 68-69; ASST PROF PLANT SCI, UNIV DEL, 70- Res: Soil fertility; plant nutrition; waste disposal; water use. Mailing Add: Dept of Plant Sci Univ of Del Newark DE 19711

LIEBLEIN, JULIUS, b New York, NY, Aug 5, 14; m 43; c 2. MATHEMATICAL STATISTICS. Educ: City Col, BS, 35; Brooklyn Col, MA, 40; American Univ, PhD(math), 53. Prof Exp: Teacher high sch, NY, 36-37; actuarial clerk, State Ins Fund, 37-39; statistician, New York Bd Ed, 42-43; econ analyst, Div Tax Res, US Dept Treas, DC, 43-47; mathematician & math statistician, Statist Eng Lab, Nat Bur Standards, 47-56; mathematician, Appl Math Lab, David Taylor Model Basin, US Navy Dept, 56-64; math statistician, Off Statist Prog, Bur Finance & Admin, US Post Off Dept, 64-68; opers res analyst, Tech Anal Div, Nat Bur Standards, 68-75; CONSULT EXTREME VALUES AND STATIST, 75- Honors & Awards: Belden Math Medal, 33. Mem: Assoc Casualty Actuarial Soc; Am Statist Asn; Opers Res Soc Am; Int Statist Inst. Res: Environmental problems; reactor siting; application of operations research to health fields. Mailing Add: 1621 E Jefferson St Rockville MD 20852

LIEBLING, RICHARD STEPHEN, b Brooklyn, NY, Aug 31, 38; m 70. MINERALOGY. Educ: Columbia Univ, BA, 60, MA, 61, PhD(mineral), 63. Prof Exp: Sr ceramist, Carborundum Co, 63-68; asst prof, 68-73, ASSOC PROF GEOL, HUNTER COL, 73- Mem: AAAS; Mineral Soc Am; Geol Soc Am. Res: Clay mineralogy of sediments. Mailing Add: Dept of Geol & Geog Hunter Col 695 Park Ave New York NY 10021

LIEBMAN, ALAN JOEL, b New York, NY, US citizen. PHYSICS, ELECTROPHOTOGRAPHY. Educ: Columbia Univ, BA, 64; Yale Univ, MS, 66, PhM, 68, PhD(appl sci), 70. Prof Exp: SCIENTIST, XEROX CORP, 69- Concurrent Pos: Adj prof, Rochester Inst Technol, 73- Mem: Am Asn Physics Teachers; Soc Photog Sci & Eng. Res: Development of electrostatic latent images; physics of fine particles; theory of phase transitions. Mailing Add: 266 Willowcrest Dr Rochester NY 14618

LIEBMAN, ARNOLD ALVIN, b St Paul, Minn, Mar 5, 31; m 55; c 4. ORGANIC CHEMISTRY, RADIOCHEMISTRY. Educ: Univ Minn, BS, 56, PhD(pharmaceut chem), 61. Prof Exp: Asst prof biochem, Loyola Univ, La, 61-63; Nat Inst Gen Med Sci res fel chem, Univ Calif, Berkeley, 63-66; asst prof chem, Sch Pharm, Univ Md, 66-68; sr chemist, 68-72, RES GROUP CHIEF, HOFFMANN-LA ROCHE, INC, 72- Mem: AAAS; fel Am Inst Chem; Am Chem Soc. Res: Heterocyclic chemistry of natural products; isotopic synthesis, heterocyclic chemistry. Mailing Add: Chem Res Dept Hoffmann-La Roche Inc Nutley NJ 07110

LIEBMAN, FREDERICK MELVIN, b New York, NY, July 26, 22; m 48; c 3. PHYSIOLOGY. Educ: NY Univ, BA, 42, PhD, 56; Univ Pa, DDS, 47. Prof Exp: Asst, 53-56, from instr to assoc prof, 56-65, PROF PHYSIOL & BIOPHYS, COL DENT, NY Univ, 65-, CHMN DEPT, 69- Concurrent Pos: Int Asn Dent Res rep, Int Cong Physiol, Buenos Aires, Arg, 59. Mem: AAAS; Am Physiol Soc; Harvey Soc; NY Acad Sci; fel Am Col Dent. Res: Peripheral circulation; control of circulation in the dental pulp and oral cavity; functional activity of the muscles of mastication. Mailing Add: Dept of Physiol & Pharmacol NY Univ Col of Dent New York NY 10010

LIEBMAN, JOEL FREDRIC, b Brooklyn, NY, May 6, 47; m 70. THEORETICAL CHEMISTRY. Educ: Brooklyn Col, BS, 67; Princeton Univ, MA, 68, PhD(chem), 70. Prof Exp: NATO fel, Depts Phys & Theoret Chem, Cambridge Univ, 70-71; Nat Res Coun & Nat Bur Stand fel, Inorg Chem Sect, Nat Bur Stand, 71-72; ASST PROF, DEPT CHEM, UNIV MD, BALTIMORE COUNTY, 72- Concurrent Pos: Ramsay hon fel, Ramsay Mem Fel Trust, 70; unofficial consult, Nat Bur Stand, 72-; unofficial consult, Argonne Nat Lab, 72-75, guest scientist, 75- Mem: Am Chem Soc; Am Phys Soc. Res: Chemical bonding theory, rules and regularities of molecular geometry and energetics; strain and resonance energy of alicyclic and aromatic hydrocarbons; noble gas and fluorine compounds; boron, carbon and nitrogen hydrides. Mailing Add: Dept of Chem Univ of Md Baltimore Co Baltimore MD 21228

LIEBMAN, JUDITH STENZEL, b Denver, Colo, July 2, 36; m 58; c 3. OPERATIONS RESEARCH. Educ: Univ Colo, BA, 58; Johns Hopkins Univ, PhD(opers res), 71. Prof Exp: Engr data anal, Convair Astronaut, Gen Dynamics, 58-59; programmer eng systs, Gen Elec Co, 63-64; programmer chem, Cornell Univ, 64-65; res asst opers res, Johns Hopkins Univ, 65-71, asst prof & health serv res scholar, 71-72; ASST PROF OPERS RES, UNIV ILL, URBANA, 72- Mem: Opers Res Soc Am; Am Inst Indust Engrs; Am Pub Health Asn. Res: Mathematical optimization; model building; applications of operations research in health and engineering. Mailing Add: 2110 Civil Eng Univ of Ill Urbana IL 61801

LIEBMAN, PAUL ARNO, b Pittsburgh, Pa, Aug 1, 33. BIOPHYSICS, PHYSIOLOGY. Educ: Univ Pittsburgh, BS, 54; Johns Hopkins Univ, MD, 58. Prof Exp: Intern internal med, Barnes Hosp, St Louis, Mo, 58-59; res assoc physiol, 63-65, asst prof, 65-69, ASSOC PROF ANAT, UNIV PA, 69- Concurrent Pos: Fel biophys, Univ Pa, 59-63. Mem: Biophys Soc; Optical Soc Am. Res: Vision; microspectrophotometry of single visual receptors; transducer mechanism of photoreceptors in vision. Mailing Add: Dept of Anat Univ of Pa Philadelphia PA 19104

LIEBMAN, SAMUEL, b New York, NY, May 3, 21; m 47; c 2. POLYMER CHEMISTRY, PHYSICAL CHEMISTRY. Educ: Brooklyn Col, BS, 42; Am Univ, MS, 60. Prof Exp: Chemist, US Sci Labs, 46-48; mgr, Valjean Corp, 48-57; res chemist, W R Grace & Co, 57-63; DEP DIR SCI DIV, SMITHSONIAN SCI INFO EXCHANGE, 63- Concurrent Pos: Mem-at-large, US Nat Comt Int Bldg Res, Nat Acad Sci, 70- Mem: AAAS; Am Inst Chem (secy, 67); Am Chem Soc. Res: Catalysis, polymerization, polymer modification and process development; information science; storage, retrieval and dissemination of on-going research. Mailing Add: Smithsonian Sci Info Exchange 1730 M St NW Washington DC 20036

LIEBMAN, SUSAN WEISS, b New York, NY, Dec 2, 47; m 69; c 1. MOLECULAR GENETICS. Educ: Mass Inst Technol, BS, 68; Harvard Univ, MA, 69; Univ Rochester, PhD(biophys), 74. Prof Exp: AM CANCER SOC FEL BIOPHYS, SCH MED & DENT, UNIV ROCHESTER, 74- Mem: AAAS; Genetics Soc Am. Res: Molecular genetics of yeast, including nonsense suppression, mutators, m-RNA processing. Mailing Add: Dept of Radiation Biol & Biophys Univ Rochester Sch Med & Dent Rochester NY 14642

LIEBNER, EDWIN J, b Chicago, Ill, July 12, 21; m 63; c 2. RADIOLOGY. Educ: Univ Ill, BS, 44, MD, 46. Prof Exp: Resident radiol, Ill Res Hosps, 53-56; from asst prof to assoc prof, 56-66, PROF RADIOL, UNIV ILL HOSP, 66-, ACTG HEAD DEPT RADIOL, UNIV ILL COL MED, 71-; DIR RADIOTHER DIV, ILL RES & EDUC HOSPS, 61- Concurrent Pos: Consult radiol, Vet Admin Hosp, Hines, 64- Mem: Am Radium Soc; Roentgen Ray Soc; Am Soc Therapeut Radiol; Radiol Soc NAm. Res: Therapeutic lymphography; refrigeration and irradiation; therapeutic pediatric radiology. Mailing Add: Dept of Radiol Col of Med Univ of Ill Chicago IL 60612

LIEBNITZ, PAUL W, b Kansas City, Mo, Jan 18, 35; m 61; c 3. MATHEMATICS. Educ: Rockhurst Col, BS, 55; Univ Kans, MA, 57, PhD(math), 64. Prof Exp: Asst prof, 61-67, ASSOC PROF MATH, UNIV MO-KANSAS CITY, 67- Mem: Am Math Soc; Math Asn Am. Res: Topology, theory of retracts. Mailing Add: Dept of Math Univ of Mo Kansas City MO 64110

LIEBOW, AVERILL ABRAHAM, b Austria, Mar 31, 11; nat US; m 46; c 3. PATHOLOGY. Educ: City Col New York, BS, 31; Yale Univ, MD, 35. Prof Exp: Prof path, Sch Med, Yale Univ, 51-68; prof path & chmn dept, 68-75, EMER PROF PATH, SCH MED, UNIV CALIF, SAN DIEGO, 75- Concurrent Pos: Res consult, Armed Forces Inst Path, 46-; mem comt path, Nat Acad Sci-Nat Res Coun, 53-63, chmn comt, 58-63; mem path adv coun, Vet Admin, 65-74 & cardiopulmonary training comt, 68-71. Mem: Soc Exp Biol & Med; Am Asn Path & Bact; Int Asn Med Mus (pres, 53); hon fel Am Col Chest Physicians. Res: Radiation effects; cancer, especially of the lungs; pulmonary circulation and disease.

LIEBSCHUTZ, ALAN MORTON, b Chicago, Ill, Jan 12, 26; m 68; c 3. PHYSICS. Educ: Purdue Univ, BS, 47, MS, 49, PhD(physics), 53. Prof Exp: Instr physics, Purdue Univ, 47-53; sr nuclear engr, Convair Div, Gen Dynamics Corp, 53-55; staff specialist, Lockheed Aircraft Corp, 55, opers res scientist, 55-57, staff scientist, 57-58, div scientist, 58-59, proj mgr, 59-62; asst mgr, Radiation Effects Res Dept, Hughes Aircraft Co, 62-68; sr scientist, Autonetics Div, NAm Rockwell Corp, 68-69, sci dir underground nuclear testing, 69; MGR ELECTRONIC & ELECTROMAGNETIC HARDNESS DEPT, VULNERABILITY & HARDNESS LAB, SCI & TECHNOL DIV, TRW SYSTS GROUP, 69- Mem: Am Phys Soc; Inst Elec & Electronics Eng; Sigma Xi. Res: Effects of nuclear radiation on space and missile parts, circuits, boxes, subsystems and systems. Mailing Add: 407 Rosarita Dr Fullerton CA 92635

LIEBSON, SIDNEY HAROLD, b New York, NY, July 9, 20; m 47; c 2. PHYSICS. Educ: City Col New York, BS, 39; Univ Mich, MS, 40; Univ Md, PhD(physics), 47. Prof Exp: Physicist, Naval Res Lab, 40-49, head electromagnetics br, 49-55; mgr res & develop, Nuclear Develop Corp Am, 55-59; asst dir physics, Armour Res Found, Ill Inst Technol, 59-60; mgr phys res, Nat Cash Register Co, 60-66; mgr xerographic technol, 66-69, sr corp planner, 69-74, MGR MFG RES & DEVELOP, XEROX CORP, 74- Mem: Fel Am Phys Soc; Inst Elec & Electronics Eng. Res: Solid state phenomena; Geiger counters; electronic circuit analysis and design; discharge mechanism of self-quenching Geiger-Müller counters; scintillation and fluorescence of organics; photoconductivity; manufacturing technologies. Mailing Add: Xerox Corp Stamford CT 06904

LIECHTY, RICHARD D, b Geneva, Wis, Oct 20, 25; m 52; c 3. SURGERY, ENDOCRINOLOGY. Educ: Yale Univ, BA, 50; Northwestern Univ, MD, 54; AM Bd Surg, dipl, 61. Prof Exp: From asst prof to assoc prof surg, Univ Iowa, 65-72; prof surg, Univ Colo, 72-74. Concurrent Pos: NIH fel cancer, 57-59 & fel surg, Univ Mich, 59. Mem: Am Col Surg; Am Thyroid Soc. Res: Cancer-endocrinology. Mailing Add: 4200 E 9th Ave Denver CO 80220

LIEDTKE, CLAUS-EBERHARD, b Leipzig, Ger, Mar 21, 42; m 73; c 1. HEALTH SCIENCES, COMPUTER SCIENCES. Educ: Tech Univ, Dipl Ing, 68, PhD(elec eng), 72. Prof Exp: Res asst, Heinrich Hertz Inst Schwingungsforschung, Berlin, 68-70 & Tech Univ, Berlin, 70-72; ASST PROF HEALTH COMPUT SCI, UNIV MINN, 73-, DIR COMPUT OPER, ELECTROCARDIOGRAM-VECTORCARDIOGRAM LAB, 75- Concurrent Pos: Consult, Vet Admin Hosp, Minneapolis, 75- Mem: Ger Info Soc; Inst Elec & Electronics Engrs; Sigma Xi. Res: Digital processing of biomedical signals such as electroencephalograms, evoked responses and vectorcardiograms; image processing for radiological applications; classification of bacteria, vectorcardiograms, evoked responses. Mailing Add: Div Health Comput Sci Box 511 Univ Minn Mayo Mem Bldg Minneapolis MN 55455

LIEDTKE, JAMES DALE, b Waco, Nebr, May 20, 37; m 61; c 2. CHEMISTRY. Educ: Univ Portland, BS, 59; Wash State Univ, PhD(chem), 64. Prof Exp: Instr chem, Whittier Col, 63-65; asst prof, 65-72, ASSOC PROF CHEM, ORE COL EDUC, 72- Concurrent Pos: NSF acad year exten grant, 66-68. Mem: Am Chem Soc. Res: Organoboron chemistry. Mailing Add: Dept of Sci & Math Ore Col of Educ Monmouth OR 97361

LIEF, FLORENCE SUSKIND, b New York, NY, July 30, 11; m 33; c 3. MEDICAL MICROBIOLOGY, VETERINARY VIROLOGY. Educ: Barnard Col, BA, 31; NY Univ, MSc, 33; Univ Pa, PhD(med microbiol), 55; Am Bd Microbiol, dipl. Prof Exp:

Asst bact, Col Med, NY Univ, 31-34; bacteriologist dept med, NY Hosp-Cornell Med Ctr, 34-40; mem res staff, Children's Hosp, Philadelphia, 54-55; asst prof microbiol, NY Med Col, 55-56; assoc virol, Dept Prev Med, Pub Health & Pediat, Sch Med, 56-59, asst prof, 59-65, assoc prof pediat, Sch Med & Microbiol, Sch Vet Med, 65-71, PROF MICROBIOL, SCH VET MED, UNIV PA, 71- Concurrent Pos: Mem res staff, Children's Hosp, Philadelphia, 56-66; consult animal influenza, WHO, 61- Mem: Fel Am Acad Microbiol; Am Soc Microbiol; Am Asn Immunol; NY Acad Sci. Res: Influenza and parainfluenza viruses of man and lower animals; human papilloma virus; viral etiology of multiple sclerosis. Mailing Add: Univ of Pa Lippincott Bldg Philadelphia PA 19103

LIEF, HAROLD ISAIAH, b New York, NY, Dec 29, 17; m 61; c 5. PSYCHIATRY. Educ: Univ Mich, AB, 38; NY Univ, MD, 42; Columbia Univ, cert psychoanal, 50. Hon Degrees: MA, Univ Pa, 71. Prof Exp: Intern, Queens Gen Hosp, Jamaica, NY, 42-43; resident psychiat, Long Island Med Col, 46-48; res asst, Col Physicians & Surgeons, Columbia Univ, 49-51; from asst prof to prof psychiat, Sch Med, Tulane Univ, 51-67; PROF PSYCHIAT, SCH MED, UNIV PA, 67-, DIR DIV FAMILY STUDY, 67-, DIR MARRIAGE COUN OF PHILADELPHIA & CTR STUDY SEX EDUC IN MED, 68- Concurrent Pos: Vis prof, Sch Med, Univ Va, 58; pres, Sex Info & Educ Coun of US, 68-70; consult, Dept Health, Educ & Welfare, 69-76, WHO, 71 & 74, AMA, 70-75 & Psychiat Educ Br, NIMH, 74-75. Mem: Fel Am Acad Psychoanal (pres, 67-68); fel Am Psychiat Asn; fel Am Col Psychiat; fel Am Col Psychoanal; Am Psychosomatic Soc. Res: Marital and sexual relations; sex education in medicine; psycho-endocrinological-pharmacologic aspects of human sexuality. Mailing Add: Div of Family Study Rm 210 4025 Chestnut St Philadelphia PA 19104

LIEGEY, FRANCIS WILLIAM, b Frenchville, Pa, Jan 4, 23; m 47; c 6. MICROBIOLOGY. Educ: St Bonaventure Univ, BS, 47, MS, 50, PhD(microbiol), 59. Prof Exp: From instr to assoc prof biol, St Bonaventure Univ, 48-64; PROF BIOL, IND UNIV PA, 64-, CHMN DEPT, 72- Mem: Am Soc Microbiol. Res: Microbial ecology of acid mine streams. Mailing Add: Dept of Biol Weyandt Hall Ind Univ of Pa Indiana PA 15701

LIEGNER, LEONARD M, b Brooklyn, NY, May 22, 19; m 41; c 2. ONCOLOGY. Educ: NY Univ, BA, 41, MD, 44; Royal Col Surgeons & Royal Col Physicians, dipl, 55; Am Bd Radiol, dipl, 56. Prof Exp: Asst vis surgeon, Col Med, NY Univ, 53-54; ASST CLIN PROF RADIOL, COL PHYSICIANS & SURGEONS, COLUMBIA UNIV, 66-; DIR RADIATION THER, ST LUKE'S HOSP CTR & WOMAN'S HOSP, 58- Concurrent Pos: Nat Found Infantile Paralysis fel, 49; Am Cancer Soc fel surg cancer, Col Med, NY Univ, 53-54; Nat Cancer Inst fel, 54-55; fel, Holt Radium Inst, 55-56; fel radiation ther, Mem Hosp Cancer & Allied Dis, New York, 56-57; Firestone spec fel, 57; res fel exp path, Sloan-Kettering Inst Cancer Res, 57; consult, St Barnabas Hosp, 60-; asst attend radiologist, Presby Hosp, New York, 69- Honors & Awards: Distinguished fel, Am Col Nuclear Med. Mem: Am Col Radiol; Radiol Soc NAm; Am Radium Soc; James Ewing Soc; Am Soc Therapeut Radiol. Res: Radiation therapy of cancer in larynx, uterus, cervix, bladder; skin reactions to supervoltage radiations; techniques of isotope administration; supervoltage chest roentgenography; radioisotope seed implants for cancer. Mailing Add: St Luke's Hosp Ctr 421 W 113 St New York NY 10025

LIEM, KAREL F, b Java, Indonesia, Nov 24, 35; m 65. VERTEBRATE MORPHOLOGY. Educ: Indonesia Univ, BSc, 57, MSc, 58; Univ Ill, PhD(zool), 61. Prof Exp: Asst prof zool, Leiden Univ, 62-64; from asst prof to assoc prof anat, Univ Ill Col Med, 64-72; HENRY BRYANT BIGELOW PROF & CUR ICHTHYOL & PROF BIOL, HARVARD UNIV, 72- Concurrent Pos: Head, Div Vert Anat, Chicago Natural Hist Mus, Ill, 65-72; mem, Comt Latimeria, Nat Acad Sci, 67; Guggenheim fel, 70-71; mem, Vis Comt, New Eng Aquarium, 74-; trustee, Cohosset Marine Biol Sta, 74-; ed, Copeia & assoc ed, J Morphol, 74- Mem: Am Soc Zool; Am Soc Ichthyol & Herpet; Soc Syst Zool; fel Zool Soc London; Netherlands Royal Zool Soc. Res: Evolution of chordate structure; functional anatomy of teleosts; morphology and hydrodynamics of air-breathing teleost blood circulations; sex reversal in teleosts; functional anatomy and evolution of African cichlid fishes. Mailing Add: Mus of Comp Zool Harvard Univ Cambridge MA 02138

LIEMOHN, HAROLD BENJAMIN, b Minneapolis, Minn, Feb 2, 35; m 57; c 2. PLASMA PHYSICS, SPACE PHYSICS. Educ: Univ Minn, BA, 56, MS, 59; Univ Wash, PhD(physics), 62. Prof Exp: Teaching asst physics, Univ Minn, 56-59; staff mem geo-astrophys, Sci Res Labs, Boeing Co, 59-63; asst prof atmospheric & space sci, Southwest Ctr Advan Studies, 63-66; staff mem, Environ Sci Res Labs, Boeing Co, 66-74; MEM STAFF, BATTELLE NORTHWEST LAB, 74- Concurrent Pos: Adj asst prof, Southern Methodist Univ, 64-65; vis assoc prof, Univ Wash, 68-; chmn & secy local arrangements, Ann Meeting, Comt Space Res, Seattle, 71; reporter particle-wave interactions, Comn V, Int Asn Geomag & Aeronomy, 71-; mem, US Comn IV, Int Union Radio Sci. Res: AAAS; Am Phys Soc; Am Geophys Union; Am Astron Soc; Am Inst Aeronaut & Astronaut. Res: Theoretical research in radiation belt physics, waves in hot plasmas, whistler propagation characteristics, radiation in magnetoplasma, atmospheric gravity waves in ionosphere and hydromagnetic waves in magnetosphere. Mailing Add: Math Bldg Battelle Northwest Lab PO Box 999 Richland WA 99352

LIEN, ARTHUR PHILIP, b Liberal, Kans, May 27, 14; m 42; c 2. ORGANIC CHEMISTRY, RESOURCE MANAGEMENT. Educ: Ottawa Univ, Kans, AB, 37; Ohio State Univ, MSc, 39, PhD(org chem), 41. Hon Degrees: DSc, Ottawa Univ, Kans, 62. Prof Exp: Asst inorg chem, Ohio State Univ, 37-39, asst org chem, 39-40, spec asst, 40-41; res chemist, Stand Oil Co, Ind, 41-46, group leader, 46-51, sect leader, 51-53, asst div dir, 53-55, div dir, 55-58, dir petrol res, 58-60, mgr res & develop, Am Oil Co Div, 60-63; dir planning, Battelle Mem Inst, 63-68; dir corp chem res lab, Allied Chem Corp, 68-72, dir sci rels, 72-73; PRES, TECHNOL TRANSFER ASSOCS, 73- Concurrent Pos: Consult petrol, petrochem & pharmaceut cos, 73- Honors & Awards: Am Chem Soc Award, 54. Mem: AAAS; Am Chem Soc; Am Inst Chem; Soc Chem Indust; emer mem Indust Res Inst. Res: Homogeneous and heterogeneous catalysis; petroleum and petrochemical products and processes; agricultural chemicals; polymers, plastics and fibers; pollution control technology; technical and business planning; international development; science administration; technology acquisition and licensing. Mailing Add: 126 Goltra Dr Basking Ridge NJ 07920

LIEN, ERIC JUNG-CHI, b Kaohsiung, Taiwan, Nov 30, 37; m 65; c 2. PHARMACEUTICAL CHEMISTRY. Educ: Taiwan Univ, BS, 60; Univ Calif, San Francisco, PhD(pharmaceut chem), 66. Prof Exp: Res assoc bio-org chem, Pomona Col, 67-68; asst prof, 68-72, ASSOC PROF PHARMACEUT & BIOMED CHEM, SCH PHARM, UNIV SOUTHERN CALIF, 72- Mem: AAAS; Am Pharmaceut Asn; Am Chem Soc; Am Asn Cols Pharm. Res: Structure-activity relationship and bio-organic chemistry; physical organic chemistry; natural products. Mailing Add: Sect of Pharm & Biomed Chem Univ of Southern Calif Univ Park Los Angeles CA 90033

LIEN, ERIC LOUIS, b Hammond, Ind, Apr 9, 46; m 69; c 3. BIOCHEMISTRY. Educ: Col Wooster, BA, 68; Univ Ill, Urbana-Champaign, MS, 71, PhD(biochem), 72. Prof Exp: Fel biochem, Sch Med, Univ Pa, 72-75; SR BIOCHEMIST, WYETH LABS, 76- Mem: AAAS; Am Chem Soc. Res: Hypothalmic releasing factors; vitamin B12 absorption. Mailing Add: Wyeth Labs Box 8299 Philadelphia PA 19101

LIEN, YEONG-CHUNG EDMUND, b Fu-Kian, China, May 20, 47; US citizen; m 71; c 1. COMPUTER SCIENCES. Educ: Nat Taiwan Univ, BS, 68; Univ Calif, Berkeley, MS, 70, PhD(comput sci), 72. Prof Exp: Asst prof, 72-76, ASSOC PROF, DEPT COMPUT SCI, UNIV KANS, 76- Mem: Asn Comput Mach; Soc Indust & Appl Math. Res: Database management; theory of computation; machine organization. Mailing Add: Dept of Comput Sci Univ of Kans Lawrence KS 66045

LIENER, IRVIN ERNEST, b Pittsburgh, Pa, June 27, 19; m 46; c 2. BIOCHEMISTRY, NUTRITION. Educ: Mass Inst Technol, BS, 41; Univ Southern Calif, PhD(biochem, nutrit), 49. Prof Exp: Instr, 49-50, from asst prof to assoc prof, 50-59, PROF BIOCHEM, UNIV MINN, ST PAUL, 59- Concurrent Pos: Guggenheim fel, Carlsberg Lab, Copenhagen, Denmark, 57. Mem: Hon fel Venezuelan Asn Advan Sci; Am Chem Soc; Am Soc Biol Chem; Am Inst Nutrit. Res: Isolation and characterization of antinutritional factors in legumes; structure and mechanism of action of proteolytic enzymes and their naturally-occurring inhibitors. Mailing Add: Dept of Biochem Col of Biol Sci Univ of Minn St Paul MN 55101

LIENGME, BERNARD V F, b London, England, Feb 13, 39; m 64; c 3. PHYSICAL CHEMISTRY. Educ: Imp Col, London, BSc, 60, dipl & PhD(phys chem), 64. Prof Exp: Res chemist, Morganite Res, London, 60-61; fel surface chem, Mellon Inst, 64-66; teaching fel Mössbauer spectros, Univ BC, 66-68; asst prof, 68-72, ASSOC PROF CHEM, ST FRANCIS XAVIER UNIV, 72- Mem: Assoc Royal Inst Chem. Res: Surface chemistry and Mössbauer spectroscopy. Mailing Add: Box 28 St Francis Xavier Univ Antigonish NS Can

LIENHARD, GUSTAV E, b Plainfield, NJ, June 21, 38; m 60; c 2. BIOCHEMISTRY. Educ: Amherst Col, BA, 59; Yale Univ, PhD(biochem), 64. Prof Exp: From asst prof to assoc prof biochem & molecular biol, Harvard Univ, 65-72; assoc prof, 72-75, PROF BIOCHEM, DARTMOUTH MED SCH, 75- Concurrent Pos: Res fel biochem, Brandeis Univ, 63-65; NSF res grant, 66- Mem: Am Chem Soc; Am Soc Biol Chem. Res: Mechanisms of enzyme action; mechanisms of transport across biological membranes; mechanistic organic chemistry of reactions that are models for enzyme catalyzed reactions. Mailing Add: Dept of Biochem Dartmouth Med Sch Hanover NH 03755

LIENK, SIEGFRIED ERIC, b Gary, Ind, Oct 16, 16; m 51; c 1. ENTOMOLOGY. Educ: Univ Idaho, BS, 42; Univ Ill, MS, 47, PhD, 51. Prof Exp: Entomologist pear psylla control, USDA, 42, Alaska Insect Control Proj, 49; assoc prof fruit invests, 50-70, PROF ENTOM, STATE AGR EXP STA, STATE UNIV NY COL AGR, CORNELL UNIV, 70-, ENTOMOLOGIST, 50- Res: Biology, ecology and control of phytophagous mites; biology and control of stone fruit insects. Mailing Add: Dept of Entom NY State Agr Exp Sta Geneva NY 14456

LIEPINS, RAIMOND, b Plavinas, Latvia, May 19, 30; US citizen; m 61; c 4. POLYMER CHEMISTRY, ORGANIC CHEMISTRY. Educ: Southern Ill Univ, BA, 54; Univ Minn, MS, 56; Kans State Univ, PhD(org chem), 60. Prof Exp: Res chemist, B F Goodrich Co, 60-64; res assoc polymer res, Univ Ariz, 64-66; SR CHEMIST, RES TRIANGLE INST, 66- Mem: AAAS; Am Chem Soc; Am Phys Soc; fel Am Inst Chem; NY Acad Sci. Res: Flame retardance; low pressure plasma applications; high temperature polymers; plasticization. Mailing Add: Camille Dreyfus Lab Res Triangle Inst PO Box 12194 Research Triangle Park NC 27709

LIER, FRANK GEORGE, b New York, NY, Feb 19, 13; m 37. BOTANY, ECOLOGY. Educ: Columbia Univ, PhD, 50. Prof Exp: Asst bot, 46-47, lectr, Sch Gen Studies, 47-50, from asst to assoc prof, 50-67, PROF BOT, SCH GEN STUDIES, COLUMBIA UNIV, 67- Mem: Bot Soc Am; Torrey Bot Club (pres, 64); Am Inst Biol Sci; NY Acad Sci; Am Bryol & Lichenological Soc. Res: Plant morphology; developmental anatomy. Mailing Add: Dept of Biol Sci Columbia Univ New York NY 10027

LIER, JOHN, b Netherlands, Feb 23, 24; US citizen; m 50; c 2. GEOGRAPHY. Educ: Clark Univ, MA, 63; Univ Calif, Berkeley, PhD(geog), 68. Prof Exp: Asst prof geog, San Francisco State Col, 65-66; asst prof, Univ Hawaii, 66-67; asst prof, 68-71, ASSOC PROF GEOG, CALIF STATE UNIV, HAYWARD, 71- Mem: Am Geog Soc; Asn Am Geog; Asn Asian Studies; Netherlands Asn Econ & Social Geog. Res: Urban morphology as related to rural economy in Saskatchewan; geography study of man's mobility. Mailing Add: Dept of Geog 25800 Hillary St Calif State Univ Hayward CA 94542

LIES, THOMAS ANDREW, b Oak Park, Ill, Jan 16, 29; m 59; c 2. ORGANIC CHEMISTRY. Educ: John Carroll Univ, BS, 49; Univ Chicago, SM, 51; Univ Wis-Madison, PhD(org chem), 58. Prof Exp: Res chemist, Wyandotte Chem Corp, 58-59; RES CHEMIST, AM CYANAMID CO, 59- Mem: Am Chem Soc. Res: Synthesis of pesticides. Mailing Add: Cherry Hill Rd RD 5 Princeton NJ 08540

LIESE, HOMER C, b New York, NY, Oct 25, 31; m 55. MINERALOGY, PETROLOGY. Educ: Syracuse Univ, BS, 53; Univ Utah, MS, 57, PhD(mineral), 62. Prof Exp: X-ray technician, Kennecott Copper Inc, 60-61; ASSOC PROF GEOL, UNIV CONN, 62- Mem: Geol Soc Am; Mineral Soc Am; Soc Appl Spectros. Res: Spectroscopy of minerals. Mailing Add: Dept of Geol Univ of Conn Storrs CT 06268

LIETH, HELMUT H F, b Steeg/Am Kürten, Ger, Dec 16, 25; m 52; c 4. BOTANY, ECOLOGY. Educ: Univ Cologne, PhD(bot), 53; Agr Univ Stuttgart, Privat Dozent, 60. Prof Exp: Sci asst bot, Agr Univ Stuttgart, 55-60, lectr, 60-64, sr lectr, 64-66; lectr ecol, Tech Univ Stuttgart, 63-66; asst prof bot, Agr Univ Stuttgart, 66; prof, Univ Hawaii, 67; assoc prof, 67-70, PROF BOT & ECOL, UNIV NC, CHAPEL HILL, 70- Concurrent Pos: Guest prof, Cent Univ Venezuela, 61 & Univ del Tolima, Colombia, 63-64; Nat Acad Sci rep, Task Force Planning Global Network Environ Monitoring, 69-70; mem bd dir, Orgn Trop Studies, 71-; guest researcher, KFA Julich/W Ger, 73-74; ecol consult, Govt of Portugal, 72-74. Mem: Fel AAAS; Am Inst Biol Sci; Ecol Soc Am; Int Soc Biometeorol; Intecol. Res: Systems analysis; primary productivity; energy exchange; phenology; geobotany; environmental biophysics. Mailing Add: Dept of Bot Univ of NC Chapel Hill NC 27514

LIETMAN, PAUL STANLEY, b Chicago, Ill, Mar 24, 34; m 56; c 3. CLINICAL PHARMACOLOGY. Educ: Western Reserve Univ, AB, 55; Columbia Univ, MD, 59; Johns Hopkins Univ, PhD(physiol chem), 68. Prof Exp: Asst prof pediat, 68-72, asst prof pharmacol, 69-72, ASSOC PROF MED, PEDIAT & PHARMACOL, MED SCH, JOHNS HOPKINS UNIV, 72- Concurrent Pos: Consult, Am Hosp Formulary Serv, 67-; investr, Howard Hughes Med Inst, 68-72. Mem: Am Soc Pharmacol & Exp Therapeut; Soc Microbiol; Soc Pediat Res; Am Acad Pediat. Res: Developmental pharmacology; anticonvulsants; antibiotics. Mailing Add: Johns Hopkins Hosp Baltimore MD 21205

LIETZ, GERARD PAUL, b Chicago, Ill, Dec 10, 37; m 64; c 4. NUCLEAR PHYSICS. Educ: DePaul Univ, BS, 59; Univ Notre Dame, PhD(nuclear physics), 64. Prof Exp: Exchange asst nuclear physics, Univ Basel, 66-67; ASST PROF PHYSICS, DePAUL UNIV, 67- Mem: Am Phys Soc; Am Asn Physics Teachers. Res: Energy levels in nuclei. Mailing Add: Dept of Physics DePaul Univ Chicago IL 60604

LIETZE, ARTHUR, b Vancouver, BC, Oct 24, 28; nat US. IMMUNOCHEMISTRY, ALLERGY. Educ: Univ BC, BA, 51; Ind Univ, PhD, 57. Prof Exp: Chemist, Pac Oceanog Group, Can, 52; asst chem, Ind Univ, 53-56; res assoc allergy, Ore State Univ, 57-60, asst prof, 60-61; asst prof physiol chem, Univ Wis, 61-63; res assoc, Palo Alto Med Res Found, 63-65; immunochemist, Immunochem Dept, Samuel Merritt Hosp, 65-70; IMMUNOCHEMIST, IMMUNO-CHEM LABS, 71- Mem: AAAS; Am Chem Soc; Am Acad Allergy. Res: Food allergy; structure of allergens and proteins. Mailing Add: 2401 Atlantic Blvd Long Beach CA 90806

LIETZKE, MILTON HENRY, b Syracuse, NY, Nov 23, 20; m 43; c 3. PHYSICAL CHEMISTRY. Educ: Colgate Univ, BA, 42; Univ Wis, MS, 44, PhD(chem), 49. Prof Exp: Asst, Univ Wis, 42-43, instr chem, 43-44; lab foreman, Tenn Eastman Corp, 44-47; asst, Univ Wis, 47-49; RES CHEMIST, OAK RIDGE NAT LAB, 49- Concurrent Pos: Prof, Univ Tenn, 63- Mem: Am Chem Soc. Res: Electrochemistry, electrodeposition; potential measurements; high temperature solution thermodynamics; corrosion research; phase studies; application of high speed computing techniques to chemical problems. Mailing Add: 4165 Towanda Trail Knoxville TN 37919

LIEUX, MEREDITH HOAG, b Morgan City, La, Nov 9, 39; m 68. PALYNOLOGY, BOTANY. Educ: La State Univ, Baton Rouge, BS, 60, PhD(bot), 69; Univ Miss, MS, 64. Prof Exp: Teacher pub schs, Lake Charles & Monroe, La, 60-71; res assoc geol, 71-72, instr bot, 72-74, ASST PROF BOT, LA STATE UNIV, BATON ROUGE, 74- Mem: Am Asn Stratig Palynologists; Bot Soc Am; Sigma Xi. Res: Holocene spore and pollen studies in the Gulf of Mexico Region; pollen morphology involving light, scanning electron and transmission electron microscopy; applied insect-pollen related studies, or melissopalynology. Mailing Add: Box 155 Dept of Bot La State Univ Baton Rouge LA 70803

LIEVENSE, STANLEY JAMES, b Holland, Mich, Nov 20, 18; m 45; c 4. FISHERIES. Educ: Univ Mich, BS, 43. Prof Exp: Dist fisheries biologist, Inst Fisheries Res, Dept Natural Resources, Mich, 45-48, dist fisheries supvr, 48-65, tech asst to chief fisheries div, 65-74; MGR NATURAL RESOURCES, MICH TRAVEL BUR, 74- Mem: Am Fisheries Soc; Outdoor Writers Am. Res: Trout lakes; lake management; fisheries gear development. Mailing Add: Mich Travel Bur 300 S Capitol Suite 102 Lansing MI 48913

LIFLAND, LEONARD, organic chemistry, textiles, see 12th edition

LIFSON, NATHAN, b Minneapolis, Minn, Jan 30, 11; m 39; c 2. PHYSIOLOGY. Educ: Univ Minn, BA, 31, PhD(physiol), 43; Columbia Univ, MD, 37. Prof Exp: Intern, San Diego County Gen Hosp, 37-38; asst, 39-42, from instr to assoc prof, 42-49, PROF PHYSIOL, UNIV MINN, MINNEAPOLIS, 49- Mem: Am Physiol Soc; Soc Exp Biol & Med. Res: Secretion of gastric juice; glycogen deposition in liver; metabolism of perfused organs; transcapillary exchange; energy and material balance; intestinal absorption and secretion. Mailing Add: 405 Millard Hall Univ of Minn Minneapolis MN 55455

LIFTON, ROBERT JAY, b New York, NY, May 16, 26; m 52; c 2. PSYCHIATRY. Educ: NY Med Col, MD, 48. Hon Degrees: ScD, Lawrence Univ, 71 & Merrimac Col, 73; DHL, Wilmington Col, Ohio, 75. Prof Exp: Intern gen med, Jewish Hosp, Brooklyn, NY, 48-49; psychiat resident, State Univ NY Downstate Med Ctr, 49-51; res assoc, Asia Found, 54; mem fac, Wash Sch Psychiat, 54-55; res psychiatrist, Walter Reed Army Inst Res, 56; res assoc psychiat & assoc E Asian studies, Harvard Univ, 56-61; Found Fund Res Psychiat assoc prof, Yale Univ, 61-67, PROF PSYCHIAT & FEL, BRANFORD COL, YALE UNIV, 67- Concurrent Pos: Asia Found fel study Chinese communist thought reform, Hong Kong & Washington, DC, 54-55; Ford Found fel, 56-57 & Found Fund Res Psychiat fel, 58-61; res assoc, Univ Tokyo, 60-61; consult, Behav Sci Study Sect, NIMH, 62-64 & Comt Invasion Privacy, NY Bar Asn, 63-64; Comt Fac Res Int Studies res fel Japanese Youth, Japan, 64-67; coordr, Group Study Psychohist Process, 66-; res training & res in cross cult psychol patterns in Far East, Ford Found; fel, Branford Col, Yale Univ, 67-; Edward W Hazen Found fel study contemp psychol or protean style, 70. Honors & Awards: Nat Bk Award in Sci, 69; NY Soc Clin Psychol Pub Serv Award, 70; William V Silverberg Mem Lect Award, Am Acad Psychoanal, 71; Karen Horney Lectr Award, 72; Soc Adolescent Psychiat Distinguished Serv Award, 72; Mt Airy Found Gold Medal Award for Excellence in Psychiat, 73. Mem: AAAS; Am Psychiat Asn; Asn Asian Studies; Am Anthrop Asn; Am Acad Psychoanal. Res: Psychohistorical studies; relationship between individual psychology and historical change, including study of Japanese youth; individual patterns in extreme historical situations, including Chinese thought reform; long-term reactions to the atomic bomb in Hiroshima; death symbolism; comtemporary psychological styles. Mailing Add: Dept of Psychiat Yale Univ Sch of Med 34 Park St New Haven CT 06519

LIGENZA, JOSEPH RAYMOND, b Chicago, Ill, June 18, 24; m 48; c 3. SOLID STATE CHEMISTRY. Educ: Ill Inst Technol, BS, 51; Columbia Univ, MA, 52, PhD(phys chem), 54. Prof Exp: Res asst chem, Cyclotron Labs, Columbia Univ, 51-54; MEM TECH STAFF PHYS CHEM, BELL TEL LABS, 54- Mem: AAAS. Res: Silicon semiconductor surface physics and chemistry; chemical reactions in plasmas; plasma oxidation of metals and semiconductors; glass physics and chemistry. Mailing Add: Bell Tel Labs 600 Mountain Ave Murray Hill NJ 07974

LIGETT, WALDO BUFORD, organic chemistry, see 12th edition

LIGGERO, SAMUEL HENRY, b Amsterdam, NY, Apr 17, 42; m 66. PHYSICAL ORGANIC CHEMISTRY. Educ: Fordham Univ, BS, 64; Georgetown Univ, PhD(chem), 69. Prof Exp: NIH fel, Princeton Univ, 69-70; sr lab supvr film develop, 70-71, RES GROUP LEADER FILM DEVELOP, POLAROID CORP, 72- Mem: The Chem Soc; AAAS; Soc Photog Scientists & Engrs; Am Chem Soc. Res: Application of physical organic chemistry principles to the development of instant color photographic transparencies employing diffusion transfer processes. Mailing Add: 69 Sheridan Rd Wellesley Hills MA 02181

LIGGETT, LAWRENCE MELVIN, b Denver, Colo, June 22, 17; m 43; c 2. ANALYTICAL CHEMISTRY, ORGANIC CHEMISTRY. Educ: Cent Col, Iowa, AB, 38; Iowa State Col, PhD(chem), 43. Prof Exp: Res chemist, Nat Defense Res Comt, Iowa State Col, 41-43; plant supt, Alkali Chlorates & Perchlorates Cardox Corp, 43-48; supvr inorg res, Wyandotte Chems Corp, 48-55; dir res, Speer Carbon Co, 55-64, vpres & tech dir, 65-67, vpres & gen mgr, Airco Speer Electronics, 67-70, pres, Airco Speer Electronics Div, 70-75, PRES, VACUUM EQUIPMENT & SYSTS, AIRCO TEMESCAL DIV, AIRCO INC, 75- Mem: Am Chem Soc; Electrochem Soc. Res: Alkali perchlorate production; nonblack pigments for rubber and paper; carbon and graphite technology; resistors; capacitors and electronic components; technical management. Mailing Add: 622 Vermont Rd St Mary's PA 15857

LIGGETT, ROBERT WINSTON, organic chemistry, see 12th edition

LIGGETT, THOMAS MILTON, b Danville, Ky, Mar 29, 44. MATHEMATICS. Educ: Oberlin Col, AB, 65; Stanford Univ, MS, 66, PhD(math), 69. Prof Exp: Asst prof, 69-73, ASSOC PROF MATH, UNIV CALIF, LOS ANGELES, 73- Concurrent Pos: Sloan fel, 73. Mem: Am Math Soc; Math Asn Am; Inst Math Statist. Res: Probability theory. Mailing Add: Dept of Math Univ of Calif 405 Hilgard Ave Los Angeles CA 90024

LIGGETT, WALTER STEWART, JR, b Abington, Pa, Aug 27, 40; m 62; c 3. STATISTICS. Educ: Rensselaer Polytech Inst, BS, 61, MS, 64, PhD(math), 67. Prof Exp: Prin engr, Submarine Signal Div, Raytheon Co, Portsmouth, 65-73; mathematician, Rand Inst, New York, 73-75; STATISTICIAN, DIV ENVIRON PLANNING, TENN VALLEY AUTHORITY, 75- Mem: Inst Math Statist; Soc Indust & Appl Math; Am Statist Asn. Res: Water and air quality; aquatic biology; radiological hygiene; sampling; data analysis; experimental design. Mailing Add: Div of Environ Planning Tenn Valley Authority Chattanooga TN 37401

LIGH, STEVE, b Canton, China, Nov 12, 37; US citizen. MATHEMATICS. Educ: Univ Houston, BS, 61; Univ Mo-Columbia, MA, 62; Tex A&M Univ, PhD(math), 69. Prof Exp: Instr math, Ohio Univ, 62-64, Houston Baptist Col, 65-66 & Tex A&M Univ, 68-69; asst prof, Univ Fla, 69-70; assoc prof, 70-72, PROF MATH, UNIV SOUTHWESTERN LA, 72- Mem: Math Asn Am; Am Math Soc. Res: Algebra; generalizations of rings; near rings. Mailing Add: Dept of Math Univ of Southwestern La Lafayette LA 70501

LIGHT, ALBERT, b Brooklyn, NY, June 19, 27; m 52; c 2. BIOCHEMISTRY. Educ: City Col New York, BS, 48; Yale Univ, PhD(biochem), 55. Prof Exp: Fel biochem, Cornell Univ, 55-57; asst res prof, Univ Utah, 57-63; assoc prof, Univ Calif, Los Angeles, 63-65; ASSOC PROF BIOCHEM, PURDUE UNIV, 65- Mem: AAAS; Am Chem Soc; Am Soc Biol Chem; Am Asn Univ Prof. Res: Protein chemistry and enzymology; protein folding; relationship of structure to function of biologically active proteins. Mailing Add: Dept of Chem Purdue Univ West Lafayette IN 47906

LIGHT, AMOS ELLIS, b Greencastle, Ind, Oct 22, 10; m 43; c 2. PHARMACOLOGY, NUTRITION. Educ: DePauw Univ, AB, 30; Syracuse Univ, MA, 32; Army Med Col, cert, 43; Am Bd Nutrit, dipl, 51; Am Bd Clin Chem, dipl, 52. Prof Exp: Asst chem, DePauw Univ, 29-30; asst, Syracuse Univ, 30-32; asst, Yale Univ, 33-34; asst, Columbia Univ, 34-36; asst, Welcome Res Labs, 36-42, biochemist, 46-63; PHARMACOLOGIST, BUR MED, US FOOD & DRUG ADMIN, 63- Concurrent Pos: Nutritionist, Spanish Nutrit Surv Team, Int Comt Nutrit Nat Defense, 58. Honors & Awards: Cosmetic Indust Buyers & Suppliers Award, Toilet Goods Asn, 54. Mem: AAAS; Am Chem Soc; fel Am Asn Clin Chem; Am Inst Nutrit; fel NY Acad Sci. Res: Biological effects of x-rays; nucleic acids and anti-metabolites; pituitary growth hormones; vitamin B complex in rats and bacteria; chronic toxicity tests; hair growth; sex hormones; radiopaques; polymyxins; diuretics; antispasmodics; busulfan cataracts. Mailing Add: 6909 Cherry Lane Annandale VA 22003

LIGHT, DONALD WILLIS, b Norton, Kans, Nov 17, 11; m 38; c 4. CHEMISTRY. Educ: Harvard Univ, BS, 32; Stanford Univ, MA, 33; Technische Hochschule Dresden, Ger, Dr Ing(chem), 34. Prof Exp: Chemist, Anaconda Copper Co, NJ, 35-37; res chemist, Am Cyanamid Co, Conn, 37-41; dir res, Ludlow Mfg & Sales Co, Mass, 41-50 & Angier Corp, 51-53; PRES & TECH CONSULT, PEPPERELL BRAIDING CO, INC, 53- Mem: AAAS; Am Chem Soc. Res: Physical and colloidal chemistry. Mailing Add: Pepperell Braiding Co Inc Pepperell MA 01437

LIGHT, IRWIN JOSEPH, b Montreal, Que, July 21, 34; m 59; c 1. PEDIATRICS. Educ: McGill Univ, BS, 55, MD, 59. Prof Exp: Intern med, Royal Victoria Hosp, 59-60, resident, 60-61; resident pediat, Montreal Children's Hosp, 61-63; from asst prof to assoc prof pediat, 65-73, from asst prof to assoc prof obstet & gynec, 68-73, PROF PEDIAT, OBSTET & GYNEC, UNIV CINCINNATI, 73-, DIR NEWBORN CLIN SERV, 73-; RES ASSOC PEDIAT, CHILDREN'S HOSP RES FOUND, 63- Concurrent Pos: Clin fel pediat, Cincinnati Gen Hosp & res fel, Univ Cincinnati, 63-65. Mem: AAAS; Soc Pediat Res; Am Pediat Soc. Res: Neonatal infectious diseases; newborn metabolism. Mailing Add: Children's Hosp Res Found Elland & Bethesda Ave Cincinnati OH 45229

LIGHT, JOHN CALDWELL, b Mt Vernon, NY, Nov 24, 34; c 3. CHEMICAL PHYSICS. Educ: Oberlin Col, BA, 56; Harvard Univ, PhD(chem), 60. Prof Exp: NSF fel, Brussels, 59-61; instr chem, 61-63, from asst to assoc prof, 63-70, dir mat res lab, 70-73, PROF CHEM, UNIV CHICAGO, 70- Concurrent Pos: Sloan fel, 66. Mem: AAAS; Am Inst Phys; NY Acad Sci. Res: Theoretical studies of elementary gas phase reactions; quantum mechanics and chemical kinetics; scattering theory. Mailing Add: Dept of Chem Univ of Chicago Chicago IL 60637

LIGHT, JOHN HENRY, b Annville, Pa, Dec 15, 24; m 50; c 3. MATHEMATICS. Educ: Lebanon Valley Col, BS, 48; Pa State Univ, MS, 50 & 57. Prof Exp: Res assoc, Ord Res Lab, Pa State Univ, 51-58, eng mech, 58-59; assoc prof, 59-74, PROF MATH, DICKINSON COL, 74- Concurrent Pos: Consult, Naval Supply Depot, Mechanicsburg, Pa, 59-62. Mem: Am Math Soc. Res: Spectroscopy; environmental testing. Mailing Add: Dept of Math Dickinson Col Carlisle PA 17013

LIGHT, KENNETH KARL, b Palmyra, Pa, Nov 6, 40; m 64; c 2. ORGANIC CHEMISTRY. Educ: Lebanon Valley Col, BS, 62; Univ Del, PhD(org chem), 66. Prof Exp: Res chemist, Mobil Oil Corp, 66-68; PROJ LEADER, INT FLAVORS & FRAGRANCES, INC, 68- Mem: Am Chem Soc. Res: Synthesis of natural products; isolation and identification of natural products; synthesis of fragrance chemicals. Mailing Add: Int Flavors & Fragrances Inc 1515 Hwy 36 Union Beach NJ 07735

LIGHT, MERLE ROBERT, animal husbandry, see 12th edition

LIGHT, ROBLEY JASPER, b Roanoke, Va, Nov 8, 35; m 60; c 1. BIOCHEMISTRY, ORGANIC CHEMISTRY. Educ: Va Polytech Inst, BS, 57; Duke Univ, PhD(org chem), 61. Prof Exp: NSF fel biochem, Harvard Univ, 60-62; instr, 62-63, asst prof, 63-67, ASSOC PROF BIOCHEM, FLA STATE UNIV, 67-, ASSOC CHMN DEPT CHEM, 74- Concurrent Pos: USPHS res career develop award, 67-72. Mem: Am Chem Soc; Am Soc Biol Chemists. Res: Lipid metabolism, structure, and function; polyketides and other secondary metabolites of microorganisms. Mailing Add: Dept of Chem Fla State Univ Tallahassee FL 32306

LIGHT, RUPERT EDWIN, b Patrick Springs, Va, June 22, 18; m 52; c 3. POLYMER CHEMISTRY. Educ: Va Polytech Inst, BS, 41; Univ Ill, MS, 46, PhD(chem), 49. Prof Exp: Res chemist polymers, Cluett, Peabody & Co, Inc, 49-50, Becco Chem Div, Food Mach & Chem Corp, 51-57 & Hooker Chem Co, 57-58; chief chemist, Chem

Rubber Prod Inc, 58-62; group leader polymer res, Lucidol Div, Pennwalt Corp, 62-72; CONSULT, 73- Mem: Am Chem Soc. Res: Polymer synthesis, characterization and modification; protective coatings. Mailing Add: 250 Clearfield Dr Williamsville NY 14221

LIGHT, TRUMAN S, b Hartford, Conn, Dec 16, 22; m 46; c 3. ANALYTICAL CHEMISTRY. Educ: Harvard Univ, SB, 43; Univ Minn, MS, 49; Univ Rome, DrChem, 61. Prof Exp: Asst prof chem, Boston Col, 49-59; staff scientist, Res & Adv Develop Div, Avco Corp, 59-64; sr res chemist, 64-72, MGR CHEM ANAL & MAT LAB, FOXBORO CO, 72- Concurrent Pos: NSF fel, Chem Inst, Univ Rome, Italy, 60-61; consult, Children's Med Ctr, Boston, Mass, 56-60 & Watertown Arsenal, 51-55. Mem: Am Chem Soc; Electrochem Soc; Soc Appl Spectros; Instrument Soc Am. Res: Instrumental methods of analysis; electrochemistry; physical chemistry; materials sciences, water quality and pollution controls. Mailing Add: 4 Webster Rd Lexington MA 02173

LIGHTBODY, ALBERT, chemistry, see 12th edition

LIGHTBODY, DAVID (BRUCE), nuclear physics, see 12th edition

LIGHTBODY, JAMES JAMES, b Detroit, Mich, Mar 1, 39; m 64; c 2. IMMUNOLOGY, BIOCHEMISTRY. Educ: Wayne State Univ, BA, 61, BS, 64, PhD(biochem), 66. Prof Exp: Res assoc biochem & NIH trainee, Brandeis Univ, 67-69; instr pediat & Swiss Nat Sci Found grant, Univ Bern, 69-70; sr res assoc immunol, Basel Inst Immunol, 70-71; res assoc, Univ Wis, 71-72; ASST PROF BIOCHEM, SCH MED, WAYNE STATE UNIV, 72-, ASSOC IMMUNOL, 73- Concurrent Pos: Vis prof, Mem Sloan-Kettering Cancer Ctr, 73; lectr, Cancer Inst, Cairo Univ, 74. Mem: AAAS; Am Asn Immunol; Transplantation Soc. Mailing Add: Dept of Biochem Wayne State Univ Sch of Med Detroit MI 48201

LIGHTFOOT, DONALD RICHARD, b Los Angeles, Calif, Aug 8, 40. BIOCHEMICAL GENETICS. Educ: Univ Redlands, BA, 62; Univ Ariz, MS, 67, PhD(biochem), 72. Prof Exp: Teacher, Philippine High Sch, Peace Corps, 62-64; res trainee biochem, Med Sch, Univ Ore, 69-71; fel, Univ Calif, Riverside, 71-74; ASST PROF BIOCHEM & NUTRIT, VA POLYTECH INST, 74- Mem: AAAS; Am Chem Soc. Res: Plant virology; tobacco mosaic virus infection process; biochemical genetics; gene titration in tobacco species; minor nucleosides; transfer RNA and messenger RNA structure and function. Mailing Add: Dept of Biochem & Nutrit Va Polytech Inst Blacksburg VA 24061

LIGHTFOOT, HAIDEH NEZAM, b Iran; m 72. MOLECULAR BIOLOGY. Educ: Univ London, BSc, 66; Univ Teheran, MS, 69; Univ Ore, PhD(microbiol), 74. Prof Exp: Res scientist microbiol, Inst Pub Health Res, Teheran, 66-69; res biochemist, Univ Calif, Riverside, 73-74; ADJ ASST PROF MICROBIOL, VA POLYTECH INST & STATE UNIV, 75- Mem: Am Soc Microbiol. Res: Translational control of protein synthesis in eucaryotes; decline of protein synthesis in maturing reticulocytes; interaction of diphtheria toxin with mammalian protein synthesizing machinery; regulation of protein synthesis through phosphorylation-dephosphorylation of ribosomal proteins and cyclic nucleotides. Mailing Add: 608 Watson Lane Blacksburg VA 24060

LIGHTHALL, HARRY, JR, mathematics, see 12th edition

LIGHTLE, PAUL CHARLES, b Globe, Ariz, Nov 17, 18; m 40; c 5. FOREST PATHOLOGY. Educ: Univ Ariz, BS, 40, MS, 47; Univ Calif, PhD, 52. Prof Exp: Agent plant path, Bur Plant Indust, USDA, 40-41, agent forest path, 41-46, plant pathologist, Bur Plant Indust, Soils & Agr Engr, 46-54, plant pathologist, Div Forest Dis Res, Calif Forest & Range Exp Sta, Forest Serv, 54-56, plant pathologist, Southern Forest Exp Sta, Miss, 56-60, proj leader, Albuquerque Dis Lab, Rocky Mt Forest & Range Exp Sta, 60-73; RETIRED. Concurrent Pos: Officer-in-charge, Field Off, Div Forest Path, USDA, Ariz, 41-47 & Calif, 52-54. Mem: Am Phytopath Soc. Res: Deterioration of wood in use; deterioration of wood in standing timber; needle diseases of conifers. Mailing Add: 2405 Nolte Dr Prescott AZ 86301

LIGHTMAN, ALAN PAIGE, b Memphis, Tenn, Nov 28, 48. THEORETICAL ASTROPHYSICS, THEORETICAL PHYSICS. Educ: Princeton Univ, AB, 70; Calif Inst Technol, MA, 73, PhD(physics), 74. Prof Exp: Res assoc physics, Calif Inst Technol, 74; res assoc astrophys, Cornell Univ, 74-75, res assoc theoret astrophys, 75-76; ASST PROF ASTRON, HARVARD UNIV, 76- Mem: Int Soc Gen Relativity & Gravitation. Res: Theoretical frameworks for analyzing modern gravitation theories; relativistic astrophysics, particularly the interactions of neutron stars and black holes with their astrophysical environment. Mailing Add: Dept of Astron Harvard Univ Cambridge MA 02138

LIGHTNER, DAVID A, b Los Angeles, Calif, Mar 25, 39. ORGANIC CHEMISTRY. Educ: Univ Calif, Berkeley, AB, 60; Stanford Univ, PhD, 63. Prof Exp: NSF fels, Stanford Univ, 63-64 & Univ Minn, 64-65; asst prof chem, Univ Calif, Los Angeles, 65-72; assoc prof, Tex Tech Univ, 72-74; ASSOC PROF CHEM, UNIV NEV, RENO, 74- Mem: AAAS; Am Chem Soc; The Chem Soc; Ger Chem Soc; NY Acad Sci. Res: Photooxidation of biological materials; mass spectrometry; synthesis and stereochemistry; circular dichroism and optical rotatory dispersion; phototherapy and jaundice. Mailing Add: Dept of Chem Univ of Nev Reno NV 89507

LIGHTNER, JAMES EDWARD, b Frederick, Md, Aug 29, 37. MATHEMATICS, EDUCATION. Educ: Western Md Col, AB, 58; Northwestern Univ, AM, 62; Ohio State Univ, PhD(math, educ), 68. Prof Exp: Teacher, Frederick County Bd Educ, Md, 58-62; instr math, 62-65, asst prof, 65-68, chmn dept, 68-73, ASSOC PROF MATH, WESTERN MD COL, 68-, DIR JANUARY TERM, 69- Concurrent Pos: Fed Liaison Rep, Western Md Col, 73- Mem: Math Asn Am; Nat Coun Teachers Math. Res: Undergraduate mathematics curricula; secondary mathematics curricula and methodology. Mailing Add: Dept of Math Western Md Col Westminster MD 21157

LIGHTNER, JERRY P, b St Edward, Nebr, May 14, 29; m 53; c 2. BIOLOGY. Educ: Wayne State Col, BS, 52; Univ Northern Colo, MA, 54, EdD(sci educ), 61. Prof Exp: Teacher high schs, Nebr, 52-53 & Mont, 54-64; exec secy, 65-70, EXEC DIR, NAT ASN BIOL TEACHERS, 70- Concurrent Pos: Mem steering comt, Biol Sci Curric Study, 66-68. Mem: AAAS; Nat Asn Biol Teachers (secy-treas, 61-64). Res: Biological science education. Mailing Add: Nat Asn Biol Teachers 11250 Roger Bacon Dr Reston VA 22090

LIGHTSTONE, ALBERT HAROLD, b Ottawa, Ont, Nov 28, 26; m 54; c 3. MATHEMATICS. Educ: Carleton Col, Can, BSc, 52; Univ NB, MA, 53; Univ Toronto, PhD(math), 55. Prof Exp: Lectr math, Univ Alta, 55-57 & Univ Calif, 57-58; asst prof, Carleton Univ, Can, 58-60; assoc prof, Univ Victoria, BC, 60-61; assoc prof, Univ Victoria, BC, 61-65; assoc prof, 65-74, PROF MATH, QUEEN'S UNIV, ONT, 74- Concurrent Pos: Can Coun vis fel, Yale Univ, 71-72. Mem: Math Asn Am. Res: Mathematical logic; foundations of mathematics; nonstandard analysis. Mailing Add: Dept of Math Queen's Univ Kingston ON Can

LIGHTY, PAUL ELLIOTT, b Madison, Wis, Apr 9, 09; m 29; c 3. CHEMICAL PHYSICS. Educ: Univ Wis, PhB(chem), 36. Prof Exp: Chem analyst, Burgess Battery Co, Ill, 30-34; res chemist, Thomas A Edison, Inc, NJ, 36-38; spectroscopist, Int Nickel Co, WVa, 38-45; chemist, Sperry Gyroscope Co, NY, 45-48; spectroscopist, Lime Crest Res Lab, 48-50; mgr eng, ITT Fed Labs, 50-67; sr mem tech staff, Standard Telecommun Labs, Harlow, Eng, 68-70; mgr advan develop, ITT Lamp Div, 70-73; RETIRED. Concurrent Pos: Consult, ITT Corp, 73- Mem: AAAS; fel Royal Hort Soc; Soc Appl Spectros (pres, 54); Am Chem Soc; Electrochem Soc. Res: Instrumental analysis of metals; spectrographic analysis of nickel; alloys and biologic materials; solid state physics; ultra high purity materials; solid state electronic devices; spectroscopy; electrochemical devices; thermionic cathode research. Mailing Add: RD 1 Lafayette NJ 07848

LIGHTY, RICHARD WILLIAM, b Freeport, Ill, Nov 8, 33; m 55; c 2. PLANT GENETICS, HORTICULTURE. Educ: Pa State Univ, BS, 55; Cornell Univ, MS, 58, PhD(genetics), 60. Prof Exp: Geneticist, Longwood Gardens, Pa, 60-67; ASSOC PROF PLANT SCI & COORD LONGWOOD PROG ORNAMENTAL HORT, UNIV DEL, 67- Mem: AAAS; Am Asn Bot Gardens & Arboretums. Res: Plant breeding; cytotaxonomy; horticultural taxonomy; floriculture. Mailing Add: Agr Hall Univ of Del Newark DE 19711

LIGON, EDGAR WILLIAM, JR, b Lawrenceville, Va, Oct 25, 12; m 40; c 3. PHARMACOLOGY. Educ: Univ Richmond, BS, 32; Duke Univ, PhD(cytol), 38. Prof Exp: Instr zool, Univ Richmond, 33-35; asst, Duke Univ, 35-38; from instr to asst prof pharmacol, Sch Med, George Washington Univ, 39-48; pharmacologist in chg, Pharmacol & Rodenticide Lab, Prod & Mkt Admin, Insecticide Div, Agr Res Admin, USDA, 48-55, head, Pharmacol & Rodenticide Sect, Pesticide Regulation Div, 55-60; pharmacologist, Hazardous Substances Act, 61-68, chief hazardous substances br, 68-71, ASST TOXICOL, DIV CHEM HAZARDS, BUR PROD SAFETY, US FOOD & DRUG ADMIN, 71- Concurrent Pos: Fel biol, City Col New York. Mem: AAAS; Am Soc Pharmacol & Exp Therapeut. Res: Toxicology. Mailing Add: Div of Chem Hazards US Food & Drug Admin Bethesda MD 20016

LIGON, JAMES DAVID, b Wewoka, Okla, Feb 2, 39; m 67; c 1. ZOOLOGY. Educ: Univ Okla, BS, 61; Univ Fla, MS, 63; Univ Mich, PhD(zool), 67. Prof Exp: Asst prof biol, Idaho State Univ, 67-68; asst prof, 68-71, ASSOC PROF BIOL, UNIV N MEX, 71- Mem: Am Ornith Union; Cooper Ornith Soc. Res: Avian ecology and behavior. Mailing Add: Dept of Biol Univ of NMex Albuquerque NM 87106

LIGUORI, VINCENT ROBERT, b Brooklyn, NY, Dec 15, 28; m 47; c 5. MARINE MICROBIOLOGY. Educ: St Francis Col, NY, BS, 51; Long Island Univ, MS, 58; NY Univ, PhD(microbiol), 67. Prof Exp: Res asst cancer chemother, Sloan-Kettering Inst Cancer Res, 55-56; supvr oncol lab, Vet Admin Hosp, NY, 56-62; res asst microbiol, New York Aquarium, 62-65; asst prof biol & marine sci, Long Island Univ, 65-66; res assoc microbiol, Osborn Labs Marine Sci, New York Aquarium, 66-71; lectr, 66-68, ASSOC PROF BIOL & DEP CHMN DEPT BIOL SCI, KINGSBOROUGH COMMUNITY COL, 71- Concurrent Pos: Lectr, Nassau County Mus Natural Hist, 65-68, Richmond Col, 67- & Queens Col, 68-; mem bd dir, Mid Atlantic Natural Sci Coun, Inc, 75-; mem, Bermuda Biol Sta Res, 75- Mem: AAAS; Am Soc Zool. Res: Biological effects of natural products and the mechanism of adhesion in marine invertebrates; role of marine microorganisms in the disease processes of marine animals. Mailing Add: Dept Biol Sci 2001 Oriental Blvd Kingsborough Community Col Brooklyn NY 11235

LIKENS, GENE ELDEN, b Pierceton, Ind, Jan 6, 35; m 56; c 3. AQUATIC ECOLOGY, LIMNOLOGY. Educ: Manchester Col, BS, 57; Univ Wis, MS, 59, PhD(zool), 62. Prof Exp: Asst zool, Univ Wis, 57-61; instr, Dartmouth Col, 61; from proj asst to res assoc, Univ Wis, 62, res assoc meteorol, 62-63; instr biol sci, Dartmouth Col, 63, asst prof, 63-66, assoc prof, 66-69; assoc prof, 69-72, actg chmn sect, 73-74, PROF ECOL & SYSTS, CORNELL UNIV, 72- Concurrent Pos: Vis lectr, Univ Wis, 63; vis assoc ecologist, Brookhaven Nat Lab, 68; NATO sr fel, Eng & Sweden, 69; Guggenheim fel, 72-73; mem, US Nat Comt Int Hydrol Decade, 66-70; mem adv panel, US Senate Comt Pub Works, 70-73; US Nat Rep, Int Limnol Socs, 70-; mem ecol adv comt & sci adv bd, Environ Protection Agency, 74-; mem comt water qual policy, Nat Acad Sci, 73- Honors & Awards: Am Motors Conserv Award, 69. Mem: Fel AAAS; Am Polar Soc; Am Soc Limnol & Oceanog (vpres, 75-76, pres-elect, 76-77); Ecol Soc Am; Int Asn Theoret & Appl Limnol. Res: Circulation in lakes using radioactive tracers; meromictic lakes; biogeochemistry and analysis of ecosystems; antarctic and arctic limnology; precipitation chemistry. Mailing Add: Sect of Ecol & Systs Div of Biol Sci Cornell Univ Ithaca NY 14853

LIKES, CARL JAMES, b Charleston, SC, Sept 11, 16; m 43. PHYSICAL CHEMISTRY. Educ: Col Charleston, BS, 37; Univ Va, PhD(phys chem), 41. Prof Exp: Instr chem, Univ Va, 41-43; asst prof, Tulane Univ, 43, asst prof, 44-46; prof & head dept, Hampden-Sydney Col, 47-52; proj supvr, Va Inst for Sci Res, 52-58; PROF CHEM, COL CHARLESTON, 58- Mem: Am Chem Soc. Res: Electrophoretic and ultracentrifugal analysis of proteins. Mailing Add: Dept of Chem Col of Charleston Charleston SC 29401

LIKINS, ROBERT CAMPBELL, b Springfield, Mo, July 1, 21; m 43; c 2. BIOCHEMISTRY. Educ: Univ Kansas City, DDS, 45, BA, 46. Prof Exp: Carnegie fel, Univ Rochester, 45-46; dent officer, Nat Inst Dent Res, 46-62, chief extramural prog br, 62-68; PROF DENT SURG, UNIV CHICAGO & DIR, WALTER G ZOLLER MEM DENT CLIN, 68- Concurrent Pos: Mem dent study sect, USPHS, 60-64. Mem: Fel AAAS; Am Col Dent; Biophys Soc; Soc Exp Biol & Med; Int Asn Dent Res. Res: Biochemistry and biophysics of calcification; calcium metabolism; research administration. Mailing Add: 1842 Hanover Lane Flossmoor IL 60422

LIKOFF, WILLIAM, b Philadelphia, Pa, Feb 5, 12; m 32; c 2. MEDICINE. Educ: Dartmouth Col, BA, 33; Hahnemann Med Col, MD, 38. Prof Exp: Assoc prof med, 45-58, prof med, head cardiovasc div & dir cardiovasc inst, 59-68, DIR CATEGORIC PROGS, HAHNEMANN MED COL, 68- Mem: fel AMA; fel Am Heart Asn; fel Am Col Physicians; distinguished fel Am Col Cardiol (pres, 67-68); fel Am Col Chest Physicians. Res: Cardiovascular diseases. Mailing Add: Dept of Med Hahnemann Med Col Philadelphia PA 19102

LIKUSKI, HENRY JOHN, b Hillcrest, Alta, Sept 27, 35; m 58; c 4. ANIMAL NUTRITION, POULTRY NUTRITION. Educ: Univ Alta, BSc, 58, MSc, 59; Univ Ill, PhD(nutrit), 64. Prof Exp: Res assoc, Vet Admin Hosp & State Univ NY Downstate Med Ctr, 63-65; researcher, 65-68, sr scientist, 68-72, GROUP LEADER, RES CTR, CAN PACKERS LTD, 72- Res: Hypothalmic regulation of food intake; nutrient requirements of livestock and poultry; nutrient composition of feeds. Mailing Add: Can Packers Ltd Res Ctr 1211 St Clair Ave W Toronto ON Can

LIKUSKI, ROBERT KEITH, b Hillcrest, Alta, Oct 16, 37; m 71; c 1. BIOMEDICAL ENGINEERING. Educ: Univ Alta, BS, 59; Univ Ill, MS, 61, PhD(elec eng), 64. Prof Exp: Asst prof elec eng, Univ Tex, Austin, 65-70; proj engr comput memories, Micro-Bit Corp, 70-76; PROJ ENGR BIOMED ENG, BERKELEY BIO-ENG, INC, 76-

Mem: Inst Elec & Electronics Engrs. Res: Development of instrumentation for medical use. Mailing Add: Berkeley Bio-Eng Inc 600 McCormick St San Leandro CA 94577

LILES, JAMES NEIL, b Akron, Ohio, Apr 25, 30; m 55; c 3. COMPARATIVE PHYSIOLOGY, INSECT PHYSIOLOGY. Educ: Miami Univ, BA, 51; Ohio State Univ, MSc, 53; PhD(insect physiol), 56. Prof Exp: Asst prof biol, Univ SC, 56-58; res assoc entom, Ohio State Univ, 58-60; from asst prof to assoc prof, 60-71, PROF ENTOM, UNIV TENN, 71- Mem: AIAS; Am Soc Zool; Geront Soc; Entom Soc Am; Am Mosquito Control Asn. Res: Aging in insects; insect nutrient utilization. Mailing Add: Dept of Zool Univ of Tenn Knoxville TN 37916

LILES, SAMUEL LEE, b Texas City, Tex, June 24, 42; m 63; c 1. PHYSIOLOGY. Educ: McNeese State Col, BS, 64; La State Univ Med Ctr, New Orleans, PhD(physiol), 68. Prof Exp: Instr, 68-70, asst prof, 70-75, ASSOC PROF PHYSIOL, LA STATE UNIV MED CTR, NEW ORLEANS, 75- Concurrent Pos: Nat Inst Neurol Dis & Stroke grant, La State Univ Med Ctr, New Orleans, 70- Mem: AAAS; Am Physiol Soc; NY Acad Sci; Soc Neurosci. Res: Regional neurophysiology; experimental dyskinesia and correlated macroelectrophysiological and microelectrophysiological events in brain. Mailing Add: 4435 Baccich St New Orleans LA 70122

LILEY, NICHOLAS ROBIN, b Halifax, Eng, Dec 17, 36; m 61. ZOOLOGY. Educ: Oxford Univ, BA, 59, DPhil(zool), 64. Prof Exp: Nat Res Coun Can fel zool, 63-65, asst prof, 65-70, ASSOC PROF ZOOL, UNIV BC, 70- Mem: Animal Behav Soc; Can Soc Zool. Res: Comparative ethology and the evolution of behavior; endocrine mechanisms in control of behavior. Mailing Add: Dept of Zool Univ of BC Vancouver BC Can

LILEY, PETER EDWARD, b Barnstaple, Eng, Apr 22, 27; m 63; c 2. PHYSICS, CHEMICAL ENGINEERING. Educ: Univ London, BSc, 51, PhD(physics), 57; Imp Col, Univ London, dipl, 57. Prof Exp: Chem engr, Brit Oxygen Eng, Ltd, 55-57; from asst prof to assoc prof, 57-72, PROF MECH ENG, PURDUE UNIV, WEST LAFAYETTE, 72- Mem: Brit Inst Physics. Res: Thermodynamic and transport properties of fluids; cryogenic engineering; high pressure. Mailing Add: Sch of Mech Eng Purdue Univ West Lafayette IN 47907

LILIEN, OTTO MICHAEL, b New York, NY, Apr 26, 24; m; c 6. GENITOURINARY SURGERY. Educ: Jefferson Med Col, MD, 49; Columbia Univ, MA, 60. Prof Exp: Lectr zool, Columbia Univ, 56-58; from asst prof urol surg to assoc prof urol, 61-67, PROF UROL, STATE UNIV NY UPSTATE MED CTR, 67-, CHMN DEPT, 63- Concurrent Pos: Nat Cancer Inst trainee, 56-58. Mem: AMA; Am Urol Asn; fel Am Col Surg. Res: Renal and cell physiology. Mailing Add: Dept of Urol State Univ of NY Upstate Med Ctr Syracuse NY 13210

LILIENFELD, ABRAHAM MORRIS, b New York, NY, Nov 13, 20; m 43; c 3. EPIDEMIOLOGY, BIOSTATISTICS. Educ: Johns Hopkins Univ, AB, 41, MPH, 49; Univ Md, MD, 44. Hon Degrees: DSc, Univ Md, 75. Prof Exp: Assoc pub health physician, State Dept Health, NY, 49-50; dir southern health dist, City Dept Health, Baltimore, 50-52; lectr public health, Sch Hyg & Pub Health, Johns Hopkins Univ, 50-52, asst prof epidemiol, 52-54; assoc prof med statist, Med Sch, Univ Buffalo, 54-58; prof pub health asmin & dir div chronic dis, Sch Hyg & Pub Health, 58-70, prof epidemiol & chmn dept, Sch Hyg & Pub Health, 70-75, UNIV DISTINGUISHED SERV PROF EPIDEMIOL, JOHNS HOPKINS UNIV, 75- Concurrent Pos: Dir, Nat Gamma Globulin Eval Ctr, 53-54, consult, NIH, 57-; chief dept statist & epidemiol, Roswell Park Mem Inst, 54-58; mem, Nat Adv Heart Coun, 62-66; staff dir, President's Comn Heart Dis, Cancer & Stroke, 64-65. Honors & Awards: Bronfman Award, Am Pub Health Asn, John Snow Award, Epidemiol Sect. Mem: AAAS; Am Epidemiol Soc; Am Pub Health Asn; Am Statist Asn; Soc Epidemiol Res. Res: Epidemiology of cirrhosis and inflammatory bowel diseases; cancer; cardiovascular diseases. Mailing Add: 3203 Old Post Dr Pikesville MD 21208

LILIENFIELD, LAWRENCE SPENCER, b New York, NY, May 5, 27; m 50; c 3. MEDICINE, PHYSIOLOGY. Educ: Villanova Col, BS, 45; Georgetown Univ, MD, 49, MS, 54, PhD, 56; Am Bd Internal Med, dipl, 57. Prof Exp: Intern med, Georgetown Univ Hosp, 49-50, from jr asst resident to sr asst resident, 50-53, asst chief cardiovasc res lab & attend physician, 56; instr med, Med Sch, 55-57, instr physiol, 56-57, from asst prof to assoc prof med, physiol & biophys, 57-64, PROF PHYSIOL & BIOPHYS, SCHS MED &DENT, GEORGETOWN UNIV, 64-, CHMN DEPT PHYSIOL, SCH MED, 63- Concurrent Pos: Am Heart Asn res fel, 57; USPHS sr res fel, 59 & res career award, 63; attend physician, DC Gen Hosp, 56 & Vet Admin Hosp, 57; estab investr, Am Heart Asn, 58; consult, USPHS, 65-; vis prof, Univ Saigon; vis prof, Univ Tel-Aviv, 67-68; assoc, Comt Int Exchange Persons, 71- Mem: AAAS; Biophys Soc; Am Physiol Soc; Soc Exp Biol & Med; AMA. Res: Transcapillary exchange; hemodynamics; blood distribution in organs; renal concentrating mechanisms; method of blood flow measurement. Mailing Add: Dept of Physiol Sch of Med Georgetown Univ Washington DC 20007

LILIENTHAL, BERNARD, b Hastings, NZ, Dec 16, 25; Australian citizen; m 49; c 4. ORAL BIOLOGY, MICROBIOLOGY. Educ: Univ Sydney, BDS, 47, BSc, 49, DDSc, 56; Oxford Univ, DPhil, 52; FRACDS, 62. Prof Exp: Res officer, Inst Dent Res, Sydney, Australia, 52-56; res chemist, Colonial Star Refining Co, 56-57; sr lectr bact, Univ Melbourne, 59-64; pvt pract, 64-71; prof oral diag & radiol, Univ Sask, 71-72; PROF ORAL BIOL, FAC DENT, DALHOUSIE UNIV, 72- Concurrent Pos: Med Res Coun Australia sr res fel, Inst Dent Res, 58; vis prof dent res, Col Dent, Univ Ill, 64. Mem: Am Acad Dent Radiol; Int Asn Dent Res; Can Asn Clin Pharmacol & Chemother. Res: Biochemistry and microbiology of oral diseases. Mailing Add: Dept of Oral Biol Dalhousie Univ Fac of Dent Halifax NS Can

LILL, GORDON GRIGSBY, b Mt Hope, Kans, Feb 23, 18; m 43; c 2. MARINE GEOLOGY. Educ: Kans State Col, BS, 40, MS, 46. Hon Degrees: DSc, Univ Miami, 66. Prof Exp: Asst chief party, State Hwy Comn, Kans, 41; asst geol, Univ Calif, 46-47; head geophys br, US Off Naval Res, 47-59, earth sci adv, 59-60; corp res adv, Lockheed Aircraft Corp, 60-64; dir proj Mohole, NSF, 64-66; sr sci adv, Lockheed Aircraft Corp, 66-70; DEP DIR, NAT OCEAN SURV, NAT OCEANIC & ATMOSPHERIC ADMIN, US DEPT COM, 70- Concurrent Pos: Pvt res, US Nat Mus; geologist, Bikini Sci Resurv, 47; mem mineral surv, Cent & Western Prov, Liberia, 49-50; consult comt geophys & geol res & develop bd, Nat Mil Estab, 47-53; chmn panel oceanog, Int Geophys Year; mem comt, Am Miscellaneous Soc, Proj Mohole, Nat Acad Sci-Nat Res Coun; vchmn, Calif Adv Comn Marine & Coastal Resources, 68-70; mem adv coun, Inst Marine Resources, Univ Calif, 69-71. Mem: Fel AAAS; fel Geol Soc Am; Am Geophys Union; Marine Technol Soc. Res: Sedimentary petrology; submarine geology. Mailing Add: 9606 Hillridge Dr Kensington MD 20795

LILLARD, DORRIS ALTON, b Thompson Station, Tenn, July 17, 36. FOOD CHEMISTRY, BIOCHEMISTRY. Educ: Middle Tenn State Univ, BS, 58; Ore State Univ, MS, 61, PhD(food sci), 64. Prof Exp: Res fel lipid autoxidation, Ore State Univ, 58-64; asst prof food flavor chem, Iowa State Univ, 64-68; ASSOC PROF FOOD SCI, UNIV GA, 68- Mem: AAAS; Am Chem Soc; Am Oil Chem Soc; Inst Food Technol; Am Meat Sci Asn. Res: Flavor chemistry of foods; autoxidation of lipids; mycotoxins in foods; food microbiology. Mailing Add: Dept of Food Sci Univ of Ga Athens GA 30602

LILLEGRAVEN, JASON ARTHUR, b Mankato, Minn, Oct 11, 38; m 64. PALEONTOLOGY. Educ: Calif State Col Long Beach, BA, 62; SDak Sch Mines & Technol, MS, 64; Univ Kans, PhD(zool), 68. Prof Exp: Instr zool, Calif State Col Long Beach, summer, 64; NSF fel paleont, Univ Calif, Berkeley, 68-69; asst prof zool, San Diego State Univ, 69-71, assoc prof, 71-74, prof, 74-75; ASSOC PROF GEOL, UNIV WYO, 76- Mem: Geol Soc Am; Paleont Soc; Soc Vert Paleont; Am Soc Mammal; Soc Study Evolution. Res: Terrestrial paleoecology; mammalian paleontology and morphology, comparative anatomy and evolution. Mailing Add: Dept of Geol Univ of Wyo Laramie WY 82070

LILLEHEI, C WALTON, b Minneapolis, Minn, Oct 23, 18; m 46; c 4. SURGERY. Educ: Univ Minn, BS, 39, MB, 41, MD, 42, MS & PhD(surg), 51; Am Bd Surg, dipl; Am Bd Thoracic Surg, dipl. Prof Exp: Intern, Minneapolis Gen Hosp, 41; asst physiol, Univ Hosp. Univ Minn, 48-59, sr surgeon res, 49-50, clin instr surg, Med Sch, 49-51, from assoc prof to prof, 51-67; chmn dept surg, 67-70, LEWIS ATTERBURY STIMSON PROF SURG, MED CTR, CORNELL UNIV, 67- Concurrent Pos: Surgeon-in-chief, New York Hosp, 67-70, attend surgeon, 70- Honors & Awards: Theobald Smith Award, AAAS, 51, Gould Mem Award, 56; Award, Am Col Chest Physicians, 52; Lasker Award, Am Pub Health Asn, 55; Modern Med Award, 57; Hektoen Gold Medal, AMA, 57, Billings Silver Medal, 72; Hunter Mem Award, Am Therapeut Soc, 58; Frederick Med Achievement Award, Int Consul Health, 58; Officer, Order of Leopold, Belg, 60; Gairdner Found Int Award, 63; Professorship Medal, Santo Tomas, 64. Mem: AAAS; Soc Thoracic Surg; fel Am Col Surg; fel Am Col Chest Physicians; fel Am Col Cardiol (pres, 66-67). Res: General and cardiovascular surgery. Mailing Add: New York Hosp Cornell Med Ctr 525 E 68th St New York NY 10021

LILLEHEI, RICHARD CARLTON, b Minneapolis, Minn, Dec 10, 27; m 52; c 4. SURGERY. Educ: Univ Minn, BA, 48, MD, 52, PhD, 60. Prof Exp: From asst prof to assoc prof, 60-66, PROF SURG, MED SCH, UNIV MINN, MINNEAPOLIS, 66- Concurrent Pos: Markle scholar, 60. Mem: Am Col Surg; Am Surg Asn; Soc Univ Surg; Am Asn Thoracic Surg. Res: Experimental surgery; organ transplantation; irreversible shock; cardiopulmonary assistance by mechanical means. Mailing Add: Mayo Box 490 Univ of Minn Hosps Minneapolis MN 55455

LILLEHOJ, EIVIND B, b Kimballton, Iowa, Aug 11, 28; m 48; c 4. PLANT PHYSIOLOGY, BIOCHEMISTRY. Educ: Iowa State Univ, BS, 60, MS, 62, PhD(plant physiol), 64. Prof Exp: NIH fel, Carlsberg Lab, Copenhagen, Denmark, 64-65; MICROBIOLOGIST, NORTHERN REGIONAL RES LAB, USDA, 65- Mem: Am Soc Plant Physiol; Am Soc Microbiol; Am Inst Biol Sci. Res: Fungal physiology; mycotoxins; fermentation; microbial products. Mailing Add: Northern Regional Res Lab USDA Peoria IL 61604

LILLELAND, OMUND, b Stavanger, Norway, Mar 12, 99; US citizen; m 34; c 2. POMOLOGY. Educ: Univ Calif, BS, 21, PhD, 34. Prof Exp: Jr pomologist, 26-30, from asst to assoc pomologist, 31-46, pomologist, 46-66, EMER POMOLOGIST, UNIV CALIF, DAVIS, 67-; CONSULT, 67- Res: Growth of fruits; thinning of deciduous tree fruits; phosphate and potash nutrition of fruit trees. Mailing Add: 40 College Park Davis CA 95616

LILLER, MARTHA HAZEN, b Cambridge, Mass, July 15, 31; m 59; c 2. ASTRONOMY. Educ: Mt Holyoke Col, AB, 53; Univ Mich, MA, 55, PhD(astron), 58. Prof Exp: Instr astron, Mt Holyoke Col, 57-59; lectr & res assoc, Univ Mich, 59-60; res assoc, 57-59, res fel, 60-69, CUR ASTRON PHOTOGS, COL OBSERV, HARVARD UNIV, 69- Concurrent Pos: Lectr, Wellesley Col, 61-63 & 67-68. Mem: Am Astron Soc; Int Astron Union. Res: Photometry; globular clusters; quasars. Mailing Add: Harvard Col Observ 60 Garden St Harvard Univ Cambridge MA 02138

LILLER, WILLIAM, b Philadelphia, Pa, Apr 1, 27; m 59; c 3. ASTRONOMY. Educ: Harvard Univ, AB, 49; Univ Mich, AM, 50, PhD(astron), 53. Prof Exp: Mem meteor exped, Harvard Univ, 47-48, supt, 52-53; asst, McMath-Hulbert Observ, Univ Mich, 52, from instr astron to assoc prof astron, 53-60; chmn dept astron, 60-66, prof, 60-70, ROBERT WHEELER WILLSON PROF APPL ASTRON, HARVARD UNIV, 70- Concurrent Pos: Guggenheim fel, 64-65. Mem: Am Astron Soc; fel Royal Astron Soc. Res: Photoelectric photometry of planetary nebulae and hot stars; investigation of x-ray sources; spectrophotometry. Mailing Add: Harvard Col Observ 60 Garden St Cambridge MA 02138

LILLEVIK, HANS ANDREAS, b Sherman, SDak, Feb 4, 16; m 46; c 4. BIOCHEMISTRY. Educ: St Olaf Col, BA, 38; Univ Minn, MS, 40, PhD(biochem), 46. Prof Exp: Instr biochem, Univ Minn, 42-44; res chemist, Minn Mining & Mfg Co, 44-45; from instr to asst prof, 46-57, ASSOC PROF CHEM & BIOCHEM, MICH STATE UNIV, 57- Concurrent Pos: Am Scand Found fel, Carlsberg Lab, Denmark, 47-48. Mem: AAAS; Am Chem Soc; Am Dairy Sci Asn. Res: Chemical properties and biological function of proteins and enzymes. Mailing Add: Dept of Biochem Mich State Univ East Lansing MI 48823

LILLEY, ARTHUR EDWARD, b Mobile, Ala, May 29, 28. ASTRONOMY. Educ: Univ Ala, BS, 50, MS, 51; Harvard Univ, PhD(radio astron), 54. Prof Exp: Physicist, Naval Res Lab, 54-57; asst prof radio astron, Yale Univ, 57-59; assoc prof, 59-63, PROF RADIO ASTRON, HARVARD UNIV, 63-; ASTRONOMER-IN-CHARGE, SMITHSONIAN ASTROPHYS OBSERV, 65- Concurrent Pos: Res Corp grant, 57-59; Sloan res fel, 58-60. Honors & Awards: Bok Prize, Harvard Univ, 58. Mem: AAAS; Int Union Radio Sci; Int Astron Union; Am Astron Soc; Am Phys Soc. Res: Spectral line and satellite radio astronomy; radio astronomical navigation techniques. Mailing Add: 89 Fletcher Rd Belmont MA 02178

LILLEY, JOHN RICHARD, b Fall River, Mass, Apr 2, 34; m 66; c 4. PLASMA PHYSICS. Educ: Univ Calif, Berkeley, AB, 56; Univ Idaho, MSc, 62. Prof Exp: Physicist, Gen Elec Co, Wash, 56-62; sr scientist, Western Div, McDonnell Douglas Astronaut Co, 62-72; STAFF MEM, LOS ALAMOS SCI LAB, 72- Concurrent Pos: Lectr, Radiol Physics Fel Prog, AEC, 61. Res: Radiation shielding; reactor physics; statistics; nonlinear optimization; experimental data analysis; thermal analysis; nuclear weapons effects; nuclear weapons design theory. Mailing Add: 2 Cherokee Lane Los Alamos NM 87544

LILLICH, THOMAS TYLER, b Cincinnati, Ohio, Sept 8, 43; m 65; c 2. MICROBIOLOGY, CELL BIOLOGY. Educ: Miami Univ, AB, 65; NC State Univ, MS, 68, PhD(microbiol), 70. Prof Exp: Asst prof, 72-75, ASSOC PROF ORAL BIOL, COL DENT, UNIV KY, 75-, ASST PROF CELL BIOL, COL MED, 73-, MEM MICROBIOL GRAD FAC, SCH BIOL SCI, 74- Concurrent Pos: NIH res fel, Univ Ky, 70-72, & Agr Res Serv contractee, 73-76. Mem: Am Soc Microbiol; Am Asn

Dent Schs (secy microbiol sect, 74-75, chmn-elect, 75-76); Int Asn Dent Res. Res: Effects of tobacco smoke components on microbial electron transport structure and function. Mailing Add: Dept of Oral Biol Col of Dent Univ of Ky Lexington KY 40506

LILLICK, LOIS CAROL, b Cincinnati, Ohio, Dec 20, 13. MEDICINE. Educ: Univ Cincinnati, AB, 33, AM, 34; Univ Mich, PhD(biol), 38; New York Med Col, MD, 53; Univ Calif, MPH, 63; Am Bd Prev Med, dipl, 64. Prof Exp: Asst bot, Univ Cincinnati, 33-35; Benedict fel physiol, Harvard Univ, 38-39; Benedict fel bot, Univ Cincinnati, 39-40; from instr to assoc prof bact, New York Med Col, 40-53, prof microbiol & chmn dept, 53-62; USPHS fel, Univ Calif, 62-63; chief chronic illness & aging unit, 63-65, asst chief bur chronic dis, 65-71, DEP DIR, CHIEF HEALTH FACIL RESOURCES DIV, STATE DEPT PUB HEALTH, CALIF, 71- Concurrent Pos: Actg dir dept bact, New York Med Col, 45-53. Mem: AAAS; Am Pub Health Asn; Am Geriat Soc; Am Cancer Soc. Res: Chronic diseases; medical and long term care. Mailing Add: 5348 Wedge Circle Fair Oaks CA 95628

LILLIE, CHARLES FREDERICK, b Indianola, Iowa, Feb 20, 36; m 65; c 3. ASTROPHYSICS. Educ: Iowa State Univ, BS, 57; Univ Wis, Madison, PhD(astrophys), 68. Prof Exp: Instr eng, NASA Flight Res Ctr, Edwards, Calif, 60-62; teaching asst, Dept Astron, Univ Wis, 62-64, res asst, Washburn Observ, 64-68, res assoc, Space Astrophys Lab, 68-70; asst prof, 70-73, ASSOC PROF PHYSICS & ASTROPHYS, DEPT PHYSICS & ASTROPHYS & LAB ATMOSPHERIC & SPACE PHYSICS, UNIV COLO, 73- Concurrent Pos: Co-investr, Apollo 17 Ultraviolet Spectrometer Exp, 72-74; team mem, Large Space Telescope Inst Definition Team High Resol Spectrograph, 73-75. Mem: Am Astron Soc; AAAS; Int Astron Union. Res: Cometary physics, planetary physics and ultraviolet stellar astronomy from rockets and space vehicles; sources of the light in the night sky: zodiacal light, integrated starlight and diffuse galactic light. Mailing Add: Lab Atmospheric & Space Physics Univ of Colo Boulder CO 80302

LILLIE, JOHN HOWARD, b Oak Park, Ill, Dec 16, 40; m 63; c 2. ANATOMY, DENTISTRY. Educ: Univ Mich, DDS, 66, PhD(anat), 72. Prof Exp: ASST PROF ANAT, SCH MED & ASST PROF, SCH DENT, UNIV MICH, ANN ARBOR, 72-, STAFF MEM, DENT RES INST, 72- Mem: Am Soc Cell Biol; Am Asn Anat. Res: Cellular control mechanisms in endocrine secretion and epithelia-connective tissue interactions; features of synthesis and control in the production of basal lamina constituents. Mailing Add: Dept of Anat Med Sci II Bldg Univ of Mich Ann Arbor MI 48104

LILLIE, RALPH DOUGALL, b Cucamonga, Calif, Aug 1, 96; m 20; c 7. PATHOLOGY, HISTOCHEMISTRY. Educ: Stanford Univ, AB, 17, MD, 20. Prof Exp: Intern, San Francisco Hosp, 19-20; pvt pract, Calif, 20; prof, Univ Md, 29-30; chief path lab, NIH, 37-48, path & pharmacol, Nat Inst Arthritis & Metab Dis, 48-53, path anat dept, Clin Ctr, 52-53 & lab path & histochem, 53-60; RES PROF PATH, LA STATE UNIV MED CTR, NEW ORLEANS, 60- Concurrent Pos: Am vpres, Int Cong Histochem & Cytochem, Ger, 64; ed, J Histochem & Cytochem, Histochem Soc, 52-64. Honors & Awards: Award, Asn Mil Surg US, 58. Mem: Histochem Soc (secy, 50-55, pres, 58-); Am Soc Exp Path; Am Asn Path & Bact; Biol Stain Comn (pres, 59-63); Int Acad Path (vpres, 43-47, pres, 48). Res: Seratonin in mast cells; nonadrenaline in adrenal; catechol in enterochromaffin; hematoxylin substitutes; tyrosine; fats; pigments; enterchromaffin; carcinoid tumors; melanosis-pseudomelanosis and intestinal iron uptake and storage; chromatin hematoxylin, iron hematoxylin mechanisms; arginine, elastin stains. Mailing Add: Dept of Path La State Univ Med Ctr New Orleans LA 70112

LILLIE, ROBERT JONES, b Rochester, Minn, Apr 15, 21; m 46; c 2. POULTRY NUTRITION. Educ: Pa State Col, BS, 44, MS, 46; Univ Md, PhD(poultry nutrit), 49. Prof Exp: Asst poultry dept, Univ Md, 45-47; poultry husbandman, Animal & Poultry Husb Res Br, 47-72, RES ANIMAL SCIENTIST, NONRUMINANT ANIMAL NUTRIT LAB, NUTRIT INST, AGR RES SERV, USDA, 72- Concurrent Pos: Mem standard diet subcomt, Nat Res Coun, 54. Honors & Awards: Award, Am Poultry Sci Asn, 50. Mem: Am Poultry Sci Asn; Am Inst Nutrit; Worlds Poultry Cong. Res: Vitamins, antibiotics, surfactants, arsenicals, unidentified factors, proteins and amino acids; pesticides; reproductive efficiency; air pollutants affecting poultry; trace minerals in swine. Mailing Add: Nonruminant Animal Nutrit Lab Nutrit Inst Bldg 200 Agr Res Ctr Beltsville MD 20705

LILLIEFORS, HUBERT W, b Reading, Pa, June 14, 28; m 53; c 2. STATISTICS. Educ: George Washington Univ, BA, 52, PhD(statist), 64; Mich State Univ, MA, 53. Prof Exp: Mathematician, Diamond Ord Fuze Labs, 53-55; sr scientist opers res, Lockheed Missile Systs Div, 55-56, opers analyst, Opers Eval Group, 56-57; mathematician opers res, Appl Physics Lab, Johns Hopkins Univ, 57-64; instr, 62-64, assoc prof, 64-67, PROF STATIST, GEORGE WASHINGTON UNIV, 67- Mem: Opers Res Soc Am; Am Statist Asn; Inst Math Statist. Res: Operations research; information theory; nonparametric statistics; statistical inference. Mailing Add: Dept of Statist George Washington Univ Washington DC 20006

LILLIEN, IRVING, b New York, NY, Feb 2, 29. ORGANIC CHEMISTRY. Educ: Univ Denver, BS, 50; Purdue Univ, MS, 52; Polytech Inst Brooklyn, PhD(org chem), 59. Prof Exp: Fel org chem, Wayne State Univ, 59-61; asst prof, Georgetown Univ, 61-62; asst prof, Univ Miami, 62-65, Sch Med, 65-67; assoc prof, Marshall Univ, 67-69; ASSOC PROF ORG CHEM, MIAMI-DADE JR COL, 69- Concurrent Pos: Air Force Off Sci & Res grant, 63-65. Mem: AAAS; Am Chem Soc; The Chem Soc. Res: Physical-organic chemistry; mechanisms of organic reactions; chemistry and conformation of small and medium size rings. Mailing Add: Dept of Chem Miami-Dade Jr Col 11011 SW 104 St Miami FL 33156

LILLING, HERBERT JEROME, organic chemistry, polymer chemistry, see 12th edition

LILLWITZ, LAWRENCE DALE, b Hinsdale, Ill, June 1, 44; m 68; c 3. INDUSTRIAL ORGANIC CHEMISTRY. Educ: Ill Benedictine Col, BS, 66; Univ Notre Dame, PhD(org chem), 70. Prof Exp: GROUP LEADER, CHEM DIV, QUAKER OATS CO, 70- Mem: Am Chem Soc. Res: Monomer synthesis; organic reaction mechanisms; homogeneous and heterogeneous catalysis. Mailing Add: John Stuart Res Lab Quaker Oats Co 617 W Main St Barrington IL 60010

LILLY, ARNYS CLIFTON, JR, b Beckley, WVa, June 3, 34; m 56; c 3. PHYSICS. Educ: Va Polytech Inst, BS, 57; Carnegie Inst Technol, MS, 63. Prof Exp: Res physicist, Gulf Res & Develop Co, Pa, 57-65; res physicist, 65-71, sr scientist, Physics Div, 71-74, ASSOC PRIN SCIENTIST, PHILIP MORRIS RES CTR, 74- Mem: Am Phys Soc. Res: Ion & electron optics; dielectric theory and experiment; electrostatics and organic conduction; space charge in insulators; thermal physics; combustion; fluid mechanics. Mailing Add: Physics Div Philip Morris Res Ctr PO Box 3-D Richmond VA 23206

LILLY, DANIEL MCQUILLAN, b Central Falls, RI, July 20, 10; m 48; c 1. PROTOZOOLOGY. Educ: Providence Col, AB, 31, MSc, 36; Brown Univ, PhD(biol), 40. Prof Exp: Instr biol, Providence Col, 32-41, asst prof, 41-42; from asst prof to prof, 46-75, EMER PROF BIOL, ST JOHN'S UNIV, NY, 75- Concurrent Pos: Scholar, Marine Biol Lab, Woods Hole. Mem: AAAS; Am Soc Zoologists; Soc Protozool (treas, 57-60); Asn Mil Surgeons US; NY Acad Sci. Res: Nutrition and growth in ciliate protozoa; interrelationships of different species of protozoa. Mailing Add: 56-10 187th St Flushing NY 11365

LILLY, DAVID J, b Washington, DC, Sept 21, 31; m 56; c 4. AUDIOLOGY. Educ: Univ Redlands, BA, 54, MA, 57; Univ Pittsburgh, PhD(audiol), 61. Prof Exp: Res assoc, Cent Inst Deaf, St Louis, Mo, 61-64; PROF AUDIOL, UNIV IOWA, 64- Concurrent Pos: NIH res fel, 61, Nat Inst Neurol Dis & Blindness trainee, 62-63; consult hearing aid res & procurement prog, Vet Admin, 66- Mem: Acoust Soc Am; Am Speech & Hearing Asn; Audio Eng Soc. Res: Experimental audiology, especially on measurements of acoustic impedance at the tympanic membrane of normal and pathologic ears; hearing aids; speech audiometry; masking; auditory adaptation; bone conduction; audiometric standards and calibration. Mailing Add: Dept of Speech Path & Audiol Univ of Iowa Iowa City IA 52242

LILLY, DOUGLAS KEITH, b San Francisco, Calif, June 16, 29; m 54; c 3. METEOROLOGY. Educ: Stanford Univ, BS, 50; Fla State Univ, MS, 54, PhD(meteorol), 59. Prof Exp: Meteorologist, Free Europe Press, 56-57; res meteorologist, US Weather Bur, 58-65; prog scientist, 65-73, SR SCIENTIST, NAT CTR ATMOSPHERIC RES, 73- Concurrent Pos: Consult, Nat Oceanic & Atmospheric Admin, 74- Honors & Awards: 2nd Half Century Award, Am Meteorol Soc, 72. Mem: Fel Am Meteorol Soc; Am Geophys Union. Res: Atmospheric convection, turbulence and mountain waves. Mailing Add: Nat Ctr for Atmospheric Res PO Box 3000 Boulder CO 80303

LILLY, FRANK, b Charleston, WVa, Aug 28, 30. GENETICS, ONCOLOGY. Educ: WVa Univ, BS, 51; Univ Paris, PhD(org chem), 59; Cornell Univ, OhD(biol PhD(biol, genetics), 65. Prof Exp: Res fel, 65-67, from asst prof to assoc prof, 67-74, PROF IMMUNOGENETICS & ONCOGENETICS, ALBERT EINSTEIN COL MED, 74- Concurrent Pos: New York City Health Res Coun career scientist award, Albert Einstein Col Med, 67-72; mem, Breast Cancer-Virus Working Group, Nat Cancer Inst, 72-; mem bd dirs, Leukemia Soc Am, 73- Mem: AAAS; NY Acad Sci; Genetics Soc Am; Transplantation Soc; Am Asn Cancer Res. Res: Oncogenetics, study of genes which influence susceptibility or resistance to oncogenic agents in mice; immunogenetics. Mailing Add: Dept of Genetics Albert Einstein Col of Med Bronx NY 10461

LILLY, JOHN HENRY, b Elk Mound, Wis, Oct 22, 07; m 32; c 3. ENTOMOLOGY. Educ: Univ Wis, BS, 31, PhD(econ entom), 39. Prof Exp: Asst, Univ Wis, 31-36, instr econ entom, 36-41, asst prof zool & econ entom, 41-46, assoc prof, 46-48; prof zool & entom, Iowa State Univ, 48-58; head dept, 58-65, PROF ENTOM, UNIV MASS, AMHERST, 58- Concurrent Pos: Sabbatical to Cambridge Univ, 66; Ford Found prog specialist entom, Univ Agr Sci, India, 67-68. Mem: Entom Soc Am. Res: Biology and control of field crop insects, particularly soil-infesting forms. Mailing Add: Dept of Entom Fernald Hall Univ of Mass Amherst MA 01002

LILLY, PERCY LANE, b Spanishburg, WVa, July 14, 27; m 51; c 4. PLANT TAXONOMY. Educ: Concord Col, BS, 50; Univ WVa, MS, 51; Pa State Univ, PhD, 57. Prof Exp: Instr biol, Salem Col, WVa, 51-53; from asst to assoc prof, 56-64, PROF BIOL, HEIDELBERG COL, 64-, CHMN DEPT, 65- Concurrent Pos: Spec field staff mem, Rockefeller Found, Colombia, 68-69. Mem: AAAS; Bot Soc Am. Res: Plant genetics and microbiology; nitrogen fixation in Azotobacter. Mailing Add: Dept of Biol Heidelberg Col Tiffin OH 44883

LILLYA, CLIFFORD PETER, b Chicago, Ill, May 23, 37; m 62; c 2. ORGANIC CHEMISTRY. Educ: Kalamazoo Col, AB, 59; Harvard Univ, PhD(chem), 64. Prof Exp: Staff assoc, 63-64, from asst prof to assoc prof, 64-73, PROF CHEM, UNIV MASS, AMHERST, 73- Concurrent Pos: Alfred P Sloan Found fel, 69-71. Mem: Am Chem Soc. Res: Photochemistry and chemistry of organotransition metal compounds. Mailing Add: Dept of Chem Univ of Mass Amherst MA 01002

LILYQUIST, MARVIN RUSSELL, b Barnum, Minn, Oct 2, 25; m 47; c 3. ORGANIC CHEMISTRY. Educ: Manchester Col, BA, 48; Univ Fla, MS, 50, PhD(org chem), 55. Prof Exp: Chemist, Fla State Rd Dept, 50-51; res assoc org fluorine chem, Univ Fla, 51-55; res chemist, Chemstrand Corp, 55-61, group leader, Chemstrand Res Ctr, 61-74, SR RES SPECIALIST MKT, MONSANTO TEXTILES CO, 74- Mem: Am Chem Soc; Fiber Soc. Res: Synthetic polymers; elastomers; fibers; characterization methods; fiber technology; organic synthesis; fluorine compounds; monomers; heterocyclics; non-woven fabric technology. Mailing Add: Monsanto Triangle Pk Develop Ctr PO Box 12274 Research Triangle Park NC 27709

LIM, DAVID J, b Seoul, Korea, Nov 27, 35; m 66; c 1. OTOLARYNGOLOGY, ELECTRON MICROSCOPY. Educ: Yonsei Univ, Korea, AB, 55, MD, 60. Prof Exp: Intern, Nat Med Ctr, Seoul, Korea, 60-61, resident otolaryngol, 61-64; res assoc, 66-67, asst prof, 67-71, ASSOC PROF OTOLARYNGOL, COL MED, OHIO STATE UNIV, 71-, DIR OTOL RES LABS, 67- Concurrent Pos: Spec fel otol res, Mass Eye & Ear Infirmary & Harvard Med Sch, 65-66; mem task force, Am Acad Opthal & Otolaryngol & Am Bd Otolaryngol, 69-72. Mem: AAAS; Electron Micros Soc Am; Acoust Soc Am; Soc Univ Otolaryngol; Am Laryngol Rhinol & Otol Soc. Res: Investigation of the ear as to the function of hearing and hearing disorders with the use of electron microscopy, autoradiography and cytochemistry. Mailing Add: Otol Res Labs Dept Otolaryngol Ohio State Univ Col of Med Columbus OH 43210

LIM, EDWARD C, b Seoul, Korea, Nov 17, 32; nat US; m 58; c 2. PHYSICAL CHEMISTRY. Educ: St Procopius Col, BS, 54; Okla State Univ, MS, 57, PhD(chem), 59. Prof Exp: Instr phys chem, Loyola Univ, Ill, 58-60, from asst prof to prof, 60-68; PROF CHEM, WAYNE STATE UNIV, 68- Mem: Am Phys Soc; Am Chem Soc. Res: Molecular electronic spectrosocpy; solid state photochemistry. Mailing Add: Dept of Chem Wayne State Univ Detroit MI 48202

LIM, HENRY S, b Pung Chun, Korea, Sept 28, 29; US citizen; m 56; c 2. ANESTHESIOLOGY. Educ: Yonsei Univ, Korea, MD, 53; Am Bd Anesthesiol, dipl, 65. Prof Exp: From instr to asst prof anesthesiol, 63-69, asst prof gynec & obstet, 65-74, ASSOC PROF ANESTHESIOL, MED SCH, JOHNS HOPKINS UNIV, 69- Mem: Fel Am Col Anesthesiol; Am Soc Anesthesiol; Int Anesthesia Res Soc; Soc Obstet Anesthesiol & Perinatol; Int Asn Study Pain. Res: Obstetric anesthesia. Mailing Add: Johns Hopkins Hosp Baltimore MD 21205

LIM, JOHNG KI, b Seoul, Korea, Feb 12, 30. GENETICS. Educ: Univ Minn, BS, 58, MS, 60, PhD(genetics), 64. Prof Exp: From asst prof to assoc prof, 63-69, PROF BIOL, UNIV WIS-EAU CLAIRE, 69- Mem: AAAS; Genetics Soc Am; Environ Mutagen Soc. Res: Chemical mutagenesis; cytogenetics. Mailing Add: Dept of Biol Univ of Wis Eau Claire WI 54701

LIM, LOUISE CHIN, b Honolulu, Hawaii, Apr 9, 22; m 53. MATHEMATICS. Educ: Univ Calif, AB, 43, MA, 44, PhD(math), 48. Prof Exp: Asst math, Univ Calif, 43-48; asst prof, 48-70, ASSOC PROF MATH, UNIV ARIZ, 70- Concurrent Pos: Res assoc, Univ Calif, 53, Ford Found fel, 53-54; res assoc, Radcliffe Col, 53-54. Mem: Am Math Soc; Math Asn Am; Asn Symbolic Logic. Res: Relation, projective and cylindric algebras. Mailing Add: Dept of Math Univ of Ariz Tucson AZ 85721

LIM, RAMON (KHE SIONG), b Cebu, Philippines, Feb 5, 33; m 61; c 3. NEUROCHEMISTRY. Educ: Univ Santo Tomas, Manila, MD, 58; Univ Pa, PhD(biochem), 66. Prof Exp: Intern, Long Island Col Hosp, NY, 59-60; USPHS trainee & fel, Univ Pa, 62-66; asst res biochemist, Ment Health Res Inst, Univ Mich, 66-69; ASST PROF NEUROSURG & BIOCHEM, UNIV CHICAGO, 69- Concurrent Pos: NIMH spec res fel, 68-69. Mem: AAAS; Am Soc Neurochem; Int Soc Neurochem; Am Soc Biol Chem; Soc Neurosci. Res: Fractionation and functional correlation of brain proteins; brain cell biology; biochemical determinants of neural mechanism and animal behavior. Mailing Add: Div of Neurosurg Univ of Chicago Hosps Chicago IL 60637

LIM, SUNG MAN, US citizen; m 68; c 2. PLANT PATHOLOGY. Educ: Seoul Univ, Korea, MS, 59; Miss State Univ, MS, 63; Mich State Univ, PhD(crop sci & plant path), 66. Prof Exp: Agronomist, Crop Exp Sta, Suwon, Korea, 60-61; res asst, Miss State Univ, 61-63 & Mich State Univ, 63-66; res assoc, 67-71, ASST PROF PLANT PATH, UNIV ILL, URBANA, 71- Mem: Am Phytopath Soc; Am Genetic Asn; Am Soc Agron; Crop Sci Soc Am. Res: Epidemics of plant diseases; genetics of host-pathogen interactions. Mailing Add: Dept of Plant Path Univ of Ill Urbana IL 61801

LIM, TECK-KAH, b Malacca, Malaysia, Dec 1, 42; m 66; c 2. THEORETICAL NUCLEAR PHYSICS. Educ: Univ Adelaide, BS, 64, PhD(nuclear physics), 68. Prof Exp: Lectr math, Univ Malaya, 68; res assoc nuclear physics, Fla State Univ, 68-70; asst prof physics, 70-75, ASSOC PROF PHYSICS, DREXEL UNIV, 75- Mem: Am Phys Soc; Sigma Xi. Res: Few-nucleon problem; direct reaction theory; medium energy physics; molecular physics. Mailing Add: Dept of Physics Drexel Univ Philadelphia PA 19104

LIM, THOMAS PYUNG KEE, b Seoul, Korea, June 1, 24; nat US; m 55; c 3. THORACIC DISEASES. Educ: Severance Union Med Col, MD, 48; Northwestern Univ, Ill, MS, 51, PhD(physiol), 53. Prof Exp: Res assoc physiol, Sch Med, Stanford Univ, 53-54 & Northwestern Univ, Ill, 54-56; mem staff, Lovelace Found, 56-61; dir cardiopulmonary lab, Tucson Med Ctr, Ariz, 61-67; mem staff, Dept Med, Syracuse Univ Hosp, 67-68; assoc prof med & physiol, Med Ctr, Univ Nebr, Omaha, 68-69; DIR CARDIOPULMONARY LAB, IMMANUEL MED CTR, 70- Mem: AAAS; Am Physiol Soc; Am Heart Asn; Am Thoracic Soc; Am Col Chest Physicians. Res: Pulmonary diseases; clinical cardiopulmonary physiology, biostatistics. Mailing Add: Immanuel Med Ctr 6901 N 72nd St Omaha NE 68122

LIM, YONG WOON, b Seoul, Korea, Oct 25, 35; m 68; c 3. SURFACE CHEMISTRY, COLLOID CHEMISTRY. Educ: Ohio Wesleyan Univ, AB, 57; Univ Dayton, MS, 63; State Univ NY Col Forestry, Syracuse, PhD(chem), 69. Prof Exp: Res chemist, Paper Res Dept, NCR Corp, 69-71, group leader anal chem, Appleton Papers Div, 71-74; RES ASSOC CHEM, TISSUE & TOWEL RES & DEVELOP, AM CONSUMER PROD, AM CAN CO, NEENAH, WIS, 74- Mem: Am Chem Soc; Tech Asn Pulp & Paper Indust. Res: Application of surface and colloid chemistry to paper machine wet-end operations; adhesive creping; morphology of cellulose and synthetic fibers; nonwovens and radfoam process; electrostatic and print bonding.

LIMA, DANIEL ANTHONY, organic chemistry, see 12th edition

LIMARZI, LOUIS ROBERT, b Chicago, Ill, Nov 27, 03; m; c 2. MEDICINE. Educ: Univ Ill, BS, 28, MD, 30, MS, 35. Prof Exp: Intern, Ill Res & Educ Hosp, Chicago, 30-31; clin asst med, 32-35, clin assoc, 35-40, from asst prof to assoc prof, 40-55, PROF MED & DIR HEMAT SECT, UNIV ILL COL MED, 55- Concurrent Pos: Resident, Ill Res & Educ Hosp, Chicago, 32-35, attend physician, 35-; investr, Midwest Coop Chemother Group, USPHS; civilian med consult, Hines Vet Admin Hosp, Ill & Surgeon Gen Off, 40-45, USPHS, 40- & Fed Civil Defense Admin, 50-; mem adv bd, Hemat Res Found, 40-; mem comt civil defense blood & blood derivatives, Ill Civil Defense Orgn, 50-; attend physician, West Side Vet Admin Hosp, 54-; hemat ed, Abstr Bioanal Tech, 61. Mem: AAAS; fel Am Soc Clin Path; fel AMA; fel Col Am Path; fel Am Col Physicians. Res: Leukemia; intermediate metabolism of leukemic leukocytes and effects of anti-leukemic agents; idiopathic throbocytopenic purpura; polycythemia vera. Mailing Add: 910 N East Ave Oak Park IL 60302

LIMBER, DAVID NELSON, b Alexandria, Va, May 25, 28; m 58; c 2. ASTRONOMY, ASTROPHYSICS. Educ: Ohio State Univ, AB & BSc, 50; Univ Chicago, PhD(astron), 53. Prof Exp: Fel astron, NSF, Princeton Univ, 53-54, Higgins fel, 56-57; asst prof, Univ Rochester, 57-58; asst prof, Univ Chicago, 58-62, assoc prof, 62-68; PROF ASTRON, UNIV VA, 68- Mem: AAAS; Am Astron Soc; Royal Astron Soc. Res: Stellar structure, stellar evolution; galaxies. Mailing Add: Leander McCormick Observ Univ of Va PO Box 3818 Charlottesville VA 22903

LIMBIRD, ARTHUR GEORGE, b East Cleveland, Ohio, Apr 10, 44; m 66; c 2. RESOURCE GEOGRAPHY. Educ: Miami Univ, AB, 66; Mich State Univ, MA, 68, PhD(geog), 72. Prof Exp: ASST PROF GEOG, BOWLING GREEN STATE UNIV, 72- Concurrent Pos: Consult, Ohio Biol Surv, 72-73. Mem: Asn Am Geogr; NZ Geog Soc. Res: Soil geography and related agricultural land use; soil as a resource—its capabilities and limitations. Mailing Add: Dept of Geog Bowling Green State Univ Bowling Green OH 43402

LIMBURG, WILLIAM W, b Buffalo, NY, Nov 9, 35; m 66. ORGANIC POLYMER CHEMISTRY. Educ: Univ Buffalo, BA, 59, MA, 62; Univ Toronto, PhD(organosilicon chem), 65. Prof Exp: From sr chemist to assoc scientist, 65-66, scientist, 67-73, SR SCIENTIST, XEROX CORP, 73- Mem: Am Chem Soc; The Chem Soc; Soc Photog Scientists & Engrs. Res: Synthesis of organometallic compounds; mechanistic and stereochemical studies of molecular rearrangements of carbon-functional silicon-containing compounds; non-silver halide imaging methods; synthesis of organic photoconductive materials; synthesis of novel polysiloxanes. Mailing Add: Xerox Corp 800 Phillips Rd Webster NY 14580

LIME, BRUCE JAMES, b Kansas City, Mo, July 31, 21; m 52; c 6. FOOD CHEMISTRY. Educ: Kans State Teachers Col Pittsburg, BS, 50; NC State Univ, MS, 68. Prof Exp: Sales rep, A J Griner Co, Mo, 50; res fel chem, Tex Citrus Comn, 50-52; chemist, 52-62, res chemist, 62-72, RES LEADER, FOOD CROPS UTILIZATION LAB, USDA, 72- Mem: Am Chem Soc; Inst Food Technol. Res: Chemistry of southern grown fruits and vegetables; utilization research of citrus, avocados and all vegetables; sugar crops. Mailing Add: 650 Bougainvillea Weslaco TX 78596

LIMERICK, JACK MCKENZIE, SR, b Fredericton, NB, July 16, 10; m 37. CHEMISTRY. Educ: Univ NB, BSc, 31, MSc, 34. Prof Exp: Res chemist, Fraser Co, 34-37; chief chemist, Bathurst Power & Paper Co, 37-41, supt control dept, 41-44, tech & res dir, Bathurst Paper Co, Ltd, 44-67, assoc dir res & develop, Consol-Bathurst Ltd, 67-71; CONSULT, PULP, PAPER & CONTAINER INDUST, IRAN, 72- & BRAZIL, 73- Concurrent Pos: Lectr, Royal Tech Inst, Sweden, 52. Honors & Awards: Award, Tech Asn Pulp & Paper Indust, 59. Mem: Am Chem Soc; Tech Asn Pulp & Paper Indust; Can Pulp & Paper Asn; fel Chem Inst Can; Pulp & Paper Res Inst Can. Res: Pulp; paper; containers. Mailing Add: Suite 903 201 Metcalfe Ave Montreal PQ Can

LIMON, PETER JACOB, particle physics, see 12th edition

LIMPEL, LAWRENCE EUGENE, b Milwaukee, Wis, Dec 6, 30; div; c 2. ENTOMOLOGY. Educ: Univ Wis, BS, 52, MS, 54, PhD(entom), 57. Prof Exp: Entomologist & Diamond Shamrock fel, 47-61, leader, 61-63, PROG DIR BIOCIDAL CHEM, DIAMOND SHAMROCK CORP, 63- Res: Insect physiology; insecticides; herbicides; fungicides. Mailing Add: Diamond Shamrock Corp 1086 N Broadway Yonkers NY 10701

LIMPER, KARL ESSLINGER, b Wichita, Kans, Mar 1, 14; m 44; c 1. MICROPALEONTOLOGY. Educ: Beloit Col, BS, 35, MS, 37; Univ Chicago, PhD(geol), 53. Prof Exp: Asst geol, Beloit Col, 35-37; instr, Miami Univ, 39-41 & Hamilton Col, 41-42; from asst to assoc prof, 46-56, chmn dept, 56-60, dean col arts & sci, 60-71, PROF GEOL, MIAMI UNIV, 56- Mem: AAAS; Paleont Soc; Geol Soc Am; Nat Asn Geol Teachers. Res: Invertebrate paleontology; geomorphology; structure. Mailing Add: Dept of Geol Shideler Hall Miami Univ Oxford OH 45056

LIMPERIS, THOMAS, b Detroit, Mich, Feb 19, 31; m; c 5. PHYSICS. Educ: Univ Mich, BS, 58, MSEE, 60. Prof Exp: Mem res staff, Univ Mich, 58-66, dir info & optical sci group, 66-68; PRES, SENSORS, INC, 68- Res: Infrared detectors; infrared spectroscopy; optical properties of materials. Mailing Add: Sensors Inc 3908 Varsity Dr Ann Arbor MI 48104

LIMPEROS, GEORGE, chemistry, see 12th edition

LIMPERT, FREDERICK ARTHUR, b Frankfort, NY, Feb 4, 21; m 44. HYDROLOGY. Educ: Wash State Univ, BSCE, 43. Prof Exp: Civil engr, Columbia Basin Proj, Wash Bur Reclamation, 46-61; HEAD HYDROL SECT & CHIEF HYDROLOGIST, BONNEVILLE POWER ADMIN, 61- Concurrent Pos: Mem, Interagency Adv Comt Water Data, 72- & Coord Coun Water Data Acquisition Methods, 74- Mem: Fel Am Soc Civil Engrs; Nat Soc Prof Engrs. Res: Use of satellite data for determining areal snow cover and cloud classification for areal precipitation. Mailing Add: Bonneville Power Admin PO Box 3621 Portland OR 97208

LIN, ADA WEN-SHUNG MA, b Canton, China, May 19, 37; m 63; c 2. ENZYMOLOGY, BIOCHEMISTRY. Educ: Nat Taiwan Univ, BS, 60; Columbia Univ, MA, 63, PhD(biochem), 66. Prof Exp: NIH grant biochem, Columbia Univ, 66-67 & traineeship, Univ Pa, 67-68; res assoc, Inst Cancer Res, Philadelphia, Pa, 69-70; RES BIOCHEMIST, CLIN RES CDIV, US NAVAL HOSP, 71- Mem: AAAS; Am Chem Soc. Res: Physical chemical identification and characterization of enzymes; enzymatic reaction mechanisms; kinetics; drug metabolism; clinical biochemistry. Mailing Add: Clin Invest Ctr US Naval Hosp Philadelphia PA 19145

LIN, BENJAMIN MING-REN, US citizen. COMPUTER SCIENCES. Educ: Taipei Inst Technol, dipl, 61; Univ Wyo, MS, 67; Univ Iowa, PhD(elec eng), 73. Prof Exp: Engr, Radio Wave Res Labs, 62-65; design & develop engr, Collins Radio Co, 67-68; engr, Addressograph Multigraph ,Corp, 68-69; PROF COMPUT SCI, MOORHEAD STATE UNIV, 73- Mem: Sigma Xi; Asn Comput Mach; Inst Elec & Electronics Engrs; Comput Soc. Res: Application of microprocessors in computer systems for small business; fault-tolerant computing systems design; computer architecture in microprogramming environment. Mailing Add: Dept of Comput Sci Moorhead State Univ Moorhead MN 56560

LIN, BOR-LUH, b Fukien, China, Mar 4, 35; m 63. MATHEMATICS. Educ: Nat Taiwan Univ, BS, 56; Univ Notre Dame, MS, 60; Northwestern Univ, PhD(math), 63. Prof Exp: Asst prof math, 63-67, assoc prof, 67-72, PROF MATH, UNIV IOWA, 72- Concurrent Pos: Vis assoc prof, Ohio State Univ, 70-71. Mem: Am Math Soc. Res: Functional analysis, general topology. Mailing Add: Dept of Math Univ of Iowa Iowa City IA 52240

LIN, CHANG KWEI, b Taiwan, Sept 11, 41; m 67; c 2. PHYCOLOGY. Educ: Chung Shim Univ, BSc, 64; Univ Alta, MSc, 68; Univ Wis-Milwaukee, PhD(bot), 73. Prof Exp: Asst cur, Acad Natural Sci, Philadelphia, 72-74; RES INVESTR LIMNOL, UNIV MICH, ANN ARBOR, 74- Mem: AAAS; Phycol Soc Am; Am Soc Limnol & Oceanog; Soc Int Limnol. Res: Physio-ecology of algae in general, with particular interest in nutrient limitation to phytoplankton growth in the Great Lakes. Mailing Add: Great Lakes Res Div Univ of Mich Ann Arbor MI 48109

LIN, CHENG SHAN, b Pingtan, China, Mar 2, 12; m 52. ENTOMOLOGY. Educ: Fukien Christian Univ, BS, 37; Ft Hays Kans State Col, MS, 50; Cornell Univ, PhD, 55. Prof Exp: Instr biol, Fukien Sci Inst, China, 34-36; prin high sch, 37-45; dir, Fukien Christian Univ, China, 45-46; instr biol, Foochow Col, 46-47; asst, Cornell Univ, 52-55; chmn natural sci div, 56-66, PROF BIOL, HUSTON-TILLOTSON COL, 56- Res: Insect biology and ecology; immature insects; biology and behavior of solitary digger wasps. Mailing Add: Dept of Biol Huston-Tillotson Col Austin TX 78702

LIN, CHE-SHUNG, b Taipei, Formosa, Oct 20, 33; Can citizen; m 64; c 1. QUANTUM CHEMISTRY. Educ: Nat Taiwan Univ, BSc, 56, MSc, 60; Univ Sask, PhD(chem), 65. Prof Exp: Fel chem, Univ Alta, 64-66 & Ind Univ, 66-67; asst prof, 67-71, ASSOC PROF CHEM, UNIV WINDSOR, 71- Mem: Am Phys Soc. Res: Low energy collisions; mathematical analysis and formalism. Mailing Add: Dept of Chem Fac of Arts & Sci Univ of Windsor Windsor ON Can

LIN, CHIA CHIAO, b Foochow, China, July 7, 16; m. APPLIED MATHEMATICS. Educ: Nat Tsing Hua Univ, China, BSc; Univ Toronto, MA, 41; Calif Inst Technol, PhD(aeronaut), 44. Prof Exp: Asst, Tsing Hua Univ, China, 37-39; from asst to res engr, Calif Inst Technol, 43-45; from asst prof appl math to assoc prof appl math, Brown Univ, 45-47; assoc prof math, 47-53, prof, 53-66, INST PROF MATH, MASS INST TECHNOL, 66- Concurrent Pos: Guggenheim fels, 54-55, 60. Mem: Nat Acad Sci; Am Astron Soc; Soc Indust & Appl Math; Am Math Soc; Am Inst Aeronaut & Astronaut. Res: Hydrodynamics; stellar dynamics; astrophysical problems; spiral structure of galaxies; density wave theory developed in great mathematical detail with predictions checked against various astronomical observations. Mailing Add: Dept of Math Mass Inst Technol Cambridge MA 02139

LIN, CHII-DONG, b Taiwan. ATOMIC PHYSICS. Educ: Nat Taiwan Univ, BS, 69; Univ Chicago, MS, 73, PhD(physics), 74. Prof Exp: FEL ASTROPHYS, CTR ASTROPHYS, HARVARD COL OBSERV, 74- Mailing Add: Ctr for Astrophys Harvard Col Observ 60 Garden St Cambridge MA 02138

LIN, CHIN-CHUNG, b Taipei, Taiwan, Oct 8, 37. BIOCHEMICAL PHARMACOLOGY. Educ: Chung Hsing Univ, Taiwan, BS, 60; Tuskegee Inst, MS, 65; Northwestern Univ, PhD(biochem), 69. Prof Exp: Sr scientist, 69-75, PRIN SCIENTIST, SCHERING CORP, 75- Concurrent Pos: Res fel biochem, Med Sch, Northwestern Univ, 69. Mem: AAAS; Am Chem Soc; Am Soc Pharmacol & Exp Therapeut; NY Acad Sci. Res: Drug metabolism and the mechanism of enzymatic hydroxylation. Mailing Add: Schering Corp Dept of Biochem 60 Orange St Bloomfield NJ 07003

LIN, CHI-WEI, b Hong Kong, May 16, 37; m 65; c 1. CANCER, BIOCHEMISTRY. Educ: Nat Taiwan Univ, BS, 61; Univ Wis-Madison, MS, 65, PhD(biochem), 69. Prof Exp: Fel cancer res, 69-71, res assoc, 71-72, ASST PROF PATH, SCH MED, TUFTS UNIV, 72- Mem: Biochem Soc; AAAS; Sigma Xi. Res: Biochemical characteristics of cancer, specifically, the studies of tumor-associated enzymes and isozymes; processes of synthesis and distribution of acid hydrolases and the biogenesis of lysosomes. Mailing Add: Tufts Cancer Res Ctr Sch of Med Tufts Univ 136 Harrison Ave Boston MA 02111

LIN, CHYI-CHYANG, b China, May 21, 33; m; c 1. GENETICS. Educ: Nat Taiwan Univ, BS, 60; Univ Man, MSc, 64, PhD(genetics), 68. Prof Exp: Cytogeneticist, Winnipeg Children's Hosp, 68-69; from instr to asst prof pediat & path, McMaster Univ, 69-73; ASSOC PROF PEDIAT & MED BIOCHEM, UNIV CALGARY, 73- Concurrent Pos: Dir cytogenetics lab, Med Ctr, McMaster Univ, 71-73. Mem: Am Genetic Asn; Can Asn Genetics & Cytol; Am Human Genetics Soc. Res: Cytogenetic studies on plants and animals; biochemical genetics; somatic cell genetics. Mailing Add: Div of Pediat & Med Biochem Univ of Calgary Fac of Med Calgary AB Can

LIN, DENIS CHUNG KAM, b Hong Kong, July 7, 44; Can citizen; m 69; c 1. ANALYTICAL CHEMISTRY, MASS SPECTROMETRY. Educ: Univ Man, BSc, 68, MSc, 70, PhD(chem), 72. Prof Exp: Fel, Univ Montreal, 72-74; STAFF CHEMIST, BATTELLE MEM INST, 74- Mem: Am Chem Soc; Am Soc Mass Spectrometry; Int Asn Forensic Toxicologists. Res: Identification and quantification of low levels of drugs and their metabolites in biological samples by mass spectrometry and other techniques; nucleic acid and protein sequencing; pyrolytic reactions. Mailing Add: Battelle Mem Inst 505 King Ave Columbus OH 43201

LIN, DUO-LIANG, b Juian, China, May 16, 30; m 63; c 1. PHYSICS. Educ: Taiwan Prov Norm Univ, BSc, 56; Tsing Hua, China, MSc, 58; Ohio State Univ, PhD(physics), 61. Prof Exp: Res assoc physics, Yale Univ, 61-64; asst prof, 64-67, ASSOC PROF PHYSICS, STATE UNIV NY BUFFALO, 67- Concurrent Pos: Sr vis, Oxford Univ, 70-71; vis prof, Nat Taiwan Univ. Mem: Am Phys Soc. Res: Problems in nuclear physics such as nuclear forces and structure; coherent radiation from a system of atoms. Mailing Add: Dept of Physics State Univ of NY Buffalo NY 14214

LIN, EDMUND CHI CHIEN, b Peking, China, Oct 28, 28; nat US. BIOCHEMISTRY. Educ: Univ Rochester, AB, 52; Harvard Univ, PhD, 57. Prof Exp: Instr biochem, 57-60, assoc, 60-63, from asst prof to assoc prof, 63-69, PROF MICROBIOL & MOLECULAR GENETICS, HARVARD MED SCH, 69-, CHMN DEPT, 73- Concurrent Pos: Vis prof, Univ Calif, Berkeley, 72; Ruby Boyer Miller fel med res, 72. Mem: Am Soc Microbiol; Am Soc Biol Chem. Res: Bacterial physiology and genetics and biochemical evolution. Mailing Add: Microbiol & Molecular Genetics Harvard Med Sch Boston MA 02115

LIN, FU HAI, b Fukien, China, Feb 15, 28; US citizen; m 56; c 5. BACTERIOLOGY, BIOCHEMISTRY. Educ: Nat Taiwan Univ, BS, 53; Univ WVa, MS, 59; Rutgers Univ, PhD(bact), 65. Prof Exp: Asst, Univ WVa, 58-59; tech asst biochem, Boyce Thompson Inst, 59-61; res asst, Rutgers Univ, 61-65; asst mem biochem, Albert Einstein Med Ctr, 65-69; sr res scientist, 69-70, ASSOC RES SCIENTIST, INST BASIC RES MENT RETARDATION, 70- Mem: AAAS; Am Soc Microbiol; Am Chem Soc. Res: Biochemistry and replication of ribonucleic acid virus; metabolism of amino acids; mechanism of the synthesis of folic acid; metabolism of ribonucleic acid and the control mechanisms of its biosynthesis; role of deoxyribonucleic acid in the replication of riobonucleic acid viruses; biochemistry and function of proteins of slow viruses. Mailing Add: Inst Basic Res Ment Retardation 1050 Forest Hill Rd Staten Island NY 10314

LIN, JAMES C H, b Macao, Aug 12, 32; m 67; c 2. GENETICS, CELL PHYSIOLOGY. Educ: Taiwan Prov Norm Univ, BS, 54; Rice Univ, MA, 60, NC State Univ, PhD(genetics), 65. Prof Exp: Lab instr zool, Nat Taiwan Univ, 55-57; res asst nuclear med, Methodist Hosp, Houston, Tex, 59-60 & Hermann Hosp, 60; asst prof, 65-70, ASSOC PROF BIOL, NORTHWESTERN STATE UNIV, 70- Mem: Genetics Soc Am. Res: Chemical mutagenesis; effects of divalent metal ions on the fecundity, hatchability and fertility of habrobracon; urate accumulation in insects; crossing over in Drosophila. Mailing Add: Dept of Biol Sci Northwestern State Univ Natchitoches LA 71457

LIN, JAMES FANG-MING, food technology, see 12th edition

LIN, JEONG-LONG, b Taichung, Formosa, Dec 17, 35; m 59; c 3. PHYSICAL CHEMISTRY. Educ: Queen's Univ, Ont, PhD(chem), 64. Prof Exp: Res assoc chem, Univ Chicago, 64-66; asst prof, 66-68, assoc prof, 68-74, PROF CHEM, BOSTON COL, 74- Mem: Am Chem Soc; Am Phys Soc. Res: Irreversible thermodynamics and electrochemistry; theory of elementary reactions. Mailing Add: Dept of Chem Boston Col Chestnut Hill MA 02167

LIN, JIANN-TSYH, b Taoyuan, Taiwan, Jan 15, 40; m 69; c 2. BIOCHEMISTRY. Educ: Chung-Hsing Univ, Taiwan, BS, 63; Univ Miss, MS, 67; Drexel Univ, PhD(biochem), 71. Prof Exp: Res fel biochem, Univ Tenn, Memphis, 71-72; Hormel fel, 72-73, res fel, 73-74, RES ASSOC BIOCHEM, HORMEL INST, UNIV MINN, 75- Mem: Am Chem Soc. Res: Diol lipid metabolism; lipids of cancerous tissues; lipid and steroid biochemistry. Mailing Add: Hormel Inst Univ of Minn Austin MN 55912

LIN, KANG, b Chen-Tu, China, Dec 17, 40; m 64; c 4. ORGANIC CHEMISTRY. Educ: Tunghai Univ, BS, 61; Univ Chicago, PhD(org chem), 66. Prof Exp: Res fel org chem, Harvard Univ, 66-68; SR RES CHEMIST, E I DU PONT DE NEMOURS & CO, INC, 68- Mem: Am Chem Soc. Res: Process development of biochemicals. Mailing Add: 25 Quartz Mill Rd Newark DE 19711

LIN, KUANG-FARN, b Taiwan, China, Feb 25, 36; m 58; c 2. POLYMER SCIENCE. Educ: Cheng Kung Univ, Taiwan, BSc, 57; NDak State Univ, MS, 63, PhD(polymers, coatings), 69. Prof Exp: Asst instr chem, Chinese Naval Acad, 57-59; supt synthetic resins, Yung Koo Paint & Varnish Mfg Co, 59-61; chemist, 63-67, res chemist, 69-73, proj leader, 73-75, SR RES CHEMIST, HERCULES RES CTR, 74- Concurrent Pos: Asst, NDak State Univ, 63 & 67. Mem: Am Chem Soc; Sigma Xi; Tech Asn Pulp & Paper Indust. Res: Structure-property relationship; adhesion, coatings and polymer synthesis. Mailing Add: Mat Sci Div Hercules Res Ctr Wilmington DE 19899

LIN, LEU-FEN HOU, b Kwangtung, China, Feb 2, 44; m 71. BIOCHEMISTRY. Educ: Nat Taiwan Univ, BS, 67; Univ Minn, PhD(biochem), 72. Prof Exp: INSTR BIOCHEM, MT SINAI SCH MED, 72- Res: Mitochondrial membranes biosynthesis. Mailing Add: Dept of Biochem Mt Sinai Sch of Med New York NY 10029

LIN, LILY, b Shanghai, China, Mar 15, 48; US citizen; m 71. PHOTOBIOLOGY. Educ: Univ Wash, BS, 70; Univ Calif, Los Angeles, PhD(biol), 75. Prof Exp: RES ASSOC MICROBIOL, UNIV ILL, URBANA, 75- Mem: AAAS; Am Soc Photobiol; Am Soc Plant Physiologists. Res: Regulation and biosynthesis in membrane development; biochemistry and physics of the photosynthetic apparatus, pigment-protein complexes of bacteria, algae, plants; electron transport, photophosphorylation, vision. Mailing Add: Dept of Microbiol Univ of Ill Urbana IL 61801

LIN, OTTO CHUI CHAU, b Kwongtang, China, Aug 8, 38; m 63; c 2. POLYMER CHEMISTRY, RHEOLOGY. Educ: Nat Taiwan Univ, BS, 60; Columbia Univ, MA, 63, PhD(phys chem), 67. Prof Exp: Res chemist, 67-69, staff chemist, 69-71, RES ASSOC, MARSHALL RES LAB, FABRICS & FINISHES DEPT, E I DU PONT DE NEMOURS & CO, INC, 71- Mem: AAAS; Am Chem Soc; NY Acad Sci; Soc Rheol. Res: Physical chemical characterization of polymers; rheological properties of polymers; sedimentation; viscometry; organic coatings; ecological impacts of polymer applications. Mailing Add: Res Lab 3500 Grays Ferry Ave E I du Pont de Nemours & Co Inc Philadelphia PA 19146

LIN, PI-ERH, b Taiwan, China, Jan 8, 38; m 63; c 2. MATHEMATICAL STATISTICS. Educ: Taiwan Norm Univ, BSc, 61; Columbia Univ, PhD(math statist), 68. Prof Exp: Consult med ctr, Columbia Univ, 67-68; asst prof, 68-74, ASSOC PROF STATIST, FLA STATE UNIV, 74- Concurrent Pos: Fla State Univ fac res grant, 71-72. Mem: Inst Math Statist; Am Statist Asn. Res: Multivariate analysis; statistical inference. Mailing Add: Dept of Statist Fla State Univ Tallahassee FL 32306

LIN, RENG-LANG, b Hsin-Chu, Taiwan, Feb 28, 37; m 65; c 2. BIOCHEMISTRY, PSYCHOPHARMACOLOGY. Educ: Nat Taiwan Univ, BS, 59, MS, 63; Okla State Univ, PhD(biochem), 69. Prof Exp: Fel, Univ Wis-Madison, 69-71; res scientist, Galesburg State Hosp, Ill, 71-75; RES SCIENTIST, ILL STATE PSYCHIAT INST, 75- Mem: Am Chem Soc; AAAS. Res: Biochemistry of mental illness; biosynthesis and metabolism of biogenic amines; biochemistry and pharmacology of psychoactive drugs. Mailing Add: Res Dept Ill State Psychiat Inst 1601 W Taylor St Chicago IL 60612

LIN, ROBERT PEICHUNG, b China, Jan 24, 42; US citizen. SPACE PHYSICS, SOLAR PHYSICS. Educ: Calif Inst Technol, BS, 62; Univ Calif, Berkeley, PhD(physics), 67. Prof Exp: Asst res physicist, 67-74, ASSOC RES PHYSICIST, SPACE SCI LAB, UNIV CALIF, BERKELEY, 74- Mem: Am Geophys Union; Am Astron Soc. Res: Solar flares, radio bursts and cosmic rays; interplanetary particles; magnetospheric processes; lunar magnetism. Mailing Add: Space Sci Lab Univ of Calif Berkeley CA 94720

LIN, SHEN, b Amoy, China, Feb 4, 31; m 56, 71; c 3. COMPUTER SCIENCE, MATHEMATICS. Educ: Univ Philippines, BS, 51; Ohio State Univ, MA, 53, PhD(math), 63. Prof Exp: Asst prof math, Univ Ohio, 59-62; lectr & res assoc, Ohio State Univ, 62-63; MEM TECH STAFF, MATH & STATIST RES CTR, BELL LABS, INC, 63- Concurrent Pos: Vis lectr, Princeton Univ, 72. Mem: AAAS; Am Math Soc; Math Asn Am; Soc Indust & Appl Math. Res: Application of computers to mathematical science. Mailing Add: Math & Statist Res Ctr Bell Labs Inc Murray Hill NJ 07974

LIN, SHENG HSIEN, b Sept 17, 37; Chinese citizen; m 70. CHEMICAL KINETICS, CHEMICAL PHYSICS. Educ: Nat Taiwan Univ, BS, 59, MS, 61; Univ Utah, PhD(chem), 64. Prof Exp: Fel chem, Columbia Univ, 64-65; from asst to assoc prof, 65-72, PROF CHEM, ARIZ STATE UNIV, 72- Concurrent Pos: A P Sloan fel, 67-69; Guggenheim fel, 71-73. Mem: Am Chem Soc. Res: Energy transfer; atomic collisions; optical rotations and the Faraday effect; reaction kinetics; high order phase transactions; magnetic properties of molecules. Mailing Add: Dept of Chem Ariz State Univ Tempe AZ 85281

LIN, SHIN, b Hong Kong, Feb 14, 45; US citizen; m 69; c 1. BIOCHEMISTRY, BIOPHYSICS. Educ: Univ Calif, Davis, BS, 65; San Diego State Univ, MS, 67; Univ Calif, Los Angeles, PhD(biol chem), 71. Prof Exp: Fel, Univ Calif, San Francisco, 71-74; ASST PROF BIOPHYS, JOHNS HOPKINS UNIV, 74- Concurrent Pos: NIH res career develop award, 76-81. Honors & Awards: Robert C Kirkwood Mem Award, San Francisco Heart Asn, 74. Res: Biochemical and biophysical studies on the molecular basis of sugar transport and cellular motility in eukaryotic cells. Mailing Add: Dept of Biophys Johns Hopkins Univ Baltimore MD 21218

LIN, SHU-REN, b Taiwan, Nov 2, 36; m 62; c 3. RADIOLOGY. Educ: Kaohsiung Med Col, Taiwan, MD, 62. Prof Exp: Intern, Deaconess Hosp, Buffalo, 64-65; surg resident, Sisters Charity Hosp, Buffalo, 65-66; radiol resident, Thomas Jefferson Univ Hosp, Philadelphia, 66-70, fel neuroradiol, 70-71; asst prof radiol, Sch Med, Univ Pa, 71-75; ASSOC PROF & CHIEF NEURORADIOLOGIST, SCH MED, UNIV ROCHESTER, 75- Mem: Asn Univ Radiologists; Am Roentgen Ray Soc; Radiol Soc NAm; Am Soc Neuroradiol. Res: Study of mechanisms of post cardiac arrest deterioration and its possible treatment; neurotoxicity of water contrast media. Mailing Add: Dept of Diag Radiol Sch Med & Dent Univ Rochester Rochester NY 14642

LIN, SHWU-YENG TZEN, b Tainan, Formosa, May 11, 34; m 60; c 3. TOPOLOGY. Educ: Nat Taiwan Univ, BSc, 58; Tulane Univ, MS, 62; Univ Fla, PhD(math), 65. Prof Exp: Asst math, Inst Math, Academia Sinica, 58-60; instr, Tulane Univ, 61-63; lectr, 64-65, asst prof, 65-71, ASSOC PROF MATH, UNIV SOUTH FLA, 71- Concurrent Pos: Reviewer, Math Rev, Am Math Soc, 68-; Zentralblatt für Mathmatik, 70- Mailing Add: Dept of Math Univ of SFla Tampa FL 33620

LIN, SIN-SHONG, b Taiwan, Oct 24, 33; m 64; c 3. HIGH TEMPERATURE CHEMISTRY. Educ: Nat Taiwan Univ, BS, 56; Nat Tsing-Hua Univ, Taiwan, MS, 58; Univ Kans, PhD(chem), 66. Prof Exp: Fel, Northwestern Univ, Evanston, 66-67; RES CHEMIST, MAT RES LAB, ARMY MAT & MECH RES CTR, 67- Mem: Am Chem Soc; Am Vacuum Soc. Res: Thermodynamics of vaporization processes; material research and development; atmospheric sampling of gases; nucleation and condensation studies by mass spectrometry. Mailing Add: Army Mat & Mech Res Ctr Watertown MA 02172

LIN, SONG-LING, b Taipei, Taiwan, Dec 31, 36; m 64; c 1. PHYSICAL PHARMACY, BIOPHARMACY. Educ: Nat Taiwan Univ, BS, 59; Univ Ill, Chicago,

MS, 64, PhD(pharm), 66. Prof Exp: Teaching asst med ctr, Univ Ill, Chicago Circle, 62-66; sr res pharmacist, Ciba Pharmaceut Co, 66-67, supvr phys pharm & stability sect, 67-72; ASST DIR PHARM DIV, AYERST LABS DIV, AM HOME PROD CORP, 72- Mem: Am Pharmaceut Asn; Am Chem Soc; Chinese Pharmaceut Asn; Acad Pharmaceut Sci; Parenteral Drug Asn (vchmn, 72-). Res: Pre-formulation investigation of drugs; kinetics and stabilization of drugs; new dosage form design and process development; in vitro and in vivo study of drug and its dosage form; pharmaceutical analysis. Mailing Add: Pharm Div Ayerst Labs Div Am Home Prod Corp Rouses Point NY 12979

LIN, SPING, b Canton, China, Sept 8, 18; nat US; m 46; c 2. NEUROCHEMISTRY, PHYSIOLOGY. Educ: Sun Yat-Sen Univ, BA, 40; Univ Minn, MS, 50, PhD(entom), 52. Prof Exp: Asst entom, Sun Yat-Sen Univ, 40-44, instr, 44-47; res fellow, 54-61, res assoc, 61-63, asst prof, 63-69, ASSOC PROF NEUROL, MED SCH, UNIV MINN, MINNEAPOLIS, 69- Concurrent Pos: US State Dept fel. Mem: AAAS; Am Soc Neurochem. Res: Neurobiology. Mailing Add: Dept of Neurol Univ of Minn Med Sch Minneapolis MN 55455

LIN, STEPHEN FANG-MAW, b Nantou, Taiwan, Aug 21, 37; m 66; c 1. PHYSICAL CHEMISTRY. Educ: Nat Taiwan Univ, BS, 60; Univ Ill, Urbana, MS, 68, PhD(phys chem), 70. Prof Exp: ASSOC PROF CHEM, NC CENT UNIV, 70- Mem: AAAS; Am Chem Soc. Res: Conformation of proteins. Mailing Add: Dept of Chem NC Cent Univ Durham NC 27707

LIN, SUE CHIN, b Taipei, China, Nov 8, 36; m 62; c 2. MATHEMATICS. Educ: Univ Calif, Berkeley, MA, 64, PhD(math), 67. Prof Exp: Asst prof math, Univ Miami, 67-69; mem, Inst Advan Study, 69-71; ASSOC PROF MATH, UNIV ILL, CHICAGO CIRCLE, 71- Mem: Am Math Soc. Res: Functional analysis. Mailing Add: Dept of Math Univ of Ill Chicago IL 60680

LIN, TSAU-YEN, b Taiwan, China, July 18, 32; m 67. BIOCHEMISTRY. Educ: Nat Taiwan Univ, BS, 55, MS, 57; Univ Calif, Berkeley, PhD(biochem), 65. Prof Exp: Instr clin chem, Kaohsiung Med Col, Taiwan, 57-58; res chemist, China Chem & Pharmaceut Co, 58-59; res asst biochem, US Naval Med Res Unit Number 2, Taiwan, 59-61; res biochemist, Univ Calif, Berkeley, 65-67, asst res biochemist, 67-68; SR RES BIOCHEMIST, MERCK INST THERAPEUT RES, MERCK SHARP & DOHME RES LABS, MERCK & CO, 69- Mem: AAAS; Am Chem Soc. Res: Biosynthesis and function of complex carbohydrates; mechanism and active site structure of enzymes related to carbohydrate metabolism; structural and functional aspects of cell wall and membrane. Mailing Add: Merck Sharp & Dohme Res Labs Rahway NJ 07065

LIN, TSUE-MING, b Ping-tung, Taiwan, June 10, 35; US citizen; m 64; c 3. IMMUNOLOGY, MICROBIOLOGY. Educ: Nat Taiwan Univ, DVM, 58, dipl pub health, 60; Tulane Univ, MS, 64; Univ Tex Med Br, PhD(microbiol), 68. Prof Exp: Teaching asst med parasitol, Col Med, Nat Taiwan Univ, 60-62; res assoc, 68-69, instr, 69-72, ASST PROF PEDIAT, MED SCH, UNIV MIAMI, 72- Concurrent Pos: Teaching asst, Taipei Med Col, 61-62; consult, Cordis Labs, 72-74. Mem: AAAS; Am Soc Parasitol; Am Soc Microbiol; Am Asn Immunol. Res: Host-parasite relationship; pathophysiology; immunology; human heart autoimmune system; trichinosis; amebiasis; human pregnancy-associated plasma proteins. Mailing Add: Nat Children's Cardiac Hosp Dept Pediat Univ Miami Med Sch Miami FL 33152

LIN, TSUNG-MIN, b Chefoo, China, Oct 8, 16; nat US; m 43; c 2. PHYSIOLOGY. Educ: Nat Tsing Hua Univ, China, BS, 38; Univ Ill, MS, 52, PhD, 54. Prof Exp: Asst physiol, Nat Tsing Hua Univ, China, 39-40; asst, Nat Chung Cheng Med Col, 40-41, instr physiol, 41-43; lectr, Nat Kweiyang Med Col, 43-46, asst prof, 46-48; sr instr, Peking Union Med Col, 48-51; asst prof clin sci, Col Med, Univ Ill, 54-58; sr pharmacologist, 56-63, RES ASSOC, RES LABS, ELI LILLY & CO, 64- Concurrent Pos: NSF deleg, Int Cong Physiol Sci, 59 & 1st Int Pharmacol Cong, 61. Mem: Am Physiol Soc; Am Soc Pharmacol & Exp Therapeut; Am Gastroenterol Asn. Res: Gastrointestinal physiology and pharmacology. Mailing Add: Res Labs Eli Lilly & Co Indianapolis IN 46206

LIN, TUNG-PO, b Fukien, China, Dec 31, 26; nat US; m 56; c 4. MATHEMATICS. Educ: Nat Cent Univ, China, BSc, 49; Mass Inst Technol, PhD(phys chem), 58. Prof Exp: Res chemist, E I du Pont de Nemours & Co, Del, 58-61; from asst prof to assoc prof math, 61-69, PROF MATH, CALIF STATE UNIV, NORTHRIDGE, 69- Concurrent Pos: Consult, IBM Corp, 61-68. Mem: Am Math Soc; Math Asn Am. Res: Functional analysis; applied mathematics. Mailing Add: 8954 Chimineas Ave Northridge CA 91324

LIN, WEI-CHING, b Taipei, China, Dec 31, 30; m 59. SPACE PHYSICS. Educ: Nat Taiwan Univ, BSc, 54; Univ Iowa, MSc, 61, PhD(physics), 65. Prof Exp: Res assoc space physics, Univ Iowa, 63-64; asst prof, Dalhousie Univ, 64-68; ASSOC PROF SPACE PHYSICS, UNIV P E I, 68- Mem: Am Geophys Union; Am Asn Physics Teachers. Res: Galactic and solar cosmic rays. Mailing Add: Dept of Physics Univ of PEI Charlottetown PE Can

LIN, YEONG-JER, b Taiwan, China, Nov 11, 36; m 66; c 2. METEOROLOGY, ATMOSPHERIC SCIENCES. Educ: Nat Taiwan Univ, BS, 59; Univ Wis-Madison, MS, 64; NY Univ, PhD(meteorol), 69. Prof Exp: Res asst meteorol, Univ Wis-Madison, 62-64; asst res scientist, NY Univ, 65-69, assoc res scientist, 69; asst prof, 69-72, ASSOC PROF METEOROL, ST LOUIS UNIV, 72- Concurrent Pos: NSF res grant, St Louis Univ, 70-72 & 72-75. Mem: Am Meteorol Soc; Am Geophys Union. Res: Dynamical and observational studies of severe local storms; numerical modelling of meso-scale circulation. Mailing Add: Dept of Earth & Atmospheric Sci St Louis Univ St Louis MO 63103

LIN, YONG YENG, b Feb 2, 33; Taiwan citizen; m 61; c 2. BIO-ORGANIC CHEMISTRY. Educ: Nat Taiwan Univ, BSc, 56; Tokyo Kyoiku Univ, MSc, 63; Tohoku Univ, Japan, PhD(org chem), 66. Prof Exp: Res assoc chem, Fla State Univ, 66-68; res assoc biochem, Univ Tex Med Br, Galveston, 68-70; res assoc chem, Univ Toronto, 70-72; RES SCIENTIST CHEM, UNIV TEX MED BR, 72- Mem: AAAS; Am Chem Soc; Sigma Xi. Res: Mechanisms of biological oxidation; enzyme models; application of enzymes and enzyme models to preparative organic chemistry; organic synthesis. Mailing Add: Div of Biochem Univ of Tex Med Br Galveston TX 77550

LIN, YOU-FENG, b Feng-Shan, Taiwan, July 31, 32; m 60; c 1. TOPOLOGY. Educ: Nat Taiwan Univ, BS, 57; Univ Fla, PhD(math), 64. Prof Exp: Asst math, Inst Math, Chinese Acad Sci, 56-59; from asst to assoc prof, 64-69, res asst prof, 65-66, PROF MATH, UNIV S FLA, 69- Mem: Am Math Soc; Math Asn Am. Res: Topological algebra; structure of topological semigroups; semigroup of measures; topology and relation-theory. Mailing Add: Dept of Math Univ of S Fla Tampa FL 33620

LIN, YU-CHONG, b Taiwan, Repub China, Apr 24, 35; m 60; c 2. PHYSIOLOGY. Educ: Taiwan Norm Univ, BS, 59; Univ NMex, MS, 64; Rutgers Univ, PhD(physiol), 68. Prof Exp: Teaching asst biol, Taiwan Norm Univ, 60-62; ASST PROF PHYSIOL, SCH MED, UNIV HAWAII, MANOA, 69- Concurrent Pos: Fel, Inst Environ Stress, Univ Calif, Santa Barbara, 68-69; physiologist consult, Cardiopulmonary Div, St Francis Hosp, 69- Mem: AAAS; Am Physiol Soc; Fedn Am Socs Exp Biol. Res: Cardiovascular research in the area of diving, exercise, and effect of environmental factors. Mailing Add: Dept of Physiol Sch of Med Univ of Hawaii Honolulu HI 96822

LINAM, JAY H, b Carey, Idaho, Mar 9, 31; m 65; c 2. ENTOMOLOGY. Educ: Univ Idaho, BS, 53; Univ Utah, MS, 57, PhD(entom, zool), 65. Prof Exp: Asst entomologist, Ecol Res Lab, Univ Utah, 58-59; mgr, Magna Mosquito Abatement Dist, Utah, 60-62; from instr to assoc prof, 65-75, PROF BIOL, UNIV SOUTHERN COLO, 75- Mem: Etom Soc Am; Am Mosquito Control Asn. Res: Taxonomy and biology of mosquitoes of Western United States. Mailing Add: Dept of Biol Univ of Southern Colo Pueblo CO 81001

LINARES, OLGA FRANCES, b Panama, Nov 10, 36; US & Panama citizen. ANTHROPOLOGY. Educ: Vassar Col, BA, 58; Harvard Univ, PhD(anthrop), 64. Prof Exp: Instr anthrop, Harvard Univ, 65; lectr, Univ Pa, 67-71; RES SCIENTIST ANTHROP, SMITHSONIAN TROP RES INST, 73- Concurrent Pos: NSF fel, 65 & 70-72; res award, Am Philos Soc, 69 & Smithsonian Inst, 75; assoc cur, Peabody Mus, Harvard Univ, 74- Mem: Am Anthrop Soc; Soc Am Archaeol; AAAS. Res: Human adaptations to the tropical forest, past and present. Mailing Add: Smithsonian Trop Res Inst Box 2072 Balboa CZ

LINCICOME, DAVID RICHARD, b Champaign, Ill, Jan 17, 14; m 41, 53; c 2. PARASITOLOGY, PHYSIOLOGY. Educ: Univ Ill, BS & MS, 37; Tulane Univ, PhD(parasitol), 41; Am Bd Med Microbiol, dipl, 65. Prof Exp: Asst zool, Univ Ill, 37; asst trop med, Sch Med, Tulane Univ, 34-41; from instr to asst prof zool, Univ Ky, 41-47; asst prof parasitol, Univ Wis, 47-49; sr res parasitologist, E I du Pont de Nemours & Co, 49-54; from asst prof to prof zool, Howard Univ, 55-70; CHMN ED BD, EXP PARASITOL, 70- Concurrent Pos: USPHS res grants, 58-68; ed, Exp Parasitol, 49-; guest scientist, Naval MEd Res Inst, 55-61; ed, Int Rev Trop Med, 60-; vis scientist, Lab Phys Biol, Nat Inst Arthritis & Metab Dis, 64-65; chmn comt exam & cert, Am Bd Med Microbiol, 72-; dir res, Am Dairy Goat Asn, mem bd dirs, 73-; ed, Trans, Am Micros Soc, 70-71. Mem: Fel AAAS; Am Soc Parasitol; Helminth Soc (secy, 60, vpres, 61, pres, 68); fel NY Acad Sci; Am Dairy Goat Asn. Res: Diagnosis of protozoan and helminthic diseases; amebiasis; taxonomy and systematics of Acanthocephala, Nematoda and Cestoda; epidemiology of tropical diseases; molecular biology of parasitism; nutritional exchange between parasite and host. Mailing Add: 7118 Cedar Ave Takoma Park MD 20012

LINCK, ALBERT JOHN, b Portsmouth, Ohio, Aug 18, 26; m 57; c 2. PLANT PHYSIOLOGY. Educ: Ohio State Univ, BSc, 50, MSc, 51, PhD(plant physiol), 55. Prof Exp: Instr plant physiol, 55-56, from asst prof to assoc prof, 56-61, asst dir, Minn Agr Exp Sta, 66-71, PROF PLANT PHYSIOL, UNIV MINN, ST PAUL, 61-, DEAN, COL AGR, 71-, ASSOC VPRES ACAD ADMIN, 72- Mem: AAAS; Am Soc Plant Physiol; Bot Soc Am; Scand Soc Plant Physiol; Am Inst Biol Sci. Res: Translocation of inorganic and organic compounds; mechanism of action of growth regulators. Mailing Add: 213 Morrill Hall Univ of Minn Minneapolis MN 55455

LINCK, RICHARD WAYNE, b Los Angeles, Calif, Apr 9, 45; m 72; c 1. CELL BIOLOGY. Educ: Stanford Univ, BA, 67; Brandeis Univ, PhD(biol), 72. Prof Exp: Fel res microtubules, Med Res Coun, Eng, 71-73; instr, 74-75, ASST PROF, DEPT ANAT, HARVARD MED SCH, 75- Concurrent Pos: Fel, Europ Molecular Biol Org, 73; res grant, NIH, 74. Mem: Am Soc Cell Biol; Biophys Soc. Res: Relation of structure to motility; biochemistry and ultrastructure of microtubule proteins in cilia and flagella; reassembly of such proteins and enzymatic interaction with accessory components; optical diffraction of electron micrographs. Mailing Add: Dept of Anat Harvard Med Sch 25 Shattuck St Boston MA 02115

LINCK, ROBERT GEORGE, b St Louis, Mo, Nov 18, 38; m 62. INORGANIC CHEMISTRY. Educ: Case Western Reserve Univ, BS, 60; Univ Chicago, PhD(chem), 63. Prof Exp: Asst prof, 66-72, ASSOC PROF CHEM, UNIV CALIF, SAN DIEGO, 72- Mem: AAAS; Am Chem Soc. Res: Rates of inorganic reactions, especially electron-transfer reactions; electronic structure and photochemistry of complex ions. Mailing Add: Dept of Chem Univ of Calif San Diego La Jolla CA 92093

LINCOLN, CHARLES ALBERT, b Rudyard, Mont, May 13, 39; m 63; c 2. THEORETICAL PHYSICS. Educ: Mont State Univ, BS, 62, MS, 64; Univ Va, DSc(eng physics), 69. Prof Exp: Instr physics, 64-66, ASST PROF PHYSICS, STATE UNIV NY COL FREDONIA, 69- Concurrent Pos: Fulbright exchange prof, Newcastle upon Tyne Polytech, Newcastle/Tyne, Eng, 73-74. Mem: Am Phys Soc; Am Asn Physics Teachers; Math Asn Am; Inst Elec & Electronics Eng. Res: Field theoretic methods in statistical mechanics and fluids; a generalized dynamical formalism of statistical mechanics; information theory and electroacoustics. Mailing Add: Dept of Physics State Univ of NY Col Fredonia NY 14063

LINCOLN, CHARLES GATEWOOD, b Bentonville, Ark, Feb 23, 14; m 38. ENTOMOLOGY. Educ: Univ Ark, BSA, 34; Cornell Univ, PhD(entom), 38. Prof Exp: Asst entom, Cornell Univ, 34-35, instr, 35-41, asst prof, 41-42; exten entomologist, Exten Serv, 42-50, head dept, 51-69, PROF ENTOM, COL AGR, UNIV ARK, 50- Concurrent Pos: Mem, Ark State Plant Bd, 51- Mem: Entom Soc Am. Res: Cotton insects. Mailing Add: Dept of Entom Agr Bldg Univ of Ark Fayetteville AR 72701

LINCOLN, JEANNETTE VIRGINIA, b Ames, Iowa, Sept 7, 15. GEOPHYSICS, SOLAR PHYSICS. Educ: Wellesley Col, BA, 36; Iowa State Univ, MS, 38. Prof Exp: Asst household equip, Iowa State Univ, 37-38, instr, 38-42; physicist, Nat Bur Standards, DC, 42-54, sect chief, Radio Warning Serv, Colo, 54-65, dep chief data serv, Inst Telecommun Sci & Aeronomy, Environ Sci Serv Admin, 65-66, dep chief data serv & chief, Upper Atmosphere Geophys, 66-70, CHIEF DATA SERV & DIR WORLD DATA CTR A, SOLAR-TERRESTRIAL PHYSICS, NAT GEOPHYS & SOLAR TERRESTRIAL DATA CTR, ENVIRON DATA SERV, NAT OCEANIC & ATMOSPHERIC ADMIN, 70- Concurrent Pos: Mem US preparatory comt study group ionospheric propagation, Int Radio Consult Comt, 59-; secy, Int Ursigram & World Days Serv, 61-, mem US Comn 3, Int Sci Radio Union, 63-, secy, Ionospheric Network Adv Group, 69-; forecasting reporter, Int Asn Geomagnetism & Aeronomy, 63-67, mem comns IV & V, 67-; mem working groups 3 & 5, Inter-Union Comn on Solar-Terrestrial Physics, 69-72; Am Geophys Union mem, Am Geophys Union-Int Sci Radio Union Bd of Radio Sci, 69-74. Honors & Awards: Gold Medal, Dept of Com, 73. Mem: Fel AAAS; Sigma Xi; Am Geophys Union; Am Astron Soc; Int Asn Geomagnetism & Aeronomy. Res: Radio propagation disturbances and forecasts; solar-terrestrial relationships; publication of solar and geophysical data; prediction of solar indices; data center management. Mailing Add: World Data Ctr A Nat Oceanic & Atmospheric Admin Boulder CO 80302

LINCOLN, KENNETH ARNOLD, b Oakland, Calif, Oct 1, 22; m 56; c 4. HIGH TEMPERATURE CHEMISTRY. Educ: Stanford Univ, AB, 44, MS, 48, PhD(phys chem), 57. Prof Exp: Phys chemist, US Naval Radio Defense Lab, 58-69; RES SCIENTIST, NASA-AMES RES CTR, 70- Mem: Am Chem Soc; Am Soc Mass

Spectrometry; Am Sci Affil; Combustion Inst. Res: Thermochemistry of the vaporization of refractory materials; thermokinetics of pulsed energy deposition; development of instrumentation combining lasers and high-speed mass spectrometry for in-situ analyses of short-lived chemical species; space flight spectrometric instrumentation. Mailing Add: 2016 Stockbridge Ave Redwood City CA 94061

LINCOLN, LEWIS LAUREN, b Canandaigua, NY, Oct 9, 26; m 49; c 6. PHOTOGRAPHIC CHEMISTRY. Prof Exp: Lab technician, 46-60, res chemist, 60-65, sr res chemist, 65-70, RES ASSOC CHEM, RES LABS, EASTMAN KODAK, 70- Mem: Am Chem Soc. Res: The study and synthesis of photographic sensitizing dyes and addenda. Mailing Add: 456 Pellet Rd Webster NY 14580

LINCOLN, RALPH ERNEST, genetics, deceased

LINCOLN, RICHARD CRIDDLE, b Boston, Mass, Nov 25, 42; c 2. APPLIED PHYSICS. Educ: Cornell Univ, BEP, 66, MS, 68, PhD(mat sci), 71. Prof Exp: Instr & res assoc mat sci, Cornell Univ, 70-71; TECH STAFF MEM APPL PHYSICS, SANDIA LABS, 71- Res: High pressure and high temperature experimental techniques; analysis of nuclear waste management systems. Mailing Add: Sandia Labs Albuquerque NM 87115

LINCOLN, RICHARD G, b Portland, Ore, Nov 1, 23; m 46; c 2. PLANT PHYSIOLOGY. Educ: Ore State Univ, BS, 49; Univ Calif, Los Angeles, PhD, 55. Prof Exp: Plant physiologist, Sugarcane Field Sta, USDA, La, 55-56; PROF BOT, CALIF STATE UNIV, LONG BEACH, 56- Mem: AAAS; Am Soc Plant Physiol. Res: Plant photoperiodism; inhibition of flowering and extraction of the flowering stimulus; rhythmic phenomena in fungi. Mailing Add: Dept of Biol Calif State Univ Long Beach CA 90804

LIND, ARTHUR CHARLES, b Chicago, Ill, May 28, 32; m 57; c 3. APPLIED PHYSICS, SPACE PHYSICS. Educ: Univ Ill, Urbana, BS, 55; Rensselaer Polytech Inst, PhD(physics), 66. Prof Exp: Physicist, Knolls Atomic Power Lab, 58-61 & Watervliet Arsenal, US Army, 63-66; assoc scientist, 66-76, SCIENTIST, McDONNELL DOUGLAS RES LABS, 76- Mem: Am Phys Soc; Inst Elec & Electronics Engrs. Res: Theoretical and experimental studies of electromagnetic scattering; propagation of electromagnetic waves in turbulent media; measurement of dielectric properties at high temperatures; nuclear magnetic resonance. Mailing Add: McDonnell Douglas Res Labs Box 516 St Louis MO 63166

LIND, BARBARA MILDRED, biochemistry, see 12th edition

LIND, CHARLES DOUGLAS, b Pittsburgh, Pa, May 12, 30; m 55; c 2. PHYSICAL CHEMISTRY. Educ: Pa State Univ, BS, 52; Univ Ore, MS, 54, PhD(chem), 56. Prof Exp: Fel chem, Univ Ore, 56-57; CHEMIST, NAVAL WEAPONS CTR, 57- Mem: Am Chem Soc. Res: Chemistry of explosives; detonation theory; combustion theory. Mailing Add: Naval Weapons Ctr Code 4541 China Lake CA 93555

LIND, DAVID ARTHUR, b Seattle, Wash, Sept 12, 18; m 45; c 4. NUCLEAR PHYSICS. Educ: Univ Wash, Seattle, BS, 40; Calif Inst Technol, MS, 43, PhD(physics), 48. Prof Exp: Jr aerodynamicist, Boeing Airplane Co, Wash, 42-43; physicist, Appl Physics Lab, Univ Wash, Seattle, 43-45; res fel physics, Calif Inst Technol, 48-50; Guggenheim fel, Nobel Inst Physics, Stockholm, 50-51; asst prof, Univ Wis, 51-56; assoc prof, 56-59, PROF PHYSICS, UNIV COLO, BOULDER, 59-, CHMN DEPT PHYSICS & ASTROPHYS, 74- Concurrent Pos: Consult off instnl prog, NSF, 63-66; physicist div res, US AEC, 69-70; mem prog adv comn, Los Alamos Meson Facil, 71-74, chmn users group, 75-76. Mem: Fel Am Phys Soc; Sigma Xi; Am Asn Physics Teachers. Res: X-rays; crystal diffraction; nuclear spectroscopy; sector focused cyclotron design; charged particle scattering; reaction studies; fast neutron spectroscopy. Mailing Add: Dept of Physics & Astrophys Univ of Colo Boulder CO 80302

LIND, EDWARD LOUIS, physical chemistry, see 12th edition

LIND, HOWARD ERIC, b Providence, RI, Feb 22, 13; m 37; c 4. BACTERIOLOGY. Educ: RI State Col, BS, 34, MS, 35; Mass Inst Technol, MPH, 37; Northwestern Univ, PhD(bact), 43; Am Bd Microbiol, dipl. Prof Exp: Asst bact, RI State Col, 34-35; lectr, RI Sch Educ, 35; asst bacteriologist, State Health Dept, RI, 35-36, sanitarian & bacteriologist, 36-37; bacteriologist, St Louis County Hosp & dir labs, County Health Dept, 37-40; sr bacteriologist, Chicago Br Lab, State Dept Health, Ill, 40-43; bacteriologist, Dow Chem Co, Mich, 45-46; res dir, Sias Lab, Brooks Hosp, Brookline, Mass, 46-71; mem staff, Maine Dept Health & Welfare, 70-71, ASST DIR, PUB HEALTH LABS, MAINE DEPT HUMAN SERV, 71- Concurrent Pos: Consult, RI Health Labs; lectr sch med, Wash Univ, 38; supvr lab, Booth Mem Hosp, 48-50; pres & dir res, Lind Labs, Inc, 56-70. Mem: Am Soc Microbiol; Am Pub Health Asn; Inst Food Technol; NY Acad Sci. Res: Evaluation of antibiotics in urinary tract infections; electronic counting of bacteria; hospital infections and phage typing; state laboratory evaluation; quality control. Mailing Add: Pub Health Labs Dept of Human Serv Augusta ME 04330

LIND, JAMES FOREST, b Fillmore, Sask, Nov 22, 25; m 50; c 5. SURGERY. Educ: Queen's Univ, Ont, MD, CM, 51; FRCS(C), 59. Prof Exp: Registr, Clatterbridge Hosp, Cheshire, Eng, 56-58; from lectr to prof surg, Univ Mann, 66-73; PROF SURG, MED CTR, McMASTER UNIV, 73- Concurrent Pos: Teaching fel anat, Queen's Univ, Ont, 53-54, fel med, 54-55, fel surg, 55-56, George Christian Hoffman award surg, 57-58; John S McEachern fel, Can Cancer Soc, 56-57; fel physiol, Mayo Found, 58-60; Markle scholar med, 60-65; asst surgeon, Dir Casualty & Dir Sect Gastroenterol, Clin Invest Unit, Winnipeg Gen Hosp, 60-73, surgeon in chief, 66-73; surg consult, Deer Lodge Hosp, 60-73. Mem: Fel Am Col Surg; Can Soc Clin Invest; Can Med Asn; Can Asn Clin Surg; Can Asn Gastroenterol. Res: Gastroeneterology. Mailing Add: Dept of Surg McMaster Univ Med Ctr Hamilton ON Can

LIND, MAURICE DAVID, b Jamestown, NY, July 25, 34; m 62; c 1. PHYSICAL CHEMISTRY, X-RAY CRYSTALLOGRAPHY. Educ: Otterbein Col, BS, 57; Cornell Univ, PhD(phys chem), 62. Prof Exp: NSF fel, 62-63; res chemist phys chem, Union Oil Co, Calif, 63-66; MEM TECH STAFF, SCI CTR/ROCKWELL INT, 66- Mem: AAAS; Am Chem Soc; Am Phys Soc; Am Crystallog Asn; Sigma Xi. Res: Crystal chemistry; crystal growth. Mailing Add: 1690 Stoddard Ave Thousand Oaks CA 91360

LIND, NIELS CHRISTIAN, b Copenhagen, Denmark, Mar 10, 30; Can citizen; m 57; c 3. APPLIED MECHANICS. Educ: Royal Tech Univ Denmark, MSc, 53; Univ Ill, PhD(theoret & appl mech), 59. Prof Exp: Design engr, Dominia Ltd, Denmark, 53-54; engr, Bell Tel Co, Can, 54-55; field engr, Drake & Merritt Co, Labrador, 55-56; design engr, Fenco, Que, 56; asst stress anal, Univ Ill, 56-57, instr, 57-58, res assoc, 58-59, asst prof theoret & appl mech, 59-60; assoc prof civil eng, 60-62, PROF CIVIL ENG, UNIV WATERLOO, 62- Concurrent Pos: Mem, Can Nat Study Group Math Higher Educ, Orgn for Econ Coop & Develop, 63-65; vis prof, Univ Laval, 69. Mem: Fel Am Acad Mech (pres, 71-72). Res: Structural mechanics; theory of design; structural reliability and optimization. Mailing Add: Dept of Civil Eng Univ of Waterloo Waterloo ON Can

LIND, OWEN THOMAS, b Emporia, Kans, June 2, 34; m 54; c 2. LIMNOLOGY, BIOLOGY. Educ: William Jewell Col, AB, 56; Univ Mich, MS, 60; Univ Mo, PhD(zool), 66. Prof Exp: Biologist, Parke, Davis & Co, Mich, 56-60; asst prof biol, William Jewell Col, 60-62; res assoc limnol, Univ Mo, 66; asst prof biol, 66-69, ASSOC PROF BIOL, BAYLOR UNIV, 69- Concurrent Pos: Co-investr, Off Water Resources Res grant, 66, prin investr thermal pollution, 71; consult, US Nat Park Serv, Guadalupe Mt Nat Park, 69-; dir, Inst Environ Studies, 71-; mem selection comt, Tyler Ecol Award, 74- Mem: Am Soc Limnol & Oceanog; Int Asn Theoret & Appl Limnol; Brit Freshwater Biol Asn. Res: Limnology of polluted waters; primary production and community metabolism in relation to water quality and eutrophication. Mailing Add: Dept of Biol Baylor Univ Waco TX 76703

LIND, ROBERT WAYNE, b Ishpeming, Mich, Aug 25, 39; m 64; c 2. THEORETICAL PHYSICS. Educ: Mich Technol Univ, BS, 61; Univ Pittsburgh, PhD(physics), 70. Prof Exp: Engr, Ford Motor Co, 63-66; res assoc physics, Syracuse Univ, 70-72; sr res assoc, Temple Univ, 72-73; res assoc, Fla State Univ, 73-74; ASST PROF PHYSICS, WVA INST TECHNOL, 74- Mem: Am Asn Physics Teachers; Int Soc Gen Relativity & Gravitation. Res: Investigation of the long range forces, gravitational and electromagnetic, and their relationship to the nature of space and time; some cosmology and mathematical physics. Mailing Add: Dept of Physics WVa Inst of Technol Montgomery WV 25136

LIND, VANCE GORDON, b Brigham City, Utah, Feb 12, 35; m 64; c 1. PHYSICS, ASTROPHYSICS. Educ: Utah State Univ, BS, 59; Univ Wis, MS, 61, PhD(elem particles), 64. Prof Exp: Eng asst, Edgerton, Germeshausen & Grier, Inc, summer, 59, res asst, 60; res assoc, Univ Mich, summer, 64; asst prof physics, 64-68, assoc prof, 68-75, PROF PHYSICS, UTAH STATE UNIV, 75- Concurrent Pos: Utah State Univ Res Found grant, 64-66; investr, NSF res grant, Utah State Univ, 66-68 & 68-71; Mem: AAAS; Am Phys Soc; Sigma Xi. Res: Elementary particle interactions, meson and nucleon interactions with nuclei, astronomy and astrophysics. Mailing Add: Dept of Physics Utah State Univ Logan UT 84321

LINDAHL, CHARLES BLIGHE, inorganic chemistry, see 12th edition

LINDAHL, CLARENCE HOMER, b Polk, Nebr, Oct 19, 06; m 30; c 2. MATHEMATICS. Educ: Nebr State Teachers Col, Kearny, BS, 29; Univ Colo, MS, 35; Iowa State Univ, PhD(math), 52. Prof Exp: Prin & teacher high sch, Nebr, 29-36, supt schs, 36-43; actg dean men & prof math, Nebr State Teachers Col, Wayne, 43-47; from asst prof to assoc prof, 47-62, PROF MATH, IOWA STATE UNIV, 62-, IN CHG INSTR, 61- Mem: Fel AAAS; Sigma Xi; Math Asn Am. Res: Overlapping pfaffians with application to utility theory. Mailing Add: 1111 Scholl Rd Ames IA 50010

LINDAHL, ROY LAWRENCE, b Los Angeles, Calif, Aug 22, 25; m 48; c 4. DENTISTRY. Educ: Univ Southern Calif, BS & DDS, 50; Univ Mich, MS, 52; Am Bd Pedodontics, dipl, 56. Prof Exp: Asst prof, 52-56, PROF PEDODONTICS, SCH DENT, UNIV NC, CHAPEL HILL, 56-, DIR CONTINUING EDUC & DENT DEMONSTR PRACT, 70- Concurrent Pos: Mem bd trustees, MC Cerebral Palsy Hosp, 57-63; consult, Womack Army Hosp, Ft Bragg, NC, 60 & Youth Fitness Comn, 60; examr, Am Bd Pedodontics, 60-67, chmn, 67. Mem: AAAS; Am Soc Dent for Children (from secy to pres, 69-73); Am Dent Asn; Am Acad Pedodontics (vpres, 62-63, pres-elect, 63-64, pres, 64-65); Int Asn Dent Res. Res: Pedodontics; pulp therapy; effective utilization of dental auxiliary personnel; problems of the handicapped patient; pre-payment dental care programs. Mailing Add: Dept of Pedodontics Univ of NC Chapel Hill NC 27514

LINDALL, ARNOLD WALFRED, b Duluth, Minn, July 11, 37; m 55; c 2. ENDOCRINOLOGY. Educ: Univ Minn, BA, 56, MD & PhD(anat, biochem), 62. Prof Exp: NIH spec fel neurochem & neuroanat, Univ Minn, 62-64; intern, Mt Sinai Hosp, Minneapolis, 64; ASST PROF ANAT, MED SCH, UNIV MINN, MINNEAPOLIS, 65-, ASST PROF MED, 70-, ASSOC PROF LAB MED, 73- Concurrent Pos: Markle scholar, 65-67; resident, Hennepin County Gen Hosp, 67-70; dir metrop ref lab, Minneapolis War Mem Blood Bank, 72- Mem: Am Asn Anat; Endocrine Soc. Res: Endocrine cytochemistry; radioimmunoassay of hormones and clinical research in endocrine disease. Mailing Add: Univ of Minn Dept of Lab Med Box 198 Mayo Minneapolis MN 55455

LINDAU, EVERT INGOLF, b Vaxjo, Sweden, Oct 4, 42. SOLID STATE PHYSICS. Educ: Chalmers Univ Technol, Sweden, Civilingenjor, 68, Technol Licentiat, 70, PhD(physics), 71, DrTechnol, 72. Prof Exp: Res asst physics, Chalmers Univ Technol, Sweden, 68-71; res scientist, Varian Assocs, 71-72; res assoc physics, 72-74, PROF PHYSICS, STANFORD UNIV, 74- Mem: Am Phys Soc; Am Vacuum Soc; Swed Soc Technol. Res: Optical and photoemission studies of the electronic structure of materials using synchrotron radiation with emphasis on surface properties; surface states, surface photoemission, physisorbtion, chemisorbtion, surface composition and catalytic activities. Mailing Add: Stanford Electronics Lab Stanford Univ Stanford CA 94305

LINDAUER, IVO EUGENE, b Grand Valley, Colo, Apr 7, 31; m 57; c 2. PLANT ECOLOGY. Educ: Colo State Univ, BS, 53, PhD(bot), 70; Univ Northern Colo, MA, 60. Prof Exp: Instr biol, Univ Northern Colo, 60-64, asst prof sci, 64-65; res assoc & teaching asst bot, Colo State Univ, 65-67; asst prof, 67-71, ASSOC PROF BOT, UNIV NORTHERN COLO, 71- Concurrent Pos: US Bur Reclamation grants, 69-72, 70-72; Tri-Univ Proj grant, NY Univ, 70; US Bur Reclamation grant, proposed Narrows Dam site, 70-72. Mem: Ecol Soc Am; Nat Asn Biol Teacher (secy-treas, 69-70); Am Inst Biol Sci. Res: Analysis of vegetational communities found along flood plains; ecological studies of river bottom ecosystems. Mailing Add: Dept of Biol Sci Univ of Northern Colo Greeley CO 80631

LINDAUER, MAURICE WILLIAM, b Millstadt, Ill, Sept 25, 24; m 46; c 3. ANALYTICAL CHEMISTRY, PHYSICAL CHEMISTRY. Educ: Wash Univ, AB, 49, AM, 53; Harvard Univ, MEd, 62; Fla State Univ, PhD, 70. Prof Exp: Res chemist, Mallinckrodt Chem Works, 52-55 & Am Zinc, Ill, 55-56; res chemist nitrogen div, Allied Chem & Dye Corp, 56-57; assoc prof anal & phys chem, 57-71, PROF CHEM, VALDOSTA STATE COL, 71- Concurrent Pos: NSF sci fac fel, 64-65. Mem: Am Chem Soc; Sigma Xi. Res: Mössbauer spectroscopy; history of chemistry. Mailing Add: Dept of Chem Valdosta State Col Valdosta GA 31601

LINDBECK, WENDELL ARTHUR, b Rockford, Ill, Sept 28, 12; m 38; c 3. ORGANIC CHEMISTRY. Educ: Beloit Col, BS, 36; Univ Wis, PhM, 37, PhD(chem), 40. Prof Exp: Teacher, Tenn Jr Col, 40-44; tech coord, Goodyear Synthetic Rubber Corp, Ohio, 44-47; assoc prof & chmn, Natural Sci Div, Univ Ill, 47-49; assoc prof phys sci & chem, 49-52, PROF PHYS SCI & CHEM, NORTHERN ILL UNIV, 52- Concurrent Pos: NSF sci fac fel, Univ Calif, Berkeley, 60-61; US

AEC grant, Argonne Nat Lab, 69-70. Mem: Am Chem Soc. Res: Synthesis of organic compounds. Mailing Add: Dept of Chem Northern Ill Univ De Kalb IL 60115

LINDBERG, DAVID SEAMAN, SR, b Merrill, Wis, July 17, 29; m 51; c 3. ACADEMIC ADMINISTRATION, HEALTH SCIENCES. Educ: Univ Wis-Stevens Point, BS, 58; Univ Fla, MEd, 69, EdD, 70. Prof Exp: Lab supvr, Marshfield Clin, Wis, 58-66; res asst clin path, Col Med, Univ Fla, 66; supvry med technologist, Lab Serv, Vet Admin Hosp, Gainesville, Fla, 67-68; asst prof health related professions, Ctr Allied Health Instrnl Personnel, Univ Fla, 71-72, asst prof med technol, Col Health Related Professions, 72-74; ASSOC PROF MED TECHNOL & ASST DEAN SCH ALLIED HEALTH PROFESSIONS, LA STATE UNIV MED CTR, NEW ORLEANS, 74- Honors & Awards: Prof Achievement Award Educ, Am Soc Med Technologists, 74, Outstanding Performance Recognition Certs, 74 & 75. Mem: Am Soc Allied Health Professions; Am Soc Clin Pathologists; Am Soc Med Technologists. Res: Continuing education in the health professions. Mailing Add: Rte 4 Box 203XL Covington LA 70433

LINDBERG, DONALD ALLAN BROR, b Brooklyn, NY, Sept 21, 33; m 57; c 3. PATHOLOGY, COMPUTER SCIENCE. Educ: Amherst Col, AB, 54; Columbia Univ, MD, 58; Am Bd Path, dipl, 63. Prof Exp: From instr to assoc prof path, Sch Med, 62-69, chmn dept info sci, Med Ctr, 68-71, PROF PATH, SCH MED, UNIV MO-COLUMBIA, 69-, DIR INFO SCI GROUP, MED CTR, 71- Concurrent Pos: Markle scholar, 64-69; mem, Comput Res & Biomath Study Sect, NIH, 67-71 & Comput Sci & Eng Bd, Nat Acad Sci, 71-73; chmn, CBX Adv Comt, Nat Bd Med Examrs, 71-74, mem, Joint CBX Comt, Nat Bd Med Examrs & Am Bd Internal Med, 74-; US rep, Comt Comput Med, Int Fedn Info Processing. Mem: Am Soc Exp Path; Soc Exp Biol & Med; Am Soc Clin Path; Col Am Path. Res: Information processing; computers in medicine; infectious diseases. Mailing Add: 605 Lewis Hall Univ of Mo Columbia MO 65201

LINDBERG, GEORGE DONALD, b Salt Lake City, Utah, Feb 9, 25; m 55; c 3. PLANT PATHOLOGY. Educ: Ariz State Univ, BS, 50; Okla State Univ, MS, 52; Univ Wis, PhD(plant path), 55. Prof Exp: Asst prof, 55-59, assoc prof, 59-70, PROF PLANT PATH, LA STATE UNIV, BATON ROUGE, 70- Mem: Am Phytopath Soc. Res: Plant virology; diseases of forage crops; abnormalities in the fungi. Mailing Add: Dept of Bot & Plant Path La State Univ Baton Rouge LA 70803

LINDBERG, HOWARD AVERY, b Chicago, Ill, May 30, 10; m 38; c 3. MEDICINE. Educ: Northwestern Univ, BS, 31, MS, 34, MD, 35; Am Bd Internal Med, dipl, 42. Prof Exp: Clin asst, 41-43, instr, 43-45, assoc, 45-48, from asst prof to assoc prof, 48-75, PROF MED, NORTHWESTERN UNIV, CHICAGO, 75- Concurrent Pos: Fel, Med Sch, Northwestern Univ, Chicago, 37-41; mem staff, Cardiovasc Renal Clin, 41-56; mem attend staff, Northwestern Mem Hosp, 42-, chief staff, 56; med dir, Jewel Co, Inc & A C Nielson Co; med dir, Peoples Gas, Light & Coke Co, 40-75; med consult, Amsted Industs & Benefit Trust Life Ins Co. Mem: Am Fedn Clin Res (pres, 48); fel Am Col Physicians; Am Soc Internal Med; NY Acad Sci. Res: Cardiology. Mailing Add: 670 N Michigan Ave Chicago IL 60611

LINDBERG, JAMES BECKWITH, b Chicago, Ill, June 5, 28; m 51; c 2. GEOGRAPHY. Educ: Denison Univ, BA, 50; Univ Mich, MBA, 52; Univ Wis, PhD(geog), 62. Prof Exp: Petrol economist, Stand Oil Co, Ind, 55-57; from instr to asst prof, 60-66, ASSOC PROF GEOG, UNIV IOWA, 66- Mem: Asn Am Geog; Regional Sci Asn. Res: Industrial location; central place systems. Mailing Add: Dept of Geog Univ of Iowa Iowa City IA 52240

LINDBERG, JAMES GEORGE, b Grand Rapids, Mich, Sept 19, 40; m 62; c 3. ORGANIC CHEMISTRY. Educ: Kalamazoo Col, BA, 62; Baylor Univ, PhD, 69. Prof Exp: Asst prof, 67-72, ASSOC PROF CHEM, DRAKE UNIV, 72- Concurrent Pos: Drake Univ Res Coun grant, 68-70. Mem: Am Chem Soc; The Chem Soc. Res: Structural studies of Grignard reagents derived from hindered ketones; nuclear magnetic resonance spectroscopic studies of steric effects; formation and structure of hydroperoxides of hindered ketones; synthesis of enolethers. Mailing Add: Dept of Chem Drake Univ Des Moines IA 50311

LINDBERG, JOHN ALBERT, JR, b New York, NY, Apr 19, 34; m 64; c 2. MATHEMATICAL ANALYSIS. Educ: Wagner Col, BA, 54; Univ Minn, MA, 57, PhD(math), 60. Prof Exp: Instr math, Univ Minn, 58-59 & Yale Univ, 60-62; from asst to assoc prof, 62-72, PROF MATH, SYRACUSE UNIV, 72- Concurrent Pos: Res fel math, Yale Univ, 68-69. Mem: AAAS; Am Math Soc; Math Asn Am; Sigma Xi. Res: Theory of algebraic extensions of Banach algebras and factorization of polynomials over such algebras; inverse producing normed extensions. Mailing Add: Dept of Math Room 200 Carnegie Syracuse Univ Syracuse NY 13210

LINDBERG, LOIS HELEN, b Scott Air Force Base, Ill, Sept 1, 32. MEDICAL MICROBIOLOGY. Educ: San Jose State Col, AB, 52; Univ Calif, MPH, 58; Stanford Univ, PhD, 67. Prof Exp: Jr microbiologist, State Dept Pub Health, Calif, 53-54; instr bact, San Jose State Col, 54-55; assoc pub health, Pub Health Lab, Univ Calif, 55-58; asst prof bact, 58-65, assoc prof biol, 65-70, PROF BIOL, SAN JOSE STATE COL, 70- Concurrent Pos: NSF sci teachers fel, 62, fel, 66 & 67; res assoc, Stanford Univ Med Sch. Mem: AAAS; Am Pub Health Asn; Am Soc Microbiol. Res: Medical microbiology as related with the pathology and immunology of streptoccal infections. Mailing Add: Dept of Biol Sci San Jose State Col San Jose CA 95114

LINDBERG, ROBERT BENJAMIN, b Grand Rapids, Mich, Dec 26, 14; m 42; c 2. MEDICAL MICROBIOLOGY. Educ: Univ Mich, BS, 35, MS, 36, PhD(bact), 50. Prof Exp: Instr bact, Univ Mich, 37-41; US Army, 42-, chief dept bact, 18th Med Gen Lab, 44-47, 406th Med Gen Lab, Japan, 50-53, Walter Reed Army Inst Res, 54-57 & Med Lab, Ger, 57-61, CHIEF DEPT MICROBIOL, INST SURG RES, US ARMY, 61- Honors & Awards: Army Res & Develop Achievement Award, US Army Res & Develop Command, 72. Mem: AAAS; Am Soc Microbiol; Am Asn Immunol; Am Pub Health Asn; Am Acad Microbiol. Res: Pathogenesis and epidemiology of salmonellosis and shigellosis; antigenic structure shigella and pathogenic fungi; pathogenic coliform bacteria; bacteriology gas gangrene; chemotherapy of burns; phage typing Pseudomonas; diagnostic methods; bacteriology of Mima group. Mailing Add: Dept of Microbiol US Army Inst of Surg Res Ft Sam Houston TX 78234

LINDBERG, ROBERT GENE, zoology, see 12th edition

LINDBERG, STEVEN EDWARD, b St Paul, Minn, Oct 17, 42; m 65; c 2. ORGANIC CHEMISTRY, POLYMER CHEMISTRY. Educ: Gustavus Adolphus Col, BS, 64; Univ Minn, Minneapolis, PhD(org chem), 69. Prof Exp: Res chemist, 69-75, RES SUPVR, RES & DEVELOP DEPT, AMOCO CHEM CORP, 75- Mem: Am Chem Soc; Soc Petrol Engrs. Res: Polyelectrolytes; tertiary oil recovery; cold flow improvers. Mailing Add: Res & Develop Dept Box 400 Amoco Chem Corp Warrenville Rd Naperville IL 60540

LINDBLOM, ROBERT O, organic chemistry, see 12th edition

LINDBURG, DONALD GILSON, b Wagner, SDak, Nov 6, 32; m 54; c 3. PHYSICAL ANTHROPOLOGY, PRIMATOLOGY. Educ: Houghton Col, BA, 56; Univ Chicago, MA, 62; Univ Calif, Berkeley, PhD(anthrop), 67. Prof Exp: Res asst primatol, Nat Ctr Primate Biol, 64-66; res anthropologist, Sch Med, Univ Calif, Davis, 69-72, asst prof anthrop, Univ Calif, Davis, 67-73; chmn & assoc prof, Ga State Univ, 73-75; ASSOC PROF ANTHROP, UNIV CALIF, LOS ANGELES, 75- Concurrent Pos: Res anthropologist, Nat Ctr Primate Biol, 66-69; NSF fel, Univ Calif, Davis, 72-75. Mem: AAAS; Animal Behav Soc; Am Anthrop Asn; Am Asn Phys Anthrop Int Primatol Soc. Res: Socio-ecology of primates. Mailing Add: Dept of Anthrop Univ of Calif Los Angeles CA 90024

LINDE, ALAN TREVOR, b Lowood, Australia, Feb 13, 38; m 60; c 3. GEOPHYSICS. Educ: Univ Queensland, BSc, 59, PhD(physics), 72. Prof Exp: Lectr physics, Univ Queensland, 62-72; STAFF MEM GEOPHYS, DEPT TERRESTRIAL MAGNETISM, CARNEGIE INST WASHINGTON, 72- Mem: Am Geophys Union; Seismol Soc Am. Res: Theoretical and observational studies of earthquake source mechanisms to determine properties of the earth's interior and hence to understand the earth's tectonic engine. Mailing Add: Carnegie Inst 5241 Branch Rd NW Washington DC 20015

LINDE, HARRY WIGHT, b Woodridge, NJ, Jan 1, 26; m 56; c 2. PHARMACOLOGY, ANESTHESIOLOGY. Educ: Tufts Col, BS, 50; Mass Inst Technol, PhD(chem), 53. Prof Exp: Sr chemist, Res Labs, Air Reduction Co, Inc, 53-56; res assoc anesthesia, Med Sch, Univ Pa, 56-64; group leader, Air Prod & Chem, Inc, 63-65; asst prof anesthesia, 65-70, asst dir anesthesia res ctr, 67-71, ASSOC PROF ANESTHESIA, MED SCH, NORTHWESTERN UNIV, CHICAGO, 70-, COORDR RES & SPONSORED PROGS, 71- Concurrent Pos: Res assoc, Col Med, Univ Ill, 55-56; mem, Comt Admis, Northwestern Univ, 67-, human subjects rev, 70- & res comt, 71-; consult res anesthesia, Vet Admin Lakeside Hosp, Chicago, 68-; assoc staff mem, Chicago Wesley Mem Hosp, 69-72 & Northwestern Mem Hosp, 72-; consult, US Naval Hosp, Great Lakes, 69- Mem: Fel AAAS; Am Chem Soc; Int Anesthesia Res Soc; Am Soc Anesthesiol; Sigma Xi. Res: Pharmacology of anesthesia; cardiovascular and respiratory physiology; gas analysis; bioanalytical chemistry. Mailing Add: Dept of Anesthesia Northwestern Univ Med Sch Chicago IL 60611

LINDE, LEONARD M, b New York, NY, June 1, 28; m 51; c 2. CARDIOLOGY, PHYSIOLOGY. Educ: Univ Calif, BS, 47, MD, 51; Am Bd Pediat, dipl & cert cardiol, 57. Prof Exp: Intern, Morrisania City Hosp, New York, 51-52; sr resident pediat, Children's Hosp, Los Angeles, 52-53 & 55-56; PROF PEDIAT & CARDIOL, SCH MED, UNIV CALIF, LOS ANGELES, 57-, PHYSIOL, 59-; CHIEF PEDIAT CARDIOL, ST VINCENT'S HOSP, LOS ANGELES, 73- Concurrent Pos: Fel pediat cardiol, Med Ctr, Univ Calif, Los Angeles, 56-57; consult, Child Cardiac Clin, Los Angeles City Health Dept, 57- & Surg Gen, US Air Force; mem, Courtesy Staff, St John's Hosp, Santa Monica, 58-; vis prof, Univ Tokyo, 65- Mem: Fel Am Acad Pediat. Res: Pediatric cardiology; cardiopulmonary physiology; clinical cardiology; cardiac catheterization. Mailing Add: Dept of Pediat & Cardiol UCLA Med Ctr Los Angeles CA 90024

LINDE, PETER FRANZ, b Berlin, Ger, June 9, 26; nat US; m 53; c 4. PHYSICAL CHEMISTRY. Educ: Reed Col, BA, 46; Univ Ore, MA, 49; Wash State Univ, PhD(chem), 54. Prof Exp: Phys chemist, Sandia Corp, 53-57; from asst prof to assoc prof, 57-66, PROF CHEM, SAN FRANCISCO STATE UNIV, 66- Mem: Am Chem Soc. Res: Electrochemistry of quaternary ammonium compounds; supporting electrolytes in polarography; shock tube measurements. Mailing Add: Dept of Chem San Francisco State Univ San Francisco CA 94132

LINDEBERG, GEORGE KLINE, b Spencer, Iowa, June 6, 30; m 54; c 2. SOLID STATE PHYSICS. Educ: St Olaf Col, BA, 52; Princeton Univ, PhD(exp physics), 57. Prof Exp: Asst physics, Princeton Univ, 56-57; PHYSICIST, MINN MINING & MFG CO, 57- Mem: Am Phys Soc. Res: Non-equilibrium electronic processes in solids; thermodynamics; physics operations research. Mailing Add: Rte 3 Hudson WI 54016

LINDEBORG, ROBERT G, b Tacoma, Wash, Oct 12, 10; m 40; c 3. BIOECOLOGY, MAMMALOGY. Educ: Univ Ill, BS, 34, MS, 35; Univ Mich, PhD(zool), 48. Prof Exp: Asst prof biol sci, Mich State Col, 46-49; asst prof, 49-54, chmn dept biol, 71-74, PROF ZOOL, N MEX HIGHLANDS UNIV, 54- Mem: Fel AAAS. Res: Physiological ecology; vertebrates. Mailing Add: Dept of Biol NMex Highlands Univ Las Vegas NM 87701

LINDEGREN, CARL ROBERT, organic chemistry, see 12th edition

LINDELL, THOMAS JAY, b Red Wing, Minn, July 22, 41; m 65; c 2. MOLECULAR PHARMACOLOGY. Educ: Gustavus Adolphus Col, BS, 63; Univ Iowa, PhD(biochem), 69. Prof Exp: ASST PROF PHARMACOL, MED CTR, UNIV ARIZ, 70- Concurrent Pos: USPHS fel biochem, Univ IWash, 68-69 & biochem, biophys & develop biol, Univ Calif, San Francisco, 69-70; assoc ed, J Life Sci. Mem: AAAS; Am Chem Soc; Sigma Xi; Am Soc Pharmacol & Exp Therapeut. Res: Control of eukaryotic transcription. Mailing Add: Dept of Pharmacol Univ of Ariz Med Ctr Tucson AZ 85724

LINDEMAN, LOUIS PAUL, physical chemistry, see 12th edition

LINDEMAN, ROBERT D, b Ft Dodge, Iowa, July 19, 30; m 54; c 5. INTERNAL MEDICINE, PHYSIOLOGY. Educ: State Univ NY Col Forestry, Syracuse Univ, BS, 52; State Univ NY, MD, 56. Prof Exp: From asst resident to asst instr internal med, State Univ NY Upstate Med Ctr, 57-60; med officer, Okla State Dept Health, 60-62; med officer geront, Baltimore City Hosps, Md, 62-66; asst prof med & prev med, 66-68, assoc prof med & physiol, 68-71, ASSOC PROF BIOSTATIST & EPIDEMIOL, MED CTR, UNIV OKLA, 69-, PROF MED & PHYSIOL, 71-, CHIEF RENAL SECT, 67- Concurrent Pos: Clin asst med, Univ Okla, 60-62; instr, Sch Med, Johns Hopkins Univ, 62-66; asst chief res staff, Oklahoma City Vet Admin Hosp, 67-; int chmn, Metab-Endocrinol Prog Comt, Vet Admin, 71; assoc ed, The Kidney, 74-; mem, US Pharmacopeia Comt Rev & chmn, Subcomt Electrolytes, Large Volume Parenterals & Renal Drugs, 75- Mem: Am Fedn Clin Res; Geront Soc; Int Soc Nephrology; Soc Exp Biol & Med; fel Am Col Physicians. Res: Renal and electrolyte problems; hypertension; renal and cardiovascular physiology; aging; trace metal metabolism. Mailing Add: Oklahoma City Vet Admin Hosp 921 NE 13th St Oklahoma City OK 73104

LINDEMAN, VERLUS FRANK, b Ashton, Ill, Sept 1, 02; m 29; c 2. ZOOLOGY. Educ: NCent Col, BA, 26; Univ Iowa, MS, 29, PhD(zool), 30. Prof Exp: Asst, Univ Iowa, 26-29, instr, 29-30; instr zool, 30-35, from asst to assoc prof, 35-45, prof, 45-69, actg chmn dept, 43-45, EMER PROF ZOOL, SYRACUSE UNIV, 69- Mem: Am Soc Zool; Soc Exp Biol & Med. Res: Phosphotases in connective tissues; 5' nucleolidase in skin and wounds; effect of ions on the crayfish heart; amphibian metamorphosis; respiration of invertebrates; nerve metabolism; acetycholine and cholinesterase activity in nervous tissue; metabolism of the retina. Mailing Add: 4135 Meadowview Dr Lake Worth FL 33460

LINDEMANN, CHARLES BENARD, b Staten Island, NY, Dec 17, 46. CELL PHYSIOLOGY, BIOPHYSICS. Educ: State Univ NY Albany, BS, 68, PhD(biol), 72. Prof Exp: Res assoc cell physiol, Pac Biomed Res Ctr, Univ Hawaii, 72-73; res assoc biophys, State Univ NY Albany, 73-74; ASST PROF PHYSIOL, OAKLAND UNIV, 74- Mem: Biophys Soc; Am Soc Cell Biol. Res: Flagellar and ciliary motility: both the mechanisms of force production and the factors which control coordination are under investigation in mammalian sperm and unicellular ciliates. Mailing Add: Dept of Biol Sci Oakland Univ Rochester MI 48063

LINDEMANN, MARTIN KARL, polymer chemistry, see 12th edition

LINDEMER, TERRENCE BRADFORD, b Gary, Ind, Feb 17, 36; m 62; c 2. HIGH TEMPERATURE CHEMISTRY, NUCLEAR CHEMISTRY. Educ: Purdue Univ, BS, 58; Univ Fla, PhD(metall eng), 66. Prof Exp: Mem res staff, Inland Steel Co, 58-61 & Solar Aircraft Co, 61-63; MEM RES STAFF, CHEM TECHNOL DIV, OAK RIDGE NAT LAB, 66- Mem: Am Ceramic Soc. Res: Thermodynamic and kinetic factors affecting reactor performance of nuclear fuels and Fission products. Mailing Add: Oak Ridge Nat Lab PO Box X Oak Ridge TN 37830

LINDEMUTH, IRVIN RAYMOND, JR, b Danville, Pa, Dec 15, 43; m 64; c 2. PLASMA PHYSICS. Educ: Lehigh Univ, BS, 65; Univ Calif, Davis, MS, 67, PhD(eng appl sci), 71. Prof Exp: COMPUT PHYSICIST, LAWRENCE LIVERMORE LAB, 71- Res: Numerical solution of multidimensional plasma models, particularly magnetohydrodynamic partial differential equations. Mailing Add: Lawrence Livermore Lab L-80 PO Box 808 Livermore CA 94550

LINDEN, DUANE B, b Toledo, Ohio, June 1, 30; m 67; c 3. PLANT GENETICS, CELL BIOLOGY. Educ: Hiram Col, AB, 52; Univ Minn, PhD(plant genetics), 56. Prof Exp: Res assoc plant genetics, Univ Minn, 56-57; asst prof genetics, Univ Fla, 57-61; assoc scientist, PR Nuclear Ctr, 61-65; assoc prof biol, 65-69, PROF BIOL, KEAN COL NJ, 69-, CHMN DEPT, 73- Mem: AAAS; Am Inst Biol Sci; Genetics Soc Am; Nat Asn Biol Teachers; Inst Soc Ethics & Life Sci. Res: Effects of radiation on biological systems; study of paramutagenic systems in maize. Mailing Add: Dept of Biol Kean Col of NJ Union NJ 07083

LINDEN, JAMES CARL, b Greeley, Colo, Sept 12, 42; m 68; c 1. PLANT BIOCHEMISTRY. Educ: Colo State Univ, BS, 64; Iowa State Univ, PhD(biochem), 69. Prof Exp: Alexander von Humboldt stipend, Bot Inst, Univ Munich, 69, fel plant biochem, 71; fel cancer res, Med Sch, St Louis Univ, 71-72; BIOCHEMIST, GREAT WESTERN SUGAR CO, 72- Mem: Am Chem Soc; Am Soc Plant Physiologists. Res: Plant cell walls are subject to chemical and physical modifications which alter solute diffusion characteristics; enzyme immobilization and microbiological fermentation are other areas of activity. Mailing Add: Great Western Sugar Co PO Box 149 Loveland CO 80537

LINDENA, SIEGFRIED JOHANNES, b Uthwerdum, WGer, July 7, 24; US citizen; m 67; c 1. ENERGY CONVERSION, MAGNETISM. Educ: Tech Univ Hannover, Dipl Ing, 52, Dr Ing(elec & electronics eng), 55. Prof Exp: Sci asst high voltage direct current transmission & energy conversion & distribution, Tech Univ Hannover, 52-55; dir res magnetics & control & supvr MS degree studies, Calor Emag EAG, Ratingen, WGer, 55-57; chief engr power conversion, Magnetic Res Corp, El Segundo, Calif, 57-59; sr scientist inverter technol, Electro Solids Corp, San Fernando, 59-63; dir res energy conversion & mil applns & mgr mil prod, Int Tel & Tel Co, Sylmar, 63-68, instr energy conversion technol, 64-68; SR STAFF SCIENTIST MIL & SPACE APPLNS & SOLAR ENERGY CONVERSION, XEROX ELECTRO OPTICAL SYSTS, PASADENA, 68- Concurrent Pos: Lectr magnetism & magnetic design, Jet Propulsion Lab, 70, on loan, res & design leader solar energy conversion & utilization demonstr, 76- Mem: Sr mem Inst Elec & Electronics Engrs. Res: Solar energy collection, conversion and utilization; energy conversion in space applications; high voltage, direct current power transmission and distribution; magnetism and continued studies to simplify design approaches. Mailing Add: 11002 Densmore Ave Granada Hills CA 91344

LINDENAUER, S MARTIN, b New York, NY, Dec 10, 32; m 56; c 4. SURGERY. Educ: Tufts Univ, MD, 57. Prof Exp: From instr to assoc prof, 64-72, PROF SURG, UNIV MICH, ANN ARBOR, 72-, ASST DEAN MED SCH, 74-; CHIEF STAFF, VET ADMIN HOSP, 74- Concurrent Pos: Chief surg serv, Vet Admin Hosp, 68-74. Mem: Am Col Surg; Asn Acad Surg; Soc Vascular Surg; Int Cardiovasc Soc; Soc Surg Alimentary Tract. Res: Vascular surgery; biliary tract surgery. Mailing Add: Dept of Surg Univ of Mich Ann Arbor MI 48104

LINDENBAUM, ARTHUR, b Camden, NJ, May 3, 16; m 43; c 3. BIOCHEMISTRY, RADIOBIOLOGY. Educ: Niagara Univ, BS, 39, MA, 42; Univ Minn, PhD, 49. Prof Exp: Paper chemist, Int Paper Co, NY, 41-42; asst, Univ Minn, 46-48; assoc biochemist, 49-70, BIOCHEMIST, ARGONNE NAT LAB, 70- Concurrent Pos: Guggenheim fel, Col Physicians & Surgeons, Columbia Univ, 60-61; adj prof, Stritch Sch Med & assoc prof, Sch Dent, Loyola Univ, Chicago; mem, Comt 30, Nat Coun Radiation Protection; chmn, Plutonium Task Group, Int Comn Radiol Protection. Mem: Am Soc Biol Chemists; Radiation Res Soc; Health Physics Soc. Res: Radioactive tracers; citrate metabolism; metal complexes; therapy of experimental metal poisoning; bone; calcification of cartilage; colloidal radioelement metabolism; autoradiography. Mailing Add: Div of Biol & Med Res Argonne Nat Lab Argonne IL 60439

LINDENBAUM, SEYMOUR JOSEPH, b New York, NY, Feb 3, 25; m 58. EXPERIMENTAL HIGH ENERGY PHYSICS. Educ: Princeton Univ, AB, 45; Columbia Univ, MA, 48, PhD(physics), 51. Prof Exp: Res assoc, Nevis Cyclotron Lab, Columbia Univ, 47-51; assoc physicist, 51-54, physicist, 54-63, SR PHYSICIST, BROOKHAVEN NAT LAB, 63-; MARK W ZEMANSKY CHAIR PHYSICS, CITY COL NEW YORK, 70- Concurrent Pos: Group leader high energy counter res group, Brookhaven Nat Lab, 54-; vis prof, Univ Rochester, 58-59; vis, Europ Orgn Nuclear Res; consult, Saclay Nuclear Res Ctr. Mem: AAAS; fel Am Phys Soc; NY Acad Sci. Res: High energy elementary particle interactions; high energy experimental techniques. Mailing Add: Physics Dept Brookhaven Nat Lab Upton NY 11973

LINDENBAUM, SIEGFRIED, b Unna, Ger, July 24, 30; nat US; m 56; c 3. PHYSICAL CHEMISTRY. Educ: Rutgers Univ, BS, 52, PhD(chem), 55. Prof Exp: Chemist, Oak Ridge Nat Lab, 55-71; ASSOC PROF PHARMACEUT CHEM, UNIV KANS, 71- Res: Physical chemistry of ion exchange; solvent extraction; separations; thermodynamics of electrolyte solutions. Mailing Add: Sch of Pharm McCollum Labs 2065 Ave A Campus W Univ of Kans Lawrence KS 66044

LINDENBERG, KATJA LAKATOS, b Quito, Ecuador, Nov 2, 41; US citizen; m 70. PHYSICAL CHEMISTRY, CHEMICAL PHYSICS. Educ: Alfred Univ, BA, 62; Cornell Univ, PhD(theoret physics), 67. Prof Exp: Res assoc & asst prof physics, Univ Rochester, 67-69; lectr chem & res chemist, 69-72, asst prof chem residence, 72-73, ASST PROF CHEM, UNIV CALIF, SAN DIEGO, 73- Concurrent Pos: Res physicist, Univ Calif, San Diego, 69-71; researcher, Oak Ridge Summer Inst Theoret Biophys, 69-75; consult, Chem Div, Oak Ridge Nat Lab, 75. Mem: Am Phys Soc. Res: Theory of stochastic processes with applications to physical and chemical systems; energy transfer in polymers and in organic solids. Mailing Add: Dept of Chem Univ of Calif at San Diego La Jolla CA 92037

LINDENBERG, RICHARD, b Bocholt, Ger, Feb 18, 11; US citizen; m 37; c 1. NEUROPATHOLOGY. Educ: Univ Berlin, MD, 44. Prof Exp: Chief resident neuropath, Kaiser-Wilhelm Inst Brain Res, 36-39; resident & dir anat lab, Neuropsychiat Hosp, Univ Frankfurt, 45-47; res neuropathologist, Sch Aviation Med, Randolph Field, Tex & Army Chem Ctr, Md, 47-51; DIR NEUROPATH & LEGAL MED, MD STATE DEPT HEALTH & MENT HYG, 51- Concurrent Pos: Clin prof path, Sch Med, Univ Md, 51-, lectr neuroanat, Dent Sch, 55-; lectr neuro-ophthal, Sch Med, Johns Hopkins Univ, 59-, lectr forensic path, Sch Hyg, 64-; consult, Greater Baltimore Med Ctr, 59- Mem: Am Asn Neuropath; Am Soc Clin Path; fel Col Am Path; AMA; World Fedn Neurol. Res: Neuropathology of head injury and circulatory disorders; neuro-ophthalmologic pathology; forensic pathology. Mailing Add: 111 Penn St Baltimore MD 21201

LINDENBLAD, GORDON ERIC, b Port Jefferson, NY, May 7, 28; m 54; c 3. BIOCHEMISTRY. Educ: Bates Col, BS, 48; Georgetown Univ, MS, 55, PhD(chem), 58. Prof Exp: Asst biol, Brookhaven Nat Lab, 49-50; res biochemist, Cancer Res Lab, Garfield Mem Hosp, 57-58; sr chemist, Radioisotope Lab, E R Squibb & Sons Div, Olin Mathieson Chem Corp, 57-58, supvr, 59-65; mgr radioisotopes & labeled compounds div, Isotopes, Inc, 65-66; asst dir radiopharmaceut dept, Neisler Labs, Inc, Union Carbide Corp, 66-69; DIR RADIOPHARMACEUT RES & DEVELOP, MALLINCKRODT, INC, 69- Concurrent Pos: Consult, Children's Hosp, DC, 57-58. Mem: Soc Nuclear Med; NY Acad Sci. Res: Medical applications of radioisotopes; clincal chemistry. Mailing Add: Mallinckrodt Inc Second & Mallinckrodt Sts St Louis MO 63160

LINDENFELD, PETER, b Vienna, Austria, Mar 10, 25; nat US; m 53; c 2. LOW TEMPERATURE PHYSICS. Educ: Univ BC, BASc, 46, MASc, 48; Columbia Univ, PhD(physics), 54. Prof Exp: Asst physics, Univ BC, 46-47; asst, Columbia Univ, 48-52, res scientist, 53; vis lectr, Drew Univ, 52-53; from instr to assoc prof, 53-66, PROF PHYSICS, RUTGERS UNIV, 66- Concurrent Pos: Dir NSF in-serv insts for high sch teachers, 64-66; regional counr NJ, Am Inst Physics, 63-71; Rutgers Res Coun fel & guest scientist fac sci, Univ Paris-South, 70-71. Mem: Fel Am Phys Soc; Am Asn Physics Teachers. Mailing Add: Dept of Physics Rutgers Univ New Brunswick NJ 08903

LINDENMAYER, ARISTID, b Budapest, Hungary, Nov 17, 25; nat US; m 58; c 1. THEORETICAL BIOLOGY. Educ: Pazmany Peter Univ, Hungary, Dipl, 48; Univ Mich, MS, 53, PhD(bot), 56. Prof Exp: Fel, Johnson Res Found, Univ Pa, 55-56, instr bot, 56-58; NSF fel, Univ London, 58-59; asst prof bot, Univ Pa, 59-62; USPHS fel, Inst Statist, NC State Col, 62-63; from asst prof to assoc prof biol, Queens Col NY, 63-68; PROF PHILOS BIOL, UNIV UTRECHT, 68- Mem: AAAS; Am Soc Plant Physiol; Bot Soc Am; Neth Soc Logic & Philos Sci (treas, 70-74); Neth Systs Res Soc (pres, 70-71). Res: Mathematical models of plant development; logical foundations of biology. Mailing Add: Theoret Biol Group Univ of Utrecht 8 Padualaan Utrecht Netherlands

LINDENMAYER, GEORGE EARL, b Port Arthur, Tex, Aug 22, 40; m 63; c 2. BIOCHEMICAL PHARMACOLOGY. Educ: Baylor Univ, BS, 62; Baylor Col Med, MD & MS, 67, PhD(pharmacol), 70. Prof Exp: Instr pharmacol, Baylor Col Med, 69-70; staff assoc cardiol, Nat Heart & Lung Inst, 70-72; asst prof pharmacol & med, Baylor Col Med, 72-74, assoc prof cell biophys & med, 74-75; ASSOC PROF PHARMACOL & MED, MED UNIV SC, 75- Concurrent Pos: Estab investr, Am Heart Asn, 73. Mem: Am Soc Pharmacol & Exp Therapeut; Int Study Group Res Cardiac Metab; Am Chem Soc; Biophys Soc; Am Heart Asn. Res: Information transfer between extracellular and intracellular environments of mammalian cells, particularly myocardial cells. Mailing Add: Dept of Pharmacol Med Univ of SC Charleston SC 29401

LINDENMEIER, CHARLES WILLIAM, b Ft Collins, Colo, Dec 2, 30; m 58; c 2. THEORETICAL PHYSICS, NUCLEAR PHYSICS. Educ: Colo State Univ, BS, 52; Cornell Univ, PhD(theoret physics), 60. Prof Exp: Sr physicist, Hanford Labs, Gen Elec Co, 60-63, mgr theoret physics, 63-64; mgr, Pac Northwest Labs, Battelle Mem Inst, 65-70, mgr math & physics res, 70-73; MGR DESIGN ANAL, LASER ENRICHMENT DEPT, EXXON NUCLEAR CO, 74- Mem: Am Phys Soc; Am Nuclear Soc. Res: Reactor physics; neutron thermalization; nuclear reactions; computer applications; laser isotope separation. Mailing Add: Exxon Nuclear Co Res & Tech 2955 George Washington Way Richland WA 99352

LINDENMEYER, PAUL HENRY, b Bucyrus, Ohio, May 4, 21; m 44; c 3. PHYSICAL CHEMISTRY, MATERIALS SCIENCE. Educ: Bowling Green State Univ, BS, 44; Ohio State Univ, PhD(chem), 51. Prof Exp: Asst, Ohio State Univ, 46-49, res assoc, Res Found, 49-51; res chemist, Visking Co Div, Union Carbide Corp, 51-53, res supvr, 53-57, mgr, Pioneering Res Dept, 57-59; mgr fiber sci, Chemstrand Res Ctr, Inc, 59-69; head mat sci lab, Boeing Sci Res Labs, 69-72, sci adv, Boeing Aerospace Co, 72-73; prog dir, Div Mat Res, NSF, 73-75; MAT RES CONSULT, 75- Honors & Awards: US Sr Scientist Award, Humboldt Found, 75. Mem: AAAS; Fiber Soc; Am Chem Soc; Am Phys Soc; Am Crystallog Asn. Res: X-ray crystallography; spectroscopy; microscopy; crystal growth and structure of high polymers; materials processing. Mailing Add: 165 Lee St Seattle WA 98109

LINDENSTRUTH, ALBERT FRANK, organic chemistry, see 12th edition

LINDER, ALLAN DAVID, b Grand Island, Nebr, Sept 27, 25; m 49; c 1. VERTEBRATE ZOOLOGY. Educ: Univ Nebr, BSc, 51; Okla State Univ, MSc, 52, PhD(zool), 56. Prof Exp: Asst prof zool, Univ Wichita, 56-59 & Southern Ill Univ, 59-60; chmn dept, 60-75, PROF ZOOL, IDAHO STATE UNIV, 60- Mem: Am Soc Ichthyologists & Herpetologists; Soc Syst Zool; Soc Vert Paleont. Res: Ichthyology, paleo-ichthyology and herpetology. Mailing Add: Idaho State Univ Box 8007 Pocatello ID 83209

LINDER, BRUNO, b Sniatyn, Poland, Sept 3, 24; nat US; m 53; c 5. THEORETICAL CHEMISTRY, CHEMICAL PHYSICS. Educ: Upsala Col, BS, 48; Univ Ohio, MS, 50; Univ Calif, Los Angeles, PhD(chem), 55. Prof Exp: Asst chem, Univ Ohio, 48-49; asst chem, Univ Calif, Los Angeles, 50-55, asst res chemist, 55; proj assoc theoret chem, Naval Res Lab, Wis, 55-57; from asst prof to assoc prof, 57-65, PROF PHYS CHEM, FLA STATE UNIV, 65- Concurrent Pos: Guggenheim fel, Inst Theoret Physics, Univ Amsterdam, 64-65; chmn chem physics prog, Fla State Univ, 71-73 & 75-; vis prof, Hebrew Univ, Jerusalem, 73. Mem: Am Chem Soc; Am Phys Soc. Res: Intermolecular forces; nuclear reactions; spectral shifts and intensities; dielectric theory; theory of adsorption. Mailing Add: Dept of Chem Fla State Univ Tallahassee FL 32306

LINDER, DONALD ERNST, b Yoakum, Tex, Oct 4, 38; m 61; c 3. ANALYTICAL CHEMISTRY. Educ: Sul Ross State Univ, BS, 61; Tex A&M Univ, MS, 64, PhD(chem), 67. Prof Exp: RES SCIENTIST, CONTINENTAL OIL CO, 66- Mem: Am Chem Soc. Res: Liquid chromatography, adsorption, liquid-liquid, ion exchange and gel permeation; large scale preparative gas-liquid chromatography; analytical distillations. Mailing Add: 2409 Cardinal Ponca City OK 74601

LINDER, ERNEST GUSTAF, b Waltham, Mass, May 16, 02; m 45; c 2. PHYSICS. Educ: Univ Iowa, BA, 25, MS, 27; Cornell Univ, PhD(physics), 31. Prof Exp: Res physicist, Mfg Co, RCA Corp, 32-42 & RCA Labs, Inc, 42-56, adminr, 56-66, mgr, 66-67, consult, 67-75; RETIRED. Mem: Fel Am Phys Soc; fel Inst Elec & Electronics Eng. Res: Photoelectricity; thermoelectricity; crystals; vapor pressure; electrical discharges; mass spectroscopy; radio; microwaves; electronics; radar; nuclear physics; energy conversion. Mailing Add: 16 Colonial Club Dr Apt 205 Boynton Beach FL 33435

LINDER, FORREST EDWARD, b Waltham, Mass, Nov 21, 06; m 31; c 2. STATISTICS. Educ: State Univ Iowa, BA, 30, MA, 31, PhD(psychol, math), 32. Prof Exp: Tech expert, Div Vital Statist, Bur of Census, 35-42, asst chief, 42-45; asst chief, Med Statist Div, US Navy, 44-46; dep chief, Nat Off Vital Statist, USPHS, 46-47; chief, Demog & Social Statist Br, UN, 47-56; dir nat health surv, USPHS, 56-60, dir, Nat Ctr Health Statist, 66-67; PROF BIOSTATIST & DIR INT PROG LABS POP STATIST, UNIV NC, CHAPEL HILL, 67- Concurrent Pos: Consult, Ford Found, India, 62 & 64 & WHO, 66 & 68-71, mem expert adv panel health statist, 67; mem policy res adv comt, Nat Inst Child Health & Human Develop, 67-71; mem res adv comt, AID, 68; consult, Pan Am Health Orgn, 68 & 69; chmn, Nat Comt Health & Vital Statist, Nat Ctr Health Statist, 69-72; mem adv comt statist policy, Off Budget & Mgt, US Exec Off, 72; mem world fertility surv steering comt, Int Statist Inst, 72; mem, Inter-Am Statist Inst; pres, Int Inst Vital Regist & Statist, 74- Honors & Awards: Distinguished Serv Award, Dept Health, Educ & Welfare, 66; Bronfman Prize, Am Pub Health Asn, 67. Mem: Fel AAAS; Int Union Sci Study Pop; fel Am Statist Asn; fel Am Pub Health Asn; Pop Asn Am. Res: Development of statistical methods for measurement of population change; public health statistics; census and vital statistics methods. Mailing Add: Dept of Biostatist Sch Pub Health Univ of NC Chapel Hill NC 27514

LINDER, HARRIS JOSEPH, b Brooklyn, NY, Jan 3, 28; m 52; c 4. ZOOLOGY. Educ: Long Island Univ, BS, 51; Conrell Univ, MS, 55, PhD(zool), 58. Prof Exp: Asst zool, Cornell Univ, 52-57; resident res assoc, Div Biol & Med, Argonne Nat Lab, 57-58; asst prof, 58-63, ASSOC PROF ZOOL, UNIV MD, COLLEGE PARK, 63- Mem: AAAS; Am Soc Zool; Am Micros Soc; Am Inst Biol Sci; Soc Study Reproduction. Res: Comparative invertebrate endocrinology; neurosecretion; autoradiographic and electrophoretic studies of reproduction in coelenterates and annelids. Mailing Add: Dept of Zool Univ of Md College Park MD 20742

LINDER, JACQUES FRANCOIS, physics, see 12th edition

LINDER, JEROME, organic chemistry, see 12th edition

LINDER, LOUIS JACOB, b East St Louis, Ill, May 10, 16; m 48; c 3. ANALYTICAL CHEMISTRY. Educ: Wash Univ, AB, 41. Prof Exp: Chemist, Eagle-Picher Lead Co, 41-44; anal chemist, Alumina & Chem Div, Res Labs, Aluminum Co Am, 46-50, res chemist, 50-72; LAB MGR, SCH SCI & TECHNOL, SOUTHERN ILL UNIV, 72- Mem: Soc Appl Spectros. Res: Analytical procedures on aluminous materials; application of optical emission spectroscopy to analysis of alumina, aluminous ores and sodium aluminate liquors; spectrographic analysis of gallium oxide and metal. Mailing Add: 7907 W Washington St Belleville IL 62223

LINDER, RAYMOND, b Grand Island, Nebr, Sept 9, 22; m 49; c 4. WILDLIFE ECOLOGY. Educ: Univ Nebr, BS, 53, PhD(zool, physiol), 64; Iowa State Univ, MS, 55. Prof Exp: Biologist, Nebr Game Comn, 55-60; instr zool, 60-62, asst prof wildlife, 64-71, ASSOC PROF WILDLIFE, SDAK STATE UNIV, 71-, LEADER SDAK COOP WILDLIFE RES UNIT, 67- Mem: Wildlife Soc; Animal Behav Soc. Res: Ecology of black-footed ferret and prairie dogs. Mailing Add: Dept Wildlife & Fisheries Sci SDak State Univ Brookings SD 57006

LINDER, REGINA, b New York, NY, June 21, 45. MICROBIOLOGY. Educ: City Col New York, BS, 67; Univ Mass, MS, 69; NY Univ, PhD(microbiol), 75. Prof Exp: ASST RES SCIENTIST MICROBIOL, SCH MED, NY UNIV, 75- Mem: Am Soc Microbiol. Res: Investigation of the interaction between a hemolytic product of a sea anemone and its phospholipid inhibitor; studies on the enzyme target of penicillin in bacterial cells. Mailing Add: Dept of Microbiol Sch of Med NY Univ New York NY 10016

LINDER, SEYMOUR MARTIN, b New York, NY, Dec 17, 25; m 55; c 2. INDUSTRIAL ORGANIC CHEMISTRY. Educ: City Col New York, BS, 46; Polytech Inst Brooklyn, MS, 49, PhD(chem), 53. Prof Exp: Jr chemist, Hoffmann-La Roche, Inc, 46-51; proj leader, Becco Chem Div, FMC Corp, 53-58 & Org Chem Div, 58-72; DIR SYNTHESIS RES, ALCOLAC, INC, 72- Mem: AAAS; Am Chem Soc; Am Inst Chem. Res: Chemistry of hydrogen peroxide and acids; organic oxidation reactions; epoxidations; hydroxylations; epoxyresins; process development; terpene and medicinal chemistry; insecticides; gas chromatography; specialty organic chemicals; functional monomers; quaternary salts; copolymerizable surfactants. Mailing Add: 1902 Tadcaster Rd Baltimore MD 21228

LINDER, SOLOMON LEON, b Brooklyn, NY, Mar 13, 29; m 53; c 3. OPTICS. Educ: Rutgers Univ, BS, 50; Wash Univ, PhD(physics), 55. Prof Exp: Mem tech staff, Bell Tel Labs, Inc, 55-62; sr group engr, McDonnell Aircraft Corp, 62-67, SR GROUP ENGR, McDONNELL DOUGLAS ASTRONAUT CO, 67- Concurrent Pos: Eve instr, Fairleigh Dickinson Univ, 59-62, univ col, Wash Univ, 63-67, Fla Technol Univ, 70-71 & univ col, Wash Univ, 75- Mem: Optical Soc Am; sr mem Inst Elec & Electronics Eng. Res: Nuclear magnetic resonance; military systems; electrooptics. Mailing Add: 14571 Coeur D'Alene Ct Chesterfield MO 63017

LINDERMAN, ROBERT G, b Crescent City, Calif, Feb 2, 39; m 61; c 1. PLANT PATHOLOGY. Educ: Fresno State Col, BA, 60; Univ Calif, Berkeley, PhD(plant path), 67. Prof Exp: Lab technician plant path, Univ Calif, Berkeley, 64-67, asst res plant pathologist, 67; res plant pathologist, Crops Res Div, Agr Res Serv, USDA, 67-74; ASSOC PROF BOT & PLANT PATH, ORE STATE UNIV, 74- Mem: Am Phytopath Soc. Res: Ecology of soil-borne fungus plant pathogens; biological control; biological effects of plant residue decomposition in soil; ornamental plant diseases. Mailing Add: Dept of Bot Ore State Univ Corvallis OR 97331

LINDFORS, KARL RUSSELL, b Saginaw, Mich, July 10, 37; m 58; c 2. PHYSICAL CHEMISTRY. Educ: Univ Mich, BS, 59; Univ Wis, PhD(phys chem), 64. Prof Exp: Spectroscopist, Tracerlab, 63-64; PROF CHEM, CENT MICH UNIV, 64- Mem: Am Phys Soc; Am Chem Soc; Sigma Xi. Res: Molecular spectroscopy; species in solution. Mailing Add: Dept of Chem Cent Mich Univ Mt Pleasant MI 48858

LINDGREN, BERNARD WILLIAM, b Minneapolis, Minn, May 13, 24; m 45; c 3. MATHEMATICS. Educ: Univ Minn, PhD(math), 49. Prof Exp: Instr math, Univ Minn, 43-44, 46-49 & Mass Inst Technol, 49-51; res mathematician, Minn-Honeywell Regulator Co, 51-53; from instr to assoc prof, 53-69, chmn dept statist, 63-73, PROF MATH, UNIV MINN, MINNEAPOLIS, 69- Res: Analysis; probability; statistics. Mailing Add: 1860 Noble Dr Golden Valley MN 55422

LINDGREN, DAVID LEONARD, b St Paul, Minn, Sept 17, 06; m 33; c 3. ENTOMOLOGY. Educ: Univ Minn, BS, 30, MS, 31, PhD(entom), 35. Prof Exp: Jr entomologist, 35-41, from asst entomologist to entomologist, 41-74, EMER ENTOMOLOGIST & LECTR, CITRUS EXP STA, UNIV CALIF, RIVERSIDE, 74- Mem: Fel AAAS; Entom Soc Am; Am Asn Cereal Chem. Res: Insecticides; citrus insects; stored product insects. Mailing Add: Dept of Entom Univ of Calif Riverside CA 92502

LINDGREN, DAVID TREADWELL, b Ipswich, Mass, Mar 1, 39; m 64; c 2. GEOGRAPHY. Educ: Boston Univ, AB, 60, AM, 62, PhD(geog), 69. Prof Exp: Intel geog, Cent Intel Agency, 64-66; chmn urban studies prog, 75-76, PROF GEOG, DARTMOUTH COL, 66- Concurrent Pos: Sabbatical leave, Int Inst Aerial Surv & Earth Sci, Neth, 73. Mem: Asn Am Geog; Am Soc Photogram. Res: The development of urban applications for remote sensing. Mailing Add: Dept of Geog Dartmouth Col Fairchild Bldg Hanover NH 03755

LINDGREN, FRANK TYCKO, b San Francisco, Calif, Apr 14, 24; m 53. BIOPHYSICS. Educ: Univ Calif, Berkeley, BA, 47, PhD(biophys), 55. Prof Exp: Res asst biophysicist, 55-56, res assoc biophysicist, 56-67, RES BIOPHYSICIST, DONNER LAB, UNIV CALIF, BERKELEY, 67- Concurrent Pos: Assoc ed, Lipids, Am Oil Chemists Soc, 66- Mem: Am Oil Chemists Soc. Res: Physical chemistry and biochemistry of blood lipids and lipo-proteins as they occur in states of health and diseases; instrumentation and engineering necessary to facilitate such investigations. Mailing Add: 108 Donner Lab Univ of Calif Berkeley CA 94720

LINDGREN, RICHARD ARTHUR, b Providence, RI, June 2, 40; m 63; c 4. NUCLEAR PHYSICS. Educ: Univ RI, BA, 62; Wesleyan Univ, MA, 64; Yale Univ, PhD(nuclear physics), 69. Prof Exp: Res assoc nuclear physics, Univ Md, College Park, 69-70; res assoc, Nat Res Coun, Nat Acad Sci, 70-71 & Univ Rochester, 71-73; RES PHYSICIST NUCLEAR PHYSICS, NAVAL RES LAB, WASHINGTON, DC, 73- Concurrent Pos: Instr, George Mason Univ, 73-75; assoc prof, Cath Univ Am, 75- Mem: Sigma Xi. Res: Nuclear structure studies using inelastic electron scattering, particularly those nuclear states excited strongly via an interaction between the electron and nuclear magnetization currents, such as giant magnetic dipole transitions. Mailing Add: Code 6632 Naval Res Lab Washington DC 20375

LINDGREN, ROBERT M, b Concord, NH, June 16, 32; m 57; c 4. PHYSICAL CHEMISTRY. Educ: Univ Maine, BS, 59, MS, 62, PhD(phys chem), 67. Prof Exp: Instr, Univ Maine, 62-64; scientist, Oxford Paper Co, Maine, 64-69; sr res prof, Plastic Coating Corp, Mass, 69-71; ASSOC PROF CHEM, HUSSON COL, 71 Mem: Am Chem Soc. Res: Molten salt chemistry; environmental science. Mailing Add: Dept of Sci Husson Col Bangor ME 04401

LINDGREN, VINCENT VICTOR, chemistry, see 12th edition

LINDGREN, WILLIAM FREDERICK, b San Mateo, Calif, Dec 23, 42; c 2. MATHEMATICS. Educ: SDak Sch Mines & Technol, BS, 64, MS, 66; Southern Ill Univ, PhD(math), 71. Prof Exp: Mathematician & analyst, Atomic Energy Div, Phillips Petrol Co, 66-67; ASSOC PROF MATH, SLIPPERY ROCK STATE COL, 71- Mem: Am Math Soc; Sigma Xi. Res: General topology, particularly abstract spaces. Mailing Add: Dept of Math Slippery Rock State Col Slippery Rock PA 16057

LINDHOLM, DALE DAVID, b Duluth, Minn, Dec 4, 31; m 53; c 2. INTERNAL MEDICINE, NEPHROLOGY. Educ: Univ Minn, Duluth, BA, 53; Univ Minn, Minneapolis, BS, 55, MD, 57. Prof Exp: Intern med, St Luke's Hosp, Duluth, 57-58; staff physician, USPHS Hosp, Rosebud, SDak, 58-59, med resident, Seattle, Wash, 59-62; res fel nephrology, Sch Med, Univ Wash, 62-64; chief res, USPHS Hosp, New Orleans, La, 64-66; from asst prof to prof med, Sch Med, Tulane Univ, 64-74, head renal sect, 65-74; PROF MED & CHMN DIV NEPHROLOGY, W VA UNIV MED CTR, 74- Concurrent Pos: Mem, Cent Clin Invest Comt, USPHS, 64-66, Nat Kidney Found, 66-; consult, USPHS Hosp, New Orleans, 66-74; chief, Hemodialysis Sect, Vet Admin Hosp, 66-74. Mem: Am Soc Artificial Internal Organs; Am Soc Nephrology; AMA; Am Heart Asn; Int Soc Nephrology. Res: Clinical and experimental investigation, training and teaching in nephrology, including renal clinico-pathological correlations, dialysis, renal transplantation and fluid-electrolyte physiology; drug therapy, especially as modified by renal insufficiency and dialysis; hemodialyzer development; investigations in uremia. Mailing Add: Div of Nephrology WVa Univ Med Ctr Morgantown WV 26506

LINDHOLM, DENNIS A, solid state physics, see 12th edition

LINDHOLM, ROBERT D, b Rockford, Ill, June 17, 40; m 62; c 2. PHYSICAL CHEMISTRY. Educ: Northern Ill Univ, BS, 63, MS, 64; Univ Southern Calif, PhD(phys chem), 69. Prof Exp: Sr res chemist, 68-73, RES ASSOC, EASTMAN KODAK CO, 73- Mem: Am Chem Soc. Res: Photochemistry of transition-metal complexes; silver halide photochemistry. Mailing Add: Res Lab Eastman Kodak Co Rochester NY 14650

LINDHOLM, ROY CHARLES, b Washington, DC, Mar 8, 37; m 65; c 2. GEOLOGY. Educ: Univ Mich, BS, 59; Univ Tex, MA, 63; Johns Hopkins Univ, PhD(geol), 67. Prof Exp: Instr, Johns Hopkins Univ, 65-66; asst prof, 67-69, ASSOC PROF GEOL, GEORGE WASHINGTON UNIV, 69- Mem: Am Asn Petrol Geol; Soc Econ Paleont & Mineral. Res: Paleozoic carbonate rocks of eastern United States; Precambrian sandstones of New Mexico; sequences of carbonate cements. Mailing Add: Dept of Geol George Washington Univ Washington DC 20006

LINDHORST, TAYLOR ERWIN, b St Louis, Mo, Aug 11, 28; m 51; c 2. MYCOLOGY. Educ: St Louis Col Pharm, BS, 51; Wash Univ, MA, 54, PhD(mycol), 67. Prof Exp: Instr pharm, 51-52, resident biol, 52-55, assoc instr, 56-59, asst prof, 59-67, assoc prof, 67-74, PROF BIOL & DEAN STUDENTS, ST LOUIS COL PHARM, 74- Mem: Am soc Pharmacog; Bot Soc Am. Res: Mycological studies concerning response and growth variation to antibiotic substances. Mailing Add: 834 Montmartre Ct St Louis MO 63141

LINDLEY, BARRY DREW, b Orleans, Ind, Jan 25, 39; m 64; c 3. PHYSIOLOGY, BIOPHYSICS. Educ: DePauw Univ, BA, 60; Western Reserve Univ, PhD(physiol), 64. Prof Exp: Asst prof, 65-68, ASSOC PROF PHYSIOL, SCH MED, CASE

WESTERN RESERVE UNIV, 68- Concurrent Pos: NSF fel neurophysiol, Nobel Inst Neurophysiol, Karolinska Inst, Sweden, 64-65; Lederle med fac award, 6770; USPHS res career develop award, 71-; mem, Physiol Study Sect, NIH, 75- Mem: Am Physiol Soc; Soc Gen Physiologists; Biophys Soc. Res: Ion and water transport; membrane permeability; irreversible thermodynamics; electrophysiology of nerve, muscle and glandular tissue. Mailing Add: Dept of Physiol Case Western Reserve Univ Sch Med Cleveland OH 44106

LINDLEY, CHARLES EDWARD, b Macon, Miss, Dec 21, 21; m 45. ANIMAL HUSBANDRY. Educ: Miss State Univ, BS, 46; Wash State Univ, MS, 48; Okla State Univ, PhD, 57. Prof Exp: Asst, Wash State Univ, 46-48, asst prof animal husb, 48-51; asst, Okla State Univ, 51-52; chmn dept animal sci, 52-69, PROF ANIMAL SCI, MISS STATE UNIV, 52-, DEAN COL AGR, 69- Mem: Am Soc Animal Sci. Res: Livestock production and animal breeding. Mailing Add: Off of Dean Col of Agr Miss State Univ Drawer AG Miss State MS 39762

LINDLEY, STANLEY BRYAN, b Bloomington, Ind, June 9, 06; m 38; c 2. PSYCHIATRY, PSYCHOLOGY. Educ: Univ Kans, AB, 27; Stanford Univ, AM, 28; Yale Univ, PhD(physiol, psychol), 33; Univ Minn, MD, 39. Prof Exp: Asst instr, Univ Ill, 28-29; asst psychol, Inst Human Rel, Yale Univ, 29-34; instr, Col Educ, Univ Minn, 34-37; asst physician, Fergus Falls State Hosp, Minn, 39-40, asst supt, 40-43; supt, Willmar State Hosp, Minn, 43-50; chief acute intensive treatment serv, Vet Admin Hosp, Knoxville, Iowa, 50-52, chief prof serv, 52-57; dir, Vet Admin Hosp, St Cloud, Minn, 57-70; dir, Vet Admin Hosp, Salisbury, NC, 70-73. Res: Experimental and educational psychology; animal behavior; muscle tonus and work output; performance of simple motor tests by psychiatric patients; aptitude tests. Mailing Add: Apt 101 623 W Fir St Fergus Falls MN 56537

LINDMAN, ERICK LEROY, JR, b Seattle, Wash, Mar 20, 38; m 63; c 4. PLASMA PHYSICS. Educ: Calif Inst Technol, BS, 60; Univ Calif, Los Angeles, MS, 63, PhD(physics), 64. Prof Exp: Res scientist, Univ Tex, Austin, 64-65, asst prof physics, 65-68; physicist, Austin Res Assocs, 68-71; STAFF MEM, LOS ALAMOS SCI LAB, 71- Mem: Am Phys Soc. Res: Collisionless shocks; plasma instabilities; numerical simulation of plasma effects. Mailing Add: T-6 Los Alamos Lab PO Box 1663 Los Alamos NM 87545

LINDMARK, RONALD DORANCE, b Leonard, Minn, May 3, 33; m 61; c 2. FOREST ECONOMICS. Educ: Univ Minn, BS, 61, MS, 63; Ohio State Univ, PhD(agr econ & rural sociol), 71. Prof Exp: Res asst forest econ, Univ Minn, 61-63; res economist, Cent States Forest Exp Sta, 63-66, res economist, Northeastern Forest Exp Sta, 66, mkt analyst, N Cent Forest Exp Sta, 66-68, from actg proj leader to proj leader, Mkt Res, 68-74, ASST DIR, INTERMT FOREST & RANGE EXP STA, US FOREST SERV, 74- Mem: Soc Am Foresters; Am Agr Econ Asn; Am Econ Asn. Res: Marketing of forest products; research administration; research planning. Mailing Add: 776 Ben Lomond Dr South Ogden UT 84403

LINDMAYER, JOSEPH, b Budapest, Hungary, May 8, 29; US citizen; m 55; c 2. SOLID STATE PHYSICS. Educ: Williams Col, MS, 63; Aachen Tech Univ, PhD, 68. Prof Exp: Scientist, Inst Measurements Tech, Hungarian Acad Sci, 55-56; scientist res ctr, Sprague Elec Co, 57-63, dept head semiconductor physics, 63-68; br mgr, Comsat Labs, Commun Satellite Corp, Clarksburg, 68-74; DIR PHYSICS LAB, DEFENSE LANGUAGE INST, 74- Concurrent Pos: Vis lectr, Yale Univ, 68. Mem: Inst Elec & Electronics Eng. Res: Semiconductor physics, electronics and devices. Mailing Add: Physics Lab Presidio of Monterey Defense Language Inst Monterey CA 93940

LINDNER, ELEK, b Budapest, Hungary, June 3, 24; US citizen; m 60; c 1. ANALYTICAL BIOCHEMISTRY, MARINE BIOLOGY. Educ: Budapest Tech Univ, Dipl Chem Eng, 46, PhD(biochem), 74. Prof Exp: Prof asst agr chem, Budapest Tech Univ, 47-48 & food chem, 48-50; chief chemist, Anal & Res Lab, Elida Cosmetic Factory, 50-51; res chemist, Res Inst, Fatty Oil Chem Indust, 51-56, Res & Develop Div, Lever Bros Co, NJ, 57-61 & Chevron Res Corp, Calif, 61-64; prod mgr, Sawyer Tanning Co, Calif, 64-65; res chemist, Paint Lab, Mare Island Naval Shipyard, 65-73; CHEMIST, NAVAL UNDERSEA CTR, 73- Concurrent Pos: Res chemist, Naval Ship Res & Develop Ctr, Annapolis, Md, 73. Mem: Am Chem Soc. Res: Chemistry and biochemistry of food and agricultural products; analytical methods; chemistry of fatty oils; detergents and surfactants; biochemistry of marine organisms. Mailing Add: Naval Undersea Ctr NUC 406 San Diego CA 92132

LINDNER, KENNETH E, b Nelson, Wis, Nov 29, 22; m 47. RADIOCHEMISTRY. Educ: Univ Iowa, MA, 53, PhD(sci educ), 66. Prof Exp: Asst prof chem, Wis State Univ, La Crosse, 56-64; instr radiation safety, Univ Iowa, 64-66; prof chem, Wis State Univ, La Crosse, 66-67, head acad affairs, Wis State Univ Syst, 67-71, CHANCELLOR, UNIV WIS-LA CROSSE, 71- Concurrent Pos: Consult, Gundersen Clin, La Crosse & boiling water reactor, Dairyland Power, 66- Mem: Am Chem Soc; Health Physics Soc. Res: Chemical methods of radiation dosimetry. Mailing Add: 1725 State St La Crosse WI 54601

LINDNER, LUTHER EDWARD, b Toledo, Ohio, Aug 6, 42; m 69; c 2. PATHOLOGY. Educ: Univ Toledo, BS, 64; Western Reserve Univ, MD, 67, Case Western Reserve Univ, PhD(exp path), 74. Prof Exp: From intern to resident, Univ Hosp, Cleveland, 67-72; fel path, Case Western Reserve Univ, 69-72; staff pathologist, William Beaumont Army Med Ctr, 72-74, chief, Anatomic Path, 74-75; ASST PROF LAB MED, UNIV NEV, 75- Concurrent Pos: Consult path, Reno Vet Admin Hosp. Mem: Am Soc Cytol. Res: Studies of anatomic changes in disease with histochemical correlations and application to diagnosis. Mailing Add: Anderson Health Sci Bldg Univ of Nev Reno NV 89507

LINDNER, MANFRED, b Chicago, Ill, Oct 21, 19; m 46; c 2. NUCLEAR CHEMISTRY. Educ: Northwestern Univ, Ill, BS, 40; Univ Calif, Berkeley, PhD(nuclear chem), 48. Prof Exp: Chemist, Hanford Eng Works, Wash, 44-46; res asst, Univ Calif, Berkeley, 46-48; asst prof chem, Wash State Col, 48-51; chemist, Calif Res & Develop Co, 51-53; SR CHEMIST, RADIOCHEM DIV, LAWRENCE LIVERMORE LAB, 53- Concurrent Pos: Rothschild fel, Weizmann Inst Sci, 62-63. Mem: AAAS; Am Phys Soc. Res: Neutron capture cross-sections; nuclear structure. Mailing Add: Lawrence Livermore Lab Livermore CA 94550

LINDNER, MILTON JEROME, fisheries, deceased

LINDON, JOHN ARNOLD, b Chicago, Ill, Mar 11, 24; m 53; c 2. PSYCHIATRY, PSYCHOANALYSIS. Educ: Univ Louisville, MD, 48; Am Bd Psychiat & Neurol, dipl, 59. Prof Exp: Psychiat trainee, Neuropsychiat Ctr, Los Angeles, 49-52; instr psychiat, Col Med Evangelists, 52-53; asst clin prof, 60-68, ASSOC CLIN PROF PSYCHIAT, SCH MED, UNIV SOUTHERN CALIF, 68- Concurrent Pos: Psychoanal fel, Southern Calif Psychoanal Inst, 50-55; pvt pract psychoanal & psychiat, 51-; mem fac, Southern Calif Psychoanal Inst, 56-, mem bd trustees, 60-63 & 68-, supv & training analyst, 69-, secy-treas, 71-; consult, Calif State Dept Ment Hyg, 58-; pres, Psychiat Res Found, 58-; contrib ed, Ann Surv Psychoanal, 59-; ed-in-chief, Psychoanal Forum, 65-; consult, Parole Outpatient Clin, Calif State Dept Corrections, 62- Honors & Awards: Clin Essay Prize, Int Psychoanal Asn, 57. Mem: AAAS; fel Am Psychiat Asn; Am Psychoanal Asn; Am Med Asn; Fedn Am Sci. Res: Early mental development; infantile growth and development; extending psychoanalytic treatment; assessing characterological change; psychoanalytic supervision. Mailing Add: Med Plaza 10921 Wilshire Blvd Los Angeles CA 90024

LINDORFER, ROBERT KARL, b St Paul, Minn, May 6, 17; m; c 6. BACTERIOLOGY. Educ: St Thomas Col, BS, 47; Univ Minn, MS, 52, PhD, 53. Prof Exp: Instr bact, 53-54, from asst prof to assoc prof, 54-65, PROF BACT, UNIV MINN, MINNEAPOLIS, 65- Mem: Am Soc Microbiol; Am Acad Microbiol; Conf Res Workers Animal Dis. Res: Factors influencing resistance to infection; acquired tolerance; staphylococcal toxins. Mailing Add: Div of Vet Microbiol & Pub Health Univ of Minn St Paul MN 55101

LINDQUIST, ANDERS GUNNAR, b Lund, Sweden, Nov 21, 42; m 66; c 2. APPLIED MATHEMATICS. Educ: Royal Inst Technol, Sweden, MS, 67, TeknL, 68, TeknD(optimization), 72. Prof Exp: Res assoc optimization, Royal Inst Technol, Sweden, 69-72, docent, 72; vis asst prof math, Univ Fla, 72-73; ASSOC PROF MATH, UNIV KY, 74- Mem: Soc Indust & Appl Math. Res: Optimal stochastic control theory and estimation. Mailing Add: Dept of Math Univ of Ky Lexington KY 40506

LINDQUIST, CLARENCE BERNHART, b Superior, Wis, Dec 21, 13; m 41; c 5. APPLIED MATHEMATICS. Educ: Wis State Univ, Superior, BEd, 37; Univ Wis, MPh, 39, PhD(math), 41. Prof Exp: Asst math, Univ Wis, 37-40; from instr to asst prof, US Naval Acad, 41-46; assoc prof, Univ Minn, Duluth, 46-53, prof math & eng & head dept, 53-57; prog & res specialist higher educ, 57-70, regional coordr grad acad progs, 70-74, CHIEF TRAINING, BUR POST-SEC EDUC, OFF EDUC, US DEPT HEALTH, EDUC & WELFARE, 74- Concurrent Pos: Consult, Conf Bd Math Sci. Mem: Fel AAAS; Am Math Soc; Math Asn Am. Res: Mathematical theory of elasticity; survey of collegiate programs in science and mathematics education; Soviet education. Mailing Add: 6008 Utah Ave NW Washington DC 20015

LINDQUIST, DAVID GREGORY, b Chicago, Ill, Feb 14, 46; m 73. ICHTHYOLOGY. Educ: Univ Calif, Los Angeles, BA, 68; Calif State Univ, Hayward, MA, 72; Univ Ariz, PhD(zool), 75. Prof Exp: ASST PROF BIOL, UNIV NC, WILMINGTON, 75- Mem: AAAS; Am Soc Ichthyologists & Herpetologists; Am Soc Zoologists; Animal Behav Soc. Res: Ethology and behavioral ecology of fishes. Mailing Add: Dept of Biol Univ of NC Wilmington NC 28401

LINDQUIST, DONALD ARTHUR, b Manhattan, Kans, Nov 22, 30; m 52; c 3. ENTOMOLOGY. Educ: Ore State Col, BS, 52; Iowa State Univ, MS, 56, PhD(entom), 58. Prof Exp: Asst insect toxicol, Iowa State Univ, 54-58; entomologist, entom res div, Agr Res Serv, USDA, 58-67; ENTOMOLOGIST, INT ATOMIC ENERGY AGENCY, 67- Mem: Entom Soc Am. Res: Insect Toxicology and physiology; systemic insecticides; cotton insects, radiation and radioisotopes as related to entomology. Mailing Add: Int Atomic Energy Agency PO Box 645 A-1011 Vienna Austria

LINDQUIST, EVERT E, b Susanville, Calif, June 26, 35; m 57; c 4. ACAROLOGY, SYSTEMATIC ENTOMOLOGY. Educ: Univ Calif, Berkeley, BS, 57, MS, 59, PhD(entom), 63. Prof Exp: RES SCIENTIST, BIOSYST RES INST, AGR CAN, 62- Concurrent Pos: Adj prof, Carleton Univ, 71-; vis lectr, Ohio State Univ Summer Acarology Prog, 72- Mem: Entom Soc Can; Acarological Soc Am. Res: Systematics of Acarina; symbiotic relationships between mites and insects; geographic distribution of arctic mites. Mailing Add: Biosyst Res Inst Can Agr Ottawa ON Can

LINDQUIST, FRANK EUGENE, b Kingsburg, Calif, Mar 19, 07; m 34; c 3. CHEMISTRY. Educ: Univ Calif, BS, 29, MS, 30, PhD(biochem), 37. Prof Exp: Jr res assoc, Stanford Univ, 37; biochemist, Med Sch, Univ Pa, 37-39; instr & Nemours Found fel, 38-39; instr biochem, Col Med, Baylor Univ, 39-42; assoc chemist, Western Regional Res Lab, Bur Agr & Indust Chem, USDA, 42-48, from chemist to sr chemist, 48-55; sr res chemist, Calif Packing Corp, 55-63, mgr prod & process res, 63-69; consult prod develop, Del Monte Corp, 69-72; RETIRED. Mem: Am Chem Soc. Res: Food chemistry and processing. Mailing Add: 382 Via Casitas Greenbrae CA 94904

LINDQUIST, JOHN RAYMOND, b Red Lodge, Mont, Mar 29, 15; m 42; c 2. CHEMISTRY. Educ: Wash State Univ, BS, 38. Prof Exp: Res chemist, Hawley Pulp & Paper Co, Ore, 38-42; asst chemist, Southern Regional Res Lab, USDA, La, 42-44, assoc chemist, Western Regional Res Lab, Calif, 44-47; sr chemist, Rexall Drug Co, 47-52, dir prod develop dept, 52-57; mgr res & develop, Thayer Labs, Revlon, Inc, 57-60; VPRES & DIR RES, ANDREW JERGENS CO, 60- Mem: Am Chem Soc; Inst Food Technol; fel Am Inst Chem; Am Pharmaceut Asn; NY Acad Sci. Res: Pulp and paper utilization of agricultural products; production of rubber from domestic plants; food products; pharmaceuticals; cosmetics; method of vulcanizing low molecular weight rubber. Mailing Add: Res Dept Andrew Jergens Co 2535 Spring Grove Ave Cincinnati OH 45214

LINDQUIST, LAWRENCE WILLARD, b Galesburg, Ill, Mar 9, 21; m 45; c 2. CULTURAL ANTHROPOLOGY, APPLIED ANTHROPOLOGY. Educ: Northern Baptist Theol Sem, ThB, 50; Northwestern Univ, MA, 51; Oxford Univ, DPhil(social anthrop), 56. Prof Exp: Instr relig studies, Brown Univ, 56-57, instr polit sci, 57-58; from asst prof to assoc prof, 58-66, PROF ANTHROP, RI COL, 66-, CHMN DEPT ANTHROP & GEOG, 70-, COORDR INT EDUC, 73- Concurrent Pos: Dept Health Educ & Welfare grant, Southeast Asian Curric Develop Proj, 69; mem, Int Coun Educ for Teaching, 69-. Mem: Fel Am Anthrop Asn. Res: Hindu and Buddhist cultures; anthropology and education; kinship and political organization; primitive religions; culture contact. Mailing Add: Dept of Anthrop & Geog RI Col Providence RI 02908

LINDQUIST, RICHARD KENNETH, b Minneapolis, Minn, Oct 2, 42; m 69. ENTOMOLOGY. Educ: Gustavus Adolphus Col, BA, 64; Kans State Univ, MS, 67, PhD(entom), 69. Prof Exp: Instr entom, Kans State Univ, 68-69; ASST PROF ENTOM, OHIO AGR RES & DEVELOP CTR, 69- Mem: Entom Soc Am. Res: Biology, ecology and control of insect and mite pests of floral and greenhouse vegetable crops. Mailing Add: Dept of Entom Ohio Agr Res & Develop Ctr Wooster OH 44691

LINDQUIST, RICHARD WALLACE, b Worcester, Mass, May 6, 33; m 57; c 2. PHYSICS. Educ: Worcester Polytech Inst, BS, 54; Princeton Univ, AM, 57, PhD(physics), 62. Prof Exp: Instr physics, Princeton Univ, 58-60; asst prof, Adelphia Univ, 60-64; res assoc, Univ Tex, 64-65; ASSOC PROF PHYSICS, WESLEYAN UNIV, 65- Mem: Am Phys Soc; Am Asn Physics Teachers. Res: General relativity; geometrodynamics; gravitational collapse; relativistic transport theory. Mailing Add: Dept of Physics Wesleyan Univ Middletown CT 06457

LINDQUIST, ROBERT HENRY, b Minneapolis, Minn, Feb 27, 28; m 50; c 2. PHYSICAL CHEMISTRY. Educ: Univ Minn, BChem, 49, MS, 50; Univ Calif,

PhD(chem), 55. Prof Exp: Res chemist, 55-60, sr res chemist, 60-64, sr res assoc, 64-75, ASST TO PRES, CHEVRON RES CO, 75- Mem: Am Chem Soc; Am Phys Soc. Res: Solid state physics; magnetic resonance; physics of ultra-fine particles; heterogeneous catalysis; reaction kinetics; synthetic fuels; alternate energy sources. Mailing Add: Chevron Res Co 576 Standard Ave Richmond CA 94802

LINDQUIST, ROBERT MARION, b Cumberland, Wis, Dec 4, 23; m; c 3. ORGANIC CHEMISTRY, PHOTOGRAPHIC CHEMISTRY. Educ: Univ Wis, BS, 44; Univ Minn, PhD, 50. Prof Exp: Res chemist photog res, Gen Aniline & Film Corp, 50-56; assoc chemist photog processes, 56-57, staff chemist, 57-62, adv chemist, 62-63, develop chemist, 63-65, SR CHEMIST, IBM CORP, 65- Honors & Awards: First Level Invention Award, IBM Corp, 62, Outstanding Contrib Award, 74. Mem: Sr mem Am Chem Soc; Sigma Xi; sr mem Soc Photog Scientists & Engrs. Res: Electrophotographic processes. Mailing Add: 4788 Briar Ridge Trail Boulder CO 80301

LINDQUIST, ROBERT NELS, b Bakersfield, Calif, Sept 29, 42; m 68. BIOCHEMISTRY, ORGANIC CHEMISTRY. Educ: Occidental Col, BA, 65; Ind Univ, PhD(chem), 68. Prof Exp: Chemist, Shankman Labs, 65; res chemist, Shell Develop Co, 68-71; ASST PROF CHEM, SAN FRANCISCO STATE UNIV, 71- Mem: AAAS; Am Chem Soc. Res: Enzyme and enzyme model reaction kinetics and mechanisms. Mailing Add: Dept of Chem San Francisco State Univ San Francisco CA 94132

LINDQUIST, WILLIAM DEXTER, b Spokane, Wash, Aug 14, 16; m 42; c 3. VETERINARY PARASITOLOGY. Educ: Univ Idaho, BS, 40, MS, 42; Johns Hopkins Univ, ScD(hyg), 49. Prof Exp: From asst prof to prof bact & pub health, Mich State Univ, 49-68; PROF PARASITOL, KANS STATE UNIV, 68- Concurrent Pos: Sr Fulbright scholar, Fed Vet Res Lab, Nigeria, 57-58; USAID sci adv, Univ Nigeria, 63-65. Mem: Am Soc Parasitol. Res: Veterinary helminthology; entomology and protozoology and helminth immunity. Mailing Add: Dept of Infectious Dis Kans State Univ Manhattan KS 66502

LINDSAY, CHARLES MCCOWN, b Fayetteville, Tenn, July 5, 32; m 55; c 4. MATHEMATICS. Educ: Univ of the South, BS, 54; Univ Iowa, MS, 57; George Peabody Col, PhD(math), 65. Prof Exp: From instr to assoc prof, 57-71, PROF MATH, COE COL, 71-, CHMN DEPT, 63-, INTERIM DEAN, 75- Mem: Am Math Soc; Math Asn Am. Res: Mathematics education. Mailing Add: Dept of Math Coe Col Cedar Rapids IA 52402

LINDSAY, DALE RICHARD, b Bunker Hill, Kans, Aug 9, 13; m 37; c 3. RESEARCH ADMINISTRATION, MEDICAL EDUCATION. Educ: Univ Kans, AB, 37, MA, 38; Iowa State Univ, PhD(entom), 43. Prof Exp: Agent, Bur Entom & Plant Quarantine, USDA, 37-39; asst biol, Iowa State Univ, 40-42, res assoc & instr, 42-43; asst sanitarian, Malaria Control War Areas, USPHS, 43-45, sr asst sanitarian, Commun Dis Ctr, 45-47, sr asst scientist, 47-48, chief & sr scientist, Field Sta, Ga, 48-53, chief prog anal sect, Div Res Grants, NIH, 53-55, from asst chief to chief, 55-63; dep to gen dir, Mass Gen Hosp, Boston, 63-65; spec asst to chancellor, Univ Calif, Davis, 65-68, asst chancellor res & health sci, 68-69; assoc comnr sci, Food & Drug Admin, 69-71; assoc dir med & allied health educ, Duke Univ, 71-74; ASST DIR SCI COORD, NAT CTR TOXICOL RES, 75- Mem: Fel AAAS; fel Am Pub Health Asn; Am Soc Trop Med & Hyg; Entom Soc Am. Res: Taxonomy of Homoptera; insect toxicology; ecological studies of Coleoptera; ecology and control of Diptera; epidemiology and ecology of enteric infections; interdisciplinary research administration; health sciences. Mailing Add: Nat Ctr for Toxicol Res Jefferson AR 72079

LINDSAY, DAVID TAYLOR, b Philadelphia, Pa, Mar 22, 35; m 59; c 1. DEVELOPMENTAL BIOLOGY. Educ: Amherst Col, BA, 57; Johns Hopkins Univ, PhD(biol), 62. Prof Exp: Asst prof, 62-69, ASSOC PROF ZOOL, UNIV GA, 69- Concurrent Pos: Nat Sci Found res grant develop biol, 63-66. Mem: AAAS; Soc Develop Biol; Am Soc Zool; NY Acad Sci. Res: Mechanisms of cellular differentiation; regulation of protein synthesis in differentiation; biological role of histone proteins; bilateral symmetry. Mailing Add: Dept of Zool Univ of Ga Athens GA 30601

LINDSAY, DELBERT W, b Blackfoot, Idaho, May 9, 24; m 49; c 3. SYSTEMATIC BOTANY. Educ: Univ Utah, BS, 46, MS, 49, PhD(biol), 61. Prof Exp: PROF BIOL, RICKS COL, 60- Mem: Bot Soc Am. Res: Ecology and climatology; the flora of Eastern Idaho. Mailing Add: Dept of Biol Ricks Col Rexburg ID 83440

LINDSAY, DOUGLAS ROME, b Port Arthur, Ont, Apr 10, 21; m 51; c 1. PLANT ECOLOGY, TAXONOMY. Educ: Queen's Univ, Ont, BA, 49; Univ Wis, MS, 51. Prof Exp: Res off, Plant Res Inst, Can Dept Agr, 49-56; PROF BIOL, LAKEHEAD UNIV, 57-, CHMN DEPT, 73- Mem: AAAS; Am Inst Biol Sci; Can Bot Asn. Res: Ecology of weedy plants; phytogeography of Northwestern Ontario; biology of boreal forest plants. Mailing Add: Dept of Biol Lakehead Univ Thunder Bay ON Can

LINDSAY, DWIGHT MARSEE, b Versailles, Ind, June 19, 21; m 43; c 2. MAMMALOGY. Educ: Hanover Col, AB, 47; Univ Ky, MS, 49; Univ Cincinnati, PhD, 58. Prof Exp: Asst zool, Univ Ky, 47-48, instr, 51-52; from instr to assoc prof, 49-63, PROF BIOL, GEORGETOWN COL, 63-, CHMN DEPT BIOL SCI, 74- Mem: Am Soc Mammalogists. Res: Endocrinology; physiology; histology; embryology. Mailing Add: Dept of Biol Georgetown Col Georgetown KY 40324

LINDSAY, EVERETT HAROLD, JR, b La Junta, Colo, July 2, 31; m 53; c 3. VERTEBRATE PALEONTOLOGY. Educ: Chico State Col, AB, 53, MA, 57; Cornell Univ, MST, 62; Univ Calif, Berkeley, PhD(paleont), 67. Prof Exp: Asst prof, 67-71, ASSOC PROF GEOL, UNIV ARIZ, 71- Mem: AAAS; Soc Vert Paleont; Am Soc Mammalogists; Paleont Soc. Res: Biostratigraphy; taxonomy and evolution of small mammal fossils. Mailing Add: Dept of Geosci Univ of Ariz Tucson AZ 85721

LINDSAY, GEORGE EDMUND, b Pomona, Calif, Aug 17, 16. PLANT TAXONOMY. Educ: Stanford Univ, BA, 51, PhD, 56. Prof Exp: Dir, Desert Bot Garden, Ariz, 39-40; admin asst, Arctic Res Lab, Off Naval Res, 52-53; exec dir, San Diego Mus Natural Hist, 56-63; EXEC DIR, CALIF ACAD SCI, 63- Mem: AAAS; Cactus & Succulent Soc Am; Int Orgn Succulent Plant Studies; Am Asn Mus; Asn Dirs Sci Mus. Res: Taxonomic botany; taxonomy and ecology of Cactaceae and xerophytic plants of Baja California and other parts of Mexico. Mailing Add: Calif Acad of Sci Golden Gate Park San Francisco CA 94118

LINDSAY, HAGUE LELAND, JR, b Ft Worth, Tex, Jan 24, 29; m 56; c 4. VERTEBRATE ZOOLOGY. Educ: Tex Christian Univ, BA, 49; Univ Tex, MA, 51, PhD(zool), 58. Prof Exp: Res scientist, Univ Tex, 54; ASSOC PROF ZOOL, UNIV TULSA, 56- Mem: AAAS; Am Fisheries Soc; Am Soc Ichthyol & Herpet. Res: Vertebrate speciation, especially with Amphibians. Mailing Add: Dept of Life Sci Univ of Tulsa Tulsa OK 74104

LINDSAY, HARRY LEE, b Cotesfield, Nebr, Sept 3, 25; m 49; c 4. VIROLOGY. Educ: Univ Nebr, BS, 50, MS, 52; Univ Wis, PhD(bact), 63. Prof Exp: Microbiologist, Gateway Chemurgic Co, 51-52 & Hiram Walker & Sons, Inc, 53-59; MICROBIOLOGIST, LEDERLE LABS, AM CYANAMID CO, 63- Mem: Am Soc Microbiol; Am Chem Soc. Res: Respiratory viruses, including rhinoviruses and influenza; herpesviruses, antiviral agents; organ cultures; tissue culture and mycoplasma; antibiotic fermentations and culture improvement. Mailing Add: Lederle Labs Am Cyanamid Co Pearl River NY 10965

LINDSAY, HUGH ALEXANDER, b Moose Jaw, Sask, Mar 5, 26; m 56; c 3. PHYSIOLOGY. Educ: Univ Western Ont, BSc, 49, MSc, 52; Univ Toronto, PhD(pharmacol), 55; WVa Univ, MD, 73. Prof Exp: Asst pharmacol, Univ Toronto, 52-55; from asst prof to assoc prof, 55-70, PROF PHYSIOL, SCH MED, WVA UNIV, 70- Mem: AAAS; NY Acad Sci; Am Physiol Soc; Pharmacol Soc Can. Res: Growth in congenital cardiovascular disease; osteoporosis. Mailing Add: Dept of Physiol & Biophys WVa Univ Sch of Med Morgantown WV 26506

LINDSAY, JACQUE K, b Rochester, NY, Dec 6, 25; m 53; c 1. ORGANIC CHEMISTRY. Educ: Hobart Col, BS, 49; Duke Univ, PhD(org chem), 57. Prof Exp: Lab asst, 45-46, chemist, Distillation Prod Industs Div, 49-53, res chemist, 57-63, RES ASSOC, EASTMAN KODAK CO, 64- Mem: Am Chem Soc. Res: Organometallic and photographic chemistry. Mailing Add: Res Labs Eastman Kodak Co Rochester NY 14604

LINDSAY, JAMES GORDON, b Minneapolis, Minn, June 16, 25; Can citizen; m 50; c 4. PHYSICAL CHEMISTRY, CERAMICS. Educ: Univ Man, BSc, 46; McMaster Univ, MSc, 48, PhD(phys chem), 51. Prof Exp: Res chemist, Aluminum Labs, Ltd, 51-56, group leader chem div, 56-58, head ceramics div, Alcan Res & Develop, Ltd, 68-71, head raw mat dept, Arvida Res Lab, 71-73, COORDR RES & DEVELOP, ALCAN INT, LTD, 73- Mem: Fel Chem Inst Can; Am Ceramic Soc. Res: Alumina precipitation and calcination; impurities in alumina; gallium extraction; development of ceramic aluminas; bauxite calcination; refractories technology. Mailing Add: Alcan Int Ltd PO Box 6090 Montreal PQ Can

LINDSAY, JAMES GORDON, JR, b Norfolk, Va, Jan 23, 41; m 62; c 1. NUCLEAR PHYSICS. Educ: Va Polytech Inst & State Univ, BS, 64, MS, 66, PhD(physics), 71. Prof Exp: Asst prof, 69-74, ASSOC PROF PHYSICS, APPALACHIAN STATE UNIV, 74- Mem: Am Asn Physics Teachers; Am Nuclear Soc; Am Soc Mech Engrs. Res: Neutron activation analysis; thermal neutron cross sections. Mailing Add: Dept of Physics Appalachian State Univ Boone NC 28607

LINDSAY, JOHN RALSTON, b Renfrew, Ont, Dec 23, 98; nat US; m 37; c 3. OTOLARYNGOLOGY. Educ: McGill Univ, MD, 25. Hon Degrees: MD, Univ Uppsala, 63. Prof Exp: From asst prof to assoc prof surg, 35-41, PROF OTOLARYNGOL, UNIV CHICAGO, 41- Concurrent Pos: Mem nat adv neurol dis & blindness coun, USPHS, 56-60; ed, Yearbk Ear, Nose & Throat, 59-69. Mem: Am Laryngol, Rhinol & Otol Soc (pres, 62); AMA; Am Laryngol Asn; Col Oto-Rhino-Laryngol Amicitiae Sacrum (pres, 67); Am Col Surg (pres, 56-57). Res: Otology and pathology of inner ear diseases; physiology of labyrinthine fluids; fenestration operation for relief of deafness due to otosclerosis. Mailing Add: Dept of Surg Univ of Chicago Chicago IL 60637

LINDSAY, JON WILLIAM, developmental biology, anatomy, see 12th edition

LINDSAY, KENNETH LAWSON, b Springfield, Ill, Aug 26, 25; m 49; c 1. ORGANIC CHEMISTRY. Educ: Univ Ill, BS, 48; Univ Minn, PhD(chem), 52. Prof Exp: Res chemist, 52-55, develop chemist, 55-61, develop assoc, 61-63, PROCESS RES SUPVR, ETHYL CORP, 63- Mem: Am Chem Soc; Am Inst Chem. Res: Applied kinetics; organometallic chemistry; chlorine chemistry. Mailing Add: Ethyl Corp PO Box 341 Baton Rouge LA 70821

LINDSAY, RAYMOND H, b Perry, Ga, Dec 9, 28; m 54; c 5. BIOCHEMISTRY, PHARMACOLOGY. Educ: Jacksonville State Col, BS, 48; Univ Ala, MS, 57, PhD(pharmacol), 61. Prof Exp: From asst prof to assoc prof pharmacol, 63-71, from asst prof to assoc prof med, 63-72, PROF MED, MED CTR, UNIV ALA, BIRMINGHAM, 72-, PROF PHARMACOL, 71-; DIR PHARMACOL RES UNIT, VET ADMIN HOSP, 71- Concurrent Pos: NIH fel physiol chem, Univ Wis, 60-62, univ fel, 62-63; dir metab res, Vet Admin Hosp, Birmingham, 65-67, asst chief radioisotope serv, 64-72. Mem: AAAS; Endocrine Soc; Am Thyroid Asn; Am Chem Soc; Am Fedn Clin Res; Am Soc Pharmacol & Exp Therapeut. Res: Biochemistry, pharmacology and physiology of thyroid function; intermediary metabolism of amino acids; pyrimidine biosynthesis. Mailing Add: Vet Admin Hosp Birmingham AL 35233

LINDSAY, RICHARD H, b Portland, Ore, Sept 24, 34; m 58; c 6. NUCLEAR PHYSICS. Educ: Univ Portland, BS, 56; Stanford Univ, MS, 58; Wash State Univ, PhD(nuclear physics), 61. Prof Exp: Teaching assoc physics, Wash State Univ, 60-61; assoc prof, 61-66, PROF PHYSICS, WESTERN WASH STATE COL, 66- Mem: Am Phys Soc; Am Asn Physics Teachers. Res: Nuclear reactions with 30 to 65 million electron volts alpha particles; reactions with 14 million electron volts neutrons; theoretical nuclear physics; instrument design. Mailing Add: Dept of Physics Western Wash State Col Bellingham WA 98225

LINDSAY, ROBERT, b New Haven, Conn, Mar 3, 24; m 52; c 3. MAGNETISM. Educ: Brown Univ, ScB, 47; Rice Inst, MA, 49, PhD(physics), 51. Prof Exp: Physicist thermodynamics sect, Nat Bur Standards, 51-53; asst prof physics, Southern Methodist Univ, 53-56; from asst prof to assoc prof, 56-65, PROF PHYSICS, TRINITY COL CONN, 65- Mem: Am Phys Soc; Am Asn Physics Teachers; Sigma Xi. Mailing Add: Dept of Physics Trinity Col Hartford CT 06106

LINDSAY, ROBERT BRUCE, b New Bedford, Mass, Jan 1, 00; m 22; c 2. PHYSICS. Educ: Brown Univ, AB & MS, 20; Mass Inst Technol, PhD(theoret physics), 24. Hon Degrees: EdD, RI Col, 59; ScD, Southeastern Mass Univ, 68. Prof Exp: Instr physics, Mass Inst Technol, 20-22; from instr to asst prof, Yale Univ, 23-30; assoc prof theoret physics, 30-36, Hazard prof, 36-?1, chmn dept, 34-54, dir ultrasonics lab, 46-56, res anal group, 48-54, dean grad sch, 54-66, EMER PROF PHYSICS, BROWN UNIV, 71- Concurrent Pos: Vis prof, Polytech Inst Brooklyn, 32, 35-36 & 41; consult, US Bur Ships, 44-47; vis lectr, US Navy Electronics Lab, 46 & 49; consult, Signal Corps, 52, Naval War Col, 52, NSF, 52 & Dept Defense, 54-57; ed-in-chief, Acoust Soc Am, 57-; mem, Div Phys Sci, Nat Res Coun, 53-57 & post doctoral fel bd, 58-60; mem adv comt physics, Nat Bur Standards, 54-57, chmn, 57-60; mem governing bd, Am Inst Physics, 56-71; pres, Asn Grad Schs, 64-65; US mem Int Comn Acoustics, 63-69; series ed, Benchmark Papers on Acoustics, Dowden, Hutchinson & Ross, Inc Publ, 70- & Benchmark Papers on Energy, 73- Honors & Awards: Gold Medal, Acoust Soc Am, 63. Mem: AAAS(vpres, 58 & 68); Am Math Soc; fel Am Phys Soc; fel Acoust Soc Am (vpres, 51-56, pres, 56-57); fel Am Acad Sci (vpres, 57-59). Res: Atomic structure; Bohr theory; quantum mechanics; self-consistent field calculus; acoustics; filtration of sound; ultrasonic transmission; underwater sound; philosophy and history of physics; science and society. Mailing Add: Dept of Physics Brown Univ Providence RI 02912

LINDSAY, ROBERT CLARENCE, b Montrose, Colo, Nov 30, 36; m 57; c 2. FOOD SCIENCE, FOOD CHEMISTRY. Educ: Colo State Univ, BS, 58, MS, 60; Ore State Univ, PhD(food sci), 65. Prof Exp: Asst prof food sci, Ore State Univ, 64-69; assoc prof, 69-74, PROF FOOD SCI, UNIV WIS-MADISON, 74- Mem: Am Chem Soc; Inst Food Technologists; Am Dairy Sci Asn; Am Soc Microbiol. Res: Flavor chemistry; sensory evaluation of food. Mailing Add: Dept of Food Sci Univ of Wis-Madison Madison WI 53706

LINDSAY, STUART, b Sacramento, Calif, Feb 28, 12; m 36; c 3. PATHOLOGY. Educ: Univ Calif, AB, 34, MD, 38; Am Bd Path, dipl, 45. Prof Exp: Intern surg, Univ Hosp, Univ Calif, 37-38, asst resident, 39-40; asst resident, San Francisco Hosp, 38-39; instr path, Sch Med, 40-45, asst prof, Schs Med & Dent, 45-50, assoc prof path & dent, Sch Dent, 49-50, from assoc prof to prof, 50-71, vchmn dept, 56-60, EMER PROF PATH, SCH MED, UNIV CALIF, SAN FRANCISCO, 71- Concurrent Pos: Resident, Univ Hosp, Univ Calif, 40-42; asst vis pathologist, Univ Hosp, Univ Calif, San Francisco Hosp & Laguna Honda Home, 41-46, vis pathologist, 46-71; pathologist, Community Hosp, San Mateo, 42-54, Dante Hosp, 42, Mills Mem Hosp, 44-50, South San Francisco Hosp, 48-50 & Sequoia Hosp, Redwood City, 50-; dir, San Mateo County Pub Health Lab, 46-49; consult, Vet Admin Hosp, Oakland & Letterman Gen Hosp, 55-72; mem coun arteriosclerosis, Am Heart Asn. Mem: AAAS; Am Soc Exp Path; fel Col Am Path; NY Acad Sci; Int Acad Path. Res: Thyroid; arteriosclerosis; amyloidosis; gargoylism. Mailing Add: Sequoia Hosp Alameda & Whipple Redwood City CA 94062

LINDSAY, WILLARD LYMAN, b Dingle, Idaho, Apr 7, 26; m 51; c 4. SOIL SCIENCE. Educ: Utah State Univ, BS, 52, MS, 53; Cornell Univ, PhD(soil sci), 56. Prof Exp: Asst, Utah State Univ, 52-53 & Cornell Univ, 53-56; soil chemist, Soils & Fertilizer Res Br, Tenn Valley Authority, 56-60; asst prof agron & asst agronomist, Exp Sta, 60-62, from assoc prof to prof, 62-70, centennial prof, 70-74, PROF AGRON, COLO STATE UNIV, 74- Concurrent Pos: Vis prof, State Agr Univ, Wageningen, 72. Mem: Soil Sci Soc Am; fel Am Soc Agron; Int Soc Soil Sci. Res: Chemical nature of phosphate reactions in soils; physicochemical equilibria of plant nutrients in soils; chemistry and availability of micronutrients to plants; equilibrium of metal chelates in soils; solubility of heavy metals in soils. Mailing Add: Dept of Agron Colo State Univ Ft Collins CO 80521

LINDSAY, WILLIAM GERMER, JR, b Cleveland, Ohio, Nov 22, 28; m 56; c 3. PHYSIOLOGY. Educ: Oberlin Col, AB, 51; Univ Pa, MS, 57, PhD(zool), 62. Prof Exp: Instr physiol, Albany Med Col, 62-66; asst prof biol, 66-69, ASSOC PROF BIOL, ELMIRA COL, 69- Mem: AAAS; Soc Study Reproduction. Res: Spermatozoa metabolism; biological limnology. Mailing Add: Dept of Biol Elmira Col Elmira NY 14901

LINDSAY, WILLIAM TENNEY, JR, b Scranton, Pa, Apr 4, 24; m 51; c 2. PHYSICAL CHEMISTRY. Educ: Rensselaer Polytech Inst, BChE, 48; Mass Inst Technol, PhD(phys chem), 52. Prof Exp: Engr, Procter & Gamble Co, 48; asst, Mass Inst Technol, 49-51, res assoc, 52-53; sr scientist, Atomic Power Div, Westinghouse Elec Corp, 53-54, supv engr, 55-59, fel engr, Res Labs, 59-64, mgr phys chem dept, 64-73, MGR PHYS & INORG CHEM DEPT, RES LABS, WESTINGHOUSE ELEC CORP, 73- Mem: Fel AAAS; Am Chem Soc; Am Phys Soc; Electrochem Soc; NY Acad Sci. Res: Electrolytic solutions; nuclear reactor coolant technology. Mailing Add: Box 167 R D 6 Irwin PA 15642

LINDSEY, ALTON ANTHONY, b Monaca, Pa, May 7, 07; m 39; c 2. PLANT ECOLOGY. Educ: Allegheny Col, BS, 29; Cornell Univ, PhD(bot), 37. Prof Exp: Asst bot, Cornell Univ, 29-33; biologist, Byrd Antarctic Exped, 33-35; asst bot, Cornell Univ, 35-37; instr bot, Am Univ, 37-40; asst prof, Univ Redlands, 40-42 & Univ NMex, 42-47; from asst prof to prof, 47-74, EMER PROF BOT, PURDUE UNIV, 74- Concurrent Pos: Botanist, Purdue Can-Arctic Permafrost Exped, 51, ecologist, Purdue Res Team, Sonoran Desert, 53-54; bot ed, Ecol, Ecol Soc Am, 57-61, managing ed, Ecol & Ecol Monogr, 72-74; dir, Ind Natural Areas Surv, 67-68. Honors & Awards: Spec Cong Medal, 35. Mem: Fel AAAS; Ecol Soc Am. Res: Indiana vegetation; flood plain ecology; ecological methodology. Mailing Add: Dept of Biol Sci Purdue Univ Lafayette IN 47907

LINDSEY, CASIMIR CHARLES, b Toronto, Ont, Mar 22, 23; m 48. ICHTHYOLOGY. Educ: Univ Toronto, BA, 48; Univ BC, MA, 50; Cambridge Univ, PhD(zool), 52. Prof Exp: Res biologist, BC Dept Game, 52-57; from asst prof to assoc prof zool, Univ BC, 57-66, cur fishes, Inst Fisheries, 52-66; PROF ZOOL, UNIV MAN, 66- Concurrent Pos: Vis prof, Univ Singapore, 62-63, Wallace mem lectr, 63; fisheries consult, Pakistan, 64 & Fiji, 71. Mem: Am Soc Ichthyologists & Herpetologists; Am Fisheries Soc; fel Royal Soc Can; Can Soc Zoologists (vpres, 75-); Can Soc Environ Biologists (vpres, 74-). Res: Meristic variation; taxonomy; zoogeography of northern freshwater fishes; comparison of tropical and temperate fisheries. Mailing Add: Dept of Zool Univ of Man Winnipeg MB Can

LINDSEY, DAVID ALLEN, b Nebraska City, Nebr, May 26, 42; m 66; c 1. GEOLOGY. Educ: Univ Nebr, BS, 63; Johns Hopkins Univ, PhD(geol), 67. Prof Exp: Geologist, US Geol Surv, Colo, 67-74, STAFF GEOLOGIST MINERAL RESOURCES, US GEOL SURV, VA, 74- Concurrent Pos: Geol Soc Am res grant, 65-66. Mem: Geol Soc Am; Am Asn Petrol Geol; Soc Econ Paleontologists & Mineralogists. Res: Precambrian glacial deposits and Tertiary alluvial conglomerates and sandstones; beryllium deposits in tuff. Mailing Add: US Geol Surv Nat Ctr Reston VA 22092

LINDSEY, DONALD LEROY, b Stockton, Kans, May 25, 37; m 61; c 3. PLANT PATHOLOGY. Educ: Ft Hays Kans State Col, BS, 59; Colo State Univ, MS, 62, PhD(plant path), 65. Prof Exp: Jr plant pathologist, Colo State Univ, 61-65, asst plant pathologist, 66-69; instr bot, Colo State Col, 66; asst prof, 69-74, ASSOC PROF PLANT PATH, N MEX STATE UNIV, 74- Mem: AAAS; Am Phytopath Soc; Sigma Xi. Res: Biological control of plant pathogens; ecology of soil fungi. Mailing Add: Dept of Bot & Entom NMex State Univ Las Cruces NM 88003

LINDSEY, DORTHA RUTH, b Kingfisher, Okla, Oct 26, 26. HEALTH SCIENCE. Educ: Okla State Univ, BS, 48; Univ Wis, MS, 56; Ind Univ, PED, 63. Prof Exp: Instr health, phys educ & recreation, Okla State Univ, 48-50; instr, Monticello Col, 51-54; instr, DePauw Univ, 54-56; PROF HEALTH, PHYS EDUC & RECREATION, OKLA STATE UNIV, 56- Concurrent Pos: Consult, Payne County Guid Ctr, 66-71; ed, Fencing Guide, Am Asn Health, Phys Educ & Recreation, 61-62. Mem: Am Asn Health, Phys Educ & Recreation; Am Corrective Ther Asn. Res: Physical education; corrective therapy; electromyographical and kinesiological analyses of muscle action; quackery in physical fitness and reducing; therapeutic exercise. Mailing Add: Dept of Health Phys Educ & Recreation Okla State Univ Stillwater OK 74074

LINDSEY, EDWARD STORMONT, b West Palm Beach, Fla, June 3, 30; m 53; c 2. MEDICINE. Educ: Tulane Univ, BS, 51, MD, 58, MMedSci, 68. Prof Exp: Intern, Charity Hosp, La, 58-59, resident surg, 59-61 & thoracic surg, 63-64; from instr to asst prof, 63-68, ASSOC PROF SURG, TULANE UNIV, 68-, DIR TRANSPLANTATION RES UNIT, 66- Concurrent Pos: Resident surg, Southern Baptist Hosp, 61-62; Nat Heart Inst spec fel, Univ Edinburgh, 64-65; consult surg, Charity Hosp La & Keesler Air Force Hosp, 65-; mem adv comt, Nat Transplant Registry, 66-67. Mem: Transplantation Soc; Am Col Surg; NY Acad Sci; Asn Advan Med Instrumentation; Am Soc Artificial Internal Organs. Res: Thoracic and vascular surgery; transplantation biology. Mailing Add: Dept of Surg Tulane Univ New Orleans LA 70112

LINDSEY, GEORGE ROY, b Toronto, Ont, June 2, 20; m 51; c 2. SYSTEMS ANALYSIS. Educ: Univ Toronto, BA, 42; Queen's Univ Ont, MA, 46; Cambridge Univ, PhD(physics), 50. Prof Exp: Defence sci officer oper res, Can Defence Res Bd, 50-53; sr oper res officer, Air Defence Command, Royal Can Air Force, 54-59; dir defence systs anal group, Can Dept Nat Defence, 59-61; oper res group leader, Antisubmarine Warfare Res Ctr, Supreme Allied Comdr, Atlantic, Italy, 61-64; sr oper res scientist, 64-67, CHIEF OPER RES ANAL ESTAB, DEPT NAT DEFENCE, 68- Concurrent Pos: Mem, Can Govt Bicult Develop Prog, 70-71 & Can Comt Int Inst Appl Systs Anal, 73-; consult, Inst Res Pub Policy, 75- Mem: Opers Res Soc Am; Am Statist Asn; Inst Strategic Studies; Can Oper Res Soc (pres, 61); Can Inst Int Affairs. Res: Military operational research; forecasting and analysis of trends for research on public policy. Mailing Add: Oper Res & Anal Estab Dept of Nat Defence Ottawa ON Can

LINDSEY, JAMES RUSSELL, b Tifton, Ga, Dec 6, 33; m 58; c 4. PATHOLOGY. Educ: Univ Ga, BS, 56, DVM, 57; Auburn Univ, MS, 60. Prof Exp: From instr to asst prof parasitol, Sch Vet Med, Auburn Univ, 57-61; from instr to asst prof path & lab animal med, Johns Hopkins Univ, 63-68; PROF COMP MED & ASSOC PROF PATH, SCH MED & DENT, UNIV ALA, BIRMINGHAM, 68- Concurrent Pos: Fel path, Johns Hopkins Univ, 61-63. Mem: Am Vet Med Asn. Res: Comparative pathology. Mailing Add: Dept of Comp Med Univ of Ala Sch of Med & Dent Birmingham AL 35294

LINDSEY, JULIA PAGE, b Pine Bluff, Ark, Dec 9, 48. MYCOLOGY, PLANT PATHOLOGY. Educ: Hendrix Col, BA, 70; Univ Ariz, MS, 72, PhD(plant path), 75. Prof Exp: Teaching asst biol, Univ Ariz, 70-72 & plant path, 72-75; SEED ANALYST SEED CERT, ARK STATE PLANT BD, 75- Mem: Mycol Soc Am. Res: A compilation of descriptive, cultural and taxonomic data concerning woodrotting basidiomycetes that decay aspen in North America; identification and taxonomy of plant pathogenic fungi. Mailing Add: 63 Lefever Lane Little Rock AR 72207

LINDSEY, MARVIN FREDERICK, b Stockville, Nebr. PLANT BREEDING. Educ: Univ Nebr, BSc, 53, MSc, 55; NC State Univ, PhD(genetics), 60. Prof Exp: Asst prof agron, Univ Nebr, 60-63; geneticist, Rockefeller Found, 64-66; asst prof agron, Univ Wis, 66-69; RES AGRONOMIST, DeKALB AgRES, INC, 70- Mem: Am Soc Agron; Crop Sci Soc Am. Res: Maize breeding and genetics. Mailing Add: DeKalb AgRes Inc DeKalb IL 60115

LINDSEY, NORMA J, b Canton, Tex, June 16, 29. CLINICAL MICROBIOLOGY. Educ: Tex Woman's Univ, BA & BS, 51; Univ Calif, MPH, 64; Colo State Univ, PhD(microbiol), 69. Prof Exp: Bacteriologist, Dallas Health Dept Lab, Tex, 51-54; microbiologist, Ariz Health Dept Labs, Phoenix, 56-65; teaching asst microbiol, Colo State Univ, 66-67; chief of microbiol, NMex Health Labs, 69-70; res microbiologist, Dept Health, Educ & Welfare, 70-73; ASST PROF & ACTG HEAD, MICROBIOL SECT, CLIN LABS, UNIV KANS MED CTR, 73- Mem: AAAS; Am Soc Microbiol; Am Pub Health Asn; NY Acad Sci; Sigma Xi. Res: Clinical and applied microbiology. Mailing Add: 6317 W 58th St Shawnee Mission KS 66202

LINDSEY, RICHARD VERNON, JR, b Mt Pulaski, Ill, Apr 20, 16; m 40; c 3. ORGANIC CHEMISTRY. Educ: Knox Col, AB, 37; Univ Ill, PhD(org chem), 41. Prof Exp: Asst chem, Univ Ill, 37-39; res chemist, 41-50, RES SUPVR, CENT RES DEPT, E I DU PONT DE NEMOURS & CO, INC, 50- Mem: Am Chem Soc. Res: Organometallic compounds; heterogeneous catalysis; homogeneous catalysis. Mailing Add: Cent Res & Develop Dept E I du Pont de Nemours & Co Inc Wilmington DE 19898

LINDSEY, WILLIAM B, b Iowa Park, Tex, July 26, 22; m 48; c 2. ORGANIC CHEMISTRY. Educ: Univ Tex, BS, 48; Ind Univ, MA, 49, PhD(chem), 54. Prof Exp: Res chemist, E I du Pont de Nemours & Co, Inc, Buffalo, 52-70, staff scientist, Film Dept, Richmond, 70-75. Res: Organic coatings; polymerization; surface phenomena; polymer stabilization; coating techniques; surface treatment for adhesion; adhesion; film extrusion and orientation; film evaluation; monomer and general organic synthesis; cellulose chemistry. Mailing Add: 1019 Eighth Ave N Clinton IA 52732

LINDSKOG, GUSTAF ELMER, b Boston, Mass, Feb 7, 03; m 34; c 2. SURGERY. Educ: Mass Agr Col, BS, 23; Harvard Univ, MD, 28; Am Bd Surg, dipl, 52 & Am Bd Thoracic Surg, eert. Prof Exp: Intern surg, Lakeside Hosp, 28-29; asst surg & path, Sch Med, Yale Univ, 29-30; asst res surgeon, obstetrician & gynecologist, New Haven Hosp, 30-32; Nat Res Coun fel, Mass Gen Hosp, 32-33; from instr to prof, 33-71, EMER PROF SURG, SCH MED, YALE UNIV, 71- Concurrent Pos: Res surgeon, New Haven Hosp, 33-34; chmn, Am Bd Surg, 57-58. Mem: Soc Univ Surg; Soc Clin Surg; Am Asn Thoracic Surg; Am Surg Asn; fel Am Col Surg. Res: Thoracic surgery and physiology. Mailing Add: Dept of Surg Yale Univ Sch Med 333 Cedar St New Haven CT 06511

LINDSLEY, CHARLES HALSEY, chemistry, see 12th edition

LINDSLEY, DAN LESLIE, JR, b Evanston, Ill, Oct 13, 25; m 47; c 4. GENETICS. Educ: Univ Mo, AB, 47, MA, 49; Calif Inst Technol, PhD(genetics), 52. Prof Exp: Nat Res Coun fel biol, Princeton Univ, 52-53; NSF fel, Univ Mo, 53-54; from assoc biologist to biologist, Oak Ridge Nat Lab, 54-67; PROF BIOL, UNIV CALIF, SAN DIEGO, 67- Concurrent Pos: NSF sr fels, Univ Sao Paulo, 60-61 & Inst Genetics, Univ Rome, 65-66; USPHS spec fel, Dept Genetics, Div Plant Indust, Commonwealth Sci & Indust Res Orgn, Canberra, Australia, 72-73. Mem: Nat Acad Sci; Genetics Soc Am (treas, 75-78). Res: Cytogenetics of Drosophila. Mailing Add: Dept of Biol Univ of Calif San Diego La Jolla CA 92037

LINDSLEY, DAVID FORD, b Cleveland, Ohio, May 18, 36; m 60; c 3. PHYSIOLOGY. Educ: Stanford Univ, BA, 57; Univ Calif, Los Angeles, PhD(anat, neurophysiol), 61. Prof Exp: Asst prof physiol, Med Sch, Stanford Univ, 63-67; ASSOC PROF PHYSIOL, MED SCH, UNIV SOUTHERN CALIF, 67- Concurrent Pos: USPHS fels, Moscow State Univ, 61-62 & Cambridge Univ, 62-63; Lederle med fac award, 64-67; visitor, Max Planck Inst Psychiat, Munich, 71 & 74-75; Guggenheim fel, 74-75. Mem: Am Physiol Soc; Am Asn Anatomists; Soc Neurosci; Int Brain Res Orgn. Res: Central nervous system neurophysiology, especially visual system, reticular formation and limbic system and brain mechanisms of attention and behavior. Mailing Add: Dept of Physiol Univ of Southern Calif Med Sch Los Angeles CA 90033

LINDSLEY, DONALD B, b Brownhelm, Ohio, Dec 23, 07; m 33; c 4. PSYCHOPHYSIOLOGY. Educ: Wittenberg Univ, AB, 29; Univ Iowa, MA, 30, PhD, 32. Hon Degrees: DSc, Brown Univ, 58, Wittenberg Univ, 59, Trinity Col, 65 & Loyola Univ Chicago, 68. Prof Exp: Instr psychol, Univ Ill, 32-33; Nat Res Coun fel physiol & neuropsychiat, Harvard Med Sch & Mass Gen Hosp, Boston, 33-35; res assoc anat, Sch Med, Western Reserve Univ, 35-38; asst prof psychol, Brown Univ, 38-46; prof, Northwestern Univ, 46-51; chmn dept psychol, 59-62, PROF PSYCHOL & PHYSIOL, UNIV CALIF, LOS ANGELES, 51- Concurrent Pos: Dir, Psychol & Neurophysiol Labs, Bradley Hosp, East Providence, RI, 38-46 & Nat Defense Res Comt Proj, Yal Yale Contract, Fla, 43-46; mem sci adv bd, US Air Force, 47-49, chmn human resources comt, 48-49; mem aviation psychol comt, Nat Res Coun, 47-49 & undersea warfare comt, 50-66; consult, USPHS, 51-54, 58-61, 65-69 & NSF, 52-54; Am Inst Biol Sci, NASA Panel, 65-; mem space sci bd, Nat Acad Sci, 67-70, chmn long-duration missions in space comt, 67-71, mem space med comt, 67-; treas, Int Brain Res Orgn, 67-71; mem sci & technol adv coun, Calif Assembly, 69-71. Honors & Awards: President's Cert Merit, 48. Mem: Nat Acad Sci; AAAS (vchmn, 53); Soc Exp Psychol; Am Physiol Soc; Am Electroencephalog Soc (pres, 65). Res: Brain function; emotion; behavior disorders; electroencephalography; neurophysiology; vision and visual perception. Mailing Add: Dept of Psychol Univ of Calif Los Angeles CA 90024

LINDSLEY, DONALD HALE, geology, see 12th edition

LINDSTEDT-SIVA, K JUNE, b Minneapolis, Minn, Sept 24, 41; m 69. BIOLOGY. Educ: Univ Southern Calif, AB, 63, MS, 67, PhD(biol), 71. Prof Exp: Asst coordr sea grant progs, Univ Southern Calif, 71; environ specialist, Southern Calif Edison Co, 71-72; asst prof biol, Calif Lutheran Col, 72-73; SCI ADV, ATLANTIC RICHFIELD CO, 73- Concurrent Pos: Consult, Jacques Cousteau, Metromedia Producers Co, 70 & Southern Calif Edison Co, 72; mem fate & effects of oil task force, Am Petrol Inst, 73-; mem opers subcomt, Marine Water Qual Comn, Water Control Fedn, 75-76; Calif Mus Found sci fel, 76. Honors & Awards: Trident Award Marine Sci, Int Rev Subaqueous Activities, Ustica, Italy, 70. Mem: Marine Technol Soc; AAAS; Sigma Xi; Am Soc Zoologists. Res: Chemoreception in aquatic animals, especially chemical control of feeding behavior in sea anemones; effects of oil on marine organisms; environmental planning in industry, implementing planning during the early stages of project development. Mailing Add: Environ Affairs Atlantic Richfield Co 515 S Flower St Los Angeles CA 90071

LINDSTROM, EUGENE SHIPMAN, b Ames, Iowa, Jan 12, 23; m 49; c 4. BACTERIOLOGY. Educ: Univ Wis, BA, 47, MS, 48, PhD(bact), 51. Prof Exp: Asst bact, Univ Wis, 46-51, AEC fel enzyme chem, 51-52; from asst prof to assoc prof, 52-64, asst dean, 66-70, PROF BACT, PA STATE UNIV, 64-, ASSOC DEAN COL SCI, 70- Concurrent Pos: NSF fel, Univ Minn, 61. Mem: AAAS; Am Soc Microbiol; Am Acad Microbiol; Am Soc Biol Chem; Brit Soc Gen Microbiol. Res: Bacterial physiology; physiology of Athiorhodaceae; physiology and ecology of photosynthetic bacteria. Mailing Add: 211 Whitmore Lab Pa State Univ University Park PA 16802

LINDSTROM, FREDERICK JOHN, b La Crosse, Wis, Dec 22, 29. ANALYTICAL CHEMISTRY. Educ: Univ Wis, BS, 51, MS, 53; Iowa State Univ, PhD(chem), 59. Prof Exp: Asst, Univ Wis, 51-53; asst, Iowa State Univ, 55-58; from asst prof to assoc prof chem, 58-69, PROF CHEM, CLEMSON UNIV, 69- Mem: Am Chem Soc. Res: Organic analytical reagents; complexometric titrations; spectrochemistry. Mailing Add: Dept of Chem Clemson Univ Clemson SC 29631

LINDSTROM, FREDRICK THOMAS, b Astoria, Ore, July 30, 40; m 64; c 2. APPLIED MATHEMATICS. Educ: Ore State Univ, BS, 63, MS, 65, PhD(appl math), 69. Prof Exp: Res asst, 64-69, asst prof, 69-74, ASSOC PROF AGR CHEM & MATH, ORE STATE UNIV, 74- Mem: Soc Indust & Appl Math; Am Math Soc; Am Statist Asn. Res: Mass transport phenomenon, especially in porous and permeable mediums; compartmental analysis and the mathematical modeling of drug distributions in mammalian tissue systems. Mailing Add: Statist Dept Ore State Univ Corvallis OR 97331

LINDSTROM, IVAR E, JR, b Milligan, nebr, Oct 15, 29; m 52; c 2. PHYSICS. Educ: Nebr Wesleyan Univ, AB, 50; Univ Ore, MA, 52, PhD(physics), 59. Prof Exp: MEM STAFF PHYSICS, LOS ALAMOS SCI LAB, UNIV CALIF, 58- Mem: Am Phys Soc. Res: Explosives, particularly initiation by shock waves; nuclear spectroscopy; solid state physics. Mailing Add: Los Alamos Sci Lab Box 1663 Los Alamos NM 87544

LINDSTROM, RICHARD EDWARD, b Bristol, Conn, June 15, 32; m 52; c 3. PHYSICAL PHARMACY. Educ: Univ Conn, BS, 55; Syracuse Univ, MS, 62, PhD(phys chem), 67. Prof Exp: Asst prof chem, US Air Force Acad, 62-66 & Salem State Col, 66-68; ASSOC PROF PHARMACEUT, UNIV CONN, 68- Concurrent Pos: Consult, Vick Chem Co, 74- Mem: Am Chem Soc; Am Pharmaceut Asn; Acad Pharmaceut Sci. Res: Thermodynamics of solution phenomena via molar volume and solubility data. Mailing Add: Sch of Pharm Univ of Conn Storrs CT 06268

LINDSTROM, RICHARD S, b Cleveland, Ohio, Mar 5, 27; m 53; c 3. HORTICULTURE. Educ: Ohio State Univ, BS, 50, MS, 51, PhD(hort), 56. Prof Exp: Instr hort, Mich State Univ, 53-56, from asst prof to assoc prof, 56-68; PROF HORT, VA POLYTECH INST & STATE UNIV, 68- Mem: Am Soc Hort Sci. Res: Physiology of floricultural plants including work with growth regulators, nutrition and photoperiodic control. Mailing Add: Dept of Hort Va Polytech Inst & State Univ Blacksburg VA 24061

LINDSTROM, TED RILEY, biophysics, physical organic chemistry, see 12th edition

LINDSTROM, WENDELL DON, b Kiron, Iowa, Feb 7, 27; m 50; c 2. MATHEMATICS. Educ: Univ Iowa, AB, 49, MS, 51, PhD(math), 53. Prof Exp: Instr math, Iowa State Univ, 53-54, asst prof, 54-58; assoc prof, 58-66, PROF MATH, KENYON COL, 66- Concurrent Pos: NSF sci fac fel, Univ Calif, Berkeley, 62-63; vis prof, Robert Col, Istanbul, 68-69. Mem: Am Math Soc; Math Asn Am. Res: Fields, rings, algebras, differential algebra; algebraic geometry. Mailing Add: Dept of Math Kenyon Col Gambier OH 43022

LINDSTRUM, ANDREW O, JR, b Galesburg, Ill, Nov 12, 14. MATHEMATICS. Educ: Harvard Univ, AB, 35; Univ Ill, AM, 36, PhD(math), 39. Prof Exp: Asst math, Univ Ill, 36-39; instr, Univ Notre Dame, 39-40; instr, Univ Wis, 40-46; instr, Northwestern Univ, 46-48; from asst prof to prof, Knox Col, 48-63; PROF MATH, SOUTHERN ILL UNIV, EDWARDSVILLE, 63- Mem: Am Math Soc; Am Meteorol Soc; Math Asn Am. Res: Theory of functions; algebra. Mailing Add: Fac of Math Studies Southern Ill Univ Edwardsville IL 62025

LINDUSKA, JOSEPH PAUL, b Butte, Mont, July 25, 13; m 36; c 2. BIOLOGY. Educ: Univ Mont, BA, 36, MA, 38; Mich State Univ, PhD, 49. Prof Exp: Asst zool, Mich State Univ, 39-40; leader, Pittman-Robertson res proj, Game Div, Mich State Conserv Dept, 41-42, game biologist, 42-43, 46-47; foreign plant quarantine inspector, Bur Entom & Plant Quarantine, USDA, La, 43-44, entomologist, Fla, 44-46; biologist, Patuxent Res Refuge, US Fish & Wildlife Serv, 47-49, asst chief br wildlife res, 49-51, chief br game mgt, 51-56, actg chief br wildlife res, 54-55; dir wildlife mgt, Remington Arms Co, Md, 56-59, pub rels & wildlife mgt, 59-66; ASSOC DIR, US BUR SPORT FISHERIES & WILDLIFE, 66- Honors & Awards: Conserv educ award, Wildlife Soc, 63; conserv serv award, US Dept Interior, 65. Mem: Wildlife Soc (pres, 65-66); Soc Exp Biol & Med; Wildlife Mgt Inst; Soil Conserv Soc Am. Res: Vertebrate ecology; insect control; biology of insects; game animal research; pheasants; ecological study of small mammals on Michigan farmland. Mailing Add: US Bur of Sport Fish & Wildlife Washington DC 20240

LINDVIG, PHILIP ERVIN, physical chemistry, see 12th edition

LINDY, LOWELL BERGA, chemistry, see 12th edition

LINE, JOHN PAUL, b Pontiac, Mich, Mar 2, 29; m 57; c 4. MATHEMATICS. Educ: Univ Mich, BS, 50, MS, 51. Prof Exp: Instr math, Oberlin Col, 55 & Univ Rochester, 55-56; asst prof, 56-62, ASSOC PROF MATH, GA INST TECHNOL, 62- Mem: Am Math Soc; Math Asn Am. Res: Integral transformations as applied to solution of boundary value problems in partial differential equations. Mailing Add: Sch of Math Ga Inst of Technol Atlanta GA 30332

LINE, LLOYD ERNEST, JR, b Talbott, Tenn, June 19, 18; m 41; c 1. PHYSICAL CHEMISTRY. Educ: Carson-Newman Col, BS, 39; Univ Tenn, MS, 41, PhD(phys chem), 51. Prof Exp: Asst, Purdue Univ, 40-41; chemist, Trojan Powder Co, 41-42; asst chemist, Geophys Lab, Carnegie Inst, 42-43; assoc res chemist, Sun Oil Co, 46; instr anal chem, Univ Tenn, 46-50; res chemist, E I du Pont de Nemours & Co, 51-52; sr scientist, Texaco Exp, Inc, 52-57, head combustion lab, 57-61, explor res lab, 61-62, from asst dir to dir res, 62-68, RES ASSOC, RICHMOND RES LABS, TEXACO INC, 68- Mem: Am Chem Soc. Res: Homogeneous catalysis; chemical separations; waxy crudes and fuels; alkylation. Mailing Add: Richmond Res Labs Texaco Inc Richmond VA 23234

LINEBACK, DAVID R, b Russellville, Ind, June 7, 34; m 56; c 3. CARBOHYDRATE CHEMISTRY. Educ: Purdue Univ, BS, 56; Ohio State Univ, PhD(org chem), 62. Prof Exp: Res chemist, Monsanto Chem Co, 56-57; fel, Univ Alta, 62-64; from instr to asst prof biochem, Univ Nebr, Lincoln, 64-69; assoc prof, 69-74, PROF GRAIN SCI & INDUST, KANS STATE UNIV, 74- Mem: Am Asn Cereal Chem; Inst Food Technol; Am Chem Soc. Res: Reaction and structure of carbohydrates; characterization of enzymes of starch hydrolysis and synthesis; cereal chemistry. Mailing Add: Dept of Grain Sci & Indust Kans State Univ Manhattan KS 66506

LINEBACK, JERRY ALVIN, b Ottawa, Kans, Oct 25, 38; m 69; c 2. GEOLOGY. Educ: Univ Kans, BS, 60, MS, 61; Ind Univ, PhD(geol), 64. Prof Exp: Asst geologist, 64-67, ASSOC GEOLOGIST, STRATIG & AREAL GEOL SECT, ILL STATE GEOL SURV, 67- Mem: Geol Soc Am; Soc Econ Paleontologists & Mineralogists. Res: Stratigraphy of Illinois Basin; geologic mapping; paleoecology and paleontology of Paleozoic and Pleistocene sediments; Pleistocene stratigraphy; geology and geochemistry of lake sediments; Pleistocene-Holocene paleo-magnetic history. Mailing Add: Ill State Geol Surv Natural Resources Bldg Urbana IL 61801

LINEBACK, NEAL GAMBILL, b Winston-Salem, NC, June 21, 40; m 64; c 1. GEOGRAPHY. Educ: E Carolina Univ, BA, 63; Univ Tenn, Knoxville, MS, 67, PhD(geog), 70. Prof Exp: Teacher geog, Henry County Sch Syst, Va, 63-65; asst prof, 69-74, ASSOC PROF GEOG, UNIV ALA, 74- Mem: Asn Am Geogr. Res: State and regional atlas production; computer map production; land use analysis; physical environmental evaluation. Mailing Add: Dept of Geog Univ of Ala PO Box 1945 University AL 35486

LINEBERGER, WILLIAM CARL, b Hamlet, NC, Dec 5, 39. CHEMICAL PHYSICS. Educ: Ga Inst Technol, BEE, 61, MSEE, 63, PhD, 65. Prof Exp: Asst prof elec eng, Ga Inst Technol, 65; res physicist atmospheric physics, Aberdeen Res & Develop Ctr, Md, 67-68; res assoc physics, Inst, 68-70, asst prof chem, Univ, 70-74, PROF CHEM, UNIV COLO, BOULDER, 74-, FEL PHYSICS, JOINT INST LAB ASTROPHYS, 71- Mem: AAAS; Am Phys Soc. Res: Negative ion structure; molecular fluorescence; ion molecule reactions; tunable lasers. Mailing Add: Dept of Chem Univ of Colo Boulder CO 80302

LINEGAR, CHARLES RAMON, b Brookville, Ind, Sept 27, 05. PHARMACOLOGY. Educ: DePauw Univ, AB, 27; Western Reserve Univ, PhD, 34. Prof Exp: Biochemist, Cleveland City Hosp, Ohio, 31-32; from instr to assoc prof pharmacol, Sch Med, Georgetown Univ, 34-39, assoc prof, Sch Med & prof, Sch Dent, 39-43; chief biol develop & control lab, E R Squibb & Sons, 43-47, dir pharmacol develop div, 47-52, dir toxicol div, 52-58, assoc dir pharmacol div, 58-63, dir toxicol & path, 63-68, sr sci adv, 68-74. Mem: Am Soc Pharmacol & Exp Therapeut; Soc Toxicol; Soc Exp Biol & Med; Pan-Am Med Asn. Res: Muscle creatine; central stimulants and depressants; autonomic drugs; curare and curare-like substances; antibiotics; toxicology; bioassay; pathology. Mailing Add: 1115 Revere Rd North Brunswick NJ 08902

LINEHAN, URBAN JOSEPH, b Brockton, Mass, Oct 13, 11; m 50; c 3. PHYSICAL GEOGRAPHY. Educ: Bridgewater State Col, BS, 33; Clark Univ, MA, 46, PhD(geog), 55. Prof Exp: Instr geog, Univ Cincinnati, 40-45; from instr to asst prof, Univ Pittsburgh, 45-48; asst prof, Cath Univ Am, 48-56; analyst, US Govt, 56-73; RETIRED. Res: Synoptic climatology of Pittsburgh, Pennsylvania; areal and temporal distribution of tornado deaths in the United States; landscapes and off-road recreation of Southwestern United States. Mailing Add: PO Box 113 Moab UT 84532

LINES, MALCOLM ELLIS, b Banbury, Eng, Apr 26, 36; m 62; c 2. THEORETICAL SOLID STATE PHYSICS. Educ: Oxford Univ, BA, 59, MA & DPhil(physics), 62. Prof Exp: Fel physics, Magdalen Col, Oxford Univ, 61-63, 65-66; MEM TECH STAFF, BELL LABS, 63-65, 66- Concurrent Pos: Consult, Atomic Energy Res Estab, Harwell, Eng, 73. Mem: Fel Brit Inst Physics; fel Phys Soc Gt Brit. Res: Statistical mechanics; magnetism; ferroelectricity. Mailing Add: Bell Labs Murray Hill NJ 07974

LINEVSKY, MILTON JOSHUA, physical chemistry, see 12th edition

LINEWEAVER, HANS, physical chemistry, see 12th edition

LINEWEAVER, JOSEPH ASHBY, reproductive physiology, see 12th edition

LINFIELD, WARNER MAX, b Hannover, Ger, Jan 8, 18; nat US; m 45; c 1. APPLIED CHEMISTRY, INDUSTRIAL ORGANIC CHEMISTRY. Educ: George Washington Univ, BS, 40; Univ Mich, MS, 41, PhD(pharmaceut chem), 43. Prof Exp: Anna Fuller Fund res fel, Northwestern Univ, 43-44; res chemist, Emulsol Corp, Ill, 44-46; group leader, E F Houghton & Co, Pa, 46-52 & Quaker Chem Prod Co, 52-55;

dir res, Soap Div, Armour & Co Ill, 55-58, tech dir grocery prod div, 58-63; vpres, Culver Chem Co, 63-65; mgr org chem res, IIT Res Inst, 65-71; RES LEADER, EASTERN REGIONAL RES CTR DIV, USDA, 71- Concurrent Pos: Assoc ed, J Am Oil Chemists Soc, 74- Mem: AAAS; Am Chem Soc; Am Oil Chemists Soc; Am Inst Chem. Res: Surface-active agents; soaps and detergents; synthesis of germicides; anti-malarial drugs; textile finishing agents; food technology. Mailing Add: 600 E Mermaid Lane Philadelphia PA 19118

LINFOOT, JOHN ARDIS, b Grand Forks, NDak, May 16, 31; m 55; c 2. MEDICINE, ENDOCRINOLOGY. Educ: Univ NDak, BA, 53, BS & MS, 55; Harvard Univ, MD, 57; Am Bd Internal Med, dipl. Prof Exp: Asst physiol & pharmacol, Univ NDak, 53-55; intern, Univ Utah Hosps, 57-58, from resident to chief resident internal med, 58-61; RES ASSOC, DONNER LAB & PHYSICIAN-IN-CHG, DONNER PAVILION, UNIV CALIF, BERKELEY, 61- Concurrent Pos: Fel metab & endocrinol, Univ Utah Hosps, 59-60; lectr, Univ NDak, 69-; dir endocrine & metab serv, Alta Bates Hosp, 70; consult, Cowell Mem Hosp & Children's Hosp East Bay. Mem: AAAS; Am Fedn Clin Res; fel Am Col Physicians; Endocrine Soc; Am Diabetes Asn. Res: Growth hormone; acromegaly; Cushing's syndrome; diabetic retinopathy; heavy particle pituitary irradiation. Mailing Add: Donner Lab Univ Calif Berkeley CA 94720

LINFORD, JOHN HERBERT, b Grimston, Eng, Nov 25, 12; Can citizen; m 40; c 2. PHYSICAL CHEMISTRY. Educ: Univ Man, BSc, 35, MSc, 37; Univ London, PhD(phys chem), 49. Prof Exp: Chemist, Man Cancer Inst, 50-54; asst prof biochem, 55-64, hon asst prof med, 65-74, ASSOC PROF MED, UNIV MAN, 74-; CHEMIST, CANCER RES, MAN CANCER FOUND, 65- Mem: Assoc Royal Inst Chem; assoc Inst Physics; Tissue Cult Asn. Res: Physical chemistry of interactions of compounds of biological importance; chemical and biological properties of alkylating agents conjugated with proteins. Mailing Add: Cancer Found 700 Bannatyne Winnipeg MB Can

LING, ALFRED SOY CHOU, b New York, NY, Mar 16, 28; m 54; c 3. NEUROCHEMISTRY, ENDOCRINOLOGY. Educ: Princeton Univ, AB, 48; Univ Ill, MSc, 50; Univ Md, PhD(pharmacol), 60, MD, 62. Prof Exp: Asst endocrinol & mammal physiol, Univ Ill, 50-53, asst endocrinol, 53-54; asst pharmacol, Sch Med, Univ Md, 55-59; intern med, Pa Hosp, Philadelphia, 62-63, resident internal med, 63-65; res assoc, 65-67, asst prof, 67-70, ASSOC PROF ENDOCRINOL & NEUROCHEM, ROCKEFELLER UNIV, 70; EXEC DIR CLIN RES, WALLACE LABS, 74- Concurrent Pos: From asst physician to assoc physician, Rockefeller Univ, 65-70, physician, 70-; assoc dir clin pharmacol, Ciba-Geigy Corp, 70-74. Mem: AAAS; Sigma Xi; NY Acad Sci; Am Diabetes Asn; fel Am Col Clin Pharmacol & Therapeut. Res: Role of thyroid gland in brain metabolism; auto-immune aspects of thyroiditis; total body x-irradiation effect on blood volume of intact and adrenalectomized animals; effect of chemo-convulsant agents on brain metabolism; use of new convulsant agent, hexafluorodiethyl ether, in therapy of mentally ill patients. Mailing Add: Wallace Labs Half Acre Rd Cranbury NJ 08512

LING, CHUNG-MEI, b Chekiang, China, May 5, 31; m 57; c 2. BIOCHEMISTRY. Educ: Nat Taiwan Univ, BS, 58; Ill Inst Technol, MS, 62, PhD(biochem), 65. Prof Exp: Teaching asst biochem & physiol, Ill Inst Technol, 60-64; res assoc biochem res, Michael Reese Res Found, Chicago, 64-65; asst prof biochem, Ill Inst Technol, 65-68; molecular biologist, 68-71, assoc res fel virol, 71-74, RES FEL, DEPT BIOCHEM & HEAD MOLECULAR BIOL LAB, ABBOTT LABS, 74- Concurrent Pos: Adj asst prof, Ill Inst Technol, 68-69. Mem: Am Chem Soc; AAAS; Sigma Xi; Am Soc Biol Chemists. Res: Molecular biology and detection of hepatitis viruses. Mailing Add: Dept of Biochem Abbott Labs North Chicago IL 60064

LING, DANIEL, b Wetherden, Eng, Mar 16, 26; Can citizen; m 58; c 2. AUDIOLOGY, COMMUNICATIONS. Educ: St John's Col, Univ York, dipl, 50; Victoria Univ, Manchester, dipl, 51; McGill Univ, MS, 66, PhD(human commun dis), 68. Prof Exp: Organizer educ deaf, Reading Educ Comt, 55-63; prin, Montreal Oral Sch Deaf, 63-66; DIR RES DEAF CHILDREN, McGILL UNIV, 66-, ASSOC PROF & DIR ORAL REHAB, SCH HUMAN COMMUN DIS, 70-; DIR, SPEECH & HEARING DIV, ROYAL VICTORIA HOSP, 70- Concurrent Pos: Can Fed Prov Health grants, McGill Univ, 66-; res assoc educ, Cambridge Univ, 55-58; asst prof audiol, Sch Human Commun Dis, McGill Univ, 68-70; hon dir, Coun Children's Audiol Rehab, Ctr Deaf Children, Mexico City, 66- Mem: Am Speech & Hearing Asn; Acoust Soc Am. Res: Communication development in deaf children; speech recognition using linear and coding amplifiers; diagnostic procedure relative to deafness. Mailing Add: 69 Nelson St Montreal PQ Can

LING, DANIEL SETH, JR, b Chicago, Ill, Oct 22, 24; m 46; c 2. THEORETICAL PHYSICS. Educ: Univ Mich, BSE(physics) & BSE(math), 44, MS, 45, PhD(physics), 48. Prof Exp: Asst prof, 48-54, ASSOC PROF PHYSICS, UNIV KANS, 54-, ASSOC PROF ASTRON, 73- Mem: AAAS; Am Phys Soc. Res: Nuclear physics. Mailing Add: Dept of Physics Univ of Kans Lawrence KS 66044

LING, DONALD PERCY, b Albany, NY, Jan 2, 12; m 40; c 1. MATHEMATICS. Educ: Amherst Col, BA, 33; Columbia Univ, MA, 38, PhD(math), 44. Prof Exp: Instr math, Phillips Acad, Andover, 35-37; instr math, Columbia Univ, 40-44, mathematician, Appl Math Group, Nat Defense Res Comt, 44-45; res mathematician, Bell Tel Labs, Inc, NJ, 45-58, dir mil anal, 58-61, exec dir mil systs res div, 61-67, vpres mil systs eng, 67-69, vpres systs res, 69-71, pres, BellComm Inc, 71; RETIRED. Mem: Nat Acad Eng; Am Math Soc; Soc Indust & Appl Math. Res: Pure mathematics; guidance techniques; systems analysis; space technology. Mailing Add: 1816 Nakomis Ct NE Albuquerque NM 87112

LING, GEORGE M, b Trinidad, WI, Apr 11, 23; Can citizen; m 55; c 2. PHARMACOLOGY. Educ: McGill Univ, BA, 43; Univ BC, MA, 57, PhD, 60. Prof Exp: Res asst pharmacol, Univ BC, 57-60, res assoc, 60-61, from asst prof to assoc prof, 61-65; PROF PHARMACOL & CHMN DEPT, UNIV OTTAWA, 65- Concurrent Pos: Asst res anatomist, Brain Res Inst, Med Ctr, Univ Calif, Los Angeles, 62-63, assoc res pharmacologist, 63-64. Mem: Am Soc Pharmacol & Exp Therapeut; Pharmacol Soc Can; Can Physiol Soc; Can Soc Chemother; Int Soc Chemother. Res: Pharmacology of central nervous system; neuro-humoral mediators and mechanisms of brain function and behavior. Mailing Add: Dept of Pharmacol Univ of Ottawa Fac of Med Ottawa ON Can

LING, GILBERT NING, b Nanking, China, Dec 26, 19; m 51; c 3. PHYSIOLOGY. Educ: Nat Cent Univ, China, BSc, 43; Univ Chicago, PhD(physiol), 48. Prof Exp: Comen fel, Univ Chicago, 48-50; instr physiol optics, Sch Med, Johns Hopkins Univ, 50-53; from asst prof to assoc prof neurophysiol, Univ Ill, 53-57; sr staff scientist, Eastern Pa Psychiat Inst, 57-62; DIR DEPT MOLECULAR BIOL, PA HOSP, 62- Concurrent Pos: Mem, Woods Hole Marine Biol Corp. Mem: Am Physiol Soc. Res: Molecular mechanisms in self function. Mailing Add: Pa Hosp Eighth & Spruce St Philadelphia PA 19107

LING, HARRY WILSON, b Painesville, Ohio, Feb 14, 27; m 53. INORGANIC CHEMISTRY. Educ: Bowling Green State Univ, AB, 50; Ohio State Univ, PhD(inorg chem), 54. Prof Exp: Res chemist, Pigments Dept, Res Div, 54-64, tech serv chemist, Sales Div, 64-67, sr res chemist, 67-69, supvr, 69-72, prod mgr, 72-74, MGR TECH SERV, E I DU PONT DE NEMOURS AND CO, INC, 74- Mem: Am Chem Soc; Fedn Soc Paint Technol; Sigma Xi. Res: Inorganic nitrogen chemistry; elemental silicon; anodic oxidation of metal substrates; electrolytic capacitors; titanium dioxide pigments; pigment colors. Mailing Add: Pigments Dept Chestnut Run Lab E I du Pont de Nemours & Co Inc Wilmington DE 19898

LING, HSIN YI, b Taiwan, Dec 5, 30; m 58; c 2. MICROPALEONTOLOGY. Educ: Nat Taiwan Univ, BS, 53; Tohuku Univ, MS, 58; Wash Univ, PhD(geol), 63. Prof Exp: Instr geol, Nat Taiwan Univ, 54-55; res engr, Res Ctr, Pan Am Petrol Corp, Okla, 60-63; res instr geol oceanog, 63-64, from res asst prof to res assoc prof, 64-74, RES PROF, DEPT OCEANOG, UNIV WASH, 74- Mem: AAAS; Soc Econ Paleontologists & Mineralogists; Paleont Soc; Paleont Res Inst; Am Quaternary Asn. Res: Palynology and geological oceanography. Mailing Add: Dept of Oceanog Univ of Wash Seattle WA 98195

LING, HUBERT, b Chungking, China, Apr 28, 42; US citizen; m 67; c 1. BIOCHEMICAL GENETICS, MYCOLOGY. Educ: Queens Col NY, BS, 63; Brown Univ, MS, 66; Wayne State Univ, PhD(biol), 69. Prof Exp: ASSOC PROF BIOL, UNIV DEL, 69- Concurrent Pos: Univ Res Found fel & Res Corp fel, Univ Del, 70-72, USPHS fel, 72-74. Mem: Bot Soc Am; Mycol Soc Am. Res: Somatic cell fusion, genetic and biochemical aspects; biology of the myxogastres, Myxomycetes; genetics of reproduction and heterokaryosis. Mailing Add: Dept of Biol Sci Univ of Del Newark DE 19711

LING, NAN-SING, biochemistry, see 12th edition

LING, ROBERT FRANCIS, b Hong Kong, Apr 21, 39; US citizen; m 63; c 1. STATISTICS. Educ: Berea Col, BA, 61; Univ Tenn, MA, 63; Yale Univ, MPhil, 68, PhD(statist), 71. Prof Exp: Asst prof math, E Tenn State Univ, 64-66; from instr to asst prof statist, Univ Chicago, 70-75; ASSOC PROF STATIST, CLEMSON UNIV, 75- Mem: Am Statist Asn; Am Math Asn; Inst Math Statist. Res: Cluster analysis; statistical computing; interactive data analysis; Bayesian statistics. Mailing Add: Dept of Math Sci Clemson Univ Clemson SC 29631

LING, RUFUS CHIN-YU, theoretical physics, see 12th edition

LING, SAMUEL CHEN-YING, b Canton, China, May 7, 29; US citizen; m 57; c 3. PHYSICS. Educ: Nat Taiwan Univ, BS, 51; Baylor Univ, MS, 53; Ohio State Univ, PhD(physics), 69. Prof Exp: Asst prof physics, Augustana Col, Ill, 59-64; asst prof, 69-74, ASSOC PROF PHYSICS, WRIGHT STATE UNIV, 74- Mem: Am Phys Soc. Res: Low energy experimental nuclear physics investigating excited states in light and medium nuclei. Mailing Add: Dept of Physics Wright State Univ Dayton OH 45431

LINGAFELTER, EDWARD CLAY, JR, b Toledo, Ohio, Mar 28, 14; m 38; c 5. CHEMISTRY. Educ: Univ Calif, BS, 35, PhD(chem), 39. Prof Exp: Assoc phys chem, 39-41, from instr to assoc prof, 41-52, assoc dean, Grad Sch, 60-68, PROF CHEM, UNIV WASH, 52- Mem: AAAS; Am Chem Soc; Am Crystallog Asn (pres, 74). Res: Colloidal electrolytes; crystal structure of paraffin-chain compounds; structure of coordination compounds; hydrogen bond. Mailing Add: Dept of Chem BG-10 Univ of Wash Seattle WA 98195

LINGANE, JAMES JOSEPH, b St Paul, Minn, Sept 13, 09; m 38; c 4. ANALYTICAL CHEMISTRY. Educ: Univ Minn, ChB, 35, PhD(chem), 38. Hon Degrees: MA, Harvard Univ, 46. Prof Exp: Asst chem, Univ Minn, 35-37, instr & Baker fel, 38-39; instr, Univ Calif, 39-41; instr, 41-44, fac instr, 44-46, assoc prof, 46-50, PROF CHEM, HARVARD UNIV, 50- Concurrent Pos: Priestley lectr, Pa State Univ, 53. Honors & Awards: Gorman Res Conf Award, Am Asn Advan Sci, 52; Fisher Award, Am Chem Soc, 58. Mem: Am Chem Soc; Am Acad Arts & Sci; hon mem Brit Soc Anal Chem. Res: Electrochemistry; electroanalysis; instrumental methods of analysis; polarographic analysis; with the dropping-mercury electrode; physiochemical methods of chemical analysis. Mailing Add: Dept of Chem Harvard Univ Cambridge MA 02138

LINGANE, PETER JAMES, b Oakland, Calif, May 12, 40; m 67; c 2. ELECTROCHEMISTRY, METALLURGY. Educ: Harvard Univ, AB, 62; Calif Inst Technol, PhD(chem), 66. Prof Exp: Asst prof chem, Univ Minn, Minneapolis, 66-70; SR CHEMIST, LEDGEMONT LAB, KENNECOTT COPPER CORP, 70- Mem: Am Chem Soc; Am Inst Mining, Metall & Petrol Eng. Res: Chemistry related to the hydrometallurgical processing of nonferrous ore minerals; kinetics and mechanisms of solution reactions with particular emphasis upon the reactions which surround electrode processes. Mailing Add: Ledgemont Lab 128 Spring St Kennecott Copper Corp Lexington MA 02173

LINGAPPA, BANADAKOPPA THIMMAPPA, b Mysore, India, Mar 19, 27; nat US; m 53; c 3. MICROBIOLOGY. Educ: Benaras Hindu Univ, BSc, 50, MSc, 52; Purdue Univ, PhD, 57. Prof Exp: Lectr mycol, Benaras Hindu Univ, 52-53; res asst, Purdue Univ, 53-57; res assoc, Univ Mich, 57-59; res assoc, Mich State Univ, 59-60, asst prof med mycol, 60; Nat Inst Sci India sr res fel, Bot Lab, Univ Madras, 61; asst prof, Mich State Univ, 61-62; from asst prof to assoc prof, 62-68, PROF BIOL, COL HOLY CROSS, 68- Concurrent Pos: Vis assoc prof, Mass Inst Technol, 68-; vis prof, Inst Gen Bot, Univ Geneva, 69-70. Mem: Mycol Soc Am; Bot Soc Am; Am Soc Microbiol. Res: Physiology of fungi; dormancy and germination of spores; self-inhibitors; morphogenesis; methane production by anaerobic fermentation of solid waste. Mailing Add: Dept of Biol Col of the Holy Cross Worcester MA 01610

LINGAPPA, YAMUNA, b Mysore, India, Dec 6, 29; nat US; m 53; c 3. MICROBIOLOGY. Educ: Mysore Univ, BSc, 49; Madras Univ, BT, 51; Purdue Univ, MS, 55, PhD, 58. Prof Exp: Res assoc, Univ Mich, 57-59 & Mich State Univ, 59-60; sci pool officer, Govt India, 61; RES ASSOC BIOL, COL HOLY CROSS, 63- Concurrent Pos: Vis scientist, Inst Bot, Univ Geneva, 69-70; instr human nutrit, Clark Univ & Worcester State Col, 74; res consult, Dept Pub Health, City Worcester; fac adv, Undergrad Res Partic Proj Methane Generation, Col Holy Cross. Mem: Bot Soc Am; Mycol Soc Am; Int Soc Human & Animal Mycol; Am Soc Microbiol. Res: Human nutrition; water quality; solid waste disposal; physiology of pathogenic fungi; microbial interactions. Mailing Add: 4 McGill St Worcester MA 01607

LINGENFELTER, RICHARD EMERY, b Farmington, NMex, Apr 5, 34; m 57; c 2. ASTROPHYSICS, COSMIC RAY PHYSICS. Educ: Univ Calif, Los Angeles, AB, 56. Prof Exp: Physicist, Lawrence Radiation Lab, Univ Calif, 57-62; assoc res geophysicist, Univ Calif, Los Angeles, 62-66, res geophysicist, Inst Geophys & Planetary Physics, 66-69, PROF IN RESIDENCE, DEPT GEOPHYS & SPACE PHYSICS, 69- & DEPT ASTRON, 74- Concurrent Pos: Fulbright res fel geophys & planetary physics, Tata Inst Fundamental Res, Bombay, India, 68-69. Mem: Am Phys Soc; Am Geophys Union. Res: Cosmic ray origins and interactions; gamma ray astronomy; solar flare particle interactions; planetology; radiocarbon variations. Mailing Add: Dept of Astron Univ of Calif Los Angeles CA 90024

LINGG, AL JOSEPH, b Mt Hope, Kans, Mar 26, 38; m 61; c 2. MICROBIOLOGY. Educ: Kans State Univ, BS, 64, MS, 66, PhD(microbiol), 69. Prof Exp: Instr, Kans State Univ, 66-68; ASST PROF MICROBIOL, UNIV IDAHO, 69- Mem: AAAS; Am Soc Microbiol. Res: Fish diseases; eutrophication; carbon turnover in natural systems; oral vaccines against fish diseases. Mailing Add: Dept of Bact Univ of Idaho Moscow ID 83843

LINGLE, JOHN CLAYTON, plant physiology, soil science, see 12th edition

LINGREL, JERRY B, b Byhalia, Ohio, July 13, 35; m 58; c 2. BIOCHEMISTRY. Educ: Otterbein Col, BS, 57; Ohio State Univ, PhD(biochem), 60. Prof Exp: From asst prof to assoc prof, 62-72, PROF BIOL CHEM, UNIV CINCINNATI, 72- Concurrent Pos: Fel biol, Calif Inst Technol, 60-62; consult, Cincinnati Vet Admin Hosp, 72- Mem: Am Chem Soc; Am Soc Biol Chemists. Res: Regulation of gene expression in animal cells; hemoglobin biosynthesis; messenger RNA. Mailing Add: Dept of Biol Chem Univ of Cincinnati Col of Med Cincinnati OH 45267

LINGREN, WESLEY EARL, b Pasadena, Calif, Aug 27, 30; m 61; c 2. PHYSICAL CHEMISTRY, OCEANOGRAPHY. Educ: Seattle Pac Col, BS, 52; Univ Wash, MS, 54, PhD(electrochem), 62. Prof Exp: Instr phys sci, Pasadena Col, 56-58; from asst prof to assoc prof, 62-68, PROF & CHMN DEPT CHEM, SEATTLE PAC COL, 68-, DIR GEN HONORS, 70- Concurrent Pos: Res assoc, US Naval Radiol Defense Lab, 63-69; NSF fel, Yale Univ, 67-68. Mem: Am Chem Soc; Sigma Xi. Res: Rates of electrode reactions; electroanalytical chemistry; oxidation states of elements in seawater oceanography. Mailing Add: Dept of Chem Seattle Pac Col Seattle WA 98119

LINHART, YAN BOHUMIL, b Prague, Czech, Oct 8, 39; US citizen. EVOLUTION, GENETICS. Educ: Rutgers Univ, New Brunswick, BS, 61; Yale Univ, MF, 63; Univ Calif, Berkeley, PhD(genetics), 72. Prof Exp: Jr specialist forest genetics, Sch Forestry, Univ Calif, Berkeley, 63-65, asst specialist, 65-66; ASST PROF BIOL, UNIV COLO, BOULDER, 71- Concurrent Pos: Res grant, Univ Colo, Boulder, 71-73. Mem: AAAS; Am Soc Naturalists; Soc Study Evolution. Res: Adaptation; population biology; reproductive biology of plants. Mailing Add: Dept of Biol Univ of Colo Boulder CO 80302

LINI, DAVID CHARLES, physical organic chemistry, see 12th edition

LINIGER, WERNER, b Berne, Switz, Dec 22, 27; m 56; c 2. NUMERICAL ANALYSIS. Educ: Swiss Fed Inst Technol, Dipl, 51; Univ Lausanne, Dr es Sc(math), 56. Prof Exp: Asst numerical anal, Univ Lausanne, 52-55; mathematician, Swiss Nat Inst Accident Ins, 55-57; mem res staff, Remington Rand Univac Div, Sperry Rand Corp, 57-59; RES STAFF MEM, MATH SCI DEPT, RES CTR, IBM CORP, 59- Concurrent Pos: Invited prof, Swiss Fed Inst Technol, 72-73. Mem: Soc Indust & Appl Math. Res: Research in numerical analysis; ordinary differential equations; applied mathematics. Mailing Add: IBM Res Ctr PO Box 218 Yorktown Heights NY 10598

LININGER, LLOYD LESLEY, b Iowa City, Iowa, Mar 13, 39. MATHEMATICS. Educ: Univ Iowa, PhD(math), 64. Prof Exp: Asst prof math, Univ Mo, 64-65; asst to Prof Montgomery, Inst Advan Study, Princeton Univ, 65-67; res instr, Univ Mich, 67-70; ASSOC PROF MATH, STATE UNIV NY ALBANY, 70- Mem: Am Math Soc. Res: Topology. Mailing Add: Dept of Math State Univ of NY Albany NY 12203

LINIS, VIKTORS, b Rostov, Russia, Sept 14, 16; m 52. MATHEMATICS. Educ: Univ Latvia, dipl, 40; McGill Univ, MSc, 51, PhD(math), 53. Prof Exp: Instr, Univ Sask, 52-54; from asst prof to assoc prof, 54-57, PROF MATH & CHMN DEPT, UNIV OTTAWA, 57- Mem: AAAS; Am Math Soc; Math Asn Am; Can Math Cong. Res: Univalent functions; convex figures. Mailing Add: Dept of Math Fac of Sci & Eng Univ of Ottawa Ottawa ON Can

LINK, ARTHUR JÜRGEN, geology, petrography, see 12th edition

LINK, BERNARD ALVIN, b Columbus, Wis, Mar 23, 41. MEAT SCIENCE. Educ: Univ Wis-Madison, BS, 62, MS, 64, PhD(meat & animal sci), 68. Prof Exp: Res asst meat sci, Univ Wis-Madison, 62-68; Welch Found fel, Tex A&M Univ, 68-72, res scientist meat chem, 70-72, res assoc biochem & biophys, 72-73; RES BIOCHEMIST, CARGILL INC, 73- Mem: Am Meat Sci Asn; Am Soc Animal Sci; Inst Food Technol. Res: Soy protein products. Mailing Add: Cargill Inc Cargill Bldg Minneapolis MN 55402

LINK, CONRAD BARNETT, b Dunkirk, NY, Mar 5, 12; m 40; c 3. HORTICULTURE. Educ: Ohio State Univ, BS, 33, MS, 34, PhD(hort), 40. Prof Exp: Hybridist, Good & Reese Co, Ohio, 34-35; asst hort, Ohio State Univ, 35-38, exten specialist, 39-40; from instr to asst prof floricult, Pa State Univ, 38-45; horticulturist, Brooklyn Bot Garden, 45-48; PROF HORT, UNIV MD, COLLEGE PARK, 48- Concurrent Pos: Mem, Nat Coun Ther & Rehab Through Hort. Mem: Fel AAAS; fel Am Soc Hort Sci; Bot Soc Am; Am Hort Soc (secy, 48-49); Int Soc Hort Sci. Res: Photoperiodism; plant anatomy, nutrition and propagation. Mailing Add: Dept of Hort Holzapfel Hall Univ of Md College Park MD 20742

LINK, GORDON LITTLEPAGE, b Charleston, WVa, Feb 9, 32; m 55; c 1. PHYSICAL CHEMISTRY. Educ: Col William & Mary, BS, 54; Univ Va, PhD(phys chem), 58. Prof Exp: MEM TECH STAFF, BELL LABS, INC, 58- Mem: Am Phys Soc; Am Chem Soc. Res: Dielectrics. Mailing Add: Bell Labs PO Box 261 Rm 1a332 Murray Hill NJ 07971

LINK, HAROLD LEONARD, chemistry, see 12th edition

LINK, JOHN CLARENCE, b Iowa, Jan 5, 08; m 36; c 3. PHYSICS. Educ: Creighton Univ, AB, 28; Cath Univ Am, MA, 29. Prof Exp: Electronic scientist, US Dept Navy, 29-64, consult, Naval Res Lab, 55-64; mgr missile projs, Aerospace Corp, 64-66; mem staff, Avco Missile Systs Div, 66-68; INDEPENDENT CONSULT ELECTROMAGNETIC REFLECTORS, 68- Honors & Awards: Distinguished Civilian Serv Award, US Navy, 55. Res: Countermeasures; electromagnetic reflectors; energy conversion. Mailing Add: 6413 Halleck St SE Washington DC 20028

LINK, KARL PAUL, b La Porte, Ind, Jan 31, 01; m 30; c 3. BIOCHEMISTRY. Educ: Univ Wis, BS, 22, MS, 23, PhD, 25. Prof Exp: Int Ed Bd fel, Univ St Andrews, 25-26, Graz Univ, 26 & Univ Zurich, 26-27; from asst prof to prof, 27-71, EMER PROF BIOCHEM, UNIV WIS-MADISON, 71- Concurrent Pos: Consult, Pabst Brewing Co. Honors & Awards: Cameron Prize, Univ Edinburgh, 53; Lasker Award, 55 & 60; Scott Medal, 59; S Kovalenko Medal, Nat Acad Sci, 67. Mem: Nat Acad Sci; AAAS; Harvey Soc. Res: Chemistry of the sugars; 4-hydroxy-coumarins; blood coagulation; anticoagulant Dicumarol and warfarin sodium for clinical use; warfarin for rodenticidal use; disease resistance in plants. Mailing Add: 1111 Willow Lane Madison WI 53705

LINK, PETER K, b Batavia, Java, Nov 7, 30; US citizen; m 57; c 2. GEOLOGY, METEOROLOGY. Educ: Univ Wis, BS, 53, MS, 55, PhD(stratig geol), 65. Prof Exp: Teaching asst, Univ Wis, 53-55; geologist, Esso Standard Inc, Libya, 57-58, party chief, 58-59, subsurface geologist, 59-60, regional geologist, 60-61; regional geologist, Humble Oil & Refining Co, Okla, 62-63; res geologist, Atlantic Richfield Co, Dallas, 65-68, sr res geophysicist, 68-70; sr res scientist, Amoco Prod Co, 70-74; CONSULT, 74- Concurrent Pos: Teaching asst, Univ Wis, 63-65. Mem: Am Asn Petrol Geologists; fel Geol Soc Am. Res: Stratigraphy; structure; tectonics; field, regional, well site, subsurface and petroleum geology; research operations; stratigraphic-seismic research exploration; exploration programs; sedimentation; photogeology; minerals and petroleum exploration. Mailing Add: 2151 S Norfolk Terr Tulsa OK 74114

LINK, RICHARD FOREST, b Eugene, Ore, July 3, 28; m 50; c 3. MATHEMATICAL STATISTICS. Educ: Univ Ore, BS, 48, Princeton Univ, MA, 51, PhD(math), 53. Prof Exp: Asst, Univ Ore, 49-50; asst, Princeton Univ, 50-54; asst, Sandia Corp, 54-55; asst prof statist, Ore State Univ, 55-60, assoc prof, 60-63; mgr statist anal, Spec Comput Systs Projs, RCA Corp, 65-67; chief statistician, Louis Harris & Assocs, Inc, 67-69; VPRES CONSULT, ARTRONIC INFO SYSTS, INC, 69- Concurrent Pos: Vis lectr, Princeton Univ, 63-72. Mem: Fel AAAS; Am Statist Asn; Inst Math Statist; Opers Res Soc Am. Res: Sampling; geological statistics; operations research; systems analysis. Mailing Add: Artronic Info Systs Inc 420 Lexington Ave Rm 2835 New York NY 10017

LINK, ROGER PAUL, b Woodbine, Iowa, Jan 24, 10; m 38; c 1. VETERINARY PHARMACOLOGY. Educ: Iowa State Univ, DVM, 34; Kans State Univ, MS, 38; Univ Ill, PhD, 51. Prof Exp: Instr vet pharmacol, Mich State Univ, 34-35; from instr to asst prof vet physiol, Kans State Univ, 35-46; from asst prof to assoc prof, 46-55, PROF VET PHYSIOL & PHARMACOL, COL VET MED, UNIV ILL, URBANA, 55-, HEAD DEPT, 61- Concurrent Pos: Vis prof, Chas Pfizer & Co, Inc, 55; panel mem antibact drugs, Nat Acad Sci-Nat Res Coun, 66; consult, Div Physician Manpower, Bur Health Manpower, Dept Health, Educ & Welfare, 67; mem revision comt, US Pharmacopoeia & Nat Formulary. Mem: Am Vet Med Asn (pres, 72); Conf Res Workers Animal Dis. Res: Renal glucose threshold in the horse and pig; growth and reproduction in cats; steroid therapy; toxicology of agricultural chemicals. Mailing Add: 1708 Pleasant St Urbana IL 61801

LINK, VERNON BENNETT, b Copper Flat, NMex, Sept 16, 09; m 35. PUBLIC HEALTH. Educ: Univ Southern Calif, AB, 32, MD, 36; Univ Mich, MPH, 44; Am Bd Prev Med, 49. Prof Exp: Intern, US Marine Hosp, USPHS, 35-36, from asst surgeon to sr surgeon, 36-52, med dir, 52-61; mem staff, Bur Commun Dis, Calif State Dept Health, 62-63; comnr health, Ulster County Dept Health, 65-70; COMNR HEALTH, DUTCHESS COUNTY DEPT HEALTH, 70- Concurrent Pos: Mem expert adv panel plague, WHO, 52-72. Mem: Am Soc Trop Med & Hyg; Sigma Xi; fel Am Pub Health Asn; Am Col Prev Med; AMA. Mailing Add: Dutchess County Dept of Health 22 Market St Poughkeepsie NY 12601

LINK, WILLIAM B, b Darke, WVa, Mar 25, 28; m 56; c 3. ANALYTICAL CHEMISTRY, ORGANIC CHEMISTRY. Educ: Shepherd Col, BS, 53. Prof Exp: Med technician, Baker Vet Ctr, Martinsburg, WVa, 55; chemist, 55-57, anal chemist, 57-62, supvry chemist, 62-63, SUPVRY ANAL RES CHEMIST, US FOOD & DRUG ADMIN, 63- Mem: Asn Off Anal Chemists; Am Chem Soc. Res: Chemistry of all color additives used in foods, drugs and cosmetics. Mailing Add: Div of Colors Technol HFF-436 US Food & Drug Admin Washington DC 20204

LINK, WILLIAM EDWARD, b Ironwood, Mich, Jan 24, 21; m 47; c 2. ANALYTICAL CHEMISTRY. Educ: Northland Col, BA, 42; Univ Wis, MS, 51, PhD, 54. Prof Exp: Asst prof chem, Northland Col, 47-52; group leader res lab, ADM Chem, 54-69, group leader res ctr, Ashland Chem Co, 69-71, MGR ANAL CHEM RES & DEVELOP DIV, ASHLAND OIL & REF CO OHIO, 71- Concurrent Pos: Ed, Off & Tentative Methods, Am Oil Chemists Soc, 71- Mem: Am Chem Soc; Am Oil Chemists Soc (pres, 75-76). Res: Organic analytical research; fats and oils chemistry; industrial fatty derivatives analysis; resin analysis. Mailing Add: 6039 Sedgwick Rd Worthington OH 43085

LINK, WILLIAM JAMES, organic chemistry, see 12th edition

LINK, WILLIAM T, nuclear physics, see 12th edition

LINKE, CHARLES EUGENE, b Sioux City, Iowa, Dec 9, 20. SPEECH PATHOLOGY. Educ: Univ Iowa, BA, 49, MA, 52, PhD, 53. Prof Exp: Coordr speech & hearing serv, State Bd Health, Del, 53-55; speech therapist, Harlem Consol Schs, Ill, 55-56; ASSOC PROF OTOLARYNGOL, SCH MED, TULANE UNIV, 56- Mem: Fel Am Speech & Hearing Asn. Res: Audiology. Mailing Add: Speech & Hearing Ctr Sch of Med Tulane Univ New Orleans LA 70112

LINKE, ERNEST GEORGE, b Barberton, Ohio, Feb 20, 28; m 51; c 4. CLINICAL BIOCHEMISTRY. Educ: Baldwin-Wallace Col, BS, 50; Ohio State Univ, MS, 52; Purdue Univ, PhD(biochem), 56. Prof Exp: Res chemist paper technol, Mead Corp, 56-59; res biochemist & proj supvr, Borden Foods Co NY, 59-70, res biochemist, Anal Dept, 70-72; resident biochemist, Dept Path, State Univ NY Upstate Med Ctr, 72-75; HEAD RADIO-IMMUNOASSAY DEPT, CLIN LABS OF NASHVILLE, 75- Mem: Am Chem Soc. Res: Analytical biochemistry, specifically food products; carbohydrate chemistry. Mailing Add: 2525 Park Plaza PO Box 70 Nashville TN 37203

LINKE, HARALD ARTHUR BRUNO, b Bautzen, Ger, Aug 18, 36; m 71. MICROBIOLOGY. Educ: Univ Berlin, BSc, 61; Univ Göttingen, MSc, 63, PhD(biochem, microbiol), 67. Prof Exp: Res assoc enzym, Univ Göttingen, 66-67; fel biochem, Rutgers Univ New Brunswick, 67-69; res microbiologist, Allied Chem Corp, 69-72; res assoc, Inst Microbiol, Rutgers Univ, 72-73; ASST PROF, DEPT MICROBIOL, NY UNIV DENT CTR, 73- Concurrent Pos: Referee, Zentralblatt fuer Bakteriologie II. Abteilung, 66- Mem: Am Chem Soc; Am Soc Microbiol; Ger Chem Soc. Res: Isolation and characterization of enzymes; utilizing isotope techniques in the study of microorganisms; biosynthesis and biodegradation of chemical and natural compounds; taxonomy of streptococci; etiology of dental caries; artificial sweeteners. Mailing Add: Dept of Microbiol NY Univ Dent Ctr 421 First Ave New York NY 10010

LINKE, WILLIAM FINAN, b Ravena, NY, Aug 5, 24; m 49; c 3. PHYSICAL CHEMISTRY. Educ: City Col New York, BS, 45; NY Univ, MS, 46, PhD(chem), 48. Prof Exp: Asst chem, NY Univ, 45-48, from instr to asst prof, 48-57; group leader phys chem, 57-59, group leader paper chem, 59-64, mgr res & develop paper & film chem, 65-67, tech dir paper chem dept, 67-70, dir res, Indust Chem & Plastics Div, 71-74, TECH DIR PAPER CHEM DEPT, AM CYANAMID CO, 74- Mem: AAAS; Am Chem Soc; Tech Asn Pulp & Paper Indust. Res: Solubilities; phase equilibria; polyelectrolytes; stability of colloids; flocculation; adsorption; mining and paper chemicals; sizing; polymers. Mailing Add: Indust Chem & Plastics Div Am Cyanamid Co 1937 W Main St Stamford CT 06904

LINKENHEIMER, WAYNE HENRY, b Pittsburgh, Pa, May 23, 28; m 52; c 2. PHYSIOLOGY. Educ: Univ Pittsburgh, BS, 50, MS, 52, PhD, 54. Prof Exp: Res assoc, Sch Med, Univ Pittsburgh, 52-54; physiologist, Lederle Labs Div, Am Cyanamid Co, 54-59, group leader physiol, Agr Div, 59-64, mgr nutrit & physiol sect, 64-70; DIR ANIMAL HEALTH LABS, E R SQUIBB & SONS, INC, 70- Mem: AAAS; Am Physiol Soc; Am Soc Animal Sci; NY Acad Sci. Res: Physiological relationships of sulfhydryl compounds in shock; erythropoietin and iron metabolism; radiotracer techniques; pharmacology of sulfonamides. Mailing Add: Squibb Agr Res Ctr E R Squibb & Sons Inc Three Bridges NJ 08887

LINKER, ALFRED, b Vienna, Austria, Nov 23, 19; US citizen; m 54; c 2. BIOCHEMISTRY, CARBOHYDRATE CHEMISTRY. Educ: City Col New York, BS, 49; Columbia Univ, PhD(biochem), 54. Prof Exp: Assoc biochem, Columbia Univ, 56-59; asst res prof biochem & path, 60-64, assoc res prof biochem, 64-72, ASSOC PROF PATH, COL MED, UNIV UTAH, 64-, RES PROF BIOCHEM, 72- Concurrent Pos: Res biochemist, Vet Admin Hosp, Salt Lake City, 60- Mem: Am Soc Biol Chemists; . AAAS. Res: Structure, function and metabolism of the glycosaminoglycans of connective tissue, including studies of heparin, heparitin sulfate, the chondroitin sulfates, hyaluronic acid, and a variety of degradative enzymes isolated from mammalian and bacterial sources. Mailing Add: Vet Admin Hosp Salt Lake City UT 84113

LINKIE, DANIEL MICHAEL, b Utica, NY, Aug 12, 40; m 64; c 2. ENDOCRINOLOGY, PHYSIOLOGY. Educ: State Univ NY Albany, BS, 62, MS, 63; Univ Mich, PhD(zool), 71. Prof Exp: From res asst to res assoc obstet & gynec, Albany Med Col, Union Univ, 63-67; asst prof reproductive med & obstet & gynec, Univ Calif, San Diego, 73-75; ASST PROF PHYSIOL & OBSTET & GYNEC, COL PHYSICIANS & SURGEONS, COLUMBIA UNIV, 75- Concurrent Pos: NIH fel, Univ Tex Health Sci Ctr, Dallas, 71-73. Mem: AAAS; Soc Study Reproduction; NY Acad Sci; Endocrine Soc. Res: Mechanism of hormone action, specifically estradiol; interaction of steroids and gonadotropins; serum binding proteins; reproductive endocrinology. Mailing Add: Col of Phys & Surg Columbia Univ 630 W 168th St New York NY 10032

LINKLETTER, GEORGE ONDERDONK, b Mineola, NY, Jan 24, 43; m 65. QUATERNARY GEOLOGY, GEOCHEMISTRY. Educ: Dartmouth Col, AB, 65, AM, 67; Univ Wash, PhD(geol), 71. Prof Exp: Geologist, US Army Cold Regions Res & Eng Lab, 64-65 & Dartmouth Victoria Island Exped, 66; asst geol, Univ Wash, 67-70; asst prof, Lafayette Col, 70-72; ASST PROF GEOL, DESERT RES INST, UNIV NEV, 72- Concurrent Pos: Geologist, US Geol Surv, 67 & 68; consult environ res & appln, 71- Mem: AAAS; Am Quaternary Asn; Geol Soc Am; Glaciol Soc; Sigma Xi. Res: Geochemistry of weathering and soil formation in Antarctica; glacial geology; geomorphology; environmental geology and geochemistry; polar atmospheric and snow and ice chemistry. Mailing Add: Desert Res Inst Univ of Nev Stead Campus Reno NV 89507

LINKSWILER, HELLEN, b Lawton, Okla, Jan 5, 12. NUTRITION. Educ: Okla Agr & Mech Col, BS, 39; Univ Wis, MS, 49, PhD(nutrit), 51. Prof Exp: Teacher high sch, Okla, 39-44; lab asst, Univ Wis, 46-51; dir human nutrit res lab, Univ Ala, 51-54; dir food & nutrit res lab, Sch Home Econ, Univ Nebr, 54-60; PROF FOOD & NUTRIT, UNIV WIS-MADISON, 60- Honors & Awards: Borden Award, Am Home Econ Asn, 71. Mem: Am Home Econ Asn; Am Dietetic Asn; Am Inst Nutrit. Res: Calcium and protein interrelationships; amino acid requirements; vitamin B6 requirements and metabolism; magnesium interrelationships and requirements. Mailing Add: Dept of Nutrit Sci Univ of Wis Madison WI 53706

LINKSZ, ARTHUR, b Hlohovec, Czech, June 23, 00; nat US; m 29; c 2. OPHTHALMOLOGY. Educ: Univ Kiel, DMSc, 25; Univ Pecs, MD, 28. Prof Exp: Instr, Eye Inst, Dartmouth Med Sch, 39-43; instr, Postgrad Med Sch, NY Univ, 44-50, asst prof ophthal, 50-55, from assoc clin prof to clin prof, 55-68; CLIN PROF OPHTHAL, NY MED COL, 68- Concurrent Pos: Head aniseikonia dept, Manhattan Eye & Ear Hosp, 44-55, assoc attend surgeon, 50-55, attend surgeon, 55-; fac mem, Lancaster Basic Sci Course, 46-; vis lectr, Sch Med, Univ Colo, 53 & 57; Edward Jackson mem lectr, Am Acad Ophthal & Otolaryngol, 58; vis prof, Med Sch, Univ Pa, 67- & Med Sch, Tulane Univ, 68-; vis lectr, Inst Visual Sci, Presby Med Ctr, San Francisco, Calif. Honors & Awards: Medal of Honor, Am Acad Ophthal & Otolaryngol, 54; Semmelweis Medal, Am Hungarian Med Asn, 54 & 68. Mem: AAAS; Am Ophthal Soc; Asn Res Vision & Ophthal; Am Acad Ophthal & Otolaryngol; NY Acad Med. Res: Physiology, refraction and motility of the eye; pharmacology of ophthalmic drugs. Mailing Add: Dept of Ophthal NY Med Col New York NY 10028

LINLEY, JOHN ROGER, b Leeds, Eng, Aug 10, 38; m 68; c 3. ENTOMOLOGY, INSECT PHYSIOLOGY. Educ: Univ London, BSc, 59, MSc, 62, PhD(med entomol), 66. Prof Exp: Entomologist, Sandfly Res Unit, Ministry Health, Jamaica, 59-64; lectr, London Sch Hyg & Trop Med, Univ London, 64-66; PROJ LEADER & CHIEF MED ENTOM, ETHOLOGY SECT, FLA MED ENTOMOL LAB, FLA DEPT HEALTH & REHAB SERV, 66- Concurrent Pos: NIH grant, Nat Inst Allergy & Infectious Dis, 69- Mem: Royal Entom Soc. Res: Medical entomology, especially biology of sandflies and mosquitoes, with special reference to behavior and physiology. Mailing Add: Fla Med Entomol Lab PO Box 520 Vero Beach FL 32960

LINMAN, JAMES WILLIAM, b Monmouth, Ill, July 20, 24; m 46; c 4. MEDICINE. Educ: Univ Ill, BS, 45, MD, 47; Am Bd Internal Med, dipl, 55, cert hemat, 74. Prof Exp: From intern to jr clin instr internal med, Univ Mich, 47-51, instr, 51-52 & 54-55, asst prof, 55-56; from asst prof to assoc prof med, Northwestern Univ, 56-65; from assoc prof to prof internal med, Mayo Grad Sch Med, Univ Minn, 65-72, consult, Div Hemat, Mayo Clin, 65-72; PROF MED & DIR OSGOOD MED CTR, UNIV ORE HEALTH SCI CTR, 72-, HEAD DIV HEMAT, 74- Mem: Fel Am Col Physicians; Soc Exp Biol & Med; Am Fedn Clin Res; Am Soc Clin Invest; Am Soc Hemat. Res: Hematology. Mailing Add: 3630 SW Shattuck Rd Portland OR 97221

LINN, BRUCE OSCAR, b East Orange, NJ, Dec 12, 29; m 51; c 3. ORGANIC CHEMISTRY. Educ: Duke Univ, BS, 52, PhD(org chem), 56. Prof Exp: Asst, Duke Univ, 52-54 & Off Naval Res, 53-54; SR CHEMIST, MERCK SHARP & DOHME RES LABS, 56- Mem: Am Chem Soc. Res: Medicinal and synthetic organic chemistry in animal health. Mailing Add: 743 Wingate Dr Bridgewater NJ 08807

LINN, CARL BARNES, b Auburn, Nebr, Feb 18, 07; m 32; c 3. ORGANIC CHEMISTRY. Educ: Univ Nebr, BS, 29, MSc, 30; Stanford Univ, PhD(chem), 34. Prof Exp: Asst, Univ Nebr, 34-35; from res chemist to group leader, Universal Oil Prod Co, 35-64; prin chemist, Midwest Res Inst, 64-68; SR RES CHEMIST, C J PATTERSON CO, 68- Mem: Fel AAAS; Am Chem Soc. Res: Grignard reagent; pyrolysis of hydrocarbons; homogeneous and heterogeneous catalytic reactions of hydrocarbons and derivatives; hydrogen fluoride technology; high pressure and high temperature technology; Friedel-Crafts reactions; carbohydrate chemistry, including catalytic condensation of sugars and derivatives with hydrocarbons and derivatives; chemistry of lactic acid. Mailing Add: 2125 Beverly Dr Prairie Village KS 66208

LINN, DEVON WAYNE, b Estherville, Iowa, Oct 9, 29; m 53; c 3. LIMNOLOGY. Educ: Mankato State Col, BA, 52; Ore State Univ, MS, 55; Utah State Univ, PhD(fishery biol, statist), 62. Prof Exp: Chemist, Mayo Clin, Minn, 52-53; res biologist, Fisheries Res Inst, Univ Wash, 55-58; asst prof biol, Dakota Wesleyan Univ, 62-64; from asst prof to assoc prof, 64-73, chmn dept, 69-73, PROF BIOL, SOUTHERN ORE STATE COL, 73- Concurrent Pos: Consult, City of Fairmont, Minn, 62-; Peace Corps vol serving as Dep to Chief Fisheries Officer, Fisheries Dept Ministry Agr & Natural Resources, Lilongwe, Malawi, EAfrica, 73-75. Mem: Am Sci Affil. Res: Physiological effects of radiation and pesticides; water pollution and abatement; environmental quality and resource management. Mailing Add: Dept of Biol Southern Ore State Col Ashland OR 97520

LINN, JAY GEORGE, JR, b Pittsburgh, Pa, Aug 15, 16; m 51; c 3. OPHTHALMOLOGY. Educ: Allegheny Col, AB, 37; Temple Univ, MD, 40. Prof Exp: From instr to asst prof, 46-63, ASSOC PROF OPHTHAL, SCH MED, UNIV PITTSBURGH, 63- Mem: Fel AAAS; AMA; Am Med Writers' Asn; Asn Res Vision & Ophthal; Am Acad Ophthal & Otolaryngol. Res: Pharmacology; neuro-ophthalmology. Mailing Add: 401 Jenkins Bldg Pittsburgh PA 15222

LINN, JOHN CHARLES, b Bellingham, Wash. COMPUTER SCIENCE, SYSTEMS THEORY. Educ: Univ Wash, BS, 68; Stanford Univ, MS, 69, PhD(elec eng), 73. Prof Exp: Res engr laser commun, Honeywell Inc, 68; instr comput sci, Stanford Univ, 72; MEM TECH STAFF COMPUT SCI, TEX INSTRUMENTS INC, 73- Mem: Inst Elec & Electronics Engrs; Asn Comput Mach. Res: Computer architecture; algorithms, memory organization; human speech and language. Mailing Add: Tex Instruments Inc Cent Res Lab MS 132 PO Box 5936 Dallas TX 75222

LINN, KURT O, b Denver, Colo, Mar 1, 30; m 56; c 1. GEOLOGY. Educ: Colo Sch Mines, GeolE, 52; Harvard Univ, MA, 58, PhD(geol), 64. Prof Exp: Geologist, Tex Gulf Sulphur Co, 52, Ressurrection Mining Co, 55-57 & Int Minerals & Chem Corp, 60-65; geologist, Tex Gulf Sulphur Co, 65-68, supvr spec projs, 68-69; sr geologist, Australian Inland Explor Co, 70-74; MGR SPEC PROJS, TEXASGULF INC, 75- Mem: Am Inst Mining & Metall Engrs; Soc Econ Geologists; Am Inst Mining, Metall & Petrol Engrs. Res: Geology of ore deposits. Mailing Add: Texasgulf Inc 200 Park Ave New York NY 10017

LINN, MANSON BRUCE, b New Ross, Ind, June 15, 08; m 32; c 2. PLANT PATHOLOGY. Educ: Wabash Col, AB, 30, Cornell Univ, PhD(plant path), 40. Prof Exp: Lab asst bot, Wabash Col, 28-29, instr, 30-32; asst plant path, Cornell Univ, 41-42; asst prof veg dis, Veg Crop Exten, 42-45, from asst prof to prof plant path, 45-74, EMER PROF PLANT PATH, UNIV ILL, URBANA, 74- Mem: AAAS; Am Phytopath Soc; Potato Asn Am. Res: Soil fungicides; vegetable and canning crop diseases. Mailing Add: 204 E Mumford Dr Urbana IL 61801

LINN, RICHARD HARRY, b Auburn, Nebr, July 23, 18; m 43; c 2. INTERNAL MEDICINE. Educ: Univ Nebr, BA, 40, MD, 43; Am Bd Internal Med, dipl, 52; Am Bd Pulmonary Dis, dipl, 55. Prof Exp: Resident internal med, USPHS Hosps, San Francisco, Calif, 47-49, resident chest dis, 49-50, dep chief med, Seattle, Wash, 50-53, chief med, New Orleans, La, 54-57, San Francisco, 57-62; chief clin res sect, Lab Med & Biol Sci Div Air Pollution, USPHS, 62-67, med officer-in-chg, USPHS Clin, Ohio, 67-69, AREA SUPVR, COLO, UTAH & IDAHO, FED EMPLOYEE HEALTH, USPHS, 69- Concurrent Pos: Asst clin prof, Sch Med, Tulane Univ, La, 54-57, Sch Med, Univ Calif, San Francisco, 61-62, Col Med, Univ Cincinnati, 64-69; vis physician, Charity Hosp, New Orleans, La, 54-57; mem, Cent Clin Invest Comt, Div Hosps, USPHS, 56-62; mem, Chronic Dis Subcomt, Nat Tuberc Asn, 62-64. Mem: Fel Am Col Physicians; fel Am Col Chest Physicians; Am Thoracic Soc. Res: Research administration; pulmonary diseases; clinical research; air pollution. Mailing Add: Bldg 40 Denver Fed Ctr Denver CO 80225

LINN, ROBERT, b Cleveland, Ohio, May 12, 26; m 56; c 2. PLANT ECOLOGY, RESOURCE MANAGEMENT. Educ: Kent State Univ, BS, 50, MA, 52; Duke Univ, PhD(bot), 57. Prof Exp: Chief naturalist, Isle Royale Nat Park, 56-63, res botanist, DC off, 63-67, from dep chief scientist to chief scientist, 67-72, chief scientist & dir off natural sci, 72-73, SR SCIENTIST ECOL, NAT PARK SERV, 73- Concurrent Pos: Adj prof biol sci, Mich Technol Univ, 73-; mem comn ecol, Int Union Conserv Nature & Natural Resources. Mem: Fel AAAS; Ecol Soc Am; Am Bryol & Lichenological Soc; Int Asn Ecol. Res: Causes of succession in spruce-fir and northern hardwood communities on Isle Royale in Lake Superior; spruce-fir, maple-birch transition areas; bryophyte succession in North Carolina Piedmont old fields. Mailing Add: Dept of Biol Sci Mich Technol Univ Houghton MI 49931

LINN, STUART MICHAEL, b Chicago, Ill, Dec 16, 40; m 67; c 1. BIOCHEMISTRY. Educ: Calif Inst Technol, BS, 62; Stanford Univ, PhD(biochem), 66. Prof Exp: Helen Hay Whitney fel, Univ Geneva, 66-68; asst prof, 68-73, ASSOC PROF BIOCHEM, UNIV CALIF, BERKELEY, 73- Concurrent Pos: Res grants, USPHS, Univ Calif, Berkeley, 68- & AEC, 70-; Guggenheim fel, 74-75. Mem: AAAS; Am Soc Microbiol; Am Soc Biol Chemists. Res: Biochemistry of nucleic acids; nucleic acid enzymes. Mailing Add: Dept of Biochem Univ of Calif Berkeley CA 94720

LINN, THOMAS ARTHUR, JR, b Colorado Springs, Colo, Jan 9, 33; m 63. ANALYTICAL CHEMISTRY, RADIOCHEMISTRY. Educ: Colo Col, BSc, 55; Univ Ill, MSc, 57; Ariz State Univ, PhD(anal chem), 68. Prof Exp: Scientist, Westinghouse Elec Corp, Idaho, 58-61; engr, Gen Elec Co, 61-62; tech staff mem, Sandia Labs, NMex, 62-64; res assoc radiochem, Radiation Ctr, Ore State Univ, 67-69; ASSOC SCIENTIST, KENNECOTT RES CTR, METAL MINING DIV, KENNECOTT COPPER CORP, 69- Concurrent Pos: Adj asst prof nuclear sci, Univ Utah, 69- Mem: Am Chem Soc; Soc Appl Spectros; Geochem Soc. Res: Methods development in chemical analysis; instrumental applications in trace element determinations; nuclear techniques applied to problems in geochemistry and analytical sciences. Mailing Add: Kennecott Res Ctr 1515 Mineral Sq Salt Lake City UT 84111

LINN, TRACY CLAUD, b Norman, Okla, May 20, 38; m 62; c 2. BIOCHEMISTRY. Educ: Univ Okla, BS, 59; Univ Minn, Minneapolis, PhD(biochem), 65. Prof Exp: NIH fel biochem, Univ Minn, St Paul, 65-66; NIH fel biochem, Univ Tex, Austin, 66-68, res assoc biochem, 68-70; ASST PROF BIOCHEM, UNIV TEX HEALTH SCI CTR DALLAS, 70- Res: Regulatory mechanisms of enzyme systems. Mailing Add: Dept of Biochem Univ of Tex Health Sci Ctr Dallas TX 75235

LINN, WILLIAM JOSEPH, b Crawfordsville, Ind, July 14, 27; m 56; c 2. ORGANIC CHEMISTRY. Educ: Wabash Col, AB, 50; Univ Rochester, PhD(chem), 53. Prof Exp: RES CHEMIST, CENT RES DEPT, E I DU PONT DE NEMOURS & CO, INC, 53- Concurrent Pos: Res assoc, Northwestern Univ, 69-70. Mem: Am Chem Soc. Res: Organometallic compounds; heterogeneous and homogeneous catalysis;

catalytic oxidation. Mailing Add: Cent Res Dept E I du Pont de Nemours & Co Inc Wilmington DE 19898

LINNA, TIMO JUHANI, b Tavastkyro, Finland, Mar 16, 37; m 61; c 3. CANCER, IMMUNOLOGY. Educ: Univ Uppsala, BMed, 59, MD, 65, PhD(histol), 67. Prof Exp: Asst prof histol, Med Sch, Univ Uppsala, 67-71; asst prof, 70-71, ASSOC PROF MICROBIOL & IMMUNOL, SCH MED, TEMPLE UNIV, 71-, ADV CLIN IMMUNOL, 72- Concurrent Pos: USPHS int res fel, Univ Minn, Minneapolis, 68-70, Univ Minn spec res fel, 70. Mem: Am Soc Exp Path; NY Acad Sci; Reticuloendothelial Soc; Swed Royal Lymphatic Soc; Am Asn Immunologists. Res: Immunobiology; experimental pathology; tumor immunology; cell kinetics. Mailing Add: Dept of Microbiol & Immunol Temple Univ Sch of Med Philadelphia PA 19140

LINNARTZ, NORWIN EUGENE, b Fischer, Tex, Apr 9, 26; m 57; c 2. FOREST SOILS. Educ: Tex A&M Univ, BS, 53; La State Univ, MF, 59, PhD(soils), 61. Prof Exp: Range mgt asst soil conserv serv, USDA, 53-54, range conservationist, 54-57; asst forestry, Sch Forestry & Agr Exp Sta, 57-60, from asst prof to assoc prof, 61-70, PROF FORESTRY, LA STATE UNIV, BATON ROUGE, 70- Mem: Fel AAAS; Ecol Soc Am; Soc Range Mgt; Soc Am Foresters; Soil Sci Soc Am. Res: Site quality relationships of forest soils; forest soil-moisture-plant relationships; forest fertilization; tree nutrition studies; forest range. Mailing Add: Sch Forestry & Wildlife Mgt La State Univ Baton Rouge LA 70803

LINNELL, ALBERT PAUL, b Canby, Minn, June 30, 22; m 44; c 5. ASTROPHYSICS. Educ: Col Wooster, AB, 44; Harvard Univ, PhD(astron), 50. Hon Degrees: MA, Amherst Col, 62. Prof Exp: From instr to prof astron, Amherst Col, 49-66; PROF ASTRON, MICH STATE UNIV, 66- Concurrent Pos: Mem adv comt, Comput Ctr, Mass Inst Technol, 60-63; mem bd dirs, Asn Univs for Res Astron, 62-65. Mem: Int Astron Union; AAAS; Am Astron Soc. Res: Instrumentation for photoelectric photometry; photometry and theory of eclipsing binaries. Mailing Add: 1918 Yuma Trail Okemos MI 48864

LINNELL, ROBERT HARTLEY, b Kalkaska, Mich, Aug 15, 22; m 50; c 4. ACADEMIC ADMINISTRATION, INSTITUTIONAL RESEARCH. Educ: Univ NH, BS, 44, MS, 47; Univ Rochester, PhD(chem), 50. Prof Exp: Instr chem, Univ NH, 47; asst prof, Am Univ Beirut, 50-52, assoc prof & chmn dept, 52-55; vpres, Tizon Chem Co, 55-58, dir, 55-62; assoc prof chem, Univ Vt, 58-61; lab dir, Scott Res Labs, 61-62; prog dir phys chem, NSF, 62-65, staff assoc planning, 65-67, staff assoc dept develop prog, 67-69; dean col letters, arts & sci, 69-70, PROF CHEM, UNIV SOUTHERN CALIF, 69-, DIR OFF INSTNL STUDIES, 70- Concurrent Pos: Grants, Res Corp, 50-54 & 58-60, NSF, 59-61, USPHS, 61-62 & Am Petrol Inst, 61-62; consult, Reheis Corp, 58-61, Tizon Chem Co, 58-62, Col Chem Consult Serv, Lake Erie Environ Studies Prog & Environ Protection Agency; grant, Exxon Educ Found, 74-76. Honors & Awards: Outstanding Achievement Award, Univ NH, 69. Mem: AAAS; Am Chem Soc; Air Pollution Control Asn; Asn Instnl Res; Am Asn Higher Educ. Res: Hydrogen bonds; air pollution, energy planning and policy research in higher education; science and public policy; science manpower; curriculum and programs, policies, financial management and information systems in higher education; student and faculty surveys. Mailing Add: Off of Instnl Studies Univ of Southern Calif Los Angeles CA 90007

LINNEMANN, ROGER E, b St Cloud, Minn, Jan 12, 31; m 51; c 5. RADIOLOGY, NUCLEAR MEDICINE. Educ: Univ Minn, Minneapolis, BA, 52, BS & MD, 56; Am Bd Radiol, cert, 64; Am Bd Nuclear Med, cert, 72. Prof Exp: Intern, Walter Reed Army Hosp, 56-57; physician, US Army, Europe, 57-61; res assoc radiobiol, Walter Reed Army Hosp, 61-62, resident radiol, 62-65; cmndg officer, Nuclear Med Res Detachment, US Army, Europe, 65-68; asst prof radiol, Univ Minn, Minneapolis, 68; radiologist, Hosp, 68-69, asst prof, 69-74, ASSOC PROF CLIN RADIOL, UNIV PA, 74-; PRES, RADIATION MGT CORP, 69- Concurrent Pos: US deleg radiation protection comt & panel experts med aspects nuclear biol & chem warfare, NATO, 65-68; Nat Res Coun James Picker Found res grant radiol, 68-69; nuclear med consult, Philadelphia Elec Co, 68-; mem ad hoc comt med aspects radiation accidents, AEC, 69- Mem: AMA; Am Col Radiol; Am Nuclear Soc; Am Pub Health Asn; Indust Med Asn. Res: Medical aspects of nuclear industry accidents; kidney function studies using isotopes; radiological health. Mailing Add: Radiation Mgt Corp Sci Ctr Bldg 2 3508 Market St Philadelphia PA 19104

LINNENBOM, VICTOR JOHN, b St Louis, Mo, Feb 10, 15; m 43; c 3. OCEANOGRAPHY. Educ: Univ Iowa, BA & MS, 38; Wash Univ, PhD(chem), 49. Prof Exp: Chemist, Monsanto Chem Co, 38-42; head radiochem sect, 49-53, head radiation effects br, 53-62, assoc supt, Radiation Div, 62-65, supt, Ocean Sci & Eng Div, 65-72, chief scientist, Off Naval Res, London, 72-74, SUPT, OCEAN SCI DIV, NAVAL RES LAB, 74- Honors & Awards: Hulburt Annual Sci Award, 63. Mem: AAAS; Am Chem Soc; Sigma Xi. Res: Oceanography; environmental effects of pollution; geochemistry. Mailing Add: Ocean Sci Div US Naval Res Lab Washington DC 20375

LINNER, EDWARD ROBERT, b Buffalo, NY, Oct 12, 99; m 31; c 1. PHYSICAL CHEMISTRY. Educ: Univ Buffalo, BS, 25; Univ Minn, PhD(biochem), 23. Prof Exp: Anal & plant chemist, Acheson Graphite Co, NY, 20-24; asst chem, Univ Wis, 25-28; instr, Lafayette Col, 28-31; instr anal chem, Univ Minn, 31-34; from instr to prof, 34-65, chmn dept chem, 59-61, EMER PROF GEN & PHYS CHEM, VASSAR COL, 65- Concurrent Pos: Adj prof, Marist Col NY, 65-69. Mem: Am Chem Soc. Res: Adsorption from solution relation to structure; relation of activation to adsorbability; activity of some organic acids; history of chemistry in the late eighteenth and early nineteenth centuries. Mailing Add: Dept of Chem Vassar Col Poughkeepsie NY 12601

LINNERUD, ARDELL CHESTER, b Whitehall, Wis, Apr 9, 31; m 56. EXPERIMENTAL STATISTICS. Educ: Wis State Univ River Falls, BS, 53; Univ Minn, MS, 62, PhD(dairy husb), 64. Prof Exp: Res asst dairy husb, Univ Minn, 57-63, consult biomet, 64; fel biomath, 64-67, asst prof statist, 67-75, ASSOC PROF STATIST, NC STATE UNIV, 75- Concurrent Pos: Statist consult, Inst for Aerobics Res, 74- Mem: Am Dairy Sci Asn; Am Soc Animal Sci. Res: Design of experiments and mathematical model building; animal science and exercise physiology. Mailing Add: Dept of Statist NC State Univ Raleigh NC 27607

LINNSTAEDTER, JERRY LEROY, b Lindale, Tex, July 25, 37; m 62; c 3. MATHEMATICS. Educ: Tex A&M Univ, BA, 59, MS, 61; Vanderbilt Univ, PhD(math), 70. Prof Exp: Instr math, Northeastern La State Col, 61-63 & Vanderbilt Univ, 67-68; assoc prof, 68-71, PROF MATH, ARK STATE UNIV, 71-, CHMN DIV MATH & PHYSICS, 68- Concurrent Pos: Prin investr NASA res grant, Ark State Univ, 69-70. Mem: Am Math Soc; Math Asn Am. Res: Multistage calculus of variations; classical analysis. Mailing Add: Div of Math & Physics Ark State Univ Drawer F State University AR 72467

LINS, THOMAS WESLEY, b Nov 24, 23; US citizen; m 69. MARINE GEOLOGY, STRUCTURAL GEOLOGY. Educ: Cornell Univ, BS, 48; Univ Kans, MS, 59, PhD(geol), 69. Prof Exp: Dist geologist, Sunray Oil Corp, 50-51; dist geologist, Monsanto Chem Co, 51-57, asst div geologist, 57-60, div geologist, 60-61, res geologist, 61-63; asst prof geol, Lamar Univ, 68-74; ASST PROF GEOL & GEOG, MISS STATE UNIV, 74- Mem: AAAS; Geol Soc Am. Res: Tectonics, structure of island arcs and trenches, recent sedimentation of the Gulf of Mexico. Mailing Add: Dept of Geol & Geog Miss State Univ State College MS 39762

LINSAY, ERNEST CHARLES, b Cleveland, Ohio, May 3, 42; m 66; c 3. ORGANIC CHEMISTRY. Educ: Yale Univ, BS, 63; Univ Wis-Madison, PhD(org chem), 68. Prof Exp: RES CHEMIST, ORGANICS DEPT, HERCULES INC, 68- Mem: Am Chem Soc. Res: Physical organic chemistry; dispersions and emulsions. Mailing Add: Hercules Res Ctr Wilmington DE 19899

LINSCHEID, HAROLD WILBERT, b Goessel, Kans, Sept, 24, 06; m 33; c 3. MATHEMATICAL ANALYSIS. Educ: Bethel Col, Kans, BA, 29; Phillips Univ, MEd, 36; Univ Okla, MA, 40, PhD, 55. Prof Exp: Prin high sch, Okla, 29-36; instr, Okla Jr Col, 36-38; instr math, Univ Okla, 38-41; instr math & physics, Bluffton Col, 41-43; army specialized training prog, Univ Nebr, 43-44; asst prof math & physics, Eastern NMex Col, 44-46; assoc prof, Col Emporia, 51-58; assoc prof, 58-75, PROF MATH, WICHITA STATE UNIV, 75- Mem: Am Math Soc; Math Asn Am. Res: Algebra; geometry; physics; electricity and magnetism. Mailing Add: 3701 E Funston Wichita KS 67218

LINSCHITZ, HENRY, b New York, NY, Aug 18, 19; m 64; c 1. PHYSICAL CHEMISTRY. Educ: City Col New York, BS, 40; Duke Univ, MA, 41, PhD(chem), 46. Prof Exp: Mem staff, Explosives Res Lab, Nat Defense Res Comt, 43; sect leader, Los Alamos Sci Lab, 43-45; fel, Inst Nuclear Studies, Univ Chicago, 46-48; from asst prof to assoc prof chem, Syracuse Univ, 48-57; assoc prof, 57-59, PROF CHEM, BRANDEIS UNIV, 59-, CHMN DEPT, 58- Concurrent Pos: Vis scientist, Brookhaven Nat Lab, 56-57; Fulbright vis prof, Hebrew Univ, Israel, 60; mem adv comt space biol, NASA, 60-61, study sect biophys & biophys chem, NIH, 62-66 & comt photobiol, Nat Res Coun, 64-69; Guggenheim fel, Weizmann Inst, 71-72. Mem: AAAS; Am Chem Soc. Res: Photochemistry; spectroscopy and luminescence of complex molecules; photobiology. Mailing Add: Dept of Chem Brandeis Univ Waltham MA 02154

LINSCOTT, DEAN L, b Blue Springs, Nebr, Mar 31, 32; m 53; c 6. AGRONOMY. Educ: Univ Nebr, BSc, 53, MSc, 57, PhD(agron), 61. Prof Exp: Instr agron, Agr Res Serv, USDA, Univ Nebr, 57-61, RES AGRONOMIST, AGR RES SERV, USDA, CORNELL UNIV, 61- Mem: AAAS; Weed Sci Soc Am; Soil Sci Soc Am; Am Soc Agron; Crop Sci Soc Am. Res: Absorption, translocation and degradation of herbicides; persistence of herbicides; vegetation management; plant protection. Mailing Add: Dept of Agron Cornell Univ Ithaca NY 14850

LINSCOTT, WILLIAM DEAN, b Bakersfield, Calif, Apr 23, 30; m 55; c 3. IMMUNOLOGY. Educ: Univ Calif, Los Angeles, BA, 51, PhD(infectious dis), 60. Prof Exp: Asst prof, 64-70, ASSOC PROF MICROBIOL, MED CTR, UNIV CALIF, SAN FRANCISCO, 70- Concurrent Pos: USPHS res fels, Labs Microbiol, Howard Hughes Med Inst, Fla, 60-62 & Div Exp Path, Scripps Clin & Res Found, Calif, 62-64. Res: Complement; immunologic unresponsiveness. Mailing Add: Dept of Microbiol Univ of Calif Med Ctr San Francisco CA 94143

LINSEY, CLARENCE WAYNE, chemistry, see 12th edition

LINSK, JACK, b New York, NY, May 20, 18; m 39; c 1. ORGANIC CHEMISTRY. Educ: City Col New York, BS, 40; Univ Iowa, MS, 41; Ohio State Univ, PhD(org chem), 48. Prof Exp: Chemist, Geo A Breon & Co, Mo, 41-45; asst chem, Ohio State Univ, 46-47; res chemist, Standard Oil Co, Ind, 48-63; tech specialist, Lockheed Propulsion Co, 63-71, RES SPECIALIST, LOCKHEED MISSILES & SPACE CO, 71- Mem: Am Chem Soc; Am Inst Aeronaut & Astronaut; Am Inst Chem. Res: Steroids; polynuclear hydrocarbons; petroleum chemistry; organic synthesis; solid propellants. Mailing Add: 1254 Klee Ct Sunnyvale CA 94087

LINSKY, CARY BRUCE, b Chicago, Ill, June 9, 42; m 68; c 2. BIOLOGICAL CHEMISTRY. Educ: Univ Wis-Madison, BS, 64; Loyola Univ, PhD(biochem), 71. Prof Exp: SR RES SCIENTIST BIOL, JOHNSON & JOHNSON, 71- Mem: Am Chem Soc; Sigma Xi. Res: Role of inflammatory response and local environment in cutaneous wound healing; cellular components of inflammation; scar formation in surgical wounds; collagen biochemistry. Mailing Add: Dept of Skin Biol Johnson & Johnson North Brunswick NJ 08902

LINSKY, JEFFREY L, b Buffalo, NY, June 27, 41; m 67; c 2. SPACE PHYSICS. Educ: Mass Inst Technol, BS, 63; Harvard Univ, AM, 65, PhD(astron), 68. Prof Exp: Res assoc astrophys, 68-69, LECTR, DEPT PHYSICS & ASTROPHYS & DEPT ASTROGEOPHYS, 69-, ASSOC PROF ADJOINT, DEPT ASTROGEOPHYS, UNIV COLO, 74-; ASTRONOMER, LAB ASTROPHYS, NAT BUR STANDARDS, 69- Concurrent Pos: Mem, Joint Inst Lab Astrophys, 68-71, fel, 71-; consult, NASA, 72- Mem: Am Astron Soc; Int Astron Union. Res: Radiative transfer; formation of spectral lines in the solar and stellar chromospheres; atmospheres of latetype stars. Mailing Add: Joint Inst for Lab Astrophys Univ of Colo Boulder CO 80302

LINSLEY, EARLE GORTON, b Oakland, Calif, May 1, 10; m 35; c 2. ENTOMOLOGY. Educ: Univ Calif, BS, 32, MS, 33, PhD(entom), 38. Prof Exp: Asst entom, Univ Calif, 33-35; agr, Univ Calif, Los Angeles, 35-37; assoc quarantine entomologist, State Dept Agr, Calif, 38-39; instr entom & jr entomologist, 39-43, asst prof & asst entomologist, 43-49, assoc prof & assoc entomologist, 49-53, chmn dept entom & parasitol, 51-59, asst dir, Agr Exp Sta, 60-63, assoc dir, 63-73, dean, Col Agr Sci, 60-73, PROF ENTOM & ENTOMOLOGIST, AGR EXP STA, UNIV CALIF, BERKELEY, 53- Concurrent Pos: Instr, Yosemite Sch Field Nat Hist, 38-41; res assoc, Calif Acad Sci, 39-; secy, Am Comn Entom Nomenclature, 43-48; ed, Pan-Pac Entomologist, 43-50; Guggenheim fel, Am Mus Natural Hist, 47-48; collabr, US Dept Interior, 53-54; res prof, Miller Inst Basic Res in Sci, 60-; mem comt insect pests, Nat Res Coun-Nat Acad Sci, 63-69; partic, Galapagos Int Sci Proj, 64; mem Orgn Trop Studies, 68- Mem: AAAS; Entom Soc Am (vpres, 46, 48, pres, 52); Ecol Soc Am; Am Soc Nat; Soc Syst Zool. Res: Systematic entomology; ecology and taxonomy of Coleoptera and Hymenoptera Apoidea; geographical distribution; mimicry and adaptive coloration; ethology of solitary bees; entomophagous Coleoptera; interrelations of flowers and insects; host specificity. Mailing Add: Div of Entom & Parasitol Univ of Calif Berkeley CA 94720

LINSLEY, ROBERT MARTIN, b Chicago, Ill, Feb 19, 30; m 54; c 3. INVERTEBRATE PALEONTOLOGY. Educ: Univ Mich, BS, 52, MS, 53, PhD(geol), 60. Prof Exp: Ford intern geol, 54-55, from instr to asst prof, 55-64, assoc prof & chmn dept, 64-71, dir natural sci course, 62-70, phys sci course, 59-64, PROF GEOL, COLGATE UNIV, 71- Concurrent Pos: Mem Paleont Res Inst. Mem: AAAS; Geol Soc Am; Paleont Soc; Soc Study Evolution. Res: Evolution, functional morphology; behavior and taxonomy of Gastropoda. Mailing Add: Dept of Geol Colgate Univ Hamilton NY 13346

LINSTER, RICHARD LEO, physics, see 12th edition

LINSTROMBERG, WALTER WILLIAM, b Beaufort, Mo, Oct 30, 12; m 43; c 2. ORGANIC CHEMISTRY. Educ: Univ Mo, AB, 37, MA, 50, PhD(chem), 55. Prof Exp: Instr chem, Univ Mo, 52-55; from asst prof to assoc prof, 55-60, PROF ORG CHEM, UNIV NEBR, OMAHA, 60- Concurrent Pos: Vis prof, Utah State Univ, 57; vis prof & res assoc, Univ Nebr, 60. Mem: Am Chem Soc. Res: Pharmaceutical chemistry. Mailing Add: Dept of Chem Univ of Nebr Omaha NE 68101

LINT, HAROLD L, b Lewiston, Idaho, Nov 27, 17; m 42; c 4. TAXONOMY. Educ: Univ Calif, Los Angeles, AB, 40, MA, 42. Prof Exp: Inspector, US Food & Drug Admin, 42-47; instr bot, 47-52, assoc prof biol sci, 52-62, PROF BIOL SCI, CALIF STATE POLYTECH UNIV, POMONA, 62- Mem: AAAS; Am Soc Plant Taxon; Am Inst Biol Sci; Int Asn Plant Taxon. Res: Genera Agastache and Juncus. Mailing Add: Dept of Biol Sci Calif State Polytech Univ Pomona CA 91766

LINTHICUM, BETTY LOUISE, plant physiology, see 12th edition

LINTNER, ANTHONY ETHELBERT, b Tolna, Hungary, Nov 27, 05; US citizen; m 40, c 2. CHEMISTRY. Educ: Inst Technol, Germany, Dipl, 30, DTech Sc, 33. Prof Exp: Chem engr, Filtex Co, Hungary, 33-38; tech dir, Teka Co, 39-45; res chemist, Sinnova Co, France, 47-48; chem consult, Argentina & Chile, 49-54; res chemist, E F Drew Co, Can, 55-56; res group leader, Calgon Corp, 57-70; CONSULT CHEMIST, 70- Mem: Am Chem Soc; fel Am Inst Chem; Am Asn Textile Chem & Colorists; Soc Cosmetic Chem. Res: Surfactants and textile chemistry. Mailing Add: 920 Berkshire Ave Pittsburgh PA 15226

LINTNER, CARL JOHN, JR, b Louisville, Ky, July 15, 17; m 48; c 2. PHARMACEUTICAL CHEMISTRY. Educ: Univ Ky, BS, 40; Univ Wis, PhD(pharmaceut chem), 50. Prof Exp: Res chemist, 50-55, SECT HEAD, UPJOHN CO, 55- Concurrent Pos: Adj prof, Fla A&M Univ, Col Pharm, 75- Honors & Awards: W E Upjohn Outstanding Achievement Award, The Upjohn Co, 60. Mem: AAAS; fel Acad Pharmaceut Sci; Am Pharmaceut Asn. Res: Organic synthesis; determination of functional groups; essential oil determination; kinetic studies; chromatography and ion exchange resins; phytochemistry; tablet coatings; ointment bases; chemistry of antibiotics; instrumentation tablet compression; pharmaceutical product stability. Mailing Add: 2125 Aberdeen Dr Kalamazoo MI 49008

LINTNER, MICHAEL ALAN, organic chemistry, see 12th edition

LINTON, EVERETT PERCIVAL, b St John West, NB, Dec 30, 06; m 36; c 3. PHYSICAL CHEMISTRY. Educ: Mt Allison Univ, BSc, 28; McGill Univ, MSc, 30, PhD(chem), 32. Prof Exp: Instr chem, Mt Allison Univ, 28-29; Royal Soc Can fel, Univ Munich, 32-33; chemist, Biol Bd Can, Halifax, NS, 34-36; instr chem, Acadia Univ, 36-41; asst phys chemist, Fisheries Res Bd, Halifax, 41-44; PROF CHEM, ACADIA UNIV, 44-, HEAD DEPT, 66- Mem: Am Chem Soc. Res: Preparation of hydrogen peroxide; measurement of dielectric constants; interaction of neutral molecules; air-drying solids; smokes; colloidal chemistry; dipole moments of amine oxides; drying and smoke curing of fish. Mailing Add: Dept of Chem Acadia Univ Wolfville NS Can

LINTON, FRED E J, b Italy, Apr 8, 38; US citizen; m 66. MATHEMATICS. Educ: Yale Univ BS, 58; Columbia Univ, MA, 59, PhD(math), 63. Prof Exp: From asst prof to assoc prof, 63-72, chmn dept, 75, PROF MATH, WESLEYAN UNIV, 72- Concurrent Pos: Res Coun res fel, Swiss Fed Inst Technol, 66-67; Izaak Walton Killam sr res fel, Dalhousie Univ, 69-70. Mem: Am Math Soc. Res: Categorical algebra, a branch of positive speculative philosophy. Mailing Add: Dept of Math Wesleyan Univ Middletown CT 06457

LINTON, HOWARD RICHARD, b Philadelphia, Pa, Nov 16, 30; m 56; c 4. INORGANIC CHEMISTRY. Educ: Western Md Col, BS, 54; Univ Pa, MS, 56, PhD(chem), 58. Prof Exp: Res chemist, 58-64, sr res chemist, 64-68, prod supvr qual control, 68-70, prod specialist-sales, 70-73, DEVELOP FEL, PIGMENTS PLANT, E I DU PONT DE NEMOURS & CO, INC, NJ, 73- Mem: Am Chem Soc. Res: Spectroscopic analysis; molecular structure; infrared, x-ray and Raman spectra; inorganic synthesis; nonaqueous chemistry. Mailing Add: Pigments Plant E I du Pont de Nemours & Co Edgemoor DE 19802

LINTON, JOE R, b Carbondale, Ill, July 4, 31; m 54; c 3. ZOOLOGY. Educ: Univ Mo, BA, 57, MA, 59, PhD(zool), 62. Prof Exp: NIH fel, Marine Lab, Univ Miami, 62-63; asst prof zool, 63-74, ASSOC PROF BIOL, UNIV S FLA, 74- Concurrent Pos: NIH grant, 64-66. Mem: AAAS; Am Soc Zool. Res: Animal physiology; endocrinology. Mailing Add: Dept of Biol Univ of SFla 4202 Fowler Ave Tampa FL 33620

LINTON, KENNETH JACK, b Muskegon, Mich, Dec 6, 35; m 57; c 3. AQUATIC ECOLOGY, BIOMETRY. Educ: Mich State Univ, BS, 60, MS, 64, PhD(limnol), 67. Prof Exp: Instr limnol, Mich State Univ, 66-67; ASSOC PROF BIOL, CLARION STATE COL, 67- Concurrent Pos: Consult, UNESCO, 72-74 & Westinghouse Environ Systs Dept, 72- Mem: AAAS; Am Fisheries Soc; Am Soc Limnol & Oceanog; Ecol Soc Am; Int Asn Theoret & Appl Limnol. Res: Application of population and community dynamics to water quality investigation. Mailing Add: Dept of Biol Clarion State Col Clarion PA 16214

LINTON, PATRICK HUGO, b Lineville, Ala, Sept 27, 25; m 52; c 4. PSYCHIATRY. Educ: Birmingham-Southern Col, AB, 49; Univ Ala, MD, 53. Prof Exp: Intern, US Naval Hosp, Jacksonville, Fla, 53-54; resident, Menninger Sch Psychiat, 54-56; staff psychiatrist, Vet Admin Hosp, Ft Lyon, Colo, 56-58; resident, Menninger Sch Psychiat, 58-59; actg chief neurol, Vet Admin Hosp, Topeka, Kans, 59-60; staff psychiatrist, Vet Admin Hosp, New Orleans, La, 6O-61; from asst prof psychiat to assoc prof, Sch Med, Univ Ala, Birmingham, 61-69, dir psychiat res & head div behav sci, 64-68, actg chmn dept psychiat, 68-69, ASSOC PROF DENT, SCH MED, UNIV ALA, BIRMINGHAM, 68-, PROF PSYCHIAT & CHMN DEPT, 69- Concurrent Pos: Actg chief psychiat serv, Vet Admin Hosp, Birmingham, Ala, 62-68; mem acad health affairs comt, Vet Admin Med Dist 14; mem, Jefferson, Blount, St Clair Ment Health Authority Bd. Mem: Am Med Asn; Am Psychiat Asn; Am Psychosom Soc. Mailing Add: Dept of Psychiatry Sch of Med Univ of Ala Birmingham AL 35294

LINTON, ROBERT WALTER, b Cape May, NJ, Dec 14, 26; m 60; c 4. INORGANIC CHEMISTRY. Educ: Pa Mil Col, BS, 50. Prof Exp: Res chemist, Pa Indust Chem Co, 51-52; asst to pres mkt res, Sindlinger Co, 52-55; res chemist, E J Houdry, 55-57; res chemist, Phila Quartz Co, 57-64, group leader silica & silicates, 64-66; res coord, 66-70, mgr lab tech serv, 70-73; MGR TECH DEVELOP & SERV, PITTSBURGH CORNING, 73- Mem: Am Chem Soc. Res: Properties and uses of synthetic silicas and soluble silicates; distillation and polymerization; market research; catalytic oxidation of hydrocarbons; insulation systems. Mailing Add: 800 Presque Isle Dr Pittsburgh PA 15239

LINTON, THOMAS LARUE, b Carlisle, Tex, July 25, 35; m 61; c 2. FISHERIES. Educ: Lamar State Col, BS, 59; Univ Okla, MS, 61; Univ Mich, PhD(fisheries), 66. Prof Exp: Res asst zool, Univ Ga, 63-65, res assoc, 65-67, asst prof, 67-70; mem staff, Div Com Sports Fisheries, NC Dept Conserv & Develop, 70-73, MEM STAFF, NC DEPT NATURAL & ECON RESOURCES, 73- Concurrent Pos: Res grants, Ga Game & Fish Comn, 65-68 & US Dept Interior, 66-; Ga rep biol comt, Atlantic State Marine Fisheries Comn, 64-66. Mem: AAAS; Am Fisheries Soc. Res: Physiology; commercial and sport fisheries; pollution ecology. Mailing Add: Dept of Natural & Econ Resources PO Box 27687 Raleigh NC 27611

LINTVEDT, RICHARD LOWELL, b Edgerton, Wis, June 23, 37; m 59; c 3. PHYSICAL INORGANIC CHEMISTRY. Educ: Lawrence Univ, BA, 59; Univ Nebr, PhD(inorg chem), 66. Prof Exp: Res chemist, Chem Div, Morton Int, 59-62; asst prof inorg chem, 66-71, ASSOC PROF CHEM, WAYNE STATE UNIV, 71- Concurrent Pos: Petrol Res Fund grant, 66-68, 70-73; Res Corp grant, 69-71. Mem: Am Chem Soc. Res: Electronic structure and bonding in inorganic coordination and chelate compounds; inorganic photochemistry; physical inorganic chemistry; magnetochemistry of transition metal complexes. Mailing Add: Dept of Chem Wayne State Univ Detroit MI 48202

LINTZ, JOSEPH, JR, b New York, NY, June 15, 21; m 44; c 3. GEOLOGY. Educ: Williams Col, AB, 42; Univ Okla, MS, 47; Johns Hopkins Univ, PhD(geol), 56. Prof Exp: Jr geologist, Gen Petrol Corp, 47-48; geologist, Pure Oil Co, 49; from asst prof to assoc prof, 51-65, PROF GEOL, MACKAY SCH MINES, UNIV NEV, RENO, 65- Concurrent Pos: Instr, Mt Lake Biol Sta, Va, 51; asst geologist, Nev Bur Mines, 51-56, assoc geologist, 56-; vis prof, Bandung Tech Inst, 59-61; Nat Acad Sci-Nat Res Coun res assoc, Manned Spacecraft Ctr, Tex, 66-67; consult, Econ Comm Asia & Far East, UN, 71 & 74; consult, Atomic Energy Comn, 71- Mem: Paleont Soc; Am Asn Petrol Geol. Res: Remote sensing of environment; petroleum possibilities and Pennsylvanian system of Nevada. Mailing Add: Mackay Sch of Mines Univ of Nev Reno NV 89507

LINZ, PETER, b Apatin, Jugoslavia, July 19, 36; US citizen. COMPUTER SCIENCE. Educ: McGill Univ, BSc, 57; Univ Mich, MS, 60; Univ Wis, PhD(comput sci), 68. Prof Exp: Res engr, Dominion Eng Ltd, 57-59; assoc programmer, IBM Corp, 63-65; staff specialist numerical anal, Comput Ctr, Univ Wis, 65-68; asst prof comput sci, NY Univ, 68-70; asst prof, 70-72, ASSOC PROF MATH, UNIV CALIF, DAVIS, 72- Res: Numerical analysis; quadrature methods; solution and applications of integral equations. Mailing Add: Dept of Math Univ of Calif Davis CA 95616

LINZER, MELVIN, b New York, NY, Aug 5, 37; m 64; c 3. PHYSICAL CHEMISTRY, ULTRASOUND. Educ: Brooklyn Col, BS, 57; Princeton Univ, MA, 59, PhD(chem), 62. Prof Exp: Res assoc chem, Princeton Univ, 61; Nat Acad Sci-Nat Res Coun fel, 61-63, PHYSICAL CHEMIST, NAT BUR STANDARDS, 63- Honors & Awards: Ross Coffin Purdy Award, Am Ceramic Soc, 75. Mem: Am Phys Soc; Am Ceramic Soc; Sigma Xi. Res: Nondestructive evaluation; ultrasound medical diagnosis; acoustic emission; magnetic resonance spectroscopy; measurement techniques for spectroscopic and materials applications; remote monitoring of laser pollutants by laser techniques; shock wave structure. Mailing Add: Inorg Chem Sect Nat Bur of Standards Div 313.01 Washington DC 20234

LINZER, ROSEMARY, b Chicago, Ill, Feb 11, 44. ORAL MICROBIOLOGY. Educ: Alverno Col, BA, 66; Northwestern Univ, MS, 71, PhD(molecular biol & biochem), 73. Prof Exp: Res technician, Biochem Dept, Northwestern Univ, 66-69, res assoc, Dept Microbiol, Med Sch, 73-75; ASST RES PROF ORAL BIOL, STATE UNIV NY BUFFALO, 75- Mem: Int Asn Dent Res; Am Soc Microbiol. Res: Oral microorganisms, specifically the purification and characterization of the serotype antigens of Streptococcus mutans with emphasis on their immunogenicity and vaccine potential. Mailing Add: Dept of Oral Biol State Univ of NY 4510 Main St Buffalo NY 14226

LINZEY, ALICIA VOGT, b Bloomingburg, NY, Jan 27, 43; m 63; c 2. MAMMALOGY. Educ: Cornell Univ, BS, 64, MS, 65. Prof Exp: Lab instr, Dept Ecol & Syst, Cornell Univ, 64-65, res assoc, 65-66; collabr mammal, Nat Park Serv, Gt Smoky Mountains Nat Park, 63-70; RES ASSOC VERT ZOOL & MAMMAL, UNIV S ALA, 68- Concurrent Pos: Ford Found fel, 65. Mem: Am Soc Mammalogists; Am Inst Biol Sci; Nat Audubon Soc. Res: Male reproductive anatomy and its use in the systematics of North American and Indo-Australian rodents; natural history and distribution of native mammals, especially of the Great Smoky Mountains National Park and Alabama. Mailing Add: Dept of Biol Sci Univ of SAla Mobile AL 36688

LINZEY, DONALD WAYNE, b Baltimore, Md, Sept 4, 39; m 63; c 2. WILDLIFE BIOLOGY, MAMMALOGY. Educ: Western Md Col, AB, 61; Cornell Univ, MS, 63, PhD(vert zool, mammal), 66. Prof Exp: Instr biol, Cornell Univ, 66-67; asst prof, 67-71, ASSOC PROF BIOL SCI, UNIV S ALA, 71-, CUR ZOOL, NATURAL HIST COLLECTIONS, 70- Mem: Am Soc Mammal; Wildlife Soc; Nat Audubon Soc; Am Inst Biol Sci; Soc Study Amphibians & Reptiles. Res: Life history; population dynamics; ecology; growth and development; systematics; state and regional surveys of mammal distribution. Mailing Add: Dept of Biol Sci Univ of S Ala Mobile AL 36688

LIONETTI, FABIAN JOSEPH, b Jersey City, NJ, Mar 3, 18; m 43; c 3. BIOCHEMISTRY. Educ: NY Univ, AB, 43, MS, 45; Rensselaer Polytech Inst, PhD(phys chem), 48. Prof Exp: From instr to assoc prof biochem, Sch Med, Boston Univ, 49-65; assoc mem, Inst Health Sci, Brown Univ, 65-68; SR INVESTR, CTR BLOOD RES, 68- Honors & Awards: Mathewson Medal, Am Inst Metals, 52. Mem: Am Soc Biol Chemists; Am Chem Soc; Cryobiol Soc. Res: Carbohydrate metabolism in human blood cells. Mailing Add: Ctr for Blood Res 800 Huntington Ave Boston MA 02115

LIOTTA, SILVESTER, physical chemistry, see 12th edition

LIOU, HORNG ING, b Hunan, China, June 21, 33; m 65; c 2. PHYSICS. Educ: Nat Taiwan Univ, BS, 60; Columbia Univ, MS, 67, PhD(physics), 68. Prof Exp: Res assoc physics, Nevis Labs, Columbia Univ, 68-74; MEM STAFF, PHYSICS DEPT, BROOKHAVEN NAT LAB, 74- Res: Neutron velocity spectroscopy. Mailing Add: Physics Dept Brookhaven Nat Lab Upton NY 11973

LIOU, JUHN G, b Taiwan, Dec 28, 39; m 65; c 2. GEOLOGY. Educ: Nat Twiwan Univ, BS, 62; Univ Calif, Los Angeles, PhD(geol), 70. Prof Exp: Teaching asst, Nat Taiwan Univ, 63-65; teaching asst, Univ Calif, Los Angeles, 65-69, NSF fel, 69-70; res assoc geochem, Manned Spacecraft Ctr, NASA, 70-72; ASST PROF GEOL, STANFORD UNIV, 72- Mem: Geol Soc Am; Am Geophys Union; Mineral Soc Am. Res: To understand, through hydrothermal experiments and field observations, the paragenesis of metamorphic minerals and their imposed physical conditions. Mailing Add: Dept of Geol Stanford Univ Stanford CA 94305

LIOY, FRANCO, b Gorizia, Italy, May 24, 32; m 66; c 1. PHYSIOLOGY. Educ: Univ Rome, MD, 56; Univ Minn, Minneapolis, PhD(physiol), 67. Prof Exp: Instr med, Univ Rome, 56-61; instr physiol, Univ Minn, Minneapolis, 66-67; from asst prof to assoc prof, 67-75, PROF PHYSIOL, UNIV BC, 75- Concurrent Pos: Mem, Sci Subcomt, Can Heart Found. Mem: Can Physiol Soc. Res: Cardiovascular physiology; coronary circulation; effect of hyperthermia on circulatory system; sympathetic control of circulation. Mailing Add: Dept of Physiol Univ of BC Vancouver BC Can

LIPA, JOHN A, b York, Eng, June 25, 43; Australian citizen. LOW TEMPERATURE PHYSICS. Educ: Univ Western Australia, BSc, 64, PhD(physics), 69. Prof Exp: Res fel, Univ Western Australia, 69; Commonwealth Sci & Indust Res Orgn overseas fel, 69-70, res assoc, 70-72, RES PHYSICIST, STANFORD UNIV, 72- Mem: Am Phys Soc. Res: Development of superconducting gyroscope for performing tests of general relativity, the the study of superfluid helium-3. Mailing Add: Dept of Physics Stanford Univ Stanford CA 94305

LIPARI, NUNZIO OTTAVIO, b Ali' Terme, Italy, Jan 1, 45. SOLID STATE PHYSICS, MOLECULAR PHYSICS. Educ: Univ Messina, Laurea Physics, 67; Lehigh Univ, PhD(physics), 70. Prof Exp: Res asst solid state physics, Lehigh Univ, 67-70; res assoc, Univ Ill, 70-72; from asst scientist to scientist physics, 72-75, SR SCIENTIST PHYSICS, WEBSTER RES LAB, XEROX CORP, 75- Mem: Am Phys Soc. Res: Optical properties of solids; electron-phonon interaction in molecular systems; excitation and impurity states in semiconductors. Mailing Add: Webster Res Lab Xerox Corp Webster NY 14580

LIPE, JOHN ARTHUR, b Los Fresnos, Tex, Aug 28, 43; m 64; c 2. HORTICULTURE, PLANT PHYSIOLOGY. Educ: Tex A&M Univ, BS, 65, MS, 68, PhD(plant physiol), 71. Prof Exp: ASST PROF HORT, TEX A&M UNIV RES & EXTEN CTR OVERTON, 71- Honors & Awards: Award, Am Soc Plant Physiol, 71. Mem: Am Soc Plant Physiol; Am Soc Hort Sci. Res: Plant growth regulation, especially with fruits; role of ethylene in fruit dehiscence. Mailing Add: Tex A&M Univ Res & Exten Ctr PO Drawer E Overton TX 75684

LIPELES, MARTIN, b New York, NY, June 22, 38; m 68; c 1. ATOMIC PHYSICS, ATMOSPHERIC CHEMISTRY. Educ: Columbia Univ, AB, 60, MA, 62, PhD(physics), 66. Prof Exp: Part-time res physicist, Radiation Lab, Columbia Univ, 62-66; MEM TECH STAFF, SCI CTR, ROCKWELL INT, 66- Mem: AAAS; Am Phys Soc; Am Chem Soc; Sigma Xi. Res: Inelastic, ion-atom collisions at low energies; physics and chemistry of photochemical aerosol formation in the atmosphere. Mailing Add: Sci Ctr Rockwell Int 1049 Camino dos Rios Thousand Oaks CA 91360

LIPETZ, JACQUES, cell biology, see 12th edition

LIPETZ, LEO ELIJAH, b Lincoln, Nebr, Aug 10, 21; m 47; c 3. BIOPHYSICS, NEUROSCIENCES. Educ: Cornell Univ, BEE, 42; Univ Calif, PhD(biophys), 53. Prof Exp: Jr elec engr, Radar Lab, Signal Corps, 42-43; mem tech staff, Bell Tel Labs, 43-36; asst physics, Univ Southern Calif, 46-47; instr biophys, Exten Div, Univ Calif, 48; from instr to asst prof ophthal, 54-60, asst prof physiol, 56-61, assoc prof biophys, 60-61, actg chmn div, 65-67, ASSOC PROF PHYSIOL, OHIO STATE UNIV, 61-, PROF BIOPHYS, 65-, CHMN DIV, 67-, CHMN DEPT, 71-, RES ASSOC, INST RES VISION, 65- Concurrent Pos: Nat Found fel, Johns Hopkins Univ, 53-54; mem, Biophys Sci Training Comt, Nat Inst Gen Med Sci, 68-70. Mem: AAAS; Biophys Soc; Am Physiol Soc; Asn Res Vision & Ophthal; Soc Neurosci. Res: Biophysics of the visual system; physical basis of behavior; electrophysiology; optics; information transfer; effect of high energy radiation. Mailing Add: Dept of Biophys Ohio State Univ Columbus OH 43210

LIPICKY, RAYMOND JOHN, b Cleveland, Ohio, May 3, 33; m 58. INTERNAL MEDICINE, PHARMACOLOGY. Educ: Ohio Univ, AB, 55; Univ Cincinnati, MD, 60. Prof Exp: From intern to resident med, Barnes Hosp, St Louis, Mo, 60-62; resident, Strong Mem Hosp, Rochester, NY, 64-65; from asst prof to assoc prof pharmacol, 66-72, from asst prof to assoc prof med, 66-73, PROF PHARMACOL, COL MED, UNIV CINCINNATI, 72-, PROF MED & DIR DIV CLIN PHARMACOL, 73- Concurrent Pos: Fel pharmacol, Univ Pa, 62-63 & Univ Cincinnati, 63-64; trainee cardiol, Strong Mem Hosp, Rochester, 65-66; mem corp, Marine Biol Lab, Woods Hole, Mass. Mem: Soc Neurosci; Biophys Soc; Am Physiol Soc; Am Soc Pharmacol & Exp Therapeut. Res: Ion transport; clinical pharmacology; membrane permeability; bioelectric potentials; hemodynamics. Mailing Add: Div of Clin Pharmacol Univ of Cincinnati Col of Med Cincinnati OH 45267

LIPIN, BRUCE REED, b New York, NY, Nov 27, 47; m 71; c 1. PETROLOGY. Educ: City Col New York. BS, 70; Pa State Univ, PhD(mineral, petrol), 75. Prof Exp: Fel, Geophys Lab, Carnegie Inst Washington, 73-74; Nat Res Coun res assoc fel, 74-75, GEOLOGIST, US GEOL SURV, 75- Mem: Mineral Soc Am. Res: Experimental study into the behavior of chrome-bearing minerals in silicate melts. Mailing Add: Stop 959 Nat Ctr US Geol Surv Reston VA 22092

LIPINSKI, CHRISTOPHER ANDREW, b Dundee, Scotland, Feb 1, 44; US citizen; m 69; c 2. ORGANIC CHEMISTRY. Educ: San Francisco State Col, BS, 65; Univ Calif, Berkeley, PhD(org chem), 68. Prof Exp: Nat Inst Gen Med Sci fel, Calif Inst Technol, 69-70; SR RES SCIENTIST, MED RES LABS, PFIZER PHARMACEUT, 71- Mem: Am Chem Soc. Res: Medicinal chemistry. Mailing Add: Med Res Labs Pfizer Pharmaceuticals Groton CT 06340

LIPKA, BENJAMIN, b New York, NY, Feb 5, 29; m 58; c 3. ORGANIC CHEMISTRY. Educ: NY Univ, BA, 49, PhD(org chem), 58. Prof Exp: Chemist, Geigy Chem Corp, 59-60; sr chemist, Allied Chem Corp, 60-67; SR DEVELOP CHEMIST, UPJOHN CO, 67- Mem: Am Chem Soc. Res: Laboratory synthesis and chemical plant production of organic compounds. Mailing Add: Upjohn Co 410 Sackett Point Rd North Haven CT 06473

LIPKE, HERBERT, b New York, NY, Jan 10, 23; m 48; c 4. BIOCHEMISTRY. Educ: Cornell Univ, BS, 47, MS, 48; Univ Ill, PhD(entom), 53. Prof Exp: Chemist, Calif Packing Corp, 48-50; asst entom, Univ Ill, 51-53, res assoc, 53-58; biochemist, Exp Zool Br, Physiol Div, Army Chem Ctr, 58-67; PROF BIOL, UNIV MASS, BOSTON, 67- Concurrent Pos: US Pub Health Serv fel, Univ Ill, 55; WHO fel, London Sch Hyg & Trop Med, 60-62. Mem: AAAS; Am Chem Soc; Am Physiol Soc; Am Soc Biol Chem; Brit Biochem Soc. Res: Biochemistry and physiology of invertebrates. Mailing Add: Dept of Biol Univ of Mass Harbor Campus Boston MA 02125

LIPKE, WILLIAM G, b Chesterton, Ind, Dec 19, 36; m 57; c 4. PLANT PHYSIOLOGY, PLANT BIOCHEMISTRY. Educ: Purdue Univ, BS, 59; Univ Nebr, MS, 62; Tex A&M Univ, PhD(plant physiol), 66. Prof Exp: From asst prof to assoc prof plant physiol, 65-74, ASSOC PROF BIOL, NORTHERN ARIZ UNIV, 74-, PLANT PHYSIOLOGIST, 65- Mem: AAAS; Am Soc Plant Physiol. Res: Plant physiology, especially mineral nutrition; weed science, especially plant enzymes. Mailing Add: Box 5640 Biol Sci Northern Ariz Univ Flagstaff AZ 86001

LIPKIN, DAVID, b Philadelphia, Pa, Jan 30, 13; m 42; c 4. ORGANIC CHEMISTRY. Educ: Univ Pa, BS, 34; Univ Calif, PhD(org chem), 39. Prof Exp: Petrol res chemist, Res & Develop Dept, Atlantic Ref Co, Pa, 34-36; res fel chem, Univ Calif, 39-42, chemist, 42-43; chemist, Manhattan Dist, Los Alamos, NMex, 43-46; assoc prof, 46-48, chmn dept, 64-70, PROF CHEM, WASH UNIV, 48- Concurrent Pos: Guggenheim fel, 55; trustee, Argonne Univs Asn, 69-71; vis res scientist, John Immes Inst, Norwich, England, 71. Honors & Awards: St Louis Award, Am Chem Soc, 70. Mem: Am Chem Soc. Res: Free radicals; organic phosphorus compounds; nucleic acids; electrochemical synthesis. Mailing Add: Dept of Chem Wash Univ St Louis MO 63130

LIPKIN, GEORGE, b New York, NY, Dec 31, 30; m 57; c 2. CELL BIOLOGY, DERMATOLOGY. Educ: Columbia Univ, AB, 52; State Univ NY Downstate Med Ctr, MD, 55. Prof Exp: From instr to assoc prof, 61-74, PROF DERMAT, MED SCH, NY UNIV, 74- Concurrent Pos: Nat Cancer Inst res grants, Dermat Found, Med Ctr, NY Univ, 67- Mem: Soc Invest Dermat; Am Fedn Clin Res; Am Acad Dermat; Harvey Soc; Am Asn Cancer Res. Res: Biology of malignant melanoma; biologic transformation of malignant cells. Mailing Add: Dept of Dermat NY Univ Med Ctr New York NY 10016

LIPKIN, HARRY JEANNOT, b New York, NY, June 16, 21; m 49; c 2. NUCLEAR PHYSICS, PARTICLE PHYSICS. Educ: Cornell Univ, BEE, 42; Princeton Univ, AM, 48, PhD(physics), 50. Prof Exp: Mem staff, Radiation Lab, Mass Inst Technol, 42-46; vis res fel reactor physics, AEC, France, 53-54; vis assoc prof physics, Univ Ill, 58-59; from assoc prof to prof, 59-71, HERBERT H LEHMAN CHAIR THEORET PHYSICS, WEIZMANN INST SCI, ISRAEL, 71- Concurrent Pos: Res physicist, Weizmann Inst Sci, 52-60, actg head dept physics, 60-61; consult, AEC, Israel, 55-58; vis lectr, Hebrew Univ, Israel, 56-58; vis prof, Univ Ill, 62-63 & Tel Aviv Univ, 65-66; vis prof, Princeton Univ, 67-68; vis scientist, Argonne Nat Lab & Nat Accelerator Lab, Ill, 71-72. Mem: Am Phys Soc; Ital Phys Soc; Phys Soc Israel; Europ Phys Soc; Israel Acad Sci & Humanities. Res: Elementary particle theory; theoretical and experimental nuclear structure; collective motion in many particle systems; Beta decay; Mössbauer effect; reactor and particle physics. Mailing Add: Dept of Nuclear Physics Weizmann Inst of Sci Rehovoth Israel

LIPKIN, LEWIS EDWARD, b New York, NY, Nov 2, 25; m 52; c 2. NEUROPATHOLOGY, COMPUTER SCIENCES. Educ: NY Univ, BA, 44; Long Island Col Med, MD, 49; Am Bd Path, dipl & cert anat path & neuropath, 52. Prof Exp: From intern med to resident path, Mt Sinai Hosp NY, 49-53; asst prof path & neuropath, State Univ NY Downstate Med Ctr, 56-62; HEAD NEUROPATH, PATH SECT, PERINATAL RES BR, NAT INST NEUROL DIS & STROKE, 62- Concurrent Pos: Asst pathologist, Kings County Hosp, 56-62; USPHS sr res fel neuropath, Mt Sinai Hosp NY, 55-56; USPHS res grant, 59-62; consult, Nat Inst Neurol Dis & Stroke, 61-62. Mem: Asn Res Nerv & Ment Dis; Int Acad Path; Am Asn Neuropath; Asn Comput Mach. Res: Computer analysis of microscopic images, especially neuropathologic material; analysis of development of central nervous system; automation of radioautography. Mailing Add: Path Sect Perinatal Res Br Nat Inst of Neurol Dis & Stroke Bethesda MD 20014

LIPKIN, MACK, b New York, NY, Mar 22, 07; m 36; c 3. INTERNAL MEDICINE, PSYCHIATRY. Educ: City Col New York, BS, 26, Cornell Univ, MD, 30; Am Bd Internal Med, dipl. Prof Exp: From intern to house physician, Univ Hosp, New York, 31-33; resident, Metrop Hosp, 33; resident, Rockland State Hosp, 33-34; from instr med to lectr psychiat, Columbia Univ, 36-52; clin asst prof med, 52-64, ASSOC PROF MED, MED COL, CORNELL UNIV, 64- Concurrent Pos: Asst attend physician, Univ Hosp, 34-46, chief div psychosomatic med, 39-46; dir, Group Health Ins, 38; asst vis physician, Goldwater Hosp, 44-46; med consult, Rehab Clin New York Hosp, 44-46; from asst attend physician to assoc attend physician, 52-; adj physician & dir, Psychosom Serv, Mt Sinai Hosp, 46-52; consult, Surgeon Gen, US Army, 48-55; sr consult, Vet Admin, 52-55; pres, Sergei Zlinkoff Found Med Educ Res, 56-; trustee, Riverside Res Inst, 67-75; vis prof, Dartmouth Med Sch, 68-69; vis prof, Med Sch, Univ Ore, 72-; trustee, NY Sch Psychiat, 74-; pres, Ruth W Dolen Found Med Educ & Res, 74- Mem: Am Col Physicians; Am Psychosom Soc; AMA; NY Acad Med. Res: Psychosomatic medicine; xiphoid syndrome; cardiovascular disease. Mailing Add: 450 Riverside Dr New York NY 10027

LIPKIN, MARTIN, b New York, NY, Apr 30, 26; m 58. GASTROENTEROLOGY, ONCOLOGY. Educ: NY Univ, AB, 46, MD, 50. Prof Exp: Instr physiol, Sch Med, Univ Pa, 53-54; from instr to asst prof, 58-64, ASSOC PROF MED, MED COL, CORNELL UNIV, 64-; ASSOC MEM, SLOAN-KETTERING INST CANCER RES, 72- Concurrent Pos: Fel physiol, Med Col, Cornell Univ, 52-53; USPHS res fel, 55-56; fel med, Med Col, Cornell Univ, 55-58; NIH res carrcareer prog award, 61-71; res collabr, Brookhaven Nat Lab, 58-72; dir gastroenterol res unit, Cornell Med Div, Bellevue Hosp, 58-68; guest investr, Rockefeller Inst, 59-60; assoc attend physician, New York Hosp, 70- & Mem Hosp, 71-; assoc prof, Grad Sch Med Sci, Cornell Univ, 71-; lectr, NY State Med Soc, 71, secy, Sect Gastroenterol & Colon & Rectal Surg, 71, vchmn, 72, chmn, 73. Mem: Fel Am Col Physicians; Am Soc Clin Invest; Am Physiol Soc; Am Soc Exp Path; Am Asn Cancer Res. Res: Proliferation and differentiation of premalignant and malignant gastrointestinal cells in man. Mailing Add: Sloan Kettering Cancer Ctr 1275 York Ave New York NY 10021

LIPMAN, HARRY JEROME, b Johnstown, Pa, Aug 4, 13; m 36; c 3. FOOD CHEMISTRY. Educ: Univ Pittsburgh, BS, 34, MS, 36, PhD(chem), 39. Prof Exp: Asst biol, Univ Pittsburgh, 34-39; bacteriologist, Hachmeister, Inc, Pa, 39-40; chemist, Ideal Soap & Chem Co, 40-43; food technologist, Closure Div, Anchor Hocking Glass Corp, Pa, 43-45; chief chemist, 45-52, dir res & develop, 52-67, VPRES, MALLET & CO, INC, 67- Mem: Am Chem Soc; Inst Food Technol; Am Oil Chem Soc. Res: Enzyme physiology; food preservation; stain technique; antisepsis; dry cleaning; development of food products. Mailing Add: 5731 Woodmont Ave Pittsburgh PA 15217

LIPMAN, JOSEPH, b Toronto, Ont, Can, June 15, 38; m 62; c 2. MATHEMATICS. Educ: Univ Toronto, BA, 60; Harvard Univ, MA, 61, PhD(math), 65. Prof Exp: Asst prof math, Queen's Univ, Ont, 65 & Purdue Univ, 66-67; vis asst prof, Columbia Univ, 67-68; from asst prof to assoc prof, 68-72, PROF MATH, PURDUE UNIV, W LAFAYETTE, 72- Mem: Am Math Soc; Can Math Cong. Res: Algebraic geometry. Mailing Add: Dept of Math Purdue Univ West Lafayette IN 47906

LIPMAN, PETER WALDMAN, b New York, NY, Apr 21, 35; m 62. GEOLOGY. Educ: Yale Univ, BS, 58; Stanford Univ, MS, 59, PhD(geol), 62. Prof Exp: GEOLOGIST, US GEOL SURV, 62- Concurrent Pos: NSF fel, Geol Inst, Tokyo, 64-65. Mem: Geol Soc Am; Mineral Soc Am; Res: Petrology and structural geology; volcanology, especially geology of calderas and related ash flows. Mailing Add: US Geol Surv Bldg 25 Denver Fed Ctr Denver CO 80225

LIPMAN, RONALD STEWART, psychopharmacology, research management, see 12th edition

LIPMANN, FRITZ (ALBERT), b Koenigsberg, Ger, June 12, 99; nat US; m 31; c 1. BIOCHEMISTRY. Educ: Univ Berlin, MD, 24, PhD(chem), 27. Hon Degrees: MD, Univ Aix-Marseille, 47; ScD, Univ Chicago, 53; DHL, Brandeis Univ, 59 & Albert Einstein Col Med, 64; DSc, Univ Paris, 66, Harvard Univ, 67 & Rockefeller Univ, 71. Prof Exp: Asst intermediary metab, Kaiser Wilhelm Inst Biol, Berlin & Heidelberg, 27-31; res assoc tissue cult & intermediary metab, Biol Inst Carlsberg Found, Copenhagen, 32-39; res assoc biochem, Med Col, Cornell Univ, 39-41; head biochem res lab & prof biol chem, Harvard Univ & Mass Gen Hosp, 41-57; PROF BIOCHEM, ROCKEFELLER UNIV, 57- Concurrent Pos: Rockefeller Found fel, Rockefeller Inst, 31-32; assoc, Harvard Med Sch, 46-49, prof, 49-57. Honors & Awards: Nobel Prize in Med & Physiol, 53; Carl Neuberg Medal, 48; Mead Johnson & Co Award, 48; Nat Medal Sci, 66. Mem: Nat Acad Sci; Am Soc Biol Chemists (pres, 60-61); Harvey Soc; Am Philos Soc; fel Royal Danish Acad. Res: Energy metabolism; sulfate activation and protein synthesis; discovery and identification of coenzyme A. Mailing Add: Rockefeller Univ New York NY 10021

LIPNER, HARRY JOEL, b New York, NY, Aug 26, 22; m 49; c 4. REPRODUCTIVE ENDOCRINOLOGY. Educ: Long Island Univ, BS, 42; Univ Chicago, MS, 47; Univ Iowa, PhD(physiol), 52. Prof Exp: Res assoc thyroid iodine trap, Univ Iowa, 52; Nat Cancer Inst res fel thyroid physiol, 52-54; instr clin path, Chicago Med Sch, 54-55; from asst prof to assoc prof, 55-65, PROF PHYSIOL, FLA STATE UNIV, 65- Concurrent Pos: NIH fel & vis prof dept anat, Harvard Med Sch, 69-70; Fulbright vis prof, Ctr Advan Biochem, Indian Inst Sci, Bangalore, India, 74-75. Mem: AAAS; Endocrine Soc; Am Soc Zool; Soc Study Reproduction; Am Physiol Soc. Res: Mechanism of ovulation. Mailing Add: Dept of Biol Sci Fla State Univ Tallahassee FL 32306

LIPNICK, ROBERT LOUIS, chemistry, see 12th edition

LIPOWITZ, JONATHAN, b Paterson, NJ, Apr 25, 37; m 60; c 2. ORGANOMETALLIC CHEMISTRY, POLYMER CHEMISTRY. Educ: Rutgers Univ, Newark, BS, 58; Univ Pittsburgh, PhD(chem), 64. Prof Exp: Fel, Pa State Univ, University Park, 64-65; res chemist, Res Labs, 65-74, sr proj chemist, 74-75, ASSOC SCIENTIST, RES DEPT, DOW CORNING CORP, 75- Mem: Am Chem Soc; Sigma Xi; Am Soc Testing & Mat. Res: Silicone flammability, mechanisms of; effects of silicones on flammability of organic polymers; silicone condensation polymerization. Mailing Add: Silicone Res Dept Dow Corning Corp Midland MI 48640

LIPOWSKI, STANLEY ARTHUR, b Warsaw, Poland, Sept 16, 05; US citizen, m 45; c 1. ORGANIC POLYMER CHEMISTRY. Educ: Warsaw Tech Univ, MS, 29; Danzig Tech Univ, PhD(chem), 31. Prof Exp: Chemist, M D Lipowski Tanning Co, Poland, 31-35; chief chemist, J B Rakower Tanning Co, 36-43; tech dir, State Tannery, 45-46; tech dir Thor Schriwer Co, Norway, 47-49; res chemist, Peter Moeller Chem Co, 50-52, div mgr, 53-54; res chemist, B D Eisendrath Co, Wis, 54-55; res chemist, Nopco Chem Co, 55-60, sr res chemist, 61-69, dir resin & polymer res & develop, Nopco Div, Diamond Shamrock Corp, 69-71; CONSULT CHEMIST & ENGR, 71- Concurrent Pos: Mem bd dirs, Nat Youth Sci Found, 65. Mem: Am Leather Chem Asn; sr mem Am Chem Soc. Res: Organic synthesis; synthetic tanning agents; surfactants; sulfanation processes; phenolic and aminoplast resins; oils and fats chemistry; complexing agents; chelation; water soluble organic polymers. Mailing Add: 25 Ashwood Dr Livingston NJ 07039

LIPOWSKI, ZBIGNIEW J, b Warsaw, Poland, Oct 26, 24; Can citizen; m 46; c 2. PSYCHIATRY. Educ: Nat Univ Ireland, MB, BCh & BAO, 53; McGill Univ, dipl, 59. Prof Exp: Demonstr psychiat, McGill Univ, 59-62, lectr, 62-65, from asst prof to assoc prof, 65-71; PROF PSYCHIAT, DARTMOUTH MED SCH, 71- Concurrent Pos: Res fel psychophysiol, Allan Mem Inst Psychiat, 57-58; Mona Bronfman Shenkman teaching fel psychiat, Harvard Univ & Mass Gen Hosp, 58-59; clin asst, Allan Mem Inst Psychiat, 59-62, from asst psychiatrist to psychiatrist, 62-71; consult psychiat, Montreal Neurol Inst, 68-71; psychiatrist, Mary Hitchcock Mem Hosp, Hanover, NH, 71- Mem: Fel Am Psychiat Asn; Am Psychosom Soc; Acad Psychosom Med; Int Col Psychosom Med; Can Med Asn. Res: Psychosomatic medicine and psychopathology related to physical illness. Mailing Add: Dept of Psychiat Dartmouth Med Sch Hanover NH 03755

LIPP, HAYDEN IVAN, b Washington, Pa, Jan 16, 43; m 63; c 2. POLYMER CHEMISTRY. Educ: Carnegie-Mellon Univ, BS, 64; Univ Ill, Urbana, MS, 66, PhD(org chem), 68. Prof Exp: RES CHEMIST, PLASTICS DEPT, E I DU PONT DE NEMOURS & CO, INC, 68-69, 71- Res: Synthesis, evaluation, and process development of engineering thermoplastics. Mailing Add: Plastics Dept E I du Pont de Nemours & Co Inc Parkersburg WV 26101

LIPP, STEVEN ALAN, b Brooklyn, NY, Jan 25, 44. INORGANIC CHEMISTRY. Educ: Brooklyn Col, BS, 65; Univ Calif, Berkeley, PhD(inorg chem), 70. Prof Exp: MEM TECH STAFF, RCA LABS, DAVID SARNOFF RES CTR, 70- Honors & Awards: Achievement Award, RCA Labs, 74, David Sarnoff Award, RCA Corp, 75. Mem: Electrochem Soc. Res: Preparation and evaluation of new cathodoluminescent materials, as well as the design and testing of enhancements for chemical milling. Mailing Add: RCA Labs David Sarnoff Res Ctr Princeton NJ 08540

LIPPA, ERIK ALEXANDER, b Minneapolis, Minn, Nov 7, 45; m 74. NUMBER THEORY. Educ: Calif Inst Technol, BS, 67; Univ Mich, MS, 68, PhD(math), 71. Prof Exp: NATO fel math, Oxford Univ, 71-72; ASST PROF MATH, PURDUE UNIV, WEST LAFAYETTE, 72- Mem: Math Asn Am; Sigma Xi. Res: Analytic number theory, specifically Siegel modular forms of several complex variables and their associated Dirichlet series. Mailing Add: Dept of Math Purdue Univ West Lafayette IN 47907

LIPPARD, STEPHEN J, b Pittsburgh, Pa, Oct 12, 40; m 64; c 2. INORGANIC CHEMISTRY, BIOPHYSICAL CHEMISTRY. Educ: Haverford Col, BA, 62; Mass Inst Technol, PhD(chem), 65. Prof Exp: NSF fel, Mass Inst Technol, 65-66; from asst prof to assoc prof, 66-72, PROF CHEM, COLUMBIA UNIV, 72- Concurrent Pos: Consult, Esso Res & Eng Co, 67-; Alfred P Sloan Found fel, 68-70; John Simon Guggenheim Mem fel, Sweden, 72. Honors & Awards: Camille & Henry Dreyfus Teacher-Scholar Award, 72. Mem: AAAS; Am Chem Soc; Am Crystallog Asn; Biophys Soc. Res: Inorganic and organometallic coordination chemistry, especially preparation, structural properties and dynamic processes of transition metal complexes; structure and function of metal ions in biology; heavy metal probes of macromolecular structure. Mailing Add: Dept of Chem Columbia Univ New York NY 10027

LIPPARD, VERNON WILLIAM, b Marlboro, Mass, Oct 4, 05; m 31; c 1. PEDIATRICS. Educ: Yale Univ, BS, 26, MD, 29. Hon Degrees: ScD, Univ Md, 55. Prof Exp: Intern pediat, New Haven Hosp, Conn, 29-30; asst res pediatrician, NY Nursery & Child's Hosp, New York, 30-31, res pediatrician, 31-32; res pediatrician, NY Hosp, 32-33; instr & assoc pediat, Med Col, Cornell Univ, 33-38; dir, Cmn Study Crippled Children, 38-39; assoc dean, Col Physicians & Surgeons, Columbia Univ, 39-46; prof pediat & dean, Sch Med, La State Univ, 46-49 & Sch Med, Univ Va, 49-52; prof, 53-71, dean 52-67, asst to pres med develop, 67-71, EMER PROF PEDIAT, SCH MED, YALE UNIV, 71-, EMER DEAN, 67- Concurrent Pos: Mem bd med consult, Oak Ridge Inst Nuclear Studies, 47-52, Brookhaven Nat Lab, 57-62 & Josiah Macy Jr Found, 67-75; mem bd dirs, Grant Found, 67-; med dir, Nat Fund Med Educ, 71-75. Mem: Soc Pediat Res; Asn Am Med Cols (pres, 54-55); Asn Am Physicians. Res: Acid-base balance and immunology in infancy; development of hypersensitiveness; medical education. Mailing Add: Sterling Hall of Med Yale Univ 333 Cedar St New Haven CT 06510

LIPPE, ROBERT LLOYD, b New York, NY, May 8, 23; m; c 2. CHEMISTRY. Educ: Yale Univ, BE, 42; Princeton Univ, MS, 47. Prof Exp: Eng trainee, Joseph E Seagram & Sons, Inc, Md, 43; jr scientist in charge radiographic res, Manhattan Dist, 45-46; chemist, Standard Varnish Works, 47-51; asst to vpres, Standard Tech Chems, Inc, 51-54; PRES, NOD HILL CHEM CORP, 53-; PRES, ARGUS PAINT & LACQUER CORP, 54- Mem: Am Chem Soc; Am Inst Chem Eng; Fedn Am Sci. Res: Radiography of explosives; organotitanium compounds; synthetic and natural resins for coatings; industrial organic coatings. Mailing Add: 310 Greenpoint Ave Brooklyn NY 11222

LIPPEL, KENNETH, b New York, NY, Feb 21, 29; m 61; c 1. BIOCHEMISTRY. Educ: City Col New York, BS, 49, MBA, 60; Univ Fla, PhD(biochem), 66. Prof Exp: NIH fel biochem, Univ Calif, Los Angeles, 66-68; asst prof dermat & biochem, Sch Med, Univ Miami, 68-70; res biochemist, Lipids Br, Human Nutrit Res Div, Agr Res Serv, USDA, 70-72; HEALTH SCI ADMINR, LIPID METAB BR, DIV HEART & VASCULAR DIS, NAT HEART & LUNG INST, NIH, 72- Concurrent Pos: Fel, Coun Arteriosclerosis, Am Heart Asn, 75- Mem: AAAS; Am Soc Biol Chem; Am Heart Asn; Am Chem Soc; Sigma Xi. Res: Regulation of fatty acid and lipid metabolism; relationship of lipoprotein metabolism to atherosclerosis; vitamin A metabolism. Mailing Add: 116 Beaumont Rd Silver Spring MD 20904

LIPPERT, ARNOLD LEROY, b Kewanee, Ill, Nov 12, 10; m 36; c 3. ORGANIC CHEMISTRY. Educ: Univ Ill, BS, 31; Johns Hopkins Univ, PhD(chem), 34. Prof Exp: Org res chemist, Exp Sta, E I du Pont de Nemours & Co, 34-35; tech asst to gen supt, Joseph Bancroft & Sons Co, 35-36, chem dir, 36-41; consult dyes, Off Prod Mgt, DC, 41-42; chief dyestuffs sect, War Prod Bd, 42-44; chem dir, Joseph Bancroft & Sons Co, 45-47, vpres, 48-63, pres, 64-65; intern, 65-66, dean col arts & sci, 66-71, DEAN, COL GRAD STUDIES, UNIV DEL, 71-, PROF TEXTILE CHEM, 73- Mem: Am Chem Soc; Am Asn Textile Chem & Colorists. Res: Textile products; commerical chemistry; cyclic diamides. Mailing Add: Col Grad Studies Univ of Del Newark DE 19711

LIPPERT, BYRON E, b Los Angeles, Calif, July 1, 29; m 52; c 3. PHYCOLOGY. Educ: Univ Ore, BS, 54, MS, 57; Ind Univ, PhD(bot), 66. Prof Exp: Instr biol, Eastern Ore Col, 56-59; asst prof, 60-69, ASSOC PROF BIOL, PORTLAND STATE UNIV, 69- Mem: Phycol Soc Am; Brit Phycol Soc; Int Phycol Soc; Bot Soc Am. Res: Morphology, life cycles and sexual reproduction in desmids. Mailing Add: Dept of Biol Portland State Univ Portland OR 97207

LIPPERT, ERNEST LAVERNE, JR, physical chemistry, see 12th edition

LIPPERT, LAVERNE FRANCIS, b Deerlodge, Mont, Sept 21, 28; m 50; c 5. PLANT PATHOLOGY. Educ: State Col Wash, BS, 50; Univ Calif, Davis, PhD(plant path), 59. Prof Exp: Asst plant path, Univ Calif, Davis, 55-58; from asst olericulturist to assoc olericulturist, 58-72, PROF VEG CROPS & OLERICULTURIST, UNIV CALIF, RIVERSIDE, 72-, VCHMN DEPT PLANT SCI, 75- Mem: Am Genetic Soc; Am Hort Soc. Res: Vegetable crops breeding, especially peppers and melons. Mailing Add: Dept of Plant Sci Univ of Calif Riverside CA 92502

LIPPINCOTT, BARBARA BARNES, b Raleigh, Ill, Oct 27, 34; m 56; c 3. MICROBIOLOGY, PLANT PHYSIOLOGY. Educ: Wash Univ, St Louis, AB, 55, MA, 57, PhD(zool), 59. Prof Exp: Jane Coffin Childs Mem Fund Med Res fel physiol genetics, Lab Physiol Genetics, Nat Ctr Sci Res, France, 59-60; RES ASSOC BIOL SCI, NORTHWESTERN UNIV, 60- Concurrent Pos: Vis scholar, Univ Calif, Berkeley, 70-71; vis scientist, Inst Bot, Univ Heidelberg, 74. Mem: Am Soc Microbiol; Sigma Xi. Res: Electron spin resonance in biological systems; crown-gall tumor formation; control mechanisms in replication, growth and development. Mailing Add: Dept of Biol Sci Northwestern Univ Evanston IL 60201

LIPPINCOTT, ELLIS RIDGEWAY, chemistry, see 12th edition

LIPPINCOTT, JAMES ANDREW, b Cumberland Co, Ill, Sept 13, 30; m 56; c 3. PLANT PHYSIOLOGY. Educ: Earlham Col, AB, 54; Wash Univ, AM, 56, PhD, 58. Prof Exp: Res assoc plant physiol & lectr bot, Wash Univ, 58-59; Jane Coffin Childs Mem Fund Med Res fel, Lab Phytotron, Nat Ctr Sci Res, France, 59-60; from asst prof to assoc prof, 60-73, PROF BIOL SCI, NORTHWESTERN UNIV, 73- Concurrent Pos: Vis assoc prof, Univ Calif, Berkeley, 70-71; vis prof, Univ Heidelberg, 74. Mem: AAAS; Am Soc Plant Physiol; Bot Soc Am; Am Soc Microbiol; Am Phytopath Soc. Res: Crown-gall tumor formation control mechanisms in replication, growth and development; tumor induction in plants by Agrobacterium tumefaciens. Mailing Add: Dept of Biol Sci Northwestern Univ Evanston IL 60201

LIPPINCOTT, SARAH LEE, b Philadelphia, Pa, Oct 26, 20. ASTRONOMY. Educ: Univ Pa, BA, 42; Swarthmore Col, MA, 50. Hon Degrees: DSc, Villanova Univ, 73. Prof Exp: Res asst astron, 42-51, res assoc, 52-72, LECTR, SPROUL OBSERV, SWARTHMORE COL, 61-, DIR, 72- Concurrent Pos: Mem Fr solar eclipse exped to Oland, Sweden, 54; partic vis prof prog, NSF-Am Astron Soc, 61-; vpres comn 26, Int Astron Union, 70-73, pres, 73-76. Mem: Am Astron Soc. Res: Parallaxes of nearby stars; double stars; stellar masses; chromosphere studies; spicules. Mailing Add: Sproul Observ Swarthmore Col Swarthmore PA 19081

LIPPINCOTT, STUART WELLINGTON, b Worcester, Mass, Nov 6, 07. PATHOLOGY. Educ: Clark Univ, AB, 29; McGill Univ, MD, 35. Prof Exp: Asst bact & physiol, Mass State Col, 29-30; asst demonstr, Path Inst, McGill Univ, 35-36; asst instr, Univ Pa & res pathologist, Univ Hosp, 37-38; demonstr path & lectr prosector, Med Sch, McGill Univ & Royal Victoria Hosp, Montreal, 38-39; Nat Cancer Inst res fel, 40-42; prof path, Sch Med, Univ Wash, 46-55; sr pathologist, Res Hosp & sr scientist, Med Res Ctr, Brookhaven Nat Lab, 55-63; prof path, Bowman Gray Sch Med, 63-65; PROF RADIOL & CHMN DEPT, MED COL VA, VA COMMONWEALTH UNIV, 65- Mem: Soc Exp Biol & Med; Am Asn Path & Bact; Am Asn Cancer Res; AMA. Res: Experimental pathology; metabolism of spontaneous mammary tumors and ultraviolet irradiation in mice; dietary production of gastric and hepatic lesions in rats; tropical diseases; serum proteins in cancer; pathology of particle radiation; turnover of labeled human serum protein fractions in neoplastic diseases. Mailing Add: Dept Radiol Radiation Biol Div Med Col of Va Health Sci Ctr Richmond VA 23219

LIPPINCOTT, WILLIAM THOMAS, physical organic chemistry, see 12th edition

LIPPITT, LOUIS, b New York, NY, Mar 19, 24; m 48; c 4. GEOPHYSICS. Educ: City Col New York, BS, 47; Columbia Univ, MA, 53, PhD(geol), 59. Prof Exp: Physicist, Columbia Univ, 47-50 & NY Univ, 51-53; geologist, Standard Oil Co Calif, 54-58; GEOPHYSICIST & RES STAFF ENGR, LOCKHEED MISSILES & SPACE CO, 58- Concurrent Pos: Instr, Allan Hancock Col, 69- Mem: Geol Soc Am; Am Geophys Union; Soc Explor Geophys. Res: Satellite systems; geophysical exploration. Mailing Add: Lockheed Missiles & Space Co Box 1506 Vandenberg AFB CA 93437

LIPPKE, HAGEN, b Yorktown, Tex, Nov 4, 36; m 58; c 3. ANIMAL NUTRITION. Educ: Tex A&M Univ, BS, 59, MS, 61; Iowa State Univ, PhD(animal nutrit), 66. Prof Exp: From asst prof to assoc prof ruminant nutrit, 66-74, ASSOC PROF ANIMAL SCI, TEX A&M UNIV, 74- Mem: Am Dairy Sci Asn; Am Soc Animal Sci. Res: Ruminant nutrition; forage utilization by cattle: forage characteristics influencing intake and digestibility. Mailing Add: Dept of Animal Sci Tex A&M Univ College Station TX 77843

LIPPMAN, ALFRED ERIC, organic chemistry, see 12th edition

LIPPMAN, GARY EDWIN, b Little Rock, Ark. MATHEMATICAL ANALYSIS. Educ: San Jose State Col, BA, 63; Univ Calif, Riverside, MA, 65, PhD(math), 70. Prof Exp: Asst prof math, Kenyon Col, 70-71; ASST PROF MATH, CALIF STATE UNIV, HAYWARD, 71- Concurrent Pos: Vis asst prof math, Univ Tenn, 72. Mem: Am Math Soc; Math Asn Am; Comput Law Soc. Res: Fourier analysis. Mailing Add: Dept of Math Calif State Univ Hayward CA 94542

LIPPMAN, MARC ESTES, b New York, NY, Jan 15, 45; m 66; c 2. ENDOCRINOLOGY, ONCOLOGY. Educ: Cornell Univ, BA, 64; Yale Univ, MD, 68. Prof Exp: House officer med, Johns Hopkins Univ, 68-70; clin assoc oncol, Nat Cancer Inst, NIH, 70-71, res assoc biochem, 71-73; fel endocrinol dep med, Med Sch, Yale Univ, 73-74; SR INVESTR ONCOL MED BR, NAT CANCER INST, NIH, 74- Concurrent Pos: Mem steering comt, Breast Cancer Task Force, 75- Mem: Endocrine Soc; Am Diabetes Asn; Am Fedn Clin Res; Tissue Cult Asn; Am Asn Cancer Res. Res: Regulation of gene expression in cells in culture by peptide and steroid hormones. Mailing Add: Nat Cancer Inst NIH Bethesda MD 20014

LIPPMANN, BERNARD ABRAM, b New York, NY, Aug 18, 14; m 38; c 1. QUANTUM MECHANICS. Educ: Polytech Inst Brooklyn, BEE, 34; Univ Mich, MS, 35; Harvard Univ, PhD(physics), 48. Prof Exp: Mem staff, Radiation Lab, Mass Inst Technol, 43-46; group leader, Submarine Sig Co, Mass, 46-48; group leader, Nat Bur Standards, 48; physicist, Naval Res Lab, 48-52; physicist, Nuclear Develop Assocs, Inc, 52-53, tech res group, 53-54; inst math sci, NY Univ, 54-57; physicist, Lawrence Radiation Lab, Univ Calif, 57-62 & Gen Res Corp, Calif, 62-69; chmn dept, 69-70, PROF PHYSICS, NY UNIV, 69- Concurrent Pos: Mem fac, Grad Sch, Univ Md, 48-52 & Univ Calif, Santa Barbara, 62-64. Mem: Fel Am Phys Soc. Res: Theory of scattering and reaction processes; laser fusion; laser isotope separation. Mailing Add: Physics Dept New York Univ 4 Washington Pl New York NY 10003

LIPPMANN, DAVID ZANGWILL, b Houston, Tex, July 6, 25; m 49. PHYSICAL CHEMISTRY. Educ: Univ Tex, BSc, 47, MA, 49; Univ Calif, Berkeley, PhD(phys chem), 53. Prof Exp: Chemist, Reaction Motors, Inc, 54-57, Fulton-Irgon Div, Lithium Corp Am, 57-61 & Proteus, Inc, 61-63; asst prof, 63-69, ASSOC PROF CHEM, SOUTHWEST TEX STATE UNIV, 69- Concurrent Pos: Consult, Proteus, Inc. Mem: Am Chem Soc. Res: Theoretical physical chemistry, especially thermodynamics and statistical mechanics; rocketry and ballistics. Mailing Add: Dept of Chem Southwest Tex State Univ San Marcos TX 78666

LIPPMANN, HEINZ ISRAEL, b Breslau, Ger, May 21, 08; nat US; m 36; c 3. MEDICINE. Educ: Univ Freiburg, BA, 26; Univ Berlin, MD, 31; Univ Genoa, MD, 33. Prof Exp: From asst prof to assoc prof, 55-68, PROF REHAB MED, ALBERT EINSTEIN COL MED, 69- Concurrent Pos: Assoc attend physician, Montefiore Hosp, Bronx, 44-; lectr, Columbia Univ, 46-62; consult, Workman's Circle Home for Aged, 51-; chief peripheral vascular clin, Sydney Hillman Health Ctr, 51-67; chief peripheral vascular clin & vis physician, Bronx Munic Hosp Ctr, 56-; chief attend physician, Brace Clin, 57-67, chief peripheral vascular clin, 57-, dir amputee ctr, 61-; consult peripheral vascular dis, Englewood Hosp, NJ, 67- & Vet Admin Hosp, East Orange, NJ; dir, Cardiac Stress Unit, Jewish Hosp & Rehab Ctr, Jersey City, 74-; attend physician & chief, Rehab Med Dept, Barnert Mem Hosp, Paterson, NJ, 75- Honors & Awards: NAm Roentgen Soc Award, 58; Gold Medal Sci Exhibit, Am Cong Rehab Med, 59. Mem: Am Heart Asn; fel Am Col Physicians; Am Cong Rehab Med; NY Acad Sci; Am Acad Phys Med & Rehab. Res: Vascular physiology; peripheral vascular diseases; prosthetics; rheumatology. Mailing Add: Dept of Med Albert Einstein Col of Med New York NY 10461

LIPPMANN, IRWIN, b New York, NY, Dec 30, 30; m 56; c 3. PHARMACEUTICAL CHEMISTRY. Educ: Rutgers Univ, BS, 52, MS, 56; Univ Mich, PhD(pharmaceut chem), 60. Prof Exp: Instr anal chem, Rutgers Univ, 55-56; Am Found Pharmaceut Educ fel, 56-59; sr res pharmacist, Squibb Inst Med Res, 59-62; ASST PROF PHARM, MED COL VA, 64-; res assoc, 62-74, ASSOC DIR BIOPHARMACEUT RES, A H ROBINS CO, INC, 74- Mem: Am Pharmaceut Asn; Am Chem Soc. Res: Biopharmaceutics-pharmacokinetics; in vitro—in vivo correlations; sustained-action drugs; physical pharmacy. Mailing Add: A H Robins Co Inc 1211 Sherwood Ave Richmond VA 23220

LIPPMANN, MORTON, b Brooklyn, NY, Sept 21, 32; m 56; c 3. ENVIRONMENTAL HEALTH. Educ: Cooper Union, BChE, 54; Harvard Univ, SM, 55; NY Univ, PhD(indust hyg), 67. Prof Exp: Asst sanit engr, USPHS, Ohio, 55-57; indust hygienist, US AEC, NY, 57-62; sr res engr, Del Electronics Corp, 62-64; assoc res scientist aerosol physiol, 64-67, asst prof, 67-70, ASSOC PROF AEROSOL PHYSIOL & DIR AEROSOL RES LAB, INST ENVIRON MED, NY UNIV, 70- Mem: Am Conf Govt Indust Hygienists; Am Indust Hyg Asn; Air Pollution Control Asn. Res: Environmental hygiene; regional deposition and bronchial clearance of inhaled particles; sampling and analysis of atmospheric particles; aerodynamic behavior of respirate aerosols. Mailing Add: NY Univ Inst of Environ Med 550 First Ave New York NY 10016

LIPPMANN, SEYMOUR A, b Brooklyn, NY, Nov 23, 19; m 45; c 3. APPLIED PHYSICS. Educ: Cooper Union, BChE, 42. Prof Exp: Group leader appl physics, Res Dept, US Rubber Co, 47-60, dept mgr phys res, 60-71, res assoc, 71-75, MGR TIRE-VEHICLE SYSTEMS LABS, UNIROYAL TIRE COMPANY, UNIROYAL INC, 75- Mem: Am Phys Soc; Soc Automotive Eng; Inst Elec & Electronics Eng; Am Soc Testing & Mat; Sigma Xi. Res: Physics of polymeric materials; transmission of noise and vibrations; design of electronic instrumentation for the study of dynamic systems and properties; perception of sound in the presence of background noise; dynamics of the human as a link in control systems. Mailing Add: 12767 Lincoln Huntington Woods MI 48070

LIPPMANN, WILBUR, b Galveston, Tex, Sept 6, 30. BIOCMEMICAL PHARMACOLOGY. Educ: Tex A&M Col, BS, 51; Univ Tex, MA, 56, PhD(biochem), 61. Prof Exp: Res biochemist, Biochem Inst, Univ Tex, 54-56, 58-62; res biochemist, Virus Inst, Univ Calif, Berkeley, 56-58; res biochemist, Univ Tex M D Anderson Hosp & Tumor Inst, 58; res biochemist, Lederle Labs, Am Cyanamid Co, NY, 62-66; head biogenic amine lab, 66-69, DIR DEPT BIOCHEM PHARMACOL, AYERST LABS, CAN, 69- Concurrent Pos: Fel, Univ Tex, 61-62. Mem: AAAS; Am Chem Soc; Am Soc Pharmacol & Exp Therapeut; Pharmacol Soc Can; NY Acad Sci. Res: Biosynthesis and mode of action of the biogenic amines; biochemical mechanisms of action of drugs with respect to cardiovascular, central nervous and gastrointestinal systems; biochemical mechanisms involved with gonadotrophin secretion. Mailing Add: 3250 Forest Hill Apt 1202 Montreal PQ Can

LIPPS, EMMA LEWIS, b Alexandria, Va, Feb 8, 19. PLANT ECOLOGY. Educ: Wesleyan Col, BA, 40; Emory Univ, MS, 49. Prof Exp: Asst biol, Agnes Scott Col, 42-43; from asst to assoc prof, 43-62, dir NSF in-serv insts, 58-61, PROF BIOL, SHORTER COL, GA, 62-, CUR EARTH SCI, MUS, 70- Concurrent Pos: Asst, Univ Tenn, 55-56; Southern Fel Fund fel, 61-62. Mem: AAAS. Res: Relationship of present and primeval forests of northwest Georgia to geology and soils; northwestern Georgia's Pleistocene fossils. Mailing Add: Dept of Biol Shorter Col Rome GA 30161

LIPPS, FRANK, meteorology, fluid dynamics, see 12th edition

LIPPS, JERE HENRY, b Los Angeles, Calif, Aug 28, 39; m 64. GEOLOGY, INVERTEBRATE PALEONTOLOGY. Educ: Univ Calif, Los Angeles, AB, 62, PhD(geol), 66. Prof Exp: Asst res geologist invert paleont, Calif Res Corp, 63-65; res geologist, Univ Calif, Los Angeles, 65-67; from asst prof to assoc prof, 67-75, PROF GEOL, UNIV CALIF, DAVIS & BODEGA MARINE LAB, 75-, RES GEOLOGIST, 67- Concurrent Pos: Res assoc, Los Angeles County Mus, 63- Mem: AAAS; Geol Soc Am; Paleont Soc; Soc Protozool; Soc Econ Paleont & Mineral. Res: Ecology of Cenozoic Foraminifera; evolutionary biology of protists. Mailing Add: Dept of Geol Univ of Calif Davis CA 95616

LIPPSCHUTZ, EUGENE J, b Buffalo, NY, Feb 16, 08; m 39; c 1. CARDIOLOGY. Educ: Georgetown Univ, BS, 30, MD, 32. Prof Exp: PROF MED, STATE UNIV NY BUFFALO, 63-, ASSOC VPRES HEALTH SCI, 70-; PHYSICIAN, BUFFALO GEN HOSP, 57- Concurrent Pos: Physician, E J Meyer Mem Hosp, 64-; fel coun clin cardiol, Am Heart Asn. Mem: Fel Am Col Physicians. Mailing Add: Fac of Health Sci 425 Kimball Tower State Univ of NY Buffalo NY 14214

LIPPSON, ROBERT LLOYD, b Detroit, Mich, Apr 18, 31; m 72; c 7. MARINE BIOLOGY. Educ: Mich State Univ, BS, 63, MS, 64, PhD(zool), 75. Prof Exp: Res biologist, Chesapeake Biol Lab, Univ Md, 68-71; fisheries biologist, 71-73, asst coordr water resources, 73-75, RES COORDR ENVIRON ASSESSMENT DIV, OXFORD LAB, NAT MARINE FISHERIES SERV, 75- Mem: Estuarine Res Fedn; Atlantic Estuarine Res Soc; Sigma Xi. Res: Population dynamics and physiological-ecology of crustacea. Mailing Add: Oxford Lab Nat Marine Fisheries Serv Oxford MA 21654

LIPS, HILAIRE JOHN, b Can, July 20, 18; m 46; c 5. CHEMISTRY. Educ: Univ BC, BA, 38, MA, 40; McGill Univ, PhD(agr chem), 44. Prof Exp: BIOCHEMIST & FOOD INFO OFFICER, NAT RES COUN CAN, 43- Mem: Am Oil Chem Soc. Res: Fats; oils; foods. Mailing Add: Nat Res Coun of Canada Ottawa ON Can

LIPSCHULTZ, FREDERICK PHILLIP, b Los Angeles, Calif, Aug 27, 37. LOW TEMPERATURE PHYSICS. Educ: Stanford Univ, BS, 59; Cornell Univ, PhD(physics), 66. Prof Exp: Res asst physics, Cornell Univ, 62-65; res assoc, Brookhaven Nat Lab, 65-67; asst prof, 67-72, ASSOC PROF PHYSICS, UNIV CONN, 72- Mem: AAAS; Am Phys Soc. Res: Thermal and acoustic properties of solids at low temperatures; fatigue in alloys; nondestructive testing of metals. Mailing Add: Dept of Physics U-46 Univ of Conn Storrs CT 06268

LIPSCHUTZ, LOUIS SANDERSON, b May 23, 04; US citizen; m 31; c 3. PSYCHIATRY, PSYCHOANALYSIS. Educ: Univ Mich, BA, 26, MD, 30; Am Bd Psychiat & Neurol, dipl. Prof Exp: Intern, Receiving Hosp, Detroit, 30-31, resident psychiatrist, 31-33, out-patient neurologist, 33-35; asst psychiatrist, Wayne County Gen Hosp, 35-36; instr psychiat, Col Med, Wayne State Univ, 36-44, clin asst prof, 45-50, prof lectr, Sch Social Work, 46-60, clin assoc prof psychiat, Col Med, 51-68, EMER CLIN ASSOC PROF PSYCHIAT, COL MED, WAYNE STATE UNIV, 69- Concurrent Pos: Sr psychiatrist, Wayne County Gen Hosp, 36-42, dir psychiat, 45-49, mem hon life staff, 68-; sr consult, Vet Admin Ment Hyg Clin, 46-63; consult psychiatrist, Wayne County Ment Health Clin, 50-67; assoc psychiatrist, Sinai Hosp Detroit, 53-69, emer assoc psychiatrist, 69-; guest instr, NY Psychoanal Inst, 57-61; training analyst, Mich Psychoanal Inst, 57-72. Mem: Am Psychoanal Asn; fel Am Col Physicians; fel Am Psychiat Asn; Am Med Asn. Res: Medicine. Mailing Add: 2243 Golfview Dr Troy MI 48084

LIPSCHUTZ, MARTIN MORRIS, mathematics, see 12th edition

LIPSCHUTZ, MICHAEL ELAZAR, b Philadelphia, Pa, May 24, 37; m 59; c 3. PHYSICAL CHEMISTRY, COSMOCHEMISTRY. Educ: Pa State Univ, BS, 58; Univ Chicago, SM, 60, PhD(phys chem), 62. Prof Exp: NSF fel, Physics Inst, Berne, 64-65; asst prof chem, 65-68, asst prof geosci, 67-68, assoc prof, 68-73, PROF CHEM & GEOSCI, PURDUE UNIV, 73- Concurrent Pos: Nininger meteorite res award, 62; Fulbright-Hays scholar, Tel Aviv Univ, 71-72; consult, NASA, 73- Mem: AAAS; Am Geophys Union; Meteoritical Soc; Int Asn Geochem & Cosmochem; Am Chem Soc. Res: Neutron activation and atomic absorption methods for trace and ultratrace analysis; geochemistry; stable isotopes in lunar samples and meteorites; cosmogenic nuclear reactions; high pressure and temperature reactions. Mailing Add: Dept of Chem Purdue Univ West Lafayette IN 47907

LIPSCHUTZ-YEVICK, MIRIAM AMALIE, mathematics, see 12th edition

LIPSCOMB, DAVID M, b Morrill, Nebr, Aug 4, 35; m 57; c 3. AUDIOLOGY. Educ: Univ Redlands, BA, 57, MA, 59; Univ Wash, PhD(audiol), 66. Prof Exp: Asst prof audiol, WTex State Univ, 60-62; asst prof, 62-64 & 66-69, assoc prof, 69-72, PROF AUDIOL & SPEECH PATH, UNIV TENN, KNOXVILLE, 72-, DIR, NOISE RES LAB, 71- Mem: Am Speech & Hearing Asn; Acoust Soc Am. Res: Effect of high intensity noise upon the eriphera; peripheral auditory mechanism. Mailing Add: Dept of Audiol & Speech Path Univ of Tenn Knoxville TN 37916

LIPSCOMB, ELIZABETH LOIS, b Hackensack, NJ, Nov 25, 27; m 49; c 1. BIOCHEMISTRY. Educ: Duke Univ, BS, 49; Univ Pittsburgh, MS, 51; NC State Univ, PhD(biochem), 73. Prof Exp: RES ASSOC BIOCHEM, NC STATE UNIV, 73- Res: Catalytic and structural characteristics of the branched-chain amino acid aminotransferase of Salmonella typhimurium; creative application of techniques to elucidate the amino acid sequence is stressed. Mailing Add: Dept of Biochem NC State Univ Raleigh NC 27607

LIPSCOMB, HARRY SHEPHERD, medicine, see 12th edition

LIPSCOMB, NATHAN THORNTON, b Jan 16, 34; US citizen; m 62; c 2. POLYMER CHEMISTRY. Educ: Eastern Ky State Col, BS, 56; Univ Louisville, PhD(phys chem), 60. Prof Exp: From asst prof to assoc prof, 60-75, PROF CHEM, UNIV LOUISVILLE, 75- Mem: Am Chem Soc; Sigma Xi. Res: Kinetics of polymerization; radiation induced polymerization; polymer characterization; polymer properties. Mailing Add: Dept of Chem Univ of Louisville Louisville KY 40208

LIPSCOMB, PAUL ROGERS, b Clio, SC, Mar 23, 14; m 40; c 2. ORTHOPEDIC SURGERY. Educ: Univ SC, BS, 35; Med Col, SC, MD, 38; Univ Minn, MS, 42; Am Bd Orthop Surg, dipl. Prof Exp: Intern, Cooper Hosp, NJ, 38-39; intern orthop surg, Mayo Found, Univ Minn, 39-42, from asst prof to prof, 49-69; PROF ORTHOP SURG & CHMN DEPT, SCH MED, UNIV CALIF, DAVIS, 69- Concurrent Pos: Consult, Mayo Clin, Minn, St Mary's Hosp & Methodist Hosp, 42-69, David Grant US Air Force Med Ctr, Travis AFB, Calif, 70- & Letterman Gen Hosp, Presidio, San Francisco, 72-; secy, Am Bd Orthop Surg, 68, pres, 71-73. Mem: Fel Clin Orthop Soc; fel AMA; fel Am Col Surg; fel Am Acad Orthop Surg; Am Orthop Asn (pres, 74-75). Res: Surgery of the hand. Mailing Add: Dept of Orthop Surg Univ of Calif Sch of Med Davis CA 95616

LIPSCOMB, ROBERT DEWALD, b Tulia, Tex, Dec 29, 17; m 43; c 2. ORGANIC CHEMISTRY. Educ: Univ Nebr, BS, 40, MS, 41; Univ Ill, PhD(org chem), 44. Prof Exp: Lab asst chem, Univ Nebr, 40-41; spec asst, Univ Ill, 41-42, investr, Off Sci Res & Develop & Nat Defense Res Comt, 42-44 & Univ Nebr, 44-45; RES CHEMIST, E I DU PONT DE NEMOURS & CO, 45- Mem: Am Chem Soc. Res: Chemistry of quinoline and benzoquinoline derivatives; amine bisulfites; organic polysulfides; free radicals; high temperature chemistry; polymerization. Mailing Add: Cent Res Dept E I du Pont de Nemours & Co Wilmington DE 19898

LIPSCOMB, WILLIAM NUNN, JR, b Cleveland, Ohio, Dec 9, 19; m 44; c 2. PHYSICAL CHEMISTRY. Educ: Univ Ky, BS, 41; Calif Inst Technol, PhD(phys chem), 46. Hon Degrees: DSc, Univ Ky, 63; MA, Harvard Univ, 59. Prof Exp: Asst prof phys chem, Univ Minn, 46-50, assoc prof & actg chief div, 50-54, prof & chief, 54-59; prof, 59-71, chmn dept chem, 62-65, ABBOTT & JAMES LAWRENCE PROF CHEM, HARVARD UNIV, 71- Concurrent Pos: Guggenheim fel, Oxford, 54-55, Cambridge, 72-73; NSF sr fel, 65-66; overseas fel, Churchill Col, Cambridge, 66; grants, Off Naval Res, 58-, Off Ord Res, 54-56, NSF, 56-65, Air Force Off Sci Res, 58-64, NIH, 58-, Upjohn Co, 58 & Adv Res Projs Agency, 61-; Frontiers in Chem lectr, 54 & 67; Four College lectr series, 63; distinguished lectr, Howard Univ & Welch Found lectr, Univ Tex, 66; Phillips lectr, Univ Okla, 67; Priestley lectr, Pa State Univ, 67; William Pyle Phillips lectr, Haverford Col, 68; Baker lectr, Cornell Univ & Coover lectr, Iowa State Univ, 69; Nieuwland lectr, Univ Notre Dame & Am Cyanamid lectr, Univ Conn, 70; invited lectr series, Univ Calif, Los Angeles, 70 & 71; Pi Alpha lectr, Emory Univ, 3M lectr, Univ Minn, Sun Oil lectr, Ohio Univ & Welch Found lectr series, 71; Mem nat comt crystallog, Nat Res Coun, 54-58, 60-63, 65-67; rev comt, Chem Div, Argonne Nat Lab, 56-65; chmn prog comt, Int Cong Crystallog, Can, 57, gen lectr, Moscow, 66; chem vis comt, Brookhaven Nat Lab, 63-65 & Calif Inst Technol, 69-74. Honors & Awards: Howe Award, Am Chem Soc, 58; Distinguished Serv Award, Am Chem Soc, 68; George Ledlie Prize, Harvard Univ, 71; Peter Debye Award, Phys Chem, Am Chem Soc, 73; Evans Award, Ohio State Univ, 74. Mem: Nat Acad Sci; Am Phys Soc; Am Chem Soc; Am Crystallog Asn (pres, 55); fel Am Acad Arts & Sci. Res: Diffraction studies of crystals and molecules of biochemical interest. Mailing Add: Dept of Chem Harvard Univ 12 Oxford St Cambridge MA 02138

LIPSETT, FREDERICK ROY, b Vancouver, BC, Can, Sept 26, 25; m 57; c 2. FORENSIC SCIENCE. Educ: Univ BC, BApSc, 48, MApSc, 51; Univ London, PhD(physics), 54. Prof Exp: SR RES OFFICER RADIO & ELEC ENG DIV, NAT RES COUN CAN, 54- Concurrent Pos: Part time lectr, Carleton Univ, Can, 62-64. Mem: Opers Res Soc Am. Res: Luminescence of organic solids; analysis and computer simulation of police patrol operations. Mailing Add: Nat Res Coun of Canada Ottawa ON Can

LIPSETT, MARIE NIEFT, b Chicago, Ill, Aug 10, 20; m 48; c 2. BIOLOGICAL CHEMISTRY. Educ: Northwestern Univ, BS, 41, MS, 43; Univ Southern Calif, PhD(biochem), 48. Prof Exp: Fel, US Pub Health Serv, Univ Calif, Los Angeles, 48-50; res assoc exp med, Univ Southern Calif, 50-53 & Sloan-Kettering Inst, 55-57; CHEMIST, NAT INST ARTHRITIS & METAB DIS, NAT INSTS HEALTH, 58- Mem: AAAS; Am Soc Biol Chem. Res: Interactions of polynucleotides and related compounds; immunochemistry of enzymes; natural thionucleotides. Mailing Add: Nat Inst of Arthritis & Metab Dis Nat Insts of Health Bethesda MD 20014

LIPSETT, MORTIMER BROADWIN, b New York, NY, Feb 20, 21; m 48; c 2. INTERNAL MEDICINE. Educ: Univ Calif, AB, 43; Univ Southern Calif, MS, 47, MD, 51; Am Bd Internal Med, dipl. Prof Exp: Res assoc, Univ Southern Calif, 46-47; intern, Los Angeles County Hosp, Calif, 51-52; resident internal med, Sawtelle Hosp, 52-54; asst mem, Sloan-Kettering Inst, 56-57; med officer, NIH, 57-59; assoc clin prof med, Sch Med, Howard Univ, 61; chief endocrinol br, Nat Cancer Inst, 65-69; assoc sci dir reprod biol, Nat Inst Child Health & Human Develop, 69-74; PROF MED, SCH MED, CASE WESTERN RESERVE UNIV, 74-; DIR, CANCER CTR, NORTHEAST OHIO, 74- Concurrent Pos: USPHS fel, Sloan-Kettering Inst, 54-56; instr, Sch Med, Cornell Univ, 56-57; sr assoc attend physician, Freedman's Hosp, 58-73; mem, Cleveland Clin; ed-in-chief, J Clin Endocrinol & Metab, 68-73. Mem: Endocrine Soc (secy-treas, 74-); Am Asn Cancer Res; Am Soc Clin Invest; Asn Am Physicians; fel Am Col Physicians. Res: Endocrinology; oncology; reproductive biology. Mailing Add: Dept of Med Sch of Med Case Western Reserve Univ Cleveland OH 44106

LIPSETT, SOLOMON GEORGE, b London, Eng, June 28, 00; m 31; c 2. PHYSICAL CHEMISTRY, INORGANIC CHEMISTRY. Educ: Univ Man, BSc, 21, MSc, 25; McGill Univ, PhD(chem), 27. Prof Exp: Instr chem, Univ Man, 22-24; asst prof, Man Agr Col, 24-25; res chemist, 27-65, pres, 65-69, CONSULT, J T DONALD & CO, LTD, 69- Mem: Am Chem Soc; fel Chem Inst Can. Res: Industrial problems involving chemistry. Mailing Add: 4970 Hingston Ave Montreal PQ Can

LIPSEY, SALLY IRENE, b Dec 31, 26; US citizen; m 48; c 3. MATHEMATICS EDUCATION. Educ: Hunter Col, AB, 47; Univ Wis, AM, 48; Columbia Univ, DEduc, 65. Prof Exp: Asst, Dept Math, Univ Wis, 47-48; high sch teacher, Bd Educ, New York, 48-49; lectr, Hunter Col, 49-53 & Barnard Col, 53-59; asst prof, Bronx Community Col, 59-65; asst prof educ, 65-71, ASST PROF MATH, BROOKLYN COL, 71- Mem: Nat Coun Teachers Math; Math Asn Am. Mailing Add: Dept of Math Brooklyn Col Brooklyn NY 11210

LIPSHITZ, STANLEY PAUL, b Cape Town, SAfrica, Nov 25, 43. APPLIED MATHEMATICS. Educ: Univ Natal, BSc, 64; Univ SAfrica, MSc, 65; Univ Witwatersrand, PhD(math), 70. Prof Exp: Vis lectr math, Univ Ariz, 67-68; ASST PROF APPL MATH, UNIV WATERLOO, 70- Mem: Am Math Soc. Res: Calculus of variations and applications to physical field theories; constrained variational problems; canonical formalisms. Mailing Add: Dept of Appl Math Univ of Waterloo Waterloo ON Can

LIPSHULTZ, LARRY I, b Philadelphia, Pa, Apr 24, 42; m 66; c 2. UROLOGY. Educ: Franklin & Marshall Col, BS, 60; Univ Pa, MD, 68. Prof Exp: Asst instr urol, Univ Pa, 73-74, instr, 74-75; ASST PROF UROL & CLIN FEL REPROD MED, UNIV TEX MED BR, HOUSTON, 75- Concurrent Pos: Res scholar, Am Urol Asn, 75-77. Mem: Am Fertil Soc; Am Soc Andrology. Res: The evaluation and diagnosis of reproductive disorders in the male, especially in the field of infertility; androgen binding protein in the human testis and epididymis and evaluation of androgen binding protein as a possible marker of sertoli cell function. Mailing Add: Univ of Tex Med Sch 6400 W Cullen Houston TX 77030

LIPSHUTZ, NELSON RICHARD, b Phila, Pa, July 14, 42; m 64; c 2. OPERATIONS RESEARCH, HIGH ENERGY PHYSICS. Educ: Univ Pa, AB, 62, MBA, 72; Univ Chicago, SM, 63, PhD(physics), 67. Prof Exp: Res assoc physics, Univ Chicago, 67; from instr to asst prof, Duke Univ, 67-70; mgt res analyst, Mgt & Behav Sci Ctr, Wharton Sch Finance & Commerce, Univ Pa, 70-72; MGT CONSULT, ARTHUR D LITTLE, INC, 72- Mem: Am Phys Soc; Inst Mgt Sci. Res: Theory of elementary particles; mathematical analysis of management decision problems; economic analysis of regulated industries. Mailing Add: Arthur D Little Inc 25 Acorn Park Cambridge MA 02140

LIPSICAS, MAX, b Tel-Aviv, Palestine, July 13, 25; m 64. CHEMISTRY, PHYSICS. Educ: Univ London, BSc, 50, BScEng, 51; Univ BC, PhD(physics), 60. Prof Exp: Physicist, Metrop-Vickers Elec Co, Eng, 51-57; asst res specialist, Dept Physics, Rutgers Univ, 61-63; assoc physicist, Brookhaven Nat Lab, 63-66; assoc prof chem, 66-69, actg chmn dept, 68-70, PROF, BELFER GRAD SCH SCI, YESHIVA UNIV, 69- Concurrent Pos: Lectr & tutor, Postgrad Eng Course, Univ Manchester, 52-55; res collabr, Brookhaven Nat Lab. Mem: Am Phys Soc. Res: Nuclear magnetic resonance, including studies of gases and liquids; critical phenomena; electronic phase transitions. Mailing Add: Belfer Grad Sch of Sci Yeshiva Univ New York NY 10033

LIPSICH, H DAVID, b Pittsburgh, Pa, Feb 20, 20; m 46; c 2. MATHEMATICS. Educ: Univ Cincinnati, MA, 45, PhD(math), 49; Princeton Univ, MA, 46. Prof Exp: From instr to assoc prof, 46-61, vprovost, 67, PROF MATH & HEAD DEPT, UNIV CINCINNATI, 61-, PROVOST UNDERGRAD STUDIES, 67- Concurrent Pos: NSF fac fel, 59-60. Mem: Am Math Soc; Math Asn Am; Asn Symbolic Logic. Res: Mathematical logic; set theory. Mailing Add: Dept of Math Univ of Cincinnati Cincinnati OH 45221

LIPSIG, JOSEPH, b Brooklyn, NY, Dec 13, 30; m 60; c 1. PHYSICAL CHEMISTRY. Educ: Brooklyn Col, BA, 50; Polytech Inst Brooklyn, PhD(phys chem), 61. Prof Exp: Res assoc, Cornell Univ, 60-62; sr res chemist, Atlantic Ref Co, 62-66; asst prof, 66-68, ASSOC PROF CHEM, STATE UNIV NY COL OSWEGO, 68- Mem: Am Chem Soc; Am Phys Soc. Res: Catalysis; surface chemistry; chemical kinetics; electrochemistry. Mailing Add: Dept of Chem State Univ of NY Oswego NY 13126

LIPSITT, DON RICHARD, b Boston, Mass, Nov 24, 27; m 53; c 2. PSYCHIATRY, PSYCHOANALYSIS. Educ: NY Univ, BA, 49; Boston Univ, MA, 50; Univ Vt, MD, 56; Boston Psychoanal Soc & Inst, cert. 69. Prof Exp: Asst, 62-65, from instr to asst prof, 65-74, ASSOC PROF PSYCHIAT, HARVARD MED SCH, 74-; CHIEF PSYCHIAT, MT AUBURN HOSP, CAMBRIDGE, 69- Concurrent Pos: Teaching fel psychiat, Harvard Med Sch, 60-62; Dept Health, Educ & Welfare res grant, 66-68; head integration clin, Beth Israel Hosp, Boston, 62-69, asst psychiat, 62-64, assoc, 64-66, asst psychiatrist, 66, dir med psychol liaison serv, 66-69; consult behav sci, Lincoln Lab, Mass Inst Technol, consult & psychiat & instr, Faulkner Workshop Chronic Complainer, Faulkner Hosp, Boston, 69-73; mem fac, Boston Psychoanal Soc & Inst, 71-72 & Sch Social Work, Simmons Col, 71-; consult, Dept Psychiat, Cambridge Hosp, 71-; assoc attend psychiatrist, McLean Hosp, 71-; ed, Int J Psychiat Med; consult, NIMH, 74-75. Mem: Fel Am Psychiat Asn. Res: Application of medical psychology to health problems in hospital and community; relationship of varieties of doctor-patient interaction to invalidism and chronicity. Mailing Add: 15 Griggs Rd Brookline MA 02146

LIPSITZ, PAUL, b York, Pa, Apr 23, 23; m 48; c 4. ORGANIC CHEMISTRY. Educ: Lebanon Valley Col, BS, 44; Univ Cincinnati, MS, 48, PhD(chem), 50. Prof Exp: Chemist, E I du Pont de Nemours & Co, 50-59; sr patent agent, Pennsalt Chems Corp, 59-69; SR PATENT AGT, PATENT & LICENCES DEPT, SUN VENTURES, INC, 69- Mem: Am Chem Soc; Am Inst Chem. Mailing Add: 205 Suffolk Rd Flourtown PA 19031

LIPSITZ, PHILIP JOSEPH, b Piketberg, SAfrica, May 17, 28; m 58; c 3. MEDICINE, PEDIATRICS. Educ: Univ Cape Town, MB, ChB, 52; Royal Col Physicians & Surgeons, dipl child health, 56; Am Bd Pediat, dipl. Prof Exp: House surgeon, Univ Cape Town, 52; house physician, Somerset Hosp, Cape Town, SAfrica, 53; resident surg house officer, Gen Hosp, Salisbury, SRhodesia, 53; Charles' house physician, St Hosp, London, Eng, 55; resident med officer, Banstead Br, Queen Elizabeth Hosp for Children, 56; house physician, Royal Hosp Sick Children, Edinburgh, Scotland, 56; registr, Prof Unit, Children's Hosp, Sheffield, Eng, 57-58; resident med officer, Red Cross War Mem Children's Hosp, Univ Cape Town, 58; consult pediat pract with hosp appointments, Southwest Africa, 62-65; asst prof pediat, Med Col Ga, 65-67, assoc prof, 67-68; assoc prof, Beth Israel Med Ctr & Mt Sinai Sch Med, 68-73; prof pediat, Health Sci Ctr, State Univ NY Stony Brook, 73-75; DIR PEDIAT, SOUTH SHORE DIV, LONG ISLAND JEWISH-HILLSIDE MED CTR, FAR ROCKAWAY, 73-, PHYSICIAN-IN-CHG NEWBORN SERV, LONG ISLAND JEWISH-HILLSIDE MED CTR, NEW HYDE PARK, 74- Concurrent Pos: Fel pediat, Sch Med, Western Reserve Univ, 58-60; clin & res fel pediat & med, Children's Hosp Med Ctr, Harvard Med Sch, 60-61. Mem: Med Asn SAfrica; Brit Med Asn; Soc Pediat Res; NY Acad Sci; AMA. Res: Physiology of the newborn. Mailing Add: LI Jewish-Hillside Med Ctr 270-05 76th Ave New Hyde Park NY 11040

LIPSKY, JOSEPH ALBIN, b Glen Lyon, Pa, Mar 31, 30; m 57; c 2. PHYSIOLOGY. Educ: Pa State Univ, BSc, 51; Ohio State Univ, MSc, 59, PhD(physiol), 61. Prof Exp: Asst prof, 61-67, ASSOC PROF PHYSIOL, COL MED, OHIO STATE UNIV, 67- Concurrent Pos: Consult to coun, Nat Bd Dent Exam, 67- Mem: AAAS; Am Physiol Soc. Res: Carbon dioxide transients and stores; hyperventilation. Mailing Add: Dept of Physiol Ohio State Univ Col of Med Columbus OH 43210

LIPSKY, SANFORD, b New York, NY, Jan 24, 30. PHYSICAL CHEMISTRY. Educ: Syracuse Univ, AB, 50; Univ Chicago, PhD(chem), 54. Prof Exp: Res assoc chem, Radiation Proj, Univ Notre Dame, 54-57; asst prof, NY Univ, 57-59; from asst prof to assoc prof, 59-67, PROF CHEM, UNIV MINN, MINNEAPOLIS, 67- Mem: Am Phys Soc; Am Chem Soc; Radiation Res Soc Am. Res: Photochemistry; photophysics; electron impact spectroscopy. Mailing Add: Dept of Chem Univ of Minn Minneapolis MN 55455

LIPSKY, SEYMOUR RICHARD, b New York, NY, Aug 5, 24; m 52; c 3. BIOCHEMISTRY, PHYSICAL CHEMISTRY. Educ: NY Univ, BA, 44; State Univ NY, MD, 49; Am Bd Internal Med, dipl, 56. Prof Exp: Fel, Univ Calif, 51-52; fel, 52-55, from instr to assoc prof, 55-72, PROF MED, SCH MED, YALE UNIV, 72-, DIR SECT PHYS SCI, 67- Concurrent Pos: Consult lunar sci subcomt & lunar & planetary prog comt, NASA, 61-62, biophys & biophys chem study sect, NIH, 64- & Nat Adv Heart Coun, Nat Heart Inst, 64-; mem postdoctoral fel comt, Nat Acad Sci, 65- Mem: Am Soc Biol Chem; Am Chem Soc; Am Soc Clin Invest. Res: Physical-chemical separation techniques; gas chromatography; mass spectrometry; ionization phenomena in gases; chemistry of membranes; analysis of planetary atmospheres; analysis of planetary surfaces for organic compounds; chemistry of macromolecules; nuclear magnetic resonance spectrometry; high performance liquid chromatography. Mailing Add: Sect of Phys Sci Yale Univ Sch of Med New Haven CT 06510

LIPSON, HERBERT GEORGE, b Boston, Mass, July 4, 25; m 51; c 3. SOLID STATE PHYSICS. Educ: Mass Inst Technol, BS, 48; Northeastern Univ, MS, 64. Prof Exp: Jr physicist metall physics, Sylvania Elec Prods, Inc, 48-50; physicist, Brookhaven Nat Lab, 51; physicist, Naval Res Lab, 51-55; physicist, Lincoln Lab, Mass Inst Technol, 55-58; PHYSICIST, AIR FORCE CAMBRIDGE RES LAB, 58- Mem: Am Phys Soc; Sigma Xi; Am Soc Testing & Mat. Res: Optical properties of solids; lattice vibrations, impurities and plasma effects in semiconductors; laser and laser window material properties. Mailing Add: 68 Aldrich Rd Wakefield MA 01880

LIPSON, JOSEPH, b New York, NY, Apr 19, 27; m 50; c 4. PHYSICS. Educ: Yale Univ, BS, 50; Univ Calif, Berkeley, PhD(physics), 56. Prof Exp: Asst res physicist, Univ Calif, Berkeley, 56-57 & Univ Alta, 57-60; assoc prof physics & geol, Univ Pittsburgh, 61-64, dir, Curric Design Group, Learning Res & Develop Ctr, 64-67; prof sci educ, Nova Univ, 67-69; dir, Sci-Math Div, Learning Res Assoc, 69-70, author & educ consult, 70-71; mem, US Comnr Educ Planning Unit, Dept Health, Educ & Welfare, 71; assoc dean grad col, Univ Ill, Chicago Circle, 71-72, assoc vchancellor acad affairs, 72-75; VPRES ACAD AFFAIRS, UNIV MID-AM, 75- Mem: Am Phys Soc; Sigma Xi; Am Educ Res Asn; Nat Sci Teachers Asn; Nat Coun Teachers Math. Res: Development of an applied science of instruction. Mailing Add: 500 Sycamore Dr Lincoln NE 68510

LIPSON, LEONARD BERGER, geophysics, see 12th edition

LIPSON, MELVIN ALAN, b Providence, RI, June 1, 36; m 61; c 4. ORGANIC CHEMISTRY. Educ: Univ RI, BS, 57; Syracuse Univ, PhD(org chem), 63. Prof Exp: Res chemist, I C I (Organics) Inc, 63 & Eltex Res Corp, 63-64; res supvr org synthesis, Wayland Chem Div, Philip A Hunt Chem Corp, RI, 64-67, res mgr, 67-69; TECH DIR, DYNACHEM CORP, 69- Mem: AAAS; Am Chem Soc; The Chem Soc. Res: Amino acids and peptides; chelating agents; photographic chemicals; surface active compounds; polymers; dyestuffs; carbohydrates; corrosion inhibitors. Mailing Add: Dynachem Corp 13000 E Firestone Blvd Santa Fe Springs CA 90670

LIPSON, RICHARD L, b Philadelphia, Pa, July 21, 31; c 2. INTERNAL MEDICINE. Educ: Lafayette Col, BA, 52; Jefferson Med Sch, MD, 56; Univ Minn, MSc, 60. Prof Exp: Res asst biophys, Mayo Clin & Mayo Found, 62-63; ASST PROF MED, COL MED, UNIV VT, 63- Concurrent Pos: Assoc dir rheumatism res unit, Univ Vt, 64-67. Honors & Awards: Arnold J Bergen Res Award, Mayo Found, 60. Mem: AMA; Am Fedn Clin Res; Am Rheumatism Asn; fel Am Col Physicians. Res: Biophysics and bionics of connective tissue, osmotic pressure, viscosity, electrogoniometry, biomechanics; fiber-optics instruments; cancer detection by fluorescence; fluorescent endoscopy; photosensitivity; rheumatology. Mailing Add: 233 Pearl St Burlington VT 05401

LIPTON, ALLAN, b New York, NY, Dec 29, 38; m 65; c 1. INTERNAL MEDICINE, ONCOLOGY. Educ: Amherst Col, BA, 59; NY Univ, MD, 63; Am Bd Internal Med, dipl, 70. Prof Exp: Intern med, Bellevue Hosp, New York, 63-64, resident, 64-65; asst prof, 71-74, ASSOC PROF MED, HERSHEY MED CTR, PA STATE UNIV, 74-, CHIEF DIV ONCOL, DEPT MED, 74- Concurrent Pos: Fel hemat, Mem Hosp, New York, 67-68, fel oncol, 68-69; Dernham fel, Salk Inst Biol Studies, 69-71. Mem: AAAS; Am Asn Cancer Res; Am Fedn Clin Res. Res: Control of growth of normal and malignant cells by serum factors. Mailing Add: Dept of Med Hershey Med Ctr Pa State Univ Hershey PA 17033

LIPTON, JAMES MATTHEW, b Aug 10, 38; US citizen; m 59; c 2. NEUROSCIENCES. Educ: Univ Colo, PhD(physiol psychol), 64. Prof Exp: Asst prof psychol, 66-73, asst prof physiol, 72, ASSOC PROF NEUROL, PHYSIOL & PSYCHOL, UNIV TEX HEALTH SCI CTR, DALLAS, 73- Concurrent Pos: USPHS fel, Neuropath Lab, Med Sch, Univ Mich, 64-66; USPHS sr fel, Inst Animal Physiol, UK, 70-71; consult neurol, Vet Admin Hosp, Dallas, 74- Mem: AAAS; Soc Neurosci; Am Physiol Soc; Am Psychol Asn. Res: Central control of body temperature. Mailing Add: Dept of Neurol Univ of Tex Health Sci Ctr Dallas TX 75235

LIPTON, MORRIS ABRAHAM, b New York, NY, Dec 27, 15; m 40; c 3. EXPERIMENTAL PSYCHIATRY, MEDICINE. Educ: City Col New York, BS, 35; Univ Wis, PhM, 37, PhD(biochem), 39; Univ Chicago, MD, 48. Prof Exp: Asst physiol, Univ Chicago, 40-41, instr, 40-45, res assoc med, 45-46, instr med, 48-51, asst prof psychiat & med, 51-54; asst prof, Med Sch, Northwestern Univ, 54-59; from assoc prof to prof, 63-65, chmn dept, 70-73, SARAH GRAHAM KENAN DISTINGUISHED PROF PSYCHIAT, SCH MED, UNIV NC, CHAPEL HILL, 65-, PROF BIOCHEM & NEUROBIOL & HEAD DIV RES DEVELOP, DEPT PSYCHIAT, 63-, DIR, BIOL SCI RES CTR, CHILD DEVELOP INST & SCH MED, 65- Concurrent Pos: NIMH res career award, 62; chief psychosom serv, Vet Admin Hosp, Chicago & asst dir prof serv, Vet Admin Res Hosp, 54-59; chmn, Psychopharmacol Study Sect, NIMH, 64, consult preclin psychopharm, 72-; chmn, Comt Res Group Advan Psychiat, 66-; chmn, Task Force Megavitamin Ther Psychiat, Am Psychiat Asn, 69-74; Mem, Coop Studies Eval Comt, Vet Admin, 71-; consult, Bur Drugs, Food & Drug Admin, 72- Mem: AAAS; Am Soc Biol Chemists; Am Col Neuropsychopharmacol (pres-elect); Fedn Am Socs Exp Biol. Res: Neuropsychopharmacology; extra pituitary effects of the hypothalamic hormones. Mailing Add: Biol Sci Res Ctr Univ of NC Sch of Med Chapel Hill NC 27514

LIPTON, SAMUEL HARRY, b Burlington, Wis, Feb 6, 21. BIOCHEMISTRY, ORGANIC CHEMISTRY. Educ: Univ Wis, BS, 42, MS, 47, PhD(biochem, org chem), 48. Prof Exp: Asst, Univ Wis, 42, 46-48; res assoc, Univ Chicago, 49-50; res chemist, Lab, Pabst Brewing Co, Wis, 50-55; biochemist, Bioferm Corp, Calif, 56; proj assoc, Univ Wis, 57-67; RES CHEMIST, NUTRITION INSTITUTE, AGR RES SERV, USDA, 68- Mem: AAAS; Am Chem Soc. Res: Nucleotides and coenzymes; toxic nitriles and lathyrism; amino acids. Mailing Add: 6980 Hanover Pkwy Greenbelt MD 20770

LIPTON, WERNER JACOB, b Ger, Oct 16, 28; nat US; m 52; c 4. PLANT PHYSIOLOGY. Educ: Mich State Univ, BS, 51, MS, 53; Univ Calif, PhD(plant physiol), 57. Prof Exp: Asst, Univ Calif, 53-57; assoc plant physiologist, 57-63, SR PLANT PHYSIOLOGIST, HORT FIELD STA, USDA, 63- Concurrent Pos: Assoc ed, Am Soc Hort Sci, 72-76. Mem: AAAS; Am Soc Hort Sci; Am Soc Plant Physiol; Am Meteorol Soc. Res: Postharvest physiology of vegetables; emphasis on effects of modified atmospheres and preharvest environmental factors. Mailing Add: US Dept of Agr PO Box 8143 Fresno CA 93727

LIPWORTH, EDGAR, b London, Eng, Dec 20, 23; m 53; c 2. ATOMIC PHYSICS. Educ: Manchester Univ, BSc, 47; Columbia Univ, MA, 49, PhD, 56. Prof Exp: Asst instr, Columbia Univ, 47-49, asst, 49-56; physicist, Lawrence Radiation Lab, Univ Calif, Berkeley, 56-61; assoc prof, 61-62, PROF PHYSICS, BRANDEIS UNIV, 62- Concurrent Pos: Lectr, Univ Calif, 59. Mem: Fel Am Phys Soc. Res: Molecular and atomic beams. Mailing Add: Dept of Physics Brandeis Univ Waltham MA 02154

LIRA, EMIL PATRICK, b Chicago, Ill, Mar 17, 34; m 58; c 4. ORGANIC CHEMISTRY. Educ: Elmhurst Col, BS, 56; Rutgers Univ, PhD(org chem), 63. Prof Exp: Chemist, Swift & Co, 56; chemist, Corn Prod Co, 58-59; res chemist, 63-69, Int Mineral & Chem Corp, 63-69, supvr org synthesis, 69-73, mgr org chem, 73-74; DIR RES, VELSICOL CHEM CORP, 74- Mem: Am Chem Soc. Res: Organic research and development with plasticizers, adhesives, polymer additives; plant growth regulators; pesticides; animal health products; synthetic sweeteners; organic processes. Mailing Add: 50 W Roxbury Ct Des Plaines IL 60018

LIS, ADAM W, b Przemysl, Poland, Jan 5, 25; US citizen. BIOCHEMISTRY. Educ: Univ Ark, BS, 49; Univ Calif, Berkeley, PhD(biochem), 60. Prof Exp: Res biochemist, Univ Calif, San Francisco, 60-62; res assoc, 63-65, asst prof, 65-67, ASSOC PROF NUCLEIC ACIDS, UNIV ORE HEALTH SCI CTR, 67-, DIR NUCLEIC ACIDS LAB, 66- Concurrent Pos: Nat Cancer Inst fel, Univ Uppsala, 62-63; ed, Physiol Chem & Physics, 67- Mem: Brit Biochem Soc; Am Chem Soc; Am Soc Cell Biol; Radiation Res Soc Am; Biophys Soc. Res: Minor components in nucleic acids and their function; body fluids analysis in malignant and metabolic diseases. Mailing Add: 7950 SW Crestline Dr Portland OR 97219

LIS, EDWARD FRANCIS, b Chicago, Ill, Apr 1, 18; m 44; c 2. PEDIATRICS. Educ: Univ Ill, BS, 39, MD, 43. Prof Exp: From asst prof to assoc prof, 52-65, PROF PEDIAT, UNIV ILL COL MED, 65-, MED DIR, CTR HANDICAPPED CHILDREN & DIR DIV SERV CRIPPLED CHILDREN, 59- Res: Handicapped children. Mailing Add: Dept of Pediat Univ of Ill Col of Med Chicago IL 60612

LIS, ELAINE WALKER, b Denver, Colo, Apr 25, 24; m 58; c 3. NUTRITION, BIOCHEMISTRY. Educ: Mills Col, AB, 45; Univ Calif, Berkeley, PhD(nutrit), 60. Prof Exp: Asst nutrit, Univ Calif, Berkeley, 56-60, USPHS fel, 60-62; lectr, Portland State Univ, 64-68; ASSOC PROF, CRIPPLED CHILDREN'S DIV, MED SCH, UNIV ORE, 68- Concurrent Pos: Consult, Crippled Children's Div, Med Sch, Univ Ore, 64-68. Mem: Am Asn Univ Prof; Am Asn Ment Deficiency; Am Soc Cell Biol; Am Home Econ Asn. Res: Metabolic approach to possible causes of retardation, emotional disturbances or other handicapping conditions with emphasis on urinary studies of ultraviolet absorbing end products of metabolism in mentally retarded children. Mailing Add: Crippled Children's Div Univ of Ore Health Sci Ctr Portland OR 97201

LIS, MARTIN, b Praha, Czech, Jan 6, 35; Can citizen; m 60; c 2. EXPERIMENTAL MEDICINE. Educ: Sch Vet Med, Brno, Czech, DVM, 60; Czech Acad Sci, CSc(animal physiol), 67; McGill Univ, PhD(exp med), 72. Prof Exp: Fel animal physiol, Lab Animal Physiol, Czech Acad Sci, 60-68; fel exp med, Inst Exp Med & Surg, Univ Montreal, 68-69; fel exp med, 69-72, sr scientist, 72-75, DIR LAB COMP ENDOCRINOL, CLIN RES INST MONTREAL, 75-; ASST PROF MED, UNIV MONTREAL, 72- Mem: Can Soc Endocrinol & Metab. Res: Pituitary and hypothalamic peptides; hormone producing tumors; hormone responsive mammary tumors; morphine-like endogenous peptides; mechanism of action of peptide hormones. Mailing Add: Clin Res Inst 110 Pine Ave W Montreal PQ Can

LISA, JOSEPH DANIEL, b Jersey City, NJ, May 8, 28; div; c 1. PHYSIOLOGY, BIOMEDICAL ENGINEERING. Educ: St Peter's Col, NJ, BA, 51; Fordham Univ, MS, 54, PhD(physiol), 58. Prof Exp: Head bioastronaut, Advan Systs Dept, 63-68, RES SCIENTIST ENVIRON PHYSIOL, RES DEPT, GRUMMAN AEROSPACE CORP, 68- Mem: Aerospace Med Asn; NY Acad Sci. Res: Cytopathological investigation; holographic techniques; screening of cancer cells. Mailing Add: Res Dept Grumman Aerospace Corp Bethpage NY 11747

LISAK, ROBERT PHILIP, b Brooklyn, NY, Mar 17, 41; m 64; c 2. NEUROLOGY, IMMUNOLOGY. Educ: NY Univ, BA, 61; Columbia Univ, MD, 65. Prof Exp: Intern med, Montefiore Hosp & Med Ctr, 65-66; res assoc immunol, Lab Clin Sci, NIMH, 66-68; jr resident med, Bronx Munic Med Ctr, Albert Einstein Col Med, 68-69; resident neurol, Hosp, 69-72, trainee allergy & immunol, 71-72, ASST PROF NEUROL, SCH MED, UNIV PA, 72-, MEM IMMUNOL GRAD GROUP, 75- Concurrent Pos: Consult neurol, Vet Admin Hosp, Philadelphia, 72-; spec consult, Nat Multiple Sclerosis Soc, 75. Mem: Am Asn Immunologists; Am Fedn Clin Res; Am Acad Neurol; AAAS. Res: Humoral and cell-mediated immunologic mechanisms involved in clinical and experimental diseases of the central and peripheral nervous system and muscle. Mailing Add: Dept of Neurol Univ of Pa Hosp Philadelphia PA 19104

LISANKE, ROBERT JOHN, SR, b New York, NY, June 7, 32; m 57; c 4. ORGANIC POLYMER CHEMISTRY. Educ: Fordham Univ, BS, 54; Notre Dame Univ, MS, 56. Prof Exp: Control chemist, Chas Pfizer & Co, 50-54; chemist, Union Carbide Plastics Co, 55, chemist, Silicones Div, Union Carbide Corp, 56-59; chemist, Hooker Chem Corp, 59-60; res chemist, Mobil Oil Corp, NJ, 60-66; org chemist, Gen Elec Co, Syracuse, 66-71, process chemist, 71-73; SR CHEMIST, PRATT & WHITNEY AIRCRAFT DIV, UNITED TECHNOLOGIES CORP, WEST PALM BEACH, 74- Mem: Am Chem Soc; Am Inst Chem Eng; Inst Elec & Electronics Eng; Am Mgt Asn. Res: Thermogravimetry; ultra-high vacuum technology; chemical kinetics; materials science; computer programming. Mailing Add: PO Box 9101 Riviera Beach FL 33404

LISANO, MICHAEL EDWARD, b Houston, Tex, Oct 6, 42; c 2. REPRODUCTIVE PHYSIOLOGY. Educ: Sam Houston State Univ, BS, 64, MS, 66; Tex A&M Univ, PhD(physiol), 70. Prof Exp: Instr biol, Hardin-Simmons Univ, 66-67; ASST PROF PHYSIOL, AUBURN UNIV, 70- Mem: Soc Study Reproduction. Res: Reproductive physiology and endocrinology of wild turkeys and white-tailed deer. Mailing Add: Dept of Zool & Entom Auburn Univ Auburn AL 36830

LISANSKY, EPHRAIM THEODORE, b Baltimore, Md, Nov 2, 12; m 37; c 2. INTERNAL MEDICINE, PSYCHOSOMATIC MEDICINE. Educ: Johns Hopkins Univ, BA, 33; Univ Md, MD, 37. Prof Exp: PROF MED & ASSOC PROF CLIN PSYCHIAT, SCH MED & LECTR, SCH SOCIAL WORK & SCH LAW, UNIV MD, BALTIMORE. Concurrent Pos: Assoc prof community dent, Sch Dent & chmn comt for continuing med educ, Sch Med, Univ Md, Baltimore. Mailing Add: 6804 Park Heights Ave Baltimore MD 21215

LI-SCHOLZ, ANGELA, b Hong Kong, Aug 15, 36; US citizen; m 66; c 2. EXPERIMENTAL PHYSICS. Educ: Manhattanville Col, BA, 56; NY Univ, MS, 57, PhD(physics), 63. Prof Exp: Jr res assoc nuclear physics, Brookhaven Nat Lab, 60-63; res assoc high energy physics, NY Univ, 63; res staff physicist, Yale Univ, 63-65, res assoc nuclear physics, 66-67; asst prof physics, City Col New York, 65-66; res assoc solid state physics, Univ Pa, 67-70; res assoc nuclear chem, Rensselaer Polytech Inst, 70-72; ASSOC PROF SCI, EMPIRE STATE COL, STATE UNIV NY, 72- Concurrent Pos: Vis prof, State Univ NY, Albany, 71- Mem: Am Phys Soc. Res: Atomic inner shell ionization phenomena; elemental analysis by electron impact ionization; nuclei spectroscopy; interaction of nuclei with electromagnetic fields in solids. Mailing Add: Empire State Col 135 Western Ave Albany NY 12203

LISCO, HERMANN, b Schul-Pforta, Ger, Oct 24, 10; nat US; m 33; c 3. PATHOLOGY. Educ: Univ Berlin, MD, 36. Prof Exp: Instr path, Johns Hopkins Univ, 36-40; instr, Harvard Univ, 40-46; sr scientist div biol & med res, Argonne Nat Lab, 46-63; lectr, 63-70, ASSOC PROF ANAT, HARVARD MED SCH, 70-, ASSOC DEAN, 67- Concurrent Pos: Finney-Howell Cancer Res Found fel, Johns Hopkins Univ, 38-40; Cancer Comn fel, Harvard Univ, 40-41; dir med div, Argonne Nat Lab, 47-52; mem comt growth, Nat Res Coun, 51-55; sci secy, Radiation Comt, UN, 57-58; sr res assoc, Cancer Res Inst, New Eng Deaconess Hosp, 63-70. Mem: Am Soc Exp Path; Am Asn Path & Bact; Am Asn Cancer Res. Res: Experimental pathology. Mailing Add: Harvard Med Sch Boston MA 02115

LISCOMBE, ERNEST A R, b Winnipeg, Man, Mar 10, 28; m 50; c 2. ENTOMOLOGY. Educ: Univ Man, BSA, 49, PhD(entom), 64; Univ Minn, MSc, 57. Prof Exp: Asst, Univ Minn, 50-51; entomologist, Pillsbury Mills, Inc, 51-57; res officer entom, Res Br, Can Dept Agr, 57-65; entomologist, Bd Grain Comnr Can, 65-69; VPRES & TECH DIR, PHOSTOXIN SALES, INC, 69- Mem: Entom Soc; Entom Soc Can. Res: Use of insecticides and fumigants for the control of insects attacking stored products; grain sanitation program in the licensed elevator system in Canada. Mailing Add: Phostoxin Sales Inc PO Box 469 Alhambra CA 91802

LISELLA, FRANK SCOTT, b Lancaster, Pa, Aug 11, 36; m 58. PUBLIC HEALTH. Educ: Pa State Teachers Col, Millersville, 57; Tulane Univ La, MPH, 61; Univ Iowa, PhD(prev med), 70. Prof Exp: Sanitarian, Pa Dept Health, 57-64; coordr, Commun Dis Control Proj, USPHS, Fla, 64-66, chief training & consultation, Pesticides Prog, Nat Commun Dis Ctr, 66-68; asst to dir div community studies, Food & Drug Admin, Ga, 69-70; asst dir div pesticide community studies, Environ Protection Agency, 70-72; health sci adv, Nat Med Audiovisual Ctr, 72-73; CHIEF, PROG DEVELOP BR, ENVIRON HEALTH SERVS DIV, CTR DIS CONTROL, GA, 73- Concurrent Pos: Adj prof, Dekalb Col, 70- Mem: Nat Environ Health Asn; Am Pub Health Asn. Res: Epidemiology of acute intoxications involving chemical agents of various types; etiology of self-induced intoxications involving medicants, pesticides and other chemical compounds and measures for prevention of repetitive episodes. Mailing Add: 3042 Hathaway Ct Chamblee GA 30341

LISH, PAUL MERRILL, b Idaho, Feb 2, 21; m 41; c 4. PHARMACOLOGY. Educ: Idaho State Univ, BS, 49; Univ Nebr, MS, 51; St Louis Univ, PhD(pharmacol), 55. Prof Exp: Instr, Sch Med, St Louis Univ, 53-55; pharmacologist, group leader & sect leader, Mead Johnson Res Ctr, Ind, 55-62, dir pharmacol, 63-67, vpres biol sci, 67-69; DIR RES & DEVELOP, PHARMACEUT GROUP, CHROMALLOY AM CORP, 69- Mem: AAAS; Am Chem Soc; Soc Exp Biol & Med; Am Soc Pharmacol & Exp Therapeut; NY Acad Sci. Res: Pharmacology of peripheral neuroeffector and cardiovascular systems; adrenergics and anti-adrenergics, anti-allergics, gastrointestinal drugs. Mailing Add: Chromalloy Am Corp Pharmaceut Group 6200 S Lindbergh St Louis MO 63123

LISK, DONALD JAMES, b Buffalo, NY, May 12, 30; m 59; c 2. SOIL CHEMISTRY. Educ: Univ Buffalo, BA, 52; Cornell Univ, MS, 54, PhD(soil chem), 56. Prof Exp: From asst prof to assoc prof, 56-66, PROF ENTOM, PESTICIDE RESIDUE LAB, CORNELL UNIV, 66- Mem: Am Chem Soc. Res: Pesticide chemistry. Mailing Add: Dept of Food Sci Cornell Univ Ithaca NY 14850

LISK, GEORGE F, b Belchertown, Mass, Mar 14, 14; m 42; c 1. ORGANIC CHEMISTRY. Educ: Clark Univ, BA, 35, MA, 37. Prof Exp: Res chemist, Nat Aniline Div, 37-40, 41-52, sect leader, 52-53, group leader, 53-57, supt chem res, 57-58, mgr res & develop, 58-60, MGR RES ADMIN, SPECIALTY CHEM DIV, ALLIED CHEM CORP, 60- Mem: Am Chem Soc. Res: Intermediates and colorants. Mailing Add: 11 Brantwood Rd Amherst NY 14226

LISK, ROBERT DOUGLAS, b Pembroke, Ont, Nov 10, 34. PHYSIOLOGY, ENDOCRINOLOGY. Educ: Queen's Univ, Ont, BA, 57; Harvard Univ, AM, 59, PhD(biol), 60. Prof Exp: From instr to assoc prof, 60-70, PROF BIOL, PRINCETON UNIV, 70- Concurrent Pos: NSF grants, 60-66; vis asst prof, Sch Med, Univ Calif, Los Angeles, 65; NIH grants, 66-71; mem panel regulatory biol, Nat Sci Found, 72-75. Mem: AAAS; Am Soc Zool; Am Asn Anat; Endocrine Soc; Am Physiol Soc. Res: Neuroendocrine mechanisms such as sites of action and interactions of hormones on central nervous system for regulation of hormone output by endocrine organs and hormone role in differentiation and triggering of behavioral sequences. Mailing Add: Dept of Biol Princeton Univ Princeton NJ 08540

LISKA, BERNARD JOSEPH, b Hillsboro, Wis, May 31, 31; m 52; c 2. FOOD TECHNOLOGY. Educ: Univ Wis, BS, 53, MS, 56, PhD(dairy & food technol), 57. Prof Exp: From asst prof to assoc prof animal sci, 52-65, dir, Food Sci Inst, 69-75, PROF FOOD SCI, PURDUE UNIV, 65-, DIR, AGR EXP STA, 75- Concurrent Pos: Sci ed, J Food Sci, 70- Mem: Inst Food Technologists; Am Dairy Sci Asn; Am Chem Soc. Res: Food bacteriology; lactic cultures; bulk handling of milk; milk quality and enzymes; pesticide residue analysis Mailing Add: Agr Exp Sta Purdue Univ West Lafayette IN 47906

LISKA, JOHN WILLIAM, JR, organic chemistry, see 12th edition

LISKA, KENNETH J, b Hinsdale, Ill, June 4, 29; m 57; c 3. CHEMISTRY, ECOLOGY. Educ: Univ Ill, BS, 51, MS, 53, PhD(med chem), 56. Prof Exp: Assoc prof pharmaceut chem, Duquesne Univ, 56-61 & Univ Pittsburgh, 61-69; assoc prof chem, US Int Univ, 69-75; INSTR CHEM, MESA COL, 75- Mem: Am Chem Soc. Res: Synthetic organic medicinal chemistry. Mailing Add: 2947 Honors Ct San Diego CA 92122

LISKEY, NATHAN EUGENE, b Live Oak, Calif, Apr 26, 37; m 57; c 3. HEALTH SCIENCE. Educ: La Verne Col, BA, 59; Ind Univ, Bloomington, MS, 61, HSD(health safety), 69. Prof Exp: Teacher pub schs, Calif, 59-65; asst prof health & phys educ, 65-68, assoc prof, 69-75, PROF HEALTH SCI, CALIF STATE UNIV, FRESNO, 75- Concurrent Pos: USPHS grant, HEW, 68-69; sex therapist, Ctr Coun & Ther, 73- Mem: Soc Sci Study Sex; Am Asn Sex Educ & Coun. Res: Physical and emotional aspects of behavior relating to accident prevention; human sexuality; sexual behavior of the aged. Mailing Add: Dept Health Sci Calif State Univ Fresno CA 93740

LISKOWITZ, JOHN W, physical chemistry, see 12th edition

LISMAN, FREDERICK LOUIS, b Wilkes-Barre, Pa, Jan 14, 39; m 62; c 3. NUCLEAR CHEMISTRY. Educ: Fairfield Univ, BS, 60; Purdue Univ, PhD(nuclear chem), 65. Prof Exp: Sr res radiochemist, Idaho Nuclear Corp, 65-70; asst prof, 70-72, ASSOC PROF CHEM, FAIRFIELD UNIV, 72-, CHMN DEPT, 75- Mem: AAAS; Am Chem Soc. Res: Fission yield determination; radiochemical separations; mass spectrometric techniques; measurement of fissionable material; mass and charge distribution in low Z fission; chemical separation techniques; energy resources. Mailing Add: 353 W Rutland Rd Milford CT 06460

LISMAN, HENRY, b Boston, Mass, July 3, 13; m 38; c 2. MATHEMATICS. Educ: Univ Boston, BS, 34, MS, 35, PhD(physics), 39. Prof Exp: Asst physics, Univ Boston, 34-35; instr, Northeastern Univ, 40-42; from assoc physicist to physicist, Sig Corps Eng Labs, NJ, 42-47; from instr to assoc prof, 47-57, PROF MATH, YESHIVA UNIV, 57- Concurrent Pos: Consult physicist, US Army Electronics Labs, 49-68. Mem: Am Phys Soc; Am Math Soc. Res: Electromagnetic wave propagation. Mailing Add: 3777 Independence Ave Bronx NY 10463

LISONBEE, LORENZO KENNETH, b Mesa, Ariz, Nov 25, 14; m 38; c 8. BIOLOGY, SCIENCE EDUCATION. Educ: Ariz State Univ, BA, 37, MA, 40, EdD, 63. Prof Exp: Teacher sci & dept chmn high schs, Ariz, 40-58; SCI CONSULT, PHOENIX HIGH SCHS, 58-; FAC ASSOC, ARIZ STATE UNIV, 63- Concurrent Pos: Consult, Am Geol Inst & Am Inst Biol Sci & Biol Sci Curriculum Study; contribr, Encyclopedia Britannica, 62 & 74. Mem: Fel AAAS; Nat Asn Res Sci Teaching; Nat Sci Teachers Asn; Nat Asn Biol Teachers. Res: Research in science teaching; desert biology. Mailing Add: 1007 W Second Pl Mesa AZ 85201

LISS, ALAN, b Pittsburgh, Pa, Sept 14, 47; m 71; c 1. MICROBIAL GENETICS, VIROLOGY. Educ: Univ Calif, Berkeley, BS, 69; Univ Rochester, PhD(microbiol), 73. Prof Exp: Fel microbiol, York Univ, 73-74; Nat Cancer Inst fel, Scripps Clin & Res Found, 74-75; ASST PROF BIOL, UNIV CONN, 75- Honors & Awards: Sigrid Juselius Found Award, 75. Mem: AAAS; Am Soc Microbiol; Sigma Xi. Res: Genetics and virology of members of the mycoplasmas; isolation of auxotrophic mutants; characterization of plasmid DNA; investigation of host-parasite interactions in mycoplasma-related infections. Mailing Add: Dept of Microbiol U-44 Univ of Conn Storrs CT 06268

LISS, LEOPOLD, b Lwow, Poland, Nov 19, 23; nat US; m 48; c 2. NEUROPATHOLOGY. Educ: Lwow Gramar Sch, Poland, BA, 41; Univ Heidelberg, MD, 50; Univ Mich, MS, 55. Prof Exp: From instr to asst prof neuropath, Univ Mich, 51-60; assoc prof, 60-64, PROF NEUROPATH, OHIO STATE UNIV, 64- Mem: Am Asn Pathologists & Bacteriologists; Am Soc Exp Path; Soc Neurosci; Am Asn Cancer Res; Asn Res Nerv & Ment Dis. Res: Clinical and experimental neuropathology; tissue cultures of nervous system. Mailing Add: Upham Hall Ohio State Univ Columbus OH 43210

LISS, MAURICE, b Boston, Mass, Dec 18, 26. BIOCHEMISTRY, BIOLOGY. Educ: Harvard Univ, AB, 49; Tufts Univ, PhD(biochem), 58. Prof Exp: Chemist, Peter Bent Brigham Hosp, 49-51; chemist, Mass Dept Pub Safety, 51-53; Am Cancer Soc res fel, enzymol, Brandeis Univ, 58-60; res assoc dermat, Sch Med, Tufts Univ, 61-63, from asst prof to assoc prof, 63-68; assoc prof, 68-73, PROF BIOL, BOSTON COL, 73- Mem: Am Chem Soc; Am Soc Biol Chem; AAAS; Am Asn Immunologists. Res: Proteins; immunology. Mailing Add: Dept of Biol Boston Col Chestnut Hill MA 02167

LISS, RAYMOND LEONARD, physical chemistry, organic chemistry, see 12th edition

LISS, ROBERT H, b Boston, Mass, Nov 2, 36; m 61; c 2. CYTOLOGY, CYTOPATHOLOGY. Educ: Tufts Univ, BS, 59; Univ Mass, MA, 61, PhD(cytol, biochem), 64. Prof Exp: ASSOC SURG, HARVARD MED SCH, 70-; HEAD ELECTRON MICROS LAB, LIFE SCI DIV, ARTHUR D LITTLE, INC, 65-; RES ASSOC CARDIOVASC SURG, CHILDREN'S MED CTR, BOSTON, 75- Concurrent Pos: Vis scientist, Tex Heart Inst, 72- Mem: AAAS; Am Soc Cell Biol; Electron Micros Soc Am; Soc Develop Biol. Res: Radioautography; electron microscopy. Mailing Add: Life Sci Div Arthur D Little Inc Acorn Park Cambridge MA 02140

LISSANT, ELLEN KERN, b St Louis, Mo, Nov 4, 22; m 47; c 3. PHYCOLOGY. Educ: Wash Univ, St Louis, AB, 44, AM, 46, PhD(bot), 68. Prof Exp: Lab instr bot, Wash Univ, St Louis, 43-45; teacher, Webster Groves High Sch, 45-46; asst bot, Wash Univ, St Louis, 46-47; asst herbarium, Stanford Univ, 47; from lectr to assoc prof, 60-73, PROF BIOL, FONTBONNE COL, 73- Mem: Bot Soc Am; Phycol Soc Am. Res: Palaeobotany; genetics; morphogenetic studies in the genus Erythrocladia Rosenvinge. Mailing Add: 12804 Westledge Lane St Louis MO 63131

LISSANT, KENNETH JORDAN, b London, Eng, Aug 6, 20; nat US; m 47; c 3. COLLOID CHEMISTRY. Educ: Ottawa Univ, Kans, AB, 41; Wash Univ, St Louis, MS, 43; Stanford Univ, PhD(chem), 47. Prof Exp: Asst chem, Ottawa Univ, Kans, 39-41; asst chem, Wash Univ, St Louis, 41-44, instr physics, 43-44; res chemist, 44-65, advan res coordr, 65-68, DIR ADVAN RES, PETROLITE CORP, 68- Concurrent Pos: Bristol-Myers fel. Mem: AAAS; Am Chem Soc. Res: Vacuum-tube titrimeters; solubilization of liquids; polymerization of unsaturates; foams; surfactants; emulsions; information retrieval; pollution abatement. Mailing Add: Petrolite Corp 369 Marshall Ave Webster Groves MO 63119

LISSEY, ALLAN, b Regina, Sask, Mar 6, 36; m 58; c 3. HYDROGEOLOGY, ENGINEERING GEOLOGY. Educ: Univ Sask, BEng, 59, MSc, 64, PhD(hydrol), 68. Prof Exp: Engr, City Regina, Sask, 59-62; sci officer groundwater res, Geol Surv Can, Ont, 64-66; res scientist, Inland Waters Br, Dept Energy, Mines & Resources, Ont, 66-69; assoc prof hydrogeol & Nat Res Coun Can grant, Brock Univ, 69-75; SR HYDROLOGIST, EBA ENG CONSULTS, LTD, ALTA, 75- Mem: Eng Inst Can; fel Geol Asn Can. Res: Hydrogeology, especially groundwater flow problems; environmental geology. Mailing Add: EBA Eng Consults Ltd 14535 118 Ave Edmonton AB Can

LISSNER, DAVID, b Rochester, NY, July 25, 31. MATHEMATICS. Educ: Mass Inst Technol, BS, 53; Cornell Univ, PhD(math), 59. Prof Exp: Design engr, NAm Aviation, Inc, 53-55; Off Naval Res fel math, Northwestern Univ, 59-60; instr, Yale Univ, 60-62; asst prof, 62-64, ASSOC PROF MATH, SYRACUSE UNIV, 64- Mem: Am Math Soc. Res: Ring theory; linear, commutative and homological algebra; algebraic geometry. Mailing Add: Dept of Math Syracuse Univ Syracuse NY 13210

LIST, ALBERT, JR, b East Orange, NJ, Nov 5, 28; m 53; c 2. PLANT PHYSIOLOGY. Educ: Univ Mass, BS, 53; Cornell Univ, MS, 58, PhD(plant physiol), 61. Prof Exp: Instr bot, Douglass Col, Rutgers Univ, 61-62, asst prof bot & biol, 62-65; fel, Univ Pa, 65-66, lectr biomet & bot, 66-67; assoc prof biol, 67-74, ASSOC PROF BIOL SCI, DREXEL UNIV, 74- Concurrent Pos: NSF res grants, 63-65, 66-

67 & 70-72; USPHS grant, 70-73. Mem: AAAS; Bot Soc Am; Am Soc Plant Physiol; Am Inst Biol Sci; Soc Develop Biol. Res: Developmental botany; control theory for plant root growth; relationships of relative elemental growth rates to bioelectric and membrane properties in roots; air pollution effects on growth. Mailing Add: Dept of Biol Sci Drexel Univ 32nd & Chestnut Sts Philadelphia PA 19104

LIST, JAMES CARL, b Paducah, Ky, July 6, 26; m 47; c 2. HERPETOLOGY. Educ: Notre Dame Univ, BS, 48, MS, 49; Univ Ill, PhD(zool), 56. Prof Exp: From instr to asst prof biol, Loyola Univ, Ill, 52-57; from instr to assoc prof, 57-66, PROF BIOL, BALL STATE UNIV, 66- Mem: Soc Syst Zool; Am Soc Ichthyol & Herpet; Soc Study Evolution; Herpetologists League; Soc Study Amphibians & Reptiles. Res: Anatomy and ecology of amphibians and reptiles. Mailing Add: Dept of Biol Ball State Univ Muncie IN 47306

LIST, ROLAND, b Frauenfeld, Switz, Feb 21, 29; m 56; c 2. ATMOSPHERIC PHYSICS. Educ: Swiss Fed Inst Technol, Dipl phys, 52, Dr sc nat(atmospheric physics), 60. Prof Exp: Sect head atmospheric ice formation, Swiss Fed Inst Snow & Avalanche Res, 52-63; assoc chmn dept physics, 69-73, PROF METEOROL, UNIV TORONTO, 63-, ADMINR NEGOTIATED DEVELOP GRANT ATMOSPHERIC DYNAMICS, 74- Concurrent Pos: Chmn working group cloud physics & weather modification, World Meteorol Orgn, 70-, chmn panel experts in weather modification of exec comt, 72-, chmn sci comt, precipitation enhancement proj, 75-; mem comn cloud physics, Int Asn Meteorol & Atmospheric Physics-Int Union Geod & Geophys, 71-; Univ Toronto sci res rep, Univ Corp Atmospheric Res, Colo, 71-, trustee, 74-, mem sci prog eval comt, 74-75; mem subcomt meteorol & atmospheric sci, Assoc Comt Geod & Geophys, Nat Res Coun, 71-74, chmn, 72-74, mem, Assoc Comt Geod & Geophys, 72-74; mem, Can Nat Comt, Nat Res Coun, 74-, mem adv comt cloud physics & weather modification, Atmospheric Environ Serv, 74- Honors & Awards: Medal, Univ Leningrad, 70. Mem: Fel Am Meteorol Soc; Am Geophys Union; Can Asn Physicists; Royal Meteorol Soc; Can Meteorol Soc. Res: Precipitation physics; hail formation; artificial growth; ice structures; aerodynamics; heat and mass transfer; rain formation; cloud dynamics; precipitation modeling; atmospheric boundary layer studies with acoustic Doppler radar; weather modification; snow, avalanches. Mailing Add: Dept of Physics Univ of Toronto Toronto ON Can

LISTER, BRADFORD CARLTON, b Marblehead, Mass, Jan 6, 47; m 72. POPULATION BIOLOGY. Educ: Tufts Univ, BA, 69; Princeton Univ, PhD(pop biol), 75. Prof Exp: ASST PROF ZOOL, UNIV MASS, 75- Res: Field studies of terrestrial vertebrates aimed at determining the effects of competition on the ecology and evolution of natural populations; the evolution of niche width, patterns of variation in insular species and comparisons of community organization. Mailing Add: Dept of Zool Morrill Sci Ctr Univ of Mass Amherst MA 01002

LISTER, CLIVE R B, b Uxbridge, Eng, Feb 8, 36. GEOPHYSICS. Educ: Cambridge Univ, BA, 59, PhD(geophys), 62. Prof Exp: Consult geophys, Saclant ASW Res Ctr, 62-63; chief geophysicist, Ocean Sci & Eng Co, 63-65; res asst prof oceanog, 65-68, asst prof, 68-73, ASSOC PROF GEOPHYS & OCEANOG, UNIV WASH, 73- Concurrent Pos: Mem, Joides Heat Flow Comt, 67-74. Mem: Am Geophys Union; Brit Astron Asn. Res: Measurement of heat flow through the ocean floor; acoustic sub-bottom profiling; thermal theory applied to geodynamics; aspects of porous convection and water penetration applied to geothermal problems. Mailing Add: Dept of Oceanog WB-10 Univ of Washington Seattle WA 98195

LISTER, EARL EDWARD, b Harvey, NB, Apr 14, 34; m 56; c 3. RUMINANT NUTRITION. Educ: McGill Univ, BS, 55, MS, 57; Cornell Univ, PhD(nutrit), 60. Prof Exp: Feed nutritionist, Ogilvie Flour Mills Ltd, 60-65; res scientist, 65-75, DEP DIR, ANIMAL RES INST, RES BR AGR CAN, OTTAWA, 75- Concurrent Pos: Assoc ed, Can J Animal Sci, 73-75, ed, 75-77. Mem: Can Soc Animal Sci; Am Dairy Sci Asn; Am Soc Animal Sci. Res: Nutritional requirements of immature ruminants; feeding and management systems for beef cows and calves; production of beef from dairy breeds of cattle. Mailing Add: Animal Res Inst Agr Can Res Br Cent Exp Farm Ottawa ON Can

LISTER, FREDERICK MONIE, b Trenton, NJ, May 9, 23; m 54; c 3. MATHEMATICS. Educ: Tufts Univ, BS, 47; Univ Mich, MA, 51; Univ Utah, PhD(math), 66. Prof Exp: Instr math, Phillips Acad, Mass, 47-49; instr, Western Wash Col Educ, 54-56, from asst prof to assoc prof, 58-67; asst prof, Chico State Col, 57-58; prof, Southern Ore Col, 67-68; assoc prof, 68-69, PROF MATH, CENT WASH STATE COL, 69- Mem: Am Math Soc; Math Asn Am. Res: Geometric topology; embeddings of 2-spheres in Euclidean 3-space. Mailing Add: Dept of Math Cent Wash State Col Ellensburg WA 98926

LISTER, MAURICE WOLFENDEN, b Tunbridge Wells, Eng, Mar 27, 14; nat Can; m 40; c 5. INORGANIC CHEMISTRY. Educ: Oxford Univ, PhD, 38, MA, 47. Prof Exp: Assoc prof, 53-62, PROF CHEM, UNIV TORONTO, 62- Res: Complex inorganic compounds; mechanisms of inorganic reactions; magnetic susceptibilities. Mailing Add: Dept of Chem Univ of Toronto Toronto ON Can

LISTER, RICHARD MALCOLM, b Sheffield, Eng, Nov 14, 28; m 53; c 4. PLANT VIROLOGY. Educ: Sheffield Univ, BSc, 49, dipl ed, 50; Cambridge Univ, dipl agr sci, 51; Imp Col Trop Agr, Trinidad, dipl, 52; St Andrews Univ, PhD, 63. Prof Exp: Plant pathologist, WAfrican Cocoa Res Inst, 52-56; plant pathologist, Scottish Hort Res Inst, 56-66; assoc prof, 66-72, PROF PLANT VIRUSES, PURDUE UNIV, W LAFAYETTE, 72- Concurrent Pos: Fel bot & plant path, Purdue Univ, 63-64; NSF res grants, 67-72; assoc ed, Virology, 71- Mem: Fel Am Phytopath Soc; Brit Asn Appl Biol. Res: Viruses and virus diseases of horticultural and agricultural plants, especially those transmitted by soil-living vectors; defectiveness and dependence in plant viruses. Mailing Add: Dept of Bot & Plant Path Life Sci Bldg Purdue Univ West Lafayette IN 47906

LISTER, ROBERT HILL, b Las Vegas, NMex, Aug 7, 15; m 42; c 2. CULTURAL ANTHROPOLOGY. Educ: Univ NMex, BA, 37, MA, 38; Harvard Univ, MA, 47, PhD(anthrop), 50. Prof Exp: From instr to prof anthrop, Univ Colo, 47-70; PROF ANTHROP, UNIV N MEX, 71-; chief archeologist, 72-74, CHIEF, CHACO CTR, NAT PARK SERV, 71- Concurrent Pos: Ford Found fel, Mex, 53-54, Univ Colo fac fels, 53-54, 55, 59 & 63-64; mem adv panel anthrop, NSF, 62-63; mem fel rev comt, NSF-Nat Acad Sci, 64-65; dir, Univ Colo Archeol Res Ctr, Mesa Verde Nat Park, 65-70. Mem: Am Anthrop Asn; Soc Am Archaeol (pres, 70-71). Res: Archeology of southwestern United States and Mexico; origin and diffusion of majolica pottery throughout world. Mailing Add: Chaco Ctr Nat Park Serv PO Box 26176 Albuquerque NM 87125

LISTERMAN, THOMAS WALTER, b Cincinnati, Ohio, Dec 21, 38; m 69; c 1. SOLID STATE SCIENCE. Educ: Xavier Univ, BS, 59; Ohio Univ, PhD(solid state physics), 65. Prof Exp: Sr res physicist, Mound Lab, Monsanto Res Corp, 65-67; asst prof, 67-72, asst provost, 71-73, asst dean sci & eng, 70-71, ASSOC PROF PHYSICS, WRIGHT STATE UNIV, 72- Mem: Am Phys Soc. Res: Electronic properties of materials; positron annihilation; cryogenics. Mailing Add: Dept of Physics Wright State Univ Dayton OH 45431

LISTGARTEN, MAX, b Paris, France, May 24, 35; Can citizen; m 63; c 3. DENTISTRY, PERIODONTOLOGY. Educ: Univ Toronto, DDS, 59; FRCD(C), 69. Hon Degrees: MA, Univ Pa, 71. Prof Exp: Intern dent, Hosp for Sick Children, Toronto, 59-60; res assoc periodont, Harvard Med Sch, 63-64; from asst prof to assoc prof, Fac Dent, Univ Toronto, 64-68; assoc prof, 68-71, PROF PERIODONT & DIR PERIODONT RES, UNIV PA, 71 Concurrent Pos: Nat Res Coun Can fel periodont, Harvard Med Sch, 60-63; US ed, J Biol Buccale, 72- Honors & Awards: Award Basic Res Periodont Dis, Int Asn Dent Res, 73. Mem: AAAS; Am Dent Asn; Am Acad Periodont; Int Asn Dent Res; Am Asn Anatomists. Res: Ultrastructural investigations of the supporting structures of teeth and associated microbial flora in health and disease. Mailing Add: Ctr for Oral Health Res Univ of Pa Philadelphia PA 19104

LISTON, DAVID M, JR, information science, see 12th edition

LISTON, THOMAS VINCENT, organic chemistry, physical chemistry, see 12th edition

LISTOWSKY, IRVING, b Vilna, Poland, Dec 21, 35; US citizen; m 63; c 3. BIOCHEMISTRY. Educ: Yeshiva Univ, BA, 57; Polytech Inst Brooklyn, PhD(org chem), 63. Prof Exp: From instr to asst prof, 65-73, ASSOC PROF BIOCHEM, ALBERT EINSTEIN COL MED, 73- Concurrent Pos: NIH career develop award, 71-76; sr investr, NY Heart Asn, 68-70. Mem: Am Soc Biol Chemists. Res: Use of physico-chemical techniques to gain insight into structure-function relationships of various biological substances, especially proteins, carbohydrates and interacting macromolecules. Mailing Add: Albert Einstein Col of Med Yeshiva Univ Bronx NY 10461

LIT, ALFRED, b New York, NY, Nov 24, 14; m 47. VISION, EXPERIMENTAL PSYCHOLOGY. Educ: Columbia Univ, BS, 38, AM, 43, PhD, 48. Prof Exp: Lectr optom, Columbia Univ, 46-48, assoc, 48-49, asst psychol, 49, asst prof optom, 49-52, assoc prof, 52-56; res psychologist, Univ Mich, 56-59; head, Human Factors Staff, Systs Div, Bendix Corp, 59-61; PROF PSYCHOL, SOUTHERN ILL UNIV, 61- Concurrent Pos: Res grant, Am Acad Optom, Columbia Univ, 49, mem psychol staff, Off Naval Res Contract, 49-56; lectr, Univ Mich, 57-58; res grants, Eye Inst, USPHS & NSF, 62; mem, Armed Forces-Nat Res Coun Comt Vision; mem, exec comt, Eng Biophysics Prog, Southern Ill Univ, 72-; mem exec comt, Molecular Sci Doctoral Prog, Southern Ill Univ, 73-; sci referee, Am J Optom, 73. Honors & Awards: Sigma Xi-Kaplan Res Award, Southern Ill Univ Found, 71. Mem: Fel AAAS; fel Optical Soc Am; fel Am Psychol Asn; fel Am Acad Optom; Asn Res Vision & Ophthal. Res: Perception; systems research on human factors. Mailing Add: Dept of Psychol Southern Ill Univ Carbondale IL 62901

LIT, JOHN WAI-YU, b Canton, China, Aug 31, 37; Can citizen; c 2. OPTICS. Educ: Univ Hong Kong, BSc, 58, dipl Ed, 61; Univ Laval, DSc(optics), 69. Prof Exp: Head physics, Diocesan Boys Sch, 61-64; teacher sci, Quebec High Sch, 64-65; fel optics, Univ Western Ont, 68-69; res assoc, 69-71, asst prof, 71-75, ASSOC PROF OPTICS, UNIV LAVAL, 75- Concurrent Pos: Consult, Commun Res Ctr, Can, 73, Bell-Northern Res, Ottawa, 73-75, Gen-Tec Inc, Quebec, 74 & Defence Res Estab, Valcartier, Can, 75-76; assoc ed, Optical Soc Am, 74-79. Mem: Can Asn Physicists; Optical Soc Am; Inst Elec & Electronics Engrs. Res: Diffraction and propagation of electromagnetic waves, fiber and integrated optics; optical instrumentation. Mailing Add: Dept of Physics Univ Laval Ste-Foy PQ Can

LITCHFIELD, CAROL DARLINE, b Cincinnati, Ohio, Oct 10, 36; m 60. MARINE MICROBIOLOGY, MICROBIAL PHYSIOLOGY. Educ: Univ Cincinnati, BS, 58, MS, 60; Tex A&M Univ, PhD(biochem), 69. Prof Exp: Res scientist oceanog, Tex A&M Univ, 60-65; from instr to asst prof, 70-75, ASSOC PROF MICROBIOL, RUTGERS UNIV, NEW BRUNSWICK, 75- Mem: Am Soc Microbiol; Am Soc Limnol & Oceanog; Int Asn Theoret & Appl Limnol. Res: Microbiology and biochemistry of marine sediments and sea water; extracellular microbial proteases and their ecological importance. Mailing Add: Dept of Microbiol Rutgers Univ New Brunswick NJ 08903

LITCHFIELD, CHARLES CARTER, b Pasadena, Calif, Feb 18, 32; m 60. BIOCHEMISTRY. Educ: Rensselaer Polytech Inst, BS, 53; Am Inst Foreign Trade, BFT, 57; Tex A&M Univ, PhD(chem), 66. Prof Exp: Chemist, Procter & Gamble Co, 53-60; from asst prof to assoc prof lipid biochem, Tex A&M Univ, 60-69; assoc prof lipid biochem, 69-73, ASSOC PROF BIOCHEM, RUTGERS UNIV, 73- Concurrent Pos: Vis scientist, Fisheries Res Bd Can, 67. Honors & Awards: Bond Award, Am Oil Chem Soc, 63 & 66. Mem: Am Chem Soc; Am Oil Chem Soc; Am Littoral Soc; Soc Chem Indust. Res: Biochemistry of lipids of marine organisms; analysis of natural fat triglyceride mixtures; gas liquid chromatography of lipids; biochemical systematics of lipids. Mailing Add: Dept of Biochem Nelson Biol Labs Rutgers Univ New Brunswick NJ 08903

LITCHFIELD, JOHN HYLAND, b Scituate, Mass, Feb 13, 29; m 66; c 1. FOOD SCIENCE, INDUSTRIAL MICROBIOLOGY. Educ: Mass Inst Technol, SB, 50; Univ Ill, MS, 54, PhD(food technol), 56. Prof Exp: Chief chemist, Searle Food Corp, Fla, 50-51; res food technologist, Swift & Co, Ill, 56-57; asst prof food eng, Ill Inst Technol, 57-60; sr food technologist, 60-61, proj leader biosci, 61-62, asst chief biosci res, 62-64, chief biochem & microbiol res, 64-67, microbiol & environ biol res, 67-68, assoc mgr life sci, Dept Chem & Chem Eng, 68-70, mgr biol & med sci sect, Columbus Labs, 70-72, SR TECH ADV, COLUMBUS LABS, BATTELLE MEM INST, 73- Concurrent Pos: Consult to food indust, 57-60. Mem: Fel AAAS; fel Am Inst Chemists; fel Am Acad Microbiol; Soc Indust Microbiol (pres, 71-72); fel Am Pub Health Asn. Res: Food processing and preservation; food, industrial, sanitary and public health microbiology; microbial biochemistry; mass cultivation of microorganisms; nutrional evaluation of food processing methods. Mailing Add: Columbus Labs Battelle Mem Inst 505 King Ave Columbus OH 43201

LITCHFORD, ROBERT GARY, b Los Angeles, Calif, June 22, 34; m 71; c 5. PARASITOLOGY. Educ: Ga Southern Col, BS, 60; Rice Univ, PhD(biol), 65. Prof Exp: Asst prof, 65-67, ASSOC PROF BIOL, UNIV TENN, CHATTANOOGA, 67- Concurrent Pos: Biol consult, Hensley & Schmidt Engrs, Inc, 74- Mem: Am Soc Parasitol; Soc Nematol; NY Acad Sci. Res: Physiology of parasitism. Mailing Add: Dept of Biol Univ of Tenn Chattanooga TN 37401

LITHERLAND, ALBERT EDWARD, b Wallasey, Eng, Mar 12, 28; Can citizen; m 56; c 2. NUCLEAR PHYSICS. Educ: Univ Liverpool, BSc, 49, PhD(physics), 55. Prof Exp: Nat Res Coun Can fel nuclear physics, Atomic Energy Can, Ltd, 53-55, sci officer, 55-60; Atomic Energy Can Ltd vis scientist, Oxford Univ, 60-61; sci officer, Atomic Energy Can, Ltd, 61-66; PROF PHYSICS, UNIV TORONTO, 66- Honors & Awards: Gold Medal, Can Asn Physicists, 71. Mem: Fel Am Phys Soc; Can Asn Physicists; Royal Soc Can. Res: Nuclear spectroscopy of light nuclei using charged particle accelerators; radiative capture of charged particles by nuclei; collective motion in light nuclei; fast neutron spectroscopy. Mailing Add: Dept of Physics Univ of Toronto Toronto ON Can

LITMAN, BURTON JOSEPH, b Boston, Mass, May 8, 35; m 58; c 2. BIOCHEMISTRY. Educ: Boston Univ, BA, 58; Univ Ore, PhD(biophys chem), 66. Prof Exp: ASST PROF BIOCHEM, SCH MED, UNIV VA, 68- Concurrent Pos: USPHS fel biochem, Sch Med, Univ Va, 66-68, Nat Eye Inst grant, 70-77; NSF grant, 74-78. Mem: AAAS; Biophys Soc; Am Soc Photobiol; Am Chem Soc; Asn Res Vision & Ophthal. Res: Structure-function relationships in biological membranes with particular emphasis on the molecular mechanism of vision. Mailing Add: Dept of Biochem Univ of Va Sch of Med Charlottesville VA 22901

LITMAN, GARY WILLIAM, b Shoemaker, Calif, June 26, 45; m 70. IMMUNOLOGY, BIOCHEMISTRY. Educ: Univ Minn, BA, 67, PhD(microbiol), 72. Prof Exp: Res asst microbiol, Univ Minn, 67-68, teaching specialist microbiol & pediat, 68-70, instr pediat & path, 70-72, asst prof path, 72; ASSOC MEM, DEPT MACROMOLECULAR BIOCHEM, SLOAN-KETTERING INST, 72-; ASSOC PROF BIOL, SLOAN-KETTERING DIV, GRAD SCH MED SCI, CORNELL UNIV, 73- Mem: AAAS; Am Asn Immunologists; NY Acad Sci; Am Soc Zoologists; Biophys Soc. Res: Evolution of immunoglobulin structure; conformation of immunoglobulins; chemical carcinogenesis. Mailing Add: Dept of Macromolec Biochem Sloan-Kettering Inst Cancer Res New York NY 10021

LITMAN, IRVING ISAAC, b Chelsea, Mass, Nov 16, 25. FOOD TECHNOLOGY. Educ: Univ Mass, BA, 49, MS, 51; Wash State Univ, PhD(food technol), 56. Prof Exp: Processed food inspector, Prod & Mkt Admin, USDA, 50-51; food technologist, Gen Prod Div, Qm Food & Container Inst, US Armed Forces, 51-53; asst dairy technologist, Wash State Univ, 53-55; jr res chemist, Univ Calif, 55-56; proj leader, Res Ctr, Gen Foods Corp, 56-62; flavor chemist, Givaudan Corp, 62-64; sect head, Durkee Famous Foods, 64-65; RES DIR, STEPAN FLAVORS & FRAGRANCES, INC, 66- Mem: The Chem Soc; Inst Food Technologists; AAAS; Sigma Xi; Flavor Chemists Soc. Res: Development of synthetic and natural flavorings for food, tobacco and pharmaceuticals. Mailing Add: 1832 Wildberry Dr Glenview IL 60025

LITOVITZ, THEODORE AARON, b New York, NY, Oct 14, 23; m 46; c 2. PHYSICS. Educ: Cath Univ, AB, 46, PhD, 50. Prof Exp: From asst prof to assoc prof, 50-59, PROF PHYSICS, CATH UNIV, 59- Concurrent Pos: Consult, Univ Hosp, Georgetown, 50-57. Mem: Am Phys Soc; Acoust Soc Am; Am Philos Soc. Res: Ultrasonic propagation and light scattering in studies of molecular motions in liquids. Mailing Add: Dept of Physics Cath Univ of Am Washington DC 20064

LITSKY, BERTHA YANIS, b Chester, Pa, Jan 2, 20; m 65; c 2. MICROBIOLOGY, HOSPITAL ADMINISTRATION. Educ: Philadelphia Col Pharm, BSc, 42; NY Univ, MPA, 64; Walden Univ, PhD(educ), 74. Prof Exp: Head dept bact, Assoc Labs of Philadelphia, 42-44; asst supvr prod, Nat Drug Co, 44-45; res bacteriologist, Univ Pa, 45-50; self-employed, Pa & NY, 50-56; head dept bact, Staten Island Hosp, NY, 56-64; environ microbiol consult, 64-74; NURSE CONSULT, BINGHAM ASSOCS FUND, NEW ENG MED CTR HOSP, 74- Concurrent Pos: Mem, Standards Comt, Asn Operating Room Nurses, 75- Mem: Am Hosp Asn; Am Soc Microbiol; Am Pub Health Asn; Inst Sanit Mgt; Royal Soc Health. Res: Environmental and clinical microbiology; control of cross-infection in hospitals; hospital sanitation, environmental microbiology and administration; antimicrobial agents, antiseptics, disinfectants and germicides; disinfection and sterilization; aseptic practices in the operating room. Mailing Add: Marshall Hall Univ of Mass Amherst MA 01002

LITSKY, WARREN, b Worcester, Mass, June 10, 24; m 65. BACTERIOLOGY. Educ: Clark Univ, AB, 45; Univ Mass, MS, 48; Mich State Univ, PhD(bact), 51. Prof Exp: Inst bact, Mich State Univ, 51; from asst res prof to res prof, 51-61, dir tech guid ctr, Environ Control, 68-73, COMMONWEALTH PROF BACT, UNIV MASS, AMHERST, 61-, DIR INST AGR & INDUST MICROBIOL, 63-, CHMN DEPT ENVIRON SCI, 72- Concurrent Pos: Ed, Newsletter, 62-66; NIH res fel marine microbiol, Oceanog Inst, Fla State Univ, 65-66; adv comt radiation preservation foods, Nat Acad Sci-Nat Res Coun, 65-67; consult, Div Water Supply & Pollution Control, Fed Water Pollution Control Admin, Environ Protection Agency & Sanit Prog, Health & Hosp Facil, Dept Health, Educ & Welfare; affiliate prof bact, Clark Univ, 66-72; mem task group microbiol aspects of raw shucked oysters, Nat Res Coun; pres, Environ Mgt Inc, 67-; mem adv coun, Mass Water Resource Res Ctr, 67-72; spec comn to study methods to reuse solid waste, Mass, 70-73; comt on environ impact statements, Lower Pioneer Valley Planning Comn, 71-; res grant study sect, Environ Protection Agency; consult, Shellfish Prog, Food & Drug Admin, 74-; mem, Mass Comn Nuclear Safety, 74-; mem, US Comt Israel Environ, 73-; Dir, Rush-Hampton Industs, 75- Honors & Awards: Sayer Prize, 51; dipl, Am Intersoc Acad Cert Sanit; Ed Award, Hosp Mgt Mag, 69. Mem: Fel AAAS; Soc Indust Microbiol; Am Soc Microbiol; fel Am Pub Health Asn; Brit Soc Appl Bact. Res: Water and sewage pollution; thermal death and vaccine production; fermentations; food and dairy bacteriology; disinfection; marine microbiology. Mailing Add: Dept Environ Sci Marshall Hall Univ of Mass Amherst MA 01002

LITSTER, JAMES DAVID, b Toronto, Ont, Can, June, 19, 38; m 65. SOLID STATE PHYSICS. Educ: McMaster Univ, BEng, 61; Mass Inst Tech, PhD(physics), 65. Prof Exp: From instr to assoc prof, 65-75, PROF PHYSICS, MASS INST TECHNOL, 75- Concurrent Pos: Fel, John Simon Guggenheim Mem Found, 71-72; lectr physics, Harvard Med Sch, 74- Mem: AAAS; Am Phys Soc. Res: Magnetism; light scattering; liquid crystals. Mailing Add: Dept of Physics Mass Inst of Technol Cambridge MA 02139

LITT, BERTRAM D, b New York, NY, Oct 6, 25; m 60; c 2. BIOSTATISTICS. Educ: NY Univ, BA, 46; Univ Mo, MA, 52. Prof Exp: Assoc res scientist, Human Eng, Col Eng, Res Div, NY Univ, 54-60; assoc proj dir, Am Orthotic & Prosthetic Asn, 60-62; statistician, Bur Med & Surg, Vet Admin, 62-66 & Bur Med, Dept Navy, 66-69; MATH STATISTICIAN, BUR DRUGS, ECON DEVELOP AGENCY, 69- Concurrent Pos: Consult, Clin Trials Comt, Rheumatoid Arthritis Found, 72-75. Mem: Am Statist Asn; Biomet Soc. Res: Design and analysis of studies, particularly clinical and animal trials and design of medical information systems. Mailing Add: 14402 Woodcrest Dr Rockville MD 20853

LITT, GERALD JOSEPH, analytical chemistry, see 12th edition

LITT, MICHAEL, b New York, NY, Apr 17, 33; m 56. BIOCHEMISTRY. Educ: Oberlin Col, BA, 54; Harvard Univ, PhD(chem), 58. Prof Exp: Instr chem, Reed Col, 58-62, assoc prof, 64-67; assoc prof biochem, 67-71, PROF BIOCHEM & MED GENETICS, MED SCH, UNIV ORE, 71- Concurrent Pos: NIH spec fel, Mass Inst Technol, 62-63; NSF fel, Auckland Univ, 66-67. Mem: Am Chem Soc; Am Soc Biol Chemists. Res: Structure of RNA; structure and function of transfer RNA. Mailing Add: Dept of Biochem Univ of Ore Med Sch Portland OR 97201

LITT, MORTIMER, b Brooklyn, NY, Sept 28, 25; m 54; c 3. IMMUNOLOGY. Educ: Columbia Univ, BA, 47; Univ Rochester, MD, 52. Prof Exp: Med house officer & asst resident physician, Peter Bent Brigham Hosp, 52-54; instr, 56-59, assoc, 60-65, asst prof, 65-71, ASSOC PROF BACT & IMMUNOL, HARVARD MED SCH, 71-, ASST DEAN TEACHING RESOURCES, 73-; ASST DIR, DEPT BACT, BOSTON CITY HOSP, 69- Concurrent Pos: Res fel, Harvard Med Sch, 54-56 & 56-59; Helen Hay Whitney Found fel, 59-63; estab investr, Am Heart Asn, 63-68. Mem: Am Asn Immunologists; Am Fedn Clin Res; NY Acad Sci. Res: Eosinophil leukocyte. Mailing Add: Harvard Med Sch Boston MA 02115

LITT, MORTON HERBERT, b Brooklyn, NY, Apr 10, 26; m 57; c 2. POLYMER CHEMISTRY. Educ: City Col New York, BS, 47; Polytech Inst Brooklyn, MS, 53, PhD(polymer chem), 56. Prof Exp: Turner & Newall res fel, Manchester Univ, 56-57; res assoc, State Univ NY Col Forestry, Syracuse, 58-60; sr scientist, Cent Res Lab, Allied Chem Corp, 60-64, assoc dir res, 65-67; ASSOC PROF POLYMER SCI, CASE WESTERN RESERVE UNIV, 67- Mem: AAAS; Am Chem Soc; NY Acad Sci; Am Phys Soc; The Chem Soc. Res: Ionic and free radical polymerization mechanisms; organo-fluorine chemistry; polymer mechanical properties; polymer electrical properties. Mailing Add: Dept of Macromolecular Sci Case Western Reserve Univ Cleveland OH 44106

LITTAU, VIRGINIA CONWAY, cell biology, see 12th edition

LITTAUER, RAPHAEL MAX, b Leipzig, Ger, Nov 28, 25; US citizen, m 50; c 2. EXPERIMENTAL HIGH ENERGY PHYSICS, ELECTRONICS. Educ: Cambridge Univ, MA & PhD(physics), 50. Prof Exp: Asst physics, Cambridge Univ, 47-50; res assoc nuclear physics, Cornell Univ, 50-54 & Synchrotron Lab, Gen Elec Co, 54-55; res assoc prof physics, 55-63, res prof, 63-65, PROF PHYSICS & NUCLEAR STUDIES, CORNELL UNIV, 65-, CHMN DEPT, 74- Mem: Am Phys Soc. Mailing Add: 1417 Slaterville Rd Ithaca NY 14850

LITTELL, ARTHUR SIMPSON, b Washington, DC, Jan 18, 25; m 48, 71; c 5. BIOSTATISTICS. Educ: Harvard Univ, AB, 44; Johns Hopkins Univ, AM, 48, ScD(biostatist), 51. Prof Exp: From instr to assoc prof biostatist, Sch Med, Western Reserve Univ, dir dept, Biomet Comput Lab, 63-70; PROF BIOMET, UNIV TEX SCH PUB HEALTH HOUSTON, 70- Concurrent Pos: Mem epidemiol & dis control study sect, Div Res Grants, NIH, 68-72. Mem: Biomet Soc; Am Statist Asn; Am Heart Asn; Am Pub Health Asn; Soc Epidemiol Res. Res: Actuarial and statistical methods in medicine and public health; applications of computers in public health; cooperative studies; cardiovascular research. Mailing Add: Univ of Tex Sch of Pub Health Box 20186 Houston TX 77025

LITTELL, RAMON CLARENCE, b Rolla, Kans, Nov 18, 42; m 66; c 1. AGRICULTURAL STATISTICS, MATHEMATICAL STATISTICS. Educ: Kans State Teachers Col, BS, 64; Okla State Univ, MS, 66, PhD(statist), 70. Prof Exp: PROF STATIST, UNIV FLA, 70- Mem: Am Statist Asn. Res: Determination of relative efficiencies of tests of hypothesis, nonparametric estimation, design of experiments and statistical computing. Mailing Add: Statist Inst of Food & Agr Sci Univ of Fla Gainesville FL 32601

LITTERIA, MARILYN, b Cleveland, Ohio, Aug 9, 31. NEUROENDOCRINOLOGY. Educ: Case Western Reserve Univ, BS, 55; Univ Calif, Berkeley, PhD(physiol), 67. Prof Exp: Lab technician physiol, Case Western Reserve Univ, 55-59, lab technician anat, 59-60; teaching asst physiol, Univ Calif, Berkeley, 61-63; res fel endocrinol, Scripps Clin & Res Found & Develop Neuroendocrinol Lab, Vet Admin Hosp, San Fernando, 67-69; res assoc reproductive biol, Univ NC, 70-71; instr anat & Sloan fel, Northwestern Univ, 71-72; RES PHYSIOLOGIST NEUROLOGY, VET ADMIN HOSP, DOWNEY, ILL, 72- Concurrent Pos: USPHS res grant neurol dis & stroke, 74-75. Mem: Endocrine Soc; Sigma Xi. Res: The role of sex steroids in central nervous system development; the role of sex steroids on brain metabolism. Mailing Add: 14 S Elmwood Waukegan IL 60085

LITTLE, A BRIAN, b Montreal, Que, Mar 11, 25; US citizen; m 49; c 6. OBSTETRICS & GYNECOLOGY. Educ: McGill Univ, BA, 48, MD, CM, 50; Royal Col Physicians & Surgeons, Can, cert obstet & gynec, 55, FRCPS(C), 57; Am Bd Obstet & Gynec, dipl, 59. Prof Exp: Intern, Montreal Gen Hosp, 50-51; asst resident & resident, Boston Lying-in-Hosp & Free Hosp Women, Boston, 51-54; asst obstet, Harvard Med Sch, 55-56, instr obstet & gynec, 56-58, tutor med sci, 57-65, assoc obstet & gynec, 58-63, asst prof, 63-65; prof, 66-72, ARTHUR H BILL PROF OBSTET & GYNEC & DIR DEPT REPRODUCTIVE BIOL, SCH MED, CASE WESTERN RESERVE UNIV, 72- Concurrent Pos: Teaching fel obstet & gynec, Harvard Med Sch, 52-54; instr, Sch Nursing, Boston Univ, 51 & 55-57; asst obstetrician outpatients, Boston Lying-in-Hosp, 55-56, asst obstetrician, 56-58, assoc, 58-59, obstetrician & gynecologist, 59-65; sr obstetrician prenatal metab div, USPHS, 55-56; assoc vis surgeon, Boston City Hosp, 55-64, assoc dir dept obstet & gynec, 58-63, dir, 63-65, vis surgeon, 65; asst surgeon, Free Hosp Women, 58-64, mem courtesy staff, 64-65; mem consult staff, Sturdy Mem Hosp, Attleboro, 61-65; chief consult, Hunt Mem Hosp, Danvers, 62-65; mem consult staff, Elliott Community Hosp, Keene, NH, 63-65; dir dept obstet & gynec, Cleveland Metrop Hosp, Ohio, 66-72; assoc obstetrician & gynecologist, Univ Hosps, Cleveland, 66-72, dir dept obstet & gynec, 72- Mem: AMA; fel Am Col Obstet & Gynec; Am Gynec Soc; Soc Gynec Invest; fel Am Col Surgeons. Res: Steroid mechanism in vivo, in vitro, primarily in reproduction. Mailing Add: Univ Hosps 2065 Adelbert Rd Cleveland OH 44106

LITTLE, ANGELA C, b San Francisco, Calif, Jan 12, 20; m 47; c 1. FOOD SCIENCE. Educ: Univ Calif, AB, 40, MS, 54, PhD(agr chem), 69. Prof Exp: Res asst food sci, 53-56, jr specialist, 56-58, from asst specialist to assoc specialist, 58-69, asst food scientist, 69-71, ASSOC FOOD SCIENTIST, UNIV CALIF, BERKELEY, 71-, LECTR FOOD SCI, 69- Mem: AAAS; Optical Soc Am; Inter-Soc Color Coun; Inst Food Technol; Am Soc Testing & Mat. Res: Colorimetry; methodology and systems relating to objective measurements; color vision. Mailing Add: Dept of Nutrit Sci Univ of Calif Berkeley CA 94720

LITTLE, CHARLES EDWARD, b Kansas City, Kans, Apr 18, 26; m 47; c 3. MATHEMATICS. Educ: Univ Kans, AB, 48; Ft Hays Kans State Col, MS, 55; Colo State Col, EdD(math educ), 64. Prof Exp: Instr pub sch, Kans, 51-60, supt, 60-61; instr math, Colo State Col, 61-63; from asst prof to assoc prof, 64-68, chmn dept math, 67-70, PROF MATH EDUC, NORTHERN ARIZ UNIV, 68-, DEAN COL ARTS & SCI, 74- Mem: Math Asn Am. Res: Training of elementary and secondary mathematics teachers; methods of instruction in mathematics at college level, particularly educational media and techniques for handling large groups; mathematics for the social and behavioral sciences; linear models and statistics. Mailing Add: Dept of Math Northern Ariz Univ Flagstaff AZ 86001

LITTLE, CHARLES GORDON, b Hunan, China, Nov 4, 24; m 54; c 5. ATMOSPHERIC PHYSICS, REMOTE SENSING. Educ: Univ Manchester, BSc, 48, PhD(radio astron), 52. Prof Exp: Jr engr, Cosmos Mfg Co, Eng, 44-46; jr physicist, Ferranti, Ltd, 46-47; asst lectr physics, Univ Manchester, 52-53; prof geophys res & dep dir geophys inst, Univ Alaska, 54-58; chief radio astron & arctic propagation sect, Nat Bur Standards, 58-60, upper atmosphere & space physics div, Boulder Labs, 60-

62, dir cent radio propagation lab, 62-65; dir inst telecommun sci & aeronomy, Environ Sci Serv Admin, 65-67, DIR WAVE PROPAGATION LAB, ENVIRON RES LABS, NAT OCEANIC & ATMOSPHERIC ADMIN, 67- Concurrent Pos: Consult, US Nat Comt for Int Geophys Year, 57-59. Mem: AAAS; Inst Elec & Electronics Eng; Royal Astron Soc; Int Union Radio Sci. Res: Remote measurement of atmosphere and ocean, using electromagnetic and acoustic waves. Mailing Add: Wave Propagation Lab Nat Oceanic & Atmospheric Admin Boulder CO 80302

LITTLE, CHARLES HARRISON ANTHONY, b Toronto, Ont, May 4, 39; m 64; c 2. TREE PHYSIOLOGY. Educ: Univ NB, BScF, 61; Yale Univ, MF, 62, PhD(tree physiol), 66. Prof Exp: RES SCIENTIST, CAN FORESTRY SERV, 66- Mem: Can Soc Plant Physiol. Res: Production, movement and transformation of carbohydrate; growth substance regulation of bud and cambial activity. Mailing Add: Maritimes Forest Res Ctr Can Forestry Serv PO Box 4000 Fredericton NB Can

LITTLE, CHARLES ORAN, b Schulenburg, Tex, July 21, 35; m 55; c 3. ANIMAL NUTRITION, AGRICULTURE. Educ: Univ Houston, BS, 57; Iowa State Univ, MS, 59, PhD(animal nutrit), 60. Prof Exp: Res asst animal nutrit, Iowa State Univ, 57-60; from asst prof to assoc prof, 60-67, indust res grants, 61-72, Agr Res Serv grant, 64-67, PROF ANIMAL SCI, UNIV KY, 67-, ASSOC DEAN RES, 69- Concurrent Pos: Assoc dir Ky Agr Exp Sta, 69-; mem & chmn, Southern Regional Res Comt, 72-75; mem, Southern Res Planning Comt, 73-75. Honors & Awards: Distinguished Nutritionist Award, Nat Distillers Feed Res Coun, 64; Outstanding Res Award, Thomas Poe Cooper & Ky Res Founds, 67. Mem: AAAS; Am Soc Animal Sci; Am Inst Nutrit. Res: Ruminant nutrition with emphasis on techniques and basic biochemical and physiological explanations of animal responses; general research administration. Mailing Add: S107 Agr Sci Ctr Univ of Ky Lexington KY 40506

LITTLE, EDWIN DEMETRIUS, b Orlando, Fla, July 2, 26. ORGANIC CHEMISTRY. Educ: Rollins Col, BS, 48; Duke Univ, AM, 53. Prof Exp: Sr res chemist, Nitrogen Div, 53-63, supvry res chemist, 63-66, res assoc, Plastics Div, 66-74, GROUP LEADER, SPECIALITY CHEM DIV, ALLIED CHEM CORP, 74- Mem: AAAS; fel Am Inst Chem; Am Chem Soc; NY Acad Sci. Res: Heterocyclic nitrogen compounds; epoxide reactions; monomer synthesis. Mailing Add: Speciality Chem Div Allied Chem Corp PO Box 1087R Morristown NJ 07960

LITTLE, ELBERT LUTHER, JR, b Ft Smith, Ark, Oct 15, 07; m 43; c 3. BOTANY, DENDROLOGY. Educ: Univ Okla, BA, 27, BS, 32; Univ Chicago, MS & PhD(bot), 29. Prof Exp: Asst prof biol, Southwestern Okla State Univ, 30-33; from asst forest ecologist to assoc forest ecologist, Ariz, 34-42, dendrologist, DC, 42-67, CHIEF DENDROLOGIST, US FOREST SERV, WASHINGTON, DC, 67- Concurrent Pos: Dendrologist, Latin Am Forest Resources Proj, Ecuador & Costa Rica, 43; botanist, For Econ Admin, Bogota, 43-45; prod specialist, US Com Co, Mexico City, 45; vis prof, Univ Andes, Venzuela, 53-54 & 60; botanist from Univ Md, Guyana, 55; mem, Int Comn Nomenclature Cultivated Plants, 56-; consult & prof, UN Mission, Inter-Am Inst Agr Sci, Costa Rica, 64-65 & 67, consult, Ecuador, 65, Nicaragua, 71; collabr, US Nat Mus, 65-; vis prof, Va Polytech Inst & State Univ, 66-67. Honors & Awards: Distinguished Serv Award, USDA, 73. Mem: Fel AAAS; Soc Am Foresters; Bot Soc Am; Am Soc Plant Taxon; Ecol Soc Am. Res: Trees of United States and tropical America; rare and endangered plants; plant taxonomy; ecology. Mailing Add: 924 20th St S Arlington VA 22202

LITTLE, ELLIS BEECHER, b Havana, Ill, Feb 25, 14; m 41; c 7. ZOOLOGY, BIOLOGY. Educ: Univ Ill, Urbana, BS, 38, MS, 44, EdD, 55. Prof Exp: Chemist, Continental Can Co, 42-46; from instr to assoc prof, 46-69, PROF ZOOL & BIOL, UNIV ILL, CHICAGO CIRCLE, 69-, ASSOC DEAN LIB ARTS & SCI, 57- Concurrent Pos: Exam consult, Nat Funeral Serv Exam Bds, 46-49, exec secy, 67-68; team mem, Nat Coun Accreditation Teacher Educ, 71 & 72. Mem: Am Asn Higher Educ. Res: The fallacy of using the correction formula in test scoring; creativity and use of the correctional formula; uncertain answers in test scoring; curiosity and man's higher nature; scholastic performance of white students from integrated high schools in college. Mailing Add: Dept of Biol Sci Univ of Ill at Chicago Circle Chicago IL 60607

LITTLE, ERNEST LEWIS, JR, b Passaic, NJ, Aug 29, 19; m 46. CHEMISTRY. Educ: Mass Inst Technol, SB, 41, PhD(org chem), 47. Prof Exp: RES CHEMIST, E I DU PONT DE NEMOURS & CO, INC, 47- Mem: Am Chem Soc. Res: Organic chemistry; organometallic compounds; synthetic rubber; use of dienes and alkyl aromatic hydrocarbons in alfin catalysis; refractory metals; polymorphism. Mailing Add: Cent Res Dept E I du Pont de Nemours & Co Inc Wilmington DE 19898

LITTLE, FRANK JAMES, JR, invertebrate zoology, biological oceanography, see 12th edition

LITTLE, GORDON RICE, b Mexico City, Mex, Oct 22, 45; US citizen; m 70; c 1. OPTICAL PHYSICS, SOLID STATE PHYSICS. Educ: Ohio State Univ, BS, 66, MS, 70, PhD(physics), 73. Prof Exp: Res assoc, Dept Physics, Ohio State Univ, 73; PHYSICIST, SYST RES LABS, 73- Res: Design and development of prototype optical measuring systems; optical Fourier transforms, propagation of visible and near infrared radiation through turbulent media. Mailing Add: 3311 Oakmont Ave Kettering OH 45429

LITTLE, HAROLD FRANKLIN, b Williamsport, Pa, June 18, 32; m 59; c 1. ENTOMOLOGY. Educ: Lycoming Col, AB, 54; Pa State Univ, MS, 56, PhD(zool), 59. Prof Exp: From asst prof to assoc prof biol, 59-63; from asst prof to assoc prof, 63-71, PROF BIOL, UNIV HAWAII, HILO, 71- CHMN DIV NAT SCI, 68-71 & 73-, CHMN BIOL DEPT, 73- Mem: AAAS; Am Inst Biol Sci; Am Soc Zool; Entom Soc Am. Res: Honey bee response to sounds and vibrations; effects of ionizing radiation on insect midgut epithelium; autoradio- graphy of insect midgut epithelial cell replacement; fine structure of insect pheromone glands. Mailing Add: Dept of Biol Univ of Hawaii Hilo HI 96720

LITTLE, HENRY NELSON, b Portland, Maine, Oct 17, 20; m 48; c 4. BIOCHEMISTRY. Educ: Cornell Univ, BS, 42; Univ Wis, MS, 46, PhD(biochem), 48. Prof Exp: Res assoc biochem, Univ Chicago, 48-49; asst prof biol, Johns Hopkins Univ, 49-51; from assoc prof chem to prof chem, 51-66, PROF BIOCHEM, UNIV MASS, AMHERST, 66- Mem: AAAS; Am Chem Soc; Am Soc Biol Chem. Res: HeHemoproteins; porphyrins; biosynthesis of chlorophyll; oxidases. Mailing Add: Dept of Biochem Univ of Mass Amherst MA 01002

LITTLE, HEWARD WALLACE, b Vancouver, BC, June 27, 16; m 43. ECONOMIC GEOLOGY. Educ: Univ BC, BASc, 38, MASc, 40; Univ Toronto, PhD(econ geol), 47. Prof Exp: Coordr Uranium Prog, 67-75, head cordilleran sect, 64-67, GEOLOGIST, GEOL SURV CAN, 47-, HEAD, URANIUM RESOURCE EVAL SECT, 75- Mem: Can Inst Mining & Metall. Res: Cordilleran and uranium geology. Mailing Add: 963 Blythdale Rd Ottawa ON Can

LITTLE, JAMES ALEXANDER, b Detroit, Mich, Dec 8, 22; Can citizen; m 53; c 2. MEDICINE, METABOLISM. Educ: Univ Toronto, MD, 46, MA, 50; FRCP(C), 52. Prof Exp: Res assoc med, 52-67, clin teacher, 54-63, dir diabetic clin, 54-70, assoc, 63-66, dir clin invest unit, 64-72, from asst prof to assoc prof, 66-74, PROF MED, ST MICHAEL'S HOSP, UNIV TORONTO, 74-, RES COORDR & SECY, RES SOC, 64-, DIR LIPID CLIN, 66-, PROJ DIR, UNIV TORONTO-McMASTER LIPID RES CLIN, 72- Concurrent Pos: Nat Res Coun Can fel biochem, Univ Toronto, 47-49, Can Red Cross fel arthritis, Sunnybrook Dept Vet Affairs Hosp, 51-52; mem, Can Nat Comn, Int Union Nutrit Sci, 71-75; dir div endocrinol, metab & nephrol, Lipid Clin, St Michael's Hosp, 70-73; mem, Exec Comt, Coun Atherosclerosis, Am Heart Asn, 74-; mem, Comt Nutrit & Cardiovasc Dis, Health Protection Br, Govt Can, 74- & Med Comt, Ont Heart Found, 65-75. Mem: Am Heart Asn; Can Cardiovasc Soc; Am Diabetes Asn; Nutrit Soc Can; Can Soc Clin Invest. Res: Relation between human atherosclerosis, plasma lipoproteins, nutrition and genetic factors; effect of insulin antibodies on diabetic complications. Mailing Add: Clin Invest Unit St Michael's Hosp Toronto ON Can

LITTLE, JAMES MAXWELL, b Commerce, Ga, Dec 16, 10; m 34; c 3. PHARMACOLOGY, PHYSIOLOGY. Educ: Emory Univ, AB, 32, MS, 33; Vanderbilt Univ, PhD(biochem), 41. Prof Exp: Asst physiol, Emory Univ, 33-35; asst biochem, Washington Univ, 35-36; instr, Vanderbilt Univ, 36-41; from asst prof to assoc prof pharmacol, 41-49, from asst prof to assoc prof physiol, 41-73, asst dean, 57-60, chmn dept pharmacol, 63-73, PROF PHARMACOL, BOWMAN GRAY SCH MED, 49- Concurrent Pos: Mem, Coun Basic Sci, Coun High Blood Pressure Res & Coun Circulation, Am Heart Asn. Mem: Fel AAAS; Am Physiol Soc; Am Soc Pharmacol & Exp Therapeut; Soc Exp Biol & Med; Am Soc Nephrology. Res: Renal function; fluid and salt balance; drug metabolism; circulation. Mailing Add: Bowman Gray Sch of Med Wake Forest Univ Winston-Salem NC 27103

LITTLE, JAMES NOEL, b Kansas City, Mo, July 3, 40; m 64; c 3. ANALYTICAL CHEMISTRY. Educ: Univ Kans, BS, 62, Mass Inst Technol, PhD(anal chem), 66. Prof Exp: Res chemist, Hercules, Inc, Del, 66-67; sr res chemist, 68-69, mgr chromatography res, 69-71, VPRES, WATERS ASSOCS, INC, 71- Mem: Am Chem Soc. Res: Separations; chromatography; polymer characterization; analytical methods development; spectroscopy. Mailing Add: Waters Assocs Inc 34 Maple St Milford MA 01757

LITTLE, JAMES W, b Leeds, NDak, June 17, 34; m 56; c 3. ORAL PATHOLOGY. Educ: Univ Ore, DMD, 58, MS, 63; Am Bd Oral Path, dipl, 66. Prof Exp: Asst prof, 63-68, PROF ORAL DIAG & ORAL MED & CHMN DEPT, COL DENT, UNIV KY, 68- Concurrent Pos: Am Cancer Soc clin fel, 60-61; USPHS fel, 61-63; USPHS res grant, 69-71; US Army med res contract, 65-; consult, USPHS Hosp, Lexington, Ky, 68- & US Army Hosp, Ft Knox, 69- Mem: Am Dent Asn; fel Am Acad Oral Path; Int Asn Dent Res; Am Cancer Soc; Am Acad Oral Med. Res: Oral diagnosis; experimental hyperparathyroidism; wound healing; histogenesis of the branchial cyst. Mailing Add: Dept of Oral Diag & Oral Med Col of Dent Univ of Ky Lexington KY 40506

LITTLE, JOHN BERTRAM, b Boston, Mass, Oct 5, 29; m 60; c 2. PHYSIOLOGY, RADIATION BIOLOGY. Educ: Harvard Univ, AB, 51; Boston Univ, MD, 55; Am Bd Radiol, dipl, 61, cert nuclear med, 61. Prof Exp: Intern med, Johns Hopkins Hosp, 55-56; resident radiol, Mass Gen Hosp, 58-61; USPHS res fel, 61-63, instr physiol, 63-65, asst prof, 65-69, ASSOC PROF RADIOBIOL, HARVARD SCH PUB HEALTH, 69- Concurrent Pos: Consult Mass Gen Hosp, 65-; Peter Bent Brigham Hosp, 68-; Nat Cancer Inst & Am Cancer Soc fels, Sch Pub Health, Harvard, 68-75, lectr, Med Sch, 68- Mem: AAAS; Am Physiol Soc; Radiation Res Soc; Health Physics Soc; Am Asn Cancer Res. Res: Cellular radiation biology with emphasis on repair mechanisms; experimental carcinogenesis. Mailing Add: Dept of Physiol Harvard Sch of Pub Health Boston MA 02115

LITTLE, JOHN CLAYTON, b Battle Creek, Mich, Jan 1, 33; m 53; c 4. ORGANIC CHEMISTRY. Educ: Univ Calif, BS, 54; Univ Ill, PhD(org chem), 57. Prof Exp: Res chemist, 57-62, proj leader, 62-64, group leader, 64-71, RES MGR, DOW CHEM USA, 71- Concurrent Pos: Mem, Mich Found Advan Res. Mem: Am Chem Soc; Sigma Xi; The Chem Soc. Res: Organic syntheses and structure-biological activity relationships; Diels-Alder reactions and synthetic methods; free radical reactions; catalytic oxidation and reduction; chlorination; cyanoethylation; pilot plant and process development studies. Mailing Add: Western Div Res Dow Chemical USA Pittsburg CA 94565

LITTLE, JOHN DUTTON CONANT, b Boston, Mass, Feb 1, 28; m 53; c 4. OPERATIONS RESEARCH. Educ: Mass Inst Technol, SB, 48, PhD(physics), 55. Prof Exp: Engr tube develop, Gen Elec Co, 49-50; asst physics, Mass Inst Technol, 51-54; from asst prof to assoc prof opers res, Case Inst Technol, 57-62; assoc prof opers res, 62-67, PROF OPERS RES & MGT, MASS INST TECHNOL, 67-, DIR OPERS RES CTR, 69- Concurrent Pos: Chmn, Mgt Decision Systs, Inc, 67- Mem: AAAS; Opers Res Soc Am; Inst Mgt Sci. Res: Applications in marketing and public systems. Mailing Add: Opers Res Ctr Mass Inst of Technol Cambridge MA 02139

LITTLE, JOHN ERNEST, b Rochester, NY, June 20, 14; m 46; c 2. BIOCHEMISTRY. Educ: Rutgers Univ, BS, 37; Columbia Univ, PhD(chem), 41. Prof Exp: Res chemist, Merck & Co, Inc, NJ, 41-45; assoc prof, 45-50, PROF BIOCHEM, UNIV VT, 50-, CHMN DEPT AGR BIOCHEM, 45- Mem: AAAS; Am Chem Soc. Res: Antibiotics; fungicides; vitamins. Mailing Add: Dept of Biochem Univ of Vt Burlington VT 05401

LITTLE, JOHN RUSSELL, JR, b Cheyenne, Wyo, Oct 23, 30; m 55; c 3. IMMUNOLOGY. Educ: Cornell Univ, AB, 52; Univ Rochester, MD, 56. Prof Exp: From asst prof to assoc prof med & microbiol, 69-73, PROF MED & MICROBIOL, SCH MED, WASHINGTON UNIV, 73- Concurrent Pos: Fel microbiol, Sch Med, Washington Univ, 62-64. Res: Medicine and lymphocyte membrane structure and function. Mailing Add: Jewish Hosp Washington Univ Sch of Med St Louis MO 63110

LITTLE, JOHN STANLEY, b Fredericton, NB, July 27, 31; m 57; c 1. ORGANIC CHEMISTRY. Educ: Univ NB, BS, 52, PhD(org chem), 55. Prof Exp: Lord Beaverbrook Overseas Scholar, Univ London, 55-56; res chemist, Can Industs, Ltd, 56-62, group leader fiber res & develop, 62-65; sect leader all process & prod res, Celanese Corp, Can, 65-67, mgr appln res, 67-70, mgr indust prod develop, Celanese Fibers Mkt Co, 70-71, dir spec prod develop, 71-73, DIR TEXTILE PROD DEVELOP, CELANESE FIBERS MKT CO, 73- Mem: Am Chem Soc; fel Chem Inst Can. Res: Synthetic fibers; alkaloids and steroids. Mailing Add: Celanese Fibers Mkt Co PO Box 1414 Charlotte NC 28232

LITTLE, JOHN WESLEY, biochemistry, molecular biology, see 12th edition

LITTLE, JOSEPH ALEXANDER, b Bessemer, Ala, Mar 16, 18; m 41; c 3. MEDICINE. Educ: Vanderbilt Univ, BA, 40, MD, 43. Prof Exp: Intern, Vanderbilt Univ Hosp, 43 & 46-47; resident, Childrens Hosp, Univ Cincinnati, 47-48, instr pediat, Col Med, Univ, 48-49; from asst prof to prof, Sch Med, Univ Louisville, 49-62; assoc prof, Sch Med, Vanderbilt Univ, 62-70; PROF PEDIAT & HEAD DEPT, SCH MED, LA STATE UNIV, SHREVEPORT, 70- Concurrent Pos: Dir outpatient dept, Childrens Hosp, 48-49, physician in chief, 56-; med dir, State Crippled Children Comn, Ky, 51-54; consult, State Dept Health, Ky, 49-51. Mem: AAAS; Am Pediat Soc; Am Acad Pediat; NY Acad Sci. Res: Pediatric cardiology. Mailing Add: Dept of Pediat La State Univ Sch of Med Shreveport LA 71130

LITTLE, LINDA WEST, b Kinston, NC, Feb 16, 37; m 58; c 2. ENVIRONMENTAL BIOLOGY. Educ: Univ NC-Greensboro, BA, 59; Univ NC-Chapel Hill, MSPH, 62, PhD(environ microbiol), 68. Prof Exp: Res asst bact, Sch Med, Univ NC-Chapel Hill, 59-61; asst prof biol, E Carolina Univ, 67-69; res assoc environ biol, 70-71, asst prof, 71-74, ASSOC PROF ENVIRON BIOL, WASTEWATER RES CTR, SCH PUB HEALTH, UNIV NC-CHAPEL HILL, 74- Concurrent Pos: Consult, Atomic Safety & Licensing Bd Panel, Nuclear Regulatory Comn, 74- Mem: Water Pollution Control Fedn; Am Water Works Asn; Am Soc Microbiol; Asn Prof Environ Eng. Res: Biological wastewater treatment; nitrification; impact of industrial and municipal wastewaters on the environment; treatment of industrial wastewaters, especially those from textile manufacture and food processing. Mailing Add: Wastewater Res Ctr Univ of NC Sch of Pub Health Chapel Hill NC 27514

LITTLE, MARTIN HENRY, electrochemistry, physical chemistry, see 12th edition

LITTLE, MAURICE DALE, b North Grove, Ind, Apr 13, 28; m 55; c 3. MEDICAL PARASITOLOGY. Educ: Purdue Univ, BS, 50; Tulane Univ, MS, 58, PhD(parasitol), 61. Prof Exp: Microbiologist, Ind State Bd Health, 53-56; from instr to asst prof, Sch Med, 63-68, ASSOC PROF PARASITOL, SCH PUB HEALTH & TROP MED, TULANE UNIV, 68- Concurrent Pos: NIH fel, Tulane Univ, 61-63. Mem: AAAS; Am Micros Soc; Am Soc Parasitol; Am Soc Trop Med & Hyg; Royal Soc Trop Med & Hyg. Res: Morphology, biology and epidemiology of Strongyloides species; zoonotic helminthiases; paragonimiasis; soil-transmitted helminths. Mailing Add: Dept of Trop Med Tulane Univ New Orleans LA 70112

LITTLE, MICHAEL ALAN, b Abington, Pa, Mar 24, 37; m 65; c 1. PHYSICAL ANTHROPOLOGY. Educ: Pa State Univ, BA, 62, MA, 65, PhD(anthrop), 68. Prof Exp: Asst prof anthrop, Ohio State Univ, 67-70; asst prof, 71-73, ASSOC PROF ANTHROP, STATE UNIV NY BINGHAMTON, 73- Concurrent Pos: Ohio State Univ res fel, Nunoa, Peru, 68; State Univ NY Binghamton res fel & grant, Nunoa, 72; vis assoc prof anthrop & sci coordr, US Int Biol Prog, Human Adaptability Component, Pa State Univ, 72-73; NSF sci equip grant, 74. Mem: AAAS; Am Asn Phys Anthrop; Soc Study Human Biol; Soc Study Social Biol; Human Biol Coun. Res: Biocultural adaptations; human biology; environmental stress; heat and cold adaptation; circadian rhythms; human populations at high altitude; ecology of savanna pastoralists. Mailing Add: Dept of Anthrop State Univ of NY Binghamton NY 13901

LITTLE, PERRY L, b Ball Ground, Ga, Aug 3, 28; m 51; c 1. NUTRITION, PHYSIOLOGY. Educ: Berry Col, BS, 50; Auburn Univ, MS, 57, PhD(path, nutrit, physiol), 66. Prof Exp: Teacher high sch Ala, 50-52 & 53-55; res asst poultry sci, Auburn Univ, 55-62; ASSOC PROF POULTRY SCI, SAM HOUSTON STATE UNIV, 62- Mem: Poultry Sci Asn; World Poultry Sci Asn. Res: Nutrition of parasites which involve poultry. Mailing Add: Dept of Agr Sam Houston State Univ Huntsville TX 77340

LITTLE, RALPH CARTER, physical chemistry, see 12th edition

LITTLE, RANDEL QUINCY, JR, b Richmond, Va, Aug 14, 27; m 49; c 3. ORGANIC CHEMISTRY. Educ: Univ Richmond, BS, 48; Univ Mich, MS, 49, PhD(org chem), 54. Prof Exp: Res chemist, Am Oil Co, Stand Oil Co, Ind, 53-60, group leader motor oil additives, 60-62, res supvr, 62-68, asst dir lubricants res, 68-74, DIR LUBRICANTS RES, AMOCO OIL CO, 74- Mem: Am Chem Soc; Sigma Xi; Am Soc Lubricating Engrs; Soc Automotive Engrs. Res: Organic reactions; motor oil additives; lubricants. Mailing Add: Amoco Oil Co PO Box 400 Naperville IL 60540

LITTLE, RAYMOND DANIEL, b Superior, Wis, Sept 12, 47; m 72. ORGANIC CHEMISTRY. Educ: Univ Wis-Superior, BS, 69; Univ Wis-Madison, PhD(org chem), 74. Prof Exp: Assoc chem, Yale Univ, 74-75; ASST PROF ORG CHEM, UNIV CALIF, SANTA BARBARA, 75- Mem: Am Chem Soc; Sigma Xi. Res: Development of new synthetic methods; total synthesis of pharmacologically active molecules and compounds of theoretical interest; mechanistic organic chemistry of thermal and phtochemical reactions; theoretical organic chemistry. Mailing Add: Dept of Chem Univ of Calif Santa Barbara CA 93106

LITTLE, ROBERT, b Jedburgh, Scotland, Mar 3, 39; m 62; c 3. PLASMA PHYSICS. Educ: Univ Glasgow, BSc, 61, PhD(physics), 64. Prof Exp: Res assoc physics, Cambridge Electron Accelerator, Harvard Univ, 64-73 & Mass Inst Technol, 73-75; PROF STAFF PHYSICS, PRINCETON PLASMA PHYSICS LAB, PRINCETON UNIV, 75- Res: Development of large tokomaks for controlled thermonuclear resarch. Mailing Add: Princeton Plasma Physics Lab Princeton Univ Princeton NJ 08540

LITTLE, ROBERT COLBY, b Norwalk, Ohio, June 2, 20; m 45; c 2. PHYSIOLOGY, MEDICINE. Educ: Denison Univ, AB, 42; Western Reserve Univ, MD, 44, MS, 48. Prof Exp: Intern, Grace Hosp, Detroit, Mich, 44-45; resident med, Crile Vet Hosp, Cleveland, 49-50; from asst prof to assoc prof physiol, Univ Tenn, 50-54, assoc prof med, 53-54; dir clin res, Mead Johnson & Co, 54-57; dir cardio-pulmonary labs, Scott & White Clin, Tex, 57-58; prof physiol, Seton Hall Col Med, 58-64, asst prof med, 59-64; prof physiol, chmn dept & asst prof med, Col Med, Ohio State Univ, 64-73; PROF PHYSIOL & MED & CHMN DEPT PHYSIOL, SCH MED, MED COL GA, 73- Concurrent Pos: USPHS res fel, Western Reserve Univ, 48-49. Mem: Am Physiol Soc; Soc Exp Biol & Med; Am Heart Asn; Am Fedn Clin Res. Res: Cardiovascular dynamics; heart sounds; clinical physiology; muscle dynamics. Mailing Add: Dept of Physiol Med Col of Ga Augusta GA 30902

LITTLE, ROBERT GREENWOOD, b Evanston, Ill, Dec 26, 42. INORGANIC CHEMISTRY, X-RAY CRYSTALLOGRAPHY. Educ: Univ Calif, Davis, BS, 64; State Univ NY Buffalo, PhD(inorg reaction mechanisms), 70. Prof Exp: Fel crystallog, State Univ NY Buffalo, 69-70; fel crystallog & instr chem, Univ Calif, Irvine, 70-72; fel crystallog, Northwestern Univ, Evanston, 72-74; ASST PROF CHEM, UNIV MD, BALTIMORE COUNTY, 74- Mem: AAAS; Am Chem Soc. Res: Stereochemical requirements and mechanisms of inorganic reactions. Mailing Add: Dept of Chem Univ of Md Baltimore County Baltimore MD 21228

LITTLE, ROBERT LEWIS, b Monticello, Miss, July 1, 29; m 53; c 2. GEOLOGY. Educ: Univ Miss, BA, 51, MS, 59; Univ Tenn, Knoxville, PhD(geol), 69. Prof Exp: Instr geol, Univ Miss, 58-59 & Univ Tenn, Knoxville, 59-69; asst prof, 69-72, ASSOC PROF, VALDOSTA STATE COL, 72- HEAD DEPT, 69- Mem: Geol Soc Am; Nat Asn Geol Teachers; Am Asn Univ Prof. Res: Areal geology; stratigraphy and structural geology. Mailing Add: Dept of Geol Valdosta State Col Box 185 Valdosta GA 31601

LITTLE, ROBERT NARVAEZ, JR, b Houston, Tex, Mar 11, 13; m 42; c 2. NUCLEAR PHYSICS, SCIENCE EDUCATION. Educ: Rice Inst, BA, 35, MA, 42, PhD(physics), 43. Prof Exp: Asst seismologist, Shell Oil Co, Tex, 36-42, asst physics, Rice Inst, 43; asst prof, Univ Ore, 43-44; testing supvr, Mil Physics Res Lab, Univ Tex, 44-48, from asst prof to assoc prof physics, 46-55 & res scientist, Nuclear Physics Lab, 55-60, PROF PHYSICS & EDUC, UNIV TEX, 55- Concurrent Pos: Chief nuclear physics, Gen Dynamics Corp, 54; mem, Comn Col Physics, 62-66; dir physics educ asst proj, Nat Univ Cent Am, 65-71; vis prof, Univ Valle, Guatemala, 71. Mem: Am Phys Soc; Am Asn Physics Teachers (pres, 70); Int Group Res on Educ Physics; Cent Am Soc Physics. Res: Neutron scattering; reactor physics; development and evaluation of physics teaching methods at all levels. Mailing Add: 3928 Balcones Dr Austin TX 78731

LITTLE, RUBY RICE, b Knoxville, Tenn, July 23, 07; m 43; c 3. BOTANY. Educ: Univ Tenn, BA, 32, MS, 34. Prof Exp: Asst bot, Univ NC, 34-35; asst mycol & plant path, Cornell Univ, 35-40; jr plant physiologist, Bur Plant Indust, USDA, 43; histologist, Cinchona mission, For Econ Admin, Colombia, 44-45; res botanist, Human Nutrit Res Div, 49-69 & lab technician weed invest, Plant Sci Div, Agr Res Serv, USDA, 69-74; RETIRED. Mem: AAAS; Inst Food Tech. Res: Histology of foods; plant anatomy. Mailing Add: Plant Sci Div USDA Rm 39 South Bldg Plant Indust Sta Beltsville MD 20705

LITTLE, SILAS, JR, b Newbury, Mass, Jan 17, 14; m 41; c 5. FORESTRY. Educ: Mass State Col, BS, 35; Yale Univ, MF, 36, PhD(silvicult), 47. Prof Exp: From jr forester to proj leader, 36-75, PRIN SILVICULTURIST, NORTHEASTERN FOREST EXP STA, US FOREST SERV, 75- Mem: Soc Am Foresters; Ecol Soc Am. Res: Silviculture; fire; forest management. Mailing Add: Northeastern Forest Exp Sta Pennington NJ 08534

LITTLE, THOMAS MORTON, b Picture Rocks, Pa, July 25, 10; m 38; c 2. BIOMETRICS. Educ: Bucknell Univ, AB, 31; Univ Fla, MS, 33; Univ Md, PhD(genetics), 43. Prof Exp: Asst entom, Cornell Univ, 31; chief hybridist, W Atlee Burpee Co, Calif, 34-41; asst geneticist, USDA, Md, 41-44; dir crop res, Basic Veg Prod, Inc, 44-49; asst prof biol & hort, Univ Nev, 49-53; exten veg crops specialist, 53-64, exten biometrician, 64-72, EMER EXTEN BIOMETRICIAN, UNIV CALIF, RIVERSIDE, 72- Concurrent Pos: Fulbright lectr, Univ Zagreb, 67-68; biometrician, Int Inst Trop Agr, Ibadan, Nigeria, 72-74. Mem: Am Math Soc; Am Soc Hort Sci; Am Statist Asn; Asn Comput Mach. Res: Design and analysis of agricultural experiments. Mailing Add: 1488 Argyle Lane Bishop CA 93514

LITTLE, WILLIAM ARTHUR, b Adelaide, SAfrica, Nov 17, 30; nat US; m 55; c 3. PHYSICS. Educ: Univ SAfrica, BSc, 50; Rhodes Univ, SAfrica, PhD, 55; Univ Glasgow, PhD, 57. Prof Exp: Nat Res Coun Can fel, Univ BC, 56-58; from asst prof to assoc prof, 58-65, PROF PHYSICS, STANFORD UNIV, 65- Concurrent Pos: Alfred P Sloan fel, 59-62; Guggenheim fel, 64-65; invited prof, Univ Geneva, 64-65; NSF sr fel, 71-72; consult, Syva Corp & RAI Inc. Mem: Fel Am Phys Soc. Res: Organic fluorescence; magnetic resonance; low temperature physics; superconductivity; phase transition; chemical physics; neural network theory. Mailing Add: Dept of Physics Stanford Univ Stanford CA 94305

LITTLE, WILLIAM ASA, b Ellenville, NY, July 14, 31; m 52; c 2. OBSTETRICS & GYNECOLOGY. Educ: Johns Hopkins Univ, AB, 51; Univ Rochester, MD, 55. Prof Exp: Intern obstet & gynec, Barnes Hosp, Washington Univ, 55-56; resident, Columbia-Presby Med Ctr, 56-61; from asst prof to assoc prof, Univ Fla, 61-66; PROF OBSTET & GYNEC & CHMN DEPT, UNIV MIAMI, 66- Concurrent Pos: Josia Macy Jr fel, 57-60; Am Cancer Soc fel, 60-61; consult, US Navy, 61-65; asst chief perinatal res, Nat Inst Neurol Dis & Stroke, 62-64, consult, 64-, Nat Inst Child Health & Human Develop, 63-; Markle scholar, 62-; adv, US Food & Drug Admin, 64- Mem: AAAS; Am Col Obstet & Gynec; Am Fertil Soc; Am Pub Health Asn; AMA. Res: Placental pathology and transfer; enzymology; fetal pharmacology; teratology; mental retardation; cancer; reproductive biology. Mailing Add: Dept of Obstet & Gynec Univ of Miami Sch of Med Miami FL 33152

LITTLE, WILLIAM FREDERICK, b Morganton, NC, Nov 11, 29; m 58; c 1. ORGANIC CHEMISTRY. Educ: Lenoir Rhyne Col, BS, 50; Univ NC, MA, 52, PhD(org chem), 55. Prof Exp: Instr chem, Reed Col, 55-56; from instr to asst prof, 56-65, asst to dean grad sch res admin, 59-62, PROF CHEM & CHMN DEPT, UNIV NC, CHAPEL HILL, 65-, VCHANCELLOR, DEVELOP & PUB SERV, 73- Concurrent Pos: Consult, Res Triangle Inst, 56-69, chmn exec comt, Bd Gov, 69- Mem: AAAS; Am Chem Soc. Res: Organic aromatic fluorine compounds; organometallic compounds, especially metallocenes. Mailing Add: Dept of Chem Univ of NC Chapel Hill NC 27515

LITTLEDIKE, E TRAVIS, veterinary medicine, biochemistry, see 12th edition

LITTLEFIELD, GAYLE, b Jonesville, SC, Mar 14, 42. ANATOMY, CYTOGENETICS. Educ: Winthrop Col, BA, 63; Med Col Ga, MS, 65, PhD(anat), 68. Prof Exp: Instr pediat, Med Col Ga, 68-69; RES SCIENTIST CYTOGENETICS, MED DIV, OAK RIDGE ASSOC UNIVS, 69- Mem: AAAS; Am Soc Human Genetics. Res: Cytogenetics; induced cytogenetic changes; chromosomal abnormalities and malignancy; birth defects. Mailing Add: Med Div Oak Ridge Assoc Univ Oak Ridge TN 37730

LITTLEFIELD, JAMES BEATON, b Pawtucket, RI, Apr 16, 19; m 45. MEDICINE. Educ: Univ Va, AB, 41; Univ Md, MD, 44; Am Bd Surg, dipl, 53; Am Bd Thoracic Surg, dipl, 59. Prof Exp: Instr thoracic surg, Sch Med, Univ Va, 56-59, asst prof surg, 59-62, assoc prof, 62-68; PROF SURG & CHMN DEPT, FAC MED, MEM UNIV NFLD, 68- Concurrent Pos: Am Trudeau Soc res fel, 57-59; Am Col Surg Kemper res scholar, 59-62; Ehler Mem lect, Albany Med Col, 62. Honors & Awards: Horsley Mem Prize, Sch Med, Univ Va, 58. Mem: Fel Am Col Surg; Am Asn Thoracic Surg; Soc Univ Surg; Am Fedn Clin Res; Can Asn Clin Surg. Res: Thoracic and cardiac surgery; cardiac and pulmonary physiology. Mailing Add: Dept of Surg Mem Univ of Nfld Fac of Med St John's NF Can

LITTLEFIELD, JOHN WALLEY, b Providence, RI, Dec 3, 25; m 50; c 3. PEDIATRICS. Educ: Harvard Med Sch, MD, 47. Prof Exp: From intern to resident med, Mass Gen Hosp, 47-50; from res fel to asst prof med, Harvard Med Sch, 54-66, tutor, 59-65, asst prof pediat, 66-69, prof, 70-73; PROF PEDIAT & CHMN DEPT, JOHNS HOPKINS UNIV, 74 Concurrent Pos: USPHS fel, Inst Enzyme Res, Univ Wis, 51; Am Cancer Soc scholar, 56-59; Guggenheim fel, 65-66; from clin & res fel to assoc physician & assoc pediatrician, Mass Gen Hosp, 54-74. Mem: Soc Pediat Res; Am Soc Human Genetics; Am Acad Pediat; Am Pediat Soc; Am Soc Biol Chemists. Res: Human genetics; molecular biology. Mailing Add: Dept of Pediat Johns Hopkins Univ Baltimore MD 21205

LITTLEFIELD, LARRY JAMES, b Ft Smith, Ark, Feb 7, 38; m 63. PLANT PATHOLOGY. Educ: Cornell Univ, BS, 60; Univ Minn, MS, 62, PhD(plant path), 64. Prof Exp: Res asst plant path, Univ Minn, 60-64; NSF res fel, Uppsala, 64-65;

from asst prof to assoc prof, 65-75, PROF PLANT PATH, N DAK STATE UNIV, 75- Concurrent Pos: NIH fel, Purdue Univ, 69-70; res fel, Oxford Univ, 73-74. Mem: Mycol Soc Am; Brit Mycol Soc; Am Phytopath Soc. Res: Histology of host-parasite relations; fungus physiology; electron microscopy of fungi and diseased plants. Mailing Add: Dept of Plant Pathol NDak State Univ Fargo ND 58102

LITTLEFIELD, NEIL ADAIR, b Santa Fe, NMex, Apr 25, 35; m 60; c 5. TOXICOLOGY. Educ: Brigham Young Univ, BS, 61; Utah State Univ, MS, 64, PhD(toxicol), 68. Prof Exp: Res assoc air pollution, Univ Utah, 66-67; staff scientist inhalation toxicol, Hazleton Lab, Inc, 67-70; pharmacologist pesticide regulation, Environ Protection Agency, 71-72; TOXICOLOGIST, NAT CTR TOXICOL RES, FOOD & DRUG ADMIN, 72- Concurrent Pos: Chmn, Interagency Task Force Inhalation Chronic Toxicity & Carcinogenesis, 74-75; mem, Food & Drug Admin Task Force Aerosol Prod, 75- Mem: Sigma Xi. Res: Investigations in concepts of long-term, low-dose exposures; extrapolation of animal toxicology data to risk-benefit in man; carcinogenesis; inhalation toxicology and inhalation carcinogenesis. Mailing Add: Nat Ctr Toxicol Res Jefferson AR 72079

LITTLEFORD, ROBERT ANTHONY, zoology, see 12th edition

LITTLEJOHN, OLIVER MARSILIUS, b Cowpens, SC, Sept 29, 24; m 48; c 2. PHARMACY. Educ: Univ SC, BS, 48 & 49; Univ Fla, MS, 51, PhD(pharm), 53. Prof Exp: Asst prof pharm & head dept, Southern Col Pharm, 53-56; prof & head dept, Univ Ky, 56-57; DEAN, SOUTHERN SCH PHARM, MERCER UNIV, 57- Mem: Fel Am Found Pharmaceut Educ. Res: Pharmaceutical preservatives. Mailing Add: Mercer Univ Southern Sch Pharm 345 Blvd NE Atlanta GA 30312

LITTLE-MARENIN, IRENE RENATE, b Pilsen, Czech, May 4, 41; US citizen; m 73; c 1. ASTROPHYSICS. Educ: Vassar Col, AB, 64; Ind Univ, MA, 66, PhD(astrophys), 70. Prof Exp: Fel astron, Ohio State Univ, 70-72; asst prof, Univ Western Ont, 72-73; teaching fel, Ferris State Col, 74; RES ASST SOLAR X-RAYS, AM SCI & ENG, 76- Mem: Am Astron Soc. Res: A search for the radioactive element technetium in long-period variable stars; a determination of the isotopic ratio of carbon in carbon stars and in G and K giants and supergiants. Mailing Add: 10 Weston Terr Wellesley MA 02181

LITTLEPAGE, JACK LEROY, b San Diego, Calif, Apr 14, 35; m 60. BIOLOGICAL OCEANOGRAPHY. Educ: San Diego State Col, BA, 57; Stanford Univ, PhD(biol), 66. Prof Exp: Asst prof, 65-71, ASSOC PROF BIOL, UNIV VICTORIA, 71- Mem: AAAS; Ecol Soc Am; Am Soc Limnol & Oceanog. Res: Physiology and ecology of marine zooplankton, especially copepods and euphausids; pollution monitoring, sanitary and mine. Mailing Add: Dept of Biol Univ of Victoria Victoria BC Can

LITTLER, MARK MASTERTON, b Athens, Ohio, Sept 24, 39; m 66. ECOLOGY, PHYCOLOGY. Educ: Ohio Univ, BS, 61, MS, 66; Univ Hawaii, PhD(bot), 71. Prof Exp: Chemist, Testing Lab, Ohio State Hwy Dept, 61-64; asst prof, 70-74, ASSOC PROF, BIOL SCI, UNIV CALIF, IRVINE, 74- Concurrent Pos: US Dept Interior Off Water Resources & Technol grant, 74 & Bur Land Mgt res contract, 75. Mem: Int Phycol Soc; Phycol Soc Am; Japanese Phycol Soc; British Phycol Soc; Am Soc Limnol & Oceanog. Res: Man's effect on marine ecosystems; taxonomy, developmental morphology and seasonal cycles of marine benthos and phytoplankton; standing stock, productivity and the physiological ecology of temperate and reef-building benthic organisms. Mailing Add: Dept of Ecol & Evolutionary Biol Univ Calif Irvine CA 92664

LITTLETON, C SCOTT, b Los Angeles, Calif, July 1, 33; m 61; c 2. CULTURAL ANTHROPOLOGY. Educ: Univ Calif, Los Angeles, AB, 57, MA, 62, PhD(anthrop), 65. Prof Exp: Asst prof, 62-68, ASSOC PROF ANTHROP, OCCIDENTAL COL, 68- CHMN, DEPT SOCIOL & ANTHROP, 74- Concurrent Pos: Instr, Exten, Univ Calif, Los Angeles, 60- vis lectr, Univ Calif, 69-, Am Coun Learned Socs grant, Greece & Brit, 72- Mem: Am Anthrop Asn; Am Folklore Soc; Int Soc Ethnol & Folklore. Res: Comparative Indo-European mythology; relationship between mythology and natural catastrophe, with special emphasis upon the eruption of Santorini. Mailing Add: Dept of Sociol & Anthrop Occidental Col Los Angeles CA 90041

LITTLEWOOD, BARBARA SHAFFER, b Buffalo, NY, Oct 8, 41; m 70. BIOCHEMISTRY, GENETICS. Educ: Univ Rochester, BA, 63; Univ Pa, PhD(biochem), 68. Prof Exp: NIH trainee, Cornell Univ, 68-70; res assoc biochem, 70-73, RES ASSOC PHYSIOL CHEM, UNIV WIS-MADISON, 73- Concurrent Pos: Lectr, dept genetics, Cornell Univ, 70 & dept biochem, Univ Wis-Madison, 72. Mem: Genetics Soc Am; Am Soc Microbiol. Res: Yeast genetics and biochemistry. Mailing Add: Dept of Physiol Chem Univ of Wis 689 Med Sci Madison WI 53706

LITTLEWOOD, ROLAND KAY, b Mendota, Ill, Nov 26, 42. COMPUTER SCIENCE, MOLECULAR BIOLOGY. Educ: Univ Ill, Urbana, BS, 64; Cornell Univ, PhD(genetics), 70. Prof Exp: NIH fel, Lab Molecular Biol, 70-72, RES ASSOC, BIOPHYS LAB, UNIV WIS-MADISON, 72- Mem: Genetics Soc Am; Am Soc Microbiol; Asn Comput Mach. Res: Application of computers in the biological sciences; biochemical genetics. Mailing Add: Biophys Lab 1525 Linden Dr Univ Wis Madison WI 53706

LITTLEWOOD, WILLIAM HERBERT, b Detroit, Mich, Apr 16, 24; m 54; c 2. SCIENCE POLICY, OCEANOGRAPHY. Educ: Univ Fla, BSc, 48; Univ Mich, MSc, 49. Prof Exp: Instr biol, Champlain Col, 49-50; phys oceanogr, US Naval Oceanog Off, Washington DC, 50-53 & 54-60; dep sci attache, Am Embassy, Stockholm, 60-65; asst to dir int orgn div, Off Int Affairs, NASA, 65-66; sci off int sci & technol affairs, Dept State, Washington, DC, 66-67; dep sci attache, Am Embassy, Tokyo, 67-70; ASSOC DIR OFF SCI & TECHNOL, AID, DEPT STATE, 70- Concurrent Pos: Fulbright scholar, Copenhagen Univ, 53-54; alt mem interagency comt oceanog, Fed Coun Sci Technol, 66-67, int comt, 70-; mem panel int prog, Int Coop Oceanog, 70-; US rep, Working Group Mutual Assistance, Intergovt Oceanog Comn, UNESCO, 70- Honors & Awards: Antarctic Medal, 55-59. Mem: AAAS; Polar Socs Australia, Japan, Norway, USA& NZ. Res: Physical and biological polar oceanography; international scientific relationships; Scandinavian and Japanese science organization; science and technology transfer to developing countries. Mailing Add: Off of Sci & Technol AID Dept of State Washington DC 20523

LITTMAN, ARMAND, b Chicago, Ill, Apr 4, 21; m 52; c 3. MEDICINE. Educ: Univ Ill, Chicago, BS, 42, MD, 43, MS, 48, PhD(physiol), 51. Prof Exp: Intern, Cook County Hosp, Chicago, 44; from clin asst to assoc prof, 46-64, PROF MED, UNIV ILL COL MED, 64-; CHIEF MED SERV, HINES VET ADMIN HOSP, 59- Concurrent Pos: Raymond B Allen instructorship award, Univ Ill, 57; US AEC travel award, 58; resident, Cook County Hosp, Chicago, 48-50; pvt pract, 52-59; attend physician, Res & Educ Hosps, 55-; prof, Cook County Grad Sch Med, 58- Mem: AMA; Am Col Physicians; Am Fedn Clin Res; Am Gastroenterol Asn. Res: Gastroenterology; physiology. Mailing Add: Med Serv Vet Admin Hosp Hines IL 60141

LITTMAN, FRED EMANUEL, chemistry, see 12th edition

LITTMAN, MAXWELL LEONARD, medical research, deceased

LITTMAN, WALTER, b Vienna, Austria, Sept 17, 29; US citizen; m 60; c 3. MATHEMATICAL ANALYSIS. Educ: Univ NY, BA, 52, PhD(math), 56. Prof Exp: Instr math, Univ Calif, Berkeley, 56-58, lectr, 58-59; asst prof, Univ Wis, 59-60; from asst prof to assoc prof, 60-66, PROF MATH, UNIV MINN, MINNEAPOLIS, 66- Concurrent Pos: Vis mem, Courant Inst Math Sci, NY Univ, 67-68; vis prof, Chalmers Technol Univ, Gothenburg, Sweden, 75. Mem: Am Math Soc. Res: Partial differential equations; functional analysis; mathematical physics. Mailing Add: Dept of Math Univ of Minn Minneapolis MN 55455

LITTON, COLE WALDAUER, solid state physics, molecular physics, see 12th edition

LITTON, GEORGE WASHINGTON, b Pennington Gap, Va, Feb 22, 10; m 38; c 2. ANIMAL HUSBANDRY. Educ: Va Polytech Inst, BS, 31, MS, 40. Prof Exp: County agr agent, Extension Serv, Va Polytech Inst & State Univ, 31-40, res & teaching beef & sheep mgt, 40-44, sheep specialist, 45-51, prof animal sci & head dept, 52-70, dir centennial progs, 70-73, EMER PROF ANIMAL SCI, VA POLYTECH INST & STATE UNIV, 73- Mem: Fel AAAS; Am Soc Animal Sci. Res: Animal breeding and management; beef, cattle and sheep. Mailing Add: Rt 3 Box 286 Blacksburg VA 24060

LITTRELL, ROBERT H, plant pathology, see 12th edition

LITVAK, AUSTIN S, b Staten Island, NY, Dec 1, 33. UROLOGY. Educ: Wagner Col, BS, 54; Univ Va, MD, 58; Am Bd Urol, dipl. Prof Exp: Intern surg, Univ Va Hosp, 58-59; NIH fel, Surg Br, Nat Cancer Inst, 59-60; USPHS surgeon, Claremore Indian Hosp, Okla, 60-61; resident urol, Hosp Univ Pa, Jefferson Med Col Serv & Philadelphia Gen Hosp, 61-63; pvt pract, 63-73; asst in surg, Div Urol, Hahnemann Med Col, 70-73; asst prof, 73-75, ASSOC PROF SURG, DIV UROL, UNIV KY MED CTR, 76- Concurrent Pos: Asst attend urol, Monmouth Med Ctr & Riverview Hosp; assoc attend & div rep, Bayshore Hosp; courtesy staff, Jersey Shore Med Ctr; asst urol staff, Hahnemann Med Col; mem urol staff, Univ Ky Med Ctr; consult urol, Vet Admin Hosp, 73-75. Honors & Awards: First Prize, Nat Clin Soc, 68. Mem: Am Urol Asn; Am Fertil Soc; AMA; fel Am Col Surgeons; fel Int Col Surgeons. Res: Embryology of the kidney; acute and chronic prostatitis; lower urinary tract infections in children and adults; diseases of the kidney, urethra and bladder in children. Mailing Add: Div of Urol Univ of Ky Med Ctr Lexington KY 40506

LITVAK, MARVIN MARK, b Newark, NJ, Oct 20, 33; m 63; c 2. THEORETICAL PHYSICS. Educ: Cornell Univ, BEngPhys, 56, PhD(theoret physics), 60. Prof Exp: Consult, Avco Corp, 55-60, sr staff mem, Avco-Everett Res Lab, 60-63; group leader, Lincoln Lab, Mass Inst Technol, 63-70; SR RADIO ASTRONOMER, SMITHSONIAN ASTROPHYS OBSERV, 70- Concurrent Pos: Lectr, Harvard Col Observ, 70- Mem: Am Phys Soc; Am Astron Soc. Res: Collision-free magnetohydrodynamic shock waves; laser-produced gas breakdown; gas laser characteristics; non-linear propagation effects of lasers; interstellar molecules and masers; gas dynamics; millimeter-wave radio astronomy and aeronomy. Mailing Add: Smithsonian Astrophys Observ 60 Garden St Cambridge MA 02138

LITWACK, GERALD, b Boston, Mass, Jan 11, 29; m 56, 73; c 1. BIOCHEMISTRY. Educ: Hobart Col, BA, 49; Univ Wis, MS, 51, PhD(biochem), 53. Prof Exp: From asst prof to assoc prof biochem, Rutgers Univ, 54-60; res assoc prof, Grad Sch Med, Univ Pa, 60-64; dir biochem, Div Cardiol, Philadelphia Gen Hosp, 60-64; PROF BIOCHEM, SCH MED, TEMPLE UNIV, 64-, SR RES INVESTR, FELS RES INST, 64- Concurrent Pos: Nat Found Infantile Paralysis fel, Biochem Lab, Univ Sorbonne, 53-54; trainee, Oak Ridge Inst Nuclear Studies, 55; Nat Inst Arthritis & Metab Dis res career develop award, 63-69; vis prof, Univ Calif, 56; hon prof, Rutgers Univ, 60-64; vis scientist, Univ London, 71 & Univ Calif, 72. Honors & Awards: Lalor Found Award, 56. Mem: Fel AAAS; Am Soc Biol Chemists; Am Asn Cancer Res; Brit Biochem Soc. Res: Enzyme regulation; hormonal control of enzyme formation and activity; actions of glucocorticoids. Mailing Add: Fels Res Inst Temple Univ Sch of Med Philadelphia PA 19140

LITWAK, ROBERT SEYMOUR, b New York, NY, Nov 25, 24; m; c 3. SURGERY. Educ: Ursinus Col, BS, 45; Hahnemann Med Col, MD, 49; Am Bd Surg, dipl, 56; Am Bd Thoracic Surg, dipl, 58. Prof Exp: Asst surg, Sch Med, Boston Univ, 52; from instr to assoc prof surg, Med Sch, Univ Miami, 56-62; ATTEND SURGEON & CHIEF DIV CARDIOTHORACIC SURG, MT SINAI HOSP, 62-; PROF SURG, MT SINAI SCH MED, 71- Concurrent Pos: Consult, Vet Admin Hosp, Coral Gables, Fla, 57- & Variety Children's Hosp, 59-; chief div thoracic & cardiovascular surg, Jackson Mem Hosp, 59-62. Mem: Fel Am Col Surg; fel Am Col Chest Physicians; fel Am Col Cardiol; fel NY Acad Sci. Res: Cardiovascular physiology; cardiac surgery. Mailing Add: Div of Cardiothoracic Surg Mt Sinai Hosp New York NY 10029

LITWIN, MARTIN STANLEY, b Florence, Ala, Jan 8, 30; m 60; c 2. SURGERY. Educ: Univ Ala, BS, 51, MD, 55; Am Bd Surg, dipl, 63. Prof Exp: Instr med physiol, Sch Med, Univ Ala, 53; intern surg, Michael Reese Hosp, Chicago, 55-56; asst, Peter Bent Brigham Hosp, Boston, 56-58, jr asst resident, 57 & 58-59, sr asst resident, 60-61, sr resident, 61-62; instr, Harvard Med Sch, 66; from asst prof to assoc prof, 66-75, PROF SURG, SCH MED, TULANE UNIV, 75- Concurrent Pos: Surg res fel, Harvard Med Sch, 56-58; surg res fel, St Mary's Hosp & Med Sch, London, 59-60; George Gorham Peters fel, Peter Bent Brigham Hosp, Boston, 59-60, Am Cancer Soc clin fel, 61-62; teaching fel, Harvard Med Sch, 64-65; registr to prof teaching unit, St Mary's Hosp, London, 59-60; clin investr, Vet Admin Hosp, West Roxbury, 64-66, consult, 66-; adj prof biomed eng, Northeastern Univ, 64-67; mem, Surgeon-Gen Adv Comt Optical Lasers, Working Group Safety Stand Use Lasers, Armed Forces Nat Res Coun Comt Vision, Ad Hoc Initial Rev Group, Nat Ctr Radiol Health & Spec Study Sect Laser & NIH session chmn, Gordon Conf Lasers Biol Med, 65-67; chief surg, Peter Bent Brigham Hosp, 65 & jr assoc surg, 65-66; vis surgeon, Charity Hosp of La, mem active staff, Touro Infirmary & consult, Keesler AFB Hosp, 66-; Nat Heart Inst investr career develop award, 68-72. Mem: Fel Am Col Surgeons; Am Burn Asn; Asn Acad Surg; Soc Univ Surgeons; Asn Mil Surgeons US. Res: Blood rheology; vascular and gastrointestinal surgery; surgical metabolism; blood transfusion and treatment of skin cancer. Mailing Add: Dept of Surg Tulane Univ Med Sch New Orleans LA 70112

LITWIN, SAMUEL, electrical engineering, see 12th edition

LITZ, LAWRENCE MARVIN, physical chemistry, see 12th edition

LITZENBERGER, SAMUEL CAMERON, b Calgary, AB, July 21, 14; nat US; m 41. AGRONOMY, FIELD CROPS. Educ: Colo State Univ, BS, 37; Mont State Univ, MS, 39; Iowa State Univ, PhD(plant path, crop breeding), 48. Prof Exp: From instr to assoc prof agron, Mont State Univ, 39-46, asst Exp Sta, 39-43, agronomist, 43-46; assoc agronomist, Exp Sta, Fla, 48-49; agronomist, Ala Exp Sta, Agr Res Admin,

USDA, 49-51; agriculturist, res adv agron & head dept, Agr Technol Serv, US Opers Mission, Int Coop Admin, Nicaragua, 51-57, agron adv, Cambodia, 58-63, agron adv, Food & Agr Off & Chief Div, USAID, Guinea, 64-67, agr res adv & chief prod and breeding, Tunisia, 67-68, chief regional cereals off, NAfrica, 68-69, Food & Agr Off, 69-70, agronomy res specialist & chief, Crops Prod Div, 70-75, SPEC CONSULT, OFF AGR, TECH ASSISTANCE BUR, AID, WASHINGTON, DC, 75- Honors & Awards: Superior Honor Award, AID, 75. Mem: Fel AAAS; fel Am Soc Agron; Am Phytopath Soc; Am Genetic Asn. Res: Inheritance and disease resistance of small grains; weed control; seed and soil improvement; crops production and improvement under tropical, sub-tropcial, temperate and sub-arctic environments; integrated programs for agricultural production and improvement of food crops for developing nations. Mailing Add: 1230 Tulip St Longmont CO 80501

LIU, C T, b Tai-Shin, Kiangsu, China, Oct 19, 31; m 70; c 3. PHYSIOLOGY, PHARMACOLOGY. Educ: Nat Taiwan Univ, BS, 56; Univ Tenn, MS, 59, PhD(physiol), 63. Prof Exp: Assoc res biologist pharmacol, Sterling-Winthrop Res Inst, 65-66; asst prof physiol, Baylor Col Med, 66-73; RES PHYSIOLOGIST, US ARMY MED RES INST INFECTIOUS DIS, 73- Concurrent Pos: USPHS trainee, 63-65. Mem: Soc Pharmacol & Exp Therapeut; Am Physiol Soc; Soc Exp Biol & Med; Am Soc Nephrology. Res: Cardiovascular and renal physiology; water, electrolyte and lipid metabolism; mechanisms of infectious diseases; effect of muscle trauma. Mailing Add: US Army Med Res Inst Infect Dis Ft Detrick Frederick MD 21701

LIU, CHAN NAO, b Peking, China, Mar 17, 10; US citizen; m 36; c 2. ANATOMY, NEUROPHYSIOLOGY. Educ: Peking Normal Univ, BS, 36; Univ Pa, PhD(anat), 49. Prof Exp: From asst prof to assoc prof, 52-60, PROF ANAT, UNIV PA, 60- Mem: Am Asn Anat. Res: Neuroanatomy and neurophysiology, with special focus on the plasticity of the central nervous system. Mailing Add: Dept of Anat Univ of Pa Philadelphia PA 19174

LIU, CHI TAN CHANG, b July 15, 17; US citizen; m 48; c 2. IMMUNOCHEMISTRY. Educ: Univ Liege, BSc, 37; Univ Brussels, MSc, 39; Univ Lyons, DSc(biochem), 44. Prof Exp: Chemist, Pharmaceut Div, Belgian Chem Union, 39-40; res chemist, Microchem Ctr, Univ Brussels, 40-41; instr chem, Nat Univ, Yunan, China, 46-48; engr-biochemist, Nat Inst Fermentation, Brussels, Belgium, 50; res chemist, Refined Prod Corp, NJ, 51-56; res assoc immunochem, NJ Col Med & Dent, 60-66; ASSOC IMMUNOCHEM, PUB HEALTH RES INST, CITY OF NEW YORK, 66- Mem: Am Chem Soc; Am Asn Immunologists. Res: Biochemistry; immunochemical methods of detection and study of narcotic drugs. Mailing Add: Apt 23B 180 Park Row New York NY 10038

LIU, CHIEN, b Canton, China, Mar 6, 21; m 47; c 4. INFECTIOUS DISEASES, VIROLOGY. Educ: Yenching Univ, BS, 42; WChina Union Univ, MD, 47; Am Bd Pediat, dipl, 64. Prof Exp: Intern, Ill Masonic Hosp, 46-47; med intern, Garfield Mem Hosp, Washington, DC, 47-48; asst med, Johns Hopkins Univ, 51-52, asst physician, Johns Hopkins Hosp, 49-52; res assoc bact & immunol, Harvard Med Sch, 52-55, assoc, 55-58, asst prof, 58; assoc prof pediat, 58-63, PROF MED & PEDIAT, SCH MED, UNIV KANS, 63- Concurrent Pos: Res fel med, Sch Med, Johns Hopkins Univ, 49-51; USPHS res career develop award, 63-; vis prof, Nat Defense Med Ctr, Taiwan, 66-67; med consult, US Naval Med Res Unit 2, 66-67. Mem: Soc Pediat Res; Am Soc Microbiol; Am Asn Immunologists; Am Acad Microbiol; Infectious Dis Soc Am. Mailing Add: Dept of Med Univ of Kans Sch of Med Kansas City KS 66103

LIU, CHONG TAN, b Shanghai, China, May 11, 36; US citizen; m 63; c 3. INORGANIC CHEMISTRY. Educ: Nat Taiwan Univ, BSc, 56; Univ Pittsburgh, PhD(inorg chem), 64. Prof Exp: Sr res chemist, Hooker Chem Corp, 64-71; SR RES CHEMIST, STAUFFER CHEM CO, 72- Mem: Am Chem Soc. Res: Water treatment; industrial chemical processes; metal finishing; plating on plastics; corrosion controls; high temperature chemistry; coordination chemistry. Mailing Add: Stauffer Chem Co ERC Dobbs Ferry NY 10522

LIU, CHUAN SHENG, b Kwanhsi, China, Jan 9, 39; m 65; c 3. THEORETICAL ASTROPHYSICS, PLASMA PHYSICS. Educ: Tunghai Univ, BS, 60; Univ Calif, Berkeley, MA, 64, PhD(physics), 68. Prof Exp: Asst prof in residence physics, Univ Calif, Los Angeles, 68-70; vis scientist, Gulf Gen Atomic, Inc, 70-71; mem, Inst Advan Study, 71-74; PROF PHYSICS, UNIV MD, 74- Mem: Am Phys Soc. Mailing Add: Dept of Physics & Astron Univ of Md College Park MD 20740

LIU, CHUI FAN, b Chunking, China, Apr 5, 30; US citizen; m 59; c 3. INORGANIC CHEMISTRY. Educ: Univ Ill, AB, 52, PhD(chem), 56. Prof Exp: Res chemist, Dow Chem Co, 56-58; instr chem, Univ Conn, 58-60; asst prof, Univ Mich, 60-65; assoc prof, 65-70, PROF CHEM, UNIV ILL, CHICAGO CIRCLE, 70- Concurrent Pos: NSF & NIH res grants, 62- Res: Structure and chemistry of coordination compounds; asymmetric synthesis. Mailing Add: Dept of Chem Univ of Ill at Chicago Circle Chicago IL 60680

LIU, CHUI HSUN, b China, Nov 5, 31; US citizen; m 62. ANALYTICAL CHEMISTRY, INORGANIC CHEMISTRY. Educ: Univ Ill, BA, 52, PhD(chem), 57. Prof Exp: From asst prof to assoc prof chem, Polytech Inst Brooklyn, 57-65; PROF CHEM, ARIZ STATE UNIV, 65- Mem: Am Chem Soc. Res: Chemistry, electrochemistry and spectroscopy in molten salts and other nonaqueous solvents; chemistry of coordination compounds; chelating agents in chemical separations and analyses. Mailing Add: Dept of Chem Ariz State Univ Tempe AZ 85281

LIU, CHUNG LAUNG, b Canton, China, Oct 25, 34; US citizen; m 60; c 1. COMPUTER SCIENCE. Educ: Cheng Kung Univ, Taiwan, BSc, 56; Mass Inst Technol, SM, 60, ScD(elec eng), 62. Prof Exp: From asst prof to assoc prof elec eng, Mass Inst Technol, 62-72; PROF COMPUT SCI, UNIV ILL, URBANA, 72- Mem: Math Asn Am; Opers Res Soc Am. Res: Theory of computation; combinatorial mathematics. Mailing Add: Dept of Comput Sci Univ of Ill Urbana IL 61801

LIU, DAVID H W, b Honolulu, Hawaii, Aug 5, 36; m 65; c 3. TOXICOLOGY. Educ: Univ Mich, BS, 58; Univ Ore, MA, 60; Ore State Univ, PhD(fisheries), 69. Prof Exp: Biologist, Hanford Atomic Prod Oper, Gen Elec Co, 62-65 & Pac Northwest Labs, Battelle Mem Inst, 65; NIH fel, Col Pharm, Wash State Univ, 69-71; sr aquatic biologist, Waste Mgt Syst Oper, Envirogenics Co, Aerojet Gen Corp, 71-73; SR AQUATIC TOXICOLOGIST, STANFORD RES INST, 73- Res: Toxicology of water pollutants. Mailing Add: Stanford Res Inst 333 Ravenswood Ave Menlo Park CA 94025

LIU, EDWIN CHIAP HENN, b Honolulu, Hawaii, Apr 11, 42; m 65. BIOCHEMISTRY. Educ: Johns Hopkins Univ, AB, 64; Mich State Univ, PhD(biochem), 71. Prof Exp: Res assoc biochem, AEC, Plant Res Lab, Mich State Univ, 71-72; ASST PROF BIOL, UNIV SC, 72- Mem: Am Chem Soc; Genetics Soc Am; Am Soc Plant Physiologists. Res: Biochemistry and developmental biology of isozymes. Mailing Add: Dept of Biol Univ of SC Columbia SC 29208

LIU, FOOK FAH, b Calcutta, India, Sept 30, 34; m 66; c 2. HIGH ENERGY PHYSICS. Educ: Presidency Col, Calcutta, India, BSc, 56; Purdue Univ, PhD(physics), 62. Prof Exp: Res assoc, High-Energy Physics Lab, Stanford Univ, 62-66; asst prof physics, Case Inst Technol, 66-68; staff physicist, Stanford Linear Accelerator Ctr, Stanford Univ, 68-70; asst prof, 70-74, ASSOC PROF PHYSICS, CALIF STATE COL, SAN BERNARDINO, 74- Mem: Am Phys Soc. Res: Photoinduced reactions at high energies; phenomenology of multibody final states; decay of unstable particles. Mailing Add: Dept of Physics Calif State Col San Bernardino CA 92407

LIU, FRANK FU-WEN, b Taiwan. POMOLOGY. Educ: Taiwan Univ, BS, 57; Cornell Univ, MS, 69, PhD(pomol), 74. Prof Exp: Horticulturist, Sino-Am Joint Comn Rural Reconstruct, 69-71; ASST PROF POMOL, CORNELL UNIV, 74- Mem: Am Soc Hort Sci; Sigma Xi. Res: Postharvest physiology with emphasis on the control mechanism of maturation and ripening of fruits and storage methods of fruits. Mailing Add: Dept of Pomol Cornell Univ Ithaca NY 14853

LIU, FRANK TSUNG YUAN, b Tongshan, China, Jan 1, 19; nat US; m 51; c 3. PHYSIOLOGY. Educ: Army Vet Col, China, DVM, 39; Univ Mo, MA, 47; Ohio State Univ, PhD(physiol), 54. Prof Exp: Asst prof physiol, Temple Univ, 54-61; assoc prof, Sch Dent, Univ Pittsburgh, 61-67; assoc prof, 68-70, PROF PHYSIOL, SCH DENT, UNIV MO-KANSAS CITY, 70-, CHMN DEPT, 68-, LECTR MED, SCH MED, 71- Mem: NY Acad Sci; Am Physiol Soc; Soc Exp Biol & Med; Int Asn Dent Res; Am Inst Biol Sci. Res: Endocrinology; salivary glands; effects of hormones on oral structures; reproduction physiology; trace elements on dental caries. Mailing Add: Dept of Physiol Sch of Dent Univ of Mo Kansas City MO 64108

LIU, FRED WEI JUI, b Canton, China, Jan 29, 26; nat US; m 61. PHYSICAL CHEMISTRY. Educ: St John's Univ, China, BS, 48; Temple Univ, MA, 50; Lehigh Univ, PhD(chem), 52. Prof Exp: Res assoc, Lehigh Leather Inst, Pa, 52-53; chief chemist, Lester Labs, Inc, 53-64; dir, Continental Consults, Inc, 64-74; pres, Continental Trading Co, 65-74; V PRES & PUB RELS OFFICER, SOUTHEAST LABS, INC, 74- Mem: Am Chem Soc; Nat Asn Corrosion Engrs. Res: Colloid or surface chemistry; detergents; cleaning and maintenance chemicals formulation; corrosion; water treatment; foreign trade; industrial chemicals. Mailing Add: Southeast Labs Inc 1490 Mecaslin St NW Atlanta GA 30309

LIU, HOUNG-ZUNG, b China, Jan 23, 31; m 69; c 2. BIOCHEMICAL GENETICS, MICROBIAL GENETICS. Educ: Taiwan Prov Col, BS, 53; NDak State Col, MS, 59; Cornell Univ, PhD(genetics, biochem, plant physiol), 64. Prof Exp: Asst cytol, Taiwan Agr Res Inst, Taipei, 54-56; asst gen genetics, Cornell Univ, 59-64; assoc prof genetics, 64-69, PROF GENETICS & CHMN DEPT BIOL SCI, COL ARTS & SCI, STATE UNIV NY COL PLATTSBURGH, 69- Concurrent Pos: NIH spec res fel, Marquette Univ, 67-68. Mem: Am Soc Microbiol; Am Chem Soc; Genetics Soc Am. Res: Enzymology; tryptophan operon mutants of Escherichia coli and indoleglycerol-phosphate synthetase. Mailing Add: Dept of Biol Sci State Univ of NY Col Plattsburgh NY 12901

LIU, HSING-JANG, b Kiang-Su, China, Dec 2, 42; m 66; c 1. CHEMISTRY. Educ: Taiwan Prov Norm Univ, BSc, 64; Univ NB, Fredericton, PhD(chem), 68. Prof Exp: Fel chem, Univ NB, Fredericton, 68-69; res assoc, Columbia Univ, 69-70; teaching & res assoc, Univ NB, Fredericton, 70-71; ASST PROF CHEM, UNIV ALTA, 71- Mem: Am Chem Soc; Chem Inst Can. Res: Natural products, isolation, identification and synthesis; development of novel synthetic methods. Mailing Add: Dept of Chem Univ of Alberta Edmonton AB Can

LIU, JIH-HUA, b Chekiang, China, Oct 25, 41; m 74. INDUSTRIAL ORGANIC CHEMISTRY. Educ: Worcester Polytech Inst, BS, 65; Univ Wis-Mailwaukee, PhD(chem), 73. Prof Exp: Chemist, Pfizer Corp, 65-68; res asst chem, Univ Wis-Milwaukee, 68-73; ASSOC SR INVESTR CHEM, SMITH KLINE CORP, 73- Mem: Am Chem Soc; Sigma Xi. Res: Synthesis of decinine, homoprostanoids and cephalosporins. Mailing Add: Smith Kline Corp 1500 Spring Garden St Philadelphia PA 19101

LIU, JOSEPH JENG-FU, b Chiangsi, China, Oct 24, 40; m 71. CELESTIAL MECHANICS, THEORETICAL MECHANICS. Educ: Cheng Kung Univ, Taiwan, BS, 62; Auburn Univ, MS, 66, PhD(celestial mech), 71. Prof Exp: Teaching asst appl mech, Cheng Kung Univ, 63-64 & Auburn Univ, 66-71; MEM RES STAFF ASTRODYN, NORTHROP SERV, INC, 71- Mem: Am Inst Aeronaut & Astronaut. Res: General and special perturbation theories, their applications in nonsingular solutions for the orbital and attitude motions of an artificial satellite perturbed by conservative and nonconservative forces. Mailing Add: Northrop Serv Inc 6025 Technology Dr PO Box 1484 Huntsville AL 35807

LIU, KANG-JEN, polymer chemistry, see 12th edition

LIU, LIU, b Shanghai, China, Aug 12, 30; m 56; c 3. SOLID STATE PHYSICS. Educ: Univ Taiwan, BS, 54; Univ Chicago, MS, 57, PhD(physics), 61. Prof Exp: From asst prof to assoc prof, 61-74, PROF PHYSICS, NORTHWESTERN UNIV, ILL, 74- Concurrent Pos: Consult, Argonne Nat Lab, 61-64; Fulbright Sr Res Scholar to France, Presidential Bd Foreign Scholars, 75-76. Mem: Am Phys Soc. Res: Band theory of semiconductors; theory of narrow-gap and zero-gap semiconductors. Mailing Add: Dept of Physics Northwestern Univ Evanston IL 60201

LIU, LUKE LOKIA, b Anhwei, China, June 11, 45. THEORETICAL PHYSICS, EXPLORATION GEOPHYSICS. Educ: Nat Taiwan Univ, BS, 66; Nat Tsing-hua Univ, MS, 68; Johns Hopkins Univ, PhD(elec eng), 72. Prof Exp: Res staff physics, Mass Inst Technol, 72-74; RES STAFF MEM GEOPHYSICS, SHELL DEVELOP CO, 74- Mem: Soc Explor Geophysicists. Mailing Add: Dept of Geophys Shell Develop Co PO Box 481 Houston TX 77001

LIU, MAW-SHUNG, b Taiwan, Feb 2, 40; m 66; c 1. MEDICAL PHYSIOLOGY. Educ: Kaohsiung Med Col, Taiwan, DDS, 64; Univ Ky, MSc, 70; Univ Ottawa, PhD(physiol), 75. Prof Exp: Staff dentist & lectr oral surg, Chinese Army Hosp, Kaohsiung Med Col Hosp, Taiwan, 64-68; intern path, Med Ctr, Univ Ky, 68-69; Med Res Coun Can fel, 70-73; Alcoholism & Drug Addiction Res Found Ont res scholar, 73-74; instr physiol, 74-76, ASST PROF PHYSIOL, SCH MED, LA STATE UNIV MED CTR, NEW ORLEANS, 76- Mem: Am Heart Asn; assoc Am Physiol Soc. Res: Myocardial lipid and carbohydrate metabolism, and metabolic alterations in shock. Mailing Add: 4516 Newlands St Metairie LA 70002

LIU, MICHAEL T H, b Hong Kong, China, Mar 1, 39; Can citizen. PHYSICAL CHEMISTRY. Educ: St Dunstan's Univ, BSc, 61; St Francis Xavier, MA, 64; Univ Ottawa, PhD(phys chem), 67. Prof Exp: Technician, Can Celanese Ltd, 61-62; group leader qual control, Chemcell Ltd, 64; Nat Res Coun fel, Univ Reading, 67-68; asst prof, 68-73, ASSOC PROF CHEM, UNIV PRINCE EDWARD ISLAND, 73- Concurrent Pos: Nat Res Coun grant-in-aid, 68-; Def Res Bd of Can grant-in-aid, 74-76. Mem: Chem Inst Can. Res: Kinetics and mechanism of chain reaction;

unimolecular reaction; carbene chemistry. Mailing Add: Dept of Chem Univ of Prince Edward Island Charlottetown PE Can

LIU, PAN-TAI, b Taipei, Taiwan, Sept 22, 41; m 66; c 1. APPLIED MATHEMATICS. Educ: Nat Taiwan Univ, BS, 63; State Univ NY, Stony Brook, PhD(appl math), 68. Prof Exp: Asst prof, 68-74, ASSOC PROF MATH, UNIV RI, KINGSTON, 74- Concurrent Pos: Vis prof, Dept Elec Eng, Nat Taiwan Univ, Taipei, Taiwan, 74-75. Mem: Am Math Soc. Res: Optimal controls; differential games; stochastic processes. Mailing Add: Dept of Math Univ of RI Kingston RI 02881

LIU, PAUL ISHEN, b Taiwan; US citizen. CLINICAL PATHOLOGY. Educ: Nat Taiwan Univ, MD, 60; St Louis Univ, PhD(path), 69. Prof Exp: Assoc prof path, Med Ctr, Univ Kans, 73-74; ASSOC DIR LAB MED, MED COL GA, 74- Mem: AMA; Col Am Path; Am Soc Clin Path; Asn Clin Sci; Am Soc Microbiol. Res: Leukemia; immunology. Mailing Add: Dept of Lab Med Med Col of Ga Augusta GA 30902

LIU, PINGHUI VICTOR, b Formosa, China, Feb 9, 24; nat US; m 59; c 2. MEDICAL MICROBIOLOGY. Educ: Tokyo Jikel-kai Sch Med, MD, 47; Tokyo Med Sch, PhD(microbiol), 57; Am Bd Med Microbiol, dipl, 62. Prof Exp: Intern, Mercy Hosp, Cedar Rapids, Iowa, 54-55; intern internal med, Louisville Gen Hosp, 55-56; from instr to assoc prof, 57-69, PROF MICROBIOL, SCH MED, UNIV LOUISVILLE, 69- Concurrent Pos: Res fel microbiol, Sch Med, Univ Louisville, 56-57; USPHS sr res fel, 59-, res career develop award, 62-; mem, Subcomt Pseudomonas & Related Organisms, Int Comt Bact Nomenclature, 63- Mem: AAAS; Am Soc Microbiol; Infectious Dis Soc Am; NY Acad Sci. Res: Pathogenesis and taxonomy of pseudomonads and related organisms, such as aeromonads and vibrios; extracellular toxins, such as hemolysin, lecithinase and protease; immunities to infections. Mailing Add: Dept of Microbiol Univ Louisville Health Sci Ctr Louisville KY 40201

LIU, ROBERT SHING-HEI, b Shanghai, China, Aug 1, 38; m 67; c 2. ORGANIC CHEMISTRY. Educ: Howard Payne Col, BS, 61; Calif Inst Technol, PhD(chem), 65. Prof Exp: Res chemist, E I du Pont de Nemours & Co, Inc, 64-68; assoc prof, 68-72, PROF CHEM, UNIV HAWAII, 72- Concurrent Pos: Alfred P Sloan fel, 70-72; John Simon Guggenheim Found fel, 74-75. Mem: Am Chem Soc; Am Soc Photobiology. Res: Photochemistry of polyenes; energy transfer processes in solutions; reaction mechanism; new geometric isomers of vitamin A and carotenoids. Mailing Add: Dept of Chem Univ of Hawaii Honolulu HI 96822

LIU, SAMUEL HSI-PEH, b Taiyuan, China, Apr 17, 34; m 61; c 2. THEORETICAL SOLID STATE PHYSICS. Educ: Taiwan Univ, BS, 54; Iowa State Univ, 58, PhD(physics), 60. Prof Exp: Assoc res mem, Res Lab, IBM Corp, 60-61, res staff mem, 61-64; assoc prof, 64-67, PROF PHYSICS, IOWA STATE UNIV, 67- Concurrent Pos: Vis prof, H C Oersted Inst, Copenhagen Univ, 71-72 & Univ Calif, Berkeley, 75-76. Mem: Fel Am Phys Soc. Res: Solid state theory; electronic and magnetic properties of metals and metallic compounds. Mailing Add: Dept of Physics Iowa State Univ Ames IA 50010

LIU, SI-KWANG, b Kwangsi, China, Dec 1, 25; m 60; c 2. VETERINARY PATHOLOGY. Educ: Vet Col Chinese Army, DVM, 49; Univ Calif, Davis, PhD(vet path), 64. Prof Exp: Sr vet res & diag, Provincial Taitung Agr Sta, China, 50-55; lectr vet path, Col Agr, Taiwan Univ, 56-59, chief path lab, Univ Vet Hosp, 56-59; res asst path & parasitol, Sch Vet Med, Univ Calif, Davis, 59-64; assoc pathologist, 64-66, cardiopulmonary pathologist, 66-69, ASST HEAD, DEPT PATH, ANIMAL MED CTR, 69-, SR STAFF MEM, 73- Concurrent Pos: Clinical asst prof in comp path, NY Med Col, 69- Mem: Am Vet Med Asn; Am Soc Parasitol; Sigma Xi; NY Acad Sci; Acad Vet Cardiol. Res: Parasitological and immunological pathology in domestic animals as well as zoo animals; comparative pathology in cardiovascular and orthopedic diseases. Mailing Add: Animal Med Ctr 510 E 62nd St New York NY 10021

LIU, SOPHIA YAN, b Hong Kong, May 23, 47; m 69. ORGANIC CHEMISTRY. Educ: Col Mt St Vincent, BS, 69; Columbia Univ, PhD(chem), 73. Prof Exp: RES CHEMIST, STAUFFER CHEM CO, EASTERN RES CTR, 73- Mem: Sigma Xi; Am Chem Soc. Res: Synthesis of novel organophosphorous compounds for a variety of industrial applications and end uses. Mailing Add: Stauffer Chem Co Eastern Res Ctr Dobbs Ferry NY 10522

LIU, STEPHEN C Y, b Hunan, China, Feb 24, 27; m 54; c 4. MICROBIOLOGY, IMMUNOLOGY. Educ: Taiwan Univ, BSc, 51, MSc, 54; Univ Minn, PhD, 57. Prof Exp: Instr plant path, Taiwan Univ, 51-54; from res asst to res assoc, Univ Minn, 54-58; res plant pathologist, Nat Res Coun, 58-62, asst mgr Chas Pfizer & Co, Inc, 62-65; from asst prof to assoc prof, 65-74, PROF MICROBIOL, EASTERN MICH UNIV, 74- Mem: AAAS; Am Phytopath Soc; Am Soc Microbiol; NY Acad Sci. Res: Genetics of bacteria; immunology; virology. Mailing Add: Dept of Biol Eastern Mich Univ Ypsilanti MI 48197

LIU, SU-CHIN CHANG, b Hsinchu, Taiwan, Apr 28, 38; m 62. BIOCHEMISTRY. Educ: Nat Taiwan Univ, BS, 61; Univ Idaho, MS, 64, PhD(agr biochem), 71. Prof Exp: Sci aide agr chem, Wash State Univ, 64-66; res asst physiol, Harvard Med Sch, 66-67; res staff chem engr, Mass Inst Technol, 67-68; res fel agr biochem, Univ Idaho, 68-70, res chemist plant physiol, US Forest Serv, Calif, 70-71; RES ASSOC DERMAT, MED SCH, STANFORD UNIV, 71- Res: Isolation and identification of epithelial cell growth stimulate factors and the mechanism of its function. Mailing Add: Dept of Dermat Stanford Univ Med Sch Stanford CA 94305

LIU, TAI-PING, b Taiwan, Repub of China, Nov 18, 45; m 73. MATHEMATICAL ANALYSIS. Educ: Nat Taiwan Univ, BS, 68; Ore State Univ, MS, 70; Univ Mich, PhD(math), 73. Prof Exp: ASST PROF MATH, UNIV MD, COLLEGE PARK, 73- Mem: Am Math Soc. Res: Nonlinear conservation laws; shock waves theory; gas dynamics. Mailing Add: Dept of Math Univ of Md College Park MD 20742

LIU, TEH-YUNG, b Tainan, Formosa, May 24, 32; nat US; m 61; c 3. BIOCHEMISTRY. Educ: Taiwan Nat Univ, BS, 55; Univ Pittsburgh, PhD(biochem), 61. Prof Exp: Res assoc biochem, Rockefeller Univ, 61-65, asst prof, 65-67; biochemist, Biol Dept, Brookhaven Nat Lab, 67-73; sect head biochem microbial struct, Nat Inst Child Health & Human Develop, 73-74, DEP DIR, DIV BACT PROD, BUR BIOLOGICS, NIH, 74- Mem: Am Soc Biol Chem. Res: Streptococcal proteinase; pneumococcal and meningococcal cell wall polysaccharides; human C-reactive protein, limulus lysate. Mailing Add: Bur of Biologics Bldg 29 Rm 425 8800 Rockville Pike Bethesda MD 20014

LIU, WEN CHIH, b Liao-Ning Province, China, Feb 19, 21; US citizen; m 60; c 3. BIOCHEMISTRY. Educ: Nat Hu-Nan Univ, BS, 44; Baylor Univ, MS, 53; Univ Wis-Madison, PhD(biochem), 58. Prof Exp: Res assoc biochem, Univ Wis, 58-59; res assoc, Univ Calif, Los Angeles, 59-60; sr res chemist, Pfizer, Inc, 60-69; SR RES INVESTR MICROBIOL, SQUIBB INST MED RES, 69- Mem: AAAS; Am Chem Soc. Res: Natural products; antibiotics; cancer chemotherapy; plant growth substances. Mailing Add: Squibb Inst for Med Res Princeton NJ 08540

LIU, YOUNG KING, b Nanking, China, May 3, 34; US citizen; m 64; c 2. BIOMECHANICS, BIOMEDICAL ENGINEERING. Educ: Bradley Univ, BS, 55; Univ Wis-Madison, MS, 59; Wayne State Univ, PhD(mech), 63. Prof Exp: Instr mech, Wayne State Univ, 6063; lectr, Univ Mich, Ann Arbor, 63-64, asst prof, 64-68; vis asst prof aeronaut & astronaut, Stanford Univ, 68-69; assoc prof, 69-72, PROF BIOMECH, TULANE UNIV, 72- Concurrent Pos: NIH spec res fel, Stanford Univ, 68-69; biophys consult, US Army Aeromed Res Lab, 72-; NIH res career develop award, 71. Mem: Am Soc Eng Educ; Orthop Res Soc; Am Asn Univ Profs; Sigma Xi; Am Acad Mech. Res: Biomechanics and physiologic basis of acupuncture. Mailing Add: Biomech Lab Tulane Univ Sch of Med New Orleans LA 70112

LIU-GER, TSU-HUEI, b Kwei-yang, Kwei-chow, Repub of China, Mar 10, 43; US citizen; m 71; c 1. THEORETICAL PHYSICS. Educ: Nat Taiwan Univ, BS, 64; Univ Ore, PhD(physics), 69. Prof Exp: Asst prof physics, Portland State Univ, 69-75; PHYSICIST, ELECTROMAGNETIC TRANSIENT PROG, BONNEVILLE POWER ADMIN, US DEPT INTERIOR, 75- Mem: Sigma Xi. Res: Electromagnetic transient studies of the power systems. Mailing Add: Bonneville Power Admin Rte EOGA PO Box 3621 Portland OR 97208

LIUIMA, FRANCIS ALOYSIUS, b Utena, Lithuania, Mar 8, 19; nat US. PHYSICS. Educ: Boston Col, MS, 50; St Louis Univ, PhD(physics), 54. Prof Exp: ASST PROF PHYSICS, BOSTON COL, 54- Mem: Am Phys Soc; Am Asn Physics Teachers. Res: Microwave spectroscopy. Mailing Add: Dept of Physics Boston Col Chestnut Hill MA 02167

LIUKKONEN, JOHN ROBIE, b Oakland, Calif, Oct 23, 42. MATHEMATICAL ANALYSIS. Educ: Harvard Univ, BA, 65; Columbia Univ, PhD(math), 70. Prof Exp: Asst prof, 70-75, ASSOC PROF MATH, TULANE UNIV, 75- Mem: Am Math Soc; Math Asn Am. Res: Representations of locally compact groups; harmonic analysis on locally compact groups. Mailing Add: Dept of Math Tulane Univ New Orleans LA 70118

LIVANT, PETER DAVID, b New York, NY, Sept 18, 48. PHYSICAL ORGANIC CHEMISTRY. Educ: City Col New York, BS, 69; Brown Univ, PhD(chem), 75. Prof Exp: Vis asst prof chem, 74-75, RES ASSOC CHEM, UNIV ILL, URBANA-CHAMPAIGN, 75- Mem: Am Chem Soc. Res: Mechanism of radical disproportionations; interactions of peroxide compounds with transition metal salts; chemically induced dynamic nuclear polarization studies of transition metal catalysis; tetracoordinate tetracovalent sulfur compounds; chemically induced dynamic nuclear polarization dependence on magnetic field strength. Mailing Add: Roger Adams Lab Univ of Ill Sch of Chem Sci Urbana IL 61801

LIVE, ISRAEL, b Austria, Apr 26, 07; nat US; m 36; c 2. VETERINARY SCIENCE. Educ: Univ Penn, VMD, 34, AM, 36, PhD(path), 40; Am Bd Microbiol, dipl. Prof Exp: Asst histopath & clin path, 34-37, from instr path to asst prof, 37-46, bact, 46-49, assoc prof, 49-53, PROF MICROBIOL, SCH VET MED, UNIV PA, 53- Concurrent Pos: Mem expert comt on brucellosis, WHO. Mem: Fel AAAS; fel Am Acad Microbiol; Am Vet Med Asn; Am Soc Microbiol; Am Pub Health Asn. Res: Diagnosis of filariasis in dogs; nature of Clostridium chauvoei aggressin; diagnosis, therapy and immunization in brucellosis; staphylococci in animals and man; serological characterization of staphylococci. Mailing Add: Dept of Microbiol Univ of Pa Sch of Vet Med Philadelphia PA 19104

LIVELY, DAVID HARRYMAN, b Indianapolis, Ind, Aug 17, 30; m 53; c 2. MICROBIOLOGY. Educ: Purdue Univ, BS, 52; Univ Tex, MA, 58, PhD, 62. Prof Exp: Mem water pollution serv team, State Bd Health, Ind, 52; bacteriologist, Biol Warfare Labs, US Army, 54-56; asst bacteriologist, Univ Tex, 56-57, res scientist, 57-60; sr microbiologist, 61-63, group leader microbiol, Chem Res Div, 63-66, asst mgr antibiotic develop, Antibiotic Mfg & Develop Div, 66-69, head microbiol res, 69-72, RES ASSOC, ELI LILLY & CO, 72- Mem: Fel AAAS; Am Chem Soc; Am Soc Microbiol; Brit Soc Gen Microbiol; Sigma Xi. Res: Bacterial endospore; production of biological active compounds by microorganisms. Mailing Add: Antibiotic Develop Div Eli Lilly & Co Indianapolis IN 46206

LIVENGOOD, SAMUEL MILLER, b Salisbury Pa, Nov 1, 17; m 41; c 3. ORGANIC CHEMISTRY. Educ: Juniata Col, BS, 38; Rutgers Univ, MS, 41, PhD(org chem), 43. Prof Exp: Instr chem, Rutgers Univ, 40-43; fel, Mellon Inst, 43-55, sr fel, 55-59; ASST DIR, RES & DEVELOP DEPT, CHEM DIV, UNION CARBIDE CORP, 59- Mem: Am Chem Soc; Soc Chem Indust. Res: Detergents and cosmetics; textile intermediates; humectants; water soluble resins; hydraulic fluids; heat transfer fluids· metalworking fluids. Mailing Add: Res & Dev Union Carbide Tarrytown Tech Ctr Tarrytown NY 10591

LIVERHANT, SOLOMON ELIESER, b Frankfurt am Main, Ger, Mar 21, 17; nat US; m 54; c 1. NUCLEAR PHYSICS, MATHEMATICS. Educ: Cambridge Univ, BSc, 41; Univ London, BSc, 42, MSc, 45. Prof Exp: Lectr appl math, Woolwich Polytech, Univ London, 44-46; instr physics, NY Univ, 46-48 & Balfour Munic Col, Israel, 48-50; vis asst prof, Colgate Univ, 51-53; PROF PHYSICS, STATE UNIV NY MARITIME COL, 53- Mem: Am Phys Soc; Am Nuclear Soc; assoc Brit Inst Physics & Phys Soc. Res: Nuclear, reactor and neutron physics. Mailing Add: Maritime Col State Univ of NY Ft Schuyler Bronx NY 10465

LIVERMAN, JAMES LESLIE, b Brady, Tex, Aug 17, 21; m 43, 59; c 5. PLANT PHYSIOLOGY, BIOCHEMISTRY. Educ: Agr & Mech Col, Univ Tex, BS, 49; Calif Inst Technol, PhD(plant physiol, bioorg chem), 52. Prof Exp: Fel plant physiol, Calif Inst Technol, 52-53; from asst prof to prof biochem, Agr & Mech Col, Univ Tex, 53-60; biochemist, AEC, 58-59, asst chief biol br, 59-60, chief 60-64; assoc dir biol div, Oak Ridge Nat Lab, 64-67, asst dir life sci, 67-69, assoc dir biomed & environ sci, 69-72; dir div biomed & environ res, US AEC, 72-75, asst gen mgr biomed & environ res & safety, 73-75, DIR, DIV BIOMED & ENVIRON RES & ASST ADMINR ENVIRON & SAFETY, US ENERGY RES & DEVELOP ADMIN, 75- Concurrent Pos: Consult agr chemist, 56-58; chmn, Gordon Conf Biochem & Agr, 61; Interim dir, Univ Tenn-Oak Ridge Grad Sch Biomed Sci, 65-66. Honors & Awards: Distinguished Serv Award, US AEC, 74. Mem: AAAS; Am Soc Plant Physiol; Am Chem Soc; Radiation Res Soc; Ecol Soc Am. Res: Cell physiology; photoperiodism; radiation in biological systems; immunology; bioengineering; policy science. Mailing Add: Asst Adminr for Environ & Safety US Energy Res & Develop Admin Washington DC 20545

LIVERMAN, THOMAS PHILLIP GEORGE, b Salzburg, Austria, June 18, 23; US citizen; m 46; c 2. MATHEMATICS. Educ: Univ Pa, MA, 48, PhD(math), 56. Prof Exp: Instr math, Univ Del, 46-48; engr, C N R Co, France, 48-49; mathematician, Appl Physics Lab, Johns Hopkins Univ, 51-58; assoc prof, 58-60, chmn dept, 71-74, PROF MATH, GEORGE WASHINGTON UNIV, 71- Mem: Am Math Soc; Soc Indust & Appl Math; Opers Res Soc Am; Math Soc France. Res: Functional analysis and applied mathematics; function theory. Mailing Add: Dept of Math George Washington Univ Washington DC 20052

LIVERMORE, ARTHUR HAMILTON, b Monroe, Wash, Aug 14, 15; m 40, 64; c 6. BIOCHEMISTRY, SCIENCE EDUCATION. Educ: Reed Col, BA, 40; Univ Rochester, MS, 42, PhD(biochem), 44. Prof Exp: Asst chem, Ore State Col, 40-41 & Univ Rochester, 43-44; res asst, Med Col, Cornell Univ, 44-46, res assoc, 46-48; from asst prof to prof chem, Reed Col, 48-65; dep dir educ, 63-74, HEAD, OFF SCI EDUC, AAAS, 74- Concurrent Pos: Guggenheim fel, Cambridge Univ, 55-56; training adv, Regional Ctr Educ in Sci & Math, 71-72. Mem: AAAS; fel Am Chem Soc; Soc Nuclear Med (secy, 54-55); Am Soc Biol Chem; fel Am Inst Chemists. Res: B vitamins in wheat products; destruction of thiamine by fresh fish tissues; chemical nature and synthesis of penicillin; chemistry of posterior pituitary hormones; yeast; effects of ionizing radiations on biochemical substances; new programs in high school and elementary science education. Mailing Add: AAAS 1515 Massachusetts Ave N W Washington DC 20005

LIVERMORE, BRIAN PAUL, b Austin, Minn, Dec 31, 47; m 70. MEDICAL MICROBIOLOGY. Educ: Univ Minn, BA, 69, PhD(microbiol), 74. Prof Exp: Fel biochem, Univ Ottawa, 74; ASST PROF BIOL, OAKLAND UNIV, 75- Mem: AAAS; Am Soc Microbiol; Sigma Xi. Res: Elucidation of metabolic pathways for lipids in the spirochetes, primarily Treponema pallidum and the exploration of the immune response to membrane antigens during syphillis. Mailing Add: Dept of Biol Sci Oakland Univ Rochester MI 48063

LIVERS, RONALD WILSON, b Barnes, Kans, Mar 20, 22; m 46; c 4. PLANT BREEDING. Educ: Kans State Univ, BS, 48, MS, 49; Univ Minn, PhD(plant genetics), 57. Prof Exp: Asst agronomist, Plains Substa, NMex State Univ, 52-58, assoc agronomist & supt, Substa, 58-62; assoc prof agron, 62-66, PROF CEREAL CROPS, FT HAYS EXP STA, KANS STATE UNIV, 66- Mem: AAAS; Am Soc Agron; Am Genetic Asn. Res: Breeding and genetics of wheat. Mailing Add: Ft Hays Exp Sta Kans State Univ Hays KS 67601

LIVERSAGE, RICHARD ALBERT, b Fitchburg, Mass, July 8, 25; m 54; c 4. DEVELOPMENTAL BIOLOGY. Educ: Marlboro Col, BA, 51; Amherst, Col, AM, 53; Princeton Univ, AM, 57, PhD(biol), 58. Prof Exp: Lab instr biol, Amherst Col, 54-55; instr, Princeton Univ, 58-60; from asst prof to assoc prof, 60-69, PROF BIOL, UNIV TORONTO, 69- GRAD SECY DEPT, 75- Concurrent Pos: Vis investr, Huntsman Marine Lab, NB, Can, 68-; res assoc, Dept Biophys, Strangeways Res Lab, Cambridge, Eng, 72. Mem: Am Soc Zool; Soc Develop Biol; Royal Can Inst. Res: In vivo and in vitro studies on the role of nerves and endocrine secretions in amphibian and fish appendage regeneration. Mailing Add: Ramsay Wright Zool Labs, Univ of Toronto Toronto ON Can

LIVESAY, GEORGE ROGER, b Ashley, Ill, Dec 9, 24. MATHEMATICS. Educ: Univ Ill, BS & MS, 48, PhD, 52. Prof Exp: Instr math, Univ Mich, 50-56; res assoc, 56-58, from asst prof to assoc prof, 58-69, PROF MATH, CORNELL UNIV, 69- Res: Topology. Mailing Add: Dept of Math Cornell Univ Ithaca NY 14850

LIVEZEY, ROBERT LEE, b Stockton, Calif, Jan 14, 20; m 46; c 3. BIOLOGY. Educ: Ore State Col, BS, 43, MS, 44; Cornell Univ, PhD(vert zool), 46. Prof Exp: Assoc prof biol, Sam Houston State Col, 46-47, prof & head dept, 47-49; asst prof, Univ Notre Dame, 49-53; from asst prof to assoc prof, 54-64, PROF BIOL, CALIF STATE UNIV, SACRAMENTO, 64- Mem: AAAS; Ecol Soc Am; Am Soc Ichthyol & Herpet. Res: Herpetology; ecology. Mailing Add: Dept of Biol Sci Calif State Univ Sacramento CA 95819

LIVIGNI, RUSSELL ANTHONY, b Akron, Ohio, July 20, 34. POLYMER CHEMISTRY. Educ: Univ Akron, BSc, 56, PhD(polymer chem), 60. Prof Exp: Res scientist polymer chem, Ford Sci Lab, 60-61; sr res chemist, 61-62, group leader polymer characterization & kinetics, 62-63, sect head, 63-69, sect head mat chem & polymer characterization, 69-75, DEPT MGR POLYMER & ANAL CHEM, GEN TIRE & RUBBER CO, 75- Mem: AAAS; Am Chem Soc. Res: Kinetics of free radical polymerization; kinetics and mechanism of anionic polymerization and copolymerization; determination of structure of block copolymers; characterization of polymer molecular weights and distribution; thermal analysis. Mailing Add: 2291 Manchester Rd Akron OH 44314

LIVINGOOD, CLARENCE SWINEHART, b Elverson, Pa, Aug 7, 11; m 47; c 5. DERMATOLOGY. Educ: Ursinus Col, BS, 32; Univ Pa, MD, 36; Am Bd Dermat, dipl, 61. Prof Exp: Asst prof dermat & syphil, Med Sch, Univ Pa, 46-48; prof dermat, Jefferson Med Col, 48-49; prof dermat & syphil, Sch Med, Univ Tex, 49-53; CHMN DEPT DERMAT, HENRY FORD HOSP, 53- Concurrent Pos: Consult, Vet Admin & Surg Gen, US Army; mem comn cutaneous dis, Armed Forces Epidemiol Bd; adv panel med sci, Dept Defense, 55-60; secy gen, Int Cong Dermat, 62; secy, Am Bd Dermat, 63- Mem: AAAS; Soc Invest Dermat (pres, 55); Am Dermat Asn; AMA; NY Acad Sci. Res: Epidemiology and treatment of cutaneous bacterial infections; topical corticosteroid therapy of cutaneous disease. Mailing Add: Dept of Dermat Henry Ford Hosp Detroit MI 48202

LIVINGOOD, JOHN JACOB, b Cincinnati, Ohio, Mar 7, 03; m 34; c 2. PHYSICS. Educ: Princeton Univ, AB, 25, MA, 27, PhD(physics), 29. Prof Exp: Instr physics, Palmer Phys Lab, Princeton Univ, 29-32; res assoc, Lawrence Radiation Lab, Univ Calif, 32-38; instr & tutor physics, Harvard Univ, 38-39, fac instr & tutor, 39-42, res assoc, Off Sci Res & Develop Proj, Radio Res Lab, 42-45; asst dir res div, Collins Radio Co, 45-52; assoc dir physics div, 52-56, dir particle accelerator div, 56-58, sr physicist, 58-68, CONSULT, ARGONNE NAT LAB, 68- Mem: Fel Am Phys Soc; fel Am Nuclear Soc. Res: Line spectroscopy; cyclotron design; high power oscillator tubes; artificial radioactivity. Mailing Add: 836 S County Line Rd Hinsdale IL 60521

LIVINGOOD, JOHN N B, b Birdsboro, Pa, June 8, 13; m 40; c 3. MATHEMATICS. Educ: Gettysburg Col, AB, 34; Univ Pa, AM, 36, PhD(math), 44. Prof Exp: Teacher high sch, 36-38; instr math, Gettysburg Col, 38-42; Rutgers Univ, 42-44; mathematician, Nat Adv Comt Aeronaut, 44-47; asst prof math, Rutgers Univ, 47-48; aeronaut res scientist, Nat Adv Comt Aeronaut 48-58 & NASA, 58-73; LECTR MATH, COL BOCA RATON, FLA, 73- Mem: Am Math Soc. Res: Theory of numbers; aeronautical research; turbine cooling; nuclear engineering. Mailing Add: 20 NW 24th St Delray Beach FL 33444

LIVINGSTON, ALBERT EDWARD, b Hartford, Conn, Feb 28, 36. MATHEMATICAL ANALYSIS. Educ: Boston Col, BA, 58, MA, 60; Rutgers Univ, MS, 62, PhD(math), 63. Prof Exp: Asst prof math, Lafayette Col, 63-67; from asst prof to assoc prof, 67-75, PROF MATH, UNIV DEL, 75- Mem: Am Math Soc; Sigma Xi. Res: Univalent and multivalent fuctions, particularly the application of methods of extreme point theory and subordination chains to extremal problems in multivalent function theory. Mailing Add: Dept of Math Univ of Del Newark DE 19711

LIVINGSTON, ARTHUR EUGENE, mathematics, deceased

LIVINGSTON, CLARK HOLCOMB, b Eau Claire, Wis, Nov 25, 20; m 47; c 2. PLANT PATHOLOGY. Educ: Colo Agr & Mech Col, BS, 51, MS, 53; Univ Minn, PhD, 66. Prof Exp: ASSOC PROF BOT & PLANT PATH, COLO STATE UNIV, 55- Mem: Am Phytopath Soc; Potato Asn Am. Res: Potato diseases, particularly physiology of disease and viruses. Mailing Add: Dept of Bot & Plant Path Colo State Univ Ft Collins CO 80521

LIVINGSTON, DANIEL ISADORE b New York City, NY, Oct 15, 19; m 56; c 2. PHYSICAL CHEMISTRY. Educ: City Col New York, BS, 41; Polytech Inst Brooklyn, PhD(phys chem), 50. Prof Exp: Dir polymer chem, Gen Latex & Chem Corp, 50-51; scientist, Polaroid Corp, 51-55; sr res engr, Ford Motor Co, Univ Mich, 55-57; sr res chemist, Continental Can Co, Univ Ill, 57-59; HEAD, POLYMER PHYSICS SECT, GOODYEAR TIRE & RUBBER CO, 59- Concurrent Pos: Assoc ed, Rubber Chem & Technol, 69-72; ed, Tire Sci & Technol, 72- Mem: Am Chem Soc; Am Phys Soc; Soc Rheology; Am Soc Test & Mat. Res: Physical chemistry of polymers; high polymer synthesis, research and development; radiation effects in polymer systems; polymer physics. Mailing Add: 731 Frank Blvd Akron OH 44320

LIVINGSTON, G E, b Rotterdam, Netherlands, Feb 1, 27; m 48; c 3. FOOD SCIENCE. Educ: NY Univ, BA, 48; Univ Mass, MS, 51, PhD(food technol), 52. Prof Exp: Chemist, Bur Chem, NY Produce Exchange, 49; from asst prof to assoc prof food technol, Univ Mass, 51-59; DIR, FOOD SCI ASSOCS, INC, 56- Concurrent Pos: Vis prof, Laval Univ, 54; vis lectr, City Col New York, 59-60; res supvr, Continental Baking Co, 59-62; mem, Adv Bd Mil Personnel Supplies, Nat Acad Sci-Nat Res Coun, 61-64, chmn, Comt Food Serv Systs, 68-71; mgr, Instnl Prod Dept, Morton Frozen Foods Co, 62-65; adj prof, Columbia Univ, 66-72; invitee, White House Conf Food, Nutrit & Health, 69; chmn, Comt Food & Nutrit & mem, Bd Sci Consults, Am Health Found, 69-; chmn, Panel VII, Nat Conf Food Protection, 71; consult, US Army Natick Labs, 71-; mem, Bd Govs, Food Update, Food & Drug Law Inst, 71-75; adj prof, Pratt Inst, 73-; mem, Food Stability Comn. Mem: Am Asn Cereal Chemists; Am Pub Health Asn; Inst Food Technologists; NY Acad Sci; Royal Soc Health. Res: Food colorimetry; prepared foods; food service systems; nutritive value. Mailing Add: Food Sci Assocs Inc 145 Palisade St Dobbs Ferry NY 10522

LIVINGSTON, JOEL R, JR, organic chemistry, see 12th edition

LIVINGSTON, KNOX W, b Atlanta, Ga, Apr 24, 19; m 48; c 1. FORESTRY. Educ: Univ SC, BS, 40; Duke Univ, MF, 48. Prof Exp: Asst forestry, 48-49, asst forester, 49-63, ASST PROF FORESTRY, AUBURN UNIV, 63- Mem: Soc Am Foresters. Res: Density, site, growth relations, especially planted southern pine; soil, site relations. Mailing Add: 856 Cary Dr Auburn AL 36830

LIVINGSTON, LUZERN GOULD, b Randolph, Wis, Apr 1, 05; m; c 4. BOTANY. Educ: Lawrence Col, BA, 29; Univ Wis, PhD(plant physiol), 33. Prof Exp: Lab instr & asst bot, Univ Wis, 29-33; Nat Res Found fel plant anat, Harvard Univ, 33-34, instr & tutor biol, 34-38; from asst prof to prof, 38-74, EMER PROF BOT, SWARTHMORE COL, 74- Concurrent Pos: Res assoc, Inst Cancer Res, Philadelphia, 45-52 & Philadelphia Acad Natural Sci, 55- Mem: Bot Soc Am. Res: Plant anatomy and cytology; plasmodesmata; biosynthesis of C-isotope compounds; diatomology. Mailing Add: Dept of Biol Swarthmore Col Swarthmore PA 19081

LIVINGSTON, MILTON STANLEY, b Brodhead, Wis, May 25, 05; m 30; c 2. NUCLEAR PHYSICS. Educ: Pomona Col, AB, 26; Dartmouth Col, MA, 28; Univ Calif, PhD(physics), 31. Prof Exp: Instr physics, Dartmouth Col, 28-29; res assoc, Univ Calif, 32-34; asst prof, Cornell Univ, 34-38; from assoc prof to prof, 38-70, EMER PROF, MASS INST TECHNOL, 70- Concurrent Pos: Proj chmn, Brookhaven Nat Lab, 46-48; dir, Cambridge Electron Accelerator, Harvard Univ, 56-67; assoc dir, Nat Accelerator Lab, Univ Ill, 67-70; consult, Los Alamos, Oak Ridge, Argonne & Brookhaven Nat Labs. Mem: Nat Acad Sci; Am Phys Soc; Fedn Am Sci. Res: Design of high energy accelerators. Mailing Add: 1005 Calle Largo Santa Fe NM 87501

LIVINGSTON, PETER MOSHCHANSKY, b Milwaukee, Wis, July 11, 34; m 57; c 1. PHYSICS, THEORETICAL CHEMISTRY. Educ: Univ Wis, BS, 56, PhD(theoret chem), 61. Prof Exp: Fel, Theoret Chem Inst, Univ Wis, 64-65; assoc prof appl physics, Cath Univ Am, 65-71; RES PHYSICIST, HEAD OPTICAL RADIATION BR, NAVAL RES LAB, WASHINGTON, DC, 70- Concurrent Pos: Consult, Inst Defense Anal, 65- Mem: Am Phys Soc. Res: Kinetic theory of gases and plasmas; light scattering in turbulent atmospheres; semiclassical scattering theory; high energy lasers. Mailing Add: Naval Res Lab Code 5560 Washington DC 20375

LIVINGSTON, RALPH, b Keene, NH, May 16, 19; m 43; c 4. CHEMICAL PHYSICS. Educ: Univ NH, BS, 40, MS, 41; Univ Cincinnati, DSc(chem), 43. Prof Exp: Chemist, metall lab, Univ Chicago, 43-45, assoc dir chem div, 65-75, CHEMIST, OAK RIDGE NAT LAB, 45- GROUP LEADER, 75- Concurrent Pos: Guggenheim fel & Fulbright scholar, France, 60-61; prof, Univ Tenn, 64- Mem: AAAS; Am Chem Soc; Am Phys Soc. Res: Radiation chemistry; chemical physics; pure quadrupole spectroscopy and electron spin resonance. Mailing Add: Chem Div Oak Ridge Nat Lab PO Box X Oak Ridge TN 37830

LIVINGSTON, ROBERT BLAIR, b Colorado Springs, Colo, July 21, 11; m 42; c 2. BOTANY. Educ: Colo Col, BS, 38; Duke Univ, MA, 41, PhD(plant ecol), 47. Prof Exp: Instr bot, Colo Col, 39; asst prof, Univ Mo, 46-50; from asst prof to assoc prof, 50-53, PROF BOT, UNIV MASS, AMHERST, 53- Mem: Ecol Soc Am. Res: Ecology of Berkshire, Massachusetts communities; flora of western New England; ecology of Juniperus species of the Northeast. Mailing Add: Dept of Bot Univ of Mass Amherst MA 01002

LIVINGSTON, ROBERT BURR, b Boston, Mass, Oct 9, 18; m 54; c 3. NEUROPHYSIOLOGY, NEUROANATOMY. Educ: Stanford Univ, AB, 40, MD, 44. Prof Exp: Intern, Stanford Hosp, 43, asst resident, 44; instr physiol, Sch Med, Yale Univ, 46-48; asst prof physiol, Sch Med & dir aeromed res unit, Yale Univ, 50-52; from assoc prof to prof physiol & anat, Univ Calif, Los Angeles, 52-56; dir basic res & sci dir, NIMH & Nat Inst Neurol Dis & Blindness, 56-60, chief lab neurobiol, NIMH, 60-63, chief gen res support br & assoc chief prog planning, Div Res Facil & Resources, NIH, 63-65; chmn dept neurosci, 65-71, PROF NEUROSCI, SCH MED, UNIV CALIF, SAN DIEGO, 65 Concurrent Pos: Nat Res Coun sr rel neurol, Int Physiol, Switz, 48-49; Gruber fel neurophysiol, Switz, France & Eng, 49-50; NIMH sr fel, Gothenburg Univ, 56; res asst, Harvard Med Sch, 47-48; asst to pres, Nat Acad Sci, 51-52; prof lectr, Univ Calif, Los Angeles, 56-59; guest prof, Univ Zurich, 71-72; assoc, Neurosci Res Prog. Mem: AAAS; Am Physiol Soc; Asn Res Nerv & Ment Dis; Am Neurol Asn; Am Asn Anatomists. Res: Mechanisms relating to higher nervous processes, perception, learning and memory; plasticity of ultrastructure of nervous system. Mailing Add: Dept of Neurosci Univ of Calif at San Diego La Jolla CA 92093

LIVINGSTON, ROBERT LOUIS, b Ada, Ohio, Nov 15, 18; div; c 4. PHYSICAL CHEMISTRY. Educ: Ohio State Univ, BS, 39; Univ Mich, MS, 41, PhD(phys chem),

43. Prof Exp: Res assoc, Univ Mich, 42-44; assoc chemist, Naval Res Lab, Washington, DC, 44-46; from asst prof to assoc prof, 46-54, asst head dept, 60-68, PROF CHEM, PURDUE UNIV, 54- Concurrent Pos: Consult, Educ Testing Serv, Princeton, NJ, 73-74. Mem: Am Chem Soc; AAAS. Res: Molecular structure by electron diffraction; electron diffraction by surface films; chemical education. Mailing Add: Dept of Chem Purdue Univ West Lafayette IN 47907

LIVINGSTON, ROBERT SIMPSON, b Summerland, Calif, Sept 20, 14; m 55; c 5. PHYSICS, RESEARCH ADMINISTRATION. Educ: Pomona Col, BA, 35; Univ Calif, MA, 41, PhD(physics), 41. Prof Exp: Asst physics, Pomona Col, 35-36; asst, Univ Calif, 36-39, res fel, Lawrence Radiation Lab, 39-43; physicist, Tenn Eastman Corp, 43-47; res supt, Carbide & Carbon Corp, 47-50; dir electronuclear div, 50-71, DIR PROG PLANNING & ANAL, OAK RIDGE NAT LAB, 71- Concurrent Pos: Consult, Nuclear Physics Panel, Physics Surv Comt, Nat Acad Sci, 69-72; chmn, Ad Hoc Comt Heavy Ion Sources, Nuclear Sci Div, Nat Acad Sci, 72-74. Mem: Fel Inst Elec & Electronics Engrs; fel Am Phys Soc; fel AAAS; Am Nuclear Soc; Sigma Xi. Res: Long range planning of scientific research and development in energy; design of isochronous cyclotrons; heavy particle accelerators; new particle accelerator methods; high intensity ion sources. Mailing Add: Oak Ridge Nat Lab PO Box X Oak Ridge TN 37830

LIVINGSTON, SAMUEL, b Philadelphia, Pa, July 14, 08; m 36; c 2. PEDIATRICS. Educ: Georgetown Univ, BS, 28; Vanderbilt Univ, MD, 34. Prof Exp: Asst, 36-39, from instr to assoc prof pediat, Sch Med, 39-73, dir epilepsy clin, Hosp, 45-73, EMER ASSOC PROF PEDIAT, JOHNS HOPKINS UNIV & EMER DIR EPILEPSY CLIN, JOHNS HOPKINS HOSP, 73-; DIR, SAMUEL LIVINGSTON EPILEPSY DIAG & TREATMENT CTR, 73- Concurrent Pos: Assoc pediat, Sinai Hosp, Baltimore, 38-, asst med & dir, Children's & Adults Epilepsy Clin, 50- Mem: Am Soc Pediat Res; Am Epilepsy Soc; AMA. Res: Convulsive disorders; management of the epileptic patient. Mailing Add: Samuel Livingston Epilepsy Diag & Treatment Ctr Baltimore MD 21202

LIVINGSTON, VIRGINIA b Meadville, Pa, Dec 28, 06; m 56; c 2. ONCOLOGY, INTERNAL MEDICINE. Educ: Vassar Col, AB, 30; NY Univ, MD, 36. Prof Exp: Dir cancer res, Bur Biol Res, Presby Hosp Br, Rutgers Univ, 49-53; internist, San Diego Health Asn, 54-63; ASSOC PROF LIFE SCI-MICROBIOL, UNIV SAN DIEGO, 69- Concurrent Pos: San Diego Biomed Soc res fel, 63-; Found grant, 67-68; Kerr Found grant; lectr, US Int Univ, 64-65. Mem: AAAS; AMA; Am Soc Microbiol. Res: Bacteriology; allergy and immunology. Mailing Add: 8492 Prestwick Dr La Jolla CA 92037

LIVINGSTON, WILLIAM CHARLES, b Santa Ana, Calif, Sept 13, 27; m 57; c 2. ASTRONOMY. Educ: Univ Calif, Los Angeles, AB, 53; Univ Calif, PhD(astron), 59. Prof Exp: Observer, Mt Wilson Observ, Carnegie Inst, 51-53; from jr astronr to assoc asstronr, 59-70, ASTRONR, KITT PEAK NAT OBSERV, 70- Mem: Am Astron Soc. Res: Solar spectroscopy; photoelectric image devices. Mailing Add: Kitt Peak Nat Observ 950 N Cherry Ave Tucson AZ 85726

LIVINGSTONE, DANIEL ARCHIBALD, b Detroit, Mich, Aug 3, 27; m 52; c 5. ECOLOGY. Educ: Dalhousie Univ, BSc, 48, MSc, 50; Yale Univ, PhD(zool), 53. Prof Exp: Field collector, NS Mus Sci, summers & demonstr biol, Dalhousie Univ, winters, 47-50; asst zool, Yale Univ, 50-53; Nat Res Coun Can fels, Cambridge Univ, 53-54 & Dalhousie Univ, 54-55; asst prof zool, Univ Md, 55-56; from asst prof to assoc prof, 56-66, PROF ZOOL, DUKE UNIV, 66- Concurrent Pos: Spec lectr biogeog, Dalhousie Univ, 54-55; limnologist, US Geol Surv, 56-63; Guggenheim fel, 60-61; mem environ biol panel, Nat Sci Found, 64-, consult Nat Sci Found Polar Prog, 74-76. Mem: Ecol Soc Am (ed, Ecol Monogr, 62-66); Am Soc Limnol & Oceanog; Am Soc Nat; Geochem Soc; Am Soc Ichthyol & Herpet. Res: Pollen analysis; history of lakes; Pleistocene geology of Alaska, Nova Scotia, East and Central Africa; geochemistry of hydrosphere; sodium cycle; coring technology; paleoecology; limnology; biogeography of African fishes. Mailing Add: Dept of Zool Duke Univ Durham NC 22706

LIVINGSTONE, FRANK BROWN, b Winchester, Mass, Dec 8, 28; m 60; c 1. PHYSICAL ANTHROPOLOGY. Educ: Harvard Univ, AB, 50; Univ Mich, MA, 55, PhD(anthrop), 57. Prof Exp: Nat Sci Found fel, 57-59; from asst prof to assoc prof, 59-68, PROF ANTHROP, UNIV MICH, ANN ARBOR, 68- Res: Human and population genetics; abnormal hemoglobin; cultural determinants of human evolution. Mailing Add: Dept of Anthrop 215 Angell Hall Univ of Mich Ann Arbor MI 48104

LIWSHITZ, MORDEHAI, b Vienna, Austria, Jan 14, 23; m 55; c 3. SPACE PHYSICS. Educ: Israel Inst Technol, BSc, 57, MSc, 59; Univ Md, PhD(physics), 64. Prof Exp: Res asst physics, Univ Md, 64; Nat Acad Sci-Nat Res Coun res associateship, Goddard Space Flight Ctr, NASA, 64-66; mem tech staff, Bellcomm, Inc, Washington, DC, 66-68, group supvr, 68-72; SUPVR SYSTS MODELING GROUP, BELL LABS, 72- Mem: Am Geophys Union; Am Phys Soc; NY Acad Sci. Res: Kinetic theory; non-equilibrium processes; Monte Carlo method; stochastic processes, system modeling. Mailing Add: Bell Labs Dept 3442 600 Mountain Ave Murray Hill NJ 07974

LIZARDI, PAUL MODESTO, b San Juan, PR, Oct 13, 45; m 67. CELL BIOLOGY, BIOCHEMISTRY. Educ: Univ PR, BS, 66; Rockefeller Univ, PhD(cell biol), 71. Prof Exp: ASST PROF BIOCHEM, ROCKEFELLER UNIV, 73- Concurrent Pos: Jane Coffin Childs Mem Fund Med Res fel biochem, Carnegie Inst Dept Embryol, 71-73. Mem: Am Soc Cell Biol. Res: Eukaryotic messenger RNA's; contol of RNA synthesis; control of messenger RNA translation; structure of messenger RNA. Mailing Add: Rockefeller Univ New York NY 10021

LJUNG, HARVEY ALBERT, b Greensboro, NC, Oct 26, 05; m 35; c 3. ANALYTICAL CHEMISTRY. Educ: Univ NC, BS, 27, MS, 28, PhD(chem), 31. Prof Exp: Prof chem, Guilford Col, 31-69, acad dean, 46-62, Dana prof, 69-71, EMER DANA PROF, 71- Concurrent Pos: Nat Defense Res Comt, 41-42; mem gen chem & qual anal subcomt, Exam Comt, Div Chem Educ, 60-71; chem consult. Mem: Am Chem Soc. Res: Chelation; electrochemistry. Mailing Add: 5314 W Friendly Ave Greensboro NC 27410

LJUNGDAHL, LARS GERHARD, b Stockholm, Sweden, Aug 5, 26; m 49; c 2. BIOCHEMISTRY, MICROBIOLOGY. Educ: Stockholm Tech Inst, BS, 45; Western Reserve Univ, PhD(biochem), 64. Prof Exp: Technician med chem, Karolinska Inst, Univ Sweden, 43-46; res chemist, Stockholm Brewery Co, 47-58; technician biochem, Case Western Reserve Univ, 58-59, sr instr, 64-66, asst prof, 66-67; from mem fac to assoc prof, 67-75, PROF BIOCHEM, UNIV GA, 75- Concurrent Pos: Alexander Von Humboldt Sr Scientist Award, 74- Mem: Am Soc Microbiol; Am Chem Soc; Brit Biochem Soc; Swedish Chem Soc; Am Soc Biol Chem. Res: Carbohydrate metabolism, carbon dioxide fixation, and one carbon metabolism inanaerobic microorganism; role of corrinoids, tetrahydrofolate derivatives and properties of enzymes in these processes. Mailing Add: Fermentation Plant Dept of Biochem Univ of Ga Athens GA 30602

LLAMAS, VICENTE JOSE, b Los Angeles, Calif, Feb 15, 44; m 66; c 1. SOLID STATE PHYSICS. Educ: Loyola Univ Los Angeles, BS, 66; Univ Mo-Rolla, MS, 68, PhD(physics), 70. Prof Exp: Asst physics, Univ Mo-Rolla, 66-68; asst prof, 70-75, ASSOC PROF PHYSICS & CHMN DEPT PHYSICS & MATH, N MEX HIGHLANDS UNIV, 75-, CO-DIR, SCI & MATH EDUC CTR, 74- Concurrent Pos: Consult, Fermi Nat Accelerater Lab, 70-, Minority Sci Educ Bibliog Proj, AAAS, 75- & NIH. Mem: Sigma Xi (secy-treas, 71-73, pres, 73-75); Am Phys Soc; Am Asn Physics Teachers; AAAS; Nat Sci Teachers Asn. Res: Surface studies of alkali halides in the infrared; atmospheric study of air pollutants. Mailing Add: Dept of Physics NMex Highlands Univ Las Vegas NM 87701

LLANO, GEORGE ALBERT, polar biology, lichenology. see 12th edition

LLAURADO, JOSEP G, b Barcelona, Catalonia, Spain, Feb 6, 27; m 58, 66; c 6. BIOMEDICAL ENGINEERING, PHYSIOLOGY. Educ: Balmes Inst, Barcelona, BA & BS, 44; Univ Barcelona, MD, 50; Drexel Univ, MS, 63. Prof Exp: Inst med, Sch Med, Univ Barcelona, 50-52; asst med res, Postgrad Med Sch, Univ London, 52-54; asst prof exp surg, Med Sch, Univ Otago, NZ, 54-57; sr endocrinologist, Med Res Labs, Chas Pfizer & Co, Inc, Conn, 59-61; assoc prof physiol, Sch Med, Univ Pa, 63-67; PROF BIOMED ENG & PHYSIOL, MARQUETTE UNIV & MED COL WIS, 67- Concurrent Pos: Brit Coun scholar, Postgrad Med Sch, Univ London, 52-54; Hite Found fel exp med, Univ Tex M D Anderson Hosp & Tumor Inst, 57-58; USPHS fel steroid biochem, Col Med, Univ Utah, 58-59; fel, Coun Adv Sci Invests, Spain, 50-52; Rockefeller vis prof, Univ Valle, Colombia, 58; partic, Nat Colloquim Theoret Biol, NASA, Colo, 65; consult, Vet Admin Hosp, Wood, Wis, 67-; US rep, Int Atomic Energy Agency Symp Dynamic Studies Radioisotopes Med, Rotterdam, 70 & Knoxville, Tenn, 74; vis prof, Polytech Univ, Barcelona, Spain, 73 & 75; vis prof, Univ Zulia, Venezuela, 74 & 75 & Univ Padua, Italy, 75. Honors & Awards: Catalan Jocs Florals Prize, Amsterdam, 74 & Caracas, 75. Mem: Fel Am Col Nutrit; Am Soc Pharmacol & Exp Therapeut; Catalan Soc Biol; Soc Math Biol; sr mem Inst Elec & Electronics Engrs. Res: Systems approach to physiopharmacological problems; compartmental analysis; sodium and potassium; isotopes in biomedicine; biomathematics; steroids. Mailing Add: Wing D-12N Vet Admin Ctr Wood WI 53193

LLEWELLYN, CHARLES ELROY, JR, b Richmond, Va, Jan 16, 22; m 48; c 3. PSYCHIATRY. Educ: Hampden-Sydney Col, BS, 43; Med Col Va, MD, 46; Univ Colo, MSc, 53; Am Bd Psychiat & Neurol, dipl, 56. Prof Exp: Instr psychiat, Med Col Va, 46-47; assoc, 55-56, asst prof, 56-63, asst dir psychiat outpatient div, 55-56, ASSOC PROF PSYCHIAT, SCH MED, DUKE UNIV, 63-, HEAD PSYCHIAT OUTPATIENT DIV, MED CTR, 56- Concurrent Pos: Partic, NIMH vis fac sem community psychiat, Lab Community Psychiat, Harvard Med Sch, 65-67; prog dir Duke study group, Interuniv Forum Educr Community Psychiat, 67-71; consult, State Dept Social Serv, NC, 55-; psychiat consult, NC Med Peer Rev Found, 75-; chief psychiat consult for Peer Rev & Qual Assurance, NC Div Ment Health Serv, 75-78. Mem: AMA; fel Am Psychiat Asn; Am Group Psychother Asn; Pan-Am Med Asn; Acad Relig & Ment Health. Res: Community mental health; individual and group psychotherapy. Mailing Add: Dept of Psychiat Duke Univ Med Ctr Durham NC 27710

LLEWELLYN, EDWARD JOHN, b London, Eng, Sept 17, 38; m 62; c 3. EXPERIMENTAL PHYSICS. Educ: Univ Exeter, BSc, 60, PhD(spectros), 63; Univ Sask, PhD, 71. Prof Exp: Ministry of Aviation res asst, Norman Lockyer Observ, Devon, Eng, 63-64; asst prof, 64-69, ASSOC PROF PHYSICS, UNIV SASK, 69- Mem: Am Geophys Union; Can Asn Physicists. Res: Photochromism; atmospheric spectroscopy; infrared aeronomy. Mailing Add: Dept of Physics Univ of Sask Saskatoon SK Can

LLEWELLYN, GERALD CECIL, b Lonaconing, Md, Feb 8, 40; m 62; c 3. BIONUCLEONICS. Educ: Frostburg State Col, BS, 62; Purdue Univ, MS, 66, PhD(bionucleonics), 70. Prof Exp: Instr biol chem, Frederick County Bd Educ, Md, 62-66; lectr biol & microbiol, Frederick Community Col, 66-67; ASST PROF BIOL EDUC, VA COMMONWEALTH UNIV, 69-, PROF BIOL, 70- Mem: Nat Sci Teachers Asn. Res: Toxicological responses of hamsters to aflatoxin B. Mailing Add: Dept Biol Va Commonwealth Univ Acad Ctr 901 W Franklin St Richmond VA 23220

LLEWELLYN, RAEBURN CARSON, b Corbin, Ky, May 29, 20; m 43; c 3. NEUROSURGERY. Educ: Univ Ala, AB, 42; Univ Va, MD, 45; Am Bd Neurol Surg, dipl, 54. Prof Exp: From asst prof to assoc prof, 55-60, PROF NEUROL SURG & CHMN DIV, SCH MED, TULANE UNIV, 60- Concurrent Pos: Attend neurol surgeon, Vet Hosp, 55-; chief neurol surg unit, Tulane Serv, Charity Hosp, 58- Mem: Am Asn Neurol Surg; Cong Neurol Surg; Am Acad Neurol Surg; Soc Neurol Surg. Res: Cerebral vascular disease; chemotherapy of advanced cancer of the brain. Mailing Add: Dept of Surg Tulane Univ Sch of Med New Orleans LA 70112

LLEWELLYN, RALPH A, b Detroit, Mich, June 27, 33; m 55; c 4. NUCLEAR PHYSICS. Educ: Rose-Hulman Inst Technol, BS, 55; Purdue Univ, PhD(physics), 62. Prof Exp: Asst prof physics, Rose-Hulman Inst Technol, 61-64, assoc prof, 64-68, prof, 68-70, chmn dept, 69-70; prof & chmn dept, Ind State Univ, Terre Haute, 70-73; exec secy, Bd on Energy Studies, Nat Acad Sci, Nat Res Coun, 73-74; PROF PHYSICS & CHMN DEPT, IND STATE UNIV, TERRE HAUTE, 74- Concurrent Pos: Mem, NSF Apparatus Develop Workshop, Rensselaer Polytech Inst, 64-65; prof physics & acting chmn dept, St Mary-of-the-Woods Col, 69-70; assoc ed, Phys Rev Lett, Am Inst Physics, 75- Mem: AAAS; Am Phys Soc; Am Asn Physics Teachers; Int Asn Gt Lakes Res; Am Soc Oceanog. Res: Environmental physics, particularly beta and gamma decay; Mössbauer effect in deep ocean sediments; energy resources, energy and public policy. Mailing Add: Dept of Physics Ind State Univ Terre Haute IN 47809

LLEWELLYN-THOMAS, EDWARD, b Salisbury, Eng, Dec 15, 17; Can citizen; m 47; c 3. MEDICINE, ENGINEERING. Educ: Univ London, BSc, 51; McGill Univ, MD, 55. Prof Exp: Res assoc, Cornell Univ, 57-59; sci officer psychophysiol, Defence Res Med Labs, Can, 58-61; med res assoc, Lakeshore Psychiat Hosp, 61-63; PROF PHARMACOL, UNIV TORONTO, 63-, ASSOC DEAN MED, 74- Mem: Inst Elec & Electronics Engrs; Brit Inst Elec Engrs; fel Royal Soc Can; fel Royal Soc Arts. Res: Biomedical engineering; psychopharmacology; human factors engineering. Mailing Add: Univ of Toronto Toronto ON Can

LLINAS, RODOLFO, b Bogota, Colombia, Dec 16, 34; m 65; c 2. NEUROBIOLOGY, ELECTROPHYSIOLOGY. Educ: Pontificial Univ Javeriana, Colombia, MD, 59;

Australian Nat Univ, PhD(neurophysiol), 65. Prof Exp: Instr neurophysiol, Nat Univ Colombia, 59; assoc prof, Univ Minn, 65-66; assoc mem neurobiol, Inst Biomed Res, AMA Educ & Res Found, 66-68, mem, 69-70; PROF PHYSIOL & BIOPHYS & HEAD DIV NEUROBIOL, UNIV IOWA, 70- Concurrent Pos: Res fel psychiat & neurosurg, Mass Gen Hosp, 59-61; fel physiol, Univ Minn, 61-63; res scholar, Australian Nat Univ, 63-65; prof lectr, Col Med, Univ Ill, 67-68, clin prof, 68-71; guest prof, Wayne State Univ, 67-74; assoc prof, Med Sch, Northwestern Univ, 67-71; mem, Neurol Sci Res Training A Comt, NIH, 71-74 & Neurol A Study Sect, 74-; consult, US Air Force Sch Aerospace Med, 72-75; Bowditch lectr, 73; assoc, Neurosci Res Prog, Mass Inst Technol, 74-; chief ed, Neurosci Jour, 75- Mem: Am Physiol Soc; Am Soc Cell Biol; Soc Neurosci; Int Brain Res Orgn; Biophys Soc. Res: Structural and functional studies of neuronal systems; synaptic transmission in vertebrate and invertebrate forms; evolution and development of the central nervous system. Mailing Add: Div of Neurobiol Univ of Iowa Oakdale IA 52319

LLOYD, CHARLES WAIT, b Catawba, Va, Jan 15, 14; m 43; c 2. ENDOCRINOLOGY. Educ: Princeton Univ, BA, 36; Univ Rochester, MD, 41; Am Bd Internal Med, dipl, 49. Prof Exp: Intern path, Med Col, Cornell Univ, 41-42; asst resident endocrinol, Med Sch, Duke Univ, 43-44; intern med, Med Sch, Yale Univ, 44-45; dir training prog physiol reproduction, Clin Res Unit & sr scientist, Worcester Found Exp Biol, 62-72; prof obstet & gynec & chief div reproductive biol, Hershey Med Ctr, Pa State Univ, 72-74; PROF CLIN PSYCHIAT & CODIR CTR STUDY HUMAN SEXUAL BEHAV, DEPT PSYCHIAT, UNIV PITTSBURGH, 74- Concurrent Pos: Res fel endocrinol, Thorndike Lab, Harvard Med Sch & Boston City Hosp, 45-46. Mem: AAAS; Am Soc Clin Invest; Endocrine Soc; Am Physiol Soc; Soc Exp Biol & Med. Res: Endocrinology of reproduction; pituitary hormones and their control; hormones and behavior. Mailing Add: Western Psychiat Inst & Clin 3811 O'Hara St Pittsburgh PA 15261

LLOYD, DAVID PIERCE CARADOC, b Auburn, Ala, Sept 22, 11; m 37, 57; c 3. PHYSIOLOGY. Educ: McGill Univ, BSc, 32; Oxford Univ, BA, 35, DPhil(physiol), 38, DSc, 61, MA, 64. Prof Exp: Demonstr physiol, Oxford Univ, 35-36; asst med res, Univ Toronto, 36-38, assoc, 38-39; from asst to assoc, Rockefeller Inst, 39-43; asst prof physiol, Sch Med, Yale Univ, 43-45; assoc mem, Rockefeller Univ, 46-49, mem, 49-70, prof physiol, 57-70; HON RES FEL, UNIV COL, UNIV LONDON, 70- Concurrent Pos: Arthur lectr, Am Mus Natural Hist, 58; mem, Neurol Study Sect, NIH, 61-65, mem, Res & Training Grants Comt, Nat Inst Neurol Dis & Blindness, 66-70; mem, Med Adv Comts, Nat Found Infantile Paralysis & United Cerebral Palsy Found; trustee, Int Poliomyelitis Cong. Mem: Nat Acad Sci; AAAS; Am Physiol Soc. Res: Synaptic transmission in sympathetic ganglia; conduction and synaptic transmission in spinal cord; excitation and inhibition; physiology of sweat glands. Mailing Add: New Cottage Greatham Pulborough Sussex England

LLOYD, DOUGLAS SEWARD, b Brooklyn, NY, Oct 16, 39. PUBLIC HEALTH. Educ: Duke Univ, AB, 61, MD, 71; Univ NC, MPH, 71. Prof Exp: COMNR, CONN STATE DEPT HEALTH, 73- Concurrent Pos: Mem courtesy staff, Hartford Hosp, 73-; lectr, Sch Med, Yale Univ, 73- & Univ Conn Health Ctr, 73- Mailing Add: Conn State Dept of Health 79 Elm St Hartford CT 06115

LLOYD, EDWIN PHILLIPS, b San Antonio, Tex, Sept 18, 29; m 54; c 1. ENTOMOLOGY. Educ: Tex A&M Univ, BS, 51, MS, 52, PhD(entom), 58. Prof Exp: RES ENTOMOLOGIST, BOLL WEEVIL RES LAB, AGR RES SERV, USDA, 56- Concurrent Pos: Sci adv, pilot boll weevil eradication exp, 71-73; adj assoc prof, Miss State Univ, 71- Honors & Awards: Superior Serv Award, USDA, 74; Res Award, Miss Entom Asn, 74. Mem: Entom Soc Am. Res: Cotton insects, specifically the boll weevil. Mailing Add: Miss State Univ PO Box 5367 State Col MS 39762

LLOYD, ELIZABETH LUKE, b Ballymoney, Northern Ireland, Jan 11, 26; m 55. BIOPHYSICS, MICROBIOLOGY. Educ: Queen's Univ, Belfast, BSc, 47; Univ London, MSc, 53; Oxford Univ, DPhil(med), 64. Prof Exp: Asst exp officer, Nat Phys Lab, Teddington, Eng, 47-54; sr physicist, St Bartholomew's Hosp, London, 54-57 & London Clin, 57-58; res physicist, Med Res Coun Bone Seeking Isotope Res Unit, Churchill Hosp, Oxford, 58-66; BIOPHYSICIST, RADIOL & ENVIRON RES DIV, ARGONNE NAT LAB, 66- Concurrent Pos: Mem, Int Comn Radiol Protection, 69- & Nat Comn Radiol Protection, 70- Mem: Fel Royal Soc Med; Orthop Res Soc; Radiation Res Soc; Health Physics Soc; Brit Bone & Tooth Soc. Res: Bone, calcium metabolism; quantitative characterization of bone; radiation effects, bone seeking isotopes, dose and damage; tumor virology and immunology. Mailing Add: Radiol & Environ Res Div Argonne Nat Lab Argonne IL 60439

LLOYD, FREDERICK A, b Chicago, Ill, Mar 13, 01; m 38; c 2. UROLOGY. Educ: Harvard Univ, AB, 22; Rush Med Col, MD, 27. Prof Exp: Chmn dept urol, Vet Admin Hosp, Hines, 45-73; assoc prof, 52-73, EMER ASSOC PROF UROL, MED SCH, NORTHWESTERN UNIV, 73- Concurrent Pos: Attend urologist, Passavant Mem Hosp, 49- Mem: Am Urol Asn; AMA; Am Col Surg. Res: Urinary surgery; experimental pathology of the urinary tract; tuberculosis of the urinary tract. Mailing Add: 948 Lee Rd Northbrook IL 60062

LLOYD, HARRIS HORTON, b Conway, Ark, Nov 14, 37; m 60; c 3. CANCER, CHEMOTHERAPY. Educ: Ouachita Baptist Univ, BA & BS, 59; Purdue Univ, PhD(phys chem), 68. Prof Exp: Chief dept chem, 406th Med Lab, US Army Med Command, Univ Japan, 64-67; res chemist, 68-72, HEAD, MATH BIOL & DATA ANAL SECT, SOUTHERN RES INST, 72- Mem: Am Asn Cancer Res; Am Sci Affil. Res: Chemical kinetics; data analysis and mathematical simulation; Pharmacokinetics; kinetics of tumor growth and cell killing; design of computer-based information management systems. Mailing Add: Dept of Chemother Southern Res Inst 2000 Ninth Ave S Birmingham AL 35205

LLOYD, HOWELL CLEVENGER, b Lima, Ohio, Nov 10, 37; m 71. GEOGRAPHY. Educ: Butler Univ, BA, 59; Northwestern Univ, MA, 61; PhD(geog), 64. Prof Exp: From instr geog to asst prof, 63-70, ASSOC PROF GEOG, MIAMI UNIV, 70- Concurrent Pos: Vis prof, European Study Ctr, Miami Univ, 69-71, consult, Bur Educ Field Serv, 71-, dir, Undergrad Ctr for Int Studies, 73- Mem: Asn Am Geog; Am Geog Soc; Nat Coun for Geog Educ; Int Studies Asn. Res: Political geography of European integration; political and cultural geography of European minorities; comparative analyses of public urban transportation developments, especially Europe and North America. Mailing Add: Dept of Geog Shideler Hall Miami Univ Oxford OH 45056

LLOYD, JAMES ARMON, b Nanticoke, Pa, June 17, 33. ENDOCRINOLOGY, ETHOLOGY. Educ: Pa State Univ, BS, 55, MS, 57; Johns Hopkins Univ, ScD(behav endocrinol), 61. Prof Exp: Res asst endocrinol behav, Univ Pa, 61-63; asst mem, Albert Einstein Med Ctr, 63-71; ASSOC PROF PROG REPRODUCTION MED & BIOL, UNIV TEX MED SCH HOUSTON, 71- Concurrent Pos: NIH res grants, Albert Einstein Med Ctr, 63-71. Mem: AAAS; Endocrine Soc; Am Soc Zoologists; Animal Behav Soc; NY Acad Sci. Res: Population studies; behavior. Mailing Add: Tex Med Ctr Univ of Tex Med Sch Houston TX 77025

LLOYD, JAMES EDWARD, b Oneida, NY, Jan 17, 33; m 58; c 2. EVOLUTIONARY BIOLOGY. Educ: State Univ NY Col Fredonia, BS, 60; Univ Mich, MA, 62; Cornell Univ, PhD(entom), 66. Prof Exp: Teacher high sch, 60; NSF res assoc syst & evolutionary biol, 66; from asst prof biol sci & entom to assoc profentom & nematol, 66-74, PROF ENTOM & NEMATOL, UNIV FLA, 74- Concurrent Pos: NSF res grant 68. Honors & Awards: Res Award, Sigma Xi, 74. Mem: Entom Soc Am; Soc Study of Evolution; Am Entom Soc; Asn for Trop Biol; Coleopterist's Soc. Res: Function of luminescence in insects; systematics in Lampyridae; animal behavior, ecology, evolution. Mailing Add: Dept of Entom & Nematol Univ of Fla Gainesville FL 32611

LLOYD, JAMES NEWELL, b Orange, NJ, Oct 20, 32; m 59; c 2. PHYSICS. Educ: Colgate Univ, BA, 54; Cornell Univ, PhD(physics), 63. Prof Exp: From instr to asst prof physics, 61-70, ASSOC PROF PHYSICS, COLGATE UNIV, 70- CHMN, DEPT PHYSICS & ASTRON, 73- Mem: AAAS; Am Phys Soc; Am Asn Physics Teachers. Res: Ferromagnetic resonance and transport properties in metals; electron paramagnetic resonance in macrocyclic ligand-metal ion compounds. Mailing Add: Dept of Physics & Astron Colgate Univ Hamilton NY 13346

LLOYD, JOEL JOSEPH, b New York, NY, Dec 6, 11; m 35; c 2. GEOLOGY, MATHEMATICS. Educ: City Col New York, BS, 45. Prof Exp: Sr geologist, Socony Vacuum Oil Co, Colombia, 45-52; chief geologist for opers, Union Oil Co Calif, 52-61; consult, Assoc Int Petrol Consult, 61-66; dir geol Soc Am bibliog, 66-69, DIR SCI INFO, AM GEOL INST, WASHINGTON, DC, 69- Concurrent Pos: Exec secy, CISTIP, Nat Acad Sci, 74- Mem: Am Asn Petrol Geol; fel Geol Soc Am; hon mem Geol Soc Costa Rica. Res: Structural geology and geomathematics; automatic data system applications to earth science information. Mailing Add: Am Geol Inst 2201 M St NW Washington DC 20037

LLOYD, JOHN EDWARD, b Munhall, Pa, Sept 28, 40; m 62; c 2. ENTOMOLOGY. Educ: Pa State Univ, BS, 62; Cornell Univ, PhD(entom), 67. Prof Exp: Asst prof entom, Pa State Univ, 67-68; asst prof, 68-72, ASSOC PROF ENTOM, UNIV WYO, 72- Mem: Entom Soc Am; Am Mosquito Control Asn. Res: Economic entomology; insect ecology; insects affecting livestock; insect toxicology. Mailing Add: Entom Sect Box 3354 Univ of Wyo Laramie WY 82071

LLOYD, JUSTIN THOMAS, b Jonesboro, Ark, Oct 31, 33; m 59; c 2. MATHEMATICS. Educ: Univ Ark, BS, 59; Tulane Univ, MS, 62, PhD(math), 64. Prof Exp: Asst prof math, Lehigh Univ, 64-67; asst prof, 67-70, ASSOC PROF MATH, UNIV HOUSTON, 70- Mem: Am Math Soc. Res: Ordered algebraic systems. Mailing Add: Dept of Math Univ of Houston Houston TX 77004

LLOYD, KENNETH OLIVER, b Denbigh, Wales, May 17, 36; US citizen; m 62; c 2. BIOCHEMISTRY, IMMUNOCHEMISTRY. Educ: Univ Wales, BSc, 57, PhD(chem), 60. Prof Exp: Res assoc microbiol, Columbia Univ, 63-68, asst prof biochem, 68-74; assoc prof, Sch Med, Tex Tech Univ, 74-75; ASSOC, SLOAN-KETTERING CANCER CTR, 75- Concurrent Pos: Fel, Washington Univ, 60-63; USPHS res career develop award, 68-73. Mem: AAAS; Am Chem Soc; Am Asn Immunologists; Soc Complex Carbohydrates. Res: Biochemistry, structure and immunochemistry of carbohydrate antigens. Mailing Add: Mem Sloan Kettering Cancer Ctr 1275 York Ave New York NY 10021

LLOYD, LEWIS EWAN, b Montreal, Que, Feb 3, 24; m 50; c 4. NUTRITION. Educ: Macdonald Col, McGill Univ, BSc, 48, MSc, 50, PhD(nutrit), 52. Prof Exp: Asst nutrit, Macdonald Col, McGill Univ, 48-52; res assoc, Cornell Univ, 52-53; from asst prof to assoc prof, Macdonald Col, McGill Univ, 53-60, prof animal sci & chmn dept, 60-67; dir sch home econ, 67-70, DEAN FAC HOME ECON, UNIV MAN, 70- Concurrent Pos: Underwood fel, Rowett Res Inst, Scotland, 58-59. Mem: Can Dietetic Asn; Can Inst Food Sci & Technol; Am Inst Nutrit; Nutrit Soc Can; Brit Nutrit Soc. Res: Utilization of food nutrients by the body. Mailing Add: Fac of Home Econ Univ of Man Winnipeg MB Can

LLOYD, MONTE, b Omaha, Nebr, July 6, 27; m 46, 69; c 4. ANIMAL ECOLOGY. Educ: Univ Calif, Los Angeles, AB, 52; Univ Chicago, PhD(zool), 57. Prof Exp: NSF fel, Bur Animal Pop, Oxford Univ, 57-59, Brit Nature Conserv res grant, 59-62; asst prof zool, Univ Calif, Los Angeles, 62-67; ASSOC PROF BIOL, UNIV CHICAGO, 67- Concurrent Pos: Ed, Ecology, 68- Mem: Soc Study Evolution; Am Soc Nat; Ecol Soc Am; Asn Study Animal Behav; Brit Ecol Soc. Res: Dynamics of animal populations and community ecology. Mailing Add: Dept of Biol Univ of Chicago 1101 E 57th St Chicago IL 60637

LLOYD, NELSON ALBERT, b Lorain, Ohio, Oct 12, 26; m 45; c 5. ANALYTICAL CHEMISTRY. Educ: Southern Methodist Univ, BS, 50, MS, 51; Okla State Univ, PhD(anal chem), 55. Prof Exp: Res chemist, Goodyear Atomic Corp, 54-56; assoc prof anal chem, Northeastern La State Col, 56-61 & Univ Ala, Tuscaloosa, 61-67; chmn Div Natural Sci, Mobile Col, Mobile, 67-70; CHIEF GEOCHEM DIV, GEOL SURV OF ALA, TUSCALOOSA, 70- Concurrent Pos: Consult, Tuscaloosa Metall Res Ctr, US Bur Mines, 63-67 & State Oil & Gas Bd, Ala, 66-67. Mem: Am Chem Soc. Res: Rock analysis, whole rock and trace metals in rock; trace substances in water, heavy metals, pesticides herbicides and various nitrogen species. Mailing Add: 209 32nd Ave E Tuscaloosa AL 35401

LLOYD, NORMAN EDWARD, b Oak Park, Ill, Feb 20, 29; m 51; c 8. BIOCHEMISTRY. Educ: Rockjurst Col, BS, 52; Kans State Col, MS, 53; Purdue Univ, PhD(biochem), 56. Prof Exp: Assoc chemist, Corn Prods Co, 56-58; cereal chemist, Int Milling Co, 58-59; supvr starch chem res, 60-64, dir sci develop, 64-69, asst res dir, 69-70, SUPVR CHEM RES, CLINTON CORN PROCESSING CO, 70- Mem: Am Chem Soc; Am Asn Cereal Chem. Res: Production and characterization of starches, sweeteners and enzymes; enzyme kinetics and immobilization; automation of analytical methods. Mailing Add: Res Dept Clinton Corn Processing Co Clinton IA 52732

LLOYD, PAUL EUGENE, physics, see 12th edition

LLOYD, PRESCOTT REES, b San Diego, Calif, Nov 12, 06; m 54; c 3. BACTERIOLOGY. Educ: Stanford Univ, BS, 32. Prof Exp: Area salesman, Standard Oil Co Calif, 32-34; in-chg qual control, H J Heinz Corp, 34-38; food technologist, Nat Canners Asn, 38-41; asst lab dir, Libby, McNeill & Libby, 41-47; actg mgr, Food Technol Div, Stanford Res Inst, 47-48; tech sales consult, Calif & Hawaiian Sugar Refining Corp, Ltd, 48-59; prod mgr, Pac Hawaiian Prod Co, 59-60, asst to vpres prod, 60-61; tech sales consult, Spreckels Sugar Co, 61-71; CONSULT, 71- Honors & Awards: Pub Award, Asn Food Indust Sanitarians, 54; Mem Year Award, Inst Food Technologists, 71. Mem: Am Soc Microbiol; Am Asn Cereal Chemists; Inst Food Technologists (treas, 50); Inst Sanit Mgt (pres, 57). Res: Industrial bacteriology; food processing methods; new product development. Mailing Add: 699 Oak Knoll Dr Ashland OR 97520

LLOYD, ROBERT AUSTIN, JR, organic chemistry, see 12th edition

LLOYD, ROBERT MICHAEL, b Los Angeles, Calif, Dec 26, 38. SYSTEMATIC BOTANY. Educ: Pomona Col, BA, 60; Claremont Grad Sch, MA, 62; Univ Calif, Berkeley, PhD(bot), 69. Prof Exp: Herbarium technician, Univ Calif, Los Angeles, 62-63; asst prof bot, Univ Hawaii, 68-71; asst prof, 71-75, ASSOC PROF BOT, OHIO UNIV, 75- Mem: AAAS; Am Inst Biol Sci; Am Fern Soc; Am Soc Plant Taxon; Asn Trop Biol. Res: Systematics and reproductive biology of ferns; systematics, morphology and evolution of the Pteridophyta; spore germination, breeding systems and genetics. Mailing Add: Dept of Bot Ohio Univ Athens OH 45701

LLOYD, RODNEY FREDERICK, organic chemistry, see 12th edition

LLOYD, RONALD MICHAEL, geology, see 12th edition

LLOYD, RUTH SMITH, b Washington, DC, Jan 25, 17; m 39; c 3. ANATOMY. Educ: Mt Holyoke Col, AB, 37; Howard Univ, MS, 38; Western Reserve Univ, PhD(anat), 41. Prof Exp: Asst physiol, Col Med, Howard Univ, 40-41; instr zool, Hampton Inst, Va, 41-42; tech asst physiol, Howard Univ, 42, from instr to asst prof anat, 42-58, ASSOC PROF ANAT, GRAD SCH, HOWARD UNIV, 58- Mem: Am Asn Anat; Am Asn Phys Anthrop. Res: Endocrinology; female sex cycle and relation of sex hormones to growth; medical genetics. Mailing Add: Dept of Anat Grad Sch Howard Univ Washington DC 20001

LLOYD, STUART PHINNEY, b Kansas City, Mo, Mar 23, 23. MATHEMATICS. Educ: Univ Chicago, SB, 43; Univ Ill, PhD(physics), 51. Prof Exp: Mem, Inst Advan Study, 51-52; res mathematician, 52-74, MEM TECH STAFF, BELL LABS INC, 74- Concurrent Pos: Vis assoc prof, Univ Chicago, 62-63. Mem: AAAS; Am Math Soc; Math Asn Am; Inst Math Statist. Res: Probability theory. Mailing Add: Bell Labs Inc Murray Hill NJ 07974

LLOYD, THOMAS BLAIR, b Reedsville, WVa, Aug 29, 21; m 44; c 3. INDUSTRIAL CHEMISTRY, COLLOID CHEMISTRY. Educ: Washington & Jefferson Col, BS, 42; Western Res Univ, MS, 46, PhD(phys chem), 48. Prof Exp: Asst prof chem, Muhlenberg Co, 48-54; chem investr, 54-66, RES SUPVR, NJ ZINC CO, 66- Mem: Am Chem Soc; Sigma Xi. Res: Industrial process research, particularly pigments, hydrometallurgy and pollution control. Mailing Add: NJ Zinc Co Palmerton PA 18071

LLOYD, TREVOR, b London, Eng, May 4, 06; m 36; c 2. GEOGRAPHY. Educ: Bristol Univ, BSc, 29, cert, 31, DSc(geog), 49; Clark Univ, PhD(geog), 40. Hon Degrees: MA, Dartmouth Col, 44; LLD, Univ Windsor, 73. Prof Exp: Geog specialist, Winnipeg Sch Bd, Man, Can, 36-38 & 40-41; asst prof geog, Carleton Col, 42; from asst prof to prof, Dartmouth Col, 42-59, chmn dept, 47-52; prof human geog, 59-74, chmn dept geog, 62-66, PROF GEOG, McGILL UNIV, 74- Concurrent Pos: Actg consul, Can Govt Greenland, 44-45; ed, Arctic, Arctic Inst NAm, 47-48; chief geog bur, Can Dept Mines & Resources, 47-48; gov, Inst Current World Affairs, 59-67 & 71-; proj officer, Royal Comn Govt Orgn, Can, 61-62; consult, Nat Film Bd, Can, 66-; consult, Kez Int, 68-; mem, Can Inst Int Affairs; res grants, Rockefeller Found, Scand, 47, Carnegie Corp, Scand & Greenland, 51 & 53, Off Naval Res, 53-56, Can Coun, Scand, 60 & 67, Carnegie Corp, Greenland, 67, Nuffield Found, UK, 67 & Can Coun Greenland res, McGill Univ, 70-73. Honors & Awards: Can Centennial Medal, 67. Mem: AAAS (chmn sect E, 66); Asn Am Geog; fel Am Geog Soc; Can Asn Geog (vpres, 55-57, pres, 58-59); fel Arctic Inst NAm. Res: Resource development and administration of arctic lands; underdeveloped areas elsewhere. Mailing Add: Dept of Geog Burnside Hall McGill Univ Box 6070 Montreal PQ Can

LLOYD, WILLIAM GILBERT, b New York, NY, July 10, 23; m 47; c 3. ORGANIC CHEMISTRY, APPLIED CHEMISTRY. Educ: Kalamazoo Col, AB, 47; Brown Univ, ScM, 50; Mich State Univ, PhD(org chem), 57. Prof Exp: Chemist, Dow Chem Co, 50-60, assoc scientist, 60-62; sr process res specialist, Lummus Co, NJ, 62-67; prof chem, Western Ky Univ, 67-74; sr res scientist, 74-75, CHIEF CHEMIST, INST MINING & MINERALS RES, UNIV KY, 75- Concurrent Pos: Dir, Larox Res Corp, 72- Mem: Fel Am Inst Chemists; AAAS; Am Chem Soc. Res: Catalysis of organic reactions; oxidations; free radical chemistry; chemistry of coal and coal-derived products; environmental chemistry. Mailing Add: Inst for Mining & Minerals Res Univ of Ky Lexington KY 40506

LLOYD, WILLIAM REESE, b Pueblo, Colo, Sept 26, 13; m 38. PHARMACY. Educ: Univ Colo, AB, 34, BS, 36; Univ NC, MS, 38; Univ Minn, PhD(pharmaceut chem), 41. Prof Exp: Asst chem, Univ NC, 36-38; asst pharm, Univ Minn, 38-41; asst prof, Univ Ga, 41-42; asst prof chem, Duquesne Univ, 42-44; biochemist, Armour Res Found, Ill, 44-45, supvr biochem res, 45-47, chemist, Int Div, 47-49; assoc prof pharm, Univ Tex, 49-60, asst dean, 52-60; TECH DIR, TEX PHARMACAL CO, 60- Mem: Am Chem Soc; Am Soc Oil Chem; Am Pharmaceut Asn. Res: Fats, oils and waxes; pharmaceutical formulation. Mailing Add: Tex Pharmacal Co PO Box 1659 San Antonio TX 78206

LLOYD, WINSTON DALE, b Pensacola, Fla, Sept 9, 29; m 58; c 3. ORGANIC CHEMISTRY. Educ: Fla State Univ, BS, 51; Univ Washington, PhD(org chem), 56. Prof Exp: Org chemist, Dow Chem Co, 56-58 & USDA, 59-62; res prof, 65-66, ASSOC PROF CHEM, UNIV TEX, EL PASO, 62- Mem: Am Chem Soc; Sigma Xi; Phytochem Soc NAm. Res: Stereochemistry of resin acids and steroids; mechanisms of organic chemical reactions; natural products; synthesis. Mailing Add: Dept of Chem Univ of Tex El Paso TX 79968

LO, CHAI-HO, biochemistry, see 12th edition

LO, CHENG FAN, b Taichung, Taiwan, Dec 14, 37; m 66; c 3. WOOD CHEMISTRY. Educ: Nat Taiwan Univ, BS, 62; Auburn Univ, MS, 66; Ore State Univ, PhD(wood chem), 70. Prof Exp: Res asst, Dept Forestry, Auburn Univ, 64-66; wood chemist, Forest Res Lab, Ore State Univ, 66-69; RES CHEMIST, BOISE CASCADE CHEM RES LAB, 69- Mem: Am Chem Soc; Tech Asn Pulp & Paper Indust; Forest Prod Res Soc. Res: By-products development in wood cellulose and lignin material; technical assistance to paper production. Mailing Add: 12309 NE Cassidy Ct Vancouver WA 98665

LO, CHIEN-PEN, b Canton, China, June 10, 14; nat US; m 42; c 3. ORGANIC POLYMER CHEMISTRY. Educ: Tsing Hua Univ, China, BS, 35; Univ Minn, PhD(org chem), 47. Prof Exp: Asst chem, Inst Chem, Acad Sinica, China, 35-42, res assoc, 42-43; asst, Univ Minn, 43-46; res chemist, 47-62, SR RES CHEMIST, ROHM & HAAS CO, 62- Mem: Fel AAAS; Am Chem Soc. Res: Organic synthesis; organic sulfur compounds; heterocycles; fungicides; synthetic resins; coatings. Mailing Add: Rohm & Haas Co Spring House PA 19477

LO, CHU-SHEK, b Hong Kong; m 69; c 1. MEDICAL PHYSIOLOGY. Educ: Nat Taiwan Univ, BS, 62; Univ Notre Dame, MS, 65; Ind Univ, PhD(physiol), 72. Prof Exp: NIH fel, Cardiovasc Res Inst, Univ Calif, San Francisco, 72-75; ASST PROF PHYSIOL, SCH MED, UNIV MD, BALTIMORE, 75- Res: Hormonal regulation of sodium transport in kidney and small intestine. Mailing Add: Dept of Physiol Sch of Med Univ of Md 660 W Redwood St Baltimore MD 21201

LO, DONALD HUNG-TAK, b Loyang, China, Dec 3, 43; Can citizen; m 68; c 2. CLINICAL CHEMISTRY. Educ: McMaster Univ, BSc, 65; McGill Univ, PhD(quantum chem), 69. Prof Exp: Chemist, Can Indust Ltd, 65; res assoc chem with comput, Univ Tex, Austin, 69-73; trainee clin chem, Erie County Lab, 73-75; CLIN CHEMIST, UNION CARBIDE CORP, 75- Concurrent Pos: Nat Res Coun Can fel, McGill Univ, 68-69; Robert A Welch fel, Univ Tex, Austin, 71-73. Mem: Am Asn Clin Chemists; Clin Radioassay Soc. Res: Radioimmunoassay method development and optimization; systematic radioimmunoassay method comparison; computer simulation of kinetic antigen-antibody interaction; the teaching of clinical chemistry via flowcharts; a more efficient clinical chemistry laboratory organization. Mailing Add: Corp Res Dept Union Carbide Corp Tarrytown NY 10591

LO, ELIZABETH SHEN, b Shanghai, China, Feb 24, 26; m 50; c 2. ORGANIC CHEMISTRY. Educ: St John's Univ, Shanghai, BS, 45; Univ Ill, MS, 47, PhD(chem), 49. Prof Exp: Univ Ill fel, 49-50; res chemist, Metalsalsts Corp, 51; J T Baker Chem Co Div, Vick Chem Co, 51-52; M W Kellogg Co, 53-57; Permacel Div, Johnson & Johnson, 57-60; staff chemist, IBM Corp, 60-63; sr res chemist, Thiokol Chem Corp, 65-70; vis fel, Princeton Univ, 71-73; MGR, MAT & CHEM PROCESS, FAIRCHILD-PMS PROD, 74- Mem: Am Chem Soc. Res: Polymer, rubber and resin chemistry; fluorocarbon polymers; liquid crystals. Mailing Add: 102 Maclean Circle Princeton NJ 08540

LO, GEORGE ALBERT, b Hong Kong, June 26, 34; US citizen; m 57; c 2. PHYSICAL CHEMISTRY, INORGANIC CHEMISTRY. Educ: Univ Ore, BA, 57, MA, 60; Wash State Univ, PhD(chem), 63. Prof Exp: MEM TECH STAFF, ROCKETDYNE DIV, ROCKWELL INT CORP, 63- Mem: Am Chem Soc; Am Inst Aeronaut & Astronaut. Res: Chemical kinetics and propulsion; chemistry of inorganic complexes. Mailing Add: Rocketdyne Div 6633 Canoga Ave Canoga Park CA 91304

LO, KWOK-YUNG, b Nanking, China, Oct 19, 47; m 73. RADIO ASTRONOMY, ASTROPHYSICS. Educ: Mass Inst Technol, SB, 69, PhD(physics), 74. Prof Exp: RES FEL RADIO ASTRON, OWENS VALLEY RADIO OBSERV, CALIF INST TECHNOL, 74- Mem: Am Astron Soc. Res: Microwave spectroscopy studies of phenomena associated with star formation; high angular resolution studies of galactic and extragalactic radio sources by interferometry and very long baseline interferometry techniques. Mailing Add: Owens Valley Radio Observ Calif Inst of Technol Pasadena CA 91125

LO, MIKE MEI-KUO, b Formosa, China, Sept 21, 36; m 67. PHYSICAL CHEMISTRY. Educ: Nat Taiwan Univ, BS, 59; Univ Ill, MS, 65, PhD(chem), 67. Prof Exp: SR RES CHEMIST, S C JOHNSON & SON, INC, 67- Mem: Am Chem Soc. Res: Microwave spectroscopy; ultrasonic impedometry; gas chromatography and mass spectroscopy. Mailing Add: S C Johnson & Son Inc 1525 Howe St Racine WI 53403

LO, THEODORE CHING-YANG, b Shanghai, China, Dec 22, 43; Can citizen; m 74. MICROBIAL GENETICS. Educ: Univ Man, BSc, 69; Univ Toronto, PhD(med biophys), 73. Prof Exp: Res fel biochem, Harvard Univ, 73-75; ASST PROF BIOCHEM, UNIV WESTERN ONT, 75- Mem: Am Soc Microbiol. Res: Molecular mechanisms for dicarboxylic acid transport in Escherichia coli K12. Mailing Add: Dept of Biochem Univ of Western Ont London ON Can

LOACH, KENNETH WILLIAM, b Portsmouth, Eng, Sept 5, 34; m 66; c 2. ANALYTICAL CHEMISTRY. Educ: Univ Auckland, BSc, 56, MSc, 58; Univ Wash, PhD(chem), 69. Prof Exp: Chemist, Ruakura Animal Res Sta, NZ, 58-60; div plant indust, Commonwealth Sci & Indust Res Orgn, Australia, 60-63; ASST PROF ANAL CHEM, STATE UNIV NY COL PLATTSBURGH, 63- Concurrent Pos: NSF grants, Tufts Univ, 71, State Univ NY Col Plattsburgh, 72-73, State Univ NY Res Found fel & grant, 72-73. Mem: AAAS; Am Chem Soc. Res: Principles of analytical chemistry; exact titration curves; nitroprusside reactions; spot-test procedures; trace analysis of natural waters; chemical information systems and computing; chemical structure codes. Mailing Add: Dept of Chem State Univ of NY Col Plattsburgh NY 12901

LOACH, PAUL A, b Findlay, Ohio, July 18, 34; m 57; c 3. BIOCHEMISTRY, PHYSICAL BIOCHEMISTRY. Educ: Univ Akron, BS, 57; Yale Univ, PhD(biochem), 61. Prof Exp: Nat Acad Sci-Nat Res Coun fel photosynthesis, Univ Calif, Berkeley, 61-63; from asst prof to assoc prof, 63-73, PROF CHEM, NORTHWESTERN UNIV, 73-, PROF BIOCHEM & MOLECULAR BIOL, 74- Concurrent Pos: Res Career Develop Award, NIH, 71-76. Mem: AAAS; AM Chem Soc; Am Soc Biol Chem; Biophys Soc; Am Soc for Photobiol. Res: Primary photochemistry of photosynthesis; chemistry of porphyrins and metalloporphyrins; biological oxidation and reduction; structure and function in photosynthetic membranes; photochemical models of photosynthesis. Mailing Add: Dept of Biochem & Molecular Biol Northwestern Univ Evanston IL 60201

LOADHOLT, CLAUDE BOYD, b Fairfax, SC, Mar 26, 40; m 63; c 2. BIOSTATISTICS. Educ: Clemson Univ, BS, 62, MS, 65; Va Polytech Inst, PhD(statist), 69. Prof Exp: Asst exp sta statistician, Clemson Univ, 65-66, asst prof exp statist, 68-70; ASSOC PROF BIOMET, MED UNIV SC, 70- Mem: Biomet Soc. Res: Statistical consultation in biological and medical research; design of experiments; statistical data processing. Mailing Add: Dept of Biomet Med Univ of SC Charleston SC 29401

LOAN, LEONARD DONALD, b London, Eng, Oct 6, 30; m 55; c 3. POLYMER CHEMISTRY. Educ: Univ Birmingham, BSc, 51, PhD(polymer chem), 54. Prof Exp: Sci off combustion chem, Royal Aircraft Estab, 54-57; chemist, Arthur D Little Res Inst, 57-59; prin sci off rubber chem, Rubber & Plastics Res Asn, 59-66; MEM TECHNOL STAFF POLYMER CHEM, BELL TEL LABS, 66- Mem: Am Chem Soc. Res: Polymer crosslinking and aging. Mailing Add: Bell Tel Labs Mountain Ave Murray Hill NJ 07974

LOAN, RAYMOND WALLACE, b Ephrata, Wash, Apr 24, 31; m 52; c 4. IMMUNOBIOLOGY. Educ: Wash State Univ, BS, 52, DVM, 58; Purdue Univ, MS, 60, PhD(animal path), 61. Prof Exp: Instr vet microbiol, Purdue Univ, 58-61; from asst prof to assoc prof, 61-68, PROF VET MICROBIOL, 68- CHMN DEPT, 69- Mem: AM Vet Med Asn; Am Asn Immunol; Am Col Vet Microbiol; Am Soc Microbiol; Conf Res Workers Animal Diseases. Res: Cell mediated immunity; immunologic aspects of avian leukosis. Mailing Add: Connaway Hall Univ of Mo-Columbia Columbia MO 65201

LOBASSO, FRANK ANTHONY, b New York, NY, Sept 29, 37; m 59; c 2. COSMETIC CHEMISTRY. Educ: Columbia Univ, BS, 59. Prof Exp: Prod supvr pharm, Reed & Carnrick Co, 61-62; MGR COSMETIC DEVELOP, WARNER-LAMBERT CO, 62- Mem: Soc Cosmetic Chemists. Mailing Add: 9110 Windview San Antonio TX 78239

LOBB, BARRY LEE, b Easton, Pa, Nov 25, 43; m 69; c 1. MATHEMATICS. Educ: Lafayette Col, BS, 65; Duke Univ, MA, 68, PhD(math), 69. Prof Exp: ASSOC PROF MATH, BUTLER UNIV, 69- Mem: Math Asn Am. Res: Topology. Mailing Add: Dept of Math Butler Univ Indianapolis IN 46208

LOBB, DONALD EDWARD, b Saskatoon, Sask, Apr 25, 40. PHYSICS. Educ: Univ Sask, BE, 61, MSc, 63, PhD(physics), 66. Prof Exp: Nat Res Coun Can overseas fel, 66-67; asst prof, 67-74, ASSOC PROF PHYSICS, UNIV VICTORIA, BC, 74- Mem: Inst Elec & Electronics Engrs; Can Asn Physicists. Res: Beam optics. Mailing Add: Dept of Physics Univ of Victoria Victoria BC Can

LOBDELL, DAVID HILL, b Erie, Pa, July 9, 30. PATHOLOGY. Educ: Kenyon Col, AB, 52; Univ Mich, MD, 56. Prof Exp: From intern to resident path, Bellevue Hosp, New York, 56-59, asst pathologist, 59-60; assoc pathologist, 60-63, DIR LABS, ST VINCENT'S MED CTR, 63-, DIR SCH MED TECHNOL, 63- Concurrent Pos: Instr, Sch Med, NY Univ, 59-61, asst clin prof path, 61-69; lectr histol, Fairfield Univ, 64-73. Mem: AMA; fel Col Am Path; fel Am Soc Clin Path. Res: Osmometry; myeloproliferative disorders. Mailing Add: Dept Path St Vincent's Med Ctr 2820 Main St Bridgeport CT 06606

LOBECK, CHARLES CHAMPLIN, b New Rochelle, NY, May 20, 26; m 54; c 4. PEDIATRICS. Educ: Hobart Col, AB, 48; Univ Rochester, MD, 52. Prof Exp: From instr to sr instr pediat, Sch Med & Dent, Univ Rochester, 55-58; from asst prof to prof, Sch Med, Univ Wis-Madison, 58-75, chmn dept, 64-74, assoc dean clin affairs, 74-75; DEAN, SCH MED, UNIV MO-COLUMBIA, 75- Res: Metabolic disease; membrane transport; cystic fibrosis. Mailing Add: Sch of Med Univ of Mo Med Ctr Columbia MO 65201

LOBENE, RALPH RUFINO, b Rochester, NY, Mar 30, 24; m 50. DENTISTRY, PERIODONTOLOGY. Educ: Univ Rochester, BS, 44; Univ Buffalo, DDS, 49; Tufts Univ, MS, 62; Am Bd Periodont, dipl, 65. Prof Exp: Res chemist, Manhattan Proj, Univ Rochester, 44-45; res chemist, Merck & Co, NJ, 45-46; intern periodont, Eastman Dent Dispensary, Rochester, NY, 4950; resident oral surg, Strong Mem Hosp, Rochester, 5051, asst dent surgeon, 53-62, instr dent & clin dent res, Sch Med & Dent, Univ Rochester, 53-62; asst prof periodont, Sch Dent, Univ Pac, 62-63; ASSOC PROF & ACAD ADMINR DENT ASST TRAINING PROG, NORTHEASTERN UNIV, 63-, DIR ADVAN EDUC, 70-, DEAN, FORSYTH SCH DENT HYGIENISTS, 75- Concurrent Pos: Res grants, Colgate Palmolive Co, 54-55 & 58-59; res grant, Sch Dent, Univ Pac, 62-63; res grant, Gen Elec Co, 62-65; staff dentist, Eastman Dent Dispensary, 53-54, res assoc, 58-62; pvt pract, 53-60; chief dent serv, State Indust & Agr Sch Boys, NY, 54-57; clin asst, Sch Dent Med, Tufts Univ, 61-62, lectr, 63-; asst mem staff, Forsyth Dent Ctr, Northeastern Univ, 63-66, lectr, 63-, head dept clin exp, Inst Res & Advan Study Dent, 66-, sr mem staff, 70-; consult, Dent Res Panel, Gen Elec Co, 63-; vis surgeon, Dept Dent, Boston City Hosp, 64- Mem: Fel AAAS; fel Am Col Dent; Am Dent Asn; Am Acad Oral Med; Am Acad Periodont. Res: Assessment of periodontal disease and evaluation of the effectiveness of therapeutic methods of treatment of periodontal disease; analytical methods for chemical analysis. Mailing Add: Forsyth Dent Ctr 140 Fenway Boston MA 02115

LOBER, PAUL HALLAM, b Minneapolis, Minn, Sept 25, 19. PATHOLOGY. Educ: Univ Minn, Minneapolis, BS, 42, MD, 44, PhD(path), 51; Am Bd Path, dipl, 52. Prof Exp: Mem fac, 51-60, PROF PATH, MED SCH, UNIV MINN, MINNEAPOLIS, 60-, SURG PATHOLOGIST, HOSPS, 51- Mem: Am Soc Cytol; fel Am Col Path; Am Asn Pathologists & Bacteriologists; Int Acad Path; Am Soc Clin Path. Res: Coronary heart disease. Mailing Add: Northwestern Hosp Dept of Path 810 E 27th St Minneapolis MN 55407

LOBKOWICZ, FREDERICK, bPrague, Czech, Nov 17, 32; US citizen; m 60; c 2. ELEMENTARY PARTICLE PHYSICS. Educ: Swiss Fed Inst Technol, 55, PhD(physics), 60. Prof Exp: Res assoc, 60-64, from asst prof to assoc prof, 64-73, PROF PHYSICS, UNIV ROCHESTER, 73- Concurrent Pos: Humboldt Found sr fel, 73-74; vis prof, Univ Munich, Ger, 73-74. Mem: Am Phys Soc; Swiss Phys Soc. Res: Muon and photon interactions. Mailing Add: Dept of Physics & Astron Univ of Rochester Rochester NY 14627

LOBL, RICHARD TOLSTOI, b Washington, DC, Sept 27, 42; m 63; c 2. ANATOMY, ENDOCRINE PHYSIOLOGY. Educ: George Washington Univ, AB, 64; Ind Univ, Indianapolis, MS, 66; Univ Calif, Los Angeles, PhD(anat), 69. Prof Exp: ASST PROF ANAT, HEALTH CTR, UNIV CONN, 70- Mem: AAAS; Soc Study Reproduction; Int Soc Psychoneuroendocrinol; Am Asn Anatomists; Sigma Xi. Res: Neuroendocrinology of reproduction; sexual differentiation; uptake, intracellular transport and metabolic effects of steroid sex hormones. Mailing Add: Dept of Anat Univ of Conn Health Ctr Farmington CT 06032

LOBL, THOMAS JAY, b Danville, Va, Oct 20, 44; m 68; c 2. PHARMACEUTICAL CHEMISTRY, REPRODUCTIVE PHYSIOLOGY. Educ: Univ NC, Chapel Hill, BS, 66; Johns Hopkins Univ, PhD(org chem), 70. Prof Exp: Res fel biochem, Calif Inst Technol, 70-73; RES ASSOC CHEM & BIOCHEM, UPJOHN CO, 73- Mem: Am Chem Soc; AAAS; Sigma Xi. Res: Regulation of male reproduction; spermatogenesis; epididymal function; mechanisms of fertilization; male contraception; chemical and biological deaminations; heterocyclic synthesis. Mailing Add: Fertil Res Upjohn Co Kalamazoo MI 49001

LOBO, ANGELO PETER, b Masindi, Uganda, May 19, 39; m 67; c 2. BIO-ORGANIC CHEMISTRY. Educ: Univ Bombay, BSc, 58; Univ Ind, Bloomington, PhD(org chem), 66. Prof Exp: Rockefeller Found spec lectr org chem, Makerere Univ Col, Univ Uganda, 66-68; res asst chem, Rensselaer Polytech Inst, 68-70; SR RES SCIENTIST, DIV LABS & RES, NY STATE DEPT HEALTH, 70- Mem: Am Chem Soc. Res: Active site studies of thrombin and other blood coagulation factors through the use of peptidic substrates and inhibitors. Mailing Add: Div of Labs & Res NY State Dept of Health Albany NY 12201

LOBO, FRANCIS X, b Aden, UAR, Oct 8, 25; US citizen; m 60; c 3. MICROBIOLOGY. Educ: Univ Bombay, BS, 47, MS, 50; Inst Divi Thomae, PhD(exp med, biol), 59; Nat Registry Microbiol, cert. Prof Exp: Technician, Path Dept, Worli Gen Hosp, India, 50; control & res microbiologist-chemist, Chemo Pharma Labs Ltd, Worli, Bombay, 50-57; assoc prof sci, 60-70, chmn dept biol, 70-74, PROF BIOL SCI, MARYWOOD COL, 70- Concurrent Pos: Consult, Radio Corp Am, 65-; NSF grant, Argonne Nat Lab, 66, resident res assoc, 68-69; fac res partic, Argonne Nat Lab, 67 & St Jude Children Res Hosp, Memphis, Tenn, 70; mem eval team, Pa Dept Educ, 71. Mem: Am Soc Microbiol; NY Acad Sci. Res: Intestinal microorganisms by enrichment culture techniques; citric acid from a cane-sugar molasses; beef brain extract in controlling staphylococcus infections; etiology of sludge formation in industrial wastes. Mailing Add: Dept of Biol Marywood Col Scranton PA 18509

LOBO, LUIZ CARLOS GALVAO, b Rio de Janeiro, Brazil, Jan 18, 34; m 60; c 3. ENDOCRINOLOGY, PHYSIOLOGY. Educ: Univ Brazil, MD, 57, DSc(med), 59. Prof Exp: Res assoc biophys, Inst Biophys, Univ Brazil, 57-66, assoc prof, Nat Sch Med, 61-66; prof endocrinol & dean sch med sci, Univ Brasilia, 67-71; HEAD LAB NUCLEAR MED, INST BIOPHYS & DIR OFF RES & DEVELOP EDUC TECHNOL FOR HEALTH SCI, FED UNIV RIO DE JANEIRO, 72- Concurrent Pos: Fr Govt fel biochem, Col de France, 59-60, fel physiol, 63; Orgn Am States fel med educ, 62; NIH res grant; mem, Pan-Am Health Orgn Study Group Goiter, 63-65; Pan-Am Health Orgn & WHO consult med educ, 67-71. Mem: Brazilian Acad Sci; Brazilian Endocrinol Soc (pres, 64-66); Brazilian Soc Nuclear Med; Brazilian Genetics Soc; Fr Soc Biol Chem. Res: Endocrinology, especially endemic goiter and cretinism; medical education, especially new educational methods and new curricula on health sciences professions. Mailing Add: Lab Nuclear Med Inst of Biophys Fed Univ of Rio de Janeiro Rio de Janeiro Brazil

LOBO, ROBERTO, physics, see 12th edition

LOBSTEIN, OTTO ERVIN, b Czech, Apr 12, 22; nat US; m 52; c 3. CLINICAL BIOCHEMISTRY. Educ: Univ London, BSc, 45; Smae Inst, Eng, MSF, 45; Northwestern Univ, PhD(biochem), 52; Am Bd Clin Chem, dipl, 55. Prof Exp: Asst res chemist, Howards & Sons, Ltd, Eng, 42-46; biochemist, Elgin State Hosp, Ill, 47-48; instr chem, Wesley & Passavant Mem Hosps, Ill, 49-51; res assoc zool, Univ Southern Calif, 52-53; med dir res, Chemtech Labs, 52-62; biochemist-owner, Lobstein Biochem Lab, Calif, 62-64; asst prof chem, Loyola Univ, Calif, 64-65; HEAD BIOCHEM DEPT, ST ELIZABETH HOSP MED CTR, 65- Concurrent Pos: Vis res prof, Univ Redlands, 59-65; secy-treas, Res Found Diseases Eye, 59-65; vis assoc prof, Purdue Univ, 68-73, adj prof, 73- Mem: Fel AAAS; Am Soc Microbiol; Soc Appl Spectros; sr mem Am Chem Soc; fel Am Asn Clin Chem. Res: Biochemical investigation of the crystalline lens of the eye, protein structure and constitution in normal and in cataract lenses of the human and other species; changes in protein with a changed electrolyte environment; clinical investigation of lysozyme in carcinomatosis. Mailing Add: St Elizabeth Hosp Med Ctr Lafayette IN 47904

LOBUE, JOSEPH, b Union City, NJ, Apr 19, 34; m 59; c 3. PHYSIOLOGY, HEMATOLOGY. Educ: St Peter's Col, NJ, BS, 55; Marquette Univ, MS, 57; NY Univ, PhD(physiol), 62. Prof Exp: From asst prof to assoc prof, 62-71, PROF BIOL, NY UNIV, 71-, CO-DIR, LAB EXP HEMAT, 67- Concurrent Pos: NIH fel, 62; Sigma Xi grant-in-aid, 64-65; Am Cancer Soc grant, 65-66; Nat Cancer Inst grant, 71-73 & 75-78; Nat Leukemia Asn grant, 74-76; assoc, Danforth Found, 68-; co-dir, Hemat Training Prog, NIH, 65-75. Honors & Awards: Christian R & Mary F Lindback Found Award, NY Univ, 65. Mem: AAAS; Soc Exp Biol & Med; Am Asn Anat; Harvey Soc; Am Soc Hemat. Res: Mechanisms controlling leukocyte and erythrocyte production and release; pathophysiology and cytokinetics of rodent and avian leukemias. Mailing Add: Lab of Exp Hemat NY Univ New York NY 10003

LOBUGLIO, ALBERT FRANCIS, b Buffalo, NY, Feb 1, 38; m 62; c 5. HEMATOLOGY, IMMUNOLOGY. Educ: Georgetown Univ, MD, 62. Prof Exp: Intern med, Presby Univ Hosp, Pittsburgh, 62-63, resident, 63-65; instr, State Univ NY Buffalo, 67-68, asst prof, 68-69; ASSOC PROF MED, OHIO STATE UNIV, 69- Concurrent Pos: Hemat fel, Thorndike Mem Lab, Boston City Hosp, 65-67; hemat consult, Vet Admin Hosp, Buffalo, 67-69, Dayton, Ohio, 69- Mem: Am Fedn Clin Res; Am Soc Hemat. Res: Tumor immunology; transplant immunology; human macrophage and lymphocyte functions. Mailing Add: Ohio State Univ Hosp 410 N Tenth Ave Columbus OH 43210

LOBUNEZ, WALTER, b Ukraine, Nov 22, 20; nat US; m 45; c 1. INDUSTRIAL CHEMISTRY. Educ: Univ Pa, MS, 52, PhD(chem), 54. Prof Exp: Res assoc immunol, Jefferson Med Col, 54-55; protein chem, Children's Hosp, Univ Penn, 55-59; sr scientist chem, Textile Res Inst, 59-60; res chemist, 60-67, SR RES CHEMIST, F M C CORP, 67- Mem: Am Chem Soc; Am Inst Chemists. Res: Chemistry of hydrocarbons; protein chemistry; chemistry of cellulose; soda ash processes; Ba and Sr processes. Mailing Add: 562 Ewing St Princeton NJ 08540

LOCALIO, S ARTHUR, b New York, NY, Oct 4, 11; m 45; c 4. SURGERY. Educ: Cornell Univ, AB, 33; Univ Rochester, MD, 36; Columbia Univ, DSc(med), 42; Am Bd Surg, dipl, 45. Prof Exp: From instr to assoc prof surg, 45-53, prof clin surg, 53-71, PROF SURG, POST-GRAD MED SCH, NY UNIV, 71-, PROF SCH MED, 62- Concurrent Pos: Consult, Riverview Hosp, Red Bank, NJ, 52, Monmouth Mem Hosp, Long Branch, NJ, 52 & St Barnabas Hosp, New York, 58; asst surgeon, Univ Hosp, 45-47, asst attend surgeon, 47-49, assoc attend surgeon, 49-52, attend surgeon, 52-, clin asst vis surgeon, 49-52, vis surgeon, 52-; Johnson & Johnson distinguished prof surg, NY Univ, 72. Mem: AMA; Am Asn Surg of Trauma; Am Gastroenterol Asn; Am Col Surg; NY Acad Sci. Res: Wound healing; surgery of gastro-intestinal disease. Mailing Add: Dept of Surg New York Univ Sch of Med New York NY 10016

LOCANTHI, DOROTHY DAVIS, b East St Louis, Ill, Apr 19, 13; m 43; c 3. ASTROPHYSICS, SCIENTIFIC BIBLIOGRAPHY. Educ: Vassar Col, BA, 33; Mills Col, MA, 34; Univ Calif, Berkeley, PhD(astron), 37. Prof Exp: Instr astron, Vassar Col, 37-38 & Smith Col, 38-39; comput astron, Princeton Univ Observ, 40-42 & Mt Wilson Observ, 42-43; physicist, US Navy, Calif Inst Technol, 43-45; optical engr, Ray Control Co, 45-46 & Nat Tech Labs, Beckman, 46-48; physicist, Off Naval Res, 50-53; res fel astron, Calif Inst Technol, 62-72; LIBR SEARCH MOLECULAR SPECTROS, JET PROPULSION LAB, NASA, 72- Concurrent Pos: Am Asn Univ Women fel, 39-40. Mem: Am Astron Soc; Int Astron Union. Res: Determining abundances of elements in atmospheres of S type stars. Mailing Add: Jet Propulsion Lab 4800 Oak Grove Dr Pasadena CA 91103

LOCASCIO, SALVADOR J, b Hammond, La, Oct 29, 33; m 54; c 3. HORTICULTURE. Educ: Southeastern La Col, BS, 55; La State Univ, MS, 56; Purdue Univ, PhD(plant physiol), 59. Prof Exp: Asst prof, 59-65, assoc horticulturist, 65-69, ASSOC PROF HORT, UNIV FLA, 65- HORTICULTURIST, 69- Mem: AAAS; Am Soc Hort Sci. Res: Fertilizer and water requirements of vegetables; strawberry culture; chemical weed control for vegetables; teaching of commercial vegetable crops and vegetable crops nutrition. Mailing Add: Dept of Veg Crops 3026 McCarty Hall Univ of Fla Gainesville FL 32601

LOCATELL, LOUIS, JR, organic chemistry, see 12th edition

LOCHHEAD, JOHN HUTCHISON, b Montreal, Que, Aug 7, 09; nat US; m 38; c 2. INVERTEBRATE ZOOLOGY. Educ: Univ St Andrew's, MA, 30; Cambridge Univ, BA, 32, Bachelor scholar, 33, PhD(zool), 37. Prof Exp: With Cambridge Univ Table, Marine Zool Sta, Naples, 34-35; sr cur, Mus Zool, Cambridge Univ, 35-38, instr zool, 36-38; fel by courtesy, Johns Hopkins Univ, 40; asst biologist, Va Fisheries Lab, 41-42; from instr to prof, 42-75, EMER PROF ZOOL, UNIV VT, 75- Concurrent Pos: Lectr, Col William & Mary, 41-42; instr, Woods Hole Marine Biol Lab, 43-55, mem corp, 44- Mem: Am Soc Zool. Res: Anatomy and physiology of Crustacea including their feeding mechanisms, locomotion, factors controlling swimming positions, responses to light, functions for the blood and related tissues, molting and reproduction. Mailing Add: 49 Woodlawn Rd London SW6 England

LOCHMAN-BALK, CHRISTINA, b Springfield, Ill, Oct 8, 07; m 47. INVERTEBRATE PALEONTOLOGY, STRATIGRAPHY. Educ: Smith Col, AB, 29, AM, 31; Johns Hopkins Univ, PhD(geol), 33. Prof Exp: Asst geol, Smith Col, 29-31; Nat Res Coun grant, 34; from instr to assoc prof, Mt Holyoke Col, 35-47; lectr phys sci, Univ Chicago, 47; lectr life sci, NMex Inst Min & Tech, 54; stratig geologist, State Bur Mines & Mineral Resources, 55-57; prof geol, 57-72, EMER PROF GEOL, NMEX INST MINING & TECHNOL, 72- Concurrent Pos: Am Geol Soc grants 36 & 47; NSF grant, 59. Mem: Fel AAAS; Paleont Soc; fel Geol Soc Am; Nat Asn Geol Teachers. Res: Cambrian paleontology and stratigraphy of the United States. Mailing Add: PO Box 1421 Socorro NM 87801

LOCHMÜLLER, CHARLES HOWARD, b New York, NY, May 4, 40; m 63; c 1. ANALYTICAL CHEMISTRY. Educ: Manhattan Col, BS, 62; Fordham Univ, MS, 64, PhD(anal chem), 68. Prof Exp: Asst prof, 69-74, ASSOC PROF CHEM, DUKE UNIV, 74- Mem: Am Chem Soc. Res: Factors effecting separation processes; nuclear magnetic resonance spectroscopy. Mailing Add: P M Gross Chem Lab Duke Univ Durham NC 27706

LOCHNER, ROBERT HERMAN, b Madison, Wis, Apr 17, 39; m 62; c 4. STATISTICS. Educ: Univ Wis-Madison, BS, 61, MS, 62 & 66, PhD(statist), 69. Prof Exp: Math analyst, A C Electronics Div, Gen Motors Corp, 62-65; asst prof, 68-74, ASSOC PROF STATIST & MATH, MARQUETTE UNIV, 74- Mem: Am Statist Asn; Inst Math Statist. Res: Statistical methods in reliability and life testing; Bayesian inference. Mailing Add: Dept of Math & Statist Marquette Univ Milwaukee WI 53216

LOCHSTET, WILLIAM A, b Port Jefferson, NY, Dec 5, 36; m 65. PHYSICS. Educ: Univ Rochester, BS, 57, MA, 60; Univ Pa, PhD(physics), 65. Prof Exp: Instr, 65-66, ASST PROF PHYSICS, PA STATE UNIV, 66- Mailing Add: Dept of Physics Pa State Univ University Park PA 16802

LOCICERO, JOSEPH CASTELLI, b Ontario Center, NY, m 37; c 2. ORGANIC CHEMISTRY. Educ: Univ Rochester, BA, 36; Pa State Univ, MS, 47, PhD(biochem), 48. Prof Exp: Chemist, Hooker Electrochem Co, NY, 37-43; sr res chemist, Nuodex Prods Co, NJ, 43-45; res chemist, Rohm and Haas Co, 48-52, sr scientist, 52-71; asst prof, 71-72, ASSOC PROF CHEM, CAMDEN COUNTY COL, 72- Mem: AAAS; Am Chem Soc. Res: Plasticizers, fungicides and insecticides; high pressure reactions; detergents; process development; ion exchange; sugar technology; halogenation; plastics. Mailing Add: 625 Devon Rd Moorestown NJ 08057

LOCK, COLIN JAMES LYNE, b London, Eng, Oct 4, 33; m 60; c 2. PHYSICAL INORGANIC CHEMISTRY, CRYSTALLOGRAPHY. Educ: Univ London, BSc, 54, PhD(inorg chem) & DIC, 63. Prof Exp: Asst exp officer reactor chem, Atomic Energy Res Estab, Eng, 54-57, sci officer, 60; develop chem, Atomic Energy Can Lab, Chalk River, 57-60; asst lectr inorg chem, Imp Col, London, 61-63; from asst prof to assoc prof, 63-73, PROF CHEM, McMASTER UNIV, 73- Mem: Fel Chem Inst Can; fel Royal Inst Chem; The Chem Soc. Res: Reactor and high temperature aqueous chemistry; physical methods of structure determination for heavy transition metal compounds. Mailing Add: Inst for Mat Res McMaster Univ Hamilton ON Can

LOCK, JAMES ALBERT, b Cleveland, Ohio, Feb 12, 48; m 72. THEORETICAL NUCLEAR PHYSICS. Educ: Case Western Reserve Univ, BS, 70, MS, 73, PhD(physics), 74. Prof Exp: Lectr, 70-73, RES ASSOC PHYSICS, CASE WESTERN RESERVE UNIV, 74- Mem: Sigma Xi. Res: Meson exchange currents; three-pion systems. Mailing Add: Dept of Physics Case Western Reserve Univ Cleveland OH 44106

LOCKARD, DERWOOD WARNER, anthropology, see 12th edition

LOCKARD, ISABEL, b Brandon, Man, June 27, 15. ANATOMY. Educ: Northwestern Univ, BS, 38; Univ Mich, MA, 42, PhD(anat), 46. Prof Exp: Speech correctionist, Pub Schs, Inc, 38-41; clin helper, Speech Clin, Univ Mich, 42, asst anat, 42-44 & 47; instr, Univ Pittsburgh, 44-45; instr, Sch Med Georgetown Univ, 47-49, asst prof, 49-52; from asst prof to assoc prof, 52-69, PROF ANAT, MED UNIV SC, 69- Res: Neuroanatomy; blood supply of central nervous system. Mailing Add: Dept of Anat Med Univ of SC Charleston SC 29401

LOCKARD, J DAVID, b Renovo, Pa, Dec 20, 29; m 51; c 4. BOTANY, SCIENCE EDUCATION. Educ: Pa State Univ, BS, 51, MEd, 55; PhD(bot), 62. Prof Exp: Dept chmn sci dept high sch, Pa, 53-56; consult sci teaching improvement prog, AAAS, 56-58; asst bot, Pa State Univ, 58-61; from asst prof to assoc prof, 61-70, PROF BOT & SCI EDUC, UNIV MD, COLLEGE PARK, 70- DIR SCI TEACHING CTR, 62- Concurrent Pos: NSF-AAAS grant, Develop & Maintain Int Clearinghouse Sci & Math Curric Develops, 62-; dir off bio educ, Am Inst Biol Sci, 66-67; dir, NSF-AID Study Improvisation Sci Teaching Mat Worldwide, 68-72; NSF grants, acad year inst sci supvrs, 69-73; dir NSF Impact Study, 74-75; rep, US Nat Comn to UNESCO, 75- Honors & Awards: Distinguished Serv to Sci Ed Award, Nat Sci Teacher's Asn, 74. Mem: AAAS (vpres, 71); Am Soc Plant Physiol; Nat Asn Biol Teachers; Nat Asn Res Sci Teaching (pres, 72-73); Nat Sci Teachers Asn. Res: Investigating physiology of fungi and fruiting mechanism; improving science teaching techniques and equipment; studying science and math curriculum developments internationally; consulting in science education; science writing. Mailing Add: Sci Teaching Ctr Univ of Md College Park MD 20742

LOCKARD, JOAN SALIBA, animal behavior, experimental psychology, see 12th edition

LOCKARD, RAYMOND G, b Patricia, Alta, Jan 1, 25; m 51; c 3. PLANT PHYSIOLOGY. Educ: Univ BC, BSA, 49; Univ Idaho, MSc, 51; Univ London, PhD, 56. Prof Exp: Plant physiologist, Can For Aid, Malaysia, 54-59, Ghana, 59-64; tech expert, Food & Agr Orgn, Philippines, 64-67; assoc prof, 67-72, PROF HORT, UNIV KY, 72- Mem: Am Soc Plant Physiol; Am Soc Hort Sci; Am Inst Biol Sci; Can Soc Plant Physiol. Res: Mineral nutriton; growth substances, particularly gibberellin. Mailing Add: Dept of Hort Univ of Ky Lexington KY 40506

LOCKARD, ROBERT BRUCE, b Long Beach, Calif, July 6, 31; m 59; c 2. ETHOLOGY. Educ: Univ Calif, Santa Barbara, BA, 55; Univ Wis, MS, 61, PhD(psychol, zool), 62. Prof Exp: From asst prof to assoc prof, 62-70, PROF PSYCHOL, UNIV WASH, 70- Concurrent Pos: NSF-NIH spec fel zool, Univ Calif, Davis, 69-70. Mem: Animal Behav Soc; Soc Study Evolution; Am Soc Mammalogists. Res: Animal behavior, especially of rodents in relation to ecological and evolutionary factors; ecology of desert rodents; activity rhythms and circadian activity of small mammals; ethology of rodents and small mammals. Mailing Add: Dept of Psychol Univ of Wash Seattle WA 98195

LOCKART, ROYCE ZENO, JR, b Marshfield, Ore, Sept 7, 28; m 51; c 3. MICROBIOLOGY. Educ: Whitman Col, AB, 50; Univ Wash, MS, 53, PhD(microbiol), 57. Prof Exp: Res fel, Nat Inst Allergy & Infectious Dis, 57-58, bacteriologist, Radiation Br, Nat Cancer Inst, 58-60; from asst prof to assoc prof microbiol, Univ Tex, 60-66; RES SUPVR, E I DU PONT DE NEMOURS & CO, INC, 66- Mem: AAAS; Am Soc Microbiol; Brit Soc Gen Microbiol. Res: Virus cell interactions, particularly animal viruses and their control by natural means and by chemicals. Mailing Add: E I du Pont de Nemours & Co Inc Wilmington DE 19898

LOCKE, BEN ZION, b New York, NY, Sept 8, 21; m 47; c 4. APPLIED STATISTICS, EPIDEMIOLOGY. Educ: Brooklyn Col, AB, 47; Columbia Univ, MS, 49. Prof Exp: Statistician, NY State Health Dept, 47-56; chief consult sect, Biomet Br, NIMH, 56-66; assoc prof eval & dir res & eval, Community Ment Health Ctr, Temple Univ, 66-67; asst chief, 67-75, CHIEF CTR EPIDEMIOL STUDIES, NIMH, 75- Mem: AAAS; Fel Am Pub Health Asn; Am Statist Asn; Soc Epidemiol Res. Res: Epidemiology of mental disorders; evaluation of programs designed to prevent and control mental disorders and promote mental health. Mailing Add: Ctr Epidemiol Studies NIMH Rm 10C-09 5600 Fishers Lane Rockville MD 20852

LOCKE, CHARLES STEPHEN, b Murfreesboro, Tenn, Dec 6, 42; m 67. STATISTICS. Educ: David Lipscomb Col, BA, 64; Ohio State Univ, MS, 67, PhD(statist), 73. Prof Exp: Instr math, Mid Tenn State Univ, 67-69; ASST PROF MATH & COMPUT SCI, UNIV SC, 73- Mem: Inst Math Statist; Am Statist Asn. Res: Tests of goodness of fit for composite hypotheses, including tests of normality; nonparametric statistics; linear statistical models. Mailing Add: Dept of Math & Comput Sci Univ of SC Columbia SC 29208

LOCKE, DAVID CREIGHTON, b Garden City, NY, Mar 1, 39; m 62. CHEMISTRY. Educ: Lafayette Col, BS, 61; Kans State Univ, PhD(chem), 65. Prof Exp: Res chemist, Esso Res & Eng Co, 65-67; NSF fel, Univ Col Swansea, Univ Wales, 67-68; asst prof, 68-72, ASSOC PROF CHEM, QUEENS COL, NY, 72- Mem: AAAS; Am Chem Soc; Air Pollution Control Asn; Int Inst Conserv Hist & Artistic Works. Res: Analytical chemistry; chemical separations; air pollution monitoring; chemistry of art conservation; chemical oceanography. Mailing Add: Dept of Chem Queens Col Flushing NY 11367

LOCKE, DAVID MILLARD, b Escanaba, Mich, Apr 7, 29. SCIENCE WRITING, CHEMISTRY. Educ: Univ Mich, BS, 51; Univ Ill, MS, 52, PhD(chem), 54. Prof Exp: Fulbright fel, Birkbeck Col, London, 54-55; res assoc, Rockefeller Inst, 56-61; staff writer, Am Chem Soc News Serv, 61-65, asst dir pub rels biol sci, Univ Chicago, 65-67; assoc ed, Encycl Britannica, 67-71; ASST PROF SCI INFO & ASST DIR SCI INFO PROG, ILL INST TECHNOL, 71- Mem: AAAS; Am Chem Soc; Nat Asn Sci Writers; Modern Lang Asn Am; Am Soc Info Sci. Res: Scientific and technical communication, writing, editing and information processing, especially in the fields of chemistry and biochemistry. Mailing Add: Sci Info Prog Ill Inst of Technol Chicago IL 60616

LOCKE, FREDERIC JOHN, b Baldwin, Ill, Feb 16, 18; m 57. CHEMISTRY. Educ: Univ Ill, BS, 38; Univ Minn, MS, 40, PhD(org chem), 46. Prof Exp: Asst, Univ Minn, 38-41; res chemist, 46-48, res group leader, 48-53, res sect leader, 53-58, proj planning specialist, 58-62, SR RES GROUP LEADER, MONSANTO CO, 63- Concurrent Pos: Consult plastic tech, for subsidiaries, 60-63. Mem: Am Chem Soc. Res: Surface coatings and thermosetting resins; organic synthesis; addition and condensation polymers; monomers and intermediates; plasticizers; stabilizers; plastics fabrication. Mailing Add: Res Dept Monsanto Co Springfield MA 01101

LOCKE, HAROLD OGDEN, b Camden, NJ, Sept 14, 31; m 59; c 2. PHYSICAL CHEMISTRY, ANALYTICAL CHEMISTRY. Educ: Wesleyan Univ, BA, 53, MA, 56; Rutgers Univ, PhD(chem), 62. Prof Exp: Res chemist, Armstrong Cork Co, 61-65; ANAL CHEMIST, GAF CORP, 65- Mem: Am Chem Soc. Res: X-ray crystallography; polymer characterization; surfactants. Mailing Add: 816 Prince St Easton PA 18042

LOCKE, JACK LAMBOURNE, b Brantford, Ont, May 1, 21; m 46; c 2. PHYSICS. Educ: Univ Toronto, BA, 46, MA, 47, PhD(physics), 49. Prof Exp: Demonstr physics, Univ Toronto, 45-47; astrophysicist, Dom Observ, 49-59, chief, Stellar Physics Div, 59-66; radio astronr, 66-70, ASSOC DIR, RADIO & ELEC ENG DIV & CHIEF ASTROPHYS BR, NAT RES COUN CAN, 70- Concurrent Pos: Officer-in-chg, Dom Radio Astrophys Observ, 59-62. Mem: Am Astron Soc; Can Astron Soc; Int Astron Union. Res: Astrophysics; radio astronomy; solar physics; molecular spectra; infrared spectrum of the atmosphere. Mailing Add: Astrophys Br Nat Res Coun of Can Ottawa ON Can

LOCKE, JOHN FLOWERS, b Winona, Miss, Nov 21, 08; m 37; c 1. BOTANY. Educ: Miss State Col, BS, 30, MS, 32; Univ Chicago, PhD, 34. Prof Exp: Instr, 30-32, from asst prof to prof, 34-74, EMER PROF BOT, MISS STATE UNIV, 74- Mem: Bot Soc Am; Soc Econ Bot. Res: Cytology, microsporogenesis and cytokinesis in the pollen mother cells of Asimina triloba. Mailing Add: Dept of Bot Miss State Univ PO Box 1568 State College MS 39762

LOCKE, JOHN FRANKLIN, b Moscow, Tenn, Nov 28, 03; m 34; c 1. MATHEMATICS. Educ: Memphis State Univ, BS, 27; Vanderbilt Univ, MA, 29; Univ Ill, PhD(math), 33. Prof Exp: Prof math, Memphis State Univ, 32-43; prof, 46-71, EMER PROF MATH, BIRMINGHAM-SOUTHERN COL, 71- Mem: Math Asn Am. Res: Analysis. Mailing Add: 1240 Greensboro Rd Birmingham AL 35208

LOCKE, KRYSTYNA KOPACZYK, b Warsaw, Poland, Dec 2, 26; m 70. BIOCHEMISTRY, ENZYMOLOGY. Educ: Wayne State Univ, BS, 53; Western Reserve Univ, MS, 56; Univ Ill, Champaign-Urbana, PhD(lipid chem), 62. Prof Exp: Res nutritionist, atherosclerosis res proj, VA Hosp, Downey, Ill, 56-58; Nat Inst Neurol Diseases & Blindness fel biochem, Ment Health Res Inst, Univ Mich, Ann Arbor, 62-64; trainee, Inst Enzyme Res, Univ Wis-Madison, 64-66; proj assoc, 66-69; RES BIOCHEMIST, BIOCHEM TOXICOL BR, DIV TOXICOL, FOOD & DRUG ADMIN, 69- Mem: Am Chem Soc. Res: Effects of environmental agents on biochemistry and ultrastructure of mitochondria. Mailing Add: Apt A-418 101 G St SW Washington DC 20024

LOCKE, LOUIS NOAH, b Stockton, Calif, Mar 14, 28; m 53; c 1. ANIMAL PATHOLOGY. Educ: Univ Calif, AB, 50, DVM, 56. Prof Exp: Vet, USPHS, 56-58; wildlife res biologist, 58-60, HISTOPATHOLOGIST, PATUXENT WILDLIFE RES CTR, US DEPT INTERIOR, 61- Mem: Wildlife Soc; Am Asn Avian Path; Am Vet Med Asn; Wildlife Disease Asn; Soc Toxicol. Res: Wildlife diseases, especially diseases and parasites of the mourning doves, waterfowl; effects of pollutants upon wild birds; lead poisoning in migratory birds. Mailing Add: Fish & Wildlife Health Lab Univ Wis Stock Pavilion 1655 Linden Dr Madison WI 53706

LOCKE, MICHAEL, b Nottingham, Eng, Feb 14, 29. INSECT PHYSIOLOGY. Educ: Cambridge Univ, BA, 52, MA, 55, PhD, 56. Prof Exp: Lectr zool, Univ WIndies, 56; assoc prof biol, Case Western Reserve Univ, 61-67; prof biol, 67-71, PROF ZOOL & CHMN DEPT, UNIV WESTERN ONT, 71- Concurrent Pos: Ed, Soc Develop Biol, 62-69; Raman prof, Univ Madras, 69. Mem: AAAS; Brit Soc Exp Biol; Can Soc Cell

Biol; Soc Develop Biol; Electron Micros Soc Am. Res: Coordination of growth in insects; insect cell development; mitochondrial replication and destruction in cell remodeling. Mailing Add: Dept of Zool Univ of Western Ont London ON Can

LOCKE, PHILIP M, b Rockford, Ill, July 12, 37; m 61; c 2. MATHEMATICS. Educ: Bluffton Col, BS, 59; Univ NH, MS, 64, PhD(math), 67. Prof Exp: Asst prof math, Mont State Univ, 67-68; asst prof, 68-74, ASSOC PROF MATH, UNIV MAINE, ORONO, 74- Mem: Math Asn Am. Res: Ordinary differential equations. Mailing Add: Dept of Math Univ of Maine Orono ME 04473

LOCKE, RAYMOND KENNETH, b Terre Haute, Ind, July 2, 40; m 70. BIOCHEMISTRY, METABOLISM. Educ: Wash Univ, BS, 65. Prof Exp: Res asst biochem, Univ Tex, Dallas, 66-67; res chemist, Div Nutrit, 68-69, Biochem & Metab Sect, Div Pesticides, 69-71 & Metab Br, Div Toxicol, 71-73, RES CHEMIST, BIOCHEM TOXICOL BR, DIV TOXICOL, FOOD & DRUG ADMIN, 73- Mem: AAAS; Am Chem Soc; Tissue Cult Asn; Fedn Am Scientists. Res: Biochemical studies of the comparative in vivo and in vitro metabolism of foreign compounds by animals, plants and man. Mailing Add: Apt A-418 101 G St SW Washington DC 20024

LOCKE, ROBERT F, b Washington, DC, July 16, 18; m 43; c 6. IMMUNOPATHOLOGY. Educ: Tex A&M Univ, DVM, 42; Univ Mich, MPH, 48; Univ Ill, PhD(immunol), 63; Am Col Lab Animal Med, dipl, 64. Prof Exp: Immunologist, USDA, 48-51; health officer, State Md, 51-60; NIH fel immunol, Univ Ill, 60-63; chief res lab animal med, sci & tech, Vet Admin Hosp, Hines, 63-67, part-time chief, 67; ADMINR, MED RES LAB, UNIV ILL MED CTR, 67-, ASSOC PROF ENVIRON HEALTH SCI, SCH PUB HEALTH, 71- Concurrent Pos: Consult & assoc prof, Col Med, Univ Ill, 64-; clin prof, Grad Sch, Loyola Univ, Ill, 65-, assoc prof, Stritch Sch Med, 67-; vis prof, Rush-Presby-St Lukes Med Sch; chief, Res Lab Animal, Sci & Tech, West Side Vet Admin Hosp, 68- Mem: AAAS; Am Vet Med Asn; Am Pub Health Asn; Am Asn Lab Animal Sci; Nat Soc Med Res. Res: Determinations of molecular parameters of immune globulins; intensive study of autoimmune diseases and developing concepts of antibody formation. Mailing Add: Dept of Prev Med Col of Med Univ of Ill Chicago IL 60612

LOCKE, SETH BARTON, b Casselton, NDak, Aug 11, 06; m 37; c 2. PLANT PATHOLOGY. Educ: Oregon State Col, BS, 33; Univ Wis, MS, 35, PhD(plant physiol, path), 37. Prof Exp: Res assoc plant path, Univ Wis, 37-38; asst plant pathologist & asst prof plant path, Univ Ark, 39-43; assoc plant pathologist, Bur Plant Indust, USDA, 43, plant pathologist, 43-44; asst plant pathologist, RI State Col, 44-45; assoc plant pathologist, 45-53, prof & plant pathologist, 53-71, EMER PROF PLANT PATH, WASH STATE UNIV, 71- Mailing Add: 389 Elk Creek Rd Chehalis WA 98532

LOCKE, STANLEY, b New York City, NY, June 18, 34; m 58; c 3. MATHEMATICS, PHYSICS. Educ: NY Univ, BME, 55, MS, 57, PhD(math), 60. Prof Exp: Res mathematician, Repub Aviation Corp, 59-60; sr res mathematician, 60-65, PROJ MATHEMATICIAN, SCHLUMBERGER-DOLL RES CTR, 65- Mem: Sigma Xi; Am Phys Soc. Res: Oil field wire-line services. Mailing Add: 17 Deerwood Ct Norwalk CT 06851

LOCKE, WILLIAM, b Morden, Man, Mar 16, 16; m 45. INTERNAL MEDICINE. Educ: Univ Man, MD, 38; Univ Minn, MS, 47; McGill Univ, DTM, 45. Prof Exp: First asst med, Mayo Clin, 47-48; sr active mem staff, Ochsner Found Hosp, 50-54, pres staff, 54-55, PARTNER, OCHSNER CLIN, 57-, HEAD SECT ENDOCRINOL & METAB, OCHSNER CLIN & FOUND HOSP, 69- Concurrent Pos: Nat Res Coun & Commonwealth Fund fel, Harvard Univ, 48-50; clin prof med, Sch Med, Tulane Univ, 69-; sr vis physician, Charity Hosp, New Orleans. Mem: AAAS; Endocrine Soc; Am Diabetes Asn; fel Am Col Physicians. Res: Metabolic diseases. Mailing Add: Ochsner Clin 1514 Jefferson Hwy New Orleans LA 70121

LOCKER, JOHN L, b Florence, Ala, Oct 11, 30; m 59; c 3. MATHEMATICS. Educ: Auburn Univ, PhD(math), 60. Prof Exp: Instr math, Auburn Univ, 56-59; mathematician, Redstone Arsenal, 54; assoc prof, 60-70, PROF MATH, UNIV OF NORTHERN ALA, 70- Concurrent Pos: Lectr, NSF-Ala Acad Sci Vis Sci Prog, 61-65. Mem: Math Asn Am. Res: Statistics; geometry. Mailing Add: Dept of Math Univ of Northern Ala Florence AL 35630

LOCKETT, M CLODOVIA, b Austin, Tex, Jan 23, 13. BIOLOGY. Educ: St Louis Univ, BS, 37, PhD, 52; De Paul Univ, MS, 47. Prof Exp: Asst prof biol, LeClerc Col, 47-49; from assoc prof to prof & dir dept, Notre Dame Col, 49-65; PROF BIOL, UNIV DALLAS, 65-, CHMN DEPT, 68- Concurrent Pos: Fel, Univ Okla, 61. Mem: AAAS. Res: Effects of drugs on iodine metabolism in the thyroid; the exchange and growth potential of phosphorus in algae cultures. Mailing Add: Dept of Biol Univ of Dallas Irving TX 75061

LOCKEY, RICHARD FUNK, b Lancaster, Pa, Jan 15, 40; m 67; c 2. ALLERGY, IMMUNOLOGY. Educ: Haverford Col, BS, 61; Univ Mich, Ann Arbor, MS, 72; Temple Univ, MD, 65. Prof Exp: Fel allergy & clin immunol, Univ Mich, 69-70; ASST PROF INT MED, COL MED, UNIV S FLA, 73- Concurrent Pos: Asst chief, Sect Allergy & Clin Immunol, Vet Admin Hosp, Tampa, 73- Mem: Fel Am Acad Allergy; fel Am Col Physicians. Res: Human hypersensitivity as it relates to specific animal allergenicity; hymenoptera antigenic specificity in relationship to human hypersensitivity; asthma induced by various pharmaceutical agents. Mailing Add: Vet Admin Hosp Sect of Allergy 13000 N 30th St Tampa FL 33612

LOCKHART, BROOKS JAVINS, b Sandyville, WVa, Feb 8, 20; m 40, 69; c 3. MATHEMATICS. Educ: Marshall Col, AB, 37; WVa Univ, MS, 40; Univ Ill, PhD(math), 43. Prof Exp: Asst instr math, Univ Ill, 40-43; instr, Univ Mich, 43-44 & 46-48; from asst prof to assoc prof, 48-55, PROF MATH, NAVAL POSTGRAD SCH, 55-, DEAN, 62- Mem: Math Asn Am; AAAS; Brit Math Asn. Res: Classical algebraic geometry; numerical analysis; programing digital computers. Mailing Add: Off of the Dean Naval Postgrad Sch Monterrey CA 93940

LOCKHART, ERNEST EARL, b Boston, Mass, Sept 10, 12; m 43. FOOD SCIENCE. Educ: Mass Inst Technol, SB, 34, MS, 35, PhD(biochem), 38. Prof Exp: Am-Scand Found fel enzyme chem, Biochem Inst, Stockholm, Sweden, 38-39; physiologist, Antarctic Serv, US Dept Interior, DC, 39-41; res assoc biol, Mass Inst Technol, 41-44, asst prof food technol, 44-51, assoc prof food chem, 51-55; sci dir, Coffee Brewing Inst, Inc, NY, 55-65; ASST TO VPRES CORP TECH DIV, COCA-COLA CO, 65- Mem: Fel AAAS; Am Chem Soc; fel Am Inst Chem; Am Soc Qual Control; Inst Food Technol. Res: Food chemistry; non-alcoholic beverages, coffee, soft drinks; nutrition; food standards and regulations; food formulation; sensory testing and food acceptance; quality control; taste, odor and flavor chemistry. Mailing Add: Coca Cola Co PO Drawer 1734 Atlanta GA 30301

LOCKHART, HAINES BOOTS, b Crawfordsville, Ind, Oct 29, 20; m 44; c 2. NUTRITION, BIOCHEMISTRY. Educ: Wabash Col, AB, 42; Univ Ill, PhD(biochem), 45. Prof Exp: Asst, Wabash Col, 41-42; chemist, Univ Ill, 42-45; from res chemist to head baby foods res div, Swift & Co, 45-66, head new foods div, 66-71; SECT MGR NUTRIT RES, JOHN STUART RES LABS, QUAKER OATS CO, 71- Mem: Am Chem Soc; Inst Food Technol. Res: Amino acids and proteins in nutrition; infant and geriatric nutrition. Mailing Add: John Stuart Res Labs Quaker Oats Co W Main St Barrington IL 60010

LOCKHART, HAINES BOOTS, JR, b Evergreen Park, Ill, Feb 4, 46; m 68; c 2. ENVIRONMENTAL CHEMISTRY, BIOCHEMISTRY. Educ: Wabash Col, AB, 67; Univ Nebr, Lincoln, MS, 69, PhD(chem), 73. Prof Exp: SR RES BIOCHEMIST, HEALTH & SAFETY LAB, EASTMAN KODAK CO, 72- Mem: Am Chem Soc. Res: Environmental impact of synthetic chemicals, their biodegradation, photodegradation and bioconcentration in aquatic organisms. Mailing Add: Health & Safety Lab Bldg 320 Eastman Kodak Co Kodak Park Rochester NY 14650

LOCKHART, JAMES ARTHUR, b Grand Rapids, Mich, June 7, 26; m 48; c 2. PLANT PHYSIOLOGY. Educ: Mich State Col, BS, 49, MS, 52; Univ Calif, Los Angeles, PhD(bot), 54. Prof Exp: Plant physiologist, Camp Detrick, Md, 51-52; NSF fel, Bot Lab, Univ Pa, 54-55; res fel biol, Calif Inst Technol, 55-60; assoc plant physiologist, Agr Exp Sta, Univ Hawaii, 60-65; assoc prof, 66-68, PROF BOT, UNIV MASS, AMHERST, 68- Mem: AAAS; Am Soc Plant Physiologists; Bot Soc Am; Soc Develop Biol; Scand Soc Plant Physiol. Res: Plant growth and development; mechanics of growth; theoretical analyses of plant growth, development, reproduction. Mailing Add: Dept of Bot Univ of Mass Amherst MA 01002

LOCKHART, LILLIAN HOFFMAN, b Columbus, Tex, Oct, 23, 30; m 51; c 3. MEDICINE. Educ: Rice Univ, BA, 51; Univ Tex Med Br, Galveston, MA, 55, MD, 57. Prof Exp: Asst prof, 63-72, ASSOC PROF PEDIAT & GENETICS, UNIV TEX MED BR GALVESTON, 72- Concurrent Pos: Fel hemat, Univ Tex Med Br, Galveston, 62-63. Mem: Am Acad Pediat. Res: Genetics; chromosome disorders. Mailing Add: Dept of Pediat Univ of Tex Med Br Galveston TX 77550

LOCKHART, LUTHER BYNUM, JR, b Atlanta, Ga, Sept 13, 17; m 51; c 4. ORGANIC POLYMER CHEMISTRY. Educ: Emory Univ, AB, 38; Univ NC, PhD(org chem), 42. Prof Exp: Res assoc, Naval Res Proj, NC, 42-43; res chemist, 43-54, head phys chem br, 54-74, HEAD ORGANIC CHEMISTRY BRANCH, US NAVAL RES LAB, 74-, ASSOC SUPT CHEM DIV, 73- Mem: Am Chem Soc; Sigma Xi. Res: Organic synthesis; polymer chemistry; composite materials; materials characterization. Mailing Add: 6820 Wheatley Court Falls Church VA 22042

LOCKHART, ROBERT JAMES, mathematics, astronomy, deceased

LOCKHART, WALLACE LYLE, toxicology, biochemistry, see 12th edition

LOCKHART, WILLIAM (CLARENCE), poultry nutrition, see 12th edition

LOCKHART, WILLIAM LAFAYETTE, b Nashville, Tenn, Oct 15, 36; m 60; c 2. INORGANIC CHEMISTRY. Educ: Tenn Technol Univ, BS, 58; Univ Miss, MS, 61; Vanderbilt Univ, PhD(inorg chem), 67. Prof Exp: Res biochemist, US Food & Drug Admin, 60-63; asst prof, 67-71, ASSOC PROF CHEM, W GA COL, 71- Mem: AAAS. Res: Kinetics and mechanisms of inorganic reactions. Mailing Add: Dept of Chem WGa Col Carrollton GA 30117

LOCKHART, WILLIAM RAYMOND, b Carlisle, Ind, Nov 25, 25; m 47; c 4. BACTERIOLOGY. Educ: Ind State Teachers Col, AB, 49; Purdue Univ, MS, 51, PhD(bact), 54. Prof Exp: Asst bact, Purdue Univ, 50-51; from asst prof to assoc prof, 54-60, chmn dept, 60-74, PROF BACT, IOWA STATE UNIV, 60- Mem: Am Soc Microbiol; fel Am Acad Microbiol; Brit Soc Gen Microbiol. Res: Physiology of bacterial growth; numerical taxonomy. Mailing Add: Dept of Bact Iowa State Univ Ames IA 50011

LOCKIE, LAURENCE DAGENAIS, pharmacy, deceased

LOCKLEY, JEANETTE ELAINE, b Dallas, Tex, Feb 13, 33; m 52; c 1. STATISTICS, MATHEMATICS EDUCATION. Educ: Wiley Col, BS, 53; Tex Southern Univ, MS, 54; Stanford Univ, MS, 69, PhD(educ), 70. Prof Exp: Instr math, Tex Southern Univ, 54-57; PROF MATH, MERRITT COL, 58-, CHMN DEPT MATH, PHYSICS, ENG & PHILOS, 74- Concurrent Pos: Asst prof & res assoc, Off Educ Res, Macalester Col, 69-70. Mem: Am Statist Asn; Am Math Soc; Am Educ Res Asn. Res: Cluster analytic techniques. Mailing Add: 6126 Plymouth Ave Richmond CA 94805

LOCKMANN, RONALD FREDERICK, b Glendale, Calif, Sept 26, 42; m 67; c 2. GEOGRAPHY. Educ: Univ Calif, Los Angeles, BA, 64, MA, 67, PhD(geog), 72. Prof Exp: ASST PROF GEOG, UNIV NEW ORLEANS, 72- Mem: Asn Am Geog. Res: Resource geography; cultural geography; environmental perception; nature and culture in western thought. Mailing Add: Dept of Geog Univ New Orleans Lakefront New Orleans LA 70122

LOCKNER, FREDERICK RUSSELL b Pomona, Calif, Dec 13, 40; m 63; c 1. ANIMAL BEHAVIOR. Educ: Univ Redlands, BS, 63; Calif State Col, Los Angeles, MA, 65; Univ Mont, PhD(zool), 68. Prof Exp: Res fel, Dept Ecol & Behav Biol, Univ Minn, 68-69; asst prof, 69-73, ASSOC PROF BIOL, CALIF STATE COL, SONOMA, 73- Concurrent Pos: Fac affil prof, Univ Mont. Mem: Animal Behav Soc; Am Soc Mammal. Res: Analyses of intraspecific communication mechanisms; physiology of avian vocalizations. Mailing Add: Dept of Biol Calif State Col Sonoma Rohnert Park CA 94928

LOCKO, BENJAMIN STEPHENSON, chemistry, see 12th edition

LOCKRIDGE, OKSANA MASLIVEC, b Czech, Sept 4, 41; US citizen. BIOCHEMICAL PHARMACOLOGY, BIOCHEMICAL GENETICS. Educ: Smith Col, BA, 63; Northwestern Univ, Ill, PhD(chem), 71. Prof Exp: Asst researcher biochem, 70-72, fel human genetics, 72-74, RES ASSOC PHARMACOL, UNIV MICH, ANN ARBOR, 74- Mem: Am Chem Soc; AAAS. Res: Pharmacogenetics; biochemical studies on genetic variants of human pseudocholinesterase. Mailing Add: Dept of Pharmacol Univ Mich Ann Arbor MI 48104

LOCKSHIN, MICHAEL DAN, b Columbus, Ohio, Dec 9, 37; m 65; c 1. RHEUMATOLOGY, IMMUNOLOGY. Educ: Harvard Col, AB, 59; Harvard Med Sch, MD, 63. Prof Exp: Intern, Second (Cornell) Div Med, Bellevue & Mem Hosp, New York, 63-64; epidemic intel serv officer, Epidemic Intel Serv, Commun Dis Ctr, 64-66; resident, Second (Cornell) Div Med, Bellevue & Mem Hosp, New York, 66-68; fel rheumatol, Columbia Presby Med Ctr, 68-70; asst prof, 70-75, ASSOC PROF MED, COL MED, CORNELL UNIV, 75- Concurrent Pos: Adj asst prof epidemiol, Sch Pub Health, Univ Pittsburgh, 65-66; assoc scientist & assoc attend physician, Hosp Spec Surg, New York Hosp, 70-; consult rheumatol, Mem Hosp, New York, 70-; mem bd dirs, Arthritis Found, 75- Mem: Am Rheumatism Asn; Am Col Physicians.

Res: Cellular immunology; clinical rheumatology. Mailing Add: 535 E 70th St New York NY 10021

LOCKSHIN, RICHARD ANSEL, b Columbus, Ohio, Dec 9, 37; m 63; c 2. PHYSIOLOGY, DEVELOPMENTAL BIOLOGY. Educ: Harvard Univ, AB, 59, AM, 61, PhD(biol), 63. Prof Exp: Asst prof physiol, Sch Med, Univ Rochester, 65-75; ASSOC PROF PHYSIOL, ST JOHN'S UNIV, NY, 75- Concurrent Pos: NSF fel, Inst Animal Genetics, Univ Edinburgh, 63-64, NIH fel, 64-65. Mem: AAAS; Soc Cell Biol; Soc Develop Biol; Gerontol Soc; Am Soc Entom. Res: Destruction of tissues during metamorphosis of insects; early developmental events in insect embryogenesis; cellular differentiation. Mailing Add: Dept of Biol Sci St John's Univ Jamaica NY 11439

LOCKWOOD, ARTHUR H, b Jan 26, 47; US citizen. MOLECULAR BIOLOGY, CELL BIOLOGY. Educ: Carleton Col, BA, 68; Albert Einstein Col Med, PhD(molecular biol), 72. Prof Exp: Res fel molecular biol, Dept Biochem Sci, Princeton Univ, 73-74; RES SCIENTIST CELL BIOL, SCH MED, NY UNIV, 75- Concurrent Pos: Arthritis Found fel, 74. Mem: AAAS; Am Soc Cell Biol; NY Acad Sci. Res: Biology of cytoplasmic microtubules; biochemistry of mitosis, cell senescence. Mailing Add: Dept of Cell Biol Sch of Med NY Univ 550 1st Ave New York NY 10016

LOCKWOOD, GEORGE WESLEY, astronomy, see 12th edition

LOCKWOOD, GRANT JOHN, b Byram, Conn, Oct 28, 31; m 56; c 4. EXPERIMENTAL ATOMIC PHYSICS, ELECTRON PHYSICS. Educ: Univ Conn, BA, 54, MS, 59, PhD(physics), 63. Prof Exp: Res asst physics, Univ Conn, 60-63; STAFF MEM, SANDIA LABS, 63- Mem: Am Phys Soc. Res: Electronic, atomic and molecular interactions to include ion-atom, ion-molecule, atom-atom and atom-molecule; interaction with surface and solids of ion beams. Mailing Add: Sandia Labs Div 5232 Albuquerque NM 87115

LOCKWOOD, JOHN ALEXANDER, b Easton, Pa, July 12, 19; m 42; c 3. PHYSICS. Educ: Dartmouth Col, AB, 41, Lafayette Col, MS, 43; Yale Univ, PhD(physics), 48. Prof Exp: From asst to instr physics, Lafayette Col, 41-44; tech supvr, Tenn Eastman Corp, 44-45; asst physics, Yale Univ, 45-46, asst instr, 46-47, asst, 47-48; from asst prof to assoc prof, 48-58, PROF PHYSICS, UNIV NH, 58-, ASSOC DIR RES, 74- Mem: AAAS; Am Phys Soc; Am Asn Physics Teachers. Res: Development of linear electron accelerators; cosmic ray; nuclear physics. Mailing Add: Dept of Physics Univ of NH Durham NH 03824

LOCKWOOD, JOHN LEBARON b Ann Arbor, Mich, May 28, 24; m 59; c 2. PLANT PATHOLOGY. Educ: Mich State Col, BA, 48, MS, 50; Univ Wis, PhD(plant path), 53. Prof Exp: Asst prof bot & plant path, Ohio Agr Exp Sta, 53-55; from asst prof to assoc prof, 55-67, PROF BOT & PLANT PATH, MICH STATE UNIV, 67- Concurrent Pos: NSF sr fel, Cambridge Univ, 70-71. Mem: Am Phytopath Soc; Sigma Xi; Am Soc Microbiol. Res: Ecology of root-infecting fungi; soybean diseases. Mailing Add: Dept of Bot & Plant Path Mich State Univ East Lansing MI 48824

LOCKWOOD, JOHN PAUL, b Bridgeport, Conn, Oct 26, 39; m 63; c 2. GEOLOGY. Educ: Univ Calif, Riverside, AB, 61; Princeton Univ, PhD(geol), 66. Prof Exp: GEOLOGIST, US GEOL SURV, 66- Concurrent Pos: Partic, Nat Acad Sci-Nat Res Coun Sci Exchange Prog with USSR, res at Geol Inst Acad Sci, Moscow, 66. Mem: Geol Soc Am; Mineral Soc Am. Res: Petrology, mineralogy and structural features of serpentinites; origin of eclogites and ultramafic xenoliths; general geology of the Sierra Nevada Mountains; circum-Pacific distribution of granite rocks; Caribbean geology; volcanic hazards; eruptive history and structure of Mauna Loa volcano, Hawaii. Mailing Add: US Geol Surv Hawaiian Volcano Observ Hawaii National Park HI 96718

LOCKWOOD, KARL LEE, b Shamokin, Pa, Aug 12, 29; m 57; c 2. ORGANIC CHEMISTRY. Educ: Muhlenberg Col, BS, 51; Cornell Univ, PhD(org chem), 55. Prof Exp: Asst prof chem, Western Md Col, 55-59; from asst prof to assoc prof, 59-70, PROF CHEM, LEBANON VALLEY COL, 70- Concurrent Pos: Vis prof, Univ Colo, 71-72. Mem: AAAS; Am Chem Soc; Nat Sci Teachers Asn. Res: Chemistry of alpha-pinene and some of its cyclobutane and cyclohexane model compounds. Mailing Add: Dept of Chem Lebanon Valley Col Annville PA 17003

LOCKWOOD, LINDA GAIL, b New York, NY, May 25, 36. ENVIRONMENTAL BIOLOGY, SCIENCE EDUCATION. Educ: Columbia Univ, BS, 60, MA, 61 & 65, PhD(bot), 69. Prof Exp: Asst prof bot & ecol, Teachers Col, Columbia Univ, 69-73; ASSOC PROF ENVIRON SCI, UNIV MASS, AMHERST, 73- Concurrent Pos: Jessie Smith Noyes Found grant environ sci educ, Teachers Col, Columbia Univ, 71-73; prof plant & soil sci & Sch Educ, Univ Mass, Amherst, 73-; co-dir, US Off Educ grant, 74-75; Univ Mass fac res grant, 74-75 & Water Resources Res Ctr grant, 74-75. Mem: Sigma Xi; AAAS; Bot Soc Am; Nat Asn Biol Teachers; Audubon Soc. Res: Influence of photoperiod and exogenous nitrogen-containing compounds on the reproductive cycles of the liverwort Cephalozia media Lindb; experimental morphology and physiological ecology; environmental biology, especially physiological ecology, aquatic systems; environmental science education, especially teacher training, history and philosophy of science. Mailing Add: Dept of Environ Sci Marshall Hall Univ of Mass Amherst MA 01002

LOCKWOOD, ROBERT GREENING b Faribault, Minn, Jan 12, 28; m 53; c 2. ORGANIC POLYMER CHEMISTRY, INDUSTRIAL ORGANIC CHEMISTRY. Educ: Carleton Col, BA, 49; Univ Minn, PhD(org chem), 53. Prof Exp: Lab instr inorg & org chem, Univ Minn, 49-53; res chemist, New Prod Develop Lab, Chem Div, Gen Elec Co, 53-54; sr chemist, 54-65, RES SPECIALIST, MINN MINING & MFG CO, 65- Mem: Am Chem Soc. Res: Organic synthesis; carboxylic acids and derivatives; condensation polymers; manufacture of alkylated aromatic hydrocarbons and polycarboxylic acids. Mailing Add: 2 Hingham Circle St Paul MN 55118

LOCKWOOD, WILLIAM GROVER, b Long Beach, Calif, May 25, 33; m 52. CULTURAL ANTHROPOLOGY. Educ: Fresno State Col, BA, 55; Univ Calif, Berkeley, PhD(anthrop), 70. Prof Exp: Asst prof, 69-74, ASSOC PROF ANTHROP, UNIV MICH, 74- Mem: Am Anthrop Asn; Am Ethnol Soc. Res: Peasant society and culture; European ethnology, especially of the Balkans; ethnic boundaries and interethnic relations; immigrant community in North American society. Mailing Add: Dept of Anthrop 221 Angell Hall Univ of Mich Ann Arbor MI 48104

LOCKWOOD, WILLIAM HOWARD b Chattanooga, Tenn, Mar 10, 05; m 29; c 3. CHEMISTRY. Educ: Univ Tenn, BS, 26, MS, 27; Johns Hopkins Univ, PhD(phys chem), 33. Prof Exp: Anal chemist, Jackson Lab, E I du Pont de Nemours & Co, 33-34, res chemist, 34-51, sr supvr patent serv, 51-57, div head patents & tech info, Elastomer Chem Dept, 57-70; RETIRED. Concurrent Pos: Consult, Alfred I du Pont Inst, Wilmington, Del, 71- Mem: AAAS; Am Ord Asn; Am Chem Soc; Soc Mil Eng; fel Am Inst Chem. Res: Catalytic oxidation; synthetic detergents; textile auxiliaries; acetylene; isocyanates; patents. Mailing Add: 28 Westover Circle Wilmington DE 19807

LOCKWOOD, WILLIAM RUTLEDGE, b Memphis, Tenn, Apr 10, 29; m; c 5. INTERNAL MEDICINE, EXPERIMENTAL PATHOLOGY. Educ: Univ Miss, BA, 49, MA, 50; Univ Tenn, Memphis, MD, 57. Prof Exp: Intern, Charity Hosp La, New Orleans, 57-58; resident med, 59-61, from instr to asst prof, 62-70, ASST PROF MICROBIOL & PATH, MED CTR, UNIV MISS, 66-, ASSOC PROF MED, 70- Concurrent Pos: USPHS fel, Med Ctr, Univ Miss, 61-64, grant, 64-67; vis instr, Washington Univ, 64, attend physician, Univ Miss Hosp, 64-; asst dean res & assoc chief staff res, Vet Admin Ctr, 69-73. Mem: Infectious Dis Soc Am; fel Am Col Chest Physicians; Am Soc Trop Med & Hyg; Am Soc Microbiol; fel Am Col Physicians. Res: Pathogenesis of acute inflammation; pharmacology of antimicrobial agents; electron microscopy. Mailing Add: Dept of Med Univ Miss Med Ctr Jackson MS 39216

LOCOCK, ROBERT A, b Toronto, Ont, Aug 14, 35; m 61. PHARMACEUTICAL CHEMISTRY, PHARMACOGNOSY. Educ: Univ Toronto, BSc, 59, MSc, 61; Ohio State Univ, PhD(pharm), 65. Prof Exp: Lectr pharmaceut chem, Univ BC, 61-62; asst pharm, Ohio State Univ, 64-65; asst prof, 65-70, ASSOC PROF PHARM, UNIV ALTA, 70-, ASSOC PROF PHARMACEUT SCI, 74- Mem: AAAS; Am Chem Soc; Am Soc Pharmacog. Res: Chemistry of natural products; phytochemistry; chemotaxonomy; alkaloids and terpenoids. Mailing Add: Fac of Pharm & Pharmaceut Sci Univ of Alta Edmonton AB Can

LOCONTI, JOSEPH DANIEL, chemistry, see 12th edition

LODATO, MICHAEL W, b Rochester, NY, June 17, 32; m 59; c 4. OPERATIONS RESEARCH. Educ: Colgate Univ, AB, 54; Univ Rochester, MS, 59; Rutgers Univ, PhD(math), 62. Prof Exp: Scientist, LFE Monterey Lab, 62-63; mem tech staff, Appl Math Dept, Mitre Corp, Mass, 63-65, head opers anal sub dept, 65-66; sr exec adv, Douglas Aircraft Corp, 66-67, mgr, Info Technol Dept, McDonnell Douglas Corp, 67-68; pres, Macro Systs Assocs, Inc, 68-70; prin bus planner, Xerox Data Systs, 70-71; V PRES INDUST SYSTS, INFORMATICS, INC, 71- Mem: Opers Res Soc Am. Res: Topology; planning, scheduling and resource allocation; orbital mechanics; production and inventory control. Mailing Add: 32038 Watergate Ct Westlake Village CA 91361

LODEN, HAROLD DICKSON, genetics, plant breeding, see 12th edition

LODER, EDWIN ROBERT, b Irvington, NJ, Feb 24, 25; m 45; c 4. ANALYTICAL CHEMISTRY. Educ: Syracuse Univ, BA, 52; Mass Inst Technol, PhD, 55. Prof Exp: Asst chem, Mass Inst Technol, 52-53, asst org microanal, 53-55; chemist, Eastman Kodak Co, 55-59, chief anal chemist, Maumee Chem, 59-62, dir res serv, 62-65, sect mgr & tech assoc, Gen Aniline & Film Co, NY, 65-66; from dep dir res to dir res, Du Bois Chem Div, W R Grace & Co, 66-70, dir res & v pres, Du Bois Chem Div, Chemed Corp, 70-72, sr v pres corp affairs, 72-73, exec v pres, 73-74, GROUP EXEC V PRES, DU BOIS CHEM DIV, CHEMED CORP, 74- Concurrent Pos: Instr, Univ Toledo, 61-62. Mem: Fel AAAS; fel Am Inst Chem; Am Chem Soc; Soc Photog Sci & Eng; Am Soc Qual Control. Res: Electrochemistry; spectroscopy; research management; statistics. Mailing Add: Res Dept Du Bois Chem Div Chemed Corp Du Bois Tower Cincinnati OH 45202

LODGE, ARTHUR SCOTT, b Liverpool, Eng, Nov 20, 22; m 45; c 3. PHYSICS. Educ: Oxford Univ, BA, 45, MA, 48, DPhil(physics), 49. Prof Exp: Jr sci officer pile design, Atomic Energy, Montreal Anglo-Can Proj, 45-46 & rheology, Brit Rayon Res Asn, 49-60; lectr math, Inst Sci & Technol, Univ Manchester, 61-63, sr lectr rheology, 63-68; PROF RHEOLOGY, UNIV WIS-MADISON, 68-, CHMN DEPT, 69- Concurrent Pos: Vis prof chem eng, Univ Wis-Madison, 65-66. Honors & Awards: Bingham Medal, Soc Rheology, 71. Mem: Soc Rheology; Brit Soc Rheology; fel Brit Inst Physics & Phys Soc. Res: Rheological properties of concentrated polymer solutions; molecular theories of their constitutive equations and stress/optical properties; experimental methods. Mailing Add: Rheology Res Ctr Univ of Wis-Madison Madison WI 53706

LODGE, JAMES PIATT, JR, b Decatur, Ill, Feb 4, 26; m 48; c 5. ATMOSPHERIC CHEMISTRY. Educ: Univ Ill, BS, 47; Univ Rochester, PhD(chem), 51. Prof Exp: Asst prof chem, Keuka Col, 50-52; chemist, Cloud Physics Lab, Univ Chicago, 52-55; chief chem res & develop sect, Robert A Taft Sanit Eng Ctr, USPHS, Columbus, Ohio, 55-61; prog scientist, Nat Ctr Atmospheric Res, 61-74; CONSULT ATMOSPHERIC CHEM, 74- Concurrent Pos: Consult, Cook Res Labs, 51-53 & Air & Indust Hyg Lab, Calif State Dept Pub Health; affil prof, La State Univ, 66-69; mem, State Air Pollution Variance Bd, Colo, 66-70; chmn, State Air Pollution Control Comn, Colo, 70- Honors & Awards: Distinguished Serv Award, Div Environ Chem, Am Chem Soc, 73; Frank A Chambers Award, Air Pollution Control Asn, 74. Mem: Fel AAAS; Am Chem Soc; Am Geophys Union; Am Meteorol Soc; Air Pollution Control Asn. Res: Air pollution and atmospheric chemistry; microchemical analysis; cloud physics; atmospheric electricity. Mailing Add: 385 Broadway Boulder CO 80303

LODGE, JAMES ROBERT, b Downey, Iowa, July 1, 25; m 47; c 2. REPRODUCTIVE PHYSIOLOGY. Educ: Iowa State Univ, BS, 52, MS, 54; Mich State Univ, PhD(dairy), 57. Prof Exp: Asst dairy, Mich State Univ, 54-57; res assoc dairy sci, 57-60, from asst prof to assoc prof, 60-69, PROF PHYSIOL, UNIV ILL, URBANA, 69- Concurrent Pos: Res fel, Nat Inst Child Health & Human Develop, 69-70. Mem: AAAS; Am Physiol Soc; Soc Study Reproduction; Am Soc Animal Sci; Am Dairy Sci Asn. Res: Physiology of reproduction and endocrinology. Mailing Add: Dept of Dairy Sci Univ of Ill Urbana IL 61801

LODGE, JOHN I, b Almonte, Ont, Oct 25, 15; m 46; c 1. PHYSICS. Educ: Queen's Univ (Ont), BA, 43, MA, 47; Univ Va, PhD, 51. Prof Exp: Instr physics, Queen's (Ont), 43-48; from asst prof to prof physics, Goucher Col, 51-64, chmn dept, 58-64; assoc prof physics, 64-66, chmn dept, 66-75, assoc dean sci, 73-75, PROF PHYSICS, TRENT UNIV, 66- Mem: Am Phys Soc; Am Asn Physics Teachers; Can Asn Physicists. Res: Nuclear physics. Mailing Add: Dept of Physics Trent Univ Peterborough ON Can

LODHI, MOHAMMED ARFIN KHAN, b Agra, India, Sept 17, 33; m 65; c 2. NUCLEAR PHYSICS. Educ: Univ Karachi, BSc, 52, MSc, 56; Univ London, DIC, 60, PhD(theoret physics), 63. Prof Exp: Lectr math, S M Col, Karachi, 52-59; from asst prof to assoc prof, 63-74, PROF PHYSICS, TEX TECH UNIV, 74- Concurrent Pos: Vis prof, Univ Frankfurt & Darmstadt Tech Univ, 70-71. Mem: Brit Inst Physics. Res: Structure of light nuclei from nuclear reactions and and scattering with special reference to shell and cluster models; nuclear structure and theories with particular emphasis on configuration interaction, nucleon-nucleon correlation and clustering effects of nucleons within the nucleus; nuclear systematics based on velocity-dependent potential; stability properties of super-heavy elements. Mailing Add: Dept of Physics Tex Tech Univ Lubbock TX 79409

LODISH, HARVEY FRANKLIN, b Cleveland, Ohio, Nov 16, 41; m 63; c 3. BIOCHEMISTRY, MICROBIOLOGY. Educ: Kenyon Col, AB, 62; Rockefeller Univ, PhD(biol), 66. Prof Exp: Am Cancer Soc fel biol, Med Res Coun Lab Molecular Biol, Eng, 66-68; asst prof, 68-71, ASSOC PROF BIOL, MASS INST TECHNOL, 71- Concurrent Pos: Res career develop award, Nat Inst Gen Med Sci, 71-75; mem panel develop biol, NSF, 72-; chmn, Gorder Conf on Animal Cells, 76. Mem: AAAS; Am Soc Microbiol; Am Chem Soc; Am Soc Biol Chemists. Res: Replication of ribonucleic acid viruses; mechanism and regulation of protein biosynthesis; synthesis of hemoglobin; differentiation of the slime mold dictyostelium discoideum; biosynthesis of cellular and viral membrane proteins. Mailing Add: Dept of Biol Mass Inst of Technol Cambridge MA 02139

LODMELL, DONALD LOUIS, b Polson, Mont, Aug 27, 39; m 63. VIROLOGY. Educ: Northwestern Univ, BA, 61; Univ Mont, MS, 63, PhD(microbiol), 67. Prof Exp: Scientist virol, Rocky Mountain Lab, NIH, 67-71; res assoc, Lab Oral Med, NIH, 71-72; SR SCIENTIST VIROL, ROCKY MOUNTAIN LAB, NIH, 72- Mem: Am Soc Microbiol; Am Asn Immunologists. Res: Immunological mechanisms of host defense against acute persistent and latent viral infections. Mailing Add: Rocky Mountain Lab Hamilton MT 59840

LODOEN, GARY ARTHUR, b Camp Rucker, Ala, May 3, 43; m 69. POLYMER CHEMISTRY. Educ: Univ NDak, BS, 65; Cornell Univ, PhD(org chem), 69. Prof Exp: Fel, Univ Iowa, 69-70; res chemist, 70-73, sr res chemist, 73, res & develop supvr, 73-75, PROCESS SUPVR, TEXTILE FIBERS DEPT, E I DU PONT DE NEMOURS & CO, INC, 75- Mem: Am Chem Soc. Res: Spandex chemistry and structure; polyester glycol synthesis and properties; development of new and novel raw materials for spandex yarns. Mailing Add: 1101 Crofton Ave Waynesboro VA 22980

LODWICK, GWILYM SAVAGE, b Mystic, Iowa, Aug 30, 17; m 47; c 4. RADIOLOGY, BIOENGINEERING. Educ: Univ Iowa, BA, 42, MD, 43. Prof Exp: Clin asst prof radiol, Univ Iowa, 52-55, assoc prof, Col Med, 55-56; actg dean sch med, 59, assoc dean, 59-64, PROF RADIOL & CHMN DEPT, SCH MED, UNIV MO-COLUMBIA, 56- Concurrent Pos: Fel, Armed Forces Inst Path, 51; Nat Inst Gen Med Sci spec fel, 67-68; chief radiol serv, Vet Admin Hosp, Iowa City, 52-55; consult, Ellis Fischel State Cancer Hosp, 59-; mem radiol training comt, Nat Inst Gen Med Sci, 66-70; consult, Jet Propulsion Lab, Calif Inst Technol, 69-73, mem, Comt Radiol, Nat Acad Sci; mem, Comt Radiol, Div Med Sci, Nat Res Coun, 70-75; dir, Mid-Am Bone Diag Ctr & Registry; vis prof, Keio Univ Sch Med, Tokyo, 74- Honors & Awards: Bronze Medal Award, Roentgen Ray Soc, 51; Sigma Xi Res Award, Univ Mo-Columbia, 72; Gold Medal, XIII Int Conf Radiol, Madrid, 73. Mem: Hon mem Portuguese Soc Radiol & Nuclear Med; AAAS; AMA; Radiol Soc NAm (3rd vpres, 74-75); fel Am Col Radiol. Res: Diagnostic radiology; diagnosis and prognosis of bone disease, computer-aided medical diagnosis; automated image analysis and pattern recognition; information systems. Mailing Add: Dept of Radiol Univ of Mo Med Ctr Columbia MO 65201

LOE, ROBERT WAYNE, b Larwill, Ind, Dec 2, 35; m 64; c 1. PLANT PHYSIOLOGY. Educ: Wabash Col, AB, 57; Columbia Univ, PhD(plant physiol), 64. Prof Exp: Instr gen biol & bot, Rutgers Univ, 61-64; NSF fel, Int Rice Res Inst, Los Banos, Phillipines, 65; sr physiologist, Honduras Div, Res Dept, 66-74, SR PLANT PHYSIOLOGIST, STANDARD FRUIT CO, 74- Mem: Am Soc Plant Physiologists; Am Chem Soc; Am Inst Biol Sci. Res: Inhibition of nicotine by analogs of nicotinic acid in tobacco roots; absorption and translocation of radioactive iron in rice; physiological studies on bananas and pineapple. Mailing Add: Standard Fruit Co PO Box 1689 Gulfport MS 39501

LOE WILLIAM CARROL, b Prescott, Ark, June 29, 31; m 55; c 3. REPRODUCTIVE PHYSIOLOGY. Educ: Univ Ark, Fayetteville, BSA, 53, MS, 59; La State Univ, Baton Rouge, PhD(physiol of reprod), 70. Prof Exp: Instr high schs, Ark, 55-57 & 60-62; instr animal sci, Ark A&M Col, 62-66; from asst prof to assoc prof animal sci, 66-75, PROF & HEAD DEPT AGR, SOUTHERN STATE COL, ARK, 75- Mem: Am Dairy Sci Asn. Res: Physiology of reproduction of cattle; dairy production. Mailing Add: Dept of Agr Southern State Col Box 1418 Magnolia AR 71753

LOEB, ARTHUR LEE, b Amsterdam, Neth, July 13, 23; nat US; m 56. CHEMICAL PHYSICS, APPLIED MATHEMATICS. Educ: Univ Pa, BSCh, 43; Harvard Univ, AM, 45, PhD(chem physics), 49. Prof Exp: Mem staff, Bur Study Coun, Harvard Univ, 45-49; actg head dept chem, Barlaeus Gym, Neth, 46; mem staff, Div Indust Coop & Lincoln Lab, Mass Inst Technol, 49-54; mem guest res staff, Univ Utrecht, 54-55; mem staff, Div Indust Coop & Lincoln Lab, Mass Inst Technol, 55-58, lectr, 56-58; from asst prof to assoc prof elec eng, Mass Inst Technol, 58-63; staff scientist, Ledgemont Lab, Kennecott Copper Corp, 63-73; SR LECTR & HON ASSOC VISUAL & ENVIRON STUDIES, HARVARD UNIV, 70- Concurrent Pos: Consult, Mass Inst Technol, 49-, Godfrey Lowell Cabot, Inc, 58 & IBM Corp, NY, 59. Mem: Am Phys Soc; Soc Indust & Appl Math; Am Crystallog Asn; NY Acad Sci; fel Am Inst Chem. Res: Mathematical crystallography; educational technology; design science; communication of two-dimensional and three-dimensional concepts and patterns. Mailing Add: Dept of Visual & Environ Studies Harvard Univ Cambridge MA 02138

LOEB, FELIX FAUST, JR, b Chicago, Ill, Jan 31, 30; m 55; c 2. PSYCHIATRY, PSYCHOANALYSIS. Educ: Univ Chicago, BA, 51; Harvard Univ, MD, 55; Am Bd Psychiat & Neurol, dipl & cert psychiat, 69. Prof Exp: Rotating intern, Res & Educ Hosps, Univ Ill, 55-56; resident psychiat, Western Psychiat Inst & Clin, Univ Pittsburgh, 56-58 & 60-61, instr psychiat, Sch Med, 61-64, asst prof, 64-68, mem fac, Pittsburgh Psychoanal Inst, 67-72, Inst, clin assoc prof psychiat, Sch Med, 68-72, actg dir psychiat educ, Sch Med, 69-72; MEM FAC, SAN DIEGO PSYCHOANAL FOUND, 73- Mem: AAAS; Am Psychiat Asn; Am Psychoanal Asn; AMA. Res: Linguistics-kinesics. Mailing Add: 1250 Prospect St La Jolla CA 92037

LOEB, GEORGE IRWIN, biophysical chemistry, see 12th edition

LOEB, HAROLD GORDON, biochemistry, see 12th edition

LOEB, LAWRENCE ARTHUR, b Poughkeepsie, NY, Dec 25, 36; m 58; c 3. CANCER, BIOCHEMISTRY. Educ: City Col New York, BS, 57; NY Univ, MD, 61; Univ Calif, Berkeley, PhD(biochem), 67. Prof Exp: Intern, Med Ctr, Stanford Univ, 61-62; res assoc biochem, Nat Cancer Inst, 62-64; res assoc zool, Univ Calif, Berkeley, 64-67; asst mem, 67-69, ASSOC MEM BIOCHEM, INST CANCER RES, 69- Concurrent Pos: Res grants, Am Cancer Soc, 67-69, Stanley C Dordick Found, 67, NIH & NSF, 69-75; assoc prof, Dept Path, Sch Med & mem biol & molecular biol grad groups, Univ Pa, 67- Mem: Am Asn Cancer Res; Fedn Am Socs Exp Biol; Am Soc Cell Biol. Res: The control of DNA replication in synchronous and malignant animal cells; fidelity of DNA synthesis; leukemia and zinc metalloenzymes; lymphocyte transformation. Mailing Add: Inst for Cancer Res 7701 Burholme Ave Philadelphia PA 19111

LOEB, LEOPOLD, b Franklin, La, July 22, 23; m 48; c 2. PHYSICAL CHEMISTRY. Educ: La State Univ, BS, 43; Tulane Univ, MS, 48. Prof Exp: Chemist, Southern Regional Res Lab, USDA, 48-55; chemist, Major Appliance Labs, Gen Elec Co, 56-64, mgr chem res, 64-71, MGR ADVAN TECHNOL PROGS, GEN ELEC CO, 71- Mem: Am Chem Soc; Am Oil Chem Soc; Am Asn Textile Chemists & Colorists. Res: Textile detergent chemistry. Mailing Add: Major Appliance Labs Gen Elec Co Appliance Park Louisville KY 40225

LOEB, MARCIA JOAN, b New York, NY, Mar 26, 33; m 53; c 2. INVERTEBRATE PHYSIOLOGY. Educ: Brooklyn Col, BA, 53; Cornell Univ, MS, 57; Univ Md, PhD(physiol), 70. Prof Exp: Nat Res Coun res assoc physiol & endocrinol of coelenterate develop, Naval Res Lab, 70-72; instr biol & physiol, Northern Va Community Col, 73; res assoc marine biol, Marine Sci Lab, Univ Col N Wales, 74; PROF LECTR PHYSIOL, AM UNIV, 75- Mem: Am Soc Zoologists; Atlantic Estuarine Res Soc; Estuarine Res Fedn. Res: Elucidation of the environmental, physiological and endocrine control of strobilation in the Chesapeake Bay sea nettle, Chrysaora quinquecirrha; associated physiological phenomena in C quinquecirrha; physiology of settlement in some marine bryzoan larvae. Mailing Add: 6920 Fairfax Rd Bethesda MD 20014

LOEB, MARILYN ROSENTHAL, b New York, NY, Feb 26, 30; m 49; c 3. BIOCHEMISTRY. Educ: Barnard Col, Columbia Univ, BA, 51; Bryn Mawr Col, MA, 55; Univ Pa, PhD(biochem), 58. Prof Exp: Res assoc biochem, Univ Pa, 58-59; res assoc, Med Col Pa, 65-68; res assoc biochem, Inst Cancer Res, 68-75; ASST RES PROF MICROBIOL, MED SCH, GEORGE WASHINGTON UNIV, 75- Mem: AAAS; Am Soc Microbiol. Res: Biochemistry of T4 interaction with cell envelope of Escherichia coli. Mailing Add: Dept of Microbiol George Washington Univ Sch of Med Washington DC 20037

LOEB, MELVIN LESTER, b New York, NY, Jan 20, 43; m 67; c 2. ORGANIC CHEMISTRY. Educ: City Col New York, BS, 64; Mass Inst Technol, SM, 67, PhD(chem), 69. Prof Exp: RES CHEMIST & GROUP LEADER, KRAFTCO CORP, 69- Mem: Am Chem Soc. Res: Amine oxide chemistry; dimer acid chemistry; homogeneous catalytic oxidations. Mailing Add: Kraftco Corp 801 Waukegan Rd Glenview IL 60025

LOEB, PETER ALBERT, b Berkeley, Calif, July 3, 37; m 58; c 3. MATHEMATICS. Educ: Harvey Mudd Col, BS, 59, Princeton Univ, MA, 61; Stanford Univ, PhD(math), 65. Prof Exp: Asst prof math, Univ Calif, Los Angeles, 64-68; from asst prof to assoc prof, 68-75, PROF MATH, UNIV ILL, URBANA, 75- Concurrent Pos: Grant, Ctr Advan Studies, 71. Mem: Am Math Soc. Res: Topology; potential theory; non-standard analysis. Mailing Add: Dept of Math Univ of Ill Urbana IL 61801

LOEB, ROBERT FREDERICK, internal medicine, deceased

LOEB, VIRGIL, JR, b St Louis, Mo, Sept 21, 21; m 50; c 4. ONCOLOGY, HEMATOLOGY. Educ: Washington Univ, MD, 44. Prof Exp: From instr to asst prof med, 51-56, asst prof path, 55-69, asst prof clin med, 56-68, ASSOC PROF CLIN MED, SCH MED, WASHINGTON UNIV, 68- Concurrent Pos: Nat Cancer Inst trainee, Sch Med, Washington Univ, 49-50, Damon Runyan res fel hemat, 50-52; dir cent diag labs, Barnes Hosp, St Louis, 52-68, attend physician, NIH grant prin investr, Southeastern Cancer Study Group, 56-; chmn, Cancer Clin Invest Res Comt, Nat Cancer Inst, 66-69, consult, 66-, mem, Polycythemia Vera Study Group & Diag Res Adv Group; mem, Clin Fel Comt, Am Cancer Soc; mem, Oncol Rev Group Vet Admin. Mem: Fel Am Col Physicians; Am Asn Cancer Res; Am Soc Clin Oncol; Am Soc Hemat; Int Soc Hemat. Res: Medical oncology. Mailing Add: 4989 Barnes Hosp Plaza St Louis MO 63110

LOEB, WALTER FRED, veterinary pathology, see 12th edition

LOEBBAKA, DAVID S, b Gary, Ind, Aug 18, 39; m 59; c 3. HIGH ENERGY PHYSICS. Educ: Calif Inst Technol, BS, 61; Univ Md, PhD(physics), 67. Prof Exp: Assoc res scientist, Univ Notre Dame, 66-68; asst prof high energy physics, Vanderbilt Univ, 68-74; ASSOC PROF PHYSICS, UNIV TENN, 74- Mem: Am Phys Soc; Am Asn Physics Teachers. Mailing Add: Dept of Phys Sci Univ of Tenn Martin TN 38237

LOEBECK, MAUDE ELVA, bacteriology, see 12th edition

LOEBEL, ARNOLD BERNARD, chemistry, see 12th edition

LOEBENSTEIN, WILLIAM VAILLE, b Providence, RI, Aug 9, 14; m 49; c 4. PHYSICAL CHEMISTRY, DENTAL RESEARCH. Educ: Brown Univ, ScB, 35, ScM, 36, PhD(phys chem), 40. Prof Exp: Lab glassblower, Eastern Sci Co, 38-40; res & control chemist, Corning Glass Works, RI, 41; res assoc phys chem, 46-51, CHEMIST, NAT BUR STANDARDS, 51- Concurrent Pos: Mem comt com stand, Commodity Stand Div, Off Tech Servs. Honors & Awards: Meritorious Serv Medal, US Dept Com, 58. Mem: AAAS; Am Chem Soc; Int Asn Dent Res. Res: Catalysis; glass technology; physical and chemical absorption of gases on solid surfaces; kinetics of adsorption from solutions; simplified techniques for the improvement in precision of surface area determinations from adsorption measurements; ortho-para conversion of liquid hydrogen; surface chemistry of teeth and dental materials. Mailing Add: Nat Bur of Stand Washington DC 20234

LOEBER, ADOLPH PAUL b Detroit, Mich, Feb 1, 20; m 45; c 4. PHYSICS. Educ: Wayne State Univ, BS, 41, MA, 49; Univ Chicago, cert, 43; Mich State Univ, PhD(physics), 54. Prof Exp: Technician, Metall Dept, Chrysler Corp, 41-42, engr, Truck Dept, 45-47; instr physics, Wayne State Univ, 47-50; asst, Mich State Univ, 50-54; proj engr, Chrysler Corp, 54-55, from asst mgr to mgr physics res, 55-61, mgr phys optics res, Missile Div, 61-64; assoc prof physics, Eastern Mich Univ, 64-66; mgr electrooptics, Missile Div, Chrysler Corp, 66-68; assoc prof, 68-71, PROF PHYSICS, EASTERN MICH UNIV, 71- Concurrent Pos: Consult, Missile Div, Chrysler Corp, 64-66. Mem: Acoust Soc Am; Indust Math Soc (secy, 57-58); Am Asn Physics Teachers; Optical Soc Am. Res: Ultrasonics; physical optics; polarized light. Mailing Add: Dept of Physics Eastern Mich Univ Ypsilanti MI 48297

LOEBL ERNEST MOSHE, b Vienna, Austria, July 30, 23; nat US; m 50; c 2. CHEMICAL PHYSICS, PHYSICAL CHEMISTRY. Educ: Hebrew Univ, MSc, 46; Columbia Univ, PhD(chem), 52. Prof Exp: Res chemist, Olamith Cement Co, 47; asst chemist, Columbia Univ, 48-50; instr, Rutgers Univ, 50-51; from instr to assoc prof, 52-63, PROF PHYS CHEM, POLYTECH INST NEW YORK, 63-, HEAD DIV, 65- Concurrent Pos: NSF fel, 63-64; lectr, Esso Res; ed phys chem series, Acad Press. Mem: AAAS; Am Chem Soc; Am Phys Soc. Res: Theoretical chemistry; quantum theory; polyelectrolytes; solid state; catalysis. Mailing Add: Dept of Chem Polytech Inst of NY 333 Jay St Brooklyn NY 11201

LOEBLICH, ALFRED RICHARD JR, b Birmingham, Ala, Aug 15, 14; m 39; c 4. MICROPALEONTOLOGY, PALYNOLOGY. Educ: Univ Okla, BS, 37, MS, 38; Univ Chicago, PhD(paleont), 41. Prof Exp: Instr geol, Tulane Univ, 40-42; assoc cur invert paleont & paleobot, US Nat Mus, 46-57; from sr res paleontologist to res assoc paleontologist, 57-61; sr res assoc, Chevron Res Co, Standard Oil Co Calif, 61-68, SR RES ASSOC, CHEVRON OIL FIELD RES CO, 68- & ADJ PROF GEOL, UNIV CALIF, LOS ANGELES, 72- Honors & Awards: Elected Corresp Mem, Societe Geologique Belgique, 74. Mem: AAAS; Soc Syst Zool; Soc Protozool; Int Phycol Soc; Int Asn Plant Taxon. Res: Stratigraphy; micropaleontology; recent and fossil foraminifera and phytoplankton; lower paleozoic phytoplankton systematics, paleoecology and biostratigraphy. Mailing Add: Dept of Geol Univ of Calif Los Angeles CA 90024

LOEBLICH, HELEN NINA (TAPPAN), b Norman, Okla, Oct 12, 17; m 39; c 4. MICROPALEONTOLOGY, PALEOECOLOGY. Educ: Univ Okla, BS, 37, MS, 39; Univ Chicago, PhD(geol), 42. Prof Exp: Asst geol, Univ geol, Univ Okla, 37-39; instr, Tulane Univ, 42-43; geologist US Geol Surv, 43-45 & 47; Guggenheim fel, 54; hon res assoc paleont, Smithsonian Inst, 54-57; lectr geol, Univ Calif, Los Angeles, 58-65; assoc res geologist, US Geol Surv, 61-63; sr lectr geol, 65-66, v chmn dept, 73-75, PROF GEOL, UNIV CALIF, LOS ANGELES, 66- Mem: Soc Econ Paleontologists & Mineralogists; Am Micros Soc; fel Geol Soc Am; Paleont Soc; Soc Protozool. Res: Micropaleontology; living and fossil foraminiferans, tintinnids, thecamoebians and organic-walled, siliceous and calcareous phytoplankton; morphology, taxonomy, ecology, primary productivity and food chains, evolution and extinctions. Mailing Add: Dept of Geol Univ of Calif Los Angeles CA 90024

LOEBLICH, KAREN ELIZABETH, b Ft Sill, Okla, Oct 10, 44; m 75. ANIMAL BEHAVIOR, ENTOMOLOGY. Educ: Univ Calif, Los Angeles, AB, 66, MA, 67; Univ Calif, Davis, PhD(zool), 73. Prof Exp: Res assoc entomol, Univ Calif, Davis, 71-72; res assoc, Univ Calif, Riverside, 72-73; lectr entomol & zool, San Francisco State Univ, 73-75; RES SCIENTIST ENTOMOL, AGR DIV, UPJOHN CO, 75- Mem: Asn Study Animal Behav; Entomol Soc Am; Am Inst Biol Sci; Soc Study Evolution; AAAS. Res: Behavior and evolution of Diptera; Drosophilidae of Hawaii; insect grooming behavior; integrated pest management, especially of cotton. Mailing Add: Insecticides & Fungicides 9730-50-1 Upjohn Co Kalamazoo MI 49001

LOEBNER, EGON EZRIEL, b Plzen, Czech, Feb 24, 24; nat US; m 50; c 3. SOLID STATE PHYSICS, RESEARCH ADMINISTRATION. Educ: Univ Buffalo, BA, 50, PhD(physics), 55. Prof Exp: Sr scientist, Sylvania Elec Prod, Inc, Gen Tel & Electronics Corp, 52-55; mem tech staff, RCA Labs, 55-61; mgr optoelectronics, HP Assocs, Hewlett-Packard Co, Calif, 61-65, head spec projs dept, Solid State Lab, 65-73, res adv, 73-74; COUNR SCI & TECHNOL AFFAIRS, US EMBASSY, MOSCOW, 74- Concurrent Pos: Mem, State Comn Radiation Protection, NJ, 59-62 & IMD Electronic Mat Comt, 66-72; lectr, Stanford Univ, 68-74; dir, Compusec Corp, 71-73; interdisciplinary lectr, Univ Calif, Santa Cruz, 72-73. Honors & Awards: Radio Corp Am Award, 59. Mem: Am Phys Soc; Inst Elec & Electronics Engrs; Optical Soc Am. Res: Solid state optoelectronic phenomena, materials, devices; image sensing, display and processing; medical, environmental instrumentation; neurophysiological networks; sensory perception, cognitive data processing; inventing, discovering, interdisciplinary transfer methodologies; computerized scientific research administration. Mailing Add: Moscow Embassy Dept of State Washington DC 20520

LOEFER, JOHN B b Forest Junction, Wis, June 14, 08; m 34; c 2. BIOLOGY. Educ: Lawrence Col, AB, 29, MS, 31; NY Univ, PhD(protozool), 33. Prof Exp: Asst zool, NY Univ, 33-35; assoc prof biol & chem, Berea Col, 35-43; instr biol, Brooklyn Col, 46; sr res biologist & head dept exp biol, Southwest Found Res & Ed, 46-53; COORDR BIOL SCI, OFF NAVAL RES, 53- Concurrent Pos: Prof, Trinity Univ, 50-53; res zoologist, Univ Calif, Los Angeles, 53-; Hancock res scholar, Univ Southern California, 53; res fel, Calif Inst Technol, 54-62; partic, 4th Int Cancer Cong, St Louis; 2nd Int Cong, Biochem, Paris & Int Physiol Cong, Montreal; mem 3rd Int Cong Microbiol, New York & 7th Int Cong, Stockholm, partic, 6th Int Cong, Rome, 8th Int Cong Cell Biol, Leiden & 10th, Paris & 15th Int Cong Zool, London. Mem: AAAS; Soc Exp Biol & Med; Am Physiol Soc; Am Soc Zool; Soc Protozool. Res: Morphology, culture and nutrition of flagellates and ciliates; population studies; pH and growth; symbiosis; acclimatization and pattern formation; malarial and helminth incidence and control; chlorophyll inhibition and action of antibiotics on protozoa and fungi; bovine fetal fluid composition; tumor growth and host resistance phenomena; serology of Tetrahymena; scientific administration. Mailing Add: 1233 Sagemont Place Altadena CA 91001

LOEFFEL, FRANK ALBERT, plant breeding, see 12th edition

LOEFFLER, ALBERT L, JR, b Mineola, NY, Oct 22, 27; m 57; c 3. MAGNETOHYDRODYNAMICS. Educ: Va Polytech Inst, BS, 49; Iowa State Univ, PhD(chem eng), 53. Prof Exp: Res engr, NASA, 54-59; res engr, 59-60, group leader, 60-74, STAFF SCIENTIST, RES DEPT, GRUMMAN AEROSPACE CORP, 74- Mem: Am Phys Soc. Res: Plasma physics; boundary layers; turbulence; heat transfer. Mailing Add: Res Dept Grumman Aerospace Corp Bethpage NY 11714

LOEFFLER, CLARENCE A, developmental biology, see 12th edition

LOEFFLER, ERWIN STANLEY, b Denver, Colo, Aug 3, 13; m 43; c 2. AGRICULTURAL CHEMISTRY. Educ: Univ Calif, AB, 38, PhD(agr chem), 43. Prof Exp: From jr chemist to assoc chemist, US Bur Ships, Mare Island, 42-45; technologist, Shell Oil Co, 45-48 & Shell Chem Corp div, 48-53; sr technologist, Agr Chem Div, Julius Hyman & Co, 53-55; sr technologist, Shell Chem Corp, 55-59, mgr agr chem dept, Tech Serv Lab, 59-65, mgr agr & pharmaceut develop dept, Prod Develop Ctr, 65-67, mgr agr & pharmaceut formulation dept, Agr Res Div, Shell Develop Co, 67-68, MGR FORMULATION RES, BIOL SCI RES CTR, SHELL CHEM CORP, SHELL DEVELOP CO, 68- Mem: AAAS; Am Chem Soc; Entom Soc Am. Res: Insect toxicology; test techniques for contact and residual type insecticides; formulation of insecticides and pharmaceuticals; use of organic sulfur compounds as blowfly repellants; surface activity of clays used in formulation and manufacture of pesticides; preparation of emulsion, dusts and wettable powders for insecticide use; development of controlled pesticide and anthelmentic drug release by means of polymeric matrix. Mailing Add: Biol Sci Res Ctr Shell Develop Co PO Box 4248 Modesto CA 95352

LOEFFLER, FRANK JOSEPH, b Ballston Spa, NY, Sept 5, 28; m 51; c 4. PHYSICS. Educ: Cornell Univ, BS, 51, PhD(physics), 57. Prof Exp: Mem staff, Princeton Univ, 57-58; from asst prof to assoc prof, 58-67, PROF PHYSICS, PURDUE UNIV, 67- Concurrent Pos: Vis prof, Univ Hamburg, 63-64 & Univ Heidelberg, 71; Mem bd trustees & chmn high energy physics comt, Argonne Univ Asn, 74- Mem: Am Phys Soc. Res: Elementary particle physics; experimental study of elementary particle interactions at high energy using electronic detection systems; investigation of alternate energy sources. Mailing Add: Dept of Physics Purdue Univ Lafayette IN 47907

LOEFFLER, HAROLD JULIUS, b Grand Junction, Colo, Nov 27, 09; m 36. FOOD SCIENCE, SCIENCE WRITING. Educ: Univ Calif, BS, 33; Stanford Univ, PhD(chem), 38. Prof Exp: Operator, Shell Chem Co, Calif, 33-35; jr res assoc salmon processing, Stanford Univ, 38-39; res fel citrus processing, Glass Container Asn, Lab Fruit & Veg Chem, USDA, 39-41, assoc chemist frozen foods, Western Regional Res Lab, Bur Agr & Indust Chem, 41-46; mgr com freezing fruits & veg, Glacier Packing Co, 46-48, pres & gen mgr processing develop, surv & sales, 48-73; RETIRED. Concurrent Pos: Mgr & partner, Sanger Frozen Foods Co, 51-71; guest lectr, Calif State Univ, Fresno, 73-75; sci columnist, La Jolla Light, 75- Mem: Am Chem Soc; Inst Food Technologists. Res: Heat and freezing processing of fruits and vegetables; processing methods and chemical changes. Mailing Add: 8234 Caminito Maritimo La Jolla CA 92037

LOEFFLER, JOSEF ERNST, b Wels, Austria, Oct 5, 19; nat US; m 54; c 2. AGRICULTURAL BIOCHEMISTRY. Educ: Univ Vienna, PhD(chem), 50. Prof Exp: Instr chem, Chem Inst, Univ Vienna, 49-50, asst prof agr chem, 50-57; res assoc, Princeton Univ, 57-59; BIOCHEMIST, BIOL SCI RES CTR, SHELL DEVELOP CO, 59- Concurrent Pos: Fulbright grant, 53; res assoc, Univ Wis, 53-54; res fel, Dept Plant Biol, Carnegie Inst, 54-55; consult, 50-53. Mem: Am Chem Soc. Res: Growth regulators; pesticides; plant biochemistry. Mailing Add: 322 N Santa Cruz Ave Modesto CA 95351

LOEFFLER, LARRY JAMES, b Beaver Falls, Pa, May 6, 32; m 57; c 2. ORGANIC CHEMISTRY. Educ: Princeton Univ, AB, 54, PhD(org chem), 61. Prof Exp: USPHS fel, Swiss Fed Inst Technol, 61-62; sr res chemist, Merck Sharp & Dohme Res Labs, Pa, 62-69; spec res fel, NIH, 69-71; asst prof, 71-74, ASSOC PROF MED CHEM, SCH PHARM, UNIV NC, CHAPEL HILL, 74- Mem: Am Chem Soc. Res: Medicinal chemistry; design and synthesis of compounds of interest as potential drugs; radioimmunoassay development. Mailing Add: 317 Wesley Dr Chapel Hill NC 27514

LOEFFLER, MARY CONSTANCE, b Pittsburgh, Pa, Oct 7, 22. PHYSICAL CHEMISTRY. Educ: Mt Mercy Col, Pa, BA, 44; Cath Univ Am, MS, 48; Carnegie Inst Technol, PhD(phys chem), 54. Prof Exp: Instr chem & physics, 45-54, asst prof chem, 55-56, ASSOC PROF CHEM, CARLOW COL, 56-, ASST TO PRES PLANNING, 71- Mem: Am Chem Soc; Am Asn Physics Teachers. Res: Biochemistry; chemical education. Mailing Add: Dept of Chem Carlow Col 3333 Fifth Ave Pittsburgh PA 15213

LOEFFLER, ORVILLE HUGO, chemistry, see 12th edition

LOEFFLER, ROBERT J, b Worcester, Mass, Oct 20, 22; m 50; c 4. BOTANY. Educ: Syracuse Univ, BA, 48; Univ Wis, MS, 50, PhD(bot, zool), 54. Prof Exp: From asst prof to prof, 54-73, EMER PROF BOT, CONCORDIA COL, MOORHEAD, MINN, 73- Mem: Bot Soc Am; Am Inst Biol Sci. Res: Pollen analysis of Spiritwood Lake, North Dakota; phytoplankton; plant anatomy and morphology. Mailing Add: 704 Eighth St So Moorhead MN 56560

LOEGERING, DANIEL JOHN, b Minn, Mar 11, 43; m 68; c 2. PHYSIOLOGY. Educ: St John's Univ, Minn, BS, 65; Univ SDak, Vermillion, MA, 67; Univ Western Ont, PhD(physiol), 70. Prof Exp: Instr physiol, Med Col Wis, 69-73; ASST PROF PHYSIOL, ALBANY MED COL, 73- Concurrent Pos: Wis Heart Asn fel, Med Col Wis, 70-72, NIH spec res fel, 72-73. Mem: Assoc mem Am Physiol Soc. Res: Reticuloendothelial system function as related to systemic host defense during circulatory shock and serum enzyme changes during shock and exercise. Mailing Add: Dept of Physiol Albany Med Col Albany NY 12208

LOEGERING, WILLIAM QUERIN, plant pathology, see 12th edition

LOEHLIN, JAMES HERBERT, b Mussoorie, India, May 23, 34; US citizen. PHYSICAL CHEMISTRY. Educ: Col Wooster, BA, 56; Mass Inst Technol, PhD(phys chem), 60. Prof Exp: Instr chem, Swarthmore Col, 60-61; from instr to asst prof, Col Wooster, 61-64; asst prof, Swarthmore Col, 64-66; asst prof, 66-69, chmn dept, 71-74, ASSOC PROF CHEM, WELLESLEY COL, 69- Concurrent Pos: Res assoc, Univ Chicago, 69-70. Mem: AAAS; Am Phys Soc; Am Crystallog Asn. Res: Crystallography; molecular structure and solids; solid state phase behavior. Mailing Add: Dept of Chem Wellesley Col Wellesley MA 02181

LOEHR, THOMAS MICHAEL, b Munich, Ger, Oct 2, 39; US citizen; m 65. CHEMISTRY. Educ: Univ Mich, Ann Arbor, BS, 63; Cornell Univ, PhD(chem), 67. Prof Exp: Asst prof chem, Cornell Univ, 67-68; asst prof chem, 68-74, ASSOC PROF CHEM, ORE GRAD CTR, 74- Concurrent Pos: NIH res grant, Ore Grad Ctr, 71-; vis lectr, Portland State Univ, 71-72; Res grant, NSF, 74- Mem: AAAS; Am Chem Soc; The Chem Soc; NY Acad Sci. Res: Structural inorganic chemistry; infrared and Raman spectroscopy; metal ion complexes; bioinorganic chemistry; molecular and electronic structure of metalloproteins; solid state vibrational spectroscopy; analytical applications of Raman spectroscopy. Mailing Add: Ore Grad Ctr 19600 NW Walker Rd Beaverton OR 97005

LOELIGER, DAVID A, b Scranton, Pa, Mar 1, 39; m 60; c 4. INORGANIC CHEMISTRY. Educ: Col Wooster, BA, 61; Univ Chicago, MS, 62, PhD(chem), 65. Prof Exp: Asst prof chem, Purdue Univ, 64-67; sr res chemist, Eastman Kodak Co, 67-72; ASSOC PROF CHEM, INT CHRISTIAN UNIV, 72-, RES CONSULT, ARCHEOL RES CTR, 75- Concurrent Pos: Missionary, Am Lutheran Church, 72- Mem: Am Chem Soc. Res: Oxidation-reduction and subtitution reactions of transition metal ions and complexes; topics of archeological-chemical interest. Mailing Add: Dept of Chem Int Christian Univ Tokyo Japan

LOENING, KURT L, b Berlin, Ger, Jan 18, 24; nat US; m 45; c 2. PHYSICAL CHEMISTRY, ORGANIC CHEMISTRY. Educ: Ohio State Univ, BS, 44, PhD(chem), 51. Prof Exp: From asst ed to sr assoc ed, Chem Abstracts, 51-63, assoc dir, Nomenclature, 63-64, DIR, NOMENCLATURE, CHEM ABSTRACTS, 64- Mem: AAAS; Am Chem Soc. Res: Acid-catalyzed esterification of organic acids; chemical nomenclature; literature. Mailing Add: 2064 Inchcliff Rd Columbus OH 43221

LOEPPERT, RICHARD HENRY, b Chicago, Ill, Mar 13, 14; m 40; c 1. ORGANIC CHEMISTRY. Educ: Northwestern Univ, BS, 35; Univ Minn, PhD(phys chem), 40. Prof Exp: Asst, Univ Minn, 35-39; res chemist, Richardson Co, Ill, 39-40; from instr to assoc prof, 40-59, PROF CHEM, NC STATE UNIV, 59- Mem: AAAS; Am Chem Soc. Res: Physical organic chemistry. Mailing Add: 1317 Rand Dr Raleigh NC 27608

LOEPPKY, RICHARD N, b Lewiston, Idaho, Aug 2, 37; m 65. PHYSICAL ORGANIC CHEMISTRY. Educ: Univ Idaho, BS, 59; Univ Mich, MS, 61, PhD(chem), 63. Prof Exp: Instr chem, Univ Mich, 63; NIH fel org chem, Univ Ill, 63-64; asst prof, 64-70, ASSOC PROF CHEM, UNIV MO-COLUMBIA, 70- Concurrent Pos: Resident vis, Bell Labs, 71-72. Honors & Awards: Kasimir Fajans Award, Univ Mich, 65. Mem: Am Chem Soc. Res: Mechanistic organic nitrogen and sulfur chemistry; application of Fourier Transform nuclear magnetic resonance to study of chemical dynamics. Mailing Add: Dept of Chem Univ of Mo Columbia MO 65201

LOERCHER, LARS, b Stuttgart, Ger, Feb 12, 31. PLANT PHYSIOLOGY. Educ:

Stuttgart Tech Univ, BS, 53; Univ Tübingen, PhD(plant physiol), 58. Prof Exp: Fel biochem, Texas A&M, 58-60; res fel plant physiol, Smithsonian Inst, 60-64; res investr, plant morphogenesis, Univ Pa, 64-69; ASST PROF BOT, RUTGERS UNIV, 69- Concurrent Pos: Instr, Univ Tübingen, 62-63. MEM: AAAS; Am Soc Plant Physiol. Res: Circadian rhythms in plants; effect of gravity and light on plant morphogenesis; phytochrome physiology; chemical growth regulators. Mailing Add: Dept of Biol Rutgers Univ 315 Penn St Camden NJ 08102

LOESCH, HAROLD CARL, b Tex, Oct 3, 26; m 45; c 4. BIOLOGICAL OCEANOGRAPHY. Educ: Agr & Mech Col, Tex, BS, 51, MS, 54, PhD(biol oceanog), 62. Prof Exp: Asst, Agr & Mech Col, Tex, 50-52; prin marine biologist, State of Ala, 52-58; res scientist, Res Found, Agr & Mech Col, Tex, 59-60; expert fisheries biol, UN Food & Agr Orgn, Guatemala, 60, Honduras, 60-61, Ecuador, 61-66 & San Salvador, 66-68; PROF MARINE SCI & IN OFF SEA GRANTS DEVELOP, LA STATE UNIV, BATON ROUGE, 68- Mem: Am Fisheries Soc; Am Soc Ichthyol & Herpet; Am Soc Limnol & Oceanog; Nat Shellfisheries Asn; fel Int Acad Fishery Sci. Res: Estuarine hydrology and biology; shrimp, lobster and inshore fishes ecology; fisheries dynamics. Mailing Add: Dept of Marine Sci La State Univ Baton Rouge LA 70803

LOESCH, JOSEPH, b Middle Village, NY, May 5, 30; m; c 3. MARINE BIOLOGY. Educ: Univ RI, BS, 65; Univ Conn, MS, 68, PhD, 69. Prof Exp: Res asst bluefish migrations, Marine Lab, Univ Conn, Noank, 65-66, asst proj leader, Conn Rivers Study, 66-69; ASST PROF MARINE SCI, COL WILLIAM & MARY & UNIV VA, 69- Concurrent Pos: Assoc marine scientist, Va Inst Marine Sci. Mem: Am Fisheries Soc; Atlantic Estuarine Res Soc. Res: Marine fisheries; general life history study of Alosa aestivalus and A. pseudoharengus in Connecticut rivers; biometrics and population dynamics of commercially important bivalves. Mailing Add: Dept of Appl Biol Va Inst of Marine Sci Gloucester Point VA 23062

LOESCHER, DOUGLAS H, solid state physics, see 12th edition

LOESCHER, WAYNE HAROLD, b Lima, Ohio, Nov 6, 42; m 67; c 1. PLANT PHYSIOLOGY. Educ: Miami Univ, BA, 64, MS, 66; Iowa State Univ, PhD(plant physiol), 72. Prof Exp: Res assoc plant physiol, Dept Agron, Iowa State Univ, 71-73; plant physiologist, Los Angeles Arboretum, Arcadia, Calif, 73-75; ASST PROF PLANT PHYSIOL & ASST HORTICULTURIST, WASH STATE UNIV, 75- Res: Plant growth and development, including mechanism of action of plant growth regulators; plant tissue culture. Mailing Add: Dept of Hort Wash State Univ Pullman WA 99163

LOESER, CHARLES NATHAN, b Cleveland, Ohio, May 20, 22; m 46. CYTOLOGY, NEUROANATOMY. Educ: Yale Univ, BS, 44; Western Reserve Univ, MD, 47. Prof Exp: Intern, St Francis Hosp, Hartford, Conn, 47-48; from instr to assoc prof anat, Western Reserve Univ, 49-62; Fulbright lectr, Nat Univ Trujillo, 62-63, prof neuroanat, 63-66; prof anat, Fac Med, Univ BC, 66-67; PROF ANAT, SCH MED, UNIV CONN, 67-, PROF DENT MED, HEALTH CTR, 71- Concurrent Pos: Teaching fel anat, Sch Med, Western Reserve Univ, 48-49. Mem: Am Asn Anat. Res: Scanning microspectrophotometry; fluorescence analysis of living cells. Mailing Add: Dept of Anat Univ of Conn Health Ctr Farmington CT 06032

LOESER, EUGENE WILLIAM, b Buffalo, NY, Nov 5, 26; m 55; c 1. MEDICINE. Educ: Univ Buffalo, MD, 52. Prof Exp: Asst neurol, Columbia Univ, 56-57; asst prof, Univ NC, 57-61; asst prof clin neurol, NY Univ, 64-71; CLIN ASSOC PROF NEUROL, RUTGERS MED SCH, 71- Mem: AMA; Asn Res Nerv & Ment Dis; Am Acad Neurol. Res: Medical neurology. Mailing Add: 10 Parrott Mill Rd Chatham NJ 07928

LOETTERLE, GERALD JOHN, b Edgar, Nebr, Sept 15, 06; m 37; c 3. GEOLOGY. Educ: Univ Nebr, AB, 31, MSc, 33; Columbia Univ, PhD(geol), 37. Prof Exp: Geologist, Shell Oil Co, Inc, Tex, 37-45; CONSULT GEOLOGIST, HUDNALL, PIRTLE & LOETTERLE, 45- Mem: Am Asn Petrol Geol; fel Am Geophys Union. Res: Micropaleontology; exploration geology; estimation of petroleum reserves; micropaleontology of the Niobrara formation in Kansas, Nebraska and South Dakota. Mailing Add: Hudnall Pirtle & Loetterle 308 E Third St Tyler TX 75706

LOEV, BERNARD, b Philadelphia, Pa, Feb 26, 28; m 54; c 3. ORGANIC CHEMISTRY, MEDICINAL CHEMISTRY. Educ: Univ Pa, BSc, 49; Columbia Univ, MA, 50, PhD(org chem), 52. Prof Exp: Instr inorg & org chem, Columbia Univ, 49-51; proj leader, Pennsalt Chem Co, 52-58; group leader, Smith Kline & French Labs, Pa, 58-66, sr investr, 66-67, from asst dir to assoc dir chem, 67-75; DIR CHEM RES, USV PHARMACEUT CORP, 75- Concurrent Pos: Mem adv bd, Index Chemicus & Intra-Sci Res Found; mem bd dirs, Int Heterocyclic Cong; ed bd, Jour Heterocyclic Chem, 72- Mem: AAAS; Am Chem Soc; Am Inst Chem; NY Acad Sci. Res: Organic synthesis; organic sulfur compounds; medicinal chemistry; nitrogen and sulfur heterocycles; natural products; central nervous system, cardiovascular, anti-arthritic and anti-ulcer areas. Mailing Add: USV Pharmaceut Corp 1 Scarsdale Rd Tuckahoe NY 10707

LOEVE, MICHEL, b Jan 22, 07; nat US; m 34; c 1. MATHEMATICS, STATISTICS. Educ: Univ Lyons, Actuaire, 36; Univ Paris, Dr es Sc(math), 41. Prof Exp: Reader, Univ London, 46-48; prof math, 48-73, EMER PROF MATH & STATIST, UNIV CALIF, BERKELEY, 73- Mem: Am Math Soc; Inst Math Statist. Res: Theory of probability; measure theory. Mailing Add: Dept of Math & Statist Univ of Calif Berkeley CA 94720

LOEVINGER, ROBERT, St Paul, Minn, Jan 31, 16; m 52; c 3. RADIOLOGICAL PHYSICS, METROLOGY. Educ: Univ Minn, BA, 36; Harvard Univ, MA, 41; Univ Calif, PhD(physics), 48. Prof Exp: Physicist, Radiation Lab, Univ Calif, 42-45, Los Alamos Sci Lab, 45-46, radiation lab, 46-48; asst physicist, Mt Sinai Hosp, 48-56; res assoc radiol, Sch Med, Stanford Univ, 57-60, asst prof, 60-65; mem staff, Div Isotopes, Int Atomic Energy Agency, 65-68; CHIEF DOSIMETRY SECT, CTR RADIATION RES, NAT BUR STANDARDS, 68- Mem: Fel Am Col Radiol; Am Asn Physicists in Med; Radiation Res Soc; Health Physics Soc; Radiol Soc N Am. Res: Radiation dosimetry and standards. Mailing Add: Radiation Physics C 210 Nat Bur of Standards Washington DC 20234

LOEVY, HANNELORE TASCHINI, b Berlin, Ger, Mar 12, 32; m 61. ANATOMY, HISTOLOGY. Educ: Univ Sao Paulo, CD, 52; Univ Ill, MS, 59, PhD(anat), 61. Prof Exp: Asst oral path, Univ Sao Paulo, 53-55; res assoc anat, Stritch Sch Med, Loyola Univ, Ill, 61-62; from instr to asst prof anat, Col Med, Univ Ill, 63-65; asst prof, Med & Dent Schs, Northwestern Univ, 65-68; asst prof pharmacol, Col Pharm, 68-73, ASSOC PROF PEDODONTICS, COL DENT, UNIV ILL MED CTR, 73- Concurrent Pos: NIH fel pharmacol, Col Med, Univ Ill, 62-63. Mem: Am Asn Anat; Int Asn Dent Res; Am Soc Human Genetics. Res: Teratology; dental pulp; histochemistry; pharmacogenetics; cytogenetics. Mailing Add: Col of Dent 801 S Paulina Univ of Ill at the Med Ctr Chicago IL 60612

LOEW, EARL RANDALL, b Allegan, Mich, Nov 17, 07; m 34; c 4. PHYSIOLOGY. Educ: Mich State Col, BS, 29; Wayne Univ, MS, 36; Northwestern Univ, PhD(physiol), 39. Prof Exp: Res bacteriologist, Col Med, Wayne Univ, 29-33, from instr to asst prof physiol, 33-41; sr pharmacologist, Parke, Davis & Co, 41-44; assoc prof pharmacol, Col Med, Univ Ill, 44-48; prof, 48-72, head dept, 48-69, EMER PROF PHYSIOL, SCH MED, BOSTON UNIV, 72- Mem: AAAS; Am Physiol Soc; Am Soc Exp Pharmacol & Therapeut; Soc Exp Biol & Med; Am Acad Arts & Sci. Res: Histamine; antihistamine drugs; autonomic and adrenergic blocking drugs. Mailing Add: 60 Hull St Newtonville MA 02160

LOEW, FREDERIC CHRISTIAN, organic chemistry, see 12th edition

LOEW, GILDA M HARRIS b New York, NY; c 4. THEORETICAL BIOLOGY, BIOPHYSICS. Educ: NY Univ, BA, 51; Columbia Univ, MA, 52; Univ Calif, Berkeley, PhD(chem physics), 57. Prof Exp: Res physicist, Lawrence Radiation Lab, Univ Calif, Berkeley, 57-62 & Lockheed Missiles & Space Co, 62-64; assoc quantum biophys, Biophys Lab, Stanford Univ, 64-66; from asst prof to assoc prof physics, Pomona Col, 66-69; RES BIOPHYSICIST & INSTR BIOPHYS, MED SCH, STANFORD UNIV, 69-, ADJ PROF GENETICS, MED CTR, 74- Concurrent Pos: Grants, NSF, 66-; NASA, 69- & NIH, 74- Mem: Biophys Soc; fel Am Phys Soc; Int Soc Magnetic Resonance. Res: Molecular orbital and crystal field quantum chemical calculations; models for protein active sites; mechanisms and requirements for specific drug action; theoretical studies related to chemical evolution of life. Mailing Add: Dept of Genetics Stanford Univ Med Sch Stanford CA 94305

LOEW, LESLIE MAX, b New York, NY, Sept 2, 47; m 70; c 1. PHYSICAL ORGANIC CHEMISTRY. Educ: City Col New York, BS, 69; Cornell Univ, MS, 72, PhD(chem), 74. Prof Exp: Res assoc chem, Harvard Univ, 73-74; ASST PROF CHEM, STATE UNIV NY BINGHAMTON, 74- Mem: Am Chem Soc; AAAS. Res: Organophosphorus chemistry; biomembranes; theoretical organic chemistry. Mailing Add: Dept of Chem State Univ of NY Binghamton NY 13901

LOEWE, WILLIAM EDWARD, b Chicago, Ill, Apr 22, 32; m 53; c 3. APPLIED PHYSICS. Educ: Univ Chicago, AB, 52; Univ Ill, BS, 53; Ill Inst Technol, MS, 59, PhD(physics), 63. Prof Exp: Reactor physicist, Savannah River Lab, E I du Pont de Nemours & Co, 53-54, Savannah River Plant, 54-57; assoc physicist, IIT Res Inst, 57-59, res physicist, 59-62, res physicist group leader, 62-63, mgr nuclear physics, 63-66; adv scientist, Nerva, Astro-nuclear Lab, Westinghouse Elec Corp, 66-67; SR PHYSICIST, LAWRENCE LIVERMORE LAB, 67- Mem: Am Nuclear Soc; Am Phys Soc; AAAS. Res: Nuclear and atomic physics of ionized media and radiation transport, especially exploiting very high speed computers. Mailing Add: 1072 Xavier Way Livermore CA 94550

LOEWEN, KENNETH LEROY, b Hillsboro, Kans, Mar 6, 27; m 57; c 4. MATHEMATICS. Educ: Tabor Col, AB, 48; Kans State Univ, MS, 50; Pa State Univ, PhD, 61. Prof Exp: Instr math, Kans State, 48-50; instr math & sci, Freeman Jr Col, 54-55; asst prof math, Tabor Col, 56-62, Westmont Col, 62-66 & Univ Okla, 66-71. Mem: Math Asn Am. Res: Mathematical logic; foundations of mathematics. Mailing Add: 1226 Barkley Norman OK 73069

LOEWENFELD, IRENE ELIZABETH, b Munich, Ger, June 2, 21; nat US. PHYSIOLOGY. Educ: Univ Bonn, PhD(zool), 56. Prof Exp: Asst ophthal, Columbia Univ, 58-61, instr, 61-62, res assoc, 62-68; asst prof, 68-71, ASSOC PROF OPHTHAL, SCH MED, WAYNE STATE UNIV, 71- Mem: Am Physiol Soc; Asn Res Vision & Ophthal. Res: Neurophysiology; neuroophthalmology; autonomic nervous system; pupillography; visual physiology. Mailing Add: Kresge Eye Inst Wayne State Univ Detroit MI 48201

LOEWENSON, RUTH BRANDENBURGER, b Zurich, Switz; US citizen; c 2. BIOMETRICS. Educ: Univ Minn, Minneapolis, BA, 59, MS, 61, PhD(biomet), 68. Prof Exp: From instr to asst prof, 65-72, ASSOC PROF NEUROL & BIOMET, SCH MED, UNIV MINN, MINNEAPOLIS, 72- Concurrent Pos: Consult statistician, Vet Admin Hosp, Minneapolis, 71- Mem: Am Statist Asn; Biomet Soc; Am Pub Health Asn; Soc Epidemiol Res. Res: Epidemiology of cerebral vascular disease; clinical studies in neurology. Mailing Add: Dept of Neurol Univ of Minn Health Sci Ctr Minneapolis MN 55455

LOEWENSTEIN, ERNEST VICTOR, b Offenbach am Main, Ger, Sept 3, 31; US citizen; m 61; c 2. OPTICS, SPECTROSCOPY. Educ: Cornell Univ, AB, 53; Johns Hopkins Univ, PhD(physics), 60. Prof Exp: PHYSICIST, OPTICAL PHYSICS LAB, AIR FORCE CAMBRIDGE RES LABS, 62- Mem: Fel Optical Soc Am. Res: Optical properties of far infrared materials; optical properties of the atmosphere; Fourier spectroscopy. Mailing Add: Optical Physics Lab Air Force Cambridge Res Labs Bedford MA 01730

LOEWENSTEIN, HOWARD, b New York, NY, Jan 1, 24; m 58; c 2. FORESTRY. Educ: Colo State Univ, BS, 52; Univ Wis, PhD(soils), 55. Prof Exp: Instr soils, Univ Wis, 55-56; asst prof silvicult, State Univ NY, Col Forestry, Syracuse, 57-58; from asst prof to assoc prof, 58-68, PROF FOREST SOILS, UNIV IDAHO, 68- Mem: Soil Sci Soc Am. Res: Forest soil-site relationships; forest fertilization; problems of tree seedling establishment; soil microbiology. Mailing Add: Col of Forestry Univ of Idaho Moscow ID 83843

LOEWENSTEIN, JOSEPH EDWARD, b Crockett, Tex, Nov 25, 37; m 58; c 2. ENDOCRINOLOGY, INTERNAL MEDICINE. Educ: Univ Tex, Austin, BA, 59; Wash Univ, MD, 63. Prof Exp: Intern internal med, Barnes Hosp, St Louis, 63-64, resident, 67-69; res assoc, Nat Cancer Inst, 64-66, mem staff, 66-67; instr med, Wash Univ, 70; asst prof, 70-73, ASSOC PROF MED, SCH MED, LA STATE UNIV, SHREVEPORT, 73-, CHIEF SECT ENDOCRINOL, 70- Concurrent Pos: Nat Inst Arthritis & Metab Dis fel metab, Wash Univ, 69-70; consult, US Vet Admin Hosp, Shreveport, 70- Mem: Endocrine Soc; fel Am Col Physicians. Res: Physiology of prolactin in humans; kinetics of iodine metabolism in thyroid. Mailing Add: Dept of Med La State Univ Med Ctr Shreveport LA 71130

LOEWENSTEIN, MATTHEW SAMUEL, b New York, NY, Dec 3, 41; m 65; c 2. GASTROENTEROLOGY. Educ: Union Col, BS, 62; Harvard Med Sch, MD, 67. Prof Exp: Chief Enteric Dis Sect, Ctr Dis Control, USPHS, 70-72; sr resident, Harvard Med Unit, Boston City Hosp, 72-73, clin fel med, Harvard Med Sch, 72-73, instr, 73-75, ASST PROF MED, HARVARD MED SCH & CLIN RES ASSOC, MALLORY GASTROENTEROL RES LAB, BOSTON CITY HOSP, 75- Concurrent Pos: Asst vis physician, Boston City Hosp, 75- Mem: Am Fedn Clin Res. Res: Clinical use of tumor markers, particularly carcinoembryonic antigen and alpha-fetoprotein. Mailing Add: Gastroenterol Res Lab 784 Massachusetts Ave Boston MA 02118

LOEWENSTEIN, MORRISON, b Kearney, Nebr, Aug 21, 15; m 39; c 3. DAIRY CHEMISTRY, NUTRITION. Educ: Univ Nebr, BS, 38; Kans State Col, MS, 40; Ohio State Univ, PhD(dairy tech), 54. Prof Exp: Asst dairy, Kans State Col, 38-39;

asst supt, Roberts Dairy Co, 39-40; instr dairy, NMex State Col, 40-41; from asst prof to assoc prof dairy, Okla Agr & Mech Col, 47-55; res dir, Crest Foods Co, Inc. 55-66, chmn bd, Sutton Crest Proteins Ltd, Can, 64-66; PROF DAIRY SCI, UNIV GA, 66- Mem: Am Dairy Sci Asn; Inst Food Technol, AAAS. Res: Development, modification and compositional control of new and improved dairy products and milk protein concentrates. Mailing Add: Dairy Sci Bldg Univ of Ga Athens GA 30602

LOEWENSTEIN, WALTER B b Gensungen, Ger, Dec 23, 26; US citizen; m 59; c 2. NUCLEAR SCIENCE. Educ: Univ Puget Sound, BS, 49; Ohio State Univ, PhD(physics), 54. Prof Exp: From asst physicist to sr physicist reactor physics, Argonne Nat Lab, 54-63, head, Fast Reactor Anal Sect, Reactor Physics Div, 63-66, mgr physics sect, Liquid Metal Fast Breeder Reactor, prog off, 66-68, assoc dir, EBR-II Proj, Argonne Nat Lab, 68-72, actg dir, 72, dir, 72-73; DIR, SAFETY & ANAL DEPT, ELEC POWER RES INST, 73- Concurrent Pos: Tech adv, US del, Int Conf Peaceful Uses Atomic Energy, Geneva, 58; mem staff, UK Atomic Energy Authority, Dounreay, Scotland, 59; mem, Int Atomic Energy Agency Symp, Vienna, 61, Europ-Am adv comt reactor physics, Atomic Energy Comn, 66-70 & adv comt reactor physics, 66- Mem: Fel Am Phys Soc; fel Am Nuclear Soc; Sigma Xi. Res: Fast reactor physics and related technology, including fast reactor design, analysis and planning of fast critical experiments, fast flux irradiation facilities and conceptual studies; reactor safety and physics technology including reactor design and test programs supporting design; conceptual studies and development of validated analytical methods for design and safety evaluation. Mailing Add: 3412 Hillview Palo Alto CA 94303

LOEWENSTEIN, WERNER RANDOLPH, b Spangenberg, Ger, Feb 14, 26; m 52; c 4. NEUROPHYSIOLOGY, BIOPHYSICS. Educ: Univ Chile, BSc(physics) & BSc(biol), 45, PhD(physiol), 50. Prof Exp: From instr to assoc prof physiol, Univ Chile, 49-57; res zoologist, Univ Calif, Los Angeles, 54-55; from asst prof to prof physiol, Col Physicians & Surgeons, Columbia Univ, 57-71; PROF PHYSIOL & BIOPHYS & CHMN DEPT, SCH MED, UNIV MIAMI, 71- Concurrent Pos: Fel neurophysiol, Sch Med & Hosp, Johns Hopkins Univ, 53-54, Kellogg Int fel physiol, 53-55; Block lectr, Univ Chicago, 60; ed, Biochem & Biophys 67-73; ed-in-chief, J Membrane Biol, 69- Mem: AAAS; Biophys Soc; Am Physiol Soc; Harvey Soc; fel NY Acad Sci. Res: Mechanisms of nerve impulse production and energy conversion at sensory nerve endings; neuro-muscular and synaptic transmission in the nervous system; excitation of the nerve cells; biophysics of cellular membranes; intercellular communication. Mailing Add: Dept of Physiol & Biophys Univ of Miami Sch of Med Coral Gables FL 33124

LOEWENTHAL, LOIS ANNE, b Middletown, Conn, Oct 31, 26. ZOOLOGY. Educ: Mt Holyoke Col, AB, 48; Brown Univ, AM, 50, PhD, 54. Prof Exp: Asst biol, Brown Univ, 48-53; res assoc zool, Mt Holyoke Col, 50-51; instr animal genetics, Univ Conn, 54-56; from instr to assoc prof zool, 57-74, ASSOC PROF EXP BIOL, UNIV MICH, ANN ARBOR, 74- Mem: AAAS; Am Soc Zool; Am Asn Anat; NY Acad Sci. Res: Histology and embryology; skin and hair growth. Mailing Add: Dept of Zool Univ of Mich Ann Arbor MI 48104

LOEWUS, FRANK A, b Duluth, Minn, Oct 22, 19; m 47; c 3. BIOCHEMISTRY. Educ: Univ Minn, BSc, 42, MSc, 50, PhD(biochem), 52. Prof Exp: Asst agr biochem, Univ Minn, 47-51; res assoc biochem, Univ Chicago, 52-55; chemist, USDA, 55-64; prof cell & molecular biol, Dept Biol, State Univ NY Buffalo, 64-75; PROF BIOCHEM & CHMN DEPT AGR CHEM, WASH STATE UNIV, 75- Mem: Phytochemical Soc NAm (pres, 75-76); AAAS; Am Chem Soc; Am Soc Biol Chem; Am Soc Plant Physiol. Res: Intermediary metabolism in plants, mechanisms of enzyme action; biochemistry of natural products. Mailing Add: Dept of Agr Chem Wash State Univ Pullman WA 99163

LOEWUS, MARY WALZ, biochemistry, physical chemistry, see 12th edition

LOEWY, ARIEL GIDEON, b Bucharest, Roumania, Mar 12, 25; US citizen; m 51; c 4. PHYSIOLOGY. Educ: McGill Univ, BSc, 45, MSc, 47; Univ Pa, PhD, 51. Prof Exp: Asst instr, Univ Pa, 47-49; NIH fel & univ res fel phys chem, Harvard Univ, 50-52; Nat Res Coun fel, Cambridge Univ, 52-53; from instr to assoc prof, 53-65, PROF BIOL, HAVERFORD COL, 65-, CHMN DEPT, 57- Concurrent Pos: Mem comn & exec comt, Undergrad Ed in Biol Sci, NSF; biol ed, Holt, Rinehart & Winston, Inc. Mem: Am Soc Biol Chem; Am Soc Cell Biol. Res: Photosynthesis; protoplasmic streaming and contract; fibrin formation; structural proteins in cellular physiology. Mailing Add: Dept of Biol Haverford Col Haverford PA 19041

LOEWY, ARTHUR DECOSTA, b Chicago, Ill, Jan 9, 43; m 71. NEUROANATOMY. Educ: Lawrence Univ, BA, 64; Univ Wis-Madison, PhD(anat), 69. Prof Exp: Res assoc & instr neuroanat, Univ Chicago, 69-71; res assoc, 71-74, sr res fel neuroanat, Mayo Grad Sch Med, Univ Minn, 74-75; ASST PROF ANAT & NEUROBIOL, SCH MED, WASHINGTON UNIV, 75- Concurrent Pos: Res assoc neuroanat, Mayo Found, 71-75. Mem: Anat Soc Gt Brit & Ireland; Am Soc Cell Biol; Am Asn Anat; Soc Neurosci. Res: Degenerative and regenerative changes in the central nervous system; spinal cord organization. Mailing Add: Dept of Anat & Neurobiol Sch of Med Washington Univ St Louis MO 63110

LOFBERG, ROBERT TOR, physical chemistry, analytical chemistry, see 12th edition

LOFFELMAN, FRANK FRED, b St. Louis, Mo, Nov 29, 25; m 45; c 1. ORGANIC CHEMISTRY. Educ: Loyola Univ (Ill), BS, 49; Notre Dame Univ, PhD(chem), 54. Prof Exp: Res chemist, 54-65, GROUP LEADER, CHEM RES DIV, AM CYANAMID CO, 65- Mem: Am Chem Soc. Res: Organic synthesis; dyestuffs; optical bleaches; medicinals. Mailing Add: Chem Res Div Am Cyanamid Co Bound Brook NJ 08805

LOFGREEN, GLEN PEHR, b St David, Ariz, Sept 28, 19; m 45; c 7. ANIMAL NUTRITION. Educ: Univ Ariz, BS, 44; Cornell Univ, MS, 46, PhD(animal nutrit), 48. Prof Exp: Asst prof animal husb, Mont State Col, 48; from asst prof to assoc prof, 48-61, PROF ANIMAL HUSB, UNIV CALIF, DAVIS, 61- Concurrent Pos: Grant, Univ Hawaii, 58-59; consult, USDA; mem subcomt beef cattle nutrit, Nat Res Coun. Honors & Awards: Am Feed Mfrs Nutrit Award, 63. Mem: Am Soc Animal Sci (vpres, 60); Am Dairy Sci Asn. Res: Nutrient requirements and feed evaluation on large domestic animals; calcium and phosphorus metabolism. Mailing Add: Dept of Animal Sci Univ of Calif Davis CA 95616

LOFGREN, CLIFFORD SWANSON, b St James, Minn, July 29, 25; m 54; c 3. ENTOMOLOGY. Educ: Gustavus Adolphus Col, BA, 50; Univ Minn, MS, 54; Univ Fla, PhD(entom), 68. Prof Exp: Entomologist, Entom Res Div, Agr Res Serv, USDA, 55-57, Plant Pest Control Div, 57-63 & Insects Affecting Man Res Lab, 63-74; ASST PROF ENTOM & ASST ENTOMOLOGIST INST FOOD AGR SCI, USDA, UNIV FLA, 74- Mem: Entom Soc Am; Am Mosquito Control Asn. Res: Methods of controlling insects of medical importance, particularly insecticides and equipment evaluation; studies on resistance, chemosterilants and biology. Mailing Add: 1321 NW 31st Dr Gainesville FL 32605

LOFGREN, EDWARD JOSEPH, b Chicago, Ill, Jan 18, 14; m 38, 68; c 3. PHYSICS. Educ: Univ Calif, AB, 38, PhD(physics), 46. Prof Exp: Asst, Univ Calif, 38-40, physicist, Lawrence Radiation Lab, 40-44 & 45-46, group leader, Los Alamos Sci Lab, 44-45; asst prof physics, Univ Minn, 46-48; group leader, 48-73, ASSOC DIR, LAWRENCE BERKELEY LAB, UNIV CALIF, 73- Concurrent Pos: With European Orgn Nuclear Res, 59; mem, High Energy Physics Adv panel, 67-70. Mem: AAAS; Am Phys Soc. Res: Elementary particle physics; accelerators for particle and heavy-ion physics and for biomedical applications. Mailing Add: Lawrence Berkeley Lab Berkeley CA 94720

LOFGREN, JAMES R, b West Point, Nebr, May 18, 31; m 62; c 2. PLANT BREEDING, GENETICS. Educ: Univ Nebr, BS, 60; NDak State Univ, MS, 62; Kans State Univ, PhD(plant breeding, genetics), 68. Prof Exp: Asst agron, NDak State Univ, 60-62; res asst, Kans State Univ, 62-67; asst prof, Northwest Exp Sta, Univ Minn, 67-71; AGRONOMIST-PLANT BREEDER, DAHLGREN & CO, INC, 71- Mem: Am Soc Agron; Genetics Soc Can. Res: Breeding and genetics of sunflowers to improve productivity and quality. Mailing Add: Dahlgren & Co Inc 1220 Sunflower St Crookston MN 56716

LOFGREN, KARL ADOLPH, b Killeberg, Sweden, Apr 1, 15; US citizen; m 42; c 2. SURGERY. Educ: Harvard Med Sch, MD, 41; Univ Minn, MS, 47; Am Bd Surg, dipl, 53. Prof Exp: Intern, Univ Minn Hosp, 41-42; resident surg, Mayo Grad Sch Med, Univ Minn, 42-44 & 46-48; resident, Royal Acad Hosp, Univ Uppsala, 49; asst to staff, Mayo Clin, 49-50; from instr to asst prof, Mayo Grad Sch Med, 51-64, ASSOC PROF SURG, MAYO MED SCH, UNIV MINN, 74-; MEM SURG STAFF, MAYO CLIN, 50-, HEAD SECT PERIPHERAL VEIN SURG, MAYO FOUND, 66- Mem: Fel Am Col Surgeons; Int Cardiovasc Soc; AMA; Sigma Xi. Res: Peripheral venous disorders. Mailing Add: 1001 Seventh Ave NE Rochester MN 55901

LOFGREN, NORMAN LOWELL, b Oroville, Calif, Dec 26, 21; m 47. CHEMISTRY. Educ: Univ Calif, BS, 43, PhD(phys chem), 49. Prof Exp: Res chemist, Radiation Lab, Univ Calif, 43-49; from asst prof to assoc prof, 49-59, PROF CHEM, CALIF STATE UNIV, CHICO, 59- Mem: Am Chem Soc. Res: Thermodynamic properties of metal halides; preparation and properties of refractories; solid state galvanic cells. Mailing Add: Dept of Chem Calif State Univ Chico CA 95926

LOFGREN, RUTH, b Huntsville, Utah, Nov 25, 16. BIOLOGY, SCIENCE EDUCATION. Educ: Univ Utah, AB, 39, AM, 40; Univ Mich, PhD(bact), 44. Prof Exp: Asst bact, Univ Mich, 40-45; from instr to asst prof, 45-53; res assoc biol, Found Integrated Educ, 53-56; asst prof biol, 56-70, ASSOC PROF SCI EDUC, BROOKLYN COL, 70- Mem: AAAS; Nat Asn Biol Teachers; Nat Sci Teachers Asn. Res: Ecology. Mailing Add: 3310 Ave H Brooklyn NY 11210

LOFLAND, HUGH B, JR, biochemistry, pathology, deceased

LOFLIN, ZACHARIAH LOWE, b McComb, Miss, Nov 11, 09; m 43; c 2. MATHEMATICS. Educ: La State Univ, BS, 31, MS, 33; Columbia Univ, PhD, 49. Prof Exp: Prof math, Southwest Jr Col, Miss, 33-36; heating engr, Doherty-Stirling, Inc, 36-40; from asst prof to prof math & chmn dept, Univ Southwestern La, 40-69; prof & head dept, La Col, 69-74; MGR LAB DIV, HOWELL CORP, HOUSTON, TEX, 74- Concurrent Pos: Asst, La State Univ, 38-40; chief engr, Miss Ord Plant, 42; mem, Natural Gas Processors Asn, pres, 62. Mem: Am Math Soc; Am Chem Soc; Am Soc Eng Educ; Math Asn Am; Soc Petrol Engrs. Res: Petroleum chemistry; partial differential equations; boundary value problems. Mailing Add: Howell Corp 2040 North Loop W Suite 204 Houston TX 77018

LOFQUIST, GEORGE W, b Brookhaven, Miss, Oct 6, 30; m 55; c 2. MATHEMATICS. Educ: Univ NC, BS, 52, MEd, 59; La State Univ, Baton Rouge, MS, 63, PhD(math), 67. Prof Exp: Instr math, La State Univ, New Orleans, 59-64 & Baton Rouge, 66-67; from asst prof to assoc prof, 67-73, PROF MATH, ECKERD COL, 73- Mem: Am Math Soc; Math Asn Am. Res: Algebra; number theory. Mailing Add: Dept of Math Eckerd Col St Petersburg FL 33733

LOFQUIST, MARVIN JOHN, b Chicago, Ill, Oct 19, 43; m 65; c 2. INORGANIC CHEMISTRY. Educ: Augustana Col, BA, 65; Northwestern Univ, PhD(inorg chem), 70. Prof Exp: Asst prof chem, Camrose Lutheran Col, 69-73; MEM FAC, DEPT CHEM, FERRIS STATE COL, 73- Mem: Am Chem Soc. Res: Kinetics and mechanisms of organometallic transition metal complexes. Mailing Add: Dept of Chem Ferris State Col Big Rapids MI 49307

LOFSTROM, JOHN GUSTAVE, b Mason, Wis, June 4, 27; m 52; c 3. ANALYTICAL CHEMISTRY. Educ: Northwestern Univ, BS, 50; Univ Wis, PhD(chem), 54. Prof Exp: Asst chem, Univ Wis, 50-52; res chemist, 53-66, SR RES CHEMIST, PHOTO PROD DEPT, E I DU PONT DE NEMOURS & CO, INC, 66- Res: Instrumental analyses. Mailing Add: 58 McGuire St Metuchen NJ 08840

LOFT, JOHN T, organic chemistry, see 12th edition

LOFTFIELD, ROBERT BERNER, b Detroit, Mich, Dec 15, 19; m 46; c 10. ORGANIC CHEMISTRY, BIOCHEMISTRY. Educ: Harvard Univ, BS, 41, MA, 42, PhD(org chem), 46. Prof Exp: Asst chem, Harvard Univ, 42-44, res assoc, 44-46; res assoc, Mass Inst Technol, 46-48; res assoc, Mass Gen Hosp, 48-56, assoc biochemist, 56-64; assoc, Harvard Med Sch, 56-60, asst prof org chem, 60-64; chmn dept, 64-71, PROF BIOCHEM, MED SCH, UNIV N MEX, 64- Concurrent Pos: Fel, Brookhaven Nat Lab, 50; Runyon fel, Medinska Nobel Inst, Stockholm, 52-53; Guggenheim fel, Med Res Coun, Cambridge, Eng, 61-62; USPHS sr res fel, Dunn Sch Path, Oxford Univ, 71-72; chief spec assistance div, Off Strategic Serv, 46-47; tutor, Harvard Univ, 48-64; instr, Marine Biol Lab, Woods Hole, 59-62; mem biochem study sect, USPHS, 64-68; mem adv comn pathogenesis of cancer, Am Cancer Soc, 64-67, mem adv comn proteins & nucleic acids, 71-74. Mem: Am Chem Soc; Am Soc Biol Chem; Am Asn Cancer Res; Biophys Soc; Am Pub Health Asn. Res: Radioactive carbon 14 techniques; organic synthesis; organic reaction mechanisms; protein synthesis. Mailing Add: Univ of NMex Med Sch Albuquerque NM 87131

LOFTHUS, ORIN MERWIN, b Orfordville, Wis, July 15, 05; m 36; c 1. ZOOLOGY. Educ: St Olaf Col, AB, 28; Univ Minn, AM, 33, PhD(physiol, zool), 41. Prof Exp: Teacher & prin high sch, SDak, 28-29; teacher, Minn, 29-31; asst zool, Univ Minn, 31-35; from assoc prof to prof biol & chmn div natural sci, Augustana Col, 35-45; assoc prof physiol, Univ SDak, 45-47, prof zool, 47-53; dean col, 53-64, prof, 53-74, EMER PROF BIOL, ST OLAF COL, 74- Concurrent Pos: Head dept, Univ SDak, 48-53. Mem: AAAS; Am Soc Zool. Res: Experimental studies on Rotifers and Stentor Ultra structure of fresh-water planarian. Mailing Add: Dept of Biol St Olaf Col Northfield MN 55057

LOFTIN, HORACE (GREELEY), b Beaufort, NC, Nov 4, 27; m 56; c 3. ECOLOGY. Educ: Duke Univ, BS, 50; Fla State Univ, MS, 56, MA, 60, PhD(biol sci), 65. Prof Exp: Biol ed, Sci Serv, 52-53 & 55-56; biologist, Fla Game & Freshwater Fish Comn, 59-61; lectr biol, Fla State Univ, 61-67, asst prof, 67-71, assoc dir CZ Branch, 61-71;

biologist-in-charge water quality surv, Panama Canal Co, 71-73, marine biologist, 72-74, ENVIRON & ENERGY OFFICER, PANAMA CANAL CO, 74- Concurrent Pos: Dir, Ctr Trop Studies, Fla State Univ, 61-71; comnr, Panamanian Nat Comn Protection Wildlife, 65-72; adv, Dept Marine Sci, Univ Panama, 66-; res assoc biol sci & anthrop, Fla State Univ, 71-; ecol consult, Panama Canal Co, 71- Mem: AAAS; Am Ornith Union; Am Soc Ichthyol & Herpet. Res: General tropical ecology; biology of tropical water quality; Central American freshwater fish distribution; North American migrant birds in the neotropics; conservation in American tropics. Mailing Add: Box 3204 Balboa CZ

LOFTON, WILLIAM MILFORD, JR, b Mendenhall, Miss, Apr 22, 04; m 31, 51. INDUSTRIAL ORGANIC CHEMISTRY. Educ: Miss Col, BA, 25; Univ NC, MA, 26, PhD(chem), 28. Prof Exp: Asst chemist, Univ NC, 25-28; prof, Guilford Col, 28-31; chemist, Texas Co, 31-32; res chemist, Pa Coal Prod Co, 33-36, chief chemist, 36-37; res chemist, Va Smelting Co, 37-38; chief chemist, Pa Coal Prod Co, 38-46; asst dir inst indust res, Univ Louisville, 46-49; dir, Res & Develop Div, Sloss-Sheffield Steel & Iron Co, 49-53; dir chem res, US Pipe & Foundry Co, 53-69, tech consult, 69-72, RETIRED. Mem: Am Chem Soc; fel Am Inst Chemists; Am Inst Chem Engrs. Res: Evaluation of lubricants; emulsification of asphalt and bituminous materials; synthesis of organic compounds; synthetic glues; synthetic and coal chemicals technology. Mailing Add: 4624 Dolly Ridge Rd Birmingham AL 35243

LOFTSGAARDEN, DON OWEN, b Big Timber, Mont, July 7, 39; m 62; c 2. MATHEMATICAL STATISTICS. Educ: Mont State Univ, BS, 61, MS, 63, PhD(math statist), 64. Prof Exp: Res engr, Autonetics Div, NAm Aviation, Inc, 62; statistician, Battelle Mem Inst, 63; instr statist, Mont State Univ, 64-65; asst prof, Western Mich Univ, 65-67; from asst prof to assoc prof, 67-75, PROF MATH, UNIV MONT, 75- Mem: Am Statist Asn; Inst Math Statist; Math Asn Am; Opers Res Soc Am. Res: Statistical inference. Mailing Add: Dept of Math Univ of Mont Missoula MT 59801

LOGAN, ALAN, b Newcastle-on-Tyne, Eng, Sept 20, 37; m 62; c 2. PALEOECOLOGY. Educ: Univ Durham, BSc, 59, PhD(paleont), 62. Prof Exp: Lectr paleont, Univ Leeds, 64-67; asst prof, 67-70, ASSOC PROF GEOL, UNIV NB, ST JOHN, 70- Concurrent Pos: Nat Res Coun fel, McMaster Univ, 62-64; vis fel, Univ Calgary. Mem: Brit Palaeontograph Soc. Res: Paleontology, paleoecology and ecology of Permian, Triassic and recent bivalves and brachiopods. Mailing Add: Dept of Geol Univ of NB Tucker Park Saint John NB Can

LOGAN, BRIAN ANTHONY, b Newcastle-upon-Tyne, Eng, Dec 22, 38; m 69. NUCLEAR PHYSICS. Educ: Univ Birmingham, BSc, 60, PhD(physics), 64. Prof Exp: Res assoc physics, Univ Birmingham, 64-65; lectr, 65-66, asst prof, 66-72, ASSOC PROF PHYSICS, UNIV OTTAWA, 72- Mem: Can Asn Physicists. Res: Nuclear physics; atomic physics investigations with polarized photon beams. Mailing Add: Dept of Physics Univ of Ottawa Ottawa ON Can

LOGAN, CHARLES DONALD, b St John, NB, May 15, 24; m 53; c 3. WOOD CHEMISTRY. Educ: Mt Allison Univ, BSc, 45; McGill Univ, PhD(org chem), 49. Prof Exp: From res chemist to sr res chemist, 49-65, asst dir, 65-74, DIR CHEM RES, ONT PAPER CO, LTD, 74- Mem: Am Chem Soc; Am Pulp & Paper Assoc; Brit Paper & Board Makers Asn; Can Res Mgt Asn; Chem Inst Can. Res: Vanillin and lignin chemistry; ion exchange chemical recovery; pulp and paper by-product utilization. Mailing Add: Ontario Paper Co Thorold ON Can

LOGAN, CHERYL ANN, b Syracuse, NY, Apr 1, 45. ANIMAL BEHAVIOR, NEUROPSYCHOLOGY. Educ: Southern Methodist Univ, BA, 67; Univ Calif, San Diego, PhD, 74. Prof Exp: ASST PROF PSYCHOL, UNIV NC, GREENSBORO, 74- Mem: Animal Behav Soc; Am Psychol Asn. Res: Conducting investigations of behavioral plasticity in invertebrate organisms; focused on interactions between habituation and sensitization in coelenterates; theoretical interests center on ecology and evolution of learning. Mailing Add: Dept of Psychol Univ of NC Greensboro NC 27412

LOGAN, DAVID MACKENZIE, b Toronto, Ont, July 23, 37; m 60; c 3. MOLECULAR BIOLOGY, BIOCHEMISTRY. Educ: Univ Toronto, BA, 60, MA, 63, PhD(med biophys), 65. Prof Exp: Res assoc biochem, NIH, 65-67; Nat Res Coun Can fel, McMaster Univ, 67-68; asst prof, 68-74, ASSOC PROF MOLECULAR BIOL, YORK UNIV, 68- Concurrent Pos: Jane Coffin Childs Mem Fund fel med res, 65-67. Mem: AAAS; Biophys Soc; Can Biochem Soc. Res: Metabolism of metabolically and physically altered RNA, particularly bacterial transfer. Mailing Add: Dept of Biol York Univ Toronto ON Can

LOGAN, GEORGE BRYAN, b Pittsburgh, Pa, Aug 1, 09; m 39; c 2. PEDIATRICS. Educ: Washington & Jefferson Col, BS, 30; Harvard Univ, MD, 34; Univ Minn, MS, 40; Am Bd Pediat, dipl, 41. Prof Exp: From instr to prof pediat, Mayo Grad Sch Med, Univ Minn, 40-73, prof pediat, Mayo Med Sch, 73-75, EMER STAFF, MAYO CLIN, 75- Concurrent Pos: Consult, Sect Pediat, Mayo Clin, 40-68, sr consult, 68-75; chmn sub-bd allergy, Am Bd Pediat, 63-66. Mem: AAAS; NY Acad Sci; AMA; Am Col Allergists; Am Acad Pediat (pres, 67-68). Res: Allergic and liver diseases in children. Mailing Add: 200 First St SW Rochester MN 55902

LOGAN, JAMES EDWARD, b Thorndale, Ont, Jan 14, 20; m 48; c 2. CLINICAL CHEMISTRY. Educ: Univ Western Ontario, BSc, 49, PhD(biochem), 52. Prof Exp: Sr res asst biochem, Univ Western Ontario, 52-54; chemist biol control labs, 54-59, sr biochemist clin labs, 59-73, chief clin chem, 73-75, ACTG DIR, BUR CLIN CHEM & HEMAT, LAB CTR DIS CONTROL, CAN DEPT NAT HEALTH & WELFARE, 75- Mem: Can Biochem Soc; Can Soc Clin Chem; Am Asn Clin Chem. Res: Chemistry of peripheral nervous system; radioisotope tracer studies; quality control and methodology; hemoglobin; evaluation of diagnostic kits and clinical laboratory instruments; radioimmunoassay; reference methods; trace element analyses. Mailing Add: Bur of Clin Chem & Hemat Lab Ctr for Dis Control Tunney's Pasture Ottawa ON Can

LOGAN, JOHN MERLE, b Pittsburgh, Pa, July 7, 34; m 61. STRUCTURAL GEOLOGY, TECTONOPHYSICS. Educ: Mich State Univ, BS, 56; Univ Okla, MS, 62, PhD(geol), 65. Prof Exp: Geologist, Shell Develop Co, 65-67; asst prof, 67-74, ASSOC PROF EXP ROCK DEFORMATION, DEPT GEOL, TEX A&M UNIV, 67- Concurrent Pos: Advan Projs Res Agency, US Dept Defense res grant, 67-; consult, Gen Motors Corp, 67-68. Mem: Assoc Geol Soc Am; assoc Am Geophys Union. Res: Experimental rock deformation as applied to structural geological problems. Mailing Add: Dept of Geol Tex A&M Univ College Station TX 77843

LOGAN, JOSEPH GRANVILLE, JR, b Washington, DC, June 8, 20; m 44; c 2. PHYSICS. Educ: DC Teachers Col, BS, 41; Univ Buffalo, PhD(physics), 55. Prof Exp: Physicist aerodyn, Nat Bur Standards, 43-47; physicist aerodyn propulsion, Cornell Aeronaut Lab, Inc, 47-57; head aerophys lab, Space Technol Labs, Inc, 57-59, mgr propulsion res dept, 59-60; dir aerodyn & propulsion lab, Aerospace Corp, 60-67; spec asst to dir res & develop, Western Div, McDonnell Douglas Astronaut Co, 67-69, mgr vulnerability & hardening develop eng, 69-72, chief engr nuclear weapons effects, Western Div, 72-74; PRES, APPL ENERGY SCI, INC, 74- Mem: Am Phys Soc; Am Inst Aeronaut & Astronaut. Res: New energy systems. Mailing Add: Appl Energy Sci Inc PO Box 36583 Los Angeles CA 90036

LOGAN, KENNETH CALVIN, chemistry, see 12th edition

LOGAN, LOWELL ALVIN, b Langley, Ark, Oct 29, 21; m 44. ECOLOGY, PLANT TAXONOMY. Educ: Henderson State Col, BS, 43; Univ Ark, MS, 47; Univ Mo, PhD(bot), 59. Prof Exp: Instr biol, Ark Polytech Col, 46-49, head dept, 49-60; assoc prof bot, La Polytech Inst, 60-62 & La State Univ, 62-65; prof, Memphis State Univ, 65-67; V PRES ACAD AFFAIRS, SOUTHERN STATE COL, ARK, 67- Mem: AAAS; Ecol Soc Am; Bot Soc Am. Res: Ecology and distribution of American Beech; local floras; ecological factors affecting vegetation in restricted habitats. Mailing Add: 504 Alice St Magnolia AR 71753

LOGAN, RALPH ANDRE, b Cornwall, Ont, Sept 22, 26; nat US; m 50; c 9. SOLID STATE PHYSICS. Educ: McGill Univ, BSc, 47, MSc, 48; Columbia Univ, PhD(physics), 52. Prof Exp: Asst physics, Columbia Univ, 49-52; MEM TECH STAFF, BELL LABS, 52- Mem: Fel Am Phys Soc; sr mem Inst Elec & Electronic Engrs. Res: Semiconductor research. Mailing Add: 179 Mills St Morristown NJ 07960

LOGAN, RICHARD FINK, b Great Barrington, Mass, June 1, 14; m 39; c 2. GEOGRAPHY, ARID LANDS. Educ: Clark Univ, BA, 36, MA, 37; Harvard Univ, MA, 48, PhD(geog), 49. Prof Exp: Instr geog, Clark Univ, 37-43; instr Yale Univ, 43-46; asst prof, Conn Col, 46-47; instr, Harvard Univ, 47-48; from asst prof to assoc prof, 48-61, PROF GEOG UNIV CALIF, LOS ANGELES, 61- Concurrent Pos: Nat Acad Sci-Nat Res Coun fel, SW Africa, 56-57, Soc Sci Res Coun fel, 61-62, NSF fel, 65; vis prof, Univ Khartoum, 64, Hebrew Univ Jerusalem, 70 & Univ Stellenbosch, 73; consult on long-range desert planning, Dept Nature Conserv, SW Africa Admin, 68-; co-ed, Geoforum, 69-, Madoqua, 69- & J Calif Anthrop, 73-; mem comn arid lands, AAAS, 72-73; v pres, Environ Policy Res Corp, 72-73; consult on environ sensitivity, SCalif Edison Co, 74- Mem: Hon mem SW Africa Sci Soc. Res: Deserts, physical conditions, utilization by different cultures, planning for future use; environmental impact studies. Mailing Add: Dept of Geog Univ of Calif Los Angeles CA 90024

LOGAN, ROBERT KALMAN, b New York, NY, Aug 31, 39. PHYSICS, FUTUROLOGY. Educ: Mass Inst Technol, BS, 61, PhD(physics), 65. Prof Exp: Res asst physics, Univ Ill, 65-67; res asst physics, 67-68, asst prof, 68-75, ASSOC PROF PHYSICS, UNIV TORONTO, 75- Res: Strong interactions of elementary particles; futures research into the planning and designing of the future; high energy and elementary particle physics. Mailing Add: Dept Physics Fac Arts & Sci Univ Toronto St George Campus Toronto ON Can

LOGAN, ROWLAND ELIZABETH, b Los Angeles, Calif, Aug 1, 23. PHYSIOLOGY. Educ: Univ Calif, AB, 45; Northwestern Univ, MS, 51, PhD(physiol), 54. Prof Exp: Instr physiol, Sch Med, WVa Univ, 54-55; instr biol, Bard Col, 56-58; ASST PROF BIOL, GETTYSBURG COL, 58- Mem: AAAS. Res: Cell metabolism; arthropod behavior. Mailing Add: Dept of Biol Gettysburg Col Gettysburg PA 17325

LOGAN, TED JOE, b Ft. Wayne, Ind, June 22, 31; m 54; c 2. INDUSTRIAL CHEMISTRY. Educ: Ind Univ, AB, 53; Purdue Univ, MS, 56, PhD(chem), 58. Prof Exp: RES CHEMIST, PROCTER & GAMBLE CO, 58- Mem: Am Chem Soc. Res: Product development & research. Mailing Add: Proctor & Gamble Co Sharon Woods Tech Ctr Cincinnati OH 45241

LOGAN, TERRY JAMES, b Georgetown, Guyana, Feb 6, 43; US citizen; m 73; c 2. SOIL CHEMISTRY. Educ: Calif Polytech State Univ, BS, 66; Ohio State Univ, MS, 69, PhD(soil sci), 71. Prof Exp: Asst prof soil chem, Ohio Agr Res & Develop Ctr, 71-72; ASST PROF SOIL CHEM, OHIO STATE UNIV, 72- Mem: Soil Sci Soc Am; Am Soc Agron; Int Soil Sci Soc; Soil Conserv Soc Am. Res: Non-point sources of pollution; phosphate chemistry of soil and sediments; land disposal of sewage sludge; erosion and sedimentation of agricultural soils. Mailing Add: Dept of Agron Ohio State Univ Columbus OH 43210

LOGEMANN, GEORGE WAHL, applied mathematics, computer science, see 12th edition

LOGEMANN, JERILYN ANN, b Berwyn, Ill, May 21, 42. SPEECH PATHOLOGY. Educ: Northwestern Univ, Chicago, BA, 63, MS, 64, PhD(speech path), 68. Prof Exp: Res assoc, 70-74, ASST PROF NEUROL & OTOLARYNGOL, NORTHWESTERN UNIV, CHICAGO, 74- Concurrent Pos: NIH fel, Northwestern Univ, Chicago, 68-70; consult, Downey Vet Admin Hosp, 73-; assoc attend staff, Northwestern Mem Hosp, 73- Mem: Am Speech & Hearing Asn; Linguistic Soc Am. Res: Speech science; laryngeal physiology; voice disorders; language disorders; language development. Mailing Add: Dept of Neurol Northwestern Univ Chicago IL 60611

LOGERFO, JOHN J, b New York, NY, Feb 12, 18; m 55. ZOOLOGY, MEDICAL TECHNOLOGY. Educ: NY Univ, BA, 42; Columbia Univ, MA, 52, EdD, 61. Prof Exp: Lab supvr med tech, Lenox Hill Hosp, New York, 41-57; chief biochemist, Clin Lab, S Shore Anal Labs, 57-58; instr biol & gen sci, 58-60, from asst prof to assoc prof, 60-69, PROF BIOL, C W POST COL, LI UNIV, 69-, DIR MED TECHNOL, 63-, CHMN DEPT HEALTH SCI & CHMN PREMED COMT, 71- Concurrent Pos: Consult biochemist, St Claires Hosp, New York, 55-58; res assoc, Community Hosp, Glen Cove, NY, 58-64, USPHS res grant, 62-64; prog dir allied health sci traineeship dept, USPHS, 67- Mem: AAAS; NY Acad Sci; Am Soc Med Technol; Am Soc Microbiol; Asn Schs Allied Health Professions. Res: Hematology; parasitology; comparative anatomy, embryology; chordate vertebrate morphology; experimental embryology. Mailing Add: Dept of Health Sci CW Post Col L I Univ Greenvale NY 11548

LOGGINS, PHILLIP EDWARDS, b Yorkville, Tenn, Feb 12, 21; m 42; c 2. ANIMAL NUTRITION. Educ: Okla State Univ, BS, 52, MS, 53. Prof Exp: Instr animal husb, Univ Fla, 53-55, from asst prof to assoc prof, 55-74, PROF ANIMAL HUSB, UNIV FLA, 74-, ANIMAL HUSBANDMAN, AGR EXP STA, 55- Mem: Am Soc Animal Sci. Res: Animal nutrition; parasitic effect on nutritional requirements; feeding requirements of animals during reproduction. Mailing Add: Dept of Animal Sci Univ of Fla Gainesville FL 32601

LOGIC, JOSEPH RICHARD, b Iron Mountain, Mich, Apr 23, 35; m 64. CARDIOVASCULAR PHYSIOLOGY, NUCLEAR MEDICINE. Educ: Marquette Univ, MD, 60, MS, 63, PhD(physiol), 64. Prof Exp: Intern med, C T Miller Hosp, St Paul, Minn, 60-61; instr physiol, Sch Med, Marquette Univ, 61-64 & 65-66; resident med, Mayo Clin, Rochester, Minn, 64-65; asst prof med, Col Med, Univ Ky, 66-69; assoc prof med & physiol, Univ Tenn, Memphis, 69-73; resident nuclear med, 74, ASSOC PROF NUCLEAR MED & MED, MED CTR, UNIV ALA, BIRMINGHAM, 74- Res: Peripheral circulatory failure; adrenergic blockade; electrolyte role in cardiac electrophysiology and contractility; myocardial nuclear medicine. Mailing Add: Div of Nuclear Med Univ of Ala Med Ctr Birmingham AL 35294

LOGIN, ROBERT BERNARD, b Brooklyn, NY, Nov 15, 42; m 71; c 1. ORGANIC CHEMISTRY, POLYMER CHEMISTRY. Educ: Brooklyn Col, BA, 66; Purdue Univ, PhD(org chem), 70. Prof Exp: Chemist paper specialties, Spring House Lab, Rohm and Haas Co, 70-73; sr chemist, 73-74, SECT HEAD FIBER SPECIALTIES, BASF-WYANDOTTE CORP, 74- Mem: Am Chem Soc; Am Asn Textile Chemists & Colorists. Res: Design of new and improved fiber processing auxiliaries and additives such as spin finish components and internal antistats. Mailing Add: BASF Wyandotte Corp Cent Res 1609 Biddle Ave Wyandotte MI 48192

LOGOTHETIS, ANESTIS LEONIDAS, b Thessaloniki, Greece, June 29, 34. POLYMER CHEMISTRY. Educ: Grinnell Col, BA, 55; Mass Inst Technol, PhD(org chem), 58. Prof Exp: Res chemist, Cent Res Dept, 56-66, Elastomers Dept, 66-72, SUPVR DEVELOP, E I DU PONT DE NEMOURS & CO, INC, 72- Mem: Am Chem Soc. Res: Synthetic organic chemistry; development of new products. Mailing Add: Elastomers Dept Louisville Works E I du Pont de Nemours & Co Inc Louisville KY 40201

LOGOTHETIS, JOHN ACHILLES, neurology, see 12th edition

LOGOTHETOPOULOS, J, b Athens, Greece, Mar 12, 18; Can citizen; m 53; c 1. MEDICINE, PHYSIOLOGY. Educ: Nat Univ Athens, MD, 41; Univ Toronto, PhD(physiol), 62. Prof Exp: From asst prof to assoc prof, 59-64, PROF MED RES, BANTING & BEST DEPT MED RES & DEPT PHYSIOL, UNIV TORONTO, 64- Concurrent Pos: Res fel, Postgrad Med Sch, Univ London, 52-56; fel med res, Banting & Best Dept Med Res, Univ Toronto, 56-59. Mem: Am Diabetes Asn; Am Soc Exp Path; Can Physiol Soc. Res: Structure and function of the thyroid and the pituitary gland; experimental diabetes; structure and function of the islets of Langerhans. Mailing Add: Banting & Best Dept Med Res Univ of Toronto Toronto ON Can

LOGSDON, CHARLES ELDON, b Mo, May 8, 21; m 48; c 3. PLANT PATHOLOGY. Educ: Univ Kansas City, AB, 42; Univ Minn, PhD, 54. Prof Exp: Res prof plant path, 53-68, plant pathologist, 53-71, PROF PLANT PATH, UNIV ALASKA, 68-, ASSOC DIR, INST AGR SCI, 71- Mem: AAAS; Am Phytopath Soc; Am Soc Microbiol; Potato Asn Am. Res: Potato and vegetable diseases. Mailing Add: Univ of Alaska Inst of Agr Sci Palmer AK 99645

LOGUE, BRUCE R, b Augusta, Ga, Oct 9, 11; m 38; c 2. MEDICINE. Educ: Emory Univ, BS, 34, MD, 37; Am Bd Internal Med, dipl, 44. Prof Exp: From instr to assoc prof med, Sch Med, 40-57, PROF MED, SCH MED & CHIEF MED SERV, UNIV HOSP, EMORY UNIV, 57- Concurrent Pos: Consult, Grady Mem Hosp; past chmn, Am Bd Cardiovasc Dis. Mem: Am Heart Asn (vpres, 64-65); Am Clin & Climat Asn; Am Fedn Clin Res (past pres); fel Am Col Physicians. Res: Cardiovascular disease. Mailing Add: Emory Univ Sch of Med Atlanta GA 30322

LOGUE, MARSHALL WOFORD, b Danville, Ky, June 4, 42. CHEMISTRY. Educ: Centre Col Ky, AB, 64; Ohio State Univ, PhD(chem), 69. Prof Exp: Res assoc chem, Univ Ill, 69-71; ASST PROF CHEM, UNIV MD, BALTIMORE COUNTY, 71- Mem: AAAS; Am Chem Soc; The Chem Soc. Res: Synthetic organic chemistry; bio-organic chemistry; pyrimidines; nucleosides. Mailing Add: Dept of Chem Univ of Md Baltimore County Baltimore MD 21228

LOGULLO, FRANCIS MARK, b Wilmington, Del, Dec 19, 39; m 62; c 3. ORGANIC POLYMER CHEMISTRY. Educ: Univ Del, BS, 61; Case Inst Technol, PhD(org chem), 65. Prof Exp: Res chemist, 65-70, SR RES CHEMIST, E I DU PONT DE NEMOURS & CO, INC, 70- Mem: Am Chem Soc. Res: Polymer chemistry; synthetic fibers; chemistry of arynes. Mailing Add: Fibers Dept Exp Sta 302 E I du Pont de Nemours & Co Wilmington DE 19898

LOH, EUGENE, solid state physics, physical metallurgy, see 12th edition

LOH, HORACE H, b Canton, China, May 28, 36; m 62; c 1. BIOCHEMISTRY, BIOCHEMICAL PHARMACOLOGY. Educ: Nat Taiwan Univ, BS, 58; Univ Iowa, PhD(biochem), 65. Prof Exp: Lectr biochem pharmacol, Univ Calif, San Francisco, 67, asst res pharmacologist, 67-68; assoc prof biochem pharmacol, Wayne State Univ, 68-70; chief, Drug Dependence Res Ctr, Mendocino State Hosp, Talmage, Calif, 71-72; RES SPECIALIST, LANGLEY PORTER NEUROPYSCHIAT INST, 72-; PROF PHARMACOL, SCH MED, UNIV CALIF, SAN FRANCISCO, 72- Concurrent Pos: Fel biochem, Univ Calif, San Francisco, 65-66. Mem: Am Chem Soc. Res: Enzyme regulatory mechanisms; mechanisms of drug tolerance; drug metabolism. Mailing Add: Dept of Pharmacol Univ of Calif Med Ctr San Francisco CA 94122

LOH, HUNG YU, b Soochow, China, July 17, 07; m 30; c 5. PHYSICS. Educ: Soochow Univ, China, BS, 31; Va Polytech Inst, MS, 43; Johns Hopkins Univ, PhD(physics), 46. Prof Exp: Asst physics, Soochow Univ, 31-34, instr, 34-37, lectr, 37-41; instr, Va Polytech Inst, 42-43, asst prof, 46; assoc prof, Soochow Univ, 46-47, prof & dean admin, 47-48; assoc prof, 48-58, PROF PHYSICS, VA POLYTECH INST & STATE UNIV, 58- Mem: Optical Soc Am. Res: Light source; interferometry; thin film; holography. Mailing Add: Dept of Physics Va Polytech Inst & State Univ Blacksburg VA 24061

LOH, JEROME WEI-PING, b China, Mar 26, 21; nat US; m 50; c 3. PATHOLOGY, HEMATOLOGY. Educ: Nat Med Col, Shanghai, China, MD, 46; Cornell Univ, cert, 49; Univ Mich, MPH, 50; Boston Univ, PhD, 54; Am Bd Path, dipl, 56; Royal Col Physicians & Surgeons Can, cert internal med, 57, cert hemat, 64, FRCP. Prof Exp: Instr path, Sch Med, Ind Univ, 54-55; asst clin pathologist, Quincy City Hosp, Mass, 55-56; pathologist in chg clin path, 56-61, CHIEF PATHOLOGIST & DIR LABS, METHODIST HOSP, 61- Concurrent Pos: Res assoc, Sch Med, Boston Univ, 56-57; asst prof, Chicago Med Sch, 60-65, clin assoc prof path, 66-; chief pathologist, Lake County Coroner's Off, 66-; consult, Dr Norman Beatty Mem & Paramore Hosps, Ind. Mem: Fel Am Col Physicians; Am Soc Clin Path; Col Am Path; AMA; Am Pub Health Asn. Res: Clinical pathology; forensic pathology; microbiology. Mailing Add: Dept of Path Chicago Med Sch Chicago IL 60612

LOH, PHILIP CHOO-SENG, b Singapore, Sept 14, 25; nat US; m 55; c 2. VIROLOGY. Educ: Morningside Col, BS, 50; Univ Iowa, MS, 53; Univ Mich, MPH, 54, PhD, 58; Am Bd Microbiol, dipl, 63. Prof Exp: Res assoc, Virus Lab, Univ Mich, 58-61; assoc prof, 61-66, PROF VIROL, UNIV HAWAII, 66- Concurrent Pos: USPH spec res fel, NIH, 67-68; Eleanor Roosevelt int cancer fel, Int Union Against Cancer, Geneva, 75. Mem: Fel AAAS; Am Asn Immunol; Am Soc Microbiol; Tissue Cult Asn; Soc Exp Biol & Med. Res: Biosynthesis and pathobiology of animal viruses at the cellular level. Mailing Add: 2552 Peter St Honolulu HI 96816

LOHER, WERNER J, b Landshut, Ger, June 27, 29; m 61; c 1. ZOOLOGY. Educ: Univ Munich, PhD(zool), 55; Univ London, PhD(entom) & DIC, 59. Prof Exp: Asst prof zoophysiol, Univ Tübingen, 60-65, privat docent, 65-67; assoc prof, 67-70, PROF ENTOM, UNIV CALIF, BERKELEY, 70- Concurrent Pos: Sr res award, Antilocust Res Ctr, Eng, 56-59; vis lectr, Glasgow Univ, 67. Mem: Brit Soc Exp Biol; Ger Zool Soc. Res: Circadian rhythms and sexual behavior in insects. Mailing Add: Dept of Entom Univ of Calif Berkeley CA 94720

LOHMAN, FRED HERMAN, analytical chemistry, see 12th edition

LOHMAN, KENNETH ELMO, b Los Angeles, Calif, Sept 11, 97; m 31. GEOLOGY. Educ: Calif Inst Technol, BS, 29, MS, 31, PhD(paleont & geol), 57. Prof Exp: Chemist, Cert Lab Prod, Inc, Calif, 24-31; geologist, US Geol Surv, 31-67; RES ASSOC, US NAT MUS, SMITHSONIAN INST, 67- Concurrent Pos: Chmn Am Comn Stratig Nomenclature, 59-61. Honors & Awards: Distinguished Serv Award, US Dept Interior, 67. Mem: AAAS; fel Geol Soc Am; hon mem Soc Econ Paleont & Mineral (vpres, 61-62); Am Asn Petrol Geol; fel Royal Micros Soc. Res: Diatoms; paleontology and stratigraphy; stratigraphic nomenclature; paleoecology; petroleum geology; photomicrography. Mailing Add: Smithsonian Inst Nat Mus of Natural History Washington DC 20560

LOHMAN, STANLEY WILLIAM, b Los Angeles, Calif, May 19, 07; m 33; c 3. GEOLOGY. Educ: Calif Inst Technol, BS, 29, MS, 38. Prof Exp: Asst mineral, Calif Inst Technol, 29-30; from jr geologist to prin geologist, Br Ground Water, Water Resources Div, US Geol Surv, 30-74, dist geologist chg ground water invests, Kans, 37-45 & Colo, 45-51, staff geologist, States in Ark-White-Red Basins, 51-56, br area chief, Rocky Mountain area, 56-59, res geologist, 59-62, staff geologist, 62-74; RETIRED. Honors & Awards: Distinguished Serv Award, US Dept Interior, 74. Mem: Fel Geol Soc Am; Am Asn Petrol Geologists; Am Geophys Union. Res: Ground water geology and hydrology. Mailing Add: 2060 S Madison Denver CO 80210

LOHMAN, TIMOTHY GEORGE, b Park Ridge, NJ, Dec 10, 40; m 61; c 4. ANIMAL NUTRITION. Educ: Univ Ill, Urbana, BS, 62, MS, 64, PhD(body compos), 67. Prof Exp: Res assoc whole-body counting, 67-69, ASST PROF BODY COMPOS ANIMALS & MAN, DEPT ANIMAL SCI & PHYS EDUC, UNIV ILL, URBANA, 69- Mem: AAAS; Am Soc Animal Sci. Res: Exercise physiology; animal body composition; atherosclerosis in relation to nutrition and genetics; physical exercise and body composition. Mailing Add: Dept of Animal Sci & Phys Educ 132 Davenport Hall Univ of Ill Urbana IL 61801

LOHMANN, KARL H, b Berlin, Ger, May 30, 24; US citizen; m 61; c 2. ORGANIC CHEMISTRY. Educ: Mass Inst Technol, BS, 50, PhD(org chem), 59. Prof Exp: Mgr textile finishing, Tintorex, Colombia, 50-55; res chemist, E I du Pont de Nemours & Co, Inc, Del, 59-61; res chemist, 61-71, GROUP LEADER, DYES APPLN, TOMS RIVER CHEM CORP, 71-, GROUP LEADER, DYES DISPERSIONS, 74- Mem: Am Chem Soc; Asn Textile Chemists & Colorists. Res: Dye and fiber research and application. Mailing Add: Dyes Res & Develop Lab Toms River Chem Corp Box 71 Toms River NJ 08753

LOHNER, DONALD J, b Brooklyn, NY, Mar 10, 39. ORGANIC CHEMISTRY. Educ: Queens Col, BS, 61; Adelphi Univ, PhD(org chem), 66. Prof Exp: Instr chem, Adelphi Univ, 64-66; res chemist, 66-76, SR RES CHEMIST, E I DU PONT DE NEMOURS & CO, !NC, 76- Mem: Am Chem Soc; Soc Photog Sci & Eng; The Chem Soc. Mailing Add: E I du Pont de Nemours & Co Photo Prod Dept Parlin NJ 08859

LOHR, DELMAR FREDERICK, JR, b Madison Co, Va, Sept 9, 34. POLYMER CHEMISTRY. Educ: Va Polytech Inst, BS, 62; Duke Univ, MA, 63, PhD(chem), 65. Prof Exp: Res org chemist, 65-70, SR RES SCIENTIST, FIRESTONE TIRE & RUBBER CO, 70- Mem: Am Chem Soc. Res: Synthesis and reactions of aromatic heterocycles, particularly those containing both nitrogen and sulfur; polymer synthesis and characterization. Mailing Add: 200 Casterton Ave Akron OH 44303

LOHR, DENNIS EVAN, b Waukegan, Ill, Jan 12, 44. PHYSICAL BIOCHEMISTRY. Educ: Beloit Col, BA, 65; Univ NC, Chapel Hill, PhD(biochem), 69. Prof Exp: Teacher chem, Peace Corps, Kenya, EAfrica, 70-71; RES ASSOC BIOCHEM, ORE STATE UNIV, 72- Res: Enzymatic investigation of the subunit structure of yeast chromatin and physical characterization of the subunits. Mailing Add: Dept of Biochem & Biophys Ore State Univ Corvallis OR 97331

LOHR, JOHN MICHAEL, b Chicago, Ill, June 21, 44; m 67. PLASMA PHYSICS. Educ: Univ Tex, Austin, BS, 66; Univ Wis-Madison, MS, 67, PhD(nuclear physics), 72. Prof Exp: RES ASSOC PLASMA PHYSICS, FUSION RES CTR, UNIV TEX, AUSTIN, 72- Mem: Am Phys Soc; Am Asn Physics Teachers. Res: Nonlinear plasma phenomena; wave coupling; beam-plasma instability. Mailing Add: Fusion Res Ctr Univ of Tex Austin TX 78712

LOHR, LAWRENCE LUTHER, JR, b Charlotte, NC, May 29, 37; m 63; c 1. Educ: Univ NC, BS, 59; Harvard Univ, AM, 62, PhD(chem), 64. Prof Exp: Res assoc chem, Univ Chicago, 63-65; res scientist, Sci Lab, Ford Motor Co, 65-68; assoc prof, 68-73, PROF CHEM, UNIV MICH, ANN ARBOR, 73- Concurrent Pos: Consult, Ford Motor Co, 68-71 & Bell Tel Labs, 69 & 72; Alfred P Sloan res fel, 69-71; vis prof & scholar, Univ Calif, Berkeley, 74-75. Mem: Am Phys Soc; Am Chem Soc. Res: Theories of chemical bonding; interpretation of electronic spectra of molecules and solids with emphasis on properties of transition metal complexes; calculation of photoionization cross-sections. Mailing Add: Dept of Chem 1040 Chem Bldg Univ of Mich Ann Arbor MI 48104

LOHR, LESTER JAY, b Mt Pleasant, Pa, Sept 26, 14; m 41; c 1. ANALYTICAL CHEMISTRY, ORGANIC CHEMISTRY. Educ: Bethany Col, BS, 40; Univ Pa, MS, 66. Prof Exp: Res chemist, Koppers Co, NJ, 40-42; fel, Mellon Inst, 42-45; anal res chemist, Gen Aniline & Film Corp, 46-52; tech assoc, Eastern Lab, E I du Pont de Nemours & Co, 52-63; ASSOC PROF ADVAN ORG & GEN CHEM, RUTGERS UNIV, CAMDEN, 63- Mem: Am Chem Soc. Res: Distillation; gas chromatography; infrared; near and far infrared; ultraviolet; microelementary and group analysis of organic compounds; organic synthesis and characterization. Mailing Add: Rutgers Univ 406 Penn St Camden NJ 08102

LOHR DE IRIZARRY, MILDRED TUCKER, b Brandy, Va, Dec 10, 05; m 40; c 2. GEOGRAPHY. Educ: Longwood Col, BA, 27; Columbia Univ, MA, 33. Prof Exp: Teacher high sch, PR, 28-38; prof hist & polit sci & chmn dept, 38-58, prof soc studies, 58-64, PROF GEOG & CHMN DEPT, INTER-AM UNIV PR, 64- Concurrent Pos: Partic seminar tour, USSR, 64. Mem: Asn Am Geog. Res: Political geography; geography of Puerto Rico; introduction to college geography. Mailing Add: Dept of Geog Inter-Am Univ of PR San German PR 00753

LOHRDING, RONALD KEITH, b Coldwater, Kans, Jan 1, 41; m 62; c 2. STATISTICAL ANALYSIS, RESEARCH ADMINISTRATION. Educ: Southwestern Col, Kans, BA, 63; Kans State Univ, MA, 66, PhD(statist), 68. Prof Exp: Instr math, St John's Col Prep Sch, PR, 63-64; consult statistician, 68-74, GROUP LEADER

STATIST SERV GROUP, LOS ALAMOS SCI LAB, UNIV CALIF, 74- Concurrent Pos: Adj prof, Univ NMex, 69-70. Mem: Am Statist Asn; Inst Math Statist. Res: Hypothesis testing; distribution-free discriminant analysis; experimental designs and cost-risk-benefit analysis of energy developments. Mailing Add: Los Alamos Sci Lab Los Alamos NM 87544

LOHRENGEL, CARL FREDERICK, II, b Kansas City, Mo, Nov 24, 39; m 71. GEOLOGY. Educ: Univ Kansas City, BS, 62; Univ Mo-Columbia, MA, 64; Brigham Young Univ, PhD(geol), 68. Prof Exp: Res assoc, Marine Inst, Univ Ga, 68-69; ASST PROF GEOL, SNOW COL, 69- Mem: Am Asn Petrol Geol; Paleont Soc; Am Asn Stratig Palynologists. Res: Palynology of the Upper Cretaceous of Utah; Upper Cenozoic and modern dinoflagellates of the Georgia coastal plain; Cretaceous stratigraphy of Utah. Mailing Add: Dept of Geol Snow Col Ephraim UT 84627

LOHRMANN, ROLF, b Bissingen-Enz, Ger, Mar 2, 30; m 60. BIO-ORGANIC CHEMISTRY. Educ: Stuttgart Tech Univ, dipl(chem), 58, Dr rer nat(chem), 60. Prof Exp: Proj assoc, Inst Enzyme Res, Univ Wis, 62-65; sr res assoc, 65-74, ASSOC RES PROF, SALK INST BIOL STUDIES, 74- Mem: Ger Chem Soc. Res: Prebiotic chemistry; molecular evolution. Mailing Add: Salk Inst for Biol Studies PO Box 1809 San Diego CA 92112

LOHSE, CARLETON LESLIE, b Minot, NDak, Nov 10, 36; m 68; c 2. ANATOMIC PATHOLOGY. Educ: Iowa State Univ, DVM, 60; Univ Sydney, PhD(anat), 71. Prof Exp: Lectr vet anat, Univ Sydney, 66-68; vet practr, Boise Animal Hosps, Inc, 71-72; lectr, 72-73, ASST PROF ANAT, SCH VET MED, UNIV CALIF, DAVIS, 73- Concurrent Pos: Res consult, Gaines Nutrit Ctr, Gen Foods Corp, 74- Mem: Am Asn Vet Anatomists; Am Vet Med Asn. Res: Gross and microscopic anatomy of the parotid gland and duct in relation to surgery on these structures; muscle growth and pathoanatomic studies of animal organs and tissues. Mailing Add: Dept of Anat Univ of Calif Sch of Vet Med Davis CA 95616

LOHUIS, DELMONT JOHN, b Oostburg, Wis, Jan 24, 14; m 37; c 3. CHEMISTRY. Educ: Carroll Col, Wis, BA, 34; Univ Wis, MS, 36. Prof Exp: Res chemist, 35-51, asst mgr res, 51-56, from assoc dir to dir, 56-60, asst to corp vpres res & develop, 60-61, dir res & develop, Milk Container Div, 61-64, asst to corp vpres res & develop, 64-70, dir res & develop, Consumer & Serv Prod, 70-71, DIR NEW PROD OFF, CORP RES & DEVELOP DEPT, AM CAN CO, 71- Mem: Am Chem Soc; fel Am Inst Chem. Res: Container construction materials; paper based consumer products; specialty chemicals. Mailing Add: Corp Res & Develop Dept Am Can Co American Lane Greenwich CT 06830

LOHWATER, A J, b Rochester, NY, Oct 20, 22; m 49; c 3. MATHEMATICS. Educ: Univ Rochester, PhD(math), 51. Prof Exp: Instr math, Univ Rochester, 43-44 & 46-47; from instr to assoc prof, Univ Mich, 49-59; prof, Rice Univ, 59-65; vprovost, 69-70, PROF MATH, CASE WESTERN RESERVE UNIV, 65-, CHMN DEPT, 68- Concurrent Pos: Fulbright res grant, Finland, 55-56; Guggenheim fel, 55-; vis prof, Univ Helsinki, 56; ed, Russian-Eng Math Dictionary, 59-; consult ed, Addison-Wesley Int Series, 60-; res assoc, Brown Univ, 61-65; exec ed, Math Rev, Am Math Soc, 61-65. Mem: NY Acad Sci; London Math Soc; Finnish Math Soc; Swiss Math Soc. Res: Conformal mapping; meromorphic and harmonic functions. Mailing Add: Dept of Math Case Western Reserve Univ Cleveland OH 44106

LOIRE, NORMAN PAUL, b St Louis, Mo, May 7, 27; m 53; c 2. ORGANIC CHEMISTRY. Educ: Shurtleff Col, BS, 51; NY Univ, PhD(chem), 60. Prof Exp: Chemist, Ciba Pharmaceut Prod Co, 52-54; res chemist, Benger Res Lab, Textile Fibers Dept, E I du Pont de Nemours & Co, 58-62; sr res chemist, Narmco Res & Develop Div, Whittaker Corp, 62-67; sr res chemist, Chemplex Co, 67-71; LAB MGR, SABER LABS, WHEELING, 71- Mem: Am Chem Soc. Res: Process development manufacturing of organic compounds; synthesis of elastomers; aromatic polyamides and heterocyclic monomers and polymers. Mailing Add: 471 Killarney Pass Circle Mundelein IL 60060

LOISELLE, JEAN-MARIE, biochemistry, see 12th edition

LOISELLE, ROLAND, b Montreal, Que, Can, July 30, 28. PLANT BREEDING. Educ: McGill Univ, BSc, 49, MSc, 51; Univ Wis, PhD(plant breeding, genetics), 55. Prof Exp: Cerealist, Cereal Crops Div, Can Dept Agr, 55-59, geneticist, Genetics & Plant Breeding Res Inst, 59-64, res sta, Res Br, 64-70, HEAD RES BR, CENT OFF PLANT GENE RESOURCES CAN, CAN DEPT AGR, 70- Concurrent Pos: Secy, Can Comt Plant Gene Resources, 70-; ed, Can J Plant Sci, 72- Mem: Genetics Soc Can; Can Soc Agron; Agr Inst Can. Res: Development of a national computerized system for the recording, storage and retrieval of informational data on more than 86,000 plant cultivars or genetic stocks maintained in the country; international and national exchange of plant genetic material. Mailing Add: Cent Off Plant Gene Resources Can Dept of Agr Ottawa ON Can

LOIZZI, ROBERT FRANCIS, b Oak Park, Ill, Oct 18, 35; m 60; c 4. PHYSIOLOGY, CELL BIOLOGY. Educ: Loyola Univ, Ill, BS, 57; Marquette Univ, MS, 60; Iowa State Univ, PhD(cell biol), 66. Prof Exp: Instr physiol, Iowa State Univ, 65-66; asst prof, 66-71, ASSOC PROF PHYSIOL, UNIV ILL COL MED, 71- Mem: AAAS; Am Soc Zool; Am Soc Cell Biol; Am Physiol Soc. Res: Cyclic nucleotides and regulation of cell proliferation and lactation in normal and neoplastic mammary gland; fine structure, cytochemistry, and physiology of crustacean hepatopancreas; gill; transepithelial transport. Mailing Add: 135 E View St Lombard IL 60148

LOK, ROGER, b Macao, Oct 19, 43; US citizen; m 70. ORGANIC CHEMISTRY. Educ: Univ Calif, Berkeley, BS, 66; Univ Washington, PhD(org chem), 71. Prof Exp: Fel, Dept Pharmacol, Yale Univ, 71-74; RES CHEMIST, EASTMAN KODAK CO, 74- Mem: Am Chem Soc. Res: Organic synthesis; preparation of dyes; enzyme immobilization; affinity chromatography; immunochemistry. Mailing Add: Kodak Park Eastman Kodak Res Labs Rochester NY 14650

LOKEN, HALVAR YOUNG, bio-organic chemistry, see 12th edition

LOKEN, KEITH I, b Sandstone, Minn, Oct 3, 29; m 57; c 3. VETERINARY MICROBIOLOGY. Educ: Univ Minn, St Paul, BS, 51, DVM, 53, PhD(vet med), 59. Prof Exp: Res fel vet med, 55-58, from instr to assoc prof, 58-71, prof vet microbiol, 71-74, PROF VET BIOL, UNIV MINN, ST PAUL, 74- Concurrent Pos: Fulbright res fel, NZ Dept Agr, Ruakura Agr Res Ctr, 65-66. Mem: Am Vet Med Asn; Wildlife Dis Asn; Am Soc Microbiol. Res: Teaching veterinary microbiology; host-parasite relationships; epidemiology of infectious diseases of animals. Mailing Add: Dept of Vet Biol Col of Vet Med Univ of Minn St Paul MN 55101

LOKEN, MERLE KENNETH, b Hudson, SDak, Jan 21, 24; m 47; c 5. NUCLEAR MEDICINE, BIOPHYSICS. Educ: Augustana Col, BA, 46; Mass Inst Technol, BS, 48, MS, 49; Univ Minn, PhD(biophys), 56, MD, 62; Am Bd Nuclear Med, cert, 72. Prof Exp: Asst physics, Mass Inst Technol, 48-49; asst prof, Augustana Col, 49-51; instr biophys, 53-56, from asst prof to assoc prof, 56-68, PROF RADIOL, UNIV MINN, MINNEAPOLIS, 68-, DIR DIV NUCLEAR MED, 64- Mem: Fel Am Col Radiol; Soc Nuclear Med; Radiol Soc NAm. Res: Clinical uses of radioisotopes; radiation dosimetry and hazards; effects of radiation on biological systems; applications of radioisotopes as tracer elements in metabolic studies of normal and cancer cells. Mailing Add: Dept of Radiol Univ of Minn Hosp Minneapolis MN 55455

LOKEN, OLAV HELGE, physical geography, see 12th edition

LOKEN, STEWART CHRISTIAN, b Montreal, Que, Feb 16, 43; m 70; c 1. EXPERIMENTAL HIGH ENERGY PHYSICS. Educ: McMaster Univ, BSc, 66; Calif Inst Technol, PhD(physics), 71. Prof Exp: Res assoc physics lab nuclear studies, Cornell Univ, 71-74; PHYSICIST, LAWRENCE BERKELEY LAB, UNIV CALIF, 74- Res: Measurement of high energy muon-nucleon scattering and experimental studies of rare muon-induced processes at high energies. Mailing Add: Bldg 50-149 Lawrence Berkeley Lab Univ of Calif Berkeley CA 94720

LOKENSGARD, JERROLD PAUL, b Saskatoon, Sask, July 30, 40; US citizen; m 65; c 2. ORGANIC CHEMISTRY. Educ: Luther Col, Iowa, BA, 62; Univ Wis, Madison, MA, 64, PhD(org chem), 67. Prof Exp: NIH fel, 67; res assoc chem, Iowa State Univ, 67, ASST PROF CHEM, LAWRENCE UNIV, 67- Mem: AAAS; Am Chem Soc. Res: Natural products synthesis, especially insect defensive compounds; diazo ketones; small strained hydrocarbons; novel organic compounds. Mailing Add: Dept of Chem Lawrence Univ Appleton WI 54911

LOKKEN, JOHN ERWIN, b Canwood, Sask, Can, Oct 22, 24; m 47; c 2. GEOPHYSICS. Educ: Univ Western Ont, BSc, 51, MSc, 52; Univ BC, PhD(physics), 56. Prof Exp: Physicist, Pac Naval Lab, Defence Res Bd Can, 55-66 & Saclant Anti Submarine Warfare Res Centre, La Spezia, Italy, 66-69; HEAD ELECTROMAGNETICS SECT, DEFENSE RES BD, DEFENSE RES ESTAB PAC, CAN, 69- Mem: Can Asn Physicists; Can Geophys Union. Res: Underwater acoustics, geomagnetics and data reduction. Mailing Add: Defense Res Estab Pac FMO Victoria BC Can

LOKKEN, ROBERT JOSEPH, organic chemistry, see 12th edition

LOKKEN, STANLEY JEROME, b Fargo, NDak, Sept 22, 31; m 65. PHYSICAL CHEMISTRY. Educ: NDak State Univ, BS, 53; Univ Calif, Berkeley, MS, 54; Iowa State Univ, PhD(phys chem), 62. Prof Exp: Res chemist, Glidden Co, 62-65 & Continental Oil Co, 65-68; ASST PROF CHEM, UNIV WIS-PLATTEVILLE, 68- Mem: Am Chem Soc. Res: Ion exchange theory and techniques; radiochemistry; solution kinetics and mechanisms of reactions; titanium chemistry; phosphate chemistry. Mailing Add: Dept of Chem Univ of Wis Platteville WI 53818

LOLLAR, ROBERT MILLER, b Lebanon, Ohio, May 17, 15; m 41; c 2. LEATHER CHEMISTRY, ENVIRONMENTAL CHEMISTRY. Educ: Univ Cincinnati, ChE, 37, MS, 38, PhD(leather chem), 40. Prof Exp: Develop chemist, Best Foods Corp, Ind, 40-41; assoc prof tanning res & assoc dir, Tanners' Coun Res Lab, 41-58; tech dir, Armour Leather Co Div, Armour & Co, 58-64, dir tech eval, 65-75; TECH DIR, TANNERS' COUN AM, UNIV CINCINNATI, 75- Concurrent Pos: Pres, Lollar & Assocs, Consults, 75- Honors & Awards: Alsop Medal, Am Leather Chemists Asn, 54. Mem: Am Chem Soc; Am Soc Qual Control; Am Leather Chemists Asn (pres, 66-68); Inst Food Technol; fifth mem World Mariculture Soc. Res: Research administration; statistical quality control; collagen chemistry; industrial biochemistry; marine biology. Mailing Add: Tanners' Coun of Am Univ of Cincinnati Location 14 Cincinnati OH 45221

LOLLEY, RICHARD NEWTON, b Blaine, Kans, May 25, 33; m 59; c 3. PHYSIOLOGY, BIOCHEMISTRY. Educ: Univ Kans, BS, 55, PhD(physiol), 61. Prof Exp: Pharmacist, Hawk Pharm, Inc, 55-56; res pharmacologist, 65-71, CHIEF, LAB DEVELOP NEUROL, VET ADMIN HOSP, 71-; ASSOC PROF ANAT, UNIV CALIF, LOS ANGELES, 70- Concurrent Pos: USPHS res fel biochem, Maudsley Hosp, Univ London, 61-62; fel neuropath, McLean Hosp & Harvard Med Sch, 62-65; asst prof anat, Univ Calif, Los Angeles, 66-70. Mem: Int Soc Neurochem; Am Soc Neurochem; Soc Neurosci; Am Asn Anat; Asn Res Vision & Ophthal. Res: Chemical and physiological investigation of retina and regions of the developing brain; quantitative histochemical studies of normal tissues and of regions of the central nervous system afflicted by inherited diseases; comparative aspects of normalcy and disease. Mailing Add: Lab Develop Neurol Vet Admin Hosp Sepulveda CA 91343

LOLY, PETER DOUGLAS, b Edmonton, Eng, Mar 7, 41; Can citizen; m 68; c 2. THEORETICAL PHYSICS, SOLID STATE PHYSICS. Educ: Univ London, BSc, 63, PhD(physics) & DIC, 66. Prof Exp: Fel, Theoret Physics Inst, Alberta, 66-68; asst prof, 68-75, ASSOC PROF PHYSICS, UNIV MAN, 75- Concurrent Pos: Travel fel, Nat Res Coun Can, 75; sabbatical, exchange prof, Lab Solid State Physics, Univ Paris-Sud, 75-76. Mem: Am Inst Physics; Can Asn Physicists; Brit Inst Physics & Phys Soc. Res: Spin waves in ferromagnets and anti-ferromagnets; light scattering; density of states; electronic susceptibilities; Fermi surface instabilities; Brillouin zone sums; lattice green functions; dimensional and anisotropic properties. Mailing Add: Dept of Physics Univ of Man Winnipeg MB Can

LOMAN, AUGUST ADRIAAN, plant pathology, biochemistry, see 12th edition

LOMAN, LAVERNE, b Stratford, Okla, June 10, 28; m 44; c 1. MATHEMATICS. Educ: Univ Okla, BS, 56, MA, 57, PhD(math educ), 61. Prof Exp: From asst to instr math, Univ Okla, 56-61; from asst prof to assoc prof, 61-65, PROF MATH, CENT STATE UNIV, OKLA, 66- Mem: Math Asn Am; Nat Coun Teachers Math. Res: Mathematics education. Mailing Add: Dept of Math Cent State Univ Edmond OK 73034

LOMANITZ, RACHEL, b Mexico City, Mex, May 23, 15; US citizen. MEDICAL MICROBIOLOGY, MEDICAL MYCOLOGY. Educ: Univ Okla, BS, 35, MS, 36, PhD(med microbiol), 64. Prof Exp: Teacher high sch, Okla, 36-38; res asst clin chem, Res Hosp, Univ Ill, 39-40; chem technician, Boiler Water Lab, Chicago, Ill, 40-41; dept librarian plant genetics, Weizmann Inst, 54-56; chem technician, Okla Med Res Found, 57; res asst med mycol, Med Ctr, Univ Okla, 60-61, co-prin investr med microbiol, 60-62; asst microbiologist, Bronx-Lebanon Hosp Ctr, Bronx, NY, 65; mem res staff med mycol, NY State Dept Health, 66; head mycol sect, Smith, Miller & Patch, Inc, 66-69; HEAD MYCOL SECT, JOHN F KENNEDY COMMUNITY HOSP, 69- Mem: Am Soc Microbiol; Mycol Soc Am. Res: Immunology of Cryptococcus neoformans; production of a delayed hypersensitivity in experimental animals; correlation of immune response and cellular components; enhancing effect of Hodgkin's serum on dissemination of experimental murine cryptococcosis. Mailing Add: John F Kennedy Med Ctr Edison NJ 08817

LOMANITZ, ROSS, b Bryan, Tex, Oct 10, 21; m 47. THEORETICAL PHYSICS. Educ: Univ Okla, BS, 40; Cornell Univ, PhD(theoret physics), 51. Prof Exp: Teaching asst physics, Univ Calif, Berkeley, 40-42, physicist, Lawrence Radiation Lab, 42-43,

LOMANITZ

teaching asst physics, 46-47; teaching asst, Cornell Univ, 47-49; assoc prof, Fisk Univ, 49; laborer, Okla, 49-54; tutor physics & math, Okla, 54-60; assoc prof physics, Whitman Col, 60-62; asst prof, 62-66, ASSOC PROF PHYSICS, NMEX INST MINING & TECHNOL, 66-, ASSOC PHYSICIST, 62- Concurrent Pos: NSF res grant, 63- Mem: Am Phys Soc; Am Geophys Union; Math Asn Am. Res: Electromagnetic isotope separator; quantum electrodynamics; superconductivity; theoretical plasma physics; theoretical ground water hydrology. Mailing Add: Dept of Physics NMex Inst of Mining & Technol Socorro NM 87801

LOMAS, HAROLD, b Stockport, Eng, Sept 10, 10; m 43; c 4. SURFACE CHEMISTRY. Educ: Univ London, BSc, 33. Prof Exp: Chief res chemist, Geigy Co, Ltd, Eng, 33-49; tech dir, Can Aniline & Extract Co, Ltd, Can, 50-59; res dir, E F Drew Co, Ltd, 59-61; sr res scientist, Ont Res Found, 61-67, asst dir, Dept Org Chem, 67-75; CHEMICAL CONSULT, 75- Mem: Faraday Soc; fel Chem Inst Can; Inst Textile Sci; Brit Oil & Colour Chem Asn. Res: Preparation, properties and applications of surface active organic chemicals and related compounds. Mailing Add: 61 Maple Ave Grand Vista Gardens Dundas ON Can

LOMAX, EDDIE, b Atlanta, Ga, Aug 12, 23; m 48. ORGANIC CHEMISTRY. Educ: Morehouse Col, BS, 48; Atlanta Univ, MS, 51. Prof Exp: Control chemist, 51-53, res chemist, 53-57, lab mgr, 57-71, asst tech dir, 71-73, TECH DIR, PURITAN CHEM CO, 73- Concurrent Pos: Sci consult, Atlanta Bd Educ. Mem: Am Chem Soc. Res: Free radical mechanism in solutions; surfactants, insecticides and disinfectants; floor polishes; cyclobutadiene series. Mailing Add: 495 Harlan Rd SW Atlanta GA 30311

LOMAX, MARGARET IRENE, b Roanoke, Va, Nov 13, 38; m 64; c 2. MOLECULAR GENETICS. Educ: Western Reserve Univ, BA, 60; Univ Mich, PhD(biol chem), 64. Prof Exp: Res assoc biol chem, Univ Mich, Ann Arbor, 64-67, instr, 67-68, ASST RES SCI, DIV BIOL SCI, UNIV MICH, ANN ARBOR, 74- Concurrent Pos: Am Cancer Soc fel, 64-66. Mem: AAAS; Am Soc Microbiol. Res: Use of DNA restriction endonucleases in construction of new bacterial genotypes. Mailing Add: Div of Biol Sci Univ of Mich Ann Arbor MI 48104

LOMAX, PETER, b Eng, May 12, 28; m 57; c 3. PHARMACOLOGY. Educ: Univ Manchester, MB & ChB, 54, MD, 64, DSc, 71. Prof Exp: Resident surgeon neurosurg, Univ Manchester, 54-57, lectr physiol, 57-61, asst prof, 61-69; assoc prof, 69-75, PROF PHARMACOL, SCH MED, UNIV CALIF, LOS ANGELES, 75- Mem: AAAS; Am Physiol Soc; Am Soc Pharmacol & Exp Therapeut; Brit Med Asn. Res: Pharmacological and immunological studies of drug abuse; role of the central nervous system in cold acclimation; effect of drugs on temperature regulation; experimental epilepsy. Mailing Add: Dept of Pharmacol Univ of Calif Sch of Med Los Angeles CA 90024

LOMBARD, DAVID BISHOP, b Lexington, Mass, June 10, 30; m 52; c 5. EXPERIMENTAL PHYSICS. Educ: Northeastern Univ, BS, 53; Pa State Univ, MS, 55, PhD(physics), 59. Prof Exp: Physicist, Lawrence Livermore Lab, Univ Calif, 59-70; mgr, Atcor, Inc, 70-71; pres, Geo-Resource Assocs, 71-72; vpres, Subcom, Inc, 72-74; prog mgr, NSF, 74; BR CHIEF, GEOTHERMAL ENERGY DIV, ELECTRONIC RESOURCES DEVELOP ADMIN, 75- Mem: Am Phys Soc; Am Nuclear Soc; Am Inst Mining, Metall & Petrol Engrs. Res: Geopressured and hot dry rock geothermal energy; neutron physics, fission-to-indium age of neutrons in water; strong shocks in high-density materials; applications of nuclear explosions to natural gas stimulation; oil shale; mining and leaching; disposal of radioactive wastes. Mailing Add: Div of Geothermal Energy Elec Resources Develop Admin Washington DC 20545

LOMBARD, LOUISE SCHERGER, b Wichita, Kans, Nov 20, 21; m 48; c 4. PATHOLOGY. Educ: Kans State Univ, DVM, 44; Univ Wis, MS, 47, PhD(path), 50. Prof Exp: Assoc vet, Morgan's Animal Hosp, 44-45; diagnostician, Corn States Serum Co, 45-46; instr path, Univ Wis, 46-50; instr, Woman's Med Col Pa, 50-51; res assoc virol, Univ Pa, 51-53, asst prof path, 53-55; biologist, Nat Cancer Inst, 55-57; assoc pathologist, Argonne Nat Labs, Ill, 57-64; assoc prof path, Stritch Sch Med, Loyola Univ Chicago, 64-69; res scientist, Univ Chicago, 69-70, assoc prof path & pharmacol, 70-73; sect head path, Abbott Labs, Abbott Park, Ill, 73-75; ASSOC PATH, ARGONNE NAT LABS, ILL, 75- Concurrent Pos: Consult pathologist, Chicago Zool Park, 60- & Argonne Nat Lab, 64-75. Mem: AAAS; Am Soc Exp Path; Wildlife Dis Asn; Am Vet Med Asn; Am Asn Cancer Res. Res: Neoplasms in animals; chemical carcinogenesis; viral oncology; radiobiology. Mailing Add: D202 Argonne Nat Lab Argonne IL 60439

LOMBARD, PORTER BRONSON, b Yakima, Wash, Feb 6, 30; m 55; c 3. HORTICULTURE. Educ: Pomona Col, BA, 52; Wash State Univ, MS, 55; Mich State Univ, PhD(hort), 58. Prof Exp: Asst horticulturist, Citrus Exp Sta, Calif, 58-63; assoc prof, 63-70, PROF HORT, ORE STATE UNIV, 70-, SUPT SOUTHERN ORE EXP STA, 63- Mem: AAAS; Am Soc Hort Sci. Res: Pear varieties; rootstocks; nutrition, pear fruit bud hardiness and water requirements. Mailing Add: Dept of Hort Ore State Univ Corvallis OR 97331

LOMBARD, RICHARD ERIC, b Brooklyn, NY, May 16, 43; m 67; c 2. MORPHOLOGY. Educ: Hanover Col, AB, 65; Univ Chicago, PhD(anat), 71. Prof Exp: Res assoc, Mus Vert Zool, Univ Calif, Berkeley, 71; res assoc, Univ Southern Calif, 71-72; ASST PROF ANAT, UNIV CHICAGO, 72- Mem: AAAS; Am Soc Ichthyologists & Herpetologists; Am Soc Zoologists; Soc Study Amphibians & Reptiles; Soc Study Evolution. Res: The evolutionary and functional morphology of major adaptive features of lower vertebrates including auditory periphery in frogs, feeding apparatus of frogs and salamanders and the vestibular system in salamanders. Mailing Add: Dept of Anat Univ of Chicago Chicago IL 60637

LOMBARD, STEWART MATTHEW, analytical chemistry, see 12th edition

LOMBARDI, MAX H, b Huanuco City, Peru, Apr 25, 32; m 61; c 3. RADIATION BIOLOGY, NUCLEAR MEDICINE. Educ: Univ Lima, BSc & DVM, 58; Cornell Univ, MSc, 61. Prof Exp: From asst prof to assoc prof biochem & nutrit, Vet Col Peru, 60-64; scientist biomed appln & consult, lectr & overall coord progs Latin Am, Oak Ridge Assoc Univs, 64-68, SR SCIENTIST & COORDR RADIATION BIOL & MED RADIOISOTOPE TRAINING PROGS, OAK RIDGE ASSOC UNIVS, 68- Mailing Add: Oak Ridge Assoc Univs PO Box 117 Oak Ridge TN 37830

LOMBARDI, PAUL SCHOENFELD, b Salt Lake City, Utah, Nov 13, 40; m 68; c 2. MICROBIOLOGY, VIROLOGY. Educ: Univ Utah, BA, 63, MA, 65; Univ Rochester, PhD(microbiol), 69. Prof Exp: Instr, 71-73, ASST PROF MICROBIOL, COL MED, UNIV UTAH, 73- Concurrent Pos: Damon Runyon Mem Fund fel, Swiss Inst Exp Cancer Res, 69-70; Am Cancer Soc fel, Univ Utah, 71-73, NIH grant, 74-76. Mem: Am Soc Microbiol. Res: Cell-virus interactions of polyoma virus in permissive cells; stru structural proteins of polyoma virions; mycoplasma viruses and their interactions with mammalian cells. Mailing Add: Dept of Microbiol Univ of Utah Col of Med Salt Lake City UT 84132

LOMBARDINI, JOHN BARRY, b San Francisco, Calif, July 2, 41; m 68; c 1. PHARMACOLOGY. Educ: St Mary's Col Calif, BS, 63; Univ Calif, San Francisco, PhD(biochem), 68. Prof Exp: Fel, Sch Med, Johns Hopkins Univ, 68-72, res assoc pharmacol, 72-73; ASST PROF PHARMACOL, SCH MED, TEX TECH UNIV, 73- Mem: Am Soc Pharmacol & Exp Therapeut. Res: Enzymatic synthesis of cysteinesulfinic acid and characterization of the co-factor requirements; formation, function and regulatory properties of S-adenosyltransferase; function of taurine as a possible neurotransmitter or modulator of nerve impulses plus its role in cardiac metabolism. Mailing Add: Dept of Pharmacol & Therapeut Sch of Med Tex Tech Univ Lubbock TX 79409

LOMBARDINO, JOSEPH GEORGE, b Brooklyn, NY, July 1, 33; m 60; c 3. ORGANIC CHEMISTRY. Educ: Brooklyn Col, BS, 54; Polytech Inst Brooklyn, PhD, 58. Prof Exp: SR RES INVESTR, PFIZER, INC, 57- Mem: Am Chem Soc; Am Inst Chemists; Inflammation Res Asn. Res: Synthetic organic medicinals; nitrogen heterocycles; anti-inflammatory drugs; immunoregulatory drugs. Mailing Add: 13 Laurel Hill Dr Niantic CT 06357

LOMBARDO, ANTHONY, b Brooklyn, NY, Jan 4, 39. ORGANIC CHEMISTRY. Educ: Queens Col, NY, BS, 61; Syracuse Univ, PhD(org chem), 67. Prof Exp: Fel, Univ Calif, Santa Barbara, 67-68; ASST PROF CHEM, FLA ATLANTIC UNIV, 68- Res: Coenzyme models; donor-acceptor complexes; kinetics; spectroscopy. Mailing Add: Dept of Chem Fla Atlantic Univ Boca Raton FL 33432

LOMBARDO, MICHAEL E, b New Rochelle, NY, May 3, 24; m 49; c 2. BIOCHEMISTRY. Educ: Columbia Univ, AB, 48; Fordham Univ, MS, 51, PhD, 53. Prof Exp: Assoc biochem, Col Physicians & Surgeons, Columbia Univ, 53-60; consult to chem indust, 60-65; sr organo-biochemist, Schwarz Biores, Inc, NY, 66-67; DIR, AUTOMATED BIOCHEM LABS, INC, 67- Concurrent Pos: Dir labs, Rampo Gen Hosp, Spring Valley, NY, 66- Mem: Am Chem Soc; Am Soc Biol Chem; Am Soc Microbiol; NY Acad Sci; Brit Biochem Soc. Res: Biochemistry of nucleic acids and steroid hormones; antimicrobial agents; clinical biochemistry. Mailing Add: Automated BioChem Labs Inc 768 N Main St Spring Valley NY 10977

LOMBARDO, PASQUALE, b Brooklyn, NY, Nov 2, 30; m 57; c 4. ANALYTICAL CHEMISTRY. Educ: City Col New York, BS, 52; Columbia Univ, MA, 53. Prof Exp: Sr res chemist pesticide chem, Allied Chem Corp, 55-66; res chemist pesticide chem, 66-73, SUPVRY CHEMIST INDUST POLLUTANTS, FOOD & DRUG ADMIN, 73- Mem: Am Chem Soc. Res: Investigation of industrial chemicals which may enter the environment and be magnified up the food chain, ultimately resulting in the contamination of human food. Mailing Add: 200 C St SW Washington DC 20204

LOMBROSO, CESARE THOMAS, b Rome, Italy, Oct 9, 17; nat US; m 43; c 3. MEDICINE. Educ: Univ Palermo, AB, 34; Univ Genoa, MD, 45; Univ Rome, PhD, 51. Prof Exp: Intern physiol, Med Univ, Genoa, 35-39; intern, G Gaslini Hosp, 45-46; pediat training, Univ Genoa, 46-48, vol asst, 48-49, asst & chief-of-ward, 49-50; instr pediat, 51-58, clin assoc, 58-63, asst clin prof, 63-66, assoc prof neurol, 66-70, PROF NEUROL, HARVARD MED SCH, 70-, CHIEF SEIZURE UNIT & CHIEF DIV NEUROPHYSIOL, HARVARD MED SCH, 66-, SR NEUROLOGIST, CHILDREN'S HOSP MED CTR, 66- Concurrent Pos: Fel med, Children's Hosp, Boston, 50-51; clin fel neurol, Mass Gen Hosp, 56-57; asst physician, Harvard Med Sch, 52-58, assoc physician, Neurol Inst, 56-68; res assoc, Vet Nat Epilepsy Ctr, Boston, 53-57; resident neurol, Mass Gen Hosp, Boston, 54-55, neuropath, 55-56; consult, Peter Bent Brigham Hosp, 57-59, assoc staff med, 59-; assoc neurol, Children's Med Ctr, 58-; mem comt 50, Epilepsy Found Am. Mem: Am Epilepsy Soc; Am Electroencephalog Soc; Asn Res Nerv & Ment Dis; Am Acad Neurol; Am Neurol Asn. Res: Neurophysiology; epilepsy, clinical and experimental studies; electroencephalography; sonar echoencephalography. Mailing Add: Children's Hosp Med Ctr 300 Longwood Ave Boston MA 02115

LOMEN, DAVID ORLANDO, bDecorah, Iowa, May 11, 37; m 61; c 1. APPLIED MATHEMATICS. Educ: Luther Col, Iowa, BA, 59; Iowa State Univ, MS, 62, PhD, 64. Prof Exp: Design specialist, Gen Dynamics/Astronaut, 63-66; from asst prof to assoc prof, 69-74, PROF MATH, UNIV ARIZ, 74- Concurrent Pos: Consult var industs. Mem: Soc Indust & Appl Math; Am Math Soc. Res: Diffusion processes in soils and physiology. Mailing Add: Dept of Math Univ of Ariz Tucson AZ 85721

LOMENICK, THOMAS FLETCHER, geology, health physics, see 12th edition

LOMMASSON, ROBERT CURTIS, b Topeka, Kans, Jan 4, 17; m 43; c 4. PLANT MORPHOLOGY. Educ: Univ Kans, AB, 38, MA, 40; Univ Iowa, PhD(bot), 48. Prof Exp: Instr bot, Univ Iowa, 46-48; from asst prof to assoc prof, 48-63, PROF BOT, UNIV NEBR, LINCOLN, 63- Mem: Bot Soc Am; Am Fern Soc (pres, 62-63); Int Soc Plant Morphol. Res: Grass leaf anatomy; fern leaf venation; flora of the great plains. Mailing Add: Sch of Life Sci Univ of Nebr Lincoln NE 68588

LOMNITZ, CINNA, b Cologne, Ger, May 4, 25; m 51; c 4. SEISMOLOGY. Educ: Univ Chile, CE, 48; Harvard Univ, MS, 50; Calif Inst Technol, PhD(geophys), 55. Prof Exp: Res fel seismol, Calif Inst Technol, 55-57; prof geophys, Univ Chile, 57-64, dir inst geophys & seismol, 58-64; assoc res seismologist, Seismog Sta, Univ Calif, Berkeley, 64-68; PROF SEISMOL & HEAD SEISMOG STA, INST GEOPHYS, NAT UNIV MEX, 68- Concurrent Pos: Consult, Geol Surv, Chile, 58-; vis assoc, Calif Inst Technol & Univ Calif, San Diego, 69- Mem: Seismol Soc Am; Am Geophys Union. Res: Geophysics of the solid earth; creep properties of rocks; viscoelasticity and internal friction in solids; seismicity; structure of the Andes; origin of earthquakes and tsunamis. Mailing Add: Inst of Geophys Nat Univ of Mex Mexico DF Mexico

LOMON, EARLE LEONARD, b Montreal, Que, Nov 15, 30; nat US; m 51; c 3. THEORETICAL PHYSICS. Educ: McGill Univ, BSc, 51; Mass Inst Technol, PhD(theoret physics), 54. Prof Exp: Res physicist, Can Defence Res Bd, 50-51 & Baird Assocs, Mass, 52-53; Nat Res Coun Can overseas res fel, Inst Theoret Physics, Denmark, 54-55; fel, Weizmann Inst, 55-56; res assoc, Lab Nuclear Studies, Cornell Univ, 56-57; assoc prof theoret physics, McGill Univ, 57-60; assoc prof physics, 60-70, PROF PHYSICS, MASS INST TECHNOL, 70- Concurrent Pos: Guggenheim Mem Found fel, 65-66; vis scientist, Los Alamos Sci Lab, 68-; proj dir, Unified Sci & Math for Elem Sch, 71- Mem: Am Phys Soc; Can Asn Physicists. Res: Nuclear and high energy physics; field theory. Mailing Add: Dept of Physics Mass Inst of Technol Cambridge MA 02139

LOMONACO, CARMINE JOSEPH, b Brooklyn, NY, Dec 23, 37; m 64; c 3. ENDODONTICS. Educ: St John's Univ, NY, BS, 59; Seton Hall Col Med & Dent, DDS, 64; NJ Col Med & Dent, cert endodontics, 69. Prof Exp: Asst prof endodontics & oral biol, 70-72, asst dean student affairs, 71-72, ASST PROF ORAL BIOL, 70-, ASSOC DEAN STUDENT AFFAIRS & ASSOC PROF ENDODONTICS, NJ DENT SCH, COL MED & DENT, 72- Concurrent Pos: Int Col Dentists fel, 75. Mem: Am Asn Endodontists. Res: A bacterial examination of endodontically involved teeth, utilizing the electrosterilization technique with and without disinfectants. Mailing Add: 6 Klimback Ct West Caldwell NJ 07006

LOMONACO, SAMUEL JAMES, JR, b Dallas, Tex, Sept 23, 39; m 68; c 1. MATHEMATICS. Educ: St Louis Univ, BS, 61; Princeton Univ, PhD(math), 64. Prof Exp: Asst prof math, St Louis Univ, 64-65 & Fla State Univ, 65-69; res mathematician, Tex Instruments, Inc, 69-71; ASSOC PROF COMPUT SCI, STATE UNIV NY ALBANY, 71- Concurrent Pos: Indust prof, Southern Methodist Univ, 69-71. Mem: Am Math Soc; Asn Comput Mach; Math Asn Am; Soc Indust & Appl Math. Res: Algebraic topology and computer science. Mailing Add: Dept of Comput Sci State Univ of NY Albany NY 12203

LOMONT, JOHN S, b Ft Wayne, Ind, Aug 26, 24. MATHEMATICAL PHYSICS. Educ: Purdue Univ, MS, 47, PhD(physics), 51. Prof Exp: Physicist theoret solid state physics, NAm Aviation, Inc, 51-52; physicist, Res Dept, Michelson Lab, Naval Ord Test Sta, 52-54; physicist, NY Univ, 54-57 & Int Bus Mach Corp, 57-60; prof math, Polytech Inst Brooklyn, 62-65; PROF MATH, UNIV ARIZ, 65- Concurrent Pos: Sabbatical, Courant Inst Math Sci, NY Univ, 71-72. Mem: Am Phys Soc; Am Math Soc. Res: Applied group theory; quantum field theory; functional analysis. Mailing Add: Dept of Math Univ of Ariz Tucson AZ 85721

LONADIER, FRANK DALTON, b Clarence, La, May 6, 32; m 59; c 3. PHYSICAL CHEMISTRY, INORGANIC CHEMISTRY. Educ: Northwestern State Col, La, BS, 54; Univ Tex, PhD(phys chem), 59. Prof Exp: Res asst, Los Alamos Sci Lab, Univ Calif, 57-58; sr res chemist, 59-61, group leader inorg & nuclear chem, 61-64, sect mgr mat eval, 64-65, sect mgr nuclear develop, 65-67, mgr nuclear prod, 67-69, MGR EXPLOSIVE TECHNOL, MOUND LAB, MONSANTO RES CORP, 69- Mem: Am Chem Soc. Res: Actinide elements, particularly uranium and plutonium; inorganic chemistry of polonium; behavior of secondary explosives; environmental pollutant abatement. Mailing Add: Mound Lab Monsanto Res Corp Miamisburg OH 45342

LONARD, ROBERT (IRVIN), b Valley Falls, Kans, June 5, 42; m 65; c 1. PLANT TAXONOMY. Educ: Kans State Teachers Col, BSE, 64, MS, 66; Tex A&M Univ, PhD(plant taxon), 70. Prof Exp: ASST PROF BIOL, PAN AM UNIV, 70- Mem: AAAS; Am Soc Plant Taxonomists; Int Asn Plant Taxonomists. Res: Flora of south Texas; grass systematics. Mailing Add: Dept of Biol Pan Am Univ Edinburg TX 78539

LONBERG-HOLM, KNUD KARL, b New York, NY, Sept 22, 31; m 52, 61; c 3. BIOCHEMISTRY. Educ: Harvard Univ, BA, 53; Univ Calif, Berkeley, PhD(biochem), 62. Prof Exp: Chemist, Hyman Labs, Fundamental Res, Inc, 59-60; BIOCHEMIST, CENT RES DEPT, E I DU PONT DE NEMOURS & CO, INC, 62- Concurrent Pos: USPHS fel, Univ Uppsala, 67-69. Res: Molecular biology; virus-cell interaction; biochemical virology. Mailing Add: Cent Res Dept E I du Pont de Nemours & Co Inc Wilmington DE 19898

LONCAREVIC, BOSKO (D), b Belgrade, Yugoslavia, Dec 11, 30; Can citizen; m 61; c 2. MARINE GEOPHYSICS. Educ: Univ Toronto, BASc, 55, MA, 58; Cambridge Univ, PhD(geophys), 61. Prof Exp: Asst sci & indust res, Cambridge Univ, 61-63; head, Marine Geophys Lab, 63-69, asst dir res, Atlantic Lab, 69-72, DIR, ATLANTIC GEOSCI CENTRE, BEDFORD INST, 72- Concurrent Pos: Mem spec study group, Int Gravity Comn, 62-67; mem gravity & seismol subcomt, Nat Res Coun Can Adv Comt Geophys, 63-69; assoc, Inst Oceanog & hon asst prof, Dalhousie Univ, 64- Mem: Am Geophys Union; Royal Astron Soc; Geol Asn Can. Res: Magnetic and gravity surveys. Mailing Add: Atlantic Geosci Centre Bedford Inst PO Box 1006 Dartmouth NS Can

LONCRINI, DONALD FRANCIS, b Springfield, Mass, Mar 24, 30; m 56; c 3. ORGANIC CHEMISTRY. Educ: Siena Col, BS, 51; Fla State Univ, PhD(org chem), 56. Prof Exp: Chemist, Gen Elec Co, 51-56, develop chemist, 56-60, advan develop chemist, Insulating Mat Dept, 60-62, specialist, NY, 63-68; dir res & develop, P D George Co, Mo, 68-72; MGR RES, MALLINCKRODT CHEM WORKS, 72- Mem: Am Chem Soc; The Chem Soc. Res: Organosilicon compounds; fluorinated amino acids; chemistry of bicylic compounds; carbonates; polycarbonates; polymers; high temperature polymers; organic synthesis and processes. Mailing Add: Res & Develop Labs Mallinckrodt Chem Works St Louis MO 63160

LONDERGAN, MARTIN CHRISTOPER, b London, Ohio, Dec 23, 09; m 42; c 3. PHYSICAL CHEMISTRY. Educ: Ohio Univ, BS, 36; Iowa State Col, PhD(phys chem), 42. Prof Exp: Lab asst, Ohio Univ, 35-37; asst, Iowa State Col, 37-42; MEM STAFF, PIGMENTS DEPT, E I DU PONT DE NEMOURS & CO, INC, 42- Mem: Am Chem Soc. Res: Photochemistry; powder metallurgy; paints; photochemistry of formation of sulfuryl chloride. Mailing Add: 601 Jackson Ave Wilmington DE 19804

LONDON, GEORGES W, high energy physics, see 12th edition

LONDON, IRVING MYER, b Malden, Mass, July 24, 18; m 55; c 2. MEDICINE. Educ: Harvard Univ, AB, 39, MD, 43. Hon Degrees: ScD, Univ Chicago, 66. Prof Exp: From instr to assoc prof med, Columbia Univ, 47-55; prof & chmn dept, Albert Einstein Col Med, 55-70; PROF MED, HARVARD UNIV & PROF MED & BIOL, MASS INST TECHNOL, 70- Concurrent Pos: Asst physician, Presby Hosp, NY, 46-52, from asst attend physician to assoc attend physician, 52-55; dir, Harvard-Mass Inst Technol prog in health sci & technol, 69-; vis prof, Albert Einstein Col Med, 70-; physician, Peter Bent Brigham Hosp; mem med fel bd & subcomt blood & related probs, Nat Acad Sci-Nat Res Coun, 55-63; subcomt intravenous alimentation, US Army, 56-60; res coun mem, Pub Health Res Inst, NY, 58-63; mem exec comt, Health Res Adv Coun, 58-63; bd sci consult, Sloan-Kettering Inst Cancer Res, 60-72, mem adv bd clin res, 72-; metab study sect mem, USPHS, 60-63, chmn, 61-63, mem bd sci coun, Nat Heart Inst, 64-68; mem panel biol sci & advan med, Nat Acad Sci, 66-67, bd med, 67-70 & exec comt, Inst Med, 70-73; mem adv comt to dir, NIH, 66-70; mem, Nat Cancer Adv Bd, 72- Honors & Awards: Smith Award, AAAS, 53. Mem: Nat Acad Sci; Am Acad Arts & Sci; Am Soc Exp Biol & Med; Am Soc Clin Invest; Am Soc Biol Chem. Res: Hemoglobin metabolism; metabolism of erythrocytes. Mailing Add: Harvard-MIT Prog Health Sci Tech 77 Massachusetts Ave Cambridge MA 02139

LONDON, JACK P, b Chicago, Ill, Feb 23, 35; m 59; c 1. BACTERIOLOGY, BIOCHEMISTRY. Educ: Univ Southern Calif, BA, 58, MS, 60, PhD(bact, biochem), 64. Prof Exp: RES MICROBIOLOGIST, NAT INST DENT RES, 66- Concurrent Pos: NIH fel, 64-66. Mem: Am Soc Microbiol. Res: Physiology of phototrophic and chemolithotrophic microorganisms; regulation and function of microbial enzyme systems. Mailing Add: Nat Inst of Dent Res Bethesda MD 20014

LONDON, JULIUS, b Newark, NJ, Mar 26, 17; m 46; c 2. METEOROLOGY. Educ: Brooklyn Col, AB, 41; NY Univ, MS, 48, PhD(meteorol), 51. Prof Exp: Meteorologist, US Weather Bur, 42; instr meteorol, US Air Force, 42-47; res assoc meteorol, NY Univ, 48-52, asst prof, 52-56, res assoc prof, 56-59, assoc prof, 59-61; chmn dept astro-geophys, 66-69, PROF ASTRO-GEOPHYS, UNIV COLO, BOULDER, 61- Concurrent Pos: Lectr, Columbia Univ, 54-55; vis prof, Pa State Univ, 55; vis scientist, High Altitude Observ, Univ Colo, 56, 57, 59 & 60; Max Planck Inst Physics, Göttingen, 58; res scientist, Nat Ctr Atmospheric Res, 61-; mem, Int Ozone Comn, Int Asn Meteorol & Atmospheric Physics, 60-, secy, Int Radiation Comn, 63-71, pres, 71-; chmn panel ozone, Nat Acad Sci-Nat Res Coun, 64-65; vis prof, Swiss Fed Inst Technol, 67; ed, Contrib Atmospheric Physics, 72- Mem: AAAS; Am Asn Univ Prof; fel Am Meteorol Soc; Royal Meteorol Soc; Am Geophys Union. Res: Atmospheric radiation; physics of the atmosphere; ozone. Mailing Add: Dept of Astro-Geophys Univ of Colo Boulder CO 80302

LONDON, MORRIS, b New York, NY, July 2, 22; m 48; c 2. BIOCHEMISTRY. Educ: City Col New York, BS, 46; Fordham Univ, MS, 48; Ohio State Univ, PhD(physiol chem), 50. Prof Exp: Res assoc biochem, Columbia Univ, 51-56; chief biochemist, North Shore Hosp, Manhasset, NY, 56-69; chief clin chem, Beth Israel Hosp, 69-71; CHIEF CLIN CHEM, BROOKDALE HOSP MED CTR, BROOKLYN, 71- Mem: Am Chem Soc; Am Asn Clin Chem; Am Soc Biol Chem. Res: Uric acid metabolism; uricase isolation; properties of prostatic acid phosphatase; clinical analyses. Mailing Add: Brookdale Hosp Med Ctr Linden Blvd at Brookdale Plaza Brooklyn NY 11212

LONDON, ROBERT ELLIOT, b Brooklyn, NY, Oct 25, 46; m 69; c 1. BIOPHYSICAL CHEMISTRY. Educ: Brooklyn Col, BS, 67; Univ Ill, MS, 69, PhD(physics), 73. Prof Exp: Fel, 73-75, STAFF MEM BIOPHYS CHEM, LOS ALAMOS SCI LAB, 75- Mem: Biophys Soc. Res: Nuclear magnetic resonance studies of biologically important molecules. Mailing Add: Group CNC-4 Los Alamos Sci Lab Los Alamos NM 87545

LONDON, S J, b New York, NY, Oct 15, 17; m 43; c 2. MEDICINE, CLINICAL PHARMACOLOGY. Educ: Univ Louisville, BA, 37, MD, 41; Am Bd Internal Med, dipl, 51. Prof Exp: Resident med, Sea View Hosp, Staten Island, NY, 46-47; resident, Halloran Vet Admin Hosp, 47-48; staff physician, Manhattan Beach Vet Admin Hosp, Brooklyn, 48-50; pvt pract internal med, 50-58; med dir, Purdue Frederick Co, NY, 58-59, vpres, 59-61, vpres & med dir, M R Thompson, Inc, 61-64; from assoc dir to dir clin res, 64-70, DIR BIOMED SCI, VICK DIV RES & DEVELOP, RICHARDSON-MERRELL, INC, 70- Mem: Fel Am Med Writers' Asn; Am Soc Clin Pharmacol & Therapeut; Am Acad Clin Toxicol. Res: Clinical pharmacology of subjective responses; respiratory pharmacology; diabetes; atherosclerosis; thyroid diseases; hepatobiliary and pancreatic diseases. Mailing Add: Vick Div Res & Develop Richardson-Merrell Inc Mt Vernon NY 10551

LONDON, WILLIAM THOMAS, b New York, NY, Mar 11, 32; m 57; c 4. INTERNAL MEDICINE, ENDOCRINOLOGY. Educ: Oberlin Col, BA, 53; Cornell Univ, MD, 57. Prof Exp: Intern med, Bellevue Hosp, 57-58, resident, Med Ctr, 58-60; res epidemiologist, Nat Inst Arthritis & Metab Dis, 62-66; assoc, 66-71, ASST PROF MED, SCH MED, UNIV PA, 71-, RES PHYSICIAN, INST CANCER RES, 66- Concurrent Pos: Fel endocrinol, Sloan-Kettering Inst, NY, 60-62; asst, Med Col, Cornell Univ, 60-62; instr, Sch Med, George Washington Univ, 64- Mem: Am Thyroid Asn; Am Asn Cancer Res. Res: Susceptibility factors to cancer; variations in host response to hepatitis B infection. Mailing Add: Inst for Cancer Res 7701 Burholme Ave Philadelphia PA 19111

LONE, MUHAMMAD ASLAM, b East Punjab, India, Jan 28, 37; m 70; c 1. EXPERIMENTAL NUCLEAR PHYSICS. Educ: Punjab Univ, West Pakistan, BSc, 58, MSc, 60; State Univ NY Stony Brook, PhD(physics), 67. Prof Exp: Lectr physics, Govt Col, Lahore, Pakistan, 60-62; fel, Ind Univ, Bloomington, 67-68; Nat Res Coun Can fel, 68-70, asst res officer physics, 70-73, ASSOC RES OFFICER PHYSICS, CHALK RIVER NUCLEAR LABS, ATOMIC ENERGY CAN LTD, 73- Mem: Can Asn Physicists; Am Phys Soc. Res: Nuclear spectroscopy by gamma ray, neutron, and charged particle induced reactions; investigation of nuclear reaction mechanism. Mailing Add: Chalk River Nuclear Labs Atomic Energy of Can Ltd Chalk River ON Can

LONERGAN, JAMES ARTHUR, physics, see 12th edition

LONERGAN, LESTER HAROLD, b Topeka, Kans, Apr 12, 05; m; c 2. HEALTH EDUCATION. Educ: Union Col, Nebr, BA, 26; Loma Linda Univ, MD, 31; Harvard Univ, MPH, 52. Prof Exp: From instr to assoc prof pharmacol, Sch Med, 31-67, assoc prof trop med, Sch Health, 67-74, EMER ASSOC PROF HEALTH EDUC, SCH HEALTH, LOMA LINDA UNIV, 74- Res: Toxicology of tobacco; health education in tobacco, alcohol and drugs of addiction; health education in developing countries. Mailing Add: Sch of Health Loma Linda Univ Loma Linda CA 92354

LONEY, ROBERT AHLBERG, b Odebolt, Iowa, June 16, 22; m 56; c 3. STRUCTURAL GEOLOGY, PETROLOGY. Educ: Univ Wash, BS, 49, MS, 51; Univ Calif, Berkeley, PhD(geol), 61. Prof Exp: Geologist, Superior Oil Co, Tex, 51-52 & Wyo, 52-54; GEOLOGIST, US GEOL SURV, 56- Mem: AAAS; Geol Soc Am; Mineral Soc Am; Ger Geol Asn. Res: Structural petrology and petrology of mafic-ultramafic complexes and associated terranes; Pacific coastal region. Mailing Add: US Geol Surv 345 Middlefield Rd Menlo Park CA 94025

LONG, ALAN JACK, b Baton Rouge, La, Oct 17, 44; m 66; c 2. FOREST ECOLOGY, FOREST GENETICS. Educ: Univ Calif, Berkeley, BS, 67, MS, 71; NC State Univ, PhD(forestry, genetics), 73. Prof Exp: Asst prof forest genetics, Pa State Univ, 73-74; RES SCIENTIST REGENERATION ECOL, WEYERHAEUSER CO, 74- Res: Technology requisite for plantation establishment and early growth of western conifers; use of clonal material in tree improvement and regeneration programs; root growth of conifer seedlings. Mailing Add: 1608 W Mellen St Centralia WA 98531

LONG, ALEXIS BORIS, b New York, NY, Sept 9, 44; m 74. CLOUD PHYSICS. Educ: Reed Col, BA, 65; Syracuse Univ, MS, 66; Univ Ariz, PhD(atmospheric sci), 72. Prof Exp: Res asst cloud physics, Inst Atmospheric Physics, Univ Ariz, 69-72; NSF fel & vis scientist, Div Cloud Physics, Commonwealth Sci & Indust Res Orgn, 72-73; res assoc cloud physics, Coop Inst Res Environ Sci, Univ Colo, 73-75; SCIENTIST & HEAD HAIL SUPPRESSION GROUP, NAT HAIL RES EXP, NAT CTR ATMOSPHERIC RES, 75- Mem: Am Meteorol Soc; Am Geophys Union; AAAS. Res: Weather modification; hail suppression; rain augmentation; precipitation measuring systems; ice crystals and hail growth; stochastic coalescence growth of cloud droplets; cloud droplet collision efficiencies; nucleation theory. Mailing Add: Nat Ctr Atmospheric Res PO Box 3000 Boulder CO 80303

LONG, ANDREW FLEMING, JR, b Amboy, WVa, Dec 20, 38. MATHEMATICS. Educ: WVa Univ, BS, 60, MS, 61; Duke Univ, PhD(math), 65. Prof Exp: Asst prof math, St Andrews Presby Col, 65-67; ASST PROF MATH, UNIV NC, GREENSBORO, 67- Mem: Math Asn Am; Am Math Soc. Res: Irreducible factorable polynomials over a finite field; number theory. Mailing Add: Dept of Math Univ of NC Greensboro NC 27412

LONG, ARTHUR OWEN, b Danbury, Conn, May 12, 21; m 44; c 5. PHYSICAL CHEMISTRY. Educ: Brown Univ, AB, 42; Univ Wis, PhD(chem), 50. Prof Exp: Asst prof chem, Univ Vt, 49-51; res assoc, US Naval Ord Lab, 51-53; ASSOC PROF CHEM, STATE UNIV NY ALBANY, 53- Mem: Am Chem Soc. Res: Reactions of glass surfaces with ions in aqueous solution; exchange of nitrogen-15 between nitrogen dioxide and nitrogen monoxide. Mailing Add: Dept of Chem State Univ of NY Albany NY 12203

LONG, AUSTIN, b Olney, Tex, Dec 12, 36; m 61; c 2. GEOCHEMISTRY. Educ: Midwestern Univ, BS, 57; Columbia Univ, MA, 59; Univ Ariz, PhD(geochem), 66. Prof Exp: Res asst geochem, Geochronol Labs, Univ Ariz, 59-63; geochemist, Smithsonian Inst, 63-68; ASSOC PROF GEOSCI, LAB ISOTOPE GEOCHEM, UNIV ARIZ, 68- Mem: Geochem Soc. Res: Pleistocene paleoclimatology; radiocarbon dating; stable isotope geochemistry. Mailing Add: Lab of Isotope Geochem Dept of Geosci Univ of Ariz Tucson AZ 85721

LONG, CALVIN H, b Myerstown, Pa, Feb 16, 27; m 54; c 2. ANALYTICAL CHEMISTRY. Educ: Univ Miami, BS, 50; Franklin & Marshall Col, MS, 56; Stanford Univ, PhD(chem), 63. Prof Exp: Chemist, Armstrong Cork Co, 50-58; res chemist, Chevron Res Co, 63-64; res group leader anal chem, 64-68, SECT MGR, KERR-McGEE CORP, 69- Mem: Am Chem Soc; Am Soc Testing & Mat. Res: Chemical equilibria; mineral benefication. Mailing Add: 4609 NW 62nd St Oklahoma City OK 73122

LONG, CALVIN LEE, b NC, Jan 27, 28; m 51; c 3. BIOCHEMISTRY. Educ: Wake Forest Col, BS, 48; NC State Col, MS, 51; Univ Ill, PhD, 54. Prof Exp: Assoc chemist biochem, Gen Food Corp, 54-57, proj leader, 57-62; res assoc, Harvard Univ, 63 & Col Physicians & Surgeons, Columbia Univ, 64-74; ASSOC PROF BIOCHEM & SURG, MED COL OHIO, 75- Mem: AAAS; Am Inst Nutrit; Am Chem Soc; NY Acad Sci. Res: Intermediary metabolism and nutritional biochemistry. Mailing Add: Dept of Surg Med Col of Ohio Toledo OH 43614

LONG, CALVIN THOMAS, b Rupert, Idaho, Oct 10, 27; m 52; c 2. MATHEMATICS. Educ: Univ Idaho, BS, 50; Univ Ore, MS, 52, PhD(math), 55. Prof Exp: Analyst, Nat Security Agency, 55-56; from asst prof to assoc prof, 56-65, PROF MATH, WASH STATE UNIV, 65-, CHMN DEPT, 70- Concurrent Pos: Educ consult, Wash State Dept Educ, 61-67 & NSF, 63-64 & 72; assoc ed, Math Mag, 64-64-69; vis prof, Univ BC, 72; consult, Educ Comn States, Nat Assessment Educ Progress, 75. Mem: Math Asn Am; Nat Coun Teachers Math. Res: Probabilistic and combinatorial number theory and other combinatorial problems. Mailing Add: Dept of Math Wash State Univ Pullman WA 99163

LONG, CEDRIC WILLIAM, b Minneapolis, Minn, Mar 4, 37. BIOCHEMISTRY, CELL BIOLOGY. Educ: Univ Calif, Los Angeles, BA, 60, MA, 62; Princeton Univ, PhD(biochem), 66. Prof Exp: Am Cancer Soc fel biochem, Univ Calif, Berkeley, 66-68; Nat Cancer Inst fel path, Med Sch, NY Univ, 68-69, instr cell biol, 69-70; sr scientist, 70-72, HEAD CELL BIOL SECT, FLOW LABS, INC, 72- Mem: AAAS; Am Soc Microbiol; Am Soc Biol Chem. Res: Genetic and biochemical aspects of mammalian cell growth, interaction of tumor viruses with cells in culture, and functional aspects of viral proteins. Mailing Add: Flow Labs Inc Cell Biol Sect 1710 Chapman Ave Rockville MD 20852

LONG, CHARLES ALAN, b Pittsburg, Kans, Jan 19, 36; m 60; c 2. ZOOLOGY. Educ: Kans State Col Pittsburg, BS, 57, MS, 58; Univ Kans, PhD(zool), 63. Prof Exp: Asst zool, Univ Kans, 59-63; instr, Univ Ill, Urbana, 63-65, asst prof zool & life sci, 65-66; asst prof biol, 66-68, ASSOC PROF BIOL, UNIV WIS-STEVENS POINT, 68-, DIR MUS NATURAL HIST, 69- Concurrent Pos: Fac fel, Univ Ill, 64. Mem: Am Inst Biol Sci; Am Soc Naturalists; Am Soc Mammal. Res: Vertebrate zoology, particularly systematics and zoogeography of mammals and their evolution; morphology and ecology; variability. Mailing Add: Dept of Biol Univ of Wis Stevens Point WI 54481

LONG, CHARLES ANTHONY, b San Antonio, Tex, Feb 22, 45. CHEMICAL PHYSICS. Educ: Carleton Col, BA, 67; Ind Univ, PhD(chem physics), 72. Prof Exp: Fel chem physics, Univ Calif, Riverside, 72-73; ASST PROF CHEM, LAKE FOREST COL, 73- Concurrent Pos: NSF res grant, 74. Mem: Am Phys Soc; Am Chem Soc. Res: Applications of lasers to chemical problems and physics of small molecules. Mailing Add: Dept of Chem Lake Forest Col Sheridan & College Rd Lake Forest IL 60045

LONG, CHARLES H, b Batesville, Ark, Dec 20, 35; m 61. ANIMAL NUTRITION. Educ: Ark State Col, BSA, 60; Mich State Univ, PhD(animal nutrit), 64. Prof Exp: Asst prof animal husb, Univ Mo, 64-67; mgr livestock nutrit res, Western Grain Co, 67-69; mgr animal nutrit, ConAgra, Inc, Ala, 69-74; ANIMAL NUTRITIONIST, SOUTHERN FARMERS ASN, 74- Mem: Am Soc Animal Sci. Res: Protein nutrition of neonatales; antibody absorption and production; feed composition; milk secretion and composition. Mailing Add: Southern Farmers Asn 824 N Palm North Little Rock AR 72114

LONG, CLARENCE SUMNER, JR, b Adairsville, Ga, June 5, 29; m 56; c 4. GEOLOGY. Educ: Tulane Univ, BS, 51; Univ Colo, PhD(geol), 66. Prof Exp: Geologist, Gulf Oil Corp, 54-57; asst geol, Univ Colo, 57-64; explor geologist, Pan Am Petrol Corp, 64-66; asst prof geol, Univ Ga, 66-69; assoc prof, 69-74, PROF GEOL, WGA COL, 74-, CHMN DEPT, 69- Concurrent Pos: Chattahoochee-Flint Planning Comn proj dir, Mineral Study Chattahoochee-Flint Area, 67-69. Mem: AAAS; Geol Soc Am; Am Asn Petrol Geol. Mailing Add: Dept of Geol WGa Col Carrollton GA 30117

LONG, CLIFFORD A, b Chicago, Ill, Apr 10, 31; m 57; c 4. MATHEMATICS. Educ: Univ Ill, BS, 54, MS, 55, PhD(math), 60. Prof Exp: From instr to assoc prof, 59-71, PROF MATH, BOWLING GREEN STATE UNIV, 71- Mem: Am Math Soc; Math Asn Am. Res: Computer graphics; numerical analysis. Mailing Add: Dept of Math Bowling Green State Univ Bowling Green OH 43403

LONG, DALE DONALD, b Louisa, Va, Jan 30, 35; m 65; c 2. EXPERIMENTAL PHYSICS. Educ: Va Polytech Inst, BS, 58, MS, 62; Fla State Univ, PhD(physics), 66. Prof Exp: Instr physics, Va Polytech Inst, 60; instr, Samford Univ, 60-62; ASST PROF PHYSICS, VA POLYTECH INST & STATE UNIV, 67- Mem: Am Phys Soc. Res: Experimental nuclear physics; experimental biophysics. Mailing Add: Dept of Physics Va Polytech Inst & State Univ Blacksburg VA 24061

LONG, DANIEL R, b Redding, Calif, June 9, 38; m 61; c 2. PHYSICS. Educ: Univ Wash, PhD(physics), 67. Prof Exp: Asst prof, 67-71, ASSOC PROF PHYSICS, EASTERN WASH STATE COL, 71- Mem: Am Phys Soc. Res: Electron impact ionization of metastable helium; experimental examination of the mass separation dependence of the gravitational constant. Mailing Add: Dept of Physics Eastern Wash State Col Cheney WA 99004

LONG, DARYL CLYDE, b Mason City, Iowa, Aug 19, 39; m 60; c 3. SOIL SCIENCE. Educ: Iowa State Univ, BS, 62, MS, 64; Univ Nebr, Lincoln, PhD, 67. Prof Exp: Instr soils, Univ Nebr, Lincoln, 64-67; ASST PROF SCI & MATH, PERU STATE COL, 67- Mem: Am Soc Agron; Soil Sci Soc Am. Res: Mechanics of soil erosion and plant removal of nutrients from soil aggregates. Mailing Add: Dept of Sci & Math Peru State Col Peru NE 68421

LONG, DAVID MICHAEL, b Shamokin, Pa, Feb 26, 29; c 6. CARDIOVASCULAR SURGERY, THORACIC SURGERY. Educ: Muhlenberg Col, BS, 51; Hahnemann Med Col, MS, 54, MD, 56; Univ Minn, PhD(physiol), 65; Am Bd Surg, dipl, 66; Bd Thoracic Surg, dipl, 67. Prof Exp: Instr surg, Univ Minn, 65; from asst prof to assoc prof, Chicago Med Sch, 65-67; from assoc prof to prof surg, Abraham Lincoln Sch Med, Univ Ill Med Ctr, 69-73, attend staff & head div cardiovasc & thoracic surg, Hosp, 67-73; CLIN ASSOC PROF RADIOL, UNIV CALIF, SAN DIEGO, 73- Concurrent Pos: Assoc prof, Cook County Grad Sch Med, 65-73; assoc attend staff, Cook County Hosp, 65-73; asst dir dept surg res, Hektoen Inst Med Res, 65-68, dir, 68-73; attend staff, W Side Vet Admin Hosp, 66-73; consult, Chicago State Tuberc Sanitarium, 67-72; pvt pract, 73- Honors & Awards: First Prize Res, Am Urol Asn, 66. Mem: AAAS; Am Asn Thoracic Surg; fel Am Col Cardiol; fel Am Col Chest Physicians; fel Am Col Surg. Res: Surgical research; physiology and morphology; cancer chemotherapy; development of the radiopaque compound perfluorocarbon. Mailing Add: Suite 221 5565 Grossmont Ctr Dr La Mesa CA 92401

LONG, DONLIN MARTIN, b Rolla, Mo, Apr 14, 34; m 59; c 3. NEUROSURGERY, ELECTRON MICROSCOPY. Educ: Univ Mo, MD, 59; Univ Minn, PhD(anat), 64. Prof Exp: Clin assoc, Surg Neurol Bd, NIH, 65-67; assoc prof neurosurg, Univ Minn Hosps, 67-73; PROF NEUROL SURG & DIR DEPT, SCH MED, JOHNS HOPKINS UNIV, 73- Concurrent Pos: Consult neurosurgeon, Vet Admin Hosp, Minneapolis, 67- Mem: AAAS; Am Asn Neurol Surg; Cong Neurol Surg; Am Asn Neuropath; Soc Neurosci. Res: Electron microscopy of normal and abnormal control nervous system. Mailing Add: Dept of Neurol Surg Johns Hopkins Univ Sch of Med Baltimore MD 21218

LONG, EARL ELLSWORTH, b Akron, Ohio, Mar 27, 19; m 41; c 4. PUBLIC HEALTH LABORATORY ADMINISTRATION. Educ: Univ Akron, BSc, 42; Univ Pa, MSc, 47. Prof Exp: Asst instr med bact, Sch Med, Univ Pa, 45-48; asst prof bact, Univ Akron, 48-49; dir labs, Akron Health Dept, 49-61; DIR LABS, GA DEPT PUB HEALTH, 61-63 & 63- Mem: Am Soc Microbiol; fel Am Pub Health Asn; Asn State & Territorial Pub Health Labs Dirs; Sigma Xi. Res: State public health laboratory administration with emphasis on implementation of rapidly changing concepts in service and research. Mailing Add: Div of Labs Ga Dept Pub Health 47 Trinity Ave SW Atlanta GA 30334

LONG, EDWARD B, b White Plains, NY, Dec 5, 27; m 52, 70; c 3. AQUATIC ECOLOGY. Educ: Hamilton Col, BA, 52; Kent State Univ, MS, 71, PhD(ecol), 75. Prof Exp: Mem staff mkt, Carbon Prod Div, Union Carbide Corp, 6-52-64, proj mgr, New Prod Mkt Develop, 64-69; CHIEF ENVIRON PLANNING, NORTHEAST OHIO AREAWIDE COORD AGENCY, 75- Mem: Am Soc Limnol & Oceanog; Ecol Soc Am; AAAS. Res: Directing planning for long-term areawide management of waste-water for Northeast Ohio. Mailing Add: NE Ohio Areawide Coord Agency 400 The Arcade Cleveland OH 44114

LONG, ERNEST CROFT, b London, Eng, Nov 5, 20; m 54; c 1. PHYSIOLOGY. Educ: Univ London, MB & BS, 53, PhD, 57. Prof Exp: Lectr physiol, St Mary's Hosp Med Sch, 54-56; assoc prof physiol & assoc pediat, Sch Med, Duke Univ, 56-66, prof community health sci, 66-73, asst dean med stud affairs, 65-67, assoc dean undergrad med educ, 67-73; FIELD STAFF MEM, ROCKEFELLER FOUND, 73- Concurrent Pos: Hanes fel pediat, Duke Hosp, 53-54; field dir, Div Int Med Educ, Asn Am Med Col, Guatemala, 71-73. Res: International health; respiration mechanics in newborn. Mailing Add: Rt 7 Box 218 Erwin Rd Durham NC 27706

LONG, ERNEST M, b Spring City, Tenn, Aug 23, 35; m; c 2. FOREST GENETICS. Educ: Univ Ark, BSF, 63; Tex A&M Univ, PhD(genetics), 72. Prof Exp: Forester, Tex Forest Serv, Tex A&M Univ, 63-65, silviculturist, 65-69, assoc geneticist, 72-75, ASST PROF FORESTRY, DEPT FOREST SERV, TEX A&M, 73- Concurrent Pos: Mem, Regional Christmas Tree Res Comt, Coop State Res Serv, USDA, 75-80. Mem: Soc Am Foresters. Res: Basic and applied research in forest genetics and tree improvement involving progeny testing, seed orchard management and biochemical genetics. Mailing Add: Dept of Forest Sci Tex A&M Univ College Station TX 77843

LONG, ESMOND RAY, b Chicago, Ill, June 16, 90; m 22; c 2. PATHOLOGY. Educ: Univ Chicago, AB, 11, PhD(path), 19; Rush Med Col, MD, 26. Hon Degrees: ScD, Univ Pa, 48. Prof Exp: Asst path, Univ Chicago, 11-13; asst, Div Plant Biol, Carnegie Inst, 14-15; from instr to prof path, Univ Chicago, 19-32; prof, 32-55, dir labs, Henry Phipps Inst, 32-35, dir, Inst, 35-55, pres, Wistar Inst, 39-42, EMER PROF PATH, UNIV PA, 55- Concurrent Pos: Trudeau fel, Saranac Lab, 20; mem med adv bd, Leonard Wood Mem Eradication Leprosy, 32-37 & 39-48, consult, 63-70; consult, Div Med Sci, Nat Res Coun, 32-55, chmn, 36-70; consult, US Indian Serv, 35-50, US Vet Admin, 46-51 & USPHS, 46-55; mem nat adv health coun, USPHS, 38-42; mem nat comn, UNESCO, 51-54; Clendening lectr, Univ Kans, 54; ed, Int J Leprosy, 64-69. Honors & Awards: Trudeau Medal, Nat Tuberc Asn, 32; Philadelphia Award, 54; Gold Headed Cane Award, Am Asn Path & Bact, 70. Mem: Nat Acad Sci; AAAS (vpres, Sect L, 36); Nat Tuberc & Respiratory Dis Asn; Am Asn Path & Bact (pres, 36); master Am Col Physicians. Res: Tuberculosis; history of medicine. Mailing Add: Apt 23B 220 Locust St Philadelphia PA 19106

LONG, FRANK WESLEY, JR, b Springfield, Ill, Aug 26, 25; m 51;c 3. ORGANIC CHEMISTRY, POLYMER CHEMISTRY. Educ: Univ Ill, BS, 46; Univ Iowa, PhD(org chem), 50. Prof Exp: Asst, Univ Iowa, 46-50; res chemist, Surfactant Group, Gen Aniline & Film Co, 50-52; proj mgr, Textile Dyeing & Finishing Lab, Philadelphia Qm Depot, Pa, 52-53; res chemist resins & plastics, Hooker Electrochem Co, 53-54, sales develop, 54-59, supvr prod develop, Hooker Chem Corp, 59-61, sect mgr, 61-64; dir prod develop, Princeton Chem Res, Inc, 64-67; PROD DIR COMMERCIAL DEVELOP, ARCO CHEM CO, ATLANTIC RICHFIELD CO, PHILADELPHIA, 67- Concurrent Pos: Adj prof, Univ Pa, 71. Mem: Am Chem Soc; Soc Plastics Engrs; Commercial Develop Asn; Am Asn Textile Chemists & Colorists. Res: Pyrrolizidines, sulfonation of camphor; surface active agents; textile dyeing and finishing agents; epoxy, polyester and urethane resins, fire retardant polymers; resins made from pyromellitic dianhydride; petrochemicals. Mailing Add: 292 W Riverside Dr Princeton NJ 08540

LONG, FRANKLIN A, b Great Falls, Mont, July 27, 10; m 37; c 2. PHYSICAL CHEMISTRY, SCIENCE POLICY. Educ: Univ Mont, AB, 31, MA, 32; Univ Calif, PhD(phys chem), 35. Prof Exp: Instr chem, Univ Calif, 35-36; instr, Univ Chicago, 36-37; from instr to assoc prof chem, 37-46, chmn dept, 50-60, dir prog sci, technol & soc, 69-73, PROF CHEM, CORNELL UNIV, 46-, LUCE PROF SCI & SOC, 69- Concurrent Pos: Res supvr, Explosives Res Lab, Nat Defense Res Comt, 42-45; consult, Ballistics Res Lab, Dept Army, 53-59; mem sci adv bd, Air Force Off Sci Res, 59-63; mem, Pres Sci Adv Comt, 61 & 64-67; asst dir, US Arms Control & Disarmament Agency, 62-63; mem bd sci & technol for int develop, Nat Acad Sci, 74-; mem, US-India Comn Educ & Cult Affairs, 74-; mem bd & consult, Inmont Corp & Exxon Corp; mem bd, Alfred P Sloan Found, Arms Control Asn & Assoc Univs,

Inc. Mem: Nat Acad Sci; AAAS; Am Chem Soc; Am Acad Arts & Sci. Res: Kinetics of solution reactions; isotopic chemistry; arms control; science and public policy. Mailing Add: 632 Clark Hall Cornell Univ Ithaca NY 14853

LONG, FRANKLIN LESLIE, b Hinesville, Ga, July 22, 18; m 43. SOIL CHEMISTRY, SOIL FERTILITY. Educ: Univ Ga, BS, 57, MS, 58; Univ Fla, PhD(soils), 60. Prof Exp: RES SOIL SCIENTIST, AGR RES SERV, USDA, 60- Mem: Am Soc Agron; Sigma Xi; Int Soc Soil Sci. Res: Soils in the Coastal Plains and Piedmont areas of United States. Mailing Add: Dept of Agron & Soils Auburn Univ Auburn AL 36830

LONG, GARY JOHN, b Binghamton, NY, Dec 3, 41; m 63; c 1. PHYSICAL INORGANIC CHEMISTRY. Educ: Carnegie-Mellon Univ, BS, 64; Syracuse Univ, PhD(chem), 68. Prof Exp: Asst prof, 68-74, ASSOC PROF CHEM, UNIV MO-ROLLA, 74- Concurrent Pos: Res assoc, Inorg Chem Lab & St John's Col, Oxford Univ, 74-75. Mem: Am Chem Soc; Am Phys Soc; The Chem Soc; Sigma Xi. Res: Transition metal inorganic coordination chemistry; Mössbauer and electronic spectroscopy; high-pressure optical and infrared spectroscopy; magnetic studies of coupled systems. Mailing Add: Dept of Chem Univ of Mo Rolla MO 65401

LONG, GEORGE DONALD, b Elizabeth, NJ, Nov 12, 29; m 52; c 2. PHYSICS. Educ: Lehigh Univ, BS, 51; Univ Pa, PhD(physics), 56. Prof Exp: Res scientist, Res Ctr, Minneapolis-Honeywell Regulator Co, 55-60, MGR RES DEPT, CORP RES CTR, HONEYWELL INC, 60- Mem: Fel Am Phys Soc. Res: Semiconductors, especially electrical properties related to energy band structure and scattering of carriers, and applied to transistors; infrared detectors; other semiconductor devices. Mailing Add: Honeywell Inc Corp Res Ctr 10701 Lyndale Ave S Bloomington MN 55420

LONG, GEORGE GILBERT, b Cincinnati, Ohio, July 12, 29; m 52; c 3. INORGANIC CHEMISTRY. Educ: Ind Univ, AB, 51; NC State Univ, MS, 53; Univ Fla, PhD(chem), 57. Prof Exp: Chemist, Ethyl Corp, 57-58; from asst prof to assoc prof, 58-70, grad adminr, 66-68, PROF CHEM, NC STATE UNIV, 70-, CHMN ANAL-INORG CHEM, DEPT CHEM, 69- Mem: Am Chem Soc. Res: Chemistry of group V metalloids-organometalloid compounds; 121-Sb Mössbauer spectroscopy, structure and syntheses; vibrational spectroscopy. Mailing Add: Dept of Chem NC State Univ Raleigh NC 27607

LONG, GEORGE LOUIS, b Atkin, Minn, Dec 20, 43; m 67; c 2. BIOCHEMISTRY, ENZYMOLOGY. Educ: Pac Lutheran Univ, BA, 66; Brandeis Univ, PhD(biochem), 71. Prof Exp: NIH trainee molecular endocrinol sch med, Univ Calif, San Diego, 71-73; ASST PROF CHEM, POMONA COL, 73- Mem: Am Chem Soc; AAAS. Res: Comparative enzymology of glycolytic enzymes; isolation, characterization and immobilization of enzymes for industrial biochemical reactors. Mailing Add: Dept of Chem Pomona Col Claremont CA 91711

LONG, HARRIET RUTH, b Buffalo, NY, Jan 23, 13. GEOGRAPHY. Educ: State Univ NY Teachers Col, Buffalo, BS, 33; Clark Univ, MA, 41, PhD, 55. Prof Exp: Asst prof geog, Miss State Col Women, 43-44 & 46-48, assoc prof, 48-49; geog educ, Genn & Co, 49-53; assoc prof, 55-67, dir liberal arts prog, 62-67, prof geog & chmn dept geog earth sci, 67-74, EMER PROF GEOG, EDINBORO STATE COL, 74- Concurrent Pos: Mem, Int Geog Union; mem exec bd, Nat Coun Geog Educ, 60-64, chmn planning comt, 62-65. Mem: Asn Am Geog; Am Geog Soc; Nat Geog Soc; Coun Geog Educ; Soc Woman Geog. Res: Urban Europe and Southeast Asia. Mailing Add: Dept of Geog Edinboro State Col Edinboro PA 16412

LONG, HENRY FOLTZ, bacteriology, see 12th edition

LONG, HOWARD CHARLES, b Seizholtzville, Pa, Dec 12, 18; m 45; c 3. PHYSICS. Educ: Northwestern Univ, BS, 41; Ohio State Univ, PhD(physics), 48. Prof Exp: Physicist, Naval Ord Lab, 42-45; instr physics, Ohio State Univ, 47-48; asst prof, Washington & Jefferson Col, 48-51; physicist, Naval Ord Lab, 51-52; assoc prof physics & chmn dept, Am Univ, 52-53; prof & chmn dept, Gettysburg Col, 53-59; chmn dept, 63-74, PROF PHYSICS, DICKINSON COL, 59- Concurrent Pos: Consult, Naval Ord Lab, 54-73. Mem: Am Phys Soc; Am Asn Physics Teachers; Audio Eng Soc. Res: Low period fluctuations in earth's magnetism; environmental noise reduction; air pollution by solid particulates; molecualr structure and infrared spectroscopy; electromagnetism. Mailing Add: Dept of Physics & Astron Dickinson Col Carlisle PA 17013

LONG, JAMES ALVIN, b Porto Alegre, Brazil, July 13, 17; US citizen; m 40; c 4. EXPLORATION GEOPHYSICS. Educ: Univ Okla, BA, 37. Prof Exp: Computer, Stanolind Oil & Gas Co, 37-40, seismologist & party chief, 40-46; party chief, United Geophys Corp, 46-48, supvr, Venezuela, Brazil & Chile, 48-54, area mgr, Southern SAm, 54-59, regional opers mgr, S & Cent Am, 59-62, sr geophysicist, 62-65, coordr digital tech, 65-66, asst chief geophysicist, 66-67, regional mgr, Latin Am, 67-73; sr geophysicist, Tetra Tech, Inc, 73-74; GEOPHYS ADV, BOLIVIAN GOVT OIL CO, SANTA CRUZ, 74- Concurrent Pos: Independent int consult geophysicist, 72- Honors & Awards: Best Paper Award, Soc Explor Geophys, 66. Mem: Soc Explor Geophysicists; Am Asn Petrol Geologists; Mex Asn Petrol Geophysicists; Mex Asn Explor Geophysicists. Res: Operations of surface seismic sources; interpretation of seismic reflections from carboniferous glaciated areas. Mailing Add: 3951 Gulf Shore Blvd N Naples FL 33940

LONG, JAMES DELBERT, b Dover, Okla, Dec 18, 39. HORTICULTURE, PLANT PHYSIOLOGY. Educ: Okla State Univ, BS, 62; Univ Md, College Park, MS, 67, PhD(hort), 69. Prof Exp: Res asst weed control, Univ Md, College Park, 64-67, instr hort, 67-68; RES BIOLOGIST AGR CHEM, E I DU PONT DE NEMOURS & CO, INC, WILMINGTON, DEL, 68- Mem: Am Soc Hort Sci; Weed Sci Soc Am. Res: Control and modification of plant growth through the use of chemicals. Mailing Add: Rte 8 Box 139 Elkton MD 21921

LONG, JAMES DUNCAN, b Rusk, Tex, Sept 23, 25. ZOOLOGY. Educ: Sam Houston State Col, BS, 48, MA, 51; Univ Tex, PhD, 57. Prof Exp: Teacher, High Sch, Tex, 48-49 & Pub Schs, 51-52; instr biol, Lamar State Col Technol, 52-53; asst, Univ Tex, 53-56; assoc prof biol & head dept, Ill Col, 56-59; assoc prof, 59-63, dir dept, 63-72, PROF BIOL, SAM HOUSTON STATE UNIV, 63- Mem: AAAS; Soc Syst Zool; Entom Soc Am; Am Mosquito Control Asn. Res: Mosquito biology. Mailing Add: Dept of Biol Sam Houston State Univ Huntsville TX 77341

LONG, JAMES EARL, b Steelton, Pa, Jan 28, 28; m 53; c 2. TOXICOLOGY, INDUSTRIAL HYGIENE. Educ: Gettysburg Col, BA, 50; Univ Pittsburgh, MS, 55, ScD(indust health), 59. Prof Exp: Toxicologist, Med Res Labs, Army Chem Ctr, Md, 50-53, aerosol chemist, Chem & Radiol Labs, 53-54; res assoc respiratory physiol, Toxicol & Indust Hyg, Grad Sch Pub Health, Univ Pittsburgh, 55-58, instr indust hyg, 58-60; res indust hygienist, Environ Health Lab, Am Cyanamid Co, 60-62, sr res environ health scientist, 62-64; indust toxicologist & hygienist, M&T Chem, Inc, NJ, 64-68; dir environ toxicol, Int Res & Develop Corp Mich, 68-69; MGR TOXICOL, MINN MINING & MFG CO, 69- Concurrent Pos: AEC fel indust hyg, Grad Sch Pub Health, Univ Pittsburgh, 54-55. Mem: AAAS; Am Indust Hyg Asn; Water Pollution Control Fedn; Am Chem Soc; Soc Toxicol. Res: Industrial hygiene sampling and instrumentation; assessment of physiological impairment by pulmonary irritants; synergistic effects of aerosols and vapors; inhalation toxicology. Mailing Add: Minn Mining & Mfg Co 3M Ctr St Paul MN 55101

LONG, JAMES FRANTZ, b Center Valley, Pa, Sept 17, 31; m 56; c 3. PHYSIOLOGY. Educ: Mich State Univ, BS, 57, MS, 59; Univ Pa, PhD(physiol), 64. Prof Exp: From instr to assoc prof physiol, Albany Med Col, 64-69; prin scientist, 69-75, RES FEL, DEPT PHARMACOL, SCHERING CORP, 75- Mem: Am Physiol Soc; Am Gastroenterol Asn. Res: Gastrointestinal physiology and pharmacology. Mailing Add: Dept of Pharmacol Schering Corp Bloomfield NJ 07003

LONG, JAMES HARVEY, JR, b Johnson City, Tenn, Sept 14, 44; m 65; c 1. CHEMISTRY, CHEMICAL ENGINEERING. Educ: Univ Tenn, BS, 65, PhD(chem), 68. Prof Exp: Chemist, Shell Chem Co, 68-71, sr chemist, Shell Develop Co, 72-73, PROCESS MGR, SHELL CHEM CO, 73- Mailing Add: 11323 Sagewillow Lane Houston TX 77034

LONG, JAMES WILLIAM, b Boise, Idaho, Aug 26, 43; m 65; c 1. BIOCHEMISTRY. Educ: Univ Wash, BS, 65; Univ Calif, Berkeley, PhD(biochem), 69. Prof Exp: Res assoc biochem, Purdue Univ, West Lafayette, 70-71, NIH res fel, 71-72, res assoc, 72-73; res assoc, Univ Ore, 73-74; ASST PROF CHEM, COL GREAT FALLS, 74- Mem: Am Chem Soc. Res: Nuclear magnetic resonance in biological studies; structure-function relationships in enzymes; mechanisms of enzyme action; enzyme model systems; role of metal ions in enzyme catalysis. Mailing Add: Dept of Chem Col of Great Falls Great Falls MT 59405

LONG, JEROME R, b Lafayette, La, May 17, 35; m 62; c 2. PHYSICS. Educ: Univ Southwestern La, BS, 56; La State Univ, MS, 58, PhD(physics), 65. Prof Exp: Res engr, Gen Dynamics/Pomona, 58-59; fel metall, Univ Pa, 65-67; asst prof physics, 67-71, ASSOC PROF PHYSICS, VA POLYTECH INST & STATE UNIV, 71- Mem: Am Phys Soc. Res: Transport effects in metals; experimental cryophysics; fermi surfaces. Mailing Add: Dept of Physics Va Polytech Inst & State Univ Blacksburg VA 24061

LONG, JOHN ARTHUR, b Kingman, Kans, July 30, 34. CYTOLOGY, ANATOMY. Educ: Univ Kans, AB, 56; Univ Wash, PhD(zool), 64. Prof Exp: Res assoc anat, Harvard Med Sch, 66-67; asst prof, 67-74, ASSOC PROF ANAT, MED CTR, UNIV CALIF, SAN FRANCISCO, 74- Mem: AAAS; Am Soc Zool. Res: Fine structure of steroid secreting cells in ovary and adrenal glands. Mailing Add: Dept of Anat Univ of Calif Med Ctr San Francisco CA 94143

LONG, JOHN EDWARD, food chemistry, see 12th edition

LONG, JOHN FREDERICK, b Napoleon, Ohio, May 30, 24; m 48; c 5. VETERINARY PATHOLOGY. Educ: Ohio State Univ, BA, 47, MSc, 48, DVM, 55, PhD(comp neuropath), 66. Prof Exp: Res asst animal sci, Ohio Agr Exp Sta, 49-50; res asst, Vet Diag Lab, State of Ohio, 55-63; res assoc comp neuropath, 63-64, NIH res fel, 64-66, instr vet path, 66-67, NIH spec res fel comp neuropath, 67-68, asst prof vet path, 68-71, ASSOC PROF VET PATH, OHIO STATE UNIV, 71- Mem: Am Vet Med Asn; Am Asn Avian Path. Res: Comparative neuropathology; viral encephalomyelitides; use of brain explant culture and germ-free animals in the study of the effects of encephalitogenic agents; demyelinating encephalomyelitides of animals. Mailing Add: 2765 Bexley Park Rd Columbus OH 43209

LONG, JOHN KELLEY, b NY, Dec 12, 21; m 48; c 3. NUCLEAR PHYSICS, NUCLEAR ENGINEERING. Educ: Columbia Univ, BS, 42; Ohio State Univ, PhD(physics), 53. Prof Exp: Chemist plastics, Hercules Powder Co, 42-45; engr, Wright Field, 47-50; physicist, Battelle Mem Inst, 52-55; physicist, Idaho Div, Argonne Nat Lab, 55-74; REACTOR ENGR, US NUCLEAR REGULATORY COMN, 74- Mem: Am Nuclear Soc. Res: Fast reactor physics; critical experiments; reactor licensing; fast reactor safety test facilities; plutonium toxicity. Mailing Add: US Nuclear Regulatory Comn Washington DC 20555

LONG, JOHN MARSHALL, mathematical statistics, see 12th edition

LONG, JOHN PAUL, b Albia, Iowa, Oct 4, 26; m 50; c 3. PHARMACOLOGY. Educ: Univ Iowa, BS, 50, MS, 52, PhD(pharmacol), 54. Prof Exp: From asst to instr pharmacol, Univ Iowa, 50-54; res assoc, Sterling-Winthrop Res Inst, 54-56; from asst prof to assoc prof, 56-62, PROF PHARMACOL, COL MED, UNIV IOWA, 62-, HEAD DEPT, 70- Mem: Am Soc Pharmacol & Exp Therapeut; Soc Exp Biol & Med. Res: Structure-activity relationships of autonomic and anesthetic agents. Mailing Add: Dept of Pharmacol Univ of Iowa Col of Med Iowa City IA 52240

LONG, JOHN VINCENT, b San Diego, Calif, Feb 18, 10; m 38; c 4. PHYSICS. Educ: Univ Calif, Los Angeles, AB, 37. Prof Exp: Lab asst physics, San Diego State Col, 32-35; serv demonstr, Ford Motor Co, Calif, 35-36; res engr, Douglas Aircraft Co, 36-37; geophysicist, Continental Oil Co, Okla, 37-40; res engr, 40, res physicist & asst dir res, 46-51, DIR RES, SOLAR DIV, INT HARVESTER CO, 51- Mem: Soc Explor Geophys; Acoust Soc Am; Soc Exp Stress Anal; Am Ceramic Soc; assoc Inst Elec & Electronics Engrs. Res: Ceramics; metallurgy; vibration and sound; high altitude research; ceramic coatings for high temperature corrosion and oxidation protection of iron, stainless steel; super alloys and refractory metals. Mailing Add: 1756 E Lexington Ave El Cajon CA 92021

LONG, JOHN WILLIAM, III, organic chemistry, see 12th edition

LONG, JOSEPH K, b Greenville, Ky, Mar 31, 37; m 65; c 1. ANTHROPOLOGY. Educ: Southern Methodist Univ, AB, 59; Univ Ky, MS, 64; Univ NC, Chapel Hill, PhD(anthrop), 73. Prof Exp: Asst anthrop, Univ Ky, 63, asst geront, 64, instr anthrop, 64-66; instr, Univ Wis-Green Bay 66-68; from instr to asst prof, Southern Methodist Univ, 71-74; ASST PROF ANTHROP, UNIV NH, 74- Concurrent Pos: Carnegie Found investr prog learning, 67; prog developer, Articulated Instrnl Media, Univ Wis-Madison, 67-68; lectr, St Norbert Col, 68; Fulbright Hays res fel, Mexico City, 68; lectr-consult, Dept Social & Prev Med, Univ West Indies, 70-71; consult, Caribbean Food & Nutrit Inst, Pan Am Health Orgn, Kingston, Jamaica, 70-71; assoc, Current Anthrop, 71-; consult, Pharmacol Res-Herbalism in Latin Am, Tex A&M Univ, 72-; mem, Parapsychol Found. Mem: Fel AAAS; fel Am Anthrop Asn; Am Soc Psychical Res; Soc Study Human Biol; Am Asn Phys Anthrop. Res: Applied anthropology; epidemiology; culture change; bio-social evolution; urban anthropology; development of instructional programs; anthropology films; medical anthropology; psychosomatic medicine; parapsychological anthropology. Mailing Add: Dept of Anthrop Univ of NH Plymouth NH 03264

LONG, JOSEPH POTE, b Baker Summit, Pa, Feb 26, 13; m 42; c 4. OBSTETRICS & GYNECOLOGY. Educ: Juniata Col, BS, 34; Univ Pa, MS, 48. Prof Exp: From demonstr to assoc prof, 48-75, CLIN PROF OBSTET & GYNEC, JEFFERSON

MED COL, THOMAS JEFFERSON UNIV, 75- Mem: AMA; Am Col Surg; Am Col Obstet & Gynec; Am Fertil Soc; NY Acad Sci. Res: Benign and malignant pelvic tumors, especially ovarian lesions; clinical research in dysmenorrhea. Mailing Add: 130 S Ninth St Philadelphia PA 19107

LONG, KEITH ROYCE, b Lincoln, Kans, Mar 17, 22; m 45; c 5. ENVIRONMENTAL HEALTH. Educ: Univ Kans, AB, 51, MA, 53; Univ Iowa, PhD, 60. Prof Exp: Asst instr bact, Univ Kans, 52-53, instr bact res, Med Ctr, 53-56; sr bacteriologist & virologist, State Hyg Lab, 56-57, instr, Inst Agr Med, 57-58, asst bact, 58-60, assoc prof hyg & prev med, Inst Agr Med, Col Med, 60-69, PROF PREV MED & ENVIRON HEALTH SCI, INSTR AGR MED & ENVIRON HEALTH, COL MED, UNIV IOWA, 69-, DIR, INST, 74-, PROF CIVIL ENG, UNIV, 70- Mem: Soc Occup & Educ Health; Soc Epidemiol Res. Res: Environmental toxicology; epidemiology; pesticides. Mailing Add: Inst Agr Med & Environ Health Univ of Iowa Col of Med Iowa City IA 52240

LONG, KENNETH MAYNARD, b Nappanee, Ind, July 10, 32; m 52; c 5. INORGANIC CHEMISTRY. Educ: Goshen Col, BS, 54; Mich State Univ, MA, 60; Ohio State Univ, PhD(chem), 67. Prof Exp: Instr, Parochial Sch, Ark, 54-56; instr, High Sch, Mich, 56-61; from instr to asst prof chem, 62-70, asst dean, 71-75, ASSOC PROF CHEM, WESTMINSTER COL, PA, 70- Mem: Am Chem Soc; Nat Sci Teachers Asn. Res: Macrocyclic complexes of transition metals; catalytic properties of transition metal complexes; kinetics. Mailing Add: Dept of Chem Westminster Col New Wilmington PA 16142

LONG, LAWRENCE WILLIAM, b Akron, Ohio, Nov 6, 42; m 69. BIOCHEMISTRY. Educ: Franklin & Marshall Col, AB, 65; Villanova Univ, PhD(chem), 71. Prof Exp: Instr biochem, Thomas Jefferson Univ, 71-73; res scientist, Stevens Inst Technol, 73-74; PROJ LEADER CHEM, ANHEUSER-BUSCH, INC, 74- Mem: Am Chem Soc; AAAS; Sigma Xi. Res: Use of yeast in food and industrial applications. Mailing Add: Anheuser-Busch Inc 721 Pestalozzi St St Louis MO 63118

LONG, LELAND TIMOTHY, b Auburn, NY, Sept 6, 40; m 70; c 1. GEOPHYSICS, SEISMOLOGY. Educ: Univ Rochester, BS, 62; NMex Inst Mining & Technol, MS, 64; Ore State Univ, PhD(geophys), 68. Prof Exp: ASST PROF GEOPHYS, GA INST TECHNOL, 68- Mem: Am Geophys Union; Seismol Soc Am; Soc Explor Geophys. Res: Earthquake seismology; regional gravity studies. Mailing Add: Sch of Geophys Sci Ga Inst Technol Atlanta GA 30332

LONG, LEON EUGENE, b Wanatah, Ind, May 4, 33; m 56; c 2. GEOCHEMISTRY. Educ: Wheaton Col, BS, 54; Columbia Univ, MA, 58, PhD(geochem), 59. Prof Exp: Geochemist, Lamont Geol Observ, Columbia Univ, 59-60; NSF fel, Oxford Univ, 60-62; from asst prof to assoc prof, 62-75, PROF GEOL, UNIV TEX, AUSTIN, 75- Mem: AAAS; fel Geol Soc Am; Am Geophys Union; Geochem Soc; Sigma Xi. Res: Isotopic age methods. Mailing Add: Dept of Geol Sci Univ of Tex Austin TX 78712

LONG, LOREN MARLIN, b Dallas, Tex, Jan 7, 14; m 43; c 3. CHEMISTRY. Educ: NTex State Col, BS, 37, MS, 38; Univ Tex, PhD(org chem), 41. Prof Exp: Res chemist & sect leader, 41-51, asst dir chem res, 51-54, DIR CHEM RES, PARKE DAVIS & CO, 54- Mem: AAAS; Am Chem Soc. Res: Anti-epileptic drugs; antibiotics; hypnotics; vitamins; improved processes for preparation of anticonvulsants; preparation of antibiotic substances; hydantoins. Mailing Add: Res & Biol Lab Parke Davis & Co PO Box 118 GPO Detroit MI 48232

LONG, LOUIS, JR, b New York, NY, Jan 3, 03; m 38; c 4. ORGANIC CHEMISTRY. Educ: Princeton Univ, BS, 23; Mass Inst Technol, SM, 25, Harvard Univ, MA, 37, PhD(org chem), 39. Prof Exp: Asst chem, Harvard Univ, 37-38; Smith, Kline & French fel, Univ Va, 39-41, Du Pont fel, 41-42; spec res assoc, Harvard Univ, 42-44; res assoc, Mass Inst Technol, 44-46; consult chemist, 46-48; pres, Concord Labs, 48-53; consult chemist, 53-55; asst chief org chem br, Pioneering Res Div, Qm Res & Eng Command, US Dept Army, 55-65, head org chem lab, US Army Natick Labs, 65-73; RETIRED. Mem: Fel AAAS; Am Chem Soc; NY Acad Sci; The Chem Soc. Res: Carcinogenic compounds; ozonization; adrenal cortex hormones; sucrose; carbohydrates; sulfur compounds. Mailing Add: Carr Rd Concord MA 01742

LONG, MARGARET ELEANOR, b Philadelphia, Pa, Mar 19, 03. CYTOCHEMISTRY. Educ: Univ Pa, BS, 25, MS, 28, PhD(cytol), 40. Prof Exp: Res technician, Wistar Inst, 26-30; technician gen zool & histol, Duke Univ, 30-32; asst histol & neurol, Ohio State Univ, 35-38; res assoc nutrit, Inst Exp Biol, Univ Calif, 38-40; instr anat & histol, Univ Miss, 42-44; instr anat, Med Col Ala, 44-46; from instr to asst prof biol, Wash Sq Col, NY Univ, 46-50; asst prof obstet & gynec, Col Physcians & Surgeons, Columbia Univ, 50-54, res assoc, 54-65, res assoc path, 65-68; RES ASSOC PATH, LENOX HILL HOSP, 68- Mem: Fel AAAS; Am Soc Zool; Am Asn Anat; Am Asn Cancer Res; Histochem Soc. Res: Human lung research; cytology of orthoptera; histology of muscle-tendon attachment; histochemistry of adrenal; gynecologic cancer research. Mailing Add: Lenox Hill Hosp Dept Path 100 E 77th St New York NY 10021

LONG, MARY JEAN, b Duluth, Minn, Mar 24, 14; m 36; c 2. MEDICAL TECHNOLOGY, BIOCHEMISTRY. Educ: Mich State Univ, BS, 35, MS, 61, PhD(path), 68. Prof Exp: Lab technician, Parke Davis & Co, 35-36; supvr clin chem, Blodgett Mem Hosp, 51-62; instr path, Mich State Univ, 62-69 & med technician, 66-68; ASST PROF PATH, SCH ALLIED HEALTH PROFESSIONS, COL MED, UNIV NEBR, OMAHA, 69- Honors & Awards: Hilkowitz Award, Am Soc Med Technol, 61. Mem: Am Soc Med Technol; NY Acad Sci; Am Soc Allied Health Professions; Am Asn Clin Chemists. Res: Clinical laboratory science; clinical chemistry. Mailing Add: Dept of Biol Univ of Nebr Omaha NE 68105

LONG, MICHAEL EDGAR, b CZ, June 22, 46; m 68. PHYSICAL CHEMISTRY. Educ: Univ Toledo, BEd, 68; Wayne State Univ, PhD(chem), 73. Prof Exp: Fel chem, Cornell Univ, 73-75; RES CHEMIST, EASTMAN KODAK CO, 75- Concurrent Pos: NIH fel, Cornell Univ, 74-75. Mem: Am Chem Soc; Am Phys Soc; Sigma Xi. Res: Molecular electronic spectroscopy and photophysical processes in organic molecules. Mailing Add: Eastman Kodak Co Res Labs Bldg 82 Rochester NY 14650

LONG, NORMAN OLIVER, b Niagara Falls, NY, June 1, 09; m 31; c 3. INORGANIC CHEMISTRY, ORGANIC CHEMISTRY. Educ: Hiram Col, BA, 32; Univ Buffalo, PhD(inorg chem), 35. Prof Exp: Asst chem, Univ Buffalo, 32-35; from instr to asst prof, SDak State Col, 35-39; prof, Wis State Teachers Col, Superior, 39-42; res chemist, Mathieson Alkali Works, NY, 42-46; prof chem & head dept, Shurtleff Col, 46-48; assoc prof, Univ Ky, 48-50; prof & head dept, Evansville Col, 50-57; prof & adv, Univ Indonesia, 57-59; prof & chmn dept, Ind Inst Technol, 59-65; prof & head dept chem, 65-75, chmn div nat sci & math, 68-73, EMER PROF CHEM, CENT METHODIST COL, 75- Mem: Am Chem Soc. Res: Administration and teaching of chemistry. Mailing Add: 710 N Church St Fayette MO 65248

LONG, PAUL EASTWOOD, JR, b Philadelphia, Pa, Oct 9, 42; m 69; c 1. METEOROLOGY, NUMERICAL ANALYSIS. Educ: Drexel Univ, BS, 65, MS, 68, PhD(physics), 70. Prof Exp: Mathematician, Philco-Ford Corp, 64-65; fel, Drexel Univ, 70-71; assoc, Nat Weather Serv, 71-73, res meteorologist, 73-74; RES METEOROLOGIST, SAVANNAH RIVER LAB, E I DU PONT DE NEMOURS & CO, INC, 74- Mem: Am Meteorol Soc; Am Inst Physics. Res: Numerical planetary boundary layer modeling. Mailing Add: 315 Homestead Lane SE Aiken SC 29801

LONG, RAYMOND CARL, b Shattuck, Okla, June 17, 39; m 59; c 4. PLANT PHYSIOLOGY. Educ: Kans State Univ, BS, 61, MS, 62; Univ Ill, Urbana, PhD(plant physiol), 66. Prof Exp: Asst prof, 66-73, ASSOC PROF CROP SCI, NC STATE UNIV, 73- Mem: Am Soc Plant Physiol; Am Soc Agron. Res: Biochemistry of growth and senescence of higher plants; diurnal variations in metabolism. Mailing Add: Dept of Crop Sci NC State Univ Raleigh NC 27607

LONG, RICHARD GENE, b Dallas, Iowa, June 28, 31; m 53; c 4. MATHEMATICS. Educ: Reed Col, BA, 52; Univ Wash, PhD, 57. Prof Exp: Actg instr math, Univ Wash, 56-57; from instr to asst prof, Wesleyan Univ, 57-65; assoc exec dir comt on educ media, Math Asn Am, 65-66, exec producer, Individual Lect Proj, Calif, 66-67; mathematician, Cambridge Conf Sch Math, Mass, 67-68; proj dir, Individual Lect Film Proj, Math Asn Am, 68-69; chmn dept, 69-73, ASSOC PROF MATH, LAWRENCE UNIV, 69- Mem: AAAS; Math Asn Am; Am Math Soc. Res: Normed linear spaces; convexity. Mailing Add: Dept of Math Lawrence Univ Appleton WI 54911

LONG, ROBERT GRANT, b Crystal Falls, Mich, Dec 19, 18; m 43; c 2. GEOGRAPHY OF LATIN AMERICA, CARTOGRAPHY. Educ: Univ Mich, BA, 41; Syracuse Univ, AM, 43; Northwestern Univ, PhD(geog), 49. Prof Exp: Geographer, US Bd Geog Names, US Dept Interior, 43-44; Soc Sci Res Coun fel, Brazil, 48; from asst prof to assoc prof, 49-66, PROF GEOG, UNIV TENN, KNOXVILLE, 66- Mem: Asn Am Geog; Am Geog Soc. Res: Regional and economic geography of Latin America, particularly Brazil and Central American republics. Mailing Add: Dept of Geog Univ of Tenn Knoxville TN 37916

LONG, ROBERT RADCLIFFE, b Glen Ridge, NJ, Oct 24, 19; m 63; c 2. METEOROLOGY. Educ: Princeton Univ, AB, 41; Univ Chicago, MS, 49, PhD, 50. Prof Exp: Meteorologist, US Weather Bur, 46-47; sr investr, hydrodyn lab, Univ Chicago, 49-51; from asst prof to assoc prof meteorol, 51-59, PROF FLUID MECH, JOHNS HOPKINS UNIV, 59- Concurrent Pos: Mem adv panel gen sci, US Secy Defense Res & Eng. Mem: Am Meteorol Soc. Res: Geophysical fluid mechanics; theoretical studies and laboratory models of geophysical phenomena; general circulation of the atmosphere; atmospheric and oceanic flow over barriers. Mailing Add: Dept of Mech Johns Hopkins Univ Baltimore MD 21218

LONG, ROBERT SIDNEY, organic chemistry, see 12th edition

LONG, ROBERT WILLIAM, b New Albany, Ind, Mar 14, 17; m 46; c 2. PHYSICAL CHEMISTRY. Educ: Ind State Univ, AB, 38; Univ Calif, Berkeley, PhD(chem), 41. Prof Exp: Res chemist, Union Oil Co Calif, 41-50; PROF CHEM, EL CAMINO COL, 50- Concurrent Pos: NSF sci fac fel, Calif Inst Technol, 61-62. Mem: Am Chem Soc. Res: Gas phase association of hydrogen and deuterium fluorides; shale oil chemistry and utilization; nitrocycloalkanes. Mailing Add: Dept of Chem El Camino Col Torrance CA 90506

LONG, ROBERT WILLIAM, JR, b Ashland, Ky, Nov 23, 27; m 53; c 4. SYSTEMATIC BOTANY. Educ: Ohio Wesleyan Univ, BA, 50; Ind Univ, MA, 52, PhD(bot), 54. Prof Exp: Asst field bot, Univ Minn, 49; asst bot, Ind Univ, 50-53; instr biol, Southern Methodist Univ, 53-54; from asst prof to assoc prof, Ohio Wesleyan Univ, 54-62; assoc prof bot & bact, 62-64, chmn dept, 62-64 & 66-71, cur herbarium, 62-64, PROF BOT & BACT & DIR HERBARIUM, UNIV S FLA, 64- Concurrent Pos: NSF res awards, 57-; ed, Plant Sci Bull, Bot Soc Am; fel, Harvard Univ, 68; comnr, Comn Undergrad Educ Biol Sci, 70-71; consult ecol, Tampa Elec Co, 74-75 & Sarasota County, Fla, 75. Mem: Am Soc Plant Taxon (treas, 65-70); Bot Soc Am; Asn Trop Biol; Int Asn Plant Taxon. Res: Cytogenetic and evolutionary systematics of plants; floristics of southern Florida; biosystematics of Ruellia and other Acanthaceae; ecology of Florida. Mailing Add: Dept of Biol Univ of S Fla Tampa FL 33620

LONG, SALLY YATES, b Moyock, NC, Nov 8, 41. EMBRYOLOGY, TERATOLOGY. Educ: Col William & Mary, BS, 63; Univ Fla, PhD(anat), 67. Prof Exp: Lectr genetics, McGill Univ, 68-70; res assoc teratology, Karolinska Inst, Sweden, 70-71; ASST PROF ANAT, MED COL WIS, 71- Concurrent Pos: NIH fel, McGill Univ, 68-70. Mem: Teratology Soc; Am Asn Anat; Europ Teratology Soc. Res: Interactions of genetic and environmental factors in causing malformations, especially cleft palate and limb defects. Mailing Add: Dept of Anat Med Col of Wis Milwaukee WI 53233

LONG, STERLING KRUEGER, bacteriology, see 12th edition

LONG, TERRILL JEWETT, b Newark, Ohio, Mar 19, 32; m 55; c 4. BOTANY. Educ: Ohio Univ, BSAg, 56; Ohio State Univ, MSc, 59, PhD(bot), 61. Prof Exp: NIH fel, Oak Ridge Nat Lab, 61-63, res assoc bot, 63-64; asst prof biol, Vanderbilt Univ, 64-65; res assoc biochem, Ohio State Univ, 65-67; asst prof, 67-70, ASSOC PROF BIOL, CAPITAL UNIV, 70- Concurrent Pos: Consult, C S Fred Mushroom Co, 66-70. Mem: Am Soc Plant Physiol; Bot Soc Am; Mycol Soc Am. Res: Physiology and biochemistry of irradiated wheat and mushrooms and related fungi. Mailing Add: Dept of Biol Capital Univ 2199 E Main St Columbus OH 43209

LONG, THEODORE ALFRED, b Miami, Okla, Jan 23, 18; m 38; c 2. ANIMAL NUTRITION. Educ: Okla State Univ, BS, 52, MS, 54, PhD, 56. Prof Exp: PROF ANIMAL NUTRIT, PA STATE UNIV, 56- Mem: AAAS; Am Soc Animal Sci. Res: General nutrition of ruminants; preservation and nutritive value of forages and grains; recycling wastes through ruminants; roughage type and the nutrition and health of ruminants. Mailing Add: 306 Animal Indust Bldg Pa State Univ University Park PA 16802

LONG, THOMAS CARLYLE, b San Bernardino, Calif, July 26, 41; m 66; c 1. GENETICS. Educ: Univ Calif, Santa Barbara, BA, 63, Univ Calif, Berkeley, MSc, 65; Univ Wales, PhD(genetics), 69. Prof Exp: Asst prof zool & bot, Univ NC, Chapel Hill, 69-75; ASSOC PROF BIOL, SWEET BRIAR COL, 75- Mem: AAAS; Genetics Soc Am. Res: Evolutionary genetics. Mailing Add: Sweet Briar Col Sweet Briar VA 24595

LONG, THOMAS ROSS, b Lexington, Ky, Nov 6, 29; m 52; c 3. SOLID STATE PHYSICS. Educ: Ohio Wesleyan Univ, BA, 51; Case Inst Technol, MS, 53, PhD(physics), 56. Prof Exp: Mem tech staff, Bell Tel Labs, Inc, NJ, 56-67, HEAD FUNDAMENTAL STUDIES DEPT, BELL LABS, OHIO, 67- Mem: AAAS; Inst Elec & Electronics Engrs; Am Phys Soc. Res: Communications device and techniques; memory and logic devices; contact physics. Mailing Add: Bell Labs 6200 E Broad St Columbus OH 43213

LONG, WALTER K, b Austin, Tex, Jan 26, 19; m 50; c 1. HUMAN GENETICS. Educ: Univ Tex, BA, 40; Harvard Univ, MD, 43. Prof Exp: Res asst cardiol, Thorndike Mem Lab, Boston City Hosp, Mass, 45-48; RES SCIENTIST HUMAN GENETICS, UNIV TEX, AUSTIN, 59-, LECTR ZOOL, 70- Concurrent Pos: Life Ins med res fel, 47-48; chief cardiovasc sect, William Beaumont Army Hosp, Ft Bliss, Tex, 51-53. Mem: Am Soc Human Genetics. Res: Relation between sulfhydryl compounds and pharmacology of organic mercurial diuretics; pentose phosphate metabolic pathway in relation to certain human diseases. Mailing Add: Dept of Zool Univ of Tex Austin TX 78712

LONG, WILLIAM ELLIS, b Minot, NDak, Aug 18, 30; m 55, 71; c 6. STRATIGRAPHY, STRUCTURAL GEOLOGY. Educ: Univ Nev, BS, 57; Ohio State Univ, MSc, 61, PhD(geol), 64. Prof Exp: Instr geol, Ohio State Univ, 63-64; explor geologist, Tenneco Oil Co, La, 64-65; from asst prof to assoc prof, 65-72, PROF GEOL, ALASKA METHODIST UNIV, 72- Concurrent Pos: Mem, US Antarctic Res Prog, NSF Geol Invest, 63-64; mem discharge prediction glacial meltwater, Off Water Res, 68-70; consult, Shelf Explor Co, 71 & Forest Oil Co, 74-75; investr potential natural landmarks in Alaska, Nat Park Serv, 71; vis lectr, Univ Canterbury, 72. Mem: AAAS; Am Asn Petrol Geol; Am Inst Prof Geol; Geol Soc Am; Glaciol Soc. Res: Stratigraphic, geologic and glaciological exploration of Gondwana sequences of Antarctica during International Geophysical Year and following years; stratigraphic and glacial geology. Mailing Add: Dept of Geol Alaska Methodist Univ Anchorage AK 99504

LONG, WILLIAM HENRY, b Decatur, Ala, Sept 20, 28; m 53; c 3. ENTOMOLOGY. Educ: Univ Tenn, BA, 52; NC State Col, MS, 54; Iowa State Col, PhD, 57. Prof Exp: From asst prof to prof entom, La State Univ, 57-65; PROF BIOL SCI, NICHOLLS STATE UNIV, 65- Concurrent Pos: Pres, Long's Pest Control, Inc, 72-; consult entom, UN Food & Agr Orgn, United Arab Republic, 73-74; entom expert, Int Atomic Energy Agency, 75; mem staff, Nuclear Ctr, Agr Col, Univ Sao Paulo, 75-76. Mem: Entom Soc Am; Am Soc Sugarcane Technologists. Res: Development of sugar cane pest management programs; agricultural and horticultural entomology; household and structural pest control. Mailing Add: PO Box 1193 Thibodaux LA 70301

LONG, WILMER NEWTON, JR, b Hagerstown, Md, Apr 24, 18; m 42; c 2. MEDICINE, OBSTETRICS & GYNECOLOGY. Educ: Juniata Col, BS, 40; Johns Hopkins Univ, MD, 43. Prof Exp: Instr gynec & obstet, Sch Med, Johns Hopkins Univ, 48-65; assoc prof, 65-67, PROF GYNEC & OBSTET, SCH MED, EMORY UNIV, 67- Concurrent Pos: Pvt pract obstet, 48-65; med officer in chg obstet & gynec, Navajo Med Ctr, Ft Defiance, Ariz, 53-55. Mem: Am Col Obstet & Gynec; AMA. Res: Diabetes in pregnancy. Mailing Add: 80 Butler St SE Atlanta GA 30303

LONGACRE, SUSAN ANN BURTON, b Los Angeles, Calif, May 26, 41; m 64; c 2. SEDIMENTARY PETROLOGY, PETROLEUM GEOLOGY. Educ: Univ Tex, Austin, BS, 64, PhD(geol), 68. Prof Exp: Res assoc III, 69-72, res assoc IV, 72-75, RES SCIENTIST I GEOL, EXPLOR & PROD RES LAB, GETTY OIL CO, 75- Mem: Am Asn Petrol Geologists; Geol Soc Am; Soc Econ Paleontologists & Mineralogists. Res: Petrology and petrography of carbonate and clastic sediments, particularly those Permian, Jurassic and Cretaceous sediments that accumulated in shallow marine to continental depositional environments. Mailing Add: Getty Oil Co 3903 Stoney Brook Houston TX 77063

LONGACRE, WILLIAM ATLAS, b Carthage, Tenn, Aug 3, 06; m 29; c 2. GEOPHYSICS. Educ: Mich Col Mining & Technol, BS & EMet, 29, MA, 41. Prof Exp: Instr math & physics, 29-37, from asst prof to prof, 37-74, head dept physics, 54-63, EMER PROF PHYSICS, MICH TECHNOL UNIV, 74-, DIR GEOPHYS, 63- Concurrent Pos: Geophys res asst, Univ Mich, 29-37; consult geophysicist, Mining & Milling Serv Co, Ltd, Can, 37, Consol Mining & Smelting Co, Ltd, 41-42 & Cleveland Cliffs Iron Co, Mich, 46-54. Mem: Soc Explor Geophys. Res: Exploration geophysics; magnetic, electrical and gravity theory. Mailing Add: Dept of Physics Mich Technol Univ Houghton MI 49931

LONGACRE, WILLIAM ATLAS, II, b Hancock, Mich, Dec 16, 37. ANTHROPOLOGY, ARCHAEOLOGY. Educ: Univ Ill, AB, 59; Univ Chicago, MA, 62, PhD(anthrop), 63. Prof Exp: Field asst, Field Mus Natural Hist, Chicago, 59-61, res asst anthrop, 61-64; vis asst prof, Univ Ill, Urbana, 64; from asst prof to assoc prof, 64-74, PROF ANTHROP, UNIV ARIZ, 74-, DIR, ARCHEOL FIELD SCH, 66- Concurrent Pos: Vis assoc prof anthrop, Yale Univ, 71-72; chmn, Southwestern Anthrop Res Group, 71-73; mem exec comt, Asn Field Archaeol, 72-; fel, Ctr Advan Study Behav Sci, 72-73. Mem: Fel AAAS; Am Anthrop Asn; Soc Am Archaeol. Res: Archaeological method and theory; new world prehistory; American southwest; changing patterns of social organization and evolution of cultural systems. Mailing Add: Dept of Anthrop Univ of Ariz Tucson AZ 85721

LONGAKER, PERRY R, b Dayton, Ohio, Aug 7, 34; m 60; c 2. PHYSICS. Educ: Heidelberg Col, BS, 56; Case Inst Technol, MS, 58, PhD(physics), 62. Prof Exp: STAFF PHYSICIST, LINCOLN LAB, MASS INST TECHNOL, 62- Mem: Am Phys Soc. Res: Experimental research in optical masers and quantum electronics. Mailing Add: Rm B-260 Lincoln Lab Mass Inst of Technol Lexington MA 02173

LONGANBACH, JAMES ROBERT, b Akron, Ohio, July 4, 42; m 66; c 2. CHEMISTRY. Educ: Univ Akron, BS, 64; Yale Univ, MS, 66, MPh, 67, PhD(chem), 69. Prof Exp: Chemist, E I du Pont de Nemours & Co, Inc, 69-71; SR CHEMIST, RES DIV, OCCIDENTAL PETROL CORP, 71- Mem: Am Chem Soc; Am Inst Chem Engr. Res: Physical-organic chemistry of coal and pyrolysis products of coal. Mailing Add: Occidental Petrol Corp 1855 Carrion Rd La Verne CA 91750

LONGENECKER, BRYAN MICHAEL, b Dover, Del, Sept 1, 42; m 63; c 2. IMMUNOLOGY, CELL BIOLOGY. Educ: Univ Mo, AB, 64, PhD(zool), 68. Prof Exp: Med Res Coun Can fel, 68-71, Nat Cancer Inst Can res grant, 71-73, NAT CANCER INST CAN RES SCHOLAR IMMUNOL, UNIV ALTA, 71- Mem: AAAS. Res: Genetic control of allo-immunocompetence and resistance to virally induced neoplasms. Mailing Add: Dept of Immunol Univ of Alta Edmonton AB Can

LONGENECKER, DONALD EUGENE, reproductive physiology, see 12th edition

LONGENECKER, HERBERT EUGENE, b Lititz, Pa, May 6, 12; m 36; c 4. BIOLOGICAL CHEMISTRY. Educ: Pa State Col, BS, 33, MS, 34, PhD(agr biol chem), 36. Hon Degrees: ScD, Duquesne Univ, 51; LLD, Loyola Univ, 63; LittD, Univ Miami, 72. Prof Exp: Asst agr & biochem, Pa State Col, 33-35, instr, 35-36; Nat Res Coun fel, Univ Liverpool, 36-37, Univ Cologne, 37-38 & Queen's Univ, Ont, 38; sr res fel & lectr chem, Univ Pittsburgh, 38-41, from asst prof to prof, 41-55, dean res natural scis, 44-55, dean grad sch, 46-55; vpres in charge Chicago Prof Cols, Univ Ill, 55-60; pres, 60-75, EMER PRES, TULANE UNIV, 75- Concurrent Pos: Mem food & nutrit bd, Nat Res Coun, 43-53, chmn comt food protection, 48-53; mem res coun, Chem Corps Adv Bd, 49-65; mem adv panel biol & chem warfare, Off Asst Secy Defense, 53-61; mem nat selection comn Fulbright student awards, 53-55, chmn, Western Europe Sect, 54-55; mem bd gov, Inst Med Chicago, 57-60; trustee, Coun Southern Univs, 60-75, Inst Defense Anal, 60-, Am Univs Field Staff, 60-74, Southwestern Res Inst, 60-69, Nat Med Fels, Inc, 65- & Alfred P Sloan Found, 71-; trustee, Nutrit Found, 61-, chmn, 65-72; mem, Coun Financial Aid to Educ, 64-71; chmn acad bd adv, US Naval Acad, 66-72; dir, Bush Found, 69-; mem panel sci & technol, US House of Rep Comt Sci & Astronaut, 70-73. Mem: AAAS; Am Oil Chem Soc (vpres, 46); fel Am Pub Health Soc; fel Am Inst Chem; fel NY Acad Sci. Res: Nutrition; fat metabolism; research administration. Mailing Add: Tulane Univ New Orleans LA 70118

LONGENECKER, JOHN BENDER, b Salunga, Pa, July 8, 30; m 54; c 2. NUTRITION, BIOCHEMISTRY. Educ: Franklin & Marshall Col, BS, 52; Univ Tex, MS, 54, PhD(biochem), 56. Prof Exp: Res biochemist, E I du Pont de Nemours & Co, Inc, Del, 56-61; group leader, Mead Johnson & Co, Ind, 61-64; PROF NUTRIT & HEAD DIV, UNIV TEX, AUSTIN, 64- Concurrent Pos: USPHS grant, 64-71; Allied Health Fel grant, 69-74. Mem: Am Chem Soc; Am Inst Nutrit; NY Acad Sci. Res: In vivo plasma amino acid studies to evaluate protein and amino acid nutrition; interrelationships among nutrients; nutritional status studies. Mailing Add: Nutrit Div Dept of Home Econ Univ of Tex Austin TX 78712

LONGENECKER, WILLIAM HILTON, b Cambridge, Md, Mar 28, 18; m 44. ORGANIC CHEMISTRY. Educ: Ohio State Univ, BA, 41; Georgetown Univ, MS, 49. Prof Exp: Chemist, Kankakee Ord Works, 42, Universal Oil Prod Co, 43, Armour & Co, 43-44, Toxicity Lab, Univ Chicago, 44, NIH, 46-49, Exp Sta, E I du Pont de Nemours & Co, Inc, 49-62, Am Petrol Inst, 62-63 & Tech Info Div, Ft Detrick, 63-70; CHEMIST, NAT AGR LIBR, USDA, 70- Mem: AAAS; Am Chem Soc; Sigma Xi; fel Am Inst Chem. Res: Systematic chemical nomenclature; chemical notation systems and machine methods of chemical documentation; bibliography compilations. Mailing Add: 11311 Cedar Lane Beltsville MD 20705

LONGERBEAM, JERROLD KAY, b Downey, Iowa, Nov 10, 22; m 47; c 3. SURGERY, PHYSIOLOGY. Educ: Univ Louisville, MD, 48. Prof Exp: From instr to assoc prof, 58-68, PROF SURG, LOMA LINDA UNIV, 68-; CHIEF SURG, RIVERSIDE COUNTY GEN HOSP, 63- Concurrent Pos: Bank of Am Giannini fel, 58-61; fel surg physiol, Grad Sch, Univ Minn, 59-63; USPHS spec fel, 61-63; asst head physician, Los Angeles County Gen Hosp, 58-59. Mem: Am Col Surg; Soc Cryobiol. Res: Total body irradiation; pathophysiology and treatment of hemorrhagic and septic shock; renal, gastric and lung homotransplantation; renal physiology. Mailing Add: Riverside Co Gen Hosp 9851 Magnolia Ave Riverside CA 92503

LONGERICH, HENRY PERRY, analytical chemistry, see 12th edition

LONGEST, WILLIAM DOUGLAS, b Pontotoc, Miss, Jan 22, 29; m 60. INVERTEBRATE ZOOLOGY. Educ: Baylor Univ, BSc, 54, MSc, 56; La State Univ, PhD(invert zool, ecol), 66. Prof Exp: Teacher, Parma High Sch, 55-56; instr biol, Northwest Jr Col, 56-59; prof natural sci, Blue Mountain Col, 59-62; instr biol, Memphis State Univ, 62-63; teaching asst zool, La State Univ, 63-65, instr, 65-66; from asst prof to assoc prof biol, 66-73, PROF BIOL, UNIV MISS, 73- Mem: Bot Soc Am; Ecol Soc Am; Am Soc Zool. Res: Taxonomy of freshwater Tricladida; foliar embryos of Kalanchoe studied in an explant medium; study of freshwater triclads in the Florida Parishes of Louisiana. Mailing Add: Dept of Biol Univ of Miss University MS 38677

LONGFELLOW, JOHN MCPHERSON, organic chemistry, see 12th edition

LONGFIELD, JAMES EDGAR, b Mt Brydges, Ont, Mar 12, 25; nat US; m 47; c 3. PHYSICAL CHEMISTRY. Educ: Univ Western Ont, BSc, 47, MSc, 48; Univ Rochester, PhD(phys chem), 51. Prof Exp: Asst, Univ Rochester, 48-50; res chemist, Res Div, 51-57, group leader eng res, 57-62, mgr eng res, 62-72, dir process eng dept, Chem Res Div, 72-74, DIR BOUND BROOK LABS, CHEM RES DIV, AM CYANAMID CO, 74- Mem: Am Chem Soc; Am Inst Chem Eng. Res: Vapor phase reactions of organic compounds; reaction kinetics; catalysis; reactor design and mechanism studies. Mailing Add: Chem Res Div Am Cyanamid Co Bound Brook NJ 08805

LONGHI, RAYMOND, b Plymouth, Mass, Nov 14, 35; m 61; c 3. INORGANIC CHEMISTRY, ORGANIC CHEMISTRY. Educ: Univ Mass, BS, 57; Dartmouth Col, MA, 59; Univ Ill, PhD(inorg chem), 62. Prof Exp: Res chemist, 62-64, sr res chemist, 64-65, res supvr, 65-69, sr supvr tech, 69-71, sr supvr res & develop, 71-74, TECH SUPT, E I DU PONT DE NEMOURS & CO, INC, 74- Mem: Am Asn Textile Chemists & Colorists; Am Chem Soc; The Chem Soc. Res: Structures of transition metal complexes; reactions of nitrogen oxide; characterization of organic compounds; textile fibers. Mailing Add: E I du Pont de Nemours & Co Inc Box 4831 Martinsville VA 24112

LONGHURST, WILLIAM MURRAY, b Oakland, Calif, Nov 12, 17; m 40; c 3. WILDLIFE MANAGEMENT. Educ: Stanford Univ, AB, 39; Univ Calif, MA, 40; Cornell Univ, PhD(zool), 42. Prof Exp: Jr econ biologist, State Div Fish & Game, Calif, 42-43; partner, Longhurst & Graham Flying Serv, Napa County Airport, Calif, 46-47; res assoc, Mus Vert Zool, 47-51, ASST SPECIALIST & WILDLIFE BIOLOGIST, AGR EXP STA, UNIV CALIF, DAVIS, 51-, LECTR, UNIV, 63- Mem: Am Soc Mammal; Ecol Soc Am; Wildlife Soc; Soc Range Mgt; Wildlife Disease Asn. Res: Relations of wildlife to agriculture, especially ecology; food habits and parasites of deer; deer management; ecology of jackrabbits; radiobiology, nutrition and ruminant physiology; coyote-livestock relationships; African wildlife. Mailing Add: Univ of Calif Hopland Field Sta Box 476 Hopland CA 95449

LONGINI, RICHARD LEON, b US, Mar 11, 13; m 37; c 2. PHYSICS, BIOENGINEERING. Educ: Univ Chicago, BS, 40; Univ Pittsburgh, MS, 44, PhD(physics), 48. Prof Exp: Physicist, Chicago TV & Res Labs, Inc, 34-35, Akay Electron Co, 35-38 & Wheelco Instruments Co, 38-41; physicist, Westinghouse Elec Corp, 41-51, sect mgr solid state electronics, 51-56, adv physicist, 56-58, sect mgr semiconductors, 58-60, consult physicist, 60-62; PROF SOLID STATE ELECTRONICS, CARNEGIE-MELLON UNIV, 62-, SUPVR MED SYSTS ENG LAB, 64- Mem: Fel Am Phys Soc; fel Inst Elec & Electronics Engr. Res: Solid state and medical electronics; data analysis and automated aids for diagnoses; education. Mailing Add: 6731 Forest Glen Rd Pittsburgh PA 15217

LONGLEY, B JACK, b Dousman, Wis, July 19, 13; m 48; c 3. SURGERY. Educ: Univ Wis, BS, 34, PhD(pharmacol), 40, MD, 42. Prof Exp: Instr, 47-49, ASSOC PROF SURG, SCH MED & ASST DIR TUMOR CLIN UNIV WIS, MADISON, 49-; ASST CHIEF SURG SERV, VET ADMIN HOSP, 50- Res: Cardiovascular research. Mailing Add: Dept of Surg Univ of Wis Sch of Med Madison WI 53705

LONGLEY, GLENN, JR, b Del Rio, Tex, June 2, 42; m 61; c 2. LIMNOLOGY. Educ: Southwest Tex State Univ, BS, 64; Univ Utah, MS, 66, PhD(environ biol), 69. Prof Exp: ASST PROF BIOL, SOUTHWEST TEX STATE UNIV, 69- Concurrent Pos: Univ res grant, Southwest Tex State Univ, 71. Mem: AAAS; Am Fisheries Soc; Am Soc Limnol & Oceanog; Entom Soc Am; Water Pollution Control Fedn. Res: Use of subterranean fauna as indicators of ground water quality; pollution biology; water pollution; heavy metals; organic wastes; pesticides; population dynamics; plankton. Mailing Add: Aquatic Sta Dept of Biol Southwest Tex State Univ Box 46 San Marcos TX 78666

LONGLEY, HERBERT J, b Tahoka, Tex, Jan 3, 26; m 52, 62; c 3. THEORETICAL PHYSICS. Educ: Univ Tex, BS, 46, PhD(physics), 52; Tex Tech Col, BS, 48. Prof Exp: Asst prof & res assoc physics, NMex Inst Mining & Technol, 52-54; staff mem, Los Alamos Sci Lab, Univ Calif, 54-71; STAFF MEM, MISSION RES CORP, 71- Concurrent Pos: Mem staff, Los Alamos Nuclear Corp, 70-71. Mem: Am Phys Soc. Res: Nuclear weapons and weapons testing; hydrodynamics, numerical solutions; radioactive waste storage; nuclear weapons effects. Mailing Add: PO Drawer 719 Santa Barbara CA 93101

LONGLEY, JAMES BAIRD, b Baltimore, Md, June 27, 20; m 44; c 4. HISTOCHEMISTRY. Educ: Haverford Col, BSc, 41; Cambridge Univ, PhD(zool), 50. Prof Exp: From asst scientist to scientist, Nat Inst Arthritis & Metab Dis, 50-60; assoc prof anat, Sch Med, Georgetown Univ, 60-62; PROF ANAT & CHMN DEPT, SCH MED, UNIV LOUISVILLE, 62- Concurrent Pos: USPHS sr res fel, 60-62; instr, Sch Med, Johns Hopkins Univ, 51-52; asst ed, J Histochem & Cytochem, Histochem Soc, 57-64, actg ed, 64-65; ed, Stain Technol, Biol Stain Comn, 73- Mem: Histochem Soc; Am Asn Anat; Am Soc Cell Biol; Biol Stain Comn. Res: Renal histochemistry, morphology and physiology. Mailing Add: Dept of Anat Univ Louisville Health Sci Ctr Box 1055 Louisville KY 40201

LONGLEY, RAYMOND IRVING, JR, chemistry, see 12th edition

LONGLEY, RICHMOND WILBERFORCE, b Paradise, NS, Oct 16, 07; m 35; c 2. METEOROLOGY, CLIMATOLOGY. Educ: Acadia Univ, BSc, 28; Harvard Univ, AM, 29, EdM, 32; Univ Toronto, MA, 40. Prof Exp: Instr, Hobart Col, 29-31; teacher & prin, High Schs, NS, 33-39; meteorologist, Meteorol Serv Can, 40-59; from asst prof to prof geog, 59-73, EMER PROF GEOG, UNIV ALTA, 73- Concurrent Pos: Adv, Educ Fund Bauru, Brazil, 73-74; sr meteorologist, Weather Bur, Pretoria, SAfrica, 74-76. Res: Elements of meteorology; theoretical and applied meteorology. Mailing Add: 11333 73rd Ave Edmonton AB Can

LONGLEY, WILLIAM JOSEPH, b Middleton, NS, May 25, 38; m 63; c 2. REPRODUCTIVE PHYSIOLOGY, ENDOCRINOLOGY. Educ: Univ Toronto, BSA, 61, MSA, 63; Univ Mass, PhD(vet animal sci), 67. Prof Exp: Lectr physiol, 67-68, asst prof, 68-73, ASST PROF PATH, MED SCH, DALHOUSIE UNIV, 73-; ENDOCRINOLOGIST, NS DEPT PUB HEALTH, 73- Mem: Can Soc Clin Chem; Soc Study Reproduction. Res: Endocrinology of the female, particularly fetal-placental function as related to steroid synthesis; clinical chemistry of various hormones including thyroid and adrenal. Mailing Add: Hormone Lab Path Inst 5788 University Ave Halifax NS Can

LONGLEY, WILLIAM WARREN, b Paradise, NS, Apr 8, 09; US citizen; m 35, 57; c 3. GEOLOGY. Educ: Acadia Univ, BS, 31; Univ Minn, MS & PhD(geol), 37. Prof Exp: Instr geol, Dartmouth Col, 35-40; from asst prof to assoc prof geol & geophys, 40-52, PROF GEOL, UNIV COLO, BOULDER, 52- Concurrent Pos: Consult, Que. Dept Mines, 36-50, Kennecott Copper Corp, 45- & Kennco Explor Ltd, 46- Mem: Fel Geol Asn Can; fel Geol Soc Am; Soc Econ Geol; Am Asn Petrol Geol; Soc Explor Geophys. Res: Photogeology; mineral deposits in pre-Cambrian shield of Canada. Mailing Add: 106 Geol Bldg Univ of Colo Boulder CO 80302

LONGLEY, WILLIAM WARREN, JR, b Hanover, NH, Aug 30, 37; m 60; c 2. PHYSICS. Educ: Univ Colo, BA, 58, PhD(physics), 63. Prof Exp: Physicist, Boulder Labs, Nat Bur Standards, 56-59; engr, Denver Div, Martin Co, 59-60, sr engr, 60-63; assoc physicist, Midwest Res Inst, 64-68; asst prof, 68-70, ASSOC PROF PHYSICS, UPPER IOWA COL, FAYETTE, 70- Concurrent Pos: Fel, Theoret Physics Inst, Univ Alta, 63-64. Mem: Fel AAAS; Am Phys Soc; Sigma Xi. Res: Computer applications; economic statistics. Mailing Add: 508 Franklin Fayette IA 52142

LONGLEY-COOK, MARK T, b Tonbridge, Eng, June 29, 43; US citizen; m 67. ACOUSTICS. Educ: Cornell Univ, BS, 65, MEng, 66; Univ Ariz, MS, 70, PhD(physics), 72. Prof Exp: Physicist, Brookhaven Nat Lab, 66 & Univ Ariz, 66-72, fel physics, Inst Atmospheric Physics, 72; PHYSICIST, ACOUSTIC ENG BR, ENVIRON ENG DIV, NAVAL AIR REWORK FACILITY, NORTH ISLAND, SAN DIEGO, 72- Concurrent Pos: Consult, Tucson Gas & Elec Co, 71-72 & Multi-Syst Assoc, 75- Mem: AAAS. Res: Community noise; aircraft noise; liquid crystals; instrumentation; analysis; thermal conductivity; programming; atmospheric electricity; diffusion in the atmosphere. Mailing Add: 312 F Ave Coronado CA 92118

LONGMIRE, CONRAD LEE, b Loyston, Tenn, Aug 23, 21; m 43; c 7. PHYSICS. Educ: Univ Ill, BS, 43; Rochester Univ, PhD(theoret physics), 48. Hon Degrees: DSc, New Eng Col Pharm, 61. Prof Exp: Mem staff, Radiation Lab, Mass Inst Technol, 43-46; instr physics, Columbia Univ, 48-49; mem staff, Los Alamos Sci Lab, 49-57, alternate leader, Theoret Div, 57-68; PRES, LOS ALAMOS NUCLEAR CORP, 68- Concurrent Pos: Vis prof, Cornell Univ, 53-54; chmn bd, Mission Res Corp, 70- Honors & Awards: E O Lawrence Award, AEC, 61; Air Force Commendation Meritorious Civilian Serv, 65. Mem: Am Phys Soc; fel Am Acad Arts & Sci. Res: Nuclear physics; energy sources weapons and rockets; plasma physics. Mailing Add: Mission Res Corp PO Drawer 719 Santa Barbara CA 93102

LONGMIRE, DENNIS B, b Dayton, Ohio, June 4, 44. RUMINANT NUTRITION. Educ: Univ Tenn, BS, 66, MS, 69, PhD(animal sci), 73. Prof Exp: Dairy res specialist, 73-74, sr ruminant nutritionist, 74-75, RUMINANT FEEDS DIR, CENT SOYA CO INC, 76- Mem: Am Dairy Sci Asn. Res: Management with emphasis on feed processing and production systems. Mailing Add: Cent Soya Co Inc 1200 N Second St Decatur IN 46733

LONGMIRE, MARTIN SHELLING, b Morristown, Tenn, Mar 6, 31. CHEMICAL PHYSICS, PHYSICAL CHEMISTRY. Educ: Univ Cincinnati, BS, 53; Mass Inst Technol, PhD(phys chem), 61. Prof Exp: Res assoc phys chem, Ohio State Univ, 61-62; res fel, Mellon Inst, 62-64; res assoc, Mass Inst Technol, 64-65, physicist, Electronics Res Ctr, NASA, Mass, 65-70; ASSOC PROF PHYSICS, WESTERN KY UNIV, 70- Mem: AAAS; Am Phys Soc; Am Inst Chem. Res: Atomic and molecular collisions; meteor physics; absorption of solar ultraviolet light by minor constituents and contaminants in the upper atmosphere. Mailing Add: Dept of Physics & Astron Western Ky Univ Bowling Green KY 42101

LONGMIRE, WILLIAM POLK, JR, b Sapulpa, Okla, Sept 14, 13; m 39; c 3. SURGERY. Educ: Univ Okla, AB, 34; Johns Hopkins Univ, MD, 38; Am Bd Surg, dipl. Prof Exp: Mem vis staff, Sapulpa City Hosp, Okla, 40-42; asst surg, Sch Med, Johns Hopkins Univ, 42-43, from instr to assoc prof, 43-48; PROF SURG & CHMN DEPT, SCH MED, UNIV CALIF, LOS ANGELES, 48- Concurrent Pos: Cushing fel exp surg, Sch Med, Johns Hopkins Univ, 39-40, Halsted fel surg path, 40; asst res, Johns Hopkins Hosp, 42-44, res, 44, surgeon, 47-48; guest prof, Univ Berlin, 52-54; consult, Air Surgeon, US Air Force, 55-75, Vet Admin Hosp, Wadsworth & Harbor County Gen Hosp, 60- & Med Corps, US Army, 60- Mem: Soc Clin Surg; Am Surg Asn; AMA; Am Col Surg (pres); hon mem Asn Surg Gt Brit & Ireland. Res: Tissue transplantation; cancer of the stomach; reconstructions of the biliary system. Mailing Add: Dept of Surg Univ of Calif Sch of Med Los Angeles CA 90024

LONGMORE, WILLIAM JOSEPH, b La Jolla, Calif, Oct 7, 31; m 53; c 4. BIOCHEMISTRY. Educ: Univ Calif, Berkeley, AB, 57; Univ Kans, PhD(biochem), 61. Prof Exp: Res assoc biochem, Scripps Clin & Res Found, 63-66; from asst prof to assoc prof, 66-73, PROF BIOCHEM, SCH MED, ST LOUIS UNIV, 73- Concurrent Pos: Nat Heart Inst fel metab res, Scripps Clin & Res Found, 61-63. Mem: AAAS; Am Chem Soc; Am Soc Biol Chem. Res: Phospholipid metabolism; control mechanisms for regulation of carbohydrate and lipid metabolism, especially in lung tissue. Mailing Add: Dept of Biochem St Louis Univ Sch of Med St Louis MO 63104

LONGMUIR, IAN STEWART, b Glasgow, Scotland, Mar 12, 22; m 49; c 4. BIOCHEMISTRY, PHYSIOLOGY. Educ: Cambridge Univ, BA, 43, MA & MB, BChir, 48. Prof Exp: Res assoc colloid sci, Cambridge Univ, 48-51; prin sci officer, Ministry Supply, Eng, 51-54; sr lectr biochem, Univ London, 54-65; PROF CHEM & BIOCHEM, NC STATE UNIV, 65- Concurrent Pos: Ed jour, Brit Polarographic Soc, 57-62; Isaac Ott fel, Univ Pa, 62-63. Mem: AAAS; Am Physiol Soc; Am Chem Soc; Am Soc Biol Chem; Aerospace Med Asn. Res: Oxygen transport in blood and tissue; inert gas metabolism. Mailing Add: Dept of Biochem NC State Univ Raleigh NC 27607

LONGNECKER, DANIEL SIDNEY, b Omaha, Nebr, June 8, 31; m 52; c 4. PATHOLOGY. Educ: State Univ Iowa, AB, 54, MD, 56, MS, 62. Prof Exp: From assoc to assoc prof path, Univ Iowa, 61-69; assoc prof, Sch Med, St Louis Univ, 69-72; PROF PATH, DARTMOUTH MED SCH, 72- Concurrent Pos: NIH spec fel, Dept Path, Univ Pittsburgh, 65-67; USPHS res grants, Univ Iowa, 67-69, St Louis Univ, 69-71 & Dartmouth Col, 75-; vis asst prof, Dept Path, Univ Pittsburgh, 65-67. Mem: Am Soc Clin Path; Int Acad Path; Soc Exp Biol & Med; Am Asn Path & Bact; Am Soc Exp Path. Res: Biochemical mechanisms of cell injury; experimental pancreatitis; carcinogenesis. Mailing Add: Dept of Path Dartmouth Med Sch Hanover NH 03755

LONGNECKER, DAVID EUGENE, b Kendallville, Ind, May 29, 39; m 63; c 3. ANESTHESIOLOGY. Educ: Ind Univ, AB, 61, MD, 64. Prof Exp: Intern, Blodgett Mem Hosp, Grand Rapids, 64-65; resident anesthesiol, Ind Univ, Indianapolis, 65-68; clin assoc, NIH, 68-70; asst prof, Univ Mo-Columbia, 70-73; ASSOC PROF ANESTHESIOL, UNIV VA, 74- Concurrent Pos: NIH spec res fel, Ind Univ, 67-68; contrib ed, Anesthesia & Analgesia, Int Anesthesia Res Soc J, 74-; res career develop award, Nat Heart & Lung Inst, 75. Mem: Am Soc Anesthesiologists; Inst Anesthesia Res Soc; Am Physiol Soc; Asn Univ Anesthetists. Res: Microcircularoty mechanisms during hemorrhagic shock; effect of anesthetics on the microcirculation during normovolemia and hypovolemia. Mailing Add: Dept of Anesthesiol Univ of Va Charlottesville VA 22903

LONGNECKER, THOMAS CHRISTOPHER, agronomy, plant physiology, see 12th edition

LONGO, FRANK JOSEPH, b Cleveland, Ohio, Nov 16, 39; m 62; c 6. CELL BIOLOGY. Educ: Loyola Univ, BS, 62; Ore State Univ, MS, 65, PhD(cell biol), 67. Prof Exp: Asst prof, 70-75, ASSOC PROF ANAT, CTR FOR HEALTH SCI, UNIV TENN, MEMPHIS, 75- Mem: AAAS; Am Soc Cell Biol; Am Asn Anat; Soc Study Reproduction. Res: Cellular and developmental biology at the fine structural and biochemical levels; comparative pronuclear development and fusion; gametogenesis and fertilization; cell division and differentiation. Mailing Add: Dept of Anat Univ of Tenn Ctr Health Sci Memphis TN 38163

LONGO, FREDERICK R, b Trenton, NJ, May 4, 30; m; c 6. PHYSICAL CHEMISTRY. Educ: Villanova Col, BA, 53; Drexel Inst, MS, 58; Univ Pa, PhD(phys chem), 62. Prof Exp: Chemist, Am Biltrite Rubber Co, 55-57; assoc prof, 57-68, PROF CHEM, DREXEL UNIV, 68-, HEAD DEPT CHEM & CHEM ENG, EVENING COL, 73- Mem: Am Chem Soc; Sigma Xi. Res: Synthesis and spectral properties of porphyrins; emphasis on conversion of solar energy into chemical potential energy. Mailing Add: Dept of Chem Drexel Univ Philadelphia PA 19104

LONGO, JOHN M, b Hartford, Conn, Nov 6, 39; m 64; c 3. INORGANIC CHEMISTRY. Educ: Univ Conn, BA, 61, PhD(inorg chem), 64. Prof Exp: Fel, Univ Stockholm, 64-65; chemist, Lincoln Lab, Mass Inst Technol, 65-70; CHEMIST, CORP RES LABS, EXXON RES & ENG CO, 70- Mem: Electrochem Soc; Am Chem Soc. Res: Preparation and characterization of solid state inorganic materials. Mailing Add: Corp Res Labs Exxon Res & Eng Co Linden NJ 07036

LONGO, JOSEPH THOMAS, b Ferndale, Mich, Jan 13, 42; m 64; c 2. SOLID STATE PHYSICS. Educ: Univ Detroit, BS, 64; Mich State Univ, MS, 66, PhD(solid state physics), 68. Prof Exp: Asst, Mich State Univ, 64-68; fel, 68-69, MEM TECH STAFF, N AM ROCKWELL SCI CTR, 69- Mem: Am Phys Soc. Res: High field magnetoresistance and Hall effect in intermetallic compounds; crystal growth, optical and device properties of narrow gap semiconductors. Mailing Add: 254 Venado Thousand Oaks CA 91360

LONGO, LAWRENCE DANIEL, b Los Angeles, Calif, Oct 11, 26; m 48; c 4. PHYSIOLOGY. Educ: Pac Union Col, BA, 49; Loma Linda Univ, MD, 54. Prof Exp: Asst prof obstet & gynec, Univ Ibadan, 59-62; asst prof, Univ Calif, Los Angeles, 62-64; lectr physiol, Univ Pa, 64-66, asst prof physiol, 66-68; assoc prof, 68-73, PROF PHYSIOL, OBSTET & GYNEC, LOMA LINDA UNIV, 73- Concurrent Pos: USPHS fel obstet & gynec, Univ Calif, Los Angeles, 59, spec fel physiol, Univ Pa, 64-66 & res career develop award, 66-68, res career develop award, Loma Linda Univ, 68- & grant, 69-; consult, Nat Inst Child Health & Human Develop, 71; corresp, Fedn Am Socs Exp Biol, 71- Mem: AAAS; NY Acad Sci; Am Physiol Soc; Soc Gynec Invest; Perinatal Res Soc. Res: Fetal and placental physiology; kinetics of placental transfer of respiratory gases; fetal oxygenation. Mailing Add: Dept Obstet Gynec & Physiol Loma Linda Univ Loma Linda CA 92354

LONGO, MICHAEL JOSEPH, b Philadelphia, Pa, Apr 7, 35; m 58; c 3. HIGH ENERGY PHYSICS. Educ: La Salle Col, BA, 56; Univ Calif, Berkeley, PhD(physics), 61. Prof Exp: NSF fel physics, Saclay Nuclear Res Ctr, France, 61-62; from asst prof to assoc prof, 62-68, PROF PHYSICS, UNIV MICH, ANN ARBOR, 68- Mem: Am Phys Soc; Sigma Xi. Res: Nucleon-nucleon and pion-nucleon interaction at high energies; K-meson decays; spark chambers and scintillation counters. Mailing Add: Dept of Physics Univ of Mich Ann Arbor MI 48104

LONGONE, DANIEL THOMAS, b Worcester, Mass, Sept 16, 32; m 54. ORGANIC CHEMISTRY. Educ: Worcester Polytech Inst, BS, 54; Cornell Univ, PhD(org chem), 58. Prof Exp: Res assoc org chem, Univ Ill, 58-59; from instr to assoc prof, 59-71, PROF ORG CHEM, UNIV MICH, ANN ARBOR, 71- Concurrent Pos: Am Chem

Soc-Petrol Res Fund int fel, 67-68; Fulbright scholar, 70-71; vis prof, Univ Cologne, 70-71; consult, Gen Motors Res Lab. Mem: Am Chem Soc. Res: Synthetic and mechanistic organic chemistry; bridged aromatic compounds; cyclophane chemistry; monomer synthesis and polymerization. Mailing Add: 3400 Chemistry Bldg Univ of Mich Ann Arbor MI 48104

LONGPRE, EDWIN KEITH, b Detroit, Mich, Mar 7, 33; m 65; c 1. SYSTEMATIC BOTANY. Educ: Univ Mich, BS, 55, MS, 56; Mich State Univ, PhD(bot), 67. Prof Exp: Instr bot, Tex Tech Col, 56-57; ASSOC PROF BOT & BIOL, WESTERN STATE COL COLO, 65- Mem: Am Soc Plant Taxon; Am Inst Biol Sci; Int Asn Plant Taxon. Res: Systematical studies in the tribe Heliantheae of the family Compositae; general cytotaxonomical studies. Mailing Add: Dept of Bot Western State Col of Colo Gunnison CO 81230

LONGREE, KARLA, b Immekeppel, Ger, Sept 7, 05; nat US. FOOD SCIENCE. Educ: Univ Berlin, DrAgr, 31; Cornell Univ, PhD(plant path), 38. Prof Exp: Asst biol, Off Soils & Forestry, Univ Berlin, 31-33; asst plant path, NY State Col Agr, Cornell Univ, 34-40, instr, State Univ NY Col Home Econ, Cornell Univ, 40-41; from assoc prof to prof foods & nutrit, Hampton Inst, 41-50; from assoc prof to prof, 50-67, EMER PROF INSTNL MGT, NY STATE COL HUMAN ECOL, CORNELL UNIV, 67- Res: Quantity foods; cooling rates and bacterial counts of foods prepared and stored in quantity. Mailing Add: 305 Highland Rd Ithaca NY 14850

LONGRIE, DEAN PAUL, b New London, Wis, Aug 30, 44; m 67; c 2. WILDLIFE ECOLOGY. Educ: Univ Wis-Milwaukee, BS, 68; Mich State Univ, MS, 70, PhD(wildlife ecol), 72. Prof Exp: Sr ecologist, Michael Baker Jr, Inc, 73-75; WILDLIFE BIOLOGIST, FED POWER COMN, 75- Concurrent Pos: Ecol Soc Am rep to bd dirs, Renewable Natural Resources Found, 75- Mem: Wildlife Soc; Ecol Soc Am; Am Inst Planners. Res: Utility right-of-way construction, operation and maintenance: evaluating effects of construction on wildlife and vegetation, and evaluating operation and maintenance methods in terms of wildlife and vegetation. Mailing Add: Fed Power Comn 825 N Capitol St NE Washington DC 20426

LONGROY, ALLAN LEROY, b Flint, Mich, May 28, 36; m 55; c 3. ORGANIC CHEMISTRY. Educ: Univ Mich, AB, 58, MS, 61, PhD(chem), 63. Prof Exp: Res fel chem, Brandeis Univ, 62-64; asst prof, Ind Univ, 64-67; asst prof, 67-69, ASSOC PROF CHEM, PURDUE UNIV, 69- Mem: Am Chem Soc. Res: Elimination reactions of beta-ketols and derivatives; epoxidation reactions; stereochemical control of epoxide reductions; ester pyrolysis; organic reaction mechanisms and kinetics. Mailing Add: Dept of Chem Purdue Univ Fort Wayne IN 46815

LONGSHORE, JOHN DAVID, b Birmingham, Ala, Mar 8, 36; m 64; c 2. PETROLOGY. Educ: Emory Univ, BA, 57; Rice Univ, MA, 59, PhD(geol), 65. Prof Exp: Teacher, Westminster Schs, Ga, 60-62; PROF GEOL, HUMBOLDT STATE UNIV, 65- Concurrent Pos: NASA res grant chem invest Medicine Lake Area, 67-69. Mem: AAAS; Am Geophys Union; Nat Asn Geol Teachers. Res: Chemistry and petrology of igneous rocks. Mailing Add: Dept of Geol Humboldt State Univ Arcata CA 95521

LONGSWORTH, LEWIS GIBSON, b Somerset, Ky, Nov 16, 04; m 29; c 3. PHYSICAL CHEMISTRY. Educ: Southwestern Col, Kans, AB, 25; Univ Kans, MA, 27, PhD(chem), 28. Prof Exp: Nat Res Coun fel, 28-30, asst phys chem, 30-39, assoc, 39-45, assoc mem, 45-49, mem, 49-70, prof chem, 57-70, EMER PROF CHEM, ROCKEFELLER UNIV, 70- Mem: Nat Acad Sci; Am Chem Soc; Harvey Soc; Electrochem Soc; fel NY Acad Sci (vpres, 44). Res: Transference phenomena in solutions of electrolytes and electrophoresis of proteins by the moving boundary method; optical methods in electrophoresis; thermal and isothermal diffusion. Mailing Add: 144-60 29th Ave Flushing NY 11354

LONGTIN, BRUCE, b North Fork, Calif, Aug 23, 13; m 53; c 6. THERMODYNAMICS. Educ: Univ Calif, BS, 35, MS, 37, PhD(chem), 38. Prof Exp: Asst chem, Univ Calif, 35-38, Shell Oil Co fel, 38-39; from instr to assoc prof, Ill Inst Technol, 39-51; from chemist to sr chemist, 51-74, STAFF CHEMIST, E I DU PONT DE NEMOURS & CO, INC, 74- Concurrent Pos: Assoc chemist, Argonne Nat Labs, 48-49. Mem: Am Chem Soc. Res: Thermodynamics of industrial processes; thermodynamic properties of solutions; chemistry, radiolysis and control of impurities in water coolant and moderator of nuclear reactors. Mailing Add: E I du Pont de Nemours & Co Savannah River Plant Aiken SC 29802

LONGWELL, ARLENE CROSBY (MAZZONE), b Buffalo, NY, Nov 26, 30; m 57, 62; c 1. GENETICS, CYTOGENETICS. Educ: Southwest Mo State Col, AB, 53; Univ Mo, PhD(genetics), 57. Prof Exp: Res assoc, Div Biol & Med Res, Argonne Nat Lab, 57-59; res assoc, Genetics Inst, Univ Lund, 59-60; res assoc path, Children's Cancer Res Found & Harvard Med Sch, 60-65; RES GENETICIST, LAB EXP BIOL, NORTH ATLANTIC COASTAL FISHERIES RES CTR, NAT MARINE FISHERIES SERV, 65- Concurrent Pos: Am Cancer Soc fel, 59-62. Mem: Bot Soc Am; Am Genetic Asn; Am Soc Cell Biol. Res: Cytogenetics, genetics and breeding of the oyster and other commercial marine species; mutagenic effects of marine pollutants on marine species. Mailing Add: North Atlantic Coastal Fisheries Res Ctr Lab Exp Biol Milford CT 06460

LONGWELL, CHESTER RAY, geology, deceased

LONGWORTH, JAMES W, b Stockton Heath, Eng, Sept 16, 38; m 65; c 2. BIOPHYSICS, CHEMICAL PHYSICS. Educ: Univ Sheffield, BSc, 59, PhD(biochem), 62. Prof Exp: USPHS fel phys chem, Univ Minn, 62-63; mem staff, Bell Tel Labs, 63-65; MEM STAFF, BIOL DIV, OAK RIDGE NAT LAB, 65- Concurrent Pos: Mem, US Nat Comt Photobiol, 73-77. Mem: Am Soc Photobiol; Biophys Soc; Brit Biochem Soc; Brit Biophys Soc; Am Soc Biol Chem. Res: Photophysics and excited state chemistry of proteins, nucleic acids and their synthetic analogues, particularly their luminescent behavior; use of optical methods to study conformation and function of proteins and nucleic acids and their complexes. Mailing Add: Oak Ridge Nat Lab Biol Div PO Box Y Oak Ridge TN 38730

LONGWORTH, RUSKIN, b Oldham, Eng, Aug 13, 27; m 57; c 4. POLYMER CHEMISTRY, POLYMER PHYSICS. Educ: Univ London, BSc, 50, PhD(chem), 56. Prof Exp: Asst, Polytech Inst Brooklyn, 52-55; chemist, Vauxhall Motors Ltd, Eng, 56-57; SR RES CHEMIST, PLASTICS DEPT, EXP STA, E I DU PONT DE NEMOURS & CO, INC, 57- Mem: Am Chem Soc; NY Acad Sci; assoc Royal Inst Chem. Res: Physical chemistry of polymers, especially rheology, solution properties and polyelectrolytes. Mailing Add: Plastics Dept Res & Develop Div E I du Pont de Nemours & Co Inc Wilmington DE 19898

LONGYEAR, JOHN MUNRO, III, b Houghton, Mich, July 30, 14; m 37; c 2. ARCHAEOLOGY, PHYSICAL ANTHROPOLOGY. Educ: Cornell Univ, AB, 36; Harvard Univ, PhD(anthrop), 40. Prof Exp: Res assoc archaeol, Peabody Mus, Harvard Univ, 40-48; from asst prof to assoc prof, 48-60, PROF ANTHROP, COLGATE UNIV, 60- Concurrent Pos: Mem staff electronics, Radiation Lab, Mass Inst Technol, 43-45; Ford Found fel, Colgate Univ, 54-55; NY State Regents fel, Cornell Univ & Mex, 67-68. Mem: AAAS; Am Anthrop Asn; Soc Am Archaeol; Mex Anthrop Asn. Res: Central American archaeology; archaeology of upstate New York. Mailing Add: Dept of Social Relations Colgate Univ Hamilton NY 13346

LONGYEAR, JUDITH QUERIDA, b Harrisburg, Pa, Sept 20, 38; m 67; c 2. PURE MATHEMATICS. Educ: Pa State Univ, BA, 62, MS, 64, PhD(math), 72. Prof Exp: Atmospheric physicist, White Sands Missile Range, 66-67; consult, Auerbach Corp, 67-68; asst prof math, Community Col Philadelphia, 68-70; John Wesley Young res assoc, Dartmouth Col, 72-74; asst prof, 74-76, ASSOC PROF MATH, WAYNE STATE UNIV, 76- Mem: Am Math Soc; Math Asn Am; Am Women in Math; Soc Indust & Appl Math. Res: Combinatorial mathematics; block designs; Hadamard matrices; transversal theory; tactical configurations. Mailing Add: 605 MacKenzie Hall Wayne State Univ Detroit MI 48202

LONKY, MARTIN LEONARD, b New York, NY, Jan 5, 44; m 66. ELECTRONIC PHYSICS, SOLID STATE PHYSICS. Educ: Rensselaer Polytech Inst, BS, 64; Univ Del, MS, 67, PhD(physics), 72. Prof Exp: Teaching asst physics, Univ Del, 64-67, res fel, 67-72; Presidential intern chem, US Army Land Warfare Lab, 72, res analyst, 72-73; sr engr electronics, 73-75, FEL ENGR PHYSICS, WESTINGHOUSE ELEC CORP, 76- Mem: Electrochem Soc; Inst Elec & Electronics Engrs. Res: Electron device physics, with emphasis on memory field effect transistors and transparent gate metal-oxide-silicon technology; device fabrication technologies. Mailing Add: Advan Technol Labs MS 3525 Westinghouse Elec Corp Baltimore MD 21203

LONNING, THOR J GRELL, physical chemistry, polymer chemistry, see 12th edition

LONSDALE, HAROLD KENNETH, b Westfield, NJ, Jan 19, 32; m 53; c 2. PHYSICAL CHEMISTRY. Educ: Rutgers Univ, BS, 53; Pa State Univ, PhD(chem), 57. Prof Exp: Staff mem, Gen Atomic Co, 59-70; prin scientist, Alza Corp, 70-72; vis scientist, Max Planck Inst Biophys, 73; vis prof, Weizmann Inst, 74; PRES, BEND RES, INC, 75- Concurrent Pos: Ed, J Membrane Sci, 75-; adj prof, Ore State Univ, 76- Mem: Am Chem Soc. Res: Transport in synthetic membranes, desalination by reverse osmosis; controlled release of biologically active agents. Mailing Add: Bend Res Inc 64550 Research Rd Bend OR 97701

LONSDALE, RICHARD ELLIS, b Stockton, Calif, Nov 27, 26; m 55; c 1. ECONOMIC GEOGRAPHY, GEOGRAPHY OF THE SOVIET UNION. Educ: Univ Calif, Los Angeles, AB, 49, MA, 53; Syracuse Univ, PhD(geog), 60. Prof Exp: Geogr, Cent Intel Agency, 53-57; asst prof geog, Harpur Col, 60-62; assoc prof, Univ NC, Chapel Hill, 62-70; PROF GEOG & CHMN DEPT, UNIV NEBR, LINCOLN, 71- Concurrent Pos: Ed, Southeastern Geogr, 65-71; consult, Battelle Mem Inst, 68-69; mem, Comn Col Geog, 68-73; consult, Res Triangle Inst, 69-70; lectr, Univ New Eng, Australia, 70; ed proc, Asn Am Geog, 72, mem publ comt, 72-75, chmn publ comt, 73-75; vis prof, Australian Royal Mil Col, 74; mem adv comt, Old West Regional Comn, 75-77; mem comn rural planning, Int Geog Union, Moscow, 74-77; chmn US-Australian Joint Seminar Geog, 75-77. Mem: Asn Am Geog; Am Asn Advan Slavic Studies. Res: Location of industry, especially problems of decentralization and economic development of smaller communities; industrial development, especially in small towns and in Socialist societies. Mailing Add: Dept of Geog Univ of Nebr Lincoln NE 68508

LONSETH, ARVID TURNER, b Bellingham, Wash, Dec 25, 12; m 35; c 2. MATHEMATICAL ANALYSIS, NUMERICAL ANALYSIS. Educ: Stanford Univ, AB, 35; Univ Calif, PhD(math), 39. Prof Exp: Asst, Univ Calif, 36-39; instr math, Ill Inst Technol, 39-40; instr, Iowa State Univ, 40-42; contract employee, Bur Ord, US Dept Navy, 42; from instr to asst prof math, Iowa State Univ, 43-44; asst prof, Northwestern Univ, 44-48; assoc prof, 48-49, chmn dept, 55-68, PROF MATH, ORE STATE UNIV, 49- Concurrent Pos: NATO vis prof, Univ Iceland, 62; guest prof, Aachen Tech Univ, 62; res prof, Math Res Ctr, Univ Wis, 68; vis prof, Colo State Univ, 70. Res: Differential and integral equations, especially approximation methods and errors. Mailing Add: Dept of Math Ore State Univ Corvallis OR 97331

LONSKI, JOSEPH, b Port Jefferson, NY, 43; c 2. DEVELOPMENTAL BIOLOGY. Educ: Cornell Univ, BS, 64; Univ Calif, Los Angeles, MA, 66; Princeton Univ, PhD(biol), 73. Prof Exp: Instr biol, Southampton Col, 66 & Princeton Univ, 71-72; ASST PROF BIOL, BUCKNELL UNIV, 72- Mem: Sigma Xi; Am Soc Plant Physiol; Soc Develop Biol. Res: Chemotaxis in the myxobacteria and cellular slime molds. Mailing Add: Dept of Biol Bucknell Univ Lewisburg PA 17837

LONTZ, JOHN FRANK, organic chemistry, biophysics, see 12th edition

LONTZ, ROBERT JAN, b Wilmington, Del, Oct 19, 36; m 62; c 2. PHYSICS. Educ: Yale Univ, BSc, 58; Duke Univ, PhD(physics), 62. Prof Exp: Asst, Physics Div, 62-64, chief gen physics br, 64-67, assoc dir, Physics Div, 67-73, DIR, PHYSICS DIV, US ARMY RES OFF, 73- Res: Paramagnetic resonance spectroscopy; lasers. Mailing Add: US Army Res Off PO Box 12211 Research Triangle Park NC 27709

LOO, TI LI, b Changsha, China, Jan 7, 18; nat US; m 51; c 3. CLINICAL PHARMACOLOGY, MEDICINAL CHEMISTRY. Educ: Tsing Hua Univ, China, BSc, 40; Oxford Univ, PhD, 47. Prof Exp: Asst pharmacol chem, Oxford Univ, 46-47; res assoc, Christ Hosp Inst Med Res, 51-54; supvry chemist, NIH, 55-65; PHARMACOLOGIST, DEPT DEVELOP THERAPEUT & PROF PHARMACOL, UNIV TEX M D ANDERSON HOSP & TUMOR INST, 65- Concurrent Pos: Fel org chem, Univ Md, 47-51. Mem: Am Chem Soc; Am Asn Cancer Res; Am Soc Pharmacol & Exp Therapeut; fel Am Pharmaceut Asn; Acad Pharmaceut Sci. Res: Pharmacology of anticancer drugs; cancer chemotherapy; metabolism of drugs; chemical structure and biological activities; pharmacodynamics. Mailing Add: M D Anderson Hosp & Tumor Inst 6723 Bertner Dr Houston TX 77030

LOO, YEN HOONG, b Honolulu, Hawaii, Dec 19, 14. BIOCHEMISTRY. Educ: Columbia Univ, BA, 37; Univ Mich, PhD(biochem), 43. Prof Exp: Res asst biochem, Univ Ill, 44-51; biochemist, Nat Heart Inst, 51-52; res biochemist, Labs, Eli Lilly & Co, 52-68; ASSOC RES SCIENTIST, NY STATE INST BASIC RES MENT RETARDATION, 68- Concurrent Pos: Fel, Univ Tex, 43-44; fel, Inst Animal Physiol & Cambridge Univ, Eng, NIH, 66-67. Mem: AAAS; Am Soc Biol Chem; Am Soc Neurochem; NY Acad Sci. Res: Neurochemistry. Mailing Add: Inst Basic Res Ment Retardation 1050 Forest Hill Rd Staten Island NY 10314

LOOFBOURROW, GUY NORMAN, b Westmoreland, Kans, Dec 26, 11; m 35; c 1. PHYSIOLOGY. Educ: Park Col, AB, 34; Univ Mich, MS, 37, PhD(zool), 46. Prof Exp: Asst zool, Univ Mich, 39-40 & 41-42 & physiol, 40-41 & 43-44; instr zool, RI State Col, 44-46; res assoc neurophysiol, Univ Minn, 46-48; asst prof, 48-63, ASSOC PROF PHYSIOL, UNIV KANS, 63- Mem: AAAS; Am Physiol Soc; NY Acad Sci. Res: Neurophysiology; neurophysiological mechanisms in behavior. Mailing Add: Dept of Physiol Univ of Kans Sch of Med Kansas City KS 66103

LOOK, DAVID C, b St Paul, Minn, Dec 19, 38; m 68; c 2. SOLID STATE PHYSICS.

Educ: Univ Minn, BPhys, 60, MS, 62; Univ Pittsburgh, PhD(physics), 65. Prof Exp: Res physicist, Aerospace Res Labs, 66-69; SR RES PHYSICIST, UNIV DAYTON, 69- Mem: Am Phys Soc; Am Sci Affil. Res: Transport properties; nuclear magnetic resonance; ion implantation; radiation damage in semiconductors. Mailing Add: Dept of Physics Univ of Dayton Dayton OH 45469

LOOK, MELVIN, organic chemistry, see 12th edition

LOOKER, JAMES HOWARD, b Bloomingburg, Ohio, Nov 24, 22; m 46; c 2. ORGANIC CHEMISTRY. Educ: Ohio State Univ, BS, 43, PhD(chem), 49. Prof Exp: From instr to assoc prof, 50-60, PROF CHEM, UNIV NEBR, LINCOLN, 60- Concurrent Pos: NIH spec fel guest prof, Univ Vienna, 63-64. Res: Flavonoid substances; diazoesters; arylserines; sulfonic esters. Mailing Add: Dept of Chem Univ of Nebr Lincoln NE 68588

LOOKER, JEROME J, b Columbus, Ohio, July 7, 35; m 57; c 3. ORGANIC CHEMISTRY. Educ: Kenyon Col, AB, 58; Univ Ill, MS, 60, PhD(org chem), 61. Prof Exp: Nat Sci Found fel, Cornell Univ, 61-62; RES CHEMIST, EASTMAN KODAK CO, 62- Res: Synthetic organic chemistry. Mailing Add: Eastman Kodak Co Rochester NY 14650

LOOMAN, JAN, b Apeldoorn, Netherlands, Oct 18, 19; Can citizen; m 58; c 2. BOTANY. Educ: Univ Wis, MSc, 60, PhD(bot, soils), 62. Prof Exp: Technician pasture res, Cent Inst Agr Res, Wageningen, Netherlands, 52-54; technician pasture res, 54-58, RES SCIENTIST, RES STA, CAN DEPT AGR, 62- Mem: Agr Inst Can; Can Bot Soc; Ecol Soc Am; Am Bryol & Lichenological Soc. Res: Pasture research; classification of plant communities, including lichens and bryophytes, particularly classification in relation to practical application. Mailing Add: Res Sta Can Dept of Agr Swift Current SK Can

LOOMANS, MAURICE EDWARD, b Wisconsin Rapids, Wis, Aug 10, 33; m 57; c 3. DERMATOLOGY. Educ: Hope Col, BA, 57; Univ Wis, MS, 59, PhD(biochem), 62. Prof Exp: RES CHEMIST, MIAMI VALLEY LABS, PROCTER & GAMBLE CO, 62- Mem: Soc Invest Dermat. Res: Keratinization; epidermal cellular control; acne. Mailing Add: Miami Valley Labs Procter & Gamble Co Cincinnati OH 45239

LOOMIS, ALBERT GEYER, b Lexington, Mo, Feb 17, 93; m 19, 32; c 3. CHEMISTRY. Educ: Univ Mo, AB, 14, AM, 15; Univ Calif, PhD(chem), 19. Prof Exp: Instr chem, Univ Ill, 19-20; asst prof, Univ Mo, 20; Nat Res Coun fel, Cryogenic Lab, US Bur Mines, 21-22, phys chemist, 21-28, chemist, Explosives Div, Exp Sta, Pa, 28-29; chief chemist, Gulf Res & Develop Co, 29-35; asst dir, Shell Develop Co, 35-42, assoc dir, 42-45; consult chemist & petrol engr, 45-48; petrol engr, US Bur Mines, 48-63; CONSULT CHEMIST, LOOMIS LABS, 63- Concurrent Pos: Lectr, George Washington Univ, 23-25; coop expert, Int Critical Tables, 25-27; sr indust fel, Mellon Inst, 29-35. Honors & Awards: Distinguished Serv Hon Award, US Dept Interior, 63; Anthony F Lucas Gold Medal Award, Am Inst Mining, Metall & Petrol Eng, 72. Mem: Am Chem Soc; Am Inst Mining, Metall & Petrol Eng. Res: Extraction of radium, vanadium and uranium from carnotite ore; liquid ammonia systems; thermodynamic properties of hydrocarbon systems at low temperatures; recovery of helium from natural gas; flame temperatures and explosives; colloid physics of clay dispersions; utilization of redwood products; petroleum production. Mailing Add: 85 Parnassus Rd Berkeley CA 94708

LOOMIS, ALDEN ALBERT, b Pittsburgh, Pa, July 22, 34; m 57; c 3. GEOLOGY, OCEANOGRAPHY. Educ: Stanford Univ, AB, 56, PhD(petrol, geol), 61. Prof Exp: Asst prof geol, San Jose State Col, 60-61; SR SCIENTIST, JET PROPULSION LAB, CALIF INST TECHNOL, 61- Concurrent Pos: Assoc prof, Calif State Col Los Angeles, 65-66; consult geoscientist, 69-; eng geologist, State of Calif, 72- Res: Space applications to oceanography and geology; igneous petrology, volcanology; metamorphic petrology; gravity and crustal structure; geology of moon and Mars; development of experiments for lunar and planetary exploration; engineering and environmental geology; mineral exploration. Mailing Add: Jet Propulsion Lab Calif Inst of Technol 4800 Oak Grove Dr Pasadena CA 91103

LOOMIS, ALFRED LEE, physics, deceased

LOOMIS, EDMOND CHARLES, b San Francisco, Calif, June 25, 21; m 48; c 4. PARASITOLOGY, ACAROLOGY. Educ: Univ Calif, Berkeley, BS, 47, PhD(parasitol), 59. Prof Exp: Sr vector control specialist, State Dept Pub Health, Calif, 52-59; malaria adv, USAID, Indonesia, 59-61; PARASITOLOGIST & LECTR, AGR EXTEN SERV, UNIV CALIF, DAVIS, 62- Concurrent Pos: Parasitologist, Long Pocket Labs, Commonwealth Sci & Indust Res Orgn, Queensland, Australia, 69-70; parasitologist, Waireka Res Sta, Ivon Watkins Dow Ltd, 74-75. Mem: Am Soc Parasitol; Entom Soc Am; Am Mosquito Control Asn. Res: Epizootiology; epidemiology; parasitic and arthropod borne diseases of animals and man; acarology; statewide agricultural sanitation program coordination. Mailing Add: Dept of Entom Univ of Calif Davis CA 95616

LOOMIS, HAROLD GEORGE, b Erie, Pa, Aug 22, 25; m 47; c 4. MATHEMATICS. Educ: Stanford Univ, BS, 50; Pa State Univ, MA, 52, PhD(math), 57. Prof Exp: Sr mathematician, Haller, Raymond & Brown, Inc, 52-55; instr math, Pa State Univ, 55-57; asst prof, Amherst Col, 57-62; mem staff, Res Corp, Syracuse Univ, 62-63; asst prof, Univ Hawaii, 63-66; MATHEMATICIAN, PAC OCEANOG LAB, NAT OCEANIC & ATMOSPHERIC ADMIN, 66- Mem: Soc Indust & Appl Math; Math Asn Am. Res: Analysis; numerical analysis; oceanography. Mailing Add: 115 Maunalua Ave Honolulu HI 96821

LOOMIS, LYNN H, b Afton, NY, Apr 25, 15; m 39; c 2. MATHEMATICS. Educ: Rensselaer Polytech Inst, BS, 37; Harvard Univ, AM, 38, PhD(math), 42. Prof Exp: Fac instr math, 41-46, assoc prof, 46-56, DWIGHT PARKER ROBINSON PROF MATH, HARVARD UNIV, 56- Mem: Am Math Soc; Math Asn Am; Am Acad Arts & Sci. Res: Abstract analysis. Mailing Add: Dept of Math Harvard Univ Cambridge MA 02138

LOOMIS, RICHARD BIGGAR, b Lincoln, Nebr, June 18, 25; m 47; c 3. ACAROLOGY, HERPETOLOGY. Educ: Univ Nebr, BSc, 48; Univ Kans, PhD(zool), 55. Prof Exp: Lab asst biol & vert zool, Univ Nebr, 47-48; asst, US Navy Chigger Proj, Univ Kans, 48-53, asst instr biol & comp anat, 53, asst, Sch Med, 53-55; PROF BIOL, CALIF STATE UNIV, LONG BEACH, 55- Concurrent Pos: Prin investr, USPHS Grant, 60-74; res assoc, Los Angeles County Mus Natural Hist. Mem: Am Soc Mammal; Am Soc Ichthyol & Herpet; Am Soc Parasitol; Am Soc Acarologists. Res: Systematics, life histories and ecology of parasitic acarines, especially trombiculid mites and their vertebrate hosts in North America; medical acarology. Mailing Add: Dept of Biol Calif State Univ Long Beach CA 90840

LOOMIS, ROBERT HENRY, b Atlanta, Ga, Nov 9, 23; m 45; c 4. ZOOLOGY, LIMNOLOGY. Educ: Univ Ga, BS, 47; Okla State Univ, MS, 51, PhD(zool), 56. Prof Exp: Instr biol, Piedmont Col, 48, Cent State Col, Okla, 51-52 & Jimma Agr Sch, Ethiopia, 52-54; asst prof, Cent State Col, Okla, 54-55; from asst prof to prof, Northeastern State Col, 55-63; prof, Parsons Col, 63-68; prof & chmn div sci, Pikeville Col, 68-75; PROF LIFE SCI, SACRAMENTO CITY COL, 75- Mem: AAAS; Am Inst Biol Sci. Res: Watershed conditions on fish populations; food habits of mesopelagic fishes; identification of photosynthetic active components of phytoplankton communities; temperature acclimation in crayfish populations. Mailing Add: Dept of Life Sci Sacramento City Col Sacramento CA 95822

LOOMIS, ROBERT MORGAN, b Mauston, Wis, Aug 31, 22; m 48; c 6. FORESTRY. Educ: Univ Mich, BS; Univ Mo, MS, 65. Prof Exp: Forester, Ochoco Nat Forest, Ore, 48-51; adminr, Ottawa Nat Forest, Mich, 51-56 & Mo Nat Forests, 56-57; fire researcher, Cent States Forest Exp Sta, Columbia Forest Res Ctr, Mo, 57-66, res forester, 66-71; RES FORESTER, NCENT FOREST EXP STA, MICH STATE UNIV, 71- Mem: Soc Am Foresters. Res: Forest fire effects, fuels and danger rating. Mailing Add: NCent Forest Exp Sta 1407 S Harrison Rd East Lansing MI 48823

LOOMIS, ROBERT SIMPSON, b Ames, Iowa, Oct 11, 28; m 51; c 2. PLANT PHYSIOLOGY. Educ: Iowa State Univ, BS, 49; Univ Wis, MS, 51, PhD(bot), 56. Prof Exp: Instr agron & jr agronomist, 56-58, asst prof & asst agronomist, 58-64, assoc prof & assoc agronomist, 64-68, dir, Inst Ecol, 69-72, assoc dean environ studies, 70-72, PROF AGRON & AGRONOMIST, UNIV CALIF, DAVIS, 68- Concurrent Pos: NIH spec fel, Harvard Univ, 63-64; NZ Nat Res Adv Coun res fel, 71. Mem: AAAS; Am Soc Plant Physiol (secy, 65-67); Am Soc Sugar Beet Technol; Am Soc Agron; Ecol Soc Am. Res: Physiology of field crops including growth and development; crop ecology with emphasis on productivity and system simulation; sugar beet production. Mailing Add: Dept of Agron & Range Sci Univ of Calif Davis CA 95616

LOOMIS, TED ALBERT, b Spokane, Wash, Apr 24, 17; m; c 2. PHARMACOLOGY, TOXICOLOGY. Educ: Univ Wash, BS, 39; Univ Buffalo, MS, 41, PhD(pharmacol), 43; Yale Univ, MD, 46. Prof Exp: Intern, US Marine Hosp, 46-47; assoc prof, 47-59, PROF PHARMACOL, SCH MED, UNIV WASH, 59- Concurrent Pos: State toxicologist, Wash, 55- Res: Pesticide and insecticide toxicology; anticoagulant agents; alcohol research; toxicological methods; mechanisms of drug action and action of toxic chemicals. Mailing Add: Dept of Pharmacol Univ of Wash Sch of Med Seattle WA 98195

LOOMIS, THOMAS CLEMENT, b Pekin, Ill, Sept 15, 24; m 51; c 8. ANALYTICAL CHEMISTRY. Educ: Bradley Univ, BS, 48; Iowa State Univ, PhD(chem), 53. Prof Exp: MEM TECH STAFF, CHEM RES DEPT, BELL LABS, INC DIV, AM TEL & TEL CO, 53- Mem: Soc Appl Spectros. Res: X-ray spectrochemical analysis; trace and micro analysis. Mailing Add: MH 1A 208 Bell Labs Inc Murray Hill NJ 07974

LOOMIS, TIMOTHY PATRICK, b Alhambra, Calif, May 25, 46. PETROLOGY, TECTONICS. Educ: Univ Calif, Davis, BS, 67; Princeton Univ, PhD(geol), 71. Prof Exp: J W Gibbs instr geol, Yale Univ, 71-73; adj asst prof, Univ Calif, Los Angeles, 73-74; ASST PROF GEOL, UNIV ARIZ, 74- Mem: Geol Soc Am. Res: Mechanisms of metamorphic reactions, dynamic igneous processes and regional tectonic implications. Mailing Add: Dept of Geosci Univ of Ariz Tucson AZ 85721

LOOMIS, WALTER DAVID, b Fayetteville, Ark, Mar 2, 26; m 52. BIOCHEMISTRY. Educ: Iowa State Univ, BS, 48; Univ Calif, PhD(comp biochem), 53. Prof Exp: Instr biochem, 53-54, from asst prof to assoc prof, 54-68, PROF BIOCHEM, ORE STATE UNIV, 68- Concurrent Pos: USPHS res career develop award, 61-67; vis researcher, Univ Col Wales, 65-66. Mem: Am Chem Soc; Am Soc Plant Physiol; Am Soc Biol Chem; Phytochem Soc NAm; Can Soc Plant Physiol. Res: Plant enzymes and proteins; terpene metabolism. Mailing Add: Dept of Biochem & Biophysics Ore State Univ Corvallis OR 97331

LOOMIS, WHEELER, b Parkersburg, WVa, Aug 4, 89; m 22; c 3. PHYSICS. Educ: Harvard Univ, AB, 10, AM, 13, PhD(physics), 17. Hon Degrees: SD, Univ Ill, Urbana, 69. Prof Exp: Instr math, Harvard Univ, 13-15; res physicist, Lamp Div, Westinghouse Elec Co, NJ, 17, Pa, 19-20; from asst prof to assoc prof physics, NY Univ, 20-29; prof & head dept, 29-57, dir control systs lab, 52-59, EMER PROF PHYSICS, UNIV ILL, URBANA, 59- Concurrent Pos: Guggenheim Found fel, Göttingen & Zurich, 28-29; assoc dir radiation lab, Mass Inst Technol, 41-46, dir Lincoln Lab & Proj Charles, 51-52; chmn personnel comt, AEC, 47; adv comt, Ballistic Res Lab, Aberdeen Proving Ground; div comt, NSF, chmn sci manpower, Nat Acad Sci; mem, Lexington Proj, 48; US Air Force Proj Paris, 54. Mem: Nat Acad Sci; AAAS (vpres sect B, 48); Am Phys Soc (vpres, 48, pres, 49); Optical Soc Am; Am Asn Physics Teachers. Res: Molecular spectra. Mailing Add: 804 W Illinois St Urbana IL 61801

LOOMIS, WILLIAM FARNSWORTH, JR, b Boston, Mass, Sept 17, 40; m 62; c 1. DEVELOPMENTAL BIOLOGY. Educ: Harvard Univ, BS, 62; Mass Inst Technol, PhD(microbiol), 65. Prof Exp: NIH fel, Brandeis Univ, 65-66; asst prof, 66-73, ASSOC PROF BIOL, UNIV CALIF, SAN DIEGO, 73- Mem: Soc Develop Biol; Am Soc Biol Chemists. Res: Cellular interactions involved in the biochemical differentiation in Dictyostelium discoideum; genetics of slime molds. Mailing Add: Univ Calif San Diego Dept Biol PO Box 109 La Jolla CA 92037

LOONEY, DUNCAN HUTCHINGS, b Muskogee, Okla, July 26, 23; m 51; c 3. PHYSICS. Educ: Purdue Univ, BS, 48; Mass Inst Technol, PhD(physics), 53. Prof Exp: Develop engr, 53-58, head dept solid state devices, 58-63, dir, Power Systs Lab, 63-74, DIR, LOOP PLANT INSTALLATION & ELEC PROTECTION LAB, BELL LABS, INC, 74- Mem: Inst Elec & Electronics Eng. Res: Solid state physics; magnetic devices; power systems. Mailing Add: Bell Labs Inc Whippany Rd Whippany NJ 07981

LOONEY, FRANKLIN SITTIG, JR, physical chemistry, see 12th edition

LOONEY, NORMAN E, b Adrian, Ore, May 31, 38; m 57; c 3. POMOLOGY, PLANT PHYSIOLOGY. Educ: Wash State Univ, BS, 60, PhD(hort), 66. Prof Exp: Sr exp aid hort, Wash State Univ, 60-62, res asst post-harvest hort, 62-66; pomologist, 66-74, HEAD POMOLOGY SECT, AGR CAN, 75- Mem: Am Soc Hort Sci; Am Soc Plant Physiol; Can Soc Hort Sci; Int Soc Hort Sci; Am Inst Biol Sci. Res: Physiology of growth, development and ripening of fruits; investigations of agroclimatology and plant growth regulators. Mailing Add: Pomology Sect Res Sta Agr Can Summerland BC Can

LOONEY, RALPH WILLIAM, b Spencer, WVa, June 30, 31; m 61; c 3. PHYSICAL CHEMISTRY, POLYMER CHEMISTRY. Educ: WVa Univ, BS, 53, MS, 54; Univ Wis, PhD(phys chem), 69. Prof Exp: Res chemist, Esso Res & Eng Co, 60-63, sr chemist, 63-66, sect head new chem intermediates, 66-72, RES ASSOC, ELASTOMERS TECHNOL DIV, EXXON CHEM CO, 72- Mem: Am Chem Soc. Res: Polymerization catalysts; kinetics of polymerization; polymer physics and ozonolysis of olefins; radiation chemistry. Mailing Add: Exxon Chem Co PO Box 45 Linden NJ 07036

LOONEY, WILLIAM BOYD, b South Clinchfield, Va, Mar 18, 22; m 55; c 2. RADIOBIOLOGY, BIOPHYSICS. Educ: Emory & Henry Col, BS, 44; Med Col Va, MD, 48; Cambridge Univ, PhD(radiobiol, biophys), 60. Prof Exp: Intern, Presby Hosp, Chicago, 48-49, asst resident, 49-50; asst prof radiol, Johns Hopkins Univ, 59-60; from asst prof to assoc prof, 61-68, PROF RADIOBIOL & BIOPHYS & DIR DIV, UNIV VA, 68- Concurrent Pos: Mem interdisciplinary prog biophys, Univ Va, 66- Mem: AAAS; Am Asn Cancer Res; Am Soc Cell Biol; Biophys Soc; Radiation Res Soc. Res: Cancer; mathematical evaluation of tumor growth curves; cell cycle and cell kinetics studies in experimental tumors; modification of tumor growth rates and cell kinetics by radiation, alone or in combination with different chemotherapeutic agents; host-tumor interaction. Mailing Add: Div of Radiobiol & Biophys Univ of Va Hosp Charlottesville VA 22901

LOOP, ANNE SCHUMACHER, public health, see 12th edition

LOOP, JOHN WICKWIRE, b Belvidere, Ill, July 23, 24; m 57; c 5. RADIOLOGY, MEDICINE. Educ: Univ Wyo, BS, 48; Harvard Med Sch, MD, 52. Prof Exp: Intern med & surg, King County Hosps, Seattle, Wash, 52-53; asst resident radiol, Univ Chicago, 53-56, instr, 57-58; assoc radiologist, Mass Inst Technol, 58-59; from instr to asst prof, 59-65, ASSOC PROF RADIOL, UNIV WASH, 65-; RADIOLOGIST-IN-CHIEF, HARBORVIEW MED CTR, SEATTLE, 71- Concurrent Pos: Swedish Govt fel, Univ Lund, 56-57; mem med radiation adv comt, US Dept Health, Educ & Welfare, 71-75. Mem: Am Soc Neuroradiol; Asn Univ Radiol; Am Col Radiol. Res: Radiological diagnosis. Mailing Add: Dept of Radiol Harborview Med Ctr Seattle WA 98104

LOOP, MICHAEL STUART, b Pittsburgh, Pa, Feb 28, 46; c 1. VISION, HERPETOLOGY. Educ: Fla State Univ, BS, 68, MS, 71, PhD(psychobiol), 72. Prof Exp: NIH fel neurol surg, Univ Va, 72-74, Sloane Found fel physiol, 74-75; VIS ASST PROF, PHYSIOL & BIOPHYS, UNIV ILL, 75- Mem: Am Soc Ichthyologists & Herpetologists; Animal Behav Soc. Res: Vertebrate visual system psychophysiology; comparative animal behavior. Mailing Add: 524 Burrill Hall Univ of Ill Urbana IL 61801

LOOS, HENDRICUS G, b Amsterdam, Neth, Dec 18, 25; nat US; m 52; c 2. ENVIRONMENTAL PHYSICS. Educ: Univ Amsterdam, Drs(math), 51; Univ Delft, ScD, 52. Prof Exp: Res engr, Nat Aeronaut Res Inst, Neth, 46-52; res fel, Calif Inst Technol, 52-55; sr engr, Propulsion Res Corp, 55-57; sr physicist, Giannini Sci Corp, 57-66; mem staff, Douglas Advan Res Lab, 66-70, sr staff scientist, McDonnell-Douglas Astronaut Co, 70-71; prof math, Cleveland State Univ, 71-74; DIR, LAGUNA RES LAB, 74- Concurrent Pos: Lectr, Univ Calif, Riverside, 63-64, assoc prof in residence, 64-70, adj prof, 70- Mem: Am Phys Soc; Am Math Soc. Res: Gauge theory; atmospheric physics; fluid mechanics; general relativity. Mailing Add: Laguna Res Lab 21421 Stans Lane Laguna Beach CA 92651

LOOS, KARL RUDOLF, b New York, NY, July 10, 39; m 65; c 3. PHYSICAL CHEMISTRY, SPECTROCHEMISTRY. Educ: Rensselaer Polytech Inst, BS, 60; Mass Inst Technol, PhD(phys chem), 65. Prof Exp: Res assoc, Inst Phys Chem, Swiss Fed Inst Technol, 65-66; SR RES CHEMIST, SHELL DEVELOP CO, 67- Mem: Am Chem Soc. Res: Vibrational spectroscopy and structure of catalysts; infrared and Raman spectroscopy. Mailing Add: Shell Develop Co PO Box 1380 Houston TX 77001

LOOSANOFF, VICTOR L, b Kiev, Russia, Oct 3, 99; nat US; m 28. FISHERIES. Educ: Univ Wash, BS, 27; Yale Univ, PhD(biol), 36. Prof Exp: Aquatic biologist, State Dept Fisheries, Wash, 27-30; chief marine biologist, Comn Fisheries, Va, 31; dir marine biol lab, 35-62, AQUATIC BIOLOGIST, US FISH & WILDLIFE SERV, 32- Concurrent Pos: Sci consult, Bingham Oceanog Lab, Yale Univ; hon prof, Rutgers Univ, 53-59; dir, Taylor Libr; adj prof, Univ of the Pac, 63-; lectr, Univ Wash, 63-; NSF grant, 66-68; consult, US Nat Marine Fisheries Serv, 66-; mem, Nat Tech Adv Comt Water Qual Criteria, 67-68; consult maricult, Lummi Indian Tribe, Wash, 68- Honors & Awards: Dept Interior Distinguished Serv Award, 65. Mem: AAAS; hon mem Nat Shellfisheries Asn (vpres, 45, pres, 47-49). Res: Marine biology in relation to ecology and physiology of oysters and other pelecypods. Mailing Add: 17 Los Cerros Dr Greenbrae CA 94904

LOOSE, LELAND DAVID, b Reading, Pa, Jan 25, 40; m 71. PHYSIOLOGY, IMMUNOLOGY. Educ: Tenn Wesleyan Col, BS, 63; ETenn State Univ, MA, 65; Univ Mo, Columbia, PhD(physiol), 70. Prof Exp: Instr physiol, Lees-McRae Col, 65-67; asst prof, Sch Med, Tulane Univ, 70-74; asst prof, 74-75, ASSOC PROF PHYSIOL, DEPT PHYSIOL & INST EXP PATH & TOXICOL, ALBANY MED COL, 75- Mem: Am Physiol Soc; Am Soc Trop Med & Hyg; NY Acad Sci; Am Soc Zool; Sigma Xi. Res: Physiological control mechanisms of immune responses; differentiation of lymphoid tissue with special reference to hormonal effects; macrophage antigen processing; calcium alterations in shock. Mailing Add: Inst Exp Path & Toxicol Albany Med Col Albany NY 12208

LOOSLI, CLAYTON G, b Marysville, Idaho, Mar 18, 05; m 33; c 2. MEDICINE. Educ: Univ Idaho, BS, 30, MS, 31; Univ Chicago, PhD(anat), 34, MD, 37. Hon Degrees: ScD, Univ Idaho, 58. Prof Exp: Asst anat, Univ Chicago, 34-35; house officer, Johns Hopkins Hosp, 37-38; from asst to prof med, Univ Chicago, 38-58, dir student health serv, 46-49, chief, Sect Prev Med, 49-58; prof med & dean sch med, 58-64, HASTINGS PROF MED, SCH MED, UNIV SOUTHERN CALIF & MED DIR, HASTINGS FOUND, 64- Concurrent Pos: Consult, Secy War, 41-43; mem, Armed Forces Epidemiol Bd, 57-63, mem comn influenza, 48-70; mem, Nat Adv Allergy & Infectious Dis Coun, 62-65; mem bd vaccine develop, Nat Inst Allergy & Infectious Dis, 66-70. Res: Air-borne infections; methods of control with glycol vapors and dust suppressive measures; epidemiology and immunological aspects of epidemic influenza and histoplasmosis; lung development. Mailing Add: Univ of Southern Calif Sch Med 2025 Zonal Ave Los Angeles CA 90033

LOOSLI, JOHN KASPER, b Clarkston, Utah, May 16, 09; m 36; c 3. ANIMAL NUTRITION. Educ: Utah State Univ, BS, 31; Colo State Univ, MS, 32; Cornell Univ, PhD(animal nutrit), 38. Prof Exp: Instr agr, Col Southern Utah, 33-35; asst animal nutrit, Cornell Univ, 35-38; agent, Bur Biol Surv, USDA, 38-39; from asst prof to prof animal nutrit, 39-74, head dept animal sci, 63-71, EMER PROF ANIMAL NUTRIT, CORNELL UNIV, 74- Concurrent Pos: Collabr, US Fish & Wildlife Serv, 39-56; vis prof, Univ Philippines, 53-54 & 66 & Univ Ibadan, 72-74; consult, US Army Vet Grad Sch, 54 & USAID, Nigeria, 61; ed, J Animal Sci, 55-58; Fulbright lectr, Univ Queensland, 60; mem comt animal nutrit, Agr Bd, Nat Res Coun; vis prof, Univ Fla, 74-, actg chmn dept animal sci, 75- Honors & Awards: Am Feed Mfrs Award Nutrit, 50; Borden Award Dairy Prod, 51; Morrison Award, 56. Mem: Am Soc Animal Sci (vpres, 59, pres, 60); Am Dairy Sci Asn (pres, 70-71); Am Inst Nutrit; Brit Soc Animal Prod. Res: Fat metabolism and requirements; vitamin requirements; lactation; mineral requirements; feed composition. Mailing Add: Dept of Animal Sci Univ of Fla Gainesville FL 62311

LOOYENGA, ROBERT WILLIAM, b NDak, Oct 21, 39; m 63; c 4. ANALYTICAL CHEMISTRY. Educ: Hope Col, AB, 61; Wayne State Univ, PhD(anal chem), 69. Prof Exp: Fel chem, Univ Wis-Milwaukee, 70; res chemist, Printing Develop Inc, 70-72; ASST PROF CHEM, SDAK SCH MINES & TECHNOL, 72- Concurrent Pos: Chemist, SDak Racing Comn, 75- Mem: Am Chem Soc; Sigma Xi. Res: Analytical research and analysis of trace metals and organics in municipal and natural waters, of selenium and tellurium, and of abused drugs; analytical separations and methods development. Mailing Add: Dept of Chem SDak Sch of Mines & Technol Rapid City SD 57701

LOPATA, EUGENE STEPHEN, b Pittsburgh, Pa, May 24, 43. POLYMER CHEMISTRY. Educ: Univ Pittsburgh, BA, 65, BS, 67; Univ Mich, PhD(chem), 73. Prof Exp: Nat Res Coun fel chem, 73-75; researcher chem, Univ Calif, Berkeley, 75-76. Concurrent Pos: Contractor, Dept Civil Eng, Univ Calif to NASA-Ames Res Ctr, 75-76. Mem: Am Chem Soc; Am Phys Soc; Fedn Am Scientists. Res: All aspects of the physical chemistry of epoxy resins; infrared absorption spectroscopy; gel permeation chromatography; nuclear magnetic resonance; thermal analysis; kinetics of polymerization and decomposition. Mailing Add: 655 S Fair Oaks Ave Apr 0-317 Sunnyvale CA 94086

LOPATIN, GEORGE, polymer chemistry, polymer physics, see 12th edition

LOPER, DAVID ERIC, b Oswego, NY, Feb 14, 40; m 66; c 3. MAGNETOHYDRODYNAMICS, APPLIED MATHEMATICS. Educ: Carnegie Inst Technol, BS, 62; Case Inst Technol, MS, 64, PhD(mech eng), 65. Prof Exp: Sr scientist, Douglas Aircraft Corp, 65-68; asst prof, 68-72, PROF MATH, FLA STATE UNIV, 72- Concurrent Pos: Nat Ctr Atmospheric Res fel, 67-68; sr vis fel, Univ Newcastle-upon-Tyne, Eng, 74-75. Mem: Am Phys Soc; Am Geophys Union; Soc Indust & Appl Math. Res: Boundary layers in rotating, stably stratified, electrically conducting fluids; evolution of the earth's core including stratification, heat transfer, solidification and particle precipitation. Mailing Add: 18 Keen Bldg Fla State Univ Tallahassee FL 32306

LOPER, GERALD D, b Brooklyn, NY, May 4, 37; m 60; c 1. NUCLEAR PHYSICS. Educ: Univ Wichita, AB, 59; Okla State Univ, MS, 62, PhD(physics), 64. Prof Exp: Asst prof, 64-67, ASSOC PROF PHYSICS, WICHITA STATE UNIV, 67-, CHMN DEPT, 66- Mem: Am Phys Soc. Res: Measurement of positron lifetimes in solids; nuclear spectroscopy; internal conversion. Mailing Add: Dept of Physics Wichita State Univ Wichita KS 67208

LOPER, GERALD MILTON, b Sykesville, Md, Jan 7, 36; m 62; c 2. AGRONOMY, BIOCHEMISTRY. Educ: Univ Md, Bsc, 58; Univ Wis, MSc, 60, PhD(agron), 61. Prof Exp: Res agronomist, USDA, SDak, 62-67; assoc prof, 69-74, PROF AGRON & PLANT GENETICS, UNIV ARIZ, 74-; RES PLANT PHYSIOLOGIST, FED HONEY BEE LAB, 67- Mem: AAAS; Am Soc Agron. Res: Effect of environment and infective organisms on the chemical composition of forages in relation to animal nutrition; attractiveness of forage legumes to honey bees; seed production and crop physiology investigations. Mailing Add: Dept of Agron Univ of Tucson Tucson AZ 85721

LOPER, JOHN C, b Hadley, Pa, June 21, 31; m 56; c 3. MICROBIOLOGY, GENETICS. Educ: Western Md Col, BA, 52; Emory Univ, MS, 53; Johns Hopkins Univ, PhD(biol), 60. Prof Exp: From instr to asst prof pharmacol, Sch Med, St Louis Univ, 60-63; from asst prof to assoc prof, 63-74, PROF MICROBIOL, COL MED, UNIV CINCINNATI, 74- Concurrent Pos: NIH res grants, 62-; NIH spec vis fel genetics, Res Sch Biol Sci, Australian Nat Univ, 70-71; Environ Protection Agency grant, 76-; mem biol comt, Argonne Nat Lab-Argonne Univ Asn, 70-73. Mem: Am Soc Microbiol; Genetics Soc Am; NY Acad Sci; Environ Mutagen Soc. Res: Bacterial genetics; resistance plasmids in pseudomonas; mutagenesis. Mailing Add: Dept of Microbiol Univ of Cincinnati Col of Med Cincinnati OH 45219

LOPES, LOUIS ALOYSIUS, mathematics, see 12th edition

LOPEZ, ANTHONY, b Chile, SAm, May 13, 19; US citizen; m 47; c 3. FOOD SCIENCE. Educ: Catholic Univ, Chile, BS, 42; Univ Mass, PhD(food tech), 47. Prof Exp: Chemist, SA Organa, Chile, 42-45; tech dir, Indust de Productos Alimenticios, 48-52; assoc res prof food technol, Univ Mass, 52-53; assoc prof, Univ Ga, 53-54; PROF FOOD SCI & TECHNOL, VA POLYTECH INST & STATE UNIV, 54- Concurrent Pos: Instr, UN Latin Am Fisheries Training Ctr, Chile, 52; lectr, Ministry Commerce, Spain, 60; consult food processing, Govt Spain, 62, 63; consult food technol, UN Food & Agr Orgn, Chile, 66 & Brazil, 69, 72 & 75; Orgn Am States in Mex, 70-74; tech ed, Food Prod Mgt, 71- Mem: Am Chem Soc; Inst Food Technol; Chilean Soc Nutrit. Res: Processing and nutritive value of fish; composition of fresh fruits and vegetables; processing of fruits and vegetables; chemical changes in processed foods during storage; food packaging; microwave irradiation of foods; effect of processing on nutritive value of foods. Mailing Add: Dept of Food Sci & Technol Va Polytech Inst & State Univ Blacksburg VA 24061

LOPEZ, ANTONIO VINCENT, b Montgomery, Ala, Apr 24, 38. PHARMACEUTICAL CHEMISTRY, PHARMACOGNOSY. Educ: Auburn Univ, BS, 59, MS, 61; Univ Miss, PhD(pharm chem), 66. Prof Exp: Chmn dept pharmaceut chem, 66-76, CHMN DIV NATURAL SCI, SOUTHERN SCH PHARM, MERCER UNIV, 76- Mem: AAAS; Am Pharmaceut Asn; Am Chem Soc. Res: Central nervous system drugs. Mailing Add: Southern Sch of Pharm Mercer Univ 345 Boulevard NE Atlanta GA 30312

LOPEZ, CARLOS, b Ponce, PR, Jan 15, 42; m 70; c 1. IMMUNOLOGY, VIROLOGY. Educ: Univ Minn, BS, 65, MS, 66, PhD(pub health), 70. Prof Exp: Res fel, Univ Minn, 70-72, asst prof path, 72-73; ASSOC MEM SLOAN-KETTERING CANCER CTR & ASST PROF BIOL, SLOAN-KETTERING DIV, SCH MED, CORNELL UNIV, 73- Concurrent Pos: NIH fel, 70-71; fel, Nat Thoracic & Respiratory Dis Asn, 71-73. Mem: Am Asn Immunologists; Am Asn Exp Pathologists; Am Soc Microbiol; AAAS. Res: Immunological resistance to virus infections; immunologic response to virus induced tumors. Mailing Add: Sloan-Kettering Cancer Ctr 425 E 68th St New York NY 10021

LOPEZ, DIANA MONTES DE OCA, b Havana, Cuba, Aug 26, 37; US citizen; m 58; c 3. MICROBIOLOGY. Educ: Univ Havana, BS, 60; Univ Miami, MS, 68, PhD(microbiol), 70. Prof Exp: Res assoc, 70-71, INSTR MICROBIOL, SCH MED, UNIV MIAMI, 71- Mem: Am Soc Microbiol; Tissue Cult Asn. Res: Virus cell interactions; biochemistry of viruses; viruses of lower vertebrates; viral immunology. Mailing Add: Dept of Microbiol Univ of Miami Sch of Med Coral Gables FL 33134

LOPEZ, GENARO, b Brownsville, Tex, Jan 24, 47; m 72. ECONOMIC ENTOMOLOGY. Educ: Tex Tech Univ, BS, 70; Cornell Univ, PhD(econ entom), 75. Prof Exp: Res asst entom, Cornell Univ, 70-75; ENTOMOLOGIST, TEX AGR EXTEN SERV, TEX A&M UNIV, 75- Mem: Entom Soc Am; Acaralogical Soc Am. Res: Bionomics, ecology and control of insects affecting man's home environment. Mailing Add: Dept of Entom Tex A&M Univ College Station TX 77843

LOPEZ, HADY, b Oriente, Cuba, Nov 2, 14. NUTRITIONAL BIOCHEMISTRY, DENTAL EPIDEMIOLOGY. Educ: Univ Havana, PharmD, 39. Prof Exp: Instr, Sch Pharm, Univ Havana, 42-45; tech asst, Nutrit Clin, Hillman Hosp, Birmingham, Ala, 48-49; tech asst, Med Sch, Univ Havana, 49-51; res asst, Dept Nutrit & Metab, Northwestern Univ, Ill, 52; prof pharm, Univ Vilanueva, Cuba, 54-61, dir dept, 56-61, tech dir, Bur Stand, 58-61; tech asst nutrit & food sci, Mass Inst Technol, 62-65, mem res staff, Div Sponsored Res, Oral Sci Labs, 65-68; RES ASSOC BIOCHEM, INST DENT RES, UNIV ALA, BIRMINGHAM, 69-, INSTR, SCH DENT, 72- Concurrent Pos: Pan Am Sanit Bur fel nutrit, Yale Univ, 46-47; instr, Sch Pub Health, Finlay Inst, Cuba, 42-46; dir, Labs FIM of Nutrit, Havana, Cuba, 53-58; dir biochem lab, Ministry Pub Health, 59-61. Honors & Awards: Pharmacol Award, Univ Havana, 41; San Esteban Conde de Canongo Award, Acad Sci, Havana, 54. Mem: Int Asn Dent Res; Am Pharmaceut Asn. Res: Effect of nutrients on caries and bone healing. Mailing Add: 1908 Southwood Rd Birmingham AL 35201

LOPEZ, RAFAEL, b Dominican Republic, Dec 15, 29; m 56; c 2. PEDIATRICS, HEMATOLOGY. Educ: Seton Hall Univ, BSc, 52; Univ PR, MD, 56. Prof Exp: ASSOC PROF PEDIAT, NEW YORK MED COL, FLOWER & FIFTH AVE HOSP, 65- Mem: Soc Study Blood; Int Soc Hemat; Am Soc Hemat; NY Acad Sci. Res: Glutathione reductase as a tool for diagnosis of riboflavin deficiency in infants, children, adolescents; malabsorption syndromes and the effect of phototherapy upon this vitamin in the newborn. Mailing Add: Apt 2E 4489 Broadway New York NY 10040

LOPEZ-ROSA, JULIO HERIBERTO, phytopathology, see 12th edition

LOPEZ-SANTOLINO, ALFREDO, b Salamanca, Spain, July 23, 31; m 62; c 2. MEDICINE, BIOCHEMISTRY. Educ: Inst Ensenanza Media, Salamanca, BS, 49; Lit Univ Salamanca, MD, 55, PhD(med sci), 58; Tulane Univ, PhD(biochem), 63. Prof Exp: Asst prof physiol med, Sch Med, Lit Univ Salamanca, 56-58; instr biochem, Cali Univ Sch Med, 58-59; asst prof internal med, Col Med & biochemist, Clin Res Ctr, Univ Iowa, 64-67; assoc prof, 67-74, PROF INTERNAL MED, MED SCH, LA STATE UNIV MED CTR, NEW ORLEANS, 74- Concurrent Pos: Mem coun atherosclerosis, Am Heart Asn. Mem: AAAS; Am Oil Chem Soc; Soc Nutrit Educ; Am Inst Nutrit; Am Soc Clin Nutrit. Res: Nutrition and metabolic diseases; metabolism of lipids and steroid hormones. Mailing Add: Dept of Med La State Univ Med Ctr New Orleans LA 70112

LOPREST, FRANK JAMES, b New York, NY, Jan 8, 29; m 60; c 5. PHYSICAL CHEMISTRY. Educ: St John's Univ, NY, BS, 50; NY Univ, MS, 52, PhD, 54. Prof Exp: Res chemist, Oak Ridge Nat Lab, 54-56; sr res chemist & supvr adv res, Reaction Motors Div, Thiokol Chem Corp, 56-65; tech assoc, Res & Develop Div, 65-67, sect mgr new imaging processes res, 67-69, MGR APPL CHEM, RES & DEVELOP & RES SERVS, INDUST PHOTO DIV, GAF CORP, 69- Mem: AAAS; Am Chem Soc; Soc Photog Scientists & Engr; Sigma Xi. Res: Heterogeneous equilibria; kinetics of liquid solid reactions; high temperature materials; physical chemistry of liquid and solid propellants; photochemistry and photoconductors, imaging systems. Mailing Add: GAF Corp Indust Photo Div Charles St Binghamton NY 13902

LOPUSHINSKY, THEODORE, b Brooklyn, NY, Oct 25, 37. ECOLOGY, PATHOLOGY. Educ: Pa State Univ, BS, 59; Univ Tenn, Knoxville, MS, 61; Mich State Univ, PhD(ecol, path), 69. Prof Exp: Asst prof natural sci, Mich State Univ, 69-70; prog rep, 70-71, ACTG DIR, MICH ASN REGIONAL MED PROGS, 72- Mem: Wildlife Soc. Res: Parasitism and disease pathologies in wildlife populations. Mailing Add: Mich Regional Med Prog 1111 Michigan Ave East Lansing MI 48823

LOPUSHINSKY, WILLIAM, b Rome, NY, July 25, 30; m 60; c 2. PLANT PHYSIOLOGY. Educ: State Univ NY, BS, 53, MS, 54; Duke Univ, PhD(plant physiol), 60. Prof Exp: Asst plant physiol, Duke Univ, 57-60, res assoc bot, 60-61; PLANT PHYSIOLOGIST, FOREST HYDROL LAB, USDA, 62- Mem: Am Soc Plant Physiol. Res: Plant water relations. Mailing Add: Forest Hydrol Lab 1133 N Western Ave Wenatchee WA 98801

LORAND, JOHN PETER, b Wilmington, Del, Dec 6, 36; m 64; c 3. PHYSICAL ORGANIC CHEMISTRY. Educ: Brown Univ, ScB, 58; Harvard Univ, PhD(org chem), 64. Prof Exp: NSF scientist, Univ Calif, Los Angeles, 64-65; asst prof org chem, Boston Univ, 65-71; asst prof, 71-73, ASSOC PROF ORG CHEM, CENT MICH UNIV, 73- Mem: Am Chem Soc. Res: Free radicals; C-H hydrogen bonding; charge-transfer complexes. Mailing Add: Dept of Chem Cent Mich Univ Mt Pleasant MI 48859

LORAND, JOYCE BRUNER, b Omaha, Nebr, Feb 3, 23; m 53; c 1. ZOOLOGY. Educ: Creighton Univ, BS, 44; Univ Iowa, MS, 45, PhD(zool), 50. Prof Exp: Asst zool, Univ Iowa, 46-50, res assoc urol, 50-52, res assoc zool, 52-53; independent investr, Marine Biol Lab, Woods Hole, 53-54; RES ASSOC CHEM, NORTHWESTERN UNIV, EVANSTON, 55- Concurrent Pos: Asst, Max Planck Inst Biol, 49; independent investr, Marine Biol Lab, Woods Hole, 56-75. Mem: AAAS; Am Soc Zool. Res: Embryology; endocrinology; sex differentiation in vertebrates; blood clotting. Mailing Add: Dept of Biochem & Molecular Biol Northwestern Univ Evanston IL 60201

LORAND, LASZLO, b Gyor, Hungary, Mar 23, 23; nat US; m 53; c 1. BIOCHEMISTRY, PHYSIOLOGY. Educ: Leeds Univ, PhD(biomolecular struct), 51; Budapest Tech Univ, absolutorium med, 48. Prof Exp: Demonstr biochem, Budapest Univ, 46-48; asst biomolecular struct, Leeds Univ, 48-52; res assoc physiol & pharmacol, Wayne State Univ, 52-53, asst prof, 53-55; asst prof, 55-57, assoc prof, 57-61, PROF CHEM, NORTHWESTERN UNIV, 61- Concurrent Pos: Beit Mem fel, Eng, 52; Lalor fac award, 57; USPHS career award, 62; mem corp, Marine Biol Lab, Woods Hole, Mass. Mem: AAAS; Am Soc Biol Chem; Soc Exp Biol & Med; Am Physiol Soc; Brit Biochem Soc. Res: Blood proteins; coagulation of blood; muscle chemistry; protein and enzyme chemistry. Mailing Add: Biochem Div Dept of Chem Northwestern Univ Evanston IL 60201

LORANGER, WILLIAM FARRAND, b Detroit, Mich, Nov 6, 25. XERORADIOGRAPHY. Educ: Denison Univ, BA, 47; Univ Ill, MS, 50, PhD(chem & x-ray diffraction), 52. Prof Exp: Asst, Anal Div, Ill State Geol Surv, 47-49; proj scientist, Wright Air Develop Div, US Air Force, Ohio, 51-54, instr physics & chem, US Mil Acad, 54-56; sales engr, X-Ray Dept, Gen Elec Co, 56-57; asst prof, Univ Fla, 57-58; tech adv indust sales, X-Ray Dept, Gen Elec Co, 58-61; prod mgr x-ray & electron optics, Picker X-Ray Corp, NY, 62-70, mkt mgr, Indust Div, Picker Corp, 70-72; new mkt res mgr, 72-73, DIR EDUC, XERORADIOGRAPHY, XEROX CORP, 73- Mem: AAAS; Am Chem Soc; Sigma Xi; Am Crystallog Asn. Res: X-ray diffraction and emission; optical methods of instrumental analysis; instrumental chemical analysis; radiography; applied x-rays. Mailing Add: Spec Prod & Systs Div Xerox Corp 125 N Vinedo Ave Pasadena CA 91107

LORBEER, JAMES W, b Oxnard, Calif, Oct 30, 31; m 64. PLANT PATHOLOGY, MYCOLOGY. Educ: Pomona Col, BA, 53; Univ Wash, MS, 55; Univ Calif, Berkeley, PhD(plant path), 60. Prof Exp: Asst bot, Univ Wash, 53-55; asst plant path, Univ Calif, Berkeley, 55-60; from asst prof to assoc prof, 60-72, PROF PLANT PATH, CORNELL UNIV, 72- Mem: Mycol Soc Am; Am Phytopath Soc; NY Acad Sci; Brit Mycol Soc. Res: Diseases of vegetable crops; epidemiology; plant disease control; biology of Botrytis; fungal genetics. Mailing Add: Dept of Plant Path Cornell Univ Ithaca NY 14850

LORBER, ARTHUR, b Cologne, Ger, June 22, 29; US citizen; m 61; c 3. INTERNAL MEDICINE, RHEUMATOLOGY. Educ: Reed Col, AB, 51; Univ Calif, Berkeley, MS, 54; Stanford Univ, MD, 56; Am Bd Internal Med, dipl, 63. Prof Exp: Clin investr, Vet Admin Ctr, 60-64; from clin instr to asst clin prof med, Univ Calif, Los Angeles, 61-68; ASSOC PROF MED, SCH MED, UNIV SOUTHERN CALIF, 67- Concurrent Pos: Southern Calif Div, Am Rheumatism Asn fel, Univ Calif, Los Angeles & Vet Admin Ctr, Los Angeles, 59-60; attend physician, Arthritis Clin, Vet Admin Hosp, Wadsworth, 62-66, assoc chief staff res, Sepulveda, 64-66, chief autoimmunol chem lab, 64-71; attend staff physician, Olive View Hosp, 69; chief rheumatology, Long Beach Vet Admin Hosp, 71-; dir rheumatology, Mem Hosp Med Ctr, Long Beach, 73- Mem: Fel Am Col Physicians; Am Rheumatism Asn. Res: Metabolic and immunological disturbances of connective tissue disorders and their broader application to human disease and aging. Mailing Add: Dept of Med Univ of Southern Calif Med Sch Los Angeles CA 90033

LORBER, MORTIMER, b New York, NY, Aug 30, 26; m 56; c 2. PHYSIOLOGY, HEMATOLOGY. Educ: NY Univ, BS, 45; Harvard Univ, DMD, 50, MD, 52. Prof Exp: Rotating intern, Univ Chicago Clins, 52-53; resident hemat, Mt Sinai Hosp, NY, 53-54, asst resident med, 57; med officer hemat res, Naval Med Res Inst, 55-56; sr asst resident med, Univ Hosp, 58, from instr to asst prof, 59-68, ASSOC PROF PHYSIOL, SCH MED & DENT, GEORGETOWN UNIV, 68- Concurrent Pos: Lederle Med Fac Award, Georgetown Univ, 60-63, USPHS res career develop award, 63-70. Mem: Am Soc Hemat; Int Soc Hemat; Sigma Xi; Int Asn Dent Res; Am Physiol Soc. Res: Splenic function; iron metabolism in Gaucher's disease; permeability of teeth. Mailing Add: Dept of Physiol & Biophys Georgetown Univ Sch Med & Dent Washington DC 20007

LORBER, STANLEY H, b New York, NY, Nov 23, 17; m 45; c 3. MEDICINE. Educ: Univ Pa, AB, 39, MD, 43; Am Bd Internal Med, dipl, 51; Am Bd Gastroenterol, dipl, 52. Prof Exp: Intern, Hosp Univ Pa, 43-44, resident med, 46-48; from assoc to assoc prof, 48-63, clin prof, 63-66, PROF MED, SCH MED, TEMPLE UNIV, 66-, HEAD DEPT GASTROENTEROL, 63- Concurrent Pos: McNeil fel gastroenterol, Univ Pa, 46-48; consult, Food & Drug Admin; team physician, Philadelphia 76ers; pres, Philadelphia GI training group. Mem: AMA; Am Gastroenterol Asn; Am Col Physicians; Am Physiol Soc; Soc Physicians Nat Basketball Asn (pres). Mailing Add: Dept of Gastroenterol Temple Univ Health Sci Ctr Philadelphia PA 19140

LORBER, VICTOR, b Cleveland, Ohio, Apr 22, 12; m 37; c 3. PHYSIOLOGY. Educ: Univ Chicago, BS, 33; Univ Ill, MD, 37; Univ Minn, PhD(physiol), 43. Prof Exp: From instr to asst prof, Med Sch, Univ Minn, 41-46; from assoc prof biochem to prof, Case Western Reserve Univ, 46-51; PROF PHYSIOL, SCH MED, UNIV MINN, MINNEAPOLIS, 52- Concurrent Pos: Career investr, Am Heart Asn, 51. Mem: Am Physiol Soc; Soc Exp Biol & Med. Res: Cardiac metabolism; ionic fluxes in heart muscle. Mailing Add: Dept of Physiol 450 Millard Hall Univ of Minn Sch of Med Minneapolis MN 55455

LORCH, EDGAR RAYMOND, b Nyon, Switz, July 22, 07; nat US; m 37, 56; c 5. MATHEMATICS. Educ: Columbia Univ, AB, 28, PhD(math), 33. Prof Exp: Asst math, Columbia Univ, 28-30, instr, 31-33; Nat Res Coun fel, Harvard Univ, 33-34; Cutting traveling fel, Univ Szeged, 34-35; instr math, Columbia Univ, 35-41; from asst prof to assoc prof, 41-48, chmn dept, 68-72, PROF MATH, COLUMBIA UNIV, 48- Concurrent Pos: Ed, Am Math Soc Bull, 41-46 & Ind Univ Math J, 66-; res mathematician, Nat Defense Res Coun, 43-45; sci adv to chief of staff, US Army, 48-49; vis prof, Carnegie Inst Technol, 49, Univ Rome, 53-54 & 66, Col of France, 58, Stanford Univ, 63, Mid East Tech Univ, Ankara, 65 & Univ Florence, 75; Fulbright lectr, Italy, 53-54 & France, 58; mem, Sec Sch Math Curric Improvement Study, 66-; vis lectr, Fordham Univ, 66-73. Mem: Am Math Soc; Math Asn Am; Math Soc France; Austrian Math Soc; Math Union Italy. Res: Linear spaces; normed rings; theory of convex bodies and integration; point set topology. Mailing Add: 445 Riverside Dr New York NY 10027

LORCH, JOAN, b Offenbach, Ger, June 13, 23; m 52; c 2. CELL BIOLOGY, PROTOZOOLOGY. Educ: Univ Birmingham, BSc, 45; Univ London, PhD(physiol), 48. Prof Exp: Nuffield fel, King's Col, Univ London, 49-52; res assoc cell biol, Ctr Theoret Biol, State Univ NY Buffalo, 63-68, res asst prof, 68-72; lectr, 71-72, ASST PROF BIOL, CANISIUS COL, 72- Mem: AAAS; Am Soc Cell Biol; Inst Soc, Ethics & Life Sci. Res: Nuclear-cytoplasmic relationships; species specificity; protozoa; bioethics. Mailing Add: Dept of Biol Canisius Col Buffalo NY 14208

LORCH, LEE (ALEXANDER), b New York, NY, Sept 20, 15; m 43; c 1. MATHEMATICS. Educ: Cornell Univ, BA, 35; Univ Cincinnati, MA, 36, PhD(math), 41. Prof Exp: Asst mathematician, Nat Adv Comt Aeronaut, 42-43; instr math, City Col New York, 46-49; asst prof, Pa State Univ, 49-50; assoc prof & chmn dept, Fisk Univ, 50-53, prof & chmn dept, 53-55; prof & chmn dept, Philander Smith Col, 55-58; vis lectr, Wesleyan Univ, 58-69; from assoc prof to prof, Univ Alta, 59-68; PROF MATH, YORK UNIV, 68- Mem: Am Math Soc; Math Asn Am; Nat Inst Sci; Can Math Cong; fel Royal Soc Can. Res: Fourier series; special functions; summability; ordinary differential equations. Mailing Add: Dept of Math York Univ Downsview ON Can

LORCH, STEVEN KALMAN, b New York, NY, Aug 21, 44; m 67; c 2. FORENSIC SCIENCE. Educ: City Col New York, BS, 66; State Univ NY Binghamton, MA, 70; Univ Md, PhD(plant physiol), 72. Prof Exp: Res assoc, Mich State Univ-AEC Plant Res Lab, 72-73; crime lab scientist, 73-75, CHIEF DRUG IDENTIFICATION UNIT, DIV CRIME DETECTION, MICH DEPT PUB HEALTH, 75- Mem: Am Soc Plant Physiologists; Bot Soc Am; Sigma Xi; Am Acad Forensic Sci. Res: Identification of controlled and prescription drugs; gas chromatographic-mass spectrometry; forensic plant identification. Mailing Add: Div of Crime Detection Mich Dept of Pub Health 3500 N Logan St Lansing MI 48914

LORD, ARTHUR E, JR, b Buffalo, NY, Apr 7, 35; m 62; c 1. PHYSICS, ENVIRONMENTAL SCIENCES. Educ: Purdue Univ, BSc, 57, MSc, 59; PhD(metall), Columbia Univ, 64. Prof Exp: Res assoc appl math, Brown Univ, 64-66, asst res prof physics, 66-68; assoc prof, 68-75, PROF PHYSICS, DREXEL UNIV, 75- Concurrent Pos: Fel, Columbia Univ, 64. Mem: Am Phys Soc; Am Soc Plant Physiologists; Nat Asn Jazz Educr; Microbeam Anal Soc. Res: Acoustic emission studies in soils, coal, pipelines and other construction materials; use of scanning electron microscope and microprobe, microwave and electron spin resonance measurements in soil and plant science. Mailing Add: 1006 Edwards Dr Springfield PA 19064

LORD, CLIFFORD SYMINGTON, geology, see 12th edition

LORD, CLIFTON FRANCIS, JR, pharmacy, see 12th edition

LORD, GEOFFREY HAVERTON, b Georgetown, Guyana, May 17, 23; nat US; m 52; c 2. PATHOLOGY. Educ: Univ Toronto, DVM, 49; Univ Wis, MS, 50, PhD, 53. Prof Exp: Asst vet sci, Univ Wis, 49-53; sr pathologist, Dept Pharmacol, 53-57, asst dir, 57-60, DIR, JOHNSON & JOHNSON RES FOUND, 60- Concurrent Pos: Actg state wildlife pathologist, Wis, 52-53; consult, Middlesex Gen Hosp, New Brunswick. Mem: Am Col Lab Animal Med; Am Vet Med Asn; Am Asn Accreditation Lab Animal Care. Res: Experimental vascular surgery; laboratory animal pathology; drug safety evaluation; wound healing; safety evaluation of devices. Mailing Add: Johnson & Johnson Res Found New Brunswick NJ 08903

LORD, JERE JOHNS, b Portland, Ore, Jan 3, 22; m 47; c 3. PHYSICS. Educ: Reed Col, AB, 43; Univ Chicago, MS, 48, PhD(physics), 50. Prof Exp: Civilian with Radiation Lab, Univ Calif, 42-46; res assoc physics, Univ Chicago, 50-52; instr, 52-62, PROF PHYSICS, UNIV WASH, 62- Mem: Am Phys Soc. Res: Cosmic ray and high energy physics. Mailing Add: Dept of Physics Univ of Wash Seattle WA 98105

LORD, JERE WILLIAMS, JR, b Baltimore, Md, Oct 12, 10; m 41, 71; c 3. SURGERY. Educ: Princeton Univ, AB, 33; Johns Hopkins Univ, MD, 37; Am Bd Surg, dipl. Prof Exp: From intern to resident surgeon, NY Hosp, 37-44; PROF CLIN SURG, POSTGRAD SCH MED, MED CTR, NY UNIV, 53- Concurrent Pos: Consult surgeon, Univ Hosp, Bellevue Hosp, Fourth Div Med Bd & Doctors Hosp & Hackensack Hosp, NJ, St Luke's Hosp, Newburgh, NY, Norwalk Hosp, Conn, 50-, Cent Suffolk Hosp, Riverhead, NY, 51-, Elizabeth Horton Mem Hosp, Middletown, NY, 54-, St Agnes Hosp, White Plains, NY, 55-, Paterson Gen Hosp, NJ, 58 & Univ Hosp; chief, Vascular Surg, Columbus Hosp, NY, 66- Mem: Am Col Surg; Am Surg Asn; James IV Asn Surg (secy, 67-75); Am Heart Asn (secy, 53-55); Int Cardiovasc Soc (treas, 53-60, vpres, 61-63). Res: Cardiovascular surgery, especially atherosclerosis; gastrointestinal surgery, particularly portal hypertension and intestinal obstruction. Mailing Add: 50 Sutton Pl S New York NY 10022

LORD, NORMAN WILLIAM, physics, systems analysis, see 12th edition

LORD, PIERRE DAVID, organic chemistry, environmental chemistry, see 12th edition

LORD, REXFORD DUNBAR, vertebrate ecology, see 12th edition

LORD, RICHARD COLLINS, JR, b Louisville, Ky, Oct 10, 10; m 43; c 4. PHYSICAL CHEMISTRY. Educ: Kenyon Col, BSc, 31; Johns Hopkins Univ, PhD(phys chem), 36. Hon Degrees: Kenyon Col, DSc, 57. Prof Exp: Instr chem & physics, Tome Sch, 31-32; asst chem, Johns Hopkins Univ, 33-36; Nat Res Coun fel, Univ Mich, 36-37 & Copenhagen Univ, 37-38; asst prof, Johns Hopkins Univ, 38-42; res assoc, Mass Inst Technol, 42-43; assoc prof, Johns Hopkins Univ, 45-46; assoc prof, 46-54, PROF CHEM, MASS INST TECHNOL, 54-, DIR SPECTROS LAB, 46- Concurrent Pos: Tech aide, Optics Div, Nat Defense Res Comt, Mass Inst Technol, 43-46, mem, 47-53, chmn panel infrared, Res & Develop Bd, 52-53; consult, E I du Pont de Nemours & Co, Inc, 48-; mem, US Nat Comt, Int Comn Optics, 56-59; comn molecular spectros, Int Union Pure & Appl Chem, 57-71, pres, 61-67; Reilly lectr, Univ Notre Dame, 58; vis prof, Univ Ga, 71. Honors & Awards: Presidential Cert Merit, 48; Pittsburgh Award Spectros, 66. Mem: AAAS; Am Chem Soc; fel Optical Soc Am (pres, 64); fel Am Acad Arts & Sci; hon mem Soc Appl Spectros. Res: Infrared spectroscopy, especially far infrared; infrared and Raman spectra of polyatomic molecules, particularly deuterium compounds; calculation of thermodynamic properties of gases and crystals from spectroscopic data; biophysical spectroscopy. Mailing Add: Dept of Chem Mass Inst of Technol Cambridge MA 02139

LORD, SAMUEL SMITH, JR, b Rockland, Maine, Apr 10, 27; m 48; c 5. ANALYTICAL CHEMISTRY. Educ: Tufts Col, BS, 47; Mass Inst Technol, PhD(anal chem), 52. Prof Exp: Res chemist, Fabrics & Finishes Dept, E I du Pont de Nemours & Co, Inc, 47-49, res chemist, Org Chem Dept, 52-57, res supvr, Elastomer Chem Dept, 57-59, div head, 59-65, supt qual control, 65-67, supt monomer area, 67-70, gen prod supt, 70-71, asst works dir, Maydown Works, Du Pont Co (UK) Ltd, 71-75, WORKS MGR, BEAUMONT WORKS, E I DU PONT DE NEMOURS & CO, INC, 75- Mem: AAAS; Am Chem Soc. Res: Polarography; coulometry; infrared and ultraviolet spectrophotometry; urethane chemistry. Mailing Add: Beaumont Works PO Box 3269 E I du Pont de Nemours & Co Inc Beaumont TX 77704

LORD, WILLIAM JOHN, b Farmington, NH, Nov 3, 21; m 47; c 1. POMOLOGY. Educ: Univ NH, BS, 43, MS, 53; Pa State Univ, PhD(hort), 55. Prof Exp: EXTEN PROF POMOL, AGR EXTEN SERV, UNIV MASS, AMHERST, 55- Mem: Am Soc Hort Sci. Res: Weed control; nutrition; growth regulators. Mailing Add: French Hall Univ of Mass Amherst MA 01002

LORDI, NICHOLAS GEORGE, b Orange, NJ, Mar 25, 30; m 61; c 3. PHARMACY. Educ: Rutgers Univ, BSc, 52 & MSc, 53; Purdue Univ, PhD(pharmaceut chem), 55. Prof Exp: From asst prof to assoc prof pharm, 57-64, PROF PHARM, RUTGERS UNIV, 64- Mem: Am Pharmaceut Asn; NY Acad Sci. Res: Polarography as applied to pharmaceutical systems; rheology of semisolids; pharmaceutical technology. Mailing Add: Col of Pharm Rutgers Univ New Brunswick NJ 08903

LORDS, JAMES LAFAYETTE, b Salt Lake City, Utah, Apr 5, 28; m 55; c 2. PLANT PHYSIOLOGY. Educ: Univ Utah, BS, 50, MS, 51, PhD(plant physiol), 60. Prof Exp: Asst bot, Univ Utah, 56-58, instr biol, 58-59; proj assoc plant path, Univ Wis, 60-62; asst prof, 62-66, ASSOC PROF MOLECULAR & GENETIC BIOL, UNIV UTAH, 66- Res: Thermophilic mechanisms; microwave interactions with biological systems. Mailing Add: Dept of Biol Univ of Utah Salt Lake City UT 84112

LÖRE, ASKELL, b Reykjavik, Iceland, Oct 20, 16; m 40; c 2. CYTOTAXONOMY, CYTOGENETICS. Educ: Reykjavik Col, BA, 37; Univ Lund, BS, 41, PhD, 42, DSc, 43. Prof Exp: Swedish Acad Sci fel, 42-45; geneticist, Res Inst, Univ Reykjavik, 42-45, dir, Inst Bot & Plant Breeding, 45-51; assoc prof bot, Univ Man, 51-56; res prof biosysts, Inst Bot, Univ Montreal, 56-63, Guggenheim Mem Found fel, 63-64; from assoc prof to prof bot, assoc curator phanerogams, mus & res assoc, Inst Arctic & Alpine Res, Univ Colo, Boulder, 64-73, chmn div pop studies, 71-73, chmn dept biol, 66-70. Concurrent Pos: Res assoc, Univ Lund, 41-45; consult, Flora Europaea, 58-; pres, Int Orgn Plant Biosyst, 60-64; pres, Int Comt Chemotaxon, 64- Mem: AAAS; Genetics Soc Am; Genetics Soc Can; Int Asn Plant Taxon; fel Icelandic Acad Sci & Letters. Res: Cytotaxonomy and geobotany of Icelandic, Scandinavian, Mediterranean and American plants; polyploidy; arctic and boreal retrieval of cytological and taxonomical data; chromosome manuals; sytogenetics; physiology of sex determination; qualitative methods in evolution and taxonomy. Mailing Add: 473 Harvard Lane Boulder CO 80303

LORE, JOHN M, JR, b New York, NY, July 26, 21; m; c 4. OTOLARYNGOLOGY, SURGERY. Educ: Col of Holy Cross, BS, 44; NY Univ, MD, 45; Am Bd Otolaryngol, dipl, 54; Am Bd Surg, dipl, 56. Prof Exp: Intern, St Vincent's Hosp, New York, 45-46, resident otolaryngol & head & neck surg, 48-50; asst resident gen surg, St Clare's Hosp, 50-52, sr resident, 54-55; asst resident surg & radiation, Mem Cancer Ctr, 52-53; asst clin prof surg & asst attend surgeon, NY Med Col, Flower & Fifth Ave Hosps, 64-66; PROF OTOLARYNGOL & CHMN DEPT, SCH MED, STATE UNIV NY BUFFALO, 66- Concurrent Pos: Fel exp surg, St Clare's Hosp, 53-54; asst vis surgeon, Metrop Hosp Ctr, New York, 64-66; dir surg, Good Samaritan Hosp, Suffern, attend surgeon, St Clare's Hosp, New York & consult surgeon, Tuxedo Mem Hosp, 65-66; head dept otolaryngol & chief combined head & neck serv, Buffalo Gen Hosp & Buffalo Children's Hosp, 66-; head dept otolaryngol & chief combined head & neck serv, E J Meyer Mem Hosp, 66-; consult, Buffalo Vet Admin Hosp, 66-; vis prof, Col Med, Baylor Univ, 67; consult, Roswell Park Mem Inst, 68-; clin consult, NY State Dept Health, 68-; consult, Deaconess Hosp, Buffalo, NY, 69; vis prof, Denver Med Ctr, 69 & Dept Otolaryngol, Bethesda Naval Med Ctr, Md, 71. Honors & Awards: Hektoen Gold Medal, AMA, 52. Mem: Fel Am Col Surg; Am Cancer Soc; fel Am Acad Ophthal & Otolaryngol; AMA; James Ewing Soc. Res: General surgery, including maxillofacial surgery and plastic surgery of the head and neck. Mailing Add: 100 High St Buffalo NY 14203

LOREE, THOMAS ROBERT, b Seattle, Wash, Feb 1, 36; m 58; c 3. SOLID STATE PHYSICS, LASERS. Educ: Willamette Univ, BA, 57; Univ Wis, MS, 60, PhD(solid state physics), 62. Prof Exp: MEM RES STAFF, LOS ALAMOS SCI LABS, 62- Concurrent Pos: Consult, Particle Technol, Inc, 73- Mem: Am Phys Soc. Res: Electronic properties of metals and semiconductors; shock waves in metals; dye laser research; laser system design; non-linear optics research and design. Mailing Add: Los Alamos Sci Labs Box 1663 Los Alamos NM 87545

LORENSEN, LYMAN EDWARD, b Lincoln, Nebr, Sept 26, 23; m 50; c 3. ORGANIC POLYMER CHEMISTRY. Educ: Univ Nebr, BS, 47; Cornell Univ, PhD(chem), 52. Prof Exp: Jr chemist, Bristol Labs, 47-48; asst org chem, Cornell Univ, 50-52; chemist, Shell Develop Co, 52-64, mem staff, Mfg Res Dept, 58-60; MEM STAFF POLYMERS & PLASTICS, LAWRENCE LIVERMORE LAB, UNIV CALIF, 64- Mem: Am Chem Soc; Sigma Xi. Res: High temperature polymers; polymers for geothermal applications; unsaturated glycols; possible precursors in biosynthesis of rubber; lubricating oil additives; silicone and epoxy polymers; filled polymers. Mailing Add: 9 Broadview Terr Orinda CA 94563

LORENSON, MARY GRACE YOUNG, biochemistry, see 12th edition

LORENTE DE NO, RAFAEL, b Zaragoza, Spain, Apr 8, 02; nat US; m 31; c 1. PHYSIOLOGY. Educ: Univ Madrid, MD, 23. Hon Degrees: MD, Univ Uppsala, 53. Prof Exp: Asst, Inst Cajal, Madrid, 21-29; head dept otolaryngol, Valdecilla Hosp, Santander, 29-31; neuroanatomist, Cent Inst for Deaf, St Louis, Mo, 31-36; assoc, 36-38, assoc mem, 38-41, mem & prof, 41-74, EMER PROF PHYSIOL, ROCKEFELLER UNIV, 74- Concurrent Pos: Lectr, Med Sch, Wash Univ, 35-36. Mem: Nat Acad Sci; Am Physiol Soc; Am Asn Anat; Am Neurol Asn; Am Acad Arts & Sci. Res: Anatomy of the central nervous system; neurophysiology. Mailing Add: Dept of Physiol Rockefeller Univ New York NY 10021

LORENTS, DONALD C, b Minn, Mar 26, 29; m 52; c 2. ATOMIC PHYSICS. Educ: Concordia Col, Moorhead, Minn, BA, 51; Univ Nebr, MA, 54, PhD(physics), 58. Prof Exp: Res physicist, Westinghouse Res Labs, 58-59; physicist, Stanford Res Inst, 59-63, chmn dept molecular physics, 63-67, head atomic & molecular collisions sect, 67-68, physicist, 69-70, SR PHYSICIST, STANFORD RES INST, 70- Concurrent Pos: Vis res physicist, Inst Physics, Aarhus Univ, 68-69. Mem: AAAS; fel Am Phys Soc. Res: Atomic and molecular collision processes with emphasis on scattering, charge transfer and excitation in ion-atom or ion-molecule collisions. Mailing Add: Dept of Molecular Physics Stanford Res Inst 333 Ravenswood Menlo Park CA 94025

LORENTZ, GEORGE G, b St Petersburg, Russia, Feb 25, 10; m 42; c 5. MATHEMATICAL ANALYSIS. Educ: Univ Leningrad, Cand, 35; Univ Tübingen, Dr rer nat (math), 44. Prof Exp: Lectr math, Univ Leningrad, 36-42 & Univ Frankfurt, 46-48; prof, Univ Tübingen, 48-49; from asst to asst prof, Univ Toronto, 49-53; prof, Wayne State Univ, 53-58; prof, Syracuse Univ, 58-69; PROF MATH, UNIV TEX, AUSTIN, 69- Concurrent Pos: Res grants, NSF & Off Sci Res. Mem: Am Math Soc; Math Asn Am; Ger Math Soc. Res: Mathematical analysis, especially approximations and expansions; summability; Birkhoff interpolation; functional analysis, especially Banach function spaces; interpolation theorems for operators. Mailing Add: Dept of Math RLM 8-100 Univ of Tex Austin TX 78712

LORENTZEN, KEITH EDEN, b Heber City, Utah, Apr 13, 21; m 47; c 6. PHYSICAL ORGANIC CHEMISTRY. Educ: Univ Utah, BA, 42, MS, 47; Pa State Univ, PhD(chem), 51. Prof Exp: Chemist, Standard Oil Co (Ind), 51-62; asst prof, 63-69, asst chmn dept, 66-70, ASSOC PROF CHEM, IND UNIV NORTHWEST, 69-, CHMN DEPT, 70- Mem: Am Chem Soc. Res: Conductivity measurements; chemistry of lubricating oils and additives; organic analytical chemistry; chromatography; polarography; Friedel-Crafts acylation of xylenes. Mailing Add: Dept of Chem Ind Univ Northwest Gary IN 46408

LORENZ, CARL EDWARD, b New York, NY, Aug 22, 33; m 56; c 3. ORGANIC CHEMISTRY. Educ: NY Univ, BA, 53, PhD(chem), 57. Prof Exp: Asst chem, NY Univ, 53-57; chemist, Plastics Dept, Exp Sta, 57-63, sr res chemist, 63-68, supvr, 68, sr supvr, 68-69, lab supt, 69-70, res lab mgr, Sabine River Works, 70-72, res mgr, Wilmington, 72-74, ASST DIR, INT DEPT, E I DU PONT DE NEMOURS & CO, INC, 74- Honors & Awards: Award, Am Inst Chem, 53. Mem: Am Chem Soc; Am Inst Chem; The Chem Soc. Res: Fluorocarbon monomer syntheses and polymerizations; high pressure hydrocarbon syntheses; heterogeneous catalysis; chemistry of anionic and radical polymerizations. Mailing Add: Int Dept DuPont Bldg E I du Pont de Nemours & Co Inc Wilmington DE 19801

LORENZ, DONALD H, b Brooklyn, NY, Oct 18, 36; m 62. ORGANIC CHEMISTRY, POLYMER CHEMISTRY. Educ: Polytech Inst Brooklyn, BS, 58, PhD(org chem), 63. Prof Exp: Asst org chem, Polytech Inst Brooklyn, 58-59, organometallics, 59-62; asst scientist chem eng res div, NY Univ, 62-63; sr polymer chemist, Tex-US Chem Co, 63-65; explor polymer chemist, Gen Aniline & Film Co, 65-70, group leader polymer synthesis, 70-74, MGR VINYL POLYMER RES, GAF CORP, 74- Mem: Am Chem Soc. Res: Organometallic chemistry; elastomers; resins; adhesives; polymers of vinyl ethers and vinyl amides; polyurethanes; fire retardants; VV & EB curable resins. Mailing Add: 12 Radel Pl Basking Ridge NJ 07920

LORENZ, DOUGLAS, b Dubuque, Iowa, July 12, 28; m 68; c 2. MICROBIOLOGY, VIROLOGY. Educ: Univ Calif, Los Angeles, BA, 53, PhD(microbiol), 63. Prof Exp: Res fel virol, Med Sch, Univ Minn, 63-64; asst prof, Sch Med, Univ NMex, 64-65; sr scientist, Life Sci Div, Melpar, Inc, Va, 65-67; res microbiologist, Div Biol Standards, NIH, 67-72, RES MICROBIOLOGIST, BUR BIOLOGICS, FOOD & DRUG ADMIN, 72- Mem: AAAS; Am Soc Microbiol. Res: Animal models for human hepatitis; herpesvirus tumors; animal testing of human vaccines; effectiveness of various commercial disinfectants on Newcastle disease virus by usedilution test; adsorption rates of Rous sarcoma virus on susceptible and non-susceptible cells. Mailing Add: Bldg 29 FDA 8800 Rockville Pike Bethesda MD 20014

LORENZ, EDWARD NORTON, b West Hartford, Conn, May 23, 17; m 48; c 3. METEOROLOGY. Educ: Dartmouth Col, AB, 38; Harvard Univ, AM, 40; Mass Inst Technol, SM, 43, ScD(meteorol), 48. Prof Exp: Asst meteorol, Mass Inst Technol, 46-48, mem staff, 48-54; vis assoc prof, Univ Calif, Los Angeles, 54-55; from asst prof to assoc prof, 55-62, PROF METEOROL, MASS INST TECHNOL, 62- Mem: Am Math Soc; Am Meteorol Soc. Res: General circulation of the atmosphere; dynamical and statistical weather prediction. Mailing Add: Dept of Meteorol Mass Inst of Technol Cambridge MA 02139

LORENZ, FREDERICK WHARTON, b Berkeley, Calif, Dec 30, 08; wid; c 1. PHYSIOLOGY. Educ: Univ Calif, BS, 31, PhD(physiol), 38. Prof Exp: Analyst, Poultry Husb Div, Univ Calif, 31-35, instr col agr, 38-43, from asst prof to prof poultry husb, 43-63, prof physiol & physiologist, 64-74, chmn dept animal physiol, 64-68, EMER PROF ANIMAL PHYSIOL, EXP STA, UNIV CALIF, DAVIS, 74- Concurrent Pos: Physiologist, White Labs, Inc, NJ, 45-46. Honors & Awards: Poultry Sci Res Prize, 45; Borden Award, Poultry Sci Asn, 60. Mem: Am Soc Zool; Endocrine Soc; Soc Exp Biol & Med; Soc Study Reproduction; Poultry Sci Asn. Res: Endocrinology; physiology of reproduction including egg formation and fertility. Mailing Add: Dept of Animal Physiol Univ of Calif Davis CA 95616

LORENZ, JOHN CLARK, b Coshocton, Ohio, June 11, 23; m 53; c 4. ORGANIC CHEMISTRY. Educ: Carnegie Inst Technol, BS, 48; Univ Ill, PhD, 51. Prof Exp: RES CHEMIST, E I DU PONT DE NEMOURS & CO INC, 51- Mem: Am Chem Soc. Res: Fluorocarbon elastomers; polyurethanes and other elastomer polymers. Mailing Add: RD 2 Box 354 Berkeley Ridge Hockessin DE 19707

LORENZ, KLAUS J, b Berlin, Ger, June 22, 36; US citizen; m 60; c 2. CEREAL CHEMISTRY. Educ: Northwestern Univ, Ill, PhB, 68; Kans State Univ, MS, 69, PhD(food sci), 70. Prof Exp: Baking technologist, Am Inst Baking, 61-65; food technologist, Nat Dairy Prod Corp, 65-68; asst prof, 70-74, ASSOC PROF FOOD SCI & NUTRIT, COLO STATE UNIV, 74- Mem: Am Asn Cereal Chem; Inst Food Technologists; Swiss Soc Food Sci & Technol. Res: Development of high-protein foods; evaluation of new cereal grain varieties. Mailing Add: Dept of Food Sci & Nutrit Colo State Univ Ft Collins CO 80521

LORENZ, MAX RUDOLPH, b Detroit, Mich, June 25, 30; m 55; c 3. PHYSICAL CHEMISTRY, PHYSICS. Educ: Rensselaer Polytech Inst, BChE, 57, PhD(phys chem), 60. Prof Exp: Res assoc, Gen Elec Res Lab, 60-63; res staff mem, Thomas J Watson Res Ctr, NY, 63-73, HEAD DEPT INORG MAT, IBM RES DIV, IBM CORP, 73- Mem: Fel Am Phys Soc; Electrochem Soc; NY Acad Sci; fel Am Inst Chem; sr mem Inst Elec & Electronics Engrs. Res: Physics and chemistry of semiconductors, especially the role and control of defects and electrical and optical properties; preparation and properties of thin film magnetic materials, garnets and amorphous bubble materials. Mailing Add: IBM Res Lab K44/281 5600 Cottle Rd San Jose CA 95193

LORENZ, OSCAR ANTHONY, b Colorado Springs, Colo, Dec 5, 14; m 47. VEGETABLE CROPS, PLANT NUTRITION. Educ: Colo State Col, BS, 36; Cornell Univ, PhD(veg crops), 41. Prof Exp: Asst hort, Colo State Col, 36-37; asst veg crops, Cornell Univ, 37-41; from instr to assoc prof, 41-55, vchmn dept, 55-64, PROF VEG CROPS, COL AGR, UNIV CALIF, DAVIS, 55-, CHMN DEPT, 64- Concurrent Pos: Mem, Agr Comn to Bermuda, 39. Honors & Awards: Vaughn Award, 42. Mem: Fel Am Soc Hort Sci; fel Am Soc Plant Physiol; fel Am Soc Agron; fel Am Potato Asn. Res: Mineral nutrition of vegetable crops; boron deficiency in table beets; soils and plant nutrient relationships; environmental factors affecting vegetable production. Mailing Add: Dept of Veg Crops Col of Agr Univ of Calif Davis CA 95616

LORENZ, PHILIP BOALT, b Dayton, Ohio, Aug 14, 20; m 46; c 3. PHYSICAL CHEMISTRY. Educ: Swarthmore Col, AB, 41; Harvard Univ, MA, 46, PhD(chem), 49. Prof Exp: Asst biol, Princeton Univ, 42-43; asst, Phys Chem, SAM Labs, Columbia Univ, 44-45; phys chemist surface chem, Petrol Res Ctr, US Bur Mines, 49-71, res chemist, Petrol Prod & Environ Res, 71-75, RES CHEMIST, BARTLESVILLE ENERGY RES CTR, ENERGY RES & DEVELOP ADMIN, 75- Mem: Am Chem Soc; Sigma Xi; Soc Petrol Engrs; Am Inst Mining & Metall Engrs. Res: Surface chemistry; electrochemistry; petroleum engineering. Mailing Add: Bartlesville Energy Res Ctr PO Box 1398 Bartlesville OK 74003

LORENZ, PHILIP JACK, b Atlanta, Ga, Apr 15, 24; m 70; c 2. ATMOSPHERIC PHYSICS. Educ: Oglethorpe Univ, BS, 49; Vanderbilt Univ, MS, 52. Prof Exp: Lab asst, Oglethorpe Univ, 48; qual control tech, Transparent Package Co, 50-51; asst prof physics, Lemoyne Col, 52-54, Ky Wesleyan Col, 54-56 & Upper Iowa Univ, 56-61; res assoc, Syracuse Univ, 63-65, vis instr, 65-66; chmn dept, 66-74, ASSOC PROF PHYSICS, UNIV OF THE SOUTH, 66- Concurrent Pos: Lab asst, Vanderbilt Univ, 52; consult physicist, Empirical Explor Co, Ky, 56; univ fel, Syracuse Univ, 58-59, Nat Sci Found fac fel, 61-63; textbook consult, J B Lippincott Co, 71-72. Mem: AAAS; Am Phys Soc; Am Asn Physics Teachers; Hist Sci Soc. Res: Atmospheric electricity in fair and foggy weather; geophysics of environmental radioactivity at sandstone sinkhole sites; history of medieval Persian and Arabic science; history of astronomy. Mailing Add: Dept of Physics Univ of the South Sewanee TN 37375

LORENZ, RALPH WILLIAM, b Waseca, Minn, Aug 19, 07; m 39; c 2. FORESTRY. Educ: Univ Minn, BS, 30, PhD(plant Physiol), 38. Prof Exp: Asst plant physiol, Univ Minn, 30-33; jr forester, US Forest Serv, 33-35; instr forestry, Univ Farm, Univ Minn, 35-38; assoc, 38-41, asst chief, 41-44, assoc chief, 44-47, assoc prof forest res, 47-55, prof, 55-73, actg head dept, 65-66, EMER PROF FORESTRY, UNIV ILL, URBANA, 73- Mem: Soc Am Foresters. Res: Forest research in planting; silviculture; dendrology; regeneration and forest management. Mailing Add: 1707 S Pleasant Urbana IL 61801

LORENZ, ROMAN R, b Breslau, Ger, July 15, 35; US citizen; m 60; c 3. ORGANIC CHEMISTRY. Educ: Rensselaer Polytech Inst, BS, 58; Univ Mich, MS, 60, PhD(med chem), 62. Prof Exp: Res org chemist, 62-69, sr res chemist & sect head, 69-74, SR RES ASSOC & SECT HEAD, STERLING-WINTHROP RES INST, 74- Mem: Am Chem Soc. Res: Synthesis of organic and medicinal compounds. Mailing Add: Sterling-Winthrop Res Inst Rensselaer NY 12144

LORENZEN, CARL JULIUS, biological oceanography, see 12th edition

LORENZEN, EVELYN JUNE, b El Reno, Okla, June 2, 21. PEDIATRICS, NUTRITION. Educ: Univ Okla, BS, 42; Cornell Univ, PhD(nutrit), 46; Univ Ill, MD, 51; Am Bd Nutrit, Dipl. Prof Exp: Instr nutrit, Cornell Univ, 43-46; res dir & res fel, Harvard Med Sch, 46-47; intern, Charity Hosp New Orleans, La, 51-52, res pediat, 52-54; asst clin prof, 54-65, ASSOC CLIN PROF PEDIAT, BAYLOR COL MED, 65- & UNIV TEX MED SCH, 75- Concurrent Pos: Pvt practice, 54- Mem: AAAS; Am Med Asn; fel Am Acad Pediat. Res: Biochemistry. Mailing Add: 6615 Travis Houston TX 77025

LORENZEN, GERALD ANDREW, mammalian physiology, biochemistry, see 12th edition

LORENZEN, JERRY ALAN, b Grand Island, Nebr, Oct 3, 44; m 67; c 1. SURFACE CHEMISTRY. Educ: Midland Lutheran Col, BS, 66; Okla State Univ, PhD(chem), 70. Prof Exp: Instr chem, Okla State Univ, 69-70; staff chemist, Mat Lab, 70-73, mgr environ technol, 73-75, ADV CHEMIST, IBM CORP, 75- Mem: Am Chem Soc. Res: Environmental chemistry; corrosion; energy transfer. Mailing Add: IBM Corp Neighborhood Rd Kingston NY 13760

LORENZETTI, OLFEO J, b Chicago, Ill, Oct 25, 36; m 62; c 3. PHARMACOLOGY, BIOCHEMISTRY. Educ: Univ Ill, Chicago, BS, 58; Ohio State Univ, MS, 62, PhD(pharmacol), 65. Prof Exp: Asst chief pharmacist, WSuburban Hosp, 58; instr pharm, Univ Ill, Chicago, 58-59; asst instr pharmacol, Ohio State Univ, 59-62; from res pharmacologist to sr res pharmacologist, Therapeut Res Labs, Dome Chem Inc Div, Miles Labs, Inc, Ind, 64-69; sr res scientist, Alcon Labs Inc, 69-71, dir immunol & biochem res, 71-72, mgr biol res, 72-74; vpres, 67-70, PRES PHARMACEUT CONSULT, INC, 70-; DIR DERMAT/PEDIAT-UROL SERV, ALCON LABS, INC, 74- Concurrent Pos: Assoc prof pharmacol, Univ Tex Health Sci Ctr, Dallas, 70-; adj prof, Tex Christian Univ, 72- Mem: AAAS; Am Chem Soc; Soc Cosmetic Chem; Am Acad Clin Toxicol; Soc Invest Dermat. Res: Pharmacodynamics; evaluations of analgesic, anti-inflammatory agents and antiglaucoma agents; development of drug screening programs; autonomic and biochemical pharmacology; topical pharmacology of eye and skin; ophthalmology; dermatology; immunology; drug metabolism; pharmacokinetics. Mailing Add: Alcon Labs Inc 6201 S Freeway Ft Worth TX 76101

LORENZO, ANTONIO V, b Vigo, Spain, July 23, 28; US citizen; m 58; c 2. NEUROPHARMACOLOGY, NEUROCHEMISTRY. Educ: Univ Chicago, BA, 56, BS, 58, PhD(pharmacol), 66. Prof Exp: From asst to assoc neurol, Children's Hosp Med Ctr, 64-68, instr pharmacol, 66-68; asst prof, 69-71, ASSOC PROF PHARMACOL, HARVARD MED SCH, 71-; DIR NEUROL RES, CHILDREN'S HOSP MED CTR, 69- Concurrent Pos: Epilepsy Found Am fel, Children's Hosp Med Ctr, 65, Nat Inst Neurol Dis & Stroke proj grant, 71-73; NIH career develop award, Harvard Med Sch, 70-75. Mem: AAAS; Am Soc Pharmacol & Exp Therapeut; NY Acad Sci; Am Soc Neurochem; Am Acad Neurol. Res: Pathophysiology of the blood, role of brain barrier, putative transmitter seizures; cerebrospinal fluid transport phenomena; cerebrospinal fluid dynamics. Mailing Add: Neurol Res Dept Children's Hosp Med Ctr Boston MA 02115

LORENZO, GEORGE ALBERT, organic chemistry, see 12th edition

LORENZO, MICHAEL ALOYSIUS, zoology, deceased

LOREY, FRANK WILLIAM, b Staten Island, NY, May 7, 29; m 51; c 3. PAPER CHEMISTRY. Educ: State Univ New York Col Forestry, Syracuse, BS, 51, MS, 52. Prof Exp: Res engr, Mead Corp, Ohio, 52-54; assoc prof pulp & paper chem & pilot plant group leader, State Univ New York Col Forestry, Syracuse, 54-66; asst to gen mgr, 66-67, corp tech dir, 67-75, VPRES, GARDEN STATE PAPER CO, GARFIELD, 75- Concurrent Pos: Develop consult, AB Kamyr, Sweden, 65. Mem: Tech Asn Pulp & Paper Indust. Res: Improved methods in pulping of wood and use of chemicals for influencing paper properties; development of processes and design of systems for deinking of waste papers. Mailing Add: 82 Dogwood Terr Ramsey NJ 07446

LORHAN, PAUL HERMAN, b Mont Clare, Pa, Apr 7, 08; m 42; c 5. ANESTHESIOLOGY. Educ: Ohio State Univ, AB, 31; Creighton Univ, MD, 35; Am Bd Anesthesiol, dipl, 41. Prof Exp: From instr to prof anesthesiol, Univ Kans, 38-58, dir anesthesia, 38-39; PROF DENT ANESTHESIOL, UNIV CALIF, LOS ANGELES, 58-; DIR ANESTHESIOL, HARBOR GEN HOSP, TORRANCE, CALIF, 58- Concurrent Pos: Consult, US Army Hosp, Ft McArthur, 58-; del, White House Conf Aging, 71; consult, US Naval Hosp, Long Beach, Calif. Mem: Am Soc Anesthesiol (vpres, 54); Int Anesthesia Res Soc; fel AMA; fel Am Col Anesthesiol; Acad Anesthesiol. Res: Anoxia in anesthesia; vasopressor drugs in anesthesia; circulatory dynamics during anesthesia; effects of methoxyflurane on renal and hepatic function; massive blood transfusions; anesthesia for the aged; evaluation of antiemetics. Mailing Add: 913 Via Mirola Palos Verdes Estates CA 90274

LORIA, ROGER MOSHE, b Antwerpen, Belgium, Apr 19, 40; US citizen; div; c 1. VIROLOGY. Educ: Bar-Ilan Univ, Israel, BS, 65; State Univ NY, Buffalo, MS, 68; Boston Univ, PhD(microvirol), 72. Prof Exp: Asst prof biochem, Mass Col Optom, 69-70; asst virol, Sch Med, Boston Univ, 68-72, instr microbiol, 72-74; ASST PROF MICROBIOL, MED COL VA, 74- Concurrent Pos: Mass Heart Asn fel, 72-74; res assoc, Sch Med, Boston Univ, 74; NIH res grant, Arthritis & Metab Dis, 74 & Heart & Lung Div, 75; Young investr develop award, Am Diabetes Asn, 75. Mem: Am Soc Microbiol; AAAS; Am Fedn Clin Res. Res: Investigation on the role of group B coxsackieviruses in diabetes, atherosclerosis and cardiovascular disease in experimental animal modes; general aspects of host-virus interaction; viral infection of the oral route. Mailing Add: Dept of Microbiol Med Col Va Box 847 Richmond VA 23298

LORIMER, JOHN WILLIAM, b Oshawa, Ont, Apr 16, 29; m 54; c 3. PHYSICAL CHEMISTRY. Educ: Univ Toronto, BA, 51, MA, 52, PhD(phys chem), 54. Prof Exp: Asst phys chem, Univ Leiden, Netherlands, 54-56; asst res officer, Atlantic Regional Lab, Nat Res Coun Can, 56-61, assoc res officer, 61; asst prof, 61-65, ASSOC PROF PHYS CHEM, UNIV WESTERN ONT, 65- Mem: The Chem Soc; fel Chem Inst Can. Res: Thermodynamics of liquids; transport in membranes; irreversible thermodynamics. Mailing Add: Dept of Chem Univ of Western Ont London ON Can

LORIMER, NANCY L, b Mishawaka, Ind, Feb 8, 47; m 72; c 1. GENETICS. Educ: Ind Univ, AB, 69; Univ Notre Dame, PhD(biol), 75. Prof Exp: Fel genetic control, Int Centre Insect Ecol & Physiol, 74-75; RES ENTOMOLOGIST, NCENT FOREST EXP STA, FOREST SERV, USDA, 75- Concurrent Pos: Consult, WHO, 73. Mem: Entom Soc Am; Genetics Soc Am. Res: Assessment of genetic variation in forest insect populations and how these variations interact with other factors to influence population dynamics. Mailing Add: USDA Forest Serv NCent Forest Exp Sta Folwell Ave St Paul MN 55101

LORINCZ, ALLAN LEVENTE, b Chicago, Ill, Oct 31, 24; m 52; c 3. DERMATOLOGY. Educ: Univ Chicago, SB, 45, MD, 47. Prof Exp: Res fel dermat, Cancer Clin, 50-51, from instr to assoc prof, 51-67, PROF DERMAT, UNIV CHICAGO, 67- Concurrent Pos: Mem dermat training grants comt, USPHS, 61-64; mem comt cutaneous syst, Div Med Sci, Nat Res Coun, 62-65; nat consult to Surgeon Gen, US Air Force, 62-; mem dermat adv comt, Food & Drug Admin, 71-72. Mem: Soc Invest Dermat; Soc Exp Biol & Med; Am Soc Dermatopath; Am Dermat Asn; Am Fedn Clin Res. Res: Psoriasis; cutaneous fungus infections; biochemistry and physiology of the skin, especially melanin chemistry and sebaceous gland control by endocrine factors; immunology. Mailing Add: Dept of Med Univ of Chicago Chicago IL 60637

LORINCZ, ANDREW ENDRE, b Chicago, Ill, May 17, 26; m 65. PEDIATRICS, BIOCHEMISTRY. Educ: Univ Chicago, PhB, 48, BS, 50, MD, 52. Prof Exp: From intern to jr asst resident pediat, Univ Chicago Clin, 52-54, jr asst resident, Rosenthal Clin, 54-55, instr, Sch Med, 56-59; from asst prof to assoc prof, Sch Med, Univ Fla, 59-68; PROF PEDIAT, ASSOC PROF BIOCHEM & DIR CTR DEVELOP & LEARNING DISORDERS, MED CTR, UNIV ALA, BIRMINGHAM, 68-, PROF DENT & ASSOC PROF PEDIAT OPTICS, 70-, ASSOC PROF SCH NURSING & ENG BIOPHYS, 71- Concurrent Pos: Res fel, Univ Chicago, 54-55 & Arthritis & Rheumatism Found res fel, 55-58; instr, La Rabida Inst, 57-59; consult ed, Am J Dis of Children, 71-; med consult, Headstart, 71- Mem: Am Chem Soc; Am Soc Pediat Res; fel Am Acad Pediat; Soc Invest Dermat; Orthop Res Soc. Res: Heritable disorders of connective tissue acid mucopolysaccharides; inborn errors of metabolism; mental retardation; biochemistry. Mailing Add: Ctr for Develop & Learning Dis Univ of Ala Birmingham AL 35233

LORING, ARTHUR PAUL, b New York, NY, May 22, 36; m 63; c 3. GEOLOGY, OCEANOGRAPHY. Educ: Columbia Univ, AB, 58; Pa State Univ, MS, 61; NY Univ, PhD(geol), 66. Prof Exp: Lectr geol, Brooklyn Col, 62-65; instr, 66-67; asst prof, Upsala Col, 67; asst prof, 67-73, ASSOC PROF GEOL, YORK COL, NY, 73- Mem: AAAS; fel Geol Soc Am; Am Soc Photogram; Asn Eng Geol; Sigma Xi. Res: Distribution and stratigraphy of planktonic foraminifera and general geologic field mapping in areas of folded and faulted sediments. Mailing Add: Dept of Geol York Col 150-14 Jamaica Ave Jamaica NY 11432

LORING, DOUGLAS HOWARD, b Concord, NH, July 25, 34; Can citizen; m 61; c 3. MARINE GEOCHEMISTRY. Educ: Acadia Univ, BSc, 54, MSc, 56; Univ Manchester, PhD(geochem), 60. Prof Exp: Tech officer, Geol Surv Can, 54-55; res fel geochem, Univ Manchester, 57-60; SCIENTIST, BEDFORD INST, 60- Concurrent Pos: Spec lectr, Dalhousie Univ, 62- Mem: Mineral Asn Can; fel Geol Asn Can. Res: Geochemistry of ancient and modern marine sediments; marine geology of the Gulf of St Lawrence. Mailing Add: Marine Ecol Lab Bedford Inst Box 1006 Dartmouth NS Can

LORING, HUBERT SCOTT, biochemistry, deceased

LORING, MARVIN F, b Jersey City, NJ, Feb 5, 23. RADIOLOGY. Educ: Chicago Med Sch, MB, 46, MD, 47. Prof Exp: ASST PROF RADIOL, MED COL, CORNELL UNIV, 58- Concurrent Pos: Brit-Am Exchange cancer res fel, Royal Cancer Hosp, London, Eng, 52-53; traveling fel, Am Cancer Soc, Radium Inst, Stockholm, Sweden, Univ Manchester, Eng & Curie Hosp, Paris, France, 53-54; asst attend radiologist, NY Hosp-Cornell Med Ctr, 58-68; med res collabr, Brookhaven Nat Lab, 61-62; asst attend radiotherapist, Mem Hosp Cancer & Allied Dis, 66-68; vis res scientist, Hadassah Med Sch, Hebrew Univ, Israel, 68-69. Mem: Am Col Radiol; Am Roentgen Ray Soc; Am Radium Soc; Radiol Soc NAm. Res: Therapeutic radiology. Mailing Add: NShore Hosp Cornell Univ Med Col Manhasset NY 11031

LORING, ROBERT DAVID, b New York, NY, July 8, 23; m 43; c 2. GEOGRAPHY, GEOGRAPHY OF LATIN AMERICA. Educ: Ohio Univ, BA, 47; Ind Univ, MA, 48. Prof Exp: ASSOC PROF GEOG, DEPAUW UNIV, 48- Concurrent Pos: Lectr, Butler Univ, 52- & Ind Univ Exten & Ind Univ-Purdue Univ, Indianapolis, 60- Mem: Asn Am Geogr; Nat Coun Geog Educ; Am Geog Soc; Can Asn Geogr; Royal Can Geog Soc. Mailing Add: Dept of Earth Sci DePauw Univ Greencastle IN 46135

LORING, WILLIAM BACHELLER, b Haileybury, Ont, Mar 4, 15; m 45; c 2. ECONOMIC GEOLOGY. Educ: Mich Col Min, BS, 40; Univ Ariz, MS, 47, PhD, 59. Prof Exp: Field geologist, Noranda Mines Co, Can, 41-42; inspector, US Dept Eng, 42-43; field engr, US Bur Mines, Mich, 43-44; party chief, Nfld Geol Surv, 44; party chief, Mining Geophys Co, Can, 44-45; mine mgr, Discovery Yellowknife Gold Mine, 46; geologist, Great Northern Explor Co, Ariz, 46-48; geologist, Eagle-Picher Mining & Smelting Co, 49-55; chief geologist, Big Indian Dist, Hidden Splendor Mining Co, 55-62; staff geologist, Atlas Minerals, 62-66; mine geologist, US Smelting, Ref & Mining Co, NMex, 66-67; dist geologist, 67-71, STAFF GEOLOGIST, CITIES SERV MINERALS CORP, WYO, 71- Mem: Am Inst Mining, Metall & Petrol Eng; Soc Econ Geol; Can Inst Mining & Metall; Int Asn Genesis Ore Deposits. Res: Ore deposits, especially controlling structures and surface indications. Mailing Add: Apt 101 1955 Haro Vancouver BC Can

LORING, WILLIAM ELLSWORTH, b Portland, Maine, Nov 6, 20; m 47; c 3. PATHOLOGY. Educ: Bowdoin Col, BS, 43; Columbia Univ, MD, 46; Am Bd Path, dipl, 54. Prof Exp: Res asst, Sch Med, Yale Univ, 51-52, instr, 52-53; asst prof, Sch Med, Univ NC, 53-56; assoc prof, Sch Med, NY Univ, 56-66; dir labs, Mercy Hosp, 66-72; CONSULT PATH & FORENSIC MED, 72- Concurrent Pos: Teaching fel path, Col Med, State Univ NY, 49-51; John Polachek Found fel, 58-61; lectr, Columbia Univ, 56-66; investr, Health Res Coun, NY, 60-65; consult, St Vincent's Hosp, Bridgeport, Conn. Mem: Am Soc Exp Path; Am Soc Clin Path; Am Thoracic Soc; Am Soc Cytol; Am Asn Path & Bact. Res: Chest pathology. Mailing Add: 7 Riverside Dr Falmouth ME 04105

LORIO, PETER LEONCE JR, b New Orleans, La, Apr 10, 27; m 57; c 6. FOREST SOILS, FOREST ECOLOGY. Educ: La State Univ, BS, 53; Duke Univ, MF, 54; Iowa State Univ, PhD(forestry soils), 62. Prof Exp: Soil scientist, Standard Fruit & Steamship Co, 54-58, chief soil scientist, 58-59; soil scientist, 62-68, PRIN SOIL SCIENTIST, FOREST INSECT RES PROJ, SOUTHERN FOREST EXP STA, US FOREST SERV, 68- Mem: Am Soc Agron; Soil Sci Soc Am; Int Soc Soil Sci. Res: Soil factors affecting pine susceptibility to bark beetles; soil water; tree rooting; rootlet pathogens; tree physiology; stand composition, age, density. Mailing Add: 2500 Shreeveport Highway Pineville LA 71360

LORIO, WENDELL JOSEPH, fisheries, wildlife biology, see 12th edition

LORKOVIC, HRVOJE RADOSLAV, b Zagreb, Yugoslavia, Nov 12, 30; m 56; c 3. MUSCULAR PHYSIOLOGY. Educ: Univ Belgrade, BA, 53; Univ Zagreb, PhD(physiol), 61. Prof Exp: Asst physiol, Inst Med Res, Yugoslavia Acad Sci & Arts, Zagreb, 56-62; asst, Univ Tübingen, 62-63; asst prof physiol & neurol, 67-74, RES SCIENTIST, MED SCH, UNIV IOWA, 74- Concurrent Pos: Wellcome res fel, Nat Inst Med Res, London, 63-64; NIH res fel, Univ Minn, Minneapolis, 64-67; NIH grant, 73-77. Mem: Am Physiol Soc. Res: Contractures of skeletal muscle; denervated muscle; excitation-contraction coupling; effects of ions and cholinergic drugs. Mailing Add: Dept of Neurol Univ of Iowa Iowa City IA 52240

LORRAIN, PAUL, b Montreal, Que, Sept 8, 16; m 44; c 4. ELECTROMAGNETISM. Educ: Univ Ottawa, BA, 37; McGill Univ, BSc, 40, MSc, 41, PhD(physics), 47. Prof Exp: Lectr physics, Sir George Williams Col, 42-43, Univ Laval, 43-46 & Inst Physics, Univ Montreal, 46; res assoc, Lab Nuclear Studies, Cornell Univ, 47-49; head dept, 57-66, PROF PHYSICS, UNIV MONTREAL, 49- Concurrent Pos: Vis prof fac sci, Univ Grenoble, France, 61-62; mem, Nat Res Coun, 60-66; vis prof, Faculty Sci, Univ Madrid, 68-69. Mem: Royal Soc Can; Am Phys Soc; Can Asn Physicists (pres, 64-65). Res: Electromagnetism and relativity. Mailing Add: Dept of Physics Univ of Montreal Case Postal 6128 Montreal PQ Can

LORSCHEIDER, FRITZ LOUIS, b Rochester, NY, Aug 27, 39; m 67; c 2. PHYSIOLOGY, ENDOCRINOLOGY. Educ: Univ Wis, Milwaukee, BSc, 63; Mich State Univ, MSc, 67, PhD(physiol, endocrinol), 69. Prof Exp: Res asst endocrinol, Radioisotope Unit, Marquette Sch Med, 63-64; ASSOC PROF PHYSIOL, FAC MED, UNIV CALGARY, 70- Concurrent Pos: NIH fel, Mich State Univ, 70. Mem: Can Physiol Soc; Can Soc Clin Invest. Res: Thyroid and reproductive physiology; antenatal diagnosis of high-risk pregnancies. Mailing Add: Div of Med Physiol Univ of Calgary Fac of Med Calgary AB Can

LORTIE, MARCEL, b Quebec, Que, Apr 15, 31; m 56; c 2. ENVIRONMENTAL SCIENCES. Educ: Laval Univ, BASc, 56; Univ Wis, PhD(plant path), 62. Prof Exp: Agr res officer forest path, Sci Serv, Can Dept Agr, 56-63; PROF FOREST PATH, LAVAL UNIV, 63-; REGIONAL DIR CAN FORESTRY, CAN DEPT ENVIRON, 70- & ENVIRON MGT SERV, QUEBEC REGION, 74- Honors & Awards: Gold Medal, Can Inst Forestry, 56. Mem: Fr-Can Asn Advan Sci; Can Inst Forestry. Mailing Add: Environ Mgt Serv PQ Region PO Box 10100 Saint Foy PQ Can

LORY, EARL CHRISTIAN, b Windsor, Colo, June 25, 06; m 33; c 2. PHYSICAL CHEMISTRY. Educ: Colo Agr & Mech Col, BS, 28; Johns Hopkins Univ, PhD(phys chem), 32. Prof Exp: From asst prof to assoc prof chem, Adams State Teachers Col, Colo, 33-38; dir res, J C Oliver Mem Res Found, St Margaret Mem Hosp, Pa, 38-42; from assoc prof to prof chem, 46-74, actg acad vpres, 68-74, EMER PROF CHEM, UNIV MONT, 74- Mem: Am Chem Soc. Res: Contact catalysis; catalytic oxidation of carbon monoxide; critical temperatures of solutions. Mailing Add: Dept of Chem Univ of Mont Missoula MT 59801

LORZ, ALBERT PROTUS, b Meadville, Pa, Sept 11, 07; m 39. PLANT CYTOGENETICS. Educ: Allegheny Col, AB, 30; Univ Va, PhD(genetics), 35. Prof Exp: Asst prof biol, Canisius Col, 35-38; prof & head dept, Seton Hall Col, 38-43; assoc hort, Purdue Univ, 45-48; prof veg crops, 48-72, assoc horticulturist, 48-49, horticulturist, 49-72, EMER PROF VEG CROPS, INST FOOD & AGR SCI, UNIV FLA, 72- Mem: Fel AAAS; fel Am Soc Hort Sci; Am Genetic Asn; Int Soc Hort Sci; Am Inst Biol Sci. Res: Cytology of polysomaty; breeding of vegetable crops; genetics of sex in spinach; varietal improvement of vegetable legumes and tomatoes for mechanical harvest. Mailing Add: 10615 Newberry Rd Gainesville FL 32601

LORZ, EMIL, organic chemistry, see 12th edition

LOS, MARINUS, b Ridderkerk, Netherlands, Sept 18, 33; m 57; c 4. CHEMISTRY. Educ: Univ Edinburgh, BSc, 55, PhD(chem), 57. Prof Exp: Res fel, Nat Res Coun Can, 58-60; res chemist, 60-71, GROUP LEADER, ORGANIC SYNTHESIS, AM CYANAMID CO, 71- Concurrent Pos: Sr res fel, Dept Pharmacol, Univ Edinburgh, 69-70. Mem: Am Chem Soc. Res: Aliphatic and aromatic chemistry, especially nitrogen heterocycles; natural products, especially alkaloids and terpenes. Mailing Add: Agr Div Am Cyanamid Co PO Box 400 Princeton NJ 08540

LOSCALZO, ANNE GRACE, b New York, NY, Sept 2, 17; m 40; c 1. MICROCHEMISTRY, ANALYTICAL CHEMISTRY. Educ: NY Univ, BA, 37, MS, 41, PhD(chem), 43. Prof Exp: Asst instr chem, Wash Square Col, NY Univ, 41-43, instr, 43-46; lectr, City Col New York, 53-58; from asst prof to assoc prof, 58-71, PROF CHEM, LONG ISLAND UNIV, 71- Mem: Am Chem Soc. Res: Educational projects to improve learning abilities of students in chemistry. Mailing Add: Dept of Chem Conolly Col Long Island Univ Brooklyn NY 11201

LOSCHE, CRAIG KENDALL, pedology, see 12th edition

LOSCHIAVO, SAMUEL RALPH, b Transcona, Man, June 28, 24; m 50; c 2. INSECT PHYSIOLOGY. Educ: Univ Man, BSc, 46, MSc, 50, PhD, 64. Prof Exp: Chemist, Man Sugar Co, 48; RES SCIENTIST, CAN DEPT AGR, 49- Concurrent Pos: Hon prof, Univ Man. Mem: Entom Soc Am; Entom Soc Can; Sigma Xi. Res: Biology, behavior and control of insects associated with stored grain and milled cereal products. Mailing Add: Can Agr Res Sta 25 Dafoe Rd Winnipeg MB Can

LOSEE, DAVID LAWRENCE, b Mineola, NY, July 19, 39; m 63; c 2. SOLID STATE PHYSICS, SEMICONDUCTORS. Educ: Cornell Univ, BEng, 62, MS, 63; Univ Ill, PhD(solid state physics), 67. Prof Exp: RES ASSOC, EASTMAN KODAK CO, 67- Mem: Am Phys Soc. Res: Physics of the noble gas solids; physics of semiconductors and semiconductor devices. Mailing Add: Res Labs Eastman Kodak Co Rochester NY 14650

LOSEE, FRED LESTER, b San Francisco, Calif, June 1, 17; m 45; c 4. BIOCHEMISTRY. Educ: Univ Calif, BS & DDS, 42; Georgetown Univ, MS, 47. Prof Exp: Asst to Prof NueKolls, Col Dent, Univ Calif, 39-42; head dept biochem, Naval Dent Sch, 48-50, head dept dent, Am Samoa, 50-51, oral surgeon, Mare Island Naval Shipyard, 51, asst chief dent div, Med Res Inst, Nat Naval Med Ctr, 52-58, dir dent res, Med Res Coun, NZ, 58-61, sr res officer, Dent Res Fac, US Naval Training Ctr, Ill, 61-65; DIR CARIES RES, EASTMAN DENT CTR, 65- Concurrent Pos: Lectr, Naval Dent Sch, 49-50 & 51-; Naval deleg subcomt biochem, Comt Dent, Nat Res Coun. Mem: AAAS; Am Chem Soc; Am Dent Asn; fel Am Col Dent; Am Inst Nutrit. Res: Organic components of the oral hard tissues; chemical and histochemical study of the dental decay process; micronutrients and their effect on calcification of bone and teeth; soil mineralcaries relationships. Mailing Add: 800 E Main St Rochester NY 14603

LOSEE, MICHAEL LEONARD, b Catskill, NY, Apr 19, 41; m 63; c 2. ORGANIC CHEMISTRY. Educ: Dartmouth Col, AB, 63; Lehigh Univ, PhD(chem), 67. Prof Exp: Sr res chemist, 67-73, RES GROUP LEADER, MONSANTO CO, 73- Mem: Am Chem Soc. Res: Bonding, synthesis and reactions of organometallic compounds; development of flame retardants for polymeric substrates. Mailing Add: 204 E Swon Webster Groves MO 63119

LOSEKAMP, BERNARD FRANCIS, b Cincinnati, Ohio, July 16, 36; m 58; c 4. POLYMER CHEMISTRY, ORGANIC CHEMISTRY. Educ: Xavier Univ, Ohio, BS, 58, MS, 61; Univ Akron, PhD(polymer chem), 66. Prof Exp: Res asst, Wm S Merrell Co, Ohio, 61; res chemist, Inst Polymer Sci, Univ Akron, 61-64; asst ed, 64-67, assoc ed, 67-69, sr indexer, 69-71, group leader, 71-72, SR ED, CHEM ABSTR SERV, OHIO STATE UNIV, 72- Mem: Am Chem Soc-Rubber Div. Res: Acenaphthene arsenicals; synthesis and characterization of polymers; polymer nomenclature; thermal polymerization. Mailing Add: Chem Abstr Serv Ohio State Univ Columbus OH 43210

LOSEY, GEORGE SPAHR, JR, b Louisville, Ky, June 30, 42; m 67; c 2. MARINE ZOOLOGY, ETHOLOGY. Educ: Miami Univ, BS, 64; Scripps Inst Oceanog, Univ Calif, PhD(marine biol), 68. Prof Exp: NIH res fel fish behav, Hawaii Inst Marine

Biol, 68-70; asst prof, 70-75, ASSOC PROF ZOOL, HAWAII INST MARINE BIOL, UNIV HAWAII, 75- Honors & Awards: Stoye Award, Am Soc Ichthyol & Herpet, 67. Mem: AAAS; Animal Behav Soc; Am Soc Ichthyol & Herpet. Res: Ethology and ecology of fish, including Hypsoblennius; symbiotic cleaner fish and mimicry in Blenniidae fish; behavioral ecology of herbivorous fish. Mailing Add: Dept of Zool Univ of Hawaii Honolulu HI 96822

LOSEY, GERALD OTIS, b Detroit, Mich, Nov 13, 30; m 63. ALGEBRA. Educ: Univ Mich, BS, 52, MS, 53, PhD(math), 58. Prof Exp: Res instr math, Princeton Univ, 57-58; instr, Univ Wis, 58-61, asst prof, 61-64; assoc prof, 64-67, PROF MATH, UNIV MAN, 67- Mem: Am Math Soc; Can Math Cong. Res: Group theory; ring theory. Mailing Add: Dept of Math Univ of Man Winnipeg MB Can

LOSHAEK, SAMUEL, polymer chemistry, see 12th edition

LOSICK, RICHARD MARC, b Jersey City, NJ, July 27, 43; m 70. MOLECULAR BIOLOGY. Educ: Princeton Univ, AB, 65; Mass Inst Technol, PhD(biochem), 69. Prof Exp: Harvard Soc fels jr fel biochem, 68-71, asst prof, 71-74, ASSOC PROF BIOCHEM & BIOL, HARVARD UNIV, 74- Honors & Awards: Camille & Henry Dreyfus Award, Camille & Henry Dreyfus Found, 73. Mem: Am Soc Biol Chemists; Am Soc Microbiol. Res: Bacterial sporulation; regulatory subunits of RNA polymers. Mailing Add: Biol Labs Harvard Univ Cambridge MA 02138

LOSIN, EDWARD THOMAS, b Racine, Wis, July 9, 23; m 50; c 2. PHYSICAL ORGANIC CHEMISTRY, ENERGY CONVERSION. Educ: Univ Ill, BS, 48; Columbia Univ, AM, 50, PhD(chem), 54. Prof Exp: Res assoc, Eng Res Inst, Univ Mich, 54-57; res chemist, Union Carbide Corp, 57-61; chem dept mgr, Isomet Corp, 61-63; sr res scientist, 63-71, mgr non-metallic mat, 71-73, SR RES SCIENTIST, ALLIS-CHALMERS CORP, 73- Mem: Am Chem Soc; The Chem Soc. Res: Reaction mechanisms of organic, stereospecific and free radical gas-phase reactions; electrical insulation materials and systems for various applications; epoxy technology; high temperature fuel gas cleanup; coal combustion of pulverized fuel in entrained-bed and fluid-bed combustors; coal beneficiation. Mailing Add: Allis-Chalmers Corp Adv Tech Ctr PO Box 512 Milwaukee WI 53201

LOSPALLUTO, JOSEPH JOHN, b New York, NY, Nov 8, 25. BIOCHEMISTRY. Educ: City Col New York, BS, 45; NY Univ, PhD, 53. Prof Exp: Chemist, Fleischman Labs, Stand Brands, Inc, 45-47; instr biochem, NY Univ, 53-58; from asst prof to assoc prof, 58-72, PROF BIOCHEM, UNIV TEX HEALTH SCI CTR, DALLAS, 72- Concurrent Pos: Res fel, Arthritis & Rheumatism Found, 56-58; res fel, Whitney Found, 58-61; sr investr, Arthritis Found, 61-66. Mem: Fel AAAS; Am Asn Immunol; Am Rheumatism Asn. Res: Proteins, especially chemistry and immunology; antibodies and connective tissue chemistry. Mailing Add: Dept of Biochem Univ of Tex Health Sci Ctr Dallas TX 75235

LOSSING, FREDERICK PETTIT, b Norwich, Ont, Aug 4, 15; m 38; c 3. CHEMICAL PHYSICS. Educ: Univ Western Ont, BA, 38, MA, 40; McGill Univ, PhD(phys chem), 42. Prof Exp: Res chemist, Shawinigan Chem, Ltd, 42-46; prin res officer, 46-69, ASST DIR DIV CHEM, NAT RES COUN CAN, 69- Mem: Fel Royal Soc Can; Royal Astron Soc Can. Res: Mass spectrometry; chemical kinetics; photochemistry; properties of free radicals; ionization processes. Mailing Add: Div of Chem Nat Res Coun Ottawa ON Can

LOSSOW, WALTER JUDAH, b New York, NY, Apr 21, 25; m 55. PHYSIOLOGY. Educ: Brown Univ, AB, 48; Univ Calif, PhD(physiol), 53. Prof Exp: Physiologist, Univ Calif, Berkeley, 53-68; LECTR PHYSIOL, LASSEN COL, 70- Mem: Am Physiol Soc. Mailing Add: Sci-Math Div Lassen Col Susanville CA 96130

LOSTAGLIO, VINCENT JOSEPH, physical chemistry, see 12th edition

LOSURDO, ANTONIO, b Spadafora, Italy, Jan 1, 43; US citizen. PHYSICAL CHEMISTRY, WATER CHEMISTRY. Educ: Syracuse Univ, BA, 65, PhD(chem), 70. Prof Exp: Res asst chem, Syracuse Univ, 66-69; NIH fel, Rutgers Univ, New Brunswick, 69-70, instr, 70-71; MEM VIS FAC CHEM, SYRACUSE UNIV, 71- Concurrent Pos: Res assoc, Ohio State Univ, 72-73, lectr chem, 73-74; res assoc chem, Clark Univ, 74-75. Mem: Am Chem Soc; NY Acad Sci; Sigma Xi. Res: Thermodynamics, transport and spectroscopic properties of simple and hydrophobic electrolytes at variuos temperatures; densimetry, calorimetry, NMR, isopiestic, analytical ultracentrifuge; electroanalytical chemistry. Mailing Add: 312 Swansea Ave Syracuse NY 13206

LOTAN, JAMES E, b Mich, Mar 20, 31; m 51; c 5. FORESTRY, ECOLOGY. Educ: La State Univ, BSF, 59; Univ Mich, MF, 61, PhD, 70. Prof Exp: Forestry technician, Southern Forest & Range Exp Sta, US Forest Serv, La, 57-59, fire control, Deerlodge Nat Forest, Mont, 59; asst forest res, Univ Mich, 60; res forester, 61-65, proj leader forest sci res, 65-74, PROG MGR, INTERMOUNTAIN FOREST & RANGE EXP STA, US FOREST SERV, 74- Mem: AAAS; Am Forestry Asn; Soc Am Foresters; Ecol Soc Am. Res: Forest ecology, fire management and silviculture; multifunctional RD&A Program integrating fire management into multiple-use planning process. Mailing Add: Intermountain Forest Exp Sta North Forest Fire Lab Drawer G Missoula MT 59801

LOTHERS, JOHN EDMOND, JR, b Wichita, Kans, Nov 25, 31; m 62; c 2. GENETICS. Educ: Okla State Univ, BS, 54; Kans State Univ, MS, 56; Univ Kans, PhD(zool, genetics), 66. Prof Exp: Teaching asst chem, Kans State Univ, 54-56; lab technologist, St Francis Hosp, Wichita, Kans, 57-58; instr chem & biol, King's Col, NY, 58-61; from asst prof to assoc prof, 66-75, PROF BIOL, COVENANT COL, TENN, 75- Mem: Am Inst Biol Sci. Res: Physiology of reproduction in mammals and mammalian genetics. Mailing Add: 201 Hardy Rd Lookout Mountain TN 37350

LOTHROP, EVERETT WINFRED, JR, b Newton, Mass, July 18, 18; m 42; c 2. TEXTILE PHYSICS, POLYMER PHYSICS. Educ: Oberlin Col, AB, 40; Northwestern Univ, PhD(physics), 49. Prof Exp: Asst physics, Northwestern Univ, 40-43, instr, 43-44, res assoc, 44-49; asst prof, Univ Kans, 49-53; res physicist, Am Viscose Corp, 53-54, HEAD CHEM GROUP, FMC CORP, 54- Mem: Am Phys Soc. Mailing Add: Chem Group FMC Corp Marcus Hook PA 19061

LOTHROP, WARREN CRAIG, b Brookline, Mass, Mar 7, 12; m 41; c 3. ORGANIC CHEMISTRY. Educ: Harvard Univ, AB, 33, MA, 35, PhD(org chem), 37. Prof Exp: Nat Tuberc fel, Yale Univ, 37-38; instr chem, Trinity Col, Conn, 38-42; tech aide, Off Sci Res & Develop, DC, 42-45; asst prof chem, Williams Col, 45-46; mem staff, Arthur D Little, Inc, 46-53, vpres, 54-60, sr vpres & dir, 60-63; vpres corporate develop, Armour & Co, 63-70, sci adv, Phoenix, Ariz, 71-73; RETIRED. Concurrent Pos: Tech aide, Nat Defense Res Comt, DC, 42-45. Mem: Am Chem Soc (asst to ed jour, 47-49); Indust Res Inst. Res: Aromatic hydrocarbons; Mills-Nixon effect; lipids of tubercle bacillus; biphenylene; high explosives; management of research. Mailing Add: Beach St RD 1 Kennebunkport ME 04046

LOTKIN, MARK MAX, b Copenhagen, Denmark, July 28, 12; nat US; m 45; c 2. MATHEMATICS. Educ: Univ Kiel, PhD(math physics), 37. Prof Exp: Instr math physics, Col Paterson, 42-43; instr math, Carleton Col, 43-44; asst prof, Wabash Col, 44-45; mathematician & math adv, Comput Lab, Ballistic Res Lab, Aberdeen Proving Ground, Md, 45-54; chief math sect, Res Div, Avco Corp, Mass, 55-59; tech proj coordr, Surface Comn Div, Radio Corp Am, 59-62; CONSULT MATH, REENTRY SYSTS DEPT, MISSILE & SPACE DIV, GEN ELEC CO, 62- Mem: Am Math Soc; Soc Indust & Appl Math; Math Asn Am; Asn Comput Mach. Res: Computing machines and procedures; numerical analysis; applied mathematics; systems analysis. Mailing Add: 362 Juniper St Marlton NJ 08053

LOTKOWSKI, WLADYSLAW, economic geography, see 12th edition

LOTLIKAR, PRABHAKAR DATTARAM, b Shirali, India, May 21, 28; US citizen; m 60; c 1. BIOCHEMISTRY, PHARMACOLOGY. Educ: Univ Bombay, BS, 50, MS, 54; Ore State Univ, PhD(biochem, pharmacol, bact), 60. Prof Exp: Asst chemist, Raptakos Brett & Co, Ltd, India, 50-55; proj assoc, McArdle Lab Cancer Res, Univ Wis, 63-65, instr, 65-66; res instr, 67-68, asst prof, 68-75, ASSOC PROF BIOCHEM, FELS RES INST, SCH MED, TEMPLE UNIV, 75-, INVESTR, 67- Concurrent Pos: Res fel oncol, McArdle Lab Cancer Res, Univ Wis, 60-63. Mem: AAAS; Am Chem Soc; Am Asn Cancer Res; Am Soc Biol Chem; NY Acad Sci. Res: Mechanisms of chemical carcinogenesis. Mailing Add: Fels Res Inst Temple Univ Sch of Med Philadelphia PA 19140

LOTSPEICH, FREDERICK BENJAMIN, b Konawa, Okla, Jan 10, 14; m 48; c 4. ENVIRONMENTAL SCIENCES, FRESH WATER ECOLOGY. Educ: State Col Wash, BS, 50, MS, 52, PhD, 56. Prof Exp: Soil scientist, US Geol Surv, 54-57; soil sci, Agr Res Serv, USDA, 57-66; SOIL SCIENTIST, ALASKA WATER LAB, ENVIRON PROTECTION AGENCY, 66- Mem: AAAS; Soil Sci Soc Am; Am Inst Biol Sci; Ecol Soc Am. Res: Geochemistry and ecology of rock weathering and soil genesis; chemical composition of soils and vegetation as indicators of mineral deposits; scientific hydrology; water pollution control in cold climates; ecological approach to water pollution problems; watershed properties and their influence on aquatic ecosystems; sedimentation effects on freshwater aquatic life. Mailing Add: Alaska Water Lab Environ Protection Agency College AK 99701

LOTSPEICH, FREDERICK JACKSON, b Keyser, WVa, Mar 12, 25; m 48; c 1. BIO-ORGANIC CHEMISTRY. Educ: WVa Univ, BS, 48, MS, 51; Purdue Univ, PhD(chem), 55. Prof Exp: Res chemist, E I du Pont de Nemours & Co, 48-50; asst org chem, WVa Univ, 50-52; asst, Purdue Univ, 52-53; asst prof chem, Simpson Col, 54-56; from asst prof to assoc prof, 56-66, PROF BIOCHEM, MED CTR, W VA UNIV, 66- Mem: Am Chem Soc. Res: Chemistry and biochemistry of S-adenosyl methionine and derivatives. Mailing Add: WVa Univ Med Ctr Morgantown WV 26506

LOTSPEICH, JAMES FULTON, b Cincinnati, Ohio, Oct 22, 22; m 60. PHYSICS. Educ: Princeton Univ, BA, 43; Univ Cincinnati, MS, 49; Columbia Univ, PhD(physics), 58. Prof Exp: Lab instr gen physics, Univ Cincinnati, 47-48; asst, Columbia Univ, 51-56; RES PHYSICIST, LABS, HUGHES AIRCRAFT CO, 56- Mem: AAAS; Am Phys Soc; Sigma Xi; NY Acad Sci. Res: Microwave spectroscopy and molecular structure; electrooptic techniques; applied laser technology; photodetection techniques. Mailing Add: Laser Dept Hughes Res Labs 3011 Malibu Canyon Rd Malibu CA 90265

LOTT, ANTONE LYMAN, II, inorganic chemistry, see 12th edition

LOTT, FRED WILBUR, JR, b Ohio, Oct 8, 17; m 41; c 3. MATHEMATICS, MATHEMATICAL STATISTICS. Educ: Cedarville Col, AB, 39; Univ Mich, MA, 46, PhD(math), 55. Prof Exp: From asst prof to assoc prof, 49-61, PROF MATH, UNIV NORTHERN IOWA, 61-, ASST VPRES ACAD AFFAIRS, 71- Concurrent Pos: Opers analyst, US Air Force, 55-64. Mem: Am Math Soc; Math Asn Am; Am Statist Asn; Inst Math Statist. Mailing Add: Dept of Math Univ of Northern Iowa Cedar Falls IA 50613

LOTT, JAMES ROBERT, b Houston, Tex, Jan 16, 24; m 42; c 4. PHYSIOLOGY, BIOPHYSICS. Educ: Univ Tex, BA, 49, MA, 51, PhD(physiol, bact), 56. Prof Exp: Med bacteriologist, Brackenridge Hosp, Austin, Tex, 55; lectr zool, Univ Tex, 55-56, res scientist, Radiobiol Lab, Balcones Res Inst, 56; instr physiol, Sch Med, Emory Univ, 56-57; from asst prof to assoc prof, 57-64, PROF BIOL, N TEX STATE UNIV, 64- Concurrent Pos: Sr res investr, AEC, 58-; NSF grant, 63-64. Mem: Am Physiol Soc; Soc Gen Physiol; Radiation Res Soc; Int Soc Biometeorol. Res: Neurophysiology; effects of x-irradiation, microwaves, and electric fields on the nervous system; effects of electric fields on cancer growth; effects of stress on heart action; endocrinology; effects of x-irradiation on the adrenal-pituitary axis; ion and water flux in root systems. Mailing Add: 1907 Locksley Lane Denton TX 76201

LOTT, JAMES STEWART, b Sarnia, Ont, Apr 10, 20; m 51; c 4. MEDICINE, RADIOLOGY. Educ: Univ Western Ont, BA, 43, MD, 46; Royal Col Physicians & Surgeons, dipl med radiother, 52 & specialist therapeut radiol, 54. Prof Exp: Instr radiol, Univ Western Ont, 52-62, assoc prof radiother & actg head dept, 62-63; assoc prof radiol, Sch Med, Johns Hopkins Univ & head div radiother, Hosp, 64-71; PROF THERAPEUT RADIOL & CHMN DEPT, QUEEN'S UNIV, ONT, 71-; DIR ONT CANCER FOUND, KINGSTON CLIN, 71- Concurrent Pos: Fel histol, Univ Western Ont, 46-47, fel path, 47-48, fel radiol, 48-49; Can Cancer Soc fel radiother, 50-51; Brit Empire Cancer Campaign exchange fel, 51-52; consult, Westminster Vet Hosp, 56-63 & St Joseph's Hosp, London, Can, 56-63; consult & radiologist in chg ther, Hackley Hosp, Muskegon, Mich, 63-64. Mem: AMA; Can Med Asn; Can Asn Radiol. Res: Radiobiology applied to radiotherapy; clinical radiotherapy applied to cancer. Mailing Add: Ont Cancer Found Kingston Clin Kingston Gen Hosp Kingston ON Can

LOTT, JOHN ALFRED, b Ger, Oct 30, 36; US citizen; m 63; c 1. ANALYTICAL CHEMISTRY. Educ: Rutgers Univ, BS, 59, MS, 61, PhD(anal chem), 65. Prof Exp: Instr chem, Rutgers Univ, 64-65; asst prof, Flint Col, Univ Mich, 65-68; ASST PROF CLIN CHEM, OHIO STATE UNIV, 68- Res: Instrumentation; methodology development; enzymology; specific-ion electrodes. Mailing Add: Rm 342 Univ Hosp Ohio State Univ 410 W Tenth Ave Columbus OH 43210

LOTT, JOHN NORMAN ARTHUR, b Summerland, BC, Jan 20, 43; m 66. PLANT ANATOMY, PLANT PHYSIOLOGY. Educ: Univ BC, BSc, 65; Univ Calif, Davis, MSc, 67, PhD(bot), 69. Prof Exp: Res asst bot, Univ Calif, Davis, 65-69; asst prof, 69-75, ASSOC PROF BIOL, McMASTER UNIV, 75- Concurrent Pos: Nat Res Coun Can res grant, 74- Mem: AAAS; Can Bot Asn; Am Soc Plant Physiol; Bot Soc Am. Res: Ultrastructure and physiological studies of developong and germinating seeds, including chloroplast development, nuclear pore formation and protein body formation. Mailing Add: Dept of Biol McMaster Univ Hamilton ON Can

LOTT, LAYMAN AUSTIN, b Ft Collins, Colo, Sept 21, 37; m 58; c 4. PHYSICS.

Educ: Colo State Univ, BS, 59, MS, 61; Iowa State Univ, PhD(physics), 65. Prof Exp: Res physicist, Rocky Flats Div, Dow Chem USA, 65-71, sr res physicist, 71-73; QUAL ENG SPECIALIST, IDAHO NAT ENG LAB, 73- Mem: Am Phys Soc; Am Soc Nondestructive Test; Sigma Xi. Res: Solid state physics; physical properties of materials; nondestructive testing; development of advanced nondestructive testing methods. Mailing Add: 701 9th St Idaho Falls ID 83401

LOTT, PETER F, b Berlin, Ger, Mar 26, 27; nat US; m 56; c 2. PHYSICAL CHEMISTRY, ANALYTICAL CHEMISTRY. Educ: St Lawrence Univ, BS, 49, MS, 50; Univ Conn, PhD(chem), 56. Prof Exp: Asst instr chem, Univ Conn, 54-56; res chemist, E I du Pont de Nemours & Co, 56; assoc prof, Univ Mo, 56-59; chemist, Pure Carbon Co, 59-60; assoc prof chem, St John's Univ, NY, 60-64; PROF CHEM & LECTR MED, UNIV MO-KANS CITY, 64- Mem: Am Chem Soc; Am Microchem Soc; Soc Appl Spectros; The Chem Soc. Res: Analytical methods development; trace and instrumental analysis; chemical kinetics; radiochemistry; physical measurements; organic reagents. Mailing Add: Dept of Chem Univ of Mo Kansas City MO 64110

LOTT, RICHARD VINCENT, b Edgerton, Mo, Nov 12, 99; m 28. POMOLOGY. Educ: Univ Mo, BS, 25, AM, 26; Univ Ill, PhD(pomol), 39. Prof Exp: Asst prof & asst horticulturist, Exp Sta, Colo Agr & Mech Col, 26-30; head dept hort, Exp Sta, Miss State Col, 30-35; assoc horticulturist in charge teaching & res, Exp Sta, NMex State Col, 35-39; from assoc prof to prof, 39-68, pomol res, Agr Exp Sta, 39-68, EMER PROF POMOL, COL AGR, UNIV ILL, URBANA, 68- Mem: Am Soc Hort Sci; Am Soc Plant Physiol. Res: Fruit physiology and effect of production practices; study of fruit quality, its characteristics and affecting factors; fruit color color, its measurement and significance; fruit pigments; fruit development, maturation and ripening; fruit morphology and anatomy. Mailing Add: 11034 Meade Dr Sun City AZ 85351

LOTT, SAM HOUSTON, JR, b New Orleans, La, Sept 22, 36; m 59. PHYSICS, HEALTH PHYSICS. Educ: La State Univ, BS, 58; Vanderbilt Univ, MS, 60, PhD(physics), 65. Prof Exp: Res assoc physics, 65-66, DIR RADIATION SAFETY OFF, VANDERBILT UNIV, 66- Concurrent Pos: Consult, 72- Mem: Am Phys Soc; Health Phys Soc; Am Asn Physicists in Med. Res: Three-color photometric study of variable stars; Zeeman & Faraday effects in high pulsed magnetic fields; calibration techniques for therapeutic machines. Mailing Add: U-0209 Med Ctr Vanderbilt Univ Nashville TN 37232

LOTZ, FREDERICK, b Hrastovac, Yugoslavia, July 16, 23; Can citizen; m 52; c 5. PHARMACOLOGY, PHYSIOLOGY. Educ: Univ Western Ont, BA, 49, MSc, 53, PhD(physiol), 57. Prof Exp: Asst prof cancer res, Cancer Res Inst, Univ Sask, 58-60; asst prof, Cancer Res Ctr, Univ BC, 60-62; assoc prof physiol, Ont Vet Col, 62-65; ASSOC PROF PHYSIOL, UNIV GUELPH, 65- Concurrent Pos: Nat Cancer Inst Can res fel, Chester Beatty Inst Cancer Res, Royal Cancer Hosp, London, Eng, 57-58. Mem: AAAS; NY Acad Sci; Can Physiol Soc; Can Soc Immunol. Res: Cardiovascular research; coagulation platelets; congenital disorders of atherosclerosis; lipid metabolism; lipid transport pharmacology. Mailing Add: Dept of Biomed Sci Univ of Guelph Guelph ON Can

LOTZ, JOHN ROBERT, b Pottsville, Pa, June 4, 18; m 47; c 1. ANALYTICAL CHEMISTRY. Educ: Pa State Univ, BS, 40, PhD(chem), 54. Prof Exp: Asst fuel technol, Pa State Univ, 40-42, instr chem, 42-44; res chemist, M W Kellogg Co, 44-45; instr chem, Pa State Univ, 47-54; fel, Mellon Inst, 54-57; ASST PROF CHEM, PA STATE UNIV, UNIVERSITY PARK, 57- Mem: Am Chem Soc. Res: Coordination compounds; gas chromatography. Mailing Add: Dept of Chem Pa State Univ University Park PA 16802

LOTZ, WILLIAM EDWARD, physiology, see 12th edition

LOU, KINGDON, b Stockton, Calif, Aug 3, 22; m 45; c 2. IMMUNOLOGY. Educ: Stanford Univ, AB, 52, AM, 56; Am Bd Bioanal, dipl. Prof Exp: From res asst to res assoc, Hyland Labs, Baxter Labs, 57-64, dir immunol dept, Res Div, 64-67; sr immunochemist, Res Div, Hoffmann-La Roche, 67-68; dir immunol, Kallestad Labs, 68-69; DIR IMMUNOL RES, ICL SCI, 70- Mem: Am Soc Microbiol; Am Asn Clin Chemists; NY Acad Sci. Res: Immunochemical diagnostic reagents; isolation and purification of serum protein constituents. Mailing Add: 21866 Michigan Lane El Toro CA 92630

LOU, MARJORIE FENG, b Tze-Chwan, China, Sept 27, 39; m 64; c 1. BIOCHEMISTRY. Educ: Nat Taiwan Univ, BS, 60; Va Polytech Inst, MS, 62; Boston Univ, PhD(biochem), 66. Prof Exp: Res assoc biochem, Howe Lab Ophthal, Harvard Med Sch, 65-67; res assoc, 67-69, RES ASSOC, ALFRED I DU PONT INST, 69- Mem: Am Chem Soc; NY Acad Sci; AAAS. Res: Control mechanism of enzymes; biosynthesis and structure of glycopeptides and glycoproteins; ninhydrin positive constituents in normal and pathological human urine. Mailing Add: Dept of Biochem Alfred I du Pont Inst Wilmington DE 19899

LOUCAS, SPIRO P, b New York, NY, Feb 26, 29; m 51; c 3. ORGANIC CHEMISTRY, BIOCHEMISTRY. Educ: Fordham Univ, BS, 57; Rutgers Univ, MS, 61, PhD(pharmaceut chem), 65. Prof Exp: Asst prof org chem, Wagner Col, 61-65; asst prof biochem, 65-67, ASST PROF PHARM CHEM, COL PHARM, COLUMBIA UNIV, 67-; OPTHALMIC BIOCHEMIST, BETH ISRAEL MED CTR, NEW YORK, 73- Concurrent Pos: Head anal drug control, Dept Pharm, Mt Sinai Hosp, New York, 61- Mem: Am Chem Soc. Res: Hyperbaric oxygenation; biochemical and biopharmaceutical studies of the effects of oxygen high pressure on physiological systems. Mailing Add: 16 Toni Ct Plainview NY 11803

LOUCH, CHARLES DUKES, b Kanpor, India, Dec 24, 25; m 56; c 1. ZOOLOGY. Educ: Col Wooster, BA, 50; Univ Wis, MS, 52, PhD(zool), 55. Prof Exp: Asst prof biol, Hope Col, 55-57; from asst prof to assoc prof, 57-68, PROF BIOL, LAKE FOREST COL, 68- Concurrent Pos: USPHS res fel, Johns Hopkins Univ, 59-60. Mem: Ecol Soc Am. Res: Vertebrate ecology, particularly relationship between population density and physiological activities of members of population. Mailing Add: Dept of Biol Lake Forest Col Lake Forest IL 60045

LOUCK, JAMES DONALD, b Grand Rapids, Mich, Dec 13, 28; m 60; c 3. MATHEMATICAL PHYSICS. Educ: Ala Polytech Inst, BS, 50; Ohio State Univ, MS, 52, PhD(physics), 58. Prof Exp: Staff mem, Los Alamos Sci Lab, 58-60; assoc res prof physics, Auburn Univ, 60-63; STAFF MEM, LOS ALAMOS SCI LAB, UNIV CALIF, 63- Mem: Am Phys Soc; AAAS. Res: Application and development of group theoretical methods in physics. Mailing Add: Los Alamos Sci Lab Univ of Calif Los Alamos NM 87545

LOUCKS, ORIE LIPTON, b Minden, Ont, Oct 2, 31; m 55; c 3. BOTANY, ECOLOGY. Educ: Univ Toronto, BSc, 53, MSc, 55; Univ Wis, PhD(bot), 60. Prof Exp: Forest ecologist, Dept Forestry, Can Govt, 55-62; from asst prof to assoc prof, 62-68, PROF BOT, UNIV WIS-MADISON, 68- Concurrent Pos: Univ Wis rep, State Bd Preserv Sci Areas, 64-; consult, Can Land Inventory, 65; coordr environ mgt progs, US/Int Biol Prog, Univ Tex, 73. Honors & Awards: George Mercer Award, Ecol Soc Am, 64. Mem: AAAS; Soc Am Foresters; Ecol Soc Am; Am Inst Biol Sci; Soc Gen Systs Res. Res: Forest ecology and stand dynamics; micrometeorology and forest hydrology; ecosystem modeling and analysis; watershed and water quality systems studies; computer simulation of land use change. Mailing Add: Dept of Bot Birge Hall Univ of Wis Madison WI 53706

LOUD, ALDEN VICKERY, b Boston, Mass, Apr 6, 25; m 50; c 4. CELL BIOLOGY, BIOPHYSICS. Educ: Mass Inst Technol, BS & MS, 51, PhD(biophys), 55. Prof Exp: Res assoc, Detroit Inst Cancer Res & asst prof biophys, Col Med, Wayne State Univ, 57-65; asst prof path, Col Physicians & Surgeons, Columbia Univ, 65-68; ASSOC PROF PATH, NY MED COL, 68- Concurrent Pos: Res fel med, Mass Gen Hosp, 51-57. Mem: Electron Micros Soc Am; Am Soc Cell Biol; Int Soc Stereology; Royal Micros Soc. Res: Intermediary steroid metabolism; quantitative electron microscopy and methods of ultrastructure research; correlation of cellular ultrastructure with metabolic function. Mailing Add: Dept of Path NY Med Col Valhalla NY 10595

LOUD, OLIVER SCHULE, b Vernal, Utah, Jan 16, 11; m 35; c 2. HISTORY & PHILOSOPHY OF SCIENCE. Educ: Harvard Univ, AB, 29; Columbia Univ, AM, 40, EdD, 43. Prof Exp: Master, Nichols Sch, NY, 29-32; instr high sch, Ohio, 32-36; teacher gen sci, Sarah Lawrence Col, 36-40; res assoc, Bur Educ Res Sci, Columbia, 39-43; asst prof physics, Antioch Col, 43-44; instr, Ohio State Univ, 44; tech supvr, Tenn Eastman Corp, Tenn, 44-45; assoc prof, 45-48, PROF PHYS SCI, ANTIOCH COL, 48- Concurrent Pos: Ford Found fel, Harvard Univ, 52-53. Mem: AAAS; Soc Social Responsibility Sci. Res: Science in general education; suggestions for teaching problems of good land use. Mailing Add: 1430 Meadow Lane Yellow Springs OH 45387

LOUD, WARREN SIMMS, b Boston, Mass, Sept 13, 21; m 47; c 3. MATHEMATICS. Educ: Mass Inst Technol, SB, 42, PhD(math), 46. Prof Exp: Instr math, Mass Inst Technol, 43-47; from asst prof to assoc prof, 47-59, PROF MATH, UNIV MINN, MINNEAPOLIS, 59- Concurrent Pos: Res engr, Mass Inst Technol, 45-47, vis fel, 55-56; guest prof, Darmstadt Tech Univ, 64-65; vis prof, Kyoto Univ, Japan, 74-75. Mem: AAAS; Am Math Soc; Soc Indust & Appl Math (ed, 61-); Math Asn Am. Res: Theory of differential equations; numerical methods of solution of differential equations; stationary solutions of Van der Pol's equation with a forcing term; nonlinear mechanics. Mailing Add: Dept of Math Univ of Minn Minneapolis MN 55455

LOUDEN, L RICHARD, b Monroe, Wash, July 8, 33; m 63. GEOCHEMISTRY. Educ: Univ Würzburg, PhD(geochem), 63. Prof Exp: Assoc prof geochem, Univ Houston, 63-64; geologist, Magnet Cove Barium Corp, 64-65, supvr, X-ray Dept, 65-67, mgr anal sect, 67-69, tech adv, 69-71, spec proj engr, 71-72, develop mgr, Dresser Pollution, Dresser Oilfield Prod Div, 72-73, prod mgr, 73-75, MKT MGR, DRESSER-SWACO, 75- Concurrent Pos: Co-worker, NASA grant, Univ Houston, 63-64. Mem: AAAS; Marine Tech Soc; Clay Minerals Soc; Ger Geol Asn; Nat Oilfield Equip Mfrs & Distribr Soc. Res: Organic geochemistry, oceanography, clay mineralogy, and x-ray analysis; new and novel equipment and chemicals for oilwell and other drilling practices. Mailing Add: 8011 Highmeadow Houston TX 77042

LOUDERBACK, ALLAN, cell physiology, biochemistry, see 12th edition

LOUDFOOT, JAMES HERBERT, b Glasgow, Scotland, Oct 28, 16; Can citizen; m 53; c 1. ORGANIC CHEMISTRY, BIOCHEMISTRY. Educ: Univ Glasgow, BSc, 39; Univ Man, PhD(biochem & org chem), 56. Prof Exp: Chemist, Imp Chem Indust Ltd, Eng, 40-45; lectr chem, S E Essex Tech Col, Eng, 45-46; lectr, 46-52, ASST PROF CHEM, UNIV MAN, 52- Mem: Am Chem Soc; The Chem Soc. Res: Synthesis and properties of 2, 4-dinitrophenyl peptides; other aspects of peptide chemistry and any field of organic chemistry which relates to biochemistry. Mailing Add: Dept of Chem Univ of Man Winnipeg MB Can

LOUDON, GORDON MARCUS, b Baton Rouge, La, Oct 10, 42; m 64; c 2. BIOCHEMISTRY, ORGANIC CHEMISTRY. Educ: La State Univ, Baton Rouge, BS, 64; Univ Calif, Berkeley, PhD(org chem), 68. Prof Exp: USPHS fel, Univ Calif, Berkeley, 69-70, lectr biochem, 70; ASST PROF CHEM, CORNELL UNIV, 70- Mem: Am Chem Soc; AAAS. Res: Mechanisms of enzyme catalysis; enzyme model systems; bioanalytical methods. Mailing Add: Dept of Chem Cornell Univ Ithaca NY 14853

LOUDON, ROBERT G, b Edinburgh, Scotland, June 27, 25; US citizen; m 55; c 3. INTERNAL MEDICINE. Educ: Univ Edinburgh, MB & ChB, 47. Prof Exp: House physician gen med, Western Gen Hosp, Edinburgh, Scotland, 47-48; sr house physician tuberc wards, City Hosp, 49-50; asst med officer, Tor-na-Dee Sanatorium, Aberdeen, 50-51; house physician, Chest Hosp, Brompton Hosp, London, Eng, 51-52; clin tutor gen med, Royal Infirmary, Edinburgh, 53-54; staff physician, South-East Kans Tuberc Hosp, Chanute, 56-60, supt, 60-61; from asst prof to assoc prof internal med, Univ Tex Southwestern Med Sch Dallas, 61-69; assoc prof med, Sch Med, George Washington Univ, 69-71; PROF INTERNAL MED, MED CTR & DIR PULMONARY DIS DIV, COL MED, UNIV CINCINNATI, 71- Concurrent Pos: Assoc med, Univ Kans, 57-61; staff physician, Woodlawn Hosp, Dallas, 61-69; chief res in respiratory dis, Vet Admin Cent Off, Washington, DC, 69-71. Mem: Am Thoracic Soc; AMA. Res: Chest diseases; tuberculosis; aerobiology. Mailing Add: Pulmonary Dis Div Univ Cincinnati Med Ctr Cincinnati OH 45267

LOUGEAY, RAY LEONARD, b Medford, Ore, Feb 9, 44; m 68. PHYSICAL GEOGRAPHY, REMOTE SENSING. Educ: Rutgers Univ, AB, 66; Univ Mich, MS, 69, PhD(phys geog), 71. Prof Exp: Lectr phys geog, Univ Mich, 69-70; ASST PROF GEOG, STATE UNIV NY COL GENESEO, 71- Mem: Asn Am Geogr; Am Geog Soc; Am Meteorol Soc; Am Soc Photogram; Glaciological Soc. Res: Applied climatology and environmental modification as a function of radiative energy balances and hydrologic water balances; Alpine periglacial environments. Mailing Add: Dept of Geog State Univ of NY Col Geneseo NY 14454

LOUGHBOROUGH, DWIGHT LOGAN, b Plains, Mont, Mar 11, 12; m 41; c 1. PHYSICS. Educ: Univ Wis, BA, 33, MA, 34, PhD(physics), 36. Prof Exp: Asst forest prod lab, US Forest Serv, 34-36; physicist, 36-50, dir phys res, Res Ctr, 50-58, prod res & develop, 58-64, SR TECH ADMINR, TIRE CO, B F GOODRICH CO, 64- Mem: Am Chem Soc; Am Phys Soc. Res: Structure and properties of high polymers; stress analysis; heat transfer; product design. Mailing Add: B F Goodrich Tire Co Akron OH 44311

LOUGHEED, EVERETT CHARLES, horticulture, plant physiology, see 12th edition

LOUGHEED, MILFORD SEYMOUR, b New Westminster, BC, Dec 13, 14; nat US; m 40; c 1. GEOLOGY. Educ: Univ BC, BASc, 40; Princeton Univ, MA, 51, PhD(geol), 53. Prof Exp: Student asst, Geol Surv Can, 35-40; geologist, Siscoe Gold Mines, 40-41; geologist, Anaconda Copper Mining Co, 42-45; geologist, Yukon

Northwest Explor Co, 46; assoc prof geol & head dept, Bates Col, 51-55; asst prof, 55-59, assoc prof, 60-65, chmn dept, 64-67, PROF GEOL, BOWLING GREEN STATE UNIV, 65- Mem: Fel AAAS; fel Geol Soc Am; Nat Asn Geol Teachers; Mineral Asn Can; Geol Asn Can. Res: Petrology and Precambrian iron. Mailing Add: Dept of Geol Bowling Green State Univ Bowling Green OH 43402

LOUGHER, EDWIN HENRY, b Greenfield, Ind, May 30, 20; m 48; c 2. PHYSICAL CHEMISTRY, RESEARCH ADMINISTRATION. Educ: Purdue Univ, BSChE, 42; Ohio State Univ, PhD(phys chem), 52. Prof Exp: Chem engr, Sinclair Ref Co, 42-43; prin chemist, 51-57, assoc chief phys chem div, 57-71, contracts adminr, 71-73, SUPVR, BUS REP, BATTELLE MEM INST, COLUMBUS LABS, 73- Mem: Am Chem Soc; Inst Elec & Electronics Eng. Res: Solid state, particularly thermoelectricity. Mailing Add: Battelle Mem Inst 505 King Ave Columbus OH 43201

LOUGHHEED, THOMAS CROSSLEY, b Sherbrooke, Que, Oct 22, 29; m 53; c 3. FOOD SCIENCE. Educ: Bishop's Univ, BSc, 49; McGill Univ, MSc, 54; Univ London, PhD(microbiol), &dipl, Imp Col, 58. Prof Exp: Res officer biochem, Can Dept Agr, 53-63; RES OFFICER, ANAL BIOCHEM, JOHN LABATT LTD, 63- Mem: Am Chem Soc; Micros Soc Can; Royal Micros Soc. Res: Applications of microscopy to research in food science. Mailing Add: Cent Res & Develop Dept John Labatt Ltd PO Box 5050 London ON Can

LOUGHLIN, TIMOTHY ARTHUR, b Bay Shore, NY, Nov 16, 42; m 65; c 2. APPLIED MATHEMATICS. Educ: State Univ NY Stony Brook, BS, 64; Rensselaer Polytech Inst, MS, 66, PhD(math), 69. Prof Exp: ASST PROF MATH, UNION COL, NY, 69- Mem: Math Asn Am. Res: Network theory; realization of matrices as impedance and admittance matrices. Mailing Add: Dept of Math Union Col Schenectady NY 12308

LOUGHMAN, BARBARA ELLEN EVERS, b Frankford, Ind, Oct 26, 40; m 62; c 2. IMMUNOBIOLOGY. Educ: Univ Ill, BS, 62; Univ Notre Dame, PhD(microbiol & immunol), 72. Prof Exp: From asst res microbiologist to assoc res microbiologist, Ames Res Lab, Miles Labs Inc, 62-71, res scientist immunol, 71-72; staff fel immunol, Nat Inst Child Health & Human Develop, 72-74; RES SCIENTIST IMMUNOL, HYPERSENSITIVITY DIS RES, UPJOHN CO, 74- Mem: AAAS; Asn Gnotobiotics; Am Asn Immunologists. Res: Cellular immunology with emphasis on regulatory mechanisms in cell-mediated and humoral immune responsiveness using controlled in vitro and in vivo systems as models for specific intervention in an immune response. Mailing Add: Hypersensitivity Dis Res Upjohn Co Kalamazoo MI 49001

LOUGHRAN, EDWARD DAN, b Canton, Ohio, June 2, 28; m 59; c 3. ANALYTICAL CHEMISTRY. Educ: Ohio State Univ, BS, 50; MS, 53, PhD(chem), 55. Prof Exp: Asst chem, Res Found, Ohio State Univ, 53-55; MEM STAFF, LOS ALAMOS SCI LAB, 55- Mem: Fel Am Inst Chem. Res: Analytical mass spectrometry; ionization and appearance potential measurements using the mass spectrometer; physical properties, modes of decomposition and radiation chemistry of chemical explosives. Mailing Add: Box 1663 Los Alamos NM 87544.

LOUGHRAN, GERARD ANDREW, SR, b Mt Vernon, NY, Sept 10, 18; m 45; c 4. ORGANIC CHEMISTRY. Educ: Fordham Univ, BS, 41; NY Univ, MS, 48. Prof Exp: Anal chemist, NY Quinine & Chem Works, 41-43; asst chem, Fordham Univ, 43-44; chemist, Am Cyanamid Co, 46-56; chemist, R T Vanderbilt Co, 56-59; chemist, 60-73, PROJ SCIENTIST, US AIR FORCE MAT LAB, 73- Mem: AAAS; fel Am Inst Chem; Am Chem Soc. Res: Petroleum and rubber chemicals; polymer chemistry; high temperature materials. Mailing Add: US Air Force Mat Lab Wright Patterson AFB OH 45433

LOUGHRIDGE, MICHAEL SAMUEL, b Jacksonville, Tex, Aug 27, 36; m 61; c 1. MARINE GEOLOGY. Educ: Rice Univ, BA, 58; Harvard Univ, MA, 61, PhD(geol), 67. Prof Exp: Grad res geologist II, Marine Phys Lab, Scripps Inst, Calif, 61-63, postgrad res geologist II, 63-64, postgrad res geologist III, 64-67, asst res geologist, 67-68; SCI STAFF ASST, OCEANOG SURV DEPT, US NAVAL OCEANOG OFF, 68- Mem: AAAS; Geol Soc Am; Am Geophys Union; assoc mem Soc Explor Geophys. Res: Studies of specialized techniques of echo sounding and the micro-topography of the sea floor; studies of fine scale magnetics of the sea floor; instrumentation for marine geology; seismic profiling; quantitative geomorphology; stream hydraulics; relationships between archaeology and geology. Mailing Add: 5009 Jamestown Rd Washington DC 20016

LOUGHRY, FRANK GLADE, b Marion Center, Pa, Apr 16, 10; m 44. SOIL CONSERVATION. Educ: Pa State Univ, BS, 31, PhD(agron, soils), 60; Ohio State Univ, MS, 34. Prof Exp: Asst agron, Ohio Agr Exp Sta, 31-33; soil scientist, USDA Soil Conserv Serv, 34-35, asst regional soil scientist, Northeastern US, 36-45, state soil scientist, Pa, 45-66; soil scientist, Pa Dept Health, 66-70; CHIEF, SOIL SCI UNIT, PA DEPT ENVIRON RESOURCES, 71- Mem: Fel AAAS; Am Soc Agron; Int Soc Soil Sci; Soil Conserv Soc Am. Res: Relation of soil morphology to aeration; soil factors affecting renovation of waste; interpretation of soil data for environmental protection; use of soil surveys in environmental programs. Mailing Add: 1105 Enterline Ct Harrisburg PA 17110

LOUGHTON, ARTHUR, b Wisbech, Eng, May 25, 31; Can citizen; m 55; c 2. HORTICULTURE. Educ: Univ Nottingham, Eng, BSc, 54, MSc, 60. Prof Exp: Hort officer res, Stockbridge House Exp Hort Sta, Ministry Agr, Fisheries & Food, Eng, 54-62, dep dir, 62-67; res scientist veg res, 67-75, DIR HORT RES, HORT EXP STA, ONT MINISTRY AGR & FOOD, 75- Mem: Can Soc Hort Sci. Res: Field vegetables: production and management of asparagus, carrots, cole crops and potatoes; evaluation of various horticultural crops as alternatives for tobacco growers; administration of total station program in fruit and vegetable research. Mailing Add: Ont Ministry of Agr & Food Hort Exp Sta Box 587 Simcoe ON Can

LOUGHTON, BARRY G, b Newcastle on Tyne, Eng, Sept 29, 35; m 62. BIOLOGY. Educ: Univ Nottingham, BSc, 58; Queen's Univ, Ont, MSc, 62, PhD(entom), 66. Prof Exp: Lectr zool, Queen's Univ, Ont, 63-66; asst prof, 66-70, ASSOC PROF BIOL, YORK UNIV, 70- Mem: NY Acad Sci; Can Zool Soc; Entom Soc Can; Can Biochem Soc. Res: Haemolymph proteins of insects, their chemical and biological properties. Mailing Add: Dept of Biol York Univ Toronto ON Can

LOUIE, RAYMOND, b Canton, China, June 22, 36; US citizen; m 62; c 1. PLANT PATHOLOGY. Educ: Univ Calif, Berkeley, BS, 59; Cornell Univ, MS, 65, PhD(plant path), 68. Prof Exp: ASSOC PROF VIROL, OHIO STATE UNIV & RES PLANT PATHOLOGIST, OHIO AGR RES & DEVELOP CTR, USDA, 67- Mem: Am Phytopath Soc. Res: Epiphytology of plant viruses; virus vector relationships; mechanical transmission of plant viruses. Mailing Add: Plant Sci Res Div Dept of Plant Path Ohio Agr Res & Develop Ctr Wooster OH 44691

LOUIE, ROBERT EUGENE, b Oakland, Calif, Aug 2, 29; m 62; c 1. VIROLOGY. Educ: Univ Calif, Berkeley, BA, 51, MA, 53, PhD(bacteriol), 63. Prof Exp: Res asst virol, Ft Detrick, Md, 54-55; RES MICROBIOLOGIST VIROL, CUTTER LABS, 61- Mem: Am Soc Microbiol; Sigma Xi. Res: Development of viral vaccines for human use; viral chemotherapy; virus-cell relationships. Mailing Add: Microbial Res Dept Cutter Labs Fourth & Parker Sts Berkeley CA 94710

LOUIS, JOHN, b Chicago, Ill, June 21, 24; m 67. HEMATOLOGY, CLINICAL PHARMACOLOGY. Educ: Univ Ill, BS, 48, MS & MD, 50. Prof Exp: Instr med, Col Med, Univ Ill, 51-65; asst prof, Stritch Sch Med, Loyola Univ, Chicago, 65-70; Prof med, Chicago Med Sch, 75; chief hematol sect, Vet Admin Hosp, Downey, Ill, 75; assoc dir, Div Hematol & Oncol, Chicago Med Sch, 75; CONSULT HEMAT & ONCOL, 70- Concurrent Pos: Consult to various hosps & Chicago State TB Sanatorium, 58-; chmn leukemia criteria comt, NIH, 61-65, leukemia task force, 62-65. Res: Clinical pharmacology of drugs relating to hematology and cancer. Mailing Add: 347 Circle Lane Lake Forest IL 60045

LOUIS, KWOK TOY, b Shanghai, China, Jan 22, 27; m 54; c 3. TEXTILE CHEMISTRY. Educ: Tex Tech Col, BS, 51. Prof Exp: Lab dir, Otto Goedecke, Inc, Tex, 53-54; develop chemist, Burlington Indust, Inc, NC, 55-56; chief chemist, United Piece Dye Works, SC, 57-61; applns chems, Ciba Chem & Dye Co, 61-62, group leader appln res & qual control, 63-64, admin mgr res & appln, Tech Appln Prod, 64-68, mgr cent lab, 68-71; DIR TECH DEPT, DYES & CHEM DIV, CROMPTON & KNOWLES CORP, NJ, 71- Mem: Am Asn Textile Chemists & Colorists; AAAS; NY Acad Sci. Mailing Add: 442 Ellis Place Wyckoff NJ 07481

LOUIS, LAWRENCE HUA-HSIEN, b Canton, China, Apr 23, 08; nat US; m 42; c 4. BIOCHEMISTRY. Educ: Univ Mich, BS, 32, MS, 33, ScD, 37. Prof Exp: Fel physiol, Univ Pa, 40-41; asst internal med, 41-46, instr biochem, 46-48, from asst prof to assoc prof, 48-69, PROF BIOCHEM, UNIV MICH, ANN ARBOR, 70- Mem: AAAS; Am Chem Soc; Am Soc Biol Chem. Res: Endocrinology and metabolism. Mailing Add: Dept of Internal Med Univ of Mich Ann Arbor MI 48104

LOUIS, THOMAS MICHAEL, b Pensacola, Fla, Dec 27, 44; m 69; c 1. REPRODUCTIVE ENDOCRINOLOGY. Educ: Va Polytech Inst & State Univ, BS, 68, MS, 71; Mich State Univ, PhD(physiol, dairy sci), 75. Prof Exp: LALOR RES FEL REPRODUCTIVE ENDOCRINOL, UNIV OXFORD, 75- Honors & Awards: Richard Hoyte Res Prize, Am Dairy Sci Asn, 75. Mem: AAAS; Am Soc Animal Sci; Sigma Xi; Soc Study Endocrinol; Soc Study Reproduction. Res: Study of the effect of prostaglandins on the male and female reproductive processes; studies include endocrinology of parturition, fetal endocrinology, endocrinology of the estrous and menstrual cycle and endocrine control of the hypothalamus and pituitary; effects of prostaglandins on sperm output and production. Mailing Add: Dept of Obstet & Gynec John Radcliffe Hosp Univ Oxford Oxford England

LOUISELL, WILLIAM HENRY, b Mobile, Ala, Aug 22, 24; m 51; c 5. PHYSICS. Educ: Univ Mich, BS, 48, MS, 49, PhD(physics), 53. Prof Exp: Mem tech staff, Bell Tel Labs, NJ, 53-66; chmn dept physics, 67-69, PROF PHYSICS & ELEC ENG, UNIV SOUTHERN CALIF, 66- Concurrent Pos: Consult, US Army Missile Command, Ala, 70-, App Math & Sci Lab, Aberdeen Proving Ground, 74- & Los Alamos Sci Lab, 75- Mem: Fel Am Phys Soc; sr mem Inst Elec & Electronics Eng. Res: Microwave and quantum electronics; nonlinear optics. Mailing Add: Dept of Physics Univ of Southern Calif Los Angeles CA 90007

LOUIS-FERDINAND, ROBERT THOMAS, pharmacology, biochemistry, see 12th edition

LOULOUDES, SPIRO JAMES, b Stratford, Conn, Mar 22, 28; m 54; c 3. ENTOMOLOGY. Educ: DePauw Univ, BA, 52; Kans State Col, MS, 55, PhD(entom), 58. Prof Exp: Asst entom, Kans State Col, 56-57, instr, 57-58; ENTOMOLOGIST INSECT PHYSIOL, INSECT PATH LAB, AGR RES CTR, AGR RES SERV, USDA, 58- Mem: Entom Soc Am; Am Chem Soc. Res: Insect physiology, pathology and biochemistry; dynamic aspects of cuticular hydrocarbon synthesis and the exotoxins of insect pathogens. Mailing Add: Insect Path Lab Bldg A Agr Res Ctr Beltsville MD 20705

LOUNSBURY, FLOYD GLENN, b Stevens Point, Wis, Apr 25, 14; m 52; c 1. ETHNOLOGY. Educ: Univ Wis, BA, 41, MA, 46; Yale Univ, PhD(anthrop), 49. Prof Exp: Asst prof anthrop, Yale Univ, 49-55, assoc prof, 55-61, PROF ANTHROP, YALE UNIV, 61- Mem: Nat Acad Sci; Ling Soc Am; Soc Am Archaeol. Res: Linguistics; Maya hieroglyphic writing. Mailing Add: Dept of Anthrop Yale Univ New Haven CT 06520

LOUNSBURY, FRANKLIN, b Chicago, Ill, May 6, 12; m 41; c 3. MEDICINE. Educ: Univ Wis, AB, 34; Northwestern Univ, MD, 39, MS, 48. Prof Exp: Asst prof, 54-65, ASSOC PROF SURG, NORTHWESTERN UNIV, CHICAGO, 65- Concurrent Pos: Attend physician, Northwestern Mem Hosp, 46- Mem: Am Col Surg. Res: Abdominal surgery, especially of the biliary tract. Mailing Add: Dept of Surg Col of Med Northwestern Univ Chicago IL 60611

LOUNSBURY, JOHN BALDWIN, physics, chemistry, see 12th edition

LOUNSBURY, JOHN FREDERICK, b Perham, Minn, Oct 26, 18; m 43; c 3. ECONOMIC GEOGRAPHY, PHYSICAL GEOGRAPHY. Educ: Univ Ill, BS, 42, MS, 46; Northwestern Univ, PhD(geog), 51. Prof Exp: Instr geog, Univ Ill, 46-49; field dir, Rural Land Classification Puerto Rico, 49-51; chmn dept earth sci, Antioch Col, 51-61; head dept geog & geol, Eastern Mich Univ, 61-69; CHMN DEPT GEOG, ARIZ STATE UNIV, 69- Concurrent Pos: Ford Found fel urban studies, Antioch Col, 58-60, Lily Found fel indust develop, 60-61; NSF comn Col Geog fel, Asn Am Geog, 63-; vis prof, Univ Ga, Mich State Univ, Wayne State Univ & Wesleyan Univ, 57-70; NSF dir, Comn Col Geog, 63-74; mem exec bd, Nat Coun Geog Educ, 68-71; dir environ-based educ proj, US Off Educ, 74-75; dir spatial anal of land use proj, NSF, 75- Honors & Awards: J Geog Award, Nat Coun Geog Educ, 61. Mem: AAAS; Asn Am Geog; Am Geog Soc. Res: Physical, economic and resource geography as it applies to land use changes and resource exploitation; impact of population growth and environmental change in selected areas of Anglo-American and Latin America. Mailing Add: Dept of Geog Ariz State Univ Tempe AZ 85281

LOUNSBURY, RICHARD WILLIAM, b Chicago, Ill, Aug 29, 17; m 54; c 4. GEOLOGY. Educ: Univ Chicago, BS, 41; Stanford Univ, PhD(geol), 51. Prof Exp: Actg instr geol, Stanford Univ, 47, 50; instr, State Col Wash, 48-49; instr, Pomona Col, 50-51; asst prof, Beloit Col, 51-53; instr, Purdue Univ, 53-56, assoc prof, 56-60, prof, 60-66, chmn dept, 54-66; PROF GEOL, MEMPHIS STATE UNIV, 68- Concurrent Pos: Fulbright scholar, 60-61. Mem: AAAS; Geol Soc Am; Am Asn Petrol Geol; Am Soc Test & Mat; Mineral Soc Gt Brit & Ireland. Res: Petrology; mineralogy; engineering geology. Mailing Add: Dept of Geol Memphis State Univ Memphis TN 38152

LOURENCO, RUY VALENTIM, b Lisbon, Portugal, Mar 25, 29; US citizen; m 60; c 2. MEDICINE, PHYSIOLOGY. Educ: Univ Lisbon, BSc, 46, MD, 51. Prof Exp:

Intern. Lisbon City Hosps. 52. resident internal med. 53-55; asst med. Sch Med. Lisbon. 56-59; from asst prof to assoc prof. NJ Col Med. 63-67; assoc prof med. 67-69. PROF MED & PHYSIOL. ABRAHAM LINCOLN SCH MED. UNIV ILL COL MED. 69-. DIR PULMONARY SECT. DEPT MED. 70- Concurrent Pos: Fel med. Cologne Univ. 57 & Columbia-Presby Med Ctr. 59-63; Lederle int fel. 59-60; Polachek Found fel. 61-63; attend physician. Newark City Hosp. 65-67; consult physician. Vet Admin Hosps. 65-; dir respiratory res. Hektoen Inst Med Res. Chicago. 67-70; attend physician. Univ Ill Hosp. 67-. dir pulmonary serv & labs. 70-; dir respiratory physiol lab. Cook County Hosp. Chicago. 67-69. dir dept pulmonary med. 69-70; mem cardio-pulmonary coun. Am Heart Asn; mem task force respiratory sci. Nat Heart & Lung Inst. 71-72; mem study sect. NIH. 72-76; chmn sci assembly. Am Thoracic Soc. 74-75. Mem: Am Physiol Soc; Am Fedn Clin Res; Am Thoracic Soc; Am Soc Clin Invest; Soc Exp Biol & Med. Res: Internal medicine; chest diseases; respiratory physiology and biochemistry; regulation of ventilation; muscles of breathing; pulmonary defense mechanisms. Mailing Add: Abraham Lincoln Sch of Med Univ of Ill Col of Med Chicago IL 60680

LOURIA, DONALD BRUCE, b New York. NY. July 11. 28; m 55; c 3. INTERNAL MEDICINE. MICROBIOLOGY. Educ: Harvard Univ. BS. 49. MD. 53. Prof Exp: From instr to assoc prof med. Col Med. Cornell Univ. 58-69; PROF PREV MED & COMMUNITY HEALTH & CHMN DEPT. NJ MED SCH. COL MED & DENT NJ. 69- Concurrent Pos: Pres. NY State Coun Drug Addiction. 65- Mem: Am Soc Clin Invest; Am Fedn Clin Res; Am Soc Microbiol; Am Col Physicians; AMA. Res: Mycology. especially fungal toxins and the pathogenesis of Candida infections; mycoplasmas and their relation to human disease. Mailing Add: Dept Prev Med & Community Health NJ Med Sch Newark NJ 07103

LOURIE, ALAN DAVID, b Boston. Mass. Jan 13. 35; m 59; c 2. ORGANIC CHEMISTRY. Educ: Harvard Univ. AB. 56; Univ Wis. MS. 58; Univ Pa. PhD(org chem). 65; Temple Univ. JD. 70. Prof Exp: Res chemist. Monsanto Co. 57-59; res chemist. Wyeth Labs. 59-60. lit chemist. 60-62. patent chemist. 62-64; patent agent chem. Smith Kline & French Labs. 64-70. patent attorney. 70-71. assoc patent counsel. 71-74. ASST DIR. PATENT DEPT. SMITHKLINE CORP. 74- Mem: Am Chem Soc. Res: Synthesis of heterocyclic compounds; medicinal chemistry. Mailing Add: 1500 Spring Garden St Philadelphia PA 19101

LOURIE, HERBERT, b St George. SC. Mar 6. 29; m 48; c 4. NEUROSURGERY. Educ: Univ SC. BS. 48; Duke Univ. MD. 52. Prof Exp: From instr to assoc prof. 60-68. PROF NEUROSURG. STATE UNIV NY UPSTATE MED CTR. 68- Concurrent Pos: Fel clin neurosurg. Wash Univ. 54. fel neurophysiol. 59; Am Heart Asn res grant. 66-67. Mem: Cong Neurol Surg; Am Asn Neurol Surg; Am Col Surg; Am Acad Neurol Surg; Soc Univ Neurosurg. Res: Hypertensive vascular disease in experimental animals; neurophysiologic investigation of coma and arousal in cat. Mailing Add: 713 E Genesee St Syracuse NY 13214

LOURIE, REGINALD SPENCER, b Brooklyn. NY. Sept 10. 88; m 31; c 3. PSYCHIATRY. Educ: Cornell Univ. BS. 30; Long Island Col Med. MD. 36; Columbia Univ. ScD(psychiat). 42. Prof Exp: Asst bact. Res Labs. New York Dept Health. 30-31; intern & resident pediat & psychiat. Long Island Col Hosp. 36-38. NY State Psychiat Inst. 38-39 & Bellevue Psychiat Hosp. 39-40; assoc res psychiat. Col Physicians & Surgeons. Columbia Univ. 42-46; instr pediat & psychiat. Sch Med. Rochester Univ. 46-48; asst prof. Col Med. George Wash Univ. 48-53. assoc clin prof pediat psychiat. 54-65. prof. 65-71. Prof. 71-74. EMER PROF CHILD HEALTH & DEVELOP & PSYCHIAT. COL MED. GEORGE WASH UNIV. 74- Concurrent Pos: Scottish Rite Fund & Markle Found fel. NY State Psychiat Inst & Hosp. 40-43; asst pediatrician. Vanderbilt Clin. 41-46; civilian with off Sci Res & Develop. 42; assoc pediatrician. Univ Hosp. George Wash Univ. 48-; dir dept psychiat. Children's Hosp & Hillcrest Children's Ctr. Washington. DC. 48-74; consult. Walter Reed Army Med Ctr. 49- & President's Panel Ment Retardation. 59-61; mem. Nat Adv Ment Health Coun. 64-68; pres. Joint Coun Ment Health Children. 65-75; lectr. Howard Univ. Am Univ & Cath Univ Am; consult. Children's Bur. Nat Bur Standards. Nat Inst Ment Health & Nat Naval Med Ctr; chmn. Human Serv Inst Children & Families. 73-74; sr consult. Psychiat Inst Wash. DC. 74- Honors & Awards: Comdr. Royal Order of Phoenix. Greece; McGAvin Award. Am Psychiat Asn; Dickinson Medal. State Univ NY. Downstate Med Sch. 64. Mem: Am Psychiat Asn; Am Orthopsychiat Asn (pres. 57-58); Am Psychoanal Asn; Am Acad Child Psychiat (secy. 61-63. pres. 63-65); Int Asn Child Psychiat (treas. 70-74). Res: Psychopathology and psychophysiology in childhood; autonomic nervous system function in children. Mailing Add: 4305 Thornapple St Chevy Chase MD 20015

LOUSTALOT, ARNAUD JOSEPH, b New Orleans. La. July 14. 13; m 37; c 3. PLANT PHYSIOLOGY. Educ: La State Univ. BS. 35. MS. 36; Cornell Univ. PhD(pomol). 39. Prof Exp: Asst. La State Univ. 35-36; asst. Cornell Univ. 36-39; agent & asst pomologist nut invests. Bur Plant Indust. Soils & Agr Eng. 43-45. assoc chemist. Fed Exp Sta. Puerto Rico. 44-46. chemist. 46-47. asst dir & sr plant physiologist. 48-53. ADMINR. COOP STATE RES SERV & PRIN PLANT PHYSIOLOGIST. USDA. 54- Mem: Am Soc Hort Sci; Am Soc Plant Physiol; Am Soc Agron. Res: Photosynthesis; mineral nutrition of plants; weed control; agricultural climatology. Mailing Add: Coop State Res Serv USDA Washington DC 20250

LOUSTAUNAU, JOAQUIN, b San Louis Potosi. Mex. Sept 17. 36; m 66. MATHEMATICS. Educ: Okla State Univ. BS. 58. MS. 60; Univ Ill. PhD(math). 65. Prof Exp: Instr math. Inst Tech & Higher Educ. Monterrey. Mex. 60-61; ASST PROF MATH. NMEX STATE UNIV. 65- Mem: Math Asn Am; Am Math Soc. Res: Functional analysis. Mailing Add: Dept of Math NMex State Univ Las Cruces NM 88001

LOUTTIT, ROBERT IRVING, b Honolulu. Hawaii. July 23. 29; m 54; c 3. EXPERIMENTAL HIGH ENERGY PHYSICS. Educ: Univ NH. BS. 52; Wash Univ. PhD(physics). 58. Prof Exp: From asst physicist to assoc physicist. 58-64. PHYSICIST. BROOKHAVEN NAT LAB. 64- Concurrent Pos: Physicist. Nuclear Res Ctr. Saclay. France. 63-64. Mem: AAAS; Am Phys Soc. Res: Bubble chamber development; neutrino interactions. Mailing Add: Accelerator Dept Brookhaven Nat Lab Upton NY 11973

LOUX, HARVEY MONROE, organic chemistry. see 12th edition

LOVAAS, ALLAN L, wildlife ecology. see 12th edition

LOVAGLIA, ANTHONY RICHARD, b San Jose. Calif. Jan 25. 23; m 44; c 3. MATHEMATICS. Educ: Univ Calif. Los Angeles. AB. 45. PhD(math). 51; Stanford Univ. MS. 48. Prof Exp: From asst prof to assoc prof. 51-60. PROF MATH. SAN JOSE STATE COL. 60- Mem: Am Math Soc; Math Asn Am. Res: Analysis. Mailing Add: Dept of Math San Jose State Col San Jose CA 95114

LOVALD, ROGER ALLEN, b Marshall. Minn. Aug 8. 38; m 57; c 2. ORGANIC POLYMER CHEMISTRY. Educ: Univ Minn. BChem. 60; Univ Wis. PhD(org chem). 65. Prof Exp: Chemist. Spring Res Lab. Rohm & Haas Co. 65-67; cent res. 67-71. sect leader. Resin Develop. 71-75. TECH DIR RESINS. GEN MILLS CHEM. INC. 75- Res: Heteroaliphatic and organic chemistry; addition and condensation polymerization; acrylics; polyamides; polyesters; polyurethanes. Mailing Add: Gen Mills Chem Inc 2010 E Hennepin Ave Minneapolis MN 55413

LOVAS, FRANCIS JOHN, b Cleveland. Ohio. July 29. 41; m 70. MOLECULAR SPECTROSCOPY. RADIO ASTRONOMY. Educ: Univ Detroit. BS. 63; Univ CAlif. Berkeley. PhD(phys chem). 67. Prof Exp: Res grant. Lawrence Radiation Lab. Univ Calif. Berkeley. 67-68; NATO fel. Phys Inst. Free Univ Berlin. 68-70; Nat Res Coun-Nat Bur Standards. Assoc. 70-72. DIR MICROWAVE SPECTRAL DATA CTR. NAT BUR STANDARDS. 72- Mem: Am Phys Soc. Res: Properties of diatomic molecules by high temperature microwave adsorption and molecular beam electric resonance techniques; microwave spectroscopy of transient molecules and molecular radio astronomy; critical evaluation of microwave spectroscopic data. Mailing Add: Optical Physics Molecular Spectros Nat Bur Standards Washington DC 20234

LOVASS-NAGY, VICTOR, b Debrecen. Hungary. Apr 25. 23; m 51; c 2. APPLIED MATHEMATICS. Educ: Budapest Tech Univ. dipl. 47. PhD(math). 49. Prof Exp: Instr math. Budapest Tech Univ. 47-49. from asst prof to assoc prof. 49-58; consult engr. Ganz Elec Works. Hungary. 60-64; reader eng math. Univ Khartoum. 64-66; PROF MATH. CLARKSON COL TECHNOL. 66- Mem: Soc Indust & Appl Math; Am Math Soc; Math Asn Am; Tensor Soc; sr mem Inst Elec & Electronics Eng. Res: Matrix theory; numerical analysis; network theory. Mailing Add: Dept of Math Clarkson Col of Technol PO Box 471 Potsdam NY 13676

LOVE, ALLAN WALTER, b. Toronto. Ont. May 28. 16; US citizen; m 46; c 3. ELECTROMAGNETISM. Educ: Univ Toronto. BA. 38. MA. 39. PhD(microwave physics). 51. Prof Exp: Res officer. Radiophys Lab. Commonwealth Sci & Indust Res Orgn. Australia. 46-48; demonstr asst. Physics Lab. Univ Toronto. 48-51; chief instrumentation. Newmont Explor Ltd. Conn & Ariz. 51-57; staff scientist. Giannini Res Lab. Wiley Electronics Co. Ariz. 57-62. mgr physics lab. Calif. 62-63; group scientist. Antenna Lab. Autonetics Div. NAm Aviation. Inc. 63; area mgr. Nat Eng Sci Co. 63-65; group scientist theoret anal. Autonetics Div. NAm Aviation. Inc. 65-71. mem tech staff. Space Div. NAm Rockwell Corp. 71-73. PROG MGR. SPACE DIV. ROCKWELL INT. 73- Mem: Inst Elec & Electronics Eng. Res: Microwave and millimeter wave physics; plasma effects and antenna theory and design; development of spacecraft antenna systems. Mailing Add: 518 Rockford Pl Corona Del Mar CA 92625

LOVE, BETHOLENE FRANCES, b Glasgow. WVa. Sept 22. 20. MEDICAL TECHNOLOGY. Educ: Western Reserve Univ. BS. 46; Univ Okla. MS. 50. Prof Exp: Sr bacteriologist. WVa Dept Health. 46-48 & 50-53; from instr to assoc prof. 55-72. PROF MED TECHNOL. W VA UNIV. 72- Concurrent Pos: Mem ed bd. Am J Med Technol. 63-74. ed. 67-70; chmn publ comt. Am Soc Med Technologists. 68-70. chmn educ sect. Sci Assembly. 70-71. Mem: Am Soc Med Technologists; Asn Schs Allied Health Professions. Res: Medical technology education; allied health education. Mailing Add: 312 Mulberry St Morgantown WV 26505

LOVE, CALVIN MILES, b Chicago. Ill. Mar 2. 37; m 60; c 3. INORGANIC CHEMISTRY. RADIOCHEMISTRY. Educ: Ill Inst Technol. BS. 59; Mich State Univ. PhD(inorg chem). 64. Prof Exp: RES SPECIALIST. MOUND LAB. MONSANTO RES CORP. 64- Mem: AAAS; Am Chem Soc. Res: Kinetics and mechanisms of inorganic oxidation-reduction reactions; plutonium separation and recovery; polonium process development; metal distillation; metal hydrides; radiation damage; thermal analysis; hydrides for hydrogen storage. Mailing Add: Mound Lab Monsanto Res Corp Miamisburg OH 45342

LOVE, CHAILLE M, b Farmerville. La. Dec 8. 11; m 45; c 4. PHYSIOLOGY. Educ: Northwestern State Col. La. AB. 32; Stanford Univ. AM. 48; Ore State Univ. PhD(physiol). 53. Prof Exp: Teacher. Pub Sch. 32-36; lectr. Univ Calif. 47; assoc prof. 48-58. PROF LIFE SCI. SACRAMENTO STATE UNIV. 58- Mem: AAAS; Int Primatol Soc. Res: Temperature regulating mechanism of the rabbit; medical technology; methods of teaching science in higher education; biological electronics. Mailing Add: Dept of Biol Calif State Univ Sacramento 6000 J St Sacramento CA 95819

LOVE, DANIEL LINDSLEY, b Portland. Ore. May 26. 28; m 50; c 2. RADIOCHEMISTRY. Educ: Reed Col. BA. 50; Univ Portland. MS. 51; Pa State Univ. PhD(fuel tech). 55. Prof Exp: Asst phys chem. Univ Tex. 51-52 & Pa State Univ. 52-55; head radiochem anal sect. US Naval Radiol Defense Labs. 55-64. appl res br. Calif. 64-69. chief nuclear chem div. Naval Ord Lab. 69-75. HEAD. NUCLEAR BR. NAVAL SURFACE WEAPONS CTR. 75- Concurrent Pos: Part-time mem fac. Montgomery Col; consult. State of Md Dept of Natural Resources. 71- Mem: Am Chem Soc. Res: Analytical chemistry; polarography; radioavtive tracers; fuel chemistry; separation and decay schemes of short-lived radionuclides; oceanography; radiation chemistry; Mössbauer spectrometry; mechanisms of carcinogenesis; lasers; fluorine chemistry; power plant siting. Mailing Add: 4416 Norbeck Rd Rockville MD 20853

LOVE, DAVID S, b Scottsbluff. Nebr. May 2. 32; m 54; c 2. ZOOLOGY. Educ: Univ Colo. BA. 54. MA. 56. PhD(zool). 62. Prof Exp: Res assoc. Inst Cell Biol. Univ Conn. 59-61; res assoc biochem. Univ Wash. 61-64. sr instr. 64-65. ASST PROF DEVELOP EMBRYONIC MUSCLE. SCH MED. CASE WESTERN RESERVE UNIV. 65- Mem: Am Cancer Soc fel. 62-64. Res: Role of the endocrine system in embryonic muscle development. Mailing Add: Dept of Anat Case Western Reserve Univ Sch of Med Cleveland OH 44106

LOVE, DAVID VAUGHAN, b St John. NB. Aug 25. 19; m 43; c 3. FOREST MANAGEMENT. Educ: Univ NB. BSc. 41; Univ Mich. MF. 46. Prof Exp: From lectr to prof. 46-72. ASST DEAN FORESTRY. UNIV TORONTO. 72- Concurrent Pos: Vpres. Conservation Coun Ont. 59-64. pres. 74-75; vchmn. Can Coun on Rural Develop. 75-; rep. Can Forestry Asn. 73. Mem: Soc Am Foresters; Can Pulp & Paper Asn; Can Inst Forestry (secy-mgr. 48-54. pres. 65-66); Ont Forestry Asn (vpres. 70-71. pres. 72-73); Can Forestry Asn (chmn-Objectives & Structure Comt. 72-73; vpres. 74-75. pres. 75). Res: Land use. Mailing Add: Fac of Forestry Univ of Toronto 203 College St Toronto ON Can

LÖVE, DORIS, b Kristianstad. Sweden. Jan 2. 18; m 40; c 2. CYTOTAXONOMY. Educ: Kristianstad Col. BA. 37; Univ Lund. BSc. 41. PhD. 43. DSc. 44. Prof Exp: Res assoc genetics. Univ Lund. 41-45; res assoc. Inst Bot & Plant Breeding. Univ Iceland. 45-51; herbarium curator. Univ Man. 51-56; assoc res prof biosyst inst. Univ Montreal. 56-63; mus assoc phytogeog & res assoc. Inst Arctic & Alpine Res & Dept Biol. Univ Colo Boulder. 64-73. Mem: Genetics Soc Am; Soc Study Evolution; Int Asn Plant Taxon; Scandinavian Asn Geneticists; fel Mendelian Soc Lund. Res: Taxonomy and geobotany of boreal. arctic and mediterranean plants; polyploidy; chromosome manuals; computer retrieval of cytological and taxonomical data; cytogenetics and physiology of sex determination; geomorphology of mountains;

Pleistocene geology and botany; dispersal mechanisms; physiology of krummholz. Mailing Add: 473 Harvard Lane Boulder CO 80303

LOVE, GEORGE M, b Lima, Ohio, Oct 5, 44; m 72. ORGANIC CHEMISTRY. Educ: DePauw Univ, BA, 66; Wake Forest Univ, MA, 68; Mich State Univ, PhD(org chem), 72. Prof Exp: Fel org chem, Rutgers Univ, 72-73; SR RES CHEMIST, MERCK INC, 73- Mem: Sigma Xi; Am Chem Soc. Res: Process research in organic chemistry. Mailing Add: Merck Inc Rahway NJ 07065

LOVE, GEORGE ROSS, b Toronto, Can, Sept 3, 17; m 43; c 5. ELECTRICITY. Educ: Univ Western Ont, BA, 41, MA, 43; Univ Toronto, PhD(physics), 48. Prof Exp: PROF PHYSICS, CARLETON UNIV, 48- Mem: Am Asn Physics Teachers; Can Asn Physicists. Mailing Add: 2321 Hillary Ave Ottawa ON Can

LOVE, HARRY SCHROEDER, JR, b Idabel, Okla, Aug 20, 27; m 52; c 2. BOTANY, ECOLOGY. Educ: Okla State Univ, BS, 52, MS, 58, PhD(bot), 71. Prof Exp: ASSOC PROF BIOL, E CENT STATE COL, 67- Mem: AAAS; Am Inst Biol Sci; Nat Asn Biol Teachers. Res: Terrestrial plant ecology, especially clonal and root-graft relationships. Mailing Add: Dept of Biol E Cent State Col Ada OK 74820

LOVE, HUGH MORRISON, b Northern Ireland, Aug 21, 26. PHYSICS. Educ: Queen's Univ, Belfast, BSc & PhD(physics), 50. Prof Exp: Asst lectr, Queen's Univ, Belfast, 46-50; lectr, Univ Toronto, 50-52; from asst prof to assoc prof, 52-65, PROF PHYSICS, QUEEN'S UNIV, ONT, 65-, ASSOC DEAN SCI, 70- Mem: Am Phys Soc. Res: Solid state physics; surface physics. Mailing Add: Dept of Physics Queen's Univ Kingston ON Can

LOVE, JAMES ALLAN, b Johnstone, Scotland, Nov 7, 42; m 67; c 2. POLYMER CHEMISTRY. Educ: Univ Strathclyde, BSc, 64, PhD(chem), 68. Prof Exp: Res chemist, Int Cellulose Res, Int Paper Co, 68-71; res dir, Fabricated Plastics Ltd, 71-72; res chemist, Com Alcohols Ltd, Int Paper Co, 72-73, RES ASSOC CHEM, INT PAPER CO, 73- Mem: Chem Inst Can; Tech Asn Pulp & Paper Indust; The Chem Soc; Forest Prod Res Soc; Plastics Inst. Res: Cellulose polymer combinations; composite wood structures; properties of fiber reinforced plastics. Mailing Add: Int Paper Co Box 797 Tuxedo Park NY 10987

LOVE, JIM, b Bathgate, Scotland, Oct 21, 38; m 62; c 2. ORGANIC CHEMISTRY. Educ: Univ Edingurgh, BSc, 60, PhD(carbohydrate chem), 63. Prof Exp: Fel, Scripps Inst, Univ Calif, 63-64 & Ohio State Univ, 64-65; res chemist, Mich, 65-67, western div, 67-74, RES SPECIALIST, DOW CHEM CO, WESTERN DIV, PITTSBURG, 74- Mem: Am Chem Soc; The Chem Soc. Res: Carbohydrate chemistry, particularly polysaccharide and mucopolysaccharide structural determination and biological activity; synthesis and biological activity of heterocyclic compounds. Mailing Add: 3320 Lancashire Place Concord CA 94520

LOVE, JOHN DAVID, b Riverton, Wyo, Apr 17, 13; m 40; c 4. GEOLOGY. Educ: Univ Wyo, BA, 33, MA, 34; Yale Univ, PhD(geol), 38. Hon Degrees: LLD, Univ Wyo, 61. Prof Exp: Asst geologist, Geol Surv Wyo, 33-37; field asst, US Geol Surv, 38; asst geologist, Shell Oil Co, Inc, 38-40, geologist, 40-42; asst geologist, 42-43, from assoc geologist to prin geologist, 43-56, supvr heavy metals br, 66-68, Northern Rocky Mts Br, 64-66, 67-69, in charge Wyo basins fuel proj, 43-56, STAFF GEOLOGIST, US GEOL SURV, 56-, SUPVR, LARAMIE OFF, ROCKY MT ENVIRON GEOL BR, 69- Concurrent Pos: Adj prof, Univ Wyo, 69-; affil prof geol, Univ Idaho, 74- Mem: AAAS; fel Geol Soc Am; Am Asn Petrol Geol; Soc Econ Paleont & Mineral. Res: Geology of fuels; uranium, vanadium and gold investigations; stratigraphic and structural geology. Mailing Add: US Geol Surv Box 3007 Univ Sta Laramie WY 82070

LOVE, LAWRENCE DUDLEY, plant ecology, soils, see 12th edition

LOVE, LEON, b New York, NY, Sept 7, 23; m 56; c 3. RADIOLOGY. Educ: City Col New York, BS, 43; Chicago Med Sch, MD, 46; Am Bd Radiol, dipl, 51. Prof Exp: Radiologist, Cook County Hosp, Chicago, 56-61; assoc prof radiol, Chicago Med Sch, 58-67, clin prof, 67-69; PROF RADIOL & CHMN DEPT, MED CTR, LOYOLA UNIV CHICAGO, 69- Concurrent Pos: Consult, Dwight Vet Admin Hosp, 56-62; dir diag radiol, Cook County Hosp, Chicago, 61-69; consult, House of Correction, Chicago, 61- & WSide Vet Admin Hosp, 62- Mem: Am Col Radiol; Radiol Soc NAm. Res: Renal angiography; radiology of the gastro-intestinal tract. Mailing Add: Dept of Radiol Loyola Univ Med Ctr Maywood IL 60153

LOVE, LINDA J CLINE, b Richmond, Mo, Oct 1, 40; m 72. ANALYTICAL CHEMISTRY. Educ: Univ Mo-Columbia, BS, 62, MA, 65; Univ Ill, Urbana, PhD(chem), 69. Prof Exp: Fel chem, Univ Fla, 69-70; asst prof anal chem, Mich State Univ, 70-72; ASST PROF ANAL CHEM, SETON HALL UNIV, 72- Mem: Am Chem Soc; Soc Appl Spectros; Optical Soc Am; Sigma Xi. Res: Development of new instrumentation and methodology in luminescence; laser-Raman, atomic, absorption-emission-fluorescence spectroscopy and automation of chemical instrumentation. Mailing Add: Dept of Chem Seton Hall Univ South Orange NJ 07079

LOVE, NORMAN DUANE, b Howell, Mich, Jan 1, 39; m 62; c 3. LOW TEMPERATURE PHYSICS. Educ: Albion Col, AB, 60; Western Mich Univ, MA, 62; Mich State Univ, PhD(physics), 67. Prof Exp: ASST PROF PHYSICS, MARYVILLE COL, 67-, DIR COMPUT SERV, 71- Concurrent Pos: Nat Sci Found comput grant, 68-73. Mem: Am Phys Soc; Am Asn Physics Teachers. Res: Effect of magnons on transport of phonons; phase boundaries in an antiferro magnetic material using calorimetric techniques. Mailing Add: Dept of Physics Maryville Col Maryville TN 37801

LOVE, ROBERT, b Glasgow, Scotland, Apr 27, 21; nat US; m 45; c 3. PATHOLOGY. Educ: Univ Glasgow, MB & ChB, 44, MD, 50. Prof Exp: House physician, Duke St Hosp, Glasgow, Scotland, 44 & Stonehouse Hosp, Lanarkshire, 45; registr, Dept Med, Royal Infirmary, Glasgow, 48; jr pathologist, Cambridge Univ, 48-50; pathologist, Sect Viral & Rickettsial Res, Lederle Labs, Am Cyanamid Co, 51-55; pathologist, Lab Path, Nat Cancer Inst, 55-60; PROF PATH, JEFFERSON MED COL, 60-; BR CHIEF PROG ANAL & FORMULATION, NAT CANCER INST, 74- Concurrent Pos: Mem health res facil sci rev comt, NIH, 67-70; ed, Carcinogenesis Abstr, 70-75; chmn, Gordon Res Conf Cancer, 71. Mem: Am Soc Exp Path; Am Asn Cancer Res. Res: Cytopathology; virus infection and neoplasia; nucleolar structure and function; cytochemistry; electron microscopy; carcinogenesis. Mailing Add: Rm 10A52 Bldg 31 Prog Anal & Formulation Br Nat Cancer Inst Bethesda MD 20014

LOVE, ROBERT ALEXANDER, b Brooklyn, NY, Apr 4, 13; wid; c 3. RADIOLOGICAL HEALTH, INDUSTRIAL MEDICINE. Educ: Brown Univ, BA, 37; Cornell Univ, MD, 42. Prof Exp: Intern med, Kings County Hosp, Brooklyn, NY, 42-43, resident, 46-47; HEAD INDUST MED, BROOKHAVEN NAT LAB, 47- Concurrent Pos: Radiation consult, Roger Williams Gen Hosp, Providence, RI, 64-; assoc prof, State Univ NY Stony Brook. Mem: Fel Indust Med Asn; AMA. Res: Treatment of radiation injuries; radiation decontamination of the human, both external and internal. Mailing Add: Med Dept Brookhaven Nat Lab Upton NY 11973

LOVE, ROBERT LYMAN, b Oswego, NY, July 28, 25; m 50; c 4. HEALTH SCIENCES. Educ: Syracuse Univ, AB, 47, MSEd, 49. Prof Exp: Teacher, Middlesex Valley Cent Sch, 49-53; from instr to assoc prof, 53-61, PROF PHYSIOL & BIOCHEM, STATE UNIV NY AGR & TECH COL ALFRED, 61-, DEAN, SCH ALLIED HEALTH TECHNOL, 75- Concurrent Pos: Adv comt Allied Health Proj, Am Asn Community & Jr Col, 73; consult, Steering Comt, Am Med Asn, 75-77. Mem: Asn Schs Allied Health Professions; Am Soc Med Technol; Am Med Record Asn; Am Asn Clin Chem. Mailing Add: State Univ of New York Agr & Tech Col Alfred NY 14802

LOVE, ROBERT MERTON, b Tantallon, Sask, Can, Jan 29, 09; nat US; m 36; c 3. CYTOGENETICS, ECOLOGY. Educ: Univ Sask, BSc, 32, MSc, 33; McGill Univ, PhD(genetics), 35. Prof Exp: Instr, Univ Sask, 30-32, asst, 32-33; asst, McGill Univ, 33-34; asst agr scientist, Cereal Div, Cent Exp Farm, Can Dept Agr, 35-40; instr agron & jr agronomist, 40-41, asst prof & asst agronomist, 41-45, assoc prof & assoc agronomist, 45-51, chmn dept agron, 59-70, PROF AGRON & AGRONOMIST, EXP STA, UNIV CALIF, DAVIS, 51-, CHMN GRAD PROG ECOL, 71- Concurrent Pos: Spec lectr, McGill Univ, 38; Can Dept Agr del, Int Genetics Cong, Scotland, 39; organizer, Cytogenetic Lab, Brazilian Ministry Agr, Rio Grande do Sul, 48-49; Fulbright res scholar, NZ & Australia, 56-57, Greece, 67; chmn range improvement adv comt, State Bd Forestry, Calif, 54-67; Rockefeller Found travel grant, 64; vis prof, Univ Ghana, 70-71 & Univ BC & Univ Guelph, 71; consult, Int Coop Admin, Govt Spain, 60, Food & Agr Orgn, Greece, 70 & Kenya Meat Comn, 71; mem panel resource technol, Nat Acad Sci-Nat Res Coun; mem ecol adv comt, Sci Adv Bd, US Environ Protection Agency, 74- Honors & Awards: Calouste Gulbenakian Award, Portugal, 67; Stevenson Award, Am Soc Agron, 52, Agronomic Serv Award, 66; Medallion Award, Am Forage & Grassland Coun, 66. Mem: Fel AAAS; Bot Soc Am; fel Am Soc Agron; Genetics Soc Am; Ecol Soc Am. Res: Cytogenetics of range forage crops and species; effect of grazing treatment on range species; interspecific hybridization of grasses; range plant improvement; ecology of grasslands. Mailing Add: 740 Miller Dr Davis CA 95616

LOVE, RUSSELL JACQUES, b Chicago, Ill, Jan 11, 31; m 61; c 2. SPEECH PATHOLOGY. Educ: Northwestern Univ, Ill, BS, 53, MA, 54, PhD(speech path), 62. Prof Exp: Speech & hearing therapist, Moody State Sch Cerebral Palsied Children, Tex, 54-56; staff clinician, Cerebral Palsy Speech Clin, Northwestern Univ, Ill, 58-61; audiologist, WSide Vet Admin Hosp, Chicago, Ill, 61-62; res speech pathologist, Vet Admin Hosp, Coral Gables, Fla, 62-64; assoc prof speech path, DePaul Univ, 64-67; asst prof, 67-69, ASSOC PROF SPEECH & LANG PATH, SCH MED, VANDERBILT UNIV, 69- Concurrent Pos: Consult speech pathologist, Michael Reese Hosp & Med Ctr, Chicago, Ill, 64-67; chief speech pathologist, Bill Wilkerson Hearing & Speech Ctr, Tenn, 67-71, consult & res speech pathologist, 71- Mem: Am Speech & Hearing Asn. Res: Organic speech and language disorders; speech and language development. Mailing Add: Hearing & Speech Sci Vanderbilt Univ Med Sch Nashville TN 37232

LOVE, SAMUEL HARRIS, b South Hill, Va, July 28, 27; m 53; c 3. MICROBIOLOGY. Educ: Univ Va, BA, 50; Miami Univ, MS, 51; Univ Pa, PhD(microbiol), 54. Prof Exp: Res assoc biochem, Mass Inst Technol, 54, fel, Nat Found Infantile Paralysis, 54-55; from instr to asst prof, 55-66, ASSOC PROF MICROBIOL & IMMUNOL, BOWMAN GRAY SCH MED, 66-, DIR BIOCHEM CORE LAB, 74- Concurrent Pos: USPHS sr res fel, 56-61; res assoc, Univ Pa, 59. Mem: AAAS; Am Soc Microbiol. Res: Purification and characterization of biologically active macromolecules. Mailing Add: Dept of Microbiol Bowman Gray Sch of Med Winston-Salem NC 27103

LOVE, THEODORE ARCEOLA, b Columbus, Ga, May 25, 09; m 43. MATHEMATICS. Educ: Talladega Col, AB, 29; Univ Mich, MA, 32; NY Univ, PhD, 51. Prof Exp: Dir instr & dean, Southern Norm Sch, 29-33; prof math & prin, Lab Schs, Ala State Col, 33-42, prof math, 47-48; prof & head dept, Tenn Agr & Indust State Col, 51-56; PROF MATH & CHMN DEPT, FISK UNIV, 56-, DIR SUMMER SESSION, 68- Mem: Am Math Soc; Math Asn Am. Res: Improved curriculum and instruction in undergraduate mathematics. Mailing Add: Dept of Math Fisk Univ Nashville TN 37203

LOVE, WARNER EDWARDS, b Philadelphia, Pa, Dec 1, 22; m 45; c 2. BIOPHYSICS. Educ: Swarthmore Col, BA, 46; Univ Pa, PhD(physiol), 51. Prof Exp: Asst instr physiol, Univ Pa, 48-49, fel biophys, Johnson Found, 51-53, assoc, 53-55; res asst physics, Inst Cancer Res, 55-56, res assoc, 56-57; from asst prof to assoc prof, 57-65, chmn dept, 72-75, PROF BIOPHYS, JOHNS HOPKINS UNIV, 65- Concurrent Pos: Phillips lectr, Haverford Col, 55; mem corp, Woods Hole Marine Lab. Mem: Soc Gen Physiol; Am Physiol Soc; Biophys Soc; Am Crystallog Asn; Am Soc Biol Chem. Res: Biological ultrastructural basis of functions; x-ray crystallography of macromolecules. Mailing Add: Thomas C Jenkins Dept of Biophys Johns Hopkins Univ Baltimore MD 21218

LOVE, WILLIAM ALFRED, b Pittsburgh, Pa, Aug 4, 32; m 57. PHYSICS. Educ: Carneige Inst Technol, BS, 54, MS, 55, PhD(physics), 58. Prof Exp: Res physicist, Carnegie Inst Technol, 58-59; fel, Nat Sci Found, European Orgn Nuclear Res, Switzerland, 59-60; from asst physicist to assoc physicsit, 60-66, PHYSICSIT, BROOKHAVEN NAT LAB, 66- Mem: Fel Am Phys Soc. Res: Particle physics. Mailing Add: Brookhaven Nat Lab Upton NY 11973

LOVE, WILLIAM F, b Houston, Tex, July 3, 25; m 51; c 3. SOLID STATE PHYSICS. Educ: Rice Inst, BS, 45, MA, 47, PhD, 49. Prof Exp: Instr physics, Randal Morgan Lab, Univ Pa, 49-52, asst prof, 52-54; from asst prof to assoc prof, 54-63, PROF PHYSICS, UNIV COLO, BOULDER, 63- Mem: Am Phys Soc. Res: Symmetry properties of crystals; galvanomagnetic properties of metals and semiconductors in high magnetic fields. Mailing Add: Dept of Physics Univ of Colo Boulder CO 80304

LOVE, WILLIAM GARY, b Meridian, Miss, Aug 16, 41; m 66; c 2. NUCLEAR PHYSICS. Educ: Univ Tenn, BS, 63, PhD(physics), 68. Prof Exp: Res assoc physics, Fla State Univ, 68-70; ASST PROF PHYSICS, UNIV GA, 70- Mem: Am Phys Soc. Res: Study of the properties of the nucleon-nucleon interaction as they are manifested in multi-nucleon systems, for example, in scattering. Mailing Add: Dept of Physics & Astron Univ of Ga Athens GA 30602

LOVE, WILLIAM JUNIOR, theoretical mechanics, applied mechanics, see 12th edition

LOVE, WILLIAM P, b Tallahassee, Fla, Apr 15, 39. MATHEMATICS EDUCATION. Educ: Fla State Univ, BS, 61, MS, 62, PhD(math educ), 69. Prof Exp: Aerospace technologist, Langley Res Ctr, NASA, 61-62; instr physics, Gibbs Jr Col, Fla, 62-65; instr math, New Col, 69-70; ASST PROF MATH, UNIV NC, GREENSBORO, 70- Mem: Math Asn Am. Res: Teacher education; computer assisted instruction. Mailing Add: Dept of Math Univ of NC Greensboro NC 27412

LOVECCHIO, FRANK VITO, b Syracuse, NY, Apr 30, 43. ANALYTICAL CHEMISTRY. Educ: Syracuse Univ, AB, 65, PhD(chem), 70. Prof Exp: Fel, Ohio State Univ, 70-73; RES CHEMIST, EASTMAN KODAK CO, 73- Mem: Am Chem Soc; Soc Photog Scientists & Engrs. Res: Reactions and mechanisms of coordination compounds, including electron transfer reactions. Mailing Add: Eastman Kodak Co Rochester NY 14650

LOVECCHIO, KAREN K, b Peekskill, NY, June 23, 42; m 68. ELECTROCHEMISTRY. Educ: State Univ NY Binghamton, BA, 64; Syracuse Univ, PhD(electrochem), 69. Prof Exp: Chemist, Syracuse Univ Res Corp, 69-70; ed, Chem Abstr Serv, 70-73; ANAL CHEMIST, EASTMAN KODAK CO, 74- Mem: Am Chem Soc; Soc Photog Scientists & Engrs. Res: Polarography; coulometry; non-aqueous titrations. Mailing Add: Eastman Kodak Co Kodak Park Bldg 34 Rochester NY 14650

LOVEJOY, DAVID ARNOLD, b Nashua, NH, Dec 12, 43; m 69. MAMMALIAN ECOLOGY. Educ: Univ Conn, BA, 65, PhD(zool, ecol), 70. Prof Exp: Asst prof, 70-71, ASSOC PROF BIOL, WESTFIELD STATE COL, 71- Mem: Am Soc Mammal; Ecol Soc Am. Res: Ecology of small mammals; Siphonapteran parasites of mammals. Mailing Add: Dept of Biol Westfield State Col Westfield MA 01085

LOVEJOY, DEREK R, b London, Eng, Jan 19, 28; Can citizen; m 53; c 3. SOLID STATE PHYSICS. Educ: Univ London, BS, 50; Univ Toronto, MA, 52, PhD(physics), 54. Prof Exp: Assoc res officer, Appl Physics Div, Nat Res Coun Can, 54-66; proj officer, Res Div, 66-72, SR TECH ADV, TECH ADV DIV, UN DEVELOP PROG, 72- Concurrent Pos: Expert thermal metrol, Nat Phys Lab Metrol Proj, Cairo, United Arab Repub, UNESCO, 64-65. Mem: Can Asn Physicists. Res: Liquid helium physics; temperature scales and measurements from very low to very high temperatures. Mailing Add: UN Develop Prog New York NY 10017

LOVEJOY, EARL MARK PAUL, b New York, NY, Oct 8, 25; m 49; c 5. STRUCTURAL GEOLOGY, GEOMORPHOLOGY. Educ: Rutgers Univ, BS, 49; Colo Sch Mines, MS, 51; Univ Ariz, PhD(geol), 64. Prof Exp: Explor geologist, Atomic Energy Comn, 53-55; pres, Sierra Mineral Ltd, 55-56; geologist, Isbell Construct Co, 56; valuation engr, US Dept Interior, 56-62; geologist, Climax Molybdenum Corp, 64; asst prof geol, Univ Ariz, 64-65; assoc prof, 65-75, PROF GEOL, UNIV TEX, EL PASO, 75- Mem: AAAS; Geol Soc Am; Am Asn Petrol Geol; Sigma Xi. Res: Basin Range Cenozoic structure and geomorphology; cordilleran geotectonics; arid lands geomorphology. Mailing Add: Dept of Geol Sci Univ of Tex El Paso TX 79912

LOVEJOY, ELWYN RAYMOND, b Nashua, NH, Aug 7, 27; m 62; c 2. POLYMER CHEMISTRY. Educ: Boston Univ, AB, 49; Univ Ill, MS, 50, PhD(chem), 53. Prof Exp: RES CHEMIST, E I DU PONT DE NEMOURS & CO, INC, 53- Mem: Sci Res Soc Am; Am Chem Soc. Res: Polymers; fluorocarbon and high temperature resins. Mailing Add: Highland Meadows RD 2 Box 308 Hockessin DE 19707

LOVEJOY, OWEN, b Paducah, Ky, Feb 11, 43; m 69. HUMAN BIOLOGY, BIOMECHANICS. Educ: Western Reserve Univ, BA, 65; Case Inst Technol, MA, 67; Univ Mass, Amherst, PhD(human biol), 70. Prof Exp: ASSOC PROF PHYS ANTHROP, KENT STATE UNIV, 69-; ASST CLIN PROF, DIV ORTHOP SURG, SCH MED, CASE WESTERN RESERVE UNIV, 70- Mem: Brit Soc Study Human Biol; Am Asn Phys Anthrop; Am Eugenics Soc. Res: Primate anatomy, biomechanics and taxonomy; human palaeontology and palaeodemography; skeletal biology. Mailing Add: Dept of Anthrop Kent State Univ Kent OH 44242

LOVEJOY, ROLAND WILLIAM, b Portland, Ore, June 18, 31; m 59; c 2. PHYSICAL CHEMISTRY. Educ: Reed Col, BA, 55; Wash State Univ, PhD(chem), 60. Prof Exp: Fel chem, Univ Wash, 59-62; asst prof, 62-67, ASSOC PROF CHEM, LEHIGH UNIV, 67- Mem: Am Phys Soc. Res: Analysis of vibration-rotation spectra of inorganic and organic molecules using infrared and Raman spectroscopy; applications of far-infrared interferometry. Mailing Add: Dept of Chem Lehigh Univ Bethlehem PA 18015

LOVELACE, ALAN MATHIESON, organic chemistry, see 12th edition

LOVELACE, C JAMES, b Holdenville, Okla, Sept 26, 34; m 72; c 2. PLANT PHYSIOLOGY, BIOCHEMISTRY. Educ: Harding Col, BS, 61; Utah State Univ, MS, 64, PhD(plant physiol), 66. Prof Exp: PROF BOT, HUMBOLDT STATE UNIV, 65- Mem: Am Soc Plant Physiol. Res: Air pollution; fluoride research in relation to enzyme reactions within plants; heavy metal toxicants; organic fluoride biosynthesis in plants; chlorophyl biosynthesis. Mailing Add: Dept of Biol Humboldt State Univ Arcata CA 95521

LOVELACE, CLAUD WILLIAM VENTON, b London, Eng, Jan 16, 34. THEORETICAL PHYSICS. Educ: Univ Capetown, BS, 54. Prof Exp: Dept Sci & Indust Res res fel, Imp Col, Univ London, 61-62, lectr physics, 62-65; sr physicist, Europ Orgn Nuclear Res, Geneva, 65-71; PROF PHYSICS, RUTGERS UNIV, NEW BRUNSWICK, 70- Res: Theoretical particle physics; strong interactions; high energy phenomenology. Mailing Add: Dept of Physics Rutgers Univ New Brunswick NJ 08903

LOVELACE, LOLLIE ROBERTA, zoology, cytology, see 12th edition

LOVELAND, DONALD WILLIAM, b Rochester, NY, Dec 26, 34. MATHEMATICS, COMPUTER SCIENCE. Educ: Oberlin Col, AB, 56; Mass Inst Technol, SM, 58; NY Univ, PhD(math), 64. Prof Exp: Mathematician & programmer, Int Bus Mach Corp, 58-59; instr math, NY Univ, 63-64, asst prof, 64-67; from asst prof to assoc prof, Carnegie-Mellon Univ, 67-73; PROF COMPUT SCI & CHMN DEPT, DUKE UNIV, 73- Mem: Am Math Soc; Asn Comput Mach; Asn Symbolic Logic; AAAS. Res: Artificial intelligence; theorem proving by computer; foundational study of the notion of random sequence; mathematical theory and hierarchies of computational complexity. Mailing Add: Dept of Comput Sci Duke Univ Durham NC 27706

LOVELAND, J WEST, physical chemistry, analytical chemistry, see 12th edition

LOVELAND, ROBERT EDWARD, b Camden, NJ, May 3, 38; m 62; c 3. BIOLOGY. Educ: Rutgers Univ, Camden, AB, 59; Harvard Univ, MA, 61, PhD(biol), 63. Prof Exp: Asst prof biol, Long Beach State Col, 63-64; asst prof zool, 64-70, ASSOC PROF ZOOL, RUTGERS UNIV, NEW BRUNSWICK, 70- Concurrent Pos: NSF sci fac fel, Univ BC, 71-72. Mem: AAAS; Am Soc Zool; Am Inst Biol Sci. Res: Distribution of marine invertebrates; chemoreception in isopods; mathematical models of biological systems. Mailing Add: Dept of Zool Rutgers Univ New Brunswick NJ 08904

LOVELAND, WALTER (DAVID), b Chicago, Ill, Dec 23, 39; m 62. NUCLEAR CHEMISTRY. Educ: Mass Inst Technol, SB, 61; Univ Wash, PhD(chem), 66. Prof Exp: Res assoc chem, Argonne Nat Lab, 66-67; res asst prof, 67-68, asst prof, 68-74, ASSOC PROF CHEM, ORE STATE UNIV, 74- Concurrent Pos: US Atomic Energy Comn res grant, Ore State Univ, 68- Mem: AAAS; Am Phys Soc; Am Chem Soc. Res: Nuclear reactions, especially fission; activation analysis; use of computers for data acquisition; environmental chemistry. Mailing Add: Radiation Ctr Ore State Univ Corvallis OR 97331

LOVELESS, CHARLES M, ecology, biometry, see 12th edition

LOVELESS, FREDERICK CHARLES, organic chemistry, see 12th edition

LOVELESS, LOYAL E, b Trent, Tex, Feb 27, 23; m 44; c 4. CLINIL CHEMISTRY. Educ: Univ Tex, BA, 46, MA, 48, PhD(biochem), 51. Prof Exp: Res scientist, Biochem Inst, Univ Tex, 48-50; res chemist, Mound Lab, Monsanto Chem Co, 51-53, group leader, Microbiol Group, 53-56, Animal & Plant Nutrit, Inorg Res Dept, 56-60 & Animal Nutrit & Animal Health, 60-62; pres, Lab Exp Biol, 62-72; VPRES & GEN MGR, SMITH KLINE CLIN LABS, SAN FRANCISCO, 72- Mem: Am Chem Soc; Am Soc Microbiol; Inst Food Technol; Am Asn Clin Chem; Asn Clin Sci. Res: Animal toxicology; forensic toxicology; clinical chemistry; new drugs; food and feed additives; agricultural pesticides. Mailing Add: 271 La Questa Woodside CA 94062

LOVELETTE, CHARLES ALAN, organic chemistry, see 12th edition

LOVELL, CALVIN, organic chemistry, see 12th edition

LOVELL, DONALD JOSEPH, b Racine, Wis, June 8, 22; m 53; c 1. OPTICS. Educ: Univ Wis, PhB & MS, 47. Prof Exp: Asst math & physics, Univ Wis, 46-47; instr physics, Norwich Univ, 47-48; physicist, Naval Res Lab, 49-52; engr, Air Arm Div, Westinghouse Elec Corp, 52-53; proj engr, Photoswitch, Inc, 54-56; head res group, Barnes Eng Co, 56-58; prin engr optics, Prod Div, Bendix Corp, 59-61; res physicist, Univ Mich, 61-68; sr res physicist, Astron Res Facility, Univ Mass, Amherst, 68-69; PROF OPTICS, MASS COL OPTOM, 69-; OPTICAL CONSULT, 73- Concurrent Pos: Lectr, Univ Conn, 57-58; consult, Inst Defense Anal, 63-69. Mem: Optical Soc Am; Soc of Photo-Optical Instrumentation Engr. Res: Infrared physics; optical instruments; history of optics. Mailing Add: Barton Road Stow MA 01775

LOVELL, EDWIN LISTER, b Vancouver, BC, Jan 18, 15; nat US; m 41; c 2. CHEMISTRY. Educ: Univ BC, BS, 35, MA, 37; McGill Univ, PhD(wood chem), 40. Prof Exp: Res assoc lignin & plastics, McGill Univ, 40-41; res assoc, Rayonier Inc, 41-48, asst dir res, 48-52, res mgr, 52-55, dir res, Rayonier Can Ltd, 55-60, res mgr, Olympic Res Div, Rayonier, Inc, 60-68, RES MGR, OLYMPIC RES DIV, ITT RAYONIER, INC, 68- Mem: Am Chem Soc; Tech Asn Pulp & Paper Indust; Can Pulp & Paper Asn. Res: Synthesis and characterization of pure high polymers; cellulose and wood chemistry. Mailing Add: Olympic Res Div ITT Rayonier Inc Shelton WA 98584

LOVELL, FREDERICK MAURICE, b Maesteg, SWales, Jan 6, 30; US citizen; m 57; c 3. X-RAY CRYSTALLOGRAPHY. Educ: Univ Wales, BSc, 52, PhD(physics, crystallog), 60. Prof Exp: Res fel chem, Univ Leeds, Eng, 59-60; res fel, Univ Sydney, Australia, 60-63; res assoc, Colo Univ, Boulder, 63-64; asst prof biochem, Columbia Univ, 64-69; asst prof physics, Trinity Univ, San Antonio, 69-70; SR X-RAY CRYSTALLOGR, LEDERLE LABS, DIV AM CYANAMID CO, 70- Mem: Am Crystallog Asn. Res: X-ray crystal structure analysis of compounds of pharmaceutical interest. Mailing Add: Lederle Labs Pearl River NY 10965

LOVELL, HAROLD LEMUEL, b Bellwood, Pa, July 13, 22; m 44; c 2. FUEL SCIENCE, MINERAL ENGINEERING. Educ: Pa State Univ, BS, 43, MS, 45, PhD(fuel tech), 52. Prof Exp: Asst chem micros, Pa State Univ, 43-44, microchem, 44-45; res chemist, Mallinckrodt Chem Works, 45-47; asst fuel technol, 47-51, res assoc spectros, 51-52, asst prof, 52-58, mineral prep, 58-64, assoc prof, 64-71, actg head dept mineral prep, 64-68, ASSOC PROF MINERAL ENG, PA STATE UNIV, 71-, DIR MINE DRAINAGE RES SECT, 68- Concurrent Pos: Consult, US Dept Com-Com Tech Adv Bd, 74-75. Mem: Am Chem Soc; Am Inst Mining, Metall & Petrol Eng; Am Soc Test Mat. Res: Mineral preparation; analytical chemistry; absorption and emission; microchemistry; coal constitution chemistry; chemical utilization of coal; mine water pollution-treatment; coal preparation. Mailing Add: 109 Mineral Industry Bldg Pa State Univ University Park PA 16802

LOVELL, JAMES BYRON, b Fallentimber, Pa, Mar 19, 27; m; c 2. ENTOMOLOGY. Educ: Pa State Univ, BS, 50; Univ Ill, MS, 55, PhD(entom), 56. Prof Exp: Entomologist, US Army Chem Ctr, Md, 50-53; asst, Univ Ill, 53-56; RES ENTOMOLOGIST, AGR DIV, AM CYANAMID CO, 56- Mem: AAAS; Entom Soc Am. Res: Insect physiology and toxicology; mode of action of insecticides; mechanism of resistance in insects. Mailing Add: Agr Div Am Cyanamid Co Princeton NJ 08540

LOVELL, JAMES EDGELEY, b Ft Atkinson, Wis, Oct 9, 23; m 46; c 2. VETERINARY ANATOMY. Educ: Iowa State Univ, DVM, 46, MS, 55, PhD(vet anat), 58. Prof Exp: Pvt vet practice, 46-51; instr vet anat, Vet Col, Iowa State Univ, 53-55, asst prof, Vet Obstet & Radiol, 55-58, assoc prof, 58-60, animal histol, Vet Med Res Inst, 60-67, head dept, 67-73, PROF VET BIOL STRUCT, COL VET MED, UNIV ILL, URBANA, 67- Concurrent Pos: NSF-NATO fel, Inst Reprod Physiol & Path, Vet Col Norway, 60-61. Mem: Am Vet Med Asn; Am Asn Vet Anat (pres); World Asn Vet Anat. Res: Histological study of canine skin; artificial insemination in swine; histochemical and ultrastructural investigations of reproductive organs of farm animals; comparison of normal and hypertrophied muscle. Mailing Add: Dept of Vet Biol Struct Univ of Ill Col of Vet Med Urbana IL 61801

LOVELL, JAMES F, b Lovell, Okla, June 9, 34; m 54; c 4. ENVIRONMENTAL BIOLOGY, SCIENCE ADMINISTRATION. Educ: Okla State Univ, BS, 56, MS, 58; Kans State Univ, PhD(biol sci), 64. Prof Exp: Teacher high sch, Okla, 56-58; assoc prof biol sci, Pa State Col, 63-66; ASSOC PROF ECOL & ENVIRON HEALTH, SOUTHWESTERN STATE UNIV, OKLA, 66-, CHMN DIV BIOL SCI, 67- Concurrent Pos: Coord basic & spec improv grants allied health professions, NIH; dir, Nat Sci Found Summer Inst Grants, 73 & 75; mem, State Pollution Control Bd, 75-; exec secy-treas, Okla Acad Sci, 71- Mem: Ecol Soc Am; Asn Schs Allied Health Professions; Am Inst Biol Sci; sci fel AAAS. Res: Ecological and net productivity studies; development of medical records administration, health care administration and nursing BS degree programs. Mailing Add: Div of Biol Sci Southwestern State Univ Weatherford OK 73096

LOVELL, RICHARD ARLINGTON, b Kentland, Ind, Aug 4, 30; m 65; c 5. BIOCHEMISTRY. Educ: Xavier Univ, Ohio, BS, 52, MS, 53; St Louis Univ, Lic Philos, 59; McGill Univ, PhD(biochem), 63. Prof Exp: Proj assoc physiol, Epilepsy Res Ctr, Univ Wis, 64-65; instr psychiat, 66-69, ASST PROF NEUROCHEM IN PSYCHIAT, UNIV CHICAGO, 69- Concurrent Pos: USPHS Psychopharmacol Res Training Prog fel psychiat, Yale Univ, 65-66; res fel, Schweppe Found, 68-71. Mem: AAAS; Am Soc Neurochem; Am Epilepsy Soc; Soc Neurosci. Res: Neurochemistry;

neuropharmacology; biochemical control mechanisms in the nervous system. Mailing Add: Pritzker Sch of Med Univ of Chicago Chicago IL 60637

LOVELL, RICHARD THOMAS, b Lockesburg, Ark, Feb 21, 34; m 63; c 2. FISHERIES. Educ: Okla State Univ, BS, 56, MS, 58; La State Univ, PhD(nutrit, biochem), 63. Prof Exp: From asst prof to assoc prof food sci, La State Univ, 63-69; assoc prof, 69-75, PROF FISHERIES & ALLIED AQUACULT, AUBURN UNIV, 75- Concurrent Pos: Consult fish cult, US AID, 72-74; columnist, Com Fish Farmer & World Aquacult, 74-; mem, Comt Animal Nutrit, Nat Res Coun-Nat Acad Sci, 74-; assoc ed, Trans Am Fisheries Soc, 75- Mem: Fel Am Inst Chemists; Am Fisheries Soc; Am Chem Soc; Inst Food Technologists. Res: Fish nutrition, especially vitamin C requirements and energy metabolism of warm water fish cultured for food; intensively cultured food fishes with geosmin-related off-flavor caused by microorganisms in the culture environment. Mailing Add: Dept of Fisheries & Allied Aquacult Auburn Univ Auburn AL 36830

LOVELL, ROBERT GIBSON, b Ann Arbor, Mich, May 13, 20; m 48; c 5. INTERNAL MEDICINE, ALLERGY. Educ: Univ Mich, MD, 44, AB, 57. Prof Exp: From instr to asst prof, 50-73, fac secy, 54-56, asst dean sch med, 57-59, CLIN PROF INTERNAL MED, SCH MED, UNIV MICH, ANN ARBOR, 73- Concurrent Pos: Consult physician, US Vet Admin Hosp, 54-55; consult, President's Comn Vet Pensions, US Air Force, 55; consult, Wayne County Gen Hosp, Mich, 59-; chmn med ed comt, St Joseph Mercy Hosp, 63- Mem: AMA; assoc Am Col Chest Physicians; fel Am Acad Allergy. Res: Use of medications and aerosol preparations in treatment of bronchial asthma. Mailing Add: 3000 Geddes Ave Ann Arbor MI 48104

LOVELL, STUART ESTES, b Seattle, Wash, Oct 8, 28; m 55; c 2. Educ: Univ Wash, BS, 53; Brown Univ, PhD(chem), 58. Prof Exp: Proj assoc chem, Univ Wis, 58-63, asst prof comput sci, 63-67, mgr computer serv, 65-75, SYSTEM ANALYST, KITT PEAK OBSERV, 75- Mem: Asn Comput Mach. Res: Systems programming; computer based systems. Mailing Add: PO Box 26732 Tucson AZ 85726

LOVELL, WILLIAM STUART, physical chemistry, see 12th edition

LOVELOCK, DAVID, b Bromley, Eng. MATHEMATICS, THEORETICAL PHYSICS. Educ: Univ Natal, BSc, 59, Hons, 60, PhD(math), 62, DSc, 74. Prof Exp: Res asst math, Univ Natal, 60-61; jr fel, Bristol Univ, 62-63, lectr, 63-69; assoc prof, 69-74, prof appl math, Univ Waterloo, Ont, 74; PROF MATH, UNIV ARIZ, 74- Concurrent Pos: Nat Res Coun Can grant, Univ Waterloo, 69-, adj prof 25appl math, 74- Mem: Am Math Soc; Tensor Soc. Res: General relativity; calculus of variations; differential geometry. Mailing Add: Dept of Math Univ of Tucson AZ 85721

LOVEN, ANDREW WITHERSPOON, physical chemistry, surface chemistry, see 12th edition

LOVENBERG, WALTER MCKAY, b Trenton, NJ, Aug 9, 34; m 58; c 2. BIOCHEMISTRY. Educ: Rutgers Univ, BS, 56, MS, 58; George Washington Univ, PhD(biochem), 62. Prof Exp: Biochemist, 59-72, trainee, 62-63, HEAD SECT BIOCHEM PHARMACOL, HYPERTENSION-ENDOCRINE BR, NAT HEART & LUNG INST, 72- Mem: Am Soc Biol Chem; Am Soc Pharmacol & Exp Therapeut; Biochem Soc; Am Soc Neurochem. Res: Enzymatic mechanisms and the chemistry of proteins involved in neurohumoral amine biosynthesis. Mailing Add: Rm 7N262 Clin Ctr Nat Inst Health Bethesda MD 20014

LOVENBURG, MERVIN FRANK, geology, marine geology, see 12th edition

LOVER, MYRON JORDON, physical chemistry, organic chemistry, see 12th edition

LOVERING, EDWARD GILBERT, b Winnipeg, Man, Oct 15, 34; m 58; c 3. PHARMACEUTICAL CHEMISTRY. Educ: Univ Man, BSc, 57, MSc, 58; Univ Ottawa, PhD(chem), 61. Prof Exp: Sci officer radiation chem, Defense Res Bd, 58-59; Nat Res Coun Can fel, Oxford Univ, 61-63; res chemist, Polymer Corp Ltd, 63-69, assoc scientist polymer chem, 69-71; RES SCIENTIST POLYMER CHEM, CAN DEPT NAT HEALTH & WELFARE, 71- Mem: Am Chem Soc; Chem Inst Can; Acad Pharmaceut Sci. Res: Correlations between bioavailability and physical properties of drug formulations; biomedical and health aspects of polymers; pharmaceutical analysis. Mailing Add: Can Health Protection Br Health & Welfare Tunney's Pasture Ottawa ON Can

LOVERING, THOMAS SEWARD, b St Paul, Minn, May 12, 96; m 19; c 1. ECONOMIC GEOLOGY. Educ: Univ Minn, EM, 22, MS, 23, PhD(econ geol), 24. Prof Exp: Instr geol, Univ Minn, 22-24; instr, Univ Ariz, 24-25; jr geologist, US Geol Surv, 25-26, from asst geologist to geologist, 26-46, br res geologist, 47-54, chief geochem explor sect, 54-58, div res scientist, 58-65; PROF GEOL & CONSULT SCI & ENG, UNIV ARIZ, 66- Concurrent Pos: Mem, Yale Exped, Nfld & Labrador, 20; prof, Univ Mich, 34-42, 46-47; consult remote sensors, NASA; AID tech adv & consult, Govt Mex, 62; UN consult, Mineral Explor Seminar, Mex City, 64; geologist, US Geol Surv, Colo; mem comt terrestrial resources & future man, Nat Acad Sci; lectr, Am Inst Mining, Metall & Petrol Eng, 65; US del, Int Geol Cong, Algiers, 52, Centenary Cong Mineral Sci & Indust, France, Paris, 55, Inst Geol Cong, Mex City, 56, Pakistan Sci Cong, West Pakistan, 57, US rep mineral resources, Econ Comn Asia & Far East, Bangkok, 54, Japanese-Am energy panel, Joint US-Japan Coop Develop & Utilization Nat Resources, 65. Honors & Awards: US Dept Interior Distinguished Serv Gold Medal, 59; Penrose Gold Medal, Soc Econ Geol, 65; Jackling Award, Am Inst Mining, Metall & Petrol Eng, 65. Mem: Nat Acad Sci; fel Geol Soc Am (past pres); fel Mineral Soc Am; Asn Explor Geochem; Soc Environ Geochem & Health. Res: Mathematics of heat conduction and model experiments; petrology of rock alteration; geochemical exploration; structures; geology of Precambrian; mining geology; rock failure; physiography; biochemistry and geochemistry of weathering; mineral economics; stable isotopes in altered rocks. Mailing Add: Dept of Geosci Univ of Ariz Tucson AZ 85721

LOVESTEDT, STANLEY ALMER, b Iliff, Colo, June 7, 13; m 40; c 3. ORAL SURGERY. Educ: Univ Southern Calif, BS & DDS, 38; Univ Minn, MS, 45; Am Bd Oral Surg, dipl. Prof Exp: Asst, Mayo Grad Sch Med, Univ Minn, 41-43, consult, 43-46, from instr to assoc prof oral surg, 60-69; consult, 43-62, head dept, 55-62, SR CONSULT, DEPT DENT & ORAL SURG, MAYO CLIN, 62-; CLIN PROF ORAL SURG, MAYO MED SCH, UNIV MINN, 69- Mem: AAAS; Am Soc Oral Surg; Am Dent Asn; Am Acad Dent Radiol; fel Am Col Dent (pres, 69). Res: Radiology; oral medicine. Mailing Add: Mayo Clin 200 First St SW Rochester MN 55901

LOVETT, EVA G, b Orange, NJ, Aug 17, 40; m 63. ORGANIC CHEMISTRY. Educ: Douglass Col, Rutgers Univ, BA, 62; Univ Rochester, PhD(chem), 66. Prof Exp: Sr chemist, Merck, Sharp & Dohme Res Lab, 66-67; RES ASSOC CHEM, WASH UNIV, ST LOUIS, 69- Mem: The Chem Soc. Res: Synthesis, degradation and mass spectroscopy of natural products, particularly purines, pyrimidines and related heterocyclic compounds. Mailing Add: Dept of Chem Wash Univ St Louis MO 63130

LOVETT, JACK R, b Logan, Utah, Nov 29, 32; m 60; c 4. ACOUSTICS, PHYSICAL OCEANOGRAPHY. Educ: Brigham Young Univ, BA, 58. Prof Exp: Physicist, US Naval Ord Test Sta, 58-67, PHYSICIST, NAVAL UNDERSEA CTR, 67- Mem: Acoust Soc Am; Am Geophys Union; Sigma Xi; Sci Res Soc NAm. Res: Underwater sound propogation and attenuation; turbulence theory and effect on propogation; instrumentation; salinity; temperature; pressure; sound velocity. Mailing Add: Naval Undersea Ctr Code 409 San Diego CA 92132

LOVETT, JAMES SATTERTHWAITE, b Fallsington, Pa, Aug 22, 25; m 46; c 1. BOTANY. Educ: Earlham Col, AB, 53; Mich State Univ, PhD, 59. Prof Exp: Fel bot, Mich State Univ, 59-60; from asst prof to assoc prof, 60-69, PROF MYCOL, PURDUE UNIV, 69-, ASSOC HEAD DEPT OF BIOL SCI, 75- Concurrent Pos: Nat Sci Found sr fel, 66-67; Europ Molecular Biol Orgn fel, 71; assoc ed, Exp Mycol, 76-79; ed bd, J of Bacteriol, 75-77. Mem: AAAS; Am Soc Microbiol; Soc Develop Biol; Bot Soc Am; Mycol Soc Am. Res: Physiology of fungi; genetic and metabolic control of development in the lower fungi, principally the aquatic Phycomycetes. Mailing Add: Dept of Biol Sci Purdue Univ West Lafayette IN 47907

LOVETT, JOHN ROBERT, b Norristown, Pa, June 17, 31; m 56; c 3. PETROLEUM CHEMISTRY, LUBRICATION ENGINEERING. Educ: Ursinus Col, BS, 53; Univ Del, MS, 55, PhD(chem), 57. Prof Exp: Res chemist polymer processes, Esso Res & Eng Co, 57-59, proj leader high energy propellants, 59-60, sr chemist, 60-61, sect head, 61-64, asst dir process res div & head govt res dept, 65-66, dir govt res lab, 66-68; dir, Enjay Additives Lab, 68-69; mgr, Additive Chem, Worldwide, Esso Res & Eng Co, 69-70; vpres paramins dept, Enjay Chem Co, 70-73; TECHNOL MGR, EXXON CHEM CO, 73- Mem: AAAS; Am Chem Soc; Soc Automotive Engrs; Am Petrol Inst. Res: Synthetic organic chemistry; research administration. Mailing Add: Exxon Chem Co PO Box 536 Linden NJ 07036

LOVETT, JOSEPH, b Columbus Co, NC, Feb 24, 33; m 57; c 2. ENVIRONMENTAL HEALTH, PUBLIC HEALTH. Educ: Wake Forest Univ, BS, 56; Univ NC, Chapel Hill, MSPH, 60; Univ Minn, Minneapolis, MS, 65, PhD(environ health, microbiol), 71. Prof Exp: Asst supt water treat, City of Raleigh, NC, 57-58; regional consult, Interstate Carrier Prog Environ Eng & Food Protection, USPHS, 60-64, CHIEF MYCOL SECT, FOOD MICROBIOL BR, DIV MICROBIOL, BUR FOODS, FOOD & DRUG ADMIN, USPHS, 66- Mem: AAAS; Am Soc Microbiol; Am Dairy Sci Asn; Am Pub Health Asn. Res: Toxic microbial metabolites in foods and the ecology of toxigenic microorganisms. Mailing Add: Food Microbiol Br Food & Drug Admin Cincinnati OH 45226

LOVETT, PAUL SCOTT, b Philadelphia, Pa, Dec 14, 40; m 64; c 1. MICROBIOLOGY. Educ: Delaware Valley Col, BS, 64; Temple Univ, PhD(microbiol), 68. Prof Exp: USPHS fel microbiol, Scripps Clin & Res Found, Calif, 68-70; asst prof, 70-74, ASSOC PROF BIOL SCI, UNIV MD, BALTIMORE COUNTY, 74- Mem: Am Soc Microbiol. Res: Microbial genetics; mechanisms of bacteriophage infection; bacillus plasmids. Mailing Add: Dept of Biol Sci Univ of Md Baltimore County Catonsville MD 21228

LOVETT, WILLIAM EDWIN, organic chemistry, see 12th edition

LOVETT-DOUST, JOHN WILLIAM, b London, Eng, July 9, 14; nat Can; m 47; c 5. PSYCHIATRY. Educ: Univ London, BSc, 39, MB & BS, 42; FRCP(C), 55. Prof Exp: Res internal med, King's Col Hosp, London, 42-43; pathophysiologist, Chem Defence Exp Sta, Eng, 44-45; sr res psychiat, Maudsley & Bethlem Royal Hosps, 47-49; sr lectr psychophys rels, Inst Psychiat, Univ London, 51-52; PROF PSYCHIAT, UNIV TORONTO, 52-; CHMN DRUG RES UNIT, 58-; DIR RES UNIT, ONT HOSP, TORONTO, 53- Concurrent Pos: Nuffield fel, Maudsley Hosp, 49-50 & Med Ctr, Cornell Univ, 50-51; consult physician, Maudsley & Bethlem Royal Hosps, 51-52 & Prov Ont, 57-; mem panel psychiat res, Defence Res Bd, Can, 56- & Psychiat Res Comt, Dept Health, Ont, 58- Mem: Am Psychosom Soc; Am Fedn Clin Res; Can Psychiat Asn; Royal Soc Med; Brit Med Asn. Res: Biological aspects of psychiatry; physiology of the emotions, psychosomatic medicine; systems of oriental medicine; biological rhythms. Mailing Add: Clarke Inst of Psychiat 250 College St Toronto ON Can

LOVETTE, MARIBETH, b Bradford, Pa, June 8, 49. OPTICS. Educ: Pa State Univ, BS(physics) & BS(math), 71, MS, 74, PhD(physics), 76. Prof Exp: RES ENGR SPECTRORADIOMETRY, GTE SYLVANIA, GEN TEL & ELECTRONICS CORP, 76- Mem: Optical Soc Am. Res: Spectroradiometry; photometry; color. Mailing Add: GTE Sylvania 100 Endicott St Danvers MA 01923

LOVINGOOD, JUDSON ALLISON, b Birmingham, Ala, July 18, 36; m 55; c 4. MATHEMATICS, ELECTRICAL ENGINEERING. Educ: Univ Ala, BSEE, 58, PhD(math), 68; Univ Minn, MS, 63. Prof Exp: Assoc engr, Martin Co, 58-59; res engr, Honeywell Inc, 59-62; aerospace engr, 62-64, dep chief astrodyn guid theory div, 64-69, chief dynamics & control div, Aero-Astrodyn Lab, 69-74, DIR SYSTS DYNAMICS LAB, MARSHALL SPACE FLIGHT CTR, NASA, 74- Concurrent Pos: Asst prof, Univ Ala, Huntsville, 68- Mem: Am Inst Aeronaut & Astronaut. Res: Optimal and adaptive control theory research applications to launch and space vehicles; mathematical research in guidance theory, control theory and celestial mechanics. Mailing Add: Dynamics & Control Div NASA-Marshall Space Flight Ctr Huntsville AL 35812

LOVINGOOD, PAUL EVANS, JR, b Sylva, NC, Dec 21, 30; m 55; c 3. GEOGRAPHY, GEOLOGY. Educ: Univ NC, AB, 56, MA, 58, PhD(geog), 63. Prof Exp: Assoc prof geog, Clarion State Col, 59-60; assoc prof, Appalachian State Teachers Col, 60-61; assoc prof, 61-73, PROF GEOG, UNIV SC, 73- Concurrent Pos: Fulbright lectr, Univ Botswana, Lesotho & Swaziland, 66-67; mem nat adv bd, Inst Cultural Exchange through Photog, 66- Mem: Asn Am Geog; African Studies Asn. Res: Relation of agricultural land use to the physical environment; patterns of land use; computer mapping and field methods. Mailing Add: Dept of Geog Univ of SC Columbia SC 29208

LOVINS, ROBERT E, b Ashgrove, Mo, Sept 25, 35; m 56; c 1. MASS SPECTROMETRY, PROTEIN BIOCHEMISTRY. Educ: Univ Calif, Riverside, AB, 58; San Jose State Col, MS, 61; Univ Calif, Davis, PhD(chem), 63. Prof Exp: Lectr, chem & res chemist, Univ Calif, 63-65; res assoc chem, Mass Inst Technol, 65-66, asst dir, Mass Spectrometry Lab, 66-69; ASSOC PROF BIOCHEM & DIR HIGH RESOLUTION MASS SPECTROMETRY CTR, UNIV GA, 69- Concurrent Pos: NIH Res Career Develop Award, 71- Mem: Am Soc Biol Chem; Am Soc Mass Spectrometry. Res: Application of mass spectrometry to the sequence analysis of heterogeneous proteins; structural heterogeneity of antibody proteins. Mailing Add: Dept of Biochem Univ of Ga Athens GA 30601

LOVRIEN, REX EUGENE, b Eagle Grove, Iowa, Jan 25, 28; m 56; c 2. PHYSICAL BIOCHEMISTRY. Educ: Univ Minn, BS, 53; Univ Iowa, PhD, 58. Prof Exp: Res assoc phys chem, Yale Univ, 58-61; asst prof, 65-69, ASSOC PROF PHYS BIOCHEM, UNIV MINN, 69- Mem: Am Chem Soc; Biophys Soc. Res:

Macromolecular biochemistry; solution physical chemistry; light energy utilization. Mailing Add: Dept of Biochem Gortner Lab Col of Biol Sci Univ of Minn St Paul MN 55101

LOVSHIN, LEONARD LOUIS, JR, b Rochester, Minn, Mar 21, 42; m 73; c 1. FISHERIES MANAGEMENT. Educ: Miami Univ, BA, 64; Univ Wis, MS, 66; Auburn Univ, PhD(fisheries), 72. Prof Exp: ASSOC PROF FISHERIES, AUBURN UNIV, 72- Concurrent Pos: USAID-Auburn Univ proj coordr, Tech Assistance Prog Fisheries Develop, Ctr Ichthyol Res, Fortaleza, Brazil, 72- Mem: Am Fisheries Soc. Res: Fish culture research dealing with Tilapias, all male hybrid tilapias, native species indigenous to Brazil, and the extension of research results to local fish farmers. Mailing Add: Dept of Fisheries Auburn Univ Auburn AL 36830

LOVVORN, ROY LEE, b Woodland, Ala, Jan 24, 10; m 36; c 3. AGRONOMY, BOTANY. Educ: Auburn Univ, BS, 31; Univ Mo, AM, 33; Univ Wis, PhD(agron), 42. Prof Exp: County agent, Mo Agr Exten Serv, 33-34; assoc agronomist, Soil Conserv Serv, USDA, 35-36, head div weed invests, 50-53; asst agronomist, Exp Sta, NC State Univ, 36-39, assoc prof agron, 39-45, prof, 45-50, dir instr, Sch Agr & Life Sci, 53-54, dir res, 55-69; ADMINR, COOP STATE RES SERV, USDA, 69- Concurrent Pos: Agent, USDA, 45-50; consult res admin, Govt India, 63-64; mem agr res needs SAm, Nat Acad Sci, 64, Latin Am Sci Bd, 67, bd sci & technol for int develop. Mem: AAAS; Am Soc Agron. Res: Pasture management; environment in relation to productivity of some pasture species in the Coastal Plain. Mailing Add: Coop State Res Serv USDA Washington DC 20250

LOW, BARBARA WHARTON, b Lancaster, Eng, Mar 23, 20; nat US; m 50. BIOCHEMISTRY. Educ: Oxford Univ, BA, 42, MA, 46, DPhil(chem), 48. Prof Exp: Res assoc, Harvard Med Sch, 48, assoc phys chem, 48-50, asst prof, 50-56; assoc prof, 56-66, PROF BIOCHEM, COL PHYSICIANS & SURGEONS, COLUMBIA UNIV, 66- Concurrent Pos: Assoc mem, Lab Phys Chem, Harvard Univ, 50-54; vis prof, Univ Strasbourg, 65; mem biophys & biophys chem study sect, USPHS, 66-69. Mem: AAAS; Am Chem Soc; Am Soc Biol Chem; Am Crystallog Asn; Am Acad Arts & Sci. Res: X-ray crystal structure of non-enzyme proteins and peptides, particularly erabutoxin, transferrin, proinsulin and oxytocin; acetylcholinesterase structure; protein-protein interactions and initiation of quasi-ordered arrays; prediction of protein conformation; methods in x-ray analysis. Mailing Add: Columbia Univ Col Physicians & Surgeons 630 W 160th St New York NY 10032

LOW, BOBBI STIERS, b Louisville, Ky, Dec 4, 42. EVOLUTIONARY BIOLOGY, ECOLOGY. Educ: Univ Louisville, BA, 62; Univ Tex, Austin, MA, 64, PhD(evolutionary zool), 67. Prof Exp: Can Med Res Coun fel physiol, Univ BC, 67-69; Commonwealth Sci & Res Orgn res assoc ecol, Univ Melbourne & Univ S Australia, 69-72; asst prof, 72-75, ASSOC PROF RESOURCE ECOL, SCH NATURAL RESOURCES, UNIV MICH, ANN ARBOR, 75- Mem: AAAS; Am Soc Naturalists; Sigma Xi; Soc Study Evolution. Res: Evolution of life history strategies; herbivorous competition; reproductive ecology in arid environments. Mailing Add: Sch of Natural Resources Univ of Mich Ann Arbor MI 48104

LOW, CHARLES JAMES, b Hamilton, Ont, July 3, 41; m 66; c 2. AQUATIC ECOLOGY. Educ: Univ Guelph, BSc, 67; Univ BC, MSc, 71, PhD(aquatic biol), 75. Prof Exp: Biologist, Sask Dept Natural Resources, 73-75; BIOLOGIST, ENVIROCON LTD, 75- Res: Taxonomy and ecology of aquatic organisms. Mailing Add: Envirocon Ltd 3374 Stephenson Point Rd Nanaimo BC Can

LOW, DONALD GOTTLOB, veterinary medicine, see 12th edition

LOW, EMMET FRANCIS, JR, b Peoria, Ill, June 10, 22; div. APPLIED MATHEMATICS. Educ: Stetson Univ, BS, 48; Univ Fla, MS, 50, PhD(math), 53. Prof Exp: Instr phys sci, Univ Fla, 50-51, physics, 51-54; aeronaut res scientist, Nat Adv Comt Aeronaut, 54-55; asst prof math, Univ Miami, 55-59; vis res scientist, Courant Inst Math Sci, NY Univ, 59-60; assoc prof math & chmn dept, Univ Miami, 60-66, actg dean col arts & sci, 66-67, prof math & assoc dean faculties, 68-72; DEAN OF COL & PROF MATH, CLINCH VALLEY COL, UNIV VA, 72- Mem: AAAS; Am Math Soc; Soc Indust & Appl Math; Math Asn Am. Res: Stress and functional analysis. Mailing Add: Clinch Valley Col Univ of Va Wise VA 24293

LOW, FRANCIS EUGENE, b New York, NY, Oct 27, 21; m 48; c 3. THEORETICAL PHYSICS. Educ: Harvard Univ, BS, 42; Columbia Univ, AM, 47, PhD(physics), 49. Prof Exp: Instr physics, Columbia Univ, 49-50; mem, Inst Advan Study, 50-52; from asst prof to assoc prof physics, Univ Ill, 52-56; prof, 57-68, KARL COMPTON PROF PHYSICS, MASS INST TECHNOL, 68-, DIR CTR THEORETICAL PHYSICS, 74- Concurrent Pos: Consult, AEC, 55-; Loeb lectr, Harvard Univ, 59; Fulbright fel, 61-62; Guggenheim fel, 61-62. Mem: Nat Acad Sci; Am Acad Arts & Sci; fel Am Phys Soc. Res: Theoretical, atomic and nuclear physics; field theory. Mailing Add: Dept of Physics Mass Inst of Technol Cambridge MA 02139

LOW, FRANK JAMES, b Mobile, Ala, Nov 23, 33; m 56; c 3. SOLID STATE PHYSICS. Educ: Yale Univ, BS, 55; Rice Univ, MA, 57, PhD(physics), 59. Prof Exp: Mem tech staff, Tex Instruments, Inc, 59-62; assoc scientist, Nat Radio Astron Observ, WVa, 62-65; prof physics, Univ Ariz, 65-66, res assoc, Lunar & Planetary Lab, 64-65, RES PROF, LUNAR & PLANETARY LAB, UNIV ARIZ, 66- Concurrent Pos: Prof space sci, Rice Univ, 66-71; adj prof, 71-; pres, Infrared Labs, Inc, Ariz, 67- Honors & Awards: H A Wilson Award, Rice Univ, 59; Helen B Warner Prize, Am Astron Soc, 68; Tex Instruments Corp Award, 76. Mem: Nat Acad Sci; Am Phys Soc; Am Astron Soc. Res: Infrared astronomy; infrared physics; cryogenic engineering. Mailing Add: Lunar & Planetary Lab Univ of Ariz Tucson AZ 85721

LOW, FRANK NORMAN, b Brooklyn, NY, Feb 9, 11. ANATOMY. Educ: Cornell Univ, AB, 32, PhD(micros anat), 36. Prof Exp: From instr to asst prof, Univ NC, 37-45; assoc, Sch Med, Univ Md, 45; assoc prof, Sch Med, WVa Univ, 46; asst prof, Sch Med, Johns Hopkins Univ, 46-49; from assoc prof to prof, Sch Med, La State Univ, 49-64; Hill res prof, 64-73, Chester Fritz Distinguished prof, 75-77, RES PROF, SCH MED, UNIV N DAK, 73- Concurrent Pos: Charlton fel anat, Med Sch, Tufts Univ, 36-37; mem, Great Plains Regional Res Rev & Adv Comt, Am Heart Asn, 72-74. Mem: Electron Micros Soc Am; Am Asn Anat; Am Asn Hist Med; Am Soc Cell Biol. Res: Transmission and scanning electron microscopy; fine structure of lung; subarachnoid space development of connective tissues. Mailing Add: Dept of Anat Univ of NDak Sch of Med Grand Forks ND 58202

LOW, HANS, b Vienna, Austria, Oct 22, 21; nat US; m 49; c 3. ORGANIC CHEMISTRY. Educ: Marietta Col, BS, 50; Purdue Univ, MS, 52; St Louis Univ, PhD(chem), 59. Prof Exp: Instr German & Latin, Marietta Col, 47-50; res chemist, 52-67, group leader lubricant additives, 67-72, STAFF TECHNOLOGIST, CHEM RES & DEV, SHELL OIL CO, 72- Mem: Am Chem Soc, NY Acad Sci. Res: Petroleum solvents and lubricants; synthetic lubricants; lubricant additives; industrial toxicology; public health. Mailing Add: Chem Res & Develop Shell Oil Co One Shell Plaza Houston TX 77002

LOW, JAMES ALEXANDER, b Toronto, Ont, Sept 22, 25; m 52; c 3. MEDICINE. Educ: Univ Toronto, MD, 49; FRCS(C). Prof Exp: Clin teacher obstet & gynec, Univ Toronto, 55-65; PROF OBSTET & GYNEC & HEAD DEPT, QUEEN'S UNIV, ONT, 65- Mem: Soc Gynec Invest; Can Soc Clin Invest; Soc Obstet & Gynec Can. Res: Perinatal foetal environment; bladder function and control. Mailing Add: Dept of Obstet & Gynec Queen's Univ Kingston ON Can

LOW, JESSOP BUDGE, b Millville, Utah, Mar 30, 14; m 37; c 5. WILDLIFE ECOLOGY. Educ: Utah State Univ, BS, 37; Iowa State Univ, MS, 39, PhD(econ, zool), 41. Prof Exp: Asst, Iowa State Univ, 37-41; asst game technician, Ill State Natural Hist Surv, 41-43; asst prof in charge dept, Utah State Univ, 43-44; wildlife mgt biologist, Utah State Fish & Game Dept, 44-45; Utah Coop Wildlife Res Unit, 45-75; RETIRED. Concurrent Pos: Ecologist, W F Sigler Assoc Consult Firm, 74- Honors & Awards: Distinguished Serv Award, Dept Interior, US Fish & Wildlife Serv, 74. Mem: Wildlife Soc; Soc Range Mgt. Res: Wildlife research; conservation of natural resources. Mailing Add: Dept of Wildlife Sci Utah State Univ Logan UT 84322

LOW, KENNETH BROOKS, JR, b New Rochelle, NY, Jan 19, 36; m 60; c 2. GENETICS. Educ: Amherst Col, BA, 58; Univ Pa, MS, 60, PhD(molecular biol), 65. Prof Exp: Asst prof radiobiol, 68-71, asst prof radiobiol & microbiol, 71-73, ASSOC PROF RADIOBIOL & MICROBIOL, YALE UNIV, 73- Concurrent Pos: USPHS fel, Med Ctr, NY Univ, 66-68. Mem: Am Soc Microbiol. Res: Molecular genetics; genetic recombination and control. Mailing Add: Radiobiol Labs Yale Univ Med Sch New Haven CT 06510

LOW, LEONE YARBOROUGH, b Cushing, Okla, Aug 27, 35; m 57; c 2. MATHEMATICAL STATISTICS. Educ: Okla State Univ, BS, 56, MS, 58, PhD(math), 61. Prof Exp: Instr math, Univ Ill, 60-64; asst prof, 64-68, ASSOC PROF MATH, WRIGHT STATE UNIV, 68- Concurrent Pos: Nat Res Coun res assoc, Wright-Patterson AFB, 67-68; consult, Systs Res Lab, 71- Mem: AAAS; Inst Math Statist; Am Statist Asn; Sigma Xi; Biometric Soc. Res: Variance component models in the analysis of variance. Mailing Add: Dept of Math Wright State Univ Colonel Glenn Hwy Dayton OH 45431

LOW, LOH-LEE, b Kuala Lumpur, Malaysia, Jan 15, 48; m 73. FISHERIES. Educ: Univ Wash, BS, 70, MS, 72, PhD(fisheries), 74. Prof Exp: Fishery biologist, Univ Wash, 74; FISHERY BIOLOGIST, NORTHWEST FISHERIES CTR, NAT MARINE FISHERIES SERV, 74- Concurrent Pos: Consult, Food & Agr Orgn, UN, 75. Res: Fisheries population dynamics; computer modelling of fisheries systems; international fisheries management. Mailing Add: NW Fish Ctr Nat Marine Fish Serv 2725 Montlake Blvd E Seattle WA 98112

LOW, MANFRED JOSEF DOMINIK, b Karlsbad, Bohemia, June 18, 28; nat US; m 65. PHYSICAL CHEMISTRY. Educ: NY Univ, BA, 52, MS, 54, PhD(phys chem), 56. Prof Exp: Asst chem, NY Univ, 52-55; res chemist, Davison Chem Co Div, W R Grace & Co, 56-58; sr chemist, Texaco, Inc, 58-61; asst prof chem, Rutgers Univ, 61-67; assoc prof, 67-72, PROF CHEM, NY UNIV, 72- Mem: Am Chem Soc; Soc Appl Spectros; NY Acad Sci. Res: Chemisorption; heterogeneous catalysis; infrared spectra of surfaces; infrared emission spectroscopy; surface chemistry and physics; Fourier transform spectroscopy. Mailing Add: Dept of Chem NY Univ 4 Washington Place New York NY 10003

LOW, MARC E, b Ada, Okla, Sept 25, 35; m 57; c 2. MATHEMATICS. Educ: Okla State Univ, BS, 58, MS, 60; Univ Ill, PhD(math), 65. Prof Exp: Instr math, 64-65, asst prof, 65-71, ASSOC PROF MATH, WRIGHT STATE UNIV, 71-, ASST DEAN, COL SCI & ENG, 73- Mem: Math Asn Am; Am Math Soc. Res: Elementary and analytic number theory. Mailing Add: Dept of Math Wright State Univ Colonel Glenn Hwy Dayton OH 45431

LOW, MORTON DAVID, b Lethbridge, Alta, Mar 25, 35; m 59; c 3. NEUROPHYSIOLOGY. Educ: Queen's Univ, Ont, MD & CM, 60, MSc, 62; Baylor Univ, PhD(physiol), 66. Prof Exp: From instr to asst prof physiol, Baylor Univ Col Med, 65-68; ASSOC PROF MED, UNIV BC, 68-; DIR EEG, VANCOUVER GEN HOSP, 68- & SHAUGNESSEY HOSP, 71- Concurrent Pos: Fel anat, Queen's Univ, Ont, 61-62; fel physiol, Baylor Univ Col Med, 63-65; Med Res Coun Can grants, Univ BC, 68-; Mr & Mrs P A Woodward's Found grants, Vancouver Gen Hosp, 69- Mem: AAAS; Am EEG Soc; Can Soc EEG (secy, 70-72, pres, 72-74); Am Epilepsy Soc. Res: Neural basis of perception and performance; brain mechanisms in maintenance and disorders of consciousness. Mailing Add: Dept of EEG Vancouver Gen Hosp Vancouver BC Can

LOW, NIELS LEO, b Copenhagen, Denmark, Dec 16, 16; nat US; m 43; c 2. MEDICINE. Educ: Med Col SC, MD, 40. Prof Exp: Clin instr pediat, Marquette Univ, 46-53; res assoc neurol, Univ Ill, 54; assoc res prof pediat, Univ Utah, 56-58; asst prof neurol, 60-67, assoc prof clin neurol, 67-75, PROF CLIN NEUROL & CLIN PEDIAT, COL PHYSICIANS & SURGEONS, COLUMBIA UNIV, 75-; DIR PEDIAT, BLYTHEDALE CHILDREN'S HOSP, VALHALLA, NY, 67- Concurrent Pos: Fel, Columbia Univ, 55 & 58-59; consult, NIH. Mem: Assoc Am EEG Soc; fel Am Acad Neurol; fel Am Acad Pediat; Am Epilepsy Soc; Int Child Neurol Asn (pres, 75-). Res: Pediatric neurology; metabolic disease affecting brain of children. Mailing Add: Columbia Univ Col Physicians & Surgeons New York NY 10027

LOW, PHILIP FUNK, b Carmangay, Alta, Oct 15, 21; nat US; m 42; c 6. SOIL CHEMISTRY. Educ: Brigham Young Univ, BS, 43; Calif Inst Technol, MS, 44; Iowa State Univ, PhD(soil chem), 49. Prof Exp: Soil scientist, USDA, 49; asst prof soil chem, 49-52, assoc prof, 52-55, PROF SOIL CHEM, PURDUE UNIV, 55- Concurrent Pos: Sigma Xi res award, Purdue Univ, 60; distinguished vis award to Australia, 68; consult, Esso Prod Res Lab & US Army Cold Regions Res & Eng Lab. Honors & Awards: Achievement Award, Soil Sci Soc Am, 63. Mem: Soil Sci Soc Am (pres-elect, 71-72, pres, 72-73); fel Am Soc Agron; Clay Minerals Soc. Res: Physical and colloidal chemistry of soils. Mailing Add: Dept of Agron Purdue Univ West Lafayette IN 47906

LOW, ROBERT BURNHAM, b Greenfield, Mass, Sept 19, 40; m 67; c 1. PHYSIOLOGY. Educ: Princeton Univ, AB, 63; Univ Chicago, PhD(physiol), 68. Prof Exp: NIH fel biol, Mass Inst Technol, 68-70; asst prof physiol, 70-74, ASSOC PROF PHYSIOL & BIOPHYS, UNIV VT, 74- Concurrent Pos: NIH & Muscular Dystrophy res grants, Univ Vt. Mem: Am Soc Cell Biol. Res: Mammalian protein turnover; physiology and biochemistry of muscle; plasma membrane turnover; macrophage response to environment. Mailing Add: Dept of Physiol & Biophys Given Bldg E-211 Univ of Vt Burlington VT 05401

LOW, ROBERT JAMES, b Cincinnati, Ohio, Oct 29, 22; m 48; c 3. GEOPHYSICS. Educ: Harvard Univ, SB, 44; Columbia Univ, MBA, 48. Prof Exp: Purchasing agent, Miller & Carrell Mfg Co, Colo, 48-49; purchasing agent refractories sales, Denver Fire Clay Co, 49-50; admin officer, High Altitude Observ, Univ Colo, 50-56, exec officer, 56-60, lectr phys sci, Univ Colo, 58-59 & 64-65, bus mgr, Nat Ctr Atmospheric Res &

asst secy-treas, Univ Corp Atmospheric Res, 60-64, actg asst dean grad sch, Univ Colo, 64-65, asst dean, 65-66, proj coord unidentified flying objects study, 66-68, spec asst to vpres acad affairs, 68-69; vpres admin, 69-74, ASST TO PRES, PORTLAND STATE UNIV, 74- Concurrent Pos: Chmn, Rocky Mt Sci Coun, 59-61; consult, State Bd Control, Fla, 62-63. Mem: Fel AAAS; Am Astron Soc; Am Meteorol Soc; Am Geophys Union; NY Acad Sci. Res: Solar physics; solar-terrestrial relationships. Mailing Add: Portland State Univ PO Box 751 Portland OR 97207

LOW, ROGER DEAN, applied mathematics, see 12th edition

LOW, WILLIAM, b Vienna, Austria, Apr 25, 22; nat Can; m 48, 70; c 9. PHYSICS. Educ: Queen's Univ, Ont, BA, 46; Columbia Univ, MA, 47, PhD, 50. Prof Exp: Tutor physics, Queen's Univ, Ont, 45-46; asst, Columbia Univ, 46-50; instr, Hebrew Univ, Israel, 50-54, lectr, 54-55; res assoc, Univ Chicago, 55-56; assoc prof, 56-61, PROF PHYSICS, HEBREW UNIV, ISRAEL, 61- Concurrent Pos: Guggenheim fel, 63-64; ed, Physics Letters; chmn Israel comt, Int Union Radio Sci; rector, Jerusalem Col Technol, Israel. Honors & Awards: Morrison Award, NY Acad Sci, 56; Israel Prize Exact Sci, 61; Rothschild Prize Physics, 64. Mem: Am Phys Soc; NY Acad Sci; Phys Soc Israel (vpres, 58-60, pres, 60-61 & 70-72); Europ Phys Soc; Int Union Pure & Appl Physics. Res: Paramagnetic resonance in solids; microwave spectroscopy in gases; quantum electronics; electron density behind shock waves; light scattering from macromolecules. Mailing Add: Microwave Div Dept of Physics Hebrew Univ Jerusalem Israel

LOWANCE, FRANKLIN ELTA, b Monroe Co, WVa, Dec 29, 07; m 31; c 1. PHYSICS. Educ: Roanoke Col, BS, 27; Duke Univ, MA, 31, PhD(physics), 35. Prof Exp: Prof eng & physics, Edinburg Col, 33-35; assoc prof math & astron, Wofford Col, 35-38; head dept physics & eng, Centenary Col, 38-42; assoc prof physics, Ga Inst Tech, 41-43, prof, 45-49; res assoc, Harvard Univ, 44; mem staff, Radiation Lab, Mass Inst Technol, 45; tech dir, Naval Civil Eng Lab, 49-53; assoc tech dir, Naval Ord Test Sta, 53-54; dir res & vpres, Air Brake Div, Westinghouse Elec Corp, 55-58; vpres eng, Crosley Div, Avco Corp, 58-60; pres, Adv Tech Corp, 60-62; CONSULT & VPRES, MERD CORP, 62- Mem: Am Phys Soc; Nat Soc Prof Eng; Am Asn Physics Teachers. Res: Microwave propagation and beacons; acoustics; ferromagnetism; magnetothermoelectricity. Mailing Add: 41 Tierra Cielo Santa Barbara CA 93105

LOWDEN, J ALEXANDER, b Toronto, Ont, Feb 21, 33; m 56; c 4. NEUROCHEMISTRY. Educ: Univ Toronto, MD, 57; McGill Univ, PhD(biochem), 64. Prof Exp: Resident pediat, Hosp Sick Children, 58-60; res assoc, Univ Toronto, 65-67; assoc scientist, 64-74, ASSOC DIR, RES INST, HOSP SICK CHILDREN, 75- ; ASSOC PROF PEDIAT, UNIV TORONTO, 67- Concurrent Pos: Fel neurochem, Montreal Neurol Inst, 61-64; Helen Hay Whitney Found fel, 63-66. Mem: AAAS; Can Biochem Soc; Int Soc Neurochem; Am Soc Neurochem; Soc Pediat Res. Res: Chemistry and metabolism of brain lipids, especially gangliosides; structure and composition of the developing brain. Mailing Add: Res Inst Hosp Sick Children 555 University Ave Toronto ON Can

LOWDEN, WILLIAM HERBERT, bio-organic chemistry, see 12th edition

LOWDER, J ELBERT, b Pinedale, Wyo, Mar 18, 40; m 64; c 3. APPLIED PHYSICS. Educ: Univ Calif, Berkeley, BS, 63, MS, 65; Univ Calif, San Diego, PhD(eng physics), 71. Prof Exp: Flight test engr, Northrop Aircraft Corp, 63-64; proj engr, Aeronutronic Div, Philco-Ford Corp, 65-68; staff mem, 71-75, ASST GROUP LEADER APPL PHYSICS, LINCOLN LAB, MASS INST TECHNOL, 75- Mem: Optical Soc Am. Res: Effects of atmospheric aerosols on propagation of laser radiation; interaction of high power laser radiation with solid surfaces; laser radar applications; passive infrared detection systems. Mailing Add: Lincoln Lab MIT 244 Wood St Lexington MA 02173

LOWDIN, PER-OLOV, b Uppsala, Sweden, Oct 28, 16; m 60; c 4. THEORETICAL PHYSICS, QUANTUM BIOLOGY. Educ: Univ Uppsala, Fil Kand, 37, Fil Mag, 39, Fil Lic, 42, Fil Dr(theoret physics), 48. Prof Exp: Lectr math & physics, 42-48, asst prof theoret physics, 48-55, assoc prof quantum chem, 55-60, head dept, 60-74, PROF QUANTUM CHEM, UNIV UPPSALA, 60-; GRAD RES PROF CHEM & PHYSICS & DIR QUANTUM THEORY PROJ, UNIV FLA, 60- Concurrent Pos: Fel, Swiss Fed Inst Technol, 46; H H Wells Phys Lab, Univ Bristol, 49; vis prof & consult, Duke Univ, Univ Chicago, Mass Inst Technol & Calif Inst Technol, 50-60; ed-in-chief, Advances in Quantum Chem, Int J Quantum Chem. Mem: Swed Royal Soc Arts & Sci; Swed Royal Soc Sci; Norweg Acad Sci & Letters; Int Soc Quantum Biol (pres, 71-72); Int Acad Quantum Molecular Sci (vpres, (vpres, 68-). Res: Theoretical physics and chemistry; quantum theory of atoms, molecules and solid state; quantum genetics and pharmacology. Mailing Add: Quantum Theory Proj 365 Williamson Hall Univ of Fla Gainesville FL 32603

LOWE, CARL CLIFFORD, b West Salem, Ohio, Jan 1, 19; m 42; c 3. PLANT BREEDING. Educ: Colo Agr & Mech Col, BS, 48; Cornell Univ, MS, 50, PhD(plant breeding), 52. Prof Exp: From asst prof to assoc prof, 52-62, PROF PLANT BREEDING, NY STATE COL AGR & LIFE SCI, CORNELL UNIV, 62-, BIOMETRY, 70- Mem: Am Soc Agron. Res: Forage crops breeding. Mailing Add: NY State Col of Agr & Life Sci Cornell Univ Plant Breeding Dept Ithaca NY 14850

LOWE, CHARLES HERBERT, JR, b Los Angeles, Calif, Apr 16, 20; m 44; c 2. ZOOLOGY. Educ: Univ Calif, Los Angeles, AB, 43, PhD(zool), 50. Prof Exp: Consult, AEC, 47-50; instr, 50-53, from asst prof to assoc prof, 53-64, PROF ZOOL, UNIV ARIZ, 64- Mem: AAAS; Am Soc Ichthyol & Herpet; Ecol Soc Am; Soc Study Evolution; Soc Syst Zool. Res: Animal and plant ecology; systematics; evolution; vertebrate zoology. Mailing Add: Dept of Zool Univ of Ariz Tucson AZ 85721

LOWE, CHARLES SAMUEL, physical chemistry, see 12th edition

LOWE, CHARLES UPTON, b Pelham, NY, Aug 24, 21; m 55; c 4. PEDIATRICS. Educ: Harvard Univ, BS, 42; Yale Univ, MD, 45. Prof Exp: From intern to asst resident pediat, Children's Hosp, Boston, 45-46; resident, Mass Gen Hosp, 47; assoc prof pediat, Sch Med, State Univ NY Buffalo, 51-55, res prof, 55-65; prof, Col Med, Univ Fla, 65-68, dir human develop ctr, 66-68; SCI DIR, NAT INST CHILD HEALTH & HUMAN DEVELOP, 68- Concurrent Pos: Nat Res Coun fel, Med Sch, Univ Minn, 48-51; Buswell fel, Sch Med, State Univ NY Buffalo, 55; ed-in-chief, Pediat Res, 65-; John F Kennedy Mem Lectr, 66; Grover Powers Mem Lectr, 69. Honors & Awards: Super Serv Award, NIH, 71; Clifford G Grulee Award, Am Acad Pediat, 71. Mem: Soc Pediat Res; Soc Exp Biol & Med; Am Soc Exp Path; Am Pediat Soc; Am Soc Clin Invest. Res: Clinical and laboratory study of nutritional disease, including celiac and cystic fibrosis of the pancreas; relationship between adrenocortical steroids and nucleic acid metabolism; inborn errors of metabolism and parenteral fluid therapy. Mailing Add: Nat Inst of Child Health & Human Develop Bethesda MD 20014

LOWE, DONALD RAY, b Sacramento, Calif, Sept 22, 42; m 64; c 1. SEDIMENTOLOGY. Educ: Stanford Univ, BS, 64; Univ Ill, Urbana, PhD(geol), 67. Prof Exp: Instr geol, Univ Ill, Urbana, 67-68; res assoc, US Geol Surv, Calif, 68-70; asst prof, 70-73, ASSOC PROF GEOL, LA STATE UNIV, BATON ROUGE, 73- Mem: Geol Soc Am; Soc Econ Paleont & Mineral; Int Asn Sedimentologists. Res: Sediment transport systems; experimental sedimentology. Mailing Add: Dept of Geol La State Univ Baton Rouge LA 70803

LOWE, EDMUND WARING, chemistry, see 12th edition

LOWE, HARRY J, b Nogales, Ariz, Dec 21, 19; m 47; c 5. ANESTHESIOLOGY. Educ: Univ Ariz, BS, 44; Johns Hopkins Univ, SM, 45, MD, 49. Prof Exp: Assoc prof biochem, Univ Tex, Southwest Med Sch, 53-56; prin res scientist, Roswell Park Mem Inst, NY, 58-62; resident anesthesiol & dir hyperbaric med, Millard Fillmore Buffalo, 62-66; prof anesthesiol & chmn dept, Pritzker Sch Med, Univ Chicago, 66-73; PROF ANESTHESIOL, UNIV SOUTHERN CALIF, 73- Concurrent Pos: Am Cancer Soc fel, Johns Hopkins Univ, 49-52. Mem: AAAS; AMA; Am Chem Soc. Res: Quantitative automated administration of volatile anesthetics in closed circuit systems; acid-base regulation of physiological ventilation during anesthesia. Mailing Add: Rancho Los Amigos Hosp 7601 E Imperial Hwy Downey CA 90242

LOWE, IRVING J, b Woonsocket, RI, Jan 4, 29; m 53; c 2. SOLID STATE PHYSICS. Educ: Cooper Union, BEE, 51; Washington Univ, St Louis, PhD(physics), 57. Prof Exp: Fel, Sloan Found & res assoc physics, Washington Univ, St Louis, 56-58; asst prof, Univ Minn, 58-62; assoc prof, 62-66, PROF PHYSICS, UNIV PITTSBURGH, 66- Mem: Am Phys Soc. Res: Experimental and theoretical studies of the structure and behavior of solids using nuclear magnetic resonance techniques. Mailing Add: Dept of Physics Univ of Pittsburgh Pittsburgh PA 15260

LOWE, JACK IRA, b Fairmount, Ga, Dec 8, 27; m 57. MARINE ECOLOGY, TOXICOLOGY. Educ: Berea Col, AB, 50; Univ Ga, MS, 55. Prof Exp: Biologist, US Fish & Wildlife Serv, 57-61 & US Bur Com Fisheries, 61-70; aquatic biologist, 70-71, dep lab dir, 71-75, ASSOC DIR TECH ASSISTANCE, ENVIRON RES LAB, ENVIRON PROTECTION AGENCY, 75- Mem: Am Fisheries Soc; Wildlife Soc; Nat Shellfisheries Asn; Atlantic Estuarine Res Soc. Res: Estuarine and coastal ecology; effects of organic pollutants on marine organisms and their environment. Mailing Add: Environ Res Lab Environ Protection Agency Gulf Breeze FL 32561

LOWE, JAMES HARRY, JR, b Vonore, Tenn, Mar 15, 31; m 55; c 3. ENTOMOLOGY. Educ: Univ Tenn, BA, 55; Ohio State Univ, MSc, 57; Yale Univ, PhD(forest entom), 66. Prof Exp: Res entomologist, Northeastern Forest Exp Sta, USDA, Conn, 59-62, insect res ecologist, 63-65; ASSOC PROF FORESTRY & ZOOL, UNIV MONT, 65- Mem: Entom Soc Am; Entom Soc Can. Res: Ecology of insects in forest communities; insect dispersal and distribution; alpine entomology; behavioral and meteorological aspects of flight of insects. Mailing Add: Sch of Forestry Univ of Mont Missoula MT 59801

LOWE, JAMES N, b Grand Forks, NDak, May 3, 36; m 61; c 3. ORGANIC CHEMISTRY. Educ: Antioch Col, BS, 59; Stanford Univ, PhD(chem), 64. Prof Exp: Asst prof chem, Smith Col, 63-65; ASST PROF CHEM, UNIV OF THE SOUTH, 65- Concurrent Pos: Am Chem Soc Petrol Res Fund grant, 64-66 & 67-69; fel, Univ Calif, Davis, 70-71. Mem: Am Chem Soc. Res: Coenzyme mechanisms. Mailing Add: Dept of Chem Univ of the South Sewanee TN 37375

LOWE, JAMES URBAN, JR, b Durham, NC, June 30, 21. PHYSICAL ORGANIC CHEMISTRY. Educ: Va State Col, BS, 42, MS, 46; Howard Univ, PhD, 63. Prof Exp: Asst prof chem, Tenn State Col, 47-52 & Ft Valley State Col, 52-56; fel, Howard Univ, 56-59, instr, 59-60; res chemist, US Govt, Md, 60-68; ASSOC PROF BIOCHEM & ASST DEAN ADMIN, MEHARRY MED COL, 69- Mem: Am Chem Soc; Sigma Xi. Res: Synthesis of 0-nitrobenzoates; aryloxyaliphatic acids; nitroguanidines; physical studies of beta diketones; nuclear magnetic resonance, ultraviolet, infrared spectroscopy of guanidines and perfluoroaromatics; longitudinal study of scholastic performance of Meharry medical students. Mailing Add: 4230 Eatons Creek Rd Route 3 Nashville TN 37218

LOWE, JANET MARIE, b Ellensburg, Wash, Jan 13, 24. MICROBIOLOGY, EMBRYOLOGY. Educ: Univ Wash, BS, 45; Univ Chicago, SM, 47. Prof Exp: Res assoc bact, Univ Chicago, 47-49; instr biol, 49-54, asst prof, 54-58, assoc prof zool, 58-74, PROF BIOL & DIR ALLIED HEALTH SCI PROG, CENT WASH STATE COL, 74- Concurrent Pos: NSF res grant, 58-60. Mem: AAAS; Am Soc Microbiol; Am Inst Biol Sci. Res: Chick embryology; bacteriology. Mailing Add: Dept of Biol Cent Wash State Col Ellensburg WA 98926

LOWE, JOHN EDWARD, b Newark, NJ, May 20, 35; m 57; c 2. VETERINARY SURGERY. Educ: Cornell Univ, DVM, 59, MS, 63. Prof Exp: Intern vet surg, 59-60, resident, 60-61, instr vet path, 61-63, asst prof vet surg, 63-68, ASSOC PROF VET SURG, NY STATE COL VET MED, CORNELL UNIV, 68-, COORD MGR, EQUINE RES PARK, 74-, ASSOC PROF, NY STATE COL AGR & LIFE SCI, 68- Mem: Am Vet Med Asn; Am Asn Equine Practitioners. Res: Endocrine control of the equine skeletal system; effect of nutrition on equine bone and joint disease; equine gastrointestinal surgery. Mailing Add: NY State Col of Vet Med Cornell Univ Ithaca NY 14850

LOWE, JOHN PHILIP, b Rochester, NY, Aug 28, 36; m 59; c 2. QUANTUM CHEMISTRY. Educ: Univ Rochester, BS, 58; Johns Hopkins Univ, MAT, 59; Northwestern Univ, PhD(quantum chem), 64. Prof Exp: Teacher high sch, NJ, 59-60; NIH fel theoret chem, Johns Hopkins Univ, 64-66; asst prof chem, 66-70, ASSOC PROF CHEM, PA STATE UNIV, UNIVERSITY PARK, 70- Concurrent Pos: Petrol Res Fund starter grant, 66-68, type AC grant, 69-71. Mem: AAAS; Am Chem Soc; Am Phys Soc. Res: Electron correlation; nuclear magnetic resonance spin-spin coupling; chemical reactivities; relations between Huckel and ab initio calculations. Mailing Add: Dept of Chem Pa State Univ University Park PA 16802

LOWE, JOHN THOMAS, physical chemistry, see 12th edition

LOWE, JOSIAH LINCOLN, b Hopewell, NJ, Feb 13, 05; m 32; c 1. MYCOLOGY. Educ: Syracuse Univ, BS, 27; Univ Mich, PhD(bot), 38. Prof Exp: Asst bot, 33-38, instr, 38-44, from asst prof to assoc prof, 44-51, prof forest bot, 52-59, RES PROF FOREST BOT, STATE UNIV NY COL ENVIRON SCI & FORESTRY, 59- Mem: AAAS; assoc Mycol Soc Am (pres, 60-61). Res: Taxonomy of lichens and polyporaceae. Mailing Add: Dept of Forest Bot & Path State Univ of NY Col of Environ Sci & Forestry Syracuse NY 13210

LOWE, KURT, b Munich, Ger, Nov 21, 05; nat US; m 40; c 1. PETROLOGY. Educ: City Col New York, BS, 33; Columbia Univ, MA, 37, PhD(petrol), 47; Am Inst Prof Geol, cert. Prof Exp: Lab asst & tutor geol, Eve Session, 33-42, tutor, 46-47, instr, 47-50, from asst to assoc prof, 50-64, chmn dept, 57-68, prof, 65-72, EMER PROF GEOL, CITY COL NEW YORK, 72- Concurrent Pos: Asst, Columbia Univ, 40-42; consult, NY, 36-42 & 47-; consult, NY Trap Rock Corp, 53- Honors & Awards: Neil

Miner Award, Nat Asn Geol Teachers, 68. Mem: Fel AAAS; fel Am Geol Soc; fel Mineral Soc Am; Geochem Soc; Nat Asn Geol Teachers (pres, 51-52). Res: Mineragraphy; optical mineralogy; structural petrology of granites; Storm King granite at Bear Mountain, New York; structure of the Palisades of Rockland County, New York. Mailing Add: 49-01 Francis Lewis Blvd Bayside NY 11364

LOWE, LAWRENCE E, b Toronto, Ont, Mar 29, 33; m 57; c 3. SOIL CHEMISTRY. Educ: Oxford Univ, BA, 54, MA, 61; McGill Univ, MSc, 60, PhD(agr chem), 63. Prof Exp: Soil chemist, Res Coun Alta, 63-66; from asst to assoc prof, 66-75, PROF SOILS, UNIV BC, 75- Mem: Can Soc Soil Sci; Int Soc Soil Sci; Soil Sci Soc Am. Res: Sulphur in soils; soil organic matter. Mailing Add: Dept of Soils Univ of BC Vancouver BC Can

LOWE, ORVILLE G, b San Gabriel, Calif, Feb 18, 25. ORGANIC CHEMISTRY. Educ: Univ Calif, Berkeley, BS, 50; Northwestern Univ, PhD(chem), 54. Prof Exp: Res chemist, Celanese Corp, 54-55 & Kelco Co, 55-58; Rask-Oerstead fel, Tech Univ Denmark, 58-59; res chemist, US Geol Surv, 60-61; chief chemist, Rachel Labs, Inc, 62-64; sr chemist, Space & Info Systs Div, NAm Aviation, Inc, 65-67; res assoc, Sch Pharm, Univ Southern Calif, 68-72; DIR, LOWE LABS, 70- Mem: AAAS; The Chem Soc. Res: Medicinal chemistry; pharmaceutical synthesis; heterocyclic compounds; uronic acids; organic sulfur compounds. Mailing Add: 3815 Los Feliz Blvd Los Angeles CA 90027

LOWE, REX LOREN, b Marshalltown, Iowa, Dec 28, 43; m 64; c 2. PHYCOLOGY. Educ: Iowa State Univ, BS, 66, PhD(phycol), 70. Prof Exp: Asst bot, Iowa State Univ, 66-69; ASST PROF BIOL, BOWLING GREEN STATE UNIV, 70- Concurrent Pos: Consult, Icthyol Assocs, 71-; collabr, US Nat Park Serv, 75-76. Mem: Phycol Soc Am; Int Phycol Soc; Am Inst Biol Sci; Brit Phycol Soc. Res: Diatom taxonomy and ecology. Mailing Add: Dept of Biol Bowling Green State Univ Bowling Green OH 43403

LOWE, RICHIE HOWARD, b Huff, Ky, Apr 9, 35; m 58; c 2. PLANT PHYSIOLOGY, BIOCHEMISTRY. Educ: Univ Ky, BS, 58, MS, 59; Ore State Univ, PhD(plant physiol), 63. Prof Exp: Res plant physiologist, 63-74, PLANT PHYSIOLOGIST, AGR RES SERV, USDA, 74- Mem: Am Soc Plant Physiol. Res: Enzymatic activity and biochemical changes associated with plant senescence and post harvest physiology; inorganic nitrogen and phosphorous metabolism. Mailing Add: Agr Res Ctr Univ of Ky Lexington KY 40506

LOWE, ROBERT FRANKLIN, JR, b Chicago, Ill, Nov 14, 41. CARDIOVASCULAR PHYSIOLOGY. Educ: Univ Wis, BS, 64, PhD(physiol), 69. Prof Exp: ASST PROF PHYSIOL, SCH MED, TULANE UNIV, 70- Concurrent Pos: NIH fel, Univ Wis-Madison, 69-70. Mem: Am Physiol Soc; Am Heart Asn; Am Fedn Clin Res. Res: Autonomic pharmacology. Mailing Add: Tulane Univ Dept Physiol 1430 Tulane Ave New Orleans LA 70116

LOWE, ROBERT PETER, b Cambridge, Eng, July 8, 35; Can citizen. PHYSICS, AERONOMY. Educ: Univ Western Ont, BSc, 57, PhD(atomic physics), 67. Prof Exp: Sci officer, Defense Res Bd, Can, 56-68; asst prof, 68-71, ASSOC PROF ATOMIC PHYSICS & AERONOMY, UNIV WESTERN ONT, 71- Mem: Am Geophys Union; Can Asn Physicists. Res: Infrared airglow; stratospheric composition; atmospheric chemistry; electronic, vibrational and rotational excitation in ion-molecular collisions. Mailing Add: Dept of Physics Univ of Western Ont London ON Can

LOWE, RONALD EDSEL, b Terre Haute, Ind, Jan 8, 35; m 55; c 2. ENTOMOLOGY. Educ: Ohio State Univ, BSc, 62; Purdue Univ, PhD(entom), 67. Prof Exp: Res asst entom, Purdue Univ, 62-66; res entomologist, 66-75, PROJ LEADER, CENT AM RES, INT PROGS DIV, AGR RES SERV, USDA, 75- Concurrent Pos: Courtesy prof, Univ Fla, 66- Mem: AAAS; Entom Soc Am; Am Mosquito Control Asn; Soc Invert Path. Res: Population dynamics of sterile-male release programs; growth regulation compounds for control of medically important insects; pathogenic microorganisms for biological control programs; ecology and epidemiology; agricultural research administration. Mailing Add: Agr Res Serv USDA PO Box 14565 Gainesville FL 32604

LOWE, WARREN, b San Francisco, Calif, June 4, 22; m 58. ORGANIC CHEMISTRY. Educ: Univ Calif, Berkeley, BS, 45. Prof Exp: Res asst, Manhattan Proj, US AEC, Univ Calif Radiation Lab, 43-45; from res chemist to sr res chemist, 45-74, SR RES ASSOC CHEM, CHEVRON RES CO, 74- Mem: AAAS; Am Chem Soc; Sigma Xi; fel Am Inst Chem. Res: Exploratory research of petroleum products and chemicals. Mailing Add: Chevron Res Co 576 Standard Ave PO Box 1627 Richmond CA 94802

LOWE-JINDE, LOUISE, cell biology, see 12th edition

LOWELL, ARTHUR IRWIN, polymer chemistry, plastics, see 12th edition

LOWELL, FRANCIS CABOT, b Boston, Mass, Aug 6, 09; m 38; c 3. MEDICINE. Educ: Harvard Univ, BS, 32; Harvard Med Sch, MD, 36; Am Bd Internal Med, dipl, 43; Am Bd Allergy, dipl, 46. Prof Exp: Intern, Boston City Hosp, 37-38; asst resident, Thorndike Mem Hosp, Boston, 38-40; from instr to assoc prof med, Sch Med, Univ, 40-58; asst prof, 58-69, ASSOC PROF, HARVARD MED SCH, 69- Concurrent Pos: Asst dean, Sch Med, Boston Univ, 48-51; mem staff, Robert Dawson Evans Mem Hosp, 44-58; mem staff, Mass Mem Hosp, 44-58; physician & chief allergy unit, Mass Gen Hosp, 58-; mem study sect, Admin Hosp, 63-; ed, J Allergy, Am Acad Allergy, 57-63. Mem: AAAS; Am Soc Clin Invest; Am Asn Immunol; fel Am Col Physicians; Am Acad Allergy (secy, 54-58, pres, 58-59). Res: Applied immunology and allergy. Mailing Add: Allergy Unit Mass Gen Hosp Boston MA 02114

LOWELL, GARY RICHARD, b Modesto, Calif, Sept 26, 42; m 68. GEOLOGY. Educ: San Jose State Col, BS, 65; NMex Inst Mining & Technol, PhD(geol), 69. Prof Exp: ASST PROF GEOL, SOUTHEAST MO STATE COL, 69- Mem: Geol Soc Am. Res: Igneous petrology; volcanology; metallic ore deposits. Mailing Add: Dept of Earth Sci Southeast Mo State Col Cape Girardeau MO 63701

LOWELL, JAMES DILLER, b Lincoln, Nebr, Aug 17, 33; m 57; c 4. PETROLEUM GEOLOGY, STRUCTURAL GEOLOGY. Educ: Univ Nebr, BSc, 55; Columbia Univ, MA, 57, PhD(geol), 58. Prof Exp: Geologist, Am Overseas Petrol Ltd, 58-65; asst prof geol, Washington & Lee Univ, 65-66; sr res specialist, Esso Prod Res Co, 66-73; explor geologist, Exxon Co, USA, 73-74; mgr geol, Northwest Explor Co, 74-76; CONSULT GEOLOGIST, 76- Concurrent Pos: Assoc, Oil & Gas Consults Int Inc, 76- Mem: Fel Geol Soc Am; Am Asn Petrol Geologists; Am Geophys Union. Res: Structural geology of sedimentary rocks. Mailing Add: 5836 S Colorow Dr Morrison CO 80465

LOWELL, ROBERT PAUL, b Chicago, Ill, Apr 10, 43; m 66; c 2. GEOPHYSICS. Educ: Loyola Univ Chicago, BS, 65; Ore State Univ, MS, 67, PhD(geophys), 72. Prof Exp: ASST PROF GEOPHYS, GA INST TECHNOL, 71- Mem: Am Geophys Union; Sigma Xi. Res: Thermal geophysics; heat transfer processes on ocean ridges; geothermal energy; mantle convection; thermoelastic phenomena. Mailing Add: Sch of Geophys Sci Ga Inst of Technol Atlanta GA 30332

LOWELL, SEYMOUR, b New York, NY, Nov 10, 31; m 56; c 2. PHYSICAL CHEMISTRY. Educ: Long Island Univ, BS, 55; Adelphi Univ, MS, 57, PhD(chem), 63. Prof Exp: Asst prof chem, US Merchant Marine Acad, 58-62; asst prof phys chem, 62-66, actg chmn dept, 66-67, assoc prof, 66-69, chmn dept chem, 67-71, PROF CHEM, C W POST COL, LONG ISLAND UNIV, 69-; PRES, QUANTACHROME CORP, 68- Mem: AAAS; Am Chem Soc; NY Acad Sci. Res: Surface chemistry and instrumentation; adsorption isotherms. Mailing Add: Dept of Chem C W Post Col Long Island Univ Greenvale NY 11548

LOWELL, SHERMAN CABOT, b Olean, NY, Aug 15, 18; m 41; c 2. PHYSICS, THEORETICAL NUMERICAL ANALYSIS. Educ: Univ Chicago, BS, 40; NY Univ, PhD(math), 49. Prof Exp: Sci liaison officer math sci, Office Naval Res, London, Eng, 49-51; from asst to assoc prof math, NY Univ, 51-57; prof math, head dept & dir grad progs math & appl sci, Adelphi Univ, 57-62; prof math & info sci, 62-66, PROF PHYSICS & COMPUT SCI, WASH STATE UNIV, 66-, MATHEMATICIAN, COMPUT CTR, 62- Concurrent Pos: Asst to sci dir, Inst Math Sci, NY Univ, 53-57; vis scientist, Lab Physics of Solids, Paris, 68 & Nat Ctr Very Low Temp, Grenoble, France, 69; consult serv bur, Int Bus Mach Corp, 59-62 & Lawrence Livermore Lab, Univ Calif, 63- Mem: AAAS; Am Meteorol Soc; Soc Indust & Appl Math; Am Phys Soc. Res: Lattice dynamics; wave propagation; numerical analysis. Mailing Add: NE 1610 Upper Dr Pullman WA 99163

LOWELL, WAYNE RUSSELL, b Palouse, Wash, Mar 21, 06; m 34; c 3. ECONOMIC GEOLOGY. Educ: Wash State Univ, BS, 36; Univ Chicago, MS, 39, PhD(geol), 42. Prof Exp: Asst geologist, State Dept Geol & Mineral Industs, Ore, 39-40; asst prof geol & actg chmn dept, Univ Mont, 42-46, from assoc prof to prof, 46-50, chmn dept, 46-50; assoc prof, 50-57, dir, Geol Field Sta, 50-62, PROF GEOL, IND UNIV, BLOOMINGTON, 57- Mem: Geol Soc Am; Soc Econ Geol; Am Inst Mining, Metall & Petrol Eng. Res: Nonmetallics. Mailing Add: Dept of Geol Ind Univ Bloomington IN 47401

LOWEN, JACK, b New York, NY, July 17, 20; m 44; c 1. ANALYTICAL CHEMISTRY. Educ: City Col New York, BS, 41. Prof Exp: Chemist, Hoosier Ord, 42; chem engr, Edgewood Arsenal, 42, phys chemist, 46-47; anal chemist, Mass Inst Technol, 48-51, Lincoln Lab, 51-62; mem staff, Sprague Elec Co, Mass, 62-65; mem staff electronics res ctr, NASA, Mass, 65-70; mgr semiconductor mat, Alpha Industs, Inc, 70-71; anal chemist, Arnold Greene Testing Labs, 71 & Forsyth Dent Ctr, Boston, 71-72; ANAL CHEMIST, US CUSTOMS, 72- Concurrent Pos: Chem consult, Consol Still & Sterilizer, Mass, 70- Mem: Am Chem Soc; Electrochem Soc. Res: Thin films; semiconductor surfaces and chemistry. Mailing Add: 121 Pine St Needham MA 02192

LOWEN, WARREN KEALOHA, b Newton, Kans, Apr 3, 21; m 45; c 2. ANALYTICAL CHEMISTRY. Educ: Univ Kans, AB, 42, MS, 47, PhD(chem), 49. Prof Exp: Anal chemist, Pan Am Refining Corp, 43-46; mem res staff, Grasselli Chem Dept, 49-52, res supvr anal res, 52-56, res sect head, Indust & Biochem Dept, 56-62, res mgr, 62-74, DIR LAB, BIOCHEM DEPT, E I DU PONT DE NEMOURS & CO, INC, 74- Mem: Am Chem Soc; Sci Res Soc Am. Res: Analytical chemistry of agricultural chemicals; ion exchange resins. Mailing Add: 611 Entwisle Ct Westminster Wilmington DE 19808

LOWENBACH, HANS, b Duisburg, Ger, Jan 31, 05; nat US; m 41; c 3. PSYCHIATRY. Educ: Univ Hamburg, MD, 29; Am Bd Psychiat & Neurol, dipl. Prof Exp: Asst, Physiol Inst, Univ Freiburg, 30-32; asst med clin, Univ Cologne, 32-33; asst, Neurophysiol Div, Kaiser Wilhelm Inst Hirnforschung, Berlin-Dahlem, 33-35; ship's surgeon, Whaling Expeds, 36-38; asst psychiat, Johns Hopkins Hosp, 39 & 40; from assoc prof to prof, 40-74, EMER PROF PSYCHIAT, SCH MED, DUKE UNIV, 74- Concurrent Pos: Nansen Found fel, Univ Oslo, 35-36; consult, Vet Admin, US Army & Off Health, Educ & Welfare, civilian with Off Sci Res & De elop & Field Info Agencies Technol, 46. Mem: AAAS; Am EEG Soc; AMA; Am Psychiat Asn; Asn Res Nerv & Ment Dis. Res: Therapy of psychiatric disorders. Mailing Add: Dept of Psychiat Duke Univ Med Ctr Durham NC 27710

LOWENGRUB, MORTON, b Newark, NJ, Mar 31, 35; m 61; c 1. MATHEMATICS. Educ: NY Univ, BA, 56; Calif Inst Technol, MS, 58; Duke Univ, PhD(math), 61. Prof Exp: Instr math, Duke Univ, 60-61; asst prof, NC State Col, 61-62; Leverhulme res fel, Glasgow Univ, 62-63; asst prof, Wesleyan Univ, 63-66; NSF fel, Glasgow Univ, 66-67; assoc prof, 67-72, PROF MATH, IND UNIV, BLOOMINGTON, 72- Concurrent Pos: Sr res fel, Sci Res Coun, Gt Brit, 73-74. Mem: Math Asn Am; Am Math Soc; Soc Indust & Appl Math. Res: Mathematical theory of elasticity. Mailing Add: Dept of Math Ind Univ Bloomington IN 47401

LOWENHAUPT, BENJAMIN, b St Louis, Mo, July 15, 18; m 50; c 3. BIOPHYSICS, PLANT PHYSIOLOGY. Educ: Iowa State Col, BS, 40; Univ Chicago, MS, 41; Univ Calif, Berkeley, PhD(plant physiol), 54. Prof Exp: Res assoc, Univ Calif, Berkeley, 54-55; res assoc, Rockefeller Inst, 56-60, NIH fel, 60-62; sr res assoc physiol, Col Med, Univ Cincinnati, 62-67; PROF BIOL, EDINBORO STATE COL, 67- Concurrent Pos: Sabbatical, Flinders Univ SAustralia, 74-75. Mem: AAAS; Am Chem Soc; Am Soc Plant Physiol. Res: Biochemistry; role of inorganic ions in biology; ion transport and excitation. Mailing Add: Dept of Biol Edinboro State Col Edinboro PA 16412

LOWENHEIM, FREDERICK ADOLPH, b New Rochelle, NY, Oct 6, 09; m 36; c 2. ELECTROCHEMISTRY. Educ: Columbia Univ, AB, 30, PhD(chem), 34. Prof Exp: Asst chem, Columbia Univ, 30-34; chemist, Comolite Corp, 34-36; res supvr, M&T Chem Inc, 36-59, res assoc, 62-66, staff asst, 66-70, res coord, 67-70; TECH WRITING & CONSULT, 70- Concurrent Pos: Lectr, Rutgers Univ, 47-48 & Stevens Inst Technol, 50-56; tech ed, Plating, 63-74. Honors & Awards: Award of Merit, Am Soc Testing & Mat, 73. Mem: AAAS; Am Chem Soc; Electrochem Soc; Am Electroplaters Soc; hon fel Am Soc Testing & Mat. Res: Tin plating; detinning; rare metals; antimony; tin chemicals; alloy deposition; corrosion. Mailing Add: 637 W Seventh St Plainfield NJ 07060

LOWENSTAM, HEINZ ADOLF, b Siemianowita, Ger, Oct 9, 12; nat US; m 37; c 3. ECOLOGY. Educ: Univ Chicago, PhD(paleont), 39. Prof Exp: Cur paleont, Ill State Mus, 40-43; assoc geologist, State Geol Surv, Ill, 43-49, geologist, 49-50; res assoc, Univ Chicago, 48-50, assoc prof, 50-52; PROF PALEOECOL, CALIF INST TECHNOL, 52- Concurrent Pos: Spec staff to aid war effort, Coal & Oil Develop, State Geol Surv, Ill, 43-45. Mem: AAAS; fel Geol Soc Am; Paleont Soc; Soc Study Evolution; assoc Soc Econ Paleont & Mineral. Res: Paleoecology; biogeochemistry; paleo-temperatures; evolution of reef ecology; impact of the evolution of life on chemical and physical processes in the oceans; minerals in hard tissue precipitates of marine invertebrates. Mailing Add: Div of Geol & Planetary Sci Calif Inst of Technol Pasadena CA 91109

LOWENSTEIN, CARL DAVID, b New York, NY, Sept 3, 34; m 65; c 1. APPLIED PHYSICS. Educ: Kent State Univ, BA, 55; Harvard Univ, SM, 56, PhD(physics), 63. Prof Exp: Res fel, Harvard Univ, 63-64; asst res physicist, 64-69, ASSOC SPECIALIST, MARINE PHYSICS LAB, UNIV CALIF, SAN DIEGO, 69- Concurrent Pos: Mem sensors comt, US Navy Deep Submergence Syst Prog, 64- Mem: Acoust Soc Am; Audio Eng Soc; Inst Elec & Electronics Eng; Math Asn Am. Res: Synthesis of directive arrays; signal processing; underwater acoustics; computer applications. Mailing Add: Marine Physics Lab Univ of Calif San Diego CA 92132

LOWENSTEIN, EDWARD, b Duisburg, Ger, May 29, 34; US citizen; m 59; c 3. ANESTHESIOLOGY, CARDIOPULMONARY PHYSIOLOGY. Educ: Univ Mich, MS, 59; Am Bd Anesthesiol, dipl. Prof Exp: Assoc anesthesia, 68-70, asst prof, 70-72, ASSOC PROF ANESTHESIA, HARVARD MED SCH, 72- Concurrent Pos: Assoc anesthetist, Mass Gen Hosp, 68-71, anesthetist, 71- Mem: Am Soc Anesthesiol. Res: Physiological effects of cardiac disease, cardiopulmonary bypass and cardiac surgery. Mailing Add: Dept of Anesthesia Mass Gen Hosp Boston MA 02114

LOWENSTEIN, JEROLD MARVIN, b Danville, Va, Feb 11, 26; m 49; c 3. NUCLEAR MEDICINE. Educ: Columbia Univ, BS, 46, MD, 53. Prof Exp: Physicist, Los Alamos Sci Lab, 46-48; instr med & radiol, Sch Med, Stanford Univ, 57-58; asst clin prof, 63-68, ASSOC CLIN PROF MED, THYROID RES, MED CTR, UNIV CALIF, SAN FRANCISCO, 68- Concurrent Pos: Nat Found fel radiobiol, 55-56; NIH res grant thyroid & cellular changes in pregnancy, 60-66, etiology of nontoxic nodular goiter, 61-66, measurement of blood digitalis levels, 62-66; dir nuclear med, Presby Med Ctr, San Francisco, 59-; partic, Galapagos Int Sci Proj, 64; dir nuclear med, Children's Hosp, 64- Mem: AMA; Soc Nuclear Med; Am Fedn Clin Res. Res: Applications of physics to medicine, especially medical uses of radioactive isotopes. Mailing Add: Thyroid Res 122 MR2 Univ of Calif Med Ctr San Francisco CA 94122

LOWENSTEIN, JOHN MARTIN, b Berlin, Ger, Oct 28, 26; m 54. BIOCHEMISTRY. Educ: Univ Edinburgh, BSc, 50; Univ London, PhD, 53. Prof Exp: Demonstr chem & biochem, Med Sch, St Thomas' Hosp, Eng, 50-53; res assoc biochem, Med Sch, Univ Wis, 53-55; Beit mem fel med res, Oxford Univ, 55-58; PROF BIOCHEM, BRANDEIS UNIV, 59- Mem: AAAS; Am Chem Soc; Am Soc Biol Chem; The Chem Soc; Brit Biochem Soc. Res: Regulated enzymes; integration and control of metabolic pathways. Mailing Add: Dept of Biochem Brandeis Univ Waltham MA 02154

LOWENSTEIN, LEAH MIRIAM, b June 17, 30; US citizen; m 54; c 3. MEDICINE, BIOCHEMISTRY. Educ: Univ Wis, BS, 50, MD, 54; Oxford Univ, DPhil, 58; Am Bd Internal Med, dipl, 63. Prof Exp: Instr med, Sch Med, Tufts Univ, 61-64; res assoc, Harvard Med Sch, 64-65, assoc, 65-68; asst prof med, 68-71, ASSOC PROF MED & BIOCHEM, SCH MED, BOSTON UNIV, 71- Concurrent Pos: Assoc med, Thorndike Mem Lab, 65-68; vis scientist, Med Sch, Univ Pa, 66-67; trustee Hampstead Med Found, 67-; assoc med, Harvard Med Sch, 69-70; mem, Nat Kidney Found; mem coun kidney in cardiovasc dis, Am Heart Asn; mem first-level rev comt, Artificial Kidney-Chronic Uremia Prog, NIH, 72-; mem-at-large, Sect N (Med Sci) Comt, AAAS, 73-; mem, Exp Models Aging Comt, Nat Inst Aging, 74- Mem: AAAS; Am Fedn Clin Res; Am Asn Study Liver Dis; Am Soc Nephrology; Am Med Soc Alcoholism. Res: Renal and hepatic metabolism; alcoholism. Mailing Add: Sch of Med Boston Univ Boston MA 02118

LOWENSTEIN, MICHAEL ZIMMER, b Hornell, NY, Oct 4, 38; m 62; c 2. PHYSICAL CHEMISTRY. Educ: Oberlin Col, AB, 60; Ariz State Univ, MS, 62, PhD(x-ray crystallog), 65. Prof Exp: Asst prof chem, 64-71, PROF CHEM, ADAMS STATE COL, 71- Concurrent Pos: AEC fac res assoc, Ariz State Univ, 70-71; consult, Citizen's Workshop, Energy Res & Develop Agency, 74-76; vis prof, Solar Energy Appln Lab, Colo State Univ, Ft Collins, 75-76. Mem: Am Chem Soc. Res: Instrumentation; energy problems; applications of solar energy to residential heating and cooling; solar thermal storage. Mailing Add: Dept of Chem Adams State Col Alamosa CO 81102

LOWENTHAL, DENNIS DAVID, b Yakima, Wash, Nov 10, 42; m 66; c 2. PLASMA PHYSICS, ELEMENTARY PARTICLE PHYSICS. Educ: Calif State Univ, Northridge, BS, 65; Univ Calif, Los Angeles, MS, 66; Univ Calif, Irvine, PhD(physics), 75. Prof Exp: Sci tech aid res & develop, Aeronutronic Div, Philco-Ford Corp, 66-69; SR SCIENTIST FUSION RES, MATH SCI NW, 75- Mem: Am Phys Soc; Optical Soc Am; AAAS. Res: Experimental search for the double beta decay of selenium 82; fusion power research. Mailing Add: Math Sci NW PO Box 1887 Bellevue WA 98009

LOWENTHAL, FRANKLIN, b Breslau, Ger, Aug 27, 38; US citizen; m 65. MATHEMATICS. Educ: City Col New York, BS, 59; Stanford Univ, MS, 62 & 63, PhD(math), 65. Prof Exp: Instr math, NY Univ, 64-65; asst prof, Univ Ore, 65-74; ASSOC PROF MATH, UNIV WIS-PARKSIDE, 74- Mem: Am Math Soc. Res: Classical analysis; subgroups and subsemigroups generated by pairs of infinitesimal transformation. Mailing Add: Dept of Math Univ of Wis-Parkside Kenosha WI 53140

LOWENTHAL, JOSEPH PHILIP, b New York, NY, July 21, 19; m 54; c 2. MICROBIOLOGY, IMMUNOLOGY. Educ: Brooklyn Col, AB, 39; Johns Hopkins Univ, ScD(microbiol), 52; Am Bd Med Microbiol, dipl. Prof Exp: Lab asst, City Dept Health, New York, 40-44; bacteriologist, Vet Admin Hosp, Ft Harrison, Ind, 47-48 & Army Med Serv Grad Sch, 52-56; chief, Microbiol Sect, Dept Biologics Res, 56-62, DEPT BIOLOGICS RES, WALTER REED ARMY INST RES, 62- Concurrent Pos: Mem comn rickettsial dis, Armed Forces Epidemiol Bd, 66- Honors & Awards: Award, Off Surgeon Gen, Dept Army, 57. Mem: AAAS; Am Soc Microbiol; NY Acad Sci. Res: Botulinal hemagglutination; bacterial enzymes; bacterial, viral and rickettsial vaccines. Mailing Add: Dept of Biologics Res Walter Reed Army Inst Res Washington DC 20012

LOWENTHAL, JULIUS, b Wussow, Ger, June 4, 17. BIOCHEMISTRY, PHARMACOLOGY. Educ: Univ Montreal, BSc, 46, MSc, 47, PhD(biochem), 50. Prof Exp: Res asst, Univ Sask, 51, res assoc, 52, asst prof physiol, 53-58, assoc prof pharmacol, 58-60; assoc prof, Univ Man, 60-64; assoc prof biochem, Queen's Univ, Ont, 65-68; PROF PHARMACOL & THERAPEUT, McGILL UNIV, 68- Res: Synthesis of compounds with vitamin K-like and vitamin E-like activity; mode of action and structure activity relation of vitamin E, vitamin K and other biologically active quinones; bioassay of vitamin K; effect of drugs on formation of plasma clotting factors. Mailing Add: Dept of Pharmacol & Therapeut McGill Univ Montreal PQ Can

LOWENTHAL, MILTON, medicine, see 12th edition

LOWENTHAL, WERNER, b Krefeld, Ger, Dec 20, 30; US citizen; m 61; c 2. PHARMACY. Educ: Albany Col Pharm, Union Univ, NY, BS, 53; Univ Mich, PhD(pharmaceut chem), 58. Prof Exp: Asst, Univ Mich, 53-55; res pharmacist, Abbott Labs, 57-61; asst prof pharm, 61-66, assoc prof, 66-71, PROF PHARM, SCH PHARM, MED COL VA, VA COMMONWEALTH UNIV, 71-, PROF EDUC PLANNING & DEVELOP, 74- Concurrent Pos: Mem revision comt, Nat Formulary; mem, US Pharmacopoeia Rev Comt, 75-80. Mem: AAAS; Am Pharmaceut Asn; Am Asn Cols Pharm; Nat Soc Performance Improv. Res: Pharmaceutical product development; drug absorption; programmed instruction; curriculum development. Mailing Add: Sch of Pharm Med Col of Va Va Commonwealth Univ Richmond VA 23298

LOWER, GERALD MALCOLM, JR, b Sheboygan, Wis, Apr 4, 45; m 65; c 1. ONCOLOGY, BIOCHEMISTRY. Educ: Univ Wis-Madison, BS, 67, MS, 69, PhD(oncol), 71. Prof Exp: ASST PROF CLIN ONCOL, UNIV HOSPS, MED SCH, UNIV WIS-MADISON, 71- Mem: AAAS. Res: Environmental carcinogenesis; occurrence and formation of environmental chemical carcinogens; metabolism and mechanism of action of chemical carcinogens. Mailing Add: Div of Clin Oncol Univ Hosp Madison WI 53706

LOWER, RICHARD ROWLAND, b Detroit, Mich, Aug 15, 29; m 53; c 5. THORACIC SURGERY, CARDIOVASCULAR SURGERY. Educ: Amherst Col, AB, 51; Cornell Univ, MD, 55; Am Bd Surg, dipl, 63, cert thoracic surg, 64. Prof Exp: Intern, King County Hosp, 55-56; asst prof surg, Stanford Univ, 62-65; PROF SURG & CHMN DIV THORACIC & CARDIAC SURG, MED COL VA, 65- Concurrent Pos: Res asst, Stanford Univ Hosps, 58-63; chief thoracic & cardiovasc surg, Palo Alto Vet Admin Hosp, 64-65. Res: Cardiac transplantation. Mailing Add: Div of Thoracic & Cardiac Surg Med Col of Va Richmond VA 23219

LOWER, STEPHEN K, b Oakland, Calif, Sept 8, 33; m 63. PHYSICAL CHEMISTRY. Educ: Univ Calif, Berkeley, BA, 55; Ore State Univ, MS, 58; Univ BC, MSc, 60, PhD(phys chem), 63. Prof Exp: Fel phys chem, Polytech Inst Brooklyn, 63-64 & Univ Calif, Los Angeles, 64-65; ASST PROF PHYS CHEM, SIMON FRASER UNIV, 65- Concurrent Pos: Nat Res Coun Can grants, 65-71, mem panel on comput assisted instruction lang, 70- Mem: AAAS; Am Chem Soc; Asn Develop Instructional Systs. Res: Electronic spectra and physical chemistry of organic solids and charge-transfer complexes, fluorescence spectroscopy; instructional systems design; computer-assisted instruction and instructional technology applied to college teaching. Mailing Add: Dept of Chem Simon Fraser Univ Burnaby BC Can

LOWER, WILLIAM RUSSELL, b La Junta, Colo, Oct 28, 30; m 71; c 2. GENETICS, ENVIRONMENTAL HEALTH. Educ: Univ Calif, Los Angeles, BA, 53; Univ Calif, Berkeley, PhD(genetics), 65. Prof Exp: Res assoc genetics of nematodes, Kaiser Found Res Inst, 64-66; res assoc, Clin Pharmacol Res Inst, 67-69; res assoc biol monitoring, Environ Health Surveillance Ctr, 70-72, ASSOC PROF COMMUNITY HEALTH & MED PRACT & BIOL, UNIV MO-COLUMBIA, 72-, GROUP LEADER, ENVIRON TRACE SUBSTANCES RES CTR, 72- Concurrent Pos: Fel, Univ Mo, 69-70. Mem: Genetics Soc Am; Soc Study Evolution. Res: Genetic effects of exposure to trace substances; biological monitoring of trace elements. Mailing Add: Dept Commun Health & Med Pract Univ of Mo Columbia MO 65201

LOWERY, CHARLES E, JR, b Austin, Tex, Sept 1, 31; m 53; c 1. MICROBIOLOGY. Educ: Univ Tex, BA, 56, MA, 58, PhD(microbiol), 65. Prof Exp: RES MICROBIOLOGIST, CHEM DIV, MILES LABS, INC, 65- Mem: Am Soc Microbiol. Res: Antibiotics; petroleum microbiology; fermentation chemistry. Mailing Add: 2725 Neff St Elkhart IN 46514

LOWERY, GEORGE HINES, JR, b Monroe, La, Oct 2, 13; m 37; c 2. ORNITHOLOGY. Educ: La State Univ, BS, 34, MS, 36; Univ Kans, PhD, 49. Prof Exp: Instr, 36-39, from asst prof to prof, 38-55, BOYD PROF ZOOL, LA STATE UNIV, BATON ROUGE, 55-, DIR MUS NATURAL SCI, 52-, MUS ZOOL, 36- Mem: Cooper Ornith Soc; Wilson Ornith Soc; Am Ornith Union. Res: Taxonomy, classification and distribution of birds and mammals; bird migration. Mailing Add: Mus of Natural Sci La State Univ Baton Rouge LA 70803

LOWERY, KIRBY, JR, b Palacios, Tex, Oct 7, 43; m 62; c 2. ORGANOMETALLIC CHEMISTRY, ORGANIC POLYMER CHEMISTRY. Prof Exp: Stephen F Austin Univ, BS, 67; Tex A&M Univ, MS, 68, PhD(chem), 71. Prof Exp: Fel chem, Ore Grad Ctr, 71; res chemist, 73-75, SR RES CHEMIST, DOW CHEM CO, 75- Res: Transition metal chemistry, organic chemistry, polymer chemistry, specifically Ziegler-type catalysis. Mailing Add: Dow Chem Co Bldg B-3827 Freeport TX 77541

LOWERY, THOMAS J, b Brooklyn, NY, Jan 21, 22; m 50; c 9. CYTOLOGY, PHYSIOLOGY. Educ: St Francis Col, NY, BS, 46; Fordham Univ, MS, 49, PhD(cytol), 55. Prof Exp: Instr zool & genetics, Stonehill Col, 50-51; instr zool, physiol & anat, Duquesne Univ, 51-53; instr biol, Fordham Univ, 53-55; assoc prof cytol & physiol, Villanova Univ, 55-62; ASSOC PROF BIOL, LA SALLE COL, 62- Mem: Am Micros Soc; Torrey Bot Club. Res: Cytological implications of relationship between the chromosome number of maternal tissue, embryo, and endosperm and its effect on producing various aberrancies such as somatoplastic sterility and neoplasm of plants. Mailing Add: Dept of Biol La Salle Col Philadelphia PA 19141

LOWEY, SUSAN, b Vienna, Austria, Jan 22, 33; nat US. PROTEIN CHEMISTRY, PHYSICAL CHEMISTRY. Educ: Columbia Univ, BA, 54; Yale Univ, PhD(chem), 58. Prof Exp: Res fel biol, Harvard Univ, 57-59; assoc prof biochem, 72-74, PROF BIOCHEM, BRANDEIS UNIV, 74-, MEM STAFF, ROSENSTIEL BASIC MED SCI RES CTR, 72-; RES ASSOC, CHILDREN'S CANCER RES FOUND, 59- Mem: Am Chem Soc. Res: Physical chemistry of muscle proteins. Mailing Add: Children's Cancer Res Found 35 Binney St Boston MA 02115

LOWIG, HENRY FRANCIS JOSEPH, b Prague, Czech, Oct 29, 04; m 49; c 2. PURE MATHEMATICS. Educ: Ger Univ, Prague, Dr rer nat, 28; Univ Tasmania, DSc, 51. Prof Exp: Privat-docent math, Ger Univ, Prague, 35-38; lectr, Univ Tasmania, 48-51, sr lectr, 51-57; from assoc prof to prof, 57-70, EMER PROF MATH, UNIV ALTA, 70- Mem: Am Math Soc; Can Math Cong. Res: Functional analysis; lattice theory; universal algebra. Mailing Add: 15212 81st Ave Edmonton AB Can

LOWITZ, DAVID AARON, b Newark, NJ, Dec 18, 28; m 53; c 4. CHEMICAL PHYSICS. Educ: Rutgers Univ, BA, 50; Pa State Univ, MS, 53, PhD(physics), 55. Prof Exp: Asst physics, Pa State Univ, 50-53, res asst, 55-56; physicist, Gulf Res & Develop Co, 56-64; res assoc & head cent res physics sect, Lord Corp, 64-67; MGR PHYSICS DIV, PHILIP MORRIS RES CTR, 67- Concurrent Pos: Am Petrol Inst fel, 55-56. Mem: Am Phys Soc; Int Soc Quantum Biol. Res: Combustion of cellulosic materials; charge transport in solids; quantum chemistry; liquid pressure-volume-temperature and pressure-temperature-Viscosity; electromagnetic wave propagation; dielectrics. Mailing Add: Philip Morris Res Ctr PO Box 26583 Richmond VA 23261

LOWKE, JOHN JAMES, b Tanunda, SAustralia, Apr 3, 34; m 63; c 2. PLASMA PHYSICS, LASERS. Educ: Univ Adelaide, BSc, 56, dipl, 58, PhD(physics), 63. Prof Exp: Lectr math, Adelaide Teachers Col, Australia, 63; sr physicist, 64-71, FEL SCIENTIST, WESTINGHOUSE RES & DEVELOP CTR, 71- Concurrent Pos: Sr Rothmans fel, Australian Nat Univ, 70. Mem: Am Phys Soc. Res: Physics of high temperature gases; physics of arcs as they occur in circuit breakers and lamps. Mailing Add: Westinghouse Res & Develop Ctr Churchill Borough Pittsburgh PA 15235

LOWMAN, BERTHA PAULINE, b Newton, NC, Mar 17, 29. MATHEMATICS. Educ: Lenoir-Rhyne Col, BS, 51; Univ Ala, MA, 52. Prof Exp: Instr math, Campbell Col, 52-53; instr sci & math, Anderson Col, 53-54; asst, Univ NC, 55; asst prof math, Hardin-Simmons Univ, 55-59, ECarolina Col, 59-60 & Elon Col, 60-62; ASST PROF MATH, WESTERN KY UNIV, 62- Mem: Math Asn Am; Nat Coun Teachers Math. Res: Number theory and algebra; geometry and history of mathematics. Mailing Add: Dept of Math Western Ky Univ Bowling Green KY 42101

LOWMAN, EDWARD WYNNE, b Orangeburg, SC, July 16, 15. PHYSICAL MEDICINE. Educ: The Citadel, BS, 36; Med Col SC, MD, 40; Univ Minn, MS, 50; Am Bd Phys Med & Rehab, dipl, 48. Prof Exp: PROF PHYS MED & REHAB & CLIN DIR, INST PHYS MED & REHAB, SCH MED, NY UNIV, 51- Concurrent Pos: Attend physician, NY Univ Hosp; mem courtesy staff, St Vincent's Hosp, 51- Mem: AMA; Am Rheumatism Asn; Am Acad Phys Med & Rehab; fel Am Col Physicians; Am Cong Rehab Med. Res: Rehabilitation; arthritis; rheumatism. Mailing Add: 400 E 34th St New York NY 10016

LOWMAN, FRANK (GEORGE), radiation ecology, see 12th edition

LOWMAN, PAUL DANIEL, JR, b Elizabeth, NJ, Sept 26, 31; m 58. ASTROGEOLOGY, PHOTOGEOLOGY. Educ: Rutgers Univ, BS, 53; Univ Colo, PhD(geol), 63. Prof Exp: AEROSPACE TECHNOLOGIST, GODDARD SPACE FLIGHT CTR, NASA, 59- Concurrent Pos: Vis lectr, US Air Force Inst Technol, 63-64; lectr, Cath Univ, 63-66; lectr, Univ Calif, Santa Barbara, 70. Honors & Awards: John C Lindsay Mem Award, Goddard Space Flight Ctr, 74. Mem: Geol Soc Am; AAAS; Am Geophys Union. Res: Planetology; lunar geology; geologic application of orbital photography; remote sensing; comparative planetology. Mailing Add: Code 922 Goddard Space Flight Ctr Greenbelt MD 20771

LOWMAN, ROBERT MORRIS, b Baltimore, Md, Dec 31, 12; m 37; c 2. RADIOLOGY. Educ: Harvard Univ, AB, 32; Univ Md, MD, 36. Hon Degrees: MA, Yale Univ, 65. Prof Exp: Instr radiol, Grad Sch Med, Univ Pa, 36-38, asst dir dept radiol, Grad Hosp, 40-45; asst, Sch Med, Boston Univ, 38-40; assoc prof, 55-62, actg chmn dept, 73, PROF RADIOL, MED SCH, YALE UNIV, 62-; DIR MEM UNIT, YALE-NEW HAVEN HOSP, 62- Concurrent Pos: Angiol Res Found honors achievement award, 64-65; attend physician, Grace-New Haven Community Hosp, 45-, dir dept radiol, Mem Unit, 45-; consult, W Haven Vet Hosp, 62-; pres, New Eng Roentgen Ray Soc, 71-72, mem exec comt & chmn exec bd, 72-74; fel, Davenport Col, Yale Univ. Mem: Fel Am Col Radiol; Radiol Soc NAm; Sigma Xi. Res: Thoracic lymphatics; embryology of the bladder; cardiac kymography; experimental coronary arteriography. Mailing Add: Mem Unit Box 1001 Yale-New Haven Med Ctr New Haven CT 06504

LOWN, JAMES WILLIAM, b Blyth, Eng, Dec 19, 34; m 62. PHYSICAL ORGANIC CHEMISTRY. Educ: Univ London, BSc, 56, PhD(org chem) & dipl, Impt Col, 59. Prof Exp: Asst lectr chem, Imp Col, Univ London, 59-61; fel, Univ Alta, 61-62, asst prof, 62-63; res chemist, Walter Reed Army Inst Res, DC, 62-63; from asst to assoc prof, 64-74, PROF CHEM, UNIV ALTA, 74- Mem: Am Chem Soc; The Chem Soc. Res: Organic reaction mechanisms; heterocyclic synthesis; antibiotics. Mailing Add: Dept of Chem Univ of Alta Edmonton AB Can

LOWNDES, DOUGLAS H, JR, b Pasadena, Calif, Jan 3, 40; m 61; c 1. LOW TEMPERATURE PHYSICS. Educ: Stanford Univ, BS, 61; Univ Colo, PhD(physics), 69. Prof Exp: Res asst solid state physics, Hewlett-Packard Assocs, Calif, 62-63; NSF fel physics, Sch Math & Phys Sci, Univ Sussex, 68-70; ASST PROF PHYSICS, UNIV ORE, 70-, ASSOC, SOLAR ENERGY CTR, 74- Mem: Am Phys Soc; Am Asn Physics Teachers. Res: Experimental techniques; thermometry; specific heat studies; de Haas-van Alphen and Fermi surface studies; magnetic linear chains; transition metal carbides. Mailing Add: Dept of Physics Univ of Ore Eugene OR 97403

LOWNDES, ROBERT P, b Derby, Eng, Dec 11, 39. PHYSICS. Educ: Univ London, BSc, 62, Queen Mary Col, PhD(exp solid state physics), 67. Prof Exp: Res assoc physics, Mass Inst Technol, 67-68; asst prof, 68-72, ASSOC PROF PHYSICS, NORTHEASTERN UNIV, 72- Mem: Am Inst Physics; Brit Inst Physics. Res: High pressure dielectric and far infrared spectroscopic studies of solids. Mailing Add: Dept of Physics Northeastern Univ Boston MA 02115

LOWNEY, EDMUND DILLAHUNTY, b Port Arthur, Tex, Nov 8, 31; m 58; c 2. DERMATOLOGY. Educ: Univ Tex, BA, 53; Yale Univ, PhD(psychol), 57; Univ Pa, MD, 60. Prof Exp: From instr to asst prof dermat, Univ Mich, Ann Arbor, 64-67; assoc prof, Med Col Va, 67-69; PROF DERMAT, UNIV HOSP,,HOSP, COL MED, OHIO STATE UNIV, 69- Mem: Soc Invest Dermat. Res: Immunology. Mailing Add: Dept of Dermat Univ Hosp Ohio State Univ Col of Med Columbus OH 43210

LOWNSBERY, BENJAMIN FERRIS, b Wilmington, Del, July 28, 20; m 50; c 1. PLANT NEMATOLOGY. Educ: Univ Del, BA, 42; Cornell Univ, PhD(plant path), 50. Prof Exp: Chemist explosives div, E I du Pont de Nemours & Co, 42-45; asst plant path, Cornell Univ, 45-50; asst plant pathologist, Conn Agr Exp Sta, 51-53; from asst to assoc nematologist, 54-67, lectr nematol, 60-70, PROF NEMATOL, UNIV CALIF, DAVIS, 70-, NEMATOLOGIST, EXP STA, 67- Concurrent Pos: Mem subcomt nematodes, Agr Bd, Nat Acad Sci-Nat Res Coun, 66-68. Honors & Awards: Stark Award, Am Soc Hort Sci, 70. Mem: Soc Nematol; Am Phytopath Soc. Res: Biology and control of nematodes parasitizing fruit and nut crops; nematode pathogenicity. Mailing Add: Dept of Nematol Univ of Calif Davis CA 95616

LOWRANCE, EDWARD WALTON, b Ogden, Utah, June 17, 08; m 35; c 2. ANATOMY. Educ: Univ Utah, AB, 30, AM, 32; Stanford Univ, PhD(biol), 37. Prof Exp: Asst zool, Stanford Univ, 32-34, Rockefeller asst exp embryol, 34-36 & 37-38; from instr to assoc prof zool, Univ Nev, 38-49; asst prof anat, Sch Med, Univ SDak, 49-50; assoc prof, 50-55, PROF ANAT, SCH MED, UNIV MO-COLUMBIA, 55- Concurrent Pos: Actg assoc prof, Sch Med, Univ Kans, 44-46; State secy, Mo State Anat Bd, 69- Mem: AAAS; Am Asn Anat; Am Micros Soc; NY Acad Sci. Res: Comparative and experimental embryology; quantitative and statistical anatomy. Mailing Add: Dept of Anat Univ of Mo Med Ctr Columbia MO 65201

LOWRANCE, WILLIAM WILSON, JR, b El Paso, Tex, May 8, 43. ORGANIC CHEMISTRY, SCIENCE POLICY. Educ: Univ NC, Chapel Hill, AB, 65; Rockefeller Univ, PhD(biochem), 70. Prof Exp: Res chemist, Tenn Eastman Co, Kingsport, 70-71; res consult, NC Dept Educ, Raleigh, 71-72; asst exec ed, J Cell Biol, New York, 72-73; resident fel, Nat Acad Sci, Washington, DC, 73-75; HARVARD UNIV RES FEL, PROG SCI & INT AFFAIRS, HARVARD UNIV, 75- Mem: AAAS. Res: National science policy; the concepts of risk and safety; ethical responsibilities of technical people; nuclear proliferation; the relations between art and science; synthetic and mechanistic organic photochemistry; cellular biochemistry. Mailing Add: Prog for Sci & Int Affairs Harvard Univ 9 Divinity Ave Cambridge MA 02138

LOWREY, ALFRED HOLLAND, chemical physics, see 12th edition

LOWREY, CHARLES BOYCE, b New Orleans, La, Mar 15, 41; m 61; c 3. PHYSICAL ORGANIC CHEMISTRY. Educ: Centenary Col, BS, 63; Univ Houston, PhD(heterocyclic chem), 68. Prof Exp: Teaching asst chem, Univ Houston, 63-66; asst prof, 66-73, ASSOC PROF CHEM, CENTENARY COL LA, 73-, ASST DEAN COL, 74- Concurrent Pos: Consult, Baifield Industs, La, 66-; water pollution consult, Ford Battery Plant, Shreveport, 68-73 & Gould Battery Plant, Shreveport, 73-75. Mem: AAAS; Am Chem Soc; The Chem Soc. Res: Synthesis and study of electronic effects in substituted benzo(b) furans and benzo(b) thiophenes. Mailing Add: Dept of Chem Centenary Col of La Box 4188 Shreveport LA 71104

LOWREY, GEORGE HARRISON, b Mansfield, Ohio, Nov 14, 17; m 43; c 3. PEDIATRICS. Educ: Univ Mich, Ann Arbor, AB, 40, MD, 43. Prof Exp: Rockefeller Found fel, Med Ctr, Univ Mich, Ann Arbor, 46-47, from asst prof to prof pediat, 49-70; PROF PEDIAT & ASSOC DEAN STUDENT AFFAIRS, SCH MED, UNIV CALIF, DAVIS, 70- Concurrent Pos: From assoc prof to prof postgrad med, Univ Mich, Ann Arbor, 56-70; sci ed, Mich Med, 66-70; adv continuing educ, Calif Med Asn, 70-73; Earl H Baxter vis prof, Ohio State Univ, 71- Mem: AMA; Am Pediat Soc; Am Acad Pediat; Am Diabetic Asn; Asn Am Med Cols. Res: Growth and development of children; endocrinology in children. Mailing Add: Off of Student Affairs Univ of Calif Sch of Med Davis CA 95616

LOWREY, ROBERT DEAN, b Washington, Pa, Aug 17, 23; m 51; c 2. PHOTOCHEMISTRY. Educ: Univ Akron, BS, 45; Purdue Univ, MS, 49, PhD(org chem), 50. Prof Exp: Asst prof chem, Univ Alaska, 50-51; proj chemist, Thiokol Chem Corp, Alaska, 51-55; SR RES SPECIALIST, VISUAL PROD DIV, MINN, MINING & MFG CO, 56- Mem: AAAS; Soc Photog Sci & Eng; Am Inst Aeronaut & Astronaut. Res: Optical properties of polymers; imaging chemistry by non silver systems; new coatings with optical clarity. Mailing Add: 28 Pheasant Lane St Paul MN 55110

LOWREY, ROBERT S, b Rome, Ga, June 9, 34; m 58; c 3. ANIMAL NUTRITION, BIOCHEMISTRY. Educ: Univ Ga, BSA, 56, MS, 57; Cornell Univ, PhD(animal nutrit), 61. Prof Exp: Asst prof radiation res, Univ Tenn, 61-64; assst prof animal sci, Coastal Plain Exp Sta, 64-70, assoc prof, 70-73, PROF ANIMAL SCI, UNIV GA, 73- Mem: Am Soc Animal Sci; Am Dairy Sci Asn; Am Inst Nutrit. Res: Nutrition of beef cattle; evaluation of chemical and in vivo procedures as predictors of forage quality; development of forage systems for beef cattle. Mailing Add: Dept of Animal Sci Univ of Ga Athens GA 30601

LOWRIE, HARMAN SMITH, b Soochow, China, June 30, 26; m 51; c 4. ORGANIC CHEMISTRY. Educ: Ohio State Univ, BSc, 48, PhD(chem), 52. Prof Exp: Res chemist med dept, Ohio State Univ, 52-54; CHEMIST, G D SEARLE & CO, 54- Mem: Am Chem Soc. Res: Organic synthesis; heterocyclics; medicinal chemistry. Mailing Add: 832 Rolling Pass Glenview IL 60025

LOWRIGHT, RICHARD HENRY, b Bethlehem, Pa, Aug 31, 40; m 66. SEDIMENTOLOGY. Educ: Franklin & Marshall Col, AB, 62; Pa State Univ, PhD(geol), 71. Prof Exp: Teacher pub sch, NY, 64-66; ASST PROF GEOL, SUSQUEHANNA UNIV, 71- Mem: Geol Soc Am; Soc Econ Paleontologists & Mineralogists. Res: Application of hydraulic equivalence studies to ancient rocks. Mailing Add: Dept of Geol Susquehanna Univ Selinsgrove PA 17870

LOWRY, BETTY JEAN RAGLE, b San Diego, Calif, May 16 34; m 57. ORGANIC CHEMISTRY. Educ: Univ Calif, Berkeley, BS, 56; Univ Wash, Seattle, PhD(org chem), 63. Prof Exp: Res chemist, E I du Pont de Nemours & Co, 62-64; res assoc synthesis psychotropic agents, Col Pharm, Univ Wash, Seattle, 65, sr res assoc, 66-70. Mem: Am Chem Soc. Res: Organic synthesis. Mailing Add: 14007-183rd Ave SE Renton WA 98055

LOWRY, CHARLES DOAK, b Chicago, Ill, June 16, 96; m 22, 61; c 4. CHEMISTRY, INFORMATION SCIENCE. Educ: Northwestern Univ, BS, 17, MS, 20; Harvard Univ, PhD(org chem), 24. Prof Exp: Analyst, Armour & Co, Ill, 19; instr org chem, Stanford Univ, 22-24; Sheldon traveling fel from Harvard Univ to Munich, 24-25; res chemist, Am Meat Inst, Ill, 25-27; chemist, Universal Oil Prod Co, 27-50; prof, Loyola Univ, Ill, 51; consult chem mgt res, 52-59; partner, Lowry-Preston Tech Abstracts Co, 60-62; partner Lowry-Cocroft Abstracts, 63-72; RETIRED. Concurrent Pos: Exec dir petrol panel, Res & Develop Bd, Defense Dept, 48. Mem: AAAS· Com Develop Asn; Am Soc Info Sci; Am Chem Soc; Am Oil Chem Soc. Res: Antioxidants; petroleum chemistry; chemical market research. Mailing Add: 321 Euclid Ave Oak Park IL 60302

LOWRY, EDWARD MACLEAN, b Dallas, Tex, Apr 9, 14; m 44; c 2. BIOLOGY. Educ: Ripon Col, AB, 36; Univ Mo, PhD(zool), 53. Prof Exp: Assoc biologist fisheries, State Conserv Comn, Mo, 47-54; asst prof biol, Meredith Col, 54-56; res asst prof zool, NC State Col, 56-59; assoc prof sci & math, 59-64, PROF BIOL, UNIV WIS-STOUT, 64-, CHMN DEPT, 71- Mem: Am Fisheries Soc. Res: Farm pond fish culture; fish growth; limnology; aquatic ecology. Mailing Add: Dept of Biol Univ of Wis-Stout Menomonie WI 54751

LOWRY, ERIC G, b Berlin, Ger, Nov 23, 16; US citizen; m 54; c 1. PHYSICAL CHEMISTRY. Educ: Univ Geneva, PhD(phys chem), 43. Prof Exp: Res chemist fluorochem, Gen Chem Div, Allied Chem Corp, 47-49; res chemist photog, Remington-Rand Div, Sperry Rand Corp, 51-58; res chemist lithography, Polychrome Corp, 59; res chemist, 59-65, CHIEF CHEMIST REPROGRAPHY, CHARLES BRUNING CO DIV, ADDRESSOGRAPH-MULTIGRAPH CORP, 65- Mem: AAAS; Am Chem Soc; Soc Photog Sci & Eng; Tech Asn Pulp & Paper Indust. Res: Reprography. Mailing Add: 73 Lewis St Middletown CT 06457

LOWRY, GEORGE GORDON, b Chico, Calif, Jan 12, 29; m 53; c 4. PHYSICAL CHEMISTRY. Educ: Chico State Col, AB, 50; Stanford Univ, MS, 52; Mich State Univ, PhD(phys chem), 63. Prof Exp: Res asst, Stanford Res Inst, 51; res chemist, Dow Chem Co, 51-62; NSF fel, 62-63; from asst prof to assoc prof chem, Claremont Men's Col, 63-68; ASSOC PROF CHEM, WESTERN MICH UNIV, 68- Mem: Am Chem Soc; The Chem Soc. Res: Polymerization kinetics and processes; copolymerization; statistical theory of kinetic chain processes; physical properties of liquids and non-ionic solutions. Mailing Add: Dept of Chem Western Mich Univ Kalamazoo MI 49008

LOWRY, GERALD LAFAYETTE, b Harrisburg, Pa, Sept 12, 28; m 49; c 3. FORESTRY, SOIL SCIENCE. Educ: Pa State Univ, BS, 53; Ore State Univ, MS, 55; Mich State Univ, PhD(forestry), 61. Prof Exp: Asst, Ore State Univ, 53-55; instr stripmine reclamation, Ohio Agr Exp Sta, Wooster, 55-61; res forester, Pulp & Paper Res Inst Can, 61-72; ASSOC PROF, STEPHEN F AUSTIN STATE UNIV, 72- Concurrent Pos: Asst prof, Ohio State Univ, 57-58; spec res asst, Mich State Univ, 58-59; vchmn forestry comt, Coun Fertilizer Appln, 61-63, chmn, 63-65. Mem: Soc Am Foresters; Soil Sci Soc Am; Can Inst Forestry. Res: Forest soil-site relationships; rehabilitation of burned and cutover lands; coal stripmine reclamation; soil chemistry, physics and fertility; tree physiology and silviculture. Mailing Add: Sch of Forestry Stephen F Austin State Univ Nacogdoches TX 75961

LOWRY, JEAN, b Indianapolis, Ind, Feb 7, 21. GEOLOGY. Educ: Pa State Univ, BS, 42; Yale Univ, PhD(geol), 51. Prof Exp: Jr economist, Off Price Admin, 42-43; jr geologist, US Geol Surv, 43-46, asst geologist, 46-49; dist geologist, State Geol Surv, Va, 49-57; from asst to assoc prof geol, 58-68, PROF GEOL, ECAROLINA UNIV, 68- Concurrent Pos: Vis prof, Concepcion Univ, 62-63. Mem: Geol Soc Am; Nat Asn Geol Teachers. Res: Stratigraphy and structure of southern Appalachians; caves. Mailing Add: 211 S Eastern St Greenville NC 27834

LOWRY, LARRY KENNETH, clinical chemistry, biochemistry, see 12th edition

LOWRY, MURRILL M, zoology, see 12th edition

LOWRY, NANCY, b Newburgh, NY, Sept 4, 38; m 61; c 3. PHYSICAL ORGANIC CHEMISTRY. Educ: Smith Col, AB, 60; Mass Inst Technol, PhD(chem), 65. Prof Exp: Res assoc chem, Mass Inst Technol, 65-66 & Amherst Col, 66-67; lectr, Smith Col, 67-69, res assoc, 69-70; asst prof, 70-74, ASSOC PROF CHEM, HAMPSHIRE COL, 74- Mem: AAAS; NY Acad Sci; Inst Soc, Ethics, & Life Sci. Res: Free radical mechanisms and stereochemistry. Mailing Add: Sch of Nat Sci & Math Hampshire Col Amherst MA 01002

LOWRY, OLIVER HOWE, b Chicago, Ill, July 18, 10; m 35; c 5. PHARMACOLOGY, BIOCHEMISTRY. Educ: Northwestern Univ, BS, 32; Univ Chicago, PhD(biochem), 37. Prof Exp: Instr biochem, Harvard Med Sch, 37-42; mem staff, Pub Health Res Inst, NY, 42-44, assoc chief, Div Physiol & Nutrit, 44-47; dean, 55-58, PROF PHARMACOL & HEAD DEPT, EDWARD MALLINCKRODT SCH MED, WASH UNIV, 47- Concurrent Pos: Commonwealth Found fel, Carlsberg Lab, Copenhagen Univ, 37. Honors & Awards: Midwest Award, Am Chem Soc, 62, Scott Award, 63. Mem: Nat Acad Sci; Am Soc Pharmacol & Exp Therapeut; Am Soc Biol Chem; Am Chem Soc; Histochem Soc. Res: Tissue electrolytes; chemistry of aging; nutrition and detection of nutritional deficiency; histochemistry; neurochemistry. Mailing Add: Edward Mallinckrodt Sch of Med Wash Univ St Louis MO 63110

LOWRY, PHILIP HOLT, b New York, NY, Feb 20, 18; m 45; c 2. OPERATIONS RESEARCH. Educ: Princeton Univ, AB, 39; Yale Univ, MA, 42, PhD(int rels), 49. Prof Exp: Meteorologist, Brookhaven Nat Lab, 47-51; opers analyst, Opers Res Off, Johns Hopkins Univ, 51-61; opers analyst, Res Anal Corp, 61-72; OPERS ANALYST, OPER ANAL DIV, GEN RES CORP, 72- Mem: Opers Res Soc Am; Am Meteorol Soc; Am Astron Soc. Res: Military operations research; impact of technology on international relations; nuclear policy and strategy. Mailing Add: Oper Anal Div Gen Res Corp McLean VA 22101

LOWRY, ROBERT JAMES, b Chelsea, Mich, Aug 26, 12; m 34; c 1. BOTANY. Educ: Univ Mich, BS, 40, MS, 41, PhD(bot), 47. Prof Exp: Res assoc, Univ Mich, 42-45; asst prof bot, Mich State Univ, 46-48; from asst prof to assoc prof, 48-58, PROF BOT, UNIV MICH, ANN ARBOR, 59- Mem: AAAS; Am Bryol & Lichenological Soc; Bot Soc Am; Am Genetic Asn; Torrey Bot Club. Res: Cytotaxonomy; electron microscopy. Mailing Add: Dept of Bot Univ of Mich Ann Arbor MI 48104

LOWRY, THOMAS HASTINGS, b New York, NY, June 16, 38; m 61; c 3. ORGANIC CHEMISTRY. Educ: Princeton Univ, AB, 60; Harvard Univ, PhD(chem), 65. Prof Exp: NIH fel chem, Mass Inst Technol, 64-65, res assoc, 65-66; ASST PROF CHEM, SMITH COL, 66- Mem: Am Chem Soc; Am Phys Soc. Res: Physical organic and free radical chemistry. Mailing Add: Dept of Chem Smith Col Northampton MA 01060

LOWRY, WALLACE DEAN, b Medford, Ore, Oct 5, 17; m 42. GEOLOGY. Educ: Ore State Univ, BS, 39, MS, 40; Univ Rochester, PhD(geol), 43. Prof Exp: Geologist, Dept Geol & Mineral Indust, Univ Ore, 42-47; geologist, Texaco, Inc, 47-49; assoc prof, 49-58, PROF GEOL, VA POLYTECH INST & STATE UNIV, 58- Mem: Fel Geol Soc Am; Am Asn Petrol Geol. Res: Late Cenozoic stratigraphy of the lower Columbia River basin; ferruginous bauxite deposits of Northwestern Oregon; silica sands of Western Virginia; porosity of sandstone reservoir rocks; role of Tertiary volcanism in tectonism; relation of silicification and dolomitization; geology of the Blue Mountains, Oregon; mechanics of Appalachian thrusting; North American geosynclines. Mailing Add: Dept of Geol Sci Va Polytech Inst & State Univ Blacksburg VA 24061

LOWRY, WILLIAM PRESCOTT, b Colon, Panama, Nov 2, 27; US citizen; m 52, 65; c 2. METEOROLOGY. Educ: Univ Cincinnati, AB, 50; Univ Wis, MS, 55; Ore State Univ, PhD(gen sci), 62. Prof Exp: Physicist, Snow, Ice & Permafrost Res Estab Corps Engrs, 51-53; res assoc meteorol, Univ Wis, 53-55; res meteorologist, Ore Forest Res Lab, 55-61; asst prof forest meteorol, Ore State Univ, 61-64, from asst prof to assoc prof biometeorol, 69-72; prof geog, Univ Ill, Urbana-Champaign, 72-76. Concurrent Pos: Meteorologist, Taft Sanit Eng Ctr, US Weather Bur, Ohio, 63; vis assoc prof, Univ Pa, 70-71. Mem: AAAS; Am Meteorol Soc. Res: Microclimatology; biometeorology; climatology; air pollution meteorology; urban climatology; climatic components of energy demand; urban effects on regional climate. Mailing Add: 3725 NW Hayes St Corvallis OR 97330

LOWRY, WILLIAM THOMAS, b Hobbs, NMex, Dec 11, 42; m 65; c 2. FORENSIC MEDICINE. Educ: E Tex State Univ, BS, 65, MS, 67; Colo State Univ, PhD(natural prod chem), 71; Am Inst Chemists, cert, 75. Prof Exp: Chemist, Fed Bur Invest, 65; res assoc biochem, Va Polytech Inst & State Univ, 71-72; spec agent, Fed Bur Invest, 72-73; TOXICOLOGIST, SOUTHWESTERN INST FORENSIC SCI, 73- Concurrent Pos: Assoc consult attend staff toxicol, Parkland Mem Hosp, 73-; instr path, Univ Tex Southwestern Med Sch, 73-75, instr path & forensic sci, 75-; adj asst prof chem, E Tex State Univ, 76- Mem: Am Acad Clin Toxicol; Am Acad Forensic Sci; Am Chem Soc; Am Inst Chemists; Am Soc Pharmacog. Res: Chemistry of respiration abnormalities in the brainstem in relation to sudden infant death syndrome, specifically chemistry of the synapse and glial cells. Mailing Add: Southwestern Inst Forensic Sci PO Box 35728 Dallas TX 75235

LOWSTUTER, WILLIAM ROBERT, b Denver, Colo, June 16, 13; m 38; c 3. ORGANIC CHEMISTRY. Educ: Allegheny Col, AB, 36; Univ Pittsburgh, PhD(org chem), 40. Prof Exp: Res chemist, Am Cyanamid Co, Conn, 40; res & develop chemist, Abrasive Co, Pa, 40-46; dir res, Atomix, Inc, Del, 46-49; res chemist applns res div, Hercules Powder Co, 49-54, res chemist sales res div, 54-58, res assoc chemist, Allegany Ballistics Lab, 58-60, tech employment mgr, 60-66; mgt consult, Warren King Assocs, Inc, Ill, 66-67; asst prof chem & chem eng & asst to head dept, Univ Ill, Urbana, 67-73; RETIRED. Mem: Am Chem Soc. Res: Development of consumer products; technical management and personnel recruiting. Mailing Add: 8718 Orient Way NE St Petersburg FL 33702

LOWTHER, GERALD EUGENE, b Lancaster, Ohio, Sept 16, 43; m 65; c 1. OPTOMETRY, VISUAL PHYSIOLOGY. Educ: Ohio State Univ, BSc, 66, OD, 67, MSc, 69, PhD(physiol optics), 72. Prof Exp: Pvt pract, 67-69; asst prof, 69-75, ASSOC PROF OPTOM & PHYSIOL OPTICS, COL OPTOM, OHIO STATE UNIV, 75- Concurrent Pos: Adv ophthal drugs adv comt, Food & Drug Admin, 73- Mem: AAAS; Am Acad Optom; Am Optom Asn. Res: Physiology of the cornea and tear film; contact lens development, fitting and patient care; optical performance of ophthalmic lenses. Mailing Add: Col of Optom Ohio State Univ Columbus OH 43210

LOWTHER, JAMES KERR, behavioral ecology, pollution, see 12th edition

LOWTHER, JOHN LINCOLN, b Burlington, Iowa, Sept 5, 43. COMPUTER SCIENCE. Educ: Univ Iowa, BA, 65, MS, 67, PhD(comput sci), 75. Prof Exp: Instr math, Southwest Minn State Col, 67-71; instr, 74-75, ASST PROF COMPUT SCI, MICH TECHNOL UNIV, 75- Mem: Asn Comput Mach; Math Asn Am; Sigma Xi. Res: Compiler writing; programming languages; computational complexity theory; computer science education. Mailing Add: Dept of Math & Comput Sci Mich Technol Univ Houghton MI 49931

LOWTHER, JOHN STEWART, b Cochrane, Ont, July 31, 25; m 53. PALEONTOLOGY, PALEOBOTANY. Educ: McGill Univ, BSc, 49, MSc, 50; Univ Mich, PhD(geol), 57. Prof Exp: From instr to asst prof, 56-72, ASSOC PROF GEOL, UNIV PUGET SOUND, 72- Mem: Paleont Soc; Bot Soc Am; Am Asn Petrol Geologists; Nat Asn Geol Teachers; Brit Paleont Asn. Res: Sedimentology; economic geology; Mesozoic paleobotany and stratigraphy; arctic paleobotany; geology education. Mailing Add: Dept of Geol Univ of Puget Sound Tacoma WA 98416

LOWY, BERNARD, b New York, NY, Feb 29, 16; m 50; c 2. MYCOLOGY. Educ: Univ Long Island, BS, 38; Univ Iowa, MS, 49, PhD(bot), 51. Prof Exp: Tech asst biol, Univ Long Island, 38-42, instr, 46-48; instr bot, Univ Iowa, 49-51; from asst to assoc prof, 51-62, PROF BOT & CUR MYCOL HERBARIUM, LA STATE UNIV, BATON ROUGE, 62- Concurrent Pos: Fulbright scholar, Peru, 58-59 & Brazil, 65-66; vis prof, Univ Tucuman, 59; Am Philos Soc grant, Mex, 62; Sigma Xi grant, Guatemala, 63; res partic, Orgn Trop Studies, Costa Rica, 64; mem numerous mycol expeds, Mex, Cent Am, SAm & West Indies, 50-74; consult ed, Revista Interam, Interam Univ, PR, 71-; chmn, Ethnomycol Sect, Int Mycol Cong, 75-77. Mem: Mycol Soc Am; Bot Soc Am; Am Bryol & Lichenological Soc; Mex Soc Mycol; Int Asn Plant Taxon. Res: Taxonomy and phylogeny of neotropical tremellaceous fungi; ethnomycology of Central America. Mailing Add: Dept of Bot La State Univ Baton Rouge LA 70803

LOWY, BERTRAM ALAN, b New York, NY, Sept 4, 23; m 50; c 2. BIOCHEMISTRY. Educ: City Col New York, BS, 47; Univ Ill, MS, 48; Cornell Univ, PhD, 53. Prof Exp: Asst, Protein Chem Div, Sloan-Kettering Inst, 48-52; res assoc biochem, Col Physicians & Surgeons, Columbia Univ, 54-55; from asst prof to assoc prof biochem in med, 55-71, PROF BIOCHEM & MED, ALBERT EINSTEIN COL MED, 71- Concurrent Pos: Res fel, Protein Chem Div, Sloan-Kettering Inst, 52-53; fel cytochem, Carlsberg Lab, Copenhagen, 53-54; USPHS career award, 59-69; guest investr, Biol Inst, Carlsberg Found, Copenhagen, 65-66. Mem: AAAS; Am Soc Biol Chem; Am Chem Soc; Harvey Soc; Brit Biochem Soc. Res: Nucleic acid and nucleotide metabolism; protein biosynthesis; erythroid tissue metabolism; synchronous cell growth. Mailing Add: Dept of Biochem Albert Einstein Col of Med Bronx NY 10461

LOWY, KARL, b Vienna, Austria, Nov 15, 04; US citizen; m 43. NEUROPHYSIOLOGY. Educ: Univ Vienna, MD, 27. Prof Exp: Intern, Allgemeines Krankenhaus, Vienna, Austria, 29-30; instr otolaryngol, Univ Vienna, 35-38; intern, Med Arts Ctr Hosp, New York, 38-39; SR RES ASSOC PSYCHOL, CTR BRAIN RES, UNIV ROCHESTER, 43-, CLIN ASSOC PROF OTOLARYNGOL, 59-, PROF PSYCHOL, 60- Concurrent Pos: Res fel psychol, Univ Rochester, 41-43. Mem: AAAS; Am Physiol Soc; AMA; fel Acoust Soc Am; Am Otol Soc. Res: Physiology of audition. Mailing Add: Ctr for Brain Res Univ of Rochester Rochester NY 14627

LOWY, MICHAEL JEFFREY, cultural anthropology, see 12th edition

LOWY, PETER HERMAN, b Vienna, Jan 3, 14; nat US; m 40; c 3. ORGANIC CHEMISTRY. Educ: Univ Vienna, Dr(chem), 36. Prof Exp: Food chemist, Rochester, NY, 40-45; res asst, 46-49, from res fel chem to sr res fel, 49-72, RES ASSOC BIOL, CALIF INST TECHNOL, 72- Mem: AAAS; Am Chem Soc; Fedn Am Socs Exp Biol; Am Soc Hemat. Res: Organic chemical synthesis, particularly of radioactive compounds; isolation and structure determination of bio-organic substances. Mailing Add: Dept of Biol Calif Inst of Technol Pasadena CA 91109

LOY, JAMES BRENT, b Borger, Tex, Feb 28, 41; m 67. PLANT SCIENCE, DEVELOPMENTAL GENETICS. Educ: Okla State Univ, BS, 63; Colo State Univ, MS, 65, PhD(genetics), 67. Prof Exp: Asst prof plant sci, 67-73, ASSOC PROF PLANT SCI, UNIV NH, 73- Concurrent Pos: Vis scholar bot, Univ Calif, Berkeley, 74-75. Mem: Am Soc Hort Sci; Am Genetic Asn; Soc Econ Bot; Am Soc Plant Physiol; Bot Soc Am. Res: Cucurbit breeding; physiological genetics of dwarfism in Citrullus lanatus; hormonal regulation of cell division in shoot meristems. Mailing Add: Dept of Plant Sci Univ of NH Durham NH 03824

LOY, ROBERT GRAVES, b Prescott, Ariz, Feb 7, 24; m 51; c 5. ANIMAL PHYSIOLOGY. Educ: Ariz State Univ, BS, 55; Univ Wis, MS, 56, PhD(physiol of reprod), 59. Prof Exp: Instr genetics, Univ Wis, 56-59; asst prof animal husb, Univ Calif, Davis, 59-66; from asst to assoc prof vet sci, Univ Ky, 66-71; agr consult, 71-74; ASSOC PROF VET SCI, UNIV KY, 74- Mem: Soc Study Reproduction; Am Soc Animal Sci; Endocrine Soc. Res: Physiology and endocrinology of reproduction in farm animals. Mailing Add: Dept of Vet Sci Univ of Ky Lexington KY 40506

LOY, WAYNE RICHARD, b Macomb, Ill, Sept 7, 21; m 47; c 1. WATER CHEMISTRY. Educ: Western Ill Univ, BS, 43, MS, 47. Prof Exp: Instr chem, Wis State Univ, Oshkosh, 47-49; assoc prof, Wis State Col & Inst Technol, 49-59; ASSOC PROF CHEM, UNIV WIS-PLATTEVILLE, 59- Mem: Am Chem Soc; Am Soc Eng Educ. Res: Radioisotope techniques in water analysis and sanitary sciences. Mailing Add: Dept of Chem Univ of Wis Platteville WI 53818

LOY, WILLIAM GEORGE, b Dawson, NMex, Oct 13, 36; m 62; c 1. GEOGRAPHY, CARTOGRAPHY. Educ: Univ Minn, Duluth, BA, 58, Univ Minn, Minneapolis, PhD(geog), 67; Univ Chicago, MS, 62. Prof Exp: Asst prof geog, 67-73, ASSOC PROF GEOG, UNIV ORE, 73- Concurrent Pos: Dir, Atlas of Ore Proj, 74-76. Mem: Asn Am Geog; Am Geog Soc; Am Soc Photogram. Res: Interpretation in archaeology; atlas cartography; geomorphology; air photo use in archaeology and schools. Mailing Add: Dept of Geog Univ of Ore Eugene OR 97403

LOYACANO, HAROLD ANTHONY, b New Orleans, La, Mar 30, 41; m 62; c 2. ICHTHYOLOGY, FISHERIES MANAGEMENT. Educ: Tulane Univ, BS, 62; La State Univ, Baton Rouge, MS, 67; Auburn Univ, PhD(fisheries mgt), 70. Prof Exp: ASST PROF ENTOM & ECON ZOOL, CLEMSON UNIV, 70- Mem: Am Fisheries Soc; World Maricult Soc; Am Soc Ichthyologists & Herpetologists. Res: The increase of fish production in ponds, cages and raceways; identification of fish species present in South Carolina. Mailing Add: Dept of Entom & Econ Zool Clemson Univ Clemson SC 29631

LOYD, COLEMAN MONROE, b Broken Bow, Nebr, Oct 3, 16; m 42; c 2. PHYSICS. Educ: Nebr State Col, BS, 39; Wayne State Univ, MA, 48; Tex A&M Univ, MS, 55. Prof Exp: High sch teacher, Nebr, 39-42; mathematician, Jam Handy Orgn, 42-45; design engr, Gen Motors Corp, 45-46; instr physics, Assoc Cols Upper NY, Utica, 46-48; asst prof, WVa Inst Technol, 48-53; from instr to asst prof, 53-74, ASSOC PROF PHYSICS, TEX A&M UNIV, 74-, COORDR, NSF TRAINING PROGS, 59- Concurrent Pos: Vis scientist, Tex Acad Sci, 59- Mem: Am Asn Physics Teachers. Res: Molecular structure of simple molecules using spectroscopic methods. Mailing Add: Dept of Physics Tex A&M Univ College Station TX 77843

LOYD, DAVID HERON, b Shreveport, La, July 3, 41; m 60; c 2. ATOMIC PHYSICS, NUCLEAR PHYSICS. Educ: Univ Tex, Austin, BS, 63, MA, 64, PhD(physics), 70. Prof Exp: ASST PROF PHYSICS, ANGELO STATE UNIV, 69- Mem: Am Phys Soc. Res: Atomic collisions. Mailing Add: Dept of Physics Angelo State Univ San Angelo TX 76901

LOYNACHAN, THOMAS EUGENE, b Oskaloosa, Iowa, Nov 18, 45; m 67; c 2. SOIL MICROBIOLOGY, SOIL FERTILITY. Educ: Iowa State Univ, BS, 68, MS, 72; NC State Univ, PhD(soil sci), 75. Prof Exp: ASST PROF AGRON, UNIV ALASKA, 75- Mem: AAAS; Soil Sci Soc Am; Am Soc Agron; Coun Agr Sci & Technol. Res: Soil testing procedures for state of Alaska; oil degradation in Arctic soils; revegetation of oil-spill areas in Arctic and sub-Arctic. Mailing Add: Agr Exp Sta PO Box AE Palmer AK 99645

LOZANO, EDGARDO A, b Tampico, Mex, Nov 20, 24; m 49; c 3. BACTERIOLOGY. Educ: Univ Tex, BA, 48; Univ Wis, MS, 54; Mont State Univ, PhD(microbiol), 65. Prof Exp: Bacteriologist vaccine prod, Agr Res Serv, 48-50; res, Am Sci Labs, 54-55; dept head prod & develop, Corn States Labs, 55-59; dir bio-prod, Philips Roxane Inc, 63-64; asst prof bact, 65-68, ASSOC PROF BACT, MONT STATE UNIV, 68- Res: Bacteriological antigens and their purification; bacterial toxins; electrophoresis; telemetry of domestic animals. Mailing Add: 1924 Sourdough Rd Bozeman MT 59715

LOZERON, HOMER ALAN, biochemistry, virology, see 12th edition

LOZIER, GERALD SCOTT, b Loudonville, Ohio, Oct 3, 30; m 54; c 2. PHYSICAL CHEMISTRY. Educ: Western Reserve Univ, BS, 52, MS, 53, PhD(chem), 56. Prof Exp: asst chem, Western Reserve Univ, 50-54; mem tech staff labs, 55-58, proj leader electrochem res & develop, 58-59, group leader battery res & develop, Semiconductor & Mat Div, 59-62, mgr, 62-63, sr engr dir energy conversion dept, 64-65, mgr thermoelec mat, 66, MEM TECH STAFF, RCA LABS, RCA CORP, 66- Honors & Awards: RCA Award, 57 & 63. Mem: Am Chem Soc; Electrochem Soc; Soc Am Archaeol. Res: Electrochemistry; primary, secondary and fuel cells; solid state problems related to physical chemistry; ultrasonics; thermoelectricity; high temperature chemistry; electrophotography; liquid crystals; color TV tube processing. Mailing Add: RCA Labs RCA Corp Princeton NJ 08540

LOZNER, EUGENE LEONARD, b Stamford, NY, Apr 29, 15; m 42; c 1. CLINICAL MEDICINE. Educ: Columbia Univ, AB, 33; Cornell Univ, MD, 37. Prof Exp: Intern, Albany Hosp, 37-38; from asst resident physician to resident physician, Boston City Hosp, 38-41; assoc path, Med Sch, George Washington Univ, 42-46; instr med, Harvard Med Sch, 46-47; assoc prof, 47-56, PROF MED, COL MED, STATE UNIV NY UPSTATE MED CTR, 56- Concurrent Pos: Fel, Thorndike Mem Lab, Boston City Hosp, Mass, 38-39; asst, Harvard Med Sch, 39-41; assoc physician, Children's Hosp, Mass, 46-47; asst, Peter Bent Brigham Hosp, 46-47; tutor, Harvard Univ, 47; dir, Clin Lab, Syracuse Univ Hosp, 47-56, attend physician, 48-; dir, Clin Labs & asst physician, Syracuse Mem Hosp, 48-56, attend physician, 56- Mem: AAAS; Am Soc Clin Invest; Soc Exp Biol & Med; Am Soc Clin Path; Am Fedn Clin Res (pres, 46). Res: Hematology; hemorrhagic diseases; vitamin C and K nutrition; thrombocytopenia; biochemistry; glucose tolerance; plasma preservation. Mailing Add: State Univ Hosp 750 E Adams St Syracuse NY 13210

LOZO, FRANK EDGAR, b Ft Worth, Tex, Aug 26, 14; m 39; c 4. GEOLOGY. Educ: Tex Christian Univ, BS, 35, MS, 37; Princeton Univ, PhD(geol), 41. Prof Exp: Instr biol & geol, Tex Christian Univ, 39-41, asst prof, 41-44; paleontologist, Shell Oil Co, 44-46, geologist, 46-48, sr geologist, Shell Develop Co, 48-63, res assoc, 63-71; CONSULT, PRESTWICK RES CTR, 71- Mem: Fel Geol Soc Am; Soc Econ Paleont & Mineral; Paleont Soc; Am Asn Petrol Geol. Res: Stratigraphy; micropaleontology; groundwater hydrology. Mailing Add: Prestwick Res Ctr 7223 Prestwick Rd Houston TX 77025

LOZZIO, BISMARCK BERTO, b Patagones, Arg, Jan 27, 31; m 55; c 1. HEMATOLOGY, IMMUNOLOGY. Educ: Bernardino Rivadavia Col, Arg, BS & BA, 49; Univ Buenos Aires, MD, 55. Prof Exp: Instr internal med & assoc gastroenterologist, Clin Univ Hosp & Inst Med Invest, Univ Buenos Aires, 55-58; assoc gastroenterologist, NIH, Buenos Aires, Arg, 58-65; res assoc, 65-67, from asst prof to assoc prof res, 68-75, RES PROF, MEM RES CTR, UNIV TENN, KNOXVILLE, 75- Concurrent Pos: Arg Nat Coun Sci & Technol res grant, NIH, Buenos Aires, 58-64; NSF grant, Univ Tenn, Knoxville, 67-69, NIH grant, 68-76, Am Cancer Soc grant, 70-75. Mem: Am Soc Hemat; Am Asn Immunol; Am Asn Cancer Res; Soc Exp Biol & Med; Am Fedn Clin Res. Res: Pathophysiology of the spleen; cell homeostasis; leukemia; antigens. Mailing Add: Univ of Tenn Mem Res Ctr 1924 Alcoa Hwy Knoxville TN 37920

LOZZIO, CARMEN BERTUCCI, b Buenos Aires, Arg, Dec 20, 31; m 55; c 1. MEDICAL GENETICS, CELL BIOLOGY. Educ: Univ Buenos Aires, physician, 55, MD, 60. Prof Exp: Physician in chg cytol, Rivadavia Hosp, Buenos Aires, 56-60; instr genetics, Univ Buenos Aires, 60-65; from res assoc to asst res prof, 65-72, ASSOC RES PROF MED GENETICS, MEM RES CTR & HOSP, UNIV TENN, KNOXVILLE, 72-, DIR BIRTH DEFECTS, CTR, 66- Concurrent Pos: Arg Asn Prog Sci Millet fel & Arg Nat Res Coun fel radiation res, Rivadavia Hosp & Arg AEC, 57-60; grants, Arg Nat Res Coun, Univ Buenos Aires, 61-65, Pan Am Union, Biol Div, Oak Ridge Nat Lab, 64, Am Cancer Soc, Univ Tenn, Knoxville, 66-71, Nat Found-March of Dimes, 66-, Physicians Med Educ & Res Found, 69-70, NIH, 69-71 & US Dept Health, Educ & Welfare, 70-74, Tenn Dept Welfare, 74-, Tenn Dept Ment Health, 74- Honors & Awards: Honor Cert, World Cong Obstet & Gynec & Int Cong Internal Med, 64. Mem: Genetics Soc Am; Genetics Soc Can; Am Asn Ment Deficiency; Am Soc Human Genetics; NY Acad Sci. Res: Studies on human genetics and cytogenetics; genetic counseling and prenatal diagnosis of hereditary disorders; experimental studies on cell culture of human diploid strains with genetic markers and the effect of antimetabolites on mammalian cell cultures. Mailing Add: Birth Defects Ctr Univ of Tenn Mem Res Ctr & Hosp Knoxville TN 37920

LU, ANTHONY Y H, b Hupei, China, Jan 12, 37; m 65; c 1. BIOCHEMISTRY. Educ: Nat Taiwan Univ, BS, 58; Univ NC, Chapel Hill, PhD(biochem), 66. Prof Exp: Fel inst sci & technol, Univ Mich, Ann Arbor, 66-70; sr biochemist, 70-74, RES FEL, RES DIV, HOFFMANN-LA ROCHE INC, 74- Mem: AAAS; Am Chem Soc; Am Soc Pharmacol & Exp Therapeut; Am Soc Biol Chemists; NY Acad Sci. Res: Basic research in biochemistry and biochemical pharmacology. Mailing Add: Res Div Hoffmann-La Roche Inc Nutley NJ 07110

LU, BENJAMIN CHI-KO, b Changchow, China, Mar 9, 32; m 62; c 2. GENETICS, CELL BIOLOGY. Educ: Taiwan Univ, BS, 55; Univ Alta, MS, 62, PhD(bot, genetics), 65. Prof Exp: Instr bot, Taiwan Univ, 58-60; fel fungal genetics, Cambridge Univ, 66-67; vis fel, Copenhagen Univ, 66; asst prof, 67-70, ASSOC PROF GENETICS, UNIV GUELPH, 70- Concurrent Pos: Nat Res Coun Can overseas fels, 65-67, res grant, Rask-Örsted Found fel & Carlsberg Found grant, 66-67. Mem: AAAS; Genetics Soc Am; Bot Soc Am; Am Soc Microbiol; Genetics Soc Can. Res: Fungal genetics; meiotic systems and fine structure of meiotic chromosomes in fungi; genetic recombination. Mailing Add: Dept of Bot & Genetics Univ of Guelph Guelph ON Can

LU, FRANK CHAO, b Hupeh, China, Mar 9, 15; nat US; m 39; c 2. PHARMACOLOGY. Educ: Cheeloo Univ, MD, 39. Prof Exp: Assoc ed, Coun on Pub, Chinese Med Asn, 40-42; sr asst pharmacol, Cheeloo Univ, 42-44, lectr, 45-47; lectr, WChina Union Univ, 44-45; pharmacologist, Food & Drug Labs, Can Dept Nat Health & Welfare, 51-60, actg head pharmacol & toxicol sect, 60-63, head, 63-65; CHIEF FOOD ADDITIVES, WHO, 65- Concurrent Pos: Res fel exp surg, McGill Univ, 47-48, med res fel pharmacol, 48-51. Mem: AAAS; Am Soc Pharmacol & Exp Therapeut; Soc Toxicol; NY Acad Sci; Can Physiol Soc. Res: Animal parasites in West China; distribution of sulfonamides in body fluids; physiology and pharmacology of coronary circulation; bioassay of drugs; cardiac glycosides; blood dyscrasias; toxicology of drugs, food additives, pesticides and contaminants. Mailing Add: WHO Food Additives Unit Ave Appia 1211 Geneva Switzerland

LU, GORDON GO, b Kiangsu, China, Jan 3, 16; nat US; m 48; c 1. PHARMACOLOGY, MEDICINE. Educ: Tung Chi Univ, MD, 41; Univ Md, PhD(pharmacol), 52. Prof Exp: From asst to asst prof pharmacol, Sch Med, Tung Chi Univ, China, 41-47; res assoc, Sch Med, Stanford Univ, 47-50; res pharmacologist, Schering Corp, 53-54; sr pharmacologist, Res Found, Johnson & Johnson, 54-57; pharmacologist, Coty, Inc, 57-59; from res assoc to asst res prof pharmacol, Sch Med, Univ Md, 59-62; assoc prof, Sch Pharm, George Washington Univ, 62-64; pharmacologist, Wallace Labs, NJ, 64-73; SCI ASST TO PRES, AFFILIATED MED RES INC, 74-, DIR LABS, 75- Mem: AAAS; Soc Exp Biol & Med; Am Soc Pharmacol & Exp Therapeut; NY Acad Sci. Res: Evaluation and safety studies of new drugs; cardiovascular, antifibrillatory, coronary vasodilators; experimental atherosclerosis; general anesthetics and immunopharmacology. Mailing Add: 5 Rocky Brook Rd Cranbury NJ 08512

LU, KUO CHIN, b Singapore, Dec 26, 17; US citizen; m 58; c 1. SOIL MICROBIOLOGY, PLANT PATHOLOGY. Educ: Nanking Univ, BS, 37; Ore State Univ, PhD(microbiol), 53. Prof Exp: Jr bacteriologist, Ore State Univ, 53-57; asst soil microbiologist, Cornell Univ, 57-59; res scientist, US Army Biol Warfare Lab, Md, 59-60; soil microbiologist, 60-67, PRIN MICROBIOLOGIST, FORESTRY SCI LAB, USDA, 67-; ASSOC PROF SOIL MICROBIOL, ORE STATE UNIV, 67- Mem: AAAS; Am Soc Microbiol; Am Phytopath Soc; Mycol Soc Am; Soil Sci Soc Am. Res: Antagonistic organisms against root-rot pathogens; biological control of forest diseases; rhizosphere association of mycorrhizal roots; influence of characteristic carbon-nitrogen ratio in decomposition of forest litters; biochemistry. Mailing Add: Forestry Sci Lab USDA 3200 Jefferson Way Corvallis OR 97331

LU, KUO HWA, b Antung, China, Jan 7, 23; US citizen; m 56; c 4. BIOSTATISTICS, GENETICS. Educ: Nat Cent Univ, China, BS, 45; Univ Minn, MS, 48, PhD(genetics), 51. Prof Exp: Agr adv, Continental Develop Found, 52-53; assoc prof appl statist, Utah State Univ, 56-60; assoc prof, 60-63, PROF BIOSTATIST, DENT SCH, UNIV ORE, 63-, PROF MED PSYCHOL, MED SCH, 71- Concurrent Pos: Eli Lilly fel, Univ Minn, 53-56; sta statistician, Utah Agr Exp Sta, 56-60; consult, Lab Nuclear Med & Radiation Biol, Univ Calif, Los Angeles, 63-, NIH fel, vis prof, 66-67; consult, Appl Math Assoc, Inc, 63-; adj prof, Portland State Univ, 67- Mem: AAAS; Biomet Soc; Am Math Soc; Am Statist Asn; Int Asn Dent Res. Res: Development and application of statistical methodology in biomedical research; statistical methods; dental public health; actuarial investigations in medical and dental insurance programs. Mailing Add: Univ of Ore Health Sci Ctr Portland OR 97201

LU, KWANG-TZU, b China. ATOMIC SPECTROSCOPY. Educ: Nat Taiwan Univ, BS, 63; Univ Chicago, PhD(physics), 71. Prof Exp: Res assoc physics, Univ Ariz, 71-72; consult, Univ Chicago, 72-73; res assoc, Imp Col London, 73-75; RES ASSOC PHYSICS, ARGONNE NAT LAB, 76- Res: Non-linear optical spectroscopy. Mailing Add: Chem Div Argonne Nat Lab Argonne IL 60439

LU, MARY KWANG-RUEY CHAO, b Liao-ning, China, Sept 6, 35; US citizen; m 61; c 2. ORGANIC CHEMISTRY, MATHEMATICS. Educ: Notre Dame Col, Ohio, BS, 59; Univ Detroit, MS, 61; Univ Tenn, Knoxville, PhD(org chem), 68. Prof Exp: Technician, Chem Lab, NY Hosp, New York, 59; chemist, US Testing Co, Inc, 61-63; asst prof chem, Morris Col, SC, 63-64; PROF CHEM & MATH, LINCOLN MEM UNIV, 68- Mem: AAAS; Am Chem Soc. Res: Organometallic chemistry. Mailing Add: Lincoln Mem Univ Harrogate TN 37752

LU, MATTHIAS CHI-HWA, b Fukien, China, Jan 3, 40; m; c 2. PHARMACY, MEDICINAL CHEMISTRY. Educ: Kaohsiung Med Col, Taiwan, BSc, 63; Ohio State Univ, PhD(med chem), 69. Prof Exp: Res asst med chem, Univ Iowa, 64-67 & Ohio State Univ, 67-69; res assoc, Col Pharm, Univ Mich, Ann Arbor, 69-71, instr, 71-72, asst prof, 72-73; ASST PROF MED CHEM, COL PHARM, UNIV ILL MED CTR, 73- Mem: Am Chem Soc. Res: Steroidogenesis and metabolisms; enzyme inhibitors, synthesis and bioassay; agents for insect growth inhibition, radiation protection and heart diseases. Mailing Add: Dept of Med Chem Col of Pharm Univ of Ill Med Ctr 833 S Wood Chicago IL 60680

LU, PHILLIP KEHWA, b Anhui, China, Oct 11, 32; m 59; c 3. ASTRONOMY, PHYSICS. Educ: Maritime Col, Taiwan, BS, 60; Wesleyan Univ, MA, 65, DSc, 69; Columbia Univ, PhD(astron & sci educ), 70. Prof Exp: Math analyst inst math, Chinese Acad Sci, 60-63; instr comput sci, Jefferson Prof Inst, 65-67; res assoc astron observ, Yale Univ, 67-70; chmn earth & space sci dept, 73-74, ASST PROF ASTRON, OBSERV, WESTERN CONN STATE COL, 70- Concurrent Pos: Consult, Bd Educ, New York, 74-75; sci educ scholar, NSF, 74-75. Mem: Fel Royal Astron Soc; Am Astron Soc; Am Phys Soc; Sigma Xi. Res: Quasistellar sources; galaxies; photometry and astrometry. Mailing Add: Western Conn State Col Observ Danbury CT 06811

LU, PONZY, b Shanghai, China, Oct 7, 42; US citizen. MOLECULAR BIOLOGY. Educ: Calif Inst Technol, BS, 64; Mass Inst Technol, PhD(biophys), 70. Prof Exp: Arthritis Found fel biophys, Max Planck Inst Biophys Chem, 70-73; Europ Molecular Biol Orgn fel genetics, Univ Geneva, 73; ASST PROF CHEM, UNIV PA, 73-

Concurrent Pos: Am Cancer Soc res grant, 73-75. Mem: AAAS; Biophys Soc; Sigma Xi. Res: Molecular compoents involved in the regulation of gene expression. Mailing Add: Dept of Chem Univ of Pa Philadelphia PA 19174

LUBAN, MARSHALL, b Seattle, Wash, May 29, 36; m 59; c 4. PHYSICS. Educ: Yeshiva Univ, AB, 57; Univ Chicago, SM, 58, PhD(theoret physics), 62. Prof Exp: Mem, Inst Advan Study, 62-63; asst prof physics, Univ Pa, 63-66; Guggenheim Mem Found fel, 66-67, chmn dept, 67-70, dean fac natural sci, 69-71, ASSOC PROF PHYSICS, BAR-ILAN UNIV, ISRAEL, 67- Concurrent Pos: Mem, Israel Coun Res & Develop, 70- Mem: Am Phys Soc. Res: Statistical mechanics and the many-body problem. Mailing Add: Dept of Physics Bar-Ilan Univ Ramat-Gan Israel

LUBAROFF, DAVID MARTIN, b Philadelphia, Pa, Feb 1, 38; m 61; c 3. IMMUNOLOGY. Educ: Philadelphia Col Pharm & Sci, BS, 61; Georgetown Univ, MS, 64; Yale Univ, PhD(microbiol), 67. Prof Exp: Assoc, 69-70, ASST PROF IMMUNOL, UNIV PA, 70- Concurrent Pos: USPHS fel, Univ Pa, 67-69. Mem: AAAS. Res: Delayed hypersensitivity reactions; transplantation immunology; immunologic tolerance; biology of immunoreactive cells. Mailing Add: Dept of Path Univ of Pa Philadelphia PA 19104

LUBARSKY, ROBERT, b New York, NY, Apr 25, 20; m 46; c 3. MYCOLOGY. Educ: Okla Agr & Mech Col, BS, 50; Univ Calif, Los Angeles, MS, 52, PhD(microbiol), 54. Prof Exp: Asst, Univ Calif, Los Angeles, 52-54; res mycologist gen med res, Vet Admin Ctr, Los Angeles, 54-56; res mycologist, Los Angeles County Hosp, 56-57; from asst prof to assoc prof, 57-66, PROF BOT & BACT, CHAFFEY COL, 66- Mem: Mycol Soc Am; Am Soc Microbiol; NY Acad Sci. Res: Coccidioidomycosis and other deep mycoses. Mailing Add: Dept of Bot Chaffey Col Alta Loma CA 91701

LUBASH, GLENN DAVID, b Jersey City, NJ, Oct 14, 29; m 52; c 1. INTERNAL MEDICINE, NEPHROLOGY. Educ: Columbia Univ, BA, 50; MD, 54. Prof Exp: From intern to asst resident internal med, Bellevue Hosp, New York, 54-57, chief resident, 58-59; from instr to asst prof, Med Col, Cornell Univ, 61-67; assoc prof, Sch Med, Univ Md, Baltimore City, 67-71; prof, Sch Med, Univ NMex, 71-73; PVT PRACT, 73- Concurrent Pos: NY Heart Asn fel nephrol, Bellevue Hosp, New York, 57-58; Am Heart Asn advan res fel, 61-63; Am Heart Asn estab invest, NY Hosp-Cornell Med Ctr, 63-68; vis res assoc, St Mary's Hosp, London, 65; consult, USPHS Hosp & Vet Admin Hosp, Baltimore, Md, 68-71; mem vol fac, Sch Med, Univ NMex, 73-; mem coun kidney & cardiovasc dis, Am Heart Asn. Mem: Am Soc Artificial Internal Organs; Am Soc Nephrol; fel Am Col Physicians; Am Heart Asn. Res: Renin-angiotension-aldosterone system in hypertension. Mailing Add: 201 Cedar SE Albuquerque NM 87106

LUBATTI, HENRY JOSEPH, b Oakland, Calif, Mar 16, 37; m 68; c 3. PHYSICS. Educ: Univ Calif, Berkeley, AB, 60, PhD(physics), 66; Univ Ill, Urbana, MS, 63. Prof Exp: Physicist, Boeing Co, Wash, 60-61; res assoc physics, Linear Accelerator Lab, Univ Paris, 66-68; asst prof, Mass Inst Technol, 68-69; assoc prof, 69-74, PROF PHYSICS, UNIV WASH, 74-, SCI DIR VISUAL TECH LAB, 69- Concurrent Pos: Vis lectr, Int Sch Physics, Erice, Sicily 68 & Herceg-Novi Int Sch, Yugoslavia, 69; Alfred P Sloan fel, 71- Mem: AAAS; fel Am Phys Soc; Pattern Recognition Soc. Res: High energy physics using visual techniques such as bubble and streamer chambers; application of pattern recognition techniques of high energy physics; phenomenology of elementary particles. Mailing Add: Visual Tech Lab Dept of Physics FM-15 Univ of Wash Seattle WA 98195

LUBAWY, WILLIAM CHARLES, b South Bend, Ind, Nov 30, 44; m 71; c 1. PHARMACOLOGY. Educ: Butler Univ, BS, 67; Ohio State Univ, MS, 69, PhD(pahrmacol), 72. Prof Exp: ASST PROF PHARMACOL, COL PHARM, UNIV KY, 72- Mem: Acad Pharmaceut Sci. Res: Metabolism of tobacco smoke compoents by isolated organ systems; inhibition of drug metabolism by hydroxylated drug metabolites. Mailing Add: Col of Pharm Univ of Ky Lexington KY 40506

LUBBERTS, GERRIT, b Oldemarkt, The Netherlands, Sept 15, 35; US citizen; m 59; c 2. SOLID STATE ELECTRONICS. Educ: Univ Rochester, 62, MS, 67, PhD, 71. Prof Exp: Technician, Case-Hoyt Corp, 56-58; technician, 58-62, res physicist, 62-71, SR RES PHYSICIST, EASTMAN KODAK CO, 71- Mem: Optical Soc Am; Am Phys Soc. Res: Physical factors which influence the structure of photographic and x-ray images; tunneling phenomena in solids; surface barrier photodetectors. Mailing Add: Eastman Kodak Co Res Lab 1669 Lake Ave Rochester NY 14650

LUBCHENCO, LULA O, b Russia, Apr 21, 15; US citizen; m 39; c 4. PEDIATRICS. Educ: Univ Denver, AB, 35; Univ Colo, MD, 39. Prof Exp: Intern pediat, Med Ctr, Univ Colo, 30-40 & Strong Mem Hosp, Rochester, NY, 40-41; from resident to chief resident, Children's Hosp, Denver, Colo, 41-44; from assoc to assoc prof, 43-69, PROF PEDIAT, MED CTR, UNIV COLO, DENVER, 69- Concurrent Pos: Res fel, Children's Hosp, Denver, 45-46; consult, Colo State Dept Pub Health, 47- & Parents Blind Children, 50-; spec consult, Children's Bur Comt Ment Retardation, 63-; mem, Children's Bur Maternity & Newborn Adv Comt, 63-; mem, Child Health & Develop-Indices Gestational Age Comt, NIH, 63-; vis assoc prof, Univ Chile, 64. Honors & Awards: Eleanor Roosevelt Mem Award, 64. Mem: Am Acad Pediat; Am Pediat Soc; Soc Pediat Res; Am Asn Ment Deficiency; AMA. Res: Newborn pediatrics. Mailing Add: Newborn Serv Div Perinatal Med Univ of Colo Med Ctr Denver CO 80220

LUBECK, AXEL JOHN, b Shanghai, China, Mar 8, 43; US citizen; m 69; c 2. ANALYTICAL CHEMISTRY. Educ: Univ Colo, BA, 64. Prof Exp: Chemist anal chem, 64-74, SUPVR, CHEM ANAL, DENVER RES CTR, MARATHON OIL CO, 74- Mem: Sigma Xi. Res: Analysis of petroleum, petroleum products and petrochemicals; gas and liquid chromatographic separations. Mailing Add: Marathon Oil Co PO Box 269 Littleton CO 80120

LUBELL, DAVID, b Brooklyn, NY, Apr 1, 32; m 60; c 3. MATHEMATICS. Educ: Columbia Univ, BS, 56; NY Univ, PhD(math), 60. Prof Exp: Benjamin Peirce instr math, Harvard Univ, 60-61; res instr, NY Univ, 61-62; sr mathematician, Systs Res Group Inc, 62-66; asst prof math, NY Univ, 66-70; assoc prof, 70-74, PROF MATH, ADELPHI UNIV, 75- Concurrent Pos: Consult, Systs Res Group Inc, 66-67 & US Air Force, 67-68; math adv, Nassau County Med Ctr, 69-72. Mem: Am Math Soc. Res: Combinatorics; biomathematics. Mailing Add: Dept of Math Adelphi Univ Garden City NY 11530

LUBELL, MARTIN S, b New York, NY, June 5, 32; m 62. SOLID STATE PHYSICS. Educ: Mass Inst Technol, SB, 54; Univ Calif, Berkeley, MA, 56. Prof Exp: Asst, Univ Calif, 55-56; res physicist, Res Labs, Westinghouse Elec Corp, 56-67; RES PHYSICIST, OAK RIDGE NAT LAB, 67- Mem: Am Phys Soc. Res: Low temperature physics; superconductivity; fusion reactor technology. Mailing Add: Oak Ridge Nat Lab Bldg 9201-2 Thermonuclear Dept Oak Ridge TN 38730

LUBENSKY, TOM C, b Kansas City, Mo, May 7, 43; m 68; c 1. THEORETICAL SOLID STATE PHYSICS. Educ: Calif Inst Technol, BS, 64; Harvard Univ, MA, 65, PhD(physics), 69. Prof Exp: NSF fel physics, Fac Sci, Orsay, France, 69-70; res asst, Brown Univ, 70-71; asst prof, 71-75, ASSOC PROF PHYSICS, UNIV PA, 75- Concurrent Pos: Sloan Found fel, 75. Res: Liquid crystals, phase transitions, cooperative phenomena in random systems and applications of the Wilson renormalization group. Mailing Add: Dept of Physics Univ of Pa Philadelphia PA 19174

LUBEROFF, BENJAMIN JOSEPH, b Philadelphia, Pa, Apr 17, 25; m 44; c 3. INDUSTRIAL CHEMISTRY. Educ: Cooper Union, BChE, 49; Columbia Univ, AM, 50, PhD(phys org chem), 53. Prof Exp: Statutory asst chem, Columbia Univ, 49-51; inst, Cooper Union, 51-53; chemist high pressure lab, Am Cyanamid Co, 53-57; head gen chem res sect, Stauffer Chem Co, 57-62; mgr process res dept, Lummus Co, 62-70; CONSULT & ED, CHEMTECH, AM CHEM SOC, 70- Honors & Awards: Cooper Medal, 49. Mem: Am Chem Soc; Am Inst Chem; Am Inst Chem Eng. Res: Research management; applied physical chemistry; high temperature and pressure processes; petrochemicals; catalysis; pesticides; analytical chemistry. Mailing Add: 19 Brantwood Dr Summit NJ 07901

LUBIN, ARDIE, b Chicago, Ill, Feb 17, 20; m 68. PSYCHOPHYSIOLOGY, BIOSTATISTICS. Educ: Univ Chicago, BS, 39; Univ London, PhD(psychol), 51. Prof Exp: Lectr psychol, Inst Psychiat, Univ London, 48-51; res psychologist, Personnel Res Br, Adj Gen Off, US Dept Army 51-52; res psychologist, Walter Reed Army Inst Res, 53-64; RES PSYCHOLOGIST, US NAVAL HEALTH RES CTR, 64- Concurrent Pos: Lectr, Am Univ, 53- & Howard Univ, 56-; vis prof, Mich State Univ, 64; Fulbright fel, Univ Leeds, 68-69; mem, Clin Psychopharmacol Res Br, NIMH, 71-73. Mem: Am Psychol Asn; Psychomet Soc; Am Soc Psychophysiol Res: fel Royal Statist Soc; Brit Psychol Soc. Res: Mathematical psychology; sleep loss and biofeedback; biorhythms. Mailing Add: US Naval Health Res Ctr San Diego CA 94152

LUBIN, ARTHUR RICHARD, b Newark, NJ, Mar 24, 47. MATHEMATICAL ANALYSIS. Educ: Mich State Univ, BS, 67; Univ Wis, MA, 68, PhD(math), 72. Prof Exp: Asst prof math, Tulane Univ, 72-73 & Northwestern Univ, 73-75; ASST PROF MATH, ILL INST TECHNOL, 75- Mem: Am Math Soc. Res: Operator theory; functional analysis; Hardy spaces. Mailing Add: Dept of Math Ill Inst of Technol Chicago IL 60616

LUBIN, CLARENCE ISAAC, b Albany, Ga, Oct 15, 00; m 54. MATHEMATICAL ANALYSIS. Educ: Univ Cincinnati, ChemE, 23; Harvard Univ, PhD(math), 29. Prof Exp: Instr math, 23-36, from asst to assoc prof, 29-56, prof, 56-71, EMER PROF MATH, COL ENG, UNIV CINCINNATI, 71-; RES & WRITING, 71- Mem: Am Math Soc; Math Asn Am. Res: Differential equations; analysis. Mailing Add: 838 Clifton Hills Terr Cincinnati OH 45220

LUBIN, GERALD I, b Brooklyn, NY, Mar 29, 28; m 54; c 4. CHILD PSYCHIATRY. Educ: Univ Notre Dame, BS, 49; Univ Tex Med Br Galveston, MD, 54; Am Bd Psychiat & Neurol, dipl & cert adult psychiat, 62, cert child psychiat, 68. Prof Exp: Intern, Queen of Angels Hosp, Los Angeles, 54-55; resident psychiat, US Vet Admin Ctr & Univ Calif, Los Angeles, 55-59; psychiatrist in charge, San Pedro Br, Los Angeles Sch Guid Clin, 59-63; asst prof psychiat & pediat, Med Sch, Univ Southern Calif, 65-75, dir training child psychiat & asst dir, Univ Affil Prog, Childrens Hosp of Los Angeles, 65-75; CHILD PSYCHIATRIST, WINDWARD CHILDREN'S SERV TEAM, WINDWARD MENT HEALTH CLIN, HAWAII, 75- Concurrent Pos: Clin instr, Univ Calif, Los Angeles, 59-63; supvr residents in adult psychiat, Los Angeles Harbor Gen Hosp, 59-63; supvr residents in child psychiat, Los Angeles County Gen Hosp & Med Ctr, Univ Southern Calif, 60-69; chief prof educ, Porterville State Hosp, 63; fel child psychiat, Mt Sinai Hosp, Los Angeles, 63-65; med dir, Culver City Guid Clin, 65; adj asst prof educ, Univ Southern Calif, 67-; attend staff psychiatrist, Cedars Mt Sinai Med Str, Children's Hosp & Los Angeles County Gen Hosp; consult psychiatrist, Porterville State Hosp, Spastic Childrens Found, Los Angeles County, State Dept Ment Hyg, Dept Voc Rehab & Los Angeles County Dept Ment Hyg. Mem: Am Acad Child Psychiat; Am Geriat Soc; Am Med Asn; Am Psychiat Asn; Am Orthopsychiat Asn. Res: Adult and child psychiatry; interdisciplinary diagnosis and treatment; psychophysiological medicine; use of psychotropic drugs; group therapy with the retarded. Mailing Add: Children's Team Windward Mental Health Clin Kaneohe HI 96744

LUBIN, JONATHAN DARBY, b Staten Island, NY, Aug 10, 36. MATHEMATICS. Educ: Columbia Univ, AB, 57; Harvard Univ, AM, 58, PhD(math), 63. Prof Exp: Instr math, Bowdoin Col, 62-63, from asst to assoc prof, 63-67; assoc prof, 67-70, PROF MATH, BROWN UNIV, 70- Concurrent Pos: Assoc prof, Inst Henri Poincare, Univ Paris, 68-69; lectr, Math Inst, Copenhagen Inst, 74-75. Mem: Am Math Soc. Res: Algebraic geometry; number theory. Mailing Add: Dept of Math Brown Univ Providence RI 02912

LUBIN, MARTIN, b NY, Mar 30, 23; m 42; c 4. CELL BIOLOGY. Educ: Harvard Univ, AB, 42, MD, 45; Mass Inst Technol, PhD(biophys), 54. Prof Exp: Res assoc biol, Mass Inst Technol, 53-54; assoc pharmacol, Harvard Med Sch, 54-57, asst prof, 57-68; PROF MICROBIOL & HEAD DIV CELL BIOL, DARTMOUTH MED SCH, 68- Concurrent Pos: USPHS sr res fel, 56-61; Lalor Found fel, 57-59; Guggenheim fel, Lab Molecular Biol, Cambridge Univ, 65-66, Commonwealth Fund fel, 66-67. Mem: Am Soc Biol Chem; Am Soc Microbiol; Biophys Soc; Am Soc Cell Biol; Soc Gen Physiol. Res: Active transport; regulation of synthesis of macromolecules; ribosome structure and function; animal cells and viruses. Mailing Add: Dept of Microbiol Dartmouth Med Sch Hanover NH 03755

LUBIN, MOSHE J, b Tel Aviv, Israel, May 24, 38; US citizen; m 61; c 1. PLASMA PHYSICS, AERODYNAMICS. Educ: Israel Inst Technol, BSc, 61; Cornell Univ, PhD(aeronaut eng), 66. Prof Exp: Asst prof plasma physics, 65-68, assoc prof, 68-74, assoc prof optics, 70-74, DIR LAB LASER ENERGETICS, UNIV ROCHESTER, 70-, PROF, DEPT MECH & AEROSCI, INST OPTICS, 74- Mem: Am Phys Soc; Am Inst Aeronaut & Astronaut; NY Acad Sci; Inst Elec & Electronics Eng. Res: Nonlinear wave propagation in plasma media; magnetoaerodynamics; electromagnetic plasma interactions; thermonuclear fusion; laser fusion; x-ray laser development; fundamental interaction of radiation with matter. Mailing Add: Lab for Laser Energetics Univ of Rochester Rochester NY 14627

LUBITZ, BETTY BAUM, b New York, NY, Oct 14, 25; m 46; c 3. ANALYTICAL CHEMISTRY, ORGANIC CHEMISTRY. Educ: Brooklyn Col, BA, 45; Univ Mich, MS, 49, PhD(org chem), 57. Prof Exp: Technician, Rockefeller Inst, 45-46; instr chem, Univ Mich, 51, res assoc radiation lab, Res Inst, 56-57, assoc res chemist, 57-60; asst prof chem, Skidmore Col, 65-66; chemist, Behr-Manning Co, 66-67; ASSOC RES CHEMIST, STERLING-WINTHROP RES INST, 68- Mem: Am Chem Soc. Res: Dielectric relaxation; oxime equilibria; gel filtration; chromatography. Mailing Add: Anal Dept Sterling-Winthrop Res Inst Columbia Turnpike Rensselaer NY 12144

LUBITZ, CECIL ROBERT, b Brooklyn, NY, Mar 18, 25; m 46; c 3. NUCLEAR PHYSICS. Educ: US Naval Acad, BS, 45; Univ Mich, MSEE, 49, PhD(physics), 60.

Prof Exp: Res assoc elec eng, Res Inst, Univ Mich, 49-54; PHYSICIST, KNOLLS ATOMIC POWER LAB, GEN ELEC CO, 60- Mem: Am Nuclear Soc. Res: Neutron cross sections for technological applications. Mailing Add: Knolls Atomic Power Lab Gen Elec Co Schenectady NY 12301

LUBKIN, ELIHU, b Brooklyn, NY, Oct 25, 33; m 62; c 2. THEORETICAL PHYSICS. Educ: Columbia Univ, AB, 54, AM, 57, PhD(physics), 60. Prof Exp: Asst theoret physics radiation lab, Univ Calif, Berkeley, 59-61; res assoc high energy group, Brown Univ, 61-63, res asst prof theoret physics, 63-66; ASSOC PROF PHYSICS, UNIV WIS-MILWAUKEE, 66- Mem: Am Phys Soc. Res: Differential geometry used to interpret the old and for new constructions in physics; interpretation of quantum mechanics; quantum measurement theory; quantum psychology. Mailing Add: Dept of Physics Univ of Wis Milwaukee WI 53211

LUBKIN, GLORIA BECKER, b Philadelphia, Pa, May 16, 33; div; c 2. PHYSICS. Educ: Temple Univ, AB, 53; Boston Univ, MA, 57. Prof Exp: Mathematician aircraft div, Fairchild Stratos Corp, 54 & Letterkenny Ord Depot, US Defense Dept, 55-56; physicist tech res group, Control Data Corp, 56-58; actg chmn dept physics, Sarah Lawrence Col, 61-62; vpres, Lubkin Assocs, 62-63; assoc ed, 63-69, SR ED, PHYSICS TODAY, AM INST PHYSICS, 70- Concurrent Pos: Consult ctr for hist & philos of physics, Am Inst Physics, 66-67; Nieman fel, Harvard Univ, 74-75. Mem: Fel Am Phys Soc. Res: Nuclear physics; physics writing and editing. Mailing Add: Am Inst of Phys 335 E 45th St New York NY 10017

LUBLIN, PAUL, b New York, NY, Sept 8, 24; m 52; c 3. PHYSICAL CHEMISTRY. Educ: NY Univ, BA, 48; Purdue Univ, MS, 49. Prof Exp: Res chemist, Pigment Div, Am Cyanamid Co, 51-53; asst res staff mem, Res Div, Raytheon Mfg Co, 53-54; appln engr, Instrument Div, Philips Electronics, 54-56; sr engr, 56-59, res engr, 59-61, adv res engr, 61-63, eng specialist, 63-67, MEM TECH STAFF & SECT HEAD, GEN TEL & ELECTRONICS LABS, INC, BAYSIDE, 67- Mem: Am Crystallog Asn; Soc Appl Spectros; Sigma Xi; Electron Probe Anal Soc Am; fel Am Inst Chem. Res: Materials analysis, applications of x-ray diffraction and spectroscopy to structure and chemical identification of materials; electron probe and scanning electron microscopy as applied to electronic materials; laboratory automation. Mailing Add: Gen Tel & Electronics Labs Inc 40 Sylvan Rd Waltham MA 02154

LUBORSKY, FRED EVERETT, b Philadelphia, Pa, May 14, 23; m 46; c 3. PHYSICAL CHEMISTRY. Educ: Univ Pa, BS, 47; Ill Inst Technol, PhD(phys chem), 52. Prof Exp: Asst chemist, Ill Inst Technol, 47-51; res assoc res lab, 51-52, phys chemist instrument dept, 52-55, physicist appl physics unit, 55-58, PHYS CHEMIST, RES & DEVELOP CTR, GEN ELEC CO, 58- Concurrent Pos: Mem div eng & indust res, Nat Acad Sci, 52-55; pres, Magnetics Soc, Inst Elec & Electronics Eng, 75-; ed-in-chief, Inst Elec & Electronics Eng Transactions Magnetics, 72-75. Mem: AAAS; Am Chem Soc; Am Phys Soc; Am Inst Chem; NY Acad Sci. Res: Nucleation and growth of sub-micron size particles; development of single domain particle permanent magnetic materials; electrochemistry; magnetism; magnetic thin films; amorphous magnetic materials; magnetic separation. Mailing Add: Res & Develop Ctr Gen Elec Co Box 8 Schenectady NY 12301

LUBORSKY, SAMUEL WILLIAM, b Philadelphia, Pa, Jan 18, 31; m 53; c 3. PHYSICAL CHEMISTRY. Educ: Univ Mich, BS, 52; Northwestern Univ, PhD, 57. Prof Exp: Fel, 57-58, BIOCHEMIST, NIH, 58- Mem: Am Chem Soc. Res: Molecular biology; biochemistry. Mailing Add: Nat Inst of Health Bethesda MD 20014

LUBRAN, MYER MICHAEL, b London, Eng, Mar 9, 15; m 44; c 3. CLINICAL PATHOLOGY. Educ: Univ London, MB, BS, 38, BSc, 43, PhD(chem), 55; FRCPath. Prof Exp: Lectr physiol, biochem & clin chem, Guy's Med Sch, London, 38-43; lectr physiol & biochem, Med Sch, Univ Birmingham, 43-44; asst clin pathologist, Emergency Health Serv, Eng, 46-48; consult pathologist, Nat Health Serv, WMiddlesex Hosp, 48-64; prof path & dir clin chem, Sch Med, Univ Chicago, 64-70; PROF PATH, UNIV CALIF, LOS ANGELES, 70-; CHIEF CLIN PATH, HARBOR GEN HOSP, 70- Concurrent Pos: Mem hosp mgt comt & chmn med staff comt, WMiddlesex Hosp, 56-58, dir cent sterile supply dept & chmn cross-infection comt, 58-64; mem exam bd, Int Med Lab Technol, Eng; examr, Royal Col Path. Mem: AAAS; Am Soc Clin Path; Asn Clin Sci (pres, 74-75); Am Asn Clin Chem; NY Acad Sci. Res: Clinical pathology; trace metals. Mailing Add: Harbor Gen Hosp 1000 W Carson St Torrance CA 90509

LUBY, ELLIOT DONALD, b Detroit, Mich, Apr 3, 24; m 50; c 3. PSYCHIATRY, LAW. Educ: Univ Mo-Columbia, BS, 47; Wash Univ, MD, 49; Am Bd Psychiat & Neurol, dipl, 57. Prof Exp: Resident psychiat, Menninger Found, 50-51; sr asst surgeon, USPHS, 51-52; resident psychiat, Yale Univ, 52-54; chief adult inpatient sect, Lafayette Clin, Detroit, 57-62, assoc dir in chg clin serv, 62; PROF LAW, WAYNE STATE UNIV, 62-, PROF PSYCHIAT, 65- Honors & Awards: Gold Medal Award, Am Acad Psychosom Med, 62. Mem: NY Acad Sci; AMA; Am Psychiat Asn; Am Psychosom Soc; fel Am Col Psychiat. Res: Psychopharmacology; model psychoses, drug induced sleep deprivation; law and psychiatry; schizophrenia.

LUBY, PATRICK JOSEPH, b Zanesville, Ohio, May 20, 30; m 56; c 4. AGRICULTURAL ECONOMICS. Educ: Univ Dayton, BA, 52; Purdue Univ, MS, 54, PhD(agr econ), 56. Prof Exp: Instr agr econ, Purdue Univ, 54-56, asst prof, 56-58; economist, 58-66, gen mgr provisions, 66-71, GEN MGR PROVISIONS & PROCUREMENT, OSCAR MAYER & CO, 71-, VPRES, 72-, CORP ECONOMIST, 74- Mem: Am Agr Econ Asn. Res: Use of statistical methods to analyze and forecast meat and livestock supplies and prices; efficient marketing of livestock and meats. Mailing Add: Oscar Mayer & Co PO Box 1409 Madison WI 53701

LUBY, ROBERT JAMES, b Kansas City, Mo, Apr 13, 28; m 51; c 8. OBSTETRICS & GYNECOLOGY. Educ: Rockhurst Col, BS, 48; Creighton Univ, MD, 52, MS, 59. Prof Exp: Intern obstet & gynec, Creighton Mem St Joseph Hosp, Omaha, 52-53, resident, 55-58; assoc prof, Col Med, Univ Nebr, 68-69; assoc dir obstet & gynec, 69-72, PROF OBSTET & GYNEC, CREIGHTON UNIV, 69-, CHMN DEPT, 72- Mem: AMA; Am Col Obstet & Gynec; Am Col Surg. Res: Nutritional aspects of infectious perinatal morbidity and mortality. Mailing Add: Dept of Obstet & Gynec Creighton Univ Omaha NE 68108

LUCANSKY, TERRY WAYNE, b Massillon, Ohio, Aug 21, 42; m 66. BOTANY. Educ: Univ SC, BS, 64, MS, 67; Duke Univ, PhD(bot), 71. Prof Exp: ASST PROF BOT, UNIV FLA, 71- Mem: Bot Soc Am; Am Fern Soc; Am Inst Biol Sci; Sigma Xi. Res: Comparative anatomical and morphological studies of tropical pteridophytes; anatomical studies of aquatic plants in relation to their ecology. Mailing Add: Dept of Bot Univ of Fla Gainesville FL 32611

LUCAS, ALEXANDER RALPH, b Vienna, Austria, July 30, 31; US citizen; m 56; c 4. CHILD PSYCHIATRY. Educ: Mich State Univ, BS, 53; Univ Mich, MD, 57. Prof Exp: Rotating intern, Univ Mich Hosp, 57-58; resident child psychiat, Hawthorn Ctr, 58-59; resident psychiat, Lafayette Clin, Detroit, 59-61; resident child psychiat, Hawthorn Ctr, 61-62; from staff child psychiatrist to sr psychiatrist, Hawthorn Ctr, Northville, Mich, 62-67; from asst prof to assoc prof psychiat, Wayne State Univ, 67-71; HEAD SECT CHILD & ADOLESCENT PSYCHIAT, MAYO CLIN, 71-; ASSOC PROF PSYCHIAT, MAYO MED SCH, 73- Concurrent Pos: Res child psychiatrist & res coordr, Lafayette Clin, Detroit, 67-71; consult, Dept of Corrections, State of Minn Dept Pub Welfare, 72 & NIMH, 74- Mem: Am Orthop Asn; Am Psychiat Asn; Am Acad Child Psychiat; Soc Prof Child Psychiat; Soc Biol Psychiat. Res: Biologic aspects of child psychiatry. Mailing Add: Mayo Clin Rochester MN 55901

LUCAS, ALFRED MARTIN, b New Albany, Ind, Oct 16, 00; m 28; c 2. ANATOMY. Educ: Wabash Col, AB, 24; Washington Univ, PhD(zool), 29. Prof Exp: Lab asst zool, Wabash Col, 21-24; instr, Washington Univ, 24-25, instr zool & histol, Dent Sch, 25-30, asst prof cytol, Sch Med, 30-34; asst prof zool, Univ Iowa, 34-35; assoc prof, Iowa State Univ, 35-44; res zoologist, Agr Res Serv, USDA, 44-70; RETIRED. Concurrent Pos: With Marine Biol Lab, Woods Hole; grant, Wash Univ, Bermuda, 30; collab avian anat proj, USDA, 70-74; chmn, Int Comt Avian Anat Nomenclature, 71-73, subcomt chmn, Nomina Anatomica Avium, 71-; res prof poultry sci, Mich State Univ, 63- Honors & Awards: Tom Newman Mem Int Award Poultry Husb Res, 73. Mem: Am Soc Zool; Am Asn Anat; Poultry Sci Asn; Am Asn Avian Path; World Asn Vet Anat. Res: Avian, gross and microscopic anatomy; hematology. Mailing Add: 6035 Grand River Dr Grand Ledge MI 48837

LUCAS, COLIN CAMERON, b Winnipeg, Man, Dec 15, 03; m 42; c 1. BIOCHEMISTRY. Educ: Univ BC, BASc, 25, MASc, 26; Univ Toronto, PhD, 36. Hon Degrees: DSc, Acadia Univ, 64. Prof Exp: Prof chem, Brandon Col, 26-27 & 29-34; asst path chem, 27-29, Banting & Best Dept Med Res, 34-35, from asst to assoc prof, 35-46, prof, 46-69, EMER PROF, BANTING & BEST DEPT MED RES, UNIV TORONTO, 69- Mem: Am Chem Soc; Am Soc Biol Chem; fel Royal Soc Can; Nutrit Soc Can; fel Chem Inst Can. Res: Oceanography; clinical micro-methods for lead; cysteine and glutathione; silicosis; blood esterases; chemical composition of royal jelly; chemotherapy; lipids; lipotropic factors; nutrition. Mailing Add: 19 Hillside Ave Wolfville NS Can

LUCAS, DAVID OWEN, b Orange, Calif, Oct 19, 42; m 63; c 2. IMMUNOLOGY. Educ: Duke Univ, BA, 64, PhD(microbiol, immunol), 69. Prof Exp: ASST PROF MICROBIOL, COL MED, UNIV ARIZ, 70- Concurrent Pos: Res fel immunol, Children's Hosp Med Ctr, Harvard Med Sch, 68-70. Mem: AAAS; Am Soc Microbiol. Res: Cellular immunology; lymphocyte metabolism. Mailing Add: Dept of Microbiol Ariz Med Ctr Univ of Ariz Tucson AZ 85724

LUCAS, DONALD BROOKS, b Hillsboro, Ore, Dec 18, 14; m 41; c 2. ORTHOPEDIC SURGERY. Educ: Univ of the Pac, AB, 38; Univ Ore, MD, 42. Prof Exp: Lectr, 50-52, from asst prof to assoc prof, 53-64, vchmn dept, 57-70, PROF ORTHOP SURG, SCH MED, UNIV CALIF, SAN FRANCISCO, 64-, CHMN DEPT, 70- Concurrent Pos: Consult, Vet Admin Hosp, San Francisco, Calif & Children's Hosp. Mem: AAAS; AMA; Am Orthop Asn; Am Acad Orthop Surg; Int Soc Orthop Surg & Traumatol. Res: Scoliosis. Mailing Add: Dept of Orthop Surg Univ of Calif Sch of Med San Francisco CA 94122

LUCAS, EDGAR ARTHUR, b Franklin, Ind, Oct 28, 33; m 60; c 2. ANATOMY, NEUROPHYSIOLOGY. Educ: Ball State Univ, BA, 61, MS, 65; Univ Calif, PhD(anat), 72. Prof Exp: Teacher, Sch, Town of Griffith, 61-62; planner admin, Rocketdyne Div, NAm Rockwell Corp, 62-63, assoc res engr, 64-65; instr, 72-74, ASST PROF ANAT, MED CTR, UNIV ARK, LITTLE ROCK, 74- Concurrent Pos: NIMH grant, 73-74. Mem: Soc Neurosci; Asn Psychophysiol Study Sleep; Am Asn Anat. Res: Biological rhythms; sleep; neuroanatomy. Mailing Add: Dept of Anat Univ of Ark Med Ctr Little Rock AR 72201

LUCAS, FRED VANCE, b Grand Junction, Colo, Feb 7, 22; m 48; c 2. PATHOLOGY. Educ: Univ Calif, AB, 42; Univ Rochester, MD, 50. Prof Exp: From asst to asst prof path, Med Sch, Univ Rochester, 51-55; assoc prof, Col Physicians & Surgeons, Columbia Univ, 55-60; PROF & CHMN DEPT, SCH MED, UNIV MO-COLUMBIA, 60-, RES ASSOC, SPACE SCI RES CTR, 64- Concurrent Pos: Vet fel path, Med Sch, Univ Rochester, 50-51, Gleeson fel, 51-52, Lilly fel, 52-53; Lederle med fac award, 54; from asst resident to chief resident, Strong Mem Hosp, Rochester, 51-54; consult, Highland Hosp, Rochester, 54-55; assoc attend pathologist, Presby Hosp, New York, 55-; consult, NIH, 66- & Vietnam med educ proj, US AID-AMA, 67- Mem: Am Soc Exp Path; Harvey Soc; Am Asn Path & Bact; Am Soc Clin Path; Col Am Path. Res: Oxidative enzymes in proliferating tissue; plasma proteins studies employing C-14; hemoglobin; activation and inactivation of human chromosomes; heterogeneity of hemin of hemoglobin; ultrastructure of normal and abnormal human endometrium. Mailing Add: Dept of Path Univ of Mo Sch of Med Columbia MO 65201

LUCAS, GENE ALLAN, b Des Moines, Iowa, Oct 15, 28; m 48; c 3. GENETICS. Educ: Drake Univ, BA, 54, MA, 58; Iowa State Univ, PhD(genetics), 68. Prof Exp: Lab instr biol, Drake Univ, 54-59, instr, 60-67; asst genetics, Iowa State Univ, 61-66; asst prof biol, 68-74, ASSOC PROF BIOL, DRAKE UNIV, 74- Mem: AAAS; Genetics Soc Am; Int Oceanog Found. Res: Pigmentation, especially of aquarium fish; pigment genetics of Siamese fighting fish; application of biological principles to world problems; race and population problems; teaching; biology and behavior of Siamese fighting fish. Mailing Add: Dept of Biol Drake Univ Des Moines IA 50311

LUCAS, GEORGE BLANCHARD, b Philipsburg, Pa, Mar 8, 15; m 40, 55; c 7. PLANT PATHOLOGY. Educ: Pa State Col, BS, 40; La State Univ, MS, 42, PhD(plant path), 46. Prof Exp: From asst prof to assoc prof, 46-63, PROF PLANT PATH, NC STATE UNIV, 63- Mem: Bot Soc Am; Mycol Soc Am; Am Phytopath Soc. Res: Tobacco diseases; genetics of fungi. Mailing Add: Dept of Plant Path NC State Univ Raleigh NC 27607

LUCAS, GEORGE BOND, b New Orleans, La, Dec 21, 24; m 62; c 2. ORGANIC GEOCHEMISTRY. Educ: Tulane Univ, BS, 48; Iowa State Col, PhD(phys & org chem), 52. Prof Exp: Res fel, Northwestern Univ, 52-53; sr res chemist res div, Redstone Arsenal, Rohm and Haas Co, 53-56; from asst to assoc prof phys chem, 56-67, PROF PHYS CHEM, COLO SCH MINES, 67- Mem: Am Chem Soc; Geochem Soc. Res: Solution kinetics; organic mechanisms; origin of petroleum. Mailing Add: Dept of Chem Colo Sch of Mines Golden CO 80401

LUCAS, GLENNARD RALPH, b Marissa, Ill, Feb 22, 16; m 41; c 2. ORGANIC POLYMER CHEMISTRY. Educ: Monmouth Col, BS, 38; Columbia Univ, PhD(phys org chem), 42. Prof Exp: Asst chem, Monmouth Col, 36-38 & Columbia Univ, 38-41; res chemist, Gen Elec Co, Mass, 42-52, process engr, NY, 52-54, supvr process eng, 54-56, mgr adv proj develop, Mass, 56-58; RES DIR, SIGNODE CORP, 58- Mem: AAAS; Am Chem Soc; Am Mgt Asn; Soc Plastics Eng. Res: Mechanisms of organic reactions; polymer studies of styrene and silicone resins; plastic and steel strapping materials; high speed paint cure. Mailing Add: Res Dept Signode Corp 3650 W Lake St Glenview IL 60025

LUCAS, HENRY LAURENCE, JR, b Pasadena, Calif, Jan 8, 16; m 45; c 5. STATISTICS. Educ: Univ Calif, Davis, BS, 37; Cornell Univ, PhD(animal nutrit), 43. Prof Exp: Supvr adv registry cow test, Univ Calif, 38; asst animal nutrit, Univ Calif, Davis, 38-39; dairy cattle feed investr, Cornell Univ, 40-43, res assoc animal nutrit, 43, instr animal physiol, 44, res assoc poultry nutrit, 44-45; from assoc prof to prof exp statist, 46-57, WILLIAM NEAL REYNOLDS DISTINGUISHED PROF STATIST, NC STATE UNIV, 57-, DIR BIOMATH PROG, 61- Concurrent Pos: Consult, Oak Ridge Nat Lab, 51-, E I du Pont de Nemours & Co, 55-, Army Res Off, 61-, Res Triangle Inst, 62-, Lovelace Clin, 65- & Idaho Fish & Game Dept, 66-; mem, Comt Feed Compos, Nat Res Coun, 46-58, prog & pub comt, Int Grassland Cong, 52, Gov Sci Adv Comt, 61-64, phys sci panel, President's Study Comt, NIH, 64, biomet & epidemiol adv comt, Nat Inst Gen Med Sci, 64-, soil pollution panel, Off Sci & Technol, 65, Comn Undergrad Educ Biol Sci, 67- & chmn math comt, Nat Acad Sci-Nat Res Coun, 67- Mem: Fel AAAS; Am Dairy Sci Asn; Am Soc Animal Sci; Soc Range Mgt; Biomet Soc. Res: Nutrition of dairy cattle and poultry; pasture and range evaluation; quantitative and theoretical biology, especially nutrition; biomathematics training and research. Mailing Add: Dept of Exp Statistics NC State Univ PO Box 5154 Raleigh NC 27607

LUCAS, JAMES ROBERT, b Mankato, Minn, Apr 26, 47. GEOLOGY. Educ: Mankato State Univ, BA, 69; Univ Iowa, MA, 73, PhD(geol), 76. Prof Exp: Instr earth sci, Providence Sch, South St Paul, Minn, 69-70; RES GEOLOGIST, IOWA GEOL SURV, 75- Concurrent Pos: Adj instr geol, Univ Iowa, 74-; prin investr, NASA grant, 75-; mem appln surv group, NASA, 76- Mem: Geol Soc Am; Am Quaternary Asn; Am Soc Photogrammetry; Sigma Xi; Asn Am Geogrs. Res: Land classification of southeastern Iowa from computer enhanced LANDSAT images; glacial geomorphology of northwestern Iowa; mapping via rock stratigraphy and remote sensing techniques; semi-quantitative analysis of clay minerals by x-ray diffraction. Mailing Add: Iowa Geol Surv 123 N Capitol St Iowa City IA 52242

LUCAS, JOHN PAUL, b Youngstown, Ohio, Nov 16, 45; m 68. MICROBIOLOGY. Educ: Univ Pittsburgh, BS, 67, MS, 69, ScD(microbiol), 73. Prof Exp: Res assoc virol, Grad Sch Pub Health, Univ Pittsburgh, 74; MICROBIOLOGIST, FOOD & DRUG ADMIN, 74- Mem: Am Soc Microbiol; Sigma Xi. Res: Methods development research in the area of cosmetic microbiology. Mailing Add: Div of Microbiol Food & Drug Admin 200 C St SW Washington DC 20204

LUCAS, JOSEPH JAMES, b Union Hill, NJ, June 5, 14; m 54; c 3. BIOMETRICS, GENETICS. Educ: Univ Calif, BS, 52, MS, 53; Wash State Univ, PhD(genetics, statist), 58. Prof Exp: Asst genetics & statist, Wash State Univ, 58-59; fel, Brown Univ, 59-61, res assoc biomet, 61-62; assoc prof, 62-70, PROF BIOMET, UNIV CONN, 70-, ASST DIR STORRS AGR EXP STA, 74- Concurrent Pos: Consult inst human biol, Brown Univ, 59-62, res unit, RI Hosp Nurses Asn, 61-62 & RI Chemother Cancer Clin, RI Hosp, 61-62. Mem: Biomet Soc; Am Genetic Asn; Am Soc Animal Sci; Am Statist Asn. Res: Application of statistical methods to biology teaching; biological statistics and experimental design. Mailing Add: Storrs Agr Exp Sta Univ of Conn Box V-10 Storrs CT 06268

LUCAS, KENNETH ROSS, b Bradford, Pa, June 4, 39; m 61; c 2. ANALYTICAL CHEMISTRY, ELECTROCHEMISTRY. Educ: Univ Pittsburgh, BS, 61; Univ Ill, MS, 64, PhD(anal chem), 66. Prof Exp: SR RES CHEMIST, FIRESTONE TIRE & RUBBER CO, 66- Mem: Am Chem Soc; Electrochem Soc. Res: Molten salt and organic electrochemistry; electro-organic synthesis; polymer morphology analysis. Mailing Add: Firestone Tire & Rubber Co 1200 Firestone Pkwy Akron OH 44317

LUCAS, LEON THOMAS, b Halifax, NC, July 30, 42; m 64; c 1. PLANT PATHOLOGY, MICROBIOLOGY. Educ: NC State Univ, BS, 64; Univ Calif, Davis, PhD(plant path), 68. Prof Exp: Res asst plant path, Univ Calif, Davis, 64-68; ASST PROF PLANT PATH, NC STATE UNIV, 68- Mem: Am Phytopath Soc. Res: Diseases of turfgrasses and forage crops in North Carolina; bacterial diseases of plants. Mailing Add: Dept of Plant Path NC State Univ Raleigh NC 27607

LUCAS, MIRIAM SCOTT, natural science, deceased

LUCAS, MYRON CRAN, b Cincinnati, Ohio, Nov 15, 46; m 73. BIOCHEMICAL GENETICS. Educ: Lewis & Clark Col, BS, 69; Wash State Univ, PhD(genetics), 74. Prof Exp: Res assoc bot, Univ Ill, Urbana, 73-75; RES ASSOC GENETICS, UNIV GA, 75- Mem: Genetics Soc Am; Am Soc Microbiol; Am Soc Plant Physiologists; AAAS. Res: Biochemical genetics of Neurospora crassa; structure and function of low molecular weight RNA; gene regulation and synthesis of messenger RNA; biosynthesis of membrane proteins and assembly of membranes. Mailing Add: 225 Country Side Dr Irving TX 75062

LUCAS, OSCAR NESTOR, b Resistencia, Arg, Aug 6, 32. HEMATOLOGY, PHYSIOLOGY. Educ: Univ Buenos Aires, Dentist, 58, DDS, 59; Univ Sask, PhD(physiol), 65. Prof Exp: Res assoc physiol, Med Sch, Univ Sask, 63-64; from asst prof to assoc prof, Med & Dent Sch, Univ Alta, 65-68; assoc prof, 68-70, PROF ORAL BIOL, DENT SCH, UNIV ORE, 70-, PROF DENT, MED SCH, 70- Concurrent Pos: Univ Buenos Aires fel, Jefferson Med Hosp, Philadelphia, Pa, 59-60, Cardeza Found fel, 60-63; affil, Div Hemat, Med Sch, Univ Ore, 71- Mem: Can Physiol Soc; Int Soc Hemat; Int Asn Dent Res. Res: Fibrinolysis; mast cell and connective tissue reparative process. Mailing Add: Dept of Oral Biol Univ of Ore Dent Sch Portland OR 97201

LUCAS, ROBERT ARMISTEAD, organic chemistry, see 12th edition

LUCAS, ROBERT CHARLES, resource geography, land economics, see 12th edition

LUCAS, ROBERT ELMER, b Malolos, Philippines, June 27, 16; m 41; c 5. SOIL SCIENCE. Educ: Purdue Univ, BSA, 39, MS, 41; Mich State Col, PhD(soil sci), 47. Prof Exp: Asst soils, Va Truck Exp Sta, 41-43 & Mich State Col, 45-46; agronomist, Wm Gehring, Inc, Ind, 46-51; assoc prof soil sci, 51-57, PROF SOIL SCI, MICH STATE UNIV, 57-, EXTEN SPECIALIST, 53- Concurrent Pos: With AID, Arg, 72-73. Mem: Soil Sci Soc Am; fel Am Soc Agron. Res: Soil fertility; management; trace elements; organic soils. Mailing Add: Dept of Crop & Soil Sci Mich State Univ East Lansing MI 48823

LUCAS, RUSSELL VAIL, JR, b Des Moines, Iowa, Nov 2, 28; m 51; c 4. PEDIATRIC CARDIOLOGY. Educ: Macalester Col, BA, 50; Wash Univ, MD, 54. Prof Exp: From intern to resident pediat, Univ Hosp, Univ Minn, Minneapolis, 54-56, resident, 58-59; from asst prof to assoc prof pediat, Med Ctr, WVa Univ, 61-66; assoc prof, 66-69, PROF PEDIAT, UNIV MINN, MINNEAPOLIS, 69- Concurrent Pos: NIH fel pediat cardiol, Univ Hosps, Univ Minn, Minneapolis, 59-61; NIH res career develop award, WVa Univ, 63-66; mem med adv bd, Coun Circulation & mem Coun Clin Cardiol, Am Heart Asn. Honors & Awards: Distinguished Achievement Award, Am Heart Asn, 66. Mem: Soc Pediat Res; Am Acad Pediat; Asn Am Med Cols; Am Fedn Clin Res; Am Heart Asn. Res: Physiology of ventricular function; pathology, physiology and natural history of congenital cardiac defects. Mailing Add: Pediat Cardiol Univ of Minn Hosps Minneapolis MN 55455

LUCAS, STEPHEN BERNARD, b Grassflat, Pa, May 29, 36; m 57; c 3. SCIENCE EDUCATION, HEALTH SCIENCES. Educ: Clarion State Col, BS, 57; Pa State Univ, MEd, 62, PhD(curric & instr), 72. Prof Exp: Instr biol pub schs, Pa, 57-65; asst prof biol, Col Misericordia, 65-69; ASSOC PROF SCI EDUC & HEALTH, EDINBORO STATE COL, 71- Honors & Awards: Pres Award, Swift Instrument Co, 75. Mem: Nat Sci Teachers Asn; Nat Asn Elem Teachers Sci; Nat Asn Res Sci Teaching. Res: Learning theory as applied to science activities in teaching in the science classroom; competency base teacher education in science. Mailing Add: Rm 309 Butterfield Hall Edinboro State Col Edinboro PA 16444

LUCAS, THOMAS RAMSEY, b Tampa, Fla, June 9, 39; m 70. MATHEMATICS. Educ: Univ Fla, BS, 61; Univ Mich, Ann Arbor, MS, 62; Ga Inst Technol, PhD(math), 70. Prof Exp: Sr engr, Martin Co, 62-65; asst prof math, 69-75, ASSOC PROF MATH, UNIV NC, CHARLOTTE, 75- Mem: Am Math Soc; Soc Indust & Appl Math. Res: Numerical analysis; approximation theory; spline theory. Mailing Add: Dept of Math Univ of NC Box 20428 Charlotte NC 28223

LUCAS, WILLIAM FRANKLIN, b Detroit, Mich, Apr 21, 33; m 57; c 4. OPERATIONS RESEARCH, APPLIED MATHEMATICS. Educ: Univ Detroit, BS, 54, MA, 56, MS, 58; Univ Mich, PhD(math), 63. Prof Exp: Instr math, Univ Detroit, 56-58 & 61-62, asst prof, 62-63; res instr, Princeton Univ, 63-65; Fulbright fel & vis asst prof econ & statist, Mid East Tech Univ, Ankara, 65-66; vis assoc prof, Math Res Ctr, Univ Wis-Madison, 66-67; mathematician, Rand Corp, 67-69; assoc prof opers res & appl math, 69-70, dir ctr appl math, 71-74, PROF OPERS RES & APPL MATH, CORNELL UNIV, 70- Concurrent Pos: Consult, Rand Corp, 69- & Educ Develop Ctr, 75-; sci exchange with USSR, US Nat Acad Sci, 76; Chautauqua lectr, AAAS, 75-76. Mem: Am Math Soc; Math Asn Am; Soc Indust & Appl Math; Opers Res Soc Am; Inst Mgt Sci. Res: Elasticity; applied mathematics; game theory. Mailing Add: Dept of Opers Res 334 Upson Hall Cornell Univ Ithaca NY 14853

LUCAS-LENARD, JEAN MARIAN, b Bridgeport, Conn, July 17, 37; m 64. MOLECULAR BIOLOGY. Educ: Bryn Mawr Col, AB, 59; Yale Univ, PhD(protein synthesis), 63. Prof Exp: USPHS fel enzymol, Inst Physiochem Biol, Paris, 63-64; guest investr protein synthesis, Rockefeller Univ, 64-65, res assoc, 65-68, asst prof, 68-70; ASSOC PROF BIOL, UNIV CONN, 70- Concurrent Pos: Estab investr, Am Heart Asn, 70-71; NIH career develop award, 71- Res: Mechanism of protein biosynthesis in eukaryotes and prokaryotes; translational control mechanisms in virus infected cells. Mailing Add: Dept of Biochem & Biophys Univ of Conn Storrs CT 06268

LUCCA, JOHN J, b Brooklyn, NY, July 12, 21; m 46; c 6. DENTISTRY. Educ: NY Univ, AB, 41; Columbia Univ, DDS, 47; Am Bd Prosthodontics, dipl. Prof Exp: From instr to assoc prof dent, 47-64, PROF PROSTHODONTICS & DIR DIV, SCH DENT & ORAL SURG, COLUMBIA UNIV, 64- Concurrent Pos: Consult, Vet Admin & USPHS; hon police surgeon & consult, New York Police Dept, 64-; consult, US Naval Dent Sch; attend, Presby Hosp & Grasslands Hosp; consult ed prosthodont, Progreso-Odonto-Stomatologique. Honors & Awards: Ewell Medal, 47. Mem: Fel Am Col Dent; fel Am Col Prosthodont; fel Int Col Dent. Res: Precision attachment; partial dentures. Mailing Add: Div of Prosthodontics Columbia Univ New York NY 10032

LUCCHESI, CLAUDE A, b Chicago, Ill, Apr 20, 29; m 54; c 2. ANALYTICAL CHEMISTRY, PHYSICAL CHEMISTRY. Educ: Univ Ill, BS, 50; Northwestern Univ, PhD, 54. Prof Exp: Asst, Northwestern Univ, 50-54; spectros group leader, Shell Develop Co, Tex, 54-56; dir anal res dept, Sherwin-Williams Co, 56-61; mgr anal & phys chem dept, Mobil Chem Co, 61-67, mgr cent coatings lab, 67-68; LECTR CHEM & DIR ANAL SERV, NORTHWESTERN UNIV, 68- Mem: Am Chem Soc; Soc Appl Spectros; Instrument Soc Am; Fedn Soc Paint Technol; Brit Soc Anal Chem. Res: General applied spectroscopy; x-ray spectroscopy; chelate chemistry; differential thermal analysis; plastics and coating characterization and analysis; trace analysis. Mailing Add: Dept of Chem Northwestern Univ Evanston IL 60201

LUCCHESI, JOHN CHARLES, b Cairo, Egypt, Sept 3, 34; US citizen; m 55; c 1. DEVELOPMENTAL GENETICS. Educ: La Grange Col, AB, 55; Univ Ga, MS, 58; Univ Calif, Berkeley, PhD(zool), 63. Prof Exp: NIH res assoc biol, Univ Ore, 63-65; from asst prof to assoc prof, 65-72, PROF ZOOL, UNIV NC, CHAPEL HILL, 72- Concurrent Pos: Vis investr, Max Planck Inst Biol, Tübingen, Ger, 69; NIH res career develop award, 70- Mem: Genetics Soc Am; Am Soc Cell Biol. Res: Cytogenetics, especially chromosome organization; biochemistry of development. Mailing Add: Dept of Zool Univ of NC Chapel Hill NC 27514

LUCCHESI, PETER J, b New York, NY, Sept 23, 26; m 49; c 2. PHYSICAL CHEMISTRY. Educ: NY Univ, AB, 49, MS, 53, PhD(chem), 54. Prof Exp: Instr chem, Adelphi Col, 52, NY Univ, 53-54 & Ill Inst Technol, 54-55; res chemist, 55-68, DIR CORP RES LAB, ESSO RES & ENG CO, 68- Mem: Am Chem Soc. Res: Radiation chemistry; heterogeneous catalysis; crystal growth and dissolution. Mailing Add: 157 Meadowbrook Dr North Plainfield NJ 07060

LUCCHITTA, BAERBEL KOESTERS, b Muenster, Ger, Oct 2, 38; US citizen; m 64; c 1. ASTROGEOLOGY. Educ: Kent State Univ, BS, 61; Pa State Univ, MS, 63, PhD(geol), 66. Prof Exp: GEOLOGIST, CTR ASTROGEOL, US GEOL SURV, 71- Mem: AAAS. Res: Dark mantles, secondary craters, basin formation, plains formation, scarps and ridges, northside and Apollo 17-site geological map of the moon; erosion, landform development, map of Ismenius Lacus, canyons and scarps of Mars; geomorphology and structural geology of earth. Mailing Add: Ctr of Astrogeol US Geol Surv 601 E Cedar Ave Flagstaff AZ 86001

LUCE, EVERETT N, b Merna, Nebr, May 14, 09; m 30; c 1. PHARMACY. Educ: Univ Colo, BS, 30. Hon Degrees: LLD, Cent Mich Univ, 57; LHD, Northwood Inst, 65. Prof Exp: Chemist, Dow Chem Co, 30-35, group leader, 35-36, asst dir e anal lab, 36-53, dir, 53-68, mgr Mich educ rels, 68-70; SPEC ASST TO PRES, DELTA COL, 70- Mem: AAAS; Am Chem Soc; hon mem Am Soc Test & Mat. Res: Analytical chemistry, also development of methods, equipment and new approaches. Mailing Add: Delta Col University Center MI 48710

LUCE, JAMES EDWARD, b Toronto, Ont, Aug 24, 35. PAPER CHEMISTRY. Educ: Univ Toronto, BASc, 56; McGill Univ, PhD(chem), 60. Prof Exp: Asst mgr basic res, CIP Res Ltd, 60-71; sci admin officer, Atomic Energy Can, Ltd, 71; GROUP DIR PAPER TECHNOL, INT PAPER CO, 72- Mem: Tech Asn Pulp & Paper Indust; Can Pulp & Paper Asn; Chem Inst Can; Brit Paper & Board Indust Fedn. Res: Application of modern instrumental techniques to diagnosis of problems encountered in papermaking, particularly in papermachines. Mailing Add: Int Paper Co PO Box 797 Tuxedo Park NY 10987

LUCE, JOHN SIDNEY, b New Orleans, La, Apr 11, 09; m 33; c 2. PLASMA PHYSICS. Hon Degrees: DSc, Auburn Univ, 61. Prof Exp: Printer, Security Lithograph Co, 26-42; co-owner, Webb Appliance Co, 38-43; engr physics, Univ Calif, Berkeley, 42-45 & Tenn Eastman Corp, 45-47; physicist, Union Carbide Nuclear Co, 47-50; head, Explor Physics Group, Oak Ridge Nat Lab, 50-59, asst dir, Thermonuclear Exp Div, 59-60; mgr res div, Aerojet-Gen Nucleonics Div, Aerojet Gen Corp, 61-68; DIR RES, BERKELEY ANAL SCI SERV, 69-; PHYSICIST, LAWRENCE LIVERMORE LAB, 72- Concurrent Pos: US del, Int Conf Peaceful Uses of Atomic Energy, Geneva, 58, Conf Plasma Physics & Controlled Nuclear Fusion Res, Salzburg, 59 & Int Atomic Energy Agency Conf, Culham, 65; mem, NASA Res Adv Comt Elec Energy Systs, 60-63; consult, US Dept Defense, 60- Mem: Fel Am Phys Soc; Am Inst Aeronaut & Astronaut; NY Acad Sci. Res: High temperature plasma physics; controlled fusion, ion sources, mass spectroscopy, accelerators and arc research. Mailing Add: 1877 Warsaw Ave Livermore CA 94550

LUCE, ROBERT GEORGE, physics, see 12th edition

LUCE, ROBERT WILLIAM, b Teaneck, NJ, July 27, 38. GEOCHEMISTRY. Educ: Dartmouth Col, BA, 60; Univ Ill, MS, 62; Stanford Univ, PhD(geochem), 69. Prof Exp: GEOLOGIST, US GEOL SURV, MENLO PARK, CALIF, 68- Mem: Geol Soc Am; Am Chem Soc; Mineral Soc Am; Clay Minerals Soc. Res: Geochemistry of mineral weathering, including kinetics, surface chemistry and clay mineral formation. Mailing Add: US Geol Surv 345 Middlefield Rd Menlo Park CA 94025

LUCE, WILLIAM GLENN, b Beaver Dam, Ky, Mar 21, 36; m 70; c 2. ANIMAL NUTRITION. Educ: Univ Ky, BS, 58; Univ Nebr, MS, 64, PhD(animal nutrit), 65. Prof Exp: Mgt trainee grocery & meat merchandising, Kroger Co, Ky, 58-60, co-mgr grocery & meat merchanidising, 60-62; asst nutrit res, Univ Nebr, 62-65; asst prof swine exten, Univ Ga, 65-68; ASST PROF SWINE EXTEN, OKLA STATE UNIV, 68- Mem: Am Soc Animal Sci. Res: Swine nutrition; cereal grain utilization and amino acid requirements. Mailing Add: Dept of Animal Sci Okla State Univ Stillwater OK 74074

LUCEY, CAROL ANN, b Johnstown, NY, Sept 16, 43; m 64; c 1. THEORETICAL PHYSICS, PHILOSOPHY OF SCIENCE. Educ: Harpur Col, BA, 65; State Univ NY Binghamton, MA, 68; Brown Univ, PhD(physics), 72. Prof Exp: ASST PROF PHYSICS, JAMESTOWN COMMUNITY COL, 73-, ACTG ASSOC DEAN INSTR, 76- Mem: Am Phys Soc; Philos Sci Asn. Res: Study of cosmological implications for elementary particle physics; consequences of the hot model and possible neutrino degeneracy in CP violations; scientific methodology. Mailing Add: Dept of Physics Jamestown Community Col Jamestown NY 14701

LUCEY, JEROLD FRANCIS, b Holyoke, Mass, Mar 26, 26; m 50; c 3. PEDIATRICS. Educ: Dartmouth Col, AB, 48; NY Univ, MD, 52. Prof Exp: Intern pediat, Bellevue Hosp, New York, 52-53; asst resident, Columbia-Presby Med Ctr, 53-55; from instr to assoc prof, 56-66, PROF PEDIAT, COL MED, UNIV VT, 66- Concurrent Pos: Bowen Brooks scholar, NY Acad Med, Bellevue Hosp, New York, 54; Meade-Johnson fel, Columbia-Presby Med Ctr, 54-55; Nat Found Infantile Paralysis res fel, Harvard Med Sch, 55-56; Markle scholar, 59-64; res fel biochem, Harvard Med Sch, 60-61; consult, Vt State Health Dept, 56-77; chmn, Nat Bd Med Exam, 68-72; mem, Am Bd Pediat Exam, 70; ed-in-chief, Pediatrics, 73- Mem: Soc Pediat Res; fel Am Acad Pediat; Am Pediat Soc; Am Soc Photobiol; Royal Soc Med. Res: Neonatal physiology; bilirubin metabolism. Mailing Add: Med Ctr Hosp of Vt 52 Overlake Park Burlington VT 05401

LUCEY, JULIANA MARGARET, b Santa Monica, Calif. NUMERICAL ANALYSIS. Educ: Univ Wash, AM, 62; Univ Ariz, MS, 72; St Louis Univ, PhD(math), 72. Prof Exp: Teacher high sch math & chmn dept, Sisters of the Holy Names of Jesus & Mary, 65; asst prof math, Holy Names Col, 65-69; teaching asst, Univ Ariz, 69-71; teaching fel, St Louis Univ, 73-74; asst prof, Holy Names Col, 74-75; INSTR MATH, WAYNE STATE UNIV, 75- Mem: Am Math Soc; Am Statist Asn. Res: Solutions of stiff ordinary differential equations by a fifth order composite multistep method; index and Lefschetz number in the structure of gratings. Mailing Add: Dept of Math Wayne State Univ Detroit MI 48202

LUCEY, ROBERT FRANCIS, b Worcester, Mass, Mar 13, 26; m 52; c 7. AGRONOMY. Educ: Univ Mass, BVA, 50; Univ Md, MS, 54; Mich State Univ, PhD(field crops), 59. Prof Exp: Asst prof agron, Univ NH, 57-61; from asst to assoc prof field crops, 61-70, PROF FIELD CROPS, NY STATE COL AGR & LIFE SCI, CORNELL UNIV, 70-, CHMN DEPT AGRON, 75- Mem: Am Soc Agron. Res: Production of field crops, especially crop-climate relationships; adaptability; plant competition. Mailing Add: NY State Col of Agr & Life Sci Dept of Agron Cornell Univ Ithaca NY 14850

LUCHINS, EDITH HIRSCH, b Poland, Dec 21, 21; nat US; m 42; c 5. MATHEMATICS. Educ: Brooklyn Col, BA, 42; NY Univ, MS, 44; Univ Ore, PhD(math), 57. Prof Exp: Govt inspector anti-aircraft dirs, Sperry Gyroscope Co, NY, 42-44; instr math, Brooklyn Col, 44-46 & 48-49; asst appl math lab, NY Univ, 46; Am Asn Univ Women res fel & res assoc math, Univ Ore, 57-58; from res assoc to assoc prof math, Univ Miami, 59-62; assoc prof, 62-70, PROF MATH, RENSSELAER POLYTECHNIC INST, 70- Mem: Math Asn Am; Am Math Soc; Soc Indust & Appl Math. Res: Banach algebras; functional analysis; mathematical psychology. Mailing Add: Dept of Math Sci Rensselaer Polytechnic Inst Troy NY 12181

LUCHSINGER, WAYNE WESLEY, b Milaca, Minn, May 8, 24; m 43; c 4. BIOCHEMISTRY. Educ: Univ Minn, BS, 51, MS, 54, PhD(biochem), 56. Prof Exp: Asst biochem, Univ Minn, 51-55; sr chemist, Kurth Malting Co, 56-58, asst dir res, 58-60; assoc prof biochem, WVa Univ, 60-66; assoc prof chem, 66-68, PROF CHEM, ARIZ STATE UNIV, 68- Mem: AAAS; Am Chem Soc; Am Soc Brewing Chem; Am Asn Cereal Chem. Res: Enzymes; barley carbohydrates; chemistry and mechanism of action of carbohydrases; carbohydrate structure. Mailing Add: Dept of Chem Ariz State Univ Tempe AZ 85281

LUCHTEL, DANIEL LEE, b Carroll, Iowa, Jan 13, 42; m 73; c 1. MICROSCOPIC ANATOMY. Educ: St Benedict's Col, Kans, BS, 63; Univ Wash, PhD(zool), 69. Prof Exp: NIH fel, 69-71, res assoc develop biol, 71-73, biol struct, 73, environ health, 74, ASST PROF ENVIRON HEALTH, UNIV WASH, 75- Concurrent Pos: Res fel, Hubrecht Lab, Utrecht, Neth, 72. Mem: AAAS; Am Inst Biol Sci; Am Soc Zool. Res: Lung ultrastructure and effects of gaseous and particulate air pollutants; lung development; pulmonary edema; alveolare macrophage cultures. Mailing Add: Dept of Environ Health Univ of Wash Seattle WA 98195

LUCIANO, DOROTHY SCHWIMMER, physiology, see 12th edition

LUCID, MICHAEL FRANCIS, b Indianapolis, Ind, Feb 23, 37; m 67; c 3. INORGANIC CHEMISTRY. Educ: Ind Univ, Bloomington, BS, 61; Purdue Univ, Lafayette, MS, 65. Prof Exp: Res chemist, 65-67, SR RES CHEMIST INORG CHEM, KERR McGEE CORP, 67- Mem: Am Chem Soc. Res: Hydrometallurgy; solvent extraction; ion exchange; solution chemistry; geochemistry; solution mining, uranium, vanaduum, copper. Mailing Add: Kerr McGee Tech Ctr PO Box 25861 Oklahoma City OK 73125

LUCIEN, HAROLD WILLIAM, organic chemistry, biochemistry, see 12th edition

LUCIER, JOHN J, b Detroit, Mich, Aug 10, 17. ORGANIC CHEMISTRY. Educ: Univ Dayton, BS, 37; Western Reserve Univ, MS, 50, PhD(org chem), 51. Prof Exp: Instr chem, 45-47 & 51-52, from asst to assoc prof, 52-63, PROF CHEM, UNIV DAYTON, 63-, CHMN DEPT, 64- Mem: AAAS; Am Chem Soc; Soc Appl Spectros; NY Acad Sci; The Chem Soc. Res: Organic synthesis; infrared spectroscopy. Mailing Add: Dept of Chem Univ of Dayton Dayton OH 45409

LUCIS, OJARS JANIS, b Latvia, Apr 2, 24; Can citizen; m 49; c 2. ENDOCRINOLOGY, ENVIRONMENTAL HEALTH. Educ: Sir George Williams Univ, BSc, 54; McGill Univ, MSc, 57, PhD(invest med), 59, MD, CM, 61, cert clin chem, 74. Prof Exp: Res asst invest med, McGill Univ, 56-60; asst prof endocrinol, Dalhousie Univ, 65-71; MED OFFICER, HEALTH & WELFARE CAN, 71-, DIV CHIEF ENDOCRINOL & METAB, 74- Concurrent Pos: Med Res Coun Can fel, McGill Univ, 62-63, res scholar, 63-65; Med Res Coun Can scholar steroid biochem, Dalhousie Univ, 65-68; asst pathologist, Prov NS Dept Pub Health, 66-68, assoc pathologist, 68-71. Mem: Can Med Asn; Can Soc Clin Chem; Endocrine Soc; Can Physiol Soc. Res: Biosynthesis and metabolism of hormones; immunochemical assays of hormones; interaction of trace elements with the cells and the mammalian organism; biosynthesis and isolation of cadmium binding proteins; pharmacology and toxicology of drugs. Mailing Add: Health & Welfare Can Health Protection Br Bur Drugs Ottawa ON Can

LUCIS, RUTA, b Rujiena, Latvia, Apr 9, 25; Can citizen; m 49; c 2. COMPARATIVE ENDOCRINOLOGY. Educ: Sir George Williams Univ, BSc, 57; McGill Univ, MS, 64, PhD(invest med), 66. Prof Exp: Res asst endocrinol, McGill Univ, 62-65; res asst, Path Inst, 66-71; clin chemist, Corp Labs, Ottawa, 72-73; CLIN CHEMIST, ANIMAL RES INST, OTTAWA, 73- Mem: NY Acad Sci. Res: Biochemistry of steroids; immunochemical assays and metabolism of hormones; environmental health. Mailing Add: 1512 Caverley St Ottawa ON Can

LUCK, CLARENCE FREDERICK, JR, b Buffalo, NY, Oct 4, 25; m 51, 69; c 5. PHYSICS. Educ: Univ Buffalo, BA, 49; Duke Univ, PhD(physics), 56. Prof Exp: Tool designer, Bell Aircraft Corp, 43-44; supvr microwave tube res, Polarad Electronics Corp, 56-60; sr staff physicist, Res Div, 60-62, mgr laser advan develop ctr, Spec Microwave Devices Oper, 62-66, PRIN ENGR, LASER ADVAN DEVELOP CTR, SPEC MICROWAVE DEVICES OPER, RAYTHEON CO, 66- Mem: Am Phys Soc; Optical Soc Am. Res: Spectroscopy; optics; photography; optical maser; laser and laser machining equipment. Mailing Add: 45 Weir Rd Waltham MA 02154

LUCK, DAVID GEORGE CROFT, b Whittier, Calif, July 26, 06; m 30; c 1. PHYSICS. Educ: Mass Inst Technol, SB, 27, PhD(physics), 32. Prof Exp: Res engr, Thomas A Edison, Inc, 29; asst physics, Mass Inst Technol, 29-32; res engr, Victor Div, Radio Corp Am, NJ, 32-42, labs div, 42-53, aviation staff engr, Eng Prod Div, 54-64; mem tech staff, Defense Res Corp, 64-69, Strategic Systs Div, Gen Res Corp, 69-74; CONSULT, 74- Honors & Awards: Ballantine Medal, Franklin Inst, 53; Achievement Awards, RCA Labs; Pioneer Award, Nat Airborne Electronics Conf. Mem: Fel Inst Elec & Electronics Eng; Am Inst Navig. Res: Magnetic materials; resistors; electronic systems for communications and specialized communication systems; fire control; air traffic control; color television; radar, especially FM radar; radio navigation; radio direction finding; navigation aides; omni-range system of air navigation. Mailing Add: 4756 Calle Camarada Santa Barbara CA 93110

LUCK, DAVID JONATHAN LEWIS, b Milwaukee, Wis, Jan 7, 29. CYTOLOGY. Educ: Univ Chicago, SB, 49; Harvard Med Sch, MD, 53. Prof Exp: From asst prof to assoc prof, 64-68, PROF CELL BIOL, ROCKEFELLER UNIV, 68- Concurrent Pos: Teaching fel, Harvard Med Sch, 57-59; fel, Rockefeller Univ, 59-64; res physician, Mass Gen Hosp, Boston, 57-59. Res: Biochemical cytology; cell structure; biochemical function. Mailing Add: Rockefeller Univ New York NY 10021

LUCK, DENNIS NOEL, b Durban, SAfrica, Dec 8, 39; m 69. MOLECULAR BIOLOGY. Educ: Univ Natal, BSc, 61, MSc, 63; Oxford Univ, DPhil(molecular biol), 66. Prof Exp: Lectr biochem, Univ Natal, 66-68; vis asst prof pharmacol, Baylor Col Med, 69; asst prof zool, Univ Tex, Austin, 70-72; asst prof, 72-75, ASSOC PROF BIOL, OBERLIN COL, 75- Mem: Brit Biochem Soc; Am Soc Zool; Am Soc Cell Biol; Sigma Xi. Res: Regulation of genetic transcription and translation; control mechanisms in endocrinology; effects of sex steroid hormones on metabolism of reproductive organs. Mailing Add: Dept of Biol Oberlin Col Oberlin OH 44074

LUCK, JAMES VERNON, b Hannibal, Mo, June 27, 06; m 38; c 2. ORTHOPEDIC SURGERY. Educ: St Louis Univ, MD, 31; Univ Iowa, MS, 37. Prof Exp: Intern, Hollywood Presby Hosp, 31-32; from intern to resident physician, Los Angeles County Gen Hosp, 32-35; assoc orthop surg, Univ Iowa, 35-39; assoc clin prof, 52-60, CLIN PROF SURG, SCH MED, UNIV SOUTHERN CALIF, 60- Concurrent Pos: Mem Nat Res Coun, Orthop Subcomt, 46-49; med dir & chief staff, Los Angeles Orthop Hosp, 55-68; vpres, Am Bd Orthop Surg, 60-61; mem, Nat Adv Coun, Voc Rehab Admin, Educ & Welfare; pres, Orthop Res Educ Found, 66-68. Honors & Awards: Award, Univ Mo Sch Med, 64 & St Louis Univ Sch Med, 65. Mem: Am Orthop Asn; AMA; Am Col Surg; Am Acad Orthop Surg (pres elect, 60, pres, 61); Int Soc Orthop Surg & Traumatol. Mailing Add: Dept of Surg Univ of Southern Calif Sch of Med Los Angeles CA 90007

LUCK, JOHN VIRGIL, b Chalmers, Ind, Jan 20, 26; m 45; c 3. MICROBIOLOGY. Educ: Purdue Univ, BS, 49, MS, 51, PhD(microbiol, biochem), 54. Prof Exp: Dir beer fermentation res, Pabst Brewing Co, 53-55; proj leader chem res, Gen Foods Corp, NY, 55-58; head biol chem dept, Armour & Co, 58-60; dir res & develop, Durkee Famous Foods, Glidden Co, Ill, 60-70; V PRES & TECH DIR, GEN MILLS, INC, MINNEAPOLIS, 70- Concurrent Pos: Mem, Res & Develop Assocs. Mem: AAAS; Am Chem Soc; Am Oil Chem Soc; Inst Food Technol; Soc Indust Microbiol. Res: Food chemistry; fats; starch; proteins emulsifiers; enzymology. Mailing Add: 5428 W Highwood Dr Edina MN 55436

LUCK, RUSSELL M, b Reading, Pa, May 11, 26; m 63; c 2. POLYMER CHEMISTRY, ORGANIC CHEMISTRY. Educ: Albright Col, BSc, 47; Bucknell Univ, MSc, 48. Prof Exp: Asst chem, Bucknell Univ, 47-48; asst prod mgr, Wyomissing Glazed Papers, Inc, 48-51; engr, Mat Eng Dept, 53-60, sr engr, Res & Develop Ctr, 60-71, FEL SCIENTIST, RES & DEVELOP CTR, WESTINGHOUSE ELEC CORP, 71- Mem: Am Chem Soc. Res: Organic and inorganic polymers for application as lubricants and electrical insulations with high temperature capabilities. Mailing Add: Westinghouse Elec Res & Develop Ctr Insul & Appl Chem Res Dept Pittsburgh PA 15235

LUCK-ALLEN, ETTA ROBENA, b Danville, Va, Dec 24, 19; div. MYCOLOGY. Educ: Howard Univ, SB, 41; Univ Iowa, MS, 42; Univ Toronto, MA, 56, PhD(bot), 58. Prof Exp: Instr biol, Fla Agr & Mech Col, 48-49; assoc prof, Tex Southern Univ, 49-54; asst, 56-59, lectr, 59-63, asst prof, 63-69, ASSOC PROF MYCOL, UNIV TORONTO, 69- Concurrent Pos: Res grant, Nat Res Coun Can, 75- Mem: Mycol Soc Am; Brit Mycol Soc; Bot Soc Am; Can Bot Asn. Res: Mycology; pathology; cultural and taxonomic studies of tremellaceous fungi; coprophilous ascomycetes; Tremellales of Ontario. Mailing Add: Dept of Bot Univ of Toronto Toronto ON Can

LUCKE, JOHN BECKER, b New York, NY, Feb 26, 08; m 37; c 3. GEOLOGY. Educ: Princeton Univ, BS, 29, AM, 32, PhD(geol), 33. Prof Exp: Geologist, Torrey, Fralich & Simmons, Pa, 29; asst geol, Princeton Univ, 30-32; prof, John Marshall Col, 33-34; asst dist geologist, Tex Co, 34-35; asst soil surv, Soil Conserv Serv, USDA, 35-36; asst prof geol, Univ WVa, 36-40; assoc prof geol & head dept geol & geog, Univ Conn, 40-48, prof geol & head dept, 48-63; prof & chmn dept, 64-73, EMER PROF GEOL & CONSULT, GRAND VALLEY STATE COL, 73- Concurrent Pos: Asst geologist, Sloan & Zook Co, Pa, 30 & State Geol Surv, WVa, 36-38; mem, Conn Geol & Natural Hist Surv Comn, 46-63, dir, 54-60; mem, Nat Geog-Woods Hole Oceanog Atlantic Exped, Mid-Atlantic Ridge, 48 & Nat Park Serv Katmai exped, 53; del, Int Geol Cong, Copenhagen, 60, Prague, 68 & Montreal, 72 & Int Asn Quaternary Res Cong, Colo, 65. Mem: AAAS; fel Geol Soc Am; Am Soc Photogram; Am Asn Petrol Geol; Nat Asn Geol Teachers (vpres, 53). Res: Geomorphology and oceanography, especially marine shorelines; photogeology. Mailing Add: The Anchorage E-24 15 Pleasant St Harwich Port MA 02646

LUCKE, ROBERT LANCASTER, b Norfolk, Va, July 22, 45. ASTROPHYSICS. Educ: Johns Hopkins Univ, BA, 68, MA, 72, PhD(physics), 75. Prof Exp: ASSOC RES SCIENTIST PHYSICS, JOHNS HOPKINS UNIV, 75- Res: Far ultraviolet albedo of the moon, thesis topic; x-ray astronomy. Mailing Add: 10608 Ridge Dr Clinton MD 20735

LUCKE, WILLIAM E, b Grand Island, Nebr, July 31, 36; m 59; c 4. ANALYTICAL CHEMISTRY. Educ: Univ Nebr, BS, 58; Ohio State Univ, PhD(chem), 63. Prof Exp: Res chemist, Olympic Res Div, Rayonier Inc, 63-69; res assoc, Cincinnati Milling Mach Co, 69-71, supvr, Cimcool Customer Lab Serv, Cincinnati Milacron Inc, 71-74, SR ANAL CHEMIST, CIMCOOL DIV, CINCINNATI, MILACRON INC, 74- Mem: AAAS; Am Chem Soc. Res: Carbohydrate, cellulose and wood chemistry; analytical chemistry of industrial working products. Mailing Add: Cincinnati Milacron Inc 4701 Marburg Ave Cincinnati OH 45209

LUCKE, WILLIAM HUNTER, b Concord, NC, July 28, 19; m 42; c 5. SURFACE PHYSICS. Educ: Hampden-Sydney Col, BS, 41; Univ Va, MS, 48, PhD(physics), 49. Prof Exp: Asst prof physics, Southern Ill Univ, 49-52 & 52-53; sr physicist, Sharples Corp, 53; PHYSICIST, US NAVAL RES LAB, 53- Mem: Am Phys Soc; assoc Inst Elec & Electronics Engrs. Res: Biological effects of ultraviolet and x-rays on haploid and polyploid yeast; characteristics of flux reversal in square loop core and magnetic amplifier circuits; thermoelectricity; solid state physics; ion implantation and its effects on surfaces. Mailing Add: Naval Res Lab 5555 Overlook Ave Washington DC 20375

LUCKE, WINSTON SLOVER, applied physics, see 12th edition

LUCKEN, KARL ALLEN, b Portland, NDak, Apr 7, 37; m 60; c 3. CROP BREEDING. Educ: Concordia Col, Minn, BA, 59; Iowa State Univ, PhD(plant breeding), 64. Prof Exp: From asst prof to assoc prof, 64-74, PROF AGRON, NDAK STATE UNIV, 74- Mem: Am Soc Agron. Res: Breeding and development of hybrid wheat; genetics of extrachromosomal variability in wheat. Mailing Add: Dept of Agron NDak State Univ Fargo ND 58102

LUCKENBACH, THOMAS ALEXANDER, b Plains, Pa, Feb 26, 33; m 59; c 6. PHYSICAL CHEMISTRY, POLYMER CHEMISTRY. Educ: Kings Col, Pa, BS, 54; Catholic Univ, PhD(phys chem), 58. Prof Exp: Sr chemist, Harris Res Labs, Inc, DC, 58-61; group leader, Toni Co Div, Gillette Corp, Ill, 61-64; res assoc, Huyck Felt Co, 64-65, asst mgr res, 65-66, res assoc, Huyck Res Ctr, 66-68, mgr chem & phys res, 68-73, mgr formex prods, Res & Develop, 73-74; TECH MGR, B F GOODRICH TIRE CO, 74- Mem: Am Chem Soc; Am Inst Chemists; Sigma Xi. Res: Chemistry of human hair; synthetic fibers; chemical and physical testing of fibers and fabrics; paper making fabrics; textile resin treatments; textile adhesives; plastics; fiberglass; tire cords; rubber compounding; tire technology. Mailing Add: B F Goodrich Tire Co 500 S Main St Akron OH 44318

LUCKENBAUGH, RAYMOND WILSON, b Hanover, Pa, Dec 29, 21; m 47; c 2. AGRICULTURAL CHEMISTRY. Educ: Gettysburg Col, AB, 43; Univ Md, PhD(chem), 52. Prof Exp: Jr chemist org synthesis, Rohm & Hass Co, 47-48; agr chemist, 52-58, res scientist, 58-62, res assoc, 62-64, RES SUPVR, E I DU PONT DE NEMOURS & CO, INC, 64- Mem: Am Chem Soc; Weed Sci Soc Am. Res: Naphthoquinones; synthesis and toxicology of agrichemicals. Mailing Add: E I du Pont de Nemours & Co Exp Sta Wilmington DE 19898

LUCKENS, MARK MANFRED, b Kiev, Russia, Apr 7, 12; US citizen; m 43; c 2. PHARMACOLOGY, TOXICOLOGY. Educ: Columbia Univ, BS, 35; NY Univ, MS, 50; Univ Conn, PhD(pharmacol, toxicol), 63; Am Bd Indust Hyg, dipl. Prof Exp: Jr chemist, Wilkow Food Prod, 28-33, chemist, 33-36; chief chemist, Technichem Labs, 37-41; inspector, Chem Warfare Serv, 41-43; dir, Emmet Tech Assocs, 48-54; toxicologist, Conn State Dept Health, 54-61; asst prof, 61-67, ASSOC PROF TOXICOL & PHARMACOL, COL PHARM, UNIV KY, 67-, DIR, INST ENVIRON TOXICOL & OCCUP HYG, 62-, MEM FAC & CO-DIR, INTERDISCIPLINARY GRAD PROG TOXICOL, 73- Concurrent Pos: Consult, Ky State Dept Health, 61-, Lexington-Fayette County Dept of Health, Ky Poison Info & Control Prog, 61-, Lab Serv, Childrens's Hosp, Louisville, Ky, 63- & Spindletop Res Ctr, 65-; Fulbright travel grant, 65-66; award, Partners-in-the-Americas, 65-66; mem, adv comt pesticides, Ky Dept Agr, 65-; mem exec bd, Am Asn Poison Control Ctrs; vis prof, Polytech Inst of Guayaquil; vis dir, Oceanog Inst Ecuador & Environ Protection Inst Ecuador. Mem: Fel AAAS; fel Am Inst Chem; fel Am Acad Indust Hyg; fel Am Acad Forensic Sci; Am Chem Soc. Res: Toxicodynamics; comparative toxicology and pharmacology; environmental, occupational, clinical, analytical, food and forensic toxicology; chemical pathology; drug action in hibernation; biorhythms; effects of psychosocial parameters on toxicity and pharmacologic action. Mailing Add: Col Pharm Univ of Ky Lexington KY 40506

LUCKETT, WINTER PATRICK, b Atlanta, Ga, Mar 23, 37. ANATOMY, EMBRYOLOGY. Educ: Univ Mo, AB, 61, MA, 63; Univ Wis-Madison, PhD(anat), 67. Prof Exp: Instr, 68-69, ASST PROF ANAT, COL PHYSICIANS & SURGEONS, COLUMBIA UNIV, 69- Mem: AAAS; Am Asn Anat; Soc Study Reproduction; Int Primatol Soc. Res: Comparative morphogenesis of the placenta and fetal membranes; comparative structure of the ovary; endocrinology of reproduction; evolution of primates. Mailing Add: Col Physicians & Surgeons Columbia Univ New York NY 10032

LUCKEY, EGBERT HUGH, b Jackson, Tenn, Jan 1, 20; m 42, 70; c 4. INTERNAL MEDICINE. Educ: Union Univ, Tenn, BS, 41; Vanderbilt Univ, MD, 44. Hon Degrees: ScD, Union Univ, Tenn, 54. Prof Exp: Intern med, New York Hosp, 44-45, asst resident, 45-46, asst resident cardiol, 48-49; from instr to assoc prof med, Med Col, 49-57, prof & chmn dept, 57-66, dean med col & assoc dean grad sch, 54-57, VPRES MED AFFAIRS, CORNELL UNIV, 66-, PRES, NEW YORK HOSP-CORNELL MED CTR, 66-, VPRES SOC NEW YORK HOSP, 66- Concurrent Pos: Asst vis physician & asst dir, Second Med Div, Bellevue Hosp Ctr, Cornell Univ, 49-50, dir, 50-54, vis physician, 50-54, vpres med bd, 52-54, secy-treas exec comt med bd, 53-54; from asst attend physician to attend physician, New York Hosp, 52-57, physician-in-chief, 57-66; consult, US Vet Admin, New York, 54, mem dean's comt, 54-57, mem spec adv group, 64-69; sci consult, Sloan-Kettering Inst, 54-57; mem comt med, W K Kellogg Found, 55-61; trustee, Cornell Univ, 57-62 & Vanderbilt Univ, 62-; med dir, Russell Sage Inst Path, 58-67, mem bd dirs, 58-; mem bd dirs, Josiah Mach Jr Found, 58-; mem heart spec proj comt, NIH, 63-67. Mem: Asn Acad Health Ctrs; Harvey Soc; master Am Col Physicians; Asn Am Physicians; Am Heart Asn. Mailing Add: 525 E 68th St New York NY 10021

LUCKEY, GEORGE WILLIAM, b Dayton, Ohio, Apr 17, 25; m 58; c 3. PHYSICAL CHEMISTRY. Educ: Oberlin Col, BA, 47; Rochester Univ, PhD(chem), 50. Prof Exp: Mem staff, Photog Theory Dept, 50-56, Appl Photog Div, 56-60, MEM STAFF, SPEC RES DEPT, EASTMAN KODAK CO, 61- Mem: Am Chem Soc; Am Phys Soc; The Chem Soc; Electrochem Soc. Res: Photochemistry; photographic theory; luminescence; processing chemistry; photographic and radiographic systems. Mailing Add: 240 Weymouth Dr Rochester NY 14625

LUCKEY, PAUL DAVID, JR, b Pittsburgh, Pa, May 18, 28; m 55; c 1. PHYSICS. Educ: Carnegie Inst Technol, BS, 49; Cornell Univ, PhD(physics), 54. Prof Exp: Res assoc physics, Cornell Univ, 53-56; mem sci res staff, 56-70, SR RES SCIENTIST PHYSICS, MASS INST TECHNOL, 70- Mem: Am Phys Soc. Res: Meson physics; photoproduction of Pi mesons; electron synchrotrons. Mailing Add: Dept of Physics Mass Inst of Technol Cambridge MA 02139

LUCKEY, ROBERT RUEL RAPHAEL, b Houghton, NY, Nov 19, 17; m 45; c 5. MATHEMATICS. Educ: Houghton Col, AB & BS, 37; NY Univ, AM, 39; Cornell Univ, PhD, 42. Prof Exp: Teacher high sch, NY, 37-39; instr, Cornell Univ, 39-42; from instr to assoc prof, 42-50, PROF MATH & PHYSICS, HOUGHTON COL, 50-, DIR PUB RELS, 54-, V PRES DEVELOP, 61- Res: Applications of Fourier integrals; analysis in mathematics. Mailing Add: Houghton Col Houghton NY 14744

LUCKEY, THOMAS DONNELL, b Casper, Wyo, May 15, 19; m 43; c 3. BIOCHEMISTRY, NUTRITION. Educ: Colo Agr Col, BS, 41; Univ Wis, MS, 44, PhD(biochem), 46. Prof Exp: Asst, Agr & Mech Col, Tex, 41-42 & Univ Wis, 42-46; asst res prof biochem, Univ Notre Dame, 46-54; PROF BIOCHEM, SCH MED, UNIV MO-COLUMBIA, 54- Concurrent Pos: NSF traveling fel, Paris Nutrit Cong, 57; Univ Mo fel, Stockholm Microbiol Cong, 58; Commonwealth res fel, 61-62; Am Inst Nutrit traveling fel, Cong, 63; dir, WCent States Biochem Conf, 64-; moderator symp gnotobiol, Int Meeting Microbiol, Moscow, 66; mem subcomt interaction of infection & nutrit, Nat Acad Sci, 72-74; nutrit consult, NASA Johnson Space Ctr, Houston; consult, McDonnell Aircraft Corp, Mygrodol Prod Inc & Gen Elec Co. Mem: AAAS; Am Chem Soc; Soc Exp Biol & Med; Am Soc Microbiol; Am Inst Nutrit. Res: Nutrition and metabolism of germ-free vertebrates; folic acid and related compounds in chick nutrition; comparative nutrition; modes of action of antibiotics; gnotobiology; thymic hormones. Mailing Add: Dept of Biochem Univ of Mo Sch of Med Columbia MO 65201

LUCKHAM, DAVID COMPTOM, b Kingston, Jamaica, Sept 7, 36. COMPUTER SCIENCE. Educ: Univ London, BSc, 56, MSc, 57; Mass Inst Technol, PhD(math logic), 63. Prof Exp: Res assoc comput sci, Mass Inst Technol, 63-65; lectr math, Univ Manchester, 65-68; res assoc comput sci, Stanford Univ, 68-70; from asst prof to assoc prof, Univ Calif, Los Angeles, 70-72; RES COMPUT SCIENTIST, STANFORD UNIV, 72- Concurrent Pos: Consult, Bolt, Beranek & Newman Inc, 63-65 & Jet Propulsion Lab, 71-; Sci Coun res grant, Univ Manchester, 65-68; lectr, Ctr Comput & Automation, Imp Col, Univ London, 67-68; Hayes sr fel, Harvard Univ, 76-77. Mem: Am Math Soc; Asn Comput Mach; Asn Symbolic Logic. Res: Theory of computation; automated proof procedures and applications to computer-aided instruction, mathematics and verification of programs; artificial intelligence. Mailing Add: Dept of Comput Sci Stanford Univ Stanford CA 94305

LUCKMAN, CYRIL EDMUND, b Capetown, SAfrica, Jan 3, 10; nat US; m 37; c 5. ZOOLOGY. Educ: Wheaton Col, BS, 37; Univ Ill, MS, 39, PhD, 56. Prof Exp: Instr biol, King's Col, Del, 39-47; asst prof, 47-53, assoc prof, 54-57, PROF ZOOL, WHEATON COL, ILL, 58- Res: Histology. Mailing Add: Dept of Biol Wheaton Col Wheaton IL 60187

LUCKMANN, FREDERICK H, b Union City, NJ, Sept 18, 10; m 45; c 3. FOOD TECHNOLOGY, ANALYTICAL CHEMISTRY. Prof Exp: Chief chemist, 34-64, asst dir, Best Foods Div, CPC Int, 64-75; RETIRED. Mem: Am Chem Soc; Am Oil Chem Soc; Inst Food Technol. Res: Vitamin A; carotene, vegetable fats and oils; analytical methods for foodstuffs; sorbic acid. Mailing Add: 805 Embree Crescent Westfield NJ 07090

LUCKMANN, WILLIAM HENRY, b Cape Girardeau, Mo, Jan 15, 26; m 49; c 5. ENTOMOLOGY. Educ: Univ Mo, BS, 49; Univ Ill, MS, 51, PhD, 56. Prof Exp: Asst entomologist, State Natural Hist Surv, Ill, 51-53 & tech develop, Shell Chem Corp, Colo, 53-54; assoc entomologist, 54-59, entomologist, 59-65, ENTOMOLOGIST & HEAD SECT ECON ENTOM, STATE NATURAL HIST SURV, ILL, 65- Mem: Entom Soc Am. Res: Ecology; biology; applied control. Mailing Add: 163 Natural Resources Bldg State Natural Hist Surv Urbana IL 61803

LUCO, JOAQUIN, b Santiago, Chile, July 18, 13; m 37; c 4. PHYSIOLOGY, NEUROPHYSIOLOGY. Educ: Univ Chile, MD, 36. Hon Degrees: Dr, Cath Univ Chile, 55 & Austral Univ, Chile, 71. Prof Exp: Fel, Cath Univ Chile, 36-37; John Simon Guggenheim Mem Found fel, 37-39; prof gen physiol, 41-42, prof biochem, 42-49, prof pharmacol, 43-49, PROF NEUROPHYSIOL, CATH UNIV CHILE, 50- Concurrent Pos: Rockefeller Found traveling fel, 44-45; Gildemeister Found fel, Nat Inst Cardiol, Mex, 49; vis prof, Med Ctr, State Univ NY, 50, Univ of the Repub, Uruguay, 54 & Ctr Advan Studies, Mex, 72; Grass Found Alexander Forbes lectr, Marine Biol Lab, Woods Hole, 60; hon prof, Fac Med, Univ of the Repub, Uruguay, 63; mem bd dirs, Chilean Comn Sci & Technol Res, 66-71; vis seminar, Int Brain Res Orgn-UNESCO, 66-; vice-rector sci invest, Austral Univ, Chile, 68-71, acad med, Fac Sci, Univ Chile, 68- & acad sci, Chilean Inst, 69- Honors & Awards: Premio Nacional de Ciencias, 75. Mem: NY Acad Sci; Chilean Biol Soc (pres, 52). Mailing Add: Lab of Neurobiol Cath Univ of Chile Santiago Chile

LUCOVSKY, GERALD, b New York, NY, Feb 28, 35; m 57; c 5. SOLID STATE PHYSICS. Educ: Univ Rochester, BS, 56, MA, 58; Temple Univ, PhD(physics), 60. Prof Exp: Mem staff solid state physics, Philco Corp, Pa, 58-65; sr scientist, Xerox

Corp, 65-67; assoc prof eng, Case Western Reserve Univ, 67-68; mgr, Photoconductor Res Br, Xerox Corp, 68-69, solid state res br, 69-70, solid state sci br, Palo Alto Res Ctr, 70-73, assoc lab mgr, Gen Sci Lab, 73-74, RES FEL, GEN SCI LAB, PALO ALTO RES CTR, XEROX CORP, 74-, LAB MGR, 75- Mem: Fel Am Phys Soc. Res: Optical properties of solids; lattice dynamics; amorphous semiconductors. Mailing Add: Xerox Palo Alto Res Ctr 3333 Coyote Hill Rd Palo Alto CA 94304

LUCY, FRANK ALLEN, b Schoolcraft, Mich, July 22, 07; m 35; c 2. PHYSICS. Educ: Stanford Univ, AB, 29, AM, 30, PhD(chem), 33. Prof Exp: Asst instr chem, Stanford Univ, 30-34; chemist, Richards Chem Works, NJ, 34-41, Pa Salt Mfg Co, 41-45 & Shell Develop Co div, Shell Oil Co, Calif, 46; assoc, Stanford Res Inst, 47; scientist, Los Alamos Sci Lab, Univ Calif, NMex, 48-58; proj scientist, Missile & Space Vehicle Dept, Gen Elec Co, Philadelphia, 59-65, consult scientist, Missile & Space Div, 65-69; PROF MATH, CUSHING JR COL, 70- Mem: Optical Soc Am. Res: Design and construction of optical instruments for laboratory use; mathematical theory of intermolecular attraction; mechanics of high-speed flows in condensed phases. Mailing Add: 420 Merwyn Rd Merion Station PA 19066

LUCY, MARY (CORCORAN), botany, see 12th edition

LUDDEN, GERALD D, b Quincy, Ill, Sept 6, 37; m 61; c 3. MATHEMATICS. Educ: St Ambrose Col, BA, 59; Univ Notre Dame, MS, 61, PhD(math), 66. Prof Exp: Lectr math, Ind Univ, 65-66; asst prof, 66-70, ASSOC PROF MATH, MICH STATE UNIV, 70- Mem: Math Asn Am; Am Math Soc; Tensor Soc. Res: Hypersurfaces of manifolds with an f-structure; submanifolds of real and complex space forms. Mailing Add: Dept of Math Mich State Univ East Lansing MI 48823

LUDDEN, THOMAS MARCELLUS, b Kansas City, Mo, Jan 16, 46; m 67; c 2. BIOPHARMACEUTICS, DRUG METABOLISM. Educ: Univ Mo-Kansas City, BS, 69, PhD(pharmacol), 73. Prof Exp: Vis res assoc pharmaceut, Ohio State Univ, 74-75; ASST PROF CLIN PHARMACOKINETICS, UNIV TEX-AUSTIN, 75- Res: Applied pharmacokinetics and new drug development. Mailing Add: 7703 Floyd Curl Dr San Antonio TX 78284

LUDEKE, CARL ARTHUR, b Cincinnati, Ohio, Sept 26, 14. PHYSICS, OCEANOGRAPHY. Educ: Univ Cincinnati, AB, 35, PhD(physics), 38. Prof Exp: Instr math, John Carroll Univ, 38-40; instr, Univ Cincinnati, 40-43, from asst prof to assoc prof mech, 42-54, prof physics, 54-72; prof, Phys Oceanog Lab, 72-75, PROF PHYSICS, RES CTR, NOVA UNIV, 75- Concurrent Pos: Consult, Gen Elec Co, 56- Mem: Int Asn Analog Comput. Res: Nonlinear mechanics; vibration analysis; shock mounts; mathematical physics; energy from the sun, sea and atmosphere. Mailing Add: 8000 N Ocean Dr Dania FL 33004

LUDEMANN, CARL ARNOLD, b Brooklyn, NY, June 21, 34; m 56; c 2. NUCLEAR PHYSICS. Educ: Brooklyn Col, BS, 56; Univ Md, PhD(nuclear physics, elec eng), 64. Prof Exp: Res assoc physics, Univ Md, 64-65; vis scientist, 64-65, PHYSICIST, ELECTRONUCLEAR DIV, OAK RIDGE NAT LAB, 65- Mem: Am Phys Soc; Am Asn Physics Teachers. Res: Neutron threshold measurements; gamma ray spectroscopy; angular correlation and nuclear reaction mechanism; nuclear structure studies. Mailing Add: 130 Newhaven Rd Oak Ridge TN 37830

LUDER, WILLIAM FAY, chemistry, see 12th edition

LUDERS, RICHARD CHRISTIAN, b Staten Island, NY, July 23, 34; m 57; c 2. ANALYTICAL CHEMISTRY. Educ: Wagner Col, BS, 56. Prof Exp: Chemist, S B Penick & Co, 56-57; chemist, Ciba Pharmaceut Co, 58-66, group supvr anal res, 66-67, supvr, 67-70, head bioanal studies, Drug Metab Sect, 70-72, SR CHEMIST, DRUG METAB DIV, CIBA-GEIGY CORP, 72- Mem: Am Chem Soc. Res: Gas liquid chromatographic analysis and methods development for pharmaceutical compounds, preparations and raw materials; blood level determinations of pharmaceutical compounds. Mailing Add: Drug Metab Div Ciba-Geigy Corp Ardsley NY 10502

LUDFORD, GEOFFREY STUART STEPHEN, b London, Eng, Feb 2, 28; nat US; m 50; c 2. APPLIED MATHEMATICS. Educ: Cambridge Univ, BA, 48, MA & PhD(math), 52, ScD(math), 62. Prof Exp: Asst appl math, Harvard Univ, 50-51; from asst prof to assoc prof math, Univ Md, 51-59, prof aeronaut engr, 59-60; prof appl math, Brown Univ, 60-61; PROF APPL MATH, CORNELL UNIV, 61- Concurrent Pos: Guggenheim fel, Harvard Univ, 57-58; NSF sr fel, Univ Paris, 68-69. Mem: Am Math Soc; Soc Indust & Appl Math; Ger Soc Appl Math & Mech; Soc Eng Sci; Soc Natural Philos. Res: Fluid mechanics; magneto-hydrodynamics; differential equations; mathematical theory of combustion. Mailing Add: Dept of Theoret & Appl Math Cornell Univ Ithaca NY 14853

LUDIN, ROGER LOUIS, b Jersey City, NJ, June 13, 44; m 66; c 2. NUCLEAR PHYSICS. Educ: Brown Univ, ScB, 66; Worcester Polytech Inst, MS, 68, PhD(physics), 69. Prof Exp: Fel, Worcester Polytech Inst, 69-71; ASSOC PROF PHYSICS, BURLINGTON COUNTY COL, 71- Mem: AAAS; Am Phys Soc; Am Asn Physics Teachers. Res: Neutron-deuteron scattering. Mailing Add: Dept of Physics Burlington County Col Pemberton NJ 08068

LUDINGTON, MARTIN A, b Detroit, Mich, Mar 7, 43; m 64; c 2. NUCLEAR PHYSICS. Educ: Albion Col, AB, 64; Univ Mich, MS, 65, PhD(physics), 69. Prof Exp: ASST PROF PHYSICS, ALBION COL, 69- Mem: Am Phys Soc; Am Asn Physics Teachers. Res: Nuclear spectroscopy. Mailing Add: Dept of Physics Albion Col Albion MI 49224

LUDINGTON, STEPHEN DEAN, b Omaha, Nebr, Apr 13, 44. GEOCHEMISTRY. Educ: Stanford Univ, BS, 67; Univ Colo, MS, 69, PhD(geol), 74. Prof Exp: Res assoc geochem, 74-75, GEOLOGIST, US GEOL SURV, 75- Mem: Mineral Soc Am; Am Geophys Union. Res: Role of fluorine and chlorine in igneous and metamorphic petrology, with emphasis on micas; thermodynamics of magmas, with reference to geothermal systems. Mailing Add: Mail Stop 959 US Geol Surv Reston VA 22092

LUDKE, JAMES LARRY, b Vicksburg, Miss, Jan 11, 42; m 65. ENVIRONMENTAL BIOLOGY. Educ: Millsaps Col, BS, 64; Miss State Univ, MS, 67, PhD(physiol), 70. Prof Exp: Res asst physiol, Miss State Univ, 70-71; RES PHYSIOLOGIST, PATUXENT WILDLIFE RES CTR, US FISH & WILDLIFE SERV, 71- Mem: Am Soc Zoologists; Sigma Xi; AAAS. Res: Study of the chronic or lethal effects of pollutants on nontarget species; emphasis on fate of chemicals, diagnostic methods and chemical interactions. Mailing Add: Patuxent Wildlife Res Ctr US Fish & Wildlife Serv Laurel MD 20811

LUDLAM, WILLIAM MYRTON, b Teaneck, NJ, Mar 31, 31; m 54; c 3. OPTOMETRY, PHYSIOLOGICAL OPTICS. Educ: Columbia Univ, BS, 53, MS, 54; Mass Col Optom, OD, 63. Prof Exp: DIR VISION RES LAB, OPTOM CTR NY, 61-; ASSOC PROF PHYSIOL OPTICS & OPTOM, STATE UNIV NY COL OPTOM, 71- Concurrent Pos: Res grants, Am Optom Found, 55-56, NY Acad Optom, 57, Optom Ctr Res Fund, 60-61 & NIH, 63- Mem: Fel AAAS; fel Am Acad Optom; Optical Soc Am; NY Acad Sci. Res: Ocular dioptric components; pathophysiology of strabismus and its remediation; ametropia and its etiology. Mailing Add: State Univ NY Col of Optom 122 E 25th St New York NY 10010

LUDLUM, DAVID BLODGETT, b Brooklyn, NY, Sept 30, 29; m 52; c 2. PHARMACOLOGY. Educ: Cornell Univ, BA, 51; Univ Wis-Madison, PhD(chem), 54; NY Univ, MD, 62. Prof Exp: Res chemist, Polychem Dept, E I du Pont de Nemours & Co, Inc, Del, 54-58; intern, 3rd & 4th Med Divs, Bellevue Hosp, 62-63; asst prof pharmacol & Am Cancer Soc fac res assoc, Sch Med, Yale Univ, 63-68; assoc prof, 68-70, PROF CELL BIOL & PHARMACOL, SCH MED, UNIV MD, BALTIMORE CITY, 70- Concurrent Pos: Markle scholar acad med, Yale Univ & Univ Md, 67-72; Nat Inst Gen Med Sci career develop award, Yale Univ, 68. Mem: Am Chem Soc; Am Soc Pharmacol & Exp Therapeut; Am Soc Biol Chem; Am Asn Cancer Res. Res: Pharmacology of alkylating agents; cancer chemotherapy; mutagenesis and carcinogenesis; molecular and clinical pharmacology. Mailing Add: Dept of Pharmacol & Exp Therapeut Univ of Md Sch of Med Baltimore MD 21201

LUDLUM, JOHN CHARLES, b Chevy Chase, Md, Feb 2, 13; m 40. GEOLOGY. Educ: Lafayette Col, BS, 35; Cornell Univ, MS, 39, PhD(struct geol), 42. Prof Exp: Mem staff, Socony Vacuum Oil Co, 35-37 & Amerada Petrol Corp, 37; from asst instr to instr geol, Cornell Univ, 37-42; from asst prof to prof, 46-72, dir ctr resource develop, 62-63, dir off res & develop, Appalachian Ctr, 63-66, from asst dean to dean grad sch, 66-72, EMER PROF GEOL, WVA UNIV, 72- Concurrent Pos: Consult, 46-62; coop econ geologist, State Geol Surv, 46- Mem: Fel Geol Soc Am; Soc Econ Geol; Am Asn Petrol Geol; Am Inst Mining, Metall & Petrol Eng. Res: Structural and economic geology of West Virginia; natural and human resources research applied toward improvement of the economy and life in West Virginia and the Appalachian highlands. Mailing Add: 612 Callen Ave Morgantown WV 26505

LUDLUM, KENNETH HILLS, b Albany, NY, Nov 16, 29; m 53; c 4. PHYSICAL CHEMISTRY. Educ: Col Educ Albany, BA, 51, MA, 52; Rensselaer Polytech Inst, PhD(phys chem), 61. Prof Exp: Chemist, Beacon Res Lab, 61-62, sr chemist, 62-65, res chemist, 65-73, SR RES CHEMIST, TEXACO INC, 73- Mem: Am Chem Soc; Catalysis Soc. Res: Reaction kinetics; catalysis and surface chemistry. Mailing Add: Texaco Res Ctr PO Box 509 Beacon NY 12508

LUDMAN, ALLAN, b Brooklyn, NY, Mar 7, 43. GEOLOGY, PETROLOGY. Educ: Brooklyn Col, BS, 63; Ind Univ, Bloomington, AM, 65; Univ Pa, PhD(geol), 69. Prof Exp: Asst prof geol, Smith Col, 69-75; ASST PROF EARTH & ENVIRON SCI, QUEENS COL, NY, 75- Concurrent Pos: Field geologist, Maine Geol Surv, 66- Mem: Geol Soc Am; Mineral Soc Am. Res: Regional geologic mapping in central Maine; low-temperature metamorphism of pelitic and calcareous rocks; tectonic evolution of northeastern New England. Mailing Add: Dept of Earth & Environ Sci Queens Col Flushing NY 11300

LUDMAN, JACQUES ERNEST, b Chicago, Ill, Nov 26, 34; m 70; c 1. SOLID STATE PHYSICS. Educ: Middlebury Col, BA, 56; Northeastern Univ, PhD(solid state physics), 73. Prof Exp: RES PHYSICIST, AIR FORCE CAMBRIDGE RES LAB, 59- Res: Injection laser development; radiation damage effects on semiconductor devices; infrared sensor physics. Mailing Add: Dept for Electronic Technol ETSD Hanscom Field Bedford MA 01731

LUDOVICI, PETER PAUL, b Pittsburgh, Pa, Aug 9, 20; m 45; c 5. BACTERIOLOGY. Educ: Washington & Jefferson Col, BS, 42; Univ Pittsburgh, MS, 49, PhD(bact), 51. Prof Exp: Res bacteriologist immunol, West Penn Hosp, 49-51; res assoc, Univ Pittsburgh, 51; res assoc obstet & gynec, Univ Mich, 51-54; instr, 54-56, asst prof, 56-63, microbiol, obstet & gynec, 63-64, microbiol & cent tissue cult facilities, 64-65; assoc prof microbiol, 65-69, PROF MICROBIOL & MED TECHNOL, UNIV ARIZ, 69- Mem: AAAS; Am Soc Microbiol; Tissue Cult Asn; Soc Exp Biol & Med. Res: Tissue culture; cancer; virology; cell transformations; pseudomonas infections. Mailing Add: 5425 E Rosewood Ave Tucson AZ 85721

LUDOWIEG, JULIO, b Trujillo, Peru, Feb 10, 24; US citizen; m 53; c 2. BIOCHEMISTRY. Educ: San Marcos Univ, Lima, BS, 49; Univ Chicago, PhD(biochem), 60. Prof Exp: Asst biochem, Univ Chicago, 51-60; RES BIOCHEMIST, MED CTR, UNIV CALIF, SAN FRANCISCO, 60- Concurrent Pos: Res grants, Sch Med, Univ Calif, San Francisco, 61-65; Arthritis Found grants, 63 & 64; prin investr, USPHS Grants, 65-72. Mem: AAAS; Am Chem Soc; Fedn Am Soc Exp Biol. Res: Biochemical research in connective tissue involving acid mucopolysaccharides and proteins; asymmetric behavior of enzymes, their specificity and mechanism of action. Mailing Add: Dept of Orthop Surg Univ of Calif Med Ctr San Francisco CA 94122

LUDUENA, FROILAN PINDARO, b Reconquista, Arg, Aug 18, 06; nat US; m 41; c 1. PHARMACOLOGY. Educ: Rosario Med Sch, MD, 31; Oxford Univ, PhD(med), 37. Prof Exp: Res assoc dept pharmacol, Stanford Univ, 34-35, assoc prof, 38-40, actg asst prof pharmacol, Sch Med, 41-42, asst prof, 43-46; res fel, Sterling-Winthrop Res Inst, 46-71; RETIRED. Concurrent Pos: Arg Asn Advan Sci fel, Stanford Univ, 40-41. Mem: AAAS; Soc Exp Biol & Med; Am Soc Pharmacol & Exp Therapeut. Res: Cacti alkaloids; sympathomimetic amines; glycols; analgesics; local anesthetics; antispasmodics; coronary dilators; autonomic system. Mailing Add: 160 Terrace Ave Albany NY 12203

LUDVIGSEN, BERNHARD (THÖGER) FRANTS (JOSEF), b Copenhagen, Denmark, Oct 3, 23; m 44; c 2. BIOCHEMISTRY, CLINICAL CHEMISTRY. Educ: Copenhagen Univ, Mag scient, 54. Prof Exp: Res chief clin chem, Med Lab, Copenhagen Univ, Denmark, 54-58; chief clin serv lab, Sask Cancer & Med Res Inst, Saskatoon, Can, 58-59; asst prof & head anesthesia res lab, Dept Anesthesia, Univ Sask, 59; dir biochem, Muscular Dystrophy Res Lab, Univ Alta Hosp, 60-63; HEAD DEPT CHEM, GREENVILLE HOSP SYST, 63- Concurrent Pos: Mem Danish Nat Comn, Int Union Biochem, 53-54; consult, Path Assocs Greenville, PA, 67- Mem: AAAS; Am Chem Soc; Am Asn Clin Chem. Res: Fibrinogen and heavy metals; chemistry methodology; hypothermia; enzymes in muscle; tissue homogenization; lead poisoning; laboratory administration; automation; computers. Mailing Add: Greenville Hosp Syst Greenville SC 29602

LUDVIK, GEORGE FRANKLIN, b Dearborn, Mich, Aug 12, 19; m 42; c 3. ENTOMOLOGY. Educ: Univ Ill, AB, 41, AM, 47, PhD(entom), 49; Am Registry Cert Entomologists, dipl. Prof Exp: Spec res asst, Ill Natural Hist Surv, 46-49; entomologist, Tenn Valley Authority, 49-54; res specialist entom, Monsanto Co, St Louis, 54-73; HEAD BIOASSAY LAB, ZOECON CORP, 73- Mem: Entom Soc Am; Am Mosquito Control Asn. Res: Insect toxicology; insecticide research and development; insect rearing; mosquito control. Mailing Add: Zoecon Corp 975 California Ave Palo Alto CA 94304

LUDWICK, ADRIANE GURAK, b Passaic, NJ, June 16, 41; m 68; c 2. ORGANIC CHEMISTRY. Educ: Rutgers Univ, New Brunswick, AB, 63; Univ Ill, Urbana, MS,

65, PhD(chem), 67. Prof Exp: Asst prof chem, Tuskegee Inst, 67-68; vis asst prof & res assoc, Univ Ill, Urbana, 68-69; asst prof, 69-74, ASSOC PROF CHEM, TUSKEGEE INST, 74- Concurrent Pos: Res assoc, Environ Sci Div, Oak Ridge Nat Lab, 74. Mem; AAAS; Am Chem Soc; Sigma Xi. Res: Interaction of metal ions with olefinic insecticides in soil systems. Mailing Add: Dept of Chem Tuskegee Inst AL 36088

LUDWICK, ALBERT EARL, b San Diego, Calif, Feb 27, 40; m 61; c 2. SOIL FERTILITY. Educ: Calif State Polytech Col, BS, 62; Univ Wis, MS, 64, PhD(soil fertil), 67. Prof Exp: Instr soil fertil, Univ Wis, 66-68, asst prof, 68-70; asst prof, 70-75, ASSOC PROF SOIL FERTIL, COLO STATE UNIV, 75- Mem: Am Soc Agron; Soil Sci Soc Am; Coun Soil Testing & Plant Anal. Res: Management of plant nutrients with emphasis on nitrogen; development of soil testing for fertilizer recommendations; calibration of testing procedures for fertilizer recommendations. Mailing Add: Dept of Agron Plant Sci Bldg Colo State Univ Ft Collins CO 80523

LUDWICK, JIMMY DONALD, radiochemistry, see 12th edition

LUDWICK, JOHN CALVIN, JR, b Berkeley, Calif, Apr 25, 22; m 50, 57. MARINE GEOLOGY. Educ: Univ Calif, Los Angeles, AB, 47; Scripps Inst Oceanog, MS, 49, PhD(oceanog), 51. Prof Exp: Res asst, Scripps Inst Oceanog, 49-50; sedimentationist, Gulf Res & Develop Co, Tex, 50, group leader, 51-52, party chief, 53-59, Pa, 59-61, sr res geologist, 61-63, sect suvpr, 63-68; SLOVER PROF & DIR INST OCEANOG, OLD DOM UNIV, 68- Mem: Soc Econ Paleont & Mineral; fel Geol Soc Am; Am Geol Inst; AAAS. Res: Marine sedimentation; sedimentary petrology; physical oceanography; environmental interpretation of ancient sediments; shoal construction; tidal current analysis; beach processes. Mailing Add: Inst of Oceanog Old Dom Univ Norfolk VA 23508

LUDWICK, LARRY MARTIN, b Jamestown, NY, Oct 15, 41; m 68; c 1. INORGANIC CHEMISTRY. Educ: Mt Union Col, BS, 63; Univ Melbourne, BSc, 65; Univ Ill, Urbana, MS, 67, PhD(inorg chem), 69. Prof Exp: Res chemist, PPG Industs, 65; asst prof, 69-73, ASSOC PROF CHEM, TUSKEGEE INST, 74- Mem: AAAS; Am Chem Soc; The Chem Soc. Res: Ozone oxidation of chemical pollutants; complexation of silver in natural waters. Mailing Add: Dept of Chem Tuskegee Inst Tuskegee Institute AL 36088

LUDWICK, THOMAS MURRELL, b Cox's Creek, Ky, Aug 2, 15. DAIRY HUSBANDRY. Educ: Eastern Ky Teachers Col, BS, 36; Univ Ky, MS, 39; Univ Minn, PhD(dairy sci, animal genetics), 42; Univ Chicago, dipl, 43; Univ Va, dipl, 43. Prof Exp: Dairy & tobacco farmer, Ky, 25-39; asst cattle breeding & physiol, Univ Minn, 39-42; asst prof dairy sci, Univ Ky, 46-48; assoc prof, 48-55, PROF DAIRY SCI, OHIO STATE UNIV, 55- Concurrent Pos: Teacher high sch, Ky, 37-38; dir, Ohio Regional Dairy Cattle Breeding Proj, 48- Mem: Am Soc Animal Sci. Res: Physiology of reproduction and milk secretion; artificial insemination; animal breeding. Mailing Add: 8071 Sawmill Rd Dublin OH 43017

LUDWIG, ARMIN K, b Toledo, Ohio, Dec 30, 30. URBAN GEOGRAPHY, GEOGRAPHY OF BRAZIL. Educ: Ball State Univ, BA, 52; Mich State Univ, MA, 54; Univ Ill, Urbana-Champaign, PhD(geog), 62. Prof Exp: From instr to assoc prof geog, Colgate Univ, 58-75; MEM FAC, DEPT GEOG, HERBERT H LEHMAN COL, 75- Concurrent Pos: Soc Sci Res Coun grant field work, Brazil, 63-64. Mem: Asn Am Geogr; Latin Am Studies Asn; Conf Latin Am Geogr. Res: Urban and rural spatial systems in Brazil and the United States; spatial ecological systems in Brazil, Caribbean and the United States. Mailing Add: Dept of Geog & Geol Herbert H Lehman Col Bronx NY 10468

LUDWIG, BERNARD JOHN, b Burlington, Vt, Nov 2, 12; m 33; c 5. ORGANIC CHEMISTRY. Educ: Univ Vt, BS, 35, MS, 36; Columbia Univ, PhD(org chem), 40. Prof Exp: Asst chemist, Exp Sta, Univ Vt, 35-37; asst instr chem, Columbia Univ, 36-38; chemist res & develop, Wallace & Tiernan, Inc, 39-47; tech dir, Dabney Pharmacol Co, Inc, 47-48; asst dir res & develop, Carter Prod, Inc, 48-64, vpres chem res, 64-73, VPRES RES & DEVELOP, WALLACE LABS DIV, CARTER-WALLACE INC, 73- Mem: AAAS; Am Chem Soc; Am Pharmaceut Asn; fel Am Inst Chemists; NY Acad Sci. Res: Ethical pharmaceuticals; muscle relaxing drugs; tranquilizing drugs; antiatherosclerosis agents; psychopharmacological agents; bacterial endotoxins; immunological agents. Mailing Add: 1159 Stockton Pl North Brunswick NJ 08902

LUDWIG, CARL EDWARD, b Manchester, NH, Sept 4, 15; m 44; c 3. BIOLOGY, ENTOMOLOGY. Educ: Mass State Col, BS, 35; Boston Univ, MA, 36, PhD(biol), 49. Prof Exp: Instr biol, Boston Univ, 39-42, 46-47; asst, Stanford Univ, 47-49; from asst prof to assoc prof biol, 49-57, PROF BIOL & CHMN DIV SCI & MATH, SACRAMENTO STATE COL, 57- Mem: AAAS; Entom Soc Am. Res: Embryology and morphology of insects, particularly Diptera; comparative anatomy and embryology of vertebrates. Mailing Add: Dept of Biol Sacramento State Col Sacramento CA 95819

LUDWIG, CHARLES HEBERLE, b Minneapolis, Minn, May 1, 20; m 56; c 2. WOOD CHEMISTRY. Educ: Macalester Col, BA, 42; Univ Wash, PhD(chem), 61. Prof Exp: Chemist, D A Dodd, Mfg Chemist, 47-55 & Univ Wash, 56-61; MEM RES STAFF, GA PAC CORP, 61- Mem: Am Chem Soc; Sigma Xi. Res: Nuclear magnetic resonance spectroscopy of lignins and lignin models; chemistry of lignosulfonates and other lignins. Mailing Add: 437 15th St Bellingham WA 98225

LUDWIG, CLAUS BERTHOLD, b Berlin, Ger, Nov 18, 24; m 54; c 2. MOLECULAR SPECTROSCOPY, ENVIRONMENTAL PHYSICS. Educ: Aachen Tech Univ, MS, 51, PhD(physics), 53. Prof Exp: Design analyst, Eng Dept, Int Harvester Corp, 53-58; sr staff scientist, Space Sci Lab, Gen Dynamics-Convair, 58-72; SCIENTIST, SCI APPLN, INC, 72- Mem: Am Phys Soc; Optical Soc Am; Combustion Inst; assoc fel Am Inst Aeronaut & Astronaut Engrs; AAAS. Res: Molecular physics; high temperature molecular spectroscopy; infrared phenomena; radiative energy transfer; optical properties of small solid particles; guiding research and development in remote sensing of air pollution; development of air pollution monitors based on optical methods. Mailing Add: 5218 Cassandra Lane San Diego CA 92109

LUDWIG, DONALD A, b New York, NY, Nov 14, 33; m 53; c 2. MATHEMATICS. Educ: NY Univ, BA, 54, MS, 57, PhD(math), 59. Prof Exp: Res assoc math, Inst Math Sci, NY Univ, 59-60; Fine instr, Princeton Univ, 60-61; asst prof, Univ Calif, Berkeley, 61-64; from assoc prof to prof, NY Univ, 64-74; PROF MATH, UNIV BC, 74- Concurrent Pos: Guggenheim fel, Tel Aviv, Rehovot, Dundee, 70-71. Mem: Am Math Soc; Soc Indust & Appl Math. Res: Partial differential equations; mathematical methods for population biology. Mailing Add: Dept of Math Univ of BC Vancouver BC Can

LUDWIG, EDWARD JAMES, b New York, NY, Apr 13, 37; m 58; c 4. NUCLEAR PHYSICS. Educ: Fordham Univ, BS, 58; Ind Univ, MS, 60, PhD(physics), 63. Prof Exp: Res fel physics, Rutgers Univ, 63-66; asst prof, 66-71, ASSOC PROF PHYSICS, UNIV NC, CHAPEL HILL, 71- Mem: Am Phys Soc. Res: Solid state detectors; nuclear reactions and scattering cross sections and polarization effects; reaction mechanisms. Mailing Add: Dept of Physics Univ of NC Chapel Hill NC 27514

LUDWIG, ERNEST HARRY, b Philadelphia, Pa, Dec 29, 15; m 48; c 2. MICROBIOLOGY. Educ: Univ Pa, BA, 37, MS, 38, PhD(microbiol), 48. Prof Exp: Asst instr bact, Univ Pa, 46-47; instr, Sch Med, Univ WVa, 47-48, from asst prof to assoc prof, 49-52; res assoc, Dept Epidemiol & Microbiol, Grad Sch Pub Health, Univ Pittsburgh, 52-56; PROF MICROBIOL, PA STATE UNIV, 56-, ASSOC DEAN, GRAD SCH, 68- Concurrent Pos: USPHS fel, Microbiol Inst, 52. Mem: AAAS; Am Soc Microbiol; fel Am Acad Microbiol; NY Acad Sci. Res: Tissue culture; virus replication in, and biology of, mammalian cells in culture. Mailing Add: 211 Kern Grad Bldg Pa State Univ University Park PA 16802

LUDWIG, FRANK ARNO, b West Reading, Pa, Jan 17, 31. ELECTROCHEMISTRY, PHYSICAL CHEMISTRY. Educ: Calif Inst Technol, BS, 53; Case Western Reserve Univ, MS, 65, PhD(phys chem), 68. Prof Exp: Proj engr, Carter Labs, Inc, 53-56; res engr, Hughes Aircraft Co, 56-57; vpres, Tech Commun, Inc, 55-58; dept mgr fuel cells, thermogalvanics, Electro-Optical Systs, Inc, 58-62; mgr org electrolyte batteries dept, Res & Develop Ctr, Whittaker Corp, 68-69; TECH SUPVR, RES LAB, FORD MOTOR CO, 69- Mem: Electrochem Soc; Am Chem Soc; AAAS. Res: Materials, corrosion, chemical and electrochemical kinetics in molten salt batteries; conception and development of new sodium-sulfur battery electrodes; electroanalytical chemistry, surface chemistry and new concepts in energy storage and conversion devices. Mailing Add: 18457 Magnolia Pkwy Southfield MI 48075

LUDWIG, FREDERIC C, b Bad Nauheim, WGer, Jan 22, 24; US citizen; m 58; c 4. EXPERIMENTAL PATHOLOGY. Educ: Univ Tübingen, MD, 49; Univ Paris, ScD(radiobiol), 58. Prof Exp: Sect chief radiation path, AEC, France, 55-59; assoc res pathologist, Med Ctr, Univ Calif, San Francisco, 58-62, lectr path, 62-65, assoc prof in residence, 65-71; PROF PATH & RADIOL SCI, COL MED, UNIV CALIF, IRVINE, 71- Concurrent Pos: Consult, Stanford Res Inst, 65- Honors & Awards: Award, Nat Inst Hyg, France, 59. Mem: Radiation Res Soc Am; Am Soc Exp Path; NY Acad Sci; Fr Asn Anat; Ger Path Soc. Res: Abscopal effects of radiation; radiation injury in blood forming organs; pathogenesis of radiation leukemia; homeostasis of white blood cells. Mailing Add: Dept of Path & Radiol Sci Univ of Calif Col of Med Irvine CA 92664

LUDWIG, FREDERICK JOHN, SR, b St Louis, Mo, June 20, 28; m 56; c 2. ANALYTICAL CHEMISTRY, ORGANIC CHEMISTRY. Educ: Washington Univ, AB, 50; St Louis Univ, PhD(chem), 53. Prof Exp: Lab asst chem, St Louis Univ, 50-53; res chemist, Uranium Div, Mallinckrodt Chem Corp, 55-59; group leader, Petrolite Corp, 59-73, RES SCIENTIST, TRETOLITE DIV, PETROLITE CORP, 73- Mem: Am Chem Soc; Sigma Xi. Res: Gas-liquid and liquid-solid chromatography; infrared spectroscopy; wax-polymers; water-treatment chemicals; nuclear magnetic resonance spectroscopy. Mailing Add: Res Lab Petrolite Corp 369 Marshall Ave St Louis MO 63119

LUDWIG, GARRY (GERHARD ADOLF), b Mannheim, Ger, Sept 4, 40; Can citizen; m 68. MATHEMATICS. Educ: Univ Toronto, BSc, 62; Brown Univ, PhD(physics), 66. Prof Exp: Asst prof, 66-72, ASSOC PROF MATH, UNIV ALTA, 72- Concurrent Pos: Nat Res Coun grants, 67-75. Mem: Am Math Soc; Am Phys Soc; Can Math Cong; Can Asn Physicists. Res: Gravitational radiation; spinor methods. Mailing Add: Dept of Math Univ of Alta Edmonton AB Can

LUDWIG, GEORGE DORING, biochemistry, biophysics, deceased

LUDWIG, GERALD W, b New York, NY, Jan 7, 30; m 51; c 3. SEMICONDUCTORS, SOLID STATE PHYSICS. Educ: Harvard Univ, AB, 50, AM, 51, PhD(chem physics), 55. Prof Exp: Physicist, 55-63, liaison scientist, 63-65, physicist, 65-71, MGR SIGNAL SEMICONDUCTOR BR, RES & DEVELOP CTR, GEN ELEC CORP, 71- Mem: Fel Am Phys Soc; sr mem, Inst Elec & Electronics Engrs; Electrochem Soc. Res: Semiconductor device physics; semiconductor materials and processing; x-ray and cathode ray phosphors; Gunn effect; electron paramagnetic resonance; transport properties of semiconductors. Mailing Add: Res & Develop Ctr Gen Elec Co PO Box 8 Schenectady NY 12301

LUDWIG, HOWARD C, b Beaver Falls, Pa, July 31, 16; m 41; c 1. CHEMICAL PHYSICS, PLASMA PHYSICS. Educ: Geneva Col, BS, 41. Prof Exp: Chem analyst, Armstrong Cork Co, 41-42; spectroscopist, Properller Div, Curtiss-Wright Corp, 42-46; res engr, Res Labs, Westinghouse Elec Corp, 46-59, FEL SCIENTIST, RES & DEVELOP CTR, WESTINGHOUSE ELEC CORP, 59- Honors & Awards: IR 100 Award, 63; Lincoln Gold Medal, Am Welding Soc, 56. Mem: Am Chem Soc; Am Welding Soc; Am Phys Soc. Res: Research and development of high pressure plasmas. Mailing Add: 159 Roberta Dr Pittsburgh PA 15221

LUDWIG, HUBERT JOSEPH, b Lincoln, Ill, July 27, 34; m 65; c 2. MATHEMATICS. Educ: Univ Ill, Urbana, BS, 56; St Louis Univ, MS, 64, PhD(math), 68. Prof Exp: Instr math, chem & eng mech, Springfield Col, Ill, 56-65; teaching asst math, St Louis Univ, 65-68; asst prof, 68-75, ASSOC PROF MATH, BALL STATE UNIV, 75- Mem: Math Asn Am; Am Math Soc. Res: Autometrized spaces; 2-metric spaces. Mailing Add: Dept of Math Sci Ball State Univ Muncie IN 47306

LUDWIG, JAMES PINSON, b Port Huron, Mich, Mar 20, 41; m 66; c 1. ECOLOGY, VERTEBRATE BIOLOGY. Educ: Univ Mich, BS, 62, MS, 65, PhD(zool), 68. Prof Exp: Res asst, Pac Proj, Smithsonian Inst, 63-64; instr biol, St Cloud State Col, 67-68; asst prof, Mackinac Col, 68-69; dir ctr environ studies, Bemidji State Col, 69-73; CONSULT ECOLOGIST, 73- Concurrent Pos: Fel limnol, Univ Minn, 69; mem, Minn Gov Environ Educ Coun; US Fish & Wildlife Serv res grant; prin investr, Off Water Resources Res Res Grant. Mem: Ecol Soc Am; Am Ornithologists Union; Cooper Ornith Soc; Wilson Ornith Soc; Int Asn Great Lakes Res. Res: Ecology of the Great Lakes, especially of gulls and terns; primary productivity of the Great Lakes. Mailing Add: RR 4 Box 228 Bemidji MN 56601

LUDWIG, JEROME HOWARD, b St Joseph, Mich, Mar 31, 33; m 57; c 2. ORGANIC CHEMISTRY. Educ: Kalamazoo Col, BA, 55; Univ Cincinnati, MS, 57, PhD(org chem), 59. Prof Exp: Res assoc, Upjohn Co, 59-62, res assoc & anal supvr, 62-63; group leader appl res & new prod develop, Emery Industs, Inc, Ohio, 63-68; TECH DIR, SYNTHETIC PROD CO, 68- Mem: Am Chem Soc; Soc Plastic Eng. Res: New product development; plastics additives; process research. Mailing Add: 17419 Fernway Rd Shaker Heights OH 44120

LUDWIG, MARTHA LOUISE, b Pittsburgh, Pa, Aug 16, 31; m 61. BIOCHEMISTRY. Educ: Cornell Univ, BA, 52, PhD(biochem), 56; Univ Calif, Berkeley, MA, 55. Prof Exp: Res fel biochem, Harvard Med Sch, 56-59; res assoc biol, Mass Inst Technol, 59-62; res fel chem, Harvard Univ, 62-67; res fel biochem, 67-68, assoc prof, 68-75, PROF BIOL CHEM, UNIV MICH, ANN ARBOR, & RES BIOPHYSICIST,

BIOPHYS RES DIV, 75- Mem: Am Chem Soc; Am Soc Biol Chemists; Biophys Soc; Am Crystallog Asn. Res: Protein crystallography; protein structure and function. Mailing Add: Dept of Biol Chem Univ of Mich Ann Arbor MI 48104

LUDWIG, OLIVER GEORGE, b Philadelphia, Pa, Nov 15, 35. PHYSICAL CHEMISTRY. Educ: Villanova Univ, BS, 57; Carnegie Inst Technol, MS, 60, PhD(quantum chem), 61. Prof Exp: Mem math lab & sr res worker theoret chem, Cambridge Univ, 61-63; asst prof chem & fac assoc comput ctr, l)niv Notre Dame, 63-68; ASSOC PROF CHEM, VILLANOVA UNIV, 68- Concurrent Pos: NSF fel, 61-63; actg chemn dept chem, Villanova Univ, 69-70. Mem: Am Chem Soc; Am Phys Soc; Asn Comput Mach. Res: Quantum chemistry; chemical applications of digital computers; development of methods for scientific computing. Mailing Add: Dept of Chem Villanova Univ Villanova PA 19805

LUDWIG, PAUL DAVID, JR, entomology, insect toxicology, see 12th edition

LUDWIG, RALPH ANTONY, b Calgary, Alta, July 12, 15; m 40; c 5. PLANT PATHOLOGY. Educ: Univ Alta, BSc, 37, MSc, 39; McGill Univ, PhD, 47. Prof Exp: From lectr to assoc prof plant path, McGill Univ, 40-51; res officer, Pesticide Res Inst, Can Dept Agr, Ont, 51-59, dir, Res Sta, NS, 59-61, res inst, Plant Res Inst, Ont, 61-65, ASST DIR-GEN ADMIN, RES BR, AGR CAN, 65- Concurrent Pos: Mem, Ont Inst Agrologists. Mem: Am Phytopath Soc; Can Phytopath Soc; Agr Inst Can; Can Bot Soc. Mailing Add: Res Br Agr Can Ottawa ON Can

LUDWIG, RICHARD ELI, b Pottstown, Pa, Oct 16, 29; m 55; c 2. ORGANIC CHEMISTRY. Educ: Ursinus Col, BS, 52; Univ Del, MS, 54, PhD, 56. Prof Exp: SR RES CHEMIST, TEXTILE RES LAB, E I DU PONT DE NEMOURS & CO, 55- Mem: Am Chem Soc; Am Inst Chemists; Tech Asn Pulp & Paper Indust. Res: Catalytic air oxidation of aromatic hydrocarbons; synthesis of trialkyl-pyrrolidine triones; epoxy resin chemistry; powder compaction; new product development of nonwoven materials; printing technology. Mailing Add: Textile Res Lab E I du Pont de Nemours & Co Wilmington DE 19898

LUDWIG, THEODORE FREDERICK, b Castlewood, SDak, July 8, 24; m 45; c 1. PROSTHODONTICS. Educ: Cent Col, Iowa, AB, 45; Ohio State DDS, 59, MSc, 63. Prof Exp: Asst prof dent, Sch Dent, WVa Univ, 63-67; asst prof, Sch Dent, Univ Iowa, 67-69; ASSOC PROF PROSTHODONTICS, COL DENT MED, MED UNIV SC, 69- Concurrent Pos: NIH grant, 62-63; mem, Carl O Boucher Prosthodontic Conf, 66- Mem: Am Dent Asn. Res: Esthetics in complete dentures; design and metals in removable partial dentures. Mailing Add: 772 Tennent St Charleston SC 29412

LUDWIG, WALTER JOHN, b Barberton, Ohio, Mar 29, 33; m 58; c 2. PHARMACY. Educ: Ohio Northern Univ, BSc, 55; Philadelphia Col Pharm & Sci, MSc, 57. Prof Exp: Pharmacist, USPHS, Phoenix Indian Hosp, 57-59; analyst nutrit & metab dis, NIH, 59-61; scientist qual control, 61-64, sr scientist, 64-66, group leader, 66-69, SUPVR PROCESS DEVELOP, MEAD JOHNSON & CO, 69- Mem: Am Pharmaceut Asn; Acad Pharmaceut Sci; Parenteral Drug Asn. Res: Pharmaceutical dosage form design and pharmaceutical process design. Mailing Add: Mead Johnson & Co 2404 Pennsylvania St Evansville IN 47721

LUDWIG, WILLIAM JACKSON, b New York, NY, May 25, 32; m 60; c 2. MARINE GEOPHYSICS. Educ: Univ Houston, BS, 56; Hokkaido Univ, PhD, 69. Prof Exp: Res asst geophys, 56-60, res scientist, 60-65, res assoc geophys, 65-69, SR RES ASSOC GEOPHYS, LAMONT-DOHERTY GEOL OBSERV, COLUMBIA UNIV, 69- Mem: Am Geophys Union; Soc Explor Geophys; Geol Soc Am; Am Asn Petrol Geologists. Res: Structure and constitution of the earth. Mailing Add: Lamont-Doherty Geol Observ Columbia Univ Palisades NY 10964

LUDWIN, ISADORE, b Malden, Mass, Feb 23, 15; m 49; c 5. ANIMAL GENETICS, PHYSIOLOGY. Educ: Univ Mass, BS, 37; Univ Wis, MS, 39; Harvard Univ, PhD(genetics), 48. Prof Exp: Statist analyst, US Dept Navy, DC, 41-42; asst prof biol, Univ Mass, 46-48; fel, Tufts Col, 48; assoc cancer biologist, Roswell Park Mem Inst, 49-51; prof sci, Calvin Coolidge Col, 57-62; prof biol, Cambridge Jr Col, 57-67; lectr, Northeastern Univ, 67-68; PVT RES & DEVELOP, 68- Mem: AAAS; Asn Advan Med Instrumentation. Mailing Add: 1073 Centre St Newton MA 02159

LUEBBE, RAY HENRY, JR, b Schenectady, NY, Mar 31, 31; m 59; c 3. PHYSICAL CHEMISTRY. Educ: Dartmouth Col, AB, 53; Univ Wis, PhD(phys chem), 58. Prof Exp: Asst phys chem, Univ Wis, 53-55; chemist, Photo Prod Dept, E I du Pont de Nemours & Co, 58-64; SCIENTIST, XEROX CORP, 64- Res: Hot atom and photo chemistry; photographic science; photopolymerization; electrophotography. Mailing Add: 825 Glenleven Crescent Mississauga ON Can

LUEBBEN, RALPH A, b Milwaukee, Wis, Feb 14, 21; m 47. ANTHROPOLOGY. Educ: Purdue Univ, BS, 43; Univ NMex, MA, 51; Cornell Univ, PhD(anthrop), 55. Prof Exp: Asst prof anthrop, Iowa State Col, 55-57; assoc prof, Grinnell Col, 57-61 & Ariz State Col, 61-63; prof, Colo Woman's Col, 63-67; PROF ANTHROP, GRINNELL COL, 67- Mem: Am Anthrop Asn; Soc Appl Anthrop; Am Ethnol Soc; Soc Am Archeol. Res: Ethnography of Navajo Indians, Spanish-speaking Americans in Rio Grande Drainage, and Germanic Europe; archeology of American southwest and northern Mexico. Mailing Add: Dept of Anthrop Grinnell Col Grinnell IA 50112

LUEBBERS, SCOTT SHERWOOD, solid state physics, see 12th edition

LUEBKE, EMMETH AUGUST, b Manitowoc, Wis, Aug 1, 15; c 2. PHYSICS. Educ: Ripon Col, BA, 36; Univ Ill, PhD(physics), 41. Prof Exp: Asst physics, Univ Ill, 36-41; group leader, Radiation Lab, Mass Inst Technol, 41-45; res assoc, Res Lab, Gen Elec Co, 45-50, mgr reactor eval, Knolls Atomic Power Lab, 50-55, gen physics, missile & space vehicle dept, 55-58, gen physicist, Gen Eng Lab, 58-63, physicist, Tempo, 63-72; ADMIN LAW JUDGE, US NUCLEAR REGULATORY COMN, 72- Concurrent Pos: Mem, Joint Liquid Metals Comt, US Navy AEC, 50-55; presiding tech mem, Atomic Safety & Licensing Bd. Mem: Fel Am Phys Soc; Am Nuclear Soc. Res: Linear accelerator; velocity spectrometer measurement of neutron cross section; microwave radar components; system design; liquid metal heat transfer; design and evaluation of reactor power plants; breeder; submarine propulsion; central station types; environmental controls. Mailing Add: 5500 Friendship Blvd Chevy Chase MD 20015

LUEBS, RALPH EDWARD, b Wood River, Nebr, Mar 21, 22; m 51; c 4. SOILS. Educ: Univ Nebr, BS, 48, MS, 52; Iowa State Univ, PhD(soil fertil), 54. Prof Exp: Asst agron, Univ Nebr, 48-49; soil scientist, Exp Sta, Agr Res Serv, USDA, 55-56, soil scientist, Ft Hays Exp Sta, Kans State Col, 56-59, soil scientist, Exp Sta, Univ Calif, Riverside, 59-75; CHIEF AGRON GROUP, WOODWARD-CLYDE CONSULTS, 75- Mem: Am Soc Agron; Soil Sci Soc Am; Sigma Xi. Res: Nitrogen availability and rainfall use efficiency for dryland crops; pathways of nitrogen originating in animal wastes. Mailing Add: 2909 West 7th Ave Denver CO 80204

LUECK, CHARLES HENRY, b St Paul, Minn, Oct 1, 28; m 55; c 6. ANALYTICAL CHEMISTRY. Educ: Col St Thomas, BS, 50; Univ Detroit, MS, 53; Wayne State Univ, PhD, 56. Prof Exp: Res chemist, 56-66, ANAL RES SUPVR, TEXTILE FIBERS DEPT, E I DU PONT DE NEMOURS & CO, 66- Mem: Am Chem Soc. Res: Spectrophotometric analysis; chemical degradation studies. Mailing Add: Textile Fibers Dept E I du Pont de Nemours & Co Old Hickory TN 37138

LUECK, JAMES DODDS, biochemistry, see 12th edition

LUECK, LESLIE MELVIN, b Chippewa Falls, Wis, July 28, 20. PHARMACY. Educ: Univ Wis, BS, 49, MS, 51, PhD, 54. Prof Exp: Asst dir qual control, 54-63, dir, 63-68, VPRES QUAL CONTROL & GOVT REGULATIONS, PARKE, DAVIS & CO, 68- Mem: Am Pharmaceut Asn. Res: Protective ointments: compressed tablet formulation; pharmacuetical formulation. Mailing Add: Parke Davis & Co Joseph Campau at the River Detroit MI 48232

LUECK, ROGER HAWKS, b Fox Lake, Wis, Dec 17, 96; m 24; c 2. RESEARCH ADMINISTRATION, CORROSION. Educ: Carroll Col, BS, 19; Univ Wis, MS, 21. Hon Degrees: DSc, Carroll Col, 43. Prof Exp: Instr chem, Univ Wis, 19-22; chemist, Am Can Co, Ill, 22-26, dist mgr, Res Dept, Calif, 26-34, mgr, Hawaiian Div, 34-35, admin mgr, Res Dept, Ill, 35-41, dir res, 41-44, mgr sales, Pac Div, 44-50, gen mgr, Res & Tech Dept, NY, 50-54, vpres res & develop, 55-62, CONSULT RES ADMIN & PACKAGING TECHNOL, AM CAN CO, 62- Concurrent Pos: Consult, Off Qm Gen, US Army, 42-44; dir, James Dole Corp, 61-; trustee, Midwest Res Inst; mem, Vis Comt, Dept Food Technol, Mass Inst Technol & Adv Coun, Col Eng, NY Univ. Mem: Am Chem Soc; AAAS; Am Mgt Asn (vpres); fel Am Inst Chemists; Inst Food Technologists (vpres, 44). Res: Kinetics of chemical reaction; corrosion of tinplate; organic coatings on metal; canning technology; metal complexes with flavones; frothy fermentation of saccharine foods; administration of industrial research. Mailing Add: 20016 Winter Lane Saratoga CA 95070

LUECKE, GLENN RICHARD, b Brian, Tex, May 19, 44; m 67; c 2. MATHEMATICAL ANALYSIS. Educ: Mich State Univ, BS, 66; Calif Inst Technol, PhD(math), 70. Prof Exp: Asst prof, 69-72, ASSOC PROF MATH, IOWA STATE UNIV, 72- Mem: Am Math Soc; Math Asn Am. Res: Study of continuous linear transformations in Hilbert space. Mailing Add: Dept of Math Iowa State Univ Ames IA 50010

LUECKE, RICHARD WILLIAM, b St Paul, Minn, July 12, 17; m 41; c 3. BIOCHEMISTRY, NUTRITION. Educ: Macalester Col, BA, 39; Univ Minn, MS, 41, PhD(biochem), 43. Prof Exp: PROF BIOCHEM, MICH STATE UNIV, 41- Concurrent Pos: Assoc prof biochem, Tex A&M Univ, 43-45; consult, Armour Res Labs, Chicago, 55-66; mem comt on animal nutrit, Nat Resources Coun, 55-65; mem food & nutrit bd, Food & Agr Orgn, UN, 60-65; consult, Merck Sharp & Dohme Res Labs, 62-69. Honors & Awards: Award, Am Soc Animal Sci, 56. Mem: Am Chem Soc; Am Inst Nutrit; Brit Nutrit Soc; Soc Exp Biol & Med. Res: Trace element metabolism in animals. Mailing Add: Dept of Biochem Mich State Univ East Lansing MI 48824

LUEDECKE, LLOYD O, b Hamilton, Mont, July 28, 34; m 57; c 2. DAIRY BACTERIOLOGY. Educ: Mont State Col, BS, 56; Mich State Univ, MS, 58, PhD(food sci), 62. Prof Exp: Asst prof dairy sci, 62-70, assoc prof & assoc dairy scientist, 70-73, ASSOC PROF FOOD SCI, WASH STATE UNIV, 73- Mem: Am Dairy Sci Asn; Inst Food Technol. Res: Heat resistance of psychrophiles; bacteriological aspects of mastitis. Mailing Add: Dept of Food Sci & Technol Clark Hall Wash State Univ Pullman WA 99163

LUEDEMAN, JOHN KEITH, b Ft Wayne, Ind, Apr 27, 41; m 63; c 4. ALGEBRA. Educ: Valparaiso Univ, BA, 63; Southern Ill Univ, Carbondale, MA, 65; State Univ NY, Buffalo, PhD(math), 69. Prof Exp: Instr math, State Univ NY, Buffalo, 67-68; asst prof, 68-72, ASSOC PROF MATH, CLEMSON UNIV, 72- Concurrent Pos: Consult math, Oconee County Sch Syst, SC, 74- Mem: Am Math Soc; Math Asn Am; Sigma Xi. Res: Ring and module theory; category theory; semigroups; topology. Mailing Add: Dept of Math Clemson Univ Clemson SC 29631

LUEDEMANN, GEORGE MILLER, mycology, plant taxonomy, see 12th edition

LUEDEMANN, LOIS W, b Chicago, Ill, Mar 22, 31; m 54; c 1. GEOCHEMISTRY, MATERIALS SCIENCE. Educ: Hunter Col, BA, 51; Syracuse Univ, MS, 53; Pa State Univ, PhD(mineral), 56. Prof Exp: Asst mineral, Pa State Univ, 52-56; res assoc chem, Syracuse Univ, 56-59; instr mat sci, 59-60, ASSOC PROF GEOL & CHEM, FAIRLEIGH DICKINSON UNIV, 60- Mem: Nat Asn Geol Teachers; Am Chem Soc; Am Ceramic Soc; Am Mineral Soc; Am Soc Metals. Res: Petrography and x-ray analysis of uranium bearing shales; crystal chemical study of hydrous vanadates; preparation and characterization of boron-hydrides; phase equilibria in alkalimetal-alkaline earth metal systems. Mailing Add: Dept of Chem Fairleigh Dickinson Univ Rutherford NJ 07070

LUEHR, CHARLES POLING, b Plentywood, Mont, Sept 27, 30. APPLIED MATHEMATICS. Educ: Ore State Col, BS, 53, MS, 56; Univ Calif, Berkeley, PhD(appl math), 62. Prof Exp: Mem prof staff, Gen Elec Co, Calif, 62-68; fel, 68-70, asst prof, 70-75, ASSOC PROF MATH, UNIV FLA, 75- Mem: AAAS; Am Math Soc; Am Phys Soc; Am Chem Soc; Soc Indust & Appl Math. Res: Methods of mathematical physics; tensor analysis with applications in physics; theory of spinors with applications in quantum mechanics and relativity theory; applications of modern differential geometry to general relativity. Mailing Add: Dept of Math Univ of Fla Gainesville FL 32611

LUEHRMANN, ARTHUR WILLETT, JR, b New Orleans, La, Mar 8, 31; m 61; c 2. SOLID STATE PHYSICS, COMPUTER SCIENCE. Educ: Univ Chicago, AB, 55, SB, 57, SM, 61, PhD(physics), 66. Prof Exp: From instr to asst prof, 65-70, ADJ ASSOC PROF PHYSICS & ASST DIR OFF ACAD COMPUT, DARTMOUTH COL, 70- Concurrent Pos: Consult, NSF Off Comput Activities, Dartmouth Col, 68-, res grant, 69-76; Fulbright lectr, Fulbright Comn, Colombia, SAm, 69. Honors & Awards: Distinguished Serv Citation, Am Asn Physics Teachers, 71. Mem: Am Asn Physics Teachers; Am Phys Soc. Res: Solid state theory; band structure; computational physics; computer-based instruction. Mailing Add: Kiewit Comput Ctr Dartmouth Col Hanover NH 03755

LUEHRS, DEAN C, b Fremont, Nebr, Aug 20, 39; m 69. INORGANIC CHEMISTRY. Educ: Mich State Univ, BS, 61; Univ Kans, PhD(chem), 65. Prof Exp: Asst prof, 65-69, ASSOC PROF CHEM, MICH STATE TECHNOL UNIV, 69- Mem: Am Chem Soc. Res: Nonaqueous solvents; electrochemistry; optical activity of inorganic compounds. Mailing Add: Dept of Chem & Chem Eng Mich Technol Univ Houghton MI 49931

LUEKING, DONALD ROBERT, b Cincinnati, Ohio, Nov 24, 46; m 73. MICROBIAL BIOCHEMISTRY. Educ: Ind Univ, Bloomington, BS, 69, PhD(microbiol), 73. Prof

Exp: Trainee microbiol, Univ Pa, 73-74, fel, 74-75; FEL MICROBIOL, UNIV ILL, URBANA, 75- Mem: Am Soc Microbiol; AAAS; Sigma Xi. Res: The use of the photosynthetic bacteria as a model system for the study of the factors involved in the regulation of membrane biosynthesis and differentiation. Mailing Add: Dept of Microbiol Burrill Hall Univ of Ill Urbana IL 61801

LUENINGHOENER, GILBERT CARL, b Hooper, Nebr; m 27; c 2. GEOLOGY. Educ: Midland Col, BS, 27; Univ Nebr, MS, 34, PhD(geol), 47. Prof Exp: Instr high sch, 27-28; instr geol, 29-46, prof astron, 29-74, chmn natural sci & math div, 42-68, prof geol, 46-70, EMER PROF GEOL, MIDLAND LUTHERAN COL, 74-, DIR WALTER D BEHLEN OBSERV & GILBERT C LUENINGHOENER PLANETARIUM, 65- Concurrent Pos: Res assoc, Mus, Univ Nebr, 50-58, planetarium consult, 58- Mem: Fel Geol Soc Am. Res: Geomorphology and sedimentation; the post-Kansas geologic history of lower Platte valley area; graphic resume of the Pleistocene of Nebraska; grinding and polishing telescope mirrors. Mailing Add: Dept of Geol Midland Lutheran Col Fremont NE 68025

LUEPSCHEN, NORMAN SIEGFRIED, b Buffalo, NY, Jan 6, 33; m 53; c 4. PLANT PATHOLOGY. Educ: Wheaton Col, Ill, BS, 54; Cornell Univ, PhD(plant path), 60. Prof Exp: Asst plant path, Cornell Univ, 54-59, exten specialist, 59-60; assoc pathologist, Mkt Qual Serv, USDA, 60-61; asst plant pathologist, 62-65, assoc plant pathologist, 65-75, PROF PLANT PATH, COLO STATE UNIV, 75- Mem: Am Phytopath Soc; Am Soc Hort Sci; Am Pomol Soc; Mycol Soc Am. Res: Tree fruit diseases; storage, transit and market diseases; antibiotics. Mailing Add: Western Slope Br Sta Colo State Univ Grand Junction CO 81501

LUESCHEN, WILLIAM EVERETT, b Springfield, Ill, Jan 29, 42; m 65; c 2. AGRONOMY. Educ: Southern Ill Univ, BS, 64; Univ Ill, MS, 66, PhD(agron), 68. Prof Exp: ASSOC PROF AGRON & AGRONOMIST, SOUTHERN EXP STA, UNIV MINN, 68- Mem: Am Soc Agron; Crop Sci Soc Am; Weed Sci Soc Am. Res: Crop production, management, physiology and weed science. Mailing Add: Southern Exp Sta Univ of Minn Waseca MN 56093

LUESSENHOP, ALFRED JOHN, b Chicago, Ill, Feb 6, 26; m 52; c 4. MEDICINE, NEUROSURGERY. Educ: Yale Univ, BS, 49; Harvard Med Sch, MD, 52. Prof Exp: Intern surg, Univ Chicago, 52-53; resident neurosurg, Mass Gen Hosp, 53-58; vis scientist, Nat Inst Neurol Dis & Blindness, 59-60; from instr to assoc prof neurosurg, 60-73, PROF SURG, SCH MED, GEORGETOWN UNIV, 73-, CHIEF DIV NEUROSURG, 65- Concurrent Pos: Teaching fel, Harvard Med Sch, 57-58; res fel neurosurg, Harvard Med Sch, 53-54; res consult, Nat Inst Neurol Dis & Stroke, 60-65, clin consult, 65-; clin consult, Vet Admin Hosp, 65-; consult, Fed Aviation Agency, 67 & Nat Naval Med Ctr, 67- Mem: Cong Neurol Surg; Am Asn Neurol Surg. Res: Cerebrovascular disease. Mailing Add: Dept of Surg Georgetown Univ Hosp Washington DC 20007

LUETSCHER, JOHN ARTHUR, JR, b Baltimore, Md, Aug 30, 13; m 34; c 2. MEDICINE. Educ: Princeton Univ, AB, 33; Johns Hopkins Univ, MD, 37. Prof Exp: Intern med, Johns Hopkins Hosp, 37-38; asst, Peter Bent Brigham Hosp, Boston, 38-40; asst resident physician, Johns Hopkins Hosp, 40-42; from instr to asst prof med, Sch Med, Johns Hopkins Univ, 42-48; assoc prof, 48-55, PROF MED, SCH MED, STANFORD UNIV, 55- Concurrent Pos: Fel, Harvard Med Sch, 38-40; mem coun high blood pressure res, Am Heart Asn. Mem: Am Soc Clin Invest; Endocrine Soc; Asn Am Physicians. Res: Clinical investigation on chemical problems, for example, plasma fractionation and use of fractions in kidney disease; aldosterone. Mailing Add: Dept of Med Stanford Univ Med Ctr Palo Alto CA 94305

LUETZELSCHWAB, JOHN WILLIAM, b Hammond, Ind, Sept 8, 40; m 63; c 2. HEALTH PHYSICS. Educ: Earlham Col, AB, 62; Washington Univ, MA & PhD(physics), 68. Prof Exp: Asst prof, 68-73, ASSOC PROF PHYSICS, DICKINSON COL, 73- Mem: Am Asn Physics Teachers; Health Physics Soc. Res: Environmental radioactivity; effects of nuclear power plants on the environment. Mailing Add: Dept of Physics & Astron Dickinson Col Carlisle PA 17013

LUETZOW, ARTHUR EDWARD, b Milwaukee, Wis, Mar 3, 38. CHEMISTRY. Educ: St Norbert Col, BS, 61; Marquette Univ, MS, 65; Ohio State Univ, PhD(chem), 71. Prof Exp: FEL, SWEDISH FOREST PROD RES LAB, 71- Concurrent Pos: Vis res assoc, Univ Mont, 72-73 & Ohio State Sch Pharm, 74-75; consult, Robert Ventre Assocs, DC, 75. Mem: Am Chem Soc; Sigma Xi. Res: Environmental chemistry, carbohydrate chemistry. Mailing Add: N41 W6366 Jackson St Cedarburg WI 53012

LUFBURROW, ROBERT ALLEN, b New Brunswick, NJ, July 8, 22; m 56; c 3. PHYSICS. Educ: Berea Col, BA, 48; Purdue Univ, MS, 50; Washington Univ, MA, 64. Prof Exp: Res assoc physics, Oceanog Inst, Woods Hole, 52-58; asst prof, 58-67, ASSOC PROF PHYSICS, ST LAWRENCE UNIV, 67- Mem: Am Asn Physics Teachers. Res: Optical physics; solar energy. Mailing Add: 29 W Main St Canton NY 13617

LUFKIN, DANIEL HARLOW, b Philadelphia, Pa, Sept 26, 30; m 51; c 3. SOLAR PHYSICS. Educ: Mass Inst Technol, BS, 52, MS, 58; Univ Stockholm, Fil lic meteorol, 64. Prof Exp: Meteorol officer, Air Weather Serv, US Air Force, DC, 53-69, dir solar forecast facility, 69-73; CONSULT, SOLAR ENERGY SCI SERV, 74-; ASST PROF ASTRON, HOOD COL, 75- Mem: Am Meteorol Soc; Optical Soc Am; Pattern recognition Soc; Int Solar Energy Soc; Am Soc Heating Refrig & Air Conditioning Engrs. Res: Application of solar energy to heating and cooling and generation of power. Mailing Add: 303 W College Terr Frederick MD 21701

LUFKIN, EDWARD GWYNNE, b Northfield, Minn, Oct 15, 35; m 61; c 2. ENDOCRINOLOGY, INTERNAL MEDICINE. Educ: Carleton Col, Minn, BA, 57; Med Sch, Northwestern Univ, Chicago, MD, 61. Prof Exp: Resident physician internal med, Vet Admin Res Hosp, Chicago, 62-65; fel endocrinol & renal dis, Med Ctr, Univ Colo, Denver, 65-66; internist, 98th Gen Hosp, APO New York, 66-69; res internist metab, US Army Med Res & Nutrit Lab, Fitzsimons Army Med Ctr, Denver, 69-74; CONSULT ENDOCRINOL & INTERNAL MED, MAYO CLIN, MAYO GRAD SCH MED, ROCHESTER, MINN, 74- Mem: Am Col Physicians; Am Fedn Clin Res; Endocrine Soc; Cent Soc Clin Res. Res: Platelet function in diabetes; development of radioimmunoassay. Mailing Add: Mayo Clin Rochester MN 55901

LUFKIN, JAMES E, b Gloucester, Mass, Mar 11, 20; m 44; c 3. ORGANIC CHEMISTRY. Educ: Univ NH, BS, 41. Prof Exp: Chemist, Explosives Dept, Eastern Lab, E I du Pont de Nemours & Co, 41-52, asst supt, Repauno Works, 52-53, spec asst, Process Sect, 53-54, dir, Carney's Point Process Lab, 54-60, Carney's Point Develop Lab, 60-70, tech progs mgr, 71-72, PROD SALES MGR, POLYMER INTERMEDIATES DEPT, EXPLOSIVES DEPT, E I DU PONT DE NEMOURS & CO, 72- Mem: Inst Food Technologists. Res: Explosives; polymer intermediates; plasticizers; dye intermediates; cellulose derivatives; smokeless powder; explosion hazards; food additives. Mailing Add: 564 Hunter St Woodbury NJ 08096

LUFT, JOHN HERMAN, b Portland, Ore, Feb 6, 27; m 49; c 3. HISTOLOGY. Educ: Univ Wash, BS, 49, MD, 53. Prof Exp: Intern, Peter Bent Brigham Hosp, Boston, Mass, 53-54; from asst prof anat to assoc prof biol struct, 56-67, PROF BIOL STRUCT, MED SCH, UNIV WASH, 67- Concurrent Pos: Nat Res Coun Rockefeller fel, Harvard Med Sch, 54-56; USPHS sr fel, 57-65. Mem: AAAS; Am Asn Anat; Electron Micros Soc Am; Int Soc Cell Biol. Res: Microscopy and electron microscopy; fixatives; basic cellular structure and function; external cell coats; ultrastructure. Mailing Add: Dept of Biol Struct Univ of Wash Med Sch Seattle WA 98195

LUFT, LUDWIG, b Lvov, Poland, Nov 9, 26; nat US; m 52; c 2. PHYSICAL CHEMISTRY. Educ: Univ Frankfurt, Dipl, 51; Univ Kans, PhD(phys chem), 56. Prof Exp: Asst, Univ Kans, 52-55; asst prof chem, Univ Miami, 55-57; res supvr, MSA Res Corp, 57-58; tech & managerial mem staff, Gen Elec Co, 58-62; sr scientist, Allied Res Assocs, 62-63; dir res, Instrumentation Lab Inc, 63; PRES, LUFT INSTRUMENTS, INC, 63- Mem: AAAS; Am Chem Soc; Soc Am. Res: Chemical engineering; automatic controls; methods development; electrochemistry. Mailing Add: Old Winter St Lincoln MA 01773

LUFT, STANLEY JEREMIE, b Turin, Italy, Sept 26, 27; US citizen; m 55; c 4. GEOLOGY. Educ: Syracuse Univ, AB, 49; Pa State Col, MS, 51. Prof Exp: Asst geol, Pa State Col, 49-51; explor geologist, NJ Zinc Co, 51-54; geologist mineral deposits, US Geol Surv, 54-56; geologist, Northern Pac Rwy Co, 56-58; prof geol & Mineral & head dept, Oriente Univ, 59-60; proj geologist, Callahan Mining Corp, 61; GEOLOGIST, US GEOL SURV, 61- Mem: Geol Soc Am; Soc Econ Geol; AAAS; Am Inst Mining, Metall & Petrol Eng. Res: Petrography and petrology of volcanic rocks; origin and movement of ash-flow sheets; geology of metallic and nonmetallic deposits; stratigraphy; Pleistocene of northern Kentucky. Mailing Add: US Geol Surv 22 Commonwealth Ave Erlanger KY 41018

LUFT, ULRICH CAMERON, b Berlin, Ger, Apr 25, 10; nat US; m 41; c 1. HUMAN PHYSIOLOGY. Educ: Univ Berlin, MD, 37. Prof Exp: Chief high altitude physiol, Aeromed Res Inst, Univ Berlin, 37-45, actg dir dept physiol, Univ, 46-47; res physiologist & assoc prof physiol, Sch Aviation Med, Air Univ, Randolph AFB, Tex, 47-54; HEAD DEPT PHYSIOL, LOVELACE FOUND, 54- Concurrent Pos: Consult human factors group, Comt Space Technol, Nat Adv Comt Aeronaut, 58; consult, Nat Acad Sci, 58 & Air Res & Develop Command, 58; mem adv bd, Off Manned Space Flight, NASA, 65-; assoc physiol, Univ NMex. Mem: AAAS; Soc Exp Biol & Med; Am Physiol Soc; Aerospace Med Asn; fel Am Col Chest Physicians. Res: Physiology of respiration and circulation; aviation medicine; acclimatization to high altitudes; cold and hot climates; physical exercise; clinical physiology. Mailing Add: Lovelace Found Dept of Physiol 5200 Gibson Blvd Albuquerque NM 87108

LUFTIG, RONALD BERNARD, b Brooklyn, NY, Dec 8, 39; m 61; c 3. MICROBIOLOGY, BIOPHYSICS. Educ: City Col New York, BS, 60; NY Univ, MS, 62; Univ Chicago, PhD(biophys), 67. Prof Exp: Asst prof microbiol, Med Ctr, Duke Univ, 69-73; SR SCIENTIST, WORCESTER FOUND EXP BIOL, 74- Concurrent Pos: NSF fel, Calif Inst Technol, 67-69; NIH res grant, Med Ctr, Duke Univ, 70-73; res grant, Worcester Found, 74-77. Mem: AAAS; Am Soc Biol Chem; Am Soc Microbiol; Am Soc Cell Biol; NY Acad Sci. Res: Viral and membrane ultrastructure; phage morphogenesis. Mailing Add: Worcester Found Exp Biol 222 Maple Ave Shrewsbury MA 01545

LUGAR, RICHARD CHARLES, b Philadelphia, Pa. CHEMISTRY. Educ: Univ Pa, BS, 62, PhD(chem), 69. Prof Exp: TEACHER ORG CHEM, DELAWARE VALLEY COL, 67- mem: Am Chem Soc. Res: Conformational analysis of alicyclic systems. Mailing Add: Dept of Chem Delaware Valley Col Doylestown PA 18901

LUGASSY, ARMAND AMRAM, b Kenitra, Morocco, July 23, 33; m 66; c 2. MATERIALS SCIENCE, PROSTHODONTICS. Educ: Toulouse Fac Med & Pharm, France, Chirurgien-Dentiste, 59; Univ Pa, DDS, 62, PhD(metall, mat sci), 68. Prof Exp: Monitor oper dent, Toulouse Fac Med & Pharm, France, 58-59; instr, Sch Dent Med, Univ Pa, 62-63; asst prof biol mat, Dent-Med Sch, Northwestern Univ, 68-71; ASSOC PROF FIXED PROSTHODONTICS, SCH DENT, UNIV OF THE PAC, 71- Concurrent Pos: Nat Inst Dent Res traineeship, Sch Metall & Mat Sci, Univ Pa, 63-68; consult, USPHS Hosp, San Francisco, Calif, 71- Mem: Am Soc Metals; Int Asn Dent Res. Res: Physical properties of calcified tissues; behavior of materials and devices in clinical applications. Mailing Add: Dept of Fixed Prosthodontics Univ of the Pac Sch of Dent San Francisco CA 94115

LUGINBILL, PHILIP, JR, entomology, see 12th edition

LUGINBUHL, ROY EMIL, b Manchester, Conn, Aug 24, 21; m 47; c 3. VIROLOGY. Educ: Univ Conn, BS, 47, MS, 52; Yale Univ, PhD, 59. Prof Exp: Asst mgr, Westbrook Lab, Eastern States Farmers Exchange, 43-44; from asst instr to assoc prof, 47-62, PROF ANIMAL DISEASES, UNIV CONN, 62- Mem: AAAS; NY Acad Sci. Res: Avian respiratory diseases; avian encephalomyelitis; eastern equine encephalomyelitis; enteric viruses of animals; avian tumor viruses. Mailing Add: Dept of Animal Dis Univ of Conn Storrs CT 06268

LUGINBUHL, WILLIAM HOSSFELD, b Des Moines, Iowa, Mar 11, 29; m 55; c 5. PATHOLOGY. Educ: Iowa State Univ, BS, 49; Northwestern Univ, MD, 53. Prof Exp: Intern, Wesley Mem Hosp, Chicago, Ill, 53-54; resident path, Children's Mem Hosp, 54-55; resident, Univ Hosps Cleveland, Ohio, 55-57; from asst prof to assoc prof, 60-67, assoc dean col, 67-70, DEAN HEALTH SCI & COL, COL MED, UNIV VT, 70-, PROF PATH, 67- Concurrent Pos: Fel, Col Med, Univ Vt, 59-60. Mem: Col Am Path; Am Soc Clin Path. Res: Gynecologic and obstetrical pathology; endometrial anatomy and physiology. Mailing Add: Off of the Dean Univ of Vt Col of Med Burlington VT 05401

LUGINBYHL, THOMAS TERENCE, b Stinnet, Tex, Feb 20, 23; m 45; c 4. OCCUPATIONAL HEALTH, INFORMATION SCIENCE. Educ: William Jewell Col, BA, 45; Univ Chicago, MBA, 63; George Washington Univ, MS, 66. Prof Exp: Chief eng & functional test, 801st Air Div, Strategic Air Command, US Air Force, 55-58, chief eng sci, Hq, 58-62, chief sci & resources br, Sci & Technol Off, Defense Intel Agency, 63-65, dep dir sci & tech info, Off Aerospace Res, 65-69, dir sci & tech info, Hq, 69-70; CHIEF TECH INFO RESOURCES, NAT INST OCCUP SAFETY & HEALTH, 71- Concurrent Pos: Guest lectr, Air Force Inst Technol, 66-70. Honors & Awards: Meritorious Serv Medal & Joint Serv Commendation, US Dept Defense, 65. Mem: Am Soc Info Sci; Am Indust Hyg Asn; Am Conf Govt Indust Hygienists. Res: Scientific and technical information systems and methods for improvement in information transfer and utilization, particularly as relates to the fields of occupational safety and health. Mailing Add: ORSD-NIOSH Rm 318 Park Bldg 5600 Fishers Lane Rockville MD 20852

LUGN, ALVIN LEONARD, b Mediapolis, Iowa, Nov 22, 95; m 21, 57; c 2. GEOLOGY. Educ: Augustana Col, AB, 16; Univ Iowa, MS, 25, PhD(geol), 27. Prof Exp: Instr sci, Upsala Col, 16-17; teacher high sch, Mich, 19; prof chem & geol &

head dept, Lenoir-Rhyne Col, 19-23; asst geol, Univ Iowa, 23-26; prof geol & biol, Midland Col, 26-27; from instr to prof, 27-62, EMER PROF GEOL, UNIV NEBR, LINCOLN, 62-; PROF EARTH SCI & CHMN DEPT GEOG & EARTH SCI, LENOIR-RHYNE COL, 62- Concurrent Pos: Geologist, Iowa Geol Surv, 24-26 & Nebr Geol Surv, 28-; spec consult, State Mus, Univ Nebr, Lincoln, 58 & Miss State Geol Surv, 53; res assoc & lectr, Stanford Univ, 57; collabr, Encycl Earth Sci, 75. Honors & Awards: Distinguished Achievement Award, Augustana Col, Ill, 64. Mem: Soc Econ Geol; fel Geol Soc Am; Soc Econ Paleont & Mineral; Am Asn Petrol Geol; Pop Asn Am. Res: Ground water geology and resources; Cretaceous, Tertiary and Pleistocene stratigraphy; sedimentary petrography; igneous petrology; origin and source of loess deposits in North America; known and new mineral and ore deposits of western North Carolina; Pleistocene geology; reframing of the Blanco Beds of west Texas. Mailing Add: 510 Seventh Ave NE Hickory NC 28601

LUGO, ARIEL EMILIO, b Mayaguez, PR, Apr 28, 43; m 64; c 2. ECOLOGY. Educ: Univ PR, Rio Piedras, BS, 63, MS, 65; Univ NC, Chapel Hill, PhD(bot), 69. Prof Exp: Lab asst cell physiol, Univ PR, 63; res asst, Rain Forest Radiation Proj, PR Nuclear Ctr, 63-65, consult seedling metab, Rain Forest Proj, 66; lab asst bot, Univ NC, 67; asst prof, Univ Fla, 67-73; asst secy, Dept Natural Resources, Commonwealth PR, 73-75; ASST PROF BOT, UNIV FLA, 75- Concurrent Pos: Researcher granite outcrop ecosyst metab, 65-69; consult, Fla Defenders Environ, Save Our Bays Asn, US Forest Serv, Am Oil Co & H W Lochner Inc, 69-; Div Biol Sci grant, 71-72; US Dept Interior grant, 71-73; Fla Dept Natural Resources grant, 72-75; US Dept Interior & State of Fla grant, 73-74; hon lectr, Univ PR, 74-75. Mem: Bot Soc Am; Am Soc Limnol & Oceanog; Ecol Soc Am; Am Inst Biol Sci; Asn Trop Biol. Res: Energetics and modeling of ecosystems. Mailing Add: Dept of Bot Univ of Fla Gainesville FL 32611

LUGO, HERMINIO LUGO, b San German, PR, June 6, 18; m 41; c 2. PLANT PHYSIOLOGY. Educ: Polytech Inst, PR, BA, 39; Cornell Univ, MS, 48, PhD, 54. Prof Exp: Teacher pub sch, PR, 41-46; instr biol & bot, Polytech Inst PR, 46-47; from asst prof to prof biol, bot & plant physiol, Col Agr, Mayaguez, 48-60; prof biol, 60-69, asst dean studies, 60-66, acad coord, Rio Piedras Campus, 66-69, PROF ECOL, UNIV PR, 69-, DIR PREMED STUDIES, 71- Concurrent Pos: Fel, Inst Ecol, Univ Ga, 68-69. Mem: Bot Soc Am; Am Soc Agr Sci. Res: Germination of vanilla seeds. Mailing Add: Dept of Ecol Univ of PR Rio Piedras PR 00931

LUGO-LOPEZ, MIGUEL ANGEL, b Mayaguez, PR, July 21, 21; m 45; c 2. SOIL PHYSICS. Educ: Univ PR, BSA, 43; Cornell Univ, MS, 45, PhD(soil sci), 50. Prof Exp: Asst, Fed Exp Sta, PR, 43-44; asst prof agron, Univ PR, 46-48, asst scientist & soil scientist, Agr Exp Sta, 48-57, assoc soil scientist & soil scientist chg, Gurabo Substa, 57-60; asst dir chg, 60-61, asst dir, Agr Exp Sta, 61-64, actg dir, 64-66, assoc dir, 66-69, dir off progs & plans & assoc dean, Col Agr Sci, 69-72, dean students, 72-74, PROF SOIL SCI & SOIL SCIENTIST, UNIV PR, MAYAGUEZ, 74- Mem: Soil Sci Soc Am; Am Soc Agron; Am Soc Agr Sci; Int Soc Soils. Res: Morphology and characterization of Puerto Rican soils; physical properties of tropical soils; soils management. Mailing Add: Box 506 Isabela Agr Exp Sta Isabela PR 00662

LUGTHART, GARRIT JOHN, JR, b Los Angeles, Calif, Feb 11, 23; m 55; c 3. ENTOMOLOGY, GENETICS. Educ: Mich State Univ, BS, 50, MS, 51; Univ Wis, PhD(entom), 59. Prof Exp: Asst prof biol, Adrian Col, 56-61; ASSOC PROF BIOL, LE MOYNE COL, NY, 61- Mem: AAAS; Entom Soc Am. Res: Biology and control of insects injurious to man. Mailing Add: Dept of Biol Le Moyne Col Syracuse NY 13214

LUH, BOR SHIUN, b Shanghai, China, Jan 13, 16; m 40; c 1. FOOD SCIENCE. Educ: Chiao Tung Univ, BS, 38; Univ Calif, MS, 48, PhD(agr chem), 52. Prof Exp: Instr, Chiao Tung Univ, 38-41; chemist, Ma Ling Canned Foods Co, Ltd, 41-46; asst, 48-51, jr specialist, Dept Food Technol, 52-56, from jr food technologist to assoc food technologist, 56-69, FOOD TECHNOLOGIST, UNIV CALIF, DAVIS, 69-, LECTR FOOD TECHNOL, 57- Concurrent Pos: Consult, Food Indust Res & Develop Inst, Hsinchu, Taiwan, 70- Mem: AAAS; Am Chem Soc; Inst Food Technol; Am Oil Chem Soc. Res: Chemistry of foods; food processing; biochemistry. Mailing Add: Dept of Food Sci Univ of Calif Davis CA 95616

LUH, JIANG, b Haining, Chekiang, China, June 24, 32; m 56; c 3. ALGEBRA. Educ: Taiwan Normal Univ, BS, 56; Univ Nebr, MS, 59; Univ Mich, PhD(math), 63. Prof Exp: Assoc prof math, Ind State Univ, 63-66 & Wright State Campus, Miami-Ohio State Univ, 66-68; assoc prof, 68-71, PROF MATH, NC STATE UNIV, 71- Mem: Am Math Soc; Math Asn Am. Res: Ring theory; semi-group theory; linear algebra. Mailing Add: 5613 Deblyn Dr Deblyn Park Raleigh NC 27612

LUHAN, JOSEPH ANTON, b Chicago, Ill, Feb 6, 01; m 44. NEUROLOGY. Educ: Northwestern Univ, BS, 27, MD, 28, MS, 31, PhD(neurol), 34; Am Bd Psychiat & Neurol, dipl, 38. Prof Exp: Clin asst neurol, Northwestern Univ, 28-34, instr, 34-35, assoc, 35-41; from assoc clin prof to clin prof neurol & psychiat, 41-54, prof, 54-73, EMER PROF NEUROL & PSYCHIAT, STRITCH SCH MED, LOYOLA UNIV CHICAGO, 73- Concurrent Pos: Practicing physician, 28-; asst dir, Psychiat Inst, Munic Court, Ill, 35-40; attend neurologist, Cook County Hosp, 39-59, consult neurologist, 59-; dir neuropath lab, 40-61, consult neuropathologist, 61-; sr attend neurologist & psychiatrist, Loretto Hosp, 41-73, dept chmn, 62-69; expert civilian consult to Surgeon Gen, US Dept Army, Percy Jones Gen Hosp, Mich, 47-50; former consult neuropsychiatrist, St Anthony's Hosp, Chicago & McNeal Mem Hosp, Berwyn, Ill; attend neurologist, Loyola Univ Chicago Hosp, 69-73; attend psychiatrist, Mem Hosp & attend neurologist, Doctors Hosp, Hollywood, Fla, 73- Mem: Fel AAAS; AMA; fel Am Psychiat Asn; Am Neurol Asn; fel Am Acad Neurol. Res: Neuropathology; clinical neurology and psychiatry. Mailing Add: 4016 Grant St Hollywood FL 33021

LUHBY, ADRIAN LEONARD, b New York, NY, Dec 21, 16; m 67; c 1. HEMATOLOGY, PEDIATRICS. Educ: Columbia Univ, AB, 38; NY Univ, MD, 43. Prof Exp: Intern path & bact, Mt Sinai Hosp, New York, 44-45, intern med & surg, 45-46; res assoc immunol, Children's Hosp, Ohio State Univ, 48-49, asst resident pediat, Hosp & instr, Univ, 49-50; from instr to assoc prof, 50-59, PROF PEDIAT, NEW YORK MED COL, 59- Concurrent Pos: Fel hemat, Children's Hosp, Boston, Mass, 46-48. Honors & Awards: Distinguished Serv Award, Cooley's Anemia Found, 63. Mem: Am Fedn Clin Res; Soc Pediat Res; Am Inst Nutrit; Am Physiol Soc; Int Soc Hemat. Res: Morphologic hematology; megaloblastic anemias; physiology, metabolism, biochemistry and nutrition of folic acid, vitamin B-12 and vitamin B-6. Mailing Add: Dept of Ped New York Med Col New York NY 10029

LUHMAN, GLADYS FINNEY, b San Francisco, Calif, July 31, 10; m 42. PHYSICS. Educ: Univ Calif, AB, 30, MA, 32, PhD(physics), 34. Prof Exp: Instr chem & math, Lone Mountain Col, 33-35; instr, 35-72, EMER INSTR PHYSICS, CITY COL, SAN FRANCISCO, 72- Mem: Am Phys Soc; Am Asn Physics Teachers. Res: Natural radioactivity; weak alpha-active sources; determination of alpha radioactivity of weak solid sources. Mailing Add: Dept of Physics City Col of San Francisco San Francisco CA 94112

LUI, YIU-KWAN, b Hong Kong, Mar 24, 37; US citizen; m 67; c 1. PHYSICAL CHEMISTRY. Educ: Chung Chi Col, Hong Kong, BS, 59; Lehigh Univ, MS, 61, PhD(phys chem), 66. Prof Exp: Res chemist, Titanium Pigment Div, NL Indust, 65-74, Indust Chem Div, 75; RES CHEMIST, ENGELHARD INDUST DIV, ENGELHARD MINERALS & CHEM CORP, 76- Mem: Am Chem Soc. Res: Colloid and surface properties of silica and alumina; heterogeneous catalysis; physical properties of rheological additives; dispersion stability; physical and surface properties of titanium dioxide pigments. Mailing Add: Engelhard Indust Div Engelhard Minerals & Chem Corp Edison NJ 08817

LUIBRAND, RICHARD THOMAS, b Detroit, Mich, Apr 13, 45. ORGANIC CHEMISTRY. Educ: Wayne State Univ, BS, 66; Univ Wis, PhD(org chem), 71. Prof Exp: Fel, Alexander von Humboldt Found, WGer, 71-72; ASST PROF ORG CHEM, CALIF STATE UNIV, HAYWARD, 72- Concurrent Pos: Cottrell res grant, Res Corp, 73. Mem: Am Chem Soc; AAAS. Res: Reaction mechanisms in organic chemistry; natural products chemistry. Mailing Add: Dept of Chem Calif State Univ Hayward CA 94542

LUICK, JACK ROGER, b Niagra Falls, NY, Jan 2, 21; m 48; c 8. PHYSIOLOGY, NUTRITION. Educ: Univ Calif, BS, 50, PhD(nutrit), 56. Prof Exp: Assoc specialist radiobiol, Univ Calif, 56-58, assoc res physiologist, 58-64; UN expert nutrit, Inst Appl Nuclear Energy in Agr, Yugoslavia, 64-65; assoc prof physiol, 65-67, PROF NUTRIT, UNIV ALASKA, 67- Concurrent Pos: Fulbright res scholar nutrit, NEng, Australia, 65-66. Mem: AAAS; Am Dairy Sci Asn; Soc Exp Biol & Med; Am Soc Animal Sci; Am Physiol Soc. Res: Metabolism of intact animals; environmental and nutritional physiology of arctic ungulates, especially reindeer, caribou, and moose; quantitative aspects of intermediary metabolism with interspecies comparisons; mineral and water metabolism. Mailing Add: Inst of Arctic Biol Univ of Alaska Fairbanks AK 99701

LUINE, VICTORIA NALL, b Pine Bluff, Ark, Apr 22, 45. NEUROCHEMISTRY. Educ: Allegheny Col, BS, 67; State Univ NY, Buffalo, PhD(pharmacol), 71. Prof Exp: Res assoc, 72-75, ASST PROF NEUROCHEM, ROCKEFELLER UNIV, 75- Mem: AAAS; Soc Neurosci. Res: Effect of hormones and drugs on enzymes and metabolites in the developing central nervous system. Mailing Add: Dept of Neurochem Rockefeller Univ New York NY 10021

LUISADA, ALDO AUGUSTO, b Florence, Italy, June 26, 01; nat US; m 31; c 1. MEDICINE, PHYSIOLOGY. Educ: Univ Florence, MD, 24. Prof Exp: Pvt docent med & clin med, Univ Padua, 29-32; asst prof med, Univ Naples, 31-35; prof, Univs Sassari & Ferrara, 35-38; instr physiol & pharmacol & lectr med, Med Sch, Tufts Univ, 43-49; from asst prof to assoc prof med, 49-60, dir div cardiol, 52-71, prof cardiovasc res & dir div, 61-72, prof med, 60-71, prof physiol & biophys, 69-71, DISTINGUISHED PROF PHYSIOL & MED, CHICAGO MED SCH, 71- Concurrent Pos: Assoc, Beth Israel Hosp, Boston, Mass, 43-49; chief cardiac clin, Mt Sinai Hosp, Chicago, 49-60, from assoc attend cardiologist to attend cardiologist, 55-71; consult, La Rabida Sanitarium, 53-57 & Hines Vet Admin Hosp, 54-; sr med officer, Oak Forest Hosp, 71-, chmn cardiol, 72- Mem: AAAS; fel Am Physiol Soc; Soc Exp Biol & Med; fel AMA; Am Heart Asn. Res: Electrokymography; phonocardiography; pulmonary circulation, intracardiac pressures; experimental and clinical studies of pulmonary edema; digitalis; heart failure. Mailing Add: 5000 S Cornell Ave Chicago IL 60615

LUISADA-OPPER, ANITA VICTORIA, b Vienna, Austria; US citizen; m 42; c 1. BIOCHEMISTRY, CLINICAL CHEMISTRY. Educ: Univ Vienna, MagPharm, 33, PhD(biochem), 34; Univ Genoa, PhD(chem), 35. Prof Exp: Res asst microanal, Inst Biol Chem, Univ Vienna, 34-36; chemist, Jedlersdorfer-Fabrik, Vienna, 36-38; res asst, Inst Cancer Res, Univ Paris, 39-40; group leader res chem, Va Carolina Chem Corp, NJ, 42-48; from res asst to res assoc chem, Mt Sinai Hosp, New York, 50-65; asst prof, 65-73, ASSOC PROF MED, DIV HEPATIC METAB & NUTRIT, COL MED & DENT NJ, 73- Concurrent Pos: Fel bact, Hosp Hotel Dieu, Paris, 38-39. Mem: AAAS; Am Chem Soc. Res: Enzyme induction; enzyme and microsomal enzyme levels in normal and damaged liver; coenzymes, lipid levels and subcellular particle fractionation used to illustrate metabolic liver damage under influence of alcohol; study of alcoholic hyaline; trace metals in biology. Mailing Add: Dept of Med Col of Med & Dent NJ East Orange NJ 07019

LUIZZO, JOSEPH ANTHONY, b Tampa, Fla, Dec 16, 26; m 51; c 3. FOOD SCIENCE. Educ: Univ Fla, BS, 50, MSA, 55; Mich State Univ, PhD(nutrit, biochem), 58. Prof Exp: Res microbiologist, Univ Fla, 50-51, asst, 54-55; dir microbiol, Nutrilite Prod, Inc, Calif, 51-53, asst to dir biol res, 53-54; asst, Mich State Univ, 55-58; asst prof biochem, 58-62, assoc prof food sci & technol, 62-69, PROF FOOD SCI, LA STATE UNIV, BATON ROUGE, 69- Mem: AAAS; Am Inst Nutrit; Inst Food Technologists; Am Inst Chemists. Res: Detection and isolation of growth factors and growth inhibitors from natural materials required by microorganisms and animals; radiation preservation of foods; development of protein supplements for deprived countries. Mailing Add: Dept of Food Sci La State Univ Baton Rouge LA 70803

LUKACH, CARL ANDREW, b Wilkes-Barre, Pa, Dec 18, 30; m 53; c 3. ORGANIC CHEMISTRY, POLYMER CHEMISTRY. Educ: Lehigh Univ, BS, 52, MS, 53; Univ Notre Dame, PhD(org chem), 57. Prof Exp: Res supvr, 56-73, RES MGR, ORGANICS DIV, HERCULES INC, 73- Mem: Am Chem Soc; Sigma Xi. Res: Polymerization and copolymerization of olefins and olefin oxides; polymerization kinetics; conformational analysis; reverse osmosis; cross-linking agents; paper chemistry. Mailing Add: 4807 Lancaster Pike Wilmington DE 19807

LUKACS, EUGENE, b Szombathely, Hungary, Aug 14, 06; US citizen; m 35. MATHEMATICS. Educ: Univ Vienna, PhD(math), 30. Prof Exp: Mathematicianm US Naval Ord Test Sta, Calif, 48-50 & Nat Bur Standards, 50-53; head statist br, Off Naval Res, 53-55; prof math & dir statist lab, Cath Univ, 55-72; PROF MATH, BOWLING GREEN STATE UNIV, 72- Concurrent Pos: From lectr to adj prof, Am Univ, 54-56; vis prof, Sorbonne, 61, 66, Swiss Fed Inst Technol, 62, Inst Technol Austria, 70 & Univ Hull, 71. Mem: Fel AAAS; fel Inst Math Statist; Am Math Soc; fel Am Statist Asn; Biomet Soc. Res: Probability theory; mathematical statistics. Mailing Add: Dept of Math Bowling Green State Univ Bowling Green OH 43403

LUKAS, DANIEL STANLEY, b Jersey City, NJ, Aug 4, 23; m 50; c 3. CARDIOLOGY. Educ: Columbia Univ, AB, 44, MD, 47. Prof Exp: From intern to instr med, New York Hosp-Cornell Med Ctr, 47-53; asst prof, Med Col, 53-57, ASSOC PROF MED, MED COL, CORNELL UNIV, 57-; CHIEF CARDIOPULMONARY SERV & ATTEND PHYSICIAN, MEM HOSP, 71- Concurrent Pos: NIH res fel, Med Col, Cornell Univ, 50; dir cardiopulmonary lab, New York Hosp-Cornell Med Ctr, 52-71; assoc attend physician, New York Hosp, 57-; assoc mem, Sloan-Kettering Inst, 71-; mem cardiovasc & renal study sect, NIH, 73-, chmn, 74- Mem: Am Soc Clin Invest; Asn Univ Cardiologists; Am Fedn Clin Res; AAAS; Sigma Xi. Res: Metabolism of digitalis steroids in man; cardiovascular physiology; pulmonary physiology. Mailing Add: Cardiopulmonary Serv Dept of Med Sloan-Kettering Cancer Ctr New York NY 10021

LUKAS, GEORGE, b Budapest, Hungary, Mar 16, 31; m 56; c 2. ORGANIC CHEMISTRY. Educ: Univ Budapest, BS, 54; Polytech Inst Brooklyn, MS, 60; Mass Inst Technol, PhD(org chem), 63. Prof Exp: Develop engr, United Pharmaceut Works, Hungary, 54-56; chemist, Avery Industs, Calif, 57; develop engr, Chas Pfizer & Co, 57-59; NIH fel, Inst Chem Natural Substances, Gif-Sur-Yvette, France, 63-64; res chemist, 64-65, res biochemist, 65-67, group leader, Biochem Dept, 67-71, MGR DRUG METAB, CIBA-GEIGY CORP, 71- Mem: Am Chem Soc; NY Acad Sci. Res: Chemistry of natural products; pharmacodynamics; absorption and disposition of drugs. Mailing Add: 6 Devoe Rd Armonk NY 10504

LUKAS, JOAN DONALDSON, b New Haven, Conn, June 19, 42; m 63; c 2. MATHEMATICS. Educ: Columbia Univ, AB, 63; Mass Inst Technol, PhD(math), 67. Prof Exp: Asst prof, 67-74, ASSOC PROF MATH, UNIV MASS,BOSTON, 74- Concurrent Pos: Vis lectr, Brandeis Univ, 71. Mem: Am Math Soc; Math Asn Am; Asn Symbolic Logic. Res: Mathematical logic; recursive function theory. Mailing Add: Dept of Math Univ of Mass Boston MA 02116

LUKASEWYCZ, OMELAN ALEXANDER, b Mostyska, Ukraine, Sept 28, 42; US citizen; m 68; c 1. IMMUNOBIOLOGY. Educ: St Joseph's Col, Pa, AB, 64; Villanova Univ, MS, 68; Bryn Mawr Col, PhD(microbiol), 72. Prof Exp: Res asst microbiol, Univ Tex, Austin, 70-72; scholar tumor immunol, Med Sch, Univ Mich, Ann Arbor, 72-75; lectr microbiol, 73-75; ASST PROF MED MICROBIOL & IMMUNOL, MED SCH, UNIV MINN, DULUTH, 75- Mem: Am Soc Microbiol; AAAS; Sigma Xi. Res: Evaluation of immunocompetent cell populations in immune mechanisms of leukemia; suppression of cellular immunity by drugs, x-irradiation and starvation; contribution of B and T cell subsets; role of macrophage. Mailing Add: Dept Med Microbiol & Immunol Univ of Minn Med Sch Duluth MN 55812

LUKASIK, STEPHEN JOSEPH, b Staten Island, NY, Mar 19, 31; m 53; c 4. PHYSICS. Educ: Rensselaer Polytech Inst, BS, 51; Mass Inst Technol, SM, 53, PhD(physics), 56. Prof Exp: Asst physics, Mass Inst Technol, 51-55; scientist, Westinghouse Elec Corp, 55-57; chief, Fluid Physics Div, Davidson Lab, Stevens Inst Technol, 57-66, assoc res prof physics, 59-66; dir nuclear test detection, Advan Res Projs Agency, 66-68, dept dir, 68-71, dir, 71-74; VPRES SYSTS DEVELOP DIV, XEROX CORP, 74- Concurrent Pos: Acoust engr, Bolt, Beranek & Newman Co, 52-55; consult, Vitro labs, Vitro Corp Am, 59-66; mem, Bd Trustees, Stevens Inst Technol & Bd Overseers, Ctr Naval Anal, Univ Rochester, 75- Honors & Awards: Ottens Res Award, 63; Distinguished Civilian Serv Award, US Dept Defense, 74. Mem: AAAS; Am Phys Soc. Res: Relaxation processes in gases and liquids; viscous boundary layer phenomena; energy dissipation processes in water waves; interaction of explosives with magnetic fields. Mailing Add: 11203 Farmland Dr Rockville MD 20852

LUKAT, ROBERT TIMON, b Louisville, Ky, July 27, 17; m 39; c 3. CHEMISTRY. Educ: Pacific Union Col, BA, 39. Prof Exp: Chemist, Hollingsworth & Whitney Pulp & Paper Co, 40-42; chief chemist supvr, Dixie Ord Works, US Dept Army, 42-45; plant chemist, Commercial Solvents Corp, 45-53; ammonia plant supt, Am Cyanamid Co, 53-56; tech dir, Southern Nitrogen Co, 56-67, vpres, 62-67, vpres & tech dir, 67-70, V PRES MFG & TECH DIR, KAISER AGR CHEM, 70- Mem: Am Chem Soc; Am Inst Chem; Am Inst Chem Eng. Res: Nitrogen fertilizer chemicals; high pressure processes; plant design and construction. Mailing Add: 12504 Bridlewood Dr Savannah GA 31406

LUKE, HERBERT HODGES, b Pavo, Ga, Feb 2, 23; m 46; c 2. PLANT PATHOLOGY. Educ: Univ Ga, BS, 50; La State Univ, MS, 52, PhD, 54. Prof Exp: Plant pathologist, Delta Br Exp Sta, USDA, Miss, 54-55, PLANT PATHOLOGIST, AGR EXP STA, UNIV FLA, USDA, 55-, PROF PLANT PATH, 70- Mem: Am Phytopath Soc. Res: Chemical nature of disease resistance in plants, particularly isolation and identification of host metabolites that inhibit pathogenesis of pathogen; chemical and genetic control of small grain diseases. Mailing Add: Agr Exp Sta Univ of Fla Gainesville FL 32603

LUKE, JAMES LINDSAY, b Cleveland, Ohio, Aug 29, 32; m 57; c 3. PATHOLOGY. Educ: Columbia Univ, BS, 56; Western Reserve Univ, MD, 60. Prof Exp: Intern Path, Yale-New Haven Hosp, 60-61; chief resident, Inst Path, Western Reserve Univ, 61-63; staff researcher, Lab Exp Path, Nat Inst Arthritis & Metab Dis, 63-65; assoc med examr forensic path, Off Chief Med Examr, New York, 65-67; prof forensic path, Sch Med, Univ Okla, 67-71; CHIEF MED EXAMR, WASHINGTON, DC, 71- Concurrent Pos: State med examr, Okla, 67-71; clin prof path, Georgetown Univ, George Washington Univ & Howard Univ, 71- Mem: Fel Am Acad Forensic Sci. Res: Epidemiological research in legal medicine; pathology of strangulation, hanging and sudden natural death; aspects of forensic pathology as related to pediatrics; experimental pathology of quantitated blunt force injury. Mailing Add: 5240 Loughboro Rd NW Washington DC 20016

LUKE, JON CHRISTIAN, b Minneapolis, Minn, Aug 10, 40. APPLIED MATHEMATICS. Educ: Mass Inst Technol, SB, 62, SM, 63; Calif Inst Technol, PhD(appl math), 66. Prof Exp: NSF fel, 66-68; asst prof math, Univ Calif, San Diego, 68-73; postdoctoral assoc, Univ Minn, 73-74; vis assoc, Calif Inst Technol, 74-75; ASST PROF MATH SCI, IND UNIV-PURDUE UNIV, INDIANAPOLIS, 75- Mem: Sigma Xi; Am Math Soc. Res: Nonlinear methods in applied mathematics; applications in nonlinear wave problems, geomorphology, economics and acoustics. Mailing Add: Dept of Math Ind Univ-Purdue Univ Indianapolis IN 46205

LUKE, OREN VICTOR, JR, organic chemistry, see 12th edition

LUKE, ROBERT A, b Rigby, Idaho, Jan 5, 38; m 64; c 4. PARTICLE PHYSICS. Educ: Utah State Univ, BS, 62, MS, 66, PhD(physics), 68. Prof Exp: Asst prof, 68-72, ASSOC PROF PHYSICS, BOISE STATE COL, 72- Mem: Am Asn Physics Teachers. Res: X-ray investigation of clay mixtures; multi-pion production in pion proton interactions. Mailing Add: 9121 Pattie Dr Boise ID 83704

LUKE, STANLEY D, b Sialkot, WPakistan, Jan 1, 28; m 52; c 5. MATHEMATICS. Educ: Univ Panjab, WPakistan, BA, 47, MA, 49; Carnegie-Mellon Univ, MS, 54; Univ Pittsburgh, PhD(math), 68. Prof Exp: Prof math, Gordon Col, WPakistan, 49-64; instr, Univ Pittsburgh, 67-68; PROF MATH, NEBR WESLEYAN UNIV, 68- Mem: Math Asn Am. Res: Mathematical analysis with special interest in summability. Mailing Add: Dept of Math Nebr Wesleyan Univ Lincoln NE 68504

LUKE, YUDELL LEO, b Kansas City, Mo, June 26, 18; m 42; c 4. MATHEMATICS. Educ: Univ Ill, BS, 39, MS, 40. Prof Exp: Asst math, Univ Ill, 40-42; res mathematician, Midwest Res Inst, 46-48, engr chg anal, 48-50, head math anal sect, 50-67, prin adv math, 67-71; PROF MATH, UNIV MO-KANSAS CITY, 71- Mem: Am Math Soc; Sigma Xi; Soc Indust & Appl Math; Math Asn Am. Res: Numerical analysis and applied mathematics. Mailing Add: Dept of Math Univ of Mo Kansas City MO 64110

LUKEHART, CHARLES MARTIN, b DuBois, Pa, Dec 21, 46; m 73. ORGANOMETALLIC CHEMISTRY. Educ: Pa State Univ, BS, 68; Mass Inst Technol, PhD(inorg chem), 72. Prof Exp: Res assoc chem, Tex A&M Univ, 72-73; ASST PROF INORG CHEM, VANDERBILT UNIV, 73- Mem: Am Chem Soc. Res: Synthesis, characterization and chemical reactivity of organometallic and coordination complexes containing transition metals. Mailing Add: Dept of Chem Vanderbilt Univ Nashville TN 37235

LUKEMEYER, JACK WARREN, b Huntingburg, Ind, Apr 4, 32; m 55; c 3. MICROBIOLOGY, BIOCHEMISTRY. Educ: DePauw Univ, AB, 57; Miami Univ, Ohio, MS, 59; Ind Univ, PhD(microbiol), 62. Prof Exp: Asst prof, 63-68, asst dean sponsored progs, Med, 70-72, ASSOC PROF PEDIAT, SCH MED, IND UNIV, 68-, ASSOC DEAN ALLIED HEALTH SCI, MED, 72- Concurrent Pos: USPHS fel, 62-63. Mem: Am Soc Allied Health Prof; Am Soc Microbiol; Asn Gnotobiotics; Sigma Xi. Res: Allied health education communicable diseases; germ-free animals; viral chemotherapy; biochemistry of premature infants and animals. Mailing Add: Dean's Off Ind Univ Sch of Med 1100 W Michigan St Indianapolis IN 46202

LUKEN, WILLIAM LOUIS, b Dayton, Ohio, Feb 15, 47; m 73. THEORETICAL CHEMISTRY. Educ: Mass Inst of Technol, BS, 69; Yale Univ, PhD(chem), 74. Prof Exp: Fel chem, Yale Univ, 74-76; ASST PROF CHEM, DUKE UNIV, 76- Res: Theory of the electronic structure of atoms and molecules; electron correlation effects; energies of electronic states of atoms and molecules; electron correlation effects; energies of electronic states of atoms and molecules; radiative transition probabilities; molecular potential energy surfaces. Mailing Add: Paul M Gross Chem Lab Duke Univ Durham NC 27706

LUKENS, FRANCIS DRING WETHERILL, b Philadelphia, Pa, Oct 5, 99; m 33; c 2. MEDICINE. Educ: Yale Univ, AB, 21; Univ Pa, MD, 25. Prof Exp: Intern, Pa Hosp, 25-27, resident med, 27-28; from instr to prof, Univ Pa, 30-66, dir, George S Cox Med Res Inst, 36-66; PROF MED, UNIV PITTSBURGH, 66- Concurrent Pos: Jacques Loeb fel med, Johns Hopkins Univ, 28-30; chief staff, Vet Admin Hosp, 66-70, staff physician, 70- Mem: Am Physiol Soc; Soc Exp Biol & Med; Am Soc Clin Invest; Endocrine Soc (pres, 65); fel AMA. Res: Diabetes mellitus; endocrine control of metabolism. Mailing Add: Apt 45 Neville House 552 N Neville St Pittsburgh PA 15213

LUKENS, LEWIS NELSON, b Philadelphia, Pa, Jan 21, 27; m 64; c 4. BIOCHEMISTRY. Educ: Harvard Univ, AB, 49; Univ Pa, PhD(biochem), 58. Prof Exp: Instr biochem, Mass Inst Technol, 56-58; Nat Res Coun res fel chem, Columbia Univ, 58-59, USPHS res fel, 59-60; asst prof biochem, Yale Univ, 64-66; ASSOC PROF BIOCHEM, WESLEYAN UNIV, 66- Mem: Am Chem Soc; Am Soc Biol Chem. Res: Protein synthesis and its control in eukaryotes, especially collagen. Mailing Add: Dept of Biol Wesleyan Univ Middletown CT 06457

LUKENS, PAUL W, JR, b Hibbing, Minn, Apr 24, 28; m 60; c 2. MAMMALOGY. Educ: Univ Minn, BS, 52, PhD(zool), 63; Tex A&M Univ, MS, 56. Prof Exp: From instr to assoc prof, 61-70, PROF ZOOL, UNIV WIS-SUPERIOR, 70- Concurrent Pos: Bd regents res grant, Univ Wis, 65-66. Mem: Am Soc Mammal. Res: Identification, interpretation and paleoecology of vertebrate faunas from archaeological sites; paleozoology; environmental conservation. Mailing Add: Dept of Biol Univ of Wis-Superior Superior WI 54880

LUKENS, RAYMOND JAMES, b Beverly, NJ, Feb 25, 30; m 54; c 5. PLANT PATHOLOGY. Educ: Rutgers Univ, BS, 54, MS, 55; Univ Md, PhD(bot), 58. Prof Exp: Asst plant pathologist, Conn Agr Exp Sta, 57-60, assoc plant pathologist, 60-69, plant pathologist, 70-75; SR PLANT PATHOLOGIST, ORTHO DIV, CHEVRON CHEM CO, 75- Mem: Soc Indust Microbiol; Am Phytopath Soc; Bot Soc Am. Res: Chemistry of fungicides; correlation between structure and activity of fungicides; fungicide screening and plant disease control. Mailing Add: Agr Res Lab Chevron Chem Co Richmond CA 94804

LUKER, WILLIAM DEAN, b Yazoo City, Miss, Sept 21, 20. ANALYTICAL CHEMISTRY, CHEMICAL ENGINEERING. Educ: La State Univ, BS, 41; Univ Wis, PhD(chem), 55. Prof Exp: Process engr, Union-Camp Corp, Ga, 41-52; technologist, E I du Pont de Nemours & Co, Inc, 55-58; res chemist, State Chem Lab, 58-65, asst state chemist, 65-67, assoc prof chem eng, 58-65, ASSOC STATE CHEMIST, MISS STATE CHEM LAB, MISS STATE UNIV, 67-, ASSOC PROF CHEM, UNIV, 65- Mem: Am Chem Soc; assoc mem Am Inst Chem Eng. Res: Fats and oils, including studies of unsaponifiable matter. Mailing Add: Dept of Chem Miss State Univ State College MS 39762

LUKERT, MICHAEL T, b Kansas City, Mo, June 28, 37; m 61; c 2. GEOLOGY, GEOCHEMISTRY. Educ: Univ Ill, BS, 60; Northern Ill Univ, MS, 62; Case Western Reserve Univ, PhD(geol), 73. Prof Exp: Instr geol, Northern Ill Univ, 62-64; from asst prof to assoc prof, 67-74, PROF GEOL, EDINBORO STATE COL, 74- Concurrent Pos: Consult, Pa Geol Surv, 75. Mem: Geol Soc Am; Geochem Soc. Res: Geochronology; igneous and metamorphic petrology; instrumental techniques in geochemistry; geostatistics. Mailing Add: Dept of Earth Sci Edinboro State Col Edinboro PA 16444

LUKERT, PHIL DEAN, b Topeka, Kans, Nov 1, 31; m 56; c 4. MICROBIOLOGY. Educ: Kans State Univ, BS, 53, DVM, 60, MS, 61; Iowa State Univ, PhD(microbiol), 67. Prof Exp: Res assoc microbiol, Kans State Univ, 60-61; res vet, Nat Animal Dis Lab, Agr Res Serv, USDA, Iowa, 61-67; ASSOC PROF MED MICROBIOL, COL VET MED, UNIV GA, 67- Mem: Am Vet Med Asn; Am Soc Microbiol; Am Asn Avian Path. Res: Animal virology, particularly pathogenesis of viral infections, identification of new pathogenic viruses and the development of new diagnostic methods for viral diseases. Mailing Add: Dept of Med Microbiol Col of Vet Med Univ of Ga Athens GA 30601

LUKES, JAMES JOSEPH, chemistry, see 12th edition

LUKES, ROBERT MICHAEL, b San Francisco, Calif, Mar 27, 23; m 49; c 6. ORANIC CHEMISTRY. Educ: Univ San Francisco, BS, 43; Univ Calif, MS, 47; Univ Notre Dame, PhD(org chem), 49. Prof Exp: Res chemist, Merck & Co, Inc, 49-53; res assoc, Res Labs, Gen Elec Co, 54-58, supvr, Insulation Lab, Locomotive & Car Equip Dept, 58-64, MGR FINISH SYSTS LAB, MAJOR APPLIANCE LABS, GEN ELEC CO, 64- Mem: Am Chem Soc; Am Electroplaters Soc; fel Am Inst Chemists; Fedn Socs Paint Technol. Res: Hydrogenation; steroid synthesis; plastics; resins; electrical insulation; surface coatings; paint; surface chemistry; electroless plating. Mailing Add: Major Appliance Labs Gen Elec Co Appliance Park 35-1117 Louisville KY 40225

LUKES, THOMAS MARK, b San Jose, Calif, Mar 28, 20; m 52; c 4. FOOD SCIENCE. Educ: San Jose State Col, BS, 47; Univ Calif, Berkeley, MS, 49. Prof Exp: Microbiologist, Real Gold Citrus, Mutual Orange Distributor, 49-51; head lab qual control, Gentry Div, Consol Food Corp, 51-62; assoc prof, 62-73, PROF FOOD PROCESSING & HEAD DEPT FOOD INDUST, CALIF POLYTECH STATE UNIV, SAN LUIS OBISPO, 73- Mem: AAAS; Am Chem Soc; Inst Food Technol.

Res: Application of evolutionary operations to the food processing industry; development of chemical methods of flavor evaluation and application of new developments in food dehydration to the industrial scale. Mailing Add: Dept of Food Indust Calif Polytech State Univ San Luis Obispo CA 93407

LUKEZIC, FELIX LEE, b Florence, Colo, May 27, 33; m 55; c 2. PLANT PATHOLOGY. Educ: Colo State Univ, BS, 56, MS, 58; Univ Calif, PhD(plant path), 63. Prof Exp: Asst plant path, Colo State Univ, 56-58; lab technician, Univ Calif, 58-63; plant pathologist, Div Trop Res, United Fruit Co, Honduras, 63-65; from asst prof to assoc prof, 65-75, PROF PLANT PATH, PA STATE UNIV, 75- Mem: Am Phytopath Soc; Am Soc Plant Physiol. Res: Physiology of plant parasitism, especially bacterial caused diseases. Mailing Add: 211 Buckhout Labs Pa State Univ University Park PA 16802

LUKIN, LARISSA SKVORTSOV, b Lvov, Poland, Aug 30, 25; nat US; m 45; c 1. PHYSIOLOGY. Educ: Col Women, Poland, BA, 44; Univ Heidelberg, Cand med, 49; Columbia Univ, PhD, 55. Prof Exp: Asst physiol, Columbia Univ, 51-55, instr, 55-56; asst prof, Ohio State Univ, 56-63; sr scientist, Hamilton Standard, 63-64; asst res physiologist, Biomech Lab, Med Ctr, Univ Calif, San Francisco, 64-66, assoc res physiologist, 66-70; ASSOC PROF PHYSIOL, SCH DENT, UNIV OF THE PAC, 70- Concurrent Pos: Vis asst prof, Stanford Univ, 65. Mem: Am Physiol Soc. Res: Cardiovascular physiology; blood volumes; cardiac outputs; energy metabolism in exercise; cardiopulmonary physiology. Mailing Add: Univ of the Pac Sch of Dent San Francisco CA 94115

LUKIN, MARVIN, b Cleveland, Ohio, Feb 12, 28; m 62; c 2. ORGANIC CHEMISTRY, BIOCHEMISTRY. Educ: Ohio Univ, BS, 49; Case Western Reserve Univ, MS, 54, PhD(org chem), 56. Prof Exp: Fel org synthesis, Mellon Inst, 56-57; fel protein chem, Albert Einstein Col Med, 57-61; res assoc immunochem, St Lukes Hosp, 61-63; staff asst, Cleveland Clin, 63-65; fel antibiotics, Case Western Reserve Univ, 66-67; asst prof, 67-75, ASSOC PROF CHEM, YOUNGSTOWN STATE UNIV, 75- Mem: Am Chem Soc; The Chem Soc. Res: Organic synthesis; peptide synthesis. Mailing Add: Dept of Chem Youngstown State Univ Youngstown OH 44503

LUKOSEVICIUS, PETRAS POVILAS, b Kulupenai, Lithuania, May 19, 20; Can citizen; m 50; c 2. AGRICULTURE, PLANT BREEDING. Educ: Univ Bonn, dipl, 48; McGill Univ, MSc, 56, PhD(plant breeding), 62. Prof Exp: Plant protection officer, 57-63, res officer, 63-65, res scientist, 65-71, RES MGR, EXP STA, CAN DEPT AGR, 71- Concurrent Pos: Mem, Tobacco Workers Conf, 64- Mem: Genetics Soc Can; Agr Inst Can; Can Soc Agron. Res: Cereal breeding, particularly disease resistance; cigar tobacco genetics and breeding, especially flavor and aroma. Mailing Add: Exp Farm Can Dept of Agr 479 Bourbonnais Montreal PQ Can

LULLA, KOTUSINGH, b Shikarpur, WPakistan, Sept 19, 35; m 61. ATOMIC PHYSICS. Educ: Univ Bombay, BSc, 57; NY Univ, MS, 60, PhD(physics), 64. Prof Exp: NASA fel physics, Univ Pittsburgh, 63-65, res assoc, 65-66, asst prof, 66-69; asst prof, Howard Univ, 69-71; ASSOC PROF & CHMN DEPT PHYSICS, NY INST TECHNOL, 71- Mem: Am Phys Soc. Res: Experimental research in the field of inter-atomic forces; plasma physics and physics of upper atmosphere. Mailing Add: Dept of Physics NY Inst of Technol Old Westburg NY 11568

LULOFF, JEROME SELIG, organic chemistry, see 12th edition

LUM, BERT KWAN BUCK, b Honolulu, Hawaii, May 9, 29; m 52; c 4. PHARMACOLOGY. Educ: Univ Mich, BS, 51, PhD(pharmacol), 56; Univ Kans, MD, 60. Prof Exp: From instr to asst prof pharmacol, Med Ctr, Univ Kans, 56-62; from asst prof to prof, Sch Med, Marquette Univ, 62-69, asst chmn dept, 64-69; PROF PHARMACOL & CHMN DEPT, SCH MED, UNIV HAWAII, MANOA, 69- Mem: AAAS; Am Soc Pharmacol & Exp Therapeut; Am Soc Clin Pharmacol & Chemother. Res: Cardiovascular and autonomic pharmacology. Mailing Add: Dept of Pharmacol Univ of Hawaii Sch of Med Honolulu HI 96822

LUM, DAVID WALKER, biochemistry, organic chemistry, see 12th edition

LUM, KIN K, b Ipoh, Malaya, Sept 4, 40; US citizen; m 65; c 2. PHOTOGRAPHIC CHEMISTRY. Educ: Hong Kong Baptist Col, BSc, 62; Baylor Univ, PhD(org chem), 66. Prof Exp: Res fel, Utah State Univ, 66-68; RES ASSOC, EASTMAN KODAK CO RES LABS, 68- Mem: Am Chem Soc; Soc Photog Sci & Eng. Res: Application of novel imaging chemistry into color image transfer systems. Mailing Add: Eastman Kodak Co Res Labs 1669 Lake Ave Rochester NY 14650

LUM, PATRICK TUNG MOON, b Honolulu, Hawaii, Nov 6, 28; m 56. ENTOMOLOGY. Educ: Earlham Col, BA, 50; Univ Ill, MS, 52, PhD(entom), 56. Prof Exp: Asst entom, Univ Ill, 54-56, res assoc, 56, USPHS res fel, 57; res biologist, Entom Res Ctr, Fla State Bd Health, 57-65; RES BIOLOGIST, STORED PROD INSECT RES & DEVELOP, USDA, 65- Mem: AAAS; Entom Soc Am; Am Mosquito Control Asn. Res: Physiology of egg hatching and reproduction in mosquitoes; pathogenecity of micro-organisms to mosquitoes; photoperiodism and circadian rhythms in insects; factors affecting growth and development in insects. Mailing Add: Stored Prod Insect Res & Develop US Dept of Agr PO Box 5125 Savannah GA 31403

LUMB, ETHEL SUE, b Huntsville, Mo, Dec 21, 16. EMBRYOLOGY, CYTOLOGY. Educ: Univ Mo, AB, 39, BS & MA, 41; Wash Univ, St Louis, PhD(zool, bot), 49. Prof Exp: Teacher high sch, Ill, 41-42; instr zool, Stephens Col, 42-46; asst, Wash Univ, St Louis, 46-48; from instr to asst prof, Univ Rochester, 49-53; from asst prof to assoc prof zool, 3-74, ASSOC PROF BIOL, VASSAR COL, 74- Concurrent Pos: Mem corp, Mt Desert Island Biol Lab. Mem: AAAS; Am Soc Zool; Am Soc Cell Biol; Soc Develop Biol; NY Acad Sci. Res: Histochemistry of nucleic acids; development of chick excretory system; embryonic excretory system; lysosomes; electron microscopy; cytochemistry. Mailing Add: Dept of Biol Vassar Col Poughkeepsie NY 12601

LUMB, GEORGE DENNETT, b London, Eng, Jan 26, 17; nat US; m 45; c 1. PATHOLOGY. Educ: Univ London, MB & BS, 39, MD, 46. Prof Exp: Assoc prof path, Univ London, 53-57; prof, Univ Tenn, 57-59; dir clin labs & pathologist, James Walker Mem Hosp, 59-65; dir, Warner-Lambert Res Inst Can, 65-69, vpres & dir, Pharmaceut Co, Fla, 69-71, dir med serv & res & develop, 71-73; VPRES MED AFFAIRS, SYNAPSE COMMUN SERV INC, 73- Concurrent Pos: Traveling res fel, Westminster Hosp, London, consult pathologist, 48-57; vis assoc prof health affairs, Sch Med, Univ NC, 60-; assoc prof, Univ Toronto, 66-71. Mem: Am Soc Exp Path; Am Asn Path & Bact; Col Am Path; Int Acad Path; Path Soc Gt Brit & Ireland. Res: Cardiac research; conduction of specialized muscle pathways in hogs; experimental production of infarcts in canine hearts. Mailing Add: Synapse Commun Serv Inc Old Greenwich CT 06870

LUMB, JUDITH RAE b Bridgeport, Conn, Mar 19, 43; m 64; c 2. IMMUNOLOGY. Educ: Univ Kans, BA, 65, MA, 66; Stanford Univ, PhD(med microbiol), 69. Prof Exp: Asst prof, 69-75, ASSOC PROF BIOL, ATLANTA UNIV, 75- Concurrent Pos: NIH career develop award, 75-80. Mem: AAAS; Reticuloendothelial Soc; Am Soc Microbiol. Res: Biochemistry of alkaline phosphatase of C57BL lymphomas; derepression of embryo functions in C57BL lymphomas; computer simulation of the development of the thymus. Mailing Add: Dept of Biol Atlanta Univ Atlanta GA 30314

LUMB, RALPH F, b Worcester, Mass, May 27, 21; m 41; c 8. PHYSICAL CHEMISTRY, NUCLEAR SCIENCES. Educ: Clark Univ, AB, 47, PhD(phys chem), 51. Prof Exp: Instr chem, Assumption Col, 47-48 & Northeastern Univ, 49-51; chief, Chem-Physics Br, Div Nuclear Mat Mgt, US AEC, 51-56; proj leader, Quantum Inc, 56-59, vpres, 59-60; dir, Western NY Nuclear Res Ctr, Inc, 60-68; pres, Advan Technol Consult Corp, 68-71; PRES, NUSAC INC, 71- Concurrent Pos: Secy, Adv Comt Uranium Standards, AEC, 53-56; mem, Adv Comt Safeguarding Spec Nuclear Mat, 67-69 & Safeguards Steering Group, Atomic Indust Forum, 66-; consult, Univ Buffalo Nuclear Reactor Proj, 56-60 & Safeguards Br, Int Atomic Energy Agency, 63- Mem: Fel AAAS; fel Am Inst Chemists; Am Nuclear Soc; Am Chem Soc. Res: Applications of nuclear energy; nuclear research; reactor design, operation and utilization. Mailing Add: 7777 Leesburg Pike Falls Church VA 22043

LUMB, ROGER H, b Union, NJ, June 29, 40; m 62. ANIMAL PHYSIOLOGY. Educ: Alfred Univ, AB, 62; Univ SC, MS, 65, PhD(biol), 67. Prof Exp: Instr biol, Univ SC, 65-67; from asst prof to assoc prof, 67-74, PROF BIOL, WESTERN CAROLINA UNIV, 74- Concurrent Pos: Consult, US Bur Indian Affairs, Cherokee Indian Schs, 68-69; panelist, NSF Undergrad Equip Grant Prog, 69. Mem: AAAS; Am Soc Zool. Res: Developmental aspects of lipid metabolism in insects. Mailing Add: Dept of Biol Western Carolina Univ Cullowhee NC 28723

LUMB, WILLIAM VALJEAN, b Sioux City, Iowa, Nov 26, 21; m 49; c 1. VETERINARY MEDICINE. Educ: Kans State Univ, DVM, 43; Tex A&M Univ, MS, 53; Univ Minn, PhD(vet med), 57. Prof Exp: From intern to resident, Angell Mem Animal Hosp, Boston, 46-48; from instr to assoc prof med & surg, Tex A&M Univ, 49-52; assoc prof clin & surg, Colo State Univ, 54-58 & surg & med, Mich State Univ, 58-60; assoc prof med, 60-63, PROF SURG & DIR SURG LAB, COL VET MED, COLO STATE UNIV, 63- Honors & Awards: Gaines Award, 65. Mem: Fel AAAS; Am Vet Med Asn; Am Asn Vet Clinicians; fel Am Col Vet Surg; NY Acad Sci. Res: Experimental surgery and anesthesiology. Mailing Add: Surg Lab Colo State Univ Ft Collins CO 80521

LUMBERS, SYDNEY BLAKE, b Toronto, Ont, Aug 6, 33. GEOLOGY, PETROLOGY. Educ: McMaster Univ, BSc, 58; Univ BC, MSc, 60; Princeton Univ, PhD(geol), 67. Prof Exp: Geologist, Ont Div Mines, Ministry Natural Resources, 62-73; CUR GEOL, ROYAL ONT MUS, 73- Concurrent Pos: Mem comt study of solid earth sci Can, Sci Coun Can, 68-69; corresp, Subcomt Precambrian Stratig, Int Union Geol Sci, 72- Mem: Geol Soc Am; Mineral Asn Can; Geol Asn Can; Sigma Xi. Res: Precambrian geology; evolution of Grenville Province of Canadian Precambrian Shield; metamorphism; petrogenesis of anorthosite suite rocks and alkalic rocks; geochronology; relationship of mineral deposits to stratigraphy, metamorphism and plutonism. Mailing Add: Dept of Mineral & Geol Royal Ont Mus 100 Queen's Park Toronto ON Can

LUMENG, LAWRENCE, b Manila, Philippines, Aug 10, 39; US citizen; m 66; c 1. MEDICINE, BIOCHEMISTRY. Educ: Ind Univ, Bloomington, BS, 60; Ind Univ, Indianapolis, MD, 64, MS, 69; Am Bd Internal Med, dipl, 70. Prof Exp: Asst prof, 71-74, ASSOC PROF MED BIOCHEM, SCH MED, IND UNIV, INDIANAPOLIS, 74- Concurrent Pos: Res & educ associateship, Vet Admin Hosp, Indianapolis, 71-73, clin investr, 73- Mem: AMA; Am Fedn Clin Res; AAAS; Am Gastroenterol Asn; Am Asn Study Liver Dis. Res: Regulation of metabolic pathways; ethanol metabolism; clinical liver diseases. Mailing Add: Dept of Med Ind Univ Med Ctr Indianapolis IN 46202

LUMER, GUNTER, b Frankfurt am Main, Ger, May 29, 29; m 61; c 3. MATHEMATICS. Educ: Univ of the Repub, Uruguay, MS, 57; Univ Chicago, PhD(math), 59. Prof Exp: Asst math, Univ Chicago, 58-59; vis asst prof, Univ Calif, Los Angeles, 59-60 & Stanford Univ, 60-61; from asst prof to assoc prof, 61-67, PROF MATH, UNIV WASH, 67- Concurrent Pos: Vis prof, Univ Grenoble, 62-64, Univ Strasbourg, 67-68 & Univ Paris, 71-72; NSF res grant, 62- Mem: Am Math Soc. Res: Functional analysis; spectral theory; semigroups; harmonic analysis; function algebras in connectio with the theory of analytic functions of one and several variables; analytic functions. Mailing Add: Dept of Math Univ of Wash Seattle WA 98105

LUMPKIN, HENRY EARL, b Ingram, Tex, Mar 1, 20; m 42; c 4. ANALYTICAL CHEMISTRY. Educ: Southwest Tex State Univ, BS, 41. Prof Exp: Res specialist, Humble Oil & Refining Co, 45-65, sect supvr, Esso Prod Res Co, 65-66, sr res specialist, 66-68, RES ASSOC, EXXON RES & ENG CO, 69- Mem: Am Chem Soc; Am Soc Testing & Mat. Res: Analytical research in mass spectrometry; development of procedures for analysis of petroleum and petrochemicals with the mass spectrometer; coal liquefaction products. Mailing Add: Anal Lab Exxon Res & Eng Co PO Box 4255 Baytown TX 77520

LUMPKIN, LEE ROY, b Oklahoma City, Okla, Sept 6, 25; m 53; c 5. DERMATOLOGY, PATHOLOGY. Educ: Univ Okla, BA, 49, MD, 53; Am Bd Dermat, dipl. Prof Exp: Intern, Tripler Gen Hosp, US Air Force, Honolulu, Hawaii, 53-54, resident dermat, Walter Reed Gen Hosp, Washington, DC, 58-61, chief dermat serv & clins, 3070th Air Force Hosp, Torrejon AFB, Spain, 61-64, chief dermat & clins, Air Force Hosp, Carswell AFB, Tex, 65-67, chief dermat serv, Wilford Hall Air Force Med Ctr, Lackland AFB, 67-72, dir residency training, 69-72; PROF DERMAT & HEAD DIV, ALBANY MED COL, 72- Concurrent Pos: Fel dermal-path, Armed Forces Inst Path, 64-65; vis lectr, US Air Force Sch Aerospace Med; clin assoc prof, Univ Tex Med Sch, San Antonio; assoc mem comn of cutaneous dis, Armed Forces Epidemiol Bd; US Air Force rep, Nat Prog Dermat. Honors & Awards: Cert of Appreciation, Strategic Air Command, 64 & Surgeon Gen Air Force, 69; James Clarke White Award, 71. Mem: Fel Am Col Physicians; fel Am Acad Dermat; Soc Air Force Physician (pres-elect, 71); AMA; fel Am Soc Dermatopath. Mailing Add: Div of Dermat Albany Med Col Albany NY 12208

LUMRY, RUFUS WORTH, b Bismarck, NDak, Nov 3, 20; m 43; c 3. PHYSICAL CHEMISTRY. Educ: Harvard Univ, AB, 42, MS, 48, PhD(chem physics), 49. Prof Exp: Res assoc, Div Eight, Nat Defense Res Comt, 42-45; Merck fel, Univ Utah, 48-50, asst prof phys chem, 50-53, asst res prof biochem, 51-53; assoc prof, 53-57, PROF PHYS CHEM, UNIV MINN, MINNEAPOLIS, 57-, DIR LAB BIOPHYS CHEM, 63- Concurrent Pos: NSF sr fel & vis prof, Lab Carlsberg, Copenhagen, 59-60; vis prof, Inst Protein Res, Osaka, Japan, 61 & Inst Biol Chem Rome, 63. Mem: Am Chem Soc; Soc Biol Chem; Sigma Xi; Biophys Soc; Photobiol Soc. Res: Biophysical chemistry; enzymes, proteins; fast reactions. Mailing Add: Sch of Chem Univ of Minn Minneapolis MN 55455

LUMSDEN, DAVID NORMAN, b Buffalo, NY, Aug 29, 35; m 63; c 2. GEOLOGY. Educ: State Univ NY, Buffalo, BA, 58, MA, 60; Univ Ill, PhD(geol), 65. Prof Exp: Res engr, Carborundum Co, 60-62; sr geologist, Pan Am Petrol Corp, 65-67; ASST PROF GEOL, MEMPHIS STATE UNIV, 67- Mem: Geol Soc Am; Am Asn Petrol Geol; Soc Econ Paleont & Mineral. Res: Study of carbonate and quartzose sedimentary rocks. Mailing Add: Dept of Geol Memphis State Univ Memphis TN 38152

LUMSDEN, RICHARD, b New Orleans, La, Apr 6, 38; m 59. CELL BIOLOGY, PARASITOLOGY. Educ: Tulane Univ, BSc, 60, MSc, 62; Rice Univ, PhD(biol), 65. Prof Exp: From asst prof to assoc prof, 65-73, PROF BIOL, TULANE UNIV, 73-, TROP MED, 74-, ANAT, 75- Concurrent Pos: Am Cancer Soc res grant, 65-66; NIH res grants, 65-68, 69-73 & 72-77, career develop award, 69-74; NSF res grant, 68-72; ed consult, J Parasitol, 75-; res contract, US Food & Drug Admin, 74-76. Mem: AAAS; Am Soc Parasitol; Am Soc Zoologists; Am Soc Trop Med & Hyg; Am Soc Cell Biol. Res: Cytology and biochemistry of parasitic helminths and host-parasite relationships. Mailing Add: Dept of Biol Tulane Univ New Orleans LA 70118

LUMSDEN, ROBERT DOUGLAS, b Washington, DC, June 21, 38; m 60; c 2. PLANT PATHOLOGY. Educ: NC State Univ, BS, 61, MS, 63; Cornell Univ, PhD(plant path), 67. Prof Exp: RES PLANT PATHOLOGIST, SOILBORNE DIS LAB, PLANT PROTECTION INST, AGR RES CTR W, USDA, 66- Mem: Am Phytopath Soc. Res: Physiology of plant diseases, including the physiology of pathogenesis and disease resistance; pathology and biological control of plant pathogens, especially soilborne plant pathogens. Mailing Add: 262 Biosci Bldg Plant Protection Inst Agr Res Ctr W USDA Beltsville MD 20705

LUMSDEN, WILLIAM WATT, JR, b Dallas, Tex, Dec 21, 20; m 45; c 2. GEOLOGY, PALEONTOLOGY. Educ: Univ Calif, Los Angeles, AB, 55, PhD, 64. Prof Exp: Asst geol, Univ Calif, Los Angeles, 55-58; from asst prof to assoc prof, 58-70, chmn dept, 58-74, PROF GEOL, CALIF STATE UNIV, LONG BEACH, 70- Concurrent Pos: Leverhulme fel for Gt Brit, Aberdeen Univ. Mem: AAAS; Paleont Soc; Soc Econ Paleont & Mineral; Am Asn Petrol Geol. Res: Invertebrate paleontology; field geology; stratigraphy. Mailing Add: Dept of Geol Calif State Univ Long Beach CA 90840

LUNA, ROBERT EARL, atmospheric physics, see 12th edition

LUNAN, KENNETH DALE, b Detroit, Mich, Nov 24, 30; m 54; c 2. BIOCHEMISTRY. Educ: Calif Inst Technol, BS, 53; Iowa State Univ, MS, 55; Univ Calif, Los Angeles, PhD(biochem), 62. Prof Exp: USPHS fel, Calif Inst Technol, 62-65; asst prof path, Univ Southern Calif, 65-66; res scientist, 66-74, SR RES SCIENTIST, STANFORD RES INST, 74- Mem: AAAS. Res: Biosynthesis of vitamins, amino acids, and terpenoids; biochemistry of nucleic acids and viral replication; general intermediary metabolism and enzyme regulation. Mailing Add: Stanford Res Inst 333 Ravenswood Dr Menlo Park CA 94025

LUNCHICK, MYRON EDWIN, theoretical mechanics, applied mechanics, see 12th edition

LUND, ARNOLD JEROME, b Clarissa, Minn, July 9, 16; m 43; c 4. BACTERIOLOGY. Educ: St Olaf Col, BA, 38; Univ Minn, PhD(bact), 49. Prof Exp: Teacher high sch, Minn, 38-40; res fel, Hormel Inst, Univ Minn, 49-52, asst prof bact, 52-64; assoc prof, 64-67, PROF MICROBIOL, MANKATO STATE COL, 67- Mem: AAAS; Am Soc Microbiol. Res: Physiology of bacterial spores; effects of low temperature on microorganisms; food microbiology. Mailing Add: Dept of Biol Mankato State Col Mankat MN 56001

LUND, CURTIS JOSEPH, b LaSita, Kans, June 8, 07; m 31; c 1. OBSTETRICS & GYNECOLOGY. Educ: Kans State Col, BS, 29; Univ Wis, MS, 30, MD, 35; Am Bd Obstet & Gynec, dipl. Prof Exp: Asst zool, Univ Wis, 29-31, instr anat, Med Sch, 31-33; intern, Cincinnati Gen Hosp, 35-36; resident obstet-gynec, Wis Gen Hosp & Lying-in Hosp, 36-39; from res asst fel to res assoc fel, Med Sch, Univ Wis, 39-43; from instr to asst prof, 43-45; assoc prof, Med Sch, Univ Minn, 45-47; prof, Med Ctr, La State Univ, 47-52; prof & chmn dept, 52-72, EMER PROF OBSTET-GYNEC, SCH MED & DENT, UNIV ROCHESTER, 72-; prof obstet & gynec, 72-75, EMER PROF OBSTET & GYNEC, UNIV MINN, MINNEAPOLIS, 75-, ACTG HEAD DEPT, 74- Concurrent Pos: Dir, Am Bd Obstet & Gynec. Mem: Am Gynec Soc; Soc Gynec Invest; Am Asn Obstet & Gynec; Am Col Obstet & Gynec. Res: Cinefluorography and cystometry in the female bladder and urethra. Mailing Add: Dept of Obstet & Gynec Univ of Minn Minneapolis MN 55455

LUND, DARYL B, b San Bernardino, Calif, Nov 4, 41; m 63; c 2. FOOD SCIENCE. Educ: Univ Wis, Madison, BS, 63, MS, 65, PhD(food sci, chem eng), 68. Prof Exp: From instr to asst prof, 67-72, ASSOC PROF FOOD SCI, UNIV WIS-MADISON, 72- Mem: Inst Food Technol; Am Soc Agr Eng. Res: Food engineering; fouling of heat exchangers; electrolysis in food processing; nutrient retention in thermal processing; blanching and drying foods. Mailing Add: Dept of Food Sci Univ of Wis 110 Babcock Hall Madison WI 53706

LUND, DONALD S, b Evanston, Ill, Sept 23, 32. PHYSICS. Educ: Northwestern Univ, BS, 54; Univ NMex, MA, 61. Prof Exp: Res asst, Univ NMex, 59-61; physicist, Nat Bur Standards, 61-62; res asst, High Altitude Observ, 62-65; PHYSICIST, WAVE PROPAGATION LAB, NAT OCEANIC & ATMOSPHERIC ADMIN, 65- Mem: Am Phys Soc; Am Geophys Union. Res: Aeronomy; radio physics and astronomy. Mailing Add: PO Box 1664 Boulder CO 80302

LUND, DOUGLAS E, b Newcastle, Nebr, Dec 12, 33; m 58; c 2. GENETICS. EMBRYOLOGY. Educ: Nebr Wesleyan Univ, BA, 58; Univ Nebr, MS, 60, PhD(zool), 62. Prof Exp: Asst prof zool, 62-68, PROF BIOL, KEARNEY STATE COL, 68- Mem: AAAS. Res: Temperature effects on early developmental stages of mammalian embryos; carbon dioxide sensitivity in Drosophila. Mailing Add: Dept of Biol Kearney State Col Kearney NE 68847

LUND, ERNEST HOWARD, b Vancouver, Wash, Apr 19, 15; m 36; c 2. GEOLOGY. Educ: Univ Ore, BS, 44; Univ Minn, PhD(geol), 50. Prof Exp: Instr geol, Univ Minn, 49-50; assoc prof, Fla State Univ, 50-57; ASSOC PROF GEOL, UNIV ORE, 57- Mem: Mineral Soc Am; Geol Soc Am. Res: Petrology. Mailing Add: Dept of Geol Univ of Ore Eugene OR 97403

LUND, EVERETT EUGENE, b Alta, Iowa, Sept 7, 07; m 32; c 3. PARASITOLOGY, PROTOZOOLOGY. Educ: Iowa State Col, BS, 29; Univ Calif, MA, 30, PhD(protozool, parasitol), 32. Prof Exp: Adj prof biol, Am Univ, Beirut, 32-35, assoc prof & chmn dept, 35-37; from asst prof to assoc prof, Alfred Univ, 37-43; biologist, Rabbit Exp Sta, US Fish & Wildlife Serv, 44-46; parasitologist, 46-53, PARASITOLOGIST, BELTSVILLE RES CTR, USDA, 53- Concurrent Pos: Instr, Exten Div, Univ Calif, 45-51; Am Soc Parasitologists rep, Agr Res Inst, 59-; Soc Protozoologists rep, Nat Res Coun, 60-63. Honors & Awards: Cert Merit & Award, USDA, 58. Mem: AAAS; Am Soc Trop Med & Hyg; Am Soc Parasitol; Am Soc Microbiol; Soc Protozool. Res: Locomotor and feeding organelles of ciliate protozoa; intestinal protozoa and helminths of man; diseases of the domestic rabbit; chemotherapy in rabbits; coccidiosis; diseases of poultry; ecological aspects of parasitism. Mailing Add: Beltsville Parasitol Lab USDA Beltsville MD 20705

LUND, HARTVIG ROALD, b Fargo, NDak, May 15, 33; m 57; c 4. AGRONOMY. Educ: NDak State Univ, BS, 55, MS, 58; Purdue Univ, PhD(agron, plant breeding), 65. Prof Exp: Res asst agron, NDak State Univ, 55-58, asst prof, 59-62; res asst, Purdue Univ, 62-65; assoc prof, 65-74, PROF AGRON, ASSOC DEAN, COL AGR & ASSOC DIR AGR EXP STA, NDAK STATE UNIV, 74- Concurrent Pos: Asst dean, Col Agr & asst dir, Agr Exp Sta, NDak State I)niv, 71-74. Mem: Am Soc Agron; Crop Sci Soc Am. Res: Rust genetics of durum wheat; chemical mutagenesis in corn; corn breeding and corn endosperm genetics. Mailing Add: Col of Agr Morrill Hall NDak State Univ Fargo ND 58102

LUND, HORACE ODIN, b Minneapolis, Minn, May 16, 08; m 40; c 3. ENTOMOLOGY. Educ: Univ Minn, AB, 31, MS, 33, PhD(entom), 36. Prof Exp: From instr to prof, 36-74, EMER PROF ENTOM, UNIV GA, 74- Concurrent Pos: Ed, J Ga Entom Soc, 65- Mem: Entom Soc Am; Am Soc Trop Med & Hyg. Res: Ecology; medical entomology; longevity and productivity in Trichogramma evanescens; filarial periodicity in the crow; oviposition in Anopheles; termite biology and control. Mailing Add: Dept of Entom Univ of Ga Athens GA 30601

LUND, JOHN EDWARD, b Detroit, Mich, Mar 16, 39; m 59; c 2. VETERINARY PATHOLOGY. Educ: Mich State Univ, BS, 62, MS & DVM, 64; Wash State Univ, PhD(vet sci), 69; Am Cl Vet Pathologists, dipl. Prof Exp: Instr path, Med Sch, Stanford Univ, 68-70; from asst prof to assoc prof, Sch Vet Sci & Med, Purdue Univ, 70-73; sr scientist, 73-74, MGR & RES ASSOC, EXP PATH SECT, BIOL DEPT, BATTELLE NORTHWEST LABS, WASH, 74- Concurrent Pos: Consult & vet pathologist, Inst Chem Biol, Univ San Francisco, 69-72. Mem: AAAS; Am Soc Vet Clin Path; Int Acad Path. Res: Hematologic diseases of animals; neutrophil kinetics in blood and bone marrow; chemical carcinogenesis. Mailing Add: Exp Path Sect Biol Dept Battelle Northwest Labs Richland WA 99352

LUND, JOHN TURNER, b Brooklyn, NY, Nov 3, 29; m 55; c 2. PHYSICAL CHEMISTRY. Educ: Brown Univ, AB, 51; Univ Wash, PhD(phys chem), 54. Prof Exp: Fel, Univ Wash, 54-55; res chemist, 55-59, res supvr, Benger Lab, 59-62, end use res, 62-63, sr supvr, Dacron Plant Technol, 63-65, tech supt, 65-68, dir, Dacron Res Lab, 68-69, TECH MGR, TEXTILE FIBERS DEPT, E I DU PONT DE NEMOURS & CO, INC, 69- Mem: Am Chem Soc; Am Phys Soc. Res: Industrial research on textile fibers. Mailing Add: Textile Fibers Dept E I du Pont de Nemours & Co Wilmington DE 19898

LUND, LANNY JACK, b Dalton, Nebr, May 1, 43; m 64; c 2. SOIL MORPHOLOGY. Educ: Univ Nebr, BS, 65, MS, 68; Purdue Univ, PhD, 71. Prof Exp: ASST PROF SOIL SCI & ASST SOIL SCIENTIST, UNIV CALIF, RIVERSIDE, 71- Mem: Am Soc Agron; Soil Conserv Soc Am. Res: Soil genesis; clay mineralogy; soil and the environment. Mailing Add: Dept of Soil Sci & Agr Eng Univ of Calif Riverside CA 92502

LUND, LILLIAN O, b Norwich, NDak, Sept 9, 07. TEXTILES. Educ: St Olaf Col, BA, 30; Univ Minn, MS, 44. Prof Exp: Teacher pub sch, Minn, 30-43; PROF TEXTILES, SDAK STATE UNIV, 44- Mem: Am Soc Testing & Mat; Am Asn Textile Technol; Am Asn Textile Chem & Colorists; Am Home Econ Asn. Res: Natural fibers; research in textiles and clothing directly related to the consumer. Mailing Add: Dept of Textiles & Clothing SDak State Univ Brookings SD 57006

LUND, LOUIS HAROLD, b Jefferson City, Mo, Mar 17, 19; m 42; c 2. CHEMICAL PHYSICS. Educ: Kans Wesleyan Univ, AB, 40; Univ Mo, AM, 43, PhD(physics), 48. Prof Exp: Instr physics, Univ Mo, 43-44; physicist, Lucas-Harold Corp, 44-45; instr math, 45-47, physics, 47-48, asst, 48-52, assoc prof, 52-55, PROF PHYSICS, UNIV MO-ROLLA, 55- Mem: Am Phys Soc. Res: Liquid structure, x-ray scattering by liquids. Mailing Add: Dept of Physics Univ of Mo Rolla MO 65401

LUND, MELVIN ROBERT, b Siren, Wis, Oct 17, 22; m 46; c 3. DENTISTRY. Educ: Univ Ore, DMD, 46; Univ Mich, MS, 54. Prof Exp: From instr to prof restorative dent, Loma Linda Univ, 53-71; PROF OPER DENT & CHMN DEPT, IND UNIV-PURDUE UNIV, INDIANAPOLIS, 71- Concurrent Pos: Fel, Claremont Grad Sch, 69-70. Mem: Int Asn Dent Res. Res: Physical research in dental materials; biologic research in dental procedures. Mailing Add: Dept of Oper Dent Ind Univ-Purdue Univ Sch of Dent Indianapolis IN 46202

LUND, PAUL K, pathology, see 12th edition

LUND, RICHARD, b New York, NY, Sept 17, 39; m 65; c 3. VERTEBRATE PALEONTOLOGY. Educ: Univ Mich, Ann Arbor, BS, 61, MS, 63; Columbia Univ, PhD(zool), 68. Prof Exp: Asst cur fossil fish, Sect Vert Fossils, Carnegie Mus, 66-69; asst prof earth & plant sci, Univ Pittsburgh, 69-74; asst prof, 74-75, ASSOC PROF BIOL, ADELPHI UNIV, 75- Concurrent Pos: Pittsburgh Found fel, Carnegie Mus, 67-69, res assoc, 69-; Pittsburgh Found fel, Univ Pittsburgh, 71-74; res assoc, WVa Geol Surv, 74. Mem: AAAS; Soc Vert Paleont; Am Soc Icthyol & Herpet; Soc Econ Paleont & Mineral. Res: Fossil fish; late Paleozoic biostratigraphy. Mailing Add: Dept of Biol Adelphi Univ Garden City NY 11530

LUND, STEVE, b Wis, Dec 3, 23; m 46; c 5. AGRONOMY. Educ: Clemson Col, BS, 49; Univ Wis, MS, 51, PhD(agron), 53. Prof Exp: Exten agronomist, Clemson Col, 53-54; from asst res specialist to assoc res specialist farm crops, Rutgers Univ, New Brunswick, 54-62, res prof, 62-75; SUPT & PROF, COLUMBIA BASIN AGR RES CTR, ORE STATE UNIV, 75- Concurrent Pos: Chmn dept soils & crops, Rutgers Univ, New Brunswick, 71-75. Mem: Am Soc Agron; Crop Sci Soc Am. Res: Cereal breeding. Mailing Add: 1201 SW 23rd St Pendleton OR 97801

LUND, WILLIAM ALBERT, JR, b Worcester, Mass, July 14, 30; m 58. ICHTHYOLOGY. Educ: Univ Mass, BS, 53, Cornell Univ, MS, 56, PhD(ichthyol), 60. Prof Exp: Fishery biologist, State Conserv Dept, Mass, 53; asst ichthyol, Cornell Univ, 55; from asst prof to assoc prof biol, 64-74, ASSOC PROF BIOL SCI, UNIV CONN, 74- Mem: AAAS; Am Soc Ichthyol & Herpet; Am Fisheries Soc; Am Soc Limnol & Oceanog; Biomet Soc. Res: Taxonomy; ecology and life history of fishes; dynamics of fish populations. Mailing Add: Dept of Biol Sci Univ of Conn Storrs CT 06268

LUND, ZANE FRANKLIN, b Wolf Point, Mont, Feb 15, 22; m 46; c 4. AGRONOMY. Educ: Ala Polytech Inst, BS & MS, 49. Prof Exp: From agronomist cotton to agronomist soils, 51-72, ACTG RES LEADER SOILS, SOIL & WATER CONSERV RES, AGR RES SERV, USDA, 72- Mem: Am Soc Agron; Sigma Xi. Res: Factors influencing rooting depth, disposal and utilization of animal waste and on conservation cropping management systems. Mailing Add: Dept of Agron Auburn Univ Auburn AL 36830

LUNDBERG, CHARLES ANDREW, JR, b Boston, Mass, June 12, 42; m 71; c 1. SYNTHETIC ORGANIC CHEMISTRY. Educ: Harvard Univ, AB, 63, AM, 65, PhD(chem), 70. Prof Exp: Res chemist, Esso Res & Eng Co, 69-70; RES CHEMIST, MERRELL-NAT LABS, 70- Concurrent Pos: Referee, J Org Chem, 73-; instr, Xavier Univ, 74- Mem: Am Chem Soc; NY Acad Sci. Res: Design and synthesis of novel substances as potential therapeutic agents. Mailing Add: Dept of Org Chem Merrell-Nat Labs Cincinnati OH 45215

LUNDBERG, GEORGE DAVID, b Pensacola, Fla, Mar 21, 33; m 56; c 3. PATHOLOGY. Educ: Univ Ala, BS, 52; Med Col Ala, MD, 57; Am Bd Path, dipl anat & clin path, 62; Baylor Univ, MS, 64. Prof Exp: Intern, Tripler Gen Hosp, Honolulu, Hawaii, 57-58; resident path, Brooke Gen Hosp, San Antonio, Tex, 58-62; chief anat path, Letterman Gen Hosp, San Francisco, Calif, 62-63, res officer, 63-64; chief path, William Beaumont Gen Hosp, El Paso, Tex, 64-67; assoc prof, 67-70, PROF PATH, SCH MED, UNIV SOUTHERN CALIF, 70-; asst dir labs, 68-73, ASSOC DIR LABS, LOS ANGELES COUNTY/UNIV SOUTHERN CALIF MED CTR, 73- Concurrent Pos: US Army Med Res & Develop Command res grants, 63-64 & 65-67. Mem: Col Am Path; Am Soc Clin Path; Am Asn Path & Bact; Int Acad Path; Am Acad Forensic Sci. Res: General pathology; laboratory computer applications; diseases produced by drugs; toxicology; drug abuse; laboratory management. Mailing Add: Dept of Path Sch of Med Univ of Southern Calif Los Angeles CA 90033

LUNDBERG, GUSTAVE HAROLD, b Fremont, Nebr, Sept 5, 01; m 35. APPLIED MATHEMATICS. Educ: Midland Col, BS, 24; Colo State Col, MA, 37; Vanderbilt Univ, MA, 32; George Peabody Col, PhD(math), 51. Prof Exp: Prof math & sci, Dana Col, 24-29; teacher high sch, Colo, 29-34; instr, Allen Acad, 34-41; prof appl math, Vanderbilt Univ, 42-67; prof, 67-72, EMER PROF MATH, AUSTIN PEAY STATE UNIV, 72-; EMER PROF APPL MATH, VANDERBILT UNIV, 67- Concurrent Pos: Res partic, Oak Ridge Nat Lab, 57; vis prof, George Peabody Col, 60 & 61; ed, J Tenn Acad Sci, 63-66, pres, Tenn Acad Sci, 69. Mem: Am Math Soc; Soc Indust & Appl Math; Am Soc Eng Educ; Math Asn Am. Res: Engineering mathematics. Mailing Add: 2001 21st Ave S Nashville TN 37212

LUNDBERG, ROBERT DEAN, b Valley City, NDak, May 30, 28; m 53; c 2. POLYMER CHEMISTRY. Educ: Harvard Univ, BA, 52, MA & PhD(phys chem), 57. Prof Exp: Chemist, Eastman Kodak Co, 52-53; chemist, Union Carbide Corp, 57-62, group leader, Res & Develop Dept, Union Carbide Chem & Plastic Co, 62-69; vpres res, Inter-Polymer Res Corp, 69-70; res assoc, 70-71, SR RES ASSOC, EXXON RES & ENG RES LAB, 71- Mem: Am Chem Soc; NY Acad Sci. Res: Synthesis of synthetic polypeptides; polymer interactions. Mailing Add: Corp Res Lab Exxon Res & Eng Co Linden NJ 07036

LUNDBERG, WALTER OSCAR PAUL, b Minneapolis, Minn, Dec 15, 10; m 42; c 4. LIPID CHEMISTRY, BIOCHEMISTRY. Educ: Johns Hopkins Univ, PhD(chem), 34. Prof Exp: Instr chem, Johns Hopkins Univ, 34-35; Hormel Res Found fel, 38-39, res chemist, 41-44, from asst prof to assoc prof physiol chem, 44-49, prof biochem, 49-75, EMER PROF BIOCHEM, UNIV MINN, MINNEAPOLIS, 75- Concurrent Pos: Res dir, Hormel Inst, 44-49, exec dir, 49-74, consult, 74-; pres, World Cong, Int Soc Fat Res & Am Oil Chem Soc, 70; first int lectr, Fats & Oils Group, Soc Indust Chem; Am Chem Soc rep, Food Protection Comt, Nat Res Coun-Liaison Panel; mem, Coun Basic Sci, Am Heart Asn. Honors & Awards: Normann Medal, Ger Soc Fat Sci, 57; Alton E Bailey Award, Am Oil Chem Soc, 67; Marques de Acapulco Medal, Span Inst Fat Res, 70; Chevreul Medal, Fr Asn Specialists in Fats & Oils, 70; Award in Lipid Chem, Am Oil Chem Soc, 75. Mem: AAAS; Am Chem Soc; Am Oil Chem Soc (ed, Lipids; secy, 61-62, vpres, 62-63, pres, 63-64); Am Soc Biol Chem; Int Soc Fat Res (pres, 68-70). Res: Physical, organic, biological and clinical chemistry of fatty acids and other lipids; lipid nutrition and metabolism; antioxidants. Mailing Add: Hormel Inst 801-16th Ave NE Austin MN 55912

LUNDBLAD, ROGER LAUREN, b San Francisco, Calif, Oct 31, 39; m 66. BIOCHEMISTRY, HEMATOLOGY. Educ: Pac Lutheran Univ, BS, 61; Univ Wash, PhD(biochem), 65. Prof Exp: Res assoc biochem, Univ Wash, 65-66; res assoc, Rockefeller Univ, 66-68; asst prof, 68-71, ASSOC PROF PATH & BIOCHEM, DENT RES CTR, UNIV NC, CHAPEL HILL, 71- Concurrent Pos: Mem coun basic sci & coun thrombosis, Am Heart Asn. Mem: AAAS; Am Chem Soc; Am Soc Biol Chem; Int Soc Thrombosis & Haemostasis. Res: Mechanism of blood coagulation; protein chemistry; proteins of the central nervous system; secretory proteins of the parotid gland. Mailing Add: Dept of Path & Biochem Dent Res Ctr Univ of NC Chapel Hill NC 27514

LUNDE, ANDERS STEEN, applied statistics, demography, see 12th edition

LUNDEEN, ALLAN JAY, b New York, NY, Aug 24, 32; m 54; c 4. ORGANIC CHEMISTRY. Educ: Southwestern Col, Kans, AB, 54; Rice Univ, PhD(chem), 57. Prof Exp: Res chemist org chem, 57-60, sr res chemist, 60-62, res group leader, 62-70, DIR EXPLOR RES, CONTINENTAL OIL CO, 70- Mem: Am Chem Soc. Res: Chemistry of mustard oil glucosides; reactions of carbonium ions; heterogenous catalysis; chemistry of organoaluminum compounds; hydrocarbon oxidation. Mailing Add: 1605 Blackard Lane Ponca City OK 74601

LUNDEEN, CARL VICTOR, JR, b Baltimore, Md, Jan 20, 43; m 65; c 2. BIOCHEMISTRY. Educ: Univ NC, Chapel Hill, AB, 65; Rockefeller Univ, PhD(life sci), 72. Prof Exp: Res assoc plant biol, Rockefeller Univ, 71-72; asst prof chem, 72-74, ASST PROF BIOL, UNIV NC, WILMINGTON, 74- Mem: AAAS; Sigma Xi. Res: Attempting to elucidate the mechanisms by which autonomous cells attain the capability for rapid growth. Mailing Add: Dept of Biol Univ of NC Wilmington NC 28401

LUNDEEN, GLEN ALFRED, b Sterling, Colo, June 7, 22; m 48; c 6. FOOD SCIENCE. Educ: Univ Calif, BS, 47; Ore State Univ, MS, 49, PhD(food technol), 52. Prof Exp: In chg qual control, Smith Canning & Freezing Co, Ore, 50; asst prof hort & asst horticulturist, Univ Ariz, 51-52; prof subtrop hort, Am Univ, Beirut, 52-54, prof & head dept food technol, 54-57; asst prof hort, Mich State Univ, 57-60 & food sci, 60-64; res chemist, Nutrit Div, Wyeth Lab, Inc, 64; assoc prof, Fresno State Col, 64-68; DIR FOOD SCI RES CTR, 68- Mem: AAAS; Am Chem Soc; Am Soc Hort Sci; Inst Food Technol. Res: Food antioxidants; oxidation-reduction potentials in food products; enzymes in fruits and vegetables; compounding nutrition foods; milk chemistry; preservation of fruit and vegetable products; world food problems; human nutrition problems. Mailing Add: 3451 E Bellaire Way Fresno CA 93726

LUNDEGARD, ROBERT JAMES, b Youngstown, Ohio, Feb 22, 27; m 51; c 1. STATISTICS, RESEARCH ADMINISTRATION. Educ: Univ Ohio, BS, 50; Purdue Univ, MS, 52, PhD(math, statist), 57. Prof Exp: Asst prof math, Syracuse Univ, 56-60; head logistics & math statist br, 60-69, DIR MATH & INFO SCI DIV, OFF NAVAL RES, 69- Mem: Am Statist Asn; Math Asn Am; Inst Math Statist; Asn Comput Mach. Res: Statistical inference; models for problems in engineering management. Mailing Add: Math & Info Sci Div Off of Naval Res Arlington VA 22217

LUNDELIUS, ERNEST LUTHER, JR, b Austin, Tex, Dec 2, 27; m 53; c 2. VERTEBRATE PALEONTOLOGY. Educ: Univ Tex, BS, 50; Univ Chicago, PhD(paleozool), 54. Prof Exp: Fulbright scholar vert paleont, Univ Western Australia, 54-55; res fel paleoecol, Calif Inst Technol, 56-57; from asst prof to assoc prof, 57-69, PROF GEOL, UNIV TEX, AUSTIN, 69- Mem: Soc Vert Paleont (secy-treas, 75-); Soc Study Evolution. Res: Pleistocene vertebrates; paleoecology; adaptive morphology; Australian marsupials. Mailing Add: Dept of Geol Sci Univ of Tex Austin TX 78712

LUNDELL, ALBERT THOMAS, b Riverside, Calif, Dec 23, 31; m 52; c 3. MATHEMATICS. Educ: Univ Utah, AB, 52, AM, 55; Brown Univ, PhD(math), 60. Prof Exp: Instr math, Brown Univ, 59-60; lectr, Univ Calif, Berkeley, 60-62; asst prof, Purdue Univ, 62-66; assoc prof, 66-69, chmn dept, 70-72, PROF MATH, UNIV COLO, BOULDER, 70- Mem: Am Math Soc. Res: Algebraic topology. Mailing Add: Dept of Math Univ of Colo Boulder CO 80302

LUNDELL, CYRUS LONGWORTH, botany, see 12th edition

LUNDELL, O ROBERT, b Revelstoke, BC, Nov 7, 31; m 56; c 2. PHYSICAL CHEMISTRY. Educ: Queen's Univ, Ont, BA, 54; Mass Inst Technol, PhD(phys chem), 58. Prof Exp: Lectr chem, Royal Mil Col, Ont, 58-61; from asst prof to assoc prof, 61-71, actg chmn dept biol, 67-68, assoc dean, 68-74, PROF CHEM, YORK UNIV, 71-, DEAN, FAC SCI, 74- Concurrent Pos: Res assoc, Mass Inst Technol, 60-62. Mem: Chem Inst Can. Res: Calorimetry and kinetics of gas phase reactions. Mailing Add: Dept of Chem York Univ 4700 Keele St Downsview ON Can

LUNDEN, ALLYN OSCAR, b Toronto, SDak, Feb 5, 31; m 55; c 3. PLANT BREEDING, PLANT GENETICS. Educ: SDak State Col, BS, 52, MS, 56; Univ Fla, PhD(plant genetics), 60. Prof Exp: Asst agronomist, SDak State Col, 55-56; asst scientist plant genetics, Univ Tenn-AEC Agr Res Lab, 59-62, assoc prof agron, 62-64; ASSOC PROF AGRON, SDAK STATE UNIV, 64- Mem: Am Soc Agron. Res: Irradiation sensitivity of plant tissues; genetic effects of ionizing and ultraviolet irradiation of plant tissues; sorghum, soybeam and oilseed crop breeding. Mailing Add: Dept of Plant Sci SDak State Univ Brookings SD 57007

LUNDGREN, DAVID LEE, b Aberdeen, Wash, Sept 28, 31; wid; c 5. MICROBIOLOGY. Educ: Ore State Univ, BS, 54; Univ Utah, MS, 61, PhD(microbiol), 67. Prof Exp: Bacteriologist, Univ Utah, 54-59, chief epizool diag lab, 59-62, chief infectious disease lab, 61-64, microbiologist, Biol Div, Dugway Proving Ground, 64-66; RES VIROLOGIST, LOVELACE FOUND, 66- Concurrent Pos: Adj prof biol, Univ NMex, 72- Mem: AAAS; Am Soc Microbiol; Radiation Research Soc; Soc Exp Biol & Med; NY Acad Sci. Res: Toxicity of inhaled radionuclides and fossil fuel pollutants; effects of inhaled radionuclides and fossil fuel pollutants on respiratory infection. Mailing Add: Inhalation Toxicol Res Inst Lovelace Found PO Box 5890 Albuquerque NM 87115

LUNDGREN, DONALD GEORGE, b Manchester, NH, Aug 29, 24; m 49; c 2. MICROBIOLOGY. Educ: St Anselm Col, BA, 49; Stetson Univ, MA, 50; Syracuse Univ, PhD(bact), 54. Prof Exp: Asst instr, 52-54, instr plant sci, 54-58, from asst prof bact to assoc prof microbiol, 58-65, PROF BIOL, SYRACUSE UNIV, 65-, CHMN DEPT BIOL, 71- Mem: AAAS; Am Soc Microbiol; Electron Micros Soc Am. Res: Bacterial physiology; industrial microbiology; electron microscopy. Mailing Add: Dept of Biol Syracuse Univ Biol Res Labs Syracuse NY 13210

LUNDGREN, HAROLD PALMER, b Minneapolis, Minn, Mar 22, 11; m 39; c2. TEXTILE CHEMISTRY. Educ: NDak Agr Col, BS, 32; Univ Minn, PhD(physiol chem), 35. Prof Exp: Asst, NDak Agr Col, 31-32; res fel, Phys Chem Inst, Univ Uppsala, 35-37; res assoc colloid chem, Univ Wis, 37-41; from assoc chemist to sr chemist, Western Regional Res Lab, Agr Res Serv, USDA, 41-53, head protein sect, 53-57, chief wool & mohair lab, 57-74; ACTG CHMN DIV TEXTILE & CLOTHING, UNIV CALIF, DAVIS, 74- Honors & Awards: H D Smith Mem Award, Am Soc Testing & Mat, 65; Olney Gold Medal Award, Am Asn Textile Chemists & Colorists, 68; Norman E Borlaug Award, World Farm Found, 75. Mem: Am Chem Soc; Am Soc Biol Chemists; Fiber Soc; Am Asn Textile Chemists & Colorists; fel Brit Textile Inst. Res: Physical chemistry of proteins and protein derivatives; synthetic and natural protein fibers. Mailing Add: Div of Textiles & Clothing Univ of Calif Davis CA 95616

LUNDGREN, J RICHARD, b Springfield, Mass, Oct 1, 42; m 64; c 2. ALGEBRA. Educ: Worcester Polytech Inst, BS, 64; Ohio State Univ, MS, 69, PhD(math), 71. Prof Exp: Proj engr, New Eng Tel, 64-67; ASST PROF MATH, ALLEGHENY COL, 71- Mem: Math Asn Am. Res: Group theory and the classification of finite simple groups; applications of group theory to algebraic coding theory. Mailing Add: 993 First St Meadville PA 16335

LUNDGREN, LAWRENCE WILLIAM, JR, b Attleboro, Mass, Mar 17, 32; div; c 2. ENVIRONMENTAL GEOLOGY. Educ: Brown Univ, AB, 53; Yale Univ, PhD(geol), 58. Prof Exp: From instr to assoc prof geol, 56-67, chmn dept geol sci, 71-74, PROF GEOL, UNIV ROCHESTER, 67- Concurrent Pos: Fulbright lectr, Finland, 67-68; NSF fac fel geog & environ eng, Johns Hopkins Univ, 76. Mem: AAAS; Geol Soc Am. Res: Structure of metamorphic rocks; geology and public policy. Mailing Add: Dept of Geol Univ of Rochester Rochester NY 14627

LUNDIN, FRANK E, JR, b Chicago, Ill, Aug 25, 28; m 49; c 4. EPIDEMIOLOGY. Educ: Manchester Col, BA, 49; Ind Univ, MD, 53; Johns Hopkins Univ, MPH, 59, DrPH, 62. Prof Exp: USPHS, 53-, intern, Hosp, Norfolk, Va, 53-54, staff physician, Hosp, Carville, La, 54-56, epidemiologist, Cancer Invest, Nat Cancer Inst, Univ Tenn, 56-58, instr, Johns Hopkins Univ, 60-61, res assoc, 61-62, res scientist, Epidemiol Br, Nat Cancer Inst, 62-67, sr epidemiologist, Occup Studies, Div Environ Health Sci, NIH, 67-71, sr epidemiologist, Epidemiol Br, Epidemiol & Biomet Br, Nat Inst Child Health & Human Develop, 71-74, DEP CHIEF, EPIDEMIOL STUDIES BR, BUR RADIOL HEALTH, FOOD & DRUG ADMIN, USPHS, 74- Mem: Soc Epidemiol Res; Am Pub Health Asn. Res: Epidemiology of cancer, especially of the cervix; epidemiology of lung cancer; leukemia and lymphoma; occupational cancer in infant ann fetal mortality and parental smoking; health effects of radiation. Mailing Add: Bur Radiol Health Food & Drug Admin 5600 Fishers Lane Rockville MD 20852

LUNDIN, ROBERT ENOR, b Boston, Mass, Mar 19, 27; m 52; c 2. NUCLEAR MAGNETIC RESONANCE. Educ: Harvard Univ, AB, 50; Univ Calif, Berkeley, PhD(chem), 55. Prof Exp: Res chemist, Res Ctr, Texaco, Inc, 55-58; RES CHEMIST, WESTERN REGIONAL RES LAB, USDA, 58- Mem: AAAS; Am Chem Soc; Am Phys Soc. Res: Nuclear magnetic resonance spectroscopy; catalysis; radiation chemistry; gaseous thermodynamics. Mailing Add: Western Regional Res Lab US Dept of Agr Berkeley CA 94710

LUNDIN, ROBERT FOLKE, b Rockford, Ill, July 20, 36; m 58. GEOLOGY, PALEONTOLOGY. Educ: Augustana Col, Ill, AB, 58; Univ Ill, MS, 61, PhD(geol), 62. Prof Exp: From asst prof to assoc prof, 62-74, res comt res grant, 66-67, PROF GEOL, ARIZ STATE UNIV, 74- Concurrent Pos: Petrol Res Fund res grants, 63-65, 66-68 & 70-72; Res Corp res grant, 70. Mem: Soc Econ Paleont & Mineral; Am Asn Petrol Geol; Geol Soc Am; Paleont Soc; Nat Asn Geol Teachers. Res: Siluro, Devonian and Mississippi ostracodes, conodonts and stratigraphy; Cenozoic stratigraphy; freshwater ostra ostracodes. Mailing Add: Dept of Geol Ariz State Univ Tempe AZ 85281

LUNDQUIST, CHARLES ARTHUR, b Webster, SDak, Mar 26, 28; m 51; c 5. SPACE SCIENCES. Educ: SDak State Univ, BS, 49; Univ Kans, PhD(physics), 53. Prof Exp: Asst prof eng res, Pa State Univ, 53-54; physicist, Tech Feasibility Study Off, Redstone Arsenal, 54-56, chief physics & astrophys sect, Army Ballistic Missile Agency, 56-60; chief physics & astrophys br, Marshall Space Flight Ctr, NASA, 60-62; asst dir sci, Smithsonian Astrophys Observ, 62-73; DIR SPACE SCI LAB, MARSHALL SPACE FLIGHT CTR, NASA, 73- Concurrent Pos: Assoc, Harvard Col Observ, 62-73. Mem: AAAS; Int Astron Union; Am Astron Soc; Am Geophys Union; Am Phys Soc. Res: Spacecraft orbital mechanics and orbit determination; space technology; classical mechanics; radiative transfer. Mailing Add: Space Sci Lab NASA Marshall Space Flight Ctr Huntsville AL 35812

LUNDQUIST, JOSEPH THEODORE, JR, analytical chemistry, electrochemistry, see 12th edition

LUNDQUIST, MARJORIE ANN, b Newport News, Va, Aug 17, 38. INDUSTRIAL HYGIENE. Educ: Randolph-Macon Woman's Col, AB, 59; Univ Va, MS, 62, PhD(physics), 65. Prof Exp: Res fel, Dept Phys & Inorg Chem, Univ Adelaide, 65-66; physicist mat sci, 67, tech prog planner lead-acid battery eng, 67-72, supvr monitoring, anal & compliance occup safety & health, 72-74, MGR INDUST HYG, GLOBE-UNION, INC, 74- Concurrent Pos: Adj asst prof, Dept Energetics, Col Eng & Appl Sci, Univ Wis-Milwaukee, 74-75. Mem: Am Indust Hyg Asn; Am Soc Testing & Mat. Res: Cigarette smoking as a source of occupational exposure to lead and other metals; biological monitoring of lead-exposed employees; correlation between airborne lead exposure and lead absorption of employee populations. Mailing Add: Globe-Union Inc 5757 N Green Bay Ave Milwaukee WI 53201

LUNDQUIST, NORMAN STANLEY, b South Bend, Wash, Apr 4, 11; m 37; c 3. DAIRY SCIENCE. Educ: State Col Wash, BS, 33, MS, 39; Univ Wis, PhD(dairy biochem), 44. Prof Exp: Instr high schs, Wash, 33-39; asst in dairy husb, Univ Wis, 40-41, instr, 41-44; assoc prof & assoc dairy husbandman, State Col Wash, 44-48; tech adv on feeds, Jack Steele, Inc, 48-49; assoc prof, 49-59, PROF ANIMAL SCI, PURDUE UNIV, 59- Mem: Fel AAAS; Am Soc Animal Sci; Am Dairy Sci Asn. Res: Relation of vitamins A, C and niacin to early calfhood diseases; effect of sulfa drugs and calfhood vaccination on blood plasma vitamin levels; effect of chlorobutanol on synthesis of B-complex vitamins by the rumen; forage utilization by the ruminant. Mailing Add: Dept of Animal Sci Purdue Univ West Lafayette IN 47906

LUNDQUIST, WILLIAM EMIL, organic chemistry, see 12th edition

LUNDSGAARDE, HENRY P, b Copenhagen, Denmark, Dec 22, 38; US citizen; m 67; c 2. ANTHROPOLOGY. Educ: Univ Calif, Santa Barbara, BA, 61; Univ Wis-Madison, MS, 63, PhD(anthrop), 66. Prof Exp: Asst anthrop, Univ Wis-Madison, 62-63; asst prof, Univ Calif, Santa Barbara, 65-70; assoc prof & chmn dept, Univ Houston, 70-72; PROF ANTHROP & CHMN DEPT, UNIV KANS, 72- Concurrent Pos: Adj res asst anthrop, Univ Ore, 64-65; NIMH grant, Univ Calif, Gilbert Islands, 66-67; fac res grants, 66-68; Am Coun Learned Socs fel, Law Sch, Harvard Univ, 69-70 & univ fel, 69-70; fac res grant, Univ Kans, 72- Mem: Fel Am Anthrop Asn; Asn Soc Anthrop Oceania; Law & Soc Asn; Am Soc Criminol. Res: Ethnology and social anthropology; comparative law and social organization; Pacific Island cultures; Gilbertese law and political development; homicide in United States. Mailing Add: Dept of Anthrop Univ of Kans Lawrence KS 66044

LUNDSTED, LESTER GORDON, b Hankison, NDak, Sept 12, 14; m 42; c 2. INDUSTRIAL ORGANIC CHEMISTRY. Educ: Wis State Teachers Col, EdB, 35; Univ Wis, PhM, 40, PhD(org chem), 42. Prof Exp: Asst, Forest Prod Lab, US Forest Serv, 40; asst chem, Univ Wis, 40-42; org chemist, Wyandotte Chem Corp, 42-45, sect head, 45-50, asst to dir res & develop, 50-53, dir chem res, 53-64, dir inorg & org chem, 64-66; DIR ORG RES & DEVELOP, BASF WYANDOTTE CORP, 66- Mem: Chem Mkt Res Asn; Am Chem Soc; Sigma Xi. Res: High pressure hydrogenation; reactions of alkylene oxides; surfactant development; polymer syntheses and applications; antistatic agents; analytical technology. Mailing Add: Cent Res & Develop Dept BASF Wyandotte Corp Wyandotte MI 48192

LUNDVALL, RICHARD, b Boxholm, Iowa, Dec 10, 20; m 41; c 3. VETERINARY MEDICINE. Educ: Iowa State Univ, DVM, 44, MS, 56. Prof Exp: From instr to assoc prof, 44-71, PROF VET MED & SURG, IOWA STATE UNIV, 71- Res: Large animal surgery; ophthalmology. Mailing Add: Vet Med Clin Iowa State Univ Ames IA 50012

LUNDY, RICHARD ALAN, b Sullivan, Ind, Aug 20, 34; m 60; c 2. HIGH ENERGY PHYSICS. Educ: Univ Chicago, BS, 56, MS, 57, PhD(physics), 62. Prof Exp: Assoc physicist, Argonne Nat Lab, 62-68, assoc dir high energy facility, 68-70; sect leader, Meson Lab, 70-75, HEAD NEUTRINO SECT, FERMI NAT LAB, 75- Mem: Am Phys Soc. Res: Elementary particles; physics instrumentation. Mailing Add: Fermi Nat Lab PO Box 500 Batavia IL 60510

LUNDY, TALMAGE E, b Andalusia, Ala, Feb 17, 17; m 48. BIOLOGY. Educ: Livingston State Col, BS, 48; Univ Ala, MA, 49, MEd, 53, PhD(biol), 62. Prof Exp: Teacher high sch, Ala, 49-50; mgr drug store, 50-51; teaching prin pub sch, 51-52, teacher & prin, 52-55, supv prin 55-57, guid counsr & teacher, 57-58 & 59-61; assoc prof biol, ECarolina Univ, 62-69; assoc prof, 69-74, PROF BIOL, PENSACOLA COL, 74- Concurrent Pos: Consult, Green-Thumb Nursery, Fla, 50-60; assoc dir, NSF Biol Inst Jr High Sch Teachers, 62-65; curriculum consult, 64-71. Mem: Nat Audubon Soc. Res: Bird banding and migratory studies; avian ecology; transfer of viruses in genus Camelia; endoparasitism of domesticated dog; ectoparasitism and endoparasitism of common house sparrow; helminthology of swine. Mailing Add: Dept of Biol Pensacola Col Pensacola FL 32503

LUNER, CHARLES, b Poland, June 1, 25; nat US; m 53; c 2. PHYSICAL CHEMISTRY. Educ: Loyola Col, Can, BS, 47; McGill Univ, PhD(phys chem), 52. Prof Exp: Res assoc polymer chem & kinetics, Ill Inst Technol, 52-54; res assoc photo chem, State Univ NY Col Forestry, Syracuse Univ, 54-56; sr technologist, Appl Res Lab, US Steel Corp, 56-63; ASSOC CHEMIST, ENVIRON STATEMENT PROJ, ARGONNE NAT LAB, 63- Mem: Am Chem Soc; Electrochem Soc. Res: Chemical kinetics; environmental impact statements; sodium technology for fast reactors. Mailing Add: Environ Statement Proj Argonne Nat Lab Argonne IL 60439

LUNER, PHILIP, b Vilno, Poland, June 1, 25; US citizen; m 51; c 2. PHYSICAL CHEMISTRY. Educ: Loyola Col, BSc, 47; McGill Univ, PhD(phys chem), 51. Prof Exp: Res chemist, Pulp & Paper Res Inst Can, 51-54; group leader, Sulfite Pulp Mfrs League, 54-57; from res assoc to assoc prof, 57-64, PROF PULP & PAPER RES, STATE UNIV NY COL ENVIRON SCI & FORESTRY, 64- Mem: Am Chem Soc; Tech Asn Pulp & Paper Indust; Can Pulp & Paper Asn. Res: Diffusion and penetration studies of pulping; chromophores in model lignin compounds; mechanical properties of fibers and paper; surface chemical properties of wood polymers. Mailing Add: State Univ NY Col Environ Sci & Forestry Syracuse NY 13210

LUNER, STEPHEN JAY, b New York, NY, Oct 2, 40; m 65; c 3. BIOPHYSICS. Educ: Calif Inst Technol, BS, 61; Univ Calif, Los Angeles, PhD(biophys), 69. Prof Exp: Res biophysicist, 68-71, asst res biophysicist, Biophys Lab, 71, ASST PROF IN RESIDENCE PEDIAT, UNIV CALIF, LOS ANGELES, 72- Concurrent Pos: NIMH trainee, Univ Calif, Los Angeles, 69-71. Mem: Biophys Soc; Sigma Xi; Am Soc Hemat. Res: Biophysics of the cell surface; electrophoresis; cell surface antigens; effects of enzymes on cell interactions; immunohematology. Mailing Add: Dept of Pediat Univ of Calif Ctr for Health Sci Los Angeles CA 90024

LUNG, BEN, b Porterville, Calif, Jan 27, 39. CELL BIOLOGY, HISTOLOGY. Educ: San Jose State Col, BA, 64; Univ Calif, Davis, MA, 66; Univ Md, PhD(histol), 72. Prof Exp: Asst histol, Univ Md, 70-71, actg asst prof histol, 73, fel develop biol, 74-75, RES AFFIL DEVELOP BIOL, DEPT ANAT, SCH MED, STANFORD UNIV, 75- Res: Ultrastructure and biology of mammalian sperm chromosomes and nuclei; somatic chromosomes and nuclei; reproductive biology. Mailing Add: Dept of Anat Stanford Univ Sch of Med Stanford CA 94305

LUNGER, PHILIP DUPONT, b Bryn Mawr, Pa, Apr 1, 35; m 59; c 3. ZOOLOGY, VIROLOGY. Educ: Colo State Univ, BS, 57; Ind Univ, MA, 59, PhD(zool), 62 Prof Exp: Res assoc biophys cytol, Rockefeller Inst, 62-66; asst prof, 66-70, ASSOC PROF ZOOL, UNIV DEL, 70- Mem: Am Soc Zoologists. Res: Cell fine structure; reptilian virus studies; virus studies. Mailing Add: Dept of Biol Sci Univ of Del Newark DE 19711

LUNGSTROM, LEON, b Lindsborg, Kans, July 22, 15; m 65. MEDICAL ENTOMOLOGY. Educ: Bethany Col, Kans, BS, 40; Kans State Univ, MS, 46, PhD(med entom), 50. Prof Exp: Entomologist, USPHS, Commun Dis Ctr, 49-52; biologist, 52-73, PROF BIOL & HEAD DEPT, BETHANY COL, KANS, 52- Concurrent Pos: NSF fac fel, Stanford Univ, 59-60 & Univ Okla, 65. Mem: AAAS. Res: Mosquitoes. Mailing Add: Dept of Biol Bethany Col Lindsborg KS 67456

LUNIN, JESSE, b New York, NY, May 11, 18; m 42; c 2. SOIL SCIENCE. Educ: Okla Agr & Mech Col, BS, 39; Cornell Univ, MS, 47, PhD, 49. Prof Exp: Soil technologist, Soil Conserv Serv, USDA, 40-42; asst, Cornell Univ, 45-49; head res sect, W Indies Sugar Co, Dom Repub, 49-56; soil scientist, Soil & Water Conserv Res Div, 56-65, chief northeast br, 65-68, chief soil chemist, 68-72, STAFF SCIENTIST, NAT PROG STAFF SOIL, WATER & AIR SCI, AGR RES SERV, USDA, 72- Concurrent Pos: Consult, Israeli Govt, 57. Mem: Am Chem Soc; Am Soc Agron; Int Soc Soil Sci. Res: Soil chemistry, especially soil-plant relationships. Mailing Add: Nat Prog Staff Soil Water & Air Sci BARC-West Agr Res Serv Beltsville MD 20705

LUNIN, MARTIN, b New York, NY, Aug 31, 17; m 47. PATHOLOGY. Educ: Okla Agr & Mech Col, BS, 38; Wash Univ, DDS, 50; Columbia Univ, MPH, 52. Prof Exp: Assoc prof path, Univ Tex Dent Br, 59-64; asst dean curriculum affairs, 69-71, assoc dean acad affairs, 71-74, PROF PATH & HEAD SCH DENT, UNIV MD, BALTIMORE, 64- Concurrent Pos: Sr consult, Univ Tex M D Anderson Hosp & Tumor Inst, 60-64; consult, Vet Admin Hosp, 62- & Children's & Lutheran Hosps, Baltimore, Md, 64- Mem: AAAS; Am Dent Asn; Am Acad Oral Path; Int Asn Dent Res. Res: Diseases of the soft and hard tissues of the head and neck. Mailing Add: 666 W Baltimore St Baltimore MD 21201

LUNK, WILLIAM ALLAN, b Johnstown, Pa, May 6, 19; m 47; c 4. ORNITHOLOGY. Educ: Univ WVa, AB, 41, MS, 46; Univ Mich, PhD(zool), 55. Prof Exp: Instr biol, Univ WVa, 46-47; preparator, 49-59, assoc cur exhibits & lectr zool, Univ, 59-64, CUR EXHIBITS, EXHIBIT MUS, UNIV MICH, ANN ARBOR, 64- Concurrent Pos: Consult, Kalamazoo Nature Ctr, Mich, 63- Mem: Cooper Ornith Soc; Wilson Ornith Soc; assoc Am Ornith Union. Res: Ornithological life history; taxonomy and distribution; fossil birds; exhibit techniques. Mailing Add: Exhibit Mus Univ of Mich Ann Arbor MI 48104

LUNN, WILLIAM HENRY WALKER, organic chemistry, see 12th edition

LUNNEY, DAVID CLYDE, b Charleston, SC. CHEMICAL INSTRUMENTATION. Educ: Univ SC, BS, 59, PhD(phys chem), 65. Prof Exp: NIH fel, Duke Univ, 66-68; asst prof, 68-73, ASSOC PROF CHEM, E CAROLINA UNIV, 73- Concurrent Pos: Pres, Serendipity Systs, Inc, 75- Mem: Am Chem Soc. Res: Chemical instrumentation and computerization. Mailing Add: Dept of Chem E Carolina Univ Greenville NC 27834

LUNSFORD, CARL DALTON, b Richmond, Va, Feb 11, 27; m 47; c 3. PHARMACEUTICAL CHEMISTRY. Educ: Univ Richmond, BS, 49, MS, 50; Univ Va, PhD(chem), 53. Prof Exp: Instr chem, Univ Va, 52-53; res chemist, 53-57, assoc dir chem res, 58, dir, 59-64, dir labs, 62-64, dir res, 64-66, asst vpres, 66-74, VPRES, A H ROBINS CO, INC, 74- Mem: AAAS; Am Chem Soc; Am Inst Chemists. Res: Medicinal and organic chemistry and development. Mailing Add: A H Robins Co Inc 1211 Sherwood Ave Richmond VA 23220

LUNSFORD, JACK HORNER, b Houston, Tex, Feb 6, 36; m 60; c 7. PHYSICAL CHEMISTRY. Educ: Tex A&M Univ, BS, 57; Rice Univ, PhD(chem eng), 62. Prof Exp: Asst prof chem eng, Univ Idaho, 61-62; asst prof chem, Sam Houston State Col, 65-66; from asst prof to assoc prof, 66-71, PROF CHEM, TEX A&M UNIV, 71- Honors & Awards: Paul H Emmett Award, Catalysis Soc, 75. Mem: Am Chem Soc. Res: Surface chemistry and heterogeneous catalysis, using modern spectroscopic techniques. Mailing Add: Dept of Chem Tex A&M Univ College Station TX 77843

LUNT, HAROLD WILLIAM, zoology, see 12th edition

LUNT, OWEN RAYNAL, b El Paso, Tex, Apr 8, 21; m 53; c 3. SOIL FERTILITY. Educ: Brigham Young Univ, AB, 47; NC State Univ, PhD(agron), 51. Prof Exp: Lectr soil chem, NC State Univ, 50; from instr to assoc prof soil sci, 51-63, actg dir lab nuclear med & radiation biol, 65-68, actg chmn dept biophys, 65-70, PROF BIOL, UNIV CALIF, LOS ANGELES, 63-, DIR LAB NUCLEAR MED & RADIATION BIOL, 68- Mem: Soil Sci Soc Am; fel Am Soc Agron; Am Soc Hort Sci; Am Nuclear Soc. Res: Soil chemistry; environmental pollution. Mailing Add: 1200 Roberto Lane Los Angeles CA 90024

LUNT, STEELE RAY, b Mammoth, Utah, Jan 5, 35; m 59; c 5. ENTOMOLOGY, GENETICS. Educ: Univ Utah, BS, 57, MS, 59, PhD(entom), 64. Prof Exp: From asst prof to assoc prof, 64-74, PROF BIOL, UNIV NEBR AT OMAHA, 74- Mem: Am Inst Biol Sci; Am Mosquito Control Asn; Entom Soc Am. Res: Control, systematics, ecology, and medical importance of mosquitoes. Mailing Add: 3853 N 100th Ave Omaha NE 68134

LUNTZ, ALAN COOPER, physical chemistry, see 12th edition

LUNTZ, JEROME D, electrical engineering, nuclear engineering, see 12th edition

LUNTZ, MYRON, b New York, NY, Jan 16, 40; m 64; c 2. RADIATION PHYSICS. Educ: City Col New York, BS, 62; Univ Conn, MS, 64, PhD(physics), 68. Prof Exp: Res asst physics, Univ Conn, 64-68, fel, 68; vis scientist, Inst Physics, Univ Aarhus, 68-69; asst prof, 69-74, ASSOC PROF PHYSICS, STATE UNIV NY COL FREDONIA, 74- Concurrent Pos: Vis assoc prof physics, Univ Del, 75- Mem: Am Phys Soc; Am Asn Physics Teachers; Sigma Xi. Res: Theoretical study of the penetration of matter by energetic charged particles, with emphasis on effects associated with the spatial distribution of energy desposition about particle tracks. Mailing Add: Dept of Physics State Univ of NY Col Fredonia NY 14063

LUOMA, JOHN ROBERT VINCENT, b Huntingdon, Pa, June 3, 38; m 61; c 2. PHYSICAL CHEMISTRY. Educ: Ohio Univ, BA & BS, 61; Purdue Univ, Lafayette, PhD(phys chem), 66. Prof Exp: Asst prof chem, NDak State Univ, 66-69; asst prof, 69-74, ASSOC PROF CHEM, CLEVELAND STATE UNIV, 74- Res: Mössbauer spectroscopy; molecular beam computations. Mailing Add: Dept of Chem Cleveland State Univ Cleveland OH 44115

LUOMALA, KATHARINE, b Cloquet, Minn, Sept 10, 07. CULTURAL ANTHROPOLOGY, ETHNOGRAPHY. Educ: Univ Calif, Berkeley, AB, 31, MA, 33, PhD(anthrop), 36. Prof Exp: Researcher, Western Mus Labs, Nat Park Serv, US Dept Interior, 36-37; Dorothy Bridgman Atkinson fel, Am Asn Univ Women, Univ Chicago, 37-38, Bishop Mus-Yale Univ fel, 38-40; res asst art, Univ Calif, 41; social sci analyst & study dir, Div Prog Survs, Bur Agr Econ, USDA, 42-44; social sci analyst & asst head, Community Anal Sect, War Relocation Authority, 44-46; from asst prof to prof, 48-73, chmn dept, 54-57, actg chmn, 60 & 64, EMER PROF ANTHROP, UNIV HAWAII, HONOLULU, 73- Concurrent Pos: Hon assoc anthrop, Bernice P Bishop Mus, 41-; assoc ed, J Am Folklore Soc, 47-52, ed, 52-53; Wenner-Gren Viking Fund fel, Gilbert Islands, 48-49; Univ Hawaii res grants, 50-62; Guggenheim Found fels, Harvard Univ, 56 & Europ Mus, 60; Finnish-Am Soc-Ford Found fel, 60; Nat Res Coun-NSF fels, 66-67; vis sr res assoc anthrop, US Nat Mus, 66-67; Am Coun Learned Socs travel grants to Europe. Mem: Fel Am Anthrop Asn; fel Am Folklore Soc; Polynesian Soc; Int Soc Folk Narrative Res (vpres, 62-). Res: Ethnography of Polynesia and Micronesia, especially Gilbert Islands; ethnography of the Dieguenо Indians, California; mythology and folklore of these and others areas. Mailing Add: Dept of Anthrop Univ of Hawaii 2424 Maile Way Honolulu HI 96822

LUOTO, LAURI, b Gardner, Mass, Aug 28, 16; m 43; c 3. VETERINARY MEDICINE, EPIDEMIOLOGY. Educ: Mich State Univ, DVM, 43; Harvard Univ, MPH, 48. Prof Exp: Pvt pract, 43-47; vet epidemiologist, USPHS, 48-50 & Rocky Mountain Lab, Mont, 50-61; exec secy, Div Res Grants, NIH, 61-65; sci adminr extramural prog, Nat Cancer Inst, 65-68; PROF MICROBIOL & ASSOC DEAN COL VET MED & BIOMED SCI, COLO STATE UNIV, 68- Mem: Fel Am Pub Health Asn; Am Vet Med Asn; NY Acad Sci. Res: Virology; tissue culture; serology; immunology; scientific administration. Mailing Add: Col of Vet Med & Biomed Sci Colo State Univ Ft Collins CO 80521

LUPAN, DAVID MARTIN, b Cleveland, Ohio, Oct 23, 45; m 68; c 2. MEDICAL MYCOLOGY. Educ: Univ Ariz, BS, 67; Univ Iowa, MS, 70, PhD(microbiol), 73. Prof Exp: ASST PROF MICROBIOL, SCH MED SCI, UNIV NEV, RENO, 73- Mem: Sigma Xi; Am Soc Microbiol; Int Soc Human & Animal Mycol; Med Mycol Soc of the Americas. Res: The mechanism of pathogenesis of fungi. Mailing Add: Sch Med Sci Univ of Nev Reno NV 98507

LUPINSKI, JOHN HENRY, b Schenectady, NY, Feb 28, 27; m 54; c 3. POLYMER CHEMISTRY. Educ: State Univ Leyden, BS, 49, MS, 53, PhD(chem), 59. Prof Exp: Res chemist, Res & Develop Ctr, 60-72, PROJ MGR, CORP RES & DEVELOP CTR, GEN ELEC CO, 72- Mem: AAAS; Am Chem Soc; Royal Neth Chem Soc. Res: Organic conductors; polymer electro-chemistry; polymer powders; electrostatics. Mailing Add: 26 Country Fair Lane Scotia NY 12302

LUPTON, CHARLES HAMILTON, JR, b Norfolk, Va, July 17, 19; m 45; c 3. MEDICINE, PATHOLOGY. Educ: Univ Va, BA, 42, MD, 44; Am Bd Path, dipl, 51. Prof Exp: Asst prof path, Sch Med, Univ Va, 51-53; assoc prof, 55-60, PROF PATH, MED CTR, UNIV ALA, BIRMINGHAM, 60-, CHMN DEPT, 61- Concurrent Pos: Consult, Vet Admin, Birmingham & Tuskegee. Mem: AAAS; Am Asn Path & Bact; Am Soc Clin Path; AMA; Col Am Path. Res: Cardiovascular diseases, especially the kidney as studied by simpler histochemical techniques. Mailing Add: Dept of Path Univ of Ala Birmingham AL 35294

LUPTON, JOHN MADISON, b Wilmington, Del, Mar 12, 15; m 40; c 3. PHYSICAL CHEMISTRY. Educ: Oberlin Col, AB, 37; Univ Del, MA, 43; Univ Amsterdam, NPhDr, 54. Prof Exp: Res chemist, 37-59, RES ASSOC, E I DU PONT DE NEMOURS & CO, INC, 59- Mem: AAAS; Am Chem Soc. Res: Polymer science, rheology, high pressure physical chemistry. Mailing Add: Plastics Dept Exp Sta E I du Pont de Nemours & Co Inc Wilmington DE 19898

LUPTON, WILLIAM HAMILTON, b Charlottesville, Va, July 25, 30. PLASMA PHYSICS. Educ: Univ Va, BA, 50; Univ Md, PhD(physics), 60. Prof Exp: Physicist, Radio Div, Nat Bur Stand, 52-55; PHYSICIST, PLASMA PHYSICS DIV, US NAVAL RES LAB, 60- Mem: AAAS; Am Phys Soc; Inst Elec & Electronics Engrs. Res: Plasma spectroscopy; high voltage and high current pulse technology; high power laser development. Mailing Add: Naval Res Lab Code 7700 4555 Overlook Ave Washington DC 20390

LUPU, CHARLES ISIAL, biochemistry, see 12th edition

LU QUI, IVAN JAMES, b Trinidad, WI; Can citizen; m; c 2. NEUROANATOMY. Educ: Carleton Univ, BSc, 56; Univ Ottawa, MSc, 66, PhD(neuroanat), 68. Prof Exp: Lectr anat, Med Sch, Univ Man, 68-69, asst prof, 69-70; ASST PROF ANAT, SCH MED, WAYNE STATE UNIV, 70- Concurrent Pos: Med Res Coun Can grant, Univ Man, 69-70. Res: Basal ganglia; light and electron microscope investigation; supra-optic nucleus light and electron microscope investigation. Mailing Add: Dept of Anat Wayne State Univ Med Sch Detroit MI 48201

LURA, RICHARD DEAN, b Kenosha, Wis, Aug 21, 45; m 68. PHYSICAL ORGANIC CHEMISTRY. Educ: Univ Wis, BS, 67; Iowa State Univ, PhD(chem), 71. Prof Exp: ASST PROF CHEM, MILLIGAN COL, 71- Concurrent Pos: Consult, R I Schattner Co, 72-75. Res: Research and devleopment of germicidal and sporicidal solutions for hospital and home use. Mailing Add: 1903 Eastwood Dr Johnson City TN 37601

LURIA, SALVADOR EDWARD, b Turin, Italy, Aug 13, 12; nat US; m 45; c 1. BACTERIOLOGY. Educ: Turin Univ, MD, 35. Prof Exp: Res fel, Curie Lab, Inst Radium, Paris, 38-40; res asst surg bact, Columbia Univ, 40-42; Guggenheim fel, Vanderbilt & Princeton Univs, 42-43; from instr to assoc prof bact, Univ Ind, 43-50; prof, Univ Ill, 50-59; prof microbiol & chmn microbiol comt, 59-64, SEDGWICK PROF BIOL, MASS INST TECHNOL, 64-, INST PROF, 70-, DIR, CTR CANCER RES, 74- Concurrent Pos: Investr, Off Sci Res & Develop, Carnegie Inst, 45-46; Jesup lectr, Columbia Univ, 50; lectr, Univ Colo, 50; Nieuwland lectr, Univ Notre Dame, 59; non-resident fel, Salk Inst, 65-; ed, Virol, 55-; sect ed, Biol Abstr, 58- Honors & Awards: Nobel Prize in Med, 69. Mem: Nat Acad Sci; Nat Inst Med; AAAS; Am Soc Nat; Am Soc Microbiol. Res: Bacterial viruses; microbial genetics; biological effects of radiation. Mailing Add: Dept of Microbiol Mass Inst of Technol Cambridge MA 02139

LURIE, ARNOLD PAUL, b Brooklyn, NY, July 22, 32; m 54; c 3. ORGANIC CHEMISTRY. Educ: NY Univ, BA, 54; Purdue Univ, PhD(org chem), 58. Prof Exp: Lab asst org chem, Purdue Univ, 54-56, fel, 58; res chemist, 58-61, sr res chemist, 61-66, INFO SCIENTIST, RES LAB, EASTMAN KODAK CO, 65-, RES ASSOC, 66- Mem: Am Chem Soc. Res: Synthetic and theoretical organic chemistry related to photographic systems; computerized handling of information. Mailing Add: Res Labs Eastman Kodak Co Rochester NY 14650

LURIE, ARON OSHER, b Johannesburg, SAfrica, Sept 9, 25; US citizen; m 52; c 2. INTERNAL MEDICINE, BIOCHEMISTRY. Educ: Univ Witwatersrand, BSc, 46, MB, BCh, 51; Univ Cape Town, PhD(endocrinol), 61; Am Bd Internal Med, dipl, 74. Prof Exp: Registr internal med, King Edward VIII Hosp, Univ Natal, 56-58; sr bursar endocrinol, Groote Schuur Hosp, Univ Cape Town, 58-60; registr, North Middlesex Hosp, London, Eng, 61-62; registr, Guys Hosp, 63-64; res assoc biol chem, Harvard Univ, 64-68; ASSOC PROF MED, SCH MED, TUFTS UNIV, 68-; DIR DEPT CLIN BIOCHEM, BOSTON CITY HOSP, 68- Concurrent Pos: Dir steroid hormone res unit, Boston City Hosp, 66-68; assoc vis physician, Boston City Hosp; teaching assoc, Newton-Wellesley Hosp, 74- Mem: AMA; Endocrine Soc; Brit Med Asn. Res: Clinical biochemistry; reproductive endocrinology; gas liquid chromatography of steroid hormones; endocrinological disturbances of malnutrition. Mailing Add: Dept of Clin Biochem Boston City Hosp Boston MA 02118

LURIE, DAN, b Tel Aviv, Israel, Dec 4, 33; US citizen; m 59; c 2. BIOSTATISTICS. Educ: Southern Methodist Univ, BS, 61, MS, 64; Tex A&M Univ, PhD(statist), 71. Prof Exp: Instr, San Antonio Col, 65-67 & Tex A&M Univ, 67-70; ASST PROF BIOMET, MED UNIV SC, 71- Concurrent Pos: Consult biostatist, Sch Aerospace Med, Brooks AFB, Tex, 64-67. Mem: Am Statist Asn; Biomet Soc. Res: Design and analysis of medical experiments; simulation method; nonparametric statistics; sampling techniques. Mailing Add: Dept of Biomet Med Univ of SC Charleston SC 29401

LURIE, FRED MARCUS, b Boston, Mass, Nov 16, 30. PHYSICS. Educ: Univ NC, Chapel Hill, BS, 52; Univ Ill, Urbana, MS, 57, PhD, 63. Prof Exp: Teaching asst physics, Univ Ill, Urbana, 57-59; from instr to asst prof, Univ Pa, 63-67; asst prof, 67-70, ASSOC PROF PHYSICS, IND UNIV, BLOOMINGTON, 70- Mem: Am Phys Soc. Mailing Add: Dept of Physics Ind Univ Bloomington IN 47401

LURIE, HARRY I, b Johannesburg, SAfrica, Jan 3, 13; m 41; c 2. PATHOLOGY. Educ: Univ Witwatersrand, BSc, 32, MB, ChB, 36, FRCPath, 74. Prof Exp: Pathologist, SAfrican Inst Med Res, Johannesburg, 39-62; lectr, Univ Witwatersrand, 46-62; PROF PATH, MED COL VA, 62- Mem: Int Soc Human & Animal Mycol; Int Acad Path. Res: Host-parasite relationship in fungal infections, particularly Sporotrichosis and Cryptococcosis. Mailing Add: Dept of Path Med Col of Va Box 817 Richmond VA 23219

LURIE, JOAN B, b New York, NY, Jan 21, 41; m 61; c 2. THEORETICAL SOLID STATE PHYSICS. Educ: Brooklyn Col, BS, 61; Rutgers Univ, MS, 62, PhD(physics), 67. Prof Exp: Mem tech staff physics res, RCA Labs, 62-66; fel appl math, Univ Col, Univ London, 67-68; syst programmer comput sci, Appl Data Res, 69-70; fel solid state physics, Rutgers Univ, 70-72; asst prof, 72-76, ASSOC PROF PHYSICS, RIDER COL, 76- Mem: Am Phys Soc; Am Asn Univ Prof. Res: Theoretical research in lattice dynamics of solid state of rare gases; computer assisted instruction, particularly in physics and mathematics. Mailing Add: Dept of Math & Physics Rider Col Lawrenceville NJ 08648

LURIE, NANCY OESTREICH, b Milwaukee, Wis, Jan 29, 24; div. ANTHROPOLOGY. Educ: Univ Wis-Madison, BA, 45; Univ Chicago, MA, 47; Northwestern Univ, PhD(anthrop), 52. Prof Exp: Instr anthrop & sociol, Univ Wis-Milwaukee, 47-49; substitute instr anthrop, Univ Colo, 50; instr anthrop & sociol, Univ Wis-Milwaukee, 51-53; AAAS grant, Nat Arch, 53-54; res assoc, Peabody Mus, 54-56; lectr anthrop, Rackham Sch, Univ Mich, 57-59, lectr, Sch Pub Health, 59-61, asst prof, 61-63; from assoc prof to prof & chmn dept, Univ Wis-Milwaukee, 63-72; CUR ANTHROP, MILWAUKEE PUB MUS, 72- Concurrent Pos: Consult & expert witness for var law firms representing Indian clients before US Indian Claims Comn, 54-; Am Philos Soc grant, 55-56; Bollingen Found grant, 58; lectr anthrop, Wayne State Univ, 60; Lichtenstern-Univ Chicago grant, Chicago Indian Conf, 61; Fulbright-Hays vis lectr, Aarhus Univ, 65-66; Wenner-Gren suppl grant Lapp res, Norway & USSR, 66; mem fac, Univ Wis-Milwaukee, 72- Honors & Awards: Co-recipient, Anisfield-Wolf Award, Sat Rev, 68. Mem: AAAS; fel Am Anthrop Asn; Am Ethnol Soc; Soc Appl Anthrop;Am Soc Ethnohist. Res: North American Indian, especially ethnology and ethnohistory; action anthropology; ethnohistory; museology. Mailing Add: Dept of Anthrop Milwaukee Pub Mus 800 W Wells St Milwaukee WI 53233

LURIE, PAUL RAYMOND, b Amsterdam, NY, Nov 18, 17; m 42; c 1. PEDIATRICS. Educ: Harvard Univ, AB, 38; Columbia Univ, MD, 42. Prof Exp: Fel cardiovasc physiol, Sch Med, Yale Univ, 48-50; from asst prof to prof pediat, Sch Med, Ind Univ, Indianapolis, 50-67; PROF PEDIAT, SCH MED, UNIV SOUTHERN CALIF, 67-; HEAD DIV CARDIOL, CHILDREN'S HOSP LOS ANGELES, 67- Mem: Soc Pediat Res; Am Pediat Soc; Int Cardiovasc Soc. Res: Pediatric cardiology; physiology, diagnosis and treatment of congenital heart disease. Mailing Add: Children's Hosp of Los Angeles 4650 Sunset Blvd Los Angeles CA 90027

LURIO, ALLEN, b Rockville Center, NY, July 11, 29. PHYSICS. Educ: Mass Inst Technol, BS, 50; Columbia Univ, MS, 52, PhD, 56. Prof Exp: Asst, Columbia Univ, 51-55; instr physics, Yale Univ, 55-57; RES PHYSICIST, IBM CORP, 57- Concurrent Pos: Adj asst prof, Columbia Univ, 64-67, adj assoc prof, 67-, mem exec comt, Div Electron & Atomic Physics, 66-69. Mem: Fel Am Phys Soc. Res: Atomic beam and optical double resonance spectroscopy; far infrared spectroscopy; ion solid interac interactions. Mailing Add: 126 Judson Ave Dobbs Ferry NY 10522

LURYE, JEROME ROBERT, b New York, NY, Nov 5, 20; m 49; c 2. MATHEMATICS. Educ: City Col New York, BEE, 42; NY Univ, PhD(math), 51. Prof Exp: Develop engr, Westinghouse Elec Corp, NJ, 42-45; jr scientist, Math Res Group, Wash Sq Col, NY Univ, 46-51, res assoc electromagnetics div, Inst Math Sci, 51-55; sr supvry scientist, TRG, Inc, 55-68; ASSOC PROF MATH, ST JOHN'S

UNIV, NY, 68- Mem: AAAS; Inst Elec & Electronics Engrs. Res: Microwave electron tubes; mathematical methods in electromagnetic theory; fluid dynamics. Mailing Add: Dept of Math St John's Univ Grand Central & Utopia Pkwys Jamaica NY 11432

LUSAS, EDMUND W, b Woodbury, Conn, Nov 25, 31; m 57; c 3. FOOD SCIENCE, FOOD TECHNOLOGY. Educ: Univ Conn, BS, 54; Iowa State Univ, MS, 55; Univ Wis, PhD(food technol), 58; Univ Chicago, MBA, 72. Prof Exp: Proj leader, Res Labs, 58-64, mgr canned foods res, 64-66, mgr pet foods res, 66-72, MGR SCI SERV, QUAKER OATS CO, 72- Mem: Sigma Xi; Inst Food Technol; Am Chem Soc; Am Dairy Sci Asn. Res: Color in canning beets; criteria for ingredient selection and stability factors in foods; human and pet food development; computerization of analytical laboratory operations; research and development administration; technical staff services management. Mailing Add: Quaker Oats Co 617 W Main St Barrington IL 60010

LUSCOMBE, HERBERT ALFRED, b Johnstown, Pa, Aug 9, 16; m 42; c 3. DERMATOLOGY. Educ: St Vincent Col, BSc, 36; Jefferson Med Col, MD, 40. Prof Exp: PROF DERMAT & HEAD DEPT, JEFFERSON MED COL, 59- Mailing Add: Dept of Dermat Jefferson Med Col Philadelphia PA 19107

LUSE, ROBERT ARTHUR, agricultural chemistry, radiation biology, see 12th edition

LUSEBRINK, THEODORE ROBERT, physical chemistry, deceased

LUSENA, CHARLES V, b Palermo, Italy, Feb 15, 19; Can citizen; m 47; c 2. MOLECULAR BIOLOGY. Educ: McGill Univ, BScA, 42, MSc, 44, PhD(biochem), 47. Prof Exp: SR RES OFFICER, DIV BIOL SCI, NAT RES COUN CAN, 47- Res: Yeast mitochondrial DNA and nucleases. Mailing Add: Div of Biol Sci Nat Res Coun 100 Sussex Dr Ottawa ON Can

LUSH, DONALD LAWRENCE, b Toronto, Ont, Jan 18, 45; c 4. FRESH WATER ECOLOGY, LIMNOLOGY. Educ: Trent Univ, BSc, 68; Univ Waterloo, MSc, 70, PhD(fresh water ecol), 74. Prof Exp: COORDR ENVIRON SERV ENVIRON MGT, BEAK CONSULTS LTD, 74- Mem: Can Soc Environ Biologists; Int Asn Theoret & Appl Limnol. Res: Defining the chemical and biological nature of natural metallorganic materials and other natural organic fresh water abiotic particulates and their role in nutrient cycling and ecosystem energetics. Mailing Add: Beak Consults Ltd 6870 Goreway Dr Mississauga ON Can

LUSH, JAY LAURENCE, b Shambaugh, Iowa, Jan 3, 96; m 23; c 2. ANIMAL BREEDING, ANIMAL GENETICS. Educ: Kans State Col, BS, 16, MS, 18; Univ Wis, PhD(genetics), 22. Hon Degrees: DAgr, Univ Uppsala, 57, Univ Giessen, 57 & Danish Vet & Agr Col, 58; LLD, Mich State Univ, 64; DSc, Univ Ill, 69, Kans State Univ, 70, Univ Wis, 70, Swiss Fed Inst Technol, 71 & Agr Univ Norway. Prof Exp: Asst genetics, Univ Wis, 19-21; animal husbandman, Exp Sta, Agr & Mech Col, Univ Tex, 21-30; prof animal breeding, 30-66, EMER PROF ANIMAL SCI & CHARLES F CURTISS DISTINGUISHED PROF AGR, IOWA STATE UNIV, 66- Concurrent Pos: Nat Res Coun fel, Danish Vet & Agr Col, 34; guest prof, Rural Univ Minais Gerais, Brazil, 41; lectr, Training Ctr, Food & Agr Orgn, India, 54 & Nat Inst Farming Technol, Buenos Aires, Arg, 66-67; vis prof, Ohio State Univ, 68. Honors & Awards: Morrison Award, 46; Borden Award, 58; Von Mathusius Medal, Ger Soc Animal Breeding, 60; Order of Merit, Italy, 65; Animal Breeding & Genetics Award, Am Soc Animal Sci, 65; Nat Medal of Sci, 68. Mem: Nat Acad Sci; Am Soc Animal Sci (pres, vpres & secy-treas); Genetics Soc Am; Norweg Acad Sci & Lett; Royal Swedish Acad Agr & Forestry. Res: Genetics of farm animals; genetics of populations. Mailing Add: Dept of Animal Sci Kildee Hall Iowa State Univ Ames IA 50010

LUSHBAUGH, CLARENCE CHANCELUM, b Covington, Ky, Mar 15, 16; m 42, 63; c 3. PATHOLOGY. Educ: Univ Chicago, BS, 38, PhD(path), 42, MD, 48. Prof Exp: Asst path, Univ Chicago, 39-42, from instr to asst prof, 42-49, pathologist, Toxicity Lab, 41-49; mem staff, Los Alamos Sci Lab, Univ Calif, 49-63; chief scientist appl radiobiol, 63-75, NASA Total Body Irradiation Proj, 64-75, CHIEF SCIENTIST EXP & HISTOCHEM PATH, MARMOSET RES COLONY, OAK RIDGE ASSOC UNIVS, 64-, DIR, RADIATION EMERGENCY ASSISTANCE CTR/TRAINING SITE, 74-, CHMN, MED & HEALTH SCI DIV, 75- Concurrent Pos: Pathologist, Los Alamos Med Ctr, NMex, 49-63; mem path study sect, NIH, 61-64; mem radiobiol adv panel, Space Sci Bd, Nat Acad Sci-Nat Res Coun, 66-72; mem adv comt space radiation effects lab, Col William & Mary, 68- Mem: Am Soc Exp Path; Soc Exp Biol & Med; Health Physics Soc; Radiation Res Soc. Res: Pathology of obstetric shock; chemotherapy of cancer; mitotic poisons; radiation damage; diagnostic radioisotopology; human radiobiology; electronic clinical pathology; primate pathology. Mailing Add: Oak Ridge Assoc Univs PO Box 117 Oak Ridge TN 37830

LUSHBOUGH, CHANNING HARDEN, b Watertown, SDak, Aug 11, 29; m 52; c 4. NUTRITION, BIOCHEMISTRY. Educ: Univ Chicago, AB, 48, AM, 52, PhD(nutrit, biochem), 56. Prof Exp: Res chemist, Res Lab, Carnation Co, Wis, 50-51; assoc biochemist & actg chief, Div Biochem & Nutrit, Am Meat Inst Found, Ill, 56-59; dir prod info, Res Ctr, Mead Johnson & Co, Ind, 59-67; vpres planning & develop, Blue Cross, NY, 67-71; assoc dir, Consumers Union US, 71-73; dir & exec secy, Citizens Comn on Science, Law & Food Supply, Rockefeller Univ, 73-75; CORP DIR, QUAL ASSURANCE, KRAFTCO CORP, ILL, 75- Concurrent Pos: Instr grad nutrit, Ill Inst Technol, 56; lectr, Univ Chicago, 57-59 & Northwestern Univ, 58-59. Mem: Am Inst Nutrit; AAAS; Am Home Econ Asn; Brit Nutrit Soc; NY Acad Sci. Res: Nutritional quality of natural proteins; effects of processing on vitamin retention; relations of dietary fat, protein and carbohydrates to atherosclerosis. Mailing Add: Kraftco Corp Kraftco Court Glenview IL 60025

LUSIS, ALDONS JEKABS, b Esslingen, Ger, June 22, 47; US citizen. MOLECULAR BIOLOGY. Educ: Wash State Univ, BS, 69; Ore State Univ, PhD(biochem), 73. Prof Exp: RES ASSOC MOLECULAR BIOL, ROSWELL PARK MEM INST, 73- Concurrent Pos: NIH fel, 74- Mem: Sigma Xi. Res: Mechanisms controlling developmental expression of enzymes in mammals; processing of mouse lysosomal enzymes. Mailing Add: Dept of Molecular Biol Roswell Park Mem Inst Buffalo NY 14263

LUSK, GRAHAM, food science, food technology, see 12th edition

LUSK, JOAN EDITH, b Teaneck, NJ, July 29, 42. BIOCHEMISTRY. Educ: Radcliffe Col, BA, 64; Harvard Univ, PhD(biol chem), 70. Prof Exp: Nat Cystic Fibrosis Res Found fel biol, Mass Inst Technol, 70-71, NIH fel biol, 71-72; ASST PROF CHEM, BROWN UNIV, 72- Mem: Am Soc Microbiol; AAAS. Res: Membrane structure and function; colicin action; transport. Mailing Add: Dept of Chem Brown Univ Providence RI 02912

LUSK, JOHN WILLIAM, b Lowden, Wash, Sept 10, 17; m 45; c 3. DAIRY SCIENCE. Educ: Wash State Univ, BSc, 41, MSc, 47; Miss State Univ, PhD(animal nutrit), 67. Prof Exp: Supvr, Dairy Herd Improv Asn, Walla Walla County, Wash, 41-43; dairy herdsman, Wash State Univ, 43-46, supt testing purebred dairy cattle, 46-47; instr dairy prod, Tenn Jr Col, 47-48; from instr to assoc prof, 48-75, PROF DAIRY SCI, MISS STATE UNIV, 75- Mem: Am Dairy Sci Asn; Am Soc Animal Sci. Res: Forage evaluation; ruminant nutrition. Mailing Add: 410 White Dr Starkville MS 39759

LUSKIN, LEO SAMUEL, b Buffalo, NY, Feb 1, 14; m 39; c 3. ORGANIC CHEMISTRY, POLYMER CHEMISTRY. Educ: Univ Mich, BSc, 36, MSc, 42. Prof Exp: Chemist coal tar prod, Barrett Div, Allied Chem Corp, 43-44; chemist org chem & polymers, 44-62, head tech writing sect, Spec Prods Dept, 62-68, PROMOTION MGR, PLASTICS INTERMEDIATES, ROHM AND HAAS CO, 68- Mem: Am Chem Soc. Res: Synthesis of organic chemicals; polymers; monomers. Mailing Add: Rohm and Haas Co Independence Mall West Philadelphia PA 19105

LUSSIER, JEAN JACQUES, medicine, deceased

LUSSIER, JEAN PAUL, b Montreal, Que, Sept 17, 17; m 43; c 7. DENTISTRY. Educ: Univ Montreal, BA, 38, DDS, 42, MS, 52; Univ Calif, PhD(endocrinol), 59; FRCD(C). Hon Degrees: DSc, McGill Univ, 72. Prof Exp: Lectr mat med, Fac Dent, Univ Montreal, 44-46, lectr physiol, Fac Med, 46-48, asst prof, 48-52; asst endocrinol, Grad Sch, Univ Calif, 52-54; assoc prof physiol & fac med, 54-58, secy, 56-58, assoc dean & dir studies, 58-62, PROF DENT, DEAN & DIR STUDIES, FAC DENT, UNIV MONTREAL, 62- Concurrent Pos: Mem exec assoc comt dent res, Nat Res Coun Can, 54-59, chmn, 64-67; mem exec comt, Med Res Coun Can, 67-70; mem coun higher educ, Minister Educ, 68-71; consult, Minister Nat Defence, Can Royal Dent Corps, 70-73, Int Dent Fedn & WHO; past pres, Asn Can Faculties Dent; mem bd trustees, Univ Montreal, 75- Honors & Awards: Award, Am Acad Oral Med, 58. Mem: Fel AAAS; fel Am Col Dent. Res: Nutrition; endocrinology; bone physiology; dental education; preventative dentistry. Mailing Add: Fac of Dent Univ of Montreal PO Box 6209 Montreal PQ Can

LUSSIER, ROGER JEAN, b Newport, RI, Apr 29, 43; m 66; c 1. INORGANIC CHEMISTRY. Educ: Univ Mass, Amherst, BS, 65; Brown Univ, PhD(inorg chem), 69. Prof Exp: NSF grant, Cath Univ Am, 69-70; RES CHEMIST, DAVISON DIV, W R GRACE & CO, 70- Mem: Am Chem Soc. Res: Heterogeneous catalysis; reaction mechanisms; homogeneous catalysis; transition metal chemistry; surface chemistry; mineral synthesis. Mailing Add: Dept 902 Washington Res Ctr W R Grace & Co 7399 Rte 32 Columbia MD 21044

LUSSKIN, ROBERT MILLER, b Dec 14, 21; m 47; c 2. ORGANIC CHEMISTRY. Educ: Harvard Univ, AB, 43; NY Univ, MS, 46, PhD(chem), 50. Prof Exp: Chemist, Spencer Kellogg & Sons, 43 & Grosvenor Labs, 45-46; with Trubek Labs, 47-55, res dir, 56-60; dir res, UOP Chem Co, Universal Oil Prod Co, 60-67; supt nonwoven lab, Kimberly-Clark Corp, Wis, 67-68, mgr basic & explor res, 68-72, mgr new concepts res, 72-75; TECH DIR, RESOURCE PLANNING ASSOCS, 75- Mem: AAAS; Am Chem Soc. Res: Business strategy development; consumer new products; polymer and fiber research; chemical intermediates; energy and materials management; pollution control. Mailing Add: Resource Planning Assocs 44 Brattle St Cambridge MA 02138

LUST, GEORGE, b Dessau, Ger, Jan 20, 38; US citizen; m 60; c 2. BIOCHEMISTRY. Educ: Univ Mass, BS, 60; Cornell Univ, PhD(biochem), 64. Prof Exp: NIH fel with Prof F Lynen, Max Planck Inst Cell Chem, Univ Munich, 66-67; biochemist, Walter Reed Army Med Ctr, 67-68; asst prof, 68-74, ASSOC PROF BIOCHEM, COL VET MED, CORNELL UNIV, 74- Mem: Am Chem Soc; Am Soc Microbiol. Res: Folic acid and vitamin B12 in methyl group metabolism; effects of infectious stress on mammalian protein metabolism; development and metabolism of connective tissue; osteoarthritis. Mailing Add: Col of Vet Med Cornell Univ Ithaca NY 14850

LUSTED, LEE BROWNING, b Mason City, Iowa, May 22, 22; m 43; c 2. RADIOLOGY. Educ: Cornell Col, BA, 43; Harvard Med Sch, MD, 50; Am Bd Radiol, dipl. Hon Degrees: DSc, Cornell Univ, 63. Prof Exp: Spec res assoc, Radio Res Lab, Harvard Univ, 43-46; from instr to asst prof radiol, Med Sch, Univ Calif, San Francisco, 55-57; asst radiologist, NIH, 57-58; from asst prof to assoc prof radiol, Sch Med, Univ Rochester, 58-60, prof biomed eng, 60-62; prof radiol, Med Sch, Univ Ore & sr scientist, Ore Primate Res Ctr, 62-68; prof radiol & chmn dept, Stritch Sch Med, Loyola Univ Chicago, 68-69; PROF RADIOL & VCHMN DEPT, UNIV CHICAGO, 69- Concurrent Pos: Chmn comt comput biol & med, Nat Acad Sci-Nat Res Coun, 58-59; consult, Strong Mem Hosp, 58-62; chmn adv comt comput res, NIH, 60-64; assoc dean prof affairs & chief of staff, Loyola Univ Hosp, Chicago, 68-69. Mem: Fel AAAS; fel Am Col Radiol; fel Inst Elec & Electronics Engrs; Roentgen Ray Soc; Radiol Soc NAm. Res: Study of medical decision making; application of signal detection theory to assess system and observer performance in radiographic diagnosis. Mailing Add: Dept of Radiol Univ of Chicago Chicago IL 60637

LUSTFIELD, CHARLES DAVENPORT, b Oak Park, Ill, June 20, 38; m 62; c 1. MATHEMATICAL ANALYSIS. Educ: Univ Ill, BS, 60; Ariz State Univ, MA, 62, PhD(math), 68. Prof Exp: Instr, 66-68, ASST PROF MATH, OHIO UNIV, 68- Mem: Am Math Soc; Math Asn Am. Res: Univalent functions. Mailing Add: Dept of Math Ohio Univ Athens OH 45701

LUSTGARTEN, CATHERINE SUE, b Gary, Ind, Sept 26, 48. VETERINARY MEDICINE. Educ: Purdue Univ, DVM, 72. Prof Exp: Assoc vet pvt pract, Hobart Animal Clin Inc, 72-73; RES VET TOXICOL, LOVELACE FOUND MED EDUC & RES, INHALATION TOXICOL RES INST, 73- Mem: Am Vet Med Asn. Res: Long-term toxic effects of inhaled particles and gasses, especially inhaled radionuclides, with emphasis on the biliary excretion of inhaled toxicants. Mailing Add: Inhalation Toxicol Res Inst Lovelace Found PO Box 5890 Albuquerque NM 87115

LUSTGARTEN, JACK ABRAHAM, b Montreal, Que, Mar 22, 34; m 61; c 3. CLINICAL CHEMISTRY. Educ: McGill Univ, BS, 57, PhD(biochem), 64; Am Bd Clin Chem, dipl, 70; Nat Registry Clin Chemists, cert. Prof Exp: Res assoc biochem, Stanford Univ, 64-65; sr fel clin chem, Univ Wash, 65-67; head clin chem, Allentown Gen Hosp, 67-69; HEAD CLIN CHEM, SINAI HOSP BALTIMORE, 69- Concurrent Pos: Consult, Sherman Labs, Baltimore, Md, 69- & Md Regional Lab, 69-71; lectr, Essex Community Col, 75- Mem: Fel Am Asn Clin Chemists. Res: Clinical chemistry methodology development. Mailing Add: Dept of Path Sinai Hosp of Baltimore Baltimore MD 21215

LUSTGARTEN, RONALD KRISSES, b New York, NY, Feb 24, 42. CHEMISTRY. Educ: Columbia Univ, AB, 62; Pa State Univ, PhD(chem), 66. Prof Exp: NIH fel, Univ Calif, Los Angeles, 66-68; res assoc & Mellon fel chem, Carnegie-Mellon Univ, 68-74; STAFF MEM, UPJOHN CO, 75- Mem: AAAS; Am Chem Soc; Fedn Am Scientists. Res: Organic mechanisms; reactive intermediates; kinetics. Mailing Add: Upjohn Co Kalamazoo MI 49001

LUSTICK, SHELDON IRVING, b Syracuse, NY, Aug 16, 34; m 70. ENVIRONMENTAL PHYSIOLOGY, VERTEBRATE ZOOLOGY. Educ: San

Fernando Valley State Col, BA, 63; Syracuse Univ, MS, 65; Univ Calif, Los Angeles, PhD(zool), 68. Prof Exp: Asst prof, 68-72, ASSOC PROF PHYSIOL, OHIO STATE UNIV, 72- Concurrent Pos: Dept of Interior res grant, 69-72 & 73-75. Mem: AAAS; Animal Behav Soc; Wildlife Soc; Am Soc Zoologists. Res: How animals adapt physiologically to environmental stress. Mailing Add: Dept of Zool Ohio State Univ Columbus OH 43210

LUSTIG, BERNARD, b Kolomea, Austria, Dec 21, 02; nat US; m 38; c 2. BIOCHEMISTRY. Educ: Univ Vienna, PhD(chem), 25. Prof Exp: Chemist, Rudolf Hosp, Vienna, 26-32; chief biochemist, Pearson Cancer Found, 33-38 & West London Hosp, 38-40; biochemist, Lawrence R Bruce Inc, 40-44, dir res, 45-58; vpres chg res, 58-68, VPRES & DIR BASIC RES, CLAIROL INC, 68- Honors & Awards: Prize, Asn Chocolate Mfrs, 30. Mem: AAAS; Am Chem Soc; Soc Exp Biol & Med; Am Asn Textile Chemists & Colorists; NY Acad Sci. Res: Chemistry and biochemistry of proteins and lipids; biochemistry of cancer; chemistry and technology of keratin fibers. Mailing Add: 38 Chester St Stamford CT 06905

LUSTIG, CLAUDE DAVID, b Berlin, Ger, June 21, 33; m 67; c 2. PHYSICS. Educ: Oxford Univ, BA, 55, MA, 58, DPhil(physics), 58. Prof Exp: Res officer, Clarendon Lab, Oxford Univ, 58-59; res assoc plasma physics, Princeton Univ, 59-60; sr sci officer, Serv Electronics Res Lab, Eng, 61-62; res staff mem, 62-68, head systs tech dept, 68-70, MGR ELECTRON & ION PHYSICS DEPT, SPERRY RAND RES CTR, 70- Mem: Am Phys Soc; Inst Elec & Electronics Eng. Res: Gas discharge displays; plasma physics, interaction of microwaves with plasmas. Mailing Add: Sperry Rand Res Ctr Sudbury MA 01776

LUSTIG, ERNEST, physical chemistry, see 12th edition

LUSTIG, HARRY, b Vienna, Austria, Sept 23, 25; nat US; wid; c 2. PHYSICS. Educ: City Col New York, BS, 48; Univ Ill, MS, 49, PhD(physics), 53. Prof Exp: Asst physics, Univ Ill, 49-53; from instr to assoc prof, 53-67, chmn dept, 65-70, exec officer PhD prog physics, 68-70, assoc dean sci, 72-75, dean col lib arts & sci, 73-74, PROF PHYSICS, CITY COL NEW YORK, 67-, DEAN SCI, 75- Concurrent Pos: Prin scientist, Nuclear Develop Corp Am, 56-61; vis res asst prof, Univ Ill, 59-60; fel, Colo Inst Theoret Physics, 60; Fulbright lectr, Univ Dublin, 64-65; vis prof, Univ Colo, 66 & Univ Wash, 67 & 69; sr officer, UNESCO, Paris, 70-72, consult, 72-75. Mem: Am Phys Soc; Fedn Am Scientists; fel NY Acad Sci. Res: Theoretical nuclear physics; Mössbauer effect; solar energy. Mailing Add: Dept of Physics City Col of New York New York NY 10031

LUSTIG, LAWRENCE KENNETH, b New York, NY, June 22, 30; m 67; c 2. EARTH SCIENCES. Educ: City Col New York, BS, 55; Univ NMex, MS, 58; Harvard Univ, PhD(geol), 63. Prof Exp: Jr geophysicist, Shell Oil Co, 57-58; geologist, Arabian Am Oil Co, 58-59 & Water Resources Div, US Geol Surv, 62-64; assoc prof geol & sr scientist, Off Arid Lands Res, Univ Ariz, 64-68; sr ed, 68-74, MANAGING ED YEARBKS, ENCYCL BRITANNICA, 74- Concurrent Pos: Res hydrologist, Water Resources Div, US Geol Surv, 64-68. Mem: Fel AAAS; Geol Soc Am; Am Geophys Union; Soc Econ Paleontologists & Mineralogists; Am Asn Petrol Geologists. Res: Quantitative geomorphology; fluvial processes & sedimentation; arid lands geomorphology and hydrology; earth-moon relations. Mailing Add: Encycl Britannica 425 N Michigan Ave Chicago IL 60611

LUSTIG, MAX, b Chicago, Ill, Apr 9, 32; m 54; c 1. INORGANIC CHEMISTRY, AIR POLLUTION. Educ: Univ Calif, Los Angeles, BS, 57; Univ Wash, PhD(inorg chem), 62. Prof Exp: Chemist, Olin Mathieson Chem Corp, 57-58; res chemist, Redstone Arsenal Res Div, Rohm and Haas Co, Ala, 62-68; asst prof chem, Memphis State Univ, 68-73; RES CHEMIST, IIT RES INST, 73- Concurrent Pos: Eve instr, Univ Ala, 63- Mem: Am Chem Soc; fel Am Inst Chemists. Res: Physical and chemical studies of boron hydrides; chemistry of non-metal compounds with oxygen and fluorine, especially peroxides and hypofluorites; free radical chemistry; organometallic compounds; air pollution studies; high vacuum techniques; reaction kinetics involving air pollutants in the troposphere and stratosphere. Mailing Add: Chem Div IIT Res Inst 10 W 35th St Chicago IL 60616

LUSTIG, STANLEY, b Brooklyn, NY, Feb 23, 33; m 60; c 2. PHYSICAL CHEMISTRY. Educ: Univ Toledo, BS, 58. Prof Exp: Chemist, Save Elec Corp, 58-59; res chemist, 59-65, res proj leader, 65-70, group leader, 70-73, TECH MGR, FILMS-PKG DIV, UNION CARBIDE CORP, 73- Mem: Am Chem Soc. Res: Plastic products and processes; crystal structure of polymers. Mailing Add: 561 Lakewood Blvd Park Forest IL 60466

LUTEN, DANIEL B, JR, b Indianapolis, Ind, Mar 15, 08; m 37; c 3. RESOURCE GEOGRAPHY. Educ: Dartmouth Col, AB, 29; Univ Calif, Berkeley, PhD(chem), 33. Prof Exp: Instr chem, Dartmouth Col, 30-31; instr, Univ Calif, Berkeley, 34-35; res chemist, Shell Oil Co, 35-36 & Shell Develop Co, 36-61; lectr, 61-75, EMER LECTR GEOG, UNIV CALIF, BERKELEY, 75- Concurrent Pos: Tech adv, Nat Resources Sect, Hq, Supreme Comdr Allied Powers, 48-50; vis prof geog, La State Univ, Baton Rouge, 70. Mem: Asn Am Geogr; Am Geog Soc; Pop Asn Am; Ecol Soc Am; Am Chem Soc. Res: Natural resources; population; conservation; criteria for decision; chemical reaction volocities. Mailing Add: Dept of Geog Univ of Calif Berkeley CA 94720

LUTES, CHARLENE MCCLANAHAN, b Grundy, Va, Feb 4, 38; m 62; c 1. GENETICS, DEVELOPMENTAL BIOLOGY. Educ: Radford Col, BS, 59; Ohio State Univ, MSc, 62, PhD(genetics), 68. Prof Exp: From instr to assoc prof, 64-75, PROF BIOL, RADFORD COL, 75- Mem: AAAS; Am Soc Zoologists; Am Inst Biol Sci. Res: Developmental genetics of wing venation patterns in drosophila melanogaster. Mailing Add: Dept of Biol Radford Col Radford VA 24141

LUTES, DALLAS D, b St Louis, Mo, July 12, 25; m 45; c 2. PLANT PATHOLOGY. Educ: La Polytech Inst, BS, 49; Univ Mo, PhD(bot), 54. Prof Exp: Instr bot, ETex State Col, 54-55; from assoc prof to prof bot, 55-74, head dept, 63-73, PROF BOT & BACT, LA TECH UNIV, 74- Mem: AAAS. Res: Disease resistance by breeding; virus transmission; seed germination affected by light; mistletoe seed germination; fern taxonomy and distribution. Mailing Add: Dept of Bot & Bact La Tech Univ Ruston LA 71270

LUTES, OLIN SILAS, b Faribault, Minn, Apr 29, 22; m 48, 71; c 3. ELECTROMAGNETISM. Educ: Carnegie Tech Univ, BS, 44; Columbia Univ, MA, 50; Univ Md, PhD(physics), 56. Prof Exp: Physicist, Sinclair Ref Co, 46-48 & Nat Bur Stand, 51-56; SR PRIN SCIENTIST, RES CTR, HONEYWELL, INC, 56- Mem: Am Phys Soc; Inst Elec & Electronics Engrs. Res: Magnetic thin flim physics. Mailing Add: Res Ctr Honeywell Corp 10701 Lyndale Ave S Bloomington MN 55420

LUTEY, RICHARD WILLIAM, b Ironwood, Mich, Feb 22, 35; m 57; c 3. INDUSTRIAL MICROBIOLOGY, PLANT PATHOLOGY. Educ: Mich State Univ, BS, 57; Univ Minn, MS, 59, PhD(plant path), 63. Prof Exp: Res asst plant path, Univ Minn, 57-59, res fel, 59-63; asst regional mgr, 63-70, regional mgr, Northern Region, 70-74, ASST REGIONAL MGR, BUCKMAN LABS, INC, 74- Mem: AAAS; Am Phytopath Soc Soc Indust Microbiol; Tech Asn Pulp & Paper Indust. Res: Control of microorganisms in industrial process water and processing systems; influence of microorganisms on quality of grain and grain products. Mailing Add: Buckman Labs Inc 1256 N McLean Blvd Memphis TN 38108

LUTEYN, JAMES LEONARD, b Kalamazoo, Mich, June 23, 48. SYSTEMATIC BOTANY. Educ: Western Mich Univ, BA, 70; Duke Univ, MA, 72, PhD(bot), 75. Prof Exp: ASSOC CUR BOT, NEW YORK BOT GARDENS, 75- Mem: Sigma Xi; Am Soc Plant Taxon; Bot Soc Am. Res: Evolution and systematics of the neotropical Ericaceae-Vacciniceae. Mailing Add: New York Bot Garden Bronx NY 10458

LUTH, WILLIAM CLAIR, b Winterset, Iowa, June 28, 34; m 53; c 3. GEOLOGY, GEOCHEMISTRY. Educ: Univ Iowa, BA, 58, MS, 60; Pa State Univ, PhD(geochem), 63. Prof Exp: Res assoc geochem, Pa State Univ, 63-65; asst prof, Mass Inst Technol, 65-68; ASSOC PROF GEOCHEM, STANFORD UNIV, 68- Concurrent Pos: Alfred P Sloan Found res fel, Mass Inst Technol, 66-67. Mem: Am Geophys Union; Geol Soc Am; Mineral Soc Am; Geochem Soc. Res: Experimental petrology; physical chemistry of the igneous and metamorphic rocks; phase equilibria in silicate-volatile systems at high pressure and temperature. Mailing Add: Dept of Geol Stanford Univ Stanford CA 94305

LUTHE, JOHN CHARLES, b Des Moines, Iowa, May 4, 49; m 72. THEORETICAL HIGH ENERGY PHYSICS. Educ: Iowa State Univ, BS, 71; Univ Wis-Madison, PhD(physics), 75. Prof Exp: SCHOLAR PHYSICS, UNIV MICH, ANN ARBOR, 75- Mem: Am Phys Soc. Res: Phenomenology of high energy interactions; two-body and inclusive hadronic interactions; lepto-hadronic interactions. Mailing Add: Dept of Physics Univ of Mich Ann Arbor MI 48104

LUTHER, CHESTER FRANCIS, b Auburn, Calif, Oct 12, 06; m 31; c 3. MATHEMATICS. Educ: Stanford Univ, AB, 28, AM, 29, PhD(math), 32. Prof Exp: From actg instr to instr math, Stanford Univ, 28-36; from assoc prof to prof, 36-46, dean, Col Liberal Arts, 40-46, James J Matthews Prof Math, 46-73, EMER PROF MATH, WILLAMETTE UNIV, 73- Concurrent Pos: NSF award, Univ Colo, 53. Mem: AAAS; Am Math Soc; Math Asn Am. Res: Theory of substitution groups. Mailing Add: Dept of Math Willamette Univ Salem OR 97301

LUTHER, EDWARD TURNER, b Nashville, Tenn, Feb 11, 28; m 55; c 2. GEOLOGY. Educ: Vanderbilt Univ, BA, 50, MS, 51. Prof Exp: From geologist to asst state geologist, 51-67, CHIEF GEOLOGIST, TENN DIV GEOL, 67- Concurrent Pos: Instr, Univ Tenn, Nashville, 55-57; fuels engr, Tenn Valley Authority, 57. Mem: Fel Geol Soc Am. Res: Areal and economic geology of various areas in Tennessee, particularly the stratigraphy and structural geology of the Cumberland Plateau; coal resources, particularly in Eastern United States. Mailing Add: Tenn Div of Geol G5 State Off Bldg Nashville TN 37219

LUTHER, HERBERT ADESLA, b Eldred, Pa, Jan 3, 10; m 37; c 3. MATHEMATICS. Educ: Univ Pittsburgh, AB, 34; Univ Iowa, MA, 35, PhD(math), 37. Prof Exp: From instr to prof, 37-75, actg head dept, 67-68, head dept, 68-70, EMER PROF MATH, TEX A&M UNIV, 75- Concurrent Pos: Mathematician, Westinghouse Res Labs, Pa, 45-46; ord res grant, 54-55; res economist, Tex Transportation Inst, 57; NASA res grant, 65-67. Mem: Am Math Soc; Math Asn Am; Soc Indust & Appl Math. Res: Iterative techniques; polynomial factorization and Runge-Kutta methods. Mailing Add: Dept of Math Tex A&M Univ College Station TX 77843

LUTHER, HERBERT GEORGE, b Brooklyn, NY, Oct 1, 14; m 38; c 4. CHEMISTRY. Educ: Cooper Union, New York, BChE, 40; NY Univ, MS, 44; Polytech Inst Brooklyn, DChE, 57. Prof Exp: With Sunshine Biscuit Co, 34-41, dir biochem labs, 41-44; asst dir tech serv, Chas Pfizer & Co, 45-52, dir agr res & develop, 52-59, sci dir agr, 59-69; PRES, LUTHER ASSOCS, 69- Concurrent Pos: Mem, Agr Res Inst; consult res & develop. Mem: Am Chem Soc; Crop Sci Soc Am; Soil Sci Soc Am; Am Soc Agron; Am Asn Cereal Chem. Res: Antibiotics; vitamins; steroids; tranquilizers; unidentified growth factors; enzymes; antioxidants; nutrition; animal health; operations research; food and feed technology. Mailing Add: Head of the River Smithtown NY 11787

LUTHER, MARVIN L, b Waterloo, Iowa, Nov 16, 34; m 59; c 3. ATOMIC PHYSICS, NUCLEAR PHYSICS. Educ: Macalester Col, BA, 57; Univ Fla, MS, 60; Va Polytech Inst, PhD(physics), 67. Prof Exp: Asst prof physics, Randolph-Macon Men's Col, 60-63; asst prof, 66-67, ASSOC PROF PHYSICS, ILL STATE UNIV, 67- Mem: AAAS; Am Asn Physics Teachers. Res: Atomic spectroscopy using collisionally excited beams provided by an accelerator. Mailing Add: Dept of Physics Ill State Univ Normal IL 61761

LUTHER, NORMAN Y, b Palo Alto, Calif, June 3, 36; m 58; c 4. PURE MATHEMATICS. Educ: Stanford Univ, BS, 58; Univ Iowa, MS, 60, PhD(math), 63. Prof Exp: Instr math, Univ Iowa, 63; NSF fel, 63-64; asst prof, 64-69, ASSOC PROF MATH, WASH STATE UNIV, 69- Concurrent Pos: Assoc prof, Albany State Col, Ga, 71-72. Mem: Inst Math Statist; Am Math Soc; Math Asn Am. Res: Probability and statistics; measure theory. Mailing Add: Dept of Math Wash State Univ Pullman WA 99163

LUTHER, WILLIAM FRANCIS, organic chemistry, see 12th edition

LUTHERER, LORENZ OTTO, physiology, see 12th edition

LUTHEY, JOE LEE, b Winslow, Ariz, Sept 21, 43. SPACE PHYSICS. Educ: Univ Calif, Berkeley, AB, 65; Univ Kans, Lawrence, PhD(physics), 70. Prof Exp: Res assoc space physics, Univ Iowa, Iowa City, 70-73; resident res assoc space physics, 73-75, CONSULT RADIATION PHYSICS, JET PROPULSION LAB, CALIF INST TECHNOL, 75- Concurrent Pos: Consult, Physics Dept, Univ Iowa, 73-74; resident res assoc, Nat Res Coun, Jet Propulsion Lab. 73-75. Mem: Am Geophys Union. Res: Test/create Jovian radiation belt models; determine x-ray and gamma-ray emission from natural and artificial satellites in the Jovian trapped electron proton belts. Mailing Add: 350 E Del Mar Blvd 223 Pasadena CA 91101

LUTHI, BRUNO, b Weinfelden, Switz, Oct 6, 31; m 63; c 3. SOLID STATE PHYSICS. Educ: Swiss Fed Inst Technol, dipl, 55, PhD(solid state physics), 59. Prof Exp: Instr physics, Univ Chicago, 60-61; res scientist, Res Lab, Int Bus Mach Corp, Switz, 62-66; assoc prof, 66-69, PROF PHYSICS, RUTGERS UNIV, NEW BRUNSWICK, 69- Mem: Am Phys Soc; Swiss Phys Soc. Res: Transport properties in metals; magnetism; spinwave research; magnetoacoustics; magnetic phase transitions; structural phase transitions; crystalline electric field effects. Mailing Add: Dept of Physics Rugers Univ New Brunswick NJ 08903

LUTHY, JAKOB WILHELM, b Staefa, Switz, Jan 31, 19; nat US; m 48; c 3. CHEMISTRY. Educ: Swiss Fed Inst Technol, MS, 44, DSc(org chem), 47. Prof Exp: Asst prof org technol, Swiss Fed Inst Technol, 46-47; chemist, Gen Aniline & Film

LUTHY

Corp, NJ, 47-48; chemist, Sandoz Chem Works, 48-51; head appln lab, Chem Div, 51-54, dir appln & promotion, 54-58, tech mgr, Dyestuff Div, 58-64, exec vpres, 64-67, PRES COLORS & CHEM DIV & DIR, SANDOZ, INC, 67- Concurrent Pos: Dir, Toms River Chem Corp, 58- Mem: Fel Am Chem Soc; Am Asn Textile Chemists & Colorists; fel Swiss Chem Soc. Res: Dyestuffs. Mailing Add: Sandoz Inc 59 Rte 10 Hanover NJ 07936

LUTMER, ROBERT F, biochemistry, see 12th edition

LUTOMIRSKI, RICHARD, plasma physics, applied physics, see 12th edition

LUTON, EDGAR FRANK, b Memphis, Tenn, Mar 3, 21; m 44; c 3. INTERNAL MEDICINE. Educ: Univ Tenn, Memphis, MD, 44; Am Bd Internal Med, dipl & cert nephrol, 74. Prof Exp: Staff physician neuropsychiat serv med teaching group, Vet Admin Hosp, Memphis, 48-49, resident internal med, 49-51, staff physician med serv, 51-59, sect chief internal med & allergy, 59-67; from asst to assoc prof, 61-74, PROF MED, CTR HEALTH SCI, UNIV TENN, MEMPHIS, 74-; SECT CHIEF ALLERGY & NEPHROLOGY, VET ADMIN HOSP, MEMPHIS, 67- Mem: Fel Am Col Physicians; Am Soc Nephrology; Int Soc Nephrology. Res: Allergy and nephrology. Mailing Add: Vet Admin Hosp 1030 Jefferson Ave Memphis TN 38104

LUTRICK, MONROE CORNEALOUS, b Grayson, La, July 22, 27; m 52; c 4. AGRONOMY, SOIL CHEMISTRY. Educ: La State Univ, BS, 51, MS, 53; Ohio State Univ, PhD(agron), 56. Prof Exp: Asst agron, La State Univ, 51-53; asst agronomist, 56-67, ASSOC SOILS CHEMIST, AGR RES CTR, UNIV FLA, 67- Mem: Am Soc Agron. Res: Soil chemistry and maximum production of field crops; utilization of liquid digested sludge on agricultural lands. Mailing Add: Agr Res Ctr Univ of Fla Jay FL 32565

LUTS, HEINO ALFRED, b Torva, Estonia, Dec 18, 19; nat US; m 54; c 4. MEDICINAL CHEMISTRY. Educ: Upsala Col, BA, 52; Univ Miss, MS, 58, PhD, 66. Prof Exp: Asst chem, Inst Therapeut Res, Warner-Hudnut, Inc, 49-52; res chemist, Wallace Labs Div, Carter Prod, Inc, 53-55, C D Smith Pharmacal Co, 55-56 & Ciba Pharmaceut Prod, Inc, 56-57; NSF asst, Univ Miss, 57-58; dir, Southern Vitamin Prod, 58-67; PROF CHEM, EASTERN KY UNIV, 67- Concurrent Pos: Proj leader, Horizons, Inc, 58-61; pres, Struct-Activity Res, Inc, 60-67; Fulbright-Hays sr lectr, Finland, 75. Mem: AAAS; Am Chem Soc; Int Pharmaceut Fedn; Am Pharmaceut Asn. Res: Analgetics; antispasmatics; tranquilizers; sedatives; metabolites; diuretics; synthetic antibiotics; vitamins. Mailing Add: Dept of Chem Eastern Ky Univ Richmond KY 40475

LUTSCH, EDWARD F, b Chicago, Ill, Nov 23, 30; m 65. ZOOLOGY. Educ: Northern Ill Univ, BS, 52; Northwestern Univ, MS, 57, PhD(biol), 62. Prof Exp: Asst prof zool, Univ Ill, Chicago, 62-68; from asst prof to assoc prof, 68-74, PROF BIOL, NORTHEASTERN ILL UNIV, 74- Mem: AAAS; Am Inst Biol Sci; Am Soc Zoologists. Res: Biological rhythms and clocks; rhythmic response of animals to pharmacological drugs; comparative physiology; animal behavior. Mailing Add: Dept of Biol Northeastern Ill Univ Chicago IL 60625

LUTSKY, BARRY NEAL, biochemistry, see 12th edition

LUTSKY, IRVING, b Paterson, NJ, June 12, 26; m 48; c 4. LABORATORY ANIMAL MEDICINE. Educ: Rutgers Univ, BS, 48; Purdue Univ, MS, 51; Univ Pa, VMD, 55; Am Col Lab Animal Med, dipl, 66. Prof Exp: Res asst poultry diseases, Purdue Univ, 49-51; staff vet, Fromm Labs, 55-58; asst prof vet sci, Med Col Wis, 60-66, assoc prof comp med, 66-72, adminr surg res lab, Allen Bradley Med Sci Lab, 61-72; vis prof comp med, 71-72, ASSOC PROF & CHMN DEPT COMP MED, SCH MED, HEBREW UNIV, 72- Mem: Am Soc Microbiol; Am Asn Lab Animal Sci; Am Vet Med Asn; Asn Gnotobiotics. Res: Infectious diseases; natural disease resistance; applied gnotobiology; occupational allergies. Mailing Add: Dept of Comp Med Sch of Med Hebrew Univ POB 1172 Jerusalem Israel

LUTT, CARL J, b Guthrie Co, Iowa, Feb 10, 21; m 45; c 2. ANATOMY, PHYSIOLOGY. Educ: Creighton Univ, BSM, 42, MD, 45. Prof Exp: Dir student health servs, 60-65, PROF BIOL & HEALTH SCI, CALIF STATE COL HAYWARD, 60- Mem: Am Col Health Asn. Mailing Add: Dept of Biol & Health Sci Calif State Col 25800 Hillary St Hayward CA 94542

LUTTERMOSER, GEORGE WILLIAM, b Detroit, Mich, Aug 16, 12; m 41; c 2. MEDICAL PARASITOLOGY. Educ: Wayne State Univ, AB, 33; Johns Hopkins Univ, ScD(parasitol), 37. Prof Exp: Asst helminth, Sch Hyg & Pub Health, Johns Hopkins Univ, 35-37; jr parasitologist, Zool Div, USDA, 37-40; Inter-Am Cult Exchange Prog fel from US State Dept to Venezuelan Ministry Health, 40-41, helminthologist, Nat Inst Hyg, 41-43; mem Venezuelan field party, Health & Sanit Div, Off Inst Inter-Am Affairs, 43-46; asst prof, Col Vet Med, Univ Pa, 46-47; sr asst scientist, USPHS, 47-49; med parasitologist, 49-62, SCIENTIST ADMINR, NIH, 62- Concurrent Pos: WHO consult to Venezuela; vis res prof, Am Univ Beirut, 61. Mem: Am Soc Parasitologists; Am Soc Trop Med & Hyg. Res: Life cycles of helminths; host-parasite relations; field control and chemotherapy of human parasitic infections; administration-scientific review of research project grant applications and development of research programs. Mailing Add: Res Grants Rev Br NIH Dept of Res Grants Bethesda MD 20014

LUTTINGER, JOAQUIN MAZDAK, b New York, NY, Dec 2, 23. THEORETICAL PHYSICS. Educ: Mass Inst Technol, BS, 44, PhD(physics), 47. Prof Exp: Swiss-Am exchange fel, 47-48; Nat Res Coun fel, 48-49; Jewett fel, Inst Advan Study, 49-50; from asst prof to assoc prof physics, Univ Wis, 50-53; assoc prof, Univ Mich, 53-57 & Higher Training College, Paris, 57-58; prof, Univ Pa, 58-60; PROF PHYSICS, COLUMBIA UNIV, 60- Mem: Nat Acad Sci; Fel Am Phys Soc. Res: Theoretical magnetism; quantum field theory; statistical mechanics; theory of solids. Mailing Add: Dept of Physics 208 Hamilton Hall Columbia Univ New York NY 10027

LUTTINGER, LIONEL, b New York, NY, Jan 30, 20; m 55; c 5. PHYSICAL CHEMISTRY, COLLOID CHEMISTRY. Educ: City Col New York, BS, 49; Polytech Inst Brooklyn, PhD(phys chem), 54. Prof Exp: Chemist, Carl Schleicher & Schuell Co, 46-51; fel, Yale Univ, 54-55; chemist, Am Cyanamid Co, 55-70; SR RES CHEMIST, PERMUTIT CO, 70- Concurrent Pos: Lectr, Univ Conn, 58, 64-68. Res: Water and waste treatment; flocculation; paper chemistry; coagulation; filtration; separations; trace metals removal; rheology; reverse osmosis; water soluble polymers; biochemistry; ultra filtration. Mailing Add: Permutit Res Ctr Ridge Rd Monmouth Junction NJ 08852

LUTTMANN, FREDERICK WILLIAM, b New Brunswick, NJ, Aug 9, 40; m 66. MATHEMATICS. Educ: Amherst Col, AB, 61; Stanford Univ, MS, 64; Univ Ariz, PhD(math), 67. Prof Exp: Assoc, Univ Ariz, 63-67; assoc prof math, Alaska Methodist Univ, 67-70; asst prof, 70-74, ASSOC PROF MATH, SONOMA STATE COL, 74- Mem: AAAS; Am Math Soc; Math Asn Am. Res: Steiner symmetrization of convex bodies; polynomial interpolation. Mailing Add: Dept of Math Sonoma State Col Rohnert Park CA 94928

LUTTON, CHARLES EDWARD, experimental pathology, internal medicine, see 12th edition

LUTTON, EDWIN SCOTT, b Munhall, Pa, July 15, 11; m 35; c 4. PHYSICAL CHEMISTRY. Educ: Swarthmore Col, AB, 32; Yale Univ, PhD(phys chem), 35. Prof Exp: Asst chem, Yale Univ, 32-34; res asst, Procter & Gamble Co, 35-39, chem engr, 39-41, res chemist, 41-74; RETIRED. Concurrent Pos: Assoc ed, Lipids, 69-74; ed, Abstr, Am Oil Chem Soc, 52-54. Honors & Awards: Award Lipid Chem, Am Oil Chem Soc, 71. Mem: AAAS; Am Chem Soc; Am Oil Chem Soc. Res: Phase chemistry; fat composition and consistency; liquid-liquid extraction; crystallization; electrolytic solutions; surface activity. Mailing Add: 5656 Ridge Ave Cincinnati OH 45213

LUTTON, JOHN DUDLEY, cell biology, cell physiology, see 12th edition

LUTTON, LEWIS MONTFORT, b Cincinnati, Ohio, July 14, 45; m 70; c 2. VERTEBRATE ZOOLOGY. Educ: Swarthmore Col, BA, 68; Cornell Univ, PhD(environ physiol), 76. Prof Exp: INSTR BIOL, ALLEGHENY COL, 74- Mem: Am Soc Mammalogists; Am Soc Zoologists; Nat Asn Biol Teachers; Fedn Am Socs Exp Biol. Res: Various aspects of the environmental physiology of small mammals, including exercise physiology, hibernation and adaptations to hypoxic environments. Mailing Add: Dept of Biol Allegheny Col Meadville PA 16335

LUTTRELL, ERIC MARTIN, b Wheeling, WVa, May 12, 41; m 63; c 2. PETROLEUM GEOLOGY. Educ: Univ Wis-Madison, BS, 62, MS, 65; Princeton Univ, PhD(geol), 68. Prof Exp: Geologist, Producing Dept, 68-69, sr geologist, 69-73, RES GEOLOGIST, RES & TECH DEPT, TEXACO INC, 73- Mem: Geol Soc Am; Soc Econ Paleontologists & Mineralogists. Res: Applications of clastic sedimentology; organic geochemistry and geologic thermometry to petroleum exploration. Mailing Add: Texaco Inc PO Box 425 Bellaire TX 77401

LUTTRELL, EVERETT STANLEY, b Richmond, Va, Jan 10, 16; m 44; c 3. MYCOLOGY, PLANT PATHOLOGY. Educ: Univ Richmond, BS, 37; Duke Univ, MA, 39, PhD(bot), 40. Prof Exp: Assoc botanist, Exp Sta, Univ Ga, 42-47; asst prof, Univ Mo, 47-49; assoc botanist, Exp Sta, 49-52, plant pathologist, 52-55, head dept plant path, 55-56, head dept plant path & genetics & chmn div plant path, 66-69, PROF PLANT PATH & BOT, UNIV GA, 66- Mem: Bot Soc Am; Mycol Soc Am; Am Phytopath Soc. Res: Morphology of Ascomycetes. Mailing Add: Dept of Plant Path Univ of Ga Athens GA 30602

LUTTRELL, GEORGE HOWARD, b Glendale, Calif, Dec 23, 41; m 64. ANALYTICAL CHEMISTRY. Educ: Univ Tex, BS, 65; Southern Methodist Univ, MS, 69; Univ Ga, PhD(chem), 75. Prof Exp: Res chemist anal, Alcon Labs, 69-72, RES CHEMIST ANAL, CTR LABS, DIV ALCON LABS, 75- Mem: Am Chem Soc. Res: Preconcentration of trace metal cations and oxyanions for analysis by x-ray fluorescence using immobilized complexing and chelating reagents. Mailing Add: Ctr Labs 35 Channel Dr Port Washington NY 11050

LUTTS, JOHN A, b Baltimore, Md, Feb 26, 32; m 67; c 3. MATHEMATICS. Educ: Spring Hill Col, BS, 57; Univ Pa, MA, 59, PhD(math), 61; Woodstock Col, Md, STL, 65. Prof Exp: From instr to asst prof math, Loyola Col, Md, 65-66; asst prof, 66-70, fac growth fel, 67, fac res grant, 70-71, ASSOC PROF MATH, UNIV MASS, HARBOR CAMPUS, 70- Mem: Am Math Soc; Math Asn Am. Res: Cultural history of mathematics; discrete mathematics. Mailing Add: Dept of Math Univ of Mass Harbor Campus Boston MA 02125

LUTWAK, LEO, b New York, NY, Mar 27, 28; m 50; c 5. ENDOCRINOLOGY, NUTRITION. Educ: City Col New York, BS, 45; Univ Wis, MS, 46; Univ Mich, PhD(biochem), 50; Yale Univ, MD, 56. Prof Exp: Biochemist med, Brookhaven Nat Lab, 50-52; clin assoc metab, Metab Dis Br, Nat Inst Arthritis & Metab Dis, 57-59, sr investr, 60-63; Jameson prof clin nutrit, Grad Sch Nutrit, Cornell Univ, 63-72; PROF MED, SCH MED, UNIV CALIF, LOS ANGELES, 72-, PROF NUTRIT, SCH PUB HEALTH, 73-; SECT CHIEF METAB, VET ADMIN HOSP, SEPULVEDA, 72- Concurrent Pos: NSF sr NASA fel, Ames Res Lab, Moffett Field, Calif, 70-71; prin investr, NASA, 63-; consult, Div Res Grants, NIH, 64-69; consult, Tompkins County Hosp, Ithaca, NY, 64-69; vis prof, Sch Med, Stanford Univ, 70-71. Mem: AAAS; Am Physiol Soc; Am Inst Nutrit; Am Soc Clin Nutrit; Endocrine Soc. Res: Isotope kinetics in metabolic bone disease; calcium, phosphorus and magnesium in human nutrition; effect of space flight on bone and muscle metabolism; obesity control; diabetes and electrolyte metabolism. Mailing Add: Sect of Metab Dept of Med Vet Admin Hosp Sepulveda CA 91343

LÜTY, FRITZ, b Essen, Ger, Apr 12, 28; m 60; c 2. SOLID STATE PHYSICS. Educ: Univ Göttingen, dipl physics, 53; Stuttgart Univ, Dr rer nat(physics), 55. Prof Exp: Asst physics, Stuttgart Univ, 53-62, dozent, Physics Inst, 64-65; vis assoc prof, Univ Ill, Urbana, 63; PROF PHYSICS, UNIV UTAH, 65- Concurrent Pos: Vis prof, Soc Advan Sci, Japan, 73. Mem: Fel Am Phys Soc; Ger Phys Soc. Res: Defects in ionic crystals; radiation damage; absorption and emission spectroscopy; field emission; magneto-optics, paraelectric and paraelastic effects; low temperature dielectric and electro-caloric studies; Raman-scattering; phase transitions. Mailing Add: Dept of Physics Univ of Utah Salt Lake City UT 84112

LUTZ, ALBERT WILLIAM, b Baltimore, Md, Sept 26, 24; m 51; c 2. AGRICULTURAL CHEMISTRY. Educ: Johns Hopkins Univ, AB, 49, MA, 50, PhD(chem), 53. Prof Exp: Assoc prof chem, Col William & Mary, 53-56; res chemist, Chemagro Corp, 56-57; res chemist, 57-59, sr res chemist, 59-69, GROUP LEADER HERBICIDES, AGR DIV, AM CYANAMID CO, 69- Mem: AAAS; Am Chem Soc. Res: Pesticides, particularly growth regulants and herbicides. Mailing Add: Agr Div Am Cyanamid Co PO Box 400 Princeton NJ 08540

LUTZ, ARTHUR LEROY, b Louisville, Ohio, Oct 22, 08; m 37; c 2. NUCLEAR PHYSICS. Educ: Capital Univ, BS, 31; Ohio State Univ, MS, 36, PhD(physics), 43. Prof Exp: High sch teacher, Ohio, 31-40; asst physics, Ohio State Univ, 40-43; PROF PHYSICS, WITTENBERG UNIV, 43- Concurrent Pos: Fac fel, NSF, 60-61. Mem: AAAS; Am Phys Soc; Am Asn Physics Teachers. Res: Radioactive isotopes; internal conversion and K-capture in the radioactive isotopes of lead and bismuth. Mailing Add: Dept of Physics Wittenberg Univ Springfield OH 45501

LUTZ, BARRY LAFEAN, b Windsor, Pa, Jan 2, 44; m 66. ASTROPHYSICS, MOLECULAR SPECTROSCOPY. Educ: Lebanon Valley Col, BS, 65; Princeton Univ, AM, 67, PhD(astrophys sci), 68. Prof Exp: Fel physics, Nat Res Coun Can, 68-70; res astronr, Lick Observ, Univ Calif, 70-71; adj asst prof, 73-74, ADJ ASSOC PROF ASTROPHYS, STATE UNIV NY STONY BROOK, 74-, SR RES ASSOC, 71- Mem: Int Astron Union; Am Astron Soc; Sigma Xi. Res: High resolution spectroscopy of the interstellar medium and of stellar and planetary atmospheres;

laboratory astrophysics; intensity measurements and long path length planetary atmospheres simulations. Mailing Add: Dept of Earth & Space Sci State Univ of NY Stony Brook NY 11794

LUTZ, BRUCE CHARLES, b London, Ont, May 16, 20; m 45; c 3. PHYSICS. Educ: Western Ont Univ, BA, 42, MA, 44; Johns Hopkins Univ, PhD, 54. Prof Exp: Instr electronics & radio, Western Ont Univ, 41-44; lectr electronics & physics, Univ Man, 45-47; instr electronics, 47-57, assoc prof elec eng, 57-62, PROF ELEC ENG, UNIV DEL, 62-, ACTG CHMN DEPT, 73- Mem: Am Inst Aeronaut & Astronaut (treas, Rocket Soc, 59-60); Inst Elec & Electronics Eng. Res: Nuclear reactor physics and engineering; plasma-microwave interaction; signal analysis. Mailing Add: Dept of Elec Eng Univ of Del Col of Eng Newark DE 19711

LUTZ, CHARLES WILLIAM, b Philadelphia, Pa, Nov 10, 29; m 53; c 2. PHYSICAL CHEMISTRY, INORGANIC CHEMISTRY. Educ: Temple Univ, AB, 53; Bryn Mawr Col, PhD(phys chem), 61. Prof Exp: Res chemist, 61-67, SR RES CHEMIST DETERGEN APPLNS, INORG CHEM DIV, FMC CORP, 67- Mem: Am Chem Soc. Res: Heterogeneous equilibria; non-stoichiometric tungsten compounds; crystal-field spectra; thermodynamics, kinetics and physical properties of polyphosphates and polyphosphoric acids; phosphate glasses and coatings; corrosion inhibition; peroxygen compounds; crystal chemistry. Mailing Add: Indust Chem Div FMC Corp Res & Develop Box 8 Princeton NJ 08540

LUTZ, DONALD ALEXANDER, b Syracuse, NY, Apr 2, 40. MATHEMATICS. Educ: Syracuse Univ, BS, 61, MS, 63, PhD(math), 65. Prof Exp: Instr math, Syracuse Univ, 65; asst prof, 65-70, ASSOC PROF MATH, UNIV WIS-MILWAUKEE, 70- Concurrent Pos: Lectr, Univ Md, 67-69; vis asst prof, Math Res Ctr, Univ Wis-Madison, 69-70. Mem: Am Math Soc. Res: Systems of linear ordinary differential equations with meromorphic coefficients; systems of linear difference equations. Mailing Add: Dept of Math Univ of Wis Milwaukee WI 53201

LUTZ, GARSON ALVIN, organic chemistry, see 12th edition

LUTZ, GEORGE JOHN, b New England, NDak, May 9, 33; m 65; c 2. PHYSICAL CHEMISTRY. Educ: Augustana Col, BA, 53; Iowa State Univ, PhD(phys chem), 62. Prof Exp: Jr chemist, Ames Lab, Univ Iowa, 55-58; resident res assoc, Argonne Nat Lab, 62-64; CHEMIST, NAT BUR STANDARDS, 64- Concurrent Pos: Sr US Scientist award, Alexander von Humboldt Found, Bonn, WGer, 74-75. Mem: Am Chem Soc; Am Nuclear Soc; Am Soc Metals. Res: Activation analysis; applications of radioactive isotopes. Mailing Add: Nat Bur of Standards Washington DC 20234

LUTZ, HAROLD JOHN, b Saline, Mich, Aug 11, 00; m 26; c 2. FORESTRY. Educ: Mich State Col, BS, 24; Yale Univ, MF, 27, PhD(forestry), 33. Prof Exp: Tech asst, US Forest Serv, 24-26, assoc silviculturist, Allegheny Forest Exp Sta, 28-29; asst forester, Conn Exp Sta, 27-28; asst prof forestry, Pa State Col, 29-31; from asst prof to prof, 33-48, Morris K Jesup prof silvicult, 48-65, Oastler prof forest ecol, 65-68, OASTLER EMER PROF FOREST ECOL, YALE UNIV, 68- Concurrent Pos: Walker-Ames prof, Univ Wash, 59; H R MacMillan lectr, Univ BC, 49; summers, US Forest Serv, 49-52 & 57 & vis prof, Univ Colo, 64-69. Honors & Awards: Soc Am Foresters Award, 57. Mem: Fel Soc Am Foresters. Res: Forest ecology and soils. Mailing Add: Rte 6 Box 346 Thomas St Allegan MI 49010

LUTZ, HARRY FRANK, b Philadelphia, Pa, Jan 30, 36; m 60; c 2. NUCLEAR PHYSICS. Educ: Univ Pa, AB, 57; Mass Inst Technol, PhD(physics), 61. Prof Exp: PHYSICIST, LAWRENCE LIVERMORE LAB, 61- Mem: Am Phys Soc. Res: Nuclear reactions and nuclear spectroscopy. Mailing Add: Lawrence Livermore Lab PO Box 808-L531 Livermore CA 94550

LUTZ, JOHN EWALD, b Benton Harbor, Mich, Sept 16, 27; m 51; c 4. ZOOLOGY. Educ: Yale Univ, BS, 50; Univ Mich, AM, 51, PhD(zool, mammal), 64. Prof Exp: Biol aide, US Fish & Wildlife Serv, Alaska, summers, 48-51, wildlife mgt biologist, 52-53; instr zool, Univ Mich, 56-57; from asst prof to assoc prof biol, Eastern Mich Univ, 57-65; from asst prof to assoc prof, 65-72, PROF BIOL, BELOIT COL, 72- Concurrent Pos: Sabbatical, Tulane Univ, 72. Mem: Am Soc Zool; Am Soc Mammal; Animal Behav Soc; Ecol Soc Am. Res: Mammalogy; ecology; animal behavior; developmental biology. Mailing Add: Dept of Biol Beloit Col Beloit WI 53511

LUTZ, JOHN GEORGE, b Brooklyn, NY, Nov 13, 00; m 31; c 2. ORGANIC BIOCHEMISTRY. Educ: Polytech Inst Brooklyn, BS, 22; Columbia Univ, MA, 24, PhD(chem), 34. Prof Exp: From instr to asst prof chem, Westminster Col, 24-30; asst, Columbia Univ, 31-33; instr, Union Col, NY, 35-37; prof & head dept, Albany Col Pharm, 37-42; from assoc prof to prof, 42-65, actg chmn dept, 42-43, chmn dept, 43-59, EMER PROF CHEM, HOFSTRA UNIV, 65- Concurrent Pos: Prof, State Univ NY Col Oneonta, 69-70. Mem: Am Chem Soc. Res: Purification of enzymes; adsorption of proteins. Mailing Add: 34 College Terr Oneonta NY 13820

LUTZ, JULIE HAYNES, b Mt Vernon, Ohio, Dec 17, 44; m 66; c 2. ASTRONOMY. Educ: San Diego State Univ, BA, 65; Univ Ill, MS, 68, PhD(astron), 72. Prof Exp: ASST PROF ASTRON, WASH STATE UNIV, 72- Mem: Am Astron Soc. Res: Planetary nebulae; stellar evolution. Mailing Add: Dept of Pure & Appl Math Wash State Univ Pullman WA 99163

LUTZ, KENNETH RUSSELL, b National City, Calif, Apr 24, 25; m 50; c 4. SPEECH PATHOLOGY, AUDIOLOGY. Educ: Pac Union Col, BA, 51; Univ Redlands, MA, 53; Univ Pittsburgh, PhD(speech path & audiol), 62. Prof Exp: Asst pharmacol, Col Med Evangelists, 51-53; speech & hearing consult, Div Spec Educ, Iowa Dept Pub Instr, 53-55; speech therapist, Monona County, Iowa, 55-57; clin asst speech path, Univ Pittsburgh, 57-58, res asst, 58-59; instr speech path, 60-61, asst prof orthod, 61-64, asst prof speech path, Col Arts & Sci, La Sierra Campus, 62-66, vis prof, 66-70, staff audiologist, Med Ctr, 70-73, ASST PROF OTOLARYNGOL, SCH MED, LOMA LINDA UNIV, 64-, ASSOC PROF SPEECH PATH, COL ARTS & SCI, LA SIERRA CAMPUS, 70- Concurrent Pos: Res fel speech path, Univ Pittsburgh, 59-60; Nat Inst Dent Res grant, 64-67; abstr ed, Cleft Palate J, 62-63, 64-66; dir speech & audiol serv, White Mem Med Ctr, 65-68; dir speech path & audiol dept, Glendale Adventist Hosp, 65-70; consult, Montrose Convalescent Hosp, 68-70 & Glen Manor Convalescent Hosp, 69-70; consult audiol, Riverside Gen Hosp, 71- Mem: AAAS; Am Cleft Palate Asn; Am Speech & Hearing Asn; Int Asn Dent Res; Sigma Xi. Res: Duration of gestation for infants with cleft palate; associated teratology with cleft palate; audible scales of nasal voice quality; intelligibility and nasality of cleft palate speakers; maladaptive tongue function; psycholinguistics. Mailing Add: Dept of Commun Loma Linda Univ La Sierra Campus Riverside CA 92505

LUTZ, PATRICK B, inorganic chemistry, see 12th edition

LUTZ, PAUL E, b Hickory, NC, June 25, 34; m 57; c 1. INVERTEBRATE ZOOLOGY, ECOLOGY. Educ: Lenoir Rhyne Col, AB, 56; Univ Miami, MS, 58; Univ NC, PhD(zool), 62. Prof Exp: Asst zool, Univ Miami, 56-58 & Univ NC, 58-61; from instr to assoc prof, 61-70, PROF BIOL, UNIV NC, GREENSBORO, 70- Concurrent Pos: Am Philos Soc grant, 64; NSF grants, 65-67 & 69-71. Mem: AAAS; Am Soc Zoologists; Ecol Soc Am. Res: Ecology and physiology of aquatic insects, especially effects of temperature and photoperiod as they affect seasonal regulation of developmental patterns. Mailing Add: Dept of Biol Univ of NC Greensboro NC 27412

LUTZ, RAYMOND PAUL, b Cleveland, Ohio, May 31, 32. PHYSICAL ORGANIC CHEMISTRY. Educ: Univ Fla, BS, 53, MS, 55; Calif Inst Technol, PhD(org chem), 62. Prof Exp: Res chemist, E I du Pont de Nemours & Co, Ky & Mich, 55-57; instr chem, Harvard Univ, 61-64, lectr, 64-65; asst prof, Univ Ill, Chicago, 65-68; asst prof, 68-69, ASSOC PROF CHEM, PORTLAND STATE UNIV, 69- Mem: Am Chem Soc; The Chem Soc. Res: Reaction mechanisms, including displacement reactions and thermal isomerizations Mailing Add: Dept of Chem Portland State Univ Portland OR 97207

LUTZ, ROBERT WILLIAM, b Mason City, Iowa, Sept 14, 37; m 56; c 4. CHEMICAL PHYSICS, COMPUTER SCIENCE. Educ: Drake Univ, BA, 62; Univ NMex, MS, 66; Ill Inst Technol, PhD(physics), 69. Prof Exp: Res asst physics, Los Alamos Sci Lab, 62-64, staff mem, 64-66; asst prof, 69-73, ASSOC PROF PHYSICS, DRAKE UNIV, 73-, DIR COMPUT SERV, 74- Mem: Am Phys Soc; Am Asn Physics Teachers; Combustion Inst; Sigma Xi; Asn Comput Mach. Res: Computer assisted instruction; computers in undergraduate curriculum. Mailing Add: Comput Ctr Drake Univ Des Moines IA 50311

LUTZ, RUTH NELLIE, nutrition, see 12th edition

LUTZ, WILSON BOYD, b Mogadore, Ohio, May 12, 27; m 50; c 2. BIOCHEMISTRY, ORGANIC CHEMISTRY. Educ: Manchester Col, BA, 50; Ohio State Univ, PhD(org chem), 55. Prof Exp: Fel biochem, Med Col, Cornell Univ, 55-57; scientist, Warner-Lambert Res Inst, 57-60, sr scientist, 60-62; asst prof, 62-64, ASSOC PROF CHEM, MANCHESTER COL, 65- Concurrent Pos: Consult, Warner-Lambert Res Inst, 63-66; guest worker, NIH, 71. Mem: Am Chem Soc. Res: Synthesis of new derivatives of hydroxylamine and substances of biological interest including melanogenic indoles. Mailing Add: Dept of Chem Manchester Col North Manchester IN 46962

LUTZKER, EDYTHE, b Berlin, Ger, June 25, 04; c 3. HISTORY OF SCIENCE, HISTORY OF MEDICINE. Educ: City Col New York, BA, 54; Columbia Univ, MA, 59. Prof Exp: Res asst hist of sci, City Col New York, 52-55; RES & WRITING, 55- Honors & Awards: Am Philos Soc Johnson Fund Grant, 64 & Penrose Fund Grant, 65; NIH grant, 66 & 68-74. Mem: AAAS; Am Asn Hist Med; Am Soc Microbiol; Int Soc Hist Med; Royal Soc Med. Res: Social history; participation by women in science and medicine; pioneers of 19th century medicine in British Empire and India. Mailing Add: 201 W 89th St New York NY 10024

LUVALLE, JAMES ELLIS, b San Antonio, Tex, Nov 10, 12; m 46; c 3. PHYSICAL CHEMISTRY, PHOTOGRAPHIC CHEMISTRY. Educ: Univ Calif, Los Angeles, AB, 36, AM, 37; Calif Inst Technol, PhD(chem), 40. Prof Exp: Instr chem, Fisk Univ, 40-41; res chemist, Res Labs, Eastman Kodak Co, 41-42, phys chemist, Kodak Res Labs, 43-53; phys chemist, Nat Defense Res Comt, Chicago, 42 & Calif Inst Technol, 42; proj dir, Tech Opers, Inc, 53-59; dir basic res, Fairchild Camera & Instrument Corp, NY, 59-63, dir res, Ill, 63-68; tech dir microstatics lab, SCM Corp, 68-69; dir physics & chem res, Smith Corona Marchant Labs, 69-70; sci coordr, Res & Develop Labs, SCM Bus Equip Div, Calif, 70-75; LAB ADMINR, CHEM DEPT, STANFORD UNIV, 75- Concurrent Pos: Vis lectr, Brandeis Univ, 57-59; vis scholar, Stanford Univ, 71-75; independent consult, 75- Mem: AAAS; Am Chem Soc; Am Phys Soc; Soc Photog Scientists & Engrs; fel The Chem Soc. Res: Photochemistry; electron diffraction; magnetic susceptibility; reaction kinetics and mechanisms; photographic theory; magnetic resonance; solid state physics; neurochemistry, chemistry of memory and learning. Mailing Add: 3580 Evergreen Dr Palo Alto CA 94303

LUX, CARL RAY, b Gadsden, Ala, Feb 27, 48; m 74. NUCLEAR CHEMISTRY. Educ: Mo Southern State Col, BS, 70; Purdue Univ, MS, 72, PhD(nuclear chem), 75. Prof Exp: RES CHEMIST RES & DEVELOP, EAGLE-PICHER INDUST INC, 75- Mem: Am Chem Soc. Res: Boron compounds for use in control rods of nuclear reactors; production of lithium for pH control in boiling water nuclear reactors. Mailing Add: Eagle-Picher Indust Inc Miami Res Labs 200 9th St NW Miami OK 74354

LUX, FRANCIS A, analytical chemistry, deceased

LUX, JOHN HERBERT, b Logansport, Ind, Feb 3, 18; m 40; c 2. CHEMISTRY. Educ: Purdue Univ, BS, 39, PhD(chem), 42. Prof Exp: Instr chem, Purdue Univ, 39-42; res & develop engr, Carbide & Carbon Chem Corp, 42-43; asst dir res & develop, Neville Co, WVa, 42-45; consult, Pittsburgh, Pa, 45-47; dir asphalt div, Witco Chem Co, 47-48, dir res, 48-50; mgr new prod develop, Chem Div, Gen Elec Co, Mass, 50-52; vpres, Shea Chem Corp, 52-55; pres, Haveg Industs, Inc, 55-66; CHMN BD & CHIEF EXEC OFFICER, AMETEK, INC, 66- Concurrent Pos: Dir, Hercules Powder Co, 64-66. Mem: AAAS; Am Inst Aeronaut & Astronaut; Am Chem Soc; Am Inst Chem Engrs; Commercial Develop Asn. Res: Vinyl resins; plasticizers; metal organic compounds; wetting agents; reactions and preparation of acetylene; phosphorus and silicone compounds; plastic materials for high temperature construction. Mailing Add: 5260 Chelsea La Jolla CA 92037

LUX, RUSSELL EUGENE, organic chemistry, see 12th edition

LUXEMBURG, WILHELMUS ANTHONIUS JOSEPHUS, b Delft, Neth, Apr 11, 29; m 55; c 2. MATHEMATICAL ANALYSIS. Educ: State Univ Leiden, BSc, 50, MSc, 53; Delft Univ Technol, PhD, 55. Prof Exp: Fel math, Queen's Univ, Can, 55-56; asst prof, Univ Toronto, 56-58; from asst prof to assoc prof, 58-62, PROF MATH, CALIF INST TECHNOL, 62- Mem: Am Math Soc; Can Math Cong; Neth Math Soc; corresp mem Royal Acad Sci Amsterdam. Res: Functional analysis, particularly measure and integration theory, Banach function space theory and theory of locally convex spaces; Riesz spaces; nonstandard analysis. Mailing Add: Dept of Math Calif Inst of Technol Pasadena CA 91125

LUXENBERG, HAROLD RICHARD, b Chicago, Ill, Feb 2, 21; m 42; c 3. APPLIED MATHEMATICS. Educ: Univ Calif, Los Angeles, BA, 42, MA, 48, PhD(math), 50. Prof Exp: Mathematician, Nat Bur Stand, 50-51; res physicist, Hughes Res & Develop Labs, 51-53; consult engr, Remington Rand, Inc, 53-55; proj consult, Litton Industs, 56-58; mgr display dept, Thompson-Ramo-Wooldridge Corp, 59-60; vpres eng & asst gen mgr, Houston Fearless Corp, 61-63; consult, Lux Assocs, 64-70; PROF COMPUT SCI, CALIF STATE UNIV, CHICO, 70- Concurrent Pos: Lectr & instr, Univ Calif, Los Angeles, 52-69. Mem: Sr mem Inst Elec & Electronics Engrs; Sigma Xi. Res: Data display; document storage and retrieval; photo-optical systems; digital computers in command and control applications. Mailing Add: 20 Sunland Dr Chico CA 95926

LUYENDYK, BRUCE PETER, b Freeport, NY, Feb 23, 43; m 67. MARINE GEOPHYSICS. Educ: San Diego State Col, BS, 65; Scripps Inst Oceanog, Univ Calif, San Diego, PhD(oceanog), 69. Prof Exp: Geophysicist, US Navy Electronics Lab, 65-66; res asst oceanog, Scripps Inst Oceanog, Univ Calif, San Diego, 65-69; fel, Woods Hole Oceanog Inst, 69-70, asst scientist, 70-73; asst prof, 73-75, ASSOC PROF GEOL SCI, UNIV CALIF, SANTA BARBARA, 75- Concurrent Pos: Mem working group marine geophys data, Comn Oceanog, Nat Acad Sci, 71; mem working group Mid-Atlantic Ridge, US Geodyn Comn, 71; ed adv, Geol Mag, 74- Mem: AAAS; Am Geophys Union; fel Geol Soc Am. Res: Geotectonics; paleomagnetism; paleoceanography. Mailing Add: Dept of Geol Sci Univ of Calif Santa Barbara CA 93106

LUYET, BASILE JOSEPH, biophysics, deceased

LUYKX, PETER (VAN OOSTERZEE), b Detroit, Mich, Dec 14, 37; m 60; c 2. CYTOLOGY. Educ: Harvard Univ, AB, 59; Univ Calif, Berkeley, PhD(zool), 64. Prof Exp: Asst prof cytol, Univ Minn, Minneapolis, 64-67; asst prof, 67-74, ASSOC PROF CYTOL, UNIV MIAMI, 67- Concurrent Pos: NIH res grants, 65-73. Mem: AAAS; Am Soc Cell Biol. Res: Ultrastructure of chromosomes and centrioles in meiosis and mitosis. Mailing Add: Dept of Biol Univ of Miami Coral Gables FL 33124

LUYTEN, JAMES REINDERT, b Minneapolis, Minn, Dec 26, 41; m 67; c 3. PHYSICAL OCEANOGRAPHY. Educ: Reed Col, AB, 63; Harvard Univ, AM, 65, PhD(chem physics), 69. Prof Exp: Res fel geophys fluid dynamics, Harvard Univ, 69-71; asst scientist, 71-75, ASSOC SCIENTIST, WOODS HOLE OCEANOG INST, 75- Mem: Am Geophys Union. Res: Theoretical and observational study of the dynamics of low frequency variability of ocean circulation; moored current meter arrays; dynamics of the Gulf Stream system. Mailing Add: Woods Hole Oceanog Inst Woods Hole MA 02543

LUYTEN, WILLEM JACOB, b Semarang, Dutch E Indies, Mar 7, 99; nat US; m 30; c 3. ASTRONOMY. Educ: Univ Amsterdam, BA, 18; State Univ Leiden, PhD(astron), 21. Hon Degrees: DSc, Case Western Reserve Univ & Univ St Andrews, 71. Prof Exp: Asst, Observ, State Univ Leiden, 20-21; fel, Lick Observ, Univ Calif, 21-22, Kellogg fel, 22-23; astronr, Harvard Observ, 23-27, asst prof astron, Univ, 27-30; from asst prof to prof, 31-75, EMER PROF ASTRON, UNIV MINN, MINNEAPOLIS, 75- Concurrent Pos: Mem, Lick Observ Calif Eclipse Exped, Ensenada, Baja, Calif, 23 & Hamburg Observ Eclipse Exped, Jokkmokk, Lapland, 27; Guggenheim fel, 28-30 & 37-38. Honors & Awards: Watson Medal, Nat Acad Sci, 64; Bruce Medal, Astron Soc of Pac, 68. Mem: Nat Acad Sci; AAAS; Am Astron Soc; Am Asn Variable Star Observers; Int Astron Union. Res: Stellar motions; nearby stars; white dwarfs; origin of the solar system. Mailing Add: Space Sci Ctr Univ of Minn Minneapolis MN 55455

LUZZATTI, LUIGI, b Rome, Italy, Sept 6, 14; US citizen. PEDIATRICS. Educ: Univ Minn, MS, 42, MD, 43. Prof Exp: Res asst pediat, Univ Minn, 40-43, intern, 42-43; resident, Mt Sinai Hosp, New York, 43-44; instr, Cornell Univ, 44-45, asst prof, 48-50; med dir & chief dept pediat, Children's Hosp of East Bay, Oakland, Calif, 51-53; from asst prof to assoc prof, 54-71, dir pediat outpatient dept, 55-63, dir pediat cytogenetic lab, 63-73, PROF PEDIAT, MED CTR, STANFORD UNIV, 57-, DIR BIRTH DEFECT CTR, 67- Concurrent Pos: Prog dir, NIH metab & hereditary dis training grant, 60-63. Mem: AAAS; Am Acad Pediat; Am Pediat Soc; Am Soc Human Genetics. Res: Chromosomal abnormalities in birth defects; computer analysis of congenital defects in children; role of clinical nurse specialist in birth defects and genetic counseling; impact of genetic counseling on family decisions. Mailing Add: Dept of Pediat Stanford Univ Sch of Med Stanford CA 94305

LUZZI, LOUIS A, b Westerly, RI, June 17, 32; m 53; c 2. PHYSICAL PHARMACY, COLLOID CHEMISTRY. Educ: Univ RI, BS, 59, MS, 62, PhD(pharm), 66. Prof Exp: Res trainee, Abbott Labs, 59-61; asst prof, 66-69, ASSOC PROF PHARM, UNIV GA, 69- Concurrent Pos: Consult, Sandoz Pharmaceut, 66-; Vet Corp Am res grant, 67-69; Abbott Labs res grant, 68-; exec secy-treas conf teachers, Am Asn Cols Pharm, 69-; Webster Co grant, 72- Mem: AAAS; Am Pharmaceut Asn; Acad Pharmaceut Sci. Res: Coacervation and interphasal polymerization as applied to microencapsulation; protein binding of drugs, especially as determined by nuclear magnetic resonance and spectrophotofluorometric probe techniques. Mailing Add: 160 Tara Pl Athens GA 30601

LUZZIO, ANTHONY JOSEPH, b Lawrence, Mass, Oct 13, 24; m 52; c 4. IMMUNOLOGY. Educ: Univ Mass, BS, 47; Kans State Col, MS, 50, PhD(microbiol), 55. Prof Exp: Bacteriologist, Wyo State Vet Lab, 47-49; asst, Univ Kans, 50-52; chemist, Hercules Powder Co, 52-54; chief, Immunol Br, US Army Med Res Lab, 55-71, res immunologist, Blood Transfusion Res Div, 71-74, IMMUNOLOGIST, LETTERMAN ARMY INST RES, 74- Concurrent Pos: Lectr, Univ Louisville, 55-74. Mem: AAAS; Am Soc Microbiol; Am Asn Immunol; Radiation Res Soc. Res: Effects of ionizing radiation on immune mechanisms; effects of arctic climates on immunity; protein degradation and alterations in antigenic specificity by exposure to ionizing rays; immune mechanisms in leishmaniasis. Mailing Add: Letterman Army Inst Res Presidio of San Francisco CA 94129

LWOWSKI, WALTER WILHELM GUSTAV, b Garmisch, Ger, Dec 28, 28; US citizen. ORGANIC CHEMISTRY. Educ: Univ Heidelberg, dipl, 54, Dr rer nat, 55. Prof Exp: Fel, Univ Calif, Los Angeles, 55-57; asst, Univ Heidelberg, 57-59; res fel chem, Harvard Univ, 59-60; asst prof, Yale Univ, 60-66; RES PROF CHEM, NMEX STATE UNIV, 66- Mem: Fel AAAS; Am Chem Soc; fel NY Acad Sci; Ger Chem Soc; The Chem Soc. Res: Reactions mechanisms; electron-deficient nitrogen intermediates; photochemistry; heteorcyclic chemistry; heteroatom rearrangements. Mailing Add: Dept of Chem NMex State Univ PO Box 3-C Las Cruces NM 88003

LYALL, DAVID, surgery, deceased

LYCAN, D RICHARD, b Sheboygan, Wis, Dec 17, 33; m 56; c 2. GEOGRAPHY. Educ: Univ Idaho, BSc, 56; George Washington Univ, MA, 61; Univ Wash, Seattle, PhD(geog), 64. Prof Exp: Asst prof geog, Univ Victoria, BC, 64-70; ASSOC PROF GEOG, PORTLAND STATE UNIV, 70- Concurrent Pos: Vis asst prof, Univ Wash, Seattle, 67. Mem: Asn Am Geog; Can Asn Geog; Regional Sci Asn. Res: Regional analysis; urban geography; quantitative methods. Mailing Add: Dept of Geog Portland State Univ Portland OR 97207

LYCKLAMA, HEINZ, b Blija, Holland, May 1, 43; Can citizen; m 64;c 3. NUCLEAR PHYSICS, COMPUTER SCIENCE. Educ: McMaster Univ, BEngr, 65, PhD(nuclear physics), 69. Prof Exp: MEM TECH STAFF, BELL TEL LABS, 69- Res: Communications research; computer science; computer operating systems research. Mailing Add: Bell Tel Labs Murray Hill NJ 07974

LYDA, STUART D, b Bridger, Mont, June 6, 30; m 53; c 5. PLANT PATHOLOGY. Educ: Mont State Col, BS, 56, MS, 58; Univ Calif, PhD(plant path), 63. Prof Exp: Lab technician, Univ Calif, 59-62; assoc prof plant path, Univ Nev, Reno, 62-67; ASSOC PROF PLANT PATH, TEX A&M UNIV, 67- Mem: AAAS; Am Phytopath Soc. Res: Fungus and plant physiology; mycology. Mailing Add: Dept of Plant Sci Tex A&M Univ College Station TX 77843

LYDDANE, RUSSELL HANCOCK, physics, see 12th edition

LYDING, ARTHUR R, b New York, NY, May 12, 25; m 57; c 1. POLYMER CHEMISTRY, ORGANIC CHEMISTRY. Educ: Cornell Univ, BA, 45; Univ Pa, BS, 48, PhD(chem), 51. Prof Exp: Instr, Cornell Univ, 44-45; control chemist, Gen Baking Co, 46; res chemist, Heyden Chem Corp, 50-52; sr res chemist, Olin Industs, 52-56, group leader polymers div, Olin Mathieson Chem Corp, Conn, 57-64, tech asst to vpres res & develop, Pkg Div, 64-69; sr res scientist, FMC Corp, 69-75; SECT LEADER INDUST CHEM DIV, NL INDUSTS, INC, HIGHTSTOWN, 75- Concurrent Pos: Asst prof, Southern Conn State Col, 64-69. Mem: Am Chem Soc. Res: Agricultural chemicals; research and development of monomers and polymers; plastics and plastics additives; oil additives; cellulose chemistry and packaging; textile stain repellents and flame retardants; fluorochemicals; emulsion polymerization; synthesis of polymers and plastics additives; coatings.

LYDOLPH, PAUL E, b Bonaparte, Iowa, Jan 4, 24; m 66; c 5. CLIMATOLOGY, GEOGRAPHY. Educ: Univ Iowa, BA, 48; Univ Wis, MS, 51, PhD(geog), 55. Prof Exp: From asst prof to assoc prof geog, Los Angeles State Col, 52-59; assoc prof, 59-62, chmn dept, 63-72, PROF GEOG, UNIV WIS-MILWAUKEE, 62- Concurrent Pos: Ford Found fel Slavic studies, Univ Calif, 56-57; consult, Encycl Britannica Films, 65 & Arid Lands Res, Univ Ariz, 66-; short term lectr, Univ Hawaii, Oxford Univ & Stockholm Sch Econ. Mem: Asn Am Geogr; Am Geog Soc; Am Asn Advan Slavic Studies. Res: Geography and climate of the Union of Soviet Socialist Republic. Mailing Add: Dept of Geog Univ of Wis Milwaukee WI 53201

LYDON, CAROL GUZE KONRAD, b St Louis, Mo, Nov 12, 35; m 59, 69; c 3. CELL BIOLOGY. Educ: Wash Univ, AB, 57; Univ Calif, Berkeley, PhD(zool), 63. Prof Exp: Res biologist, Naval Biol Labs, Univ Calif, Berkeley, 63-64; instr biol, Brandeis Univ, 64-65 & Univ Mass, Boston, 65-66; fel bact, Univ Calif, Los Angeles, 66-67; asst prof, 67-72, ASSOC PROF BIOL, CALIF STATE COL, DOMINGUEZ HILLS, 72- Concurrent Pos: Res assoc med genetics, Harbor Gen Hosp, Med Sch Campus, Univ Calif, Los Angeles, 74- Mem: AAAS; Sigma Xi; Am Soc Human Genetics. Res: Physiology of mitotic cells; bacteriophage genetics; blue-green algae mutant production; human genetics; studies of human trisomic cells in culture. Mailing Add: Dept of Biol Calif State Col Dominguez Hills CA 90747

LYDY, DAVID LEE, b Elwood, Ind, Apr 27, 36; m 59; c 3. SURFACE CHEMISTRY. Educ: Ind Univ, AB, 58; Univ Ill, PhD(inorg chem), 63. Prof Exp: Res chemist, 63-68, SECT HEAD, MIAMI VALLEY LABS, PROCTER & GAMBLE CO, 68- Mem: Am Chem Soc. Res: Detergency; new opportunities research. Mailing Add: Miami Valley Labs Procter & Gamble Co Cincinnati OH 45239

LYE, ROBERT GLEN, b Kimberley, BC, Sept 21, 26; m 51; c 3. SOLID STATE PHYSICS. Educ: Univ BC, BASc, 50, MASc, 52; Univ Minn, PhD(elec eng), 57. Prof Exp: Jr develop engr, Consol Mining & Smelting Co, Can, 51-53; res physicist, Res Labs, Nat Carbon Co Div, Union Carbide Corp, 57-65; mem staff, Res Inst Advan Studies, Martin Co, 65-69; head physics dept & assoc dir, Res Inst Advan Studies, 69-74, CORP SCIENTIST, MARTIN MARIETTA CORP, 74- Mem: AAAS; Am Phys Soc; Can Asn Physicists. Res: Physical electronics; electronic, thermal and optical properties of crystals. Mailing Add: Martin Marietta Corp 1450 S Rolling Rd Baltimore MD 21227

LYERLA, TIMOTHY ARDEN, b Long Beach, Calif, Mar 5, 40; m 64; c 1. DEVELOPMENTAL GENETICS. Educ: Univ Calif, Davis, BA, 63; San Diego State Col, MA, 67; Pa State Univ, University Park, PhD(zool), 70. Prof Exp: NIH fel, Northwestern Univ, Ill, 70-71; ASST PROF BIOL, CLARK UNIV, 71- Mem: AAAS; Am Soc Zoologists; Soc Develop Biol. Res: Development of pigment patterns in amphibians; cell regression in vertebrate embryogenesis; isozymes in development. Mailing Add: Dept of Biol Clark Univ Worcester MA 01610

LYERLY, LARRY ALEXANDER, b Crescent, NC, Sept 1, 34; m 60. ANALYTICAL CHEMISTRY. Educ: Catawba Col, AB, 58; Fla State Univ, MS, 61. Prof Exp: RES CHEMIST, R J REYNOLDS TOBACCO CO, 61- Mem: Am Chem Soc. Res: Analytical instrumentation. Mailing Add: 5800 Phillips Bridge Rd Lewisville NC 27023

LYERLY, PAUL JUNIOR, b Granite Quarry, NC, Feb 24, 18; m 42; c 2. AGRONOMY. Educ: NC State Col, BS, 38; Iowa State Col, MS, 40, PhD(plant breeding), 42. Prof Exp: Asst, Iowa State Col, 38-41, corn investr, 41-42; agronomist, Exp Sta, 42-45, cotton breeder, 45-47, supt, 47-59, res coordr, Trans-Pecos Area, 59-72, PROF & RESIDENT DIR, AGR RES CTR, TEX A&M UNIV, EL PASO, 72-, RES COORDR, TRANS-PECOS AREA, 74- Mem: AAAS; Am Soc Agron; Am Soc Plant Physiol; Am Phytopath Soc; Weed Sci Soc Am. Res: Weed control; cotton breeding and production. Mailing Add: Agr Res Ctr 10601 North Loop Tex A&M Univ Box 454 Rte 2 El Paso TX 79927

LYFORD, JOHN H, JR, b Chicago, Ill, July 10, 28; m 51; c 6. ECOLOGY. Educ: Carleton Col, BA, 50; Ore State Univ, MS, 62, PhD(bot), 66. Prof Exp: Pub sch teacher, Wash, 55-62; res biologist, Ore State Game Comn, 63-65; asst prof, 65-72, ASSOC PROF BIOL, ORE STATE UNIV, 72- Mem: AAAS; Ecol Soc Am; Am Soc Limnol & Oceanog; Phycol Soc Am. Res: Trophic structure of aquatic communities; estuarine productivity. Mailing Add: Dept of Gen Sci Ore State Univ Corvallis OR 97331

LYFORD, SIDNEY JOHN, JR, b Exeter, NJ, Jan 20, 37; m 61; c 3. ANIMAL NUTRITION, BIOCHEMISTRY. Educ: Univ NH, BS, 58; NC State Univ, MS, 60, PhD(animal nutrit), 64. Prof Exp: ASSOC PROF ANIMAL NUTRIT, UNIV MASS, AMHERST, 63- Mem: Am Dairy Sci Asn; Am Soc Animal Sci; Sigma Xi. Res: Mechanism of action of certain natural inhibitors and of volatile fatty acid absorption from the ruminant stomach; pectin degradation; nutritive evaluation of byproduct materials as animal feedstuffs. Mailing Add: Dept of Vet & Animal Sci Stockbridge Hall Univ of Mass Amherst MA 01002

LYGRE, DAVID GERALD, b Minot, NDak, Aug 10, 42; m 66. BIOCHEMISTRY. Educ: Concordia Col, Moorhead, Minn, BA, 64; Univ NDak, PhD(biochem), 68. Prof Exp: Am Cancer Soc fel, Case Western Reserve Univ, 68-70; asst prof chem & res corp grant, 70-73, ASSOC PROF CHEM, CENT WASH STATE COL, 73- Mem: Am Chem Soc; Sigma Xi. Res: Enzymology of carbohydrate metabolism; development of experiments for instructional use. Mailing Add: Dept of Chem Cent Wash State Col Ellensburg WA 98926

LYJAK, ROBERT FRED, b Detroit, Mich. MATHEMATICS. Educ: Wayne State Univ, BS, 51; Univ Mich, Ann Arbor, MA, 53, PhD(math), 60. Prof Exp: Assoc mathematician, Res Inst, Univ Mich, 53-56, instr math, Univ, 56-58, mathematician,

Res Inst, 59-62; res mathematician, Conduction Corp, 62-63 & Res Inst, Univ Mich, 63-66; assoc prof, 66-69, chmn dept, 67-70, PROF MATH, UNIV MICH-DEARBORN, 69- Mem: Am Math Soc. Res: Transformation groups; mathematical models of stochastic systems. Mailing Add: Dept of Math Univ of Mich Dearborn MI 48128

LYKE, EDWARD BONSTEEL, b Boston, Mass, Nov 9, 37; m 62; c 2. CYTOLOGY, INVERTEBRATE ZOOLOGY. Educ: Miami Univ, BA, 59; Univ Wis-Madison, MS, 62, PhD(zool), 65. Prof Exp: From asst prof to assoc prof, 65-73, PROF BIOL SCI, CALIF STATE UNIV, HAYWARD, 73- Mem: AAAS; Am Soc Zoologists; Am Inst Biol Sci; Marine Biol Asn UK. Res: Invertebrate cytology and histology; spermatogenesis and oogenesis; morphology of bioluminescent systems. Mailing Add: Dept of Biol Sci Calif State Univ Hayward CA 94542

LYKKEN, GLENN IRVEN, b Grafton, NDak, Jan 27, 39; m 64; c 2. PHYSICS. Educ: Univ NDak, BS, 61; Univ NC, MS, 64, PhD(physics), 66. Prof Exp: Asst physics, Univ NDak, 61-62 & Univ NC, 62-65; asst prof, 65-70, ASSOC PROF PHYSICS, UNIV NDAK, 70- Concurrent Pos: Vis prof, Univ NC, 69-70. Res: Thin films; superconducting tunneling; neutron activation analysis. Mailing Add: Dept of Physics Univ of NDak Grand Forks ND 58201

LYKKEN, LOUIS, chemistry, deceased

LYKOS, PETER GEORGE, b Chicago, Ill, Jan 22, 27; m 50; c 3. PHYSICAL CHEMISTRY. Educ: Northwestern Univ, BS, 50; Carnegie Inst Technol, PhD(chem), 55. Prof Exp: Instr chem, Carnegie Inst Technol, 54-55; from instr to assoc prof chem, 55-64, dir comput ctr & comput sci dept, 64-71, PROF CHEM, ILL INST TECHNOL, 64- Concurrent Pos: Consult, Solid State Sci Div, Argonne Nat Lab, 58-67; consult, Dept Radiation Ther, Michael Reese Hosp, 66-70; pres, Four Pi, Inc, 66-; mem-at-large & chmn comt comput in chem, Nat Acad Sci-Nat Res Coun, 68-74; prog dir, Off Comput Activities, NSF, 71-73; co-chmn, Nat Resource Comput in Chem Proposal Develop Team, Argonne Univs Asn-Argonne Nat Lab, 74-; chmn comput & soc comt, Asn Comput Mach, 73-75; chmn comput in chem comt, Am Chem Soc, 73-77. Mem: Asn Comput Mach; Am Chem Soc. Res: Quantum chemistry; computational chemistry; computer simulation of liquid solutions. Mailing Add: 316 N Ridgeland Ave Oak Park IL 60302

LYLE, GLORIA GILBERT, b Atlanta, Ga, Aug 7, 23; m 47. ORGANIC CHEMISTRY. Educ: Vanderbilt Univ, BA, 44; Emory Univ, MS, 46; Univ NH, PhD, 58. Prof Exp: Instr chem, Hollins Col, 46-47; res assoc, McArdle Lab Cancer Res, Univ Wis, 47-49; from instr to assoc prof, 51-74, PROF CHEM, UNIV NH, 74- Concurrent Pos: USPHS res fel, 58-59; vis assoc prof, Univ Va, 73-74. Mem: Am Chem Soc; NY Acad Sci; The Chem Soc. Res: Organic synthesis; natural products; optical rotatory dispersion and circular dichroism; stereochemistry. Mailing Add: Dept of Chem Parsons Hall Univ of NH Durham NH 03824

LYLE, JAMES ALBERT, b Lexington, Ky, Sept 19, 16; m 48; c 1. BOTANY. Educ: Univ Ky, BS, 40; NC State Col, MS, 46; Univ Minn, PhD(plant path), 53. Prof Exp: Jr plant pathologist, Exp Sta, Univ Hawaii, 46-47; from asst plant pathologist to assoc plant pathologist, 47-54, PROF BOT & MICROBIOL & HEAD DEPT, AUBURN UNIV, 64- Mem: Am Phytopath Soc; Am Inst Biol Sci. Res: Fungus ecology and plant disease control, especially peanuts. Mailing Add: Dept of Bot & Microbiol Auburn Univ Auburn AL 36830

LYLE, RICHARD EDWARD, organic chemistry, see 12th edition

LYLE, ROBERT EDWARD, b Atlanta, Ga, Jan 26, 26; m 47. ORGANIC CHEMISTRY. Educ: Emory Univ, BA, 45, MS, 46; Univ Wis, PhD(chem), 49. Prof Exp: Asst prof chem, Oberlin Col, 49-51; from asst prof to assoc prof, 51-57, PROF CHEM, UNIV NH, 57- Concurrent Pos: USPHS spec fel, 65-66; vis prof, Univ Va, 73-74; adj prof chem, Bowdoin Col, 75- Mem: AAAS; Am Chem Soc; The Chem Soc. Res: Heterocyclic compounds; stereochemistry; synthetic medicinals. Mailing Add: Dept of Chem Parsons Hall Univ of NH Durham NH 03824

LYLE, WILLIAM MONTGOMERY, b Summerside, PEI, Oct 4, 13; m 56; c 3. OPTOMETRY. Educ: Col Optom Ont, dipl, 38, OD, 58; Ind Univ, Bloomington, MS, 63, PhD(physiol optics), 65. Prof Exp: Pvt pract optom, 38-60; res assoc physiol optics, Ind Univ, 60-62, lectr, 62-65; asst prof optom, Col Optom Ont, 65-67; chief path sect, 67-74, assoc prof, 67-70, PROF OPTOM, UNIV WATERLOO, 70-, DIR CLINS, 74- Concurrent Pos: Med Res Coun Can grant, Univ Waterloo, 70-71; pres, Asn Schs Optom Can, 71- Mem: AAAS; Am Acad Optom; Am Soc Human Genetics; Royal Soc Health. Res: Side effects of drugs; inheritance of astigmatism; intraracial differences in refraction. Mailing Add: Sch of Optom Univ of Waterloo Waterloo ON Can

LYLES, GEORGE ROBERT, b Flint, Mich, Oct 27, 31; m 51; c 3. ANALYTICAL CHEMISTRY. Educ: Ark State Col, BS, 51; La State Univ, MS, 58. Prof Exp: Anal chemist water pollution, Water Co, Ark, 51-52; anal chemist & mgr, 55-65, PRES, KEMTECH LABS, 65- Mem: Am Chem Soc. Res: Qualitative and quantitative micro analytical chemistry. Mailing Add: 944 W Lakeview Dr Baton Rouge LA 70810

LYLES, SANDERS TRUMAN, b Reeves, La, May 24, 07; m 46; c 4. BACTERIOLOGY. Educ: Rice Univ, BA, 30, MA, 31; Southwestern Baptist Sem, ThM, 36, ThD, 49; Univ Tex, PhD(bact, biochem), 55. Prof Exp: From instr to assoc prof, 46-59, res scientist, 52-54, PROF BIOL, TEX CHRISTIAN UNIV, 59- Mem: Am Soc Microbiol. Res: Epidemiology and antibiotic resistance of staphylococcus; biochemical studies of blood serum. Mailing Add: Dept of Biol Tex Christian Univ Ft Worth TX 76129

LYMAN, CHARLES PEIRSON, b Brookline, Mass, Sept 23, 12; m 41; c 5. BIOLOGY. Educ: Harvard Univ, AB, 36, MA, 39, PhD(biol), 42. Prof Exp: Asst physiol, 42, asst cur, Mus Comp Zool, 45-50, fel anat, Med Sch, 46-48, res assoc, 48-62, assoc cur, Mus Comp Zool, 50-57, res assoc, 58-68, asst prof anat, Med Sch, 62-67, ASSOC PROF ANAT, HARVARD MED SCH, 67-, CUR MAMMAL, MUS COMP ZOOL, 68- Mem: AAAS; Am Physiol Soc; Am Soc Mammal; Am Soc Zoologists; Am Acad Arts & Sci. Res: Hibernation and temperature regulation in mammals; physiological ecology. Mailing Add: Mus of Comp Zool Harvard Univ Cambridge MA 02138

LYMAN, DONALD JOSEPH, b Chicago, Ill, Nov 5, 26; m 48; c 2. POLYMER CHEMISTRY, BIOMATERIALS. Educ: Univ Nev, BS, 49; Univ Del, MS, 51, PhD(chem), 52. Prof Exp: Asst chem, Univ Del, 50-52; res chemist high polymers, E I du Pont de Nemours & Co, 52-61; sr polymer chemist, Stanford Res Int, 61-64, head biomed polymer res, 64-69; prof mat sci, 69-74, RES ASSOC PROF SURG, UNIV UTAH, 69-, PROF BIOENG, 74- Concurrent Pos: Lectr, Dept Mat Sci, Stanford Univ, 64-68; chmn, Gordon Conf Sci & Technol Biomat; mem coun thrombosis, Am Heart Asn; mem comt surv mat sci & eng, Nat Acad Sci; mem eval panel polymer div, Nat Bur Standards, 73-76. Honors & Awards: Am Soc Artificial Internal Organs Award, 69. Mem: AAAS; Am Chem Soc; Am Soc Artificial Internal Organs; NY Acad Sci; Soc Biomat. Res: Synthetic polymers and polymer intermediates; mechanisms of polymerization; structure-property relationships of polymers; biomedical polymers; implants for artificial organs and reconstruction surgery. Mailing Add: Dept of Bioeng 2086 Merrill Eng Bd Univ of Utah Salt Lake City UT 84112

LYMAN, ERNEST MCINTOSH, b Berlin, Ger, Sept 23, 10; m 34; c 5. PHYSICS. Educ: Pomona Col, BA, 31; Dartmouth Col, MA, 33; Univ Calif, PhD(physics), 38. Prof Exp: Assoc physics, Univ Ill, 38-40; mem staff, Radiation Lab, Mass Inst Technol, 40-45; assoc prof, 45-51, assoc head dept, 70-73, PROF PHYSICS, UNIV ILL, URBANA, 51- Mem: Fel Am Phys Soc; Am Asn Physics Teachers. Res: Nuclear physics; beta-ray spectra; cyclotron design; betatron design problems; electronics; high energy physics; electrical breakdown in high vacuum. Mailing Add: Dept of Physics Univ of Ill Urbana IL 61801

LYMAN, FRANK LEWIS, b Springfield, Ill, Nov 6, 21; m 47; c 6. TOXICOLOGY. Educ: Swarthmore Col, AB, 43; Hahnemann Med Col, MD, 46. Prof Exp: Intern, WJersey Hosp, Camden, NJ, 47; physician, coach & instr biol, William Penn Col, 47-48; pvt pract, Iowa, 48-55; staff pediatrician, US Naval Hosp, Beaufort, SC, 55-57; assoc med dir, Mead Johnson & Co, 57-60; assoc dir, Med Dept, Geigy Chem Corp, 60-61, asst to med dir, 61-63; dir indust med, Ciba-Geigy Corp, 63-76; CONSULT TOXICOL, 76- Concurrent Pos: Instr, Seton Hall Col, 60-62. Mem: AAAS; AMA; Soc Toxicol; Am Acad Clin Toxicol; Am Asn Poison Control Ctrs. Res: Dietary management of phenylketonuria; toxicology of fluorescent whitening agents. Mailing Add: 48A Long Beach Blvd North Beach NJ 08008

LYMAN, HARVARD, b San Francisco, Calif, Sept 25, 31. PLANT PHYSIOLOGY, MOLECULAR BIOLOGY. Educ: Univ Calif, Berkeley, BA, 53; Univ Wash, MS, 57; Brandeis Univ, PhD(biol), 60. Prof Exp: Asst biol, Univ Wash, 55-57; instr, Brooklyn Col, 60-62; vis scientist biochem, Brookhaven Nat Lab, 62-63; asst prof biol, Brooklyn Col, 63-65; asst scientist microbiol, Brookhaven Nat Lab, NY, 65-67, assoc scientist, Med Dept, 67-68; ASSOC PROF BIOL, STATE UNIV NY STONY BROOK, 68- Concurrent Pos: NIH res grant, 63-65; NSF travel grant, 64, grant, 70-72. Mem: AAAS; Am Soc Plant Physiologists; Soc Protozoologists; Am Soc Cell Biologists; Biophys Soc. Res: Biosynthesis and inheritance of cellular organelles; development, physiology and differentiation of algae and fleshy and unicellular fungi. Mailing Add: Dept of Biol State Univ of NY Stony Brook NY 11794

LYMAN, JOHN, b Berkeley, Calif, Oct 28, 15; m 46; c 2. CHEMICAL OCEANOGRAPHY. Educ: Univ Calif, BS, 36; Univ Calif, Los Angeles, MS, 51, PhD(oceanog), 58. Prof Exp: Chemist, Union Oil Co, Calif, 37; asst oceanog, Scripps Inst Oceanog, Univ Calif, 37-41; oceanogr, Hydrographic Off, Navy Dept, 46-51, dir div oceanog, 51-59; assoc prog dir oceanog, NSF, 59-63, prog dir, 63-64; oceanog coordr, Bur Commercial Fisheries, 64-66; consult oceanogr, 66-68; prof oceanog & marine sci coordr, Univ NC, Chapel Hill, 68-73; CONSULT OCEANOGR, 73- Concurrent Pos: Oceanog ed, Trans, Am Geophys Union, 47-58; mem atomic safety & licensing bd panel, Nuclear Regulatory Comn, 72- Mem: AAAS; fel Am Geophys Union. Res: Chemistry of sea water; environmental chemistry. Mailing Add: 404 Clayton Rd Chapel Hill NC 27514

LYMAN, JOHN TOMPKINS, b Berkeley, Calif, May 25, 32; m 59; c 3. BIOPHYSICS. Educ: Univ Calif, AB, 54 & 58, PhD(biophys), 65. Prof Exp: Res asst, 59-65, BIOPHYSICIST, LAWRENCE BERKELEY LAB, 65- Mem: Am Asn Physicists in Med. Res: Radiation physics; radiation therapy; radiobiology; heavy charged-particle radiation dosimetry; radiobiology and radiotherapy. Mailing Add: Lawrence Berkeley Lab Bldg 55 Univ of Calif Berkeley CA 94720

LYMAN, ONA RUFUS, b Jamaica, Vt, Nov 18, 30; m 54; c 3. PHYSICS. Educ: Univ Vt, BA, 52. Prof Exp: Jr engr, Sprague Elec Co, 52-54; PHYSICIST, TERMINAL BALLISTICS LAB, BALLISTICS RES LAB, ABERDEEN PROVING GROUND, 56- Res: Neutron shielding; combustion; interaction of laser beams with materials; blast and fragment protection for industrial workers. Mailing Add: 303 Carter St Aberdeen MD 21001

LYMAN, RICHARD LEE, b Gilroy, Calif, Apr 7, 27; m 51; c 3. NUTRITION, BIOCHEMISTRY. Educ: Univ Calif, Berkeley, BA, 49, PhD(nutrit), 57; Univ Wis, MS, 51. Prof Exp: Sr lab technician poultry husb, Univ Calif, Berkeley, 51-53, res asst, 54-57; chemist, Western Utilization Labs, Calif, 57-58; from asst prof to assoc prof, 58-67, PROF NUTRIT, UNIV CALIF, BERKELEY, 67-, BIOCHEMIST, EXP STA, 73- Mem: AAAS; Am Inst Nutrit; Soc Exp Biol & Med. Res: Endocrine and diet relationships to lipid and phospholipid metabolism; physiological effects of amino acid deficiency and imbalances with emphasis on gastrointestinal function. Mailing Add: Dept of Nutrit Sci Univ of Calif Berkeley CA 94720

LYMAN, WILLIAM CHESTER, JR, b Hilton, NY, Aug 28, 21; m 47; c 4. ORGANIC CHEMISTRY. Educ: Univ Mich, BS, 48. Prof Exp: Res chemist, Nat Aniline Div, Allied Chem Corp, 49-51; res chemist, Distillation Prod Industs Div, 51-60, res admin, 60-67, tech asst to vpres mkt, 67-69, COORDR ORG CHEM, EASTMAN KODAK CO, 69- Mem: Am Chem Soc. Res: Fats and oils; dye intermediates; vitamins; amino acids. Mailing Add: Eastman Kodak Co 343 State St Rochester NY 14650

LYMAN, WILLIAM RAY, b Stratton, Vt, May 30, 20; m 44; c 4. CHEMISTRY. Educ: Univ Vt, BS, 41; Mass Inst Technol, PhD(org chem), 47; Columbia Univ, AM, 47. Prof Exp: Asst chem, Columbia Univ, 41-44; jr chemist, Tenn Eastman Corp Div, Eastman Kodak Co, 44-46; res chemist, Resinous Prod & Chem Co, 47-48; res chem-chemist, 48-66, lab head, 66-73, PROJ LEADER, ROHM AND HAAS CO, 73- Mem: Am Chem Soc. Res: Pesticide residue analysis; fate of pesticides in plant and animal systems and in the environment. Mailing Add: Rohm & Haas Co Springhouse PA 19477

LYMN, RICHARD WESLEY, b Flushing, NY, July 26, 44; m 70; c 1. BIOPHYSICS. Educ: Johns Hopkins Univ, BA, 64; Univ Chicago, PhD(biophys), 70. Prof Exp: USPHS fel biophys, Univ Chicago, 70-71; Brit-Am fel, Am Heart Asn, Lab Molecular Biol, Cambridge, Eng, 71-74; SR STAFF FEL BIOPHYS, PHYS BIOL LAB, NAT INST ARTHRITIS METAB & DIGESTIVE DIS, NIH, 74- Mem: Biophys Soc. Res: Detailed molecular mechanism of muscle contraction, using a combination of x-ray diffraction techniques and physiological measurements to characterize steps in the development of tension. Mailing Add: Lab of Phys Biol Bldg 6 Rm 104 NIH Bethesda MD 20014

LYNCH, BENJAMIN LEO, b Omaha, Nebr, Dec 29, 23; m 56; c 6. ORAL SURGERY. Educ: Creighton Univ, BS, 45, DDS, 47, MA, 53; Am Bd Oral Surg, dipl. Prof Exp: From asst instr to assoc prof oral surg, 48-53, dir dept, 54-55 & 60-67, dean sch dent, 54-61, assoc prof oral surg, 54-55, PROF ORAL SURG, SCH DENT, CREIGHTON UNIV, 57-, COORDR GRAD & POSTGRAD STUDIES, 67- Concurrent Pos: Pres dent staff, Children's Mem Hosp, 52 & 59; pvt pract, 65-; mem,

Omaha-Douglas County Health Bd, 66-68, vpres, 67, pres, 68; mem bd dir, Nebr Blue Cross, 68-; bd mem, Nebr Dent Serv Corp, 71-, pres, 74-75; guest lectr, Walter Reed Army Inst Res, DC; consult, Vet Hosp & Strategic Air Command Hq, Omaha, Nebr & Jenny Edmundson Hosp, Council Bluffs, Iowa. Mem: Am Soc Oral Surg; Am Dent Asn; fel Am Col Dent. Res: Dental education. Mailing Add: 128 Swanson Prof Bldg 8601 W Dodge Rd Omaha NE 68114

LYNCH, BRIAN MAURICE, b Melbourne, Australia, Jan 20, 30; m 56; c 2. PHYSICAL ORGANIC CHEMISTRY. Educ: Univ Melbourne, BSc, 52, PhD(chem), 56. Prof Exp: Fel & vis prof cancer chemother, NMex Highlands Univ, 56-57; asst prof org chem, St Francis Xavier Univ, Can, 57-58; res officer chem, Div Coal Res, Commonwealth Sci & Indust Res Orgn, Australia, 58-59; asst prof phys chem, Mem Univ Nfld, 59-62; assoc prof, 62-68, PROF ORG CHEM, ST FRANCIS XAVIER UNIV, 68-, CHMN DEPT, 72- Concurrent Pos: Nat Res Coun Can sr res fel, Australian Nat Univ, 68-69. Mem: Am Chem Soc; fel Chem Inst Can; fel The Chem Soc; fel Royal Inst Chem. Res: Physical organic chemistry of nitrogen-heterocyclic compounds; substitutions, syntheses, ring-fission reactions of heterocyclic compounds; proton magnetic resonance spectra of aromatic compounds. Mailing Add: Dept of Chem St Francis Xavier Univ Antigonish NS Can

LYNCH, CAROL BECKER, b New York, NY, Dec 3, 42; m 67. BEHAVIORAL GENETICS. Educ: Mt Holyoke Col, AB, 64; Univ Mich, MA, 65; Univ Iowa, PhD(zool), 71. Prof Exp: NSF fel, Inst Behav Genetics, Univ Colo, 72-73; ASST PROF BIOL, WESLEYAN UNIV, 73- Mem: Beahv Genetics Asn; Genetics Soc Am; Animal Behav Soc; Soc Study Evolution. Res: Study of genetic and environmental influences on behavioral thermoregulation and its physiological correlates in mice; comparisons of behavioral heterosis in laboratory and wild populations of house mice. Mailing Add: Dept of Biol Wesleyan Univ Middletown CT 06457

LYNCH, CHARLES ANDREW, organic chemistry, see 12th edition

LYNCH, CHARLES THEODORE, inorganic chemistry, analytical chemistry, see 12th edition

LYNCH, CORNELIUS JAMES, b Lowell, Mass, Oct 19, 31; m 63; c 2. BIOSTATISTICS, OPERATIONS RESEARCH. Educ: Cath Univ Am, AB, 57; Georgetown Univ, MA, 58; Am Univ, PhD(statist), 72. Prof Exp: Mathematician, Naval Res & Develop Ctr, 57-64; sr analyst opers res, Res Anal Corp, 65-73; chief biostatistician, Litton-Bionetics, Inc, 73-74; CHIEF BIOSTATISTICIAN, ENVIRO CONTROL, INC, 74- Mem: Am Statist Asn; Biomet Soc. Res: Development of a less hazardous cigarette; national and international epidemiology of tobacco-related diseases; cost-effective approaches toward reduction in risk from maritime transport of hazardous cargoes. Mailing Add: Enviro Control Inc 1530 E Jefferson St Rockville MD 20852

LYNCH, DAN K, b San Francisco, Calif, Aug 6, 20; m 53; c 1. INDUSTRIAL CHEMISTRY. Educ: Principia Col, BS, 42; Stanford Univ, MA, 44. Prof Exp: Instr, Principia Col, 44-48; anal chemist, Monsanto Co, Mo, 48-53, res chemist, Org Chem Div, 53-63, res specialist, 63-71; GROUP LEADER PLANT PROCESS TECHNOL, WM G KRUMMRICH PLANT, MONSANTO INDUST CHEM CO, 71- Mem: Am Chem Soc; fel Am Inst Chemists. Res: Plant process improvement and maintenance research. Mailing Add: Wm G Krummrich Plant Monsanto Indust Chem Co Sauget IL 62201

LYNCH, DANIEL MATTHEW, b Detroit, Mich, June 28, 21. PLANT ECOLOGY. Educ: Univ Detroit, AB, 43; Mich State Univ, MS, 48; Wash State Univ, PhD(bot), 52. Prof Exp: Asst bot, Mich State Univ, 47-48 & Wash State Univ, 48-52; from instr to assoc prof, 54-65, PROF BIOL, ST EDWARD'S UNIV, 65- Mem: AAAS; Ecol Soc Am; Bot Soc Am. Res: Ecology of the southwestern grasslands and woodlands. Mailing Add: Div of Phys & Biol Sci St Edward's Univ Austin TX 78704

LYNCH, DARREL LUVENE, b Dewey, Okla, Feb 6, 21; m 49; c 4. ORGANIC CHEMISTRY, SOIL MICROBIOLOGY. Educ: Univ Ill, PhD(agron), 53; Univ Del, MS, 57. Prof Exp: Instr & asst soil biol, Univ Ill, 48-52; asst prof agron, Univ Del, 52-58; asst prof soil sci, Univ Alta, 58-60; assoc prof chem, Ga Southern Col, 60-62; assoc prof, 62-66, PROF BIOL SCI, NORTHERN ILL UNIV, 66- Mem: Am Soc Microbiol. Res: Nitrogen fixation of Rhizobia and nodulation; soil organic matter; soil polysaccharides; morphology and nutrition studies with algae; ultrastructure studies with the Actinoplanaceae; pigment production in bacteria. Mailing Add: 306 Dresser Rd DeKalb IL 60115

LYNCH, DAVID DEXTER, theoretical physics, see 12th edition

LYNCH, DAVID WILLIAM, b Rochester, NY, July 14, 32; m 54; c 3. SOLID STATE PHYSICS. Educ: Rensselaer Polytech Inst, BS, 54; Univ Ill, MS, 55, PhD(physics), 58. Prof Exp: Fulbright fel, Pavia, Italy, 58-59; from asst prof to assoc prof, 59-66, PROF PHYSICS, IOWA STATE UNIV, 66- Concurrent Pos: Sr physicist, Ames Lab, US ERDA, 66- Mem: AAAS; fel Am Phys Soc; Optical Soc Am. Res: Optical properties of solids, including use of synchrotron radiation and modulation-spectroscopy. Mailing Add: Dept of Physics Iowa State Univ Ames IA 50011

LYNCH, DON MURL, b Delano, Calif, Feb 19, 34. ORGANIC CHEMISTRY. Educ: Fresno State Col, AB, 60; Univ Calif, Berkeley, PhD(org chem), 64. Prof Exp: SR RES CHEMIST, ABBOTT LABS, 64- Mem: Am Chem Soc. Res: Synthesis of potential pharmaceuticals and agricultural chemicals; isolation, structure and synthesis of natural products. Mailing Add: Abbott Labs North Chicago IL 60064

LYNCH, DONALD FRANCIS, b Seattle, Wash, June 1, 31. GEOGRAPHY OF ALASKA & NORTHERN LANDS, RESOURCE GEOGRAPHY. Educ: Yale Univ, BA, 52, PhD(Russian studies & geog), 65. Prof Exp: Asst prof geog, Dartmouth Col, 61-63; analyst, Res Anal Corp, Va, 63-66; group engr, Ryan Aeronaut Co, Calif, 66-70; assoc prof, 70-75, PROF GEOG & HEAD DEPT, UNIV ALASKA, FAIRBANKS, 75- Concurrent Pos: Res assoc, Mineral Indust Res Lab, Univ Alaska, 74- Mem: Asn Am Geog. Res: Geography of Arctic coal; urban geography; historical geography of Russian America. Mailing Add: Box 81676 College AK 99701

LYNCH, DONALD WALTON, forestry, see 12th edition

LYNCH, EDWARD CONOVER, b Fayette, Mo, Feb 24, 33; m 55; c 4. INTERNAL MEDICINE, HEMATOLOGY. Educ: Wash Univ, BA, 53, MD, 56. Prof Exp: From intern to asst resident med, Barnes Hosp, St Louis, Mo, 56-58; from assoc resident to chief resident, Strong Mem Hosp, Rochester, NY, 58-60; from instr to assoc prof, 62-72, assoc dean, Med Sch, 71-74, PROF MED, BAYLOR COL MED, 72-, DEAN STUDENT AFFAIRS, 74- Concurrent Pos: Adj assoc prof biomed eng, Rice Univ, 71-73, adj prof, 73- Mem: Am Col Physicians; Am Fedn Clin Res; Southern Soc Clin Invest; Am Soc Hemat; Am Soc Artificial Internal Organs. Res: Effects of physical forces on erythrocytes and blood rheology; internal distribution of iron in various anemias. Mailing Add: Baylor Col of Med Tex Med Ctr Houston TX 77025

LYNCH, EUGENE JOSEPH MICHAEL, b New York, NY, July 1, 27; m 50; c 3. PHYSICS. Educ: Cornell Univ, BA, 50; Duke Univ, PhD(physics), 59. Prof Exp: Asst physics, Duke Univ, 53-54, instr, 58-59; res physicist, Tonawanda Labs, Linde Div, Union Carbide Corp, 59-69; proj leader, New Ventures Div, 69-73, SR SCIENTIST, XEROX CORP, 73- Mem: Am Phys Soc; Asn Comput Mach. Res: Theoretical physics; cryogenics; liquid helium; superconductivity; heat transfer; solid state physics; electronics; statistics; inertial guidance; color science; computer graphics; computer interfacing; systems design. Mailing Add: W147 Xerox Corp Xerox Sq Rochester NY 14644

LYNCH, EVA MARIA, b Philadelphia, Pa, Mar 13, 16. MEDICINE. Educ: Chestnut Hill Col, AB, 44; Inst Divi Thomae, MS, 46, PhD(exp med), 57. Prof Exp: Assoc prof & res prof biol, 46-71, PROF BIOL & RES & CHMN DEPT BIOL, CHESTNUT HILL COL, 71- Concurrent Pos: Cert-radiol health, US Dept Health, Educ & Welfare, 56-57; fel, Purdue Univ, 63 & Am Univ, 64; consult educ progs off, Goddard Space Flight Ctr, NASA. Mem: AAAS; Am Soc Microbiol; NY Acad Sci; Am Inst Biol Sci. Res: Virology; philosophy of science. Mailing Add: Dept of Biol Chestnut Hill Col Philadelphia PA 19118

LYNCH, FRANCIS WATSON, b Winona, Minn, June 21, 06; m 31; c 3. DERMATOLOGY. Educ: Univ Minn, BS, 28, MB, 29, MD, 30, MS, 33; Am Bd Dermat, dipl, 36. Prof Exp: Mem fac, 33-57, head dept dermat, 57-71, prof, 57-74, EMER PROF DERMAT, MED SCH, UNIV MINN, MINNEAPOLIS, 74- Concurrent Pos: Pvt pract, 33-; chief dermat, Ancker City & County Hosps, 55-57, chief staff, 57, hon consult staff, 58-; chief staff, St Joseph's Hosp, 59; vpres, Am Bd Dermat, 57, pres, 58, adv, 60- Honors & Awards: Finnerud Award, Dermat Found, 72. Mem: Soc Invest Dermat (vpres, 52); Am Cancer Soc; Am Dermat Asn (pres, 64); hon mem Am Acad Dermat (vpres, 50, pres, 60); fel AMA. Mailing Add: Ste 206 Kellogg Sq St Paul MN 55101

LYNCH, G PAUL, b Leominster, Mass, Sept 17, 25; m 62; c 2. PHYSIOLOGY. Educ: Univ Mass, BS, 50; Univ Idaho, MS, 56; Cornell Univ, PhD(animal physiol, biochem), 62. Prof Exp: Asst prof animal sci, Univ WVa, 56-59; res asst, Cornell Univ, 59-62; ANIMAL PHYSIOLOGIST, AGR RES SERV, USDA, 62- Mem: Fel AAAS; Am Dairy Sci Asn; Am Soc Animal Sci; Sigma Xi. Res: Methods of determining the body composition of live animals; Pb and Cl toxicity in livestock; nitrogen and protein metabolism in young ruminants. Mailing Add: Ruminant Nutrition Lab Bldg 200 Agr Res Serv US Dept of Agr Beltsville MD 20705

LYNCH, GERALD, physics, see 12th edition

LYNCH, GERALD JOHN, b New York, NY, Nov 10, 34; m 55; c 2. NUTRITION, FOOD TECHNOLOGY. Educ: NY Univ, BA, 57; Fairleigh Dickinson Univ, MBA, 67. Prof Exp: Chemist, Gen Foods Corp, NJ, 57-61; chemist, Merck & Co, Inc, 62-64; sr chemist, Nat Starch & Chem Corp, 64-65; proj leader basic res, Res Ctr, Fair Lawn, 66, group leader cracker prod develop, 66-71, head dept nutrit prod develop, 71-73, MGR FOOD SERV DIV, NABISCO, INC, 73- Mem: Inst Food Technol; Am Asn Cereal Chem. Res: Bakery; snacks; coffee; institutional products, restaurants, vending, schools; meat products; product development from basic research to plant production and monitoring; technical service for developed products. Mailing Add: 262 Clinton Ave North Plainfield NJ 07063

LYNCH, HARRY JAMES, b Glenfield, Pa, Jan 18, 29; m 63. NEUROENDOCRINOLOGY. Educ: Geneva Col, BS, 57; Univ Pittsburgh, PhD(biol), 71. Prof Exp: Clin chemist, Western Pa Hosp, Pittsburgh, 55-66; sr tech fel res asst, Univ Pittsburgh, 66-71; NIH fel, 71-73, res assoc, Lab Richard Wurtman, 74-75, LECTR, LAB NEUROENDOCRINE REGULATION, DEPT NUTRIT & FOOD SCI, MASS INST TECHNOL, 75- Mem: Endocrine Soc; Am Asn Clin Chemists; Am Inst Biol Sci; AAAS; Am Soc Zoologists. Res: Neuroendocrine regulation exemplified by the pineal gland of vertebrate animals; pineal gland function as evidenced by melatonin biosynthesis and excretion; physiological, pharmacological, and environmental factors that influence pineal function. Mailing Add: 37-307 Lab Neuroendocrine Reg Dept of Nutrit & Food Sci MIT Cambridge MA 02139

LYNCH, HARVEY LEE, b Minden, Nebr, July 17, 39; m 69. EXPERIMENTAL HIGH ENERGY PHYSICS. Educ: Mass Inst Technol, SB, 61; Stanford Univ, PhD(physics), 66. Prof Exp: Vis scientist, Europ Orgn Nuclear Res, 66-68; res assoc, 68-73, ASST PROF, STANFORD LINEAR ACCELERATOR CTR, 73- Res: Study of electron positron annihilation at high energy. Mailing Add: Stanford Linear Accelerator Ctr Stanford CA 94305 19

LYNCH, HENRY T, b Lawrence, Mass, Jan 4, 28; m 51; c 3. MEDICAL GENETICS, INTERNAL MEDICINE. Educ: Univ Okla, BS, 51; Univ Denver, MA, 52; Univ Tex, MD, 60. Prof Exp: Intern, St Mary's Hosp, Evansville, Ind, 60-61; resident internal med, Col Med, Univ Nebr, 61-64; asst prof biol & asst internist, M D Anderson Hosp & Tumor Inst, Tex, 66-67; assoc prof, 67-71, PROF PREV MED & PUB HEALTH, SCH MED, CREIGHTON UNIV, 71-, CHMN DEPT, 67- Concurrent Pos: USPHS sr clin cancer trainee, Eppley Cancer Inst, Nebr, 64-66. Mem: Am Soc Human Genetics; Am Soc Clin Oncol; Am Asn Cancer Res. Res: Cancer genetics. Mailing Add: Dept of Prev Med & Pub Health Creighton Univ Sch of Med Omaha NE 68178

LYNCH, JAMES CARLYLE, b Clifton Hill, Mo, Mar 1, 42; m 65; c 1. NEUROPHYSIOLOGY. Educ: Univ Mo, AB, 64; Stanford Univ, MA & PhD(neurol sci), 71. Prof Exp: INSTR PHYSIOL, SCH MED, JOHNS HOPKINS UNIV, 74- Concurrent Pos: Nat Inst Neurol Dis & Stroke neurophysiol training grant, Dept Physiol, Sch Med, Johns Hopkins Univ, 71-73. Mem: Soc Neurosci. Res: Central neural mechanisms of sensation, perception and motor control. Mailing Add: Dept of Physiol Johns Hopkins Univ Sch of Med Baltimore MD 21205

LYNCH, JOHN BROWN, b Akron, Ohio, Feb 5, 29; m 50; c 2. PLASTIC SURGERY. Educ: Vanderbilt Univ, BS, 49; Univ Tenn, MD, 52; Am Bd Surg & Am Bd Plastic Surg, dipl. Prof Exp: Internship, John Gaston Hosp, Tenn, 53-54; resident surg, Univ Tex Med Br, Galveston, 56-59, res plastic surg, 59-62, from instr to assoc prof, 62-73; PROF PLASTIC SURG, VANDERBILT UNIV SCH MED, 73-, CHMN DEPT, 73- Concurrent Pos: Nat consult plastic surg to Surgeon Gen, USAF, 74-; mem, FDA Adv Panel, HEW, Gen Surg & Plastic Surg Devices, 74- Mem: AMA; Am Soc Plastic & Reconstructive Surgeons; Am Asn Plastic Surg; fel Am Col Surg; Plastic Surg Res Coun. Res: Pathophysiological aspects of burns and laboratory projects related to congenital anomalies. Mailing Add: Vanderbilt Univ Hosp Rm S-2221 Nashville TN 37232

LYNCH, JOHN DOUGLAS, b Collins, Iowa, July 30, 42; m 64; c 1. ZOOLOGY, HERPETOLOGY. Educ: Univ Ill, Urbana, BA, 64, MS, 65; Univ Kans, PhD(zool), 69. Prof Exp: Asst prof zool, 69-73, ASSOC PROF ZOOL, UNIV NEBR, LINCOLN, 73- Mem: Am Soc Ichthyol & Herpet; Soc Systs Zool; Soc Study Evolution; Soc Study Amphibians & Reptiles. Res: Systematics and zoogeography of leptodactyloid frogs; ecology of anuran larvae; evolution in simple ecosystems. Mailing Add: Sch of Life Sci Univ of Nebr Lincoln NE 68508

LYNCH, JOHN EDWARD, b Taunton, Mass, Feb 3, 23; m 46; c 2. BACTERIOLOGY, PARASITOLOGY. Educ: Providence Col, BS, 49; Mich State Univ, MS, 50, PhD(bact), 52; Am Bd Med Microbiol, dipl. Prof Exp: Res bacteriologist, Chas Pfizer & Co, Inc, 52-56, head parasitol lab, 57-60; res microbiologist, Hoffmann-La Roche, Inc, 60-61; res virologist, Chas Pfizer & Co, Inc, 61-63, mgr dept bact & parasitol, 63-70, ASST DIR DEPT PHARMACOL, PFIZER, INC, 71- Mem: Am Soc Microbiol; Soc Exp Biol & Med; fel Am Acad Microbiol; fel Royal Soc Trop Med & Hyg. Res: Chemotherapy of infectious diseases. Mailing Add: Dept of Pharmacol Pfizer Inc Groton CT 06340

LYNCH, KENNETH MERRILL, pathology, deceased

LYNCH, LEO, mathematical statistics, see 12th edition

LYNCH, MAURICE ALEXANDER, JR, inorganic chemistry, see 12th edition

LYNCH, MAURICE PATRICK, b Boston, Mass, Feb 24, 36; m 65; c 2. BIOLOGICAL OCEANOGRAPHY. Educ: Harvard Col, AB, 57; Col William & Mary, MA, 65, PhD(marine sci), 72. Prof Exp: Assoc marine scientist, 71-73, sr marine scientist & head dept spec progs, 73-75, ASST DIR & HEAD DIV BIOL OCEANOG, VA INST MARINE SCI, 75-; ASSOC PROF MARINE SCI, COL WILLIAM & MARY & UNIV VA, 76- Concurrent Pos: Asst prof marine sci, Col William & Mary & Univ Va, 71-75; adj prof earth sci, Va State Col, 74- Mem: Am Inst Biol Sci; Marine Technol Soc; Am Soc Zoologists; Am Fisheries Soc; Am Soc Limnol & Oceanog. Res: Management of marine and estuarine resources with special emphasis on management-research interactions and communications; physiology of marine and estuarine organisms with special emphasis on development of physiological conditions indices. Mailing Add: Va Inst of Marine Sci Gloucester Point VA 23062

LYNCH, PETER JOHN, b Minneapolis, Minn, Oct 22, 36; m 64; c 2. DERMATOLOGY. Educ: Univ Minn, Minneapolis, BS, 59, MD, 61. Prof Exp: Clin instr dermat, Univ Minn, Minneapolis, 65-66; from asst prof to assoc prof, Univ Mich, Ann Arbor, 70-73; assoc prof, 73-75, PROF DERMAT & CHIEF DIV, UNIV ARIZ, 75- Concurrent Pos: Consult, Wayne County Gen Hosp, Eloise, Mich, 68-73, Vet Admin Hosp, Ann Arbor, Mich, 71-73, Vet Admin Hosp, Tucson, Ariz & Pima County Gen Hosp, Tucson, Ariz, 74- Mem: AAAS; Am Acad Dermat; Soc Invest Dermat; Am Dermat Asn; Asn Am Med Cols. Res: Clinical subjects in diseases of the skin. Mailing Add: Div of Dermat Univ of Ariz Med Ctr Tucson AZ 85724

LYNCH, PETER ROBIN, b Philadelphia, Pa, July 18, 27; m 53; c 3. PHYSIOLOGY. Educ: Univ Miami, BS, 50; Temple Univ, MS, 54, PhD, 58. Prof Exp: From instr to assoc prof physiol, 58-70, PROF PHYSIOL & RADIOL, TEMPLE UNIV, 70- Mem: Am Physiol Soc. Res: Cardiovascular and radiologic physiology; rheology. Mailing Add: Dept of Physiol Temple Univ Med Sch Philadelphia PA 19140

LYNCH, RICHARD WALLACE, b Ft Leavenworth, Kans, June 17, 39; m 62; c 3. CHEMICAL PHYSICS, CHEMICAL ENGINEERING. Educ: Univ Calif, Berkeley, BS, 62; Univ Ill, MS, 64, PhD(chem eng), 66. Prof Exp: Tech staff mem chem physics, 66 & 68-71, supvr appl mat sci div, 71-73, SUPVR CHEM TECHNOL DIV, SANDIA LABS, 73- Mem: AAAS; Am Inst Chem Engrs; Am Phys Soc. Res: Nuclear waste solidification; process development. Mailing Add: Orgn 5824 Sandia Labs Albuquerque NM 87115

LYNCH, ROBERT EMMETT, b Chicago, Ill, Feb 5, 32; m 55; c 3. APPLIED MATHEMATICS. Educ: Cornell Univ, BEngPhys, 54; Harvard Univ, MA, 59, PhD(appl math), 63. Prof Exp: Sr res mathematician, Res Labs, Gen Motors Corp, 61-64; asst prof math & res mathematician, Univ Tex, Austin, 64-66, assoc prof, 66-67; ASSOC PROF MATH, PURDUE UNIV, 67- Mem: Am Math Soc; Math Asn Am; Soc Indust & Appl Math. Res: Numerical analysis, particularly numerical solution of partial differential equations and applied mathematics. Mailing Add: Div of Math Sci Purdue Univ West Lafayette IN 47906

LYNCH, THOMAS ALOYSIUS, pharmacology, see 12th edition

LYNCH, THOMAS FRANCIS, b Minneapolis, Minn, Feb 25, 38; m 61; c 3. ARCHAEOLOGY, ANTHROPOLOGY. Educ: Cornell Univ, BA, 60; Univ Chicago, MA, 62, PhD(anthrop), 67. Prof Exp: Archaeologist, Idaho State Univ Mus, 63-64; instr anthrop, 64-65, asst prof, 65-69, assoc prof, 69-74, PROF ANTHROP, CORNELL UNIV, 74-, CHMN ARCHAEOL CONCENTRATION, 71-, CHMN DEPT ANTHROP, 74- Mem: AAAS; Am Archaeol; Prehist Soc; Inst Andean Studies; Am Quaternary Asn. Res: Archaeology of Stone Age man, especially in Andean South America. Mailing Add: Dept of Anthrop Col Arts & Sci Cornell Univ Ithaca NY 14853

LYNCH, THOMAS JOHN, b Quincy, Mass, Mar 3, 41; m 69. POLYMER CHEMISTRY. Educ: Boston Col, BS, 62; Mass Inst Tech, PhD(org chem), 66. Prof Exp: Res chemist, 66-71, SR RES CHEMIST, GULF RES & DEVELOP CO, 71- Mem: AAAS; Am Chem Soc; The Chem Soc. Res: Properties and synthesis of specialty polymers for lubricants, flow modifiers and surface active agents; synthesis and degradation of functional fluids; polymer process development; polymer catalyst research and development. Mailing Add: Gulf Res & Develop Corp Chemical Row Orange TX 77630

LYNCH, VINCENT DE PAUL, b Niagara Falls, NY, May 27, 27; m 54; c 4. PHARMACOLOGY. Educ: Niagara Univ, BS, 50; St John's Univ, NY, BS, 54; Univ Conn, MS, 56, PhD(pharmacol), 59. Prof Exp: Asst pharmacol, Univ Conn, 56-58; from asst prof to assoc prof, 58-66, chmn dept pharmacog, pharmacol & allied sci, 61-73, PROF PHARMACOL, ST JOHN'S UNIV, NY, 66-, DIR DIV TOXICOL, 69-, CHMN DEPT PHARMACEUT SCI, 73- Concurrent Pos: Res fel pharmacol, Univ Conn, 54-56; consult, NY State Off Drug Abuse Serv & NY State Assembly Ment Health Comt Drug Abuse Adv Coun. Mem: Int Soc Psychoneuroendocrinol; Int Narcotic Enforcement Off Asn; Sigma Xi. Res: Neuropharmacology; toxicology; drug abuse. Mailing Add: Dept of Pharmaceut Sci St John's Univ Jamaica NY 11439

LYNCH, WESLEY CLYDE, b Vancouver, Wash, Feb 28, 44; m 65. NEUROPSYCHOLOGY. Educ: Univ Hawaii, BA, 67; Hollins Col, MA, 68; Univ NMex, PhD(exp psychol), 72. Prof Exp: Fel physiol psychol, Rockefeller Univ, 71-73, asst prof, 73-75; VIS ASST FEL PHYSIOL PSYCHOL, JOHN B PIERCE FOUND LAB, 75- Concurrent Pos: Adj asst prof, Rockefeller Univ, 75-; res assoc psychol, Yale Univ, 75- Mem: Am Psychol Asn; AAAS; Sigma Xi. Res: Psychological and physiological bases of motivation, reward and learning. Mailing Add: John B Pierce Found Lab 290 Congress Ave New Haven CT 06519

LYNCH, WILLIAM C, b Cleveland, Ohio, Apr 27, 37; m 61; c 4. MATHEMATICS, COMPUTER SCIENCE. Educ: Case Univ, BS, 59; Univ Wis, MS, 60, PhD(math), 63. Prof Exp: Actg instr numerical anal, Univ Wis, 62-63, asst prof, 63; from asst prof to assoc prof, 63-74, PROF COMPUT ENG, CASE WESTERN RESERVE UNIV, 74- Concurrent Pos: Vis prof, Comput Lab, Univ Newcastle, 70-71. Mem: AAAS; Asn Comput Mach; Am Math Soc. Res: Mathematical linguistics; design, construction, measurement and modelling of operating systems. Mailing Add: Dept of Comput & Info Sci Case Western Reserve Univ Cleveland OH 44106

LYND, JULIAN QUENTIN, b Joplin, Mo, Feb 11, 22; m 43; c 2. SOIL SCIENCE. Educ: Univ Ark, BS, 43; Mich State Univ, MS, 47, PhD(soil sci), 48. Prof Exp: Asst prof soil sci, Mich State Univ, 48-51; assoc prof, 52-57, PROF AGRON, OKLA STATE UNIV, 57- Mem: Fel Am Soc Agron; Soil Sci Soc Am; Int Soc Soil Sci; Am Soc Microbiol; Mycol Soc Am. Res: Soil microbiology; induced antibiosis to carcinogenic mycotoxins and biopathway of biotoxin degradation. Mailing Add: Dept of Agron Okla State Univ Stillwater OK 74074

LYND, LANGTRY EMMETT, b Can, Feb 8, 19; nat US; m 42; c 2. CHEMICAL METALLURGY, MINERALOGY. Educ: Univ Man, BSc, 41; Rutgers Univ, MS, 55, PhD(geol), 57. Prof Exp: Mgr raw mat sect, Res & Develop Dept, Titanium Pigment Div, Sayreville, 48-72, SR RES SCIENTIST CENT RES LAB, NL INDUSTS, INC, HIGHTSTOWN, NJ, 72- Mem: Geol Soc Am; Am Inst Mining, Metall & Petrol Eng. Res: Preparation and evaluation of concentrates for titanium dioxide pigment processes; utilization of titaniferous magnetite; treatment of industrial plant wastes; titanium geology and mineralogy; petrography and mineragraphy. Mailing Add: NL Industs Inc PO Box 420 Hightstown NJ 08520

LYNDE, RICHARD ARTHUR, b Orange, NJ, Apr 12, 42; m 61; c 2. INORGANIC CHEMISTRY. Educ: Hamilton Col, AB, 64; Iowa State Univ, PhD(inorg chem), 70. Prof Exp: Asst prof, 70-73, CHMN DEPT CHEM, MONTCLAIR STATE COL, 73-, ASSOC PROF, 75- Mem: Am Chem Soc. Res: Elucidation of the stoichiometry, structure and bonding of compounds formed by the post-transition and transition metals in unusual oxidation states. Mailing Add: Dept of Chem Montclair State Col Upper Montclair NJ 07043

LYNDON, ROGER CONANT, b Calais, Maine, Dec 18, 17. MATHEMATICS. Educ: Harvard Univ, AB, 39, MA, 41, PhD(math), 46. Prof Exp: Sci liaison officer, US Off Naval Res, Eng, 46-48; from instr to asst prof math, Princeton Univ, 48-53; from asst prof to assoc prof, 53-58, PROF MATH, UNIV MICH, ANN ARBOR, 58- Concurrent Pos: Vis assoc prof, Univ Calif, 56-57; mathematician, Inst Defense Anal, Princeton Univ, 59-60; vis prof, Queen Mary Col, Univ London, 60-61 & 64-65. Mem: Am Math Soc; Math Asn Am; Asn Symbolic Logic; London Math Soc. Res: Abstract algebra; group theory; mathematical logic. Mailing Add: Dept of Math Angell Hall Univ of Mich Ann Arbor MI 48104

LYNDRUP, MARK LEROY, b Traverse City, Mich, Aug 29, 39; m 63; c 2. PHYSICAL BIOCHEMISTRY. Educ: Trinity Col, Conn, BS, 61; Northwestern Univ, PhD(chem), 66. Prof Exp: Res fel, Inst Phys Chem, Sweden, 66-68; res chemist, Western Utilization Res & Develop Div, Agr Res Serv, USDA, 68-69; asst prof chem, Trinity Col, 69-70; asst prof chem, Lebanon Valley Col, 70-73; ASST PROF CHEM, MONTCLAIR STATE COL, 73- Mem: Am Chem Soc. Res: Macromolecular chemistry; solution properties of water soluble polymers; characterization of egg white proteins; computers interfacing. Mailing Add: Dept of Chem Montclair State Col Montclair NJ 07043

LYNDS, BEVERLY T, b Shreveport, La, Aug 19, 29; m 54; c 1. ASTRONOMY. Educ: Centenary Col, BS, 49; Univ Calif, PhD(astron), 55. Prof Exp: Res assoc astron, Nat Radio Astron Observ, Green Bank, WVa, 60-62; asst prof astron, Univ Ariz & asst astronomer, Steward Observ, 62-65, assoc prof astron Univ Ariz & assoc astronomer, Steward Observ, 65-71; ASST DIR, KITT PEAK NAT OBSERV, 71-, ASTRONOMER, 74- Concurrent Pos: Consult, Astron Adv Panel, NSF, 75- & NSF Sci & Technol Policy Off, Adv Group Sci Progs, 75-; councilor, Am Astron Soc, 74-77. Mem: AAAS; Am Astron Soc; Int Astron Union. Res: Interstellar medium; galactic structure; composition of galaxies. Mailing Add: Kitt Peak Nat Observ PO Box 26732 Tucson AZ 85726

LYNDS, CLARENCE ROGER, b Kirkwood, Mo, July 28, 28; m 54; c 1. ASTRONOMY. Educ: Univ Calif, AB, 52, PhD(astron), 55. Prof Exp: Asst, Lick Observ, 52; astron, Univ Calif, 53-54, jr res astronr & assoc astron, 55-58; Nat Res Coun Can fel, Dom Astrophys Observ, Can, 58-59; asst astronr, Nat Radio Astron Observ, 59-61; from asst to assoc astronr, 61-68, ASTRONR, KITT PEAK NAT OBSERV, 68- Mem: Nat Acad Sci; Am Astron Soc; Royal Astron Soc; Int Astron Union. Res: Photometry & spectroscopy of quasi-stellar objects and galaxies; observational cosmology; optical interferometry. Mailing Add: Kitt Peak Nat Observ PO Box 26732 Tucson AZ 85726

LYNE, EVERETT, b Fall River, Mass, June 21, 06; m 31; c 1. PHYSIOLOGY. Educ: NY Univ, BS, 29, MS, 31, PhD(biol, physiol), 45. Prof Exp: Asst biol, Wash Sq Col, NY Univ, 28-30, instr sci, Sch Educ, 30-36; from instr to asst prof biol, Hofstra Col, 36-45, chmn dept biol, 36-41, dir summer sch, 37-41, admin asst dean, 37-39; from asst prof to prof, 45-75, EMER PROF BIOL, NY UNIV, 75- Mem: AAAS; Nat Asn Biol Teachers; Nat Asn Res Sci Teaching; Nat Sci Teachers Asn. Res: Cytology; developmental anatomy; fatigue and its effect on the growth picture during early development. Mailing Add: Dept of Sci Educ Rm 685 Ed Bldg Sch of Educ NY Univ New York NY 10003

LYNE, LEONARD MURRAY, SR, b Riverhurst, Sask, Aug 20, 19; m 46; c 2. PAPER CHEMISTRY. Educ: Queen's Univ Ont, BSc, 42, MSc, 46. Prof Exp: Chemist, Int Nickel Co, 42-43; res chemist, Dom Plywoods, Ltd, 45-46; res chemist, E B Eddy Co, 46-56, res mgr, 56-62; head printability, Pulp & Paper Res Inst, 62-65; ASST RES DIR PULP & PAPER, ONT PAPER CO, 65- Mem: Tech Asn Pulp & Paper Indust; Chem Inst Can; Can Pulp & Paper Asn. Res: Fundamental and applied research of pulp and paper. Mailing Add: Ontario Paper Co Thorold ON Can

LYNESS, WARREN IRL, organic chemistry, biochemistry, see 12th edition

LYNGE, WALTER CLARENCE, mathematics, see 12th edition

LYNK, EDGAR THOMAS, b Kansas City, Mo, Aug 26, 41. LASERS. Educ: Yale Univ, BS, 63, MS, 65, PhD(physics), 70. Prof Exp: Assoc prof physics, Southern Univ, 69-74; STAFF PHYSICIST, RES & DEVELOP CTR, GEN ELEC CO, 74- Mem: Am Phys Soc. Res: Atomic excitation cross sections; computerized tomography for medical imaging. Mailing Add: Res & Develop Ctr Gen Elec Co PO Box 8 Schenectady NY 12301

LYNN, EDWARD JOSEPH, b New York, NY, Apr 16, 38; c 1. PSYCHIATRY. Educ: Brooklyn Col, BS, 62; Albert Einstein Col Med, MD, 65; Mich State Univ, MA, 72; Am Bd Psychiat & Neurol, cert, 71. Prof Exp: Intern med, Univ Rochester, 65-66, resident psychiat, 66-69, instr, 68-69; from asst prof to assoc prof psychiat, Mich State Univ, 69-75; dir, St Lawrence Hosp Community Ment Health Ctr, 72-75; ASSOC

PROF PSYCHIAT, UNIV NEV, 75-; CHIEF MENTAL HEALTH SERV, VET ADMIN HOSP, RENO, NEV, 75- Concurrent Pos: Lectr continuing educ prog, Mich State Univ, 70-71. Mem: Am Group Psychother Asn; Am Psychiat Asn; Asn Am Med Cols; AMA. Res: Manic depressive psychosis and the use of lithium; amphetamine abuse; nitrous oxide as a psychedelic and therapeutic agent; pathological grief; suicide; selection of therapists. Mailing Add: Dept of Psychiat Univ of Nev Reno NV 89507

LYNN, HUGH BAILEY, b Verona, NJ, Aug 13, 14; m 40; c 3. SURGERY. Educ: Princeton Univ, AB, 36; Columbia Univ, MD, 40. Prof Exp: Assoc surg, Newark Babies Hosp, 52-53; assoc prof surg & chief sect pediat surg, Sch Med, Univ Louisville, 53-60; HEAD SECT PEDIAT SURG, MAYO CLIN, 61-, PROF SURG, MAYO GRAD SCH MED, UNIV MINN, 71- Concurrent Pos: Teaching fel, Harvard Univ, 51-52; surgeon-in-chief, Children's Hosp, Louisville, Ky, 53-60. Mem: Fel Am Col Surg; Am Acad Pediat. Mailing Add: Mayo Clin 200 First St SW Rochester MN 55901

LYNN, JOHN R, b Dallas, Tex, Mar 8, 30; m 54; c 5. OPHTHALMOLOGY. Educ: Rice Univ, BA, 51; Univ Tex, MD, 55. Prof Exp: Res assoc, Univ Iowa Hosps, 61-63; from asst prof to assoc prof, 63-70, PROF SURG, UNIV TEX HEALTH SCI CTR DALLAS, 70-, CHMN DEPT OPHTHAL, 63- Concurrent Pos: Nat Inst Neurol Dis & Blindness spec fel, Univ Iowa Hosps, 61-63 & Eye Clin, Univ Tübingen, 62-63. Mem: AMA; Am Acad Ophthal & Otolaryngol; Asn Res Vision & Ophthal. Res: Methods of clinical perimetry; acute visual function effects by raising the intraocular pressure; threshold, summation and visual acuity of accentric scotomatous areas during phototopic, mesopic and scotopic adaptations. Mailing Add: Dept of Opthal Univ of Tex Health Sci Ctr Dallas TX 75235

LYNN, JOHN WENDELL, b New York, NY, Mar 23, 25; m 46; c 3. ORGANIC CHEMISTRY. Educ: Yale Univ, BS, 48, PhD(chem), 51. Prof Exp: Res chemist & proj leader, Org Chem Res Dept, 51-55, group leader, 55-60, res assoc, 60-61, asst dir res & develop, 61-69, mgr new mkt develop, 69-70, dir technol, Fibers & Fabrics, 69-72, new venture mgr, Chem & Plastics, 72-73, ASSOC DIR RES & DEVELOP, CHEM & PLASTICS, UNION CARBIDE CORP, 73- Mem: Am Chem Soc; Electrochem Soc. Res: Nitrogenous substances; vinyl monomers; organic synthesis; synthetic fibers; vinyl fabrics; nonwovens; thermoplastic M & E resins; phenolic resins. Mailing Add: 209 Lynn Lane Westfield NJ 07090

LYNN, JOSEPH ALDEN, b Dallas, Tex, Dec 31, 39; m 61; c 3. PATHOLOGY, ELECTRON MICROSCOPY. Educ: Southern Methodist Univ, BS, 63; Univ Tex Med Br Galveston, MA, 63, MD, 64. Prof Exp: Intern path, Univ Chicago Hosps & Clins, 64-65; resident, Med Ctr, Baylor Univ, 65-66, assoc staff pathologist & dir electron micros sect, 67-69; surg pathologist, US Naval Hosp, San Diego, 69-71; staff pathologist & dir electron micros sect, Med Ctr, Baylor Univ, 71-73, assoc prof path, Dent Col, 71-73; CHIEF DEPT PATH, WESTGATE HOSP & MED CTR, 73- Concurrent Pos: Fel, Med Ctr, Baylor Univ, 66-67; Am Cancer Soc clin fels, 66-68, adv clin fel, 68-; assoc pathologist, Ford, Lynn & Assocs, 73- Mem: AAAS; AMA; Am Asn Anat; Electron Micros Soc Am; Col Am Path. Res: Ultrastructural pathology; human neoplasia, diagnostic criteria and viral etiology; experimental renal hypertension and toxemia of pregnancy; diabetes and atherosclerosis. Mailing Add: Dept of Path Westgate Hosp & Med Ctr Denton TX 76201

LYNN, JOSEPH THOMAS, physics, see 12th edition

LYNN, KELVIN GIDEON, b Rapid City, SDak, Feb 2, 48. SOLID STATE PHYSICS, METALLURGY. Educ: Univ Utah, BS, 70 & 71, PhD(mat sci), 74. Prof Exp: ASSOC PHYSICIST, BROOKHAVEN NAT LAB, 74- Concurrent Pos: Resident visitor, Tech Staff, Bell Labs, NJ, 75- Mem: Am Phys Soc; Am Inst Metall Engrs; Am Soc Metals. Res: Use of positron annihilatron in defect and defect-free solids. Mailing Add: Physics Dept Bldg 510 Brookhaven Nat Lab Upton NY 11973

LYNN, MELVYN STUART, b London, Eng, July 7, 37; m 60; c 3. MATHEMATICS. Educ: Oxford Univ, BA, 58, MA, 65; Univ Calif, Los Angeles, MA, 60, PhD(math), 62. Prof Exp: Res asst numerical anal, Univ Calif, Los Angeles, 59-60, res mathematician, 60-62; res mathematician, Calif Res Corp, 62; jr fel math, Nat Phys Labs, Eng, 62-63, sr sci off, 63-64; staff mem, IBM Sci Ctr, Calif, 64-65, mgr res dept, Tex, 65-68, mgr ctr, 68-71; PROF MATH SCI & DIR INST COMPUT SERV & APPLN, RICE UNIV, 71- Mem: Soc Indust & Appl Math; Asn Comput Mach. Res: Numerical solution of systems of linear and integral equations; biomathematics; computer science. Mailing Add: Inst for Comput Serv & Appln Rice Univ Houston TX 77001

LYNN, MERRILL, b New Columbia, Pa, Nov 20, 30; m 57. POLYMER CHEMISTRY. Educ: Bucknell Univ, BS, 56; Univ Fla, PhD(chem), 61. Prof Exp: Res chemist, Esso Res & Eng Co, 61-69; SR RES CHEMIST, CORNING GLASS WORKS, 70- Mem: AAAS; Am Chem Soc; Am Inst Chem; NY Acad Sci. Res: Bonding to glass surfaces; glass reinforced plastics; polymer modifications; coating resins; immobilized enzymes. Mailing Add: 2920 Olcott Rd Big Flats NY 14814

LYNN, RALPH BEVERLEY, b Penetanguishene, Ont, Aug 24, 21; m 44; c 4. SURGERY. Educ: Queen's Univ, Ont, MD, CM, 45; FRCS(E), 48; FRCS, 49; Royal Col Physicians & Surgeons Can, cert, 57; FRCS(C), 65. Prof Exp: Jr intern, Kingston Gen Hosp, 44-46; sr intern surg, Royal Victoria Hosp, Montreal, Que, 46-47; sr registr, Post-Grad Med Sch, Univ London, 47-48; clin tutor, Royal Infirmary, Edinburgh, Scotland, 48-49; asst lectr surg, Post-Grad Med Sch, Univ London, 49-50 & 52-54; sr registr, Southampton Chest Hosp, Eng, 54-55; from asst prof to assoc prof surg, Univ Sask, 55-58; assoc prof, 58-62, PROF SURG, SCH MED, QUEEN'S UNIV, ONT, 62-; HEAD CARDIOTHORACIC UNIT, KINGSTON GEN HOSP, 58- Concurrent Pos: Nat Res Coun Can scholar, Western Reserve Univ, 50-51; traveling fel, Post-Grad Med Fedn, Johns Hopkins Univ, 51-52; Markle scholar, Univ Sask, 55-57; surgeon, Cleveland City Hosp, Ohio, 50-51; consult, Hotel Dieu & Can Forces Hosp, 58- & Dept Vet Affairs, 58-; fel coun clin cardiol, Am Heart Asn, 65. Mem: Asn Thoracic Surg; fel Am Col Surg; fel Am Col Chest Physicians; NY Acad Sci; Can Thoracic Soc. Res: Thoracic, cardiovascular and peripheral vascular surgery. Mailing Add: Dept of Surg Queen's Univ Sch of Med Kingston ON Can

LYNN, RAYMOND J, b Bitner, Pa, Oct 23, 28; m 58; c 3. MEDICAL MICROBIOLOGY. Educ: Univ Pittsburgh, BS, 52, MS, 53; Univ Pa, PhD(med microbiol), 56. Prof Exp: Asst biol, Univ Pittsburgh, 52-53; res investr microbiol, Univ Pa, 53-56, res microbiologist, 56-57; res assoc microbiol, Sch Med, Univ Pittsburgh, 58-60, instr, 60-61; from asst prof to assoc prof, 61-70, PROF MICROBIOL, SCH MED, UNIV SDAK, VERMILLION, 70- Concurrent Pos: Secy-treas, SDak Bd Examr in Basic Sci, 71-; rep Dak Affil, regional rev comt, Am Heart Asn, 75-78. Mem: AAAS; Am Pub Health Asn; Am Soc Microbiol; Soc Exp Biol & Med; NY Acad Sci. Res: Immunology of the Mycoplasmataceae; role of L-forms in sequelae disease states; immunochemistry of streptococcal L-forms and relation of such antigens to rheumatic fever and acute glomerular nephritis. Mailing Add: Dept of Microbiol Univ of SDak Sch of Med Vermillion SD 57069

LYNN, ROBERT THOMAS, b Coleman, Tex, Jan 15, 31; m 54; c 2. ANIMAL BEHAVIOR, ECOLOGY. Educ: Fla State Univ, BA, 56, MA, 57; Univ Okla, PhD(zool), 63. Prof Exp: Instr biol, Austin Col, 57-59; asst prof, Emory & Henry Col, 63-64; assoc prof, Presby Col, SC, 64-67; ASSOC PROF BIOL SCI, SOUTHWESTERN OKLA STATE UNIV, 67- Mem: AAAS; Ecol Soc Am; Animal Behavior Soc; Am Ornith Union; Wilson Ornith Soc. Res: Ecology and behavior of birds and lizards. Mailing Add: 1208 N Indiana Weatherford OK 73096

LYNN, ROGER YEN SHEN, b Shanghai, China, Jan 18, 41. APPLIED MATHEMATICS. Educ: Cheng Kung Univ, Taiwan, BS, 61; Brown Univ, MS, 64; NY Univ, PhD(math), 68. Prof Exp: Lectr math, Univ Ind, Bloomington, 68-69, asst prof, 69-71; ASST PROF MATH, VILLANOVA UNIV, 71- Mem: Am Math Soc; Soc Indust & Appl Math; NY Acad Sci. Res: Asymptotic solutions of differential equations. Mailing Add: Dept of Math Villanova Univ Villanova PA 19085

LYNN, THOMAS NEIL, JR, b Ft Worth, Tex, Feb 14, 30; m 52; c 3. MEDICINE, PREVENTIVE MEDICINE. Educ: Univ Okla, BS, 51, MD, 55. Prof Exp: From intern to asst resident med, Barnes Hosp, St Louis, 55-57; clin assoc, Nat Heart Inst, Md, 57-59; chief res, 59-61, instr, 61-63, asst prof prev med, 61-64, asst prof med, 63-69, assoc prof prev med & pub health, 64-69, vchmn dept, 63-69, PROF FAMILY PRACT, COMMUNITY MED & DENT, & CHMN DEPT, MED CTR, UNIV OKLA, 69-, ACTG DEAN COL MED, 74- Mem: AAAS; Asn Teachers Prev Med; Am Acad Family Physicians; Am Fedn Clin Res; Asn Am Med Cols. Res: Epidemiology of coronary artery disease; psycho-social aspects of dependence and rehabilitation; ballistocardiography and electrocardiography. Mailing Add: Dept of Community Health Univ of Okla Med Ctr Oklahoma City OK 73104

LYNN, WARREN CLARK, b Satanta, Kans, Dec 4, 35; m 60; c 3. SOIL SCIENCE. Educ: Kans State Univ, BS, 57, MS, 58; Univ Calif, PhD(soil sci), 64. Prof Exp: SOIL SCIENTIST, SOIL SURV INVEST UNIT, USDA, 63- Mem: Soil Sci Soc Am; Clay Minerals Soc; Int Peat Soc. Res: Properties of cat clays or acid sulfate soils; clay minerals in relation to soil properties; organic soils. Mailing Add: Soil Surv Invest Unit US Dept of Agr Fed Bldg Lincoln NE 68508

LYNN, WILLIAM GARDNER, b Washington, DC, Dec 26, 05; m 33; c 2. ZOOLOGY. Educ: Johns Hopkins Univ, AB, 28, PhD(zool), 31. Prof Exp: From asst to instr to assoc zool, Johns Hopkins Univ, 28-42; from assoc prof to prof biol, 42-74, head dept, 58-63, EMER PROF BIOL, CATH UNIV AM, 74- Concurrent Pos: Fel Rockefeller Found, Yale Univ, 39-40; Fulbright scholar, Univ Col, WIndies, 52-53. Mem: Am Soc Naturalists. Res: Anatomy of reptiles; amphibian metamorphosis. Mailing Add: Dept of Biol Cath Univ of Am Washington DC 20017

LYNN, WILLIAM SANFORD, b Clarendon, Va, June 14, 22; m 49; c 4. MEDICINE. Educ: Ala Polytech Inst, BS, 43; Columbia Univ, MD, 46. Prof Exp: Instr biochem, Univ Pa, 52-55; from asst prof to assoc prof, 56-64, PROF MED & BIOCHEM, DUKE UNIV, 64- Concurrent Pos: Markle scholar, 55-59. Mem: Am Soc Biol Chem; Am Soc Clin Invest. Res: Biochemistry; oxidative phosphorylation and hormone action. Mailing Add: Dept of Biochem Duke Univ Durham NC 27706

LYNN, YEN-MOW, b Shanghai, China, Jan 17, 35; m 64; c 3. APPLIED MATHEMATICS. Educ: Nat Taiwan Univ, BS, 55; Calif Inst Technol, MS, 57, PhD, 61. Prof Exp: From asst res scientist to assoc res scientist, Courant Inst Math Sci, NY Univ, 60-64; assoc prof, Ill Inst Technol, 64-67; assoc prof, 67-74, PROF MATH, UNIV MD BALTIMORE COUNTY, 74- Concurrent Pos: Consult, Ames Res Ctr, NASA, 66; consult, Ballistic Res Lab, US Army, 69- Mem: AAAS; Am Math Soc; Soc Indust & Appl Math; Math Asn Am; Am Phys Soc. Res: Magneto-gas-dynamics; plasma physics; partial differential equations; rotating fluids; aerodynamics. Mailing Add: Dept of Math Univ of Md Baltimore County Baltimore MD 21228

LYNTON, ERNEST ALBERT, b Berlin, Ger, July 17, 26; nat US; m 53; c 2. ACADEMIC ADMINISTRATION, LOW TEMPERATURE PHYSICS. Educ: Carnegie Inst Technol, BS, 47, MS, 48; Yale Univ, PhD(physics), 51. Prof Exp: Asst, Off Naval Res, Yale Univ, 48-50; AEC fel, Univ Leiden, 51-52; from asst prof to prof physics, Rutgers Univ, 52-74, dean Livingston Col, 65-74; SR VPRES ACAD AFFAIRS, UNIV MASS, 74-, COMMONWEALTH PROF PHYSICS, 75- Concurrent Pos: Vis prof, Univ Grenoble, 59-60; mem, Comn Higher Educ, Mid States Asn, 70-75. Mem: Fel Am Phys Soc. Res: Low temperature physics helium 3 and helium 4 mixtures; superconductors; dilute metallic alloys; thermal conductivity. Mailing Add: Dept of Physics Univ of Mass 1 Washington Mall Boston MA 02108

LYNTS, GEORGE WILLARD, b Edgerton, Wis, July 26, 36; m 59; c 2. GEOLOGY, PALEONTOLOGY. Educ: Univ Wis, BS, 59, MS, 61, PhD(geol), 64. Prof Exp: USPHS fel, Columbia Univ, 64-65; asst prof, 65-71, ASSOC PROF GEOL, DUKE UNIV, 71- Concurrent Pos: Mem, Cushman Found Foraminifera Res. Mem: AAAS; Am Soc Limnol & Oceanog; Soc Econ Paleont & Mineral; Paleont Soc; Protozool Soc. Res: Biology and ecology of the Foraminifera; application of quantitative techniques to the solution of geological and biological problems; micropaleontology and paleoecology of the oceans. Mailing Add: Dept of Geol Duke Univ Box 6665 Col Sta Durham NC 27708

LYON, CAMERON KIRBY, b Islampur, India, July 23, 23; US citizen; m 48; c 3. ORGANIC CHEMISTRY. Educ: Col Wooster, BA, 47; Northwestern Univ, PhD(chem), 52. Prof Exp: Chemist, Jackson Lab, E I du Pont de Nemours & Co, 51-59; CHEMIST, WESTERN REGIONAL RES LAB, USDA, ALBANY, 59- Mem: Am Chem Soc; Am Oil Chem Soc. Res: Polymers; urethanes; fats and oils; oilseed proteins. Mailing Add: 5 North Lane Orinda CA

LYON, DAVID N, b Altoona, Kans, Apr 15, 19; m 42; c 2. PHYSICAL CHEMISTRY. Educ: Univ Mo, MA, 42; Univ Calif, PhD(chem), 48. Prof Exp: Res assoc, 48-51, asst res chemist, 51-53, assoc res chemist, 53-59, res chem engr, 59-65, lectr chem eng, 57-65, asst dean, Col Chem, 69-72, PROF CHEM ENG, COL CHEM, UNIV CALIF, BERKELEY, 65- Mem: AAAS; Am Chem Soc; Am Inst Chem Eng; Sigma Xi. Res: Chemical thermodynamics; cryogenic engineering. Mailing Add: Col of Chem Univ of Calif Berkeley CA 94720

LYON, DONALD WILKINSON, b Manchester, Eng, Aug 6, 16; nat US; m 42; c 3. INORGANIC CHEMISTRY. Educ: Ohio Wesleyan Univ, BA, 37; Ohio State Univ, PhD(inorg chem), 41. Prof Exp: Res chemist, 41-54, tech supvr, 54-62, ADMIN SUPVR, PIGMENTS DEPT, E I DU PONT DE NEMOURS & CO, INC, 62- Mem: Am Chem Soc. Res: Titanium dioxide. Mailing Add: 110 Banbury Dr Windsor Hills Wilmington DE 19803

LYON, DUANE EDGAR, b Muskegon, Mich, Mar 12, 39; m 61; c 2. FOREST PRODUCTS. Educ: Univ Mich, BS, 62, MS, 63; Univ Calif, Berkeley, PhD(forest prod), 75. Prof Exp: Asst technologist, Dept Wood Technol, Wash State Univ, 63-66; asst specialist, Forest Prod Lab, Univ Calif, 66-73; ASST PROF FOREST PROD, MISS FOREST PROD LAB, MISS STATE UNIV, 73- Mem: Forest Prod Res Soc; Soc Wood Sci & Technol. Res: Development and characterization of composite engineering materials made wholly or in part from wood. Mailing Add: Miss Forest Prod Lab PO Drawer FP Mississippi State MS 39762

LYON, EDWARD SPAFFORD, b Chicago, Ill, Feb 26, 26; m 51; c 11. GENITOURINARY SURGERY. Educ: Univ Chicago, PhB, 48, SB, 50, MD, 53. Prof Exp: Intern, Univ Hosps, 53-54, resident surg, 54-56, resident urol, 56-59, asst prof, Univ, 59-65, ASSOC PROF UROL, UNIV CHICAGO, 65- Res: Urolithiasis. Mailing Add: Dept of Urol Univ of Chicago Chicago IL 60637

LYON, GORDON FREDERICK, b London, Eng, May 10, 22; Can citizen; m 43; c 1. PHYSICS. Educ: Univ Sask, BA, 56, MA, 58, PhD(physics), 61. Prof Exp: Instr physics, Univ Sask, 56-62; from asst prof to assoc prof, 62-69, PROF PHYSICS, UNIV WESTERN ONT, 69- Concurrent Pos: Mem comn 6, Int Union Geod & Geophys-Int Asn Geomag & Aeronomy, 63-; mem subcomt aeronomy, Nat Res Coun Can, 66- Mem: Am Geophys Union; Am Asn Physics Teachers; Can Asn Physicists. Res: Radio physics of the upper atmosphere; scattering of radio waves by ionospheric inhomogeneities; ionospheric absorption; travelling ionospheric disturbances; ionospheric electron content utilizing beacon satellites; associated geophysical phenomena; aurora. Mailing Add: Dept of Physics Univ of Western Ont London ON Can

LYON, HARVEY WILLIAM, b Chicago, Ill, May 20, 20; m 46; c 3. DENTAL RESEARCH. Educ: Marquette Univ, BS, 42, DDS, 45; Georgetown Univ, MS, 51, PhD(anat), 56. Prof Exp: Head dept periodontia, Naval Dent Sch, Md, 51-52, consult, 55-59, res officer, Naval Med Res Inst, 52-53, 54-59, dent officer, 62, Head dent res dept, 62-64, head res br, Dent Div, Bur Med & Surg, 64-65, dent proj officer, Off Naval Res, 64-65; sr res assoc, 65-69, secy coun dent res & dir dent res info ctr, 68-70, head electron optics facil, 70, dir clin studies, 69-74, SECY COUN DENT RES, RES INST, AM DENT ASN, 74- Concurrent Pos: Guest scientist, Nat Bur Stand, 49-51; Navy rep dent study sect, NIH, 58-59; res collabr, Brookhaven Nat Lab, 59; consult, Naval Dent Res Inst, Ill, 67-; res assoc, Col Dent, Northwestern Univ, Ill, 69-75, mem exec comt, Div Med Sci, Nat Res Coun, 69-74. Mem: AAAS; fel Am Col Dent; Int Asn Dent Res (asst secy-treas, 69-70); Electron Probe Anal Soc Am; Int Dent Fedn. Res: Dental pathology, physiology and anatomy; dental materials; electron microprobe analysis; scanning electron microscopy. Mailing Add: Am Dent Asn Res Inst 211 E Chicago Ave Chicago IL 60611

LYON, IRVING, biochemistry, biophysics, see 12th edition

LYON, JOHN BLAKESLEE, JR, b Auburn, NY, Mar 17, 25; m 48; c 2. BIOCHEMISTRY. Educ: Hamilton Col, AB, 50; Brown Univ, ScM, 52, PhD(biol), 54. Prof Exp: Asst biol, Brown Univ, 50-52; Life Ins Med Res Fund fel biochem, 54-56, from instr to assoc prof, 56-70, PROF BIOCHEM, EMORY UNIV, 70- Concurrent Pos: Lederle Med Fac award, Emory Univ, 56-59, USPHS sr res fel, 59. Mem: Am Soc Biol Chem. Res: Regulatory mechanisms of metabolism; glycogen metabolism; vitamin B-six. Mailing Add: Dept of Biochem Emory Univ Atlanta GA 30322

LYON, LEONARD JACK, b Sterling, Colo, Oct 31, 29; m 56; c 2. WILDLIFE ECOLOGY, FOREST ECOLOGY. Educ: Colo State Univ, BS, 51, MS, 53; Univ Mich, PhD(wildlife mgt), 60. Prof Exp: Res biologist & proj leader pheasant habitat, Colo Game & Fish Dept, 55-62; WILDLIFE BIOLOGIST & PROJ LEADER FOREST WILDLIFE HABITAT, FORESTRY SCI LAB, INTERMT FOREST & RANGE EXP STA, US FOREST SERV, 62- Concurrent Pos: Res assoc, Mont State Univ, 65- Mem: AAAS; Wildlife Soc; Am Inst Biol Sci; Ecol Soc Am. Res: Forest seral ecology; wildlife habitat. Mailing Add: Forestry Sci Lab US Forest Serv Intermt Forest & Range Exp Sta Missoula MT 59801

LYON, LUTHER LAWRENCE, JR, b Greensburg, Kans, Oct 28, 18; m 42. PHYSICAL INORGANIC CHEMISTRY. Educ: Southwestern Col AB, 39; Univ Kans, MS, 41; Univ Southern Cal, PhD(chem), 44. Prof Exp: Asst lab instr, Univ Kans, 39-41; assoc prof chem, Univ Wichita, 46-58, prof, 58-60; alternate group leader, Los Alamos Sci Lab, 60-73; RETIRED. Concurrent Pos: Dir, Res Found, Univ Wichita, 51-54. Mem: Am Chem Soc. Res: Inorganic carbon chemistry; surface area and adsorption measurements; diffusion of gases. Mailing Add: 3007 Woodland Ave Los Alamos NM 87544

LYON, PATRICIA JEAN, b Seattle, Wash, Jan 9, 31; m 70. ANTHROPOLOGY. Educ: Univ Calif, Berkeley, BA, 52, PhD(anthrop), 67. Prof Exp: Res anthropologist, Am Mus Natural Hist Exped, Peruvian Montana, 60-61; prof sociol, Nat Univ Huamanga, Peru, 61; prof anthrop, Nat Univ Cuzco, 63-64; actg instr, Univ Calif, Berkeley, 66-67, actg asst prof, 67-68; asst prof, Wash Univ, 68-70; RES ASSOC, INST ANDEAN STUDIES, 70-, CO-ED NAWPA PACHA, 75- Concurrent Pos: Lectr, Ministry Pub Health, Cuzco, Peru, 63; Wenner-Gren Found Mus res fel, R H Lowie Mus Anthrop, Univ Calif, Berkeley, 70-71; asst ed, Nawpa Pacha, Inst of Andean Studies, 70-75. Mem: AAAS; Inst Andean Studies (vpres, 68, secy-treas, 75); Am Anthrop Asn; Soc Am Archaeol; Am Folklore Soc. Res: Ethnology of South America; South American folklore; African folklore; Andean culture history; Andean archaeology; culture change; comparative religion; ethnolinguistics; ethnomusicology. Mailing Add: Inst of Andean Studies PO Box 9307 Berkeley CA 94709

LYON, RICHARD HALE, b Marquette, Mich, Nov 15, 20; m 44; c 3. MICROBIOLOGY, BIOCHEMISTRY. Educ: Univ Minn, BA, 47, MS, 62, PhD(microbiol, biochem), 65. Prof Exp: City bacteriologist, Sioux City Dept Health, Iowa, 48-49; bacteriologist, Vet Admin Ctr, Sioux Falls, SDak, 49-54; RES MICROBIOLOGIST, BACT RES LAB, VET ADMIN HOSP, MINNEAPOLIS, 54- Mem: Am Soc Microbiol; fel Am Inst Chem; Am Thoracic Soc. Res: Microbial physiology, specifically as it pertains to metabolic differences in the mycobacteria and to the relationship of these differences to drug susceptibility, taxonomy and virulence. Mailing Add: Bact Res Lab Infectious Dis Sect Vet Admin Hosp 54th St & 48th Ave S Minneapolis MN 55417

LYON, RICHARD KENNETH, b Cleveland, Ohio, Dec 22, 33; m 68. PHYSICAL CHEMISTRY. Educ: Col William & Mary, BS, 55; Harvard Univ, PhD(phys chem), 60. Prof Exp: Chemist, 60-64, sr chemist, Cent Basic Res Lab, 64-67, res assoc, 67-75, SR RES ASSOC, EXXON RES & ENG CO, 75- Mem: Am Phys Soc. Res: Chemical reaction kinetics; unimolecular reaction theory; cage effect in solution and gas phase; gas phase detonations and shock waves; radiation and high pressure chemistry; laser isotope separation. Mailing Add: Exxon Res & Eng Co Linden NJ 07036

LYON, ROBERT LYNDON, b Dolgeville, NY, Apr 17, 27; m 56; c 3. FOREST ENTOMOLOGY, INSECT TOXICOLOGY. Educ: Syracuse Univ, BS, 53, MS, 54; Univ Calif, Berkeley, PhD(insect toxicol), 61. Prof Exp: Res entomologist, 53-72, SUPVRY RES ENTOMOLOGIST & PROJ LEADER, INSECTICIDE EVAL PROJ, PAC SOUTHWEST FOREST & RANGE EXP STA, US FOREST SERV, 72- Mem: Entom Soc Am; Entom Soc Can. Res: Development of safe, selective, nonpersistent and effective chemical insecticides and techniques to manage forest insect populations and protect forest resource values with minimal adverse effects on the environment. Mailing Add: Pac Southwest Forest & Range Exp US Forest Serv 1960 Addison St Berkeley CA 94701

LYON, RONALD JAMES PEARSON, b Northam, WAustralia, Jan 15, 28; US citizen; m 61; c 4. GEOLOGY, MINERALOGY. Educ: Univ Western Australia, BS, 48, Hons, 49; Univ Calif, Berkeley, PhD(geol), 54. Prof Exp: Geologist, Lake George Mines, Captains Flat, NSW, 49-51; Goewey res fel geol, Univ Calif, Berkeley, 51-54; res off mining, Commonwealth Sci Res Orgn, Australia, 54-56; geochemist, Kennecott Res Ctr, Utah, 56-59; sr geochemist, Stanford Res Inst, 59-63; Nat Acad Sci sr fel geol, Ames Res Ctr, NASA, 63-65; assoc prof, 65-71 PROF MINERAL EXPLOR, STANFORD UNIV, 72- Concurrent Pos: Chmn geol panel, Nat Acad Sci summer study on space appln, Woods Hole, Mass, 67-69; consult planetary atmospheres, NASA, 68-70; consult & prin assoc, Earth Satellite Corp, 70-; mem remote sensing group, Int Hydrol Decade, Nat Acad Sci, 72- Honors & Awards: Photog Interpretation Award, Am Soc Photogram, 72. Mem: AAAS; Soc Econ Geol; Am Soc Photogram. Res: Use of airborne geophysical techniques and remote sensing in exploration for mineral deposits; recognition of rock and soil materials using ERTS & SKYLAB spectral data. Mailing Add: Dept of Appl Earth Sci Stanford Univ Stanford CA 94305

LYON, WALDO (KAMPMEIER), b Los Angeles, Calif, May 19, 14; m 37; c 2. PHYSICS. Educ: Univ Calif, Los Angeles, AB, 36, MA, 37, PhD(physics), 41. Prof Exp: Asst physics, Univ Calif, Los Angeles, 40-41; chief scientist, Arctic Submarine Res, US Navy Electronics Lab, 41-66, DIR, ARCTIC SUBMARINE LAB, NAVAL UNDERSEAS RES & DEVELOP CTR, 66- Concurrent Pos: Sr scientist, Wave Measurement Group, Bikini atom bomb tests, 46; lectr, Univ Calif, Los Angeles, 48-49; physicist, Submarine Opers, US Navy-Byrd Antarctic exped, 46-47; chief scientist, US-Can Aleutian exped, 49, Beauford Sea expeds, 51-54; sr scientist, Transpolar Submarine Exped, 57-75. Honors & Awards: Distinguished Civilian Serv Award, US Navy, 55 & 58; US Dept Defense, 56; Am Soc Naval Engrs Gold Medal Award, 59; President's Distinguished Fed Civilian Serv Award, 62. Mem: Fel AAAS; fel Am Phys Soc; fel Acoust Soc Am; fel Arctic Inst NAm; Am Soc Naval Eng. Res: Ocean-cryology and physics of sea ice; underwater acoustics. Mailing Add: 1330 Alexandria Dr San Diego CA 92107

LYON, WILLIAM D, b Chicago, Ill, Sept 3, 36; m 58; c 3. THEORETICAL CHEMISTRY. Educ: Univ Ill, BS, 58; Univ Wis, PhD(phys chem), 67. Prof Exp: Asst chem, Univ Wis, 61-66; res assoc, Univ Minn, 66-67; ASST PROF CHEM, UNIV AKRON, 67- Mem: Am Phys Soc. Res: Molecular quantum mechanics; perturbation theory of small molecules using pseudopotentials to calculate wave functions. Mailing Add: Dept of Chem Univ of Akron Akron OH 44325

LYON, WILLIAM FRANCIS, b Mt Gilead, Ohio, Jan 24, 37; m 62; c 4. ECONOMIC ENTOMOLOGY. Educ: Ohio State Univ, BSc, 59, MSc, 62, PhD(entom), 69. Prof Exp: County exten agent, Ohio Coop Exten Serv, 59-61; surv entomologist, Ohio Agr Res & Develop Ctr, 62-64; exten entomologist, Ohio Coop Exten Serv, 66-72; Plant protection entomologist, Makerere Univ, Uganda, 72-73; pest mgt entomologist, Univ Nairobi, Kenya, 73-74; EXTEN ENTOMOLOGIST, OHIO COOP EXTEN SERV, 74- Mem: Entom Soc Am; Am Inst Biol Sci; E African Acad. Res: Identification and control of household, livestock, poultry and stored grain insects. Mailing Add: 1735 Neil Ave Ohio State Univ Columbus OH 43210

LYON, WILLIAM GRAHAM, b Chelsea, Mass, Apr 29, 44; m 65; c 2. PHYSICAL CHEMISTRY. Educ: Univ Mich, BS, 66, MS, 68, PhD(chem), 73. Prof Exp: Fel phys chem, Univ Mich, 73-74; FEL PHYS CHEM, ARGONNE NAT LAB, 74- Mem: Assoc Sigma Xi. Res: Thermodynamics of the solid state. Mailing Add: Chem Div Argonne Nat Lab 9700 SCass Ave Argonne IL 60439

LYON, WILLIAM SOUTHERN, JR, b Pulaski, Va, Jan 25, 22; m 46; c 2. RADIOCHEMISTRY. Educ: Univ Va, BS, 43; Univ Tenn, MS, 68. Prof Exp: Chemist, E I du Pont de Nemours & Co, WVa, 43-44 & Wash, 44-45; lab foreman, Tenn Eastman Corp, 45-47; chemist, 47-62, GROUP LEADER RADIOCHEM, OAK RIDGE NAT LAB, 62- Concurrent Pos: Consult, Thai Atomic Energy for Peace Lab, Bangkok, 66-; mem sci comt 25, Nat Coun Radiation Protection, 67-; assoc ed, Radiochem-Radioanal Lett, 70-; regional ed, J Radioanal Chem, 71- Mem: Am Chem Soc; Am Nuclear Soc. Res: Trace element analysis; new energy sources; nuclear decay schemes; specialized radioactivity measurements. Mailing Add: Anal Chem Div Oak Ridge Nat Lab PO Box X Oak Ridge TN 37830

LYONS, DON CHALMERS, b Jackson, Mich, May 5, 99; m 23; c 3. BACTERIOLOGY. Educ: Univ Mich, DDS, 21, MS, 32; Mich State Univ, PhD(bact), 35; Am Bd Oral Med, dipl. Prof Exp: Instr oral surg, Univ Mich, 21, asst, Univ Hosp, 23-25; DIR, LYONS RES LAB, 37- Concurrent Pos: Practicing oral surgeon, 35-; res assoc, Mich State Univ & Univ Mich, 23-26; Fulbright prof, Univ Lima, 62, Univ Guenca, 64, Cent Univ Ecuador, 64 & Univ Tehran, 67; mem staff, Mercy, Sheldon & Marlin Community Hosps; chief dent staff, W A Foote Hosp; mem bd, Beth Moser Ment Clin, Mich; chmn, Am Bd Oral Med. Mem: Fel AAAS; Sigma Xi; Am Dent Asn; fel Am Pub Health Asn; fel Am Acad Oral Med (pres). Res: Antibiosis associated with biochemical activities of bacteria; oral surgery. Mailing Add: Lyons Res Lab 420 W Michigan Ave Jackson MI 49201

LYONS, DONALD HERBERT, b Buffalo, NY, Feb 28, 29; m 51; c 3. PHYSICS. Educ: Univ Buffalo, BA, 49; Univ Pa, MA, 51, PhD(physics), 54. Prof Exp: Staff scientist, Lincoln Lab, Mass Inst Technol, 56-61 & Sperry Rand Res Ctr, 61-63; res prof physics, Inst Solid State Physics, Univ Tokyo, 63-64; staff scientist, Sperry Rand Res Ctr, Mass, 64-66; assoc prof, 66-68, chmn dept, 67-68 & 70-72, PROF PHYSICS, UNIV MASS, BOSTON, 68- Concurrent Pos: Fulbright grant, 63-64. Mem: Am Phys Soc. Res: Theoretical magnetism; communication theory; theoretical nuclear physics. Mailing Add: Dept of Physics Univ of Mass 100 Arlington St Boston MA 02116

LYONS, ERWIN JOHN, b Madison, Wis, July 12, 09; m 32; c 2. GEOLOGY. Educ: Univ Wis, BA, 35, PhD(geol), 47. Prof Exp: Jr geologist, Emergency Conserv, Wis, 35; geologist, Wis Geol Surv, 36-39; instr mineral, Johns Hopkins Univ, 39-41; prof geol, Wis Inst Technol, 41-48; geologist, US Geol Surv, 43-58; ASSOC PROF GEOL, UNIV KY, 58- Res: Coal investigations and engineering geology. Mailing Add: Dept of Geol Bowman Hall Univ of Ky Lexington KY 40506

LYONS, EUGENE T, b Yankton, SDak, May 6, 31. PARASITOLOGY. Educ: SDak State Univ, BS, 56; Kans State Univ, MS, 58; Colo State Univ, PhD(parasitol), 63. Prof Exp: Asst prof, 58-60 & 63-70, ASSOC PROF PARASITOL, UNIV KY, 70- Mem: Am Soc Parasitol; Wildlife Dis Asn. Res: Parasites of jackrabbits, fur seals, horses, sheep and cattle. Mailing Add: Dept of Vet Sci Agr Exp Sta Univ of Ky Lexington KY 40506

LYONS, GEORGE D, b New Orleans, La, Jan 19, 28; m 54; c 5. OTOLARYNGOLOGY. Educ: Southeastern La Col, BS, 50; La State Univ, New Orleans, MD, 54. Prof Exp: From clin instr to clin assoc prof, 58-70, assoc prof, 70-71, PROF OTOLARYNGOL & HEAD DEPT, SCH MED, UNIV NEW

ORLEANS, 71- Concurrent Pos: Mem, Soc Acad Chmn Otolaryngol, 72- Honors & Awards: Recognition Award, AMA. Mem: Fel Am Laryngol, Rhinol & Otol Soc; fel Am Acad Facial Plastic & Reconstruct Surg; fel Am Col Surg; fel Pan-Am Soc Otolaryngol. Res: Regional plastic surgery; otology. Mailing Add: Dept of Otorhinolaryngol Univ of New Orleans Sch of Med New Orleans LA 70112

LYONS, HAROLD, b New York, NY, Mar 27, 19; m 41; c 3. ANALYTICAL CHEMISTRY, MOLECULAR PATHOLOGY. Educ: City Col New York, BS, 45; Okla State Univ, MS, 49, PhD(chem), 51. Prof Exp: Chemist, Climax Rubber Co, 37-41; res chemist, Ruberoid Co, 45-48; sr res chemist, Gen Elec Co, 51-52 & Pa Salt Mfg Co, 52-55; lab mgr, Koppers Co, Inc, 55-58; assoc prof, 58-60, PROF CHEM, SOUTHWESTERN AT MEMPHIS, 60-; PROF PATH, MED UNITS, UNIV TENN, MEMPHIS, 63- Mem: AAAS; Am Chem Soc. Res: Instrumental analysis; polymer physical chemistry; analytical biochemistry; biophysical chemistry. Mailing Add: Dept of Chem Southwestern at Memphis Memphis TN 38112

LYONS, HAROLD, b Buffalo, NY, Feb 16, 13; m 37; c 2. PHYSICS. Educ: Univ Buffalo, BA, 33; Univ Mich, MA, 35, PhD(physics), 39. Prof Exp: Lab instr physics, Univ Buffalo, 33-34; Rackham asst nuclear physics, Univ Mich, 37-39; res physicist, Naval Res Lab, DC, 39-41; from asst physicist to chief physicist, Nat Bur Standards, 41-55; head atomic physics dept, Res Labs, Hughes Aircraft Co, 55-60; vpres & mgr quantum electronics div, Electro-Optical Systs, Inc, 61-62; res theoret physicist anat, Space Biol Lab, Brain Res Inst & Instr Laser Quantum Electronics, 66-74, INSTR LASERS, EXTEN, UNIV CALIF, LOS ANGELES, 74- Concurrent Pos: Consult physicist, Aerospace Indust, 62-; mem panel radiating systs, Res & Develop Bd; mem Comn I, radio measurements & standards, Int Sci Radio Union. Honors & Awards: Gold Medal, US Dept Commerce, 49, Commemorative Award, 73; Flemming Award, US Govt, 49; Super Accomplishment Award, Nat Bur Standards, 49; Cert Merit, Franklin Inst, 58. Mem: Fel Am Phys Soc; Sigma Xi; fel Inst Elec & Electronics Eng. Res: Brain research, memory and learning using spectroscopic methods; biophysics; lasers; quantum electronics. Mailing Add: 1101 El Medio Ave Pacific Palisades CA 90272

LYONS, HAROLD ALOYSIUS, b Brooklyn, NY, Sept 14, 31; m 40; c 8. MEDICINE. Educ: St Johns Univ, NY, BS, 35; Long Island Col Med, MD, 40; Am Bd Internal Med, dipl, 50. Prof Exp: Intern, Brooklyn Hosp, 40-41; asst clin prof med, Long Island Col Med, 46-50; asst prof, Col Med, Georgetown Univ, 50-52; clin asst prof, 52-53, assoc prof, 53-56, PROF MED, COL MED, STATE UNIV NY DOWNSTATE MED CTR, 56-, DIR PULMONARY DIS DIV & CARDIOVASC LAB, 70- Concurrent Pos: Consult, Hosps, 52; dir, Pulmonary Dis Div, Kings County Hosp Ctr, 53-; med adv, Security Serv Admin, USPHS, 62- Mem: AAAS; Am Thoracic Soc; Harvey Soc; fel AMA; Am Heart Asn. Res: Internal medicine; cardiopulmonary physiology; pulmonary diseases; biomedical engineering; mechanics of respiration; ventilation-perfusion relationships. Mailing Add: Dept of Med Pulmonary Div State Univ NY Downstate Med Ctr Brooklyn NY 11203

LYONS, HAROLD DWIGHT, b Elizabethton, Tenn, Mar 13, 28; m 52; c 3. SYNTHETIC ORGANIC CHEMISTRY. Educ: Carson-Newman Col, BS, 49; Univ Ala, MS, 51, PhD(org chem), 53. Prof Exp: Group leader res & develop, Phillips Petrol Co, 52-59; assoc prof org chem, Carson-Newman Col, 59-60; sr res chemist, Gulf Oil Corp, Kans, 60-66; SR RES CHEMIST, DEERING MILLIKEN RES CORP, 66- Mem: Am Chem Soc. Res: Polymer chemistry and emulsions; polyolefin technology; textile chemistry. Mailing Add: Deering Milliken Res Corp PO Box 1927 Spartanburg SC 29304

LYONS, JAMES EDWARD, b Montpelier, Vt, Oct 20, 37; m 63; c 2. ORGANIC CHEMISTRY, ORGANOMETALLIC CHEMISTRY. Educ: Boston Col, BS, 59; Purdue Univ, MS, 61; Univ Calif, Davis, PhD(org chem), 68. Prof Exp: Chemist, Res & Develop Ctr, Gen Elec Co, 62-64; RES CHEMIST, SUN OIL CO, 68- Mem: AAAS; Am Chem Soc. Res: Mechanisms and synthetic applications of transition metal catalyzed reactions in organic and organometallic systems. Mailing Add: Res & Develop Div Sun Oil Co Marcus Hook PA 19061

LYONS, JAMES MARTIN, b Livermore, Calif, Oct 9, 29; m 52; c 2. PLANT PHYSIOLOGY. Educ: Univ Calif, Berkeley, BS, 51; Univ Calif, Davis, MS, 58, PhD(plant physiol), 62. Prof Exp: Asst plant physiologist, Univ Calif, Riverside, 62-66, vchmn dept veg crops, 64-66, asst prof, 65-66, assoc prof, chmn dept & assoc plant physiologist, 66-70; chmn dept veg crops, 70-73, PROF VEG CROPS & PLANT PHYSIOLOGIST, UNIV CALIF, DAVIS, 70-, ASSOC DEAN, COL BIOL & AGR SCI, 73- Mem: AAAS; Am Soc Hort Sci; Am Soc Plant Physiol; Weed Sci Soc Am; Int Soc Hort Sci. Res: Biochemistry and physiology of fruit ripening; senescence and growth regulators in vegetable crops. Mailing Add: Col of Biol & Agr Sci Univ of Calif Davis CA 95616

LYONS, JOHN BARTHOLOMEW, b Quincy, Mass, Nov 22, 16; m 45; c 5. GEOLOGY. Educ: Harvard Univ, AB, 38, AM, 39, PhD(geol), 42. Prof Exp: Geologist, US Geol Surv, 41-45; asst prof, 46-52, PROF GEOL, DARTMOUTH COL, 52- Mem: Fel Geol Soc Am; Mineral Soc Am. Res: Petrology; structural geology; glaciology. Mailing Add: Dept of Earth Sci Dartmouth Col Hanover NH 03755

LYONS, JOHN WINSHIP, inorganic chemistry, physical chemistry, see 12th edition

LYONS, JOSEPH F, b Wappingers Falls, NY, Nov 27, 20; m 46; c 6. PETROLEUM CHEMISTRY. Educ: Fordham Univ, BS, 41; Purdue Univ, MS, 48, PhD(chem), 50. Prof Exp: Chemist, 41-46, 50-53, group leader, 53-60, res supvr, 60-73, ASST MGR, TEXACO, INC, 73- Mem: Am Chem Soc; Sigma Xi. Mailing Add: Texaco Res Ctr Box 509 Beacon NY 12508

LYONS, JOSEPH PAUL, b Ardmore, Pa, Dec 9, 47; m 70; c 1. OPERATIONS RESEARCH, PUBLIC HEALTH ADMINISTRATION. Educ: Bloomsburg State Col, BA, 70; Johns Hopkins Univ, ScD, 75. Prof Exp: Syst analyst ment health, Pa Off Ment Health, 70-71; SCIENTIST ALCOHOLISM, RES INST ALCOHOLISM, 75- Concurrent Pos: Nat Inst Ment Health trainee, Johns Hopkins Univ, 71-75; assoc consult, Elliott Assocs, 71-74; admin consult, Md Dept Ment Hyg, 73; asst clin prof, Dept Psychiat, Sch Med & adj asst prof, Dept Indust Eng, Sch Eng, State Univ NY Buffalo, 75- Mem: Oper Res Soc Am; AAAS; Asn Ment Health Admin. Res: Problem oriented record and its application to alcoholism service delivery; treatment planning in both in-patient and out-patient settings and systems design for delivery of alcoholism services. Mailing Add: Res Inst on Alcoholism 1021 Main St Buffalo NY 14203

LYONS, KENNETH BRENT, b St Louis, Mo, Aug 31, 46; m 68; c 1. SOLID STATE PHYSICS. Educ: Univ Okla, BS, 68, MS, 69; Univ Colo, PhD(physics), 73. Prof Exp: MEM RES STAFF, BELL TEL LABS, 73- Mem: Am Phys Soc. Res: Raman and Brillouin light scattering, with emphasis on non-equilibrium phenomena and surface effects. Mailing Add: 1A139 Bell Tel Labs 600 Mountain Ave Murray Hill NJ 07974

LYONS, MARGARET S, b Edinburgh, Scotland, Sept, 4, 22; US citizen; m 49; c 4. PHYSICAL CHEMISTRY. Educ: Smith Col, AB, 43; Wesleyan Univ, MA, 45; Univ Wis, PhD(chem), 49. Prof Exp: Lectr, Albertus Magnus Col, 56-64; asst prof Southern Conn State Col, 64-67; vis asst prof, Conn Col, 68-71; MEM STAFF, DEPT PHYSICS, HOPKINS GRAMMAR SCH, 71- Mem: Am Asn Physics Teachers. Mailing Add: 7 Ridgewood Terr North Haven CT 06473

LYONS, PAUL CHRISTOPHER, b Cambridge, Mass, Oct 1, 38; m 63; c 4. MINERALOGY, PETROLOGY. Educ: Boston Univ, AB, 63, AM, 64, PhD(geol), 69. Prof Exp: Pub sch teacher, Mass, 64-68; instr, 68-69, ASST PROF PHYS SCI, BOSTON UNIV, 69- Concurrent Pos: Res grants, Boston Univ, 71-72 & Mineral Soc Gt Britain. Mem: AAAS; Geol Soc Am; Mineral Soc Am. Res: Geology of granites, eastern Massachusetts; Pennsylvanian plant megafossils of New England. Mailing Add: Div of Sci Boston Univ Boston MA 02215

LYONS, PETER BRUCE, b Hammond, Ind, Feb 23, 43; m 63; c 2. PLASMA PHYSICS. Educ: Univ Ariz, BS, 64; Calif Inst Technol, PhD(physics), 69. Prof Exp: STAFF MEM, LOS ALAMOS SCI LAB, 69- Mem: Am Phys Soc; Inst Elec & Electronics Eng. Res: X-ray interactions and dosimetry; high intensity monoenergetic x-ray generation; x-ray and nuclear detectors and instrumentation; low energy nuclear physics; accelerator instrumentation; low energy nuclear physics; accelerator technology; plastic scintillators; fiber optics; plasma diagnostics. Mailing Add: Los Alamos Sci Lab Group MS-410 Los Alamos NM 87545

LYONS, PETER FRANCIS, b Philadelphia, Pa, Nov 29, 42; m 68; c 1. PHYSICAL CHEMISTRY, POLYMER SCIENCE. Educ: Villanova Univ, BS, 64; Princeton Univ, MA, 67, PhD(chem), 70. Prof Exp: RES CHEMIST, TEXTILE FIBERS DEPT, E I DU PONT DE NEMOURS & CO, INC, 68- Mem: Am Chem Soc. Res: Physical chemistry of polymeric systems including work on degradation, strength mechanisms and viscosity theory. Mailing Add: Textile Fibers Dept E I du Pont de Nemours & Co Inc Wilmington DE 19898

LYONS, PHILIP AUGUSTINE, b Lancashire, Eng, May 26, 16; US citizen; m 49; c 4. PHYSICAL CHEMISTRY. Educ: La Salle Col, BA, 37; Univ Wis, PhD(chem), 48. Prof Exp: From instr to assoc prof, 48-65, PROF CHEM, YALE UNIV, 65- Concurrent Pos: Consult, Audiotape Corp; vis prof, Univ Islamabad, WPakistan, 71. Mem: Am Chem Soc. Res: Raman spectra; nonaqueous solutions; diffusion in liquids; Soret effect; critical solution phenomena. Mailing Add: Dept of Chem Yale Univ New Haven CT 06520

LYONS, RICHARD BERNARD, b Corvallis, Ore, Aug 23, 34; m 57; c 4. CELL BIOLOGY, MEDICAL GENETICS. Educ: Univ Ore, BS, 57, MS & MD, 60. Prof Exp: Intern, Med Sch Hosps & Clins, Univ Ore, 60-61, from instr to asst prof anat, Med Sch, 61-66; asst prof anat & exp med, 64-66; WITH ARCTIC HEALTH RES CTR, USPHS, 66-; ASSOC PROF MED SCI, UNIV ALASKA, 71- Mem: Am Chem Soc; Am Soc Cell Biol; Am Soc Zool; Am Heart Asn; Am Geriat Soc. Res: Chemical embryology; human cytogenetics; developmental and cell biology; biology of isozymes; anatomy and physiology of whales. Mailing Add: Arctic Health Res Ctr USPHS College AK 99701

LYONS, WALTER ANDREW, b Brooklyn, NY, June 14, 43. METEOROLOGY, AIR POLLUTION. Educ: St Louis Univ, BS, 64; Univ Chicago, MS, 65, PhD(meteorol), 70. prof meteorol, Univ Wis-Milwaukee, 69-76; DIR, WSTP WEATHER SERV, 76- Concurrent Pos: Mem energy budget panel, Int Field Year for Great Lakes, 68-73; chmn comn on meterol aspects of air pollution, Am Meteorol Soc, 73-75; adj prof dept mech eng, Univ Minn, Minneapolis, 76- Mem: Am Meteorol Soc; Am Geophys Union; Air Pollution Control Asn; Int Asn Gt Lakes Res. Res: Air pollution meteorology, especially the effect of small scale features, such as cities and lakes, on the regional transport of pollutants; study using aircraft, balloons, ground networks, photography and satellites. Mailing Add: Hubbard Broadcasting 3415 University Ave SE Minneapolis MN 55414

LYRENE, PAUL MAGNUS, b Ala, Apr 16, 46. PLANT BREEDING. Educ: Auburn Univ, BS, 68; Univ Wis, MS, 70, PhD(plant breeding), 74. Prof Exp: ASST PROF AGRON, UNIV FLA EXP STA, 74- Mem: Am Soc Agron; Crop Sci Soc Am; Am Soc Sugarcane Technologists. Res: Sugarcane variety improvement; sugarcane tissue culture for variety improvement; surgarcane cytogenetics; inheritance studies in sugarcane. Mailing Add: Sugarcane Field Sta Star Rte Box 8 Canal Point FL 33438

LYS, JEREMY EION ALLEYNE, b Dannevirke, NZ, Apr 17, 38; m 68. HIGH ENERGY PHYSICS. Educ: Univ Canterbury, BSc, 58, MSc, 60; Oxford Univ, PhD:physics), 64; Mitchell Col, New South Wales, DipEd, 74. Prof Exp: Res fel physics, Univ Liverpool, 63-65; res assoc physics, Univ Mich, 66-72; sch teacher sci, De la Salle Col, New South Wales, 72-74; RES ASSOC PHYSICS, FERMI NAT ACCELERATOR LAB, 75- Res: Strong interactions in high energy physics using the bubble chamber technique. Mailing Add: Fermi Nat Accelerator Lab PO Box 500 Batavia IL 60510

LYSENKO, MICHAEL GEORGE, b Brandon, Man, Nov 20, 17; nat US; m 45; c 4. PARASITOLOGY. Educ: Univ Man, BA, 39, dipl, 40; Iowa State Univ, PhD, 50. Prof Exp: Lectr sci, St John's Col, Can, 43-45; lectr zool, Univ Man, 45-47; asst prof, 50-55, ASSOC PROF MED MICROBIOL & ZOOL, MED SCH, UNIV WIS-MADISON, 55- Mem: AAAS; Am Soc Parasitol; Am Soc Trop Med & Hyg; Soc Protozool. Res: Immunity in parasitic infections; physiology and metabolism of protozoa. Mailing Add: Dept of Med Microbiol Med Sch Univ of Wis Madison WI 53706

LYSER, KATHERINE MAY (MRS E SHOUBY), b Berkeley, Calif, May 11, 33; m 65. NEUROEMBRYOLOGY; ONCOLOGY. Educ: Oberlin Col, AB, 55, Radcliffe Col, MA, 57, PhD(biol), 60. Prof Exp: Instr zool, Oberlin Col, 57-58; NSF fel exp embryol, Col France, 60-61; res fel, Med Col, Cornell Univ, 61-62, instr anat, 62-64; asst prof, Sch Med & Dent, Georgetwon Univ, 64-65; ASSOC PROF BIOL SCI, HUNTER COL CITY UNIV NY, 65- Concurrent Pos: Part-time fac mem, Sarah Lawrence Col, 61-62; USPHS res grants, Med Col, Cornell Univ, 63 & Hunter Col, 65-70; United Cerebral Palsy Res & Educ Found grant, Cornell Univ & Georgetown Univ, 64-65; guest investr, P A Weiss Lab, Rockefeller Univ, 67-70. Mem: Tissue Cult Asn; Am Soc Zool; Am Asn Anat; Soc Neurosci; Soc Develop Biol. Res: Factors controlling differentiation in the embryonic nervous system and in human neuroblastomas; cytology of normal nervous system and neural tumors. Mailing Add: Box 1030 Dept of Biol Sci Hunter Col 695 Park Ave New York NY 10020

LYSIAK, RICHARD JOHN, b Chicago, Ill, Dec 29, 28; m 53; c 2. PHYSICS. Educ: Aeronaut Univ, Chicago, BSAE, 50; Tex Christian Univ, BA, 59, MA, 60, PhD(physics), 63. Prof Exp: Sr aerosyst engr, Gen Dynamics/Ft Worth, 54-61; teaching fel, 61-63, asst prof, 63-67, ASSOC PROF PHYSICS, TEX CHRISTIAN UNIV, 67-, CHMN DEPT, 69- Res: Quantum electronics; optics; random noise theory. Mailing Add: Dept of Physics Tex Christian Univ Fort Worth TX 76129

LYSNE, PETER C, b Milwaukee, Wis, July 20, 39; m 62; c 2. APPLIED PHYSICS. Educ: Grinnell Col, BA, 61; Ariz State Univ, PhD(physics), 66. Prof Exp: STAFF MEM, SHOCK PHYSICS RES, SANDIA LABS, 66- Mem: Am Phys Soc. Res: Thermodynamics and its relation to shock physics; shock propagation in solid, liquid and porous media; shock-wave induced depolarization of ferroelectrics Mailing Add: Sandia Labs Albuquerque NM 87115

LYSYJ, IHOR, b Tarnow, Poland, Apr 13, 29; nat US; m 57; c 2. ANALYTICAL CHEMISTRY, ENVIRONMENTAL TECHNOLOGY. Educ: Ukrainian Tech Inst, Ger, MS, 50. Prof Exp: Anal chemist, Ex-Lax, Inc, NY, 52-54; dir res, Gaston Johnston Corp, 54-56; anal chemist, Cent Res Lab, Food Mach & Chem Corp, 56-60; res scientist, Ethicon, Inc, 60-61; PRIN SCIENTIST, ROCKETDYNE DIV, ROCKWELL INT CORP, 61- Mem: Am Chem Soc. Res: Microchemistry; chemical detection and sensing technology; environmental quality monitoring systems and networks. Mailing Add: Rocketdyne Div Rockwell Int Corp Canoga Park CA 91304

LYTLE, CARL DAVID, b Millersburg, Ohio, Jan 28, 41; m 63; c 2. BIOPHYSICS. Educ: Kent State Univ, BS, 63; Cornell Univ, MS, 65; Pa State Univ, PhD(biophys), 68. Prof Exp: Res biophysicist, Bur Radiol Health, USPHS, 68-70, chief path studies sect, 70, chief path studies sect, Environ Protection Agency, 70-71, chief mult environ stresses br, 71-75, MEM STAFF, BUR RADIOL HEALTH, FOOD & DRUG ADMIN, 75- Concurrent Pos: Adj prof, George Washington Univ, 71- Mem: AAAS; Biophys Soc; Am Soc Microbiol; Radiation Res Soc. Res: Radiation virology, oncology; host cell reactivation; radiation enhanced transformation; ultraviolet carcinogenesis. Mailing Add: Bur of Radiol Health Food & Drug Admin 5600 Fishers Lane Rockville MD 20852

LYTLE, CHARLES FRANKLIN, b Crawfordsville, Ind, May 13, 32; m 55; c 5. ZOOLOGY. Educ: Wabach Col, AB, 53; Ind Univ, MA, 58, PhD(zool), 59. Prof Exp: Asst zool, Ind Univ, 53-55, 57-58, res assoc, 59-60; asst prof, Tulane Univ, 60-62; res analyst, US Govt, 63-64; from asst prof to assoc prof zool, Pa State Univ, 64-69; assoc prof, 69-72, PROF ZOOL, NC STATE UNIV, 72-, TEACHING COORDR BIOL SCI, 69- Concurrent Pos: Fel embryol, Ind Univ, 59; consult, US Dept Army, 62-63; consult, Educ Testing Serv, 69- Mem: Fel AAAS; Am Soc Zool; Am Inst Biol Sci; Sigma Xi; Asn Southeastern Biologists. Res: Invertebrate zoology; cellular structure and function in invertebrate development; systematics and ecology of Hydrozoa; biological education; instructional television. Mailing Add: Dept of Zool N C State Univ PO Box 5577 Raleigh NC 27607

LYTLE, CHARLES WILLIAM, mathematics, see 12th edition

LYTLE, ERNEST JAMES, JR, b Eastlake, Fla, June 28, 13; m 41; c 3. MATHEMATICAL STATISTICS. Educ: Univ Fla, BS, 35, MA, 40, PhD(math), 56. Prof Exp: Teacher pub sch, Fla, 35-39, prin, 39-41; instr math, Univ Fla, 46-50, 52-53, asst, 53-55, instr, 55-56; assoc engr, Mil Prods Div, Int Bus Mach Corp, NY, 56-57; assoc res prof, Statist Lab, Univ Fla, 57-59; sr mem staff statist & math, Radiation, Inc, Fla, 59-61; sr staff engr, Guid & Control Sect, Systs Dept, Martin Co, 61-68; chmn dept, 74-75, prof, 68-75, EMER PROF MATH SCI, FLA TECHNOL UNIV, 75- Mem: AAAS; Math Asn Am; Am Statist Asn; Inst Math Statist. Res: Applied statistics to systems analysis. Mailing Add: 2107 Whitehall Dr Winter Park FL 32789

LYTLE, FARREL WAYNE, b Cedar City, Utah, Nov 10, 34; m 54; c 4. SOLID STATE PHYSICS. Educ: Univ Nev, BS, 56, MS, 58. Prof Exp: Chemist, US Bur Mines, 55-58; sr basic res scientist, Boeing Sci Res Labs, 60-74; PRES, EXAFS CO, 74- Mem: AAAS; Am Phys Soc; fel Am Inst Chem. Res: X-ray physics, spectroscopy and diffraction; x-ray astronomy; radiation chemistry; cryogenics; structural inorganic chemistry; amorphous structures; structure of catalysts. Mailing Add: 10815 24th Ave S Seattle WA 98168

LYTLE, FRED EDWARD, b Lewisburg, Pa, Jan 13, 43; m 67; c 1. CHEMISTRY. Educ: Juniata Col, BS, 64; Mass Inst Technol, PhD(chem), 68. Prof Exp: Asst prof, 68-74, ASSOC PROF CHEM, PURDUE UNIV, WEST LAFAYETTE, 74- Honors & Awards: Merck Co Found Fac Develop Award, 69. Mem: Am Chem Soc; Soc Appl Spectros. Res: Time resolved spectroscopy; trace analysis; raman absorption spectroscopy; computer interpretation, storage and revival of analytical data. Mailing Add: Dept of Chem Purdue Univ West Lafayette IN 47907

LYTLE, IVAN M, b San Francisco, Calif, Oct 3, 24; c 4. PHYSIOLOGY. Educ: Univ Tulsa, BS, 49, Univ Calif, Davis, MA, 56, PhD(physiol), 58. Prof Exp: NIH fel biochem, Univ Utah, 58-59, asst prof biol, 59-62, assoc prof & chmn dept, 62-69, prof, 69-70; assoc prof, 70-72, PROF BIOL, UNIV ARIZ, 72-, DEPT HEAD, 75- Concurrent Pos: Mem nurse sci comt, NIH, 68-71. Mem: Am Soc Zool; Brit Soc Study Fertility; Soc Study Reproduction. Res: Hormone control of ovulation; placental hormones. Mailing Add: Dept of Gen Biol Univ of Ariz Tucson AZ 85721

LYTLE, JACK RUSSELL, statistics, see 12th edition

LYTLE, JAMES BERT, b Little Rock, Ark, May 2, 32; m 58; c 2. PHYSIOLOGY, RADIATION BIOLOGY. Educ: Ark Agr & Mech Col, BS, 58; Univ Ark, MS, 60; Tex A&M Univ, PhD(radiation biol), 64. Prof Exp: Res asst microbiol, Univ Tex M D Anderson Hosp & Tumor Inst, 60-61; asst prof biol, Maryville Col, 64-67; CHMN BIOL-HEALTH SCI DIV, COLUMBUS COL, 67- Concurrent Pos: Res partic, Sect Path & Physiol, Oak Ridge Nat Lab, 66. Res: Mitotic activity of fetal and adult hemopoietic organs in the rat; kinetics of the cell cycle in adult and fetal intestinal epithelium in the rodent. Mailing Add: Div of Biol-Health Sci Columbus Col Columbus GA 31907

LYTLE, LOY DENHAM, b Glendale, Calif, Apr 8, 43; m 74; c 2. PSYCHOPHARMACOLOGY, NEUROSCIENCES. Educ: Univ Calif, Santa Barbara, BA, 66; Princeton Univ, PhD(psychol), 70. Prof Exp: NIMH fel neuropharmacol, 70-72, ASST PROF PSYCHOPHARMACOL, MASS INST TECHNOL, 72- Concurrent Pos: Alfred P Sloan fel neurosci, 75. Mem: Am Soc Pharmacol & Exp Therapeut; Nutrit Soc; Int Soc Develop Psychobiol; Neurosci Soc; AAAS. Res: Effects of drugs on physiological and behavioral development; diet and drug induced changes in behavior; effects of drugs on brain and peripheral neurotransmitters. Mailing Add: Dept of Nutrit & Food Sci Bldg 37 Rm 327 Mass Inst of Technol Cambridge MA 02139

LYTLE, RAYMOND ALFRED, b Spartanburg, SC, Sept 23, 19; m 44; c 4. MATHEMATICS. Educ: Wofford Col, BS, 40; Univ Va, MA, 46; Univ Ga, PhD, 55. Prof Exp: Instr math, Univ Va, 42-46; adj prof, 46-56, ASSOC PROF MATH, UNIV SC, 56- Concurrent Pos: Researcher, Univ Ga, 52- Mem: Am Math Soc; Math Asn Am. Res: Topology. Mailing Add: Dept of Math Univ of SC Columbia SC 29208

LYTTLE, NELSON EDWARDS, chemistry, research administration, see 12th edition

LYTTON, BERNARD, b London, Eng, June 28, 26; m; c 4. UROLOGY. Educ: Univ London, MB, BS, 48; FRCS, 55. Prof Exp: House officer med & surg, London Hosp, 55-61; from asst prof to assoc prof urol, 62-71, PROF UROL, SCH MED, YALE UNIV, 71-, CHIEF SECT UROL, 67- Concurrent Pos: Brit Empire Cancer res fel surg, Univ Hosp, King's Col, Univ London, 61-62; USPHS grant; resident surg, Royal Victoria Hosp, McGill Univ, 57-58; attend, Yale-New Haven Hosp, 62-; consult, West Haven Vet Admin Hosp, 62- & Hartford Hosp & Hosp of St Raphael, 68- Mem: AAAS; fel Am Col Surg; Soc Pelvic Surg; Am Asn Genito-Urinary Surg; Asn Univ Urol. Res: Immunologic aspects of cancer; delayed hypersensitivity response to autogenous tumor extracts; effects of pressure on infections; renal responses to alterations in bladder pressure; compensatory renal growth in parabiotic animals and effects of hemodialysis. Mailing Add: 789 Howard Ave New Haven CT 06510

LYZNICKI, EDWARD PETER, JR, b Chicago, Ill. ORGANIC CHEMISTRY. Educ: Notre Dame Univ, BS, 66; Kans State Univ, PhD(chem), 71. Prof Exp: NIH fel, Dept Chem, Univ SC, 72-73; res chemist, Verona Corp, Div Mobay Chem, 73-75; SR RES CHEMIST, PIGMENTS DIV, CHEMETRON CORP, 75- Mem: Am Chem Soc. Res: Organic synthesis; structure determinations; mechanisms of organic reactions; process industrial development. Mailing Add: 300 Fallenleaf Lane Holland MI 49423

M

MA, CHIN WAH, b Shanghai, China. THEORETICAL NUCLEAR PHYSICS. Educ: Univ Calif, Berkeley, PhD(physics), 69. Prof Exp: Res assoc physics, Yale Univ, 69-70; physicist, Univ Calif, Davis, 70-72; vis asst prof, Ind Univ, 72-74; ASST PROF PHYSICS, TEX A&M UNIV, 74- Mem: Am Physics Soc; AAAS. Res: Nuclear structure and nuclear reaction study; nuclear collective excitation; heavy-ion and pion interact with nucleus. Mailing Add: Cyclotron Inst Tex A&M Univ College Station TX 77843

MA, CYNTHIA SANMAN, b Hong Kong, May 16, 40; US citizen; m 66; c 2. STATISTICS, COMPUTER SCIENCE. Educ: Siena Col, BS, 62; Fla State Univ, MS, 66, PhD(systs anal), 69. Prof Exp: Statistician, Community Studies Inst, Kansas City, Mo, 62-63; res asst statist, Fla State Univ, 63-66, comput programmer & analyst, Dept Physics, 66-67, systs design analyst, Systs Planning & Develop Ctr, 67-69; ASSOC PROF DATA PROCESSING & STATIST, BALL STATE UNIV, 69- Concurrent Pos: Consult, 70- Mem: Am Statist Asn; Asn Systs Mgt; Asn Comput Mach. Res: Design of computer-based systems for finance analysis and teaching; design and alocation resource model for planning; use of graphics for information systems. Mailing Add: Dept of Finance & Mgt Ball State Univ Muncie IN 47304

MA, LAURENCE JUN-CHAO, b China, June 12, 37; US citizen; m 64; c 1. GEOGRAPHY. Educ: Nat Taiwan Univ, BA, 60; Kent State Univ, MA, 68; Univ Mich, PhD(geog), 71. Prof Exp: ASST PROF GEOG, UNIV AKRON, 71- Mem: Am Geog Soc; Asn Am Geogr; Asn Asian Studies. Res: Non-western urbanism; cultural geography; geography of China. Mailing Add: Dept of Geog Univ of Akron Akron OH 44325

MA, NANCY SHUI FONG, Hong Kong citizen. CYTOGENETICS. Educ: Mt St Vincent Col, Can, BSc, 64; Northeastern Univ, MS, 68; Boston Col, PhD(biol), 73. Prof Exp: ASSOC PATH, HARVARD UNIV, 73- Concurrent Pos: Cytogenetics consult, Pathobiol Inc, 75- Res: Cytogenetic studies of the New World monkeys. Mailing Add: New Eng Regional Primate Res Ctr Harvard Med Sch Southborough MA 01772

MA, PANG-FAI, b Hong Kong, Mar 2, 39; m 66; c 2. BIOCHEMISTRY. Educ: Chung Chi Col, Hong Kong, BSc, 62; Fla State Univ, PhD(chem), 68. Prof Exp: Instr chem, Literary Col, Hong Kong, 62-63; res assoc biochem, Fla State Univ, 68-69; asst prof, 69-72, ASSOC PROF CHEM, BALL STATE UNIV, 72- Concurrent Pos: Ind Delaware County Cancer Soc res grant, Ball State Univ, 71-73, 73- Mem: AAAS; Am Chem Soc; Am Soc Zool. Res: Comparative biochemistry and developmental studies of enzymes; biomedical research. Mailing Add: Dept of Chem Ball State Univ Muncie IN 47306

MA, ROBERTA MOHLING, biochemistry, see 12th edition

MA, SHANG-KENG, b Chungkin, China, Sept 24, 40. PHYSICS. Educ: Univ Calif, Berkeley, BA, 62, PhD(physics), 66. Prof Exp: Asst res physicist, 66-67, asst prof, 67-71, ASSOC PROF PHYSICS, UNIV CALIF, SAN DIEGO, 71- Concurrent Pos: Mem, Inst Advan Study, 68-69 & 70-; Sloan fel, 71- Mem: Am Phys Soc. Res: Statistical mechanics; many body theory. Mailing Add: Dept of Physics Univ of Calif at San Diego La Jolla CA 92037

MA, TE HSIU, b Hopei, China, Aug 24, 24; nat US; m 60; c 2. CYTOGENETICS. Educ: Cath Univ Peking, BS, 48; Nat Taiwan Univ, MS, 50; Univ Va, PhD(biol), 59. Prof Exp: Asst cytologist, Sugarcane Cytol, Taiwan Sugar Exp Sta, 50-55; from asst prof to assoc prof biol, Emory & Henry Col, 59-64; asst prof, Western Ill Univ, 64-66; geneticist, Radiation Biol Lab, Smithsonian Inst, Wash, DC, 66-69; assoc prof, 69-73, PROF BIOL SCI, WESTERN ILL UNIV, 73- Concurrent Pos: Res partic, Oak Ridge Nat Lab, 63-64; res consult, Oak Ridge Inst Nuclear Studies, 63-; Atomic Energy Comn res grant, 65; Environ Protection Agency res grant, 71-73; radiation safety officer, Western Ill Univ, 71- Mem: AAAS; Am Genetics Soc; Am Soc Cell Biol; NY Acad Sci. Res: Cytogenetics of sugarcane, corn and Tradescantia; radiation effects on chromosomes of Tradescantia and Vicia; air pollutant effects on chromosomes of Tradescantia; radioactive pollutant effect on fish. Mailing Add: Dept of Biol Sci Western Ill Univ Macomb IL 61455

MA, TSU SHENG, b Canton, China, Oct 15, 11; nat US; m 42; c 2. MICROCHEMISTRY, ORGANIC CHEMISTRY. Educ: Tsing Hua Univ, China, BS, 31; Univ Chicago, PhD(chem), 38. Prof Exp: Instr in charge microchem lab, Univ Chicago, 38-46; prof, Peking Univ, 46-49; sr lectr microchem, Univ Otago, NZ, 49-51; asst prof chem, NY Univ, 51-54, assoc prof, 54-58; PROF CHEM, CITY UNIV NEW YORK, 58- Concurrent Pos: Vis prof, Tsinghua Univ, Peking, 47, Lingnan Univ, 49, NY Univ, 54-60; Fulbright-Hays lectr, China, Japan, Hong Kong, India, Malaya, Australia & NZ, 61-62, Korea, Thailand, Hong Kong & Iran, 68-69; Am specialist, Bur Educ & Cult Affairs, US State Dept, Ceylon, Burma, Thailand, Hong Kong & Philippines, 64; mem comn reagents & reactions, Int Union Pure & Appl Chem, 64-69; ed, Mikrochimica Acta, 65-; vis prof, Univ Singapore, 75-76. Mem: AAAS; Am Inst Chem; Am Chem Soc; Soc Appl Spectros; Am Microchem Soc. Res: Synthetic drugs; medicinal plants; microchemical analysis; microtechniques in organic chemistry; small scale experiments for teaching general and organic chemistry. Mailing Add: Dept of Chem City Univ of New York Brooklyn NY 11210

MA, WAI-SAI, b Canton, China, Sept 13, 43; m 72; c 1. IMMUNOCHEMISTRY. Educ: Univ Cincinnati, BS, 66; State Univ NY Albany, PhD(biochem), 71. Prof Exp: Fel immunochem, Kidney Dis Inst, NY State Dept Health, 71-73; RES SCIENTIST IMMUNOCHEM, DOME LABS DIV, MILES LABS, INC, 74- Mem: Am Chem

Soc; AAAS. Res: Immediate hypersensitivity with emphasis in the physicochemical and biological properties of purified allergens, standardization of allergenic extracts and alternate method to hyposensitization immunotherapy. Mailing Add: Dome Labs Div Miles Labs Inc 400 Morgan Lane West Haven CT 06516

MA, ZEE-MING, b Shanghai, China, Feb 4, 42; m 64; c 2. HIGH ENERGY PHYSICS. Educ: Southwestern Univ, Tex, BS, 62; Duke Univ, PhD(physics), 67. Prof Exp: From instr to asst prof, 67-74, ASSOC PROF, PHYSICS, MICH STATE UNIV, 74- Mem: Am Phys Soc. Res: Experimental high energy physics. Mailing Add: Dept of Physics Mich State Univ East Lansing MI 48823

MAACK, ARTHUR CHARLES, bacteriology, see 12th edition

MAAG, DALE D, b Wray, Colo, July 25, 10; m 36; c 2. BIOCHEMISTRY. Educ: Colo State Univ, BS, 35; Univ Mich, MS, 38; Univ Colo, PhD, 59. Prof Exp: Asst prof chem, Univ Denver, 46; from asst prof to prof, 47-74, chmn dept, 60-69, EMER PROF CHEM, COLO STATE UNIV, 74- Mem: Am Chem Soc. Res: Effect of selenium bearing plants on livestock. Mailing Add: Dept of Chem Colo State Univ Ft Collins CO 80521

MAAG, THEODORE AUGUSTUS, b Lansing, Mich, Feb 3, 31; m 51; c 4. VIROLOGY, IMMUNOLOGY. Educ: Mich State Univ, BS, 54, MS, 60; Univ Ga, DVM, 62, PhD, 66. Prof Exp: Regional field res rep, Lederle Labs Div, Am Cyanamid Corp, 53-54; serviceman, Strain Poultry Farms, Inc, 54-58; res asst avian microbiol, Univ Ga, 58-60, fel oncogenic virol, 62-64; Nat Cancer Inst fel avian tumor viruses, Duke Univ, 64-66; res fel, Merck Inst Therapeut Res, Merck Sharp & Dohme Res Labs, NJ, 66-70; PRES & CHIEF RES EXEC, MAAG & EASTERBROOKS, INC, 71- Mem: AAAS; Am Vet Med Asn; Am Soc Microbiol; Am Asn Avian Pathologists; Tissue Cult Asn. Res: Manufacturing veterinary vaccines; contract research in animal diseases. Mailing Add: Maag & Easterbrooks Inc 1805 Tapawingo Dr Gainesville GA 30501

MAAG, URS RICHARD, b Winterthur, Switz, Jan 20, 38; m 65; c 2. STATISTICS. Educ: Swiss Fed Inst Technol, Dipl Math, 61; Univ Toronto, MA, 62, PhD, 65. Prof Exp: From lectr to assoc prof math, 64-73, ASSOC PROF INFO SCI, UNIV MONTREAL, 73- Mem: Can Math Cong; Am Statist Asn; Inst Math Statist; Statist Sci Asn Can. Res: Nonparametric statistics, robust methods in multivariate analysis; applications to genetics. Mailing Add: Dept of Info Sci Univ of Montreal Box 6128 Montreal PQ Can

MAAN, SHIVCHARAN SINGH, b Karnal, Panjab, India, Jan 11, 26; m 55; c 2. GENETICS, PLANT BREEDING. Educ: Univ Panjab, India, BSc, 48; Indian Agr Res Inst, IARI, 50; Kans State Univ, MS, 59, PhD(genetics, biochem), 61. Prof Exp: Res asst wheat breeding, Indian Agr Res Inst, 51-57; asst cytogenetics, Kans State Univ, 57-61; instr wheat cytogenetics, Univ Nebr, 61-63; from asst prof to assoc prof wheat cytogenetics, 63-73, PROF AGRON, NDAK STATE UNIV, 63- Mem: Am Soc Agron; Crop Sci Soc Am. Res: Wheat cytogenetic research; wheat breeding. Mailing Add: Dept of Agron NDak State Univ Fargo ND 58102

MAAR, JAMES RICHARD, b Wellsville, NY, Oct 7, 43; m 68. MATHEMATICAL STATISTICS, STATISTICAL ANALYSIS. Educ: Eckerd Col, BS, 65; Brown Univ, ScM, 67; George Washington Univ, PhD(math statist), 73. Prof Exp: MATHEMATICIAN, US DEPT DEFENSE, FT GEORGE G MEADE, 67- Concurrent Pos: Asst prof lectr mgt sci, Col Gen Studies, George Washington Univ, 73- Mem: Sigma Xi; Inst Math Statist; Am Statist Asn; Math Asn Am. Res: Multivariate data analysis, including cluster analysis, discriminat analysis, statistical computing and graphical data analysis; new counterexamples to plausible but false statements in probability theory and mathematical statistics. Mailing Add: 3906 Walt Ann Dr Ellicott City MD 21043

MAAS, EUGENE VERNON, b Jamestown, NDak, Dec 18, 36; m 61; c 3. PLANT PHYSIOLOGY. Educ: Jamestown Col, BS, 58; Univ Ariz, MS, 61; Ore State Univ, PhD(soils), 66. Prof Exp: Res assoc soil physics, Univ Ariz, 61; asst soils, Ore State Univ, 61-66; plant physiologist, Mineral Nutrit Lab, 66-68, PLANT PHYSIOLOGIST, AGR RES SERV, US DEPT AGR, US SALINITY LAB, 68- Concurrent Pos: Nat Acad Sci-Nat Res Coun resident res associateship, 66-68. Mem: Am Soc Plant Physiol; Am Soc Agron. Res: Ion absorption and transport in plants; environmental physiology of plants; tolerance to salts and air pollution. Mailing Add: US Salinity Lab PO Box 672 Riverside CA 92502

MAAS, JAMES WELDON, b St Louis, Mo, Oct 26, 29; m 53; c 1. PSYCHIATRY. Educ: Wash Univ, BA, 50, MD, 54. Prof Exp: Intern med, Grady Mem Hosp, Atlanta, Ga, 54-55; resident psychiat, Cincinnati Gen Hosp, Ohio, 55-56 & 58-60; chief sect psychosom med, NIMH, 60-66; prof psychiat, Univ Ill Col Med, 66-72, PROF PSYCHIAT, SCH MED, YALE UNIV, 72- Concurrent Pos: Dir res, Ill State Psychiat Inst, 66-72. Mem: AAAS; Am Psychiat Asn; Am Psychosom Soc. Res: Relationship between biology and behavior; brain chemistry and behavior; biochemistry of synapse; biogenic amines in brain; neurochemistry. Mailing Add: Dept of Psychiat Yale Univ Sch Med New Haven CT 06510

MAAS, JOHN LEWIS, b Detroit, Mich, Aug 13, 40; m 62; c 3. PLANT PATHOLOGY. Educ: Mich State Univ, BS, 62; Univ Wash, MS, 64; Ore State Univ, PhD(plant path), 68. Prof Exp: RES PLANT PATHOLOGIST, FRUIT LAB, PLANT GENETICS & GERMPLASM INST, AGR RES CTR, NORTHEAST REGION, US DEPT AGR, 68- Mem: Am Phytopath Soc; Mycol Soc Am. Res: Fungus diseases of small fruit crops; etiology and control of Phytophthora fragariae root rot and Botrytis cinerea fruit rot of strawberry. Mailing Add: Plant Genetics & Germplasm Inst Agr Res Ctr NE Region US Dept of Agr Beltsville MD 20705

MAAS, KEITH ALLAN, b Burlington, Wis, Apr 7, 36; m 63; c 2. PHOTOGRAPHIC CHEMISTRY. Educ: Mass Inst Technol, SB, 58; Univ Vt, MS, 60; Univ Calif, Davis, PhD(chem), 63. Prof Exp: Res chemist, tech rep & col rels rep, E I du Pont de Nemours & Co, Inc, 63-71; SR STAFF SCIENTIST, PHOTOGRAPHIC SYSTS, TECHNICOLOR GRAPHIC SERVS, INC, 72- Mem: Am Chem Soc; Soc Photog Scientists & Engrs; Am Soc Photogram. Res: Photographic materials and processes; techniques of image presentation. Mailing Add: EROS Data Ctr Sioux Falls SD 57198

MAAS, PETER, b Evanston, Ill, Apr 9, 39; m 64; c 2. SOLID STATE PHYSICS, BIOPHYSICS. Educ: Mass Inst Tech, BS, 62; Stanford Univ, MS, 64; Univ Colo, PhD(physics), 69. Prof Exp: Asst engr, Lockheed Missile & Space Co, 62-63, grad study scientist, 63-65, scientist, 64-65; asst physics, Univ Colo, Boulder, 65-67, solid state physics, 67-69, res assoc & instr biophys, Med Ctr, 69-70; LECTR APPL PHYSICS, UNIV STRATHCLYDE, 70- Mem: Asn Comput Mach; Am Asn Physics Teachers. Res: Molecular physics; computer uses in science. Mailing Add: Dept of Appl Physics Univ of Strathclyde 107 Rottenrow Glasgow Scotland

MAAS, WERNER KARL, b Kaiserslautern, Ger, Apr 27, 21; nat US; m 60; c 3. MOLECULAR GENETICS. Educ: Harvard Univ, BA, 43; Columbia Univ, PhD(zool), 48. Prof Exp: Asst zool, Columbia Univ, 43-45; mem staff, Med Col, Cornell Univ, 49-54; from asst prof pharmacol to assoc prof microbiol, 54-63, PROF MICROBIOL, SCH MED, NY UNIV, 63-, ADV GRAD DEPT, 64-, CHMN DEPT BASIC MED SCI, 75- Concurrent Pos: Vis investr, Mass Gen Hosp, 52-53; dir honors prog, Sch Med, NY Univ, 58-61, co-dir, USPHS Genetics Training Grant, 61-68, dir, 68-75, dir, Microbiol Training Grant, 64-69; mem, NIH Genetics Training Grants Comt, 61-65, mem study sect microbial chem, 68-72, chmn, 70-72; mem staff, Univ Brussels, 63; mem test comt microbiol, Nat Bd Med Exam, 71-75. Mem: Am Soc Microbiol; Am Soc Biol Chem; Genetics Soc Am. Res: Microbial genetics and physiology, with emphasis on regulatory mechanisms, especially of protein synthesis; amino acid permeases; polyamine metabolism; genetics of extrachromosomal elements. Mailing Add: Dept of Microbiol NY Univ Sch of Med New York NY 10016

MAASBERG, ALBERT THOMAS, b Bronx, NY, Feb 10, 15; m 41; c 2. CHEMISTRY, CHEMICAL ENGINEERING. Educ: State Univ NY, BS, 36. Prof Exp: Res dir cellulose prod dept, 46-50, asst prod mgr, 50-52, mgr, 52-54, tech dir plastics prod dept, 54-56, dir res & develop, Midland Div, 56-63, DIR CONTRACT RES, DEVELOP & ENG, DOW CHEM USA, 63- Mem: Am Chem Soc; Sigma Xi; Tech Asn Pulp & Paper Indust; Am Inst Chem Eng. Res: Pulp and paper manufacturing; cellulose ethers; water soluble polymers; chemicals and plastics process and product; government and industrial contract research and development administration. Mailing Add: 566 Bldg Dow Chem USA Midland MI 48640

MAASEIDVAAG, FRODE, b Stavanger, Norway, July 30, 37; US citizen; m; c 2. BIOENGINEERING, ELECTROPHYSIOLOGY. Educ: Univ Mich, BSEE, 63, MS, 66, PhD(bioeng), 69. Prof Exp: Instr, Univ Mich, Ann Arbor, 68-72, asst prof ophthal, 72-74; ASSOC PROF OPHTHAL & BIOENG, UNIV MO-COLUMBIA, 74- Concurrent Pos: Res engr, Environ Res Inst Mich, 73- Mem: AAAS; Inst Elec & Electronics Eng; Soc Neurosci. Res: Clinical electrophysiology; biomedical instrumentation; diagnostic ultrasound. Mailing Add: Univ of Missouri 141 E E Bldg Columbia MO 65201

MAASKE, CLARENCE ALFRED, b Mayville, Wis, Oct 3, 08; m 37; c 3. PHYSIOLOGY. Educ: Univ Wis, BA, 30, PhD(physiol), 37. Prof Exp: Instr physiol, Sch Med, Univ Wis, 35-37; instr, Sch Med, Loyola Univ, 37-41, assoc, 41-42; chmn dept physiol, 52-70, assoc prof physiol & pharmacol, 45-50, PROF PHYSIOL, SCH MED, UNIV COLO, DENVER, 50-, DIR MED STUDENT ADV PROG, 70- Concurrent Pos: Consult, Fitzsimmons Gen Hosp, 46-60, Vet Admin Hosp, 48-62 & Nat Jewish Hosp, 52- Mem: Fel AAAS; Am Physiol Soc; Soc Exp Biol & Med. Res: Chemical mediation; cardiovascular and respiratory physiology; aviation physiology. Mailing Add: Dept of Physiol Univ of Colo Sch of Med Denver CO 80220

MAASS, ALFRED ROLAND, b Plymouth, Wis, Apr 14, 18; m 47; c 3. BIOCHEMISTRY. Educ: Antioch Col, BS, 42; Univ Wis, MS, 47, PhD(biochem), 50. Prof Exp: Consult, Argonne Nat Labs, 51; sr res biochemist, 51-56, group leader, Biochem Sect, 56-57, asst sect head, 57-62, sect head, 62-67, assoc dir biochem, 67-74, DIR DEVELOP PROJ, SMITH, KLINE & FRENCH LABS, 74- Concurrent Pos: Chmn civilian adv comt radiation safety, City Philadelphia, 64-73. Mem: Fel AAAS; Health Physics Soc; Am Chem Soc; Am Acad Neurol; NY Acad Sci. Res: Isotope tracers; drug metabolism; enzymes; neurobiochemistry; gastric and renal physiology; anti-hypertensive, diuretic, uricosuric and anti-lipemic agents. Mailing Add: 415 Cornell Ave Swarthmore PA 19081

MAASS, GEORGE JOSEPH, b New York, NY, Jan 6, 38; m 61; c 2. PHYSICAL CHEMISTRY. Educ: Fordham Univ, BS, 61; Iowa State Univ, PhD(phys chem), 67. Prof Exp: Res scientist paper chem, Union-Camp Res Labs, 67-70; RES SCIENTIST, SCOTT PAPER CO, 70- Mem: Am Chem Soc. Res: Low temperature magnetic properties; molecular quantum mechanics; thermodynamic and surface properties of fibers; penetration and structure of porous media. Mailing Add: Scott Paper Co Scott Plaza Philadelphia PA 19113

MAASS, WOLFGANG SIEGFRIED GÜNTHER, b Helsinki, Finland, Oct 23, 29; m 60; c 2. PLANT PHYSIOLOGY, PLANT BIOCHEMISTRY. Educ: Univ Tübingen, Dr rer nat, 57. Prof Exp: Asst bot, Univ Tübingen, 57; sci collabr, Max-Planck Inst Protein & Leather Res, 58-60; fel biol, Dalhousie Univ, 60-62; asst res officer, 62-65, ASSOC RES OFFICER, ATLANTIC REGIONAL LAB, NAT RES COUN CAN, 66- Concurrent Pos: Mem, Plant Phenolics Group NAm. Mem: Can Soc Plant Physiol; Ger Bot Soc. Res: Taxonomy and distribution of Sphagnum; chemical taxonomy; biochemical aspects of cellular development; biosynthesis of pulvinic acid derivatives and phenolic compounds in general. Mailing Add: Nat Res Coun Can Atlantic Regional Lab 1411 Oxford Halifax NS Can

MAASSAB, HUNEIN FADLO, b Damascus, Syria, June 11, 28; nat US; m 59; c 2. EPIDEMIOLOGY, VIROLOGY. Educ: Univ Mo, BA, 50, MA, 52; Univ Mich, MPH, 55, PhD(epidemiol sci), 56; Am Bd Med Microbiol, dipl. Prof Exp: Res assoc, 57-60, from asst prof to assoc prof, 60-72, PROF EPIDEMIOL, UNIV MICH, ANN ARBOR, 72- Mem: AAAS; Am Asn Immunol; Brit Soc Gen Microbiol; Tissue Cult Asn; Soc Exp Biol & Med. Res: Metabolism of infection; host-virus interaction; immunology; tissue culture; tumors; biology of myxoviruses. Mailing Add: Dept of Epidemiol Univ of Mich Sch of Pub Health Ann Arbor MI 48109

MAATMAN, RUSSELL WAYNE, b Chicago, Ill, Nov 7, 23; m 48; c 5. PHYSICAL CHEMISTRY. Educ: Calvin Col, AB, 46; Mich State Col, PhD(chem), 50. Prof Exp: Asst prof chem, DePauw Univ, 49-51; sr technologist, Socony-Mobil Oil Co, 51-58; assoc prof chem, Univ Miss, 58-63; PROF CHEM, DORDT COL, 63- Mem: Am Chem Soc; Am Sci Affiliation. Res: Catalysis; solution-solid reactions; ion solvation. Mailing Add: Dept of Chem Dordt Col Sioux Center IA 51250

MABEN, JERROLD WILLIAM, b Detroit, Mich, Feb 17, 29; m 56; c 3. SCIENCE EDUCATION, SCIENCE WRITING. Educ: Wayne State Univ, BA, 50, BS, 51, MS, 54; Ohio State Univ, PhD, 71. Prof Exp: Teaching asst physics, Wayne State Univ, 48-51; teacher pub schs, Mich, 51-56; instr & coordr student teaching & sci & math educ, Mich State Univ, 56-58, coordr sci & math teaching ctr, 58-63; assoc prof sci & math educ & dir sci educ progs, Univ Akron, 63-70; CHMN SCI EDUC, LEHMAN COL, 71- Concurrent Pos: Researcher, Governor's Comn Educ Finance, Mich, 53-57; res assoc educ planning, Wayne State Univ, 54-56; co-dir inst col & high sch teachers phys sci & math, Mich State Univ, 58-59, dir NSF Traveling Sci & Math Teacher Prog, 59-61, dir, In-Serv Inst Earth Sci Teachers, Univ Akron, 69-70; lectr sci, Agency Int Develop Latin Am Prog, Univ Akron, 63-65, dir aerospace & conserv sci educ workshops, 65-66, NASA seminar & workshop space-oriented sci & math, 66-69; ed, Discovering Sci News, 68-71; res assoc, Educ Resources Info Ctr Sci & Math & Environ Educ, 69-71; consult Bd of Educ, Pontiac, Mich, 61-63 & Cleveland, Ohio, 65-66; consult, Govt Guyana, SA, Ministry Educ, Agency Int Develop, 68-69; consult math, NY Bd Educ, 73-; mem bd dir, Coun Elem Sci Int, 73- & Asn Educ Teachers Sci, 74-; dir, Viking Student Proj, NASA, 74-75; consult film ed, Sci Screen Report, Allegro Film Prods, 71- & What Is Science Series, Prentice Hall Media Prods, 75-; mem membership Comt, Nat Asn Res Sci Teaching, 75- Mem: Fel AAAS; Nat Asn Res Sci Teaching; Nat Sci Teachers Asn; Am Educ Res Asn; NY Acad Sci. Res:

Science curriculum development for science courses kindergarten through twelfth grade, community-technical colleges and university general education science; coordination of national science teaching study; competency-based science education; environmental education. Mailing Add: 33 Midbrook Lane Old Greenwich CT 06870

MABEY, WILLIAM RAY, b Los Angeles, Calif, Oct 16, 41. PHYSICAL ORGANIC CHEMISTRY. Educ: Univ Calif, Riverside, BA, 65; San Diego State Univ, MS, 68; Univ Ore, PhD(chem), 72. Prof Exp: PHYS ORG CHEMIST, STANFORD RES INST, 72- Mem: Am Chem Soc; AAAS. Res: Environmental chemistry; kinetics and mechanisms of hydrolysis, oxidation and photochemistry in environmen; persistence and fate of chemicals in environment. Mailing Add: Stanford Res Inst 333 Ravenswood Menlo Park CA 94025

MABIE, CURTIS PARSONS, b Memphis, Tenn, Feb 26, 32; m 59; c 1. DENTAL MATERIALS, MICROSCOPY. Educ: Western Reserve Univ, BA, 54; Univ Mich, MS, 58. Prof Exp: Res geologist, US Bur Mines, 58-61; anthracologist, Appl Res Ctr, US Steel Corp, 61-63; res microscopist, IRC, 63-65; res engr, Norton Co, 65-66; ceramic technologist, NL Indust Inc, 66-67; CHIEF SCIENTIST, CERAMICS DIV, AM DENT ASN, NAT BUR STANDARDS, 68- Mem: Am Ceramic Soc; Am Soc Testing & Mat. Res: Dental porcelains, investments, fillers, and cements; microscopy of dental materials and biological calcifications. Mailing Add: Bldg 224 Rm A164 Nat Bur of Standards Washington DC 20234

MABIS, ALTON JOHN, b Edgerton, Ind, Oct 6, 20; m 43; c 3. PHYSICAL CHEMISTRY. Educ: Capital Univ, BS, 42; Ohio State Univ, PhD(chem), 47. Prof Exp: Chem lab asst, Capital Univ, 40-42; asst chem, Ohio State Univ, 42-43, 45-46; res scientist, Manhattan Proj, Columbia Univ, 43-45; res scientist, Carbide & Carbon Chem Corp, NY, 45; res chemist, 47-71, COORDR PHD RECRUITING, MIAMI VALLEY LABS, PROCTER & GAMBLE CO, 71- Concurrent Pos: Bd Regents, Capital Univ, 72- Mem: Am Chem Soc. Res: X-ray diffraction; crystal structure; liquid crystals. Mailing Add: Procter & Gamble Co Miami Valley Labs PO Box 39175 Cincinnati OH 45247

MABROUK, AHMED FAHMY, b Cairo, UAR, Sept 30, 23; m 54; c 3. AGRICULTURAL CHEMISTRY. Educ: Univ Cairo, BSc, 45, MSc, 50; Ohio State Univ, PhD(lipid chem), 54. Prof Exp: Chemist, Ministry Agr, Egypt, 45-46; instr chem, Fac Agr, Univ Cairo, 46-48, instr agr indust, 48-51; fel agr chem, Ohio State Univ, 54-55; lectr & asst prof chem, Univ Cairo, 55-58, lectr, Grad Sch, 56-58; res org chemist, Am Meat Inst Found, Univ Chicago, 58-61; prin org chemist, Northern Utilization Res & Develop Div, USDA, 61-65; HEAD FLAVOR CHEM GROUP, FOOD LAB, US ARMY NATICK LABS, 65- Concurrent Pos: Fel physiol chem, Ohio State Univ, 56; lectr to postgrads, Agr Schs, Egypt. Consult, Tahreer Prov Authority, Egypt, 55-57 & Nat Serv Coun, 56-57. Abstractor, Chem Abstr, 55-65. Honors & Awards: Sci Dir Silver Key Res Award, 68. Mem: Am Oil Chemists Soc; Am Chem Soc; Inst Food Technologists; The Chem Soc; Brit Soc Chem Indust. Res: Chromatography of organic compounds; gel permeation chromatography; gas chromatography; ultrafiltration; fat and flavor chemistry; heterogeneous and homogenous hydrogenation of fats and organic compounds; chemical kinetics of oxidative rancidity; antioxidants; isolation and synthesis of naturally occurring compounds. Mailing Add: 9 Wildewood Terr Framingham MA 01701

MABRY, TOM JOE, b Commerce, Tex, June 6, 32; m 54, 71; c 1. ORGANIC CHEMISTRY, PHYTOCHEMISTRY. Educ: ETex State Univ, BS & MS, 53; Rice Univ, PhD(chem), 60. Prof Exp: NIH fel chem, Org Chem Inst, Univ Zurich, 60-61; res scientist, 62, from asst prof to prof phytochem, 63-73, PROF BOT, UNIV TEX, AUSTIN, 73- Concurrent Pos: Guggenheim fel, Univ Freiburg, 71. Mem: Am Chem Soc; The Chem Soc; Bot Soc Am; Phytochem Soc NAm (vpres, 65, pres, 66-67). Res: Natural products chemistry; biochemical systematics; molecular evolution. Mailing Add: Dept of Bot Univ of Tex Austin TX 78712

MACADAM, DAVID LEWIS, b Philadelphia, Pa, July 1, 10; m 38; c 4. PHYSICS. Educ: Lehigh Univ, BS, 32; Mass Inst Technol, PhD, 36. Prof Exp: Sr res assoc, Res Labs, Eastman Kodak Co, 36-75; RETIRED. Concurrent Pos: Mattiello mem lectr, Fedn Socs Paint Technol, 65; Hurter & Driffield mem lectr, Royal Photog Soc, 66; chmn tech comt colorimetry, Int Comn Illum; del, Inter-Soc Color Coun; ed jour, Optical Soc Am, 64- Honors & Awards: Godlove Award, Inter-Soc Color Coun, 63; Lomb Medal, Optical Soc Am, 40. Mem: Fel Optical Soc Am (pres, 62). Res: Optics; color photography; influence of color contrast on visual acuity; spectroradiometry; color television; photographic image structure. Mailing Add: 68 Hammond St Rochester NY 14615

MACALADY, DONALD LEE, b Shamokin, Pa, Apr 19, 41; m 64; c 2. PHYSICAL INORGANIC CHEMISTRY, ENVIRONMENTAL CHEMISTRY. Educ: Pa State Univ, BS, 63; Univ Wis-Madison, PhD(chem), 69. Prof Exp: Jr engr, H R B Singer, Inc, 63; asst prof chem, Grinnell Col, 68-70; asst prof, 70-74, ASSOC PROF CHEM, NORTHERN MICH UNIV, 74- Concurrent Pos: NSF Fac fel sci, Rosenstiel Sch of Marine & Atmospheric Sci, Univ Miami, 75-76. Mem: AAAS; Am Chem Soc; Water Pollution Control Fedn. Res: Nuclear quadrupole resonance; aquatic chemistry. Mailing Add: Dept of Chem Northern Mich Univ Marquette MI 49855

MACALPIN, ARCHIE JUSTUS, geology, see 12th edition

MACALPINE, GORDON MADEIRA, b Bozeman, Mont, Feb 23, 45; m 67. ASTROPHYSICS. Educ: Earlham Col, AB, 67; Univ Wis, PhD(astron), 71. Prof Exp: Mem, Inst Advan Study, 71-72; ASST PROF ASTRON, UNIV MICH, ANN ARBOR, 72- Mem: Am Astron Soc. Res: Detailed theoretical and observational investigation of the emission-line regions in quasi-stellar objects and Seyfert galaxies. Mailing Add: Dept of Astron Univ of Mich Ann Arbor MI 48109

MACALUSO, ANTHONY, SR, b New Orleans, La, Oct 4, 39; m 61; c 3. CHEMISTRY. Educ: Loyola Univ, BS, 61; Tulane Univ, MS, 63, PhD(chem), 65. Prof Exp: SR CHEMIST, RES & TECH DEPT, TEXACO INC, 67- Mem: Am Chem Soc. Res: Exploratory research in organic and petroleum chemistry. Mailing Add: 4135 42nd St Port Arthur TX 77640

MACALUSO, MARY CHRISTELLE, b Lincoln, Nebr, July 9, 31. ANATOMY. Educ: Col St Mary, Nebr, BS, 56; Univ Notre Dame, MS, 61; Univ Nebr, PhD(anat), 66. Prof Exp: From instr to assoc prof, 61-73, PROF BIOL, COL ST MARY, NEBR, 73-, CHMN DEPT, 67- Mem: Nat Asn Biol Teachers. Res: Sex education; ultrastructure of the corpus luteum of pregnancy and the corpus luteum of lactation in Swiss mice during the first nineteen days postpartum. Mailing Add: Dept of Biol Col of St Mary Omaha NE 68124

MACALUSO, PAT, b New York, NY, Aug 29, 16; m 41; c 2. PHYSICAL CHEMISTRY, COMPUTER SCIENCE. Educ: City Col NY, BS, 39; Polytech Inst Brooklyn, MS, 46. Prof Exp: Chemist indust & consumer specialties, Foster D Snell, Inc, 40-42, group leader, 42-45, acct exec, 45-47; group leader agr chem formulations, 47-54, head prod develop & res sect, 54-65, head polymer prod sect, 65-67, MGR INFO SERV, EASTERN RES CTR, STAUFFER CHEM CO, 67- Mem: Am Chem Soc; Am Soc Testing & Mat; NY Acad Sci; Asn Comput Mach. Res: Polymers, elastomers; polyvinyl chloride; industrial chemical applications; chemical specialties; chemistry of sulfur; physical chemistry of proteins; computer applications in chemical research. Mailing Add: 9 Church Court White Plains NY 10603

MACANDER, RUDY F, organic chemistry, see 12th edition

MACARTHUR, DONALD M, b Detroit, Mich, Jan 7, 31; m 62; c 2. ENVIRONMENTAL MANAGEMENT, HEALTH SCIENCES. Educ: Univ St Andrews, BSc, 54; Univ Edinburgh, PhD(x-ray crystallog), 57. Prof Exp: Lectr chem, Univ Conn, 57-58; from sr scientist to mgr chem & life sci res ctr, Melpar Inc, 58-66; dep dir chem & mat, Off Secy of Defense, 66, dep dir res & technol, 66-70; PRES & CHIEF EXEC OFFICER, ENVIRO CONTROL, INC, 70- Concurrent Pos: Consult, Off Water Resources, Dept of Interior, DC, 65-66; contrib ed, Am Ord Mag, 65-66; consult, Off Dir Defense Res & Eng; mem bd, Diversitron, Inc; Module Systs & Develop, Inc; pres, Dynamac Inc, 75- Mem: AAAS; Am Chem Soc; Am Water Works Asn; fel Am Inst Chem. Res: Air quality and water quality trends; health effects of air pollution; hazardous materials transportation and associated risk analysis; health research in cancer and relationship of nutrition to cancer. Mailing Add: 5313 Albermarle St NW Washington DC 20016

MACARTHUR, JOHN DUNCAN, b Toronto, Ont, Apr 13, 36; m 59; c 2. NUCLEAR PHYSICS. Educ: Univ Western Ont, BSc, 58; McMaster Univ, PhD(nuclear physics), 62. Prof Exp: Asst prof, 62-71, ASSOC PROF PHYSICS, QUEEN'S UNIV, ONT, 71- Mem: Can Asn Physicists; Am Asn Physics Teachers. Res: Low energy nuclear physics. Mailing Add: Dept of Physics Queen's Univ Kingston ON Can

MACARTHUR, JOHN WOOD, b Chicago, Ill, Sept 1, 22; m 47; c 5. PHYSICS. Educ: Univ Toronto, BA, 45; Rensselaer Polytech Inst, PhD(physics), 53. Prof Exp: MEM FAC PHYSICS, MARLBORO COL, 48- Concurrent Pos: With res div, Mass Inst Technol, 57-58; with lunar & planetary lab, Univ Ariz, 66-67. Mem: Am Phys Soc; Am Asn Physics Teachers; Am Geophys Union. Res: Astronomy; geophysics; population biology. Mailing Add: Dept of Physics Marlboro Col Marlboro VT 05344

MACARTHUR, KENNETH WILLIAM, b Milwaukee, Wis, Dec 24, 12; m 41; c 3. ENTOMOLOGY. Educ: Wis State Col, Milwaukee, BE, 36; Univ Mich, MS, 50. Prof Exp: From asst cur to assoc cur invert zool, 37-60, cur div entom & parasitol, 60-70, CUR DEPT INVERT ZOOL, MILWAUKEE PUB MUS, 70- Concurrent Pos: Lectr, Marquette Univ, 43-44. Mem: Entom Soc Am. Mailing Add: Milwaukee Pub Mus 800 W Wells St Milwaukee WI 53233

MACARTHUR, NORMAN CURRIE, organic chemistry, polymer chemistry, see 12th edition

MACARTHUR, ROBERT HELMER, biology, deceased

MACAULAY, WESLEY CLAUDE, b East Dudswell, Que, Oct 28, 09; m 43; c 2. PHARMACEUTICAL CHEMISTRY. Educ: Univ Sask, PhC, 33, BSP, 36; Univ Ont, PhC, 42; Purdue Univ, MS, 42; Univ Montreal, DPharm, 55. Prof Exp: From instr to asst prof pharm, Ont Col Pharm, 37-39; from instr to asst prof, 39-46, PROF & DEAN PHARM, UNIV SASK, 46- Mem: Am Pharmaceut Asn; Can Pharmaceut Asn; Can Soc Hosp Pharmacists; fel Chem Inst Can. Res: Manufacture of compressed tablets; preparations intended for parenteral administration; allergenic extracts. Mailing Add: Col of Pharm Univ of Sask Saskatoon SK Can

MACAVOY, THOMAS COLEMAN, b Jamaica, NY, Apr 24, 28; m 52; c 4. ORGANIC CHEMISTRY. Educ: Queens Col, NY, BS, 50; St Johns Univ, MS, 52; Univ Cincinnati, PhD(chem), 57. Prof Exp: Anal chemist, Chas Pfizer & Co, Inc, 53-54; sr chemist, 57-61, mgr electronic res, 61-64, dir phys res, 64-66, gen mgr & vpres, Electronic Prod Div, 66-69, gen mgr tech prod group, 69-71, PRES & DIR, CORNING GLASS WORKS, 71- Concurrent Pos: Chmn bd, Cormedics Corp, NJ & Corhart Refractories Co, Ky. Res: Composition and properties of glass: complex phosphates and phosphate glasses; ion exchange equilibria; high temperature inorganic chemistry; growth of single crystals of refractory compounds. Mailing Add: Corning Glass Works Houghton Park Corning NY 14830

MACBETH, ROBERT ALEXANDER, b Edmonton, Alta, Aug 26, 20; m 49; c 4. SURGERY. Educ: Univ Alta, BA, 42, MD, 44; McGill Univ, MSc, 47, dipl surg, 52; FRCS(C), 52. Prof Exp: Assoc prof, Univ Alta, 57-60, prof surg & head dept, 60-75, dir surg serv, Univ Hosp, 60-75; PROF SURG, DALHOUSIE UNIV, 75-, ASSOC DEAN, 75- Concurrent Pos: Res fel endocrinol, McGill Univ, 46-47, teaching fel anat, 47-48; Nuffield Found traveling fel surg, Brit Postgrad Med Sch, 50-51; consult, Dept Vet Affairs, Col Mewburn Pavillion, Edmonton, 57-75; dir med educ, Prov of New Brunswick, 75- Mem: Am Surg Asn; fel Am Col Surg; Soc Univ Surg; Can Med Asn; Can Soc Clin Invest. Res: Metabolic adaption to cold; magnesium metabolism; serum and tissue glycoproteins in relation to malignant and inflammatory disease. Mailing Add: Dept of Med Educ 310 St John Gen Hosp St John NB Can

MACCABE, JEFFREY ALLAN, b Oakland, Calif, Jan 30, 43; m 71. DEVELOPMENTAL BIOLOGY. Educ: Univ Calif, Davis, BS, 64, PhD(genetics), 69. Prof Exp: Res assoc, State Univ NY Albany, 69-71; ASST PROF ZOOL, UNIV TENN, KNOXVILLE, 72- Mem: AAAS; Am Genetic Asn; Am Soc Zool; Soc Develop Biol; NY Acad Sci. Res: Morphogenesis of the vertebrate limb. Mailing Add: Dept of Zool Univ of Tenn Knoxville TN 37916

MACCABEE, BRUCE SARGENT, b Rutland, Vt, May 6, 42; m 70. THERMODYNAMICS, ELECTROOPTICS. Educ: Worcester Polytech Inst, BS, 64; Am Univ, MS, 67, PhD(physics), 70. Prof Exp: Res assoc physics, Am Univ, 67-72; RES PHYSICIST, WHITE OAK LAB, NAVAL SURFACE WEAPONS CTR, 72- Concurrent Pos: Consult, Nat Invests Comt Aerial Phenomena, 66-, Tracor, Inc, 70-71, Compackager Corp, 70-73 & Sci Appln, Inc, 73-74. Mem: Am Phys Soc; AAAS; Am Optical Soc. Res: Critical phenomena, light scattering, critical equation of state of fluids; electrooptical imaging systems; forward looking infrared imaging systems. Mailing Add: White Oak Lab Naval Surface Weapons Ctr Silver Spring MD 20910

MACCABEE, HOWARD DAVID, biophysics, radiological physics, see 12th edition

MACCALLUM, CRAWFORD JOHN, b New York, NY, May 28, 29; m 51; c 5. RADIATION PHYSICS. Educ: Princeton Univ, BA, 51; Univ NMex, PhD(physics), 62. Prof Exp: STAFF MEM, SANDIA CORP, 57- Concurrent Pos: Fulbright lectr, Univ Cairo, 64-65. Mem: Am Phys Soc. Res: Radiation physics; analytical methods electron penetration in solids; gamma ray astronomy. Mailing Add: Sandia Corp 5231 Sandia Base Albuquerque NM 87115

MACCALLUM, DONALD KENNETH, b Los Angeles, Calif, Apr 13, 39; m 62; c 2. ANATOMY, HISTOLOGY. Educ: Pomona Col, BA, 61; Univ Southern Calif, MS,

64, PhD(anat), 66. Prof Exp: Staff mem exp path, Walter Reed Army Inst Res, 66-68; asst prof, 69-72, ASSOC PROF ANAT, DENT SCH, UNIV MICH, ANN ARBOR, 72-, ASST PROF ANAT, MED SCH & STAFF MEM CYTOL, DENT RES INST, 69- Concurrent Pos: Nat Inst Dent Res fel, Case Western Reserve Univ, 68-69; asst prof lectr, Sch Med, George Washington Univ, 67-68. Mem: Am Asn Anat; Am Soc Cell Biol. Res: Oral mucosal ultrastructure; epithelial-connective tissue interaction; experimental cytology. Mailing Add: Dept of Anat Univ of Mich Med Sch Ann Arbor MI 48104

MACCAMY, RICHARD C, b Spokane, Wash, Sept 26, 25; m 49; c 3. MATHEMATICS. Educ: Reed Col, AB, 49; Univ Calif, Berkeley, PhD(math), 55. Prof Exp: Res engr, Univ Calif, Berkeley, 50-54, res mathematician, 54-56; from asst prof to assoc prof, 56-64, PROF MATH, CARNEGIE-MELLON UNIV, 64-, ASSOC CHMN DEPT, 74- Concurrent Pos: Air Force Off Sci Res fel, 56-66; Soc Naval Archit & Marine Engrs fel, 57-63; Boeing Sci Res Lab grant, 63-64. Mem: Am Math Soc. Res: Partial differential and integral equations; fluid dynamics; elasticity; electromagnetic theory. Mailing Add: Dept of Math Carnegie-Mellon Univ Pittsburgh PA 15213

MACCANNELL, KEITH LEONARD, b Transcona, Man, Jan 8, 34. CLINICAL PHARMACOLOGY. Educ: Univ Man, BSc & MD, 58, PhD(pharmacol), 63; FRCPS(C). Prof Exp: Vis asst prof pharmacol & internal med, Sch Med, Emory Univ, 64-66; assoc prof pharmacol & asst prof internal med, Univ BC, 67-69; PROF PHARMACOL & THERAPEUT & CHMN DEPT, FAC MED, UNIV CALGARY, 69- Concurrent Pos: Can Found for Advan Therapeut fel, Emory Univ, 64-66; Markle scholar & Med Res Coun Can scholar, 66- Mem: Am Fedn Clin Res; Pharmacol Soc Can; Can Med Asn. Res: Cardiovascular physiology-pharmacology. Mailing Add: Dept of Pharmacol & Therapeut Univ of Calgary Fac of Med Calgary AB Can

MACCANON, DONALD MOORE, b Norwood, Iowa, June 17, 24; m 46; c 2. CARDIOVASCULAR PHYSIOLOGY, PHARMACOLOGY. Educ: Drake Univ, BA, 48; Univ Iowa, MS, 51, PhD(physiol), 53. Prof Exp: Asst, Drake Univ, 42-43; asst physiol, Univ Iowa, 50-51, res assoc, 53-54; asst prof physiol & pharmacol, Sch Med Sci, Univ SDak, 54-60; asst prof physiol & pharmacol & chief exp cardiol, Chicago Med Sch, 60-61, from assoc prof to prof cardiovasc res, 61-71, assoc dir, 62-71, from assoc prof to prof physiol, 64-71; health scientist adminr, Training Grants & Awards Br, Nat Heart & Lung Inst, 71-72, actg head postdoctoral sect, Fel Br, Nat Inst Gen Med Sci, 72-73; health scientist adminr, Cardiac Functions Br, Div Heart & Vascular Dis, Nat Heart & Lung Inst, 73-74, CHIEF MANPOWER BR, DIV HEART & VASCULAR DIS, NAT HEART & LUNG INST, 74- Concurrent Pos: Life Ins res fel physiol, Univ Iowa, 53-54; NIH career prog awardee, 61-71; consult, Physiol Fels Rev Panel, USPHS, 65-68, Anesthesiol Training Comt, NIH, 68-70 & Chicago Heart Res Comt, 69-70. Honors & Awards: Morris L Parker Award Meritorious Res, 64; Res Award, Interstate Postgrad Med Asn, 67. Mem: Fel AAAS; fel Am Col Cardiol; Am Physiol Soc; Am Heart Asn. Res: Cardiac function; pulmonary circulation; hemodynamics; cardiac vibrations. Mailing Add: 401 Nina Pl Rockville MD 20852

MACCARTHY, HUBERT REAGH, b Eng, June 22, 11; Can citizen; m 53; c 1. ENTOMOLOGY. Educ: Univ BC, BA, 50; Univ Calif, PhD(entom), 53. Prof Exp: Student entomologist, 48, asst plant path, 49, asst entomologist, 50-55, HEAD, INSECT SECT, CAN DEPT AGR, 55- Mem: Am Entom Soc; Am Phytopath Soc. Res: Insect transmission of plant virus diseases. Mailing Add: 6660 NW Marine Dr Vancouver BC Can

MACCARTY, COLLIN STEWART, b Rochester, Minn, Sept 20, 15; m 40; c 3. NEUROSURGERY. Educ: Dartmouth Col, AB, 37; Johns Hopkins Univ, MD, 40; Univ Minn, MS, 44; Am Bd Neurol Surg, dipl, 72. Prof Exp: From instr to prof neurosurg, Mayo Grad Sch Med, Univ Minn, 61-73, PROF NEUROSURG, MAYO MED SCH, 73-, MEM DEPT NEUROSURG, MAYO CLIN, 46-, CHMN, 63- Concurrent Pos: Pres-elect med staff, Mayo Clin, 65, pres, 66; secy cong affairs, Liaison & Admin Coun & chmn prog comt, World Fedn Neurosurg Socs, 65-69; nat consult neurol surg, US Air Force, 72- Mem: Am Col Surg; Soc Neurol Surg; Neurosurg Soc Am (vpres, 54, pres, 59); Am Asn Neurol Surg (vpres, 6S-66, pres, 70-71). Mailing Add: Mayo Clin Dept of Neurol Surg 200 First St SW Rochester MN 55901

MACCHESNEY, JOHN BURNETTE, b Glen Ridge, NJ, July 8, 29; m 53; c 1. SOLID STATE CHEMISTRY. Educ: Bowdoin Col, BA, 51; Pa State Univ, PhD, 59. Prof Exp: MEM TECH STAFF, BELL LABS, INC, 59- Res: Preparation and properties of electronic materials. Mailing Add: Bell Labs Inc 600 Mountain Ave Murray Hill NJ 07971

MACCHI, I ALDEN, b Bologna, Italy, Feb 21, 22; nat US; m 53; c 1. ENDOCRINOLOGY. Educ: Clark Univ, BA, 47, MA, 50; Boston Univ, PhD(endocrinol), 54. Prof Exp: Mem res staff, Worcester Found Exp Biol, 50-54; asst prof physiol, Clark Univ, 54-56; from asst prof to assoc prof 56-64, exec asst res, Biol Sci Ctr, 56-67, PROF BIOL, BOSTON UNIV, 64-, ACTG CHMN DEPT, COL LIB ARTS, 74- Concurrent Pos: Lalor Found fel, 55; vis lectr & Dept Sci & Indust Res Eng sr res fel, Univ Sheffield, 62-63. Mem: Am Physiol Soc; Am Soc Zool; Endocrine Soc; Soc Exp Biol & Med; Am Diabetes Asn. Res: Comparative aspects of corticosteroid biogenesis; regulation of adrenocortical and pancreatic endocrine secretion; transplantation of adrenal and endocrine pancreas. Mailing Add: 52 Roundwood Rd Newton MA 02164

MACCINI, JOHN ANDREW, b Boston, Mass, July 9, 28; m 61; c 4. GEOLOGY. Educ: Boston Univ, BA, 52, MA, 54; Ohio State Univ, PhD(earth sci educ), 69. Prof Exp: Sr tech officer, Nfld Geol Surv, 53; construct & soils engr, Thompson & Lichtner Co, Brookline, Mass, 54-58; teacher earth sci, Lincoln-Sudbury Regional High Sch, Mass, 58-66; teaching assoc geol, Ohio State Univ, 66-69; assoc prof geol & sci educ, Univ Md, College Park, 69-71; PROG MGR SCI EDUC, NSF, 71- Concurrent Pos: Regional dir skylab student proj, NASA, 70-71. Mem: Nat Sci Teachers Asn; Nat Asn Geol Teachers; Nat Asn Res Sci Teaching. Res: Improvement of undergraduate science instruction with specific focus on audio-visual tutorial laboratory development for undergraduate geology courses. Mailing Add: Div of Sci Educ Educ Directorate Resources Improv NSF 1800 G St Washington DC 20550

MACCLINTOCK, COPELAND, b Princeton, NJ, Dec 3, 30; m 56; c 2. INVERTEBRATE PALEONTOLOGY. Educ: Franklin & Marshall Col, BS, 54; Univ Wyo, MA, 57; Univ Calif, Berkeley, PhD(paleont), 64. Prof Exp: Field asst, US Geol Surv, 54-55; res asst, 63-65, RES ASSOC, PEABODY MUS, YALE UNIV, 65- Concurrent Pos: NASA res grant, 65-68. Mem: Geol Soc Am; Paleont Soc. Res: Microstructure and growth of fossil and recent mollusk shells; relationship between shell structures and classification, phylogeny and ecology of mollusks. Mailing Add: Peabody Mus Yale Univ New Haven CT 06520

MACCLUER, JEAN WALTERS, b Columbus, Ohio, Mar 30, 37. HUMAN GENETICS, POPULATION GENETICS. Educ: Ohio State Univ, BSc, 59; Univ Mich, MSc, 63, PhD(human genetics), 68. Prof Exp: Res asst mech eng, Battelle Mem Inst, 59-60; elec eng, Antenna Lab, Ohio State Univ, 60-62; res assoc human genetics, Univ Mich, 68-71; res assoc anthrop, 71-72, asst prof, 72-74, ASSOC PROF BIOL, PA STATE UNIV, UNIVERSITY PARK, 74- Concurrent Pos: Mem, Arteriosclerosis Res Ctrs Adv Comt, Nat Heart & Lung Inst, 72-74. Mem: Am Soc Human Genetics; Am Soc Naturalists; Pop Asn Am; Soc Study Evolution; Soc Study Social Biol. Res: Population genetics; genetic demography; computer simulation. Mailing Add: Dept of Biology Pa State Univ University Park PA 16802

MACCOLL, ROBERT JOSEPH, b Brooklyn, NY, Mar 27, 42; m 63; c 3. PHYSICAL CHEMISTRY, BIOCHEMISTRY. Educ: Queens Col, NY, BA, 63; Univ Miss, MS, 67; Adelphi Univ, PhD(phys chem), 69. Prof Exp: Fel, 69-70, res scientist, 70-72, SR RES SCIENTIST, DIV LABS & RES, NY STATE DEPT HEALTH, 72- Concurrent Pos: Adj asst prof microbiol, Albany Med Col, 74- Mem: Am Chem Soc. Res: Protein chemistry; protein subunit studies; virus assembly; aggregation of C-phycocyanin. Mailing Add: Div of Labs & Res NY State Dept of Health Albany NY 12201

MACCOLLOM, GEORGE BUTTERICK, b Boston, Mass, June 10, 25; m 53; c 4. ENTOMOLOGY. Educ: Univ Mass, BS, 50; Cornell Univ, PhD, 54. Prof Exp: Entomologist, Exp Sta, 54-66, PROF ENTOM, UNIV VT, 66- Mem: Entom Soc Am. Res: Biology and control of apple insects; detection and measurement of environmental contamination by insecticides. Mailing Add: Dept of Soil Sci Univ of Vt Burlington VT 05401

MACCONNACHIE, HUGH JOHN, b New Glasgow, NS, Dec 29, 20; m 42; c 3. RESTORATIVE DENTISTRY. Educ: Dalhousie Univ, DDS, 53; Ind Univ, MSD, 67. Prof Exp: PROF OPER DENT, DALHOUSIE UNIV, 65- Mem: Int Asn Dent Res; Can Dent Asn. Res: Restorative dental materials. Mailing Add: Fac of Dent Dalhousie Univ Halifax NS Can

MACCONNELL, JOHN GRIFFITH, b Chicago, Ill, Oct 14, 42. NATURAL PRODUCTS CHEMISTRY. Educ: Univ Ill, Urbana, BS, 64; Univ Mich, Ann Arbor, MS, 68, PhD(org chem), 69. Prof Exp: Res assoc org chem, Univ Ga, 69-70, NY State Col Forestry, Syracuse Univ, 70-72 & Univ Ga, 72-73; asst prof chem, Dalton Jr Col, 73-74; instr psychiat & Chem, Emory Univ, 74-75; SR CHEMIST, MERCK SHARPE & DOHME RES LABS, 75- Mem: AAAS; Am Chem Soc; The Chem Soc; NY Acad Sci. Res: Isolation and characterization of complex natural products; structure determination via spectrometric methods. Mailing Add: Dept of Appl Microbiol & Nat Prod Merck Sharpe & Dohme Res Labs Rahway NJ 07065

MACCONNELL, WILLIAM PRESTON, b NB, Can, June 15, 18; nat US; m 43; c 3. FORESTRY. Educ: Univ Mass, BS, 43; Yale Univ, MF, 48. Prof Exp: Assoc prof, 48-63, PROF FORESTRY, UNIV MASS, AMHERST, 63- Mem: Soc Am Foresters; Am Soc Photogram. Res: Forest management; aerial photogrammetry. Mailing Add: Dept of Forestry Univ of Mass Amherst MA 01002

MACCORMICK, ALASDAIR JOHN, b Glasgow, Scotland, Feb 13, 36; Brit citizen. MEDICAL STATISTICS. Educ: Univ Edinburgh, BS, 57; Univ Wis, MS, 64 & 66, PhD(statist), 70. Prof Exp: Exp officer grade 3 elec eng, Dept Supply, Australian Govt, 60-62; proj assoc statist, Inst Environ Studies, 70-72, ASST PROF STATIST, DEPT OPHTHAL, MED SCH, UNIV WIS-MADISON, 72- Mem: Fel Royal Statist Soc; Am Statist Asn. Res: Diabetic retinopathy; diabetes; clinical controlled trials; natural history of disease; epidemiology. Mailing Add: Dept of Ophthal Univ Wis Hosps 1300 University Ave Madison WI 53706

MACCOSS, MALCOLM, b Cleator, Eng, June 2, 47; m 71; c 1. BIO-ORGANIC CHEMISTRY. Educ: Univ Birmingham, Eng, BSc, 68, PhD(chem), 71. Prof Exp: Fel chem, Univ Alta, 72-74, res assoc, 74-76; ASST SCIENTIST BIO-ORG CHEM, ARGONNE NAT LAB, 76- Mem: The Chem Soc; Am Chem Soc. Res: Chemistry and biochemistry of nucleic acids, nucleosides and nucleotides; synthesis of nucleic acid components of potential biological and medicinal interest. Mailing Add: Div of Biol & Med Res Argonne Nat Lab Argonne IL 60439

MACCOY, CLINTON VILES, b Brookline, Mass, Mar 27, 05; m 36; c 2. ECOLOGY. Educ: Harvard Univ, AB, 28, AM, 29, PhD(biol), 34. Prof Exp: Cur fishes & mammals, Secy & ed, Boston Soc Natural Hist, 29-39; asst prof zool, Mass State Col, 39-44; from assoc prof to prof, 44-70, EMER PROF BIOL, WHEATON COL, MASS, 70-; DIR NORWELL LABS, 70- Concurrent Pos: Asst biologist, State Fish & Game Dept, NH, 39; res assoc, Woods Hole Oceanog Inst, 55-64. Mem: Am Soc Mammal; Am Soc Ichthyol & Herpet; Am Soc Limnol & Oceanog; Int Soc Limnol. Res: Limnology; oceanography, especially estuarine areas; intertidal in-fauna, especially meiofauna. Mailing Add: Norwell Labs 77Winter St Norwell MA 02061

MACCRACKEN, ELLIOTT B, b Winnetka, Ill, May 24, 11; m 56; c 3. SCIENCE EDUCATION, MATHEMATICS. Educ: Ore State Col, BS, 32; Columbia Univ, MA, 41; Stanford Univ, EdD, 53. Prof Exp: Teacher pub schs, Ore, 34-36, high schs, 36-42; engr, Watson Labs, 46; from instr to assoc prof, 46-55, PROF SCI & CHMN SCI & MATH DIV, SOUTHERN ORE COL, 55- Mem: Am Asn Physics Teachers; Inst Elec & Electronics Engrs; Am Asn Univ Profs. Res: Preparation of teachers. Mailing Add: 645 Glenwood Ashland OR 97520

MACCRACKEN, MICHAEL CALVIN, b Schenectady, NY, May 20, 42; m 67; c 2. ATMOSPHERIC PHYSICS, AIR POLLUTION. Educ: Princeton Univ, BSE, 64; Univ Calif, Davis, MS, 66, PhD(appl sci), 68. Prof Exp: Physicist, 68-74, DEP DIV LEADER ATMOSPHERIC & GEOPHYS SCI, LAWRENCE LIVERMORE LAB, UNIV CALIF, 74- Mem: Am Meteorol Soc; Am Geophys Union; Am Quaternary Asn; Arctic Inst NAm. Res: Numerical simulation of processes governing the global climate and the factors causing climatic change, and regional air quality for land use planning and control strategy assessment. Mailing Add: Lawrence Livermore Lab PO Box 808 Livermore CA 94550

MACCREADY, PAUL BEATTIE, JR, b New Haven, Conn, Sept 29, 25; m 57; c 2. AIR POLLUTION. Educ: Yale Univ, BS, 47; Calif Inst Technol, MS, 48, PhD(aeronaut), 52. Prof Exp: Meteorol consult, Salt River Valley Water Users' Asn, 50-51; pres, Meteorol Res, Inc, 51-70 & Atmospheric Res Group, 58-70; PRES, AERO VIRONMENT INC, 71-, CONSULT, ADV COMT REACTOR SAFEGUARDS, 73- Concurrent Pos: Res asst, Calif Inst Technol, 52-53; consult, President's Adv Comt Weather Control, 56-57. Mem: AAAS; assoc fel Am Inst Aeronaut & Astronaut; Am Geophys Union; Am Meteorol Soc; Weather Modification Asn. Res: Instrumentation development in aeronautics and atmospheric science; basic and applied studies in turbulence and diffusion; cloud physics, cloud electrification and weather modification. Mailing Add: Aero Vironment Inc 145 Vista Ave Pasadena CA 91107

MACCREADY, ROBERT ALVAN, b Fredericton, NB, May 8, 02; US citizen; m 48; c 4. PUBLIC HEALTH. Educ: Dartmouth Col, BS, 25; Harvard Univ, MD, 32; Am Bd Path, cert clin microbiol, 60. Prof Exp: Teacher pub sch, NJ, 25-28; practicing physician, Mass, 34-41; epidemiologist, State Dept Pub Health, Mass, 41-46, asst dir, Div Commun Dis, 46-54, Biol Labs, Inst Labs, 54-57, dir, DIV DIAG LABS, 57-68;

RETIRED. Concurrent Pos: Instr, Med Sch, Harvard Univ, 46-49, assoc, 49-58, Sch Pub Health, 58-64, lectr, 64-69; lectr Simmons Col, 53-64; adj prof microbiol, Windham Col, 69-72; consult State Univ NY, Buffalo, 70- Mem: Am Soc Microbiol; fel Am Med Asn; fel Am Pub Health Asn. Res: Public health diagnostic bacteriology and epidemiology; surveillance studies of outbreaks, especially of enteric disease; screening for metabolic disorders; laboratory administration; effectiveness of newborn screening programs for phenylketonuria and other inborn errors of metabolism in the prevention of mental retardation. Mailing Add: RFD 1 Box 386 Jaffrey NH 03452

MACCREARY, DONALD, b Literberry, Ill, Nov 18, 02; m 32; c 1. ENTOMOLOGY. Educ: Iowa Wesleyan Col, BS, 29; Univ Md, MS, 30. Prof Exp: Fel, Crop Protection Inst, 30-32, asst entomologist, 32-41, assoc entomologist, 41-50, assoc res prof, 50-58, res prof, 58-71, EMER PROF ENTOM, UNIV DEL, 71- Mem: Entom Soc Am; Mosquito Control Asn. Res: Bionomics and control of agricultural insects and insects affecting man and animals; general entomology. Mailing Add: Dept of Entomol Univ of Del Newark DE 19711

MACDIARMID, ALAN GRAHAM, b Masterton, NZ, Apr 14, 27; m 54; c 4. INORGANIC CHEMISTRY. Educ: Univ NZ, BSc, 48, MSc, 50; Univ Wis, MS, 52, PhD(chem), 53; Cambridge Univ, PhD(chem), 55. Prof Exp: Asst lectr chem, St Andrews Univ, 55; from instr to assoc prof, 55-64, PROF CHEM, UNIV PA, 64- Concurrent Pos: Sloan fel, Univ Pa, 59-63. Mem: Am Chem Soc; The Chem Soc. Res: Preparation and characterization of simple inorganic derivatives of silicon hydrides, organosilicon compounds, derivatives of sulfur nitrides, fluorides of silicon, sulfur and phosphorus; ultra high pressure synthesis. Mailing Add: Dept of Chem Univ of Pa Philadelphia PA 19104

MACDIARMID, WILLIAM DONALD, b Arcola, Sask, June 22, 26; m 53; c 4. INTERNAL MEDICINE, HUMAN GENETICS. Educ: Univ Sask, BA, 47; Univ Toronto, MD, 49; Am Bd Internal Med, dipl, 67; FRCPS(C) & cert, 67. Prof Exp: Instr internal med, Col Med, Univ Utah, 64-66, from asst prof to assoc prof, 66-69; prof med fac med, Univ Man, 69-75; PROF & CHMN MED, FAC MED, MEM UNIV NFLD, 75- Concurrent Pos: NIH training grant, Univ Utah, 60-62; Neuromuscular Found fel, Univ London, 62-64; consult, Winnipeg Children's Hosp, 69- & Winnipeg Gen Hosp, 70-; physician-in-chief, St Boniface Gen Hosp; sr consult, Janeway Child Health Ctr, St Clare's Mercy Hosp. Mem: AAAS; fel Am Col Physicians; Genetics Soc Can; Am Soc Human Genetics. Mailing Add: Dept of Med Gen Hosp St John's NF Can

MACDONALD, ALAN ANGUS, b Hutchinson, Kans, June 27, 42. ORGANIC CHEMISTRY. Educ: Harvey Mudd Col, BS, 63; Univ Wash, PhD(org chem), 68. Prof Exp: Res chemist, 68-75, TECH AST TO DIR, RICHMOND RES CTR, STAUFFER CHEM CO, 75- Mem: Am Chem Soc. Res: Agricultural chemistry; insecticides, herbicides, plant growth regulators. Mailing Add: Richmond Res Ctr Stauffer Chem Co 1200 S 47th St Richmond VA 94804

MACDONALD, ALASTAIR DAVID, b Victoria, BC, July 19, 43; m 68; c 2. BOTANY, MORPHOLOGY. Educ: Univ Victoria, BC, BSc, 65; McGill Univ, PhD(bot), 72. Prof Exp: Lectr, 69-71, ASST PROF BIOL, LAKEHEAD UNIV, 71- Mem: AAAS; Can Bot Asn; Bot Soc Am; Am Inst Biol Sci. Res: Developmental floral morphology; theoretical morphology. Mailing Add: Dept of Biol Lakehead Univ Thunder Bay ON Can

MACDONALD, ALEX BRUCE, b Anaconda, Mont, Mar 7, 34; m 59; c 2. IMMUNOLOGY, BIOCHEMISTRY. Educ: Carroll Col, Mont, AB, 56; Mich State Univ, PhD(biochem), 67. Prof Exp: Chemist, Anaconda Aluminum Co, Mont, 56-57; biochemist, Abbott Labs, 59-63; asst biochem, Mich State Univ, 63-67; res assoc, 69-70, asst prof microbiol, 70-75, ASSOC PROF IMMUNOL, SCH PUB HEALTH, HARVARD UNIV, 75- Concurrent Pos: NIH fel immunol, Med Sch, Univ Ill, 67-69. Mem: Am Chem Soc; Fedn Am Socs Exp Biol; Am Asn Immunol. Res: Immunological response to infectious agents, including chlamylia and virus studies on possible competition between substances in immune response leading to detrimental effects in the host. Mailing Add: Dept of Microbiol Harvard Sch of Pub Health Boston MA 02115

MACDONALD, ALEXANDER, JR, b Quincy, Mass, Oct 29, 36; m 62; c 1. ANALYTICAL CHEMISTRY. Educ: Northeastern Univ, BS, 59; Univ Iowa, MS & PhD(chem), 63. Prof Exp: Asst prof chem, Mich State Univ, 65-67; sr chemist, 67-71, res fel, 71-74, RES GROUP CHIEF, HOFFMAN-LA ROCHE, INC, 75- Mem: Am Chem Soc. Res: Chromatography; aerosols; automated analysis; residue analysis; veterinary drug metabolism. Mailing Add: Hoffman-La Roche Inc Nutley NJ 07110

MACDONALD, ALEXANDER DANIEL, b Sydney, NS, Apr 8, 23; m 46; c 4. PHYSICS. Educ: Dalhousie Univ, BSc, 45, MSc, 47; Mass Inst Technol, PhD(physics), 49. Prof Exp: Res assoc physics, Mass Inst Technol, 48-49; from asst prof to prof, Dalhousie Univ, 49-60; sr specialist, Microwave Physics Lab, Gen Tel & Electronics Lab, Inc, 60-62; prof appl math & head div, Dalhousie Univ, 62-65; sr mem res labs, 65-73, DIR, ELECTRONIC & COMMUN SCI LAB, PALO ALTO LABS, LOCKHEED MISSILES & SPACE CO, 73- Concurrent Pos: Res scientist, Defence Res Bd, Can, 49-52; mem nat comn electronics, Int Union Radio Sci, 52-; specialist, Sylvania Elec Prod Co, 56-57. Mem: Fel Am Phys Soc. Res: Physical electronics; ultrasonics; theoretical physics; computer science. Mailing Add: 3056 Greer Rd Palo Alto CA 94303

MACDONALD, BURNS, b Baltimore, Md, Feb 10, 28. EXPERIMENTAL PHYSICS. Educ: Univ Calif, Berkeley, BA, 49, MA, 51, PhD(physics), 64. Prof Exp: Res assoc physics, Lawrence Radiation Lab, Univ Calif, 64-65; asst prof, Va Polytech Inst, 65-70; PHYSICIST, LAWRENCE BERKELEY LAB, 71- Mem: Am Phys Soc. Res: Medium energy physics. Mailing Add: Lawrence Radiation Lab 80-226 Univ of Calif Berkeley CA 94720

MACDONALD, CAROLYN TROTT, b Iowa City, Iowa, June 23, 41; m 62; c 3. THEORETICAL CHEMISTRY, APPLIED MATHEMATICS. Educ: Univ Minn, BS, 61; Univ Ore, MA, 63 & 65; Brown Univ, PhD(chem), 68. Prof Exp: Asst prof chem, Moorhead State Col, 68-74; ASST PROF PHYS SCI, UNIV MO-KANSAS CITY, 74- Mem: AAAS; Am Chem Soc; Am Phys Soc; Am Asn Physics Teachers; Math Asn Am. Res: Mathematical model for kinetics of polypeptide synthesis on polyribosomes; formulation of mathematical model for chemical/physical/biological processes. Mailing Add: Dept of Chem Univ of Mo Kansas City MO 64110

MACDONALD, DAVID HOWARD, b Cleveland, Ohio, Sept 23, 34; m 56; c 3. PLANT NEMATOLOGY, POMOLOGY. Educ: Purdue Univ, BS, 56; Cornell Univ, MS, 62, PhD(pomol), 66. Prof Exp: From asst prof to assoc prof plant nematol, 66-74, ASSOC PROF PLANT PATH, UNIV MINN, ST PAUL, 74- Mem: Am Phytopath Soc; Soc Nematol. Res: Chemical and physical factors affecting rate of build-up of plant parasitic nematodes. Mailing Add: Dept of Plant Path Univ of Minn St Paul MN 55101

MACDONALD, DAVID J, b San Diego, Calif, May 14, 32; m 62; c 2. PHYSICAL CHEMISTRY, INORGANIC CHEMISTRY. Educ: Calif Inst Technol, BS, 53, MS, 54; Univ Calif, Los Angeles, PhD(phys chem), 60. Prof Exp: Chem engr, Apache Powder Co, 54-55; res asst, Univ Calif, Los Angeles, 60; asst prof chem, Univ Nev, Reno, 63-68; proj leader, 68-69, RES CHEMIST, RENO METALL RES CTR, US BUR MINES, 69- Mem: AAAS; Am Chem Soc; The Chem Soc. Res: Kinetics of substitution reactions in coordination compounds; molten-salt chemistry; metallurgical chemistry, primarily liquid-liquid extraction. Mailing Add: Reno Metall Res Ctr US Bur Mines 1605 Evans Ave Reno NV 89505

MACDONALD, DIGBY DONALD, b Thames, NZ, Dec 7, 43; m 67; c 3. PHYSICAL CHEMISTRY, ELECTROCHEMISTRY. Educ: Univ Auckland, BSc, 65, Univ Calgary, PhD(chem), 69. Hon Degrees: MSc, Univ Auckland, 66. Prof Exp: Res officer chem, Atomic Energy Can, Ltd, 69-72; lectr, Victoria Univ Wellington, 73-74; HON ASSOC & PROF CHEM, UNIV CALGARY, 75-; SR RES ASSOC, ALBERTA SULPHUR RES, LTD, 75- Mem: Electrochem Soc. Res: The electrochemical and thermodynamic behavior of metals in high temperature/high pressure solutions; corrosion; theoretical electrochemistry; effect of pressure on reactions in solution; liquid mixtures; sulphur; structure of water. Mailing Add: 5808 66th Ave NW Calgary AB Can

MACDONALD, DONALD CRAIG, physics, see 12th edition

MACDONALD, DONALD LAURIE, b Toronto, Ont , Nov 16, 22; nat US; m 47; c 3. BIOCHEMISTRY. Educ: Univ Toronto, BA, 44, MA, 46, PhD(chem), 48. Prof Exp: Asst, Univ Calif, 48-50, instr, 50-52, asst prof biochem, 52-57; vis scientist, Nat Inst Arthritis & Metab Dis, 57-60, chemist, 60-61; USPHS fel, Heidelberg, 61-62; prof biochem, 62-67, PROF BIOCHEM & BIOPHYS, ORE STATE UNIV, 67- Concurrent Pos: USPHS career develop award, 62-67; fel, Harvard Med Sch, 70-71. Mem: Am Chem Soc; Am Soc Biol Chem; The Chem Soc. Res: Degradation of monosaccharides; sugar phosphates; glycoproteins. Mailing Add: Dept of Biochem & Biphhysics Ore State Univ Corvallis OR 97331

MACDONALD, DONALD MACKENZIE, b Newark, NJ, July 26, 28; Can citizen; m 60, c 2. CELLULOSE CHEMISTRY. Educ: Mt Allison Univ, BSc, 49; Univ NB, Fredericton, MSc, 51, PhD(org chem), 53. Prof Exp: Nat Res Coun Can fel, Nat Res Coun Labs, 53-54; defense sci serv officer org & polymer chem, Defense Res Bd Can, 54-60; res assoc cellulose chem, Can Int Paper Res Ltd, 60-69; RES CHEMIST, INT PAPER CO, 69- Mem: Am Chem Soc. Res: Chemistry of viscose process and cellulose derivatives. Mailing Add: Corporate Res Ctr Int Paper Co Box 797 Tuxedo Park NY 10987

MACDONALD, DUNCAN ROSS, b Hamilton, Ont, Oct 15, 28; m 59; c 2. FOREST ENTOMOLOGY. Educ: Univ Toronto, BScF, 52; Univ Mich, MF, 57. Prof Exp: Res off forest entom, Forest Res Lab, Can Dept Forestry, NB, 52-68, head forest entom sect, BC, 68- 69, PROG MGR FOREST PROTECTION RES, PAC FOREST RES CENTER, DEPT ENVIRON, CAN FORESTRY SERV, 69-, DEP DIR, 73- Mem: Can Inst Forestry; Can Entom Soc. Res: Aerial spraying against spruce budworm; effects of pesticides on population dynamics of the budworm, associated species and the forest. Mailing Add: Pac Forest Res Ctr Dept of Environ Can Forestry Serv 506 Burnside Rd W Victoria BC Can

MACDONALD, ELEANOR JOSEPHINE, b West Somerville, Mass, Mar 4, 09; c 1. EPIDEMIOLOGY. Educ: Radcliffe Col, AB, 28. Prof Exp: Statistician, Mass Dept Pub Health, 30-35, epidemiologist, 35-40; res statistician, Div Cancer Res, Conn Dept Health, 41-48; prof epidemiol, epidemiologist & head dept, 48-74, EMER PROF EPIDEMIOL, UNIV TEX M D ANDERSON HOSP, 74- Concurrent Pos: Vis lectr res methods, Dent Sch, Tufts Univ, 33-43; lectr soc sci, Regis Col, Mass, 36-38; consult, Nat Adv Cancer Coun, 44-46; consult, Mem Hosp, New York, 47-57; asst clin prof, Sch Med, Yale Univ, 48-60; prof biostatist, Postgrad Sch Med, Univ Tex, 48-63; mem dir adv coun, Univ Tex, M D Anderson Hosp & Tumor Inst, 48-59, 63-65, mem adv comt biomath & sci computation, 62-63, chmn educ comt, 63-65, consult epidemiologist, Tex Dept Health, 49-63; consult statistician, Tex Cancer Coord Coun, 49-63; chmn definitions comt, End Results Eval Sect, Cancer Chemother Nat Serv Ctr, 58-59; statist consult, Dept Pediat, Baylor Univ Col Med, 58-62; prof biostatist, Univ Tex Grad Sch Biomed Sci, Houston, 63-65. Honors & Awards: Myron Gordon Award, 8th Int Pigment Cell Growth Conf, 72; Outstanding Serv Award, Am Cancer Soc, 73. Mem: AAAS; Am Asn Cancer Res; Pub Health Cancer Asn Am (pres, 58, secy-treas, 51-57); Am Pub Health Asn; Biomet Soc. Mailing Add: Univ of Tex Syst Cancer Ctr M D Anderson Hosp & Tumor Inst Houston TX 77030

MACDONALD, ETTA MAE, b Houston, Tex, Nov 26, 19. MICROBIOLOGY. Educ: Univ Tex, BS & BA, 40, MD, 49; Univ Wis, MS, 43, PhD(pharmacol), 47. Prof Exp: Tutor pharm, Univ Tex, 40-41; asst, Univ Wis, 41-47; res assoc prev med, 47-49, asst prof bact & parasitol, 50-55, intern, Med Br, Hosp, 49-50, ASSOC PROF MICROBIOL, UNIV TEX MED BR GALVESTON, 55- Concurrent Pos: Fel, Oak Ridge Inst Nuclear Studies, 52; Smith-Mundt grant, Helsinki, Finland, 52-53. Mem: AAAS; Am Soc Parasitol; Am Soc Trop Med & Hyg; Wildlife Dis Asn; Mycol Soc Am. Res: Biological fermentations; chemotherapy, infection and immunity of Trichomonas; infections and immunity of Filariasis; medical mycology. Mailing Add: Dept of Microbiol Univ of Tex Med Br Galveston TX 77551

MACDONALD, EVE LAPEYROUSE, b Baton Rouge, La, Jan 2, 29; m 50; c 1. DEVELOPMENTAL BIOLOGY, CELL BIOLOGY. Educ: Wellesley Col, AB, 50; Bryn Mawr Col, MA, 65, PhD(develop biol), 67. Prof Exp: Teaching asst, Bryn Mawr Col, 64-66; asst prof, 67-75, ASSOC PROF BIOL, WILSON COL, 75-, DIR INST ELECTRON MICROS, 75- Mem: AAAS; Am Soc Zool; Am Inst Biol Sci; Soc Develop Biol; Electron Microscope Soc Am. Res: Development of the amphibian eye at different temperatures; cellular differentiation in Rana pipiens and Xenopus laevis; ultrastructure of differentiating cells. Mailing Add: Dept of Biol Wilson Col Chambersburg PA 17201

MACDONALD, FRANK WHITEMORE, sanitary engineering, see 12th edition

MACDONALD, GORDON ANDREW, b Boston, Mass, Oct 15, 11, m 38; c 4. GEOLOGY. Educ: Univ Calif, Los Angeles, AB, 33, AM, 34; Univ Calif, PhD, 38. Prof Exp: Asst geol sci, Univ Calif, 36-38; asst geologist, Shell Oil Co, 38-39; from asst geologist to geologist, US Geol Surv, 39-58; SR PROF GEOL & GEOPHYSICS, UNIV HAWAII, 58- Concurrent Pos: Asst prof, Univ Southern Calif, 47-48; dir, Hawaiian Volcano Observ, 51-56. Mem: Geol Soc Am; Mineral Soc Am; Am Geophys Union; Int Asn Volcanology. Res: Volcanology; ground-water geology; petrography and petrology of Hawaiian volcanoes; geologic mapping in Hawaii and California; prediction and control of volcanic eruptions and alleviation of volcanic disasters. Mailing Add: Inst of Geophysics Univ of Hawaii Honolulu HI 96822

MACDONALD, GORDON J, b Staten Island, NY, May 18, 34; m 56; c 2. PHYSIOLOGY, ANATOMY. Educ: Rutgers Univ, BS, 55, MS, 58, PhD, 61. Prof

Exp: Res asst dairy sci, Rutgers Univ, 57-58; trainee endocrinol, Univ Wis, 61-63; res fel physiol, Sch Dent Med, Harvard Univ, 63-64, res asst, 64-68; res assoc anat, New Eng Regional Primate Res Ctr, Harvard Med Sch, 68-, asst prof, Lab Human Reproduction & Reproductive Biol, 69-73; ASSOC PROF ANAT, RUTGERS MED SCH, COL MED & DENT NJ, 73- Mem: Am Soc Zool; Endocrine Soc; Soc Study Reproduction; NY Acad Sci; Int Soc Res Reprod. Res: Role of pituitary gonadotropins in ovarian function; hormonal control of mammary gland growth and lactation; endocrinology reproduction. Mailing Add: Dept of Anat Rutgers Med Sch-CMDNJ Piscataway NJ 08854

MACDONALD, GORDON JAMES FRASER, b Mexico City, Mex, July 30, 29; nat US; m 50, 69; c 4. GEOPHYSICS. Educ: Harvard Univ, AB, 50, AM, 52, PhD(geophys), 54. Prof Exp: From asst prof to assoc prof geol, Mass Inst Technol, 54-58; prof geophys, Univ Calif, Los Angeles, 58-68, assoc dir inst geophys & planetary sci, 60-68, chmn dept planetary & space sci, 65-66; prof geophys, Univ Calif, Santa Barbara, 68-74; PROF ENVIRON STUDIES & POLICY & PROF EARTH SCI, DARTMOUTH COL, 74-; MEM, COUN ENVIRON QUAL, EXEC OFF PRESIDENT, 70- Concurrent Pos: Mem space sci bd, Nat Acad Sci, 62-70; mem adv panel weather modification, NSF, 64-67; with Woods Hole Oceanog Inst, 64-; mem, US-Japen Comt Sci Coop, Dept State, 65-67, consult, 67-70; mem, President's Sci Adv Comt, 65-69; vchmn, Environ Studies Bd, 69, chmn, 70; mem sci & technol comt, NASA; vchancellor res & grad affairs, Univ Calif, Santa Barbara, 68-70; ed, Rev Geophys, Am Geophys Union, 62-70; chmn, Nat Res Coun Natural Resources & mem ex officio Coord Comt Air Qual Studies. Honors & Awards: James B Macelwane Award, Am Geophys Union, 65. Mem: Nat Acad Sci; AAAS; Am Acad Arts & Sci; Am Philos Soc; Geophys Union. Res: Rotation of the earth; physics of interior of planets; use of computers in geophysics. Mailing Add: Dept of Earth Sci Dartmouth Col Hanover NH 03755

MACDONALD, HAROLD CARLETON, b Englishtown, NS, Sept 18, 30; m 55; c 5. GEOLOGY. Educ: State Univ NY, Binghampton, BA, 60; Univ Kans, MS, 62, PhD(geol), 69. Prof Exp: Explor geologist, Sinclair Oil & Gas Co, 62-65; res assoc geol, Ctr Res, Univ Kans, 69-70; asst prof, Ga Southern Col, 70-71; assoc prof geol, Univ Ark, Fayetteville, 71-73; ASSOC DIR, ARK WATER RESOURCES RES CTR, 73- Mem: Geol Soc Am; Am Asn Petrol Geol; Sigma Xi; Am Soc Photogram; Am Geophys Union. Res: Geoscience evaluation of side-looking airborne imaging radars; geological remote sensing, geohydrology in carbonate terrains. Mailing Add: Dept of Geol Univ of Ark Fayetteville AR 72701

MACDONALD, HUBERT C, JR, b Detroit, Mich, Aug 3, 41; m 67; c 4. PHYSICAL CHEMISTRY, ANALYTICAL CHEMISTRY. Educ: Wheeling Col, BS, 63; Univ Mich, MS, 65, PhD(chem), 69. Prof Exp: RES SCIENTIST, KOPPERS CO, INC, 69- Mem: Am Chem Soc; Electrochem Soc. Res: Electrochemistry; x-ray spectroscopy; inorganic chemistry; catalyst testing; gas chromatography. Mailing Add: Koppers Co Inc 440 College Park Dr Monroeville PA 15146

MACDONALD, JACK ROBERT, physics, nuclear physics, see 12th edition

MACDONALD, JAMES CAMERON, b Can, Nov 3, 26; m 57. BIOCHEMISTRY. Educ: Univ Alta, BSc, 49; Univ Wis, MSc, 51, PhD(physiol chem), 53. Prof Exp: Asst, Univ Wis, 49-53; Anna Fuller Fund fel, Univ Sheffield, 53-54 & Oxford Univ, 54-55; Nat Res Coun Can fel, McMaster Univ, 55-56; with Gerber Baby Foods Co, 56-57; from asst res officer to assoc res officer, 57-64, SR RES OFFICER MICROBIOL, PHYSIOL & BIOCHEM, PRAIRIE REGIONAL LAB, NAT RES COUN CAN, 64- Mem: Am Chem Soc; Chem Inst Can. Res: Microbiology. Mailing Add: Prairie Regional Lab Nat Res Coun Can Saskatoon SK Can

MACDONALD, JAMES REID, b St Helena, Calif, Aug 22, 18; m 41, 57, 69; c 3. VERTEBRATE PALEONTOLOGY, GEOLOGY. Educ: Univ Calif, BA, 40, MS, 47, PhD(paleont), 49. Prof Exp: Lab asst mus paleont, Univ Calif, 45-46, lab technician, 46-49; from asst prof to assoc prof geol, SDak Sch Mines & Technol, 49-57, cur, 49-57; occup analyst, Idaho Dept Hwy, 57-58; res assoc, Am Mus Natural Hist, 58-60; spec projs technician, Idaho Dept Hwy, 60-61; dep dir admin, State of Idaho, 61-62; assoc prof geol, Univ Idaho, 62; sr cur vert paleont, Los Angeles County Mus Natural Hist, 62-69; instr geog, Calif State Polytech Col, San Luis Obispo, 70; PROF GEOL, FOOTHILL COL, 70- Concurrent Pos: Adj prof, Univ Southern Calif, 65-69. Mem: Geol Soc Am; Soc Vert Paleont; Paleont Soc; Soc Study Evolution; Am Soc Mammal. Res: Pliocene faunas of Nevada, California and South Dakota; Oligocene and Miocene mammals of South Dakota and adjacent regions. Mailing Add: Dept of Geol Foothill Col 12345 El Monte Rd Los Altos Hills CA 94022

MACDONALD, JAMES ROBERT, b Toronto, Ont, May 24, 36; m 68; c 2. EXPERIMENTAL PHYSICS. Educ: Univ Toronto, BA, 58; McMaster Univ, MSc, 64, PhD(physics), 66. Prof Exp: Teacher, North York Bd Educ, 58-62; res asst, McMaster Univ, 63-66; Nat Res Coun Can fel, Niels Bohr Inst, Copenhagen, 66-68; from asst prof to assoc prof, 68-75, PROF PHYSICS, KANS STATE UNIV, 75- Mem: Am Phys Soc; Can Asn Physicists; Am Asn Physics Teachers. Res: Atomic collision studies pertaining to interactions of energetic particles with matter. Mailing Add: Dept of Physics Kans State Univ Manhattan KS 66502

MACDONALD, JAMES ROSS, b Savannah, Ga, Feb 27, 23; m 46; c 3. SOLID STATE PHYSICS. Educ: Williams Col, BA, 44; Mass Inst Technol, SB, 44, SM, 47; Oxford Univ, DPhil(physics), 50, DSc, 67. Prof Exp: Asst, Radar & Electronics Lab, Mass Inst Technol, 43-44, asst elec eng, 46-47; physicist, Armour Res Found, Ill Inst Technol, 50-52; assoc physicist, Argonne Nat Lab, 52-53; res physicist semiconductor & solid state physics, Tex Instruments Inc, 53-74, dir solid state physics res, 55-61, dir physics res lab, 61-63, dir cent res labs, 63-72, asst vpres corp res & eng, 67, vpres, 68-74, vpres corp res & develop, 73-74; W R KENAN JR PROF PHYSICS, DEPT PHYSICS & ASTRON, UNIV NC, CHAPEL HILL, 74- Concurrent Pos: Adj assoc prof, Southwestern Med Sch, Univ Tex, 54, adj prof, 71-74; chmn, Numerical Data Adv Bd, Nat Acad Sci, 70-74, mem, Comt Motor Vehicle Emissions, 71-73, chmn, 73-74; mem coun, Nat Acad Eng, 71-74. Honors & Awards: Achievement Award, Inst Elec & Electronics Engrs, 62, Merit Serv Award, 74; Achievement Award, Inst Radio Engrs. Mem: Nat Acad Sci; Nat Acad Eng; fel Am Phys Soc; fel Inst Elec & Electronics Engrs. Res: Semiconductors; space-charge; electrolyte double layer; data analysis; equations of state. Mailing Add: 308 Laurel Hill Rd Chapel Hill NC 27514

MACDONALD, JOHN ALAN, b Honolulu, Hawaii, Oct 25, 41; m 65; c 2. COMPARATIVE PHYSIOLOGY. Educ: Stanford Univ, AB, 63; Univ Tex, Austin, PhD(zool), 70. Prof Exp: Field res asst zool, US Antarctic Res Prog, McMurdo Sound, 63-64; asst, Univ Tex, Austin, 65-70; NIH traineeship neurophysiol, Univ Calif, Los Angeles, 70-72; LECTR BIOL, UNIV AUCKLAND, 72- Mem: Am Soc Zool; Physiol Soc NZ. Res: Neurosensory physiology of invertebrates and lower vertebrates; chemoreception; signal interaction and changes of excitability in the peripheral nervous system. Mailing Add: Dept of Zool Univ of Auckland Private Bag Auckland New Zealand

MACDONALD, JOHN BARFOOT, b Toronto, Ont, Feb 23, 18; m 67; c 5. MICROBIOLOGY. Educ: Univ Toronto, DDS, 42; Univ Ill, MS, 48; Columbia Univ, PhD(bact), 53; Univ Man, LLD, 62; Univ BC, DSc, 67. Prof Exp: Lectr prev dent, Univ Toronto, 42-44, instr bact, 46-47; res asst, Univ Ill, 47-48; asst prof, Univ Toronto, 49-53, assoc prof bact & chmn div dent res, 53-56, prof, 56; prof microbiol, Sch Dent Med, Harvard Univ, 56-62, dir postdoctoral studies, 60-62; pres, Univ BC, 62-67; prof higher educ, Univ Toronto, 68; EXEC DIR, COUN ONT UNIVS, 68- Concurrent Pos: Charles Tomes lectr, Royal Col Surgeons, Eng, 62; dir, Forsyth Dent Infirmary, 56-62, consult, 62-; consult, Univ BC, 55-56; consult, Dent Med Sect, Corporate Res Div, Colgate-Palmolive Co, 58-62; mem bd, Banff Sch Advan Mgt, 62-67, chmn, 66-67; consult, Donwood Found, Toronto, 67-, Sci Coun Can & Can Coun Support Res in Can Univs, 67-69 & Addiction Res Found, Toronto, 68; mem dent study sect, NIH, 61-65. Mem: AAAS; Am Soc Microbiol; NY Acad Sci; Can Ment Health Asn; Int Asn Dent Res (pres, 68). Res: Ecology of mucous membranes; mixed anaerobic infections. Mailing Add: 130 St George St Suite 8039 Toronto ON Can

MACDONALD, JOHN CAMPBELL FORRESTER, b Kelowna, BC, Sept 24, 20; m 50; c 3. PHYSICS. Educ: Univ BC, BA, 41, MA, 48; Univ Toronto, PhD(exp physics), 51. Prof Exp: Physicist radiol physics, Toronto Gen Hosp, 51-57; asst prof, 61-71, PROF PHYSICS, UNIV WESTERN ONT, 71-; SR PHYSICIST, ONT CANCER FOUND, VICTORIA HOSP, 57- Concurrent Pos: Clin teacher fac med, Univ Toronto, 54-57; mem, Can Sci Mission to USSR, 66. Mem: Can Asn Physicists; Brit Inst Radiol. Res: Physics applied to medicine, especially applications of ionizing radiation in treatment of cancer. Mailing Add: Ontario Cancer Found Victoria Hosp London ON Can

MACDONALD, JOHN CHISHOLM, b Boston, Mass, Mar 17, 33; m 64; c 2. ANALYTICAL CHEMISTRY. Educ: Boston Col, BS, 55, MS, 57; Univ Va, PhD(chem), 62. Prof Exp: AEC fel & res assoc chem, Pa State Univ, 62-63; sr res chemist, Monsanto Res Corp, Mass, 63-66; from asst prof to assoc prof, 66-75, PROF CHEM, FAIRFIELD UNIV, 75- Concurrent Pos: NIH spec res fel, Med Lab, Sch of Med, Yale Univ, 72-73. Mem: AAAS; Am Chem Soc. Res: Chemical analysis and instrumentation; computers in chemistry; science for society. Mailing Add: 250 Strobel Rd Trumbull CT 06611

MACDONALD, JOHN JAMES, b New Glasgow, NS, Oct 31, 25; m 52; c 7. ELECTROCHEMISTRY. Educ: St Francis Xavier Univ, Can, BSc, 45; Univ Toronto, MA, 47, PhD(chem), 51. Prof Exp: Demonstr, Univ Toronto, 45-49; assoc prof chem, 49-60, dean sci, 60-70, ACAD VPRES, ST FRANCIS XAVIER UNIV, 70- Concurrent Pos: Researcher, Ottawa Univ, 59-60; chmn bd gov, Atlantic Inst Educ, 70; mem, Maritime Prov Higher Educ Comn, 74- Honors & Awards: Centennial Medal, 67. Mem: AAAS; Chem Inst Can; Comp Educ Soc; Can Soc Study Higher Educ. Res: Electrochemical kinetics. Mailing Add: St Francis Xavier Univ Antigonish NS Can

MACDONALD, JOHN LAUCHLIN, b Woodstock, NB, Oct 7, 38; US citizen; m 63; c 2. MATHEMATICS. Educ: Harvard Univ, AB, 59; Univ Chicago, MS, 61, PhD(math), 65. Prof Exp: Humboldt fel math, Math Seminar, Frankfurt, 65-66; asst prof, 66-74, ASSOC PROF MATH, UNIV BC, 74- Concurrent Pos: Nat Res Coun grant, 66-68. Mem: Am Math Soc; Can Math Cong. Res: Category theory. Mailing Add: Dept of Math Univ of BC Vancouver BC Can

MACDONALD, JOHN MACKILLOP, petroleum chemistry, organic chemistry, see 12th edition

MACDONALD, JOHN MARSHALL, b Dunedin, NZ, Nov 9, 20; m 52; c 3. PSYCHIATRY, CRIMINOLOGY. Educ: Univ Otago, MD, 46; Univ London, dipl, 50. Prof Exp: House physician, New Plymouth Gen Hosp, 46; asst physician, Belmont Hosp, London, Eng, 48-49; asst physician, Royal Edinburgh Hosp, 49-51; instr psychiat, 51-54, asst prof, 54-61, assoc prof, 61-71, PROF PSYCHIAT, SCH MED, UNIV COLO MED CTR, DENVER, 71-, DIR FORENSIC PSYCHIAT, MED CTR, 60- Concurrent Pos: Med consult, Colo State Hosp, 62- & Fitzsimons Gen Hosp, US Army, 64-; med adv, Social Security Admin, 64- Mem: Fel Am Psychiat Asn. Res: Crime. Mailing Add: 2205 E Dartmouth Circle Englewood CO 80110

MACDONALD, KENNETH, b Fennimore, Wis, Nov 3, 09; m 33; c 2. HYGIENE, PREVENTIVE MEDICINE. Educ: Park Col, AB, 33; Univ Iowa, MS, 36, PhD(hyg, prev med), 39. Prof Exp: From instr to asst prof, 39-50, ASSOC PROF HYG & PREV MED, COL MED, UNIV IOWA, 50- Mem: AAAS; Am Soc Trop Med & Hyg; Am Soc Parasitol; Am Pub Health Asn; Royal Soc Trop Med & Hyg. Res: Parasitic diseases; trichinosis; amebiasis; preventive treatment; nutritional requirements of parasitic protozoa; host defensive mechanisms against infection. Mailing Add: 259-A Med Labs Univ of Iowa Col of Med Iowa City IA 52240

MACDONALD, MALCOLM DUNCAN, b Kimberley, BC, May 23, 27; m 49; c 4. CYTOGENETICS. Educ: Univ Alta, BSc, 50; Univ Minn, PhD(plant genetics), 59. Prof Exp: Asst entomologist, Sci Serv Lab, 49-51, asst cytogeneticist, 51-59, cytogeneticist, 59-71, CORN BREEDER, AGR CAN RES STA, 71- Res: Cytogenetics of winter-wheat; inheritance of resistance to winter injury. Mailing Add: Agr Can Res Sta Lethbridge AB Can

MACDONALD, MICHAEL RAYMOND, b Sydney Mines, NS, Feb 27, 12; m 45; c 1. PREVENTIVE MEDICINE. Educ: St Francis Xavier Univ, BA, 32, BSc, 44; McGill Univ, MD & CM, 38; Univ Toronto, DPH, 43; Royal Col Physicians & Surgeons Can, cert pub health. Prof Exp: Gen pract, NS, 39-42; div med health officer, Dept Pub Health, NS, 43-51; asst adminr, 51-66, ADMINR, VICTORIA GEN HOSP, HALIFAX, 66-; ASST PROF PREV MED, DALHOUSIE UNIV, 64- Concurrent Pos: Registr-secy, Prov Med Bd NS, 58- Honors & Awards: Can Centennial Medal, 67. Mem: Can Pub Health Asn; Can Col Health Serv Execs. Res: Hospital administration. Mailing Add: 30 Armshore Dr Halifax NS Can

MACDONALD, NOEL CHARLES, b San Francisco, Calif, Dec 31, 40; m 63; c 2. SURFACE PHYSICS. Educ: Univ Calif, Berkeley, BS, 63, MS, 65, PhD(elec eng), 67. Prof Exp: Actg asst prof elec eng, Univ Calif, Berkeley, 67-68; mem tech staff scanning electron micros, Sci Ctr, NAm Rockwell Corp, Calif, 68-70; VPRES, PHYS ELECTRONICS INDUST, INC, 70- Concurrent Pos: Coordr & lectr, Univ Calif, Los Angeles, 70. Honors & Awards: Victor G Macres Award, Microbeam Anal Soc, 73. Mem: AAAS; Inst Elec & Electronics Eng; Am Phys Soc; Electron Micros Soc Am; Microbeam Anal Soc. Res: Microelectronic applications of scanning electron microscopy; computer controlled and time resolved scanning electron microscopy; microscopic Auger electron microscopy; surface physics techniques. Mailing Add: Phys Electronics Ind Inc 6509 Flying Cloud Dr Eden Prairie MN 55343

MACDONALD, NORMAN SCOTT, b Boston, Mass, Jan 16, 17; m 51; c 2. BIOLOGICAL CHEMISTRY. Educ: Western Reserve Univ, AB, 38; Ohio State Univ, MSc, 40, PhD(org chem), 42. Prof Exp: Asst chem, Ohio State Univ, 38-42; res assoc, Mass Inst Technol, 42-43; asst prof, Occidental Col, 46-48; assoc prof biophys, 49-67, PROF RADIOL, SCH MED, UNIV CALIF, LOS ANGELES, 67-, DIR

BIOMED CYCLOTRON FACIL, 71- Mem: Am Chem Soc; Am Soc Biol Chem; Am Soc Nuclear Med; Radiation Res Soc; Health Physics Soc. Res: Bone physiology; metabolism of calcium and radioisotopes; applications of radioisotopes in nuclear medicine; measurement of radioactivity in human body. Mailing Add: 900 Veteran Ave Los Angeles CA 90024

MACDONALD, PAUL WILLIAM, plant physiology, biology, see 12th edition

MACDONALD, RICHARD ANNIS, b Manistee, Mich, July 23, 28; m 54; c 3. MEDICINE, PATHOLOGY. Educ: Albion Col, AB, 51; Boston Univ, MD, 54. Prof Exp: Intern med, Boston City Hosp, Mass, 54-55, resident path, Mallory Inst Path, 56-59; instr, Harvard Med Sch, 59-60, assoc, 60-62, asst prof, 62-65; prof, Sch Med, Univ Colo, 66-69; prof, Sch Med, Boston Univ, 69-71; prof & chmn dept, Sch Med, Univ Mass, 71-73; PROF PATH, SCH MED, BOSTON UNIV, 73- Concurrent Pos: Fel path, Harvard Med Sch, 57-59; Am Heart Asn sr res fel, Mallory Inst Path, Boston City Hosp, 58-59. Mem: AAAS; Am Asn Path & Bact; Soc Exp Biol & Med; Am Soc Exp Path; Am Fedn Clin Res. Res: Experimental pathology; liver and hematological diseases; nutrition. Mailing Add: Dept of Path Norwood Hosp Norwood MA 02062

MACDONALD, ROBERT NEAL, b Mansfield, Ohio, Nov 2, 16; m 41; c 3. ORGANIC CHEMISTRY. Educ: Oberlin Col, AB, 38; Yale Univ, PhD(org chem), 41. Prof Exp: CHEMIST, E I DU PONT DE NEMOURS & CO, 41- Mem: Am Chem Soc; Sigma Xi. Res: Synthesis of high polymers. Mailing Add: E I du Pont de Nemours & Co Wilmington DE 19898

MACDONALD, RODERICK, JR, b Charleston, SC, Oct 16, 26; m 51; c 5. OPHTHALMOLOGY. Educ: Davidson Col, BS, 47; Med Col SC, MD, 50. Prof Exp: Resident ophthal, New Orleans Eye, Ear, Nose & Throat Hosp, La, 52-53 & 55-56; instr ophthal, Sch Med, Tulane Univ, 56-57; exec dir dept ophthal & from asst prof to assoc prof ophthal, 57-65, actg chmn dept, 65, assoc pharmacol, 69, assoc dean, 69-70, PROF OPHTHAL & CHMN DEPT, SCH MED, UNIV LOUISVILLE, 65-, VDEAN, 70- Concurrent Pos: Mem vision res training comt, NIH, 68-71. Mem: AAAS; Am Col Surg; Am Ophthal Soc; Royal Soc Health; Contact Lens Asn Ophthal. Res: External diseases of the eye; ocular immunology. Mailing Add: 301 E Walnut St Louisville KY 40202

MACDONALD, RODERICK PATTERSON, b Detroit, Mich, Nov 9, 24. BIOCHEMISTRY. Educ: Mich State Univ, BS, 47; Univ Detroit, MS, 49; Wayne State Univ, PhD(physiol chem), 52; Am Bd Clin Chem, dipl; Can Bd Clin Chem, dipl. Prof Exp: Asst chem, Univ Detroit, 47-49; spec instr, Mortuary Sch, Wayne State Univ, 49-51; supvr chem lab, 52-59, RES ASSOC, HARPER HOSP, 52-, DIR CLIN CHEM, 59-; ASST PROF PATH, WAYNE STATE UNIV, 64- Concurrent Pos: Ed-in-chief, Stand Methods Clin Chem, Vol 6. Honors & Awards: McLean Award, 52. Mem: Am Chem Soc; Am Asn Clin Chem (secy, 65-68); NY Acad Sci; Can Soc Clin Chem. Res: Metabolism and physiology of bone; thyroid diseases and hypometabolism; ultramicro clinical methods; general clinical chemistry; clinical enzymology. Mailing Add: Dept of Path Wayne State Univ Sch of Med Detroit MI 48202

MACDONALD, RONALD NEIL ANGUS, b Calgary, Alta, Jan 6, 35; m 62; c 3. HEMATOLOGY, ONCOLOGY. Educ: Univ Toronto, BA, 55; McGill Univ, MD, 59; FRCP(C). Prof Exp: Demonstr med, McGill Univ, 65-66, lectr, 66-67, asst prof, 67-71, asst dean, Fac Med, 67-68, assoc dean, 68-70; exec dir, Prov Cancer Hosp Serv, 71-75; PROF, FAC MED, UNIV ALTA, 71-, DIR DIV ONCOLOGY, 76- Concurrent Pos: Sir Edward Beatty scholar, McGill Univ, 63-64, Nat Cancer Inst fel, 64-65. Mem: AAAS; Am Soc Clin Oncology; Can Soc Hemat; Am Col Physicians. Res: Health care delivery for cancer patients. Mailing Add: Dr W W Cross Inst Edmonton AB Can

MACDONALD, ROSEMARY A, b Leamington Spa, Eng, Oct 7, 30; m 65. SOLID STATE PHYSICS. Educ: St Andrews Univ, BSc, 54; Oxford Univ, DPhil(solid state physics), 59. Prof Exp: Lectr physics, Somerville Col, Oxford Univ, 57-59; res assoc, Univ Md, 59-61; NATO res fel, Bristol Univ, 61-62; lectr, Sheffield Univ, 62-64; PHYSICIST, NAT BUR STAND, 64- Mem: Am Phys Soc; Brit Inst Physics. Res: Lattice dynamics; point defects; phase transitions; molecular dynamical calculations of nonequilibrium behavior of solids and liquids. Mailing Add: Heat Div Nat Bur of Stand Washington DC 20234

MACDONALD, RUSSELL EARL, b NB, Can, Feb 18, 28; m 59; c 3. BACTERIOLOGY. Educ: Acadia Univ, BA, 50, MA, 52; Univ Mich, PhD(bact), 57. Prof Exp: Asst bact, Acadia Univ, 50-52; asst, Univ Mich, 53-55, instr & res assoc, 56-57; asst prof, 57-62, ASSOC PROF BACT, CORNELL UNIV, 62- Concurrent Pos: Jane Coffin Childs Mem Fund Med Res fel, 64-65; vis scientist, Ames Res Ctr, 74-75. Mem: Am Soc Microbiol. Res: Microbial ecology; molecular biology; control mechanisms; ribosomes; membrane transport. Mailing Add: 410 Stocking Hall Cornell Univ Ithaca NY 14850

MACDONALD, STEWART FERGUSON, b Toronto, Ont, Aug 17, 13; m 45; c 5. ORGANIC CHEMISTRY. Educ: Univ Toronto, BA, 36, MA, 37; Munich Tech Univ, Dr rer nat(chem), 39. Prof Exp: Res asst med res, Univ Toronto, 39-40; chemist, Welland Chem Works, 40-42; asst prof med res, Univ Toronto, 42-48; Wellcome fel, Cambridge Univ, 48-52; from assoc res officer to sr res officer, 52-64, PRIN RES OFFICER, NAT RES COUN CAN, 64- Mem: Am Chem Soc; The Chem Soc; fel Royal Soc Can. Res: Pyrroles; porphyrins. Mailing Add: Div of Pure Chem Nat Res Coun Can Ottawa ON Can

MACDONALD, TIMOTHY, b Portland, Maine, May 16, 36; m 61; c 2. GENETICS, BIOCHEMISTRY. Educ: Univ Hawaii, BA, 66, PhD(horticult), 69. Prof Exp: Res assoc biochem, Brown Univ, 68; res assoc genetics, Univ Hawaii, 68-69; AEC res grant biol, Brookhaven Nat Lab, 69-70; asst prof, 70-74, ASSOC PROF BIOL, LOWELL TECHNOL INST, 74- Mem: AAAS; Am Chem Soc. Res: Development Developmental, biochemical and isoenzymic genetics; clinical applications of isoenzymes. Mailing Add: Dept of Biol Sci Lowell Technol Inst Lowell MA 01854

MACDONALD, WALTER CHARLTON, b Regina, Sask, Nov 1, 30; m 62; c 2. GASTROENTEROLOGY. Educ: Univ BC, MD, 55; FRCPS(C). Prof Exp: Resident med, Shaughnessy Hosp, Vancouver & Royal Victoria Hosp, Montreal, 56-59; instr, 64-68, ASST PROF MED, UNIV BC, 68- Concurrent Pos: Fel gastroenterol, Univ Wash, 60-62; attend physician, Shaughnessy Hosp, Vancouver, 64-; asst cytologist, Vancouver Gen Hosp, 63-, attend physician, 64- Mem: Am Gastroenterol Asn; Can Asn Gastroenterol; Can Soc Clin Invest. Res: Environmental factors in the pathogenesis of gastric cancer; early diagnosis and natural history of gastric cancer. Mailing Add: 6875 Angus Dr Vancouver BC Can

MACDONALD, WILLIAM, b Salem, Ohio, Nov 25, 27; m 51, 65; c 4. NUCLEAR PHYSICS. Educ: Univ Pittsburgh, BS, 50; Princeton Univ, PhD(physics), 55. Prof Exp: Asst, Princeton Univ, 50-54; theoret physicist, Radiation Lab, Univ Calif, 54-55; vis prof physics, Univ Wis, 55-56; from asst prof to assoc prof, 56-63, PROF PHYSICS, UNIV MD, COLLEGE PARK, 63- Concurrent Pos: Consult, Lockheed Missiles & Space Lab, 56-65; NATO fel, Univ Paris, 62-63. Mem: Am Phys Soc; Sigma Xi; Am Asn Physics Teachers. Res: Structure of light nuclei; nuclear reaction theory; plasma physics; space physics. Mailing Add: Dept of Physics & Astron Univ of Md College Park MD 20742

MACDONALD, WILLIAM DAVID, b Chatham, Ont, Feb 28, 37; m 70. STRUCTURAL GEOLOGY. Educ: Univ Western Ont, BSc, 59; Princeton Univ, PhD(geol), 65. Prof Exp: Asst prof geol, Villanova Univ, 64-65; asst prof, 65-70, ASSOC PROF GEOL, STATE UNIV NY BINGHAMTON, 70- Mem: Geol Soc Am; Am Geophys Union. Res: Structural geology, tectonics and paleomagnetism of Latin America and Caribbean regions; computer applications in geology and geophysics; orientation analysis. Mailing Add: Dept of Geol Sci State Univ of NY Binghamton NY 13901

MACDONALD, WILLIAM E, JR, b Columbus, Ohio, Nov 21, 16; m 40. PHARMACOLOGY, TOXICOLOGY. Educ: Emory Univ, AB, 39; Univ Fla, MS, 51; Univ Miami, PhD(pharmacol), 61; Am Bd Indust Hyg, dipl. Prof Exp: Formulator, Biscayne Chem Co, 39-40; spec officer, Seaboard Air Line RR, 40-41; chemist, Fla State Bd Health, 41-46, indust hyg chemist, 46-55; instr, Sch Med, Univ Miami, 55-63, asst prof pharmacol & toxicol, 63-75, assoc dir, Res & Teaching Ctr Toxicol, 73-75; STAFF TOXICOLOGIST, TRACOR JITCO, INC, 75- Concurrent Pos: Am Cancer Soc instnl grants, 64-66. Mem: Soc Toxicol; Am Chem Soc; Conf Govt Indust Hygienists; Am Indust Hyg Asn. Res: Pharmacology, toxicology and industrial hygiene of silicones, nitroolefins, benzene, food additives including drugs, pesticides and anticorrosives; bladder carcinogens; subliminal toxicology of pesticides. Mailing Add: Tracor Jitco Inc 1776 E Jefferson St Rockville MD 20852

MACDONNELL, DONALD R, b Springfield, Mass, Apr 8, 14; m 41. PHARMACY, CHEMISTRY. Educ: Mass Col Pharm, BS, 39, MS, 47. Prof Exp: Prof serv rep, Armour & Co, Ill, 40-41; explosives inspector, Ala Ord Works, US Civil Serv, 41-42; anal chemist, E L Patch, Mass, 45-47; from jr pharmaceut chemist to sr pharmaceut chemist, Smith Kline & French Labs, 47-58, group leader, 58-61, from asst sect head to sect head, Menley & James Div, 61-65, dept head, 65-68, dir pharmaceut develop, 68-74, CONSULT, MENLEY & JAMES DIV, SMITH KLINE CORP, 74- Mem: AAAS; Am Pharmaceut Asn; Inst Food Technologists; Soc Cosmetic Chem; NY Acad Sci. Res: Development of medicinal dosage forms; sustained release medication for twelve-hour action; improved enteric coatings; extraction and purification of alkaloids; sterile products; medicated cosmetics; suppositories; stability testing; biological availability of drugs in animals and humans. Mailing Add: Smith Kline Corp 1500 Spring Garden St Philadelphia PA 19101

MACDONNELL, JOHN JOSEPH, b Springfield, Mass, Mar 28, 27. MATHEMATICS. Educ: Univ Boston, AB, 50, MA, 51; Cath Univ Am, PhD(math), 57. Prof Exp: From instr to assoc prof, 60-69, ASSOC PROF MATH, COL OF THE HOLY CROSS, 69- Mem: Math Asn Am; Am Math Soc. Res: Convergence theorems for Dirichlet type series. Mailing Add: Dept of Math Col of the Holy Cross Worcester MA 01610

MACDORAN, PETER FRANK, b Los Angeles, Calif, Jan 2, 41; m 64; c 2. GEODESY. Educ: Calif State Univ, Northridge, BS, 64; Univ Calif, Santa Barbara, MS, 66. Prof Exp: ARIES PROJ MGR RADIO GEOD, JET PROPULSION LAB, CALIF INST TECHNOL, 68- Honors & Awards: NASA Except Sci Achievement Award, 70. Mem: Am Geophys Union; Inst Elec & Electronics Engrs. Res: Research and development of radio interferometry for applications to problems of geodesy and earth physics. Mailing Add: 4800 Oak Grove Dr Pasadena CA 91103

MACDOUGALL, DANIEL, biochemistry, see 12th edition

MACDOUGALL, DUNCAN PECK, b College Station, Tex, Apr 13, 09; m 42; c 1. PHYSICAL CHEMISTRY. Educ: Pomona Col, AB, 29; Univ Calif, PhD(chem), 33. Prof Exp: Instr chem, Univ Calif, 33-34 & Harvard Univ, 34-37; asst prof, Clark Univ, 37-41; phys chemist, US Bur Mines, Pa, 41-43; assoc dir res, Nat Defense Res Comt, Explosives Res Lab, Carnegie Inst Technol, 43-46; chief explosives div, Naval Ord Lab, 46-48; div leader, 48-70, ASST DIR WEAPONS, LOS ALAMOS SCI LAB, 70- Honors & Awards: Medal of Merit, 48. Res: Cryogenics below one degree absolute; vibrations of polyatomic molecules; high explosives chemistry and physics; nuclear weapons. Mailing Add: 1984 Peach St Los Alamos NM 87544

MACDOUGALL, EDWARD BRUCE, b Sault Ste Marie, Ont, Aug 17, 39; m 60; c 3. ECONOMIC GEOGRAPHY. Educ: Univ Toronto, BScF, 61, MScF, 62, PhD(geog), 67. Prof Exp: Lectr geog, Univ Toronto, 66-67, asst prof, 67-69; asst prof regional planning, Univ Pa, 69-71, assoc prof landscape archit & regional planning, 71-74; ASSOC PROF GEOG, FAC FORESTRY, UNIV TORONTO, 74- Mem: Sigma Xi; Asn Am Geog; Asn Comput Mach. Res: Computer-assisted regional planning; rural land use; quantitative methods in spatial analysis. Mailing Add: Fac of Forestry Univ of Toronto Toronto ON Can

MACDOUGALL, JOHN D, nuclear physics, solid state physics, see 12th edition

MACDOUGALL, JOHN DOUGLAS, b Toronto, Ont, Mar 9, 44; m 68; c 1. GEOCHEMISTRY, METEORITICS. Educ: Univ Toronto, BSc, 67; McMaster Univ, MSc, 68; Univ Calif, San Diego, PhD(earth sci), 72. Prof Exp: Asst res geologist, Univ Calif, Berkeley, 72-74; ASST PROF EARTH SCI, SCRIPPS INST OCEANOG, UNIV CALIF, SAN DIEGO, 74- Mem: AAAS; Am Geophys Union; Meteoritical Soc. Res: Fission track geochronology; chemical exchange between seawater and oceanic rocks; manganese nodule growth; origin and evolution of meteorites; processes on the lunar regolith. Mailing Add: Geol Res Div Scripps Inst of Oceanog La Jolla CA 92093

MACDOUGALL, JOHN TAYLOR, b Brandon, Man, Apr 16, 14; m 41; c 2. SURGERY. Educ: Univ Alta, BA, 33; McGill Univ, MD & CM, 37; FRCS(E), 47; FRCS(C), 50. Prof Exp: Lectr, 55-60, from asst prof to assoc prof, 60-71, PROF SURG, UNIV MAN, 71- Concurrent Pos: Asst surg, Winnipeg Gen Hosp, 51-; surgeon, Grace Hosp, 51-59, pres med staff, 57-59, chief dept surg, 58-60; consult, Deerlodge Hosp, 51-63, chief dept surg, 63-; consult surgeon, St Boniface Hosp. Mem: Fel Am Col Surg; Can Asn Clin Surg. Res: Oesophageal and peripheral vascular surgery. Mailing Add: Abbott Clin 2740 Osborne St N Winnipeg MB Can

MACDOUGALL, ROBERT DOUGLAS, b McVille, NDak, Jan 2, 22; m 61; c 2. PETROLEUM GEOLOGY. Educ: Univ Mont, BA, 49; Univ Minn, Minneapolis, MS, 52. Prof Exp: Field geologist, Arabian Am Oil Co, Saudi Arabia, 52-56, subsurface geologist, 56-59; consult, 59-62; GEOLOGIST, US GEOL SURV, 62- Mem: Am Asn Petrol Geol; fel Royal Geog Soc. Res: Subsurface geology. Mailing Add: 646 Oak St Mandeville LA 70448

MACDOWALL, FERGUS D H, b Victoria, BC, Dec 19, 24; m 49; c 3. PLANT

PHYSIOLOGY. Educ: McGill Univ, BSc, 46; Univ Minn, MS, 47; Stanford Univ, PhD(biol), 54. Prof Exp: Instr plant physiol, Univ Mich, 49-51; RES SCIENTIST, CHEM & BIOL RES INST, CAN DEPT AGR, 52- Concurrent Pos: AEC fel, Univ Minn, 63-65, lectr, 65. Mem: Am Soc Plant Physiol; Bot Soc Am; Am Inst Biol Sci; Am Soc Photobiol; Int Solar Energy Soc. Res: Photobiological mechanisms; chloroplasts; growth; ozone damage; frost resistance; tobacco; wheat. Mailing Add: Chem & Biol Res Inst Agr Can Ottawa ON Can

MACDOWELL, DENIS W H, b Belfast, Northern Ireland, Jan 26, 24; US citizen; m 49; c 4. ORGANIC CHEMISTRY. Educ: Queen's Univ, Belfast, BSc, 45, MSc, 47; Mass Inst Technol, PhD(org chem), 55. Prof Exp: Res chemist, Linen Indust Res Asn, 45-48; instr, Univ Toronto, 48-51; NIH fel org chem, Ohio State Univ, 55-57; lectr, Univ Toronto, 57-59; PROF CHEM, WVA UNIV, 59- Mem: Am Chem Soc. Res: Polycyclic aromatic hydrocarbon derivatives; thiophene chemistry. Mailing Add: Dept of Chem WVa Univ Morgantown WV 26506

MACDOWELL, JOHN FRASER, b Oct 10, 32; US citizen; m 57; c 5. GEOLOGY, CHEMISTRY. Educ: Univ Mich, BS, 58, MS, 59. Prof Exp: Chemist, 59-61, sr chemist, 61-64, res mgr glass ceramics, 64-66, DIR CHEM RES, CORNING GLASS WORKS, 66- Concurrent Pos: Mem panel ceramic processing, Mat Adv Bd, Nat Acad Sci-Nat Res Coun, 65- Mem: Am Chem Soc; Am Ceramic Soc; Mineral Soc Am; Brit Soc Glass Technol. Res: Devitrification of glass and ceramic materials formed in this manner; controlling nucleation to create unique ceramics. Mailing Add: Corning Glass Works Sullivan Park Corning NY 14830

MACDOWELL, ROBERT W, b Detroit, Mich, Dec 11, 24; m 44; c 3. MATHEMATICS. Educ: Oberlin Col, AB, 48; Univ Mich, AM, 49, PhD(math), 53. Prof Exp: From instr to asst prof math, Univ Rochester, 51-57; res assoc, Cornell Univ, 57-58; from assoc prof to prof, Antioch Col, 58-70; PROF MATH, HIRAM COL, 70-, VPRES, COL & DEAN, 74- Concurrent Pos: Sci fac fel, 57-58. Mem: Am Math Soc; Math Asn Am; Asn Symbolic Logic. Res: Model theory in mathematical logic; logic of quantum mechanics. Mailing Add: Dept of Math Hiram Col Hiram OH 44234

MACDOWELL, SAMUEL WALLACE, b Recife, Brazil, Mar 24, 29; m 53; c 3. PHYSICS. Educ: Univ Recife, BS, 51; Univ Birmingham, PhD(physics), 58. Prof Exp: Instr physics, Brazilian Ctr Phys Res, 54-56; instr, Princeton Univ, 59-60; from assoc prof to prof, Brazilian Ctr Phys Res, 60-63; mem, Inst Advan Study, 63-65; assoc prof, 65-67, PROF PHYSICS, YALE UNIV, 67- Concurrent Pos: Sci secy, Tenth Int Conf High Energy Physics, Univ Rochester, 60; prof, Cath Univ Rio de Janiero, 62-63. Mem: Fel Am Phys Soc; Brazilian Acad Sci. Res: Strong and weak interactions of elementary particles; s-matrix theory; analytic properties of scattering amplitudes and form factors; group theory and symmetries of interactions; gauge theories; spontaneous symmetry breaking. Mailing Add: Dept of Physics Yale Univ New Haven CT 06520

MACE, ARNETT C, JR, b Hackers Valley, WVa, Nov 18, 37; m 62; c 2. FOREST HYDROLOGY. Educ: WVa Univ, BS, 60; Univ Ariz, MS, 62, PhD(forest hydrol), 68. Prof Exp: Res forester forest hydrol, Res Br, US Forest Serv, 62-64; instr, Univ Ariz, 64-67; from asst prof to assoc prof, 67-74, head forest biol dept, 72-74, PROF & HEAD FOREST RESOURCES DEPT, UNIV MINN, 74- Mem: Soc Am Foresters; Am Geophys Union; Am Water Resources Asn. Res: Applied and basic research related to forest land management practices and policies, with particular emphasis on water resources problems related to forest land management. Mailing Add: Univ of Minn 110 Green Hall St Paul MN 55108

MACE, KENNETH DEAN, b Pekin, Ill, Jan 29, 26; m; c 1. MICROBIOLOGY, PLANT PHYSIOLOGY. Educ: Univ Ark, BS, 51, MS, 57, PhD(plant physiol), 64. Prof Exp: Bacteriologist, Pepsi Cola Co, NY, 51-54; asst epidemiol, Univ Ark, 54-55 & bact, 55-57; bact chemist, Pepsi Cola Co, NY, 57-59; asst bot & bact, Univ Ark, 59-64; from asst prof to assoc prof, 64-71, PROF BIOL, STEPHEN F AUSTIN STATE UNIV, 71- Concurrent Pos: Dept Health, Educ & Welfare grants, 71-73. Mem: AAAS; Bot Soc Am; NY Acad Sci; Am Soc Microbiol. Res: Cyclic adenosine monophosphate relationships early in infection by New Castle disease virus. Mailing Add: Dept of Biol Stephen F Austin State Univ Nacogdoches TX 75961

MACE, ROBERT, physics, see 12th edition

MACEDO, PEDRO BUARQUE DE, b Copenhagen, Denmark, July 16, 38; m 59; c 3. PHYSICS. Educ: George Washington Univ, BS, 59; Cath Univ Am, PhD(physics), 63. Prof Exp: Physicist, Nat Bur Standards, 63-70; PROF CHEM ENG & MAT SCI, CATH UNIV AM, 70- Concurrent Pos: Co-dir, Vitreous State Lab, 70- Mem: Acoust Soc Am; Am Ceramic Soc. Res: Liquid and glassy states involving mechanical relaxations; electrical relaxation; optical properties characterizing microstructure; glass technology. Mailing Add: Dept of Chem Eng & Mat Sci Cath Univ of Am Washington DC 20017

MACEK, ANDREJ, b Zagreb, Yugoslavia, Oct 24, 26; nat US; m 56; c 2. PHYSICAL CHEMISTRY. Educ: Georgetown Univ, BS, 50; Cath Univ Am, MS, 51, PhD(phys chem), 53. Prof Exp: Asst, Cath Univ Am, 51-53, res assoc, 53-54; asst prof, Lafayette Col, 54-55; res assoc, US Naval Ord Lab, 55-60; phys chemist, Atlantic Res Corp, 60-69, chief kinetics & combustion group, 69-74; proj leader flame & combustion res, Nat Bur Stand, 74-75; COMBUSTION SCIENTIST, ENERGY RES & DEVELOP ADMIN, 75- Concurrent Pos: Adj prof, Am Univ, 58- Mem: Am Chem Soc; Combustion Inst. Res: Combustion; chemical kinetics and thermodynamics. Mailing Add: 5515 Eastbourne Dr Springfield VA 22151

MACEK, JOSEPH, b Rapid City, SDak, July 4, 37; m 64; c 2. PHYSICS. Educ: SDak State Col, BS, 60; Rensselaer Polytech Inst, PhD(physics), 64. Prof Exp: Nat Res Coun fel, Nat Bur Stand, 64-66; Oxford-Harwell fel, Atomic Energy Res Estab, UK, 66-68; from asst prof to assoc prof, 68-74, PROF PHYSICS, UNIV NEBR, LINCOLN, 74- Mem: Am Phys Soc. Res: Atomic physics. Mailing Add: Behlen Lab of Physics Univ of Nebr Lincoln NE 68508

MACEK, ROBERT JAMES, b Rapid City, SDak, July 14, 36; m 62; c 2. ELEMENTARY PARTICLE PHYSICS, EXPERIMENTAL NUCLEAR PHYSICS. Educ: SDak State Univ, BS, 58; Calif Inst Technol, PhD(physics), 65. Prof Exp: Res assoc physics, Univ Pa, 66-69; GROUP LEADER & MEM STAFF MEDIUM ENERGY PHYSICS DIV, LOS ALAMOS SCI LAB, 69- Mem: AAAS; Am Phys Soc; Sigma Xi. Res: Weak interactions and rare decay modes of mesons; pion production and absorption reactions on nuclei; nuclear instrumentation; particle beam optics and beam instrumentation. Mailing Add: MP-13 MS838 Los Alamos Sci Lab Los Alamos NM 87545

MACEK, THOMAS JOSEPH, b Newark, NJ, Oct 25, 17; m 42; c 2. PHARMACEUTICAL CHEMISTRY. Educ: Rutgers Univ, BS, 38, PhD(physiol, biochem), 50; Univ Fla, MS, 40. Prof Exp: Control chemist, Burroughs Wellcome & Co, Inc, 40-41; res assoc, Merck & Co, Inc, 42-56, from asst mgr to mgr, 50-56, mgr pharmaceut res, Merck Sharp & Dohme Res Labs, 56-60, dir pharmaceut res & develop, 60-70; dir rev, US Pharmacopeia, 70-72; SR VPRES, REGULATORY AFFAIRS & QUAL ASSURANCE, CHEM MFG, TECHNICON INSTRUMENTS CORP, 72- Concurrent Pos: Expert comt on stand for pharmaceuticals, WHO, 71-72. Mem: Am Chem Soc; Am Pharmaceut Asn (1st vpres, 64-65); Acad Pharmaceut Sci (pres, 70-71); Int Pharmaceut Fedn. Res: Drug extraction; stabilization of pharmaceuticals; new product development; technical service; vitamins; antibiotics; adrenocortical hormones; narcotics; diuretics; synthetic therapeutics; drug standards; clinical chemistry; hematology; automated analysis; industrial management; drug and diagnostics regulations. Mailing Add: Chem Mfg Technicon Instruments Corp Tarrytown NY 10591

MACELROY, ROBERT DAVID, b Boston, Mass, May 14, 39. EXOBIOLOGY. Educ: Univ Mass, BS, 61, MA, 63, PhD, 66. Prof Exp: Teaching assoc, Univ Mass, 61-63; fel, Oak Ridge Inst Nuclear Studies, 63-66; Nat Acad Sci-Nat Res Coun resident res assoc microbial physiol, Ames Res Ctr, 66-69, sr investr, 69-73, tech asst, Washington, DC, 73-74, MEM STAFF, NASA, WASHINGTON, DC, 74- Concurrent Pos: Prin sci expert, Roman Cath Univ, Nijmegen, 70-71. Mem: AAAS; Am Soc Microbiol. Res: Microbial ribosomes and enzymology; regulatory mechanisms of microbial metabolism; ultraviolet effects on bacterial physiology; autotrophic metabolism; physiology of thermophilic bacteria. Mailing Add: Code SL 400 Maryland Ave SW Washington DC 20546

MACERO, DANIEL JOSEPH, b Revere, Mass, Nov 19, 28; m 52; c 2. ANALYTICAL CHEMISTRY. Educ: Mass Inst Technol, BS, 51; Univ Vt, MS, 53; Univ Mich, PhD(chem), 58. Prof Exp: Instr chem, Univ Mich, 56-57; asst prof, 57-64, actg chmn, 70-71, ASSOC PROF ANAL CHEM, SYRACUSE UNIV, 64-, VCHMN CHEM, 70- Concurrent Pos: Res assoc, Univ NC, Chapel Hill, 67-68. Mem: AAAS; Am Chem Soc. Res: Kinetics of electron transfer; electrochemistry of transition metal complexes; computer assisted instruction. Mailing Add: Dept of Chem Syracuse Univ Syracuse NY 13210

MACEWAN, DOUGLAS W, b Ottawa, Ont, Nov 11, 24; m; c 4. RADIOLOGY. Educ: McGill Univ, BSc, 48, MD, CM, 52; FRCPS(C), 58. Prof Exp: Asst radiologist, Montreal Children's Hosp, 58-63; asst prof radiol, McGill Univ, 63-65; PROF RADIOL, UNIV MAN, 66-; CHIEF RADIOLOGIST, HEALTH SCI CENTRE, 66- Concurrent Pos: Radiologist, Montreal Gen Hosp, 63-65; mem, Int Comt Radiol Info. Mem: Can Med Asn; Can Asn Radiol; fel Am Col Radiol; Radiol Soc NAm. Res: Renal disease; standards, administration and costs of radiology. Mailing Add: Health Sci Centre 700 William Ave Winnipeg MB Can

MACEWEN, JAMES DOUGLAS, b Detroit, Mich, July 31, 26; m 47; c 1. TOXICOLOGY. Educ: Wayne State Univ, BS, 49, PhD(pharmacol, physiol), 62; Univ Mich, Ann Arbor, MPH, 57; Am Bd Indust Hyg, dipl, 64. Prof Exp: Chemist, Water Bd, City of Detroit, Mich, 49, asst indust hyg, Bur Indust Hyg, 51-56; chemist, USPHS Hosp, Detroit, 50-51; asst prof & res assoc toxicol, Wayne State Univ, 56-63; dir environ health, Toxic Hazards Res Unit, Systemed Corp, 63-74; DIR ENVIRON HEALTH, TOXIC HAZARDS RES LAB, UNIV CALIF, DAYTON, OH, 74- Mem: Soc Toxicol; Am Indust Hyg Asn; Am Pub Health Asn; NY Acad Sci. Res: Inhalation toxicology; environmental analysis and control. Mailing Add: Toxic Hazards Res Lab Univ of Calif Dayton OH 45431

MACEY, ROBERT IRWIN, b Minneapolis, Minn, Sept 22, 26; m 56; c 2. PHYSIOLOGY, BIOPHYSICS. Educ: Univ Minn, BA, 47; Univ Chicago, PhD(math biol), 54. Prof Exp: Instr chem & physiol, George Williams Col, 49-51; instr math, Ill Inst Technol, 53-54; res assoc physiol, Aeromed Lab, Univ Ill, 55-57, asst prof, Col Med, 57-60; from asst prof to assoc prof, 60-69, PROF PHYSIOL, UNIV CALIF, BERKELEY, 69- Concurrent Pos: Asst, Univ Chicago, 53-54; consult, Rand Corp, 63 & NIH, 65. Mem: Biophys Soc; Biomet Soc; Am Physiol Soc. Res: Theoretical biophysics; membrane transport; kidney; circulation. Mailing Add: Dept of Physiol Univ of Calif Berkeley CA 94720

MACEY, WADE THOMAS, b Mt Airy, NC, Jan 13, 36; m 58; c 3. MATHEMATICS. Educ: Guilford Col, BS, 60; Fla State Univ, MS, 62, PhD(math educ), 70. Prof Exp: Instr math, Oxford Col, Emory Univ, 62-65; ASSOC PROF & HEAD, DEPT MATH, PFEIFFER COL, 67- Mem: Math Asn Am. Res: Effect of prior instruction of selected topics of logic on the understanding of the limit of a sequence. Mailing Add: Dept of Math Pfeiffer Col Misenheimer NC 28109

MACFADDEN, CLIFFORD HERBERT, b Salem, Mich, Jan 30, 08; m 34. GEOGRAPHY. Educ: Univ Mich, AB, 37, MA, 39, PhD(geog), 48. Prof Exp: Asst geog, Univ Mich, 37-42; carographic engr, US War Dept, 42-46; lectr geog, Univ Calif, Los Angeles, 46-48, from asst prof to assoc prof, 48-58, chmn dept, 61-66, dir, Univ Calif Study Ctr, Chinese Univ Hong Kong, 72-74, PROF GEOG, UNIV CALIF, LOS ANGELES, 58- Mem: Asn Am Geog; Nat Coun Geog Educ. Res: Southern Asia and Pacific; post-colonial structuring; industrial-economic geography; regional geography. Mailing Add: Dept of Geog Univ of Calif Los Angeles CA 90024

MACFADDEN, DONALD LEE, b Port Deposit, Md, Dec 12, 26; m 53; c 4. PHYSIOLOGY, BIOLOGY. Educ: Univ Del, BS, 53, MS, 55; Univ Kans, PhD(physiol), 59. Prof Exp: Asst prof nutrit, Univ Mass, 60-63; prof physiol, 63-73, PROF BIOL, KING COL, 73-, HEAD DEPT, 63- Concurrent Pos: Sterling-Winthrop Res Corp res grant, 61-64. Mem: AAAS; Am Inst Biol Sci. Res: Mode of action of antibiotics in growth stimulation; studies on histological and immunological mechanisms involved in homograft rejections with intent of reducing severity of this response. Mailing Add: Dept of Biol King Col Bristol TN 37620

MACFADDEN, KENNETH ORVILLE, b Philadelphia, Pa, Sept 5, 45; m 68; c 2. ANALYTICAL CHEMISTRY. Educ: Juniata Col, BS, 66; Georgetown Univ, PhD(phys chem), 72. Prof Exp: Fel kinetics, Univ Calgary, 71-73; asst prof chem, Stockton State Col, 73-75; ANAL CHEMIST, AIR PROD & CHEM CO, 75- Mem: Am Chem Soc; AAAS; Sigma Xi. Res: Mass spectrometric analysis and methods development. Mailing Add: Air Prod & Chem Co Box 427 Marcus Hook PA 19061

MACFADYEN, DONALD JOHN, b Eden, Man, May 27, 29; m 51; c 4. NEUROLOGY. Educ: Univ Sask, BA, 49, MSc, 61; Univ Toronto, MD, 53. Prof Exp: Lectr, Univ Sask, 63-64, lectr neuropath, 63-65, asst prof neurol, 64-65; ASSOC PROF NEUROL, UNIV BC, 65-, HEAD DIV, 70- Concurrent Pos: Res fel cancer, Nat Cancer Inst Can, 54-55; teaching fel neurol, Harvard Med Sch, 62-63; head sub-dept neurol, Vancouver Gen Hosp, 65-; chief neurologist, Shaughnessy Hosp, Can Dept Vet Affairs, Vancouver, 66- Mem: Am Acad Neurol; Can Neurol Soc; Can Soc Electroencephalog. Res: Muscle pathology; end-plate morphology in neuropathies and myopathies; central nervous system tissue regeneration. Mailing Add: Dept of Neurol Univ of BC Vancouver BC Can

MACFADYEN, DOUGLAS ARCHIBALD, b Toronto, Ont, Feb 20, 05; nat US; m 31, 49; c 1. BIOCHEMISTRY, PATHOLOGY. Educ: Univ Toronto, BSc, 26, MA, 27, MD, 31. Prof Exp: Demonstr biochem, Univ Toronto, 26-28; asst path, Rockefeller Inst, 31-35, asst chem, 36-40, assoc, 40; chief biochem, Alfred I du Pont Inst, 40-45;

prof biochem, Univ Ill, 45-59; prof path & dir exp prog med educ, Sch Med, Ind Univ, Bloomington, 59-73; RETIRED. Concurrent Pos: Vis asst prof, Univ Pa, 40-45; chmn dept biochem, Presby Hosp, 45-59; trustee, Am Bd Path, 51-; consult, Surgeon Gen, US Army. Mem: AAAS; Am Soc Biol Chem; Harvey Soc; fel NY Acad Sci. Res: Monometric and gasometric analysis; spectrophotometry; transplantability of animal tissues; histochemical processes. Mailing Add: RR 4 Lucan Rd Bloomington IN 47401

MACFADYEN, JOHN ARCHIBALD, JR, b Scranton, Pa, July 10, 22; m 46; c 3. GEOLOGY. Educ: Williams Col, BA, 48; Lehigh Univ, MS, 50; Columbia Univ, PhD(geol), 62. Prof Exp: Asst geol, Lehigh Univ, 48-50 & Columbia Univ, 50-52; from instr to prof geol, 52-74, EDWARD BRUST PROF GEOL & MINERAL, WILLIAMS COL, 74-, CHMN DEPT, 68- Concurrent Pos: Geologist, Vt Geol Surv, 51-54. Mem: Geol Soc Am; Soc Rheol; Nat Asn Geol Teachers; Am Geophys Union. Res: Structural geology; deformation and flow of solids. Mailing Add: Dept of Geol Williams Col Williamstown MA 01267

MACFARLAND, HAROLD NOBLE, b London, Eng, July 30, 17; m 43; c 2. INDUSTRIAL HYGIENE, TOXICOLOGY. Educ: Univ Toronto, BA, 41, MA, 42, PhD(physiol hyg), 49; Am Bd Indust Hyg, dipl. Prof Exp: Lectr indust hyg, Univ Toronto, 47-49; chief toxicol sect, Defence Res Med Labs, Dept Nat Defence, 49-52; asst chief occup health labs, Dept Nat Health & Welfare, 52-54, head res group, Div Occup Health, 54-59, sr scientist biol unit, 59-62; sr toxicologist, Hazleton Labs, Inc, 62-65; dir inhalation div & vpres & dir, Resources Res, Inc, 65-68; prof natural sci, York Univ, 68-72; vpres, Bio-Res Labs Ltd & Eco-Res Ltd, 72-74; DIR TOXICOL, MED DEPT, GULF OIL CORP, 75- Concurrent Pos: Mem comt on toxic dusts & gases, Am Nat Stand Inst & toxicol res comt, Am Petrol Inst. Mem: Am Indust Hyg Asn; Soc Toxicol; Air Pollution Control Asn; NY Acad Sci. Res: Experimental toxicology, particularly on air-borne toxicants; research and development on air quality criteria and standards; design of inhalation studies and equipment; related university teaching. Mailing Add: Gulf Bldg PO Box 1166 Pittsburgh PA

MACFARLANE, CONSTANCE IDA, b Charlottetown, PEI. BIOLOGY. Educ: Dalhousie Univ, BA, 29, MSc, 32. Hon Degrees: DSc, Acadia Univ, 75; LLD, Dalhousie Univ, 75 & Univ PEI, 75. Prof Exp: Asst bot, Dalhousie Univ, 30-32; head dept sci, High Sch, Can, 32-33 & 39-46; lectr bot & dean women's residence, Univ Alta, 46-48; spec lectr biol, Victoria Col, 48-49; spec lectr, Acadia Univ, 49-61, res assoc, 61-65, assoc prof, 65-70; INDUST CONSULT, 70- Concurrent Pos: Vprin, Mt Allison Sch Girls, Can, 42-44, prin, 55-56; marine phycologist & dir seaweeds div, NS Res Found, 49-70; consult, Indust Develop Br, Fisheries Serv, Environ Can, 68-75. Mem: Int Phycol Soc; Brit Phycol Soc; Can Bot Asn; Bot Soc Am. Res: Cultivation of commercial algae; recovery of plants after harvesting; phenology. Mailing Add: 1101 Wellington St Halifax NS Can

MACFARLANE, JOHN O'DONNELL, b Valparaiso, Ind, Apr 16, 20; m 43; c 4. MEDICAL MICROBIOLOGY, ONCOLOGY. Educ: Purdue Univ, BS, 41; Univ Cincinnati, PhD(bact), 48. Prof Exp: Bacteriologist, Virus Div, E R Squibb & Sons, NJ, 42-43 & Antibiotic Div, Upjohn Co, Mich, 43-46; bacteriologist, Eli Lilly & Co, 48-53, head biol develop, 53-58, asst dir, 58-63; tech assoc, 63-64, from sr virologist to prin virologist, 64-67, HEAD LIFE SCI SECT, MIDWEST RES INST, 67- Mem: AAAS; Am Soc Microbiol; NY Acad Sci; Sigma Xi; Am Inst Biol Sci. Res: Viral, phage and microbial agents as indicators of pollution and application of enzymes for medicine, toxicology and monitoring. Mailing Add: 9220 Roe Ave Prairie Village KS 66207

MACFARLANE, JOHN T, b Hamilton, Ont, Nov 23, 23; m 46; c 5. PHYSICS. Educ: McMaster Univ, BA, 44; Univ Montreal, MSc, 53. Prof Exp: Asst prof physics, Sir George Williams Univ, 45-51; lectr physics & math, Univ Col, Ethiopia, 51-54; res scientist, Can Armament Res & Develop Estab, Que, 54-58; assoc prof physics, Univ Col, Haile Sellassie, 58-63, dean sci, 59-63, prin, 61-63; prof physics, Univ Libya, 63-64; head dept, Univ Col Sci Educ, Ghana, 64-65; assoc prof, Mt Allison Univ, 65-69; vrector, Nat Univ Rwanda, 69-72; assoc prof, 72-75, PROF PHYSICS, MT ALLISON UNIV, 75- Concurrent Pos: Asst lectr, Univ Montreal, 47-49; mem, Nat Comts Educ in Ethiopia, 60-63. Mem: Inst Elec & Electronics Engrs; Am Asn Physics Teachers; Fr Phys Soc. Res: Systems analysis; gas laser physics. Mailing Add: Dept of Physics Mt Allison Univ Sackville NB Can

MACFARLANE, MALCOLM DAVID, b Cambridge, Mass, Sept 26, 40. CLINICAL PHARMACOLOGY, MEDICAL RESEARCH. Educ: New Eng Col Pharm, BS, 62; Georgetown Univ, PhD(pharmacol), 67. Prof Exp: Instr pharmacol, Kirksville Col, 67-69; asst prof, Univ Southern Calif, 69-74; DIR RES, INST RES, MEYER LABS, 74- Concurrent Pos: Consult, Kirksville Osteop Hosp & Still-Hildreth Hosp, 67-69, Rom-Amer Pharmaceuts, Ltd, LAC/USC Med Ctr, 69-74, Calif Dept Consumer Affairs, 72-74 & Superior Court of Calif, 72-74. Mem: Fel Am Col Clin Pharmacol; Am Pharmaceut Asn; Am Soc Pharmacol & Exp Therapeut; Am Geriat Soc; NY Acad Sci. Res: Pharmacology of the autonomic nervous system; cardiovascular and renal pharmacology; microvascular renal physiology; narcotic analgesics and treatment of addiction; neuropsychopharmacology; clinical nutrition; clinical and geriatric pharmacology. Mailing Add: Meyer Labs Inst of Res 1900 W Commercial Blvd Ft Lauderdale FL 33309

MACFARLANE, ROBERT, JR, b Brooklyn, NY, Aug 26, 30; m 52; c 5. PHYSICAL CHEMISTRY. Educ: Brown Univ, ScB, 52; Yale Univ, PhD(phys chem), 56. Prof Exp: Sr res chemist, Chem Div, US Rubber Co, 56-65; from res chemist to sr res chemist, Imp Oil Enterprises, Ltd, Esso Res & Eng Co, 65-74; GROUP LEADER, ALLIED CHEM CORP, 74- Concurrent Pos: Lectr grad sch, Brooklyn Polytech Inst, 69-70. Mem: Am Chem Soc; Soc Rheol; NY Acad Sci; Soc Plastics Engrs. Res: Emulsion stability; controlled aglomeration of colloids; mechanics of liquid-liquid mixing; molecular structure versus physical properties of polymers; polymerization kinetics and mechanisms; thermally stable polymers; techniques of polymer characterization. Mailing Add: 25 Linda Pl Fanwood NJ 07023

MACFARLANE, ROGER MORTON, b Dunedin, NZ, Oct 25, 38; m 59; c 2. SOLID STATE PHYSICS. Educ: Univ Canterbury, BSc, 60, PhD(solid state physics), 64. Prof Exp: Asst lectr physics, Univ Canterbury, 64-65; res assoc, Stanford Univ, 65-68; STAFF MEM, IBM RES LAB, 68- Concurrent Pos: Sci Res Coun sr vis fel, Oxford Univ, 74. Mem: Am Phys Soc. Res: Laser spectroscopy and studies of phase transitions in magnetic solids and organic solids. Mailing Add: IBM Res Lab San Jose CA 95193

MACFARLANE, RONALD DUNCAN, b Buffalo, NY, Feb 21, 33; m 56; c 2. BIOPHYSICAL CHEMISTRY. Educ: Univ Buffalo, BA, 54; Carnegie Inst Technol, MS, 57, PhD(chem), 59. Prof Exp: Res fel nuclear chem, Lawrence Radiation Lab, Univ Calif, Berkeley, 59-62; asst prof chem, McMaster Univ, 62-64, assoc prof, 65-67; PROF NUCLEAR CHEM, TEX A&M UNIV, 67- Concurrent Pos: Guggenheim fel, 69-70. Mem: Am Chem Soc; Am Phys Soc. Res: Mass spectroscopy of biomolecules; pattern recognition. Mailing Add: Cyclotron Inst Tex A&M Univ College Station TX 77843

MACFARLANE, SIDLEY KERR, b Scotland, Jan 11, 09; nat US; m 40; c 2. GEOGRAPHY. Educ: Univ Syracuse, AB, 38, PhD, 60; Clark Univ, MA, 41. Prof Exp: Asst prof geog, Middlebury Col, 41-45; asst prof, 46-50, dean, Burrstone Campus, 50-74, EMER ASSOC PROF GEOG & EMER DEAN COL, BURRSTONE CAMPUS, UTICA COL, 74- Mailing Add: Dept of Geog Utica Col Utica NY 13502

MACGEE, JOSEPH, b Edinburgh, Scotland, Nov 24, 25; nat US; m 47; c 1. BIOCHEMISTRY, ANALYTICAL CHEMISTRY. Educ: Univ Calif, PhD(comp biochem), 54. Prof Exp: Lab technician res & develop, Merck & Co, Inc, 44-50; asst bact, Univ Calif, 52-53; res anal biochemist, Procter & Gamble Co, 55-61; asst prof, 61-70, ASSOC PROF BIOL CHEM & EXP MED, COL MED, UNIV CINCINNATI, 70-; RES BIOCHEMIST, VET ADMIN HOSP, 61- Concurrent Pos: USPHS fel, Univ Ill, 54-55. Res: Analytical biochemistry. Mailing Add: Basic Sci Lab Vet Admin Hosp 3200 Vine St Cincinnati OH 45220

MACGILLIVRAY, ARCHIBALD DEAN, b Vancouver, BC, Dec 28, 29; m 57; c 3. APPLIED MATHEMATICS. Educ: Univ BC, BASc, 55; Calif Inst Technol, MS, 57, PhD(aeronaut), 60. Prof Exp: Instr math, Calif Inst Technol, 60-61; mathematician, Lincoln Lab, Mass Inst Technol, 61-62, instr math, 62-64; asst prof, 64-67, ASSOC PROF MATH, STATE UNIV NY BUFFALO, 67- Concurrent Pos: Sr scientist, Jet Propulsion Labs, Calif Inst Technol, 60-61; consult, Lincoln Lab, 62-64. Res: Asymptotic expansions of solutions of differential equations. Mailing Add: Dept of Math State Univ of NY Buffalo NY 14214

MACGILLIVRAY, M ELLEN, b Fredericton, NB, Nov 26, 25; m 48. ENTOMOLOGY. Educ: Univ NB, BA, 47; Univ Mich, MSc, 51; State Univ Leiden, DSc, 58. Prof Exp: Sr agr asst, Div Entom, 47-48, tech officer, 48-54, res officer, 54-59, RES SCIENTIST, RES BR, CAN DEPT AGR, 59- Mem: Entom Soc Can (pres-elect, 75-76); Can Soc Zool; Entom Soc Am. Res: Systematics and biology of aphids of the Atlantic provinces of Canada and the New England states; potato pest management. Mailing Add: Res Sta Can Dept of Agr Box 20280 Fredericton NB Can

MACGILLIVRAY, MARGARET HILDA, b San Fernando, Trinidad, WI, Aug 30, 30; Can citizen; m 57; c 3. ENDOCRINOLOGY, IMMUNOLOGY. Educ: Univ Toronto, MD, 56. Prof Exp: From lectr to asst prof, 64-68, ASSOC PROF PEDIAT ENDOCRINOL, CHILDREN'S HOSP & MED SCH, STATE UNIV NY BUFFALO, 68- Concurrent Pos: AEC grant biol, Calif Inst Technol, 60-61; USPHS grant endocrinol & metab, Mass Gen Hosp, 61-64. Mem: Soc Pediat Res; Endocrine Soc. Res: The role of growth hormone in dwarfism of childhood; the mechanism of action of growth hormone in protein synthesis and cellular multiplication. Mailing Add: Children's Hosp 219 Bryant St Buffalo NY 14222

MACGINITIE, HARRY DUNLAP, b Lynch, Nebr, Mar 29, 96; m 35; c 2. PALEOBOTANY, STRATIGRAPHY. Educ: Fresno State Col, AB, 26; Univ Calif, PhD(paleont, geol), 35. Prof Exp: From instr to prof phys sci, 27-60, chmn div natural sci, 47-60, EMER PROF PHYS SCI, HUMBOLDT STATE UNIV, 60- Concurrent Pos: Assoc, Mus Paleont, Univ Calif, Berkeley, 61- Mem: Fel Geol Soc Am. Res: Physical sciences; Tertiary paleobotany; paleoclimatology; early Eocene Wind River flora of northwestern Wyoming. Mailing Add: 3747 Young Ave Napa CA 94558

MACGRAW, FRANK MOSS, b San Francisco, Calif, Aug 31, 23; m 43; c 2. GEOGRAPHY. Educ: Stanford Univ, BA, 51, MA, 58, EdD, 64. Prof Exp: Head dept & curriculum specialist, High Sch, Calif, 50-66; assoc prof, 66-74, PROF GEOG, SOUTHERN ORE COL, 74-, COORDR GEOG, 66- Concurrent Pos: Consult high sch dist, 64-66; mem prog learning comt, Nat Coun Soc Studies, 66- Mem: Asn Am Geog. Res: Feed-back mechanisms in teacher training; simulation. Mailing Add: Dept of Geog Southern Ore Col Ashland OR 97520

MACGREGOR, ALEXANDER HAMILTON, b Birmingham, Eng, Oct 8, 29; m 58; c 3. MEDICAL RESEARCH. Educ: Royal Col Physicians, Edinburgh, LRCP, 56; Royal Col Surgeons, Edinburgh, LRCS, 56; FRCS(C), 72. Prof Exp: Assoc dir med res, William S Merrell Co Div, Richardson-Merrell, Inc, 65-69; dir clin res, Ames Co Div, 69-73, DIR MED RES, AMES CO DIV, MILES LABS, 73- Mem: Fel Am Col Obstet & Gynec. Res: Reproductive endocrinology. Mailing Add: Miles Labs Ames Co Div 1127 Myrtle St Elkhart IN 46514

MACGREGOR, CAROLYN HARVEY, b Cambridge, Mass, Nov 7, 42; m 65. PROTEIN CHEMISTRY. Educ: Univ Va, BS, 64, PhD(biol), 69. Prof Exp: NIH fel, 69-72, RES ASSOC MICROBIOL, MED SCH, UNIV VA, 72- Res: Organization and function of E coli membrane proteins; regulation of nitrate reductase system in E coli; cell membrane proteases. Mailing Add: Dept of Microbiol Univ of Va Med Sch Charlottesville VA 22904

MACGREGOR, DUGAL, b Fernie, BC, Mar 7, 25; m 45; c 4. FOOD SCIENCE. Educ: Univ BC, BSA, 50; Ore State Col, MS, 52, PhD(bact), 55. Prof Exp: Instr bact, Ore State Col, 52-55; serologist, 55-57, FOOD TECHNOLOGIST, CAN DEPT AGR, 57- Concurrent Pos: Prof nutrit & food sci & head dept, Univ Ghana, 71-73. Mem: Can Inst Food Technologists; Inst Food Technologists. Res: Fruit and vegetable processing; chemistry and product development. Mailing Add: Fruit Processing Lab Can Dept of Agr Res Sta Summerland BC Can

MACGREGOR, IAN DUNCAN, b Calcutta, India, Jan 5, 35; Can citizen; m 56; c 4. PETROLOGY, GEOCHEMISTRY. Educ: Univ Aberdeen, BSc, 57; Queen's Univ, Ont, MSc, 60; Princeton Univ, PhD(geol), 64. Prof Exp: Sr asst, Geol Surv Can, Ottawa, 57, party chief, 58-59; fel exp petrol, Geophys Lab, Washington, DC, 64-65; assoc prof high pressure exp petrol, Southwest Ctr Advan Studies, Tex, 65-69; chmn dept, 69-74, PROF GEOL, UNIV CALIF, DAVIS, 69- Concurrent Pos: Chmn petrol panel, Joint Oceanog Insts Deep Earth Sampling, 66- Mem: Am Geophys Union; Mineral Soc Am. Res: Field geology; experimental petrology. Mailing Add: Dept of Geol Univ of Calif Davis CA 95616

MACGREGOR, IAN ROBERTSON, b Andover, Eng, Mar 7, 17; nat US. CHEMISTRY. Educ: Univ Cincinnati, AB, 41, PhD(org chem), 45. Prof Exp: Lab asst chem, Univ Cincinnati, 41-44, instr, 44-45, 47-48, from asst prof to assoc prof, 48-60, assoc dean admin, 60-61; financial vpres, 61-65, PROF CHEM, UNIV AKRON, 65-, VPRES PLANNING, 67- Mem: Am Chem Soc. Res: Organic medicinal chemistry; organic synthesis; polynuclear compounds; synthetic organic coatings; fluorene and thiaxanthene derivatives. Mailing Add: 3459 Granger Rd Akron OH 44313

MACGREGOR, JOHN MALCOLM, b Carson City, Mich, May 5, 08; m 42; c 3. SOIL SCIENCE. Educ: Univ Alta, BSc, 36, MSc, 38; Univ Minn, PhD(soils), 42. Prof Exp: Soil survr, Univ Alta, 36-40; assoc soils, Armour Res Found, Ill, 42-43; from asst prof to assoc prof, 43-55, prof, 56-74, EMER PROF SOILS, UNIV MINN, ST PAUL, 74- Concurrent Pos: NSF fel, 60. Mem: Am Soc Agron; Soil Sci Soc Am. Res: Solonetz soils of western Canada; fertility and rarer element content of Minnesota soils; effect of fertilizers on plant growth; soil compaction for airports. Mailing Add: 1795 N Fairview Ave St Paul MN 55113

MACGREGOR, MALCOLM HERBERT, b Detroit, Mich, Apr 24, 26; m 49; c 3. PHYSICS. Educ: Univ Mich, BA, 49, MS, 50, PhD(physics), 54. Prof Exp: PHYSICIST, LAWRENCE LIVERMORE LAB, UNIV CALIF, 53- Concurrent Pos: NATO fel, Inst Theoret Physics, Denmark, 60-61; lectr, Univ Calif, 63; consult, Appl Radiation Corp, 56-, Gen Atomic Div, Gen Dynamics Corp, 62-69 & Sci Applns, Inc, 69- Mem: Fel Am Phys Soc; Italian Phys Soc. Res: Neutron scattering; two-nucleon problem; elementary particle structure. Mailing Add: 4651 Almond Circle Livermore CA 94550

MACGREGOR, RONAL ROY, b Hayward, Calif, July 4, 39; m 65; c 2. BIOCHEMISTRY. Educ: Calif State Col, Long Beach, BS, 64; Ind Univ, PhD(biochem), 68. Prof Exp: Res assoc biochem, 68-70, RES CHEMIST, VET ADMIN HOSP, KANS CITY, MO, 70- Concurrent Pos: Instr biochem, Sch Dent, Univ Mo, Kans City, 68-70, asst prof, 70- Mem: Endocrine Soc. Res: The biosynthesis of proparathormone and its conversion to parathormone, the packaging and secretion of parathormone; the mechanisms of these processes and their control. Mailing Add: Vet Admin Hosp Calcium Res 4801 Linwood Blvd Kansas City MO 64128

MACGREGOR, THOMAS HAROLD, b Jersey City, NJ, June 21, 33; m 56; c 1. MATHEMATICS. Educ: Lafayette Col, AB, 54; Univ Pa, MA, 56, PhD(math), 61. Prof Exp: Instr math, Col South Jersey, Rutgers Univ, 58-59; prof, Lafayette Col, 59-67; PROF MATH, STATE UNIV NY ALBANY, 67-, CHMN DEPT, 75- Concurrent Pos: NSF grants, 62-64, 65-67 & 68-71. Mem: Am Math Soc; Math Asn Am. Res: Complex analysis, specifically the theory of univalent functions. Mailing Add: Dept of Math State Univ of NY Albany NY 12203

MACH, EDWARD EUGENE, physical organic chemistry, analytical chemistry, see 12th edition

MACH, GEORGE ROBERT, b Cedar Falls, Iowa, July 23, 28; m 52; c 3. MATHEMATICS. Educ: Univ Northern Iowa, BA, 50; Univ Iowa, MS, 51; Purdue Univ, PhD(math), 63. Prof Exp: From instr to assoc prof, 54-67, PROF MATH, CALIF POLYTECH STATE UNIV, SAN LUIS OBISPO, 67- Mem: Math Asn Am. Res: Mathematics education. Mailing Add: Dept of Math Calif Polytech State Univ San Luis Obispo CA 93407

MACH, MARTIN HENRY, b New York, NY, Feb 10, 40; div; c 1. PHYSICAL ORGANIC CHEMISTRY. Educ: City Col New York, BS, 61; Clark Univ, MA, 65; Univ Calif, Santa Cruz, PhD(chem), 73. Prof Exp: Assoc scientist chem, Polaroid Corp, Mass, 65-69; MEM TECH STAFF CHEM, AEROSPACE CORP, 73- Mem: Am Chem Soc. Res: Analytical organic chemistry using interfaced, computerized gas chromatography-mass spectrometry, including forensic science, fuel synthesis and analysis, and lubrication phenomena. Mailing Add: Aerospace Corp Bldg 130-108 PO Box 92957 Los Angeles CA 90009

MACHACEK, MARIE ESTHER, b Cedar Rapids, Iowa, Sept 12, 47; m 67; c 1. THEORETICAL HIGH ENERGY PHYSICS. Educ: Coe Col, BA, 69; Univ Mich, MS, 70; Univ Iowa, PhD(physics), 73. Prof Exp: Teaching asst physics, Univ Iowa, 70-71; hon fel physics, Univ Wis-Madison, 73-74; jr fel, Mich Soc Fellows, Univ Mich, 74-76; HON RES FEL, HARVARD UNIV, 76- Concurrent Pos: Lectr physics, Univ Wis-Madison, 74. Mem: Am Phys Soc. Res: Melosh transformation; nonleptonic weak decays; unified theories of weak, electromagnetic and strong interactions such as nonabelian gauge theories, quark models and applications to new narrow resonance phenomena. Mailing Add: Lyman Lab Physics Harvard Univ Cambridge MA 02138

MACHACEK, MILOS, b Prague, Czech, Sept 11, 32; US citizen; m 52; c 2. ATOMIC PHYSICS. Educ: Univ Tex, BS, 58, MA, 60, PhD(physics), 64. Prof Exp: Staff res engr, AC Electronics Div, Mass, 64-69, STAFF RES ENGR, DELCO ELECTRONICS, GEN MOTORS CORP, 69- Concurrent Pos: Russ physics translr, Consult Bur, Inc, 67- Mem: Am Phys Soc. Res: Computational atomic and molecular physics; magnetic resonance; quantum electronics; underwater acoustics. Mailing Add: Delco Electronics 6767 Hollister Ave Goleta CA 93017

MACHADO, EMILIO ALFREDO, b Buenos Aires, Arg, Feb 12, 27. EXPERIMENTAL PATHOLOGY. Educ: Mariano Moreno Col, Arg, BD, 44; Univ Buenos Aires, Physician, 52, MD(path), 53. Prof Exp: Instr path, Univ Buenos Aires, 53-61, chief lab path, Clin Univ Hosp, 54-60; chief lab exp path, NIH, Inst Gastroenterol, Buenos Aires, 63-70; assoc investr exp path, Univ Buenos Aires, 70-73; RES ASSOC PROF EXP PATH, MEM RES CTR, UNIV TENN, KNOXVILLE, 73- Mem: AAAS; Reticuloendothelial Soc; Int Acad Path; NY Acad Sci. Res: Structural and functional research on heterotransplanted malignant tumors in animal models. Mailing Add: Mem Res Ctr Univ of Tenn 1924 Alcoa Hwy Knoxville TN 37920

MACHAMER, HAROLD EUGENE, b Tower City, Pa, Jan 29, 22; m 51; c 3. MICROBIOLOGY. Educ: Pa State Univ, BS, 42, MS, 48, PhD(bact, chem), 50. Prof Exp: Res microbiologist, Tex Co, 50-52; res microbiologist, 52-54, lab dir antibiotic develop, 54-60, asst dir microbiol res dept, 60-62, dir antibiotic & microbial tech res dept, 63-66, DIR BIOL RES & DEVELOP, PARKE, DAVIS & CO, 66- Mem: AAAS; Am Chem Soc; Am Soc Microbiol; Soc Indust Microbiol; NY Acad Sci. Res: Microbiological fermentation; microbial physiology. Mailing Add: Biol Res & Develop Dept Parke Davis & Co Detroit MI 48232

MACHATTIE, LESLIE BLAKE, b Weihwei, China, Jan 29, 17; m 42; c 2. METEOROLOGY. Educ: Dalhousie Univ, BSc, 37, MSc, 39; Univ Toronto, MA, 40. Prof Exp: Meteorologist forecasting, Meteorol Serv Can, 41-51; meteorologist, Irish Meteorol Serv, Eire, 52-53; RES SCIENTIST METEOROL, CAN FORESTRY SERV, 54- Concurrent Pos: Assoc ed, J Appl Meteorol, Am Meteorol Soc, 62-67; counr, Royal Meteorol Soc, Can Br, 63-66; Rapporteur Agrotopoclimatol, Comn Agr Meteorol World Meteorol Orgn, 67-71, mem, Working Group on appln of meteorol to forestry, 75-78. Mem: Can Meteorol Soc; Am Meteorol Soc; Royal Meteorol Soc. Res: The meteorology and climatology of forest fire occurrence, behavior and control; cumulus cloud seeding for fire suppression; topoclimatology. Mailing Add: Forest Fire Res Inst Dept of Environ Ottawa ON Can

MACHATTIE, LLOYD ELLIOT, b Changte, China, Jan 8, 15; m 52; c 3. PHYSICS. Educ: Dalhousie Univ, BSc, 36, MSc, 38; Univ Va, PhD(physics), 41. Prof Exp: Res physicist, Univ Toronto, 41-43; physicist, Appl Physics Lab, Johns Hopkins Univ, 43-45; res physicist, Univ Toronto, 45-48; physicist, Comput Devices Can, Ltd, Ottawa, 49-51; DEFENCE SCI SERV OFFICER, DEFENCE & CIVIL INST ENVIRON MED, 51- Mem: Can Asn Physicists (secy, 47, treas, 48). Res: Electronics; environmental stress instrumentation. Mailing Add: Defence & Civil Inst Environ Med PO Box 2000 Downsview ON Can

MACHATTIE, LORNE ALLISTER, b Anyang, China, Apr 7, 25; Can citizen; m 52; c 3. BIOPHYSICS, GENETICS. Educ: Univ Toronto, BA, 49; Univ Western Ont, MSc, 55; Univ Buffalo, PhD(physiol), 61. Prof Exp: Instr physiol, Sch Med, Univ Buffalo, 61; res assoc, Johns Hopkins Univ, 65-67; lectr biol chem, Harvard Univ, 67-69; ASSOC PROF BIOPHYS, UNIV TORONTO, 69- Concurrent Pos: Teaching fel physiol, Sch Med, Univ Buffalo, 53-61; fel biophys, Johns Hopkins Univ, 61-65; Damon Runyon cancer res fel, 62-64; USPHS career develop award, 65-67. Mem: Fel AAAS; Biophys Soc. Res: Electron microscopy; correlation of molecular with genetic organization of bacteriophage chromosomes. Mailing Add: Dept of Med Genetics Univ of Toronto Sch of Med Toronto ON Can

MACHEAK, MERLIN EDWARD, b Lamont, Iowa, May 27, 17; m 45; c 4. VETERINARY MEDICINE, MICROBIOLOGY. Educ: Univ Iowa, BA, 42; Iowa State Univ, DVM, 50; Mont State Univ, MS, 61. Prof Exp: Vet med officer, 50-61, head anaerobic prod, 61-67, CHIEF VET, BACT LAB, VET SERV LABS, ANIMAL & PLANT HEALTH INSPECTION SERV, USDA, 67- Concurrent Pos: Mem adv panel microbiol attributes & procedures, US Pharmaceopeia, 67- Mem: Am Vet Med Asn; US Animal Health Asn; Am Inst Biol Sci; Wildlife Dis Asn. Res: Immunology of veterinary biological products of bacterial origin, particularly those containing clostridia; sterilization and sterility testing of pharmaceuticals and biological products. Mailing Add: Vet Serv Lab Box 844 Ames IA 50010

MACHEL, ALBERT R, b Dallas, Tex, Apr 17, 20; m 60; c 1. ANALYTICAL CHEMISTRY, INORGANIC CHEMISTRY. Educ: NTex Teachers Col, BS, 46; Univ Tex, MA, 49; Tex A&M Univ, PhD(chem), 58. Prof Exp: Jr chemist, Helium Plant, US Bur Mines, Tex, 42-44; instr chem, Hardin Jr Col, 46-48, asst prof, 50-51; asst prof, Tex Col Arts & Indust, 52-55 & ETex State Col, 57-58; asst prof, 58-64, ASSOC PROF CHEM, STEPHEN F AUSTIN STATE UNIV, 64- Mem: Am Chem Soc. Res: Organic chemistry; biochemistry. Mailing Add: Dept of Chem Stephen F Austin State Univ Box 6164 Nacogdoches TX 75961

MACHELL, GREVILLE, b Blackburn, Eng, Nov 19, 29; m 52; c 4. ORGANIC CHEMIS7RY, TEXTILE CHEMISTRY. Educ: Univ London, BSc, 48, Hons, 50, PhD(org chem), 52. Prof Exp: Res chemist, Brit Rayon Res Asn, 55-59 & Brit Celanese Ltd, 60-61; res chemist, Deering Milliken Res Corp, 61-64, dept mgr radiation chem, 64-67, mgr tech opers div, 67-69, gen mgr decorative fabrics div, 69-75, DIR DEVELOP, DECORATIVE FABRICS DIV, DEERING MILLIKEN INC, 75- Mem: Am Chem Soc; The Chem Soc. Res: Reaction mechanisms; alkaline degradation of carbohydrates; chemical modification of textile fibers; radiation-initiated graft polymerization. Mailing Add: Decorative Fabrics Div Deering Milliken Elm City Plant La Grange GA 30240

MACHEMER, PAUL EWERS, b Romney, WVa, Jan 30, 19; m 41; c 3. ANALYTICAL CHEMISTRY. Educ: Princeton Univ, AB, 40; Univ Pa, MS, 43, PhD(chem), 49. Prof Exp: From chemist to chief chemist, Anal Lab, Warner Co, 41-44; from jr chemist to sr shift supvr, Manhattan Proj, Carbide & Carbon Chem Co, 44-45; from asst prof to assoc prof, Villanova Univ, 49-55; from asst prof to assoc prof, 55-67, PROF CHEM, COLBY COL, 67- Concurrent Pos: Vis prof, Rosemont Col, 51-52. Mem: AAAS; Am Chem Soc; Electrochem Soc. Res: Inorganic analytical chemistry; instrumental analysis. Mailing Add: Dept of Chem Colby Col Waterville ME 04901

MACHIA, BOLLERA MUDDAPPA, b Coorg, India, Aug 8, 33; US citizen; m 66; c 2. PLANT PHYSIOLOGY. Educ: Univ Poona, BSc, 60; Kans State Univ, MS, 62, PhD(plant physiol), 65. Prof Exp: Instr food technol, Kans State Univ, 62-65; fel plant nutrit, Univ Calif, Berkeley, 65-67; res plant physiologist, Univ Calif, Riverside, 67-68; PROF BIOCHEM & CELL PHYSIOL, ST MARY'S UNIV, TEX, 68- Concurrent Pos: Consult, Wilco Peanut Co, Tex, 72- Mem: Am Chem Soc; Am Soc Plant Physiologists; Sigma Xi; Am Inst Biol Sci; AAAS. Res: Environmental factors, particularly plant nutrient elements and key enzymes of nitrogen metabolism for increased protein synthesis in rice plants and evaluation of the quality of rice protein as affected by this treatment. Mailing Add: Dept of Life Sci St Mary's Univ San Antonio TX 78284

MACHIELE, DELWYN EARL, b Zeeland, Mich, Dec 30, 38; m 62; c 2. ORGANIC CHEMISTRY. Educ: Hope Col, AB, 60; Univ Ill, Urbana, PhD(org chem), 64. Prof Exp: RES ASSOC, EASTMAN KODAK CO, 64- Mem: Am Chem Soc. Mailing Add: Res Lab Eastman Kodak Co 1669 Lake Ave Rochester NY 14650

MACHIN, J, b Herne Bay, Eng, June 29, 37; m 61; c 3. COMPARATIVE PHYSIOLOGY. Educ: Univ London, BSc, 59, PhD(zool), 62. Prof Exp: Asst prof, 62-66, ASSOC PROF ZOOL, UNIV TORONTO, 66- Concurrent Pos: Vis assoc prof, Univ Wash. Honors & Awards: T H Huxley Award, Zool Soc London, 62. Mem: AAAS; Am Physiol Soc; Marine Biol Asn UK; Brit Soc Exp Biol; Zool Soc London. Res: Comparative physiology of terrestrial moist skinned animals, particularly physics and physiology of evaporation; water transport in insects; osmotic regulation in cells. Mailing Add: Dept of Zool Univ of Toronto Toronto ON Can

MACHLEDER, WARREN HARVEY, b New York, NY, Aug 2, 43; m 67; c 1. ORGANIC CHEMISTRY. Educ: NY Univ, BA, 64; Ind Univ, Bloomington, PhD(chem), 68. Prof Exp: Res chemist, 68-74, PROJ LEADER, ROHM AND HAAS CO, 74- Mem: Am Chem Soc. Res: Allene oxidations; petroleum additives, antiwear agents and carburetor detergents. Mailing Add: Rohm and Haas Co Norristown Rd Springhouse PA 19477

MACHLIN, LAWRENCE JUDAH, b New York, NY, June 24, 27; m 53; c 3. NUTRITIONAL BIOCHEMISTRY. Educ: Cornell Univ, BS, 48, MNS, 49; Georgetown Univ, PhD(biochem), 54. Prof Exp: Poultry nutritionist, USDA, 49-50, biochemist, 50-53, biochemist, Agr Res Serv, 53-56; biochemist, Monsanto Co, 56-73; CHIEF RES GROUP, HOFFMANN-LA ROCHE, 73- Concurrent Pos: Lectr, Washington Univ, 69-72. Mem: Am Inst Nutrit; Endocrine Soc; Soc Nutrit Educ; Am Soc Animal Sci; NY Acad Sci. Res: Regulation and function of growth hormone and insulin in farm animals; biochemical function and nutritional role of amino acids, fatty acids, antioxidants, ascorbic acid, tocopherols and retinoids. Mailing Add: Hoffmann-La Roche Inc 340 Kingsland St Nutley NJ 07110

MACHLIS, LEONARD, b Seattle, Wash, Apr 13, 15; m 46; c 4. PLANT PHYSIOLOGY. Educ: State Col Wash, BS, 37; Univ Hawaii, MS, 39; Univ Calif, PhD(plant physiol), 43. Prof Exp: Res assoc, Nat Defense Res Comt, Nat Bur Stand, Washington, DC, 44-45; instr bot, Univ Ill, 45-46; from instr to assoc prof bot, 46-59, res prof, Miller Inst Basic Sci, 60-61 & 67-68, asst chancellor educ develop, 68-70, actg chmn dept bot, 73-74, PROF BOT, UNIV CALIF, BERKELEY, 59-, CHMN DEPT, 62-68 & 74- Concurrent Pos: Guggenheim Found fel, 57-58; ed, Ann Rev Plant Physiol, 58-72; consult, Sci Facil Sect, NSF, 64-68; mem cell biol study sect, NIH, 65-68. Mem: Am Soc Plant Physiol; Bot Soc Am; Scand Soc Plant Physiol. Res: Physiology of fungi and algae. Mailing Add: Dept of Bot Univ of Calif Berkeley CA 94720

MACHLIS, SAMUEL, b New York, NY, May 25, 07; m 39; c 1. TEXTILE

CHEMISTRY. Educ: NY Univ, BS, 30, MS, 31, PhD(biochem), 35. Prof Exp: Asst, Wash Sq Col, NY Univ, 31-35; asst res, Benton & Bowles, 35-36; chief chemist, Sedley Chem Co, 41-71; VPRES, ANSCOTT CHEM INDUST, INC, 71- Concurrent Pos: Dir res, O D Chem Corp, 44-46; res dir, Emtec Res Assocs, 47-50; pres, Stamford Chem Co, 50- Mem: Assoc Am Chem Soc; assoc Am Oil Chem Soc. Res: Synthetic detergents; emulsifiers; anti-caking of synthetics; packaging synthetics; role of glutathiene in the metabolism of yeast. Mailing Add: 64 Avimore Dr New Rochelle NY 10804

MACHLOWITZ, ROY ALAN, b Brooklyn, NY, Apr 16, 21; m 49; c 2. ANALYTICAL BIOCHEMISTRY, VIROLOGY. Educ: Brooklyn Col, AB, 41. Prof Exp: Jr chemist, US Naval Boiler & Turbine Lab, 42-45, chemist, Naval Air Exp Sta, 45-51; res assoc antibiotics, 51-56 & virus & tissue cult res, 56-64, SR RES BIOCHEMIST, MERCK SHARP & DOHME RES LABS, 64- Mem: AAAS; Am Chem Soc. Res: Purification of antibiotics; purification and assay of viruses. Mailing Add: 520 Laverock Rd Glenside PA 19038

MACHLUP, STEFAN, b Vienna, Austria, July 1, 27; nat US; m 61, 71; c 2. PHYSICS. Educ: Swarthmore Col, BA, 47; Yale Univ, MS, 49, PhD, 52. Prof Exp: Asst, Yale Univ, 49-51; mem tech staff, Bell Tel Labs, Inc, 52-53; res assoc physics, Univ Ill, 53-55; mem sci staff, Van der Waals Lab, Amsterdam, 55-56; asst prof, 56-61, ASSOC PROF PHYSICS, CASE WESTERN RESERVE UNIV, 61- Concurrent Pos: Consult, Res Ctr, Clevite Corp, 57-62, Educ Serv Inc, Auckland, NZ, 61, Ibadan, Nigeria, 62 & Watertown, Mass, 64-65; NSF sci fac fel, Univ Liverpool, 62-63; consult, Mass Inst Technol, 68, 69. Mem: AAAS; Am Phys Soc; Am Asn Physics Teachers. Res: Theory of solids; imperfections in metals; polaron mobility; fluctuations and irreversible processes; noise in semiconductors; transport processes in dense gases; underwater sound scattering. Mailing Add: Dept of Physics Case Western Reserve Univ Cleveland OH 44106

MACHNE, XENIA, b Trieste, Italy, Sept 28, 21. PHYSIOLOGY. Educ: Univ Padua, MD, 46. Prof Exp: Asst physiol, Univ Parma, 47-48; asst, Univ Bologna, 48-51; res assoc physiol, Univ Ore, 56-58; asst prof pharmacol, Univ Ill, 58-62; assoc prof physiol, Sch Med, Tulane Univ, 62-71; PROF PHARMACOL, UNIV MINN, MINNEAPOLIS, 71- Concurrent Pos: Brit Coun scholar, Univ Col, Univ London, 51-53; Fulbright fel, Univ Calif, Los Angeles, 53-55. Mem: Soc Neurosci; Am Physiol Soc. Res: Neurophysiology; neuropharmacology; electrophysiological methods; microelectrode techniques. Mailing Add: Dept of Pharmacol Univ of Minn Minneapolis MN 55455

MACHOVER, MAURICE, b New York, NY, Dec 5, 31; m 64. MATHEMATICAL ANALYSIS. Educ: Brooklyn Col, BS, 56; Columbia Univ, MA, 58; NY Univ, MS, 60, PhD(math), 63. Prof Exp: Lectr math & physics, Brooklyn Col, 56-58; asst math, NY Univ, 58-63, asst res scientist, 63; asst prof math, Fairleigh Dickinson Univ, 64-65, assoc prof & chmn dept, 65-67; ASSOC PROF MATH, ST JOHN'S UNIV, NY, 67- Concurrent Pos: Asst physics, Columbia Univ, 56-58. Mem: Am Math Soc; Math Asn Am; Asn Symbolic Logic; Soc Indust & Appl Math. Res: Eigenfunction expansions as applied to self adjoint differential and integral equations and their extentions to non-self adjoint problems. Mailing Add: Dept of Math St John's Univ Jamaica NY 11432

MACHT, MARTIN BENZYL, b Baltimore, Md, Aug 31, 18; m 39. PHYSIOLOGY. Educ: Johns Hopkins Univ, AB, 39, PhD(neurophysiol), 42, MD, 45. Prof Exp: Asst instr psychol, Johns Hopkins Univ, 39-42, instr physiol, Med Sch, 42-45; intern & asst resident med, Jewish Hosp, Cincinnati, Ohio, 45-46; physiologist & sta surgeon, Climatic Res Lab, Mass, 46-48; instr pharmacol, 48-54, ASST PROF PHYSIOL & ASST CLIN PROF MED, COL MED, UNIV CINCINNATI, 54- Concurrent Pos: Chief res, Jewish Hosp, 48-49, dir med educ, 49-50; co-dir internal med, Rollman Receiving Hosp. Mem: Am Physiol Soc; assoc Am Psychol Asn; AMA; Am Diabetes Asn; Am Geriat Soc. Res: Neurophysiology; localization of function in central nervous system; temperature regulation; peripheral blood flow; frost bite; neural basis of emotion; decerebrate preparations in the chronic state; psychopharmacology; psychosomatic medicine; renal disease. Mailing Add: Dept of Physiol Univ of Cincinnati Col of Med Cincinnati OH 45221

MACHTINGER, LAWRENCE ARNOLD, b St Louis, Mo, Mar 11, 36; m 64; c 1. MATHEMATICS, EDUCATION. Educ: Wash Univ, BSChE, 59, AM, 63, PhD(math), 65. Prof Exp: Asst prof math, Webster Col, 64-65; asst prof, St Louis Univ, 65-66; asst prof, Ill Inst Technol, 66-72; dir sci & math proj, ASSOC PROF MATH, PURDUE UNIV, NCENT CAMPUS, 72- Concurrent Pos: Dir, Curric Improv & Teacher Training Inst Proj; consult, high sch math; consult, Madison Proj, Syracuse Univ & Webster Col, 64-72; mem math adv coun, State of Ind, 74- Mem: Am Math Soc; Math Asn Am; Nat Coun Teachers Math. Res: Group theory and ordered rings. Mailing Add: 2930 Alexander Crescent Flossmoor IL 60422

MACHUSKO, ANDREW JOSEPH, JR, b Hiller, Pa, Dec 31, 37; m 62; c 2. MATHEMATICS. Educ: Calif State Col, Pa, BS, 59; Univ Ga, MA, 64, PhD(math), 68. Prof Exp: Teacher high sch, Ohio, 59-62; instr math, Univ Ga, 67-68; asst prof, Univ Tenn, Knoxville, 68-70; ASSOC PROF MATH, CALIF STATE COL, PA, 70- Mem: Am Math Soc; Math Asn Am. Res: Topology; algebra. Mailing Add: Dept of Math Calif State Col California PA 15419

MACIAG, THOMAS EDWARD, b Bayonne, NJ, Nov 19, 46; m 74. MOLECULAR BIOLOGY. Educ: Rutgers Univ, BA, 68; Univ Pa, PhD(molecular biol), 75. Prof Exp: Fel biochem & biophys sch med, Univ Pa, 75-76. Mem: AAAS; Am Chem Soc. Res: Enzyme technology; extracellular proteolytic enzymes and maintenance of homeostasis; diseases involving cell proliferation. Mailing Add: 4321 Osage Ave Philadelphia PA 19104

MACIAG, WILLIAM JOHN, JR, b Rome, NY, May 28, 36; m 58; c 2. MICROBIOLOGY. Educ: Univ Buffalo, BS, 57; Syracuse Univ, MS, 59, PhD(microbiol), 63. Prof Exp: Asst instr microbiol, Syracuse Univ, 59-63; res microbiologist, 63-72, SR RES MICROBIOLOGIST, STINE LAB, E I DU PONT DE NEMOURS & CO, INC, 72- Mem: AAAS; Soc Indust Microbiol; Am Soc Microbiol. Res: Pharmaceutical drug research; intermediary metabolism; autotrophic mechanisms. Mailing Add: Stine Lab E I du Pont de Nemours & Co Inc Newark DE 19711

MACIAK, GEORGE M, b Poland, Sept 14, 21; nat US; m 47; c 2. MICROCHEMISTRY. Educ: Swiss Fed Inst Technol, ChemEng, 45. Prof Exp: Anal chemist, Exp Labs, Bally Shoe Ltd, Switz, 47-49; microanalyst, Swiss Fed Inst Technol, 49-50 & Micro-Tech Labs, Ill, 50-52; microanalyst, 52-67, SR MICROANAL CHEMIST, LILLY RES LABS, 67- Mem: Am Chem Soc; AAAS; Am Microchem Soc; Asn Swiss Microchemists. Res: Erythromycin; anti-cancer agents; new procedures for microanalysis; automation and computerization of micro analytical procedures. Mailing Add: Lilly Res Labs Div Eli Lilly & Co Indianapolis IN 46206

MACIAS, EDWARD S, b Milwaukee, Wis, Feb 21, 44; m 67; c 2. NUCLEAR CHEMISTRY, AIR POLLUTION. Educ: Colgate Univ, AB, 66; Mass Inst Technol, PhD(nuclear chem), 70. Prof Exp: Asst prof, 70-76, ASSOC PROF CHEM, WASHINGTON UNIV, 76- Concurrent Pos: Consult, Argonne Nat Lab, 67, US Dept Transp, 75-76 & Meteorol Res Inc, 75-76; vis scientist, Lawrence Livermore Lab, Univ Calif, 71. Mem: AAAS; Sigma Xi; Am Phys Soc; Am Chem Soc. Res: Nuclear structure studies; atomic structure studies via x-ray spectroscopy; atmospheric aerosol physics and chemistry studies. Mailing Add: Dept of Chem Washington Univ St Louis MO 63130

MACIEL, GARY EMMET, b Niles, Calif, Jan 18, 35; m 56; c 1. PHYSICAL CHEMISTRY. Educ: Univ Calif, Berkeley, BS, 56; Mass Inst Technol, PhD(chem), 60. Prof Exp: Res asst chem, Mass Inst Technol, 59-60, NSF fel, 60-61; from asst prof to prof, Univ Calif, Davis, 61-70; PROF CHEM, COLO STATE UNIV, 71- Mem: AAAS; Am Chem Soc; Am Phys Soc. Res: Nuclear magnetic resonance and its application to chemical problems, particularly using the less common nuclei; relationships between coupling constants and chemical shifts and molecular structure; ion cyclotron resonance studies on ion-molecule reactions. Mailing Add: Dept of Chem Colo State Univ Ft Collins CO 80521

MACINKO, GEORGE, b Nesquehoning, Pa, Jan 19, 31; m 54; c 2. GEOGRAPHY, ENVIRONMENTAL SCIENCES. Educ: Univ Idaho, BA, 53; Univ Mich, MA, 57, PhD(geog), 61. Prof Exp: Instr geog & geol, Univ Idaho, 58-61; from asst prof to assoc prof geog, Univ Del, 61-67; assoc prof, 67-68, PROF GEOG, CENT WASH STATE COL, 68- Concurrent Pos: NSF sci fac fel, Univ Mich, 65-66; vis prof geog, Dartmouth Col, 69-70; consult, NH Off Planning, 70 & Boulder Area Growth Study Comn, 72-73. Mem: AAAS; Asn Am Geog; Am Inst Biol Sci. Res: Environmental management; philosophical bases for planning land use; population-resource relationships. Mailing Add: Dept of Geog Cent Wash State Col Ellensburg WA 98926

MACINNES, DAVID FENTON, JR, b Abington, Pa, Mar 19, 43. INORGANIC CHEMISTRY. Educ: Earlham Col, BA, 65; Princeton Univ, MA, 70, PhD(chem), 72. Prof Exp: Teacher chem, Westtown Sch, Pa, 70-73; ASST PROF CHEM, GUILFORD COL, 73- Mem: Am Chem Soc. Res: Transition metal complexes; x-ray crystallography; chemical education; computers and chemistry. Mailing Add: Dept of Chem Guilford Col Greensboro NC 27410

MACINNES, JAMES WILLIAM, molecular biology, neurobiology, see 12th edition

MACINNIS, AUSTIN J, b Virginia, Minn, Mar i5, 31; m 57; c 3. PARASITOLOGY, BIOCHEMISTRY. Educ: Concordia Col, Moorhead, Minn, BA, 57; Fla State Univ, MS, 59, PhD(parasitol), 63. Prof Exp: Asst, Fla State Univ, 57-59; NIH scholar parasitol, Rice Univ, 63-65; from asst prof to assoc prof zool, 65-73, PROF ZOOL, UNIV CALIF, 73- Concurrent Pos: NSF foreign travel award, 65- Mem: Fel AAAS; Am Soc Parasitol; Am Soc Trop Med & Hyg; Am Soc Zool; fel Royal Soc Trop Med & Hyg. Res: Behavior, physiology and biochemistry of parasitism. Mailing Add: Dept of Zool Univ of Calif Los Angeles CA 90024

MACINNIS, MARTIN BENEDICT, b Big Pond, NS, Aug 16, 25; m 53; c 4. CHEMISTRY, PHYSICS. Educ: St Francis Xavier Univ, BSc, 46; Col of the Holy Cross, MS, 53. Prof Exp: Prof chem, Loyola Col Montreal, 46-52; from engr to sr engr, Sylvania Elec Prod Inc, 53-60, develop engr, 60-63, advan develop engr, GTE Sylvania Inc, 63-74, head dept chem develop, 69-74, SECT HEAD CHEM DEVELOP & PROCESS ENG, GTE SYLVANIA INC, 74- Mem: Am Chem Soc. Res: Chemistry of tungsten, rhenium, molybdenum, rare earths, tantalum and niobium; solvent extraction; pyrometallurgy; hydrometallurgy; chemical vapor deposition; germanium and silicon; sugar chemistry; organic phosphors and fire retardants. Mailing Add: Chem & Metall Div GTE Sylvania Inc Towanda PA 18848

MACINTOSH, FRANK CAMPBELL, b Baddeck, NS, Dec 24, 09; m 38; c 5. PHYSIOLOGY. Educ: Dalhousie Univ, BA, 30, MA, 32; McGill Univ, PhD(physiol), 37. Hon Degrees: LLD, Univ Alta, 64 & Queen's Univ, Can, 65; MD, Univ Ottawa, 74. Prof Exp: Instr biol, Dalhousie Univ, 30-31, biochem, 31-32 & pharmacol, 32-33; demonstr physiol, McGill Univ, 36-37; mem res staff, Med Res Coun Gt Brit, 38-49; JOSEPH MORLEY DRAKE PROF PHYSIOL, McGILL UNIV, 49- Concurrent Pos: Treas, Int Union Physiol Sci, 62-68; mem, Sci Coun Can, 66-71; ed, Can J Physiol & Pharmacol, 68-72. Mem: Am Physiol Soc; Am Soc Pharmacol & Exp Therapeut; Royal Soc Can; Can Physiol Soc (pres, 60-61); Brit Physiol Soc. Res: Pharmacologically active constituents of tissues. Mailing Add: Dept of Physiol McGill Univ Montreal PQ Can

MACINTYRE, BRUCE ALEXANDER, b Oak Park, Ill, Sept 10, 42; m 65; c 2. PHYSIOLOGY. Educ: Carroll Col, Wis, BS, 63; Ind Univ, Bloomington, PhD(physiol), 68. Prof Exp: Asst prof, 68-73, ASSOC PROF BIOL, CARROLL COL, WIS, 73- Mem: AAAS; Am Inst Biol Sci. Res: Human thermoregulatory control mechanisms; cellular enzymology, including carbohydrate metabolism of choriocarcinoma cells. Mailing Add: 100 N East Ave Waukesha WI 53186

MACINTYRE, FERREN, b Seattle, Wash, Sept 15, 30. CHEMICAL OCEANOGRAPHY. Educ: Univ Calif, Riverside, BA, 60; Mass Inst Technol, PhD(phys chem), 65. Prof Exp: Asst prof oceanog, Univ Calif, San Diego, 66-69; RES ASSOC, MARINE SCI INST, UNIV CALIF, SANTA BARBARA, 69- Concurrent Pos: Guggenheim fel, 70; res fel, Univ Melbourne, 73-76. Res: Surface chemistry of the oceans; chemical interchange between ocean and atmosphere; marine environmental chemistry; physical chemistry of seawater. Mailing Add: Marine Sci Inst Univ of Calif Santa Barbara CA 93106

MACINTYRE, GILES T, b Bridgeton, NJ, Oct 6, 26; m 59; c 3. VERTEBRATE ZOOLOGY, VERTEBRATE PALEONTOLOGY. Educ: Columbia Univ, BS, 55, MA, 57, PhD(zool), 64. Prof Exp: Asst zool, Columbia Univ, 56-57, lectr, 57-59; lectr biol, 62-64, instr, 64-65, asst prof, 66-68, ASSOC PROF BIOL, QUEENS COL, NY, 69- Mem: Am Soc Mammal; Soc Syst Zool; Soc Study Evolution; Soc Vert Paleont. Res: Mesozoic mammals; quasi-mammals and therapsid reptiles; carnivore evolution and systematics; basicranial osteology and associated anatomy; functional anatomy of teeth and jaws; adaptive radiation. Mailing Add: Dept of Biol Queens Col Flushing NY 11367

MACINTYRE, MALCOLM NEIL, b Honolulu, Hawaii, Aug 12, 19; m 41; c 4. CYTOGENETICS. Educ: Univ Mich, BA, 48, MA, 50, PhD(zool), 55. Prof Exp: From instr to sr instr, 54-57 & asst prof to assoc prof, 57-67, PROF ANAT, CASE WESTERN RESERVE UNIV, 67- Concurrent Pos: Mem anat sci training comt, NIH, 62-, chmn, 65- Mem: AAAS; Teratol Soc; Am Asn Anat; Am Soc Zool; Soc Human Genetics. Res: Human reproduction and development; human reproductive failure; association of chromosomal abnormalities with congenital malformations; amniotic fluid cultures; prenatal genetic evaluation; genetic counseling. Mailing Add: Dept of Anat Case Western Reserve Univ Cleveland OH 44106

MACINTYRE, THOMAS MARTIN, b NS, Can, June 11, 14; m 42; c 4.

AGRICULTURAL CHEMISTRY, ANIMAL NUTRITION. Educ: St Francis Xavier Univ, BSc, 37; McGill Univ, BScA, 39, MSc, 41. Prof Exp: Prof chem, St Francis Xavier Univ, 42-49; res officer nutrit, 49-59, SUPT, NAPPAN EXP FARM, CAN DEPT AGR, 59- Mem: Agr Inst Can; Can Soc Animal Prod. Res: Animal management. Mailing Add: Can Dept of Agr Nappan Exp Farm Nappan NS Can

MACINTYRE, WALTER MACNEIL, chemistry, see 12th edition

MACINTYRE, WILLIAM G, chemical oceanography, see 12th edition

MACINTYRE, WILLIAM JAMES, b Cannan, Conn, Nov 26, 20; m 47; c 2. NUCLEAR MEDICINE, MEDICAL PHYSICS. Educ: Western Reserve Univ, BS, 43, MA, 47; Yale Univ, MS, 48, PhD(physics), 50. Prof Exp: Res asst, Atomic Energy Med Res Proj, 47, res assoc, 49-51, sect chief radiation physics, 51-58, sr instr radiol, 50-52, asst prof, 52-63, from asst prof to assoc prof biophys, 58-71, PROF BIOPHYS, SCH MED, CASE WESTERN RESERVE UNIV, 71-; PHYSICIST, CLEVELAND CLIN, 72- Concurrent Pos: Lectr, Mid East Regional Ctr Arab Countries, Egypt, 64; mem US nat comt med physics, Int Atomic Energy Agency, 66-72; mem coun cardiovasc radiol, Am Heart Asn. Mem: Biophys Soc; Radiol Soc NAm; Am Asn Physicists in Med; Soc Nuclear Med (pres, 76-77); Am Physiol Soc. Res: Techniques of radionuclide and functional scanning; application of radionuclide techniques to organ dynamic measurements; applications of computers to imaging and dynamic studies; measurement and analysis of electrical parameters of the brain. Mailing Add: 3108 Huntington Rd Shaker Heights OH 44120

MACIOLEK, JOHN A, b Milwaukee, Wis, Nov 2, 28; m 56; c 3. LIMNOLOGY. Educ: Ore State Univ, BS, 50; Univ Calif, Berkeley, MA, 54; Cornell Univ, PhD(limnol), 61. Prof Exp: Fishery biologist, Eastern Fish Nutrit Lab, NY, 56-61 & Sierra Nevada Aquatic Res Lab, Calif, 61-65, LEADER HAWAII COOP FISHERY RES UNIT, US FISH & WILDLIFE SERV, 66- Concurrent Pos: NSF-Am Soc Limnol & Oceanog travel grant, XIV Cong, Int Soc Limnol, Austria, 59 & XVI Cong, Poland, 65; assoc prof, Univ Hawaii, 67-; assoc zoologist, Hawaii Inst Marine Biol, 69-; actg chmn, Hawaii State Natural Area Reserves Syst Comn, 73- Mem: Am Soc Limnol & Oceanog; Int Soc Limnol; Sigma Xi. Res: Stream ecology; diadromous and euryhaline fauna of Oceania; inshore marine fishes; anchialine pools; aquatic zoogeography. Mailing Add: Hawaii Coop Fishery Res Unit 2538 The Mall Univ of Hawaii Honolulu HI 96822

MACIOR, LAZARUS WALTER, b Yonkers, NY, Aug 26, 26. BOTANY. Educ: Columbia Univ, AB, 48, MA, 50; Univ Wis, PhD(bot, zool), 59. Prof Exp: Asst bot, Columbia Univ, 48-50; head dept biol, St Francis Col, Wis, 60-62; instr, Marquette Univ, 62-64; asst prof biol, Loras Col, 65-67; asst prof, 67-68, ASSOC PROF BIOL, UNIV AKRON, 68- Concurrent Pos: Res assoc, Inst Arctic & Alpine Res, Univ Colo, 66-67 & Inst Polar Studies, Ohio State Univ, 71-73. Mem: Fel AAAS; Am Soc Plant Taxon; Bot Soc Am; Ecol Soc Am; Nat Asn Biol Teachers. Res: Plant morphology; insect-flower pollination relationships; evolution. Mailing Add: Dept of Biol Univ of Akron Akron OH 44325

MACIVER, DONALD STUART, b Cambridge, Mass, Oct 3, 27; m 47; c 2. INDUSTRIAL CHEMISTRY. Educ: Univ Boston, AB, 52; Univ Pittsburgh, PhD(phys chem), 57. Prof Exp: Sect head catalysis, Gulf Res & Develop Co, 57-64; mgr chem res, Eastern Res Ctr, NY, 64-67, dir, Western Res Ctr, Richmond, Calif, 67-74, dir, St Gabriel Plant, La, 72-74, VPRES & GEN MGR, STAUFFER CHEM CO OF WYO, 74- Mem: Am Chem Soc. Res: Adsorption; catalysis; resonance spectroscopy; fuel cell technology; petroleum processing; petrochemicals; chemical processes; agricultural chemicals; caustic-chlorine manufacture. Mailing Add: One Sally Ann Rd Orinda CA 94563

MACK, CHARLES LAWRENCE, JR, b Cleveland, Ohio, July 20, 26; m 50; c 4. APPLIED PHYSICS. Educ: Harvard Univ, BSc, 48; Univ Pa, MSc, 53. Prof Exp: Instr physics, Univ Pa, 49-50, instr med, Sch Med, 52-53; mem staff physics, Lincoln Lab, Mass Inst Technol, 53-57; tech consult, Supreme Hq, Allied Powers Europe, Paris, 57-59; mem staff, Mitre Corp, Bedford, Mass, 59; MEM STAFF PHYSICS, LINCOLN LAB, MASS INST TECHNOL, 59- Res: Conversion of solar to electrical energy by low temperature fluids in gravity-driven engines. Mailing Add: 7 Parker St Lexington MA 02173

MACK, CLINTON OLMSTED, b Upland, Ind, Sept 29, 07; m 35; c 2. HUMAN PHYSIOLOGY, DEVELOPMENTAL BIOLOGY. Educ: Ohio State Univ, BS, 28; Northwestern Univ, MA, 32; Western Reserve Univ, PhD(zool), 42. Prof Exp: Teacher high sch, Ohio, 28-30; asst chem, Northwestern Univ, 30-32; prof, Lees Jr Col, Ky, 34-38; prof biol, Cleveland Bible Col, 38-40; assoc prof, Asbury Col, 40-42; prof, Cascade Col, 42-43; from asst prof phys sci to prof biol, 43-74, EMER PROF BIOL, WHEATON COL, ILL, 74- Mem: AAAS. Res: Longevity, reproductivity and growth of Daphnia magna when reared on commerical yeast. Mailing Add: 486 E Exchange St Sycamore IL 60178

MACK, HARRY JOHN, b Gatesville, Tex, Mar 18, 26; m 55; c 4. HORTICULTURE. Educ: Tex A&M Univ, BS, 50, MS, 52; Ore State Univ, PhD(hort), 55. Prof Exp: Instr hort, Tex A&M Univ, 50-51; from instr to assoc prof, 55-69, PROF HORT, ORE STATE UNIV, 69- Mem: Am Soc Hort Sci; Am Sci Affil. Res: Vegetable crops physiology; population density; mineral nutrition; irrigation and water relations; growth regulators. Mailing Add: Dept of Hort Ore State Univ Corvallis OR 97331

MACK, HARRY PATTERSON, b Toledo, Ohio, Jan 9, 24; m 45; c 6. ANATOMY. Educ: Univ Md, MD, 48. Prof Exp: Intern, Univ Md Hosp, 48-50, from instr to assoc anat, Sch Med, 50-52, from asst prof to assoc prof, 54-60; mem staff Christ Hosp Inst Med Res, 60-64, sr clin pharmacologist, Sterling-Winthrop Res Inst, NY, 64-66; assoc dir med res, Mead Johnson Res Ctr, Ind, 67-68, dir cardiovasc res, 68-71; MGR CLIN RES, PROCTER & GAMBLE MIAMI VALLEY LABS, 71- Concurrent Pos: Lectr, Sch Med, Univ Cincinnati, 60-64, adj prof anat, 67- Mem: AAAS; Am Asn Anat; AMA; Biol Stain Comn. Res: Porphyrin metabolism and pharmacology; radiobiology and photosensitization; growth and differentiation; experimental and comparative pathology; clinical, experimental and therapeutic pharmacology. Mailing Add: Miami Valley Labs PO Box 39175 Cincinnati OH 45247

MACK, IRVING, b Vilna, Poland, Apr 15, 19; nat US; m 59; c 2. INTERNAL MEDICINE. Educ: Univ Chicago, BS, 39, MD, 42; Am Bd Internal Med, dipl, 52; Am Bd Pulmonary Dis, dipl, 62. Prof Exp: Intern, Cook County Hosp, 42-43; from resident to sr resident internal med, Michael Reese Hosp, 44-45; asst med, Chicago Med Sch, 49-51, instr, 51-52, assoc, 52-54, from asst prof to assoc prof, 54-73; CLIN PROF MED, PRITZKER SCH MED, UNIV CHICAGO, 73- Concurrent Pos: Sr res fel, Dept Cardiovasc res, Michael Reese Hosp, 45-46; adj attend physician, Chest Dept, Michael Reese Hosp, 47-54, assoc attend physician, 54-59, attend physician, 59-73, sr attend physician, 73-, assoc cardiovasc res, 50-57, chief Thursday Chest Clin, 52-70, adj attend physician, Winfield Hosp, 47-54, from assoc attend physician to attend physician, 54-63; consult, Herrick House Rheumatic Fever, 49-54. Mem: AAAS; Am Psychosom Soc; Am Thoracic Soc; AMA; Am Heart Asn. Res: Pulmonary physiology; cor pulmonale; clinical cardiology; cardiac arrhythmias; electrocardiography; psychosomatic interrelationships; clinical pulmonary disease. Mailing Add: 5490 South Shore Dr Chicago IL 60615

MACK, JAMES PATRICK, b Newark, NJ, Dec 9, 39; m 68; c 1. CELL BIOLOGY. Educ: Monmouth Col, NJ, BS, 62; William Paterson Col NJ, MA, 66; Columbia Univ, EdD(cell biol), 71. Prof Exp: Teacher biol, Shore Regional High Sch, NJ, 62-66; asst prof cell physiol, Monmouth Col, NJ, 68-69; teacher biol chem, Lakewood High Sch, NJ, 71-74; RES SCIENTIST CELL BIOL, LAMONT DOHERTY GEOL OBSERV, COLUMBIA UNIV, 69-; ASST PROF CELL PHYSIOL, MONMOUTH COL, NJ, 74- Concurrent Pos: Teacher biol, Jersey City State Col, 70-73 & biol & chem, Ocean County Col, NJ, 71-; res scientist cell biol, Creedmore Inst, Queens Village, NY, 74- Honors & Awards: Am Chem Soc Award, 73. Mem: Sigma Xi; Am Chem Soc; Am Inst Biol Sci. Res: Determining the role of vitamin A, outside the visual cycle, incellular membranes; effect of vitamin E and hydrocortisone on aging cells. Mailing Add: Dept of Biochem Monmouth Col West Long Branch NJ 07764

MACK, JOHN EDWARD, physics, astrophysics, see 12th edition

MACK, JOHN ELDON, mathematics, see 12th edition

MACK, JULIUS L, b Gadsden, SC, June 14, 30; m 58; c 3. CHEMICAL PHYSICS. Educ: SC State Col, BS, 52; Howard Univ, MS, 57, PhD(phys chem), 65. Prof Exp: Res chemist, Naval Ord Sta, 56-71; chmn div natural sci, 71-73, PROF CHEM, FED CITY COL, 71-, DEAN SCH NATURAL, APPL & HEALTH SCI, 73- Mem: AAAS; Am Chem Soc; Am Phys Soc. Res: Study of structural and thermodynamic properties of high temperature molecules; infrared spectroscopy of matrix isolated species and mass and infrared spectra of hot vapors; vibrational spectra of unstable species. Mailing Add: Sch of Natural Appl & Health Sci Fed City Col Washington DC 20001

MACK, LAWRENCE LLOYD, b Springfield, Mo, Dec 10, 42; m 66; c 2. PHYSICAL CHEMISTRY, BIOCHEMISTRY. Educ: Middlebury Col, AB, 65; Northwestern Univ, PhD(phys chem), 69. Prof Exp: Fel, Univ Calif, Berkeley, 69-70, instr chem & NIH fel, 70; asst prof chem, St Lawrence Univ, 70-72; ASSOC PROF CHEM, BLOOMSBURG STATE COL, 72- Mem: Am Chem Soc; Sigma Xi. Res: Development and employment of physical methods to study solution behavior of biopolymers; biophysical chemistry and biophysics; computer assisted instruction. Mailing Add: Dept of Chem Bloomsburg State Col Bloomsburg PA 17815

MACK, PAULINE BEERY, b Norborne, Mo, Dec 19, 91; m 23. CHEMISTRY. Educ: Univ Mo, AB, 13; Columbia Univ, AM, 19; Pa State Col, PhD(biol & phys chem, physics), 32. Hon Degrees: ScD, Western Col Women & Moravian Col Women, 50. Prof Exp: Teacher high schs, Mo, 13-19; from instr to assoc prof chem, Pa State Col, 19-35, prof household chem, 35-52, dir mass studies human nutrit, Ellen H Richards Inst, 40-52; DIR RES INST, TEX WOMAN'S UNIV, 53- Concurrent Pos: Dir mass studies human nutrit, State Dept Health, Pa, 35-52. Honors & Awards: Garvan Medal, Am Chem Soc, 50; Astronauts' Silver Snoopy Award, 70. Mem: AAAS; fel Soc Res Child Develop; Am Chem Soc; Am Soc Testing & Mat; fel Am Pub Health Asn. Res: Human nutrition; chemistry of calcium in relation to evaluation of mineral density in bones from x-rays. Mailing Add: Res Inst Tex Woman's Univ PO Box 23895 Denton TX 76204

MACK, REX CHARLES, b Whitewater, Wis, Oct 25, 20; m 46; c 1. PHYSICS. Educ: Whitewater State Col, BEd, 42; Ohio State Univ, PhD(physics), 53. Prof Exp: Physicist, Rand Corp, 56-57; chief scientist, Hughes Aircraft Co, 57-65; sci adv, Off Dep Chief Staff Res & Develop, Hq, Dept Air Force, 65-67; dir develop planning, Hughes Aircraft Co, 67-68; consult, Rex C Mack Assocs, 68-70; PROF SYSTS MGT, UNIV SOUTHERN CALIF, 70- Res: Nuclear physics; space technology; systems analysis. Mailing Add: 300 Ivy Terr Fallbrook CA 92028

MACK, RICHARD BRUCE, b South Paris, Maine, Sept 18, 28; m 54. APPLIED PHYSICS. Educ: Colby Col, BA, 51; Harvard Univ, MS, 57, PhD, 64. Prof Exp: PHYSICIST ELECTROMAGNETIC THEORY, AIR FORCE CAMBRIDGE RES CTR, 51- Mem: Inst Elec & Electronics Engrs; Sigma Xi. Res: Antennas; scattering of electromagnetic waves; radar systems; propagation of waves in plasmas. Mailing Add: Air Force Cambridge Res Ctr Laurence G Hanscom Field Bedford MA 01731

MACK, RICHARD NORTON, b Providence, RI, July 31, 45. PLANT ECOLOGY. Educ: Western State Col Colo, BA, 67; Wash State Univ, PhD(bot), 71. Prof Exp: Instr bot, Wash State Univ, 70; ASST PROF BIOL SCI, KENT STATE UNIV, 71- Mem: Ecol Soc Am; Brit Ecol Soc; Am Quaternary Asn. Res: Mineral cycling by terrestrial plants; quaternary vegetation history of Pacific Northwest. Mailing Add: Dept of Biol Sci Kent State Univ Kent OH 44242

MACK, ROBERT EMMET, b Morris, Ill, Mar 26, 24; m 51; c 5. INTERNAL MEDICINE. Educ: Univ Notre Dame, BS, 46; St Louis Univ, MD, 48. Prof Exp: From intern to resident internal med, St Louis Univ, 48-52, from instr to asst prof, 53-61; chief med serv, Hutzel Hosp, 61-66, med dir, 65-66, dir, 66-70; asst prof, 61-67, PROF INTERNAL MED, SCH MED, WAYNE STATE UNIV, 67-; PRES, HUTZEL HOSP, 70- Concurrent Pos: Chief, Med Serv, Vet Admin Hosp, 56-61; consult, St Louis City Hosp. Mem: Am Thyroid Asn; Am Physiol Soc; Endocrine Soc; Am Col Physicians; Am Fedn Clin Res. Res: Thyroid gland; cardiac output and coronary blood flow; radioisotopes. Mailing Add: Hutzel Hosp 432 E Hancock Ave Detroit MI 48201

MACK, STANLEY ZANER, b Swift Current, Sask, May 29, 24; m 52; c 4. PHYSICS. Educ: Univ Toronto, BASc, 48, MA, 49, PhD(physics), 51. Prof Exp: Tech officer, Geod Surv Can, 48-49 & Dom Observ, 49-53; sci res officer, Armament Res & Develop Estab, 53-56, Hq, 56-60, Theoret Studies Group, 60-62 & Pac Naval Lab, 62-69, mem planning staff, 69-75, ACTG CHIEF PLANS, DEFENCE RES BD CAN, 75- Mem: AAAS; World Future Soc; Asn Sci & Technol Community Can. Res: Geophysics. Mailing Add: 2112 Thistle Crescent Ottawa ON Can

MACK, WALTER NOEL, b Marcellus, Mich, May 22, 11; m 40; c 1. BACTERIOLOGY. Educ: Mich State Univ, BS, 39, MS, 40; Univ Calif, PhD(comp path), 47. Prof Exp: Bacteriologist viruses, State Dept Health, Mich, 39-40, serologist, 40-41; technician, Sch Pub Health, Univ Mich, 41-42; asst, Hooper Found, Univ Calif, 43-47; from asst prof to assoc prof, 47-56, PROF BACT, MICH STATE UNIV, 56- Concurrent Pos: Consult, US Biol Warfare Labs, 58-59. Mem: AAAS; Am Soc Microbiol; Am Asn Immunol; NY Acad Sci. Res: Viruses and virus diseases. Mailing Add: Dept of Microbiol & Pub Health Mich State Univ East Lansing MI 48823

MACKAL, ROY PAUL, b Milwaukee, Wis, Aug 1, 25; m 63; c 1. BIOCHEMISTRY. Educ: Univ Chicago, BS, 49, PhD(biochem), 53. Prof Exp: From instr to asst prof, 53-64, ASSOC PROF BIOCHEM, UNIV CHICAGO, 64-, RES ASSOC, 53-, UNIV SAFETY COORDR, 74- Mem: Am Asn Phys Anthrop; Am Soc Biol Chem; AAAS.

Res: Virus; bacteriophage; chemical physical anthropology; synthesis of DNA; safety and environmental health. Mailing Add: 9027 S Oakley Ave Chicago IL 60620

MACKANESS, GEORGE BELLAMY, b Sydney, Australia, Aug 20, 22; m 45; c 1. IMMUNOLOGY. Educ: Univ Sydney, MB & BS, 45; Univ London, dipl clin path, 48; Oxford Univ, MA, 49, DPhil(path), 53. Prof Exp: Resident med, Sydney Hosp, 45-46; resident path, Kanematsu Inst, Univ Sydney, 46-47; demonstr, Oxford Univ, 48-53; asst prof, Australian Nat Univ, 54-62; prof microbiol, Univ Adelaide, 62-65; DIR, TRUDEAU INST, 65- Concurrent Pos: USPHS traveling fel, Rockefeller Univ, 59-60; USPHS res grant, Trudeau Inst, 65-; consult, USPHS, 66-; adj prof, Sch Med, NY Univ, 68-; consult, Nat Acad Med, 69-71 & US Armed Force Epidemiol Br, 69- Mem: AAAS; Am Soc Microbiol; Am Thoracic Soc; Am Asn Immunol. Res: Cellular aspects of immunology; resistance to infectious disease; antitumor immunity. Mailing Add: Trudeau Inst PO Box 59 Saranac Lake NY 12983

MACKAUER, MANFRED, b Wiesbaden, Ger, June 3, 32; m 59; c 1. ENTOMOLOGY. Educ: Univ Frankfurt, Drphilnat, 59. Prof Exp: Res asst parasitol, Univ Frankfurt, 59-61; res scientist, Res Inst, Can Dept Agr, 61-67; PROF BIOL SCI, SIMON FRASER UNIV, 67-, CHMN INT BIOL PROG PROJ, 68- Concurrent Pos: Comn Int de Lutte Biol, Zurich study grant, 61; mem taxon sect, Orgn Int de Lutte Biol, 58- Mem: Entom Soc Am; Soc Syst Zool; Entom Soc Can; Ger Soc Appl Entom; Ger Entom Soc. Res: Bionomics, phylogeny and taxonomy of parasitic Hymenoptera; host specificity of hymenopterous parasites of aphids; nonchemical controls of pest insects, especially aphids. Mailing Add: Dept of Biol Sci Simon Fraser Univ Burnaby BC Can

MACKAY, ALBERT GEORGE, b Peacham, Vt, Oct 5, 07; m 32; c 3. SURGERY. Educ: Univ Vt, BS, 29, MD, 32; Am Bd Surg, dipl. Hon Degrees: DSc, Univ Vt, 72. Prof Exp: Instr anat, 33-35, instr path, 35-37, from instr to assoc prof surg, 39-42, chmn dept, 42-70, PROF SURG, COL MED, UNIV VT, 42- Concurrent Pos: Attend surgeon and chief surg, Mary Fletcher Hosp, Burlington, Vt, 42-; attend surgeon, Degoesbriand Mem Hosp, 44-; consult surgeon, Hosps, 55-; mem staff, Med Ctr, Hosp of Vt, Burlington; attend surgeon, Fanny Allen Hosp, Consol Porter Hosp; Brightlook Hosp, Heaton Hosp & Vt State Hosp. Honors & Awards: Distinguished Serv Award, Vt Med Soc, 69. Mem: Am Thoracic Soc; Am Geriat Soc; AMA; Asn Am Med Cols; fel Am Col Surg. Res: Surgical diseases of the stomach, duodenum, pancreas and biliary tract. Mailing Add: Dept of Surg Univ of Vt Col of Med Burlington VT 05401

MACKAY, BRUCE, b Edinburgh, Scotland, May 18, 30; m 66. PATHOLOGY. Educ: Univ Edinburgh, MB & ChB, 56, PhD(anat), 61. Prof Exp: Lectr anat, Univ Edinburgh, 61; asst prof, Univ Iowa, 61-63; from asst resident to assoc resident surg, Vancouver Gen Hosp, BC, 63-65, from assoc resident to resident path, 65-67; from instr to asst prof, Univ Wash, 67-69; ASST PATHOLOGIST, DEPT ANAT PATH & ASST PROF PATH, UNIV TEX M D ANDERSON HOSP & TUMOR INST HOUSTON, 69-, ASST PROF PATH, UNIV TEX GRAD SCH BIOMED SCI HOUSTON, 70- Concurrent Pos: Resident orthop, St Paul's Hosp, Vancouver, BC, 65; chief resident path, King County Hosp, Seattle, Wash, 68-69. Mem: Fel Col Am Path; Am Asn Path & Bact; Am Asn Anat; Int Acad Path. Res: Anatomical pathology; electron microscopy, particularly of tumors. Mailing Add: Dept of Path M D Anderson Hosp & Tumor Inst Houston TX 77025

MACKAY, COLIN FRANCIS, b Waterbury, Conn, Sept 21, 26. PHYSICAL CHEMISTRY. Educ: Univ Notre Dame, BS, 50; Univ Chicago, PhD(chem), 56. Prof Exp: From asst prof to assoc prof, 56-68, PROF CHEM, HAVERFORD COL, 68- Mem: AAAS; Am Chem Soc. Res: Atomic reactions; chemistry of highly reactive species. Mailing Add: Dept of Chem Haverford Col Haverford PA 19041

MACKAY, DONALD ALEXANDER-MORGAN b Gt Brit, Feb 8, 26; nat US; m 52; c 4. FOOD CHEMISTRY. Educ: Oxford Univ, BA, 48; Yale Univ, PhD(chem), 54. Prof Exp: Proj leader, Evans Res & Develop Corp, NY, 53-56, from assoc dir res to dir res, 56-61; mgr corp res, Gen Foods Corp, 61-64; asst to vpres technol, Coca-Cola Co, 64-66; vpres res & develop, Evans Res & Develop Corp, 66-69; dir spec projs, Snell Div, Booz Allen & Hamilton Inc, 69-73; VPRES RES & DEVELOP, LIFE SAVERS/SQUIBB CORP, 73- Mem: Am Chem Soc; NY Acad Sci. Res: Chromatography; trace analysis; odor measurement; flavor, sulfur and keratin; reaction mechanisms; biogenesis of natural products; product development; consumer studies; new product planning; research organization; dental research; nutrition. Mailing Add: 135 Deerfield Lane Pleasantville NY 10570

MACKAY, DONALD CYRIL, b St Stephen, NB, Can, Aug 10, 23; m 51; c 2. SOIL CHEMISTRY, PLANT CHEMISTRY. Educ: McGill Univ, BSc, PhD(agr chem), 54; Cornell Univ, MS, 49. Prof Exp: Soil surveyor & lectr chem, Nova Scotia Agr Col, 45-48, soil specialist, 49-51; plant nutritionist, 54-61, HEAD SOIL SCI SECT, RES STA, CAN DEPT AGR, 61- Mem: Soil Sci Soc Am; Am Soc Agron; Can Soc Soil Sci; Agr Inst Can; Int Soc Soil Sci. Res: Diagnosis of plant nutrient status; determination of optimum soil management and fertilizer practices for crop production. Mailing Add: Soil Sci Sect Res Sta Can Dept of Agr Lethbridge AB Can

MACKAY, DONALD DOUGLAS, b Lorne, NS, Apr 29, 08; m 36; c 2. CHEMISTRY. Educ: Acadia Univ, BSc, 29. Prof Exp: Chemist, Aluminum Co Can, Ltd, 29-31; chemist, McColl-Frontenac Oil Co, 31-35; chemist, Aluminum Co Can, Ltd, 35-36, head tech control, Alumina Plant, 36-37, chief chemist, Demerara Bauxite Co, Ltd Div, 38, from supvr to asst supt, Chem Div, Aluminum Co, 39-42, chem engr, 42-45, chief alumina div, Raw Mat Dept, Aluminium Labs, Ltd Div, 45-49, head, 49-66, vpres div, 59-66, dir, 65-66; managing dir, Indian Aluminium Co, Ltd, Calcutta, 66-69; exec vpres raw mat, Alcan Aluminium Ltd & pres, Alcan Ore Ltd, Montreal, Que, 69-73; ALUMINUM INDUST ADV & CONSULT, 73- Concurrent Pos: Dir & vpres, Fluoresqueda, SA, Mex, 57-66; dir, Nfld Fluorspar, Ltd, Nfld & Southeast Asia Bauxites, Ltd, Singapore, 58-66; dir, Alcan Queensland Pty, Ltd, 64-66. Mem: Am Chem Soc; Chem Inst Can. Res: Investigation and acquisition of raw material deposits throughout the world for the aluminum industry, especially bauxite and fluorspar; appraisal of economics of producing alumina and fluoride materials from these deposits. Mailing Add: PO Box 210 Mansonville PQ Can

MACKAY, FRANCIS PATRICK, b Waterbury, Conn, July 12, 29. ORGANIC CHEMISTRY. Educ: Univ Notre Dame, BS, 51; Col Holy Cross, MS, 52; Pa State Univ, PhD(chem). Prof Exp: Res chemist, E I du Pont de Nemours & Co, Inc, 56-58; asst prof, 58-70, ASSOC PROF CHEM, PROVIDENCE COL, 70-, CHMN DEPT, 73- Mem: Am Chem Soc; Sigma Xi. Res: Preparation and reactions of organosilicon compounds; reactions of metal chelate compounds. Mailing Add: Dept of Chem Providence Col Providence RI 02918

MACKAY, IAN FRANCIS STUART, b Vina del Mar, Chile, June 14, 12; m 38; c 1. HUMAN PHYSIOLOGY. Educ: Royal Col Surgeons, Ireland, Lic, 36; Univ Manchester, MSc, 39, PhD(physiol), 42. Prof Exp: Student demonstr anat, Royal Col Surgeons, Ireland, 34-36; demonstr & asst lectr human physiol, Univ Manchester, 36-45; lectr exp physiol, Univ Sheffield, 45-47; reader exp pharmacol, Univ Durham, 47-49; prof physiol, Univ West Indies, 49-62; PROF CLIN PHYSIOL, SCH MED, UNIV PR, SAN JUAN, 62- Mem: Am Physiol Soc; fel Am Col Cardiol; Brit Physiol Soc. Res: Respiratory physiology; nutritional studies; pancreatic secretion; clinical cardiovascular problems. Mailing Add: Gen Clin Res Ctr Univ of PR Sch of Med San Juan PR 00905

MACKAY, JOHN KELVIN, b Kirkland Lake, Ont, Jan 4, 38; m 63; c 1. SURFACE CHEMISTRY. Educ: Queens Univ, Ont, BSc, 59; McGill Univ, PhD(chem), 67. Prof Exp: Teaching asst chem, McGill Univ, 59-64; tech officer organometallic chem, petrochem & polymer lab, ICI Ltd, Eng, 64-66; from res chemist to sr res chemist, Res Ctr, Hooker Chem Corp, NY, 66-69, proj coordr electrochem, 69-70, mgr electrochem res, 70-71, mgr chem res, 71-73, mgr chem-electrochem res, Hooker Chem & Plastics Corp, 73-74; dir venture res, Oxy Metal Industs Int, Switz, 74-75; MGR FUTURES RES & DEVELOP, HOOKER CHEM & PLASTICS CORP, 75- Mem: Am Chem Soc; Electrochem Soc. Res: Analytical chemistry and automation; surface treatment systems; energy storage and conversion devices; management. Mailing Add: Res Ctr Hooker Chem & Plastics Corp Grand Island NY 14072

MACKAY, JOHN ROSS, b Formosa, Dec 31, 15; Can citizen; m 44; c 2. GEOMORPHOLOGY, CARTOGRAPHY. Educ: Clark Univ, BA, 39; Boston Univ, MA, 41; Univ Montreal, PhD, 49. Prof Exp: Asst prof geog, McGill Univ, 46-49; from asst prof to assoc prof, 49-57, PROF GEOG, UNIV B C, 57- Concurrent Pos: Chmn Can nat comt, Int Geog Union, 64- Mem: Am Geog Soc; Geol Soc Am; Asn Am Geog; fel Royal Soc Can; Can Asn Geog. Res: Arctic geomorphology, expecially glaciation and patterned ground; ground ice; deltas. Mailing Add: Dept of Geog Univ of B C Vancouver BC Can

MACKAY, JOHN WARWICK, b Can, June 1, 23; m 48; c 4. SOLID STATE PHYSICS. Educ: Univ Sask, BSc, 45; Purdue Univ, MS, 49, PhD(physics), 53. Prof Exp: Asst physics, Nat Res Coun Can, 45-46; asst, 46-53, from asst prof to assoc prof, 53-66, PROF PHYSICS, PURDUE UNIV, 66- Concurrent Pos: Consult, NASA, 63- Mem: Am Phys Soc. Res: Microwave electron accelerators; radiation damage in semiconductors. Mailing Add: Dept of Physics Purdue Univ West Lafayette IN 47906

MACKAY, JOHNSTONE SINNOTT, b Schenectady, NY, Dec 27, 13; m 40; c 2. CHEMISTRY. Educ: Cornell Univ, BCh, 35; Columbia Univ, PhD(chem eng), 39. Prof Exp: Res chemist, Catalytic Develop Co, 39-42 & Am Cyanamid Co, 42-52; res chemist, Pittsburgh Chem Co, 52-58, mgr gen res, 58-60, dir res & develop, 60-64, dir res, Pittsburgh Chem Div, US Steel Corp, 64-67, SR RES CONSULT, USS CHEM DIV, US STEEL CORP, 67- Honors & Awards: Young Auth Prize, Electrochem Soc, 41. Mem: AAAS; Am Chem Soc; Electrochem Soc; Am Inst Chem Eng; NY Acad Sci. Res: Semiconductors; catalysts and catalytic processes; high pressure nitrogen chemistry; acrylonitrile and coal chemicals; blast furnace technology; protective coatings. Mailing Add: 139 Altadena Dr Pittsburgh PA 15228

MACKAY, KENNETH DONALD, b Detroit, Mich, July 18, 42; m 64; c 2. ORGANIC CHEMISTRY. Educ: Univ Mich, BS, 64; Univ Minn, PhD(chem), 69. Prof Exp: Sr res chemist, Gen Mills, Inc, 68-74, group leader, Gen Mills Chem, Inc, 74-75, RES ASSOC, GEN MILLS CHEM, INC, 75- Mem: Am Chem Soc; Am Inst Mining, Metall & Petrol Engrs. Res: Conformational analysis; physical organic chemistry; organic synthesis; solvent extraction. Mailing Add: Gen Mills Chem Inc 2010 E Hennepin Ave Minneapolis MN 55413

MACKAY, KENNETH PIERCE, JR, b Detroit, Mich, Feb 7, 39; m 63; c 1. METEOROLOGY, AIR POLLUTION. Educ: Univ Mich, BSE, 61, MS, 65; Univ Wis, PhD(meteorol), 70. Prof Exp: Asst res meteorologist, Univ Mich, 61-65; res asst meteorol, Univ Wis, 65-69; asst prof, 69-75, ASSOC PROF METEOROL, SAN JOSE STATE UNIV, 75- Mem: Am Meteorol Soc; Air Pollution Control Asn. Res: Boundary Layer Winds. Mailing Add: Dept of Meteorol San Jose State Univ San Jose CA 95192

MACKAY, LOTTIE ELIZABETH BOHM, b Vienna, Austria, June 7, 27; US citizen; m 52; c 4. SCIENCE WRITING, SCIENCE EDUCATION. Educ: Vassar Col, AB, 47; Yale Univ, PhD(org chem), 52. Prof Exp: Sr res chemist, Standard Oil Develop Corp, 52-53 & Burroughs-Wellcome Co, 53-55; sci consult, Hudson Inst, 67; sci & math ed AV educ mat, Educ AV Corp, 68-70; SCI & MATH ED AV EDUC MAT, PRENTICE-HALL MEDIA, INC, 70- Mem: Nat Sci Teachers Asn; Nat Coun Teachers Math; Am Asn Physics Teachers. Mailing Add: Prentice-Hall Media Inc 150 White Plains Rd Tarrytown NY 10591

MACKAY, RALPH STUART, b San Francisco, Calif, Jan 3, 24; m 60. BIOPHYSICS, ELECTRONICS. Educ: Univ Calif, Berkeley, AB, 44, PhD(physics), 49. Prof Exp: Asst physics, Univ Calif, Berkeley, 44-48, elec eng, 47-49, lectr, 49-52, electron microscopist, 46-51, asst prof elec eng, 52-57, dir res & develop lab, Med Ctr, San Francisco, 54-58, assoc res biophysicist, 54-57, lectr, 55-60, assoc clin prof exp radiol & assoc res physicist, 58-60, res biophysicist, Berkeley, 60-67, assoc clin prof optom, 60-62, clin prof, 62-64, biophysicist, Space Sci Lab, 63, lectr med physics, 63-67; PROF BIOL, BOSTON UNIV & PROF SURG, MED CTR, 67- Concurrent Pos: Guggenheim fel, Karolinska Inst, Sweden, 57-58; Fulbright fel, Cairo Univ, 60, bis prof, 60-61; sr scientist, Galapagos Inst Sci Proj, 64; US ed, Ultrasonics, 65-; Erskine fel, Univ Canterbury, NZ, 69; consult, Dep Proj Off, US Naval Radiol Defense Lab, 56; chronic uremia consult, Nat Inst Arthritis & Metab Dis, 66-69; mem bd dir, Biotronics, Inc, 60-67, mine adv comt, Nat Acad Sci-Nat Res Coun, 61-, bio-instrumentation adv coun, Am Inst Biol Sci, 65-71 & comt emergency med serv, Nat Res Coun, 68-70; distinguished visitor, US Antarctic Prog, 70; mem bd gov, Int Inst Med Electronics & Biol Eng, Paris, 70- Honors & Awards: Apollo Award, 62. Mem: Fel Inst Elec & Electronics Eng; Am Inst Biol Sci; Undersea Med Soc; Biomed Eng Soc. Res: Medical engineering; biology. Mailing Add: Dept of Biol Boston Univ Boston MA 02215

MACKAY, RAYMOND ARTHUR, b New York, NY, Oct 30, 39; m 66; c 1. PHYSICAL CHEMISTRY. Educ: Rensselaer Polytech Inst, BS, 61; State Univ NY Stony Brook, PhD(phys chem), 66. Prof Exp: Guest res assoc nuclear eng dept, Brookhaven Nat Lab, 63-64, res assoc chem, 66; res chemist phys res lab, US Army Edgewood Arsenal, Md, 67-69; asst prof, 69-74, ASSOC PROF CHEM, DREXEL UNIV, 74- Concurrent Pos: Asst prof exten div, Univ Del, 67-68; consult, Edgewood Arsenal, Md, 70-76. Mem: AAAS; Am Chem Soc. Res: Catalysis of organic reactions by metal complexes; liquid crystals; charge-transfer complexes; reactions in microemulsions. Mailing Add: Dept of Chem Drexel Univ Philadelphia PA 19104

MACKAY, ROSEMARY JOAN, b Eng, July 18, 36; m 62. FRESH WATER ECOLOGY. Educ: Univ London, BSc, 57, Dipl, 58; McGill Univ, MSc, 68, PhD(biol), 72. Prof Exp: High sch teacher biol, Govt Kenya, 58-61; high sch teacher biol, St Mary's Acad, Winnipeg, 61-62; engr water anal, Winnipeg Water Works, 62-63; technician med, Montreal Gen Hosp, 63-65; res fel entom, Royal Ont Mus, 73-74; ASST PROF ZOOL, UNIV TORONTO, 74- Concurrent Pos: Fel, McGill Univ, 72-73; Nat Res Coun Can fel, 73. Mem: NAm Benthol Soc; Ecol Soc Am; Entom Soc Am; Entom Soc Can; Brit Ecol Soc. Res: Ecology of aquatic insects, especially trichoptera, with emphasis on resource-partitioning among closely related species;

investigating ecology of sympatric hydropsychidae in Ontario streams. Mailing Add: Dept of Zool Univ of Toronto Toronto ON Can

MACKAY, VIVIAN LOUISE, b Columbus, Ohio, Jan 8, 47. MOLECULAR BIOLOGY, BIOCHEMICAL GENETICS. Educ: Capital Univ, BS, 68; Case Western Reserve Univ, PhD(microbiol), 72. Prof Exp: Fel biochem, Univ Calif, Berkeley, 72-74; ASST RES PROF MICROBIOL, WAKSMAN INST MICROBIOL, RUTGERS UNIV, 74- Mem: Am Soc Microbiol; AAAS; Genetics Soc Am. Res: Biochemical and genetic investigations of cellular regulatory system in yeast that controls sexual conjugation, meiosis, genetic recombination, and DNA repair. Mailing Add: Waksman Inst of Microbiol Rutgers Univ New Brunswick NJ 08903

MACKAY, WILLIAM CHARLES, b Innisfail, Alta, Nov 29, 39; m 63; c 3. ENVIRONMENTAL PHYSIOLOGY, COMPARATIVE PHYSIOLOGY. Educ: Univ Alta, BSc, 61, BEd, 65, MSc, 67; Case Western Reserve Univ, PhD(biol), 71. Prof Exp: Res asst physiol, Yale Univ, 70-71; asst prof zool, 71-75, ASSOC PROF ZOOL, UNIV ALTA, 75- Mem: AAAS; Am Soc Zool; Can Soc Zool. Res: Comparative and environmental physiology of osmotic and ionic regulatory mechanisms. Mailing Add: Dept of Zool Univ of Alta Edmonton AB Can

MACKEL, DONALD CHARLES, b Madison, SDak, Nov 29, 27; m 52; c 3. MEDICAL MICROBIOLOGY. Educ: Univ Fla, BS, 50, MS, 51; Tulane Univ, MPH, 65. Prof Exp: Bacteriologist, Fla State Bd Health, 50-52 & Armed Forces Epidemiol Bd, Korea, 51; lab officer airborne & enteric dis, Ctr Dis Control, La, 52-53, asst, Enteric Dis Invest Unit, 53-57, bacteriologist, Phoenix Field Sta, 57-64, asst chief, Biophys Sect, 65-68, asst chief, Microbiol Control Sect, 68-73, DEP CHIEF, EPIDEMIOL INVESTS LAB BR, USPHS, 73- Mem: Am Soc Microbiol; Sigma Xi; Royal Soc Health. Res: Infectious diseases of animals; epidemiology of enteric diseases of man and animal; hospital infections. Mailing Add: 1742 Timothy Dr NE Atlanta GA 30329

MACKELLAR, ALAN DOUGLAS, b Detroit, Mich, Sept 3, 36; m 60; c 2. PHYSICS. Educ: Univ Mich, BSE(physics) & BSE(math), 58; Tex A&M Univ, PhD(physics), 66. Prof Exp: Nuclear engr, Oak Ridge Nat Lab, 60-61; instr physics, Tex A&M Univ, 61-63; fel, Oak Ridge Nat Lab, 63-65; instr, Mass Inst Technol, 65-67; mem fac, Dept Physics, Rice Univ, 67-68; asst prof, 68-72, ASSOC PROF PHYSICS, UNIV KY, 72- Res: Theoretical nuclear physics, especially scattering theory and nuclear many body problem; reactor physics and radiation shielding. Mailing Add: Dept of Physics Univ of Ky Lexington KY 40506

MACKELLAR, DONALD GORDON, b Wilkes-Barre, Pa, May 28, 22; m 45; c 2. TOXICOLOGY, ORGANIC CHEMISTRY. Educ: Cornell Univ, BA, 43. Prof Exp: Chemist, Manhattan Proj, Linde Air Prod Co, 43-46; res chemist, Inorg Chem Div, NY, 46-57; supvr prepilot lab, Md, 57-62; mgr org prod & processes, Inorg Div, 62-70, MGR PROD SAFETY & TOXICOL, INDUST CHEM DIV, FMC CORP, 70- Mem: Am Chem Soc; Am Oil Chem Soc; Inst Food Technol. Res: Development of new products and processes using peroxygens; investigation of toxic and physical hazards of products and intermediates. Mailing Add: FMC Corp Box 8 Princeton NJ 08540

MACKELLAR, INGEBORG, nutrition, see 12th edition

MACKELLAR, WILLIAM JOHN, b Detroit, Mich, July 14, 35; m 62; c 4. ANALYTICAL CHEMISTRY. Educ: Concordia Col, Moorhead, Minn, BA, 65; Wayne State Univ, PhD(chem), 70. Prof Exp: ASST PROF CHEM, CONCORDIA COL, MOORHEAD, MINN, 69- Concurrent Pos: Chem analyst, Ctr Environ Studies, Tri Col Univ, 74 & Lower Sheyenne River Basin Study, NDak Water Resources Res Inst, 75- Mem: Am Chem Soc. Res: Coordination complex formation reactions and mechanisms; chemical instrumentation; electroanalytical determination of chemical pollutants. Mailing Add: Dept of Chem Concordia Col Moorhead MN 56560

MACKELVIE, ROBIN MAXWELL, b Cape Town, SAfrica, Apr 17, 23; Can citizen; m 55; c 2. BACTERIOLOGY. Educ: Univ BC, BSc, 62, PhD(bact physiol), 65. Prof Exp: SCIENTIST, HALIFAX LAB, FISHERIES & MARINE SERV, ENVIRON CAN, 65- Mem: Brit Soc Gen Microbiol. Res: Fish diseases. Mailing Add: Halifax Fisheries & Marine Serv Halifax Lab PO Box 429 Halifax NS Can

MACKENTHUN, KENNETH MARSH, b Bushong, Kans, May 15, 19; m 42; c 1. AQUATIC BIOLOGY. Educ: Col Emporia, BA, 41; Univ Ill, MA, 46. Prof Exp: Fisheries biologist, Wis Conserv Dept, 46-49; pub health biologist & chief aquatic nuisance control, Wis Comt Water Pollution, 49-62; aquatic biologist, Water Pollution Control Admin, Robert A Taft Sanit Eng Ctr, 62-69, chief tech studies br, Div Tech Support, 69-71; DIR DIV APPL TECHNOL, OFF WATER PROGS, ENVIRON PROTECTION AGENCY, 71- Mem: Am Water Works Asn; Water Pollution Control Fedn. Res: Water pollution biology and control; wastes and sewage treatment; algae and aquatic weed control; aquatic ecology. Mailing Add: Div Appl Technol Off Water Progs Environ Protection Agency Washington DC 20460

MACKENZIE, ANGUS FINLEY, b Elrose, Sask, Can, Oct 2, 32; m 55; c 3. SOIL CHEMISTRY. Educ: Univ Sask, BSA, 54, MSc, 57; Cornell Univ, PhD(soil chem), 59. Prof Exp: Asst prof soil chem, Ont Agr Col, 59-62; from asst prof to assoc prof, 62-72, PROF SOIL SCI, MACDONALD COL, McGILL UNIV, 72-, HEAD DEPT, 66- Mem: Am Soc Agron; Can Soc Soil Sci; Int Soc Soil Sci. Res: Chemistry of the major forms of nutrients in the soil; soil pollution; organic matter transformations. Mailing Add: Dept of Soil Sci Macdonald Col of McGill Univ Ste Anne de Bellevue PQ Can

MACKENZIE, CHARLES AUGUSTUS, organic chemistry, see 12th edition

MACKENZIE, CHARLES WESTLAKE, III, b New York, NY, Jan 25, 46; m 69; c 1. BIOCHEMISTRY. Educ: Univ Pac, BS, 68; Univ Southern Calif, PhD(biochem), 74. Prof Exp: Res chemist, Curtis Nuclear Corp, Los Angeles, 69-70; from teaching asst to res asst biochem, Univ Southern Calif, 71-74; FEL PHARMACOL, UNIV MINN, 74- Concurrent Pos: Fel, USPHS, NIH, 75- Res: Mechanism of action and roles of cyclic nucleotides; biochemical control mechanisms in animal cells; mechanism of action of hormones. Mailing Add: Dept of Pharmacol Univ of Minn 105 Millard Hall Minneapolis MN 55455

MACKENZIE, CLYDE LEONARD, JR, b Oak Bluffs, Mass, June 4, 31; m 65; c 1. MARINE BIOLOGY, MARINE ECOLOGY. Educ: Univ Mass, BS, 55; Col William & Mary, MA, 58. Prof Exp: Res asst marine biol, Va Fisheries Lab, 55-58; asst biologist, Biol Lab, US Bur Com Fisheries, Conn, 58-61, fishery biologist, Lab Exp Biol, Nat Marine Fisheries Serv, US Dept Com, 61-72; oyster consult, Prov Dept Fisheries, PEI, Can, 72-73; FISHERY BIOLOGIST, NAT MARINE FISHERIES SERV, US DEPT COM, 73- Concurrent Pos: Oyster consult, Miss Marine Conserv Comn, 75. Mem: AAAS; Ecol Soc Am; Nat Shellfisheries Asn. Res: Predator-prey relationships and food chain systems in benthic communities; biology of gastropods, echinoderms and crustaceans; development of oyster culture. Mailing Add: Sandy Hook Lab Nat Marine Fisheries Serv Highlands NJ 07732

MACKENZIE, CORTLANDT JOHN GORDON, b Toronto, Ont, Sept 6, 20; m 45; c 3. PUBLIC HEALTH, EPIDEMIOLOGY. Educ: Queens Univ, Ont, MD, CM, 51; Univ Toronto, DPH, 56; Royal Col Physicians Can, cert, 61. Prof Exp: Dir health units, Peace River, 54-55, West Kootenay & Selkirk, 56-59 & Cent Vancouver Island, 59-63; from asst prof to assoc prof prev med, 63-71, PROF & DEPT HEAD, HEALTH CARE & EPIDEMIOL, UNIV BC, 71- Concurrent Pos: Res fel, Univ BC, 61-62; mem main bd exam, Med Coun Can, 66, chmn prev med exam comt, 72-; vpres, Family Planning Fedn Can, 71-72; consult family planning & birth control, Dept Nat Health & Welfare, 71-72, mem med adv group health of immigrants, 72-; Can deleg, Int Planned Parenthood Western Hemisphere, 72; chmn adv comt pub health option, Brit Col Inst Technol; mem, Pollution Control Bd BC & Traffic Injury Res Found Can; chmn, Royal Comn Herbicides & Pesticides; mem test comt, Med Coun Can. Mem: Fel Royal Soc Health; Can Med Asn; Can Pub Health Asn; Can Asn Teachers Social & Prev Med (secy); Asn Teachers Prev Med. Mailing Add: Dept of Health Care & Epidemiol Univ of BC Vancouver BC Can

MACKENZIE, COSMO GLENN, b Baltimore, Md, May 22, 07; m 36; c 2. BIOCHEMISTRY. Educ: Johns Hopkins Univ, AB, 32, ScD(biochem), 36. Prof Exp: Asst prof biochem, Sch Hyg & Pub Health, Johns Hopkins Univ, 38-42; from asst prof to assoc prof, Med Col, Cornell Univ, 46-50; chmn dept, 50-73, prof, 50-75, EMER PROF BIOCHEM, SCH MED, UNIV COLO, 75- Concurrent Pos: USPHS fel, Johns Hopkins Univ, 36-38. Mem: Am Soc Biol Chem; Soc Exp Biol & Med; Harvey Soc; Am Inst Nutrit; fel AAAS. Res: Vitamin E; antioxidants; antithyroid action of thioureas and sulfonamides; biochemistry of one-carbon compounds; s-amino acids and enzymes; regulation of lipid metabolism and lipid accumulation in cultured mammalian cells. Mailing Add: Dept of Biochem Univ of Colo Sch of Med Denver CO 80220

MACKENZIE, DAVID BRINDLEY, b Victoria, BC, May 1, 27; nat US; m 54; c 4. GEOLOGY. Educ: Calif Inst Technol, BS, 50; Princeton Univ, PhD(geol), 54. Prof Exp: Geologist, Am Overseas Petrol Ltd, 53-57; res geologist, Ohio Oil Co, 57-63; MGR GEOL DEPT, DENVER RES CTR, MARATHON OIL CO, 63- Mem: Geol Soc Am; Soc Econ Paleont & Mineral; Am Asn Petrol Geol. Res: Sedimentology; stratigraphy; petroleum geology. Mailing Add: Denver Res Ctr Marathon Oil Co PO Box 269 Littleton CO 80121

MACKENZIE, DAVID ROBERT, b Beverly. Mass, Oct 19, 41; m 63; c 3. PLANT PATHOLOGY, PLANT BREEDING. Educ: Univ NH, BS, 64; Pa State Univ, MS, 67, PhD(plant path), 70. Prof Exp: Res asst plant path, Pa State Univ, 64-70; mem field staff, Rockefeller Found, 70-73; plant breeder, Asian Veg Res & Develop Ctr, 73-74; ASST PROF PLANT PATH, PA STATE UNIV, UNIVERSITY PARK, 74- Mem: Am Phytopath Soc. Res: Potato breeding and pathology for the development of improved potato cultivars; basic epidemiological investigations through genetic modeling. Mailing Add: Dept of Plant Path Buckhout Lab Pa State Univ University Park PA 16802

MACKENZIE, DONALD HERSHEY, b Quincy, Mass, June 23, 09; m 34; c 3. INORGANIC CHEMISTRY. Educ: Northeastern Univ, BChE, 31, BS, 32; Univ Boston, EdM, 35. Prof Exp: Instr chem eng, Northeastern Univ, 31-42, asst prof chem, 42-45, dean, Lincoln Inst, 45-64, assoc dean, Col Eng, 64-74; RETIRED. Mem: Am Soc Eng Educ; Inst Elec & Electronics Engrs. Res: Semi-micro qualitative analysis. Mailing Add: 78 Kemper St Wollaston MA 02170

MACKENZIE, DONALD ROBERTSON, b Man, Can, Dec 9, 21; m 49; c 3. PHYSICAL CHEMISTRY. Educ: Queen's Univ, Ont, MA, 44; Univ Toronto, PhD, 50. Prof Exp: Asst res officer, Atomic Energy Can, Ltd, 50-58; CHEMIST, DEPT APPL SCI, BROOKHAVEN NAT LAB, 58- Mem: Am Chem Soc; Fedn Am Sci; AAAS; The Chem Soc. Res: Radiation, fluorine and nuclear chemistry; chemical kinetics of high temperature reactions in coal hydrogenation. Mailing Add: Dept of Appl Sci Brookhaven Nat Lab Upton NY 11973

MACKENZIE, FREDERICK THEODORE, b Garwood, NJ, Mar 17, 34; m 60; c 2. GEOCHEMISTRY, SEDIMENTOLOGY. Educ: Upsala Col, BS, 55; Lehigh Univ, MS, 59, PhD(geol), 62. Prof Exp: Geologist, Shell Oil Co, 62-63; staff geochemist & mem corp, Bermuda Biol Sta Res, 63-65; res fel geol, Harvard Univ, 65, vis scholar, 65-66, asst prof geol, 67-69, assoc prof geol sci, 70-72, PROF GEOL SCI, NORTHWESTERN UNIV, EVANSTON, 72-, CHMN DEPT, 70- Concurrent Pos: Vis lectr, Lehigh Univ, 63-65; comt mem, Nat Acad Sci, 72-74; res fel, Univ Brussels, 74. Mem: Int Asn Cosmochem & Geochem; Am Asn Petrol Geologists; AAAS; Soc Econ Paleont & Mineral; Geol Soc Am. Res: Control of the chemical composition of seawater; history of the oceans from a chemical and sedimentologic approach; chemical cycles of the elements and man's contributions. Mailing Add: Dept of Geol Sci Northwestern Univ Evanston IL 60201

MACKENZIE, GEORGE HENRY, b Bishop Auckland, Eng, Feb 11, 40; m 70. APPLIED PHYSICS. Educ: Univ Birmingham, BSc, 61, PhD(physics), 65. Prof Exp: Res assoc physics, Cyclotron Lab, Mich State Univ, 65-68; res assoc physics, 68-75, RES PHYSICIST, TRI-UNIV MESON FACIL PROJ, UNIV BC, 75- Mem: Can Asn Physicists. Res: Cyclotron beam dynamics and diagnostics; linear and nonlinear optics calculations for beam transport systems; design and commissioning of beam lines. Mailing Add: TRIUMF Proj Univ of BC Vancouver BC Can

MACKENZIE, GLENN S, b Brooklyn, NY, July 7, 38; m 63; c 2. GEOPHYSICS. Educ: Univ Calif, Los Angeles, AB, 60; Univ Calif, San Diego, MS, 62, PhD(earth sci), 65. Prof Exp: From res geophysicist to sr res geophysicist, 65-70, sr res assoc, Calif, 70-74, SR RES ASSOC, CHEVRON OIL FIELD RES CO, 74- Mem: Am Geophys Union; Seismol Soc Am; Soc Explor Geophys. Res: Solid earth geophysics; oceanography; computer applications. Mailing Add: Explor Dept Chevron Oil Field Res Co Box 36487 Houston TX 77036

MACKENZIE, INNES KEITH, b Stornoway, Scotland, Nov 16, 22; Can citizen; m 46; c 5. PHYSICS. Educ: Univ Western Ont, BSc, 48, MSc, 49; Univ BC, PhD(physics), 53. Prof Exp: Sci officer physics, Can Defence Res Bd, 53-60; from assoc prof to prof, Dalhousie Univ, 60-67; chmn dept, 67-70, PROF PHYSICS, UNIV GUELPH, 67- Concurrent Pos: Res grants, 61-; mem, Can Assoc Comt Space Res, 61- Res: Positron annihilation; metal defects; physics in archaeology. Mailing Add: Dept of Physics Univ of Guelph Guelph ON Can

MACKENZIE, JAMES STEWART, physical chemistry, see 12th edition

MACKENZIE, JAMES W, b Cleveland, Ohio, Oct 17, 25; m 50; c 1. THORACIC SURGERY. Educ: Univ Mich, BS, 48, MD, 51. Prof Exp: Resident surg, Univ Mich, 52-53, 55-58, resident thoracic surg, 58-60, instr surg, 60-62; from asst prof to prof, Univ Mo, 62-69; dean, 71-75, PROF SURG & CHMN DEPT, RUTGERS MED SCH COL MED & DENT NJ, 69- Concurrent Pos: Chief sect thoracic & cardiovasc surg, Sch Med, Univ Mo, 62-69; consult, Ellis Fischel State Cancer Hosp, 64-69. Mem: Am Col Surg. Res: Cardiac surgery. Mailing Add: Rutgers Med Sch Col of Med & Dent of NJ New Brunswick NJ 08903

MACKENZIE, JULIA BUZZ, b Colorado Springs, Colo, Apr 5, 11; m 36; c 2. BIOCHEMISTRY. Educ: Colo Col, AB, 32; Johns Hopkins Univ, ScD(biochem), 39. Prof Exp: From instr to assoc prof, Sch Hyg & Pub Health, Johns Hopkins Univ, 39-43; res assoc biochem, Med Col, Cornell Univ, 48-50; asst prof pediat res, 51-54, from asst prof to assoc prof biochem, 54-67, PROF BIOCHEM, MED SCH, UNIV COLO, DENVER, 67- Mem: Am Inst Nutrit; Am Soc Biol Chem. Res: Effect of vitamin E deficiency in the rat and rabbit; toxicity of thiourea in the weanling and adult rat; effect of various drugs on thyroid gland in rats and other species. Mailing Add: Dept of Biochem Univ of Colo Med Sch Denver CO 80220

MACKENZIE, KENNETH ROSS, b Portland, Ore, June 15, 12; m 37; c 3. PHYSICS. Educ: Univ BC, BA, 35, MA, 37; Univ Calif, Berkeley, PhD(nuclear physics), 40. Prof Exp: Res assoc, Berkeley, 38-46, assoc prof, 47-52, PROF PHYSICS, UNIV CALIF, LOS ANGELES, 53-, CHMN DEPT, 74- Mem: Am Phys Soc. Res: Nuclear and plasma physics. Mailing Add: Dept of Physics Univ of Calif Los Angeles CA 90024

MACKENZIE, KENNETH VICTOR, b Brandon, Man, Aug 29, 11; nat US; m 33, 68; c 4. PHYSICS, OCEANOGRAPHY. Educ: Univ Wash, Seattle, BS, 34, MS, 36. Prof Exp: Sr lab asst, Mat Testing, Ore State Hwy Dept, 36-41; head physicist, Puget Sound Magnetic Degaussing Range, US Navy, 41-44; assoc physicist, Appl Physics Lab, Univ Wash, Seattle, 44-46; physics group leader, US Navy Electronics Lab, 46-51, sect head, Oceanog Br, 51-55, head shallow water acoust processes sect, 55-61, exchange scientist from US Off Naval Res to HM Underwater Weapons Estab, Eng, 61-62, head deep submergence group, 62-67, supvry physicist, Acoust Propagation Div, Ocean Sci Dept, Naval Undersea Ctr, 67-73; SR PHYSICIST & PRIMARY ADV, US NAVAL OCEANOG OFF, 73- Concurrent Pos: Consult allied govts & defense indust. Honors & Awards: Cert Merit, Off Sci Res, 45; Cert Except Serv, Bur Ord, 45; Super Achievement Award, Navy Electronics Lab, 60; Commendation, Trieste Searches USS Thresher, US Navy Unit, 63; Super Achievement Award, Inst Navig, 71. Mem: Fel Acoust Soc Am; Am Geophys Union; Am Phys Soc; Marine Technol Soc (co-founder); Sigma Xi. Res: Underwater acoustics; deep submergence research; physical oceanography; instrumentation; analysis. Mailing Add: US Naval Oceanog Off Washington DC 20373

MACKENZIE, MALCOLM R, b Oakland, Calif, Jan 15, 35; m 59; c 3. INTERNAL MEDICINE, IMMUNOLOGY. Educ: Univ Calif, Berkeley, AB, 56; Univ Calif, San Francisco, MD, 59. Prof Exp: Intern & resident, Univ Calif Hosps, San Francisco, 59-62; asst res physician, Div Hemat, Univ Calif, 66-67; asst prof med, Univ Calif, San Francisco, 67-68; asst prof, Univ Cincinnati, 68-70; ASSOC PROF MED, SCH MED, UNIV CALIF, DAVIS, 70- Concurrent Pos: USPHS fel physiol chem, Univ Wis, 62-64; Am Cancer Soc fel, Div Hemat, Univ Calif, 65-66; Am Cancer Soc scholar, 66-68. Mem: Am Fedn Clin Res; Am Rheumatism Asn; Am Soc Hemat; Am Asn Immunol. Res: Structure and function of human IgM antibodies; origin of human lymphocyte malignancies; human multiple myxedema. Mailing Add: Dept of Internal Med Univ of Calif Sch of Med Davis CA 95616

MACKENZIE, NEIL MITCHILL, b Westport, Conn, Oct 2, 11; m 36; c 4. ORGANIC CHEMISTRY. Educ: Yale Univ, AB, 34, PhD(org chem), 37. Prof Exp: Lab asst chem, Yale Univ, 35-36; res chemist, Tex Co, NY, 36; res chemist, Calco Chem Div, Am Cyanamid Co, 37-44, asst chief chemist, Res Dept, 44-47 & Develop Dept, 47-52, chief develop chemist, Azo Dye Dept, Bound Brook Labs, 52-56, group leader res div, 56-58, asst to dir process develop, 58-59, group leader res div, 59-64, staff asst to dir res & develop, 64-67, bus mgr res & develop, Org Chem Div, 67-74; RETIRED. Mem: Am Chem Soc; fel Am Inst Chemists. Res: Azo Dyes and azo dye intermediates with emphasis on metal complex compounds of azo dyes; sesquiterpenes in birchtar oil; rubber chemicals and elastomers. Mailing Add: 515 Watchung Rd Bound Brook NJ 08805

MACKENZIE, RICHARD STANLEY, b Detroit, Mich, Dec 28, 33; m 57; c 2. DENTISTRY. Educ: Univ Mich, DDS, 58, MS, 60; Univ Pittsburgh, PhD(psychol), 65. Prof Exp: Assoc prof dent behav sci & head dept, Univ Pittsburgh, 65-68, dir educ res, Sch Dent, 66-69, prof higher educ, Sch Dent Med, 68-69; dir, Off Dent Educ, 69-74, PROF DENT EDUC, COL DENT & COL EDUC, UNIV FLA, 69-, CHMN DEPT, 75- Concurrent Pos: Mem USPHS adv comt, Dent Health Res & Educ, 68-72; mem oral biol training comt, Vet Admin Hosps, DC, 69-72, mem res serv merit rev bd, 72-75; USPHS career develop award, 70-75; consult, WHO-Pan Am Health Orgn, 71- & Am Dent Asn, 72-; chmn adv comt, Educ Testing Serv, 73- Mem: AAAS; Am Psychol Asn; Am Educ Res Asn. Res: Analysis and evaluation of clinical judgment. Mailing Add: 3715 SE 37th St Gainesville FL 32601

MACKENZIE, ROBERT DOUGLAS, b Chicago, Ill, Aug 18, 28; m 52; c 4. BIOCHEMISTRY. Educ: Univ Cincinnati, BS, 52; Mich State Univ, MS, 54, PhD(biochem), 57. Prof Exp: Asst, Mich State Univ, 53-57, res assoc, 57; res biochemist, 57-63, HEAD HEMAT SECT, PHARMACOL DEPT, MERRELL-NAT LABS DIV, RICHARDSON-MERRELL, INC, 63- Concurrent Pos: Adj assoc prof, Univ Cincinnati, 67-69; mem coun thrombosis, Am Heart Asn. Mem: AAAS; Am Chem Soc; Soc Exp Biol & Med; Am Soc Pharmacol & Exp Therapeut. Res: Animal and lipid metabolism; blood coagulation; radiobiochemistry; toxicology; nutrition. Mailing Add: Merrell-Nat Labs 110 Amity Rd Cincinnati OH 45215

MACKENZIE, ROBERT EARL, b Calif, Mar 17, 20; m 50; c 4. MATHEMATICS. Educ: Calif Inst Technol, BS, 42; Princeton Univ, MA, 48, PhD(math), 50. Prof Exp: Physicist, US Naval Ord Lab, 42-45; from instr to asst prof math, 50-61, asst chmn dept, 62-67, ASSOC PROF MATH, IND UNIV, BLOOMINGTON, 61- Concurrent Pos: Ed, J Math & Mech, 56-62 & 71- Mem: Am Math Soc. Res: Modern algebra and algebraic number theory. Mailing Add: RR 12 Bloomington IN 47401

MACKENZIE, SCOTT, JR, b Sedalia, Mo, Mar 10, 20; m 47; c 3. CHEMISTRY. Educ: Univ Pa, BS, 42; Univ Ill, MS, 44, PhD(org chem), 47. Prof Exp: Res chemist, E I du Pont de Nemours & Co, Va, 44-46; fel, Gen Mills Co, Univ Minn, 47-48; instr chem, Columbia Univ, 48-51; from asst prof to assoc prof, 51-66, PROF CHEM, UNIV R I, 66- Mem: Am Chem Soc. Res: Organic chemistry; carbanions; ultraviolet spectroscopy. Mailing Add: Dept of Chem Univ of R I Kingston RI 02881

MACKENZIE, WALTER CAMPBELL, b Glace Bay, NS, Aug 17, 09; m 38; c 3. MEDICINE. Educ: Dalhousie Univ, BSc, 29, MD, CM, 33; Univ Minn, MS, 37. Prof Exp: Prof surg & chmn dept, Med Sch, 49-60, dean fac med, 59-74, dir surg servs, Univ Hosp, 49-60, EMER PROF SURG, UNIV ALTA, 74- Concurrent Pos: Moynihan lectr, Royal Col Surgeons, Eng, 54; Sommer mem lectr, Univ Portland, 59; Banks mem lectr, Univ Liverpool, 59; sr consult, Dept Vet Affairs for Northern Alta, Can, 48-; fel, Royal Col Physicians & Surg Can. Mem: Am Surg Asn; Asn Am Med Cols; Am Col Surg; Asn Med Cols Can. Res: Gastrointestinal physiology; tissue transplantation; wound healing. Mailing Add: Fac of Med Univ of Alta Edmonton AB Can

MACKEY, BRUCE ERNEST, b Akron, Ohio, Feb 9, 39; m 61; c 2. BIOMETRICS, PLANT BREEDING. Educ: Univ Akron, BS, 61; Cornell Univ, MS, 64, PhD(plant breeding), 66. Prof Exp: Statistician, Biomet Serv, Md, 66-69, biometrician, Biomet Serv, Calif, 69-71, BIOMETRICIAN, WESTERN REGIONAL RES LAB, AGR RES SERV, USDA, 71- Res: Applications of quantitative genetics to plant breeding research; design and analysis of agricultural research data. Mailing Add: Western Regional Res Lab Agr Res Serv 800 Buchanan St Berkeley CA 94710

MACKEY, GEORGE WHITELAW, b St Louis, Mo, Feb 1, 16; m 60; c 1. MATHEMATICS. Educ: Rice Inst Technol, BA, 38; Harvard Univ, AM, 39, PhD(math), 42; Oxford Univ, MA, 66. Prof Exp: Instr math, Ill Inst Technol, 42-43 & Harvard Univ, 43-44; oper analyst, Off Field Servs, Off Sci Res & Develop, 44; assoc res mathematician, Appl Math Group, Columbia Univ, 44-45; from fac instr to prof math, 45-69, LANDON T CLAY PROF MATH & THEORETICAL SCI, HARVARD UNIV, 69- Concurrent Pos: Guggenheim fel, 49-50, 61-62, 70-71; George Eastman vis prof, Oxford Univ, 66-67; vis prof, Tata Inst Fundamental Res, India, 70-71. Mem: Nat Acad Sci; Am Math Soc (vpres, 64-65); Am Acad Arts & Sci; Am Philos Soc. Res: Abstract analysis; infinite dimensional representations of locally compact groups and applications to quantum mechanics and other branches of mathematics. Mailing Add: Dept of Math Harvard Univ Cambridge MA 02138

MACKEY, HENRY JAMES, b Vicksburg, Miss, Nov 25, 35; m 59; c 4. SOLID STATE PHYSICS. Educ: La State Univ, BS, 57, MS, 59, PhD(physics), 63. Prof Exp: Res assoc physics, La State Univ, 63-64; from asst prof to assoc prof, 64-69, PROF PHYSICS, N TEX STATE UNIV, 69- Mem: Am Phys Soc. Res: Electron, phonon transport phenomena in metals; Fermi surface mappings; surface scattering contribution to electrical resistivity. Mailing Add: Dept of Physics N Tex State Univ Denton TX 76203

MACKEY, JAMES E, b Tupelo, Miss, Feb 4, 40; m 65; c 3. SOLID STATE PHYSICS. Educ: Tulane Univ, BS, 62; Univ Miss, MS, 65, PhD(physics), 69. Prof Exp: Asst prof, 68-74, ASSOC PROF PHYSICS, HARDING COL, 74- Mem: Am Asn Physics Teachers. Res: Nonlinear properties of solids; ultrasonic second-harmonic generation in solids. Mailing Add: Dept of Physics Harding Col Box 582 Searcy AR 72143

MACKEY, JAMES P, b Akron, Ohio, Feb 28, 30; m 50; c 1. BIOLOGY. Educ: Univ Akron, BS, 51; Ohio State Univ, MS, 54; Univ Ore, PhD(biol), 57. Prof Exp: From asst prof to assoc prof, 57-74, PROF BIOL, SAN FRANCISCO STATE UNIV, 74- Mem: Ecol Soc Am; Soc Study Evolution; Am Soc Ichthyol & Herpet. Res: Vertebrate ecology; herpetology; intraspecific variation of tree frogs. Mailing Add: Div of Natural Sci San Francisco State Univ San Francisco CA 94132

MACKEY, JOHN LINN, b St James, Mo, Dec 1, 34; m 56; c 2. PHYSICAL CHEMISTRY, INORGANIC CHEMISTRY. Educ: Southeast Mo State Col, BS, 56; Iowa State Univ, PhD(chem), 60. Prof Exp: Fel chem, Ames Lab, Atomic Energy Comn, 60-61; asst prof, Univ Mo, Rolla, 61-63; from asst prof to assoc prof, 63-74, PROF CHEM, AUSTIN COL, 74- Concurrent Pos: Robert A Welch Found res grant, 65-70; vis prof, Univ Newcastle, 70-71. Mem: Am Chem Soc. Res: Chemistry of rare earth elements; chemical applications of the Mössbauer effect. Mailing Add: Dept of Chem Austin Col Sherman TX 75090

MACKEY, KAREN ETHEL, b Philadelphia, Pa, Jan 13, 45. COMPUTER SCIENCE. Educ: Pa State Univ, BS, 65, MS, 68, PhD(comput sci), 73. Prof Exp: Instr, Pa State Univ, 71 & 73; ASST PROF COMPUT SCI, STATE UNIV NY BINGHAMTON, 73- Mem: Asn Comput Mach; Am Math Soc; Asn Women in Math. Res: Adaptive operating systems. Mailing Add: Sch of Advan Technol State Univ of NY Binghamton NY 13901

MACKEY, MICHAEL CHARLES, b Kansas City, Kans, Nov 16, 42; m 64. BIOPHYSICS. Educ: Univ Kans, BA, 63; Univ Wash, PhD(physiol), 68. Prof Exp: Res assoc electrocardiol, Sch Med, Univ Okla, 63-64; biophysicist, Phys Sci Lab, Div Comput Res & Technol, NIH, 69-71; ASST PROF PHYSIOL, McGILL UNIV, 71- Mem: Biophys Soc. Res: Mathematical modeling of the relation between transport of charged and uncharged molecular species through excitable membranes; excitable membrane structure and function. Mailing Add: Dept of Physiol McGill Univ Montreal PQ Can

MACKI, JACK W, b Mullan, Idaho, June 16, 39; m 62; c 2. MATHEMATICS. Educ: Univ Idaho, BS, 60; Calif Inst Technol, PhD(math), 64. Prof Exp: Staff mem, Los Alamos Sci Lab, 64-65; from asst prof to assoc prof, 66-75, PROF MATH, UNIV ALTA, 75- Mem: Am Math Soc. Res: Ordinary differential equations. Mailing Add: Dept of Math Univ of Alta Edmonton AB Can

MACKIE, GEORGE OWEN, b Louth, Eng, Oct 20, 29; m 56; c 5. ZOOLOGY. Educ: Oxford Univ, BA, 53, MA & DPhil, 56. Prof Exp: Lectr zool, Univ Alta, 56-58, from asst prof to prof, 58-68; chmn dept biol, 71-74, PROF BIOL, UNIV VICTORIA, BC, 68- Concurrent Pos: Nat Res Coun Can overseas fel, 63-64; assoc ed, Can J Zool, 71- Mem: Am Soc Zool; Soc Exp Biol & Med; Can Soc Zool; Brit Soc Exp Biol. Res: Invertebrate neurobiology, structure, especially Coelenterata. Mailing Add: Dept of Biol Univ of Victoria Victoria BC Can

MACKIE, RICHARD JOHN, b Foster City, Mich, July 6, 33; m 57; c 4. WILDLIFE MANAGEMENT, ECOLOGY. Educ: Mich State Univ, BS, 58; Wash State Univ, MS, 60; Mont State Univ, PhD(wildlife mgt), 65. Prof Exp: Res biologist, Mont Fish & Game Dept, 60-65, res coordr, 65-66; from asst prof to assoc prof entom, fisheries & wildlife, Univ Minn, St Paul, 66-70; ASSOC PROF WILDLIFE MGT, MONT STATE UNIV, 70- Mem: Wildlife Soc; Ecol Soc Am; Soc Range Mgt. Res: Reproductive cycle of chukar; range ecology and relations of mule deer, elk and cattle; big game habitat relationships; big game range survey techniques; browse plant ecology. Mailing Add: Dept of Biol Mont State Univ Bozeman MT 59715

MACKIE, WALLACE ZILISCH, soil chemistry, see 12th edition

MACKIEWICZ, JOHN STANLEY, b Waterbury, Conn, July 12, 30; m 57; c 1. PARASITOLOGY. Educ: Cornell Univ, BS, 53, MS, 54, PhD(parasitol), 60. Prof Exp: Asst med entom & parasitol, Cornell Univ, 54-57, 59-60, instr, 57-59; NIH fel parasitol, Switz, 60-61; from asst prof to assoc prof biol, 61-68, prof biol sci, 68-73, DISTINGUISHED TEACHING PROF BIOL SCI, STATE UNIV NY ALBANY, 73- Concurrent Pos: Vis assoc prof, Univ Tenn, 67-68. Mem: AAAS; Am Soc Parasitol; Am Soc Mammal; Am Ornith Union; Am Micros Soc. Res: Parasites of freshwater fish; Cestoidea; Caryophylidea; conservation. Mailing Add: Dept of Biol Sci State Univ of NY Albany NY 12222

MACKIN, ROBERT JAMES, JR, animal breeding, animal genetics, see 12th edition

MACKIN, ROBERT JAMES, JR, b Little Rock, Ark, Dec 4, 25; c 6. PHYSICS. Educ: Yale Univ, BE, 49; Calif Inst Technol, MS, 51, PhD(physics), 53. Prof Exp: Res assoc, Calif Inst Technol, 53-54 & Nuclear Physics Br, Off Naval Res, 54-56; res assoc, Thermonuclear Exp Div, Oak Ridge Nat Lab, 56-59, group leader, 59-62; mgr physics sect, Space Sci Div, 62-67, mgr, Lunar & Planetary Sci Sect, 67-69, MGR SPACE SCI DIV, JET PROPULSION LAB, CALIF INST TECHNOL, 69- Concurrent Pos: Traveling lectr, Oak Ridge Inst Nuclear Studies, 58-60. Mem: AAAS; Am Phys Soc. Res: Plasma and interplanetary physics; controlled fusion; planetary science. Mailing Add: Jet Propulsion Lab Space Sci Div Calif Inst of Technol Pasadena CA 91103

MACKINNEY, ARCHIE ALLEN, JR, b St Paul, Minn, Aug 16, 29; m 55; c 3. HEMATOLOGY. Educ: Wheaton Col, Ill, BA, 51; Univ Rochester, MD, 55; Am Bd Internal Med, dipl, 62. Prof Exp: Resident med, Univ Wis Hosps, 55-59; clin assoc hemat, Nat Inst Arthritis & Metab Dis, 59-61; clin investr, 61-64, from asst prof to assoc prof med, 64-74, PROF MED, SCH MED, UNIV WIS-MADISON, 74-; CHIEF HEMAT, VET ADMIN HOSP, 64- Concurrent Pos: Vet admin res grants, 61-64; NIH res grants, 64-67 & 70-75; chief nuclear med, Vet Admin Hosp, 64-74; chmn, Wis Hemat Study Group. Mem: Am Fedn Clin Res; Cent Soc Clin Res; Am Soc Hemat. Res: Cell proliferation in vitro; leukocytes in health and disease. Mailing Add: Dept of Med Univ of Wis Sch of Med Madison WI 53706

MACKINNEY, ARLAND LEE, b Hendersonville, NC, Nov 29, 31; m 55; c 3. NUCLEAR PHYSICS, RESEARCH ADMINISTRATION. Educ: NC State Col, BS, 53; Ind Univ, MS, 55; Mass Inst Technol, SM, 67. Prof Exp: Physicist, Knolls Atomic Power Lab, AEC, 55-58; physicist, 58-60, supvr, 60-64, sect chief, 64-67, tech adv to mgr physics lab, 67-68, from asst mgr to mgr qual control dept, 68-71, SPEC ASST NUCLEAR OPERS TO DIV V PRES, BABCOCK & WILCOX CO, 71- Mem: Am Nuclear Soc. Res: Nuclear reactor physics; planning, operation and analysis of critical experiments. Mailing Add: Babcock & Wilcox Co Mount Vernon IN 47620

MACKINNEY, GORDON, b London, Eng, May 28, 05; m 28; c 1. AGRICULTURAL CHEMISTRY. Educ: Univ Toronto, BSA, 26; Univ Calif, PhD(plant physiol), 33. Prof Exp: Asst soil chemist, United Fruit Co, Honduras, 27-30; Nat Res Coun fel plant biol, Carnegie Inst, Stanford, 33-35; from instr to asst prof food technol, 36-43, from assoc prof to prof, 44-72, EMER PROF FOOD TECHNOL, UNIV CALIF, BERKELEY, 72- Concurrent Pos: Guggenheim Mem fel, Birmingham & London, 47; consult, Hazardous Mat Adv Comt-Herbicides, 72-73 & Orgn Am States, Campinas, Brazil, 73. Mem: AAAS; Am Chem Soc; Am Soc Plant Physiol; Fr Acad Agr; cor mem Bordeaux Nat Acad Sci, Lit & Arts. Res: Natural coloring matters; food deterioration; measurement of color. Mailing Add: Dept of Nutrit Sci Univ of Calif Berkeley CA 94720

MACKINNEY, HERBERT WILLIAM, b London, Eng, Feb 22, 07; nat US; m 44; c 1. POLYMER CHEMISTRY, ENVIRONMENTAL CHEMISTRY. Educ: Swiss Fed Inst Technol, dipl, 28; McGill Univ, MSc, 33, PhD(cellulose chem), 35. Prof Exp: Chemist, Lever Bros Ltd, Eng, 28-31; res chemist, Can Int Paper Co, Ont, 31-32; res assoc, Macdonald Col, McGill Univ, 36; res chemist, Kendall Co, 36-39, Sylvania Indust Corp, 39-40 & Bakelite Corp Div, Union Carbide Corp, 41-58; staff chemist, Int Bus Mach Corp, 58-61, adv chemist, 61-66, sr chemist, IBM Corp, 66-72; CONSULT, 72- Honors & Awards: Honor Scroll, Am Inst Chemists, 57. Mem: Am Chem Soc; fel Am Inst Chemists. Res: Polymers and plastics; adhesives; environmental deterioration of materials; plastic laminates. Mailing Add: 740 S Alton Way Denver CO 80231

MACKINNON, JUAN ENRIQUE, b Montevideo, Uruguay, June 24, 04; m 34; c 5. MEDICAL MYCOLOGY, INSECT TOXICOLOGY. Educ: Univ of the Repub, Uruguay, MD, 33. Prof Exp: Microbiologist, 69-74, EMER PROF, FAC MED, UNIV OF THE REPUB, URUGUAY, 74- Honors & Awards: Rhoda Benham Award, Med Mycol Soc of the Americas, 70. Mem: Am Acad Microbiol; Brit Mycol Soc; Royal Soc Trop Med & Hyg; Int Soc Human & Animal Mycol (pres, 58-62); Soc Dermat & Syphil Uruguay. Res: Pathogenic yeasts; causal organisms of mycetoma; morphology of pathogenic fungi in the tissues; poisonous spiders; necrotic arachnidism; pathogenesis of South American Blastomycosis; effect of ambient temperature and of climate on the mycoses; ecology of pathogenic fungi. Mailing Add: Juan Benito Blanco 768 Montevideo Uruguay

MACKIW, VLADIMIR NICOLAUS, b Stanislawiw, Western Ukraine, Sept 4, 23; nat US; m 51; c 3. INORGANIC CHEMISTRY, PHYSICAL CHEMISTRY. Educ: Univ Breslau, dipl, 44; Univ Erlangen, dipl, 46. Prof Exp: Chemist, Lingman Lake Mines, 48 & Prov Bur Mines, Man, 49; chemist, Ont, 49-50, chief res chemist, 50-51, dir res, Alta, 52-55, dir res & develop div, 55-68, DIR & MEM BD, SHERRITT GORDON MINES, LTD, 64-, V PRES, 67-, V PRES TECHNOL & CORP DEVELOP, 68- Concurrent Pos: Mem, Nat Res Coun Can. Honors & Awards: Jules Garnier Prize, Fr Metall Soc, 66; Int Nickel Co Medal, Can Inst Mining & Metall, 66; R S Jane Mem Lect Award, Chem Inst Can, 67. Mem: Fel Chem Inst Can; Can Inst Mining & Metall; Am Inst Mining, Metall & Petrol Eng. Res: Extraction from ores by chemical methods; powder metallurgy; inorganic chemicals; kinetics and thermodynamics of inorganic reactions. Mailing Add: Sherritt Gordon Mines Ltd 25 King St W PO Box 28 Toronto ON Can

MACKLEM, PETER TIFFANY, b Kingston, Ont, Oct 4, 31; m 54; c 5. PULMONARY PHYSIOLOGY, EXPERIMENTAL MEDICINE. Educ: Queen's Univ, Ont, BA, 52; McGill Univ, MD, CM, 56; FRCPS(C), 62. Prof Exp: Fel, Royal Victoria Hosp, Montreal, Que, 60-61, res fel, 61-63; Meakins Mem fel, 63-64; McLaughlin traveling res fel, Sch Pub Health, Harvard Univ, 64-65; from asst prof to assoc prof, 65-71, PROF EXP MED, McGILL UNIV, 71-; DIR, MEAKINS-CHRISTIE LABS, ROYAL VICTORIA HOSP, 71- Concurrent Pos: Watson scholar, McGill Univ, 61-63; asst physician, Royal Victoria Hosp, 67-71, sr physician, 71- Mem: Am Physiol Soc; Am Soc Clin Invest; Can Med Asn; Can Soc Clin Invest. Res: Mechanical properties of lungs; relationship between lung structure and function; airway dynamics. Mailing Add: Meakins-Christie Labs 3775 University St Montreal PQ Can

MACKLER, BRUCE, b Philadelphia, Pa, May 23, 20; m 49; c 2. PEDIATRICS. Educ: Temple Univ, MD, 43. Prof Exp: Intern, Temple Univ Hosp, 43-44; resident physician pediat, Willard Parker Hosp, New York, 44; resident physician, Univ Iowa, 46-47; resident physician, Children's Hosp, Univ Cincinnati, 47-48, res assoc, 50-53, asst prof pediat, Univ, 53-54; asst prof enzyme chem, Univ Wis, 55-57; assoc prof, 57-61, PROF PEDIAT, SCH MED, UNIV WASH, 61- Concurrent Pos: USPHS fel res found, Children's Hosp, Univ Cincinnati, 48-50; fel, Inst Enzyme Res, Univ Wis, 53-55; estab investr, Am Heart Asn, 55-60. Mem: Am Soc Biol Chem; Am Soc Pediat Res. Res: Carbohydrate metabolism; metalloflavoproteins; electron transport systems. Mailing Add: Dept of Pediat Univ of Wash Seattle WA 98105

MACKLER, SAUL ALLEN, b New York, NY, Dec 9, 13; m 40; c 3. SURGERY. Educ: Columbia Univ, BS, 33; Univ Chicago, MD, 37. Prof Exp: Intern, Michael Reese Hosp, 38-39, resident surg, 40; ASSOC PROF THORACIC SURG, MED SCH, UNIV CHICAGO, 47-; PROF, COOK COUNTY GRAD SCH MED, 46- Concurrent Pos: Fel thoracic surg, Barnes Hosp, St Louis, 41 & 42; attend, assoc & consult thoracic surgeon, Hosps, 46-; chmn dept surg, Michael Reese Hosp, 62-64. Mem: Fel AMA; Soc Thoracic Surg; fel Am Col Surg; fel Int Col Surg; fel Am Col Chest Physicians. Res: Physiology and disease of the esophagus and mediastinum; cancer of the esophagus; disease of the heart and great vessels; injuries of the chest. Mailing Add: 55 E Washington St Chicago IL 60602

MACKLES, LEONARD, b New York, NY, Jan 17, 29; m 54; c 2. COSMETIC CHEMISTRY, PHARMACEUTICAL CHEMISTRY. Educ: Long Island Univ, BS, 51. Prof Exp: Org chemist, Colloids Inc, 53-56, head org chemist, Arlen Chem Corp, 56-58; tech dir, Chemclean Corp, 58-61; head chemist, Schenley Res Inst, 61-63, asst dir res, 63-65; sr res scientist prod develop, 65-73, PRIN RES INVESTR CONCEPT DEVELOP, PROD DIV, BRISTOL MYERS CO, 73- Mem: Fel AAAS; Am Chem Soc; fel Am Inst Chemists; NY Acad Sci; Brit Soc Chem Indust. Res: Development of consumer products in the fields of pharmaceuticals, toiletries and household specialties. Mailing Add: 311 E 23rd St New York NY 10010

MACKLIN, JOHN WELTON, b Ft Worth, Tex, Dec 11, 39. INORGANIC CHEMISTRY, SPECTROSCOPY. Educ: Linfield Col, BA, 62; Cornell Univ, PhD(inorg chem), 68. Prof Exp: ASST PROF CHEM, UNIV WASH, 68- Mem: Am Chem Soc. Res: Spectroscopic measurements, particularly Raman, applied to elucidation of structural characteristics of inorganic solids, liquids and solutions. Mailing Add: Dept of Chem BG-10 Univ of Wash Seattle WA 98195

MACKLIN, JUNE, b Bryant, Ind, June 8, 25. ANTHROPOLOGY. Educ: Purdue Univ, BS, 46; Univ Chicago, MS, 53; Univ Pa, PhD(anthrop), 63. Prof Exp: Teacher, Bryant High Sch, Ind, 46-48; staff mem, Nat Conf Christians & Jews, 51-53; instr sociol & anthrop, 56-58 & 59-63, from asst prof to assoc prof, 63-72, co-chmn dept, 70-73, PROF SOCIOL & ANTHROP, CONN COL, 72- Concurrent Pos: NIMH grants field res, Mex, 65-68, NIMH grant, 75; vis assoc prof, Dept Psychiat, Col Med & Dent NJ, NJ Med Sch, 75-76. Honors & Awards: Nicholas Salgo Outstanding Teacher Award, Salgo-Noren Found, Inc, 65. Mem: Fel Am Anthrop Asn. Res: Medical anthropology, especially in Mexico and Latin America; social and cultural change; people of Mexican descent living in the United States. Mailing Add: Dept of Sociol & Anthrop Conn Col New London CT 06320

MACKLIN, MARTIN, b Raleigh, NC, Aug 27, 34; div; c 2. PHYSICAL BIOLOGY, BIOMEDICAL ENGINEERING. Educ: Cornell Univ, BME, 57, MIE, 58; Case Western Reserve Univ, PhD(biomed eng), 67. Prof Exp: Instr mech eng, Cornell Univ, 56-58; sr engr, Hamilton Standard Div, United Aircraft Corp, 58-61; prod planning specialist, Moog Servocontrols, Inc, 61-62; staff specialist, Thompson-Ramo-Wooldridge, Inc, 62-65; asst prof, 67-72, ASSOC PROF BIOMED ENG, CASE WESTERN RESERVE UNIV, 72- Concurrent Pos: Established investr, Am Heart Asn, 69-74. Mem: AAAS; Biophys Soc; Am Soc Zoologists; Soc Gen Physiologists. Res: Control of growth and tissue properties as mediated by inorganic ions; transport of ions and electrogenesis in coelenterate epithelia and intestinal villi; fetal electrocardiography. Mailing Add: Dept of Biomed Eng Case Western Reserve Univ Cleveland OH 44106

MACKLIN, PHILIP ALAN, b Richmond Hill, NY, Apr 13, 25; m 53; c 3. QUANTUM MECHANICS, ASTRONOMY. Educ: Yale Univ, BS, 44; Columbia Univ, MA, 49, PhD(physics), 56. Prof Exp: Physicist, Carbide & Carbon Chem Corp, Tenn, 46-47; res scientist, AEC, Columbia Univ, 49-51; instr physics, Middlebury Col, 51-54, actg chmn dept, 53-54; from asst prof to assoc prof, 54-61, PROF PHYSICS, MIAMI UNIV, 61-, CHMN DEPT, 72- Concurrent Pos: Vis prof, Univ NMex, 57-68; physicist, Los Alamos Sci Labs, 60-62. Mem: AAAS; Am Phys Soc; Am Asn Physics Teachers. Res: Beta and gamma spectroscopy; interpretation of quantum mechanics. Mailing Add: 211 Oakhill Dr Oxford OH 45056

MACKLIN, RICHARD LAWRENCE, b Jamaica, NY, Dec 24, 20; m 45; c 4. NUCLEAR PHYSICS. Educ: Yale Univ, BS, 41, PhD(org chem), 44. Prof Exp: Lab asst, Yale Univ, 41-44; chemist, Indust Labs, Carbide & Carbon Chem Co, 44-48, sr physicist, 48-52; SR PHYSICIST, OAK RIDGE NAT LAB, 52- Mem: Fel AAAS; fel Am Phys Soc; Int Asn Geochem & Cosmochem. Res: Radioactivity; nuclear data; neutron capture experiments; nuclear physics instrumentation; astrophysics and cosmology experiments; nuclear safety. Mailing Add: Physics Div Oak Ridge Nat Lab PO Box X Oak Ridge TN 37831

MACKNIGHT, FRANKLIN COLLESTER, b Louisville, Ky, June 30, 09; m 32; c 1. PETROLEUM GEOLOGY. Educ: Univ Chicago, PhB, 32, PhD, 38. Prof Exp: Asst geologist, Pure Oil Co, 37 & Ill State Geol Surv, 39-40; asst exam phys sci, Bd Exams, Chicago, 40-41; asst prof geol, Mt Union Col, 41-43; asst prof geog, Fenn Col, 43-44; geologist, Tex Co, 44-47; prof geol & geog, Evansville Col, 47-51; chief geologist, S C Yingling Oil Opers, 51-52; asst prof geol, Univ Pittsburgh, 53-61; assoc prof gen studies, 61-63 & sci, 63-66, PROF GEN SCI, STATE UNIV NY COL BUFFALO, 66- & GEOSCI, 72- Concurrent Pos: Consult, 52- Mem: AAAS; Soc Study Evolution; Paleont Soc; Philos Sci Asn; Asn Am Geog. Res: Carboniferous paleobotany; oil field structure and exploration; history and philosophy of science; evolution; phylogeny; aesthetics; professional career counseling. Mailing Add: Dept of Gen Sci State Univ of NY Col Buffalo NY 14222

MAC KNIGHT, WILLIAM JOHN, b New York, NY, May 5, 36; m 67. PHYSICAL CHEMISTRY. Educ: Univ Rochester, BS, 58; Princeton Univ, MA, 63, PhD(phys chem), 64. Prof Exp: Res assoc, Princeton Univ, 64-65; from asst prof to assoc prof, 65-74, PROF CHEM, UNIV MASS, AMHERST, 74- Mem: Am Chem Soc; Am Phys Soc. Res: Physical chemistry of high polymers; sulfur chemistry. Mailing Add: Dept of Chem Univ of Mass Amherst MA 01002

MACKO, EDWARD, b Philadelphia, Pa, Jan 5, 19; m 44; c 3. PHARMACOLOGY. Educ: La Salle Col, BA, 41. Prof Exp: Instr pharmacol, Med Sch, Temple Univ, 44-48; sr pharmacologist & asst sect head, 62-71, asst dir pharmacol, 71-74, ASSOC DIR PHARMACOL, SMITH KLINE & FRENCH LABS, 74- Mem: Am Physiol Soc; Am Chem Soc; Am Asn Lab Animal Sci. Res: Pharmacological screening; central nervous system; antiobesity. Mailing Add: 1530 Spring Garden St Philadelphia PA 19101

MACKOWIAK, ELAINE DECUSATIS, b Hazleton, Pa, Apr 28, 40; m 64; c 2. PHARMACOLOGY, PHARMACEUTICAL CHEMISTRY. Educ: Temple Univ, BS, 62, MS, 65; Thomas Jefferson Univ, PhD(pharmacol), 74. Prof Exp: Asst chief pharmacist, Holy Redeemer Hosp, Meadowbrook, Pa, 62-63; lectr radiol health, Sch Dist Philadelphia, 64-68; instr pharmaceut chem, 64-72, ASST PROF PHARMACEUT CHEM, TEMPLE UNIV, 72- Concurrent Pos: Actg chmn dept pharmaceut chem, Temple Univ, 74-75. Mem: Am Pharmaceut Asn; Health Physics Soc. Res: Melanin formation; human autopsied samples for research; enzyme purification; pharmacy manpower, especially women and pharmaceutical education. Mailing Add: Temple Univ Sch of Pharm 3307 N Broad St Philadelphia PA 19140

MACKOWIAK, ROBERT CARL, b Hazleton, Pa, May 13, 38; m 64; c 2. CARDIOVASCULAR PHYSIOLOGY, INTERNAL MEDICINE. Educ: Univ Pa, AB, 60; Thomas Jefferson Univ, MD, 64. Prof Exp: Intern med, Methodist Hosp, Philadelphia, 64-65; instr physiol, 65-66, asst prof, 67-70, ASSOC PROF PHYSIOL &

ASST PROF MED, THOMAS JEFFERSON UNIV, 71-, ASSOC DEAN, 73- Concurrent Pos: Resident med & cardiol, Mercy Cath Med Ctr, Philadelphia, 68-70, consult cardiol, 71- Mem: Am Physiol Soc; Am Fedn Clin Res; Bioeng Soc; Aerospace Med Asn; Am Heart Asn. Res: Cardiac electrophysiology and mechanics; autonomic reflexes; noninvasive measures of cardiac function; techniques for evaluating medical education and clinical competence. Mailing Add: Off of the Dean Thomas Jefferson Univ Philadelphia PA 19107

MACKSEY, HARRY MICHAEL, b Detroit, Mich, Feb 20, 47; m 68. SEMICONDUCTORS. Educ: Univ Mich, Ann Arbor, BS, 68; Univ Ill, Urbana, MS, 70, PhD(physics), 72. Prof Exp: Fel physics, Univ Ill, Urbana, 72-73; MEM TECH STAFF, TEX INSTRUMENTS, INC, 73- Mem: Am Phys Soc; Inst Elec & Electronics Engrs. Res: Development of high power gallium arsenide field-effect transistors for microwave amplification. Mailing Add: Tex Instruments Inc 13500 N Cent Expressway MS 118 Dallas TX 75222

MACLACHLAN, DONALD STUART, b Kenmore, Ont, Can, Sept 1, 23; m 48; c 3. PLANT PATHOLOGY. Educ: McGill Univ, BSc, 48, MSc, 49; Univ Wis, PhD, 52. Prof Exp: Chief seed potato cert, 46-66, DIR PLANT PROTECTION DIV, AGR CAN, 66- Mem: Potato Asn Am; Can Phytopath Soc (secy-treas, 54); Agr Inst Can. Res: Seed potato certification; plant quarantine. Mailing Add: Plant Protection Div Agr Can Sir John Carling Bldg Ottawa ON Can

MACLACHLAN, GORDON ALISTAIR, b Saskatoon, Sask, June 30, 30; m 59; c 2. PLANT BIOCHEMISTRY. Educ: Univ Sask, MA, 54; Univ Man, PhD(biochem), 56. Prof Exp: Nat Res Coun res fel plant physiol, Imp Col, London, 56-58, sci officer, Res Inst Plant Physiol, 58-59; asst prof bot, Univ Alta, 59-62; assoc prof, 62-69, chmn dept, 70-74, PROF BIOL, McGILL UNIV, 70- Concurrent Pos: Assoc ed, Can J Biochem; mem grants comt cell biol, Nat Res Coun Can. Mem: Can Soc Plant Physiol; Am Soc Phytochem; Am Soc Plant Physiol; Can Soc Cell Biol. Res: Metabolism of growing plants. Mailing Add: Dept of Biol McGill Univ Montreal PQ Can

MACLACHLAN, JAMES ANGELL, JR, b Cambridge, Mass, May 18, 38; m 60; c 2. HIGH ENERGY PHYSICS. Educ: Univ Mich, AB, 59; Yale Univ, MS, 62, PhD(physics), 68. Prof Exp: Consult programmer, Yale Comput Ctr, 69; PHYSICIST, FERMI NAT ACCELERATOR LAB, 69- Mem: Am Phys Soc. Res: Hadron physics, especially small angle elastic and inelastic scattering of mesons, nucleons, and hyperons; meson form factors; hyperon decays. Mailing Add: Fermi Nat Accelerator Lab PO Box 500 Batavia IL 60510

MACLACHLAN, JAMES CRAWFORD, b Detroit, Mich, Jan 13, 23; m 50. GEOLOGY. Educ: Wayne State Univ, AB, 48; Princeton Univ, MA, 51, PhD(geol), 52. Prof Exp: Consult geologist, Ministerio de Minas e Hidrocarburos, Caracas, Venezuela, 49-51; geologist, Mineral Deposits Br, US Geol Surv, 52-53 & Fuels Br, 53-56; mem explor projs group, Phillips Petrol Co, 57-62; consult, Shallow Well Explor Co, 62-67, secy-treas, 65-67; asst prof, 67-68, ASSOC PROF GEOL, METROP STATE COL, 68-, CHMN DEPT EARTH SCI, 70- Concurrent Pos: Consult geologist, 62- Mem: Geol Soc Am; Am Asn Petrol Geol; Am Ord Asn. Res: Stratigraphy and sedimentation of Paleozoic and Precambrian rocks of Eastern Colorado and Wyoming; mineralogy and petrology. Mailing Add: Dept of Earth Sci Metrop State Col Denver CO 80204

MACLACHLAN, JAMES DANIEL, b Boston, Mass, June 12, 43; m 71. POLYMER CHEMISTRY, ORGANIC CHEMISTRY. Educ: Tufts Univ, BS, 65; Purdue Univ, West Lafayette, PhD(org chem), 70. Prof Exp: RES CHEMIST, ELASTOMERS LAB, E I DU PONT DE NEMOURS & CO, INC, 69- Mem: Am Chem Soc. Res: Polymer chemistry related to elastomer preparation and/or modification, specifically fluorocarbon polymers. Mailing Add: Du Pont Elastomers Lab Chestnut Run Wilmington DE 19898

MACLANE, GERALD ROBINSON, mathematics, see 12th edition

MACLANE, SAUNDERS, b Norwich, Conn, Aug 4, 09; m 33; c 2. MATHEMATICS. Educ: Yale Univ, PhB, 30; Univ Chicago, MA, 31; Univ Göttingen, PhD(math), 34. Hon Degrees: MA, Harvard Univ, 42; DSc, Purdue Univ, 65, Yale Univ, 69 & Coe Scol, 74; LLD, Glasgow Univ, 71. Prof Exp: Pierce instr math, Harvard Univ, 34-36; instr, Cornell Univ, 36-37; instr, Univ Chicago, 37-38; from asst prof to prof, Harvard Univ, 38-47; prof, 47-63, MAX MASON DISTINGUISHED SERV PROF MATH, UNIV CHICAGO, 63- Concurrent Pos: Guggenheim fel, Swiss Fed Inst Technol & Columbia Univ, 47-48; mem exec comt, Int Math Union, 54-58; vis prof, Univ Heidelberg, 58, Univ Frankfurt, 60 & Tulane Univ, 69; mem coun, Nat Acad Sci, 59-62 & 69-72; Fulbright fel, Australian Nat Univ, 69; mem, Nat Sci Bd, 74- Honors & Awards: Chauvenet Prize, Math Asn Am, 41; Distinguished Serv Award, Math Asn Am, 75. Mem: Nat Acad Sci (vpres, 73-77); Am Math Soc (vpres, 46-48, pres, 73-74); Am Philos Soc (vpres, 68-71); Math Asn Am (vpres, 48, pres, 50); Am Acad Arts & Sci. Res: Algebra; topology; algebraic topology; logic; category theory. Mailing Add: Dept of Math Univ of Chicago Chicago IL 60637

MACLAREN, MALCOLM DONALD, b Tarrytown, NY, Aug 5, 36. COMPUTER SCIENCE. Educ: Harvard Univ, AB, 58, MA, 60, PhD(math), 62. Prof Exp: Mem staff math, Boeing Sci Res Labs, Wash, 60-64; from asst mathematician to assoc mathematician, Argonne Nat Lab, 64-72; MGR ADVAN LANG SYSTS DEVELOP, CAMBRIDGE INFO SYSTS LAB, HONEYWELL, INC, 74- Concurrent Pos: Vchmn comt X3J1, Am Nat Standards Inst, 71-72. Mem: Am Math Soc; Asn Comput Mach; Soc Indust & Appl Math. Res: Computation and programming, especially design, definition and implementation of languages, operating systems and similar software; formal theory of languages and programming. Mailing Add: Cambridge Info Systs Lab Honeywell Inc 575 Technol Square Cambridge MA 02139

MACLAREN, RICHARD OLIVER, b Missoula, Mont, Sept 4, 24; m 48; c 5. PHYSICAL CHEMISTRY. Educ: Univ Ore, BA, 49, MA, 50; Univ Wash, PhD(chem), 54. Prof Exp: Res chemist, Olin Mathieson Chem Corp, 54-55, pilot plant supvr, 55-56, group leader, 56-57, head thermodyn sect, 57-60; chief chem sect, United Tech Ctr, United Aircraft Corp, 60-63, MGR COMBUSTION RES & DEVELOP BR, CHEM SYSTS DIV, UNITED TECHNOL CORP, 63- Mem: Am Chem Soc. Res: Physical and inorganic chemistry involving high temperature processes; electrochemistry; solid propellant combustion, ignition and propellant chemistry; chemistry of fluorine compounds; chemical propulsion concetps. Mailing Add: 1031 Pinenut Ct Sunnyvale CA 94087

MACLAREN, WALTER ROGERS, b Yokohama, Japan, Dec 7, 10; US citizen; m 42; c 5. ALLERGY. Educ: Queen's Univ, Ont, 33; Harvard Univ, MD, 38; Am Bd Allergy & Immunol, dipl, 74. Prof Exp: Asst, Mass Gen Hosp, 39-41; lectr physiol, 43-47, asst prof med, 48-56, ASSOC PROF MED, SCH MED, UNIV SOUTHERN CALIF, 56-; CHIEF, ADULT ALLERGY DEPT, LOS ANGELES COUNTY-UNIV SOUTHERN CALIF MED CTR, 52- Concurrent Pos: Bradford fel, Mass Gen Hosp & Harvard Med Sch, 39-41; aeromed dir, Lockheed Aircraft Corp, 43-46; pvt pract, 47- Mem: Fel Am Acad Allergy; fel Am Col Allergists; fel Am Asn Clin Immunol & Allergy; Am Thoracic Soc; Pan-Am Med Asn. Res: Allergy and immunology. Mailing Add: 94 N Madison Ave Pasadena CA 91101

MACLATCHY, CYRUS SHANTZ, b Galt, Ont, May 24, 41; m 63; c 3. PLASMA PHYSICS. Educ: Acadia Univ, BS, 64; Univ BC, MS, 66, PhD(physics), 70. Prof Exp: ASST PROF PHYSICS, ACADIA UNIV, 70- Res: Use of Langmuir probes in diagnostics of diffusion plasmas. Mailing Add: Dept of Physics Acadia Univ Wolfville NS Can

MACLAUCHLAN, DONALD WELLS, b Bedegue, PEI, June 8, 05; m 38. PHYSICAL CHEMISTRY. Educ: Mt Allison Univ, BSc, 29; McGill Univ, PhD(chem), 37. Prof Exp: Lab technician, Atlantic Sugar Ref Co, NB, 23-24; instr chem, McGill Univ, 35-37; lectr physics, Univ NB, 40-42; prof chem, 42-74, dean men, 43-60, assoc dean arts & sci, 66-69, EMER PROF CHEM, MT ALLISON UNIV, 74-, EXEC DIR ALUMNI, 69- Res: Properties of hydrogen peroxide. Mailing Add: Box 91 Sackville NB Can

MACLAUGHLIN, DOUGLAS EARL, b Indiana, Pa, Nov 18, 38; m 67. SOLID STATE PHYSICS. Educ: Amherst Col, BA, 60; Univ Calif, Berkeley, PhD(physics), 66. Prof Exp: NATO fel physics, Atomic Energy Res Estab, Harwell, Eng, 66-67; res assoc, Lab Physique Solides, Fac Sci, Orsay, France, 67-69; asst prof, 69-73, ASSOC PROF PHYSICS, UNIV CALIF, RIVERSIDE, 73- Mem: Am Phys Soc. Res: Use of nuclear magnetic resonance and relaxation to investigate electronic structure of metals at low temperatures, particularly in the superconducting state. Mailing Add: Dept of Physics Univ of Calif Riverside CA 92502

MACLAURY, DONALD WAYNE, b Waterbury, Conn, Feb 28, 05; m 46; c 1. POULTRY BREEDING. Educ: Cornell Univ, BS, 37; Iowa State Col, MS, 40, PhD(poultry breeding), 55. Prof Exp: Instr poultry husb, Iowa State Col, 37-38; asst, Exp Sta, 40-52, assoc, 52-54, poultry husbandman, 54-60, prof poultry sci, 60-66, prof animal sci, 66-74, EMER PROF ANIMAL SCI, UNIV KY, 74- Mem: Fel AAAS; Genetics Soc Am; Poultry Sci Asn; Am Statist Asn; World Poultry Sci Asn. Res: Poultry incubation, breeding and growth. Mailing Add: 106 Brigadoon Pkwy Lexington KY 40503

MACLAURY, MICHAEL RISLEY, b Hays, Kans, Dec 28, 43; m 66. ORGANOMETALLIC CHEMISTRY. Educ: Antioch Col, BS, 67; Stanford Univ, PhD(inorg chem), 74. Prof Exp: Res asst nuclear med, Stanford Med Ctr, 70-72; STAFF CHEMIST, GEN ELEC CORP RES & DEVELOP, 74- Concurrent Pos: Fel chem catalysis, NSF US/USSR Exchange Prog, 74. Mem: Am Chem Soc. Res: Mechanisms of organometallic reactions as they relate to possible catalytic reactions; the role of inorganic materials in flame retarding polymers. Mailing Add: Gen Elec Res & Develop Ctr PO Box 8 Bldg K-1 Rm 5A10 Schenectady NY 12301

MACLAY, CHARLES WYLIE, b Fannettsburg, Pa, Oct 17, 29; m 50; c 2. MATHEMATICS. Educ: Shippensburg State Col, BS, 56; Univ Va, MEd, 59, EdD(math educ), 68. Prof Exp: Teacher math, Seaford Spec Schs, Del, 56-58; instr, Shippensburg State Col, 60-62; teacher, Abington Schs, 62-69; PROF MATH, EAST STROUDSBURG STATE COL, 69- Mem: Math Asn Am; Nat Coun Teachers Math. Res: Learning sequences and sequencing in math; curriculum constructing and testing. Mailing Add: East Stroudsburg State Col East Stroudsburg PA 18301

MACLAY, WILLIAM NEVIN, b Belleville, Pa, Dec 30, 24; m 49; c 5. PHYSICAL CHEMISTRY, RESEARCH ADMINISTRATION. Educ: Juniata Col, BS, 47; Yale Univ, PhD(phys chem), 50. Prof Exp: Assoc prof chem, Davis & Elkins Col, 50-51; res chemist, B F Goodrich Co, 51-59; mgr latices res, 59-65, mgr polystyrene res, 62-65, asst mgr plastics res, 65-67, mgr commercial develop, 67-68, asst mgr res, 68, VPRES & DIR RES, KOPPERS CO INC, 68- Mem: Am Chem Soc. Res: Surface and colloid chemistry; high polymer latices; polystyrene molding resins; copolymerization; graft polymers. Mailing Add: 539 Greenleaf Dr Monroeville PA 15146

MACLEAN, BONNIE KUSESKE, b St Cloud, Minn, Jan 23, 42; m 64; c 2. ENTOMOLOGY. Educ: Gustavus Adolphus Col, BA, 63; Purdue Univ, MS, 65, PhD(entom), 72. Prof Exp: Instr biol, Youngstown State Univ, 68-69; RES & WRITING, 69- Mem: Sigma Xi; Entom Soc Am. Res: Effects of petroleum hydrocarbons on DNA synthesis in the mosquito Culex pipiens. Mailing Add: 280 E Western Reserve Rd Poland OH 44514

MACLEAN, DAVID BAILEY, b Summerside, PEI, July 15, 23; m 45; c 7. ORGANIC CHEMISTRY. Educ: Acadia Univ, BSc, 42; McGill Univ, PhD(chem), 46. Prof Exp: Res chemist, Dom Rubber Co, 46-49; assoc prof indust chem, NS Tech Col, 49-54; assoc prof chem, 54-60, PROF CHEM, McMASTER UNIV, 60- Mem: Am Chem Soc; Chem Inst Can. Res: Isolation and structure of alkaloids; mass spectrometry of organic compounds. Mailing Add: Dept of Chem McMaster Univ Hamilton ON Can

MACLEAN, DAVID BARKER, analytical chemistry, see 12th edition

MACLEAN, DAVID BELMONT, b Cleveland, Ohio, Sept 15, 41; m 64; c 1. INSECT ECOLOGY, INSECT TAXONOMY. Educ: Heidelberg Col, BS, 63; Purdue Univ, MS, PhD(entom, biometry), 69. Prof Exp: ASST PROF BIOL, YOUNGSTOWN STATE UNIV, 68- Mem: Am Inst Biol Sci; Entom Soc Am; Entom Soc Can. Res: Interrelationships of insect and plant communities; symbiotic relationship of ambrosia beetles and associated fungi; systematics of Polistes; cytotaxonomic techniques. Mailing Add: Dept of Biol Youngstown State Univ Youngstown OH 44503

MACLEAN, DAVID CAMERON, b New Rochelle, NY, Dec 8, 33; m 56; c 3. PLANT PHYSIOLOGY. Educ: Univ Conn, BS, 60; Mich State Univ, MS, 62, PhD(plant physiol), 65. Prof Exp: Res assoc plant physiol, Mich State Univ, 61-65; PLANT PHYSIOLOGIST, BOYCE THOMPSON INST PLANT RES, 65- Concurrent Pos: Consult to several aluminum & phosphate corps, 65-; mem panel on nitrogen oxides & med & biol effects of environ pollutants, Nat Res Coun, Nat Acad Sci, 73- Honors & Awards: Dow Award, Am Soc Hort Sci, 70. Mem: AAAS; Am Soc Plant Physiol; Am Soc Hort Sci; NY Acad Sci; Sigma Xi. Res: Physiology of senescence in plant tissues; effects of air pollutants on vegetation; plant growth and development. Mailing Add: Boyce Thompson Inst Plant Res 1086 N Broadway Yonkers NY 10701

MACLEAN, DONALD ISADORE, b Norwalk, Conn, Nov 25, 29. PHYSICAL CHEMISTRY. Educ: Boston Univ, AB, 53; LicPhil, Weston Col, 54; Catholic Univ, PhD(phys chem), 58. Prof Exp: Humboldt res fel, Univ Göttingen, 62-65; from asst prof to assoc prof phys chem, Boston Col, 66-73; VPRES ACAD AFFAIRS, CREIGHTON UNIV, 73- Concurrent Pos: Adj prof, Boston Col, 73-; mem bd trustees, St Louis Univ, 75- Mem: Am Chem Soc; Combustion Inst; Am Asn Higher Educ. Res: Fast reaction kinetics; combustion chemistry; mass, electron spin resonance and optical spectroscopy. Mailing Add: Off of VPres Acad Affairs Creighton Univ Omaha NE 68178

MACLEAN, LLOYD DOUGLAS, b Calgary, Alta, June 15, 24; m 54; c 5. SURGERY. Educ: Univ Alta, BSc, 43, MD, 59; Univ Minn, PhD(surg), 57; FRCPS(C). Prof Exp: From instr to assoc prof, Univ Minn, 56-65; PROF SURG, McGILL UNIV, 65-, CHMN DEPT, 68- Concurrent Pos: Chief surg serv, Ancker Hosp, St Paul, 57-65; surgeon-in-chief, Royal Victoria Hosp, Montreal, 65- Mem: Am Physiol Soc; Soc Exp Biol & Med; Soc Univ Surgeons; AMA; Am Surg Asn. Res: Blood flow to heart; intestinal blood flow in shock due to hemorrhage and endotoxin and cardiac excitability; shock and transplantation. Mailing Add: Dept of Surg Royal Victoria Hosp Montreal PQ Can

MACLEAN, PAUL DONALD, b Phelps, NY, May 1, 13; m 42; c 5. NEUROPHYSIOLOGY. Educ: Yale Univ, BA, 35, MD, 40. Prof Exp: Intern, Johns Hopkins Hosp, 40-41; asst resident med, New Haven Hosp, Conn, 41-42; asst path, Sch Med, Yale Univ, 42; clin instr med, Med Sch, Washington Univ, 46-47; USPHS res fel psychiat, Harvard Med Sch & Mass Gen Hosp, 47-49; asst prof physiol, Sch Med, Yale Univ, 49-51, from asst prof to assoc prof psychiat, 51-56, assoc prof physiol, 56; chief limbic integration & behav sect, NIH, 57-71, CHIEF LAB BRAIN EVOLUTION & BEHAV, NIMH, 71- Concurrent Pos: Dir EEG lab, New Haven Hosp, Conn, 51-52; attend physician, Grace-New Haven Hosp, 53-56; NSF sr res fel, Switz, 56-57; Salmon lectr award, NY Acad Med, 66; Hinks lectr award, Ont Ment Health Found, 69; Mider lectr award, NIH, 72; mem bd trustees, Prof Percival Bailey Educ Proj. Honors & Awards: Distinguished Res Award, Asn Res Nerv & Ment Dis, 64; Salmon Medal Distinguished Res in Psychiat, 66; Superior Serv Award, US Dept Health, Educ & Welfare, 67; Spec Award, Am Psychopath Asn, 71; Karl Spencer Lashley Award, Am Philos Soc, 72. Mem: Am Electroencephalog Soc; Am Neurol Asn; Am Asn Hist Med; Soc Neurosci; Pavlovian Soc NAm. Res: Forebrain mechanisms of species-typical and emotional behavior. Mailing Add: Lab of Brain Evolution & Behav NIMH Bldg 110-NIHAC Bethesda MD 20014

MACLEAN, STEPHEN FREDERICK, JR, b Los Angeles, Calif, Jan 18, 43; m 67; c 1. ECOLOGY. Educ: Univ Calif, Santa Barbara, BA, 64; Univ Calif, Berkeley, PhD(ecol), 69. Prof Exp: Actg asst prof zool, Univ Mont, 69-70; asst prof, Univ Ill, 70-71; NSF res grant Arctic Alaska, 71-72, asst prof, 71-74, ASSOC PROF BIOL SCI, UNIV ALASKA, 74- Concurrent Pos: NSF res grant Arctic Alaska, Univ Mon, 70-71; prog integrator, US Int Biol Prog Tundra Biome, 71- Mem: Ecol Soc Am; Brit Ecol Soc; Arctic Inst NAm; Am Ornithologists Union; Cooper Ornith Soc. Res: Population ecology and energetics of arctic birds, mammals and insects; systems analysis of tundra ecosystem. Mailing Add: Dept of Biol Sci Univ of Alaska College AK 99701

MACLEAN, WALLACE H, b PEI, Jan 10, 31; m 60; c 2. GEOLOGY, GEOCHEMISTRY. Educ: Colo Sch Mines, GeolE, 55; McGill Univ, MSc, 64, PhD(geol), 68. Prof Exp: Mine geologist, United Keno Hill Mines, Can, 55-57; econ geologist, Ministry Petrol & Mineral Resources, Saudi Arabia, 57-62; prof assoc, 67-70, asst prof, 70-74, ASSOC PROF GEOL SCI, McGILL UNIV, 74- Honors & Awards: Waldemar Lindgren Citation Award, 69. Mem: Geol Asn Can; Mineral Asn Can; Can Inst Mineral Metall. Res: Phase equilibria work in oxide-sulfide-silicate systems; petrogenesis of sulfide magmatic ore bodies; field research in massive volcanogenic sulfide deposits. Mailing Add: Dept of Geol Sci McGill Univ PO Box 6070 Montreal PQ Can

MACLEAN, WILLIAM PLANNETTE, III, b Bainbridge, Md, Sept 20, 43; m 66. BIOLOGY. Educ: Princeton Univ, BA, 65; Univ Chicago, PhD(evolutionary biol), 69. Prof Exp: Asst prof, 69-75, ASSOC PROF BIOL, COL VIRGIN ISLANDS, 75- Mem: Am Soc Ichthyologists & Herpetologists; Soc Study Evolution. Res: Patterns of adaptation in squamates; phylogeny of squamates; island biogeography. Mailing Add: Col of the Virgin Islands St Thomas VI 00801

MACLEAY, RONALD E, b Buffalo, NY, Dec 3, 35; m 60; c 4. SYNTHETIC ORGANIC CHEMISTRY. Educ: St Bonaventure Univ, BS, 57, MS, 59; Univ Buffalo, PhD(org chem), 65. Prof Exp: Res & develop chemist, 59-61, sr res chemist, 64-66, res group leader nitrogen chem, 66-75, SR GROUP LEADER RES, LUCIDOL DIV, PENNWALT CORP, 75- Mem: Am Chem Soc. Res: Reduction of carbonyl compounds with lithium tetrakis-aluminate; synthesis of azo compounds for free radical initiation of polymerization. Mailing Add: Lucidol Div Pennwalt Corp 1740 Military Rd Buffalo NY 14240

MACLELLAN, CHARLES ROGER, b New Glasgow, NS, May 8, 23; m 49; c 2. ENTOMOLOGY. Educ: McGill Univ, BSc, 50; Queen's Univ Ont, MA, 54. Prof Exp: RES SCIENTIST & ECOLOGIST, RES BR, CAN DEPT AGR, 50- Concurrent Pos: Res fel entom, Commonwealth Sci & Indust Res Orgn, Australia, 62-64. Mem: Entom Soc Can; Ecol Soc Australia; Japanese Soc Appl Entom & Zool; Japanese Soc Pop Ecol; Entom Soc Australia. Res: Ecological research and population dynamics of tortricid pests of pome fruits and the natural enemies which attack them. Mailing Add: Res Sta Kentville NS Can

MACLENNAN, DONALD ALLAN, b San Turse, PR, Mar 27, 36; m 67; c 1. ATOMIC PHYSICS. Educ: Univ Calif, Berkeley, BS, 59, PhD(physics), 66; Case Western Reserve Univ, MBA, 72. Prof Exp: Engr, Atomic Power & Equip Div, 59, 60, res scientist, Res & Develop Ctr, 66-68, group leader discharge eng, 68-70, MGR FLUORESCENT LAMP ENG, LAMP DIV, GEN ELEC CO, 70- Mem: Am Phys Soc. Res: Experimental atomic and molecular physics; gaseous electronics; surface physics. Mailing Add: Gen Elec Co Lamp Div Sect 1300 Nela Park East Cleveland OH 44112

MACLEOD, ALASTAIR WILLIAM, b Vancouver, BC, Aug 26, 16; m 54. PSYCHIATRY. Educ: Glasgow Univ, BSc, 38, MB, ChB, 41, DPH, 42. Prof Exp: Asst dir dept psychol med, York Clin, Guy's Hosp, London, Eng, 45-50; from asst prof to assoc prof psychiat, McGill Univ, 52-72; from asst dir to assoc dir, 57-69, EXEC DIR, MENT HYG INST, 69-; PROF PSYCHIAT, McGILL UNIV, 59- Concurrent Pos: Mem attend staff, Royal Victoria Hosp, 52-, sr psychiatrist attend staff, 71-; consult ment health, Protestant Sch Bd Gtr Montreal, 59-; training analyst, Can Inst Psychoanal; psychiat consult, Montreal Gen, Jewish Gen & Queen Elizabeth Hosps, Montreal, 66-; spec lectr, Dept Health & Social Med & Sch Social Work, McGill Univ. Mem: Fel Am Psychiat Asn; Can Psychoanal Soc (pres, 53-55). Res: Aftereffects of early maternal deprivation, their recognition and response to treatment. Mailing Add: Ment Hyg Inst 3690 Peel St Montreal PQ Can

MACLEOD, CHARLES FRANKLYN, b Halifax, NS, Jan 15, 24; m 51; c 3. ECOLOGY. Educ: McGill Univ, BSc, 48; Univ BC, MA, 50; Univ Minn, PhD(ecol, statist), 59. Prof Exp: Teaching asst zool, NC State Col, 51-53; res officer small mammal ecol, Can Dept Forestry, 56-62; asst prof, 62-66, actg chmn dept, 65-66, chmn dept, 66-69, ASSOC PROF BIOL, SIR GEORGE WILLIAMS CAMPUS, CONCORDIA UNIV, 66- Mem: Ecol Soc Am; Am Inst Biol Sci; Am Soc Mammal; Can Soc Zoologists. Res: Small mammal population studies. Mailing Add: Dept Biol Sci Sir Geo Wms Campus Concordia Univ Montreal PQ Can

MACLEOD, DON PUTNAM, b Inverness, NS, Aug 6, 31; m 53; c 4. ELECTROPHYSIOLOGY. Educ: Dalhousie Univ, BSc, 53, MSc, 55, PhD(biol sci), 59. Prof Exp: Nat Heart Found fel pharmacol, Dalhousie Univ, 59-62; lectr pharmacol, Univ Alta, 62-64, Nat Heart Found res assoc, 63-64, asst prof, 64-67; assoc prof physiol & biophys, 67-71, PROF PHYSIOL & BIOPHYS, DALHOUSIE UNIV, 71- Concurrent Pos: Nat Heart Found fel, Univ Alta, 62-63 & sr res fel, 64-67; mem, Int Study Group Cardiac Metab. Mem: Can Physiol Soc; Pharmacol Soc Can. Res: Cardiovascular pharmacology; analysis of tissues for morphine; hydrocarbon; adrenalin arrhythmias, antiarrhythmic activity of rauwolfia alkaloids; effect of glucose on ventricular electrophysiology; relationship of glucose to potassium transport; ionic basis for cardiac electrical activity; hemicholinium and cardiac arrhythmias. Mailing Add: Dept of Physiol & Biophys Dalhousie Univ Halifax NS Can

MACLEOD, DONALD IAIN ARCHIBALD, b Glasgow, Scotland, Oct 2, 45; m 74. VISION. Educ: Univ Glasgow, MA, 67; Cambridge Univ, PhD(exp psychol), 74. Prof Exp: Res assoc vision, Inst Perception, Soesterberg, 67-68; res assoc zool, Cambridge Univ, 72; vis asst prof psychobiol, Fla State Univ, 72-74; ASST PROF PSYCHOL, UNIV CALIF, SAN DIEGO, 74- Mem: Psychonomic Soc; Asn Res Vision & Ophthal. Res: Retinal mechanisms in human vision; human color vision. Mailing Add: Dept of Psychol Univ of Calif San Diego La Jolla CA 92093

MACLEOD, DONALD RICHARD EASON, b Tamsui, Taiwan, Feb 20, 10; Can citizen; m 40; c 5. MEDICAL MICROBIOLOGY. Educ: Univ Toronto, BA, 32, MD, 35, BSc, 38, DPH, 46. Prof Exp: Res assoc, Connaught Med Res Labs, 45-56, res mem, 56-70, asst dir, Connaught Labs Ltd, 70-75; RETIRED. Concurrent Pos: Assoc microbiol, Sch Hyg, Univ Toronto, 46-48, from asst prof to assoc prof, 48-70. Mem: Can Med Asn; Can Pub Health Asn; Can Soc Microbiol; Can Soc Immunol. Res: Poliomyelitis vaccine; antibody response, duration, combined antigens; herpesvirus simiae, passive immunity; oral poliovirus vaccine; antibody response, virus excretion and neurovirulence after human passage; serological surveys. Mailing Add: 22 Austin Crescent Toronto ON Can

MACLEOD, ELLIS GILMORE, b Washington, DC, Sept 3, 28. EVOLUTIONARY BIOLOGY, ENTOMOLOGY. Educ: Univ Md, BS, 55, MS, 60; Harvard Univ, PhD(biol), 64. Prof Exp: Fel evolutionary biol, Harvard Univ, 64-66; asst prof, 66-69, ASSOC PROF ENTOM, UNIV ILL, URBANA, 69- Concurrent Pos: NSF grant syst biol, 69- Mem: Royal Entom Soc London; AAAS. Res: Speciation and higher levels of evolution of insects, including studies of phylogeny deduced from the fossil record, and such comparative studies of contemporary species as behavior, ecology and chromosome cytology. Mailing Add: Dept of Entom Univ of Ill Urbana IL 61801

MACLEOD, GUY FRANKLIN, b Lowell, Mass, June 22, 97; m 24; c 3. AGRICULTURE, ENTOMOLOGY. Educ: Mass Col, BS, 20; Cornell Univ, PhD(entom), 30. Prof Exp: Asst, NY Exp Sta, Univ Geneva, 20-24; exten entomologist, Pa State Col, 24-29; prof entom, NY State Col Agr & entomologist, Exp Sta, 30-39; prof entom & entomologist, Exp Sta, Univ Calif, 39-42; chief chem div war food admin, USDA, 42-45; tech vpres & dir, Sunland Industs, 45-58; mgr pub rels & tech serv, Western Agr Dept, Niagara Chem Div, FMC Corp, 58-62; consult, Union Carbide Corp, Calif, 62-69; dir agr educ study, 69-71, CONSULT TO VPRES AGR SCI, UNIV CALIF, BERKELEY, 71- Concurrent Pos: Consult, Tex Gulf Sulfur, 36; trustee, Mosquito Abatement Dist; pres, Insecticide Inst. Mem: AAAS; fel Am Inst Chemists; entom Soc Am. Res: Insect control through chemicals; insecticides; herbicides; fertilizers; chemical and engineering problems of manufacturing; insect and plant physiology. Mailing Add: Rm 323 Univ Hall Univ of Calif Berkeley CA 94720

MACLEOD, JOHN, b Edinburgh, Scotland, Nov 29, 05; nat US; m 34; c 2. PHYSIOLOGY. Educ: NY Univ, AB, 34, MS, 37; Cornell Univ, PhD, 41. Prof Exp: Asst physiol, 38-39, res assoc anat, 41-46, from asst prof to prof, 46-72, EMER PROF ANAT, MED COL, CORNELL UNIV, 72- Honors & Awards: Lasker Award, 45. Mem: AAAS; Am Physiol Soc; Am Fertil Soc; Harvey Soc; Am Asn Anat. Res: Physiology of human spermatozoa, reproduction and sterility; tissue metabolism. Mailing Add: Dept of Anat Cornell Univ Col of Med New York NY 10021

MACLEOD, JOHN CAMERON, aquatic biology, pollution, see 12th edition

MACLEOD, JOHN CAMPBELL, b Fenelon Falls, Ont, Feb 24, 11; m 40; c 3. FORESTRY. Educ: Univ NB, BScF, 34. Prof Exp: With tech control & sales dept, Howard Smith Paper Mills, Ltd, 34-39, forester, Woodlands Div, 45-46; res officer fire control, Fed Govt, Ottawa, 46-51; head fire protection sect, Res Div, Dept Forestry Can, 51-61, assoc dir, Forest Res Br, 61-65, assoc dir prog coord fire, 65-72, DIR PETAWAWA FOREST EXP STA, CAN FORESTRY SERV, 73- Mem: Can Inst Forestry. Res: Forest research administration. Mailing Add: Can Forestry Serv Chalk River ON Can

MACLEOD, LLOYD BECK, b Apr 27, 30; Can citizen; m 55; c 3. AGRONOMY. Educ: McGill Univ, BSc, 52, MSc, 53; Cornell Univ, PhD(soil sci), 62. Prof Exp: Res officer agron, Res Br, Can Dept Agr, 53-59, head soils & plant nutrit sect, 62-65, res scientist, PEI, 65-67, head soils & plant nutrit sect, 67-70, DIR RES STA, AGR CAN, 70- Mem: Am Soc Agron; Soil Sci Soc Am; Can Soc Soil Sci; Can Soc Agron; Int Soc Soil Sci. Res: Soil fertility and plant nutrition of forage species; nutrient competition; aluminum tolerance; effect of potassium on utilization of ammonium and nitrate sources of nitrogen by forage and cereal crops. Mailing Add: Can Agr Res Sta PO Box 1210 Charlottetown PE Can

MACLEOD, MICHAEL CHRISTOPHER, b Allentown, Pa, Feb 7, 47; m 69; c 1. MOLECULAR BIOLOGY. Educ: Calif Inst Technol, BS, 69; Univ Ore, PhD(molecular biol), 74. Prof Exp: Fel biol, Univ Ore, 74-75; FEL GENE REGULATION, BIOL DIV, OAK RIDGE NAT LAB, 75- Res: Purifying tyrosine aminotransferase mRNA from rat liver; cDNA copies will be used to quantitate changes in mRNA concentration during steroid-mediated induction and deinduction of the enzyme. Mailing Add: Biol Div Oak Ridge Nat Lab Oak Ridge TN 37830

MACLEOD, ROBERT ANGUS, b Athabasca, Alta, July 13, 21; m 48; c 6. MICROBIOLOGY. Educ: Univ BC, BA, 43, MA, 45; Univ Wis, PhD(biochem), 49. Prof Exp: Instr chem, Univ BC, 45-46; asst prof biochem, Queen's Univ Ont, 49-52; biochemist, Fisheries Res Bd, Can, 52-60; assoc prof agr bact, 60-64, PROF MICROBIOL, MACDONALD COL, McGILL UNIV, 64-, CHMN DEPT, 74- Concurrent Pos: Mem marine sci ctr, McGill Univ. Honors & Awards: Harrison Prize, Royal Soc Can, 60; Award, Can Soc Microbiol, 73. Mem: Fel Royal Soc Can; Am Soc Biol Chem; Am Soc Microbiol; Can Soc Microbiol. Res: Nutrition and metabolism of marine bacteria; function of inorganic ions in bacterial metabolism; microbial biochemistry. Mailing Add: Dept of Microbiol MacDonald Col McGill Univ Ste Anne de Bellevue PQ Can

MACLEOD, ROBERT MEREDITH, b Newark, NJ, May 14, 29; m 51; c 4. BIOCHEMISTRY. Educ: Seton Hall Univ, BS, 52; NY Univ, MS, 56; Duke Univ, PhD(biochem), 59. Prof Exp: Res biochemist, Schering Corp, NJ, 48-56; instr biochem, Sch Med, Duke Univ, 59-60; asst prof biochem in internal med, 60-66, chmn div biomed eng, 64-65, dir div cancer studies, 69-72, assoc prof, 66-73, PROF INTERNAL MED, SCH MED, UNIV VA, 73- Concurrent Pos: Am Heart Asn res fel, 59-60; USPHS res grant, 64-, career develop award, 65-71; cancer travel fel, WHO, 68. Mem: AAAS; Am Physiol Soc; Am Fedn Clin Res; Am Asn Cancer Res; Int Soc Neuroendocrinol. Res: Hormonal control of biochemical mechanisms which regulate normal and neoplastic growth. Mailing Add: Dept of Internal Med Univ of Va Sch of Med Charlottesville VA 22901

MACLEOD, STEPHEN CLAIR, b Bonshaw, PEI, Sept 19, 34; m; c 4. OBSTETRICS & GYNECOLOGY. Educ: Dalhousie Univ, MD, FRCS, 64. Prof Exp: Pvt pract, NS, 59-60; Can Res Coun fel, Dalhousie Univ, 60-61, asst resident obstet & gynec, 61-62, chief resident, 62-63; chief resident surg, Women's Col Hosp, Toronto, 63-64; McLaughlin traveling fel, Royal Women's Hosp, Melbourne, 64-65; lectr, 66-68, asst prof, 68-70, ASSOC PROF OBSTET & GYNEC, DALHOUSIE UNIV, 71- Concurrent Pos: Mem staff, Victoria Gen Hosp & Halifax Infirmary, 64-; mem staff, Grace Maternity Hosp, 64-, pres med staff, 71-72; Can Life Ins res fel, 67-72; mem adv comts birth control pill & res, Dept Health & Welfare Can; mem ad hoc comt fertil control, Int Develop Res Coun; mem comt pituitary gonadotrophins, Med Res Coun. Honors & Awards: Killam Award, 64. Mem: Fel Am Col Obstet & Gynec; Soc Obstet & Gynaec Can; AMA. Res: Female endocrinology and infertility; control of hypothalamic pituitary ovarian function in the human female; antinatal monitoring of high risk pregnancies. Mailing Add: 546 Young Ave Halifax NS Can

MACLINN, WALTER ARNOLD, chemistry, see 12th edition

MACLULICH, DUNCAN ALEXANDER, b Toronto, Ont, July 2, 09; m 42; c 3. WILDLIFE BIOLOGY. Educ: Univ Toronto, BScF, 31, PhD, 37. Prof Exp: Asst biol, Univ Toronto, 31-36; bacteriologist, Mt Sanitarium, Hamilton, Ont, 36-37; ranger, Algonquin Prov Park, 38-40; navig instr, Royal Can Air Force, 40-43, in exp flying, Empire Cent Flying, 43-45, chief proj engr, Exp Proving Estab, 46-50, dep dir, Instrument & Elec Develop, 50-60; from assoc prof to prof, 62-75, chmn dept, 62-75, EMER PROF BIOL, WILFRID LAURIER UNIV, WATERLOO, ONT, 75- Concurrent Pos: Ed, Ont Bird Banding J. Mem: Fel AAAS; Am Soc Parasitol; Soc Am Foresters; Am Soc Mammal; Wildlife Soc. Res: Wildlife population; parasitology; aerodynamics; instruments; nature photography; management; research on white pine and on animal populations. Mailing Add: 26 Stewart St Strathroy ON Can

MACLURE, KENNETH CECIL, b Montreal, Que, Oct 14, 14; m 49 49; c 4. NUCLEAR PHYSICS. Educ: McGill Univ, BSc, 34, MSc, 50, PhD(nuclear physics), 52. Prof Exp: Royal Can Air Force, 39-67, navigator, Navig Inst & Navig Staff Off, 39-44; officer chg test & develop, Empire Air Navig Sch, Eng, 44-46, dir arctic res, Defence Res Bd, Ont, 47-48, chief proj engr, Exp & Proving Estab, 53-54, dir armament eng, Air Force Hqs, 54-58, air attache, Can Embassy, Poland, 58-61, mem res staff, Pac Naval Lab, Defence Res Bd, BC, 61-67, mem res staff, Defense Serv Sci Off, 67-75, chief defence res staff, London, 71-75, SCI ADV TO DEP CHIEF OF DEFENCE STAFF, NAT DEFENCE HQS, OTTAWA, 75- Concurrent Pos: Defence Serv Sci off, Defence Res Estab Pac, 67-71. Mem: Am Inst Navig; assoc fel Can Aeronaut & Space Inst; fel Arctic Inst NAm; fel Brit Inst Navig; Royal Inst Navig, London (vpres, 74-75). Res: Military applications of nuclear physics; air navigation; arctic research. Mailing Add: 16 Birch Ave Ottawa ON Can

MACMAHAN, HORACE ARTHUR, JR, b Freeport, Maine, Aug 13, 28; m 65; c 1. EARTH SCIENCE. Educ: Univ Maine, BA, 54; Univ Utah, MSEd, 63; Univ Colo, EdD(sci educ), 67. Prof Exp: Staff asst, Earth Sci Curric Proj, Univ Colo, 63-64; assoc prof earth sci, State Univ NY Col Oneonta, 67-68; assoc prof sci educ, Weber State Col, 68-69; ASSOC PROF GEOG & GEOL, EASTERN MICH UNIV, 69- Mem: AAAS; Nat Asn Geol Teachers; Nat Coun Geog Educ; Nat Sci Teachers Asn; Am Educ Res Asn. Res: Developed composite paleogeographic maps of North America for each geologic time period; determination of the most effective mode of presenting map concepts in geology. Mailing Add: Dept of Geog & Geol Eastern Mich Univ Ypsilanti MI 48197

MACMAHON, BRIAN, b Eng, Aug 12, 23; m 48; c 4. EPIDEMIOLOGY. Educ: Univ Birmingham, MB, ChB & DPH, 49, PhD, 52, MD, 55; Harvard Univ, SM, 53. Prof Exp: Lectr social med, Univ Birmingham, 53-54; from assoc prof to prof environ med & community health, State Univ NY Downstate Med Ctr, 55-58; PROF EPIDEMIOL & HEAD DEPT, SCH PUB HEALTH, HARVARD UNIV, 58- Mem: Am Pub Health Asn; Am Epidemiol Soc. Res: Epidemiology of noninfectious diseases. Mailing Add: Dept of Epidemiol Harvard Univ Sch of Pub Health Boston MA 02115

MACMAHON, HAROLD EDWARD, b Aylmer, Ont, Mar 30, 01; US citizen; m 34; c 4. PATHOLOGY, BACTERIOLOGY. Educ: Univ Western Ont, MD, 25; Am Bd Path, dipl & cert path anat. Hon Degrees: BA, Univ Western Ont, 22, ScD, 48. Prof Exp: Intern, Montreal Gen Hosp, 25-26; asst, Boston City Hosp, 26-29; asst path, Univ Hamburg, 29-30; prof path & chmn dept, 30-71, EMER PROF PATH, SCH MED & SCH DENT MED, TUFTS UNIV, 71- Concurrent Pos: Instr, Harvard Med Sch, 28-29; asst, Univ Berlin, 31-32; pathologist-in-chief, Tufts-New Eng Med Ctr Hosps; consult pathologist, Mt Auburn Hosp, New Eng Med Ctr Hosps, Carney Hosp, Lynn Hosp, Leonard Morse Hosp, Malden Hosp & Cape Cod Hosp; consult, Armed Forces Inst Path, Boston Vet Hosp & USPHS Hosp, Boston; vis prof, Med Sch, Univ Mass, 71- Mem: Royal Col Physicians; Am Asn Path & Bact; Am Med Asn; Int Acad Path; Ger Path Soc. Res: Pathology of the heart, lungs, liver, kidneys and blood vessels. Mailing Add: 19 Hubbard Park Cambridge MA 02138

MACMAHON, JAMES A, b Dayton, Ohio, Apr 7, 39; m 63. ECOLOGY, VERTEBRATE ZOOLOGY. Educ: Mich State Univ, BS, 60; Univ Notre Dame, PhD(biol), 64. Prof Exp: Asst dir biol, Dayton Mus Natural Hist, 63-64; from asst prof to assoc prof, Univ Dayton, 64-71; assoc prof zool, 71-74, PROF BIOL, UTAH STATE UNIV, 74-, ASST DIR US INT BIOL PROG-DESERT BIOME, 71- Mem: Soc Study Evolution; Soc Syst Zool; Ecol Soc Am; Am Soc Zoologists; Am Soc Ichthyologists & Herpetologists. Res: Water relations of amphibians; energy exchange in plant and animal populations; evolution and ecology of reptiles and amphibians; biology of arachnids. Mailing Add: Dept of Biol Utah State Univ Logan UT 84321

MACMANUS, JOHN PATRICK, b Dublin, Ireland, July 15, 43; Can citizen. BIOCHEMISTRY. Educ: Nat Univ Ireland, BSc, 65; Univ Lancaster, Eng, PhD(biochem), 68. Prof Exp: Fel biochem, 68-69, RES OFFICER BIOCHEM, NAT RES COUN CAN, 69- Mem: Biochem Soc. Res: Mechanism of control of the initiation of DNA synthesis, and the role of hormones, cyclic-nucleotides and ions therein. Mailing Add: Div of Biol Sci Nat Res Coun of Can Ottawa ON Can

MACMASTERS, WILLIAM JOSEPH, b Girard, Ohio, Nov 15, 16; m 45. PLANT PHYSIOLOGY. Educ: Ohio State Univ, BS, 49, MS, 50, PhD(bot), 53. Prof Exp: Asst plant physiol, Ohio State Univ, 49-50, head bot asst, 50-51, asst plant physiol, 51-53; asst prof biol & chem, Heidelberg Col, 53-56; ASSOC PROF BIOL, MIAMI UNIV, 56- Mem: Bot Soc Am; Am Soc Plant Physiol. Res: Photoperiodism; water relations. Mailing Add: Dept of Zool Miami Univ Oxford OH 45056

MACMILLAN, BRUCE GREGG, b Cincinnati, Ohio, Apr 19, 20; m 45; c 2. SURGERY. Educ: Col Wooster, BA, 42; Univ Cincinnati, MD, 45; Am Bd Surg, dipl, 54. Prof Exp: From instr to assoc prof, 52-64, SHRINE PROF SURG, COL MED, UNIV CINCINNATI, 64-; CHIEF STAFF, SHRINERS BURNS INST, 67- Concurrent Pos: Mem active staff, C R Holmes & Children's Hosps, 52-; attend surgeon, clinician & dir dept surg photog, Cincinnati Gen Hosp, 52-; consult surg res unit, Brooke Army Med Ctr, Ft Sam Houston, Tex. Mem: AMA; Am Col Surgeons; Am Asn Surg of Trauma; Int Soc Burn Injuries; Am Surg Asn. Res: Thermal trauma; surgical infections; audio-visual education. Mailing Add: 202 Goodman St Cincinnati OH 45219

MACMILLAN, DONALD BASHFORD, mathematics, see 12th edition

MACMILLAN, DUNCAN ROBERT, b Frederickton, NB, Jan 3, 31; m 60; c 4. PEDIATRIC ENDOCRINOLOGY. Educ: McGill Univ, BSc, 52, MD, CM, 55; FRCP(C). Prof Exp: Res fel pediat endocrinol, Nuffield Inst Child Health, Eng, 60-62; clin & res fel, Univ Iowa, 62-64, instr pediat, 64-65; from asst prof to assoc prof pediat, 65-71, asst dean clin affairs & assoc in psychiat, 70-72, PROF PEDIAT, SCH MED, UNIV LOUISVILLE, 71- Concurrent Pos: Nat Inst Child Health & Human Develop res fel, 67-69. Mem: Am Diabetes Asn; Royal Soc Med; fel Am Acad Pediat. Res: Influence of growth hormone and chorionic somatomammotropin on prenatal and postnatal growth of twins and single births; medical education. Mailing Add: Dept of Pediat Sch of Med Univ Louisville Health Sci Ctr Louisville KY 40202

MACMILLAN, FRANCIS SOPHUS KILMER, biochemistry, see 12th edition

MACMILLAN, JAMES DAVID, microbiology, biochemistry, see 12th edition

MACMILLAN, JAMES G, organic chemistry, see 12th edition

MACMILLAN, WILLIAM HOOPER, b Boston, Mass, Oct 21, 23; m 48; c 3. PHARMACOLOGY. Educ: McGill Univ, BA, 48; Yale Univ, PhD, 54. Prof Exp: From instr to assoc prof, 54-63, chmn dept, 62-63, PROF PHARMACOL, COL MED, UNIV VT, 64-, DEAN GRAD COL, 63-69 & 71- Concurrent Pos: USPHS fel, Oxford Univ, 58-59; Ford Found sci adv, Haile Selassie I Univ, 69-71; educ consult, World Bank, 72-74; consult, New Eng Asn Schs & Cols, 72- Mem: Sigma Xi; AAAS; Am Soc Pharmacol & Exp Therapeut; NY Acad Sci. Res: Autonomic pharmacology; mechanism of drug action; graduate studies in biomedical sciences. Mailing Add: Grad Col Univ of Vt Burlington VT 05401

MACMILLEN, RICHARD EDWARD, b Upland, Calif, Apr 19, 32; m 53; c 2. PHYSIOLOGICAL ECOLOGY, VERTEBRATE BIOLOGY. Educ: Pomona Col, BA, 54; Univ Mich, MS, 56; Univ Calif, Los Angeles, PhD(zool), 61. Prof Exp: Res zoologist, Univ Calif, Los Angeles, 59-60; from instr to assoc prof zool, Pomona Col, 60-68; ASSOC PROF BIOL, UNIV CALIF, IRVINE, 68- Concurrent Pos: NSF res grants, 61-72; Fulbright advan res grant zool & comp physiol, Monash Univ, Australia, 66-67; partic subprog Hawaiian terrestrial biol, Int Biol Prog, Nat Res Coun, 71-74; mem, Int Hibernation Info Exchange. Mem: Fel AAAS; Am Soc Zool; Ecol Soc Am; Am Soc Mammal; Australian Soc Mammal. Res: Physiological ecology of thermo-regulation and water economy in terrestrial vertebrates. Mailing Add: Dept of Pop & Environ Biol Univ of Calif Irvine CA 92664

MACMORINE, HILDA MILDRED GRACE, b Kars, Ont, June 26, 16. ORGANIC CHEMISTRY. Educ: Univ Toronto, BA, 38, MA, 43, PhD(chem), 55. Prof Exp: From res asst to res assoc, Connaught Med Res Labs, Univ Toronto, 40-61, res mem, Univ, 61-71, res mem Connaught Labs, Ltd, 61-71, asst dir, 71-73, DIV DIR HUMAN BIOL, CONNAUGHT LABS, LTD, 73-, ASST PROF MICROBIOL, SCH HYG, UNIV TORONTO, 71- Concurrent Pos: Can Pub Health grant improved methods viral vaccine prod, 71- Mem: Can Pub Health Asn; fel Chem Inst Can; Am Soc Microbiol; Tissue Cult Asn; NY Acad Sci. Res: Animal cell nutrition and improved methods for viral vaccines. Mailing Add: Connaught Labs Ltd 1755 Steeles Ave West Willowdale ON Can

MACMULLAN, RALPH AUSTIN, zoology, deceased

MACMULLEN, CLINTON WILLIAM, b St Cloud, Minn, May 16, 09; m 33; c 1. ORGANIC CHEMISTRY. Educ: Univ Minn, BChE, 30, PhD(org chem), 35. Prof Exp: Group leader, Rohm & Haas Co, Pa, 35-43; tech dir, Cowles Chem Co, 43-52; sect chief, Olin Industs, Inc, 52-53, mgr res, Org Chem Div, Olin Mathieson Chem Corp, 53-60, dir applns res, Chem Div, 60-62, tech adv to corp vpres res & develop, 62-65, tech dir chem div, 65-69, tech dir new venture identification, Olin Corp, 69-72; CONSULT, 72- Mem: Am Chem Soc; Am Oil Chem Soc; Sigma Xi; Soc Plastics Eng; fel Am Inst Chemists. Res: Petrochemicals; polymers; lubricants; plastics; sodium metasilicate; organosilicon compounds; surface active agents; bactericides; fungicides. Mailing Add: 73 Jesswig Dr Hamden CT 06517

MACNAB, ROBERT MARSHALL, b Barnsley, Eng, Feb 3, 40. BIOPHYSICS. Educ: Univ St Andrews, Scotland, BSc, 62; Univ Calif, Berkeley, PhD(chem), 69. Prof Exp: Technologist petrochem, Brit Petrol Co, 62-65; res assoc biochem, Univ Calif, Berkeley, 70-73; ASST PROF BIOPHYS, YALE UNIV, 73- Res: Bacterial chemotaxis as a model for chemo-mechanical and sensory transduction. Mailing Add: Molecular Biophys & Biochem Yale Univ Box 1937 Yale Sta New Haven CT 06520

MACNAIR, RICHARD NELSON, b Newton, Mass, Oct 19, 29; m 60; c 1. ORGANIC CHEMISTRY. Educ: Middlebury Col, AB, 52; Univ Del, PhD(org chem), 60. Prof Exp: Chemist, Arthur D Little, Inc, 60-63; sr chemist, Tracerlab Div, Lab for Electronics, Inc, 64; res chemist, 64-69, SUPVRY RES CHEMIST, US ARMY NATICK LABS, 69- Concurrent Pos: Mem, Int Oceanog Found; comt chmn 52nd nat meeting printing & advertising, Am Inst Chemists. Mem: Am Chem Soc; Sigma Xi; Am Inst Chemists. Res: Heterocyclic chemistry; graft polymerization; chemical protective clothing; activated carbon sorption-desorption; carbon fibers and fabrics. Mailing Add: ClothingEquip & Mat Eng Lab US Army Natick Develop Ctr Natick MA 01760

MACNAMARA, EDLEN EVERETT, pedology, environmental sciences, deceased

MACNAMARA, THOMAS E, b Airdrie, Scotland, May 23, 29; US citizen; m; c 5. ANESTHESIOLOGY. Educ: Glasgow Univ, MB, ChB, 52. Prof Exp: Instr anesthesia, Med Ctr, Georgetown Univ, 57-60; asst, Mass Gen Hosp & Harvard Med Sch, 60-62; PROF ANESTHESIA & CHMN DEPT, MED CTR, GEORGETOWN UNIV, 63- Concurrent Pos: Lectr, Bethesda Naval Med Ctr, 63-75, consult, 63-; vpres fac senate, Georgetown Univ, 67-71, pres, 73-75; chief anesthesia dept, NIH Clin Ctr, Bethesda, Md, 75-; consult, DC Gen & Vet Admin Hosps & Charles Town Gen Hosp, Ranson, WVa. Mem: Am Soc Anesthesiol; Brit Med Asn; Royal Soc Med; Int Anesthesia Res

Soc. Mailing Add: Dept of Anesthesia Georgetown Univ Med Ctr Washington DC 20007

MACNAMEE, JAMES K, b Philadelphia, Pa, May 15, 16; m 50. VETERINARY PATHOLOGY. Educ: Auburn Univ, BS, 36, DVM, 37, MS, 42. Prof Exp: Prof & chmn dept path & parasitol, Col Vet Med, Auburn Univ, 42-47; prof path, Col Vet Med, Univ Ga, 47-49; comp pathologist, Med Res Labs, Edgewood Arsenal, Md, 49-55 & Biol Res Labs, Ft Detrick, Md, 55-58; dir diag lab, DOD, Schofield Barracks, Hawaii, 58-62; dir vet sect, Sixth Army Med Lab, Ft Baker, Calif, 62-65; HEALTH SCIENTIST ADMINR, NIH, 65- Mem: Am Soc Lab Med; Am Asn Zool Parks & Aquariums; Avicult Soc Am; Am Asn Vet Toxicologists; AAAS. Res: Psittacine birds, particularly respiratory diseases, reproduction and nutritional problems. Mailing Add: 5415 Wilson Lane Bethesda MD 20014

MACNAUGHTON, EARL BRUCE, b Maple, Ont, Aug 29, 19; m 43; c 2. PHYSICS. Educ: Univ Toronto, BA, 41, MA, 46, PhD(physics), 48. Prof Exp: Instr physics, Univ Toronto, 41-44; prof, Ont Agr Col, 48-56, head dept, 56-65; prof & head dept, Wellington Col, 65-70, assoc dean sci, 66-70, DEAN COL PHYS SCI, UNIV GUELPH, 70- Mem: Am Phys Soc; Am Asn Physics Teachers; Can Asn Physicists. Res: Molecular spectroscopy; electronics and instrumentation. Mailing Add: Col of Phys Sci Univ of Guelph Guelph ON Can

MACNAUGHTON, WILLIAM NORMAN, b Vegreville, Alta, Apr 8, 18; m 44; c 4. ANIMAL BREEDING, GENETICS. Educ: Univ Alta, BSc, 41, MSc, 48; Iowa State Col, PhD(animal breeding), 56. Prof Exp: Asst animal husb, Exp Sta, Lethbridge, Alta, 40-46, animal husbandman, 46-47, geneticist, 47-51, animal husbandman, Exp Farm, Brandon, Man, 51-60, dir res admin & mgt, Res Sta, Melfort, Sask, 60-66, DIR RES ADMIN & MGT, RES STA, CAN DEPT AGR, BRANDON, 66- Mem: Agr Inst Can; Can Soc Animal Sci. Res: Breeding, physiology, management and production in beef cattle, swine, poultry, barley, forage crops, corn, soil-plant relationships and weed control. Mailing Add: Res Sta Can Dept of Agr Brandon MB Can

MACNEAL, RICHARD HENRI, applied mechanics, see 12th edition

MACNEIL, JOSEPH H, b Sydney, NS, Apr 13, 31; m 57; c 5. FOOD SCIENCE. Educ: McGill Univ, BSA, 55; Mich State Univ, MS, 58, PhD(food sci), 61. Prof Exp: Asst agr rep, NS Dept Agr, 52-53; field serv rep, Maritime Coop Serv, 55-57; res asst poultry mkt, Mich State Univ, 57-61; asst prof poultry sci, Univ Conn, 61-62; sr res assoc food res, Lever Bros Co, NJ, 62-64; asst prof poultry sci, 64-71, assoc prof food sci, 71-73, PROF FOOD SCI, PA STATE UNIV, UNIVERSITY PARK, 73- Concurrent Pos: Fulbright scholar, 74-75. Mem: Inst Food Technol; Poultry Sci Asn. Res: Processing technology of poultry meat, quality maintenance, sensory evaluation techniques, flavor changes and nutritional status of food. Mailing Add: Dept of Food Sci Pa State Univ University Park PA 16802

MACNEILL, ARTHUR EDSON, b Waltham, Mass, July 14, 12; m 41; c 2. PHYSIOLOGY, BIOMEDICAL ENGINEERING. Educ: Harvard Univ, AB, 33, MD, 37. Prof Exp: Asst physician, Hitchcock Clin, Hanover, NH, 38-41; instr anat & secy, Dartmouth Med Sch, 44-45, res assoc physiol sci, 46-50; dir labs, Chronic Dis Res Inst, Sch Med, State Univ NY Buffalo, 50-56, lectr med, 51-57, from asst res prof to assoc res prof surg, 57-64; DIR DIALYSIS RES INST, 64- Concurrent Pos: Physician, Univ Fla, 46-47; asst physiol, Mary Hitchcock Mem Hosp, Hanover, NH, 46-50; consult surv Ger med & biol res, US Dept State, 50; mem res staff, Buffalo Gen Hosp, 51-57, head dept therapeut eng, 58-64, res consult, Dept Surg, 58-64; chief clin invest consult, Poliomyelitis Respirator Ctr, Buffalo, 54-55; Buswell res fel, State Univ NY Buffalo, 57-63; res adv, Children's Hosp, 57-66; consult hosps, indust & sci corp, 64-; consult med assoc, Crotched Mountain Rehab Ctr, NH. Honors & Awards: Cert of Merit, AMA, 49; Seaman Award, Asn Mil Surgeons US, 49; First Award Sci Res, Med Soc NY, 53, Redway Prize, 63. Mem: AAAS; AMA; Am Heart Asn; Sigma Xi; Am Soc Nephrol. Res: Applied physiology; blood dialysis, oxygenation, pumping and rheometry; therapeutic computation by diffusion methods; bedside care of patients; devices for treatment of blood stream and transfusion blood. Mailing Add: Dialysis Res Inst Box 334 Sunapee NH 03782

MACNEILL, BLAIR HILTON, mycology, plant pathology, see 12th edition

MACNEILL, IAN B, b Regina, Sask, Dec 12, 31; m 52; c 3. MATHEMATICS, STATISTICS. Educ: Univ Sask, BA, 62; Queen's Univ, Ont, MA, 65; Stanford Univ, PhD(statist), 69. Prof Exp: Asst prof math, Univ Toronto, 66-71; ASSOC PROF APPL MATH, UNIV WESTERN ONT, 71- Res: Time series analysis; statistical ecology. Mailing Add: Dept of Appl Math Univ of Western Ont London ON Can

MACNEILL, RUPERT HEATH, b PEI, Can, July 13, 14; m 39; c 1. PHYSICS, GEOLOGY. Educ: Acadia Univ, BSc, 50, MSc, 51. Prof Exp: Lectr geol, 51-52, from asst prof to assoc prof, 52-68, EDWIN DAVID KING ASSOC PROF GEOL & DIR EXTEN, SUMMER SCH & STUDENT ASSISTANCE, 68- Concurrent Pos: Geologist, NS Res Found, summers 49-75, consult, 75-; mem meteorite comt, Nat Res Coun, 67-; chmn, Curric Comt Geol, NS Dept of Educ, 70- Honors & Awards: Hunt Prize, 49; Mining Soc NS Prize, 51. Mem: Nat Asn Geol Teachers; Geol Asn Can; Mineral Asn Can; Geol Soc Am; Soc Explor Geophys. Res: Pleistocene of Nova Scotia; supervision of student research. Mailing Add: PO Box 340 Wolfville NS Can

MACNEILLE, HOLBROOK MANN, mathematics, deceased

MACNERNEY, JOHN SHERIDAN, b New York, NY, Jan 10, 23; m 45. MATHEMATICAL ANALYSIS. Educ: Univ Tex, BA, 48, PhD(math), 51. Prof Exp: Instr pure math, Univ Tex, 48-51 & Northwestern Univ, 51-52; from asst prof to prof, Univ NC, 52-67; PROF PURE MATH, UNIV HOUSTON, 67- Mem: AAAS; Am Math Soc; Math Asn Am; Sigma Xi. Res: Analytic functions; continued fractions; Hilbert spaces; integral equations. Mailing Add: Dept of Math Univ of Houston Houston TX 77004

MACNICHOL, EDWARD FORD, JR, b Toledo, Ohio, Oct 24, 18; m 40; c 2. BIOPHYSICS. Educ: Princeton Univ, AB, 41; Johns Hopkins Univ, PhD, 52. Prof Exp: Mem staff, Radiation Lab, Mass Inst Technol, 41-43, assoc group leader, 43-44, group leader, 45-46; asst, Johnson Found, Univ Pa, 46-48; asst biophys, Johns Hopkins Univ, 49-51, from instr to prof, 52-68; dir, Nat Inst Neurol Dis & Stroke, 68-73; PRIN SCIENTIST, LAB SENSORY PHYSIOL & ASST DIR, MARINE BIOL LAB, WOODS HOLE, 73- Concurrent Pos: Vis prof, Venezuelan Inst Sci Res, 57; guest scientist, US Naval Res Inst, 57-60; actg dir, Nat Eye Inst, 68-69; mem bd dirs, Deafness Res Found, 73- Mem: Am Phys Soc; Biophys Soc; Am Physiol Soc; Inst Elec & Electronics Engrs; NY Acad Sci. Res: Neurophysiology of retina and other sensory systems; instrumentation for biological research. Mailing Add: Lab of Sensory Physiol Marine Biol Lab Woods Hole MA 02543

MACNINTCH, JOHN EDWIN, b Moncton, NB, Nov 7, 35; m 58; c 2. BIOCHEMISTRY. Educ: McGill Univ, BScAgr, 58; Purdue Univ, MS, 63, PhD(biochem), 65. Prof Exp: Nat Heart Asn fel cardiovasc training prog, Bowman Gray Sch Med, 65-66; sr res biochemist, Biochem Dept, 66-72, sr res biochemist, Pharmacol Dept, Fibrinolytic Res & Develop, 72-74, ASST DIR RES PLANNING, BRISTOL LABS INC, 74- Res: Atherosclerosis and general cardiovascular disease biochemistry; lipid biochemistry; biochemistry of drug addiction; administration, library and information services; drug evaluations; planning. Mailing Add: Bristol Labs Inc PO Box 657 Syracuse NY 13201

MACOMBER, HILLIARD KENT, physics, see 12th edition

MACOMBER, JAMES DALE, b Albany, Calif, June 23, 39; m 63; c 3. CHEMICAL PHYSICS. Educ: Univ Calif, Berkeley, BS, 60; Mass Inst Technol, PhD(phys chem), 65. Prof Exp: Physicist, Pitman-Dunn Res Labs, Frankford Arsenal, Pa, 65-67; asst prof, 67-70, chmn phys chem div, Chem Dept, 74-76, ASSOC PROF CHEM, LA STATE UNIV, BATON ROUGE, 70- Concurrent Pos: NIH res grant, 67-70. Mem: Optical Soc Am; Inst Elec & Electronics Engrs. Res: Interaction of laser radiation with matter; nonlinear optical phenomena; spectroscopy at high electromagnetic field strengths; optical analogs of transient nuclear magnetic resonance effects. Mailing Add: Dept of Chem La State Univ Baton Rouge LA 70803

MACOMBER, RICHARD WILTZ, b Chicago, Ill, June 6, 32; m 57. PALEONTOLOGY, STRATIGRAPHY. Educ: Northwestern Univ, BS, 54, MS, 59; Harvard Univ, AM, 63; Univ Iowa, PhD(paleont), 68. Prof Exp: Geologist, Bear Creek Mining Co, 57-58; geologist, US Geol Surv, 62-63; cur geol, Northwestern Univ, 65-66; from instr to asst prof, 66-71, ASSOC PROF EARTH SCI, LONG ISLAND UNIV, 71- Mem: Geol Soc Am; Int Palaeont Union; Brit Palaeont Asn; Paleont Soc. Res: Ordovician brachiopod paleontology; Ordovician stratigraphy of North America and Europe. Mailing Add: Dept of Physics Long Island Univ Brooklyn NY 11201

MACOMBER, ROGER STARK, chemistry, see 12th edition

MACON, NATHANIEL, b Durham, NC, Nov 15, 26; m 53; c 4. NUMERICAL ANALYSIS, SYSTEMS SCIENCE. Educ: Univ NC, BA, 46, MA, 48, PhD(math), 50. Prof Exp: Instr math, Univ NC, 48-50; asst prof, Auburn Univ, 51-56; math appln specialist, Gen Elec Co, Ohio, 56-57; assoc prof math, Auburn Univ, 57-59, prof & dir, Comput Lab, 59-65; mem staff, Inst Defense Anal, Arlington, Va, 65-67; adj prof, 66-67, PROF MATH, AM UNIV, 67- Concurrent Pos: Fulbright student, Univ Amsterdam, 50-51; consult, Air Proving Ground Ctr, Eglin AFB, Fla, 58-62 & NASA, 60-65; chief comput sect, SHAPE Tech Ctr, Neth, 62-64; consult, Fed Preparedness Agency, 68- & Leasco Systs & Res Corp, 68-72; tech prog chmn, Spring Joint Comput Conf, Am Fedn Info Processing Socs, 72. Mem: Am Math Soc; Math Asn Am; Asn Comput Mach; Int Coun Comput Commun (vpres, 72-); Soc Appl Learning Technol. Res: Numerical analysis; interpolation; approximation; continued fractions; computer conferencing and applications. Mailing Add: 6104 Namakagan Rd Bethesda Md 20016

MACPEEK, DONALD LESTER, b Andover, NJ, Apr 4, 28; m 50; c 3. INDUSTRIAL ORGANIC CHEMISTRY. Educ: Rensselaer Polytech Inst, BS, 49, MS, 51, PhD(chem), 52. Prof Exp: Asst, Rensselaer Polytech Inst, 49-52; res chemist, Union Carbide Chem Co, 52-67, tech mgr oxidation prod opers group, Chem & Plastics Develop Div, 67-69, technol mgr aldehydes, alcohols & plasticizer intermediates opers group, 69-71 & acrolein, acrylic acid & acrylate esters opers group, 72-74, ASSOC DIR RES & DEVELOP DEPT, CHEM & PLASTICS DIV, UNION CARBIDE CORP, 74- Mem: Am Chem Soc. Res: Oxidation of organic compounds, especially hydrocarbons and aldehydes; hydroformylation; specialty organic chemicals; process development; management of research and development. Mailing Add: 1518 Village Dr South Charleston WV 25309

MACPHAIL, DONALD DOUGALD, b Dayton, Ohio, Mar 13, 22; m 50. GEOGRAPHY. Educ: Mich State Univ, BS, 47; Univ Mich, MA, 49, PhD(geog), 53. Prof Exp: Dist chief rural land classification prog, Div Agr Econ, Univ PR, 50-51; asst prof geog, Western Wash Col, 52-56; from asst prof to assoc prof & chmn dept, 56-60, PROF GEOG, UNIV COLO, BOULDER, 60- Concurrent Pos: Fulbright lectr, Inst Geog, Univ Chile, 59-60; fac res fel, Univ Colo, 60. Mem: Asn Am Geogr; Am Geog Soc. Res: Rural land classification and planning; regional geography of Latin America; cartography and field mapping. Mailing Add: Dept of Geog Univ of Colo Boulder CO 80304

MACPHAIL, MORAY ST JOHN, b Kingston, Ont, May 27, 12; m 39; c 1. MATHEMATICS. Educ: Queen's Univ Ont, BA, 33; McGill Univ, MA, 34; Oxford Univ, DPhil, 36. Prof Exp: From instr to prof math, Acadia Univ, 37-47; vis lectr, Queen's Univ Ont, 47-48; assoc prof, 48-53, dir sch grad studies, 60-63, dean fac grad studies, 63-69, PROF MATH, CARLETON UNIV, 53- Concurrent Pos: Instr, Princeton Univ, 41-42. Mem: Am Math Soc; Math Asn Am; fel Royal Soc Can; Can Math Cong. Res: Analysis; theory of series. Mailing Add: Dept of Math Carleton Univ Ottawa ON Can

MACPHEE, ALBERT WILLIAM, b Gore, NS, Sept 13, 17; m 43; c 6. ENTOMOLOGY. Educ: McGill Univ, BSc, 40, MSc, 47, PhD(entom), 60. Prof Exp: Res entomologist, 45-62, HEAD ENTOM SECT, RES STA KENTVILLE CAN DEPT AGR, 62- Concurrent Pos: Agr Inst Can scholar, Univ Calif, 47-48. Mem: Prof Inst Pub Serv Can; Agr Inst Can; Entom Soc Can. Res: Ecology; population dynamics; cold-hardiness of arthropods. Mailing Add: Res Sta Can Dept Agr Kentville NS Can

MACPHEE, CRAIG, b San Francisco, Calif, Nov 3, 18; m 42; c 3. ZOOLOGY. Educ: Univ BC, BA, 47, MA, 49; Univ Wash, PhD(zool), 54. Prof Exp: Asst prof biol, Eastern Wash Col Educ, 54-57; from asst prof to assoc prof, 57-66, PROF FISHERY MGT, UNIV IDAHO, 66- Mem: Am Soc Limnol & Oceanog; Am Soc Ichthyol & Herpet; Am Fisheries Soc; Am Inst Fishery Res Biol. Res: Limnology; ecology. Mailing Add: Col of Forestry Univ of Idaho Moscow ID 83843

MACPHEE, HUGH JOHN, physics, see 12th edition

MACPHEE, KENNETH ERSKINE, b Toronto, Ont, Sept 17, 26; m 49; c 3. ORGANIC CHEMISTRY, POLYMER CHEMISTRY. Educ: Acadia Univ, BSc, 47; Univ Oxford, BSc, 51, DPhil(org chem), 53. Prof Exp: Res scientist polymer chem, Res Lab, Dom Rubber Co, 53-64, group leader, 64-74; MEM STAFF, UNIROYAL LTD RES LABS, 74- Mem: Chem Inst Can. Res: Organic synthesis; pharmacological action of chemicals; elastomers and plastics. Mailing Add: Uniroyal Ltd Res Labs Guelph ON Can

MACPHERSON, A, b Campbeltown, Scotland, Sept 28, 22. GEOGRAPHY. Educ: Univ Edinburgh, MA, 50. Prof Exp: Asst lectr geog, Univ Edinburgh, 51-54, from lectr to sr lectr, 56-65; lectr, Aberdeen Univ, 54-56; head dept, 65-71, prof urban climatol, 65-74, PROF GEOG, SIMON FRASER UNIV, 74- Concurrent Pos: Vis prof, Univ BC, 58-59. Mem: Royal Scottish Geog Soc; fel Royal Meteorol Soc; Am

Geog Soc; Asn Am Geog; Can Asn Geog. Res: Land use problems in upland areas of United Kingdom; air pollution in relation to weather types. Mailing Add: Dept of Geog Simon Fraser Univ Burnaby BC Can

MACPHERSON, ANDREW HALL, b London, Eng, June 2, 32; Can citizen; m 57; c 3. ECOLOGY, ZOOGEOGRAPHY. Educ: Carleton Univ, BS, 54; McGill Univ, MSc, 57, PhD, 67. Prof Exp: Asst cur zool, Nat Mus Can, 57-58; res wildlife biologist, Can Wildlife Serv, 58-64, regional supvr wildlife biol, 64-67; sci adv, Sci Secretariat, Privy Coun Off, 67-68 & Sci Coun Can, 68-69; res supvr mammal, 69-70, dir western region, 70-74, DIR-GEN ENVIRON MGT SERV, WESTERN & NORTHERN REGION ENVIRON, CAN WILDLIFE SERV, 74- Concurrent Pos: Mem, Tech Comt for Caribou Preservation, 64-69; chmn, Polar Bear Group and mem, Survival Serv Comn, Int Union Conserv of Nature & Natural Resources, 70-72. Honors & Awards: Centennial Medal, 67. Mem: Soc Syst Zool; Am Soc Mammal; fel Arctic Inst NAm (gov, 72-76). Res: Taxonomy of Laridae; ecology and population dynamics of Alopex; zoogeography of Arctic mammals; ecology and population processes of Rangifer. Mailing Add: Can Wildlife Serv 818 10025 Jasper Ave Edmonton AB Can

MACPHERSON, CATHERINE FRANCES CONWAY, b Lowell, Mass, June 11, 14; m 39; c 2. IMMUNOCHEMISTRY, PSYCHIATRY. Educ: Mt Allison Univ, NB, Can, BSc, 35; Dalhousie Univ, MSc, 37; Columbia Univ, PhD, 45. Prof Exp: Demonstr, Dalhousie Univ, 36-38; res assoc, McGill Univ, 46-48, res assoc, Univ Clin, Montreal Gen Hosp, 58-60; res assoc, Montreal Neurol Inst, 60-63; assoc prof psychiat, McGill Univ, 63-72; ASSOC PROF PSYCHIAT, UNIV WESTERN ONT, 72- Concurrent Pos: Fel, McGill Univ, 43-46. Mem: Am Soc Biol Chem; Am Asn Immunol; Am Soc Neurochem. Res: Protein denaturation; estimation of brain antigens by quantitative immunochemical methods; immunochemical approaches to studies of demyelinating disease and behavior. Mailing Add: Dept of Psychiat Univ of Western Ont Hosp London ON Can

MACPHERSON, COLIN ROBERTSON, b Aberdeen, Scotland, Sept 2, 25; m 49; c 4. PATHOLOGY. Educ: Univ Cape Town, MB, ChB, 46, MMed & MD, 54. Prof Exp: Asst lectr path, Univ Cape Town, 48-50, lectr, 50-55; asst renal physiol, Post-Grad Sch Med, Univ London, 55-56; from asst prof to prof bact & path, Ohio State Univ, 56-75, vchmn dept path, 60-72, actg chmn dept, 72-75; DIR DIV LAB MED, UNIV CINCINNATI, 75- Concurrent Pos: Consult, Vet Admin Hosp, Dayton, Ohio, 65- Res: Immuno-hematology; laboratory screening procedures. Mailing Add: Cincinnati Gen Hosp 234 Goodman St Cincinnati OH 45267

MACPHERSON, CULLEN H, b San Mateo, Calif, Dec 6, 27; m 51; c 3. BIOPHYSICS. Educ: San Jose State Col, AB, 49; Stanford Univ, MA, 51. Prof Exp: Mgr reproducing components div, Electro-Voice Inc, 54-56; biophysicist, Tektronix, Inc, 56-61; PRES & CHMN BD, ARGONAUT ASSOCS, INC, 59- Concurrent Pos: Scientist, Dept Neurophysiol, Ore Regional Primate Res Ctr, 64- Mem: AAAS; Am Asn Physics Teachers; Acoust Soc Am; Audio Eng Soc; Inst Elec & Electronics Eng. Res: Limited energy measurements in biological systems; constant current neuronal stimulation; stereo vector electrocardiography. Mailing Add: Argonaut Assocs Inc 2677 NW Westover Rd Portland OR 97210

MACPHERSON, HERBERT GRENFELL, b Victorville, Calif, Nov 2, 11; m 37; c 2. NUCLEAR SCIENCE, NUCLEAR ENGINEERING. Educ: Univ Calif, AB, 32, PhD(physics), 37. Prof Exp: Jr meteorologist, USDA, 36-37; res physicist, Nat Carbon Co, 37-50, asst dir res, 50-56; with Oak Ridge Nat Lab, 56-60, assoc dir reactor prog, 60-63, asst lab dir, 63-64, dep dir, 64-70; PROF NUCLEAR ENG, UNIV Tenn, 70- Concurrent Pos: Consult, Oak Ridge Nat Lab, 70- & US AEC, 72-74; actg dir, Inst Energy Anal, 74-75, consult, 75- Mem: Am Nuclear Soc; Am Phys Soc; Am Soc Metals. Res: Fundamentals of the carbon arc; high temperature properties of carbon and graphite; heavy particles in cosmic radiation; nuclear reactor technology; safety of nuclear reactors; energy policy. Mailing Add: 102 Orchard Circle Oak Ridge TN 37830

MACPHERSON, L W, b Glasgow, Scotland, June 1, 20; Can citizen; m 45; c 4. MICROBIOLOGY, PUBLIC HEALTH. Educ: Univ Edinburgh, dipl vet state med, 49, PhD(virol), 55. Prof Exp: Vet officer, Ministry Agr, 49-52; fel microbiol, Can Dept Nat Defence, 52-55; res officer virol & head virus sect, Animal Dis Res Inst, Can Dept Agr, 55-57; actg head dept microbiol, 57-68, PROF MICROBIOL & VET PUB HEALTH, SCH HYG, UNIV TORONTO, 57- Concurrent Pos: Consult, Res Inst, Hosp Sick Children, Toronto, 59-; mem adv comt bact warfare, Can Dept Nat Defence, 62-64. Mem: Fel Royal Soc Health; Royal Col Vet Surg. Res: Zoonotic diseases, particularly myxoviruses and enteroviruses. Mailing Add: Dept of Microbiol Sch of Hyg Univ of Toronto Toronto ON Can

MACPHERSON, LLOYD BERTRAM, b Annapolis Royal, NS, July 6, 13; m 45; c 3. BIOCHEMISTRY. Educ: Acadia Univ, BSc, 34; Univ Toronto, PhD(path chem), 49. Hon Degrees: DSc, Acadia Univ, 73. Prof Exp: Asst, Banting Inst, Univ Toronto, 46-49, asst prof, 49-52; from asst prof to assoc prof biochem, 52-63, from asst dean fac med to assoc dean fac med, 58-71, PROF BIOCHEM, DALHOUSIE UNIV, 63-, DEAN FAC MED, 71- Honors & Awards: Mem, Order of Brit Empire, 44. Res: Chemistry of phospholipids and inositol; lipid metabolism. Mailing Add: Off of Dean of Fac Med Dalhousie Univ Halifax NS Can

MACPHERSON, RODERICK IAN, b St Thomas, Ont, Feb 22, 35; m 57; c 5. RADIOLOGY, PEDIATRICS. Educ: Univ Man, BSc & MD, 58; Royal Col Physicians Can, cert radiol, 63; FRCP(C), 64. Prof Exp: Asst radiologist, Montreal Children's Hosp, Que, 64-65; asst radiologist, Royal Victoria Hosp, Montreal, 65; asst radiologist, Shaughnessy Hosp, Vancouver, BC, 65-67; assoc radiologist, Children's Hosp Winnipeg, 67-69; asst prof, 67-69, ASSOC PROF RADIOL, UNIV MAN, 69-, ASSOC PROF PEDIAT, 71-; DIR DEPT RADIOL, CHILDREN'S HOSP WINNIPEG, 69- Concurrent Pos: Instr, McGill Univ, 65 & Univ BC, 65-67. Mem: Can Asn Radiol; Am Roentgen Ray Soc; Soc Pediat Radiol. Res: Diagnostic radiology; clinical, pathologic and radiologic aspects of pediatric chest diseases, renal diseases and skeletal diseases. Mailing Add: Dept of Radiol Children's Hosp of Winnipeg Winnipeg MB Can

MACPHERSON, WALTER EVERETT, b Wadsworth, Nev, Dec 19, 99; m 28; c 1. INTERNAL MEDICINE. Educ: Pac Union Col, AB, 22; Loma Linda Univ, MD, 24. Prof Exp: From instr to prof physiol, 26-36, from assoc prof to prof internal med, 36-67, dean Loma Linda div, 35-36, Los Angeles div, 36-42, pres col, 42-48 & 51-54, dean med sch, 54-62, vpres med affairs, 54-67; EMER PROF & EMER VPRES, SCH MED, LOMA LINDA UNIV, 67- Concurrent Pos: Dir educ, White Mem Med Ctr, 67-71. Mem: Am Soc Internal Med; fel AMA; fel Am Col Physicians. Res: Etiology of pneumonia; diagnosis of hepatic disease. Mailing Add: 661 Glenmore Blvd Glendale CA 91206

MACPHILLAMY, HAROLD BELDING, organic chemistry, see 12th edition

MACQUARRIE, IAN GREGOR, b Hampton, PEI, July 6, 33; m 57; c 4. BOTANY, PLANT PHYSIOLOGY. Educ: Dalhousie Univ, BSc, 57, MSc, 58; Univ London, PhD(plant physiol), 61. Prof Exp: Asst prof biol, Dalhousie Univ, 61-65 & St Dunstan's Univ, 65-67; asst prof, 67-74, ASSOC PROF BIOL, UNIV PEI, 74- Mem: Can Biol Asn; Can Soc Wildlife & Fishery Biol. Res: Upland game; fragile habitats; sand dune systems; hedgerows. Mailing Add: Dept of Biol Univ of PEI Charlottetown PE Can

MACQUEEN, JOHN THOMAS, physical chemistry, see 12th edition

MACQUEEN, ROBERT MOFFAT, b Memphis, Tenn, Mar 28, 38; m 60; c 2. SOLAR PHYSICS. Educ: Southwestern at Memphis, BS, 60; Johns Hopkins Univ, PhD(atmospheric sci), 68. Prof Exp: Actg asst prof physics, Southwestern at Memphis, 61-63; instr astron, Goucher Col, 64-66; consult, Appl Physics Lab, Johns Hopkins Univ, 66; SR STAFF SCIENTIST, HIGH ALTITUDE OBSERV, NAT CTR ATMOSPHERIC RES, 67- Concurrent Pos: Lectr, Dept Astrogeophys, Univ Colo, Boulder, 69- Honors & Awards: Exceptional Sci Achievement Medal, NASA, 74. Mem: Am Asn Physics Teachers; Am Geophys Union; Optical Soc Am; Am Astron Soc. Res: Structure and evolution of the solar electron corona; infrared thermal emission of the interplanetary medium; temperature structure of solar photosphere and corona. Mailing Add: Nat Ctr for Atmospheric Res High Altitude Observ Box 3000 Boulder CO 80303

MACQUEEN, ROGER WEBB, b Toronto, Ont, Nov 5, 35; m 59; c 4. GEOLOGY. Educ: Univ Toronto, BA, 58, MA, 60; Princeton Univ, MA & PhD(geol), 65. Prof Exp: Geologist, V C Illing & Partners, Eng, 60-62; RES SCIENTIST, INST SEDIMENTARY & PETROL GEOL, GEOL SURV CAN, 65- Concurrent Pos: Spec lectr, Dept Geol, Univ Calgary, 67-68; vis assoc prof, Erindale Col, Univ Toronto, 71-72; chmn, Am Comn Stratig Nomenclature, 74-75. Mem: AAAS; Geol Soc Am; Soc Econ Paleontologists & Mineralogists; fel Geol Asn Can; Can Soc Petrol Geologists. Res: Regional geology; sedimentology; geochemistry of sedimentary rocks; base metals in sedimentary rocks. Mailing Add: Inst Sediment & Petrol Geol Geol Surv Can 3303 33rd St NW Calgary AB Can

MACQUILLAN, ANTHONY M, b London, Eng, Feb 3, 28; Can citizen; m 53; c 4. MICROBIAL PHYSIOLOGY. Educ: Univ BC, BSA, 56, MSc, 58; Univ Wis, PhD(bact), 62. Prof Exp: Nat Res Coun Can fel, 61-63; asst prof, 63-72, ASSOC PROF MICROBIOL, UNIV MD, COLLEGE PARK, 72- Mem: AAAS; Am Soc Microbiol. Res: Yeast DNA repair systems and Mitochondrial biogenesis; regulation of permeability systems in yeast. Mailing Add: Dept of Microbiol Univ of Md College Park MD 20742

MACQUOWN, WILLIAM CHARLES, JR, b Pittsburgh, Pa, Sept 8, 15; m 39; c 2. GEOLOGY. Educ: Univ Rochester, BS, 38, MS, 40; Cornell Univ, PhD(struct geol), 43. Prof Exp: Asst geol, Univ Rochester, 38-40 & Cornell Univ, 40-43; petrol geologist, Magnolia Petrol Co, 43-46; asst prof geol, Univ Ky, 47-48; dist geologist, Coop Ref Asn, 48-50; geologist, Denver Area, Deep Rock Oil Corp, 50-54; chief geologist, Sohio Petrol Co, Okla, 54-57, mgr, 57-60; petrol consult, 60-61; PROF GEOL, UNIV KY, 61- Mem: AAAS; fel Geol Soc Am; Am Asn Petrol Geol; Soc Econ Paleont & Mineral. Res: Stratigraphy and carbonate petrology. Mailing Add: Dept of Geol Univ of Ky Lexington KY 40506

MAC RAE, ALFRED URQUHART, b New York, NY, Apr 14, 32; m 67; c 2. PHYSICS. Educ: Syracuse Univ, BS, 54, PhD(physics), 60. Prof Exp: HEAD EXPLOR SEMICONDUCTOR TECHNOL DEPT, BELL LABS, 60- Mem: Fel Am Phys Soc; sr mem Am Vacuum Soc; sr mem Inst Elec & Electronics Engrs. Res: Low energy electron diffraction; surfaces; infrared photoconductivity; fluctuations in solids; ion implantation; silicon integrated circuit processing. Mailing Add: Bell Labs Murray Hill NJ 07974

MACRAE, DONALD, b Stornoway, Scotland, June 12, 16; m 42; c 5. NEUROLOGY. Educ: Glasgow Univ, MB & ChB, 40, MD, 52; FRCPS, 47. Prof Exp: From asst prof to assoc prof, 54-63, PROF NEUROL, UNIV CALIF, SAN FRANCISCO, 63- Mem: Am Acad Neurol; Am Neurol Asn; Royal Soc Med. Res: Epilepsy; multiple sclerosis; myasthenia gravis; psychophysiological disorders. Mailing Add: Dept of Neurol Univ of Calif Med Ctr San Francisco CA 94143

MACRAE, DONALD ALEXANDER, b Halifax, NS, Feb 19, 16; m 39; c 3. ASTRONOMY. Educ: Univ Toronto, BA, 37; Harvard Univ, PhD(astron), 43. Prof Exp: Asst, Univ Toronto, 37-38; res astronr, Univ Pa, 41-42; instr, Cornell Univ, 42-44; res physicist, Manhattan Proj, Tenn, 45-46; asst prof astron, Case Inst Technol, 46-53; assoc prof, 53-55, PROF ASTRON, UNIV TORONTO, 56-, CHMN DEPT ASTRON & DIR DAVID DUNLAP OBSERV, 65- Concurrent Pos: Mem bd trustees, Univs Space Res Asn, 69-; mem bd dirs, Can-France-Hawaii Telescope Corp, 73- Mem: Royal Soc Can; Royal Astron Soc Can; fel Royal Astron Soc; Can Astron Soc; Am Astron Soc. Res: Galactic and radio astronomy. Mailing Add: David Dunlap Observ Univ of Toronto Richmond Hill ON Can

MACRAE, EDITH KRUGELIS, b Waterbury, Conn, Jan 24, 19; m 50; c 1. ANATOMY. Educ: Bates Col, BS, 40; Columbia Univ, MA, 41, PhD(zool), 46. Prof Exp: Instr, Vassar Col, 45-47; Donner Found fel, Carlsberg Lab, Denmark, 47-48, Am Cancer Soc fel, 48-49; res assoc, Yale Univ, 49-51; instr biol, Mass Inst Technol, 51-53; asst res zoologist, Univ Calif, Berkeley, 54-56; from asst prof to prof anat, Univ Ill Med Ctr, 57-72, actg head dept, 69-70; PROF ANAT, MED SCH, UNIV NC, CHAPEL HILL, 72- Concurrent Pos: Guggenheim fel, 64-65. Mem: AAAS; Am Soc Zoologists; Am Soc Anatomists; Am Soc Cell Biol; Electron Micros Soc Am. Res: Histology; cytochemistry; fine structure. Mailing Add: Dept of Anat Univ of NC Med Sch Chapel Hill NC 27514

MACRAE, HERBERT F, b Middle River, NS, Mar 30, 26; m 55; c 4. BIOCHEMISTRY, ANIMAL SCIENCE. Educ: McGill Univ, BSc, 54, MSc, 56, PhD(biochem), 60. Prof Exp: Chemist, Can Dept Nat Health & Welfare, 60-61; asst prof biochem, MacDonald Col, McGill Univ, 61-67, assoc prof, 67-70, prof animal sci, 70-72, chmn dept, 67-72; PRIN, NS AGR COL, 72- Mem: Can Biochem Soc; Agr Inst Can; Can Soc Animal Sci. Res: Estrogens of avian species; milk proteins; organophosphorus insecticides; skeletal muscle enzymes and proteins. Mailing Add: NS Agr Col Truro NS Can

MACRAE, NEIL D, b Scotstown, Que, Oct 9, 35. GEOLOGY. Educ: Queen's Univ, Ont, BSc, 61; McMaster Univ, MSc, 63, PhD(geol), 66. Prof Exp: Res assoc petrol, Univ Minn, 65-66; fel, 66-68, asst prof, 68-72, ASSOC PROF GEOL, UNIV WESTERN ONT, 72- Res: Geochemistry and petrology of silicate-oxide-sulfide relations in mafic and ultramafic intrusions and skarns. Mailing Add: Dept of Geol Univ of Western Ont London ON Can

MACRAE, PATRICK DANIEL, b Calgary, Alta, Apr 26, 28; m 58. DENTISTRY. Educ: Univ Alta, DDS, 60, cert pediat dent, 67. Prof Exp: Assoc prof, 60-73, PROF PEDIAT DENT, UNIV ALTA, 73- Concurrent Pos: Dir dent serv, Glenrose Hosp,

Edmonton, Alta, 66- Mem: Can Dent Asn. Res: Sociology of health care. Mailing Add: Dept of Pediat Dent Univ of Alta Fac of Dent Edmonton AB Can

MACRAE, ROBERT E, b Detroit, Mich, May 1, 34; m 57; c 1. MATHEMATICS. Educ: Univ Chicago, AB, 53, SM, 56, PhD(math), 61. Prof Exp: Staff mem, Los Alamos Sci Lab, 57-59; NSF fel, 61-62; Ritt instr math, Columbia Univ, 62-65; asst prof, Univ Mich, 65-67; assoc prof, 67-72, PROF MATH, UNIV COLO, BOULDER, 72- Concurrent Pos: Consult, Inst Defense Anal, 65 & Math Rev, 65- Mem: Am Math Soc. Res: Commutative ring theory; homological algebra. Mailing Add: Dept of Math Univ of Colo Boulder CO 80302

MACRAKIS, MICHAEL S, b Crete, Greece, Nov 21, 24; US citizen; m 53; c 3. ELECTROMAGNETICS, PLASMA PHYSICS. Educ: Athens Polytech Inst, Dipl eng, 51; Mass Inst Technol, MS, 53; Harvard Univ, PhD(appl physics), 58. Prof Exp: Staff mem, Lincoln Lab, Mass Inst Technol, 58-65; res fel, Harvard Univ, 65-66; br chief phys electronics, NASA, 66-70; CONSULT ENG & APPL PHYSICS, 70- Concurrent Pos: Consult, Sylvania Elec Prod, Inc, 57-58; vis prof, Brown Univ, 67; sr res assoc, Elec Eng Dept, Mass Inst Technol, 72-75; fel, John F Kennedy Sch Govt, Harvard Univ, 75- Mem: Am Phys Soc. Res: Electromagnetism; plasma physics; surface phenomena; energy policy. Mailing Add: 24 Fieldmont Rd Belmont MA 02178

MACRI, ALFRED ROGER, b New York, NY, June 28, 32. CHEMISTRY, GEOLOGY. Educ: Queens Col, BS, 55; NMex State Univ, MA, 58; NY Univ, PhD(sci educ), 63. Prof Exp: Asst sci, NY Univ, 59-63; ASSOC PROF CHEM & GEOL, STATEN ISLAND COMMUNITY COL, 63- Mem: Am Chem Soc; Nat Sci Teachers Asn. Res: Atomic structure. Mailing Add: Staten Island Community Col Dept of Sci 715 Ocean Terr Staten Island NY 10301

MACRI, FRANK JOHN, b Portland, Maine, Jan 12, 23; m 48; c 4. PHARMACOLOGY. Educ: Univ Maine, BA, 43; George Washington Univ, BS, 50; Georgetown Univ, PhD(pharmacol), 53. Prof Exp: Anal chemist, Naval Powder Factory, 43-45, org chemist, 46-48; biochemist, NIH, 48-50; pharmacologist, Irwin, Neisler & Co, 51-53; chief pharmacologist, Eaton Labs, Inc, 53-55; HEAD SECT PHARMACOL, NAT EYE INST, BETHESDA, 55- Mem: Am Soc Pharmacol & Exp Therapeut; Asn Res Vision & Ophthal; NY Acad Sci. Res: Pharmacodynamics of autonomic nervous system agents; effect of hormones on vascular physiology and pharmacology; curaremimetics and central nervous system depressants; structure-activity relationships in biological processes; pharmacology of the eye.

MACSWAIN, JOHN WINSLOW, entomology, deceased

MACSWAN, IAIN CHRISTIE, b Ocean Falls, BC, Apr 15, 21; US citizen; m 43; c 3. PLANT PATHOLOGY. Educ: Univ BC, BSA, 42, MSA, 62. Prof Exp: Prin res asst, Dom Lab Plant Path, Univ BC, 46-47; asst prof & plant pathologist, BC Dept Agr, 47-55; from asst prof to assoc prof, 55-67, PROF PLANT PATH, ORE STATE UNIV, 67-, EXTEN PLANT PATHOLOGIST, 61- Mem: Am Phytopath Soc; Can Phytopath Soc. Res: Extension pathology; diseases of horticultural crops. Mailing Add: 1089 Cordley Hall Ore State Univ Corvallis OR 97331

MACSWEEN, JOSEPH MICHAEL, b Antigonish, NS, Mar 1, 33; m 57; c 4. IMMUNOLOGY. Educ: St Francis Xavier Univ, BSc, 52; Dalhousie Univ, MD, 57, MSc, 69; FRCP(C), 70. Prof Exp: Resident med, Dalhousie Univ, 66-69; fel clin immunol, Montreal Gen Hosp, 69-70; Med Res Coun res fel immunol, Walter & Eliza Hall Inst Med Res, Australia, 71-72; lectr med, 72-74, asst prof, 74-75, ASSOC PROF MED, DALHOUSIE UNIV, 75- Concurrent Pos: Regional coordr myeloma chemother nat trial, Nat Cancer Inst Can, 74- Mem: Can Soc Clin Invest; Can Soc Immunol; AAAS; Am Fedn Clin Res. Res: Lymphocyte-macrophage interactions; characterization of macrophage migration inhibition factor; determination of T lymphocyte function in chronic lymphocytic leukemia. Mailing Add: Camp Hill Hosp Robie St Halifax NS Can

MACTAVISH, JOHN N, b Detroit, Mich, May 23, 39; m; c 1. ENVIRONMENTAL SCIENCES, GEOLOGY. Educ: Bowling Green Univ, BS, 61, MA, 63; Case Western Reserve Univ, PhD(geol), 71. Prof Exp: Asst prof geol, Col Arts & Sci, 68-71, ASST PROF ENVIRON STUDIES, ASST TO THE DEAN & DIR ENVIRON STUDIES PROG, WILLIAM JAMES COL, GRAND VALLEY STATE COLS, 71- Mem: Geol Soc Am; Paleont Soc; Nat Asn Geol Teachers. Res: Invertebrate paleontology. Mailing Add: William James Col Grand Valley State Cols Allendale MI 49401

MACUR, GEORGE J, b Chicago, Ill, Feb 22, 33; m 60; c 2. PHYSICAL CHEMISTRY. Educ: DePaul Univ, BS, 54, MS, 57; Ill Inst Technol, PhD(chem), 65. Prof Exp: Chemist, IIT Res Inst, 60-65 & Argonne Nat Lab, 65-68; CHEMIST, IBM CORP, 68- Res: Surface chemistry and technology; analytical chemistry. Mailing Add: IBM Corp 1701 North St Endicott NY 13760

MACURDA, DONALD BRADFORD, JR, b Boston, Mass, Aug 8, 36; m 61; c 2. PALEONTOLOGY, MARINE BIOLOGY. Educ: Univ Wis, BS, 56, PhD(geol), 63. Prof Exp: From instr to asst prof, 63-68, ASSOC PROF GEOL & MINERAL, UNIV MICH, ANN ARBOR, 68-, ASSOC CUR INVERT PALEONT, MUS PALEONT, 63- Concurrent Pos: NSF grants, Univ Mich. Mem: Am Asn Petrol Geologists; Geol Soc Am; Brit Paleont Asn; Paleont Soc; Soc Econ Paleontologists & Mineralogists. Res: The study of the funcitonal morphology and ecology of modern and ancient crinoids, invertebrate. Mailing Add: Mus of Paleont Univ of Mich Ann Arbor MI 48104

MACVICAR, MARGARET LOVE AGNES, b Hamilton, Ont, Nov 20, 43; US citizen. MATERIALS SCIENCE. Educ: Mass Inst Technol, SB, 64, ScD(metall, mat sci), 67. Prof Exp: NATO fel, Univ Cambridge, 67-68, Am Asn Univ Women Marie Curie fel, 68-69; from instr to asst prof physics, 69-74, ASSOC PROF PHYSICS, MASS INST TECHNOL, 74- Concurrent Pos: Dir, Sprague Elec Co, Mass; consult, Lilly Endowment, Inc, Ind, 73, IBM Corp, NY, 73-, Danforth Found, Mo, 74- & Utah State Univ, 75; mem bd dirs, Oral Educ Ctr of Southern Calif, Los Angeles; trustee, Carnegie Found for Advan Teaching; mem vis comt, Gen Motors Inst; mem educ comt, Boston Mus Sci; mem adv comt on hist of physics, Am Inst Physics, 73-; mem, Carnegie Coun for Policy Studies in Higher Educ, 75- Mem: Am Inst Physics; Am Vacuum Soc; AAAS. Res: Superconductivity, especially of thin films, refractory metals, single crystals, tunnelling and barrier fabrication; deaf education; education reform. Mailing Add: Rm 8-201 Mass Inst of Technol Cambridge MA 02139

MACVICAR, ROBERT WILLIAM, b Princeton, Minn, Sept 28, 18; m 49; c 2. CHEMISTRY. Educ: Univ Wyo, BA, 39; Okla State Univ, MS, 40; Univ Wis, PhD(biochem), 46. Prof Exp: Assoc prof biochem, Okla State Univ, 46-49, prof & head dept, 49-57, dean grad sch, 53-64, vpres, 57-64; vpres acad affairs, Southern Ill Univ Syst, 64-68; chancellor, Southern Ill Univ, Carbondale, 68-70; PRES, ORE STATE UNIV, 70- Mem: Am Chem Soc; Am Soc Biol Chemists; Am Inst Nutrit. Res: Nitrogen metabolism of plants and animals; nutrition and metabolism of large animals. Mailing Add: Off of the Pres Ore State Univ Corvallis OR 97331

MACVICAR-WHELAN, PATRICK JAMES, b St John's, Nfld, Oct 19, 38; m 65; c 1. ATOMIC PHYSICS, PLASMA PHYSICS. Educ: St Francis Xavier Univ, BSc, 59; Dalhousie Univ, MSc, 61; Univ BC, PhD(physics), 65. Prof Exp: Asst prof physics, St Francis Xavier Univ, 64-65; asst prof math, St Mary's Univ NS, 65-66; Off Naval Res grant, Univ Calif, Berkeley, 66-70; ASSOC PROF PHYSICS, GRAND VALLEY STATE COL, 70- Mem: Am Phys Soc; Math Asn Am; Am Geophys Union; Am Asn Physics Teachers. Res: Systems science; clustering; atomic collisions; electrical breakdown; acoustics; low temperature physics; applied mathematics; physical systems in relation to those in other disciplines. Mailing Add: Dept of Physics Grand Valley State Col Allendale MI 49401

MACWHERTER, JOHN BAIRD, b Alton, Ill, Aug 21, 23; m 45; c 3. MATHEMATICS. Educ: US Mil Acad, BS, 45; Columbia Univ, MA, 57; Prof Exp: Asst prof math, US Mil Acad, 53-56 & US Air Force Acad, 56-58; proj scientist, Air Force Cambridge Res Lab, 58-60, assoc prof math, US Air Force Acad, 61-65, prof & head dept, 65-66, asst dean & fac exec, 66-67, chief scientist, Europ Off Aerospace Res, Brussels, 67-69; asst to pres, 69-71, DIR BUS SERVS, JAMES FORD BELL TECH CTR, GEN MILLS INC, 71- Mailing Add: James Ford Bell Tech Ctr 9000 Plymouth Rd Minneapolis MN 55427

MACWILLIAM, EDGAR ALEXANDER, b McAdam, NB, Nov 6, 20; m 44; c 3. PHYSICAL CHEMISTRY. Educ: Mt Allison Univ, BSc, 41; Univ Toronto, MA, 42, PhD(phys chem), 44. Prof Exp: RES CHEMIST, EASTMAN KODAK CO, 46- Mem: Am Chem Soc; Soc Photog Sci & Eng. Res: Electrochemistry of silver halides; improved photographic emulsions. Mailing Add: Eastman Kodak Res Labs 1669 Lake Ave Rochester NY 14650

MACWILLIAMS, DALTON CARSON, b Winnipeg, Man, Mar 22, 28; US citizen; m 54; c 4. POLYMER CHEMISTRY. Educ: Univ Alta, BSc, 50; Univ Minn, PhD(anal phys chem), 55. Prof Exp: From chemist to sr res chemist, 55-75, SR RES SPECIALIST, DOW CHEM, USA, WESTERN DIV RES, 72-, PROJ MGR, 75- Mem: Am Chem Soc. Res: Synthesis, rheology, interfacial properties and applications of water soluble polymers; physical organic chemistry; crystal growth; secondary oil recovery; wet end paper chemistry. Mailing Add: Dow Chem USA 2800 Mitchell Dr Walnut Creek CA 94598

MACWILLIAMS, DONALD GRIBBLE, b El Paso, Tex, Mar 14, 22; m 46; c 7. CHEMISTRY. Educ: US Mil Acad, BS, 44; Ohio State Univ, MS, 52; Rensselaer Polytech Inst, PhD(chem), 67. Prof Exp: US Army, 40-, chief radio br, Chem Corps Sch, 52-53, from instr to prof chem, 56-65, prof physics & chem & head dept, 65-67, PROF CHEM & HEAD DEPT, US MIL ACAD, US ARMY, 67- Mem: Am Chem Soc. Res: Primary military research; training in nuclear defense. Mailing Add: Dept of Chem US Mil Acad West Point NY 10996

MACWILLIAMS, FLORENCE JESSIE, b Stoke-on-Trent, Eng, Jan 4, 17; US citizen; m 41; c 3. MATHEMATICS. Educ: Cambridge Univ, BA, 39, MA, 40; Harvard Univ, PhD, 62. Prof Exp: Tech asst comput prog, Bell Tel Labs, Inc, 56-58, assoc mem tech staff, 58-61, res mathematician, 61-64; Dept Sci & Indust Res vis fel, Cambridge Univ, 64-65; MEM TECH STAFF, BELL LABS, 65- Mem: Am Math Soc; Math Asn Am. Res: Algebraic theory of error correcting codes; group theory; use of computer in above fields. Mailing Add: Bell Labs 2C377 Murray Hill NJ 07971

MACWOOD, GEORGE EUGENE, b New York, NY, Feb 16, 08; m 43; c 6. CHEMISTRY. Educ: Columbia Univ, AB, 30, BS, 32, ChE, 33, PhD(chem), 36. Prof Exp: Cutting traveling fel, Univ Leiden, 36, Nat Res Coun fel, 37; instr chem, Ohio State Univ, 38-42; asst dir radiation lab, Univ Calif, 43-46; ASSOC PROF CHEM, OHIO STATE UNIV, 46- Mem: Am Chem Soc. Res: Raman spectra; thermodynamics and cryogenics. Mailing Add: Dept of Chem Ohio State Univ Columbus OH 43210

MACY, JOSIAH, JR, b New York, NY, Dec 17, 24; m 45; c 4. MATHEMATICAL BIOPHYSICS. Educ: Mass Inst Technol, SB, 49, PhD(math), 54. Prof Exp: Asst math, Mass Inst Technol, 52-54; mem staff opers res off, Johns Hopkins Univ, 54-58; from asst prof to assoc prof physiol & math biophys, Albert Einstein Col Med, 58-67; PROF PHYSIOL, BIOPHYS & BIOMATH & DIR DIV BIOPHYS SCI, MED CTR, UNIV ALA, BIRMINGHAM, 67- Concurrent Pos: Mem adv comt comput & res, NIH, 63-64, comput res study sect, 64-67 & biomed library review comt, Nat Libr Med, 72- Mem: AAAS; Am Math Soc; Soc Indust & Appl Math; Biophys Soc; Biomet Soc. Res: Mathematical models in nervous system; communication; information and applied probability theory; sensory biophysics. Mailing Add: Div of Biophys Sci Med Ctr Univ of Ala Birmingham AL 35294

MACY, RALPH WILLIAM, b McMinnville, Ore, July 6, 05; m 31; c 1. PARASITOLOGY. Educ: Linfield Col, BA, 29; Univ Minn, MA, 31, PhD(zool), 34. Prof Exp: Asst zool, Univ Minn, 29-33; from instr to prof biol, Col St Thomas, 34-42; prof biol, Reed Col, 42-55, head dept, 43-55; prof, 55-72, exec off dept, 58-65, EMER PROF BIOL, PORTLAND STATE UNIV, 72- Concurrent Pos: Lectr, Med Sch, Univ Ore, 43-45, prof exten div, 44-45; USPHS grants, 49-50 & 64-67; chief investr res proj, Off Naval Res, 50-59; trustee, Ore Mus Sci & Indust, 53-59 & Northwestern Sci Asn, 56-59; NSF grants, 60-64 & 66-72; del, Int Cong Trop Med & Malaria, Lisbon, 58, Int Cong Parasitol, Rome, 64 & DC, 70; guest investr, Univ Helsinki, 61, US Naval Med Res Unit, Cairo, 61-62 & Inst Trop Med, Lisbon, 62 & 69. Honors & Awards: Citation, Ore Acad Sci. Mem: Fel AAAS; charter fel Am Acad Microbiol; Am Micros Soc; Am Soc Parasitol; Wildlife Dis Asn. Res: Biology of helminths and trematodes; entomology. Mailing Add: Dept of Biol Portland State Univ Portland OR 97207

MACY, SPENCER MICHAEL, mathematical physics, see 12th edition

MADACSI, DAVID PETER, b Youngstown, Ohio, Jan 28, 44; m 70; c 2. EXPERIMENTAL SOLID STATE PHYSICS, MAGNETIC RESONANCE. Educ: Youngstown Univ, BS, 65; Univ Dayton, MS, 67; Univ Conn, PhD(physics), 72. Prof Exp: Res physicist, Ames Lab USAEC, Iowa State Univ, 67-68; res assoc solid state physics, Inst Mat Sci, Univ Conn, 72-73; ASST PROF PHYSICS, PA STATE UNIV, SHENANGO VALLEY CAMPUS, 73- Mem: Am Phys Soc; Am Asn Physics Teachers. Res: Defects in solids, especially paramagnetic impurities, radiation induced defects and color centers; magnetic, optical properties of solids, utilizing electron paramagnetic resonance and optical spectroscopy; solid state materials, particularly crystal growth and characterization; photo-catalytic hydrogen production. Mailing Add: Pa State Univ Shenango Valley Campus Sharon PA 16146

MADAN, DEVENDRA K, pharmacy, physical pharmacy, see 12th edition

MADAN, RABINDER NATH, b New Delhi, India, Mar 11, 35; m 69. THEORETICAL PHYSICS. Educ: Univ Delhi, BSc, 54, MSc, 56; Princeton Univ, MA, 64, PhD(physics), 67. Prof Exp: Lectr physics & math, Ramjas Col, Univ Delhi, 56-57; sr sci asst, Nat Phys Lab, Univ New Delhi, 57-62; res assoc theoret physics, Univ Calif, Santa Barbara, 67-68; res assoc & lectr, Univ Mass, Amherst, 68-69;

ASSOC PROF PHYSICS, NC A&T STATE UNIV, 69- Concurrent Pos: Fulbright travel award, Inst Int Educ, 62; prin investr, NASA res grant, 67-; NSF res grant theoret physics, 73-75. Mem: Am Phys Soc. Res: Electron-hydrogen scattering theory; amplitude-phase method in non-relativistic quantum mechanics; Regge Daughter Trajectories in the Bethe-Salpeter equation; spectral line satellites in alkalis; Eikonal-Glauber exchange amplitudes. Mailing Add: 5314 Bennington Dr Greensboro NC 27410

MADAN, STANLEY KRISHEN, b Lahore, Pakistan, May 1, 22; US citizen; m 58; c 1. INORGANIC CHEMISTRY. Educ: Forman Christian Col, Lahore, Pakistan, BSc, 45; Punjab Univ, Pakistan, MA, 50, MSc, 54; Univ Ill, Urbana, MS, 57, PhD(inorg chem), 60. Prof Exp: Res assoc irrig, Punjab Irrig Res Inst, Lahore, Pakistan, 45-48; demonstr physics, Forman Christian Col, 48-55; asst chem, Univ Ill, Urbana, 57-60; from asst prof to assoc prof, 60-67, PROF CHEM, STATE UNIV NY BINGHAMTON, 67- Mem: Am Chem Soc. Res: Synthesis and structure determination of new coordination compounds; kinetics and substitution reactions of triaminotriethylamine metallic complexes. Mailing Add: Dept of Chem State Univ of NY Binghamton NY 13901

MADANSKY, ALBERT, b Chicago, Ill, May 16, 34; m 56; c 3. STATISTICS. Educ: Univ Chicago, AB, 52, MS, 55, PhD(statist), 58. Prof Exp: Mathematician, Rand Corp, 57-65; vpres, Interpub Group of Co, Inc, Calif, 65-68; pres, Dataplan, Inc, NY, 68-71; prof comput sci & chmn dept, City Col New York, 71-74; PROF COMPUT SCI, GRAD SCH BUS, UNIV CHICAGO & DIR, CTR FOR MGT PUB & NONPROFIT ENTERPRISE, 74- Concurrent Pos: Fel, Ctr Advan Study Behav Sci, 61-62. Mem: Fel Am Statist Asn; fel Inst Math Statist; Economet Soc. Res: Multivariate analysis; mathematical models in the social sciences. Mailing Add: Grad Sch of Bus Univ of Chicago Chicago IL 60637

MADANSKY, LEON, b Brooklyn, NY, Jan 11, 23; m 47; c 2. NUCLEAR PHYSICS. Educ: Univ Mich, BS, 42, MS, 44, PhD(physics), 48. Prof Exp: From asst prof to assoc prof, 48-58, chmn dept, 65-68, PROF PHYSICS, JOHNS HOPKINS UNIV, 58- Concurrent Pos: Res physicist, Brookhaven Nat Lab, 52-53; NSF sr fel, 61, 69. Mem: Fel Am Phys Soc. Res: Atomic properties of elementary particles; radiative effects in elementary particles; particle counters; nuclear spectroscopy; radiative effects in beta-decay; meson-nuclear scattering. Mailing Add: Dept of Physics Johns Hopkins Univ Baltimore MD 21218

MADAPPALLY, MATHEW MATHAI, b Elikulam, Kerala, India, June 19, 38; m 72. NUTRITIONAL BIOCHEMISTRY. Educ: Univ Kerala, BSc, 59; Univ Baroda, MSc, 65; Univ Ill, Urbana, PhD(nutrit, biochem), 70. Prof Exp: Tutor biochem & biophys, St John's Med Col, Bangalore, India, 65-66; fel biochem, Sch Med, Univ Nebr, Omaha, 70, instr, 70-71; res assoc develop biochem, Inst Develop Res, Children's Hosp Res Found, 71-75; BIOCHEMIST RES & DEVELOP, COULTER DIAGNOSTICS INC, 75- Mem: Biochem Soc; Am Chem Soc. Res: Applied research in the field of clinical chemistry. Mailing Add: Coulter Diagnostics Inc 740 W 83rd St Hialeah FL 33014

MADARAS, RONALD JOHN, b Summit, NJ, Dec 18, 42; m 65; c 1. EXPERIMENTAL HIGH ENERGY PHYSICS. Educ: Cornell Univ, BEP, 65, MS, 65; Harvard Univ, PhD(physics), 73. Prof Exp: Res physicist, Lab Accelerateur Lineaire, Orsay, France, 72-75; RES PHYSICIST, LAWRENCE BERKELEY LAB, UNIV CALIF, 75- Mem: Am Phys Soc. Res: Electron-positron colliding beam physics. Mailing Add: Lawrence Berkeley Lab Univ of Calif Berkeley CA 94720

MADARIAGA, RAUL IVAN, b Santiago, Chile, Mar 29, 44; m 70; c 1. SEISMOLOGY. Educ: Univ Chile, CE, 67; Mass Inst Technol, PhD(geophys), 71. Prof Exp: Researcher, Dept Geophys, Univ Chile, 71-73; RES ASSOC, DEPT EARTH & PLANETARY SCI, MASS INST TECHNOL, 73- Mem: Am Geophys Union; Seismol Soc Am. Res: Study of the dynamics of earthquake source mechanism; numerical solution of elastic wave propagation. Mailing Add: Rm 54-514 Mass Inst of Technol Cambridge MA 02139

MADDAIAH, VADDANAHALLY THIMMAIAH, b Mysore, India, Nov 10, 29; m 53; c 2. BIOCHEMISTRY. Educ: Univ Mysore, BSc, 53, MSc, 54; Univ Ariz, PhD(biochem), 63. Prof Exp: Lectr chem, Univ Mysore, 54-60; NIH fel biol, Univ Calif, San Diego, 63-64; NIH fel biochem, Univ Alta, 64-66; sr res biochemist, Res & Develop Labs, Can Packers Ltd, 66-67; res assoc physiol & biophys, Sch Med, Univ Louisville, 67-69; RES BIOCHEMIST, NASSAU COUNTY MED CTR, 69-; ASSOC PROF PEDIAT, HEALTH SCI CTR, STATE UNIV NY STONY BROOK, 72- Res: Enzymology; metabolism; polypeptide hormones; control mechanisms. Mailing Add: Dept of Pediat Nassau County Med Ctr East Meadow NY 11554

MADDEN, DAVID LARRY, b Willow Hill, Ill, Aug 10, 32; m 53; c 3. VETERINARY MICROBIOLOGY. Educ: Kans State Univ, BS, 56, DVM & MS, 58; Purdue Univ, PhD(vet microbiol), 63. Prof Exp: Instr vet microbiol, Sch Vet Sci & Med, Purdue Univ, 58-63; resident microbiol, Lab Bact Dis, Nat Inst Allergy & Infectious Dis, 63-68, off assoc dir, Nat Inst Child Health & Human Develop, 68-70, RES MICROBIOLOGIST INFECTIOUS DIS BR, NAT INST NEUROL DIS & STROKE, 70- Concurrent Pos: Instr NIH Grad Sch Prog, 67- Mem: AAAS; Am Vet Med Asn; Am Asn Avian Path; Poultry Sci Asn. Res: Etiology of neurological diseases; concerned with isolation of etiological agent and immunological response of the host. Mailing Add: Nat Inst Neurol Dis & Stroke NIH Infectious Dis Br Bldg 36 Rm 5D-04 Bethesda MD 20014

MADDEN, GEORGE D, b Palestine, Tex, Nov 14, 34; m 57; c 3. HORTICULTURE, PLANT PHYSIOLOGY. Educ: Tex Col Arts & Indust, BS, 57; Tex A&M Univ, MS, 63, PhD, 67. Prof Exp: Farm foreman, Citrus Ctr, Tex Col Arts & Indust, 57-59; asst pomol, Tex A&M Univ, 59-62, res technician, 62-64; RES HORTICULTURIST, CROPS RES DIV, AGR RES SERV, USDA, 64- Mem: Am Soc Hort Sci. Res: Chemical fruit thinning; various techniques of propagation of the pecan; breeding and development of new varieties of pecan; pruning, growth regulators and rootstock and interstock studies with pecan. Mailing Add: US Pecan Field Sta PO Box 579 Brownwood TX 76801

MADDEN, HANNIBAL HAMLIN, JR, b New York, NY, Oct 5, 31; m 64; c 2. PHYSICS. Educ: Williams Col, Mass, BA, 52, MA, 54; Brown Univ, PhD(physics), 59. Prof Exp: Asst physics, Williams Col, 52-54; asst, Brown Univ, 54-57, assoc, 58-59, instr, 59-60; asst prof, 60-67, ASSOC PROF PHYSICS, WAYNE STATE UNIV, 67- Concurrent Pos: Fulbright traveling grant, Hannover Tech Univ, 71-72. Mem: Am Phys Soc; Am Asn Physics Teachers; Am Vacuum Soc. Res: Solid state and low temperature physics; solid state surface phenomena; low energy electron diffraction; transport properties of liquid helium II. Mailing Add: Dept of Physics Wayne State Univ Detroit MI 48202

MADDEN, JOHN WILLIAM, b Glendale, Calif, June 3, 33; m 63; c 2. SURGERY, BIOLOGY. Educ: Yale Univ, BS, 55; Harvard Univ, MD, 58. Prof Exp: Intern & resident gen surg, Mass Gen Hosp, Boston, 58-61; instr anat, Univ Pa, 61-63; intern & resident gen surg, Mass Gen Hosp, Boston, 63-65; resident plastic surg, NC Mem Hosp, Chapel Hill, 65-67; from instr to asst prof surg, Sch Med, Univ NC, Chapel Hill, 67-69; assoc prof, 69-73, PROF SURG, UNIV ARIZ, 73- Concurrent Pos: NIH fel, Sch Med, Univ Pa, 61-63; instr, Sch Med, Univ NC, Chapel Hill, 65-66; staff surgeon, Med Ctr, Univ Ariz, 69-; consult, Tucson Vet Admin Hosp, Davis-Monthan Air Force Hosp & Hand Rehab Ctr, Univ Ariz. Mem: Soc Develop Biol; Soc Exp Biol & Med; Soc Univ Surgeons; fel Am Col Surgeons; Am Soc Surg of Hand. Res: Biology and biochemistry of wound healing; pharmacological alteration of scar formation; kinesiology of hand function; constructive surgery. Mailing Add: Dept of Surg Univ of Ariz Med Ctr Tucson AZ 85724

MADDEN, RICHARD M, b Waubay, SDak, Apr 1, 28; m 54; c 1. DENTISTRY. Educ: Northern State Col, BS, 53; Univ Minn, DDS, 57; Univ Iowa, MS, 62; Am Bd Endodont, dipl. Prof Exp: Pvt pract, Minn, 57-60; assoc prof dent, Univ Iowa, 62-70; assoc prof, 70-74, PROF ENDODONT & CHMN DEPT, UNIV TEX DENT BR, HOUSTON, 74- Mem: Am Asn Endodont; Int Asn Dent Res; Am Dent Asn. Res: Endodontics; aerosols. Mailing Add: Dept of Endodont Univ of Tex Dent Br Houston TX 77025

MADDEN, ROBERT E, b Oak Park, Ill, Sept 16, 25; c 4. THORACIC SURGERY, CARDIOVASCULAR SURGERY. Educ: Univ Ill, BS, 50, MD, 52. Prof Exp: Am Cancer Soc fel, Hammersmith Hosp, London, 58-59; PROF SURG, NEW YORK MED COL, 71- Honors & Awards: Borden Res Award, Univ Ill, 52. Mem: Am Col Surgeons; Am Soc Clin Oncol; Am Asn Cancer Educ; Soc Surg Oncol. Res: Dissemination of cancer via lymphatics and blood stream; clinical chemotherapy studies; demonstrations of cellular immunology in clinical cancer. Mailing Add: New York Med Col Flower & Fifth Ave Hosps New York NY 10029

MADDEN, ROBERT PHYFE, b Schenectady, NY, Dec 20, 28; m 50; c 3. PHYSICS. Educ: Univ Rochester, BS, 50; Johns Hopkins Univ, PhD(physics), 56. Prof Exp: Physicist, Lab Astrophys & Phys Meteorol, Johns Hopkins Univ, 53-58; physicist & sect chief, US Army Eng Res & Develop Labs, Ft Belvoir, Va, 58-61; CHIEF FAR ULTRAVIOLET PHYSICS SECT, OPTICAL PHYSICS DIV, NAT BUR STANDARDS, 61- Mem: Fel Am Phys Soc; fel Optical Soc Am. Res: Atomic and molecular spectroscopy; optical instrumentation; surface physics; thin films. Mailing Add: Div Phys A257 Nat Bur of Standards Washington DC 20234

MADDEN, ROBERT WILLIAM, physics, see 12th edition

MADDEN, SIDNEY CLARENCE, b Fresno, Calif, Oct 27, 07; m 33; c 4. PATHOLOGY. Educ: Stanford Univ, AB, 30, MD, 34. Prof Exp: Intern, Johns Hopkins Hosp, 33-34; asst resident path, Strong Mem Hosp, Univ Rochester, 34-40; asst path, Sch Med & Dent, Univ Rochester, 34-37, from instr to assoc prof, 37-45; prof & chmn dept, Sch Med, Emory Univ, 45-48; head div, Brookhaven Nat Lab, 48-51, sr physician, Lab Hosp, 49-51; chmn dept, 51-70, PROF PATH, SCH MED, UNIV CALIF, LOS ANGELES, 51- Concurrent Pos: Dir labs, Park Ave Hosp, 40-45; pathologist, Emory Hosp, 45-48 & Eggleston Hosp for Children, 45-48; consult, Vet Admin Ctr, Los Angeles, 51-, Los Angeles County Harbor Gen Hosp, 52- & Armed Forces Inst Path, 55-65; sr adv, Atomic Bomb Casualty Comn, 58-66. Honors & Awards: Smith Award, AAAS, 43. Mem: AAAS; Am Soc Exp Path (pres, 52); Soc Exp Biol & Med; fel AMA; Am Asn Path & Bact (pres, 63). Res: Nucleic acid and protein metabolism after tissue injury; chemical carcinogenesis; immunology of neoplasia. Mailing Add: Dept of Path Univ of Calif Sch of Med Los Angeles CA 90024

MADDEN, STEPHEN JAMES, JR, b Newton, Mass, June 8, 36; m 58; c 2. COMPUTER SCIENCE, GEODESY. Educ: Mass Inst Technol, BS, 59, MS, 62, PhD(appl math), 66. Prof Exp: Staff mathematician, Instrumentation Lab, 59-61, instr math, 65-66, assoc dir, Measurement Systs Lab, 66-74, SECT CHIEF, C S DRAPER LAB, MASS INST TECHNOL, 74- LECTR, DEPT AERONAUT & ASTRONAUT, 68- Mem: Am Math Soc; Soc Indust & Appl Math; Inst Elec & Electronics Engrs; Am Geophys Union. Mailing Add: 5 Constitution Rd Lexington MA 02173

MADDEN, THEODORE RICHARD, b Boston, Mass, Mar 14, 25; m 74; c 1. GEOPHYSICS. Educ: Mass Inst Technol, BS, 49, PhD(geophys), 61. Prof Exp: From lectr to assoc prof, 52-64, PROF GEOPHYS, MASS INST TECHNOL, 64- Res: Geoelectricity and geomagnetism, inversion theory, atmospheric gravity waves. Mailing Add: Dept of Earth & Planetary Sci Rm 54-614 Mass Inst of Technol Cambridge MA 02139

MADDEN, THOMAS M, b Indianapolis, Ind, Mar 29, 22; m 63; c 2. OPTOMETRY. Educ: Southern Col Optom, OD, 53; Ind Univ, MS, 61. Prof Exp: Pvt pract, 54-59; asst prof, 63-68, ASSOC PROF OPTOM, IND UNIV, BLOOMINGTON, 68- Res: Accommodative-convergence relationships; clinical optometry. Mailing Add: Sch of Optom Ind Univ Bloomington IN 47401

MADDIN, CHARLES MILFORD, b Vernon, Tex, Sept 7, 27; m 48; c 1. ANALYTICAL CHEMISTRY. Educ: Univ Tex, BA, 50, PhD(chem), 53. Prof Exp: Res scientist, Univ Tex, 51-53; lab group leader, 53-64, SECT SUPVR, DOWELL DIV, DOW CHEM USA, 64- Concurrent Pos: Lectr, Univ Tulsa, 64-65. Mem: Am Chem Soc. Res: Instrumental analysis; chemical separations. Mailing Add: 7705 E 25th Pl Tulsa OK 74129

MADDING, GARY DEAN, organic chemistry, see 12th edition

MADDISON, SHIRLEY EUNICE, b SAfrica, Apr 12, 25; US citizen; m 50. IMMUNOLOGY, PARASITOLOGY. Educ: Univ Capetown, SAfrica, BSc, 44, PhD(microbiol), 62. Prof Exp: Lab technician microbiol, Prov Health Serv, Natal, SAfrica, 45-54; asst instr bact, Med Sch, Univ Natal, 54-56; sr tech officer parasitol, Amoebiasis Res Inst, Coun Sci & Indust Res, Natal, 56-64; USPHS int fel, Ctr Dis Control, Atlanta, Ga, 64-66; guest researcher, Biochem Inst, Uppsala, Sweden, 66; vis scientist & fel parasitol, 67-71, actg chief parasitic immunochem unit, 72-73, CHIEF PARASITIC IMMUNOCHEM BR, CTR DIS CONTROL, USPHS, 73- Concurrent Pos: Adv ad hoc study group parasitic dis, US Army Med Res & Develop Command, 74-; consult to Surgeon Gen, US Army, 74- Mem: Fel Am Acad Microbiol; Am Soc Trop Med & Hyg; AAAS; Reticuloendothelial Soc; Am Asn Immunologists. Res: Immunology of parasitic diseases with emphasis on the immune mechanisms and antigens related to protective immunity in schistosomiasis; immunologic response to parasitic infections in terms of immunoglobulin subclasses. Mailing Add: Bldg 5 Rm SB-15 Ctr Dis Control 1600 Clifton Rd NE Atlanta GA 30333

MADDOCK, MARSHALL, b Glendale, Calif, Sept 2, 28; m; c 3. GEOLOGY. Educ: Univ Calif, AB, 49, PhD, 55. Prof Exp: PROF GEOL, SAN JOSE STATE COL, 55- Mem: Geol Soc Am; Nat Asn Geol Teachers. Res: Mineralogy; petrology; field geology. Mailing Add: Dept of Geol San Jose State Col San Jose CA 95114

MADDOCKS, ROSALIE FRANCES, b Lewiston, Maine, Aug 27, 38. GEOLOGY, MICROPALEONTOLOGY. Educ: Univ Maine, Orono, BA, 59; Univ Kans, MA, 62, PhD(geol), 65. Prof Exp: Res assoc, Smithsonian Inst, 65-67; ASSOC PROF GEOL,

UNIV HOUSTON, 67- Mem: AAAS; Paleont Soc; Soc Syst Zoologists; Int Paleont Union. Res: Systematics, ecology and evolution of living and fossil Ostracoda. Mailing Add: Dept of Geol Univ of Houston Houston TX 77004

MADDOX, BILLY HOYTE, b Pahokee, Fla, July 2, 32; m 53; c 4. MATHEMATICS. Educ: Troy State Univ, BS, 53; Univ Fla, MEd, 57; Univ SC, PhD(math), 65. Prof Exp: Instr math, Troy State Univ, 57-60; prof & chmn dept, Presby Col SC, 64-66; assoc prof, 66-71, PROF MATH, ECKERD COL, 71- Concurrent Pos: Reviewer, Zentralblatt für Mathematics, 66-; partic, NSF Summer Inst Comput-Oriented Calculus, Fla State Univ, 70. Honors & Awards: Nat Defense Serv Medal; Europ Occup Medal. Mem: Am Math Soc; Math Asn Am. Res: Algebra; ring and module theory, specifically absolutely pure modules; application of mathematics to evaluation in higher education. Mailing Add: Dept of Math Eckerd Col St Petersburg FL 33733

MADDOX, JOHN NOLEN, industrial chemistry, nuclear chemistry, see 12th edition

MADDOX, JOSEPH VERNARD, b Montgomery, Ala, Apr 14, 38; m 71. INSECT PATHOLOGY. Educ: Auburn Univ, BS, 59, MS, 61; Univ Ill, PhD(entom), 66. Prof Exp: ASST PROF AGR ENTOM & ASSOC ENTOMOLOGIST, ILL NATURAL HIST SURV, UNIV ILL, URBANA, 66- Mem: AAAS; Entom Soc Am; Soc Invert Path. Res: Insect pathology, especially microsporidian diseases of insects. Mailing Add: Sect of Econ Entom Ill Nat Hist Surv Univ of Ill Urbana IL 61801

MADDOX, V HAROLD, JR, b New York, NY, Sept 25, 23; m 49; c 3. ORGANIC CHEMISTRY. Educ: Carnegie Inst Technol, BS, 49; Rutgers Univ, MS, 52, PhD(chem), 53. Prof Exp: Lab technician, Reichhold Chem, 41-43; asst chem, Rutgers Univ, 49-53; from assoc res chemist to res chemist, 53-60, res leader org chem, 60-62, dir lab, 62-70, SECT DIR, ORG CHEM, PARKE, DAVIS & CO, 70- Mem: Am Chem Soc; NY Acad Sci; The Chem Soc. Res: Synthesis in the chrysene series; synthesis of sernyl; synthetic organic medicinals; process research. Mailing Add: Parke Davis & Co Bldg 48 Joseph Campau Ave Detroit MI 48232

MADDOX, WILLIAM EUGENE, b Owensboro, Ky, Aug 3, 37; m 63. NUCLEAR PHYSICS. Educ: Murray State Univ, BS, 62; Ind Univ, MS, 64, PhD(physics), 68. Prof Exp: Asst prof, 67-69, ASSOC PROF PHYSICS, MURRAY STATE UNIV, 69- Mem: Am Phys Soc. Res: Experimental nuclear reaction physics. Mailing Add: Dept of Physics Murray State Univ Murray KY 42071

MADDUX, WILLIAM STERLING, physiological ecology, see 12th edition

MADDY, KEITH THOMAS, b Knoxville, Iowa, Oct 28, 23; m 46; c 6. VETERINARY MEDICINE, PUBLIC HEALTH. Educ: Iowa State Col, DVM, 45; Univ Calif, MPH, 54; Am Bd Vet Pub Health & Am Col Vet Microbiol, dipl. Prof Exp: Asst prof, Univ Nev, 45-47; vet meat inspector, USDA, 48-51; epidemiologist, Nat Commun Dis Ctr, 54-60, scientist, Lab Med & Biol Sci, Div Air Pollution, 60-62; scientist adminstr, Nat Inst Allergy & Infectious Dis, 62-64 & Nat Heart Inst, 64-69, chief pulmonary dis bd, Nat Heart & Lung Inst, 69-71; STAFF TOXICOLOGIST AGR CHEM, CALIF DEPT FOOD & AGR, 71- Mem: AAAS; Am Vet Med Asn; Am Pub Health Asn; Conf Pub Health Vets; Soc Epidemiol Res. Res: Fungal infections occurring in man and animals; coccidioidomycosis; toxicology. Mailing Add: 1413 Notre Dame Davis CA 95616

MADDY, KENNETH HILTON, b Cleveland, Ohio, May 31, 23; m 46; c 3. BIOCHEMISTRY. Educ: Pa State Univ, BS, 44; Univ Wis, MS, 48; Pa State Univ, PhD(biochem), 52. Prof Exp: Res biochemist, Monsanto Co, Mo, 52-54, develop biochemist, 54-58, proj mgr, 58-60, mgr feed chem develop, 60-66, mgr new proj develop, 66-68, mgr life sci, New Enterprise Div, 68-69, Dir, Computerized Tech Dept, 69-70; pres, Maddy Assocs Inc, 71-75; MGR PROD DEVELOP, LONZA INC, 75- Mem: Am Chem Soc; Am Soc Animal Sci; Poultry Sci Asn; Animal Nutrit Res Coun; fel Am Inst Chemists. Res: Animal and human nutrition, specifically amino acids; dietary energy and vitamins; computer program design and development for animal production and human nutrition. Mailing Add: Lonza Inc 16 Godwin Lane St Louis MO 63124

MADER, ADOLF b Tübingen, Ger, Dec 25, 34; m 65. ALGEBRA. Educ: Univ Tübingen, MS, 61; NMex State Univ, PhD(math), 64. Prof Exp: Asst prof math, Univ Idaho, 64-65; from asst prof to assoc prof, 65-74, PROF MATH & CHMN DEPT, UNIV HAWAII, 74- Mem: Am Math Soc; Math Asn Am. Res: Abelian group theory. Mailing Add: Dept of Math Univ of Hawaii Honolulu HI 96822

MADER, CHARLES LAVERN, b Dewey, Okla, Aug 8, 30; m 60; c 1. PHYSICAL OCEANOGRAPHY. Educ: Okla State Univ, BS, 52, MS, 54. Prof Exp: Mem staff, 55-74, ASST GROUP LEADER, LOS ALAMOS SCI LAB, 74- Concurrent Pos: Consult, Defense Stand Labs, Melbourne, Australia, 69; mem fac, Hawaii Inst Geophys, Univ Hawaii, 72-73. Mem: Am Chem Soc; fel Am Inst Chemists; Am Phys Soc; Combustion Inst. Res: Explosives; thermodynamics; hydrodynamics; equation of state; the study of detonation chemistry and physics; chemically reactive fluid dynamics. Mailing Add: Los Alamos Sci Lab PO Box 1663 Los Alamos NM 87544

MADER, DONALD LEWIS, b Baltimore, Md, Nov 7, 26; m 50; c 3. FOREST SOILS, FOREST HYDROLOGY. Educ: Syracuse Univ, BS, 50; Univ Wis, MS, 54, PhD(soils), 56. Prof Exp: From asst prof to assoc prof forestry, 56-70, PROF FORESTRY, UNIV MASS, AMHERST, 70- Mem: AAAS; Soc Am Foresters; Soil Sci Soc Am; Ecol Soc Am; Am Inst Biol Sci. Res: Forest soils and ecology; watershed management. Mailing Add: Dept of Forestry & Wildlife Mgt Univ of Mass Amherst MA 01002

MADER, ERICH OTTO, b Neumark, Ger, Mar 21, 05; nat US; m 32; c 1. PLANT PATHOLOGY. Educ: Univ Geisenheim, BS, 26; Cornell Univ, PhD(plant path), 36. Prof Exp: Asst plant path, Univ Berlin-Dahlem, 26-27; asst & instr, Cornell Univ, 29-36; res dir, Yoder Bros, Ohio, 37-43; assoc plant physiologist, Bur Plant Indust, USDA, 43-44; hon fel, Univ Minn, 45-46, asst prof plant path, 46-48; sci adv, Andrews Nursery, 48-50; PVT BIOL RES, 50- Concurrent Pos: Sci adv, Butler Co Mushroom Farms, Pa, 37-43. Mem: AAAS; Am Soc Plant Physiol; Bot Soc Am; Am Phytopath Soc. Res: Physiology of fungi; parasitism; fungicides. Mailing Add: 1840 Berkeley Ave St Paul MN 55105

MADER, IVAN JOHN, b Iowa, Dec 29, 23; m 46; c 4. MEDICINE. Educ: Cornell Col, 41-43; Wayne State Univ, MS, 49, MD, 51; Am Bd Internal Med, dipl, 61. Prof Exp: Chief med & dir med educ, 62-68, CHIEF OF STAFF, WILLIAM BEAUMONT HOSP, 68- Concurrent Pos: Jr assoc, Detroit Gen Hosp, Mich, 56-; consult, Vet Admin, 57-; adj asst prof, Col Med, Wayne State Univ, 58- Mem: AMA; fel Am Col Physicians; Am Fedn Clin Res. Res: Renal disease. Mailing Add: William Beaumont Hosp 3535 W 13 Mile Rd Royal Oak MI 48072

MADER, WILLIAM JOHN, b Martins Ferry, Ohio, June 6, 11; m 42; c 3. CHEMISTRY. Educ: Western Reserve Univ, AB, 33; Ohio State Univ, MSc, 35. Prof Exp: Anal res chemist, Am Cyanamid Co, 35-42; chief anal res, Merck & Co, Inc, 46-51, from asst mgr to mgr, 51-55; dir, Anal Res Div, Ciba Pharmaceut Prods, Inc, 55-60, dir qual control div, 60-68, dir drug standards lab, Am Pharmaceut Asn Found, 68-71; dir qual control, 71-74, VPRES CONTROL, ALZA CORP, 74- Concurrent Pos: Mem revision comt, US Pharmacopeia, 75-80. Honors & Awards: Powers Award, 66. Mem: AAAS; Am Chem Soc; Am Pharmaceut Asn. Res: Analytical research on pharmaceuticals, especially antibiotics, vitamins and steroids; solubility of organic compounds. Mailing Add: Alza Corp 950 Page Mill Rd Palo Alto CA 94304

MADERA-ORSINI, FRANK, b Mayaguez, PR, Sept 29, 16; m 45; c 3. BIOCHEMISTRY. Educ: Ohio Univ, BS, 39; Univ Mo, MS, 56, PhD(biochem), 59. Prof Exp: Biochemist clin chem, St Joseph Mercy Hosp, 58-64; DIR CLIN BIOCHEM & RES, WEST SUBURBAN HOSP, 64- Concurrent Pos: Consult, Ment Health State Hosp, Ill; consult, Peoples Community Hosps, Wayne, Mich, 58-64. Mem: Am Chem Soc; Am Asn Clin Chem; Am Soc Cell Biol; NY Acad Sci. Res: Clinical and cell chemistry; aldosterone, antagonists and immunology; iron metabolism. Mailing Add: Biochem Dept West Suburban Hosp 518 N Austin Blvd Oak Park IL 60302

MADERSON, PAUL F A, b Kent, Eng, Dec 19, 38; m 61; c 2. DEVELOPMENTAL ANATOMY. Educ: Univ London, BSc, 60, PhD(zool), 62, DSc(zool), 72. Prof Exp: Asst lectr zool, Univ Hong Kong, 62-65; lectr, Univ Calif, Riverside, 65-66; NIH res assoc dermat, Mass Gen Hosp, Harvard Med Sch, 66-68; from asst prof to assoc prof, 68-73, PROF BIOL, BROOKLYN COL, 73- Concurrent Pos: Asst Prof, Boston Univ, 67-68; prof path, Univ Ark Med Ctr, 75- Mem: AAAS; Zool Soc London; Anat Soc Gt Brit & Ireland; Am Soc Zoologists; Am Soc Naturalists. Res: Anatomy and cell dynamics squamate epidermis; evolutionary morphology; regenerative phenomena in lower vertebrates. Mailing Add: Dept of Biol Brooklyn Col Brooklyn NY 11210

MADEY, RICHARD, b Brooklyn, NY, Feb 23, 22; m 51; c 6. NUCLEAR PHYSICS. Educ: Rensselaer Polytech Inst, BEE, 42; Univ Calif, Berkeley, PhD(physics), 52. Prof Exp: Elec engr, Allen B DuMont Labs, 43-44; physicist, Lawrence Radiation Lab, Univ Calif, 47-53; assoc physicist, Brookhaven Nat Lab, 53-56; from scientist to sr scientist, Repub Aviation Corp, 56-61, chief staff scientist mod physics, 61-62, chief appl physics res, 63-64; prof physics, Clarkson Col Technol, 65-71, chmn dept, 65; PROF PHYSICS & CHMN DEPT, KENT STATE UNIV, 71- Concurrent Pos: Guest scientist, Nevis Cyclotron Lab, Columbia Univ, 55 & 76, Brookhaven Nat Lab, 63 & 74, Foster Radiation Lab, McGill Univ, 67 & 68, Nat Res Coun Can, 68-70, Nuclear Struct Lab, Univ Rochester, 70 & Lawrence Berkeley Lab, Univ Calif, 71 & 72; consult, Ross Radio Corp, 52-53; Wcoast Electronics Lab, Willys Motors Co, 54-55, Kaiser Aircraft & Electronics Corp, 55-56, US AEC, 65-75 & Energy Res & Develop Admin, 75-; prin investr contracts & grants, US Air Force, US AEC, NSF, Energy Res & Develop Admin, NASA & NIH; guest scientist, Univ Md Cyclotron Lab, 73-75, Ohio Univ Accelerator Lab, 75 & Ind Univ Cyclotron Facil, 76. Honors & Awards: Army-Navy E Award, 43; Letter of Commendation, British Admiralty, 46; Naval Ordn Develop Award, Naval Ordn Lab, 46; Award, Am Nuclear Soc, 65. Mem: Fel AAAS; Am Nuclear Soc; Am Phys Soc; sr mem Inst Elec & Electronics Eng; fel NY Acad Sci. Res: Nuclear, space and environmental sciences; interaction of radiation with matter and with biological systems; transport of fluids through porous media; nuclear instrumentation. Mailing Add: 215 Overlook Dr Kent OH 44240

MADEY, THEODORE EUGENE, b Wilmington, Del, Oct 24, 37; m 60; c 4. SURFACE PHYSICS, SURFACE CHEMISTRY. Educ: Loyola Col Md, BS, 59; Univ Notre Dame, PhD(physics), 63. Prof Exp: Nat Res Coun res assoc, 63-65, PHYSICIST, NAT BUR STANDARDS, 65- Concurrent Pos: Vis scientist, Tech Univ Munich, 73. Mem: Fel Am Phys Soc; Am Vacuum Soc. Res: Physics and chemistry of solid surfaces; kinetics of adsorption on single crystal surfaces; interaction of slow electrons with adsorbed species; reactions at surfaces; surface standards; catalysis. Mailing Add: Nat Bur of Standards Washington DC 20234

MADHAV, R, b Bangalore, India, Jan 14, 38; m 67; c 1. ORGANIC CHEMISTRY. Educ: Univ Madras, BS, 57; Univ Delhi, MS, 60, PhD(chem), 65. Prof Exp: Res asst agr chem, Agr Res Inst, India, 57-58; res assoc chem, Univ Delhi, 65-68; res chemist, Carnegie-Mellon Univ, 68-71, res chemist, Mellon Inst, 71-75; RES BIOCHEMIST MICROBIOL, SCH MED, UNIV PITTSBURGH, 75- Concurrent Pos: USDA fel, Univ Delhi, 65-68; NIH fel, Carnegie-Mellon Univ, 68-71, Walter Reed Army Hosp fel, Mellon Inst, 71-75. Mem: Am Chem Soc; Nat Geog Soc. Res: Natural products chemistry; oxygen heterocycles; synthesis of nitrogen heterocycles; synthesis of fatty acids; synthesis of spin labeled compounds, analogs of coenzymes; preparation of substrates on polymer supports for studying enzyme properties. Mailing Add: Dept of Microbiol Sch of Med Univ of Pittsburgh Pittsburgh PA 15213

MADHOSINGH, CLARENCE, b Trinidad, WI, Sept 19, 31; m 59; c 1. MYCOLOGY, BIOCHEMISTRY. Educ: Univ BC, BSA, 54, MA, 58; Univ Western Ont, PhD(adv mycol), 59. Prof Exp: Res scientist, Plant Res Inst, Can Dept Agr, 60-67; RES SCIENTIST, AGR RES BR, CHEM & BIOL RES INST, AGR CAN, 67- Concurrent Pos: Fel, Biochem Inst, Univ Uppsala, 71-72. Mem: Asn Trop Biol; Can Bot Soc; Can Soc Microbiol; Can Phytopath Soc. Res: Comparative biochemistry of the fungi, involving a study of serological relationships of proteins significant in the taxonomy of wood-rotting fungi, sterol metabolism and tolerance and metabolism of fungicides. Mailing Add: Chem & Biol Res Inst Res Br Agr Can Carling Ave Ottawa ON Can

MADIGOSKY, WALTER MYRON, b Derby, Conn, Dec 15, 33; m 62. PHYSICS. Educ: Fairfield Univ, BS, 55; Univ Del, MS, 57; Catholic Univ, PhD(physics), 63. Prof Exp: RES PHYSICIST, US NAVAL SURFACE WEAPONS CTR, 57- Concurrent Pos: Sci officer phys acoust, Off Naval Res, 75-76. Mem: Acoust Soc Am. Res: Physical acoustics; liquid state; ultrasonics; underwater acoustics. Mailing Add: US Naval Surface Weapons Ctr White Oak Silver Spring MD 20910

MADIN, STEWART HARVEY, b Sheffield, Eng, Apr 3, 18; nat US; m 43; c 3. ANIMAL PATHOLOGY, VIROLOGY. Educ: Univ Calif, AB, 40, PhD, 60; Agr & Mech Col Tex, DVM, 43. Prof Exp: Asst histopath, Agr & Mech Col Tex, 42-43; assoc, Exp Sta, 43-44, assoc vet sci, 46-49, prin pathologist, Off Naval Res, 49-51, res pathologist, Naval Biol Lab, 51-57, asst sci dir, 57-60, from actg sci dir to dir, 60-68, PROF PUB HEALTH, EXP PATH & MED MICROBIOL, UNIV CALIF, BERKELEY, 61-; PROF EPIDEMIOL & PREV MED, SCH VET MED, UNIV CALIF, DAVIS, 67- Concurrent Pos: Lectr, Univ Calif, Berkeley, 50-61; chmn adv comt foot-and-mouth dis, Nat Acad Sci-Nat Res Coun, 67-; consult, Pan-Am Health Orgn, 67-; consult-at-large, Nat Cancer Inst; chmn consults, Agr Res Serv; mem, Navy Sr Scientist's Coun; mem foot-and-mouth dis res comt, Nat Acad Sci; mem, US Foot-and-Mouth Dis Comn II; mem cell cult comt, Nat Cancer Inst; mem, Comt Vesicular Dis & Animal Virus Classification. Mem: Am Vet Med Asn; US Animal Health Asn; NY Acad Sci. Res: Pathology of infectious diseases; vesicular viruses of domestic animals; pathogenesis of viral diseases; experimental pathology; tissue culture; comparative medicine of terrestrial and marine animals and oncology. Mailing Add: 3510 Life Sci Bldg Univ of Calif Sch of Pub Health Berkeley CA 94720

MADISON, BERNARD L, b Rocky Hill, Ky, Aug 1, 41. MATHEMATICS. Educ:

Western Ky Univ, BS, 62; Univ Ky, MS, 64, PhD(math), 66. Prof Exp: Asst prof, 66-71, ASSOC PROF MATH, LA STATE UNIV, BATON ROUGE, 71- Mem: Am Math Soc. Res: Topology. Mailing Add: Dept of Math La State Univ Baton Rouge LA 70803

MADISON, CAROLINE RABB, b Charleston, SC, Mar 25, 21; m 44; c 1. ZOOLOGY, GENETICS. Educ: Univ Ga, BS, 42; Ohio State Univ, MS, 49, PhD(genetics), 52. Prof Exp: Asst radiotracers, Am Cyanamid Co, 42-46; from asst prof to assoc prof, 62-69, PROF BIOL, KEAN COL NJ, 69- Mem: AAAS; Genetics Soc Am; Am Genetic Asn; Am Inst Biol Sci. Res: Embryology. Mailing Add: Dept of Biol Kean Col of NJ Union NJ 07083

MADISON, DALE MARTIN, b Urbana, Ill, Oct 20, 42; m 66; c 2. ANIMAL BEHAVIOR, VERTEBRATE BIOLOGY. Educ: Univ Md, BSc, 65, MSc, 68, PhD(ethology), 71. Prof Exp: Res specialist ethology, Univ Wis, 70-71, res assoc, 71-73; ASST PROF ETHOLOGY, McGILL UNIV, 73- Mem: AAAS; Animal Behav Soc; Am Soc Mammalogists; Am Soc Ichthyologists & Herpetologists; Sigma Xi. Res: Space utilization and social organization in vertebrates, especially small mammals; chemical communication and scent marking in vertebrates; vertebrate orientation systems, homing, migration, navigation, and sensory mechanisms of orientation. Mailing Add: Dept of Biol McGill Univ Montreal PQ Can

MADISON, DON HARVEY, b Pierre, SDak, Jan 4, 45; m 66; c 1. ATOMIC PHYSICS. Educ: Sioux Falls Col, BA, 67; Fla State Univ, MS, 70, PhD(physics), 72. Prof Exp: Instr physics, Fla State Univ, 72; res assoc, Univ NC, 72-74; ASST PROF PHYSICS, DRAKE UNIV, 74- Mem: Am Phys Soc; Am Asn Physics Teachers. Res: Theoretical aspects of collisions between charged particles and atoms. Mailing Add: Dept of Physics Drake Univ Des Moines IA 50311

MADISON, JAMES AMBROSE, b Woodstock, Ill, Jan 12, 28; m 54; c 4. GEOLOGY. Educ: Univ NC, BS, 51, MS, 55; Wash Univ, PhD(earth sci), 68. Prof Exp: ASSOC PROF EARTH SCI, DePAUW UNIV, 54-, HEAD DEPT, 68- Concurrent Pos: Lectr, JESSI Prog & NSF Summer Inst Jr High Sch Teachers, 59-70; vis lectr high schs, Ind. Mem: Nat Asn Geol Teachers. Res: Mineralogy; geophysics; engineering geology. Mailing Add: Dept of Earth Sci DePauw Univ Greencastle IN 46135

MADISON, JAMES THOMAS, b Adrian, Minn, Oct 23, 33; m 58; c 1. BIOLOGICAL CHEMISTRY. Educ: Colo State Univ, BS, 55; Univ Utah, PhD(biol chem), 62. Prof Exp: Res assoc biochem, Cornell Univ, 62-64; RES CHEMIST, PLANT, SOIL & NUTRIT LAB, AGR RES SERV, USDA, 64- Concurrent Pos: NIH fel, 62-64. Mem: AAAS; Am Chem Soc; Am Soc Biol Chemists. Res: Chemistry and nucleotide sequence of soluble ribonucleic acids; protein synthesis. Mailing Add: Plant Soil & Nutrit Lab Tower Rd Ithaca NY 14853

MADISON, JOHN HERBERT, JR, b Burlington, Iowa, Jan 16, 18; m 42; c 4. HORTICULTURE. Educ: Oberlin Col, BA, 42; Cornell Univ, PhD(plant physiol), 53. Prof Exp: Instr, Cornell Univ, 52-53; from asst prof landscape hort & asst horticulturist to assoc prof & assoc horticulturist, 53-71, PROF LANDSCAPE HORT & HORTICULTURIST, UNIV CALIF, DAVIS, 71- Concurrent Pos: Fel, Univ Edinburgh, 69-70; res assoc, Univ BC, 75. Mem: Bot Soc Am; Am Soc Plant Physiologists; Am Soc Agron; Soil Sci Soc Am. Res: Monocotyledon biology, soil-water-plant relations of amended soils. Mailing Add: Dept of Environ Hort Univ of Calif Davis CA 95616

MADISON, JOHN JOSEPH, physical chemistry, see 12th edition

MADISON, KENNETH MENEFEE, b Albert Lea, Minn, Apr 5, 15; m 36; c 2. MICROBIOLOGY. Educ: George Washington Univ, AB, 37, MA, 38; Johns Hopkins Univ, PhD(zool), 42. Prof Exp: With US Treasury Dept, 36-41; res physiologist, Joseph E Seagram & Sons, Inc, Ky, 41-43; sr chemist, Schenley Res Inst, Inc, Ind, 43-44, chief bacteriologist, 44-45; fel, Mellon Inst, 45-46, sr fel, 46-50; from asst prof to assoc prof, 50-61, PROF MICROBIOL, UNIV ILL, CHICAGO CIRCLE, 61- Concurrent Pos: Vis lectr, Oriental Inst, Univ Chicago, 60-62. Mem: Am Soc Microbiol; Soc Study Evolution. Res: Nucleus; synthesis by microorganisms; origin of life; primitive evolution; history of science; Chinese philosophy. Mailing Add: Dept of Biol Sci Univ of Ill Chicago IL 60680

MADISON, LEONARD LINCOLN, b New York, NY, Feb 11, 20; div; c 2. ENDOCRINOLOGY, METABOLISM. Educ: Ohio State Univ, BA, 41; Long Island Col Med, MD, 44. Prof Exp: From asst prof to assoc prof, 52-64, PROF INTERNAL MED, UNIV TEX HEALTH SCI CTR, DALLAS, 64- Concurrent Pos: Mem gen med study sect, NIH, 62-66. Mem: Am Soc Clin Invest; Asn Am Physicians; Endocrine Soc; Am Diabetes Asn. Res: Cerebral, carbohydrate and hepatic metabolism; diabetes mellitus. Mailing Add: Dept of Internal Med Univ of Tex Health Sci Ctr Dallas TX 75235

MADISON, VINCENT STEWART, b Adrian, Minn, Feb 10, 43; m 70; c 1. BIOPHYSICAL CHEMISTRY. Educ: Univ Minn, Minneapolis, BChem, 65; Univ Ore, PhD(chem), 69. Prof Exp: Instr chem, Nat Univ Trujillo, Peru, 70; fel biochem, Univ Calif Med Ctr, 70-71, Sch Med, Harvard, 71-74; ASST PROF MED CHEM, UNIV ILL MED CTR, 75- Concurrent Pos: NIH fel, 71-73. Res: Molecular forces which determine peptide conformation and function; determination of conformation by circular dichroism and nuclear magnetic resonance spectroscopy. Mailing Add: Dept of Chem Col of Pharm Univ of Ill 833 S Wood St Chicago IL 60680

MADISON, WILLIAM LEON, b Westerly, RI, Dec 28, 19; m 47; c 3. PHARMACEUTICAL CHEMISTRY. Educ: RI Col Pharm, BS, 42; Purdue Univ, MS, 48, PhD(pharmaceut chem), 49. Prof Exp: Anal chemist, Burroughs Wellcome & Co, 42-43; sr res scientist, Bristol Labs, Inc, 49-53; mgr, Pharmaceut Develop Dept, 53-60, asst dir res & dir pharm res, 60-71, ASSOC DIR CLIN RES & SERV, McNEIL LABS, INC, 71- Mem: Am Chem Soc; Am Pharmaceut Asn. Res: Biological research with radioactive isotopes; pharmaceutical development. Mailing Add: McNeil Labs Inc Camp Hill Rd Ft Washington PA 19034

MADISSOO, HARRY, b Paide, Estonia, July 4, 24; US citizen; m 64; c 2. VETERINARY TOXICOLOGY. Educ: Hannover Vet Col, DVM, 60. Prof Exp: Res asst physiol, Med Col, Cornell Univ, 52-57; from res scientist to sr res scientist toxicol, Squibb Inst Med Res, 60-64; res scientist, 64-65, DIR TOXICOL, BRISTOL LABS, 65- Mem: AAAS; Am Vet Med Asn; Ny Acad Sci; Soc Toxicol. Res: Kidney physiology; veterinary pharmacology; toxicology of drugs, natural products and cosmetics. Mailing Add: Dept of Toxicol Bristol Labs Po Box 657 Syracuse NY 13201

MADOC-JONES, HYWEL, b Cardiff, Wales, Nov 7, 38; m 67; c 2. RADIOTHERAPY. Educ: Oxford Univ, BA, 60, MA, 66; Univ London, PhD(biochem), 65; Univ Chicago, MD, 73. Prof Exp: Fel cell biol, Ont Cancer Inst & Dept Med Biophys, Univ Toronto, 65-67; res asst prof radiation & cell biol, Mallinckrodt Inst Radiol, Sch Med, Wash Univ, 67-68, asst prof, 68-69; res assoc radio, Univ Chicago, 69-73, Intern, Univ Chicago Hosps, 73-74; FEL, M D ANDERSON HOSP & TUMOR INST, 74- Mem: Can Soc Cell Biol. Res: Radiosensitivity of tumor cells in tissue culture; mechanisms of action of cytotoxic drugs; control of cell division. Mailing Add: Dept of Radiother M D Anderson Hosp & Tumor Inst Houston TX 77025

MADOFF, MILTON, b Detroit, Mich, July 22, 24; m 48; c 2. ORGANIC CHEMISTRY. Educ: Wayne Univ, BS, 44, PhD(chem), 51; Univ Southern Calif, MS, 46. Prof Exp: Res chemist, Ingram Res Inst, 46-47; org res chemist, Edsel B Ford Inst Med Res, 51-55; self employed, 55-56; PROG DEVELOP MGR, ROCKETDYNE DIV, ROCKWELL INT CORP, 56- Mem: Am Chem Soc. Res: Polymers; organic synthesis; rocket propellants. Mailing Add: Advan Progs Rocketdyne Div Rockwell Int Corp 6633 Canoga Ave Canoga Park CA 91304

MADOFF, MORTON A, b Clinton, Mass, Oct 20, 27; m 53; c 3. INFECTIOUS DISEASES, IMMUNOLOGY. Educ: Tulane Univ, BA, 51, MD, 55. Prof Exp: Asst med, New Eng Ctr Hosp, 58-60, asst physician, 60-64; chief infectious dis serv, Lemuel Shattuck Hosp, 64-69; dir biol lab, 67-70, dir, Bur Adult & Maternal Health Servs, 70-72, SUPT, STATE LAB INST, MASS DEPT PUB HEALTH, 72- Concurrent Pos: Nat Found res fel, 58-60; Nat Inst Allergy & Infectious Dis res career develop award, 62-67; prof, Sch Med, Tufts Univ. Mem: AAAS; Am Asn Immunologists; Infectious Dis Soc Am; Am Fedn Clin Res; Soc Exp Biol & Med. Res: Cell membranes; staphylococcal toxins; microbial exotoxins. Mailing Add: State Lab Inst 375 South St Jamaica Plain MA 02130

MADOLE, RICHARD FRANK, geomorphology, photogeology, see 12th edition

MADOR, IRVING LESTER, physical chemistry, see 12th edition

MADORE, BERNADETTE, b Barnston, Que, Jan 24, 18; US citizen. BOTANY, BACTERIOLOGY. Educ: Univ Montreal, AB, 42, BEd, 43; Catholic Univ, MS, 49, PhD(ecol, bot), 51. Prof Exp: Instr math, Marie Anne Col, Can, 43-44, Dean, 52-75, PROF BIOL, ANNA MARIA COL, 49-, VPRES, 75- Mem: AAAS; Am Soc Microbiol; Nat Asn Biol Teachers. Res: Ecology. Mailing Add: Dept of Biol Anna Maria Col Paxton MA 01612

MADOW, LEO, b Cleveland, Ohio, Oct 18, 15; m 42; c 2. PSYCHIATRY, NEUROLOGY. Educ: Western Reserve Univ, BA, 37, MD, 42; Ohio State Univ, MA, 38. Prof Exp: Prof neurol, 56-64, PROF PSYCHIAT & CHMN DEPT, MED COL PA, 64- Concurrent Pos: Training analyst, Philadelphia Psychoanal Inst. Mailing Add: Med Col of Pa Dept of Psychiat 3300 Henry Ave Philadelphia PA 19129

MADOW, WILLIAM GREGORY, b New York, NY, Feb 22, 11; m 42. MATHEMATICAL STATISTICS. Educ: Columbia Univ, AB, 32, MA, 33, PhD(math), 38. Prof Exp: Asst math adv, Pop Div, US Bur Census, 40-41; actg chief, Statist Sect, Div Health & Disability Studies, Social Security Bd, 42; statist adv, Rationing Dept, Off Price Admin, 43-44; sampling specialist, US Bur Census, 44-46; vis prof statist, Univ Sao Paulo, 46-48; prof math statist, Univ NC, 48-49; prof, Univ Ill, 49-57; CONSULT PROF STATIST, STANFORD UNIV, 57-, STAFF SCIENTIST STATIST, STANFORD RES INST, 57- Concurrent Pos: Fund Advan Educ fel, Inst Advan Study, NJ, 53-54; assoc ed, Ann Math Statist, Inst Math Statist, 53-55; chmn, Comt Math Training Social Scientists, Soc Sci Res Coun, 53-61; assoc ed, J Am Statist Asn, 51-53 & 57-63; mem panel statist consults, US Bur Census, 55-68; fel, Ctr Advan Study Behav Sci, 56-57; mem res adv comt, Calif Dept Ment Hyg, 61-66; mem census adv comt, Am Statist Asn, 73-75, chmn, 75. Mem: Fel AAAS; fel Am Statist Asn; fel Inst Math Statist; Biomet Soc; Economet Soc. Res: Sample survey design and analysis; analysis of quasi-experiments; mathematical models in social science. Mailing Add: 700 New Hampshire Ave NW Washington DC 20037

MADRAZO, ALFONSO A, b Tabasco, Mex, Mar 6, 31; US citizen; m 60; c 2. PATHOLOGY. Educ: Nat Univ Mex, MD, 54; Am Bd Path, dipl, 65. Prof Exp: From instr to asst prof path, Col Med, Seton Hall Univ, 63-67; attend pathologist, St Vincents Hosp & Med Ctr, New York, 67-73; DIR LABS, CHRIST HOSP, JERSEY CITY, NJ, 73- Concurrent Pos: Res assoc path, Mt Sinai Sch Med, New York, 69-; clin assoc prof path, NJ Col Med, 73-; consult & attend pathologist, St Vincents Hosp & Med Ctr, 75- Mem: Fel Int Acad Path; fel Am Soc Cell Biol; fel Am Soc Clin Path; fel Col Am Pathologists. Res: Effects of radiation on human tissues, especially the kidney parenchyma and blood vessels. Mailing Add: Christ Hosp Jersey City NJ 07306

MADSEN, DAVID CHRISTY, b Chelsea, Mass, May 3, 43; m 67; c 2. PHYSIOLOGY, BIOCHEMISTRY. Educ: Merrimack Col, BA, 64; Univ Mass, Amherst, MA, 67, PhD(zool), 72. Prof Exp: Fel microbiol, 72-73, ASST FAC FEL MICROBIOL, LOBUND LAB, UNIV NOTRE DAME, 73- Mem: AAAS; Am Asn Univ Prof; Asn Gnotobiotics. Res: Bile acid and cholesterol metabolism; role of dietary factors and of microflora of the intestine. Mailing Add: Lobund Lab Univ of Notre Dame Notre Dame IN 46556

MADSEN, FRED CHRISTIAN, b Rocksprings, Wyo, Dec 9, 45; m 73; c 1. ANIMAL NUTRITION. Educ: Austin Peay State Univ, BS, 68; Univ Tenn, Knoxville, MS, 72, PhD(animal nutrit), 74. Prof Exp: Res assoc nutrit & physiol, Comp Animal Res Lab, 74-76; TECH SERV COORDR, SYNTEX AGRIBUS, INC, 76- Mem: Am Soc Animal Sci. Res: Factors affecting insulin secretion and metabolism in the ruminant animal; energy supplementation and the relationship to mineral utilization and physical characteristics in the gastrointestinal tract. Mailing Add: Syntex Agribus Inc PO Box 1246 SSS Springfield MO 65805

MADSEN, HAROLD F, b San Jose, Calif, Mar 31, 21; m 45; c 3. ENTOMOLOGY. Educ: San Jose Col, AB, 43; Univ Calif, PhD, 49. Prof Exp: Exten entomologist, Univ Calif, 49-52, from asst entomologist to assoc entomologist, 52-64; RES SCIENTIST, AGR CAN, 64- Mem: AAAS; Entom Soc Am; Entom Soc Can. Res: Pest management of insect pests of deciduous fruits. Mailing Add: Res Sta Agr Can Summerland BC Can

MADSEN, JAMES HENRY, JR, b Salt Lake City, Utah, July 28, 32; m 56; c 2. GEOLOGY, VERTEBRATE PALEONTOLOGY. Educ: Univ Utah, BS, 57, MS, 59. Prof Exp: CUR DINOSAUR LAB, UTAH MUS NATURAL HIST, UNIV UTAH, 59-, ASST RES PROF GEOL & GEOPHYS SCI, UNIV, 69- Mem: Paleont Soc; Soc Vert Paleont. Res: Stratigraphic and field geology; vertebrate paleontology, especially Upper Jurassic Carnosaurs. Mailing Add: Dinosaur Lab Dept Geol & Geophys Univ Utah Salt Lake City UT 84112

MADSEN, KENNETH OLAF, b Lavoye, Wyo, May 30, 26; m 50; c 3. NUTRITIONAL BIOCHEMISTRY. Educ: Univ Wyo, BS, 50; Univ Wis, MS, 53, PhD(biochem), 58. Prof Exp: From instr to assoc prof, 58-70, PROF BIOCHEM, DENT BR, UNIV TEX, HOUSTON, 70- Mem: AAAS; Soc Exp Biol & Med; Int Asn Dent Res. Res: Bull semen metabolism; fluorine metabolism and toxicity; trace mineral nutrition; oral biology and dental research, especially dental caries in rats; redox substances in nutrition; biology of the cotton rat. Mailing Add: Dept of Biochem Univ of Tex Dent Br Houston TX 77025

MADSEN, LOUIS LINDEN, b Salt Lake City, Utah, Sept 29, 07; m 32; c 8. ANIMAL NUTRITION. Educ: Utah State Col, BS, 30; Cornell Univ, PhD(animal nutrit), 34. Prof Exp: Asst animal nutrit, Cornell Univ, 30-34; Nat Res Coun fel & asst path, Columbia Univ, 34-35; asst chem, Exp Sta, Mich State Col, 36-37; assoc animal nutrit, Bur Animal Indust, USDA, 37-39, nutritionist, 39-45, head cattle res sect, Animal & Poultry Husb Res Br, Agr Res Serv, 53-55; head dept animal husb, Utah State Agr Col, 45-50, pres col, 50-53; dir inst agr sci, 55-65, dean col agr, 65-74, EMER DEAN COL AGR, WASH STATE UNIV, 74- Concurrent Pos: Mem comt animal nutrit, Nat Res Coun, 45-51; mem, Wash Bd Natural Resources & Wash Soil & Water Conserv Comt. Mem: Am Soc Animal Sci; Am Inst Nutrit. Res: Pathology and histology of nutritional diseases; purified and synthetic diet work; vitamin A in beef cattle nutrition; vitamin B complex in swine nutrition; trace elements in cattle nutrition; nutrition in range cattle and sheep. Mailing Add: Col of Agr Wash State Univ Pullman WA 99163

MADSEN, MILTON ANDREW, b Manti, Utah, Apr 21, 12. ANIMAL HUSBANDRY. Educ: Utah State Agr Col, BS, 34, MS, 39; Univ Wis, PhD, 58. Prof Exp: From instr to asst prof, 35-46, ASSOC PROF ANIMAL SCI, UTAH STATE UNIV, 47- Mem: Am Soc Animal Sci. Res: Sheep and wool production. Mailing Add: Dept of Animal Sci Utah State Univ Logan UT 84321

MADSEN, NEIL BERNARD, b Grande Prairie, Alta, Feb 8, 28; m 52; c 2. BIOCHEMISTRY. Educ: Univ Alta, BSc, 50, MSc, 52; Wash Univ, PhD, 55. Prof Exp: Instr biol chem, Wash Univ, 55-56; fel biochem, Oxford Univ, 56-57; res off, Microbiol Res Inst, Can Dept Agr, 57-62; assoc prof, 62-69, PROF BIOCHEM, UNIV ALTA, 69- Concurrent Pos: Med Res Coun Can vis scientist, Oxford Univ, Eng, 72-73; chmn sci policy comt, Can Fedn Biol Soc, 73-75. Mem: Am Soc Biol Chemists; Can Biochem Soc (past pres). Res: Enzyme chemistry; structure-function relationships in glycogen phosphorylase as determined by protein chemistry, kinetics and x-ray crystallography; biological control of glycogen metabolism. Mailing Add: Dept of Biochem Univ of Alta Edmonton AB Can

MADSEN, PAUL O, b Denmark, July 25, 27; US citizen; m 55; c 3. UROLOGY. Educ: Copenhagen Univ, MD, 52; Univ Heidelberg, DrMed, 58. Prof Exp: Intern surg, Genesee Hosp, Rochester, NY, 53-54; resident surg & urol, Gen Hosp, Buffalo, 55-57; resident, Buffalo Gen Hosp & Roswell Park Mem Inst, 59-61; vis urologist, Univ Hamburg, 61-62; assoc prof, 68-71, PROF UROL, UNIV WIS-MADISON, 71-; CHIEF UROL, VET ADMIN HOSP, 62- Mem: Int Soc Urol; Int Soc Surg; Am Urol Asn; Soc Univ Urologists. Res: Treatment of urinary tract infections; experimental pyelonephritis and hydronephrosis; prostatic tissue cultures; prostatilis. Mailing Add: Vet Admin Hosp Dept of Urol 2500 Overlook Terr Madison WI 53705

MADSEN, VICTOR ARVIEL, b Idaho Falls, Idaho, Feb 14, 31; m 55; c 2. THEORETICAL NUCLEAR PHYSICS. Educ: Wash Univ, BS, 53, PhD(physics), 61. Prof Exp: Res assoc physics, Case Inst Technol, 61-63; asst prof, 63-66, ASSOC PROF PHYSICS, ORE STATE UNIV, 66- Concurrent Pos: Physicist, Niels Bohr Inst, Copenhagen & Atomic Energy Res Estab, Eng, 69-70; consult, Los Alamos Sci Lab, Group P9, 75-76. Mem: Am Phys Soc. Res: Theory of nuclear inelastic scattering; core-polarization effects; charge-exchange reactions; photocapture. Mailing Add: Dept of Physics Ore State Univ Corvallis OR 97331

MADSEN, WILLIAM, b Shanghai, China, Dec 26, 20; US citizen; m 45; c 2. ANTHROPOLOGY. Educ: Stanford Univ, BA, 46; Univ Calif, Berkeley, PhD(anthrop), 55. Prof Exp: Teaching asst anthrop, Univ Calif, Berkeley, 51-55; from asst prof to assoc prof, Univ Tex, Austin, 55-63; res assoc, Inst for Study Human Probs, Stanford Univ, 63; prof, Purdue Univ, 64-66; PROF ANTHROP, UNIV CALIF, SANTA BARBARA, 66- Concurrent Pos: Hogg Found Ment Health fel, 57-62 & Ctr Advan Study Behav Sci, 62-63. Mem: AAAS; fel Am Anthrop Asn. Res: Acculturation; Meso-America; culture and health. Mailing Add: Dept of Anthrop Univ of Calif Santa Barbara CA 93106

MADSON, JAMES MERLE, plasma physics, see 12th edition

MADSON, WILLARD HEGLAND, b Canton, Minn, Mar 22, 04; m 33; c 4. CHEMISTRY. Educ: St Olaf Col, BA, 26; Univ Wis, MS, 27, PhD(chem), 31. Prof Exp: Asst instr chem, Univ Wis, 26-27, 30-31, instr, Exten Div, Milwaukee, 27-30; instr inorg chem, Univ Ill, 31-32, spec res assoc, 32-33; res chemist, Pigments Dept, E I du Pont de Nemours & Co, Md, 33-43, res div head, 43-44, head pigment dept, 44-48, head white pigments, Sales Serv Div, 48-58, mgr trade sales paints, Tech Serv, 58-66, consult, 66-67; asst prof, 67-70, ASSOC PROF CHEM, COL STEUBENVILLE, 70- Concurrent Pos: Tech consult, Zinc Inst, 69- Mem: Am Chem Soc. Res: Colloidal chemistry; titanium; dioxides; pigments; surface coatings. Mailing Add: Dept of Chem Col of Steubenville Steubenville OH 43954

MAECK, JOHN VAN SICKLEN, b Shelburne, Vt, Mar 10, 14; m 39; c 2. OBSTETRICS & GYNECOLOGY. Educ: Univ Vt, BS, 36, MD, 39; Am Bd Obstet & Gynec, dipl, 49. Prof Exp: Asst clin instr path, Yale Univ, 39; intern, Lenox Hill Hosp, 40-41; resident obstet & gynec, Woman's Hosp, New York, 47-48; from instr to assoc prof, 48-56, PROF OBSTET & GYNEC, COL MED, UNIV VT, 56-, CHMN DEPT, 50-, CHIEF SERV, MED CTR HOSP, 67- Concurrent Pos: Chief serv, Mary Fletcher Hosp, 50-66 & DeGoesbriand Mem Hosp, 64-66; dir, examr & first vpres, Am Bd Obstet & Gynec, Am Bd rep to Am Bd Med Specialists. Mem: Fel Am Col Surgeons; fel Am Col Obstet & Gynec; Am Fertil Soc; Am Gynec Soc; Asn Profs Gynec & Obstet. Res: Carcinoma of the cervix and endometrium; toxemia of pregnancy; estrogen metabolism; early detection of carcinoma; rubella and pregnancy; computerized problem-oriented medical records. Mailing Add: Given Bldg Univ of Vt Col of Med Burlington VT 05401

MAEHL, RONALD CHARLES, b St Louis, Mo, Jan 18, 48; m 69. SPACE PHYSICS. Educ: Univ Mo-Rolla, BS, 69; Washington Univ, MA, 71, PhD(physics), 74. Prof Exp: Nat Acad Sci-Nat Res Coun res assoc cosmic ray physics, NASA Goddard Space Flight Ctr, 74-76; RES PHYSICIST PLANETARY PHYSICS, LAB SPACE PHYSICS, AIKEN INDUSTS, 76- Res: Cosmic ray astrophysics, isotopic, chemical and energy spectra of galactic cosmic rays; planetary physics, specifically the structure of the ionospheric regions of the Earth and Venus. Mailing Add: 4419 Jupiter Rockville MD 20853

MAEHR, HUBERT, b Schlins, Austria, Feb 25, 35. NATURAL PRODUCTS CHEMISTRY. Educ: State Univ Agr & Forestry, Austria, Dipl Ing, 60; Rutgers Univ, PhD(chem microbial prod), 64. Prof Exp: Res assoc chem antibiotics, Inst Microbiol, Rutgers Univ, 64-65; RES CHEMIST DIV CHEM RES, HOFFMANN-LA ROCHE INC, 66- Mem: Am Chem Soc. Res: Separations and structure elucidations of natural products. Mailing Add: Chem Res Div Hoffmann-La Roche Inc Nutley NJ 07110

MAELAND, ARNULF JULIUS, physical chemistry, see 12th edition

MAENCHEN, GEORGE, b Moscow, Russia, Apr 27, 28; nat US; m 47; c 3. PHYSICS. Educ: Univ Calif, AB, 50, PhD(physics), 57. Prof Exp: PHYSICIST, LAWRENCE LIVERMORE LAB, 57- Mem: Am Phys Soc. Res: Nuclear and general physics. Mailing Add: Lawrence Livermore Lab Univ of Calif PO Box 808 Livermore CA 94550

MAENDER, OTTO WILLIAM b Brooklyn, NY, Jan 18, 40; m 63; c 2. CHEMISTRY. Educ: Rochester Inst Technol, BS, 63; Iowa State Univ, MS, 65; Univ Ga, PhD(org chem), 69. Prof Exp: SR RES CHEMIST, MONSANTO CO, 69- Res: Electron spin resonance study of oxidative processes; antidegradants; vulcanization. Mailing Add: Monsanto Co 260 Springside Dr Akron OH 44313

MAENGWYN-DAVIES, GERTRUDE DIANE, b Paris, France, Dec 28, 10; nat US; m 46. PHARMACOLOGY. Educ: Univ Vienna, MPharm, 37; Johns Hopkins Univ, PhD(biochem), 52. Prof Exp: Asst pharmacog, Univ Vienna, 37-38; res chemist, Warwick Chem Co, 42-43; head analyst fermentation, Overly Biochem Res Found, 43-44; asst, NY Univ, 44-45; res chemist, Quaker Oats Co, 46; from instr to asst prof ophthal, Med Sch, Johns Hopkins Univ, 52-55; assoc res prof pharmacol, Sch Med, George Washington Univ, 55-56; assoc prof, 56-63, PROF PHARMACOL, SCHS MED & DENT, GEORGETOWN UNIV, 63-, PROF PHYSIOL & BIOPHYS, 75- Concurrent Pos: Guest worker, Lab Clin Sci, NIMH, 70- Mem: Fel AAAS; Am Soc Biol Chemists; Am Soc Pharmacol & Exp Therapeut; Soc Exp Biol & Med; Brit Pharmacol Soc. Res: Enzyme and pharmacological kinetics; biochemistry; autonomics; catecholamines biosynthesis and release; stress. Mailing Add: 15205 Tottenham Terr Silver Spring MD 20906

MAENZA, RONALD MORTON, b New York, NY, May 8, 36; m 60; c 3. PATHOLOGY. Educ: Columbia Univ, BA, 57, MD, 61; Am Bd Path, dipl, 66. Prof Exp: Intern med, Bronx Munic Hosp, 61-62; resident path, Columbia Presby Hosp, NY, 62-65; resident, Englewood Hosp, NJ, 65-66; asst prof, 68-73, ASSOC PROF PATH, SCHS MED & DENT MED, UNIV CONN, 73- Concurrent Pos: Consult, Vet Admin Hosp, Newington, Conn, 68-; teaching affil, Hartford Hosp, 68- Mem: Int Acad Path; Am Asn Path & Bact. Res: Chemical carcinogenesis in tissue culture. Mailing Add: Dept of Path Univ of Conn Sch of Med Farmington CT 06032

MAERKER, GERHARD, b Bernburg, Ger, Nov, 4, 23; nat US; m 51; c 2. ORGANIC CHEMISTRY. Educ: Philadelphia Col Pharm, BS, 51; Temple Univ, MA, 52, PhD(chem), 57. Prof Exp: Res chemist, Allied Chem Corp, 52-58; res chemist, 58-71, CHIEF ANIMAL FAT PROD LAB, EASTERN REGIONAL RES CTR, USDA, 71- Mem: AAAS; Am Chem Soc; Am Oil Chem Soc. Res: Synthesis, properties and reactions of chemical derivatives of fats and fatty acids, particularly small ring heterocyclic derivatives such as epoxides and aziridines. Mailing Add: Eastern Regional Res Ctr USDA Philadelphia PA 19118

MAERKER, RICHARD ERWIN, nuclear physics, see 12th edition

MAEROV, SIDNEY BENJAMIN, b Calgary, Alta, Feb 25, 27; US citizen; m 53; c 3. ORGANIC CHEMISTRY. Educ: Univ Alta, BSc, 49; Univ Wash, PhD(org chem), 54. Prof Exp: From res chemist to sr res chemist, 54-62, RES ASSOC, DACRON RES LAB, E I DU PONT DE NEMOURS & CO, INC, 62- Mem: Am Chem Soc. Res: Polymer synthesis, especially polyesters; photochemistry of dyes and polymers; basic fiber studies; organic coatings. Mailing Add: Du Pont Exp Sta E I Du Pont de Nemours Wilmington DE 19898

MAESTRE, MARCOS FRANCISCO, b San Juan, PR, June 20, 32; US citizen; m 58; c 2. BIOPHYSICS. Educ: Univ Mich, BS, 54; Wayne State Univ, MS, 58; Yale Univ, PhD(biophysics), 63. Prof Exp: Res assoc, 63-65, ASSOC RES CHEMIST, SPACE SCI LABS, UNIV CALIF, BERKELEY, 65-, LECTR MED PHYSICS, 73- Concurrent Pos: USPHS fel, 63-65. Mem: Biophys Soc. Res: X-ray studies on virus structure; theory of the birefringence of nucleic adids; electric birefringence of bacteriophage; optical rotatory dispersion of viruses; viral nucleic acids and protein components. Mailing Add: Space Sci Labs Univ of Calif Berkeley CA 94720

MAESTRELLO, LUCIO, b Legnago, Italy, Mar 4, 28; US citizen; m 58; c 4. ACOUSTICS. Educ: Galileo-Ferraris Inst, Dipl eng, 50; Univ Southampton, DPhil(acoustics), 76. Prof Exp: Res asst aerodynamics, Imp Col, Univ London, 53-54; res assoc acoustics, Inst Aerophys, Univ Toronto, 56-57; res engr, Boeing Co, 58-70; HEAD AEROACOUSTICS SECT, NASA-LANGLEY RES CTR, 70- Res: Jet and boundary layer noise. Mailing Add: Aeroacoustics Sect NASA-Langley Res Ctr Hampton VA 23665

MAESTRONE, GIANPAOLO, b Urgnano, Italy, Jan 31, 30; US citizen; m 56; c 3. VETERINARY MEDICINE, MICROBIOLOGY. Educ: Univ Milan, DVM, 51; Am Col Vet Microbiol, dipl, 67. Prof Exp: Asst prof infectious dis, Vet Sch, Univ Milan, 51-56; res assoc microbiol, Animal Med Ctr, NY, 57-61; sr res microbiologist, Squibb Inst Med Res, NJ, 61-66; SR RES MICROBIOLOGIST, HOFFMANN-LA ROCHE, INC, 66- Mem: AAAS; Am Col Lab Animal Med; Am Vet Med Asn; Am Soc Vet Parasitol; assoc felNY Acad Med. Res: Diagnosis of infectious genital diseases of cattle; chemotherapy of infectious diseases and parasites; immunofluorescence applied to diagnosis of leptospiral and viral diseases; antibiotic sensitivity of clinical isolates; epidemiology of leptospirosis; experimental infections. Mailing Add: Hoffmann-La Roche Inc Nutley NJ 07100

MAFARACHISI, BOAZ AMON, b Rusape, Rhodesia, May 25, 36; Brit subj; m 67; c 2. NUTRITION. Educ: Union Col, Ky, BA, 65; Purdue Univ, Lafayette, MC, 67; Univ Mass, PhD(nutrit), 74. Prof Exp: NUTRITIONIST, KELLOGG CO, 73- Mem: Nutrit Today Soc; Inst Food Technol; Sigma Xi. Res: Evaluation, establishment, improvement and maintenance of nutritional quality of company food products; cooperation of current, nutritional knowledge and government regulations into new and existing Kellogg's ready-to-eat cereals. Mailing Add: Kellogg Co Battle Creek MI 49017

MAFFETT, ANDREW L, b Port Royal, Pa, Oct 23, 21; m 43. MATHEMATICS. Educ: Gettysburg Col, AB, 43; Univ Mich, MA, 48. Prof Exp: From instr to asst prof math, Gettysburg Col, 47-54; res assoc, Univ Mich, 51-52, res engr, Radiation Lab, 54-56, asst head lab, 56-57; sr mathematician, Bendix Aviation Corp, 58-59; res mathematician, Inst Sci & Technol, Univ Mich, 60-61; assoc head sr anal staff, Conductron Corp, 61-67; sr scientist, KMS Indust, Inc, 67-69; INDEPENDENT CONSULT, 69-; ADJ PROF MATH, UNIV MICH-DEARBORN, 70- Concurrent Pos: US mem comn VI, Int Sci Radio Union. Res: Electromagnetic boundary value problems arising from both the radiation and scattering of enrgy from various geometric shapes and configurations. Mailing Add: 2250 N Zeeb Rd R D 1 Dexter MI 48130

MAFFLY, LEROY HERRICK, b Berkeley, Calif, Nov 26, 27; m 52; c 3. MEDICINE. Educ: Univ Calif, AB, 49, MD, 52. Prof Exp: Intern med, Univ Calif Hosp, 52-53, asst resident, 53-54; resident, Herrick Mem Hosp, 54-55; res fel, Harvard Med Sch & Mass Gen Hosp, 57-59; res fel, Sch Med, Univ Calif, San Francisco, 59-61; from asst prof to assoc prof, 61-70, PROF MED, SCH MED, STANFORD UNIV, 70- Concurrent Pos: NSF fel, 57-58; USPHS fel, 58-60; Am Heart Asn fel, 60-61; estab investr, Am Heart Asn, 61-66, mem exec comt, Coun Kidney in Cardiovasc Dis, 68-, chmn comt, 71-73, mem cardiovasc A res study comt, 74-; mem adv comt renal dialysis ctrs, State of Calif, 66-70; mem gen med B study sect, NIH, 67-71; staff physician, Vet Admin Hosp, Palo Alto, Calif, 68-; mem sci adv bd, Nat Kidney Found, 70- Mem: Biophys Soc; Am Physiol Soc; Am Fedn Clin Res; Am Soc Clin Invest. Res: Transport processes and permeability of biological membranes; nephrology; computer-assisted education. Mailing Add: Dept of Med Stanford Univ Sch of Med Stanford CA 94305

MAGA, JOSEPH ANDREW, b New Kensington, Pa, Dec 25, 40; m 64; c 2. FOOD SCIENCE, BIOCHEMISTRY. Educ: Pa State Univ, BS, 62, MS, 64; Kans State Univ, PhD(food sci), 70. Prof Exp: Proj leader dairy prod, Borden Foods Co, 64-66; group leader simulated dairy prod, Cent Soya Co, 66-68; asst prof, 70-74, ASSOC PROF FOOD SCI, COLO STATE UNIV, 74- Mem: Am Dairy Sci Asn; Inst Food Technologists. Res: Flavor aspects of foods, especially high protein foods, including composition, chemistry and preferences. Mailing Add: Dept of Food Sci & Nutrit Colo State Univ Ft Collins CO 80521

MAGALHAES, HULDA, b New York, NY, Apr 9, 14. ZOOLOGY. Educ: Douglass Col, Rutgers Univ, BS, 35; Mt Holyoke Col, Univ, MA, 37; Duke Univ, PhD(zool), 44. Prof Exp: Asst physiol, Mt Holyoke Col, 35-37; instr, Woman's Med Col Pa, 37-40; asst zool, Duke Univ, 41-42, instr, 43-46; asst prof physiol & hyg, 46-49, assoc prof physiol, 49-54, PROF ZOOL, BUCKNELL UNIV, 54- Concurrent Pos: Vis prof, Pa State Univ, 64. Mem: AAAS; Am Soc Zool; Am Asn Lab Animal Sci; Am Genetic Asn; Soc Study Reproduction. Res: Anatomy, cytogenetics and teratology of the golden hamster. Mailing Add: Dept of Biol Bucknell Univ Lewisburg PA 17837

MAGARIAN, EDWARD O, b East St Louis, Ill, Oct 3, 35; m 59; c 2. PHARMACEUTICAL CHEMISTRY. Educ: Univ Miss, BA, 58, PhD(pharmaceut chem), 64. Prof Exp: Asst prof, Univ RI, 64-67; assoc prof, Col Pharm, Univ Ky, 67-73; ASSOC PROF PHARMACEUT CHEM, COL PHARM, NDAK STATE UNIV, 73- Mem: Am Chem Soc; Am Pharmaceut Asn; NY Acad Sci. Res: Anticonvulsants; enzyme inhibitors. Mailing Add: Dept of Pharmaceut Chem NDak State Univ Fargo ND 58102

MAGARIAN, ELIZABETH ANN, b Orlando, Fla, July 13, 40. ALGEBRA. Educ: Asbury Col, AB, 60; Fla State Univ, MS, 61, PhD(math), 68. Prof Exp: Instr math, La State Univ, New Orleans, 61-64; asst prof, 68-73, ASSOC PROF MATH, STETSON UNIV, 73- Mem: Am Math Soc; Math Asn Am. Res: Commutative ring theory. Mailing Add: Dept of Math Stetson Univ DeLand FL 32720

MAGARIAN, ROBERT ARMEN, b East St Louis, Ill, July 27, 30; m 50; c 4. MEDICINAL CHEMISTRY, ORGANIC CHEMISTRY. Educ: Univ Miss, BS, 56, BSPh, 60, PhD, 66. Prof Exp: NIH fel, Col Pharm, Univ Kans, 66-67; asst prof med chem, St Louis Col Pharm, 67-70; ASSOC PROF MED CHEM, COL PHARM, UNIV OKLA, 70- Concurrent Pos: NSF grant, St Louis Col Pharm, 69-70; dir & secy, Pharmaceut Consult, Inc, Okla, 71- Mem: AAAS; Am Chem Soc; The Chem Soc; Am Pharmaceut Asn; Acad Pharmaceut Sci. Res: Synthetic medicinal chemistry; relation of molecular structure to biological activity; synthetic estrogens and anti-Parkinson agents. Mailing Add: Col of Pharm Univ of Okla Norman OK 73069

MAGARVEY, RAYMOND HALLIDAY, b Parker's Cove, NS, Dec 27, 12; m 37; c 2. PHYSICS, PHYSICAL METEOROLOGY. Educ: Acadia Univ, BSc, 49; McGill Univ, MSc, 50. Prof Exp: From asst prof to assoc prof, 51-56, PROF PHYSICS, ACADIA UNIV, 56- Concurrent Pos: Physicist, NS Res Found, 51- Mem: Am Geophys Union; Can Asn Physicists; Can Meteorol Soc; Royal Meteorol Soc. Res: Fluid dynamics; physical meteorology. Mailing Add: Dept of Physics Acadia Univ Wolfville NS Can

MAGASANIK, BORIS, b Kharkoff, Russia, Dec 19, 19; nat US; m 49. MICROBIOLOGY. Educ: City Col NY, BS, 41; Columbia Univ, PhD(biochem), 48. Hon Degrees: MA, Harvard Univ, 58. Prof Exp: Res asst biochem, Columbia Univ, 48-49; Ernst fel bact & immunol, Harvard Univ, 49-51, assoc, 51-53, from asst prof to assoc prof, 53-60; PROF MICROBIOL, MASS INST TECHNOL, 60-, HEAD DEPT BIOL, 67- Concurrent Pos: Markle scholar, Harvard Univ, 51-56; Guggenheim fel, Pasteur Inst, Paris, 59; ed, Biochem & Biophys Res Commun, 64- Mem: AAAS; Am Chem Soc; Am Soc Biol Chem; Am Soc Microbiol; Am Acad Arts & Sci. Res: Microbial physiology and biochemistry. Mailing Add: Dept of Biol Mass Inst of Technol Cambridge MA 02139

MAGAT, EUGENE EDWARD, b Kharkov, Russia, July 8, 19; nat US; m 45; c 3. CHEMISTRY. Educ: Mass Inst Technol, BS, 43, PhD(org chem), 45. Prof Exp: Res chemist, Carothers Res Lab, Exp Sta, 45-50, res assoc, 50-52, res supvr, 52-62, res fel, 62-64, RES MGR, TEXTILE FIBERS DEPT, PIONEERING RES LAB, E I DU PONT DE NEMOURS & CO, INC, 64- Mem: Am Chem Soc. Res: Condensation polymers; textile chemistry; radiation chemistry; fiber technology. Mailing Add: Pioneering Res Lab E I du Pont de Nemours & Co Inc Wilmington DE 19898

MAGDE, DOUGLAS, b Rochester, NY, Feb 12, 42. CHEMICAL PHYSICS. Educ: Boston Col, BS, 63; Cornell Univ, MS, 68, PhD(physics), 70. Prof Exp: Res assoc biophys, Cornell Univ, 70-72; res assoc chem physics, Wash State Univ, 72-74; ASST PROF CHEM, UNIV CALIF, SAN DIEGO, 74- Mem: Am Asn Physics Teachers. Res: Concentration correlation analysis, a new kinetic probe of biophysical processes; picosecond flash photolysis and spectroscopy in the study of fast molecular dynamics. Mailing Add: Dept of Chem B-014 Univ of Calif at San Diego La Jolla CA 92093

MAGDER, JULES, b Toronto, July 17, 34; m 63; c 2. PHYSICAL CHEMISTRY, INORGANIC CHEMISTRY. Educ: Univ Toronto, BA, 58, PhD(inorg chem), 61. Prof Exp: Proj supvr, Horizons, Inc, 61-64; proj mgr pigments & chem div, Glidden Co, 64-65; dir res, Princeton Chem Res, Inc, 65-71; vpres, 72-73, PRES, PRINCETON ORGANICS, INC, 73- Mem: Am Chem Soc; Am Ceramic Soc. Res: Heterogeneous catalysis; inorganic polymers; polymer technology; composites; inorganic foams; refractory and structural clay materials; cement and concrete materials. Mailing Add: 385 Walnut Lane Princeton NJ 08540

MAGDOFF, FREDERICK ROBIN, b Washington, DC, Apr 5, 42; c 1. SOIL CHEMISTRY, ENVIRONMENTAL SCIENCE. Educ: Oberlin Col, BA, 63; Cornell Univ, MS, 65, PhD(soil sci), 69. Prof Exp: Res scientist soils, Soil Conserv Div, Israeli Ministry Agr, 69-71; fel, Univ Wis, 72-73; ASST PROF SOILS, UNIV VT, 73- Mem: Am Soc Agron; Soil Sci Soc Am. Res: Land disposal of human and agricultural wastes; soil fertility. Mailing Add: Dept of Plant & Soil Sci Univ of Vt Burlington VT 05401

MAGDOFF-FAIRCHILD, BEATRICE, biophysics, see 12th edition

MAGE, MICHAEL GORDON, b New York, Aug 17, 34; m 55; c 3. IMMUNOLOGY. Educ: Cornell Univ, AB, 55; Columbia Univ, DDS, 60. Prof Exp: USPHS fel microbiol, Columbia Univ, 60-62; res immunochemist, Nat Inst Dent Res, 62-66, RES IMMUNOCHEMIST, NAT CANCER INST, 66- Mem: AAAS; Am Chem Soc; Am Asn Immunologists; Am Pub Health Asn. Res: Protein chemistry; antibody structure and specificity; immunoglobulin structure; mammalian cell separation. Mailing Add: 7008 Wilson Lane Bethesda MD 20034

MAGEAU, RICHARD PAUL, b Flushing, NY, Oct 24, 41; m 64; c 3. MICROBIOLOGY, IMMUNOLOGY. Educ: Univ Conn, BA, 63; Univ Md, MS, 66, PhD(microbiol), 68. Prof Exp: Res fel & asst microbiol, Univ Md, 63-68; from asst prof to assoc prof, Kans State Col Pittsburgh, 68-74; RES MICROBIOLOGIST, SCI SERV, MICROBIOL STAFF, MED MICROBIOL GROUP, ANIMAL, PLANT & HEALTH INSPECTION SERV, USDA, 74- Mem: Am Soc Microbiol; NY Acad Sci; AAAS. Res: Radioimmunoassay and other immunological methods of staphylococcal enterotoxin quantitation and detection; laboratory methods development. Mailing Add: USDA APHIS Rm 101 Bldg 318 ARC-West Beltsville MD 20705

MAGEE, ADEN COMBS, III, b Dimmitt, Tex, Dec 8, 30; m 56; c 2. NUTRITION. Educ: Tex A&M Univ, BS, 53; NC State Univ, MS, 57, PhD(animal nutrit), 60. Prof Exp: Asst animal nutrit, NC State Univ, 55-60; from asst prof to assoc prof, 60-68, PROF NUTRIT, UNIV N C, GREENSBORO, 68- Mem: AAAS; Am Chem Soc; Biomet Soc; Am Inst Nutrit; Am Dietetic Asn. Res: Mineral metabolism and toxicities; mineral interrelationships. Mailing Add: Sch of Home Econ Univ of N C Greensboro NC 27412

MAGEE, CHARLES BRIAN, b Detroit, Mich, Sept 26, 26; m 55; c 7. PHYSICAL CHEMISTRY, INORGANIC CHEMISTRY. Educ: Univ Detroit, BS, 50; Purdue Univ, PhD(phys chem), 54. Prof Exp: Sr engr, Aircraft Nuclear Propulsion Dept, Gen Elec Co, Ohio, 55-59; proj scientist, Booz Allen Appl Res, Inc, Ill, 59-61; from asst prof to prof metall, Univ Denver, 61-75, actg chmn dept, 65-66, RES METALLURGIST, DENVER RES INST, UNIV DENVER, 61- Mem: Am Chem Soc; Am Nuclear Soc. Res: Molecular structure; compounds of saline hydrides and transition-metals; thermodynamic properties of intermetallic compounds; thermionic and photoelectric emission of metals and inorganic compounds; nuclear reactor materials. Mailing Add: Div Metall & Mat Sci Denver Res Inst Univ Denver Denver CO 80210

MAGEE, DONAL FRANCIS, b Aberdeen, Scotland, June 4, 24; nat US; m 50; c 5. PHYSIOLOGY, PHARMACOLOGY. Educ: Oxford Univ, BA, 44, MA, BM & BCh, 48; Univ Ill, PhD(physiol), 52. Prof Exp: Instr & res assoc clin sci, Univ Ill, 48-51; from asst prof to prof pharmacol, Sch Med, Univ Wash, 51-65; PROF PHYSIOL & CHMN DEPT PHYSIOL & PHARMACOL, CREIGHTON UNIV, 65- Concurrent Pos: Guggenheim Mem Found fel, Rowett Inst, Scotland, 59-60. Mem: Soc Exp Biol & Med; Am Physiol Soc; Am Soc Pharmacol & Exp Therapeut; Am Gastroenterol Asn; Brit Med Asn. Res: Gastrointestinal tract, especially physiology and pharmacology of pancreas and biliary canal and stomach. Mailing Add: Dept of Physiol & Pharmacol Creighton Univ Omaha NE 68131

MAGEE, DONALD WALLACE, physical chemistry, see 12th edition

MAGEE, ELLINGTON MCFALL, b San Augustine Co, Tex, Nov 30, 29; m 50. PHYSICAL CHEMISTRY. Educ: Univ Tex, BS, 53; Univ Wis, PhD(chem), 56. Prof Exp: Chemist, Humble Oil & Refining Co, Tex, 56-59; sr chemist, Esso Res & Eng Co, 59-62, res assoc, 62-68; SR RES ASSOC, GOVT RES LABS, EXXON RES & ENG CO, 68- Mem: Am Chem Soc. Res: Gas phase oxidation; oxidation of hydrocarbons; aromatic raw materials, reactions and separations; coal gasification and liquefaction; environmental evaluation. Mailing Add: Govt Res Labs Exxon Res & Eng Co PO Box 8 Linden NJ 07036

MAGEE, GORDON RICHEY, b London, Ont, May 23, 02; m 31; c 1. MATHEMATICS. Educ: Univ Western Ont, BA, 25; Univ Chicago, MS, 26, PhD(math), 33. Prof Exp: From instr to prof & head dept, 27-67, part-time prof, 67-73, EMER PROF MATH, UNIV WESTERN ONT, 73- Concurrent Pos: Examr in chief, Ont Dept Educ, 44-46, 50-52, 56-58 & 62-63. Mem: Am Math Soc; Math Asn Am; Royal Astron Soc Can (secy-treas, 31-33, pres, 43); Can Math Cong. Res: Conjugate nets of ruled surfaces in a congruence. Mailing Add: Dept of Math Univ of Western Ont London ON Can

MAGEE, JOHN FRANCIS, b Bangor, Maine, Dec 3, 26; m 49; c 3. MATHEMATICS. Educ: Bowdoin Col, AB, 46; Harvard Univ, MBA, 48; Univ Maine, AM, 53. Prof Exp: Mem staff financial anal, Johns-Manville Co, 49-50; dir res, Opers Res Group, 50-59, head, 59-62, vpres mgt serv div, 63-68, mem corp tech staff, 68-69, exec vpres & dir, 69, chief operating officer, 71, pres, 72-74, CHIEF EXEC OFFICER, ARTHUR D LITTLE INC, 74- Concurrent Pos: Pres, Inst Mgt Sci, 71-72. Mem: Opers Res Soc Am (pres, 66-67). Res: Operations research; marketing; financial planning and policy; managerial controls. Mailing Add: Arthur D Little Inc 25 Acorn Park Cambridge MA 02140

MAGEE, JOHN LAFAYETTE, b Franklinton, La, Oct 28, 14; m 48; c 3. PHYSICAL CHEMISTRY. Educ: Miss Col, AB, 35; Vanderbilt Univ, MS, 36; Univ Wis, PhD(chem), 39. Prof Exp: Nat Res Coun fel, Princeton Univ, 39-40, res assoc, 40-41; res physicist, B F Goodrich Co, 41-43; group leader, Los Alamos Sci Lab, 43-45 & Naval Ord Testing Sta, Calif, 45-46; sr scientist, Argonne Nat Lab, 46-48; from asst prof to assoc prof, 48-53, assoc dir radiation lab, 54-71, head dept chem, Univ, 67-71, PROF CHEM, UNIV NOTRE DAME, 53-, DIR RADIATION LAB, 71- Concurrent Pos: Observer, Proj Bikini, 46; mem weapons systs eval group, US Dept Defense, 55-57. Mem: AAAS; Am Chem Soc; Am Phys Soc; Radiation Res Soc; The Chem Soc. Res: Photochemistry; chemical reaction rate theory; radiation chemistry; application of statistical mechanics and quantum mechanics to chemistry. Mailing Add: Radiation Lab Univ of Notre Dame Notre Dame IN 46556

MAGEE, JOHN ROBERT, b Bristol, RI, Aug 27, 16; m 49; c 5. TEXTILE CHEMISTRY. Educ: Brown Univ, AB, 39. Prof Exp: Chem supt, Nat Dairy Prod Corp, 39-48; gen supt, Va-Carolina Chem Corp, 48-58; asst mgr spec prod, Charles Pfizer & Co, Inc, 58; staff assoc res & develop, 58-60, sect head nylon develop, 60-61, MGR DEVELOP OPERS & TECH SERV, CHEMSTRAND DIV, MONSANTO CO, 62- Concurrent Pos: Consult, Charles Pfizer & Co, Inc & Textile Res Inst, NJ, 58. Mem: Am Chem Soc; Am Asn Textile Technol. Res: Melt and wet spun man-made fibers, especially nylon, polyester, acrylics, protein and copolymer blends. Mailing Add: 3844 Dunwoody Dr Pensacola FL 32503

MAGEE, JOHN STOREY, JR, b Baltimore, Md, Mar 23, 31; m 53; c 3. INORGANIC CHEMISTRY, PHYSICAL CHEMISTRY. Educ: Loyola Col, Md, BS, 53; Univ Del, PhD(inorg chem), 61. Prof Exp: Chemist, Davison Chem Corp Div, 53-55, res chemist, Res Div, 61-63, res chemist, Davison Chem Corp Div, 63-71, RES DIR, PETROL CATALYST DEPT, DAVISON CHEM CORP DIV, W R GRACE & CO, 71- Mem: Am Chem Soc. Res: Preparation of catalysts for heterogeneous reactions;

study of compounds containing metals in unusual oxidation states; preparation and evaluation of petroleum catalysts. Mailing Add: 1195 Hoods Mill Rd Cooksville MD 21723

MAGEE, KENNETH RAYMOND, b Gardner, Ill, July 30, 26; m 48; c 4. NEUROLOGY. Educ: Purdue Univ, BS, 46; Univ Chicago, MD, 48, MS, 49; Univ Mich, MA, 53; Am Bd Psychiat & Neurol, dipl, 55. Prof Exp: Intern, 49-50, resident, 50-53, from instr to assoc prof, 53-65, PROF NEUROL, UNIV MICH, ANN ARBOR, 65- Concurrent Pos: Consult, Vet Admin Hosp, Ann Arbor, 53, Northville State Psychiat Hosp, 57- & Wayne County Gen Hosp, Eloise, 58-; clin asst prof, Georgetown Univ & clin assoc, Nat Inst Neurol Dis & Blindness, 54-56; mem ment health comn, State of Mich, 63, mem ment health adv coun, 64-; mem med adv bd, Myasthenia Gravis Found Am, 63-; mem neurol sci res training comt, Nat Inst Neurol Dis & Blindness, 63-68. Mem: Am Neurol Asn; Am Acad Neurol; Soc Clin Neurol (pres, 55-56); assoc Am Asn Neurol Surg; Asn Res Nerv & Ment Dis. Res: Neuromuscular diseases; Parkinson's disease; headache; epilepsy. Mailing Add: 614 Riverview Ann Arbor MI 48104

MAGEE, LYMAN ABBOTT, b Bogalusa, La, Apr 10, 26; m 57; c 2. MICROBIOLOGY. Educ: La Col, BS, 46; La State Univ, MS, 54, PhD(bact), 58. Prof Exp: Sr analyst, Cities Serv Refining Corp, La, 47-48; asst prof chem, La Col, 48-51, dean men, 51; Hite fel cancer res, M D Anderson Hosp & Tumor Inst, Univ Tex, 58-59; Nat Inst Allergy & Infectious Diseases fel & instr microbiol, Sch Med, 60-61, assoc prof biol, Univ, 61-65, PROF BIOL, UNIV MISS, 65-, CHMN DEPT, 71-, ASSOC MICROBIOL, SCH MED, 63- Concurrent Pos: Am Soc Microbiol Pres fel virol, Commun Dis Ctr, USPHS & Southern Res Inst, Ga, 65. Mem: Am Soc Microbiol; fel Am Acad Microbiol. Res: Cancer viruses; metabolism of exanthem viruses; microbial decomposition of pesticides. Mailing Add: Dept of Biol Univ of Miss University MS 38677

MAGEE, MICHAEL JACK, b Coleman, Tex, Feb 16, 46; m 69; c 1. COMPUTER SCIENCES. Educ: Univ Tex, BA, 68, MA, 72, PhD(comput sci), 75. Prof Exp: Assoc programmer, Int Bus Mach Corp, 68-71; ASST PROF COMPUT SCI, UNIV WYO, 75- Res: Artificial intelligence; pattern recognition; computational linguistics. Mailing Add: Dept of Comput Sci Univ of Wyo Laramie WY 82070

MAGEE, PAUL TERRY, b Los Angeles, Calif, Oct 26, 37; m 64; c 2. GENETICS, BIOCHEMISTRY. Educ: Yale Univ, BS, 59; Univ Calif, Berkeley, PhD(biochem), 64. Prof Exp: Am Cancer Soc fel, Lab Enzymol, Nat Ctr Sci Res, Gif-sur-Yvette, France, 64-66; from asst prof to assoc prof microbiol, 66-74, ASSOC PROF HUMAN GENETICS, SCH MED, YALE UNIV, 74- Res: Regulation of gene expression; developmental biology. Mailing Add: Dept of Human Genetics Yale Univ New Haven CT 06510

MAGEE, PHILIP STEWART, b Kecoughtan, Va, Dec 9, 26; m 49, 71; c 7. PHYSICAL ORGANIC CHEMISTRY. Educ: Univ Southern Calif, BS, 49, MS, 52; Univ Calif, Los Angeles, PhD(phys org chem), 55. Prof Exp: Res chemist, Calif Res Corp, Standard Oil Co Calif, 55-61, sr res chemist, 61-62, sr res chemist, Ortho Div, 62-66, group supvr pesticide res, 66-68; SUPVR ORG SYNTHESIS, ORTHO DIV, CHEVRON CHEM CO, 68- Concurrent Pos: Lectr, Univ Calif Exten, Berkeley, 56-60 & 73-; ed, Topics in Sulfur Chem, 73- Res: Solvolysis studies in neighboring group systems; radiotracer studies of inorganic exchange reactions; halogenation, oxydation and hydrogenation of petroleum based hydrocarbons; synthesis of new pesticides; inventor-monitor and orthene insecticides; simulated metabolism studies; quantitative structure-activity relations. Mailing Add: Chevron Chem Co 940 Hensley St Richmond CA 94804

MAGEE, RICHARD JOSEPH, b Hempstead, NY, Mar 15, 27; m 52; c 3. ORGANIC CHEMISTRY. Educ: St John's Univ, NY, BS, 48; Columbia Univ, PhD(chem), 54. Prof Exp: Asst prof chem, St John's Univ NY, 52-55; guest chemist, Brookhaven Nat Labs, 55; res chemist, Stamford Labs, Am Cyanamid Co, 55-59; group leader chem, 59-66, mgr org synthesis, 66-69, DIR CHEM RES & DEVELOP, AM CYANAMID AGR CTR, 69- Concurrent Pos: Lectr, Univ Conn, 58-59. Mem: AAAS; Am Chem Soc. Res: Organic chemistry of phosphorus; pesticides; animal feed and health products. Mailing Add: Am Cyanamid Agr Ctr Princeton NJ 08540

MAGEE, STEVE CARL, b Tylertown, Miss, Sept 19, 42; m 66. BIOCHEMISTRY. Educ: La State Univ, BS, 65, MS, 67, PhD(biochem), 71. Prof Exp: NIH fel biochem, Okla State Univ, 71-75; WITH RES & DEVELOP LAB, SIGMA CHEM CO, 75- Res: Enzyme structure-function relationships; effects of metabolites on kinetics of enzymic reactions; comparative enzymology. Mailing Add: Res & Develop Lab Sigma Chem Co 3500 DeKalb St St Louis MO 63178

MAGEE, THOMAS ALEXANDER, b Brookhaven, Miss, Apr 19, 30; m 61; c 1. ORGANIC CHEMISTRY. Educ: Tulane Univ, BS, 52, MS, 55; Univ Tenn, PhD(chem), 57. Prof Exp: CHEMIST, T R EVANS RES CTR, DIAMOND SHAMROCK CORP, 57- Mem: Am Chem Soc. Res: Organometallic chemistry of the transition elements; pesticide chemistry. Mailing Add: T R Evans Res Ctr Diamond Shamrock Corp PO Box 348 Painesville OH 44077

MAGEE, WAYNE EDWARD, b Big Rapids, Mich, Apr 11, 29; m 51; c 3. VIROLOGY, BIOCHEMISTRY. Educ: Kalamazoo Col, BA, 51; Univ Wis, MS, 53, PhD(biochem), 55. Prof Exp: Res scientist microbiol, Upjohn Co, 55-60, proj leader biochem, 60-63, proj leader virol res, 63-66, sr res scientist, 67-71, sr scientist exp biol, 71; prof life sci, Ind State Univ, Terre Haute, 71-74; PROF & DIR DIV ALLIED HEALTH & LIFE SCI, UNIV TEX, SAN ANTONIO, 75-, PROF BIOCHEM, HEALTH SCI CTR, 75- Concurrent Pos: Mem adj staff, Dept Biol, Western Mich Univ, 68-70, adj prof, 71; adj prof microbiol, Terre Haute Ctr Med Educ, Sch Med, Ind Univ, 72-74. Mem: AAAS; Am Chem Soc; Am Soc Biol Chem; Am Soc Microbiol; NY Acad Sci. Res: Nitrogen fixation; nucleic acid biochemistry; biochemistry of virus infection; mechanism of action of therapeutic agents and interferon. Mailing Add: Div of Allied Health & Life Sci Univ of Tex San Antonio TX 78285

MAGEE, WILLIAM LOVEL, b Ft William, Ont, Mar 24, 29; m 55; c 2. BIOCHEMISTRY. Educ: Univ Western Ont, BSc, 52, MSc, 54, PhD(biochem), 57. Prof Exp: Nat Multiple Sclerosis Soc overseas fel biochem, Guy's Hosp Med Sch, London, Eng, 58-61; res asst, 61-62, lectr, 62-63, asst prof, 63-68, ASSOC PROF BIOCHEM, HEALTH SCI CTR, UNIV WESTERN ONT, 68- Mem: AAAS; Can Biochem Soc. Res: Biochemistry of brain and peripheral nerve; biochemistry of demyelination; phospholipid chemistry and metabolism. Mailing Add: Dept of Biochem Health Sci Ctr Univ of Western Ont London ON Can

MAGEE, WILLIAM THOMAS, b San Antonio, Tex, Apr 22, 23; m 48; c 3. ANIMAL BREEDING. Educ: Tex A&M Univ, BS, 47; Iowa State Univ, MS, 48, PhD(animal husb), 51. Prof Exp: Asst animal husbandman, Tex A&M Univ, 53-55; from asst prof to assoc prof, 55-65, PROF ANIMAL BREEDING, MICH STATE UNIV, 65- Mem: Am Soc Animal Sci; Am Genetic Asn. Res: Evaluation of the effects of selection and mating systems on performance traits in beef cattle and swine. Mailing Add: Dept of Animal Husb Mich State Univ East Lansing MI 48823

MAGEL, BERNARD, chemistry, see 12th edition

MAGELI, ORVILLE LEONARD, organic chemistry, see 12th edition

MAGENHEIMER, JOHN JOSEPH, b Brooklyn, NY, Nov 27, 44; m 67; c 1. Educ: Fairfield Univ, BS, 66; Cath Univ Am, PhD(phys chem), 71. Prof Exp: TECH MGR, RAYBESTOS-MANHATTAN INC, 70- Mem: Am Chem Soc. Res: Chemical kinetics, photochemistry, colloid chemistry and adsorption chemistry; surface chemistry. Mailing Add: Corp Res Lab Raybestos-Manhattan Inc Stratford CT 06484

MAGER, MILTON, b New York, NY, Dec 13, 20; m 47; c 3. BIOCHEMISTRY. Educ: NY Univ, BA, 43; Rensselaer Polytech Univ, MS, 49; Boston Univ, PhD(biochem), 54. Prof Exp: Instr chem, Albany Col Pharm, 46-49; biochemist, Qm Res & Eng Ctr, 53-56, chief biochem sect physiol br, 56-61; chief biochem lab, 61-67, dir biochem & pharmacol lab, 67-75, DIR HEAT RES DIV, US ARMY RES INST ENVIRON MED, 75- Concurrent Pos: Asst, Sch Med, Boston Univ, 52-55, instr, 55-59, asst res prof, 59-65, assoc res prof, 65- Mem: AAAS; Am Chem Soc; Am Asn Clin Chemists; Am Physiol Soc. Res: Biochemical responses of man to environmental stress; temperature regulation; blood and tissue enzymes; lipids; methodology. Mailing Add: Heat Res Div US Army Res Inst of Environ Med Natick MA 01760

MAGERLEIN, BARNEY JOHN, b Columbus, Ohio, Nov 11, 19; m 44; c 4. MEDICINAL CHEMISTRY. Educ: Capital Univ, BS, 41; Ohio State Univ, PhD(org chem), 46. Prof Exp: Asst org chem, Ohio State Univ, 41-44, Off Sci Res & Develop contract, Univ Res Found, 44-45; chemist, 46-71, DISTINGUISHED RES SCIENTIST, UPJOHN CO, 71- Concurrent Pos: Vis scholar, Univ Calif, Los Angeles, 60-61. Mem: Am Chem Soc; Sigma Xi. Res: Organic synthesis; synthetic studies of morphine series; steroids and pteridines; antibiotics. Mailing Add: Infectious Dis Res Upjohn Co Kalamazoo MI 49007

MAGERLEIN, JOHN HAROLD, b Kalamazoo, Mich, Aug 1, 47. EXPERIMENTAL SOLID STATE PHYSICS. Educ: Kalamazoo Col, BA, 69; Univ Mich, MS, 71, PhD(physics), 75. Prof Exp: FEL PHYSICS, BELL LABS, AM TEL & TEL CO, 75- Mem: Am Phys Soc. Res: Josephson effect; liquid helium. Mailing Add: Bell Labs 600 Mountain Ave Murray Hill NJ 07974

MAGGENTI, ARMAND RICHARD, b San Jose, Calif, Feb 15, 33; m 63; c 2. PLANT NEMATOLOGY. Educ: Univ Calif, Berkeley, BS, 54, PhD, 59. Prof Exp: From asst nematologist to nematologist, 58-74, LECTR NEMATOL & CHMN DEPT, UNIV CALIF, DAVIS, 74- Concurrent Pos: Fulbright fels, Pakistan, 65, Iraq, 65-66. Mem: Soc Nematol. Res: Taxonomy and morphology of soil and freshwater nematodes; nematode parasites of fish. Mailing Add: Dept of Nematol Univ of Calif Davis CA 95616

MAGGIO, EDWARD THOMAS, b Brooklyn, NY, Mar 28, 47. PROTEIN CHEMISTRY, CLINICAL BIOCHEMISTRY. Educ: Polytech Inst Brooklyn, BS, 68; Univ Mich, MS, 69, PhD(biol chem), 73. Prof Exp: Fel biochem dept pharmaceut chem, Univ Calif, San Francisco, 73-74, NIH fel pharmaceut chem, 74-75; group leader enzym, 75, MGR BIOCHEM SECT, SYVA RES INST, 75- Mem: Am Chem Soc. Res: Enzymology in clinical and pharmaceutical chemistry; enzyme kinetics; chemical modification of proteins; physical biochemistry. Mailing Add: Syva Res Inst 3221 Porter Dr Palo Alto CA 94304

MAGGIO, FRANCIS XAVIER, b Brooklyn, NY, Jan 5, 25; m 56; c 4. CHEMISTRY. Educ: LI Univ, BS, 48. Prof Exp: Res chemist, Am Alkyd Industs, 49-51, Crown Oil Chem Co, 51-54 & Interchem Corp, 52-57; chief chemist, Mitchell-Rand Corp, 57-60; sr proj scientist, Aerovox Corp, 60; sr res chemist, Am Potash & Chem, 60-65; MGR POLYMER RES, SWEDLOW, INC, GARDEN GROVE, 65- Mem: Am Inst Chem; Am Chem Soc. Res: Phosphorus-boron polymers; epoxy chemistry; free radical polymerization of methylmethacrylates. Mailing Add: 1555 Sandalwood Dr Brea CA 92621

MAGGIO, THOMAS EDWARD, organic chemistry, medicinal chemistry, see 12th edition

MAGGIO-CAVALIERE, MARY, b White Castle, La, Aug 15, 40; m 71; c 1. CLINICAL PHARMACOLOGY. Educ: Loyola Univ, La, BS, 63; Tulane Univ, MS, 65, PhD(pharmacol), 68. Prof Exp: Asst prof pharmacol, Northeast La State Col, 68-69; ASST DIR, MED DIV, CIBA-GEIGY CORP, 69- Mem: Am Pharmaceut Asn; Am Col Clin Pharmacol; Am Soc Clin Pharmacol & Therapeut; NY Acad Sci. Res: Nerve-muscle; acetylcholine-acetylcholinesterase; toxicology; drug metabolism. Mailing Add: 25 Seven Oaks Dr Summit NJ 07901

MAGGIOLO, ALLISON, organic polymer chemistry, textile chemistry, see 12th edition

MAGGIORA, GERALD M, b Oakland, Calif, Aug 11, 38; m 63; c 2. MOLECULAR BIOPHYSICS. Educ: Univ Calif, Davis, BS, 64, PhD(biophys), 68. Prof Exp: Res assoc, 68-69, USPHS fel, 69-70, asst prof, 70-74, ASSOC PROF BIOCHEM, UNIV KANS, 74- Mem: Am Chem Soc; Am Phys Soc; Int Soc Quantum Biol. Res: Molecular quantum mechanics; quantum biochemistry; photosynthesis; vision; electromagnetic properties of large molecules; interaction of light with matter. Mailing Add: Dept of Biochem Univ of Kans Lawrence KS 66044

MAGID, ANDY ROY, b St Paul, Minn, May 4, 44; m 66. MATHEMATICS. Educ: Univ Calif, Berkeley, BA, 66; Northwestern Univ, PhD(math), 69. Prof Exp: J F Ritt asst prof math, Columbia Univ, 66-72; ASSOC PROF MATH, UNIV OKLA, 72- Concurrent Pos: Vis assoc prof math, Univ Ill, 75-76. Mem: Am Math Soc. Res: Commutative algebra; Galois theory; algebraic geometry. Mailing Add: Dept of Math Univ of Okla Norman OK 73069

MAGID, LINDA JENNY, b Omaha, Nebr, Dec 13, 46; m 69. PHYSICAL ORGANIC CHEMISTRY. Educ: Rice Univ, BA, 69; Univ Tenn, Knoxville, PhD(chem), 73. Prof Exp: Instr chem, 73-74, ASST PROF CHEM, UNIV TENN, KNOXVILLE, 74- Mem: Sigma Xi. Res: Thermodynamics of micellization and micellar solubilization in aqueous and nonaqueous solvents; micellar catalysis. Mailing Add: Dept of Chem Univ of Tenn Knoxville TN 37916

MAGID, LOUIS, b Montreal, Que, Feb 10, 11; nat US; m 46; c 1. PHARMACEUTICAL CHEMISTRY. Educ: Univ Fla, BS, 31, MS, 32, PhD(pharm), 34. Prof Exp: Asst, Univ Fla, 27-32; head pharmaceut res dept, Wm S Merrell Co, 34-46; chief pharmaceut res div, Wm R Warner & Co, 46-50; dir pharmaceut res lab, 50-60, DIR PROD DEVELOP DEPT, HOFFMANN-LA ROCHE, INC, 60- Mem: Am Pharmaceut Asn. Res: Pharmaceutical research and development; cosmetics; vitamin products. Mailing Add: Prod Develop Dept Hoffmann-La Roche Inc Nutley NJ 07110

MAGID, RONALD, b Brooklyn, NY, Dec 19, 38; m 60, 69; c 2. ORGANIC CHEMISTRY. Educ: Yale Univ, BS, 59, MS, 60, PhD(chem), 64. Prof Exp: Asst prof, Rice Univ, 64-70; asst prof, 70-72, ASSOC PROF CHEM, UNIV TENN, KNOXVILLE, 72- Mem: Am Chem Soc; The Chem Soc; Swiss Chem Soc. Res: Orbital symmetry rules; mechanisms of organolithium reactions; mechanisms of reactions of allylic compounds; synthesis of strained compounds. Mailing Add: Dept of Chem Univ of Tenn Knoxville TN 37916

MAGIDSON, OSCAR, b London, Eng, July 16, 20; nat US; m 44; c 1. MEDICINE. Educ: Univ Leeds, MB & ChB, 43, MD, 45; FRCP, 69. Prof Exp: Assoc clin prof, 56-67, CLIN PROF MED, SCH MED, UNIV SOUTHERN CALIF, 67- Concurrent Pos: Dir cardiopulmonary lab, St Vincent's Hosp, Los Angeles, 58-72; pvt pract, 72-; fel coun clin cardiol, Am Heart Asn. Mem: Fel Am Col Cardiol; fel Am Col Chest Physicians; overseas fel Brit Cardiac Soc; fel Am Col Physicians. Res: Cardiology. Mailing Add: 1127 Wilshire Blvd Los Angeles CA 90017

MAGIE, ALLAN RUPERT, b Umatilla, Fla, July 21, 36; m 61; c 3. PUBLIC HEALTH. Educ: Univ Calif, BA, 58, PhD(physiol), 63; Loma Linda Univ, MPH, 71. Prof Exp: Assoc prof biol, Pac Union Col 63-65; prof & dean sci & technol, Mountain View Col, Philippines, 65-70; asst prof environ health, 71-73, ASSOC PROF ENVIRON & TROP HEALTH, SCH HEALTH, LOMA LINDA UNIV, 73-, COORDR DOCTOR HEALTH SCI PRROG, 72- Concurrent Pos: Dir external MPH prog, Western Consortium Schs Pub Health, 73- Mem: Am Pub Health Asn; Sigma Xi. Res: Human health effects of ambient air pollutants, including fetal development, birth anomalies, respiratory disease and hospitalization; animal diseases transferrable to humans. Mailing Add: Dept of Environ & Trop Health Loma Linda Sch of Health Loma Linda CA 92354

MAGIE, ROBERT OGDEN, b Madison, NJ, July 30, 06; m 33; c 2. PLANT PATHOLOGY. Educ: Rutgers Univ, BS, 29; Univ Wis, MS, 30, PhD(plant path), 34. Prof Exp: Agt, Dutch elm dis invest, Bur Plant Indust, USDA, 35; Crop Protection Inst fel, NY Agr Exp Sta, Geneva, 35-36, res assoc, 36-40; asst prof plant path, Cornell Univ, 40-45; PROF PLANT PATH, INST FOOD & AGR SCI, AGR RES & EDUC CTR, UNIV FLA, 45- Mem: Am Phytopath Soc; Am Soc Hort Sci; Int Soc Hort Sci; Int Soc Plant Path. Res: Diseases of cut flowers crops; gladiolus flower and corm production; control of Botrytis and Fusarium diseases; caladium tuber production. Mailing Add: Agr Res & Educ Ctr Inst of Food & Agr Sci Univ Fla Bradenton FL 33505

MAGILL, CLINT WILLIAM, b Washington, DC, Sept 15, 41; m 65. GENETICS. Educ: Univ Ill, BS, 63; Cornell Univ, PhD(genetics), 69. Prof Exp: NIH fel biochem genetics, Univ Minn, 67-69; asst prof, 69-75, ASSOC PROF GENETICS, TEX A&M UNIV, 75- Mem: Genetics Soc Am. Res: Mutation; reverse mutation; genetic complementation; amino acid uptake in Neurospora crassa. Mailing Add: Genetics Sect Tex A&M Univ College Station TX 77843

MAGILL, JANE MARY (OAKES), b Hamilton, Ont, Sept 30, 40; m 65. BIOCHEMISTRY, GENETICS. Educ: Univ Western Ont, BSc, 63; Cornell Univ, PhD(genetics), 68. Prof Exp: Asst scientist genetics, Univ Minn, 68-69; instr, 70-71, ASST PROF BIOCHEM, TEX A&M UNIV, 71- Mem: Genetics Soc Am; Am Chem Soc. Res: Utilization of exogenous purines and pyrimidines. Mailing Add: Dept of Biochem Tex A&M Univ College Station TX 77843

MAGILL, KENNETH DERWOOD, JR, b Duncansville, Pa, Oct 21, 33; m 52; c 2. MATHEMATICS. Educ: Shippensburg State Col, BS, 56; Pa State Univ, MA, 60, PhD(math), 63. Prof Exp: Teacher, Central Cove Schs, 56-57; instr math, Pa State Univ, 58-63; from asst prof to assoc prof, 63-67, chmn dept, 67-70, PROF MATH, STATE UNIV NY BUFFALO, 67- Concurrent Pos: Vis prof, Univ Leeds, 68; vis mem, Inst Advan Studies, Australian Nat Univ, 70; vis prof, Univ Fla. Mem: Math Asn Am; Am Math Soc; London Math Soc. Res: Topology; rings of continuous functions; semigroups. Mailing Add: Dept of Math State Univ of NY Buffalo NY 14214

MAGILL, ROBERT EARLE, b Ft Worth, Tex, May 8, 47; m 69; c 2. BRYOLOGY, TAXONOMY. Educ: Sul Ross State Univ, BS, 69, MS, 71; Tex A&M Univ, PhD(bot), 75. Prof Exp: CURATORIAL TRAINEE, MO BOT GARDEN, 75- Concurrent Pos: Sigma Xi res grant, 73. Mem: Am Bryol & Lichenological Soc; Brit Bryol Soc; Am Fern Soc. Res: Taxonomy and phytogeography of bryophytes; use of scanning electron microscope in systematic studies of mosses. Mailing Add: Mo Bot Garden Dept of Bot 2315 Tower Grove Ave St Louis MO 63110

MAGILL, THOMAS PLEINES, b Philadelphia, Pa, May 24, 03; m 43. MICROBIOLOGY, IMMUNOLOGY. Educ: Johns Hopkins Univ, AB, 25, MD, 30. Prof Exp: Instr med, Sch Med, Johns Hopkins Univ, 33-35; asst, Rockefeller Inst, NY, 35-36; mem staff, Rockefeller Found, 36-38; from asst prof to assoc prof bact & immunol, Med Col, Cornell Univ, 38-48; prof microbiol & immunol & chmn dept, Long Island Col Med, 48-50; chmn dept, 50-70, prof, 50-73, EMER PROF MICROBIOL & IMMUNOL, COL MED, STATE UNIV NY DOWNSTATE MED CTR, 73- Mem: Am Soc Immunologists (pres, 53); Soc Exp Biol & Med; Am Soc Microbiol; Harvey Soc. Res: Infectious diseases; filterable viruses and rickettsia; variation of influenza virus. Mailing Add: 140 Main St East Hampton NY 11937

MAGILTON, JAMES HENRY, b Rockford, Iowa, July 15, 14; m 46; c 2. VETERINARY ANATOMY. Educ: Iowa State Univ, BS, 46, MS, 64, PhD(vet med), 66. Prof Exp: Gen practr vet med, Vet Clin, David City, Nebr, 46-61; from asst prof to assoc prof, 66-69, PROF VET MED, IOWA STATE UNIV, 69- Mem: Am Vet Med Asn; Int Soc Biometeorol; Am Asn Anatomists; World Asn Vet Anatomists. Res: Thermoregulation of the brain. Mailing Add: Dept of Anat Physiol & Pharmacol Iowa State Univ Col of Vet Med Ames IA 50010

MAGIN, GEORGE BENEDICT, JR, b Chattanooga, Tenn, Feb 22, 24; m 51; c 7. INORGANIC CHEMISTRY, GEOCHEMISTRY. Educ: Univ Chattanooga, BS, 46, MS, 52; Am Univ, BS, 56. Prof Exp: Instr chem, Rutgers Univ, 47-48; assoc, George Washington Univ, 48-52; chemist, Res Sect, Geol Div, US Geol Surv, 52-53, chemist mineral synthesis, 53-57, staff gen hydrol br, Water Resources Div, 57-58, res chemist & chief anal methods proj, 58-59; radiochemist res & admin, Div Isotopes Develop, AEC, 59-62, radiohydrologist & oceanogr, 62-67, process control engr, Div Oper Safety, 67-68, pollution control engr, 68-72; OCCUP HEALTH & SAFETY SPECIALIST, ENERGY RES & DEVELOP ADMIN, 72- Mem: Am Chem Soc; Geochem Soc; Mineral Soc Am. Res: Inorganic chemistry of rare earths; mineral synthesis; chemistry of water and its relations to its aquifier; tracer applications of radioisotopes; nuclear instrumentation in oceanography; pesticide chemistry; toxicology and occupational health and safety. Mailing Add: Rm E-162 Div of Occup Safety Energy Res & Develop Admin Washington DC 20545

MAGIN, RALPH WALTER, b Belleville, Ill, Oct 22, 37; m 61; c 2. ORGANIC CHEMISTRY, POLYMER CHEMISTRY. Educ: Univ Ill, BS, 59; Mass Inst Technol, PhD(org chem), 63. Prof Exp: Assignment, Univ Ariz, 63-65; sr res chemist, Org Chem Div, Monsanto Co, 66-68 & Monsanto Indust Chem Co, Mo, 68-75; RES SPECIALIST, MONSANTO POLYMERS & PETROCHEM, 75- Mem: AAAS; Am Chem Soc. Res: Polyelectrolytes, both bioactive and watersoluble, and their effect on environment; polyester condensation polymers. Mailing Add: Monsanto Polymers & Petrochem 730 Worcester St Indian Orchard MA 01151

MAGINNES, EDWARD ALEXANDER, b Ottawa, Ont, Can, Apr 19, 33; m 64; c 1. HORTICULTURE. Educ: McGill Univ, BS, 56; Cornell Univ, MS, 60, PhD(floricult), 64. Prof Exp: Res officer horticult, Can Dept Agr, 56-60; asst prof, 64-72, EXTEN SPECIALIST & ASSOC PROF HORT SCI, UNIV SASK, 72- Mem: Am Soc Hort Sci. Res: Culture of horticultural crops; influence of photoperiod and temperature on flowering, light quality and plant growth; nutrition of floriculture crops; growth retardants; moisture stress and plant growth. Mailing Add: Dept of Hort Univ of Sask Saskatoon SK Can

MAGINNIS, RICHARD L, microbiology, deceased

MAGISTRO, ANGELO JOSEPH, organic chemistry, see 12th edition

MAGLADERY, JOHN WILLIAM, b Can, Oct 10, 11; nat US; m 48; c 2. NEUROLOGY. Educ: Univ Toronto, MD, 35; Oxford Univ, DPhil, 37; FRCP & FRCP(C). Prof Exp: ASSOC PROF NEUROL MED, JOHNS HOPKINS UNIV, 46- Mem: Am Soc Physiol; Am Neurol Asn. Res: Neurophysiology. Mailing Add: Dept of Psychiat & Neurol Johns Hopkins Hosp Baltimore MD 21205

MAGLEBY, KARL LEGRANDE, b Provo, Utah. NEUROPHYSIOLOGY, BIOPHYSICS. Educ: Univ Utah, BS, 66; Univ Wash, PhD(physiol, biophys), 70. Prof Exp: NIH neurophys training grant, Univ Wash, 70-71; asst prof, 71-75, ASSOC PROF PHYSIOL & BIOPHYS, SCH MED, UNIV MIAMI, 75- Concurrent Pos: NIH res grant, Univ Wash, 72- Res: Synaptic transmission; mechanism of transmitter release; transmitter-receptor interaction. Mailing Add: Dept of Physiol & Biophys Univ of Miami Sch of Med Miami FL 33152

MAGLICH, BOGDAN, b Sombor, Yugoslavia, Aug 5, 28; m 59; c 2. EXPERIMENTAL NUCLEAR PHYSICS, PARTICLE PHYSICS. Educ: Univ Belgrade, dipl physics, 51; Univ Liverpool, MSc, 55; Mass Inst Technol, PhD(physics), 59. Prof Exp: Res asst with Boris Kidric, Inst Nuclear Sci, Univ Belgrade, 51-54; res asst, Synchrotron Lab, Mass Inst Technol, 57-59; res physicist, Lawrence Radiation Lab, Univ Calif, 59-62; sr staff mem, Europ Orgn Nuclear Res, Geneva, Switz, 62-67; vis prof physics, Univ Pa, 67-69; PROF PHYSICS, RUTGERS UNIV, 69- Mem: Am Phys Soc. Res: Nuclear reactions and instrumentation development using accelerators 100 kev, 1.5 Mev, 10 Mev and 150 Mev cyclotrons; photon reactions at 300 Mev; pi-meson scattering and polarization at 3 bev cosmotron. Mailing Add: Dept of Physics Rutgers Univ New Brunswick NJ 08903

MAGLIO, M MARTIN, organic chemistry, see 12th edition

MAGLIO, VINCENT JOSEPH, b New York, NY, Oct 2, 42. PALEOBIOLOGY, VERTEBRATE PALEONTOLOGY. Educ: City Univ New York, BA, 65; Harvard Univ, MA, 67, PhD(biol), 71. Prof Exp: Dir paleont, Harvard Kenya Expeds, 68; coordr paleont prog, Nat Mus Kenya Lake Rudolf Res Expeds, 70-71; DIR PALEONT, PRINCETON KENYA EXPEDS, 72- Concurrent Pos: Asst prof & dir nat hist mus, Princeton Univ, 71- Mem: Am Soc Mammal; Soc Study Evolution; Soc Syst Zool; Soc Vert Paleont. Res: Patterns of vertebrate evolution; origin and evolution of extant faunas; evolution of African mammals. Mailing Add: Dept of Geol 320 Guyot Hall Princeton Univ Princeton NJ 08540

MAGLIULO, ANTHONY RUDOLPH, b Brooklyn, NY, July 23, 31; m 67; c 5. MICROBIOLOGY, BIOCHEMISTRY. Educ: Brooklyn Col, BS, 55; Long Island Univ, MS, 61; St John's Univ, PhD(microbiol), 68 Prof Exp: Lab technician, Maimonides Hosp, Brooklyn, 55-56; res microbiol, Armed Forces Inst Path, 56-58; res asst, New York Hosp, 58-60; microbiologist chg clin microbiol, Misericordia Hosp, 60-61 & 62-63; lectr microbiol, Queens Col, NY, 61-62; teaching asst biol, St John's Univ, 63-65; lectr & instr, 65-68, asst prof, 68-73, ASSOC PROF BIOCHEM, JOHN JAY COL CRIMINAL JUSTICE, 73- Concurrent Pos: Instr nursing sci, Roosevelt Hosp, 68-69; coordr nursing chem & lectr nursing sci, Hunter Col, 68- Mem: AAAS; fel Am Inst Chemists; Am Chem Soc; Nat Sci Teachers Asn. Res: Isolation of new and unique microbial lipids; environmental science; forensic microbiology and chemistry. Mailing Add: Dept of Sci John Jay Col 445 W 59th St New York NY 10019

MAGLIVERAS, SPYROS SIMOS, b Athens, Greece, Sept 6, 38; US citizen; m 62; c 2. MATHEMATICS. Educ: Univ Fla, BEE, 61, MA, 63; Univ Birmingham, PhD(math), 70. Prof Exp: Teaching asst math, Univ Fla, 61-62, interim instr, 62-63; instr, Fla Presby Col, 63-64; systs analyst, Inst Social Res, Univ Mich, Ann Arbor, 65-68; res fel math, Univ Birmingham, 68-70; asst prof, 70-74, ASSOC PROF MATH, STATE UNIV NY COL OSWEGO, 74- Concurrent Pos: Consult, Nat Broadcasting Co, 67-68; consult, Ctr Human Growth & Develop, Univ Mich, Ann Arbor, 68, res assoc, 71. Mem: Am Math Soc; Math Asn Am; Edinburgh Math Soc; London Math Soc. Res: Finite groups. Mailing Add: Dept of Math State Univ of NY Col Oswego NY 13126

MAGNANI, ARTHUR, b Mark, Ill, Oct 3, 10; m 38; c 1. ORGANIC CHEMISTRY. Educ: DePauw, AB, 34; Univ Wis, PhD(org chem), 37. Prof Exp: Du Pont fel, Univ Wis, 37-38; res chemist, Nat Aniline & Chem Co, 38-41, Soya Prod Div, Glidden Co, 41-54, Julian Labs, 54-64 & Smith Kline & French Labs, 63-75; RETIRED. Mem: Fel Am Chem Soc. Res: Hormones; steroids. Mailing Add: 234 Brydon Rd Wynnewood Philadelphia PA 19151

MAGNARELLA, PAUL J, US citizen. CULTURAL ANTHROPOLOGY, SOCIAL ANTHROPOLOGY. Educ: Univ Conn, BS, 59; Fairfield Univ, MA, 62; Harvard Univ, AM, 69, PhD(anthrop), 71. Prof Exp: ASST PROF ANTHROP, UNIV VT, 71- Concurrent Pos: Assoc fel, Ctr Mid E Studies, Harvard Univ, 71-72; mem, Coun Anthrop & Educ, 69- Mem: Am Anthrop Asn; Mid E Studies Asn NAm. Res: Tradition and change in Turkey; American towns; educational institutions as sociocultural systems; college student culture. Mailing Add: Dept of Anthrop Univ of Vt Burlington VT 05401

MAGNARELLI, LOUIS ANTHONY, b Syracuse, NY, Mar 27, 45; m 69. ENTOMOLOGY. Educ: State Univ NY Col Oswego, BS, 67; Univ Mich, MS, 68; Cornell Univ, PhD(entom), 75. Prof Exp: Teacher biol, WGenesee Cent Sch Dist, NY, 68-71; asst entom, Cornell Univ, 71-75, experimentalist, 75; ASST SCIENTIST ENTOM, CONN AGR EXP STA, 75- Mem: Entom Soc Am; Am Mosquito Control Asn. Res: Blood feeding, sugar feeding, and ovarian studies of mosquitoes, deer flies and horse flies. Mailing Add: Dept of Entom Conn Agr Exp Sta PO Box 1106 New Haven CT 06504

MAGNELL, KENNETH ROBERT, b Detroit, Mich, July 27, 38; m 63; c 2.

INORGANIC CHEMISTRY. Educ: Wayne State Univ, BS, 62, MS, 66; Univ Minn, Minneapolis, PhD(inorg chem), 70. Prof Exp: ASST PROF CHEM, CENT MICH UNIV, 70- Mem: AAAS; Am Chem Soc. Res: Non-aqueous solvent equilibrium studies. Mailing Add: Dept of Chem Cent Mich Univ Mt Pleasant MI 48859

MAGNER, DESMOND, pathology, deceased

MAGNIEN, ERNEST, b Ger, Mar 28, 25; nat US; m 49; c 2. ORGANIC CHEMISTRY. Educ: City Col New York, BS, 49; Polytech Inst Brooklyn, MS, 58. Prof Exp: Org chemist, Gane & Ingram, 49-53; res chemist, Burroughs-Wellcome, Inc, 53-60; SR RES CHEMIST, USV PHARMACEUT CORP, 60- Mem: Am Chem Soc. Res: Isolation of alkaloids and natural products; pharmaceutical compounds; organic synthesis; heterocyclic chemistry. Mailing Add: 67-37 166th St Flushing NY 11365

MAGNIN, ETIENNE NICOLAS, b Valloire, France, Feb 8, 22; Can citizen; m 68. ICHTHYOLOGY. Educ: Univ Grenoble, Lic es Sc, 52; Univ Nancy, DES, 57; Univ Paris, Doct Etat Ichthyol, 62. Prof Exp: Prof zool, Univ Lyon, 55-64; PROF ICHTHYOL, UNIV MONTREAL, 64- Concurrent Pos: Consult, Nat Coun Fisheries, France, 57-64, Dept Fisheries, Que, Can, 62-64 & Conserv Dept, Wis, 63. Mem: Can Soc Zoologists; Am Fisheries Soc; Int Acad Fishery Scientists. Res: Freshwater benthos. Mailing Add: Dept of Biol Univ of Montreal PO Box 6128 Montreal PQ Can

MAGNO, MICHAEL GREGORY, b Newark, NJ, Aug 9, 42; m 65; c 3. PHYSIOLOGY. Educ: Rutgers Univ, New Brunswick, BS, 64, PhD(physiol), 69. Prof Exp: Instr physiol, Albany Med Col, 68-73; res fel, 73-75, RES ASSOC, CARDIOVASC-PULMONARY DIV, HOSP UNIV PA, 75- Mem: Am Physiol Soc. Res: Capillary exchange in the lung; cardio-pulmonary adjustments to hypoxic in birds and mammals. Mailing Add: Cardiovasc-Pulmonary Div Hosp of the Univ of Pa Philadelphia PA 19104

MAGNO, RICHARD, b Newark, NJ, May 5, 44; m 68; c 1. SOLID STATE PHYSICS. Educ: Stevens Inst Technol, BS, 66; Rutgers Univ, PhD(physics), 74. Prof Exp: FEL PHYSICS, UNIV ALTA, 74- Mem: Sigma Xi; Am Phys Soc. Res: Inelastic electron tunneling is being used to study metal-insulator-metal junctions whose barriers contain organic molecules or were formed in an organic vapor glow discharge. Mailing Add: Dept of Physics Univ of Alta Edmonton AB Can

MAGNUS, ARNE, b Oslo, Norway, Aug 17, 22; nat US; m 50; c 3. MATHEMATICS. Educ: Univ Oslo, Cand Real, 52; Washington Univ, PhD(math), 53. Prof Exp: Instr math, Univ Kans, 52-54; asst prof, Univ Nebr, 54-56; asst prof, Univ Colo, 56-65, prof, 65-66; PROF MATH, COLO STATE UNIV, 66- Mem: Math Asn Am; Am Math Soc. Res: Analytic function of one and several variables; analytic theory of continued fractions; iteration. Mailing Add: Dept of Math & Comput Sci Colo State Univ Ft Collins CO 80521

MAGNUS, GEORGE, b Ganister, Pa, Jan 16, 30; m 54; c 4. PLASTICS CHEMISTRY. Educ: Franklin & Marshall Col, BS, 52; Univ Pittsburgh, PhD(org chem), 56. Prof Exp: Chemist, Union Carbide Chem Co, 56-71; chemist, 71-74, TECHNOL MGR URETHANES, STEPAN CHEM CO, 74- Mem: Am Chem Soc. Res: Applied research and product development in urethane foams; solid and microcellular urethane elastomers; applied research in spandex fibers; development of new poly-e-caprolactone and polyadipate polyols for solid and microcellular urethane elastomers. Mailing Add: Chem Dept Stepan Chem Co Edens Expressway & Winnetka Northfield IL 60094

MAGNUSON, EUGENE ROBERT, b Emerson, Nebr, Dec 5, 33; m 60. ORGANIC CHEMISTRY. Educ: Univ Nebr, BS, 55, MS, 58; Kans State Univ, PhD(org chem), 66. Prof Exp: Res chemist, Standard Oil Co Div, Am Oil Co, Ind, 60 & Allis-Chalmers Mfg Co, Wis, 63-69; PROF CHEM, MILWAUKEE SCH ENG, 69- Mem: Am Chem Soc; Soc Plastics Eng. Res: Insulation of electrical components as applied to motors, generators and transformers by encapsulation using various thermal and electrical resin systems. Mailing Add: Dept of Chem Milwaukee Sch of Eng 1025 N Milwaukee St Milwaukee WI 53201

MAGNUSON, GUSTAV DONALD, b Chicago, Ill, Aug 22, 26; m 50; c 4. SOLID STATE PHYSICS, ATOMIC PHYSICS. Educ: Univ Chicago, PhB, 49, BS, 50; Univ Ill, MS, 52, PhD(physics), 57. Prof Exp: Asst, Univ Ill, 53-57; sr staff scientist, Gen Dynamics-Convair, Calif, 57-66; res assoc prof aerospace eng & eng physics, Univ Va, 66-69; res scientist, Atomic Physics Lab, Gulf Energy & Environ Systs, 69-73; RES PHYSICIST, IRT CORP, 73- Concurrent Pos: Asst, Anderson Phys Lab, Ill, 53-55. Mem: AAAS; Am Phys Soc; Am Asn Physics Teachers. Res: Molecular physics; radiation damage; surface-particle interactions; atomic collisions. Mailing Add: 2386 Caminito Agrado San Diego CA 92107

MAGNUSON, HAROLD JOSEPH, b Halstead, Kans, Mar 31, 13; m 35; c 2. OCCUPATIONAL MEDICINE. Educ: Univ Southern Calif, AB, 34, MD, 38; Johns Hopkins Univ, MPH, 42; Am Bd Prev Med, dipl. Prof Exp: Intern med & surg, Los Angeles County Hosp, 37-39; instr internal med, Sch Med, Univ Southern Calif, 39-41; asst surgeon, USPHS, 41-44, sr asst surgeon, 44-46, surgeon, 47-49, sr surgeon, 49-53, med dir, 53-62, chief oper res sect venereal dis prog, 55-56, chief div occup health, 56-62; dir inst indust health & chmn dept, Sch Pub Health, 62-69, PROF INDUST HEALTH, SCH PUB HEALTH & PROF INTERNAL MED, MED SCH, UNIV MICH, ANN ARBOR, 62-, ASSOC DEAN SCH PUB HEALTH, 69- Concurrent Pos: Spec consult, USPHS, 40-41; instr, Johns Hopkins Univ, 43-45; res prof, Univ NC, 45-55; spec lectr, George Washington Univ, 59-62; vpres, Permanent Comn & Int Asn Occup Health; vchmn occup med, Am Bd Prev Med. Honors & Awards: Bronze Hektoen Medal, AMA, 55; William S Knudsen Award, Indust Med Asn, 70. Mem: Fel AAAS; fel Am Col Physicians; fel Indust Med Asn; fel Am Pub Health Asn; fel AMA. Res: Public health; internal medicine. Mailing Add: Sch of Pub Health Univ of Mich Ann Arbor MI 48104

MAGNUSON, JAMES ANDREW, b Oak Park, Ill, Oct 21, 42; m 64. BIOCHEMISTRY, BIOPHYSICS. Educ: Stanford Univ, BS, 64; Calif Inst Technol, PhD(chem), 68. Prof Exp: Res fel org chem, Mellon Inst, 67-68; asst prof chem & molecular biophys, 68-75, ASSOC PROF CHEM & BIOPHYS, WASH STATE UNIV, 75- Mem: Am Chem Soc; Am Soc Biol Chemists. Res: Application of nuclear magnetic resonance spectroscopy to the study of ion binding and small molecular binding to bio-organic molecules; use of both wide-line and high resolution techniques. Mailing Add: Dept of Chem 531 Fulmer Hall Wash State Univ Pullman WA 99163

MAGNUSON, JOHN JOSEPH, b Evanston, Ill, Mar 8, 34; m 59; c 2. HYDROBIOLOGY. Educ: Univ Minn, BS, 56, MS, 58; Univ BC, PhD(zool), 61. Prof Exp: Chief tuna behav prog, Biol Lab, Bur Commercial Fisheries, US Fish & Wildlife Serv, 61-67; from asst prof to assoc prof, 68-74, PROF ZOOL, UNIV WIS-MADISON, 74- Concurrent Pos: Mem affil grad fac, Univ Hawaii, 63-67; bk rev ed, Am Fisheries Soc, 70-; prog dir ecol, NSF, 75-76. Mem: Am Fisheries Soc; Am Soc Ichthyol & Herpet; Animal Behav Soc; Ecol Soc Am; Am Inst Biol Sci. Res: Behavioral ecology of fishes; locomotion of scombrids; ecological interactions among fishes and invertebrates; crayfish ecology; dissolved oxygen and water temperature as factors influencing fish distribution; ecology of Great Lakes. Mailing Add: Dept of Zool Lab of Limnol Univ of Wis Madison WI 53706

MAGNUSON, VINCENT RICHARD, b Laurel, Nebr, May 5, 42; m 62; INORGANIC CHEMISTRY, CRYSTALLOGRAPHY. Educ: Univ Nebr, BS, 63; Univ Wis, MS, 65; Univ Ill, PhD(chem), 68. Prof Exp: ASSOC PROF CHEM, UNIV MINN, DULUTH, 68- Mem: Am Chem Soc; Am Crystallog Asn. Res: Chemistry and structural properties of organometallic complexes of Group II and III metals as determined by x-ray crystallography. Mailing Add: Dept of Chem Univ of Minn Duluth MN 55812

MAGNUSON, WINIFRED LANE, b Brady, Tex, Oct 12, 35; m 58. INORGANIC CHEMISTRY. Educ: McMurry Col, AB, 59; Univ Kans, PhD(chem), 63. Prof Exp: Asst prof chem, McMurry Col, 63-69; CHMN DEPT CHEM, KY WESLEYAN COL, 69- Concurrent Pos: Petrol Res Fund grant, 63-64; Res Corp grant, 65-66; Robert A Welch Found grant, 66-69. Mem: Am Chem Soc. Res: Inorganic reactions in molten salts; reaction of metal carbonyls. Mailing Add: Dept of Chem Ky Wesleyan Col Owensboro KY 42301

MAGNUSSON, LAWRENCE BERSELL, b Moline, Ill, Jan 3, 19; m 42; c 4. PHYSICAL INORGANIC CHEMISTRY. Educ: Augustana Col, AB, 41; Univ Calif, PhD(phys chem), 49. Prof Exp: Chemist, Chem Warfare Serv, US Dept Army, 42-43, metall lab, Univ Chicago, 43-46, Radiation Lab, Univ Calif, Berkeley, 46-49 & Argonne Nat Lab, 49-71; CHEMIST, BETZ LABS, 71- Concurrent Pos: Res fel, City Univ, London, 64-65. Mem: Am Chem Soc. Res: Complex ions; electrolytes; water treatment. Mailing Add: 431 Merion Dr Newtown PA 18940

MAGOFFIN, JAMES EDWARD, b Buffalo, NY, Dec 31, 10; m 37; c 1. PHYSICAL CHEMISTRY. Educ: Cornell Univ, BChem, 32, PhD(phys chem), 36. Prof Exp: From asst prof to assoc prof chem, Univ NC, 36-40; dir res, Thompson & Co, 40-41; chief chemist, Hydroquinone Div, 42-45, mgr chem sales develop, Tenn Eastman Corp, 45-52; sales mgr chem div, Eastman Chem Prod, Inc, 53-57, vpres, 58-66, group vpres, 66-69, sr vpres & dir, 69-73; PRES, EASTMAN CHEM PROD, INC, 73- Mem: Am Chem Soc; Am Inst Chem Engrs; fel Am Inst Chemists; Soc Chem Indust. Res: Catalysis; electrokinetics; antioxidants. Mailing Add: 1433 Linville St Kingsport TN 37664

MAGOFFIN, ROBERT LOUIS, b Ft Worth, Tex, Dec 2, 22; m 46; c 4. EPIDEMIOLOGY, MICROBIOLOGY. Educ: Tex Christian Univ, BA, 44; Univ Tex, MD, 47; Univ Minn, MPH, 54. Prof Exp: Intern internal med, Univ Minn Hosps, 47-48; USPHS res fel infectious dis, Dept Med, Univ Minn, 48-50; pub health med officer, Bur Commun Dis, 53-57, pub health med officer, Virus Lab, 58-61, ASST CHIEF VIRUS LAB, CALIF STATE DEPT PUB HEALTH, 62- Concurrent Pos: Lab instr microbiol, Univ Minn, 49-50, teaching asst epidemiol, 54. Mem: AAAS; Am Pub Health Asn; Am Epidemiol Soc. Res: Diagnosis and prevention of infectious diseases, especially viral infections of central nervous system; viral respiratory infections; respiratory infections; respiratory virus vaccines. Mailing Add: Viral & Rickettsial Dis Lab Calif State Dept of Pub Health Berkeley CA 94704

MAGORIAN, THOMAS R, b Charleston, WVa, Nov 26, 28; m 50; c 3. GEOPHYSICS. Educ: Univ Chicago, PhD(geol), 52. Prof Exp: Res geologist, Ohio River Div Labs, US Corps Engrs, 50-52, Pure Oil Co, 52-53 & Shell Develop Co, 56-62; consult petrol explor, Tex, 62-63; prin geophysicist, Cornell Aeronaut Lab, 63-73; CHIEF GEOPHYSICIST, ECOL & ENVIRON, INC, 73- Concurrent Pos: Mem joint environ effects prog, Environ Characterization Comt, US Dept Defense, 65-; consult, Ltd War Lab & Ballistics Res Lab, US Army, 65-; mem air staff, US Air Force, 66-; dir earth sci prog, Rosary Hill Col, 70-, prof earth sci, 71-74. Mem: Fel AAAS; Am Asn Petrol Geol; Am Geophys Union; Geol Soc Am; Ecol Soc Am. Res: Geomorphology; biophysics; stratigraphic seismology; hydrology; plant ecology and paleocology; operations research; origin of oil; geotechnical analysis of energy systems. Mailing Add: Box D Ecol & Environ Inc Buffalo NY 14225

MAGOSS, IMRE V, b Nagykata, Hungary, July 19, 19; US citizen; m 44; c 2. UROLOGY. Educ: Pazmany Peter Univ, Hungary, MD, 43; Am Bd Urol, dipl, 60. Prof Exp: Assoc cancer res urologist, Roswell Park Mem Inst, 55-60; from asst prof to assoc prof urol surg, 60-71, PROF SURG, STATE UNIV NY BUFFALO, 71- Mem: AMA; Am Urol Asn; fel Am Col Surg. Res: Histochemical changes of prostatic carcinoma during progression. Mailing Add: Dept of Surg State Univ of NY Sch of Med Buffalo NY 14222

MAGOUN, HORACE WINCHELL, b Philadelphia, Pa, June 23, 07; m 31; c 3. NEUROANATOMY. Educ: Univ RI, BS, 29; Syracuse Univ, MS, 31; Northwestern Univ, PhD(anat), 34. Hon Degrees: DSc, Northwestern Univ, 59 & Univ RI, 60; LHD, Wayne State Univ, 65. Prof Exp: Instr neurol, Inst Neurol, Northwestern Univ, 34-37, asst prof, Inst & Sch Med, 37-40, assoc prof, Sch Med, 40-43, prof microanat, 43-50; prof anat, Sch Med, 50-62, chmn dept, 50-55, Salmon mem lectr, 56, dean grad div, 62-72, EMER DEAN GRAD DIV, UNIV CALIF, LOS ANGELES, 72-, EMER PROF PSYCHIAT, SCH MED, 74- Concurrent Pos: Rockefeller fel, Sch Med, Johns Hopkins Univ, 39-40; staff mem, Nat Res Coun, 72-74. Honors & Awards: Jacoby Award, Am Neurol Asn, 56; Borden Award, 61; Passano Award, 63; Lashley Prize, Am Philos Soc, 70; Order Sacred Treasure, Japan, 71. Mem: Nat Acad Sci; Am Asn Anat (pres, 64); Am Neurol Asn. Res: Neurophysiology; manpower study in neurosciences. Mailing Add: 427 25th St Santa Monica CA 90402

MAGOVERN, GEORGE JEROME, b Brooklyn, NY, Nov 17, 23; m; c 5. THORACIC SURGERY, CARDIOVASCULAR SURGERY. Educ: Marquette Univ, MD, 47; Am Bd Surg, dipl, 55; Am Bd Thoracic Surg, dipl, 60. Prof Exp: Intern, Kings County Hosp, 49; resident gen surg, St Vincent's Hosp, NY, 50; resident, Brooklyn Vet Admin Hosp & Kings County Hosps, 50-53; instr surg, Col Med, State Univ NY, 53-54; resident thoracic & cardiovasc surg, George Washington Univ Hosp, 56-58; clin asst prof, 58-64, CLIN ASSOC PROF SURG & DIR SURG RES LAB, SCH MED, UNIV PITTSBURGH, 64- Concurrent Pos: Staff physician, Presby-Univ Hosp, 58-; staff physician, Allegheny Gen Hosp, 58-59, dir div surg, 69-; consult staff physician, St Margaret's, Mercy & Columbia Hosps, 58-; Health Res & Serv Found grant, NIH grant & Am Heart Asn grant. Mem: AMA; Asn Thoracic Surg; Am Surg Asn; Am Col Chest Physicians; fel Am Col Surgeons. Res: Lung transplantation; sutureless prosthetic heart valve; total heart replacement. Mailing Add: Dept of Surg Allegheny Gen Hosp Pittsburgh PA 15212

MAGRAM, SIDNEY JEROME, physical chemistry, see 12th edition

MAGRANE, JOHN KEARNS, JR, b Holyoke, Mass, Apr 5, 13; m 38; c 2. ORGANIC CHEMISTRY. Educ: Amherst Col, AB, 34; Rutgers Univ, MSc, 36, PhD(org chem), 39. Prof Exp: From asst to instr chem, Rutgers Univ, 34-40; res chemist, Catalin Corp Am, 40-41, dir res, 41-44; chief chemist, New Eng Tape Co, Inc, 44-52; group leader plastics & resins sect, 52-57, mgr, 57-59, mgr process chem res dept, 59-64, mgr org

& polymer res, Indust Chem Div, 65-66, dir res & develop, 66-71, COORDR PATENTS & LICENSES, INDUST CHEM & PLASTICS DIV, AM CYANAMID CO, 71- Mem: Am Chem Soc; Sigma Xi. Res: Synthetic resins. Mailing Add: 54 Pinner Lane Stamford CT 06903

MAGRUDER, NORMAN DAVID, animal nutrition, biochemistry, see 12th edition

MAGRUDER, SAMUEL ROSSINGTON, b Kevil, Ky, June 10, 06; m 43; c 1. HUMAN ANATOMY. Educ: Univ Ky, BS, 30; Univ Cincinnati, MA, 31, PhD(zool), 34. Prof Exp: Lab instr, Med Col, Cornell Univ, 34-36, asst anat, 36-38; from instr to assoc prof & actg chmn dept, 38-71, lectr, 73-74, EMER ASSOC PROF ANAT, SCH MED, TUFTS UNIV, 71- Concurrent Pos: Mem corp, Marine Biol Lab, Woods Hole. Mem: AAAS; Am Soc Zoologists; Am Asn Anat. Res: Vertebrate embryology; gastropod anatomy and life history; histological techniques; anatomy of nervous system. Mailing Add: Rte 4 Box 177 Magruder Rd Kevil KY 42053

MAGRUDER, WILLIS JACKSON, b Lentner, Mo, Aug 7, 35; m 54; c 3. CHEMISTRY, SCIENCE EDUCATION. Educ: Northeast Mo State Col, BS, 57; State Col Iowa, MA, 61; Colo State Col, EdD(chem, sci educ), 66. Prof Exp: High sch teacher, Mo, 58-60; instr gen & org chem, Fullerton Jr Col, 61-64; from asst prof to assoc prof sci educ, 64-67, PROF SCI & SCI EDUC, NORTHEAST MO STATE COL, 67- Concurrent Pos: Mem sci curric comt, Sec Sch Chem & Phys, State Dept Educ, 65-67; dir, Northeast Mo Regional Sci Fair, 69. Mem: Am Chem Soc; Nat Sci Teachers Asn. Res: Organolithium chemistry. Mailing Add: Dept of Sci Northeast Mo State Col Kirksville MO 63501

MAGUDER, THEODORE LEO, JR, b Meriden, Conn, Oct 14, 39; m 64; c 3. WILDLIFE BIOLOGY, ECOLOGY. Educ: Fairfield Univ, BS, 61; St John's Univ, MS, 63; State Univ NY Col Forestry, Syracuse Univ, PhD(forest zool), 68. Prof Exp: ASST PROF BIOL, UNIV HARTFORD, 68-, EDUC DIR ENVIRON CTR, GREAT MT FOREST, 73- Concurrent Pos: NSF Col Sci Improv Prog grant prin investr, Woodcock Study, Conn, 70 & US Fish & Wildlife Serv, 71-73. Mem: Wildlife Soc; Sigma Xi. Res: Wildlife biology including the study of woodchuck, woodcock, crow and seagull population dynamics; development of environmental education programs in ecology and field biology. Mailing Add: Dept of Biol Univ of Hartford West Hartford CT 06117

MAGUE, JOEL TABOR, b New Haven, Conn, Nov 23, 40; m 64; c 2. INORGANIC CHEMISTRY, ORGANOMETALLIC CHEMISTRY. Educ: Amherst Col, BA, 61; Mass Inst Technol, PhD(chem), 65. Prof Exp: NIH fel, Imp Col, Univ London, 65-66; asst prof, 66-71, ASSOC PROF CHEM, TULANE UNIV, 71- Mem: AAAS; Am Chem Soc; The Chem Soc. Res: Organometallic complexes of the platinum metals; synthetic and structural studies. Mailing Add: Dept of Chem Tulane Univ New Orleans LA 70118

MAGUIRE, BASSETT, b Alabama City, Ala, Aug 4, 04. SYSTEMATIC BOTANY. Educ: Univ Ga, BS, 26; Cornell Univ, PhD, 38. Prof Exp: Instr bot, Univ Ga, 27-29; asst, Cornell Univ, 29-31; from asst prof to assoc prof, Utah State Univ, 31-43; from cur to head cur, 43-68, coordr trop res, 53-68, asst dir, 68-69, dir bot, 69-71 & 73-75, NATHANIEL LORD BRITTON DISTINGUISHED SR CUR, NY BOT GARDEN, 61-, SR SCIENTIST & EMER DIR, 75- Concurrent Pos: Botanist, NY Conserv Dept, 30-31; aquatic biologist, US Bur Fish, 32 & 34; cur, Intermountain Herbarium, 32-43; field agt, USDA, 34-35, range examr, Soil Conserv Serv, 35; instr, Cornell Univ, 37-38; non-resident prof, Utah State Univ, 43-; adj prof, Columbia Univ, 64 & Herbert H Lehman Col, 69-; mem adv comt, Cary Arboretum; consult, Nat Bulk Carriers, Inc, 55-, Eli Lilly & Co, 55-, Tex Instruments, Inc, 64- & US Army Edgewood Arsenal, 64-; consult, Cent Insts Sci & rector, Univ Brasilia, 65; consult, Nat Bot Garden, Dom Repub, 74- With exped Northern SAm, 25, leader, 44-45 & 48-49, dir, 50-75. Honors & Awards: Sarah Gildersleeve Fife Mem Award, 52; David Livingstone Centenary Medal, Am Geog Soc, 65. Mem: AAAS; hon fel Asn Trop Biol (pres, 64-65); hon mem Dom Bot Soc; Newcomen Soc; NY Acad Sci. Res: Phytogeography, South America; flora of Guayana; taxonomy of the Clusiaceae, Theaceae, Rapateaceae, Abolbodaceae, Gentianaceae; discoverer of Serrania de la Neblina, highest point in Brazil. Mailing Add: 120-24 Dreiser Loop Bronx NY 10475

MAGUIRE, BASSETT, JR, b Birmingham, Ala, Aug 30, 27; m 50; c 2. ECOLOGY. Educ: Cornell Univ, AB, 53, PhD(zool), 57. Prof Exp: From instr to asst prof, 57-64, ASSOC PROF ZOOL, UNIV TEX, AUSTIN, 64- Concurrent Pos: Ed, Ecol Modeling & Eng, 74- Mem: AAAS; Ecol Soc Am; Am Soc Limnol & Oceanog; Am Micros Soc; Am Soc Naturalists. Res: Mechanisms of community structure determination; internal niche structure; ecological and ecosystem analysis; physiological ecology. Mailing Add: Dept of Zool Univ of Tex Austin TX 78712

MAGUIRE, HENRY C, JR, b New York, NY, May 4, 28; m 53; c 3. IMMUNOLOGY, DERMATOLOGY. Educ: Princeton Univ, BA, 49; Columbia Univ, 49-50; Univ Chicago, MD, 54; Univ Pa, dipl, 61. Prof Exp: Asst instr dermat, Sch Med, Univ Pa, 58-61, instr, 61, assoc, 61-64, asst prof, 64-67, asst instr, Div Grad Med, 58-61, instr, 61-63, assoc, 63-65, asst prof, 65-67; assoc prof med, 67-75, ASSOC PROF MICROBIOL, HAHNEMANN MED COL, 69-, PROF MED, 75- Concurrent Pos: Guest investr, Rockefeller Univ, 66-67; chief investr, NIH res grant, 68-; investr, Inst Cancer Res, 72-75. Mem: AAAS; Am Fedn Clin Res; fel Am Col Physicians; Am Asn Immunologists; Am Acad Dermat. Res: Delayed hypersensitivity; tumor immunology; immunological adjuvants; hair growth.

MAGUIRE, JAMES DALE, b Chelan, Wash, Sept 16, 30; m 55; c 4. AGRONOMY. Educ: Wash State Univ, 52; Iowa State Univ, MS, 57; Ore State Univ, PhD(crops), 68. Prof Exp: Settler assistance agt irrig eng, US Bur Reclamation, 54-56; instr seed technol, Wash State Univ, 57-66; res assoc seed physiol, Ore State Univ, 66-67; ASSOC PROF AGRON, WASH STATE UNIV, 67- Concurrent Pos: Wash Crop Improv Asn fel, Wash State Univ, 68-70; vis prof, Univ Nottingham, 71- Mem: Am Soc Agron; Asn Off Seed Analysts. Res: Seed physiology studies dealing with seed vigor evaluation and determination of metabolic processes involved in dormancy, germination and seedling growth. Mailing Add: 201 Johnson Hall Dept of Agron Wash State Univ Pullman WA 99163

MAGUIRE, KEITH DEAN, b Manchester, Eng, May 11, 34; m 62; c 2. INORGANIC CHEMISTRY. Educ: Vanderbilt Univ, PhD(inorg chem), 62. Prof Exp: Res fel inorg chem, Lawrence Radiation Lab, Univ Calif, 62-63; sr res chemist, Pennsalt Chem Corp, 63-68; res assoc, Youngstown Sheet & Tube Co, 68-71; MGR NEW PROD PLANNING, FIBERCAST CO, 71- Mem: Am Chem Soc. Res: Synthetic inorganic, coordination and metallo-organic chemistry; fiberglass reinforced plastics. Mailing Add: Fibercast Co Box 968 Sand Springs OK 74063

MAGUIRE, MARJORIE PAQUETTE, b Pearl River, NY, Sept 2, 25; m 50; c 2. CYTOGENETICS. Educ: Cornell Univ, BS, 47, PhD(cytol), 52. Prof Exp: Instr bot, Cornell Univ, 52-53; res assoc, Genetics Found, 57-60, res scientist, 60-75, ASSOC PROF ZOOL, UNIV TEX, AUSTIN, 75- Concurrent Pos: NIH career develop award, 65. Mem: Genetics Soc Am; Am Genetic Asn; Bot Soc Am; Am Soc Cell Biol. Res: Chromosome mechanisms of synapsis; crossing over and disjunction. Mailing Add: Dept of Zool Univ of Tex Austin TX 78712

MAGUIRE, MILDRED MAY, b Leetsdale, Pa, May 7, 33. PHYSICAL CHEMISTRY. Educ: Carnegie-Mellon Univ, BS, 55; Univ Wis, MS, 60; Pa State Univ, PhD(chem), 67. Prof Exp: Chemist, Koppers Co, 55-58 & Am Cyanamid Co, 60-63; PROF CHEM, WAYNESBURG COL, 67- Mem: Am Chem Soc; Am Phys Soc. Res: Gas phase electron spin resonance; electron spin resonance of transition metal complexes. Mailing Add: Dept of Chem Waynesburg Col Waynesburg PA 15370

MAH, RAYMOND W, organic chemistry, see 12th edition

MAH, ROBERT A, b Fresno, Calif, Oct 28, 32. MICROBIOLOGY. Educ: Univ Calif, Davis, AB, 57, MA, 58, PhD(microbiol), 63. Prof Exp: Asst prof biol, San Fernando Valley State Col, 62-64; from asst prof to assoc prof microbiol, Univ NC, Chapel Hill, 64-71; PROF PUB HEALTH, UNIV CALIF, LOS ANGELES, 71- Mem: AAAS; Am Soc Microbiol; Soc Protozoologists. Res: Nutrition, physiology and metabolism of anaerobic bacteria and protozoa; microbial physiology and ecology of anaerobic habitats; cycling of matter by microbes. Mailing Add: Div of Environ & Nutrit Sci Univ of Calif Sch of Pub Health Los Angeles CA 90024

MAHA, GEORGE EDWARD, b Elgin, Ill, Feb 15, 24; m 53; c 5. MEDICINE, PHARMACOLOGY. Educ: Univ Notre Dame, BS, 50; St Louis Univ, MD, 53. Prof Exp: Rotating intern, Mt Carmel Mercy Hosp, Detroit, 54; resident med, Cochrane Vet Admin Hosp, St Louis, 55; fel, St Louis Univ Hosp, 56; Nat Heart Inst fel cardiol, Med Ctr, Duke Univ, 56-58; internist & cardiologist, pvt pract, 58-59; asst dir clin res, Lilly Labs Clin Res, 59-61; internist & cardiologist, Hosp Pract, 61-64; rev officer, Div New Drugs, US Food & Drug Admin, 64-66; EXEC DIR CLIN RES, MERCK SHARP & DOHME RES LABS, 66- Concurrent Pos: Clin asst prof med, Jefferson Med Col, 66- Mem: AAAS; Am Heart Asn; Am Col Cardiologists; Am Diabetes Asn; Am Soc Clin Pharmacol & Therapeut. Res: Internal medicine; clinical pharmacology. Mailing Add: Merck Sharp & Dohme Res Labs West Point PA 19486

MAHADEVA, MADHU NARAYAN, b Kallidaikurichi, India, Apr 27, 30; US citizen; m 57; c 2. ZOOLOGY. Educ: Univ Madras, MA, 51, MS, 52; Univ Calif, Los Angeles, PhD(zool), 56. Prof Exp: Lectr biol & head dept, Harward Col, Ceylon, 50-51; lect & head dept, Dharmaraja, Kandy & Zahira Cols, Ceylon, 52-53; res officer, Fisheries Res Sta, Govt of Ceylon, 53-61; head dept biol, Jamshedpur Coop Col, India, 61-67, Govt of India Univ Grants Comn fac res grant, 66-67; actg assoc prof zool & Acad Senate res grant, Univ Calif, Los Angeles, 67-68; ASSOC PROF BIOL, UNIV WIS-OSHKOSH, 68- Concurrent Pos: Teaching asst, Univ Calif, Los Angeles, 54-55; hon head dept biol, Navalar Hall, Tamil Univ Movement, Ceylon, 56-61; NSF instnl grant, Univ Wis, 70-71, fac res grants from Pres/chancellor, 72-73 & 74-75. Mem: AAAS; Am Fisheries Soc. Mailing Add: Dept of Biol Univ of Wis Oshkosh WI 54901

MAHADEVAN, PARAMESWAR b India, Apr 23, 26; m 58; c 2. ATOMIC PHYSICS, MOLECULAR PHYSICS. Educ: Univ Kerala, BSc, 44, MSc, 46; Univ London, PhD(physics), 58. Prof Exp: Lectr physics, Univ Kerala, 47-54; fel elec eng, Univ Fla, 59-60; consult physics, Gen Dynamics/Convair, 60-66; res scientist, Douglas Advan Res Labs, 67-74; STAFF SCIENTIST, AEROSPACE CORP, EL SEGUNDO, 74- Mem: AAAS; Am Phys Soc; fel Brit Inst Physics. Res: Atomic and molecular collision processes; measurement of interaction crossections; phenomena accompanying the impact of ions and atoms on clean metallic surfaces, such as sputtering, secondary electron emission and ion reflection; chemistry and physics of the earth's ionosphere; space science.

MAHADEVIAH, INALLY, b Hoovinamoda, India, May 25, 28; m 60; c 3. NUCLEAR CHEMISTRY, INORGANIC CHEMISTRY. Educ: Univ Mysore, BSc, 50, MSc, 54; Univ Cincinnati, PhD(chem), 63. Prof Exp: Lectr chem, Univ Mysore, 50-56; asst prof, Ft Hays Kans State Col, 60-61; from asst prof to assoc prof, 61-74, PROF CHEM, YOUNGSTOWN STATE UNIV, 74- Concurrent Pos: Res assoc, Univ Col, Univ London, 69-70. Mem: Am Chem Soc. Mailing Add: Dept of Chem Youngstown State Univ Youngstown OH 44555

MAHAFFEY, KATHRYN ROSE, b Johnstown, Pa, Dec 24, 43. NUTRITION, EXPERIMENTAL PATHOLOGY. Educ: Pa State Univ, BS, 64; Rutgers Univ, MS, 66, PhD(nutrit), 68. Prof Exp: NIH fel endocrine pharmacol, Sch Med, Univ NC, Chapel Hill, 67-69; res assoc, Sch Med, Univ NC, Chapel Hill, 69-70, asst prof path, 70-72; PROJ MGR LEAD CONTAMINATION OF FOODS, FOOD & DRUG ADMIN, 72-, ASST TO DIR DIV NUTRIT, 75- Concurrent Pos: Lectr, Sch Home Econ, Univ NC, Greensboro, 68-71; prog coordr, Univs Assoc Res & Educ in Path, Inc, Bethesda, Md, 71-72; adj asst prof community med, Sch Med, Georgetown Univ, 72-74. Mem: Am Inst Nutrit; Am Soc Exp Path. Res: Influence of nutritional status on susceptibility to toxic substances, particularly heavy metals; pediatric nutrition. Mailing Add: Div of Nutrit Bur of Foods Food & Drug Admin 200 C St SW Washington DC 20204

MAHAJAN, DAMODAR K, b Pilode, Maharastra; m 54; c 2. BIOCHEMISTRY, ENDOCRINOLOGY. Educ: Univ Poona, BSc(Hons), 51, MSc, 53, PhD(biochem), 64; Univ Utah, dipl steroid biochem, 61. Prof Exp: Pop Coun New York fel, Inst Steroid Biochem, Univ Utah, 60-62, Am Cancer Soc fel, Dept Biochem, 62-63; sr res asst steroid biochem, Cancer Res Inst, Bombay, India, 63-65, res officer, 65-69; instr steroid hormones, Dept Med & dir, Intermountain Regional Lab, Univ Utah, 69-72; asst prof hormones, Dept Obstet & Gynec, Pa State Univ, 72-75; ASST PROF HORMONES, DEPT REPRODUCTIVE BIOL, CASE WESTERN RESERVE UNIV, 75- Concurrent Pos: Teacher biochem, Univ Bombay, 68. Mem: AAAS; Endocrine Soc. Res: Metabolism of steroid hormones in health and endocrine disorders; inhibitory effect of progesterone, steroidogenic ability of different cells of the ovary; bioconversion of steroids by reproductive organs; radioimmunoassay of steroids; steroid binding proteins. Mailing Add: Dept of Reproductive Biol Case Western Reserve Univ Cleveland OH 44106

MAHAJAN, KISHAN PAUL, b Dina Nagar, India, Apr 20, 37; m 64; c 3. BIOCHEMISTRY, ORGANIC CHEMISTRY. Educ: Panjab Univ, India, BSc, 59, MSc, 61; Duquesne Univ, PhD(biochem), 65. Prof Exp: Lectr chem, DAV Col, Amritsar, 60-61; res assoc biochem, Duquesne Univ, 65-68; asst prof, 68-71, ASSOC PROF CHEM, STATE UNIV NY COL OSWEGO, 71- Mem: AAAS; fel Am Inst Chemists; Am Chem Soc. Res: Enzyme structure; mechanism of action of enzymes. Mailing Add: Dept of Chem State Univ NY Oswego NY 13126

MAHAJAN, SATISH CHANDER, b Chandigarh, India, Dec 6, 35; m 62; c 2. REPRODUCTIVE PHYSIOLOGY, ENDOCRINOLOGY. Educ: Punjab Univ, India, BVSc, 57; Agra Univ, MVSc, 60; Rutgers Univ, PhD(animal sci), 65. Prof Exp: Exten officer vet sci, Block Develop Area, 57-58; res asst animal sci, Indian Vet Res Inst Izatnagar, 60-61; PROF BIOL, LANE COL, 65- Concurrent Pos: Ford Found fel, Univ Wis-Madison, 65-66. Mem: AAAS; Am Soc Animal Sci. Res: Study of factors

MAHAJAN

which affect fertilization and survival of embryo in various animals. Mailing Add: Dept of Biol Lane Col Jackson TN 38301

MAHALL, BRUCE ELLIOTT, b Springfield, Mass, Apr 21, 46. PLANT ECOLOGY. Educ: Dartmouth Col, AB, 68; Univ Calif, Berkeley, PhD(bot), 74. Prof Exp: Fel physiol plant ecol, Carnegie Inst Washington, 74-75; ASST PROF PHYSIOL PLANT ECOL, UNIV CALIF, SANTA BARBARA, 75- Mem: Ecol Soc Am; Brit Ecol Soc; Am Soc Plant Physiologists; AAAS. Res: Environmental and physiological restrictions of plant distributions and dimensions of plant niches, emphasizing energy flow. Mailing Add: Dept of Biol Sci Univ of Calif Santa Barbara CA 93106

MAHAN, ARCHIE IRVIN, b Portland, Maine, Sept 1, 09; m 41; c 1. PHYSICS. Educ: Friends Univ, AB, 31; Johns Hopkins Univ, PhD(physics), 40. Prof Exp: Jr instr physics, Univ Kans, 31-32; jr instr optics, Johns Hopkins Univ, 34-38; instr physics, Georgetown Univ, 38-41; physicist, US Naval Ord Lab, 41-42, sr physicist, 45-53, div chief & optics dir, 53-55; assoc physicist, US Naval Gun Factory, 42-45; sr res physicist, 55-62, PRIN PHYSICIST, APPL PHYSICS LAB, JOHNS HOPKINS UNIV, 62- Concurrent Pos: Lectr, Grad Sch, Georgetown Univ, 52-54 & Am Inst Physics, 61-; mem, Nat Acad Sci Postdoctoral Eval Panel, 69- Mem: Fel AAAS; fel Am Phys Soc; fel Optical Soc Am (treas, 60-). Res: Infrared; optical design; interferometry; roof prisms; diffraction by telescopes, plane-parallel plates, cones and circularly symmetric apertures; radiation fields; radome boresight error; astronomical refraction; spontaneous and stimulated emission in cylinders; optical properties of cylinders; boundary value problems. Mailing Add: Appl Physics Lab Johns Hopkins Univ Silver Spring MD 20910

MAHAN, BRUCE HERBERT, b New Britain, Conn, Aug 17, 30. PHYSICAL CHEMISTRY. Educ: Harvard Univ, AB, 52, PhD(chem), 56. Prof Exp: From instr to assoc prof, 56-66, chmn dept, 68-71, PROF CHEM, UNIV CALIF, BERKELEY, 66-, PRIN INVESTR MAT RES, 73- Concurrent Pos: Vis fel, Joint Inst Lab Astrophys, Univ Colo, Boulder, 72. Honors & Awards: Calif Sect Award, Am Chem Soc, 68. Mem: Nat Acad Sci; Am Chem Soc. Res: Gas phase reaction kinetics, particularly ionic and electronic collision phenomena. Mailing Add: Dept of Chem Univ of Calif Berkeley CA 94720

MAHAN, DONALD CLARENCE, b East Chicago, Ind, May 28, 38; m 62; c 2. ANIMAL NUTRITION. Educ: Purdue Univ, West Lafayette, BS, 60, MS, 65; Univ Ill, Urbana, PhD(nutrit), 69. Prof Exp: Asst prof, Univ, 69-75, ASSOC PROF ANIMAL SCI, OHIO STATE UNIV & OHIO AGR RES DEVELOP CTR, 75- Mem: Am Soc Animal Sci; Am Inst Nutrit; Sigma Xi. Res: Protein nutrition in reproducing swine; selenium and vitamin E nutrition in swine and sheep; calcium and phosphorus nutrition in swine; management and nutrition interrelationship with swine. Mailing Add: Dept of Animal Sci Ohio Agr Res Develop Ctr Wooster OH 44691

MAHAN, GERALD DENNIS, b Portland, Ore, Nov 24, 37; m 65; c 2. THEORETICAL SOLID STATE PHYSICS. Educ: Harvard Univ, AB, 59; Univ Calif, Berkeley, PhD(physics), 64. Prof Exp: Physicist, Gen Elec Res & Develop Ctr, 64-67; assoc prof physics, Univ Ore, 67-73; PROF PHYSICS, IND UNIV, BLOOMINGTON, 73- Concurrent Pos: Alfred P Sloan res fel, 68-70. Mem: Fel Am Phys Soc. Res: Theory of optical and transport phenomena in solids. Mailing Add: Dept of Physics Ind Univ Bloomington IN 47401

MAHAN, HAROLD DEAN, b Ferndale, Mich, June 11, 31; m 54; c 5. ORNITHOLOGY, ECOLOGY. Educ: Wayne State Univ, BA, 54; Univ Mich, MS, 57; Mich State Univ, PhD(zool), 64. Prof Exp: From instr to prof biol, Cent Mich Univ, 57-72, dir environ interpretation ctr, 71-72; DIR, CLEVELAND MUS NATURAL HIST, 72- Mem: Am Ornithologists Union; Wilson Ornith Soc; Animal Behav Soc. Res: Growth and temperature regulation in birds; animal behavior. Mailing Add: Cleveland Mus of Natural Hist Wade Oval University Circle Cleveland OH 44106

MAHAN, JOHN ELMER, b Cortland, Ohio, Aug 1, 15; m 39; c 2. ORGANIC CHEMISTRY. Educ: Ohio Univ, BS, 37; Ohio State Univ, PhD(org chem), 41. Prof Exp: Fel, Univ Ill, 41-43, instr org chem, 43; sr sect chief res div, 43-56, mgr chem & rockets br, 56-59, asst dir res, 59-64, dir chem res, 64-69, SR SCIENTIST, PHILLIPS PETROL CO, 69- Mem: AAAS; Am Chem Soc. Res: Determination of structure of naturally occurring organic compounds; development of liquid and solid rocket propellants; synthesis and production of petrochemicals. Mailing Add: 526 E 16th St Bartlesville OK 74003

MAHAN, KENT IRA, b Springfield, Mo, June 7, 42; m 65; c 1. PHYSICAL CHEMISTRY, RADIOCHEMISTRY. Educ: Southwest Mo State Col, BS, 64; Columbia Univ, PhD(phys chem), 69. Prof Exp: ASST PROF CHEM, SOUTHERN COLO STATE COL, 69- Mem: Am Chem Soc. Res: Gas kinetics; hot atom chemistry; radiation chemistry; recoil tritium-hydrocarbon hot atom reactions. Mailing Add: Dept of Chem Southern Colo State Col Pueblo CO 81005

MAHANEY, WILLIAM C, b Utica, NY, May 17, 41. PHYSICAL GEOGRAPHY. Educ: Syracuse Univ, BA, 65; Ind Univ, MA, 67; Univ Colo, PhD(geog), 70. Prof Exp: Asst prof geog, Univ NDak, 70-71; asst prof, Brock Univ, summer 71; asst prof, 71-74, ASSOC PROF GEOG, YORK UNIV, 74- Mem: AAAS; Brit Soc Soil Sci; Glaciol Soc; Geol Soc Am; Am Quaternary Asn. Res: Pedology; quaternary stratigraphy. Mailing Add: 1 Bathgate Dr Toronto ON Can

MAHANTHAPPA, KALYANA T, b Hirehalli, India, Oct 29, 34; m 61; c 2. ELEMENTARY PARTICLE PHYSICS. Educ: Univ Mysore, BSc, 54; Univ Delhi, MSc, 56; Harvard Univ, PhD(physics), 61. Prof Exp: Res assoc physics, Univ Calif, Los Angeles, 61-63; asst prof, Univ Pa, 63-66; assoc prof, 66-69, fac res fel, 70-71, PROF PHYSICS, UNIV COLO, BOULDER, 69- Concurrent Pos: Mem, Inst Advan Study, 64-65; vis scientist, Int Ctr Theoret Physics, Trieste, 70-71. Mem: AAAS; fel Am Phys Soc. Res: Quantum field theory; elementary particle physics; quantum electro-dynamics; gauge theories and symmetries as applied to weak, electromagnetic and strong interactions. Mailing Add: Dept of Physics Univ of Colo Boulder CO 80302

MAHANTI, SUBHENDRA DEB, b Cuttack, India, Sept 24, 45; m 72; c 2. THEORETICAL SOLID STATE PHYSICS, BIOPHYSICS. Educ: Utkal Univ, India, BSc, 61; Univ Allahabad, MSc, 63; Univ Calif, Riverside, PhD(physics), 68. Prof Exp: Res asst physics, Univ Allahabad, 63-64 & Univ Calif, Riverside, 64-68; mem tech staff physics, Bell Tel Lab, Murray Hill, NJ, 68-70; ASST PROF PHYSICS, MICH STATE UNIV, 70- Mem: Am Phys Soc. Res: Structural, electronic and magnetic phase transitions; electron nuclear interactions in solids; energy and charge transport in biological systems. Mailing Add: Dept of Physics Mich State Univ East Lansing MI 48823

MAHAR, J MICHAEL, b Portland, Ore, Aug 21, 29; m; c 1. ANTHROPOLOGY. Educ: Reed Col, BA, 53; Cornell Univ, PhD, 66. Prof Exp: From lectr to assoc prof, 58-69, PROF ORIENTAL STUDIES, UNIV ARIZ, 69- Concurrent Pos: Fulbright-Hays res fel, India, 68-69; mem ed staff, J Asian Studies, 72- Mem: Am Oriental Soc. Res: Inter-village social organization in North India. Mailing Add: Dept of Oriental Studies Univ of Ariz Tucson AZ 85721

MAHARD, RICHARD HAROLD, b Lawton, Mich, July 5, 15; m. GEOMORPHOLOGY. Educ: Eastern Mich Univ, AB, 35; Columbia Univ, MA, 41, PhD, 49. Prof Exp: Asst geomorphol, Columbia Univ, 36-38, asst geol, 38-41; from instr to assoc prof geol, 41-55, chmn dept geol & geog, 51-61 & 70-73, PROF GEOL, DENISON UNIV, 55- Concurrent Pos: Lectr, Barnard Col, 37-38; lectr, Columbia Univ Exten, 37-38, vis prof, 56 & 61; with US Geol Surv, 45-; vis assoc prof, Stanford Univ, 52; vis prof, Univ Mich, 58. Mem: Fel AAAS (secy, Sect E, 60-68, chmn, 68-70); fel Geol Soc Am; fel Asn Am Geogr. Res: Geomorphic history of Verde Valley, Arizona; origin and nature of intrenched meanders; regional geography of the United States. Mailing Add: Dept of Geol & Geog Denison Univ Granville OH 43023

MAHARRY, DAVID EDWIN, b Zanesville, Ohio, Dec 12, 43; m 65; c 2. THEORETICAL PHYSICS. Educ: Muskingum Col, BS, 65; Univ Kans, MS, 69, PhD(physics), 71. Prof Exp: Asst prof, 71-72, asst acad dean & registrar, 72-74, asst prof, 74-76, ASSOC PROF PHYSICS, FRANKLIN COL, 76- Res: Mathematical models and computer simulation of social, biological and physical science. Mailing Add: Dept of Physics Franklin Col Franklin IN 46131

MAHAVIER, WILLIAM S, b Houston, Tex, July 30, 30; m; c 2. MATHEMATICS. Educ: Univ Tex, BS, 51, PhD(math), 57. Prof Exp: Physicist, US Air Missile Test Ctr, Calif, 51-52; mathematician, Defense Res Lab, Univ Tex, 54-57; instr math, Ill Inst Technol, 57-59; asst prof, Univ Tenn, 59-64; from asst prof to assoc prof, 64-70, PROF MATH, EMORY UNIV, 70- Mem: Am Math Soc; Math Asn Am. Mailing Add: Dept of Math Emory Univ Atlanta GA 30322

MAHDY, MOHAMED SABET, b Heliopolis, Egypt, Oct 29, 30; Can citizen; m 51; c 2. VIROLOGY, IMMUNOLOGY. Educ: Cairo Univ, BSc, 50; Univ Pittsburgh, MPH, 60, DScHyg(virol, microbiol), 63. Prof Exp: Res asst tissue cult & virol, US Naval Med Res Unit 3, 52-59; investr virol & head sect, Microbiol Unit, Med Res Inst, Egypt, 63-65; head lab virus res, 65-71, res scientist, 65-68, asst to chief virologist, 68-71, RES SCIENTIST & HEAD LAB IMMUNOGLOBULIN ASSAY, LAB SERV BR, ONT DEPT HEALTH, 71- Concurrent Pos: Guest investr, US Naval Med Res Unit 3, 63-64; lectr, Sch Hyg, Univ Toronto, 68-71, asst prof, 71-72; consult microbiol labs, Can Ctr Inland Waters, 72- Mem: Am Soc Microbiol; Can Soc Microbiol; Brit Soc Gen Microbiol; Can Pub Health Asn. Res: Immune and non-specific responses to viral infections; assay of immunoglobulins to identify viral infections; health hazards of environmental pollution with viruses. Mailing Add: Ont Dept of Health Lab Serv Br PO Box 9000 Terminal A Toronto ON Can

MAHENDRAPPA, MUKKATIRA KARIAPPA, b Coorg, India, Sept 20, 41; Can citizen; m 61; c 2. FOREST SOILS. Educ: Karnatak Univ, India, BSc, 61; Utah State Univ, MS, 63, PhD(soil chem), 66. Prof Exp: Res soil scientist, Univ Calif, Riverside, 66; RES SCIENTIST, CAN FOREST SERV, CAN DEPT ENVIRON, 66- Mem: Am Soc Agron; Soil Sci Soc Am; Can Soc Soil Sci; Int Soc Soil Sci. Res: Forest fertilization; nitrogen transformations; nutrition of forest nurseries and trees; organic matter decomposition. Mailing Add: Can Forest Serv Can Dept of Environ PO Box 4000 Fredericton NB Can

MAHENDROO, PREM P, b Indore, India, July 4, 30; m 62; c 2. SOLID STATE PHYSICS, CHEMICAL PHYSICS. Educ: Agra Univ, BSc, 50; Panjab Univ, MA, 55; Univ Tex, PhD(physics), 60. Prof Exp: Physicist, Div Acoust, Nat Phys Lab, India, 51-56; from asst prof to assoc prof, 60-74, PROF PHYSICS, TEX CHRISTIAN UNIV, 74- Concurrent Pos: Consult, Gen Dynamics Corp, Tex, 61- & Alcon Labs, 65-66; vis prof, Univ Nottingham, 74. Mem: Fel AAAS; Am Phys Soc; Am Chem Soc. Res: Acoustics and ultrasonics; nuclear magnetic and electron spin resonance; biophysics. Mailing Add: 2900 Covert Ft Worth TX 76133

MAHER, FRANK THOMAS, b East St Louis, Ill, Sept 23, 09; m 41; c 1. MEDICINE. Educ: Univ Ill, BS, 37, MS, 38, PhD(pharmacol), 41, MD, 47. Prof Exp: Asst instr mat med, Col Pharm, Univ Ill, 37-41, instr pharmacol & pharmacog, 41-43, asst prof pharmacol, 43-48, prof pharmacol & pharmacog & head dept, 48-53, asst dean pharm, 49-53; from assoc prof to prof pharmacol, Mayo Grad Sch Med, Univ Minn, 62-74; RETIRED. Concurrent Pos: Res assoc, Mayo Clin, 47-48, consult biochem, 53-58, consult clin path, 53-71, consult, Dept Lab Med Sect Diag Nuclear Med, Div Nephrology & Dept Internal Med, 71-74; mem coun circulation, Am Heart Asn. Mem: AMA; Soc Nuclear Med; Am Heart Asn; Am Soc Exp Path; Am Col Physicians. Res: Clinical pathology; renal functions; diagnostic radioactive isotopes. Mailing Add: 815 Tenth St SW Rochester MN 55901

MAHER, GEORGE GARRISON, b Washington, DC, Apr 17, 19; m 43; c 5. BIO-ORGANIC CHEMISTRY. Educ: NDak Agr Col, BSc, 41, MSc, 49; Ohio State Univ, PhD(org chem), 54. Prof Exp: Chemist, Weldon Springs Ord Works, 41-42; chief chemist, Ky Ord Works, 43-44; chemist, Tenn Eastman Corp, 44-45; asst agr chemist, NDak Agr Col, 45-49; res assoc, Res Found, Ohio State Univ, 49-54; sr res chemist, Clinton Corn Processing Co, 54-64; PRIN RES CHEMIST, NORTHERN REGIONAL RES LAB, USDA, 64- Mem: AAAS; Am Chem Soc; Sigma Xi. Res: Carbohydrate chemistry; starch derivatives; sugars and syrups; physical characterization; enzymology; propellant and explosive chemistry; research and development, chemically, of agriculturally derived commodities. Mailing Add: Northern Regional Res Lab 1815 N University St Peoria IL 61604

MAHER, JAMES VINCENT, b New York, NY, Aug 25, 42; m 66; c 2. PHYSICS. Educ: Univ Notre Dame, BS, 64; Yale Univ, MS, 65, PhD(physics), 69. Prof Exp: Appointee, Argonne Nat Lab, 68-70; ASST PROF PHYSICS, UNIV PITTSBURGH, 70- Mem: Am Phys Soc. Res: Nuclear reactions and spectroscopy; nuclear reactions induced by heavy projectiles. Mailing Add: Dept of Physics Univ of Pittsburgh Pittsburgh PA 15213

MAHER, JOHN CHARLES, b Lincoln, Nebr, Mar 24, 14; m 38; c 2. GEOLOGY. Educ: Univ Nebr, AB, 35, MSc, 37. Prof Exp: Asst instr geol, Univ Nebr, 35-36; field geologist, Nebr Geol Surv, 36-37; petrol geologist, Shell Petrol Corp, 37-38; dist geologist, US Geol Surv, 38-44, supv geologist, 44-55, regional supvr, 55-57; chief stratig sect, Pure Oil Co, 57-60, chief stratigr, 60-62; RES GEOLOGIST, US GEOL SURV, 62- Mem: Geol Soc Am; Am Asn Petrol Geologists. Res: Stratigraphy; petroleum, marine and groundwater geology. Mailing Add: US Geol Surv Menlo Park CA 94025

MAHER, JOHN FRANCIS, b Hempstead, NY, Aug 3, 29; m 53; c 5. MEDICINE, NEPHROLOGY. Educ: Georgetown Univ, BS, 49, MD, 53. Prof Exp: Intern med, Boston City Hosp, 53-54; resident, Georgetown Univ Hosp, 56-58, Nat Inst Arthritis & Metab Dis fel nephrol, Univ & Hosp, 58-60, from instr to assoc prof med, Sch Med, 60-69; prof med & dir div nephrol & clin res ctr, Univ Mo-Columbia, 69-74; PROF MED & DIR DIV NEPHROL, UNIV CONN, FARMINGTON, 74- Concurrent Pos: Ed, J Am Fedn Clin Res, 67-69. Mem: Am Fedn Clin Res; Am Soc

Artificial Internal Organs; Am Col Physicians; Am Soc Nephrol; Am Heart Asn. Res: Hemodialysis, kinetics; toxic nephropathy; dialysis of poisons; fluid and electrolyte homeostasis; transplantation immunology; uremia; renal pathology. Mailing Add: Dept of Med Univ of Conn Health Ctr Farmington CT 06032

MAHER, JOHN THOMAS, b Milford, Mass, Sept 17, 32; m 56; c 4. PHYSIOLOGY. Educ: Boston Col, BS, 53; Boston Univ, PhD(physiol), 71. Prof Exp: Instr anat, physiol & chem, Milford Hosp Sch Nursing, 54-55 & 58-59; RES PHYSIOLOGIST, US ARMY RES INST ENVIRON MED, 59- Mem: Am Col Sports Med; Am Physiol Soc; Asn Mil Surgeons US; Sigma Xi. Res: Cardiovascular physiology; environmental physiology and exercise; heat acclimatization; electrophysiology of the heart; cardiac dynamics. Mailing Add: US Army Res Inst of Environ Med Natick MA 01760

MAHER, LOUIS JAMES, JR, b Iowa City, Iowa, Dec 18, 33; m 56; c 3. PALYNOLOGY, QUATERNARY GEOLOGY. Educ: Univ Iowa, BA, 55, MS, 59; Univ Minn, Minneapolis, PhD(geol), 61. Prof Exp: NATO fel, Cambridge Univ, 61-62; from asst prof to assoc prof, 62-70, PROF GEOL, UNIV WIS-MADISON, 70- Mem: AAAS; Geol Soc Am; Ecol Soc Am. Res: Quaternary palynology in central and western United States. Mailing Add: Dept of Geol & Geophys Univ of Wis Madison WI 53706

MAHER, MICHAEL JOHN, b Napa, Calif, Oct 28, 28; m 54; c 2. COMPARATIVE ENDOCRINOLOGY. Educ: Univ Calif, Los Angeles, AB, 52, PhD(zool), 58. Prof Exp: Asst, Univ Calif, Los Angeles, 53-57; res scientist zool, Barnard Col, Columbia Univ, 58-59; res fel anat, Albert Einstein Col Med, 59-60, instr, 60-62; from asst prof to assoc prof zool, 62-69, ASSOC PROF PHYSIOL & CELL BIOL, UNIV KANS, 69- Mem: AAAS; Am Soc Zool. Res: Reptilian physiology; vertebrates; role of the thyroid gland in metabolism, especially cold blooded vertebrates; role of pancreatic and adrenal hormones in carbohydrate metabolism of reptiles. Mailing Add: Dept of Physiol & Cell Biol Univ of Kans Lawrence KS 66044

MAHER, PHILIP KENERICK, b Catonsville, Md, Dec 13, 30; m 56; c 4. PHYSICAL CHEMISTRY. Educ: Randolph-Macon Col, BS, 52; Cath Univ Am, MS, 55, PhD(phys chem), 56. Prof Exp: Sr res chemist, W R Grace & Co, Md, 56-60, res supvr, 60-62, mgr, 62-68, dir, 68-75; PRES, CATALYTIC ENVIRON SERV INC, 75- Mem: Am Chem Soc; Am Inst Chem Eng; Am Inst Chemists; Am Mgt Asn; Catalysis Soc. Res: Physical and surface chemistry of inorganic materials, including catalysis, adsorption and desorption phenomena; thermal and hydrothermal reactions of inorganic materials such as silicates, aluminate and phosphates; catalyst preparation, regeneration and rejuvenation. Mailing Add: Catalytic Environ Serv Inc 2 Village Square Baltimore MD 21210

MAHER, ROBERT FRANCIS, b Eldora, Iowa, July 14, 22; m 57; c 4. ANTHROPOLOGY. Educ: Univ Wis, BS, 48, MS, 50, PhD(anthrop), 58. Prof Exp: Instr anthrop, Univ Wis-Milwaukee, 53-54 & DePauw Univ, 56-57; from asst prof to assoc prof, 57-65, chmn dept, 67-71 & 73-74, PROF ANTHROP, WESTERN MICH UNIV, 65- Concurrent Pos: NIMH res grant, 59-60; Fulbright sr researcher, 60-61. Mem: Am Anthrop Asn; Soc Am Archaeol; Ethnol Soc. Res: Archaeology and ethnology of the Southwest Pacific. Mailing Add: Dept of Anthrop Western Mich Univ Kalamazoo MI 49001

MAHER, STUART WILDER, b Knoxville, Tenn, Aug 11, 18; m 43; c 2. GEOLOGY. Educ: Univ Tenn, AB, 46, MS, 48. Prof Exp: Asst geol, Univ Tenn, 45-46, instr, 46-48; geologist, US Geol Surv, 48-51; res assoc, Univ Tenn, 51-54; sr geologist, 54-57, prin geologist, 57-66, CHIEF GEOLOGIST, TENN STATE DIV GEOL, 66- Concurrent Pos: Asst prof, Univ Tenn, Knoxville, 71. Mem: AAAS; Am Inst Prof Geologists; Int Asn Genesis Ore Deposits. Res: Economic geology. Mailing Add: Tenn Div of Geol 4711 Old Kingston Pike Knoxville TN 37919

MAHER, VERONICA MARY, b Detroit, Mich, Feb 20, 31. CANCER, MOLECULAR BIOLOGY. Educ: Marygrove Col, BS, 51; Univ Mich, MS, 58; Univ Wis, PhD(molecular biol), 68. Prof Exp: Res assoc radiol, Sch Med, Yale Univ, 68-69; asst prof biol, Marygrove Col, 69-70; RES SCIENTIST CARCINOGENESIS, MICH CANCER FOUND, 70- Concurrent Pos: Mem carcinogenesis contract comt, Nat Cancer Inst, NIH, 75-, sci adv, Carcinogenesis Prog, Div Cancer Cause & Prev, 75- Mem: Am Asn Cancer Res; Am Soc Microbiol; Environ Mutagen Soc; Tissue Cult Asn. Res: Mutagenic and carcinogenic action of physical and of chemical carcinogenic agents in human cells in culture and the effect of DNA repair on this interaction. Mailing Add: Mich Cancer Found 110 E Warren Ave Detroit MI 48201

MAHER, WILLIAM J, b Brooklyn, NY, Mar 10, 27. VERTEBRATE ECOLOGY. Educ: Purdue Univ, BS, 51; Univ Mich, MS, 53; Univ Calif, Berkeley, PhD(zool), 61. Prof Exp: Lectr biol, San Francisco State Col, 61-62; actg asst prof, Univ Calif, Santa Barbara, 62-63; from asst prof to assoc prof, 63-74, PROF BIOL, UNIV SASK, 74- Mem: AAAS; Ecol Soc Am; Am Ornith Union; Arctic Inst NAm. Res: Predator-prey relationships; competition between closely related predators; adaptability of developmental rates to arctic environments. Mailing Add: Dept of Biol Univ of Sask Saskatoon SK Can

MAHESH, VIRENDRA B, b Khanki Punjab, India, Apr 25, 32; m 55; c 3. ORGANIC CHEMISTRY, ENDOCRINOLOGY. Educ: Patna Univ, BSc, 51; Univ Delhi, MSc, 53, PhD(org chem), 55; Oxford Univ, DPhil(biol sci), 58. Prof Exp: J H Brown Mem fel physiol, Sch Med, Yale Univ, 58-59; from asst res prof to assoc res prof, 59-66, prof, 66-70, REGENTS PROF ENDOCRINOL, MED COL GA, 70-, DIR, CTR POP STUDIES, 71-, CHMN DEPT ENDOCRINOL, 72- Honors & Awards: Rubin Award, Am Soc Study Sterility, 63; Billings Silver Medal, 65. Mem: Am Fertil Soc; Sigma Xi; Endocrine Soc; Int Soc Neuroendocrinol; Int Asn Clin Res Human Reproduction. Res: Isolation, secretion, biosynthesis and metabolism of various steroid hormones; mechanism of hormone action; control of gonadotropin secretion; ovulation; reproductive physiology. Mailing Add: Dept of Endocrinol Med Col of Ga Augusta GA 30902

MAHGOUB, AHMED, b Alexandria, Egypt, Aug 20, 41; US citizen. PHARMACOLOGY. Educ: Univ Alexandria, BS, 64, MS, 67; Univ NC, PhD(pharmacol), 71. Prof Exp: Fel pharmacol, Med Sch, Northwestern Univ, 71-73; sr biologist cell biol, Hoffmann-La Roche Inc, NJ, 73-76; ASST PROF PHARMACOL, SCH MED, UNIV PR, SAN JUAN, 76- Res: Studies of the mechanism of actions of parathyroid hormone, thyrocalcitonin, and cholecalciferol derivatives on bone culture and bone cells. Mailing Add: Dept of Pharmacol Univ of PR Sch of Med San Juan PR 00936

MAHLBERG, PAUL GORDON, b Milwaukee, Wis, Aug 1, 28; m 54; c 2. PLANT ANATOMY, CELL BIOLOGY. Educ: Univ Wis, BS, 50, MS, 51; Univ Calif, Berkeley, PhD, 58. Prof Exp: Drug salesman, Pitman-Moore Co, Wis, 53-54; asst, Univ Calif, 55-58; instr bot, Univ Pittsburgh, 58-65; ASSOC PROF BOT, IND UNIV, BLOOMINGTON, 65- Mem: AAAS; Bot Soc Am. Res: Lacticifer ontogeny and physiology in normal and abnormal tissues; lysosome origin, development and function in plant cells. Mailing Add: Dept of Plant Sci Ind Univ Bloomington IN 47401

MAHLE, NELS H, b Highland Park, Mich, June 23, 43. ANALYTICAL CHEMISTRY. Educ: Eastern Mich Univ, BS, 66; Northern Ill Univ, PhD(anal chem), 74. Prof Exp: Res assoc anal chem, Purdue Univ, 72-74; SR ANAL CHEMIST, ANAL LABS, DOW CHEM CO, 74- Mem: AAAS; Am Soc Mass Spectrometry; Am Chem Soc. Res: Mass spectrometry; laboratory automation; gas chromatography; computer analysis of spectrometry data; environmental analysis. Mailing Add: Anal Labs Dow Chem Co Midland MI 48640

MAHLER, DAVID BERNARD, dental materials, see 12th edition

MAHLER, HENRY RALPH, b Vienna, Austria, Nov 12, 21; nat US; m 48; c 3. BIOCHEMISTRY. Educ: Swarthmore Col, AB, 43; Univ Calif, PhD(org chem), 48. Prof Exp: Sr chemist, Tex Res Found, 48-49; res assoc, Univ Wis, 49-50; asst prof enzyme res, 50-55; from assoc prof to prof, 55-66, RES PROF CHEM, IND UNIV, BLOOMINGTON, 66- Concurrent Pos: Travel award, NSF, 55 & Rockefeller Found, 57; vis prof, Univ Sao Paulo, 57; vis investr & mem corp, Marine Biol Lab, Woods Hole, 60-; vis prof, Lab Biol Genetics, Nat Ctr Sci Res, France & Univ Paris, 62-63 & 69-70; NIH res career award, 62-, res career investr, 66-, mem biochem study sect, 67-71. Mem: Am Chem Soc; Am Soc Biol Chemists; Am Soc Cell Biologist; Brit Biochem Soc; Am Soc Microbiol. Res: Function, structure, genetics and biosynthesis of cell organelles and their constituents, especially of mitochondria and nerve cell synaptic membranes. Mailing Add: Dept of Chem Ind Univ Bloomington IN 47401

MAHLER, KURT, b Krefeld, Ger, July 26, 03. PURE MATHEMATICS. Educ: Univ Frankfurt, PhD(math), 29; Univ Manchester, DSc, 40. Prof Exp: Asst lectr to prof math, Univ Manchester, 38-63; prof, Inst Adv Studies, Australian Nat Univ, 63-68 & Ohio State Univ, 68-72; PROF MATH, AUSTRALIAN NAT UNIV, 72- Honors & Awards: deMorgan Medal, London Math Soc. Mem: Am Math Soc; Math Asn Am; London Math Soc; fel Royal Soc; fel Australian Acad Sci. Res: Number theory. Mailing Add: Dept of Math Inst of Adv Studies Australian Nat Univ Canberra Australia

MAHLER, RICHARD JOSEPH, b New York, NY, Mar 4, 34; m 60; c 2. INTERNAL MEDICINE, ENDOCRINOLOGY. Educ: NY Univ, BA, 55; New York Med Col, MD, 59. Prof Exp: Am Diabetes Asn metabolic res fel, New York Med Col, 62-63; NY Acad Med Glorney-Raisbeck traveling fel, Univ Durham, 63-64; from instr to assoc prof med, New York Med Col, 64-71; assoc dir dept metab & endocrinol, City of Hope Med Ctr, Duarte, Calif, 71-72; DIR METAB & ENDOCRINOL, EISENHOWER MED CTR, 72- Concurrent Pos: Am Diabetes Asn res & develop award, 66-68; NSF res grant, 66-69; New York City Health Res Coun res grants, 66-70, career scientist award, 68-71. Mem: Am Fedn Clin Res; Am Diabetes Asn; Asn Am Med Cols; Endocrine Soc; Am Physiol Soc. Res: Hormonal and non-hormonal influences on insulin action, the influence of these factors on its mechanism and its relationship to diabetes mellitus. Mailing Add: Eisenhower Med Ctr 39000 Bob Hope Dr Palm Desert CA 92260

MAHLER, ROBERT JOHN, b Los Angeles, Calif, Mar 17, 32; m 59; c 2. SOLID STATE PHYSICS. Educ: Univ Calif, Los Angeles, BA, 55; Univ Colo, PhD(physics), 63. Prof Exp: Oceanogr, US Navy Hydrographic Off, 57-58; asst nuclear magnetic resonance, Univ Colo, 58-63; physicist, 63-68, chief solid state electronics sect, 68-71, res physicist, 71-74, CHIEF OFF PROG DEVELOP, NAT BUR STANDARDS, 74- Concurrent Pos: Res assoc, Univ Colo, Boulder, 63-64, lectr, 64- Mem: Am Phys Soc. Res: Investigations of phonon-nuclear spin system interactions using pulsed nuclear magnetic resonance techniques; infrared detectors; measurement techniques. Mailing Add: Div 277 Nat Bur of Standards Boulder CO 80302

MAHLER, WALTER, b Vienna, Austria, May 7, 29; nat US; m 65; c 1. INORGANIC CHEMISTRY, ORGANIC CHEMISTRY. Educ: Monmouth Col, BS, 50; Univ Southern Calif, PhD(inorg chem), 58. Prof Exp: Res chemist, Am Dent Asn, Ill, 54; RES CHEMIST, CENT RES DEPT, EXP STA, E I DU PONT DE NEMOURS & CO, INC, 58- Mem: Am Chem Soc. Res: Turbine power fluids; high temperature organic chemistry; low temperature inorganic chemistry; inorganic fibers. Mailing Add: Cent Res Dept Exp Sta E I du Pont de Nemours & Co Inc Wilmington DE 19898

MAHLER, WILLIAM FRED, b Iowa Park, Tex, Aug 30, 30; m 55; c 2. PLANT TAXONOMY. Educ: Midwestern Univ, BS, 55; Okla State Univ, MS, 60; Univ Tenn, Knoxville, PhD(bot), 68. Prof Exp: From instr to asst prof bot, Hardin-Simmons Univ, 60-66; asst prof, 68-74, ASSOC PROF BOT, SOUTHERN METHODIST UNIV, 74-, CUR HERBARIUM, 71- Concurrent Pos: Ed & publ, Sida, Contrib to Bot, 71- Mem: Int Asn Plant Taxon; Am Soc Plant Taxon; Am Bryol & Lichenological Soc. Res: Floristic studies and pollen morphology in relation to taxonomic concepts. Mailing Add: Herbarium Southern Methodist Univ Dallas TX 75222

MAHLMAN, BERT H, b Bismarck, NDak, Nov 2, 22; m 48; c 3. POLYMER CHEMISTRY, MATERIALS SCIENCE. Educ: Univ Minn, BChE, 49. Prof Exp: Chemist, Hercules Powder Co, 49-61; RES CHEMIST, HERCULES, INC, 61- Mem: Am Chem Soc. Res: Polymer and coatings research and development. Mailing Add: Res Ctr Hercules Inc Wilmington DE 19899

MAHLMAN, GEORGE WILLIAM, b Buffalo, NY, Sept 21, 19; m 43; c 2. PHYSICS. Educ: Univ Mich, BSE, 41; Mass Inst Technol, ScD(physics), 48. Prof Exp: Assoc physicist, Res Lab, Linde Co Div, Union Carbide Corp, 48-52; sr physicist, Photoswitch Div, Electronics Corp Am, 52-55; adv develop engr, Semiconductor Div, Sylvania Elec Prod, Inc, 55-56; physicist, Res Lab, Hughes Aircraft Co, 57-61; adv engr, Westinghouse Astroelectronics Lab, 62-63; res scientist, Northrop Nortronics, 63-67; sr tech specialist, Autonetics Div, NAm Rockwell Corp, 67-71; design specialist, 71-73, CONSULT PHYSICIST, POMONA DIV, GEN DYNAMICS CORP, 73- Mem: Am Phys Soc. Res: Physical electronics; highcurrent arcs; infrared; photoconductors; semiconductor physics. Mailing Add: 2000 Port Albans Circle Newport Beach CA 92660

MAHLMAN, HARVEY ARTHUR, b La Crosse, Wis, Aug 8, 23; m 45; c 2. ANALYTICAL CHEMISTRY, INORGANIC CHEMISTRY. Educ: Univ Minn, BChem & BBA, 49; Univ Tenn, PhD, 56. Prof Exp: Chemist, Clinton Lab, Tenn, 44-46; asst, Gen Elec Co, 49-52; chemist, Oak Ridge Nat Lab, 52-73; nuclear staff power resources specialist, Fla Power & Light, Miami, 73-74; RADIATION WASTE & RADIATION CHEM STAFF, TENN VALLEY AUTH, 74- Concurrent Pos: NIH spec fel, Inst Radium, Paris, 63-64. Mem: AAAS; Am Chem Soc; Radiation Res Soc. Res: Radiochemical and neutron activation analysis; radiation chemistry. Mailing Add: 7006 Downing Dr Knoxville TN 37919

MAHLMAN, JERRY DAVID, meteorology, see 12th edition

MAHLSTEDE, JOHN PETER, b Cleveland, Ohio, June 5, 24; m 48; c 4.

MAHLSTEDE

ORNAMENTAL HORTICULTURE. Educ: Miami Univ, Ohio, BS, 47; Mich State Univ, MS, 48, PhD(ornamental hort), 51. Prof Exp: From asst prof to assoc prof, 51-57, asst dir, 65-66, actg assoc dir, 66-67, head dept hort, 61-65, PROF HORT, IOWA STATE UNIV, 57-, ASSOC DIR, AGR & HOME ECON EXP STA, 67- Honors & Awards: Colman Award, Am Asn Nueserymen, 58. Mem: Am Soc Hort Sci (pres, 71-72); Int Plant Propagators Soc (pres, 65). Res: Nursery management; plant propagation; morphological subjects; packaging and propagation. Mailing Add: 104 Curtiss Hall Iowa State Univ Ames IA 50010

MAHLUM, DANIEL DENNIS, b Wolf Point, Mont, May 5, 33; m 56; c 2. BIOCHEMISTRY. Educ: Whitworth Col, Wash, BS, 55; Univ Idaho, MS, 58; Univ Wis, PhD(biochem), 62. Prof Exp: Res technician agr chem, Univ Idaho, 55-57, asst agr chemist, 57-58; biol scientist, Hanford Labs, Gen Elec Co, 61-63, sr scientist, 63-65; SR SCIENTIST, BATTELLE-NORTHWEST, 65- Concurrent Pos: Co-chmn, Conf Early Nutrit & Environ Influences upon Behav Develop. Mem: Am Inst Nutrit; Soc Toxicol. Res: Radionuclide metabolism and toxicity; lipid metabolism; nutrition. Mailing Add: Dept of Biol Battelle-Northwest Richland WA 99352

MAHMOUD, HORMOZ MASSOUD, b Teheran, Iran, May 2, 18; nat US; m 54. THEORETICAL PHYSICS. Educ: Univ Tehran, EE, 40; Ind Univ, MS, 49, PhD(physics), 53. Prof Exp: Res assoc theoret physics, Ind Univ, 53-54 & Cornell Univ, 54-56; asst prof physics & assoc physicist, Ames Lab, Iowa State Univ, 56-60; assoc prof, 60-70, PROF PHYSICS, UNIV ARIZ, 70- Mem: Am Phys Soc. Res: Field theory; elementary particles. Mailing Add: Dept of Physics Univ of Ariz Tucson AZ 85721

MAHMOUD, IBRAHIM YOUNIS, b Baghdad, Iraq, Sept 9, 33; m 65; c 4. HERPETOLOGY. Educ: Ark Agr & Mech Col, BS, 53; Univ Ark, MA, 55; Univ Okla, PhD(zool), 60. Prof Exp: Prof biol, Northland Col, 60-63; assoc prof, 63-68, PROF BIOL, UNIV WIS-OSHKOSH, 68- Concurrent Pos: Bd regents res grants, Univ Wis, 63-65, 67-69 & 70-71. Mem: AAAS; Am Soc Zool; Am Soc Ichthyol & Herpet. Res: Reptilian physiology with special emphasis on steroid metabolism of corpus luteum of turtles. Mailing Add: Dept of Biol Univ of Wis Oshkosh WI 54901

MAHNCKE, HENRY ELMORE, b Puyallup, Wash, Jan 9, 11; m 36; c 2. INDUSTRIAL CHEMISTRY. Educ: State Col Wash, BS, 31, MS, 32; Brown Univ, PhD(phys chem), 36. Prof Exp: Res chemist, Calco Chem Co, NJ, 36-37; develop engr, Birdseye Elec Co, Mass, 37-38; chemist, Woods Hole Oceanog Inst, 38-39; res chemist, Gen Chem Co, NY, 39-43; mgr phys chem sect, Westinghouse Elec Co, 43-55; asst mgr eastern res div, Rayonier, Inc, 55-59; SUPVR CHEM SECT, SKF INDUSTS, INC, 59- Mem: Am Chem Soc; Soc Lubrication Engrs. Res: Ultraviolet absorption spectra; photochemistry; potentiometric analysis; printing ink pigments; mechanism of lubrication process; production of fine chemicals; silver plating processes; metal cleaning; cellulose chemistry. Mailing Add: SKF Industs Inc Eng Res Ctr King of Prussia PA 19406

MAHOLICK, LEONARD THOMAS, b Coaldale, Pa, Apr 18, 21; m 43; c 5. PSYCHIATRY. Educ: Univ Md, BS, 44, MD, 46. Prof Exp: Intern, Hosp, Emory, Ga, 46-47; resident, Lawson Vet Admin Hosp & Emory, Ga, 47-48; asst chief neuropsychiat serv, Oliver Gen Hosp, Augusta, Ga, 48-50; fel psychiat, Austin Riggs Ctr, Mass, 50; dir, Savannah-Chatham County Ment Health Clin, 51-52; med dir, Bradley Ctr, Inc, 55-73; SUPVR PSYCHOTHER, DEPT CLIN PSYCHOL, GA STATE UNIV, 73-; PSYCHOTHERAPIST, ATLANTA PSYCHIAT CLIN, 73- Concurrent Pos: Asst clin prof, Med Col Ga, 49-50; pvt pract, 52-; consult, Muscogee County Pub Health Dept, 52-53 & State Div Voc Rehab, Ga, 57; spec consult, NIMH, 62-66; consult, Wiregrass Ment Health Ctr, Dothan, Ala, 64-73, Voc Rehab Social Security, 65-, Ment Health Clin, Troup County Pub Health Dept, 66-70, Child Guid Clin, Columbus Health Dept, 70-73, Aftercare Clin, 70-73, Ga Asn Pastoral Care, 73-75, Cent De Kalb Ment Health Ctr, 74-75 & Coosa Ment Health Ctr, Rome, Ga, 74-; teacher behav sci & psychother, Family Pract Residents, Med Ctr, Columbus, Ga, 72-73; teacher clin psychol interns, Bradley Ctr, Inc, 72-73. Mem: Fel Royal Soc Health; fel Am Psychiat Asn; Am Soc Group Psychother & Psychodrama; fel Am Pub Health Asn; Insts Relig & Health. Res: Psychotherapy; community mental health planning and programming; mental health training; community and clinic surveys. Mailing Add: Atlanta Psychiat Clin 2905 Peachtree Rd NE Atlanta GA 30305

MAHON, JOHN HAROLD, b Barbados, BWI, July 7, 22; US citizen; m 46; c 3. AGRICULTURAL BIOCHEMISTRY. Educ: McGill Univ, BSc, 48, MSc, 49, PhD(agr biochem), 53. Prof Exp: Res chemist & actg head food chem sect, Food & Drug Directorate, Dept Nat Health & Welfare, Can, 49-58; group leader food technol & biol, Calgon Corp, 58-63, mgr res & develop dept, 63-67; dir res & develop, 67-72, VPRES RES & DEVELOP, CALGON CORP DIV, MERCK & CO INC, 72- Mem: Inst Food Technologists; Am Chem Soc; Soc Chem Indust; Mfg Chemists Soc; Indust Res Inst. Res: Development of specialty chemicals for industrial water treatment; activated carbon; paper chemicals. Mailing Add: Calgon Corp Div Merck & Co Inc PO Box 1346 Pittsburgh PA 15230

MAHON, WILLIAM A, b Airdrie, Scotland, Aug 20, 29; m 56; c 4. CLINICAL PHARMACOLOGY. Educ: Harvard Univ, SM, 62; FRCP(C), 60. Prof Exp: Sr resident med, St Joseph's Hosp, London, Ont, 56-57; teaching fel path, Univ Western Ont, 57-58; asst resident, Lemuel Shattuck Hosp, 58-59, chief resident, 59-60; assoc, Sch Med, Tufts Univ, 60-62; from asst prof to assoc prof pharmacol & med, Univ Alta, 62-66; asst prof therapeut & assoc med, 66-68, ASSOC PROF PHARMACOL, UNIV TORONTO, 66-, ASSOC PROF MED, 68- Mem: Am Fedn Clin Res; Can Soc Pharmacol. Res: Cardiovascular pharmacology in man and animals; distribution of antibiotics in normal man and in man with renal failure; methods for studying drug action in man. Mailing Add: Toronto Gen Hosp Toronto ON Can

MAHONEY, BERNARD LAUNCELOT, JR, b Boston, Mass, Nov 1, 36; m 65; c 2. ANALYTICAL CHEMISTRY, PHYSICAL CHEMISTRY. Educ: Boston Col, BS, 58, MS, 60; Univ NH, PhD(phys chem), 67. Prof Exp: From asst prof to assoc prof, 65-74, CHMN DEPT, MARY WASHINGTON COL, 71-, PROF CHEM, 74- Mem: Am Chem Soc; Soc Appl Spectros; fel Am Inst Chem. Res: Spectroscopy; photochemistry; chemical kinetics. Mailing Add: Dept of Chem Mary Washington Col Fredericksburg VA 22401

MAHONEY, CHARLES LINDBERGH, b Geneva, NY, Mar 18, 28; m 57; c 2. ENVIRONMENTAL SCIENCES. Educ: Colo State Univ, BS, 53; State Univ NY Col Environ Sci & Forestry, MS, 55, PhD(ecol, forest zool), 65. Prof Exp: From asst prof to prof biol, State Univ NY Col Geneseo, 55-68; ASSOC PROF RESOURCE CONSERV, COL FORESTRY & NATURAL RESOURCES, COLO STATE UNIV, 68- Concurrent Pos: Mem Yellowstone field res exped, Atmospheric Sci Res Ctr, State Univ NY Albany, 63-64; res scientist, Bahamas Marine Surv, Off Naval Res, 67; dir, Pingree Park Campus, Colo State Univ, 68-74; staff mem, Coop Sci Prog, Univ Sopron, Hungary, 74; consult atmospheric monitoring, Stillwater Platinum Mines, Mont, 76; res scientist, NOWCAST Weather Satellite Prog, NASA-Colo State Univ, 76. Mem: AAAS; Soc Am Foresters; Am Inst Biol Sci; Sigma Xi; Conserv Educ Asn. Res: Soil insects as indicators of use patterns in recreation areas. Mailing Add: Col Forestry & Natural Resources Colo State Univ Ft Collins CO 80523

MAHONEY, CLARENCE LYNN, organic chemistry, see 12th edition

MAHONEY, COLETTE, b Jamaica, NY, July 19, 26. BIOLOGY. Educ: Fordham Univ, MA, 52, PhD(biol), 61. Prof Exp: Teacher private sch, 52-57; instr biol, Marymount Col, Va, 57-60, assoc prof, NY, 61-68, PRES, MARYMOUNT MANHATTAN COL, 68- Mem: AAAS; Nat Asn Biol Teachers. Res: Cytology; effects of some anti-metabolites on cell division. Mailing Add: Off of the Pres Marymount Manhattan Col New York NY 10021

MAHONEY, EARLE BARNES, b Penn Yan, NY, July 3, 09; m 37; c 4. SURGERY. Educ: Hobart Col, BS, 30; Univ Rochester, MD, 34. Hon Degrees: DSC, Hobart Col, 57. Prof Exp: Intern surg, Strong Mem Hosp, NY, 34-35, asst resident, 35-36, assoc resident surg & instr orthop surg, 36-37, resident surgeon, 38-39, from asst surgeon to surgeon, 39-74; from instr to prof, 38-74, EMER PROF SURG, SCH MED & DENT, UNIV ROCHESTER, 74- Concurrent Pos: Nat Res Coun fel, Cincinnati Gen Hosp, 37-38; consult, Genesee Hosp, Rochester, 52- Mem: Soc Univ Surg (past pres); Soc Clin Surg; Soc Vascular Surg (past pres); Am Asn Cancer Res; Am Asn Thoracic Surg. Res: Cardiovascular surgery; surgical shock; venous thrombosis; cardiac surgery; protein metabolism. Mailing Add: Box 1138 Nantucket MA 02554

MAHONEY, FRANCIS JOSEPH, b Boston, Mass, Mar 18, 36; m 70; c 2. PHYSICS, BIOLOGY. Educ: Col Holy Cross, BS, 57; Univ Rochester, MS, 58; Harvard Univ, MS, 60; Mass Inst Technol, PhD(nuclear eng), 68. Prof Exp: Asst radiation physics, Harvard Univ, 58-60, asst radiation physics & biol, 60-61; physicist, Cambridge Nuclear Corp, 61-64; res asst nuclear eng, Mass Inst Technol, 64-67; res physicist, US Army Natick Labs, 67-71; grants assoc, NIH, 71-72; PROG DIR FOR RADIATION, NAT CANCER INST, 72- Concurrent Pos: Radiation physicist, Avco Corp, 66-67. Mem: AAAS; Am Phys Soc; Am Nuclear Soc; Radiation Res Soc. Res: Radiation physics and biology; nuclear physics and engineering. Mailing Add: Nat Cancer Inst NIH Bethesda MD 20014

MAHONEY, JOAN MUNROE, b Providence, RI, May 13, 38; m 71; c 1. BIOCHEMISTRY. Educ: Chatham Col, BS, 60; St Lawrence Univ, MS, 62; State Univ NY Upstate Med Ctr, PhD(biochem), 69. Prof Exp: Instr chem labs, St Lawrence Univ, 60-62; USPHS training grant, Inst Enzyme Res, Univ Wis, 68-69; res assoc mitochondrial metab, Sch Med, Ind Univ, Indianapolis, 69-71; instr, 71-74, ASST PROF BIOCHEM, TERRE HAUTE CTR MED EDUC, IND STATE UNIV, 74- Mem: AAAS; Am Chem Soc; Biophys Soc. Res: Mitochondrial metabolism; membranes and their control of processes; enzyme regulation. Mailing Add: Terre Haute Ctr for Med Educ Ind State Univ Terre Haute IN 47809

MAHONEY, JOHN ARTHUR, chemical engineering, see 12th edition

MAHONEY, RICHARD THEODORE, b Mishawaka, Ind, July 21, 43; m 63; c 2. REPRODUCTIVE BIOLOGY, CHEMICAL PHYSICS. Educ: Purdue Univ, West Lafayette, BS, 65; Univ Calif, San Diego, PhD(chem), 70. Prof Exp: Proj specialist, 70, asst prog officer, 70-72, PROG OFFICER POP, FORD FOUND, 72- Mem: AAAS; Soc Study Reproduction; Int Soc Study Reproduction; Soc Study Fertil; Am Fertil Soc. Res: Monitoring of grants supporting research on new methods of human fertility control. Mailing Add: Pop Off Ford Found 320 E 43rd St New York NY 10017

MAHONEY, ROBERT PATRICK, b Poughkeepsie, NY, Dec 27, 34; m 61; c 5. MICROBIAL PHYSIOLOGY, ELECTRON MICROSCOPY. Educ: State Univ NY Co, New Paltz, BS, 58; Syracuse Univ, MS, 61, PhD(microbiol), 64. Prof Exp: Asst prof, 64-68, ASSOC PROF BIOL, SKIDMORE COL, 68- Concurrent Pos: NSF fac fel & guest investr, Biol Dept, Woods Hole Oceanog Inst, 70-71. Mem: AAAS; Am Soc Microbiol; Electron Micros Soc Am; NY Acad Sci. Res: Physiology and electron microscopy of autotrophic bacteria; iron and sulfur oxidizing species. Mailing Add: Dept of Biol Skidmore Col Saratoga Springs NY 12866

MAHONEY, THOMAS JOSEPH, physical chemistry, organic chemistry, see 12th edition

MAHONY, DAVID EDWARD, b St John, NB. MICROBIOLOGY. Educ: Acadia Univ, BSc, 62; Dalhousie Univ, MSc, 64; McGill Univ, PhD(bact), 67. Prof Exp: Lectr, 67-69, ASST PROF MICROBIOL, FAC MED, DALHOUSIE UNIV, 69- Mem: Can Soc Microbiol; Am Soc Microbiol. Res: Bacteriophages and bacteriocins of Clostridium perfringens. Mailing Add: Dept of Microbiol Dalhousie Univ Fac of Med Halifax NS Can

MAHONY, JOHN DANIEL, b New York, NY, Jan 15, 31. NUCLEAR CHEMISTRY. Educ: St John's Univ, NY, 51; Univ Conn, MS, 53; Univ Calif, Berkeley, PhD(nuclear chem), 65. Prof Exp: Asst prof chem, Marquette Univ, 65-67; asst prof, 67-71, ASSOC PROF CHEM, MANHATTAN COL, 71- Concurrent Pos: Vis res collabr, State Univ NY Stony Brook, 67-72. Mem: Am Chem Soc. Res: Nuclear reactions and nuclear fission; track detectors; activation analysis; interdisciplinary chemical education. Mailing Add: Dept of Chem Manhattan Col Bronx NY 10471

MAHOWALD, ANTHONY P, b Albany, Minn, Nov 24, 32; m 71. DEVELOPMENTAL BIOLOGY. Educ: Spring Hill Col, BS, 58; Johns Hopkins Univ, PhD(biol), 62. Prof Exp: Res fel biol, Johns Hopkins Univ, 62-66; asst prof, Marquette Univ, 66-70; asst mem & vis scientist, Inst Cancer Res, 70-72; ASSOC PROF ZOOL, IND UNIV, BLOOMINGTON, 72- Mem: AAAS; Am Soc Zoologists; Am Soc Cell Biologists; Soc Develop Biol. Res: Differentiation of ultrastructure during embryogenesis; polar granules of Drosophila. Mailing Add: Dept of Zool Ind Univ Bloomington IN 47401

MAHOWALD, MARK EDWARD, b Albany, Minn, Dec 1, 31; m 54; c 5. MATHEMATICS. Educ: Univ Minn, BA, 53, MA, 54, PhD(math), 55. Prof Exp: Sr engr, Gen Elec Co, 56-57; asst prof math, Xavier Univ, Ohio, 57-59 & Syracuse Univ, 59-63; chmn dept, 72-75, PROF MATH, NORTHWESTERN UNIV, EVANSTON, 63- Concurrent Pos: Sloan fel, 65-67. Mem: Am Math Soc; Math Asn Am. Res: Homotopy theory; algebraic topology; topological groups. Mailing Add: 500 Greenleaf Wilmette IL 60091

MAHOWALD, THEODORE AUGUSTUS, b St Cloud, Minn, June 22, 30; m 54; c 7. BIOCHEMISTRY. Educ: St John's Univ, Minn, BA, 52; St Louis Univ, PhD(biochem), 57. Prof Exp: USPHS fel, Enzyme Inst, Univ Wis, 57-60; mem staff, Scripps Clin & Res Found, Univ Calif, 60-62; asst prof biochem, Med Col, Cornell Univ, 62-69; ASSOC PROF BIOCHEM, MED COL, UNIV NEBR AT OMAHA, 69- Mem: AAAS; Am Soc Biol Chemists; Am Chem Soc. Res: Metabolism of bile acids; mechanism and sites of enzyme action; metabolism of methionine Mailing Add: 3347 S 114th Ave Omaha NE 68144

MAHR, TIBOR G, physical chemistry, surface chemistry, see 12th edition

MAHRT, JEROME L, b Colon, Nebr, Dec 20, 37; m 60; c 3. PARASITOLOGY, ZOOLOGY. Educ: Utah State Univ, BS, 60, MS, 63; Univ Ill, PhD(vet parasitol), 66. Prof Exp: Asst prof, 66-71, ASSOC PROF ZOOL, UNIV ALTA, 71- Mem: Am Soc Parasitologists; Soc Protozoologists. Res: Protozoan parasitology with specialization in the avian and mammalian Coccidia and blood Protozoa of birds. Mailing Add: Dept of Zool Univ of Alta Edmonton AB Can

MAI, WILLIAM FREDERICK, b Greenwood, Del, July 23, 16; m 41; c 3. PLANT PATHOLOGY. Educ: Univ Del, BS, 39; Cornell Univ, PhD(plant path), 45. Prof Exp: Asst prof, 46-52, PROF PLANT PATH, CORNELL UNIV, 52- Mem: Fel Am Phytopath Soc; Soc Nematol (vpres, 68, pres, 69); Am Inst Biol Sci; Potato Asn Am; Soc Europ Nematol. Res: Plant pathogenic and soil inhabiting nematodes; diseases of plants caused by nematodes. Mailing Add: Dept of Plant Path Cornell Univ Ithaca NY 14850

MAIBACH, HOWARD I, b New York, NY, July 18, 29. DERMATOLOGY. Educ: Tulane Univ, AB, 50, MD, 55; Am Bd Dermat, dipl, 61. Prof Exp: Asst instr, Sch Med, Univ Pa, 58-61; from asst prof to assoc prof, 61-71, PROF DERMAT, SCH MED, UNIV CALIF, SAN FRANCISCO, 71-, VCHMN DIV DERMAT, 62- Concurrent Pos: USPHS fel, Pa Hosp, 59-61; lectr, Grad Sch Med, Univ Pa, 60-61; pvt pract, 61-; mem staff, Herbert C Moffitt Hosps, Univ Calif, 61-; consult, Sonoma State Hosp, Eldridge, 62-64, Calif State Pub Health Serv, Stanford Res Inst, Calif Med Facil, Vacaville, Letterman & San Francisco Gen Hosps, 62- & Vet Admin Hosp, San Francisco, 63- Mem: AAAS; Soc Invest Dermat; Am Acad Dermat; fel Am Col Physicians; Am Fedn Clin Res. Mailing Add: Dept of Dermat Univ of Calif Sch of Med San Francisco CA 94143

MAIBENCO, EDWARD PAUL, neuroanatomy, see 12th edition

MAIBENCO, HELEN CRAIG, b Scotland, June 9, 17; nat US; m 57; c 2. ANATOMY. Educ: Wheaton Col, Ill, BS, 48; De Paul Univ, MS, 50; Univ Ill, PhD, 56. Prof Exp: Instr pharmacol, Presby Hosp Sch Nursing, 50-54; asst, Univ Ill Col Med, 54-56, from instr to prof anat, 56-73; PROF ANAT, RUSH MED COL, 73- Mem: AAAS; Am Soc Zoologists; Endocrine Soc; Am Asn Anatomists. Res: Physiology of connective tissues; endocrine relationships; morphological changes associated with aging; connective tissues of the female reproductive system. Mailing Add: Dept of Anat Rush Med Col 1725 W Harrison St Chicago IL 60612

MAICKEL, ROGER PHILIP, b Floral Park, NY, Sept 8, 33; m 56; c 2. BIOCHEMISTRY, PHARMACOLOGY. Educ: Manhattan Col, BS, 54; Georgetown Univ, MS, 57, PhD(chem), 60. Prof Exp: Chemist, Lab Chem Pharmacol, Nat Heart Inst, 55-60, biochemist, 60-63, sect head biochem function, 63-65; assoc prof, 65-69, PROF PHARMACOL, IND UNIV, BLOOMINGTON, 69-, CHIEF MED SCI SECT, INST RES PUB SAFETY, 70-, DIR PHARMACOL, UNIV, 71- Concurrent Pos: Exec ed, Life Sci, 65-69; Am Chem Soc tour lectr, 65- Mem: Fel AAAS; Am Chem Soc; Am Soc Pharmacol & Exp Therapeut; fel Am Inst Chemists; NY Acad Sci. Mailing Add: Dept of Pharmacol Ind Univ Bloomington IN 47401

MAIDANIK, GIDEON, b Safad, Israel, July 3, 25; US citizen. PHYSICS. Educ: Univ Manchester, BSc, 54, MSc, 55; Brown Univ, PhD(physics), 59. Prof Exp: Asst physics, Brown Univ, 56-58; sr scientist, Brit Oxygen Res & Develop Ltd, 58-60; Sr scientist, Bolt Beranek & Newman, Inc, Mass, 60-66; SR RES SCIENTIST TECH ADV, NAVAL SHIP RES & DEVELOP CTR, 66- Mem: Fel Acoust Soc Am; Brit Inst Physics & Phys Soc. Res: Nuclear physics, especially nucleon scattering; acoustics, especially structural vibration and radiation; sonar systems; hydroacoustics. Mailing Add: Code 1902 Ship Acoust Dept Naval Ship Res & Develop Ctr Bethesda MD 20034

MAIELLO, JOHN MICHAEL, b New York, NY, Oct 5, 42. MYCOLOGY, ENVIRONMENTAL BIOLOGY. Educ: Hunter Col, BA, 65; Rutgers Univ. PhD(mycol), 72. Prof Exp: ASST PROF MYCOL, RUTGERS UNIV, NEWARK, 72- Concurrent Pos: Dir & cur mycol cult collection, Rutgers Univ, 72- Mem: Sigma Xi; Mycol Soc Am; Am Phytopath Soc; Am Inst Biol Sci. Res: Morphology and physiology of pycnidial development in the sphaeropsidales; pathology of economically important plants; air pollutants injurious to vegetation in New Jersey. Mailing Add: Dept of Bot Rutgers Univ Newark NJ 07102

MAIENSCHEIN, FRED (CONRAD), b Belleville, Ill, Oct 28, 25; m 48; c 2. NUCLEAR PHYSICS. Educ: Rose Polytech Inst, BS, 45; Ind Univ, MS, 48, PhD(physics), 49. Prof Exp: Sr physicist, Nuclear Energy Propulsion for aircraft div, Fairchild Engine & Airplane Corp, 49-51; sr physicist, 51-66, DIR NEUTRON PHYSICS DIV, OAK RIDGE NAT LAB, 66- Mem: Am Phys Soc; Sigma Xi; fel Am Nuclear Soc. Res: Neutron and reactor physics; shielding. Mailing Add: 838 W Outer Dr Oak Ridge TN 37830

MAIENTHAL, E JUNE, b Paris, Mo, Oct 4, 28; m 49. CHEMISTRY. Educ: Millikin Univ, BS, 50. Prof Exp: Control chemist, Beatrice Food Co, Decatur, Ill, 50-51; chemist, Lincoln Lab, Decatur, Ill, 51-52; chemist, 52-62, RES CHEMIST, NAT BUR STANDARDS, 62- Honors & Awards: Spec Serv Award, Nat Bur Standards, 62, Outstanding Performance Award, 70 & 74; Silver Medal Award, Dept Com, 70. Mem: Am Chem Soc; Am Inst Chemists; The Chem Soc; Am Soc Testing & Mat. Res: Development of voltammetric methods of analysis of metals, ores, lunar rocks, biological materials, standard reference materials and environmental samples, particularly for trace elements. Mailing Add: A227 Chem Bldg Div Anal Chem Nat Bur of Standards Washington DC 20234

MAIENTHAL, MILLARD, b Decatur, Ill, Feb 25, 16; m 49. ORGANIC CHEMISTRY. Educ: Millikin Univ, BS, 38; Univ Mo, PhD(chem), 49. Prof Exp: Chemist, Ill Powder Mfg Co, 41-43; instr, Univ Mo, 48-49; asst prof chem, Millikin Univ, 49-52; org chemist, Ord Corps, Diamond Ord Fuze Lab, US Dept Army, 52-60; ORG CHEMIST, US FOOD & DRUG ADMIN, WASHINGTON, DC, 60- Mem: Am Chem Soc; Am Soc Testing & Mat; Sigma Xi. Res: Synthesis of cyclic amines; aliphatic nitro compounds; silanes; reduction of oximes; thermal analysis, purification and synthesis of drugs. Mailing Add: 10116 Bevern Lane Potomac MD 20854

MAIER, CHARLES ROBERT, b Leoti, Kans, Oct 9, 28; m 56; c 3. BOTANY. Educ: Kans State Teachers Col, Emporia, BS, 53, MS, 55; Ore State Col, PhD(plant path), 59. Prof Exp: Asst, Kans State Col, 54-55; from asst prof to assoc prof plant path, NMex State Univ, 58-68; PROF BIOL, WAYNE STATE COL, 68- Mem: Am Phytopath Soc; Mycol Soc Am; Nat Sci Teachers Asn. Res: Soil microbiology; control of diseases of cotton; antibiotic chemotherapy of hop downy mildew; organismal pollution of irrigation water. Mailing Add: Dept of Biol Wayne State Col Wayne NE 68787

MAIER, EMANUEL, b Essen, Ger, Jan 4, 16; US citizen; m 40; c 2. PHYSICAL GEOGRAPHY. Educ: City Col New York, AB, 36, MEd, 42; NY Univ, PhD(Ger lit), 53; Clark Univ, PhD(geog), 61. Prof Exp: PROF GEOG & CHMN DEPT, BRIDGEWATER STATE COL, 63- Concurrent Pos: Max Planck Inst fel, Munich, Ger, 71-72. Mem: AAAS; Asn Am Geog. Res: Territoriality. Mailing Add: Dept of Geog Bridgewater State Col Bridgewater MA 02324

MAIER, EUGENE ALFRED, b Tillamook, Ore, May 7, 29; m 52; c 4. MATHEMATICS. Educ: Univ Ore, BA, 50, MA, 51, PhD(math), 54. Prof Exp: From asst prof to assoc prof & chmn dept, 55-61; assoc prof, 61-70, PROF MATH, UNIV ORE, 70- Mem: Am Math Soc; Math Asn Am. Res: Number theory; mathematics education. Mailing Add: Dept of Math Univ of Ore Eugene OR 97403

MAIER, EUGENE JACOB RUDOLPH, b Washington, DC, Sept 1, 31; m 59; c 3. GEOPHYSICS. Educ: Mass Inst Technol, BS, 53; Carnegie Inst Technol, MS, 59, PhD(meson physics), 62. Prof Exp: Physicist, Appl Physics Lab, Johns Hopkins Univ, 56; PHYSICIST, GODDARD SPACE FLIGHT CTR, NASA, GREENBELT, 62- Mem: AAAS; Am Phys Soc; Am Geophys Union. Res: Structure and direct measurements of ionosphere; measurements of interplanetary medium; spacecraft experiment instrumentation; high energy meson physics; energetic particle experiments. Mailing Add: 1173 River Bay Rd Annapolis MD 21401

MAIER, GEORGE D, b Chicago, Ill, July 24, 30. BIOCHEMISTRY. Educ: Cornell Col, BA, 53; Iowa State Univ, MS, 56, PhD(biochem), 62. Prof Exp: Asst biochemist, Am Meat Inst Found, Chicago, 57-59; fel, Mass Gen Hosp, Boston, 62-63; fel, Sch Med, Western Reserve Univ, 63-65; ASST PROF CHEM, COLBY COL, 65- Res: Protein primordial synthesis; thiamine and renin chemistry. Mailing Add: Dept of Chem Colby Col Waterville ME 04910

MAIER, JOHN, b Royersford, Pa, Mar 3, 12; m 42; c 3. PUBLIC HEALTH. Educ: Harvard Univ, BA, 34, MD, 38. Prof Exp: Mem staff, 47-55, asst dir, 55-57, assoc dir, 57-73, DIR, ROCKEFELLER FOUND, 73- Res: Malaria chemotherapy and control; insecticides. Mailing Add: Rockefeller Found 1133 Ave of Americas New York NY 10036

MAIER, JOHN G, b St Louis, Mo, Aug 23, 26; m 52; c 3. MEDICINE, RADIOBIOLOGY. Educ: St Louis Univ, MD, 52; Univ Colo, MS, 59; Univ Rochester, PhD(radiation biol), 63. Prof Exp: Chief radioisotope clin, Walter Reed Gen Hosp, US Army, 59-60, chief radiation biol, Walter Reed Army Inst Res, 62-63, chief radiation ther, 64-69, chief dept radiol, Walter Reed Gen Hosp, Washington, DC, 70-72; CHIEF RADIATION THER DIV, GEORGE WASHINGTON UNIV HOSP, 72- Concurrent Pos: Assoc prof, Sch Med, Georgetown Univ & George Washington Univ, 64- Mem: Am Col Radiol; AMA; Soc Nuclear Med; Radiol Soc NAm; Am Soc Therapeut Radiol. Res: Nuclear medicine; therapeutic radiology. Mailing Add: Radiation Ther Div George Washington Univ Hosp Washington DC 20037

MAIER, MARY LOUISE, b Elizabeth, NJ, July 21, 26. INORGANIC CHEMISTRY, PHYSICAL CHEMISTRY. Educ: St John's Univ, NY, BS, 59; Univ Detroit, MS, 66; Univ Mich, PhD(chem), 70. Prof Exp: High sch teacher, 59-66; instr chem, St Joseph's Col, NY, 70-71, ASST PROF CHEM, MEDGAR EVERS COL, 71- Concurrent Pos: Sci adv, Am Chem Soc. Mem: AAAS; Am Chem Soc. Res: Synthesis and structural analysis of paramagnetic titanium and uranium compounds through electron spin resonance measurements; remedial chemistry teaching. Mailing Add: Medgar Evers Col 1150 Carroll St Brooklyn NY 11225

MAIER, ROBERT HAWTHORNE, b New York, NY, Oct 26, 27; m 52; c 3. PLANT PHYSIOLOGY, ENVIRONMENTAL MANAGEMENT. Educ: Univ Miami, BS, 51; Univ Ill, MS, 52, PhD(soil & plant chem), 54. Prof Exp: Asst agron, Univ Ill, 51-54; from asst prof to prof agr chem & soils, Univ Ariz, 56-67, asst dean grad col, 66-67; asst chancellor & prof chem, 67-70, vchancellor & prof environ sci, 70-75, PROF SCI, ENVIRON CHANGE & ENVIRON ADMIN, UNIV WIS-GREEN BAY, 75- Concurrent Pos: Am Coun Educ fel acad admin, Univ NC, 65-66. Mem: Fel AAAS; fel Am Inst Chemists; fel Am Soc Agron; fel Soil Sci Soc Am; Am Chem Soc. Res: Cellular physiology and biochemistry of metals, metal chelates and chelating agents; analytical chemistry of biological material; chemistry of soil-plant-human relationships. Mailing Add: Dept of Sci Environ Change & Environ Admin Univ of Wis Green Bay WI 54302

MAIER, SIEGFRIED, b Stuttgart, Ger, Apr 22, 30; US citizen; m 54; c 3. MICROBIOLOGY. Educ: Capital Univ, BS, 58; Ohio State Univ, MSc, 60, PhD(microbiol), 63. Prof Exp: Asst prof bact, 63-68, ASSOC PROF MICROBIOL, OHIO UNIV, 68- Concurrent Pos: AEC res partic, Argonne Nat Lab, 69-70. Mem: Am Soc Microbiol; Can Soc Microbiol. Res: Physiology and structure of Beggiatoaceae; anaerobic sporulation in bacillus. Mailing Add: Dept of Zool & Microbiol Ohio Univ Athens OH 45701

MAIER, THOMAS O, b Rochester, NY, June 28, 45; m 69; c 2. INORGANIC CHEMISTRY. Educ: Mass Inst Technol, BS, 67; Univ Ill, Urbana, PhD(chem), 71. Prof Exp: SR RES CHEMIST, EASTMAN KODAK CO, 71- Res: Use of transition metal elements in photographic processes. Mailing Add: Eastman Kodak Co Kodak Park Rochester NY 14650

MAIER, VINCENT PAUL, biochemistry, see 12th edition

MAIER, WILLIAM BRYAN, II, chemical physics, see 12th edition

MAILLIE, HUGH DAVID, b Chester, Pa, Nov 2, 32; m 58; c 3. RADIOBIOLOGY, HEALTH PHYSICS. Educ: La Salle Col, BA, 54; Univ Rochester, MS, 56, PhD(radiation biol), 63. Prof Exp: From instr to asst prof, 62-70, ASSOC PROF RADIATION BIOL, UNIV ROCHESTER, 70-, DIR HEALTH PHYSICS DIV, 72- Mem: AAAS; Health Physics Soc. Res: Radiation dosimetry and its application to the understanding of biological effects in man and laboratory animals. Mailing Add: Health Physics Div Univ of Rochester Rochester NY 14642

MAILMAN, DAVID SHERWIN, b Chicago, Ill, June 29, 38; m 61; c 4. PHYSIOLOGY. Educ: Univ Chicago, BS, 58; Univ Ill, PhD(physiol), 64. Prof Exp: Fel biophys, Univ Md, 62-64; asst prof, 64-70, ASSOC PROF BIOL, UNIV HOUSTON, 70- Concurrent Pos: Adj assoc prof physiol, Univ Tex Med Sch Houston, 73- Mem: AAAS; Am Soc Zoologists; Soc Exp Biol & Med; NY Acad Sci; Sigma Xi. Res: Salt and water transport regulation by physical forces and hormones; membrane physiology. Mailing Add: Dept of Biol Univ of Houston Houston TX 77004

MAILMAN, RICHARD BERNARD, b New York, NY, Feb 6, 45. TOXICOLOGY, NEUROPHYSIOLOGY. Educ: Rutgers Univ, New Brunswick, BS, 68; NC State Univ, MS, 72, PhD(physiol, toxicol), 74. Prof Exp: Res assoc toxicol, NC State Univ, 74-75; FEL NEUROBIOL, SCH MED, UNIV NC, CHAPEL HILL, 76- Mem: Am Chem Soc; Sigma Xi; AAAS. Res: Metabolism of xenobiotics by hepatic mixed function oxidases; mode of action and fate of neuropharmacological agents. Mailing Add: 1915 Glenwood Ave Raleigh NC 27608

MAIN, ALEXANDER RUSSELL, b Pincher Creek, Alta, Oct 28, 25; m 51; c 2. BIOCHEMISTRY, ENZYMOLOGY. Educ: Queen's Univ, Ont, BA, 50, MA, 52; Cambridge Univ, PhD(biochem), 59. Prof Exp: Chemist, Can Fed Mines Br, 52-54 & Can Dept Health & Welfare, 54-63; assoc prof, 63-67, PROF ENZYME KINETICS, N C STATE UNIV, 67- Concurrent Pos: NIH grant. Mem: Brit Biochem Soc. Res: Pesticide toxicology; purification of the detoxication enzyme, paraoxonase; irreversible inhibition kinetics and the measurement of the affinity and phosphorylation constants of organophosphate inhibitors; purification of horse serum cholinesterase. Mailing Add: Dept of Biochem N C State Univ PO Box 5126 Raleigh NC 27607

MAIN, CHARLES EDWARD, b Triadelphia, WVa, July 25, 33; m 54; c 2. PLANT PATHOLOGY, PLANT PHYSIOLOGY. Educ: WVa Univ, BS, 59, MS, 61; Univ Wis, PhD(plant path), 64. Prof Exp: Fel plant path, Univ Wis, 64; RES PLANT PATHOLOGIST, AGR RES SERV, USDA, 64-; ASST PROF PLANT PATH, N C STATE UNIV, 64- Mem: AAAS; Am Phytopath Soc; Am Soc Plant Physiol. Res: Biochemical and physiological effects of bacterial and fungal phytopathogens on plants; chemical nature of disease resistance. Mailing Add: Dept of Plant Path N C State Univ PO Box 5126 Raleigh NC 27607

MAIN, FREDERIC HALL, b Webster, Mass, Aug 1, 20; m 48; c 5. ECONOMIC GEOLOGY. Educ: Dartmouth Col, AB, 42; Columbia Univ, MA, 48, PhD(geol), 55. Prof Exp: Miner, Callahan Mining Co, 42; asst to supt, Vt Copper Co, 42-44; geologist, US Geol Surv, 44-46 & Eagle-Picher Co, 47-48; geologist, NJ Zinc Co, 49-60, planning engr, 60-65, asst to pres, 65-70, vpres explor, 71-74; PRES, NJ ZINC EXPLOR CO, 74- Mem: Soc Econ Geologists; Am Inst Mining, Metall & Petrol Engrs. Mailing Add: NJ Zinc Explor Co 65 E Elizabeth Ave Bethlehem PA 18018

MAIN, JAMES HAMILTON PRENTICE, b Biggar, Scotland, June 7, 33; m 61; c 2. ORAL PATHOLOGY. Educ: Univ Edinburgh, BDS, 55, PhD(path), 64. Prof Exp: Lectr oral path, Univ Edinburgh, 61-66, sr lectr & consult, 66-69; PROF ORAL PATH, UNIV TORONTO, 69- Concurrent Pos: USPHS int res fel, NIH, 64-65; USPHS res grant, Univ Edinburgh, 66-69; Nat Cancer Inst Can res grant, Univ Toronto, 70-; mem nat educ comt, Nat Cancer Inst Can, 70-; head dent, Sunnybrook Hosp, 71- Honors & Awards: Colgate Prize, Int Asn Dent Res, 66; Clarke Prize Cancer Res, 68. Mem: AAAS; Am Acad Oral Path; Royal Col Path; Int Asn Dent Res; Royal Col Dentists Can. Res: Induction of neoplasia; epithelio-mesenchymal interactions; developmental aspects of oncogenesis; biological testing of dental materials; salivary gland tumors. Mailing Add: Dept of Oral Path Univ of Toronto Fac of Dent Toronto ON Can

MAIN, ROBERT ANDREW, b Billings, Mont, Sept 23, 23; m 61; c 4. ZOOLOGY, LIMNOLOGY. Educ: Univ Calif, Santa Barbara, AB, 48; Univ Wash, MS, 53; Univ Mich, PhD(zool), 61. Prof Exp: Asst zool, Univ Mich, 53-57, instr, 61; instr, Univ NH, 58-61; asst prof biol, Va Polytech, 61-62, Western Ill Univ, 62-64 & Tex Woman's Univ, 64-66; assoc prof, 66-70, PROF BIOL, CALIF STATE UNIV HAYWARD, 70- Mem: AAAS; Am Micros Soc; Am Soc Limnol & Oceanog; Ecol Soc Am. Res: Freshwater zooplankton and bottom fauna; life histories of planktonic copepods; biological illustration. Mailing Add: Dept of Biol Sci Calif State Univ Hayward CA 94542

MAIN, STEPHEN PAUL, b Iowa City, Iowa, Aug 26, 40; m 63. BOTANY, AQUATIC ECOLOGY. Educ: Valparaiso Univ, BS, 62, MALS, 65; Ore State Univ, PhD(bot), 72. Prof Exp: Teacher biol, Crescent-Iroquois Community High Sch, Ill, 62-63; lab instr, Valparaiso Univ, 63-65; teacher, Santiam High Sch, Ore, 65-69; ASST PROF BIOL, WARTBURG COL, 71- Mem: Sigma Xi; AAAS; Phycol Soc Am; Ecol Soc Am; Bot Soc Am. Res: Ecology and taxonomy of diatoms in rivers, marshlands, and marine shoreline systems. Mailing Add: Dept of Biol Wartburg Col Waverly IA 50677

MAIN, WILLIAM FRANCIS, b Fresno, Calif, July 2, 21; m 43; c 3. PHYSICS. Educ: Fresno State Col, AB, 43. Prof Exp: Electronic scientist, US Naval Res Lab, 43-55; mgr radar & data link dept, Missile & Space Div, Lockheed Aircraft Corp, 55-56, mgr electronic res, 56-65; dir electronics, Lockheed Missiles & Space Co, Calif, 65-66, asst gen mgr electronics, 66-69, asst chief engr space systs, 69-72, asst prog mgr, 72-74; PRES, PAN DATA CORP, 74- Mem: AAAS; Am Phys Soc; Sigma Xi; Inst Elec & Electronics Engrs. Res: Radar; communications and data processing systems. Mailing Add: 14495 Miranda Rd Los Altos Hills CA 94022

MAINE, FRANCIS WILLIAM, materials science, see 12th edition

MAINEN, EUGENE LOUIS, b Baltimore, Md, May 9, 40; m 66; c 1. ORGANIC CHEMISTRY. Educ: Univ Md, BS, 63; Univ Iowa, MS, 65, PhD(org chem), 68. Prof Exp: Sr res chemist, Cent Res Labs, 67-69 & Photog Prod Div, 69-71, tech serv specialist, 71-74, CUSTOMER SERV SUPVR, PHOTOG PROD DIV, MINN MINING & MFG CO, 74- Mem: Am Chem Soc; Soc Photog Scientists & Engrs; Sigma Xi. Res: High performance polymers; emulsion technology in photographic sciences. Mailing Add: Photog Prod Div Bldg 223-2SE 3M Ctr St Paul MN 55119

MAINES, MAHIN D, b Arak, Iran, July 31, 41; m 62; c 2. PHARMACOLOGY. Educ: Ball State Univ, BS, 64, MA, 67; Univ Mo, PhD(pharmacol), 70. Prof Exp: Res assoc pharmacol, Univ Mo, 70-71; NIH fel, Univ Minn, 71-73; res assoc, 73-75, ASST PROF PHARMACOL, ROCKEFELLER UNIV, 75- Concurrent Pos: Irma T Hirschl Trust career scientist award, 76-80. Mem: Am Soc Pharmacol & Exp Therapeut; Am Soc Biol Chemists; Sigma Xi. Res: Biosynthesis and degradation of cellular heme and hemoproteins, with emphasis on the investigation of the mechanisms by which these activities are regulated. Mailing Add: Dept of Metab-Pharmacol Rockefeller Univ New York NY 10021

MAINIER, ROBERT, b Pittsburgh, Pa, Oct 27, 25; m 50; c 6. ANALYTICAL CHEMISTRY. Educ: Univ Pittsburgh, BS, 49. Prof Exp: Chemist, Barrett Co, Allied Chem & Dye, 49-50; fel anal chem, Mellon Inst, 50-53; LAB GROUP MGR ABSORPTION SPECTROS, RES DEPT, KOPPERS CO INC, 53- Mem: Am Chem Soc; Soc Appl Spectros; Coblentz Soc. Res: Utilization of infrared, ultraviolet and nuclear magnetic resonance spectroscopy for characterization and analyses of commercial products. Mailing Add: Koppers Co Inc Res Dept 440 College Park Dr Monroeville PA 15146

MAINLAND, GORDON BRUCE, b Elmhurst, Ill, May 22, 45; m 67. THEORETICAL HIGH ENERGY PHYSICS. Educ: Cornell Univ, BS, 67; Univ Tex, PhD(theoret high energy physics), 71. Prof Exp: Fel theoret high energy physics, Univ Tex, 72; scholar, Sch Theoret Physics, Dublin Inst Advan Studies, 72-74; fel, 74-75, ASST PROF THEORET HIGH ENERGY PHYSICS, OHIO STATE UNIV, 75- Res: Field theory, gauge theory, weak interactions, symmetries. Mailing Add: Dept of Physics Ohio State Univ 174 W 18th Ave Columbus OH 43210

MAINS, GILBERT JOSEPH, b Clairton, Pa, Apr 20, 29; m 51; c 2. PHYSICAL CHEMISTRY. Educ: Duquesne Univ, BS, 51; Univ Calif, PhD(chem), 54. Prof Exp: Fulbright fel, Cambridge Univ, 54-55; from asst prof to assoc prof chem, Carnegie Inst Technol, 55-65; prof, Univ Detroit, 65-68, Poetker prof physics & chem, 68-71, chmn dept chem, 65-68; PROF CHEM & HEAD DEPT, OKLA STATE UNIV, 71- Concurrent Pos: Fel, Lawrence Radiation Lab, Univ Calif, 59-60. Mem: AAAS; Am Chem Soc; Am Phys Soc; Radiation Res Soc. Res: Photochemistry; radiation chemistry; chemical kinetics; elementary reactions in photolysis, radiolysis and pyrolysis; reactions of free radicals; excited molecules and ions. Mailing Add: 301 E Redbud Dr Stillwater OK 74074

MAINSTER, MARTIN ARON, b Toronto, Ont, June 30, 42; US citizen; m 65; c 1. OPHTHALMOLOGY, MATHEMATICAL PHYSICS. Educ: NC State Univ, BS, 63, PhD(physics), 69; Univ Tex Med Br Galveston, MD, 75. Prof Exp: Sr res scientist, Life Sci Div, Technol Inc, 68-70, prin res scientist, 70-71, mgr biomath anal, 71-75; intern internal med, Univ Tex Med Br Galveston, 75-76; RESIDENT OPHTHAL, SCOTT & WHITE MEM INST, 76- Mem: AAAS; Am Phys Soc; Asn Res Vision & Ophthal; Optical Soc Am; AMA. Res: Mathematical and physical analysis of problems in physiological optics and ophthalmology; mathematical and digital computer modeling of biological systems. Mailing Add: 446 Calumet Pl San Antonio TX 78209

MAIO, DOMENIC ANTHONY, b Washington, DC, June 22, 35; m 58; c 3. PHYSIOLOGY, AEROSPACE MEDICINE. Educ: Georgetown Univ, BS, 56; George Washington Univ, MS, 57; Tex A&M Univ, PhD(physiol), 68. Prof Exp: US Air Force, 58-, res scientist, US Air Force Sch Aerospace Med, 63-71, staff officer, Aerospace Biotechnol, Off Dep Chief Staff Res & Develop, Hq US Air Force, 71-74, SPEC ASST INT RES & DEVELOP, OFF ASST SECY AIR FORCE FOR RES & DEVELOP, WASHINGTON, DC, 74- Concurrent Pos: Liaison rep to appl physiol study group, NIH. Mem: Assoc fel Aerospace Med Asn. Res: Altitude and hyperbaric physiology. Mailing Add: 5306 Stonington Dr Fairfax VA 22030

MAIO, JOSEPH JAMES, b Priest River, Idaho, July 29, 29. MICROBIOLOGY, BIOCHEMISTRY. Educ: Univ Wash, BS, 55, MS, 57, PhD(microbiol), 61. Prof Exp: NIH fel tissue cult, Univ Pavia, 61-63; NIH trainee, 64-66, fel cell biol, 66-67, asst prof, 67-72, ASSOC PROF CELL BIOL, ALBERT EINSTEIN COL MED, 72- Concurrent Pos: NIH career develop award, Albert Einstein Col Med, 69-74. Res: Host-induced modification in bacteriophage; predatory fungi; enzymology and active transport processes of tissue culture cells; mammalian cytogenetics; nucleic acids of mammalian cells. Mailing Add: Dept of Cell Biol Albert Einstein Col of Med Bronx NY 10461

MAIORANA, VIRGINIA CATHERINE, b Hagerstown, Md, Aug 16, 47; m 74. EVOLUTIONARY BIOLOGY. Educ: Univ Md, BS, 69; Univ Calif, Berkeley, MA, 71, PhD(zool), 74. Prof Exp: Actg asst prof biol, Univ Calif, Berkeley, 74; RES ASSOC BIOL, UNIV CHICAGO, 74- Concurrent Pos: Asst prof biol, Mundelein Col, 75-76. Mem: Soc Study Evolution; Ecol Soc Am; Animal Behav Soc; Am Soc Ichthyologists & Herpetologists. Res: Empirical and theoretical investigations on the evolution of life history and behavior within and among phylogenetic groups and their relations to community structure and function. Mailing Add: Univ of Chicago Dept of Biol 1103 E 57th St Chicago IL 60637

MAIR, ROBERT DIXON, b Tide Head, NB, Feb 11, 21; nat US; m 43; c 4. PHYSICAL CHEMISTRY. Educ: Univ NB, BSc, 41; Brown Univ, ScM, 43, PhD(chem), 49. Prof Exp: Chemist, Polymer Corp, 43-46; res chemist anal methods develop, 48-58, sr res chemist, 58-71, RES SCIENTIST ANAL METHODS DEVELOP, HERCULES INC, 71- Mem: Am Chem Soc; Sigma Xi; Am Soc Testing & Mat. Res: Infrared spectroscopy; molecular structure of benzene; catalysis; organic peroxide analysis; polymer fractionation; thermal analysis; odor and flavor research. Mailing Add: Hercules Res Ctr Hercules Inc Wilmington DE 19899

MAIRE, FREDERICK WIRTH, b Seattle, Wash, Dec 5, 20; m 44; c 3. PHYSIOLOGY. Educ: Northwestern Univ, BS & BM, 48, MD, 49; Univ Wash, MS, 54. Prof Exp: Resident internal med, Virginia Mason Hosp, Seattle, 50-51; clin resident, 51-53; physician II, Ment Health Res Inst, 56-59; chief med, Ranier Sch Mentally Retarded, 59-61; dept neurophysiol & electroencephalog, Ment Health Res Inst, 61-67; clin dir, 67-69, CHIEF, STATE HOME & TRAINING SCH, 69- Mem: Am Physiol Soc; AMA; NY Acad Sci. Res: Neurophysiology; animal and human behavior; mechanisms responsible for mental illness. Mailing Add: State Home & Training Sch Wheat Ridge CO 80033

MAIRHUBER, JOHN CARL, b Rochester, NY, Dec 14, 22; m 46; c 2. MATHEMATICS. Educ: Univ Rochester, BS, 42, MS, 50; Univ Pa, PhD(math), 59. Prof Exp: Instr math, Univ Rochester, 56-58; from asst prof to assoc prof, Univ NH, 58-64; prof, Univ Richmond, 64-68; PROF MATH & HEAD DEPT, UNIV MAINE, ORONO, 68- Mem: Am Math Soc; Math Asn Am. Res: Complex variables; theory of numbers and approximations. Mailing Add: Dept of Math 304 Shibles Hall Univ of Maine Orono ME 04473

MAISCH, WELDON FREDERICK, b Pana, Ill, Jan 19, 35; m 60; c 2. FOOD MICROBIOLOGY, INDUSTRIAL MICROBIOLOGY. Educ: Ill Wesleyan Univ, BS, 57; Univ Ill, MS, 60, PhD(microbiol), 67. Prof Exp: Asst plant bacteriologist, A E Staley Mfg Co, 59-61; sr scientist, Mead Johnson & Co Div, Bristol-Meyers Co, 66-68; assoc res scientist, 68-71, sr res scientist, 71-74, DIR RES, HIRAM WALKER & SONS, INC, 74- Mem: Am Soc Microbiol; Am Chem Soc; NY Acad Sci. Res: Microbial metabolism and fermentation; animal feedstuffs. Mailing Add: Res Dept Hiram Walker & Sons Inc Peoria IL 61601

MAISCH, WILLIAM GEORGE, b Philadelphia, Pa, Feb 15, 29. PHYSICAL CHEMISTRY. Educ: Univ Pa, BS, 51; Brown Univ, PhD(chem), 56. Prof Exp: Res assoc phys chem, Eng Exp Sta, Univ Ill, 55-57; asst prof, Inst Molecular Physics, Univ Md, 57-63; RES CHEMIST, US NAVAL RES LAB, 63- Mem: Am Chem Soc; Am Phys Soc. Res: Optical properties of magnetic materials; molecular and crystal structure; high pressure spectroscopy. Mailing Add: Code 6452M US Naval Res Lab Washington DC 20375

MAISEL, HERBERT, b Brooklyn, NY, Sept 22, 30; m 57; c 2. COMPUTER SCIENCE, STATISTICS. Educ: City Col New York, BS, 51; NY Univ, MS, 52; Cath Univ Am, PhD(math), 64. Prof Exp: Anal statistician, Develop & Proof Serv, Aberdeen Proving Ground, 52-56, chief statist sect, 56-58; chief methodology & reliability div, Off Naval Inspector Ord, Washington, DC, 58-59; mathematician, Oper Math Br, Off Qm Gen, 59-62; tech chief modeling div, US Army Strategy & Tactics Group, 62-63; assoc prof, 63-75, PROF COMPUT SCI, GEORGETOWN UNIV, 75-, DIR COMPUT CTR, 63- Concurrent Pos: Consult, Social Security Admin, 66-73, Nat Bur Standards, 68-72 & Baltimore Housing Authority, 72-73; app to spec study group for suppl security income prog, Dept Health, Educ & Welfare, 75-76. Mem: AAAS; Am Statist Asn; Asn Comput Mach. Res: Simulation and other stochastic applications of computers; application of numerical and statistical methods to problems in physical and life sciences; teaching of computer science. Mailing Add: Acad Comput Ctr Georgetown Univ Washington DC 20057

MAISSEL, LEON I, b Cape Town, SAfrica, May 31, 30; US citizen; m 56; c 2. PHYSICS, COMPUTER SCIENCES. Educ: Cape Town Univ, BSc, 49, MSc, 51; Univ London, PhD(physics), 55. Prof Exp: Res physicist, Philco Corp, Pa, 56-60; SR PHYSICIST, IBM CORP, 60- Mem: Am Vacuum Soc. Res: Thin films, particularly cathodic sputtering, and their application; computer design, particularly array logic. Mailing Add: IBM Corp Systs Prod Div Dept C14 Bldg 704 Poughkeepsie NY 12602

MAITLEN, ELDON GENE, b West Point, Ind, Feb 20, 26; m 54; c 4. PHYSIOLOGY. Educ: Purdue Univ, BSA, 50, MS, 51, PhD(plant physiol), 54. Prof Exp: Pathologist, Niagara Chem Div, 54-59, res biologist, 59, supvr res & develop, Jackson Lab, Miss, 59-62, prod mgr insecticides res & develop, 62-66, asst dir res & develop, 66-73, DIR DEVELOP, AGR CHEM DIV, FMC CORP, 73- Res: Fungus physiology with respect to host-parasite relationships; all phases of development of plant protection agents and plant growth regulators; direction of field testing and product development, embracing toxicology, metabolism, residue determinations, environmental impact, federal registration and state recommendations. Mailing Add: 26 N Lakeside Dr W Birchwood Lakes Medford NJ 08055

MAITRA, SHYAMAL KUMAR, b Lucknow, India; m 76. BIOCHEMISTRY. Educ: Lucknow Univ, BSc, 64, MSc, 66; Georgetown Univ, PhD(chem), 73. Prof Exp: FEL, DEPT BIOCHEM, UNIV CALIF, BERKELEY, 73- Mem: Sigma Xi; AAAS. Res: Structure, function, relationship and biosynthesis of multiple forms of methyl mannose polysaccharides in mycobacterium smegmatis; characterization of mannans of mycobacteria; studies on endo alpha mannanase of bacillus sp TN 31. Mailing Add: Dept of Biochem Univ of Calif Berkeley CA 94720

MAITRA, UMADAS, b Jalpaiguri, India. BIOCHEMISTRY, MOLECULAR BIOLOGY. Educ: Univ Calcutta, BSc, 56, MSc, 58; Univ Mich, Ann Arbor, PhD(biol chem), 63. Prof Exp: Jane Coffin Childs Found Med Res fel, 63-65, from instr to asst prof, 65-72, ASSOC PROF DEVELOP BIOL & CANCER, ALBERT EINSTEIN COL MED, 72- Concurrent Pos: Am Heart Asn estab investr, Albert Einstein Col Med, 67-72, Am Cancer Soc fac res award, 72-77. Mem: Am Soc Biol Chemists. Res: Enzymatic synthesis of biologically active macromolecules. Mailing Add: Dept of Develop Biol Albert Einstein Col of Med Bronx NY 10461

MAIZEL, BENJAMIN LEO, b Riga, Latvia, Dec 21, 07; nat US; m 36. ORGANIC CHEMISTRY. Educ: Univ Chicago, BS, 28, PhD(org chem), 32. Prof Exp: Res chemist, Kimberly Clark Co, Wis, 28-30; in chg org res, Pacini Labs, Ill, 32-34; vpres & treas in chg prod & res, Vi-Co Prods Co, 34-49; pres & treas, Maizel Labs, 49-72; PRES & TREAS, BENHILL CORP, 71- Mem: Am Chem Soc; NY Acad Sci. Res: Extraction of yeast; oxidation of sterols; base exchange resins; parenteral solutions. Mailing Add: 2935 W Bryn Mawr Ave Chicago IL 60659

MAIZELL, ROBERT EDWARD, b Baltimore, Md, Aug 3, 24; m 54; c 2. INDUSTRIAL CHEMISTRY. Educ: Loyola Col, BS, 45; Columbia Univ, BS, 47, MS, 49, DLS, 57. Prof Exp: Chemist, Manhattan Proj, 45-46; reference asst, Sci & Technol Div, New York Pub Library, 47-48; teaching asst sci lit, Columbia Univ, 50; chg tech info serv, Olin Mathieson Chem Corp, NY, 50-58; NSF doc res proj dir, Am Inst Physics, NY, 58-60; supvr tech info serv, 60-65, mgr, 65-72, tech mgr, 72-74, MGR BUS & SCI INFO SERV, OLIN CORP, 74- Concurrent Pos: Chmn continuing educ comt, Olin Corp, 69-75; mem, Adv Coun, Smithsonian Scientists Info Exchange, 74-; chmn, Subcomt On-line Serv, Mfg Chemists Asn, 75- Mem: Fel Am Inst Chemists; Am Chem Soc; Sigma Xi. Res: Industrial processes for chemicals; technical and marketing intelligence; technological forecasting; audio cassettes for disseminating chemical information. Mailing Add: Olin Corp Res Ctr 275 Winchester Ave New Haven CT 06504

MAJARAKIS, JAMES DEMETRIOS, b Chicago, Ill, Oct 11, 15. SURGERY. Educ: Univ Chicago, BS, 36, MD, 40; Univ Ill, MS, 46; Am Bd Surg, dipl, 48. Prof Exp: From asst prof to assoc prof, 46-72, PROF SURG, UNIV ILL COL MED, 72- Concurrent Pos: Surg assoc attend, Cook County Hosp, 45-55; from sr surgeon to pres med staff, mem bd gov & mem exec comt, Henrotin Hosp, 52-; consult, Cancer Prev Ctr, Chicago, 52- & Univ Ill Res Hosp, 53-; sr surg consult, Westside Vet Admin Hosp, 55- Mem: Am Geriat Soc; Soc Nuclear Med; Asn Mil Surgeons US; Am Thyroid Asn; fel Am Col Surgeons. Res: Cancer, especially of the thyroid, breast and rectum. Mailing Add: Suite 808 30 N Michigan Ave Chicago IL 60602

MAJCHROWICZ, EDWARD, b Stryj, Poland, Mar 18, 20; m 56; c 2. BIOCHEMISTRY, BIOCHEMICAL PHARMACOLOGY. Educ: Univ Birmingham, BSc, 48; McGill Univ, PhD(biochem), 59. Prof Exp: Asst chemist, A Guinness, Son & Co, Eng, 49-56; res assoc biochem, McGill Univ & McGill Montreal Gen Hosp Res Inst, 56-59, fel neurochem, 59-60; res assoc biochem, Med Sch, Univ Va, Charlottesville, 61-62, asst prof, 62-63; asst prof, Sch Med, Univ NC, Chapel Hill, 63-67; sr res scientist, Squibb Inst Med Res, 67-68; res scientist, Nat Ctr Prev & Control Alcoholism, NIMH, 68-71, HEAD BIOCHEM PROG, NAT INST ALCOHOL ABUSE & ALCOHOLISM, ALCOHOL & DRUG ABUSE MENT HEALTH ADMIN, 71- Mem: AAAS; Am Chem Soc; NY Acad Sci; Am Soc Pharmacol & Exp Therapeut; Am Soc Neurochem. Res: Neurochemistry, metabolism and metabolic effects of aliphatic alcohols, aldehydes and fatty acids; biogenic amine metabolism; biological aspects of mental diseases; biochemistry and microbiology of fermentation processes. Mailing Add: Nat Inst Alcohol Abuse & Alcoholism ADAMHA Washington DC 20032

MAJDE, JEANNINE ADKINS, b Chicago, Ill, Feb 17, 40; m 65. INFECTIOUS DISEASES, IMMUNOPATHOLOGY. Educ: Univ Chicago, BS, 64; Univ Notre Dame, PhD(microbiol), 70. Prof Exp: USPHS fel immunol, La Rabida Res Inst, Univ Chicago, 70-72; MICROBIOLOGIST, OFF NAVAL RES, 72- Concurrent Pos: Adj asst prof microbiol, Stritch Sch Med, Loyola Univ Chicago, 75- Mem: Am Soc Microbiol. Res: Immunopathology of chronic viral diseases; role of occult viruses in autoimmune diseases; cellular immunity; chemotherapy of viral infections. Mailing Add: Off of Naval Res 536 S Clark St Chicago IL 60605

MAJER, JAROSLAV, b Czech, July 25, 27; m 55; c 1. BIOCHEMISTRY, ORGANIC CHEMISTRY. Educ: Charles Univ, Prague, BS, 50, MS, 52; Czech Acad Sci, PhD(microbial chem), 62. Prof Exp: Res assoc, Czech Acad Sci, 52-68; res assoc, La Rabida Children's Hosp & Res Ctr, Univ Chicago, 68-70; ASST PROF BIOCHEM, SCH MED, NORTHWESTERN UNIV, CHICAGO, 70- Mem: Am Chem Soc; Am Soc Microbiol. Res: Chemistry and biochemistry of antibiotics; metabolism of antibiotics in man. Mailing Add: Dept of Biochem Northwestern Univ Chicago IL 60611

MAJERUS, PHILIP W, b Chicago, Ill, July 10, 36; m 57; c 4. HEMATOLOGY, MEDICINE. Educ: Washington Univ, MD, 61. Prof Exp: Res assoc biochem, NIH, 63-66; from asst prof to assoc prof, 66-71, PROF BIOCHEM & MED, SCH MED, WASHINGTON UNIV, 71- Concurrent Pos: NSF res grants, 66-70; Am Cancer Soc fac res award, 66-74; NIH res grants, 66-79; mem biochem fel rev comt, NIH, 69-71, mem hemat study sect, 74-78. Mem: Fel Am Col Physicians; Am Asn Physicians; Am Soc Biol Chemists; Am Soc Clin Invest. Res: Structure and function of human blood platelets. Mailing Add: Dept of Biochem & Med Washington Univ Sch of Med St Louis MO 63110

MAJEWSKI, ROBERT FRANCIS, b Chicago, Ill, Oct 1, 27; m 61; c 3. ORGANIC CHEMISTRY. Educ: Univ Ill, BS, 51; Univ Notre Dame, PhD(org chem), 55. Prof Exp: Chemist, Armour & Co labs, 51; sr chemist res & develop, 54-60, group leader, 60-68, sect leader chem res, 68-72, SR PRIN INVESTR, MEAD JOHNSON RES CTR, 72- Mem: AAAS; Am Chem Soc; Sigma Xi. Res: Unsaturated lactones; medicinal chemistry in endocrine, central nervous and cardiovascular systems; development of organic chemical processes. Mailing Add: Mead Johnson Res Ctr Mead Johnson & Co Evansville IN 47721

MAJEWSKI, THEODORE E, b Boonton, NJ, July 5, 25; m 53; c 6. ORGANIC CHEMISTRY. Educ: Syracuse Univ, BA, 51; Univ Del, MS, 53, PhD(org chem), 60. Prof Exp: Res chemist, Dow Chem Co, 57-64, proj leader benzene chem, 64-66, org process titled specialist, 66-69; RES SCIENTIST, PHILIP MORRIS, INC, 69- Mem: Am Chem Soc. Res: Brominated salicylanilides; salicylanilides; aniline; biphenyl; nonyl phenol; chloromethylation of aromatics; bromination of aromatic compounds; flavor technology on cigaret; taste research on cigaret; menthol research. Mailing Add: 2330 Devenwood Rd Bon Air VA 23235

MAJKOWSKI, RICHARD FRANCIS, b Detroit, Mich, Apr 18, 31. PLASMA PHYSICS, SPECTROSCOPY. Educ: Univ Detroit, BS, 54, MS, 56. Prof Exp: Instr physics, Univ Detroit, 54-55; res physicist, 55-60, SR RES PHYSICIST, GEN MOTORS RES LABS, 60- Concurrent Pos: Instr, Univ Detroit, 56-58 & Detroit Inst Technol, 60- Mem: Optical Soc Am; Am Asn Physics Teachers. Res: Atomic spectroscopy; diagnostics of plasmas by spectroscopic methods, especially research in and development of spectral line profiles and their dependence on plasma parameters; holography; dimensional measurements from holographic images. Mailing Add: Gen Motors Res Labs GM Tech Ctr Warren MI 48090

MAJMUNDAR, HASMUKHRAI HIRALAL, b Baroda, India, Nov 18, 32; m 62; c 2. GEOCHEMISTRY, GEOLOGY. Educ: Univ Baroda, India, BSc, 55; Banaras Hindu Univ, MSc, 57; Univ Nancy, PhD(geochem, mineral), 61. Prof Exp: Asst prof geol, Univ Baroda, India, 57-64; Nat Acad Sci-Nat Res Coun resident res assoc geochem, Goddard Space Flight Ctr, NASA, 64-66; Nat Res Coun Can fel, spec lectr & head geochem labs, Dalhousie Univ, 66-68; prof geol, Appalachian State Univ, 68-70; MGR GEOCHEM SECT, DIV MINES & GEOL, STATE OF CALIF, 70- Concurrent Pos: Consult geologist agate mines & limestone & dolomite opers, Gujarat State, India, 57-64. Mem: Geochem Soc; Mineral Asn Can. Res: Development of procedures for major, minor and trace element determination in geological samples; computer data collection. Mailing Add: Geochem Sect Calif Div of Mines & Geol San Francisco CA 94111

MAJOR, CHARLES WALTER, b Framingham, Mass, Jan 31, 26; m 51; c 3. PHYSIOLOGY. Educ: Dartmouth Col, AB, 48; Univ Tenn, MS, 54, PhD(zool), 57. Prof Exp: Res fel, Nat Cancer Inst, 57; instr physiol, Sch Med, Univ Rochester, 57-59; from asst prof to assoc prof, 59-71, PROF ZOOL, UNIV MAINE, 71- Mem: AAAS; Am Phsyiol Soc. Res: Comparative physiology and toxicology in marine ecosystems. Mailing Add: Dept of Zool Univ of Maine Orono ME 04473

MAJOR, FOUAD GEORGE, atomic physics, see 12th edition

MAJOR, JACK, b Salt Lake City, Utah, Mar 15, 17; m 47; c 3. PLANT ECOLOGY. Educ: Utah State Agr Col, BS, 42; Univ Calif, PhD(soil sci), 53. Prof Exp: Range researcher, US Forest Serv, Utah, 42-49; range weed control researcher, 53-60, from asst prof to assoc prof bot, 55-71, PROF BOT, UNIV CALIF, DAVIS, 71- Mem: Ecol Soc Am; Am Bryol & Lichenol Soc; Brit Ecol Soc; Brit Soc Soil Sci. Res: Plant community and soil relationships; California vegetation; vegetation near Atlin Lake, British Columbia; vegetation of Teton and Gros Ventre Ranges, Wyoming. Mailing Add: Dept of Bot Univ of Calif Davis CA 95616

MAJOR, JOHN KEENE, b Kansas City, Mo, Aug 3, 24; m 70; c 3. EXPERIMENTAL NUCLEAR PHYSICS, ACADEMIC ADMINISTRATION. Educ: Yale Univ, BS, 43, MS, 47; Univ Paris, DrSc(physics), 51. Prof Exp: Lab asst physics, Sloane Physics Lab, Yale Univ, 43-44, instr, 52-55; mem sci staff, Div War Res, Columbia Univ, 44; sci consult, Sonar Anal Group, Oceanog Inst, Woods Hole, 46-47; instr physics & chem, Am Community Sch, Paris, 48-49; res assoc, Nat Ctr Sci Res, Lab Nuclear Chem, Col France, 51; assoc prof physics, Western Reserve Univ, 55-57, Perkins Prof, 57-66, chmn dept, 55-60 & 61-64; staff assoc, sci develop eval group, Div Instnl Prog, NSF, 64-67, Univ Sci Develop Sect, 67-68; prof physics & dean, Grad Sch Arts & Sci, Univ Cincinnati, 68-71 & NY Univ, 71-73; vis scholar, Alfred P Sloan Sch Mgt, Mass Inst Technol, 73-74; PROF PHYSICS, NORTHEASTERN ILL UNIV, 74- Concurrent Pos: Fel, Curie Lab, Inst Radium, Paris, 51; res asst, Edwards St Lab, Yale Univ, 52-55; consult, Am Inst Physics, 59; NSF fel, Lab Tech Physics, Munich Tech Univ, 60-61; consult, Reuter-Stokes Electronic Components, Inc, 63-66 & NSF, 68-69; vpres acad affairs, Northeastern Ill Univ, 74-75. Mem: AAAS; Am Phys Soc; Am Asn Physics Teachers; Fedn Am Scientists. Res: Experimental solid state physics; Mössbauer effect; instrumentation; acoustics; academic planning; resource allocation in higher education; college and university organization and governance; program evaluation. Mailing Add: 740 Webster Ave Chicago IL 60614

MAJOR, RANDOLPH THOMAS, b Columbus, Ohio, Dec 23, 01; m 28; c 4. CHEMISTRY. Educ: Univ Nebr, AB, 22, MS, 24; Princeton Univ, PhD(org chem), 27. Hon Degrees: DSc, Univ Nebr, 49. Prof Exp: Teacher high sch, Nebr, 22-23; asst chem, Univ Nebr, 23-24; asst, Princeton Univ, 24-25, instr & res assoc, 27-30; dir pure res, Merck & Co, Inc, 30-36, res & develop, 36-37, vpres & sci dir, 47-53, sci vpres, 53-56, sci adv, 56-67; prof chem, Univ Va, 67-70; RES PROF CHEM, UNIV CONN, 70- Concurrent Pos: Mem, Adv Coun, US Army Qm Corps, 45-47; mem & chmn, Comt Chem Warfare, Res & Develop Bd, 48-52; sci adv, Merck Inst Therapeut Res, 57-74. Mem: AAAS; Am Chem Soc; NY Acad Sci. Res: Medicinal, biological and organic chemistry. Mailing Add: Dept of Chem Univ of Conn Storrs CT 06268

MAJOR, ROBERT WAYNE, b Newark, Ohio, Sept 5, 37; m 63; c 2. PHYSICS. Educ: Denison Univ, BS, 58; Iowa State Univ, MS, 60; Va Polytech Inst, PhD(physics), 66. Prof Exp: Instr physics, Denison Univ, 60-61; asst prof, The Citadel, 61-62; from asst prof to assoc prof, 66-74, PROF PHYSICS, UNIV RICHMOND, 74- Concurrent Pos: Oak Ridge Assoc Univs fac res fel, 74. Mem: AAAS; Am Asn Physics Teachers. Res: Laser-modulated optical absorption in II-VI crystals; polarization modulation analysis of color center excited states. Mailing Add: Dept of Physics Univ of Richmond Richmond VA 23173

MAJOR, SCHWAB SAMUEL, JR, b Windsor, Mo, July 2, 24; m 51; c 3. PHYSICS. Educ: Wichita State Univ, BA, 49; Kans State Univ, MS, 53, PhD(physics), 67. Prof Exp: Elec engr, Derby Oil Refinery, Kans, 51; staff engr, Boeing Airplane Co, 51-53; instr physics, Southwestern Col, Kans, 53-55; asst prof, Midland Col, 55-59; ASSOC PROF PHYSICS, UNIV MO-KANSAS CITY, 59- Concurrent Pos: Res assoc,

Midwest Res Inst, Mo, 62-64. Mem: Am Phys Soc; Am Asn Physics Teachers; Optical Soc Am. Res: Applied quantum statistical mechanics; development of physics science techniques to model building for socio-economic interactions in political milieu. Mailing Add: Dept of Physics Univ of Mo Kansas City MO 64110

MAJORS, PAUL ALEXANDER, b Dawson Springs, Ky, Oct 16, 12; m 36; c 1. BACTERIOLOGY. Educ: Western Ky State Col, BS, 36; Univ Ky, MS, 38. Prof Exp: Clin bacteriologist, Christ Hosp, Cincinnati, Ohio, 38-47; clin bacteriologist labs, 47-49, dir, 49-69, dir microbiol & proprietary clin dept, 69-71, V PRES & TECH DIR, HILL TOP RES, INC, 71- Mem: Am Soc Microbiol. Res: Industrial microbiology, especially germicide evaluation, antibacterial properties of textiles and test method developments; clinical studies, especially antiperspirants, skin irritation and sensitization. Mailing Add: Hill Top Res Inc PO Box 138 Miamiville OH 45147

MAJORS, RIAS HILTON, b Montgomery, Ala, Mar 8, 24; m 43; c 2. ANIMAL SCIENCE. Educ: Ala Polytech Univ, BS, 48, MS, 50; Univ Ga, PhD(animal sci), 65. Prof Exp: Chief, Animal Husb Sub-unit, Commun Dis Ctr, 50, Animal Husb Unit, Virus & Rickettsia Sect, 50-59, actg chief, Sci Serv Sect, Ctr Dis Control, 59-60, chief, 60-72, dep chief, Off Res Grants, 72-73, CHIEF, REAL PROPERTY & COMMUN MGT BR, CTR DIS CONTROL, USPHS, 73- Mem: AAAS; Sigma Xi. Res: Physiology of reproduction of farm and laboratory animals; gnotobiotic and axenic laboratory animal production. Mailing Add: Real Prop & Commun Mgt Br Ctr for Dis Control Atlanta GA 30333

MAJORS, RONALD E, b Ellwood City, Pa, Apr 10, 41; m 71. ANALYTICAL CHEMISTRY. Educ: Fresno State Col, BS, 63; Purdue Univ, PhD(anal chem), 68. Prof Exp: Res asst, Purdue Univ, 63-68; res chemist, Celanese Res Co, NJ, 68-71; res chemist, Varian Aerograph, 71-73, prod mgr, Varian European opers, 73-75, APPLNS MGR, VARIAN ASSOCS, 75- Mem: Am Chem Soc; The Chem Soc; Soc Anal Chem. Res: Liquid and gas chromatography applications; column technology. Mailing Add: Varian Instrument Div 611 Hansen Way Palo Alto CA 94303

MAJUMDAR, DEBAPRASAD, b Calcutta, India, Dec 10, 41; m 71. PHYSICS. Educ: Univ Calcutta, BS, 61, MS, 63; Univ Pa, MS, 66; Univ Mich, Ann Arbor, MS, 73; State Univ NY Stony Brook, PhD(physics), 69. Prof Exp: Res assoc physics, Syracuse Univ, 69-71; res assoc, Univ Mich, Ann Arbor, 71-73, assoc res scientist nuclear eng, Phoenix Mem Lab, 73-74; ASSOC NUCLEAR ENGR REACTOR SAFETY, BROOKHAVEN NAT LAB, 74- Mem: Am Phys Soc; Am Nuclear Soc; Int Asn Hydrogen Energy. Res: Liquid metal fast breeder reactor safety analysis; production of hydrogen, solar and nuclear, and its use as energy. Mailing Add: 343 S Country Rd Brookhaven Hamlet NY 11719

MAJUMDAR, SAMIR RANJAN, b Chittagong, Pakistan, Nov 26, 36; m 62; c 2. APPLIED MATHEMATICS. Educ: Univ Calcutta, BA, 56, MA, 58; Jadavpur Univ, India, PhD(fluid mech), 63; Univ London, PhD(mech of continuous medium), 65. Prof Exp: Lectr math, Jadavpur Univ, 59-62; asst prof, Univ Ariz, 65-69; ASSOC PROF MATH, UNIV CALGARY, 69- Concurrent Pos: Assoc fel, Inst Math & Appln, UK, 65. Mem: Am Math Soc; Soc Natural Philos; Calcutta Math Soc. Res: Hydrodynamics, especially slow motion of viscous liquids. Mailing Add: Dept of Math Statist & Comput Sci Univ of Calgary Calgary AB Can

MAJUMDER, SANAT KUMER, b Khulna, India, Nov 1, 29; m 57; c 2. BOTANY, ECOLOGY. Educ: Univ Calcutta, BSc, 49, MSc, 51; Univ NH, PhD(bot), 58. Prof Exp: Asst bot, Univ NH, 55-58; res assoc biol, Brookhaven Nat Lab, 58-59; plant physiologist, Cent Rice Res Inst, Cuttack, India, 60-62; fel hort, Univ Hawaii, 62-64; asst prof biol, St Louis Univ, 64-67; from asst prof to assoc prof biol sci, Smith Col, 67-71; ASSOC PROF BIOL, WESTFIELD STATE COL, 72- Concurrent Pos: Indian Coun Agr Res grant, 61-62. Mem: Am Inst Biol Sci; Am Soc Plant Physiol; Nat Wildlife Soc; Sigma Xi. Res: Ionizing and nonionizing radiation on pollen grain; morphogenetic response of plants to growth regulators; stress physiology; response of plant growth to electric and magnetic fields. Mailing Add: Dept of Biol Westfield State Col Westfield MA 01085

MAK, STANLEY, b Canton, China, Sept 20, 33; Can citizen; m; c 1. VIROLOGY. Educ: Univ Sask, BSc, 58, MSc, 59; Univ Toronto, PhD(biophys), 62. Prof Exp: Asst prof biol, Queen's Univ, Ont, 62-68; assoc prof, 68-74, PROF BIOL, McMASTER UNIV, 74- Mem: Can Asn Physicists; Can Soc Cell Biol; Am Soc Microbiol. Res: Molecular biology of animal virus infection; transcription cellular transformation of oncogenic viruses. Mailing Add: Dept of Biol McMaster Univ Hamilton ON Can

MAKAR, BOSHRA HALIM, b Sohag, Egypt, Sept 23, 28; m 60; c 2. PURE MATHEMATICS, MATHEMATICAL ANALYSIS. Educ: Univ Cairo, BSc, 47, MSc, 52, PhD(math anal), 55. Prof Exp: Lectr math, Univ Cairo, 48-55, from asst prof to assoc prof, 55-65; vis assoc prof, Am Univ, Beirut, 66; assoc prof, Mich Tech Univ, 66-67; PROF MATH, ST PETER'S COL, NJ, 67- Concurrent Pos: Egyptian Govt sci exchange mission, Moscow State Univ, 63-64. Mem: AAAS; Math Asn Am; Am Math Soc. Res: Functions of a complex variable; functional analysis. Mailing Add: Dept of Math St Peter's Col Jersey City NJ 07306

MAKAREM, ANIS H, b Rasel-Metn, Lebanon, Dec 21, 33; US citizen; m 66; c 2. BIOCHEMISTRY. Educ: Concord Col, BS, 57; Univ Calif, San Francisco, MS, 64, PhD(biochem), 65. Prof Exp: Fel nutrit, Univ Calif, Berkeley, 65-66; res training fel clin chem, Med Sch, Yale Univ, 66-67, instr, 67-68; asst dir endocrinol, 68-69, ASST DIR CHEM, BIOSCI LABS, VAN NUYS, 69- Mem: Am Asn Clin Chem. Res: Radio immunoassays for quantitation of hormones and vitamins; disc and agarose gel electrophoresis for serum proteins; lipoproteins and hemoglobins. Mailing Add: 29395 Hillrise Dr Agoura CA 91301

MAKAREWICZ, JOSEPH CHESTER, b Attleboro, Mass, Aug 5, 47; m 71. LIMNOLOGY. Educ: Southeastern Mass Univ, BS, 69; Cornell Univ, PhD(aquatic ecol), 75. Prof Exp: Res asst ecol, Cornell Univ, 69-71; instr pop biol, Southeastern Mass Univ, 71-72; instr biol, Bristol Community Col, 72; res asst ecol, Hubbard Brook Ecosyst Study, 72-74; ASST PROF BIOL, STATE UNIV NY COL BROCKPORT, 74- Concurrent Pos: Edmund Niles Huyck Preserve fel, 75. Mem: AAAS; Am Soc Limnol & Oceanog; Int Asn Theoret & Appl Limnol; Sigma Xi. Res: Niche division of zooplankton populations, production and nutrient cycling in lakes. Mailing Add: State Univ of NY 215 Lennon Hall Brockport NY 14420

MAKEMSON, JOHN CHRISTOPHER, b San Francisco, Calif, Sept 20, 42; m 67; c 1. MICROBIOLOGY. Educ: San Francisco State Col, BA, 64, MA, 66; Wash State Univ, PhD(bact), 70. Prof Exp: ASST PROF BIOL, AM UNIV BEIRUT, 70- Mem: AAAS; Am Soc Microbiol. Res: Autolysis; bioluminescence; chemical oceanography. Mailing Add: Dept of Biol Am Univ of Beirut Beirut Lebanon

MAKENS, ROYAL FRANCIS, b Minneapolis, Minn, Mar 8, 01; m 30; c 2. NUCLEAR SCIENCE. Educ: Univ Minn, BS, 26; Mich Col Mining, MS, 30; Univ Iowa, PhD(phys chem), 38. Prof Exp: Teacher high sch, Minn, 26-29; instr chem, Mich Col Mining & Technol, 30-37, from asst prof to assoc prof, 38-47; prof, Tex Col Arts & Indust, 47-48; prof, Mich Col Mining & Technol, 48-62, chmn dept chem & chem eng, 56-60, dir nuclear eng, 60-62; nuclear engr, Idaho Opers Off, AEC, 62, chief, Reactor Technol Br, 62-65, educ officer & tech specialist, 65-72; CONSULT, 72- Concurrent Pos: Partic, Nuclear Test Series, Nev, 57. Mem: Am Chem Soc; Am Nuclear Soc; Am Soc Eng Educ; Am Inst Chem Eng; fel Inst Nuclear Eng. Res: Frother activity in ore flotation; xanthates; reaction kinetics of ethyl formate; dithiophosphates; isotope exchange kinetics; polarography; autoradiography; nuclear fuel processing; radiological health. Mailing Add: 1701 Grandview Dr Idaho Falls ID 83401

MAKER, HOWARD SMITH, b New York, NY, Oct 29, 30; m 65; c 2. NEUROCHEMISTRY, NEUROLOGY. Educ: NY Univ, BA, 52; State Univ NY, MD, 56; Am Bd Psychiat & Neurol, dipl, 64. Prof Exp: Fel neurochem, Harvard Med Sch, 62-65; fel, Mt Sinai Hosp, 65-66, res asst, 66-69; asst prof, 70-73, ASSOC PROF NEUROL, MT SINAI SCH MED, 73- Concurrent Pos: Assoc attend neurologist, Mt Sinai Hosp, 73-; consult, Heart-Lung Group, NIH, 74; attend neurologist, Bergen Pines Hosp, NJ, 74 & Bronx Vet Admin Hosp, 75- Mem: Fel Am Acad Neurol; Am Soc Neurochem; Int Soc Neurochem; AAAS; Am Histol Soc. Res: Biochemistry of the brain; myelination and demyelination in tissue culture; human brain enzymes as effected by disease of basal ganglia; brain development. Mailing Add: Dept of Neurol Mt Sinai Sch Med New York NY 10027

MAKER, PAUL D, physics, see 12th edition

MAKER, PHILIP T, b Barre, Vt, Oct 25, 07; m 32; c 2. MATHEMATICS. Educ: Brown Univ, AB, 29, AM, 32; Univ Ill, PhD(math), 38. Prof Exp: Instr math, Rutgers Univ, 38-42; asst prof, Duke Univ, 42-45; from assoc prof to prof, 57-74, EMER PROF MATH, BOSTON UNIV, 74- Mem: Am Math Soc. Res: Real variable and measure. Mailing Add: Dept of Math Boston Univ Boston MA 02215

MAKHLOUF, GABRIEL MICHEL, b Haifa, Israel, June 11, 29; m 60; c 3. MEDICINE, PHYSIOLOGY. Educ: Univ Liverpool, MB, ChB, 53; Univ Edinburgh, PhD(med), 65; FRCP, 72. Prof Exp: Sr res asst gastroenterol, Univ Edinburgh, 62-65; asst prof med, Tufts Univ, 66-68; assoc prof, Med Col, Univ Ala, Birmingham, 68-70; assoc prof, 70-72, PROF MED, MED COL VA, 72-, CO-DIR GASTROENTEROL RES, 70- Concurrent Pos: Consult, Lemuel Shattuck Hosp, Boston, 66-68 & Med Ctr, Univ Ala, 68- Mem: AAAS; Am Gastroenterol Asn; Am Fedn Clin Res; Am Physiol Soc; Biophys Soc. Res: Exocrine physiology, particularly gastric physiology; membrane transport kinetics of sensory phenomena. Mailing Add: Div of Gastroenterol Med Col of Va Richmond VA 23298

MAKI, ARTHUR GEORGE, JR, b Portland, Ore, Nov 24, 30; m 66. PHYSICAL CHEMISTRY. Educ: Univ Wash, Seattle, BS, 53; Ore State Col, PhD(phys chem), 60. Prof Exp: PHYSICIST, NAT BUR STANDARDS, 58- Mem: AAAS; Am Phys Soc; Optical Soc Am. Res: Molecular structure, particularly infrared spectroscopy. Mailing Add: Infrared Spectros Sect Nat Bur of Standards Washington DC 20234

MAKI, AUGUST HAROLD, b Brooklyn, NY, Mar 18, 30; m 52; c 4. BIOPHYSICAL CHEMISTRY. Educ: Columbia Univ, AB, 52; Univ Calif, PhD(chem), 57. Prof Exp: Instr chem, Harvard Univ, 57-60, asst prof, 60-64; from assoc prof to prof, Univ Calif, Riverside, 64-74, PROF CHEM, UNIV CALIF, DAVIS, 74- Concurrent Pos: Guggenheim fel, 70-71; assoc ed, Photochem & Photobiol, 75- Mem: AAAS; Am Phys Soc; Am Chem Soc. Res: Studies of molecular paramagnetism, principally by electron paramagnetic resonance and electron-nuclear double resonance; applications to free radicals, transition metal complexes and phosphorescent and ground state triplets. Mailing Add: Dept of Chem Univ of Calif Davis CA 95616

MAKI, LEROY ROBERT, b Astoria, Ore, May 27, 27; m 51; c 5. BACTERIOLOGY. Educ: State Col Wash, BS, 51; Univ Wis, PhD(bact), 55. Prof Exp: From asst prof to assoc prof, 55-65, PROF MICROBIOL, UNIV WYO, 65- Mem: AAAS; Am Soc Microbiol. Res: Pulmonary emphysema of cattle; taxonomy of fresh-water bacteria; bacterially induced ice nucleation. Mailing Add: Div of Microbiol & Vet Med Univ of Wyo Laramie WY 82070

MAKI, TENHO EWALD, b Ottertail Co, Minn, Mar 29, 07; m 36; c 2. FOREST SOILS, FOREST HYDROLOGY. Educ: Univ Minn, BS, 30, MS, 31, PhD, 51. Prof Exp: Field asst, Cloquet Forest Exp Sta, US Forest Serv, 30, Lakes States Forest Exp Sta, 31, Southern Forest Exp Sta, 31-32, Intermountain Forest & Range Exp Sta, 32-34, jr forester, Southern Forest Exp Sta, 34-36, from asst forester to assoc forester, Intermountain Forest & Range Exp Sta, 36-40, from forester to sr forester, Northeastern Forest Exp Sta, 40-45, officer chg, Gulfcoast Br, Southern Forest Exp Sta, 45-51; prof forest mgt, 51-55, CARL ALWIN SCHENCK PROF FOREST MGT, SCH FOREST RESOURCES, NC STATE UNIV, 55- Concurrent Pos: Head dept forest mgt, NC State Univ, 58-70, dir, Spec Field Inst Forest Biol, 60; Fulbright res scholar, Helsinki, 65-66; consult, Res Triangle Inst, 72- Mem: AAAS; Soc Am Foresters; Soil Sci Soc Am; Am Inst Biol Sci. Res: Watershed management; hydrology; forest soil amelioration; site quality evaluation; land use effects on water attributes; stream channelization effects on forest ecosystems. Mailing Add: Sch of Forest Resources NC State Univ Raleigh NC 27607

MAKIELSKI, SALLY KIMBALL, b Ft Defiance, Ariz, Nov 23, 38; m 63. POPULATION BIOLOGY, ENVIRONMENTAL HEALTH. Educ: Columbia Univ, BA, 60, MA, 61, PhD(zool), 65; Univ Va, MUP, 68. Prof Exp: Asst prof biol, Loyola Univ, La, 70-72; health planner, New Orleans Area Health Planning Coun, 72-75; SPEC ASST TO DIR URBAN ENERGY STUDIES, INST HUMAN RELS, LOYOLA UNIV, LA, 75- Concurrent Pos: Fac res grant, Loyola Univ, La, 71-72. Mem: AAAS; Am Soc Cell Biol; Lepidop Soc; Pop Asn Am; Am Soc Planning Offs. Res: Biological approach to the study of urban systems. Mailing Add: Inst of Human Rels Loyola Univ New Orleans LA 70118

MAKIN, EARLE CLEMENT, JR, b Maple Shade, NJ, Nov 13, 17; m 43; c 2. CHEMISTRY. Prof Exp: Lab technician, United Gas Improv Co, Pa, 37-39, asst chemist, 39-40, chemist, 40-41, group leader, 41-44, asst off mgr, 44-45; asst lab mgr, Co-polymer Corp, 45-46, lab mgr, 46-48; group leader, Lion Oil Co Div, Monsanto Co, 48-65, sr group leader, Hydrocarbons & Polymers Div, 65-74, SR PROCESS SPECIALIST, MONSANTO POLYMERS & PETROCHEMS CO, 74- Mem: Am Chem Soc; Am Inst Chem Eng. Res: Chemical and physical analysis of synthetic rubber; nitriles; terpenes; separation and purification processes; distillation, extraction, adsorption, clathrate and complex chemistry of hydrocarbons; chemistry of high molecular weight hydrocarbons and polymers; development of new separation procedures for petrochemicals. Mailing Add: Monsanto Co PO Box 1311 Texas City TX 77590

MAKINODAN, TAKASHI, b Hilo, Hawaii, Jan 19, 25; m 54. IMMUNOLOGY. Educ: Univ Hawaii, BS, 48; Univ Wis, MS, 50, PhD(zool, biochem), 53. Prof Exp: Asst serol, Univ Wis, 50-53; res assoc immunohemat, Mt Sinai Med Res Found, Ill, 53-54; NIH fel, 54-55; assoc biologist, Biol Div, Oak Ridge Nat Lab, 55-56, biologist, 56-57, head immunol group, 57-72; CHIEF CELLULAR & COMP PHYSIOL BR,

GERONT RES CTR, BALTIMORE CITY HOSPS, 72- Concurrent Pos: NSF sr fel, 61-62; mem microbiol fel rev comt, NIH, 67-70; prof, Grad Sch Biomed Sci, Univ Tenn, 68-72; dir training prog, Nat Inst Child Health & Human Develop, 68-72; mem adv panel regulatory biol prog, NSF, 71-73; mem adv panel, Lobund Inst, Notre Dame, 71-73; mem pub info comt, Fedn Am Socs Exp Biol, 74- Mem: AAAS; Int Soc Hemat; Geront Soc (vpres, 74-75); Am Soc Microbiol; Am Asn Immunologists. Res: Radiation immunology; mechanism of antibody formation; aging of the immune system. Mailing Add: Geront Res Ctr Baltimore City Hosp Baltimore MD 21224

MAKINS, REES THOMAS, chemistry, see 12th edition

MAKLEY, TORRENCE ALOYSIUS, JR, b Dayton, Ohio, Jan 11, 18; c 7. OPHTHALMOLOGY. Educ: Univ Dayton, BS, 40; Wash Univ, MD, 43; Northwestern Univ, 47. Prof Exp: Intern, Barnes Hosp, St Louis, Mo, 43-44; resident, 48-51, assoc prof, 52-59, chmn dept, 63-72, PROF OPHTHAL, OHIO STATE UNIV, 59-63 & 72- Concurrent Pos: Fel ophthal path, Armed Forces Inst Path, 47-48; attend staff, Mt Carmel Hosp, 55; consult staff, Children's Hosp. Mem: AMA; Am Acad Ophthal & Otolaryngol; Asn Res Ophthal. Mailing Add: Dept of Ophthal Ohio State Univ Hosps Columbus OH 43210

MAKMAN, MAYNARD HARLAN, b Cleveland, Ohio, Oct 6, 33; m 59; c 2. PHARMACOLOGY, BIOCHEMISTRY. Educ: Cornell Univ, BA, 55; Case Western Reserve Univ, MD & PhD(pharmacol), 62. Prof Exp: Asst prof, 64-70, ASSOC PROF BIOCHEM & PHARMACOL, ALBERT EINSTEIN COL MED, 70- Concurrent Pos: NIH spec fel, 65-66, career develop award, 66-71. Mem: AAAS; Am Soc Biol Chemists; Endocrine Soc; Am Soc Pharmacol & Exp Therapeut; Asn Res Vision & Ophthal. Res: Hormone action; control of hormone receptors in normal and malignant cells; biochemical influence of catecholamines, steroids, cyclic adenosine monophosphate and related drugs on lymphoid and cultured cells, lens, retina and brain. Mailing Add: Dept of Biochem Albert Einstein Col of Med Bronx NY 10461

MAKOFSKE, WILLIAM JOSEPH, nuclear physics, see 12th edition

MAKOWSKI, EDGAR LEONARD, b Milwaukee, Wis, Oct 27, 27; m 52; c 6. OBSTETRICS & GYNECOLOGY. Educ: Marquette Univ, BS, 51, MD, 54. Prof Exp: Intern, Evangelical Deaconess Hosp, Milwaukee, Wis, 54-55; resident obstet & gynec, Univ Minn, 55-59, from instr to assoc prof, 59-66; assoc prof, 66-69, PROF OBSTET & GYNEC, MED CTR, UNIV COLO, DENVER, 69- Concurrent Pos: Fel physiol, Sch Med, Yale Univ, 63-64. Res: Reproductive and fetal physiology. Mailing Add: Dept of Obstet & Gynec Univ of Colo Med Ctr Denver CO 80220

MAKOWSKI, GARY GEORGE, b Wausau, Wis, Aug 22, 45; m 70. MATHEMATICS, STATISTICS. Educ: Univ Iowa, BA, 66, MS, 67, PhD(statist), 71. Prof Exp: ASST PROF MATH & STATIST, MARQUETTE UNIV, 70- Mem: Inst Math Statist; Am Statist Asn; Am Math Soc; Biomet Soc. Res: Convergence rates of statistics and regression; probability. Mailing Add: Dept of Math & Statist Marquette Univ Milwaukee WI 53233

MAKOWSKI, MIECZYSLAW PAUL, b Warsaw, Poland, Jan 15, 22; US citizen; m 45; c 2. ELECTROCHEMISTRY, PHYSICAL CHEMISTRY. Educ: Western Reserve Univ, BA, 57, MS, 61, PhD(electrochem), 64. Prof Exp: Asst mgr plastics technol, Smith-Phoenix Mfg Co, Ohio, 50-55; res chemist, Clevite Corp, 55-61; res asst electrochem, Western Reserve Univ, 61-62; sr res chemist, 62-64, mgr chem & polymers sect, 64-74, ASSOC DIR EXPLOR DEVELOP, GOULD LABS, GOULD, INC, 74- Concurrent Pos: Mem, Frontiers in Chem Lecture Series Comt, Case Western Reserve Univ, 72- Mem: Am Chem Soc; Electrochem Soc. Res: Electrode kinetics; hydrogen electrode; fuel cell electrode structure; electrodeposition; electroless deposition of metals; surface area studies; applied polymer research; supervision of electron microprobe; x-ray diffraction; electron microscope; spectrograph. Mailing Add: Gould Labs Gould Inc 540 E 105th St Cleveland OH 44108

MAKRIDES, ALKIS CHRYSANTHOU, physical chemistry, see 12th edition

MAKSOUDIAN, Y LEON, b Beirut, Lebanon, Oct 30, 33; US citizen; m; c 3. MATHEMATICS. Educ: Calif State Polytech Col, BS, 57; Univ Minn, Minneapolis, MS, 61, PhD, 70. Prof Exp: Instr math, Westmont Col, 57-58, Northwestern Col, Minn, 58-62 & Univ Minn, Minneapolis, 62-63; from asst prof to assoc prof, 63-72, PROF MATH, CALIF STATE POLYTECH UNIV, SAN LUIS OBISPO, 72- Mem: Am Statist Asn; Math Asn Am. Res: Probability and statistics. Mailing Add: Dept of Comput Sci & Statist Calif State Polytech Univ San Luis Obispo CA 93407

MAKSUD, MICHAEL GEORGE, b Chicago, Ill, Mar 26, 32; m 59; c 2. EXERCISE PHYSIOLOGY. Educ: Univ Ill, Urbana, BS, 55; Syracuse Univ, MA, 57; Mich State Univ, PhD(phys educ, physiol), 65. Prof Exp: Instr phys educ, Univ Ill, Chicago, 59-63; from asst prof to assoc prof, 65-72, PROF PHYS EDUC, UNIV WIS-MILWAUKEE, 67-, DIR EXERCISE PHYSIOL LAB, 65- Concurrent Pos: Clin assoc, Med Col Wis, 66-; consult physiol, Res Serv, Wood Vet Admin Ctr, 67- Mem: Am Physiol Soc; Am Col Sports Med; Am Asn Health, Phys Educ & Recreation. Res: Physiological basis of performance; nutritional effects of biochemical adaptation. Mailing Add: Exercise Physiol Lab Univ of Wis Milwaukee WI 53201

MAKSYMIUK, BOHDAN, b Stanyslaviv, Ukraine, Sept 17, 26; US citizen; m 58; c 2. ENTOMOLOGY, FORESTRY. Educ: Univ Mich, BSF, 53, MF, 55; Univ Md, PhD(entom), 65. Prof Exp: Res entomologist, US Forest Serv, Washington, DC, 55-65, Forestry Sci Lab, Ore, 65-70, PRIN ENTOMOLOGIST & RES LEADER, FORESTRY SCI LAB, US FOREST SERV, ORE, 70- Mem: Entom Soc Am; Soc Am Foresters; Soc Invert Path. Res: Aerial application of pesticides, environmental contamination, spray formulations, aircraft dispersal equipment, kinetics and physics of sprays; spray behavior, deposition and assessment; biological and chemical insecticides; ecological principles; insect pathology; antimicrobial substances in plants; biological insect control. Mailing Add: Forestry Sci Lab Pac NW Forest & Range Exp Sta Corvallis OR 97331

MAKSYMOWICH, ROMAN, b Kaminka, Ukraine, Oct 15, 24; nat US; m 51; c 3. BOTANY. Educ: Univ Pa, MS, 56, PhD, 59. Prof Exp: Asst bot, Univ Pa, 52-54 & 55-57, lab instr, 54-55, asst instr biol, 57-58; from instr to assoc prof, 59-65, PROF BIOL, VILLANOVA UNIV, 66- Concurrent Pos: NSF res grants, 59-67. Mem: Bot Soc Am; Am Soc Plant Physiol. Res: Plant growth and development; quantitative analysis of cell division and biosynthesis of DNA during leaf development of Xanthium pennsylvanicum; regulation of Xanthium shoot development with gibberellic acid. Mailing Add: Dept of Biol Villanova Univ Villanova PA 19085

MALACARA, DANIEL, b Leon, Mex, June 7, 37; m 64; c 4. OPTICS. Educ: Univ Mex, BSc, 61; Univ Rochester, MSc, 63, PhD(optics), 65. Prof Exp: Asst prof astron, Tonantzinla Observ, 64-66; from assoc prof to prof optics, Univ Mex, 66-72; PROF & TECH DIR, NAT INST ASTROPHYS OPTICS & ELECTRONICS, 72- Concurrent Pos: Mem admis comn, Mex Acad Sci & acad judgement comn, Physics Inst, Univ Mex, 69-71. Honors & Awards: Sci Instrumentation Award, Mex Acad Sci, 68. Mem: Optical Soc Am; Int Astron Union; Int Comn Optics; Mex Acad Sci. Res: Optical testing and design of instruments and components; interferometry. Mailing Add: Nat Inst Astrophys Optics & Electronics Apartado Postal 216 Puebla Mexico

MALACHESKY, PAUL ANTHONY, analytical chemistry, electrochemistry, see 12th edition

MALACINKSI, GEORGE M, b Norwood, Mass, Nov 25, 40; m 65; c 2. DEVELOPMENTAL BIOLOGY, BIOCHEMISTRY. Educ: Boston Univ, AB, 62; Univ Ind, MA, 64, PhD(microbiol), 66. Prof Exp: USPHS fel biochem & develop biol, Univ Wash, 66-68; asst prof, 68-74, ASSOC PROF ZOOL, IND UNIV, BLOOMINGTON, 74- Concurrent Pos: Res assoc, Univ Zurich, Switz, 74-75. Mem: AAAS; Am Chem Soc; Soc Develop Biol. Res: Biochemical and molecular basis of the regulatory mechanisms which control the ordered sequence of events which characterize the various stages in the developmental cycle of various animals. Mailing Add: Dept of Zool Ind Univ Bloomington IN 47401

MALAHOFF, ALEXANDER, b Moscow, USSR, Feb 7, 39; m 62. GEOLOGY, GEOPHYSICS. Educ: Univ NZ, BS, 60; Victoria Univ, NZ, MSc, 62; Univ Hawaii, PhD(geophys), 65. Prof Exp: Sci officer, Dept Sci & Indust Res, NZ, 59-60; asst geophys, Univ Wis, 63-64; asst geophys, Univ Hawaii, 64-65, asst geophysicist & asst prof geosci, 65-69, asst prof oceanog, 66-69, assoc prof geosci & oceanog, 69-71; PROG DIR MARINE GEOL & GEOPHYS PROG, OFF NAVAL RES, 71- Mem: Am Geophys Union; Soc Explor Geophys; Royal Soc NZ; Geol Soc NZ. Res: Solid earth geophysics; marine geophysical studies of the Pacific Ocean crust. Mailing Add: Off of Naval Res Dept of the Navy Sci & Technol Div Code 483 Arlington VA 22217

MALAIYANDI, MURUGAN, b Madurai, India, Apr 1, 23; Can citizen; m 43; c 2. ORGANIC CHEMISTRY, ANALYTICAL CHEMISTRY. Educ: Univ Madras, BA, 46; Univ Mysore, BSc, 49, MSc, 50; Univ Toronto, PhD(org chem), 60. Prof Exp: Res scholar org chem, Central Col, India, 50-51, lectr gen chem, 51-52; lectr, Intermediate Col, 52-56, demonstr org chem, Univ Toronto, 56-59, res fel, 59-60; res assoc, Univ Pittsburgh, 61-63; fel, Univ BC, 63-65; RES SCIENTIST, DEPT AGR, GOVT CAN, 65- Mem: Am Chem Soc. Res: Metabolism of precursors in biological systems; photochemistry; circular dichroism of steroids; analysis and metabolism of pesticides, drugs and vitamins. Mailing Add: Anal Serv Sect Plant Prod Bldg Can Dept of Agr Ottawa ON Can

MALAKER, DONALD LOUIS, physics, see 12th edition

MALAMED, SASHA, b New York, NY, May 6, 28; m 56; c 1. CELL BIOLOGY. Educ: Univ Pa, BA, 48, MS, 50; Columbia Univ, PhD(zool), 55. Prof Exp: Asst zool, Univ Pa, 49-50; asst, Columbia Univ, 50-54; res assoc, Univ Iowa, 54-55 & Columbia Univ, 55-56; USPHS res fel physiol, Western Reserve Univ, 56-58; instr anat, Albert Einstein Col Med, 58-59, asst prof, 59-67; assoc prof, 67-74, PROF ANAT, COL MED & DENT NJ, RUTGERS MED SCH, 74- Concurrent Pos: Vis lectr, Cornell Col, 55; instr, Hunter Col, 55; Lederle med fac award, 61-64. Mem: Am Soc Cell Biol; Am Asn Anatomists; Biophys Soc; Am Soc Zoologists; Am Physiol Soc. Res: Mitochondrial structure and function; ultrastructural and steroidogenic relationships of adrenocortical cells. Mailing Add: Col of Med & Dent of NJ Rutgers Med Sch Piscataway NJ 08854

MALAMUD, DANIEL F, b Detroit, Mich, June 5, 39; m 61; c 2. CELL BIOLOGY. Educ: Univ Mich, BS, 61; Western Mich Univ, MA, 62; Univ Cincinnati, PhD(zool), 65. Prof Exp: Instr biol, Univ Cincinnati, 65-66; asst prof path, Temple Univ, 68-69; ASST BIOLOGIST, MASS GEN HOSP, 69-; ASST PROF PATH, HARVARD MED SCH, 70- Concurrent Pos: USPHS res fel, Fels Res Inst, Sch Med, Temple Univ, 66-68. Mem: Am Inst Biol Sci; Am Soc Cell Biol; Soc Develop Biol; NY Acad Sci; Am Soc Zool. Res: Cell culture; autoradiography; control of desoxyribonucleic acid synthesis and cell proliferation; adenyl cyclase and glycogen metabolism. Mailing Add: Surg Serv Mass Gen Hosp Boston MA 02114

MALAMUD, ERNEST ILYA, b New York, NY, May 8, 32. PHYSICS. Educ: Univ Calif, Berkeley, AB, 54; Cornell Univ, PhD(physics), 59. Prof Exp: Res assoc, Cornell Univ, 59-60; privat docent, Univ Lausanne, 61-62; Ford fel, Europ Orgn Nuclear Res, 63; guest prof, Univ Heidelberg, 64; from asst prof to assoc prof physics, Univ Ariz, 64-66; vis assoc prof, Univ Calif, Los Angeles, 66-67, assoc prof in residence, 67-68; physicist, 68-70, head, Main Ring Accelerator Sect, 70-71, HEAD INTERNAL TARGET SECT, FERMILAB, 72- Mem: Inst Elec & Electronics Engrs; Am Phys Soc; Sigma Xi. Res: High energy physics; designing, constructing and commissioning of Fermilab 400 GEV main accelerator; initiating Soviet-American joint experiment on pp scattering. Mailing Add: Fermilab PO Box 500 Batavia IL 60510

MALAMUD, HERBERT, b New York, NY, June 28, 25; m 51; c 3. MEDICAL PHYSICS. Educ: City Col, New York, BS, 49; Univ Md, MS, 52; NY Univ, PhD(physics), 57. Prof Exp: Sr engr, Physics Labs, Sylvania Elec Prod Co, 57-59; specialist res engr, Repub Aviation Corp, 59-64; res sect head, Sperry Gyroscope Co, 64-65; dir physics res, Radiation Res Corp, 65-67; vpres, Plasma Physics Corp, 67-70; SR PHYSICIST, DEPT NUCLEAR MED, QUEEN'S HOSP CTR, JAMAICA, 70- Mem: AAAS; Am Phys Soc; Soc Nuclear Med; Am Asn Physicists in Med. Res: Nuclear medicine; plasma and atomic physics. Mailing Add: 30 Wedgewood Dr Westbury NY 11590

MALAMUD, NATHAN, b Kishinev, Russia, Jan 28, 03; nat US; m 30; c 2. NEUROPATHOLOGY. Educ: McGill Univ, MD, 30. Prof Exp: Asst neuropathologist & instr psychiat, Med Sch, Univ Mich, 34-45; prof, 46-71, EMER PROF NEUROPATH IN RESIDENCE, SCH MED, UNIV CALIF, SAN FRANCISCO, 71-; NEUROPATHOLOGIST, LANGLEY PORTER NEUROPSYCHIAT INST, 46- Concurrent Pos: Consult, Armed Forces Inst Path, 44-, Letterman Army Med Ctr, 46-, Oakland Naval Hosp, 51-, Nat Inst Neurol Dis & Stroke, 55- & Vet Admin Hosps, Martinez & San Francisco, 60-75. Mem: Am Asn Neuropath (vpres, 58-59); Am Psychiat Asn; Am Acad Neurol. Res: Cerebral palsy; mental retardation; geriatric disorders; chronic alcoholism; encephalitis; radiation; epilepsy; heredodegenerative disorders. Mailing Add: 240 Dept of Path Univ of Calif Med Sch San Francisco CA 94143

MALAMUD, WILLIAM, b Kishinev, Russia, May 5, 96; nat US; m 27; c 3. PSYCHIATRY. Educ: McGill Univ, MD, 21. Hon Degrees: DSc, Boston Univ, 60. Prof Exp: Asst physician, Foxboro State Hosp, 22-23; resident neurologist, Mt Sinai Hosp, New York, 23-24; neuropathologist, Foxboro State Hosp, 26-29; assoc prof psychiat, Sch Med, Univ Iowa, 29-31, prof, 31-39; clin dir, Worcester State Hosp, Mass, 39-48; chief psychiat & neurol serv, Mass Mem Hosp, 46-58; prof dir, Nat Asn Ment Health, 58-70; PROF PSYCHIAT, SCH MED, BOSTON UNIV, 70- Concurrent Pos: Lipman fel, France, Switz, Ger, Eng & Austria, 24-26; attend psychiatrist, St Vincent's Hosp, Mass, 58- Mem: Soc Exp Biol & Med; fel AMA; fel Am Psychiat Asn (secy, 54-58, pres elect, 58-59, pres, 59-60); Am Neurol Asn; fel Am Acad Arts & Sci. Res: Psychosomatic medicine; psychopathology;

MALAMUD

neuropathology; geriatrics. Mailing Add: Dept of Psychiat Boston Univ Sch of Med Boston MA 02118

MALAN, RODWICK LAPUR, b Du Quoin, Ill, Sept 7, 16; m 43; c 6. ORGANIC CHEMISTRY. Educ: Univ Ariz, BS & MS, 41; Univ Colo, PhD(org chem), 46. Prof Exp: Lab asst, Univ Ariz, 40-41; asst, llniv Colo, 41-44; chemist, Res Labs, 44-49, Film Emulsion Div, 50-56, TECH ASSOC, FILM EMULSION DIV, EASTMAN KODAK CO, 57- Concurrent Pos: Lectr eve div, Rochester Inst Technol, 48-68. Mem: Am Chem Soc. Res: Hemicelluloses and pectic materials in corn leaves; pyridine chemistry; photographic chemicals for the color processes; chemical emulsions. Mailing Add: Eastman Kodak Co Kodak Park Bldg 30 Rochester NY 14650

MALANGA, CARL JOSEPH, b New York, NY, Aug 26, 39; m 66; c 1. CELL PHYSIOLOGY, PHARMACOLOGY. Educ: Fordham Univ, BS, 61, MS, 67, PhD(biol sci), 70. Prof Exp: Instr biol sci labs, Col Pharm, Fordham Univ, 64-67, instr anat, physiol & pharmaceut, 67-70; asst prof, 70-73, ASSOC PROF THERAPEUT, SCH PHARM, WVA UNIV, 73- Concurrent Pos: Chmn curriculum comt, Fac Senate, WVa Univ. Mem: AAAS; Am Pharmaceut Asn; Am Inst Biol Sci; Am Soc Zoologists; Am Soc Pharmacol & Exp Therapeut. Res: Effects of serotonin, catecholamines and drugs on ciliary activity and energy metabolism. Mailing Add: WVa Univ Sch of Pharm Morgantown WV 26506

MALANIFY, JOHN JOSEPH, b Troy, NY, Apr 26, 34; m 56; c 2. NUCLEAR PHYSICS. Educ: Rensselaer Polytech Inst, BS, 55, PhD(physics), 64. Prof Exp: Staff mem nuclear physics, Los Alamos Sci Lab, Univ Calif, 64-66 & Oak Ridge Nat Lab, 66-69; STAFF MEM P-11, LOS ALAMOS SCI LAB, 69- Mem: Am Nuclear Soc; Am Phys Soc. Res: Direct nuclear reaction mechanism and the nucleon-nucleon problem, especially polarization; delayed neutrons and gamma rays from fission; x-ray fluorescence; muonic atoms; neutron time of flight. Mailing Add: Los Alamos Sci Lab Los Alamos NM 87544

MALARKEY, EDWARD CORNELIUS, b Girardville, Pa, Dec 7, 36; m 65; c 1. OPTICAL PHYSICS. Educ: La Salle Col, AB, 58; Mass Inst Technol, PhD(phys chem), 63. Prof Exp: FEL PHYSICIST, APPL SCI GROUP, SYSTS DEVELOP DIV, WESTINGHOUSE ELEC CORP, 63- Concurrent Pos: Assoc prof, Anne Arundel Community Col, Md, 67. Res: Laser development and imaging; laser resonator design and development; gas discharge analysis and computer modeling; optical emission spectroscopy and analysis. Mailing Add: Westinghouse Advan Technol Lab MS 3714 Box 1521 Baltimore MD 21203

MALASHOCK, EDWARD MARVIN, b Omaha, Nebr, Mar 27, 23; m 44; c 3. UROLOGY. Educ: Univ Nebr, BA, 43, MD, 46; Am Bd Urol, dipl, 56. Prof Exp: Resident urol surg, Beth Israel Hosp, New York, 50-53; clin asst, 53-55, assoc, 55-57, asst prof, 57-61, ASSOC PROF UROL, COL MED, UNIV NEBR AT OMAHA, 61-, ASSOC, PHYS MED & REHAB, 58- Concurrent Pos: Chief urol sect, Tenth Gen Hosp, Manila, 48-49. Mem: AMA; Am Urol Asn; Am Col Surgeons. Res: Urological surgery; neurogenic bladder dysfunction; new drugs as related to urological problems. Mailing Add: Dept of Urol Univ of Nebr Col of Med Omaha NE 68131

MALASPINA, ALEX, b Athens, Greece, Jan 4, 31; nat US; m 54; c 4. NUTRITION, FOOD TECHNOLOGY. Educ: Mass Inst Technol, BS, 52, SM, 53, PhD(food tech), 55. Prof Exp: Asst, Mass Inst Technol, 53-55; coordr, New Prod Dept, Chas Pfizer & Co, 55-61; mgr qual control dept, 61-70, VPRES QUAL CONTROL & DEVELOP DEPT, COCA-COLA EXPORT CORP, 70- Res: Quality control and new product development on carbonated beverages and protein drinks. Mailing Add: Qual Control & Develop Dept Coca-Cola Export Corp Atlanta GA 30313

MALAWISTA, STEPHEN E, b New York, NY, Apr 4, 34; m 69. INTERNAL MEDICINE. Educ: Harvard Univ, AB, 54; Columbia Univ, MD, 58; Am Bd Internal Med, dipl, 65; Am Bd Dermat, dipl, 66. Prof Exp: Intern internal med, Yale-New Haven Med Ctr, 58-59, asst resident, 59-60; clin assoc, Nat Inst Arthritis & Metab Dis, 60-62; asst resident, Yale-New Haven Med Ctr, 62-63, NIH spec fel, 63-66; from asst prof to assoc prof, 66-75, PROF MED, SCH MED, YALE UNIV, 75-, CHIEF RHEUMATOL, 67- Concurrent Pos: Asst attend physician, Yale-New Haven Med Ctr, 66-69, attend physician, 69-; attend physician, Vet Admin Hosp, West Haven, 66-69, consult rheumatology, 69-; investr, Arthritis Found, 66-70; consult, Gaylord Hosp, 68-; Nat Inst Arthritis & Metab Dis career res develop award, 70-75. Mem: AAAS; fel Am Col Physicians; Am Fedn Clin Res; Am Rheumatism Asn; Am Soc Clin Invest. Res: Rheumatic diseases; gout; inflammation; phagocytosis; cell division. Mailing Add: Dept of Internal Med Yale Univ Sch of Med New Haven CT 06510

MALBICA, JOSEPH ORAZIO, b Brooklyn, NY, Apr 6, 25; m 47; c 4. BIOCHEMISTRY. Educ: Brooklyn Col, BS, 49; Fordham Univ, MS, 54; Rutgers Univ, PhD(biochem), 67. Prof Exp: Chemist, Hoffman-La Roche Inc, NJ, 54-65; instr physiol & biochem Rutgers Univ, 66-67; res biochemist, Hess & Clark Div, Richardson-Merrell, Inc, 67-69; SUPVR BIOCHEM-PHARMACOL, ICI US, INC, 69- Mem: Am Chem Soc; Am Soc Pharmacol & Exp Therapeut; NY Acad Sci; Sigma Xi. Res: Biosynthesis of natural and unnatural products of pharmacology; fermentation; isolation and purification of natural products; drug metabolism; anti-inflammatories; muscle relaxants; drug-related studies with cellular organelles; ion-transport; drug kinetics. Mailing Add: Biomed Dept ICI US Inc Wilmington DE 19897

MALBON, WENDELL ENDICOTT, b Norfolk, Va, July 18, 18; m 42; c 1. MATHEMATICS. Educ: Univ Va, BCh, 41, MA, 52, PhD(math), 55. Prof Exp: Asst math, Univ Va, 50-54, from instr to assoc prof, 54-69; PROF MATH, OLD DOM UNIV, 69- Mem: Am Math Soc; Math Asn Am. Res: Point set topology; quasi-compact mappings; application of topology to theory of functions of a complex variable. Mailing Add: Dept of Math Old Dominion Univ Norfolk VA 23508

MALBROCK, JANE C, b NJ. MATHEMATICAL ANALYSIS. Educ: Montclair State Col, BA, 64; Pa State Univ, MA, 66, PhD(math anal), 71. Prof Exp: ASST PROF MATH & COMPUT SCI, KEAN COL NJ, 71- Mem: Am Math Soc; Math Asn Am. Res: Approximation theory. Mailing Add: Dept of Math & Comput Sci Kean Col of NJ Union NJ 07083

MALCHICK, SHERWIN PAUL, b St Paul, Minn, Aug 13, 29; m 51; c 3. ORGANIC CHEMISTRY. Educ: Univ Minn, BA, 49; Univ Rochester, PhD(chem), 52. Prof Exp: Lab asst, Univ Rochester, 49-51, asst, 51-52; chemist, Stand Oil Co, Ind, 52-59, group leader explor res, 59-60; tech specialist, Nalco Chem Co, 60-61, group leader, 61-66, sect head, Corp Res, 66-67, tech mgr corp res, 67-69, tech dir com develop, 69-71, mgr textile chems, 71-73; DIR RES, PIGMENTATION DIV, CHEMETRON CORP, 73- Mem: Federated Socs Coating Technol; Am Asn Textile Chem & Colorists; Am Chem Soc; Tech Asn Pulp & Paper Indust. Res: Organic synthesis; polymerization; polymer application; additives; coordination complexes; paper chemicals; colloid chemistry; textile chemistry; surface chemistry; organic pigments and dispersions. Mailing Add: Pigments Div Chemetron Corp 491 Columbia Ave Holland MI 49423

MALCOLM ALEXANDER RUSSELL, b Providence, RI, June 28, 36; m 64. BIOCHEMICAL GENETICS, GENETIC TOXICOLOGY. Educ: Univ RI, BS, 64, MS, 70, PhD(biophys), 76. Prof Exp: Chemist, Elec Boat Div, Gen Dynamics Corp, 64-67 & USPHS, 67-69; chemist, 69-72, RES CHEMIST, ENVIRON RES LAB, US ENVIRON PROTECTION AGENCY, 72- Concurrent Pos: Panel mem subcomt environ mutagenesis, Comt Coord Toxicol & Related Progs, Dept Health, Educ & Welfare, 74- Mem: Am Chem Soc; Biophys Soc; Environ Mutagen Soc; Sigma Xi; Tissue Cult Asn. Res: Development and application of in vitro mammalian cell methods in relation to investigations of the mutagenic properties of chemicals. Mailing Add: US Environ Protection Agency South Ferry Rd Narragansett RI 02882

MALCOLM, DAVID ROBERT, b Green Lake, Wis, Jan 11, 26; m 49. ENTOMOLOGY. Educ: Minn State Teachers Col, Winona, BS, 49; State Col Wash, MS, 51, PhD(zool), 54. Prof Exp: Asst zool, State Col Wash, 49-53, res assoc entom, 50-51; instr, Iowa State Col, 54; from instr to prof, Portland State Col, 54-69; PROF BIOL & CHMN DIV SCI, PAC UNIV, 69- Concurrent Pos: Asst dean grad studies, Portland State Col, 67-69. Mem: AAAS; Entom Soc Am; Soc Syst Zool; Am Inst Biol Sci. Res: Biology and taxonomy of Chelonethida. Mailing Add: Div of Sci Pac Univ Forest Grove OR 97116

MALCOLM, JANET MAY, b Bronx, NY, Mar 25, 25; m 46; c 3. OPERATIONS RESEARCH, RESOURCE MANAGEMENT. Educ: Rutgers Univ, New Brunswick, BS, 45; Northwestern Univ, MS, 46; Columbia Univ, PhD(phys chem), 51. Prof Exp: Lectr, Hunter Col, 48-49; res instr phys chem, Univ Miami, 51-53, Sch Med, 57-59; prof chem, Univ El Salvador, 62; asst prof opers res & info sci & dir opers res prog, Am Univ, 70-75; PVT CONSULT, 75- Mem: Fel AAAS; Asn Comput Mach; Opers Res Soc Am; Inst Mgt Sci; Am Chem Soc. Res: Rank sum statistics; physical chemistry of blood serum; refrigerant desiccants; kinetics; fluid dynamics analogs to linear, quadratic and separable convex programming; indigenous marketing system in Ghana and Nigeria; socioeconomic system modelling. Mailing Add: 1607 Kirby Rd McLean VA 22101

MALCOLM, JOHN LOWRIE, b Westfield, NJ, July 30, 20; m 46; c 3. SOILS. Educ: Rutgers Univ, BSc, 43, MSc, 45, PhD(soils), 48. Prof Exp: Assoc soil chemist, Subtrop Exp Sta, Univ Fla, 48-59; soils adv, US Opers Mission, Int Coop Admin, El Salvador, 59-63, USAID, India, 63-69; soils specialist, EAsia Bur/Tech Off, AID, DC, 70, PROJ MGR FERTILIZER SPECIALIST, TECH ASSISTANCE BUR/OFF AGR, AID, 70- Concurrent Pos: Proj mgr, Food & Agr Orgn, UN, Ghana, 69- Mem: AAAS; Am Chem Soc; Soil Sci Soc Am. Res: Soil and analytical chemistry; agronomy; vegetable and sub-tropical fruit production. Mailing Add: 1607 Kirby Rd McLean VA 22101

MALCOLM, MICHAEL ALEXANDER, b Denver, Colo, Feb 7, 45; m 73. COMPUTER SCIENCES, NUMERICAL ANALYSIS. Educ: Univ Denver, BSME, 66, MS, 68; Stanford Univ, MS, 70, PhD(comput sci), 73. Prof Exp: ASST PROF COMPUT SCI, UNIV WATERLOO, 73- Concurrent Pos: Mem prog lang comt, Can Standards Asn, 75- Mem: Asn Comput Mach; Soc Indust & Appl Math. Res: Portable programming techniques; portable systems software; mathematical software. Mailing Add: Dept of Comput Sci Univ of Waterloo Waterloo ON Can

MALCOLM, RICHARD EVELYN REGINALD, b Ottawa, Ont, July 20, 41; m 64. PHYSIOLOGY. Educ: Univ Ottawa, BSc, 64, MSc, 66; McGill Univ, PhD(neurophysiol), 71. Prof Exp: Res officer nuclear med, Royal Can Air Force, Inst Aviation Med, 66-67; RES OFFICER PHYSIOL, DEFENCE & CIVIL INST ENVIRON MED, 70- Mem: Asn Res Otolaryngol; Barany Soc. Res: Physiology of the inner ear; orientation instrumentation for aircraft and submarines; modelling of neurological processes. Mailing Add: Defense & Civil Inst Environ Med 1133 Shepard Ave W PO Box 2000 Downsview ON Can

MALCUIT, ROBERT JOSEPH, b Fredericksburg, Ohio, Feb 11, 36. PETROLOGY, ASTROGEOLOGY. Educ: Kent State Univ, BS, 68, MS, 70; Mich State Univ, PhD(geol), 73. Prof Exp: ASST PROF GEOL, DENISON UNIV, 72- Mem: Geol Soc Am; Am Geophys Union; AAAS; Sigma Xi. Res: Igneous and metamorphic petrology; zircons as petrogenetic indicators; geologic evidence relating to origin and evolution of Earth-Moon system; origin of massif-type anorthosite; origin of continents. Mailing Add: Dept of Geol & Geog Denison Univ Granville OH 43023

MALDACKER, THOMAS ANTON, b New York, NY, Apr 3, 46. ANALYTICAL CHEMISTRY. Educ: Fordham Univ, BS, 67; Purdue Univ, MS, 69, PhD(anal chem), 73. Prof Exp: SR SCIENTIST RES & DEVELOP, SANDOZ INC, 73- Mem: Am Chem Soc. Res: Development of liquid and gas chromatographic techniques for the analysis of drug substance and dosage, degradation and by-products. Mailing Add: Sandoz Inc Rt 10 East Hanover NJ 07936

MALDE, HAROLD EDWIN, b Reedsport, Ore, July 9, 23; m 54; c 2. GEOLOGY. Educ: Willamette Univ, AB, 47. Prof Exp: GEOLOGIST, US GEOL SURV, 51- Concurrent Pos: Affil prof, Univ Idaho, 68-; mem Colo consult comt, Nat Register Hist Places, 72-; mem study comt potential rehab lands surface mined for coal in western US, Nat Acad Sci, 73, mem paleoanthrop deleg to People's Repub of China, 75, mem oil shale environ adv panel, 76- Honors & Awards: Kirk Bryan Award, Geol Soc Am, 70. Mem: Fel Geol Soc Am; Soc Am Archaeol; fel AAAS; Am Quaternary Asn. Res: Cenozoic and Quaternary geology; geomorphology; environmental geology; stratigraphy and paleo-geomorphology applied to early man; technical photography. Mailing Add: US Geol Surv Mail Stop 913 Box 25046 Denver Fed Ctr Denver CO 80225

MALDONADO, JORGE EUSEBIO, b Cartago, Colombia, Nov 12, 35; m 59; c 2. HEMATOLOGY, ELECTRON MICROSCOPY. Educ: Col San Francisco Javier Pasto, Columbia, BA, 51; Pontif Univ Javeriana, MD, 58; Univ Minn, PhD(med), 68. Prof Exp: Fel, Mayo Grad Sch Med, Mayo Clin & Mayo Found, 61-65, asst prof, Mayo Med Sch, 68-73, ASSOC PROF MED, MAYO MED SCH, 73-; CONSULT HEMAT, MAYO CLIN, 65- Mem: Am Asn Cancer Res; Am Fedn Clin Res; Am Soc Cell Biol; Am Soc Hemat; Int Soc Hemat. Res: Biopathology and electron microscopy of hematologic disorders, primarily neoplastic; ultrastructure and cell biology of hematopoietic cells in preleukemic states. Mailing Add: Mayo Clin 200 First St SW Rochester MN 55901

MALDONADO, JUAN RAMON, b Holguin, Cuba, May 6, 38; US citizen; m 62; c 2. APPLIED PHYSICS. Educ: Univ Havana, MSc, 61; Univ Md, College Park, PhD(exp solid state physics), 68. Prof Exp: Elec engr, CMQ TV, Havana, 57-61; instr physics-math, Univ Havana, 60-61; supvr electronics, Univ Md, College Park, 62-65, res asst solid state physics, 65-68; MEM TECH STAFF PHYSICS, BELL TEL LABS, 68- Mem: Am Inst Physics; Inst Elec & Electronic Engrs; Sigma Xi. Res: X-ray fluorescence systems for special applications; x-ray lithography and integrated circuit technology. Mailing Add: Bell Tel Labs Mountain Ave Rm 2C240 Murray Hill NJ 07974

MALE, CAROLYN JOAN, b Troy, NY, June 26, 39. MICROBIOLOGY. Educ:

Cornell Univ, BS, 61; Univ Rochester, MS, 64, PhD(microbiol), 68. Prof Exp: USPHS fel, Med Sch, Univ Rochester, 67-69; USPHS fel, 69-70, res assoc microbiol, 70-71, INSTR MICROBIOL, MED SCH, UNIV COLO, DENVER, 71- Mem: Am Soc Microbiol. Res: Bacteriophage biochemistry, attachment, injection. Mailing Add: Dept of Microbiol Univ of Colo Med Sch Denver CO 80220

MALECHA, SPENCER R, b Chicago, Ill, Nov 13, 43; m 71. GENETICS. Educ: Loyola Univ, Ill, BS, 65; Univ Hawaii, MS, 68, PhD(genetics), 71. Prof Exp: Asst zool, 66-68, asst genetics, 68-69, ASST PROF GENETICS, SCH MED, UNIV HAWAII, MANOA, 71- Concurrent Pos: Ford Found fel, Univ Chicago, 72-73. Mem: AAAS; Soc Study Evolution; Am Soc Nat; Nat Asn Biol Teachers. Res: Ecological genetics; genetic variation in natural populations. Mailing Add: Dept of Genetics Univ of Hawaii Sch of Med Honolulu HI 96822

MALECHEK, JOHN CHARLES, b San Angelo, Tex, Aug 6, 42. RANGE SCIENCE, ECOLOGY. Educ: Tex Tech Univ, BS, 64; Colo State Univ, MS, 66; Tex A&M Univ, PhD(range sci), 70. Prof Exp: ASST PROF RANGE SCI, UTAH STATE UNIV, 70- Mem: Am Soc Animal Sci; Soc Range Mgt; Wildlife Soc. Res: Nutritional relationships of wild and domestic herbivores on rangelands. Mailing Add: Dept of Range Sci Utah State Univ Logan UT 84321

MALEENY, ROBERT TIMOTHY, b Staten Island, NY, Jan 1, 31; m 54; c 2. CHEMISTRY. Educ: Wagner Col, BS, 52; St John's Univ, MS, 54. Prof Exp: NSF res asst chem, St John's Univ, 52-53; asst tech dir, Dodge & Olcott, 53-66; tech dir, Globe Extracts, 66-68; vpres tech dir, Major Prod, 68-70 & Aromatics, Int, 70-74; DIR RES, MONSANTO FLAVOR/ESSENCE, MONSANTO CO, 74- Mem: Am Chem Soc; AAAS; Soc Flavor Chemists; Soc Cosmetic Chemists. Res: Development of flavors, fragrances and unique aromatic chemicals, including malodor counteractants. Mailing Add: 16 Mohawk Dr Ramsey NJ 07446

MALEK, EMILE ABDEL, b El Mansura, Egypt, Aug 22, 22; US citizen; m 54; c 3. MEDICAL PARASITOLOGY. Educ: Cairo Univ, BSc, 43, MSc, 47; Univ Mich, Ann Arbor, PhD(parasitol), 52. Prof Exp: Teaching asst zool, Cairo Univ, 43-47; scientist schistosomiasis control, Ministry of Health, Egypt, 52-53; from lectr to reader parasitol, Univ Khartoum, 53-59; from asst prof to assoc prof, 59-73, PROF PARASITOL, MED SCH & SCH PUB HEALTH, TULANE UNIV, 74- Concurrent Pos: NIH res career award, 62-; consult, WHO & Pan Am Health Orgn, 56, 61-64, 66 & 74, Peace Corps, 67 & USPHS, 74-; mem expert adv panel, WHO, 64-, scientist parasitic dis, 67-69. Mem: Am Soc Parasitol; Am Malacol Soc; Am Soc Trop Med & Hyg; Royal Soc Trop Med & Hyg. Res: Epidemiology and control of schistosomiasis; medical malacology; snail-transmitted helminthiases. Mailing Add: Dept of Trop Med Tulane Univ Med Ctr New Orleans LA 70112

MALEK, RICHARD BARRY, b Westfield, NJ, July 6, 36; m 63; c 3. PLANT NEMATOLOGY. Educ: Univ Maine, BS, 58; Rutgers Univ, MS, 60, PhD(plant nematol), 65. Prof Exp: Res asst entom, Rutgers Univ, 58-64; asst prof plant path, SDak State Univ, 64-68; asst prof, 68-75, ASSOC PROF NEMATOL, UNIV ILL, URBANA, 75- Mem: Soc Nematol; Helminthol Soc Wash; Europ Soc Nematologists; Orgn Trop Am Nematologists. Res: Nematode diseases of plants and their control. Mailing Add: Dept of Plant Path Univ of Ill Urbana IL 61801

MALENFANT, ARTHUR LEWIS, b Wakefield, RI, May 17, 37; m 57; c 3. ANALYTICAL CHEMISTRY. Educ: Univ RI, BS, 60; Mass Inst Technol, PhD(anal chem), 67. Prof Exp: Chemist, Corning Glass Works, NY, 60-63; res asst anal chem, Mass Inst Technol, 63-67; dir tech serv, 67-74, VPRES RES & DEVELOP, INSTRUMENTATION LAB, INC, 74- Mem: Am Chem Soc; Soc Appl Spectros; Asn Advan Med Instrumentation; Am Asn Clin Chemists. Res: Analytical techniques and methods to promote new approaches in biomedical instrumentation. Mailing Add: Instrumentation Lab Inc 113 Hartwell Ave Lexington MA 02173

MALENKA, BERTRAM JULIAN, b New York, NY, June 8, 23; m 48; c 2. THEORETICAL HIGH ENERGY PHYSICS. Educ: Columbia Univ, AB, 47; Harvard Univ, AM, 49, PhD(physics), 51. Prof Exp: Res fel, Harvard Univ, 51-54; asst prof physics, Washington Univ, 54-56; assoc prof, Tufts Univ, 56-60; assoc prof, 60-62, PROF PHYSICS, NORTHEASTERN UNIV, 62- Concurrent Pos: Adv, Harvard Univ & Mass Inst Technol, 54-; consult, Arthur D Little Inc, 59- & Am Sci & Eng, 59- Mem: Am Phys Soc; NY Acad Sci; Ital Phys Soc. Res: Theoretical nuclear physics; scattering theory at high energies; elementary particles; accelerator theory and design. Mailing Add: Dept of Physics Northeastern Univ Boston MA 02115

MALES, JAMES ROBERT, b Noblesville, Ind, Sept 28, 45; m 72; c 1. ANIMAL SCIENCE. Educ: Pa State Univ, BS, 67; Mich State Univ, MS, 69; Ohio State Univ, PhD(ruminant nutrit), 73. Prof Exp: Res assoc ruminant nutrit, Okla State Univ, 73-74; ASST PROF ANIMAL SCI, SOUTHERN ILL UNIV, CARBONDALE, 74- Mem: Am Soc Animal Sci. Res: Maximizing meat production from low quality roughages with minimum protein supplementation; beef cattle performance as it is related to maximum efficiency of beef production. Mailing Add: Dept of Animal Indust Southern Ill Univ Carbondale IL 62901

MALETSKY, EVAN M, b Pompton Lakes, NJ, June 9, 32; m 54; c 4. MATHEMATICS. Educ: Montclair State Col, BA, 53, MA, 54; NY Univ, PhD(math educ), 61. Prof Exp: PROF MATH, MONTCLAIR STATE COL, 57- Mem: Math Asn Am; Nat Coun Teachers Math. Res: Training of mathematics teachers and mathematics curriculum changes in the junior and senior high school. Mailing Add: 34 Pequannock Ave Pompton Lakes NJ 07442

MALETTE, WILLIAM GRAHAM, b Springfield, Mo, Mar 27, 22; m 45; c 2. SURGERY. Educ: Drury Col, 40-42; Washington Univ, MD, 53. Prof Exp: Physician, US Air Force, 53-58, chief exp surg, US Air Force Sch Aviation Med, 58-61, chief unit II surg & chief vascular surg serv, Air Force Hosp, Tex, 61-63; from asst prof to assoc prof surg, Med Ctr, Univ Ky, 63-75, assoc dean Vet Admin affairs, 71-75; DIR EMERGENCY MED SERV, KERN MED CTR, 75- Concurrent Pos: Chief surg serv, Vet Admin Hosp, Lexington, Ky, 63-75, chief staff, Univ Div, 71-75; participating nat surg consult & mem surg res comt, Vet Admin, Washington, DC; consult, USPHS Hosp, Lexington. Mem: Fel Am Col Chest Physicians; fel Am Col Cardiol; Aerospace Med Asn; Am Soc Artificial Internal Organs; fel Am Col Surgeons. Res: Cardiovascular surgery; tissue transplantation; aerospace physiology. Mailing Add: Kern Med Ctr Bakersfield CA 93305

MALEWITZ, THOMAS DONALD, b Holland, Mich, Apr 13, 29; m 57; c 3. ANATOMY. Educ: Hope Col, AB, 51; Univ Kans, MA, 53; Mich State Univ, PhD(anat, histol), 56. Prof Exp: Asst instr biol, Univ Kans, 52-53; instr anat, Mich State Univ, 56-57; asst prof anat & physiol, Col Pharm, Univ Fla, 57-61 & anat, Woman's Med Col Pa, 61-66; asst prof, 66-67, ASSOC PROF BIOL, VILLANOVA UNIV, 67- Mem: Am Asn Anat; Am Soc Cell Biol. Res: Human and animal histology; animal pathology; parasitology. Mailing Add: Dept of Biol Villanova Univ Villanova PA 19085

MALEY, FRANK, b Brooklyn, NY, July 26, 29; m 53; c 3. BIOCHEMISTRY. Educ: Brooklyn Col, BS, 52; Univ Wis, MS, 53, PhD, 56. Prof Exp: USPHS res fel, Sch Med, NY Univ, 56-58; sr res scientist, 58-61, assoc res scientist, 61-69, DIR DEVELOP BIOCHEM, DIV LAB & RES, NY STATE DEPT HEALTH, 69- Concurrent Pos: From asst prof to assoc prof, Albany Med Col, 58-61, adj prof, 70- Mem: Am Soc Biol Chem; Am Chem Soc. Res: Nucleotide interconversions and nucleic acid metabolism; one carbon and hexosamine metabolism; chemical synthesis of hexosamine derivatives; glycoprotein structure and biosynthesis; regulation of enzyme activity and synthesis; isolation of phage induced enzymes. Mailing Add: Div of Lab & Res NY State Dept of Health Albany NY 12201

MALEY, GLADYS FELDOTT, b Aurora, Ill, June 7, 26; m 53; c 3. BIOCHEMISTRY. Educ: NCent Col, Ill, BA, 48; Univ Wis, MS, 50, PhD, 53. Prof Exp: Am Heart Asn res fel, Sch Med, NY Univ, 56-58; adv res fel, 58-60, Am Heart Asn estab investr, 60-65, USPHS res career develop award, 65-70, ASSOC RES SCIENTIST, NY STATE DEPT HEALTH, 70- Concurrent Pos: Asst prof, Albany Med Col, Union Univ, NY, 59-65. Mem: AAAS; Am Soc Biol Chem. Res: Nucleotide interconversions; nucleic acid metabolism; structure and function of regulatory proteins. Mailing Add: Div of Labs & Res NY State Dept of Health Albany NY 12201

MALEY, MARTIN PAUL, b Seattle, Wash, Aug 27, 36; m 66. PHYSICS. Educ: Yale Univ, BS, 56; Rice Univ, MA, 63, PhD(physics), 65. Prof Exp: Fel, Los Alamos Sci Lab, 65-68; ASST PROF PHYSICS, RENSSELAER POLYTECH INST, 68- Res: Low temperature physics; magnetism; ferromagnetic resonance; microwave ultrasonics. Mailing Add: Dept of Physics Rensselaer Polytech Inst Troy NY 12181

MALHIOT, ROBERT JOSEPH, b Chicago, Ill, Nov 21, 26; m 53; c 3. PHYSICS. Educ: Denver Univ, BS, 49; Ill Inst Technol, MS, 54, PhD(physics), 57. Prof Exp: From asst prof to assoc prof, 57-70, PROF PHYSICS, ILL INST TECHNOL, 70- Concurrent Pos: Actg chmn dept physics, Ill Inst Technol, 62-68, chmn dept, 68-70. Mem: AAAS; Am Phys Soc; Fedn Am Scientists; Am Asn Physics Teachers; Sigma Xi. Res: Gravitation and general relativity. Mailing Add: Dept of Physics Ill Inst of Technol Chicago IL 60616

MALHOTRA, ASHWANI, b Lahore, WPakistan, July 6, 43; m 71; c 1. ENZYMOLOGY, CARDIOVASCULAR PHYSIOLOGY. Educ: Univ Delhi, BS, 62, MS, 64, PhD(org chem), 69. Prof Exp: Res asst chem, Univ Conn, 69-70; Fogarty Int vis fel cancer res, Nat Cancer Inst, 70-72; fel, New Eng Inst, 72-73; RES ASSOC CARDIOL, MONTEFIORE HOSP & MED CTR, ALBERT EINSTEIN COL MED, 73- Mem: Am Chem Soc. Res: Cardiovascular physiology and biochemistry with special emphasis on excitation-contraction coupling of contractile proteins and protein synthesis. Mailing Add: Div of Cardiol Dept of Med Montefiore Hosp & Med Ctr Bronx NY 10467

MALHOTRA, OM PARKASH, b Multan City, Pakistan, Sept 1, 26; m 55; c 2. PHYSIOLOGY. Educ: Panjab Agr Univ, BVSc, 49; Kans State Univ, MS, 57; Univ Ill, Urbana, PhD(vet physiol), 62. Prof Exp: Vet asst surgeon, Indian Govt, 49-56; instr vet med, Col Vet Med, Univ Ill, Urbana, 60-63, asst prof, 63; asst res prof path, Med Ctr, Univ Kans, 63-67; ASST PROF EXP PATH, INST PATH, SCH MED, CASE WESTERN RESERVE UNIV, 67-; RES CHEMIST, MED RES SERV, VET ADMIN HOSP, 67- Concurrent Pos: Asst investr, US Army Med Dept, 57-61, prin investr, 61-62; co-investr, Nat Heart Inst fel, 62-63 & NIH spec fels, 63-; res assoc, USPHS, 63-64, co-prin investr, 66- Mem: Conf Res Workers Animal Dis; Am Soc Vet Physiol & Pharmacol. Res: Blood coagulation; protein chemistry. Mailing Add: Inst of Path Case Western Reserve Univ Cleveland OH 44106

MALHOTRA, SUDARSHAN KUMAR, b Bhera, India, June 20, 33; m 63; c 2. CELL BIOLOGY. Educ: Oxford Univ, DrPhil(biol sci), 60. Prof Exp: Sr studentship, Royal Comn for Exhib of 1851, 60-62; res fel biol, Calif Inst Technol, 63-65, sr res fel, 66-67; prof zool, Univ Alta, 67-71; dean life sci, Jawaharlal Nehru Univ, New Delhi, 71-72; PROF ZOOL & DIR BIOL SCI ELECTRON MICROS, UNIV ALTA, 72- Concurrent Pos: Res fel, New Col, Oxford Univ, 61-63. Mem: AAAS; Am Soc Cell Biol; Soc Exp Biol & Med; Can Soc Cell Biol; Can Soc Zoologists. Res: Cytology of nerve cells; structure, function and biogenesis of cellular membranes and membranous organelle. Mailing Add: 12916-63rd Ave Edmonton AB Can

MALHOTRA, SUDARSHAN KUMAR, b Pasrur, Pakistan, Oct 11, 35; m 61; c 2. ORGANIC CHEMISTRY, PESTICIDE CHEMISTRY. Educ: Univ Punjab, BSc, 56; Columbia Univ, MA, 58, PhD(org chem), 62. Prof Exp: Staff scientist, Worcester Found Exp Biol, Mass, 61-65; res chemist, East Res Lab, Dow Chem Co, Mass, 65-67, sr res chemist, 67, Western Div Res Lab, 67-71, res specialist, 71-75, SR RES SPECIALIST ORG RES, DOW CHEM CO, 75- Mem: Am Chem Soc. Res: Stereochemistry; reaction mechanism; chemistry of enolate anions; enols; enamines; chemistry of natural products; insect growth regulators; insect pheremones; pyrethroids; organic phosphate and carbamates; plant growth regulators. Mailing Add: Org Res Dow Chem Co Walnut Creek CA 94565

MALICH, CHARLES WILSON, b Somerville, Tex, Feb 4, 19; m 51; c 2. BIOPHYSICS. Educ: Rice Inst, BA, 40, PhD(physics), 47; Univ Minn, MA, 42. Prof Exp: Assoc prof physics, Univ Alaska, 42-44; sr observer & physicist, Carnegie Inst, 44-46; instr physics, Univ Pa, 47-48; physicist, Naval Res Lab, 48-58; assoc prof physics, Southern Ill Univ, 58-59; physicist, Nat Inst Arthritis & Metab Dis, 59-62; RES SCIENTIST, AMES RES CTR, NASA, 62- Mem: Am Phys Soc; Biophys Soc; Genetics Soc Am. Res: Ionospheric and nuclear physics; radiation biology. Mailing Add: NASA MS 39-4 Ames Res Ctr Moffett Field CA 94035

MALICK, DONALD, mathematics, see 12th edition

MALICK, JEFFREY BEVAN, b Brooklyn, NY, Nov 14, 42; m 62; c 3. NEUROPHARMACOLOGY. Educ: Rutgers Univ, BA, 65, MS, 68; NY Univ, PhD(psychobiol), 73. Prof Exp: Res asst neuropharmacol, Lederle Labs, Div Am Cyanamid, 62-67; supvr neurophysiol, Union Carbide Corp, 67-69; sr res scientist neuropharmacol, Schering Corp, 69-74; SR RES PHARMACOLOGIST, ICI US INC, 74- Mem: Am Soc Pharmacol & Exp Therapeut; Soc Neurosci; Am Psychol Asn; Int Soc Res Aggression; NY Acad Sci. Res: Psychopharmacology; neurochemical and neuroanatomical substrates of behavior and drug action; aggressive behavior; depression; alcoholism and drug abuse. Mailing Add: ICI US Inc Biomed Res Dept Pharmacol Sect Wilmington DE 19897

MALICKI, CAROL ANN, b New York, NY, Apr 6, 49. BIOCHEMISTRY. Educ: Long Island Univ, BS, 70; Fairleigh Dickinson Univ, MS, 76. Prof Exp: Res asst obesity, 70-71, SCIENTIST PLATELETS & ATHEROSCLEROSIS, CIBA-GEIGY CORP, 71- Mem: Am Chem Soc; Sigma Xi. Res: Role of platelets in atherosclerosis; lipid-platelet interactions; vascular injury and platelet function; phospholipidases, prostaglandins and platelets; drug effects. Mailing Add: Ciba-Giegy Corp 444 Saw Mill River Rd Ardsley NY 10502

MALIK, ASRAR BARI, b Lahore, Pakistan, Dec 1, 45; Can citizen. CARDIOVASCULAR PHYSIOLOGY. Educ: Univ Western Ont, BSc(Hons), 68; Univ Toronto, MSc, 69, PhD(physiol), 71. Prof Exp: Demonstr physiol, Univ Toronto, 68-71, demonstr histol, 69-70; instr physiol, Wash Univ, 71-73; ASST PROF PHYSIOL, ALBANY MED COL, 73- Concurrent Pos: Assoc staff surg, Jewish Hosp St Louis, 72-73. Mem: Am Physiol Soc; Biophys Soc. Res: Regional circulation; regulation of pulmonary circulation; cardiac hypertrophy. Mailing Add: Dept of Physiol Albany Med Col Albany NY 12208

MALIK, DHARAM DEV, b Bhainswal Kalan, India, June 20, 31; m 51; c 2. PHYSIOLOGY, GENETICS. Educ: Punjab Univ, BVSc, 56; Tex A&M Univ, MS, 62, PhD(physiol), 64. Prof Exp: Instr reprod physiol, Col Vet Med, India, 56-60; assoc prof, Punjab Agr Univ, India, 64-67, grant, 65-67; PROF BIOL, JARVIS CHRISTIAN COL, 67- Mem: Poultry Sci Asn. Res: Sex hormones and their effects on growth and reproduction; water metabolism; environmental physiology of birds and mammals; economical rations of poultry. Mailing Add: Dept of Biol Jarvis Christian Col Hawkins TX 75765

MALIK, FAZLEY BARY, b Bankura, India, Aug 16, 34. ATOMIC PHYSICS, NUCLEAR PHYSICS. Educ: Univ Calcutta, BS, 53; Univ Dacca, MS, 55; Univ Göttingen, Dr rer nat(physics), 58. Prof Exp: Res assoc physics, Max Planck Inst Physics, 59-60; sr sci officer, Pakistan AEC, 60-69; ASSOC PROF PHYSICS, IND UNIV, BLOOMINGTON, 68- Concurrent Pos: AID res fel, Princeton Univ, 60-63; vis assoc prof, Fordham Univ, 61-62; asst prof, Yale Univ, 64-68; consult, US AEC, 66-72; vis prof, Comn Physics, Switz, 71-72; adv, Bangladesh Planning Comn, 72. Mem: Am Phys Soc. Res: Theory of atomic, molecular and nuclear structure; reactions and many body aspects. Mailing Add: Dept of Physics Ind Univ Bloomington IN 47401

MALIK, JIM GORDEN, b Elyria, Ohio, Oct 5, 28; m 53; c 3. PHYSICAL CHEMISTRY, INORGANIC CHEMISTRY. Educ: Wabash Col, AB, 50; Mich State Univ, PhD, 54. Prof Exp: Instr, Univ Minn, Duluth, 54-55, asst prof, 55-56; asst prof chem, Knox Col, 56-57; from asst prof to assoc prof, San Diego State Col, 57-64; prof, Sonoma State Col, 64-65; PROF CHEM, SAN DIEGO STATE UNIV, 65- Res: Preparation of inorganic coordination compounds; atomic and molecular structure; interferometric measurements; chemical education. Mailing Add: Dept of Chem San Diego State Univ San Diego CA 92182

MALIK, JOHN STANLEY, b Geddes, SDak, Sept 3, 20; m 54; c 1. EXPERIMENTAL PHYSICS. Educ: Kans State Teachers Col, AB, 42; Univ Mich, MS, 47, PhD, 50. Prof Exp: PHYSICIST, LOS ALAMOS SCI LAB, 50- Concurrent Pos: Sci adv, Opers Off, Energy Res & Develop Admin, Nev, 69- Mem: AAAS; Am Phys Soc. Res: Nuclear weapon testing and phenomenology; diagnostics; gamma rays; electromagnetic pulse; nuclear weapons test hazards evaluation. Mailing Add: Los Alamos Sci Lab Los Alamos NM 87545

MALIK, VEDPAL SINGH, b Sunna, India, Jan 1, 42; Can citizen; m 72; c 1. MEDICAL RESEARCH. Educ: Indian Agr Res Inst, New Delhi, MSc, 64; Dalhousie Univ, PhD(biol), 70. Prof Exp: Teaching asst biol, Dalhousie Univ, 65-68; res assoc microbiol, Sherbrooke Med Ctr, Can, 70, Tex Med Ctr, Houston, 70-71 & Mass Inst Technol, 71-72; RES SCIENTIST MICROBIOL, UPJOHN CO, 72- Concurrent Pos: Res assoc, Nat Cancer Inst, 72. Mem: Am Soc Microbiol. Res: Physiology and genetics of industrial organisms; novel microbial metabolites. Mailing Add: Infectious Dis Res Uphohn Co Kalamazoo MI 49001

MALIN, JOHN MICHAEL, b Cleveland, Ohio, July 9, 42; m 67; c 2. INORGANIC CHEMISTRY, PHOTOCHEMISTRY. Educ: Univ Calif, Berkeley, BS, 63; Univ Calif, Davis, PhD(chem), 68. Prof Exp: NIH fel inorg chem, Stanford Univ, 68-70; Nat Acad Sci fel, Inst Chem, Univ Sao Paulo, Brazil, 70-73; vis asst prof, 73-74, ASST PROF INORG CHEM, UNIV MO, COLUMBIA, 74- Mem: Sigma Xi; Am Chem Soc. Res: Studies of the synthesis, spectra, structures and reactivity, both thermal and photoinduced, of transition metal complexes. Mailing Add: Dept of Chem Univ of Mo Columbia MO 65201

MALIN, MURRAY EDWARD, b New York, NY, June 25, 27; m 52; c 2. PHYSICAL CHEMISTRY. Educ: City Col New York, BS, 48; Harvard Univ, AM & PhD(phys chem), 51. Prof Exp: Mem staff, Los Alamos Sci Lab, 51-56; vpres res & technol, Res & Develop Div, Avco Corp, 56-57, tech opers, Avco Space Systs Div, 57-68; mgr spec systs, 68-71, prog mgr, 71-75, DIV VPRES, POLAROID CORP, 75- Concurrent Pos: Mem, Adv Comt Fluid Mech, NASA, 65- Mem: Am Phys Soc; Am Chem Soc; fel Am Inst Aeronaut & Astronaut. Res: High temperature; thermodynamics; theory and structure of detonation waves; space and missile physics. Mailing Add: Polaroid Corp 1 Upland Rd Norwood MA 02062

MALIN, SHIMON, b Ramat-Gan, Israel, July 7, 37; m 60; c 3. THEORETICAL PHYSICS. Educ: Hebrew Univ, Jerusalem, MSc, 61; Univ Colo, Boulder, PhD(physics), 68. Prof Exp: ASST PROF PHYSICS, COLGATE UNIV, 68- Mem: Am Phys Soc; Am Asn Physics Teachers. Res: Group theory and its applications to high-energy physics and general relativity; foundations of quantum mechanics. Mailing Add: Dept of Physics & Astron Colgate Univ Hamilton NY 13346

MALINA, MARSHALL ALBERT, b Chicago, Ill, July 11, 28; m 55; c 2. RESOURCE MANAGEMENT. Educ: Ill Inst Technol, BS, 49; Northwestern Univ, MBA, 69. Prof Exp: Qual control chemist, Capitol Chem Co, 49-50; prod mgr, Hamilton Industs, 50-52; mgr, Gen Anal Sect, 52-68, mgr qual control, 68-69, dir qual control & tech serv, 69-70, dir res, 71-74, MGR CORP DEVELOP, VELSICOL CHEM CORP, 74- Mem: Am Chem Soc; Am Soc Testing & Mat; fel Asn Off Anal Chem; Com Develop Asn; Soc Plastics Eng. Res: Agricultural pesticides; polymers; polymer additives; plasticizers; benzoic acid and derivatives; pollution control; statistics; licensing, acquisitions, capital budgeting and planning, corporate planning. Mailing Add: Velsicol Chem Corp 341 E Ohio Chicago IL 60611

MALINA, ROBERT MARION, b Brooklyn, NY, Sept 19, 37. PHYSICAL ANTHROPOLOGY. Educ: Manhattan Col, BS, 59; Univ Wis, MS, 60, PhD(phys educ), 63; Univ Pa, PhD(anthrop), 68. Prof Exp: Asst prof, 67-71, ASSOC PROF ANTHROP, UNIV TEX, AUSTIN, 71- Mem: Am Asn Health, Phys Educ & Recreation; Am Asn Phys Anthrop; Soc Study Human Biol; Human Biol Coun. Res: Human growth and development; motor development; growth and nutrition in Mexico and Central America; growth and athletic performance. Mailing Add: Dept of Anthrop Univ of Tex Austin TX 78712

MALINAUSKAS, ANTHONY PETER, b Ashley, Pa, Mar 24, 35; m 57; c 6. PHYSICAL CHEMISTRY. Educ: Kings Col, Pa, BS, 56; Boston Col, MS, 58; Mass Inst Technol, PhD(phys chem), 62. Prof Exp: Mem res staff, 62-73, CHIEF, CHEM DEVELOP SECT, OAK RIDGE NAT LAB, 73- Mem: AAAS; Am Chem Soc; Am Phys Soc; Am Nuclear Soc. Res: Thermal transpiration; transport properties of gases; nuclear safety; fission product chemistry and transport; nuclear fuel reprocessing. Mailing Add: Oak Ridge Nat Lab Oak Ridge TN 37830

MALINDZAK, GEORGE STEVE, JR, b Cleveland, Ohio, Jan 3, 33; m 59; c 4. MEDICAL PHYSIOLOGY, ENVIRONMENTAL HEALTH. Educ: Western Reserve Univ, AB, 56; Ohio State Univ, MSc, 58, PhD(physiol, biophys), 61. Prof Exp: Consult comput, NIH, 61-62; res assoc physiol, Ohio State Univ; from instr to assoc prof, Bowman Gray Sch Med, 62-73, dir comput ctr, 64-65; RES PHYSIOLOGIST, CLIN STUDIES DIV, ENVIRON PROTECTION AGENCY, UNIV NC, CHAPEL HILL, 73- Concurrent Pos: Partic, LINC eval prog, Bowman Gray Sch Med-Mass Inst Technol, 63-65; sr res investr, NC Heart Asn, 65-75, res comt, 68-69 & 71-; consult NASA, 67-; adj assoc prof biomath & bioeng, Dept Physiol & Surg, Univ NC, Chapel Hill, 73- Mem: Sigma Xi; Am Soc Pharmacol & Exp Therapeut; Asn Comput Mach (secy-treas, 73-77). Res: Physics and mathematics of circulation; mechanical properties of arteries; transmission line characteristics of the arterial system; pressure pulse propagation of the arterial system; computer applications to biomedical research; coronary vascular control indicator; dilution mathematical analysis; time series analysis of cardiovascular events; atherosclerosis research; environmental pathophysiology of cardiopulmonary system. Mailing Add: Clin Studies Div Environ Protection Agency Univ of NC Mason Farm Rd Chapel Hill NC 27514

MALING, GEORGE CROSWELL, JR, b Boston, Mass, Feb 24, 31; m 60; c 2. PHYSICS. Educ: Bowdoin Col, AB, 54; Mass Inst Technol, SB & SM, 54, EE, 58, PhD(physics), 63. Prof Exp: Mem res staff, Mass Inst Technol, 63-65; SR PHYSICIST, IBM CORP, 65- Concurrent Pos: Consult, pvt pract, 59-65 & Indust Acoustics Co, 63-65. Mem: Fel Inst Elec & Electronics Engrs; fel Acoust Soc Am; Inst Noise Control Eng (secy, 71-74, pres, 75). Res: Physical acoustics; large amplitude wave propagation; instabilities in inhomogeneous media; radiation phenomena; noise control. Mailing Add: 62 Timberlane Dr Poughkeepsie NY 12603

MALING, HARRIET MYLANDER, b Baltimore, Md, Oct 2, 19; m 43; c 4. PHARMACOLOGY. Educ: Goucher Col, AB, 40; Radcliffe Col, AM, 41, PhD(med sci physiol), 44. Prof Exp: Asst pharmacol, Harvard Med Sch, 44-45, instr, 45-46; asst prof, Sch Med, George Washington Univ, 51-52, asst res prof, 52-54; PHARMACOLOGIST, NAT HEART & LUNG INST, BETHESDA, 54- Mem: AAAS; Soc Exp Biol & Med; Am Soc Pharmacol; NY Acad Sci. Res: Autonomic and cardiovascular drugs. Mailing Add: 406 N Taylor Ave Annapolis MD 21401

MALING, JOHN E, physics, see 12th edition

MALININ, THEODORE I, b Krasnodar, USSR, Sept 13, 33; US citizen; m 60; c 4. PATHOLOGY, EXPERIMENTAL SURGERY. Educ: Concord Col, BS, 55; Univ Va, MS, 58, MD, 60. Prof Exp: Fel path, Johns Hopkins Univ, 60-61; pathologist, Nat Cancer Inst, 61-64; asst prof path, Sch Med, Georgetown Univ, 64-69; clin assoc prof, 69-70, PROF SURG & PATH, SCH MED, UNIV MIAMI, 70- Concurrent Pos: Guest scientist, Tissue Bank, Naval Med Res Inst, 64-70; ed, Cryobiol, 64-70; consult, Bur Health Manpower, USPHS, 66-67; mem staff, Vet Admin Hosp, Miami, 70- Mem: AAAS; Am Asn Path & Bact; Soc Cryobiol; Am Soc Exp Path; Path Soc Gt Brit & Ireland. Res: Organ perfusion and preservation; experimental myocardial infarction; hemorrhagic shock; bone marrow preservation and transplantation; biological behavior of neoplastic tissue. Mailing Add: Dept of Surg Univ of Miami Sch of Med Miami FL 33152

MALINOW, MANUEL R, b Buenos Aires, Arg, Feb 27, 20; m 52; c 3. CARDIOLOGY. Educ: Univ Buenos Aires, MD, 45. Prof Exp: Res fel cardiovasc med, Michael Reese Hosp, Chicago, 45-46; dir dir res dept & electrocardiologist, Hosp Ramos Mejia, Buenos Aires, 47-57; chief res physiol, Buenos Aires Med Sch, 56-63; PROF MED, MED SCH, UNIV ORE, 64-; CHMN CARDIOVASC DIS, ORE REGIONAL PRIMATE RES CTR, 64- Concurrent Pos: Chief cardiol serv, Munic Inst Radiol & Physiother, Arg, 50-53; chief sect atherosclerosis, Nat Acad Med, Buenos Aires, 60-63, physician, Hosp, 62-63. Honors & Awards: Paul D White Prize, Arg Soc Cardiol, 54 & Gold Medal, 60; Ciba Found Award, 59; Rafael M Bullrich Prize, Nat Acad Med, Buenos Aires, 59; Gold Medal, Inter-Am Cong Cardiol, Brazil, 60; Sesquicentenary Prize, Arg Med Asn; Gold Medal & Malenky Prize, Arg Comt for Weizmann Inst, 60. Mem: AAAS; Arg Biol Soc; Royal Soc Med; Uruguay Soc Cardiol; French Soc Atherosclerosis. Res: Experimental cardiology; atherosclerosis; blood cholesterol; cardiovascular diseases; exercise and electrocardiography. Mailing Add: Ore Regional Primate Res Ctr 505 NW 185th Beaverton OR 97005

MALINOWSKI, EDMUND R, b Mahanoy City, Pa, Oct 16, 32; m 58; c 2. PHYSICAL CHEMISTRY. Educ: Pa State Univ, BS, 54; Stevens Inst Technol, MS, 56, PhD(phys chem), 61. Prof Exp: Res assoc, Nuclear Magnetic Resonance Lab, 60-63, from asst prof to assoc prof, 63-70, PROF CHEM, STEVENS INST TECHNOL, 70- Mem: Am Chem Soc. Res: Nuclear magnetic resonance; dipole moments; conformation of organic molecules; structure of water and electrolyte solutions; applications of factor analysis to chemistry. Mailing Add: Dept of Chem Stevens Inst of Technol Hoboken NJ 07030

MALINOWSKI, HENRY JOHN, b Philadelphia, Pa, Sept 12, 45; m 73. HEALTH SCIENCES. Educ: Philadelphia Col Pharm & Sci, BSc, 68, MSc, 71, PhD(pharmaceut), 73. Prof Exp: ASST PROF PHARM, UNIV MO-KANSAS CITY, 73- Mem: Am Pharmaceut Asn. Res: Pharmaceutical dosage form design, evaluation and testing; design of in vitro testing procedures for solid-dosage forms, including computer controlled systems. Mailing Add: Univ of Mo Sch of Pharm 5005 Rockhill Rd Kansas City MO 64110

MALINS, DONALD CLIVE, b Lima, Peru, May 19, 31; US citizen; m 62; c 3. BIOCHEMISTRY. Educ: Univ Wash, BA, 53; Univ Seattle, BS, 56; Univ Aberdeen, PhD(biochem), 67. Prof Exp: Org chemist, Tech Lab, US Bur Com Fisheries, 56-62, res chemist, 62-66, Food Sci Pioneer Res Lab, 66-71, prog dir, Pioneer Res Unit, Northwest Fisheries Ctr, 71-74, DIR, ENVIRON CONSERV DIV, NORTHWEST FISHERIES CTR, NAT MARINE FISHERIES SERV, 74- Concurrent Pos: Lectr, Univ Wash, 68-74, affil assoc prof, 74-; res prof, Seattle Univ, 71- Honors & Awards: US Dept Interior Achievement Awards, 56-, Dept Com Spec Achievement Award, 74. Mem: AAAS; Am Chem Soc; Am Soc Biol Chem. Res: Synthesis and metabolism of lipids; chemistry of bioacoustics; chemistry of nitrogen compounds; biochemistry of toxic metals and polynuclear hydrocarbons in marine biological systems. Mailing Add: Environ Conserv Div NW Fisheries Ctr NMFS Seattle WA 98102

MALIS, LEONARD I, b Philadelphia, Pa, Nov 23, 19; m; c 2. NEUROSURGERY. Educ: Univ Va, MD, 43. Prof Exp: Attend neurosurgeon, 51-70, NEUROSURGEON IN CHIEF & DIR DEPT, MT SINAI HOSP, 70-; PROF NEUROSURG & CHMN DEPT, MT SINAI SCH MED, 70- Concurrent Pos: Res collabr, Med Dept, Brookhaven Nat Lab, 56-70; consult neurosurgeon, Beth Israel Med Ctr, NY. Mem: Am Acad Neurol Surg; Soc Neurol Surg; Am Asn Neurol Surg; Am Physiol Soc; Cong Neurol Surg. Res: Neurophysiology. Mailing Add: Dept of Neurosurg Mt Sinai Sch of Med New York NY 10029

MALITSON, HARRIET HUTZLER, b Richmond, Va, June 30, 26; m 51; c 2. ASTRONOMY. Educ: Goucher Col, AB, 47; Univ Mich, MS, 51. Prof Exp: Jr physicist, Nat Bur Standards, 47-49 & 51-52 & US Naval Res Lab, 52-57; ASTRONOMER, GODDARD SPACE FLIGHT CTR, NASA, 60- Concurrent Pos: Mem, Comn V, Int Union Radio Sci & Comn 10, Int Astron Union. Mem: Am Astron Soc; fel Royal Astron Soc. Res: Solar physics; solar-terrestrial relationships; radio and space astronomy. Mailing Add: Code 693 NASA Goddard Space Flight Ctr Greenbelt MD 20771

MALITZ, SIDNEY, b Brooklyn, NY, Apr 20, 23; m 45; c 2. PSYCHIATRY, PSYCHOPHARMACOLOGY. Educ: Univ Chicago, MD, 46; Columbia Univ, cert psychoanal med, 59; Am Bd Psychiat & Neurol, dipl, 53. Prof Exp: Resident psychiatrist, NY State Psychiat Inst, 48-51, sr resident psychiatrist, 51-52; asst, 55-57, assoc, 57-59, asst clin prof, 60-65, assoc prof, 65-69, vchmn dept psychiat, 72-75, PROF CLIN PSYCHIAT, COL PHYSICIANS & SURGEONS, COLUMBIA UNIV, 70-, ACTG CHMN DEPT, 75-; ACTG DIR, NY STATE PSYCHIAT INST, 75- Concurrent Pos: Sr res psychiatrist, NY State Psychiat Inst, 54-56, actg chief psychiat res, 56-65, chief dept biol psychiat, 65-72, dep dir, Inst, 72-75; asst vis psychiatrist, Francis Delafield Hosp, 54-; asst attend psychiatrist, Vanderbilt Clin & Presby Hosp, 56-58, assoc attend psychiatrist, 58-71, attend psychiatrist, Presby Hosp, 71-, actg dir psychiat serv, 75; asst examr, Am Bd Psychiat & Neurol, 57- Mem: AAAS; fel Am Psychiat Asn; AMA; Am Psychopath Asn; Asn Res Nerv & Ment Dis. Res: Psychopharmacology and psychoanalysis. Mailing Add: NY State Psychiat Inst 722 W 168th St New York NY 10032

MALKEMUS, JOHN DAVID, b Louisville, Ky, Sept 6, 13; m 39; c 5. ORGANIC CHEMISTRY. Educ: De Paul Univ, BS, 34, MS, 36; Northwestern Univ, PhD(org chem), 39. Prof Exp: Procter & Gamble Co fel, Northwestern Univ, 39-40; res chemist, Colgate-Palmolive-Peet Co, NJ, 40-46; dir, Prods Appln Div, Jefferson Chem Co, Inc, 49-57, res assoc, 57-59; TECH DIR, SPECIALTY CHEMS DIV, REICHHOLD CHEMS, INC, 59-, ASST GEN MGR, 64- Mem: Am Chem Soc; Soc Plastics Engrs. Res: Petroleum derivatives; soaps and detergents; ethylene oxide derivatives; polycarbonates; polyurethanes; telomers; organic peroxides; alkylation; sulfonation; chlorination; polymerization. Mailing Add: 4603 Crestway Dr Austin TX 78731

MALKEVITCH, JOSEPH, b Brooklyn, NY, May 24, 42. MATHEMATICS. Educ: Queens Col, NY, BS, 63; Univ Wis-Madison, MS, 65, PhD(math), 69. Prof Exp: Teaching asst, Univ Wis-Madison, 64-68; asst prof, 68-74, ASSOC PROF MATH, YORK COL, NY, 74- Mem: AAAS; Am Math Soc; Math Asn Am. Res: Convex polytopes; graph theory; arrangements of curves; euclidean geometry. Mailing Add: Dept of Math York Col Jamaica NY 11451

MALKIEL, SAUL, b Boston, Mass, Dec 28, 12; m 45; c 3. IMMUNOCHEMISTRY, ALLERGY. Educ: Clark Univ, AB, 34; Boston Univ, MA, 36, PhD(chem), 42, MD, 44. Prof Exp: Asst med sci, Sch Med, Boston Univ, 35-37, asst chem, 37-44; asst path, Yale Univ, 44-45; vis investr, Rockefeller Inst, 45-48; asst prof med, Med Sch, Northwestern Univ, 48-54; ASSOC PATH, HARVARD MED SCH, 54-, RES ASSOC, SIDNEY FARBER CANCER CTR INC, 63-; ASSOC PROF MED, MED SCH, UNIV MASS, 71- Concurrent Pos: Fel, Sch Med, Boston Univ, 45; Nat Res Coun fel med sci, Rockefeller Inst, 45-46, Am Cancer Soc sr fel, 46-48; lectr, Univ Pa, 48; assoc, Peter Bent Brigham Hosp, 54-63; assoc, Children's Hosp, 63-; mem corp, Marine Biol Lab, Woods Hole. Mem: AAAS; Am Asn Immunologists; Soc Exp Biol & Med; Am Acad Allergy (pres, 70-71); Reticuloendothelial Soc. Res: Immunochemistry of hemocyanins, hemoglobin, viruses and allergic reactions; effects of stress on experimental asthma; clinical and experimental studies with the antihistamines; effect of cortisone on antibody production; fractionation and adjuvant studies on Bordetella pertussis; allergens from ragweed pollen; immuno-competence of the leukemic cell; tumor viruses; immune response to malarial infection. Mailing Add: Sidney Farber Cancer Ctr Inc 35 Binney St Boston MA 02115

MALKIN, HAROLD MARSHALL, b San Francisco, Calif, Oct 9, 23; m 49; c 4. BIOCHEMISTRY. Educ: Univ Calif, AB, 47, MA, 49; Univ Chicago, MD, 51. Prof Exp: Nat Found Infantile Paralysis res fel, Univ Calif, 51-53; Am Cancer Soc fel, Univ Brussels, 53-54; dir, Malkin Med Lab, 54-72, PRES & MED DIR, MALKIN MED LAB/SOLANO LABS, 72-; CLIN INSTR HISTOL, SCH MED, STANFORD UNIV, 56- Concurrent Pos: Summer asst, Univ Calif, 51-; intern path, Sch Med, Stanford Univ, 61-62. Mem: AAAS; Am Chem Soc; Am Asn Clin Chemists; Nat Sci Teachers Asn. Res: Cellular physiology as related to structure; nucleic acid and protein metabolism; cancer; clinical chemistry. Mailing Add: PO Box 905 Palo Alto CA 94302

MALKIN, IRVING, b Cleveland, Ohio, Dec 28, 25; m 47; c 3. INORGANIC CHEMISTRY, ELECTROCHEMISTRY. Educ: Western Reserve Univ, BS, 54; Case Inst Technol, MS, 60. Prof Exp: Chief chemist, I Schumann & Co, 50-56; res & process control inorg chem, Precision Metalsmiths Inc, 56-59; GROUP LEADER INORG CHEM & ELECTROCHEM, DIAMOND SHAMROCK CORP, 59- Mem: Am Chem Soc. Res: Catalytically active electrode surfaces; electrochemical processes; heterogeneous catalysis; corrosion inhibiting metal coatings; controlled release technology. Mailing Add: T R Evans Res Ctr PO Box 348 Diamond Shamrock Corp Painesville OH 44077

MALKIN, LEONARD ISADORE, b New York, NY, Dec 17, 36; m 59; c 2. BIOCHEMISTRY. Educ: NY Univ, AB, 57; Univ Calif, San Francisco, PhD(biochem), 62. Prof Exp: Resident res assoc biochem, Western Regional Res Lab, USDA, Calif, 62-63; NIH fel, Brown Univ, 63-65 & Mass Inst Technol, 65-68; asst prof biochem, Dartmouth Med Sch, 68-73; ASSOC PROF BIOCHEM, SCH MED, WAYNE STATE UNIV, 73- Mem: AAAS; Am Soc Biol Chemists. Res: Protein synthesis in animal cells; production and utilization of messenger RNA, both endogenous and viral, in animal systems. Mailing Add: Dept of Biochem Wayne State Univ Sch of Med Detroit MI 48201

MALKIN, MARTIN F, b Newark, NJ, June 28, 37; m 60; c 1. RESEARCH ADMINISTRATION. Educ: Univ Mich, BS, 59; Brooklyn Col, MA, 65; NY Univ, PhD(biol), 68. Prof Exp: Res assoc biochem, Rockefeller Univ, 67-69; sr res scientist, Merck Inst Therapeut Res, 69-72, res fel, Dept Basic Animal Sci Res, Merck & Co, 72-74, PROJ COORDR, DEPT PROJ PLANNING & MGT, MERCK & CO, 74- Mem: AAAS; Am Inst Biol Sci. Res: Management of development research projects in animal and human health areas. Mailing Add: Merck & Co Inc Dept of Proj Planning & Mgt Rahway NJ 07065

MALKIN, MYRON SAMUEL, physics, see 12th edition

MALKIN, RICHARD, b Chicago, Ill, Mar 25, 40; m 60; c 3. BIOCHEMISTRY. Educ: Antioch Col, BS, 62; Univ Calif, Berkeley, PhD(biochem), 67. Prof Exp: Fel biochem, Univ Gothenburg, 67-69; res biochemist, LECTR CELL PHYSIOL, UNIV CALIF, BERKELEY, 71-, ASSOC BIOCHEMIST, 75- Concurrent Pos: NATO fel, 67-68; Am Cancer Soc fel, Mem: Am Chem Soc; Am Soc Photobiol; Am Soc Plant Physiol; Biophys Soc. Res: Photosynthesis; investigations of electron transport; biological applications of electron paramagnetic resonance spectroscopy. Mailing Add: Dept of Cell Physiol Univ of Calif Berkeley CA 94720

MALKIN, WILLIAM, b New York, NY, Aug 19, 08; m 47; c 4. METEOROLOGY. Educ: Univ Minn, Minneapolis, BS, 31 & 41; NY Univ, MS, 65. Prof Exp: Meteorologist, US Weather Bur, 37-67; DIR RES & DEVELOP, ABBOTT SCI CO, 70- Concurrent Pos: Lectr, Northern Ill Univ, 67-69. Mem: Fel AAAS; Am Meteorol Am Geophys Union. Res: Chemical and medical meteorology; air pollution; hydrometeorology; water and air quality and pollution; environmental quality; aviation meteorology; environmental impact studies and statements for such industries as chemical plants and power generating plants. Mailing Add: Abbott Sci Co 1446 Oakview Dr McLean VA 22101

MALKINSON, FREDERICK DAVID, b Hartford, Conn, Feb 26, 24; m 49; c 3. MEDICINE. Educ: Harvard Univ, DMD, 47, MD, 49. Prof Exp: From instr to assoc prof dermat, Sch Med, Univ Chicago, 54-68; CHMN DEPT DERMAT, RUSH-PRESBY-ST LUKE'S MED CTR, 68-, PROF DERMAT, 71- Concurrent Pos: Assoc prof oral med, Zoller Dent Clin, Univ Chicago, 67-68, res assoc, Sect Dermat, Dept Med, 68-; prof med & dermat, Univ Ill, 68-70; ed, Yearbk of Dermat. Mem: AAAS; Soc Invest Dermat; Am Acad Dermat; Am Dermat Asn; Radiation Res Soc. Res: Percutaneous absorption; adrenal steroid effects on skin; radiation effects on skin. Mailing Add: Presby-St Luke's Hosp 1753 W Congress Pkwy Chicago IL 60612

MALKUS, WILLEM VAN RENSSELAER, b Brooklyn, NY, Nov 19, 23; m 48; c 2. PHYSICS. Educ: Univ Chicago, PhD(physics), 50. Prof Exp: Asst prof natural sci, Univ Chicago, 50-51; phys oceanogr, Woods Hole Oceanog Inst, 51-60; prof geophys, Univ Calif, Los Angeles, 60-67, prof geophys & math, 67-69; PROF APPL MATH, MASS INST TECHNOL, 69- Concurrent Pos: Prof appl math, Mass Inst Technol, 59-60; Guggenheim fel, Cambridge Univ & Univ Stockholm, 71-72. Mem: Nat Acad Sci; Am Phys Soc; fel Am Acad Arts & Sci. Res: Fluid dynamics. Mailing Add: Dept of Math Mass Inst of Technol Cambridge MA 02139

MALLAMS, ALAN KEITH, b Johannesburg, SAfrica, June 11, 40. NATURAL PRODUCTS CHEMISTRY. Educ: Univ Witwatersrand, BSc, 62, PhD(chem), 64; Univ London, PhD(chem), 67. Prof Exp: Res off org chem, African Exlosives & Chem Industs, SAfrica, 64; Exhibit 1851 fel, Queen Mary Col, Univ London, 64-66, res asst, 66-67; sr scientist, Med Res Div, 67-70, prin scientist, 70-72, SECT HEAD, ANTIBIOTICS & ANTIINFECTIVES CHEM RES, SCHERING CORP, 72- Mem: Am Chem Soc; The Chem Soc. Res: Synthesis and structural elucidation of carotenoids, antibiotics, carbohydrates and natural products of medicinal interest; antibiotics research. Mailing Add: Res Div Schering Corp Bloomfield NJ 07003

MALLAMS, JOHN THOMAS, b Ashland, Pa, Aug 29, 23; m 45; c 3. MEDICINE, RADIOLOGY. Educ: Temple Univ, MD, 46; Am Bd Radiol, dipl. Prof Exp: Intern, US Naval Hosp, Philadelphia, 46-47, resident radiol, 47-48, radiologist, Parris Island, SC, 48-49, chief radiol, Beaufort, 49-50; asst prof radiol, Baylor Col Med, 52-54, mem attend staff & dir irradiation ther & tumor clins & dir, Sammons Res Div, Med Ctr, 54-68, prof radiother, Col Dent, 66-68; prof clin radiol, Med Sch, Yale Univ, 68-70; PROF RADIOL & CHMN DEPT, MARTLAND HOSP UNIT, COL MED & DENT NJ, 70-, ASSOC DEAN PATIENT SERV, 71- Concurrent Pos: Fel, Robert Packer Hosp & Guthrie Clin, Sayre, Pa, 50-51; Am Cancer Soc fel clin radiation ther, Frances Delafield Hosp, New York, 51-52; assoc radiologist, Jefferson Davis Hosp, Houston & attend radiologist, Vet Admin Hosp, 52-54; prof, Univ Tex Southwest Med Sch, 66-68; vis prof, Med Sch, Yale Univ, 67-68; Am Col Radiol consult, State of NJ, 70; consult, Conn Hosp Planning Comn, Branford, 71. Mem: Fel Am Col Radiol; AMA; Am Radium Soc; Am Roentgen Ray Soc; Radiol Soc NAm. Res: Radiotherapy. Mailing Add: Martland Hosp Col of Med & Dent of NJ Newark NJ 07107

MALLAN, JEAN M, organic chemistry, see 12th edition

MALLARY, EUGENE COBB, b Macon, Ga, Jan 22, 12; m 47; c 4. NUCLEAR PHYSICS. Educ: Univ Ga, BS, 34, MS, 37; Ohio State Univ, PhD(nuclear physics), 49. Prof Exp: Instr physics, Ga Inst Technol, 37-39; from instr to asst prof physics, US Mil Acad, 44-48; mem sci staff, Los Alamos Sci Lab, 50-52; physicist, Eval Staff, Air War Col, 52-53; US sci adv, Allied Air Force Cent Europe, 54-57; dir aeronaut res labs, Wright-Patterson Air Force Base, Ohio, 58-61; PROF PHYSICS, CALIF STATE UNIV, CHICO, 61- Mem: AAAS; Am Phys Soc. Res: New isotopes and cross-section measurements by cyclotron; criticality measurements with fissionable materials; effects and implication of all nuclear weapons; research direction. Mailing Add: Dept of Physics Calif State Univ Chico CA 95926

MALLAY, JAMES FRANCIS, b Morristown, NJ, Dec 8, 36; m 60; c 2. NUCLEAR PHYSICS. Educ: Lafayette Col, BS, 59; Mass Inst Technol, MS, 61. Prof Exp: Assoc nuclear engr, 61-62, consult, Nuclear Reactor Dynamics, 63-64, sr nuclear engr, 65-69, supvr safety anal, 69-71, mgr licensing, 71-75, MGR LMFBR COMPONENTS, BABCOCK & WILCOX CO, 75- Concurrent Pos: Lectr, Grad Sch, Lynchburg Col, 67-71. Mem: Am Nuclear Soc. Res: Heat conduction; reactor control theory and reactor safety licensing; space-dependent reactor kinetics. Mailing Add: Babcock & Wilcox Co 570 S Hawkins Rd Akron OH 44320

MALLEN, MARIO SALAZAR, b Mexico City, Mex, Apr 30, 13; m 39; c 1. MEDICINE, HISTORY OF MEDICINE. Educ: Nat Univ Mex, MD, 35. Prof Exp: Chief allergy dept, Gen Hosp, Mex, 38-71; PROF HIST & PHILOS OF MED, SCH MED, NAT UNIV MEX, 50- Concurrent Pos: Chief immunol & bact, Inst Cardiol, 43-56; dir, Fed Dist Med Serv, Mex Ministry Health, 54-56; dir, Med Ctr, Mexico City, 54-58; consult, Mex Ministry Health, 59-; hon consult, Gen Hosp, Mex, 71-; pvt pract. Mem: Hon fel Am Col Allergy; Mex Allergy Soc (pres); Mex Dermat Soc (pres); Mex Soc Hist Med (vpres); Mex Acad Med. Res: Immunology applied to clinical medicine; history of science. Mailing Add: Montes Urales 765 Mexico DF Mexico

MALLER, OWEN, b Brooklyn, NY, Jan 27, 30; m 52; c 5. PSYCHOPHYSIOLOGY. Educ: Univ Ill, BS, 52, PhD(psychol), 64. Prof Exp: USPHS fel, 63-65; res assoc, Duke Univ, 65-66 & Vet Admin Hosp, Coatesville, Pa, 66-68; MEM, MONELL CHEM SENSES CTR, UNIV PA & VET ADMIN HOSP, PHILADELPHIA, 68- Concurrent Pos: Res assoc physiol, Med Sch, Univ Pa, 68. Mem: AAAS; Am Psychol Asn; Psychonomics Soc; Sigma Xi. Res: chemoreception and food habits in human and subhuman populations. Mailing Add: Monell Chem Senses Ctr Univ of Pa Philadelphia PA 19104

MALLERY, CHARLES HENRY, b Southampton, NY, June 3, 43; m 66; c 2. DEVELOPMENTAL PHYSIOLOGY, BIOCHEMISTRY. Educ: Univ Ga, BS, 65, PhD(bot, biochem), 70. Prof Exp: Res fel, Lab Quant Biol, 70-72, ASST PROF BIOL, UNIV MIAMI, 72- Concurrent Pos: Chmn biol task force, Fla Dept Educ, 73-; NSF inst grant, 74; consult, Dade County Dept Educ, 74- Mem: Am Soc Plant Physiologists; Scand Soc Plant Physiol; Bot Soc Am; Am Inst Biol Sci. Res: Biochemical development in the establishment of a mature eucaryotic plant cell;

control macromolecule synthesis of DNA, RNA and protein-ion regulating mechanisms; compartmental analysis during cell development. Mailing Add: Dept of Biol Univ of Miami Coral Gables FL 33124

MALLERY, OTTO TOD, b Philadelphia, Pa, Dec 22, 11; m 31, 54; c 6. INDUSTRIAL HEALTH. Educ: Yale Univ, AB, 35; Univ Pa, MD, 39. Prof Exp: From instr to assoc prof internal med, Univ Mich Hosp, 42-56; MED DIR, EMPLOYERS INS WAUSAU, 57-, VPRES, 72- Concurrent Pos: Res asst, Simpson Mem Inst, Univ Mich, 41-44, chief clin labs, 45-51, dir inst indust health, 51-56. Mem: Fel Am Pub Health Asn; Indust Med Asn; AMA; Am Acad Occup Med. Res: Hematology; internal and industrial medicine; occupational health. Mailing Add: Employers Ins of Wausau 2000 Westwood Wausau WI 54401

MALLETT, GORDON EDWARD, b Lafayette, Ind, Nov 30, 27; m 50; c 4. MICROBIOLOGY. Educ: Purdue Univ, BS, 49, MS, 52, PhD(bact), 56. Prof Exp: Bacteriologist, US Army Res & Develop Labs, Md, 56-57; sr microbiologist, Eli Lilly & Co, 57-69, head fermentation prod res dept, 66-69, dir res, Lilly Res Ctr, Ltd, Eng, 69-75, DIR CORP QUAL ASSURANCE, ELI LILLY & CO, IND, 75- Mem: Am Chem Soc; Am Soc MIcrobiol; NY Acad Sci. Res: Bacterial cell structure; microbiological conversions of steroids, alkaloids, antibiotics; microbiological conversion mechanisms. Mailing Add: Eli Lilly & Co 307 E McCarty St Indianapolis IN 46206

MALLETT, RUSSELL LLOYD, b Seattle, Wash, Nov 2, 35; m 61; c 2. APPLIED MECHANICS, DYNAMICS. Educ: Mass Inst Technol, BS, 58, PhD(appl math), 70; Stanford Univ, MS, 66. Prof Exp: Res engr, Boeing Co, 58-65; asst prof appl mech, 70-75, SR RES ASSOC APPL MECH, STANFORD UNIV, 75- Mem: Am Inst Aeronaut & Astronaut; Am Soc Mech Engrs; Soc Indust & Appl Math; Sigma Xi. Res: Rigid body, structural and continuum dynamics, shell theory and nonlinear continuum mechanics; development of efficient modeling in computational techniques; cable dynamics, metal forming analysis, and finite element procedures. Mailing Add: Dept of Mech Eng Div Appl Mech Stanford Univ Stanford CA 94305

MALLETT, WILLIAM ROBERT, b Painesville, Ohio, Sept 12, 32; m 57; c 2. ANALYTICAL CHEMISTRY, PETROLEUM CHEMISTRY. Educ: Miami Univ, BA, 61, MS, 63; Rensselaer Polytech Inst, PhD(energy transfer), 66. Prof Exp: Res chemist, 66-68, SR RES CHEMIST, UNION OIL CO CALIF, 68- Concurrent Pos: Vis res chemist, Maruzen Oil Co, Japan, 73-75. Res: Fuels research; gasoline, especially development of new gasoline additives for tomorrow's high performance internal combustion engines and methods for optimizing gasoline volatility to insure proper engine operation in all climates and seasons. Mailing Add: Union Oil Res Ctr PO Box 76 Brea CA 92621

MALLETTE, JOHN M, b Houston, Tex, Aug 6, 32; m 59; c 3. ENDOCRINOLOGY, EXPERIMENTAL EMBRYOLOGY. Educ: Xavier Univ, BS, 54; Tex Southern Univ, MS, 58; Pa State Univ, PhD(zool), 62. Prof Exp: Res technician anat, Dent Br, Univ Tex, 57-58; instr biol, Tex Southern Univ, 58-59; asst zool, Pa State Univ, 59-62; assoc prof biol, 62-64, PROF BIOL SCI & CHMN GRAD CURRIC, TENN STATE UNIV, 64- Concurrent Pos: Dir undergrad res participation prog, NSF, 64-66; grants assoc, NIH, 67; dir allied health, Meharry Med Col & Tenn State Univ. Honors & Awards: Knight of St Gregory, 71. Mem: AAAS; Am Soc Zool; Nat Inst Sci. Res: Growth of trypsin; dissociated glands in vitro and vivo; teratogenic effects of drugs in avian embryos. Mailing Add: Ctr of Allied Health Tenn State Univ Nashville TN 37203

MALLETTE, MANNEY FRANK, b Leon, Iowa, May 28, 17; m 42; c 2. BIOCHEMISTRY. Educ: Iowa State Col, BS, 40; Columbia Univ, PhD(chem), 45. Prof Exp: Fel, Cornell Univ, 45-47; instr chem, Univ Wyo, 47-48; from asst prof to assoc prof biochem, Sch Hyg & Pub Health, Johns Hopkins Univ, 48-55; assoc prof, 55-60, PROF BIOCHEM, PA STATE UNIV, 60-, ACTG HEAD DEPT BIOCHEM & BIOPHYS, 75- Concurrent Pos: Vis prof, Univ BC, 63 & Univ Hull, 70; actg head dept biochem, Pa State Univ, 73-75. Mem: Fel AAAS; Am Soc Microbiol; Am Chem Soc; Am Soc Biol Chem. Res: Enzyme synthesis; natural products; sulfur metabolism; immunochemistry. Mailing Add: Dept of Biochem & Biophys Pa State Univ University Park PA 16802

MALLEY, ARTHUR, b Chicago, Ill, Jan 7, 31; m 61. IMMUNOLOGY. Educ: San Francisco State Col, BA, 53, BS, 57; Ore State Univ, PhD(biochem), 61. Prof Exp: Fel immunochem, Calif Inst Technol, 61-63; from asst prof to assoc prof, 63-71, PROF BACT, ORE REGIONAL PRIMATE RES CTR, 71- Mem: AAAS; Am Chem Soc; Transplantation Soc; Am Asn Immunologists. Res: Isolation of antigens and antibodies involved in various allergic diseases, and the regulation of antibody formation. Mailing Add: Ore Regional Primate Res Ctr 505 NW 185th Ave Beaverton OR 97005

MALLI, GULZARI LAL, b Lehlian, India, Feb 12, 38. QUANTUM CHEMISTRY, CHEMICAL PHYSICS. Educ: Univ Delhi, BSc, 58; McMaster Univ, MSc, 60; Univ Chicago, MS, 63, PhD(chem physics), 64. Prof Exp: Mem res staff physics, Yale Univ, 64-65; asst prof theoret chem, Univ Alta, 65-66; asst prof, 66-69, ASSOC PROF CHEM, SIMON FRASER UNIV, 69- Concurrent Pos: Vis fel, Mellon Inst, 67-68. Mem: Am Phys Soc. Res: Quantum mechanics of atoms and molecules; tensor operators in atomic and molecular spectroscopy; relativistic many-electron atomic and molecular self-consistent field theory; electron correlation in many-electron systems. Mailing Add: Dept of Chem Simon Fraser Univ Burnaby BC Can

MALLICK, GEORGE THEODORE, JR, solid state physics, see 12th edition

MALLIN, MORTON LEWIS, b Feb 13, 26; m 60; c 2. MICROBIOLOGY, BIOCHEMISTRY. Educ: Phila Col Pharm, BS, 50; Hahnemann Med Col, MS, 52; Cornell Univ, PhD(bact), 56. Prof Exp: Res fel biochem, McCollum-Pratt Inst, Johns Hopkins Univ, 56-57; NIH res fel, Brandeis Univ, 57-59; res assoc, May Inst Med Res, Jewish Hosp, Cincinnati, 59-65; assoc prof, 65-71, PROF MICROBIOL & CHMN DEPT, OHIO NORTHERN UNIV, 71- Mem: Am Soc Microbiol. Res: Biochemical problems related to hypertension; oxidative phosphorylation in animal mitochondria and bacterial particles; phosphorous metabolism in anaerobic bacteria. Mailing Add: Dept of Microbiol Ohio Northern Univ Ada OH 45810

MALLING, HEINRICH VALDEMAR, b Copenhagen, Denmark, Apr 21, 31; m 55, 68; c 7. GENETICS, MICROBIOLOGY. Educ: Univ Copenhagen, BSc, 51, MSc, 53, PhD(genetics), 57. Prof Exp: Lectr genetics, Univ Copenhagen, 57; res staff mem, Leo Pharmaceut Prod, Denmark, 53-58; fel genetics, Univ Copenhagen, 58-61, lectr, 61-63; mem res staff, Biol Div, Oak Ridge Nat Lab, 63-72; MEM MUTAGENESIS BR, NAT INST ENVIRON HEALTH SCI, 72- Mem: Genetics Soc Am; Danish Microbiol Soc; Danish Biol Soc; Danish Biochem Soc; Scand Soc Cell Biol. Res: Botany; cytology; mutation induction; enzyme production by microorganisms; induction of cancer; antincoplastic activity of chemicals. Mailing Add: Mutagenic Br Nat Inst of Environ Health Sci Research Triangle Park NC 27709

MALLINSON, GEORGE GREISEN, b Troy, NY, July 4, 18; m 43, 54; c 4. SCIENCE ADMINISTRATION, SCIENCE EDUCATION. Educ: NY State Col Teachers Albany, BA, 38, MA, 41; Univ Mich, PhD(sci educ, statist), 47. Prof Exp: Teacher high schs, NY, 37-42; dir sci educ, Iowa State Teachers Col, 47-48; from assoc prof to prof exp psychol & statist, 48-53, PROF SCI EDUC & RES METHODOLOGY, WESTERN MICH UNIV, 54-, DEAN GRAD STUDIES, 56- Concurrent Pos: Burke Aaron Hinsdale scholar, Univ Mich, 47-48; actg dir, Grad Div, Western Mich Univ, 54-55, dir grad studies, 55-56; dir, NSF summer insts, 58-, in-serv insts, 59-, sec sci training progs, 59-; Nat Defense Educ Act grant, 63-66; ed, Sch Sci & Math, 57-; mem, Coun & Coop Comt, Teaching Sci & Math, AAAS, 49-; chmn coun, Cent States Univs, Inc, 65-66, pres, Bd Dirs, 70-73. Mem: AAAS; Nat Asn Res Sci Teaching (pres, 53-54); Nat Sci Teachers Asn. Res: Scientific manpower; factors related to achievement in science; basic hydrothermal methods for amelioration of taconite ore. Mailing Add: Off of the Dean Grad Col Western Mich Univ Kalamazoo MI 49001

MALLIOS, WILLIAM STEVE, statistics, see 12th edition

MALLIS, ARNOLD, b New York, NY, Oct 15, 10; m 43; c 2. ENTOMOLOGY. Educ: Univ Calif, BS, 34, MS, 39. Prof Exp: Field aide entom, Bur Entom & Plant Quarantine, USDA, 38; entomologist, Univ Calif, Los Angeles, 39-42; asst entomologist malaria control, USPHS, 43; res fel, Hercules Powder Co, Del, 44; entomologist, Gulf Res & Develop Co, 45-68; EXTEN ENTOMOLOGIST, PA STATE UNIV, 68- Mem: AAAS; Entom Soc Am; Am Mosquito Control Asn. Res: Household and livestock insects and insecticides; taxonomy and control of ants; history of entomology. Mailing Add: Exten Entomology Pa State Univ University Park PA 16802

MALLISON, GEORGE FRANKLIN, b Suffolk Co, NY, May 31, 28; m 53, 69; c 3. ENVIRONMENTAL HEALTH, PUBLIC HEALTH. Educ: Cornell Univ, BCE, 51; Univ Calif, MPH, 57. Prof Exp: Asst sanit engr, Cornell Univ, 51; from jr asst to asst sanit engr & asst to chief tech develop labs, Tech Br, USPHS, 51-53, sr asst sanit engr, Phoenix Field Sta, 54-57, sanit engr & asst to chief, Tech Br, 57-59, asst chief & sr sanit engr, 59-63, chief microbiol control sect, Ctr Dis Control, 63-74, SANIT ENGR DIR, CTR DIS CONTROL, USPHS, 66-, ASST DIR BACT DIS DIV, 74- Concurrent Pos: Mem, Conf Fed Environ Engrs; consult environ epidemiol, US Environ Protection Agency, 71- Mem: Water Pollution Control Fedn; Am Water Works Asn; Am Pub Health Asn; Am Soc Microbiol; Asn Practr Infection Control. Res: Control of nosocomial infections; hospital and environmental sanitation; water supply; epidemiologic aspects and control of environmental contamination; sanitary engineering; microbiology; solid and other waste disposal; vector control. Mailing Add: Ctr for Dis Control US Pub Health Serv Atlanta GA 30333

MALLMANN, VIRGINIA H, b Letts, Iowa, Feb 2, 18; m 45, 57. MICROBIOLOGY. Educ: Mich State Univ, BS, 54, MS, 55, PhD(microbiol), 60. Prof Exp: Instr, 60-64, asst prof, 64-69, ASSOC PROF VETERINARY IMMUNOL, MICH STATE UNIV, 69- Concurrent Pos: Partic & res contract support, Animal & Plant Health Inspection Serv & Agr Res Serv, USDA, 60-75; res grants, Am Thoracic Soc-Nat Tuberc Asn, 65-67; Mich Tuberc & Respiratory Dis Asn, 62-71. Mem: Am Soc Microbiol; fel Am Pub Health Asn; Am Thoracic Soc; NY Acad Sci. Res: Immunobiology; cellular immunity; allergenicity and pathogenicity of mycobacteria. Mailing Add: Dept of Microbiol Mich State Univ East Lansing MI 48823

MALLONEE, JAMES EDGAR, b Frederick, Md, June 20, 15; m 43; c 2. INDUSTRIAL ORGANIC CHEMISTRY. Educ: Col William & Mary, BS, 35; Univ Va, PhD(org chem), 40. Prof Exp: Control chemist, Solvay Process Co, Va, 36-37; res chemist, Jackson Lab, 40-43, chemist, Louisville Works, 43-47, res chemist, Fine Chem Div, 47-52, PROCESS SUPVR, CHAMBERS WORKS, E I DU PONT DE NEMOURS & CO, 52- Mem: Am Chem Soc. Res: Morphine chemistry; neoprene intermediates; industrial organic chemical research and development. Mailing Add: 45 Shellburne Dr Wilmington DE 19803

MALLORY, BOB FRANKLIN, b Blackwell, Okla, June 11, 32; m 61; c 4. GEOLOGY, PALEONTOLOGY. Educ: Wichita State Univ, BA, 61; Univ Mo-Columbia, PhD(geol), 68. Prof Exp: Asst prof, 68-70, ASSOC PROF EARTH SCI, NORTHWEST MO STATE COL, 70- Mem: Geol Soc Am; Soc Econ Paleont & Mineral. Res: Environmental geology; paleoecology. Mailing Add: Dept of Earth Sci Northwest Mo State Col Maryville MO 64468

MALLORY, CLELIA WOOD, b Brooklyn, NY, Feb 9, 38; m 65. ORGANIC CHEMISTRY. Educ: Bryn Mawr Col, BA, 59, MA, 60, PhD(chem), 63. Prof Exp: RES ASSOC CHEM, BRYN MAWR COL, 63- Mem: Am Chem Soc. Res: Photochemistry; nuclear magnetic resonance spectroscopy. Mailing Add: Dept of Chem Bryn Mawr Col Bryn Mawr PA 19010

MALLORY, FRANK BRYANT, b Omaha, Nebr, Mar 17, 33; m 51, 65; c 4. ORGANIC CHEMISTRY. Educ: Yale Univ, BS, 54; Calif Inst Technol, PhD(chem), 58. Prof Exp: From asst prof to assoc prof, 57-69, PROF CHEM, BRYN MAWR COL, 69- Concurrent Pos: Guggenheim fel, 63-64; Sloan res fel, 64-68; NSF sr fel, 70-71; vis assoc, Calif Inst Technol, 63-64; vis prof, State Univ NY, Albany, 67 & Yale Univ, 68; vis fel, Cornell Univ, 70-71. Mem: Am Chem Soc. Res: Preparative and mechanistic aspects of organic photochemistry; nuclear magnetic resonance spectroscopy; biosynthetic transformations in protozoans. Mailing Add: Dept of Chem Bryn Mawr Col Bryn Mawr PA 19010

MALLORY, HERBERT DEAN, b Wayland, Iowa, Oct 3, 23; div; c 3. PHYSICAL CHEMISTRY. Educ: Univ Iowa, BS, 46, PhD(phys chem, Raman spectra), 50. Prof Exp: Res chemist, US Naval Ord Lab, Md, 49-56; mem staff, Sandia Corp, NMex, 56-59; head, Explosives Detonation Br, 59-69, HEAD, DETONATION MECHANISMS BR, NAVAL WEAPONS CTR, CHINA LAKE, 69- Mem: Sigma Xi; Am Chem Soc; Combustion Inst. Res: Shock and detonation phenomena; explosive devices. Mailing Add: 125 Desert Candles Dr Ridgecrest CA 93555

MALLORY, KENNETH BRANDT, b London, Eng, Apr 27, 26; US citizen; m 49; c 4. APPLIED PHYSICS. Educ: Harvard Univ, AB, 45; Stanford Univ, MS, 50, PhD(physics), 55. Prof Exp: Res assoc microwaves, Stanford Univ, 50-59, staff mem, Linear Accelerator Ctr, 60-65, GROUP LEADER INSTRUMENTATION & CONTROL, LINEAR ACCELERATOR CTR, STANFORD UNIV, 65- Mem: AAAS. Res: Microwave measurements; design of linear accelerators; on-line computer control. Mailing Add: Stanford Linear Accelerator Ctr Stanford CA 94305

MALLORY, THOMAS E, b Alhambra, Calif, July 2, 40. PLANT MORPHOLOGY, PLANT CYTOLOGY. Educ: Univ Redlands, BS, 62; Univ Calif, Davis, MS, 65, PhD, 68. Prof Exp: Asst prof, 68-72, ASSOC PROF BIOL, CALIF STATE UNIV, FRESNO, 72- Mem: Bot Soc Am. Res: Plant morphogenesis; morphogenesis of lateral roots; mechanisms of action of herbicides. Mailing Add: Dept of Biol Calif State Univ Fresno CA 93710

MALLORY, VIRGIL STANDISH, b Englewood, NJ, July 14, 19; m 46; c 4. GEOLOGY. Educ: Oberlin Col, AB, 46; Univ Calif, MA, 48, PhD(paleont), 52. Prof Exp: Lectr paleont, Univ Calif, 50-51, consult, Mus Paleont, 52; from asst prof to assoc prof, 53-61, PROF GEOL, UNIV WASH, 61-, CUR INVERT PALEONT, THOMAS BURKE WASH STATE MUS, 62- Concurrent Pos: Consult, 54-; mem, Gov Comn Petrol Laws, Wash, 56-57; chmn, Geol & Paleont Div, Thomas Burke Wash State Mus, 61-62; geologist, US Geol Surv, 63-; ed invert paleont, Quaternary Res, 70-; mem, Abstr Rev Comt, Geol Soc Am, 75-76 & Minerals Mus Adv Comt, 75. Mem: Fel AAAS; fel Geol Soc Am; Int Paleont Union; Soc Econ Paleont & Mineral; Am Asn Petrol Geol. Res: Biostratigraphy of west coast ranges, especially Lower Tertiary; west coast Lower Tertiary foraminifera; Lower Tertiary and molluscan paleontology; Pacific coast structural geology. Mailing Add: Paleont Div Burke Wash State Mus Univ of Wash Seattle WA 98105

MALLORY, WILLAM R, b Dudley, Mo. THEORETICAL PHYSICS, OPTICS. Educ: Univ Ill, Urbana-Champaign, BS, 59; Syracue Univ, PhD(physics), 70. Prof Exp: Physicist, Gen Elec Co, 59-67; ASST PROF PHYSICS & ASTRON, UNIV MONT, 70- Mem: Am Phys Soc; Am Asn Physics Teachers. Res: Theoretical optics; quantum and statistical optics. Mailing Add: Dept of Physics & Astron Univ of Mont Missoula MT 59801

MALLORY, WILLIAM WYMAN, b New Rochelle, NY, Apr 19, 17; m 42; c 2. GEOLOGY. Educ: Columbia Univ, BA, 39, MA, 46, PhD(geol), 48. Prof Exp: Asst geol, Columbia Univ, 40-41; asst geologist, Phillips Petrol Co, Okla, 43-45, explor geologist, 46-55, supvr explor projs, 55-57; petrol consult, Ball Assocs, 58; mem, Paleotectonic Map Proj, 59-75, MEM SURV OIL & GAS RESOURCES OF US, US GEOL SURV, 75- Concurrent Pos: Ed-in-chief, Geol Atlas Rocky Mountains Region, Rocky Mountain Asn Geologist, 72. Mem: Fel Geol Soc Am; Am Asn Petrol Geologists; Sigma Xi. Res: Stratigraphy and tectonics of Colorado and western United States; continental framework and petroleum exploration in western United States. Mailing Add: US Geol Surv Fed Ctr Denver CO 80225

MALLOTT, I FLOYD, b Chicago, Ill, Oct 20, 22; m 50; c 4. PSYCHIATRY. Educ: Manchester Col, AB, 44; Loyola Univ, Ill, MD, 51. Prof Exp: Fel psychiat, 52-55, from asst instr to asst prof, 55-65, CLIN ASSOC PROF PSYCHIAT, SCH MED, UNIV PITTSBURGH, 65-; SR PSYCHIATRIST, OFF REGIONAL PROGRAMMING, WESTERN PSYCHIAT INST & CLIN, 73- Concurrent Pos: Asst chief outpatient serv, Western Psychiat Inst & Clin, 55-57, chief outpatient serv, 57-72, coordr educ, 72-73; consult, Magee-Woman's Hosp, Pittsburgh, 63- & Presby-Univ Hosp, 63-. Mem: AAAS; assoc mem Am Psychoanal Asn; AMA; Asn Am Med Cols; Am Psychiat Asn. Res: Clinical psychiatry and psychiatric education. Mailing Add: 3700 Fifth Ave Pittsburgh PA 15213

MALLOV, SAMUEL, b New York, NY, Apr 19, 19; m 43; c 2. PHARMACOLOGY. Educ: City Col New York, BS, 39; NY Univ, MS, 41; Syracuse Univ, PhD(biochem), 52. Prof Exp: Chemist, Labs, Westinghouse Elec Co, 43-44; org chemist, US Bur Mines, 44-45; res chemist, Coal Lab, Carnegie Inst Technol, 45-48; asst, 48-52, from instr to assoc prof, 53-70, PROF PHARMACOL, COL MED, STATE UNIV NY UPSTATE MED CTR, 71- Concurrent Pos: Am Heart Asn fel, State Univ NY Upstate Med Ctr, 52-53. Mem: AAAS; Am Physiol Soc; Am Soc Pharmacol & Exp Therapeut. Res: Cardiovascular, alcohol, lysosomes. Mailing Add: 210 DeForest Rd Syracuse NY 13214

MALLOW, JEFFRY VICTOR, b New York, NY, June 28, 43; m 70. ATOMIC PHYSICS. Educ: Columbia Univ, AB, 64; Northwestern Univ, MS, 66, PhD(physics, astron), 70. Prof Exp: Fel physics, Hebrew Univ Jerusalem, 70-71; res assoc, Northwestern Univ, 71-74; ASST PROF PHYSICS, OAKLAND UNIV, 74- Concurrent Pos: Res consult, Northwestern Univ, 74- Mem: AAAS

MALLOWS, COLIN LINGWOOD, b Great Sampford, Eng, Sept 10, 30; m 56; c 3. MATHEMATICAL STATISTICS. Educ: Univ London, BSc, 51, PhD(statist), 53. Prof Exp: Asst lectr statist, Univ Col London, 55-57, lectr, 58-60; res assoc statist tech res group, Princeton Univ, 57-58; mem tech staff, 60-69, DEPT HEAD, BELL LABS, 69- Concurrent Pos: Adj assoc prof, Columbia Univ, 60-64. Mem: Fel Inst Math Statist; fel Am Statist Asn; Math Asn Am; Bernoulli Soc; Royal Statist Soc. Res: Data analysis, especially informal and graphical methods; algebraic coding theory. Mailing Add: Bell Labs Inc Murray Hill NJ 07974

MALLOY, ALFRED MARCUS, b Pittsburgh, Pa, Oct 6, 03; m 43; c 1. PHYSICAL CHEMISTRY, ELECTROCHEMISTRY. Educ: Carnegie Inst Technol, BS, 25; Mich State Univ, MS, 27. Prof Exp: Asst electrochem, Mich State Univ, 25-27; foundry chemist, Cadillac Motorcar Co, Mich, 27-29; res chemist, Behr-Manning Corp, NY, 30-35; metall observer, Carnegie-Ill Steel Corp, Pa, 36-37; chemist, Bur Tests, Allegheny County, 37-40; chem engr, Bur Aeronaut, DC, 40-59; head mat protection sect, Bur Naval Weapons, 59-72, PROJ OFFICER, NAVAL AIR SYSTS COMMAND, 65-, MAT BR ENGR, 72- Concurrent Pos: Mem, Working Group Aircraft Camouflage, NATO, France, 52. Mem: Nat Asn Corrosion Engrs; fel Am Inst Chemists. Res: Environmental behavior of aircraft and missile material on land, sea and air; corrosion; thin free films; organic coatings; camouflage and visibility; adhesion; surface effects; biomedical techniques; air and water pollution control. Mailing Add: AIR-52033 Naval Air Systs Command Rm 1000 Jefferson Plaza Bldg 2 Washington DC 20360

MALLOY, THOMAS BERNARD, JR, b El Campo, Tex, Aug 20, 41; m 68; c 2. MOLECULAR SPECTROSCOPY. Educ: Univ St Thomas, Tex, BA, 64; Tex A&M Univ, PhD(chem), 70. Prof Exp: Res assoc chem, Mass Inst Technol, 70-71; asst prof, 71-74, ASSOC PROF PHYSICS & CHEM, MISS STATE UNIV, 74- Mem: Am Phys Soc; Am Chem Soc. Res: Spectroscopy of molecules with low-frequency, large-amplitude vibrational modes; microwave spectroscopy; far-infrared spectroscopy; laser Raman spectroscopy. Mailing Add: Dept of Physics Miss State Univ Mississippi State MS 39762

MALLOY, THOMAS PATRICK, b Chicago, Ill, July 28, 41; m 64; c 2. ORGANIC CHEMISTRY. Educ: Ill Inst Technol, BS, 65; Loyola Univ, PhD(org chem), 70. Prof Exp: Sr res scientist, De Soto, Inc, 70-71; consult org chem, Bernard Wolnak & Assocs, 71-74; RES CHEMIST, UNIVERSAL OIL PROD, 74- Mem: Am Chem Soc; Inst Food Technologists. Res: Physical organic, mainly molecular, rearrangements; mechanisms of reaction and organic synthesis. Mailing Add: 605 Burr Oak Dr Lake Finch IL 60047

MALLOZZI, PHILIP JAMES, b Norwalk, Conn, Feb 12, 37; m 61; c 1. PHYSICS. Educ: Harvard Univ, BA, 60; Yale Univ, MS, 62, PhD(physics), 64. Prof Exp: Instr physics, Yale Univ, 64-66; mem tech staff, 66-70, DIR LASER APPLNS CTR, COLUMBUS LABS, BATTELLE MEM INST, 70- Mem: Am Phys Soc. Res: Plasma physics; laser generated plasmas; laser physics; astrophysics. Mailing Add: Battelle Mem Inst Columbus OH 43201

MALM, DONALD E G, b Tallant, Okla, June 3, 30; m 57; c 1. TOPOLOGY. Educ: Northwestern Univ, BS, 52; Brown Univ, AM, 54, PhD(math), 59. Prof Exp: Instr math, Rutgers Univ, 57-59; vis lectr, Royal Holloway Col, Univ London, 59-60; asst prof, State Univ NY Stony Brook, 60-62; asst prof, 62-65, ASSOC PROF MATH, OAKLAND UNIV, 65- Mem: AAAS; Am Math Soc; Math Asn Am. Res: Algebraic topology. Mailing Add: Dept of Math Oakland Univ Rochester MI 48063

MALM, NORMAN R, b Boulder, Colo, June 9, 31; m 55; c 3. PLANT BREEDING, AGRONOMY. Educ: Colo State Univ, BS, 54; Univ Ill, MS, 56, PhD(agron), 60. Prof Exp: Agronomist, NMex State Univ, 61-68 & Univ Nebr, Lincoln, 68-69; agronomist, 69-72, ASSOC PROF AGRON & COTTON BREEDER, NMEX STATE UNIV, 72- Mem: Am Soc Agron; Crop Sci Soc Am. Res: Cotton breeding research for high quality fiber, disease resistance and insect resistance. Mailing Add: Dept of Agron NMex State Univ Las Cruces NM 88003

MALMBERG, EARL WINTON, b Bernadotte Twp, Minn, Oct 18, 19; m 49; c 5. CHEMISTRY. Educ: Gustavus Adolphus Col, BA, 40; Univ Nebr, MA, 42, PhD(org chem), 47. Prof Exp: Res chemist, Nat Defense Res Comt, Calif Inst Technol, 43-46 & Am Cyanamid Co, 47-48; Du Pont fel, Univ Minn, 48-49; from asst prof to assoc prof chem, Ohio State Univ, 49-59; res chemist, 59-69, sr res scientist, 69-74, CHIEF SCIENTIST CHEM, SUN OIL CO, 74- Mem: Am Chem Soc; Soc Petrol Engrs. Res: Electrochemistry; mechanism of oxidation of organic compounds; chromatography; cis-trans isomerizations; chemical reactions in shock tubes; Athabasca tar sands; petroleum recovery; petroleum source bed geochemistry. Mailing Add: Sun Oil Co 503 N Central Expressway Richardson TX 75080

MALMBERG, JOHN HOLMES, b Gettysburg, Pa, July 5, 27; m 52; c 2. PLASMA PHYSICS. Educ: Ill State Univ, BS, 49; Univ Ill, MS, 51, PhD(physics), 57. Prof Exp: Mem staff plasma physics, Gen Atomic Div, Gen Dynamics Corp, 57-69; PROF PHYSICS, UNIV CALIF, SAN DIEGO, 67- Mem: AAAS; Am Phys Soc. Res: Experimental plasma physics; fundamental properties of waves; controlled thermonuclear research; development of seismic prospecting systems; physics of elementary particles. Mailing Add: Dept of Physics Univ of Calif at San Diego La Jolla CA 92037

MALMBERG, MARJORIE SCHOOLEY, b Estherville, Iowa, Aug 20, 21; m 45; c 5. PHYSICAL CHEMISTRY. Educ: Wellesley Col, AB, 42; Univ Md, College Park, PhD(phys chem), 67. Prof Exp: Jr chemist, Nat Bur Standards, 42-45, chemist, 48-55; US Army grant & res assoc, Univ Md, College Park, 67-70; Nat Inst Gen Med Sci-Nat Inst Arthritis & Metab Dis spec res fel nuclear magnetic resonance, Nat Bur Standards, 70-72, guest worker, 72-74; ENG ANALYST, NUS CORP, 74- Mem: Am Chem Soc; Health Phys Soc. Res: Molecular microdynamics in fluids; structure of liquids; relaxation phenomena in biological molecules and polymers; infrared, raman and nuclear magnetic resonance spectroscopy; light scattering; radiological environmental monitoring.

MALMBERG, PAUL ROVELSTAD, b New Haven, Conn, Apr 15, 23; m 44; c 5. SOLID STATE ELECTRONICS. Educ: Thiel Col, BS, 44. Prof Exp: Fel life preservers, Mellon Inst, 46-48; instr physics, Univ Pittsburgh, 50-51; physicist, Westinghouse Res Labs, 51-60; Int Atomic Energy Agency UN vis prof, Tsing Hua Univ, Taiwan, 60-61; mgr sci instrumentation, 61-66, fel engr, 66-68, mgr advan circuit fabrication technol, 68-70, mgr electron beam fabrication technol, 70-72, mgr electron imaging technol, Elec Sci Div, 72-75, FEL ENGR THIN FILM DEVICES, WESTINGHOUSE RES LABS, 75- Mem: Am Phys Soc; Inst Elec & Electronic Engrs; Electrochem Soc; Soc Info Display; Fedn Am Scientists. Res: Advanced integrated circuits and solid state devices made by electron and ion beam techniques; thin film devices and systems for information processing and display and for signal transduction. Mailing Add: Thin Film Devices Westinghouse Res Labs Pittsburgh PA 15235

MALMBERG, PHILIP RAY, b Norwood, Mass, Oct 13, 20; m 56; c 3. NUCLEAR PHYSICS, SOLID STATE PHYSICS. Educ: Ill State Norm Univ, BEd, 40; Univ Iowa, MS, 44, PhD(physics), 55. Prof Exp: Jr engr, Res Lab, Sylvania Elec Prod, Inc, 44-46; res assoc nuclear physics, Univ Iowa, 50-53; PHYSICIST, NAVAL RES LAB, 55- Mem: Sigma Xi; Am Phys Soc. Res: Radiation damage by charged particles; ion implantation; materials analysis; development of specialized equipment. Mailing Add: 6818 Farmer Dr Oxon Hill MD 20022

MALMGREN, RICHARD AXEL, b St Paul, Minn, Dec 31, 21; m 46; c 2. PATHOLOGY. Educ: Wagner Col, BS, 42; Cornell Univ, MD, 45; Am Bd Path, dipl, 57. Prof Exp: Intern, Grasslands Hosp, 45-46; head serol unit, Biol Sect, Nat Cancer Inst, 48-53, med officer chg cancer invest unit, Tenn, 53-56, head cytopath serv, 56-72; HEAD CYTOPATH SERV, DEPT PATH, MED CTR, GEORGE WASHINGTON UNIV, 74- Mem: Am Soc Cytol; AMA. Res: Cancer immunology; cytology; pathologic physiology of cancer. Mailing Add: 1686 N Harbor Ct Annapolis MD 21401

MALMON, ARTHUR GERALD, b Milwaukee, Wis, Oct 29, 29; m 58; c 2. PHYSICS. Educ: Univ Wis, BS, 51, MS, 52, PhD(physics), 57. Prof Exp: Asst physics, Univ Wis, 53-54 & 55-57; res fel, NIMH, 57-58; res assoc biol, Mass Inst Technol, 58; res physicist, Opers Res, Inc, 58-61; res physicist, Nat Inst Allergy & Infectious Dis, 61-67; INDEPENDENT CONSULT, 67- Mem: Inst Elec & Electronics Eng; Am Phys Soc. Res: Small angle x-ray scattering; biomolecular structure; operations research; charged particle effects; automated microbiology; image analysis. Mailing Add: PO Box 30211 Bethesda MD 20014

MALMQUIST, CARL PHILLIP, b St Paul, Minn, Mar 10, 31; m 56; c 2. PSYCHIATRY. Educ: Univ Minn, BA, 52, MD, 58, MS, 61. Prof Exp: Resident psychiat & child psychiat, Univ Minn & Columbia Med Ctr, 58-63; training dir, Hennepin County Gen Hosp, Minneapolis, 63-65; assoc prof psychiat, Med Sch, Univ Mich, 65-66; assoc prof, Inst Child Develop, Univ Minn, Minneapolis, 66-70, prof child psychiat & dir dept, Univ, 70-71, PROF LAW & CRIMINAL JUSTICE, UNIV MINN, MINNEAPOLIS, 71- Concurrent Pos: Psychiat consult, Minneapolis Dist Ct, 66-; mem task force on right to treatment, Am Psychiat Asn, 72-73; chmn comt on psychiat & law, Group for the advan of Pschiat, 74- Mem: Am Psychiat Asn; Am Orthopsychiat Asn; Acad Child Psychiat; Am Col Psychiat. Res: Depression; childhood psychopathology; psychiatry and law; depression and acting out behaviors. Mailing Add: Dept of Criminal Justice Univ of Minn 6600 France Ave S Minneapolis MN 55435

MALMQUIST, WINSTON, virology, see 12th edition

MALMSTADT, HOWARD VINCENT, b Marinette, Wis, Feb 17, 22; m 47; c 3. CHEMISTRY. Educ: Univ Wis, BS, 43, MS, 48, PhD(chem), 50. Prof Exp: Res assoc chem, Univ Wis, 50-51; from instr to asst prof chem, 51-57, assoc prof anal chem, 57-61, PROF CHEM, UNIV ILL, URBANA, 61- Concurrent Pos: Guggenheim fel, 60; mem adv bd, NIH, 72-76. Honors & Awards: Chem Instrumentation Award, Am Chem Soc, 63; Educ Award, Instrument Soc Am, 70; Analytical Chemistry Fischer Award, Am Chem Soc, 76. Mem: Am Chem Soc; Soc Appl Spectros; Instrument Soc

Am; Optical Soc Am. Res: Clinical/analytical methodology; short-time phenomena in sparks, laser plumes and flames; applied spectroscopy; spectrochemical methods; automatic titrations; atomic and molecular absorption spectrometry; time-resolved spectroscopy; automation; reaction-rate methods. Mailing Add: Dept of Chem & Chem Eng Univ of Ill 47 Roger Adams Lab Urbana IL 61801

MALMSTROM, VINCENT HERSCHEL, b Evanston, Ill, Mar 6, 26; m 51; c 2. GEOGRAPHY. Educ: Univ Mich, AB, 47, MA, 48, PhD(geog), 54. Prof Exp: Instr geog, Mich Col Mining & Technol, 53-54; asst prof, Bucknell Univ, 55-56; asst prof, Middlebury Col, 56-57; asst prof, Univ Minn, Duluth, 57-58; from asst prof to assoc prof, 58-70, PROF GEOG, MIDDLEBURY COL, 70- Concurrent Pos: Vis prof, Univ Vt, 59-62; Asn Am Geogrs partic fel, Int Geog Cong, Sweden, 60; mem, US Nat Comt, Int Geog Union, 62-66. Mem: Asn Am Geogrs. Res: Regional geography of Northern Europe; economic geography; climatology. Mailing Add: Dept of Geog Middlebury Col Middlebury VT 05753

MALO, SALVADOR ALEJANDRO, b Mexico City, May 4, 41; m 64; c 2. SURFACE PHYSICS, EXPERIMENTAL ATOMIC SPECTROSCOPY. Educ: Nat Univ Mex, Physics Title, 66; Univ London, PhD(physics), 72. Prof Exp: Exp officer, Nat Inst Nuclear Energy, 61-64; asst officer radiation dosimetry, Int Atomic Energy Agency, 64-67; res physics, 70-74, HEAD DIV PHYS CHEM SURFACES, MEX INST PETROL, 74- Concurrent Pos: Exp officer, Nat Inst Nuclear Energy, 70-71; assoc prof, Nat Univ Mex, 71-75, titular prof, 75- Mem: Mex Soc Physics (vpres, 73-74); Nat Acad Sci Invest, Mex. Res: Matrix isolation spectroscopy and physical chemistry of surfaces in relation to catalysis; photoelectron spectroscopy, auger spectroscopy and low energy electron diffraction. Mailing Add: Mex Inst Petrol Apdo Postal 14-805 Mexico 14 D F Mexico

MALO, SIMON E, b Cuenca, Ecuador, May 30, 33; US citizen; m; c 2. HORTICULTURE, NEMATOLOGY. Educ: Univ Fla, BS, 57, MS, 60, PhD(nematol, hort), 64. Prof Exp: ASSOC HORTICULTURIST, AGR RES & EDUC CTR, UNIV FLA, 64-, ASSOC PROF HORT, 72- Mem: Am Soc Hort Sci; Soc Nematol. Res: Tropical fruits, especially avocado, mango, lime, litchi and guava. Mailing Add: Agr Res & Educ Ctr Univ of Fla 18905 SW 280th St Rt 1 Homestead FL 33030

MALOFSKY, BERNARD MILES, b New York, NY, Oct 7, 37; m 64; c 2. ORGANIC POLYMER CHEMISTRY. Educ: Calif Inst Technol, BS, 59; Univ Wash, PhD(org chem), 64. Prof Exp: Res chemist, Textile Fibers Dept, E I du Pont de Nemours & Co, Inc, 64-70; from res & develop chemist to technol mgr, 72-74, MGR PROD DEVELOP, LOCTITE CORP, 74- Mem: Am Chem Soc. Res: Anaerobic adhesives and sealants, particularly thermal resistance, cure systems, structural adhesives of high peel and impact strength, primers, ultraviolet curing adhesives, powdered metal and casting impregnation. Mailing Add: Loctite Corp 705 N Mountain Rd Newington CT 06111

MALONE, CHARLES R, b Sweetwater, Tex, May 9, 38; m 64; c 1. ECOLOGY. Educ: Tex Tech Col, BS, 63, MS, 65; Rutgers Univ, PhD(zool), 68. Prof Exp: AEC fel ecol, Oak Ridge Nat Lab, 68-70, res scientist, 70-71; PROF ASSOC, ENVIRON STUDIES BD, NAT ACAD SCI-NAT ACAD ENG, 71- Concurrent Pos: Chmn, Prog Comt, Ecol Soc Am, 72-74; mem Pub Responsibilities Comt, Am Inst Biol Sci, 74- Mem: AAAS; Ecol Soc Am; Am Inst Biol Sci; Brit Ecol Soc. Res: Impacts of pollutants on ecosystems; environmental management; science advisory process. Mailing Add: Environ Studies Bd Nat Acad of Sci Washington DC 20418

MALONE, CREIGHTON PAUL, b Beaver City, Nebr, May 3, 33; m 57; c 2. PHYSICAL CHEMISTRY. Educ: Univ Colo, BA, 58, PhD(phys chem), 62. Prof Exp: Asst phys chem, Univ Colo, 58-62; res chemist, Eng Dept, 62-65, sr res chemist, 65-69, SR RES CHEMIST, TEXTILE FIBERS DEPT, E I DU PONT DE NEMOURS & CO, INC, 69- Mem: Am Chem Soc. Res: Magnetic susceptibility of small particles; infrared adsorption and reflection spectroscopy; liquid chromatography; polymer physical chemistry; textile physical chemistry. Mailing Add: Textile Res Lab Chestnut Run E I du Pont de Nemours & Co Inc, Wilmington DE 19898

MALONE, JAMES ROBERT, organic chemistry, fuel technology, see 12th edition

MALONE, JOSEPH JAMES, b St Louis, Mo, Sept 9, 32; m 60; c 4. ALGEBRA. Educ: St Louis Univ, BS, 54, MS, 58, PhD(math), 62. Prof Exp: Instr math, Rockhurst Col, 60-62; asst prof, Univ Houston, 62-67; from assoc prof to prof, Tex A&M Univ, 67-71; PROF MATH & HEAD DEPT, WORCESTER POLYTECH INST, 71- Mem: Am Math Soc; Math Asn Am; Soc Indust & Appl Math. Res: Abstract algebra; groups; near rings. Mailing Add: 45 Adams St Westboro MA 01581

MALONE, LEO JACKSON, JR, b Wichita, Kans, July 24, 38; m 64; c 3. INORGANIC CHEMISTRY. Educ: Univ Wichita, BS, 60, MS, 62; Univ Mich, PhD(inorg chem), 64. Prof Exp: From asst prof to assoc prof, 64-73, PROF CHEM, ST LOUIS UNIV, 73- Mem: AAAS; Am Chem Soc. Res: Chemistry of carbon-monoxide-borane. Mailing Add: Dept of Chem St Louis Univ St Louis MO 63156

MALONE, LINDA CATRON, b Joplin, Mo, May 13, 44; m 73. STATISTICS. Educ: Emory & Henry Col, BS, 66; Univ Tenn, MS, 68; Va Polytech Inst & State Univ, PhD(statist), 75. Prof Exp: Instr math, John Tyler Community Col, Chester, Va, 68-71 & statist, Va Polytech Inst & State Univ, 72-75; ASST PROF STATIST, MISS STATE UNIV, 75- Mem: Am Statist Asn. Res: Applied statistics; linear models; analysis of variance; biased estimation. Mailing Add: Dept of Comput Sci & Statist Miss State Univ Mississippi State MS 39762

MALONE, MARVIN HERBERT, b Fairbury, Nebr, Apr 2, 30; m 52; c 2. PHARMACOLOGY. Educ: Univ Nebr, BS, 51, MS, 53, PhD(pharmacol, pharmaceut sci), 58. Prof Exp: Asst prof pharmacol, Univ Nebr, 51-53; asst pharmacodyn, Squibb Inst Med Res, 53-56; asst pharmacol, Univ Nebr, 56-58; asst prof pharmacol, Col Pharm, Univ NMex, 58-60; assoc prof, Sch Pharm, Univ Conn, 60-69; PROF PHYSIOL & PHARMACOL, UNIV OF THE PAC, 69- Concurrent Pos: Consult, Drug Plant Lab, Univ Wash, 60-64, Res Path Assocs, Md, 67-70, Amazon Natural Drug Co, NJ, 67-70 & Imp Chem Indust US Inc, Del, 68-; ed, Wormwood Rev, 61- & Am J Pharm Educ, 75- Honors & Awards: Mead Johnson Labs Award, 64. Mem: AAAS; Am Pharmaceut Asn; Am Soc Pharmacol & Exp Therapeut; Am Soc Pharmacog; Acad Pharmaceut Sci. Res: Screening and assay of natural products; pharmacodynamics of psychotropic and autonomic agents; biometrics; pharmacology of inflammation and antiinflammation. Mailing Add: Sect of Physiol-Pharmacol Univ of the Pac Stockton CA 95211

MALONE, MICHAEL JOSEPH, b Portland, Maine, Apr 28, 30; m 57; c 1. NEUROLOGY, NEUROCHEMISTRY. Educ: Boston Col, AB, 51; Georgetown Univ, MD, 56. Prof Exp: Resident neurol, Boston Vet Admin Hosp, Boston Univ, 60-63; Nat Inst Neurol Dis & Stroke spec fel neurochem, Harvard Med Sch, 63-65; res assoc, Mass Gen Hosp, 65-69; lectr, Boston City Hosp, 69-70; prof, Med Sch, George Washington Univ, 70-75; PROF NEUROL & PEDIAT & DIR NEUROL RES, BOSTON CITY HOSP, 75- Concurrent Pos: Vet Admin clin investr, Harvard Med Sch, 65-68; asst prof, Sch Med, Boston Univ, 67-70; Nat Inst Neurol Dis & Stroke res grant, Boston Univ, 68-71, career develop award, 69; chief neurol, Children's Hosp, Washington, DC, 70-75; consult, Walter Reed Army Med Ctr, Washington, DC, 71-, US Naval Hosp, Bethesda, Md, 71- & NIH, 71- Mem: Am Acad Neurol; Am Soc Neurochem; Int Soc Neurochem; Soc Neurosci; NY Acad Sci. Res: Biochemistry of maturation; biochemical pathology of nervous tissue. Mailing Add: 45 Greenhill Rd Sudbury MA 01776

MALONE, PHILIP GARCIN, b Louisville, Ky, Jan 12, 41. GEOCHEMISTRY. Educ: Univ Louisville, BA, 62; Ind Univ, Bloomington, MA, 64; Case Western Reserve Univ, PhD(geol), 69. Prof Exp: Nat Acad Sci-Nat Res Coun res assoc, Smithsonian Inst, 69-70; asst prof, 70-74, ASSOC PROF GEOL, WRIGHT STATE UNIV, 74- Mem: AAAS; Sigma Xi. Res: Chemistry and mineralogy of biologically precipitated materials. Mailing Add: Dept of Geol Wright State Univ Dayton OH 45431

MALONE, ROBERT CHARLES, b Wichita, Kans, Mar 14, 45; m 68. THEORETICAL PHYSICS. Educ: Washington Univ, BA, 67; Cornell Univ, MS, 70, PhD(theoret physics), 73. Prof Exp: STAFF PHYSICIST THEORET PHYSICS, LOS ALAMOS SCI LAB, 72- Mem: Am Phys Soc. Res: Hydrodynamics, energy transfer and atomic processes related to the behavior of laser-irradiated targets. Mailing Add: T-DOT MS 210 Los Alamos Sci Lab Los Alamos NM 87545

MALONE, THOMAS ELLIS, zoology, cell biology, see 12th edition

MALONE, THOMAS FRANCIS, b Sioux City, Iowa, May 3, 17; m 42; c 6. METEOROLOGY. Educ: SDak Sch Mines & Technol, BS, 40; Mass Inst Technol, ScD, 46. Hon Degrees: DEng, SDak Sch Mines & Technol, 62; DHL, St Joseph Col, 65. Prof Exp: Asst, Mass Inst Technol, 41-42, from asst prof to assoc prof meteorol, 43-55; dir weather res ctr, Travelers Ins Co, Conn, 55-57, dir res, 57-64, second vpres, 64-66, vpres & dir res, 66-67, sr vpres, 68-70; dean grad sch, Univ Conn, 70-73; DIR, HOLCOMB RES INST, BUTLER UNIV, 73- Concurrent Pos: With Off Naval Res, 50-53; ed, Compendium Meteorol; mem adv panel sci & technol, Comt Sci & Astronaut, US House of Rep; mem geophys res bd & comt water, Nat Acad Sci, chmn comt atmospheric sci, 62-68, dep foreign secy, 68-73, chmn bd int orgn & progs, 69-, mem space applns bd, 72-; secy-gen comt atmospheric sci, Int Union Geod & Geophys, 65-68; chmn, Nat Motor Vehicle Safety Adv Coun, 67-69; secy-gen sci comt probs of environ, Int Coun Sci Unions, 70-; mem, Nat Adv Comt Oceans & Atmosphere, 71-75. Honors & Awards: Losey Award, Int Aerospace Sci, 60; Brooks Award, Am Meteorol Soc, 64; Abbe Award, 68. Mem: Nat Acad Sci; fel AAAS; fel Am Meteorol Soc (secy, 57-60, pres, 60-62); fel Am Geophys Union (vpres, 60-61, pres, 61-64, secy int partic, 64-72); Int Coun Sci Unions (vpres, 70-72). Res: Applied meteorology; synoptic climatology. Mailing Add: Holcomb Res Inst Butler Univ Indianapolis IN 46208

MALONE, WILLIAM MAXTON, b New Hanover Co, NC, July 8, 36. ORGANIC CHEMISTRY. Educ: Middle Tenn State Col, BS, 59; Univ Ala, PhD, 68. Prof Exp: Res fel org chem under Dr Paul R Story, Univ Ga, 68-69; res chemist, US Plywood-Champion Papers, Inc, 69-71, GROUP LEADER, N L INDUSTS, INC, 71- Mem: Am Chem Soc. Res: Flame retardants for plastics; mechanism studies of gas phase ozonolysis plus gas phase photolysis of ozonides; synthesis and development of fire-retardant polyurethanes; theoretical studies of polymer physical and solution properties. Mailing Add: Cent Res Labs Chem Prod Dept N L Industs Inc Hightstown NJ 08520

MALONE, WINFRED FRANCIS, b Revere, Mass, Feb 10, 35; m 75. ENVIRONMENTAL HEALTH. Educ: Univ Mass, Amherst, BS, 57, MS, 61; Rutgers Univ, New Brunswick, MS, 64; Univ Mich, Ann Arbor, PhD(environ health sci), 70. Prof Exp: Asst prof environ health, Lowell Technol Inst, 64-66; res & develop officer environ health, NIH, 69-72; sci adv, Hazardous Mat Adv Comt, Environ Protection Agency, 72-73, staff dir, Sci Adv Bd, 73-74; PROG DIR, PREV CANCER CONTROL PROG, NAT CANCER INST, 74- Mem: Health Physics Soc; Am Indust Hyg Asn; Royal Soc Health; NY Acad Sci; Am Pub Health Asn. Mailing Add: Nat Cancer Inst Blair Rm 614 Bethesda MD 20014

MALONEY, CLIFFORD JOSEPH, b Wheelock, NDak, Mar 25, 10; m 42; c 1. STATISTICS. Educ: NDak Agr Col, BS, 34; Univ Minn, MA, 37; Iowa State Col, PhD(statist), 48. Prof Exp: Instr math, NDak Agr Col, 35-41; instr math, Iowa State Col, 41-42; statistician, Bur Agr Econ, USDA, 42-46; instr math, Iowa State Col, 46, res assoc statist method, 46-47; chief statist br, Chem Corps, Ft Detrick, Md, 47-58, Biol Labs, Biomath Div, 58-62; CHIEF BIOMET SECT, DIV BIOL STAND, NIH, 62- Mem: AAAS; Biomet Soc; Math Asn Am; Am Statist Asn; Inst Math Statist. Res: Biometrics; computing; retrieval. Mailing Add: Div of Biol Stand Nat Insts of Health Bethesda MD 20014

MALONEY, DANIEL EDWIN, b Jericho, Vt, Feb 9, 26; m 53; c 7. POLYMER CHEMISTRY. Educ: St Michael's Col, BS, 47; Univ Notre Dame, MS, 49, PhD(org chem), 51. Prof Exp: Res chemist, 51-61, sr res chemist, 61-67, RES ASSOC, PLASTICS DEPT, E I DU PONT DE NEMOURS & CO, 67- Mem: Am Chem Soc. Res: Plastics; polyolefins. Mailing Add: Highland Meadows RD 2 Box 304 Hockessin DE 19707

MALONEY, JAMES EUGENE, b Rollette, NDak, July 11, 35. OPERATIONS RESEARCH, FOREST ECONOMICS. Educ: Univ Calif, Berkeley, BSc, 62, MF, 63, PhD(quant methods), 72. Prof Exp: Researcher opers res, Univ Calif, Berkeley, 63-72; res scientist & proj leader opers res, Forest Fire Res Inst, Ottawa, Ont, 72-76; MGR, ECON & STATIST DIV, CAN POST OFF, 76- Concurrent Pos: Consult & prof, Univ Toronto, 72-74 & Can Comt on Forest Fire Control, 72-76. Mem: AAAS; Soc Am Foresters; Can Inst Forestry. Res: Development and application of complex postal systems using advanced operations research models, economic principles, statistical techniques. Mailing Add: Post Off Opers Res Br Campbell Bldg Ottawa ON Can

MALONEY, JAMES VINCENT, JR, b Rochester, NY, June 30, 25; m 57; c 4. SURGERY. Educ: Univ Rochester, MD, 47; Am Bd Surg, dipl, 55; Am Bd Thoracic Surg, dipl, 57. Prof Exp: Lectr biol, Sampson Col, 47; surg house officer, Johns Hopkins Univ Hosp, 47-48, from asst res to res surgeon, 50-55, instr, Sch Med, 54-55; res fel physiol, Sch Pub Health, Harvard Univ, 48-50; from asst prof to assoc prof, 55-65, PROF SURG, SCH MED, UNIV CALIF, LOS ANGELES, 65-, CHIEF DIV THORACIC SURG, 59- Concurrent Pos: Consult, Vet Admin Hosp, Los Angeles & Harbor County Gen Hosp, Torrance, 57-; Markle scholar, 58- Mem: Soc Univ Surgeons; Soc Clin Surgeons; Am Physiol Soc; AMA; Am Fedn Clin Res. Res: Surgery and physiology, especially the cardiorespiratory system. Mailing Add: Dept of Thoracic Surg Univ of Calif Sch of Med Los Angeles CA 90024

MALONEY, JOHN FRANCIS, JR, food science, see 12th edition

MALONEY, JOHN P, b Omaha, Nebr, Dec 9, 29. MATHEMATICAL ANALYSIS. Educ: Iowa State Univ, BS, 58; Georgetown Univ, MA, 62, PhD(math), 65. Prof Exp: Prod engr, Western Elec Co, 58-59; elec engr, US Govt, 58-63; instr math,

Georgetown Univ, 63-65; asst prof, Univ Nebr, Lincoln, 65-67; asst prof, 67-69, ASSOC PROF MATH, UNIV NEBR, OMAHA, 69- Mem: Am Math Soc; Math Asn Am; Soc Indust & Appl Math. Res: Integral equations. Mailing Add: Dept of Math & Comput Sci Univ of Nebr Omaha NE 68132

MALONEY, MARY ADELAIDE, b Quincy, Mass, Apr 30, 18. BIOLOGY. Educ: Col St Elizabeth, AB, 40. Prof Exp: Lab technician, Quincy City Hosp, Mass, 40-43; lab asst, Thorndike Mem Lab, Harvard Med Sch, 43-47; hematologist, Med Nutrit Lab, Univ Chicago, 47-51; chief lab asst hepatitis surv group from Harvard Univ, Off Sci Res & Develop, Japan, 51-52; asst biologist, Argonne Nat Lab, 52-64; res assoc biologist, 64-71, RES BIOLOGIST, LAB RADIOBIOL, SCH MED, UNIV CALIF, SAN FRANCISCO, 71- Mem: AAAS; Sigma Xi. Res: Radiobiology; hematologic problems. Mailing Add: Lab of Radiobiol Univ of Calif Sch of Med San Francisco CA 94122

MALONEY, THOMAS EDWARD, b Niagara Falls, NY, Sept 7, 23; m 46; c 3. ENVIRONMENTAL SCIENCES. Educ: Univ Buffalo, BA, 49, MA, 53. Prof Exp: Res biologist, Robert A Taft Sanit Eng Ctr, USPHS, 51-65; chief plankton res sect, Nat Marine Water Qual Lab, Fed Water Pollution Control Admin, 65-68, chief physiol control br, Nat Eutrophication res prog, 68-71, dep chief prog & chief physiol control br, Environ Protection Agency, 71-72, CHIEF, EUTROPHICATION & LAKE RESTORATION BR, CORVALLIS ENVIRON RES LAB, US ENVIRON PROTECTION AGENCY, 72- Concurrent Pos: Chmn, Plankton Subcomt, Biol Methods Comt, Environ Protection Agency, Biostimulation Joint Task Group & Phytoplankton Subcomt, Standard Methods for Examination Water & Wastewater, 73- Honors & Awards: Superior Serv Award, Dept Health, Educ & Welfare, 65. Mem: AAAS; Am Soc Limnol & Oceanog; Water Pollution Control Fedn; Am Phycol Soc; Am Inst Biol Sci. Res: Culturing of algae; algal physiology; chemical control of algal growth; environmental requirements of planktonic organisms; coordination and review of research and development programs to provide for control of accelerated eutrophication and development of lake restoration technology. Mailing Add: Environ Protection Agency 200 SW 35th St Corvallis OR 97330

MALONEY, THOMAS J, b Arlington, Mass, Nov 16, 22; m 48; c 4. CULTURAL ANTHROPOLOGY. Educ: Northeastern Univ, BS, 48; Harvard Univ, STB, 52; Washington Univ, AM, 56, PhD(anthrop & sociol), 66. Prof Exp: Res engr, Gen Aniline & Film Corp, 48; chem engr, Eng Exp Sta, Univ Colo, 48-49; chem engr, Aircraft Gas Turbine Div, Gen Elec Co, 51-52; minister, Unitarian Church, Davenport, Iowa, 52-53, Unitarian Church, Quincy, Ill, 53-56 & Unitarian Church, Boulder, Colo, 57-62; staff asst tech recruiting, Bettis Atomic Power Div, Westinghouse Elec Corp, 56-57; asst prof sociol & anthrop, NMex Highlands Univ, 62-67; assoc prof, Ripon Col, 67-69; ASSOC PROF ANTHROP, SOUTHERN ILL UNIV, EDWARDSVILLE, 69- Concurrent Pos: Lectr sociol, Univ Pittsburgh, 56-57; instr anthrop, Univ Colo, 57-59; res assoc, Kirschner Assocs, NMex, 68-69. Mem: Fel AAAS; fel Am Anthrop Asn; Am Ethnol Soc; Latin Am Studies Asn. Res: Human ecology and habitat disruption in Costa Rica; cultural ecology of cattle ranching in northern New Mexico; labor-intensive industries in developing nations. Mailing Add: Dept of Anthrop Southern Ill Univ Edwardsville IL 62026

MALONEY, WILLIAM FARLOW, b Minneapolis, Minn, June 20, 19; m 43; c 4. MEDICINE. Educ: Univ Minn, BBA, 41, MS, 43, MD, 46. Prof Exp: Scholar, Trudeau Sch Tuberc, 49; Nat Heart Inst trainee, 52-53; instr internal med, Univ Minn, 53-55, asst prof, 55-57; assoc prof med & dean, Sch Med, Med Col Va, 57-63; assoc dir, Asn Am Med Cols, 63-66; prof med & dean, Sch Med, 66-74, PROF HEALTH PLANNING & HEALTH PLANNING ADMINR, TUFTS UNIV, 74- Concurrent Pos: Asst dean, Col Med Sci, Univ Minn, 53-57; lectr, Med Sch, Northwestern Univ, 63-66. Res: Cardiovascular and pulmonary disease; medical education. Mailing Add: Tufts Univ Medford MA 02155

MALONEY, WILLIAM THOMAS, b Warren, Ohio, Dec 12, 35; m 58; c 2. MAGNETISM. Educ: Case Western Reserve Univ, BS, 57, MS, 58; Harvard Univ, AM, 61, PhD(appl physics), 64. Prof Exp: Asst appl physics, Harvard Univ, 60-63, lectr & res fel, 63-65; RES STAFF MEM, SPERRY RAND RES CTR, 65- Mem: Inst Elec & Electronics Engrs; Am Phys Soc; Optical Soc. Res: Optical signal processing; optical pattern recognition; optical memories; magnetic recording. Mailing Add: 119 Willis Rd Sudbury MA 01776

MALOOF, FARAHE, b Boston, Mass, Jan 24, 21; m 47; c 2. INTERNAL MEDICINE, ENDOCRINOLOGY. Educ: Harvard Univ, BA, 42; Tufts Univ, MD, 45. Prof Exp: Asst, 54-60, asst prof med, 66-72, ASSOC PROF MED, HARVARD MED SCH, 72-; PHYSICIAN, MASS GEN HOSP, 73-, CHIEF THYROID UNIT, 66- Concurrent Pos: USPHS fel, Mass Gen Hosp, 49-51; asst med, Mass Gen Hosp, 54-60, assoc physician, 65-73; sr res assoc, Grad Dept Biochem, Brandeis Univ, 58-64, adj assoc prof, 64- Mem: Am Thyroid Asn; Am Fedn Clin Res; Endocrine Soc; Am Soc Clin Invest; Am Col Physicians. Res: Biochemistry and pharmacology of the thyroid. Mailing Add: Thyroid Unit Mass Gen Hosp Boston MA 02114

MALOOF, GILES WILSON, b San Bernardino, Calif, Jan 4, 32; m 58; c 2. MATHEMATICS, GEOPHYSICS. Educ: Univ Calif, Berkeley, BA, 53; Univ Ore, MA, 58; Ore State Univ, PhD(math), 62. Prof Exp: Engr, Creole Petrol Corp, 53-54; engr, US Navy Ord Res Labs, 58-59; instr math, Ore State Univ, 61-62, asst prof, 62-68, res assoc geophys oceanog, 63-68; PROF MATH & HEAD DEPT, BOISE STATE COL, 68-, DEAN GRAD SCH, 70- Mem: Soc Indust & Appl Math; Am Math Soc; Math Asn Am. Res: Numerical filtering as applied in the interpretation of geophysical data; nonlinear functional analysis applied to integral equations. Mailing Add: Dept of Math Boise State Col 1907 Campus Dr Boise ID 83707

MALOTKY, LYLE OSCAR, b New London, Wis, Apr 14, 46; m 68; c 1. POLYMER CHEMISTRY. Educ: Augsburg Col, BA, 68; Univ Akron, PhD(polymer sci), 73. Prof Exp: CHEMIST, NAVAL EXPLOSIVE ORD DISPOSAL FACIL, 73- Mem: Am Chem Soc. Res: Applications of polymeric materials; explosive analysis and detection. Mailing Add: Naval Explosive Ord Disposal Facil Code 5031 Indian Head MD 20640

MALOUF, CARLING I, b Fillmore, Utah, June 9, 16; m 41; c 4. ANTHROPOLOGY. Educ: Univ Utah, BS, 39, MS, 40; Columbia Univ, PhD, 56. Prof Exp: From asst prof to assoc prof anthrop, 48-64, PROF ANTHROP, UNIV MONT, 64-, CHMN DEPT, 74- Concurrent Pos: Mem, Northwestern Anthrop Conf. Mem: Fel Am Anthrop Asn; Soc Am Archaeol. Res: Archaeology of Utah, Nevada, Arizona, and Montana; ethnology of Utah, New Mexico, Montana, and the Near East; modern Indian affairs in the United States; applied anthropology. Mailing Add: Dept of Anthrop Univ of Mont Missoula MT 59801

MALOY, JOHN OWEN, b Orange, NJ, Feb 7, 32; div; c 1. HIGH ENERGY PHYSICS, SPACE PHYSICS. Educ: Univ Ariz, BS, 54; Calif Inst Technol, PhD(physics), 61. Prof Exp: Group leader systs anal, Jet Propulsion Lab, Calif Inst Technol, 60-61, res fel physics, Synthrotron Lab, 61-63, sr res fel, 63-67; mgr advan develop div, Analog Technol Corp, 67-71; chief scientist, Beckman Instruments Inc, 71-74; RES STAFF PHYSICIST, UNIV SOUTHERN CALIF, 74-; STAFF SCIENTIST, BALL BROS RES CORP, 74- Concurrent Pos: Consult, Electro-Optical Systs, Inc, Calif, 63-66, Beckman Instruments, Inc, 74-, Jet Propulsion Lab, Calif Inst Technol, 76- & Melcon, 76- Mem: AAAS; Am Phys Soc; Am Geophys Union. Res: Photoproduction of pi mesons; accelerator physics and technology; development of radio frequency acceleration system; solar and planetary science; space science instrumentation systems design; instrument program management. Mailing Add: 4591 Green Tree Lane Irvine CA 92715

MALOY, JOSEPH T, b Mt Pleasant, Pa, Apr 19, 39; m 70; c 1. ANALYTICAL CHEMISTRY. Educ: St Vincent Col, BA, 61; Univ Tex, Austin, MA, 67, PhD(chem), 70. Prof Exp: Teacher, Mt Pleasant area schs, Pa, 61-65; asst prof, 70-75, ASSOC PROF CHEM, WVA UNIV, 70- Mem: Am Chem Soc; Electrochem Soc; Sigma Xi. Res: Electrochemistry; electroanalytical techniques; computer applications; electrogenerated chemiluminescence; photoluminescence. Mailing Add: Dept of Chem WVa Univ Morgantown WV 26506

MALOY, OTIS CLEO, JR, b Coeur d'Alene, Idaho, Jan 19, 30; m 53. PLANT PATHOLOGY. Educ: Univ Idaho, BS, 51, MS, 55; Cornell Univ, PhD, 58. Prof Exp: Asst plant path, Cornell Univ, 55-58; forest pathologist, US Forest Serv, 58-59; res forester, Potlatch Forests, Inc, 59-63; EXTEN PLANT PATHOLOGIST, WASH STATE UNIV, 63- Mem: Am Phytopath Soc. Res: Ecology and physiology of soil microorganisms; root rots; diseases of forest trees. Mailing Add: Dept of Plant Path Wash State Univ Pullman WA 99163

MALSBERGER, RICHARD GRIFFITH, b Philadelphia, Pa, Jan 12, 23; m 44; c 1. BIOLOGY. Educ: Lehigh Univ, BA, 48, MS, 49, PhD(bact), 58. Prof Exp: Mem staff, Biol Prod Dept, Merck Sharp & Dohme, 50-53, res assoc virol, 53-57, mgr control, 57-59; from asst prof to assoc prof, 59-66, PROF BIOL, LEHIGH UNIV, 66- Mem: AAAS; Am Inst Biol Sci; Tissue Cult Asn; Am Soc Microbiol. Res: Viral vaccines, immunology and multiplication; viral diseases of freshwater fishes; immunology. Mailing Add: Dept of Biol Lehigh Univ Bethlehem PA 18015

MALSKY, STANLEY JOSEPH, b New York, NY, July 15, 25; m 65; c 2. RADIOLOGICAL PHYSICS, MEDICAL PHYSICS. Educ: NY Univ, BSc, 46, MA, 50, MSc, 53, PhD, 63. Prof Exp: Nuclear physicist, US Navy, 50-54; asst chief radiother, Vet Admin Hosp, Bronx, 54-73; PRES, RADIOL PHYSICS ASSOCS, INC, 73- Concurrent Pos: Asst prof, NY Univ, 59-63; res collab, Med Div, Brookhaven Nat Lab, 60-67; co-dir & prof radiol sci, Manhattan Col, 63-74; chief physicist, Fordham Hosp, 71-; res prof radiol, Sch Med, NY Univ, 74- Honors & Awards: James Picker Award Res Dosimetry. Mem: Fel AAAS; fel Am Pub Health Asn; Royal Soc Health; Am Asn Physicists Med; Health Physics Soc. Res: Solid state dosimetry. Mailing Add: Radiol Physics Assocs Inc 809 Hartsdale Rd White Plains NY 10607

MALT, RONALD A, b Pittsburgh, Pa, Nov 12, 31; m 51; c 3. SURGERY, MOLECULAR BIOLOGY. Educ: Wash Univ, AB, 51; Harvard Univ, MD, 55; Am Bd Surg, dipl, 62; Bd Thoracic Surg, dipl, 63. Prof Exp: Intern surg, Mass Gen Hosp, 55-56, resident, 58-62; asst, 62-64, instr, 64-67, assoc, 67-68, from asst prof to assoc prof, 68-75, PROF SURG, HARVARD MED SCH, 75-; CHIEF GASTROENTEROL SURG, MASS GEN HOSP, 70- Concurrent Pos: USPHS spec res fel biol, Mass Inst Technol, 62-63, fel, Sch Advan Study, 63-64; Am Heart Asn estab investr, 63-68; res assoc, Mass Inst Technol, 62-64, from asst surgeon to surgeon, Mass Gen Hosp, 62-; assoc surgeon, Shriners Burns Inst, 67- Mem: Am Surg Asn; Am Soc Clin Invest; Soc Develop Biol; Am Soc Cell Biol; Am Physiol Soc. Res: Regeneration; molecular events in renal, hepatic, and enteric growth and neoplasia; liver, biliary and portal-system surgery; replantation of limbs. Mailing Add: Mass Gen Hosp Boston MA 02114

MALTENFORT, GEORGE GUNTHER, b Landsberg, Ger, Aug 13, 13; US citizen; m 46; c 1. CHEMISTRY. Educ: Northwestern Univ, BS, 34. Prof Exp: Chemist, Transparent Package Co, 34-42 & 45-46; chemist, 46-58, TECH DIR, TECH SERV LAB, CONTAINER CORP AM, 58- Concurrent Pos: Dir, Res & Develop Assocs, 65-68 & 74-76; mem packaging comt, Nat Acad Sci-Nat Res Coun, 74-77. Honors & Awards: Medal, Tech Asn Pulp & Paper Indust, 67. Mem: Am Chem Soc; Am Soc Qual Control; Tech Asn Pulp & Paper Indust; Am Soc Test & Mat. Res: Packaging, sampling, statistics and development of test methods and instruments. Mailing Add: Tech Serv Lab Container Corp of Am Carol Stream IL 60187

MALTER, MARGARET QUINN, b Philadelphia, Pa, May 12, 26; m 51; c 2. ORGANIC CHEMISTRY. Educ: Bryn Mawr Col, BA, 47, MA, 48, PhD, 52. Prof Exp: From instr to asst prof, Drexel Inst, 52-57; from asst prof to assoc prof, 57-69, PROF CHEM, IMMACULATA COL, PA, 69- Mem: AAAS; Am Chem Soc. Res: Reaction mechanisms. Mailing Add: Dept of Chem Immaculata Col Immaculata PA 19345

MALTESE, GEORGE J, b Middletown, Conn, June 24, 31; m 56; c 2. MATHEMATICS. Educ: Wesleyan Univ, BA, 53; Yale Univ, PhD(math), 60. Prof Exp: NATO fel, Univ Göttingen, 60-61; instr math, Mass Inst Technol, 61-63; from asst prof to assoc prof, Univ Md, College Park, 63-74; PROF MATH, UNIV MÜNSTER, GER, 74- Concurrent Pos: Vis prof, Univ Frankfurt, 66-67 & 70-71 & Univ Palermo, 70-71. Mem: Am Math Soc; Austrian Math Soc; Dutch Math Soc. Res: Functional analysis with emphasis on Banach algebra theory and the spectral theory of linear operators. Mailing Add: Math Inst Univ Münster Roxeler Str 64 44 Münster West Germany

MALTZ, HENRY, organic chemistry, inorganic chemistry, see 12th edition

MALTZ, ROBERT, mathematics, see 12th edition

MALTZEFF, EUGENE M, b Khabarovsk, Russia, Oct 31, 12; US citizen; m 46; c 2. FISHERIES. Educ: U Wash, BS, 39. Prof Exp: Aquatic biologist, Fish & Wildlife Serv, US Bur Com Fisheries, 44-48, fishery res biologist, 42-57, fishery biologist, 58-68, foreign fisheries analyst, Pac Northwest Region I, 68-70, foreign fisheries analyst, Nat Marine Fisheries Serv, 70-74; CONSULT MARINE AFFAIRS, 75- Mem: Am Fisheries Soc; Am Inst Fishery Res Biol. Res: Pacific salmon; stream improvement; Indian fisheries; foreign fishing. Mailing Add: 4501 Stanford Ave NE Seattle WA 98105

MALUEG, KENNETH WILBUR, b Appleton, Wis, Apr 19, 38; m 64; c 2. LIMNOLOGY. Educ: Univ Wis, BS, 60, MS, 63, PhD(zool), 66. Prof Exp: SUPVR RES AQUATIC BIOLOGIST, CORVALLIS ENVIRON RES LAB, ENVIRON PROTECTION AGENCY, 66- Mem: Am Soc Limnol & Oceanog; Sigma Xi; Amer Inst Biol Sci; Int Asn Theoret & Appl Limnol. Res: Lake restoration; eutrophication control. Mailing Add: Corvallis Environ Res Lab 200 SW 35th St Corvallis OR 97330

MALVEAUX, FLOYD J, b Opelousas, La, Jan 11, 40; m 65; c 4. MICROBIOLOGY, MICROBIAL PHYSIOLOGY. Educ: Creighton Univ, BS, 61; Loyola Univ, La, MS, 64; Mich State Univ, PhD(microbiol), 68. Prof Exp: Instr soil microbiol, Mich State

Univ, 68; ASST PROF MICROBIOL, COL MED, HOWARD UNIV, 68- Mem: AAAS; Am Soc Microbiol. Res: Characterization of enzymes and extracellular proteins of pathogenic bacteria as these products relate to virulence; physiology of microorganisms associated with plaque formation and periodontal disease. Mailing Add: Dept of Microbiol Howard Univ Col of Med Washington DC 20001

MALVEN, PAUL VERNON, b Annapolis, Md, Oct 24, 38; m 63; c 2. NEUROENDOCRINOLOGY, REPRODUCTIVE PHYSIOLOGY. Educ: Univ Ill, Urbana, BS, 60; Cornell Univ, PhD(animal physiol), 64. Prof Exp: NIH fel, Univ Calif, Los Angeles, 64-65; from asst to assoc prof animal sci, 66-72, PROF ANIMAL SCI, PURDUE UNIV, 72- Mem: Am Physiol Soc; Endocrine Soc; Am Dairy Sci Asn; Am Soc Animal Sci; Soc Study Reproduction. Res: Neuroendocrinology of reproduction. Mailing Add: Dept of Animal Sci Purdue Univ West Lafayette IN 47906

MALVILLE, JOHN MCKIM, b San Francisco, Calif, Apr 24, 34; m 60; c 2. ASTROPHYSICS. Educ: Calif Inst Technol, BS, 56; Univ Colo, PhD(astrophys), 61. Prof Exp: Res assoc astron, Univ Mich, 62-63, asst prof, 63-65; mem sr staff solar physics, High Altitude Observ, 65-70; asst dean col arts & sci, 69-70, assoc prof, 70-73, PROF ASTRO GEOPHYS, UNIV COLO, BOULDER, 73- Concurrent Pos: Am Astron Soc vis prof, 64- Mem: AAAS; Am Astron Soc; Int Astron Union. Res: Solar physics; radio astronomy; auroral, atomic and molecular physics; interstellar medium; philosophy of science; science education. Mailing Add: Dept of Astro Geophys Univ of Colo Boulder CO 80302

MALVIN, RICHARD L, b Aug 19, 27; US citizen; m; c 2. PHYSIOLOGY. Educ: McGill Univ, BSc, 50; NY Univ, MS, 54; Univ Cincinnati, PhD(physiol), 56. Prof Exp: Res assoc, 56-57, from instr to assoc prof, 57-67, PROF PHYSIOL, MED SCH, UNIV MICH, 67- Concurrent Pos: Lederle med fac award, 59-62; USPHS career develop award, 62-72. Mem: AAAS; Am Physiol Soc; Am Soc Nephrology. Res: Renal physiology; salt and water balance; control of secretion of renin and antidiuretic hormone. Mailing Add: Dept of Physiol 7730 Med Sci II Univ of Mich Ann Arbor MI 48104

MALVITZ, DOLORES MARIE, b Dearborn, Mich, Oct 17, 41. PUBLIC HEALTH, DENTAL HYGIENE. Educ: Western Mich Univ, BA, 68; Univ Mich, cert dent hyg, 61, MPH, 72, PhD(dent pub health), 74. Prof Exp: Instr dent hyg, 68-70, res assoc community dent, 73, ASST PROF DENT PUB HEALTH, UNIV MICH, 74- Concurrent Pos: Mem, Adv Comt Develop Dent Health Unit Elem Schs, Div Nutrit Educ, Nat Dairy Coun, 73-74; consult, Coun Dent Educ, Am Dent Asn, 75- Mem: Am Dent Hygienists Asn; Am Asn Dent Schs; Am Pub Health Asn; Soc Pub Health Educ. Res: Delivery of dental care; applications of behavioral and social sciences in dentistry, dental education and programs of dental health education. Mailing Add: M5515 Sch of Pub Health Univ of Mich Ann Arbor MI 48109

MALY, EDWARD J, b Troy, NY, Nov 10, 42; m 67. ECOLOGY, EVOLUTION. Educ: Univ Rochester, BS, 64; Princeton Univ, PhD(biol), 68. Prof Exp: Asst prof biol, Tufts Univ, 68-75; ASST PROF BIOL, CONCORDIA UNIV, 75- Mem: AAAS; Ecol Soc Am; Am Inst Biol Sci. Res: Population dynamics; predator-prey interactions and population growth rates; life histories and diversity of fresh-water animals. Mailing Add: Dept of Biol Sci Concordia Univ 1455 De Maisonneuve Blvd W Montreal PQ Can

MALYA, GOVINDA P A, b Cochin, India, July 17, 40; m 70; c 1. CLINICAL BIOCHEMISTRY, PATHOLOGICAL CHEMISTRY. Educ: M S Univ, Baroda, India, 65; Univ Miss, MS, 67; Drexel Univ, PhD(biochem), 70. Prof Exp: Chemist, Navaratna Pharmaceut Labs, India, 62-63; fel biol sci, Drexel Univ, 70-71; dir clin biochem, Edgewater Hosp, Chicago, 72; chief clin biochem, Mercy Cath Med Ctr, Darby, Pa, 72-74; res biochemist, 74-76, SUPVR CLIN PATH, ICI-US, INC, 76- Mem: Am Soc Clin Chemists; Am Chem Soc; Sigma Xi. Res: Determination and interpretation of clinical biochemistry and hematology tests in biological fluids to assess the safety of new pharmaceutical drugs under development. Mailing Add: ICI-US Inc Bio Med Res Dept Concord Pike & Murphy Rd Wilmington DE 19897

MALZAHN, RAY ANDREW, b Ft Madison, Iowa, July 8, 29; m 53; c 2. ORGANIC CHEMISTRY, ACADEMIC ADMINISTRATION. Educ: Gustavus Adolphus Col, BA, 51; Univ NDak, MS, 53; Univ Md, PhD(org chem), 62. Prof Exp: Assoc prof chem, 63-67, dean col arts & sci, 67-71, PROF CHEM, WTEX STATE UNIV, 67-, VPRES ACAD AFFAIRS, 71- Concurrent Pos: Fel, Univ Ariz, 61-63. Mem: Am Chem Soc. Res: Pyrolysis of allyl and propargyl ethers; polymerization of monomers derived from natural products; synthesis of arylsilanes containing carboxyl groups. Mailing Add: Dept of Chem Box 727 WTex State Univ Canyon TX 79016

MALZAHN, RONALD C, b Merrill, Wis, Mar 2, 14; m 44; c 3. BACTERIOLOGY. Educ: Univ Wis, BS, 40; Pa State Univ, MS, 42, PhD(bact), 50. Prof Exp: Bacteriologist, Com Solvents Corp, 46-48; asst prof bact, NDak Agr Col, 50-51; SR BACTERIOLOGIST, GRAIN PROCESSING CORP, 51- Mem: AAAS; Am Chem Soc; Am Soc Microbiologists. Res: Industrial microbiology. Mailing Add: 31 Colony Dr Muscatine IA 52761

MAMANTOV, GLEB, b Karsava, Latvia, Apr 10, 31; US citizen; m 56; c 3. INORGANIC CHEMISTRY, ANALYTICAL CHEMISTRY. Educ: La State Univ, BS, 53, MS, 54, PhD(chem), 57. Prof Exp: Res chemist electrochem dept, E I du Pont de Nemours & Co, 57-58; instr & res assoc chem, Univ Wis, 60-61; from asst to assoc prof, 61-71, PROF CHEM, UNIV TENN, KNOXVILLE, 71- Concurrent Pos: Consult, Oak Ridge Nat Lab, 62-; NATO sr fel, Ger, 71. Mem: AAAS; Am Chem Soc; The Chem Soc; Electrochem Soc. Res: Electrochemistry and chemistry in molten salts; fluorine chemistry; electroanalytical chemistry. Mailing Add: Dept of Chem Univ of Tenn Knoxville TN 37916

MAMAY, SERGIUS HARRY, b Akron, Ohio, May 20, 20; m 53; c 2. PALEOBOTANY. Educ: Univ Akron, BS, 44; Washington Univ, MA, 48, PhD(bot), 50. Prof Exp: Asst bot, Washington Univ, 49-50; Guggenheim fel, Cambridge Univ, 50-51; PALEOZOIC PALEOBOTANIST, US GEOL SURV, 51- Mem: Geol Soc Am; Paleont Soc; Bot Soc Am. Res: Permian floras of southwestern United States; American coal ball floras. Mailing Add: W-301 US Nat Mus Washington DC 20242

MAMELAK, JOSEPH SIMON, b Lodz, Poland, Dec 14, 23; US citizen; m 59; c 2. COMPUTER SCIENCE, APPLIED MATHEMATICS. Educ: McGill Univ, BS, 45, MS, 46; Univ Pittsburgh, PhD(math), 49. Prof Exp: Asst prof math, Univ WVa, 53-56; sr analyst opers res, Univac, Sperry Rand Corp, 56-59; mgr sci appln, RCA, 59-62, proj mgr automated design, 62-65; head dept, 65-71, PROF MATH, COMMUNITY COL PHILADELPHIA, 66- Concurrent Pos: Consult, City Philadelphia, 65-; mem, var med & educ insts, 66-; mem, Am Stand Inst, 64- Mem: Am Math Soc; Am Statist Asn; Asn Comput Mach; Math Asn Am; Can Math Cong. Res: Water pollution models using computer simulation; automated circuit design and integrated circuit layout using computers; computer utilization in schools. Mailing Add: 70 Knollwood Dr Cherry Hill NJ 08034

MAMER, ORVAL ALBERT, b Humboldt, Sask, May 2, 40; m 63; c 3. CHEMISTRY. Educ: Univ Windsor, BSc, 62, PhD(org chem), 66. Prof Exp: Fel, Wayne State Univ, 66-68; Nat Res Coun Can fel, 68-69; ASST PROF EXP MED, ROYAL VICTORIA HOSP, McGILL UNIV, 69- Concurrent Pos: Sessional lectr dept chem, Univ Windsor, 67-68. Mem: Am Chem Soc; Chem Inst Can. Res: Bio-medical mass spectrometry; inherited and acquired metabolic disorders; metabolite identification; high temperature flash vacuum thermolytic studies. Mailing Add: Royal Victoria Hosp McGill Univ Montreal PQ Can

MAMET, BERNARD LEON, b Brussels, Belg, Feb 7, 37; m 63; c 1. GEOLOGY. Educ: Free Univ Brussels, LSc, 57, PhD(stratig), 62; French Petrol Inst, cert eng geol, 59; Univ Calif, Berkeley, MA, 60. Prof Exp: Researcher, Royal Inst Natural Sci, Belg, Brussels, 56-61; asst researcher, Nat Found Sci Res, Belg, 63-65; asst prof, 65-67, ASSOC PROF GEOL, UNIV MONTREAL, 67- Mem: Geol Soc Belg; Geol Soc France; Belg Soc Geol, Paleont & Hydrol; Am Asn Petrol Geologists; Soc Econ Paleontologists & Mineralogists. Res: Carboniferous microfacies. Mailing Add: Univ of Montreal PO Box 6128 Montreal PQ Can

MAMIYA, RICHARD T, b Honolulu, Hawaii, Mar 8, 25; m 50; c 8. THORACIC SURGERY, CARDIOVASCULAR SURGERY. Educ: Univ Hawaii, BSc, 50; St Louis Univ, MD, 54. Prof Exp: From instr to sr instr surg, Sch Med, St Louis Univ, 59-61, dir sect surg, Cochran Vet Admin Hosp, 59-61; assoc prof, 67-70, PROF SURG, SCH MED, UNIV HAWAII, MANOA, 70-, CHMN DEPT, 67- Concurrent Pos: Consult, US Army Tripler Gen Hosp, 68- Mem: AMA; Am Col Surg. Mailing Add: Rm 252 Harkness Hall Univ of Hawaii Honolulu HI 96813

MAMMANO, NICHOLAS J, physical chemistry, see 12th edition

MAMMEN, EBERHARD F, b Carolinensiel, Ger, July 13, 30; m 55; c 3. PHYSIOLOGY, PHARMACOLOGY. Educ: Univ Giessen, MD, 56. Prof Exp: Instr med, Univ Marburg, 60-62; from asst prof to assoc prof physiol & pharmacol, 62-67, prof pharmacol, 67-70, PROF PHYSIOL, COL MED, WAYNE STATE UNIV, 67-, PROF PHARM OF PATH, 70-, ACTG DEAN PHARM, 72- Mem: Am Physiol Soc; Soc Exp Biol & Med. Res: Physiology of blood coagulation; pathogenesis of hemophilia A; purification and properties of factor VIII; inhibitors of blood coagulation. Mailing Add: Wayne State Univ Col of Med 1400 Chrysler Expressway Detroit MI 48207

MAMOLA, KARL CHARLES, b Greenport, NY, May 23, 42; m 63; c 1. SOLID STATE PHYSICS. Educ: State Univ NY, Stony Brook, BS, 63; Fla State Univ, MS, 65; Dartmouth Col, PhD(physics), 73. Prof Exp: From instr to asst prof physics, 65-69, asst prof, 72, ASSOC PROF PHYSICS, APPALACHIAN STATE UNIV, 72- Mem: Am Phys Soc; Am Asn Physics Teachers; Nat Sci Teachers Asn; Sigma Xi. Res: Point deflects in crystals at high pressures using optical absorption and magnetic resonance spectroscopy. Mailing Add: Dept of Physics Appalachian State Univ Boone NC 28608

MAMPE, CHARLES DOUGLASS, b Abington, Pa, Oct 13, 38; m 59; c 1. ENTOMOLOGY. Educ: Iowa State Univ, BS, 60; NDak State Univ, MS, 62; NC State Univ, PhD(entom), 65. Prof Exp: Tech mgr, Nat Pest Control Asn, 65-74; TECH DIR, WESTERN INDUSTS, 74- Mem: Entom Soc Am. Res: Biology, ecology and population dynamics use of insecticides and rodenticides in pest control. Mailing Add: Western Industs 475 Prospect Ave West Orange NJ 07052

MAMRAK, SANDRA ANN, b Cleveland, Ohio, Sept 8, 44. COMPUTER SCIENCES. Educ: Notre Dame Col, BS, 67; Univ Ill, Urbana-Champaign, MS, 73, PhD(comput sci), 75. Prof Exp: COMPUT SPECIALIST, NAT BUR STAND, 75-; ASST PROF COMPUT SCI, OHIO STATE UNIV, 75- Mem: Asn Comput Mach. Res: Statistical and simulation methodologies for performance evaluation of computer systems, with emphasis on time sharing systems and computer networks. Mailing Add: Dept of Comput & Info Sci Ohio State Univ Columbus OH 43210

MAN, EUGENE HERBERT, b Scranton, Pa, Dec 14, 23; m 45; c 4. MARINE GEOCHEMISTRY. Educ: Oberlin Col, AB, 48; Duke Univ, PhD(chem), 52. Prof Exp: Res chemist chem dept, E I du Pont de Nemours & Co, 51-58, textile fibers dept, 58-60, supvr nylon tech div, 60-61, sr supvr, 61-62; coord res, 62-66, DEAN RES COORD, UNIV MIAMI, 66- Concurrent Pos: Vis investr, Scripps Inst Oceanog, Univ Calif, San Diego, 71-72. Mem: AAAS; Am Chem Soc; Sci Res Soc Am; fel Am Inst Chem. Res: Academic research administration; organic chemistry of ocean sediments; geochemistry of amino acids in deep-sea drilling cores. Mailing Add: Off of Res Coord Univ of Miami PO Box 248293 Coral Gables FL 33124

MAN, EVELYN BROWER, b Lawrence, NY, Oct 7, 04. CLINICAL CHEMISTRY. Educ: Wellesley Col, AB, 25; Yale Univ, PhD(physiol chem), 32. Prof Exp: Instr chem, Conn Col, 25-27; technician sch med, Yale Univ, 28-29, asst, 29-30, Am Asn Univ Women fel, 33-34, from instr to asst prof psychiat, 34-50, res assoc med, 50-61; assoc mem, 61-71, EMER ASSOC MEM, INST HEALTH SCI, BROWN UNIV, 71- Mem: AAAS; Endocrine Soc; Am Thyroid Asn; Am Chem Soc; Am Soc Biol Chem. Res: Lipemia and iodemia in thyroid diseases; pregnancy and infancy. Mailing Add: Route 2 Box 270-A North Stonington CT 06359

MANAHAN, STANLEY E, analytical chemistry, environmental chemistry, see 12th edition

MANAKER, ROBERT ANTHONY, b Avenel, NJ, Feb 28, 18; m 53; c 1. MICROBIOLOGY. Educ: Rutgers Univ, BS, 50, PhD(microbiol), 53. Prof Exp: Merck-Waksman res fel, Rutgers Univ, 53-54, instr microbiol, Inst Microbiol, 54-55, asst prof virol, 55-56; res microbiologist, 56-72, CHIEF VIRAL BIOL BR, NAT CANCER INST, 72- Mem: AAAS; Am Asn Cancer Res; Am Soc Microbiol. Res: Virus-tumor relationships. Mailing Add: 5305 Baltimore Ave Chevy Chase MD 20015

MANAKKIL, THOMAS JOSEPH, b Gothuruthy, India, Dec 31, 33. PHYSICS. Educ: Univ Kerala, BSc, 53; Univ Saugar, MSc, 58; NMex State Univ, MS, 65, PhD(physics), 67. Prof Exp: Instr physics, Sacred Heart Col, India, 53-55 & 57-58; asst prof, 67-71, ASSOC PROF PHYSICS, MARSHALL UNIV, 71- Concurrent Pos: NSF Mem: Am Phys Soc; Am Asn Physics Teachers; Inst Fundamental Studies Asn. Res: Magnetic resonance; crystal field studies; nuclear physics; interactions of gamma photons with fibers. Mailing Add: Dept of Physics Marshall Univ Huntington WV 25701

MANASEK, FRANCIS JOHN, b New York, NY, July 22, 40. CELL BIOLOGY, DEVELOPMENTAL BIOLOGY. Educ: NY Univ, AB, 61; Harvard Univ, DMD, 66. Prof Exp: Fel anat, Harvard Med Sch, 66-68; vis investr develop biol, Carnegie Inst Washington, 68-69; from instr to asst prof anat, Harvard Med Sch, 69-74; ASSOC PROF ANAT, UNIV CHICAGO, 74- Concurrent Pos: Res assoc path & cardiol, Children's Hosp, 69-74; NIH res career develop award, 71. Honors & Awards: Fel, Med Found, Inc, 69. Mem: Am Asn Anatomists; Am Soc Cell Biol; Soc Develop

Biol. Res: Cell biology of developing cardiac muscle particularly synthesis of structural macromolecules and their role in myocardial morphogenesis. Mailing Add: 1025 E 57th St Chicago IL 60637

MANASEVIT, HAROLD MURRAY, b Bridgeport, Conn, Nov 1, 27; m 53; c 3. PHYSICAL INORGANIC CHEMISTRY, ORGANOMETALLIC CHEMISTRY. Educ: Univ Ohio, BS, 50; Pa State Univ, MS, 51; Ill Inst Technol, PhD(phys inorg chem), 59. Prof Exp: Chemist, Armour Res Found, 51-55; inorg chemist, US Borax Res Corp, 59-60; sr res engr, Autonetics Div, NAm Aviation, Inc, 60-61, res specialist, 61-63, sr tech specialist, 63-67, mem tech staff, NAM Rockwell, Inc, 67-70, MEM TECH STAFF, ROCKWELL INT, 70- Honors & Awards: 1975 Electronics Div Award, Electrochem Soc. Mem: AAAS; Am Chem Soc; Electrochem Soc; Sigma Xi. Res: Thin film growth of metals and semiconductors on semiconductors and insulating substrates; epitaxy; gas phase acid-base reactions; metalorganics; diboron compounds. Mailing Add: 1855 Janette Lane Anaheim CA 92802

MANASSE, ROBERT JAMES, b New York, NY, Apr 22, 40; m 70. MICROBIOLOGY. Educ: Columbia Col, AB, 61; Columbia Univ, MA, 63, PhD(biol sci), 67. Prof Exp: Lab asst microbiol, Barnard Col, Columbia Univ, 64-66; ASST MICROBIOLOGIST, BOYCE THOMPSON INST, 66- Mem: AAAS; Am Soc Microbiol; Brit Soc Gen Microbiol; Can Soc Microbiol. Res: Crown gall; plant tissue culture. Mailing Add: Boyce Thompson Inst 1086 N Broadway Yonkers NY 10701

MANASSE, ROGER, b New York, NY, Apr 9, 30; m 52; c 2. PHYSICS. Educ: Mass Inst Technol, BS, 50, PhD(physics), 55. Prof Exp: Mem staff instrumentation lab, Mass Inst Technol, 50-52, Lincoln Lab, 54-59; subdept head radar dept, Mitre Corp, 59-60, assoc dept head, 60-64, dept head, 64-67; staff mem, Gen Res Corp, 67-70; independent consult, 70-71; vpres & chief scientist, Spectra Res Systs, 72-73; INDEPENDENT CONSULT, 74- Concurrent Pos: Mem, Air Force Sci Adv Bd, 69- Mem: Inst Elec & Electronics Eng; Sigma Xi. Res: Radar systems analysis and measurement theory. Mailing Add: 234 Canon Dr Santa Barbara CA 93105

MANASTER, ALFRED B, b Chicago, Ill, May 25, 38; m 60; c 3. MATHEMATICAL LOGIC. Educ: Univ Chicago, BS, 60; Cornell Univ, PhD(math), 65. Prof Exp: Res assoc math, Cornell Univ, 65; instr, Mass Inst Technol, 65-67; asst prof, 67-71, ASSOC PROF MATH, UNIV CALIF, SAN DIEGO, 71- Mem: Am Math Soc; Math Asn Am; Asn Symbolic Logic. Res: Recursive function theory. Mailing Add: Dept of Math Univ of Calif San Diego PO Box 109 La Jolla CA 92037

MANATT, STANLEY L, b Glendale, Calif, July 13, 33; m 58; c 4. ORGANIC CHEMISTRY. Educ: Calif Inst Technol, BS, 55, PhD(chem, physics), 59. Prof Exp: Wis Alumni Res Found fel, Univ Wis, 58-59; sr scientist, Jet Propulsion Lab, Calif Inst Technol, 59-64, res specialist, 65-66, mem tech staff, 66-70, asst mgr biosci & planetology sect, 70-73, staff scientist, Sci Data Anal Sect & Chem Dept, 73-74, STAFF SCIENTIST, SCI DATA ANAL SECT, JET PROPULSION LAB, CALIF INST TECHNOL, 75- Concurrent Pos: Vis prof, Inst Org Chem, Univ Cologne, 74-75; Am ed, Org Magnetic Resonance, 68-; Alexander von Humboldt Award, 74-75. Mem: AAAS; Am Chem Soc. Res: Nuclear magnetic resonance; polymers; propellant and fluorocarbon chemistry; gas chromatography techniques; spacecraft material problems; extraterrestrial life detection; analytical chemistry; small-ring compounds; theoretical calculations on aromatic molecules; polypeptide synthesis; steric effects in organic molecules. Mailing Add: Jet Propulsion Lab Calif Inst of Technol 1201 E California Blvd Pasadena CA 91109

MANCALL, ELLIOTT L, b Hartford, Conn, July 31, 27; m; c 2. NEUROPATHOLOGY, NEUROLOGY. Educ: Trinity Col, BS, 48; Univ Pa, MD, 52; Am Bd Psychiat & Neurol, dipl & cert neurol, 59. Prof Exp: Asst prof neurol, Jefferson Med Col, 53-64, assoc prof, 64-65; PROF MED & NEUROL, HAHNEMANN MED COL, 65- Concurrent Pos: Fulbright fel, Nat Hosp Neurol Dis, London, 54-55; teaching gel neuropath, Harvard Med Sch, 56-57; mem vis fac, Sch Med, Emory Univ, 64; vis lectr, US Naval Hosp, 64-; consult, Valley Forge Gen Army Hosp, 67-74 & Pennhurst State Sch & Hosp, 68-73; chief neurol serv, Philadelphia Gen Hosp, 69-74. Mem; Fel Am Acad Neurol; Am Neurol Asn; Am Asn Neuropath; Pan-Am Med Asn; Asn Res Nerv & Ment Dis. Res: Neurology and neuropathology of metabolic diseases of the nervous system. Mailing Add: 230 N Broad St Philadelphia PA 19107

MANCERA, OCTAVIO, b Oaxaca, Mex, Mar 28, 19; m 50; c 4. ORGANIC CHEMISTRY. Educ: Nat Univ Mex, MS, 43; Oxford Univ, DPhil(chem), 46. Prof Exp: Lectr org chem, Nat Univ Mex, 47-50; res asst steroid chem, SA group leader, 53-56, asst res dir, 57-59, process res dir, 60-61, prod dir, 62-63, dir opers, 64-68, GEN MGR CHEM DIV, SYNTEX, 68- Res: Chemistry of penicillin and natural products; steroid chemistry. Mailing Add: Apartado Postal 517 Cuernavaca Mexico

MANCHE, EMANUEL PETER, b New York, NY, Apr 30, 31; m 61; c 3. CHEMISTRY. Educ: City Col New York, BS, 56; Brooklyn Col, MA, 59; Rutgers Univ, NB, PhD(chem), 65. Prof Exp: Res chemist, Am Chicle Co, 56-57; lectr chem, Brooklyn Col, 58-62, sch gen studies, 58-59; teaching asst, Rutgers Univ, 62-64, instr, 65; advan res engr, Gen Tel & Electronics Res Labs, 65-68; asst prof chem, 68-72, ASSOC PROF CHEM, YORK COL, CITY UNIV NEW YORK, 73- Concurrent Pos: Instr univ col, Rutgers Univ, 64-; invited lectr, United Hosps, Newark, 72-73. Mem; Am Chem Soc; Electron Micros Soc Am. Res: Chemical instrumentation; thermal methods of analysis including thermogravimetry and differential thermal analysis; thermoluminescent dating. Mailing Add: Dept of Natural Sci York Col City Univ of New York Jamaica NY 11451

MANCHEE, ERIC BEST, b Toronto, Oct 16, 18; m 45; c 3. GEOPHYSICS, SCIENCE ADMINISTRATION. Educ: Univ Toronto, BASc, 49, MA, 51. Prof Exp: Geophysicist, Calif Standard Co, Alta, 51-59, dist geophysicist, 59-62; head array seismol sect, 62-74, head spec proj, 74-75, BR PROG OFFICER, EARTH PHYSICS BR, DEPT ENERGY, MINES & RESOURCES, 75- Mem: Seismol Soc Am; Can Asn Physicists; Can Geophys Union. Res: Exploration and array seismology; earthquake-explosion differentiation. Mailing Add: Earth Physics Br Dept of Energy Mines & Resources Ottawa ON Can

MANCHESTER, DONALD FRASER, b Fairville, NB, Apr 18, 20; m 45; c 7. ORGANIC CHEMISTRY. Educ: Queen's Univ, Ont, BSc, 43, MSc, 49; McGill Univ, PhD(chem), 52. Prof Exp: MGR PROCESS RES, RES CTR, ABITIBI PAPER CO, LTD, 51- Mem: Tech Asn Pulp & Paper Indust; Can Pulp & Paper Asn; Chem Inst Can. Res: Polymerization of hydrogen cyanide; nitration of unsubstituted diazocycloalkines; oxidation and methylation of cellulose; recovery of pulping reagents from sulphite spent liquors; brightening of high yield chemical and mechanical pulps; pulping processes; environmental research and waste treatment. Mailing Add: Abitibi Paper Co Ltd Res Ctr Sheridan Park ON Can

MANCHESTER, J STEWART, b NB, May 22, 16; m 41; c 1. RADIOLOGY. Educ: Dalhousie Univ, MD & CM, 41. Prof Exp: Assoc prof, 53-55, PROF RADIOL & CHMN DEPT, DALHOUSIE UNIV, 55-; HEAD DEPT DIAG RADIOL, VICTORIA GEN HOSP, HALIFAX, 55- Concurrent Pos: Chief diag radiol & assoc therapeut radiol, Victoria Gen Hosp, 53-55; consult, Tri-Serv Hosp, Dept Nat Defense, Halifax, NS, 55-, Halifax Childrens Hosp, 61- & Camp Hill Hosp, Dept Vet Affairs, Halifax, 63- Mem: Fel Am Col Radiol; Can Asn Radiol; Can Med Asn. Res: Gastrointestinal radiology; biliary system with particular reference to abnormal physiology. Mailing Add: Dept of Radiol Victoria Gen Hosp Halifax NS Can

MANCHESTER, KENNETH EDWARD, b Winona, Minn, Mar 22, 25; m 46; c 2. SURFACE CHEMISTRY. Educ: San Jose State Col, AB, 49; Stanford Univ, MS, 50, PhD(thermochem), 55. Prof Exp: Fel chem, Stanford Univ, 52-55; chemist surface chem, Shell Develop Co, 55-62; sect head semiconductor chem, 62-63, dept head, 63-69, DIR SEMICONDUCTOR RES & DEVELOP, SPRAGUE ELEC CO, 69- Mem: Am Chem Soc; Am Asn Contamination Control; Am Inst Mining, Metal & Petrol Eng. Res: Energetics of liquid-liquid and liquid-solid interfaces; interaction of energetic ions or electrons with solid substrates. Mailing Add: Sprague Elec Co 115 NE Cutoff Worcester MA 01606

MANCI, ORLANDO J, JR, aeronautical engineering, see 12th edition

MANCILLA, RAFAEL, biochemistry, parasitology, see 12th edition

MANCINELLI, ALBERTO L, b Rome, Italy, Nov 22, 31; m 62. PLANT PHYSIOLOGY. Educ: Univ Rome, Dr rer nat(bot), 54. Prof Exp: Vol asst prof bot, Univ Rome, 54-64; asst prof, 64-67, ASSOC PROF PLANT PHYSIOL, COLUMBIA UNIV, 67- Concurrent Pos: Ital Nat Res Coun fel, 56-59; Ital Nat Comt Nuclear Energy fel, 59-62; NATO fel, 62-63; NSF res grant, 65-72. Mem: AAAS; Am Soc Plant Physiol; Am Inst Biol Sci; Japanese Soc Plant Physiol; Bot Soc Am. Res: Metabolism during seed germination; reactions controlling light responses in plant growth and development, particularly phytochrome controlled and high energy reaction controlled responses. Mailing Add: Dept of Biol Sci Columbia Univ New York NY 10027

MANCINI, ROBERT EDWARD, b New York, NY, Feb 28, 45. PHARMACOLOGY, TOXICOLOGY. Educ: Seton Hall Univ, BA, 66, MS, 69; Thomas Jefferson Univ, PhD(pharmacol), 73. Prof Exp: Instr, 72-73, ASST PROF PHYSIO-PHARM, PHILADELPHIA COL OSTEOP MED, 73- Mem: AAAS; Am Col Clin Pharmacol; Am Osteop Asn. Res: Interaction of drugs and solvents establishing biochemical toxicological parameters to importance of these interactions. Mailing Add: Dept of Physio-Pharm Philadelphia Col Osteop Med Philadelphia PA 19131

MANCINI, ROBERT EUSEBIO, b Buenos Aires, Arg, Sept 26, 16; m 49; c 2. MEDICINE. Educ: Univ Buenos Aires, MD, 48. Prof Exp: Instr histol, Inst Gen Anat & Embryol, Med Sch, Univ Buenos Aires, 40-46; chief lab histol, Nat Inst Endocrinol, 48-55; lectr, 55-57, vdean, 58-69, PROF HISTOL, INST GEN ANAT & EMBRYOL, MED SCH, UNIV BUENOS AIRES, 56-, DIR CTR STUDIES REPROD, 65-, PROF EMBRYOL, 69- Concurrent Pos: Guggenheim Found fel, Columbia Univ, 51, res assoc, 52; consult ed, Endocrinology & J d'Histochimie, France, 64; mem, Career for Investr, Nat Res Coun, Buenos Aires, Arg. Honors & Awards: Nat Res Coun Arg Award, 63; Gold Medal, Arg Soc Endocrinol & Metab, 57. Mem: Histochem Soc; Am Asn Anat; Endocrine Soc; Am Fertil Soc; Arg Soc Endocrinol & Metab. Res: Histology; biochemistry; experimental endocrinology. Mailing Add: Dept of Histol Med Sch Univ of Buenos Aires Buenos Aires Argentina

MANCLARK, CHARLES ROBERT, b Rochester, NY, June 22, 28; m 53; c 2. MICROBIOLOGY, IMMUNOLOGY. Educ: Calif Polytech State Univ, BS, 53; Univ Calif, Los Angeles, PhD(bact), 63. Prof Exp: Asst prof microbiol, Calif State Univ, Long Beach, 61-64; asst res bacteriologist, Univ Calif, Los Angeles, 63-65, asst prof microbiol, Col Med, Univ Calif, Irvine, 65-67; res microbiologist, Lab Bact Prod, Div Biologics Stand, NIH, 67-72; RES MICROBIOLOGIST, DIV BACT PROD, BUR BIOLOGICS, FOOD & DRUG ADMIN, 72- Concurrent Pos: Mem subcomt taxon of vibrios, Int Comt Bact Nomenclature. Mem: AAAS; Am Soc Microbiol; Mycol Soc Am; Am Soc Indust Microbiol; Int Asn Biol Stand. Res: Immunity and the immune response; diagnostic bacteriology; delayed hypersensitivity; the immune basis of infertility and sterility; site of antibody formation; host-parasite relationships in pertussis. Mailing Add: Div of Bact Prod Bur Biologics Food & Drug Admin Bethesda MD 20014

MANCUSI, MICHAEL D, b New York, NY, Aug 31, 40; m 61; c 3. COMPUTER SCIENCE. Educ: Univ Notre Dame, BS, 62; Univ Iowa, PhD(nuclear physics), 66. Prof Exp: Res asst nuclear physics, Univ Iowa, 63-66; US AEC fel, Oak Ridge Nat Lab, 66-68; MEM TECH STAFF, BELL TEL LABS, 68- Mem: Am Phys Soc; Pattern Recognition Soc; Inst Elec & Electronics Eng. Res: Low mass nuclear structure physics and small computer systems applications; computer graphics and pattern recognition. Mailing Add: Room 4F625 Bell Tel Labs Holmdel NJ 07733

MANCUSO, JOSEPH J, b Hibbing, Minn, Dec 9, 33; m 57; c 4. ECONOMIC GEOLOGY. Educ: Carleton Col, BA, 55; Univ Wis, MS, 57; Mich State Univ, PhD(geol), 60. Prof Exp: Instr geol, Mich State Univ, 58-60; asst prof, 60-71, PROF GEOL, BOWLING GREEN STATE UNIV, 71- Mem: Am Inst Prof Geol; Soc Econ Geol. Res: Economic geology, mineralogy and stratigraphy of the Lake Superior iron formations; Precambrian geology of Wisconsin, Michigan and Minnesota. Mailing Add: Dept of Geol Bowling Green State Univ Bowling Green OH 43402

MANCUSO, RICHARD VINCENT, b Rochester, NY, Nov 4, 38; m 64; c 2. NUCLEAR PHYSICS. Educ: St Bonaventure Univ, BS, 60; State Univ NY Buffalo, PhD(physics), 65. Prof Exp: Teaching asst physics, State Univ NY Buffalo, 61-65; Nat Acad Sci-Nat Res Coun res assoc nuclear physics, Van De Graaff Br, Naval Res Lab, DC, 67-69; asst prof, 69-74, ASSOC PROF PHYSICS, STATE UNIV NY COL BROCKPORT, 74- Mem: Am Phys Soc. Res: Gamma ray spectroscopy; level structures of medium weight nuclei; charged particle reactions and reaction mechanisms; application of nuclear techniques to non-nuclear problems. Mailing Add: Dept of Physics State Univ of NY Col Brockport NY 14420

MANCUSO, VINCENT J, b New York, NY, Dec 14, 39; m 65; c 1. MATHEMATICS. Educ: Fordham Univ, BS, 61; Rutgers Univ, MS, 63, PhD(math), 65. Prof Exp: Asst prof, 65-71, ASSOC PROF MATH, ST JOHN'S UNIV, 71- Mem: Am Math Soc. Res: Point set topology. Mailing Add: Dept of Math St John's Univ Jamaica NY 11358

MANDAL, ANIL KUMAR, b West Bengal, India, Nov 12, 35; m 64; c 2. CARDIOVASCULAR DISEASES. Educ: Univ Calcutta, MB, BS, 59; Am Bd Internal Med, dipl, 72. Prof Exp: Med officer, Inst Postgrad Med Educ & Res, Calcutta, 63-66; registr, R G Kar Med Col, Calcutta, 66-67; lectr path, Univ Edinburgh, 68-69; instr med, Univ Ill, Chicago Circle, 71-72; asst prof, 72-75, ASSOC PROF MED, COL MED, UNIV OKLA, 75- Concurrent Pos: Consult physician, Vet Admin Hosp, Muskogee, Okla, 72-; asst physician, Okla Med Res Found, 72-; attend nephrologist, Vet Admin Hosp & Univ Hosp, Oklahoma City, 75- Honors & Awards: Recognition Award, AMA, 69, 70, 71 & 72. Mem: Fel Am Col Physicians; Am Fedn

Clin Res; Am Soc Nephrol; Electron Micros Soc Am; Sigma Xi. Res: Pathological study by light, electron and fluorescence microscopy of kidney in experimental renal disease and hypertension. Mailing Add: 921 NE 13th St Oklahoma City OK 73104

MANDARINO, JOSEPH ANTHONY, b Chicago, Ill, Apr 20, 29; m 56; c 4. MINERALOGY. Educ: Mich Col Mining & Technol, BS, 50, MS, 51; Univ Mich, PhD(mineral), 58. Prof Exp: Asst prof mineral, Mich Col Mining & Technol, 57-59; assoc cur, 59-65, CUR DEPT MINERAL, ROYAL ONT MUS, 65- Concurrent Pos: Mem, Joint Comt Powder Diffraction Standards; Nat Res Coun Can sr res fel, Fr Bur Geol Mines Res, 68-69. Mem: Mineral Soc Am; Mineral Asn Can; Mineral Soc Gt Brit & Ireland; Fr Soc Mineral & Crystallog. Res: Crystal optics; crystallography; descriptive mineralogy. Mailing Add: Dept of Mineral & Geol Royal Ont Mus 100 Queen's Park Toronto ON Can

MANDAVA, NAGABHUSHANAM, b Bhushanagulla, India, Oct 14, 34; m 57; c 3. BIO-ORGANIC CHEMISTRY. Educ: Univ Andhra, India, BSc, 55; Banaras Hindu Univ, MSc, 57; Indian Inst Sci, Bangalore, PhD(chem), 62. Prof Exp: Res assoc, Okla State Univ, 63-65, State Univ NY Stony Brook, 65-66 & Laval Univ, 66-68; chemist, Plant Sci Res Div, 68, RES CHEMIST, PLANT PHYSIOL INST, AGR RES SERV, USDA, 68- Concurrent Pos: Nat Res Coun Can fel, 66-68. Mem: Fel Am Inst Chem; Am Chem Soc; Int Plant Growth Workers Soc. Res: Organophosphorus compounds; plant hormones, lipids, steroids, carbohydrates, alkaloids, pesticides and heterocyclic compounds; application of specroscopy and computers to structural and stereochemical problems; bioassays and tracer techniques. Mailing Add: Plant Physiol Inst Agr Res Serv USDA 101 West Bldg Beltsville MD 20705

MANDEL, BENJAMIN, b New York, NY, Aug 17, 13; m 46; c 2. VIROLOGY. Educ: City Col New York, BS, 40; NY Univ, MS, 48, PhD(microbiol), 51. Prof Exp: Assoc mem div infectious dis, Pub Health Res Inst New York, Inc, 52-67; res assoc prof, 66-71, RES PROF MICROBIOL, COL MED, NY UNIV, 71-; MEM DIV INFECTIOUS DIS, PUB HEALTH RES INST NEW YORK, INC, 67- Mem: AAAS; Am Soc Microbiol; fel NY Acad Sci. Res: Fundamentals of virus-cell relationship. Mailing Add: Pub Health Res Inst New York Inc 455 First Ave New York NY 10016

MANDEL, BENJAMIN J, b Poland, Sept 1, 12; US citizen; m 37; c 2. APPLIED STATISTICS. Educ: City Col New York, BS, 34; George Washington Univ, MA, 38; Goteborg Sch Econ & Bus Admin, Sweden, Ekonomie Licentiate, 67. Prof Exp: Jr asst statistician, Social Security Admin, 38-44, chief statist div, 44-60; asst dir statist standards & opers, US Off Educ, 60-62; dir off statist prog, US Post Off Dept, 62-70; STATIST SCI CONSULT, 70- Concurrent Pos: Prof statist, Univ Baltimore, 46-70, chmn dept, 48-70, EMER emer prof, 70-; vis prof mgt & statist, Dept Agr Grad Sch, 62-; lectr statist for mgt, Bur Training, US Civil Serv Comn, 65-; vis lectr statist, NSF, 71- Honors & Awards: Dir Citation Outstanding Contrib Mgt, Dept Health, Educ & Welfare, 59; Superior Accomplishment Award, US Post Off Dept, 67, Meritorious Serv Award, 68 & Superior Accomplishment Award, 70. Mem: Am Statist Asn; Am Soc Qual Control; Am Asn Retired Teachers; Smithsonian Assocs; Am Mgt Asn. Res: Extension of statistical theory and techniques to new areas of management, administration, accounting, auditing, inspection and quality assurance. Mailing Add: 6101 16th St NW Washington DC 20011

MANDEL, EDWARD H, b NY, June 17, 21; m; c 2. DERMATOLOGY. Educ: Cornell Univ, BA, 42; New York Med Col, MD, 45. Prof Exp: Fel med mycol, Skin & Cancer Unit, NY Univ-Bellevue Med Ctr, 48-51; PROF DERMAT & CHMN DEPT, NEW YORK MED COL, 68- Mem: Fel Am Col Physicians; Am Acad Dermat; Soc Invest Dermat. Mailing Add: Dept Dermat New York Med Col Metrop Hosp 1901 First Ave New York NY 10029

MANDEL, FREDERIC, b Teheran, Iran, Apr 6, 41. THEORETICAL CHEMISTRY, BIOPHYSICAL CHEMISTRY. Educ: Univ Calif, Berkeley, BA, 65; Univ Kans, PhD(chem), 72. Prof Exp: Fel chem, Rice Univ, 70-71; instr, Univ Rochester, 71-73; fel chem, Rice Univ, 73-75; NIH TRAINEE BIOPHYS, BAYLOR COL MED, 75- Res: Theory of liquids and macromolecules and the application of these theories to the studies of enzymes and transport across cell membranes. Mailing Add: Dept of Cell Biophys Baylor Col of Med Tex Med Ctr Houston TX 77025

MANDEL, HAROLD GEORGE, b Berlin, Ger, June 6, 24; nat US; m 53; c 2. PHARMACOLOGY. Educ: Yale Univ, BS, 44, PhD(org chem), 49. Prof Exp: Asst & lab instr chem, Yale Univ, 42-44, lab instr org chem, 47-49; res assoc, 49-50, asst res prof, 50-52, assoc prof, 52-58, PROF PHARMACOL, SCH MED, GEORGE WASHINGTON UNIV, 58-, CHMN DEPT, 60- Concurrent Pos: Advan Commonwealth Fund fel, Molteno Inst, Eng, 56 & Pasteur Inst, France, 57; travel award, Int Pharmacol Cong, Prague, 63 & Helsinki, 75; Commonwealth Fund sabbatical leave, Univ Auckland & Univ Med Sci, Thailand, 64; Am Cancer Soc Eleanor Roosevelt Int fel, Chester Beatty Res Inst, London, 70-71; lectr, US Naval Dent Sch, 59-61, 71-75, Wash Hosp Ctr, 60-66, US Army Dent Sch, 72-75 & Holy Cross Hosp, 72-74; consult, Fed Aviation Agency, 61-62; mem biochem comt, Cancer Chemother Nat Serv Ctr, 58-61, med adv comt, Therapeut Res Found, Inc, 62-, pharmacol & exp therapeut B study sect, USPHS, 63-68, comt probs drug safety, Nat Acad Sci-Nat Res Coun, 65-71 & 72-76, mem drug metab workshop progs, NY Univ, 66, George Washington Univ, 67 & Univ Calif, 68; mem cancer chemother comt, Int Union Against Cancer, 66, res comt, Children's Hosp, Washington, DC, 69- & sci adv comt, Registry Tissue Reactions to Drugs, 70-; mem cancer chemother training progs, Poland, 68, Curacao, 71, Ger, 73, Belg, 73 & Arg, 73; mem chemother comt, Am Cancer Soc, 69-73; mem cancer res training comt, Nat Cancer Inst, 71-73, mem cancer spec prog adv comt, 74-; consult, Roswell Park Inst, Buffalo, 72-74; consult toxicol & mem toxicol adv comt, Food & Drug Admin, 75-; mem merit rev bd, Vet Admin, 75- Honors & Awards: Abel Award, Am Soc Pharmacol & Exp Therapeut, 58. Mem: AAAS; Am Soc Pharmacol & Exp Therapeut (secy, 61-63, pres, 73-74); Am Chem Soc; Am Soc Biol Chem; Am Asn Cancer Res. Res: Drug metabolism; mechanism of action of antimetabolites and other anti-cancer drugs; action of growth inhibitory drugs. Mailing Add: Dept of Pharmacol George Washington Univ Sch of Med Washington DC 20037

MANDEL, IRWIN D, b New York, NY, Apr 9, 22; m 44; c 3. PREVENTIVE DENTISTRY, ORAL BIOLOGY. Educ: City Col New York, BS, 42; Columbia Univ, DDS, 45. Prof Exp: Res asst dent & oral surg, 46-50, instr, 50-57, from asst clin prof to clin prof, 57-69, dir, Lab Clin Res, 60-69, PROF DENT & ORAL SURG, COLUMBIA UNIV, 69-, DIR DIV PREV DENT, 69- Concurrent Pos: USPHS grant, 64-; Health Res Coun City of New York career scientist award, 69-; consult & chmn, Oral Biol & Med Study Sect, Coun Dent Therapeut, 68- Mem: Fel AAAS; Am Dent Asn; Int Asn Dent Res. Res: Plaque, calculus and periodontal disease; salivary composition and relation to oral and systemic disease; effect of pharmacologic agents on salivary composition. Mailing Add: Sch of Dent & Oral Surg Columbia Univ New York NY 10032

MANDEL, JOHN, b Antwerp, Belgium, July 12, 14; nat US; m 38; c 2. MATHEMATICAL STATISTICS. Educ: Univ Brussels, BS, 35, MS, 37; Eindhoven Technol Univ, PhD(appl statist), 65. Prof Exp: Res chemist, Soc Belge De Recherches, Belgium, 38-40; anal & develop chemist, Foster D Snell, Inc, 41-43; res chemist, B G Corp, 44-47; gen phys scientist, 47-48, anal statistician, 48-57, MATH STATISTICIAN, NAT BUR STANDARDS, 58- Honors & Awards: Silver Medal for Meritorious Serv, US Dept Commerce, 57, Gold Medal Award, 73. Mem: Fel Am Statist Asn; Inst Math Statist; fel Am Qual Control; fel Royal Statist Soc. Res: Statistical design of experiments; statistical analysis of data obtained in physical and chemical experimentation; development of statistical techniques for the physical sciences; interlaboratory testing; statistical evaluation of measuring processes. Mailing Add: 10710 Lombardy Rd Silver Spring MD 20901

MANDEL, JOHN HERBERT, b New York, NY, Mar 11, 25; m 50; c 3. MEDICAL MICROBIOLOGY. Educ: City Col New York, BS, 47; Univ Calif, Berkeley, MA, 49. Prof Exp: Chief diag serol sect & res bacteriologist, US Army Grad Sch, Walter Reed Army Med Ctr, 50-51; chief serol & hemat sect, FDR Vet Admin Hosp, Montrose, NY, 51-55; DIR, LEHIGH VALLEY LABS, INC, 55- Concurrent Pos: Bioanalyst dir, Am Bd Bioanalysts; specialist-microbiologist, Nat Registry Microbiologists; instr microbiol, Pa State Univ, 69-74. Mem: Am Soc Microbiol; Am Asn Bioanalysts; Am Pub Health Asn; Royal Soc Health; Int Soc Human & Animal Mycol. Res: Rapid isolation and identification of pathogenic microorganisms from biological materials, food and water; identification and significance of yeasts isolated from clinical specimens. Mailing Add: Lehigh Valley Labs Inc Microbiol Dept 1740 Allen St Allentown PA 18105

MANDEL, LAZARO J, b Lima, Peru, Oct 13, 40; US citizen; m 63. PHYSIOLOGY, BIOPHYSICS. Educ: Mass Inst Technol, BS, 61, MS, 62; Univ Pa, PhD(biomed eng), 69. Prof Exp: USPHS fel, Yale Univ Med Sch, 69-72; ASST PROF PHYSIOL, MED CTR, DUKE UNIV, 72- Mem: AAAS; Biophys Soc; Am Physiol Soc. Res: Active and passive transport; biological energy conversion; properties of biological membranes. Mailing Add: Dept of Physiol Duke Univ Med Ctr Durham NC 27710

MANDEL, LEWIS RICHARD, b Brooklyn, NY, Nov 13, 36; m 60; c 3. BIOCHEMISTRY, PHARMACOLOGY. Educ: Columbia Univ, BS, 58, PhD(biochem), 62. Prof Exp: Asst prof pharmacol, Col Pharm, Columbia Univ, 62-64; sr res biochemist, 64-67, sect head, 67-71, asst dir, 71-74, DIR BIOCHEM, MERCK SHARP & DOHME RES LABS, 74- Concurrent Pos: NIH Res Grant, 63-64. Mem: Soc Neurosci; Int Soc Neurochem; NY Acad Sci; Am Soc Biol Chem; Am Soc Pharmacol & Exp Therapeut. Res: Lipid metabolism; prostaglandins; transmethylation reactions. Mailing Add: Dept of Biochem Merck Sharp & Dohme Res Labs Rahway NJ 07065

MANDEL, MANLEY, b Philadelphia, Pa, July 10, 23; m 52; c 3. MICROBIOLOGY, MOLECULAR BIOLOGY. Educ: Brooklyn Col, BA, 43; Mich State Univ, MS, 47, PhD(bact), 52. Prof Exp: Asst biol, Brooklyn Col, 46; guest investr microbiol, Haskins Labs, 46 & 52; asst bact, Univ Calif, 48-50; instr, Univ Mass, 52-53, asst prof, 53-63; res assoc, Brandeis Univ, 63; assoc prof biol, 63-66, assoc biologist, 63-66, PROF BIOL, UNIV TEX GRAD SCH BIOMED SCI, HOUSTON, 66-, CHIEF SECT MOLECULAR BIOL, UNIV TEX MD ANDERSON HOSP & TUMOR INST, 63-, BIOLOGIST, 66- Concurrent Pos: Mem molecular biol study sect, NIH, 66-70, sci review comt health related facil, 70-74; mem bd trustees, Am Type Cult Collection, 73-76 & Nucleic Acid & Protein Biosynthesis Adv Comt, Am Cancer Soc, 70-75; consult ed, McGraw-Hill Encycl Sci & Technol, 74- Mem: Am Soc Microbiol; Biophys Soc; Genetics Soc Am; Brit Soc Gen Microbiol. Res: Molecular and genetic relations of protists; role of modification in viral DNA-host interactions. Mailing Add: M D Anderson Hosp & Tumor Inst Univ of Tex 6723 Bertner Ave Houston TX 77025

MANDEL, MORTON, b Brooklyn, NY, July 6, 24; m 52; c 2. MOLECULAR BIOLOGY. Educ: City Col New York, BCE, 44; Columbia Univ, MS, 49, PhD(physics), 57. Prof Exp: Instr civil eng, Stevens Inst Technol, 48; asst microwave components, Columbia Univ, 52-56; mem tech staff solid state devices, Bell Tel Labs, 56-57; res assoc paramagnetic resonance, Stanford Univ, 57-61, asst prof physics, 59-61; eng specialist, Gen Tel & Electronics Lab, 61-63; res assoc genetics, Sch Med, Stanford Univ, 63-64; USPHS fel microbial genetics, Karolinska Inst, Sweden, 64-66; assoc prof biophys, 66-69, chmn dept biochem & biophys, 71-72, PROF BIOPHYS, SCH MED, UNIV HAWAII, 69- Concurrent Pos: Consult, Fairchild Semiconductor Corp, 57-58, Hewlett-Packard Co, 58 & Rheem Semiconductor Corp, 59; vis prof, Worcester Found Exp Biol, 72-73. Mem: Am Phys Soc; Genetics Soc Am; Biophys Soc; Sigma Xi. Res: Nuclear magnetic resonance studies on protein-nucleic acid interactions; bacterial transformation and transfection; carcinogenesis in colon cancer. Mailing Add: Dept of Biochem & Biophys Univ of Hawaii Sch of Med Honolulu HI 96822

MANDEL, ZOLTAN, b Czech, July 18, 24; US citizen; m 57; c 1. TEXTILE CHEMISTRY. Educ: Western Reserve Univ, BS, 51, MS, 52, PhD(org chem), 55. Prof Exp: Res chemist, Diamond Alkali Co, 54-55; res chemist, 55-63, sr res chemist, 63-65, res assoc, 65-72, SR RES CHEMIST, E I DU PONT DE NEMOURS & CO, INC, 72- Mem: Am Chem Soc. Res: Synthetic fibers. Mailing Add: 4013 Greenmount Rd Longwood Wilmington DE 19810

MANDELBAUM, HUGO, b Sommerhausen, Bavaria, Oct 18, 01; nat US; m 31; c 5. PHYSICAL OCEANOGRAPHY. Educ: Univ Hamburg, Dr rer nat(geophys), 34. Prof Exp: Teacher high sch, Hamburg, 30-38; prin, Jewish Day Sch, Mich, 40-48; prof geol, 48-71, EMER PROF GEOL, WAYNE STATE UNIV, 71- Mem: AAAS; Soc Explor Geophys; Siesmol Soc Am; Am Geophys Union. Res: Tides and tidal currents; air-sea boundary problems; sedimentation; the astronomical foundation of the Jewish calendar. Mailing Add: Givath Beth Hakerem Block 1 Number 37 Jerusalem 96268 Israel

MANDELBAUM, ISIDORE, b New York, NY. SURGERY. Educ: NY Univ, AB, 48; State Univ NY Downstate Med Ctr, MD, 52. Prof Exp: From instr to asst prof, 61-68, ASSOC PROF SURG, SCH MED, IND UNIV, INDIANAPOLIS, 68- Concurrent Pos: Dazian fel path, Mt Sinai Hosp, New York, 54-55; Nat Heart Inst fel, Ind Univ, Indianapolis, 61-62; consult, Vet Admin Hosp, Indianapolis, 61- Mem: Am Asn Thoracic Surg; Soc Vascular Surg; Int Cardiovasc Soc; Am Soc Clin Invest; Am Col Surg. Res: Cardiothoracic surgery; cardiovascular and pulmonary research. Mailing Add: Dept of Surg Ind Univ Sch of Med Indianapolis IN 46202

MANDELBERG, HIRSCH I, b Baltimore, Md, Apr 16, 34; m 58; c 2. PHYSICS. Educ: Johns Hopkins Univ, BE, 54, PhD(physics), 60. Prof Exp: Res physicist res inst adv study div, Martin Co, 56-60; RES PHYSICIST, US DEPT DEFENSE, 60- Concurrent Pos: Visitor, Univ Col, Univ London, 67-68; assoc prof, Univ Col, Univ Md, 61-66. Mem: Am Phys Soc. Res: Atomic collisions; optics; optical propagation; gaseous electronics; lasers; electron collisions; optical spectroscopy; laser applications. Mailing Add: 6800 Pimlico Dr Baltimore MD 21209

MANDELBROT, BENOIT, b Warsaw, Poland, Nov 20, 24; m 55; c 2. APPLIED MATHEMATICS. Educ: Polytech Sch, Paris, Engr, 47; Calif Inst Technol, MS, 48;

Univ Paris, PhD(math), 52. Prof Exp: Mathematician, Philips Electronics, Paris, 50-53; mem staff sch math, Inst Adv Study, NJ, 53-54; assoc, Inst Henri Poincare, Paris, 54-55; asst prof math, Univ Geneva, 55-57; jr prof, Lille & Polytech Sch, Paris, 57-58; mathematician, 58-74, IBM FEL, IBM CORP, 74- Concurrent Pos: Vis prof, Harvard Univ, 62-64; inst lectr, Mass Inst Technol, 64-68 & 74-; Guggenheim fel, 69; staff mem, Nat Bur Econ Res, 69-; Trumbull lectr & vis prof, Yale Univ, 70; vis prof, Albert Einstein Col Med, 71-; lectr, Col France, 73; vis prof, Downstate Med Ctr, State Univ NY, 74. Honors & Awards: S S Wilks Lectr, Princeton Univ & A Wald Lectr, Columbia Univ, 74. Mem: Fel Inst Math Statist; fel Economet Soc; fel Inst Elec & Electronics Eng; Math Soc France. Res: Theory of stochastic processes, especially its applications; thermodynamics, noise and turbulence; natural languages; astronomy; geomorphology; commodity and security prices; self-similar or sporadic chance phenomena; fractals. Mailing Add: Thomas J Watson Res Ctr PO Box 218 Yorktown Heights NY 10598

MANDELCORN, LYON, b Montreal, Que, June 27, 26; nat US; m 55; c 3. PHYSICAL CHEMISTRY. Educ: NY Univ, BA, 47; McGill Univ, PhD(phys chem), 51. Prof Exp: Res fel photochem, Nat Res Coun Can, 51-53; res assoc microcalorimetry, Univ Montreal, 53-54; res engr, 54-64, fel scientist, 64-68, ADV SCIENTIST, WESTINGHOUSE ELEC CORP RES LABS, 68- Mem: Am Chem Soc; Electrochem Soc; Inst Elec & Electronics Eng. Res: Electrochemistry; free radicals; clathrates; dielectrics and insulation; research on development and behavior and properties of dielectrics for high voltage home equipment. Mailing Add: Westinghouse Res & Develop Ctr Beulah Rd Churchill Borough Pittsburgh PA 15235

MANDELES, STANLEY, b Brooklyn, NY, Dec 8, 24; m 47; c 2. BIOCHEMISTRY. Educ: NY Univ, BA, 47; Univ Chicago, PhD, 53. Prof Exp: Res assoc & instr biochem, Univ Chicago, 53-56; assoc chemist, Western Util Res Lab, USDA, 56-61; assoc res biochemist, Univ Calif, Berkeley, 61-71; chmn dept chem, 71-75, PROF CHEM, DOUGLASS COL, RUTGERS UNIV, NEW BRUNSWICK, 71- Mem: Am Soc Biol Chemists; Am Chem Soc; Brit Biochem Soc. Res: Biochemistry of growth; differentiation, especially role of proteins and nucleic acids. Mailing Add: Dept of Chem Douglass Col Rutgers Univ New Brunswick NJ 08903

MANDELKER, MARK, b Milwaukee, Wis, July 18, 33. MATHEMATICS. Educ: Marquette Univ, BS, 55; Harvard Univ, AM, 59; Univ Rochester, PhD(math), 66. Prof Exp: Instr math, State Univ NY Stony Brook, 62-65; asst prof, Univ Kans, 66-69; asst prof, 69-70, ASSOC PROF MATH, NMEX STATE UNIV, 70- Concurrent Pos: NSF grants, Univ Kans, 69-70, NMex State Univ, 71-73. Mem: Fel AAAS; Am Math Soc. Res: Rings of continuous functions; constructive mathematics. Mailing Add: Dept of Math Sci NMex State Univ Las Cruces NM 88001

MANDELKERN, LEO, b New York, NY, Feb 23, 22; m 46; c 3. POLYMER CHEMISTRY, BIOPHYSICS. Educ: Cornell Univ, AB, 42, PhD(chem), 49. Prof Exp: Res assoc chem, Cornell Univ, 49-52; phys chemist, Nat Bur Standards, 52-62; PROF CHEM & BIOPHYS, FLA STATE UNIV, 62- Concurrent Pos: Vis prof, Cornell Univ, 67; consult, NIH, 70- Honors & Awards: Fleming Award, 59; Polymer Award, Am Chem Soc, 75. Mem: AAAS; Am Chem Soc; fel Am Phys Soc; Biophys Soc. Res: Physical chemistry of high polymers; biophysics and macromolecules. Mailing Add: 1503 Old Fort Dr Tallahassee FL 32301

MANDELKERN, MARK ALAN, b New York, NY, Jan 28, 43; m 69. ELEMENTARY PARTICLE PHYSICS, MEDICAL PHYSICS. Educ: Columbia Univ, AB, 63; Univ Calif, Berkeley, PhD(physics), 67; Univ Miami, MD, 75. Prof Exp: Asst prof physics, 68-73, ASSOC PROF PHYSICS, UNIV CALIF, IRVINE, 73- Concurrent Pos: Researcher, Saclay Nuclear Res Ctr, France, 70-71. Mem: Fr Phys Soc. Res: Strong interactions; physiology. Mailing Add: Dept of Physics Univ of Calif Irvine CA 92664

MANDELL, ALAN, b New Bedford, Mass, Feb 26, 26; m 46; c 2. SCIENCE EDUCATION, BIOLOGY. Educ: Holy Cross Col, BS, 46; Univ Va, MEd, 56; Univ NC, DEd, 66. Prof Exp: Prof biol, Frederick Col, 61-65, chmn dept, 65-67; PROF SCI EDUC, OLD DOMINION UNIV, 67- Concurrent Pos: Consult surplus property div, US Off Educ, 56. Mem: Sci Teachers Asn. Mailing Add: Sch of Educ Old Dominion Univ Norfolk VA 23508

MANDELL, ARNOLD J, b Chicago, Ill, July 21, 34; c 2. PSYCHIATRY, NEUROCHEMISTRY. Educ: Stanford Univ, BA, 54; Tulane Univ, MD, 58. Prof Exp: Resident psychiat, Sch Med, Univ Calif, Los Angeles, 59-62, chief resident, 62-63, from asst prof to assoc prof, 63-68; assoc prof psychiat, human behav & psychobiol, Univ Calif, Irvine, 68-69; prof psychiat & chmn dept, 69-74, PROF PSYCHIAT & CO-CHMN DEPT, UNIV CALIF, SAN DIEGO, 75- Concurrent Pos: NIMH career teacher award, Univ Calif, Los Angeles, 62-67; referee, Science, Psychopharmacologica, Community Behav Sci & Am J Psychiat, 64-; mem res comt, Calif Interagency Coun Drug Abuse, 69-; ctr study narcotic addiction & drug abuse, NIMH, 69-, ad hoc sci adv bd biol res, 71-, consult, Lab Clin Sci; mem adv bd, Jerusalem Ment Health Ctr, Israel, 71-; ad hoc sci adv bd, President's Spec Actg Off Drug Abuse Prev, 71-; staff psychiatrist, Vet Admin Hosp, San Diego, Calif, 72-; consult, Ill State Psychiat Res Inst. Honors & Awards: A E Bennett Award Res Biol Psychiat, 62. Mem: AAAS; Am Inst Chem; Am Psychiat Asn; Soc Biol Psychiat; Am Col Psychiat. Res: Neurochemical and biochemical correlates of behavior in animals and man. Mailing Add: Dept of Psychiat Sch of Med Univ of Calif San Diego La Jolla CA 92093

MANDELL, JOSEPH DAVID, microbiology, see 12th edition

MANDELL, LEON, b Bronx, NY, Nov 19, 27; m 59, 71; c 2. ORGANIC CHEMISTRY. Educ: Polytech Inst Brooklyn, BS, 48; Harvard Univ, MA, 49, PhD(chem), 51. Prof Exp: Sr chemist, Merck & Co, Inc, 51-53; from asst to assoc prof org chem, 55-64, PROF ORG CHEM, EMORY UNIV, 64- Concurrent Pos: Consult, Schering Corp, 58- & Houdry Chem & Process Co, 64- Mem: Am Chem Soc; The Chem Soc. Res: Synthetic methods in organic chemistry; nuclear magnetic resonance spectroscopy; mechanisms of organic reactions; natural product chemistry. Mailing Add: Dept of Chem Emory Univ Atlanta GA 30322

MANDELL, PAUL IRVING, b Brooklyn, NY, Mar 22, 35; m 59; c 3. ECONOMIC GEOGRAPHY, ECONOMIC DEVELOPMENT. Educ: Univ Chicago, AB, 58, MA, 62; Columbia Univ, PhD(geog, Latin Am Studies), 69. Prof Exp: Asst prof geog, Food Res Inst, Stanford Univ, 66-73; MEM FAC, CTR MARINE AFFAIRS, SCRIPPS INST OCEANOG, 73- Concurrent Pos: Consult, Brazilian Inst Agr Reform, 66; dir, Stanford in Mex Prog, 69-70. Mem: Asn Am Geog; Am Econ Asn; Am Agr Econ Asn; Latin Am Studies Asn; Conf Latin Am Geog. Res: Analysis of contemporary and historical agricultural development in Latin America, particularly Brazil; development of theory and quantitative techniques for the analysis of agricultural development. Mailing Add: Scripps Inst Oceanog Ctr Marine Affairs T-16 Box 152 La Jolla CA 92037

MANDELL, ROBERT BURTON, b Alhambra, Calif, Nov 13, 33; m 59; c 2. OPTOMETRY, VISUAL PHYSIOLOGY. Educ: Los Angeles Col Optom, OD, 56; Ind Univ, MS, 58, PhD(physiol optics), 62. Prof Exp: Asst prof, 62-68, assoc prof, 68-73, PROF OPTOM, UNIV CALIF, BERKELEY, 73- Concurrent Pos: USPHS grant, 64-66. Mem: AAAS; Optical Soc Am; Am Acad Optom; Am Optom Asn. Res: Corneal contour; contact lenses. Mailing Add: Dept of Optom Univ of Calif Berkeley CA 94720

MANDELL, WALLACE, b Brooklyn, NY, Jan 25, 28; c 3. PSYCHOLOGY, PUBLIC HEALTH. Educ: City Col New York, BS, 48; Yale Univ, MS, 51; NY Univ, PhD(social psychol), 54; Johns Hopkins Univ, MPH, 59. Prof Exp: Res asst, Yale Univ, 49-51; res asst, NY Univ, 51-53; res scientist, Human Resources Res Inst, US Air Force, 54; personnel psychologist, Adj Gen Corps, 54-56; ment health res consult, Tex State Dept Health, 56-60; dir res, Wakoff Res Ctr, Staten Island Ment Health Soc, NY, 60-68; assoc prof, 68-71, PROF MENT HYG, SCH HYG & PUB HEALTH, JOHNS HOPKINS UNIV, 71- Concurrent Pos: Consult, New York City Coun Alcoholism, 66; mem, New York City Task Force on Corrections, 67; consult, Baltimore County Bur Ment Health, 69, Md Training Prog Community Psychiat, 69 & Spring Grove State Hosp, Md, 71; mem training rev comt & prev progs, Nat Inst Alcohol Abuse & Alcoholism, 73-; chmn, Md State Adv Coun Drug Abuse, 73-; mem, Baltimore Area Coun Alcoholism, 74- Mem: Fel Am Pub Health Asn; Am Psychol Asn. Res: Prevention of mental disorders through epidemiologic studies and the planning and evaluation of programs for the care of the mentally ill with special emphasis on the development of deviant behavior such as alcoholism and drug abuse. Mailing Add: Sch of Hyg & Pub Health Johns Hopkins Univ Baltimore MD 21205

MANDELS, GABRIEL RAPHAEL, b Schenectady, NY, Mar 18, 15; m 42; c 2. MICROBIAL PHYSIOLOGY, PLANT PHYSIOLOGY. Educ: Cornell Univ, BS, 37, PhD(plant physiol), 41. Prof Exp: Asst bot, Cornell Univ, 37-41, instr, 41-42 & 45-46; physiologist biol labs, US Army Qm Res & Develop Labs, 46-51, physiologist & chief biol br, 51-57; dir cent res labs, United Fruit Co, Mass, 57-62; assoc dir life sci, Pioneering Res Lab, 62-74, CHIEF FOOD MICROBIOL & NUTRIT DIV, FOOD SCI LAB, US ARMY NATICK DEVELOP CTR, 74- Mem: AAAS; Am Soc Plant Physiol; Bot Soc Am; Am Chem Soc; Am Soc Microbiol. Res: Physiology of fungi; cellular physiology; physiology and biochemistry of fungus spores. Mailing Add: US Army Natick Develop Ctr Kansas St Natick MA 01760

MANDELS, MARY HICKOX, b Rutland, Vt, Sept 12, 17; m 42; c 2. MICROBIAL PHYSIOLOGY. Educ: Cornell Univ, BS, 39, PhD(plant physiol), 47. Prof Exp: Microbiologist pioneering res div, Qm Res & Eng Ctr, 55-62, MICROBIOLOGIST, FOOD LAB, US ARMY NATICK DEVELOP CTR, 62- Mem: Bot Soc Am; Am Soc Microbiol. Res: Fungal enzymes; cellulose saccharification; fermentation technology. Mailing Add: US Army Natick Develop Ctr Kansas St Natick MA 01760

MANDELSTAM, PAUL, b Boston, Mass, Apr 18, 25; m. BIOCHEMISTRY, INTERNAL MEDICINE. Educ: Harvard Univ, AB, 44, AM, 46, MD, 50, PhD, 53; Am Bd Internal Med, dipl, 57. Prof Exp: Intern, Med Serv, Beth Israel Hosp, Boston, 50-51, asst resident, 52-53; from asst prof to assoc prof, 60-73, PROF MED, COL MED, UNIV KY, 73- Concurrent Pos: Res fel, Sch Med, Yale Univ, 55-57; Nat Found res fel, Sch Med, Wash Univ, 57-59; Nat Inst Neurol Dis & Blindness spec trainee, 59-60; asst physician, New Haven Hosp, 55-57. Mem: Fel Am Col Physicians; Am Soc Gastrointestinal Endoscopy; Cent Soc Clin Res; Am Gastroenterol Asn; Am Physiol Soc. Res: Active transport; gastroenterology. Mailing Add: Univ of K Col of Med Lexington KY 40506

MANDELSTAM, STANLEY, b Johannesburg, SAfrica, Dec 12, 28. THEORETICAL PHYSICS. Educ: Univ Witwatersrand, BSc, 51; Cambridge Univ, BA, 54; Univ Birmingham, PhD(math physics), 56. Prof Exp: Asst math physics, Univ Birmingham, 56-57; Boese fel physics, Columbia Univ, 57-58; asst res physics, Univ Calif, Berkeley, 58-60; prof math physics, Univ Birmingham, 60-63; PROF PHYSICS, UNIV CALIF, BERKELEY, 63- Mem: Fel Royal Soc. Res: Theoretical physics of elementary particles. Mailing Add: Dept of Physics Univ of Calif Berkeley CA 94720

MANDEVILLE, CHARLES EARLE, b Dallas, Tex, Sept 3, 19; m 43; c 2. PHYSICS. Educ: Rice Univ, BA, 40, MA, 41, PhD(physics), 43. Prof Exp: Mem staff radiation lab, Mass Inst Technol, 43-45; instr physics, Rice Univ, 45-46; physicist, Bartol Res Found, 46-53, asst dir, 53-59; prof physics & head dept, Univ Ala, 59-61, res physicist, 61; prof physics, Kans State Univ, 61-67; head dept, 67-75, PROF PHYSICS, MICH TECHNOL UNIV, 67-, DIR SPEC PROJ, COMT ENERGY RES, 75- Concurrent Pos: Vis lectr, Philadelphia Col Osteop Med, 50-68; consult, US Naval Ord Test Sta, 54-60, Curtiss-Wright Corp, 56-60, res ctr, Babock & Wilcox Co, 58-60, US Army Rocket & Guided Missile Agency, Redstone Arsenal, 59-61, US Naval Radiol Defense Lab, 62-64, Kaman Nuclear Corp, 64-67 & Commonwealth-Edison Co, 73-74. Mem: AAAS; fel Am Phys Soc. Res: Nuclear, experimental and solid state physics; primary disintegrations; energies of gamma rays; coincidence experiments; neutron scattering; luminescence; biophysics. Mailing Add: Dept of Physics Mich Technol Univ Houghton MI 49931

MANDICS, PETER ALEXANDER, b Budapest, Hungary, May 29, 37; US citizen; m 68; c 2. ATMOSPHERIC PHYSICS. Prof Exp: Univ Colo, Boulder, BS, 62; Mass Inst Technol, SM, 63; Stanford Univ, PhD(elec eng), 71. Prof Exp: Instr & res asst elec eng, Univ Colo, Boulder, 63-65; PHYSICIST, WAVE PROPAGATION LAB, NAT OCEANIC & ATMOSPHERIC ADMIN, US DEPT COM, 71- Concurrent Pos: Nat Res Coun res assoc atmospheric physics, Wave Propagation Lab, Nat Oceanic & Atmospheric Admin, US Dept Com, 71-73. Mem: Inst Elec & Electronic Engrs; Int Union of Radio Sci. Res: Atmospheric propagation of radio, optical and acoustic waves; development of acoustic echo sounding techniques for the remote sensing of the atmosphere; investigation of tropical marine boundry layer. Mailing Add: Wave Propagation Lab NOAA ERL R45x2 325 Broadway Boulder CO 80302

MANDL, INES, b Vienna, Austria, Apr 19, 17; nat US; m 36. BIOCHEMISTRY. Educ: Nat Univ Ireland, dipl, 44; Polytech Inst Brooklyn, MS, 47, PhD(chem), 49. Prof Exp: Res chemist, Res Labs, Interchem Corp, 45-49; res assoc, 49-55, assoc, 55-56, asst prof biochem, 56-72, ASSOC PROF REPROD BIOCHEM, COL PHYSICIANS & SURGEONS, COLUMBIA UNIV, 72- Concurrent Pos: Mem coun basic sci & coun cardiopulmonary dis, Am Heart Asn; dir gynec labs, Delafield Hosp; ed-in-chief, Connective Tissue Res-An Int Jour. Mem: AAAS; Am Soc Biol Chem; Am Chem Soc; Am Asn Cancer Res; Am Thoracic Soc. Res: Enzymes; proteins; amino acids; carbohydrates; proteolytic enzyme inhibitors; alpha-antitrypsin; emphysema; respiratory distress syndrome; enzymes of bacterial origin and their medical applications; microstructure of collagen and elastin. Mailing Add: 166 W 72nd St New York NY 10023

MANDL, PAUL, b Vienna, Austria, Feb 9, 17; nat US; m 50. APPLIED MATHEMATICS. Educ: Univ Toronto, BA, 45, MA, 48, PhD(math), 51. Prof Exp: Jr res officer aerodyn, Nat Res Coun Can, 45-48, from asst to assoc res officer, 48-60, sr res officer, 60-66; PROF MATH, CARLETON UNIV, 66- Concurrent Pos: Lectr univ exten dept, McGill Univ, 58-61 & Carleton Univ, 60-61, vis assoc prof math, 64-65, part-time lectr wing theory, Grad Div, 65-; consult lab unsteady aerodynamics, Nat Aeronaut Estab; vis prof mechanics of fluids, Univ Manchester, Eng, 74. Mem:

Can Aeronaut Inst; Can Math Cong; Can Soc Mech Eng. Res: Theoretical fluid mechanics; aerodynamics; rheology; diffraction of shockwaves by solid obstacles. Mailing Add: Dept of Math Carleton Univ Ottawa ON Can

MANDL, RICHARD H, b New York, NY, Oct 20, 34; m 56; c 4. BIOLOGY. Educ: NY Univ, BA, 66. Prof Exp: Res assoc environ biol, 52-69, ASST PLANT PHYSIOLOGIST, BOYCE THOMPSON INST PLANT RES, 69- Mem: AAAS. Res: Environmental biology; development of new food products; methods for protein, amino acid, iodo-amino acid, organic phosphate and fluoride analysis; effects of air pollutants on plants. Mailing Add: Boyce Thompson Inst for Plant Res 1086 N Broadway Yonkers NY 10701

MANDLE, ROBERT JOSEPH, b New York, NY, May 18, 19; m 43; c 4. MICROBIOLOGY. Educ: Lebanon Valley Col, BS, 42; Univ Pa, PhD, 51. Prof Exp: Asst microbiol, Rockefeller Inst, NJ, 45-50; instr, Univ Del, 50-51; asst prof, 51-65, PROF MICROBIOL, JEFFERSON MED COL, 65- Mem: Am Soc Microbiol. Res: Physiology of micro-organisms; infections and resistance. Mailing Add: Dept of Microbiol Jefferson Med Col 1020 Locus St Philadelphia PA 19107

MANDLIK, JAYANT V, b Dapoli, India, Aug 6, 30; m 71. ORGANIC CHEMISTRY, PHARMACOLOGY. Educ: Karnatak Univ, India, BS, 53, MS, 55; Univ Poona, PhD(org chem), 64. Prof Exp: Instr chem, NWadia Col, India, 55-63, asst prof, 63-69; res assoc pharmacol, Univ Ill Med Ctr, 69-73; CHEMIST, CHICAGO BD HEALTH, 73- Concurrent Pos: Univ Grants Comn India grants, NWadia Col, 63-64 & 68-69. Mem: Am Chem Soc. Res: Synthesis of organic compounds of biological importance; toxicology; drug metabolism. Mailing Add: 5652 W Sunnyside Chicago IL 60630

MANDRA, YORK T, b New York, NY, Nov 24, 22; m 46. GEOLOGY. Educ: Univ Calif, AB, 47, MA, 49; Stanford Univ, PhD(geol), 58. Prof Exp: Asst paleont, Univ Calif, 49-50; instr geol, 50-58, from asst to assoc prof, 58-65, head geol sect & chmn dept, 60-67, PROF GEOL, SAN FRANCISCO STATE UNIV, 65- Concurrent Pos: NSF fac fel & vis prof, Univ Aix Marseille, 60; res assoc, Calif Acad Sci, 66-; NSF res grants, 67- Mem: Fel AAAS; fel Geol Soc Am; Paleont Soc; Nat Asn Geol Teachers; Soc Econ Paleont & Mineral. Res: Micropaleontology, especially stratigraphic and paleoecologic aspects of Mesozoic and Cenozoic silicoflagellates. Mailing Add: 8 Bucareli Dr San Francisco CA 94132

MANDULA, BARBARA BLUMENSTEIN, b New York, NY, Dec 19, 41; m 63. BIOCHEMISTRY. Educ: City Col New York, BS, 62; Brandeis Univ, PhD(biochem), 69. Prof Exp: Res biochemist, City Hope Nat Med Ctr, 68-69; res assoc biol, Princeton Univ, 69-71; res assoc, Sch Med, Univ Southern Calif, 71-73; RES ASSOC DERMAT, MASS GEN HOSP, 74- Mem: AAAS. Res: Regulation of protein synthesis; metabolism of photosensitizing drugs; structure of ribosomes; evolution of biological macromolecules. Mailing Add: Dept of Dermat Mass Gen Hosp Boston MA 02114

MANDULA, JEFFREY ELLIS, b New York, NY, July 23, 41; m 63. PHYSICS. Educ: Columbia Col, NY, AB, 62; Harvard Univ, AM, 64, PhD(physics), 66. Prof Exp: NSF fel, Harvard Univ, 66-67; res fel physics, Calif Inst Technol, 67-69; mem natural sci, Inst Advan Study, 69-70; asst prof theoret physics, Calif Inst Technol, 70-74; ASSOC PROF APPL MATH, MASS INST TECHNOL, 74- Mem: AAAS; Am Phys Soc. Res: Theoretical elementary particle physics; quantum field theory. Mailing Add: Dept of Math Mass Inst of Technol Cambridge MA 02139

MANDULEY, ILMA MORELL, b Holquin, Cuba, Sept 19, 29. MATHEMATICS. Educ: Univ Havan, PhD(math), 53. Prof Exp: From asst prof to prof math, Polytech Inst Holquin, Cuba, 55-61; instr, 61-64, ASST PROF MATH, GUILFORD COL, 64- Concurrent Pos: Instr, Friends Sch, Cuba, 52-61. Mem: AAAS; Math Asn Am. Res: Euclidean and projective geometry; differential analysis; ordinary and partial differentiation equations. Mailing Add: Dept of Math Guilford Col Greensboro NC 27410

MANDY, WILLIAM JOHN, b Lackawanna, NY, Mar 12, 33; m 59; c 3. IMMUNOBIOLOGY, IMMUNOGENETICS. Educ: Elmhurst Col, 58; Univ Ill, Urbana, PhD(microbiol), 63; Univ Houston, MS, 62. Prof Exp: From asst prof to assoc prof, 65-73, PROF IMMUNOL, UNIV TEX, AUSTIN, 73- Concurrent Pos: NIH fel immunochem, Sch Med, Univ Calif, 63-65; USPHS career develop award, 66- Mem: AAAS; Am Soc Microbiol; Am Soc Immunol. Res: Structure of the gamma globulin molecule and nature of naturally occurring antiglobulin factors. Mailing Add: Dept of Microbiol Univ of Tex Austin TX 78712

MANERA, PAUL ALLEN, b Clovis, Calif, Nov 11, 30; m 59; c 2. HYDROGEOLOGY. Educ: Fresno State Col, BA, 55; Ariz State Univ, MA, 63 & 71, PhD(geol), 76. Prof Exp: Mining geologist, Holly Minerals Corp, 55-56; geohydrologist, Samuel F Turner & Assocs, 57-62; CONSULT GEOHYDROLOGIST, MANERA & ASSOCS, INC, 62- Mem: Geol Soc Am; Am Geophys Union; Am Water Resources Asn; Soc Explor Geophysicists; Europ Soc Explor Geophysicists. Res: Hydrologic impact of the Central Arizona Project and the Salt River Project on the ground water reservoir in Maricopa County. Mailing Add: 5251 N 16th St Suite 302 Phoenix AZ 85016

MANERI, CARL C, b Cleveland, Ohio, Jan 25, 33; m 53; c 3. MATHEMATICS. Educ: Case Western Reserve Univ, BS, 54; Ohio State Univ, PhD(math), 59. Prof Exp: Res fel & instr math, Univ Chicago, 60-62; asst prof, Syracuse Univ, 62-65; chmn dept, 68-71, ASSOC PROF MATH, WRIGHT STATE UNIV, 65- Mem: Am Math Soc; Math Asn Am. Res: Non-associative algebra and combinatorial analysis. Mailing Add: Dept of Math Wright State Univ Dayton OH 45431

MANERY, JEANNE FOREST, b Chesley, Ont, July 6, 08; m 38; c 2. PHYSIOLOGY, BIOCHEMISTRY. Educ: Univ Toronto, BA, 32, MA, 33, PhD(chem embryol), 35. Prof Exp: Asst physiol sch med dent, Univ Rochester, 35-36, instr, 37-39; Nat Res Coun fel biol sci, Dept Biol Chem, Harvard Med Sch, 36-37; demonstr biochem, 39-48, spec lectr zool, 44-45, from asst to assoc prof, 48-65, PROF BIOCHEM, UNIV TORONTO, 65- Concurrent Pos: Res scientist defense res bd, Dept Nat Defense Can, 50-51. Mem: Am Physiol Soc; Soc Gen Physiol; Can Physiol Soc; Can Biochem Soc; Brit Biochem Soc. Res: Chemical embryology; tissue electrolytes; shock; leucocytes; arctic research; mitochondria; muscle; biomembranes of muscle and erythrocytes. Mailing Add: Dept of Biochem Univ of Toronto Toronto ON Can

MANES, COLE, b Coleman, Tex, Mar 1, 30; m 56; c 5. DEVELOPMENTAL BIOLOGY. Educ: Stanford Univ, BA, 50, MD, 57; Univ Colo, Boulder, PhD(develop biol), 68. Prof Exp: ASST PROF PEDIAT & ANAT, MED CTR, UNIV COLO, 68- Mem: AAAS; Soc Develop Biol. Res: Genetic expression during early mammalian embryogenesis. Mailing Add: Univ of Colo Med Ctr Dept Pediat 42 E Ninth Ave Denver CO 80220

MANES, MILTON, b New York, NY, Oct 14, 18; m 45; c 2. PHYSICAL CHEMISTRY. Educ: City Col New York, BS, 37; Duke Univ, PhD(phys chem), 47. Prof Exp: From lab asst to jr chemist, US Food & Drug Admin, 37-41; phys chemist, US Bur Mines, 47-52; sr chemist & mgr statist design group, Koppers Co, Inc, 52-58; supvr phys chem res, Pittsburgh Chem Co, 59-64; sr fel & head adsorption fel, Mellon Inst, 64-67; PROF CHEM, KENT STATE UNIV, 67- Concurrent Pos: Vis prof, Cornell Univ, 64. Mem: AAAS; Am Chem Soc. Res: Adsorption; near equilibrium thermodynamics and kinetics. Mailing Add: Dept of Chem Kent State Univ Kent OH 44242

MANGAN, GEORGE FRANCIS, JR, b Lowell, Mass, Feb 4, 25; m 48. BIOCHEMISTRY. Educ: St Anselm's Col, BA, 50; Georgetown Univ, MS, 54, PhD, 56. Prof Exp: Res chemist, US Fish & Wildlife Serv, 50-51; res chemist, USDA, 51-53; res biochemist, Walter Reed Army Inst Res, 53-57; proj leader, US Fish & Wildlife Serv, 57-58; sr chemist, Ionics, Inc, 58-61; chief br org & biol chem, Off Saline Water, 61-65, water resources scientist, Off Water Resources Res, 65-74, CHIEF PHYS CHEM DIV, OFF WATER RES & TECHNOL, US DEPT INTERIOR, 74- Mem: AAAS; Am Chem Soc; Am Geophys Union. Res: Water resources planning and management; urban water resources research; contract and grant research program administration. Mailing Add: Off of Water Res & Technol US Dept Interior 19th & C St NW Washington DC 20240

MANGAN, JERROME, b Columbus, Ohio, Nov 18, 34; m 70; c 2. DEVELOPMENTAL BIOLOGY, GENETICS. Educ: Univ Cincinnati, BA, 60, MS, 63; Brown Univ, PhD(biol), 66. Prof Exp: NSF fel, Albert Einstein Col Med, 66-67; asst prof biol, Univ Chicago, 67-70; asst prof, 70-72, ASSOC PROF BIOL, FRESNO STATE COL, 72- Concurrent Pos: Am Cancer Soc & NIH grants, Univ Chicago, 67-68; NSF grant, Univ Chicago & Fresno State Col, 68-72. Mem: AAAS; Am Soc Microbiol. Res: The nature of control of biochemical processes in bacteria and during development of higher organisms. Mailing Add: Dept of Biol Fresno State Col Fresno CA 93710

MANGANIELLO, LOUIS O J, b Waterbury, Conn, June 6, 15; m 50; c 1. NEUROSURGERY. Educ: Harvard Univ, AB, 37; Univ Md, MD, 42; Augusta Law Sch, JD, 67; Am Bd Neurol Surg, dipl. Prof Exp: Fel neurosurg, Sch Med, Univ Md, 46-47; asst resident, Univ Md Hosp, 47-48, chief resident, 49-50; asst resident, Baltimore City Hosp, 48-49; instr neuroanat & neurosurg, Sch Med, Univ Md, 50-51; ASSOC PROF NEUROSURG, MED COL GA, 51- Concurrent Pos: Consult, Hosps, Ga. Mem: Am Asn Neurol Surg; Am Asn Cancer Res; AMA; Am Psychiat Asn; fel Am Col Surg. Res: Cancer detection and therapy; porphyrin metabolism. Mailing Add: 656 Milledge Rd Augusta GA 30904

MANGASARIAN, OLVI LEON, b Baghdad, Iraq, Jan 12, 34; US citizen; m 59; c 3. APPLIED MATHEMATICS. Educ: Princeton Univ, BSE, 54, MSE, 55; Harvard Univ, PhD(appl math), 59. Prof Exp: Mathematician, Shell Develop Co, 59-67; lectr math programming, Univ Calif, Berkeley, 65-67; assoc prof comput sci, 67-69, chmn dept, 70-73, PROF COMPUT SCI, UNIV WIS-MADISON, 69- Mem: Am Math Soc; Soc Indust & Appl Math; Asn Comput Mach. Res: Development and use of theory and computational methods of mathematical programming in various fields of applied mathematics such as operations research, optimal control theory and numerical analysis. Mailing Add: Dept of Comput Sci Univ of Wis 1210 W Dayton St Madison WI 53706

MANGAT, BALDEV SINGH, b Ludhiana, India, May 7, 35; m 60; c 2. ENTOMOLOGY, ZOOLOGY. Educ: Univ Punjab, India, MSc, 58; Univ Wis, PhD(entom), 65. Prof Exp: Instr entom, Punjab Agr Col & Res Inst, 58-60; assoc prof biol, Alcorn Agr & Mech Col, 65-66; assoc prof, 66-69, PROF BIOL, ALA A&M UNIV, 69- Mem: AAAS; Entom Soc Am. Res: Biology and physiology of corn earworm, Heliothis zea. Mailing Add: Dept of Biol Ala A&M Univ Normal AL 35762

MANGAT, BHUPINDER SINGH, b Brit citizen. PLANT BIOCHEMISTRY & PLANT PHYSIOLOGY. Educ: Univ London, BSc, 64, dipl biochem & dipl, Chelsea Col, 65, PhD(plant biochem), 68. Prof Exp: Res assoc biochem, Univ Col Swansea, Wales, 65-68; fel, Univ Toronto, 69-71; Nat Res Coun Can fel & instr biol, Queen's Univ, Ont, 71-72; asst prof, Univ NB, Fredericton, 72-75; MEM FAC, DEPT BIOL, CONCORDIA UNIV, LOYOLA CAMPUS, 75- Concurrent Pos: Mem, Biol Coun Can. Mem: Can Soc Plant Physiologists. Res: Role of tree nucleotides in plant cell metabolism; photosynthesis; photorespiration. Mailing Add: Dept of Biol Concordia Univ Montreal PQ Can

MANGE, ARTHUR P, b St Louis, Mo, Jan 28, 31; m 60; c 3. GENETICS. Educ: Cornell Univ, BEngPhys, 54; Univ Wis, MS, 58, PhD(genetics), 63. Prof Exp: Instr biol, Case Western Reserve Univ, 62-64; asst prof zool, 64-70, ASSOC PROF ZOOL, UNIV MASS, AMHERST, 70- Concurrent Pos: Sr res assoc, Univ Wash, 71-72. Mem: AAAS; Am Soc Human Genetics; Genetics Soc Am. Res: Population structure of human isolates; Drosophila genetics. Mailing Add: Dept of Zool Univ of Mass Amherst MA 01002

MANGE, FRANKLIN EDWIN, b St Louis, Mo, Feb 12, 28; m 54; c 4. ORGANIC CHEMISTRY. Educ: Mass Inst Technol, SB, 48; Univ Ill, PhD(org chem), 51. Prof Exp: Res chemist, 51-57, group leader, 57-63, sect mgr, 63-67, RES DIR, TRETOLITE DIV, PETROLITE CORP, 67- Mem: Am Chem Soc. Res: Polymer synthesis; surfactants; waxes; demulsification; flocculation; petroleum chemistry; water treatment. Mailing Add: 18 Granada Way St Louis MO 63124

MANGE, PHILLIP WARREN, b Kalamazoo, Mich, June 5, 25; m 51; c 2. PLANETARY ATMOSPHERES: SPACE PHYSICS. Educ: Kalamazoo Col, AB, 49; Pa State Univ, MS, 52, PhD(physics), 54. Prof Exp: Asst prof eng res, Ionosphere Res Lab, Pa State Univ, 54-55; admin asst to gen secy, Spec Comt, Int Geophys Year, Belgium, 55-57, prog officer, US Nat Comt, Nat Acad Sci, DC, 57-59; PHYSICIST, SPACE SCI DIV, US NAVAL RES LAB, 59- Mem: AAAS; Am Phys Soc; Am Geophys Union; Am Astron Soc. Res: Structure of high atmosphere; ultraviolet environment in the solar system. Mailing Add: Code 7101 US Naval Res Lab Washington DC 20375

MANGEL, MARGARET, b Tell City, Ind, May 13, 12. FOOD SCIENCE, NUTRITION. Educ: Ind Univ, AB, 32, cert, 34; Univ Chicago, MS, 40, PhD(food chem), 51. Prof Exp: Teacher high sch, Ind, 34-39; instr home econ, 40-43, from asst prof to assoc prof, 43-51, chmn dept, 55-60, dir sch home econ, 60-73, PROF HOME ECON, UNIV MO-COLUMBIA, 51-, DEAN COL HOME ECON, 73- Concurrent Pos: Chmn res comt, Nat Coun Adminrs Home Econ, 72-73. Mem: Fel AAAS; Am Home Econ Asn (coun prof develop); Am Chem Soc; Am Dietetic Asn; Am Pub Health Asn. Res: Food quality; utilization studies; biological pigment and enzyme assay; bio and microbiological assay; dietary and nutritional status studies. Mailing Add: Col of Home Econ Univ of Mo 113 Gwynn Hall Columbia MO 65201

MANGELSDORF, PAUL CHRISTOPH, b Atchison, Kans, July 20, 99; m 23; c 2. ECONOMIC BOTANY, GENETICS. Educ: Kans State Univ, BSc, 21, LLD, 61; Harvard Univ, MS, 23, ScD(genetics), 25. Hon Degrees: DSc, Park Col, 60, St Benedict's Col, 65, Univ NC, 75. Prof Exp: Asst geneticist, State Agr Exp Sta, Conn,

21-27; agronomist exp sta, Agr & Mech Col Tex, 27-36, agronomist & asst dir, 36-40, vdir, 40; prof bot, 40-62, asst dir bot mus, 40-45, dir bot mus, 45-68, chmn inst exp & appl bot, 46-68, Fisher prof natural hist, 62-68, EMER FISHER PROF NATURAL HIST, HARVARD UNIV, 68-: LECTR BOT, UNIV NC, CHAPEL HILL, 68- Concurrent Pos: Hon prof, San Carlos Univ, Guatemala, 56 & Nat Sch Agr, Peru, 59; with USDA, 44; mem, Rockefeller Agr Comn, Mex, 41, bd consults for agr, Rockefeller Found, 43-45 & 56-58; mem Nat Res Coun, 43-; mem basic res group, US Dept Defense, 53. Mem: Nat Acad Sci; AAAS; Am Soc Nat (pres, 51); Am Genetics Soc Am (vpres, pres, 55); Soc Econ Bot (pres, 62-63). Res: Genetics of maize; origin of cultivated plants. Mailing Add: 510 Caswell Rd Chapel Hill NC 27514

MANGELSDORF, PAUL CHRISTOPH, JR, b New Haven, Conn, Jan 31, 25; m 49; c 4. MARINE GEOCHEMISTRY. Educ: Swarthmore Col, BA, 49; Harvard Univ, PhD(chem physics), 55. Prof Exp: Instr, Univ Chicago, 55-57, asst prof chem, 57-60; RES ASSOC PHYS CHEM, WOODS HOLE OCEANOG INST, 60-; ASSOC PROF PHYSICS, SWARTHMORE COL, 61- Mem: Am Chem Soc; Am Phys Soc; Am Soc Limnol & Oceanog; Am Geophys Union; Am Asn Physics Teachers. Res: Fluid dynamics; transport properties in liquids; thermodynamics of electrolyte solutions; chemistry of sea water. Mailing Add: Dept of Physics Swarthmore Col Swarthmore PA 19081

MANGELSON, FARRIN LEON, b Levan, Utah, May 12, 12; m 36; c 9. BIOCHEMISTRY, NUTRITION. Educ: Univ Wash, BS, 38; Utah State Univ, MS, 50, PhD, 63. Prof Exp: Teacher high schs, Utah, 36-41; chemist, Remington Arms Co, Inc Div, E I du Pont de Nemours & Co, 42-43; asst state chemist, Utah, 44-47; asst chem, Exp Sta, Utah State Univ, 47-49, res instr, 49-51; from instr to assoc prof, 51-67, PROF CHEM, SNOW COL, 67-, CHMN DIV PHYS SCI & MATH, 69- Concurrent Pos: Mem res staff, Univ Calif, Berkeley, 64. Mem: AAAS; Am Chem Soc. Res: Kidney function in cattle as affected by ingestion of inorganic fluorides; effect of animal fats and proteins on blood serum cholesterol level in humans; molecular size of myosin; human nutrition. Mailing Add: Dept of Chem Snow Col Ephraim UT 84627

MANGELSON, NOLAN FARRIN, b Nephi, Utah, Jan 17, 36; m 61; c 5. PHYSICAL CHEMISTRY, NUCLEAR PHYSICS. Educ: Utah State Univ, BS, 61; Brigham Young Univ, MS, 63; Univ Calif, Berkeley, PhD(chem), 68. Prof Exp: Res fel, Nuclear Physics Lab, AEC, Univ Wash, 67-69; ASST PROF CHEM, BRIGHAM YOUNG UNIV, 69- Mem: Am Chem Soc; Am Phys Soc. Res: Nuclear reactions and spectroscopy; x-ray fluorescence. Mailing Add: Dept of Chem Brigham Young Univ Provo UT 84601

MANGER, CHARLES WALTER, b Newark, NJ, Dec 22, 13; m 47; c 4. PHYSICAL CHEMISTRY. Educ: Seton Hall Univ, BS, 48; Newark Col Eng, MS, 53. Prof Exp: Res technician, 36-52, res chemist, 52-67, SR RES CHEMIST, E I DU PONT DE NEMOURS & CO, INC, 67- Mem: Am Chem Soc; Electron Micros Soc Am. Res: Colored pigments; electronmicroscopy; x-ray diffraction; ultraviolet, visible and infrared spectrophotometry; surface chemistry; crystal structure. Mailing Add: Photo Prod Dept E I du Pont de Nemours & Co Inc Parlin NJ 08859

MANGER, GEORGE EDWARD, b Baltimore, Md, Aug 17, 02; m 39; c 1. GEOLOGY. Educ: Johns Hopkins Univ, AB, 23, PhD(paleont), 29. Prof Exp: Teacher, Md Pub Schs, 23-25; paleontologist & sedimentary petrographer, Creole Petrol Corp, Venezuela, 30-31; asst geologist, Mene Grande Oil Co, 35-43; geologist, Gulf Oil Corp, Tex, 43-44 & Gulf Refining Co, 44-48; geologist, US Geol Surv, 48-75; RETIRED. Mem: Fel Geol Soc Am; Soc Econ Paleontologists & Mineralogists; Soc Explor Geophys; Am Asn Petrol Geologists; Am Inst Mining, Metall & Petrol Engrs. Res: Orogeny in eastern Venezuela; mineral analysis, sands; geology of San Quintin Bay; stratigraphy of Tertiary of Maracaibo Lake Basin, Venezuela; micropaleontology; sedimentary petrology; stratigraphy; core analysis; physical and related properties of uranium-bearing rocks. Mailing Add: 3710 Woodbine St Chevy Chase MD 20015

MANGER, MARTIN C, b Bethlehem, Pa, Sept 20, 37; m 63. ORGANIC CHEMISTRY, ENVIRONMENTAL SCIENCES. Educ: Muhlenberg Col, BS, 59; St Lawrence Univ, MS, 61; Rutgers Univ, MS, 65; Sheffield Univ, PhD(chem), 68. Prof Exp: Res chemist, E R Squibb & Sons, Inc, 64-65; assoc prof, 68-73, PROF CHEM, ALA A&M UNIV, 73- Concurrent Pos: Sci consult, Asn Educ & Prof Opportunities Found prog for gifted children, 68- Mem: Am Chem Soc; The Chem Soc; Am Inst Chem. Res: Natural products; conformational inversion of bridged biphenyls; pollution technology; biological pigments. Mailing Add: Dept of Chem Ala A&M Univ Normal AL 35762

MANGER, WILLIAM MUIR, b Greenwich, Conn, Aug 13, 20; m 64; c 4. MEDICINE. Educ: Yale Univ, BS, 44; Columbia Univ, MD, 46; Univ Minn, PhD(med), 58; Am Bd Internal Med, dipl, 57. Prof Exp: Intern med, Columbia Presby Med Ctr, 46-47, resident, 49-50; instr, Col Physicians & Surgeons, Columbia Univ, 57-66; asst clin prof, Med Ctr, NY Univ, 68-75; ASSOC, COL PHYSICIANS & SURGEONS, COLUMBIA UNIV, 66-; ASSOC CLIN PROF, MED CTR, NY UNIV, 75- Concurrent Pos: Dir, Manger Res Found, 58-; asst physician, Presby Hosp, 57-66, asst attend, 66-70; clin asst vis physician, Columbia Div, Bellevue Hosp, 64-68; asst attend, Dept Med, NY Univ Med Ctr, 69-; trustee, Found Res in Med & Biol, 70-; consult, Southampton Hosp, 71-; trustee, Found Res in Microbiol & Human Genetics, 75-; mem cent adv comt sect circulation, Am Heart Asn, fel coun circulation, mem coun high blood pressure res; ed, Am Lect in Endocrinol. Honors & Awards: Meritorious Res Award, Mayo Found, 55. Mem: Am Physiol Soc; Soc Pharmacol & Exp Therapeut; fel Am Col Clin Pharmacol; fel Acad Psychosom Med; Soc Exp Biol & Med. Res: Chemical quantitation of epinephrine and norepinephrine in plasma and relationship of these pressor amines to hypertension, circulatory shock and mental disease. Mailing Add: 8 E 81st St New York NY 10028

MANGHAM, JESSE ROGER, b Plains, Ga, Nov 18, 22; m 43; c 5. ORGANIC CHEMISTRY. Educ: Univ Ga, BS, 43; Ohio State Univ, MS, 46, PhD(org chem), 48. Prof Exp: Asst, Ohio State Univ, 43-44 & 46-48; res chemist, Va-Carolina Chem Corp, 49-51, group leader org chem, 51-53, sect leader, 53-54; proj leader, 54-67, sr appln chemist, 67-71, APPLN RES ASSOC, ETHYL CORP, 70- Mem: Am Chem Soc; Sigma Xi. Res: Organometallics of aluminum, magnesium and boron; organophosphorus chemistry; brominated chemicals; chlorinated solvents. Mailing Add: Res & Develop Dept Ethyl Corp PO Box 341 Baton Rouge LA 70821

MANGHNANI, MURLI HUKUMAL, b Karachi, WPakistan, Apr 4, 36; m 62; c 2. GEOPHYSICS, GEOCHEMISTRY. Educ: Jaswant Col, India, BS, 54; Indian Sch Mines & Appl Geol, Dhanbad, BS, 57; Bihar Univ, MS, 58; Mont State Univ, PhD(geochem, geol, geophys), 62. Prof Exp: Fel geophys & NSF res grant, Univ Wis, 62-63; asst prof, 64-69, ASSOC PROF GEOPHYS & GEOCHEM, UNIV HAWAII, 69-, GEOPHYSICIST, HAWAII INST GEOPHYS, 74- Concurrent Pos: Asst geophysicist, Hawaii Inst Geophys, 63-74. Mem: Am Geophys Union; Soc Explor Geophys; Geochem Soc; Geol, Mining & Metall Soc India. Res: Gravity; seismology; high pressure and temperature laboratory experimentation of rock materials believed to form the lower crust and upper mantle of the earth. Mailing Add: Hawaii Inst of Geophys Univ of Hawaii 2525 Correa Rd Honolulu HI 96822

MANGIN, WILLIAM PATRICK, anthropology, social anthropology, see 12th edition

MANGLITZ, GEORGE RUDOLPH, b Washington, DC, Aug 26, 26; m 53; c 5. ENTOMOLOGY. Educ: Univ Md, BS, 51, MS, 52; Univ Nebr, PhD(entom), 62. Prof Exp: Asst entomologist, United Fruit Co, Guatemala, 51; assoc prof entom, 65-73, RES ENTOMOLOGIST, USDA, UNIV NEBR, LINCOLN, 52-, PROF ENTOM, UNIV, 73- Mem: AAAS; Entom Soc Am; Am Inst Biol Sci. Res: Field crop insects, particularly plant resistance to insects. Mailing Add: Forage Insect Lab USDA Univ of Nebr Lincoln NE 68583

MANGO, FRANK DONALD, b San Francisco, Calif, Dec 31, 32; m 59; c 2. ORGANIC CHEMISTRY, INORGANIC CHEMISTRY. Educ: San Jose State Col, BS, 59; Stanford Univ, PhD(chem), 63. Prof Exp: ORG CHEMIST, SHELL DEVELOP CO, 63- Mem: Am Chem Soc. Res: Transition metal catalysis. Mailing Add: 9086 Broadway Terr Oakland CA 94611

MANGOLD, DONALD JACOB, organic chemistry, see 12th edition

MANGRAVITE, JOHN A, organic chemistry, see 12th edition

MANGUM, BILLY WILSON, b Mize, Miss, Dec 8, 31; m 63; c 1. LOW TEMPERATURE PHYSICS, SOLID STATE PHYSICS. Educ: Univ Southern Miss, BA, 53; Tulane Univ, MS, 55; Univ Chicago, PhD(phys chem), 60. Prof Exp: NSF fel physics, Clarendon Lab, Oxford, 60-61; actg sect chief low temperature physics, 67-68, PHYSICIST, NAT BUR STANDARDS, 61- Mem: Am Phys Soc. Res: Magnetism at very low temperatures; cooperative phenomena; electron-lattice interactions. Mailing Add: 8202 Post Oak Rd Rockville MD 20854

MANGUM, CHARLOTTE P, b Richmond, Va, May 19, 38. INVERTEBRATE ZOOLOGY, COMPARATIVE PHYSIOLOGY. Educ: Vassar Col, AB, 59; Yale Univ, MS, 61, PhD(biol), 63. Prof Exp: Res assoc biol, Yale Univ, 63; NIH res fel zool, Bedford Col, London, 63-64; from asst prof to assoc prof biol, 64-74, PROF BIOL, COL WILLIAM & MARY, 74- Concurrent Pos: Assoc, Va Inst Marine Sci, 64-; vis investr, Marine Biol Lab, Mass, 66, instr, 69-73; NSF res grants, 66-75; Col William & Mary fac res fel, 73; mem, Corp Marine Biol Lab, Woods Hole; lectr, Univ Aarhus, Denmark, 74. Mem: AAAS; Am Soc Zool; Marine Biol Asn UK; Brit Soc Exp Biol. Res: Comparative physiology of respiratory pigments; temperature adaptation. Mailing Add: Dept of Biol Col of William & Mary Williamsburg VA 23185

MANGUM, JOHN HARVEY, b Rexburg, Idaho, Apr 16, 33; m 57; c 3. BIOCHEMISTRY. Educ: Brigham Young Univ, BS, 57, MS, 59; Univ Wash, PhD(biochem), 63. Prof Exp: Res assoc biochem, Scripps Clin & Res Found, 62-63; from asst prof to assoc prof, 63-74, PROF CHEM, BRIGHAM YOUNG UNIV, 74- Mem: Am Chem Soc. Res: Enzymology; one-carbon metabolism; methionine biosynthesis; virus-induced acquisition of metabolic function; folate mediated reactions in brain metabolism. Mailing Add: Dept of Chem Brigham Young Univ Provo UT 84601

MANGUS, MARVIN D, b Altoona, Pa, Sept 13, 24; m 50; c 2. GEOLOGY. Educ: Pa State Univ, BS, 45, MS, 46. Prof Exp: Geologist, US Geol Surv, 47-58; sr geologist, Guatemalan Atlantic Corp, 58-60; surface geologist, Atlantic Refining Co, Pa, 60-65, sr surface geologist, Alaska, 65-69; CONSULT GEOLOGIST, CALDERWOOD & MANGUS, 69- Mem: Am Polar Soc; Am Asn Petrol Geol; Arctic Inst NAm; fel Geol Soc Am; Am Inst Prof Geol. Res: Surface geologic mapping and regional geologic studies in Alaska, Central America, Canada and Bolivia; geological mapping in arctic Alaska, the British Mountains of northern Yukon Territory, the arctic islands of Canada and the Northwest Territories of Canada; geological well site; environmental studies of Alaska. Mailing Add: 1045 E 27th Ave Anchorage AK 99504

MANHART, JOSEPH HERITAGE, b Greencastle, Ind, Mar 26, 30; m 57; c 2. ORGANIC POLYMER CHEMISTRY. Educ: DePauw Univ, AB, 52; Ohio State Univ, PhD(org chem), 60. Prof Exp: SR RES SCIENTIST, ALCOA TECH CTR, ALUMINUM CO AM, 60- Mem: Am Chem Soc; Sci Res Soc Am. Res: Anodizing electrolytes; fire retardant fillers; ultraviolet curable coatings. Mailing Add: Alcoa Center PA 15069

MANHAS, MAGHAR SINGH, b Kothe Manhasan, India, Aug 17, 22; m 53; c 5. ORGANIC CHEMISTRY. Educ: Punjab Univ, India, BSc, 43; Allahabad Univ, MSc, 45, DPhil, 50. Hon Degrees: MEng, Stevens Inst Technol, 74. Prof Exp: Asst prof chem, BR Col, Agra, India, 50-52 & Univ Saugar, 52-60; from asst prof to assoc prof, 61-70, PROF CHEM, STEVENS INST TECHNOL, 70- Concurrent Pos: Res assoc, Stevens Inst Technol, 60-61, Ottens res award, 68. Mem: Am Cjem Soc; fel The Chem Soc; Sigma Xi. Res: Heterocyclic chemistry; medicinal chemistry; stereochemistry. Mailing Add: Dept of Chem & Chem Eng Stevens Inst of Technol Hoboken NJ 07030

MANHEIM, FRANK T, b Leipzig, Ger, Oct 14, 30; US citizen; m 61; c 2. GEOCHEMISTRY. Educ: Harvard Univ, AB, 51; Univ Minn, MSc, 53; Univ Stockholm, Fil Lic, 61, DSc(geochem), 74. Prof Exp: Geochemist, Geol Surv, Sweden, 61-62; fel & res asst strontium isotopes, Yale Univ, 63; res geologist, US Geol Surv, 64-73; CHMN DEPT MARINE SCI, UNIV S FLA, 74- Concurrent Pos: Mem, Nat Acad Sci Comt on USSR & Eastern Europe, 75-77. Mem: AAAS; Am Geophys Union; Geochem Soc; Swedish Geol Soc. Res: Geochemistry of recent and fossil sediments; marine resources; chemistry of ground and natural waters; suspended matter in ocean waters; aquaculture marine policy and scientific communications. Mailing Add: Dept of Marine Sci Univ of SFla St Petersburg FL 33701

MANHEIM, JEROME HENRY, b Chicago, Ill, July 16, 23; m 46; c 3. MATHEMATICS. Educ: Univ Ill, BS, 46, MS, 47; Columbia Univ, PhD, 61. Prof Exp: Instr math, Univ Conn, 47-49 & Chicago City Jr Col, Wilson Br, 50-52; from engr to sr engr microwave develop, Fed Telecommun Labs, Int Tel & Tel Co, 52; instr math, Cooper Union, 53-55 & 56-57, admin asst, 55-56, asst prof, 57-58; asst prof, Montclair State Col, 58-61; assoc prof, Wagner Col, 62-64; prof & head dept, North Mich Univ, 64-67; fel ctr study high educ, Univ Mich, Ann Arbor, 67-68; dean col lib arts & sci, Bradley Univ, 68-71; dean sch letters & sci, 71-73, PROF MATH, CALIF STATE UNIV, LONG BEACH, 73- Concurrent Pos: Consult, Eng Design & Develop Co, 53-56. Mem: Math Asn Am. Res: Teaching of mathematics; history of analysis and number theory; cancellable numbers; mirror multiplcation. Mailing Add: 41 Neapolitan Lane E Long Beach CA 90803

MANHEIMER, WALLACE MILTON, b New York, NY, Feb 10, 42; m 65; c 3. PLASMA PHYSICS. Educ: Mass Inst Technol, BS, 63, PhD(physics), 67. Prof Exp: Prof physics, Mass Inst Technol, 68-70; PHYSICIST, NAVAL RES LABS, 70- Mem:

MANHEIMER

Am Phys Soc. Res: Turbulence theory; laser plasma interaction; relativistic beams; controlled thermonuclear fusion. Mailing Add: Code 7750 Naval Res Labs Washington DC 20390

MANHOLD, JOHN HENRY, JR, b Rochester, NY, Aug 20, 19; m 52; c 1. PATHOLOGY. Educ: Univ Rochester, BA, 40; Harvard Univ, DMD, 44; Wash Univ, MA, 56. Prof Exp: Instr oral path, Med & Dent Schs, Tufts Univ, 47-48, dir cancer teaching prog, 48-50; asst prof gen & oral path, Sch Dent, Wash Univ, 55-56; assoc prof path, 56-57, PROF PATH & ORAL DIAG & DIR DEPT, COL MED & DENT NJ, 57- Mem: AAAS; Am Dent Asn; Am Psychol Asn; AMA; fel Am Col Dent. Res: Psychosomatics; oral diagnosis; tissue metabolism. Mailing Add: Col of Med & Dent of NJ Newark NJ 07103

MANI, INDER, b India, Feb 15, 28; m 55; c 2. POLYMER CHEMISTRY. Educ: Agra Univ, BS, 46, MS, 48, LLB, 51; Univ Fla, PhD(chem), 66. 'PXAsst prof chem, Meerut Col, Agra Univ, 48-63; res assoc Argonne Nat Lab, 66-68; res chemist, 68-73, RES SPECIALIST, DOW CHEM CO, 73- Res: Gamma-radiolysis and pulsed-radiolysis of organic compounds in the vapor and liquid state; radiation curing of organic coatings; preparation, characterization and applications of polymer latexes. Mailing Add: Dow Chem Co Bldg 1712 Midland MI 48640

MANI, RAMA I, b Madras, India, Apr 10, 27. ORGANIC CHEMISTRY, BIOLOGICAL CHEMISTRY. Educ: Univ Bombay, BSc, 47, MSc, 51, PhD(chem), 61. Prof Exp: Res scientist chem, Indian Coun Med Res, India, 51-53; res scientist chem, Coun Sci & Indust Res, 57-60; res assoc, Stanford Univ, 60-62 & Univ Southern Calif, 62-63; res assoc biol chem, Vanderbilt Univ & Meharry Med Col, 63-65; ASSOC PROF CHEM, TENN STATE UNIV, 65- Mem: Am Chem Soc. Res: Organic and medicinal chemistry; radiation research; chemical and biological studies of proteins and chemical allergenic materials inducing hypersensitivity in human and animal tissues. Mailing Add: Dept of Chem Tenn State Univ Nashville TN 37203

MANI, SRINIVASA BALASUBRA, b Madras, India, Oct 8, 39; m 70. APPLIED ANTHROPOLOGY, CULTURAL ANTHROPOLOGY. Educ: Univ Delhi, BSc, 59, MSc, 61; Cornell Univ, MA, 66; Syracuse Univ, PhD(anthrop), 70. Prof Exp: Tech asst anthrop, Nat Mus India, 62-63; ASSOC PROF ANTHROP, SLIPPERY ROCK STATE COL, 70- Concurrent Pos: Sr fel pop, Policy Inst, Syracuse Univ, 69-70; consult, SAsian Prog, Slippery Rock State Col, 70- Res: Family planning and population information, education and communication; diffusion of innovations in peasant societies; systems approach to family planning communication and adoption. Mailing Add: Dept of Sociol & Anthrop Slippery Rock State Col Slippery Rock PA 16057

MANIAN, ALBERT ARDASHES, b Charlestown, Mass, Dec 30, 18; m 63; c 2. PSYCHOPHARMACOLOGY. Educ: Mass Col Pharm, BS, 43, MS, 48; Purdue Univ, PhD(pharmacol), 54. Prof Exp: Control chemist, E L Patch Co, 47; pharmacist, New Eng Hosp, 49-51; pharmacologist, US Dept Agr, 55-61; ASST CHIEF, PHARMACOL SECT, PSYCHOPHARMACOL RES BR, NIMH, 61- Honors & Awards: Cert Merit, US Dept Agr, 59 & 60; Superior Work Performance Award, US Dept Health, Educ & Welfare, 70. Mem: AAAS; Am Soc Pharmacol & Exp Therapeut; Am Soc Pharmacog. Res: Chemical and pharmacological investigations of plant products; preclinical pharmacological evaluation of psychoactive drugs; mechanisms of drug action. Mailing Add: Pharmacol Sect NIMH Psychopharmacol Res Br Rockville MD 20852

MANIAR, ATISH CHANDRA, b Unjha, India, Jan 21, 26; Can citizen; m 51; c 4. MICROBIOLOGY, PUBLIC HEALTH. Educ: Univ Bombay, BSc, 48, MSc, 52, PhD(microbiol), 56, DSc(microbiol), 69. Prof Exp: Bacteriologist, Caius Res Lab, Bombay, India, 48-55; res bacteriologist, Alembic Chem, India, 55; bacteriologist, Glaxo Labs, Bombay, 55-59; bacteriologist, Hindustan Antibiotics Ltd, India, 61-63; bacteriologist, Winnipeg Gen Hosp, 64-67; BACTERIOLOGIST, PROV LAB, MAN, 67- Concurrent Pos: Nat Res Coun Can fel antibiotics, Can Commun Dis Ctr, Ottawa, 59-61; Royal Soc Health fel, Med Col, Winnipeg, 71; lectr, St Xavier's Col, India, 48-52; mem bd studies & fac sci, Univ Bombay, 57-59; lectr, Univ Man, 64- Mem: Am Soc Microbiol; Can Pub Health Asn; Indian Asn Microbiol; Royal Soc Health. Res: Mode of action of antibiotics; aminoacidopathy in newborn babies; staphylococcal toxins. Mailing Add: Prov Lab Med Col Bldg Emily & Bannatyne Winnipeg MB Can

MANIATIS, GEORGE MARINOS, b Athens, Greece, Dec 25, 34; m 61; c 3. BIOCHEMISTRY, CELL BIOLOGY. Educ: Nat Univ Athens, MD, 59, DrMed, 63; Mass Inst Technol, PhD(biochem), 69. Prof Exp: Res assoc biochem, Mass Inst Technol, 67-69 & 70-71; ASST PROF HUMAN GENETICS & DEVELOP, COL PHYSICIANS & SURGEONS, COLUMBIA UNIV, 71- Concurrent Pos: USPHS trainee, Tufts Univ, Boston, 62-64; Med Found fel, Mass Inst Technol, 69-70. Mem: AAAS; Am Soc Hemat; Am Fedn Clin Res. Res: Ontogeny of erythropoiesis; regulation of hemopoiesis; biochemical genetics; hematology. Mailing Add: Col of Physicians & Surgeons Columbia Univ New York NY 10032

MANIATIS, THOMAS PETER, b Denver, Colo, May 8, 43; m 68. MOLECULAR BIOLOGY. Educ: Univ Colo, Boulder, BA, 65, MA, 67; Vanderbilt Univ, PhD(molecular biol), 71. Prof Exp: NIH fel, Harvard Univ, 71-73; European molecular biol org res fel, Med Res Coun Molecular Biol, Cambridge, Eng, 73-74; res assoc biol, Harvard Univ, 74-75; mem sr scientific staff, Cold Spring Harbor Lab, 75-76; ASST PROF BIOCHEM & MOLECULAR BIOL, HARVARD UNIV, 76- Res: Chromosome structure and the control of gene expression. Mailing Add: Biol Labs Harvard Univ 16 Divinity Ave Cambridge MA 02138

MANILOFF, JACK, b Baltimore, Md, Nov 6, 38; m 60; c 2. BIOPHYSICS, MICROBIOLOGY. Educ: Johns Hopkins Univ, BA, 60; Yale Univ, MS, 64, PhD(biophys), 65. Prof Exp: Res assoc chem, Brown Univ, 64-66; asst prof, 66-71, ASSOC PROF MICROBIOL, UNIV ROCHESTER, 71- Concurrent Pos: NIH res career develop award, 70. Mem: AAAS; Biophys Soc; Am Soc Microbiol. Res: Molecular and cellular biology of mycoplasma cells and their viruses; theoretical aspects of biological processes. Mailing Add: Dept of Microbiol Univ of Rochester Rochester NY 14642

MANINGER, RALPH CARROLL, b Harper, Kans, Dec 24, 18; m 42; c 3. PHYSICS. Educ: Calif Inst Technol, BS, 41. Prof Exp: Res engr & group leader, Off Sci Res & Develop Proj, Columbia, 41-45; contract physicist, Taylor Model Basin, US Dept Navy, DC, 45, sect head, US Navy Electron Lab, Calif, 45-48; physicist, Vitro Corp Am, 48-51, from asst dir to dir phys res & develop, 51-53; tech dir, Precision Tech, Inc, 53-57, br mgr librascope div, Gen Precision, Inc, 57-62; head eng res div electronics dept, 62-68, dep head electronics eng dept, 68-71, head environ studies, 71-74, HEAD TECHNOL APPLN GROUP, LAWRENCE LIVERMORE LAB, UNIV CALIF, 75- Concurrent Pos: Mem adv comt, Statewide Air Pollution Res Ctr, Univ Calif, 71-73. Mem: AAAS; Acoust Soc Am; Am Ord Asn; Inst Elec & Electronics Eng; NY Acad Sci. Res: Under-water acoustics and electronics; non-linear vibrations; solid state radiation detectors and electron devices; high speed pulse circuity; microwave generation and propagation; initiation of explosives; fast reactions in solids; quantum electronics; environmental systems research. Mailing Add: 146 Roan Dr Danville CA 94526

MANION, JAMES J, b Butte, Mont, May 17, 22; m 46; c 3. ECOLOGY. Educ: Univ Portland, BS, 48; Univ Notre Dame, MS, 50, PhD(zool), 52. Prof Exp: Asst prof biol & head dept, St Mary's Col, 52-58; assoc prof biol sci, 58-71, PROF BIOL, CARROLL COL, 71-, CHMN DIV NATURAL SCI & MATH, 58- Concurrent Pos: Dean fac, Carroll Col, 54-71, acad vpres, 66-71. Mem: AAAS. Res: Ecology of amphibians. Mailing Add: Div of Natural Sci & Math Carroll Col Helena MT 59601

MANION, JERALD MONROE, b Beebe, Ark, Sept 24, 40; m 59; c 2. ORGANIC CHEMISTRY. Educ: Harding Col, BS, 62; Univ Miss, PhD(chem), 65. Prof Exp: PROF CHEM & CHMN DEPT, UNIV CENT ARK, 65- Mem: Am Chem Soc. Res: Gas phase kinetics of reverse Diels-Alder reactions. Mailing Add: Dept of Chem Univ of Cent Ark Conway AR 72032

MANION, MARLOW WILLIAM, b Garrett, Ind, Mar 26, 02; m 32; c 1. MEDICINE. Educ: Univ Ind, BS, 24, MD, 26; Am Bd Otolaryngol, dipl, 30. Prof Exp: Intern, Harper Hosp, Detroit, 26-27, resident otolaryngol, 27-30; from assoc to assoc prof, 30-51, chmn dept, 51-62, PROF OTOLARYNGOL, SCH MED, IND UNIV, INDIANAPOLIS, 51- Mem: Am Laryngol, Rhinol & Otol Soc; Am Acad Ophthal & Otolaryngol; Am Broncho-Esophagol Asn; AMA. Res: Otolaryngology. Mailing Add: Dept of Otolaryngol Ind Univ Sch of Med Indianapolis IN 46202

MANIOTIS, JAMES, b Detroit, Mich, Aug 17, 29; m 55; c 1. MYCOLOGY. Educ: Wayne State Univ, AB, 52, MS, 57; Univ Iowa, PhD(bot), 60. Prof Exp: Res assoc virol, Child's Res Ctr, Mich, 56-57; asst bot, Univ Iowa, 57-60; instr bot, Univ Tex, 60-61; asst prof biol, Wayne State Univ, 61-65; assoc prof bot, 65-69, ASSOC PROF BIOL, WASH UNIV, 69- Mem: AAAS; Mycol Soc Am; Bot Soc Am; Genetics Soc Am; Am Inst Biol Sci. Res: Biochemical-genetical bases for pathogenicity in ringworm fungi; biology of membrane fusion in slime molds. Mailing Add: Dept of Biol Wash Univ St Louis MO 63130

MANIRE, GEORGE PHILIP, b Roanoke, Tex, Mar 25, 19; m 43; c 2. MICROBIOLOGY. Educ: NTex State Col, BS, 40, MS, 41; Univ Calif, Berkeley, PhD(bact), 49. Prof Exp: Instr bact, Univ Tex Southwestern Med Sch, 49-50; from asst prof to assoc prof, 50-59, asst vchancellor health sci, 65-66, PROF BACT, SCH MED, UNIV NC, CHAPEL HILL, 59-, CHMN DEPT, 66-, KENAN PROF, 71- Concurrent Pos: Fulbright scholar, Serum Inst, Copenhagen, 56; China Med Bd Alan Gregg fel, Virus Inst, Kyoto, Japan, 63-64; USPHS spec fel, Lister Inst, London, 71-72; mem health sci advan award comt, NIH, 67-71 & chmn, 69-71. Mem: Am Asn Immunol; Am Acad Microbiol; Infectious Dis Soc Am; Am Soc Microbiol; Soc Exp Biol & Med. Res: Mechanisms of pathogenesis of microorganisms, especially diseases of Chlamydia. Mailing Add: Dept of Bact & Immunol Univ of NC Sch of Med Chapel Hill NC 27514

MANIS, MERLE E, b St Ignatius, Mont, Aug 20, 34; m 61; c 4. MATHEMATICS. Educ: Univ Mont, BA, 60, MA, 61; Univ Ore, PhD(math), 66. Prof Exp: From instr to asst prof, 62-73, ASSOC PROF MATH, UNIV MONT, 73- Concurrent Pos: NSF res contract, 67-68. Mem: Am Math Soc. Res: Ring theory; valuation theory; D K Harrison's theory of primes. Mailing Add: Dept of Math Univ of Mont Missoula MT 59801

MANIS, WALLACE EUGENE, b Lewiston, Mont, May 20, 13; m 49; c 3. HORTICULTURE. Educ: Univ Mont, AB, 35; Univ Mich, AM, 36. Prof Exp: Rubber planter res dept, Firestone Plantations Co, Liberia, 40-41; rubber procurement technician, Rubber Develop Corp, Brazil, 42; adv western hemisphere rubber develop prog, Rubber Invests, USDA, Brazil, 43, pathologist, US-Costa Rican Coop Rubber Plant Prog, Costa Rica, 43-44, adv rubber plantation develop prog in Uraba, Colombia, 44-46, in charge US-Costa Rican Coop Rubber Plant Exp Sub-Sta, Costa Rica, 47-54; hybrid seed & ornamental plant prod, Linda Vista Ltd, Costa Rica, 55-57; plant breeder, Rubber Res Inst Ceylon, AID, Ceylon, 58-60, rubber res officer, Ministry Agr & Natural Resources, Western Nigeria, AID, 61-63; res horticulturist in charge US plant introd sta, New Crops Res Br, 64-73, AGR RES OFFICER, INT PROG DIV, AGR RES SERV, USDA, 74- Concurrent Pos: Adj asst prof, Univ Miami, 66, adj prof, 67-; consult spec fund natural rubber develop proj, UN, Thailand & Burma, 63. Mem: Soc Econ Bot. Res: Hevea rubber research, breeding and plantation development; plant introduction; tropical and sub-tropical fruit and ornamental crops; cold tolerance studies in sub-tropical fruits; development of research projects on specific crops that have potential as opium poppy replacement in northern Thailand.

MANISCALCO, IGNATIUS ANTHONY, b New York, NY, June 25, 44; m 67; c 2. ORGANIC CHEMISTRY, BIOCHEMISTRY. Educ: Manhattan Col, BS, 65; Fordham Univ, PhD(org chem), 71. Prof Exp: Instr chem, Univ Va, 70-71; ASST PROF CHEM, SPRINGFIELD COL, 71- Concurrent Pos: Res fel, Univ Va, 70-71. Mem: Am Chem Soc. Res: Heterocyclic organic chemistry. Mailing Add: Dept of Chem Springfield Col Box 1656 Springfield MA 01109

MANJARREZ, VICTOR M, b Los Angeles, Calif, June 13, 33; m 66. MATHEMATICS. Educ: Spring Hill Col, BS, 57; Harvard Univ, MA, 58, PhD(math), 63. Prof Exp: Instr math, Cath Univ Am, 65-67; asst prof, Univ Houston, 67-71; ASSOC PROF MATH, CALIF STATE COL, HAYWARD, 71- Mem: Am Math Soc; Math Asn Am. Res: Interpolation and approximation in the complex domain; topological vector spaces. Mailing Add: Dept of Math Calif State Col Hayward CA 94542

MANKA, CHARLES K, b Flemington, Mo, Sept 28, 38; m 61; c 1. PLASMA PHYSICS, SPECTROSCOPY. Educ: William Jewell Col, AB, 60; Univ Ark, MS, 64, PhD(plasma physics), 65. Prof Exp: Asst prof, 65-68, PROF PHYSICS & DIR DEPT, SAM HOUSTON STATE UNIV, 68- Res: Temperatures of exploding wires and other transient plasma formed in vacuum; spectroscopy of these plasmas and plasma acceleration by pulse methods. Mailing Add: Dept of Physics Sam Houston State Univ Huntsville TX 77340

MANKAU, REINHOLD, b Chicago, Ill, July 22, 28; m 54; c 2. NEMATOLOGY, SOIL BIOCHEMISTRY. Educ: Univ Ill, BS, 51, MS, 53, PhD(plant path), 56. Prof Exp: Res asst, Univ Ill, 54-56; Fulbright res fel, India, 56-57; asst nematologist, 58-63, ASSOC NEMATOLOGIST & ASSOC PROF NEMATOL, UNIV CALIF, RIVERSIDE, 63- Concurrent Pos: Fulbright res fel, India, 64-65. Mem: Am Phytopath Soc; Mycol Soc Am; Soc Nematol; Indian Phytopath Soc; Soc Europ Nematologists. Res: Biological control of plant-parasitic nematodes; soil biology and biochemistry; nematode-soil microbial relationships. Mailing Add: Dept of Nematol Univ of Calif Riverside CA 92502

MANKAU, SAROJAM KURUDAMANNIL, b Kottayam, India, June 5, 30; nat US; m 54; c 1. PARASITOLOGY, NEMATOLOGY. Educ: Univ Madras, BS, 49; Univ Ill, PhD(zool), 56. Prof Exp: Asst, Univ Ill, 53-56; instr biol, Univ Redlands, 58-59; res assoc plant nematol, Citrus Exp Sta, Univ Calif, Riverside, 59-60, asst prof zool, Univ, 60-63, res assoc nematol, 59-68; asst prof biol, 68-72, ASSOC PROF BIOL, CALIF STATE UNIV, SAN BERNADINO, 72- Mem: Am Soc Parasitol. Res: Helminthology; biology of soil nematodes and invertebrates. Mailing Add: Div Natural Sci Calif State Univ 5500 State College Pkwy San Bernadino CA 92407

MANKIEWICZ, EDITH MARION, b Leipzig, Ger, May 16, 10; Can citizen; m 33; c 2. MICROBIOLOGY. Educ: Univ Leipzig, MSc, 33; Univ Lyons, MD, 38; Univ Montreal, MD, 59. Prof Exp: Asst prof path, Univ Lyons, 37-39; physician in chief, Children's Hosp, Tullins, France, 39-41; chmn dept microbiol, French Univ, Shanghai, China, 41-46; res assoc bact, McGill Univ, 47-51; DIR LABS, ROYAL EDWARD CHEST HOSP, MONTREAL, 51-; ASST PROF BACT, McGILL UNIV, 62- Concurrent Pos: Guest lectr, Univ Montreal; consult, Lakeshore Gen Hosp, 65-; mem reference labs, WHO; cert specialist microbiol, Col Physicians & Surgeons Prov Que, 60 & vpres comt accreditation. Mem: Can Med Asn; Can Asn Med Bact; Am Tuberc Asn; NY Acad Sci; Am Pub Health Asn. Mailing Add: Royal Edward Chest Hosp 3650 St Urbain Montreal PQ Can

MANKIN, CHARLES JOHN, b Dallas, Tex, Jan 15, 32; m 53; c 3. GEOLOGY. Educ: Univ Tex, BS, 54, MA, 55, PhD(geol), 58. Prof Exp: Instr geol, Univ Tex, 56-57; asst prof, Calif Inst Technol, 58-59; asst prof, 59-63, DIR SCH GEOL & GEOPHYS, UNIV OKLA, 64-; DIR OKLA GEOL SURV, 67- Concurrent Pos: Assoc prof, Univ Okla, 63-66, actg dir sch geol & geophys, 63-64; mem, Gulf Univ Res Consortium-Energy Coun, 74- & Nat Petrol Coun-Exec Adv Coun Nat Gas Surv, 75. Mem: Asn Am State Geol (pres, 75-76); Am Asn Petrol Geol; Am Inst Prof Geol; Geol Soc Am; Soc Econ Paleont & Mineral. Res: Sedimentary petrology and geochemistry; clay mineralogy. Mailing Add: Sch Geol & Geophys Univ of Okla 830 Van Vleet Oval Room 163 Norman OK 73069

MANKIN, CLEON J, b Holdenville, Okla, July 20, 15; m 44; c 3. PLANT PATHOLOGY, MYCOLOGY. Educ: NMex Highlands Univ, BS, 38; NMex State Univ, MS, 50; Wash State Univ, PhD(plant path), 53. Prof Exp: Chief clerk hospitalization unit, Vet Admin, 45-48; asst plant path, Wash State Univ, 50-53; asst prof, SDak State Univ, 53-59; plant pathologist, USDA, 59-60; assoc prof plant path, 60-67, PROF PLANT PATH, SDAK STATE UNIV, 67- Mem: Am Phytopath Soc; Mycol Soc Am. Res: Diseases of forage grasses and field crops; taxonomy and genetics of microorganisms. Mailing Add: Dept of Plant Path SDak State Univ Brookings SD 57007

MANKIN, WILLIAM GRAY, b Memphis, Tenn, Sept 2, 40. ASTROPHYSICS. Educ: Southwestern Univ, Memphis, BS, 62; Johns Hopkins Univ, PhD(physics), 69. Prof Exp: Res asst astron, Univ Mass, 67-69; sci visitor, 69-71, STAFF SCIENTIST, HIGH ALTITUDE OBSERV, NAT CTR ATMOSPHERIC RES, 71- Concurrent Pos: Lectr, Univ Mass, Amherst & Smith & Mt Holyoke Cols, 68-69. Mem: Am Astron Soc; Optical Soc Am; Am Asn Physics Teachers. Res: Solar infrared radiation; chromospheric temperature structure; thermal radiation of F corona; infrared astrophysics; planetary atmospheres. Mailing Add: High Altitude Observ Boulder CO 80302

MANKINEN, CARL BELL, genetics, see 12th edition

MANKOVITZ, RALPH, b New York, NY, Dec 25, 33; m 56; c 2. CELL BIOLOGY. Educ: City Col New York, BS, 56; Univ Calif, MS, 60; Wash Univ, PhD(molecular biol), 67. Prof Exp: Fel, Dept Med Biophys, Univ Toronto, 68-71 & Med Res Coun, 69-71; RES SCHOLAR, NAT CANCER INST CAN & ASST PROF, DEPT PATH, QUEEN'S UNIV, 71- Res: Genetics of somatic mammalian cells with the long range goal of developing tools for the study of the molecular biology of growth, differentiation and cancer. Mailing Add: Dept of Path Richardson Lab Queen's Univ Kingston ON Can

MANLEY, CHARLES HOWLAND, b Acushnet, Mass, Feb 27, 43; m 65; c 1. FOOD CHEMISTRY. Educ: Southeastern Mass Univ, BS, 64; Univ Mass, Amherst, MS, 68, PhD(food sci), 69. Prof Exp: Sr food technologist, Nestle Co, 69-71; sr res chemist, Givaudan Corp, 71-74; GROUP LEADER TEA RES, THOMAS J LIPTON INC, 74- Mem: Am Chem Soc; Inst Food Technologists; Sigma Xi; Am Inst Chemists. Res: Chemistry of flavor components in natural foods; isolation and characterization of volatile and non-volatile constitutents of tea by the use of chromatographic and chemical techniques. Mailing Add: Thomas J Lipton Inc 800 Sylvan Ave Englewood Cliffs NJ 07632

MANLEY, EMMETT S, b Jackson, Tenn, Nov 6, 36; m 58; c 2. PHARMACOLOGY, PHYSIOLOGY. Educ: Univ Tenn, BS, 59, PhD(pharmacol, physiol), 63. Prof Exp: Asst physiol, Bowman Gray Sch Med, 64-65; from instr to asst prof, 65-72, ASSOC PROF PHARMACOL, MED UNITS, UNIV TENN, MEMPHIS, 72- Concurrent Pos: USPHS trainee, 64. Mem: Am Soc Pharmacol & Exp Therapeut. Res: Cardiovascular pharmacology; catecholamines; hypercapnia; acid-base balance and drug response; blood flow determination; hemorrhagic shock; myocardial function. Mailing Add: Univ of Tenn Med Units Memphis TN 38163

MANLEY, JOHN HENRY, b Harvard, Ill, July 21, 07; m 35; c 2. NUCLEAR PHYSICS. Educ: Univ Ill, BS, 29; Univ Mich, PhD(physics), 34. Prof Exp: Instr physics, Univ Mich, 31-33; lectr, Columbia Univ, 34-37; assoc, Univ Ill, 37-41, asst prof, 41-45; assoc prof, Wash Univ, 46-47; assoc dir, Los Alamos Sci Lab, 47-51; prof physics & exec officer, Univ Wash, 51-57; res adv, 57-72, CONSULT, LOS ALAMOS SCI LAB, 72- Concurrent Pos: Res assoc, Univ Chicago, 42-43; scientist, Los Alamos Sci Lab, 43-46; asst dir res, AEC, DC, 47; Guggenheim fel, 54; US State Dept fel, 58. Mem: Fel AAAS; fel Am Phys Soc. Res: Biophysics; nuclear physics; electron microscopy. Mailing Add: Rte 1 Box 417 Espanola NM 87532

MANLEY, LEO WILLIAM, b Auburn, NY, Sept 5, 17; m 43; c 3. CHEMISTRY, MATHEMATICS. Educ: Syracuse Univ, AB, 40. Prof Exp: Chemist, Am Cyanamid Co, NJ, 40-41; chemist, 41-48, res assoc, 48-52, from asst supvr to supvr, 52-63, tech dir, 63-72, mgr prod res & develop, 64-72, MGR LUBRICANTS & ADDITIVES RES & TECH SERV, RES DEPT, MOBIL RES & DEVELOP CORP, 72- Concurrent Pos: Consult, Dept Defense, 52- & Am, Brit & Can Tripartite Meetings, 56-; mem various comts, Am Petrol Inst, 63- Mem: Am Chem Soc; Soc Automotive Eng; Coord Res Coun; Am Soc Test & Mat. Res: New and improved fuels, lubricants and other products derived from petroleum. Mailing Add: Mobil Res & Develop Corp Res Dept Paulsboro NJ 08066

MANLEY, LILLIAN C, b Birmingham, Ala, Oct 22, 32; m 53; c 2. BIOLOGY, ACADEMIC ADMINISTRATION. Educ: Univ Ala, BS, 53, MS, 58, PhD(biol), 61. Prof Exp: Asst, Southern Res Inst, Birmingham, Ala, 53-54; teaching fel biol, Univ Ala, 57-58; instr high sch, Ala, 58-59; assoc prof biol & phys sci, Livingston State Col, 59-63; prof biol & chmn div sci, Judson Col, 63-70; EXEC DIR, ALA CONSORTIUM FOR DEVELOP HIGHER EDUC, 70- Concurrent Pos: Mem int fel awards comt, Am Asn Univ Women, 62-68; mem, Gov Comt 100 & Comn Status of Women, 62-66; bd trustees, Troy State Univ, 67-70; chmn task force III, Ala Educ Study Comn, 67-69; vchmn, Ala Comn Higher Educ, 69-71; mem adv comt on coop progs, Am Asn Higher Educ, 72-75; mem bd vis, Air Univ, 74-76; mem adv comt, Univ Without Walls Prog, Univ Ala. Mem: AAAS; Am Asn Higher Educ; Am Coun of Educ; Sigma Xi. Res: Physiology and histology of mammalian thymus gland. Mailing Add: Ala Consortium Develop Higher Ed Progs Off PO Box 338 Demopolis AL 36732

MANLEY, ROCKLIFFE ST JOHN, b Kingston, Jamaica, Mar 26, 25; Can citizen; m 58; c 1. POLYMER CHEMISTRY. Educ: McGill Univ, BSc, 50, PhD(phys chem), 53; Uppsala Univ, DSc, 56. Prof Exp: Nat Res Coun Can fel, 53-55; RES ASSOC CHEM, McGILL UNIV, 58-; SR SCIENTIST, PULP & PAPER RES INST CAN, 58- Concurrent Pos: Secy-treas, Can High Polymer Forum, 67-69, prog chmn, 69-71, chmn forum, 71-73. Mem: AAAS; Fiber Soc; Am Phys Soc; Chem Inst Can. Res: Macromolecular science; flow properties of model disperse systems; polymer solution properties; morphology of crystalline polymers; polymer crystallization; molecular morphology and biosynthesis of cellulose. Mailing Add: Dept of Chem McGill Univ Montreal PQ Can

MANLEY, STEPHEN ALEXANDER, b Sydney, NS, Dec 6, 45; m 68. FOREST GENETICS. Educ: Univ NB, BScF, 67, MScF, 69; Yale Univ, MPhil, 71, PhD(forest genetics), 75. Prof Exp: Forest geneticist, Environ Can, Can Forestry Serv, 69-75; FOREST GENETICIST BREEDING, DEPT AGR & FORESTRY, 75- Mem: Can Tree Improve Asn. Res: Elucidation of the population and genetics structure of tree species and maximization of growth potential of locally adapted races. Mailing Add: Dept of Agr & Forestry Charlottetown PE Can

MANLEY, THOMAS CLINTON, b Ithaca, NY, Feb 15, 11; m 40; c 2. ELECTROCHEMISTRY. Educ: Cornell Univ, BChem, 32, MChem, 33; Rutgers Univ, PhD(phys chem), 38. Prof Exp: Pilot plant engr, 40, res engr, 41-47, asst dir res, 48-49, DIR RES, WELSBACH CORP, 49- Mem: Am Chem Soc. Res: Electric discharges in gases; chemical reactions in electric discharges; ozone production properties; reactions of ozone; absorption of gases. Mailing Add: Welsbach Corp Ozone Systs Div 3340 Stokely St Philadelphia PA 19129

MANLY, DONALD G, b Cleveland, Ohio, Oct 7, 30; m 52; c 2. ORGANIC CHEMISTRY. Educ: Brown Univ, ScB, 52; Lehigh Univ, MS, 54, PhD(org chem), 56. Prof Exp: Proj leader chem, Quaker Oats Co, 56-57, group leader, 57-63, sect leader chem res, 63-65; res mgr chem, Glyco Chem Inc, 65-69; assoc dir res, 69-71, DIR CORP RES, AIR PROD & CHEM INC, 71- Mem: Am Chem Soc; Am Oil Chem Soc; Soc Cosmetic Chem; Indust Res Inst. Res: Heterogeneous catalysis; heterocyclic compounds; fluorine chemistry; enzyme technology; pollution control. Mailing Add: Air Prod & Chem Inc PO Box 538 Allentown PA 18105

MANLY, JETHRO OATES, b NC, Jan 21, 14; m 41; c 3. PHYCOLOGY. Educ: Col William & Mary, BS, 37; Duke Univ, PhD, 53. Prof Exp: Instr biol, Col William & Mary, 46-49; instr zool, Duke Univ, 49-50, bot, 52-55; PROF BIOL, PFEIFFER COL, 55- Mem: Fel AAAS. Res: Taxonomy and distribution of marine diatoms. Mailing Add: Dept of Biol Pfeiffer Col Misenheimer NC 28109

MANLY, KENNETH FRED, b Cincinnati, Ohio, July 12, 41; m 62; c 1. VIROLOGY. Educ: Calif Inst Technol, BS, 64; Mass Inst Technol, PhD(microbiol), 69. Prof Exp: Sr cancer res scientist, 71-74, ASSOC CANCER RES SCIENTIST, ROSWELL PARK MEM INST, 74- Concurrent Pos: Am Cancer Soc fel, Mass Inst Technol, 69-71; Nat Cancer Inst grant, 72-75. Mem: AAAS; Am Soc Microbiol. Res: Biochemistry of tumor viruses; etiology of cancer. Mailing Add: Med Viral Oncol Dept Roswell Park Mem Inst Buffalo NY 14263

MANLY, MARIAN LEFEVRE, b Rochester, NY, Aug 24, 11; m 39; c 3. DENTAL RESEARCH. Educ: Univ Rochester, BA, 33, MS, 35, PhD(biochem), 39. Prof Exp: Res asst biochem, Sch Med, Univ Rochester, 33-39, lab asst chem, Col Arts & Sci, 36-37; co-owner, 53-64, SPEC PROJS RES, WESTWOOD RES LAB, INC, 64- Mem: Int Asn Dent Res. Res: Research method development in biochemistry and in evaluation of home oral products. Mailing Add: Westwood Res Lab Inc 543 High St Westwood MA 02090

MANLY, PHILIP JAMES, b Cincinnati, Ohio, Apr 12, 44; m 67; c 3. HEALTH PHYSICS. Educ: Mass Inst Technol, BS, 67; Rensselaer Polytech Inst, MS, 71. Prof Exp: Shift radcon dir, 71-72, head tech div, 72-74, sr health physicist, 74, HEAD TRAINING DIV HEALTH PHYSICS, PEARL HARBOR NAVAL SHIPYARD, 74- Concurrent Pos: Owner, Health Physics Assocs, 74- Mem: Health Physics Soc; assoc mem Sigma Xi. Res: Measurement of radiation and radionuclides in the environment; practical application of radiation protection principles. Mailing Add: 228 Plum St Wahiawa HI 96786

MANLY, RICHARD SAMUEL, b Malta, Ohio, May 31, 11; m 39; c 3. BIOCHEMISTRY. Educ: Antioch Col, BS, 33; Oberlin Col, MA, 34; Univ Rochester, PhD(biochem), 38; Am Inst Chem, cert. Prof Exp: Univ Wis Alumni Res Found fel, Univ Rochester, 39; res chemist, Procter & Gamble Co, Ohio, 39-45; asst prof, 45-47, prof dent & res dir, 47-62, chmn dept dent sci, 62-64, RES PROF DENT, SCH DENT MED, TUFTS UNIV, 64-; PRES, WESTWOOD RES LAB, 53- Concurrent Pos: Mem biochem subcomt, Nat Res Coun, 52-54; mem dent study sect, NIH, 56-59; co-chmn, Gordon Conf Calcium Phosphates, 64. Mem: Fel AAAS; Am Chem Soc; fel Am Inst Chem; Am Dent Asn; Int Asn Dent Res (vpres, 64, pres-elect, 65, pres, 66). Res: Carbohydrate metabolism; physical properties of teeth; oral physiology; glycolysis inhibition; dental plaque composition; enzymes and physical properties; toothbrush and dentifrice abrasion; tooth enamel solution rate; porosity and membrane potentials. Mailing Add: Westwood Res Lab Inc 543 High St Westwood MA 02090

MANN, ALAN EUGENE, b New York, NY, Sept 19, 39; m 63; c 1. PHYSICAL ANTHROPOLOGY, PRIMATOLOGY. Educ: Univ Pittsburgh, BA, 61; Univ Calif, Berkeley, MA & PhD(anthrop), 68. Prof Exp: Teaching asst phys anthrop, Columbia Univ, 63-64; asst, Univ Calif, Berkeley, 66-68, actg asst prof, 68-69; ASST PROF ANTHROP, UNIV & ASST CUR PHYS ANTHROP, MUS, UNIV PA, 69- Mem: Am Asn Phys Anthrop; Am Anthrop Asn; Royal Anthrop Inst Gt Brit & Ireland. Res: Analysis of hominid evolution, with emphasis on the reconstruction of behavior. Mailing Add: Dept of Anthrop Univ of Pa Philadelphia PA 19174

MANN, ALFRED KENNETH, b New York, NY, Sept 4, 20; m 46; c 4. PHYSICS. Educ: Univ Va, AB, 42, MS, 46, PhD(physics), 47. Prof Exp: Instr physics, Columbia Univ, 47-49; from asst prof to assoc prof, 49-57, PROF PHYSICS, UNIV PA, PHILADELPHIA, 57- Concurrent Pos: Fulbright fel, 55-56; NSF sr fel, 62-63. Mem: Fel Am Phys Soc. Res: Mass spectroscopy; molecular beams; photonuclear reactions; electrodynamics; elementary particle physics. Mailing Add: Dept of Physics Univ of Pa Philadelphia PA 19104

MANN, BENJAMIN MICHAEL, b Philadelphia, Pa, Apr 17, 48; m 74. TOPOLOGY. Educ: Univ Calif, Los Angeles, BA, 70; Stanford Univ, MS, 71, PhD(math), 75. Prof Exp: LECTR MATH, RUTGERS UNIV, NEW BRUNSWICK, 7S- Mem: Am Math

Soc. Res: Corbordism; infinite loop space theory; surgery; homotopy theory. Mailing Add: Dept of Math Hill Ctr Rutgers Univ New Brunswick NJ 08903

MANN, BRUCE JAMESON, b Wadsworth, Ohio, Sept 10, 38; c 1. HEALTH PHYSICS, NUCLEAR ENGINEERING. Educ: Ashland Col, BS, 60; Univ Calif, Berkeley, MS, 64 & 72. Prof Exp: Health physicist, Div Radiol Health, USPHS, Dept Health, Educ & Welfare, 61-66, Sandia Labs, 67 & Health & Safety Off, Univ Calif, Los Angeles, 67-69; assoc engr, Dept Nuclear Eng, Univ Calif, Berkeley, 72; CHIEF EVAL BR, OFF AIR & HAZARDOUS MAT, OFF RADIATION PROGS, US ENVIRON PROTECTION AGENCY, 73- Concurrent Pos: Eng assoc, Teknekron Inc, 72-73; panel mem, Fusion Res & Develop Subpanel, US AEC, 73. Mem: Health Physics Soc; Am Nuclear Soc; AAAS. Res: Environmental radiological effects of nuclear energy technologies; mathematical modeling of environmental radioactivity transport uptake and dosimetry. Mailing Add: US Environ Protection Agency PO Box 15027 Las Vegas NV 89114

MANN, CHARLES E, b Colorado Springs, Colo, May 5, 26; m; c 1. ANTHROPOLOGY. Educ: Mexico City Col, BA, 57, MA, 58; Stanford Univ, PhD(anthrop), 65. Prof Exp: Res asst anthrop, Stanford Univ, 62; from asst prof to assoc prof, San Diego State Col, 62-67; assoc prof, 67-69, PROF ANTHROP, UNIV OF THE AMERICAS, 69-, CHMN DEPT, 67- Mem: Am Anthrop Asn; Soc Appl Anthrop; Mex Soc Anthrop. Res: Linguistics, culture change, social organization, archaeology and ethnology, especially in Mesoamerica. Mailing Add: Dept of Anthrop Apdo Postal 507 Univ of the Americas Puebla Mexico

MANN, CHARLES KENNETH, b Fairmont, WVa, Jan 2, 28; m 57; c 2. ANALYTICAL CHEMISTRY. Educ: George Washington Univ, BS, 50, MS, 52; Univ Va, PhD(chem), 55. Prof Exp: Anal chemist, Nat Bur Standards, 50-52; instr chem, Univ Tex, 55-58; from asst prof to assoc prof, 58-68, PROF CHEM, FLA STATE UNIV, 68- Mem: Am Chem Soc. Res: Organic electrochemistry; electroanalytical chemistry. Mailing Add: Dept of Chem Fla State Univ Tallahassee FL 32306

MANN, CHARLES ROY, b New York, NY, Mar 27, 41. MATHEMATICAL STATISTICS, APPLIED STATISTICS. Educ: Polytech Inst Brooklyn, BS, 61; Mich State Univ, MS, 63; Univ Mo, PhD(statist), 69. Prof Exp: Instr math, Univ Maine, 63-64; asst prof statist, George Washington Univ, 69-73; HEAD STATIST DIV, GROUP OPERS INC, 73- Concurrent Pos: Consult, US Info Agency, 71-, Am Asn RR, 71- & Dept Justice, 75- Mem: Inst Math Statist; Am Statist Asn; Math Asn Am. Res: Bayesian statistics; density estimation; data analysis; legal applications of statistics. Mailing Add: Group Opers Inc 2025 Eye St NW Washington DC 20006

MANN, CHRISTIAN JOHN, b Junction City, Kans, Oct 16, 31; m 61; c 4. GEOLOGY. Educ: Univ Kans, BS, 53, MS, 57; Univ Wis, PhD(geol), 61. Prof Exp: Geologist, Gulf Oil Corp, 53 & Calif Oil Co, 57-64; sr earth scientist, Hazleton Nuclear Sci Corp, 64-65; asst prof geol, 65-69, ASSOC PROF GEOL, UNIV ILL, URBANA, 69- Mem: AAAS; fel Geol Soc Am; Int Asn Math Geol; Am Asn Petrol Geol. Res: Mesozoic and Paleozoic stratigraphy; quantitative geology; quantitative analysis of cycles in geology; nature of geologic data; data enhancement; regional stratigraphic synthesis. Mailing Add: Dept of Geol Univ of Ill Urbana IL 61801

MANN, DAVID EDWIN, JR, b Johnson City, Tenn, Feb 13, 22; m 50; c 3. PHARMACOLOGY. Educ: Harvard Univ, BS, 44; Purdue Univ, MS, 48, PhD(physiol), 51. Prof Exp: Asst prof physiol & pharmacol, Sch Pharm, 50-54, assoc prof pharmacol & chmn dept, 54-60, PROF PHARMACOL, SCH PHARM & SCH DENT, TEMPLE UNIV, 60- Honors & Awards: Lindback Award, 66. Mem: AAAS; Am Pharmaceut Asn. Res: Teratology; toxicology; carcinogenesis. Mailing Add: Dept of Pharmacol Temple Univ Philadelphia PA 19140

MANN, DAVID JACOB, b Patchoque, NY, Nov 6, 19; m 44; c 2. ORGANIC CHEMISTRY. Educ: Univ Long Island, BS, 41; Univ Ill, MS, 47, PhD(chem), 50. Prof Exp: Res chemist, Atlantic Res Corp, 50-53; res chemist, Reaction Motors, Inc, 53-54, sect supvr, 54-56, dept mgr, Reaction Motors Div, Thiokol Chem Corp, 56-61, res dir, 61-68; dir res & develop, Resins & Plastics Div, Escambia Chem Corp, 68-72; PRES, BUSINESS MGT SYSTEMS, 72- Mem: Am Chem Soc; Am Inst Aeronaut & Astronaut; Am Inst Chemists. Res: Chemistry of rocket propellants; organic, inorganic, organometallic and polymer synthesis; synthesis of rocket fuels and oxidizers; propellant ingredient manufacture; propellant formulation and evaluation; liquid and solid rocket testing; thermochemistry; chemical materials; explosives; combustion. Mailing Add: PO Box 14124 Orlando FL 32807

MANN, DENNIS KEITH, b Mt Carmel, Ill, Nov 11, 29; m 54; c 4. MEDICAL MICROBIOLOGY. Educ: Univ Ill, Urbana, BS, 51, DVM, 60, MS, 64, PhD(vet med sci), 67; Tulane Univ, MPH, 63. Prof Exp: Chemist, USI Chem Corp, Ill, 54-56; instr vet microbiol, Col Vet Med, Univ Ill, Urbana, 60-61, res assoc, 61-62; chief lab develop, State Hyg Lab, Univ Iowa, 67-71; assoc prof med microbiol, Sch Med, Southern Ill Univ, 72-75; PROF MICROBIOL & CHMN DEPT, SCH MED, MARSHALL UNIV, 75- Concurrent Pos: NIH fel, Col Vet Med, Univ Ill, Urbana, 64-67; adj assoc prof, Sangamon State Univ, 73-75. Mem: Am Vet Med Asn; Am Pub Health Asn; Am Soc Microbiol; Am Soc Trop Med & Hyg; AAAS. Res: Epidemiology, diagnosis and control of zoonotic diseases. Mailing Add: 1316 Mallory Court Huntington WV 25701

MANN, DIANA WITHERSPOON, b Teaneck, NJ, Sept 6, 35. NEUROPHYSIOLOGY. Educ: Conn Col, AB, 57; Brown Univ, PhD(biol, neurophysiol), 73. Prof Exp: Res technician renal physiol, Univ Pa, 57-58; teaching asst biochem, Conn Col, 63-65 & biol & med sci, Brown Univ, 65-67; RES ASSOC NEUROPHYSIOL, INST MARINE BIOMED RES & VIS ASST PROF BIOL, UNIV NC, WILMINGTON, 74- Mem: Sigma Xi; Soc Neurosci; Undersea Med Soc; assoc mem Am Physiol Soc. Res: Cellular level and biophysical studies of the mechanisms of synaptic transmission with emphasis on longer term aspects; effects of high pressures on neuronal physiology, particularly the synaptic transmission process. Mailing Add: Inst of Marine Biomed Res 7205 Wrightsville Ave Wilmington NC 28401

MANN, EDWARD CULLEE, b Laredo, Tex, Nov 21, 23; m 56. MEDICINE. Educ: Tulane Univ, BA, 46, MD, 50; Am Bd Obstet & Gynec, dipl, 61. Prof Exp: Intern, Walter Reed Gen Hosp, DC, 50-51; resident psychiat, Johns Hopkins Hosp, 51-54, instr, Johns Hopkins Univ, 53-54; resident obstet & gynec, NY Lying-In-Hosp, 54-58; from asst prof to clin assoc prof obstet & gynec, Med Col, Cornell Univ, 58-69; ASST PROF OBSTET & GYNEC, LA STATE UNIV MED CTR, NEW ORLEANS, 69- Concurrent Pos: Commonwealth res fel, Johns Hopkins Univ, 54-58; asst attend obstetrician & gynecologist, NY Lying-In-Hosp, 58-69; consult, NIH, 58-; mem staff, Earl K Long Mem Hosp, Baton Rouge, La, 69- Honors & Awards: Found Prize, Am Asn Obstet & Gynec, 58; Bronze Medal, Am Roentgen Ray Soc, 59; Award Radiol Soc NAm, 59. Mem: Am Fertil Soc; Am Psychiat Asn; fel Am Col Obstet & Gynec. Res: Psychosomatic aspects of obstetrics and gynecology; physiology of the uterine isthmus. Mailing Add: Earl K Long Mem Hosp 5825 Airline Hwy Baton Rouge LA 70805

MANN, ELTON W, b Oketo, Kans, Nov 9, 14; m 53; c 4. MICROBIOLOGY, PLANT PATHOLOGY. Educ: NMex State Univ, BS, 62; Univ Ariz, PhD(plant path), 66. Prof Exp: Asst prof biol, Eastern NMex Univ, 65-69; MGR BIOL SCI, HERSHEY RES LABS, HERSHEY FOOD CORP, 69- Concurrent Pos: Am Cancer Soc grant, 67-69; Nat Cotton Coun grant, 68-69. Mem: Soc Indust Microbiol; Am Inst Biol Sci. Res: Cancer; leukemia; tobacco mosaic virus; diagnosis and control of rabies; antifungal antibiotics; damping-off diseases of cotton, peanuts, wheat and corn; control of Tricophyton and other dermatophytes. Mailing Add: Briar Crest 21 Townhouse Apts Hershey PA 17033

MANN, GEORGE VERNON, b Lehigh, Iowa, Sept 15, 17; m 47. BIOCHEMISTRY, NUTRITION. Educ: Cornell Col, BA, 39; Johns Hopkins Univ, DSc(biochem), 42, MD, 45. Prof Exp: Asst chemist, State Health Dept, Md, 40-41; intern med, Johns Hopkins Hosp, 44-45; intern, Peter Bent Brigham Hosp, 46; asst prof nutrit, Harvard Univ, 49-55, asst med, 50-58; asst prof med, 58-68, ASSOC PROF BIOCHEM, SCH MED, VANDERBILT UNIV, 58-, ASSOC PROF MED, 68- Concurrent Pos: Nutrit Found res fel, Sch Pub Health, Harvard Univ, 47-49; asst resident, New Eng Deaconess Hosp, 46; asst, Peter Bent Brigham Hosp, 47-48; estab investr, Am Heart Asn, 54-62; asst dir Framingham Heart Study, USPHS, 55-58, consult, 58-; career investr, Nat Heart Inst, 62- Mem: AAAS; Am Heart Asn; Am Inst Nutrit; NY Acad Sci. Res: Atherosclerosis-cardiovascular diseases; epidemiology. Mailing Add: Div of Nutrit Vanderbilt Univ Sch of Med Nashville TN 37203

MANN, GODFREY EMILE, biological chemistry, organic chemistry, deceased

MANN, HARRY MILTON, physics, see 12th edition

MANN, HENRY BERTHOLD, b Vienna, Austria, Oct 27, 05; nat US; m 35; c 1. MATHEMATICS. Educ: Univ Vienna, PhD(math), 35. Prof Exp: Carnegie fel, Columbia Univ, 42-43; instr math, Bard Col, 43-44; from assoc prof to prof, Ohio State Univ, 46-64; prof & mem math res ctr, Univ Wis, Madison, 64-70; PROF MATH, UNIV ARIZ, 70- Honors & Awards: Cole Prize, Am Math Soc, 46. Mem: Am Math Soc; fel Inst Math Statist. Res: Elementary number theory; additive number theory; algebraic number theory; group theory; mathematical statistics. Mailing Add: Dept of Math Univ of Ariz Tucson AZ 85721

MANN, JACINTA, b Pinckneyville, Ill, May 13, 25. APPLIED STATISTICS. Educ: Southern Ill Univ, BS, 46; Univ Wis, MS, 47, PhD(educ measurement & statist), 58. Prof Exp: Statistician, Univ Wis, 48-50; sec sch teacher math, Pa, 52-56; res asst, Univ Wis, 56-57; asst prof educ, Seton Hill Col, 58-61, from asst dir admis to dir admis, 59-67; asst to pres, Scripps Col, 67-68; acad dean, 68-71, assoc prof-at-large, 71-72, PROF-AT-LARGE, SETON HILL COL, 72- Concurrent Pos: Am Coun Educ acad admin internship prog fel, Scripps Col, 67-68; NSF comput grant, Holland Col, 72. Mem: Nat Coun Measurement Educ; Am Statist Asn. Mailing Add: Seton Hill Col Greensburg PA 15601

MANN, JAMES, b Paterson, NJ, Nov 29, 13; m 42; c 4. MEDICINE. Educ: Univ Ill, AB, 35; Wash Univ, MD, 40; Am Bd Psychiat & Neurol, dipl. Prof Exp: From asst prof to assoc prof, 54-63, PROF PSYCHIAT, SCH MED, BOSTON UNIV, 63- Concurrent Pos: Dir psychiat, Briggs Clin, Boston State Hosp, 49-52, dir psychiat, Hosp, 52-59; mem fac, Sch Social Work, Smith Col, 51-58; mem, Boston Psychoanal Inst, 53-, training analyst, 62-, dean, 71-73; vis prof, Hadassah Med Sch, Hebrew Univ Jerusalem, 55-56, Inst Living, Hartford, Conn, 58 & Grad Sch, Brandeis Univ, 59. Mem: AAAS; fel Am Psychiat Asn; Am Psychoanal Asn. Res: Psychotherapy of schizophrenia; time limited psychotherapy; group psychotherapy; dynamics of teaching. Mailing Add: 20 Locke Rd Waban MA 02168

MANN, JAMES EDWARD, b Oceanside, NY, May 30, 38; m 63; c 2. ALGOLOGY. Educ: NTex State Univ, BA, 62, MA, 64; Univ Tex, Austin, PhD(algal physiol), 68. Prof Exp: ASST PROF BIOL, UNIV HOUSTON, 68- Mem: Phycol Soc Am. Res: Biochemical aspects of algal physiology, including lipid metabolism, nitrogen fixation and photosynthesis. Mailing Add: Dept of Biol Univ of Houston Houston TX 77004

MANN, JAMES EDWARD, JR, b Bluefield, WVa, Nov 17, 36; m 62; c 3. APPLIED MATHEMATICS, APPLIED MECHANICS. Educ: Va Polytech Inst, BS, 59; Harvard Univ, SM, 60, PhD(eng), 64. Prof Exp: Res engr, Esso Prod Res Co, 63-65; asst prof appl math, 65-68, ASSOC PROF APPL MATH, UNIV VA, 68- Mem: Soc Indust & Appl Math; Math Asn Am. Res: Wave propagation phenomena; problems that arise in electrodynamics, acoustics and water waves. Mailing Add: 2224 Greenbriar Dr Charlottesville VA 22901

MANN, JOHN ALLEN, b NJ, Sept 28, 21; m 43; c 3. GEOLOGY. Educ: Princeton Univ, BA, 43, MA, 49, PhD(geol), 50. Prof Exp: Geologist, Stand Oil Co, Calif, 50-54, dist geologist, 54-64; geol res supvr, Chevron Res Co, 64-68, sr staff geologist, Chevron Oil Field Res Co, 68, div geologist, Sotex Div, Chevron Oil Co, 68-71, SR STAFF GEOLOGIST-SEM COORDR, WESTERN DIV, CHEVRON OIL CO, 71- Mem: Geol Soc Am; Am Asn Petrol Geologists; Nat Asn Geol Teachers. Res: Petroleum exploration. Mailing Add: Chevron Oil Co PO Box 599 Denver CO 80201

MANN, JOHN B, food science, microbiology, see 12th edition

MANN, JOHN FRANCIS, JR, b Brooklyn, NY, Mar 21, 21; m 49. GEOLOGY. Educ: Univ Southern Calif, MS, 47, PhD(geol), 51. Prof Exp: Field geologist, Frontier Refining Co, 43-44; res geologist, US Geol Surv, 46-47; asst, Univ Southern Calif, 47-49, from asst prof to assoc prof geol, 51-58; ground water geologist, State Geol Surv, Ill, 49-51; CONSULT GROUND WATER GEOLOGIST, 51-; CONSULT, LOS ANGELES DEPT WATER & POWER, 57- & METROP WATER DIST SOUTHERN CALIF, 67- Concurrent Pos: Vis assoc prof, Univ Southern Calif, 58- Mem: Geol Soc Am; Am Asn Petrol Geol; Am Water Works Asn; Am Geophys Union; hon mem Nat Water Well Asn (vpres, 58-60). Res: Ground water geology. Mailing Add: 945 Reposado Dr La Habra CA 90631

MANN, JOSEPH BIRD, (JR), b Kearny, NJ, Dec 1, 23; m 45; c 3. CHEMICAL PHYSICS. Educ: Union Univ, NY, BS, 44; Mass Inst Technol, PhD(phys chem), 50. Prof Exp: Asst photochem, Cabot Solar Energy Fund, Mass Inst Technol, 48-50; MEM STAFF, LOS ALAMOS SCI LAB, 50- Mem: Fel AAAS; Am Inst Chem. Res: Atomic structure studies; properties of actinide and superheavy elements; relativistic Hartree-Fock calculations. Mailing Add: 2551 35th St Los Alamos NM 87544

MANN, JULIAN ADIN, JR, surface chemistry, theoretical chemistry, see 12th edition

MANN, KENNETH CLIFFORD, b Swift Current, Sask, Oct 2, 10; m 44. PHYSICS. Educ: Univ Sask, BSc, 31, BEd, 33; Univ Toronto, MA, 36, PhD(physics), 38. Prof Exp: Asst prof physics, Univ C, 39-41; res physicist, Nat Res Coun Can, 41-45; prof physics, Univ BC, 46-75; MEM STAFF, JET PROPULSION LAB, 75- Concurrent

Pos: Prod engr, Res Enterprises, Ltd, 44-45; radar rep, Brit Admiralty Tech Mission, 44-45. Mem: AAAS; Am Phys Soc; Can Asn Physicists. Res: Investigation of disintegration schemes of radioactive isotopes by beta ray spectroscopy. Mailing Add: Jet Propulsion Lab 4800 Oak Grove Dr Pasadena CA 91103

MANN, KENNETH GERARD, b Floral Park, NY, Jan 1, 41; m 64; c 3. BIOCHEMISTRY. Educ: Manhattan Col, BS, 63; Univ Iowa, PhD(biochem), 67. Prof Exp: Fel, Univ Iowa, 67-68; NIH fel, Duke Univ, 68-70; asst prof biochem, 70-75, ASSOC PROF BIOCHEM, UNIV MINN, ST PAUL, 75-; ASSOC PROF HEMATOL, MAYO MED SCH, MAYO FOUND, 74- Concurrent Pos: NIH res grant, Univ Minn, St Paul, 71-74, Dreyfus teacher grant, 71-76; estab investr, Am Heart Asn, 74-79. Mem: Am Soc Biol Chemists; Am Chem Soc; Sigma Xi; Am Soc Hematol; Am Heat Asn. Res: Protein chemistry; blood clotting. Mailing Add: Hematol Res Mayo Clin Rochester MN 55901

MANN, KENNETH H, b Dovercourt, Eng, Aug 15, 23; m 46; c 3. ECOLOGY. Educ: Univ London, BSc, 49, DSc(zool), 66; Univ Reading, PhD(zool), 53. Prof Exp: Asst lectr zool, Univ Reading, 49-51, lectr, 51-64, reader, 64-67; sr biologist, Marine Ecol Lab, Bedford Inst, Fisheries Res Bd Can, 67-72; PROF BIOL & CHMN DEPT, DALHOUSIE UNIV, HALIFAX, NS, 72- Concurrent Pos: Mem productivity freshwater subcomt, Brit Nat Comt Int Biol Prog, 64-67; consult, London Anglers Asn, 58-64; ed, J Animal Ecol, 66-67; mem, Can Comt Man & Biosphere, 73- Mem: Am Soc Limnol & Oceanog; Brit Ecol Soc; Brit Freshwater Biol Asn; Ecol Soc Am. Res: Functioning of aquatic ecosystems; primary and secondary productivity in coastal, estuarine and fresh waters; dynamics of marine food chains. Mailing Add: Biol Dept Dalhousie Univ Halifax NS Can

MANN, KINGSLEY M, b Prince Albert, Sask, Nov 12, 19; nat US; m 48; c 3. SCIENCE WRITING. Educ: Univ Alta, BSc, 44; Univ Sask, MSc, 46; Univ Wis, PhD(biochem), 49. Prof Exp: Lectr, radio-carbon lab, Univ Ill, 49-50; biochemist, 50-75, CONSULT, SCIENCE WRITER & ED, UPJOHN CO, 75- Concurrent Pos: Vis scientist, NIH, 58-59. Res: Steroids; antimetabolites; antiviral chemotherapy. Mailing Add: 5240 Maple Ridge Dr Kalamazoo MI 49001

MANN, LARRY N, b Philadelphia, Pa, Aug 21, 34; m 59; c 3. TOPOLOGY. Educ: Univ Pa, BS, 55, MA, 56, PhD(math), 59. Prof Exp: Mathematician, Radio Corp Am, 58-60; lectr math, Univ Va, 60-61, asst prof, 61-63; mem, Inst Defense Anal, 63-64; asst, Inst Advan Study, 64-65; assoc prof, 65-70, PROF MATH, UNIV MASS, AMHERST, 70- Concurrent Pos: Assoc, Off Naval Res, 60-61. Mem: Am Math Soc. Res: Applications of algebraic and differential topology to topological transformation groups. Mailing Add: Dept of Math Univ of Mass Amherst MA 01002

MANN, LEONARD ANDREW, b Cleveland, Ohio, June 30, 15. PHYSICS. Educ: Univ Dayton, BS, 37; Ohio State Univ, MSc, 45; Carnegie Mellon Univ, PhD(physics), 54. Prof Exp: Teacher high sch, Ohio, 37-48; assoc prof physics, 54-62, assoc dean, 56-61, chmn dept, 55, PROF PHYSICS, UNIV DAYTON, 62-, DEAN COL ARTS & SCI, 61- Mem: AAAS; Am Phys Soc; Am Soc Eng Educ; Am Asn Physics Teachers. Res: Meson scattering; beta spectrometry. Mailing Add: Univ of Dayton Col of Arts & Sci Dayton OH 45409

MANN, LESLIE BERNARD, b Granger, Wash, Oct 19, 19; m 44; c 3. NEUROLOGY. Educ: La Sierra Col, BS, 44; Loma Linda Univ, MD, 45. Prof Exp: Intern, Los Angeles County Hosp, 44-45; intern, 50, DIR ELECTROENCEPHALOG LAB, WHITE MEM MED CTR, 50-; ASSOC PROF NEUROL, COL MED, LOMA LINDA UNIV, 58- Concurrent Pos: Chief neuromed serv, White Mem Med Ctr, 52-62; assoc prof, Univ Calif, Irvine-Calif Col Med, 66-69; consult, Los Angeles County Gen Hosp, Glendale Adventist Hosp, Mem Hosp Glendale, Rancho Los AmigosHosp & Lincoln Hosp. Mem: Am Acad Cerebral Palsy; Am Acad Neurol; Am Electroencephalog Soc; Am Epilepsy Soc; AMA. Res: Pediatric neurology; electroencephalography; epilepsy. Mailing Add: Neurosci Med Group 1710 Brooklyn Ave Suite 121 Los Angeles CA 90033

MANN, LEWIS THEODORE, JR, b New York, NY, Aug 5, 25; m 57; c 3. CLINICAL BIOCHEMISTRY, IMMUNOCHEMISTRY. Educ: Mass Inst Technol, SB, 46; Columbia Univ, AM & Phd(org chem), 51. Prof Exp: Lectr biochem, Harvard Med Sch, 58-66; asst prof immunol, 68-72, ASSOC PROF RADIOL, SCH MED, UNIV CONN, 72- Concurrent Pos: Res fel path, Harvard Med Sch, 56-58; Nat Inst Gen Med Sci spec fel, McIndoe Mem Res Inst, East Grinstead, Eng, 66-67 & Inst for Exp Immunol, Copenhagen, 67-68. Mem: AAAS; Am Chem Soc; Am Asn Clin Chem; Soc Nuclear Med. Res: Radioimmunoassay and non-radio immune assay techniques; natural inhibitors of proteolytic enzymes in mammalian serum. Mailing Add: Dept of Radiol Univ of Conn Health Ctr Farmington CT 06032

MANN, LLOYD GODFREY, b Sterling, Mass, July 2, 22; m 59; c 3. EXPERIMENTAL NUCLEAR PHYSICS. Educ: Worcester Polytech, BS, 44; Univ Ill, MS, 47, PhD(physics), 50. Prof Exp: Mem staff radiation lab, Mass Inst Technol, 44-46; instr physics, Stanford Univ, 50-53; MEM STAFF, LAWRENCE LIVERMORE LAB, UNIV CALIF, 53- Mem: Am Phys Soc. Res: Nuclear energy levels and decay schemes. Mailing Add: Lawrence Livermore Lab Univ of Calif Box 808 Livermore CA 94550

MANN, MARSHALL JOFFREE, physiology, cytology, see 12th edition

MANN, MICHAEL DAVID, b Gold Beach, Ore, May 20, 44; m 66; c 2. NEUROPHYSIOLOGY. Educ: Univ Southern Calif, BA, 66; Cornell Univ, PhD(neurobiol & behav), 71. Prof Exp: Ford Found fel, Cornell Univ, 71; USPHS fel, Univ Wash, 71-73; ASST PROF PHYSIOL & BIOPHYSICS, UNIV NEBR MED CTR, OMAHA, 73- Mem: AAAS; Am Physiol Soc; Soc Neurosci. Res: Somatosensory system, especially in the cerebral cortex, with a view to understanding the role of the system in controlling and modulating behavior; evolution and development of the central nervous system. Mailing Add: Dept of Physiol & Biophysics Univ of Nebr Med Ctr Omaha NE 68105

MANN, NANCY ROBBINS, b Chillicothe, Ohio, May 6, 25; m 49; c 2. STATISTICAL ANALYSIS. Educ: Univ Calif, Los Angeles, BA, 48, MA, 49, PhD(biostatist), 65. Prof Exp: Mathematician inst numberical anal, Nat Bur Standards, 49-50; sr physicist, NAm Rockwell Corp, 62-70, mem tech staff & proj develop engr, 70-74, sr scientist, 74-75, PROJ MGR RELIABILITY & STATIST, ROCKETDYNE, ROCKWELL SCIENCE CTR, ROCKWELL INT, 75- Concurrent Pos: Consult, US Army Mat Command & Missile Command; mem adv comt, US Census Bur, 72-74. Mem: Fel Am Statist Asn; Inst Math Statist. Res: Point and interval estimation theory; order statistics; statistical methods in reliability. Mailing Add: Rockwell Int Sci Ctr Rockwell Int PO Box 1085 Thousand Oaks CA 91360

MANN, RALPH WILLARD, b Robinson, Ill, July 12, 16; m 44; c 2. PHYSICS. Educ: DePauw Univ, AB, 38; Wash Univ, St Louis, MS, 40. Prof Exp: Asst physics, Wash Univ, St Louis, 39-41; physicist, Naval Ord Lab, 41-45; sr res geophysicist, Humble Oil & Refining Co, 46-54, res specialist, 54-63, sr res specialist geophys, Esso Prod Res Co, Tex, 64-71; environ health specialist, Tex Air Pollution Control Serv, 71-73, MEM STAFF, TEX AIR CONTROL BD, 73- Mem: Soc Explor Geophys; Asn Comput Mach; Inst Elec & Electronics Eng. Res: Conductivity of liquid dielectrics; underwater ordnance; geophysical research and instrumentation; underwater gravity meter; seismic prospecting methods and apparatus; digital computer applications. Mailing Add: 5013 Westview Dr Austin TX 78731

MANN, RICHARD HENRY, physical chemistry, see 12th edition

MANN, ROBERT ALEXANDER, theoretical physics, see 12th edition

MANN, ROBERT LESLIE, b Decatur, Ind, Oct 2, 22; m 45; c 1. AGRICULTURAL CHEMISTRY. Educ: Ind Univ, BS, 45; Univ Minn, PhD(biochem), 49. Prof Exp: From sr scientist biochem to head dept, Eli Lilly & Co, 49-63, dir chem res, 63-66 & plant sci res, 66-67, EXEC DIR AGR CHEM & PLANT SCI RES, ELI LILLY & CO, 67- Mem: Am Soc Biol Chemists; Am Chem Soc; AAAS. Res: Synthesizing, evaluating and developing agricultural chemicals for use in plants and animals. Mailing Add: Lilly Res Labs Eli Lilly & Co Greenfield IN 46140

MANN, ROGER HUNTINGTON, physical organic chemistry, see 12th edition

MANN, STANLEY JOSEPH, b Worcester, Mass, Sept 18, 32; m 69; c 2. GENETICS, BIOLOGY. Educ: Clark Univ, AB, 53, MA, 58; Brown Univ, PhD(biol), 61. Prof Exp: Cancer res scientist, Springville Labs, Roswell Park Mem Inst, 61-63, sr cancer res scientist, 63-66; asst prof anat in dermat, 66-69, ASSOC PROF ANIMAL GENETICS IN DERMAT, TEMPLE UNIV, 69- Concurrent Pos: Asst res prof biol, State Univ NY Buffalo, 64-66. Mem: Genetics Soc Am; Am Genetic Asn. Res: Mammalian genetics; biology of the skin; phenogenetics of hair mutants in the house mouse; morphology and development of normal and abnormal mammalian hair follicles. Mailing Add: Skin & Cancer Hosp of Philadelphia Temple Univ Health Sci Ctr Philadelphia PA 19140

MANN, THURSTON (JEFFERSON), b Lake Landing, NC, June 22, 20; m 45; c 3. GENETICS. Educ: NC State Univ, BS, 41, MS, 47; Cornell Univ, PhD(genetics, plant breeding), 50. Prof Exp: From asst prof to assoc prof agron, 49-53, in charge agron teaching, 53-55, prof crop sci, 55-64, prof genetics & head dept, 64-73, PROF GENETICS & CROP SCI, NC STATE UNIV, 73- Concurrent Pos: Vis agronomist, Coop State Res Serv, USDA, 74. Mem: Am Soc Agron; Am Genetic Asn; Genetics Soc Am. Res: Tobacco genetics; interspecific hybridization and breeding procedures; inheritance of alkaloids in Nicotiana. Mailing Add: Dept of Genetics NC State Univ Raleigh NC 27607

MANN, VIRGIL IVOR, geology, deceased

MANN, WALLACE VERNON, JR, b Pembroke, Mass, Mar 17, 30; m 52; c 4. DENTISTRY. Educ: Williams Col, BA, 51; Tufts Univ, DMD, 55; Univ Ala, MS, 62. Prof Exp: From instr to prof dent, Sch Dent, Univ Ala, Birmingham, 62-74, chmn dept periodont, 65-74, asst dean sch, 66-74; DEAN DENT, UNIV MISS MED CTR, 74- Concurrent Pos: Sr res trainee, Sch Dent, Univ Ala, Birmingham, 62-63; NIH career develop award, 63-66; dir clin res training grant for DMD/PhD prog, Sch Dent Univ Ala, Birmingham, 67-74. Mem: Am Dent Asn; Am Asn Dent Schs; fel Am Col Dentists; Int Asn Dent Res. Res: Physiology and biochemistry of periodontal tissues. Mailing Add: Sch of Dent Univ of Miss Med Ctr Jackson MS 39216

MANN, WILFRID BASIL, b London, Eng, Aug 4, 08; m 38; c 3. PHYSICS. Educ: Univ London, BSc, 30, PhD(physics), 34, DSc, 51. Prof Exp: Lectr physics, Imp Col, Univ London, 33-46; Nat Res Coun Can atomic energy proj, 46-48; attache, Brit Embassy, DC, 48-51; CHIEF RADIOACTIVITY SECT, NAT BUR STANDARDS, 51- Concurrent Pos: Sci liaison officer, Brit Commonwealth Sci Off, DC, 43-45; sci adv, UK Del, UN AEC, 46-51; adj prof, Am Univ, 61-68; ed, Int J Appl Radiation & Isotopes, 65-; mem, Fed Radiation Coun, 69-70; dep chief, Appl Radiation Div, Ctr Radiation Res, 74-; chmn sci 18A, Nat Coun Radiation Protection & Measurements, 72-; NAm ed, Int J Nuclear Med & Biol, 73- Honors & Awards: Medal of Freedom, 48; Gold Medal, US Dept Com, 58. Mem: Am Phys Soc; Brit Inst Physics & Phys Soc (ed, Reports Progress Physics, 41-46). Res: Radioactivity standardization; microcalorimetry. Mailing Add: Radioactivity Sect Nat Bur of Standards Washington DC 20234

MANN, WILLIAM RICHARD, b Battle Creek, Mich, Apr 29, 16; m 40; c 1. DENTISTRY. Educ: Univ Mich, DDS, 40, MS, 42. Hon Degrees: DSc, Mich State Univ, 73. Prof Exp: Clin instr dent, Sch Dent, 40-42, from instr to assoc prof, 42-55, assoc dir, W K Kellogg Found Inst Grad & Post Grad Dent, 52-62, PROF DENT, SCH DENT, UNIV MICH, ANN ARBOR, 55-, DEAN SCH & DIR W K KELLOGG FOUND INST GRAD & POSTGRAD DENT, 62- Concurrent Pos: Mem dent res adv comt, Med Res & Develop Bd, Off Surgeon Gen, Dept Army, 54-62; consult, Univs Costa Rica & Antioquia, Colombia, 57; dir sect dent educ, surv & dent, Am Coun Educ, 58-60; mem dent training comt, Nat Inst Dent Res, 61-63; mem expert comt dent health, WHO, Switz, 62, mem expert adv panel dent health, 62-; mem adv comt dent student training, USPHS, 63-66; mem dent adv comt, W K Kellogg Found, 63-67, mem Latin Am study comt, 64-65 & mem Latin Am adv comt, 66-67; mem med adv bd, Food & Drug Admin, Dept Health, Educ & Welfare, 65-70. Mem: Am Dent Asn; Am Asn Dent Schs (pres, 70-71); fel Am Col Dent. Res: Dental education, particularly teacher-training programs; dental materials. Mailing Add: Univ of Mich Sch of Dent Ann Arbor MI 48104

MANN, WILLIAM ROBERT, b Honea Path, SC, Sept 21, 20; m 47; c 3. APPLIED MATHEMATICS. Educ: Univ Rochester, AB, 41; Univ Calif, PhD(math), 49. Prof Exp: Instr, 49-50, from asst prof to assoc prof, 50-60, PROF MATH, UNIV NC, CHAPEL HILL, 60- Mem: Am Math Soc. Res: Nonlinear boundary value problems; iterative techniques. Mailing Add: Dept of Math Univ of NC Chapel Hill NC 27515

MANNELL, WILLIAM ARNOLD, b Belmont, Ont, June 30, 21; m 46; c 2. TOXICOLOGY. Educ: Univ Western Ont, BSc, 49, PhD(biochem), 52. Prof Exp: Sr res assoc biochem, Univ Western Ont, 52-54; chemist food & drug labs, 54-65, head food additives & pesticides sect, Pharmacol Eval Div, Food & Drug Directorate, 65-67, chief div toxicol, Food Adv Bur, 67-74, CHIEF TOXICOL EVAL DIV, BUR CHEM SAFETY, FOOD DIRECTORATE, HEALTH PROTECTION BR, DEPT NAT HEALTH & WELFARE, 74- Concurrent Pos: Mem, Expert Adv Panel Food Additives, WHO, 68-78. Mem: Soc Toxicol; Pharmacol Soc Can; Can Asn Res Toxicol. Res: Biochemical studies of Wallerian degeneration; nucleic acids; toxicological evaluation of food additives, food contaminants and pesticides. Mailing Add: Food Directorate Health Protection Br Ottawa ON Can

MANNER, GEORG KARL, b Vienna, Austria, Oct 26, 30; m 60; c 1. BIOCHEMISTRY, PHARMACOLOGY. Educ: Univ Vienna, PhD(chem), 57. Prof Exp: Res assoc biochem, Mass Inst Technol, 57-64; asst prof biochem surg, 64-71, ASST PROF BIOCHEM, ALBERT EINSTEIN COL MED, 64-, ASST PROF ANESTHESIA, 71- Concurrent Pos: Head anesthesiol res lab, Montefiore Hosp &

Med Ctr, 71- Mem: AAAS; NY Acad Sci; Tissue Cult Asn. Res: Protein synthesis, regulatory mechanisms, neurotransmitters. Mailing Add: Anesthesiol Res Lab Montefiore Hosp & Med Ctr Bronx NY 10467

MANNER, HAROLD WALLACE, b Brooklyn, NY, July 31, 25; m 45; c 2. BIOLOGY. Educ: John Carroll Univ, BS, 49; Northwestern Univ, MS, 50, PhD, 52. Prof Exp: Nat Arthritis & Rheumatism Found fel, 52-53; asst prof biol, Utica Col, 53-57, assoc prof, 57-62, prof & chmn div sci & math, 62-69; prof biol & chmn dept, St Louis Univ, 69-72; PROF BIOL & CHMN DEPT, LOYOLA UNIV CHICAGO, 72- Concurrent Pos: Vis asst prof, Kenyon Col, 52-53; res assoc, NSF, 53-55 & NIH, 60-62. Mem: AAAS; Am Soc Zool. Res: Regeneration of Salamander limbs; influence of steroids on morphogenesis; effect of aquatic pollutants on teleost embryogenesis. Mailing Add: Dept Biol Loyola Univ Chicago 6525 N Sheridan Rd Chicago IL 60626

MANNER, RICHARD JOHN, b Buffalo, NY, Mar 22, 20; m 43; c 4. MEDICAL RESEARCH. Educ: Rensselaer Polytech Inst, BS, 41; Univ Rochester, MD, 51. Prof Exp: Physicist, Eastman Kodak Co, 41-42, electronics develop engr, 46-47; intern med, Rochester Gen Hosp, 51-52, preceptorship internal med, 52-55; pvt pract, NY, 55-60; assoc dir clin res, Mead Johnson Res Ctr, 60-65; staff physician, Wyeth Labs, 65-67; dir med affairs, Warren-Teed Pharmaceut Inc, Columbus, 67-75; DIR PHARMACEUT RES, ROSS LABS, COLUMBUS, 75- Concurrent Pos: Consult, Rochester Inst Technol, 53-54 & Stecher-Traung Lithograph Corp, 54-60. Mem: AMA; Am Occup Med Asn; Inst Elec & Electronics Eng; Am Soc Clin Pharmacol & Therapeut; Am Col Gastroenterol. Res: Electronic instrumentation; gastrointestinal pharmacology and instrumentation; blood lipid research. Mailing Add: 2177 Castle Crest Dr Worthington OH 43085

MANNERING, GILBERT JAMES, b Racine, Wis, Mar 9, 17; m 39, 69; c 3. PHARMACOLOGY, BIOCHEMISTRY. Educ: Univ Wis, BS, 40, MS, 43, PhD(biochem), 44. Prof Exp: Sr biochemist, Parke, Davis & Co, 44-50; consult, Chem Dept, 406th Med Gen Lab, Tokyo, Japan, 50-54; from asst prof to assoc prof pharmacol & toxicol, Univ Wis, 54-62; PROF PHARMACOL, MED SCH, UNIV MINN, MINNEAPOLIS, 62- Concurrent Pos: Consult, Wis State Crime Lab, 54-62; spec consult, Interdept Comt Nutrit Nat Defense, NIH, Ethiopia, 58; mem toxicol study sect, USPHS, 62-65, mem pharmacol-toxicol rev comt, 65-67 & pharmacol study sect, 68-69; mem comt probs drug safety, Nat Acad Sci-Nat Res Coun, 65-71. Mem: Am Soc Pharmacol & Exp Therapeut. Res: Biochemical pharmacology; drug metabolism; toxicology. Mailing Add: 1865 N Fairview Ave St Paul MN 55113

MANNERING, JERRY VINCENT, b Custer City, Okla, June 14, 29; m 53; c 3. AGRONOMY. Educ: Okla State Univ, BS, 51; Purdue Univ, MS, 56, PhD, 67. Prof Exp: Asst agronomist, Univ Idaho, 56-58; soil scientist, Agr Res Serv, USDA, 58-67; EXTEN AGRONOMIST, PURDUE UNIV, 67- Mem: Am Soc Agron; Soil Sci Soc Am; Int Soil Sci Soc; Soil Conserv Soc Am. Res: Soil erosion; waste management; efficient water use; soil management for crop production. Mailing Add: Dept of Agron Purdue Univ West Lafayette IN 47907

MANNERS, IAN ROBERT, b Chester, Eng, July 15, 42; m 67; c 1. GEOGRAPHY. Educ: Oxford Univ, BA, 64, MA, 68, DPhil(geog), 69. Prof Exp: Lectr geog, Columbia Univ, 68-69, asst prof, 69-72; ASST PROF GEOG, UNIV TEX, AUSTIN, 72- Concurrent Pos: Consult, Natural Resources Authority, Govt Jordan, 66-67; mem, Comn Col Geog, 70-74, co-chmn, Panel Environ Educ, 70-74; assoc ed, Human Ecol, 71-; mem comt environ studies, Asn Am Geogr, 75- Mem: AAAS; Asn Am Geog; Am Geog Soc; Inst Brit Geog. Res: Ecological and socio-economic aspects of agricultural change in the Middle East; application of ecological systems analysis to man-environment interactions; environmental perception. Mailing Add: Dept of Geog Univ of Tex Austin TX 78712

MANNERS, ROBERT ALAN, b New York, NY, Aug 21, 13; m 43, 55; c 4. ANTHROPOLOGY. Educ: Columbia Univ, BS, 35, MA, 39, PhD(anthrop), 50. Prof Exp: Instr anthrop, Univ Rochester, 50-52; from lectr to assoc prof, Brandeis Univ, 52-61, PROF ANTHROP, BRANDEIS UNIV, 61- Concurrent Pos: Ford Found & Univ Ill fels, Africa, 57-58; Res Inst Study Man fel, Africa, 57-58; NSF fel, Africa, 61-62; vis prof, Inter-cult Ctr Doc & Ibero-Am Univ, 69-70; consult, Ford Found, 70-71; gen ed, Am Anthropologist, 73-75. Mem: Fel Am Anthrop Asn; fel Am Ethnol Soc; fel African Studies Asn. Res: African and Caribbean ethnology; American-Indian ethnohistory; method and theory in anthropology. Mailing Add: Dept of Anthrop Brandeis Univ Waltham MA 02154

MANNEY, THOMAS RICHARD, b El Paso, Tex, Dec 20, 33; m 56; c 3. BIOPHYSICS. Educ: Western Wash State Col, BA, 57; Univ Calif, Berkeley, PhD(biophys), 64. Prof Exp: Teaching asst, Western Wash State Col, 56-58; res asst, Univ Calif, Berkeley, 58-59, biophysicist, Donner Lab, 59-60; staff biologist, Oak Ridge Nat Lab, 64-65; asst prof microbiol, Case Western Reserve Univ, 65-71; ASSOC PROF PHYSICS & BIOL, KANS STATE UNIV, 71- Concurrent Pos: USPHS res career develop award, 67-70; vis assoc prof, Univ Calif, 70-71. Honors & Awards: Chem Achievement Award, Western Wash State Col, 55, Physics Achievement Award, 56. Mem: AAAS; Genetics Soc Am; Am Soc Microbiol. Mailing Add: Dept of Physics Kans State Univ Manhattan KS 66502

MANNI, PETER EMIL, medicinal chemistry, see 12th edition

MANNING, ARMIN WILLIAM, b Milwaukee, Wis, Aug 22, 13; m 40; c 2. NUCLEAR PHYSICS. Educ: Valparaiso Univ, BA, 36; Concordia Sem, BD, 37; Univ Mich, MA, 38; Fordham Univ, PhD, 57. Prof Exp: Prof math & mod physics, Concordia Collegiate Inst, 38-56; co-chmn dept physics, 56-68, PROF PHYSICS, VALPARAISO UNIV, 56-, CHMN DEPT, 68- Mem: AAAS; Am Asn Physics Teachers; Am Geophys Union. Res: Reactor physics and radioactivity. Mailing Add: Dept of Physics Valparaiso Univ Valparaiso IN 46383

MANNING, CHARLES RICHARDSON, analytical chemistry, see 12th edition

MANNING, CLEO WILLARD, b Woodhull, Ill, Oct 10, 15; m 38; c 5. PLANT BREEDING. Educ: Ill Wesleyan Univ, BS, 40; Agr & Mech Col, Univ Tex, MS, 42; Iowa State Univ, PhD, 54. Prof Exp: Agent, Bur Plant Indust, Soils & Agr Eng, USDA, 41-45; asst agronomist, Agr & Mech Col, Univ Tex, 41-45, agronomist, 45-48, botanist, 48, agronomist & assoc prof, 48-51; PLANT BREEDER IN CHARGE, STONEVILLE PEDIGREED SEED CO, 51- Mem: AAAS; Am Inst Biol Sci; Am Genetic Asn; Genetics Soc Am; Am Soc Agron. Res: Genetics and breeding of cotton and oats; soybeans; plant exploration; interspecific relationships in Gossypium. Mailing Add: PO Box 213 Stoneville MS 38776

MANNING, DAVID TREADWAY, b Santa Monica, Calif, Sept 19, 28; m 56; c 2. BIO-ORGANIC CHEMISTRY, AGRICULTURAL CHEMISTRY. Educ: Calif Inst Technol, BS, 51, PhD(chem), 55. Prof Exp: Res proj chemist, 54-62, res scientist, 62-75, SR RES SCIENTIST, UNION CARBIDE CORP, 75- Mem: AAAS; Am Chem Soc; NY Acad Sci; Plant Growth Regulator Working Group; Am Soc Plant Physiologists. Res: Nitrosation reactions; organic reactions of nitrosyl chloride; chemistry of oximes; nitrogen-containing heterocyclic compounds; pesticide chemistry and formulants; plant growth regulators. Mailing Add: Res & Develop Dept Union Carbide Corp PO Box 8361 South Charleston WV 25303

MANNING, DEAN DAVID, b Grand Junction, Colo, Oct 11, 40; m 66. IMMUNOLOGY. Educ: Colo State Univ, BS, 62, MS, 64; Mont State Univ, PhD(microbiol), 72. Prof Exp: Fel microbiol, Mont State Univ, 72-74; ASST PROF MED MICROBIOL, MED SCH, UNIV WIS-MADISON, 75- Concurrent Pos: NIH fel, 73-74. Mem: Am Soc Microbiol. Res: Control of the immune reponse with particular reference to heavy chain isotype suppression. Mailing Add: Dept of Med Microbiol Univ of Wis Med Sch Madison WI 53706

MANNING, GEORGE WILLIAM, b Toronto, Ont, Aug 4, 11; m 40; c 3. MEDICINE. Educ: Univ Toronto, BA, 35, MA, 36, MD, 40, PhD(physiol), 48; FRCP, 48; FRCP(C). Prof Exp: Asst med, Banting Inst, Toronto, 35-37; sr intern, Toronto Gen Hosp, Can, 45-46; clin asst cardiac dept, London Hosp, 46-47; from instr to assoc prof, 47-62, PROF MED & DIR HEART UNIT, UNIV WESTERN ONT, 62- Concurrent Pos: Pvt pract, 48-; dir cardiovasc unit, Victoria Hosp, 60-70 & Univ Hosp, London, Ont, 70-; consult cardiologist, Westminster Hosp; with Nat Res Coun Can. Mem: Am Physiol Soc; fel Aerospace Med Asn; fel Am Col Cardiol; fel Am Col Physicians; Can Physiol Soc. Res: Experimental coronary occulsion; decompression sickness; clinical cardiology and electrocardiography. Mailing Add: 9 Harrison Crescent London ON Can

MANNING, GERALD STUART, b New York, NY, Dec 9, 40; m 64. BIOPHYSICAL CHEMISTRY. Educ: Rice Univ, BA, 62; Univ Calif, San Diego, PhD(phys chem), 65. Prof Exp: NATO fel, Univ Brussels, 65-66; Nat Sci Found fel, Rockefeller Univ, 66-67, asst prof chem, 67-69; assoc prof, 69-75, PROF CHEM, RUTGERS UNIV, 75- Honors & Awards: Alfred P Sloan fel, 70-72. Res: Polyelectrolytes; biopolymer conformation; transport across membranes. Mailing Add: Sch of Chem Rutgers Univ New Brunswick NJ 08903

MANNING, HAROLD EDWIN, b Huntsville, Ala, Mar 18, 35; m 66; c 4. INDUSTRIAL ORGANIC CHEMISTRY. Educ: Auburn Univ, BS, 58; Trinity Univ, MS, 62. Prof Exp: Chemist, 62-66, res chemist, 66-67, HEAD RES GROUP, PETRO-TEX CHEM CORP, 67- Mem: Am Chem Soc; Catalysis Soc. Res: Heterogeneous vapor-phase catalysis; reaction mechanisms and surface chemistry. Mailing Add: Petro-Tex Chem Corp 8600 Park Place Blvd Houston TX 77017

MANNING, HERBERT LEE, b Brooklyn, NY, July 25, 31; m 69. MICROBIOLOGY, WATER POLLUTION. Educ: City Col New York, BS, 60; Syracuse Univ, MS, 64; Univ Md, PhD(microbiol), 71. Prof Exp: Res asst med, Res Found, Upstate Med Ctr, State Univ NY, 65; asst res biologist, Virol Dept, Sterling-Winthrop Res Inst, Rensselaer, NY, 65-67; MICROBIOLOGIST, ENVIRON MONITORING & SUPPORT LAB, US ENVIRON PROTECTION AGENCY, 71- Mem: Sigma Xi; Am Soc Microbiol. Res: Evaluating, improving and developing methodology to detect enteric pathogens from water. Mailing Add: Environ Monitoring & Support Lab US Environ Protection Agency Cincinnati OH 45268

MANNING, IRWIN, b Brooklyn, NY, Mar 7, 29; m 64; c 2. THEORETICAL PHYSICS. Educ: Mass Inst Technol, BS, 51, PhD(physics), 55. Prof Exp: Res assoc & asst prof, Syracuse Univ, 55-57; res assoc, Univ Wis, 57-59; res physicist, 59-73, SUPVRY RES PHYSICIST, US NAVAL RES LAB, 73- Concurrent Pos: Mem ad hoc panel on the use of accelerators to study irradiation effects, Nat Acad Sci, 74. Mem: Am Phys Soc. Res: Atomic scattering at high energies; phenomena associated with the penetration of matter by energetic particles such as ion implantation, sputtering, neutron radiation damage. Mailing Add: US Naval Res Lab Washington DC 20375

MANNING, JAMES ARTHUR, b Calcutta, India, Mar 21, 25; m 48; c 4. PEDIATRICS, PEDIATRIC CARDIOLOGY. Educ: Columbia Univ, AB, 44, MD, 48; Am Bd Pediat, dipl, 53, cert pediat cardiol, 62. Prof Exp: Intern, Lakeside Hosp, Cleveland, 48-49; asst resident pediat, New York Hosp, 49-50; asst resident, Univ Rochester, 50-51; chief pediatrician, US Naval Hosp, Bainbridge, Md, 51-53; sr asst, Congenital Cardiac Clin, Johns Hopkins Hosp, 54-55; from instr to assoc prof, 55-67, PROF PEDIAT, SCH MED & DENT, UNIV ROCHESTER & STRONG MEM HOSP, 67- Concurrent Pos: Fel pediat, Univ Rochester, 50-51; fel pediat cardiol, Johns Hopkins Hosp, 53-54; consult pediat cardiol, Genesee Hosp & Rochester Gen Hosp, 57-; consult, Pediat Cardiac Clin, Jefferson County, 62- & Orleans County Br, Genesee Valley Heart Asn, 69-; chmn coun cardiovasc dis in the young, Am Heart Asn, 73- Mem: Am Pediat Soc; Am Acad Pediat; Asn Europ Pediat Cardiol; AMA. Mailing Add: Dept of Pediat Univ of Rochester Sch Med & Dent Rochester NY 14627

MANNING, JAMES HARVEY, b Hancock, Mich, Aug 13, 40; m 66; c 3. PAPER CHEMISTRY. Educ: Mich Technol Univ, BS, 62; Lawrence Univ, MS, 64, PhD(paper technol), 67. Prof Exp: Prod engr paper, Kimberly-Clark Corp, 61, res chem tissue, 62; qual control engr bd, Westvaco, 63; proj engr pulping, 64, res scientist cellulose, 67-72, GROUP DIR NONWOVENS, INT PAPER CO, 72- Concurrent Pos: Secy, Dissolving Pulp Comt, Tech Asn Pulp & Paper Indust, 72-74. Mem: Tech Asn Pulp & Paper Indust; NY Acad Sci. Res: Direction of a group responsible for the development of new nonwoven products, with emphasis on wet-laid nonwoven products and processes. Mailing Add: Int Paper Co PO Box 797 Tuxedo Park NY 10987

MANNING, JAMES JOSEPH, b New York, NY, Sept 14, 06; m 46; c 3. PHYSICAL CHEMISTRY, SPECTROCHEMISTRY. Educ: Cooper Union, BS, 31; NY Univ, MS, 38; Fordham Univ, PhD(phys chem), 44. Prof Exp: Researcher, Cooper Union, 32-35; chief physicist & chemist, New York City Police Crime Lab, 38-57, sr scientist & dir lab, 57-67, dir, 67-72; CONSULT, 72- Concurrent Pos: Instr, counter-intel sch, US Army & City Col New York; private practice. Res: Chemical application of spectroscopy; methods of identification; ultraviolet and infrared regions of spectra; instrumental and microanalytical methods of analysis and identification; scientific methods and techniques applied to criminal investigations; technical investigations of industrial accidents, disasters and explosions; recording devices; horse identification system. Mailing Add: 198 Broadway Rm 1105 New York NY 10038

MANNING, JAMES MATTHEW, b Boston, Mass, Jan 3, 39; m 64; c 2. BIOCHEMISTRY. Educ: Boston Col, BS, 60; Tufts Univ, PhD(biochem), 66. Prof Exp: Nat Sci Found fel biochem, Univ Rome, 66-67; res assoc, 67-69, asst prof, 69-72, ASSOC PROF BIOCHEM, ROCKEFELLER UNIV, 72- Res: Protein and synthetic peptide chemistry; mechanism of enzyme action, especially of pyridoxal phosphate enzymes; amino acid metabolism and methods for the determination of amino acids; collagen biosynthesis; chemical aspects of hemoglobinopathies. Mailing Add: Rockefeller Univ York Ave & 66th St New York NY 10021

MANNING, JARUE STANLEY, b Indiana, Pa, Sept 25, 34; m 60; c 1. VIROLOGY, BIOPHYSICS. Educ: San Francisco State Col, BA, 62; Univ Calif, Berkeley,

PhD(biophys), 69. Prof Exp: Nat Cancer Inst fel, Univ Calif, Berkeley, 69-70; ASST PROF MICROBIOL, SCH VET MED & ASST RES VIROLOGIST, COMP ONCOL LAB, UNIV CALIF, DAVIS, 71-, ASST PROF BACTERIOL, 73- Honors & Awards: Hektoen Gold Medal, Am Med Asn, 68. Mem: AAAS; Am Soc Microbiol; Biophys Soc. Res: Animal virology, including oncogenic viruses; cell-virus interaction, including virus-induced cytopathology and transformation; mechanisms of viral replication; characterization of viral components; viral immunity, including host defense mechanisms. Mailing Add: Dept of Vet Microbiol Sch of Vet Med Univ of Calif Davis CA 95616

MANNING, JEROME EDWARD, b Minneapolis, Minn, Dec 31, 40; m 62; c 2. ACOUSTICS. Educ: Mass Inst Technol, SB, 62, SM, 63, ScD(mech eng), 65. Prof Exp: Sr scientist, Bolt Beranek & Newman, Inc, Mass, 65-68; MEM STAFF, CAMBRIDGE COLLABORATIVE, INC, 68- Concurrent Pos: Lectr, Mass Inst Technol, 67-69. Mem: Acoust Soc Am. Res: Sound induced vibration; noise; random vibrations.

MANNING, JERRY EDSEL, b Redland, Calif, Oct 19, 44; m 67; c 2. BIOCHEMISTRY. Educ: Univ Utah, BS, 66, PhD(biochem), 71. Prof Exp: Res fel biol, Univ Utah, 71-72 & chem, Calif Inst Technol, 73-74; ASST PROF MOLECULAR BIOL, UNIV CALIF, IRVINE, 75- Concurrent Pos: Jane Coffin Childs fel, 73; Petrol Res Fund res grant, 74; NIH & Res Corp res grants, 75. Mem: Am Soc Cell Biol. Res: Molecular mechanisms that govern the regulation of genetic activity in the eukaryotic genome. Mailing Add: Dept of Molecular Biol & Biochem Univ of Calif Irvine CA 92664

MANNING, JOHN CRAIGE, b Detroit, Mich, Jan 5, 20; m 46; c 2. GEOLOGY. Educ: Univ Idaho, BS, 42; Stanford Univ, PhD, 51. Prof Exp: Geologist, US Geol Surv & US Army Corps Engrs, 46-47; field geologist, Amerada Petrol Corp, 49; instr & asst prof geol, Stanford Univ, 50-54; res geologist, Shell Oil Co, 54-56; chief geologist, Philadelphia Oil Co, 56; vpres & tech dir, Ranney Method West Corp, 56-57; pres, Hydro Develop, Inc, 61-70; PROF EARTH SCI, CALIF STATE COL, BAKERSFIELD, 70- Concurrent Pos: Consult geologist, 57-; prof, Fresno State Col, 58-70. Mem: Am Asn Petrol Geol; Am Inst Mining, Metall & Petrol Eng. Res: Economic geology, especially engineering ground water and mining geology; development of ground water and industrial mineral deposits. Mailing Add: 2512 Spruce St Bakersfield CA 93301

MANNING, JOHN PAUL, b Cleveland, Ohio, Oct 1, 34; m 56; c 5. HISTOCHEMISTRY, ANATOMY. Educ: Med Col Va, PhD(anat, histochem), 63; Salem Col, WVa, BS, 64. Prof Exp: Design engr atomic power div, Newport News Shipbuilding & Dry Dock Co, 55-57; instr anat, Med Col Va, 62-63; asst prof, Univ SDak, 63-64; scientist, Warner-Lambert Res Inst, 64-67, sr scientist, 67-70; asst dir clin lab, 70-71, SCI & ADMIN COORDR & DIR LAB INFO SYST PROJ, ST BARNABAS MED CTR, 71- Concurrent Pos: Asst clin prof, NJ Col Med & Dent, Newark, 71- Mem: Histochem Soc; Am Asn Anat; Am Heart Asn; Am Soc Zool; Am Asn Clin Chem. Res: Human anatomy; drug and hormonal related biochemical and histochemical enzyme changes of uterus, liver, kidney, pubic symphysis and brain; experimental technique and methodology in histochemistry and biochemistry. Mailing Add: Dept of Path St Barnabas Med Ctr Livingston NJ 07039

MANNING, JOHN RANDOLPH, b Norristown, Pa, Aug 24, 32; m 60; c 1. SOLID STATE PHYSICS, METALLURGY. Educ: Ursinus Col, BS, 53; Univ Ill, MS, 54, PhD(physics), 58. Prof Exp: Asst physics, Univ Ill, 57, res assoc, 58; physicist, 58-67, CHIEF, METAL PHYSICS SECT, NAT BUR STAND, 67- Mem: Am Phys Soc; Am Inst Mining, Metall & Petrol Engrs; Am Soc Metals. Res: Diffusion in solids; kinetic processes and defects in metals. Mailing Add: Metall Div Nat Bur of Stand Washington DC 20234

MANNING, JOHN W, b New Orleans, La, Nov 14, 30; m 54; c 6. PHYSIOLOGY. Educ: Loyola Univ, La, BS, 51; Tulane Univ, MS, 55; Loyola Univ Chicago, PhD(physiol), 58. Prof Exp: Vis scientist, Karolinska Inst, Sweden, 63-64; from asst prof to assoc prof physiol, 64-70, assoc prof anat, 67-71, PROF PHYSIOL, EMORY UNIV, 70- Concurrent Pos: USPHS fel, Emory Univ, 58-61, Am Heart Asn Advan res fel, 61-65; guest referee, Am J Physiol, 70-; vis prof, Shinshu Univ, Matsumoto, Japan, 72. Mem: AAAS; Am Physiol Soc; Am Asn Anat; Soc Neurosci. Res: Central nervous system regulation of cardiovascular activity; central response and transmission small cutaneous afferents; energetics of cardiac muscle. Mailing Add: Dept of Physiol Emory Univ Atlanta GA 30322

MANNING, MAURICE, b Loughrea, Ireland, Apr 10, 37; m 65; c 3. BIOCHEMISTRY, CHEMISTRY. Educ: Nat Univ Ireland, BSc, 57, MSc, 58, DSc, 74; Univ London, PhD(chem), 61. Prof Exp: Res assoc biochem, Med Col, Cornell Univ, 61-64; res assoc, Rockefeller Univ, 64-65; asst prof, McGill Univ, 65-69; assoc prof, 69-73, PROF BIOCHEM, MED COL OHIO, 73- Concurrent Pos: Fulbright travel grant, 61-64. Mem: Am Soc Biol Chem; Can Biochem Soc; The Chem Soc; Am Chem Soc. Res: Solid phase peptide synthesis; study of structure-function relationships and phylogeny of oxytocin and vasopressin. Mailing Add: Dept of Biochem Med Col of Ohio Toledo OH 43614

MANNING, MONIS JOSEPH, b Allentown, Pa, Mar 7, 31; m 57; c 3. PHOTOGRAPHIC CHEMISTRY, ANALYTICAL CHEMISTRY. Educ: Pa State Univ, BS, 53; Univ Cincinnati, MS, 58, PhD(org chem, molecular spectros), 60. Prof Exp: Chemist, Arthur D Little, Inc, 60-66; CHEMIST, POLAROID CORP, 66- Mem: AAAS; Am Chem Soc. Res: Application of chemical and physical science to photographic problem-solving; thermal analysis; analytical spectrophotometry; chromatographic analysis; specialty coatings formulation; adhesives; spectral properties of plastics; battery electrodes. Mailing Add: Polaroid Corp 784 Memorial Dr Cambridge MA 02139

MANNING, PETER BURNAM, food science, see 12th edition

MANNING, PHIL RICHARD, b Kans City, Mo, May 14, 21; m 48; c 2. INTERNAL MEDICINE. Educ: Univ Southern Calif, AB, 45, MD, 48. Prof Exp: Intern, Los Angeles County Hosp, 47-48; resident internal med, Vet Admin Hosp, Van Nuys & Long Beach, Calif, 48-50; from instr to assoc prof, 54-64, dir postgrad div, 53-59, PROF MED, SCH MED, UNIV SOUTHERN CALIF, 64-, ASSOC DEAN POSTGRAD DIV, 59- Concurrent Pos: Fel, Mayo Clin, 50-52. Res: Medical problems in pregnancy; evaluation of teaching techniques in postgraduate medical education; development of the community hospital as an intramural teaching center. Mailing Add: Postgrad Div Univ of South Calif Sch of Med Los Angeles CA 90033

MANNING, RAYMOND B, b Brooklyn, NY, Oct 11, 34; m 57; c 3. INVERTEBRATE ZOOLOGY, MARINE BIOLOGY. Educ: Univ Miami, BS, 56, MS, 59, PhD(marine sci), 63. Prof Exp: Res instr, Inst Marine Sci, Univ Miami, 59-63; assoc curator Crustacea, Div Marine Invert, 63-65, curator in charge div Crustacea, 65-67, chmn dept invert zool, 67-71, CURATOR DIV CRUSTACEA, SMITHSONIAN INST, 71- Res: Systematics and biology of decapod and stomatopod Crustacea. Mailing Add: Dept of Crustacea Smithsonian Inst Washington DC 20560

MANNING, ROBERT JOSEPH, b Kansas City, Kans, Jan 12, 20; m 49; c 4. PHYSICAL CHEMISTRY, ORGANIC CHEMISTRY. Educ: St Benedict's Col, Kans, BS, 43; Univ Kansas City, MS, 48. Prof Exp: Phys chemist, US Naval Ord Test Sta, 48-53; sr chemist, 53-63, prod line mgr, 63-73, PRIN CHEMIST, BECKMAN INSTRUMENTS, INC, 73- Mem: Am Chem Soc; Optical Soc Am; Soc Appl Spectros (pres-elect, 76, pres, 77). Res: Ultraviolet and infrared absorption spectroscopy, especially the near infrared; reflectance spectroscopy. Mailing Add: Beckman Instruments Inc 2500 Harbor Blvd Fullerton CA 92634

MANNING, ROBERT THOMAS, b Wichita, Kans, Oct 16, 27; m 49; c 3. MEDICINE, BIOCHEMISTRY. Educ: Univ Wichita, AB, 50; Univ Kans, MD, 54; Am Bd Internal Med, dipl. Prof Exp: Intern, Kans City Gen Hosp, 54-55; resident, Med Ctr, Univ Kans, 55-58, instr internal med, 58-59, assoc, 59-62, asst prof, 62-64, assoc prof internal med & biochem, 64-69, prof med & assoc dean, 69-71; PROF INTERNAL MED, ASSOC PROF BIOCHEM & DEAN, EASTERN VA MED SCH, 71- Concurrent Pos: Nat Inst Arthritis & Metab Dis fel, 56-58; chief first med serv, Univ Kans Med Ctr, 62-71; consult, Hampton Vet Admin Hosp, 71-; nat consult, US Air Force, 73- Mem: Fel Am Col Physicians; Am Fedn Clin Res; Cent Soc Clin Res; Am Asn Study Liver Dis; AMA. Res: Liver disease; biometrics. Mailing Add: Eastern Va Med Sch 600 Gresham Dr Norfolk VA 23507

MANNING, WALTER H, JR, b Atlantic City, NJ, Mar 9, 25; m 49; c 1. PHYSICS. Educ: Duke Univ, BS, 45; Pa State Univ, MS, 49. Prof Exp: Physicist, Electromagnetic Sect, Naval Ord Lab, Md, 50-53, sr physicist & chief systs sect, Electronics & Missile Systs Br, 53-56; chief missile br, Air Force Missile Test Ctr, Fla, 56-58, chief res br, 58-59, proj scientist, Proj Space Track, 58-62, chief space surveillance div, 59-62, aerospace sci div, 62-66; ORGANIZER-DIR, RANGE MEASUREMENTS LAB, PATRICK AFB, 66- Honors & Awards: Invention Award, Naval Ord Lab, 57; Commendation, US Dept Defense, 60; Outstanding Achievement Award, Space Cong Canaveral Coun Tech Socs, 74. Mem: Sigma Xi; Am Inst Aeronaut & Astronaut; Am Phys Soc. Res: Electron physics; electromagnetic propogation; orbital mechanics; solid state physics; image processing research; research and development of precision and microwave pointing and tracking systems; underwater optics research. Mailing Add: 411 Melbourne Ave Indialantic FL 32903

MANNING, WILLIAM JOSEPH, b Grand Rapids, Mich, June 13, 41; m 69; c 2. MICROBIOLOGY, PLANT PATHOLOGY. Educ: Mich State Univ, BS, 63; Univ Del, MS, 65, PhD, 68. Prof Exp: ASSOC PROF PLANT PATH, SUBURBAN EXP STA, UNIV MASS, WALTHAM, 68- Mem: AAAS; Am Phytopath Soc; Am Inst Biol Sci. Res: Ecology of soil-borne fungi that cause root diseases of plants; interactions between air pollutants and biological incitants of plant diseases; air pollution effects on economic plants. Mailing Add: 20 Spring St Lexington MA 02173

MANNING, WINSTON MARVEL, b Washington, DC, Mar 26, 09; m 35; c 2. NUCLEAR CHEMISTRY. Educ: Am Univ, BA, 30; Brown Univ, MS, 31, PhD(phys chem), 33. Prof Exp: Fel chem, Brown Univ, 33-34; res asst chem & bot, Univ Wis, 34-36, res assoc limnol, 36-41; phys chemist plant biol, Carnegie Inst Technol, 41-43; from chemist to div dir metall lab, Univ Chicago, 43-46; dir chem div, Argonne Nat Lab, 46-65, assoc lab dir, 65-71, actg lab dir, 67, sr resident consult, 71-72; RETIRED. Concurrent Pos: Mem subcomt radiochem, Nat Res Coun, 47-50, div chem & chem technol, 48-55; lectr, Wayne State Univ, 51; mem comt sr reviewers; Atomic Energy Comn, 54-65; US del, Int Conf Peaceful Uses Atomic Energy, Geneva, 55; actg Am ed, Inorg & Nuclear Chem, 57-58. Mem: Am Chem Soc; fel Am Nuclear Soc. Res: Photochemistry; limnology; photosynthesis; chlorophyll and carotenoid pigments; chemical and nuclear properties of transuranium elements; environmental chemistry. Mailing Add: 5524 Carpenter St Downers Grove IL 60515

MANNION, JOHN JOSEPH, b Huddersfield, Eng, June 9, 41; Irish citizen; m 66. GEOGRAPHY. Educ: Univ Col, Dublin, BA, 63, MA, 65; Univ Toronto, PhD(geog), 71. Prof Exp: Asst prof, 69-74, PROF GEOG, MEM UNIV, NFLD, 74- Mem: Can Asn Geog. Res: Atlantic migrations and the transfer of culture, northeast North America; geography and ethnicity; evolution of rural settlement pattern, Newfoundland and Maritimes, Canada; Irish immigration and settlement in 19th century eastern Canada. Mailing Add: Dept of Geog Mem Univ of Nfld St John's NF Can

MANNION, WILLIAM A, physical chemistry, see 12th edition

MANNIS, FRED, b Boston, Mass, May 3, 37; m 61; c 2. PHYSICAL CHEMISTRY. Educ: Harvard Univ, AB, 58; Mass Inst Technol, PhD(phys chem), 63. Prof Exp: Nat Sci Found fel, Univ Col, N Wales, 63; chemist, Cent Res Dept, 63-73, SR RES CHEMIST, PLASTICS DEPT, RES & DEVELOP DIV, E I DU PONT DE NEMOURS & CO, INC, 73- Res: Mössbauer effect; nuclear magnetic resonance; heterogeneous catalysis; materials research; electrochemistry; cation exchange membranes; optical fibers. Mailing Add: E I du Pont de Nemours & Co Inc 1007 Market St Wilmington DE 19898

MANNO, BARBARA REYNOLDS, b Columbus, Ohio, Mar 16, 36; m 68. PHARMACOLOGY, TOXICOLOGY. Educ: Otterbein Col, BS, 57; Ind Univ, Indianapolis, MS, 68, PhD(pharmacol), 70. Prof Exp: Asst prof pharmacol, Sch Pharm, Auburn Univ, 70-71; asst prof, 71-74, ASSOC PROF PHARMACOL, MED SCH, LA STATE UNIV, SHREVEPORT, 74-; PHARMACOLOGIST, VET ADMIN HOSP, 71- Concurrent Pos: Assoc prof clin med technol, La Tech Univ, 74- Mem: AAAS; Soc Toxicol; Am Acad Forensic Sci; Am Asn Clin Chem; fel Am Acad Forensic Sci. Res: Cardiac lipid metabolism; cardiovascular actions of marihuana; analytical toxicology. Mailing Add: Dept Pharmacol & Therapeut La State Univ Med Sch Shreveport LA 71130

MANNO, JOSEPH EUGENE, b Warren, Pa, May 5, 42; m 68. TOXICOLOGY, PHARMACOLOGY. Educ: Duquesne Univ, BS, 65, MS, 67; Ind Univ, Indianapolis, PhD(toxicol), 70. Prof Exp: Asst prof pharmacol & toxicol, Sch Pharm, Auburn Univ, 70-71; asst prof, 71-74, ASSOC PROF PHARMACOL & TOXICOL & CHIEF, SECT TOXICOL, SCH MED, LA STATE UNIV, SHREVEPORT, 74- Mem: Soc Toxicol; Am Acad Forensic Sci. Res: Cardiovascular actions of marihuana and its chemical components; analytical toxicology. Mailing Add: Dept of Pharmacol & Therapeut La State Univ Med Sch Shreveport LA 71101

MANNY, BRUCE ANDREW, b Dayton, Ohio, May 24, 44. LIMNOLOGY, AQUATIC ECOLOGY. Educ: Oberlin Col, AB, 66; Rutgers Univ, MS, 68; Mich State Univ, PhD(bot), 71. Prof Exp: Nat Sci Found res assoc limnol, W K Kellogg Biol Sta, Mich State Univ, 71-73; PROJ LEADER, GREAT LAKES FISHERY LAB, US FISH & WILDLIFE SERV, 73- Concurrent Pos: Pres, Environ Mgt & Design, Inc, 73- Honors & Awards: Nat Sci Found travel Award, Int Cong Limnol, Leningrad, 71. Mem: Ecol Soc Am; Am Soc Limnol & Oceanog; Int Asn Great Lakes Res; Int Asn Theoret & Appl Limnol; Am Fisheries Soc. Res: Ecological interactions of organic and inorganic nitrogen compounds in lake and stream metabolism; nitrogen cycle; ecosystem eutrophication. Mailing Add: Great Lakes Fishery Lab 1451 Green Rd Ann Arbor MI 48105

MANO, KOICHI, b Numazu, Japan, Nov 12, 20; m 45; c 2. THEORETICAL PHYSICS. Educ: Tokyo Univ Lit & Sci, BS, 45; Calif Inst Technol, PhD(physics), 55. Prof Exp: Lectr, Kyoto Univ, 50-55, asst prof, 55-58; scientist, Comstock & Wescott, Inc, Mass, 58-59; scientist, Wentworth Inst, Boston, Mass, 59-64; PHYSICIST, AIR FORCE CAMBRIDGE RES LABS, 64- Mem: Am Phys Soc; Sigma Xi. Res: Quantum mechanics; mathematical physics; theory of atomic structure; wave propagation through random media. Mailing Add: 334 Gray St Arlington MA 02174

MANOCHA, MANMOHAN SINGH, b Sheikhupura, India, Feb 25, 35; Can citizen; m 63; c 1. MYCOLOGY, PLANT PATHOLOGY. Educ: Punjab Univ, India, BSc, 55, MSc, 57; Indian Agr Res Inst, New Delhi, PhD(mycol, plant path), 61. Prof Exp: Coun Sci & Indust Res fel, Indian Agr Res Inst, New Delhi, 61-63; Can Dept Agr grant, Univ Sask, 63-65; fel, Nat Res Coun Can, 65-66; from asst prof to assoc prof, 66-75, PROF BIOL, BROCK UNIV, 75- Concurrent Pos: Alexander von Humboldt-Stiftung fel, Inst Plant Path, Univ Göttingen, WGer, 71-72. Mem: Can Phytopath Soc; Can Soc Cell Biol; Indian Phytopath Soc; Mycol Soc Am. Res: Fine structure and physiology of microorganisms; study of host-parasite interaction at cellular and molecular level; high resolution autoradiography associated with biochemical studies of diseased tissue. Mailing Add: Dept of Biol Sci Brock Univ St Catharines ON Can

MANOCHA, SOHAN LALL, b Sultan Pur Lodhi, India, Aug 12, 36; m 64; c 2. HISTOCHEMISTRY, NEUROANATOMY. Educ: Punjab Univ, India, BSc, 56, MSc, 57, PhD(biol), 61. Prof Exp: Lectr zool, Govt Col, Rupar, India, 61-62; Ont Cancer Res Found fel, Queen's Univ, Ont, 62-64; res assoc histochem, 64-67, from asst prof to assoc prof neurohistochem, 67-75, CHMN DIV EXP NUTRIT, YERKES PRIMATE RES CTR, EMORY UNIV, 75- Mem: Histochem Soc; Am Asn Anat; Soc Neurosci. Res: Fields of cytology, cytogenetics, histology, histochemistry, neuroanatomy and experimental nutrition using biological material related to reproductive system, skin, nervous system and biology of malnutrition; alteration of the nervous system under the impact of experimental dietary deficiency of protein in the diets of pregnant female and multigenerational study of its impact. Mailing Add: Yerkes Primate Res Ctr Emory Univ Atlanta GA 30322

MANOHAR, RAMPURKAR, b Indore, India, July 11, 25; m 49; c 3. NUMERICAL ANALYSIS. Educ: Agra Univ, BSc, 46, MSc, 48; Aligarh Muslim Univ, MSc, 54; Univ Göttingen, Dr rer nat(fluid mech, appl math), 57. Prof Exp: Lectr math, Holkar Col, 48-51; instr appl math, Eng Col, Aligarh Muslim Univ, 51-54, reader math, 58-59; res fel, Univ Göttingen & Max Planck Inst Aerodyn, 54-57; assoc prof appl sci, Punjab Eng Col, 59-62; vis assoc prof math, Math Res Ctr, US Army, Univ Wis, 62-64; assoc prof, 64-68, PROF MATH, UNIV SASK, 68- Mem: Soc Indust & Appl Math; Am Acad Mech; Am Math Soc; Can Math Cong; Indian Math Soc. Res: Fluid mechanics, boundary layer theory and viscous flows. Mailing Add: Dept of Math Univ of Sask Saskatchewan SK Can

MANOHARAN, A CHELVANAYAKAM, b Colombo, Ceylon, Jan 20, 33; m 63; c 1. PHYSICS, MATHEMATICS. Educ: Ceylon Univ, BSc, 55; Cambridge Univ, MA, 62; Brandeis Univ, PhD(physics), 63. Prof Exp: Asst lectr physics, Ceylon Univ, 55-56; instr, Princeton Univ, 62-65; asst prof physics & math, Lowell Technol Inst, 65-66; asst prof physics, Univ Wis-Milwaukee, 66-69; ASSOC PROF PHYSICS & MATH, OKLAHOMA CITY UNIV, 69-, CHMN DEPT PHYSICS, 70- Mem: Am Phys Soc; Am Math Soc. Res: Quantum field theory; mathematical physics. Mailing Add: Dept of Math & Physics Oklahoma City Univ Oklahoma City OK 73106

MANOHARAN, ARTHUR, b Tiruchirapalli, India, Sept 3, 28; m 54. OCCUPATIONAL HEALTH, PUBLIC HEALTH. Educ: Madras Univ, MB, BS, 51; London Univ, dipl, 57; Columbia Univ, DPH, 60; Royal Col Physicians & Surgeons Eng, dipl indust health, 60. Prof Exp: Instr microbiol, Stanley Med Col, Madras Univ, 51-52; med officer, Green Hosp, Manipay, Ceylon, 52-54; health officer, City Health Dept, Singapore, 54-59; health officer, Ministry of Health, Malaysia, 60-65; lectr pub health, Univ Singapore, 65-67; assoc prof int health, Sch Pub Health, Univ Hawaii, 67-69; expert occup health & Int Labour Orgn Adv to Korean Govt, 69; assoc prof pub health, Sch Med, Boston Univ, 70-75; PROF PUB HEALTH, SCH PUB HEALTH, UNIV HAWAII, HONOLULU, 75- Concurrent Pos: Cholera consult, WHO, Nepal & Afghanistan, 66. Mem: Soc Occup Med Singapore; Am Pub Health Asn. Res: Health problems in industry; population problems in developing countries; medical education. Mailing Add: Sch of Pub Health Univ of Hawaii 1890 E West Rd Honolulu HI 96822

MANON, JOHN JOSEPH, physical chemistry, see 12th edition

MANOOGIAN, ARMEN, b Galt, Ont, June 16, 34; m 64; c 2. PHYSICS. Educ: McMaster Univ, BA, 59; Fresno State Col, MSc, 61; McGill Univ, MS, 63; Univ Windsor, PhD(physics), 66. Prof Exp: Asst prof, 66-70, ASSOC PROF PHYSICS, UNIV OTTAWA, 70- Mem: Can Asn Physicists. Res: Paramagnetic resonance of transition metal impurities in crystals. Mailing Add: Dept of Physics Univ of Ottawa Ottawa ON Can

MANOS, CONSTANTINE T, b White Plains, NY, Jan 2, 33; m 71; c 2. SEDIMENTOLOGY. Educ: City Col New York, BS, 58; Univ Ill, MS, 60, PhD(geol), 63. Prof Exp: Res asst, State Geol Surv, Ill, 58-63; asst prof, Plattsburgh, 63-64, New Paltz, 64-66, assoc prof, 66-70, chmn dept geol sci, 69-72, PROF GEOL, STATE UNIV N Y, NEW PALTZ, 70- Mem: Fel Geol Soc Am; Soc Econ Paleont & Mineral; Am Geol Inst; Nat Asn Geol Teachers. Res: Heavy mineral analysis; sedimentation; stratigraphy; heavy mineral, thin section analysis of Eocene flysch sediments of northwest Greece. Mailing Add: Dept Geol Sci State Univ Col New Paltz NY 12561

MANOS, NICHOLAS EMMANUEL, b Modesto, Calif, Dec 19, 16; m 49; c 5. MATHEMATICS. Educ: Univ Calif, BA, 39, MA, 40. Prof Exp: Meteorologist, US Weather Bur, 46-49; statistician, USPHS, 50-73; assoc dir, ASTHO Health Progs Reporting Syst, 74-75; PVT CONSULT STATIST, PUB HEALTH, AIR POLLUTION, OCCUP HEALTH AND METEOROLOGY, 75- Concurrent Pos: Asst prof epidemiol & environ health, Med Sch, George Washington Univ, 68- Mem: Am Statist Asn; Am Pub Health Asn. Res: Statistics; public health; air pollution; occupational health; mathematics; meteorology. Mailing Add: 9847 Singleton Dr Washington DC 20034

MANOS, PHILIP, b Thessaloniki, Greece, May 8, 28; US citizen; m 57; c 2. ORGANIC POLYMER CHEMISTRY. Educ: Univ Thessaloniki, BS, 51; Boston Univ, PhD(chem), 61. Prof Exp: USPHS fel, Boston Univ, 59-60; res chemist, 60-68, to sr res chemist, 60-75, RES ASSOC ORG CHEM DEPT, E I DU PONT DE NEMOURS & CO, INC, 75- Mem: Am Chem Soc. Res: Synthetic and mechanistic studies in organic chemistry; polymer and membrane chemistry. Mailing Add: E I du Pont de Nemours & Co Inc Org Chem Dept Jackson Lab Deepwater NJ 19898

MANOS-HODGE, GEORGIA EVANGELINE, physiology, biology, see 12th edition

MANOUGIAN, EDWARD, b Highland Park, Mich, Apr 11, 29; m 62; c 2. THEORETICAL BIOLOGY. Educ: Wayne State Univ, BS, 51; Univ Mich, MD, 55. Prof Exp: NIH fel math, Univ Calif, Berkeley, 60-62; RES ASSOC BIOMED, DONNER LAB, UNIV CALIF, BERKELEY, 62- Mem: AAAS; Am Math Soc. Res: Endocrinology; biological rhythms. Mailing Add: Donner Lab Univ of Calif Berkeley CA 94720

MANOUGIAN, MANOUG N, b Jerusalem, Palestine, Apr 29, 35; m 60; c 1. MATHEMATICS. Educ: Univ Tex, Austin, BA, 60, MA, 64, PhD(math), 68. Prof Exp: Instr math, Haigazian Col, Lebanon, 60-62, asst prof, 64-66; ASSOC PROF MATH, UNIV S FLA, 68-, CHMN DEPT, 74- Concurrent Pos: Nat Sci Found grant, Univ S Fla, 71. Mem: AAAS; Am Math Soc; Math Asn Am; Soc Indust & Appl Math. Res: Analysis; differential and integral equations. Mailing Add: Dept of Math Univ of SFla Tampa FL 33620

MANOWITZ, MILTON, biochemistry, see 12th edition

MANOWITZ, PAUL, biochemistry, see 12th edition

MANRING, EDWARD RAYMOND, b Springfield, Ohio, Mar 21, 21; m 40; c 3. PHYSICS. Educ: Ohio Univ, BS, 44, MS, 48, PhD(physics), 52. Prof Exp: Res physicist, Monsanto Chem Co, 48-51; physicist, Geophys Res Directorate, Upper Air Observ, NMex, 52-60; head, Observational Physics Group, Geophys Corp Am, 60-66; PROF PHYSICS, NC STATE UNIV, 66- Mem: Am Phys Soc. Res: Nuclear physics; radio frequency spectroscopy; night sky intensity. Mailing Add: 1601 Dixie Trail Raleigh NC 27607

MANS, RUSTY JAY, b Newark, NJ, Sept 30, 30; m 52; c 5. BIOCHEMISTRY, ENZYMOLOGY. Educ: Univ Fla, BS, 52, MS, 54, PhD(biochem), 59. Prof Exp: Res assoc, Enzyme Group, Biol Div, Oak Ridge Nat Lab, 59-61, biochemist, 61-64; assoc prof biochem genetics, Univ Md, 64-69; prof immunol, med microbiol & radiol, Radiation Biol Lab, 69-72, PROF BIOCHEM, UNIV FLA, 72- Concurrent Pos: AEC res contract, 65- Mem: Am Soc Biol Chem; NY Acad Sci; Am Soc Plant Physiol. Res: Biosynthesis of proteins and nucleic acids; mechanism of eukaryotic transcription; mechanism of control of light induced nucleic acid synthesis in higher plants. Mailing Add: Dept of Biochem Univ of Fla 334 Nuclear Sci Ctr Gainesville FL 32610

MANSBERGER, ARLIE ROLAND, JR, b Turtle Creek, Pa, Oct 13, 22; m 46; c 3. SURGERY. Educ: Univ Md, MD, 47. Prof Exp: Res fel surg, Univ Md, Baltimore City, 49-50, asst, Sch Med, 53-56, instr, 56-59, from asst prof to prof surg, 59-74, dir clin res, Shock-Trauma Unit, 62-66, chief clin adv, 66-71, head div gen surg, Univ Md Hosp, 71-74; PROF SURG & CHMN DEPT, MED COL GA, 73- Concurrent Pos: Consult surgeon, Montebello State Hosp, Baltimore, 56-73; consult surgeon, Div Voc Rehab, State of Md, 58-73; actg chmn dept surg, Univ Md Hosp, 70-71; examr, Am Bd Surg, 73-; chief consult, Vet Admin Hosp, Augusta, Ga, 73- Mem: AMA; fel Am Col Surg. Res: Ammonia metabolism in surgical diseases; biochemical and metabolic factors in shock. Mailing Add: Dept of Surg Med Col of Ga Augusta GA 30902

MANSELL, GLENN WILLIAM, organic chemistry, see 12th edition

MANSELL, ROBERT SHIRLEY, b Roswell, GA, Apr 28, 38; m 65; c 3. SOIL PHYSICS. Educ: Univ Ga, BSA, 60, MS, 63; Iowa State Univ, PhD(agron), 68. Prof Exp: Fertilizer sales rep, Int Mineral & Chem Corp, Ga, 61; res asst soil fertility, Dept Agron, Univ Ga, 61-63; res asst soil physics, Iowa State Univ, 63, res assoc, 63-68; asst prof, 68-73, ASSOC PROF SOIL PHYSICS, UNIV FLA, 73- Mem: AAAS; Soil Sci Soc Am; Am Soc Agron; Am Geophys Union. Res: Movement of chemicals and water throug water-unsaturated soil; water movement in layered soils; soil and water pollution. Mailing Add: G149 McCarty Hall Dept of Soil Sci Univ of Fla Gainesville FL 32611

MANSFIELD, ARTHUR WALTER, b London, Eng, Mar 29, 26; m 57; c 3. MARINE BIOLOGY. Educ: Cambridge Univ, BA, 47, MA, 51; McGill Univ, PhD, 58. Prof Exp: Meteorologist, Falkland Islands Dependencies Surv, South Georgia Island, 51, base leader & biologist, S Orkneys, 52-53; demonstr zool, McGill Univ, 54-56; SCIENTIST & DIR, ARCTIC BIOL STA, FISHERIES & MARINE SERV, ENVIRONMENT CAN, 56- Concurrent Pos: Lectr, McGill Univ, 64-65. Res: Arctic marine biology, principally marine mammals. Mailing Add: Arctic Biol Sta PO Box 400 Fisheries & Marine Serv St Anne de Bellevue PQ Can

MANSFIELD, CHARLES FREDERIC, III, geology, see 12th edition

MANSFIELD, JOHN E, b Cleveland, Ohio, July 2, 38; m 68; c 1. ELEMENTARY PARTICLE PHYSICS, THEORETICAL PHYSICS. Educ: Univ Detroit, AB, 60; St Louis Univ, MS & PhL, 63; Harvard Univ, AM, 66, PhD(physics), 70. Prof Exp: Res fel physics, Univ Notre Dame, 68-71; RES PHYSICIST, SCI APPLNS, INC, 71- Mem: Am Phys Soc. Res: Internal symmetries; bootstraps; S-matrix theory; hydrodynamics; statistical physics, thermodynamics. Mailing Add: Sci Applns Inc 1651 Old Meadow Rd McLean VA 22101

MANSFIELD, JOHN MICHAEL, b Louisville, Ky, Nov 5, 45; m 66; c 1. IMMUNOLOGY, MICROBIOLOGY. Educ: Miami Univ, BA, 67, MA, 69; Ohio State Univ, PhD(microbiol), 71. Prof Exp: NSF trainee, 71; fel microbiol & immunol, 71-73, ASST PROF MICROBIOL & IMMUNOL, SCH MED, UNIV LOUISVILLE, 73- Mem: AAAS; Am Soc Microbiol; Am Soc Trop Med & Hyg. Res: Tumor immunobiology; immunopathology of experimental African trypanosomiasis; cellular immunology; cytogenetics. Mailing Add: Dept of Microbiol & Immunol Univ of Louisville Sch of Med Louisville KY 40201

MANSFIELD, JOSEPH VICTOR, b Chicago, Ill, Mar 9, 07; m 35; c 1. ORGANIC CHEMISTRY. Educ: Iowa State Col, BS, 31; Univ Chicago, PhD(org chem), 42. Prof Exp: Chemist, Chicago Steel Co, 28-29; instr physiol chem, Chicago Med Sch, 31-36; pres & res dir, Mansfield Photo Res Labs, 35-42; chief specialist photog, Air Ctr, Photog Sch, US Navy, Fla, 42-43, chief photog mate, Photo Sci Lab, Va, 43-45; dir educ, Chicago Sch Photog, 45-46; asst prof, 46-58, ASSOC PROF CHEM, UNIV ILL, CHICAGO CIRCLE, 58- Mem: AAAS; Am Chem Soc; Photog Soc Am; Am Inst Chem; Royal Photog Soc. Res: Photographic chemical products; analytical chemistry. Mailing Add: 505 N Lake Shore Dr Apt 1806 Chicago IL 60611

MANSFIELD, KEVIN THOMAS, b Yonkers, NY, Mar 26, 40; m 62; c 3. ORGANIC POLYMER CHEM. Educ: Fordham Univ, BS, 62; Ohio State Univ, MS, 65, PhD(org chem), 67. Prof Exp: Res chemist, Silicones Res Ctr, Union Carbide Corp, NY, 67-69; prod develop chemist, 69-74, develop chemist, Basel, Switzerland, 74-75, GROUP LEADER PLASTICS DEVELOP, PLASTICS & ADDITIVES DIV, CIBA-GEIGY CORP, 75- Mem: Am Chem Soc. Res: Process development; epoxies and other polymers of commercial interest. Mailing Add: Ciba-Geigy Facilities Toms River NJ 08753

MANSFIELD, LARRY EVERETT, b Seattle, Wash, Sept 8, 39; m 62. MATHEMATICS. Educ: Whitman Col, AB, 61; Univ Wash, PhD(math), 65. Prof Exp: ASST PROF MATH, QUEENS COL, NY, 65- Mem: Am Math Soc; Math Asn Am. Res: Differential geometry. Mailing Add: Dept of Math Queens Col Flushing NY 11367

MANSFIELD, LOIS E, b Portland, Maine, Jan 2, 41. NUMERICAL ANALYSIS. Educ: Univ Mich, BS, 62; Univ Utah, MS, 66, PhD(math), 69. Prof Exp: Vis asst prof comput sci, Purdue Univ, 69-70; vis asst prof math, Univ Utah, 73-74; asst prof, 70-74, ASSOC PROF COMPUT SCI, UNIV KANS, 74- Concurrent Pos: Mem, Adv Panel, Comput Sci Sect, NSF, 75- Mem: Am Math Soc; Soc Indust & Appl Math; Asn Comput Mach. Res: Numerical solution of partial differential equations, approximation theory. Mailing Add: Dept of Comput Sci Univ of Kans Lawrence KS 66045

MANSFIELD, MANFORD EDWARD, veterinary medicine, see 12th edition

MANSFIELD, MAYNARD JOSEPH, b Marietta, Ohio, Jan 28, 30; m 53; c 2. TOPOLOGY. Educ: Marietta Col, BA, 52; Purdue Univ, MS, 54, PhD(math), 56. Prof Exp: Asst math, Purdue Univ, 52-54, statist, Statist Lab, 54-55, instr math, 56-57; asst prof, Washington & Jefferson Col, 57-60, assoc prof, 60-63; assoc prof, 63-65, PROF MATH, INDIANA-PURDUE UNIVS, FT WAYNE, 65-, CHMN DEPT MATH SCI, 63- Mem: Am Math Soc; Math Asn Am. Res: Abstract topological spaces. Mailing Add: Dept of Math Sci Indiana-Purdue Univs Ft Wayne IN 46805

MANSFIELD, TOM, b Montreal, Que, Mar 5, 11; nat US; m 42; c 2. BACTERIOLOGY. Educ: Univ BC, BA, 35. Prof Exp: Bacteriologist, Am Can Co, 35-43; chief canning sect, Food Mach & Chem Corp, 43-57, asst to gen mgr, Int Mach Corp Div, 57-59, dir res & develop, 59-70, TECH DIR, FMC MACH INT, FMC CORP, 70- Mem: AAAS; Inst Food Technol. Res: Equipment for dehydrated, frozen and canned foods plants. Mailing Add: 1240 Glenn Ave San Jose CA 95125

MANSFIELD, VICTOR NEIL, b Norwalk, Conn, Mar 7, 41; m 68; c 2. ASTROPHYSICS. Educ: Dartmouth Col, BA, 63, MS, 65; Cornell Univ, PhD(astrophys), 72. Prof Exp: Res assoc astron, Cornell Univ, 71-73; ASST PROF PHYSICS & ASTRON, COLGATE UNIV, 73- Concurrent Pos: Vis asst prof astron, Cornell Univ & vis scientist, Nat Astron & Ionosphere Ctr, 75-76. Mem: Int Astron Union; Am Astron Soc; AAAS. Res: Theoretical astrophysics, especially interstellar medium, cosmology and X-ray astronomy. Mailing Add: Dept of Physics & Astron Colgate Univ Hamilton NY 13346

MANSFIELD, WALTER O, JR, physics, deceased

MANSINHA, LALATENDU, b Orissa, India, July 2, 37; m 63. GEOPHYSICS, APPLIED MECHANICS. Educ: Indian Inst Technol, Kharagpur, BSc, 57, MTech, 59; Univ BC, PhD(geophys, physics), 63. Prof Exp: Fel geophys, Rice Univ, 62-65; vis lectr, 65-66, asst prof, 66-69, ASSOC PROF GEOPHYS, UNIV WESTERN ONT, 69- Concurrent Pos: NASA-Nat Res Coun sr res assoc, Goddard Space Flight Ctr, Md, 70-71. Mem: Am Geophys Union; Seismol Soc Am; Soc Explor Geophys. Res: Source mechanism of earthquakes; determination of fracture velocity in elastic media; fracture in anisotropic media; Chandler wobble. Mailing Add: Dept of Geophys Univ of Western Ont London ON Can

MANSKE, RICHARD HELMUTH (FRED), b Berlin, Ger, Sept 14, 01; Can citizen; m 24; c 2. NATURAL PRODUCTS CHEMISTRY. Educ: Queen's Univ, Ont, BSc, 23, MSc, 24; Univ Manchester, PhD(org chem), 26, ScD, 37. Hon Degrees: DSc, McMaster Univ, 60; LLD, Queen's Univ, Ont, 67. Prof Exp: Res chemist, Gen Motors Corp, 26-27; Eli Lilly & Co res fel, Yale Univ, 27-29, Sterling fel, 29-30; assoc res chemist, Nat Res Coun Can, 30-43; dir res, Dom Rubber Co, Ltd, 43-66; RETIRED. Concurrent Pos: Adj prof, Waterloo Univ, 67- Honors & Awards: Medal, Chem Inst Can, 59; Morley Medal, Am Chem Soc, 72. Mem: Fel Royal Soc Can; hon fel Chem Inst Can (pres, 63-64); The Chem Soc. Res: Synthesis, constitution and degraduation of alkaloids; isolation of alkaloids and other plant products; synthesis of phyto-hormones; indol compounds; quinolines. Mailing Add: Waterloo ON Can

MANSKI, WLADYSLAW J, b Lwow, Poland, May 15, 15; US citizen; m 41; c 2. MICROBIOLOGY, IMMUNOCHEMISTRY. Educ: Univ Warsaw, PhM, 39; Univ Wroclaw, DSc, 51. Prof Exp: Instr anal chem, Inst Inorg & Phys Chem, Univ Warsaw, 36-39; instr, Inst Chem, Univ Lublin, 44-45; head chem lab, Inst Microbiol, Univ Wroclaw, 45-49; Rockefeller fel, US, Denmark & Sweden, 49-50; head immunochem, Inst Immunol & Exp Ther, Polish Acad Sci, 51-55, head macromolecular biochem, Inst Biochem & Biophys & head biochem lab, State Inst Hyg, 55-57; from asst prof to assoc prof, 58-75, PROF MICROBIOL, ASSIGNED TO OPHTHALMOL, COL PHYSICIANS & SURGEONS, COLUMBIA UNIV, 75- Mem: AAAS; Am Asn Immunol; Am Chem Soc; Harvey Soc; Brit Biochem Soc. Mailing Add: Columbia Univ 630 W 168th St New York NY 10032

MANSON, DONALD JOSEPH, b Chewelah, Wash, Dec 10, 30; m 67; c 3. PHYSICS, ELECTRONICS. Educ: St Louis Univ, BA, 59, PhL, 60, MS, 63, PhD(physics), 67. Prof Exp: Fel physics, St Louis Univ, 66-67, asst prof, 67-69; asst prof radiol, Med Ctr, Univ Mo-Columbia, 69-71, assoc prof radiol sci & elec eng, 71-75; VPRES RES, ALPHA ELECTRONICS LABS, 76- Concurrent Pos: Res asst biomed comput labs, Washington Univ, 67-69; assoc investr, Space Sci Res Ctr, 70-; res coordr, Nat Inst Gen Med Sci grant, 70-; tech proj dir, USPHS grant, 71- Mem: Am Phys Soc; Am Asn Physics Teachers; Asn Comput Mach; Inst Elec & Electronics Engrs; Am Asn Physicists in Med. Res: Computer diagnosis in radiology; computer applications in medicine; image analysis; ecological systems analysis; computer modeling of biological systems; energy conservation; consumer electronics. Mailing Add: Alpha Electronics Labs 2302 Oakland Columbia MO 65201

MANSON, EARLE LOWRY, JR, b Alexandria, Va, Feb 28, 48; m 72. MOLECULAR PHYSICS. Educ: The Citadel, BS, 70; Duke Univ, PhD(physics), 75. Prof Exp: RES ASSOC PHYSICS, DUKE UNIV, 76- Res: Millimeter and submillimeter wave spectra and molecular constants for the cuprous halides.

MANSON, JOHN ALEXANDER, b Dundas, Ont, Can, Aug 4, 28; m 51; c 4. PHYSICAL CHEMISTRY, POLYMER SCIENCE. Educ: McMaster Univ, BSc, 49, MSc, 51, PhD(chem), 56. Prof Exp: Asst, McMaster Univ, 47-50, 51-55; instr chem, Royal Mil Col Can, 50-51; assoc res engr, Eng Res Inst, Univ Mich, 56-57; sr res chemist, Cent Res Labs, Air Reduction Co, Inc, 57-59, sect head chem res, 59-61, supvr, 61-66; assoc prof, 66-70, PROF CHEM, LEHIGH UNIV, 70-, DIR POLYMER LAB, MAT RES CTR, 66- Concurrent Pos: Res assoc, Phoenix Proj, Univ Mich, 56-57; Airco sr staff award, Macromolecular Res Ctr, Strasbourg, France, 64-65; consult, Indust & Govt Labs. Mem: Am Chem Soc; Chem Inst Can. Res: Polymerization; physical and engineering properties of high polymers and composite systems; permeability; fracture phenomena. Mailing Add: Coxe Lab Bldg 32 Lehigh Univ Bethlehem PA 18015

MANSON, LIONEL ARNOLD, b Toronto, Ont, Dec 24, 23; nat US; m 45; c 3. IMMUNOBIOLOGY, MOLECULAR BIOLOGY. Educ: Univ Toronto, BA, 45, MA, 47; Wash Univ, PhD(biol chem), 49. Prof Exp: Nat Res Coun fel med sci, Western Reserve Univ, 49-50, from instr to sr instr microbiol, 50-54; res asst prof, Sch Med, Univ Pa, 54-66, assoc prof, 66-73, res assoc, Wistar Inst, 54-57, assoc mem, 58-65, PROF MICROBIOL, SCH MED, UNIV PA, 74-, MEM, WISTAR INST, 65- Concurrent Pos: Mem grad group microbiol, Grad Sch Arts & Sci, Univ Pa, 55-, molecular biol, 65-, immunology, 71-, chmn immunology, 74-; sr Fulbright scholar, France, 63-64; fel, Nat Cancer Inst Israel, 71-72. Mem: Sigma Xi; Am Soc Microbiol; Am Asn Biol Chemists; Am Asn Immunol; Am Chem Soc. Res: Transplantation immunobiology; molecular biology and biochemical genetics of humoral and cell-mediated immunity; structure, function and biosynthesis of cellular membranes; cell membrane differentiation; interaction of sub-cellular organelles in macromolecular biosynthesis; molecular biology of the immune response induced by tumor-specific and transplantation antigens; biochemical analysis of subcellular organelles. Mailing Add: Wistar Inst of Anat & Biol 36th & Spruce St Philadelphia PA 19104

MANSON, STEVEN TRENT, b Brooklyn, NY, Dec 12, 40; m 68; c 1. ATOMIC PHYSICS, CHEMICAL PHYSICS. Educ: Rensselaer Polytech Inst, 61; Columbia Univ, MA, 63, PhD(physics), 66. Prof Exp: Nat Acad Sci-Nat Res Coun res assoc, Nat Bur Standards, 66-68; asst prof, 68-71, ASSOC PROF PHYSICS, GA STATE UNIV, 71- Concurrent Pos: Consult, Oak Ridge Nat Lab, 68-69; consult, Argonne Nat Lab & Pac Northwest Labs, Battelle Mem Inst, 73- Mem: Am Phys Soc; Brit Inst Physics. Res: Theoretical atomic collisions; photoionization; generalized oscillator strengths; angular distribution of ionized electrons; penetration of charged particles into matter. Mailing Add: Dept of Physics Ga State Univ Atlanta GA 30303

MANSON-HING, LINCOLN ROY, b Georgetown, Guyana, May 20, 27; US citizen; m 49; c 3. DENTISTRY. Educ: Tufts Univ, DMD, 48; Univ Ala, MS, 61. Prof Exp: From asst prof to assoc prof, 56-58, PROF DENT, SCH DENT, UNIV ALA, BIRMINGHAM, 68-, CHMN DEPT DENT RADIOL, 62- Concurrent Pos: USPHS grants, 58-72; consult, State Ala Cleft Palate Clins, 57-65 & Vet Admin Hosp; Fulbright-Hays lectr, UAR, 64-65. Mem: AAAS; Int Asn Dent Res; Am Dent Asn; Am Acad Dent Radiol. Res: Dental radiology. Mailing Add: Dept of Dent Radiol Univ of Ala Med Ctr Birmingham AL 35233

MANSOUR, A MAHER, b Kallin, Egypt, July 20, 28; m 66; c 2. MOLECULAR BIOLOGY, ENDOCRINOLOGY. Educ: Cairo Univ, BSc, 50; Cambridge Univ, PhD(endocrinol), 57. Prof Exp: Lectr physiol & endocrinol, Ain Shams Univ, Cairo, 58-63; res assoc molecular biol, Temple Univ, 63-65; MED RES SCIENTIST BIOCHEM RES, EASTERN PA PSYCHIAT INST, 65- Concurrent Pos: Pop Coun fel, Rutgers Univ, 62-64; res assoc, Bur Biol Res, Rutgers Univ, 62-63. Mem: AAAS; Endocrine Soc; Am Chem Soc; Am Soc Cell Biol. Res: Intracellular mechanism of hormone action and the relationship between mitochondrial and nuclear RNA synthesis. Mailing Add: Eastern Pa Psychiat Inst Philadelphia PA 19129

MANSOUR, AGNES MARY, b Detroit, Mich, Apr 10, 31. BIOCHEMISTRY. Educ: Mercy Col, Mich, BS, 53; Catholic Univ, MS, 58; Georgetown Univ, PhD(biochem), 64. Prof Exp: Instr chem, 58-60, chmn med assocs div, 60-61, chmn dept natural sci, 64-70, PRES, MERCY COL, MICH, 71- Concurrent Pos: Educ coordr med assocs, Mt Carmel Mercy Hosp, Detroit, 58-61, instr, Nurse Anesthetist Prog, 59-61; on leave, Am Coun Educ fel, Univ Ky, 70-71; consult, Div Allied Health Manpower, Dept Health, Educ & Welfare, 72- Mem: Am Chem Soc; AAAS; Am Soc Med Technologists; Am Sch Allied Health Prof; Am Asn Higher Educ. Res: Health Science. Mailing Add: Off of the Pres Mercy Col 8200 W Outer Dr Detroit MI 48219

MANSOUR, TAG ELDIN, b Belkas, Egypt, Nov 6, 24; nat US; m 55; c 3. PHARMACOLOGY, BIOCHEMISTRY. Educ: Cairo Univ, BVSc, 46; Univ Birmingham, PhD(pharmacol), 49, DSc(biochem), 74. Prof Exp: Lectr, Cairo Univ, 50-51; Fulbright instr physiol, Sch Med, Howard Univ, 51-52; sr instr & res assoc pharmacol, Med Sch, Western Reserve Univ, 52-54; res assoc, Sch Med, La State Univ, 54-56, from asst prof to assoc prof, 56-61; assoc prof, 61-66, PROF PHARMACOL, SCH MED, STANFORD UNIV, 66- Concurrent Pos: Res fel pharmacol, Univ Birmingham, 49-50; Commonwealth Fund fel, 65; vis prof, Univ Wis, 69-70; consult, WHO, 70; mem study sect pharmacol, USPHS, 72-75. Mem: AAAS; Am Soc Pharmacol & Exp Therapeut; Am Soc Biol Chem. Res: Molecular and biochemical pharmacology; enzyme regulation; action of drugs on enzyme systems; regulation of cellular metabolism; chemotherapy of helminthiasis; physiology and biochemistry of parasitic helminths. Mailing Add: Dept of Pharmacol Stanford Univ Sch of Med Stanford CA 94305

MANSPEIZER, WARREN, b New York, NY, July 16, 33; m 62; c 4. GEOLOGY. Educ: City Col New York, BS, 56; WVa Univ, MS, 58; Rutgers Univ, PhD(geol), 63. Prof Exp: Assoc dean acad affairs, 70-71, ASSOC PROF GEOL, RUTGERS UNIV, NEWARK, 69-, CHMN DEPT, 71- Concurrent Pos: Scientist, Nat Sci Found-Moroccan Study Group, 72-75. Mem: Geol Soc Am; Am Asn Petrol Geol; Nat Asn Geol Teachers; Soc Econ Paleontol & Mineral. Res: Paleoflow structures in Triassic lava flows, Eastern North America; stratigraphy of the Triassic basalts in Morocco and Eastern North America. Mailing Add: Dept of Geol Rutgers Univ 195 University Ave Newark NJ 07102

MANTAI, KENNETH EDWARD, b Jamaica, NY, Oct 19, 42; m 62; c 2. PLANT PHYSIOLOGY. Educ: Univ Maine, BS, 64; Ore State Univ, PhD(plant physiol), 68. Prof Exp: Res fel plant physiol, Carnegie Inst, Washington, 68-69; res assoc, Brookhaven Nat Lab, 69-71; ASST PROF BIOL, STATE UNIV NY COL FREDONIA, 71- Concurrent Pos: NSF res grant, 76. Mem: Am Soc Plant Physiologists; Int Asn Great Lakes Res; Phycol Soc Am; Am Soc Limnol & Oceanog; AAAS. Res: Physiology of the green alga Cladophora glomerata in response to environmental conditions found in Lake Erie. Mailing Add: Dept of Biol State Univ NY Col Fredonia NY 14063

MANTEI, ERWIN JOSEPH, b Benton Harbor, Mich, Nov 1, 38. GEOCHEMISTRY. Educ: St Joseph's Col, Ind, BS, 60; Univ Mo, MS, 62, PhD(geochem), 65. Prof Exp: Asst prof earth sci, 65-68, assoc prof geochem, 68-70, assoc prof geol, 70-72, PROF GEOL, SOUTHWEST MO STATE UNIV, 72- Concurrent Pos: Dr Carl Hasselman stipend, Mineral Inst, Univ Heidelberg, 71-72. Mem: AAAS; Geochem Soc; Mineral Soc Am. Res: Trace element distribution in minerals and rocks associated with ore deposits. Mailing Add: Dept of Geol Southwest Mo State Univ Springfield MO 65802

MANTEI, KENNETH ALAN, b Los Angeles, Calif, Nov 22, 40; m 67; c 2. PHYSICAL CHEMISTRY. Educ: Pomona Col, BA, 62; Ind Univ, Bloomington, PhD(chem), 67. Prof Exp: Res scientist, Univ Calif, Los Angeles, 67-68; ASSOC PROF CHEM, CALIF STATE COL, SAN BERNARDINO, 68-, CHMN DEPT, 75- Mem: Am Chem Soc. Res: Kinetics and mechanism of gas-phase reactions; flash photolysis. Mailing Add: Dept of Chem Calif State Col San Bernardino CA 92407

MANTEL, LINDA HABAS, b New York, NY, May 12, 39; m 66. COMPARATIVE

PHYSIOLOGY. Educ: Swarthmore Col, BA, 60; Univ Ill, Urbana, MS, 62, PhD(physiol), 65. Prof Exp: Res fel living invert, Am Mus Natural Hist, 65-68; asst prof, 68-76, ASSOC PROF BIOL, CITY COL NEW YORK, 76- Concurrent Pos: Nat Inst Child Health & Human Develop fel, 65-66; res assoc, Am Mus Natural Hist, 68- Mem: AAAS; Am Inst Biol Sci; Am Soc Zool; Asn Women Sci. Res: Adaptations of animals to their environment, particularly invertebrates; comparative physiology of salt and water balance, particularly in crustaceans; neuroendocrine control of adaptive mechanisms. Mailing Add: Dept of Biol City Col of the City Univ New York NY 10031

MANTELL, GERALD JEROME, b US, May 11, 23; m 48; c 4. ORGANIC POLYMER CHEMISTRY. Educ: Queen's Univ, Ont, BSc, 45; NY Univ, PhD(chem), 49. Prof Exp: Res chemist, E I du Pont de Nemours & Co, 50-58; mgr applns res, Spencer Chem Co, 58-63; mgr, Gulf Oil Corp, 63-66; dir polymer res, 66-69, dir polymer & applns res & develop, 69-71, DIR RES & APPLNS DEVELOP, PLASTICS DIV, AIR PROD & CHEM, INC, 71- Mem: Am Chem Soc; Tech Asn Pulp & Paper Indust; Am Asn Textile Chemists & Colorists; Soc Plastics Engrs. Res: Elastomers; free radical reactions; textile chemicals; polymers; adhesive and paper applications; polyvinyl chloride polymers and uses. Mailing Add: Air Prods & Chem Inc PO Box 538 Allentown PA 18105

MANTEUFFEL, THOMAS ALBERT, b Woodstock, Ill, Nov 15, 48; m 72. NUMERICAL ANALYSIS. Educ: Univ Wis-Madison, BS, 70; Univ Ill, Urbana, MS, 72, PhD(math), 75. Prof Exp: ASST PROF MATH, EMORY UNIV, 75- Mem: Am Math Soc; Soc Indust & Appl Math. Res: Solutions of large sparse nonsymmetric linear systems. Mailing Add: Dept of Math Emory Univ Atlanta GA 30322

MANTHEI, ROLAND WILLIAM, b Kenosha, Wis, Apr 11, 17; m 49; c 2. PHARMACOLOGY. Educ: Univ Chicago, BS, 49, PhD(pharmacol), 53. Prof Exp: Instr pharmacol, Univ Chicago, 53-54; from asst prof to assoc prof, 54-68, PROF PHARMACOL, JEFFERSON MED COL, 68- Mem: Am Soc Pharmacol & Exp Therapeut. Res: Biochemical pharmacology; cancer chemotherapy. Mailing Add: Dept of Pharmacol Jefferson Med Col Philadelphia PA 19107

MANTHEY, ARTHUR ADOLPH, b New York, NY, June 7, 35; m 62. PHYSIOLOGY. Educ: Dartmouth Col, BA, 57; Columbia Univ, PhD(physiol), 65. Prof Exp: Fel physiol, Col Physicians & Surgeons, Columbia Univ, 65; instr med, 66-70, asst prof, 70-74, ASSOC PROF PHYSIOL & BIOPHYS, CTR HEALTH SCI, UNIV TENN, MEMPHIS, 74- Res: Electrophysiology of muscle and nerve. Mailing Add: Dept of Physiol & Biophys Univ of Tenn Ctr Health Sci Memphis TN 38163

MANTHEY, JOHN AUGUST, b Akron, Ohio, Mar 22, 25; m 49; c 5. PLANT PHYSIOLOGY. Educ: Kent State Univ, BS, 49; Univ Wyo, MS, 52. Prof Exp: Bacteriologist, Cleveland City Hosp, Ohio, 50-51; anal res chemist, Strong, Cobb & Co, Inc, 51-54; biochemist, Res Labs, 54-63, ANAL BIOCHEMIST, ELI LILLY & CO, INC, 63- Concurrent Pos: Lectr, Butler Univ, 58- Mem: Am Soc Plant Physiol. Res: Subcellular fractionation techniques in plant and animal tissues; mechanisms of metabolism of agrichemicals and therapeutic drug agents. Mailing Add: Dept of Agr Biochem Eli Lilly & Co Inc Indianapolis IN 46206

MANTHY, ROBERT SIGMUND, b Chicago, Ill, Apr 4, 39. FOREST ECONOMICS. Educ: Mich State Univ, BS, 60, MS, 63, PhD(forest econ), PhD(forest Econ), 64. Prof Exp: Proj leader, Govt Asst Progs, Northeastern Forest Exp Sta, US Forest Serv, 63-66; PROF FOREST ECON & POLICY, MICH STATE UNIV, 66- Concurrent Pos: Consult, Pub Land Law Rev Comn, 69, McClure Oil Corp, Mich, 72, US Forest Serv, Div Forest Econ, 72-73, US Forest Serv, Div Spec Progs & Policy, 72- & Forest Sci Dept, Utah State Univ, 72; sr assoc, Greentree Assoc-Resource Econ Consults, 73- Mem: AAAS; Soc Am Foresters. Res: Economics and scarcity of natural resources. Mailing Add: Dept of Forestry Mich State Univ East Lansing MI 48824

MANTIK, DAVID WAYNE, b Milan, Wis, Oct 13, 40. BIOPHYSICS. Educ: Univ Wis, BS, 62, PhD(physics), 67; Univ Ill, MS, 63. Prof Exp: NIH fel, Stanford Univ, 67-69; asst prof physics, Univ Mich-Flint, 69-72; MED STUDENT, MED SCH, UNIV MICH, 72- Res: Protein structure and function; physical mechanisms of energy transfer in biology. Mailing Add: 7437 Byron Ave Detroit MI 48202

MANTIS, HOMER THEODORE, b Reading, Pa, Aug 13, 17; m 40; c 4. PHYSICS, METEOROLOGY. Educ: Lehigh Univ, BS, 38; NY Univ, PhD(physics, meteorol), 50. Prof Exp: Res assoc meteorol, NY Univ, 48-50; from res assoc meteorol to prof mech eng, 50-74, PROF PHYSICS, UNIV MINN, MINNEAPOLIS, 74- Mem: Am Meteorol Soc. Res: Atmospheric dynamics. Mailing Add: 2352 Buford St Paul MN 55108

MANUCK, BARBARA ANN, b Brooklyn, NY, Sept 11, 48. PHYSICAL BIOCHEMISTRY. Educ: City Univ New York, BA, 69; Harvard Univ, MA, 73, PhD(chem), 76. Prof Exp: RES ASSOC, UNIV PITTSBURGH, 75- Mem: Am Chem Soc. Res: Application of nuclear magnetic resonance to the study of the physical properties of biochemical systems; specifically, nuclear magnetic resonance is used to study the structure-function relationships and protein-membrane interactions for bacterial and mammalian transport proteins. Mailing Add: 378 Crawford Hall Univ of Pittsburgh Pittsburgh PA 15260

MANUDHANE, KRISHNA SHANKAR, b Bombay, India, Aug 20, 27; m 54; c 2. INDUSTRIAL PHARMACY. Educ: Univ Poona, BSc, 49; Univ Bombay, BS, 51, MS, 54; Univ Md, PhD(pharm), 67. Prof Exp: Res chemist, Unichem Labs, Bombay, 54-55; pharmaceut chemist, Bd Experts, Drugs Control Admin, 55-59, tech officer, 59-64; head prod develop, Smith, Miller & Patch, Inc, 67-69; scientist, Parke Davis & Co, 69-71; dir tech serv-qual control, Cord Labs, 71-75; DIR PROD DEVELOP, ICN PHARMACEUT, INC, 75- Res: Research associated with development of new pharmaceutical products, improvement of marketed products, scale up, trouble shooting, improving yields and reducing costs. Mailing Add: 7839 Stonehill Dr Cincinnati OH 45230

MANUEL, OLIVER K, b Wichita, Kans, Oct 13, 36; m 60; c 5. NUCLEAR CHEMISTRY, GEOCHEMISTRY. Educ: Kans State Col, Pittsburg, BS, 59; Univ Ark, MS, 62, PhD(chem), 64. Prof Exp: Nat Sci Found fel, Univ Calif, Berkeley, 64; from asst prof to assoc prof, 64-73, PROF CHEM, UNIV MO-ROLLA, 73- Mem: AAAS; Am Chem Soc; Am Geophys Union. Res: Noble gas mass spectrometry to study chronology of solar system; geochemistry of tellurium and the halogens; charge interactions as fundamental basis for all forces in nature. Mailing Add: Dept of Chem Univ of Mo Rolla MO 65401

MANUEL, THOMAS ASBURY, b Austin, Tex, Jan 3, 36; m 58; c 2. RUBBER CHEMISTRY. Educ: Ohio Wesleyan Univ, BA, 57; Harvard Univ, AM, 58, PhD(chem), 61. Prof Exp: Res chemist, Cent Basic Res Lab, Esso Res & Eng Co, 60-63, sr chemist, 63-66, proj leader, Enjay Polymer Labs, 66-67, res assoc & sect head, 67-73, mgr, European Elastomers Tech Serv, Esso Chem Europe, Inc, 74-75, MGR NEW ELASTOMERS, ELASTOMERS TECHNOL DIV, EXXON CHEM CO, 75- Mem: Am Chem Soc. Res: Metal carbonyls; organometallic compounds; coordination compounds; polymers; elastomers. Mailing Add: Elastomers Technol Div Exxon Chem Co PO Box 45 Linden NJ 07036

MANUELIDIS, ELIAS EMMANUEL, b Constantinople, Turkey, Aug 15, 18; US citizen; m 66; c 2. PATHOLOGY, NEUROPATHOLOGY. Educ: Univ Munich, MD, 42. Hon Degrees: MA, Yale Univ, 64. Prof Exp: Sci asst path, Univ Munich, 43-46; lab dir, Ger Res Inst Psychiat, Max Planck Inst, 46-49; lab dir, Hosp Int Refugee Orgn, 49-50; neuropathologist, US Army Europ Command, 98th Gen Hosp, 50-51; from instr to assoc prof neuropath, 51-64, PROF PATH, SCH MED, YALE UNIV, 64-, PROF NEUROL, 72-, CUR BRAIN TUMOR REGISTRY, 58- Concurrent Pos: USPHS spec fel, 66-67; vis lectr, Harvard Med Sch, 66-67; consult, Fairfield, Norwich & Norwalk Hosps & NIH. Mem: AAAS; Am Acad Neurol; Am Asn Path & Bact; Am Asn Neuropath; NY Acad Sci. Res: Encephalitides and tumors; tissue cultures and transplantation of brain tumors; electron microscope utilizing transplanted tumors. Mailing Add: Dept of Path Yale Univ Sch of Med New Haven CT 06510

MANUS, LOUIS JOHN, b Lynden, Wash, Nov 13, 07; m 33; c 3. DAIRY SCIENCE. Educ: Wash State Col, BS, 32, PhD(dairy sci), 47; WVa Univ, MS, 34. Prof Exp: Dairy plant supt, WVa Univ, 34-36, instr dairy mfg, 36-41; asst prof dairy mfg sci, 41-62, assoc prof animal sci, 62-71, assoc prof food sci & technol, 71-73, EMER ASSOC PROF FOOD SCI & TECHNOL, WASH STATE UNIV, 73- Mem: Am Dairy Sci Asn; Inst Food Technol. Res: Powdered milk; lipase; Babcock fat test; nonfat milk solids tests; fresh, soft type cheese. Mailing Add: 4524 106th Pl NE Marysville WA 98270

MANUWAL, DAVID ALLEN, b South Bend, Ind, Oct 13, 42; m 68. WILDLIFE ECOLOGY. Educ: Purdue Univ, BS, 66; Univ Mont, MS, 68; Univ Calif, Los Angeles, PhD(zool), 72. Prof Exp: Biologist, Pt Reyes Bird Observ, Bolinas, Calif, 71; ASST PROF WILDLIFE SCI, COL FOREST RESOURCES, UNIV WASH, 72- Concurrent Pos: Consult, Ctr Northern Studies, Wolcott, Vt, 73; Nat Wildlife Fedn fel, 73 & 75. Mem: Ecol Soc Am; Am Ornithologists Union; Wildlife Soc. Res: Timing and synchrony of reproduction and the social structure of seabird populations; impact of timber management on forest bird communities. Mailing Add: Col of Forest Resources Wildlife Sci Group Univ of Wash Seattle WA 98195

MANVILLE, JOHN FIEVE, b Victoria, BC, Mar 18, 41; m 65; c 3. WOOD CHEMISTRY. Educ: Univ BC, BSc, 64, PhD(chem), 68. Prof Exp: RES SCIENTIST, FOREST PROD LAB, CAN FORESTRY SERV, 68- Mem: Chem Inst Can. Res: Extractives. Mailing Add: Forest Prod Lab 6620 NW Marine Dr Vancouver BC Can

MANVILLE, RICHARD HYDE, mammalogy, deceased

MANWILLER, ALFRED, b Elwood, Iowa, Aug 31, 16; m 42; c 4. PLANT BREEDING. Educ: Iowa State Univ, BS, 39, MS, 40; Pa State Univ, PhD(plant breeding), 44. Prof Exp: ASSOC AGRONOMIST, PEE DEE EXP STA, CLEMSON UNIV, 46-, ASSOC PROF AGRON, 72- Mem: Am Soc Agron. Res: Plant genetics and physiology. Mailing Add: Pee Dee Exp Sta Box 271 Florence SC 29501

MANWILLER, FLOYD GEORGE, b Bailey, Iowa, May 8, 34; m 59; c 2. FOREST PRODUCTS. Educ: Iowa State Univ, BS, 61, PhD(wood tech, plant cytol), 66. Prof Exp: Res assoc wood anat, Iowa State Univ, 64-65; WOOD SCIENTIST, SOUTHERN FOREST EXP STA, US FOREST SERV, 66- Concurrent Pos: Affil mem grad fac, La State Univ, 74- Honors & Awards: Wood Award, Forest Prod Res Soc & Wood & Wood Prod Mag, 66. Mem: Forest Prod Res Soc; Soc Wood Sci & Technol. Res: Wood anatomy; physical, mechanical and chemical properties of wood. Mailing Add: 417 Edgewood Dr Pineville LA 71360

MANYAN, DAVID RICHARD, b Providence, RI, Nov 9, 36; m 65; c 2. BIOCHEMISTRY. Educ: Bowdoin Col, AB, 58; Univ RI, MS, 65, PhD(biochem), 67. Prof Exp: Asst chemist, Metals & Controls Div, Tex Instruments, Inc, 59-60; NIH fel dermat, Sch Med, Univ Miami, 67-69, res scientist, 69-71, instr med, 71-72; dir, Am Heart Asn, 72-73; vis investr, Howard Hughes Med Inst, Univ Miami, 73-75; ASST PROF CHEM, ST FRANCIS COL, 75- Mem: AAAS; Am Chem Soc; fel Am Inst Chem; Am Soc Clin Invest; NY Acad Sci. Res: Enzymology, protein synthesis and drug effects in mitochondria; biochemistry of certain blood dyscrasias. Mailing Add: Ctr for Life Sci St Francis Col Biddeford ME 04005

MANYIK, ROBERT MICHAEL, b San Francisco, Calif, June 11, 28; m 52; c 2. PETROLEUM CHEMISTRY. Educ: Univ Calif, BS, 49; Duke Univ, PhD(org chem), 54. Prof Exp: RES CHEMIST, UNION CARBIDE CORP, 53- Mem: Am Chem Soc. Res: Organo-metallic reagents; olefins polymerization; homogeneous catalysis. Mailing Add: 1146 Summit Dr St Albans WV 25177

MANZ, BRUNO JULIUS, b Dortmund, Ger, June 26, 21; US citizen; m 53; c 4. THEORETICAL PHYSICS, MATHEMATICS. Educ: Univ Mainz, Dipl, 53; Aachen Tech Univ, Dr rer Nat(physics), 57. Prof Exp: Asst, Inst Theoret Physics, Aachen Tech Univ, 53-55; phys adv, Div Gas Turbines, Siemens Schuckert Werke, Ger, 55-57; specialist, Aeroballistics Lab, Army Ballistic Missile Agency, Redstone Arsenal, Ala, 57-59; res physicist, Air Force Missile Develop Ctr, 59-65, chief opers res div, Off Res Anal, Air Force Off Aerospace Res, 65-71, asst for study support, Kirtland AFB, 71-74, CHIEF OPERS RES DIV, DIRECTORATE OF AEROSPACE STUDIES, DEPT OF US AIR FORCE, 74- Concurrent Pos: Adj prof, Univ NMex. Honors & Awards: Award for Meritorious Civil Serv, Dept of Air Force, 72. Res: Thermodynamics of irreversible processes; absorption and dispersion of sound in crystals and fluids; theory of shock waves; relativity; electromatic wave propagation; geometrical probability; military operations. Mailing Add: 1004 Casa Grande Ct Albuquerque NM 87112

MANZELLI, MANLIO ARTHUR, b New Market, NJ, Mar 19, 17; m 47; c 1. ENTOMOLOGY. Educ: Rutgers Univ, BS, 39, MS, 41, PhD, 48. Prof Exp: Res entomologist, Va-Carolina Chem Corp, 47-54, sect leader, 55-57, sr specialist plant chem, Merck & Co, Inc, 57-60; self-employed, 60-64; res scientist & leader tech info facility, 64-67, sr scientist, 67-73, ENTOMOLOGIST, PHILIP MORRIS RES CTR, 73- Mem: AAAS; Entom Soc Am; Am Inst Biol Sci. Res: Biological control of insects; plant growth regulators; medical entomology; insect growth regulators; stored products entomology; information retrieval and storage. Mailing Add: 3412 Lochinvar Dr Bon Air VA 23235

MANZER, FRANKLIN EDWARD, b Maine, Feb 28, 32; m 54; c 4. PLANT PATHOLOGY. Educ: Univ Maine, BS, 55; Iowa State Col, PhD(plant path), 58. Prof Exp: From asst prof to assoc prof, 58-66, PROF PLANT PATH, UNIV MAINE, ORONO, 66- Mem: Am Phytopath Soc; Potato Asn Am. Res: All phases of potato disease; remote sensing. Mailing Add: 307 Deering Hall Univ of Maine Orono ME 04473

MANZER, JAMES IVAN, b Regina, Sask, Nov 6, 21; m 49; c 3. MARINE BIOLOGY, FISHERIES. Educ: Univ Man, BS, 44; Univ BC, MA, 49. Prof Exp:

From sci asst to sr scientist, 45-63, SCIENTIST, PAC BIOL STA, 63- Concurrent Pos: Scientist, Int Fur Seal Invest, Can, 52-54; sci adv to Can, Int NPac Fisheries Comn, 56-65. Res: Marine life, population biology of Pacific salmon; migration and distribution of northern fur seal; production of aquatic resources through fertilization; ecological systems. Mailing Add: Pac Biol Sta Nanaimo BC Can

MANZER, LEO ERNEST, b Timmins, Ont, Apr 4, 47; m 70; c 1. ORGANOMETALLIC CHEMISTRY. Educ: Univ Waterloo, BSc, 70; Univ Western Ont, PhD(chem), 73. Prof Exp: RES CHEMIST, E I DU PONT DE NEMOURS & CO, INC, 73- Res: Organometallic chemistry of the transition metals; homogeneous catalysis; activation of organic molecules by coordination. Mailing Add: Cent Res & Develop Dept Exp Sta E I du Pont de Nemours & Co Wilmington DE 19898

MAO, HO-KWANG, b Shanghai, China, June 18, 41; m 68. GEOCHEMISTRY. Educ: Taiwan Nat Univ, BS, 63; Univ Rochester, MS, 66, PhD(geol sci), 68. Prof Exp: Res assoc geochem, Univ Rochester, 67-68; RES FEL, CARNEGIE INST GEOPHYS LAB, 68- Mem: AAAS; Am Geophys Union. Res: High-pressure geochemistry. Mailing Add: Carnegie Inst Geophys Lab 2801 Upton St NW Washington DC 20008

MAO, JAMES CHIEH HSIA, b China, Apr 3, 28; m 58; c 1. BIOCHEMISTRY. Educ: Taiwan Nat Univ, BS, 52; Univ Wis, MS, 59, PhD(biochem), 63. Prof Exp: SR BIOCHEMIST, ABBOTT LABS, 63- Mem: AAAS; Am Chem Soc. Res: Metabolism of erythromycin; mode of action of antibiotics.

MAPES, WILLIAM HENRY, b Jonesboro, Ark, Dec 30, 39; m 64; c 2. ANALYTICAL CHEMISTRY. Educ: Carnegie Inst Technol, BS, 61; Univ Wash, PhD(chem), 66. Prof Exp: CHEMIST, MAJOR APPLIANCE LABS, GEN ELEC CO, 69- Mem: Am Chem Soc; The Chem Soc. Res: Air quality; solution thermodynamics; absorption refrigeration systems. Mailing Add: Maj Appliance Labs GE Co Appliance Park Louisville KY 40225

MAPLE, CLAIR GEORGE, b Glenwood, Ind, Mar 17, 16; m 42. APPLIED MATHEMATICS. Educ: Earlham Col, AB, 39; Univ Cincinnati, MA, 40; Carnegie Inst Technol, DSc(math), 48. Prof Exp: Instr math, WVa Inst Technol, 40-41; instr, Ohio State Univ, 41-44; res scientist & instr, Carnegie Inst Technol, 46-48; assoc prof, NTex State Col, 48-49; assoc prof, 49-55, PROF MATH, IOWA STATE UNIV, 55-, DIR COMPUT CTR, 63-, ASST PROG DIR, MATH & COMPUT SCI DIV, AMES LAB, ATOMIC ENERGY COMN, 67- Mem: AAAS; Am Math Soc; Math Asn Am. Res: Partial differential equations; computer science. Mailing Add: Computation Ctr Iowa State Univ Ames IA 50010

MAPLE, TELFORD GRANT, b Seattle, Wash, May 17, 19; m 49. PHYSICAL CHEMISTRY. Educ: Univ Wash, BS, 41; Mass Inst Technol, PhD(chem), 49. Prof Exp: Asst chem, Mass Inst Technol, 46-47, mem staff, Transistor & Solid State Group, Lincoln Lab, 52-55; res assoc bact & immunol, Harvard Med Sch, 49-52; crystal engr, Microwave Assoc, Inc, 55-56; chief engr, Thermosen, Inc, 56-59; sr physicist & mgr solid state mat & devices sect, Labs, Columbia Broadcasting Syst, Inc, 59-61; sr scientist, Res Dept, Ampex Corp, 61-62; prin scientist, Raytheon Co, 62-67; STAFF SCIENTIST, LOCKHEED PALO ALTO RES LABS, 67- Concurrent Pos: English transl consult, Gmelin Inst, 57- Mem: Am Chem Soc; Am Phys Soc; Inst Elec & Electronics Engrs; Electrochem Soc. Res: Thin films; integrated circuits; thermodynamic properties of phosphorus and germanium; phosphorus and luminescent films. Mailing Add: 1225-649 Vienna Dr Sunnyvale CA 94086

MAPLE, WILLIAM ROBERT, b Princeton, NJ, Dec 2, 24; m 52; c 3. FORESTRY. Educ: Rutgers Univ, BS, 49; Duke Univ, MF, 51. Prof Exp: Jr forester, US Forest Serv, 51-52, asst dist ranger, 52-53, ranger dist timber mgt asst, 53-55, res forester, 55-75; RETIRED. Mem: Soc Am Foresters. Res: Role of fire in longleaf pine regeneration and stand management practices and its effect on site quality. Mailing Add: 204 Spring Dr Brewton AL 36426

MAPLES, WILLIAM PAUL, b Jefferson City, Mo, Sept 1, 29; m 49; c 2. PARASITOLOGY. Educ: George Peabody Col Teachers, BS, 53, MA, 56; Univ Ga, PhD(parasitol), 66. Prof Exp: Soil engr, State Tenn Hwy Dept, 53-54; teacher, Davidson County Bd Educ, Tenn, 54-56; asst prof physics & zool, WGa Col, 56-59; res asst parasitol, Univ Ga, 59-63, res scholar, Southeastern Coop Wildlife Dis Study, 63-64, res assoc, 64-66, asst dir lab serv, 66-67; assoc prof, 67-70, PROF BIOL, W GA COL, 70- Mem: Am Soc Parasitol; Am Micros Soc. Res: Physiology, life history, embryology and taxonomy of helminths; pathobiology. Mailing Add: 128 Belva St Carrollton GA 30117

MAPLES, WILLIAM ROSS, b Dallas, Tex, Aug 7, 37; m 58; c 2. PHYSICAL ANTHROPOLOGY, PRIMATOLOGY. Educ: Univ Tex, Austin, BA, 59, MA, 62, PhD(anthrop), 67. Prof Exp: Mgr, Darajani Primate Res Sta, Kenya, 62-63; mgr, Southwest Primate Res Ctr, Nairobi, Kenya, 64-65; asst prof anthrop, Western Mich Univ, 66-68; asst prof, 68-72, ASSOC PROF ANTHROP, UNIV FLA, 72-, CHMN DEPT SOC SCI & ASSOC CUR PHYS ANTHROP, FLA STATE MUS, 73- Concurrent Pos: Western Mich Univ Fac Res Fund grant path of the Kenya baboon, 67-68, Univ Fla biomed sci grant, Kenya, 69-70, NSF grant, Kenya, 69-71, Univ Fla biomed sci grant, Kenya, 72-73. Mem: Am Asn Phys Anthrop; fel Am Anthrop Asn. Res: Primate taxonomy and primate behavior, particularly as related to adaptation; forensic identification of human skeletal remains. Mailing Add: Dept of Soc Sci Fla State Mus Gainesville FL 32611

MAPLESDEN, DOUGLAS CECIL, b Kent, Eng, Oct 30, 19; m 40; c 4. VETERINARY MEDICINE. Educ: Ont Vet Col, DVM, 50; Ont Agr Col, MSA, 57; Cornell Univ, PhD(animal nutrit), 59. Prof Exp: Private practice, Ont, Can, 50-51; herd vet, Flat Top Ranch, Tex, 51-53; from assoc prof to prof in charge ambulatory clinic, Ont Vet Col, 53-60; vet dir, Stevenson, Turner & Boyce, 60-63; dir animal health res, Ciba Pharmaceut Co, 62-69; DIR ANIMAL HEALTH RES & DEVELOP, E R SQUIBB & SONS, INC, 69- Res: Nutrition; infectious agents. Mailing Add: Squibb Agr Res Ctr E R Squibb & Sons Inc Three Bridges NJ 08887

MAPLETON, ROBERT ALLAN, b San Francisco, Calif, June 25, 10; m 43; c 2. MATHEMATICAL PHYSICS. Educ: Purdue Univ, BS, 48; Harvard Univ, MA, 49, ME, 55; Queen's Univ, Belfast, PhD, 66. Prof Exp: Res physicist, Lab for Electronics, Inc, Mass, 50-52; res physicist, US Air Force Cambridge Res Labs, 52-74; RETIRED. Mem: Am Phys Soc; Sigma Xi. Res: Scattering of heavy particles by atomic systems; elastic wave propagation in solid media. Mailing Add: RR1 Box 59a Milbridge ME 04658

MAPOTHER, DILLON EDWARD, b Louisville, Ky, Aug 22, 21; m 46; c 3. ACADEMIC ADMINISTRATION, PHYSICS. Educ: Univ Louisville, BS, 43; Carnegie-Mellon Univ, DSc(physics), 49. Prof Exp: Engr, Res Lab, Westinghouse Elec Co, 43-46; from instr to assoc prof, 49-59, PROF PHYSICS, UNIV ILL, URBANA-CHAMPAIGN, 59-, DIR OFF COMPUT SERV, 71- Concurrent Pos: Consult, govt labs, indust & univs, 54- Sloan fel, 57-61; Guggenheim fel, 60-61; vis prof, Cornell Univ, 60-61. Mem: Fel Am Phys Soc; Am Asn Physics Teachers; Asn Comput Mach; Sigma Xi. Res: Experimental physics of solids; low temperature physics; superconductivity; calorimetry; magnetic phase transitions; thermodynamics. Mailing Add: 808 S Foley Ave Champaign IL 61820

MAPP, FREDERICK EVERETT, b Atlanta, Ga, Oct 12, 10; m 63; c 2. ZOOLOGY. Educ: Morehouse Col, BS, 32; Atlanta Univ, MS, 34; Harvard Univ, MA, 42; Univ Chicago, PhD(zool), 50. Prof Exp: Instr high sch, Ga, 33-40; prof biol & head dept, Knoxville Col, 44-46; lectr, Roosevelt Col, 48-50; prof & head dept, Tenn Agr & Indust State Col, 51-52; PROF BIOL, MOREHOUSE COL, 52-, CHMN DEPT, 62- Mem: AAAS; Am Soc Zool; Am Micros Soc; NY Acad Sci. Res: Regeneration; experimental morphology; transplantation; tissue culture. Mailing Add: 703 Waterford Rd NW Atlanta GA 30318

MAQUET, JACQUES, b Brussels, Belg, Aug 4, 19; US citizen; m 46, 70; c 2. CULTURAL ANTHROPOLOGY, SOCIAL ANTHROPOLOGY. Educ: Cath Univ Louvain, LLD, 46, DPhil, 48; Univ London, PhD(anthrop), 52. Prof Exp: Anthropologist, Inst Sci Res Cent Africa, 49-51, head, Res Ctr, Rwanda & Burundi, 52-57; prof anthrop, Univ of the Congo, 57-60; dir studies, Ecole Pratique de Hautes Etudes, Univ Paris, 61-68; prof anthrop, Case Western Reserve Univ, 68-70; PROF ANTHROP, UNIV CALIF, LOS ANGELES, 71- Concurrent Pos: Vis prof, Northwestern Univ, 56; ed, Jeune Afrique, 58-60; Univ Congo to, Univ NMex, 58 & African Studies Ctr of Moscow, Leningrad & New Delhi, 59; vis prof, Univ Brussels, 63-68 & Harvard Univ, 64; UNESCO consult, Repub Congo, Brazzaville, 64-65; vis prof, Univ Montreal, 65; Wenner-Gren Found grant, Univ Paris, 65-68; res dir, Ctr Cult Anthrop, Univ Brussels, 66-68; vis prof, Univ Pittsburgh, 67. Honors & Awards: Emile Waxweiler Award, Royal Acad Belg, 61; African Art Award, First World Festival of Negro Arts, Dakar, Senegal, 66. Mem: Fel Am Anthrop Asn; Int African Inst; Int Inst Differing Civilizations; Int Asn Fr Speaking Sociol; Fr Soc Men of Letters. Res: Cultural analysis; societal analysis; aesthetic anthropology; sociology of knowledge; symbolic communication. Mailing Add: Dept of Anthrop Univ of Calif Los Angeles CA 90024

MARA, RICHARD THOMAS, b New York, NY, Mar 18, 23; m 46; c 1. PHYSICS. Educ: Gettysburg Col, AB, 48; Univ Mich, MS, 50, PhD(physics), 53. Prof Exp: From asst prof to prof, 53-70, SAHM PROF PHYSICS, GETTYSBURG COL, 70-, CHMN DEPT, 58- Mem: Am Phys Soc. Res: Molecular structure; classical field theory. Mailing Add: Rte 6 Box 234A Gettysburg PA 17325

MARABLE, JAMES HOLLEY, b Memphis, Tenn, Dec 10, 25; m 48; c 4. MATHEMATICAL PHYSICS. Educ: Univ Colo, BS, 51; Univ Tenn, PhD, 67. Prof Exp: Physicist, Oak Ridge Nat Lab, 52-57; lectr reactor anal, Oak Ridge Sch Reactor Tech, 57-65; head EMP group, 72-74, PHYSICIST, OAK RIDGE NAT LAB, 65- Mem: AAAS; Am Nuclear Soc; Am Phys Soc. Res: Low energy nuclear and neutron physics; reactor theory; shielding; quantum mechanical three body problem; electromagnetic theory; reactor sensitivity studies; effects of neutrals in plasmas. Mailing Add: Oak Ridge Nat Lab PO Box X Oak Ridge TN 37830

MARABLE, NINA LOUISE, b Wilmington, NC, July 26, 39. FOOD CHEMISTRY. Educ: Agnes Scott Col, BA, 61; Emory Univ, MS, 63; Mt Holyoke Col, PhD(chem), 67. Prof Exp: Instr chem, Mary Baldwin Col, 66-67; asst prof, Sweet Briar Col, 67-69; ASST PROF CHEM, VA POLYTECH INST & STATE UNIV, 69- Mem: Am Chem Soc; Inst Food Technologists. Res: Protein nutritive quality; processing effects on food quality; methods of protein quality evaluation. Mailing Add: Dept of Human Nutrit & Foods Va Polytech Inst & State Univ Blacksburg VA 24061

MARADUDIN, ALEXEI ALEXEI, b San Francisco, Calif, Dec 14, 31; m 54; c 2. THEORETICAL SOILD STATE PHYSICS. Educ: Stanford Univ, BS, 53, MS, 54; Bristol Univ, PhD(physics), 56. Prof Exp: Res assoc physics, Univ Md, 56-57, res asst prof, 57-58, asst res prof, Inst Fluid Dynamics & Appl Math, 58-60; physicist, Westinghouse Res Labs, 60-65; chmn dept, 68-71, PROF PHYSICS, UNIV CALIF, IRVINE, 65- Concurrent Pos: Consult, Semiconductor Br, US Naval Res Lab, 58-60, Los Alamos Sci Lab, 65- & Gen Atomic Div, Gen Dynamics Corp, 65-71. Mem: Am Phys Soc. Res: Lattice dynamics; electronic properties of solids; statistical mechanics. Mailing Add: Dept of Physics Univ of Calif Irvine CA 92664

MARAGOUDAKIS, MICHAEL E, b Myrthios, Greece, Aug 4, 32; m 68; c 1. BIOCHEMICAL PHARMACOLOGY. Educ: Nat Univ Athens, BS, 56; Ore State Univ, MS, 61, PhD(biochem), 64. Prof Exp: Res asst biochem, Ore State Univ, 61-63, instr, 64; res assoc, Albert Einstein Med Ctr, 64-66; Ciba fel, Ciba Pharmaceut Co, 66-67, sr biochemist, 67-69, head biochem pharmacol, 69-72, MGR BASIC BIOCHEM, CIBA-GEIGY CORP, 72- Mem: Am Chem Soc; Am Soc Biol Chem; Am Soc Pharmacol & Exp Therapeut. Res: Intermediary metabolism and enzymology; mode of action of drugs at the molecular level; diabetic microanolopathy; basement membrane biosynthesis. Mailing Add: Ciba-Geigy Corp Ardsley NY 10502

MARAMAN, GRADY VANCIL, b Covington Co, Ala, Feb 19, 36; m 56; c 2. BIOMATHEMATICS, PHYSIOLOGY. Educ: Auburn Univ, BS, 62; Med Col Va, PhD(physiol), 70. Prof Exp: Aerospace technologist, Langley Res Ctr, NASA, 62-73; asst prof biol & math & chmn dept math & eng, Mid Ga Col, 73-76; SANITARY ENGR, US AIR FORCE RESERVES, ROBBINS AFB, 76- Mem: AAAS; Aerospace Med Asn; Am Inst Biol Sci. Res: Behavioral effects of alcohol on performance. Mailing Add: Box 303 Cochran GA 31014

MARAMOROSCH, KARL, b Vienna, Austria, Jan 16, 15; nat US; m 38; c 1. VIROLOGY, ENTOMOLOGY. Educ: Warsaw Tech Univ, MA, 38; Columbia Univ, PhD(bot, plant path), 49. Prof Exp: Lectr biol & animal breeding, Agr Sch Rumania, 45-46; from asst to assoc, Rockefeller Inst, 49-60; entomologist, Boyce Thompson Inst Plant Res, 60-63, prog dir, 63-74; PROF MICROBIOL, RUTGERS UNIV, NEW BRUNSWICK, 74- Concurrent Pos: Vis prof, State Agr Univ, Wageningen, 53; del, Int Cong Microbiol, Italy, 53, Sweden, 58, Can, 62, USSR, 66, Mex, 70; vis prof, Cornell Univ, 57; Lalor sr fel, 57; del, Int Cong Plant Protection, Ger, 57; virologist, Food & Agr Orgn, UN, Philipine Islands, 60, world-wide coconut dis surv, 63; del, Int Cong Entom, Austria, 60, Eng, 64, USSR, 68; coordr, US-Japan Virus-Vector Conf, Japan, 65; coordr, Invertebrate Tissue Culture, Tokyo, 74; mem exec comt, Int Comn Virus Nomenclature, 65; mem, Leopoldina Acad, 71-; food & fiber panel, Nat Acad Sci, 66; consult, US State Dept, Agency Int Develop, India, 67 & Int Rice Res Inst, Philippines, 67; vis prof, Rutgers Univ, 67-68; consult, All India Cent Rice Improv Proj, Hyderabad, & Ford Found, Nigeria, 71; del, II Int Conf Virol, Hungary, 71, Madrid, 75; Fulbright distinguished prof, Yugoslavia, 72; vis prof, Fordham Univ, 73; pres, Int Conf Comp Virol, 69 & 73; 4th Int Conf Invert Tissue Culture, Can, 75. Honors & Awards: Morrison Prize, NY Acad Sci, 51; Campbell Award, AAAS, 58. Mem: AAAS; Harvey Soc; Soc Develop Biol; Am Phytopath Soc; NY Acad Sci (recording secy, 60-62, vpres, 62-63). Res: Plant pathology; insect transmission of viruses and mycoplasma-like agents. Mailing Add: Inst of Microbiol Rutgers Univ New Brunswick NJ 08903

MARAN, STEPHEN PAUL, b Brooklyn, NY, Dec 25, 38; m 71. ASTROPHYSICS. Educ: Brooklyn Col, BS, 59; Univ Mich, MA, 61, PhD(astron), 64. Prof Exp:

Astronomer-in-charge remotely controlled telescope, Kitt Peak Nat Observ, Ariz, 64-69; proj scientist for orbiting solar observs, 69, mgr, Oper Kohovtek, 73-74, HEAD ADVAN SYSTS & GROUND OBSERVATIONS BR, GODDARD SPACE FLIGHT CTR, NASA, 70- Concurrent Pos: Assoc ed, Earth & Extraterrestrial Sci, 70-; ed, Astrophys Letters, 75- Mem: Am Astron Soc; Am Phys Soc; Royal Astron Soc; Int Astron Union; Am Geophys Union. Res: Pulsars; gaseous nebulae; solar research satellites; comets. Mailing Add: NASA Goddard Space Flight Ctr Code 683 Greenbelt MD 20771

MARANDA, ELLI KONGAS, b Tervola, Finland, Jan 11, 32; m 63; c 2. ANTHROPOLOGY. Educ: Univ Helsinki, MA, 54, Mag Phil, 55. Prof Exp: From asst prof to assoc prof, 70-73, PROF ANTHROP, UNIV BC, 73- Concurrent Pos: Fel, Radcliffe Inst, 65-67, scholar, 67-68, fel, 71-; Can Coun leave fel, 75. Mem: Am Folklore Soc; for fel Am Anthrop Asn; Asn Social Anthrop Oceania; Can Soc Sociol & Anthrop; Can Ethnol Soc. Res: Ethnography of women in Melanesia; analysis of oral literature in its living context, including structural and stylistic and riddle analysis; myth in context-return to field, Malaita, Solomon Islands. Mailing Add: Dept of Anthrop & Sociol Univ of BC Vancouver BC Can

MARANDA, PIERRE, b Quebec, Can, Mar 27, 30; m 63; c 2. ANTHROPOLOGY, ETHNOLOGY. Educ: Laval Univ, BA, 49; Univ Montreal, MA, 53; Col Immaculate Conception, LPh, 55; Harvard Univ, MA, 66, PhD(anthrop), 66. Prof Exp: Asst prof classics, Col Garnier, Univ Laval, 55-58; tutor anthrop, Harvard Univ, 64-66; PROF ANTHROP & SOCIOL, UNIV BC, 69- Concurrent Pos: Fourth pilot res grant, Lab Social Rels, Harvard Univ, 64-69, Milton Fund res grant anthrop, 66-67; NIH fel & res grant, 66-68; fel oceanic ethnol, Harvard Univ, 66-69; assoc prof, Ecole Pratique des Hautes Etudes, Univ Paris, 68; for prof, Col of France, 75; vis res prof, Laval Univ, 75-76. Mem: Can Sociol & Anthrop Asn (pres, 72-73); fel Am Anthrop Asn. Res: Computer research in semantic analysis; social structure. Mailing Add: Dept of Anthrop & Sociol Univ of BC Vancouver BC Can

MARANS, NELSON SAMUEL, b Washington, DC, June 5, 24; m 54; c 3. CHEMISTRY. Educ: George Washington Univ, BS, 44; Pa State Univ, MS, 47, PhD, 49. Prof Exp: Asst anal develop, Allegany Ballistics Lab, George Washington Univ, 44-45; fel & lectr, De Paul Univ, 49-50; res assoc chem invest, Allegany Ballistics Lab, Hercules Powder Co, 50-54; group leader, Org Group, Mineral Benefication Lab, Columbia Univ, 54-55; sr scientist, Westinghouse Elec Corp, 55-57; RES SUPVR, RES DIV, W R GRACE & CO, 57- Mem: AAAS; Am Chem Soc. Res: Radiation chemistry; polymers; cyanide chemistry; amino acids; organosilicon chemistry. Mailing Add: 12120 Kerwood Rd Silver Spring MD 20904

MARANTZ, LAURENCE BOYD, b Los Angeles, Calif, July 8, 35; m 62; c 3. ORGANIC CHEMISTRY. Educ: Calif Inst Technol, BS, 57; Univ Calif, Los Angeles, PhD(org chem), 62. Prof Exp: Res chemist, Rocket Power Res Lab, Maremont Corp, 62-64, head org sect, 64-66; mat res & develop engr, Douglas Aircraft Missile & Space Systs Div, McDonnell Douglas Corp, 66-67; spec mem staff, Marquardt Corp, Calif, 67, spec mem advan tech staff, CCI-Marquardt Corp, 67-71, chief chemist, Med Systs Div, Van Nuys, 71-73, dir chem, CCI Life Systs, 73-75, DIR SCI SERV, CCI LIFE SYSTS, CCI CORP, 75- Mem: AAAS; Am Chem Soc. Res: Organo fluorine compounds; phosphorous polymers; high temperature thermodynamics; explosives and rocket propellants; ion exchange systems; polymers; instrumental design and operation; electrochemistry; medical equipment. Mailing Add: 3447 Alana Dr Sherman Oaks CA 91403

MARANVILLE, JERRY WESLEY, b Hutchinson, Kans, Sept 21, 40; m 67; c 1. AGRONOMY. Educ: Colo State Univ, BS, 62, MS, 64; Kans State Univ, PhD(agron), 67. Prof Exp: Asst prof, 67-73, ASSOC PROF AGRON, UNIV NEBR, LINCOLN, 73- Mem: Am Soc Plant Physiol; Am Inst Biol Sci; Crop Sci Soc Am; Am Soc Agron. Res: Crop physiology and protein biochemistry; development of grain sorghum strains which are high in mineral uptake and utilization efficiency. Mailing Add: 102 Kiesselbach Lab Dept Agron Univ of Nebr Lincoln NE 68503

MARANVILLE, LAWRENCE FRANK, b Utica, Kans, Jan 6, 19; m 45; c 3. PULP CHEMISTRY. Educ: Phillips Univ, BA, 40; State Col Wash, MS, 42; Univ Chicago, PhD(phys chem), 49. Prof Exp: Anal chemist, Aluminum Co Am, 42-45, res chemist, 49-51; res chemist, Olympic Res Div, Rayonier Inc, 51-59, group leader, 60-71, group leader, ITT Rayonier Inc, 71-72, ASST TO RES SUPVR, ITT RAYONIER INC, 72- Mem: AAAS; Am Chem Soc; Tech Asn Pulp & Paper Indust. Res: Raman spectroscopy; medium strong electrolytes; electrical conductivity of molten salts; lignin and tannin chemistry; ultraviolet and infrared spectrophotometry; chemical cellulose; pulp chemistry. Mailing Add: 1128 Harvard Ave Shelton WA 98584

MARASCIA, FRANK JOSEPH, b New York, NY, Aug 6, 28; m 51; c 3. ORGANIC CHEMISTRY. Educ: Bucknell Univ, BS, 52; Univ Maine, MS, 54; Univ Del, PhD(chem), 58. Prof Exp: Control chemist, M W Kellogg Co, Pullman, Inc, 52; asst to sales & tech supvr, Southern Dist, Bound Brook Labs, Am Cyanamid Co, 54, res chemist, 54-55; res supvr org chem dept, Jackson Labs, NJ, 55-57, res chemist, 57-63, asst to sales, 64, tech supvr southern dist, 64-65, res supvr, Jackson Labs, 65-66, sales supvr southern district, NC, 66-67, asst mgr chem, Org Chem Div, 67-70, asst dir tech lab, 70-73, TECH MGR INDUSTRIALS, ORG CHEM DEPT, DYES & CHEM TECH DIV, E I DU PONT DE NEMOURS & CO, INC, 73- Mem: AAAS; Am Asn Textile Chem & Colorists; Am Chem Soc; Tech Asn Pulp & Paper Indust. Res: Synthetic organic chemistry; sulfur and nitrogen compounds; textile chemicals in intermediates; paper chemicals. Mailing Add: 504 Windsor Dr Newark DE 19711

MARASPIN, LYNO EVELINO, b Pola, Italy, Mar 4, 38; US citizen. HUMAN ANATOMY, ENDOCRINOLOGY. Educ: ECarolina Univ, AB, 64; Bowman Gray Sch Med, MS, 66; Col Med & Dent, NJ, PhD(anat), 69. Prof Exp: Asst prof anat, Med Col Wis, 69-74; ASST PROF ANAT, COL MED & DENT NJ, 74- Mem: AAAS; Am Asn Anat; Am Soc Zool; Soc Study Reprod. Res: Electron microscopic examination of the cellular elements of the endometrium. Mailing Add: Dept of Anat Col Med & Dent NJ Newark NJ 07103

MARATHAY, ARWIND SHANKAR, b Bombay, India, Dec 11, 33; m 63; c 1. OPTICAL PHYSICS. Educ: Univ Bombay, BSc, 54; Imp Col, Univ London, dipl, 56; Univ London, MSc, 57; Boston Univ, PhD(physics), 63. Prof Exp: Fel physics, Boston Univ, 63-64; sr scientist optical physics, Tech Opers, Mass, 64-69; ASSOC PROF OPTICAL SCI, UNIV ARIZ, 69- Concurrent Pos: Fel mech eng, Univ Pa, 66-67; consult, McCown Labs, Ariz & Tech Opers, Mass, 69- Mem: Optical Soc Am; Am Asn Physics Teachers. Res: Physical optics; coherence theory; partial polarization; quantum coherence theory; electro-optic light modulators and scanners. Mailing Add: Dept of Optical Sci Univ of Ariz Tucson AZ 85721

MARAVETZ, LESTER L, b Cresco, Iowa, May 6, 37; m 62; c 4. AGRICULTURAL CHEMISTRY. Educ: Loras Col, BS, 59; Creighton Univ, MS, 61; Loyola Univ, PhD(org chem), 65. Prof Exp: Res chemist, Esso Res & Eng Co, 65-70; SR RES CHEMIST, AGR CHEM RES, MOBIL CHEM CO, 70- Mem: Am Chem Soc Mailing Add: 843 Carleton Rd Westfield NJ 07090

MARAVOLO, NICHOLAS CHARLES, b Chicago, Ill, Dec 4, 40. PLANT MORPHOGENESIS. Educ: Univ Chicago, BS, 62, MS, 64, PhD(bot), 66. Prof Exp: Asst prof, 66-75, ASSOC PROF BOT, LAWRENCE UNIV, 75- Concurrent Pos: Consult, State of Wis, 72- & East-Cent Wis Regional Planning Comt, 73- Mem: Bot Soc Am; Am Asn Plant Physiol; Am Bryol & Lichenological Soc. Res: Plant growth and development; biochemical changes associated with differentiation in lower green plants; hormonal physiology of development in bryophytes. Mailing Add: Dept of Biol Lawrence Univ Appleton WI 54911

MARBARGER, JOHN PORTER, b Palmyra, Pa, Sept 2, 16; m 43; c 3. PHYSIOLOGY. Educ: Lebanon Valley Col, BS, 38; Johns Hopkins Univ, PhD(zool), 41. Prof Exp: From instr to asst prof physiol, Col Med, Univ Vt, 46-48; assoc prof, 48-55, dir aeromed lab, 58-70, PROF PHYSIOL, UNIV ILL COL MED, 55-, DIR, RES RESOURCES CTR, 70-, ASSOC DEAN, GRAD COL, 72- Concurrent Pos: Ed, J Aerospace Med. Honors & Awards: Tuttle Award, 54. Mem: AAAS; Am Physiol Soc; Soc Exp Biol & Med; Am Inst Aeronaut & Astronaut; fel Aerospace Med Asn. Res: Stress physiology, especially aviation and cardiovascular physiology. Mailing Add: Res Resources Ctr Univ of Ill Col of Med Chicago IL 60612

MARBLE, ALEXANDER, b Troy, Kans, Feb 2, 02; m 30; c 1. MEDICINE. Educ: Univ Kans, AB, 22, AM, 24; Harvard Univ, MD, 27. Prof Exp: Moseley traveling fel, Austria, Ger & Eng, 31-32; from asst clin prof to clin prof, 55-68, EMER CLIN PROF MED, HARVARD MED SCH, 68-; PRES, JOSLIN DIABETES FOUND, INC, 68- Honors & Awards: Banting Medal, Am Diabetes Asn, 59. Mem: AAAS; Am Soc Clin Invest; Soc Exp Biol & Med; Endocrine Soc; Am Diabetes Asn (pres, 58-59). Res: Diabetes mellitus and carbohydrate metabolism. Mailing Add: 131 Laurel Rd Chestnut Hill MA 02167

MARBLE, DUANE F, b Seattle, Wash, Dec 10, 31; m 57; c 2. GEOGRAPHY. Educ: Univ Wash, BA, 53, MA, 56, PhD(geog), 59. Prof Exp: Asst prof real estate, Univ Ore, 59; asst prof regional sci, Univ Pa, 60-63; assoc prof geog, Northwestern Univ, 63-67, dir acad progs, Transp Ctr, 65-67, assoc dir, Ctr, 68-73, prof geog, Northwestern Univ, Evanston, 67-73; PROF GEOG, STATE UNIV NY BUFFALO, 73- Concurrent Pos: Corresp mem, Comn Quant Methods, Int Geog Union, 68- & Comn Geog Data Sensing & Processing, 68-; consult, ed, Geog Anal, 68- & Perspectives in Geog, 70-; mem, Hwy Res Bd, Nat Acad Sci-Nat Res Coun. Mem: AAAS; Regional Sci Asn (secy, 61-64); Asn Comput Mach; Oper Res Soc Am; Urban & Regional Info Systs Asn. Res: Individual travel behavior and the spatial structure of cities; computer applications to geographic research and teaching with special reference to geographic information systems. Mailing Add: Dept of Geog State Univ of NY Buffalo Amherst NY 14226

MARBLE, HOWARD BENNETT, JR, b Shelburne Falls, Mass, June 14, 23; m 48; c 5. DENTISTRY, ORAL SURGERY. Educ: Tufts Univ, DMD, 47. Prof Exp: ASSOC PROF ORAL SURG, SCH DENT, MED COL GA, 69- Concurrent Pos: Chief dent serv, Vet Admin Hosp, Augusta, Ga, 69- Mem: Am Soc Oral Surg; Int Asn Oral Surg; Int Asn Dent Res. Res: Bone graft substitutes; bone healing. Mailing Add: Sch of Dent Med Col of Ga Augusta GA 30902

MARBLE, VERN L, b Tremonton, Utah, Sept 27, 28; m 51; c 5. AGRONOMY, FIELD CROPS. Educ: Utah State Univ, BS, 51, MS, 53; Univ Calif, PhD(plant physiol), 59. Prof Exp: Asst agron, Univ Calif, 54-57; res agronomist soil fertility, Calif Spray-Chem Corp, 57-58; EXTEN AGRONOMIST, UNIV CALIF, DAVIS, 58- Concurrent Pos: Mem Nat Alfalfa Improv Conf; consult, Int Agr Develop & Nat Agr Exten Systs; consult, Food & Agr Orgn, UN & Govt of Argentina, 73-75. Honors & Awards: Seedsman of the Year, Pa Seedsmen Asn, 73; Merit Cert, Am Forage & Grasslands Coun. Mem: Am Soc Agron; Crop Sci Soc Am; Am Forage & Grasslands Coun; Am Soc Plant Physiol. Res: Physiology of crop production; forage crop production; factors affecting alfalfa seed production. Mailing Add: Agron & Range Sci Exten Univ of Calif Davis CA 95616

MARBURG, STEPHEN, b Frankfurt, Ger, July 16, 33; US citizen; m 56; c 2. ORGANIC CHEMISTRY. Educ: City Col New York, BS, 55; Harvard Univ, MA, 57, PhD(org chem), 60. Prof Exp: Chemist, Hoffman-La Roche, Inc, 55; NIH fel biochem, Brandeis Univ, 59-61; NIH fel phys org chem, Mass Inst Technol, 61-62; asst prof chem, Boston Univ, 62-65; sr res chemist, 65-74, RES FEL, MERCK SHARP & DOHME INC, 74- Mem: Am Chem Soc. Res: Synthetic organic and physical organic chemistry; organic fluorine chemistry. Mailing Add: Merck Sharp & Dohme Inc Rahway NJ 07065

MARBURGER, JOHN HARMEN, III, b New York, NY, Feb 8, 41; m 65; c 2. THEORETICAL PHYSICS. Educ: Princeton Univ, BA, 62; Stanford Univ, PhD(appl physics), 66. Prof Exp: Physicist, Goddard Space Flight Ctr, NASA, 62-63; asst prof physics & elec eng, 66-69, assoc prof, 69-72, chmn dept physics, 72-75, PROF PHYSICS & ELEC ENG, UNIV SOUTHERN CALIF, 75- Concurrent Pos: Consult, Lawrence Livermore Labs, 72- Mem: AAAS; Am Phys Soc. Res: Nonlinear optics; self-focusing and stimulated inelastic scattering of light; quantum electronics; interaction of radiation with plasmas; nonlinear field theory. Mailing Add: Dept of Physics Univ of Southern Calif Los Angeles CA 90007

MARBURGER, RICHARD EUGENE, b Detroit, Mich, May 26, 28; m 50; c 2. PHYSICS. Educ: Wayne State Univ, BS, 50, MS, 52, PhD, 62. Prof Exp: Physicist, Res Labs, Gen Motors Corp, 52-53, res physicist, 55-58, sr res physicist, 58-69; mem staff, 69-70, dir, Sch Arts & Sci, 70-72, dean acad affairs, 72-75, PROF PHYSICS, LAWRENCE INST TECHNOL, 75-, VPRES ACAD AFFAIRS, 75- Mem: Am Phys Soc; Am Asn Physics Teachers. Res: Phase transformations in solids; residual stress analysis by x-ray diffraction; metal physics; x-ray diffraction techniques; Mössbauer effect. Mailing Add: Lawrence Inst of Technol 21000 W 10 Mile Rd Southfield MI 48075

MARCALI, KALMAN, chemistry, see 12th edition

MARCANTEL, EMILY LAWS, b Columbus, Ohio, Apr 5, 43; m 67; c 2. GEOLOGY. Educ: Ohio State Univ, BA, 66, PhD(geol), 75; La State Univ, MS, 68. Prof Exp: Geologist, Spec Proj Br, US Geol Surv, 68-70; GEOLOGIST, EXPLOR & PROD RES LAB, GETTY OIL CO, 75- Mem: Geol Soc Am. Res: Carbonate petrology; conodont biostratigraphy and paleoecology; regional geology of the mid-continent Silurian, western United States Permian and Gulf Coast Cretaceous systems. Mailing Add: Explor & Prod Res Lab Getty Oil Co 3903 Stoney Brook Houston TX 77042

MARC-AURELE, JULIEN, b June 20, 29; Can citizen; m 55; c 3. PHYSIOLOGY. Educ: Univ Montreal, AB, 50, MD, 55. Prof Exp: Resident res collab, Brookhaven Nat Lab, 60-62; assoc dir clin res dept, Hotel-Dieu Hosp, 62-66; ASSOC PROF MED, UNIV MONTREAL, 70-; DIR, DIV NEPHROLOGY, DEPT MED, SACRE-COEUR HOSP, 75- Concurrent Pos: Res fel nephrology, Georgetown Univ Hosp, 58-60; Can Heart Found fel, 58-62; George Strong fel, 60; Med Res Coun Can & Que Heart Found grants, 62-; lectr, Univ Montreal, 63-69; hon lectr, McGill Univ, 65- Mem: Am Fedn Clin Res; Am Soc Artificial Internal Organs; Can Med Asn; Can Soc

Clin Invest; Am Soc Nephrology. Res: Nephrology; renal physiology in relation to human and experimental arterial hypertension; mechanisms of action of hormones in the kidney. Mailing Add: Dept of Med Sacre-Coeur Hosp 5400 W Govin Blvd Montreal PQ Can

MARCEAU, GILLES, b Quebec, Que, Aug 30, 19; m 48; c 8. ANATOMY, SURGERY. Educ: Laval Univ, BA, 40, MD, 45; FRCPS(C), 52. Prof Exp: Researcher arterial grafts, Salpetriere Hosp, Paris, 51-53, Fleming Inst, Eng, 53, Banting Inst, Ont, Can, 53 & Naval Med Res Inst, Md, 53; assoc prof, 55-69, PROF ANAT, FAC MED, LAVAL UNIV, 69- Concurrent Pos: Med Res Coun Can grants, 58-65; surgeon, Dept Vet Affairs, Ste-Foy Hosp, Que, 59- Mem: Fel Am Col Surg; Can Asn Anat. Res: Elastase; gross anatomy; arterial grafts; arteriosclerosis. Mailing Add: Dept of Anat Laval Univ Fac of Med Quebec PQ Can

MARCELLI, JOSEPH F, b Schenectady, NY, Nov 22, 26; m 52. ORGANIC CHEMISTRY. Educ: Rensselaer Polytech Inst, BS, 48, MS, 50, PhD, 57. Prof Exp: Asst, Rensselaer Polytech Inst, 49-51, res assoc, 53-56; from instr to assoc prof, 56-59, PROF CHEM, HUDSON VALLEY COMMUNITY COL, 59-, CHMN DEPT, 58-, DEAN HEALTH & PHYS SCI, 71- Concurrent Pos: Actg chmn, Dept Chem, Hudson Valley Community Col, 57-58, dir phys sci div, 67-70, dean sci & actg dean arts, 70-71. Mem: AAAS; fel Am Inst Chem; Am Pub Health Asn; Am Chem Soc; Health Physics Soc. Mailing Add: Hudson Valley Community Col 80 Vandenburgh Ave Troy NY 12180

MARCELLINI, DALE LEROY, b Oakland, Calif, Mar 19, 37; m 58; c 2. HERPETOLOGY, ANIMAL BEHAVIOR. Educ: San Francisco State Univ, BA, 64, MA, 66; Univ Okla, PhD(zool), 70. Prof Exp: Instr biol, Calif State Univ, Hayward, 71-74; RES CUR, NAT ZOOL PARK, SMITHSONIAN INST, 74- Mem: Sigma Xi; Animal Behav Soc; Am Soc Ichthyologists & Herpetologists. Res: Behavior and ecology of Gekkonid lizards, expecially acoustic behavior; breeding and captive maintenance of vertebrates. Mailing Add: Nat Zool Park Off of Animal Mgt Smithsonian Inst Washington DC 20560

MARCH, ANDREW LEE, b New Haven, Conn, Oct 14, 32; m 54; c 3. GEOGRAPHY. Educ: Swarthmore Col, BA, 53; Syracuse Univ, MA, 59; Univ Wash, PhD(geog), 64. Prof Exp: Instr geog, Ohio State Univ, 63-64; asst prof, Columbia Univ, 64-69; asst prof anthrop, Queen's Col, NY, 69-73; VIS ASST PROF GEOG & CHINA HUMANITIES, UNIV DENVER, 73- Mem: Asn Asian Studies. Res: Chinese geographic thought; geographic theories about China; Chinese traditional medicine; biogeography. Mailing Add: Dept of Geog Univ of Denver Denver CO 80210

MARCH, BERYL ELIZABETH, b Port Hammond, BC, Aug 30, 20; m 46; c 1. POULTRY NUTRITION. Educ: Univ BC, BA, 42. Prof Exp: Asst chem, Univ BC, 43; res asst, Can Fishing Co, 44-47; instr, 47-59, res assoc, 59-62, from asst prof to assoc prof, 62-70, PROF POULTRY SCI, UNIV BC, 70- Honors & Awards: Nutrit Res Award, Am Feed Mfg Asn. Mem: AAAS; Poultry Sci Asn; Nutrit Soc Can; Agr Inst Can; Soc Exp Biol & Med. Res: Physiology. Mailing Add: Dept of Poultry Sci Univ of BC Vancouver BC Can

MARCH, JERRY, b Brooklyn, NY, Aug 1, 29; m 54; c 3. ORGANIC CHEMISTRY. Educ: Brooklyn Col, MA, 53; Pa State Univ, PhD(chem), 57. Prof Exp: From asst prof to assoc prof, 56-68, PROF CHEM, ADELPHI UNIV, 68- Concurrent Pos: Vis prof, Univ Strasbourg, 67-68. Mem: Am Chem Soc; The Chem Soc. Res: Organic synthesis; organometallic compounds; aromaticity. Mailing Add: Dept of Chem Adelphi Univ Garden City NY 11530

MARCH, LOUIS CHARBONNIER, b Philadelphia, Pa, Jan 17, 41. ORGANIC CHEMISTRY. Educ: Drexel Univ, BS, 63; Univ Pa, PhD(org chem), 67. Prof Exp: Res fel bio-org chem, Harrison Dept Surg Res, Med Sch, Univ Pa, 67-70 & org chem, Dept Chem, 70-73; SR CHEMIST, CHEM DEVELOP, WYETH LABS, DIV AM HOME PROD CORP, 73- Mem: Am Chem Soc. Res: Synthesis of prostaglandins and their isolation from natural sources. Mailing Add: Wyeth Labs Chem Develop 611 E Nield St West Chester PA 19380

MARCH, RALPH BURTON, b Oshkosh, Wis, Aug 5, 19; m 42; c 3. ENTOMOLOGY. Educ: Univ Ill, AB, 41, MA, 46, PhD(entom, chem), 48. Prof Exp: From jr entomologist to assoc entomologist, Citrus Exp Sta, 48-57, dean grad div, Univ, 61-69, head div toxicol & physiol, Dept Entom, 69- 72, PROF ENTOM, UNIV CALIF, RIVERSIDE, 61-, ENTOMOLOGIST, CITRUS EXP STA, 57- Mem: AAAS; Entom Soc Am; Am Chem Soc. Res: Physiological, biochemical and toxicological studies on the mode of action of insecticides; relation of chemical structure to insecticidal activity; resistance of insects to insecticides. Mailing Add: Div of Toxicol & Physiol Dept of Entom Univ of Calif Riverside CA 92502

MARCH, RAYMOND EVANS, b Newcastle upon Tyne, Eng, Mar 13, 34; m 58; c 3. PHYSICAL CHEMISTRY. Educ: Leeds Univ, BSc, 57; Univ Toronto, PhD(phys chem), 61. Prof Exp: Res chemist, Johnson & Johnson Ltd, Can, 61-62; res assoc & res fel, McGill Univ, 62-65; asst prof, 65-68, ASSOC PROF CHEM, TRENT UNIV, 68- Mem: Chem Inst Can. Res: Gas phase reaction kinetics and chemiluminescence. Mailing Add: Dept of Chem Trent Univ Peterborough ON Can

MARCH, RICHARD PELL, b Medford, Mass, May 1, 22; m 46; c 3. FOOD SCIENCE. Educ: Univ Mass, BS, 44; Cornell Univ, MS, 48. Prof Exp: From instr to assoc prof, 48-65, PROF DAIRY INDUST, CORNELL UNIV, 65- Concurrent Pos: Chmn, Northeastern Dairy Practices Comt, 70- Mem: Am Dairy Sci Asn; Int Asn Milk, Food & Environ Sanit (secy-treas, 70-). Res: Dairy industry extension in the field of milk and milk handling on farms; processing in fluid milk plants. Mailing Add: Dept of Food Sci Cornell Univ Col of Agr Ithaca NY 14853

MARCH, ROBERT HENRY, b Yarmouth, NS, Can, July 30, 37; m 63; c 2. PHYSICS. Educ: Dalhousie Univ, MSc, 60; Oxford Univ, DPhil(physics), 65. Prof Exp: Asst prof, 65-68, ASSOC PROF PHYSICS, DALHOUSIE UNIV, 68- CHMN DEPT, 69- Mem: Can Asn Physicists. Res: Low temperature physics. Mailing Add: Dept of Physics Dalhousie Univ Halifax NS Can

MARCH, ROBERT HERBERT, b Chicago, Ill, Feb 28, 34; m 53; c 1. PHYSICS. Educ: Univ Chicago, AB, 52, SM, 55, PhD(physics), 60. Prof Exp: Lectr physics, Midwest Univ Res Asn, 60-61; from instr to assoc prof, 61-71, PROF PHYSICS, UNIV WIS-MADISON, 71- Concurrent Pos: Vis scientist, Europ Orgn Nuclear Res, 65, 67, Fermi Nat Accelerator Lab, 71- & Stanford Linear Accelerator Ctr, 75- Honors & Awards: Sci Writing Award, Am Inst Physics-US Steel Found, 71 & 75. Mem: Am Phys Soc. Res: Experimental high-energy physics; science writing. Mailing Add: Dept of Physics Univ of Wis Madison WI 53706

MARCHALONIS, JOHN JACOB, b Scranton, Pa, July 22, 40; m 69. BIOCHEMISTRY, IMMUNOLOGY. Educ: Lafayette Col, AB, 62; Rockefeller Univ, PhD(biochem), 67. Prof Exp: Am Cancer Soc fel, Walter & Eliza Hall Inst Med Res, Melbourne, 67-68; asst prof med sci, Brown Univ, 69-70; sr lectr, 70-73, ASSOC PROF MOLECULAR IMMUNOL & HEAD LAB, WALTER & ELIZA HALL INST MED RES, MELBOURNE, 73- Concurrent Pos: Consult, Miriam Hosp, Providence, RI, 69-70; assoc, dept microbiol, Monash Univ Med Sch, Melbourne, 75- Mem: AAAS; Am Asn Immunol; Australian Asn Immunol; Australian Biochem Soc; NY Acad Sci. Res: Molecular and cellular basis of immunological specificity. Mailing Add: Walter & Eliza Hall Inst Med Res PO Box Royal Melbourne Hosp Parkville Australia

MARCHAND, ALAN PHILIP, b Cleveland, Ohio, May 23, 40. PHYSICAL ORGANIC CHEMISTRY. Educ: Case Western Reserve Univ, BS, 61; Univ Chicago, PhD(org chem), 65. Prof Exp: Instr phys chem, Huston-Tillotson Col, 63-65; NIH fel phys org chem, Univ Calif, Berkeley, 65-66; asst prof, 66-70, ASSOC PROF ORG CHEM, UNIV OKLA, 70- Concurrent Pos: Sr res fel, Fulbright-Hays grant, Inst Chem, Univ Liege, Belg, 72-73. Mem: Am Chem Soc; The Chem Soc. Res: Reactions of carbethoxycarbene with carbon-halogen bonds; nuclear magnetic resonance of rigid bicyclic systems; application of molecular orbital theory to problems in organic chemistry; organometallic chemistry; enzyme-substrate interactions. Mailing Add: Dept of Chem Univ of Okla 620 Parrington Oval Norman OK 73069

MARCHAND, E ROGER, b Palo Alto, Calif, June 17, 36; m 59; c 2. ANATOMY. Educ: San Diego State Col, BS, 63; Univ Calif, Los Angeles, PhD(anat), 68. Prof Exp: ASST PROF NEUROSCI & ACAD ADMINR, OFF LEARNING RESOURCES, SCH MED, UNIV CALIF, SAN DIEGO, 68- Mem: AAAS; Soc Neurosci; Am Asn Anatomists. Mailing Add: Dept of Neurosci Univ Calif Sch of Med San Diego La Jolla CA 92038

MARCHAND, ERICH WATKINSON, b Hartford, Conn, July 7, 14; m 41; c 3. OPTICS. Educ: Harvard Univ, AB, 36; Univ Wash, MS, 41; Univ Rochester, PhD(math), 52. Prof Exp: Instr math, Univ Rochester, 43-49; PHYSICIST OPTICS, RES LABS, EASTMAN KODAK CO, 49- Mem: Optical Soc Am; Math Asn Am. Res: Mathematics. Mailing Add: 192 Seville Dr Rochester NY 14617

MARCHAND, JEAN-PAUL, b Murten, Switz, Mar 25, 33. MATHEMATICAL PHYSICS. Educ: Univ Bern, Dipl math, 58; Univ Geneva, DrSc(physics), 63. Prof Exp: Asst physics, Univ Bern, 58-60; asst, Univ Geneva, 60-63, res assoc, 63-67; asst prof physics & math, 67-68, math, 68-69, ASSOC PROF MATH, UNIV DENVER, 69- Concurrent Pos: Lectr, Univ Bern, 65-66. Res: Mathematical foundations of quantum mechanics and statistical mechanics; formal scattering theory. Mailing Add: Dept of Math Univ of Denver Denver CO 80210

MARCHAND, MARGARET O, b Shorncliffe, Man, Can, Oct 17, 25; US citizen; m 57; c 3. MATHEMATICS, STATISTICS. Educ: Univ Man, BA, 45; Univ Minn, MA, 48, PhD(math), 50. Prof Exp: Assoc prof math, Southwest Mo State Col, 50-52; statistician, Man Cancer Res Inst, Can, 52-56; asst prof math, Bemidji State Col, 56-57, 58-59, 65-66; asst prof, Univ Denver, 57-58; instr corresp dept, Univ Minn, 59-66; asst prof math, Lakehead Univ, 66-68; prof, Wis State Univ-Superior, 68-71, ASSOC PROF MATH, ADRIAN COL, 71- Mem: Am Math Soc; Math Asn Am; Sigma Xi. Mailing Add: Dept of Math Adrian Col Adrian MI 49221

MARCHANT, COSMO, b Hull, Eng, Feb 3, 13; m 41; c 5. INDUSTRIAL CHEMISTRY. Educ: Univ Toronto, BA, 35, MA, 36, PhD, 41. Prof Exp: Asst chem, Univ Toronto, 35-38, Banting & Best dept med res, 38-41; lectr chem, Queen's Univ, Can, 41-42; res chemist, Res & Develop Dept, Can Industs, Ltd, 42-44, patent attorney, Legal Dept, 44-48, develop chemist, Chems Dept, 48-52; mgr, Res & Develop Dept, Chem Develops Can, Ltd, 52-58, vpres & asst gen mgr, 58-59; mgr, New Prod Develop Dept, Domar Ltd, 59-61, dir cent develop dept, 61-71, MGR PLANNING, DOMTAR FINE PAPERS LTD, 71- Concurrent Pos: Lectr, Univ Exten, Univ Toronto, 38-41. Mem: Fel Chem Inst Can; Soc Chem Indust; Chem Mkt Res Asn; Patent & Trademark Inst Can; Can Pulp & Paper Asn, Indust Develop Sect (chmn, 75-). Res: Sulphanilamide derivatives; growth factors for yeast; explosives; surface active agents, pulp and paper, construction materials; industrial chemicals. Mailing Add: Domtar Fine Papers Ltd PO Box 7211 Montreal PQ Can

MARCHANT, DOUGLAS J, b Malden, Mass, Dec 31, 25; m 55; c 5. OBSTETRICS & GYNECOLOGY. Educ: Tufts Univ, BS, 47, MD, 51; Am Bd Obstet & Gynec, dipl, cert gynec oncol, 74. Prof Exp: Assoc prof, 65-75, PROF OBSTET & GYNEC, SCH MED, TUFTS UNIV, 75-, DIR GYNECOLOGIC ONCOL, TUFTS NEW ENG MED CTR, 74- Concurrent Pos: Consult, St Margaret's Hosp, Dorchester, Mass, 57- & Boston City Hosp, 58-; sr gynecologist, New Eng Med Ctr Hosp, 67-; consult, Sturdy Mem Hosp, Attleboro, Mass, Choate Hosp, Wouborn & Lemuel Shattuck Hosp, Boston. Mem: AMA; Am Col Obstet & Gynec; Am Col Surg; Soc Gynec Oncol; Am Asn Obstet & Gynec. Res: The transport of urine; ureteral activity; immunology of pregnancy; gynecologic oncology. Mailing Add: Dept of Obstet & Gynec Tufts Univ Boston MA 02111

MARCHESI, VINCENT T, b New York, NY, Sept 4, 35; m 59; c 3. BIOCHEMISTRY, PATHOLOGY. Educ: Yale Univ, BA, 57, MD, 63; Oxford Univ, PhD(exp path), 61. Prof Exp: From intern to resident path, Wash Univ, 63-65; res assoc cell biol, Rockefeller Univ, 65-66; staff assoc, Nat Cancer Inst, 66-68, CHIEF SECT CHEM PATH, NAT INST ARTHRITIS, METAB & DIGESTIVE DIS, 68- Mem: Am Soc Cell Biol; Histochem Soc; NY Acad Sci. Res: Inflammation, blood vessel permeability and the biochemical properties of cell surfaces; physical and chemical properties of cell membranes. Mailing Add: Bldg 4 Nat Inst of Arthritis Metab & Digestive Dis Bethesda MD 20014

MARCHESSAULT, ROBERT HENRI, b Montreal, Que, Sept 16, 28; m 52; c 6. PHYSICAL CHEMISTRY, POLYMER CHEMISTRY. Educ: Univ Montreal, BSc, 50; McGill Univ, PhD(phys chem), 54. Prof Exp: Res chemist, Am Viscose Corp, 56-59, res assoc, 59-61; assoc prof polymer & phys chem, State Univ NY Col Forestry, Syracuse Univ, 61-65, prof, 65-69; PROF CHEM & DIR DEPT, UNIV MONTREAL, 69- Concurrent Pos: Fel, Univ Uppsala, 55; distinguished res fel, State Univ NY & vis prof, Univ Strasbourg, 67-68. Mem: AAAS; Am Chem Soc; Am Phys Soc; Tech Asn Pulp & Paper Indust. Res: Physical chemical studies on natural and synthetic polymers, especially solid state characterization by electromagnetic scattering techniques; synthesis and property studies of graft and block copolymers; kinetics and mechanisms of polymeric reactions. Mailing Add: Dept of Chem Univ of Montreal PO Box 6128 Montreal PQ Can

MARCHESSAULT, VICTOR HENRI, b West Shefford, Que, Nov 16, 29; c 4. PEDIATRICS, MICROBIOLOGY. Educ: Univ Montreal, BA, 50, MD, 55; FRCP(C), 59. Prof Exp: Lectr pediat, McGill Univ, 62-66; prof & head dept, Univ, 66-71, HEAD DIV CLIN SCI, HOSP CTR, UNIV SHERBROOKE, 71- Concurrent Pos: Asst physician, Montreal Children's Hosp, 59-66. Mem: Can Pediat Soc (exec secy, 64-); Am Pediat Soc; Can Soc Clin Invest. Res: Infectious disease in virology associated with measles; mumps; rubella vaccines. Mailing Add: Clin Sci Div Univ of Sherbrooke Hosp Ctr Sherbrooke PQ Can

MARCHETTA, FRANK CARMELO, b Utica, NY, Apr 28, 20; m 49; c 3. MEDICINE. Educ: Univ Buffalo, MD, 44; Am Bd Surg, dipl, 54. Prof Exp: Resident gen surg, Deaconess Hosp, Buffalo, 47-50; resident, 50-51, assoc surgeon, 51-54, ASSOC CHIEF CANCER RES HEAD & NECK SURG, ROSWELL PARK MEM INST, 54-; CLIN ASSOC PROF ORAL PATH, DENT SCH & RES ASSOC PROF SURG, MED SCH, STATE UNIV NY BUFFALO, 59- Mem: Fel Am Col Surg; Soc Head & Neck Surgeons. Res: Cancer. Mailing Add: 192 High Park Blvd Eggertsville NY 14226

MARCHETTE, NYVEN JOHN, b Murphys, Calif, June 26, 28; m 50; c 1. VIROLOGY, RICKETTSIAL DISEASES. Educ: Univ Calif, BA, 50, MA, 53; Univ Utah, PhD, 60. Prof Exp: Bacteriologist, Ecol Res Lab, Univ Utah, 55-61, res microbiologist, 60-61; asst res microbiologist, Hooper Found, Univ Calif, San Francisco, 61-69, from asst res prof rickettsiology to assoc res prof virol, Univ, 63-70, assoc res microbiologist, Med Ctr, 69-70; assoc prof, 70-74, PROF TROP MED, SCH MED, UNIV HAWAII, MANOA, 74- Concurrent Pos: Fel, Inst Ctr Med Res & Training, Med Ctr, Univ Calif, San Francisco, 61-63; fel, Inst Med Res, Kuala Lumpur, 61-69; chief arbovirus res lab, Fac Med, Univ Malaya, 61-69. Mem: AAAS; Am Soc Microbiol; Am Soc Trop Med. Res: Ecology infectious diseases, such as virology, rickettsiology; immunology; immunopathology; pathogenesis of virus infections. Mailing Add: Univ of Hawaii Sch of Med Leahi Hosp Honolulu HI 96816

MARCHETTI, ALFRED PAUL, b Bakersfield, Calif, Feb 16, 40; m 63; c 3. PHYSICAL CHEMISTRY. Educ: Univ Calif, Riverside, BA, 61, PhD(phys chem, spectros), 66; Univ Calif, Berkeley, MS, 63. Prof Exp: NIH fel, Univ Pa, 66-69; sr res chemist, 69-75, RES ASSOC, EASTMAN KODAK CO, 75- Mem: Am Chem Soc; Sigma Xi. Res: Electronic spectroscopy of organic molecules and crystals; Stark and Zeeman effect in molecules; exciton theory and energy transfer processes in organic crystals. Mailing Add: Res Labs Eastman Kodak Co Kodak Park Bldg 81 Rochester NY 14650

MARCHETTI, MARCO ANTHONY, b New York, NY, Feb 15, 36; m 58; c 4. PLANT PATHOLOGY. Educ: Pa State Univ, BS, 57; Iowa State Univ, MS, 59, PhD(plant path), 62. Prof Exp: Res plant pathologist, US Army Biol Ctr, Ft Detrick, 62-71; res plant pathologist, Epiphytology Res Lab, 71-73, RES PLANT PATHOLOGIST, PLANT DIS RES LAB, USDA, 73- Mem: Am Phytopath Soc; Mycol Soc Am; Sigma Xi. Res: Rice diseases. Mailing Add: Plant Dis Res Lab US Dept of Agr PO Box 1209 Frederick MD 21701

MARCHI, RAYMOND PAUL, b Daly City, Calif, Mar 27, 37; m 59; c 3. PHYSICAL CHEMISTRY. Educ: Univ Calif, Berkeley, BS, 58; Purdue Univ, PhD(phys chem), 63. Prof Exp: NIH fel chem, Univ Utah, 62-63; chem physicist, Stanford Res Inst, 63-68; OPERS ANALYST, BDM SCI SUPPORT LAB, BDM SERVS CO, 68- Mem: Am Chem Soc. Res: Atomic physics; collision phenomena; liquid theory. Mailing Add: BDM Sci Support Lab BDM Serv Co PO Box 416 Ft Ord CA 93941

MARCHIN, GEORGE LEONARD, b Kansas City, Kans, July 12, 40. MOLECULAR BIOLOGY. Educ: Rockhurst Col, AB, 62; Univ Kans, PhD(microbiol), 67. Prof Exp: Res assoc microbiol, Purdue Univ, 67-68, NIH fel, 68-70; asst prof, 70-75, ASSOC PROF BIOL, KANS STATE UNIV, 75- Concurrent Pos: Comnr, Adv Lab Comn, Kans State Bd Health, 70-; dir allied health prog, Div Biol, Kans State Univ, 70-; NIH res grant, 72-75, 75-78. Mem: AAAS; Am Soc Microbiol; NY Acad Sci. Res: Enzymology; bacterial physiology. Mailing Add: Div of Biol Kans State Univ Manhattan KS 66502

MARCHINTON, ROBERT LARRY, b New Smyrna Beach, Fla, Mar 3, 39; m 64; c 1. WILDLIFE ECOLOGY, ETHOLOGY. Educ: Univ Fla, BSF, 62, MS, 64; Auburn Univ, PhD(zool), 68. Prof Exp: Mgr, Loxahatchee Refuge, US Fish & Wildlife Serv, Fla, 62; wildlife biologist, Fla Game & Fresh Water Fish Comn, 64; asst prof, 67-73, ASSOC PROF WILDLIFE ECOL, SCH FOREST RESOURCES, UNIV GA, 73- Concurrent Pos: Ga Forest Res Coun & McIntire-Stennis grant, Univ Ga, 68-75, Southeastern Coop Wildlife Dis Study Group Contract, 71-72. Mem: Soc Am Foresters; Am Soc Mammal; Wildlife Soc; Animal Behav Soc; Wildlife Dis Asn. Res: Radiotelemetric studies of the behavioral ecology of large vertebrates. Mailing Add: Sch of Forest Resources Univ of Ga Athens GA 30602

MARCHIORO, THOMAS LOUIS, b Spokane, Wash, Aug 1, 28; c 7. SURGERY. Educ: Gonzaga Univ, BS, 51; St Louis Univ, MD, 55. Prof Exp: Intern, St Mary's Group of Hosps, Mo, 55-56; asst in surg, Sch Med, Univ Colo, 59-60, from instr to assoc prof, 60-67; assoc prof, 67-69, PROF SURG, SCH MED, UNIV WASH, 69- Concurrent Pos: Clin investr, Denver Vet Admin Hosp, Colo, 62-65; consult, Children's Orthop, Vet Admin, USPHS & Harborview Hosps, Seattle, Wash, 67- Mem: Soc Univ Surg; Asn Acad Surg (secy, 67-70, pres, 74); Soc Vasc Surg; Am Heart Asn; Am Soc Transplant Surgeons (pres-elect, 76). Res: Transplantation. Mailing Add: Dept of Surg Univ of Wash Seattle WA 98195

MARCHISOTTO, ROBERT, b New York, NY, Nov 9, 29; m 52; c 3. PHARMACEUTICAL CHEMISTRY. Educ: Long Island Univ, BS, 52; Purdue Univ, MS, 54, PhD, 56. Prof Exp: Group leader, Johnson & Johnson Res Ctr, 56-61; dir int labs, Int Div, Bristol-Myers Co, 61-65; dir sci info, Vick Div Res, Richardson-Merrell Inc, 65-68; dir res & develop, Pharmaco Div, Schering Corp, 68-71; assoc, Res Corp, 71-75; EXEC DIR, PURDUE ASSOCS, 71- Mem: AAAS; Am Chem Soc; Am Pharmaceut Asn; Soc Cosmetic Chem; Licensing Execs Soc. Res: Patent and licensing administration; research and development management; technical communications; food and drug laws; pharmaceutical dosage forms; dermatologicals and cosmetics; emulsions and other colloidal systems; antiseptics; antibiotics and preservatives; natural and synthetic hydrocolloids. Mailing Add: 2 Darby Rd East Brunswick NJ 08816

MARCHOK, ANN CATHERINE, developmental biology, see 12th edition

MARCIAL, VICTOR A, b San Juan, PR, Feb 23, 24; m 57; c 5. RADIOTHERAPY. Educ: Univ PR, BS, 44; Harvard Univ, MD, 49. Prof Exp: PROF RADIATION THER, SCH MED, UNIV PR, SAN JUAN, 58-, ASSOC DIR MED PROGS, PR NUCLEAR CTR, 67- Concurrent Pos: Training fel radiation ther cancer, Nat Cancer Inst, 51-53; Am Cancer Soc fel, Europe, 53-54; dir cancer control, PR Dept Health, 55-65; consult, Vet Admin Hosp, San Juan, 56- & PR Cancer Control Prog, 65-; dir radiother div, PR Nuclear Ctr, 58-67; chmn, Interagency Comt Tobacco & Health, 64- Mem: AAAS; Asn Am Med Cols; Am Radium Soc (treas, 71-73, pres, 74-75); fel Am Col Radiol; Radiol Soc NAm. Res: Radiation therapy and cancer. Mailing Add: PO Box 20581 Rio Piedras PR 00928

MARCIAL-ROJAS, RAUL ARMANDO, b Cayey, PR, July 13, 25; US citizen; m 49; c 4. PATHOLOGY. Educ: Univ PR, BS, 45, MPH, 71; Marquette Univ, MD, 49. Prof Exp: PROF PATH & LEGAL MED & CHMN DEPT, SCH MED & DENT, UNIV PR, SAN JUAN, 60-, DIR DEPT PATH, MED CTR, 66- Concurrent Pos: Dir, Inst Legal Med, Univ & Commonwealth of PR, 60-; consult, Vet Admin Hosp, 60- & Armed Forces Inst Path, 60- Honors & Awards: Quevedo-Baez Award, PR Med Asn, 53. Mem: Am Soc Cytol; Am Acad Forensic Sci; Am Col Physicians; Col Am Path; Am Soc Clin Path. Res: Neoplastic parasitic diseases and forensic pathology. Mailing Add: Dept of Path & Legal Med Univ of PR Sch of Med San Juan PR 00905

MARCIANI, DANTE JUAN, b Lima, Peru, Feb 23, 39; m 69; c 3. BIOCHEMISTRY. Educ: San Marcos Univ, BS, 62, ScD(biol), 63; Univ Colo, Boulder, PhD(biochem), 70. Prof Exp: Asst prof, San Marcos Univ, 62-65; res asst, Univ Colo, Boulder, 65-70; vis fel, 71-73, vis assoc, 73-76, RES CHEMIST, LAB BIOCHEM, NAT CANCER INST, 76- Res: Cell membrane alterations in transformed cells. Mailing Add: 13520 Walnutwood Lane Germantown MD 20767

MARCINIAK, EWA, b Lvov, Poland, Apr 3, 26. HEMATOLOGY. Educ: Wroclaw Univ, MD, 51. Prof Exp: Fac mem exp path, Sch Med, Wroclaw Univ, 52-64; res assoc physiol, Sch Med, Wayne State Univ, 64-69; asst prof med, 69-72, ASSOC PROF MED, MED CTR, UNIV KY, 72- Concurrent Pos: Rockefeller Found fel, 61-62. Mem: Am Physiol Soc; Am Soc Hemat; Int Soc Thrombosis & Haemostasis. Res: Blood coagulation including studies of disorders in rare haemorrhagic and thromboembolic diseases; studies on clotting mechanism; studies on biological activities, biochemical structure and physical properties of prothrombin; Factor X, Antithrombin III. Mailing Add: Dept of Med Univ of Ky Med Ctr Lexington KY 40506

MARCINKOWSKY, ARTHUR ERNEST, b Moosehorn, Man, Nov 8, 31; m 59; c 4. PHYSICAL CHEMISTRY. Educ: Univ Man, BSc, 55, MSc, 58; Rensselaer Polytech Inst, PhD(phys chem), 61. Prof Exp: Sci teacher, Foxwarren Collegiate, Man, 52-53; lectr phys chem, Royal Mil Col, Ont, 57-58; res assoc & fel, Rensselaer Polytech Inst, 58-61; develop chemist, Chem & Plastics Div, Union Carbide Can, Ltd, 61-63; res assoc, Chem Div, Oak Ridge Nat Lab, 64-67; RES SCIENTIST, TECH CTR, UNION CARBIDE CORP, 67- Honors & Awards: Sci Award, Am Chem Soc, 72. Mem: Am Chem Soc; Catalysis Soc; The Chem Soc. Res: Physical chemistry of electrolyte solutions in aqueous, non-aqueous and mixed solvent media; polymerization of olefins via stereospecific catalysis; water purification and desalination; heterogeneous catalysis; chromatography; industrial separations; process development. Mailing Add: Tech Ctr Union Carbide Corp PO Box 8361 South Charleston WV 25303

MARCO, GINO JOSEPH, b Leechburg, Pa, Dec 19, 24; m 51; c 5. BIOCHEMISTRY, ORGANIC CHEMISTRY. Educ: Carnegie Inst Technol, BS, 50; Univ Pittsburgh, MS, 52, PhD(biochem), 56. Prof Exp: Res biochemist agr chem, Monsanto Chem Co, 56-60, proj leader animal nutrit & biochem, 60, group leader chem biol & animal feed res, Agr Chem Div, Res Dept, Monsanto Co, 60-66, group leader biochem of pesticide metab & resudues, 66-69; MGR METAB INVESTS, AGR BIOCHEM DEPT, GEIGY AGR CHEM DIV, CIBA-GEIGY CORP, 69- Mem: Fel AAAS; Am Chem Soc; Sigma Xi; fel Am Inst Chemists. Res: Animal biochemistry; especially in ruminant metabolism, physiology and biochemistry; organic synthesis of agricultural and radioactive chemicals; process development; development of analytical methods and metabolic information in plants, animals, fish and environment systems for use in submitting pesticide petitions. Mailing Add: 1502 Burlwood Dr Greensboro NC 27410

MARCO, PHILIP JOSEPH, b Boston, Mass, Sept 20, 14. PSYCHIATRY. Educ: Boston Col, AB, 38; Middlesex Univ, MD, 43; Univ Mo, MD, 66, MSPH, 71. Prof Exp: Dir out-patient dept, Malcolm Bliss Ment Health Ctr, 59-61; asst prof psychiat, Med Sch, Univ Mo-Columbia, 61, dir out-patient dept, Med Ctr, 61-69, ASSOC PROF PSYCHIAT, MED SCH, UNIV MO-COLUMBIA, 67-; CHIEF PSYCHIATRY, HARRY TRUMAN MEM VET HOSP, 73- Concurrent Pos: Instr psychiat, Wash Univ, 59-60; fel forensic psychiat, Univ Calif, Los Angeles, 69-70; mem staff, Springfield Med Ctr, Calif Dept Corrections, Terminal Island, 69-70; psychiat consult, Fed Bur Prisons, 69-; dir forensic clin, Dept Psychiat, Univ Mo, 70; forensic psychiatrist, Boone County Circuit Court, 70 & US Western Dist, Circuit Court, Kansas City, 74- Mem: AAAS; Am Med Asn; Fel Am Psychiat Asn (pres, 75-76); Am Acad Forensic Sci; Am Soc Clin Hypnosis. Res: Forensic psychiatry; hypnotherapy; psychosomatic medicine; psychopharmacology of depressive states. Mailing Add: Dept of Psychiat Med Ctr Univ of Mo Columbia MO 65201

MARCONI, GARY G, b Columbus, Ohio, Aug 31, 44; m 66; c 1. NATURAL PRODUCTS CHEMISTRY. Educ: Univ Dayton, BS, 66; Case Western Reserve Univ, PhD(chem), 70. Prof Exp: SR BIOCHEMIST, ELI LILLY & CO, 70- Mem: Am Chem Soc. Res: Isolation and purification of natural products produced in fermentations. Mailing Add: 307 McCarty St Indianapolis IN 46206

MARCOTTE, FRANK BASIL, physical chemistry, see 12th edition

MARCOTTE, RONALD EDWARD, b Taunton, Mass, Aug 27, 39; m 64; c 1. PHYSICAL CHEMISTRY. Educ: Univ Fla, BS, 62, PhD(phys chem), 68. Prof Exp: Ohio State Univ Res Found, vis res assoc, Aerospace Res Labs, Wright-Patterson AFB, 68-70; asst prof, 70-74, ASSOC PROF CHEM, TEX A&I UNIV, 74- Mem: Am Chem Soc. Res: Formation and decay of reactive intermediates, especially scavenger studies as well as kinetic studies of ionic and free radical intermediates using fast flow microwave discharge and mass spectrometric techniques. Mailing Add: Dept of Chem Tex A&I Univ Kingsville TX 78363

MARCOU, RENE JOSEPH, b Winslow, Maine, Jan 10, 06; m 40; c 4. MATHEMATICS. Educ: Colby Col, BS, 28; Mass Inst Technol, PhD(math), 45. Prof Exp: Asst physics, Colby Col, 26-27; asst, Mass Inst Technol, 28-30, instr, 30-32; asst prof, Boston Col, 34-44; math physicist, Raytheon Mfg Co, Mass, 44-45; from assoc prof to prof, 46-56, DIR SPACE RES ANAL LAB, BOSTON COL, 56-, RES PROF MATH, 71- Concurrent Pos: Consult, Ultrasonic Corp, 46; upper air res, Air Force Cambridge Res Ctr, 52. Mem: AAAS; Am Math Soc; Math Asn Am; Math Soc France. Res: Rectilinear congruences in Euclidean four space; space research, especially rocket and satellite aspect. Mailing Add: Space Data Analysis Lab Boston Col Chestnut Hill MA 02167

MARCOUX, JULES E, b Charny, Que, Jan 26, 24; m 55; c 6. PHYSICS. Educ: Laval Univ, BA, 47, BASc, 52; Univ Toronto, MA, 54, PhD(physics), 56. Prof Exp: Nat Res Coun Can fel, 56-57; prof physics, Royal Mil Col, Que, 57-62; prof, Laval Univ, 62-64; PROF PHYSICS, ROYAL MIL COL, QUE, 64- Mem: Am Asn Physics Teachers. Res: Physical constants of rare gases in the liquid and solid states. Mailing Add: Dept of Physics Royal Mil Col St Jean PQ Can

MARCOUX, MICHAEL ORAN, geophysics, see 12th edition

MARCOWITZ, STEWART, b Chicago, Ill, Feb 9, 34; m 56; c 2. OSTEOPATHY. Educ: Univ Chicago, AB & SB, 55, MS, 56, PhD(physics), 60; Chicago Col Osteop Med, DO, 75. Prof Exp: Res asst physics, Fermi Inst Nuclear Studies, 56-60; asst physicist, Argonne Nat Lab, 60-67, assoc physicist, 67-75; RESIDENT INTERNAL MED, MOUNT SINAI HOSP, 75- Concurrent Pos: Plan Comnr, Village of Palos Park, 67-71. Mem: Am Phys Soc. Res: Magnet design; beamology; particle accelerator design; radiation monitoring; shielding design; particle detector design; Cerenkov counter design. Mailing Add: 8036 W 126th St Palos Park IL 60464

MARCUM, JAMES BENTON, b Cedar Co, Mo, June 25, 38; m 64; c 3. ANIMAL GENETICS, CYTOGENETICS. Educ: Univ Mo-Columbia, BSAgr, 60, PhD(animal genetics), 69; Cornell Univ, MS, 61; Midwestern Baptist Theol Sem, MDiv, 65. Prof Exp: Lectr animal breeding, Univ Libya, 69-71; ASST PROF ANIMAL GENETICS, UNIV MASS, AMHERST, 71- Mem: Am Soc Animal Sci; Am Genetic Asn; Genetics Soc Can; AAAS; Sigma Xi. Res: Chromosomal identification and abnormalities in domestic animals; relationships between cytogenetics and reproductive biology; freemartin syndrome. Mailing Add: Dept of Vet & Animal Sci Univ of Mass Amherst MA 01002

MARCUS, AARON JACOB, b Brooklyn, NY, Nov 6, 25; m 55; c 3. INTERNAL MEDICINE, HEMATOLOGY. Educ: Univ Va, BA, 48; New York Med Col, MD, 53. Prof Exp: CHIEF HEMAT SECT, NEW YORK VET ADMIN HOSP, 58-; PROF MED, MED COL, CORNELL UNIV, 74- Concurrent Pos: NIH res fel, Montefiore Hosp, 56-58; attend physician, New York Hosp, 74- Mem: Am Soc Clin Invest; Asn Am Physicians; Am Physiol Soc; Am Soc Hemat. Res: Hemostasis, coagulation and thrombosis; biochemistry and physiology of blood platelets. Mailing Add: New York Vet Admin Hosp 408 First Ave New York NY 10010

MARCUS, ABRAHAM, b New York, NY, Oct 26, 30; m 55; c 4. BIOCHEMISTRY. Educ: Yeshiva Univ, BA, 50; Univ Buffalo, AM, 54, PhD, 56. Prof Exp: Asst, Univ Buffalo, 52-54; biochemist, Agr Mkt Serv, Plant Indust Sta, USDA, 58-67; mem staff biol div, 67, assoc mem, 67-71, SR MEM, INST CANCER RES, 71- Concurrent Pos: USPHS res fel, Univ Chicago, 56-58; mem staff biophys, Weizmann Inst Sci, 64-65; vis prof, Bar-Ilan Univ, Israel, 70-71. Mem: Am Chem Soc; Am Soc Biol Chem; Am Soc Plant Physiol. Res: Metabolic pathways as ascertained by enzymatic studies; metabolic control of growth and development. Mailing Add: Inst for Cancer Res 7701 Burholme Ave Philadelphia PA 19111

MARCUS, ALLAN H, b New York, NY, July 14, 39. STATISTICS, ENVIRONMENTAL SCIENCES. Educ: Case Western Reserve Univ, BS, 61; Univ Calif, Berkeley, MA, 63, PhD(statist), 65. Prof Exp: Asst prof math, Case Western Reserve Univ, 64-67; mem staff, Bellcomm, Inc, DC, 67-68; assoc prof statist & earth & planetary sci, Johns Hopkins Univ, 68-73; ASSOC PROF MATH, UNIV MD, BALTIMORE COUNTY, 73- Concurrent Pos: Consult, Rand Corp, 63-65; fel, Statist Lab, Cambridge Univ, 65-66; exec secy, Power Plant Siting Adv Comt, State of Md, 73- Mem: Am Statist Asn. Res: Applied statistics; urban transportation and environmental sciences; environmental health; biomathematics; air pollution; traffic noise. Mailing Add: Dept of Math Univ of Md Baltimore Co 5401 Wilkens Ave Baltimore MD 21228

MARCUS, ANTHONY MARTIN, b London, Eng, June 21, 29; Can citizen; div; c 1. PSYCHIATRY. Educ: Cambridge Univ, BA, 52, MA, 56; Univ London, LMS, 56; McGill Univ, dipl psychiat, 62; Royal Col Physicians & Surgeons Can, spec cert, 62; Am Bd Psychiat & Neurol, dipl, 63. Prof Exp: From instr to asst prof, 62-70, actg head dept, 70-72, ASSOC PROF PSYCHIAT, UNIV BC, 70-, DIR DIV FORENSIC PSYCHIAT, 67- Concurrent Pos: Res fel psychiat, Montreal Gen Hosp, Que, 61-62; Can Penitentiary Serv res grant study dangerous sexual offenders; assoc attend staff, Vancouver Gen Hosp. Mem: Am Psychiat Asn; Nat Coun Crime & Delinq; Can Psychiat Asn; Brit Med Asn; fel Royal Soc Med. Res: Forensic psychiatry; clinical psychiatry and psychopathology; teaching. Mailing Add: Dept of Psychiat Univ of BC Vancouver BC Can

MARCUS, ARNOLD DAVID, b Brooklyn, NY, Feb 25, 28; m 56; c 3. PHARMACY. Educ: Brooklyn Col Pharm, BS, 46; Purdue Univ, MS, 49; Univ Wis, PhD(pharm), 54. Prof Exp: From asst prof to assoc prof pharmaceut chem, 49-54; asst prof pharm, Rutgers Univ, 54-57; res assoc, Merck Sharp & Dohme Res Labs, Pa, 57-64, mgr pharmaceut res, 64-66; asst dir res, 66-67, dir res coord, 67-69, ASSOC DIR RES & DEVELOP, BRISTOL MYERS CO, 69- Mem: AAAS; Am Chem Soc; Am Pharmaceut Asn. Res: Kinetics; biopharmaceutics; pharmaceutical research and development. Mailing Add: Bristol Myers Co Hillside NJ 07205

MARCUS, BERNARD P, physics, see 12th edition

MARCUS, CAROL JOYCE, b New York, NY, Aug 13, 43; c 1. BIOCHEMISTRY. Educ: Cornell Univ, BS, 65; Duke Univ, PhD(biochem), 72. Prof Exp: Instr, 72-73, ASST PROF BIOCHEM, UNIV TENN CTR HEALTH SCI, MEMPHIS, 73- Mem: NY Acad Sci. Res: Enzyme kinetics; metabolic regulation of carbohydrate metabolism; molecular basis of beryllium toxicity. Mailing Add: Dept of Biochem Univ Tenn Ctr for the Health Sci Memphis TN 38163

MARCUS, CAROL SILBER, b New York, NY, July 2, 39; m 58; c 2. RADIATION BIOLOGY, BIOPHYSICS. Educ: Cornell Univ, BS, 60, MS, 61, PhD(biochem, genetics), 63. Prof Exp: Vis res scientist, Lab Biol Med, Netherlands, 63-64; asst res chemist, Lab Nuclear Med & Radiation Biol, Univ Calif, Los Angeles, 65-67; ASST PROF RADIATION BIOL, RADIOPHARM PROG SCH PHARM, UNIV SOUTHERN CALIF, 69- Concurrent Pos: Instr, Santa Monica City Col Exten, 66-69; Pfeiffer Found fel, Sch Pharm, Univ Southern Calif, 69-70; consult, Gen Elec Co, 70-72, Innotek, Inc, 70-73 & XMI Assocs, Inc, 71-73; adv & consult, Radiobiol for Nuclear Med Technol Training Prog, Los Angeles City Col, 71-73; consult radiopharmaceut, Food & Drug Admin, 71-76. Mem: AAAS; Soc Nuclear Med; Biophys Soc. Res: Nuclear dentistry as applied to periodontal disease; use of semiconductor microprobe radiation detectors and short lived radionuclides for biomedical applications. Mailing Add: Radiopharm Prog Sch of Pharm Univ of Southern Calif Los Angeles CA 90033

MARCUS, DAVID, b New York, NY, Feb 17, 32; m 60. PHARMACY. Educ: Columbia Univ, BS, 53, MS, 55; Univ Fla, PhD(pharm), 59. Prof Exp: Asst pharm, Univ Fla, 57-58; sr res pharmacist, 59-64, labeling mgr, 64-74, REGULATORY PROJ DIR, DRUG REGULATORY AFFAIRS, E R SQUIBB & SONS, INC, 74- Concurrent Pos: Fel, Am Found Pharmaceut Educ. Mem: Am Pharmaceut Asn; NY Acad Sci. Res: Effects of medicinal agents on the blood; theology and suspension; food, drug and cosmetic law. Mailing Add: 153 Dunhams Corner East Brunswick NY 08816

MARCUS, DONALD M, b New York, NY, Dec 10, 30; m 58; c 3. INTERNAL MEDICINE, IMMUNOCHEMISTRY. Educ: Princeton Univ, BA, 51; Columbia Univ, MD, 55. Prof Exp: Intern internal med, Presby Hosp, New York, 55-56, asst resident, 56-57; assoc resident, Strong Mem Hosp, Rochester, 59-60; assoc med, 63-64, asst prof, 64-70, assoc prof med & microbiol, 70-75, PROF MED, MICROBIOL & IMMUNOL, ALBERT EINSTEIN COL MED, 75-, DIR, DIV RHEUMATOLOGY & IMMUNOL, 73- Concurrent Pos: Helen Hay Whitney Found fel, 60-63; career scientist, Health Res Coun, New York, 63- Mem: Am Asn Immunol; Am Chem Soc; Am Rheumatism Asn; Am Soc Clin Invest. Res: Immunochemistry of human blood group antigens; blood group and cell membrane antigens; glycosphingolipids; tumor antigens; mechanism of hapten-antibody interactions. Mailing Add: Albert Einstein Col of Med 1300 Morris Park Ave Bronx NY 10461

MARCUS, ERICH, b Halle, Ger, June 1, 27; nat US; m 51; c 2. SYNTHETIC ORGANIC CHEMISTRY, TEXTILE CHEMISTRY. Educ: Univ Minn, PhD(org chem), 56. Prof Exp: Res chemist, 56-63, res scientist, 63-68, SR RES SCIENTIST, UNION CARBIDE CORP, 68- Mem: Am Chem Soc. Res: Synthetic organic chemistry in area of textile chemistry. Mailing Add: 1505 Knob Rd Charleston WV 25314

MARCUS, FRANK I, b Haverstraw, NY, Mar 23, 28; m 57; c 3. CARDIOLOGY, INTERNAL MEDICINE. Educ: Columbia Univ, BA, 48; Tufts Univ, MS, 51; Boston Univ, MD, 53. Prof Exp: Intern med, Peter Bent Brigham Hosp, Boston, 53-54, asst resident, 56-57; clin fel, Georgetown Univ Hosp, 58-59, chief med resident, 59-60; from instr to assoc prof med, Georgetown Univ, 60-68; PROF MED & CHIEF CARDIOL, ARIZ MED CTR, UNIV ARIZ, 69- Concurrent Pos: Mass Heart Asn res fel cardiol, Peter Bent Brigham Hosp, 57-58; Markle scholar, Georgetown Univ, 60-65, NIH career develop award, 65-68; chief cardiol, Georgetown Univ Med Serv Div, DC Gen Hosp, 60-68; fel coun clin cardiol, Am Heart Asn, 65-; consult, Vet Admin & Davis-Monthan AFB Hosps, Tucson, 69- Mem: Am Fedn Clin Res; Am Heart Asn; fel Am Col Physicians; Asn Univ Cardiol; Am Soc Pharmacol & Exp Therapeut. Res: Cardiovascular research; pharmacology; digitalis; metabolism. Mailing Add: Ariz Med Ctr Cardiol Sect Univ of Ariz Tucson AZ 85724

MARCUS, GEORGE JACOB, b Toronto, Ont, Mar 17, 33; m 56; c 2. REPRODUCTIVE BIOLOGY. Educ: Univ Toronto, BA, 56, PhD(biochem), 61. Prof Exp: Res assoc biodynamics, Weizmann Inst, 63-68; asst prof pop dynamics, Sch Hyg, Johns Hopkins Univ, 68-72; asst prof reproductive biol, Dept Anat & Lab Human Reproduction & Reproductive Biol, Harvard Med Sch, 72-74; RES SCIENTIST, ANIMAL RES INST, RES BR, CAN DEPT AGR, 75- Concurrent Pos: Pop Coun med fel, Weizmann Inst, 61-63; asst ed, Biol of Reproduction, 70-74. Res: Biochemical aspects of nidation; decidual induction; preimplantation embryonic development. Mailing Add: Animal Res Inst Agr Can Ottawa ON Can

MARCUS, JOSEPH, b Cleveland, Ohio, Feb 27, 28; c 2. CHILD PSYCHIATRY. Educ: Hadassah Med Sch, Hebrew Univ, MD, 58; Western Reserve Univ, BSc, 63. Prof Exp: Resident psychiat, Ministry of Health, Israel, 58-61; actg head child psychiat, Ness Ziona Rehab Ctr; sr psychiatrist, Lasker Dept Child Psychiat, Hadassah Hosp, 62-64, consult, Tel Hashomer Govt Hosp, 65-66; res assoc, Israel Inst Appl Social Res, 66-69; assoc dir, Jerusalem Infant & Child Develop Ctr, 69-70; head dept child psychiat, Eytanim Hosp, 70-72; dir child psychiat & develop, Jerusalem Ment Health Ctr, 72-75; PROF PSYCHIAT, UNIV CHICAGO & DIR UNIT RES CHILD PSYCHIAT & DEVELOP, 75- Mem: Am Acad Child Psychiat; Soc Res Child Develop. Res: Development of infants of parents with serious mental diseases, especially behavioral, neurological, physiological and biochemical aspects. Mailing Add: Dept of Psychiat Box 151 950 E 59th St Chicago IL 60637

MARCUS, JOYCE, b Los Angeles, Calif, Sept 20, 48; m 73. ANTHROPOLOGY. Educ: Univ Calif, Berkeley, BA, 69; Harvard Univ, MA, 71, PhD(anthrop), 74. Prof Exp: Lectr anthrop, 73-74, RES SCIENTIST, MUS ANTHROP, UNIV MICH, ANN ARBOR, 73-, ASST PROF ANTHROP, 74- Concurrent Pos: Dir grants, Nat Endowment Humanities, 75-77. Mem: AAAS; Am Anthrop Asn; Soc Am Archaeol. Res: Origins and evolution of meso-american writing systems; political organization of chiefdoms and states in Mexico, Guatemala and south to Panama. Mailing Add: Mus of Anthrop Univ Mus Bldg Univ of Mich Ann Arbor MI 48104

MARCUS, JULES ALEXANDER, b Coytesville, NJ, May 10, 19; m 42; c 4. PHYSICS. Educ: Yale Univ, BS, 40, MS, 44, PhD(physics), 47. Prof Exp: Instr physics, Yale Univ, 42-44; res physicist, Appl Physics Lab, Johns Hopkins Univ, 44-46; asst physics, Yale Univ, 46-47; fel, Inst Study Metals, Univ Chicago, 47-49; from asst prof to assoc prof, 49-61, PROF PHYSICS, NORTHWESTERN UNIV, 61- Mem: Fel Am Phys Soc; Am Asn Physics Teachers. Res: Low temperature solid state physics; de Haas-van Alphen effect; experimental determination of Fermi surfaces; Overhauser spin-density-waves in chromium; galvanomagnetic effects in metals. Mailing Add: Dept of Physics Northwestern Univ Evanston IL 60201

MARCUS, LEON, b New York, NY, May 1, 30; m 63; c 2. MICROBIAL GENETICS, MOLECULAR BIOLOGY. Educ: Univ Calif, Los Angeles, AB, 51; Univ Calif, Davis, PhD(microbiol), 61; Med Ctr, Loyola Univ, Ill, MD, 75. Prof Exp: Trainee microbiol, Univ Wis, 61-64; assoc res scientist, Kaiser Found Res Inst, 64-65; from asst prof to assoc prof microbiol, Stritch Sch Med, Loyola Univ, Ill, 65-75; MEM STAFF, DEPT PEDIAT, CEDARS OF LEBANON HOSP, LOS ANGELES, CALIF, 75- Concurrent Pos: Am Soc Microbiol pres fel, 63; fel microbiol chem, Hadassah Med Ctr, Hebrew Univ Jerusalem, 71-72. Mem: AAAS; Am Soc Microbiol; Brit Soc Gen Microbiol. Res: Microbial physiology; kinetics of induction of enzymes; properties of enzymes; analysis of ribonucleic acid in yeast; characterization of polysomes in yeast and azotobacter; structure-function of nitrogen fixation in azotobacter. Mailing Add: Dept of Pediat Cedars of Lebanon Hosp PO Box 54265 Los Angeles CA 90054

MARCUS, LESLIE F, b Los Angeles, Calif, Oct 22, 30; m 58; c 1. BIOMETRY, PALEONTOLOGY. Educ: Univ Calif, Berkeley, BA, 51, MA, 59, PhD(paleont), 62. Prof Exp: From asst prof to assoc prof statist, Kans State Univ, 60-67; assoc prof, 67-70, PROF BIOL, QUEENS COL, NY, 70- Concurrent Pos: Vis asst prof, Univ Kans, 63-64; NSF sci fac fel, Columbia Univ, 66-67. Mem: AAAS; Soc Study Evolution; Soc Syst Zool; Paleont Soc; Biomet Soc. Res: Vertebrate paleontology; statistical methods application to study of natural selection in fossils; geographic variation; morphology; numerical classification; multivariate statistics. Mailing Add: Dept of Biol Queens Col Flushing NY 11367

MARCUS, MARVIN, b Albuquerque, NMex, July 31, 27; m 65; c 2. ALGEBRA. Educ: Univ Calif, Berkeley, BA, 50, PhD(math), 54. Prof Exp: Res assoc, Univ Calif, Berkeley, 53-54; assoc prof math, Univ BC, 54-60, 61-62; res mathematician, Numerical Anal Sect, Nat Bur Standards, 60-61; chmn dept, 63-68, PROF MATH, UNIV CALIF, SANTA BARBARA, 62- Concurrent Pos: Fulbright grant, 54; consult, US Naval Test Sta, 55; Nat Res Coun fel, 56-57; NSF grant, 58-59, 63-66; vis distinguished prof, Univ Islamabad, W Pakistan, 70; dir, Inst Interdisciplinary Applns of Algebra and Combinatorics, 73- Mem: Am Math Soc; Math Asn Am; Soc Indust & Appl Math; Soc Tech Commun. Res: Linear and multilinear algebra. Mailing Add: Dept of Math Univ of Calif Santa Barbara CA 93106

MARCUS, MELVIN GERALD, b Seattle, Wash, Apr 13, 29; m 53; c 4. PHYSICAL GEOGRAPHY. Educ: Univ Miami, BA, 56; Univ Colo, MS, 57; Univ Chicago, PhD(geog), 63. Prof Exp: Res asst climat, Lab Climat, 58-59; from instr to asst prof geog, Rutgers Univ, 60-64; from asst prof to prof, Univ Mich, Ann Arbor, 64-73, chmn dept, 67-71; PROF GEOG, ARIZ STATE UNIV, 74-, DIR, CTR ENVIRON STUDIES, 74- Concurrent Pos: Res asst, Am Geog Soc, 57-58; sr scientist, Icefield

Ranges Res Proj, Arctic Inst NAm, 64-71; vis lectr, Univ Colo, 67; consult, High Sch Geog Proj, Boulder, 67; chmn, Comn Col Geog, 68-71; vis prof, Univ Canterbury, 72. Mem: Asn Am Geog; Am Geog Soc; Glaciol Soc; Arctic Inst NAm. Res: Glaciological and climatological work, particularly in Alpine regions; physical geography to include geographic education and urban environments; environmental education. Mailing Add: Ctr Environ Studies Ariz State Univ Tempe AZ 85281

MARCUS, MELVIN L, b Milwaukee, Wis, July 22, 40; m 62; c 4. CARDIOLOGY. Educ: Univ Wis, Milwaukee, BS, 62, MD, 66. Prof Exp: From intern to resident internal med, Bronx Munic Hosp Ctr, Albert Einstein Col Med, 66-69; fel cardiol, NIH, 69-71; cardiologist, US Army Med Corps, Walson Army Hosp, 71-73; ASST PROF MED, UNIV IOWA, 73- Concurrent Pos: Fel coun circulation, Am Heart Asn. Honors & Awards: Irving S Wright Award, Stroke Coun, Am Heart Asn, 75. Mem: Fel Am Col Cardiol; fel Am Heart Asn; Am Fedn Clin Res. Res: Regional coronary blood flow; segmental ventricular function in the presence of ischemia; neural control of regional cerebral blood flow; progression and regression of ventricular hypertrophy. Mailing Add: Cardiovasc Div Dept of Internal Med Univ of Iowa Hosp Iowa City IA 52242

MARCUS, MICHAEL BARRY, b Brooklyn, NY, Mar 5, 36; m 64; c 3. MATHEMATICS. Educ: Princeton Univ, BSE, 57; Mass Inst Technol, MS, 58, PhD(math), 65. Prof Exp: Staff mem math & electronics, Rand Corp, 58-67; asst prof, 67-73, ASSOC PROF MATH, NORTHWESTERN UNIV, 73- Concurrent Pos: Asst, Mass Inst Technol, 62-65; NSF vis prof, Westfield Col, Univ London, 70-71; vis mem, Courant Inst, NY Univ. Mem: Am Math Soc. Res: Probability theory; analysis. Mailing Add: Dept of Math Northwestern Univ Evanston IL 60201

MARCUS, PAUL MALCOLM, b New York, NY, Feb 4, 21. MATHEMATICAL PHYSICS. Educ: Columbia Univ, BA, 40; Harvard Univ, MA, 42, PhD(chem physics), 43. Prof Exp: Mem staff, Radiation Lab, Mass Inst Technol, 43-46; fel, Nat Res Coun, 46-47, res assoc physics, 47-48; sci liaison officer, Off Naval Res, US Govt, Eng, 49-50; res asst prof physics, Univ Ill, 50-52; lectr, Carnegie Inst Technol, 52-53, res physicist, 53-58, asst prof physics, 58-59; PHYSICIST, T J WATSON RES CTR, IBM CORP, 59- Mem: Fel Am Phys Soc. Res: Low temperature and solid state physics; radiation theory. Mailing Add: IBM Res Ctr Yorktown Heights NY 10598

MARCUS, PHILIP IRVING, b Springfield, Mass, June 3, 27; m 54; c 3. VIROLOGY. Educ: Univ Southern Calif, BS, 50; Univ Chicago, MS, 53; Univ Colo, PhD(microbiol, biphys), 57. Prof Exp: Lab asst infrared studies bacteria, Univ Chicago, 51-52, lab asst med & gen microbiol & microbiologist, 52-53, asst steroid enzyme induction, 53-54; asst biophys, Med Ctr, Univ Colo, 54-57, instr, 57-59, asst prof, 59-60; asst prof microbiol & immunol, Albert Einstein Col Med, 60-62, assoc prof, 62-66, prof, 66-69; head microbiol sect, 69-74, PROF BIOL, MICROBIOL SECT, UNIV CONN, 69- Concurrent Pos: USPHS sr res fel, 60-65, res career develop awardee, 65-69; on leave from Albert Einstein Col Med to Salk Inst, 67-68; mem sci bd, Damon Runyon Mem Fund Cancer Res, 69-73; ed, J Cellular Physiol, 69- Mem: AAAS; Biophys Soc; Am Soc Microbiol; Am Soc Cell Biol; NY Acad Sci. Res: Single-cell cloning techniques for mammalian cells; host-cell animal virus interactions; mechanism of cell-killing by viruses; viral inhibition; cell surfaces; viral hemadsorption; viral interference; interferon action. Mailing Add: Microbiol Sect U-44 Univ of Conn Storrs CT 06268

MARCUS, PHILIP SELMAR, b New York, NY, Jan 30, 36; m 66; c 1. MATHEMATICS. Educ: Univ Chicago, AB, 56, BS, 58, MS, 59; Ill Inst Technol, PhD(math), 68. Prof Exp: Instr math, De Paul Univ, 62-66; mem fac, Shimer Col, 66-70, dir Shimer-in-Oxford prog, 68-69, chmn dept nat sci, 67-70; ASST PROF MATH, IND UNIV, SOUTH BEND, 70- Concurrent Pos: Mgr & vpres, Midwest Chamber Orchestra, 75- Mem: AAAS; Math Asn Am; Am Math Soc. Res: Probability; geometry; mathematics education. Mailing Add: Dept of Math Ind Univ South Bend IN 46615

MARCUS, ROBERT BORIS, b Chicago, Ill, Nov 26, 34; m 57; c 2. PHYSICAL CHEMISTRY. Educ: Univ Chicago, BS, 56, SM, 58; Univ Mich, PhD(phys chem), 62. Prof Exp: Fel phys chem, Univ Mich, 61-62, instr, 62-63, mem tech staff, 63-67, SUPVR STRUCT ANAL GROUP, BELL LABS, 67- Mem: Electron Micros Soc; Am Vacuum Soc. Res: Microstructure and electrical properties of materials; microelectronics. Mailing Add: 2C-174 Bell Labs Murray Hill NJ 07974

MARCUS, ROBERT BROWN, b Phila, Pa, Dec 1, 18; m 42; c 2. PHYSICAL GEOGRAPHY. Educ: Pa State Teachers Col, W Chester, BS, 40; Univ Fla, MA, 53, EdD(geog), 56. Prof Exp: Teacher high sch, NC, 41-42; head sci dept & master chem & physics, Pennington Sch, NJ, 46-51; from instr to assoc prof, 54-68, PROF PHYS SCI & GEOG, UNIV FLA, 68- Mem: Asn Am Geog; Nat Coun Geog Educ; Int Geog Union. Res: Utilization of water and natural resources. Mailing Add: 102 Bryan Hall Univ of Fla Gainesville FL 32611

MARCUS, ROBERT TOBY, b Brookline, Mass, Dec 18, 46; m 74; c 1. COLOR SCIENCE. Educ: Rensselaer Polytech Inst, BS, 68, PhD(chem), 74. Prof Exp: Asst systs eng, Int Bus Mach Corp, Providence, RI, 68-69; res asst color sci, Rensselaer Polytech Inst, 69-74; SR RES PHYSICIST, COATINGS & RESINS DIV, PPG INDUSTS, INC, 74- Mem: Inter-Soc Color Coun; Fedn Socs Coatings Technol; Optical Soc Am. Res: Computer color control systems; instrumental color difference evaluation; standardization and comparison of color-measuring instrumentation. Mailing Add: PPG Industs Res & Develop Ctr 151 Colfax St Springdale PA 15144

MARCUS, RUDOLPH ARTHUR, b Montreal, Que, July 21, 23; nat US; m 49; c 3. PHYSICAL CHEMISTRY. Educ: McGill Univ, BSc, 43, PhD(phys chem), 46. Prof Exp: Jr res officer photochem, Nat Res Coun Can, 46-49; res assoc theoret chem, Univ NC, 49-51; asst prof phys chem, Polytech Inst Brooklyn, 51-54, assoc prof, 54-58, prof, 58-64; PROF PHYS CHEM, UNIV ILL, URBANA, 64- Concurrent Pos: Temp mem, Courant Inst Math Sci, 60-61; NSF sr fel, 60-61; Sloan fel, 60-63; vis sr scientist, Brookhaven Nat Lab, 62-64; Henry Werner lectr, Univ Kans; coun mem, Gordon Res Conf, 65-68, chmn bd trustees, 68-69; Venable lectr, Univ NC; Seydel-Wooley lectr, Ga Inst Technol; Foster lectr, State Univ NY Buffalo; mem adv coun chem dept, Princeton Univ, 72-; sr Fulbright-Hays Scholar, Fulbright Program, 72, 73; mem, Nat Res Coun-Nat Acad Sci Climatic Impact Comt, Panel Atmospheric Chem, 75-; chmn, Nat Res Coun-Nat Acad Sci Comt, Kinetics of Chem Reactions, 75-; mem, Review Comt, Radiation Lab, Univ Notre Dame, 75-; vis prof theoret chem, Oxford Univ, 75-76. Honors & Awards: Sr US Sci Award, Alexander von Humboldt Found, 76. Mem: Nat Acad Sci; Am Chem Soc; Am Phys Soc; Am Acad Arts & Sci. Res: Theoretical and experimental chemical kinetics; electron transfer, electrode and unimolecular reactions; semiclassical theory of reactive and nonreactive collisions. Mailing Add: Dept of Chem Noyes Chem Lab Univ of Ill Urbana IL 61801

MARCUS, RUDOLPH JULIUS, b Frankfurt, Ger, Mar 30, 26; nat US. PHYSICAL CHEMISTRY, Educ: Wayne State Univ, BS, 48; Univ Utah, PhD, 54. Prof Exp: Chemist, Sun Oil Co, 48-49; phys chemist, Stanford Res Inst, 54-64; CHEMIST, OFF NAVAL RES, PASADENA, CALIF, 64- Mem: Fel AAAS; fel Am Inst Chem; Am Chem Soc; Solar Energy Soc; NY Acad Sci. Res: Statistical thermodynamics; photosynthesis; solar energy; fluorescence and phosphorescence measurements; experimental design; interactive computer applications. Mailing Add: Off of Naval Res 1030 E Green St Pasadena CA 91106

MARCUS, SANFORD M, b New York, NY, Mar 18, 32; m 59; c 1. PHYSICS. Educ: Brooklyn Col, BS, 54; Columbia Univ, MS, 57; Univ Pa, PhD(physics), 64. Prof Exp: Engr, Radio Corp Am, 57-59; PHYSICIST, E I DU PONT DE NEMOURS & CO, 64- Mem: Am Phys Soc. Res: Superconductivity, especially experimental work by means of tunneling; solid state devices; transport properties of metals. Mailing Add: Cent Res Dept, Exp Sta E I du Pont de Nemours & Co Wilmington DE 19898

MARCUS, SHELDON H, b Chicago, Ill, Feb 6, 38; c 2. RESEARCH ADMINISTRATION. Educ: Univ Ill, BSc, 59; Ill Inst Technol, PhD(chem), 65. Prof Exp: From asst proj chemist to sr proj chemist, 64-69, group leader, 69-71, RES SUPVR, AMOCO CHEM CORP, 71- Concurrent Pos: Com develop proj mgr, Amoco Chem Corp, 70-73. Mem: Am Chem Soc. Res: Aromatic acids; fiber and film intermediates; oxidation; process development; industrial chemicals. Mailing Add: Res & Develop Dept Amoco Chem Corp PO Box 400 Naperville IL 60540

MARCUS, STANLEY, b New York, NY, Jan 20, 16; m 39; c 2. MICROBIOLOGY, IMMUNOLOGY. Educ: City Col New York, BA, 37; Univ Mich, MS, 39, PhD(microbiol), 42. Prof Exp: PROF MICROBIOL, COL MED, UNIV UTAH, 49- Concurrent Pos: Nat Inst Allergy & Infectious Dis res career award, 61. Mem: AAAS; Am Soc Microbiol; Am Asn Immunol; Soc Exp Biol & Med; Reticuloendothelial Soc. Res: Mechanisms of specific and nonspecific resistance to infectious and neoplastic disease; theory of testing; pyrogen tests; nontoxic enteric vaccines; standardization of mycotic sensitins; proficiency testing as basis of evaluation surveys. Mailing Add: Dept of Microbiol Univ of Utah Col of Med Salt Lake City UT 84132

MARCUS, STEPHEN, b New York, NY, Dec 27, 39; m 70; c 1. LASERS. Educ: Rensselaer Polytech Inst, BS, 61; Columbia Univ, MA, 63, PhD(physics), 68. Prof Exp: Res assoc elec eng, Cornell Univ, 67-69; staff scientist laser physics, United Aircraft Res Labs, 69-70; MEM STAFF, LINCOLN LAB, MASS INST TECHNOL, 70- Mem: Am Phys Soc. Res: Gas lasers, primarily pulsed and continuous wave carbon dioxide lasers and their applications to laser radar systems. Mailing Add: Mass Inst Technol Lincoln Lab 244 Wood St Lexington MA 02173

MARCUSE, DIETRICH, b Koenigsberg, Ger, Feb 27, 29; m 59; c 2. PHYSICS. Educ: Free Univ Berlin, Dipl phys, 54; Karlsruhe Tech Univ, DrIng, 62. Prof Exp: Mem tech staff, Siemens & Halske, Ger, 54-57; MEM TECH STAFF, BELL LABS, 57- Concurrent Pos: Adj assoc prof, Univ Utah. Mem: Fel Inst Elec & Electronics Eng; Optical Soc Am. Res: Circular electric waveguide; microwave masers; light communications. Mailing Add: Crawford Hill Lab Bell Labs Box 400 Holmdel NJ 07733

MARCUVITZ, NATHAN, b Brooklyn, NY, Dec 29, 13; m 46; c 2. MATHEMATICAL PHYSICS. Educ: Polytech Inst Brooklyn, BEE, 35, MEE, 41, DEE, 47. Prof Exp: Develop engr, Radio Corp Am, 35-40; mem staff, Radiation Lab, Mass Inst Tech, 42-46; asst prof elec eng, Polytech Inst Brooklyn, 46-49, assoc prof, 49-51, prof, 51-65, dir microwave res inst, 57-61, vpres res & actg dean, Grad Ctr, 61-63, prof electrophys, 61-65, dean res & grad ctr, 64-65, inst prof, 65-66; prof appl physics, NY Univ, 66-73; PROF APPL PHYSICS, POLYTECH INST NEW YORK, 73- Concurrent Pos: Asst dir res, Defense Res & Eng, Dept Defense, DC, 63-64; Gordon MacKay vis prof, Harvard Univ, 71. Mem: Am Phys Soc; fel Inst Elec & Electronics Eng. Res: Electromagnetics; plasma dynamics; nonlinear and turbulent wave phenomena. Mailing Add: Dept of Elect Eng & Electrophys Polytech Inst of New York Rte 110 Farmingdale NY 11735

MARCY, WILLARD, b Newton, Mass, Sept 27, 16; m 38; c 2. ORGANIC CHEMISTRY, CHEMICAL ENGINEERING. Educ: Mass Inst Technol, SB, 37, PhD(org chem), 49. Prof Exp: Asst supt, Am Sugar Ref Co, 37-42; res assoc org chem, Mass Inst Technol, 46-49; chem engr, Res & Develop Div, Am Sugar Ref Co, 49-58, head process develop, 58-64; dir patent progs, 64-67, VPRES PATENTS, RES CORP, 67- Mem: AAAS; Am Chem Soc; fel Am Inst Chem; Inst Food Technol; NY Acad Sci. Res: Carbohydrates; war gases; sugar refining; sugar by-products; patent administration. Mailing Add: Res Corp 405 Lexington Ave New York NY 10017

MARCZYNSKA, BARBARA MARY, b Cracow, Poland; US citizen; m 56; c 1. IMMUNOLOGY, GENETICS. Educ: Acad Med Cracow, MS, 56, PhD(immunol, genetics), 62. Prof Exp: From instr to assoc prof genetics & embryol, Med Sch, Cracow, 56-64; res asst, 65-72, ASST PROF VIROL, RUSH-PRESBY-ST LUKE'S MED CTR, 72- Mem: Am Soc Microbiol. Res: Virological and immunological aspects of virus-induced oncogenic transformation in non-human primates. Mailing Add: Dept of Microbiol Rush-Presby-St Luke's Med Ctr Chicago IL 60612

MARCZYNSKI, THADDEUS JOHN, b Poznan, Poland, Nov 30, 20; m 56; c 2. PHARMACOLOGY, NEUROPHYSIOLOGY. Educ: Cracow Acad Med, MD, 51, DMSc, 59. Prof Exp: Res asst pharmacol, Cracow Acad Med, 54-59, asst prof, 62-64; asst prof, 64-68, assoc prof, 68-73, PROF PHARMACOL, UNIV ILL COL MED, 73- Concurrent Pos: Rockefeller Found & Brain Res Inst fel, Univ Calif, Los Angeles, 61-62; NIH res grant, Univ Ill Col Med, 66-72. Mem: AAAS; Soc Neurosci; Biofeedback Soc; Am Soc Pharmacol & Exp Therapeut. Res: Pharmacology and electrophysiology of the central nervous system; application of computers to the analysis of information transmission coding in neuronal pathways; positive reinforcement and sensory imput. Mailing Add: Dept of Pharmacol Univ of Ill Col of Med Chicago IL 60680

MARDELLIS, ANTHONY, b Neuville-sur-Saone, France, July 17, 20, wid. MATHEMATICS. Educ: Univ Calif, Berkeley, BA, 50, MA, 52. Prof Exp: Asst math, Univ Calif, Berkeley, 51-55; instr, Loyola, Calif, 55-56; from asst prof to assoc prof, 56-70, chmn dept, 63-67, PROF MATH, CALIF STATE UNIV, LONG BEACH, 70- Mem: Am Math Soc; Math Asn Am; Math Soc France; Sigma Xi. Res: Picard-Vessiot theory; differential algebra. Mailing Add: Dept of Math Calif State Univ Long Beach CA 90840

MARDEN, MORRIS, b Boston, Mass, Feb 12, 05; m 32; c 2. MATHEMATICS. Educ: Harvard Univ, AB, 25, AM, 27, PhD(math), 28. Prof Exp: Instr math, Harvard Univ, 25-27; Nat Res Coun fel, Uni- Wis, Princeton Univ, Univ Zurich & Univ Paris, 28-30; from asst prof to prof, 30-64, distinguished prof, 64-75, chmn dept, 57-61, 63-64, EMER DISTINGUISHED PROF MATH, UNIV WIS-MILWAUKEE, 75- Concurrent Pos: Invited lectr, Math Inst, Polish Acad Sci, 58-62, Math Insts, Univs Jerusalem, Haifa & Tel Aviv, 62 & 68, Greece, 62, Japan, India, Spain & Eng, 64, Mex, Peru, Chile, Argentina, Uruguay & Brazil, 65, Budapest, Belgrade, Goteburg, 67, Montreal, 67 & 70, Finland & NZ, 70 & Australia, 71; consult, Allis-Chalmers Co, 48-60; asst ed, Bull, Am Math Soc, 42-45; vis distinguished prof math, Calif Poly State Univ, San Luis Obispo, 75-76. Mem: Fel AAAS; Am Math Soc; Math Asn Am; London Math Soc; Soc Indust & Appl Math. Res: Zeros of polynomials; entire and potential function; functions of a complex variable. Mailing Add: Dept of Math Univ of Wis Milwaukee WI 53201

MARDEN, PHILIP AYER, b Newport, NH, Oct 31, 11; m 50. MEDICINE. Educ: Dartmouth Col, AB, 33; Univ Pa, MD, 36. Prof Exp: From instr to assoc prof, 39-58, chmn dept, 59-73, PROF OTOLARYNGOL, SCH MED, UNIV PA, 58- Concurrent Pos: Consult, Vet Admin Hosp, Philadelphia, Pa. Mem: Am Acad Ophthal & Otolaryngol. Res: Streptomycin toxicity; osteitis fibrosa cystica. Mailing Add: Dept of Otolaryngol Univ of Pa Sch of Med Philadelphia PA 19104

MARDER, HERMAN LOWELL, b New York, NY, Mar 3, 31; m 55; c 4. ORGANIC POLYMER CHEMISTRY, RESEARCH ADMINISTRATION. Educ: State Univ NY, BS, 54, MS, 57, PhD(chem), 59. Prof Exp: Res chemist, E I du Pont de Nemours & Co, Inc, 58-61; sect head, Colgate Palmolive Co, 61-66; dir res & develop, Boyle Midway Div, Am Home Prod Corp, 66-69; VPRES RES & DEVELOP, INT PLAYTEX CO, 69- Mem: Am Chem Soc. Res: Paper and textile chemistry; adhesives and surface chemistry. Mailing Add: Int Playtex Corp 215 College Rd Paramus NJ 07652

MARDER, STANLEY, b Philadelphia, Pa, Aug 21, 26; m 53; c 3. INFORMATION SCIENCE, SYSTEMS ANALYSIS. Educ: Univ Pa, BA, 50; Columbia Univ, PhD(physics), 58. Prof Exp: Res physicist, Carnegie Inst Technol, 56-60; staff mem, Inst Defense Anal, 60-73; res physicist, 73-74, DIR, WASHINGTON OFF, ENVIRON RES INST MICH, ARLINGTON, VA, 74- Mem: Inst Elec & Electronic Engrs; Am Econ Asn; Am Phys Soc. Res: Signal processing; image evaluation; radar system analysis. Mailing Add: 9608 McAlpine Rd Silver Spring MD 20901

MARDINEY, MICHAEL RALPH, JR, b Brooklyn, NY, Dec 16, 34; m 60; c 3. IMMUNOLOGY, INTERNAL MEDICINE. Educ: Hamilton Col, AB, 56; Seton Hall Col Med & Dent, MD, 60; Am Bd Allergy & Immunol, dipl, 74. Prof Exp: Intern med, Kings County Hosp Ctr, Brooklyn, 60-61; resident med, Col Med, Baylor Univ, 61-62; clin assoc, Immunol Br, Nat Cancer Inst, 65-67, HEAD IMMUNOL & CELL BIOL SECT, BALTIMORE CANCER RES CTR, NAT CANCER INST, 67- Concurrent Pos: Res fel, Exp Path Div, Scripps Clin & Res Found, Univ Calif, 62-65; instr, Sch Med, Johns Hopkins Univ & physician, Allergy & Infectious Dis Clin, Johns Hopkins Hosp; physician to med staff, Good Samaritan Hosp; staff physician, Howard County Gen Hosp, South Baltimore Gen Hosp & Lutheran Hosp, 70- Mem: Transplantation Soc; Am Soc Exp Path; Am Asn Immunol; Am Asn Cancer Res; Am Acad Allergy. Res: Immunopathology; tumor immunology; allergy. Mailing Add: Baltimore Cancer Res Ctr 3100 Wyman Park Dr Baltimore MD 21211

MARDON, DAVID NORMAN, b Syracuse, NY, Mar 11, 37; m 59; c 3. MICROBIOLOGY. Educ: Syracuse Univ, AB, 62, PhD(microbiol), 68. Prof Exp: Res assoc microbiol, Oak Ridge Assoc Univs, 67-68, NIH res fels, 68-70; ASST PROF HEALTH SCI DIV, MED COL VA, VA COMMONWEALTH UNIV, 70- Concurrent Pos: Brown-Hazen Fund res grant, Va Commonwealth Univ, 70- Mem: Am Soc Microbiol; Int Soc Human & Animal Mycol. Mailing Add: Dept of Microbiol Med Col of Va Va Commonwealth Univ Richmond VA 23219

MARE, CORNELIUS JOHN, b Middleburg, SAfrica, Aug 27, 34; m 60; c 4. VETERINARY MICROBIOLOGY. Educ: Pretoria Univ, BVSc, 57; Iowa State Univ, PhD(vet microbiol), 65. Prof Exp: Private practice, 57-58; vet diagnostician, Allerton Diag Lab, SAfrica, 58-59; res virologist, Onderstepoort Vet Res Inst, 59-62, sr res virologist, 65-67; res assoc microbiol, Vet Med Res Inst Iowa, 62-65; assoc prof virol, 67-72, PROF VET MICROBIOL, IOWA STATE UNIV, 72- Concurrent Pos: Vis prof, Plum Island Animal Dis Ctr, NY, 73 & EAfrican Vet Res Inst, Nairobi, Kenya, 74. Mem: Conf Res Workers Animal Dis; US Animal Health Asn; Am Vet Med Asn; Am Soc Microbiol; Wildlife Dis Asn. Res: Viruses of domestic animals and man, expecially viruses of the herpes virus group. Mailing Add: Dept Vet Microbiol Iowa State Univ Ames IA 50010

MAREK, JERRY WILLIAM, b Brenham, Tex, Feb 24, 13. BIOCHEMISTRY, SCIENCE EDUCATION. Educ: Texas A&M Univ, BS, 35, MS 37. Prof Exp: Sci aide biochem, USDA, Tex, 36-37; res asst, Univ Wis, 37-39; instr high sch, Tex, 40-41; head dept chem, Col Emporia, 41-42; dept sci, Blinn Col, 46-47; instr chem, NC State Col, 47-51; res scientist, Plant Res Inst, Univ Tex, 51-52; instr chem, Univ Miss, 52-55; assoc prof chem & physics & head dept, 55-57, head dept sci, 56-66, PROF SCI, LANDER COL, 57-, AUDIOVISUAL COORDR, LEARNING LAB, 71- MEDIA CTR, 73- Mem: AAAS; Am Chem Soc; Asn Educ Commmun & Technol. Res: Carbohydrate analyses; nitrogen fractionation; plant tissue analysis; enzyme studies related to potato diseases and discoloration; microbiology; organic synthesis related to cellulose degradation; study of UFO activity in SC related to gold deposits and nuclear installations. Mailing Add: Sci Div Lander Col Greenwood SC 29646

MAREN, THOMAS HARTLEY, b New York, NY, May 26, 18; m 41; c 3. PHARMACOLOGY. Educ: Princeton Univ, AB, 38; Johns Hopkins Univ, MD, 51. Prof Exp: Res chemist, Wallace Labs, Carter Prods, Inc, NJ, 38-40, group leader, 41-44; chemist, Sch Hyg & Pub Health, Johns Hopkins Univ, 44-46, instr pharmacol, Med Sch, 46-51; pharmacologist, Chemother Dept, Res Div, Am Cyanamid Co, 51-54, group leader, 54-55; PROF PHARMACOL & THERAPEUT & CHMN DEPT, COL MED, UNIV FLA, 55- Concurrent Pos: Investr, Mt Desert Island Biol Lab, 53- Mem: Am Soc Pharmacol & Exp Therapeut. Res: Renal ocular, cerebrospinal and electrolyte pharmacology and physiology; carbonic anhydrase and its inhibitors; chemotherapy of infectious diseases; comparative pharmacology. Mailing Add: Univ of Fla Col of Med Gainesville FL 32610

MARENGO, NORMAN PAYSON, b New York, NY, Feb 21, 13; m 39; c 2. BOTANY. Educ: NY Univ, BS, 36, MA, 39, MS, 42, PhD(biol), 49. Prof Exp: Asst ed, NY Univ, 36-39, instr, 39-43, biol, 46-48; instr biol, Lafayette Col, 48-49, asst prof, 49-50; asst prof biol, Hofstra Col, 50-54; teacher sci, Cent High Sch, Merrick, 54-55; asst prof biol & gen sci, 55-57, assoc prof, 57-60, dir div sci, 61-67, chmn dept biol, 63-67, PROF BIOL, C W POST COL, L I UNIV, 60- Mem: AAAS; Bot Soc Am; Am Fern Soc; Torrey Bot Club; NY Acad Sci. Res: Developmental genetics; botanical cytology; microscopical technique. Mailing Add: Dept of Biol C W Post Col Long Island Univ Greenvale NY 11548

MARETZKI, ANDREW, b Berlin, Ger, Feb 23, 26; nat US; m 57; c 2. BIOCHEMISTRY. Educ: Univ Cincinnati, BS, 52; Pa State Univ, MS, 58, PhD, 60. Prof Exp: Asst res biochemist, Parke, Davis & Co, 52-53, res asst biochem, Pa State Univ, 55-60; res biochemist, Kitchwan Res Labs, 60-61; assoc scientist, Nuclear Ctr & Sch Med, Univ Puerto Rico, 61-65; assoc biochemist, 66-69, BIOCHEMIST, EXP STA, HAWAIIAN SUGAR PLANTERS ASN, 69- Mem: AAAS; Am Chem Soc; Am Soc Plant Physiol; Sigma Xi. Res: Structure of antibiotics; toxins; plant enzyme systems; membrane transport. Mailing Add: Hawaiian Sugar Planters Asn Exp Sta 1527 Keeaumoku St Honolulu HI 96822

MARETZKI, THOMAS WALTER, b Berlin, Ger, Sept 3, 21; nat US; m 51. ANTHROPOLOGY. Educ: Univ Hawaii, BA, 51; Yale Univ, PhD, 57. Prof Exp: Instr anthrop, Univ Conn, 57-61; from asst prof to prof, 61-70, chmn dept, 65-70, PROF ANTHROP & PSYCHIAT, SCH MED, UNIV HAWAII, 70- Concurrent Pos: Field researcher, Univ Pittsburgh, 60-61; assessment coordr, Peace Corps Training, Hilo, Hawaii, 62-65, prin investr, Peace Corps Impact Study, Philippines, 64-66; sr specialist, Inst Advan Studies, East West Ctr, 65-66; consult debriefing, AID Training, Far East Training Ctr, Univ Hawaii, 66-67; co-prin investr, Cult & Ment Health Proj, Soc Sci Res Inst, Nat Inst Ment Health, 67- Mem: Fel Am Anthrop Asn; fel Asn Asian Studies; fel Soc Appl Anthrop. Res: Culture and personality; socialization, religion and health in Okinawa; impact of Peace Corps in the Philippines. Mailing Add: Dept of Anthrop Univ of Hawaii Honolulu HI 96822

MAREZIO, MASSIMO, b Rome, Italy, Aug 25, 31; US citizen. CRYSTALLOGRAPHY. Educ: Univ Rome, Dr(chem), 54, Lib Doc, 65. Prof Exp: Ital Atomic Energy Comn fel physics, Univ Chicago, 59-60, res assoc, 60-63; mem tech staff, Bell Tel Labs, Inc, 63-73; RES DIR, NAT CTR SCI RES, FRANCE, 73- Mem: Am Crystallog Asn. Res: Inorganic crystal chemistry, solid state physical and high pressure chemistry. Mailing Add: Lab des Rayons X Nat Ctr for Sci Res Grenoble France

MARFEY, SVIATOPOLK PETER, b Kobaki, Poland, June 1, 25; nat US; m 64; c 1. ORGANIC CHEMISTRY, BIOCHEMISTRY. Educ: Wayne State Univ, BS, 49, MS, 53, PhD(chem), 55. Prof Exp: Res asst chem, Princeton Univ, 55-56; res assoc, Rockefeller Inst, 56-59; res assoc, Harvard Univ, 59-67; ASSOC PROF BIOL SCI, STATE UNIV NY ALBANY, 67- Concurrent Pos: Dir biochem res lab, Mass Eye & Ear Infirmary, 59-67. Mem: AAAS; Am Chem Soc; Biophys Soc. Res: Chemistry of proteins and nucleic acids; structure and function of cellular membranes. Mailing Add: Dept of Biol Sci State Univ of NY 1400 Washington Ave Albany NY 12203

MARG, ELWIN, b San Francisco, Calif, Mar 23, 18; m 42; c 1. VISION, NEUROSCIENCES. Educ: Univ Calif, AB, 40, PhD(physiol optics), 50. Prof Exp: Instr optom, 50-51, asst prof, 51-56, assoc prof, 56-62, Miller res prof, 67-68, PROF OPTOM & PHYSIOL OPTICS, UNIV CALIF, BERKELEY, 62- Concurrent Pos: NSF sr fel, Nobel Inst Neurophysiol, Karolinska Inst, Sweden, 57; Guggenheim fel, Madrid, 64; res assoc neurosci, Mt Zion Hosp & Med Ctr, San Francisco, 69- Honors & Awards: Apollo Award, Am Optom Asn, 62. Mem: Am Physiol Soc; Optical Soc Am; Asn Res Vision & Ophthal; Am Acad Optom; Soc Neurosci. Res: Neurophysiology of visual system and brain; automated eye examination; phosphene visual prosthesis; diagnosis and prognosis by single neuron responses from the brain in neurosurgery; visual acuity and development in infants. Mailing Add: Sch of Optom Univ of Calif Berkeley CA 94720

MARGACH, CHARLES BOYD, b Utica, NY, Aug 11, 12; m 37; c 2. OPTOMETRY. Educ: Northern Ill Col Optom, OD, 48; Pac Univ, BS, 50, MS, 51. Prof Exp: Prof optom, Col Optom, Pac Univ, 60-72; pvt pract optom, 72-74; PROF OPTOM, SOUTHERN CALIF COL OPTOM, 74- Concurrent Pos: Mem sect res & clin assoc, Optom Exten Prog Found, 62-, ed lit & res rev, 75-; ed, J Am Optom Asn, 65-66; contrib ed, Optical J & Rev Optom, 75- Mem: Fel Am Acad Optom. Res: Electronic magnification for low-visioned readers; psycholinguistics; altered states of consciousness. Mailing Add: Southern Calif Col of Optom 2001 Associated Rd Fullerton CA 92631

MARGALIT, NEHEMIAH, b Haifa, Israel, Nov 27, 38; US citizen; m 65; c 2. CHEMISTRY. Educ: Univ Pa, BS, 63; Drexel Univ, MS, 65, PhD(chem), 69. Prof Exp: SR SCIENTIST BATTERIES, TECHNOL CTR, ESB INC, YARDLEY, PA, 68- Mem: Electrochem Soc; Am Chem Soc; Carbon Soc; AAAS. Res: Chemistry and technology of electrochemical cells with emphasis on non-aqueous systems, including chemical and electrochemical behavior of electrodes and solutions, intracell interactions and stability of systems. Mailing Add: 18 Calicobush Rd Levittown PA 19057

MARGANIAN, VAHE MARDIROS, b Jlala, Lebanon, May 28, 38; US citizen; m 62; c 3. INORGANIC CHEMISTRY. Educ: San Francisco State Col, BS, 60; Clemson Univ, MS, 64, PhD(inorg chem), 66. Prof Exp: Teaching & res fel chem, Clemson Univ, 62-66; NSF res fel inorg chem, Univ Mass, Amherst, 66-67, from asst prof to assoc prof, 67-74, PROF CHEM, BRIDGEWATER STATE COL, 74- Mem: Am Chem Soc; Sigma Xi. Res: Synthesis and structural studies of oxo-compounds with tellurium IV halides; characterization of the products of cadmium II halides with N-bases; P-NMR of platinum II hydride systems. Mailing Add: Dept of Chem Bridgewater State Col Bridgewater MA 02324

MARGARETTEN, WILLIAM, b Brooklyn, NY, Sept 19, 29. PATHOLOGY. Educ: NY Univ, AB, 50; Northwestern Univ, MS, 51; State Univ NY, MD, 55; Am Bd Path, dipl, 67. Prof Exp: Asst prof path, Columbia Univ, 66-67; from asst prof to assoc prof, 67-75, PROF PATH, UNIV CALIF, SAN FRANCISCO, 75- Concurrent Pos: Mem coun thrombosis, Am Heart Asn. Mem: AAAS; Am Asn Path & Bact; Am Soc Exp Path. Res: Coagulation; endotoxin; inflammation. Mailing Add: Dept of Path Univ of Calif San Francisco CA 94110

MARGARIS, ANGELO, b Worcester, Mass, Dec 3, 21. MATHEMATICS. Educ: Cornell Univ, BEE, 43, PhD(math), 56; Syracuse Univ, MA, 51. Prof Exp: Elec engr, Fed Tel & Radio Corp, 46-50; instr math, Oberlin Col, 54-57; from asst prof to assoc prof, Ohio State Univ, 57-68; PROF MATH, SOUTHWESTERN AT MEMPHIS, 68- Mem: Am Math Soc; Asn Symbolic Logic. Res: Mathematical logic and foundations. Mailing Add: Dept of Math Southwestern at Memphis 2000 N Pkwy Memphis TN 38112

MARGAZIOTIS, DEMETRIUS JOHN, b Athens, Greece, Oct 14, 38; m 67. NUCLEAR PHYSICS. Educ: Univ Calif, Los Angeles, BA, 59, MA, 61, PhD(physics), 66. Prof Exp: From instr to assoc prof, 64-73, PROF PHYSICS, CALIF STATE UNIV, LOS ANGELES, 73- Mem: Am Phys Soc. Res: Few nucleon problem; nuclear structure. Mailing Add: Dept of Physics Calif State Univ Los Angeles CA 90032

MARGEN, SHELDON, b Chicago, Ill, May 9, 19; m 44; c 4. HUMAN NUTRITION. Educ: Univ Calif, AB, 38, MA, 39, MD, 43. Prof Exp: USPHS sr res fel, 47-48; res assoc, US Metab Unit, 47-50, clin instr med, Sch Med, 48-56, lectr soc res, Sch Soc Welfare, 56-62, assoc res biochemist, 52-60, res biochemist, 60-62, nutritionist, Agr Exp Sta, 62-70, chmn dept nutrit sci, 70-74, PROF HUMAN NUTRIT, UNIV CALIF, BERKELEY, 62- Concurrent Pos: Schering fel, 48-49; Damon Runyon fel, Nat Res Coun, 49-51; res assoc, Inst Metab Res, Alameda, 50-52. Mem: Endocrine Soc; Am Fedn Clin Res; Am Med Asn; Am Inst Nutrit; Am Soc Clin Nutrit. Res: Energy and general protein metabolism; protein turnover; human nutrition, experimental and programatic; hormone effects on intermediate metabolism; water and electrolyte regulation. Mailing Add: Dept of Nutrit Sci Univ of Calif Berkeley CA 94720

MARGERISON, RICHARD BENNETT, b Phila, Pa, Feb 24, 32; m 53; c 4. ORGANIC CHEMISTRY. Educ: Lehigh Univ, BS, 53, MS, 55; Univ Va, PhD(chem),

57. Prof Exp: Asst, Lehigh Univ, 53-55; res chemist med chem, Wallace & Tiernan, Inc, 57-58; sr chemist develop res, Ciba Pharmaceut Prods, Inc, 58-67, mgr process res & develop, Ciba Agrochem Co, 67-70, mgr chem mfg, 70-74, DIR CHEM MFG, PHARMACEUT DIV, CIBA GEIGY CORP, 74- Mem: AAAS; NY Acad Sci. Res: Preparation of nitrogen and sulfur aliphatic, aromatic and heterocyclic compounds as medicinal agents; substituted piperazines, diphenyl sulfides, gem-diphenyl compounds; medium size heterocyclic rings; sulfonamides; ureas. Mailing Add: 556 Morris Ave Summit NJ 07901

MARGERUM, DALE WILLIAM, b St Louis, Mo, Oct 20, 29; m 53; c 3. INORGANIC CHEMISTRY, ANALYTICAL CHEMISTRY. Educ: Southeast Mo State Col, BA, 50; Iowa State Univ, PhD, 55. Prof Exp: Chemist, Ames Lab, Iowa State Univ, 52-53; from instr to assoc prof, 54-65, PROF CHEM, PURDUE UNIV, 65- Concurrent Pos: NSF sr fel, Max Planck Inst Phys Chem, Gottingen, 63-64; vis prof, Univ Kent, Canterbury, 70; adv bd, Res Corp, 73- Mem: AAAS; Am Chem Soc. Res: Coordination chemistry; bio-inorganic; kinetics; fast reactions in solution; analytical applications of kinetics; inorganic-analytical studies of environmental solution chemistry. Mailing Add: Dept of Chem Purdue Univ West Lafayette IN 47907

MARGERUM, JOHN DAVID, physical chemistry, see 12th edition

MARGETTS, EDWARD LAMBERT, b Vancouver, BC, Mar 8, 20; m 41; c 2. MEDICINE, PSYCHIATRY. Educ: Univ BC, BA, 41; McGill Univ, MD & CM, 44; FRCP(C); FRCPsychiat. Prof Exp: Psychiatrist, Royal Victoria Hosp, Montreal, 49-55; specialist psychiatrist, Kenya Govt & med supt, Mathari Hosp, Nairobi, 55-59; PROF PSYCHIAT & LECTR HIST OF MED, UNIV BC, 64- Concurrent Pos: Asst to dir, Allan Mem Inst Psychiat, 49-51; chief serv, Shaughnessy Vet Hosp, Vancouver, BC, 64-70; Ment Health Unit, WHO, Switz, 70-72. Mem: Am Psychiat Asn; Am Asn Hist Med; Am Anthrop Asn; Can Psychiat Asn; Royal Micros Soc. Res: Ethnic, cultural and international psychiatry; history of medicine; archaeology and anthropology applied to medicine. Mailing Add: Dept Psychiat Univ BC Vancouver BC Can

MARGOLIASH, EMANUEL, b Cairo, Egypt, Feb 10, 20; m 44; c 2. MOLECULAR BIOLOGY, PROTEIN CHEMISTRY. Educ: American Univ, Beirut, BA, 40, MA, 42, MD, 45. Prof Exp: Res fel exp path, Hebrew Univ, Israel, 45-48, sr asst, 49-51, lectr & actg head, Cancer Res Labs, Hadassah Med Sch, 54-58; res assoc biochem, Molteno Inst, Cambridge Univ, 51-53; res assoc, Nobel Inst, Sweden, 58, Univ Utah, 58-60 & Montreal Res Inst, McGill Univ, 60-62; head protein sect, Abbott Labs, 62-71; PROF BIOCHEM & MOLECULAR BIOL, NORTHWESTERN UNIV, 71- Concurrent Pos: Prof lectr, Univ Chicago, 64-71. Mem: Nat Acad Sci; Am Chem Soc; Am Soc Biol Chem; Can Biochem Soc; Brit Biochem Soc. Res: Structure-function relations of heme proteins; molecular evolution and immunology; energy conservation mechanisms. Mailing Add: Dept of Biochem & Molecular Biol Northwestern Univ Evanston IL 60201

MARGOLIN, ABRAHAM STANLEY, biology, deceased

MARGOLIN, BARRY HERBERT, b New York, NY, Jan 8, 43; m 69. MATHEMATICAL STATISTICS, APPLIED STATISTICS. Educ: City Col New York, BS, 63; Harvard Univ, MA, 64, PhD(statist), 67. Prof Exp: Instr educ statist, Harvard Univ, 66-67; asst prof statist, 67-72, ASSOC PROF STATIST, YALE UNIV, 72- Concurrent Pos: Consult, Consumers Union, 67- & IBM Co, 69-70. Mem: Am Statist Asn; Inst Math Statist. Res: Data analysis; design and analysis of experiments; categorical data; contingency tables; statis- tical evaluation of computer system performance. Mailing Add: Dept of Statist Yale Univ New Haven CT 06520

MARGOLIN, ESAR GORDON, b Omaha, Nebr, Mar 17, 24; m 56; c 2. INTERNAL MEDICINE. Educ: Univ Nebr, BA, 45, MD, 47. Prof Exp: From asst prof to assoc prof, 58-69, clin prof, 69-72, PROF MED, COL MED, UNIV CINCINNATI, 72-; DIR DEPT INTERNAL MED, JEWISH HOSP, 59- Concurrent Pos: Fel med, Harvard Univ, 53-55. Mem: Am Col Physicians; Am Soc Nephrology; Int Soc Nephrology; Am Heart Asn; Am Soc Artificial Internal Organs. Res: Kidney and electrolytes. Mailing Add: Dept of Med Jewish Hosp Cincinnati OH 45229

MARGOLIN, PAUL, b New York, NY, Aug 31, 23; m 46. GENETICS, MICROBIOLOGY. Educ: NY Univ, BS, 47; Ind Univ, PhD(genetics), 55. Prof Exp: Asst zool, Ind Univ, 50-52; res fel, Calif Inst Technol, 55-56; USPHS fel, Univ Edinburgh, 57-58 & Brookhaven Nat Lab, 58; geneticist, Biol Lab, Long Island Biol Asn, 58-62; sr staff investr, Cold Spring Harbor Lab Quant Biol, NY, 62-66; MEM & CHIEF DEPT GENETICS, PUB HEALTH RES INST OF CITY OF NEW YORK, INC, 66- Concurrent Pos: Res prof, Dept Microbiol, Sch Med, NY Univ; mem sci adv comt virol & cell biol, Am Cancer Soc, 69-73. Mem: AAAS; Genetics Soc Am; Am Soc Microbiol; Am Inst Biol Sci. Res: Bacterial genetics; molecular biology. Mailing Add: Dept of Genetics 455 First Ave Pub Health Res Inst of New York New York NY 10016

MARGOLIN, SOLOMON, b Philadelphia, Pa, May 16, 20; m 47; c 4. PHARMACOLOGY, ENDOCRINOLOGY. Educ: Rutgers Univ, BSc, 41, MSc, 43, PhD(physiol, biochem), 45. Prof Exp: Asst, Rutgers Univ, 43-45; consult, 46-47; res biologist, Silmo Chem Co, 47-48; res biologist, Schering Corp, 48-52, dir pharmacol res, 52-54; chief pharmacologist, Maltbie Labs Div, Wallace & Tiernan, Inc, 54-56; chief pharmacologist, Wallace Labs, Carter-Wallace, Inc, 56-60, dir biol res, 60-64, vpres, 64-68; PRES, AFFILIATED MED RES, INC, 68- Mem: AAAS; Am Soc Animal Sci; Endocrine Soc; Am Chem Soc; Soc Exp Biol & Med. Res: Antihistamines; anticholinergics; cholecystographic media; sedative-hypnotics; tranquilizers; muscle relaxants; adrenal hormones; cardiovascular agents. Mailing Add: Affiliated Med Res Inc PO Box 57 Princeton NJ 08540

MARGOLIN, SYDNEY GERALD, b New York, NY, Apr 25, 09; m 41; c 2. PSYCHIATRY. Educ: Columbia Univ, BS, 30, MA, 31; State Univ NY Downstate Med Ctr, MD, 36; Am Bd Psychiat & Neurol, dipl, 43. Prof Exp: Mem fac, Columbia Univ, 30-32; assoc attend psychiatrist, Mt Sinai Hosp, 46-55; chief div psychosom med, Med Ctr, 55-58, PROF PSYCHIAT, SCH MED, UNIV COLO, DENVER, 55-, DIR HUMAN BEHAV LAB & UTE INDIAN PROJ, 56- Concurrent Pos: Abrahamson fel neurol, Mt Sinai Hosp, NY, 39-40; Josiah Macy Jr Found grant, Columbia Univ, 41-42; psychiatrist in chg, Male Med Serv, Worcester State Hosp, Mass, 41; mem fac, NY Psychoanal Inst, 46-55. Mem: AAAS; fel Am Col Physicians; Am Psychiat Asn; Am Psychoanal Asn; Am Psychosom Soc (pres, 52-53). Res: Psychoanalysis; psychophysiology; neuropsychology of consciousness; psychosomatic medicine; ethno-psychiatry; behavioral science. Mailing Add: 4375 S Lafayette Englewood CO 80110

MARGOLIS, BERNARD, b Montreal, Que, Aug 15, 26; m 54. PHYSICS. Educ: McGill Univ, BSc, 47, MSc, 49; Mass Inst Technol, PhD, 52. Prof Exp: Instr physics, Mass Inst Technol, 53-54; instr, Columbia Univ, 54-57, res physicist, 57-59; assoc prof physics, Ohio State Univ, 59-61; assoc prof math physics, 61-63, PROF PHYSICS, McGILL UNIV, 63- Mem: Am Phys Soc. Res: Theoretical physics. Mailing Add: Dept of Physics McGill Univ Montreal PQ Can

MARGOLIS, DAVID, plant physiology, see 12th edition

MARGOLIS, FRANK L, b Brooklyn, NY, Jan 21, 38; m 61; c 3. NEUROCHEMISTRY. Educ: Antioch Col, BS, 59; Columbia Univ, PhD(biochem), 64. Prof Exp: USPHS trainee biochem, Columbia Univ, 64-65; fel, Lab Comp Physiol, Univ Paris, 65-66; asst res microbilogist, Sch Med, Univ Calif, Los Angeles, 66-69; res assoc, 69-71, asst mem, 71-74, ASSOC MEM, ROCH INST MOLECULAR BIOL, 74- Concurrent Pos: Adj prof, City Univ New York, 71- Mem: AAAS; Am Soc Neurochem; Int Soc Neurochem; Am Soc Biol Chemists; Soc Neurosci. Res: Regulation of mammalian gene expression; biochemistry of brain regions associated with specific sensory function, especially olfaction. Mailing Add: Roche Inst of Molecular Biol Nutley NJ 07110

MARGOLIS, GEORGE, b Montgomery, WVa, Dec 12, 14; m 50; c 4. PATHOLOGY. Educ: Johns Hopkins Univ, AB, 36; Duke Univ, MD, 40. Hon Degrees: MA, Dartmouth Col, 66. Prof Exp: Intern & jr asst, Med Sch, Duke Univ, 40-41, asst resident & sr asst, 41-43, resident & instr, 43-44, assoc path, 47-51, from asst prof to prof, 51-59; prof path & chmn med, Med Col Va, 59-63; PROF PATH, DARTMOUTH MED SCH, 63- Concurrent Pos: Vol neuropath, Montefiore Hosp, 48; consult, USPHS, 56-64. Mem: AAAS; Am Asn Path & Bact; Am Asn Neuropath; Int Acad Path; Int Brain Res Orgn. Res: Neuropathology; virology; cerebrovascular disease. Mailing Add: Rennie Rd Lyme Center NH 03769

MARGOLIS, JACK SELIG, b Los Angeles, Calif, Mar 9, 32. CHEMICAL PHYSICS, SPECTROSCOPY. Educ: Univ Calif, Los Angeles, AB, 54, PhD(physics), 60. Prof Exp: Engr, Collins Radio Co, 54-55; asst physics, Univ Calif, Los Angeles, 55-60; mem tech staff, Sci Ctr, NAm Aviation, Inc, 60-64; lectr physics, Univ Calif, Santa Barbara, 64-65, asst prof, 65-66; MEM TECH STAFF, JET PROPULSION LAB, CALIF INST TECHNOL, 66- Concurrent Pos: Asst, Scripps Inst, Univ Calif, 55-56. Mem: Am Phys Soc; Am Astron Soc. Res: Spectroscopy of the earth's atmosphere; theoretical rare earth and molecular spectroscopy; induced Raman effect; charge transfer complexes; atmospheric radiation. Mailing Add: 1381 Cheviotdale Dr Pasadena CA 91105

MARGOLIS, LEO, b Montreal, Que, Dec 18, 27. PARASITOLOGY, FISH PATHOLOGY. Educ: McGill Univ, BSc, 48, MSc, 50, PhD(parasitol), 52. Prof Exp: Asst parasitol, McGill Univ, 49-52; asst zool, Macdonald Col, 50-51; from assoc scientist to prin scientist, Fisheries Res Bd Can, 52-67, head exp biol & path group, 67-73, HEAD SALMON ENHANCEMENT, AQUACULT & FISH HEALTH SECT, PAC BIOL STA, 73- Concurrent Pos: Co-chmn, Can Comt Fish Dis, 70-73; assoc ed, Can J Zool, 71-; mem comt Biol & Res, Int North Pac Fisheries Comn, 71- Mem: Am Soc Parasitol; Wildlife Dis Asn; Can Soc Zool; fel Royal Soc Can; Am Fisheries Soc. Res: Parasites of fish and marine mammals; diseases of fish. Mailing Add: Pac Biol Sta Dept of Environ Fisheries & Marine Serv Nanaimo BC Can

MARGOLIS, MAXINE LUANNA, b New York, NY, Aug 2, 42; m 70; c 1. ANTHROPOLOGY. Educ: NY Univ, BA, 64; Columbia Univ, PhD(anthrop), 70. Prof Exp: Asst prof, 69-74, ASSOC PROF ANTHROP, UNIV FLA, 74- Mem: Fel Am Anthrop Asn; Latin Am Studies Asn; Am Anthrop Asn. Res: Brazilian ethnology; plantation societies; cultural ecology; cross-cultural study of frontiers; sex roles. Mailing Add: Dept of Anthrop Univ of Fla Gainesville FL 32611

MARGOLIS, PHILIP MARCUS, b Lima, Ohio, July 7, 25; m 59. PSYCHIATRY. Educ: Univ Minn, BA, 46, BS, 47, BM, 48, MD, 49; Am Bd Psychiat & Neurol, dipl. Prof Exp: Harvard fel psychiat, Univ Minn, 49-53, instr, Med Sch, Univ Minn, 53-56; from asst prof to assoc prof, Sch Med, Univ Chicago, 56-66; prof psychiat, Med Sch, 66-71, PROF COMMUNITY MENT HEALTH, SCH PUB HEALTH, UNIV MICH, ANN ARBOR, 71-; DIR, WASHTENAW COUNTY MENT HEALTH SERV, 66- Concurrent Pos: Clin fel, Mass Gen Hosp, Boston, 52-53; consult, Vet Admin Hosp, Minneapolis, Minn, 53-54 & Family Serv Agency, St Paul, 54-56; chief psychiat in-patient serv, Billings Hosp, Univ Chicago Clins, 56-66; consult, Child & Family Serv, Chicago, 57-60 & State Psychiat Inst, 60-66; sr psychiat consult, Peace Corps, 61-66; chmn, Consult & Eval Serv Bd, Am Psychiat Asn, consult, Contact Surv Bd, 69- Mem: Am Psychiat Asn; AMA; Am Orthopsychiat Asn; World Fedn Ment Health; Int Asn Social Psychiat. Res: Social and community psychiatry; preventive psychiatry; crisis therapy; suicide studies; consultation process; inpatient psychosocial issues. Mailing Add: 228 Riverview Dr Ann Arbor MI 48104

MARGOLIS, RENEE KLEINMANN, b Paris, France, Oct 31, 38; US citizen; m 59. PHARMACOLOGY, NEUROCHEMISTRY. Educ: Univ Chicago, BS, 60, PhD(pharmacol), 66. Prof Exp: Res scientist, NY State Res Inst Neurochem & Drug Addiction, 66-68; instr pharmacol, Mt Sinai Sch Med, 68-70; asst prof, 70-74, ASSOC PROF PHARMACOL, STATE UNIV NY DOWNSTATE MED CTR, 74- Mem: Am Soc Pharmacol & Exp Therapeut; Am Soc Neurochem; Int Soc Neurochem; Soc Complex Carbohydrates; Brit Biochem Soc. Res: Glycoproteins and glycosaminoglycans of nervous tissue. Mailing Add: Dept of Pharmacol State Univ NY Downstate Med Ctr Brooklyn NY 11203

MARGOLIS, RICHARD URDANGEN, b Pittsburgh, Pa, Sept 7, 37; m 59. PHARMACOLOGY, BIOCHEMISTRY. Educ: Univ Chicago, BS, 59, PhD(pharmacol), 63, MD, 66. Prof Exp: Res assoc pharmacol, Univ Chicago, 63-66; from instr to asst prof, 66-71, ASSOC PROF PHARMACOL, SCH MED, NY UNIV, 71- Mem: NY Acad Sci; Brit Biochem Soc; Int Soc Neurochem; Am Soc Pharmacol & Exp Therapeut; Soc Neurosci. Res: Brain lipid metabolism; blood-brain barrier and cerebrospinal fluid; mucopolysaccharides and glycoproteins of nervous tissue. Mailing Add: Dept of Pharmacol NY Univ Sch of Med New York NY 10016

MARGOLIS, SAM AARON, b Cambridge, Mass, Nov 17, 33; m 60; c 2. BIOCHEMISTRY, MOLECULAR BIOLOGY. Educ: Boston Univ, AB, 55, PhD(biochem), 63; Univ RI, MS, 57. Prof Exp: Staff scientist, Worcester Found Exp Biol, 63-64; pharmacologist, Food & Drug Admin, 66-68; staff fel biochem, Nat Inst Allergy & Infectious Dis, 68-69, sr staff fel, 69-70; sr staff fel biochem, Nat Cancer Inst, 70-72; RES CHEMIST, NAT BUR STAND, 72- Concurrent Pos: Fel, Inst Enzyme Res, Univ Wis, 64-66; instr, Sch Med, Boston Univ, 63-64. Mem: AAAS; Am Chem Soc; NY Acad Sci. Res: Protein hormones and antihormones, characterization and isolation; association of metabolic pathways with biological membranes; modification of viral growth and reproduction by natural and synthetic substances. Mailing Add: 5902 Roosevelt St Bethesda MD 20034

MARGOLIS, SIMEON, b Johnstown, Pa, Mar 29, 31; m 54; c 3. BIOCHEMISTRY. Educ: Johns Hopkins Univ, BA, 53, MD, 57, PhD(lipoprotein struct), 64. Prof Exp: From intern to asst resident, Johns Hopkins Hosp, 57-59; res assoc biochem, Nat Heart Inst, 59-61; resident med, Johns Hopkins Hosp, 64-65; asst prof med & physiol chem, 65-68, ASSOC PROF MED, SCH MED, JOHNS HOPKINS UNIV, 68- Concurrent Pos: Fel biochem, Sch Med, Johns Hopkins Univ, 61-64; Nat Heart Inst res grant, 65-; mem metab study sect, USPHS; mem coun on arteriosclerosis, Am Heart Asn. Mem: Am Diabetes Asn; Endocrine Soc; Am Soc Clin Invest; Am Soc Biol Chem; Am Chem Soc. Res: Lipid biochemistry; regulation of lipid biosynthesis; metabolism of isolated hepatocytes; structure of human serum lipoproteins; role of serum lipoproteins in atherosclerosis. Mailing Add: Dept of Med Johns Hopkins Univ Sch of Med Baltimore MD 21205

MARGOLIUS, HARRY STEPHEN, b Albany, NY, Jan 29, 38; m 64; c 2. CLINICAL PHARMACOLOGY. Educ: Union Univ, BS, 59; Albany Med Col, PhD(pharmacol), 63; Univ Cincinnati, MD, 68. Prof Exp: From intern to resident med, Harvard Med Serv II & IV, Boston City Hosp, 68-70; res assoc pharmacol, Exp Therapeut Br, Nat Heart & Lung Inst, 70-72, sr clin investr hypertension res, Hypertension-Endocrine Br, 72-74; ASSOC PROF PHARMACOL & ASST PROF MED, MED UNIV SC, 74-, PROG DIR, GEN CLIN RES CTR, 74- Concurrent Pos: Attend physician, Clin Ctr, NIH, 70-74; attend physician, Med Univ SC Hosp, Charleston County Hosp & Vet Admin Hosp, 74-; mem, Hypertension Task Force, Nat Heart & Lung Inst, 75-77; Nat Heart & Lung Inst res grant, 75-78; ad hoc reviewer, NSF, 75- & Cardiovasc & Renal Study Sect, Nat Heart & Lung Inst, 76; Burroughs-Wellcome scholar clin pharmacol, 76; mem med adv bd, Coun High Blood Pressure Res, Am Heart Asn. Mem: AAAS; Am Fedn Clin Res; Am Soc Pharmacol & Exp Therapeut; Am Heart Asn. Res: Studies of the regulation of the kallikrein-kinin system and its role in renal function and the pathogenesis of hypertensive diseases using isolated cell suspensions and cultures, whole animals and clinical investigation. Mailing Add: Dept of Pharmacol Med Univ of SC 80 Barre St Charleston SC 29401

MARGOSHES, MARVIN, b New York, NY, May 23, 25; m 55; c 4. ANALYTICAL CHEMISTRY. Educ: Polytech Inst Brooklyn, BS, 51; Iowa State Col, PhD(phys chem), 53. Prof Exp: Asst, Inst Atomic Res, Iowa State Col, 50-53; res fel med, Harvard Med Sch, 54-56, res assoc, 56-57; res assoc spectrochem anal sect, Nat Bur Standards, 57-69; proj dir, Dunn Anal Instruments Div, Block Eng, Inc, 69-70; TECH DIR, TECHNICON INSTRUMENT CORP, 71- Concurrent Pos: Ed, Atomic Spectra Sect, Spectrochimica Acta, 66- Mem: Am Chem Soc; Soc Appl Spectros (pres, 74), Sigma Xi. Res: Analytical spectroscopy; clinical chemistry. Mailing Add: Technicon Instrument Corp Tarrytown NY 10591

MARGRAVE, JOHN LEE, b Kansas City, Kans, Apr 13, 24; m 50; c 2. PHYSICAL INORGANIC CHEMISTRY, FLUORINE CHEMISTRY. Educ: Univ Kans, BS, 48, PhD(chem), 50. Prof Exp: Atomic Energy Comn fel, Univ Calif, 51-52; from instr to prof chem, Univ Wis, 52-63; chmn dept chem, 67-72, PROF CHEM, RICE UNIV, 63-, DEAN ADVAN STUDIES & RES, 72- Concurrent Pos: Sloan res fel, 57-58; Guggenheim fel, 61; pres, Marchem, Inc, 70-; mem bd trustees, Ctr Res, Inc, Univ Kans, 71-75; consult, Nat Bur Standards, Argonne Nat Lab, Lawrence Radiation Lab, Oak Ridge Nat Lab, NASA & private indust; vpres bd dir, Rice Ctr for Community Design & Res, 72-; dir bd of dirs, Gulf Universities Res Consortium, 74- Honors & Awards: IR 100 award, 70; Southwest Regional Award, Am Chem Soc, 73. Mem: Nat Acad Sci; AAAS; Am Phys Soc; Am Ceramic Soc; Am Chem Soc. Res: High temperature chemistry and thermodynamics; fluorine chemistry; optical and mass spectroscopy; synthetic inorganic, plasma and high pressure chemistry; ESCA. Mailing Add: Dept of Chem Rice Univ Houston TX 77001

MARGRAVE, THOMAS EWING, JR, b Langley Field, Va, Nov 15, 38; m 64; c 4. ASTRONOMY. Educ: Univ Notre Dame, BS, 61; Rensselaer Polytech Inst, MS, 63; Univ Ariz, PhD(astron), 67. Prof Exp: Physicist, US Naval Avionics Facil, 61; aerospace technologist, NASA Manned Spacecraft Ctr, 63; asst prof astron, Georgetown Univ, 67-69; asst prof, 69-73, ASSOC PROF ASTRON, UNIV MONT, 73- Concurrent Pos: NSF sci equip grants, Univ Mont, 70-72 & 75-77; Univ Mont Found res grants, 71-72 & 73-75. Mem: Am Astron Soc. Res: Model stellar atmospheres; solar line profiles; photoelectric photometry. Mailing Add: Dept of Physics & Astron Univ of Mont Missoula MT 59801

MARGULES, DAVID LAWRENCE, physiological psychology, see 12th edition

MARGULIES, GABRIEL, b Bucharest, Rumania, Jan 4, 31; nat US; m 53; c 2. MATHEMATICS. Educ: Univ Paris, BA, 49, Sorbonne, CES, 50; Univ Wash, Seattle, BS, 53; Ind Univ, MA, 54, PhD(math, mech), 58. Prof Exp: Assoc math, Grad Inst Math & Mech, Ind Univ, 53-58; asst prof, Fla State Univ, 58-59, res dir, Grad Ctr, Eglin AFB, 59-61; sr math specialist, Space Div, West Develop Labs, Philco Corp, 61-66; SR STAFF SCIENTIST, LOCKHEED PALO ALTO RES LAB, 72- Res: Differential geometry; tensor analysis; analytical mechanics and dynamics; applied mathematics; gyrodynamics and rigid body mechanics. Mailing Add: Dept 52-56 B-201 Lockheed Palo Alto Res Lab Palo Alto CA 94304

MARGULIES, MAURICE, b Brooklyn, NY, Feb 9, 31; m 67; c 3. BIOCHEMISTRY, PLANT PHYSIOLOGY. Educ: Brooklyn Col, BA, 52; Yale Univ, MS, 53, PhD(microbiol), 57. Prof Exp: Res assoc biol, Haverford Col, 57; McCollum Pratt fel, Johns Hopkins Univ, 57-59; BIOCHEMIST, RADIATION BIOL LAB, SMITHSONIAN INST, 59- Concurrent Pos: Lectr, George Washington Univ, 64-67; res fel, Harvard Univ, 69-70. Mem: AAAS; Am Soc Plant Physiol; Am Soc Biol Chem; Am Soc Cell Biol; Am Chem Soc. Res: Chloroplast biochemistry-protein synthesis, synthesis of chloroplast membranes, photosynthesis, electron transport. Mailing Add: Radiation Biol Lab Smithsonian Inst 12441 Parklawn Rockville MD 20852

MARGULIES, MILTON, b Toronto, Ont, Nov 27, 30; m 59; c 4. ROENTGENOLOGY. Educ: Univ Toronto, MD, 56. Prof Exp: From instr to assoc prof radiol, State Univ NY Downstate Med Ctr, 61-69; ASSOC, LANKENAU HOSP, 69- Concurrent Pos: Consult, Vet Admin Hosp, Brooklyn, 67-69; clin assoc prof radiol, Jefferson Med Col, Thomas Jefferson Univ, 74- Mem: Am Col Radiol; Radiol Soc NAm. Mailing Add: 1212 Green Tree Lane Narberth PA 19072

MARGULIES, SEYMOUR, b Jaslo, Poland, Oct 3, 33; US citizen; m 59; c 2. EXPERIMENTAL HIGH-ENERGY PHYSICS. Educ: Cooper Union, BEE, 55; Univ Ill, MS, 56, PhD(physics), 62. Prof Exp: Nat Acad Sci-Nat Res Coun res fel, Max Planck Inst Nuclear Physics, Ger, 61-63; res assoc nuclear & high energy physics, Nevis Labs, Columbia Univ, 63-65; asst prof, 65-69, ASSOC PROF HIGH-ENERGY PHYSICS, UNIV ILL, CHICAGO CIRCLE, 69- Concurrent Pos: Res grant, co-prin investr, NSF, 73- Mem: Am Phys Soc; Sigma Xi. Res: Mössbauer effect; nuclear spectroscopy; nuclear disintegrations following capture of negative pi-mesons; strong interactions of elementary particles, particularly multiparticle production and high transverse momentum reactions. Mailing Add: Dept of Physics Univ of Ill Chicago Circle Chicago IL 60680

MARGULIES, WILLIAM GEORGE, b New York, NY, Oct 31, 40; m 64; c 2. MATHEMATICS. Educ: State Univ NY Col Long Island, BS, 62; Brandeis Univ, MS, 64, PhD(math), 67. Prof Exp: Asst prof math, Wash Univ, 66-69; ASST PROF MATH, CALIF STATE UNIV, LONG BEACH, 69- Concurrent Pos: NSF grant, 70-72. Mem: Math Asn Am; Am Math Soc; Soc Indust & Appl Math. Res: Analysis, partial differential equations; least action principle. Mailing Add: Dept of Math Calif State Univ Long Beach CA 90840

MARGULIS, ALEXANDER RAFAILO, b Belgrade, Yugoslavia, Mar 31, 21; nat US; m 46. RADIOLOGY. Educ: Harvard Med Sch, MD, 50. Prof Exp: Intern, Henry Ford Hosp, Detroit, 50-51; resident radiol, Univ Mich Hosps, 51-53; jr clin instr, Univ Mich, 53-54; from instr to asst prof radiol, Univ Minn, 54-57; vis assoc, Duke Univ, 58-59; from asst prof to prof, Mallinckrodt Inst Radiol, Sch Med, Wash Univ, 59-63; PROF RADIOL & CHMN DEPT, UNIV CALIF, SAN FRANCISCO, 63- Concurrent Pos: Mem comt radiol, Nat Acad Sci-Nat Res Coun, 64-; consult, Off Surgeon Gen, 67-71, Vet Admin Hosp, Ft Miley & Letterman Gen Hosp, San Francisco & Oak Knoll Naval Hosp, Oakland. Mem: AMA; fel Am Col Radiol; Am Roentgen Ray Soc; Asn Univ Radiol (past pres); Soc Gastrointestinal Radiol (pres-elect, 72). Res: Gastroenterology and arteriography. Mailing Add: Dept of Radiol Univ of Calif San Francisco CA 94143

MARGULIS, LYNN, b Chicago, Ill, Mar 5, 38; m 66; c 4. CELL BIOLOGY, EVOLUTION. Educ: Univ Wis, MS, 60; Univ Calif, Berkeley, PhD(genetics), 65. Prof Exp: Lectr & res assoc biol, Brandeis Univ, 63-64; asst prof, 66-71, ASSOC PROF BIOL, BOSTON UNIV, 71- Honors & Awards: Dimond Award, Bot Soc Am, 75. Mem: Soc Protozool; Am Inst Biol Sci; Am Soc Cell Biol; Am Soc Microbiol; Soc Study Origin Life. Res: Origin and evolution of eukaryote cells; cytoplasmic genetics; microtubules and kinetosomes; evolution of biochemical pathways in cells; morphogenesis in protozoans; spirochetes of termites. Mailing Add: Dept of Biol Boston Univ Boston MA 02215

MARGULIS, THOMAS N, b New York, NY, Sept 7, 37. STRUCTURAL CHEMISTRY. Educ: Mass Inst Technol, BS, 59; Univ Calif, Berkeley, PhD(chem), 62. Prof Exp: Asst prof chem, Brandeis Univ, 62-67; assoc prof, 67-75, PROF CHEM, UNIV MASS, BOSTON, 75- Mem: Am Crystallog Asn; Am Chem Soc. Res: Crystal and molecular structure by x-ray diffraction; small ring compounds; structural chemistry of drugs. Mailing Add: Dept of Chem Univ of Mass Boston MA 02125

MARHENKE, KARL, analytical chemistry, see 12th edition

MARIA, NARENDRA LAL, b Chamba, India, Apr 22, 28; m 57; c 1. APPLIED MATHEMATICS. Educ: Panjab Univ, India, BA, 48, MA, 49; Univ Calif, Berkeley, PhD(appl Math), 68. Prof Exp: Lectr math, Panjab Univ, India, 50-51, sr lectr, 51-59, asst prof, 59-65; teaching assoc, Univ Calif, Berkeley, 65-67; vis lectr, 67-68, assoc prof, 68-70, PROF MATH, STANISLAUS STATE COL, 70-, CHMN DEPT, 70- Mem: Am Math Soc. Res: Partial differential equations; analysis. Mailing Add: Dept of Math Stanislaus State Col Turlock CA 95350

MARIANELLI, ROBERT SILVIO, b Wilmington, Del, Dec 17, 41; m 61; c 2. INORGANIC CHEMISTRY. Educ: Univ Del, BA, 63; Univ Calif, Berkeley, PhD(chem), 66. Prof Exp: Asst prof, 66-71, ASSOC PROF CHEM, UNIV NEBR, LINCOLN, 71- Mem: AAAS; Am Chem Soc; The Chem Soc. Res: The chemistry of metalloporphyrins and related compounds. Mailing Add: Dept of Chem Univ of Nebr Lincoln NE 68508

MARIANI, ELIO PAUL, pharmaceutical chemistry, see 12th edition

MARIANI, HENRY A, b Medford, Mass, Sept 13, 24. BIOCHEMISTRY, PHYSICAL CHEMISTRY. Educ: Boston Col, AB, 47, Tufts Univ, MS, 49. Prof Exp: Instr chem, St Anselm's Col, 49-50; asst prof org chem & biochem, Merrimack Col, 52-60; chmn dept sci, Medford Pub Schs, Mass, 60-62; ASSOC PROF BIOCHEM & PHYS CHEM, BOSTON STATE COL, 62- Mem: AAAS; Am Chem Soc. Res: Cell membranes and transport-photosynthesis. Mailing Add: Dept of Chem Boston State Col Boston MA 02115

MARIANI, JOHN, mathematics, theoretical physics, see 12th edition

MARIANO, PATRICK S, b Passaic, NJ, Aug 31, 42. CHEMISTRY. Educ: Fairleigh Dickinson Univ, BSc, 64; Univ Wis, PhD(chem), 69. Prof Exp: NIH fel, Yale Univ, 68-70; ASST PROF CHEM, TEX A&M UNIV, 70- Mem: Am Chem Soc; The Chem Soc. Res: Organic chemistry; photochemistry; synthetic chemistry. Mailing Add: Dept of Chem Tex A&M Univ College Station TX 77843

MARICICH, TOM JOHN, b Anacortes, Wash, Dec 20, 38; m 64; c 3. ORGANIC CHEMISTRY. Educ: Univ Wash, BS, 61; Yale Univ, MS, 63, PhD(chem), 65. Prof Exp: Chemist, Shell Develop Co, Calif, 65-67; asst prof org chem, N Dak State Univ, 67-70, assoc prof, 70-75; ASST PROF CHEM, CALIF STATE UNIV, LONG BEACH, 75- Mem: Am Chem Soc. Res: Reactive organic intermediates, nitrenes; sulfur-nitrogen functional groups and heterocycles; non-benzenoid aromatic compounds; ylids. Mailing Add: Dept of Chem Calif State Univ Long Beach CA 90840

MARICK, LOUIS, b Butte, Mont, May 16, 03; m 30; c 4. PHYSICS, ENGINEERING. Educ: Mont Col Mineral Sci & Technol, EM, 25; Univ Wash, MS 27; Univ Wis, PhD(physics), 34. Prof Exp: Assoc physics, Univ Wash, Seattle, 27-29; asst physicist, Air Corps, US Army, 29-30; asst physics, Univ Wis, 30-34; res physicist, US Rubber Co, 34-39, mgr conductive rubber develop, 39-45, develop & sales mech div, 45-48, tire eng res, 48-53, asst mgr, testing dept, 53-55, spec asst to dir develop, 55-56, mgr passenger tire eng & design, 56-57, asst mgr passenger tire develop, 57-58, asst prod mgr develop dept, 59-61, mgr sales serv develop dept, 61-63, mgr eng standards, 63-66, Uniroyal Inc, 66-68; CONSULT, MAT DIV, US ARMY TANK-AUTOMOTIVE COMMAND, 68- Mem: Am Phys Soc; Soc Automotive Eng; Soc Indust & Appl Math; Am Soc Testing & Mat. Res: Physical properties of rubber and textiles; tire physics, engineering, testing and development. Mailing Add: Mat Div US Army Tank-Automotive Command Warren MI 48090

MARICLE, DONALD L, electrochemistry, see 12th edition

MARICONDI, CAROLYN WOOD, inorganic chemistry, see 12th edition

MARICONDI, CHRIS, b Oct 13, 41; US citizen; m 70. INORGANIC CHEMISTRY. Educ: WVa Univ, AB, 64; Univ Pittsburgh, PhD(chem), 69. Prof Exp: Asst prof, 69-75, ASSOC PROF CHEM, PA STATE UNIV, McKEESPORT, 75- Mem: Am Chem Soc. Res: Molecular structure. Mailing Add: Dept of Chem Pa State Univ McKeesport PA 15132

MARICQ, HILDEGARD RAND, b Rakvere, Estonia, Apr 23, 25; US citizen; m 48; c 3. PSYCHIATRY. Educ: Free Univ Brussels, Cand, 49, MD, 53. Prof Exp: Intern, Jersey City Med Ctr, 55-56; resident psychiat, Essex County Overbrook Hosp, Cedar Grove, NJ, 57-61; resident, Vet Admin Hosp, Lyons, NJ, 61-62, res assoc, 62-63, clin investr, 63-65, dir microcirc lab, 67-69, sr psychiatrist, 67-73, dir schizophrenia res sect, 69-73; RES ASSOC, DEPT MED, COL PHYSICIANS & SURGEONS, COLUMBIA UNIV, 73- Concurrent Pos: Res fel psychiat, Col Physicians &

Surgeons, Columbia Univ, 65-67; res assoc, Dept Psychiat, Rutgers Med Sch, 67-71, res asst prof, 71-73. Mem: AAAS; AMA; Am Psychiat Asn; Microcirc Soc; Soc Psychophysiol Res. Res: Somatic research in schizophrenia; microcirculation; human genetics; psychophysiology; microcirculation in connective tissue diseases; peripheral circulation. Mailing Add: Dept of Med Columbia Univ Col of Phys & Surg New York NY 10032

MARICQ, JOHN, b Anderlecht, Belg, Sept 14, 22; US citizen; m 48; c 3. ORGANIC CHEMISTRY. Educ: Free Univ Brussels, Lic en Sc, 48, Dr en Sc, 51. Prof Exp: Res chemist, Pharmaceut Div, Belgian Union Chem, 50-54; sr chemist, 54-74, TECH FEL, TECH DEVELOP DEPT, HOFFMANN-LA ROCHE, INC, 74- Mem: Am Chem Soc. Res: Synthetic organic chemistry; research and development of new drugs, vitamins, carotenoids and aromatics. Mailing Add: Tech Develop Dept Hoffmann-La Roche Inc Nutley NJ 07110

MARIEB, ELAINE NICPON, b Northhampton, Mass, Apr 5, 36; m 58; c 2. ANATOMY, PHYSIOLOGY. Educ: Westfield State Col, BSEd, 64; Mt Holyoke Col, MA, 66; Univ Mass, Amherst, PhD(cell biol), 69. Prof Exp: Instr zool, anat, physiol & embryol, Springfield Col, 66-67; asst prof, 69-74, ASSOC PROF BOT, ANAT, PHYSIOL & MICROBIOL, HOLYOKE COMMUNITY COL, 74- Mem: AAAS; Am Soc Zool; Sigma Xi. Res: Kinetic studies on the synthesis of sRNA in yeast; species and tissue variations in transfer RNA populations. Mailing Add: 99 Cherry St Feeding Hills MA 01030

MARIELLA, RAYMOND PEEL, b Philadelphia, Pa, Sept 5, 19; m 43; c 4. ORGANIC CHEMISTRY. Educ: Univ Pa, BS, 41; Carnegie Inst Technol, MS, 42, DSc(org chem), 45. Prof Exp: Asst, Carnegie Inst Technol, 41-44, instr, 44, res chemist, 44-45; Eli Lilly & Co fel, Univ Wis, 45-46; instr chem, Northwestern Univ, 46-49, asst prof, 49-51; assoc prof, 51-55, chmn dept, 51-70, PROF CHEM, LOYOLA UNIV CHICAGO, 55-, DEAN GRAD SCH, 69- Concurrent Pos: Mem, Gov Sci Adv Coun, Ill; exec comt, Coun Grad Schs, 71-74; exec comt mem, Midwestern Asn Grad Schs, 72-; ed, annual Proc, Midwestern Asn Grad Schs, 72-; assoc vpres res, Loyola Univ, Chicago, 74- Mem: AAAS; Am Chem Soc. Res: Synthesis of new pyridine compounds; hyperconjugations; ultraviolet absorption spectra; small ring synthesis; synthesis of carcinolytic substances. Mailing Add: Grad Sch Loyola Univ 820 N Michigan Ave Chicago IL 60611

MARIEN, DANIEL, b New York, NY, Aug 19, 25; m 59; c 3. GENETICS, ZOOLOGY. Educ: Cornell Univ, BS, 49; Columbia Univ, MA, 51, PhD, 56. Prof Exp: From instr to assoc prof, 53-70, PROF BIOL, QUEENS COL, NY, 70- Mem: Genetics Soc Am; Soc Study Evolution. Res: Population genetics and evolution; bird taxonomy. Mailing Add: Dept of Biol Queens Col Flushing NY 11367

MARIENFELD, CARL J, b Chicago, Ill, July 11, 17; m 43; c 4. PEDIATRICS, PREVENTIVE MEDICINE. Educ: Lake Forest Col, BA, 38; Univ Ill, MD, 43; Johns Hopkins Univ, MPH, 60; Am Bd Pediat, dipl, 50. Prof Exp: Resident pediat, Cook County Hosp, Chicago, Ill, 43-45; from instr to assoc prof pediat, Col Med, Univ Ill, 47-57; dir interdiv health related res, 65-68, dir environ health surveillance ctr, 68-75, PROF COMMUNITY HEALTH & MED PRACT, SCH MED, UNIV MO-COLUMBIA, 61-, PROF PEDIAT, 74-; DIR, MATERNAL & CHILD HEALTH & CRIPPLED CHILDREN'S SERV, MO DIV HEALTH, 75- Concurrent Pos: USPHS fel cardiol, Univ Ill, 48-49; assoc dir, Children's Heart Sta, Cook County Hosp, 52-57; consult, USPHS, 62-, mem, Dis Control Study Sect, 64-68; chmn subcomt young cardiac, Am Heart Asn, 64-; partic, White House Conf Health, 65; mem ment retardation res & training comt, Nat Inst Child Health & Human Develop, 67-71; mem subcomt geochem & health, Nat Acad Sci, 70-72. Mem: AAAS; AMA; Am Fedn Clin Res; Genetics Soc Am; fel Am Pub Health Asn. Res: Chronic disease epidemiology; rheumatic fever etiology and clinical management; environmental health and comparative medicine; trace substances in health research. Mailing Add: Rt 1 Ashland MO 65010

MARIER, GUY, b Que, Sept 6, 20; m 49; c 3. PHARMACOLOGY. Educ: Sem of Que, Can, BA, 40; Laval Univ, BSc, 44, PhD(pharmacol), 47. Prof Exp: Res scientist & supt, Northern Labs, Defence Res Bd, Can, 47-56; dir sci dept, Poulenc, Ltd, 56-67; VPRES & DIR MED SERV, BIO-RES LABS LTD, 67- Mem: Can Pharmacol Soc; Can Fedn Biol Soc; Can Soc Chemother; Can Asn Res Toxicol (pres). Res: Pharmaceuticals and biologicals; toxicology; clinical trials on new drugs; liaison with government agencies. Mailing Add: 2380 Charles-Gill Montreal PQ Can

MARIK, JAN, b Ungvar, USSR, Nov 12, 20; m 48; c 1. MATHEMATICAL ANALYSIS. Educ: Univ Prague, RNDr(math), 49. Prof Exp: Asst math, Prague Tech Univ, 48-50; grant, Czech Acad Sci, 50-52, sci worker, 52-53; asst, Prague Univ, 53-56, docent, 56-60, prof, 60-69; vis prof, 69-70, PROF MATH, MICH STATE UNIV, 70- Res: Surface integral and non-absolute convergent integrals in Euclidean spaces; representation of functionals by integrals; oscillatory properties of differential equations of second order. Mailing Add: Dept of Math Mich State Univ East Lansing MI 48823

MARIMONT, ROSALIND BROWNSTONE, b New York, NY, Feb 3, 21; m 51; c 2. APPLIED MATHEMATICS. Educ: Hunter Col, BA, 42. Prof Exp: Physicist electronics, Nat Bur Stand, 42-51, electronic scientist digital comput design, 51-60; MATHEMATICIAN, NIH, 60- Mem: AAAS; Am Women Math; Classification Soc. Res: Applications of linear algebra to biological problems including compartmental analysis and classification schemes; mathematical modeling of biological systems, particularly human visual and auditory systems. Mailing Add: 11512 Yates St Wheaton MD 20902

MARINACCIO, LAWRENCE, physics, see 12th edition

MARINE, IRA WENDELL, b Washington, DC, Apr 15, 27; m 53; c 4. GEOLOGY, HYDROLOGY. Educ: St John's Col, Md, BA, 49; Univ Utah, PhD(geol), 60. Prof Exp: Geologist, US Geol Surv, 51-71; RES ASSOC, E I DU PONT DE NEMOURS & CO, INC, 71- Concurrent Pos: Teaching assoc, Univ SC. Mem: Geol Soc Am; Am Asn Petrol Geol; Am Geophys Union; Am Water Well Asn; Seismol Soc Am. Res: Ground water geology and hydrology. Mailing Add: 1002 Hitchcock Dr Aiken SC 29801

MARINE, WILLIAM MURPHY, b Cleveland, Ohio, Oct 21, 32; c 4. PREVENTIVE MEDICINE, INTERNAL MEDICINE. Educ: Emory Univ, BA, 53, MD, 57; Univ Mich, MPH, 63; Am Bd Internal Med, dipl, 65. Prof Exp: From intern to resident med, NY Hosp-Cornell Med Ctr, 57-59; mem staff, Epidemic Intel Serv Kansas City Field Sta, 59-61; resident med, Grady Mem Hosp, Atlanta, Ga, 61-62; trainee epidemiol, Univ Mich, 62-64; from asst prof to assoc prof prev med, 64-70, PROF PREV MED & COMMUNITY HEALTH, SCH MED, EMORY UNIV, 70- Concurrent Pos: Milbank Mem Fund fac fel, 65; med consult, Southeastern Region, Job Corps, 73- Mem: Am Epidemiol Soc; AMA; Am Fedn Clin Res; Am Pub Health Asn; Asn Teachers Prev Med (secy-treas, 74). Res: Epidemiology and immunology of respiratory virus infections, especially influenza; evaluation of health care delivery. Mailing Add: Dept Prev Med Community Health Emory Univ Sch of Med Atlanta GA 30303

MARINELARENA, RAFAEL, b Rio Piedras, PR, Sept 20, 23; m 51. BACTERIOLOGY. Educ: Ind Univ, AB, 46, MA, 47; Univ Mich, PhD(bact), 50. Prof Exp: Asst bact, Ind Univ, 46-47; asst, Univ Mich, 48-50; assoc, Agr Exp Sta, 50-53, from asst prof to assoc prof, Sch Med, 53-62, PROF BACT, SCH MED, UNIV PR, SAN JUAN, 62-, HEAD DEPT, 69- Mem: AAAS; Am Soc Microbiol; NY Acad Sci; Latin Am Soc Microbiol; Am Inst Biol Sci. Res: Metabolism of leukocytes; fermentations; medical bacteriology; viruses. Mailing Add: Fac of Microbiol Univ of PR Sch of Med San Juan PR 00905

MARINER, ALLEN SHAN, b Newark, NJ, May 1, 25; m 61. PSYCHIATRY. Educ: Swarthmore Col, BA, 45; NY Univ, MD, 48; Am Bd Psychiat & Neurol, dipl, 55. Prof Exp: Pvt pract psychiat, 54-62; DIR, ONT COUNTY MENT HEALTH CTR, 62- Concurrent Pos: Clin instr, State Univ NY Upstate Med Ctr, 62-65; consult, social agencies, Calif, 54-62. Mem: Am Psychiat Asn; Acad Psychother. Res: Psychotherapy; educational preparation for mental health work. Mailing Add: 120 N Main St Canandaigua NY 14424

MARINER, THOMAS, b Blasdell, NY, Nov 17, 13; m 38; c 3. PHYSICS. Educ: Boston Univ, AB, 35; Princeton Univ, PhD(physics), 47. Prof Exp: Asst, Princeton Univ, 37-40; res physicist, Am Cyanamid Co, Conn, 40-48; MGR PHYS RES UNIT, ARMSTRONG CORK CO, 48- Concurrent Pos: Exten instr, New Haven State Teachers Col, 44-45. Mem: AAAS; fel Acoust Soc Am; Am Phys Soc; Am Chem Soc; Am Asn Physics Teachers. Res: Mass spectroscopy; ionization potentials; electronics; acoustical materials; architectural acoustics; vibration damping; thermal insulators. Mailing Add: Longenecker Rd RR 1 Box 3 Mount Joy PA 17552

MARINETTI, GUIDO V, b Rochester, NY, June 26, 18; m 42; c 2. BIOCHEMISTRY. Educ: Univ Rochester, BS, 50, PhD(biochem), 53. Prof Exp: Res biochemist, West Regional Res Lab, USDA, 53-54; from instr to assoc prof, 54-66, PROF BIOCHEM, SCH MED & DENT, UNIV ROCHESTER, 66- Concurrent Pos: Lederle med fac award, 55-56; Nat Heart Inst grants, 55- Mem: AAAS; Am Chem Soc; Am Soc Biol Chem. Res: Biochemistry of phosphatides and other lipids; biosynthesis of phosphatides and neutral glycerides and regulatory or control mechanisms in this process; the topology and function of lipids in cellular membranes; hormone action on cell membranes; hormone binding to receptors on cell membranes. Mailing Add: Dept of Biochem Univ of Rochester Sch Med & Dent Rochester NY 14627

MARINI, JAMES LOUIS, organic chemistry, neurochemistry, see 12th edition

MARINI, MARIO ANTHONY, b Ascoli Piceno, Italy, Oct 18, 25; nat US; m 52; c 3. BIOCHEMISTRY. Educ: St Michael's Col, BS, 49; Wayne State Univ, MS, 52, PhD, 55. Prof Exp: Res assoc, Univ Minn, 55-58; asst prof, 60-65, ASSOC PROF BIOCHEM, MED SCH, NORTHWESTERN UNIV, 65- Concurrent Pos: NIH fel biochem, Cornell Univ, 58-60. Mem: Am Chem Soc; Fedn Am Socs Exp Biol. Res: Clinical calorimetry; enzymatic mechanisms; structure-function in biological polymers; potentiometric and thermal analysis of polymers; specificity and biological significance of soluble ribonucleic acid. Mailing Add: Dept of Biochem Northwestern Univ Med Sch Chicago IL 60611

MARINO, ANDREW ANTHONY, b Philadelphia, Pa, Jan 12, 41; m 65; c 4. BIOPHYSICS. Educ: St Joseph's Col, Pa, BS, 62; Syracuse Univ, MS, 65, PhD(physics), 68. Prof Exp: RES PHYSICIST, VET ADMIN HOSP, SYRACUSE, 63- Concurrent Pos: Assoc prof, State Univ NY Upstate Med Ctr, 65- Mem: AAAS; Am Inst Physics; Orthop Res Soc. Res: Electric and magnetic properties of biological tissue; growth control systems. Mailing Add: Vet Admin Hosp Irving Ave Syracuse NY 13210

MARINO, JOSEPH PAUL, b Hazleton, Pa, Apr 20, 42; m 67; c 2. ORGANIC CHEMISTRY. Educ: Pa State Univ, BS, 63; Harvard Univ, AM, 65, PhD(chem), 67. Prof Exp: NIH fel, Harvard Univ, 67-69; asst prof, 69-74, ASSOC PROF CHEM, UNIV MICH, ANN ARBOR, 74- Mem: Am Chem Soc. Res: Sulfur chemistry; ylides; synthesis of natural products; heterocyclic chemistry. Mailing Add: Dept of Chem Univ of Mich Ann Arbor MI 48104

MARINO, LAWRENCE LOUIS, b Belleville, Ill, Nov 1, 30; m 53; c 3. HYDRODYNAMICS, NUCLEAR PHYSICS. Educ: Purdue Univ, BS, 52; Univ Calif, MA, 58, PhD, 59. Prof Exp: Systs analyst, NAm Aviation, Inc, 55; asst, Univ Calif, 57-59; staff scientist, Gen Dynamics/Convair, 59-66; PHYSICIST, LAWRENCE LIVERMORE LAB, UNIV CALIF, 66- Mem: AAAS; Am Phys Soc. Res: Atomic and molecular beams; nuclear moments; atomic collisions; hydrodynamics; nuclear and plasma physics. Mailing Add: Lawrence Livermore Lab Univ of Calif PO Box 808 Livermore CA 94550

MARINO, ROBERT ANTHONY, b Positano, Italy, Feb 19, 43; US citizen; m 67; c 1. PHYSICS. Educ: City Col York, BS, 64; Brown Univ, PhD(physics), 69. Prof Exp: Res assoc physics, Brown Univ, 69-70; ASST PROF PHYSICS, HUNTER COL, CITY UNIV NEW YORK, 70- Concurrent Pos: Consult, US Army Res Off, 70- Mem: Am Phys Soc; Am Asn Physics Teachers. Res: Nitrogen-14 nuclear quadrupole resonance; hydrogen bond studies. Mailing Add: Dept of Physics Hunter Col 695 Park Ave New York NY 10021

MARIN-PADILLA, MIGUEL, b Jumilla, Spain, July 9, 30; nat US; m 58; c 2. PATHOLOGY. Educ: Univ Granada, BS, 49, MD, 55; Educ Coun Foreign Med Grads, cert, 60; Am Bd Path, dipl & cert anat path, 65. Prof Exp: Teaching fel path, Sch Med, Boston Univ, 60-62 & Harvard Med Sch, 61-62; from instr to assoc prof, 62-75, PROF PATH, DARTMOUTH MED SCH, 75- Concurrent Pos: Consult, Vet Admin Hosp, White River Junction, Vt, 64- Mem: Teratol Soc; Am Asn Anat; Soc Neurosci. Res: Development pathology; neurohistology; human and experimental teratology. Mailing Add: Dept of Path Dartmouth Med Sch Hanover NH 03755

MARINSKY, JACOB A, inorganic chemistry, see 12th edition

MARINUS, MARTIN GERARD, b Amsterdam, Neth, June 22, 44; m 70; c 2. MICROBIAL GENETICS. Educ: Univ Otago, NZ, BSc, 65, PhD(microbiol), 68. Prof Exp: Fel genetics, Yale Univ, 68-70; vis fel microbiol, Free Univ, Amsterdam, Neth, 70-71; instr pharmacol, Col Med & Dent NJ, Rutgers Med Sch, 71-74; ASST PROF PHARMACOL, MED SCH, UNIV MASS, 74- Mem: Am Soc Microbiol. Res: Function of methylated bases in nucleic acids. Mailing Add: Dept Pharmacol Med Sch Univ Mass 55 Lake Ave N Worcester MA 01605

MARIO, ERNEST, b Clifton, NJ, June 12, 38; m 61; c 3. PHYSICAL PHARMACY. Educ: Rutgers Univ, BS, 61; Univ Rhode Island, MS, 63, PhD(enzyme kinetics), 65. Prof Exp: Instr pharm, Univ RI, 64-66; dept head anal develop, Strasenburgh LabsDiv, Wallace-Tiernan Inc, 66-69, dir qual control, 69-71, dir US pharmaceut prods, 71-73, dir prod, 73-75, VPRES MFG US, SMITH KLINE & FRENCH LABS,

75- Mem: Am Pharmaceut Asn. Mailing Add: Smith, Kline & French Labs 1500 Spring Garden St Philadelphia PA 19101

MARION, ALEXANDER PETER, b New York, NY, Apr 24, 15; m 43. PHYSICAL CHEMISTRY. Educ: City Col New York, BS, 36, MS, 39; NY Univ, PhD(chem), 44. Prof Exp: Lectr asst, 37-41, tutor, 41-43, from instr to assoc prof, 43-64, PROF CHEM, QUEENS COL, NY, 64- Concurrent Pos: Designer, Microchem Serv, 42-48. Mem: AAAS; Am Chem Soc. Res: Chemical kinetics; teaching aids; electronic laboratory apparatus. Mailing Add: Dept of Chem Queens Col Flushing NY 11367

MARION, GERMAIN BERNARD, animal physiology, see 12th edition

MARION, GILES MICHAEL, b Potsdam, NY, Mar 15, 43. FOREST SOILS, SOIL CHEMISTRY. Educ: Syracuse Univ, BS, 65, MS, 68; Univ Calif, Berkeley, PhD(soil sci), 74. Prof Exp: Res assoc soil chem, Univ Ariz, 72-74; RES SPECIALIST FOREST SOILS, WEYERHAEUSER CO, 74- Mem: Soc Am Foresters; Am Soc Agron; Soil Sci Soc Am; AAAS. Res: Analytical chemistry, particularly the development of soil and tissue chemical tests for diagnosing tree nutritional problems. Mailing Add: Weyerhaeuser Co Tech Ctr PO Box 188 Longview WA 98632

MARION, JAMES EDSEL, b Cana, Va, May 30, 35; m 57; c 2. FOOD SCIENCE, NUTRITION. Educ: Berea Col, BS, 57; Univ Ky, MS, 59; Univ Ga, PhD(nutrit), 62. Prof Exp: Res asst poultry nutrit, Univ Ky, 57-59; res asst poultry nutrit, Univ Ga, 59-62, asst food technologist, Ga Exp Sta, 62-67, assoc food scientist & head food sci dept, 67-69; asst dir res, 69-72, DIR RES, GOLD KIST RES CTR, 72- Mem: AAAS; Am Inst Nutrit; Inst Food Technol; Oil Chem Soc; Poultry Sci Asn. Res: Feed and nutrition; plant breeding; product development. Mailing Add: Gold Kist Res Ctr 2230 Industrial Blvd Lithonia GA 30058

MARION, JERRY BASKERVILLE, b Mobile, Ala, Dec 19, 29; m 52; c 2. NUCLEAR PHYSICS. Educ: Reed Col, BA, 52; Rice Univ, MA, 53, PhD(physics), 55. Prof Exp: NSF fel, Calif Inst Technol, 55-56; instr physics, Univ Rochester, 56-57; physicist, Los Alamos Sci Lab, Univ Calif, 57; PROF PHYSICS, UNIV MD, COLLEGE PARK, 57- Concurrent Pos: Sr staff scientist, Convair Div, Gen Dynamics Corp, Calif, 60-61; Guggenheim fel, Calif Inst Technol, 65-66; consult, Oak Ridge Nat Lab, 58-70; Grumman Aircraft Eng Co, 62-67; mem nuclear data group, Nat Acad Sci, 57-61; subcomt nuclear struct, Nat Acad Sci-Nat Res Coun, 59-69. Mem: Am Phys Soc. Res: Experimental low-energy nuclear physics; theory of nuclear structure and reactions. Mailing Add: Dept of Physics & Astron Univ of Md College Park MD 20472

MARION, LEO (EDMOND), b Ottawa, Ont, Mar 22, 99; m 33. NATURAL PRODUCTS CHEMISTRY. Educ: Queen's Univ, Can, BSc, 26; McGill Univ, MSc, 27, PhD, 29. Hon Degrees: DSc, Laval Univ, 54, Univ Ottawa, 58, Univ Montreal, 61, Queen's Univ, Can, 61, Univ BC, 63, Royal Mil Col Can, 65, Carleton Univ, 65, McGill Univ, 66 & Univ Poznan, 67; LLD, Univ Toronto, 62, Univ Sask, 68; DUniv, Sorbonne, 62; DCL, Bishop's Univ, Can, 66. Prof Exp: Res chemist, labs, Nat Res Coun Can, 29-52, dir div pure chem, 52-63, sr dir coun, 60-63; sci vpres, 63-65; dean fac pure & appl sci, 65-69, EMER DEAN, UNIV OTTAWA, 69- Concurrent Pos: Ed-in-chief, Can J Res, 47-65; ed, Can J Chem, 52-63; pres org div, Int Union Pure & Appl Sci, 63-65, mem bur, 65-69. Honors & Awards: Medal, French-Can Asn Advan Sci, 48; Medal, Chem Inst Can, 56; Medal, Prof Inst Pub Serv Can, 59; Jecker Prize, French Acad Sci, 63; Companion Order Can, 67; Order Brit Empire; Montreal Medal, Chem Inst Can, 69. Mem: Am Chem Soc; fel Royal Soc Can (pres, 64-65); Chem Inst Can (pres, 61-63); French-Can Asn Advan Sci; fel Royal Soc. Res: Condensation reactions; isolation of alkaloids and determination of their structure and synthesis; Lupin alkaloids; lycopodium and delphinium alkaloids; biogenesis of alkaloids. Mailing Add: Fac of Pure & Appl Sci Univ of Ottawa Ottawa ON Can

MARION, STEPHEN PAUL, chemistry, deceased

MARION, WILLIAM W, b Hillsville, Va, Feb 3, 30; m 54; c 4. FOOD SCIENCE. Educ: Berea Col, BS, 53; Purdue Univ, MS, 55, PhD(food technol), 58. Prof Exp: Instr poultry husb, Purdue Univ, 55-58; from asst prof to prof animal sci, 58-74, chmn dept poultry sci, 68-71, PROF FOOD TECHNOL & HEAD DEPT, IOWA STATE UNIV, 74- Mem: Inst Food Technol; Am Oil Chem Soc; Poultry Sci Asn; Am Inst Nutrit. Res: Structure and composition of muscle lipids; post-mortem biochemical changes in muscle. Mailing Add: Dept Food Technol Iowa State Univ Ames IA 50011

MARISCAL, RICHARD NORTH, b Los Angeles, Calif, Oct 4, 35; m 74; c 2. MARINE BIOLOGY, INVERTEBRATE ZOOLOGY. Educ: Stanford Univ, AB, 57, MA, 61; Univ Calif, Berkeley, PhD(zool), 66. Prof Exp: Asst entom, Univ Calif, Berkeley, 60-61, asst zool, 61-64, lectr, 66; fac asst, Te Vega & Int Indian Ocean Expeds, Hopkins Marine Sta, Stanford Univ, 64-65; NIH fel, Lab Quant Biol, Univ Miami, 67-68; asst prof, 68-72, ASSOC PROF BIOL SCI, FLA STATE UNIV, 72- Mem: AAAS; Am Soc Zool; Ecol Soc Am; Asn Trop Biol; NY Acad Sci. Res: Morphology and ecology of the Entroprocta; coelenterate nematocyst physiology, biochemistry and morphology; symbiosis between sea anemones, fishes and crustaceans; chemical control of feeding in corals and other coelenterates; invertebrate behavior and ecology. Mailing Add: Dept of Biol Sci Fla State Univ Tallahassee FL 32306

MARISCOTTI, MARIO ALBERTO JUAN, nuclear physics, see 12th edition

MARK, DANIEL LEE, b Des Moines, Iowa, Dec 6, 43; m 67; c 1. PARASITOLOGY, NEMATOLOGY. Educ: Drake Univ, BA, 68, MA, 70; Univ Ill, Urbana, PhD(zool), 74. Prof Exp: Teaching assoc vet parasitol, Col Vet Med, Univ Ill, Urbana, 73-75; ASST PROF BIOL, KNOX COL, ILL, 75- Mem: Am Soc Parasitologists. Res: Influence of microenvironmental factors on free-living stages of parasitic nematodes; biological indicators of fresh water pollution. Mailing Add: Dept of Biol Knox Col Galesburg IL 61401

MARK, EARL LARRY, b Ogden, Utah, Dec 13, 40; m 62; c 4. PHYSICAL CHEMISTRY. Educ: Weber State Col, BS, 65; Univ Idaho, PhD(phys chem), 70. Prof Exp: Res chemist, Amalgamated Sugar Co, 70-73; dir res, Water Refining Co, 73-74; RES MKT SPECIALIST, BLACK CLAWSON CO, 74- Mem: Am Chem Soc. Res: Ion exchange; surface adsorption; use of radiotracers in adsorption studies; activated carbon adsorption; liquid-solid separation. Mailing Add: 8716 Meadowlark Franklin OH 45005

MARK, HANS MICHAEL, b Mannheim, Ger, June 17, 29; nat US, m 51; c 2. PHYSICS. Educ: Univ Calif, AB, 51; Mass Inst Technol, PhD(physics), 54. Prof Exp: Asst, Mass Inst Technol, 52-54, res assoc, 54-55; jr res physicist, Univ Calif, 55-56, physicist, Lawrence Radiation Lab, 56-58; asst prof physics, Mass Inst Technol, 58-60; assoc prof nuclear eng, Univ Calif, Berkeley, 60-66, prof, 66-69, chmn dept, 64-69, physicist, Lawrence Radiation Lab, 60-69; leader exp physics div, 60-64, DIR, AMES RES CTR, NASA, 69- Concurrent Pos: Lectr, Dept Appl Sci, Univ Calif, Davis, 69-73; consult, Inst Defense Anal, US Army, DC; consult prof sch eng, Stanford Univ, 73-; consult, US Air force. Honors & Awards: Distinguished Serv Medal, NASA, 72. Mem: Fel Am Phys Soc; Am Geophys Union; Am Nuclear Soc; fel Am Inst Aeronaut & Astronaut. Res: Nuclear and atomic physics; nuclear instrumentation; astrophysics. Mailing Add: NASA-Ames Res Ctr Moffett Field CA 94035

MARK, HAROLD WAYNE, b Chanute, Kans, May 2, 49. PHYSICAL ORGAINC CHEMISTRY. Educ: Univ Kans, BS, 71; Northwestern Univ, PhD(chem), 75. Prof Exp: CHEMIST, PHILLIPS PETROL CO, 75- Mem: AAAS; Am Chem Soc; Sigma Xi. Res: Carbonium ion chemistry and nucleophilic substitution reactions. Mailing Add: Phillips Petrol Co Phillips Res Ctr Bartlesville OK 74004

MARK, HARRY BERST, JR, b Camden, NJ, Feb 28, 34; m 60; c 3. ELECTROCHEMISTRY, ANALYTICAL CHEMISTRY. Educ: Univ Va, BA, 56; Duke Univ, PhD(electrochem), 60. Prof Exp: Assoc, Univ NC, 60-62; fel, Calif Inst Technol, 62-63; from asst prof to assoc prof chem, Univ Mich, Ann Arbor, 63-70; PROF CHEM, UNIV CINCINNATI, 70- Concurrent Pos: Vis prof, Free Univ Brussels, 70; cong legis Counr, Am Chem Soc, 74- Mem: AAAS; Am Chem Soc; Electrochem Soc; NY Acad Sci; Am Inst Chem. Res: Heterogeneous electron transfer kinetics; electrical double layer phenomena; electroanalytical techniques; neutron activation analysis; kinetic methods for the analysis of closely related mixtures; bioelectrochemistry; environmental analysis methods. Mailing Add: Dept of Chem Univ of Cincinnati Cincinnati OH 45221

MARK, HERBERT, b Jersey City, NJ, June 10, 21; m 45; c 3. MEDICINE, CARDIOLOGY. Educ: Columbia Univ, AB, 42; Long Island Col Med, MD, 45; Am Bd Internal Med, dipl, 53. Prof Exp: Resident med, Montefiore Hosp, New York, 48-49; resident, Vet Admin Hosp, Bronx, 49-50; pvt pract, 51-64; assoc prof med, NY Med Col, 64-67; asst prof prev med & med, Albert Einstein Col Med, 67-69; clin assoc prof med, NJ Col Med, 69-72, prof med, 72-75; CHIEF, MED SERV, VET ADMIN HOSP, BRONX, NY, 75- Concurrent Pos: Fel cardiol, Montefiore Hosp, New York, 48-49, mem staff, Montefiore Hosp, assoc attend physician, 61-67, attend physician, 67-; attend physician, Vet Admin Hosp, Bronx; assoc chief med, chief cardiol & assoc attend physician, Bird S Coler Hosp, 64-67; assoc attend physician, Flower & Metrop Hosps, 64-; dir ambulatory serv, Montefiore-Morrisania Affiliation, 67-69; chief med, Jersey City Med Ctr, 69-75; adj attend physician, Med Serv, Mt Sinai Hosp, New York, 75- Mem: Am Fedn Clin Res; fel Am Col Physicians; fel Am Col Cardiol; Am Heart Asn. Res: Vectorcardiography; congenital heart disease. Mailing Add: Bronx Vet Admin Hosp 130 W Kingsbridge Rd Bronx NY 10468

MARK, HERMAN FRANCIS, b Vienna, Austria, May 3, 95; nat US; m 22; c 2. PHYSICAL CHEMISTRY. Educ: Univ Vienna, PhD, 21, Dr rer nat, 56. Hon Degrees: EngD, Univ Leige, 49; PhD, Uppsala Univ, 42, Lowell Technol Inst, 57 & Munich Tech Univ, 60. Prof Exp: Instr physics & phys chem, Univ Vienna, 19-21; instr org chem, Univ Berlin, 21-22; from res fel to group leader, Kaiser Wilhelm Inst, Dahlem, 22-26; res chemist, I G Farben-Indust, 27-28, group leader, 28-30, asst res dir, 30-32; prof chem, Univ Vienna, 32-38; adj prof org chem, 40-42, prof, 42-46, dir polymer res inst, 46-70, EMER DEAN, POLYTECH INST BROOKLYN, 70- Concurrent Pos: Assoc prof, Karlsruhe Tech Inst, 27-32; tech consult, US Navy, Qm Corps, US Army; NSF ed, J Polymer Sci, J Appl Polymer Sci; Series on Highpolymer, Rev in Polymer Sci, Resins, Rubbers, Plastics & Natural & Synthetic Fibers; chmn tech comt wood chem, Food & Agr Orgn, UN; chmn comn macromolecules, Int Union Pure & Appl Chem; chmn comt macromolecules, Nat Res Coun; vpres in-chg proj res, Am Comt, Weizmann Inst, Israel & Gov Inst; chmn, Gordon Res Conf Macromolecules & Textiles. Mem: Nat Acad Sci; AAAS; Am Chem Soc; fel Am Phys Soc; Soc Rheol. Res: Use of x-rays and electrons in the synthesis, characterization, reactions and properties of natural and synthetic macromolecules. Mailing Add: Polytech Inst of Brooklyn 333 Jay St Brooklyn NY 11201

MARK, J CARSON, b Lindsay, Ont, July 6, 13; US citizen; m 35; c 6. MATHEMATICS, MATHEMATICAL PHYSICS. Educ: Univ Western Ont, BA, 35; Univ Toronto, PhD(math), 38. Prof Exp: Instr math, Univ Man, 38-43; scientist, Montreal Lab, Nat Res Coun Can, 43-45; scientist, Los Alamos Sci Lab, 45-46, MEM STAFF, THEORET PHYSICS DIV, LOS ALAMOS SCI LAB, UNIV CALIF, 56-, DIV LEADER, 47- Concurrent Pos: Mem, Sci Adv Bd, US Air Force; sci adv, US Deleg, Conf Experts Means of Detection Nuclear Explosions, Geneva, 58. Mem: Am Math Soc; Am Phys Soc. Res: Finite group theory; transport theory; hydrodynamics; neutron physics. Mailing Add: Los Alamos Sci Lab PO Box 1663 Los Alamos NM 87544

MARK, JAMES EDWARD, b Wilkes-Barre, Pa, Dec 14, 34; m 64; c 2. POLYMER CHEMISTRY. Educ: Wilkes Col, BS, 57; Univ Pa, PhD(phys chem), 62. Prof Exp: Res chemist, Rohm & Haas Co, 55-56; res asst, Stanford Univ, 62-64; asst prof chem, Polytech Inst Brooklyn, 64-67; from asst prof to assoc prof, 67-72, PROF CHEM, UNIV MICH, ANN ARBOR, 72- Concurrent Pos: Consult, Mechrolab, Inc, Calif, 63-64; vis prof, Stanford Univ, 73-74; spec res fel, NIH, 75-76; lectr short course prog, Am Chem Soc, 73- Mem: AAAS; Am Chem Soc; Am Phys Soc; NY Acad Sci. Res: Statistical properties of chain molecules; elastic properties of polymer networks. Mailing Add: Dept of Chem Univ of Mich Ann Arbor MI 48104

MARK, JAMES WAI-KEE, b Calcutta, India, Aug 29, 43; m 69. ASTROPHYSICS, APPLIED MATHEMATICS. Educ: Univ Calif, Berkeley, BS, 64; Princeton Univ, PhD(astrophys), 68. Prof Exp: Res assoc, Plasma Physics Lab, Princeton Univ, 68-69; CLE Moore instr, 69-70, ASST PROF, APPL MATH, MASS INST TECHNOL, 70- Concurrent Pos: Vis scientist theoret astrophysics, Kitt Peak Nat Observ, 75-76. Mem: Am Astron Soc; Int Astron Union; Soc Indust & Appl Math. Res: Plasma and stellar dynamics; density waves and spiral structure in galaxies; structure, dynamics and evolution of galaxies and rotating stars. Mailing Add: Dept of Math Mass Inst Technol Cambridge MA 02139

MARK, LESTER CHARLES, b Boston, Mass, July 16, 18; m 46; c 2. MEDICINE. Educ: Univ Toronto, MD, 41; Am Bd Anesthesiol, dipl, 52. Prof Exp: Intern, Jewish Mem Hosp, 41-43; asst resident surg, Grace Hosp, New Haven, Conn, 43; resident anesthesiol, Hosp Spec Surg, New York, 47-48; clin instr, Col Med, State Univ NY, 52-53; assoc, 53-54, from asst prof to assoc prof, 54-65, PROF ANESTHESIOL, COL PHYSICIANS & SURGEONS, COLUMBIA UNIV, 65- Concurrent Pos: Res fel, Res Serv, NY Univ-Bellevue Med Ctr & Goldwater Mem Hosp, 48-51; Am Heart Asn res fel, 49-51; travel award, Int Cardiol Cong, Paris, 50; Guggenheim fel, 60-61; Macy fac scholar, Switz, 74-75; asst adj anesthesiologist, Jewish Mem Hosp, New York, 47-52; asst clin vis anesthesiologist, Jewish Mem Hosp, New York, 47-52; asst clin vis anesthesiologist, Goldwater Mem Hosp, 48-50; dir anesthesiol, Brunswick Gen Hosp, Amityville, & anesthesiologist, SNassau Communities Hosp, Oceanside, 51-53; dir anesthesiol, Freeport Hosp, assoc vis anesthesiologist, Kings County Hosp, Brooklyn & anesthesiologist, Vet Admin Hosp, Northport, 52-53; from asst attend anesthesiologist to assoc attend anesthesiologist, Presby Hosp, New York, 53-65, attend anesthesiologist, 65-, collab med & exp pharmacol, 59-; Fulbright res prof, Denmark, 60-61; actg consult, WHO Anaesthesia Ctr, Copenhagen, 60-61; consult, Coun Drugs, AMA, 62-; mem adv comt respiratory & anesthetic drugs, Food & Drug

Admin, 66-70, mem, Over-The-Counter Hypnotics, Tranquillizers & Sleep-Aids Rev Panel, 72-; mem pharmacol-toxicol rev comt, Nat Inst Gen Med Sci, 68-70, prog comt, 70-72, chmn, 71-72; mem prof adv bd, Found Thanatology, 68-, exec comt, 74-; China Med Bd vis prof, Sapporo Med Col, Japan, 67; guest scientist, Med Dept & vis attend physician, Med Res Ctr, Brookhaven Nat Lab, 68-71. Honors & Awards: Hiroshima Univ Medal, 67; Distinguished Serv Award, NY State Jour Med, 67. Mem: AAAS; Am Soc Anesthesiol; Am Soc Pharmacol & Exp Therapeut; sr mem Asn Univ Anesthet; fel Am Col Anesthesiol. Res: Barbiturates; drug metabolism and distribution; mechanisms of drug action in man; thanatology; hypnosis; acupuncture. Mailing Add: Col of Physicians & Surgeons Columbia Univ New York NY 10032

MARK, PETER HERMAN, b Mannheim, Ger, Apr 8, 31; US citizen; m 55; c 1. SURFACE PHYSICS, MATERIALS SCIENCE. Educ: Harvard Univ, BA, 53; NY Univ, PhD(physics), 58. Prof Exp: Res scientist, Polaroid Corp, 58-62; vis lectr physics, Munich Tech Univ, 61-62; mem tech staff, RCA Labs, Inc, 62-66; mem fac, 66-67, assoc prof, 67-72, PROF ELEC ENG, PRINCETON UNIV, 72- Concurrent Pos: Ed, J Vacuum Sci & Technol, 75-; mem bd dir, Int Rectifier Corp, 75- Mem: Am Vacuum Soc; Am Phys Soc. Res: Insulator physics; electronic and surface properties of semiconductors and insulators. Mailing Add: Dept of Elec Eng Princeton Univ Princeton NJ 08540

MARK, ROBERT VINCENT, b Jamaica, NY, Dec 22, 42. ORGANIC CHEMISTRY. Educ: St John's Univ, NY, BS, 64, MS, 66, PhD(org chem), 71. Prof Exp: Instr, 70-73, ASST PROF GEN & ORG CHEM, STATE UNIV NY AGR & TECH COL FARMINGDALE, 73- Concurrent Pos: NSF traineeship, 70; res assoc, Long Island Jewish-Hillside Med Ctr, 73-74; consult, Pall Corp, 75- Mem: Am Chem Soc; Am Inst Chem. Res: Preparation and mass spectral characteristics of small ring heterocyclic compounds. Mailing Add: Dept of Chem State Univ of NY Agr & Tech Col Farmingdale NY 11735

MARK, ROGER G, b Boston, Mass, June 4, 39; m 66; c 4. ELECTRICAL ENGINEERING, INTERNAL MEDICINE. Educ: Mass Inst Technol, BS, 60, PhD(elec eng), 66; Harvard Med Sch, MD, 65. Prof Exp: Intern & resident internal med, Harvard Med Serv-Boston City Hosp, 65-67; med officer, Spec Weapons Defense, US Air Force, 67-69; instr med, Harvard Med Sch, 69-72; asst prof, 69-72, ASSOC PROF ELEC ENG, MASS INST TECHNOL, 72-; ASST PROF MED, HARVARD MED SCH, 72- Res: Biomedical instrumentation; medical care delivery systems; cardiovascular physiology. Mailing Add: Rm 36-789 Mass Inst Technol Cambridge MA 02139

MARK, SHEW-KUEY, b China, Aug 8, 36; Can citizen. EXPERIMENTAL NUCLEAR PHYSICS. Educ: McGill Univ, BSc, 60, MSc, 62, PhD(nuclear physics), 65. Prof Exp: Nat Res Coun Can fel, Univ Man, 65-66; from asst prof to assoc prof, 66-75, PROF PHYSICS, McGILL UNIV, 75-, DIR FOSTER RADIATION LAB, 71- Mem: Can Asn Physicists. Res: Nuclear reactions; spectroscopy; structural studies. Mailing Add: Foster Radiation Lab McGill Univ Montreal PQ Can

MARK, VICTOR, b Marosvasarhely, Hungary, Feb 9, 21; nat US, m 55; c 6. ORGANIC CHEMISTRY. Educ: Polytech Budapest, Hungary, Dipl, 44; Northwestern Univ, PhD(chem), 55. Prof Exp: Res chemist insecticides, Arzola, Hungary, 44-46, Atox, 46-48; consult, Montecatini, Italy, 49-50, Bombrini, 50-51; res chemist, Universal Oil Prod Co, 51-53, Union Oil Co Calif, 55-56 & Monsanto Chem Co, Mo, 57-62; res scientist, Pennsalt Co, 63; sr res assoc, Hooker Chem Corp, 64-71; SPECIALIST, GEN ELEC CO, 71- Mem: Am Chem Soc; NY Acad Sci. Res: Chloro- carbons; phosphorus chemistry; reaction mechanisms; phosphorus and proton nuclear magnetic resonance spectroscopy; organic polymer chemistry. Mailing Add: Plastics Dept Gen Elec Co Mt Vernon IN 47620

MARKAKIS, PERICLES, b Cassaba, Turkey, Mar 3, 20; nat US; m 53; c 3. FOOD SCIENCE. Educ: Univ Salonika, Greece, BS, 42 & 49; Univ Mass, MS, 52, PhD(food technol), 56. Prof Exp: Instr food sci, Univ Salonika, Greece, 42-50; asst res prof food technol, Univ Mass, 55-56; sr food technologist, DCA Food Industs, Inc, 56-57; res food technologist, Univ Calif, 57-59; from asst prof to assoc prof, 59-69, PROF FOOD SCI, MICH STATE UNIV, 70- Mem: Am Chem Soc; Inst Food Technologists; Am Soc Plant Physiol. Res: Chemistry and processing of fruits and vegetables; irradiation preservation of foods. Mailing Add: Dept of Food Sci Mich State Univ East Lansing MI 48823

MARKARIAN, DERAN, genetics, see 12th edition

MARKEES, DIETHER GAUDENZ, b Basel, Switz, Oct 16, 19; nat US, div; c 1. ORGANIC CHEMISTRY. Educ: Univ Basel, Dr phil, 46. Prof Exp: Res fel med chem, Univ Va, 47-48; res assoc E R Squibb & Sons, 49-53 & Amherst Col, 53-58; from asst prof to assoc prof, 58-68, PROF CHEM, WELLS COL, 68- Mem: Am Chem Soc; Swiss Chem Soc. Res: Medicinal chemistry; chemistry of heterocycles; synthetic organic chemistry. Mailing Add: Dept of Chem Wells Col Aurora NY 13026

MARKELL, EDWARD KINGSMILL, b Brooklyn, NY, Apr 14, 18; m 53; c 2. PARASITOLOGY, TROPICAL MEDICINE. Educ: Pomona Col, BA, 38; Univ Calif, PhD(zool), 42; Stanford Univ, MD, 51. Prof Exp: Asst zool, Univ Calif, 38-41; intern, Stanford Univ Hosps, 50-51; asst prof infectious dis, Sch Med, Univ Calif, Los Angeles, 51-58; MEM DEPT INTERNAL MED, KAISER FOUND MED CTR, 58- Concurrent Pos: Markle scholar, 52-57; clin assoc prof prev med, Sch Med, Stanford Univ, 61-70, clin prof, 70- Mem: AAAS; Am Micros Soc; Am Soc Parasitol; Am Soc Trop Med & Hyg; AMA. Res: Parasitic diseases of man; filariasis. Mailing Add: Kaiser Found Med Ctr Oakland CA 94611

MARKER, DAVID, b Atlantic, Iowa, Mar 20, 37; m 66; c 2. THEORETICAL PHYSICS. Educ: Grinnell Col, BA, 59; Pa State Univ, MS, 62, PhD(physics), 66. Prof Exp: From asst prof to assoc prof, 65-72, assoc dean nat sci, 73-74, PROF PHYSICS, HOPE COL, 72-, PROVOST, 74- Mem: Am Phys Soc; Sigma Xi. Res: Theoretical high energy physics; calculation of nucleon-nucleon bremsstrahlung cross sections; analytic approximation theory. Mailing Add: Off of Provost Hope College Holland MI 49423

MARKER, LEON, b Lancaster, Pa, Jan 6, 22; m 53; c 3. PHYSICAL CHEMISTRY. Educ: Temple Univ, AB, 47; Univ Utah, PhD(phys chem), 51. Prof Exp: Res chemist, Gen Res Labs, Olin Industs, 52-57, res chemist, Film Res Dept, Olin Mathieson Chem Co, Conn, 57-63; HEAD POLYMER PHYSICS SECT, RES & DEVELOP DIV, GEN TIRE & RUBBER CO, 63- Mem: AAAS; Am Chem Soc; Soc Rheol. Res: Polymer physics; rheology; physical chemistry of polymers; chemical kinetics; kinetics of electrode reactions. Mailing Add: Res & Develop Div Gen Tire & Rubber Co Akron OH 44309

MARKERT, CLEMENT LAWRENCE, b Las Animas, Colo, Apr 11, 17; m 40; c 3. DEVELOPMENTAL GENETICS, ENZYMOLOGY. Educ: Univ Colo, BA, 40; Univ Calif, MS, 42; Los Angeles, Johns Hopkins Univ, PhD(biol), 48. Prof Exp: Merck & Co, Inc & Nat Res Coun fels, Calif Inst Technol, 48-50; from asst prof to assoc prof zool, Univ Mich, 50-57; prof biol, John Hopkins Univ, 57-65; chmn dept, 65-71, PROF BIOL, YALE UNIV, 65-, DIR CTR REPRODUCTIVE BIOL, 74- Concurrent Pos: Managing ed, J Exp Zool, 63-; trustee, Bermuda Biol Sta, 59-; panelist, NSF, 59-63; co-chmn, Develop Biol Cluster, President's Biomed Res Panel, 75; council mem, Am Cancer Soc, 75- Mem: Nat Acad Sci; Genetics Soc Am; Soc Develop Biol (pres, 63-64); Am Inst Biol Sci (pres, 66). Res: Mammalian reproductive physiology; cellular differentiation; developmental genetics; enzymology. Mailing Add: Dept of Biol Yale Univ New Haven CT 06520

MARKESBERY, WILLIAM RAY, b Florence, Ky, Sept 30, 32; m 58; c 3. NEUROLOGY, NEUROPATHOLOGY. Educ: Univ Ky, BA, 60, MD, 64. Prof Exp: Resident neurol, Col Physicians & Surgeons, Columbia Univ, 65-67, instr, 68-69, asst neurologist, Vanderbilt Clin, Columbia Presby Med Ctr, 68-69; asst prof path & neurol, Sch Med & Dent, Univ Rochester, 69-72; ASSOC PROF NEUROL & PATH, UNIV KY, 72- Concurrent Pos: USPHS-NIH-Nat Inst Neurol Dis & Blindness spec fel, Col Physicians & Surgeons, Columbia Univ, 67-69; USPHS res grant, Univ Rochester Med Ctr, 69-70; assoc neurologist, Strong Mem Hosp, 69-, asst pathologist, 70- Mem: Am Acad Neurol; Am Asn Neuropath. Res: Ultrastructural studies of human central nervous system tumors, neurological degenerative disorders and muscle diseases. Mailing Add: Dept of Neurol Univ of Ky Med Ctr Lexington KY 40506

MARKEY, SANFORD PHILIP, b Cleveland, Ohio, June 15, 42; m 66; c 1. ORGANIC CHEMISTRY, PHARMACOLOGY. Educ: Bowdoin Col, AB, 64; Mass Inst Technol, PhD(chem), 68. Prof Exp: Instr pediat, Med Sch, Univ Colo, 69, asst prof, 69-74, asst prof pharmacol, 71-74; RES SCIENTIST PHARMACOL, NIMH, 74- Concurrent Pos: NIH grant mass spectrometry, Med Sch, Univ Colo, 70-74; assoc ed, Org Mass Spectrometry, 72-74. Mem: Am Chem Soc; Am Soc Mass Spectrometry. Res: Mass spectrometry applied to clinical research. Mailing Add: Lab of Clin Sci Nat Inst Ment Health Bethesda MD 20014

MARKGRAF, JOHN HODGE, b Cincinnati, Ohio, Mar 16, 30; m 57; c 2. ORGANIC CHEMISTRY. Educ: Williams Col, BA, 52; Yale Univ, MS, 54, PhD, 57. Prof Exp: Chemist, Procter & Gamble Co, 58-59; from asst prof to assoc prof, 59-69, PROF CHEM, WILLIAMS COL, 69- Mem: Am Chem Soc. Res: Physical organic studies of heterocyclic systems. Mailing Add: Dept of Chem Williams Col Williamstown MA 01267

MARKHAM, ARLEIGH HOLDEN, b Sparta, Wis, July 9, 16; m 41, 72; c 2. LOW TEMPERATURE PHYSICS, URBAN RESEARCH AND DEVELOPMENT. Educ: Univ Wis, BA, 38, PhD(physics), 57. Prof Exp: Mgr personnel & Admin, Gen Physics Dept, Res Lab, Gen Elec Co, 57-62, liaison scientist, Res & Develop Ctr, 62-68; adminstr prog residences in eng practice, Am Soc Eng Educ, 68-71, assoc dir projs & fed rels, 71-73; REGIONAL MGR, PUB TECHNOL, INC, 73- Mem: Am Phys Soc; Am Asn Physics Teachers; Am Soc Eng Educ; Am Soc Testing & Mat. Res: Investigation of mechanisms for transfer and diffusion of technological innovation among urban governments. Mailing Add: Public Technol Inc 1140 Connecticut Ave NW Washington DC 20036

MARKHAM, CHARLES G, b Las Cruces, NMex, Feb 28, 20; m 43; c 3. GEOGRAPHY, METEOROLOGY. Educ: Univ Calif, Los Angeles, BA, 50; Colo State Col, MA, 64; Univ Calif, Berkeley, PhD(geog), 67. Prof Exp: From asst prof to assoc prof, 67-74, PROF GEOG, FRESNO STATE COL, 74- Mem: Am Meteorol Soc; Asn Am Geogrs. Res: Drought and distribution of rainfall. Mailing Add: Dept of Geog Fresno State Col Fresno CA 93726

MARKHAM, CHARLES HENRY, b Pasadena, Calif, Dec 24, 23; m 45, 71; c 5. NEUROLOGY, NEUROPHYSIOLOGY. Educ: Stanford Univ, BS, 47, MD, 51; Am Bd Psychiat & Neurol, dipl, 59. Prof Exp: Teaching fel neurol, Harvard Med Sch, 54-55; from instr to assoc prof, 56-71, PROF NEUROL, SCH MED, UNIV CALIF, LOS ANGELES, 71- Concurrent Pos: Consult, Wadsworth Vet Admin Hosp, Los Angeles, 56- Mem: Am Epilepsy Soc; Am Neurol Asn; Am Acad Neurol. Res: Vestibular, brain-stem and basal ganglia physiology; Parkinson's disease and other movement disorders. Mailing Add: Dept of Neurol Univ of Calif Sch of Med Los Angeles CA 90024

MARKHAM, ELIZABETH MARY, b New Haven, Conn, Oct 12, 29. MATHEMATICS. Educ: St Joseph Col, Conn, BA, 51; Univ Notre Dame, MS, 60, PhD(math), 64. Prof Exp: Teacher high sch, Conn, 54-59; instr math, St Joseph Col, Conn, 64-65; teacher, Our Lady of Mercy Acad, 65-66; asst prof, 66-68, ASSOC PROF MATH & CHMN DEPT, ST JOSEPH COL, CONN, 68- Concurrent Pos: Dir & instr, NSF in serv inst high sch math teachers, 66-69; instr, Cent Conn State Col, 68; NSF consult, US Agency Int Develop Inst High Sch Math Teachers, Ramjas Col, Delhi Univ, 68; mem, Conn State Adv Comt Math, 69- Mem: Math Asn Am; Am Math Soc. Res: Foundations of geometry; transformation geometry. Mailing Add: Dept of Math St Joseph Col West Hartford CT 06117

MARKHAM, JAMES J, b Oreland, Pa, Aug 23, 28; m 52; c 8. ANALYTICAL CHEMISTRY, OCEANOGRAPHY. Educ: Villanova Univ, BS, 50; Temple Univ Minn, PhD(chem), 58. Prof Exp: Res chemist, Whitemarsh Res Lab, Pa Salt Mfg Co, 50-51; from asst prof to assoc prof, 56-67, PROF CHEM, VILLANOVA UNIV, 67-, ASSOC DEAN SCI, 68- Concurrent Pos: USPHS vis res fel, Dept Inorg & Struct Chem, Univ Leeds, 65-66. Mem: Franklin Inst; Am Chem Soc; The Chem Soc. Res: Instrumentation; marine chemistry; water supply and pollution control. Mailing Add: Dept of Chem Villanova Univ Villanova PA 19085

MARKHAM, JORDON JEPTHA, b Samokov, Bulgaria, Dec 25, 16; US citizen; m 43; c 2. PHYSICS. Educ: Beloit Col, BS, 38; Syracuse Univ, MS, 40; Brown Univ, PhD(physics), 46. Prof Exp: Res physicist, Div War Res, Columbia Univ, Conn, 42-45, NY, 45; fel, Clinton Labs, Tenn, 46-47; instr physics, Univ Pa, 47-48; asst prof, Brown Univ, 48-50; physicist, Appl Physics Lab, Johns Hopkins Univ, 50-53; physicist, Zenith Radio Corp, 53-60; sci adv physics res, IIT Res Inst, 60-62, PROF PHYSICS, ITT RES INST, ILL INST TECHNOL, 62- Mem: Am Phys Soc. Res: Oceanographic effect on underwater sound; theory of imperfections in ionic crystals; absorption of sound; second order acoustic fields; spectroscopy of solids; color centers. Mailing Add: 1528 Tyrell Ave Park Ridge IL 60068

MARKHAM, M CLARE, b New Haven, Conn, Aug 12, 19. PHYSICAL CHEMISTRY. Educ: St Joseph Col, Conn, AB, 40; Cath Univ Am, PhD(chem), 52. Prof Exp: Lab instr chem, St Joseph Col, Conn, 42-45; teacher, Sacred Heart High Sch, 45-49; chmn nat sci div, 75-77, PROF CHEM, ST JOSEPH COL, CONN, 52- Concurrent Pos: Res grants, Res Corp & Sigma Xi; consult, US Air Force contracts solar energy conversion, 60-64; sci fac fel with M Calvin, Univ Calif, Berkeley, 67-68; counr, Am Chem Soc, 68-71 & 74-76; coop US scientist, Indian Inst Technol, Madras, 74-76. Mem: AAAS; Am Chem Soc. Res: Photochemical reactions; surface reactions; factors influencing energy transfer between chlorophylls and carotenoids, or oxidant-reductant pairs. Mailing Add: Dept of Chem St Joseph Col 1678 Asylum Ave West Hartford CT 06117

MARKHAM, THOMAS LOWELL, b Apex, NC, Jan 2, 39. ALGEBRA. Educ: Univ NC, Chapel Hill, BS, 61, MA, 64; Auburn Univ, PhD(math), 67. Prof Exp: Asst prof math, Univ NC, Charlotte, 67-68; ASST PROF MATH, UNIV SC, 68- Mem: Am Math Soc; Math Asn Am. Res: Linear algebra. Mailing Add: Dept of Math Univ of SC Columbia SC 29208

MARKIEWITZ, KENNETH HELMUT, b Breslau, Ger, May 18, 27; nat US; m 57; c 3. POLYMER CHEMISTRY. Educ: City Col New York, BS, 51; Columbia Univ, MA, 54, PhD(chem), 57. Prof Exp: Chemist, Schwarz Labs, Inc, NY, 51-52; sr chemist, Atlas Powder Co, 57-63; res chemist, 63-67, SR RES CHEMIST, ICI AMERICA INC, 68- Mem: Am Chem Soc. Res: Isolation of natural products; poison ivy; synthesis of alkenyl phenols; carbohydrates; amines; conformational analysis and structural determinations; polymer synthesis. Mailing Add: ICI America Inc Wilmington DE 19899

MARKING, RALPH H, b Holmen, Wis, Jan 24, 35; m 63. INORGANIC CHEMISTRY. Educ: Wis State Univ, La Crosse, BS, 57; Univ Minn, PhD(inorg chem), 65. Prof Exp: Assoc prof, 63-74, PROF CHEM, UNIV WIS-EAU CLAIRE, 74- Mem: Am Chem Soc; Am Asn Physics Teachers. Res: Thermochemistry and thermodynamics; molecular structure. Mailing Add: Dept of Chem Univ of Wis Eau Claire WI 54701

MARKIW, ROMAN TEODOR, b Tarnopol, Ukraine, June 25, 23; US citizen; m 50. BIOCHEMISTRY, ORGANIC CHEMISTRY. Educ: Univ Conn, BA, 54, PhD(biochem), 66; Rensselaer Polytech, MS, 55. Prof Exp: Biochemist, Vet Admin Hosp, 57-62; USPHS grants, Yale Univ, 65-68; res chemist, 68-72, CHIEF BIOCHEM RES LAB, VET ADMIN CTR, 72- Mem: Am Chem Soc; NY Acad Sci. Res: Isolation and identification of peptides in biological fluids; chemical reactions of polynucleotides and derivatives. Mailing Add: Vet Admin Ctr Martinsburg WV 25401

MARKLAND, ALAN COLIN, b Bolton, Eng, Aug 26, 29; US citizen; m 54; c 3. SURGERY, UROLOGY. Educ: Cambridge Univ, MB, ChB, 53, MA, 54. Prof Exp: Researcher, Mass Gen Hosp, Boston, 60-64; asst prof urol, Univ Iowa Hosp, 64; assoc prof, 64-71, PROF UROL, UNIV MINN HOSP, MINNEAPOLIS, 71- Concurrent Pos: USPHS fel, Univ Leeds, 62-63; consult surgeon, Minneapolis Vet Admin Hosp, 65. Mem: Am Fertil Soc; fel Am Col Surg; fel Am Acad Pediat; Soc Univ Urol; Soc Pediat Urol. Res: Urological surgery; pediatric urology; gender identity; neurologic vesical dysfunction. Mailing Add: Univ of Minn Health Sci Ctr 412 Union St SE Minneapolis MN 55455

MARKLAND, FRANCIS SWABY, JR, b Philadelphia, Pa, Jan 15, 36. BIOCHEMISTRY. Educ: Pa State Univ, BS, 57; Johns Hopkins Univ, PhD(biochem), 64. Prof Exp: Asst prof biochem, Sch Med, Univ Calif, Los Angeles, 66-73; ASSOC PROF BIOCHEM, SCH MED, UNIV SOUTHERN CALIF, 74- Concurrent Pos: NIH fel, Sch Med, Univ Calif, Los Angeles, 64-66 & career develop award, 68-73. Mem: Am Soc Biol Chem; Am Chem Soc; Sigma Xi. Res: Structure of proteins and relation of structure to function in enzymes; biochemistry of blood coagulation; receptor proteins for steroid hormones. Mailing Add: Cancer Res Inst Univ Southern Calif Sch of Med Los Angeles CA 90033

MARKLAND, WILLIAM R, b Brooklyn, NY, Jan 3, 19; m 42; c 3. COSMETIC CHEMISTRY. Educ: Middlebury Col, AB, 41. Prof Exp: Lab supvr, Hercules Powder Co, 42-45; chief chemist, John H Breck Inc, 45-57; res group leader hair prep, Revlon, Inc, 57-58; res mgr hair & makeup prods, Chesebrough-Pond's Inc, 58-71; CONSULT COSMETICS & TOILETRIES, 71- Mem: Am Chem Soc; fel Am Inst Chem; Soc Cosmetic Chem. Res: Surfactant and shampoo chemistry; physical and chemical behavior of the hair; transparent microemulsions of mineral oil and water; cosmetic colors and pigments. Mailing Add: 38 High St Clinton CT 06413

MARKLE, CARROLLE ANDERSON, b Ashfield, Mass, Jan 8, 08; m 53. PLANT ANATOMY. Educ: Mass State Col, BS, 32, MS, 35; Cornell Univ, PhD(plant morphol, anat), 40. Prof Exp: Instr bot, Mass State Col, 32-36; asst, Cornell Univ, 36-38, instr, 38-39; instr biol, Adelphi Col, 39-40, actg chmn dept, 40-41, chmn, 41-44; asst prof, Sweet Briar Col, 45-49; from asst prof to prof, 49-73, chmn dept, 59-64, EMER PROF BIOL, EARLHAM COL, 73- Concurrent Pos: Coordr nursing prog, Reid Mem Hosp, 51-57. Mem: AAAS; Bot Soc Am. Res: Taxonomy and morphology of Ranales; floral anatomy of Liliales; science and biology teaching; personnel and vocational guidance; nursing curriculum. Mailing Add: Norton Hill Rd Ashfield MA 01330

MARKLE, GERALD E, b Detroit, Mich, Feb 17, 14. MATHEMATICS. Educ: Univ Detroit, BS, 36, MS, 38; Univ Mich, MA, 40; Wayne State Univ, PhD, 54. Prof Exp: From instr to assoc prof math, Univ Detroit, 38-57, prof & dir comput lab, 57-63, vchmn dept math, 53-63; PROF ENG & DIR COMPUT CTR, UNIV SANTA CLARA, 63-, DIR DEPT APPL MATH, 67- Concurrent Pos: Lectr, Wayne State Univ, 46-47 & Boston Univ, 49; consult, IBM Corp, 57-58; NSF fac fel, Stanford Univ, 59-60. Mem: Math Asn Am; Soc Indust & Appl Math; Asn Comput Mach. Res: Numerical analysis; applied mathematics; computers. Mailing Add: 2461 Boxwood Dr San Jose CA 95128

MARKLE, H CHESTER, JR, b Brookville, Pa, Mar 31, 26. PHYSICAL CHEMISTRY. Educ: Franklin & Marshall Col, BS, 49; Carnegie Inst Technol, MS, 51, PhD(chem), 55. Prof Exp: From instr to asst prof chem, Chatham Col, 54-61; ASSOC PROF CHEM, SWEET BRIAR COL, 61- Mem: Am Chem Soc. Res: Chemical kinetics; thermodynamics. Mailing Add: Dept of Chem Sweet Briar Col Sweet Briar VA 24595

MARKLEIN, BERNARD C, inorganic chemistry, see 12th edition

MARKLEY, FRANCIS LANDIS, b Philadelphia, Pa, July 20, 39; m 65; c 1. THEORETICAL PHYSICS. Educ: Cornell Univ, BEP, 62; Univ Calif, Berkeley, PhD(high energy physics), 67. Prof Exp: Physicist, Lawrence Radiation Lab, Univ Calif, 67; NSF res fel theoret physics, Univ Md, 67-68; asst prof physics, Williams Col, 68-74; MEM TECH STAFF, COMPUTER SCI CORP, 74- Mem: AAAS; Am Asn Physics Teachers. Res: Scattering theory; quantum theory; statistical mechanics; theoretical mechanics; estimation theory. Mailing Add: Computer Sci Corp 8728 Colesville Rd Silver Spring MD 20910

MARKLEY, JOHN LUTE, b Denver, Colo, Mar 6, 41; m 66, 75. PHYSICAL BIOCHEMISTRY, PROTEIN CHEMISTRY. Educ: Carleton Col, BA, 63; Harvard Univ, PhD(biophys), 69. Prof Exp: Res chemist, Merck Inst Therapeut Res, 67-68, sr res chemist, 68-69; USPHS sr fel biophys, Chem Biodynamics Lab, Univ Calif, Berkeley, 70-71; asst prof, 72-76, ASSOC PROF CHEM, PURDUE UNIV, WEST LAFAYETTE, 76- Concurrent Pos: USPHS res career develop award, Nat Heart & Lung Inst, 75. Mem: AAAS; Am Chem Soc; Am Soc Biol Chem; Int Soc Magnetic Resonance. Res: Structure-function relationships in biological macromolecules; applications of nuclear magnetic resonance spectroscopy to the study of local environments of groups; proteinases and their inhibitors, nucleases, glycoproteins, electron transport proteins. Mailing Add: Dept of Chem Purdue Univ West Lafayette IN 47907

MARKLEY, KEHL, III, b Pennsburg, Pa, Sept 11, 23; m 52; c 4. PHYSIOLOGY. Educ: Pa State Univ, BS, 43; Univ Pa, MD, 47. Prof Exp: Instr pharmacol, Univ Pa, 48-49; asst resident med, NY Hosp, 49-51; MED DIR, NAT INST ARTHRITIS, METAB & DIGESTIVE DIS, 51- Res: Physiology of traumatic shock; resistance to infection; biochemistry of lipids; transplantation of skin grafts; immunologic functions of the lymphocyte; biochemistry and immunology of phytohemagglutinin. Mailing Add: Rm B1-27 Bldg 4 Nat Inst of Health Bethesda MD 20014

MARKLEY, LOWELL DEAN, b Mishawaka, Ind, Aug 27, 42; m 62; c 2. ORGANIC CHEMISTRY. Educ: Manchester Col, BA, 64; Purdue Univ, PhD(org chem), 69. Prof Exp: Res chemist, 68-72, RES SPECIALIST, AG-ORGANICS DEPT, DOW CHEM CO, 72- Mem: Am Chem Soc. Res: Synthesis of biologically active organic compounds including pharmaceuticals and agricultural products; synthesis of agricultural products. Mailing Add: Ag-Organics Dept Dow Chem Co Bldg 9001 Midland MI 48640

MARKLEY, MAX C, chemistry, see 12th edition

MARKLEY, RICHARD WILLIAM, physics, see 12th edition

MARKLEY, WILLIAM A, JR, b Sinking Springs, Pa, Aug 24, 25; m 54; c 4. MATHEMATICS. Educ: Bucknell Univ, BS, 49; Univ Pittsburgh, MLitt, 56, PhD(math), 68. Prof Exp: From asst prof to assoc prof, 56-70, PROF MATH, MT UNION COL, 70- Mem: NY Acad Sci. Res: Analysis; summability of infinite series. Mailing Add: 1435 Robinwood Rd Alliance OH 44601

MARKO, ARTHUR MYROSLAW, b Krydor, Sask, May 28, 25; m 49; c 2. MEDICINE, BIOCHEMISTRY. Educ: Univ Sask, BA, 46; Univ Toronto, MD, 49, PhD(biochem), 52. Prof Exp: Asst res officer, Atlantic Regional Labs, Can, 54-55; chief res biochemist, Dept Pub Health, Sask, 55-56; asst prof biochem, res assoc pediat & admin asst to dean med, Univ Sask, 56-60, assoc prof biochem & lectr pediat, 60-61; asst dir, 61-65, DIR BIOL & HEALTH DIV PHYSICS DIV, ATOMIC ENERGY CAN, LTD, 65- Concurrent Pos: Nat Res Coun Can res fel, Eng, 52-54; vis consult, Univ Sask Hosp, 58- Mem: Am Chem Soc; Health Physics Soc; Can Biochem Soc; Can Soc Clin Chem; Can Physiol Soc. Res: Molecular biology; effects of radiations on living matter; histones; radiation protection. Mailing Add: Biol & Health Div Atomic Energy of Can Ltd Chalk River ON Can

MARKO, JOHN ROBERT, b Bayonne, NJ, Jan 28, 38; m 65; c 1. SOLID STATE PHYSICS. Educ: Mass Inst Technol, BS, 59; Syracuse Univ, MS, 63, PhD(physics), 67. Prof Exp: Instr physics, Univ BC, 67-68, asst prof, 68-74; MEM STAFF, PAC REGION, MARINE SCI DIRECTORIATE, 74- Mem: Am Phys Soc. Res: Magnetic resonance; spin-lattice relaxation and electron spin resonance in semiconductors; transport properties of heavily doped semiconductors. Mailing Add: Pac Reg Marine Sci Directoriate 512 Fed Bldg Victoria BC Can

MARKOFF, ELLIOTT LEE, b Baltimore, Md, Nov 27, 32; m 55; c 4. PSYCHIATRY, PSYCHOANALYSIS. Educ: Harvard Univ, AB, 54; Univ Pa, MD, 58; Univ Calif, Los Angeles, MS, 64. Prof Exp: Intern, USPHS Hosp, Staten Island, NY, 58-59; resident psychiat, Lexington, Ky, 59-60; staff psychiatrist, Ft Worth, Tex, 60-61; resident psychiat, Neuropsychiat Inst, Univ Calif, Los Angeles, 61-64; pvt pract, 64-69; sr consult drug abuse, Los Angeles County Dept Ment Health, 69-70; ASST PROF PSYCHIAT & ASSOC DIR GRAD EDUC IN PSYCHIAT, SCH MED, UNIV SOUTHERN CALIF, 70- Concurrent Pos: USPHS fel social & community psychiat, Univ Calif, Los Angeles, 64-65, lectr psychiat, Sch Med, 64-69; consult, Calif Rehab Ctr, 64 & Los Angeles County Ment Health Dept, 65-70; pvt pract, 69-; consult, White House Drug Abuse Task Force, 70 & Spec Action Off for Drug Abuse Prev, 71-73; mem tech adv comt drug abuse, Los Angeles County Bd Supvr, 73-; chmn res, eval & adv panel, Los Angeles County Drug Abuse Task Force, 74-; mem, Methadone Med Adv Bd, Los Angeles County Dept Health Serv, 74- Mem: AAAS; fel Am Psychiat Asn; assoc mem Am Psychoanal Asn. Res: Psychiatric education; drug dependency, development of models for local school and local community response to drug dependence and drug abuse; organization and delivery of psychiatric program services and consultations. Mailing Add: Univ of Southern Calif Rm 311 1237 N Mission Rd Los Angeles CA 90033

MARKOFSKY, SHELDON, organic chemistry, see 12th edition

MARKOS, CHARLES S, organic chemistry, see 12th edition

MARKOS, HARRY GEORGE, animal breeding, animal genetics, see 12th edition

MARKOVETZ, ALLEN JOHN, b Aberdeen, SDak, Apr 17, 33. MICROBIOLOGY, BIOCHEMISTRY. Educ: Univ SDak, BA, 57, MA, 58; PhD(bact), 61. Prof Exp: NIH fel microbial metab, 61-62, from instr to assoc prof, 62-73, PROF MICROBIOL, UNIV IOWA, 73- Concurrent Pos: Spec res fel, NIH, Dept Biochem, Univ Calif, Davis, 69-70; Career develop award, NIH, 72-76. Mem: Am Soc Microbiol; Am Chem Soc. Res: Microbial physiology and metabolism; microbial hydrocarbon and ketone metabolism; microbial-insect interactions. Mailing Add: Dept of Microbiol Univ of Iowa Iowa City IA 52241

MARKOVITZ, ALVIN, b Chicago, Ill, May 30, 29; m 52; c 4. MICROBIOLOGY. Educ: Univ Ill, BS, 50, MS, 52; Univ Wash, PhD, 55. Prof Exp: Fel, Nat Heart Inst, 55-57; instr, La Rabida Inst, 57-59, from asst prof to assoc prof, 59-74, PROF MICROBIOL, UNIV CHICAGO, 74- Concurrent Pos: Assoc prof, La Rabida Inst, 64-67. Mem: AAAS; Am Soc Biol Chem; Am Soc Microbiol; Genetics Soc Am. Res: Regulation of protein and capsular polysaccharide synthesis in bacteria. Mailing Add: Dept of Microbiol Univ of Chicago Chicago IL 60637

MARKOVITZ, HERSHEL, b McKeesport, Pa, Oct 11, 21; m 49; c 3. PHYSICAL CHEMISTRY. Educ: Univ Pittsburgh, BS, 42; Columbia Univ, AM, 43, PhD(phys chem), 49. Prof Exp: Mathematician, Kellex Corp, 43-45; asst, Columbia Univ, 45-49; fel, 49-51, sr fel, 51-56, SR FEL, FUNDAMENTAL RES GROUP, SYNTHETIC RUBBER PROPERTIES, MELLON INST, 56-, PROF MECH & POLYMER SCI, CARNEGIE-MELLON UNIV, 67- Concurrent Pos: Lectr, Univ Pittsburgh, 56-58; vis lectr, Johns Hopkins Univ, 58-59; Fulbright lectr, Weizmann Inst, 64-65; asst ed, J Polymer Sci, 65-68, assoc ed, 69-; mem gov bd, Am Inst Physics, 70-72; adj prof, Univ Pittsburgh, 72- Honors & Awards: Bingham Medal, Soc Rheol, 67. Mem: Am Chem Soc; Am Phys Soc; Soc Rheol (vpres, 67-69, pres, 69-71); Soc Natural Philos (treas, 65-66). Res: Physics of polymers; continuum mechanics; rheology. Mailing Add: Carnegie-Mellon Univ Mellon Inst 4400 Fifth Ave Pittsburgh PA 15213

MARKOVITZ, MARK, b Rosario, Argentina, June 3, 38; US citizen. ORGANIC CHEMISTRY, POLYMER CHEMISTRY. Educ: City Col New York, 58; NY Univ, PhD(thiophene chem), 63. Prof Exp: Res fel, Thiophene Chem, NY Univ, 58-62;

CHEMIST, MAT & PROCESSES LAB, GEN ELEC CO, 62- Am Chem Soc. Res: Electrical insulating materials; organo-metallic polymers; thermosetting resins; thiophene chemistry. Mailing Add: Mat & Processes Lab Gen Elec Co Lge Steam Turbine-Generator Div Schenectady NY 12345

MARKOWITZ, ABRAHAM SAM, b New York, NY, July 12, 21; m 48; c 3. IMMUNOLOGY. Educ: NY Univ, BA, 48; Univ Southern Calif, MS, 50, PhD(bact), 52. Prof Exp: Asst prof bact, San Diego State Col, 52-54; from asst prof to assoc prof, 55-68, PROF MICROBIOL, UNIV ILL COL MED, 68-; HEAD EXP IMMUNOL, HEKTOEN INST MED RES, 58-; CHMN, DIV IMMUNOL, COOK COUNTY HOSP, 74- Mem: Am Asn Immunol; Am Soc Cell Biol; NY Acad Sci; Transplantation Soc; Int Soc Nephrology. Res: Heterophile antigens and antibodies; autogenous hypersensitivity; immunochemistry. Mailing Add: Hektoen Inst for Med Res 629 S Wood St Chicago IL 60612

MARKOWITZ, DAVID, b Paterson, NJ, Mar 24, 35; m 61; c 3. SOLID STATE PHYSICS, BIOLOGICAL PHYSICS. Educ: Mass Inst Technol, BS, 56; Univ Ill, PhD(physics), 63. Prof Exp: Res assoc physics, Rutgers Univ, 63-65; asst prof, 65-73, ASSOC PROF PHYSICS, Univ Conn, 73- Concurrent Pos: Adj prof, New Eng Inst, 70-; vis res scientist, Univ Sussex, 71-72. Mem: Am Phys Soc. Mailing Add: Dept of Physics Univ of Conn Storrs CT 06268

MARKOWITZ, HAROLD, b New York, NY, Sept 1, 25; m 53; c 4. IMMUNOCHEMISTRY. Educ: City Col New York, BS, 47; Columbia Univ, MA, 52, PhD(biochem), 53; Univ Utah, MD, 58. Prof Exp: Res assoc immunochem, Columbia Univ, 52-53; res fel med, Univ Utah, 53-59, intern, 58-59, instr, 59-61; asst prof microbiol, 62-68, ASSOC PROF MICROBIOL, MAYO GRAD SCH MED, UNIV MINN, 68-, CONSULT, MAYO CLIN, 61- Concurrent Pos: NIH fel, 59-60; assoc attend physician, Salt Lake County Gen Hosp, Utah, 59-61. Mem: Am Asn Immunol; Am Soc Microbiol; Soc Exp Biol & Med; NY Acad Sci; Cent Soc Clin Res. Res: Carbohydrate chemistry; trace metal and copper metabolism; antigens of erythrocytes and pathogenic fungi; immunohematology; phytohemagglutinins. Mailing Add: Dept of Lab Med Mayo Clinic 200 First St SW Rochester MN 55902

MARKOWITZ, JOSEPH MORRIS, b Scranton, Pa, Feb 15, 25; m 49; c 2. PHYSICAL CHEMISTRY. Educ: Bucknell Univ, BS, 47; Pa State Univ, MS, 52; Univ Mich, PhD(phys chem), 58. Prof Exp: Instr chem, Wilkes Col, 47-50; sr scientist, 55-62, FEL SCIENTIST, WESTINGHOUSE ELEC CORP, 62- Mem: AAAS. Res: X-ray diffraction and ESCA studies of solid surfaces; aqueous metallic corrosion; diffusional processes in liquids; solid state, especially phase equilibria, diffusion and self-diffusion in ceramics; thermal diffusion in metals. Mailing Add: Bettis Atomic Power Lab Westinghouse Elec Corp PO Box 79 West Mifflin PA 15122

MARKOWITZ, MELVIN MYRON, b New York, NY, Sept 17, 46. PHYCOLOGY. Educ: City Col New York, BS, 68; Univ Ill, Urbana, MS, 70, PhD(bot), 75. Prof Exp: Lectr bot, Univ Ill, Urbana, 75- Mem: Am Phycol Soc; Brit Phycol Soc. Res: Ultrastructure and histochemical investigations of the reproductive cells of algae. Mailing Add: Dept of Bot Univ of Ill Urbana IL 61801

MARKOWITZ, MILTON, b New York, NY, June 6, 18; c 4. PEDIATRICS. Educ: Syracuse Univ, AB, 39, MD, 43. Prof Exp: Asst pediat, Sch Med, Johns Hopkins Hosp, 48-49, instr, 50-55, asst prof, Sch Med, Univ, 55-62, dir pediat rheumatic clins, Children's Med & Surg Ctr, 61-69; assoc pediatrician-in-chief, Sinai Hosp, Baltimore, 63-69; PROF PEDIAT & HEAD DEPT, SCH MED, UNIV CONN HEALTH CTR, 69- Concurrent Pos: Pvt pract, 49-52; dir streptococcal dis res lab, Sinai Hosp, Baltimore, 60-69; assoc prof pediat, Sch Med, Johns Hopkins Univ, 62-69. Mem: Fel Am Acad Pediat; Am Pediat Soc. Res: Rheumatic heart disease. Mailing Add: Dept of Pediat Univ of Conn Health Ctr Farmington CT 06032

MARKOWITZ, SAMUEL SOLOMON, b Brooklyn, NY, Oct 31, 31; m 58; c 3. NUCLEAR CHEMISTRY. Educ: Rensselaer Polytech, BS, 53; Princeton Univ, MA, 55, PhD, 57. Prof Exp: Jr res assoc nuclear chem, Brookhaven Nat Lab, 55-57; NSF fel, Univ Birmingham, 57-58; from asst prof to assoc prof chem, Univ Calif, Berkeley, 58-72, mem staff, Lawrence Berkeley Lab, 58-64, PROF CHEM, UNIV CALIF, BERKELEY, 72-, SR SCIENTIST, LAWRENCE BERKELEY LAB, 64- Concurrent Pos: Imp Chem Industs hon fel, Univ Birmingham, 57-58; NSF sr fel, fac sci, Univ Paris, 64-65; vis prof, Weizmann Inst Sci, Israel, 73-74. Mem: AAAS; Am Chem Soc; Am Phys Soc; Sigma Xi. Res: Nuclear reactions at billion-electron-volt energies; fission and spallation; meson-induced reactions; nuclear activation analysis by He-3-induced reactions; radiochemistry; chemical fate of atoms produced via nuclear transformations. Mailing Add: Dept of Chem & Lawrence Lab Univ of Calif Berkeley CA 94720

MARKOWITZ, WILLIAM, b Poland, Feb 8, 07; nat US; m 43; c 1. ASTRONOMY. Educ: Univ Chicago, BS, 27, MS, 29, PhD(astron), 31. Prof Exp: Instr math, Pa State Col, 31-32; astronr, US Naval Observ, 36-66, dir time serv, 53-66; prof physics, Marquette Univ, 66-68, Wehr prof, 68-72; ADJ PROF, NOVA UNIV, 72-; ED, GEOPHYS SURV, 72- Honors & Awards: US Navy Conrad Medal, 67; Distinguished Civilian Serv Award, 67. Mem: Int Astron Union; Int Union Geod & Geophys; Am Astron Soc; Am Geophys Union. Res: Time and frequency; variations in earth rotation; secular motion of pole; SI units. Mailing Add: Dept of Physics Nova Univ 8000 N Ocean Dr Dania FL 33004

MARKOWSKI, HENRY JOSEPH, b Worcester, Mass, July 1, 29; m 54; c 5. ORGANIC POLYMER CHEMISTRY. Educ: Providence Col, BS, 52. Prof Exp: Anal chemist, Nitrogen Div, Allied Chem Corp, 52-56; mgr, Markowski's Bakery, RI, 56-57; develop chemist, Lowe Brothers Paint Co, 57-60 & Hysol Corp, 60-64; develop chemist, Insulation Mat Dept, Gen Elec Corp, 64-68; mgr resin develop, P D George Paint & Varnish Co, 68-70; tech dir, Windecker Res, Tex, 70-71; SR RES CHEMIST, CARBOLINE CO, 71- Mem: Am Chem Soc. Res: Polymer synthesis and research; urethane elastomers; epoxy coatings, organo-metallic polymers; resins, coatings, and adhesives used for electrical insulation and corrosion resistance. Mailing Add: Carboline Co 350 Hanley Industrial Ct St Louis MO 63144

MARKS, ALFRED FINLAY, b Yorktown Heights, NY, Sept 13, 32; c 4. AGRICULTURAL CHEMISTRY. Educ: Iowa State Univ, BS, 55. Prof Exp: Res chemist, Am Cyanamid Corp, 55-67; sr res chemist, Esso Res & Eng Corp, 67-70; GROUP LEADER BIOCHEM FORMULATIONS, DIAMOND SHAMROCK CORP, 70- Mem: Am Chem Soc. Res: Development of formulations and dosage forms of pesticides, pharmaceuticals and animal health products, including the investigation of formulation-bioavailability relationships. Mailing Add: Diamond Shamrock Corp T R Evans Res Ctr PO Box 348 Painesville OH 44077

MARKS, ASHER, b Atlanta, Ga, Sept 5, 26; m 49; c 2. MEDICINE. Educ: Emory Univ, MD, 50; Am Bd Internal Med, dipl, 59; Am Bd Pulmonary Dis, dipl, 69. Prof Exp: From intern to jr asst resident med, Boston City Hosp, 50-52; from instr to asst prof, 57-63, ASSOC PROF MED, MED SCH, UNIV MIAMI, 63- Concurrent Pos: Fel chest dis, Boston City Hosp, 54-56; res fel, Med Sch, Univ Miami, 56-58; consult, Vet Admin Hosp, Coral Gables, Fla, 57-; Southeast Fla Tuberc Hosp, Lantana, 58- Mem: Am Thoracic Soc; Am Heart Asn; Am Col Physicians; Am Col Chest Physicians. Res: Disease of chest; cardio-pulmonary physiology. Mailing Add: 1150 NW 14th St Miami FL 33136

MARKS, BERNARD HERMAN, b Cleveland, Ohio, Apr 21, 21; m 43; c 1. PHARMACOLOGY, BIOCHEMISTRY. Educ: Ohio State Univ, BA, 42, MD, 45, MA, 50. Prof Exp: Instr pharm & biochem, Ohio State Univ, 48-53, asst prof pharmacol, 54-56, from assoc prof to prof pharm, 57-73, chmn dept, 63-73; PROF PHARMACOL & CHMN DEPT, WAYNE STATE UNIV, 74- Mem: AAAS; Am Soc Pharmacol & Exp Therapeut. Res: Cellular, cardiovascular and endocrine pharmacology; digitalis; radio-labeled drugs. Mailing Add: Dept of Pharmacol Wayne State Univ Sch of Med Detroit MI 48201

MARKS, BURTON STEWART, b New York, NY, Oct 23, 24; m 48; c 3. POLYMER CHEMISTRY. Educ: Univ Miami, Fla, BS, 48, MS, 50; Polytech Inst Brooklyn, PhD(chem), 55. Prof Exp: Sr res chemist, Hooker Chem Co, 55-58; adv scientist, Continental Can Co, Inc, Ill, 58-62; STAFF SCIENTIST, LOCKHEED-PALO ALTO RES LABS, 62- Concurrent Pos: Adj prof, Niagara Univ, 57-58 & Roosevelt Univ, 60-62. Mem: Am Chem Soc; The Chem Soc. Res: Organic and polymer synthesis and chemistry; monomers; polymerization; resins; coatings; adhesives; foams; composite structures; ceramics; carbides. Mailing Add: Lockheed-Palo Alto Res Labs 3251 Hanover St Palo Alto CA 94304

MARKS, CHARLES, b Kremenchug, Russia, Jan 28, 22; US citizen; m 49; c 4. MEDICINE. Educ: Univ Cape Town, BA, 42, MD, 45; Marquette Univ, MS, 65; FRCS; FRCP; Tulane Univ, PhD, 73. Prof Exp: Hunterian prof surg, Royal Col Surgeons, Eng, 56; assoc prof, Sch Med, Marquette Univ, 63-67; clin prof, Sch Med, Case Western Reserve Univ, 67-71; PROF SURG, SCH MED, UNIV NEW ORLEANS, 71- Concurrent Pos: Consult surg, Salisbury, Rhodesia, 53-65, Vet Admin Hosp & Milwaukee County Gen Hosp, 63-, St Francis Hosp & Mt Sinai Hosp, Milwaukee, Wis, 64-; dir div surg, Mt Sinai Hosp, Cleveland, Ohio, 67-71; attend surgeon, Charity Hosp, Touro Infirmary & Hotel Dieu Hosp, 71-; consult cardiovasc surg, E Jefferson Gen Hosp, New Orleans, La, 71- Mem: AMA; Am Col Chest Physicians; fel Am Col Surg; fel Am Col Cardiol; fel Royal Soc Med. Res: Hepatocyte synthesis measured with tritiated thymidine autoradiography in response to variable hepatic blood flow; haemodynamic effects of protal hypertension. Mailing Add: 1680 State St New Orleans LA 70118

MARKS, CHARLES FRANCIS, b Codroy, Nfld, Oct 23, 38; m 68. NEMATOLOGY, PLANT PATHOLOGY. Educ: Macdonald Col, McGill Univ, BSc, 59; Ont Agr Col, MSA, 63; Univ Calif, Riverside, PhD(plant path), 67. Prof Exp: RES SCIENTIST, CAN DEPT AGR, 67-, PROG LEADER NEMATOL, 73- Mem: AAAS; Soc Nematol. Res: Use and mode of action of nematicides. Mailing Add: Vineland Res Sta PO Box 185 Vineland Station ON Can

MARKS, DARRELL L, b Mountain Home, Idaho, July 23, 36; m 55; c 4. PHYSICS. Educ: Northwest Nazarene Col, AB, 58; Mass Inst Technol, MS, 59; Ore State Univ, PhD(biophys), 66. Prof Exp: Mem fac, 59-75, CHMN DIV MATH & NATURAL SCI, NORTHWEST NAZARENE COL, 75- Concurrent Pos: Mem curric comn, Idaho State, 75-81. Mem: AAAS; Am Asn Physics Teachers. Res: Mass spectroscopy; helium-uranium dating; science education; new methods for teaching large lab classes. Mailing Add: Dept of Physics Northwest Nazarene Col Nampa ID 83651

MARKS, DENNIS WILLIAM, b Madison, Wis, Nov 5, 44; m 68. ASTROPHYSICS. Educ: Fordham Univ, BS, 66; Univ Mich, Ann Arbor, PhD(astron), 70. Prof Exp: Fel & asst prof, David Dunlap Observ, Univ Toronto, 70-71; asst prof, 71-75, ASSOC PROF ASTRON & PHYSICS, VALDOSTA STATE COL, 75- Mem: Am Astron Soc. Res: Internal differential rotation of stars; viscosity of gases and radiation; stellar structure and evolution. Mailing Add: Dept of Phys & Astron Valdosta State Col Valdosta GA 31601

MARKS, EDWIN POTTER, cell biology, see 12th edition

MARKS, GAYTON CARL, b Winsted, Conn, Jan 10, 21; m 46; c 4. BOTANY. Educ: Valparaiso Univ, BA, 55; Univ Mich, MS, 60. Prof Exp: Teacher sci, Munster Pub Schs, Munster, Ind, 56-62; INSTR BIOL, VALPARAISO UNIV, 62- Mem: Bot Soc Am. Res: Systematics of the local flora. Mailing Add: 237 Kolling Rd Schererville IN 46375

MARKS, GERALD SAMUEL, b Cape Town, SAfrica, Feb 13, 30; m 55; c 2. ORGANIC CHEMISTRY, PHARMACOLOGY. Educ: Univ Cape Town, BSc, 50, MSc, 51; Oxford Univ, DPhil(org chem), 54. Prof Exp: Res chemist, SAfrican Inst Med Res, 55-56; res assoc porphyrin biosynthesis, Univ Chicago, 57-59; assoc prof pharmacol, Univ Alta, 62-69; HEAD DEPT PHARMACOL, QUEEN'S UNIV, ONT, 69- Concurrent Pos: Nat Res Coun Can fel, 56-57; Brit Empire Cancer Campaign fel, Dept Chem Path, St Mary's Hosp, London, 60-62. Mem: Pharmacol Soc Can; The Chem Soc. Res: Lipids of tubercle bacilli; glycoprotein chemistry; porphyrin biosynthesis; chemistry of adrenergic receptor. Mailing Add: Dept of Pharmacol Queen's Univ Kingston ON Can

MARKS, HENRY CLAY, chemistry, see 12th edition

MARKS, HENRY L, b Waynesboro, Va, Sept 6, 35; m 59; c 1. ANIMAL GENETICS. Educ: Va Polytech Inst, BS, 58, MS, 60; Univ Md, PhD, 67. Prof Exp: Res geneticist, 60-67, RES GENETICIST, SOUTH REGIONAL POULTRY BREEDING PROJ, AGR RES SERV, USDA, 67- Mem: Poultry Sci Asn; World Poultry Sci Asn. Res: Design and test animal breeding systems for increasing production by genetic selection. Mailing Add: Room 107 Livestock-Poultry Bldg Univ of Ga Athens GA 30601

MARKS, JAMES FREDERIC, b Pittsburgh, Pa, Dec 18, 28; m 59; c 1. PEDIATRIC ENDOCRINOLOGY. Educ: Princeton Univ, AB, 50; Harvard Med Sch, MD, 54. Prof Exp: Res fel pediat endocrinol, Children's Hosp Pittsburgh, 59-61; asst prof, 61-68, ASSOC PROF PEDIAT, UNIV TEX HEALTH SCI CTR DALLAS, 68- Concurrent Pos: Vis prof, AMA-Vietnam Educ Proj, 71. Mem: Soc Pediat Res; Am Fedn Clin Res. Res: Thyroid, adrenal and growth problems in children; uric acid metabolism; cytogenetics. Mailing Add: Dept of Pediat Univ of Tex Health Sci Ctr Dallas TX 75235

MARKS, JAY GLENN, b Los Angeles, Calif, Aug 7, 16; m 42; c 3. GEOLOGY, PALEONTOLOGY. Educ: Stanford Univ, BA, 38, MA, 41, PhD(geol), 51. Prof Exp: Geologist & paleontologist, Int Ecuadorean Petrol Co, 41-46; actg instr field geol, Stanford Univ, 47-48; sr macropaleontologist, Creole Petrol Corp, Venezuela, 48-55, supvr, Geol Lab, 55-56, regional geologist, 56-57; geologist, Exxon Co, US, 57-75; RETIRED. Concurrent Pos: Mem, Paleont Res Inst. Mem: Fel Geol Soc Am; Soc Econ Paleont & Mineral; Am Asn Petrol Geol. Res: Tertiary stratigraphy and paleontology of California and Alaska; Paleozoic paleontology of Alaska. Mailing Add: 3900 S Clermont St Englewood CO 80110

MARKS, LEON JOSEPH, b Providence, RI, Nov 30, 25; m 56; c 2. INTERNAL MEDICINE, ENDOCRINOLOGY. Educ: Brown Univ, AB, 44; Johns Hopkins Univ, MD, 48; Am Bd Internal Med, dipl, 56. Prof Exp: Intern med, Jewish Hosp, Brooklyn, NY, 48-49; jr resident, Kings County Hosp, 49-50; sr resident, Montefiore Hosp, 50-51; staff physician & dir steroid res lab, 52-73, chief outpatient serv & ambulatory health care, 52-75, ASSOC CHIEF STAFF AMBULATORY HEALTH CARE, BOSTON VET ADMIN HOSP, 75- Concurrent Pos: Milton res fel, Harvard Univ, 51-52; res fel pediat, Mass Gen Hosp, 51-52; clin instr, Sch Med, Tufts Univ, 61-66, sr clin instr, 66-68, asst prof, 68- Mem: Endocrine Soc; Am Geriat Soc; AMA; fel Am Col Physicians; Am Fedn Clin Res: Res: Metabolism and endocrinology, especially adrenal steroid biochemistry and physiology as applied clinically to medicine, surgery and psychiatry; cancer of the prostate. Mailing Add: Vet Admin Hosp Boston MA 02130

MARKS, LOUIS SHEPPARD, b New York, NY, Dec 13, 17; m 44; c 4. ENTOMOLOGY, MATHEMATICAL BIOLOGY. Educ: City Col New York, BS, 39; Fordham Univ, MS, 51, PhD(entom), 54. Prof Exp: Statist analyst, US Dept Com, 40-42; prof chem & sanit sci, Am Acad, 46-51; instr biol, Fordham Univ, 51-54, from asst prof to assoc prof, 54-65; prof & head dept, Pace Col, 65-66; PROF BIOL & CHMN DEPT, ST JOSEPH'S COL, PA, 66- Concurrent Pos: Smith Mundt vis prof, Nat Univ Mex, 56; vis prof, Hunter Col, 57; fel, Harvard Univ, 62-64; fel, NC State Univ, 63 & Williams Col, 65. Mem: AAAS; Soc Syst Zool; Am Inst Biol Sci; Am Soc Zoologists; Soc Study Evolution. Res: Systematic entomology; taxonomy; morphology and zoogeography of the Lepidoptera; vertebrate coronary circulation; mathematical and evolutionary biology; zoological bibliography; chordate morphology. Mailing Add: Dept of Biol St Joseph's Col Philadelphia PA 19131

MARKS, LUTHER WHITFIELD, III, b Fairfax, Okla, Nov 2, 26; m 48; c 4. PHYSICS. Educ: Cent State Univ, Okla, BS, 49; Univ Okla, MS, 51, PhD(physics), 55. Prof Exp: Asst physics, Univ Okla, 49-55; instr, 55, from asst prof to assoc prof, 55-58, PROF PHYSICS, CENT STATE UNIV, OKLA, 58-, CHMN DEPT, 64- Mem: Am Asn Physics Teachers. Res: Electrode processes, particularly analysis of decay of activation overpotential. Mailing Add: Dept of Physics Cent State Univ Edmond OK 73034

MARKS, MELVIN ISSAC, b Montreal, Que, July 30, 40; m 73; c 2. INFECTIOUS DISEASES, PEDIATRICS. Educ: McGill Univ, BSc, 61, MD, CM, 65. Prof Exp: Intern pediat, Montreal Gen Hosp, 65-66; from jr resident to sr resident, Montreal Children's Hosp, 66-68; fel infectious dis, Univ Colo, 68-69, instr, Med Ctr, 69-70; asst physician, 70-73, ASSOC PHYSICIAN, MONTREAL CHILDREN'S HOSP, 73-, DIR INFECTIOUS DIS SERV, 70- Concurrent Pos: Examr pediat, Royal Col Physicians & Surgeons, 74- Mem: Am Soc Microbiol; Can Soc Clin Invest; Am Fedn Clin Res; Soc Pediat Res; Infectious Dis Soc Am. Res: Pathogenesis and therapy of herpes simplex infections; in vitro activity, pharmacokinetics and efficacy of antimicrobial drugs. Mailing Add: Dept of Infectious Dis Montreal Children's Hosp 2300 Tupper St Montreal PQ Can

MARKS, MORTON, b Vineland, NJ, Oct 26, 18; m 47; c 6. NEUROLOGY, PSYCHIATRY. Educ: Temple Univ, AB, 39, MD, 43; Am Bd Psychiat & Neurol, dipl & cert neurol, 51, cert psychiat, 52. Prof Exp: Clin asst neurol, 49-51, instr, 51-52, dir neurol res, Inst Phys Med & Rehab, Med Ctr, 50-62, ASST PROF CLIN NEUROL, COL MED, NY UNIV, 52- Concurrent Pos: Asst vis physician, Goldwater Mem Hosp, New York, 56-; mem med adv bd, Nat Multiple Sclerosis Soc, 58. Mem: AMA; Am Psychiat Asn; Am Fedn Clin Res; Am Acad Neurol. Res: Research in rehabilitation of chronic neurological patients and degenerative neurological diseases. Mailing Add: 566 First Ave New York NY 10016

MARKS, NEVILLE, b Dublin, Ireland, Apr 10, 30. NEUROBIOLOGY. Educ: Univ London, MSc, 55, PhD(neurochem), 59. Prof Exp: Lectr neurochem, Inst Psychiat, Univ London, 57-59; fel biochem, Northwestern Univ, 59-60; neurochem, Ment Health Res Inst, Univ Mich, 60-61; sr res scientist, 61-68, assoc res scientist, 68-70, PRIN RES SCIENTIST, NY STATE RES INST NEUROCHEM & DRUG ADDICTION, 70- Concurrent Pos: Ed, Res Methods Neurochem, 72-75, assoc ed, Neurochem Res, 75; consult, Vet Admin Hosp, East Orange, 71-75. Mem: Am Acad Neurol; Am Soc Neurochem; Int Soc Neurochem; Am Chem Soc; Am Soc Biol Chemists. Res: Protein breakdown and turnover in brain; purification of catabolic enzymes; myelin turnover in experimental demyelination; formation and breakdown of hormonal peptides. Mailing Add: NY State Inst of Neurochem Ward's Island New York NY 10035

MARKS, PAUL A, b New York, NY, Aug 16, 26; m 53; c 3. INTERNAL MEDICINE, BIOCHEMISTRY. Educ: Columbia Univ, AB, 45, MD, 49. Prof Exp: Res fel med, Med Col, Cornell Univ, 49; intern med, Presby Hosp, New York, 50, asst resident, 51; fel, Col Physicians & Surgeons, Columbia Univ, 52-53; assoc investr, Nat Inst Arthritis & Metab Dis, 53-55; instr med, Sch Med, George Washington Univ, 54-55; instr, 55-56, assoc, 56-57, from asst prof to assoc prof, 57-67, chmn dept human genetics & develop, 69-70, dean fac med & vpres in chg med affairs, 70-73, prof med, 67-74, FRODE JENSEN PROF MED, COL PHYSICIANS & SURGEONS, COLUMBIA UNIV, 74-, DIR HEMAT TRAINING, 61-, PROF HUMAN GENETICS & DEVELOP, 69-, VPRES HEALTH SCI & DIR CANCER RES CTR, 73- Concurrent Pos: Commonwealth Fund fel, 61-62; vis scientist, Lab Cellular Biochem, Pasteur Inst, 61-62; consult, Vet Admin Hosp, 62-; mem adv panel develop biol, NSF, 64-; Swiss-Am Found lectr & award in med res, 65; ed-in-chief, J Clin Invest, Am Soc Clin Invest, 67-71; mem adv panel hemat training grants prog, NIH, 69-, chmn hemat training grants comn, 71-73; trustee, Roosevelt Hosp & St Luke's Hosp, 70-; mem div med sci, Nat Res Coun, 72-, chmn exec comt, 73-; Carl R Moore lectr, Sch Med, Wash Univ, 73; honors prog lectr, Sch Med, NY Univ, 73; mem jury, Albert Lasker Awards, 74-; mem adv comt, XV Int Cong Hemat, Israel, 74; consult ed, Blood Cells, 74-; mem rev comt blood dis & blood resources panel, Nat Res & Demonstration Ctr, Nat Heart & Lung Inst, 74; mem, Dartmouth Med Sch Conf on Health Systs & President's Biomed Res Panel, 75; Frontiers in Biol Sci lectr, Case Western Reserve Univ Sch Med, 75; Sci Coun adv to bd dirs, Radiation Effects Res Found, Japan, 75. Honors & Awards: Charles Janeway Prize, 49; Joseph Mather Smith Prize, 59; Stevens Triennial Prize, Columbia Univ, 60. Mem: Nat Acad Sci; Nat Inst Med; Am Soc Clin Invest (pres, 71-72); Soc Exp Biol & Med; Harvey Soc (treas, 67-70, pres, 73-74). Res: Cellular development; protein synthesis; human genetics; hematology. Mailing Add: Col of Physicians & Surgeons Columbia Univ New York NY 10032

MARKS, RICHARD HENRY LEE, b Richmond, Va, Nov 23, 43; m 66; c 2. BIOCHEMISTRY. Educ: Univ Richmond, BS, 65; Ind Univ, Bloomington, PhD(biol chem), 69. Prof Exp: USPHS fel, Univ Calif, Santa Barbara, 69-70, univ fel, 71-72; ASST PROF BIOCHEM, COL MED & DENT NJ, 72- Mem: NY Acad Sci; Am Chem Soc. Res: Structure-function relationships in proteins, especially metalloproteins; oxidation-reduction of metalloproteins. Mailing Add: Dept of Biochem Col of Med & Dent of NJ Newark NJ 07103

MARKS, RONALD LEE, b Jersey Shore, Pa, May 23, 34; m 71; c 3. INORGANIC CHEMISTRY. Educ: Lock Haven State Col, BS, 56; Pa State Univ, MEd, 59, EdD(chem educ), 66. Prof Exp: PROF CHEM, INDIANA UNIV, PA, 59- Mem: Am Chem Soc. Res: Coordination compounds of molybdenum IV; new high school curriculum in chemistry education. Mailing Add: Dept of Chem Indiana Univ of Pa Indiana PA 15701

MARKS, SANDY COLE, JR, b Wilmington, NC, Nov 16, 37; m 62; c 2. ANATOMY. Educ: Washington & Lee Univ, BS, 60; Univ NC, DDS, 64; Johns Hopkins Univ, PhD(anat), 68. Prof Exp: Res officer, Dent Res Dept, Naval Med Res Inst, Bethesda, Md, 68-70; asst prof anat, Med Sch, Univ Mass, 70-71; assoc prof pedodont, Sch Dent, Univ NC, Chapel Hill, 71-73; ASSOC PROF ANAT, UNIV MASS MED SCH, 73- Mem: AAAS; Int Asn Dent Res; Am Asn Anatomists. Res: Bone metabolism; calcium homeostasis. Mailing Add: Dept of Anat Univ of Mass Med Sch Worcester MA 01605

MARKS, SIDNEY, b Chicago, Ill, June 28, 18; m 46; c 2. PATHOLOGY, BIOSTATISTICS. Educ: Univ Ill, BS, 38, MD, 42; Univ Idaho, MS, 61; Univ Calif, Los Angeles, PhD, 70. Prof Exp: Pathologist, Vet Admin Hosp, Albuquerque, NMex, 47-48, Kadlec Hosp & Biol Lab, Gen Elec Co, Wash, 50-53 & Kadlec Methodist Hosp, Richland, 53-65; assoc prof surg & internal med, Sch Med, Univ Md, 70-71; COORDR HUMAN STUDIES & BIOSTATIST, DIV BIOMED & ENVIRON RES, ENERGY RES & DEVELOP ADMIN, 71- Concurrent Pos: Consult, Biol Lab & Occup Health Oper, Gen Elec Co, Wash, 53-65, Hanford Environ Health Found, Richland, 65-68 & Radiobiol Lab, Univ Calif, Davis, 66-70. Mem: Col Am Pathologists; Biomet Soc; Math Asn Am; Am Statist Asn; Soc Epidemiol Res. Res: Radiation epidemiology and pathology; discriminant analysis; clinical data processing and statistical analysis. Mailing Add: Div of Biomed & Environ Res Energy Res & Develop Admin Washington DC 20545

MARKS, STUART A, b Wilmington, NC, Apr 28, 39; m 64; c 2. ANTHROPOLOGY, WILDLIFE RESEARCH. Educ: NC State Univ, BS, 61; Mich State Univ, MS, 64; Univ London, res cert anthrop, 65-67; Mich State Univ, PhD(animal ecol, anthrop), 68. Prof Exp: Conserv aid, Mich Dept Conserv, 61-62; game biologist, Alaska Dept Fish & Game, 62; instr conserv wildlife mgt, Mich State Univ, 63-64, instr animal ecol, 65; asst prof zool, Okla State Univ, 68-70; ASSOC PROF ANTHROP, ST ANDREWS PRESBY COL, 70- Concurrent Pos: Soc Sci Res Coun grant, 73. Mem: AAAS; Wildlife Soc; Ecol Soc Am; Fel Royal Anthrop Inst Gt Brit & Ireland; fel Am Anthrop Asn. Res: Hunting and the role of hunter in different societies; belief systems and their influence on use of resources; history of conservation; determination of sex and age in large mammals; interdisciplinary studies. Mailing Add: Dept of Behavior Sci St Andrews Presby Col Laurinburg NC 28352

MARKS, TOBIN JAY, b Washington, DC, Nov 25, 44. CHEMISTRY. Educ: Univ Md, College Park, BS, 66; Mass Inst Technol, PhD(chem), 70. Prof Exp: Asst prof, 70-74, ASSOC PROF CHEM, NORTHWESTERN UNIV, ILL, 74- Mem: Am Chem Soc. Res: Inorganic and organometallic chemistry; structural chemistry in solution; catalysis. Mailing Add: Dept of Chem Northwestern Univ Evanston IL 60201

MARKS, WILLIAM B, neurophysiology, see 12th edition

MARKSON, RALPH JOSEPH, b Feb 25, 31; US citizen; m 67; c 1. ATMOSPHERIC PHYSICS. Educ: Reed Col, BA, 56; Pa State Univ, MA, 67; State Univ NY Albany, PhD(atmospheric sci), 74. Prof Exp: Res engr, Convair Astronauts, 56-58; physicist, self employed, 58-65; res assoc atmospheric elec, State Univ NY Albany, 67-74; RES ASSOC ATMOSPHERIC PHYSICS, MASS INST TECHNOL, 74- Concurrent Pos: Dir, Airborne Res Assoc Inc & mem, Comt Atmospheric & Space Elec, Am Geophys Union, 74-; mem, Subcomt Global Circuit & Fair Weather Elec, Int Comn Atmospheric Elec & Subcomt Planetary & Space Problems Atmospheric Elec & secy, Subcomt Appl Atmospheric Elec, 76- Mem: Am Geophys Union; Am Meteorol Soc; Am Inst Aeronaut & Astronaut; AAAS. Res: Atmospheric electrical global circuit; thundercloud electrification; extra-terrestrial modulation of atmospheric electricity; use of atmospheric space charge as an air tracer; remote thermal detection for soaring; maritime convection and fog. Mailing Add: 46 Kendal Common Rd Weston MA 02193

MARKSTEIN, GEORGE HENRY, b Vienna, Austria, June 22, 11; nat US; m 37; c 1. APPLIED PHYSICS. Educ: Vienna Tech Univ, Ing, 35, PhD(appl physics), 37. Prof Exp: Res physicist, Allgem Gluhlampenfabrics AG, Austria, 37-38; asst seismologist, Shell Petrol Co, Colombia, 39-40, prod supt, Plastic Molding Plant, 42-43; prod supt, Globe Soc Ltd, 44-46; res physicist, Cornell Aeronaut Lab, Inc, 46-50, head combustion sect, 50-56, prin physicist, 56-71; PRIN RES SCIENTIST, FACTORY MUTUAL RES CORP, 71- Mem: AAAS; Am Phys Soc; Am Inst Aeronaut & Astronaut; Combustion Inst. Res: Combustion; fluid dynamics; reaction kinetics; fire research; radiative energy transfer. Mailing Add: Factory Mutual Res Corp 1151 Boston-Providence Turnpike Norwood MA 02062

MARKUNAS, PETER CHARLES, b Chicago, Ill, Nov 5, 11; m 41; c 3. ANALYTICAL CHEMISTRY. Educ: Shurtleff Col, BS, 34; Univ Ill, MS, 37, PhD(anal chem), 40. Prof Exp: Asst, Univ Ill, 37-40; res chemist, Nat Distillers & Chem Corp, 40-41 & Com Solvents Corp, 41-51; dir anal res, R J Reynolds Industs, Inc, 51-72; RETIRED. Mem: AAAS; Am Chem Soc. Res: Instrumentation methods of analysis; chromatography; development of methods for analysis of nitroparaffins and derivatives; penicillin; bacitracin; hexachlorocyclohexanes; complex cations in microanalysis; titrimetry in nonaqueous solvents; functional group analysis. Mailing Add: 2425 Westchester Blvd Springfield IL 62704

MARKUS, GABOR, b Budapest, Hungary, June 8, 22; nat US; m 64; c 3. BIOCHEMISTRY. Educ: Univ Budapest, MD, 47; Stanford Univ, PhD, 50. Prof Exp: Estab investr, Am Heart Asn, 60-63; assoc res prof, State Univ NY Buffalo, 63-67, chmn dept biochem, Roswell Park Div, 67-71; assoc cancer res scientist, 63-67, PRIN CANCER RES SCIENTIST, ROSWELL PARK MEM INST, 67-; RES PROF BIOCHEM, STATE UNIV NY BUFFALO, 67- Mem: AAAS; Am Soc Biol Chemists; Soc Exp Biol & Med. Res: Protein structure and conformations; enzyme regulation; biochemistry of fibrinolysis. Mailing Add: 430 Starin Ave Buffalo NY 14216

MARKUS, HELENE BABAD, b Paris, France, Mar 21, 40; US citizen; m 67; c 2. BIOCHEMISTRY. Educ: City Univ New York, BS, 61; Univ Calif, Berkeley, PhD(biochem), 65. Prof Exp: NIH fel, Dept Biol Chem, Hebrew Univ, Jerusalem, 65-67; res fel fat metab, Harvard Med Sch, 67-70; researcher biochem, Fels Res Inst, Temple Univ, 70-72; ASST PROF PHYSIOL CHEM, PHILADELPHIA COL OSTEOP MED, 72- Mem: AAAS. Res: Ketone body metabolism in starvation and diabetes. Mailing Add: Dept of Physiol Chem Philadelphia Col of Osteop Med Philadelphia PA 19139

MARKUS, LAWRENCE, b Hibbing, Minn, Oct 13, 22; m 50; c 2. MATHEMATICS. Educ: Univ Chicago, BS, 42, MS(meteorol) & MS(math), 47; Harvard Univ, PhD(math), 51. Prof Exp: Instr meteorol, Univ Chicago, 42-44, res meteorologist, Atomic Energy Proj, 44; instr math, Harvard Univ, 51-52 & Yale Univ, 52-55; lectr, Princeton Univ, 55-57; from asst prof to assoc prof, 57-60, assoc head control sci ctr, 61-63, PROF MATH, UNIV MINN, MINNEAPOLIS, 60-, DIR CONTROL SCI CTR, 65- Concurrent Pos: Fulbright fel, Paris, France, 50-51; Guggenheim fel, Univ Lausanne, 63-64; Nuffield prof, Univ Warwick, 68-69, dir control theory ctr, 70-; course dir, Int Ctr Theoret Physics, 74; lectr, Int Math Cong; prin lectr, Iranian Math Soc, 75. Mem: Am Math Soc; Math Asn Am. Res: Ordinary differential equations; control theory; differential geometry; cosmology. Mailing Add: Dept of Math Univ of Minn Minneapolis MN 55455

MARKUS, RICHARD LOUIS, organic chemistry, see 12th edition

MARKWELL, DICK ROBERT, b Muskogee, Okla, Feb 20, 25; m 49; c 4. ELECTROCHEMISTRY. Educ: Univ Wichita, BS, 48, MS, 50; Univ Wis, PhD(chem), 56. Prof Exp: Res & develop coordr, Off Chief Res & Develop, Hq, US Army, 65-67; assoc prof chem, San Antonio Col, 67-74; CHEMIST, CORPUS CHRISTI DEPT OF HEALTH, 75- Mem: Am Chem Soc. Mailing Add: 1406 Haskin Dr San Antonio TX 78209

MARKWORTH, ALAN JOHN, b Cleveland, Ohio, July 13, 37; c 3. PHYSICS. Educ: Case Inst Technol, BSc, 59; Ohio State Univ, MSc, 61, PhD(physics), 69. Prof Exp: PRIN PHYSICIST, BATTELLE-COLUMBUS, 66- Mem: Am Asn Physics Teachers; Am Inst Mining, Metall & Petrol Engrs. Res: Theory of phase transformations; computer simulation studies of kinetic processes in solids and liquids. Mailing Add: Metal Sci Group Battelle-Columbus 505 King Ave Columbus OH 43201

MARLAND, GREGG (HINTON), b Oak Park, Ill, Sept 16, 42; m 63; c 3. GEOCHEMISTRY. Educ: Va Polytech Inst & State Univ, BS, 64; Univ Minn, PhD(geol), 72. Prof Exp: Asst prof geochem, Ind State Univ, Terre Haute, 70-75; ASSOC SCIENTIST, INST FOR ENERGY ANAL, 75- Mem: AAAS; Geochem Soc; Soc Environ Geochem & Health. Res: Aqueous geochemistry; phase equilibria; environmental geochemistry, energy options and environmental implications. Mailing Add: Inst for Energy Anal Oak Ridge Assoc Univs Oak Ridge TN 37830

MARLATT, ABBY LINDSEY, b Manhattan, Kans, Dec 5, 16. NUTRITION. Educ: Kans State Univ, BS, 38; Univ Calif, cert, 40, PhD(animal nutrit), 47. Prof Exp: Asst home econ, Univ Calif, 40-45; from assoc prof to prof foods & nutrit, Kans State Univ, 45-56; vis prof home econ, Beirut Col Women, 53-54; dir col, 56-63, PROF NUTRIT & FOOD SCI, COL HOME ECON, UNIV KY, 63- Concurrent Pos: Consult, Ky State Col, 68-70. Mem: AAAS; Am Home Econ Asn; Am Dietetic Asn. Res: Human nutrition; nutrient interrelationships; nutritional status and dietary surveys; pynidoxine requirements. Mailing Add: Dept of Nutrit & Food Sci Univ of Ky Col of Home Econ Lexington KY 40506

MARLATT, ROBERT BRUCE, b Cleveland, Ohio, July 18, 20; m 46; c 3. PLANT PATHOLOGY. Educ: Univ Ariz, PhD(plant path), 52. Prof Exp: Asst plant pathologist, State Dept Agr, Calif, 52; assoc plant pathologist, Univ Ariz, 52; assoc plant pathologist, Subtrop Exp Sta, 64-70, PROF PLANT PATH & PLANT PATHOLOGIST, AGR RES & EDUC CTR, UNIV FLA, 70- Mem: Am Phytopath Soc. Res: Diseases of tropical ornamentals. Mailing Add: Agr Res & Educ Ctr Univ of Fla Homestead FL 33030

MARLATT, WILLIAM EDGAR, b Kearney, Nebr, June 5, 31; m 56; c 2. ATMOSPHERIC SCIENCE. Educ: Nebr State Col Kearney, BA, 56; Rutgers Univ, MS, 58, PhD(soil physics), 61. Prof Exp: Res asst forestry, US Forest Serv, 54-55; res asst meteorol, Rutgers Univ, 56-58, res assoc, 58-61, asst prof, 61; from asst prof to assoc prof, 61-69, head dept watershed sci, 70-74, assoc dean, Grad Sch, 67-68, PROF ATMOSPHERIC SCI, COLO STATE UNIV, 69- Concurrent Pos: Consult, Nat Bur Stand, 64-67; mem, Colo Natural Resource Ctr Coun, 66-71; consult, Martin Marietta Co, 67-; mem, Int Biol Prog Biometeorol Panel, Nat Res Coun-Nat Acad Sci, 68-70; sr scientist, Environ Resources Assocs, Inc, 69-70; consult, Manned Spaceflight Ctr, NASA, 70 & Colspan Environ Systs, Inc, 70-71; Int Biol Prog mem, Nat Adv Comt Aerobiol, 70-; consult, Thorne Ecol Found, 71-; chmn educ comt, Colo Environ Res Ctr, 71- Mem: AAAS; Am Meteorol Soc; Am Astronaut Soc; Am Geophys Union. Res: Remote sensing of atmosphere and earth surface, environment quality, interaction of climate and environment. Mailing Add: 3611 Richmond Dr Ft Collins CO 80521

MARLBOROUGH, DAVID IAN, b Henfield, Eng, May 2, 41; m 73. BIOCHEMISTRY. Educ: London Univ, BSc, 62; Univ Exeter, PhD(org chem), 66. Prof Exp: Fel, Harvard Med Sch, 66-68; Sci Res Coun fel, Imp Col, Univ London, 68-69; sr res fel structure & biophys of L-asparaginase, Microbiol Res Estab, Salisbury, Eng, 70-73; ASSOC SCIENTIST PEPTIDE HORMONE STRUCT, PAPANICOLAOU CANCER RES INST, 73-; RES ASST PROF, SCH MED, UNIV MIAMI, 73- Mem: The Chem Soc. Res: Study of polypeptide and protein conformations in solution via spectroscopic techniques including circular dichroism, nuclear magnetic resonance spectroscopy and infrared spectroscopy; protein structure. Mailing Add: Papanicolaou Cancer Res Inst 1155 NW 14th St Miami FL 33136

MARLBOROUGH, JOHN MICHAEL, b Toronto, Ont, Aug 1, 40. ASTRONOMY. Educ: Univ Toronto, BSc, 52, MA, 63; Univ Chicago, PhD(astron), 67. Prof Exp: Lectr, 67, asst prof, 67-70, ASSOC PROF ASTRON, UNIV WESTERN ONT, 70- Concurrent Pos: Vis scientist, Dominion Astrophys Observ, Victoria, BC, 73-74. Mem: Am Astron Soc; Can Astron Soc; Royal Astron Soc. Res: Stellar interiors and evolution; early-type stars with extended atmospheres; radiative transfer; gas dynamics. Mailing Add: Dept of Astron Univ of Western Ont London ON Can

MARLER, PETER, b London, Eng, Feb 24, 28; m 53; c 3. ZOOLOGY. Educ: Univ London, BSc, 48, PhD(bot), 52; Cambridge Univ, PhD(zool), 54. Prof Exp: Res fel, Jesus Col, Cambridge Univ, 54-56; from asst prof to prof zool, Univ Calif, Berkeley, 57-66; PROF ZOOL, ROCKEFELLER UNIV, 66- Concurrent Pos: Guggenheim fel, 64-65; dir, Inst Res Animal Behav. Mem: Nat Acad Sci; fel AAAS; fel Am Acad Arts & Sci; Animal Behav Soc (pres, 69-70); Am Soc Zoologists. Res: Animal behavior, particularly the function and evolution of animal communication systems, especially vocalizations of birds, monkeys and apes; ontogenic basis of bird song. Mailing Add: Dept of Zool Rockefeller Univ New York NY 10021

MARLETT, JUDITH ANN, b Toledo, Ohio, June 20, 43. NUTRITION. Educ: Miami Univ, BS, 65; Univ Minn, PhD(nutrit), 72. Prof Exp: Therapeut dietician, Minneapolis Vet Admin Hosp, 66-67; res fel nutrit, Sch Pub Health, Harvard Univ, 73-74; ASST PROF NUTRIT, UNIV WIS-MADISON, 75- Mem: Sigma Xi. Res: Role of dietary fiber in human nutrition and in the human gastrointestinal tract; role of the gastrointestinal tract in lipid metabolism. Mailing Add: Dept of Nutrit Sci Univ of Wis 1270 Linden Dr Madison WI 53706

MARLEY, GERALD C, b Lovington, NMex, Nov 11, 38; m 59. MATHEMATICS. Educ: Eastern NMex Univ, BSc, 59; Tex Tech Col, MSc, 61; Univ Ariz, PhD(math), 67. Prof Exp: Res engr, Gen Dynamics/Astronaut, 61; lectr math, Univ Ariz, 67; from asst prof to assoc prof, 67-74, PROF MATH, CALIF STATE UNIV, FULLERTON, 74- Mem: Am Math Soc; Math Asn Am; Am Sci Affil. Res: Subdivisions of Euclidean space by convex bodies. Mailing Add: Dept of Math Calif State Univ Fullerton CA 92634

MARLEY, JAMES ALOYSIUS, b Philadelphia, Pa, July 16, 33; m 59; c 3. INORGANIC CHEMISTRY, ENGINEERING MANAGEMENT. Educ: St Joseph's Col, BS, 55; Univ Pa, PhD(inorg chem), 60. Prof Exp: Res chemist, Corning Glass Works, 59-64, mgr solid state chem dept, 64-68, mgr solid state res, 68-72, DEPT MGR, RES & DEVELOP LAB, SIGNETICS CORP, 72- Mem: Am Phys Soc; Am Ceramic Soc; Electrochem Soc. Res: Crystal growth; materials characterization; ion implantation studies in semiconductors. Mailing Add: Res & Develop Lab Signetics Corp 811 E Arques Ave Sunnyvale CA 94086

MARLIN, CLIFTON BOYD, b Dorsey, Miss, Oct 24, 20; m 45; c 1. FOREST ECONOMICS. Educ: Miss State Univ, BS, 43; Duke Univ, MF, 49. Prof Exp: Instr forest econ, Miss State Univ, 49-50; forester, Miss Forestry Comn, 51-52, asst forest mgt dir, 52-53, forest mgt dir, 53-56, state forester, 56-61; asst prof forest mensuration, 61-73, ASSOC PROF FOREST ECON & MENSURATION, LA STATE UNIV, BATON ROUGE, 73- Concurrent Pos: Mem nat exec comt, Coop Forest Fire Prev, Smokey Bear Prog, 57-60. Mem: Soc Am Foresters. Res: Forest resource and production economics. Mailing Add: 5822 Clematis Dr Baton Rouge LA 70808

MARLIN, JOE ALTON, b Naylor, Mo, July 3, 35; m 60; c 2. APPLIED MATHEMATICS. Educ: Southeast Mo State Col, BS, 58; Univ Mo-Columbia, MA, 60; NC State Univ, PhD(math), 65. Prof Exp: Mem tech staff, Bell Tel Labs, 60-63; instr, 64-66, asst prof, 66-68, ASSOC PROF MATH, NC STATE UNIV, 68- Mem: Am Math Soc; Math Asn Am. Res: Oscillatory and asymptotic behavior of systems of ordinary differential equations which represent equations of motion for mechanical systems. Mailing Add: Dept of Math NC State Univ Raleigh NC 27607

MARLIN, ROBERT LEWIS, b Bronx, NY, June 28, 37; m 59; c 2. INFORMATION SCIENCE, RESEARCH ADMINISTRATION. Educ: Syracuse Univ, AB, 58, MPA, 62. Prof Exp: Asst psychologist, NY State Dept Ment Hyg, 57-59; asst scientist exp psychol, Sterling-Winthrop Res Inst, 59-60, asst scientist statist, 60-62; asst to dir new prod develop, Winthrop Labs, Sterling Drug, Inc, 62-65; coordr med affairs, Knoll Pharm Co, 65-68; coordr prod develop, Schering Corp, 68-69; clin res assoc, Sandoz Pharm Co, 69-71; dir res admin, Leo Winter Assocs, 71-72; clin res assoc med, Sandoz Pharm Co, 72-75; BIOMED CONSULT, 75- Concurrent Pos: Ed newslett, NJ Acad Sci, 70- Mem: AAAS; Drug Info Asn (secy, 68-69, vpres, 70-71); Biomet Soc; NY Acad Sci; Am Statist Asn. Mailing Add: 8 Biscay Dr Parsippany NJ 07054

MARLOW, KEITH WINTON, b Madison, Kans, Nov 14, 28; m 51; c 3. NUCLEAR PHYSICS. Educ: Kans State Univ, BS, 51; Univ Md, PhD, 66. Prof Exp: Physicist, 51-67, head reactors br, 67-70, CONSULT, RADIATION TECHNOL DIV, NAVAL RES LAB, 71- Concurrent Pos: Physicist, Inst Nuclear Physics Res, Amsterdam, 67-68. Mem: AAAS; Am Phys Soc; Sigma Xi. Res: Experimental study of nuclear structure, principally by investigating decay of radioactive nuclides; development of instrumentation for detecting low levels of radioactivity. Mailing Add: Naval Res Lab Code 6603M Washington DC 20375

MARLOW, WILLIAM HENRY, b Waterloo, Iowa, Nov 26, 24; m 48; c 5. OPERATIONS RESEARCH, MATHEMATICS. Educ: St Ambrose Col, BS, 47; Univ Iowa, MS, 48, PhD(math), 51. Prof Exp: Instr math, Univ Iowa, 48-51; res assoc, Logistics Res Proj, 51-56, prin investr, 56-69, PROF OPERS RES & DIR INST MGR SCI & ENG, GEORGE WASHINGTON UNIV, 69-, CHMN DEPT OPERS RES, 71- Concurrent Pos: Assoc res mathematician, Univ Calif, Los Angeles, 54-55. Mem: Opers Res Soc Am; Am Math Soc; Math Asn Am; Soc Indust & Appl Math; Inst Mgt Sci. Res: Mathematical methods and numerical procedures in operations research and management science; logistics; systems effectiveness. Mailing Add: Sch of Eng & Appl Sci George Washington Univ Washington DC 20052

MARLOWE, EDWARD, b New York, NY, May 5, 35; m 59; c 2. PHARMACEUTICAL CHEMISTRY. Educ: Columbia Univ, BS, 56, MS, 58; Univ Md, PhD(pharm), 62. Prof Exp: Res assoc pharm res & develop, Merck Sharp & Dohme Res Lab, 62-64; sr scientist, Ortho Pharmaceut Corp, 64-67; dir res & develop, Whitehall Labs Div, Am Home Prod, NJ, 67-72; VPRES RES & DEVELOP, RES & TECH DIV, PLOUGH INC, 72- Concurrent Pos: Prof pharmaceut, Univ Tenn, 73- Mem: Acad Pharmaceut Sci; Soc Cosmetic Chemists; Am Pharmaceut Asn. Res: Product development, pharmacology, toxicology and photobiology; analytical development; exploratory product design research; drug absorption. Mailing Add: 6368 Kirby Oaks Dr Memphis TN 38138

MARLOWE, GEORGE ALBERT, JR, b Detroit, Mich, May 25, 25; m 53; c 1. HORTICULTURE. Educ: George Washington Univ, BS, 49, MS, 50; Univ Md, PhD(hort), 55. Prof Exp: Exten specialist hort, Univ Ky, 56-62; assoc prof hort, Ohio State Univ, 62-65; exten specialist veg crops, Univ Calif, Davis, 65-69; PROF HORT & HORTICULTURIST, INST FOOD & AGR SCI, UNIV FLA, 69- Concurrent Pos: Chmn, Dept Veg Crops, Univ Fla, 69-72. Mem: Am Soc Hort Sci; Am Phys Soc. Res: Precision production technology; crop nutrition. Mailing Add: Dept of Veg Crops Univ of Fla Gainesville FL 32601

MARLOWE, JAMES IRVIN, b Southport, NC, Sept 23, 32; m 57; c 4. MARINE GEOLOGY. Educ: Fla State Univ, BS, 57; Univ Ariz, PhD(econ geol), 61. Prof Exp: Geologist, US Geol Surv, 57; part-time independent geologist, 58-60; geologist, NJ Zinc Co, 60-62; marine geologist, Atlantic Oceanog Lab Bedford Inst Oceanog, 62-70; assoc prof geol, Miami-Dade Community Col, 70-73; HEAD MARINE GEOL & GEOPHYS, DAMES & MOORE CONSULT EARTH SCI, 73- Concurrent Pos: Independent consult geol, 65-73. Mem: Soc Econ Paleont & Mineral; AAAS. Res: Mineral deposits in carbonate rocks; stratigraphy of the continental slope; marine diagenesis in carbonate rocks; offshore mineral deposits. Mailing Add: PO Box 190 Sarasota FL 33578

MARLOWE, THOMAS JOHNSON, b Fairview, NC, Sept 15, 17; m 45; c 4. ANIMAL GENETICS, ANIMAL PHYSIOLOGY. Educ: NC State Univ, BS, 40, MS, 49; Okla State Univ, PhD(animal genetics & physiol), 54. Prof Exp: High sch teacher, 40-42; training specialist & asst supvr, Vet Admin, NC, 46-48; asst county agt & livestock specialist, Va Agr Exten Serv, 49-50; instr animal husb, Miss State Univ, 51-52; res asst, Okla State Univ, 52-54; assoc prof, 54-64, PROF ANIMAL HUSB, VA POLYTECH INST & STATE UNIV, 64- Concurrent Pos: Mem, Coun Agr Sci & Technol, Va Acad Sci. Mem: Am Inst Biol Sci; Am Genetic Asn; Am Soc Animal Sci (secy-treas, 71-74, pres, 75-76). Res: Beef cattle performance testing; heritability of economic traits; genetics and pathology of hereditary dwarfism cattle; effectiveness of selection in beef cattle; cytogenetics; evaluation of sire and dam breed for crossbreeding. Mailing Add: Dept of Animal Sci Va Polytech Inst & State Univ Blacksburg VA 24061

MARMER, WILLIAM NELSON, b Philadelphia, Pa, July 19, 43; m 73; c 1. ORGANIC CHEMISTRY, LIPID CHEMISTRY. Educ: Univ Pa, AB, 65; Temple Univ, PhD(chem), 71. Prof Exp: Nat Res Coun-Agr Res Serv res assoc org chem, 70-72, res scientist, Fats & Proteins Res Found, Inc, 72-75, RES CHEMIST, EASTERN REGIONAL RES CTR, USDA, 75- Mem: Am Chem Soc; Am Oil Chemists' Soc; Sigma Xi. Res: Amine oxides; epoxides; O-acylhydroxylamines; fatty acid derivatives; acylations; fabric treatment; mixed anhydrides; lime soap dispersing agents; depot fat analysis. Mailing Add: Eastern Regional Res Ctr USDA 600 E Mermaid Lane Philadelphia PA 19118

MARMET, PAUL, b Levis, Que, May 20, 32; m 59; c 2. ATOMIC PHYSICS, MOLECULAR PHYSICS. Educ: Laval Univ, BSc, 56, DSc(physics), 60. Prof Exp: Asst molecular physics, Commonwealth Sci & Indust Res Orgn, Australia, 60-61; from asst prof to assoc prof, 61-67, PROF PHYSICS, LAVAL UNIV, 67- Concurrent Pos: Nat Res Coun Can fel, 60-61, grant, 61-, mem adv comt physics, 70-73; Defence Res Bd Can grants, 63-64 & 66-75. Honors & Awards: Herzberg Medal, Can Asn Physicists, 71. Mem: Can Asn Physicists; Royal Astron Soc Can; Fr-Can Asn Advan Sci; fel Royal Soc Can. Mailing Add: Dept of Physics Laval Univ Ste-Foy PQ Can

MARMO, FREDERICK FRANCIS, b Boston, Mass, Oct 25, 20; m 43; c 3. AERONOMY, ENVIRONMENTAL CHEMISTRY. Educ: Boston Univ, AB, 49; Harvard Univ, MS, 51, PhD(chem physics), 53. Prof Exp: Chief chem physics br, US Air Force Cambridge Res Ctr, 53-58; mgr chem physics dept, GCA Corp, Mass, 58-61, dir space sci lab, 61-66, dir space sci opers, 66-67, tech dir & vpres tech div, 67, vpres & dir res, 67-72; PRIN SCIENTIST, DEPT OF TRANSP, 72- Honors & Awards: Fermi Outstanding Scientist Award & Superior Performance & Outstanding Achievement Award, Air Force Cambridge Res Ctr, 58. Mem: Am Geophys Union; Am Meteorol Soc; Am Inst Aeronaut & Astronaut. Res: Planetary physics; environmental sciences; photochemistry; design of satellite probes for global monitoring. Mailing Add: 138 Main St Wakefield MA 01880

MARMOR, JUDD, b London, Eng, May 1, 10; US citizen; m 38; c 1. PSYCHIATRY. Educ: Columbia Univ, AB, 30, MD, 33. Hon Degrees: LHD, Hebrew Union Col, 72. Prof Exp: Pvt pract psychiat, psychoanal & neurol, 37-65; dir psychiat, Cedars-Sinai Med Ctr, 65-72; FRANZ ALEXANDER PROF PSYCHIAT, SCH MED, UNIV SOUTHERN CALIF, 72- Concurrent Pos: Clin prof, Univ Calif, Los Angeles, 52-; training analyst, Southern Calif Psychoanal Inst, 53-; mem fac psychiat & law for judiciary, Univ Southern Calif, 64-69; mem fac, Conn State Dept Ment Health, 65-67; mem homosexuality task force, NIMH, 67-69, mem social probs res rev comn, 70-72; lectr, Women's Bur, US Dept Labor, 68. Mem: Group Advan Psychiat (pres, 73-75); Am Orthopsychiat Asn; Am Psychoanal Asn; life fel Am Psychiat Asn (pres, 75-76); Am Acad Psychoanal (pres, 65-66). Res: Social and community psychiatry; psychotherapy; problems of sexual deviancy. Mailing Add: 2025 Zonal Ave Los Angeles CA 90033

MARMOR, ROBERT SAMUEL, organic chemistry, see 12th edition

MARMOR, SOLOMON, b New York, NY, Feb 25, 26; m 54; c 2. ORGANIC CHEMISTRY. Educ: City Col New York, BS, 48; Syracuse Univ, PhD(org chem), 52. Prof Exp: Res chemist, Becco Chem Div, FMC Corp, 52-56; asst prof chem, Utica Col, Syracuse Univ, 56-62; assoc prof, NMex Highlands Univ, 62-66, head dept, 64-66; assoc prof, 66-70, coordr interdept progs, 68-70, chmn dept, 68-71, actg dean sch natural sci & math, 73-74, PROF CHEM, CALIF STATE COL, DOMINGUEZ HILLS, 70- Mem: AAAS; Am Chem Soc; fel Am Inst Chemists. Res: Epoxides; hypochlorous acid reactions; hydrogen peroxide oxidations of organic compounds. Mailing Add: Dept of Chem Calif State Col Carson CA 90747

MARMUR, JULIUS, b Byelostok, Poland, Mar 22, 26; Can citizen; m 58; c 2. MOLECULAR BIOLOGY, BIOCHEMISTRY. Educ: McGill Univ, BS, 46, MS, 47; Iowa State Col, PhD(bact physiol), 51. Prof Exp: Mem staff, NIH, 51-52, Rockefeller Inst, 52-54, Pasteur Inst, Paris, 54-55 & Inst Microbiol, Rutgers Univ, 55-56; res assoc chem, Harvard Univ, 56-60; asst prof biochem, Brandeis Univ, 60-61, assoc prof, 61-63; PROF BIOCHEM, ALBERT EINSTEIN COL MED, 63-, ACTG CHMN DEPT BIOCHEM & PROF GENETICS, 74- Mem: Am Soc Biol Chem; Am Soc Microbiol. Res: Biological and physical-chemical properties of bacterial and viral nucleic acids. Mailing Add: Dept of Biochem Albert Einstein Col of Med New York NY 10461

MAROIS, PAUL HENRI, b Montreal, Que, Feb 22, 19; m 48; c 6. VETERINARY MICROBIOLOGY. Educ: Univ Montreal, DVM, 40, LSc, 44, MSc, 45. Prof Exp: Vet, 45-46, res assoc microbiol, 46-70, ASSOC DIR, INST ARMAND-FRAPPIER, 70- Concurrent Pos: Res asst, Fac Med, Univ Montreal, 46-70, res assoc, 70-, lectr, Fac Vet Med, 46- Mem: Can Vet Med Asn; Can Soc Microbiologists; Am Col Vet Med; Am Asn Lab Animal Sci; Can Asn Lab Animal Sci. Res: Veterinary virology and avian and mammal viruses; pathogenicity; antigenicity; vaccine production; zoonoses. Mailing Add: 531 Blvd des Prairies Laval-des-Rapides City of Laval PQ Can

MAROIS, ROBERT LEO, b Troy, NY, Apr 27, 35; m 61; c 4. PHARMACOLOGY. Educ: Siena Col, NY, BS, 64; Albany Med Col, PhD(pharmacol), 69. Prof Exp: Res assoc, State Univ NY Albany, 68-69; ASSOC PROF PHARMACOL, ALBANY COL PHARM, 69- Concurrent Pos: USPHS fel, State Univ NY Albany, 68-69. Res: Cardiovascular and neuromuscular pharmacology. Mailing Add: Dept of Biol Sci Albany Col of Pharm Albany NY 12208

MARON, MELVIN EARL, b Bloomfield, NJ, Jan 23, 24; m 48; c 2. INFORMATION SCIENCE. Educ: Univ Nebr, BS, 45, BA, 47; Univ Calif, Los Angeles, PhD(philos), 51. Prof Exp: Instr, Univ Calif, Los Angeles, 51-52; tech engr, Int Bus Mach Corp, 52-55; mem tech staff, Ramo-Wooldridge Corp, 55-59; mem sr res staff, Rand Corp, 59-66; PROF LIBRARIANSHIP, UNIV CALIF, BERKELEY, 66- Mem: AAAS; Asn Comput Mach; Philos Sci Asn. Res: Philosophy; computer sciences; cybernetics; mechanized literature searching and data retrieval. Mailing Add: 63 Ardilla Rd Orinda CA 94563

MARON, SAMUEL HERBERT, b Warsaw, Poland, May 28, 08; nat US; m 36; c 1. PHYSICAL CHEMISTRY, POLYMER SCIENCE. Educ: Case Western Reserve Univ, BS, 31, MS, 33; Columbia Univ, PhD(phys chem), 38. Prof Exp: Instr chem, 31-37, from asst prof to assoc prof phys chem, 37-45, PROF PHYS CHEM, CASE WESTERN RESERVE UNIV, 45- Concurrent Pos: Dir res proj, Off Rubber Reserve, 43-56, mem, Latex Adv Comt. Mem: AAAS; Am Chem Soc; Am Inst Chem Eng; Soc Rheol. Res: Phase rule; solution kinetics; thermodynamics of gases and solutions; colloid chemistry of latex; synthetic rubber; polymerization; physical chemistry of high polymers; rheology. Mailing Add: Div of Macromolecular Sci Case Western Reserve Univ Cleveland OH 44106

MARONDE, ROBERT FRANCIS, b Calif, Jan 13, 20; m 42; c 3. INTERNAL MEDICINE, CLINICAL PHARMACOLOGY. Educ: Univ Southern Calif, BA, 41, MD, 44; Am Bd Internal Med, dipl, 51. Prof Exp: Resident med, Los Angeles County Gen Hosp, 46-48; asst prof physiol, 48-49, from asst clin prof to assoc clin prof med, 49-63, assoc prof med & pharmacol, 63-68, PROF MED & PHARMACOL, SCH MED, UNIV SOUTHERN CALIF, 68-, CHIEF CLIN PHARMACOL SECT, 70- Mem: Am Soc Clin Pharmacol & Therapeut. Res: Medical computer applications for drug utilization review. Mailing Add: Dept of Med Univ Southern Calif Sch of Med Los Angeles CA 90033

MARONEY, SAMUEL PATTERSON, JR, b Wilmington, Del, Feb 3, 26; m 51; c 3. ZOOLOGY. Educ: Wesleyan Univ, BA, 50; Univ Del, MA, 53; Duke Univ, PhD(zool), 57. Prof Exp: Asst prof, 56-62, ASSOC PROF BIOL, UNIV VA, 62- Mem: Fel AAAS; Am Soc Zoologists; Am Physiol Soc; Am Soc Cell Biol. Res: Cell physiology; permeability and hemolysis of amphibian erythrocytes. Mailing Add: Dept of Biol Gilmer Hall Univ of Va Charlottesville VA 22903

MARONEY, WILLIAM, b Fargo, NDak, June 9, 08. CHEMISTRY. Educ: NDak Agr Col, BS, 29; Univ Calif, PhD(chem), 33. Prof Exp: Chemist, Com Solvents Corp, 29-30; instr chem, Univ Calif, 30-35; head dept, Col St Theresa, 35-36; from instr to prof, 36-74, chmn dept, 38-67, EMER PROF CHEM, UNIV SAN FRANCISCO, 74- Res: Photochemistry; photochemical equilibria; kinetics of reaction in solution; catalytic hydrazine reductions. Mailing Add: Dept of Chem Univ of San Francisco San Francisco CA 94117

MARONI, GUSTAVO PRIMO, b Merlo, Arg, Nov 20, 41; m 74. DEVELOPMENTAL GENETICS, BIOCHEMICAL GENETICS. Educ: Univ Buenos Aires, Lic, 67; Univ Wis, PhD(zool), 72. Prof Exp: Res assoc genetics, Dept Zool, Univ NC, 73-74 & Inst Genetics, Univ Colgne, 74-75; ASST PROF GENETICS, DEPT ZOOL, UNIV NC, CHAPEL HILL, 75- Mem: Genetics Soc Am; AAAS. Res: Regulation of sex-linked gene activity in Drosophila; control of alcohol dehydrogenase and other gene-enzyme systems. Mailing Add: Dept of Zool Wilson Hall Univ of NC Chapel Hill NC 27514

MAROTTA, CHARLES ANTHONY, b New York, NY, Apr 12, 45. MOLECULAR BIOLOGY, PSYCHIATRY. Educ: City Col New York, BS, 65; Duke Univ, MD, 69; Yale Univ, MPhil, 72, PhD(molecular biophys & biochem), 75. Prof Exp: NIH fel molecular biophys & internal med, Med Sch, Yale Univ, 69-73, fel molecular biophys & internal med, fel internal med, 71-72, fel med res, Clin Res Training Prog, 71-73, res fel human genetics, 72-73; CLIN FEL PSYCHIAT, HARVARD MED SCH, 73-, RES FEL, 75- Concurrent Pos: Fel med, Yale-New Haven Hosp, 70-73; resident psychiat, Mass Gen Hosp, 73- Honors & Awards: Physician Recognition Award, AMA, 72; Ethel B Dupont-Warren Award & William F Milton Fund Award, Harvard Med Sch, 75. Mem: AAAS; AMA. Res: Molecular genetics of hemoglobin diseases; molecular psychobiology; clinical research in psychiatry; neurobiology. Mailing Add: Res 4 Mass Gen Hosp Boston MA 02114

MAROTTA, SABATH FRED, b Chicago, Ill, Aug 26, 29. PHYSIOLOGY. Educ: Univ Ill, MS, 53, PhD(physiol), 57. Prof Exp: Res assoc animal sci, 57-58, from instr to assoc prof physiol, Col Med, 58-70, res assoc, Aeromed Lab, Med Ctr, 58-60, asst dir, 60-64, assoc dean grad col, 75, PROF PHYSIOL, MED CTR, UNIV ILL, 70-, ASSOC DIR RES RESOURCES CTR, 75- Concurrent Pos: Univ Ill adv, Chiengmai Proj, Thailand, 64-66. Mem: Aerospace Med Asn; Am Physiol Soc; Soc Exp Biol & Med. Res: Neuroendocrinology; biologic rhythms; role of the adrenal cortex in the adaption to environmental stresses. Mailing Add: Res Resources Ctr Univ of Ill Med Ctr Chicago IL 60680

MAROUSKY, FRANCIS JOHN, b Shenandoah, Pa, Oct 28, 35; m 59; c 5. HORTICULTURE. Educ: Pa State Univ, BS, 57; Univ Md, MS, 64; Va Polytech Inst, PhD(hort), 67. Prof Exp: Res asst, Univ Md, 62-64; instr, Va Polytech Inst, 64-67; RES HORTICULTURIST, AGR RES & EDUC CTR, AGR RES SERV, USDA, UNIV FLA, 67- Mem: Am Soc Hort Sci; Int Soc Hort Sci. Res: Market quality and post harvest aspects and senescence of cut flowers. Mailing Add: Agr Res & Educ Ctr Agr Res Serv USDA Univ of Fla Bradenton FL 33505

MAROV, GASPAR J, b Unije, Yugoslavia, Jan 3, 20; US citizen; m 46. SUGAR CHEMISTRY. Educ: City Col New York, BS, 42; Columbia Univ, MA, 50. Prof Exp: Asst food chem, Columbia Univ, 49-50; anal chemist, Thomas J Lipton, Inc, NJ, 50-51; res chemist, 51-57, CHIEF CONTROL CHEMIST, PEPSI-COLA CO, 57- Mem: Am Chem Soc; Sugar Indust Technologists; Soc Soft Drink Technologists; Am Water Works Asn. Res: Determination of solids in sugar solutions, syrups and carbonated beverages; measurement of sugar color. Mailing Add: Prod Control Lab Pepsi-Cola Co 4600 Fifth St Long Island City NY 11101

MAROVITZ, WILLIAM F, b Salt Lake City, Utah, Dec 10, 41; m 64. ANATOMY, OTOLARYNGOLOGY. Educ: Univ Calif, BA, 62, PhD(human anat), 66. Prof Exp: From asst prof to assoc prof anat in otolaryngol, Sch Med, Wash Univ, 66-73; DIR RES, DEPT OTOLARYNGOL, ASSOC PROF OTOLARYNGOL & ASSOC PROF ANAT, MT SINAI SCH MED, 73- Concurrent Pos: Res grant deafness, Mt Sinai Sch Med, 73-; prin investr, Nat Inst Neurol Dis & Blindness grant, 66-73; consult, Nat Inst Neurol Dis & Blindness report on human commun & its disorders, 68; actg prof anat & chmn dept, Israel Inst Technol, 72. Mem: AAAS; Am Asn Anat; NY Acad Sci. Res: Male reproductive endocrinology; developmental anatomy and biochemistry of temporal bone and its contents. Mailing Add: Dept of Otolaryngol Mt Sinai Sch of Med New York NY 10028

MARPLE, DENNIS NEIL, b Storm Lake, Iowa, Oct 31, 45; m 66; c 2. ANIMAL PHYSIOLOGY, ANIMAL SCIENCE. Educ: Iowa State Univ, BS, 67, MS, 68; Purdue Univ, PhD(physiol), 71. Prof Exp: NIH fel swine physiol, Meat & Animal Sci Dept, Univ Wis, 71-73; ASST PROF ANIMAL & DAIRY SCI, AUBURN UNIV, 73- Mem: Am Soc Animal Sci; Am Meat Sci Asn; Sigma Xi. Res: Endocrinological interactions and their regulation of growth in meat animals; study of animal physiology and meat quality. Mailing Add: Dept Animal & Dairy Sci Auburn Univ Auburn AL 36830

MARPLE, DUDLEY TYNG FISHER, physics, see 12th edition

MARPLE, LELAND WARREN, b Los Angeles, Calif, June 24, 34; m 56; c 3. ANALYTICAL CHEMISTRY. Educ: Occidental Col, BA, 56; Mass Inst Technol, PhD(anal chem), 60. Prof Exp: Fel chem, Ames Lab, USAEC, 60-63; asst prof chem, Iowa State Univ, 63-67; MEM STAFF, SYNTEX RES CORP, 67- Mem: Soc Cosmetic Chemists; Am Chem Soc. Res: Oxidation of ethylene oxide base polymers. Mailing Add: Syntex Res Corp 3401 Hillview Ave Palo Alto CA 94304

MARPLE, ROBERT P, b Woodston, Kans, Dec 6, 16; m 41; c 2. GEOGRAPHY. Educ: Ft Hays Kans State Col, AB, 48, MS, 49; Univ Nebr, PhD(geog), 58. Prof Exp: From instr to asst prof hist, Ft Hays Kans State Col, 49-56, from assoc prof to prof geog, 58-62; assoc prof, 62-65, chmn dept, 69-71, PROF GEOG, CENT MICH UNIV, 65- Mem: Asn Am Geogr; Nat Coun Geog Educ. Res: Historical geography of the United States, especially the Great Plains; teaching of geography in elementary schools. Mailing Add: Dept of Geog Cent Mich Univ Mt Pleasant MI 48859

MARQUARDT, CHARLES LAWRENCE, b Chicago, Ill, Dec 12, 36; m 63; c 3. EXPERIMENTAL SOLID STATE PHYSICS. Educ: DePaul Univ, BS, 60, MS, 63; Cath Univ Am, PhD, 72. Prof Exp: Gen physicist, 63-65, SOLID STATE RES PHYSICIST, US NAVAL RES LAB, 65- Mem: AAAS; Am Phys Soc; Sigma Xi. Res: Electromagnetic theory; ionic transport phenomena; radiation defects in solids; lunar sample analysis; laser damage in semiconductors; photochromic glasses. Mailing Add: Code 6440 US Naval Res Lab Washington DC 20390

MARQUARDT, DAWN NILAN, b Avoca, Iowa, May 6, 14; m 41; c 2. ORGANIC CHEMISTRY. Educ: Grinnell Col, BA, 35; Univ Iowa, PhD(org chem), 40. Prof Exp: Instr org chem, Univ Nebr, 40-41; res chemist, Gen Elec Co, Mass, 41-42; lab dir, Nebr Defense Corp, 42-45; res chemist, Firestone Tire & Rubber Co, 45-49; asst prof chem, Iowa State Teachers Col, 49-54; assoc prof, 54-55, PROF CHEM & CHMN DEPT, UNIV NEBR AT OMAHA, 55- Concurrent Pos: Consult, Charles Schneider Co; staff assoc, Adv Coun Col Chem, 68-69. Mem: Am Chem Soc; Sigma Xi. Res: Organic synthesis; high polymers; synthetic rubber; ring closure of chloramines. Mailing Add: Dept of Chem Univ of Nebr at Omaha Omaha NE 68101

MARQUARDT, DONALD WESLEY, b New York, NY, Mar 13, 29; m 52; c 2. STATISTICS, MATHEMATICS. Educ: Columbia Univ, AB, 50; Univ Del, MA, 56. Prof Exp: Res engr & mathematician, Exp Sta, 53-57, res proj engr & sr mathematician, 57-64, consult supvr, 64-72, field mgr, 70-73, CONSULT MGR, E I DU PONT DE NEMOURS & CO, INC, 72- Concurrent Pos: Assoc ed, Technometrics, 74- Honors & Awards: Youden Prize, Am Soc Qual Control, 74. Mem: Fel Am Statist Asn; Am Soc Qual Control; Soc Indust & Appl Math; Sigma Xi; Asn Comput Mach. Res: Statistics of nonlinear models; biased estimation; strategy of experimentation; smooth regression; mixture models and experiments; computer algorithms; applications in engineering, physical and biological sciences. Mailing Add: Eng Dept E I du Pont de Nemours & Co Inc Wilmington DE 19898

MARQUARDT, FRITZ-HANS, organic chemistry, chemical engineering, see 12th edition

MARQUARDT, HANS WILHELM JOE, b Berlin, Ger, Aug 28, 38; m 74. PHARMACOLOGY, CANCER. Educ: Univ Cologne, MD, 64. Prof Exp: Instr pharmacol, Univ Cologne, 64-68; vis investr, Div Pharmacol, Sloan-Kettering Inst Cancer Res, 68-70; vis investr cancer res, McArdle Lab Cancer Res, Univ Wis-Madison, 70-71; assoc, 71-74, ASSOC MEM, SLOAN-KETTERING INST CANCER RES, 74- Concurrent Pos: Asst prof pharmacol, Grad Sch Med Sci, Cornell Univ, 71-74, assoc prof, 75-; NIH-USPHS res career develop award, 75. Mem: Ger Pharmacol Soc; Europ Asn Cancer Res; Am Soc Pharmacol & Exp Therapeut; Am Asn Cancer Res. Res: Pharmacology and toxicology of antitumor agents and chemical carcinogens; chemical carcinogenesis and mutagenesis in tissue culture. Mailing Add: Sloan-Kettering Inst Cancer Res 410 E 68th St New York NY 10021

MARQUARDT, ROLAND PAUL, b Tulare, SDak, Sept 2, 13. ANALYTICAL CHEMISTRY, ORGANIC CHEMISTRY. Educ: Huron Col, SDak, AB, 35; Univ SDak, AM, 36. Prof Exp: Chemist, 39-58, ANAL RES SPECIALIST CHEM, DOW CHEM CO, 58- Mem: Fel Am Inst Chemists; Am Chem Soc; Sigma Xi. Res: Industrial analytical research; analytical method development, including fundamental methods for unsaturation in organic compounds and for residue analysis of pesticides, especially phenoxy acid herbicides. Mailing Add: 1212 Baldwin St Midland MI 48640

MARQUARDT, RONALD RALPH, b Bassano, Alta, May 24, 35; m 59; c 2. BIOCHEMISTRY, AVIAN PHYSIOLOGY. Educ: Univ Sask, BSA, 58; Univ Alta, MSc, 61; Wash State Univ, PhD(animal sci), 65. Prof Exp: Asst dist agriculturist, Alta Dept Agr, 58-59; asst animal sci, Univ Alta, 59-61; asst, Wash State Univ, 61-65, res assoc biochem, 65-67; ASSOC PROF ANIMAL SCI, UNIV MAN, 67- Mem: Am Chem Soc; Can Biochem Soc; Can Nutrit Soc; Can Soc Animal Sci. Res: Purification and characterization of avian glycolytic enzymes; sex hormones control mechanisms; influence of diet on synthesis and degradation of hepatic enzymes. Mailing Add: Dept of Animal Sci Univ of Man Winnipeg MB Can

MARQUARDT, WILLIAM CHARLES, b Ft Wayne, Ind, Oct 9, 24; m 48; c 3. PROTOZOOLOGY, PARASITOLOGY. Educ: Northwestern Univ, BS, 48; Univ Ill, MS, 50, PhD(zool), 54. Prof Exp: Asst, Col Vet Med, Univ Ill, 52-54; from asst prof to assoc prof parasitol, Mont State Col, 54-61; assoc prof biol, DePaul Univ, 61-62; assoc prof parasitol, Univ Ill, 62-66; PROF ZOOL, COLO STATE UNIV, 66- Concurrent Pos: Consult, Thorne Ecol Inst, Colo, 72- Mem: Am Soc Parasitologists; Soc Protozoologists (asst treas, 67-70, pres, 74); Am Soc Zoologists; Am Soc Trop Med & Hyg. Res: Transmission and host-parasite relationships in parasitic protozoa and helminths. Mailing Add: Dept of Zool & Entom Colo State Univ Ft Collins CO 80523

MARQUARDT, WILLIAM HARRISON, b Tampa, Fla, Oct 2, 46. ANTHROPOLOGY. Educ: Fla State Univ, BA, 68; Univ Ky, MA, 71; Wash Univ, St Louis, PhD(anthrop), 74. Prof Exp: ASST PROF ANTHROP, UNIV MO, 74- Mem: Am Anthrop Asn; Soc Am Archaeol; Am Ethnol Soc. Res: Theories of culture process; archaeological theory and method; quantitative approaches to anthropological data; North American archaeology. Mailing Add: Dept of Anthrop Univ of Mo Columbia MO 65201

MARQUART, JOHN R, b Benton Harbor, Mich, Feb 3, 33. PHYSICAL CHEMISTRY. Educ: Univ Ariz, BS, 55; Univ Ill, MS, 61, PhD(phys chem), 63. Prof Exp: Chem test officer, Dugway Proving Ground, US Army Chem Corps, Utah, 57-58; res assoc physics, Argonne Nat Lab, 62; chemist, Shell Develop Co, 63-68; assoc prof, 68-74, PROF CHEM, MERCER UNIV, 74- Mem: Am Chem Soc; Sigma Xi. Res: Mass spectroscopy; thermodynamics; solution behavior; high temperature gas phase kinetics. Mailing Add: Dept of Chem Mercer Univ Macon GA 31207

MARQUART, PHILIP BUTLER, b Milton, Wis, July 13, 99. MEDICINE. Educ: Univ Wis, AB, 23, MA, 25; Harvard Univ, MD, 31. Prof Exp: Psychometrist, Psychiat Field Serv, Wis State Bd Control, 26-27; student instr neural anat, Med Sch, Harvard Univ, 30-31; intern, Henry Ford Hosp, 31-32; ward physician, Northern State Hosp, 32-34; psychiat res, Psychopath Hosp, Galveston, Tex, 35-36; ward officer, Ment Hyg Div, Marine Hosp, USPHS, Galveston, 39; instr, Tex Christian Univ, 40-41; prof psychol & col psychiatrist, Wheaton Col, 45-58; prof psychol & col counr, Tenn Temple Col, 58-71; prof psychol, Independent Baptist Col, 71-75. Mem: Am Med Asn; Am Psychiat Asn. Res: Physiological studies of schizophrenia; post-monucleosis psychoneurosis; integration of a Christian understanding of the fields of psychiat and psychology. Mailing Add: PO Box 4042 Little Rock AR 72204

MARQUET, LOUIS C, b Philadelphia, Pa, May 9, 36; m 58; c 4. ATOMIC PHYSICS, SPECTROSCOPY. Educ: Carnegie Inst Technol, BS, 58; Univ Calif, Berkeley, MA, 60, PhD(physics), 64. Prof Exp: Asst prof physics, Univ Ariz, 65-67; staff scientist, 67-75, LEADER APPL RADIATION GROUP, LINCOLN LAB, MASS INST TECHNOL, 74- Mem: Optical Soc Am. Res: High power/laser technology and applications, especially propagation in the atmosphere. Mailing Add: Appl Radiation Group Lincoln Lab Mass Inst of Technol PO Box 73 Lexington MA 02173

MARQUEZ, ERNEST DOMINGO, b Tranquillity, Calif, Nov 13, 38; m 74. BIOLOGICAL CHEMISTRY. Educ: Calif State Univ, Fresno, BA, 62, MA, 68; Univ Southern Calif, PhD(microbiol), 72. Prof Exp: USPHS fel, Scripps Inst Oceanog, 71-73; ASST PROF MICROBIOL, COL MED, HERSHEY MED CTR, PA STATE UNIV, 73- Mem: Am Soc Microbiol; AAAS. Res: Analyses of malignant cell surfaces transformed by herpesviruses using biochemical, immunological and biological methods. Mailing Add: Dept of Microbiol Col Med Pa State Univ Hershey PA 17033

MARQUEZ, JOSEPH A, b New York, NY, Nov 5, 30; m 53; c 4. BIOCHEMISTRY. Educ: City Col New York, BS, 57; Fairleigh Dickinson Univ, MAS, 70. Prof Exp: Lab asst steroid identification & isolation, Columbia Univ, 54-56; lab asst natural prod isolation, 56-58, res asst antibiotic isolation, 58-61, res assoc, 62-67, res scientist, 67-68, sr scientist, 68-75, MGR ANTIBIOTIC DEPT, SCHERING CORP, 75- Mem: AAAS; Am Chem Soc; Am Soc Microbiol; NY Acad Sci; Am Inst Biol Sci. Mailing Add: Schering Corp 60 Orange St Bloomfield NJ 07003

MARQUIS, DAVID ALAN, b Pittsburgh, Pa, Jan 16, 34; m 73; c 3. FOREST ECOLOGY. Educ: Pa State Univ, BS, 55; Yale Univ, MF, 63, PhD(forest ecol), 73. Prof Exp: Res forester forest ecol, Laconia, NH, 57-65, staff asst forest mgt, Upper Darby, Pa, 65-69, RES PROJ LEADER FOREST ECOL, NORTHEASTERN FOREST EXP STA, WARREN, PA, 69- Mem: Soc Am Foresters. Res: Conduct research on ecological factors affecting regeneration and growth of hardwood forests. Mailing Add: Forestry Sci Lab PO Box 928 Warren PA 16365

MARQUIS, DAVID MALEY, b Yonkers, NY, Apr 14, 29; m 55. INDUSTRIAL ORGANIC CHEMISTRY. Educ: Stanford Univ, BS, 50; Harvard Univ, MA, 54, PhD(org chem), 55. Prof Exp: Res chemist, Jackson Lab, E I du Pont de Nemours & Co, 56-58; vpres, Minerals Refining Co, 58-61; SR RES ASSOC, CHEVRON RES CO, STANDARD OIL CO CALIF, 61- Mem: Am Chem Soc. Res: Organic and inorganic fluorine chemistry; high temperature reactions; reaction mechanisms; petrochemicals; synthetic detergents; process development. Mailing Add: 32 Brookdale Ct Lafayette CA 94549

MARQUIS, EDWARD THOMAS, b South Bend, Ind, July 10, 39; m 61; c 6. ORGANIC CHEMISTRY. Educ: Ind Univ, AB, 61; Univ Tex, PhD(org chem), 67. Prof Exp: Res chemist, 66-68, SR RES CHEMIST, JEFFERSON CHEM CO, 68- Mem: Am Chem Soc; Sigma Xi. Res: Hydrocarbon oxidations and reductions; aromatic and aliphatic isocyanates and their amine precursors; reactions and synthetic use of phosgene. Mailing Add: Jefferson Chem Co PO Box 4128 Austin TX 78751

MARQUIS, MARILYN GRACE ALDER, physical chemistry, see 12th edition

MARQUIS, NORMAN RONALD, b Laconia, NH, Jan 3, 36; m 57; c 3. PHYSIOLOGY, BIOCHEMISTRY. Educ: Univ NH, BA, 59, MS, 60; Univ Mich, PhD(physiol), 65. Prof Exp: Res fel, Harvard Med Sch, 65-67, teaching fel, 66-67; group leader biochem, 67-70, PRIN INVESTR, MEAD JOHNSON RES CTR, 70- Concurrent Pos: Adj prof, Univ Evansville, 74- Mem: Am Soc Biol Chemists; Soc Exp Biol & Med; Am Physiol Soc; Am Heart Asn; NY Acad Sci. Res: Functions of carnitine in lipid metabolism; interrelationships of thrombosis, fibrinolysis and atherosclerosis; role of prostaglandins and cyclic adenosine monophosphate in platelet aggregation and thrombosis. Mailing Add: Dept of Biol Res Mead Johnson Res Ctr Evansville IN 47721

MARQUIS, RICHARD JACK, b Mineral Wells, Tex, Jan 7, 10; m 42; c 2. PHYSICS. Educ: NTex State Teachers Col, BA, 30; Univ Tex, MA, 35. Prof Exp: Sound engr, Interstate Circuit, Inc, 31-33; tutor, Univ Tex, 34-35; sound engr, Interstate Circuit, Inc, 35-36; high sch teacher, Tex, 36-39; teacher, Arlington State Col, 39-41; spec res assoc, Nat Defense Res Coun, Harvard Univ, 41-45; vis assoc prof physics, Southern Methodist Uhiv, 46; head dept, 46-67, PROF PHYSICS, UNIV TEX, ARLINGTON, 46-, ASST DEAN SCI & MATH, 67- Mem: Acoust Soc Am. Res: Communication; electronics; audiometry; hearing aids. Mailing Add: Dept of Physics Univ of Tex Arlington TX 76019

MARQUIS, ROBERT E, b Sarnia, Ont, Jan 21, 34; US citizen; m 57; c 3. MICROBIAL PHYSIOLOGY. Educ: Wayne State Univ, BS, 56; Univ Mich, MS, 58, PhD(bact), 61. Prof Exp: NATO fel, Univ Edinburgh, 61-62, NSF fel, 62-63; from sr instr to asst prof, 63-70, ASSOC PROF MICROBIOL, SCH MED, UNIV ROCHESTER, 70- Concurrent Pos: NIH fel, Scripps Inst Oceanog, Univ Calif, San Diego, 70-71. Mem: AAAS; Am Soc Microbiol; Brit Soc Gen Microbiol. Res: Bacterial physiology; physical structure of bacterial plasma membranes; basic studies of microbial barophysiology; investigation of the physiology of bacteria in dental plaque. Mailing Add: Dept of Microbiol Univ of Rochester Rochester NY 14642

MARQUISEE, JOSEPH ALFRED, physics, physical chemistry, see 12th edition

MARQUISEE, MARK, biochemistry, molecular biology, see 12th edition

MARQUISS, ROBERT W, b Pine Bluffs, Wyo, Sept 4, 30; m 56; c 2. RANGE SCIENCE. Educ: Univ Wyo, BS, 56, MS, 57; Univ Ariz, PhD(plant sci), 67. Prof Exp: Res assoc range sci, Univ Ariz, 61-65; asst prof range mgt, 65-71, ASSOC PROF AGR, FT LEWIS COL, 71- Concurrent Pos: Asst prof range sci, Colo State Univ, San Juan Basin Br, 65-71. Mem: Soc Range Mgt. Mailing Add: Ft Lewis Col Durango CO 81301

MARR, ALLEN GERALD, b Tulsa, Okla, Apr 24, 29; m 48, 70; c 6. MICROBIOLOGY. Educ: Univ Okla, BS, 48, MA, 49; Univ Wis, PhD(bact), 52. Prof Exp: Proj assoc bact, Univ Wis, 52; instr, 52-54, from asst prof to assoc prof, 54-63, PROF BACT, UNIV CALIF, DAVIS, 63-, DEAN GRAD STUDIES & RES, 70- Concurrent Pos: Ed, J Bact. Mem: Am Soc Microbiol. Res: Growth and division of bacteria; microbial physiology. Mailing Add: Dept of Bact Univ of Calif Davis CA 95616

MARR, DAVID HENRY, chemistry, see 12th edition

MARR, ELEANOR B, b Walden, Colo, June 29, 99. ORGANIC CHEMISTRY. Educ: Univ Denver, AB, 22, AM, 23; Columbia Univ, PhD(org chem), 35. Prof Exp: High sch teacher, 23-28; instr chem, Wilson Col, 29-32; from instr to prof, 35-70, EMER PROF CHEM, HUNTER COL, 70-; CONSULT CHEMIST, 35- Mem: Fel AAAS; Am Chem Soc; Am Soc Info Sci; fel Brit Chem Soc; fel Am Inst Chemists. Res: Synthetic organic chemistry; reaction mechanisms; chemical documentation. Mailing Add: 114 Medford Leas Medford NJ 08055

MARR, HAROLD EVERETT, III, b Manila, Philippines, Oct 3, 39; US citizen; m 65;

c 4. PHYSICAL CHEMISTRY. Educ: Calif Inst Technol, BS, 62; Univ Md, College Park, PhD(chem), 70. Prof Exp: Chemist, Hercules Powder Co, Utah, 62-63, Melpar, Inc, Va, 63-64 & Univ Md, College Park, 64-65; RES CHEMIST, METALL RES CTR, US BUR MINES, 68- Concurrent Pos: Lectr, Prince George's Community Col, Md, 72-73. Mem: Am Chem Soc; Soc Appl Spectros. Res: X-ray spectroscopy; single crystal x-ray diffraction; computer processing of analytical data; energy dispersion x-ray analysis. Mailing Add: Metall Res Ctr US Bur of Mines College Park MD 20740

MARR, JAMES JOSEPH, b Hamilton, Ohio, Oct 21, 38; m 63; c 5. INFECTIOUS DISEASES, INTERNAL MEDICINE. Educ: Xavier Univ, Ohio, BS, 59; Johns Hopkins Univ, MD, 64; St Louis Univ, MS, 68; Am Bd Internal Med, dipl, 72, cert infectious dis, 74. Prof Exp: Am Cancer Soc fel & instr microbiol, Sch Med, St Louis Univ, 67-69; asst prof internal med, 70-75, asst prof path, 73-75, asst prof microbiol, 71-75, ASSOC PROF INTERNAL MED & PATH, SCH MED, WASH UNIV, 75- Concurrent Pos: Fel trop med, USPHS-La State Univ Int Ctr Med Res & Training, Costa Rica, 72; med dir, Microbiol Labs, Barnes Hosp, St Louis, 73-; consult, Vet Admin Hosp, St Louis, 72-, St Louis Childrens Hosp, 74- & Jewish Hosp, St Louis, 75- Mem: Am Soc Microbiol; Am Fedn Clin Res; fel Am Col Physicians; Infectious Dis Soc Am. Res: Metabolic regulation in microorganisms and its relationship to the pathogenesis of intracellular infections in man. Mailing Add: Dept of Med Sch of Med Wash Univ 4550 Scott Ave St Louis MO 63110

MARR, JOHN DOUGLAS, b Denver, Colo, Dec 3, 00; m 29. EXPLORATION GEOPHYSICS, EXPLORATION GEOLOGY. Educ: Univ Colo, BS, 26; Colo Sch Mines, MSc, 31, DSc(geophys eng), 32. Prof Exp: Party chief seismic, Independent Explor Co, Houston, 33-36; supvr opers interpretation, Seismic Explor Inc, Houston, 37-49, vpres, 49-63, asst mgt, Ray Geophys Div, Mandrel Indust Inc, 63, vpres & mgr data processing & interpretation, 64-65, vpres, 66; PETROL EXPLOR CONSULT, 67- Mem: Am Asn Petrol Geologists; Am Inst Mining, Metall & Petrol Engrs; Soc Explor Geophysicists; fel Geol Soc Am. Res: Application of advanced concepts in geological and geophysical exploration for hydrocarbon deposits; specializing in stratigraphic exploration and direct hydrocarbon detection. Mailing Add: 803 Old Lake Rd Houston TX 77057

MARR, JOHN MAURICE, b Jefferson City, Mo, June 15, 20; m 49. MATHEMATICS. Educ: Cent Mo State Col, BS, 41; Univ Mo, MA, 48; Univ Tenn, PhD, 53. Prof Exp: Instr math, Mo Sch Mines, 46-47; from asst instr to instr, Univ Mo, 47-49; U asst, Univ Tenn, 49-53; from asst prof to assoc prof, 53-62, PROF MATH, KANS STATE UNIV, 62- Mem: Am Math Soc; Math Asn Am. Res: Topology; convexity. Mailing Add: Dept of Math Kans State Univ Manhattan KS 66504

MARR, JOHN WINTON, b Lamesa, Tex, May 16, 14; m 42, 56; c 3. BOTANY. Educ: Tex Tech Col, BS, 36; Univ Minn, PhD(bot), 42. Prof Exp: Asst, Tex Tech Col, 34-36; asst, Northwestern Univ, 36-37; asst, Univ Minn, 37-42; from instr to assoc prof, 44-65, dir, Inst Arctic & Alpine Res, 51-67, PROF ENVIRON, POP & ORGANISMIC BIOL, UNIV COLO, BOULDER, 65- Concurrent Pos: Field asst, Univ Minn Bot Exped, ECoast Hudson Bay, 39; leader, Colo Exped Ungava, 48; vpres & trustee, Thorne Ecol Inst, 66- Mem: AAAS; Ecol Soc Am. Res: Ecology of vegetation transition areas, and of trees; tree growth-layer analysis; character of and plant reaction to the winter environment in Colorado mountains; tundra ecosystems of the Thule, Greenland region; ecosystems of Colorado. Mailing Add: Dept of Biol Univ of Colo Boulder CO 80302

MARR, PAUL DONALD, b San Francisco, Calif, Dec 31, 28; m 54; c 5. URBAN GEOGRAPHY, REGIONAL PLANNING. Educ: Univ Calif, Berkeley, AB, 51, MA, 55, PhD(geog), 67. Prof Exp: Location analyst, Bank of Am, Calif, 54-56; geogr, Stanford Res Inst, 56-60; asst prof geog, Univ Calif, Davis, 64-69, asst res geogr, Inst Govt Affairs, 67-68; ASSOC PROF GEOG, STATE UNIV NY ALBANY, 69- Concurrent Pos: Group leader policy studies, New York Sea Grant Inst, 75- Mem: AAAS; Asn Am Geogr; Am Inst Planners; Regional Sci Asn; Am Soc Planning Officers. Res: Coastal zone planning and management; environmental planning; urban and regional planning. Mailing Add: Dept of Geog State Univ of NY Albany NY 12222

MARR, ROBERT B, b Quincy, Mass, Mar 25, 32; m 54; c 3. PHYSICS. Educ: Mass Inst Technol, SB, 53; Harvard Univ, MA, 55, PhD(physics), 59. Prof Exp: Res assoc theoret physics, 59-61, assoc physicist, 61-64, physicist, 64-69, SR PHYSICIST, BROOKHAVEN NAT LAB, ASSOC UNIVS, INC, 69-, CHMN APPL MATH DEPT, 75- Mem: Am Phys Soc. Res: Applied mathematics; computers; applications in physical and biological sciences; theoretical physics. Mailing Add: Dept of Appl Math Brookhaven Nat Lab Assoc Univs Inc Upton NY 11973

MARRA, ALAN A, b Summitville, Ohio, Mar 28, 15; m 51; c 3. WOOD TECHNOLOGY. Educ: State Univ NY Col Forestry, Syracuse Univ, BS, 40, MS, 42; Univ Mich, PhD(wood technol), 55. Prof Exp: Proj engr wood propellers, Eng & Res Corp, 42-43; tech dir, Pluswood, Inc, 43-44; lab supvr wood adhesives, Monsanto Chem Co, 44-48; res engr wood prod, Univ Mich, Ann Arbor, 49-53, from instr to prof wood sci & technol, 53-68, prof archit, 68-72; MEM FAC FORESTRY, UNIV MASS, AMHERST, 72- Concurrent Pos: Chmn, Gordon Res Conf Adhesion, 59; mem, Conf Sci Wood Adhesion, NSF; Soc Wood Sci & Technol vis scientist; consult, Koppers Co, Simpson Timber Co, Alpha Res Div, Certain-Teed Prod Co. Mem: Soc Wood Sci & Technol; Forest Prod Res Soc; Am Chem Soc; Am Soc Testing & Mat. Res: Wood adhesion and impregnation; glued products; structurizing with wood. Mailing Add: Dept of Forestry Univ of Mass Amherst MA 48104

MARRA, DOROTHEA CATHERINE, b Brooklyn, NY, Jan 23, 22; m 47; c 1. SURFACE CHEMISTRY, COLLOID CHEMISTRY. Educ: Brooklyn Col, BA, 43. Prof Exp: Anal chemist, Matam Corp, 43-44; res chemist, Foster D Snell Inc, 44-69; VPRES, OMAR RES, INC, NEW YORK, 69- Mem: AAAS; Sigma Xi; Soc Cosmetic Chemists; fel Am Inst Chemists. Res: Creation and development of new products, specifically in cosmetics, toiletries and pharmaceuticals. Mailing Add: 107 Fernwood Rd Summit NJ 07901

MARRA, EDWARD FRANCIS, b Poughkeepsie, NY, May 25, 16. PREVENTIVE MEDICINE. Educ: Trinity Col, Conn, BS, 45; Boston Univ, MD, 50; Harvard Univ, MPH, 55. Prof Exp: From instr to assoc prof prev med, Sch Med, Boston Univ, 53-60; PROF SOCIAL & PREV MED & HEAD DEPT & ASSOC PROF MED, SCH MED, STATE UNIV NY BUFFALO, 60- Concurrent Pos: Asst, Mass Mem Hosps, 52-53, from asst vis physician to vis physician, 53-59, chief serv, 59-60. Mem: Fel Am Pub Health Asn; Asn Teachers Prev Med; AMA; Int Epidemiol Asn. Res: Evaluation of medical care; control of infectious disease; development of teaching programs in social and preventive medicine in medical schools. Mailing Add: Dept of Social & Prev Med State Univ of NY Sch of Med Buffalo NY 14214

MARRA, MICHAEL DOMINICK, b Brooklyn, NY, Dec 4, 22; m 47; c 1. BIOCHEMISTRY, BIOANALYSIS. Educ: Brooklyn Col, BA, 44; Rutgers Univ, MS, 56; Am Bd Bioanalysts, dipl, 69; Jackson State Univ, PhD(biochem), 74. Prof Exp: DIR CLIN LAB, SUMMIT MED GROUP, 46- Concurrent Pos: Res consult, Summit Testing Lab, 50-; Warner-Lambert res grant, 71- Mem: Am Asn Clin Chemists; fel Am Inst Chemists; Am Asn Bioanalysts. Res: Microbiology; immunology. Mailing Add: 107 Fernwood Rd Summit NJ 07901

MARRACK, DAVID, b Sawbridgeworth, Eng, Dec 25, 22; m 49; c 3. MEDICINE, PATHOLOGY. Educ: Univ London, MB, 42, MB, BS, 47, MD, 53. Prof Exp: Clin resident, London Hosp, Eng, 47-48, clin resident path, 48-49; chem pathologist & tutor clin path, Westminster Hosp & Univ London, 54-58; sr lectr chem path, Inst Neurol, Nat Hosp Nerv Dis, Queen Square, London, 58-61; assoc prof path, Univ Tex M D Anderson Hosp & Tumor Inst, 61-69; ASSOC PROF PATH, TEX MED CTR, BAYLOR COL MED, 69- Concurrent Pos: Med Res Coun scholar chem path, Postgrad Med Sch, Univ London, 51-53; Postgrad Fedn Univ London traveling fel, 53-54. Mem: Assoc Am Soc Clin Path; NY Acad Sci; Royal Col Path; Brit Biochem Soc; Brit Soc Immunol. Res: Human metabolic diseases; automation of data retrieval in clinical chemistry. Mailing Add: Dept of Path Tex Med Ctr Baylor Col of Med Houston TX 77025

MARRANZINO, ALBERT PASQUALE, b Denver, Colo, Oct 5, 27; m 50; c 5. GEOCHEMISTRY. Educ: Regis Col, BS, 49. Prof Exp: Phys sci aide, Geol & Petrol Br, 51-55, chemist, Geochem Explor Sect, 55-57, chemist, Mineral Deposits Br, 58-60, chemist, Geochem Explor & Minor Elements Br, 60-65, chief field serv sect mobile & anal chem, 65-71, adv, Geochem Labs, Off Int Geol, 71-73, DEP REGIONAL GEOLOGIST, CENT REGION, US GEOL SURV, 74- Honors & Awards: Meritorious Serv Award, US Dept Interior, 66. Mem: Am Chem Soc; Soc Appl Spectros; Asn Explor Geochem. Res: Geochemical prospecting basin and range province; mobile spectrographic techniques as applied to geochemical exploration. Mailing Add: US Geol Surv Bldg 25 Federal Ctr Denver CO 80225

MARRARO, ROBERT V, b Brooklyn, NY, May 24, 30; m 56; c 5. MEDICAL MICROBIOLOGY. Educ: Colby Col, Maine, BA, 51; Columbia Univ, BS, 55; Ariz State Univ, MS, 64; Ohio State Univ, PhD(microbiol), 71. Prof Exp: Chief microbiol br, US Air Force Hosp, Wiesbaden, Ger, 65-67, US Air Force Med Ctr, Wright-Patterson AFB, Ohio, 67-68 & US Air Force Sch Aerospace Med, Brooks AFB, Tex, 71-74; CHIEF MICROBIOL BR, WILFORD HALL US AIR FORCE MED CTR, LACKLAND AFB, TEX, 74- Concurrent Pos: Consult microbiol, Surg Gen, US Air Force, Europe, 65-67; clin instr path, Univ Tex Health Sci Ctr, San Antonio, & exam proctor, Registry Am Med Technologists, 72-; consult microbiol, Bac-Data, Med Info Systs Inc, 75. Honors & Awards: Fisher Award Med Technol, Fisher Sci Co, 72. Mem: Am Soc Microbiol; Am Pub Health Asn; Am Asn Bioanalysts; Am Med Technologists; NY Acad Sci. Res: Investigation into the etiology of chronic, recurrent diseases of the human genitourinary tract, especially the role of cell wall-deficient forms of microorganisms in these syndromes. Mailing Add: 12511 El Domingo San Antonio TX 78233

MARRAZZI, AMEDEO S, b New York, NY, Feb 6, 05; m; c 2. NEUROPHARMACOLOGY, NEUROPHYSIOLOGY. Educ: NY Univ, MD, 28. Prof Exp: Instr biol & comp anat, City Col New York, 28; intern, Bellevue Hosp, 28-30, Herter fel pharmacol, Bellevue Hosp Med Col, 31, instr pharmacol & therapeut, NY Univ & Bellevue Hosp Med Col, 31-35; from instr to asst prof pharmacol, Col Med, NY Univ, 35-43; prof & head dept Sch Med, Loyola Univ Chicago, 43-44; prof & chmn dept, Col Med, Wayne Univ, 44-48; chief toxicol br, Chem Corps Med Lab, Army Chem Ctr, Md, 48-51 & Clin Res Div, 51-56, asst sci dir, Med Directorate, Chem Warfare Labs, 56; prof physiol & pharmacol, Sch Med, Univ Pittsburgh, 56-64; Hill prof neuropharmacol, Col Med, Univ Minn, Minneapolis, 64-69; prof pharmacol, Sch Med, Univ Mo-Columbia, 69-76, chief neuropharmacol, Mo Inst Psychiat, 69-76. Concurrent Pos: Practicing physician, NY, 30-35; investr, Col Physicians & Surgeons, Columbia Univ, 33-34 & Rockefeller Inst, 42; Dazian Found fel, Univ Chicago, 42; consult, Receiving Hosp, Detroit, Mich, 45-48; lectr, Sch Med, Univ Md, 49- & Med Sch Univ Tex, 62-; dir, Vet Admin Res Labs Neuropsychiat, Pittsburgh, Pa, 56-64; Koch Mem lectr, Univ Pittsburgh, 57; liaison mem disaster comt, Nat Res Coun, 52-56; mem exec comt, Coop Chemother Studies in Psychiat, Vet Admin, 56-60; mem study sect pharmacol & exp ther, NIH, 59-63; mem sci & adv bd, Int Inst Comprehensive Med, 64-; mem Am Schizophrenia Found, 64-; consult, Vet Admin, Minneapolis, Minn, 65-69. Mem: Am Soc Pharmacol & Exp Therapeut; Am Physiol Soc; Soc Biol Psychiat(pres, 63); Am Psychiat Asn; Soc Neurosci. Res: Action potentials in localization and quantitative study of drug action, correlating with chemical structure; physiology and pharmacology of the autonomic and central nervous systems; interaction of adjacent nerve cells and fibers; cerebral homeostasis; psychopharmacology; mechanism of hallucination; experimental psychiatry. Mailing Add: 962 Lochmoor Grosse Pointe Woods MI 48236

MARRAZZI, MARY ANN, b Ann Arbor, Mich, Dec 22, 45. NEUROPHARMACOLOGY, NEUROCHEMISTRY. Educ: Univ Minn, BA, 66; Wash Univ, PhD(pharmacol), 72. Prof Exp: NIH fel pharmacol, Sch Med, Wash Univ, 72-74; vis investr neuropharmacol, Inst Psychiat, Univ Mo, 74; ASST PROF PHARMACOL, SCH MED, WAYNE STATE UNIV, 74- Mem: Soc Neurosci. Res: Hypothalamic glucoreceptors in central nervous system regulation of metabolic homeostasis and appetite; relation to possible insulin central nervous system actions; prostaglandin assay, metabolism and role in nervous system; microchemical methodology. Mailing Add: Dept of Pharmacol Sch of Med Wayne State Univ Detroit MI 48201

MARRELLO, VINCENT, b Belsito, Italy, Apr 20, 47; Can citizen; m 72; c 1. SOLID STATE PHYSICS. Educ: Univ Toronto, BASc, 70; Calif Inst Technol, MS, 71, PhD(elec eng), 74. Prof Exp: Fel appl physics, Calif Inst Technol, 74-75; MEM RES STAFF APPL PHYSICS, IBM RES LAB, 75- Mem: Am Phys Soc; Sigma Xi. Res: Device physics; material physics; optical and electrical properties of amorphous materials. Mailing Add: IBM Res Lab 5600 Cottle Rd San Jose CA 95193

MARRIAGE, LOWELL DEAN, b New Rockford, NDak, June 28, 23; m 52; c 3. WILDLIFE CONSERVATION. Educ: Ore State Univ, BS, 48. Prof Exp: Aquatic biologist, Fish Comn Ore, 48-56, water resources analyst, 56-60, asst dir, 60-62, regional fisheries biologist, Soil Conserv Serv, 62-71, REGIONAL BIOLOGIST, SOIL CONSERV SERV, USDA, 71- Mem: Am Fisheries Soc; Wildlife Soc; Am Inst Fishery Res Biol; Soil Conserv Soc Am. Res: Shellfish management and research; water projects effects on fisheries and wildlife populations; anadromous fisheries biology and management; water quality and wildlife habitat management in water development projects. Mailing Add: Soil Conserv Serv 510 Fed Off Bldg 511 NW Broadway Portland OR 97209

MARRIAGE, PAUL BERNARD, b London, Ont, Dec 15, 42; m 65; c 3. PLANT BIOCHEMISTRY, PLANT PHYSIOLOGY. Educ: Univ Western Ont, BSc, 64, PhD(bot), 69. Prof Exp: RES SCIENTIST, RES STA, CAN DEPT AGR, 68- Mem: Weed Sci Soc Am. Res: Effect of herbicides on plant physiology and biochemical processes as related to agricultural problems. Mailing Add: Res Sta Can Dept of Agr Harrow ON Can

MARRIOTT, HENRY JOSEPH LLEWELLYN, b Hamilton, Bermuda, June 10, 17; nat US; m 51; c 3. CARDIOLOGY. Educ: Oxford Univ, BA, 41, MA, 43, BM, BCh, 44. Prof Exp: House physician, St Mary's Hosp London, 44, resident med officer, Sir Alexander Fleming's Penicillin Res Unit, 45; resident, King Edward Hosp, Bermuda, 45-46; from asst to asst prof med, Med Sch, Univ Md, 48-53, assoc prof med & head div phys diag, 53-62, head div arthritis, 56-59; dir med educ & cardiol ctr, Tampa Gen Hosp, Fla, 62-65; dir clin res, Rogers Heart Found, 65-70; CLIN PROF PEDIAT, COL MED, UNIV FLA, 70-, CLIN PROF MED, 72-; DIR CORONARY CARE, ST ANTHONY'S HOSP, 70- Concurrent Pos: Fel med, Johns Hopkins Hosp, 46-47; chief EKG dept, Mercy Hosp, Baltimore, Md, 54-62; consult, Vet Admin Hosp, Bay Pines, Fla, 63-; clin prof med, Sch Med, Emory Univ, 66-70. Mem: AMA; Am Heart Asn; fel Am Col Cardiol; fel Am Col Physicians; Brit Med Asn. Res: Electrocardiography and clinical cardiology. Mailing Add: St Anthony's Hosp St Petersburg FL 33705

MARRIOTT, LAWRENCE FREDERICK, b Browns, Ill, Dec 18, 13; m 38; c 1. AGRONOMY, SOILS. Educ: Univ Ill, BS, 35; Univ Wis, MS, 53, PhD(soils), 55. Prof Exp: Asst soil exp fields, Univ Ill, 35-42; self employed, 46-51; asst soils, Univ Wis, 51-55; asst prof, 55-59, ASSOC PROF SOIL TECHNOL, PA STATE UNIV, 59- Mem: Fel AAAS; Am Soc Agron; Soil Sci Soc Am. Res: Soil problems and fertilization of grasslands; disposal and utilization of dairy manure by land application. Mailing Add: Dept of Agron Pa State Univ University Park PA 16802

MARRON, MICHAEL THOMAS, b Jan 31, 43; US citizen; m 66. THEORETICAL CHEMISTRY, PHYSICAL CHEMISTRY. Educ: Univ Portland, BS, 64; Johns Hopkins Univ, MA, 65, PhD(theoret chem), 69. Prof Exp: Res assoc, Theoret Chem Inst, Univ Wis, 69-70; ASST PROF CHEM, UNIV WIS-PARKSIDE, 70- Mem: AAAS; Am Phys Soc. Res: Molecular quantum mechanics; theoretical kinetics; biological effects of extremely low frequency electromagnetic radiation; application of computer technology to chemical problems. Mailing Add: Div of Sci Univ of Wis-Parkside Kenosha WI 53140

MARRONE, MICHAEL JOSEPH, b Lewistown, Pa, July 19, 37; m 61; c 4. SOLID STATE PHYSICS. Educ: Univ Notre Dame, BS, 59; Univ Pittsburgh, MS, 61; Cath Univ Am, PhD(physics), 71. Prof Exp: RES PHYSICIST, US NAVAL RES LAB, 61- Mem: Am Phys Soc; Sigma Xi. Res: Optical properties of solids; optical absorption and emission; radiation effects in solids; magneto-optics. Mailing Add: Code 6440 US Naval Res Lab Washington DC 20390

MARROQUIN DE LA FUENTE, JORGE SAUL, b Monterrey, Mex, Oct 6, 35; m 61; c 5. PLANT TAXONOMY, PLANT ECOLOGY. Educ: Univ Neuvo Leon, Biologist, 59; Northeastern Univ, MS, 69, PhD, 72. Prof Exp: Field entomologist, Distribuidora Shell de Mex, 60; field botanist, Nat Inst Forest Res, Mex, 61, head trop exp sta, Escarcega, 63-64; chmn sch biol sci, Univ Nuevo Leon, 64-67; asst biol, Northeastern Univ, 67-70; mem fac biol sci, Univ Nuevo Leon, 70-73; MEM FAC BIOL SCI, GRAD SCH, ANTONIO NARRO UNIV, 73-, HEAD DEPT BOT, 75- Concurrent Pos: Bank of Mex study grant, 69-70; mem bd gov, Univ Nuevo Leon, 71-75. Mem: Bot Soc Am; Am Soc Plant Taxon; Inst Asn Plant Taxon; Am Inst Biol Sci; Soc Econ Botanists. Res: Vegetation of arid lands; deciduous forest ecology and succulent plants in Mexico; Berberideceae of Mexico. Mailing Add: Grad Sch Antonio Narro Univ Buenavista Saltillo Coahuila Mexico

MARRS, BARRY LEE, b Newark, NJ, Sept 23, 42; m 66; c 2. MICROBIOLOGY, BIOCHEMISTRY. Educ: Williams Col, BA, 63; Western Reserve Univ, PhD(biol), 68. Prof Exp: NSF fel, Univ Ill, Urbana, 67-69; Am Cancer Soc fel, Stanford Univ, 69-71; res assoc microbiol, Ind Univ, Bloomington, 71-72; asst prof, 72-75, ASSOC PROF BIOCHEM, SCH MED, ST LOUIS UNIV, 75- Mem: Am Soc Microbiol. Res: Regulation of membrane formation and genetics of photosynthetic bacteria. Mailing Add: Dept of Biochem St Louis Univ Sch of Med St Louis MO 63104

MARRS, ROSCOE EARL, b Schenectady, NY, Oct 21, 46; m 74. NUCLEAR PHvSICS. Educ: Cornell Univ, AB, 68; Univ Wash, MS, 69, PhD(physics), 75. Prof Exp: RES FEL PHYSICS, CALIF INST TECHNOL, 75- Mem: Am Phys Soc. Res: Experiments in nuclear physics related to electromagnetic transitions and weak interactions in light nuclei. Mailing Add: Kellogg Lab Calif Inst of Technol Pasadena CA 91125

MARRUS, RICHARD, b Brooklyn, NY, Sept 14, 32. ATOMIC PHYSICS, NUCLEAR PHYSICS. Educ: NY Univ, BS, 54; Univ Calif, Berkeley, MA, 56, PhD(physics), 59. Prof Exp: Asst, Univ, 54-56, res physicist, Lawrence Radiation Lab, 56-66, assoc prof, Univ, 66-71, PROF PHYSICS, UNIV CALIF, BERKELEY, 71- Concurrent Pos: Guggenheim fel, 70-71. Mem: Am Phys Soc. Res: Atomic beam magnetic resonance spectroscopy; optical pumping. Mailing Add: Dept of Physics Univ of Calif Berkeley CA 94720

MARSAGLIA, GEORGE, b Denver, Colo, Mar 12, 25; m 54; c 1. MATHEMATICS, COMPUTER SCIENCE. Educ: Colo Agr & Mech Col, BSc, 47; Ohio State Univ, MA, 48, PhD(math), 50. Prof Exp: Instr math, Ohio Univ, 48; asst prof, Univ Mont, 51-53, dir statist lab, 52-53; res assoc math statist, Univ NC, 53-54; vis lectr, Okla State Univ, 54-55; Fulbright vis prof, Univ Rangoon, 55-56; mem staff, Sci Res Labs, Boeing Co, Wash, 56-70; PROF COMPUT SCI & DIR SCH COMPUT SCI, McGILL UNIV, 70- Concurrent Pos: Vis lectr, Univ NC, 53-54; consult, Westinghouse Elec Corp, 56; from lectr to vis prof, Univ Wash, 59-70. Res: Probability and measure theory; stochastic processes; mathematical statistics; biomathematics; computer sciences. Mailing Add: Sch of Comput Sci McGill Univ Montreal PQ Can

MARSALIS, SULA JOHNSON, b Meridian, Miss, Nov 18, 23; m 43; c 2. CHEMISTRY. Educ: Univ Southern Miss, BS, 50, EdD(biol, chem), 71; Univ Miss, MS, 58. Prof Exp: Teacher, Meridian City Schs, Miss, 50-53, Lincoln County Schs, 53-54 & Brookhaven High Sch, 54-56; instr chem & head dept, Copiah-Lincoln Jr Col, 56-58; from asst prof to assoc prof, 58-72, PROF CHEM, MISS STATE UNIV WOMEN, 72- Concurrent Pos: Consult, Oak Ridge Nat Lab, 66- Mem: Am Chem Soc. Res: Science education; organic reaction mechanisms. Mailing Add: Dept of Chem Miss State Univ for Women Columbus MS 39701

MARSCHKE, CHARLES KEITH, b St Paul, Minn, Mar 9, 41; m 65; c 5. BIOCHEMISTRY, MICROBIOLOGY. Educ: Univ Minn, Minneapolis, BA, 65, MS, 69, PhD(biochem), 71. Prof Exp: RES SCIENTIST, UPJOHN CO, 70- Mem: Am Soc Microbiol. Res: Metabolism, enzymology, bacterial sporulation, fermentation and production of extracellular substances by microorganisms. Mailing Add: 202 Boston Kalamazoo MI 49002

MARSDEN, BRIAN GEOFFREY, b Cambridge, Eng, Aug 5, 37; m 64; c 2. CELESTIAL MECHANICS, PLANETARY SCIENCES. Educ: Oxford Univ, BA, 59, MA, 63; Yale Univ, PhD(astron), 66. Prof Exp: Res asst astron, Yale Univ Observ, 59-65; ASTRONR, SMITHSONIAN ASTROPHYS OBSERV, 65- Concurrent Pos: Lectr, Harvard Univ, 66-; dir cent bur astron telegrams, Int Astron Union, 68- Honors & Awards: Merlin Medal, Brit Astron Asn, 65. Mem: Am Astron Soc; Royal Astron Soc; Brit Astron Asn; Int Astron Union. Res: Orbits of comets, minor planets and natural satellites; celestial mechanics; astrometry; physics of comets. Mailing Add: Smithsonian Astrophys Observ 60 Garden St Cambridge MA 02138

MARSDEN, DAVID HENRY, b Dighton, Mass, Jan 31, 21; m 43; c 5. AGRICULTURAL MICROBIOLOGY. Educ: Univ Mass, BS, 43, MS, 48; Harvard Univ, PhD(biol), 52. Prof Exp: Asst prof res, Univ Mass, 47-54; plant pathologist, Eastern States Farmers' Exchange, 54-64; PROF MGR CHEM, AGWAY, INC, 64- Mem: Am Phytopath Soc (treas, 64-67); Am Inst Biol Sci; Weed Sci Soc Am. Res: Pesticides; farm chemicals. Mailing Add: Agway Inc Syracuse NY 13201

MARSDEN, HALSEY M, b July 25, 33; US citizen; m 62; c 2. ZOOLOGY, RESEARCH ADMINISTRATION. Educ: Univ Conn, BS, 55; Univ Mo, MA, 57, PhD(zool), 63. Prof Exp: NIH fel animal behav reproduction, Jackson Lab, 63-65; res biologist, Primate Ecol Sect, Nat Inst Neurol Dis & Stroke, 65-69 & Behav Systs Sect, Lab Brain Evolution & Behav, NIMH, 69-72; RES BIOLOGIST, DEVELOP NEUROL BR, NAT INST NEUROL & COMMUNICATIVE DIS & STROKE, 72- Mem: AAAS; Wildlife Soc; Am Ornithologists Union; Animal Behav Soc; Int Soc Res Aggression. Res: Ecology; animal behavior; mammalian reproduction; behavioral-environmental systems; child and human development; primatology. Mailing Add: Develop Neurol Br Nat Inst Neurol & Commun Dis & Stroke Bethesda MD 20014

MARSDEN, JAMES G, b St Louis, Mo, Dec 12, 25; m 48; c 6. CHEMISTRY. Educ: St Louis Univ, BS, 48. Prof Exp: Chemist, Linde Co, 48-56 & Silicones Div, 56-67, GROUP LEADER RES & DEVELOP, CHEM & PLASTICS DIV, UNION CARBIDE CORP, 67- Res: Synthesis and properties of organo-functional silanes; reinforced and filled composites. Mailing Add: Union Carbide Corp Tarrytown Tech Ctr PO Box 65 Tarrytown NY 10591

MARSDEN, JERROLD ELDON, b Ocean Falls, BC, Aug 17, 42; m 65; c 1. MATHEMATICS. Educ: Univ Toronto, BSc, 65; Princeton Univ, PhD(math), 68. Prof Exp: Instr math, Princeton Univ, 68; lectr, 68-69, asst prof, 69-72, ASSOC PROF MATH, UNIV CALIF, BERKELEY, 72- Concurrent Pos: Asst prof, Univ Toronto, 70-71. Res: Mathematical physics; global analysis; hydrodynamics; quantum mechanics; nonlinear Hamiltonian systems. Mailing Add: Dept Math Evans Hall Univ of Calif Berkeley CA 94720

MARSDEN, JOAN C, invertebrate zoology, physiology, see 12th edition

MARSDEN, RALPH WALTER, b Sumner, Wis, Apr 11, 11; m 57; c 2. GEOLOGY. Educ: Univ Wis, PhB, 32, PhM, 33, PhD(geol), 39. Prof Exp: Asst instr geol, Univ Wis, 36-39; geologist, Philippine Geol Surv, 39-40, chief geol surv div, 40-45; assoc prof geol, Univ Okla, 46-47; geologist, Jones & Laughlin Steel Corp, Pa, 45-46 & 47-51; mgr geol invest, Oliver Iron Mining Div, US Steel Corp, Minn, 51-64, mgr geol invest iron ore Pa, 64-67; chmn dept, 67-74, PROF GEOL, UNIV MINN, DULUTH, 67- Concurrent Pos: Intern, Cebu, Santo Tomas & Los Banos Camps, Philippines, 42-45. Mem: AAAS; fel Geol Soc Am; Soc Econ Geologists; Mining & Metall Soc Am; Am Inst Mining, Metall & Petrol Engrs (vpres, 69). Res: Mineral deposits; economic and mining geology; geology of iron ores including their origin, extent of reserves and resources. Mailing Add: Dept of Geol Univ of Minn Duluth MN 55812

MARSH, ALBERT WILLIAM, b Crookston, Minn, July 29, 13; m 40; c 2. SOIL SCIENCE, IRRIGATION. Educ: Univ Minn, BChE, 35, MS, 38; Ore State Col, PhD(soils), 42. Prof Exp: Asst soils, Univ Minn, 37-38; jr soil surveyor, Soil Conserv Serv, USDA, 38-40; asst soils, Ore State Col, 40-42, asst prof irrig, 42-45, asst prof soils, 42-55; soil scientist, US Salinity Lab, 55-56; AGRICULTURIST, UNIV CALIF, RIVERSIDE, 56- Concurrent Pos: Consult, Land & Water Develop Div, Food & Agr Orgn, UN, 63-64 & Food & Agr Orgn, 73 & 75; consult, Harza Eng Co, 75. Honors & Awards: Man of Year, Sprinkler Irrig Asn, 71. Mem: Sprinkler Irrig Asn; AAAS; Am Soc Agron; Soil Sci Soc Am. Res: Irrigation methods, particularly drip and sprinkler; control of soil salinity; frequency and amount of irrigation. Mailing Add: Soil Sci & Agr Eng Dept Univ of Calif Riverside CA 92502

MARSH, ALICE GARRETT, b Berrien Center, Mich, Feb 20, 08; m 27; c 2. NUTRITION, FOODS. Educ: Emmanuel Missionary Col, BS, 29; Univ Nebr, MS, 38. Prof Exp: Instr, Hinsdale Acad, Ill, 28-30; dietitian, Hindsdale Sanitarium & Hosp, 30-36; instr foods & nutrit, Union Col, 37-39; asst, Human Nutrit Lab, Univ Nebr, 39-44; instr, Pub Sch, 45-47; instr, Union Col, 47, asst prof foods & nutrit, 47-48, assoc prof home econ, 48-50; assoc prof, 59-59, PROF HOME ECON, ANDREWS UNIV, 59- Concurrent Pos: Instr, Univ Nebr, 40-44, mem staff, State & Fed Res Exp Sta. Mem: Am Dietetic Asn; Am Home Econ Asn. Res: Human nutrition, especially response of blood serum lipids to a controlled diet; animal nutrition; effect of food supplementary proteins upon successive generations of animals. Mailing Add: Dept of Home Econ Andrews Univ Berrien Springs MI 49104

MARSH, BENJAMIN BRUCE, b Petone, NZ, Nov 15, 26; m 52; c 2. MEAT SCIENCE, MUSCULAR PHYSIOLOGY. Educ: Univ NZ, BSc, 46, MSc, 47; Cambridge Univ, PhD(biochem), 51. Prof Exp: Chemist, Fats Res Lab, Wellington, NZ, 47; biochemist, Low Temperature Res Sta, Cambridge, Eng, 47-51 & Dominion Lab, Wellington, NZ, 51-57; biochemist & dep dir, Meat Indust Res Inst, Hamilton, NZ, 57-71; PROF MUSCLE BIOL & MEAT SCI & DIR LAB, UNIV WIS-MADISON, 71- Honors & Awards: Distinguished Meats Res Award, Am Meat Sci Asn, 70. Mem: Am Meat Sci Asn; Am Soc Animal Sci; Inst Food Technologists. Res: Early postmortem muscle metabolism; rigor mortis; meat quality; muscular contraction and relaxation; effects of muscle shortening on meat tenderness. Mailing Add: Muscle Biol & Meat Sci Lab Univ of Wis Madison WI 53706

MARSH, BRUCE BURTON, b Dickinson Center, NY, Aug 8, 34; m 60; c 3. NUCLEAR PHYSICS. Educ: State Univ NY Albany, BS, 56; Univ Rochester, PhD(physics), 62. Prof Exp: Assoc prof, 62-64, PROF PHYSICS, STATE UNIV NY ALBANY, 64- Mem: Am Phys Soc; Am Asn Physics Teachers. Res: Low energy experimental nuclear physics; physics education. Mailing Add: Dept of Physics State Univ of NY Albany NY 12222

MARSH, BRUCE DAVID, b Munising, Mich, Jan 4, 47; m 70. GEOLOGY. Educ: Mich State Univ, BS, 69; Univ Ariz, MS, 71; Univ Calif, Berkeley, PhD(geol), 74. Prof Exp: Geophysicist, Anaconda Co, 69-70, geologist, 70-71; ASST PROF EARTH & PLANETARY SCI, JOHNS HOPKINS UNIV, 74- Concurrent Pos: Geophysicist, US Geol Surv, 75- Mem: Am Geophys Union; Geol Soc Am; Soc Explor Geophysicists; Europ Asn Explor Geophysicists. Res: Plate tectonics, continental drift, and seafloor spreading, and the physics and chemistry of the generation and evolution of magma within the earth as associated with these processes. Mailing Add: Dept of Earth & Planetary Sci Johns Hopkins Univ Baltimore MD 21218

MARSH, CONNELL LEROY, b Pagosa Springs, Colo, June 9, 18; m 42; c 2. BIOCHEMISTRY. Educ: Univ Nebr, BS, 49, MS, 51, PhD(chem), 53. Prof Exp: Asst, USPHS Proj, 49-53, from asst biochemist to assoc biochemist, Dept Animal

Path & Hyg, 53-71, from assoc prof to prof vet sci, 58-71, PROF ORAL BIOL, COL DENT, UNIV NEBR-LINCOLN, 71- Concurrent Pos: Mem bd adv, Lincoln Med Res Found, 64-70, pres, 70-; consult, Beckman Instrument Co, Inc. Mem: Conf Res Workers Animal Dis; Am Asn Clin Chem. Res: Enzymology of metazoan parasites; colostrum absorption in young animals; leukemia in man and cattle; biochemical aspects of diseases in domestic animals; oral biology; immunology; research on human antiproteases. Mailing Add: Dept of Oral Biol Univ of Nebr Col of Dent Lincoln NE 68503

MARSH, DAVID GEORGE, b London, Eng, Mar 29, 40. IMMUNOGENETICS, BIOCHEMISTRY. Educ: Univ Birmingham, BSc, 61; Cambridge Univ, PhD(biochem), 64. Prof Exp: ASST PROF MED, SCH MED, JOHNS HOPKINS UNIV, 69-, ASST PROF MICROBIOL, 72- Concurrent Pos: USPHS fel, Calif Inst Technol, 66-69; USPHS res grant, Johns Hopkins Univ, 70- & res career develop award, 71- Mem: Am Acad Allergy; Am Asn Immunol; NY Acad Sci. Res: Immunochemistry and genetics of immediate hypersensitivity. Mailing Add: Good Samaritan Hosp 5601 Loch Raven Blvd Baltimore MD 21239

MARSH, DAVID PAUL, b Seattle, Wash, Dec 10, 34; m 58; c 3. NUCLEAR PHYSICS. Educ: DePauw Univ, BA, 57; Univ Calif, Berkeley, PhD(physics), 62. Prof Exp: Asst prof physics, Univ Hawaii, 62-63; asst prof, 63-69, ASSOC PROF PHYSICS, UNIV NEV, RENO, 69- Mem: AAAS; Am Geophys Union; Am Asn Physics Teachers. Mailing Add: Dept of Physics Univ of Nev Reno NV 89507

MARSH, DONALD CHARLES BURR, b Jackson, Mich, July 20, 26. NUMBER THEORY. Educ: Univ Ariz, BS, 47, MS, 48; Univ Colo, PhD(math), 54. Prof Exp: Instr math, Univ Ariz, 48-50; asst prof, Tex Tech Col, 54-55; from instr to assoc prof, 55-66, PROF MATH, COLO SCH MINES, 66- Concurrent Pos: Asst to dir, Nat Number Theory Inst, Univ Colo, 59; ed, Aristocrat Dept, The Cryptogram. Mem: Math Asn Am; Am Cryptogram Asn (pres, 68-70). Res: Heuristics; cryptanalysis; number theory. Mailing Add: Dept of Math Colo Sch of Mines Golden CO 80401

MARSH, DONALD JAY, b New York, NY, Aug 5, 34; m 55; c 2. PHYSIOLOGY, BIOMEDICAL ENGINEERING. Educ: Univ Calif, Berkeley, AB, 55; Univ Calif, San Francisco, MD, 58. Prof Exp: NIH fel, 59-63; from asst prof to assoc prof physiol, Sch Med, NY Univ, 63-71; PROF BIOMED ENG, SCH ENG, UNIV SOUTHERN CALIF, 71- Concurrent Pos: NIH spec fel, 70-71; mem sci adv coun, Southern Calif Kidney Found, 75- Mem: Am Physiol Soc; Biomed Eng Soc; Am Soc Nephrol; Biophys Soc; Soc Gen Physiologists. Res: Renal physiology, mechanism of hypertonic urine formation; regulation of glomerular filtration and proximal tubule reabsorption; dynamics of organ level regulation of glucose metabolism. Mailing Add: Olin Hall 500 Univ of Southern Calif Los Angeles CA 90007

MARSH, FRANK DENNIS, organic chemistry, see 12th edition

MARSH, FRANK LEWIS, b Aledo, Ill, Oct 18, 99; m 27; c 2. ECOLOGY. Educ: Emmanuel Missionary Col, Andrews Univ, AB, 27, BS, 29; Northwestern Univ, MS, 35; Univ Nebr, PhD(bot), 40. Prof Exp: Instr sci & math, Hinsdale Acad, Ill, 29-34; asst zool, Northwestern Univ, 34-35; from instr to prof biol, Union Col, Nebr, 35-50; prof & head dept, Emmanuel Missionary Col, Andrews Univ, 50-58; researcher, Geo-Sci Res Inst, 58-64; prof 64-71, EMER PROF BIOL, ANDREWS UNIV, 71- Res: Ecological entomology and botany; hyperparasitism; origin of species; hybridization. Mailing Add: 216 Hillcrest Dr Berrien Springs MI 49103

MARSH, FREDERICK LEON, b Richmond, Va, Dec 20, 35; m 64. ANALYTICAL CHEMISTRY, PHYSICAL CHEMISTRY. Educ: Blackburn Univ, AB, 58; Univ Minn, PhD(anal chem), 65. Prof Exp: Trainee microchem, Northern Util Res & Develop Br, Agr Res Serv, USDA, 57, microchemist, 58; instr chem, Univ Toledo, 64-65; SR RES ELECTROCHEMIST, GOULD LABS, GOULD INC, 65- Mem: AAAS; Am Chem Soc; Electrochem Soc. Res: Microanalytical chemistry; electrochemistry; electroanalytical chemistry. Mailing Add: 1523 Windemere Dr Minneapolis MN 55421

MARSH, GAYLE G, b St Cloud, Minn, July 29, 27; m 64; c 1. NEUROPSYCHOLOGY. Educ: San Francisco State Col, BA, 49; Columbia Univ, MA, 56, PhD(clin psychol), 62. Prof Exp: Res assoc psychol, Foreign Physicians Training Study-Int Inst Educ, 62-64; asst prof med psychol, 64-73, ASSOC RES PSYCHOLOGIST, NEUROPSYCHIAT INST, UNIV CALIF, LOS ANGELES, 73- Mem: Am Psychol Asn; Int Neuropsychol Soc. Res: Effects of brain damage on intellectual, adaptive and emotional behavior. Mailing Add: Neuropsychiat Inst Univ of Calif Los Angeles CA 90024

MARSH, GLENN ANTHONY, b Chicago, Ill, Dec 20, 24; m 46; c 4. PHYSICAL CHEMISTRY, CORROSION. Educ: Ill Inst Technol, BS, 45; Northwestern Univ, MS, 46. Prof Exp: Res fel corrosion, Ill Inst Technol, 46-48; sr res chemist, Pure Oil Co, Ill, 48-56, proj technologist, 56-60, res assoc & head corrosion res sect, 60-65; SUPVR CORROSION SECT, UNION OIL CO CALIF, 66- Concurrent Pos: Ed, Corrosion Div, Electrochem Soc, 62-65. Honors & Awards: Willis Rodney Whitney Award, Nat Asn Corrosion Engrs, 71. Mem: Nat Asn Corrosion Engrs; Electrochem Soc. Res: Corrosion mechanisms, measurements and preventive methods. Mailing Add: Res Ctr Union Oil Co of Calif Box 76 Brea CA 92621

MARSH, HOWARD STEPHEN, b New York, NY, Feb 4, 42; m 68; c 1. APPLIED PHYSICS. Educ: Rensselaer Polytech Inst, BS, 63; Cornell Univ, PhD(physics), 69. Prof Exp: Asst physics, Cornell Univ, 63-65, asst, Lab Atomic & Solid State Physics, 64-69; MEM TECH STAFF, MITRE CORP, 69- Mem: Am Phys Soc; NY Acad Sci. Res: Signal transmission and detection; advanced interdisciplinary technology; military systems; information and control systems. Mailing Add: Mitre Corp Bedford MA 01730

MARSH, JAMES ALEXANDER, JR, b Wilson, NC, Dec 8, 40. MARINE ECOLOGY. Educ: Duke Univ, BS, 63; Univ Ga, PhD(zool), 68. Prof Exp: Fel environ sci, Univ NC, Chapel Hill, 68-70; asst prof marine sci, 70-74, ASSOC PROF MARINE SCI, MARINE LAB, UNIV GUAM, 74- Mem: Ecol Soc Am; Am Soc Limnol & Oceanog; Am Inst Biol Sci; AAAS. Res: Coral reef ecology; primary productivity; nutrient and energy cycling in tropical marine ecosystems. Mailing Add: Univ of Guam Marine Lab PO Box EK Agana GU 96910

MARSH, JOHN LEE, b Washington, DC, Feb 25, 16; m 43, 72; c 6. INFORMATION SCIENCE. Educ: Univ Ill, PhD(org chem), 41. Prof Exp: Res chemist, Hooker Electrochem Co, NY, 41-43; res chemist, 43-57, SR INFO SCIENTIST, CIBA-GEIGY PHARMACEUT CO, 57- Mem: Am Chem Soc; NY Acad Sci; Am Soc Info Sci. Res: Processing of scientific data and information by classical and computer techniques; communication of research knowledge; chemical nomenclature; molecular notations; published and proprietary scientific information. Mailing Add: 108 Beekman Rd Summit NJ 07901

MARSH, JOHN MACCLENAHAN, b Brooklyn, NY, Nov 23, 31; m 59; c 3. BIOCHEMISTRY. Educ: City Col New York, ScB, 53; Brown Univ, ScM, 55, PhD(biochem), 58. Prof Exp: Trainee steroid biochem, Worcester Found Exp Biol, 58-59; from asst prof to assoc prof endocrinol, 59-75, PROF BIOCHEM, SCH MED, UNIV MIAMI, 75- Concurrent Pos: Dept Health, Educ & Welfare res career develop award, 67- Mem: AAAS; Endocrine Soc; Am Soc Biol Chemists; Soc Study Reprod; Soc Gynec Invest. Res: Mechanism of hormone action; gonadotrophic control of steroidogenesis; receptors of gonadotropins; adenyl cyclase; mechanism of cyclic adenosine monophosphate action in endocrine tissues; role of prostaglandins in ovarian function; action of gonadotropin on prostaglandins; mechanism of prostaglandin action. Mailing Add: Dept of Biochem Univ of Miami Sch of Med Miami FL 33152

MARSH, JULIAN BUNSICK, b New York, NY, Jan 21, 26; m 48; c 1. BIOCHEMISTRY. Educ: Univ Pa, MD, 47. Prof Exp: Intern, Episcopal Hosp, Philadelphia, 47-48; NIH fel biochem, Dept Res Med, Univ Pa, 48-50, instr res med, 50-51, assoc biochem, Grad Sch Med, 52-53, asst prof, 53-59, assoc prof, Sch Med & Grad Sch Med, 59-63, prof, Grad Sch med, 63-65, prof biochem & chmn dept, Sch Dent Med, 65-75; PROF PHYSIOL & PROF BIOCHEM & CHMN DEPT, MED COL PA, 75- Concurrent Pos: Guggenheim Mem fel, Nat Inst Med Res, Eng, 60-61. Mem: AAAS; Am Soc Biol Chemists; Soc Exp Biol & Med. Res: Action of insulin and other hormones; carbohydrate, chromoprotein and lipoprotein metabolism; experimental nephrosis. Mailing Add: Dept of Physiol & Biochem Med Col of Pa Philadelphia PA 19129

MARSH, LELAND C, b Lyons, NY, Nov 19, 28; m 53; c 4. BOTANY. Educ: Syracuse Univ, BS, 51, PhD(bot), 62. Prof Exp: Asst prof bot & biol, Marshall Univ, 57-60; assoc prof, State Univ NY Col Buffalo, 60-65; PROF BIOL, STATE UNIV NY COL OSWEGO, 65-, CHMN DEPT BOT & PHYSIOL, 71- Mem: Bot Soc Am; Soc Study Evolution. Res: Botanical research of Typha species, including ecological, genetic and systematic studies; industrial uses of Typha. Mailing Add: Dept of Biol State Univ of NY Oswego NY 13126

MARSH, MAX MARTIN, b Indianapolis, Ind, Feb 25, 23; m 41; c 4. PHYSICAL CHEMISTRY, ANALYTICAL CHEMISTRY. Educ: Ind Univ, BS, 47. Prof Exp: Anal chemist, 47-56, head anal, Res Dept, 56-61, RES ADV, RES LABS, ELI LILLY & CO, 66-; INDUST PROF CHEM, IND UNIV, 71- Concurrent Pos: Dir phys chem res div, Eli Lilly & Co, 67-69; mem sci adv bd, Indianapolis Ctr Advan Res. Mem: AAAS; Am Chem Soc; NY Acad Sci. Res: Optical analytical techniques; molecular structure-activity relationships. Mailing Add: Lilly Res Labs Eli Lilly & Co Indianapolis IN 46206

MARSH, MICHAEL PIERCE, b Port Limon, Costa Rica, Sept 12, 32; US citizen; m 58; c 3. ECOLOGY, ANIMAL BEHAVIOR. Educ: Fresno State Col, BA, 53; Univ Calif, Berkeley, MA, 58, PhD(zool), 62. Prof Exp: Lectr zool, Univ Sydney, 61-67; fel behav, Mich State Univ, 67-68; ASSOC PROF BIOL, CENTRE COL, 68- Concurrent Pos: Australian Res Grants Comt res grants, 66-67. Mem: AAAS; Ecol Soc Am; Animal Behav Soc; Am Soc Mammal. Res: Population ecology of small mammals, especially effects of the food, density and social behavior on population dynamics. Mailing Add: Div of Sci & Math Centre Col Danville KY 40422

MARSH, NAT HUYLER, b Ft Worth, Tex, Aug 1, 14; m 42; c 3. ORGANIC CHEMISTRY. Educ: Rice Inst, BA, 38, MA, 40, PhD(org chem), 42. Prof Exp: Res chemist, Humble Oil & Ref Co, Tex, 42-45; res chemist, Am Cyanamid Co, 45-54, mgr synthetic fibers dept, 54-56, mgr synthetic fibers dept, Santa Rosa Plant, Fla, 56-60, dir res & develop, Fibers Div, 60-65, sr managing dir, Cyanamid, Japan, Ltd, 67-68; PRES, NIHON MILLIPORE LTD, 68- Mem: Am Chem Soc; Brit Soc Chem Indust. Res: Petroleum and nitrogen chemistry; guanidines and triazines; cracked gasoline and reaction of bromine with alcohols. Mailing Add: Nihon Millipore Ltd 4-15 1 Chome Shiroganedai Minato-Ku Tokyo 108 Japan

MARSH, PAUL BRUCE, b Niagara Falls, NY, Nov 21, 14; m 41; c 2. PLANT PHYSIOLOGY. Educ: Univ Rochester, AB, 37, MS, 39; Cornell Univ, PhD(plant path), 42. Prof Exp: PLANT PATHOLOGIST, BELTSVILLE AGR RES CTR, USDA, 42- Mem: Am Chem Soc; Fiber Soc; Am Phytopath Soc; Mycol Soc Am; Soc Indust Microbiol. Res: Microbial physiology; microbiology of natural fibers and seeds. Mailing Add: Beltsville Agr Res Ctr USDA Beltsville MD 20705

MARSH, PAUL MALCOLM, b Fresno, Calif, Nov 7, 36; m 65; c 2. ENTOMOLOGY. Educ: Univ Calif, Davis, BS, 58, MS, 60, PhD(entom), 64. Prof Exp: Lab technician, Univ Calif, Davis, 61-63; RES ENTOMOLOGIST, SYST ENTOM LAB, AGR RES SERV, USDA, 64- Mem: Entom Soc Am; Am Entom Soc. Res: Systematic entomology; taxonomy and biology of parasitic wasps of the family Braconidae. Mailing Add: Syst Entom Lab USDA US Nat Mus Washington DC 20560

MARSH, RICHARD EDWARD, b Jackson, Mich, Mar 6, 22; m 47; c 4. PHYSICAL CHEMISTRY. Educ: Calif Inst Technol, BS, 43; Univ Calif, Los Angeles, PhD(phys chem), 50. Prof Exp: Fel struct of metals, 50-51, fel struct of proteins, 51-55, sr fel, 55-74, RES ASSOC STRUCTURE OF PROTEINS, CALIF INST TECHNOL, 74- Concurrent Pos: Instr, Univ Calif, Los Angeles, 53. Mem: Am Crystallog Asn. Res: Crystal structure analysis; molecular structure; structure of biologic molecules. Mailing Add: Dept of Chem Calif Inst of Technol Pasadena CA 91109

MARSH, RICHARD FLOYD, b Portland, Ore, Mar 3, 39; m 59; c 5. VETERINARY VIROLOGY, VETERINARY PATHOLOGY. Educ: Wash State Univ, BS, 61, DVM, 63; Univ Wis-Madison, MS, 66, PhD(vet sci), 68. Prof Exp: Res veterinarian, Kellogg Co, Battle Creek, Mich, 63-64; NIH trainee vet sci, Univ Wis-Madison, 64-66, NIH spec fel, 66-68; vet officer, Nat Inst Neurol Dis & Stroke, NIH, USPHS, 68-70; RES ASSOC VET SCI, UNIV WIS-MADISON, 70- Mem: AAAS; Am Soc Microbiol. Res: Development and study of animal models of human disease, especially persistent virus infections of the central nervous system. Mailing Add: Dept of Vet Sci Univ of Wis Madison WI 53706

MARSH, RICHARD HAYWARD, b Detroit, Mich, Jan 6, 40; m 61; c 4. ANALYTICAL CHEMISTRY. Educ: Univ Mich, BS, 61; Wayne State Univ, MS, 64, PhD(anal chem), 67. Prof Exp: Sanit chemist, Detroit Water Dept, 61-63; teaching asst, Wayne State Univ, 64-67; SR RES SCIENTIST, FORD MOTOR CO, 67- Mem: Asn Comput Mach; Sigma Xi. Res: Neutron activation analysis; radiochemistry; environmental chemistry; computer science. Mailing Add: 2050 Glen Iris Milford MI 48042

MARSH, RICHARD RILEY, b Pittsburgh, Kans, Feb 17, 06; m 32; c 3. HUMAN BIOLOGY, FOOD SCIENCE. Educ: Baker Univ, BA, 27; Kans State Teachers Col, BS, 28; Kans State Col, MS, 31; Univ Kans, PhD(zool), 47. Prof Exp: Lab asst, Baker Univ, 26-27, head dept biol, 37-41; asst instr, Kans State Teachers Col, 27-28; clin bacteriologist, Smith Clin, 28-29; high sch instr, 29-30; asst, Kans State Col, 30-31; instr, Sch Educ, Univ Kans, 31-32; high sch instr & instr, Western State Col, Colo, 32-36; supt schs, High Sch Dist, Colo, 36-37; asst head res labs, Swift & Co, Ill, 41-44; head res lab, St Louis Independent Packing Co, 44-52; asst prof dairy technol,

Dept Food Technol & Univ Exten Div, 52-55, assoc prof food technol, 55-60, assoc dir honors prog, 60-67, prof biol, 60-72, dir honors prog, 67-72, asst to dir sch life sci, 68-72, EMER PROF BIOL, UNIV ILL, URBANA, 72- Concurrent Pos: Lectr, Univ Col, Washington Univ, 43-52. Res: Endocrinology of anterior pituitary and of corpus luteum; nerve fibers in Cambarus; erythropoiesis in the pig embryo. Mailing Add: Dept of Biol Univ of Ill Urbana IL 61801

MARSH, ROBERT CECIL, b Lexington, Ky, Feb 27, 44; m 65; c 2. MOLECULAR BIOLOGY. Educ: Western Ky Univ, BS, 65; Vanderbilt Univ, PhD(molecular biol), 71. Prof Exp: Res assoc biochem, Soc Molecular Biol Res, Stockheim, Ger, 71-75; res assoc, Princeton Univ, 75-76; ASST PROF BIOCHEM, UNIV TEX, DALLAS, 76- Mem: Am Chem Soc. Res: Mechanism of action of elongation factors in protein biosynthesis; regulation of DNA replication and gene expression by bacteriophage T4. Mailing Add: 417 Boone Ave Winchester KY 40391

MARSH, TERRENCE GEORGE, b Winnipeg, Man, Jan 12, 41; US citizen; m 65; c 2. ENVIRONMENTAL BIOLOGY. Educ: Earlham Col, AB, 63; Ore State Univ, MS, 65; Univ Ky, PhD(zool), 69. Prof Exp: Instr biol, Asbury Col, 68-69; ASST PROF BIOL, N CENT COL, ILL, 69- Mem: AAAS; Nat Speleol Soc. Res: urban biology; insect ecology. Mailing Add: Dept of Biol N Cent Col Naperville IL 60540

MARSH, WALTON HOWARD, b Bay City, Mich, Mar 26, 19; m 44; c 2. BIOCHEMISTRY. Educ: Columbia Col, AB, 40; Polytech Inst New York, MS, 43; Case Western Reserve Univ, PhD(biochem), 51. Prof Exp: Asst polymer res, Am Cyanamid Co, 40-43, asst pharm, 45-46; med res biochemist and clin biochemist, Vet Admin Hosp, 51-54; PROF PATH, STATE UNIV NY DOWNSTATE MED CTR, 54-; CHIEF BIOCHEMIST, KINGS COUNTY HOSP, BROOKLYN, 54- Concurrent Pos: Instr, Case Western Reserve Univ, 51-54. Res: Differentiation and regeneration metabolism. Mailing Add: Dept of Path State Univ NY Downstate Med Ctr Brooklyn NY 11203

MARSH, WILLIAM ERNEST, b New Brunswick, NJ, Nov 22, 39; m 62; c 2. MATHEMATICAL LOGIC. Educ: Dartmouth Col, AB, 62, MA, 65, PhD(math), 66. Prof Exp: Asst prof math & chmn dept, Talledega Col, 66-69; asst prof, 69-74, ASSOC PROF MATH, HAMPSHIRE COL, 74- Mem: Asn Symbolic Logic; Am Math Soc. Res: Model theory; foundations of mathematics; mathematical linguistics; automata theory. Mailing Add: Dept of Math Hampshire Col Amherst MA 01002

MARSHAK, ALFRED GEORGE, molecular biology, see 12th edition

MARSHAK, HARVEY, b Brooklyn, NY, Nov 9, 27; div; c 2. NUCLEAR PHYSICS. Educ: Univ Buffalo, BA, 50; Univ Conn, MA, 52; Duke Univ, PhD(physics), 55. Prof Exp: Res assoc, Duke Univ, 54-55; assoc physicist, Brookhaven Nat Lab, 55-62; PHYSICIST, NAT BUR STANDARDS, 62- Mem: Fel Am Phys Soc. Res: Nuclear orientation and spectroscopy; neutron physics; low temperature physics. Mailing Add: Physics Bldg B 128 Nat Bur of Standards Washington DC 20234

MARSHAK, ROBERT EUGENE, b New York, NY, Oct 11, 16; m 43; c 2. THEORETICAL PHYSICS, ASTROPHYSICS. Educ: Columbia Col, AB, 36; Cornell Univ, PhD(physics), 39. Prof Exp: From instr to prof physics, Univ Rochester, 39-50, Harris prof & chmn dept, 50-64, distinguished univ prof, 64-70; PRES, CITY COL NEW YORK, 70- Concurrent Pos: Physicist, Radiation Lab, Mass Inst Technol, 42-43 & Dept Sci & Indust Res Gt Brit, 43-44; dep group leader, Los Alamos Sci Lab, 44-46; mem, Inst Advan Study, 48; Guggenheim fel & prof, Sorbonne, 53-54; Guggenheim fel & guest prof, Ford Found, Europ Orgn Nuclear Res, Switz, 60-61, Yugoslavia, Israel & Japan, 67-68; vis prof, Columbia Univ, Univ Mich, Harvard Univ, Cornell Univ & Tata Inst Fundamental Res, India; Niels Bohr vis prof, Madras Univ, 63, Yalta Int Sch, Carnegie-Mellon Univ & Univ Tex; Nobel lectr, Sweden; Solvay Cong, 67; trustee, Atoms for Peace Awards. Secy, High Energy Physics Comn, Int Union Pure & Appl Physics, 57-63; chmn, Int Conf High Energy Physics, 60; mem, Nat Acad Sci Adv Comt Soviet Union & Eastern Europe, 63-66, head deleg to Poland, 64 & Yugoslavia, 65; mem, US Mission to Soviet Labs, 60; chmn vis physics comn, Brookhaven Nat Lab, 65; vis physics comn, Carnegie Inst, 66-70; mem, Sloan Fel Comt, 67-73, chmn, 72-73; mem sci coun, Int Ctr Theoret Physics, Trieste, 67-75; mem, US-Japan Sci Comt, 68-72; mem exec comt, Nat Comn, UNESCO, 71-73; mem coun, Nat Acad Sci, 71-74. Honors & Awards: Morrison Prize, NY Acad Sci, 40. Mem: Nat Acad Sci; fel AAAS; Am Acad Arts & Sci; Am Phys Soc; Fedn Am Scientists (chmn, 47-48). Res: Energy sources of stars; atomic nuclei; neutron diffusion; elementary particles. Mailing Add: Off of Pres City Col of New York New York NY 10031

MARSHAK, ROBERT REUBEN, b New York, NY, Feb 23, 23; m 48; c 3. VETERINARY MEDICINE. Educ: Cornell Univ, DVM, 45. Hon Degrees: Dr Vet Med, Univ Bern, 68. Prof Exp: Pvt pract & clin invest, 45-56; chmn dept clin studies, 61-73, dir, Bovine Leukemia Res Ctr, 66-75, PROF MED, SCH VET MED, UNIV PA, 56-, DEAN SCH VET MED, 73- Concurrent Pos: Mem comt vet med sci, Nat Acad Sci, 74- Mem: AAAS; Am Asn Cancer Res; Am Pub Health Asn; fel NY Acad Sci; Am Col Vet Internal Med (pres, 75-76). Res: Bovine leukemia; metabolic diseases of cattle. Mailing Add: Sch of Vet Med Univ of Pa Philadelphia PA 19174

MARSHALEK, EUGENE RICHARD, b New York, NY, Jan 17, 36; m 62; c 2. NUCLEAR PHYSICS. Educ: Queen's Col, NY, BS, 57; Univ Calif, Berkeley, PhD(nuclear struct), 62. Prof Exp: NSF fel nuclear physics, Niels Bohr Inst, Copenhagen, Denmark, 62-63; res assoc physics theory group, Brookhaven Nat Lab, 63-65; asst prof, 65-69, ASSOC PROF PHYSICS, UNIV NOTRE DAME, 69- Res: Nuclear structure theory; nuclear theory, particularly collective effects in atomic nuclei. Mailing Add: Dept of Physics Univ of Notre Dame Notre Dame IN 46556

MARSHALL, ALBERT WALDRON, b Portland, Ore, Aug 3, 28; m 51; c 2. MATHEMATICS. Educ: Univ Ore, BS, 51; Univ Wash, PhD(math), 58. Prof Exp: Actg asst prof statist, Stanford Univ, 58-60; staff mem, Inst Defense Anal, 60-61 & Boeing Sci Res Labs, 61-70; vis prof math, Univ Wash, 70-71; prof statist & math, Univ Rochester, 71-75; PROF MATH, UNIV BC, 75- Mem: Am Math Soc; Math Asn Am; fel Inst Math Statist. Res: Probability theory; inequalities; reliability theory. Mailing Add: Dept of Math Univ of BC Vancouver BC Can

MARSHALL ANNE (CORINNE), b Zanesville, Ohio, Oct 8, 04. BIOLOGICAL STRUCTURE. Educ: Denison Univ, BS, 25; Ohio State Univ, AM, 28, PhD(entom), 39. Prof Exp: Instr zool, McCook Jr Col, 29-32, La State Univ, 32-35 & Ohio State Univ, 35-39; from assoc prof to prof biol & head dept sci, 39-69, EMER PROF BIOL, UNIV WIS-STOUT, 69- Mem: AAAS; Sigma Xi. Res: Trichoptera. Mailing Add: 2950 Fairway Lane Zanesville OH 43701

MARSHALL, ARVLE EDWARD, b Canyon, Tex, Dec 24, 37; m 60; c 2. VETERINARY MEDICINE, VETERINARY NEUROLOGY. Educ: Tex Tech Col, BS, 60; Tex A&M Univ, DVM, 64; Univ Mo-Columbia, PhD(vet neurol), 71. Prof Exp: Asst vet pract, Tex, 64; vet meat inspector, Meat Inspection Div, USDA, Tex, 64-65; instr vet med & surg, Okla State Univ, 65-67; Nat Inst Child Health & Human Develop fel vet anat, Univ Mo-Columbia, 67-71; ASST PROF VET MED, UNIV ILL, URBANA, 71- Mem: Am Vet Med Asn; Am Asn Vet Neurol. Res: Maturation of brainstem control of circulation and nervous system control of circulation in the pig. Mailing Add: Col of Vet Med Univ of Ill Urbana IL 61801

MARSHALL, BILLY JACK, b Denison, Tex, Apr 28, 35; m 57; c 2. SOLID STATE PHYSICS, LOW TEMPERATURE PHYSICS. Educ: Austin Col, BA, 58; Rice Univ, MA, 60, PhD(physics), 62. Prof Exp: Designer, Corps Engrs, Perrin AFB, 55-56; engr, Hughes Aircraft Co, 56-57; engr, Ling-Temco-Vought, Inc, 57-58; asst prof physics, Arlington State Col, 62-65; PROF PHYSICS, TEX TECH UNIV, 65-, CHMN DEPT, 72- Concurrent Pos: Welch Found grant, Arlington State Col, 62-65 & Tex Tech Univ, 65-74. Mem: Am Inst Physics. Res: Electron-phonon studies in superconductors at very low temperatures (10 degrees Kelvin - 0.08 degrees Kelvin). Mailing Add: Dept of Physics Tex Tech Univ Lubbock TX 79409

MARSHALL, CARTER LEE, b New Haven, Conn, Mar 31, 36; c 2. PREVENTIVE MEDICINE. Educ: Harvard Univ, BA, 58; Yale Univ, MD, 62, MPH, 64; Am Bd Prev Med, dipl, 70. Prof Exp: Proj dir, Conn Dept Health, New Haven, 64-65; asst prof prev med, Sch Med, Univ Kans Med Ctr, Kansas City, 67-69; assoc prof community med, 69-75, PROF COMMUNITY MED & MED EDUC, MT SINAI SCH MED, 76-, ASSOC DEAN, 74- Concurrent Pos: Fel epidemiol & pub health, Yale Univ, 64-65; mem, Nat Med Found, 70; consult, New York Health Serv Admin, 70-71; dean health affairs, City Univ New York, 72-74. Mem: AAAS; fel Am Pub Health Asn; Asn Teachers Prev Med. Res: Development of health manpower. Mailing Add: Dept of Community Med Mt Sinai Sch of Med New York NY 10029

MARSHALL, CHARLES EDMUND, b Bredbury, Eng, Jan 9, 03; nat US; m 32; c 1. COLLOID CHEMISTRY, SOIL SCIENCE. Educ: Univ Manchester, BSc, 24, MS, 25; Univ London, PhD(agr chem), 27. Prof Exp: Asst lectr agr chem, Leeds Univ, 28-36; vis assoc prof, 35-36, from assoc prof to prof, 36-73, assoc dean grad sch, 65-66, actg dean, 66-67, EMER PROF SOILS, UNIV MO-COLUMBIA, 73- Concurrent Pos: NSF sr res fel, Imp Col, Univ London, 60-61. Honors & Awards: Hoblizelle Award, 51. Mem: AAAS; Am Chem Soc; Soil Sci Soc Am (vpres, 45, pres, 46); fel Am Soc Agron; fel Mineral Soc Am. Res: Colloid chemistry and mineralogy of clays; electrochemistry of membranes; mineral nutrition of plants; soil formation processes. Mailing Add: Dept of Agron Univ of Mo Columbia MO 65202

MARSHALL, CHARLES LOUIS, b New York, NY, Sept 28, 12; m 39; c 2. CHEMISTRY. Educ: Fordham Univ, BS, 34, MS, 39. Prof Exp: Teacher, Sch Dist NY, 34-35; instr, Fordham Univ, 35-36; teacher, Sch Dist NY, 36-40; res chemist, United Fruit Co, 46; dep de-classification officer, Tenn, 46-48, asst chief br, Washington, DC, 48, actg chief, 48-49, dep dir, Off of Classification, 49-55, DIR DIV CLASSIFICATION, ENERGY RES & DEVELOP ADMIN, WASHINGTON, DC, 55- Honors & Awards: Distinguished Serv Award, AEC, 74. Mem: Assoc Am Chem Soc; Am Nuclear Soc; fel Am Inst Chemists. Res: Organic synthesis; organic micro analysis; vitamins plant chemistry; burette clamp. Mailing Add: 4308 Lynbrook Dr Bethesda MD 20014

MARSHALL, CHARLES WHEELER, b Syracuse, NY, Oct 20, 06; m 39; c 2. ORGANIC CHEMISTRY. Educ: Univ Chicago, BS, 31, MS, 33, PhD(biochem), 49. Prof Exp: Chemist, Edwal Labs, 38-39; chief control chemist, Lakeside Labs, 39-43; res chemist, Off Sci Res & Develop, Chicago, 43-45 & G D Searle & Co, 49-67; SCI WRITER, 67- Mem: Am Chem Soc. Res: Synthesis of steroids related to adrenal cortical hormones and steroids with new pharmacological properties. Mailing Add: 230 Baseline Rd Apt B6 South Haven MI 49090

MARSHALL, CLIFFORD DANIEL, organic chemistry, see 12th edition

MARSHALL, CLIFFORD WALLACE, b New York, NY, Mar 11, 28; m 55. APPLIED MATHEMATICS, APPLIED STATISTICS. Educ: Hofstra Col, BA, 49; Syracuse Univ, MA, 50; Polytech Inst Brooklyn, MS, 55; Columbia Univ, PhD, 61. Prof Exp: Instr math, Polytech Inst Brooklyn, 50-57; mem staff, Inst Defense Anal, 58-59; prin dynamics engr, Repub Aviation Corp, 59-60; from instr to assoc prof, 60-68, PROF MATH, POLYTECH INST NEW YORK, 68- Concurrent Pos: Consult, Urban Inst, 72- Mem: Am Math Soc; Soc Indust & Appl Math; assoc Opers Res Soc Am; Math Asn Am; Am Statist Asn. Res: Combinatorial theory; finite graph theory; probability; time series analysis and forecasting. forecasting. Mailing Add: Dept of Math Polytech Inst of New York Brooklyn NY 11201

MARSHALL, DAVID JONATHAN, b Montreal, PQ, June 26, 28; m 60; c 3. MEDICINAL CHEMISTRY, RESEARCH ADMINISTRATION. Educ: McGill Univ, BSc, 49; Mass Inst Technol, PhD(chem), 53. Prof Exp: Res fel chem, Harvard Univ, 53-54; res chemist, 54-65, group leader steroids, 65-69, asst dir admin, 69-71, DIR ADMIN SERV, AYERST LABS DIV, AM HOME PROD CORP, 72- Mem: Chem Inst Can. Mailing Add: Ayerst Labs PO Box 6115 Montreal PQ Can

MARSHALL, DAVID L, biochemistry, see 12th edition

MARSHALL, DELBERT ALLAN, b Topeka, Kans, July 22, 37; m 64. ANALYTICAL CHEMISTRY. Educ: Kans State Teachers Col, BS, 59; Kans State Univ, MS, 65, PhD(anal chem), 68. Prof Exp: High sch teacher, Kans, 61-63; instr chem, Mo Valley Col, 63-64; asst prof, 67-71, ASSOC PROF CHEM, FT HAYS KANS STATE COL, 71- Mem: Am Chem Soc; Soc Appl Spectros; Coblentz Soc; The Chem Soc. Res: Chemistry of metal chelates; atomic absorption spectroscopy. Mailing Add: Dept of Chem Ft Hays Kans State Col Hays KS 67601

MARSHALL, DONALD D, b Woodland, Calif, Aug 8, 34; m 64; c 2. INORGANIC CHEMISTRY, ANALYTICAL CHEMISTRY. Prof Exp: Anal chemist, US Bur Mines, Nev, 58-60; asst prof chem, Southern Ore Col, 65-66; from asst prof to assoc prof, 66-73, PROF CHEM, CALIF STATE COL, SONOMA, 73- Mem: Am Chem Soc. Res: Determination of stability constants of inorganic compounds in aqueous solutions; water and air pollution; computer applications in chemistry. Mailing Add: Dept of Chem Calif State Col Sonoma Rohnert Park CA 94928

MARSHALL, DONALD IRVING, b Houston, Tex, Jan 22, 24; m 48; div; c 3. PLASTICS CHEMISTRY. Educ: Sam Houston Col, BS, 44; Univ Tex, MA, 46, PhD(chem), 48. Prof Exp: Develop assoc, Plastics Div, Union Carbide Corp, 48-58; sr res engr, 58-65, res leader, 65-71, SR STAFF ENGR, WESTERN ELEC CO, 71- Mem: Am Chem Soc; Soc Rheol; Soc Plastics Engrs; Plastics Inst Am (treas, 68-71). Res: Rheology; material characterization; extrusion, molding and calendering processes. Mailing Add: Western Elec Co 2000 Northeast Expwy Norcross GA 30071

MARSHALL, DONALD JAMES, b Marlboro, Mass, Apr 14, 33; m 54; c 1. PHYSICS. Educ: Mass Inst Technol, BS, 54, PhD(geophys), 59; Calif Inst Technol, MS, 55. Prof Exp: Dir res, Nuclide Corp, 58-74, GEN MGR & PUB RELS OFFICER, ALLOYD GEN VACUUM CORP, NUCLIDE CORP, 74- Mem: Am Soc Testing & Mat; NY Acad Sci. Res: Mass spectrometry; ion physics; electron beam technology. Mailing Add: Alloyd Gen Vacuum Corp 916 Main St North Acton MA 01720

MARSHALL, EDWARD EUGENE, chemistry, see 12th edition

MARSHALL, EDWIN RANDOLPH, b Jodie, WVa, Dec 3, 12; m 43; c 2. CHEMISTRY. Educ: Va Mil Inst, BS, 33; Columbia Univ, AM, 39, PhD(org chem), 42. Prof Exp: Res chemist, Pigments Dept, E I du Pont de Nemours & Co, Md, 41-42, Del, 46-75; RETIRED. Mem: AAAS; Am Chem Soc. Res: Synthetic organic chemistry; lactones; development of incendiary munitions; flame proofing of cellulose materials; titanium metal; potassium titanate; pigments. Mailing Add: 1905 Longcame Dr Graylyn Crest III Wilmington DE 19803

MARSHALL, ERNEST (ROY), b Bonner Springs, Kans, May 25, 21; m 47; c 6. HORTICULTURE. Educ: Purdue Univ, BS, 48; Cornell Univ, PhD, 51. Prof Exp: Asst veg crops, Cornell, 48-51; asst dir res, Grange League Fedn, Soil Bldg Serv, 51-56; sr fel, Boyce Thompson Inst, 56-57; dir, Res Farm, Union Carbide Corp, 57-59, mgr agr chem, Union Carbide Int Co, 59-67, int bus mgr, Union Carbide Corp, 67-68, MGR AGR CHEM, UNION CARBIDE CORP, 68- Mem: Weed Sci Soc Am; Entom Soc Am. Res: Chemical weed control; entomology; nematology; plant pathology; chemical formulations; management of agricultural chemicals. Mailing Add: Union Carbide Corp Agr Chem PO Box 1906 Salinas CA 93901

MARSHALL, FRANKLIN NICK, b Chicago, Ill, July 5, 33; m 55. PHARMACOLOGY. Educ: Univ Iowa, BS, 57, MS, 59, PhD(pharmacol), 61. Prof Exp: From pharmacologist to sr pharmacologist, Pitman-Moore Div, Dow Chem USA, 61-65, proj leader, 65-67, group leader, Dow Human Health Res Labs, 67-68, asst head dept pharmacol, Dow Human Health Res Labs, Ind, 68-72, HEAD DEPT PHARMACOL, DOW LEPETIT RES & DEVELOP LABS, 72-, ASSOC SCIENTIST, 73- Mem: AAAS; Am Soc Pharmacol & Exp Therapeut. Res: Autonomic neuromuscular and renal pharmacology; pharmacology of antibiotics and anesthetic agents; lipid metabolism in neoplasms; blood coagulation and fibrinolysis. Mailing Add: Dow Lepetit Res & Develop Labs 1701 Bldg Midland MI 48640

MARSHALL, FRED TAYLOR, b Chicago, Ill, Apr 2, 13; m 42; c 3. ORGANIC CHEMISTRY. Educ: Univ Ill, BS, 35. Prof Exp: Analyst, 35-38, chemist, Cent Res Dept, Ohio, 38-47, asst to res dir, 47-64, SR RES CHEMIST, MONSANTO CO, 64- Mem: Am Chem Soc. Res: Development of evaluation methods for plastics; impregnation of textile fibers; paper chemicals. Mailing Add: Monsanto Co 800 N Lindbergh Blvd St Louis MO 63166

MARSHALL, FREDERICK J, b Detroit, Mich, Aug 14, 20; m 46; c 7. ORGANIC CHEMISTRY. Educ: Univ Detroit, BS, 41, MS, 43; Iowa State Col, PhD(org chem), 48. Prof Exp: RES CHEMIST, ELI LILLY & CO, 48- Mem: Am Chem Soc; NY Acad Sci. Res: Pharmaceuticals; antibiotic structure; radioactive carbon synthesis. Mailing Add: Drug Metab Res Group Eli Lilly & Co Res Labs Indianapolis IN 46206

MARSHALL, FREDERICK JAMES, b Vancouver, BC, Feb 11, 25; m 48; c 4. HISTOLOGY. Educ: Univ Ore, DMD, 49; Univ Ill, MS, 59; Am Bd Endodont, dipl. Prof Exp: Assoc prof histol & endodont, Fac Dent, Univ Man, 59-65; assoc prof oper dent, Sch Dent, Univ Pittsburgh, 65-67; prof endodont & head dept, Col Dent, Ohio State Univ, 67-72; PROF ENDODONT & CHMN DEPT, DENT SCH, UNIV ORE HEALTH SCI CTR, 72- Concurrent Pos: Consult, Vet Admin Hosp, Portland, Ore; vpres, Am Asn Dent Schs. Mem: Fel Am Col Dentists; Am Dent Asn; Am Asn Endodont; Int Asn Dent Res; assoc Can Dent Asn. Res: Endodontic culturing techniques; root canal medications; computer-assisted instruction for the diagnosis of toothache; electron microscopy of dentin. Mailing Add: Dent Sch Dept of Endodont Univ of Ore Health Sci Ctr Portland OR 97201

MARSHALL, GARLAND ROSS, b San Angelo, Tex, Apr 16, 40; m 59; c 4. BIOCHEMISTRY. Educ: Calif Inst Technol, BS, 62; Rockefeller Univ, PhD(biochem), 66. Prof Exp: From instr to asst prof, 66-72, ASSOC PROF PHYSIOL & BIOPHYS, SCH MED, WASH UNIV, 72- Concurrent Pos: Fel, Oxford Univ, 66; res assoc, Comput Systs Lab, 68-; estab investr, Am Heart Asn, 70-, mem coun high blood pressure res; guest investr, Massey Univ, NZ, 75- Mem: Biophys Soc; Am Chem Soc; Am Heart Asn. Res: Solid phase peptide synthesis; conformation of peptides and small proteins; endocrinology. Mailing Add: Dept of Physiol & Biophys Wash Univ Sch of Med St Louis MO 63110

MARSHALL, GLORIA A, anthropology, see 12th edition

MARSHALL, GRAYSON WILLIAM, JR, b Baltimore, Md, Feb 12, 43; m 70. BIOMATERIALS, DENTAL RESEARCH. Educ: Va Polytech Inst & State Univ, BS, 65; Northwestern Univ, PhD(materials sci), 72. Prof Exp: Res assoc materials sci, Design & Develop Ctr, Northwestern Univ, 72-73; instr, 73-74, ASST PROF BIOMATERIALS, DENT SCH, NORTHWESTERN UNIV, CHICAGO, 74- Concurrent Pos: Nat Inst Dent Res fel, Dent Sch, Northwestern Univ, 72-73, spec dent res award, 75. Honors & Awards: Res Prize, Am Asn Dent Res, 74. Mem: Fel AAAS; Int Asn Dent Res; Electron Micros Soc Am; Am Inst Mining, Metall & Petroleum Eng; Am Soc Metals. Res: Use of metals, polymers and ceramics in dentistry and surgery; scanning electron microscopy of enamel, dentin and bone; corrosion resistance of new alloys, amalgams. Mailing Add: Dept of Biomaterials Northwestern Univ Dent Sch Chicago IL 60611

MARSHALL, HAROLD GENE, b Evansville, Ind, May 7, 28; m 53; c 2. PLANT BREEDING, PLANT GENETICS. Educ: Purdue Univ, BS, 52; Kans State Col, MS, 53; Univ Minn, PhD(plant genetics), 59. Prof Exp: Asst agron, Kans State Col, 52-53; asst, Univ Minn, 56-58; RES AGRONOMIST, OAT SECT, USDA, PA STATE UNIV, UNIVERSITY PARK, 59-, ADJ PROF PLANT BREEDING, 74- Mem: AAAS; Am Soc Agron. Res: Nature of winter hardiness of winter oats and the development of winter-hardy varieties. Mailing Add: Dept of Agron Pa State Univ Col of Agr University Park PA 16802

MARSHALL, HAROLD GEORGE, b Bedford, Ohio, May 17, 29; m 51; c 3. MARINE BIOLOGY. Educ: Baldwin-Wallace Col, BS, 51; Western Reserve Univ, MS, 53, PhD(biol), 62. Prof Exp: Biologist, Nat Dairy Labs, 51-52; instr, Cleveland City Schs, Ohio, 52-58, chmn sci dept, Bedford, 58-62; instr, Western Reserve Univ, 62-63; from asst prof to assoc prof, 63-69, NASA res grant, 70, PROF BIOL & CHMN DEPT, OLD DOM UNIV, 69- Concurrent Pos: NSF res grants, 64-74; environ consult, 72-75; NASA res grants, 74-75. Mem: AAAS; Am Soc Limnol & Oceanog; Phycol Soc Am; Int Phycol Soc; NY Acad Sci. Res: Spatial distribution and ecology of coccolithophores and other phytoplankton off the southeast coast of the United States and Caribbean Sea. Mailing Add: Dept of Biol Old Dom Univ Norfolk VA 23508

MARSHALL, HARRY BORDEN, b Listowel, Ont, July 15, 09; m 36; c 3. CHEMISTRY. Educ: Univ BC, BA, 29, MA, 31; McGill Univ, PhD(chem), 34. Prof Exp: Demonstr chem, Univ BC, 29-31; res chemist, Dow Chem Co, Mich, 35-37; res fel, Ont Res Found, 37-46; asst dir dept chem, 46-57, dir, 58-63; assoc res dir, 62-70, RES DIR, DOMTAR LTD, 70- Concurrent Pos: Howard Smith Paper Mills fel, 35-36; mem res comt, Can Chem Producers Asn & Nat House Builders Asn, 65. Honors & Awards: Montreal Medal, Chem Inst Can, 61. Mem: Tech Asn Pulp & Paper Indust; Can Pulp & Paper Asn; Chem Inst Can. Res: Research administration in pulp and paper, construction materials and chemicals. Mailing Add: 3 Sunset Ave Senneville PQ Can

MARSHALL, HENRY PETER, b Altoona, Pa, May 12, 24; m 51; c 3. PHYSICAL ORGANIC CHEMISTRY. Educ: Pa State Univ, BS, 47; Univ Calif, Los Angeles, PhD(chem), 52. Prof Exp: Fel chem, Fla State Univ, 52-53; res chemist, Celanese Corp, 53-56 & Stanford Res Inst, 56-58; SR STAFF SCIENTIST MAT, LOCKHEED MISSILES & SPACE CO, INC, 58- Mem: Am Chem Soc; Am Inst Physics; Sigma Xi. Res: Study of chemical structural aging effects of non-metallics; identification and kinetic measurements of chemical processes occurring in non-metallics, principally polymers, in all types of environments. Mailing Add: Palo Alto CA

MARSHALL, J HOWARD, III, b San Francisco, Calif, Feb 6, 36. PHYSICS, ELECTRONICS. Educ: Calif Inst Technol, BS, 57, PhD(high energy physics), 65. Prof Exp: Sr res engr, Jet Propulsion Lab, Calif Inst Technol, 62-65; vpres prod, 65-66, chmn bd & vpres advan planning, 66-71, CHMN BD & VPRES TECHNOL, ANALOG TECHNOL CORP, 71- Mem: AAAS; sr mem Inst Elec & Electronics Engrs; NY Acad Sci; Am Inst Aeronaut & Astronaut. Res: Electronic and system design of instrumentation for spaceborne and earthbound applications involving nuclear physics, high-energy physics, mass spectroscopy, gas chromatography and infrared and ultraviolet radiation; corporate management. Mailing Add: Analog Technol Corp 3410 Foothill Blvd Pasadena CA 91107

MARSHALL, JACK STANTON, b Topeka, Kans, Apr 28, 29; m 53; c 7. LIMNOLOGY. Educ: Univ Colo, BA, 54, MA, 56; Univ Mich, PhD(zool), 61. Prof Exp: Asst res limnologist, Great Lakes Res Div, Inst Sci & Technol, Univ Mich, 61-62, assoc res limnologist, 62-63; res biologist, Savannah River Lab, E I du Pont de Nemours & Co, SC, 63-69; aquatic ecologist, AEC, Md, 69-71; ECOLOGIST, ARGONNE NAT LAB, 71- Mem: Am Soc Limnol & Oceanog; Ecol Soc Am. Res: Limnology trace elements; biogeochemistry; population dynamics; radiation ecology. Mailing Add: Argonne Nat Lab 9700 S Cass Ave Argonne IL 60439

MARSHALL, JAMES ARTHUR, b Oshkosh, Wis, Aug 7, 35; m 57. ORGANIC CHEMISTRY. Educ: Univ Wis, BS, 57; Univ Mich, PhD(chem), 60. Prof Exp: USPHS fel org chem, Stanford Univ, 60-62; from asst prof to assoc prof, 62-68, PROF CHEM, NORTHWESTERN UNIV, EVANSTON, 68- Concurrent Pos: Sloan Found fel, 66-; Seidel Wooley lectr, Ga Inst Technol, 70, Am-Swiss Found lectr, 71-72; consult, Ortho Res Found & Givaudan. Mem: Am Chem Soc; The Chem Soc. Res: Synthetic organic chemistry related to natural products; stereochemistry and organic reaction mechanisms. Mailing Add: Dept of Chem Northwestern Univ Evanston IL 60201

MARSHALL, JAMES DALE, b Clarendon, Tex, Dec 18, 33; m 58; c 1. ZOOLOGY, ENTOMOLOGY. Educ: Tex Tech Col, BS, 60; Cornell Univ, PhD(entom), 64. Prof Exp: Asst prof, 64-68, ASSOC PROF BIOL, COL OF IDAHO, 68-, CHMN DEPT, 71- Mem: Entom Soc Am. Res: Ecology of insects; systematics of Coleoptera; evolutionary theory. Mailing Add: Dept of Biol Col of Idaho Caldwell ID 83605

MARSHALL, JAMES JOHN, b Edinburgh, Scotland, July 7, 43; m 66; c 2. BIOCHEMISTRY, MEDICAL RESEARCH. Educ: Univ Edinburgh, BSc, 65; Heriot-Watt Univ, PhD(appl biochem), 69. Prof Exp: Res assoc biochem, Sch Med, Univ Miami, 69-71; fel, Royal Holloway Col, Univ London, 71-72; asst prof biochem, 73-75, ASST PROF BIOCHEM & MED, SCH MED, UNIV MIAMI, 75-; DIR LAB BIOCHEM RES, HOWARD HUGHES MED INST, 73- Mem: Biochem Soc; Am Soc Microbiol; Am Chem Soc; Am Soc Biol Chemists; Am Asn Cereal Chemists. Res: Structure and mechanism of action of glycoside hydrolases; structure, function and metabolism of polysaccharides; glycoproteins, especially structure, function and synthesis; naturally occurring enzyme inhibitors. Mailing Add: Howard Hughes Med Inst PO Box 520605 Biscayne Annex Miami FL 33152

MARSHALL, JAMES LAWRENCE, b Denton, Tex, May 19, 40; m 63; c 1. ORGANIC CHEMISTRY. Educ: Ind Univ, Bloomington, BS, 62; Ohio State Univ, PhD(org chem), 66. Prof Exp: NIH fel org chem, Univ Colo, 66-67; asst prof chem, 67-71, ASSOC PROF CHEM, NTEX STATE UNIV, 71- Mem: Am Chem Soc. Res: Proton and carbon magnetic resonance studies of carbon-13 labeled compounds; conformational analysis; small polycyclic compounds; ammonia-metal reductions of aromatic compounds; nuclear magnetic resonance studies of small polycyclic compounds. Mailing Add: Dept of Chem NTex State Univ Denton TX 76203

MARSHALL, JAMES R, b Lawrence, Mass, Dec 25, 33; m 58; c 2. PHARMACEUTICS, HEALTH SCIENCES. Educ: Mass Col Pharm, BS, 59, MS, 61, PhD, 64. Prof Exp: Res chemist, Gen Foods Corp, 57-59; res asst, Mass Col Pharm, 60; teacher, Mary Brooks Sch Nursing, 60-62; head pharmaceut prod develop, Astra Pharm Prod, 63-66; head skin prod develop group, Colgate-Palmolive Co, 66-68; mgr pharmaceut prod develop, Pharmaceut Div, Pennwalt Corp, NY, 68-74; DIR RES & DEVELOP, JOHNSON & JOHNSON BABY PROD CO, 74- Concurrent Pos: Retail pharmacist, 59-63. Mem: Am Soc Testing & Mat; Am Chem Soc; Am Pharmaceut Asn; Soc Cosmetic Chemists; Int Pharmaceut Fedn. Mailing Add: Johnson & Johnson Baby Prod Co US Rte 1 North Brunswick NJ 08903

MARSHALL, JAMES TILDEN, JR, b Canadian, Tex, July 30, 45; m 68; c 2. FOOD SCIENCE, DAIRY SCIENCE. Educ: Tex Tech Univ, BS, 68, MS, 69; Mich State Univ, PhD(food sci), 74. Prof Exp: Res asst dairy sci, Tex Tech Univ, 68-69 & food sci, Mich State Univ, 69-73; ASST PROF DAIRY & FOOD SCI, MISS STATE UNIV, 74- Mem: Am Dairy Sci Asn; Inst Food Technologists; Am Oil Chemists Soc. Res: Chemical and physical properties of dairy and food products; formulation and processing of marketable products from dairy and other food by-products. Mailing Add: Dept of Dairy Sci Miss State Univ Mississippi State MS 39762

MARSHALL, JEAN MCELROY, b Chambersburg, Pa, Dec 31, 22. PHYSIOLOGY. Educ: Wilson Col, AB, 44; Mt Holyoke Col, MA, 46; Univ Rochester, PhD, 51. Prof Exp: Instr physiol, Mt Holyoke Col, 46-47; from instr to asst prof, Sch Med, Johns Hopkins Univ, 51-60; asst prof, Harvard Med Sch, 60-66; assoc prof, 66-69, PROF BIOL & MED SCI, BROWN UNIV, 69- Concurrent Pos: Res fel pharmacol, Oxford Univ, 54-55; mem physiol study sect, NIH, 67-71 & eng in biol & med training comt, 71-73; mem physiol testing comt, Nat Bd Med Examrs. Mem: Am Physiol Soc; Soc Gen Physiol; Am Soc Pharmacol & Exp Therapeut; Soc Reprod Biol. Res: Electrical and mechanical properties of cardiac and smooth muscle. Mailing Add: Div of Biol & Med Sci Brown Univ Providence RI 02912

MARSHALL, JOE TRUESDELL, JR, b Paris, France, Feb 15, 18; US citizen; m 42; c 3. ORNITHOLOGY. Educ: Univ Calif, AB, 39, PhD(zool), 48. Prof Exp: From asst prof to prof zool & cur birds, Univ Ariz, 49-63; MED BIOLOGIST, WALTER REED ARMY INST RES, 64- Concurrent Pos: Mem Calif exped to El Salvadore, 41-42; Guggenheim Mem fel, 51; mem Pac Sci Bd, Pac Island rat ecology & coral atoll ecol, 55-56; NSF grants, 57-62. Mem: Cooper Ornith Soc; fel Am Ornith Union.

Res: Ecology; taxonomy; behavior; ecology of arboviruses. Mailing Add: SEATO Med Res Lab APO San Francisco CA 96346

MARSHALL, JOHN CLIFFORD, b Whitewater, Wis, Jan 23, 35; m 57; c 2. ANALYTICAL CHEMISTRY. Educ: Luther Col, Iowa, BA, 56; State Univ Iowa, MS, 58, PhD(chem), 60. Prof Exp: Instr chem, State Univ Iowa, 60; fel, Univ Minn, 60-61; from asst prof to assoc prof, 61-74, PROF CHEM, ST OLAF COL, 74- Concurrent Pos: NSF res grant, St Olaf Col, 62-64, Petrol Res Fund grant, 64-67; res assoc, Argonne Nat Lab, 68-69; NSF fac res fel sci, Univ NC, 69-70. Mem: Am Chem Soc. Res: Flame spectroscopy and computer applications in chemistry. Mailing Add: Dept of Chem St Olaf Col Northfield MN 55057

MARSHALL, JOHN DEAN, JR, b Pittsburgh, Pa, Sept 22, 24; m 50; c 5. MICROBIOLOGY. Educ: Univ Pittsburgh, BS, 49, MS, 50; Univ Md, College Park, PhD(microbiol), 62. Prof Exp: Med Serv Corps, US Army, 50-, med bacteriologist, McGee Hosp, Pittsburgh, Pa, 50, chief bact, Serol & Blood Bank Sect, Madigan Army Hosp, Tacoma, Wash, 50-53, chief anaerobic bact, 406th Med Gen Lab, Tokyo, Japan, 53-56, asst chief, Dept Bact & Immunol, Armed Forces Inst Path, Washington, DC, 56-59, chief, Microbiol Div, Army Med Unit, Ft Detrick, Md, 62-65, chief, Infectious Dis Lab, Army Med Res Team, Walter Reed Army Inst Res, Washington, DC, 65-66, chief, Microbiol Div, Army Med Res Inst Infectious Dis, 66-72, comndr, US Army Res & Develop Gen Purpose (Far E), 72-75, CHIEF CUTANEOUS INFECTION DIV, LETTERMAN ARMY INST RES, 75- Concurrent Pos: Lectr, Howard Univ, 63-71; assoc mem comn on immunization, Armed Forces Epidemiol Bd, Washington, DC, 66-71. Mem: NY Acad Sci; Am Soc Microbiol; Soc Exp Biol & Med; Infectious Dis Soc Am; Wildlife Dis Asn. Res: Immunology and epidemiology of infectious diseases. Mailing Add: Letterman Army Inst of Res Presidio of San Francisco CA 94129

MARSHALL, JOHN HART, b Chicago, Ill, Feb 14, 25; m 55; c 2. PHYSICS. Educ: Harvard Col, AB, 45; Mass Inst Technol, PhD(physics), 52. Prof Exp: Mem res staff, Radioactivity Ctr, Mass Inst Technol, 52-55; assoc physicist, 55-67, SR BIOPHYSICIST, RADIOL PHYSICS DIV, ARGONNE NAT LAB, 67- Mem: AAAS; Am Phys Soc; Radiation Res Soc; Orthop Res Soc. Res: Theory of alkaline earth metabolism; long term effects of radioactivity; microscopic metabolism of calcium in bone; theory of the induction of bone cancer by alpha radiation. Mailing Add: Radiol & Environ Res Div Argonne Nat Lab 9700 S Cass Argonne IL 60439

MARSHALL, JOHN ROMNEY, b Los Angeles, Calif, Apr 27, 33; m 56; c 4. OBSTETRICS & GYNECOLOGY. Educ: Univ Pa, MD, 58. Prof Exp: Intern, Los Angeles Gen Hosp, 58-59; instr pharmacol, Sch Med, Univ Pa, 59-60; resident obstet & gynec, George Washington Univ Hosp, 60-63, asst clin prof, 63-69; PROF OBSTET & GYNEC & VCHMN DEPT, SCH MED, UNIV CALIF, LOS ANGELES, 70-; CHMN DEPT, HARBOR GEN HOSP, 70- Concurrent Pos: Resident, DC Gen Hosp, 60-63; sr investr, Nat Cancer Inst, 63-69; consult, Long Beach Naval Hosp, Calif, 71-; mem hon fac staff, Mem Hosp Med Ctr, Long Beach, Calif, 71- Mem: Fel Am Col Obstet & Gynec; Soc Study Reprod; Endocrine Soc; Soc Gynec Invest; Am Fertility Soc. Res: Clinical pharmacology of reproductive biology. Mailing Add: Harbor Gen Hosp 1000 W Carson Torrance CA 90509

MARSHALL, JOHN STEWART, b Welland, Ont, July 18, 11; m 40; c 2. PHYSICS, METEOROLOGY. Educ: Queen's Univ, Ont, BA, 31, MA, 33; Cambridge Univ, PhD(physics), 40. Prof Exp: From jr res physicist to asst res physicist, Ottawa Labs, Nat Res Coun Can, 39-43; oper res scientist, Can Army Oper Res Group, 44-45; from asst prof to prof physics, 45-60, chmn dept meteorol, 60-64, MACDONALD PROF PHYSICS & PROF METEOROL, MACDONALD PHYSICS LAB, McGILL UNIV, 60- Mem: Am Phys Soc; Can Meteorol Soc; fel Am Meteorol Soc; Can Asn Physicists (pres, 50); fel Royal Soc Can. Res: Radar weather; cloud physics. Mailing Add: McGill Radar Weather Observ Macdonald Col PO Box 241 Montreal PQ Can

MARSHALL, JOHN U, b Bearsden, Scotland, Sept 22, 38; Can citizen; m 63; c 1. GEOGRAPHY. Educ: Univ Toronto, BA, 61, PhD(urban geog), 68. Univ Minn, MA, 65. Prof Exp: Asst prof geog, Brock Univ, 66-68; asst prof, 68-71, ASSOC PROF GEOG, YORK UNIV, 71- Mem: Can Asn Geog; Asn Am Geogr; Regional Sci Asn. Res: Urban and population geography; statistical methods for spatially arrayed data. Mailing Add: Dept of Geog York Univ Downsview ON Can

MARSHALL, JOSEPH ANDREW, b New Brunswick, NJ, Mar 25, 37; m 58; c 2. ANIMAL BEHAVIOR. Educ: Univ Md, BS, 60, PhD(zool), 66. Prof Exp: USPHS fel, 66-68; asst prof, 68-73, ASSOC PROF BIOL, WVA UNIV, 73- Mem: Animal Behav Soc; Am Soc Ichthyol & Herpet; assoc mem Ecol Soc Am. Res: Acoustical behavior of fishes. Mailing Add: Dept of Biol WVa Univ Morgantown WV 26506

MARSHALL, KENNETH CHENERY, b Maplewood, Mo, Sept 24, 17; m 41; c 3. ORTHODONTICS. Educ: Wash Univ, DDS, 40; Univ Mich, MS, 47. Prof Exp: From instr to assoc prof, 47-62, CLIN PROF ORTHOD, ST LOUIS UNIV, 62-, DIR DEPT, 47- Mem: Am Dent Asn; Am Asn Orthod. Res: Growth and development of the human face. Mailing Add: Dept of Orthod St Louis Univ St Louis MO 63104

MARSHALL, KNEALE THOMAS, b Filey, Eng, Feb 13, 36; US citizen; m 64. OPERATIONS RESEARCH. Educ: Univ London, BSc, 58; Univ Calif, Berkeley, MS, 64, PhD(opers res), 66. Prof Exp: Metallurgist, Beaverlodge Oper, Eldorado Mining & Refining Ltd, 58-60, chief metallurgist, 60-62; mem tech staff, Bell Tel Labs, NJ, 66-68; from asst prof to assoc prof, 68-75, PROF OPERS RES, NAVAL POSTGRAD SCH, 74- Concurrent Pos: Consult, Res Proj in Higher Educ, Univ Calif, 68-70; assoc ed, Opers Res, 70- & Soc Indust & Appl Math J Appl Math, 71- Mem: Opers Res Soc Am; Soc Indust & Appl Math; Inst Mgt Sci. Res: Stochastic models of congested systems; theory of manpower and budget planning. Mailing Add: Dept of Opers Res Code 55 MT Naval Postgrad Sch Monterey CA 93940

MARSHALL, LAURISTON CALVERT, b Canton, China, June 27, 02; m 49; c 3. PHYSICS. Educ: Park Col, AB, 23; Univ Calif, PhD(physics), 29. Prof Exp: Collabr, Bur Plant Indust, USDA, 28-31, physicist, 31-37; assoc prof elec eng, Univ Calif, Berkeley, 37-45, prof, 45-54, dir microwave power lab, 46-52, staff mem, Lawrence Radiation Lab, 46-54; dir res, Link-Belt Co, 52-59; assoc tech dir microwave power lab, Varo, Inc, Tex, 59-61; chief off sci personnel & dir mat res lab, Grad Res Ctr Southwest, 61-67; prof physics, 67-75, EMER PROF PHYSICS, SOUTHERN ILL UNIV, CARBONDALE, 75- Concurrent Pos: Nat Res Coun fel, Princeton Univ, 29-31; div head, Radiation Lab, Mass Inst Technol, 41-43, dir Brit Br, Radiation Lab, 43-44; Guggenheim fel, 50-51; consult, Lawrence Radiation Lab, Univ Calif, 54-56. Honors & Awards: Cert Merit & Presidential Citation, 48. Mem: Fel AAAS; fel Inst Elec & Electronics Engrs; fel Am Phys Soc. Res: Microwaves high power; radio physics; particle accelerators; high energy, electron and solid state physics; gaseous conduction; biophysics; environmental control plant reproduction and growth; greenhouse air conditioning, evolution of atmospheres; Earth and planets; origins of life. Mailing Add: Dept of Physics Southern Ill Univ Carbondale IL 62901

MARSHALL, LAWRENCE MARCELLUS, b Pittsburgh, Pa, Mar 31, 10; m 39; c 3. BIOCHEMISTRY. Educ: Duquesne Univ, BS, 32, MS, 40; Wayne State Univ, PhD(biochem), 49. Prof Exp: Instr biol, Clark Col, 37-39; asst prof gen & org chem, Agr Mech & Normal Col, Ark, 40-44; asst physiol chem, Col Med, Wayne State Univ, 45-46; from asst prof to assoc prof biochem, 49-58, PROF BIOCHEM, MED SCH, HOWARD UNIV, 58- Concurrent Pos: Org chemist, Taft Sanit Eng Ctr, USPHS, 56-57. Honors & Awards: Lederle Med Fac Award, 54. Mem: Am Chem Soc; Am Soc Biol Chemists. Res: Physico-chemical approaches to the separation and identification of organic compounds in mixtures; study of the tricarboxylic acid cycle in physiological systems by isolation and identification of intermediates of the cycle; isotope effects during countercurrent distribution. Mailing Add: Dept of Biochem Howard Univ Med Sch Washington DC 20001

MARSHALL, LOUISE HANSON, b Perrysburg, Ohio, Oct 2, 08; m 34; c 2. NEUROSCIENCES, MAMMALIAN PHYSIOLOGY. Educ: Vassar Col, MA, 32; Univ Chicago, PhD(physiol), 35. Prof Exp: Asst physiol, Vassar Col, 30-32, instr, 36-37; asst, Univ Chicago, 34-35; physiologist, NIH, 43-65; prof assoc, Div Med Sci, Nat Acad Sci-Nat Res Coun, 65-75; ADMIN ANALYST, BRAIN RES INST, UNIV CALIF, 75- Concurrent Pos: Managing ed, Exp Neurol, 75- Mem: Am Physiol Soc; Soc Neurosci. Res: Circulatory and renal response to plasma expanders; peripheral circulation; neuroscience administration. Mailing Add: Brain Res Inst Univ of Calif Los Angeles CA 90024

MARSHALL, LYNNOR BEVERLY, b Melbourne, Australia, Mar 11, 43; m 65; c 2. BIOCHEMISTRY. Educ: Univ Melbourne, BSc, 63, BEd, 67; Monash Univ, Australia, PhD(biochem), 72. Prof Exp: Sr sci teacher chem, Victorian Educ Dept, 65; res asst biochem, Monash Univ, 66, sr teaching fel, 67-70; fel biochem, Sch Med, Stanford Univ, 71; sr res chemist, 72-74, PROD MGR PEPTIDES, BECKMAN INSTRUMENTS, INC, 75- Mem: Am Chem Soc; AAAS; Australian Biochem Soc. Res: Chemistry and biology of biologically active peptides. Mailing Add: Beckman Instruments Inc 1117 California Ave Palo Alto CA 94304

MARSHALL, MARYAN LORRAINE, b New Haven, Conn, Jan 18, 40. PHYSICAL CHEMISTRY. Educ: Conn Col, BA, 60; Yale Univ, PhD(phys chem), 65. Prof Exp: Instr chem, Randolph-Macon Woman's Col, 64-66, asst prof, 66-72; assoc prof, 72-75, PROF CHEM, CENT VA COMMUNITY COL, 75- Mem: AAAS; Am Chem Soc; Nat Sci Teachers' Asn. Mailing Add: Dept of Chem Cent Va Community Col Wards Rd S Lynchburg VA 24502

MARSHALL, MAUD ALICE, b East Lansing, Mich, July 15, 07. ORGANIC CHEMISTRY. Educ: Radcliffe Col, AB, 28; Oxford Univ, PhD(org chem), 33. Prof Exp: Res chemist, Boston City Hosp, 28-29; chemist, Robert Brigham Hosp, 29-31; asst, Exp Sta, Mass State Col, 33-34; from instr to prof, 34-75, EMER PROF CHEM, WHEATON COL, MASS, 75- Res: Stereochemistry of narcotine; antimalarials. Mailing Add: 13 W Main St Norton MA 02766

MARSHALL, NELSON, b Yonkers, NY, Dec 16, 14; m 40; c 4. BIOLOGICAL OCEANOGRAPHY. Educ: Rollins Col, BS, 37; Ohio State Univ, MS, 38; Univ Fla, PhD(biol), 41. Prof Exp: Asst, Univ Fla, 39-41; from instr to asst prof zool, Univ Conn, 41-45; asst prof & fisheries biologist, Marine Lab, Univ Miami, 45-46; assoc prof, Univ NC, 46-47; prof biol, Col William & Mary, 47-51, dean, 49-51, dir, Va Fisheries Lab, 47-50; assoc dir oceanog inst, Fla State Univ, 52-54; vis investr, Bingham Oceanog Lab, Yale Univ, 54-55; dean col lib arts, Alfred Univ, 55-59; dir int ctr marine resource develop, 72-75, PROF OCEANOG, GRAD SCH OCEANOG, UNIV RI, 59-, PROF MARINE AFFAIRS, 75- Concurrent Pos: Hon mem bd trustees, Rollins Col. Mem: Fel AAAS; Am Soc Limnol & Oceanog; Ecol Soc Am; hon mem Atlantic Estuarine Res Soc; Nat Shellfisheries Asn. Res: Estuarine and coral reef ecology; higher education for marine resource development in developing countries. Mailing Add: Grad Sch of Oceanog Univ of RI Kingston RI 02881

MARSHALL, NORMAN BARRY, b Brooklyn, NY, Oct 3, 26; m 52; c 3. PHYSIOLOGY. Educ: Long Island Univ, BS, 49; Clark Univ, MA, 52; Harvard Univ, PhD(med sci), 56. Prof Exp: Instr physiol, Med Ctr, Duke Univ, 56-58, assoc, 58-59; res assoc nutrit, 59-61, head lipid metab sect, 61-67, mgr dept biochem, 67-68, MGR HYPERSENSITIVITY DIS RES, UPJOHN CO, 68- Mem: Am Physiol Soc. Res: Regulation of food intake; endocrine control of metabolism. Mailing Add: Upjohn Co 301 Henrietta St Kalamazoo MI 49001

MARSHALL, NORTON LITTLE, b Washington, DC, Dec 30, 27. BOTANY. Educ: Pa State Univ, BS, 49; Univ Md, MS, 52, PhD(bot), 55. Prof Exp: Asst pathologist, Trop Res Dept, United Fruit Co, Honduras, 49-50; asst, Univ Md, 50-54, instr bot, 55-56; pathologist, Res Dept, Firestone Plantations Co, Liberia, 56-58; from asst prof to assoc prof, 58-66, PROF BOT, AUBURN UNIV, 66- Mem: Bot Soc Am; Am Phytopath Soc. Res: Plant pathology; microbiology. Mailing Add: Dept of Bot Auburn Univ Auburn AL 36830

MARSHALL, PHILIP RICHARD, b Decatur, Ind, Nov 13, 26; m; c 4. PHYSICAL CHEMISTRY. Educ: Earlham Col, AB, 49; Purdue Univ, MS, 51, PhD, 54. Prof Exp: Asst, Purdue Univ, 49-51; res engr, Battelle Mem Inst, 51-52; asst, Purdue Univ, 52-53; from instr to asst prof, Albion Col, 53-58; from asst prof to assoc prof chem, Cornell Col, 58-65; prof & dean, Lycoming Col, 65-69; asst prog dir, NSF, 69; dean acad affairs, 69-70, VPRES ACAD AFFAIRS, EASTERN WASH STATE COL, 70- Res: Metal-ammonia solutions; kinetics of heterogeneous reactions; properties of mixed solvents. Mailing Add: 1011 Gary St Cheney WA 99004

MARSHALL, RICHARD, b Monett, Mo, Aug 9, 22; m 51; c 2. BIOCHEMISTRY. Educ: Okla State Univ, BS, 48, MS, 50; Univ Wis, PhD, 55. Prof Exp: Biochemist, Biol Div, Hanford Works Res Dept, Gen Elec Co, 50-51; assoc chemist, Res Dept, Corn Prod Co, Ill, 55-57; sr biochemist, Grain Processing Corp, Iowa, 57-59 & Cent Res Dept, Minn Mining & Mfg Co, 59-60; res dir, Producers Creamery Co, 60-68; CLIN BIOCHEMIST, ST JOHN'S HOSP, 68- Mem: Fel Am Inst Chemists; Am Asn Clin Chemists; Am Chem Soc. Res: Reducing of nitrates, oxidation of molecular tritium, isomerization of hexoses and fumaric acid production by microorganisms; biosynthesis of citrulline; production and application of mold amylases; insect metabolism; submerged mushroom fermentation; removal of radionuclides from milk. Mailing Add: St John's Hosp Lab 1235 E Cherokee St Springfield MO 65804

MARSHALL, RICHARD ALLEN, b Madisonville, Tex, Aug 25, 35; m 59; c 2. POLYMER CHEMISTRY. Educ: Rice Univ, BA, 57; Ohio State Univ, PhD(org chem), 62. Prof Exp: Res chemist, Baytown Res & Develop Div, Esso Res & Eng Co, 62-64, sr res chemist, 64-68; mem staff, Chem Div, Vulcan Mat Co, Kans, 68-74; SR RES CHEMIST, GOODYEAR RES, 74- Mem: Am Chem Soc. Res: Charge transfer complexes; exploratory polymers and polymerization processes; process and exploratory research in chlorinated organics; polyvinyl chloride polymerization. Mailing Add: Goodyear Res 142 Goodyear Blvd Akron OH 44316

MARSHALL, RICHARD BLAIR, b Melrose, Mass, July 25, 28; m 53; c 5. PATHOLOGY. Educ: Boston Univ, BA, 49, MD, 55. Prof Exp: Intern, Detroit Receiving Hosp, 55-56; resident path, Henry Ford Hosp, Detroit, 56-60; from asst prof to prof path, Univ Tex Med Br Galveston, 64-75; PROF PATH, BOWMAN GRAY SCH MED, 75-, DIR ANAT PATH, 75- Mem: Am Soc Clin Path; Int Acad Path. Res: Cytochemical and ultrastructural studies of human endocrine pathology. Mailing Add: Dept of Path Bowman Gray Sch of Med Winston-Salem NC 27103

MARSHALL, ROBERT HERMAN, b Decatur, Ill, June 26, 25. INORGANIC CHEMISTRY. Educ: Ill State Normal Univ, 47, MS, 50; Univ Ill, PhD(inorg chem), 54. Prof Exp: Res chemist, Ethyl Corp, 54-58; assoc prof chem, La Polytech Inst, 58-60; assoc prof, 60-67, actg chmn dept, 70-71, PROF CHEM, MEMPHIS STATE UNIV, 67- Mem: Am Chem Soc. Res: Organometallic compounds. Mailing Add: Dept of Chem Memphis State Univ Memphis TN 38152

MARSHALL, ROBERT JAMES, b Ballymena, Northern Ireland, May 5, 26; m 57; c 3. CARDIOLOGY, CARDIOVASCULAR PHYSIOLOGY. Educ: Queen's Univ Belfast, MB, BCh, 48, MD, 52; FRCP(I), 67; FRCP, 74. Prof Exp: Asst lectr physiol & path, Queen's Univ Belfast, 50-51, tutor med, 51-57; res assoc physiol, Mayo Grad Sch Med, Univ Minn, 58-61; from assoc prof to prof med, 61-68, PROF PHYSIOL, SCH MED, W VA UNIV, 68- Concurrent Pos: Res fel med, Alfred Hosp, Melbourne, Australia, 57-58; fel coun clin cardiol, Am Heart Asn, 59-, mem adv bd, coun circulation, 62-; consult cardiovasc med, Radcliffe Infirmary, Oxford, Eng, 74- Mem: Am Physiol Soc; Am Soc Clin Invest; fel Am Col Physicians; fel Am Col Cardiol; Am Fedn Clin Res. Res: Physiology of peripheral circulation; control of cardiac output; regulation of pulmonary circulation; cardiovascular pharmacology. Mailing Add: Dept of Med WVa Univ Sch of Med Morgantown WV 26506

MARSHALL, ROBERT T, b Halltown, Mo, July 27, 32; m 53; c 4. MICROBIOLOGY, FOOD SCIENCE. Educ: Univ Mo, BS, 54, MS, 58, PhD(food microbiol), 60. Prof Exp: From instr to asst prof dairy microbiol, 60-65, assoc prof dairy microbiol & mfrs, 65-70, PROF FOOD SCI & NUTRIT, UNIV MO-COLUMBIA, 70- Mem: Am Soc Microbiol; Int Asn Milk, Food & Environ Sanit; Am Dairy Sci Asn; Inst Food Technol. Res: Biology of psychrotrophic bacteria and their enzymes. Mailing Add: 203 Eckles Hall Univ of Mo Columbia MO 65201

MARSHALL, ROSEMARIE, b Medford, Ore, Jan 28, 43. BACTERIOLOGY, BIOSTATISTICS. Educ: Univ Wash, BS, 64; Iowa State Univ, MS, 66, PhD(bact), 68. Prof Exp: NIH fel, Retina Found, Harvard Med Sch, 68-70; head dept bact, Grays Harbor Hosp, 70-71; ASST PROF BIOL, GA SOUTHERN COL, 71-, MEM INST ANTHROPODOLOGY & PARISTOL, 74- Mem: AAAS; Am Soc Microbiol. Res: Clinical bacteriology; turnover of macromolecules in vivo in differentiating systems; systems analysis of differentiating systems. Mailing Add: Dept of Biol Ga Southern Col Statesboro GA 30458

MARSHALL, RUSH PORTER, plant pathology, deceased

MARSHALL, SAMSON A, b Chicago, Ill, Oct 25, 24. SOLID STATE PHYSICS. Educ: Ill Inst Technol, BS, 50; Univ Mich, MS, 51; Cath Univ, PhD(physics), 56. Prof Exp: Physicist, Nat Bur Standards, 51-53, Naval Ord Lab, 53-56 & Armour Res Found, 56-65; PHYSICIST, ARGONNE NAT LAB, 65- Mem: AAAS; Am Phys Soc. Res: Microwave and radiofrequency spectroscopy of solids and gases. Mailing Add: SSS Div Argonne Nat Lab Argonne IL 60439

MARSHALL, SAMUEL WILSON, b Dallas, Tex, Sept 8, 34; m 56; c 3. PHYSICS. Educ: Va Mil Inst, BS, 55; Tulane Univ, MS, 63, PhD(physics), 65. Prof Exp: Jr res engr, Prod Res Div, Humble Oil & Refining Co, 55-56, 59-60; from asst prof to assoc prof physics, Colo State Univ, 65-75; MEM STAFF, NAVAL RES LAB, 75- Mem: Am Phys Soc. Res: Mössbauer effect; acoustic scattering. Mailing Add: Code 8160 Naval Res Lab Washington DC 20375

MARSHALL, SIDNEY PAUL, b Greenville, Fla, Oct 2, 16; m 40; c 1. DAIRY HUSBANDRY. Educ: Univ Fla, BSA, 38; Okla Agr & Mech Col, MS, 39; Univ Minn, PhD(dairy prod), 45. Prof Exp: Asst prof animal nutrit, Univ Fla, 42-45; assoc prof dairy sci, Clemson Col, 45-47; assoc prof, 47-67, PROF DAIRY SCI, UNIV FLA, 67-, DAIRY NUTRITIONIST, AGR EXP STA, 58- Mem: Am Soc Animal Sci; Am Dairy Sci Asn. Res: Nutrition of cattle; investigations on the chemical composition, biological value and feeding value of shark meal; postpartum development of bovine stomach compartments; evaluation of pastures for dairy cattle nutrition; antibiotics in calf nutrition. Mailing Add: Dept of Dairy Sci Univ of Fla Gainesville FL 32603

MARSHALL, THEODORE, b Chicago, Ill, Dec 31, 27; m 54; c 3. PLASMA PHYSICS. Educ: Ill Inst Technol, BS, 51; Cath Univ Am, PhD(physics), 62. Prof Exp: Scientist, US Naval Ord Lab, 55-62; sr consult scientist, Res & Develop Div, Avco Corp, 62-64; mgr exp physics, Parametrics Inc, Mass, 65; assoc prof elec eng, Univ RI, 65-70; PROF PHYSICS & CHMN DEPT, SUFFOLK UNIV, 70- Mem: Am Phys Soc. Res: Propagation of electromagnetic waves in ionized gas; physics of fluids; high temperature properties of gases. Mailing Add: Dept of Physics Suffolk Univ Boston MA 02114

MARSHALL, THOMAS BALL, b Upland, Pa, May 30, 22; m 48; c 5. POLYMER CHEMISTRY. Educ: Princeton Univ, AB, 42; Cornell Univ, PhD(org chem), 50. Prof Exp: Chemist, Mil Explosives Dept, Wabash River Ord Works, Ind, 42-44, res chemist, Nylon Res Div, Exp Sta, 50-53, res supvr, Dacron Res Div, 53-57, tech rep, Textile Fibers Dept, Indust Merchandising Div, 57-59, merchandising supvr, 59-60, mkt asst, 60-61, lycra prod asst, Prod Div, 61-63, venture planning specialist, Develop Dept, 63-70, DEVELOP MGR, FILM DEPT, E I DU PONT DE NEMOURS & CO, INC, 70- Mem: Am Chem Soc. Res: Synthetic fibers and films. Mailing Add: Film Dept Exp Sta E I du Pont de Nemours & Co Inc Wilmington DE 19898

MARSHALL, THOMAS C, b Cleveland, Ohio, Jan 29, 35; m 64; c 1. PHYSICS. Educ: Case Inst Technol, BS, 57; Univ Ill, MS, 58, PhD(physics), 60. Prof Exp: Asst prof elec eng, Univ Ill, 61-62; from asst prof to assoc prof, 62-70, PROF ELEC ENG, COLUMBIA UNIV, 70- Mem: Am Phys Soc. Res: Plasma and atomic physics; microwave scattering and radiation from plasmas; lasers; shock waves; plasma stability; toroidal containment experiments. Mailing Add: Plasma Lab Columbia Univ New York NY 10027

MARSHALL, THOMAS HANSON, biochemistry, see 12th edition

MARSHALL, VICTOR FRAY, b Culpeper, Va, Sept 1, 13; m 42; c 3. SURGERY. Educ: Univ Va, MD, 37; Am Bd Urol, dipl, 45; Washington & Lee Univ, DSc, 75. Prof Exp: Intern surg, New York Hosp, 37-38; asst, 38-42, from instr to assoc prof surg & urol, Med Col, 42-57, prof urol in surg, 57-70, JAMES J COLT PROF UROL IN SURG, MED COL, CORNELL UNIV, 70- Concurrent Pos: Asst resident surgeon, New York Hosp, 38-40, from asst resident urologist to resident urologist, 40-43, attend surgeon, 47-; attend surgeon-in-chg, Dept Urol, James Buchanan Brady Found, 49-; asst attend surgeon, Mem Hosp, 46-52, assoc attend surgeon, 52-; assoc attend surgeon urol serv, Mem Ctr for Cancer, 59- Mem: Soc Pelvic Surg; Clin Soc Genitourinary Surg; Am Surg Asn; fel AMA; Am Urol Asn. Res: Treatment of genitourinary cancer; mechanism of urinary control; surgical technique; pediatric urology. Mailing Add: New York Hosp 525 E 68th St New York NY 10021

MARSHALL, VINCENT DEPAUL, b Washington, DC, Apr 5, 43; m 65; c 2. MICROBIAL PHYSIOLOGY. Educ: Northeastern State Col, BS, 65; Univ Okla, MS, 67, PhD(microbiol), 70. Prof Exp: From res asst to res assoc microbiol, Univ Okla, 65-70; res assoc biochem, Univ Ill, 70-73; res scientist fermentation microbiol, Fermentation Prod Div, 73-74, res head microbial control, Prod Control Div, 75, RES SCIENTIST CANCER RES, EXP BIOL DIV, UPJOHN CO, 75- Concurrent Pos: NIH fel, Univ Ill, 71-73. Mem: Am Soc Microbiol. Res: Microbial metabolism and transformation of antibiotics, amino acids and terpenes. Mailing Add: Upjohn Co 301 Henrietta St Kalamazoo MI 49001

MARSHALL, WALTER LINCOLN, b Princeton, NJ, May 6, 25; m 50; c 3. PHYSICAL CHEMISTRY. Educ: Princeton Univ, AB, 46; Harvard Univ, PhD(chem), 50. Prof Exp: Chemist, Mat & Processes Lab, 50-53, supvr appl res & insulation mat, 53-58, mgr chem & elec insulation, 58-67, MGR MAT & PROCESSES LAB, LARGE STEAM TURBINE & GENERATOR DEPT, GEN ELEC CO, SCHENECTADY, 67- Concurrent Pos: Mem, Nat Res Coun, 53. Mem: Am Chem Soc. Res: Electrical insulation of large rotating electrical apparatus. Mailing Add: Box 292 Garnsey Rd Delanson NY 12053

MARSHALL, WILLIAM DEFORREST, b Halifax, NS, Mar 15, 46; m 68; c 2. PESTICIDE CHEMISTRY. Educ: Univ NB, BSc, 67; McMaster Univ, PhD(org chem), 73. Prof Exp: Jr chemist anal, Domtar Res Ctr, Domtar Ltd, 67-68; RES SCIENTIST PESTICIDES, CHEM & BIOL RES INST, AGR CAN, 73- Res: Analytical methodology, persistence and decomposition of selected pesticides in agricultural produce soils and waters; identification of metabolites/transformation products of these agents in the environment. Mailing Add: Chem & Biol Res Inst Agr Can K W Neatby Bldg CEF Ottawa ON Can

MARSHALL, WILLIAM E, JR, biochemistry, see 12th edition

MARSHALL, WILLIAM HAMPTON, b Montreal, Que, Apr 20, 12; nat US; m 37; c 2. WILDLIFE MANAGEMENT. Educ: Univ Calif, BS, 33; Univ Mich, MF, 35, PhD(wildlife mgt), 42. Prof Exp: Foreman, US Forest Serv, Calif, 33 & Ark, 34, asst conservationist, Mass, 35; asst, Univ Mich, 34-35; instr wildlife mgt, Utah State Col, 36; jr biologist, US Fish & Wildlife Serv, Idaho, 36-43; area supvr, War Food Admin, 43-44, wage control off, 44-45; from assoc prof to prof econ zool, 45-70, from assoc dir to dir, Lake Itasca Forestry & Biol Sta, 55-70, PROF WILDLIFE MGT, UNIV MINN, ST PAUL, 70- Mem: Wildlife Soc; Am Soc Mammal; Wilson Ornith Soc; Ecol Soc Am; Am Ornith Union. Res: Ecology and management of woodcock and grouse. Mailing Add: Dept Entom Fish & Wildlife Univ of Minn St Paul MN 55108

MARSHALL, WILLIAM JOSEPH, b Pittsburgh, Pa, Apr 10, 29; m 56; c 3. PHYSICAL CHEMISTRY. Educ: Univ Pittsburgh, BS, 51; Carnegie Inst Technol, MS, 55, PhD, 56. Prof Exp: Res supvr, E I du Pont de Nemours & Co, Inc, 56-68, res mgr pigments dept, 68-69, asst lab dir, 69-74, TECH SUPT, EDGE MOOR LAB, E I DU PONT DE NEMOURS & CO, INC, 74- Mem: Am Chem Soc. Res: Pigment technology; solid state chemistry. Mailing Add: Edge Moor Lab E I du Pont de Nemours & Co Inc Edge Moor DE 19809

MARSHALL, WILLIAM LEITCH, b Columbia, SC, Dec 3, 25; m 49; c 2. PHYSICAL CHEMISTRY. Educ: Clemson Univ, BS, 45; Ohio State Univ, PhD(phys org chem), 49. Prof Exp: Asst chem, Ohio State Univ, 45-46; from chemist to sr chemist, 49-57, group leader, 57-74, SR STAFF SCIENTIST, OAK RIDGE NAT LAB, 75- Concurrent Pos: Guggenheim fel, 56-57; mem, Org Comt, First Int Cong High Temperature Aqueous Electrolytes, Eng, 73 & Int Asn Properties of Steam Working Group, 75- Mem: AAAS; Geochem Soc; Sigma Xi; Am Chem Soc. Res: Solubilities in aqueous electrolyte systems from 0 to 374 degrees centigrade, application of theory, water desalination; electrical conductance of aqueous electrolytes to 800 degrees centigrade and 4000 atmospheres; chemistry of aqueous homogeneous reactors; uranium and thorium salt systems, effect of pressure on elastic constants of quartz; constitution of Grignard-type reagents. Mailing Add: Chem Div Oak Ridge Nat Lab Oak Ridge TN 37830

MARSHALL, WINSTON STANLEY, b Nashville, Tenn, Jan 16, 37; m 61; c 3. MEDICINAL CHEMISTRY. Educ: Vanderbilt Univ, AB, 59; Wayne State Univ, PhD(org chem), 63. Prof Exp: Sr org chemist, 63-69, res scientist med chem, 69-72, res assoc, 72-73, HEAD ORG CHEM, LILLY RES LABS, ELI LILLY & CO, 73- Mem: AAAS; Am Chem Soc. Res: Drug design in the areas of arthritis, especially relating non-steroidal anti-inflammatory agents, and of antipsychotic agents. Mailing Add: Chem Res Div Lilly Res Labs Indianapolis IN 46206

MARSHECK, WILLIAM JOHN, b Baltimore, Md, Mar 26, 42; m 63; c 3. MICROBIOLOGY. Educ: Univ Pittsburgh, BS, 64; Rutgers Univ, PhD(microbiol), 69. Prof Exp: RES INVESTR MICROBIOL, G D SEARLE & CO, 68- Mem: Am Soc Microbiol; Am Soc Indust Microbiol. Res: Microbiological fermentations; general bacteriology. Mailing Add: Dept of Microbiol Searle & Co PO Box 5110 Chicago IL 60680

MARSHO, THOMAS V, b Cleveland, Ohio, Mar 15, 40; m 63; c 3. PLANT BIOCHEMISTRY. Educ: Case Western Reserve Univ, BA, 61; Miami Univ, MA, 63; Univ NC, PhD(plant physiol), 68. Prof Exp: NIH fel photosynthesis, Res Inst Advan Studies, 68-70; ASST PROF BIOL, UNIV MD, BALTIMORE COUNTY, 70- Honors & Awards: William Chambers Coker Res Award. Mem: AAAS; Am Soc Plant Physiol. Res: Photosynthetic electron transport; photosynthetic regulation. Mailing Add: Dept of Biol Sci Univ Md Baltimore Co 5401 Wilkins Ave Baltimore MD 21228

MARSI, KENNETH LARUE, b Los Banos, Calif, Dec 13, 28; m 55; c 4. PHYSICAL ORGANIC CHEMISTRY. Educ: San Jose State Col, AB, 51; Univ Kans, PhD(org chem), 55. Prof Exp: Instr chem, Univ Kans, 54; sr res chemist, Sherwin-Williams Co, 55-57; from asst prof to assoc prof chem, Ft Hays Kans State Col, 57-61; from asst prof to assoc prof, 61-70, PROF CHEM, CALIF STATE COL LONG BEACH, 70-, CHMN DEPT, 75- Concurrent Pos: Petrol Res Fund grant, 59-61; NIH spec fel, Rutgers Univ, 67-68; NSF grants, 68-76. Mem: Am Chem Soc; Sigma Xi. Res: Synthesis and stereochemistry of reactions of organophosphorus compounds; ring closure reactions of compounds leading to phosphorus and nitrogen heterocycles. Mailing Add: Dept of Chem Calif State Univ Long Beach CA 90840

MARSICANO, FENIX R, celestial mechanics, fluid dynamics, see 12th edition

MARSLAND, DOUGLAS ALFRED, b Brooklyn, NY, Feb 17, 99; m 24. PHYSIOLOGY. Educ: NY Univ, BS, 22, PhD(physiol), 34; Columbia Univ, MA, 38. Prof Exp: Teacher, Silver Bay Sch Boys, NY, 22-24; from instr to prof, 24-63, res prof, 63-69, EMER RES PROF BIOL, WASH SQUARE COL, NY UNIV, 69- Concurrent Pos: Guggenheim fels, 51-52 & 59-60; Fulbright fel, 59-60; trustee, Marine

Biol Lab, Woods Hole, secy bd trustees, 65-69, emer trustee, 69-; Bermuda Biol Sta. Honors & Awards: Cleveland Award, AAAS, 41. Mem: AAAS; Am Soc Nat; Am Soc Zool; Soc Gen Physiol; fel NY Acad Sci. Res: Cell membranes; narcosis; physiological effects of high pressure; protoplasmic contractility and streaming; amoeboid movement; cell division; pigmentary effectors; bioluminescence; anti-mitotic effects of heavy water and colchicine; electron microscopy; fine structure of microtubules. Mailing Add: Marine Biol Lab Woods Hole MA 02543

MARSTERS, ROGER WESTCOTT, b Albany, NY, Sept 16, 18; m 41; c 2. BIOCHEMISTRY, IMMUNOHEMATOLOGY. Educ: State Univ NY Albany, AB, 39; Cornell Univ, MA, 42; Case Western Reserve Univ, PhD(clin biochem), 48. Prof Exp: Biochemist, Norwich Pharmacal Co, NY, 42-45; from instr to asst prof clin biochem, Sch Med, Case Western Reserve Univ, 48-64; ASSOC HEAD CLIN PATH & CLIN CHEMIST, ST LUKE'S HOSP, 64-; ADJ PROF CHEM, CLEVELAND STATE UNIV, 70- Concurrent Pos: Dir maternity Rh lab, Univ Hosps Cleveland, 47-55; consult immunohematologist, 50- Mem: AAAS; Am Asn Clin Chem; Am Acad Forensic Sci; Am Asn Blood Banks. Res: Clinical chemistry; Rh factor immunization; forensic hematology; laboratory instrument development. Mailing Add: St Luke's Hosp Dept Clin Path 11311 Shaker Blvd Cleveland OH 44104

MARSTON, ALFRED LAWRENCE, b Albany, NY, Apr 6, 18; m 41; c 2. PHYSICAL CHEMISTRY. Educ: Colgate Univ, AB, 40; Johns Hopkins Univ, PhD(chem), 43. Prof Exp: Res chemist, Gen Aniline & Film Corp, Pa, 43-44; fel, Mellon Inst, 44-45, head spectros div, Dept Res Chem Physics, 46-47; sr fel & spec instrumentation fel, 48-49; res chemist, E I du Pont de Nemours & Co, Inc, 50-53, res supvr, Savannah River Lab, 53-76; RETIRED. Mem: Am Chem Soc; Coblentz Soc. Res: Ultraviolet, visible, infrared, Raman and mass spectroscopy; instrumental methods of process control; analytical chemistry; radiochemistry. Mailing Add: Rte 2 Box 99A Lake Toxaway NC 28747

MARSTON, NORMAN LEE, b Hartman, Colo, Jan 8, 37; m 74; c 2. ENTOMOLOGY. Educ: Colo State Univ, BS, 58; Kans State Univ, MS, 62, PhD(entom), 65. Prof Exp: Instr entom, Kans State Univ, 65-66; asst prof, Univ Wyo, 66-67; RES ENTOMOLOGIST, BIOL CONTROL INSECTS RES LAB, USDA, 67- Mem: Entom Soc Am. Res: Insect ecology and taxonomy. Mailing Add: Biol Control of Insects Res Lab USDA PO Box A Columbia MO 65201

MARSTON, ROBERT QUARLES, b Toano, Va, Feb 12, 23; m 46; c 3. MEDICINE. Educ: Va Mil Inst, BS, 43; Med Col Va, MD, 47; Oxford Univ, BSc, 49. Prof Exp: House officer med, Johns Hopkins Hosp, 49-50; asst resident, Vanderbilt Univ, 50-51; asst resident, Med Col Va, 53-54, from asst prof to assoc prof, 54-61, asst dean, 59-61; vchancellor & dean sch med, Univ Miss, 61-66; assoc dir, Regional Med Prog, NIH, 66-68, adminstr, Health Serv & Ment Health Admin, 68, dir, NIH, 68-73; scholar-in-residence, Univ Va, Charlottesville, 73-74; PRES, UNIV FLA, 74- Concurrent Pos: Markle scholar, Med Col Va, 54-59; mem staff, Armed Forces Spec Weapons Proj, NIH, 51-53, chmn int fels rev panel, 64-66; asst prof, Univ Minn, 58-59; consult rev comt, Div Hosp & Med Facil, Dept Health, Educ & Welfare, 61-66. Mem: Am Cancer Soc; Am Heart Asn; fel Am Pub Health Asn; hon mem Nat Med Asn; hon mem Am Hosp Asn. Res: Infectious diseases; medical administration. Mailing Add: Off of the Pres Univ of Fla Gainesville FL 32601

MARSZALEK, DONALD STANLEY, geology, biology, see 12th edition

MARTAN, JAN, b Prague, Czech, Feb 13, 14; US citizen. REPRODUCTIVE BIOLOGY. Educ: Univ Chicago, MS, 60; Univ Ore, PhD(biol), 63. Prof Exp: Instr biol, Univ Ore, 63; fel Zool, Univ Mich, 63-64; from asst prof to assoc prof, 64-74, PROF ZOOL, SOUTHERN ILL UNIV, 74- Mem: AAAS; Am Asn Anat; Soc Study Reprod; Am Soc Zool; Histochem Soc. Res: Cytological and cytochemical aspects of the cells in the male genital tract and accessory sex glands. Mailing Add: Dept of Zool Southern Ill Univ Carbondale IL 62901

MARTEL, FERNAND, biology, see 12th edition

MARTEL, RENE R, b Montreal, Que, May 20, 30; m 57; c 3. PHARMACOLOGY. Educ: Univ Montreal, DVM, 56; McGill Univ, PhD, 60. Prof Exp: Mem staff pharmacol, Charles E Frosst & Co, Can, 61-63 & Bristol Labs, Can, 63-69; MEM STAFF PHARMACOL, AYERST LAB, 69- Mem: Pharmacol Soc Can. Res: Inflammation. Mailing Add: Ayerst Lab 1025 Laurentien Blvd St Laurent PQ Can

MARTEL, ROBERT WILLIAM, physical chemistry, see 12th edition

MARTEL, WILLIAM, b New York, NY, Oct 1, 27; m 55; c 4. RADIOLOGY. Educ: NY Univ, BS, 50, MD, 53. Prof Exp: PROF RADIOL, UNIV MICH, ANN ARBOR, 66- Mem: AMA; Am Roentgen Ray Soc; Radiol Soc NAm. Mailing Add: Dept of Radiol Univ of Mich Ann Arbor MI 48104

MARTELL, ARTHUR EARL, b Natick, Mass, Oct 18, 16; m 44, 65; c 8. CHEMISTRY. Educ: Worcester Polytech Inst, BS, 38; NY Univ, PhD(chem), 41. Hon Degrees: DSc, Worcester Polytech Inst, 62. Prof Exp: Asst, NY Univ, 38-40; instr chem, Worcester Polytech Inst, 41-42; from asst prof to prof, Clark Univ, 42-61, chmn dept, 59-61; prof & chmn dept, Ill Inst Technol, 61-66; DISTINGUISHED PROF CHEM & HEAD DEPT, TEX A&M UNIV, 66- Concurrent Pos: Res fel, Univ Calif, 49-50; Guggenheim fel, Univ Zurich, 54-55; NSF sr fel, 59-60; fel, Sch Advan Studies, Mass Inst Technol, 59-60; NIH fel, Univ Calif, Berkeley, 64-65; ed, J Coord Chem. Mem: AAAS; Am Chem Soc; fel Am Acad Arts & Sci; fel NY Acad Sci; Am Soc Biol Chemists. Res: Amino acid synthesis; potentiometry; physical and chemical properties, stabilities and catalytic effects of metal chelate compounds. Mailing Add: Dept of Chem Tex A&M Univ College Station TX 77843

MARTELL, EDWARD A, b Spencer, Mass, Feb 23, 18; m 42; c 4. RADIOCHEMISTRY, NUCLEAR GEOCHEMISTRY. Educ: US Mil Acad, BS, 42; Univ Chicago, PhD(nuclear chem), 50. Prof Exp: Prog dir, Armed Forces Spec Weapons Proj, Washington, DC, 50-54; res assoc, Enrico Fermi Inst Nuclear Studies, Univ Chicago, 54-56; group leader atmospheric radioactivity & fallout, Geophys Res Div, Air Force Cambridge Res Lab, Mass, 56-62; RES SCIENTIST, NAT CTR ATMOSPHERIC RES, 62- Concurrent Pos: Secy, Int Comn Atmospheric Chem & Global Pollution Int Asn Meteorol & Atmospheric Physics, 63-71, pres, 75- Mem: Fel AAAS; Am Geophys Union; Health Phys Soc; Sigma Xi. Res: Natural radioactivity; discovery of indium-115 beta negative activity; radiation and fallout effects of nuclear explosions; nuclear meteorology; upper atmosphere composition with rocket samplers; radioactive aerosols; environmental and biomedical transport of alpha emitters. Mailing Add: Nat Ctr for Atmospheric Res PO Box 3000 Boulder CO 80303

MARTELL, MICHAEL JOSEPH, JR, b Minneapolis, Minn, May 20, 32. INDUSTRIAL CHEMISTRY, PHARMACY. Educ: Univ Minn, BS, 54, PhD(pharmaceut chem), 58. Prof Exp: NIH fel, Univ Ill, 59-60; res chemist, Lederle Labs Div, 60-70, mgr prod develop, 70-75, DIR MED PROD & PROCESS DEVELOP, CYANAMID INT, AM CYANAMID CO, 75- Mem: Am Chem Soc. Res: Tetracycline antibiotics; alkaloids; biopharmaceutics as pertains to product development. Mailing Add: Int Res & Develop Am Cyanamid Co Pearl River NY 10965

MARTELLOCK, ARTHUR CARL, b Detroit, Mich, Jan 7, 28; m 49; c 3. POLYMER CHEMISTRY, ORGANIC CHEMISTRY. Educ: Wayne State Univ, AB, 51; Rutgers Univ, PhD(org chem), 57. Prof Exp: Chemist, Silicone Prod Dept, Gen Elec Co, 56-68, specialist silicone rubber develop, 68-70; SR SCIENTIST & UNIT MGR, MAT SECT, DEVELOP DEPT, XEROX CORP, 70- Mem: Fel Am Inst Chemists; Am Chem Soc. Res: Materials development; polymer molecular structure and rheology; silicone polymer synthesis; characterization and degradation kinetics. Mailing Add: Mat Sect Develop Dept Xerox Corp Xerox Sq Bldg 147 Rochester NY 14440

MARTEN, GORDON C, b Wittenberg, Wis, Sept 14, 35; m 61; c 1. AGRONOMY. Educ: Univ Wis, BS, 57; Univ Minn, MS, 59, PhD(agron), 61. Prof Exp: RES AGRONOMIST, AGR RES SERV, USDA, 61- Concurrent Pos: From asst prof to assoc prof agron & plant genetics, Univ Minn, St Paul, 61-71, prof, 71-; mem, Am Grassland Coun, 65- Mem: Crop Sci Soc Am; fel Am Soc Agron. Res: Techniques for evaluating forage crops; nutritive value of forage crops as influenced by agronomic practices; cattle and sheep grazing management; effects of ecological factors on forage quality. Mailing Add: 316 Agron & Plant Genetics Bldg Univ of Minn St Paul MN 55108

MARTEN, JAMES FREDERICK, b Liverpool, Eng, Sept 11, 31; m 53; c 2. BIOCHEMISTRY. Educ: Royal Inst Chem, ARIC, 52; Univ Leeds, PhD, 56. Prof Exp: Sr sci off, UK Atomic Energy Auth, 55-57; mgr, Borax Consol Res Labs, 57-58; tech mgr, Technicon Instruments Co Ltd, 59-64, gen mgr, Technicon Controls Inc, NY, 64-67, tech coordr, Technicon Corp, 67-69; vpres & dir mkt, Biomed Sci Inc, 69-70; vpres med develop, Damon Corp, Boston, 70-73; SECY & DIR, DELKA CORP, MASS, 73- Concurrent Pos: Dir & secy, Dakton Ltd, BWI, 74-; dir, Armendaris Corp, Mo, 75- & Medi Inc, Mass, 75- Res: Conception and design of automated medical diagnostic instrumentation.

MARTENS, CHRISTOPHER SARGENT, b Akron, Ohio, Jan 11, 46; m 68; c 2. MARINE CHEMISTRY. Educ: Fla State Univ, BS, 68, MS, 69, PhD(chem oceanog), 72. Prof Exp: Res technician Antarctic sediment chem, Fla State Univ, 68-69; res chemist & partic guest scientist, Lawrence Livermore Radiation Lab, 71-72; res staff marine chem, Dept Geol & Geophys, Yale Univ, 72-74; ASST PROF GEOL & ASST PROF MARINE SCI, MARINE SCI PROG, UNIV NC, CHAPEL HILL, 74- Mem: AAAS; Am Soc Limnol & Oceanog; Am Geophys Union; Geochem Soc. Res: Chemical processes in organic-rich coastal environments, particularly, bacterially catalyzed oxidation-reduction reactions involving dissolved gas sources and transports; tracer studies of material exchanges between coastal sediments, water and atmosphere. Mailing Add: Marine Sci Prog 12-5 Venable Hall Univ of NC Chapel Hill NC 27514

MARTENS, DAVID CHARLES, b Shawano, Wis, Apr 17, 33; m 57; c 2. SOIL SCIENCE. Educ: Univ Wis, BS, 60, MS, 62, PhD(soil sci), 64. Prof Exp: Asst prof, 64-68, ASSOC PROF SOIL SCI, VA POLYTECH INST & STATE UNIV, 68- Mem: Am Soc Agron; Soil Sci Soc Am; Soil Conserv Soc. Res: Micronutrient chemistry of soil; diagnosis of chemical factors of soil responsible for abnormal plant growth; by-product disposal. Mailing Add: Dept of Agron Va Polytech Inst & State Univ Blacksburg VA 24061

MARTENS, EDWARD JOHN, b Evergreen Park, Ill, July 31, 38; m 59; c 2. NUCLEAR PHYSICS. Educ: Mass Inst Technol, BS, 61, MS, 65, PhD(physics), 67. Prof Exp: Instr physics, Northeastern Univ, 67-69; sr scientist, Am Sci & Eng, Inc, 69-71; instr, 71-74, ASST PROF INDUST ARTS, FITCHBURG STATE COL, 74- Mem: Am Phys Soc. Res: X-ray astronomy. Mailing Add: Dept of Indust Arts Fitchburg State Col Fitchburg MA 01420

MARTENS, JACOB LOUIS, b Anchor, Ill, Apr 18, 09; m 39; c 1. PLANT MORPHOLOGY. Educ: Ind Cent Col, AB, 30; Ind Univ, AM, 32, PhD(plant morphol), 38. Prof Exp: Asst & tutor bot, Ind Univ, 30-38; head dept biol & phys educ, Piedmont Col, 38-39, head dept biol, 40-47; from assoc prof to prof, 47-74, EMER PROF BIOL, ILL STATE UNIV, 74- Mem: Am Asn Biol Teachers; Bot Soc Am; Am Inst Biol Sci; Am Genetic Asn. Res: Role of plants in science education. Mailing Add: Dept of Biol Ill State Univ Normal IL 61761

MARTENS, JOHN WILLIAM, b Desalaberry, Man, July 31, 34; m 59; c 4. PLANT PATHOLOGY. Educ: Univ Man, BSc, 62; Univ Wis-Madison, PhD(plant path, mycol), 65. Prof Exp: RES SCIENTIST CEREAL RUSTS, RES BR, AGR CAN, 65- Concurrent Pos: Head plant path sect, Plant Breeding Sta, Njoro, Kenya, under Can Int Develop Agency, 71-72; vis res scientist, Dept Sci & Indust Res, Christchurch, NZ, 75-76. Mem: Can Phytopath Soc; Am Phytopath Soc; Sigma Xi. Res: Physiologic specialization in cereal rusts; host resistance; collection, preservation and utilization of wild avena species. Mailing Add: Can Agr Res Sta 25 Dafoe Rd Winnipeg MB Can

MARTENS, LESLIE VERNON, b Peoria Heights, Ill, Oct 15, 38; m 61; c 4. DENTISTRY. Educ: Loyola Univ Chicago, DDS, 63; Univ Minn, Minneapolis, MPH, 69. Prof Exp: Pvt pract, Ill, 63; lectr prev dent, Sch Dent, 68-69, asst prof maternal & child health, Sch Pub Health, 69-70, asst prof prev med, Sch Dent, 69-71, ASSOC PROF HEALTH ECOL & ASSOC CHMN DIV, SCH DENT, UNIV MINN, MINNEAPOLIS, 71- Concurrent Pos: Consult, Cambridge State Hosp & Sch for Ment Retarded, 68-; lectr, Schs Nursing & Pharm, Univ Minn, Minneapolis, 70- & Normandale State Jr Col, 71-; consult, Minneapolis Pub Schs, 72- & USPHS, 73- Mem: Am Dent Asn; Int Asn Dent Res; Am Pub Health Asn; Am Soc Prev Dent; Behav Sci in Dent Res. Res: Preventive dentistry; health manpower; dental epidemiology; health education; health behavior. Mailing Add: Div of Health Ecol Univ of Minn Sch of Dent Minneapolis MN 55455

MARTENS, ROBERT IVAN, chemistry, see 12th edition

MARTENS, TED FRANK, organic chemistry, see 12th edition

MARTENS, VERNON EDWARD, b St Louis, Mo, Aug 15, 12; m; c 7. PATHOLOGY. Educ: St Louis Univ, BS, 35, MD, 37; Am Bd Path, dipl, 48. Prof Exp: Intern, St Louis City Hosp, Mo, 37-38; resident med, US Naval Hosps, Chelsea, Mass, 38-39, pathologist, Norman, Okla, 44-45, asst pathologist, Philadelphia, Pa, 45-47, pathologist, 50-51; dir labs, US Navy Med Sch, 51-58; DIR LABS, WASHINGTON HOSP CTR, 58- Concurrent Pos: Fel path, Hosp Univ Pa, 47-50, instr, 50-51; assoc clin prof, Sch Med, George Washington Univ, 63, 65 & 66. Mem: AMA; Am Soc Clin Path; Asn Clin Sci (pres, 57-58); Col Am Path; Int Acad Path. Res: Clinical and anatomical pathology. Mailing Add: Washington Hosp Ctr 110 Irving St NW Washington DC 20010

MARTENS, WILLIAM STEPHEN, b Pittsburgh, Pa, June 14, 35. ENVIRONMENTAL CHEMISTRY. Educ: Rutgers Univ, BS, 56, PhD(anal & inorg chem), 60. Prof Exp: Sr res chemist, Int Minerals & Chem Corp, 60-62 & Agr Div,

Allied Chem Corp, 62-69; consult, State of Va Health Dept, 69-70; SECT LEADER ENVIRON ENHANCEMENT, US NAVAL SURFACE WEAPONS CTR, 70- Mem: Am Chem Soc. Res: Sodium polyphosphate analyses; ion exchange; phosphate rock and wet-process phosphoric acid; inorganic polymers; air pollution detector development; environmental assessment, enhancement and control; air, water and solid waste pollution abatement; incineration technology. Mailing Add: US Naval Weapons Ctr Code DG-30 Dahlgren VA 22448

MARTH, ELMER HERMAN, b Jackson, Wis, Sept 11, 27; m 57. FOOD MICROBIOLOGY, DAIRY MICROBIOLOGY. Educ: Univ Wis, BS, 50, MS, 52, PhD(bact), 54. Prof Exp: Asst bact, Univ Wis, 49-54, proj assoc, 54-55, instr, 55-57; bacteriologist, Kraftco Corp, 57-59, from res bacteriologist to sr res bacteriologist, 59-63, group leader bact, 63-66, assoc mgr microbiol, 66; assoc prof, 66-71, PROF FOOD SCI & BACT, UNIV WIS-MADISON, 71- Concurrent Pos: Ed, J Milk & Food Technol, Int Asn Milk, Food & Environ Sanit, 67-; chmn, Intersoc Coun Stand Methods Exam Dairy Prods, 72-; WHO travel fel, 75. Honors & Awards: Pfizer Award, Am Dairy Sci Asn, 75. Mem: Am Soc Microbiol; Am Dairy Sci Asn; Inst Food Technol; Int Asn Milk, Food & Envi- ron Sanit; Coun Biol Ed. Res: Microbiology of dairy and food products; psychrotrophic bacteria; mycotoxins; dairy starter cultures; fermentations; microbiology of animal feeds; manufacturing of buttermilk and cottage cheese; fate of pathogenic bacteria in foods. Mailing Add: Dept of Food Sci Univ of Wis-Madison Madison WI 53706

MARTH-SNADER, ELLA CAROLYN, b Alton, Ill, Aug 21, 09; m 56. MATHEMATICS. Educ: Harris Teachers Col, AB, 30; St Louis Univ, MS, 35, PhD(math), 44. Prof Exp: Teacher pub schs, Mo, 30-43; instr math, Southeastern Mo State Col, 44; asst prof, Harris Teachers Col, 45-47, prof & dean women, 47-52; prof & chmn div math & bus ed, DC Teachers Col, 52-56; assoc prof, Chicago Teachers Col, 56-59; specialist elem math, US Off Ed, 59; PROF MATH, DC TEACHERS COL, 59-, CHMN DIV MATH, 62-65 & 72- Concurrent Pos: Consult pub schs, Mo, 45-49. Mem: Am Math Soc; Math Asn Am. Res: Mathematical analysis; analytic geometry; theory of numbers; history of transcendental numbers. Mailing Add: 3701 Connecticut Ave N W Washington DC 20008

MARTI, KURT, b Berne, Switz, Aug 18, 36; m 63; c 3. COSMOCHEMISTRY. Educ: Univ Berne, MSc, 63, PhD(geophys), 65. Prof Exp: Res chemist, 65-67, asst res chemist, 67-68, asst prof, 69-74, ASSOC PROF COSMOCHEM, UNIV CALIF, SAN DIEGO, 74- Concurrent Pos: NASA grant, Univ Calif, San Diego, 71-; prin investr, Lunar Sample Anal, 72- Mem: AAAS; Am Geophys Union; Meteoritical Soc. Res: Isotopic and nuclear cosmochemistry; origin and history of the moon and the solar system; products of extinct elements. Mailing Add: Dept of Chem Univ of Calif San Diego La Jolla CA 92037

MARTIG, ROBERT C, dairy science, see 12th edition

MARTIGNOLE, JACQUES, b Carcassonne, France, Oct 11, 39; m 62. GEOLOGY. Educ: Univ Toulouse, Lic es Sci, 61, Dr 3rd Cycle, 64, Dr Univ, 68, DSc, 75. Prof Exp: Nat Coun Arts Can fel, 64-66, lectr geol, 66-68, asst prof, 68-72, ASSOC PROF GEOL, UNIV MONTREAL, 72- Mem: Geol Asn Can; Asn Study Deep Zones Earth's Crust. Res: Precambrian geology; igneous and metamorphic petrology; structural geology. Mailing Add: Dept of Geol Univ of Montreal Montreal PQ Can

MARTIGNONI, MAURO EMILIO, b Lugano, Switz, Oct 30, 26; nat US; m 53; c 2. VIROLOGY, INVERTEBRATE PATHOLOGY. Educ: Swiss Fed Inst Technol, dipl ing agr, 50, PhD(microbiol, entom), 56. Prof Exp: Asst entom, Swiss Fed Inst Technol, 50 & 52, entomologist, Swiss Forest Res Inst, 53-56; from asst insect pathologist to assoc insect pathologist & lectr invert path, Univ Calif, Berkeley, 56-62, assoc prof, 63-65; CHIEF MICROBIOLOGIST, US FOREST SERV, 65-; PROF ENTOM, ORE STATE UNIV, 65- Concurrent Pos: Consult entomologist, Food & Agr Orgn, UN, Rome, Italy, 52-53; USPHS grant, 58-64; mem trop med & parasitol study sect, NIH, 64-65; consult med zool dept, US Naval Med Res Unit 3, 66-; mem, Int Comt Nomenclature Viruses, 66-; consult, NASA, 66-67; vis scientist insect virol, Agr Res Coun, Littlehampton, Gt Brit, 72-73. Honors & Awards: Kern Award & Silver Medal, Swiss Fed Inst Technol, 57. Mem: AAAS; Entom Soc Am; Tissue Cult Asn; Am Soc Microbiol; Soc Invert Path. Res: Insect pathology, especially viral diseases of insects; pathologic physiology; epizootiology of insect diseases; insect tissue culture. Mailing Add: Forestry Sci Lab US Forest Ser 3200 Jefferson Way Corvallis OR 97331

MARTIN, AARON JAY, b Lancaster, Pa, June 2, 28; m 52; c 1. ANALYTICAL CHEMISTRY. Educ: Franklin & Marshall Col, BS, 50; Pa State Col, MS, 52, PhD(anal chem), 53. Prof Exp: AEC res asst, Pa State Univ, 51-53; res chemist, E I du Pont de Nemours & Co, Inc, 53-58, res supvr, 58-59; dir res, F&M Sci Corp, 59-65, mgr res & eng, F&M Sci Div, Hewlett Packard Co, 65-69; PRES, MARLABS, INC, 69- Mem: AAAS; Instrument Soc Am; Am Chem Soc. Res: Polarographic behavior of organic compounds; analytical instrumentation. Mailing Add: Marlabs Inc Rte 3 Box 116 Kennett Square PA 19348

MARTIN, ABRAM VENABLE, b Clinton, SC, Apr 20, 12; m 42; c 2. MATHEMATICS. Educ: Presby Col, SC, AB, 36; Duke Univ, PhD(math), 40. Prof Exp: Instr math, Duke Univ, 39-40 & Univ Mich, 40-41; physicist, Metall Lab, Chicago, 43-46; instr math, Mass Inst Technol, 46-47; asst prof, Univ Nev, 47-49 & San Jose Col, 49-51; vis asst prof, Univ Calif, 51-54; assoc prof, Univ NMex, 54-60; prof, State Univ NY Stony Brook, 60-63; sr lectr, Univ Ibadan, 63-65; prof, Union Col, NY, 65-66; lectr, Univ Birmingham, 66-68; PROF MATH, GRAND VALLEY STATE COL, 68- Concurrent Pos: Am-Swiss Found sci exchange grant, Univ Lausanne, 59-60. Mem: Math Asn Am. Res: Topology, real variable theory. Mailing Add: Dept of Math Grand Valley State Col Allendale MI 49401

MARTIN, ALBERT, JR, b Camden, NJ, Dec 10, 16; m 39; c 3. MICROBIOLOGY. Educ: Univ Pittsburgh, BS, 42, MS, 46, PhD(genetics), 47; Registered, Nat Registry Microbiologists, 71. Prof Exp: Asst, Univ Pittsburgh, 45-46; prof biol, Mt Mercy Col, Pa, 46-53; MICROBIOLOGIST, VET NEUROPSYCHIAT HOSP, 53- Concurrent Pos: Lectr, Univ Pittsburgh, 46-54; asst prof, Dent Sch, 65- Honors & Awards: Darbaker Award, 59. Mem: AAAS; Am Soc Microbiologists; Genetics Soc Am; Am Soc Human Genetics; Sigma Xi. Res: Genetics of Habrobracon juglandis; genetics; bacteriology; tissue culture. Mailing Add: 4625 Fifth Ave Apt 100 Pittsburgh PA 15213

MARTIN, ALBERT BYRON, b Harlem, Ga, Dec 18, 15; m 44; c 2. PHYSICS. Educ: Univ Wyo, BS, 37, MS, 39; Yale Univ, PhD(physics), 47. Prof Exp: Res assoc, Radiation Lab, Mass Inst Technol, 41-42; sr physicist, Oak Ridge Nat Lab, 47-48; chief, Reactor Physics Sect, Atomic Energy Res Dept, NAm Aviation, Inc, 49-59, vpres, Atomics Int Div, 59-68, MGR SPACE SYSTS PROGS, ATOMICS INT DIV, ROCKWELL INT CORP, 68- Mem: Am Phys Soc; Am Nuclear Soc. Res: Reactor development; critical experiments; radar development; intermetallic diffusion; single crystal growth. Mailing Add: Atomics Int Div Rockwell Int Corp 8900 De Soto Canoga Park CA 91304

MARTIN, ALBERT EDWIN, b Mifflintown, Pa, Nov 25, 31; m 53; c 2. ANALYTICAL CHEMISTRY, PHARMACEUTICAL CHEMISTRY. Educ: Franklin & Marshall Col, BS, 53; Univ Calif, Los Angeles, MS, 56; Univ NC, Chapel Hill, PhD(chem), 59. Prof Exp: Sr res chemist, Chas Pfizer & Co, Inc, Conn, 59-62; mgr, 62-74, DIR ANAL RES, A H ROBINS CO, INC, 74- Concurrent Pos: Spec lectr, Va Commonwealth Univ, 67-70; mem revision comt, US Pharmacopeia XIX; consult, Gov Mgt Study Comn, Va. Mem: Am Chem Soc; Am Pharmaceut Asn; Acad Pharmaceut Sci. Res: Complex solution analysis; instrumental and electrochemical techniques; analytical chemistry of organic compounds; pharmaceutical dosage formulations. Mailing Add: 1211 Sherwood Ave Richmond VA 23220

MARTIN, ALBERT ERSKINE, JR, b Rome, Ga, Jan 22, 19; m; c 1. APPLIED PHYSICS. Educ: Tulane Univ, BS, 45, MS, 46. Prof Exp: Instr physics, Tulane Univ, 45-47 & 53-57; instr, Brown Univ, 47-50; res physicist, Southern Regional Res Lab, USDA, 50-57; sr physicist, Courtaulds, Inc, 57-58, textile physics sect head, 59-62; MGR APPL RES & SERV, FIRESTONE SYNTHETIC FIBERS CO, 62- Concurrent Pos: Lectr, Spring Hill Col, 57-59. Mem: Am Soc Testing & Mat; Fiber Soc; Am Asn Textile Technol; Soc Plastics Engrs. Res: Fiber physics; x-ray diffractometry; ultrasonics; electronics. Mailing Add: 4221 Stratford Rd Richmond VA 23225

MARTIN, ALEXANDER ROBERT, b Can, Oct 12, 28; m 51; c 3. NEUROPHYSIOLOGY. Educ: Univ Man, BSc, 51, MSc, 53; Univ London, PhD(biophys), 55; Yale Univ, MA, 68. Prof Exp: Asst biophys, Univ Col, Univ London, 53-55; from instr to assoc prof physiol, Col Med, Univ Utah, 57-66; prof, Yale Univ, 66-70; PROF PHYSIOL & CHMN DEPT, SCH MED, UNIV COLO, DENVER, 70- Concurrent Pos: Bronfmann fel neurophysiol, Montreal Neurol Inst, 55-57. Mem: Am Physiol Soc; Brit Physiol Soc. Res: Synaptic transmission. Mailing Add: Dept of Physiol Univ of Colo Sch of Med Denver CO 80220

MARTIN, ALFRED, b Pittsburgh, Pa, May 1, 19; m 46; c 2. PHYSICAL MEDICINAL CHEMISTRY. Educ: Philadelphia Col Pharm, BS, 42; Purdue Univ, MS, 48, PhD, 50. Prof Exp: From asst prof to assoc prof pharm, Temple Univ, 50-55; from assoc prof to prof, Sch Pharm, Purdue Univ, 55-66; prof, Sch Pharm, Med Col Va, 66-68; prof phys med chem & dean, Sch Pharm, Temple Univ, 68-72; PROF & DIR, DRUG DYNAMICS INST, COL PHARM, UNIV TEX AUSTIN, 73- Concurrent Pos: Pfeiffer mem res fel, Ctr Appl Wave Mech, France, 62-63; indust consult, 62- Honors & Awards: Ebert Medal, Am Pharmaceut Asn, 66, Achievement Award, 67. Mem: AAAS; Am Chem Soc; Am Pharmaceut Asn; fel Acad Pharmaceut Sci. Res: Application of physical chemical principles to pharmacy and medicinal chemistry. Mailing Add: Drug Dynamics Inst Col of Pharm Univ of Tex Austin TX 78712

MARTIN, ALICE OPASKAR, genetics, see 12th edition

MARTIN, ALLAN ERNEST, chemistry, see 12th edition

MARTIN, ARLENE PATRICIA, b Binghamton, NY, June 30, 26. BIOCHEMISTRY. Educ: Cornell Univ, BA, 48, MNutritS, 52; Univ Rochester, PhD(biochem), 57. Prof Exp: Fel, Sch Med & Dent, Univ Rochester, 57-58, instr biochem, 58-65; asst prof radiol, Jefferson Med Col, 65-67, asst prof biochem, 67-68; assoc prof, 68-74, PROF PATH & BIOCHEM, SCH MED, UNIV MO-COLUMBIA, 74- Mem: Fel AAAS; Am Chem Soc; NY Acad Sci. Res: Isolation, characterization and function of enzymes concerned with biological oxidation, especially respiratory enzymes, including components of the peroxidase, succinoxidase and pyridine nucleotide oxidase systems; structure-function relationships of mitochondria. Mailing Add: Dept of Path Univ Mo Sch of Med Columbia MO 65201

MARTIN, ARNOLD R, b Missoula, Mont, Mar 6, 36; m 59; c 4. PHARMACEUTICAL CHEMISTRY. Educ: Wash State Univ, BS, 59, MS, 61; Univ Calif, San Francisco, PhD(pharm chem), 64. Prof Exp: Actg asst prof pharm, 64-65, asst prof, 65-69, ASSOC PROF PHARM, WASH STATE UNIV, 69- Concurrent Pos: Mem, Am Found Pharmaceut Educ. Mem: AAAS; Am Chem Soc; Am Pharmaceut Asn; Acad Pharmaceut Sci. Res: Medicinal chemistry; phenothiazine tranquilizers; stereochemical and conformational studies; aminotetralin as analgesics; adrenergic blocking agents. Mailing Add: Col of Pharm Wash State Univ Pullman WA 99163

MARTIN, ARTHUR FRANCIS, b Elkins, WVa, Feb 5, 18; m 53; c 3. CHEMISTRY. Educ: Ursinus Col, AB, 38; Mass Inst Technol, PhD(org chem), 42. Hon Degrees: ScD, Ursinus Col, 63. Prof Exp: Res chemist, Exp Sta, 41-42, sr chemist, Va, 43, asst leader, Cellulose Prod Group, Del, 44, chief chemist & head lab, Cellulose Plant, Va, 45-49, mgr, Va Cellulose Res Div, Exp Sta, 49-53, spec assignment, Argonne Nat Lab, 54-55, actg mgr, Phys Chem Res Div, Res Ctr, 56, sr res chemist, Phys Chem Res Div, 57-58 & Appl Math Div, 59-63, mgr, 63-67, mgr, Opers Res Div, 67-71, SR FINANCIAL ANALYST, HERCULES, INC, 71- Mem: Am Chem Soc; Tech Asn Pulp & Paper Indust. Res: Cellulose and cellulose products; research administration; applied mathematics; operations research. Mailing Add: New Enterprise Dept Hercules Inc Wilmington DE 19899

MARTIN, ARTHUR WESLEY, JR, b Nanking, China, Dec 13, 10; US citizen; m 31, 58; c 2. PHYSIOLOGY. Educ: Col Puget Sound, BS, 31; Stanford Univ, PhD(physiol), 36. Prof Exp: Asst physiol, Stanford Univ, 34-35; asst microbiol, Marine Sta, Johns Hopkins Univ, 35-36, instr physiol, 36-37; from instr to assoc prof, 37-50, exec off dept, 48-63, PROF ZOOL, UNIV WASH, 50- Concurrent Pos: Dir prog regulatory biol, Div Biol & Med Sci, NSF, 58-59. Mem: AAAS; Am Physiol Soc; Soc Exp Biol & Med; Am Soc Zool. Res: Comparative circulatory physiology; cellular metabolism; muscle atrophy; anesthetics; excretory processes in molluscs. Mailing Add: Dept of Zool Univ of Wash Seattle WA 98105

MARTIN, ARTHUR WESLEY, III, b Palo Alto, Calif, July 5, 35; m 58; c 3. THEORETICAL PHYSICS. Educ: Harvard Univ, AB, 57; Stanford Univ, MS, 59, PhD(particle physics), 62. Prof Exp: Res assoc physics, Argonne Nat Lab, 62-64; asst prof, Stanford Univ, 64-67; assoc prof, Rutgers Univ, 67-69; ASSOC PROF PHYSICS, UNIV MASS, BOSTON, 69- Mem: Am Phys Soc. Res: Elementary particle physics; dispersion theory; general relativity. Mailing Add: Dept of Physics Univ of Mass Harbor Campus Boston MA 02125

MARTIN, ASHLEY MARVIN, III, b Memphis, Tenn, May 5, 43; m 66; c 2. NUCLEAR PHYSICS, OPTICAL PHYSICS. Educ: Memphis State Univ, BS, 65; Fla State Univ, PhD(physics), 70. Prof Exp: Asst prof, 70-71, ASSOC PROF PHYSICS & HEAD DEPT, ATHENS COL, ALA, 71- Mem: Am Phys Soc. Res: Holographic non-destructive testing; acousto-optical non-destructive testing. Mailing Add: Dept of Physics Athens Col Athens AL 35611

MARTIN, BERNARD LOYAL, b Whittier, Calif, Jan 1, 28; m 55; c 3. MATHEMATICS. Educ: Cent Wash State Col, BA, 55, MEd, 57; Ore State Univ, MS, 64, PhD(math), 66. Prof Exp: Instr high schs, Wash, 55-59, chmn dept math, 56-

59; from instr to asst prof, 59-66, assoc prof & asst dean arts & sci, 66-69, PROF MATH & DEAN ARTS & SCI, CENT WASH STATE COL, 69- Mem: Math Asn Am. Res: Mathematics education. Mailing Add: Dept of Math Cent Wash State Col Ellensburg WA 98926

MARTIN, BILLY JOE, b Talpa, Tex, May 24, 33; m 55; c 2. CELL BIOLOGY, CYTOCHEMISTRY. Educ: Univ Southern Miss, BS, 62, MS, 63; Rice Univ, PhD(biol), 70. Prof Exp: Asst prof biol, William Carey Col, 63-66; fel, Inst Pathobiol, Med Univ SC, 70-71, asst prof path, 71-73, asst prof, Sch Dent, 72-73; ASST PROF BIOL, UNIV SOUTHERN MISS, 73- Concurrent Pos: Mem, Grad Fac, Med Univ SC, 72-73. Mem: AAAS; Am Soc Zoologists; Electron Micros Soc Am; Am Soc Cell Biol. Res: Ultrastructure of cells specialized for electrolyte transport; dynamics of cell transport; cytochemistry of cell surface; use of lectins as cytochemical tools. Mailing Add: Univ Southern Miss Box 477 Southern Sta Hattiesburg MS 39401

MARTIN, BRUCE DOUGLAS, b Rochester, NY, Apr 8, 34; m 57; c 2. PHARMACEUTICAL CHEMISTRY. Educ: Albany Col Pharm, BS, 55; Univ Ill, MS, 59, PhD(pharmaceut chem), 62. Prof Exp: From asst prof to assoc prof, 61-68, PROF PHARMACEUT CHEM, DUQUESNE UNIV, 68-, DEAN SCH PHARM, 71- Concurrent Pos: Fulbright lectr, Univ Sci & Technol, Ghana, 68-69. Mem: Am Chem Soc; Am Pharmaceut Asn; The Chem Soc. Res: Organic synthesis of potential antiradiation compounds; sulfonamides; large ring compounds. Mailing Add: Sch of Pharm Duquesne Univ Pittsburgh PA 15219

MARTIN, BRUCE EUGENE, physical chemistry, see 12th edition

MARTIN, CARROLL JAMES, b Indianola, Iowa, June 1, 17; m 48; c 5. MEDICINE. Educ: Univ Iowa, BS & MD, 40; Am Bd Internal Med, dipl, 55. Prof Exp: Instr med, 50-56, clin asst prof, 56-60, clin assoc prof med physiol & biophys, 60-71, CLIN PROF MED PHYSIOL & BIOPHYS, MED SCH, UNIV WASH, 71-, DIR INST RESPIRATORY PHYSIOL, VIRGINIA MASON RES CTR, 68- Mem: Am Physiol Soc; fel Am Col Physicians; Am Thoracic Soc; Am Fedn Clin Res. Res: Respiratory physiology; distribution of ventilation and blood flow. Mailing Add: Inst of Respiratory Physiol Virginia Mason Res Ctr Seattle WA 98101

MARTIN, CHARLES EVERETT, b Moscow Mills, Mo, Nov 7, 29; m 52; c 3. VETERINARY PHYSIOLOGY. Educ: Univ Mo, BS & DVM, 58; Purdue Univ, Lafayette, MS, 67. Prof Exp: Practitioner, Green Hills Animal Hosp, 58-65; from instr to asst prof vet med & surg, Purdue Univ, Lafayette, 65-67; from asst prof to assoc prof, 67-73, PROF VET MED & SURG, UNIV MO-COLUMBIA, 73-, CHMN DEPT, 74- Concurrent Pos: Mem, NCent Res Comt, 64 & 68- & Nat Pork Producers Res Coord Comt, 70-; Mo Pork Producers & Agr Exp Sta grants, Univ Mo-Columbia, 70. Mem: Am Vet Med Asn; Am Asn Equine Practitioners; Am Asn Swine Practitioners; Am Col Theriogenology; Am Soc Study Breeding Soundness. Res: Bovine, equine and swine reproduction; physiology, endocrinology and pathology of lactation failure in swine. Mailing Add: Dept of Med & Surg Univ of Mo Sch of Vet Med Columbia MO 65201

MARTIN, CHARLES FRANKLIN, b Spokane, Wash, Dec 3, 17; m 41; c 3. PHARMACEUTICAL CHEMISTRY. Educ: State Col Wash, BS, 41, MS, 43, PhD(chem), 50. Prof Exp: From instr to assoc prof pharm, 43-59, PROF PHARMACEUT CHEM, COL PHARM, WASH STATE UNIV, 59- Concurrent Pos: Researcher, Chem Inst, Zurich, 57-58. Mem: AAAS; Am Pharmaceut Asn; Am Chem Soc. Res: Synthesis and anticonvulsant evaluations of certain aminopiperidones and piperidinediones. Mailing Add: Col of Pharm Wash State Univ Pullman WA 99163

MARTIN, CHARLES J, b New Castle, Pa, Dec 5, 21; m 45; c 4. BIOCHEMISTRY. Educ: Univ Pittsburgh, BS, 44, PhD(chem), 51. Prof Exp: Asst, Western Pa Hosp, 49-51; instr path, Western Reserve Univ, 51-53, sr instr biochem, 53-54; res assoc, Sch Med, Univ Pittsburgh, 54-57, asst res prof, 57-63; res assoc prof enzymol & hypersensitivity, 63-67, asst dean acad affairs, 68-72, asst to the pres acad affairs, Univ Health Sci-Chicago Med Sch, 72-75, PROF BIOCHEM, UNIV HEALTH SCI-CHICAGO MED SCH, 67- Mem: AAAS; Am Chem Soc; Am Soc Biol Chemists; Am Calorimetry Conf; NY Acad Sci. Res: Mechanism of enzyme action; protein modifications, calorimetry of biological systems. Mailing Add: Univ Hlth Sci-Chicago Med Sch 2020 W Ogden Ave Chicago IL 60612

MARTIN, CHARLES JOHN, b Sloatsburg, NY, Apr 3, 35; m 59; c 2. APPLIED MATHEMATICS. Educ: Union Col, NY, BS, 56; Mich State Univ, MS, 57; Rensselaer Polytech Inst, PhD(math), 61. Prof Exp: Instr math, Union Col, NY, 58-59; res asst, Rensselaer Polytech Inst, 59-61; sr staff scientist, Res & Advan Develop Div, Avco Corp, 61-66; from assoc prof to prof math, Mich State Univ, 66-75; PROF MATH & HEAD DEPT, WESTERN CAROLINA UNIV, 75- Concurrent Pos: NASA res grant, Mich State Univ, 67-71. Mem: Am Soc Mech Eng. Res: Mechanics. Mailing Add: Dept of Math Western Carolina Univ Cullowhee NC 28723

MARTIN, CHARLES K, b Fayetteville, Tenn, Aug 4, 35; m 56; c 7. MATHEMATICS. Educ: Univ Fla, BA, 62, PhD(math), 66. Prof Exp: Instr math, Univ Fla, 65-66; asst prof, Va Polytech Inst & State Univ, 66-72; ASSOC PROF MATH, GA STATE UNIV, 72-, DIR GRAD STUDIES, 75- Mem: Am Math Soc; Math Asn Am. Res: Theory of rings, algebras and homological algebras; philosophy. Mailing Add: Dept of Math Ga State Univ University Plaza Atlanta GA 30303

MARTIN, CHARLES LOUIS, b Massey, Tex, Dec 2, 93; m 20; c 1. RADIOLOGY. Educ: Univ Tex, EE, 14; Harvard Univ, MD, 19. Prof Exp: Res physician radiol, Mass Gen Hosp, 19-20; from assoc prof to prof, Baylor Univ, 20-43; from prof to clin prof, 43-66, EMER CLIN PROF RADIOL, UNIV TEX HEALTH SCI CTR, DALLAS, 66- Concurrent Pos: Radiologist, Baylor Hosp, 20-40, consult, 43; radiologist, Gaston Episcopal Hosp, 40-, mem bd dirs, 60; dir, Martin X-ray & Radium Clin, 40-66, consult, 66-75; consult, Parkland Hosp, 43- & Vet Admin Hosps, 43- Honors & Awards: Janeway Medal, Am Radium Soc, 49. Mem: Am Roentgen Ray Soc (2nd vpres, 23, secy, 25-27, vpres, 40, pres, 52); Am Radium Soc (vpres, 36, treas, 38-42, pres, 47); fel AMA; fel Am Col Radiol (vpres, 48). Res: Radium therapy; biological effects of radiation; technic and dosage determination in treatment of cancer with radiation; statistical studies showing that radiation can cure certain types of cancer. Mailing Add: 4605 Watauga Rd Dallas TX 75209

MARTIN, CHARLES WELLINGTON, JR, b Omaha, Nebr, Apr 28, 33; m 59; c 3. PETROLOGY. Educ: Dartmouth Col, AB, 54; Univ Wis, MS, 59, PhD(geol), 62. Prof Exp: From asst prof to assoc prof, 60-71, PROF GEOL, EARLHAM COL, 71- Mem: Geol Soc Am; Nat Asn Geol Teachers. Res: Petrology; structural and regional geology of western Connecticut Highlands. Mailing Add: Dept of Geol Earlham Col Richmond IN 47374

MARTIN, CHARLES WILLIAM, b Kansas City, Mo, July 16, 43; m 68; c 2. ORGANIC CHEMISTRY. Educ: Univ Pa, BS, 65; Univ Kans, PhD(chem), 73. Prof Exp: Res chemist process res, 71-72, sr res chemist amine chem, 72-75, RES SPECIALIST AMINE CHEM, DOW CHEM CO, 75- Concurrent Pos: Instr, Saganaw Valley Col, 75. Mem: Am Chem Soc. Res: Development of new amine products and applications with emphasis on the gas processing industry. Mailing Add: Dow Chem Co B1605 Bldg Freeport TX 77541

MARTIN, CHRISTOPHER MICHAEL, b New York, NY, Sept 25, 28; m 54; c 3. MEDICINE. Educ: Harvard Univ, AB, 49, MD, 53. Prof Exp: Intern med, Boston City Hosp, 53-54, asst resident, 56-57; res fel, Thorndike Mem Lab, Boston City Hosp & Harvard Med Sch, 57-59; res fel, Med Found Metrop Boston, Inc, 58-59; from asst prof to assoc prof, Seton Hall Col Med & Dent, 59-65; prof med & pharmacol, Sch Med, Georgetown Univ, 65-70; PROF MED, JEFFERSON MED COL, 70-; SR DIR, MED AFFAIRS, MERCK, SHARP & DOHME RES LABS, 70- Concurrent Pos: Dir, Georgetown Med Div, DC Gen Hosp. Mem: Am Soc Pharmacol & Exp Therapeut; Am Soc Clin Pharmacol & Therapeut; Am Asn Immunol; Infectious Dis Soc Am; Am Soc Microbiol. Res: Infectious diseases; immunology; virology; chemotherapy; virus synthesis; clinical pharmacology; carcinogens. Mailing Add: Med Affairs Merck Sharp & Dohme Res Labs West Point PA 19486

MARTIN, CONSTANCE RIGLER, b Brooklyn, NY, Dec 31, 23; m 43, 71; c 2. ENDOCRINOLOGY. Educ: Long Island Univ, BS, 44; Univ Iowa, PhD(physiol), 51. Prof Exp: Res assoc physiol & pharmacol, NY Med Col, 50-51; sr physiologist, Creedmoor Inst Psychobiol Studies, 51-53; from instr physiol & pharmacol to asst prof physiol & pharmacol, NY Med Col, 53-57; from asst prof to assoc prof biol, Long Island Univ, 59-63; asst prof physiol, 63-66, assoc prof biol sci, 66-76, PROF BIOL SCI, HUNTER COL, 76- Concurrent Pos: Am Cancer Soc Res grant, 65-68. Mem: AAAS; Am Physiol Soc; Endocrine Soc; Soc Study Reproduction; Electron Micros Soc Am. Res: Thymus gland function; reproduction physiology; biological rhythms; electrolyte metabolism. Mailing Add: Dept of Biol Sci Hunter Col 695 Park Ave New York NY 10021

MARTIN, DANIEL S, b Brooklyn, NY, Oct 29, 21; m 47; c 4. SURGERY. Educ: NY Univ, MD, 44. Prof Exp: Resident surg, Col Physicians & Surgeons, Columbia Univ, 49-55, instr, 55-58; assoc prof surg, Sch Med, Univ Miami, 58-68; chmn dept surg, Cath Med Ctr, 68-72. Concurrent Pos: Res fel path & cancer, New Eng Deaconess Hosp, Boston, 48-49; Damon Runyon cancer res fel, 49-50; fel, Dazian Found Med Res, 51; resident & attend, Presby Hosp, Columbia-Presby Med Ctr, 55-58; attend, Jackson Mem Hosp, Miami, 58-68; attend, Hosp Holy Family, St John's, St Joseph's, St Mary's & Mary Immaculate Hosps, New York, 68- Honors & Awards: Mead Johnson Award, 55. Mem: AAAS; Am Geriat Soc; Harvey Soc; Soc Exp Biol & Med; Am Asn Cancer Res. Res: Shock; membrane oxygenator; general surgery; cancer chemotherapy. Mailing Add: 60 Haven Ave Apt 30A New York NY 10032

MARTIN, DANIEL WILLIAM, b Georgetown, Ky, Nov 18, 18; m 41; c 4. PHYSICS. Educ: Georgetown Col, AB, 37; Univ Ill, MS, 39, PhD(physics), 41. Prof Exp: Asst instr, Univ Ill, 37-41; acoust develop engr, Radio Corp Am, 41-49; supvr engr, Acoust Res, Baldwin Piano Co, 49-57, res dir, 57-70, res & eng dir, D H Baldwin Co, 70-74, CHIEF ENGR, D H BALDWIN CO, 74- Concurrent Pos: Instr, Purdue Univ, 41-46; ed, Audio Trans, 54-56; asst prof, Univ Cincinnati, 65- Mem: Fel Acoust Soc Am; fel Audio Eng Soc (exec vpres, 63-64, pres, 64-65); fel Inst Elec & Electronics Engrs. Res: Acoustics of piano, organ, brass wind instruments, auditoriums; sound powered telephones; aircraft intercommunication; microphones; loudspeaker enclosures; reverberation simulation; analog-to-digital encoders; optoelectronics; audio systems. Mailing Add: D H Baldwin Co 1801 Gilbert Ave Cincinnati OH 45202

MARTIN, DANNY BERNARD, limnology, see 12th edition

MARTIN, DAVID GLENN, organic chemistry, see 12th edition

MARTIN, DAVID LEE, b St Louis, Mo, May 30, 41; m 66; c 2. BIOCHEMISTRY. Educ: Univ Minn, St Paul, BS, 63; Univ Wis-Madison, MS, 65, PhD(biochem), 68. Prof Exp: Asst prof, 68-72, ASSOC PROF CHEM, UNIV MD, COLLEGE PARK, 72- Concurrent Pos: Vis scientist, Armed Forces Radiobiol Res Inst, Md, 75- Mem: AAAS; Am Chem Soc; Biochem Soc. Res: Membrane transport of small molecules; neurotransmitter metabolism. Mailing Add: Dept of Chem Univ of Md College Park MD 20742

MARTIN, DAVID P, b New Holland, Pa, Jan 12, 42; m 75; c 2. AGRONOMY. Educ: Goshen Col, BA, 66; Mich State Univ, MS, 70, PhD(crop sci), 72. Prof Exp: Res assoc agron, Mich State Univ, 72-73; ASST PROF AGRON, OHIO STATE UNIV, 73- Mem: Am Soc Agron; Crop Sci Soc Am; Int Turfgrass Soc. Res: Turfgrass management; ecology; pest control. Mailing Add: Dept of Agron Ohio State Univ Columbus OH 43210

MARTIN, DAVID WILLIAM, b Chicago, Ill, Mar 7, 42; m 64; c 2. METEOROLOGY. Educ: Univ Wis, BS, 64, MS, 66, PhD(meteorol), 68. Prof Exp: Meteorologist, Aerophys Br, Phys Sci Lab, Redstone Arsenal, 68-69; asst scientist, 70-75, ASSOC SCIENTIST METEOROL, SPACE SCI & ENG CTR, UNIV WIS, 75- Concurrent Pos: Satellite meteorologist, Global Atmospheric Res Prog Atlantic Trop Exp, Dakar, Senegal, 74. Mem: Am Meteorol Soc; Int Asn Aerobiol. Res: Understanding the structure and behavior of convective systems, particularly tropical; applications of meteorological satellites, including wind measurements, rain estimation from image data and forecasting. Mailing Add: Space Sci & Eng Ctr Univ of Wis 1225 W Dayton St Madison WI 53706

MARTIN, DAVID WILLIS, b Philadelphia, Pa, Sept 19, 27; m 50; c 6. PHYSICS. Educ: Univ Mich, BS, 50, MS, 51, PhD(physics), 57. Prof Exp: Res assoc, Univ Mich, 53-54; resident student assoc nuclear spectros, Argonne Nat Lab, 54-56, res assoc, 56-57; from asst prof to assoc prof physics, 57-65, PROF PHYSICS, GA INST TECHNOL, 65- Concurrent Pos: Consult, Oak Ridge Nat Lab, 65-71. Mem: Am Phys Soc. Res: Ion-molecule reactions in gases at thermal energies; ionization and charge-transfer cross sections in gases at high energies; nuclear spectroscopy. Mailing Add: Sch of Physics Ga Inst of Technol Atlanta GA 30332

MARTIN, DEAN FREDERICK, b Woodburn, Iowa, Apr 6, 33; m 56; c 6. INORGANIC CHEMISTRY. Educ: Grinnell Col, AB, 55; Pa State Univ, PhD(chem), 58. Prof Exp: NSF fel chem, Univ Col, London, 58-59; from instr to asst prof inorg chem, Univ Ill, 59-64; assoc prof, 64-69, PROF INORG CHEM, UNIV S FLA, 69- Concurrent Pos: USPHS career develop award, Nat Inst Gen Med Sci, 69- Mem: AAAS; Am Chem Soc; The Chem Soc. Res: Coordination chemistry; marine chemistry. Mailing Add: Dept of Chem Univ of SFla Tampa FL 33620

MARTIN, DEWAYNE, b Wausau, Wis, July 17, 35; m 57; c 2. PETROLOGY. Educ: Univ Wis, BS, 57, MS, 60. Prof Exp: Asst prof, 61-65, ASSOC PROF GEOL, MINOT STATE COL, 65-, HEAD DEPT PHYS SCI, 69- Concurrent Pos: Co-dir, NASA Regional Space Sci Prog, 73-74. Mem: Nat Asn Geol Teachers; Nat Asn Sci Teachers. Res: Development of earth science program for teacher preparation; petrography and petrology of aerolites and siderolites. Mailing Add: Dept of Phys Sci Minot State Col Minot ND 58701

MARTIN, DIAL FRANKLIN, b Gilmer, Tex, Apr 30, 15; m 39. ENTOMOLOGY. Educ: Agr & Mech Col Tex, BS, 39, MS, 42; Iowa State Col, PhD(entom), 50. Prof Exp: From instr to asst prof entom, Agr & Mech Col Tex, 39-42, from assoc prof to prof, 46-57; leader pink bollworm invests, Entom Res Div, Agr Res Serv, 57-65, asst chief cotton insects res br, 65-68, DIR BIOENVIRON INSECT CONTROL RES LAB, US DELTA STATES AGR RES CTR, AGR RES SERV, USDA, 68- Mem: AAAS; Entom Soc Am; Ecol Soc Am; Am Inst Biol Sci. Res: Cotton insect research; economic entomology. Mailing Add: Bioenviron Insect Control Lab Agr Res Serv USDA Stoneville MS 38776

MARTIN, DON STANLEY, JR, b Indianapolis, Ind, Feb 19, 19; m 49; c 3. PHYSICAL INORGANIC CHEMISTRY. Educ: Purdue Univ, BS, 39; Calif Inst Technol, PhD(chem), 44. Prof Exp: Asst, Nat Defense Res Comt, Calif Inst Technol, 41-42; res assoc, Northwestern Univ, 42-44; assoc scientist, Manhattan Dist, 44-46; from asst prof to assoc prof, 46-55, chemist, AEC, 46-66, PROF CHEM, IOWA STATE UNIV, 55-, SECT CHIEF AMES LAB, ENERGY RES & DEVELOP ADMIN, 66- Mem: Am Chem Soc; Am Phys Soc. Res: Chemistry of the platinum elements; chemical kinetics; absorption spectra of solutions and crystals of coordination compounds; radiochemistry; applications of radioactive materials; inorganic chemistry. Mailing Add: Dept of Chem Iowa State Univ Ames IA 50010

MARTIN, DONALD BECKWITH, b Philadelphia, Pa, July 24, 27; m 56; c 4. MEDICINE. Educ: Haverford Col, AB, 50; Harvard Univ, MD, 54. Prof Exp: From intern to asst resident med, Mass Gen Hosp, 54-56, resident, 58, chief resident, 59; Med Found Boston fel, Nat Heart Inst, 60-61; Fulbright res scholar, Nat Ctr Sci Res, France, 62; from instr to asst prof med, Harvard Med Sch, 63-71; asst in med, 63-71, assoc physician, 71-73, PHYSICIAN, DIABETES UNIT, MASS GEN HOSP, 73-; ASSOC PROF MED, HARVARD MED SCH, 71- Concurrent Pos: Res fel, Harvard Med Sch & Peter B Brigham Hosp, 56-58, USPHS fel, 57-58; assoc ed, Metabolism, 68 & Diabetes; Guggenheim fel, Univ Geneva, 74-75. Mem: Endocrine Soc; fel Am Col Physicians; Am Diabetes Asn; NY Acad Sci; Royal Col Med. Res: Academic medicine; diabetes mellitus; intermediate metabolism; glucose transport in mammalian systems. Mailing Add: Diabetes Unit Mass Gen Hosp Boston MA 02146

MARTIN, DONALD CLAYTON, b Gardner, Mass, Feb 15, 08; m 39; c 2. EXPERIMENTAL PHYSICS. Educ: La State Univ, BS, 29, MS, 31; Cornell Univ, PhD(physics), 36. Prof Exp: Instr physics, La State Univ, 29-31, 34-36 & High Sch, Ala, 36-37; assoc prof, Southeastern La Col, 37-43 & Marshall Col, 43-44; physicist, Receiving Tube Div, Raytheon Mfg Co, Mass, 44-46; prof physics, Marshall Univ, 46-74, chmn dept, 51-74; RETIRED. Mem: AAAS; Am Asn Physics Teachers. Res: Atomic spectra; analysis of the spectrum of selenium II. Mailing Add: 1911 Parkview St Huntington WV 25701

MARTIN, DONALD EFFON, b Three Oaks, Mich, Apr 21, 18; m 42; c 2. METEOROLOGY. Educ: Western Mich Univ, BA, 42; Univ Chicago, MA, 53; Univ Stockholm, PhD(meteorol), 59. Prof Exp: PROF METEOROL, ST LOUIS UNIV, 67- Mem: Am Meteorol Soc. Res: Dynamic modeling; applied research in weather forecasting; severe storms research. Mailing Add: Dept of Meteorol St Louis Univ St Louis MO 63103

MARTIN, DONALD JAMES, organic chemistry, see 12th edition

MARTIN, DONALD RAY, b Marion, Ohio, Oct 21, 15; m 39; c 2. INORGANIC CHEMISTRY. Educ: Otterbein Col, AB, 37; Western Reserve Univ, MS, 40, PhD(inorg chem), 41. Prof Exp: Lectr chem, Cleveland Col, Western Reserve Univ, 40-42, lab mgr, Naval Res Proj, 41-43; from instr to asst prof chem, Univ Ill, 43-51; head chem metall br, Metall Div, US Naval Res Lab, 51-52; lab mgr, Govt Res, Mathieson Chem Corp, 52-56, mgr chem res, Aviation Div, Olin Mathieson Chem Corp, 56-57, assoc dir fuels res, Energy Div, 57-60; dir res, Libbey-Owens-Ford Glass Co, 60-61; dir tech develop, Harshaw Chem Co, 61-63, dir chem res, 63-67, vpres res & develop div, Kewanee Oil Co, 67-68; PROF CHEM & CHMN DEPT, UNIV TEX, ARLINGTON, 69- Concurrent Pos: Res chemist, E I du Pont de Nemours & Co, Inc, 41; mem chem adv comt, Air Force Off Sci Res, 55-60; trustee, Otterbein Col, 62-72. Honors & Awards: Distinguished Sci Achievement Award, Otterbein Col, 70. Mem: AAAS; Am Chem Soc; Electrochem Soc; The Chem Soc; Am Electroplaters Soc. Res: Coordination compounds of boron halides; boron and silicon hydrides; reactions with hydrogen fluoride; fluoroborates; gaseous halides; corrosion; hafnium; inorganic nomenclature; glass; electroplating; color in compounds; gold compounds. Mailing Add: 3311 Cambridge Dr Arlington TX 76013

MARTIN, DONALD STOVER, b Johnstown, Pa, Aug 1, 04; m 27; c 1. BACTERIOLOGY. Educ: Johns Hopkins Univ, AB, 25; Univ Rochester, MD, 30; Columbia Univ, MPH, 46, DPH, 50. Prof Exp: Asst physiol, Univ Rochester, 26-27, asst bact, 31-32, intern, Strong Mem Hosp, 30-31; instr med & bact, Duke Univ, 32-34, assoc, 34-38, from asst prof to assoc prof bact, 36-50; assoc prof, Sch Med, Emory Univ, 52-75; RETIRED. Concurrent Pos: Res assoc, Sch Trop Med & Rockefeller Found fel, Columbia Univ, 45-46; dean, Univ PR, 50-52; chief training br, Bact Sect, 52-57, Ctr Dis Control, USPHS, 57-75. Mem: Am Soc Microbiol; Am Soc Trop Med & Hyg; Am Soc Parasitol; AMA; Am Pub Health Asn. Res: Mycotic diseases; immunology of mycotic diseases; epidemiology of mycotic disease. Mailing Add: 369 S Paseo Altar Green Valley AZ 85614

MARTIN, DOUGLAS LEONARD, b London, Eng, Nov 11, 30. METAL PHYSICS, THERMAL PHYSICS. Educ: Univ London, BSc, 51, PhD(physics), 54, DSc, 70. Prof Exp: Nat Res Coun Can fel, 54-55; sci officer physics, Royal Aircraft Estab, Eng, 55-56; from asst res officer to assoc res officer, 57-64, SR RES OFFICER PHYSICS, NAT RES COUN CAN, 64- Mem: Brit Inst Physics; Can Asn Physicists. Res: Solid state physics; calorimetry; cryogenics. Mailing Add: Div of Physics Nat Res Coun Can Ottawa ON Can

MARTIN, DUNCAN WILLIS, b Durango, Colo, Mar 26, 31; m 64; c 2. PHYSIOLOGY, BIOPHYSICS. Educ: Univ NMex, BS, 55, MS, 56; Univ Ill, Urbana, PhD(physiol), 62. Prof Exp: Asst biol, Univ NMex, 55-56; asst physiol, Univ Ill, 56-60, res asst, 60-62, USPHS trainee, 62; fel, Marine Biol Lab, Woods Hole, 62; fel biophys, Harvard Univ, 62-65; from asst prof to assoc prof, 65-75, PROF ZOOL, UNIV ARK, 75- Concurrent Pos: Vis assoc prof physiol, Yale Univ Sch Med, 74-75. Mem: Am Soc Zool; Am Physiol Soc. Res: Membrane physiology; active transport of ions; energy requirements of active transport systems. Mailing Add: Dept of Zool Univ of Ark Fayetteville AR 72701

MARTIN, EARL CHIAFULLO, organic chemistry, see 12th edition

MARTIN, EDGAR J, b Brno, Czech, Nov 10, 08; US citizen; m 52. TROPICAL MEDICINE, PHARMACOLOGY. Educ: Ger Univ, Prague, MD, 33; Inst Trop Med, Antwerp, Belg, cert, 38. Prof Exp: Res asst biochem, Ger Univ, Prague, 33-35; med officer, Mining Co & Health Serv, Belg Congo, 38-46; res assoc malaria parasites, Inst Microbiol, Univ Montreal, 47-48; res assoc & lectr pharmacol & radiation biol, Univ Toronto, 49-53; med officer, Health Serv, Govt of Am Samoa, 55-57; consult, Environ Health Br, 62-66, MED OFFICER, FOOD & DRUG ADMIN, US DEPT HEALTH, EDUC & WELFARE, 68- Concurrent Pos: NIH grant, Lab Comp Biol, Kaiser Found Res Inst, Calif, 60-64; mem, Galapagos Int Res Proj, Univ Calif & Darwin Found, 64; assoc res physiologist, Univ Calif, Berkeley, 64-67; free lance consult, 64-68. Mem: Am Soc Trop Med & Hyg. Res: Parasitology; drug antigenicity; tropical diseases; kinetics of dialysis; erythrocyte development; radon and radium metabolism; pulmonary gas exchange; pharmacology, chemistry and antigenicity of coelenterate toxins. Mailing Add: Food & Drug Admin Bur of Drugs 5600 Fishers Lane Rockville MD 20852

MARTIN, EDWARD EUGENE, b Roodhouse, Ill, July 28, 26; m 47; c 3. AGRICULTURAL CHEMISTRY. Educ: Northeast Mo State Teachers Col, AB & BS, 49; Univ Mo, MS, 52. Prof Exp: Analyst, Univ Mo, 49-52; feed chemist, Moorman Mfg Co, 52-53; nutrit res chemist, Com Solvents Corp, 53-56; chief chemist, Cent Soya Co, 56-58, prod supvr, 58-60; LAB MGR, RALSTON PURINA CO, 60- Mem: Am Chem Soc; Animal Nutrit Res Coun. Res: Feed research. Mailing Add: Ralston Purina Co 835 S Eighth St St Louis MO 63199

MARTIN, EDWARD SHAFFER, b Terre Haute, Ind, Jan 14, 39. PHYSICAL CHEMISTRY. Educ: DePauw Univ, BA, 60; Northwestern Univ, PhD(chem), 67. Prof Exp: Lectr chem, Ind Univ, South Bend, 65-66, asst prof, 66-72; scientist, 72-74, SR SCIENTIST, ALCOA LABS, ALUMINUM CO AM, 74- Mem: Am Chem Soc; AAAS. Res: Kinetics and thermodynamics applied to the production of high purity alumina and anhydrous aluminum choride. Mailing Add: Alcoa Labs Aluminum Co of Am Alcoa Ctr PA 15069

MARTIN, EDWARD WILLIFORD, b Sumter, SC, Nov 29, 29; m 57; c 3. EMBRYOLOGY. Educ: Fisk Univ, AB, 50; Ind Univ, MA, 52; Univ Iowa, PhD, 62. Prof Exp: Actg head dept biol, Fayetteville State Teachers Col, 52; asst prof zool, 52-65, ASSOC PROF BIOL, PRAIRIE VIEW AGR & MECH COL, 65- Mem: AAAS; Nat Inst Sci; Am Soc Zoologists; Am Inst Biol Sci. Res: Synergic action and individual actions of streptomycin and aureomycin on Brucella abortus and Brucella melentensis. Mailing Add: Dept of Biol Prairie View Agr & Mech Col Prairie View TX 77445

MARTIN, EDWIN PERRY, animal ecology, see 12th edition

MARTIN, ELDEN WILLIAM, b Frankfort, Kans, Feb 2, 32; m 55; c 4. ANIMAL PHYSIOLOGY, ECOLOGY. Educ: Kans State Univ, BS, 54, MS, 59; Univ Ill, PhD(zool, ecol), 65. Prof Exp: Res grant & instr physiol, 63-65, asst prof, 65-69, ASSOC PROF PHYSIOL, BOWLING GREEN STATE UNIV, 69- Concurrent Pos: Frank M Chapman Fund & Marcia Brady Tucker travel awards, 62; Peavey Co res grant, 65-68; fac res grant, Bowling Green State Univ, 65-67; Frank M Chapman Mem Fund grant, 68-69; NSF int travel grant, 70; mem working group granivorous birds, Int Biol Prog; Int Ornith Cong. Mem: AAAS; Am Ornith Union; Wilson Ornith Soc; Am Soc Zool; Int Union Physiol Sci. Res: Physiology and physiological ecology of vertebrate animals; temperature regulation, nutrition and bioenergetics of birds and other animals; effects of gaseous pollutants on the physiology of birds. Mailing Add: Dept of Biol Sci Bowling Green State Univ Bowling Green OH 43403

MARTIN, ERIC WENTWORTH, b Kamloops, BC, Dec 6, 12; nat US; m 40; c 1. MEDICAL COMMUNICATION. Educ: Philadelphia Col Pharm, BSc, 42; Univ Pa, MS, 48, PhD(org & physiol chem), 49. Prof Exp: Pharmacist, Can, 28-36, pharmaceut chemist, 36-39; asst prof, Philadelphia Col Pharm, 49-52; sr res investr, Univ Pa, 52-56; ed, J Am Pharmaceut Asn, 56-59; exec ed, Pfizer Spectrum, Chas Pfizer & Co, Inc, 59-60; dir med commun, Lederle Labs, Am Cyanamid Corp, Pearl River, NY, 60-72; dep asst dir med commun, Bur Drugs, 73-75, DIR, PROF COMMUN, FOOD & DRUG ADMIN, 75- Concurrent Pos: Adj prof biomed commun, Columbia Univ; assoc dir, LaWall & Harrisson Consult Lab, 49-52; ed-in-chief, Remington's Pract Pharm, 42-65, Remington's Pharmaceut Sci, 65-70, Husa's Pharmaceut Dispensing, 59-71, Dispensing of Medication, 71-75 & Hazards of Medication, 71; gen chmn drug info symposia, Drug Info Asn, 73- Mem: Fel AAAS; fel Am Med Writers Asn (pres, 70-71); Drug Info Asn (founder & first pres); fel Int Acad Law & Sci. Res: Medical writing; biochemistry; communication; pharmacology. Mailing Add: PO Box 635 Woodcliff Lake NJ 07675

MARTIN, ETHELBERT COWLEY, b Liverpool, Eng, Nov 1, 10; nat US; m 41; c 3. ENTOMOLOGY. Educ: Univ Toronto, BS, 33; Cornell Univ, MS, 38, PhD(entom), 57. Prof Exp: Lectr apicult, Ont Agr Col, 33-38; lectr entom, Univ Man, 39-42; prov apiarist, Man Dept Agr, 45-50; from asst prof to prof entom, Mich State Univ, 50-75; MEM NAT PROG STAFF CROP POLLINATION, BEES & HONEY, AGR RES SERV, USDA, 75- Concurrent Pos: Tech adv, Man Coop Honey Producers, 45-50; sci adv, Univ Nigeria, 61-63; entom develop consult, Gadjah Mada Univ, Jogjakarta & Agr Inst, Bogor, Indonesia, 71. Mem: Entom Soc Am; Bee Res Asn; Sigma Xi. Res: Physical properties and fermentation of honey; nectar and pollen collecting by bees; pollination of fruit and seed crops. Mailing Add: Nat Prog Staff Rm 412 Bldg 005 Agr Res Serv USDA ARC West Beltsville MD 20705

MARTIN, EUGENE CHRISTOPHER, b Evansville, Ind, Dec 17, 25. ORGANIC POLYMER CHEMISTRY. Educ: Evansville Col, BA, 49; De Paul Univ, MS, 51; Univ Ky, PhD(chem), 54. Prof Exp: From assoc chemist to chemist, Am Oil Co, 54-60; sr res chemist, Southwest Res Inst, 60-71; RES CHEMIST, NAVAL WEAPONS CTR, 71- Mem: Am Chem Soc. Res: Polymer synthesis and modification of polyethylene and polybutadiene, cellulose derivatives, polyurethanes, polyureas and polypeptides for commerical use and biomedical applications; polymer synthesis and modification for membrane separation processes; gelation of liquids; microencapsulation of liquids and solids; organic synthesis. Mailing Add: Michelson Labs Naval Weapons Ctr Code 6058 China Lake CA 93555

MARTIN, FANT W, wildlife management, see 12th edition

MARTIN, FLORENCE THOMAS, biology, chemistry, see 12th edition

MARTIN, FRANCIS HALL, b New Orleans, La, Nov 13, 42; m 74. PHYSICAL BIOCHEMISTRY. Educ: Harvard Univ, BA, 64, PhD(biophys), 70. Prof Exp: Asst prof chem, Miles Col, 69-73; ASST PROF CHEM, SWARTHMORE COL, 74- Res: RNA structure and function; mechanisms of RNA virus replication; RNA-protein recognition. Mailing Add: Dept of Chem Swarthmore Col Swarthmore PA 19081

MARTIN, FRANCIS W, b Minneapolis, Minn, Mar 7, 11; m 44; c 3. PHYSICAL CHEMISTRY. Educ: Univ Minn, BChem, 33, PhD(phys chem), 38. Prof Exp: Instr inorg chem, Univ Mont, 38-39; res assoc, Battelle Mem Inst, 39-40; res chemist, Corning Glass Works, 40-42; res assoc, Radiation Lab, Mass Inst Technol, 42-46; RES ASSOC, CORNING GLASS WORKS, 46- Mem: AAAS; Am Chem Soc; Am Ceramic Soc; Brit Soc Glass Technol. Res: Glass; glass ceramics. Mailing Add: 101 Hornby Dr Painted Post NY 14870

MARTIN, FRANK BURKE, b Cleveland, Ohio, Mar 21, 37; m 61; c 3.

MATHEMATICS. Educ: St Mary's Col, Minn, BA, 58; Iowa State Univ, MS, 66, PhD(statist), 68. Prof Exp: Instr math, St Mary's Col, Minn, 60-63; teaching asst statist, Iowa State Univ, 63-65, res assoc, 65-67; asst prof, 67-68, ASSOC PROF STATIST & DIR STATIST CTR, UNIV MINN, ST PAUL, 68-, EXP STA STATISTICIAN, 67- Mem: Biomet Soc; Am Statist Asn. Res: Data analysis; life sciences; catagorical data; design; statistics; clinical trials. Mailing Add: 726 Lincoln Ave St Paul MN 55105

MARTIN, FRANK ELBERT, b Warrensburg, Mo, Nov 21, 13. SOLID STATE PHYSICS. Educ: Univ Mo, AB, 34, PhD(physics), 63; Univ Ill, MS, 56. Prof Exp: Instr sr high sch, Mo, 38-42; instr physics, Little Rock Jr Col, 42-43; from instr to asst prof, Cent Mo State Col, 43-58; physicist, Metall Div, US Naval Res Lab, DC, 45-54; asst instr math, Univ Ill, 55-56; instr physics, Univ Mo, 58-59; 60-62; from asst prof to assoc prof, 62-67, PROF PHYSICS, CENT MO STATE COL, 67- Concurrent Pos: NSF equip grant, Cent Mo State Col. Mem: AAAS; Am Phys Soc; Sigma Xi; Inst Elec & Electronics Eng; Am Asn Physics Teachers. Res: Elastic constants; electric contact transients; low-carbon steel dilatometry; photoelectric emission; electronic structure of semiconductors; cryogenics; charge carriers and thermal conductivity in crystalline solids. Mailing Add: 123 W South St Warrensburg MO 64903

MARTIN, FRANK GARLAND, b New Orleans, La, Oct 9, 32; m 55; c 3. EXPERIMENTAL STATISTICS. Educ: Okla State Univ, BS, 54, MS, 55; NC State Univ, PhD(exp statist), 59. Prof Exp: Sr scientist, Bettis Atomic Power Lab, Westinghouse Elec Corp, Pa, 58-62; res statistician, Stamford Res Lab, Am Cyanamid Co, Conn, 62-64; ASSOC PROF STATIST, UNIV FLA, 64- Concurrent Pos: Consult, Fla Agr Exp Sta. Mem: Am Statist Asn; Biomet Soc. Mailing Add: Dept of Statist Univ of Fla Gainesville FL 32601

MARTIN, FRANK GENE, b Clarksville, Tenn, Mar 15, 38; m 59; c 2. PHARMACOLOGY. Educ: Univ Tenn, BSPh, 59, MS, 65, PhD(pharmacol), 69. Prof Exp: Teaching fel pharmacol, Univ Tenn, 63-68; asst prof, 68-72, ASSOC PROF, SCH PHARM, UNIV KANS, 72-, CHMN DEPT PHARM PRACT, 73- Mem: AAAS; Sigma Xi. Res: Autonomic pharmacology, especially release and degradation of transmitters. Mailing Add: Dept of Pharm Pract Univ of Kans Sch Pharm Lawrence KS 66045

MARTIN, FRANK LIONEL, b Montreal, Que, May 16, 15; nat US; m 41; c 4. METEOROLOGY. Educ: Univ BC, BA, 36, MA, 38; Univ Chicago, PhD(math), 41. Prof Exp: Instr physics & math, Pa State Col, 40-41; from meteorologist to area chief meteorologist, Northwest Airlines, Minn, 42-47; assoc prof, 47-55, PROF METEOROL, NAVAL POSTGRAD SCH, 55- Concurrent Pos: Consult, Fleet Numerical Weather Facility, 61-65 & Lab Electronics, 62-64; vis scientist, Nat Ctr Atmospheric Res, Colo, 63-64; consult, US Navy Proj FAMOS, 66; Nat Acad Sci sr res assoc, Goddard Space Flight Ctr, Md, 68-69. Mem: Am Meteorol Soc; Royal Meteorol Soc. Res: Satellite meteorology; retrieval of temperature profiles from multichannel carbon dioxide fine-scale radiometers; radiational budget calculations of the earth-atmosphere system, with cloud-amount parameterization at gridpoints. Mailing Add: Dept of Meteorol Naval Postgrad Sch Monterey CA 93940

MARTIN, FRANK STEPHEN, b Salem, Ind, Oct 8, 10; m 37; c 2. CHEMISTRY. Educ: Kans State Col, BS, 33; Iowa State Col, PhD(chem), 38. Prof Exp: Asst chem, Iowa State Col, 33-35, instr, 35-38; instr phys chem, Fordham Univ, 38-39; develop chemist, 39-53, MGR PROVIDENCE LABS, US RUBBER CO, 53- Mem: Am Chem Soc. Res: Raman spectra studies of organo-mercury compounds; compound, process and product development; rubber products; natural rubber; synthetic rubbers and resins. Mailing Add: 11 Belmont Rd Cranston RI 02910

MARTIN, FRANK WINSTEAD, b St Louis, Mo, Sept 19, 22; m 51; c 2. BOTANY. Educ: St Louis Col Pharm, BS, 49; Wash Univ, AM, 53, PhD(bot), 58. Prof Exp: Instr biol & pharmacog, St Louis Col Pharm, 50-56; from assoc prof to prof pharmacog, Northeast La State Col, 56-68; dir bur financial aid & res, 68-72, VPRES RES & PLANNING, NORTHWESTERN STATE UNIV, 72- Concurrent Pos: Comnr, La Higher Educ Asst Comn Bd, 74-; mem, Nat Coun Univ Res Adminr. Mem: AAAS; Asn Inst Res. Mailing Add: Res & Planning Northwestern State Univ Natchitoches LA 71457

MARTIN, FRANKLIN WAYNE, b Salt Lake City, Utah, Apr 14, 28; m 56; c 4. GENETICS. Educ: Okla Baptist Univ, BS, 48; Univ Calif, PhD(genetics), 60. Prof Exp: Sr lab technician, Univ Calif, 54-60; asst horticulturist, Western Wash Exp Sta, 60-61; mem staff, Fed Exp Sta, 61-71, DIR, MAYAGUEZ INST TROP AGR, AGR RES SERV, PR, 71- Mem: Genetics Soc Am; Am Genetic Asn; Bot Soc Am; Am Soc Hort Sci; Soc Econ Bot. Res: Genetics, physiology and evolution of systems of incompatability and sterility in plants; genetics, breeding and development of tropical root and tuber crops; introduction and development of little known tropical fruits and vegetables. Mailing Add: Mayaguez Inst of Trop Agr PO Box 70 Mayaguez PR 00708

MARTIN, FRED ELI, organic chemistry, see 12th edition

MARTIN, FREDDIE ANTHONY, b Raceland, La, Nov 17, 45; m 69; c 3. PLANT PHYSIOLOGY, CROP PHYSIOLOGY. Educ: Nicholls State Col, BS, 66; Cornell Univ, MS, 68, PhD(veg crops), 70. Prof Exp: ASST PROF PLANT PHYSIOL, LA STATE UNIV, BATON ROUGE, 71- Mem: Am Soc Plant Physiologists; Crop Sci Soc Am; Am Soc Sugarcane Technologists; Am Soc Agronomists. Res: Determination of the biochemical and physiological basis for differences in yielding ability that are known to exist among genetic strains of Saccharum. Mailing Add: Dept of Plant Path La State Univ Baton Rouge LA 70803

MARTIN, FREDERICK JOHNSON, b York, Maine, Sept 6, 15; m 43; c 3. PHYSICAL CHEMISTRY. Educ: Bates Col, BS, 37; Mass Inst Technol, PhD(phys chem), 41. Prof Exp: Chemist, S D Warren Pulp & Paper Co, Maine, 40-41; phys chemist, Nat Defense Res Comt Explosives Res Lab, Pa, 42-45; sect leader, Manhattan Dist, Los Alamos Sci Lab, 45; phys chemist, Res Found, Ohio State Univ, 45-46 & M W Kellogg Co, NJ, 46-54; PHYS CHEMIST, RES LAB, GEN ELEC CO, 54- Mem: AAAS; Am Chem Soc; Combustion Inst. Res: Explosives; propellants; flames and detonations; flame retardant polymers; combustion of coal-derived fuels. Mailing Add: 2504 Peters Lane Schenectady NY 12309

MARTIN, FREDERICK N, b Brooklyn, NY, July 24, 31; m 54; c 2. AUDIOLOGY. Educ: Brooklyn Col, BA, 57, MA, 58; City Univ New York, PhD(speech), 68. Prof Exp: Speech therapist, Lenox Hill Hosp, 57-58; audiologist, Ark Rehab Serv, 58-60 & Bailey Ear clin, 60-66; lectr, Speech & Hearing Ctr, Brooklyn Col, 66-68, asst prof audiol, 68; from asst prof to assoc prof, 68-74, PROF AUDIOL, UNIV TEX, AUSTIN, 74- Mem: Am Speech & Hearing Asn. Res: Clinical audiology and normal audition. Mailing Add: Speech & Hearing Clin Univ of Tex Austin TX 78712

MARTIN, FREDERICK WIGHT, b Boston, Mass, Feb 16, 36; m 65. EXPERIMENTAL PHYSICS. Educ: Princeton Univ, AB, 57; Yale Univ, MS, 58, PhD(physics), 64. Prof Exp: From physicist to sr physicist, Ion Physics Corp, High Voltage Eng Corp, 63-66; asst prof atomic & solid state physics, Aarhus Univ, Denmark, 66-68; res assoc atomic physics, Univ Ky, 68-69, asst prof, 69-70; ASST PROF PHYSICS & ASTRON, UNIV MD, COLLEGE PARK, 70- Mem: Am Phys Soc. Res: Penetration of high energy particles in matter; single atomic collisions involving electron capture or loss or x-ray production by heavy ions; channeling, ion implantation and radiation damage in solids; microscopy. Mailing Add: Dept of Physics Univ of Md College Park MD 20742

MARTIN, GENE ELLIS, b Omaha, Nebr, Apr 4, 26; m 50; c 2. GEOGRAPHY. Educ: Univ Wash, BA, 49, MA, 52; Syracuse Univ, PhD(geog), 55. Prof Exp: Prof geog, Univ Chile, 54; asst prof, Univ Ga, 55-56; from asst prof to assoc prof, Univ Ore, 56-68, prof, 68-73, head dept, 69-72; SCH BEHAV & SOCIAL SCI, CALIF STATE UNIV, CHICO, 73- Concurrent Pos: Fulbright prof, Univ Buenos Aires, Arg, 59; adv planning, US AID, Guatemala, 66-68. Mem: Asn Am Geogr. Res: Latin American geography. Mailing Add: Sch of Behav & Social Sci Calif State Univ Chico CA 95926

MARTIN, GEOFFREY, chemistry, see 12th edition

MARTIN, GEOFFREY JOHN, b London, Eng, Mar 9, 34; m 65; c 2. GEOGRAPMY. Educ: London Sch Econ, BSc, 56; Univ London, PGCE, 57; Univ Fla, MA, 58. Prof Exp: Asst prof geog, Eastern Mich Univ, 59-65; assoc prof, Southern Conn State Col, 65-69, PROF GEOG, SOUTHERN CONN STATE COL, 69- Concurrent Pos: Nat Coun Geog Educ grant, 61; Am Coun Learned Socs grants, 69-70 & 72-73; NSF grant, 72; mem, Geog Thought Comn, Int Geog Union. Mem: Asn Am Geog; cor mem Int Geog Union. Res: History of geographic thought with special reference to the post-Darwinian period. Mailing Add: Dept of Geog Southern Conn State Col New Haven CT 06515

MARTIN, GEORGE C, b San Francisco, Calif, Sept 15, 33; m 53; c 2. POMOLOGY. Educ: Calif State Polytech Col, BS, 55; Purdue Univ, MS, 60, PhD(plant physiol), 62. Prof Exp: Res asst plant physiol & hort, Purdue Univ, 58-62; res plant physiologist, Crops Res Div, Agr Res Serv, USDA, Wash, 62-67; assoc pomologist, 67-73, POMOLOGIST, UNIV CALIF, DAVIS, 73- Honors & Awards: J H Gourley Award in Pomol. Mem: Am Soc Hort Sci; Am Soc Plant Physiol; Phytochem Soc NAm. Res: Chemical thinning; mechanism of fruit set, dormancy and rest; use of chemicals to aid mechanical harvest of fruit. Mailing Add: Dept of Pomol Univ of Calif Davis CA 95616

MARTIN, GEORGE CARLYLE, b Philadelphia, Tenn, Nov 10, 09; m 42; c 3. GEOLOGY. Educ: Univ Tenn, AB, 31, MA, 32; Ohio State Univ, PhD(geol), 40. Prof Exp: Asst geogr, Tenn Valley Authority, 34; instr geol, Univ Tenn, 35-38; PROF GEOG, E CAROLINA UNIV, 46- Res: Physical geography; physiography; soil conservation. Mailing Add: Dept of Geog ECarolina Univ Greenville NC 27834

MARTIN, GEORGE EDWARD, b Batavia, NY, July 3, 32; m 69. GEOMETRY. Educ: State Univ NY Albany, AB, 54, MA, 55; Univ Mich, PhD(math), 64. Prof Exp: Asst prof math, Univ RI, 64-66; asst prof, 66-70, ASSOC PROF MATH, STATE UNIV NY ALBANY, 70- Mem: Am Math Soc; Math Asn Am. Res: Geometry, specializing in projective planes. Mailing Add: Dept of Math State Univ of NY Albany NY 12222

MARTIN, GEORGE FRANKLIN, JR, b Englewood, NJ, Feb 20, 37; m 60; c 2. NEUROANATOMY. Educ: Bob Jones Univ, BS, 60; Univ Ala, MS, 63, PhD(anat), 65. Prof Exp: From instr to assoc prof, 65-73, PROF ANAT, COL MED, OHIO STATE UNIV, 73- Concurrent Pos: NIH res grants, 65-75. Mem: AAAS; Soc Neurosci; Am Asn Anat; Pan-Am Asn Anat. Res: Determining the various connections and functions of motor systems in primitive mammals. Mailing Add: Dept of Anat Ohio State Univ Col of Med Columbus OH 43210

MARTIN, GEORGE LLOYD, b Valencia, Pa, Dec 10, 19; m 42; c 3. INORGANIC CHEMISTRY. Educ: Tarkio Col, BA, 41; Ohio State Univ, PhD(chem), 47. Prof Exp: Asst chem, Ohio State Univ, 41-43; instr, Muskingum Col, 43-44; lab supvr, Mallinckrodt Chem Works, 44-46, group leader, 47-55, sect supvr uranium div, 55-57; asst dir chem dept, Nat Res Corp, 57-58, tech dir metals div, 58-59; asst prof chem, Univ Ala, 59-61; asst gen mgr metals div, Nat Res Corp, Mass, 61-63, gen mgr res div, 63-68; vpres, Norton Res Corp, 68-71; mgr high temperature chem, Kawecki Berylco Industs, 71-74; GEN MGR, M & R REFRACTORY METALS, WINSLOW, NJ, 74- Mem: Am Chem Soc; Electrochem Soc. Res: Inorganic separations; liquid-liquid extraction; ion exchange; preparation of high purity metals and compounds. Mailing Add: M & R Refractory Metals Winslow NJ 08095

MARTIN, GEORGE MONROE, b New York, NY, June 30, 27; m 52; c 4. EXPERIMENTAL PATHOLOGY. Educ: Univ Wash, BS, 49, MD, 53. Prof Exp: Eleanor Roosevelt Int Cancer Res fel, Inst Molecular Biol, Paris & fel med, surg & gynec, Montreal Gen Hosp, 53-54; asst resident path, Univ Chicago, 54-55, asst, 55-56, instr, 56-57; from asst prof to assoc prof path, 60-68, dir cytogenetics lab, Hosp, 64-68, dir med scientist training prog, 70-73, PROF PATH, UNIV WASH, 68-, ATTEND PATHOLOGIST, 59-, ADJ PROF GENETICS, 75- Concurrent Pos: Consult, Firlands Sanitarium, Seattle, Wash, 57-59 & Northern State Hosp, Sedro Woolley, 59-63; mem path B Study sect, NIH, 66-70, mem adult develop & aging res comn, 73-, chmn aging res rev comt, 75- Mem: Am Soc Exp Path; Am Soc Human Genetics; Tissue Cult Asn; Genetics Soc Am; assoc Am Asn Path & Bact. Res: Mammalian cell culture; somatic cell and human biochemical genetics; cell senescence. Mailing Add: Dept of Path Univ of Wash Seattle WA 98195

MARTIN, GEORGE REILLY, b Boston, Mass, Jan 20, 33. PHARMACOLOGY. Educ: Colgate Univ, AB, 55; Univ Rochester, PhD, 59. Prof Exp: Asst, Atomic Energy Proj, Univ Rochester, 55-58; guest worker, Nat Heart Inst, 58-59, res assoc, Nat Inst Dent Res, 59-67, CHIEF CONNECTIVE TISSUE SECT, NAT INST DENT RES, 67- Mem: Am Chem Soc; Biophys Soc; Am Soc Biol Chem. Res: Structure and metabolism of collagen and elastin; chemistry of connective tissue. Mailing Add: Connective Tissue Sect Nat Inst of Dent Res Bethesda MD 20014

MARTIN, GERALD CHARLES, JR, b Berrien Center, Mich, Feb 21, 37; m 65. NUCLEAR CHEMISTRY. Educ: Western Mich Univ, BS, 59; Univ Notre Dame, MS, 62, PhD(nuclear chem), 64. Prof Exp: NUCLEAR CHEMIST, BOILING WATER REACTOR SYSTS DEPT, VALLECITOS NUCLEAR CTR, GEN ELEC CO, 64- Mem: Am Nuclear Soc; Am Soc Testing & Mat; Am Nat Stand Inst. Res: Absolute measurements of radioactivities and neutron flux. Mailing Add: Vallecitos Nuclear Ctr Gen Elec Co Vallecitos Rd Pleasanton CA 94566

MARTIN, GLENN ELLIS, b Eureka, Mont, May 22, 47; m 70. PHYSICAL CHEMISTRY. Educ: NDak State Univ, BS, 69, PhD(phys chem), 74. Prof Exp: SR CHEMIST, INMONT CORP, 74- Mem: Am Chem Soc. Res: Synthesizing polymers to be used for industrial coatings with emphasis in automobile coatings. Mailing Add: Inmont Corp 6125 Industrial Pkwy Whitehouse OH 43571

MARTIN, GORDON EUGENE, b San Diego, Calif, Aug 22, 25; m 49; c 4. PHYSICS,

ENGINEERING. Educ: Univ Calif, Berkeley, BS, 47; Univ Calif, Los Angeles, MS, 51; San Diego State Col, MA, 61; Univ Tex, PhD, 66. Prof Exp: Electronic scientist, US Navy Electronics Lab, 47-52, supvry physicist, 54-74, SR RES PHYSICIST, US NAVAL UNDERSEAS CTR, 74- Concurrent Pos: Contract instr, San Diego State Col, 57-59. Mem: Acoust Soc Am; Inst Elec & Electronics Engrs; NY Acad Sci. Res: Underwater electroacoustic arrays, including near field and radiation impedance phenomena and effects of scattering, especially on optimum array design; wave propagation in viscoelastic piezoelectric solids of crystalline and polycrystalline forms. Mailing Add: US Naval Underseas Ctr San Diego CA 92132

MARTIN, GORDON MATHER, b Brookline, Mass, Mar 2, 15; m 40; c 3. PHYSICAL MEDICINE. Educ: Nebr Wesleyan Univ, AB, 36; Univ Nebr, MD, 40; Univ Minn, MS, 44. Prof Exp: Asst prof phys med, Sch Med, Univ Kans, 44-47; from asst prof to assoc prof, Mayo Grad Sch Med, 47-73, PROF PHYS MED, MAYO MED SCH, UNIV MINN, 73- Concurrent Pos: Consult, Mayo Clin, 47-73, sr consult phys med & rehab, 73- Mem: Am Cong Rehab Med; Am Acad Phys Med & Rehab; AMA. Res: Clinical research. Mailing Add: Mayo Clin 200 First St SW Rochester MN 55901

MARTIN, GORDON WYATT, b Newberg, Ore, Aug 1, 37; m 59; c 2. PARASITOLOGY, DEVELOPMENTAL BIOLOGY. Educ: Portland State Col, BS, 61; Ore State Univ, MA, 64, PhD(zool), 65. Prof Exp: Asst prof biol, Seattle Pac Col, 65-70; NSF sci fac res fel, Tulane Univ, 70-71; asst prof zool, Univ Mont, 71-72; vpres, Edutramix Ltd, 72-74; PRES, BASILEA PRODS, 75- Concurrent Pos: Res grant, Nat Inst Allergy & Infectious Dis, 67-70. Mem: Am Soc Parasitol; Am Micros Soc. Res: Blood and tissue fluid development in amphibia; invertebrate immunoserology; trematode development; biological programed instruction. Mailing Add: Basilea Prods PO Box 214 Newberg OR 97132

MARTIN, HANS CARL, b Winnipeg, Man, Nov 20, 37; m 66; c 1. MICROMETEOROLOGY. Educ: Univ Man, BSc, 58; Univ Western Ont, MSc, 61, PhD(physics), 66. Prof Exp: Res scientist meteorol physics, Commonwealth Sci & Indust Res Orgn, Australia, 66-68; RES SCIENTIST METEOROL PHYSICS, CAN METEOROL SERV, DEPT ENVIRON, 69- Mem: Am Meteorol Soc; Royal Meteorol Soc. Res: Air-sea interactions; energy exchange processes near the surface of the earth both on land and over water. Mailing Add: 4905 Dufferin St Downview ON Can

MARTIN, HAROLD ROLAND, b White Co, Ind, Nov 11, 19; m 52; c 3. MEDICINE. Educ: Purdue Univ, BS, 42; Ind Univ, MD, 44. Prof Exp: First asst & spec asst psychiat, Mayo Clin, 50-52; from asst to clin dir, Inst Living, 52-54; from asst prof to assoc prof neurol & psychiat, Col Med, Univ Nebr, 54-60; CONSULT PSYCHIATRIST, MAYO CLIN, 60-, ASSOC PROF PSYCHIAT, MAYO MED SCH, UNIV MINN, 70- Concurrent Pos: Clin dir, Adult Inpatient Serv, Nebr Psychiat Inst, 54-; consult, Vet Admin Hosp, Omaha, Nebr, 55 & State Div Rehabil Servs, Nebr, 59. Mem: AAAS; AMA; fel Am Psychiat Asn. Res: Psychiatry; rehabilitation. Mailing Add: Dept of Psychiat Mayo Clin 200 First St SW Rochester MN 55901

MARTIN, HELEN EASTMAN, b Alameda, Calif, May 15, 06. INTERNAL MEDICINE. Educ: Pomona Col, AB, 27; Univ Southern Calif, MD, 34. Hon Degrees: LLD, Univ Southern Calif, 72. Prof Exp: Asst path, 36-37, asst med, 37-38, instr physiol, 38-42, instr med, 42-45, from asst prof to prof, 45-68, EMER PROF MED, SCH MED, UNIV SOUTHERN CALIF, 68- Mem: Am Med Asn; Am Diabetes Asn; Am Fedn Clin Res. Res: Diabetes; fluid and electrolyte metabolism. Mailing Add: 413 Scott Pl Pasadena CA 91103

MARTIN, HERBERT LLOYD, b Somerville, Mass, Dec 7, 21; m 51; c 6. NEUROLOGY. Educ: Boston Univ, BS, 47, MD, 50. Prof Exp: Teaching fel neurol, Montreal Neurol Inst, McGill, 57-58; assoc prof clin neurol, 58-69, PROF NEUROL, UNIV VT, 69-, CHMN DEPT, 71- Mem: AMA; Am Acad Neurol; Am Epilepsy Soc; Am Asn Res Nerv & Ment Dis; Asn Am Med Cols. Mailing Add: Dept of Neurol Univ of Vt Burlington VT 05401

MARTIN, HORACE F, b Azores, Jan 11, 31; US citizen; m 54; c 7. BIOCHEMISTRY, ANALYTICAL CHEMISTRY. Educ: Providence Col, BS, 53; Univ RI, MS, 55; Boston Univ, PhD(biochem), 61; Brown Univ, MA(ad eundum), 67, MD, 75. Prof Exp: Sr res chemist, Monsanto chem Co, 57-59, group leader life sci, Monsanto Res Corp, 61-63; BIOCHEMIST, RI HOSP, PROVIDENCE, 63-; ASSOC PROF MED SCI, BROWN UNIV, 66- Concurrent Pos: Consult, Monsanto Chem Co, 63-64. Mem: Am Chem Soc; Asn Comput Mach. Res: Automation and analytical procedures; clinical pathology; normal values; instrumental methods of analysis. Mailing Add: 879 Mineral Spring Ave Pawtucket RI 02860

MARTIN, HUGH JACK, JR, b San Diego, Calif, Sept 1, 26; m 50; c 6. HIGH ENERGY PHYSICS. Educ: Calif Inst Technol, BS, 51, PhD(physics), 56. Prof Exp: Res assoc, 55-57, from asst prof to assoc prof, 57-65, PROF PHYSICS, IND UNIV, BLOOMINGTON, 65- Res: Experimental high energy, bubble chamber and spark chamber physics; pattern recognition and computer applications. Mailing Add: Dept of Physics Col Arts & Sci Ind Univ Bloomington IN 47405

MARTIN, IRVING, b Brooklyn, NY, Sept 18, 12; m 45; c 2. CHEMISTRY. Educ: City Col New York, BS, 34; Polytech Inst Brooklyn, MS, 52. Prof Exp: Biochemist, Res Labs, Mt Sinai Hosp, New York, 35-37; jr chemist, Res & Develop Div, Edgewood Arsenal, 41-44; chemist, 45-60, SR CHEMIST, RES DEPT, NAT STARCH & CHEM CORP, 61- Mem: AAAS; Am Chem Soc. Res: Biochemistry; starch; polymers; chemical literature. Mailing Add: 166 De Lacy Ave North Plainfield NJ 07060

MARTIN, JACK, b Tuscaloosa, Ala, Aug 11, 27; m 57; c 4. PSYCHIATRY. Educ: Univ Ala, BS, 49; Vanderbilt Univ, MD, 53. Prof Exp: Intern, Charity Hosp, New Orleans, La, 53-54; resident physician gen psychiat, Cincinnati Gen Hosp, 54-56, res fel child psychiat, Cincinnati Gen Hosp & Child Guid Home, 56-58; from instr to asst prof, 58-63, CLIN PROF PSYCHIAT, UNIV TEX HEALTH SCI CTR DALLAS, 63- Concurrent Pos: Med dir & pres, Shady Brook Schs. Res: Child development; clinical child psychiatry. Mailing Add: 3508 Beverly Dr Dallas TX 75205

MARTIN, JACK E, b Bogard, Mo, June 4, 31; m 55; c 3. NUTRITION, BIOCHEMISTRY. Educ: Univ Mo, BS, 53, MS, 60; Univ Fla, PhD(nutrit), 63. Prof Exp: Prod supvr biochem, Monsanto Co, Mo, 63-67; asst res nutritionist, Ralston Purina Co, 67-69; nutritionist, Ceres Land Co, 69-71; NUTRITIONIST, STERLING NUTRIT SERV, INC, 71- Mem: Am Soc Animal Sci. Res: Effect of mineral nutrition on cellulose digestion in ruminants. Mailing Add: Rte 3 Sterling CO 80751

MARTIN, JAMES CULLEN, b Dover, Tenn, Jan 14, 28; m 51; c 5. ORGANIC CHEMISTRY. Educ: Vanderbilt Univ, BA, 51, MS, 52; Harvard Univ, PhD(chem), 56. Prof Exp: From instr to assoc prof, 56-65, PROF ORG CHEM, UNIV ILL, URBANA, 65- Concurrent Pos: Sloan Found fel, 62-66; Guggenheim Mem Found fel, 65-66; assoc mem, Ctr Advan Study, Univ Ill, Urbana, 71-72. Mem: Am Chem Soc. Res: Mechanisms of organic reactions; free-radical reactions; synthesis of compounds expected to show unusual physical properties or reactivity; compounds of hypervalent sulfur; sulfuranes. Mailing Add: Dept of Chem Univ of Ill Urbana IL 61801

MARTIN, JAMES CUTHBERT, b Wilson, NC, May 8, 27; m 66. INDUSTRIAL ORGANIC CHEMISTRY. Educ: Univ NC, BS, 47. Prof Exp: From res chemist to sr res chemist, 48-67, res assoc, 67-72, SR RES ASSOC, TENN EASTMAN CO, 72- Mem: Am Chem Soc. Res: Chemistry of ketenes; small ring compounds; new polymer systems; applied organic chemistry; exploratory research in catalysis; new product development. Mailing Add: Res Labs Tenn Eastman Co Kingsport TN 37662

MARTIN, JAMES D, b Michigan City, Ind, Nov 24, 34; m 55; c 4. ORGANIC CHEMISTRY. Educ: Purdue Univ, BS, 56, MS, 58. Prof Exp: Chief propellant chem, Thiokol Chem Corp, 64-66; head propellant chem, 66-67, chief solid propellant chem, 67, MGR, SOLID PROPELLANT DEPT, ATLANTIC RES CORP, 67- Mem: Am Chem Soc; Am Inst Aeronaut & Astronaut. Res: Solid propellant research and development; binder chemistry; high energy and fuel rich propellants; controllable and slurry propellants. Mailing Add: Atlantic Res Corp Alexandria VA 22314

MARTIN, JAMES EDWIN, b Waynesboro, Ga, Oct 31, 23; m 46; c 3. PHYSICS. Educ: Mercer Univ, AB, 43; Univ Ga, MS, 48. Prof Exp: Instr physics, Univ Ga, 49-51; physicist, Oak Ridge Nat Lab, 51-53; head dept physics & math, Norman Col, 53-57; asst prof physics, Mercer Univ, 57-58; assoc prof, 58-66, dir-admis, 66-70, PROF PHYSICS & ASTRON, VALDOSTA STATE COL, 66- Mem: Am Phys Soc. Res: Nuclear physics; cosmic radiation; electronics. Mailing Add: Dept of Physics Valdosta State Col Valdosta GA 31601

MARTIN, JAMES FRANKLIN, b Harrison Co, WVa, Nov 24, 24; m 46. MICROBIOLOGY. Educ: WVa Univ, AB, 52, MS, 56; Nat Registry Microbiologists, registered. Prof Exp: Res microbiologist, Ralph M Parsons Co, Ft Detrick, Md, 54-55 & Lambert Pharmacal Co, St Louis, Mo, 55-56; scientist microbiol, 56-72, SR SCIENTIST MICROBIOL, WARNER-LAMBERT RES INST, 72- Mem: Am Soc Microbiol; Soc Indust Microbiol. Res: Pharmaceutical research and development of synthetic and natural antimicrobial agents; research and development of disinfection agents and systems. Mailing Add: Warner-Lambert Res Inst Mt Tabor Rd Morris Plains NJ 07950

MARTIN, JAMES FRANKLIN, b St Mary's, WVa, Mar 20, 17; m 42; c 2. MEDICINE, RADIOLOGY. Educ: Marietta Col, AB, 38; Western Reserve Univ, MD, 42. Prof Exp: Teaching fel radiol, Western Reserve Univ Hosp, 47-48, demonstr radiol, Western Reserve Univ, 48, from instr to sr instr, 48-50; from asst prof to assoc prof, 50-61, prof radiol, 61-75, PROF MED SONICS, ASST RADIOL, BOWMAN GRAY SCH MED, 75- Concurrent Pos: Physician, NC Baptist Hosp. Mem: Radiol Soc NAm; Am Roentgen Ray Soc; AMA; fel Am Col Radiol. Res: Clinical radiology. Mailing Add: Dept of Radiol Bowman Gray Sch of Med Winston-Salem NC 27103

MARTIN, JAMES GRUBBS, organic chemistry, see 12th edition

MARTIN, JAMES HAROLD, b Collinwood, Tenn, Oct 12, 31; m 54; c 2. DAIRY MICROBIOLOGY. Educ: Univ Tenn, BSc, 57; Ohio State Univ, MSc, 58, PhD(microbiol), 63. Prof Exp: Res asst dairy tech, Ohio State Univ, 57-58, res assoc, 60-63, asst prof, 63-65; instr dairy mfg, Miss State Univ, 58-60; from asst prof to assoc prof dairy microbiol, Univ Ga, 65-72; PROF DAIRY SCI & HEAD DEPT, SDAK STATE UNIV, 72- Mem: Am Soc Microbiol; Am Dairy Sci Asn; Inst Food Technol. Res: Physiology and metabolism of microorganisms common to milk, especially as related to bacterial spores, starter and spoilage organisms. Mailing Add: Dept of Dairy Sci SDak State Univ Brookings SD 57006

MARTIN, JAMES HAROLD, b Clarksville, Tex, May 12, 38; m 61; c 3. ANATOMY. Educ: ETex State Univ, BS, 61; Baylor Univ, MS, 66, PhD(anat), 68. Prof Exp: DIR ELECTRON MICROS LABS, MED CTR, BAYLOR UNIV, 62-, ASST PROF ANAT, 68- Concurrent Pos: Nat Inst Arthritis & Metab Dis grant, Med Ctr, Baylor Univ, 69-72. Mem: Am Asn Anat. Res: Calcification, calcium, phosphate, ion movement and homeostasis in tissues and cells. Mailing Add: Dept of Path Baylor Univ Med Ctr Dallas TX 75226

MARTIN, JAMES HENRY, III, b New Orleans, La, Mar 31, 43; m 66; c 2. VERTEBRATE PHYSIOLOGY. Educ: Univ Va, BA, 65; Univ Richmond, MS, 67; Univ Tenn, PhD(zool), 70. Prof Exp: Instr anat & physiol, Univ Tenn, Knoxville, 68; NIH fel & instr physiol, Med Col Va, Va Commonwealth Univ, 71-73; asst prof natural sci, 73-76, ASSOC PROF BIOL, J S REYNOLDS COMMUNITY COL, 76- Concurrent Pos: Vis lectr, Math & Sci Ctr, Richmond, Va, 72- Mem: AAAS; Am Soc Ichthyologists & Herpetologists; Am Soc Zoologists. Res: Physiology and biophysics of muscle contraction; transport across epithelial membranes. Mailing Add: Dept of Biol J S Reynolds Community Col Richmond VA 23228

MARTIN, JAMES JOHN, JR, b Paterson, NJ, Feb 3, 36; m 54; c 5. OPERATIONS RESEARCH, SYSTEMS ANALYSIS. Educ: Univ Wis, BA, 55; US Naval Postgrad Sch, MS, 63; Mass Inst Technol, PhD(opers res), 65. Prof Exp: US Navy, 57-, dir strategic retaliatory div, Off Asst Secy Defense Systs Anal, 65-73, OPERS RES SPECIALIST, US NAVY, 65-, SPEC ASST, OFF ASST SECY DEFENSE ATOMIC ENERGY, 73- Mem: Opers Res Soc Am; Mil Opers Res Soc. Res: Statistical decision theory. Mailing Add: 123 N Hickory Rd Sterling VA 22170

MARTIN, JAMES MILTON, b Waxahachie, Tex, May 15, 14; m 41; c 3. GEOPHYSICS. Educ: Univ Okla, BS, 38; Rensselaer Polytech Inst, MS, 46. Prof Exp: Seismic explor, Geophys Serv, Inc, 34-35 & 37-39; seismol & seismic prospecting, Magnolia Petrol Co, 40-41; tech eval, 42 & 47-57, chief underwater eval dept, 57-73, PROJ MGR TORPEDOES, US NAVAL ORD LAB, MD, 73- Honors & Awards: Superior Civil Serv Award, Dept Navy, 72. Mem: Soc Explor Geophys; Am Geophys Union. Res: Technical evaluation of naval ordnance; geophysics. Mailing Add: 314 Williamsburg Dr Silver Spring MD 20901

MARTIN, JAMES PAXMAN, b Cowley, Wyo, Sept 22, 14; m 37; c 2. SOIL MICROBIOLOGY. Educ: Brigham Young Univ, BS, 38; Rutgers Univ, PhD(soil microbiol), 41. Prof Exp: Asst soil microbiol, NJ Exp Sta, 38-41; coop agent, Soil Conserv Serv, USDA & NJ Exp Sta, 41-43; asst prof bact & asst soil microbiologist exp sta, Univ Idaho, 43-45; from asst chemist to assoc chemist, 45-57, CHEMIST, CITRUS RES CTR, UNIV CALIF, RIVERSIDE, 57-, PROF SOIL SCI, 61- Mem: Fel AAAS; Am Soc Microbiol; Soil Sci Soc Am; fel Am Soc Agron. Res: Contribution of synthesized microbial products to soil aggregation and humus; citrus replant problem; influence of pesticides on soil properties. Mailing Add: Dept of Soil Sci & Agr Eng Univ of Calif Riverside CA 92502

MARTIN, JAMES TILLISON, b Bluefield, WVa, Aug 10, 46. BEHAVIORAL PHYSIOLOGY. Educ: WVa Univ, AB, 67; Univ Conn, MS, 71; Univ Munich, PhD(zool), 74. Prof Exp: Asst pharmacologist, RMI Inst Pharmacol, State Univ Utrecht, 73-74; ASSOC ANIMAL SCI, UNIV MINN, ST PAUL, 74- Concurrent

Pos: Translr, French Nat Mus Natural Hist, 73-; Nat Inst Child Health & Human Develop fel, 74- Res: Behavioral and endocrine factors involved in the domestication process; neuroendocrine basis of emotional behavior and adrenosteroid influences on reproductive function. Mailing Add: Dept of Animal Sci Univ of Minn St Paul MN 55108

MARTIN, JEROME, b Bisbee, Ariz, Jan 12, 02; m 28; c 2. CHEMISTRY. Educ: Univ Colo, BS, 24; Univ Calif, PhD(phys chem), 28. Prof Exp: Res chemist, Com Solvents Corp, 27-39, res dir, 39-57, sci dir, 57-67; CONSULT, 67- Mem: AAAS; Am Chem Soc; Math Asn Am. Res: Catalysis; organic chemistry; antibiotics. Mailing Add: 1334 S Center St Terre Haute IN 47802

MARTIN, JERRY JUNIOR, b Darwin, Okla, Oct 28, 30; m 53; c 4. ANIMAL NUTRITION, ANIMAL PHYSIOLOGY. Educ: Okla State Univ, BS, 57, MS, 59, PhD(animal nutrit & physiol), 61. Prof Exp: Instr animal sci, Murray State Agr Col, 61-67; ASSOC PROF BIOL & ANIMAL SCI, PANHANDLE STATE COL, 67-, CHMN DIV AGR, 73- Mem: Am Inst Biol Sci; Am Soc Animal Sci. Res: High moisture content feed. Mailing Add: Dept of Biol & Animal Sci Panhandle State Col Goodwell OK 73939

MARTIN, JERRY ROY, microbiology, biochemistry, see 12th edition

MARTIN, JOEL JEROME, b Jamestown, NDak, Mar 27, 39. SOLID STATE PHYSICS. Educ: SDak Sch Mines & Technol, BS, 61, MS, 63; Iowa State Univ, PhD(physics), 67. Prof Exp: AEC fel, Ames Lab, Iowa State Univ, 67-69; asst prof physics, 69-74, ASSOC PROF PHYSICS, OKLA STATE UNIV, 74- Mem: Am Phys Soc. Res: Thermal and electrical properties of semiconductors; thermal conductivity of metals. Mailing Add: Dept of Physics Okla State Univ Stillwater OK 74074

MARTIN, JOHN BENNETT, JR, agronomy, chemistry, see 12th edition

MARTIN, JOHN DAVID, b Chicago, Ill, Nov 8, 39; m 62; c 2. NUCLEAR PHYSICS, ATMOSPHERIC PHYSICS. Educ: Va Mil Inst, BS, 61; Col William & Mary, MA, 63; Univ Fla, PhD(physics), 67. Prof Exp: Nuclear res officer physics, McClellan Cent Labs, McClellan Air Force Base, 67-70; PHYSICIST, TELEDYNE ISOTOPES, 70- Mem: Am Phys Soc. Res: Research in measurement of stable and radioactive trace gases in the atmosphere; development of nuclear counting techniques for measurement of environmental-level fission and activation radioisotopes. Mailing Add: Teledyne Isotopes 50 Van Buren Ave Westwood NJ 07675

MARTIN, JOHN ELMSLIE, b Terre Haute, Ind, Dec 17, 42; m 67; c 1. INORGANIC CHEMISTRY. Educ: DePauw Univ, BA, 64; Univ Ark, PhD(inorg chem), 69. Prof Exp: Fel, Univ Southern Calif, 68-70 & Univ Perugia, Italy, 70-71; res assoc radiation chem, Argonne Nat Lab, 71-73; SR CHEMIST INORG/ANAL CHEM, COLLINS RADIO GROUP, ROCKWELL INT, 73- Mem: Am Chem Soc; Electrochem Soc. Res: Analytical and chemical processing techniques for metal oxide semiconductor device fabrication. Mailing Add: Collins Radio Group Rockwell Int 4311 Jamboree Rd Newport Beach CA 92663

MARTIN, JOHN F, b Youngstown, Ohio, Oct 29, 23; m 43; c 2. ANALYTICAL CHEMISTRY. Educ: Youngstown Col, BS, 51; Chemist, 51-56, SR RES CHEMIST & HEAD VACUUM-FUSION GAS ANAL LAB, RES CTR, US STEEL CORP, 56- Concurrent Pos: US Steel Corp indust fel, 67-68; res assoc, Nat Bur Standards, 67-68. Res: Gases in metals by vacuum fusion; gas chromatography and mass spectrometry; carbon in steel; analysis of surface gases and films. Mailing Add: Res Ctr US Steel Corp Monroeville PA 15146

MARTIN, JOHN H, b Gosport, Ind, June 30, 16; m 42; c 4. ENVIRONMENTAL PHYSICS. Educ: Ind Univ, BS, 38; Wash Univ, PhD(physics), 49. Prof Exp: Asst physicist, US Signal Corps, 41-43 & Navy Div War Res, Univ Calif, 43-46; asst prof physics, Forman Christian Col, Lahore, Pakistan, 50-54; assoc physicist, 55-67, sr physicist & assoc div dir, 67-72, DIR SOLAR ENERGY PROGS, ARGONNE NAT LAB, 72- Concurrent Pos: Prof, Col Petrol & Minerals, Dhahran, Saudi Arabia, 68-70. Mem: AAAS; Am Phys Soc; Sigma Xi. Res: Underwater acoustics; cosmic rays; nuclear physics; atomic collisions; high energy particle accelerators; self sufficient residential life-style. Mailing Add: Argonne Nat Lab HEF-360 9700 S Cass Ave Argonne IL 60439

MARTIN, JOHN HARVEY, b Chambersburg, Pa, Jan 20, 32; m 58; c 3. INTERNAL MEDICINE, RHEUMATOLOGY. Educ: Gettysburg Col, BA, 54; Temple Univ, MD, 58; Mayo Grad Sch Med, Univ Minn, MS, 62; Am Bd Internal Med, dipl, 65, cert med, 74. Prof Exp: From asst to staff med, Mayo Clin, 64-65; from instr to asst prof, 66-70, ASSOC PROF MED, SCH MED, TEMPLE UNIV, 70- Concurrent Pos: Fel rheumatol, Temple Univ, 65-66; P S Hench scholar, Mayo Grad Sch Med, Univ Minn, 66. Mem: Fel Am Col Physicians; Am Rheumatism Asn; Am Asn Clin Res. Res: Clinical research. Mailing Add: Dept of Med Sect Rheumatol Temple Univ Hosp Philadelphia PA 19140

MARTIN, JOHN HOLLAND, b Old Lyme, Conn, Feb 27, 35; m 69; c 2. OCEANOGRAPHY, POLLUTION BIOLOGY. Educ: Colby Col, BA, 59; Univ RI, MS, 64, PhD(oceanog), 66. Prof Exp: Assoc scientist, PR Nuclear Ctr, Univ PR, Mayagüez, 66-69; sr scientist, Hopkins Marine Sta, Stanford Univ, 69-70, NSF fel, 70-71, asst prof oceanog, 71-72; asst prof, 72-75, ASSOC PROF BIOL, CALIF STATE UNIV, SAN FRANCISCO, 75- Concurrent Pos: Corresp, Sci Comt Prob Environ, 75-; partic, Am-Soviet Agreement Protection Environ Workshop, 76. Mem: Am Soc Limnol & Oceanog. Res: Trace elements in sea water and marine organisms; heavy metal pollution in the marine environment. Mailing Add: Moss Landing Marine Labs Moss Landing CA 95039

MARTIN, JOHN KENNETH, b Eng, Apr 11, 18; m 41; c 4. PEDIATRICS. Educ: Univ London, MB, Bs, 42; Royal Col Physicians & Surgeons, Can, cert, 50. Prof Exp: Pediatrician, Winnipeg Clin, Man, 50-57; prof pediat & head dept, Univ Alta, 57-71; MED DIR, GLENDALE LODGE HOSP, 71- Concurrent Pos: Instr, Univ Man, 50-57. Mem: Can Paediat Soc; Can Med Asn; Royal Soc Med; Brit Med Asn. Res: Pediatric neurology; handicapped children. Mailing Add: Glendale Lodge Hosp 4464 Markham St Victoria BC Can

MARTIN, JOHN LEE, b Houston, Tex, Nov 19, 23; m 48; c 4. BIOCHEMISTRY. Educ: Southern Methodist Univ, BS, 49; Univ Ark, MS, 53; Tex A&M Univ, PhD(biochem, nutrit), 56. Prof Exp: Instr chem, Colo State Univ, 55-57, asst prof, 57-59; prof & head dept, Baker Univ, 59-60; assoc prof, Colo State Univ, 60-67; PROF CHEM, METROP STATE COL, 67- Concurrent Pos: Affiliate prof, Colo State Univ, 67- Mem: AAAS; Am Chem Soc; Am Inst Nutrit. Res: Selenium-sulfur interrelationships; selenium enhancement of the immune response. Mailing Add: Dept of Chem Metrop State Col Denver CO 80204

MARTIN, JOHN PERRY, JR, b Dunbar, Pa. PHYSICAL CHEMISTRY, ANALYTICAL CHEMISTRY. Educ: Carnegie Inst Technol, BS, 47, MS, 55, PhD(chem), 62. Hon Degrees: MHL, Davis & Elkins Col. Prof Exp: Asst chem, Metals Res Lab, Carnegie Inst Technol, 47-50; chemist, Dunbar Corp, 50-52; chief chemist, Duraloy Co, 52-59; asst prof, 62-65, from actg chmn to chmn dept, 62-68, ASSOC PROF CHEM, DAVIS & ELKINS COL, 65- Concurrent Pos: Consult, Pa Wire Glass Co, 50-52. Mem: Am Chem Soc; Am Inst Chemists; Sigma Xi. Res: X-ray, ultraviolet visible and infra-red spectroscopy; hydrogen bonding of secondary amines with polar organic compounds and of amine complexes. Mailing Add: Dept of Chem Davis & Elkins Col Elkins WV 26241

MARTIN, JOHN ROBERT, b Lancaster, Pa, Dec 6, 23; m 45; c 3. ANALYTICAL CHEMISTRY. Educ: Goshen Col, AB, 44; Pa State Col, MS, 49, PhD(chem), 50. Prof Exp: Res chemist, 50-51, supvr, 51-64, DIV HEAD, E I DU PONT DE NEMOURS & CO, INC, 64- Mem: Am Chem Soc. Res: Application of ion exchange to analytical chemistry; analysis of fluoro compounds; functional group analysis; microanalysis. Mailing Add: 213 Hullihen Dr Newark DE 19711

MARTIN, JOHN SAMUEL, b Philadelphia, Pa, Oct 5, 43. PHYSIOLOGY, BIOPHYSICS. Educ: Temple Univ, AB, 65; Woman's Med Col, MS, 68; Thomas Jefferson Univ, PhD, 73. Prof Exp: Instr biol, Holy Family Col, 71-72; instr, 72-74, ASST PROF PHYSIOL, SCH DENT, TEMPLE UNIV, 74- Concurrent Pos: Smith Kline & French Labs fel, Dept Physiol & Biophys, Sch Dent, Temple Univ, 72-74. Mem: Am Fedn Clin Res. Res: Gastrointestinal and autonomic physiology; upper airway physiology; membrane phenomena. Mailing Add: Dept of Physiol & Biophys Temple Univ Sch of Dent Philadelphia PA 19140

MARTIN, JOHN SCOTT, b Toronto, Ont, Sept 1, 34; m 57. PHYSICAL CHEMISTRY. Educ: Univ Toronto, BA, 56; Columbia Univ, PhD(chem), 62. Prof Exp: Fel chem, Nat Res Coun Can, 61-63; asst prof, 63-69, ASSOC PROF CHEM, UNIV ALTA, 69- Mem: Chem Inst Can. Res: Hydrogen bonding and ionic solvation processes by nuclear magnetic resonance; structure and spectra of bihalide ions; computer analysis of nuclear magnetic resonance spectra of symmetric molecules. Mailing Add: Dept of Chem Univ of Alta Edmonton AB Can

MARTIN, JOHN WALTER, JR, b Hagerstown, Md, Oct 31, 22; m 43; c 4. PHARMACEUTICAL CHEMISTRY. Educ: Bridgewater Col, BA, 47; Med Col Va, BS, 49; Univ NC, PhD(pharmaceut chem), 52. Prof Exp: From asst prof to assoc prof pharmaceut chem, Col Pharm, Butler Univ, 52-61; PROF CHEM, BRIDGEWATER COL, 61- Concurrent Pos: Fulbright lectr, Cairo Univ, 65-66; fel, Univ London, 70-71. Mem: Am Chem Soc; AAAS. Res: Derivatives of amino acids and nitrogen heterocycles. Mailing Add: Dept of Chem Bridgewater Col Bridgewater VA 22812

MARTIN, JOSEPH POURCHER, planetary science, particle physics, see 12th edition

MARTIN, JULIA MAE, b Snow Hill, Md, Nov 9, 24. BIOCHEMISTRY. Educ: Tuskegee Inst, BS, 46, MS, 48; Pa State Univ, PhD(biochem), 63. Prof Exp: Instr chem, Tuskegee Inst, 48-49; from instr to asst prof, Fla Agr & Mech Univ, 49-59; assoc prof, Tuskegee Inst & res assoc, Carver Res Found, 63-66; PROF CHEM, SOUTHERN UNIV, BATON ROUGE, 66-, ACTG DEAN GRAD SCH & A&M COL, 74- Mem: Am Chem Soc; AAAS; fel Am Inst Chemists; NY Acad Sci; Nat Inst Sci. Res: Biochemical abnormalities of red blood cells of patients with hemolytic disorders. Mailing Add: Dept of Chem Southern Univ Box 9608 Baton Rouge LA 70813

MARTIN, JULIO MARIO, b Salta, Arg, Sept 16, 22; m 53; c 3. PHYSIOLOGY. Educ: Nat Univ La Plata, MD, 50. Prof Exp: Res asst, Inst Biol & Exp Med, Univ Arg, 51-55; assoc prof physiol, Nat Univ La Plata, 56-60; res fel exp path, Wash Univ, 61-63; asst prof, Univ Toronto, 63-71; asst scientist, 63-70, ASSOC SCIENTIST, RES INST, HOSP SICK CHILDREN, 70-; ASSOC PROF PHYSIOL, UNIV TORONTO, 71- Concurrent Pos: Squibb res fel, 51-53. Mem: Am Diabetes Asn; Can Physiol Soc; Arg Med Asn; Arg Physiol Soc. Res: Experimental diabetes; pancreatic islets development; insulin synthesis and release; neuroendocrine control of insulin secretion; relationship between growth hormone and beta-cells activity. Mailing Add: Res Inst 555 University Ave Toronto ON Can

MARTIN, KATHRYN HELEN, b Hartford, Conn, Oct 5, 40. HISTOPHYSIOLOGY. Educ: Cath Univ Am, AB, 62; Cornell Univ, PhD(zool), 73. Prof Exp: ASST PROF ZOOL, STATE UNIV NY COL OSWEGO, 72- Mem: Sigma Xi; Am Soc Mammalogists; AAAS. Res: Histophysiological analysis of reproductive and stress phenomena in small mammals. Mailing Add: Dept of Zool State Univ NY Col Piez Hall Oswego NY 13126

MARTIN, KENNETH EDWARD, b Cheney, Kans, Apr 5, 44; m 65; c 1. MATHEMATICS. Educ: St Benedict's Col, Kans, BA, 64; Ind Univ, Bloomington, MA, 66; Univ Notre Dame, PhD(math), 70. Prof Exp: Asst prof, 70-74, ASSOC PROF MATH, GONZAGA UNIV, 74- Mem: Math Asn Am. Res: Group theory and its generalizations; algebraic number theory. Mailing Add: Dept of Math Gonzaga Univ Spokane WA 99258

MARTIN, KENNETH JOHN, b New York, NY, July 16, 31; m 53; c 5. CLINICAL CHEMISTRY, ANALYTICAL CHEMISTRY. Educ: Univ St Louis, BS, 53; Univ Wis, PhD(electrochem), 60. Prof Exp: Res chemist, Redstone Res Labs, Ala, 60-68, tech liaison, Micromedic Systs, MICROMEDIC SYSTS, INC, ROHM AND HAAS CO, 74- Mem: Am Chem Soc; Am Asn Clin Chem. Res: Clinical chemistry; electroanalytical chemistry. Mailing Add: Micromedic Systs Inc NL Statenplein 56 Dordrecht Netherlands

MARTIN, KENNETH ROBERT, b Baltimore, Md, Apr 6, 21. GEOGRAPHY. Educ: Md State Teachers Col, Towson, BS, 43; Univ Wis, MS, 49, PhD(geog), 56. Prof Exp: Instr geog, Hunter Col, 55-56; PROF GEOG, WESTERN ILL UNIV, 56- Mem: Asn Am Geogr. Res: Regional geography of Europe; agricultural geography; physiography. Mailing Add: Dept of Geog & Geol Western Ill Univ Macomb IL 61455

MARTIN, KENNETH ROGER, organic chemistry, see 12th edition

MARTIN, LARRY DEAN, b Bartlett, Nebr, Dec 8, 43; m 67; c 2. VERTEBRATE PALEONTOLOGY. Educ: Univ Nebr, BS, 66, MS, 69; Univ Kans, PhD(biol), 73. Prof Exp: ASST PROF SYSTS & ECOL, UNIV KANS & ASST CUR VERT PALEONT, MUS NATURAL HIST, 72- Concurrent Pos: Res affil, Univ Nebr State Mus, 72-; mem, US Nat Working Group Neogene Quaternary Boundry, 74- Mem: Soc Vert Paleont; Am Soc Mammalogists; Am Quaternary Asn. Res: Fossil history of certain birds, rodents and Saber-toothed cats with emphasis on functional morphology; relationship between climatic history and vertebrate extinctions. Mailing Add: Mus of Natural Hist Univ of Kans Lawrence KS 66045

MARTIN, LAURENCE ROBBIN, physical chemistry, see 12th edition

MARTIN, LEROY BROWN, JR, b Elkin, NC, June 6, 26; m 61; c 3. MATHEMATICS, OPERATIONS RESEARCH. Educ: Wake Forest Col, BS, 49; NC

State Univ, MS, 52; Harvard Univ, MS, 53, PhD(appl math), 58. Prof Exp: Appl sci rep, Int Bus Mach Corp, 55-56, spec rep, Serv Bur Corp, 56-59, asst mgr planning & develop, 59-61; from asst prof to assoc prof math, 61-68, PROF COMPUT SCI, DIR COMPUT CTR & ASST PROVOST, NC STATE UNIV 68- Mem: AAAS; Am Asn Comput Mach; Inst Mgt Sci; Soc Indust & Appl Math. Res: Mathematical optimization; numerical analysis; management science. Mailing Add: Comput Ctr Box 5445 NC State Univ Raleigh NC 27607

MARTIN, LESTER W, b Edwards, Mo, Aug 15, 23; m 49; c 5. PEDIATRIC SURGERY. Educ: Univ Mo, BS, 44, BSc, 47; Harvard Med Sch, MD, 49; Am Bd Surg, dipl, 57. Prof Exp: From asst prof to assoc prof, 57-72, PROF SURG, COL MED, UNIV CINCINNATI, 72-; DIR PEDIAT SURG, CHILDREN'S HOSP, 57- Mem: Affil fel Am Acad Pediat; fel Am Col Surg; AMA; Brit Asn Paediat Surg. Res: Various aspects of surgery of infancy and childhood; esophageal anomalies and Hirschsprung's disease. Mailing Add: Children's Hosp 240 Bethesda Ave Cincinnati OH 45229

MARTIN, LLOYD MILO, b Bayard, Nebr, May 24, 12; m 38; c 1. ENTOMOLOGY. Prof Exp: Asst, 36-37, laborer, 37-38, from preparator to asst preparator, 38-43, curatorial asst, 43-44, from asst cur to assoc cur, 44-69, EMER CUR & RES ASSOC ENTOM, LOS ANGELES COUNTY MUS NATURAL HIST, 69- Mem: Soc Syst Zool; Lepidop Soc (pres, 72). Res: Lepidoptera of southwestern United States, especially families Hesperiidae, Acontiiae and Notodontidae. Mailing Add: 2063 Kachina Dr Prescott AZ 86301

MARTIN, LOREN GENE, b Danville, Ill, Oct 31, 42; m 67; c 1. PHYSIOLOGY, ENDOCRINOLOGY. Educ: Ind Univ, AB, 64, PhD(physiol), 69. Prof Exp: Air Force grant high altitude physiol, Ind Univ, 69-70; from instr to asst prof physiol, Sch Med, Temple Univ, 70-73; ASST PROF PHYSIOL, PEORIA SCH OF MED, COL MED, UNIV ILL, PEORIA, 73- Concurrent Pos: Lectr, Community Col Philadelphia, 70-72, Gwyned-Mercy Col, 72-73 & Eureka Col, 74-; mem, Int Study Group Res Cardiac Metab. Mem: Am Physiol Soc. Res: Environmental physiology; endocrine adaptations. Mailing Add: Peoria Sch of Med Univ of Ill Col of Med Peoria IL 61606

MARTIN, LOUIS NORBERT, b Baton Rouge, La, July 29, 42; m 68; c 1. IMMUNOLOGY. Educ: Tulane Univ, BS, 66, PhD(microbiol), 72. Prof Exp: ASSOC SCIENTIST IMMUNOL, DELTA REGIONAL PRIMATE CTR, TULANE UNIV, 73- Res: Ontogeny of immune function in chickens; oncogenic herpes viruses in non-human primates; function of immunoglobulin D. Mailing Add: Delta Regional Primate Res Ctr Tulane Univ Covington LA 70433

MARTIN, M CELINE, b Huntington, Ind, Oct 14, 09. ECOLOGY, BIOLOGY. Educ: Ind Univ, BS, 40; Loyola Univ, Ill, MEd, 47; Univ Notre Dame, MS, 48, PhD(biol), 63. Prof Exp: Assoc prof, 48-60, PROF BIOL, ST FRANCIS COL, IND, 63-, CHMN DEPT, 48- Mem: Nat Asn Biol Teachers; Ecol Soc Am; Am Inst Biol Sci. Res: Life history and ecology of the American wild geranium; effect of anesthesia on the protein levels of blood plasma. Mailing Add: Dept of Biol St Francis Col 2701 Spring St Ft Wayne IN 46808

MARTIN, MALCOLM MENCER, b Vienna, Austria, Dec 10, 20; US citizen; m 62; c 3. PEDIATRIC ENDOCRINOLOGY. Educ: Univ Durham, MB, BS, 45, MD, 52; Am Bd Internal Med, dipl & cert, 66; FRCP, 72. Prof Exp: Resident, Postgrad Sch Med, Univ London, 48-50, first asst, Diabetic & Metab Unit, King's Col Hosp, 50-53, registr, Med Unit, 53-56; physician, Out-Patient Dept, Harriet Lane Home, Johns Hopkins Hosp, 57; asst med, Peter Bent Brigham Hosp, Boston, 57-59; from asst prof to assoc prof pediat, 59-67, PROF PEDIAT & MED, SCH MED, GEORGETOWN UNIV, 67- Concurrent Pos: Ministry Educ fel, Eng, 48-49; Lund res fel, Brit Diabetes Asn, 50-51; King's Col res grant, 52-55; Leverhulme res fel, Inst Clin Res, Middlesex Hosp Med Sch, 56; NIH spec res fel, 57-59; Lederle fac award, 62-65; consult endocrinol & metab dis, Childrens' Convalescent Hosp, DC, 63; mem acad staff, Childrens' Hosp, DC, 63; mem Worcester Found, 66. Mem: AAAS; Endocrine Soc; Am Fedn Clin Res; Am Soc Human Genetics; Am Diabetes Asn. Res: Endocrinology and metabolism, particularly as related to growth and development. Mailing Add: Dept of Pediat Georgetown Univ Med Ctr Washington DC 20007

MARTIN, MARGARET EILEEN, b Albright, WVa, Oct 17, 15. BIOCHEMISTRY. Educ: WVa Wesleyan Col, BS, 40; Georgetown Univ, MS, 54, PhD(chem), 58. Prof Exp: Teacher pub schs, WVa, 35-43; med technician, Emergency Hosp, Wash, DC, 47-49 & St Elizabeth's Hosp, 49-52; chemist, US Food & Drug Admin, 52-54, Phys Biol Lab, Nat Inst Arthritis & Metab Dis, 54, Nat Heart Inst, 54-57 & Food Qual Lab, Agr Res Ctr, USDA, 57-62; CHEMIST, LAB BR, ST ELIZABETH'S HOSP, DEPT HEALTH, EDUC & WELFARE, WASH, DC, 62- Concurrent Pos: Mem spec ment health res neurochem sect, Nat Inst Ment Health, 70. Mem: Am Chem Soc; Am Inst Chemists; Am Soc Microbiol; AAAS. Res: Basic biochemistry; mental illness; toxicology; clinical chemistry. Mailing Add: 4006 Rickover Rd Silver Spring MD 20902

MARTIN, MARGARET ELIZABETH, b New York, NY, May 6, 12. STATISTICS. Educ: Barnard Col, AB, 33; Columbia Univ, MA, 34, PhD(econ), 42. Prof Exp: Economist, Div Placement & Unemployment Ins, NY State Dept Labor, 38-42; anal statistician, Statist Stand Div, US Bur Budget, 43-67; asst chief, Statist Policy Div, Off Mgt & Budget, 67-72; EXEC DIR, COMT NAT STATIST, NAT ACAD SCI-NAT RES COUN, 73- Concurrent Pos: Exec secy, President's Comt Appraise Employment & Unemployment Statist, 61-62; Nat Acad Sci rep, Int Statist Inst Biennial Meeting, Warsaw, 75. Honors & Awards: Dir Except Serv Award, US Bur Budget, 68. Mem: AAAS; fel Am Statist Asn; Am Econ Asn; Int Statist Inst; Pop Asn Am. Res: Statistical planning and coordination with application especially in demographic and economic statistics; confidentiality of statistical records. Mailing Add: Comt Nat Statist Nat Acad Sci-Nat Res Coun Washington DC 20418

MARTIN, MARGARET PEARL, b Duluth, Minn, Apr 22, 15. FOREST BIOMETRY. Educ: Univ Minn, BA, 37, MA, 39, PhD(math), Prof Exp: Instr biostatist, Univ Minn, 40-41 & Columbia Univ, 42-45; statist consult, Health Dept, NY, 45; asst prof biostatist, Univ Minn, 45-46; from asst prof to assoc prof prev med, Vanderbilt Univ, 47-58; assoc prof biostatist, Sch Hyg & Health, Johns Hopkins Univ, 59-64; asst prof biomet, 67-68, PRIN BIOMETRICIAN, UNIV MINN, ST PAUL, N CENT FOREST EXP STA, USDA, 68- Mem: Fel Am Statist Asn; Biomet Soc; Inst Math Statist. Res: Applied statistics; biological fields. Mailing Add: 1366 Selby Ave St Paul MN 55104

MARTIN, MARILYNN KAY, b Niagara Falls, NY, Oct 19, 42; div. ANTHROPOLOGY. Educ: State Univ NY Buffalo, BA, 64, MA, 66, PhD(anthrop), 70. Prof Exp: ASST PROF ANTHROP, UNIV CALIF, SANTA BARBARA, 69- Mem: Am Ethnol Soc; Am Anthrop Asn. Res: Kinship and social organization; cultural evolution; role of women; Africa. Mailing Add: Dept of Anthrop Univ of Calif Santa Barbara CA 93106

MARTIN, MARK WAYNE, b Twin Falls, Idaho, June 17, 30; m 52; c 5. GENETICS. Educ: Univ Idaho, BS, 52; Cornell Univ, MS, 54, PhD(plant breeding), 59. Prof Exp: Geneticist, Agr Res Serv, USDA, Utah State Univ, 59-67, GENETICIST, RES & EXTEN CTR, AGR RES SERV, USDA, Wash, 67- Mem: Am Soc Hort Sci; Am Phytopath Soc. Res: Breeding vegetables; disease resistance, especially resistance to virus diseases. Mailing Add: USDA Res & Exten Ctr Prosser WA 99350

MARTIN, MARTIN CLAUDE, b Mulgrave, NS, May 27, 21; m 49; c 4. EXPERIMENTAL SOLID STATE PHYSICS. Educ: St Francis Xavier Univ, Can, BSc, 45; Univ Western Ont, MSc, 51; Univ Alta, PhD, 56. Prof Exp: Lectr physics, Univ Man, 46-49; lectr, Univ Alta, 51-53, asst prof, 53-56; sci officer, Defense Res Bd, 56-57; asst prof, 57-62, ASSOC PROF PHYSICS, CLARKSON COL TECHNOL, 62- Mem: Am Phys Soc; Am Asn Physics Teachers; Can Asn Physicists. Res: Solid state physics. Mailing Add: Dept of Physics Clarkson Col of Technol Potsdam NY 13676

MARTIN, MICHAEL, b Vallejo, Calif, Jan 16, 43; m 67; c 1. POLLUTION BIOLOGY. Educ: Univ Calif, Davis, BS, 65; Sacramento State Col, MA, 67; Univ Southern Calif, PhD(biol), 72. Prof Exp: Instr physiol & biol, Glendale Community Col, 71; head instr ichthyol & freshwater ecol, NAm Sch Conserv & Ecol, 71, dean, 71-73; ASSOC WATER QUAL BIOLOGIST, CALIF DEPT FISH & GAME, 73- Mem: AAAS; Am Soc Ichthyol & Herpet; Am Fisheries Soc. Res: Vertebrate biology, morphology, systematics and ecology of fishes; marine pollution; heavy metal toxicity. Mailing Add: 2201 Garden Rd Monterey CA 93940

MARTIN, MICHAEL MCCULLOCH, b Junction City, Kans, Mar 21, 35; m 65; c 1. BIOLOGICAL CHEMISTRY. Educ: Cornell Univ, AB, 55; Univ Ill, PhD(org chem), 58. Prof Exp: NSF res fel, Mass Inst Technol, 58-59; from instr to assoc prof chem, 59-70, assoc prof zool, 68-70, PROF CHEM & ZOOL, UNIV MICH, ANN ARBOR, 70- Concurrent Pos: Consult, Socony-Mobil Oil Co, 59-64; res chemist, Entom Div, USDA, 64; Sloan Found fel, 66-68. Mem: AAAS; Am Chem Soc; Am Soc Zoologists; Am Inst Biol Sci; Soc Study Evolution. Res: Molecular aspects of ecological interactions and physiological adaptations; chemistry of insect natural products. Mailing Add: Dept of Chem Univ of Mich Ann Arbor MI 48104

MARTIN, MONROE HARNISH, b Lancaster, Pa, Feb 7, 07; m 32; c 1. APPLIED MATHEMATICS. Educ: Lebanon Valley Col, BS, 28; Johns Hopkins Univ, PhD(math), 32. Hon Degrees: DSc, Lebanon Valley Col, 58. Prof Exp: Nat res fel, Harvard Univ, 32-33; instr math, Trinity Col, 33-36; from asst prof to prof, 36-68, from actg head dept to head dept, 42-53, from actg dir to dir, Inst Fluid Dynamics & Appl Math, 53-68, res prof, 68-71, EMER RES PROF MATH, UNIV MD, COLLEGE PARK, 71- Concurrent Pos: Mem, US Nat Comt Theoret & Appl Math, 53-56; exec secy, Div Math, Nat Acad Sci-Nat Res Coun, 55-57 & 58-59, chmn comt appl math, 58-59, mem, 60-61; Guggenheim fel, 60; hon lectr, Univ St Andrews, 60; consult, Naval Ord Lab. Mem: Am Math Soc; Math Asn Am. Res: Matrices; dynamics; ergodic theory; mathematical theory of the flow of a compressible fluid; partial differential equations; the flow of a viscous fluid. Mailing Add: R R 2 Box 64 Denton MD 21629

MARTIN, MURRAY JOHN, b Regina, Sask, June 22, 35; m 63; c 1. NUCLEAR PHYSICS. Educ: Univ Sask, BA, 56, MA, 58; McMaster Univ, PhD(physics), 63. Prof Exp: Instr physics, Univ Sask, 55-57, lectr math, 57-59; res assoc physics, Nat Acad Sci, 63; RES ASSOC PHYSICS, OAK RIDGE NAT LAB, 64- Mem: Am Phys Soc; Can Asn Physicists. Res: Low-energy theoretical nuclear spectroscopy; nuclear data compilation. Mailing Add: Oak Ridge Nat Lab PO Box X Oak Ridge TN 37830

MARTIN, NATHANIEL FRIZZEL GRAFTON, b Wichita Falls, Tex, Oct 10, 28; m 54; c 2. MATHEMATICS. Educ: NTex State Col, BS, 49, MS, 50; Iowa State Col, PhD(math), 59. Prof Exp: Asst math, NTex State Col, 49-50; instr, Midwestern Univ, 50-52; asst, Iowa State Col, 55-58, instr, 58-59; from instr to asst prof, 59-64, ASSOC PROF MATH, UNIV VA, 64-, ASST CHMN DEPT, 74- Concurrent Pos: NSF fac fel & res assoc, Univ Calif, Berkeley, 65-66; vis lectr, Copenhagen Univ, 69-70; consult ed, McGraw-Hill Book Co, 72- Mem: Am Math Soc; Math Asn Am. Res: Analysis; real function theory and measure theory, particularly differentiation of set functions; ergodic theory; entropy; isomorphisms of dynamical systems. Mailing Add: Dept of Math Univ of Va Charlottesville VA 22903

MARTIN, NED HAROLD, b New Brunswick, NJ, May 18, 45. BIO-ORGANIC CHEMISTRY. Educ: Denison Univ, AB, 67; Duke Univ, PhD(org chem), 72. Prof Exp: Chemist, Res Triangle Inst, 69-70; ASST PROF ORG CHEM, UNIV NC, WILMINGTON, 72- Mem: Am Chem Soc; AAAS. Res: Synthesis of potential anti-cancer agents; alkaloid biosynthesis; iridium complexes as models for oxygenase enzymes; novel photochemical syntheses of natural products. Mailing Add: Dept of Chem Univ of NC Wilmington NC 28401

MARTIN, NORMAN MARSHALL, b Chicago, Ill, Jan 16, 24; m 50; c 3. MATHEMATICS. Educ: Univ Chicago, MA, 47; Univ Calif, Los Angeles, PhD(philos), 52. Prof Exp: Instr philos, Univ Ill, 50-51 & Univ Calif, Los Angeles, 52-53; res assoc, Willow Run Res Ctr, Mich, 53-55; mem tech staff, Space Tech Labs, Thompson-Ramo-Wooldridge, Inc, 55-59, head logic tech group, 59-61; mem tech staff & bd dirs, Logicon, Inc, 61-65, treas, 62-65; assoc prof, 66-68, res scientist, Comput Ctr, 66-71, PROF PHILOS & COMPUT SCI, UNIV TEX, AUSTIN, 68-, PROF ELEC ENG, 74- Concurrent Pos: Lectr, Univ Calif, Los Angeles, 57-65; consult, Logicon, Inc, 65- Mem: Am Math Soc; Math Asn Am; Asn Symbolic Logic; Asn Comput Mach; Am Philos Asn. Res: Systems, organization and logical design of digital computing equipment; missile guidance system engineering; applications of digital equipment; mathematical logic, especially many-valued logic; philosophy of language; switching theory. Mailing Add: 4423 Crestway Dr Austin TX 78731

MARTIN, PAUL CECIL, b Brooklyn, NY, Jan 31, 31; m 57; c 3. THEORETICAL PHYSICS. Educ: Harvard Univ, AB, 51, PhD(physics), 54. Prof Exp: NSF res fel, Univ Birmingham, 55 & Inst Theoret Physics, Denmark, 56; from asst prof to assoc prof, 56-64, chmn dept, 72-75, PROF PHYSICS, HARVARD UNIV, 64- Concurrent Pos: Res fel, Sloan Found, 59-62; vis prof, Ecole Normale Superieure, 63 & 66; ed, J Math Physics, 63-66; Guggenheim Found fel, 65 & 71; ed, Ann Physics, 68-; vis prof, Univ Paris, 71; consult, Brookhaven Nat Lab. Mem: Fel Am Acad Arts & Sci. Res: Quantum theory of fields; physics of solids and fluids; statistical mechanics. Mailing Add: Dept of Physics Harvard Univ Cambridge MA 02138

MARTIN, PAUL SCHULTZ, b Allentown, Pa, Aug 22, 28; m 50; c 3. ECOLOGY. Educ: Cornell Univ, BA, 51; Univ Mich, MA, 53, PhD, 56. Prof Exp: Rackham fel, Univ Mich & Yale Univ, 55-56; Nat Res Coun Can res fel, Univ Montreal, 56-57; res assoc palynol, 57-62, assoc prof, 62-68, PROF PALYNOL & CHIEF SCIENTIST, PALEOENVIRON STUDIES, GEOCHRONOL LABS, UNIV ARIZ, 68- Concurrent Pos: Guggenheim fel, 65-66. Mem: AAAS; Soc Study Evolution; Am Soc Nat; Ecol Soc Am. Res: Pleistocene biogeography; pollen stratigraphy; faunal extinction and its causes. Mailing Add: Dept of Geosci Univ of Ariz Tucson AZ 85721

MARTIN, PETER WILSON, b Glasgow, Scotland, Jan 7, 38; m 65. NUCLEAR

PHYSICS. Educ: Glasgow Univ, BSc, 60, PhD(nuclear physics), 64. Prof Exp: Asst lectr physics, Univ Glasgow, 64-65; asst prof, 65-70, ASSOC PROF PHYSICS, UNIV BC, 70- Res: Fast neutron scattering; charged particle reactions; nuclear orientation. Mailing Add: Dept of Physics Univ of BC Vancouver BC Can

MARTIN, RICHARD BLAZO, b Winchendon, Mass, July 1, 17; m 41; c 4. CHEMISTRY, SCIENCE ADMINISTRATION. Educ: Clark Univ, AB, 39, AM, 40, PhD(chem), 49. Prof Exp: From instr to asst prof chem, Clark Univ, 46-53; chemist, Res Br, Oak Ridge Opers, AEC, 53-57, chief, 57-59, dep dir lab & univ div, 59-72, asst br chief, Waste Mgt Br, Res & Tech Support Div, 72-73, PHYS SCIENTIST, CLASSIFICATION & TECH SUPPORT BR, OAK RIDGE OPERS, US ENERGY RES & DEVELOP ADMIN, 73- Mem: Am Chem Soc; Am Nuclear Soc; Sigma Xi. Res: Radiochemical processing; isotopic separations; reactions of aliphatic diazo compounds with alicyclic ketones; synthesis of alicyclic ketones; spectrophotometric analysis. Mailing Add: Res & Tech Sup Div Oak Ridge Ops US Energy Res & Develop Admin PO Box E Oak Ridge TN 37830

MARTIN, RICHARD GORDON, b York, Maine, June 9, 18; m 43; c 3. ONCOLOGY, SURGERY. Educ: Bates Col, BS, 40; Temple Univ, MD, 44; Am Bd Surg, dipl, 52. Prof Exp: Intern med, US Navy Hosp, Philadelphia, 44-45; resident surg, Md Gen Hosp, Baltimore, 46-50; instr, Med Sch, 50-51, asst surgeon, 51-53, from asst surgeon to assoc surgeon, 54-57, SURGEON & CHIEF SECT GEN SURG, UNIV TEX M D ANDERSON HOSP & TUMOR INST, 67- Mem: AMA; Soc Surg Alimentary Tract; Am Col Surg; James Ewing Soc. Res: Soft tissue and gastrointestinal tumors. Mailing Add: Univ of Tex M D Anderson Hosp & Tumor Inst Houston TX 77025

MARTIN, RICHARD HADLEY, JR, b Worcester, Mass, May 15, 24; m 46; c 3. POLYMER CHEMISTRY. Educ: Worcester Polytech Univ, BS, 45; Princeton Univ, MS, 47. Prof Exp: Res chemist, Plastics Div, 47-58, RES SPECIALIST, HYDROCARBONS & POLYMERS DIV, MONSANTO CO, 58- Mem: Am Chem Soc; Soc Plastics Eng. Res: Polymerization, processing and analytical characterization of high polymers. Mailing Add: 57 Brewster St Springfield MA 01119

MARTIN, RICHARD HAROLD, b Bluefield, WVa, Feb 16, 39; m 61; c 2. GEOLOGY. Educ: WVa Univ, BS, 62, MS, 64, PhD(geol), 67. Prof Exp: Field geologist, NY State Natural Gas Corp, 62 & 64; asst prof geol, 66-69, ASSOC PROF GEOL & ASST ACAD DEAN, CAMPBELL COL, 69- Mem: Geol Soc Am; Soc Econ Paleont & Mineral; Am Asn Petrol Geol. Res: Analysis of recent sedimentary environments to facilitate the interpretation of ancient sedimentary rocks. Mailing Add: Box 281 Buies Creek NC 27506

MARTIN, RICHARD HARVEY, b LaPorte, Ind, Aug 27, 32; m 54; c 2. MEDICINE, PHYSIOLOGY. Educ: Johns Hopkins Univ, 50-53; Univ Rochester, MD, 57. Prof Exp: Intern, Strong Mem Hosp, 57-58; resident, Univ Wash, 58-59 & 61-62, asst med, Div Cardiol, 62-65; asst prof med, 65-68, assoc prof med & physiol, 68-73, PROF MED, SCH MED, UNIV MO-COLUMBIA, 73-, DIR CORONARY CARE UNIT, 69-, DIR DIV CARDIOL, 70- Concurrent Pos: Fel, Coun Clin Cardiol, Am Heart Asn; res fel med, Div Cardiol, Univ Wash, 62-64; res fel physiol & biophys, 64-65. Mem: Sigma Xi; Asn Advan Med Instrumentation; Am Fedn Clin Res; fel Am Col Physicians. Res: Hemodynamic and clinical observations in acute myocardial infarction; ventricular aneurysm; left ventricular function; effect of atrial systole in man. Mailing Add: Div of Cardiol C7A Univ of Mo Med Ctr Columbia MO 65201

MARTIN, RICHARD HUGO, b Hanover, Pa, Aug 16, 36; m 59; c 2. PHYSICAL CHEMISTRY. Educ: Gettysburg Col, AB, 58; Pa State Univ, PhD(chem), 65. Prof Exp: Res assoc fel ion molecule res, Pa State Univ, 66-67; sr res chemist, Chem Div, 67-69, sr res chemist, Ecusta Paper Div, 69-74, res assoc, Fine Paper & Film Group, 74-75, SR RES ASSOC, FINE PAPER & FILM GROUP, OLIN CORP, 75- Mem: Am Chem Soc; Sigma Xi. Res: Energetics and kinetics of organic reactions; homogeneous and heterogeneous catalysis; reaction mechanisms and substituent effects; vapor phase synthesis; photochemical synthesis; instrumental measurements of dynamic systems. Mailing Add: Fine Paper & Film Group Olin Corp Pisgah Forest NC 28768

MARTIN, RICHARD MCFADDEN, b Somerville, Tenn, Aug 19, 42; m 64; c 2. PHYSICS. Educ: Univ Tenn, Knoxville, SB, 64; Univ Chicago, MS, 66, PhD(physics), 69. Prof Exp: Mem tech staff, Bell Tel Labs, 69-71; SCIENTIST, XEROX PALO ALTO RES CTR, 71- Mem: Am Phys Soc. Res: Solid state physics, mainly lattice dynamics of insulators, interaction of light with insulators and semiconductors. Mailing Add: Xerox Palo Alto Res Ctr 3333 Coyote Hill Rd Palo Alto CA 94304

MARTIN, RICHARD MCKELVY, b Los Angeles, Calif, Jan 26, 36; m 54; c 3. PHYSICAL CHEMISTRY. Educ: Univ Calif, Riverside, BA, 59; Univ Wis, PhD(phys chem), 63. Prof Exp: NIH fel, Harvard Univ, 63-65; asst prof, 65-71, ASSOC PROF CHEM, UNIV CALIF, SANTA BARBARA, 71- Mem: AAAS; Am Chem Soc; Am Phys Soc. Res: Photochemistry; molecular beams; electronic energy transfer. Mailing Add: Dept of Chem Univ of Calif Santa Barbara CA 93106

MARTIN, ROBERT ALLAN, b Paterson, NJ, Dec 3, 29; m 54; c 3. ORGANIC CHEMISTRY. Educ: Rutgers Univ, BS, 50; Princeton Univ, MA, 52, PhD(org chem), 55. Prof Exp: Group leader, Olin Mathieson Chem Corp, 54-55, res chemist, Squibb Inst Med Res, 55-57; group leader, Alco Chem Corp, 57-61; sr chemist, Riegel Textile Corp, 61-62; res chemist, 62-68, GROUP LEADER CHEM & PLASTICS, UNION CARBIDE CORP, TARRYTOWN, 68- Mem: Am Chem Soc. Res: Acrylic emulsion polymers; textile finishing resins; water-soluble polymers; cellulose derivatives; rubber accelerators and antioxidants; organic coatings; alkyd resins; polyurethanes; computer programming; solvent technology. Mailing Add: 275 Russet Rd Stamford CT 06903

MARTIN, ROBERT ALLEN, b New York, NY, Feb 19, 44; m 69. VERTEBRATE PALEONTOLOGY. Educ: Hofstra Univ, BA, 65; Tulane Univ, La, MS, 67; Univ Fla, PhD(zool), 69. Prof Exp: Asst prof biol, SDak Sch Mines & Technol, 69-72; ASST PROF BIOL, FAIRLEIGH DICKINSON UNIV, 72- Concurrent Pos: Sigma Xi grant, SDak Sch Mines & Technol, 70-71; NSF grants, 70-72. Mem: AAAS; Soc Vert Paleont; Am Soc Mammal; Am Quaternary Asn. Res: Mammalian evolution, ecology, anatomy, and systematics. Mailing Add: Dept of Biol Fairleigh Dickinson Univ Madison NJ 07940

MARTIN, ROBERT BRUCE, b Chicago, Ill, Apr 29, 29; m 53. BIOPHYSICAL CHEMISTRY. Educ: Northwestern Univ, BS, 50; Univ Rochester, PhD(phys chem), 53. Prof Exp: Asst prof chem, Am Univ Beirut, 53-56; res fel, Calif Inst Technol, 56-57 & Harvard Univ, 57-59; from asst prof to assoc prof chem, 59-65, chmn dept, 68-71, PROF CHEM, UNIV VA, 65- Concurrent Pos: NIH spec fel, Oxford Univ, 61-62; prog dir molecular biol sect, NSF, 65-66. Mem: AAAS; Am Chem Soc; Am Soc Biol Chemists. Res: Structure, equilibrium and mechanism investigations of systems with biological interest; metal ion interactions; nuclear magnetic resonance studies; optical activity of transition metal ion complexes and disulphides. Mailing Add: Dept of Chem Univ of Va Charlottesville VA 22901

MARTIN, ROBERT EDWARD, JR, b Flint, Mich, Jan 9, 31; m 53; c 3. FORESTRY. Educ: Marquette Univ, BS, 53; Univ Mich, BS, 58, MF, 59, PhD(forestry), 63. Prof Exp: Res forester, Southern Forest Fire Lab, US Forest Serv, Ga, 60-63; from asst prof to prof wood technol & forest fires, Va Polytech Inst & State Univ, 63-71; PROF FOREST RESOURCES, COL FOREST RESOURCES, UNIV WASH, 71-; RES PHYSICIST, US FOREST SERV, 71- Concurrent Pos: NSF-Soc Wood Sci & Technol vis scientist wood sci, Univ Minn, 71, Univ Idaho & Washington Univ, 72. Mem: AAAS; Forest Prod Res Soc; Soc Wood Sci & Technol; Soc Am Foresters. Res: Forest fire physics, effects and behavior; forest fire and fuel management; bark structure, properties and utilization. Mailing Add: Col of Forest Resources Univ of Wash Seattle WA 98195

MARTIN, ROBERT EUGENE, b Monterey, Tenn, July 19, 30; m 56; c 2. ANIMAL ECOLOGY. Educ: Tenn Tech Univ, BS, 52; Univ Tenn, MS, 59, PhD(zool), 63. Prof Exp: From asst prof to assoc prof, 63-72, PROF BIOL, TENN TECHNOL UNIV, 72- Mem: Ecol Soc Am; Am Fisheries Soc. Res: Fish and insect population dynamics; biometrics; fish scale structure and development. Mailing Add: Dept of Biol Box 52A Tenn Technol Univ Cookeville TN 38501

MARTIN, ROBERT FRANCOIS CHURCHILL, b Ottawa, Ont, Nov 3, 41; m 63; c 3. GEOLOGY. Educ: Univ Ottawa, Ont, BSc, 63; Pa State Univ, MS, 66; Stanford Univ, PhD(geol), 69. Prof Exp: Res assoc geol, Stanford Univ, 68-70; asst prof, 70-74, ASSOC PROF GEOL, McGILL UNIV, 74- Mem: Mineral Soc Am; Mineral Asn Can; Swiss Soc Mineral & Petrog. Res: Igneous and metamorphic mineralogy and petrology. Mailing Add: Dept of Geol Sci McGill Univ Montreal PQ Can

MARTIN, ROBERT FREDERICK, analytical chemistry, see 12th edition

MARTIN, ROBERT FREDERICK, b Weehawken, NJ, Nov 10, 38; m 60; c 3. VERTEBRATE ZOOLOGY, ECOLOGY. Educ: Fairleigh Dickinson Univ, BS, 60; Univ Tex, Austin, MA, 64, PhD(zool), 69. Prof Exp: Res sci asst zool, Univ, 64-65, CUR VERT, TEX MEM MUS, UNIV TEX, AUSTIN, 69- Concurrent Pos: Lectr zool, Univ Tex, Austin, 75- Mem: Soc Study Evolution; AAAS; Am Soc Ichthyologists & Herpetologists; Am Ornithologists Union. Res: Reproductive ecology of reptiles and birds; anuran morphology and evolution. Mailing Add: Cur of Vert Tex Mem Mus 24th & Trinity Austin TX 78705

MARTIN, ROBERT LAWRENCE, b Washington, DC, Nov 18, 33; m 55. MAMMALOGY, ZOOLOGY. Educ: Univ Maine, BS, 56; Kans State Univ, MS, 59; Univ Conn, PhD, 71. Prof Exp: Teacher high sch, Maine, 56-57; from instr to asst prof anat & biol, State Univ NY Col Plattsburgh, 61-64; collabr, Great Smoky Mountain Nat Park, Nat Park Serv, Tenn, 64-65; assoc prof, 66-71, PROF MAMMAL & BIOL, UNIV MAINE, FARMINGTON, 71- Concurrent Pos: Res assoc, Mt Washington Observ, NH, 68-74; ed, Bat Res News, 70-; res assoc, Univ Conn Paraguayan Exped, 73; mem, Univ Conn Chaco Exped, 74 & 75; mem, Ind Bat Recovery Team, US Fish & Wildlife Serv, 75- Mem: AAAS; Am Soc Mammal; NY Acad Sci; Mammal Soc Brit Isles; fel Zool Soc London. Res: Comparative vertebrate anatomy; bat studies; mammalian natural history. Mailing Add: Cape Cod Hill New Sharon ME 04955

MARTIN, ROBERT LEONARD, b Seattle, Wash, July 14, 19; m 46; c 4. PHYSICS. Educ: Reed Col, BA, 41; Univ Mich, MS, 47, PhD(physics), 56. Prof Exp: Res systs analyst, Willow Run Res Lab, Univ Mich, 50-52; asst prof physics, Reed Col, 56-62; assoc prof, 62-74, PROF PHYSICS, LEWIS & CLARK COL, 74-, CHMN DEPT, 63- Mem: Am Asn Physics Teachers; Am Phys Soc. Res: Solid state; logic of physics; optical and electrical properties of ionic solids, especially alkali halides. Mailing Add: Dept of Physics Lewis & Clark Col Portland OR 97219

MARTIN, ROBERT O, b Honolulu, Hawaii, Jan 8, 31; m 56; c 3. BIOCHEMISTRY, ORGANIC CHEMISTRY. Educ: Univ San Francisco, BS, 56; Univ Calif, Berkeley, PhD(biochem), 59. Prof Exp: USPHS fel chem, Kings Col, Newcastle-on-Tyne, 59-61; chemist, Lawrence Radiation Lab, Univ Calif, 61-65; from asst prof to assoc prof biochem, 65-74, PROF BIOCHEM, UNIV SASK, 74- Concurrent Pos: Can Med Res Coun vis scientist, Neurol Inst, London, 72-73. Mem: Am Chem Soc; fel Chem Inst Can; Can Fedn Biol Sci. Res: Alkaloid structures and biosynthesis; neurochemistry; diabetes and other disorders of carbohydrate metabolism. Mailing Add: Dept of Biochem Univ of Sask Saskatoon SK Can

MARTIN, ROBERT PAUL, b Hartford, Conn, Mar 10, 43; m 67; c 2. MATHEMATICAL ANALYSIS. Educ: Cent Conn State Col, BS, 65, MS, 68; Univ Md, MA, 72, PhD(math), 73. Prof Exp: Teacher math, Washington Jr High Sch, New Britain, Conn, 65-66; instr, Northwestern Community Col, 66-69; INSTR MATH, UNIV PA, 73- Mem: Am Math Soc; Math Asn Am. Res: Non-abelian harmonic analysis and representation theory of lie groups. Mailing Add: Dept of Math Univ of Pa Philadelphia PA 19174

MARTIN, ROBERT WILLIAM, chemistry, see 12th edition

MARTIN, ROGER CHARLES, b Janesville, Wis, June 12, 31; m 64. GEOLOGY. Educ: Univ Calif, Los Angeles, AB, 53; Univ Idaho, MS, 57; Victoria Univ Wellington, PhD(geol), 63. Prof Exp: Geologist, NZ Geol Surv, 59-60 & 63-64; geologist, Calif Dept Water Resources, 64-68; geologist, Earth Resources Opers, NAm Rockwell Corp, 68-73; GEOLOGIST, CALIF STATE LANDS DIV, 73- Concurrent Pos: Geol consult, NZ Forest Prod, Ltd, 61-64. Mem: Geol Soc Am; Am Geophys Union; Asn Eng Geologists. Res: Application of remote sensing to geoscience problems; discrimination of kimberlite and other lithologic bodies by thermal infrared and other multispectral sensors; geothermal research and development administration; exploration and environmental problems; methods of geothermal exploration, including underwater technology and methods. Mailing Add: State Lands Div 100 Oceangate Suite 300 Long Beach CA 90802

MARTIN, RONALD ALLEN, organic chemistry, see 12th edition

MARTIN, RONALD LAVERN, b Devereaux, Mich, Sept 13, 22; m 49; c 6. NUCLEAR PHYSICS. Educ: US Naval Acad, BS, 44; Mich State Univ, MS, 48, Univ Chicago, PhD(physics), 52. Prof Exp: Res assoc physics, Univ Chicago, 52-53 & Cornell Univ, 53-56; mem staff, Bell Tel Labs, 56-59; sr scientist, TRG, Inc, 59-62; assoc dir particle accelerator div, 62-67, DIR ACCELERATOR DIV, ARGONNE NAT LAB, 67- Mem: Fel Am Phys Soc. Res: High energy nuclear physics; accelerators; laser development; nonlinear properties of ferrites at microwave frequencies. Mailing Add: Accelerator Div Argonne Nat Lab Argonne IL 60439

MARTIN, RONALD LEROY, b Beloit, Wis, Sept 14, 32; m 53; c 3. ANALYTICAL CHEMISTRY. Educ: Beloit Col, BS, 53; Univ Wis, MS, 55, PhD, 57. Prof Exp: Res chemist, Standard Oil Co (Ind), 57-75; DIR ADDITIVES EVAL & FORMULATION DIV, AMOCO CHEM CO, 75- Mem: Am Chem Soc. Res: Gas chromatography; spectrochemical analysis; lubricant development. Mailing Add: 1013 Summit Hills Lane Naperville IL 60540

MARTIN, ROY JOSEPH, JR, b Lutcher, La, Jan 3, 43; m 67; c 2. NUTRITION, BIOCHEMISTRY. Educ: Univ Southwestern La, BS, 64; Univ Fla, MS, 65; Univ Calif, Davis, PhD(nutrit), 70. Prof Exp: Asst prof nutrit, 70-74, ASSOC PROF ANIMAL NUTRIT, PA STATE UNIV, UNIVERSITY PARK, 74- Mem: Am Dairy Sci Asn; Am Soc Animal Sci. Res: Metabolic regulation of growth and development; effects of early nutritional experiences. Mailing Add: Dept of Nutrit Pa State Univ University Park PA 16802

MARTIN, RUFUS RUSSELL, b Decatur, Ga, Mar 3, 36; m 61; c 3. INTERNAL MEDICINE, INFECTIOUS DISEASES. Educ: Yale Univ, AB, 56; Med Col Ga, MD, 60. Prof Exp: NIH fel infectious dis, 65-67; from asst prof to assoc prof med, Sch Med, Ind Univ, Indianapolis, 67-71; assoc prof, 71-75, PROF MED & MICROBIOL, BAYLOR COL MED, 75- Concurrent Pos: Attend physician, Vet Admin Hosp, Houston, 71- Mem: Fel Am Col Physicians; Am Fedn Clin Res; NY Acad Sci; Infectious Dis Soc Am; Am Thoracic Soc. Res: Staphylococcal immunology; histamine; lysosomes; leukocytes and inflammation; pulmonary macrophages and smoking. Mailing Add: Dept of Med Baylor Col of Med Houston TX 77025

MARTIN, RUSSELL JAMES, b Beaumont, Tex, May 15, 39; m 64; c 2. EPIDEMIOLOGY. Educ: Tex A&M Univ, BS, 61, DVM, 63; Univ Mich, MPH, 66. Prof Exp: Epidemiol intel serv officer, Ctr Dis Control, USPHS, 63-65, regional pub health vet, 66-72, chief pub health vet, 72-75, COMMUN DIS EPIDEMIOLOGIST, ILL DEPT PUB HEALTH, 75- Concurrent Pos: Asst prof vet pub health, Col Vet Med, Univ Ill, 68-72, assoc prof, 72-; clin assoc prof prev med, Peoria Sch Med, Univ Ill, 75- Mem: Am Pub Health Asn; Conf Pub Health Vets (pres, 76-77); Am Bd Vet Pub Health; Am Vet Med Asn. Res: Zoonotic diseases that occur naturally in the United States, particularly delineating the epidemiology of this group. Mailing Add: 219 Wild Rose Lane Rochester IL 62563

MARTIN, SAMUEL CLARK, b McNeal, Ariz, Apr 16, 16; m 44; c 2. RANGE CONSERVATION. Educ: Univ Ariz, BS, 42, MS, 47, PhD, 64. Prof Exp: Range conservationist, Southwestern Forest & Range Exp Sta, 42-49, range conservationist, Cent States Forest Exp Sta, 49-55, PRIN RANGE SCIENTIST, ROCKY MOUNTAIN FOREST & RANGE EXP STA, US FOREST SERV, 55- Mem: Soc Range Mgt. Res: Grazing management; noxious plant control; range revegetation. Mailing Add: Rocky Mtn Forest & Range Exp Sta PO Box 4460 Tucson AZ 85717

MARTIN, SAMUEL PRESTON, III, b East Prairie, Mo, May 2, 16; m 70; c 3. INTERNAL MEDICINE, COMMUNITY HEALTH. Educ: Wash Univ, MD, 41. Hon Degrees: MA, Univ Pa, 71. Prof Exp: Am Col Physicians fel, Rockefeller Inst, 48-49; Markle Found fel, Duke Univ, 50-55; prof internal med & chmn dept, Univ Fla, 56-62, provost health affairs, 62-69; vis prof health econ, Harvard Univ, 69-71; PROF COMMUNITY MED, COL MED & PROF HEALTH CARE SYSTS, WHARTON SCH, UNIV PA, 71-; EXEC DIR, LEONARD DAVIS INST, 74- Concurrent Pos: Dir, Fla Regional Med Prog, 67-68; mem comt accreditation, US Off Educ, 68-70; Commonwealth & USPHS fels, Harvard Univ & London Sch Hyg, 69-71; dir, SmithKline Corp, Philadelphia, 72- Honors & Awards: Order of Leopold, Belg. Mem: Asn Am Physicians; Am Asn Immunol; Am Col Physicians; Am Fedn Clin Res (vpres, 54); Am Pub Health Asn. Res: Immunology microbiology; medical economics. Mailing Add: Dept of Community Med Univ of Pa Philadelphia PA 19104

MARTIN, SARAH SMITH, b Columbia, SC, Nov 24, 43; m 68. PHARMACOLOGY, BIOCHEMISTRY. Educ: Skidmore Col, BA, 65; Univ Rochester, PhD(pharmacol), 71. Prof Exp: Instr biol, Univ Rochester, 71-72; FEL REPRODUCTIVE ENDOCRINOL, UNIV MICH, ANN ARBOR, 72- Concurrent Pos: Sr teaching fel biochem, Monash Univ, Australia, 70-71. Mem: Australian Biochem Soc. Res: Glycoprotein synthesis by isolated mitochondria; biochemical correlates to cytoplasmic inheritance of drug resistance factors by mitochondrial genome; biogenesis of mitochondria; biochemical mechanisms of hormone action. Mailing Add: Reproductive Endocrinol Unit Univ of Mich Dept Obstet & Gynec Ann Arbor MI 48104

MARTIN, SEELYE, b Northampton, Mass, Sept 22, 40. OCEANOGRAPHY. Educ: Harvard Univ, BA, 62; Johns Hopkins Univ, PhD(mech), 67. Prof Exp: Res assoc oceanog, Mass Inst Technol, 67-69; RES ASST PROF OCEANOG, UNIV WASH, 69- Mem: Am Geophys Union. Res: Internal waves in the ocean; desalination of sea ice. Mailing Add: Dept of Oceanog MSB-108 Univ of Wash Seattle WA 98105

MARTIN, STANLEY BUEL, b Tulsa, Okla, Oct 21, 27; m 51; c 4. PHYSICAL CHEMISTRY. Educ: San Jose State Col, AB, 50. Prof Exp: Chemist, US Naval Radiol Defense Lab, 50-61, supvry chemist, 61-66; prin res chemist, URS Corp, 66-69; MGR FIRE RES PROG, STANFORD RES INST, 69- Mem: Am Chem Soc; Combustion Inst; Soc Fire Protection Engrs. Res: Thermal radiation transport and effects; nuclear weapons effects; fire phenomenology; transient heat conduction in solids; thermal decomposition of organic solids; ignition processes; kinetics of pyrolysis; reactions in unsteady-state systems; combustion and fire protection research. Mailing Add: Fire Res Prog Stanford Res Inst 333 Ravenswood Ave Menlo Park CA 94025

MARTIN, STANLEY MORRIS, b Ottawa, Ont, Oct 26, 20; m 44; c 3. BIOCHEMISTRY, MICROBIOLOGY. Educ: Univ Toronto, BSA, 44; Univ Wis, MS, 48, PhD(bact), 50. Prof Exp: Asst res officer, 50-55, assoc res officer, 55-62, SR RES OFFICER, NAT RES COUN CAN, 62- Mem: Can Soc Microbiol; Int Asn Plant Tissue Cult; Am Chem Soc. Res: Fermentation biochemistry; production of metabolites by plant cell cultures; enzymes of plant cell cultures; large-scale cultivation of bacteria. Mailing Add: Nat Res Coun Div of Biol Sussex Dr Ottawa ON Can

MARTIN, STEPHEN FREDERICK, b Albuquerque, NM, Feb 8, 46; m 70. SYNTHETIC ORGANIC CHEMISTRY. Educ: Univ NMex, BS, 68; Princeton Univ, MA, 70, PhD(org chem), 72. Prof Exp: Alexander von Humboldt Found fel org chem, Univ Munich, 72-73; NIH fel, Mass Inst Technol, 73-74; ASST PROF ORG CHEM, UNIV TEX, AUSTIN, 74- Mem: Am Chem Soc. Res: Design and development of new synthetic methods; chemistry and total synthesis of natural products, particularly alkaloids and terpenes; heterocyclic chemistry. Mailing Add: Dept of Chem Univ of Tex Austin TX 78712

MARTIN, STEPHEN GEORGE, b Eagle Grove, Iowa, Sept 20, 41; m 65; c 2. BEHAVIORAL ECOLOGY. Educ: Univ Wis-Madison, BS, 64; MS, 67; Ore State Univ, PhD(zool), 70. Prof Exp: Asst prof zool, Colo State Univ, 70-73; VPRES, ECOL CONSULT, INC, 73- Honors & Awards: A Brazier Howell Award, Cooper Ornith Soc, 70. Mem: AAAS; Ecol Soc Am; Am Ornith Union; Cooper Ornith Soc; Wilson Ornith Soc. Res: Habitat structure and vertebrate social systems; adaptations for niche diversification in vertebrates; analysis of environmental impact; applied ecology; structural analysis of vertebrate communication systems. Mailing Add: Ecol Consult Inc PO Box 1057 Ft Collins CO 80522

MARTIN, SUSAN SCOTT, b Paducah, Ky, May 18, 38. PLANT CHEMISTRY. Educ: Univ Colo, BA, 60; Utah State Univ, MS, 68; Univ Calif, Santa Cruz, PhD(biol), 73. Prof Exp: Res chemist, Dept Wildlife Resources, Utah State Univ, 60-68; PLANT PHYSIOLOGIST, CROPS RES LAB, AGR RES SERV, USDA, 74- Mem: Phytochem Soc NAm; Bot Soc Am. Res: Biochemical aspects of plant-pathogen interaction, including pathogen-produced toxins and phytoalexins; physiological factors affecting sugarbeet quality and sucrose production; resin chemistry and chemical ecology of the leguminous genus Hymenaea. Mailing Add: USDA Agr Res Serv Crops Res Lab Colo State Univ Ft Collins CO 80523

MARTIN, TELLIS ALEXANDER, b Hickory, NC, May 20, 19; m 49; c 4. ORGANIC CHEMISTRY. Educ: Berea Col, BA, 42; Univ Va, MS, 45, PhD(org chem), 48. Prof Exp: Asst chem, Berea Col, 40-42, Univ Va, 42-44 & Off Sci Res & Develop, 44-48; res chemist, Gen Aniline & Film Corp, 48-53; sr chemist, 53-63, res assoc, 63-70, sr investr, 70-73, PRIN INVESTR, MEAD JOHNSON & CO, 73- Mem: AAAS; Am Chem Soc. Res: Synthesis of benzalacetophenones; aminoketones and amino alcohols of phenyl, quinoline and phenylquinoline series for possible use as antimalarials, cancer agents and antitubercular drugs; phthalocyanines; phenanthridines; biphenylsulfones; carbohydrates; steric hindrance; ring-chain tautomerism; cysteines; sulfa drugs; hypnotics; betalactams and aspartic acids; anti-inflammatory, fibrinolytic and mucolytic agents. Mailing Add: Chem Res Mead Johnson & Co Pennsylvania Ave Evansville IN 47721

MARTIN, TERENCE EDWIN, b Adelaide, Australia, Apr 28, 41; m 63; c 2. CELL BIOLOGY, MOLECULAR BIOLOGY. Educ: Univ Adelaide, BSc, 62; Cambridge Univ, PhD(biochem), 66. Prof Exp: Univ fel, Univ Chicago, 66-68, Am Cancer Soc fel, Univ Wash, 69-70, ASST PROF BIOL, UNIV CHICAGO, 71- Concurrent Pos: USPHS res grant, Univ Chicago, 71- Mem: Brit Biochem Soc; Am Soc Cell Biol. Res: Control of gene expression in eukaryotic cells; nucleic acid synthesis and metabolism; control of protein synthesis; molecular basis of cancer. Mailing Add: Dept of Biol Univ of Chicago Chicago IL 60637

MARTIN, TERRY JOE, b Baxter Springs, Kans, Dec 28, 47; m 66; c 2. PLANT PATHOLOGY. Educ: Kans State Col, Pittsburg, BS, 70; Kans State Univ, MS, 71; Mich State Univ, PhD(plant path), 74. Prof Exp: ASST PROF PLANT PATH, FT HAYS BR EXP STA, KANS STATE UNIV, 74- Mem: Am Phytopath Soc. Res: Genetic control of parasite-host interactions and the development of wheat cultivars resistant to leaf rust, stem rust, wheat streak mosaic virus, and soilborne wheat mosaic virus. Mailing Add: Ft Hays Br Exp Sta Hays KS 67601

MARTIN, THOMAS GEORGE, III, b Boston, Mass, Jan 14, 31; m 51; c 2. HEALTH PHYSICS. Educ: Northeastern Univ, BS, 58; Am Bd Health Physics, dipl, 65. Prof Exp: Head radiol safety dept, Controls for Radiation Inc, 58-61, head labs, 61-62; radiation chemist, Mass Inst Technol, 62-63; RADIATION PROTECTION OFFICER, NATICK DEVELOP CTR, US ARMY, 63- Concurrent Pos: Consult tech points of contact, Interdept Comt on Radiation Preservation of Food, 64- Mem: Health Physics Soc; Sigma Xi; Conf Radiol Health; NY Acad Sci. Res: Induced radioactivity in food sterilized by ionizing radiation; health physics problems associated with particle accelerators; trace metals in foods; activation analysis. Mailing Add: 588 Winter St Framingham MA 01701

MARTIN, THOMAS WARING, b Cumberland, Md, July 24, 25; m 47; c 4. PHYSICAL CHEMISTRY. Educ: Franklin & Marshall Col, BS, 50; Northwestern Univ, PhD(chem), 54. Prof Exp: Res fel photochem, Nat Res Coun Can, 54-55; instr chem, Williams Col, Mass, 55-57; from asst prof to assoc prof, 57-66, chmn dept, 67-73, PROF CHEM, VANDERBILT UNIV, 66- Mem: AAAS; Am Chem Soc; The Chem Soc; Royal Inst Chem. Res: Photochemistry; electron spin resonance and magneto-chemical effects; biochemical oxidoreduction and model systems; chemical kinetics and catalysis; mass spectrometry. Mailing Add: Dept of Chem Box 1506/B Vanderbilt Univ Nashville TN 37235

MARTIN, TRUMAN GLEN, b Wortham, Tex, May 24, 28; m 50; c 3. ANIMAL GENETICS. Educ: Tex A&M Univ, BS, 49; Iowa State Univ, MS, 51, PhD(animal breeding & nutrit), 54. Prof Exp: Asst dairy husb, Iowa State Univ, 49-51, asst animal breeding, 53-55; from asst prof to assoc prof, 55-63, PROF ANIMAL BREEDING, PURDUE UNIV, 63- Mem: Am Dairy Sci Asn; Am Soc Animal Sci. Res: Crossbreeding and selection of dairy and beef cattle; factors affecting body composition of swine, sheep and cattle. Mailing Add: Dept of Animal Sci Purdue Univ West Lafayette IN 47907

MARTIN, VIRGINIA LORELLE, b Mount Olive, NC, Nov 29, 39. BIOLOGY, PARASITOLOGY. Educ: Wake Forest Col, BS, 61; Emory Univ, MS, 63, PhD(biol), 67. Prof Exp: Asst prof, 66-72, ASSOC PROF BIOL, QUEENS COL, NC, 72- Mem: AAAS; Am Soc Parasitol; Am Inst Biol Sci. Res: fluorescent antibody immunodiagnosis of parasitic diseases; taxonomy and life cycle of Spirorchiidae. Mailing Add: Dept of Biol Queens Col Charlotte NC 28207

MARTIN, WALTER EDWIN, b DeKalb, Ill, Jan 14, 08; m 34; c 4. PARASITOLOGY, ZOOLOGY. Educ: Northern Ill State Teachers Col, BEd, 30; Purdue Univ, MS, 32, PhD(parasitol, zool), 37. Prof Exp: Asst biol, Purdue Univ, 30-33, instr, 34-37; from asst prof to prof, DePauw Univ, 37-47; from assoc prof to prof zool, 47-74, head dept, 48-54, chmn div biol sci, 49-56, head dept biol, 54-58, EMER PROF BIOL SCI, UNIV SOUTHERN CALIF, 74- Concurrent Pos: Mem sci exped, Honduras, 33; mem staff & scholar, Marine Biol Lab, Woods Hole, 36; vis prof, Univ Hawaii, 56-57, Univ Neuchatel, 63-64 & Univ Queensland, 70-71. Mem: AAAS; Am Soc Parasitol; Am Soc Zoologists; Am Micros Soc. Res: Morphology, taxonomy and life cycles of trematodes; embryology; radioactive isotope absorption by embryonic tissues. Mailing Add: Dept of Biol Univ of Southern Calif Los Angeles CA 90007

MARTIN, WAYNE DUDLEY, b Watertown, Ohio, Nov 22, 20; m 53; c 2. GEOLOGY. Educ: Marietta Col, BS, 48; Univ WVa, MS, 50; Univ Cincinnati, PhD(geol), 55. Prof Exp: Instr geol, Bowling Green State Univ, 51-52; from instr to assoc prof, 52-69, PROF GEOL, MIAMI UNIV, 69- Mem: Geol Soc Am; Soc Econ Paleontologists & Mineralogists; Nat Asn Geol Teachers; Am Asn Petrol Geologists. Res: Petrology of the Cincinnatian Series limestone; sedimentary facies of the Dunkard Basin. Mailing Add: Dept of Geol Miami Univ Oxford OH 45056

MARTIN, WAYNE HOLDERNESS, b Manchester, Ohio, Mar 15, 31; m 54; c 6. POLYMER CHEMISTRY. Educ: Ohio State Univ, BS, 52, PhD(chem), 58. Prof Exp: Instr chem, Ohio State Univ, 57-58; from chemist to SR CHEMIST, PLASTICS DEPT, E I DU PONT DE NEMOURS & CO, INC, 58- Mem: Am Chem Soc. Res: High polymers; plastics research, development and analysis; air pollution. Mailing Add: Plastics Dept Washington Works E I du Pont de Nemours & Co Inc Parkersburg WV 26101

MARTIN, WESTON JOSEPH, b Church Point, La, Jan 15, 17; m 48; c 5. PLANT PATHOLOGY. Educ: Southwestern La Inst, BS, 37; La State Univ, MS, 39; Univ Minn, PhD(plant path), 42. Prof Exp: Asst bot exp sta, La State Univ, 37-39; asst plant path, Univ Minn, 39-42; from assoc pathologist to pathologist, USDA, 42-47; assoc pathologist, Agr Exp Sta, 47-54, plant pathologist, 54-58, PROF PLANT PATH, LA STATE UNIV, BATON ROUGE, 58- Mem: Fel AAAS; Am Phytopath Soc; Int Soc Trop Root Crops. Res: General plant pathology; nematology; sweet potato diseases. Mailing Add: Dept of Plant Path La State Univ Baton Rouge LA 70803

MARTIN, WILLARD JOHN, b Minneapolis, Minn, May 29, 15; m 42; c 5. PHYSICAL CHEMISTRY. Educ: Univ Minn, BS, 37; Cornell Univ, PhD(phys chem), 41. Prof Exp: Lab asst, Cornell Univ, 37-41; asst prof in-chg chem dept, Univ Maine, 46-49; PROF CHEM, SDAK SCH MINES & TECHNOL, 49- Mem: Am Chem Soc. Res: X-ray diffraction; atomic and molecular structure; general chemistry; computer science. Mailing Add: Dept of Chem SDak Sch Mines & Technol Rapid City SD 57701

MARTIN, WILLIAM BUTLER, JR, b Winchendon, Mass, Aug 31, 23; m 50; c 3. PHYSICAL ORGANIC CHEMISTRY, SPECTROSCOPY. Educ: Clark Univ, AB, 48, AM, 49; Yale Univ, PhD(org chem), 53. Prof Exp: Lab instr chem, Clark Univ, 47-49 & Yale Univ, 49-50; fel, Hickrill Res Fedn, NY, 52-53; from asst prof to assoc prof, 53-63, PROF CHEM, UNION COL (NY), 63- Concurrent Pos: NIH spec res fel, Sch Advan Studies, Mass Inst Technol, 59-61; res assoc, Swiss Fed Inst Technol, 67-68 & Univ Basel, 74-75. Mem: AAAS; Am Chem Soc; Fedn Am Sci; NY Acad Sci. Res: Photochemistry; biochemistry; synthetic organic chemistry; electron spin resonance in synthesized paracyclophanes. Mailing Add: Dept of Chem Union Col Schenectady NY 12308

MARTIN, WILLIAM C, b Hamilton, Ill, Dec 17, 29; m 55; c 1. AGRONOMY. Educ: Univ Ill, BS, 54, MS, 57, PhD(agron), 60. Prof Exp: Seed analyst & inspector, Ill Crop Improv Asn, Inc, 50-60; res agronomist, Morton Chem Co Div, Morton Int, Inc, 60-67, supvr biol res, 67-70; supvr planning & prod, 70-74, SUPVR NEW PROD EVAL, NOR-AM AGR PROD, INC, 74- Mem: Am Soc Agron; Am Phytopath Soc; Weed Sci Soc. Res: Agricultural chemicals, including fungicides, bactericides, insecticides, nematicides, soil fumigants, grain and space fumigants, herbicides and growth regulators; plant breeding, genetics and physiology. Mailing Add: Nor-Am Agr Prod Inc 1275 Lake Ave Woodstock IL 60098

MARTIN, WILLIAM CLARENCE, b Dayton, Ky, Nov 27, 23; m 47; c 3. PLANT TAXONOMY. Educ: Purdue Univ, BS, 50; Ind Univ, MA, 56, PhD(bot), 58. Prof Exp: From asst prof to assoc prof, 58-71, PROF BIOL, UNIV NMEX, 71- Mem: AAAS; Am Soc Plant Taxon. Res: Floristics; genetics; plant geography. Mailing Add: Dept of Biol Univ of NMex Albuquerque NM 87131

MARTIN, WILLIAM CLYDE, b Cullman, Ala, Nov 27, 29; m 59; c 2. ATOMIC SPECTROSCOPY, ATOMIC PHYSICS. Educ: Univ Richmond, BS, 51; Princeton Univ, MA, 53, PhD(physics), 56. Prof Exp: Instr physics, Princeton Univ, 55-57; physicist, 57-62, CHIEF SPECTROS SECT, NAT BUR STANDARDS, 62- Mem: Am Phys Soc; Optical Soc Am; Am Astron Soc; Int Astron Union. Res: Optical atomic spectroscopy; atomic structure. Mailing Add: A167 Physics Bldg Nat Bur of Standards Washington DC 20234

MARTIN, WILLIAM DAVID, b Anaconda, Mont, June 24, 42; m 64; c 4. ANATOMY. Educ: Carroll Col, Mont, AB, 64; Creighton Univ, MS, 66; Univ Minn, Minneapolis, PhD(vet anat), 72. Prof Exp: Instr vet anat, Univ Minn, 66-72; instr, 72-74, ASST PROF ANAT, UNIV KY, 75- Mem: Am Soc Zoologists. Res: Muscle histochemistry and ultrastructure. Mailing Add: Dept of Anat Univ of Ky Med Ctr Lexington KY 40506

MARTIN, WILLIAM E, soils, deceased

MARTIN, WILLIAM EUGENE, ecology, see 12th edition

MARTIN, WILLIAM GERALD, b Ottawa, Ont, Mar 18, 19; m 47; c 3. PHYSICAL CHEMISTRY, BIOCHEMISTRY. Educ: Carleton Univ, BSc, 52; McGill Univ, MSc, 55, PhD(chem), 58. Prof Exp: Examr, Inspection Bd UK & Can, 40-47; res officer, 55-69, SR RES OFFICER, NAT RES COUN CAN, 69- Mem: Can Biochem Soc. Res: Structure and function of biological membranes, lipoproteins and proteins. Mailing Add: Div of Biol Sci Nat Res Coun Can Sussex Dr Ottawa ON Can

MARTIN, WILLIAM GILBERT, b Shreveport, La, June 15, 31; m 53; c 3. NUTRITION, BIOCHEMISTRY. Educ: La Polytech Inst, BS, 56; NC State Univ, MS, 58, WVa Univ, PhD(biochem), 63. Prof Exp: Res asst animal nutrit, NC State Univ, 56-58; res asst agr biochem, 58-60, from instr to assoc prof, 60-74, AGR BIOCHEMIST & PROF AGR BIOCHEM, WVA UNIV, 74- Mem: Am Inst Nutrit; Am Chem Soc; Soc Exp Biol & Med; Poultry Sci Asn; Am Soc Animal Sci. Res: Amino acid and mineral metabolism; sulfur metabolism. Mailing Add: Comt of Agr Biochem WVa Univ Morgantown WV 26505

MARTIN, WILLIAM HARRY, b Bloomington, Ill, July 26, 23; m 46; c 2. ORGANIC CHEMISTRY, TEXTILES. Educ: Bradley Univ, BS, 47; Inst Textile Technol, MS, 49, PhD(chem), 51. Prof Exp: Res assoc chem div, Inst Textile Technol, 51-53, head org chem sect & chmn comt acad studies, 53-54, supvr pilot plant, 54-55, tech admin asst, 55-56, head chem div, 55-59, assoc dir res, 59-65, dir res, 65-67; prof textile chem & assoc dir textile res ctr, Tex Tech Col, 67-68; VPRES RES & DEVELOP, SPRINGS MILLS, INC, 68- Mem: Am Chem Soc; Am Asn Textile Chem & Colorists. Res: Surface active agents; chemical modification of cellulose; chemistry of textile fibers and finishing operations. Mailing Add: Springs Mills Inc Ft Mill SC 29715

MARTIN, WILLIAM HAYWOOD, III, b Bath Springs, Tenn, Nov 29, 38; m 65; c 2. PLANT ECOLOGY, FOREST ECOLOGY. Educ: Tenn Polytech Inst, BS, 60; Univ Tenn, Knoxville, MS, 66, PhD(bot), 71. Prof Exp: ASSOC PROF BIOL SCI, EASTERN KY UNIV, 69- Concurrent Pos: Inst grants, Eastern Ky Univ, 71 & 73. Mem: AAAS; Ecol Soc Am; Sigma Xi. Res: Relationships among plant components of forests and soil and geologic, topographic parameters; relationship of plant communities and populations to climatic, soil, geologic, topographic and biotic factors; major areas of interest and expertise in forest, natural and cultivated grassland ecosystems. Mailing Add: Dept of Gen Studies Cent Univ Col Eastern Ky Univ Richmond KY 40475

MARTIN, WILLIAM MACPHAIL, b Heatherdale, PEI, Aug 16, 19; m 45; c 4. NUCLEAR PHYSICS. Educ: Queen's Univ, Ont, BSc, 41; McGill Univ, PhD(physics), 51. Prof Exp: Jr res assoc physics, Chalk River Labs, 45-46; from asst prof to assoc prof, Queen's Univ, Ont, 51-55; assoc prof, 55-63, PROF PHYSICS, McGILL UNIV, 63-, ASST CHMN DEPT, 70- Mem: Can Asn Physicists; Am Phys Soc. Res: Radioactivity; nuclear reactions and isomerism. Mailing Add: Dept of Physics McGill Univ Montreal PQ Can

MARTIN, WILLIAM PAXMAN, b American Fork, Utah, July 15, 12; m 37; c 3. SOIL MICROBIOLOGY. Educ: Brigham Young Univ, AB, 34; Iowa State Col, MS, 36, PhD(soil bact), 37. Prof Exp: Instr soil microbiol, Univ Ariz, 37-40, asst prof, 40-45; forest ecologist, Southwestern Forest & Range Exp Sta, US Forest Serv, 45-48; prof agron exp sta, Ohio State Univ, 48-54; PROF SOILS & HEAD DEPT, INST AGR, UNIV MINN, ST PAUL, 54- Concurrent Pos: Asst soil microbiologist exp sta, Univ Ariz, 37-45; asst res chemist, Soil Conserv Serv, USDA, 41-44, soil chemist, 44-45, microbiologist, Regional Salinity Lab, Calif, 45; mem comn zero tolerance & residue regulation pesticides, Nat Acad Sci-Nat Res Coun, 64-65; mem agr libr network comt, Inter-Univ Commun Coun Educ Commun, 68-69; bd mem, Agron Sci Found, 71-74. Mem: Fel AAAS; fel Am Soc Agron (pres, 75); Soil Sci Soc Am (vpres, 65, pres, 66); fel Soil Conserv Soc Am; Am Inst Biol Scientists. Res: General soils; soil and water conservation; plant ecology. Mailing Add: Dept of Soil Sci Univ of Minn St Paul MN 55101

MARTIN, WILLIAM RANDOLPH, b Knoxville, Tenn, Apr 19, 22; m 49; c 2. MICROBIOLOGY. Educ: Univ Tenn, BA, 47, MS, 50; Univ Tex, PhD(bact), 55. Prof Exp: Res asst biophys, Oak Ridge Nat Lab, 47-48; res asst bact, Univ Tex, 51-52, res scientist, 52-55; assoc bacteriologist, Am Meat Inst Found, 55-57; from instr to asst prof microbiol, 57-64, ASSOC PROF MICROBIOL, UNIV CHICAGO, 64- Concurrent Pos: USPHS career develop award, 60-; Guggenheim fel, 65-66; vis investr, Inst Microbiol, Göttingen, Ger, 65-66. Mem: Am Soc Microbiol; NY Acad Sci; Brit Soc Gen Microbiol. Res: Microbial metabolism; filamentous fungi and mechanisms of cellular resistance to anti-tumor drugs. Mailing Add: Dept of Microbiol Univ of Chicago Chicago IL 60637

MARTIN, WILLIAM ROBERT, b Toronto, Ont, Nov 25, 16; m 47; c 3. FISH BIOLOGY. Educ: Univ Toronto, BA, 38, MA, 39; Univ Mich, PhD(zool), 48. Prof Exp: Scientist-in-chg groundfish invests, Fisheries Res Bd Can, 45-63, asst chmn, 63-68, dep chmn opers, 69-71, dir-gen prog integration & develop, Fisheries & Marine Serv, 71-73, SR SCI ADV, FISHERIES RES BD, CAN DEPT ENVIRON, 73- Concurrent Pos: Exec secy, Int Comn Northwest Atlantic Fisheries, NB, 51-52. Mem: Am Fisheries Soc; Can Soc Zool; Am Inst Fishery Res Biologists. Mailing Add: Fisheries Res Bd Can Dept of Environ Ottawa ON Can

MARTIN, WILLIAM ROBERT, b Aberdeen, SDak, Jan 30, 21; m 49; c 3. PHARMACOLOGY. Educ: Univ Chicago, BS, 48; Univ Ill, MS & MD, 53. Prof Exp: Intern, Cook County Hosp, Chicago, Ill, 53-54; from instr to asst prof pharmacol, Univ Ill, 54-57; neuropharmacologist, 57-63, DIR ADDICTION RES CTR, NAT INST DRUG ABUSE, 63- Concurrent Pos: Adj assoc prof, Univ Ky, 62, adj prof, Sch Med, 71; mem expert adv panel drug dependence, WHO, 65; prof pharmacol, Univ Ill Col Med, 67. Mem: AAAS; Am Soc Pharmacol & Exp Therapeut; Am Soc Clin Pharmacol & Therapeut; Am Col Neuropsychopharmacol; Soc Neurosci. Res: Neuropharmacology; clinical pharmacology; drug addiction. Mailing Add: NIDA Addiction Res Ctr PO Box 12390 Lexington KY 40511

MARTIN, WILLIAM ROYALL, JR, b Raleigh, NC, Sept 3, 26; m 52; c 2. ORGANIC CHEMISTRY, BUSINESS ADMINISTRATION. Educ: Univ NC, AB, 48, MBA, 63; NC State Univ, BS, 52. Prof Exp: Chemist, Am Cyanamid Co, 52-54; plant chemist, Dan River Mills, Inc, 54-56; group leader, Union Carbide Corp, 56-59; res assoc & head appl chem res, NC State Univ, 59-63; tech dir, 63-73, EXEC DIR, AM ASN TEXTILE CHEMISTS & COLORISTS, 74- Concurrent Pos: US deleg & secy meeting subcomt color-fastness tests, Int Orgn Standardization, NC, 64, Würzburg, Ger, 68, US deleg tech comt textiles, London, 65, 70 & 75, secy meetings subcomt color fastness & color measurement & subcomt dimensional stability, Newton, Mass, 71 & Paris, 74; mem comt textiles, Pan Am Standards Comn, Montevideo, 66; spec lectr & adj asst prof, NC State Univ, 66-. Mem: AAAS; Am Chem Soc; Am Inst Chem; Fiber Soc; Soc Dyers & Colourists. Res: Business and technical administration. Mailing Add: Am Asn Textile Chemists & Colorists PO Box 12215 Research Triangle Park NC 27709

MARTIN, WILLIAM SONDERMAN, b Mineola, NY, May 27, 35; m 63; c 1. SOLID STATE PHYSICS. Educ: Williams Col, BA, 57; Harvard Univ, MA, 60; Univ Calif, Berkeley, PhD(physics), 65. Prof Exp: Spectroscopist, Arthur D Little, Inc, 60-61; from physicist to liaison scientist, Indust Group, Res & Develop Ctr, Gen Elec Co, 65-75; RETIRED. Mem: Am Phys Soc. Res: Lasers, especially high power glass lasers; far-infrared spectroscopy, especially Fourier spectroscopy for the region of 10 to 100 wave numbers; superconductivity, especially thin films. Mailing Add: 390 Half Mile Rd Southport CT 06490

MARTIN, WILLIAM TED, b Springdale, Ark, June 4, 11; m 38; c 4. MATHEMATICS. Educ: Univ Ark, AB, 30; Univ Ill, MA, 31, PhD(math), 34. Prof Exp: Asst math, Univ Ill, 32-33; Nat Res Coun fel math, Princeton Univ & sch math, Inst Advan Study, 34-36; from instr to asst prof, Mass Inst Technol, 36-43; prof & chmn dept, Syracuse Univ, 43-46; prof math, 46-73, exec officer dept, 46-47, head, 47-68, chmn fac, 69-71, dir div study & res in educ, 73-75, PROF EDUC & MATH, MASS INST TECH, 73- Concurrent Pos: Res assoc, Princeton Univ, 40-41; mem exec comt, Div Math & Phys Sci, Nat Res Coun, 47-48; ed, Math Surv, Am Math Soc, 50-52, Bulletin, 51-56; mem, Inst Advan Study, 51-52; trustee, Oklahoma City Univ, 66- Mem: AAAS (vpres, 51, chmn sect A, 51); Am Math Soc (vpres, 49-50, treas, 65-73); Math Asn Am; Am Acad Arts & Sci. Res: Several complex variables; integration in function space. Mailing Add: Div Study & Res Educ Rm 20C-109A Mass Inst of Technol Cambridge MA 02139

MARTIN, YVONNE CONNOLLY, b St Paul, Minn, Sept 13, 36; m 63; c 2. MEDICINAL CHEMISTRY, MOLECULAR PHARMACOLOGY. Educ: Carleton Col, BA, 58; Northwestern Univ, PhD(biochem), 64. Prof Exp: Res asst, 58-60, sr pharmacologist, 64-70, assoc res fel, 70-74, RES FEL, ABBOTT LABS, 74- Concurrent Pos: Mem ed adv bd, J Med Chem, 71-75 & Med Res Series. Mem: AAAS; Am Chem Soc. Res: Structure-activity relationships; biochemical pharmacology, including drug metabolism and drug kinetics. Mailing Add: Dept 463 Abbott Labs Abbott Park North Chicago IL 60064

MARTINDALE, ROBERT WARREN, b New Castle, Ind, Jan 18, 18; m 43; c 4. MEDICAL ADMINISTRATION. Educ: Wabash Col, AB, 41. Prof Exp: Hosp adminr, US Army & US Air Force, 41-51, chief adminr, Surgeon's Off, Hq Air Res & Develop Command, 51-53, chief opers sect, Off Surgeon Gen, Air Force, Washington, DC, 53-56, chief manpower & orgn br, 53-56, res adminr, Sch Aviation Med, Randolph AFB, Tex, 57-60, dir plans & progs, Prin Proj Off & Exec Off, Air Force Sch Aerospace Med, Brooks AFB, 60-64, hosp adminr, Air Force Hosp, Clark AFB, Philippine Islands, 64-66, exec officer & dir support serv, Air Force Sch Aerospace Med, 66-67, chief of staff, Hq Aerospace Med Div, 67; DEP DIR ADMIN, MARINE BIOMED INST, UNIV TEX MED BR GALVESTON, 68- Mem: Armed Forces Mgt Asn; Am Hosp Asn; Am Soc Oceanog. Mailing Add: Marine Biomed Inst Univ of Tex Med Br 200 University Blvd Galveston TX 77550

MARTINDALE, WILLIAM EARL, b Nashville, Ark, Sept, 4, 23; m 50; c 3. BIOCHEMISTRY. Educ: Henderson State Teachers Col, BA, 47; Univ Ark, MS, 49; Univ Ala, MS, 57, PhD(biochem), 62. Prof Exp: Biochemist, Thayer Vet Admin

Hosp, Nashville, Tenn, 50-54; asst chief radioisotope serv, Birmingham Vet Admin Hosp, 54-62; asst prof chem, Miss State Univ, 62-64; PROF CHEM & CHMN DEPT, BELMONT COL, 64- Concurrent Pos: Vis prof org chem, Trevecca Col, 69- Res: Carbohydrate metabolism in thyroid tissues; synovial permeability in arthritis; chronic vitamin B-6 deficiency in rat; folic acid deficiency in chicks. Mailing Add: Dept of Chem Belmont Col Nashville TN 37203

MARTINEAU, BERNARD, b Chateauguay, Que, Dec 6, 21; m 51; c 2. MEDICAL BACTERIOLOGY. Educ: Bourget Col, BA, 43; Univ Montreal, MD, 49. Prof Exp: From lectr to assoc prof med bact, 40-70, PROF MICROBIOL & IMMUNOL, UNIV MONTREAL, 70- Concurrent Pos: Bacteriologist, Clin Lab, Hospital Ste-Justine, 54-, dir microbiol lab, 72- Mem: Can Med Asn; Asn Fr Speaking Physicians Can. Res: Clinical bacteriology; virology; mycology; neonatal virus infection, especially cytomegalovirus. Mailing Add: Dept of Med Univ of Montreal Montreal PQ Can

MARTINEAU, PERRY CYRUS, b Boise, Idaho, Jan 25, 18; m 41; c 4. PATHOLOGY, PHARMACOLOGY. Educ: Idaho State Univ, BS, 39; Univ Mich, MD, 43, MS, 48; Am Bd Path, dipl & cert clin path, 49, cert anat path, 50. Prof Exp: Intern, Henry Ford Hosp, Detroit, 43-44, asst resident surg, 44-45, resident path, 45-48; from asst prof to assoc prof, Col Med, Wayne State Univ, 50-65; assoc prof path, 66-72, CLIN PROF PHARMACOL, UNIV PITTSBURGH, 70-, CLIN PROF PATH, 72- Concurrent Pos: Dir labs & pathologist, City Health Dept, Detroit, 50-65; assoc, Detroit Receiving Hosp, 51-65; consult pathologist, Northville State Hosp, 59-; dir lab & chief pathologist, McKeesport Hosp, Pa, 66-; consult pathologist, Mayview State Hosp, 70. Mem: Fel AAAS; Am Soc Clin Path; fel Am Pub Health Asn; fel Col Am Path; Int Acad Path. Res: Anatomical, oral and surgical pathology. Mailing Add: 1100-5 Salk Hall Univ of Pittsburgh Pittsburgh PA 15261

MARTINEAU, ROBERT JEAN, b Woonsocket, RI, Mar 29, 40. THEORETICAL PHYSICS, ELECTRONICS. Educ: Providence Col, BS, 62; Rensselaer Polytech Inst, PhD(physics), 66; Stanford Univ, MS, 75. Prof Exp: From asst prof to assoc prof physics, Providence Col, 66-73; RES ENGR, HONEYWELL RADIATION CTR, 75- Mem: Inst Elec & Electronics Engrs; Am Phys Soc. Res: Physics of infrared detectors; general relativity; quantum field theory. Mailing Add: Honeywell Radiation Ctr 2 Forbes Rd Lexington MA 02173

MARTINEK, GEORGE WILLIAM, b Chicago, Ill, Apr 23, 32. GENETICS. Educ: Concordia Teachers Col, Ill, BS, 53; Los Angeles State Col, MA, 60; Univ Calif, Los Angeles, PhD(bot), 68. Prof Exp: Teacher, Trinity Lutheran Sch, 53-58; instr biol, Concordia Teachers Col, Ill, 58-62; from asst prof to assoc prof biol, 67-75, PROF BIOL, CALIF STATE POLYTECH UNIV, POMONA, 75- Mem: AAAS; Genetics Soc Am. Res: Genetics of Chlamydomonas reinhardi; recombination. Mailing Add: Dept of Biol Sci Calif State Polytech Univ Pomona CA 91768

MARTINEK, JOHN JOEL, b Milwaukee, Wis, Jan 20, 42; m 65; c 2. ANATOMY. Educ: Wis State Univ-Whitewater, BEd, 65; Tulane Univ, PhD(anat), 69. Prof Exp: Asst prof anat, Col Med, Ohio State Univ, 69-74; MEM FAC, ANAT-HISTOL DEPT, GRINNELL COL, 74- Concurrent Pos: NIH grants, Col Med, Ohio State Univ, 69-70 & 71-72. Mem: Am Asn Anat. Res: Electron microscopy of normal and abnormal human placentas. Mailing Add: Anat-Histol Dept Grinnell Col Grinnell IA 50112

MARTINEK, ROBERT GEORGE, b Chicago, Ill, Nov 25, 19; m 52. CLINICAL CHEMISTRY. Educ: Univ Ill, BS, 41 & 45, MS, 43; Univ Southern Calif, PharmD, 54; Am Bd Bioanal, cert. Prof Exp: Pharmacist & pharmaceut chemist, Bates Labs, Inc, 45-47; chemist res lab, Diversey Corp, Victor Chem Works, 47-50; assoc chemist, AMA, 50-55; sr chemist, Mead Johnson & Co, 55-56; clin chemist, Butterworth Hosp, Grand Rapids, Mich, 56-58, Iowa Methodist Hosp, 58-62 & Chicago Bd Health, 62-65; CLIN CHEMIST, DEPT PUB HEALTH, ILL, 65- Concurrent Pos: Assoc ed, J Am Med Technologists, 64-; consult, Abel Labs & Thornburg Labs, 65; consult, Lab-Line Instruments, 71-, bd dirs, 73-; lectr dept prev med & community health, Col Med, Univ Ill Med Ctr; ed consult, Med Electronics & Data; mem clin chem adv bd, Ctr Dis Control, USPHS, Atlanta, Ga, 74- Mem: AAAS; AMA; Am Pharmaceut Asn; fel Am Inst Chemists. Res: Pharmaceutical and detergent chemistry; vitamin assay; sympathomimetic amines; clinical chemistry methodology. Mailing Add: 4736 N Tripp Ave Chicago IL 60630

MARTINELLI, ERNEST A, b Lucca, Italy, Dec 15, 19; nat US; m 46; c 3. NUCLEAR PHYSICS. Educ: Univ Calif, BS, 41, PhD(physics), 50. Prof Exp: Mem staff radiation lab, Mass Inst Technol, 42-45; instr physics, Stanford Univ, 50-51; physicist radiation lab, Univ Calif, Berkeley, 51-52, Livermore, 52-56; physicist, Aeronutronics Systs, Inc, 56-57 & Rand Corp, 57-71; SR STAFF MEM, R&D ASSOCS, 71- Mem: Am Phys Soc. Res: Nuclear weapon effects; weapon systems. Mailing Add: R&D Assocs PO Box 9695 Marina del Rey CA 90291

MARTINELLI, LOUIS CARL, b Oroville, Calif, Sept 2, 37; m 59; c 3. PHARMACEUTICAL CHEMISTRY. Educ: Univ Calif, San Francisco, PharmD, 63, PhD(pharmaceut chem), 68. Prof Exp: Asst prof med chem, Sch Pharm, Univ Ga, 68-74; ASSOC PROF CLIN PHARM & COORDR, SCH PHARM, WVA UNIV, 74- Mem: Am Chem Soc; Am Pharmaceut Asn. Res: Synthesis and study of physical-chemical properties of organic molecules potentially useful as drugs in humans. Mailing Add: Sch of Pharm WVa Univ Morgantown WV 26506

MARTINELLI, MARIO, JR, b Covington, Va, May 7, 22; m 53; c 1. METEOROLOGY, FORESTRY. Educ: Univ Chicago, BS, 44; Duke Univ, MF, 48; State Univ NY, PhD, 56. Prof Exp: Forester, Southern Pine Lumber Co, Tex, 48-49; instr & asst, Purdue Univ, 49-54; RES METEOROLOGIST, ROCKY MOUNTAIN FOREST & RANGE EXP STA, US FOREST SERV, 54- Mem: Am Meteorol Soc; Glaciol Soc. Res: Watershed management of alpine areas, especially late-lying snow-beds and avalanche research. Mailing Add: Rocky Mtn Forest & Range Exp Sta 240 W Prospect St Ft Collins CO 80521

MARTINEZ, ALBERTO MAGIN, b Matanzas, Cuba, Oct 12, 43; US citizen. ORGANIC CHEMISTRY, PHOTOGRAPHY. Educ: Univ Ill, Chicago Circle, BS, 67, MS, 69, PhD(org chem), 72. Prof Exp: SR RES CHEMIST CHEM & PHOTOG, EASTMAN KODAK CO RES LABS, 72- Mem: Am Chem Soc. Res: Study of properties of light sensitive photographic materials, synthetic aspects of compounds used in such materials; properties of small ring alicyclic organic compounds. Mailing Add: Eastman Kodak Co Res Labs 1669 Lake Ave Rochester NY 14650

MARTINEZ, JOSEPH DIDIER, b White Castle, La, Oct 19, 15; m 45. ENVIRONMENTAL GEOLOGY, ENVIRONMENTAL ENGINEERING. Educ: La State Univ, BS, 37, MS, 52, PhD(geol), 59. Prof Exp: Elec engr, Brown Co, NH, 46-47; consult engr, La, 48-49; sr res geophysicist, Humble Oil & Refining Co, Tex, 52-57, sr geologist, 59-61; assoc prof geol, Northern Ill Univ, 61-64; vis lectr, Univ Wis, 64-65; PROF ENVIRON ENG & DIR INST SALINE STUDIES, LA STATE UNIV, BATON ROUGE, 65-, DIR, INST ENVIRON STUDIES, 72- Concurrent Pos: Lectr, Rice Univ, 57-58. Mem: Geol Soc Am; Am Asn Petrol Geol; Soc Econ Paleont & Mineral; Am Geophys Union. Res: Geology of evaporites; environmental geology; studies of heavy metals in plants and animal skeletons; environmental use of the sub-surface. Mailing Add: La State Univ PO Drawer J D University Sta Baton Rouge LA 70803

MARTINEZ, LUIS OSVALDO, b Havana, Cuba, Dec 27, 27; US citizen; m 55; c 3. RADIOLOGY. Educ: Inst Sec Educ, BS, 47; Univ Havana, MD, 54. Prof Exp: From instr to asst prof, 65-68, clin asst prof, 68-70, ASSOC PROF RADIOL, SCH MED, UNIV MIAMI, 70- Concurrent Pos: Counr, Interam Col Radiol, 70-79; chief, Div Diag Radiol, Mt Sinai Med Ctr, 70-, prog dir, Diag Radiol Residency Prog, 70- & assoc dir radiol, 70- Honors & Awards: Recognition Awards, AMA, 71-74; Gold Medal, Interam Col Radiol, 75. Mem: Fel Am Col Radiol; Am Roentgen Ray Soc; Radiol Soc NAm; Am Asn Univ Radiologists; Soc Gastrointestinal Radiologists. Res: Clinical evaluation of contrast media for intravenous cholangiography. Mailing Add: Mt Sinai Med Ctr 4300 Alton Rd Miami Beach FL 33140

MARTINEZ, MARGARET YARNALL, b West Grove, Pa, Dec 26, 20; m 46; c 3. HUMAN ANATOMY, PHYSIOLOGY. Educ: Univ Pa, AB, 43; Columbia Univ, MA, 44. Prof Exp: Teacher, Tatnall Sch, 62-67; ASSOC PROF BIOL, WEST CHESTER STATE COL, 67- Mem: AAAS; Am Women Sci Asn. Res: Cetaceans; whaling. Mailing Add: Dept of Biol West Chester State Col West Chester PA 19380

MARTINEZ, MARIO GUILLERMO, JR, b Havana, Cuba, Mar 6, 24; US citizen; m 49; c 3. ORAL PATHOLOGY. Educ: Univ Havana, DDS, 47; Univ Ala, Birmingham, DMD, 64, MS, 68; Am Bd Oral Path, dipl. Prof Exp: From instr to asst prof oral path, Sch Dent, Univ Havana, 49-59, prof, 59-60; from asst prof to assoc prof, 67-75, PROF PATH, MED CTR, UNIV ALA, BIRMINGHAM, 71-, DIR CLIN CANCER TRAINING PROG, SCH DENT, 70-, DIR DIV ORAL PATH, MED CTR, 71-, SR SCIENTIST COMPREHENSIVE CANCER CTR, 75- Concurrent Pos: Consult, Vet Admin Hosp, Birmingham, Ala. Mem: Int Asn Dent Res; fel Am Acad Oral Path; fel Am Col Dent; Am Dent Asn; Am Asn Cancer Educ. Res: Ultrastructure of giant cell lesion of the jaws; oral oncology. Mailing Add: Div of Oral Path Univ of Ala Med Ctr Birmingham AL 35294

MARTINEZ, NILDA, b Puerto Rico. ORGANOMETALLIC CHEMISTRY. Educ: Hunter Col, BA, 70; Mass Inst Technol, PhD(inorg chem), 75. Prof Exp: MEM RES STAFF, T J WATSON RES LAB, IBM CORP, 75- Mem: Sigma Xi. Res: Synthesis of new organometallic compounds which exhibit interesting electronic solid state properties. Mailing Add: T J Watson Res Lab IBM Corp Box 218 Yorktown Heights NY 10598

MARTINEZ, RAFAEL JUAN, b Santurce, PR, Feb 28, 27; div; c 3. BACTERIAL PHYSIOLOGY. Educ: Univ Southern Calif, AB, 52, PhD, 56. Prof Exp: From asst prof to assoc prof bact, 61-69, chmn dept, 71-73, PROF BACT, UNIV CALIF, LOS ANGELES, 69- Concurrent Pos: Fulbright fel, 70-71; NIH res grants bact & mycol, 73-77. Mem: Am Soc Microbiol; Brit Soc Gen Microbiol. Res: Biochemistry of pathogenesis; host-parasite interactions. Mailing Add: Dept of Bact Univ of Calif Los Angeles CA 90024

MARTINEZ-CARRION, MARINO, b Felix, Spain, Dec 2, 36; US citizen; m 57; c 2. BIOCHEMISTRY. Educ: Univ Calif, Berkeley, BA, 59, MA, 61, PhD(comp biochem), 64. Prof Exp: NIH fel biochem, Rome, 64-65; from asst prof to assoc prof chem, 65-74, PROF CHEM, UNIV NOTRE DAME, 74- Concurrent Pos: NIH career develop award, 72. Mem: Am Soc Biol Chemists; Am Chem Soc. Res: Mechanisms of enzyme action; active center of pyridoxal dependent enzymes; isoenzymes; nuclear magnetic resonance of enzyme-substrate interaction; neuroreptors; neurochemistry; membrane research. Mailing Add: Dept of Chem Univ of Notre Dame Notre Dame IN 46556

MARTINEZ-LOPEZ, JORGE IGNACIO, b Santurce, PR, Oct 5, 26; m 50; c 4. INTERNAL MEDICINE, CARDIOLOGY. Educ: La State Univ, MD, 50. Prof Exp: Intern, Arecibo Dist Hosp, PR, 50-51; physician, Elizabeth, La, 53-54; resident internal med, Charity Hosp, New Orleans, La, 54-57; from instr to assoc prof, 57-69, PROF MED, LA STATE UNIV MED CTR, NEW ORLEANS, 69- Concurrent Pos: Vis physician, Charity Hosp, New Orleans, 57-64, sr vis physician, 64-, dir, Cardiol Dept, 60-; cardiologist, Heart Sta, Hotel Dieu Hosp, 63-71, head, 71-74, mem consult staff, 75- Mem: Fel Am Heart Asn; fel Am Col Chest Physicians; fel Am Col Physicians; fel Am Col Cardiol. Res: Clinical cardiology; cardiac catheterization and other special diagnostic procedures; electrocardiography. Mailing Add: Dept of Med La State Univ Med Ctr New Orleans LA 70112

MARTINEZ-MALDONADO, MANUEL, b Yauco, PR, Aug 25, 37; m 59; c 4. INTERNAL MEDICINE, NEPHROLOGY. Educ: Univ PR, San Juan, BS, 57; Temple Univ, MD, 61. Prof Exp: Intern, St Charles Hosp, Toledo, Ohio, 61-62; resident internal med, Vet Admin Hosp & Sch Med, Univ PR, San Juan, 62-65; USPHS fel, Univ Tex Southwestern Med Sch Dallas, 65-67; Lederle Labs int fel, 66-67; instr, Univ Tex Southwestern Med Sch Dallas, 67-68; from asst prof to prof, Baylor Col Med, 68-73; PROF MED & PHYSIOL, SCH MED, UNIV PR, 72-, ACTG CHMN DEPT PHYSIOL, 74- Concurrent Pos: Dir chronic dialysis unit, Parkland Mem Hosp, Dallas, Tex, 67-68; attend physician, Ben Taub Gen Hosp, Houston, 68-73 & Methodist Hosp, Houston, 69-73; assoc chief staff for res, Vet Admin Hosp, San Juan, 73, chief med serv, 74- Mem: Am Soc Clin Invest; Cent Soc Clin Res: Am Fedn Clin Res; Am Soc Nephrology; Am Physiol Soc. Res: Renal physiology; electrolyte metabolism; biochemistry of transport. Mailing Add: Med Serv Vet Admin Hosp San Juan PR 00936

MARTINEZ NADAL, NOEMI G, b Aug, 15, 17; US citizen; m 37, 53; c 2. ORGANIC CHEMISTRY, BIOCHEMISTRY. Educ: New Rochelle Col, BS, 37; Univ Paris, DSc, 54. Prof Exp: Chemist, Fed Exp Sta, 37-46; from instr to assoc prof chem, 54-70, chemist in chg chem & biochem invests, Res Ctr, 54-67, NSF grant dir, Undergrad Res Prog, 63-66, PROF CHEM, UNIV PR, MAYAGÜEZ, 70-, DIR RES, BIOCHEM LABS, FAC ENG, 68- Concurrent Pos: Consult, Essential Oil Indust & Fomento Labs. Mem: Am Soc Eng Educ; Asn Off Anal Chemists; Am Soc Agr Chemists; Am Inst Chem Eng; fel Am Inst Chemists. Res: Essential oil chemistry; antibiotics. Mailing Add: Biochem Res Labs Univ of PR Fac of Eng Mayagüez PR 00708

MARTINEZ-PICO, JOSE LUIS, b Coamo, PR, July 16, 18; m 48; c 6. PHYSICAL CHEMISTRY. Educ: Univ PR, BS, 39; Univ Mich, MS, 47; Carnegie Inst Technol, MS, 61, PhD(chem), 62. Prof Exp: From instr to assoc prof chem, 40-62, dean sch arts & sci, 67-71, PROF CHEM, UNIV PR, MAYAGUEZ, 61-, DEAN STUDIES, 71- Concurrent Pos: Consult, Univ Nicaragua, 65- Mem: Am Chem Soc. Res: Use of nuclear magnetic resonance and mass spectra in analysis of petroleum. Mailing Add: Box 1381 Mayaguez PR 00708

MARTINI, CATHERINE MARIE, b New York, NY, July 7, 24. PHYSICAL CHEMISTRY. Educ: Hunter Col, BA, 46; Univ Pa, MS, 48. Prof Exp: Tutor chem, Hunter Col, 47-49; asst, 49-53, res assoc, 53-60, res chemist, 61-66, SR RES

CHEMIST, STERLING-WINTHROP RES INST DIV, STERLING DRUG, INC, 66- Mem: Am Chem Soc; Coblentz Soc. Res: Infrared, ultraviolet and nuclear magnetic resonance spectroscopy of organic molecules. Mailing Add: Sterling-Winthrop Res Inst Rensselaer NY 12144

MARTINI, IRENEO PETER, b Dec 14, 35; Can citizen; m 62; c 2. SEDIMENTOLOGY. Educ: Univ Florence, DrGeolSci, 61; McMaster Univ, PhD(geol), 66. Prof Exp: From geologist to sr geologist, Shell Can Ltd, 66-69; ASSOC PROF SEDIMENTOLOGY, UNIV GUELPH, 69- Mem: Soc Econ Paleontologists & Mineralogists; Int Asn Sedimentol. Res: Sedimentary geology; sedimentology of Recent and Pleistocene clastic sediments and ancient sedimentary rocks; fabric of soils and sediments; analysis of hydrocarbon potentials of selected regions. Mailing Add: Dept of Land Resource Sci Univ of Guelph Guelph ON Can

MARTINO, FRANK, physics, see 12th edition

MARTINO, JOSEPH PAUL, b Warren, Ohio, July 16, 31; m 57; c 3. OPERATIONS RESEARCH, SCIENCE ADMINISTRATION. Educ: Miami Univ, AB, 53; Purdue Univ, MS, 55; Ohio State Univ, PhD(math), 61. Prof Exp: With US Air Force, 53-75, proj engr, Wright Air Develop Ctr, 55-58, mathematician, Air Force Off Sci Res, 60-62, staff scientist, 63-67, opers analyst, Res & Develop Field Univ, Bangkok, Thailand, 62-63, chief tech anal div, Air Force Off Res Anal, 68-71, staff scientist, Avionics Lab, Wright-Patterson AFB, 72-73, dir eng standardization, Defense Electronics Supply Ctr, 73-75; RES SCIENTIST, RES INST, UNIV DAYTON, 75- Concurrent Pos: Consult, Nat Coun Cath Men, 67- Mem: AAAS; Opers Res Soc Am; Inst Elec & Electronics Eng; Am Inst Aeronaut & Astronaut; Inst Mgt Sci. Res: Application of operations research to problems of technological change, with emphasis on technological forecasting. Mailing Add: 819 N Maple Ave Fairborn OH 45324

MARTINS, DONALD HENRY, b Poplar Bluff, Mo, July 31, 45; m 69; c 1. ASTROPHYSICS. Educ: Univ Mo, Columbia, BS, 67, MS, 69; Univ Fla, PhD(astron), 74. Prof Exp: Res assoc astron, Nat Res Coun, Johnson Space Ctr, 74-76; NASA RES ASSOC ASTRON, HOUSTON BAPTIST UNIV, 76- Mem: Am Astron Soc; Am Inst Physics. Res: Active in photoelectric photometry of variable stars, and in electrographic techniques in astronomy, especially for cluster photometry and surface photometry of galaxies. Mailing Add: Astrophys Sect TN 23 Johnson Space Ctr Houston TX 77058

MARTINS DA SILVA, MAURICIO, b Rio de Janeiro, Brazil, Dec 14, 16; m 43; c 3. PEDIATRICS. Educ: Pinto Ferreira Col Brazil, BS, 33; Univ Brazil, MD, 39; Am Bd Pediat, dipl, 55; Univ Minn, MPH, 57. Prof Exp: Instr biol, Univ Brazil, 35 & 36; intern pediat, Univ Minn Hosps, 41-42, resident, 42-43; resident tuberc, Sea View Hosp, Staten Island, NY, 43; Am Acad Pediat res fel pediat, Yale Univ, 43-44; head pediat serv, Sao Zacharias Hosp, Rio de Janeiro, 49-54; asst prof pediat & res assoc bact & immunol, Univ Minn, 54-57; regional adv poliomyelitis, Pan-Am Sanit Bur, 58-61, dep chief off res coord, 61-65, CHIEF DEPT RES DEVELOP & COORD, PAN-AM HEALTH ORGN, WHO, 66- Concurrent Pos: Pan-Am Sanit Bur fel, Univ Minn Hosps, 41-42; asst dir in-patient serv, Elizabeth Kenny Inst, Minneapolis, Minn, 55-57. Honors & Awards: Adolfo Lutz Medal, Brazil, 65. Mem: Am Acad Pediat; Am Pub Health Asn; Brazilian Pediat Soc. Res: Planning and coordination of biomedical research; neurotropic viral diseases, live oral poliovirus vaccine; water and electrolyte metabolism. Mailing Add: 5101 River Rd Ste 1613 Washington DC 20016

MARTINSEN, DAVID LINNEBACH, physiological ecology, radiation biology, see 12th edition

MARTINSEN, JAMES STANLEY, microbiology, see 12th edition

MARTINSON, CHARLIE ANTON, b Orchard, Colo, Sept 15, 34; m 57; c 4. PLANT PATHOLOGY. Educ: Colo State Univ, BS, 57, MS, 59; Ore State Univ, PhD(plant path), 64. Prof Exp: Asst prof plant path, Cornell Univ, 63-68; ASSOC PROF, IOWA STATE UNIV, 68- Concurrent Pos: Consult, Corn Prod Syst, Inc, 72- Mem: Fel AAAS; Am Phytopath Soc. Res: Root diseases of economically important crops; physiology of plant disease; corn diseases; role of toxins in pathogenesis; international programs in corn production and plant disease control. Mailing Add: Dept of Bot & Plant Path Iowa State Univ Ames IA 50011

MARTINSON, HAROLD GERHARD, b Hartford, Conn, Sept 9, 43; m 68; c 2. MOLECULAR BIOLOGY. Educ: Augsburg Col, BA, 65; Univ Calif, Berkeley, PhD(molecular biol), 71. Prof Exp: Fel biol, Univ Lethbridge, 71-73 & biochem, Univ Calif, San Francisco, 73-75; ASST PROF CHEM, UNIV CALIF, LOS ANGELES, 75- Res: Chromosome structure and chemistry; control of gene expression in eucaryotes. Mailing Add: Dept of Chem Univ of Calif 405 Hilgard Ave Los Angeles CA 90024

MARTINSON, TOM L, b Portland, Ore, Nov 7, 41; m 65; c 1. ECONOMIC GEOGRAPHY, GEOGRAPHY OF LATIN AMERICA. Educ: Univ Ore, BA, 63; Univ Kans, PhD(geog), 69. Prof Exp: Vis asst prof geog, Univ Colo, 66-67; from asst prof to assoc prof, 67-75, dir, Inst Int Studies, 72-75, PROF GEOG, BALL STATE UNIV, 75- Mem: Asn Am Geog; Am Geog Soc; Latin Am Studies Asn. Res: Impact of highway development on subsistence agriculture in central America. Mailing Add: Dept of Geog Ball State Univ Muncie IN 47306

MARTINSONS, ALEKSANDRS, b Russia, Nov 30, 12; US citizen; m 39; c 2. ELECTROCHEMISTRY. Educ: Univ Mich, MS, 55. Prof Exp: Sr res chemist, Am Potash & Chem Corp, 57-60; SR RES CHEMIST, CHEM DIV, PPG INDUSTS, INC, 60- Mem: Electrochem Soc. Res: Overvoltage; electro-winning and metal deposition; thin film coatings. Mailing Add: Chem Div PPG Indust Inc Barberton Tech Ctr Barberton OH 44203

MARTIRE, DANIEL EDWARD, b New York, NY, June 3, 37; m 61. PHYSICAL CHEMISTRY. Educ: Stevens Inst Technol, BE, 59, MS, 60, PhD(chem), 63. Prof Exp: Instr chem, Stevens Inst Technol, 62-63; NSF fel, Cambridge Univ, 63-64; from asst prof to assoc prof, 64-74, PROF CHEM, GEORGETOWN UNIV, 75- Mem: Am Chem Soc. Res: Thermodynamics and statistical mechanics of liquid crystals and liquid mixtures; theory of gas-liquid and liquid-liquid chromatography; charge-transfer and hydrogen-bond complex formation. Mailing Add: Dept of Chem Georgetown Univ Washington DC 20057

MARTIS, KENNETH CHARLES, b Toledo, Ohio, Dec 5, 45. GEOGRAPHY. Educ: Univ Toledo, BEd, 68; San Diego State Univ, MA, 70; Univ Mich, PhD(geog), 76. Prof Exp: Teaching asst geog, San Diego State Univ, 68-70 & Univ Mich, 72-75; ASST PROF GEOG, WVA UNIV, 75- Concurrent Pos: Teaching intern, Lilly Endowment Inc, 75-76. Mem: Asn Am Geogrs; Am Geog Soc; Nat Coun Geog Educ; Nat Geog Soc. Res: Legislative behavior with respect to natural resources policy in the United States Congress, especially the mapping of legislative votes and the spatial aspects of legislative voting behavior. Mailing Add: Dept of Geol & Geog WVa Univ Morgantown WV 26506

MARTNER, SAMUEL (THEODORE), b Prairie du Chien, Wis, Apr 20, 18; m 42; c 2. GEOPHYSICS. Educ: Univ Calif, BA, 40, Calif Inst Technol, MS, 46, PhD(geophys), 49. Prof Exp: From prod engr to prod control mgr, Los Angeles Shipbldg & Drydock Co, 41-43; ship supvr & asst to repair gen mgr, Todd Shipyards Corp, 43-45; geologist, Standard Oil Co Calif, 46; geologist, Stanolind Oil & Gas Co Div, Standard Oil Co, Ind, 47-48, seismic interpreter, 48, asst party chief, 48-49, party chief, 49-50, tech group supvr, 51-52, res group supvr, 52-58, res sect supvr, Pan Am Petrol Corp, 58-64, div geophysicist, 65-67, asst chief geophysicist, 67-68, geophys res dir, 68-71, RES CONSULT, AMOCO PROD CO, 71- Mem: Seismol Soc Am; Soc Explor Geophys; Geol Soc Am; Am Geophys Union; Inst Elec & Electronics Eng. Res: Petroleum; earth sciences. Mailing Add: Res Ctr Amoco Prod Co PO Box 591 Tulsa OK 74102

MARTOF, BERNARD STEPHEN, b Rices Landing, Pa, Aug 21, 20; m 42; c 4. ZOOLOGY. Educ: Waynesburg Col, BS, 42; WVa Univ, MS, 47; Univ Mich, PhD(zool), 51. Prof Exp: Asst zool, Univ Mich, 47-50, instr, 50-51, from asst prof to prof, Univ Ga, 51-63; head dept zool, 63-67, PROF ZOOL, NC STATE UNIV, 63- Mem: AAAS; Soc Syst Zool; Am Soc Zool; Soc Study Amphibians & Reptiles; Am Soc Naturalists. Res: Vertebrate zoology; ecology, behavior, life histories, geographic distribution, variation and taxonomy of amphibians and reptiles; sociobiology, especially analysis of vocalization and reproductive behavior. Mailing Add: Dept of Zool NC State Univ Raleigh NC 27607

MARTON, JOHN PETER, b Budapest, Hungary, May 13, 33; Can citizen. SOLID STATE ELECTRONICS. Educ: Sci Univ, Budapest, BSc, 54; Univ Western Ont, PhD(solid state), 69. Prof Exp: DIR RES & DEVELOP, WELWYN CAN LTD, 63-; ASSOC PROF ENG PHYSICS, McMASTER UNIV, 71- Mem: AAAS; Asn Prof Engrs Ont; Inst Elec & Electronics Engrs. Res: Solid state plasmas in thin films; electroless Ni-P deposition and properties; optical communication; cooperative physical phenomena in biological cells. Mailing Add: Dept of Eng Physics McMaster Univ Hamilton ON Can

MARTON, JOSEPH, b Budapest, Hungary, Mar 5, 19; US citizen; m 49; c 1. PHYSICAL ORGANIC CHEMISTRY, WOOD CHEMISTRY. Educ: Pazmany Peter Univ, Hungary, BS & MS, 42, PhD(org chem), 43. Prof Exp: Asst prof org chem, Budapest Tech Univ, 45-47, lectr, 49-56; chemist, Arzola Chem Co, 47-49; res supvr, Res Inst Indust Org Chem, 49-56; res fel wood chem, Chalmers Univ Technol, Sweden, 56-60; res assoc wood & phys org chem, Charleston Res Lab, Westvaco Corp, SC, 60-66, res assoc, Phys Org & Surface Chem, 66-72, SR RES ASSOC, LAUREL RES CTR, WESTVACO CORP, 72- Mem: AAAS; Am Chem Soc; Tech Asn Pulp & Paper Indust; Swedish Chem Soc; Tech Asn Graphic Arts. Res: Chemistry and reaction of polymeric compounds; analysis and chemistry of fiber surfaces; colloid and surface chemistry of pulp and papermaking; printability of paper; forest improvement, lignin chemistry. Mailing Add: Laurel Res Ctr Westvaco Corp Johns Hopkins Rd Laurel MD 20810

MARTON, LADISLAUS LASZLO, physics, see 12th edition

MARTON, OLIVER L, b Smyrna, Turkey, Jan 26, 04; nat US; m 38; c 1. COSMETIC CHEMISTRY. Educ: Graz Univ, PhD(chem), 26. Prof Exp: Chemist, Friedlander Labs, Ger, 26-31; consult chemist, Berkshire Knitting Mills, 31-34; chief chemist, Soc Textile Improv, 34-35; chief chemist & plant mgr cosmetics, Raymond Chem Co, Inc, 35-42; chief chemist, 42-50, chief perfumer, 47-71, CONSULT PERFUMER-CHEMIST, SHULTON, INC, 71- Concurrent Pos: Guest lectr, NY Univ, 49-50. Mem: Am Chem Soc; Soc Cosmetic Chem; NY Acad Sci; Fr Soc Perfumers; Am Soc Perfumers (pres, 59). Res: Cosmetics; toiletries; soaps; perfumes; aromatics; flavors; odor evaluations; olfactory research; quinine derivatives; textiles; fibers; plastics; fats; oils; waxes; detergents. Mailing Add: 1059 Inwood Terr Ft Lee NJ 07024

MARTON, RENATA, b Krakow, Poland, July 27, 10; US citizen; m 38; c 2. CHEMISTRY. Educ: Univ Jagello, Poland, MS, 34, PhD, 36. Prof Exp: Asst chem, Univ Jagello, Poland, 34-38; res asst cellulose derivatives, Inst Chem, Sorbonne, 38-43; from researcher to tech mgr, Chem Indust, Hungary, 45-50; head dept, Pulp & Paper Res Inst, 50-56; Rockefeller Found fel, Austrian Wood Res Inst, 56-57; from asst prof to assoc prof pulp & paper res, 57-68, PROF PULP & PAPER RES, STATE UNIV NY COL FORESTRY, 68- Mem: Tech Asn Pulp & Paper Indust. Res: Fundamentals of pulp and papermaking fibers; morphology and nature of coloring materials in wood. Mailing Add: Dept of Forestry State Univ of NY Col of Forestry Syracuse NY 13210

MARTONOSI, ANTHONY, b Szeged, Hungary, Nov 7, 28; US citizen; m 59; c 3. BIOCHEMISTRY. Educ: Univ Szeged, MD, 53. Prof Exp: Asst prof physiol, Univ Szeged, 54-57; Nat Acad Sci res fel biochem, Mass Gen Hosp, Boston, 57-59; assoc, Retina Found, 59-63, asst dir, 64-65; PROF BIOCHEM, SCH MED, ST LOUIS UNIV, 65- Concurrent Pos: USPHS grant, 59-; estab investr, Am Heart Asn, 61-66; NSF grant, 63- Mem: Am Soc Biol Chem; Biophys Soc. Res: Biochemistry of muscle contraction; contractile proteins; structure and function of membranes. Mailing Add: Dept of Biochem St Louis Univ Sch of Med St Louis MO 63104

MARTORELL, LUIS FELIPE, b Yabucoa, PR, June 20, 09; m 38; c 1. ENTOMOLOGY. Educ: Univ PR, BSA, 28; Ohio State Univ, MS, 34, PhD(entom), 43. Prof Exp: Agronomist, Ministry Agr, Venezuela, 32-33; chief div entom, Emergency Relief Admin, PR, 34; from asst to the technician to asst forester, US Forest Serv, PR, 34-36; asst entomologist, Agr Exp Sta, 36-43, assoc entomologist, 43-50, entomologist, 50-73, head dept entom, 53-73, EMER PROF ENTOM, UNIV PR, 73- Concurrent Pos: Prof zool & entom & sub-dir, Sch Agr, Maracay, Venezuela, 32; with educ gardening sect, El Mundo, 60-71. Res: Forest entomology; sugarcane insects; parasite introduction; chemical control of insect pests; cacao, papaya and rice insects in West Indies; termite taxonomy, genus Nasutitermes (New World species). Mailing Add: Dept of Entom Agr Exp Sta Univ of PR Rio Piedras PR 00928

MARTS, MARION ERNEST, b Olympia, Wash, July 20, 15; m 39; c 3. GEOGRAPHY. Educ: Univ Wash, BA, 37, MA, 44; Northwestern Univ, PhD(geog), 50. Prof Exp: From instr to prof geog, 48-63, vis provost, 63-69, PROF GEOG & URBAN PLANNING, UNIV WASH, 63-, ADJ PROF ENVIRON STUDIES, 73-, DEAN SUMMER QUARTER, 69- Concurrent Pos: Econ geogr, US Bur Reclamation, 49-51; Fulbright res grant, Italy, 54. Mem: Asn Am Geogr. Res: Water resource utilization; economic geography of western United States. Mailing Add: Dept of Geog Univ of Wash Seattle WA 98105

MARTSOLF, J DAVID, b Beaver Falls, Pa, Nov 26, 32; m 55; c 2. AGRICULTURAL METEOROLOGY. Educ: Univ Fla, BSA, 54, MSA, 62; Univ Mo-Columbia, PhD(atmospheric sci), 66. Prof Exp: Asst county agr agt, Agr Exten Serv, Univ Fla, 58-62, asst prof hort, 62-64; res asst surface energy balance study, Univ Mo-Columbia, 64-66; ASSOC PROF AGR CLIMAT, PA STATE UNIV, UNIVERSITY PARK, 66- Mem: Am Meteorol Soc; Am Soc Hort Sci. Res: Frost protection of citrus and other tree crops; modification of advective heat transport; general energy balance of vegetative canopies. Mailing Add: Dept of Hort Pa State Univ University Park PA 16802

MARTT, JACK M, b Ashland, Ky, Nov 9, 22; m 48; c 2. INTERNAL MEDICINE, CARDIOLOGY. Educ: Univ Mo, BS, 43; Wash Univ, MD, 46. Prof Exp: Asst prof internal med, Col Med, Univ Iowa, 55-56; from asst prof to assoc prof, Sch Med, Univ Mo-Columbia, 56-69, dir cardiopulmonary lab, Med Ctr, 65-69; CARDIOLOGIST, SCOTT & WHITE CLIN, 69- Concurrent Pos: Fel coun clin cardiol, Am Heart Asn, 63. Mem: Fel Am Col Physicians. Res: Cardiology research primarily in atherosclerosis. Mailing Add: Dept of Med Scott & White Clin 2401 S 31st Temple TX 76501

MARTUS, JOSEPH ARMAND, b Boston, Mass, Mar 12, 09. HISTORY OF CHEMISTRY. Educ: Boston Col, AB, 33, MA, 34; Col Holy Cross, MS, 36; Clark Univ, PhD(chem), 52. Prof Exp: Instr chem, Boston Col, 36-37, Cranwell Prep Sch, 42-44 & St George's Col, Jamaica, BWI, 44-47; from instr to prof, 47-74, chmn dept, 62-74, EMER PROF CHEM, COL HOLY CROSS, 74- Res: Chemical education; chelation. Mailing Add: Gonzaga: Jesuit Ctr for Renewal Monroe NY 10950

MARTY, ROBERT JOSEPH, b Evanston, Ill, July 6, 31; m 62; c 2. FOREST ECONOMICS. Educ: Mich State Univ, BS, 54; Duke Univ, MF, 55; Harvard Univ, MPA, 59; Yale Univ, PhD(forestry), 62. Prof Exp: Res forester economics, Forest Serv, USDA, 55-65, chief forest econ br, 65-67; assoc prof forestry, 67-71, PROF FORESTRY, MICH STATE UNIV, 71- Mem: Soc Am Foresters; Econ Asn. Res: Timber production economics; economics of public, natural resource programs and policies. Mailing Add: Dept of Forestry Mich State Univ East Lansing MI 48823

MARTY, ROGER HENRY, b Sterling, Ohio, Oct 16, 42; m 64; c 2. TOPOLOGY. Educ: Kent State Univ, BS, 64; Pa State Univ, MS, 66, PhD(math), 69. Prof Exp: Asst prof math, 69-73, ASSOC PROF MATH, CLEVELAND STATE UNIV, 73- Mem: Am Math Soc. Res: Set-theoretic topology; general topology; set-theory. Mailing Add: Dept of Math Cleveland State Univ Cleveland OH 44115

MARTY, WAYNE GEORGE, b LuVerne, Iowa, Feb 14, 32; m 54; c 3. PARASITOLOGY. Educ: Westmar Col, BA, 53; Univ Iowa, MS, 59, PhD(zool), 62. Prof Exp: Teacher high sch, Iowa, 55-57; PROF BIOL, WESTMAR COL, 59- Concurrent Pos: NIH fel malariology, 69-70. Mem: Am Soc Parasitol; Nat Asn Biol Teachers. Res: Experimental infections of parasites in abnormal hosts, specifically Trichinella Spiralis in chickens. Mailing Add: RR 2 LeMars IA 51031

MARTZ, BILL L, b Anderson, Ind, Jan 17, 22; m 48; c 3. MEDICINE. Educ: DePauw Univ, AB, 44; Ind Univ, MD, 45; Am Bd Internal Med, dipl. Prof Exp: Asst med serv, 53-55, from asst prof to assoc prof, 55-67, PROF MED, SCH MED, IND UNIV, INDIANAPOLIS, 67-; DIR CLIN INVEST, DOW CHEM CO, 74- Concurrent Pos: Assoc med serv, Indianapolis Gen Hosp, 53-55, mem vis staff, Med Serv, 56-; mem coun high blood pressure res, Am Heart Asn; res physician, Marion County Gen Hosp, 51-60, dir, Lilly Lab Clin Res, 60-72; chief med, Kansas City Gen Hosp, 72-74. Mem: AMA; Am Col Physicians; Am Fedn Clin Res; Am Col Cardiol; Am Soc Clin Pharmacol & Therapeut. Res: Cardiovascular renal disease and clinical pharmacology. Mailing Add: Dow Chem Co PO Box 68511 Indianapolis IN 46268

MARTZ, CARL D, b Quincy, Ohio, Nov 22, 13; m 39; c 3. MEDICINE. Educ: DePauw Univ, AB, 36; Ind Univ, MD, 40; Am Bd Orthop Surg, dipl. Prof Exp: From intern to resident, 40-44, from assoc to clin prof, 45-63, PROF ORTHOP SURG, SCH MED, IND UNIV, INDIANAPOLIS, 63- Concurrent Pos: Consult orthopedist, Crippled Children's Serv, Ind State Welfare Dept, 45-50, Voc Rehab Admin, 50-55 & Muscatatuck State Sch, 55; chmn emergency med serv, Ind State Adv Comt. Mem: AAAS; Clin Orthop Soc; AMA; Am Neurol Asn; Am Col Surg. Res: Biomechanics; cerebral palsy; orthopedic rehabilitation. Mailing Add: Dept of Orthop Surg Ind Univ Sch of Med Indianapolis IN 46200

MARTZ, DOWELL EDWARD, b Livonia, Mo, Sept 29, 23; m 50; c 3. PHYSICS. Educ: Union Col, Nebr, BA, 50; Vanderbilt Univ, MS, 53; Colo State Univ, PhD, 68. Prof Exp: Physicist, US Naval Ord Lab, 53-61; assoc prof physics, Pac Union Col, 61-62; sr res physicist, Calif Inst Technol, 62-64; assoc prof physics, 64-74, PROF PHYSICS, PAC UNION COL, 74- Concurrent Pos: Consult, Calif Inst Technol, 61- & Ames Res Lab, NASA, 64-65. Mem: Optical Soc Am; Sigma Xi; Am Inst Physics. Res: Infrared optics, detectors, astronomy and space research; application to long wavelength. Mailing Add: Dept of Physics Pac Union Col Angwin CA 94508

MARTZ, FREDRIC A, b Columbia City, Ind, May 21, 35; m 59; c 5. ANIMAL NUTRITION, DAIRY SCIENCE. Educ: Purdue Univ, BS, 57, MS, 59, PhD(dairy sci), 61. Prof Exp: Instr dairy sci, Purdue Univ, 60-61; from asst prof to assoc prof, 61-73, PROF DAIRY SCI & FORAGE LIVESTOCK RES COORD-AGR, UNIV MO-COLUMBIA, 73- Concurrent Pos: Consumer Coop Asn grant, 64-65, 67-71 & 73; NIH grant, 69; vis assoc prof, Cornell Univ, 71-72. Honors & Awards: Award of Merit, Gamma Sigma Delta, 73; Res Award, NSF, 75. Mem: Am Inst Nutrit; Am Dairy Sci Asn; Am Soc Animal Sci. Res: Digestibility of feedstuffs for ruminant; regulation of food intake in ruminant; recycling of fibrous wastes through ruminant feeds; forage utilization. Mailing Add: Dept of Dairy Husb 214 Eckles Hall Univ of Mo Columbia MO 65201

MARTZ, HARRY FRANKLIN, JR, b Cumberland, Md, June 16, 42; m 64; c 2. STATISTICS, OPERATIONS RESEARCH. Educ: Frostburg State Col, BS, 64; Va Polytech Inst, PhD(statist), 68. Prof Exp: Asst prof, 67-70, ASSOC PROF INDUST ENG & STATIST, TEX TECH UNIV, 70- Concurrent Pos: NASA grants, 69-71. Mem: AAAS; Inst Math Statist; Am Inst Indust Eng; Am Statist Asn; Opers Res Soc Am. Res: Empirical Bayes decision theory; reliability theory; trajectory estimation and filter theory; stochastic processes. Mailing Add: Dept of Indust Eng Tex Tech Univ Lubbock TX 79409

MARUCA, ROBERT EUGENE, b Buckhannon, WVa, Nov 25, 41; m 62; c 2. INORGANIC CHEMISTRY. Educ: WVa Wesleyan Col, BS, 63; Cornell Univ, PhD(chem), 66. Prof Exp: NIH fel chem, Ind Univ, 66-68; asst prof, Miami Univ, 68-72, ASSOC PROF CHEM, ALDERSON-BROADDUS COL, 72-, CHMN DIV NATURAL & APPL SCI, 75- Res: Chemistry of the mixed compounds of boron, silicon and nitrogen. Mailing Add: Dept of Natural Sci Alderson-Broaddus Col Philippi WV 24616

MARUCCI, AMERICO ALVIN, b Orange, NJ, July 19, 23; m; c 2. IMMUNOCHEMISTRY. Educ: Rutgers Univ, BS, 44; Johns Hopkins Univ, ScM, 51, ScD, 54. Prof Exp: Res biochemist immunochem, USDA, Denmark, 54-56; from asst prof to assoc prof microbiol, 56-74, PROF MICROBIOL, COL MED, STATE UNIV NY UPSTATE MED CTR, 74- Concurrent Pos: Consult, Syracuse Bur Labs, 60- Mem: Am Asn Immunol; Am Soc Microbiol. Res: Immunochemistry of staphylococcal hemolysins; mechanism and measurement of antigen-antibody reactions; enzyme-antienzyme studies; mechanism of complement action. Mailing Add: Dept of Microbiol 766 Irving Ave State Univ of NY Upstate Med Ctr Syracuse NY 13210

MARUCCI, PHILIP EDWARD, b Orange, NJ, Apr 11, 16; m 51. ENTOMOLOGY. Educ: Rutgers Univ, BS, 36. Prof Exp: Entom field aide, USDA, 38-40, jr entomologist, 40-41, entomologist, 45-47; res assoc, Rutgers Univ, 47-49, entomologist, USDA, 49-51; assoc res specialist, NJ Exp Sta, 51-59, res specialist, 59, exten specialist, 64-70; EXTEN SPECIALIST CRANBERRY & BLUEBERRY CULT, CRANBERRY & BLUEBERRY RES LAB & RES PROF ENTOM, RUTGERS UNIV, NEW BRUNSWICK, 70- Mem: Entom Soc Am. Res: Pollination; biological control of insects; insect vectors of plant diseases; biology and control of fruit insects; medical entomology. Mailing Add: Cranberry & Blueberry Res Lab Rutgers Univ New Lisbon NJ 08064

MARULL, JOSE DOMINGO, soils, see 12th edition

MARULLO, NICASIO PHILIP, b Apr 13, 30; US citizen; m 54; c 2. ORGANIC CHEMISTRY. Educ: Queen's Col, NY, BS, 52; Polytech Inst Brooklyn, PhD(chem), 61. Prof Exp: NIH fel & res assoc chem, Calif Inst Technol, 60-61; from asst prof to assoc prof, 61-74, PROF CHEM, CLEMSON UNIV, 74- Mem: Am Chem Soc. Res: Organic reaction mechanisms; rate processes by nuclear magnetic resonance; coordination compounds of alkali metal salts. Mailing Add: Dept of Chem Clemson Univ Clemson SC 29631

MARUSYK, RAYMOND GEORGE, b Yellowknife, NT, Mar 19, 42; m 66. VIROLOGY. Educ: Univ Alta, BSc, 65, MSc, 67; Karolinska Inst, Sweden, Fil dr(virol), 72. Prof Exp: Asst prof biochem, 72-73, ASST PROF MED BACT, UNIV ALTA, 73- Mem: Can Soc Microbiologists; Am Microbiol Soc; NY Acad Sci; Tissue Cult Asn. Res: Structural and functional relationships of viral capsid components. Mailing Add: Dept of Med Bact Univ of Alta Fac of Med Edmonton AB Can

MARUYAMA, GEORGE MASAO, b Las Animas, Colo, June 15, 18; m 46; c 2. CHEMISTRY. Educ: Western State Col Colo, BA, 41; Univ Wis, MS, 48; Am Bd Bioanal, dipl, 67; Am Bd Clin Chem, dipl, 68. Prof Exp: BIOCHEMIST & ASST DIR LAB, MED ASSOCS CLIN, 48- Mem: AAAS; Am Asn Bioanalysts (pres-elect, 65-66, pres, 66-67); Am Chem Soc; Am Asn Clin Chem; fel Am Inst Chemists. Res: Clinical chemistry. Mailing Add: 1650 Atlantic St Dubuque IA 52001

MARUYAMA, HITOSHI, b Kobe, Japan, Feb 1, 29; m 62; c 2. BIOCHEMISTRY. Educ: Int Christian Univ, Tokyo, BA, 57; Va Polytech Inst, MS, 61, PhD(biochem), 64. Prof Exp: Fel dept biochem, Sch Med, NY Univ, 64-65; res biochemist, Res Inst, St Joseph Hosp, Lancaster, Pa, 65-68; clin biochemist & dir clin chem, 68-73, ADMINR PATH LAB, WASHINGTON COUNTY HOSP ASN, 73- Mem: AAAS; Am Chem Soc; Am Inst Biol Scientists; Japanese Biochem Soc. Res: Control mechanism of metabolism; carbohydrates metabolism in humans. Mailing Add: Path Lab Washington County Hosp Asn Hagerstown MD 21740

MARUYAMA, KOSHI, b Sapporo, Hokkaido, Japan, Feb 19, 32; m 61; c 3. PATHOLOGY, VIROLOGY. Educ: Hokkaido Univ, MD, 57, DMSc, 62. Prof Exp: Res staff mem path, Nat Inst Leprosy Res, Tokyo, Japan, 62-65; res staff mem, Nat Cancer Res Inst, Tokyo, 65-67; asst prof, 67-74, ASSOC PROF VIROL, UNIV TEX M D ANDERSON HOSP & TUMOR INST, 74-; CHIEF SECT VIRAL LEUKEMIA-LYMPHOMA STUDIES, 68-, ASSOC VIROL, 70- Concurrent Pos: Leukemia Soc Am scholar, 68- Mem: AAAS; Am Asn Cancer Res; NY Acad Sci; Am Soc Microbiol. Res: Internal medicine; cancer research; leprology; viral oncology. Mailing Add: Univ of Tex M D Anderson Hosp & Tumor Inst Houston TX 77025

MARUYAMA, MAGOROH, b Tokyo, Japan, Apr 2, 29; US citizen; m 66; c 1. ANTHROPOLOGY. Educ: Univ Calif, Berkeley, BA, 51; Univ Lund, PhD(philos & anthrop), 59. Prof Exp: Lectr & asst prof psychol & US Off Educ fel, Inst Human Develop, Univ Calif, Berkeley, 60-62; NIMH res fels, Human Probs Inst, Stanford Univ, 62-64 & Inst Study Crime & Delinq, Calif, 65-66; assoc prof psychol, San Francisco State Univ, 66-68; NIMH fels, Ctr Study Violence, Brandeis Univ, 67-69 & Soc Sci Res Inst, Univ Hawaii, 70-71; vis prof comput sci, Antioch Col, 71-72; consult, Water Resources Inst, Corps Engrs, Washington, DC & Stanford Res Inst, 72-73; PROF SYSTS SCI, PORTLAND STATE UNIV, 73- Concurrent Pos: Consult, Calif State Dept Pub Health, 60-63, Off Econ Opportunity, 66-67, Navajo Community Col, 69, Nat Bur Standards, 71 & Can Ministry of State for Urban Affairs, 74; res social scientist, Calif State Dept Ment Hyg, 63-64; res consult, Univ Calif, Davis, 70. Mem: AAAS; Am Anthrop Asn; World Future Soc. Res: Cultural futuristics and heterogenization; symbiotization of heterogeneity; endogenous research and polyocular anthropology; paradigmatology and cross-paradigmatic communication; mutual causal systems; extraterrestrial community design. Mailing Add: Dept of Systs Sci Portland State Univ Portland OR 97207

MARUYAMA, YOSH, b Pasadena, Calif, Apr 30, 30; m 54; c 4. RADIOTHERAPY, RADIOBIOLOGY. Educ: Univ Calif, Berkeley, AB, 51; Univ Calif, San Francisco, MD, 55. Prof Exp: Intern, San Francisco Hosp, Calif, 55-56; residency, Mass Gen Hosp, Boston, 58-61; James Picker advan acad fel, Stanford Univ, 62-64, traveling fel, Eng, France, Scand, 64; from asst prof to assoc prof radiol, Col Med Sci, Univ Minn, 64-70, admin dir div radiother, 68-70; PROF RADIATION MED & CHMN DEPT, COL MED, UNIV KY, 70- Concurrent Pos: Consult, Vet Admin Hosp. Mem: AAAS; Radiation Res Soc; Am Asn Cancer Res; Radiol Soc NAm; Am Soc Therapeut Radiol. Res: Radiation medicine, tumor cell biology; mouse leukemia; immunoradiobiology. Mailing Add: Dept of Radiation Med Univ of Ky Med Ctr Lexington KY 40506

MARVEL, CARL SHIPP, b Waynesville, Ill, Sept 11, 94; m 33; c 2. ORGANIC POLYMER CHEMISTRY. Educ: Ill Wesleyan Univ, AB & MS, 15; Univ Ill, MA, 16, PhD(org chem), 20. Hon Degrees: DSc, Ill Wesleyan Univ, 46 & Univ Ill, 63; hon Dr, Cath Univ Louvain, 70. Prof Exp: Instr chem, 20-21, assoc, 21-23, from asst prof to prof, 23-53, res prof, 53-61, EMER RES PROF, UNIV ILL, 61-; PROF CHEM, UNIV ARIZ, 61- Concurrent Pos: Mem Bd Coord Malaria Studies, 44-46; chmn panel synthesis antimalarial drugs, Nat Res Coun, 44-46, mem mat adv bd, 54-64, chmn, 62-64; mem, Nat Adv Health Coun, 45-47; chmn, Int Union Chem Comn Encyclop Compendia, 47-50; chmn adv panel, NSF, 52-54; collabr southern utilization res br, Agr Res Serv, USDA, 54-56; mem sci adv bd, Robert A Welch Found, Tex, 71- Honors & Awards: Nichols Medal, 44, Gibbs Medal, 50, Priestley Medal, 56, Witco Award, 64, Madison Marshall Award, 66 & Borden Found Award, 73, Am Chem Soc; Gold Medal, Am Inst Chem, 55; Int Award, Soc Plastics Eng, 64; Perkin Medal, Soc Chem Indust, 65; Distinguished Serv Award, Air Force Mat Lab, 66; Air Force Systs Command Award, 66; Chem Pioneer Award, 67; John R Kuebler Award, Alpha Chi Sigma Fraternity, 70. Mem: Nat Acad Sci; AAAS; Am Chem Soc (pres, 45); Am Acad Arts & Sci; fel NY Acad Sci. Res: Synthetic organic chemistry; chemistry of high polymers; polynes; organometallic compounds; free radicals; rearrangements; natural products; hydrogen bonding; synthesis of polymers, especially those with high thermal stability; heat stable organic polymers. Mailing Add: Dept of Chem Univ of Ariz Bldg 37 Rm 618 Tucson AZ 85721

MARVEL, JOHN THOMAS, b Champaign, Ill, Sept 14, 38; m 61; c 3. ORGANIC CHEMISTRY, BIOCHEMISTRY. Educ: Univ Ill, AB, 59; Mass Inst Technol, PhD(chem), 64. Prof Exp: Res assoc agr biochem, Univ Ariz, 64-65, asst agr biochemist, 65-68; sr res chemist, 68-72, sr res group leader, Monsanto Agr Prod Co, 72-75, MGR RES, MONSANTO AGR PROD CO, MONSANTO CO, 75- Mem: AAAS; Am Chem Soc; The Chem Soc; NY Acad Sci. Res: Synthesis of carbohydrates and nucleic acids; nuclear magnetic resonance spectroscopy; mass spectrometry; photochemistry; pesticide metabolism; synthesis of herbicides and plant growth regulators. Mailing Add: 800 N Lindbergh Blvd St Louis MO 63131

MARVEL, MASON E, b Brewton, Ala, Dec 11, 21; m 45; c 3. HORTICULTURE, PLANT PATHOLOGY. Educ: Univ Mass, BS, 50; Va Polytech Inst, MS, 52; WVa Univ, PhD, 70. Prof Exp: Instr hort, WVa Univ, 51-56; tech rep, Calif Chem Corp, 56-57; from asst prof to assoc prof veg crops, 57-70, PROF VEG CROPS, UNIV FLA, 70- Concurrent Pos: Chief party, Contract Team to Nat Agr Ctr, Saigon, SVietnam, 70-72, asst dir tech assistance, Int Progs Ctr Trop Agr Inst, 72-75, team leader, Pulse Prod Prog, Near East Found, Ethiopia, 75-77; consult, Hanover Brands Inc, 73-75. Mem: Am Soc Hort Sci; Asn Univ Dirs Int Agr Progs. Res: Tropical vegetable crops production and marketing. Mailing Add: 3026 McCarty IFAS Univ of Fla Gainesville FL 32611

MARVELL, ELLIOT NELSON, b New Bedford, Mass, Sept 10, 22; m 44; c 2. ORGANIC CHEMISTRY. Educ: Brown Univ, BS, 43; Univ Ill, PhD(org chem), 48. Prof Exp: From instr to assoc prof org chem, 48-61, PROF ORG CHEM, ORE STATE UNIV, 61- Concurrent Pos: NSF fel, 56-57; Petrol Res Fund int award, 65-66. Mem: Am Chem Soc; The Chem Soc; Swiss Chem Soc. Res: Molecular rearrangements; synthesis of alicyclic molecules; pericyclic reactions; sesquiterpene synthesis. Mailing Add: Dept of Chem Ore State Univ Corvallis OR 97330

MARVIN, DANIEL EZRA, JR, b East Stroudsburg, Pa, Apr 25, 38; m 58; c 2. COMPARATIVE PHYSIOLOGY. Educ: East Stroudsburg State Col, BS, 60; Ohio Univ, MS, 62; Va Polytech Inst, PhD(zool), 67. Prof Exp: Instr zool, Ohio Univ, 61-62; asst prof biol, Radford Col, 62-67, dean natural sci, 67-70; assoc dir, 70-72, DIR, STATE COUN HIGHER EDUC, VA, 72- Concurrent Pos: Mem, Nat Adv Coun Exten & Continuing Educ, 75-78. Mem: AAAS. Res: Taxonomy and behavior of fireflies; effect of hypoxic stress on cardiac and respiratory function in freshwater fish; behavior of plethedon salamanders. Mailing Add: State Coun Higher Educ Life of Va Bldg Tenth Floor Richmond VA 23219

MARVIN, DONALD ARTHUR, b New York, NY, June 25, 34; m 61. MOLECULAR BIOLOGY. Educ: Yale Univ, BS, 56; Univ London, PhD(physics), 60. Prof Exp: Fel molecular biol, Max Planck Inst Virus Res, Tübingen, Ger, 60-65; from asst prof to assoc prof, Yale Univ, 66-75; STAFF MEM, EUROP MOLECULAR BIOL LAB, 76- Res: Structure and function of biologic systems on molecular level. Mailing Add: Europ Molecular Biol Lab Postfach 10.2209 Heidelberg West Germany

MARVIN, HENRY HOWARD, JR, b Lincoln, Nebr, Mar 9, 23; m 44; c 2. PHYSICAL CHEMISTRY. Educ: Univ Nebr, BA, 47; Univ Wis, PhD(chem), 50. Prof Exp: Res assoc phys chem, Res Lab, Gen Elec Co, 50-53, liaison scientist chem, 53-55, personnel adminr, 56-58, mgr solid state chem, 59-61, mgr eng capacitor dept, 62-64, mgr lighting res lab, Ohio, 64-69, gen mgr, High Intensity Quartz Lamp Dept, 69-75; DIR DIV SOLAR ENERGY, ENERGY RES & DEVELOP ADMIN, 75- Mem: Am Chem Soc; Electrochem Soc. Mailing Add: Washington DC

MARVIN, HORACE NEWELL, b Camden, Del, Apr 20, 15; m 40; c 5. ANATOMY, ENDOCRINOLOGY. Educ: Morningside Col, BA, 36; Univ Wis, MA, 38, PhD(zool), 41. Hon Degrees: DSc, Morningside Col, 69. Prof Exp: Asst, Morningside Col, 32-36; asst zool, Univ Wis, 36-41; asst, Dept Genetics, Carnegie Inst, 41-42; from instr to asst prof anat, Med Sch, Univ Ark, 42-48, head dept biol res, Univ Tex M D Anderson Hosp Cancer Res, 48-50; assoc prof anat, 50-59, prof & head dept, 59-67, ASSOC DEAN, COL MED, UNIV ARK FOR MED SCI, LITTLE ROCK, 65- Concurrent Pos: Vis prof, Univ Lagos, 63. Mem: Am Asn Anat; Asn Am Med Cols. Res: Medical student performance; curriculum. Mailing Add: Off of the Dean Univ Ark for Med Sci Little Rock AR 72201

MARVIN, JAMES WALLACE, b Norwalk, Conn, Apr 22, 09; m 34; c 4. BOTANY. Educ: Univ Vt, BS, 32, MS, 33; Columbia Univ, PhD(bot), 39. Prof Exp: Lab asst, Columbia Univ, 33-39; from instr to prof bot, 39-75, asst botanist, 39-45, botanist, 45-47, plant physiologist, Exp Sta, 47-75, EMER PROF BOT, UNIV VT, 75- Concurrent Pos: Chmn dept bot, Univ Vt, 44-64; Guggenheim fel, 53-54; mem, State Vt Environ Bd. Mem: AAAS; Bot Soc Am; Am Soc Plant Physiol; Soc Develop Biol; Torrey Bot Club. Res: Plant microclimate; physiology of sugar maples. Mailing Add: Dept of Bot Marsh Life Sci Bldg Univ of Vt Burlington VT 05401

MARVIN, PHILIP ROGER, b Troy, NY, May 1, 16; m 42. SOLID STATE SCIENCE. Educ: Rensselaer Polytech Inst, BS, 37; Ind Univ, DCS, 51; La Salle Col, LLB, 54. Prof Exp: Engr, Gen Elec Co, 37-42; dir chem & metall eng, Bendix Aviation Corp, NY, 43-44; dir res & develop, Milwaukee Gas Specialty Co, 45-52; vpres & dir, Commonwealth Eng Co, Ohio, 52-54 & Am Viscose Corp, 54-56; mgr res & develop div, Am Mgt Asn, 56-64; pres, Clark, Cooper, Field & Wohl, 64-65; dean prof develop, 65-73, PROF DEVELOP & ADMIN, UNIV CINCINNATI, 73- Concurrent Pos: Lectr, Bridgeport Eng Inst, 37-43, Jr Col Conn, 40-41 & war training prog, Yale Univ, 41-44; lectr, US Air Force Inst Technol, 53-; consult, NASA, 66- Mem: AAAS; Am Inst Aeronaut & Astronaut; Am Defense Preparedness Asn; Inst Elec & Electronics Eng. Res: Bio-mechanics; physics of the solid state; metallurgy of electrical steel and beryllium copper; x-ray diffraction; design of electronic controls; thermoelectric phenomena; rectification phenomena. Mailing Add: Mail Code 115 Univ of Cincinnati Cincinnati OH 45221

MARVIN, ROBERT SIDNEY, b St Paul, Minn, Oct 4, 17; m 42; c 5. PHYSICAL CHEMISTRY, RHEOLOGY. Educ: Univ Minn, BChem, 39, MS, 42; Univ Wis, PhD(chem), 49. Prof Exp: Proj assoc, Univ Wis, 49; chemist, Nat Bur Standards, 49-56, physicist, 56-75; RETIRED. Concurrent Pos: Vis prof, Kyoto Univ, 61-62; chmn, Int Comt Rheology, 72-76. Mem: Am Chem Soc; Soc Rheology (pres, 66-67); fel Am Phys Soc. Res: Viscoelastic properties of polymers. Mailing Add: 11700 Stony Creek Rd Potomac MD 20854

MARVIN, URSULA BAILEY, b Bradford, Vt, Aug 20, 21; m 52. MINERALOGY, METEORITICS. Educ: Tufts Univ, BA, 43; Harvard Univ, MS, 46, PhD, 69. Prof Exp: Asst silicate chem, Univ Chicago, 47-50; mineralogist, Union Carbide Ore Co, NY, 53-58; instr mineral, Tufts Univ, 58-61; GEOLOGIST, SMITHSONIAN ASTROPHYS OBSERV, 61-, COORDR FED WOMEN'S PROG, 73- Concurrent Pos: Assoc, Harvard Col Observ, 65-; lectr, Tufts Univ, 68-69 & Harvard Univ, 74- Mem: AAAS; Geol Soc Am; Meteoritical Soc (vpres, 73-74, pres, 75-76); Am Geophys Union; Int Asn Geochem & Cosmochem; Int Astron Union. Res: Mineralogy and petrology of meteorites and lunar samples; history of geology. Mailing Add: Smithsonian Astrophys Observ 60 Garden St Cambridge MA 02138

MARWIN, RICHARD MARTIN, b Minneapolis, Minn, Dec 10, 18; m 42; c 1. MEDICAL MICROBIOLOGY, MEDICAL MYCOLOGY. Educ: Univ Minn, BA, 41, MS, 43, PhD(bact), 47; Am Bd Microbiol, dipl, 62. Prof Exp: Teaching asst med bact, Med Sch, Univ Minn, 41-44, instr, 45-48; assoc prof, 48-51, chmn dept, 48-62, PROF BACT, MED SCH, UNIV NDAK, 51- Concurrent Pos: Chief lab bact & blood bank, Univ Minn Hosps, 45-47. Mem: Am Soc Microbiol; Mycol Soc Am. Res: Culture media modification for growth of pathogenic bacteria; effects of chemicals on pathogenic fungi of man. Mailing Add: Dept of Microbiol Univ NDak Sch of Med Grand Forks ND 58201

MARWITT, JOHN PAUL, b Cleveland, Ohio, July 14, 37; m 74. ETHNOLOGY, ARCHAEOLOGY. Educ: Fla State Univ, BS, 66; Univ Utah, PhD(anthrop), 71. Prof Exp: Staff archeologist, Can Dept Northern Affairs, 64-66; staff archeologist, Univ Utah, 66-68, asst dir archeol res, 68-71; asst prof anthrop, 71-75, ASSOC PROF ANTHROP, UNIV AKRON, 75- Concurrent Pos: Vis scholar, Colombian Inst Anthrop, 71; Univ Akron Res Coun grant fieldwork in Colombia, 72-73; vis asst prof, Univ Utah, 73-74; Nat Geog Soc res grants, 73 & 74. Honors & Awards: Soc Sigma Xi Award, 71. Mem: Am Anthrop Asn; Soc Am Archaeol. Res: Prehistory of the United States, Great Basin; prehistory and ethnology of lowland South America; the anthropology of war. Mailing Add: Dept of Sociol Univ of Akron Akron OH 44325

MARX, DONALD HENRY, b Ocean Falls, BC, Oct 3, 36; US citizen; m 57; c 5. PLANT PATHOLOGY, SOIL MICROBIOLOGY. Educ: Univ Ga, BSA, 61, MS, 62; NC State Univ, PhD(plant path), 66. Prof Exp: PLANT PATHOLOGIST, FORESTRY SCI LAB, SOUTHEASTERN FOREST EXP STA, 62- Concurrent Pos: Adv, Int Union Forest Res Orgn, 63. Honors & Awards: USDA Superior Serv Award, 71. Mem: AAAS; Am Phytopath Soc; Am Inst Biol Scientists. Res: Mycorrhizae of conifers and hardwoods; ecology and parasitism of soil-borne organisms; reforestation of adverse sites by use of specific mycorrhizae. Mailing Add: Forestry Sci Lab Carlton St Southeastern Forest Exp Sta Athens GA 30602

MARX, EGON, b Cologne, Ger, Apr 4, 37; m 65; c 2. ELECTROMAGNETICS. Educ: Univ Chile, EE, 59; Calif Inst Technol, PhD(physics), 63. Prof Exp: Assoc investr physics, Univ Chile, 63-65, independent investr, 65; asst prof, Clarkson Col Technol, 65-67; asst prof, Drexel Univ, 67-72; PHYSICIST, HARRY DIAMOND LABS, 72- Res: Field theory; relativistic quantum mechanics: electromagnetic waves; computer simulation. Mailing Add: Electromagnetic Effects Lab Harry Diamond Labs Adelphi MD 20783

MARX, GEORGE DONALD, b Antigo, Wis, Apr 30, 36; m 64; c 2. AGRICULTURE, ANIMAL PHYSIOLOGY. Educ: Univ Wis-River Falls, BS, 58; SDak State Univ, MS, 60; Univ Minn, Minneapolis, PhD(animal sci), 64. Prof Exp: Farm planner, Soil Conserv Serv, USDA, 57-58; asst dairy sci, SDak State Univ, 58-60; asst animal sci, Univ Minn, Minneapolis & St Paul, 60-64; from instr to ASSOC PROF AGR, AGR EXP STA, UNIV MINN TECH COL, CROOKSTON, 64- Mem: Am Dairy Sci Asn; Am Soc Animal Sci; Int Asn Immunity. Res: Animal management. Mailing Add: Agr Exp Sta Univ of Minn Tech Col Crookston MN 56716

MARX, GERALD ALVIN, b Milwaukee, Wis, Mar 7, 30; m 65. AGRONOMY, PLANT PATHOLOGY. Educ: Univ Wis, BS, 53, MS, 56, PhD(agron), 59. Prof Exp: PLANT BREEDER, DEPT SEED & VEG SCI, AGR EXP STA, CORNELL UNIV, 59- Mem: AAAS; Am Soc Agron; Am Genetic Asn; Crop Sci Soc Am; Am Phytopath Soc. Res: Vegetable crop breeding and genetics; breeding for disease resistance. Mailing Add: Dept of Seed & Veg Sci Cornell Univ Agr Exp Sta Geneva NY 14456

MARX, GERTIE F, b Frankfurt am Main, Ger, Feb 13, 12; US Citizen; m 40. ANESTHESIOLOGY. Educ: Univ Bern, MD, 37. Prof Exp: From asst attend to assoc atten anesthesiologist, Beth Israel Hosp, New York, 43-55; from asst prof to assoc prof, 55-70, PROF ANESTHESIOL, ALBERT EINSTEIN COL MED, 70-; ATTEND ANESTHESIOLOGIST, BRONX MUNIC HOSP CTR, 55- Concurrent Pos: Attend anesthesiologist, Bronx Vet Admin Hosp, 66-72, consult, 72- Mem: Fel Am Soc Anesthesiol; AMA; fel NY Acad Med; NY Acad Sci; assoc fel, Am Col Obstetricians & Gynecologists. Res: Obstetric anesthesia. Mailing Add: Dept of Anesthesiol Albert Einstein Col of Med Bronx NY 10461

MARX, HYMEN, b Chicago, Ill, June 27, 25; m 50; c 2. HERPETOLOGY. Educ: Roosevelt Univ, BS, 49. Prof Exp: Asst cur, Div Amphibians & Reptiles, 50-64, assoc cur, 65-73, CUR, FIELD MUS NATURAL HIST, 73-, HEAD DIV, 70- Concurrent Pos: Consult, US Naval Med Res Unit, Egypt, 53; vis scientist, NSF Field Mus, 67-71; lectr, Univ Chicago, 73- Mem: Am Soc Ichthyol & Herpet; Soc Study Amphibians & Reptiles. Res: Reptiles; systematics; North Africa and Southwestern Asia herpetology; zoogeography of Old World reptiles; phyletic character analysis; phylogeny of vipers; phylogenetic theory. Mailing Add: Field Mus Natural Hist Roosevelt Rd & Lakeshore Dr Chicago IL 60605

MARX, JAMES JOHN, JR, b Paris, Tex, Dec 17, 44; m 73. IMMUNOPATHOLOGY. Educ: St Vincent Col, BA, 66; WVa Univ, MS, 70, PhD(microbiol), 72. Prof Exp: RES SCIENTIST IMMUNOL, MARSHFIELD MED FOUND, 73- Concurrent Pos: Investr grants, Am Lung Asn, 74 & Nat Heart & Lung Inst, 75; lectr, Sch Med Technol, St Joseph's Hosp, 74- Mem: Am Acad Allergy; Am Lung Asn; Am Soc Microbiol; NY Acad Sci. Res: Basic immunologic mechanisms involved in occupational diseases. Mailing Add: Marshfield Med Found 510 NSt Joseph Ave Marshfield WI 54449

MARX, JEAN LANDGRAF, biochemistry, see 12th edition

MARX, JOHN NORBERT, b Columbus, Ohio, Oct 31, 37. ORGANIC CHEMISTRY. Educ: St Benedict's Col, Kans, BS, 62; Univ Kans, PhD(org chem), 65. Prof Exp: Fel org chem, Cambridge Univ, 65-66 & Johns Hopkins Univ, 66-67; asst prof, 67-73, ASSOC PROF ORG CHEM, TEX TECH UNIV, 73- Mem: Am Chem Soc; Am Inst Chemists; The Chem Soc. Res: Structural determination and synthesis of natural products, especially terpenes and steroids; new synthetic methods; stereochemistry; cyclohexadienone rearrangements; migrations of electronegative groups. Mailing Add: Dept of Chem Tex Tech Univ Lubbock TX 79409

MARX, JOSEPH VINCENT, b Joplin, Mo, Mar 19, 43; m 67. CLINICAL BIOCHEMISTRY. Educ: Johns Hopkins Univ, AB, 65; State Univ NY Upstate Med Ctr, PhD(biochem), 69. Prof Exp: From assoc scientist to sr scientist biochem, 69-73, PRIN SCIENTIST APPL SCI, ORTHO DIAGNOSTICS, INC, 74- Mem: Am Asn Clin Chemists; Am Chem Soc; Soc Cryobiol; NY Acad Sci. Res: Applied research and development including techniques in protein separation and purification, clinical enzymology, blood coagulation and clinical hematology. Mailing Add: Ortho Diagnostics Inc Rte 202 Raritan NJ 08869

MARX, MICHAEL, b Stuttgart, Ger, Nov 10, 33; US citizen; m 63; c 3. ORGANIC CHEMISTRY. Educ: Dartmouth Col, BA, 54; Columbia Univ, MA, 64, PhD(chem), 66. Prof Exp: Chemist, Lederle Labs Div, Am Cyanamid Co, 54-63; fel, Stanford

Univ, 66-67; chemist, 67-69, dept head, Synthetic Org Chem, 69-74, ASST DIR, INST ORG CHEM, SYNTEX RES CTR, 74- Mem: Am Chem Soc; The Chem Soc. Res: Synthetic methods; synthesis and transformations of steroids and terpenoid natural products; synthesis of medicinal agents. Mailing Add: Inst Org Chem Syntex Res Ctr Hillview Ave Palo Alto CA 94304

MARX, MORRIS LEON, b New Orleans, La, May 21, 37; m 60; c 1. MATHEMATICS. Educ: Tulane Univ, BS, 59, MS, 63, PhD(math), 64. Prof Exp: Asst prof, 66-69, ASSOC PROF MATH, VANDERBILT UNIV, 69-, DIR GRAD SCH MATH, 70-, DIR TEACHER EDUC, 73- Mem: Am Math Soc; Math Asn Am. Res: Topological analysis. Mailing Add: Dept of Math Vanderbilt Univ Nashville TN 37203

MARX, PAUL CHRISTIAN, b Los Angeles, Calif, Jan 21, 29; m 66; c 1. PHYSICAL CHEMISTRY. Educ: Univ Calif, Los Angeles, BS, 51; Northwestern Univ, PhD(chem), 55. Prof Exp: Res chemist, Stand Oil Co Calif, 54-57; sr res chemist, Gillette Co, 57-61; mem tech staff high temperature chem, Aerospace Corp, 61-71; consult, 71-73; MGR ADVAN DEVELOP, FORTIN LAMINATING CORP, 73- Mem: Am Chem Soc; Am Electrolaters Soc. Res: Catalytic chemical plating of metals and alloys, electroplating, electroforming of thin metallic foils, electrogeochemistry. Mailing Add: 9907 Rathburn Ave Northridge CA 91324

MARX, PRESTON AUGUST, JR, b New Orleans, La, Dec 1, 43; m 67; c 1. VIROLOGY. Educ: La State Univ, New Orleans, BS, 66, MS, 67, PhD(microbiol), 69. Prof Exp: Instr microbiol, Med Ctr, Univ New Orleans, 69-72; fel, St Jude Children's Res Hosp, 72-74; ASST PROF MICROBIOL, THOMAS JEFFERSON UNIV, 74- Mem: AAAS; Am Soc Microbiol. Res: Combined bacterial and viral infections; biosynthesis of viruses; biochemistry of RNA viruses. Mailing Add: Thomas Jefferson Univ Philadelphia PA 19107

MARX, STEPHEN JOHN, b New York, NY, Nov 23, 42; m 74. ENDOCRINOLOGY. Educ: Yale Univ, BA, 64; Johns Hopkins Univ, MD, 68; Am Bd Internal Med, cert, 74; Am Bd Endocrinol, cert, 76. Prof Exp: Med intern, Mass Gen Hosp, 68-69, med resident, 69-70 & 72-73; clin assoc, 70-72, SR INVESTR METAB DIS, NAT INST ARTHRITIS & DIGESTIVE DIS, NIH, 73- Mem: Endocrine Soc; Am Fedn Clin Res; AAAS. Res: Mechanism of action of hormones, hormone-receptor interaction, adenylate cyclase regulation, disorders of calcium metabolism, diagnosis and treatment of disorders of parathyroid gland. Mailing Add: Bldg 10 Rm 90-20 NIH Bethesda MD 20014

MARX, WALTER, b Karlsruhe, Ger, June 26, 07; nat US; m 54; c 1. BIOCHEMISTRY. Educ: Karlsruhe Tech Univ, Dipl Ing, 31, DIng, 33. Prof Exp: Instr, Inst Phys Chem, Karlsruhe Tech Univ, 30-33; Justus Liebig res fel, Kaiser Wilhelm Inst Med Res, Heidelberg, 33-34; Isadore Hernsheim fel, 34-36; fel, Mt Sinai Hosp, 34-37; res assoc, Sch Med, Duke Univ, 37-39; res assoc, Univ Calif, 39-44; res fel, Calif Inst Technol, 45-46; from asst prof to assoc prof, 46-54, PROF BIOCHEM, SCH MED, UNIV SOUTHERN CALIF, 54- Concurrent Pos: USPHS spec fel, 61-62. Mem: AAAS; Am Soc Biol Chemists. Res: Glycosaminoglycans; proteoglycans; heparin biosynthesis; sulfate metabolism; cholesterol metabolism; pituitary hormones. Mailing Add: Dept of Biochem Univ Southern Calif Sch of Med Los Angeles CA 90033

MARX, WALTER S, JR, photography, see 12th edition

MARY, NOURI Y, b Baghdad, Iraq, June 25, 29; m 58; c 2. PHARMACOGNOSY. Educ: Univ Baghdad, PhC, 51; Ohio State Univ, MSc, 53, PhD(pharm, pharmacog), 55. Prof Exp: Asst prof pharmacog, Col Pharm, Univ Baghdad, 56-60, assoc prof & actg dean, 60-61; vis res scientist, Sch Pharm, Univ Calif, San Francisco, 61-63; fel, Univ Conn, 63-65; assoc prof pharmacog, 65-72, PROF PHARMACOG, BROOKLYN COL PHARM, LONG ISLAND UNIV, 72- Mem: Am Pharmaceut Asn; Am Soc Pharmacog; Acad Pharmaceut Sci; Am Asn Col Pharm. Res: Chemical and biochemical studies of natural products. Mailing Add: Dept Biol Sci Brooklyn Col Pharm LI Univ 600 Lafayette Ave Brooklyn NY 11216

MARYANOFF, BRUCE ELIOT, b Philadelphia, Pa, Feb 26, 47; m 71. SYNTHETIC ORGANIC CHEMISTRY, MEDICINAL CHEMISTRY. Educ: Drexel Univ, BS, 69, PhD(org chem), 72. Prof Exp: Fel phys org chem, Princeton Univ, 72-74; SR SCIENTIST ORG CHEM, McNEIL LABS, INC, JOHNSON & JOHNSON, 74- Mem: Am Chem Soc; The Chem Soc; Sigma Xi. Res: Synthesis of biologically active compounds; applications of nuclear magnetic resonance spectroscopy and dynamic nuclear magnetic resonance; new synthetic reactions and processes; isoquinoline and indole alkaloids; sterochemistry and asymmetric synthesis. Mailing Add: McNeil Labs Inc Camp Hill Rd Ft Washington PA 19036

MARZETTI, LAWRENCE ARTHUR, b Mt Vernon, Ohio, Apr 17, 17; m 42; c 5. APPLIED STATISTICS. Educ: Morehead State Col, AB, 39. Prof Exp: Opers head 1940 census, Pop Div, Bur Census, 40-42, surv statistician, 46-51, asst budget officer, Budget Off, 52-56, chief overseas consult foreign census & statist, Int Statist Progs, 56-70, tech adv statist legis, Subcomt Census & Statist, Post Off & Civil Serv Comt, US House Rep 92nd Cong, 71-73; spec asst, Econ Census Staff, Bur Census, 73-74; STATIST CONSULT, 75- Concurrent Pos: Census consult, Tech Coop Admin, Amman, Jordan, 52. Honors & Awards: Meritorious Award, Dept Com, 52. Mem: Am Statist Asn; Soc Int Develop. Res: Foreign census methodology; mid-decade census; census confidentiality; national vote registration and election practice; international statistical consultation. Mailing Add: 4121 25th Ave SE Washington DC 20031

MARZLUF, GEORGE A, b Columbus, Ohio, Sept 29, 35; m 60; c 4. GENETICS, BIOCHEMISTRY. Educ: Ohio State Univ, BSc, 57, MS, 60; Johns Hopkins Univ, PhD(genetics), 64. Prof Exp: NSF fel biochem genetics, Sch Med, Univ Wis, 64-66; asst prof biol, Marquette Univ, 68-69; assoc prof biochem, 70-75, PROF BIOCHEM, OHIO STATE UNIV, 75- Concurrent Pos: NIH grants biochem genetics, 67-71, 71-76 & career develop award, 75-80. Mem: AAAS; Genetics Soc Am; Am Soc Microbiol. Res: Developmental and biochemical genetics; synthesis, allosteric control and turnover of enzymes and permeases in higher organisms; control of differential gene action and morphogenesis. Mailing Add: Dept of Biochem Ohio State Univ Columbus OH 43210

MARZLUFF, WILLIAM FRANK, organic chemistry, see 12th edition

MARZLUFF, WILLIAM FRANK, JR, b Washington, DC, May 7, 45; m 66; c 2. BIOCHEMISTRY. Educ: Harvard Univ, AB, 67; Duke Univ, PhD(biochem), 71. Prof Exp: NIH fel biol, Johns Hopkins Univ, 71-74; ASST PROF CHEM, FLA STATE UNIV, 74- Res: Chromosome structure; histone chemistry. Mailing Add: Dept of Chem Fla State Univ Tallahassee FL 32306

MARZOLF, GEORGE RICHARD, b Columbus, Ohio, Dec 13, 35; m 58; c 2. LIMNOLOGY. Educ: Wittenberg Col, AB, 57; Univ Mich, MS, 61, PhD(zool), 62. Prof Exp: From asst prof zool to assoc prof biol, 62-75, assoc dir div biol, 73-75, PROF BIOL, KANS STATE UNIV, 75- Concurrent Pos: Vis prof zool, Univs Wis, Okla & Ore, 66, 67 & 75. Mem: Int Asn Theoret & Appl Limnol; Am Soc Limnol & Oceanog; Ecol Soc Am; Am Micros Soc. Res: Reservoir limnology; plankton; benthos; trophic structure. Mailing Add: Div of Biol Kans State Univ Manhattan KS 66502

MARZOLF, JOHN GEORGE, b Buffalo, NY, Dec 24, 32. SOLID STATE PHYSICS. Educ: Fordham Univ, AB, 56, MA, 58, MS, 59; Johns Hopkins Univ, PhD(physics), 63. Prof Exp: Physicist, Nat Bur Stand, 63-64; res inst natural sci, Woodstock Col, Md, 64-67; ASST PROF PHYSICS, LE MOYNE COL, 67- Concurrent Pos: Prin investr, NASA res grant, 64-68. Mem: Am Phys Soc; Am Asn Physics Teachers. Res: Single crystals with the Mössbauer effect and x-rays; fundamental aspects of the Mössbauer effect; angle measurement with optical interferometry. Mailing Add: Dept of Physics Le Moyne Col Syracuse NY 13214

MARZULLI, FRANCIS NICHOLAS, b New York, NY, Feb 2, 17; m 45; c 2. PHARMACOLOGY, TOXICOLOGY. Educ: St Peters Col, BS, 37; Johns Hopkins Univ, MA, 40, PhD(physiol), 41. Prof Exp: Aquatic biologist, US Fish & Wildlife Serv, 41-43; toxicologist, Dugway Proving Ground, Utah, 46-47; toxicologist & physiologist, Army Chem Res & Develop Labs, 47-63; chief dermal toxicity br, Pharmacol Div, 63-73, spec assignment, Med Ctr, Univ Calif, San Francisco, 73-75, SR SCIENTIST, FOOD & DRUG ADMIN, 75- Concurrent Pos: Exchange scientist, Chem Defense Exp Estab, Eng, 60-61. Mem: Soc Invest Dermat; Soc Exp Biol & Med; Soc Cosmetic Chem; Soc Toxicol; Asn Res Vision & Ophthal. Res: Environmental, skin and eye physiology. Mailing Add: 8044 Park Overlook Dr Bethesda MD 20034

MARZZACCO, CHARLES JOSEPH, b Philadelphia, Pa, May 1, 42; m 64; c 1. PHYSICAL CHEMISTRY. Educ: Temple Univ, AB, 64; Univ Pa, PhD(chem), 68. Prof Exp: Grant, Princeton Univ, 68-69, instr chem, 69-70; asst prof, NY Univ, 70-73; ASST PROF CHEM, RI COL, 73- Mem: Am Chem Soc. Res: Photophysical and photochemical properties of azines and ketones. Mailing Add: Dept of Phys Sci RI Col Providence RI 02908

MASAITIS, CESLOVAS, b Kaunas, Lithuania, Mar 2, 12; nat US; m 40; c 1. MATHEMATICS. Educ: Vytauto Didziojo Univ, Lithuania, MA, 37; Univ Tenn, Knoxville, PhD(math), 56. Prof Exp: Asst astron, Vytauto Didziojo Univ, 37-40; instr, Univ Vilnius, 40-44; instr math, Nazareth Col, Ky, 50-52, Univ Ky, 52-53 & Univ Tenn, Knoxville, 53-56; mathematician, 56-63, RES MATHEMATICIAN, BALLISTIC RES LABS, ABERDEEN PROVING GROUND, 63- Concurrent Pos: Lectr, Univ Del, 57-64; res assoc surg, Univ Md, Baltimore, 64-71, res asst prof, 70-71, asst prof, 71-72; consult, Shock-Trauma Ctr, Univ Md Hosp, 72- Honors & Awards: Sustained Superior Performance Award, US Army, 62, Army Materiel Command Res & Develop Achievement Awards, 69 & 73, Army Ord Kent Award, 70. Mem: Math Asn Am. Res: Numerical analysis; approximations; application of system theory; optimization. Mailing Add: Ballistic Res Labs Aberdeen Proving Ground MD 21005

MASAKI, BEVERLY WONG, b Denver, Colo, June 20, 38; m 67. PHARMACEUTICAL CHEMISTRY, CLINICAL PHARMACOLOGY. Educ: Univ Southern Calif, PharmD, 62, PhD(pharmaceut chem), 67. Prof Exp: Res assoc health care delivery & respiratory dis, Univ Southern Calif, 67-68, asst prof pharm, Los Angeles County Med Ctr, 68-69, ASST PROF PHARMACOL, SCH PHARM, LAC/USC MED CTR, UNIV SOUTHERN CALIF, 69- Mem: Am Asn Cols Pharm; Am Soc Hosp Pharmacists; Am Pharmaceut Asn; NY Acad Sci. Res: Accidental ingestions of the pediatric age group; adverse drug reactions. Mailing Add: Sch of Pharm LAC/USC Med Ctr Univ of Southern Calif Los Angeles CA 90033

MASAMUNE, SATORU, b Fukuoka, Japan, July 24, 28; m 56; c 2. ORGANIC CHEMISTRY. Educ: Tohuku Univ, Japan, AB, 52; Univ Calif, PhD(chem), 57. Prof Exp: Proj assoc chem, Univ Wis, 56-59, lectr, 59-61; fel, Mellon Inst, 61-64; assoc prof chem, 64-67, PROF CHEM, UNIV ALTA, 67- Concurrent Pos: Ed, Organic Syntheses Inc, 71- Mem: Am Chem Soc; The Chem Soc; fel Royal Soc Can; Japan Chem Soc. Res: Organic synthesis of biologically important compounds, chemistry of cyclic pi-electron and strained systems. Mailing Add: Dept of Chem Univ of Alta Edmonton AB Can

MASANI, PESI RUSTOM, b Bombay, India, Aug 1, 19. MATHEMATICS. Educ: Univ Bombay, BSc, 40; Harvard Univ, MA, 42, PhD(math), 46. Prof Exp: Teaching fel math, Harvard Univ, 43-45; mem, Inst Advan Study, Princeton, NJ, 46-48; sr res fel, Tata Inst Fundamental Res, Bombay, 48-49; prof math & head dept, Inst Sci, Bombay, 49-59; vis lectr, Brown Univ, 59-60; prof math, Ind Univ, Bloomington, 60-72; PROF MATH, UNIV PITTSBURGH, 72- Concurrent Pos: Vis lectr, Harvard Univ & Mass Inst Technol, 57-58; vis prof, Math Res Ctr, Univ Wis, 65-66 & Statist Lab, Cath Univ Am, 66-67; vis researcher, Battelle Seattle Res Ctr, 69-70. Mem: Am Math Soc; Soc Indust & Appl Math; Math Asn Am. Res: Noncommutative analysis, specifically the factorization of operator-valued functions; prediction and filter theory of stationary stochastic processes; Hilbert spaces, specifically isometric flows, spectral integrals and vector-valued measures; ordinary linear differential systems. Mailing Add: Dept of Math Univ of Pittsburgh Pittsburgh PA 15260

MASARACCHIA, JOSEPH PHILIP, organic chemistry, see 12th edition

MASARACCHIA, RUTHANN ADELE, biochemistry, endocrinology, see 12th edition

MASAT, ROBERT JAMES, b Greeley, Colo, Sept 6, 28; m 64; c 2. COMPARATIVE PHYSIOLOGY, BIOCHEMISTRY. Educ: Univ Portland, BSc, 54; Wash State Univ, MSc, 58; St Louis Univ, PhD(biol), 64. Prof Exp: Instr biol, Rockhurst Col, 58-61, head dept, 59-61; res assoc cryobiol, Am Found Biol Res, 62; res assoc biochem, Med Ctr, La State Univ, 64, Nat Heart Inst fel, 64-66; Am Found Biol Res, 66-68; assoc prof biol, 68-73, PROF BIOL, ST AMBROSE COL, 73-, CHMNDIV NATURAL & MATH SCI, 75- Mem: AAAS; Am Soc Zool; Am Chem Soc; Am Physiol Soc. Res: Comparative biochemistry and physiology of plasma proteins; physiology of biological cyclic phenomena of hibernation and migration; active transport of substances across living membranes; physiology of hypothermia. Mailing Add: Dept of Biol St Ambrose Col Davenport IA 52803

MASCARENHAS, JOSEPH PETER, b Nairobi, Kenya, Nov 19, 29; m 60; c 2. DEVELOPMENTAL BIOLOGY. Educ: Univ Poona, BSc, 52, MSc, 54; Univ Calif, Berkeley, PhD(plant physiol), 62. Prof Exp: Res officer, Parry & Co, India, 53-56; instr biol, Amherst Col, 62-63; Res Corp Brown-Hazen Fund grant & instr bot, Wellesley Col, 63-64, asst prof, 64-67; res assoc biol, Mass Inst Technol, 67-68; assoc prof, 69-74, PROF BIOL, STATE UNIV NY ALBANY, 74- Concurrent Pos: NSF grants, Wellesley Col, Mass Inst Technol & State Univ NY Albany, 65-; vis asst prof, Mass Inst Technol, 66-67; res found fel & grant, State Univ NY Albany, 69-74. Mem: AAAS; Int Soc Develop Biol; Soc Develop Biol; Bot Soc Am; Am Soc Plant Physiol. Res: Molecular control of plant development. Mailing Add: Dept of Biol Sci State Univ NY 1400 Washington Ave Albany NY 12222

MASCHERONI, P LEONARDO, b Tucuman, Arg, July 20, 35; US citizen; m 67; c 2. THEORETICAL PHYSICS. Educ: Univ Cuyo, Arg, BS, 62; Univ Calif, Berkeley, PhD(physics), 68. Prof Exp: Res assoc & lectr physics, Temple Univ, 68-70, vis asst prof chem & physics, 70-71; res scientist assoc physics, Ctr Statist Mechanics & Thermodynamics, 71-74; RES SCIENTIST ASSOC PHYSICS, FUSION RES CTR, UNIV TEX, AUSTIN, 74- Concurrent Pos: Consult laser-plasma res, Tex Tech Univ, 75. Mem: Am Phys Soc; AAAS. Res: Current research in plasma physics; laser interaction with matter and laser fusion; turbulent heating of a plasma; fusion research in Tokomak systems. Mailing Add: Fusion Res Ctr Univ of Tex Austin TX 78712

MASCIANTONIO, PHILIP (X), b Monongehela City, Pa, Mar 14, 29; m 50; c 5. PHYSICAL CHEMISTRY, ORGANIC CHEMISTRY. Educ: St Vincent Col, BS, 50; Carnegie Inst Technol, MS, 57, PhD(chem), 60. Prof Exp: Asst chemist, Robertshaw-Fulton Div, Nat Roll & Foundry Co, 50, chief chemist, 51; prod supvr dyestuffs, Pittsburgh Chem Co, 51-55; sr res chemist, Appl Res Lab, 55-67, sect supvr, 67-69, div chief chem, 69-74, asst dir environ control, 74-75, DIR ENVIRON CONTROL, US STEEL CORP, 75- Concurrent Pos: Instr, Carnegie Inst Technol, 62-64; abstractor, Chem Abstr, 63-70. Mem: Am Chem Soc; Water Pollution Control Fedn; Air Pollution Control Asn. Res: Chemistry, properties and structure of coal; process research of chemicals polymer properties and development of air and water pollution abatement systems. Mailing Add: Box 353 RD 1 Jeannette PA 15644

MASCIO, AFEWORK ASGHEDOM, b Asmara, Ethiopia, Sept 15, 45. IMMUNOLOGY. Educ: Haile Selassie Univ, BS, 68; Colo State Univ, MS, 70; Pa State Univ, PhD(immunol), 75. Prof Exp: Res asst microbiol, Naval Med Res Unit, Ethiopia, 70-71; med rep, Ciba-Geigy Corp, Ethiopia, 71-72; ASST PROF ZOOL, DREW UNIV, 75- Mem: Am Soc Microbiol; AAAS. Res: Suppression of cell-mediated immunity by murine leukemia viruses; tumor immunology; immunocompetence; immunosuppression; RNA tumor viruses. Mailing Add: Dept of Zool Box HS-40 Drew Univ Madison NJ 07940

MASCIOLI, ROCCO LAWRENCE, b Mt Carmel, Pa, May 7, 28; m 54; c 7. ORGANIC CHEMISTRY. Educ: Bucknell Univ, BS, 52; Univ Pa, MS, 54, PhD(chem), 57. Prof Exp: Res chemist, Houdry Process Corp, 56-57, proj dir air prod & chem, 67-69, sect head appln res & develop, 69-71, ASST DIR CHEM ADDITIVES DIV, RES & DEVELOP, HOUDRY LABS, 71- Mem: Am Chem Soc. Res: Nitrogen chemistry; catalysis. Mailing Add: Air Prod & Chem Box 427 Marcus Hook PA 19063

MASCOLI, CARMINE CHARLES, b Waterbury, Conn, Jan 28, 28; m 53; c 5. VIROLOGY. Educ: Col Holy Cross, BS, 49; Univ Conn, MS, 53; Ohio State Univ, PhD(bact), 56. Prof Exp: Clin bacteriologist, Meriden Hosp, Conn, 52-53; asst bact, Ohio State Univ, 53-55; virologist & immunologist, Eli Lilly & Co, 56-60; from asst prof to assoc prof microbiol, Sch Med, WVa Univ, 60-64; from res fel to sr res fel, Merck Inst Therapeut Res, Pa, 64-70; dir qual control, Nat Drug Co, 70-71; dir qual control & dir res, Merrell Nat Labs, 71-76; CORP DIR MICROBIOL, RES & DEVELOP DEPT, BAXTER-TRAVENOL LABS, 76- Concurrent Pos: Instr, Butler Univ, 58-60. Mem: AAAS; Am Soc Microbiol; Am Asn Immunol. Res: Epidemiology; viral immunology. Mailing Add: Baxter-Travenol Labs Morton Grove IL 60053

MASEK, GEORGE EDWARD, b Norfolk, Va, Feb 10, 27; m 55; c 1. PHYSICS. Educ: Stanford Univ, PhD(physics), 56. Prof Exp: Res assoc physics, Hansen Lab, Stanford Univ, 55-56; instr, Princeton Univ, 56-57; from asst prof to prof, Univ Wash, 57-65; assoc prof, 65-67, PROF PHYSICS, UNIV CALIF, SAN DIEGO, 67- Res: Elementary particle and high energy physics. Mailing Add: Dept of Physics Univ of Calif at San Diego La Jolla CA 92037

MASELLI, JAMES MICHAEL, b Pottsville, Pa, Mar 29, 35; m 61; c 2. INORGANIC CHEMISTRY. Educ: Lafayette Col, AB, 57; Univ Pa, PhD(inorg chem), 61. Prof Exp: Univ fel, Harvard Univ, 61-62; sr res chemist, 63-66, res supvr, 66-68, mgr, 68-75, DIR, W R GRACE CO, CLARKSVILLE, 75- Mem: Am Chem Soc. Res: Inorganic synthesis and catalysis. Mailing Add: 6413 Amherst Ave Columbia MD 21046

MASELLI, JOHN ANTHONY, b New York, NY, Feb 18, 28; m 48; c 2. FOOD CHEMISTRY. Educ: City Col New York, BS, 47; Fordham Univ, MS, 49, PhD(chem), 52. Prof Exp: Instr chem, Fordham Univ, 48-52; chemist, Fleischmann Labs, Standard Brands, Inc, 52-57, mgt staff asst, 57-59, dept dir, 59-62, dir res, Fleischmann Mfg Div, 62-64, prod develop mgr M&M Candies, NJ, 64-67; vpres opers, 67-70, PRES, OZ FOOD CORP, 70-, FOUNDER, LUBIN-MASELLI LABS, 73- Mem: Am Soc Bakery Eng; Am Asn Cereal Chem; Am Asn Candy Technol; Inst Food Technol; Am Inst Chem. Res: Food, cereal and yeast chemistry; candy technology. Mailing Add: 1004 Sheridan Rd Wilmette IL 60091

MASER, MORTON D, b Hagerstown, Md, Nov 24, 34; m 55; c 2. CELL BIOLOGY. Educ: Univ Pa, AB, 55; Univ Pittsburgh, PhD(biophys), 62. Prof Exp: Res asst, Mellon Inst, 56-58, res assoc, 58-60, jr fel, 60-62, fel, 63; res assoc, Marine Biol Lab, Woods Hole, 62; res assoc, Biol Labs, Harvard Univ, 62-64, lectr, 64-66; dir cell biol lab, Millard Fillmore Hosp, Buffalo, 66-70; asst prof biol, Erie County Community Col, 70; assoc prof continuing educ, Northeastern Univ, 70-73; ASSOC PROF PATH, CREIGHTON UNIV, 73- Mem: AAAS; Electron Micros Soc Am; Am Soc Cell Biol. Res: Electron microscopical techniques; pathogenetic mechanisms. Mailing Add: Dept of Path Creighton Univ 2500 California St Omaha NE 68178

MASERICK, PETER H, b Washington, DC, Feb 8, 33; m 56; c 4. MATHEMATICS. Educ: Univ Md, BS, 55, MA, 57, PhD(math), 60. Prof Exp: NSF fel math, Univ Wis, 63-64; asst prof, 64-71, ASSOC PROF MATH, PA STATE UNIV, UNIVERSITY PARK, 71- Res: Functional analysis; convexity. Mailing Add: Dept of Math Pa State Univ University Park PA 16802

MASERJIAN, JOSEPH, b Albany, NY, Feb 10, 29; m 53; c 4. SEMICONDUCTORS, MICROELECTRONICS. Educ: Rensselaer Polytech Inst, BS, 52; Univ Southern Calif, MS, 55; Calif Inst Technol, PhD(mat sci), 66. Prof Exp: Mem tech staff, Semiconductor Div, Hughes Aircraft Co, 52-60; res specialist, 60-66, SUPVR, SEMICONDUCTOR TECHNOL GROUP, JET PROPULSION LAB, CALIF INST TECHNOL, 66- Mem: Am Phys Soc. Res: Semiconductor technology; thin films; interface physics; reliability of semiconductor devices; solar energy conversion. Mailing Add: Jet Propulsion Lab Calif Inst of Technol 4800 Oak Grove Dr Pasadena CA 91103

MASHBURN, LOUISE TULL, b Wayne, Pa, Aug 27, 30; m 58. BIOCHEMISTRY. Educ: Westhampton Col, BA, 52; Duke Univ, PhD(biochem), 61. Prof Exp: Fel biochem, Univ Del, 61-63, res assoc biochem, 63-64; RES ASSOC BIOCHEM, RES INST, HOSP JOINT DIS, 64- Concurrent Pos: Asst prof, Mt Sinai Sch Med; USPHS spec res fel, 65-67, grant, 66-75; Am Cancer Soc grant, 68-74; Leukemia Soc Am scholar, 71-76. Mem: Am Chem Soc; Am Soc Biol Chemists; NY Acad sci; Am Asn Cancer Res. Res: Biochemistry of malignant diseases; amino acid metabolism; enzymology in therapy; tissue culture. Mailing Add: Res Inst Hosp Joint Dis 1919 Madison Ave New York NY 10035

MASHBURN, THOMPSON ARTHUR, JR, b Morganton, NC, Oct 9, 36; m 58. BIOCHEMISTRY. Educ: Univ NC, AB, 56; Duke Univ, PhD(org chem), 61. Prof Exp: Res chemist, Plastics Dept, E I du Pont de Nemours & Co, Inc, 60-62, res chemist, Org Chem Dept, 62-64; RES ASSOC BIOCHEM, RES INST SKELETOMUSCULAR DIS, HOSP JOINT DIS, NEW YORK, 64- Concurrent Pos: Advan res fel, Am Heart Asn, Inc, 65-69, estab investr, 69-74; res asst prof biochem, Mt Sinai Sch Med, 73- Mem: AAAS; Am Soc Biol Chemists; Am Chem Soc. Res: Structure, chemistry and biochemistry of connective tissue and mucopolysaccharides; structure and chemistry of natural highpolymers; sugar chemistry. Mailing Add: Res Inst Hosp for Joint Dis 1919 Madison Ave New York NY 10035

MASHIMO, PAUL AKIRA, b Osaka, Japan, Oct 25, 26; m 55; c 2. ORAL MICROBIOLOGY. Educ: Osaka Dent Univ, Japan, DDS, 48; Kyoto Med Univ, Japan, PhD(microbiol), 55. Prof Exp: Instr oral surg, Osaka Dent Univ, 48-50, lectr microbiol, 50-53, from asst prof microbiol to assoc prof pub health, 53-65; prof res assoc oral biol, 66, asst res prof, 66-67, asst prof oral biol, 68-69, ASSOC PROF ORAL BIOL, SCH DENT, STATE UNIV NY BUFFALO, 70-, MEM FAC GRAD SCH, 69- Concurrent Pos: Louise C Ball fel, Sch Dent & Oral Surg, Columbia Univ, 56-58; Japan Soc fel, 56-67. Mem: Am Soc Microbiol; Int Asn Dent Res; NY Acad Sci. Res: Oral microbiology; immunology; dentistry. Mailing Add: Dept of Oral Biol State Univ of NY Sch of Dent Buffalo NY 14214

MASI, ALFONSE THOMAS, b New York, NY, Oct 29, 30; m 60; c 4. EPIDEMIOLOGY, INTERNAL MEDICINE. Educ: City Col New York, BS, 51; Columbia Univ, MD, 55; Johns Hopkins Univ, MPH, 61, DrPH(epidemiol), 63. Prof Exp: Intern med, Osler Serv, Johns Hopkins Hosp, 55-56; sr asst surgeon, Commun Dis Ctr, USPHS, 56-58; asst resident, Johns Hopkins Hosp, 58-59; assoc resident, Med Ctr, Univ Calif, Los Angeles, 59-60; res fel epidemiol, Sch Hyg & Pub Health, Johns Hopkins Univ, 60-63, asst prof epidemiol, 63-65; instr med, Sch Med, 63-67, assoc prof epidemiol, Sch Hyg & Pub Health, 65-67; prof med & prev med & chief sect rheumatol, Dept Med, 67-72, DIR DIV CONNECTIVE TISSUE DIS, UNIV TENN CTR HEALTH SCI, MEMPHIS, 72- Concurrent Pos: Consult, Radiol Health Res Br, USPHS, 63-67, Nat Inst Arthritis & Metab Dis spec fel, 63-66; res geog epidemiol sect, Vet Admin, 64-66; mem subcomt epidemiol use of hosp data, Nat Comt Health & Vital Statist, 65-69; consult, US Food & Drug Admin, 65-70; sr investr, Arthritis Found, 66-71; Russell L Cecil fel award, 70-71; mem arthritis training grant comt, Nat Inst Arthritis & Metab Dis, 71-73. Mem: Am Fedn Clin Res; Am Rheumatism Asn; fel Am Pub Health Asn; fel Am Col Physicians. Res: Application of epidemiologic methods, such as population studies, community-wide hospital surveys and case-control investigations to research in chronic diseases in order to define better their causes and pathogenesis. Mailing Add: Div of Connective Tissue Dis Univ of Tenn Ctr for Hlth Sci Memphis TN 38163

MASI, JOSEPH FRANCIS, b Fairfax Co, Va, Sept 27, 15; m 40; c 2. PHYSICAL CHEMISTRY. Educ: Am Univ, BA, 38; Va Polytech Inst, MS, 39; Univ NC, PhD(phys chem), 45. Prof Exp: Instr, Va Polytech, 39-42; instr, Univ NC, 44; assoc chemist, Nat Bur Standards, Washington, DC, 44-48, chemist, 48-54; head phys chem, Callery Chem Co, 54-59, res adminr, Propulsion Div, 59-62, chief, Propulsion Div, 62-70, CHIEF ENERGETICS DIV, AIR FORCE OFF SCI RES, 70- Concurrent Pos: Lectr, Univ Md, 68-69. Mem: Am Chem Soc; Am Inst Aeronaut & Astronaut; Combustion Inst; Sigma Xi; Am Inst Chemists. Res: Propulsion; thermodynamics; calorimetry; electrolytic solutions. Mailing Add: Air Force Off of Sci Res Energetics Div 1400 Wilson Blvd Arlington VA 22209

MASIH, SHABIR ZAHOOR, b Pendra Road, India, Apr 6, 34; m 62; c 4. BIOPHARMACEUTICS. Educ: Gujarat Univ, India, BPharm, 59; Univ Pittsburgh, MS, 70; Univ Alta, PhD(biopharmaceut), 75. Prof Exp: Lectr pharm, Christian Med Col, Vellore, India, 59-60; chief pharmacist, Missions Tablet Indust, Bangarapet, India, 60-68; teaching asst pharm, Univ Pittsburgh, 68-70 & Univ Alta, 70-74; ASST PROF INDUST PHARM, MASS COL PHARM, 74- Concurrent Pos: Consult pharmacist, Zemmer Moor Kirk Labs, Oakmont, Pa, 68-74 & New Eng Nuclear, Boston, 74-75. Res: Analytical procedure for determination of drugs in dosage forms and body fluids; routine quality control testing of pharmaceutical dosage forms; drug stability testing; drug bioavailability testing; formulation development. Mailing Add: Mass Col of Pharm 179 Longwood Ave Boston MA 02115

MASING, ULV, b Abja, Estonia, May 6, 27; Can citizen; m 60; c 2. GEOGRAPHY. Educ: Clark Univ, BA, 60; Univ Alta, MSc, 62; Univ Fla, PhD(geog), 64. Prof Exp: Asst prof geog, Univ Calgary, 64-68; assoc prof, State Univ NY Col Oswego, 68-69; PROF GEOG, EDINBORO STATE COL, 69- Concurrent Pos: Assoc prof, ECarolina Univ, 67-68. Mem: Asn Am Geogr; Am Geog Soc; Can Asn Geogr. Res: Latin American cultural geography, especially land use, settlement and colonization. Mailing Add: Dept of Geog Edinboro State Col Edinboro PA 16412

MASKAL, JOHN, b Garfield, NJ, Sept 27, 18; m 43; c 2. PHYSICAL CHEMISTRY. Educ: Syracuse Univ, BS, 40; Mich State Univ, MS, 42. Prof Exp: Supvr anal lab, Evansville Ord Plant, 42-44; engr mat testing, Chrysler Corp, 44-45; high sch teacher, Mich, 46-47; mgr plating plant, Ludington Plating Co, 47-50; supvr, Denham Mfg Co, 50-51; supvr res lab, 51-60, DIR GEN LAB, DOW CHEM USA, 60- Mem: Tech Asn Pulp & Paper Indust; fel Am Inst Chemists; Am Chem Soc. Res: Analytical and physical chemical research involving inorganic chemical manufacture. Mailing Add: Gen Lab Dow Chem USA Ludington MI 49431

MASKEN, JAMES FREDERICK, b Frederick, Md, Apr 4, 27; m 59; c 3. PHYSIOLOGY, BIOCHEMISTRY. Educ: NY Univ, BA, 53; Colo State Univ, MS, 60, PhD(physiol), 65. Prof Exp: Lab asst pharmacol, Wm R Warner Co, 50-52 & Nepera Chem Co, 52-53; res technician, Surg Dept, Sinai Hosp, Baltimore, Md, 54-55; lab technician, Biochem Div, Toni Co, 55; res asst endocrinol, 55-62; from instr to asst prof physiol, 62-69, ASSOC PROF PHYSIOL, COLO STATE UNIV, 69- Concurrent Pos: Vis assoc prof, Univ Calif, 70-71. Mem: AAAS; Am Physiol Soc; Can Physiol Soc; Soc Study Reproduction. Res: Reproductive physiology; neuroendocrinology. Mailing Add: Dept of Physiol & Biophys Colo State Univ Ft Collins CO 80521

MASKER, WARREN EDWARD, b Honesdale, Pa, July 8, 43. MOLECULAR BIOLOGY. Educ: Lehigh Univ, BS, 65; Univ Rochester, PhD(physics), 70. Prof Exp: Fel, Univ Rochester, 69-71; fel, Stanford Univ, 71-73; fel, Med Sch, Harvard Univ, 73-74; RES ASSOC, BIOL DIV, OAK RIDGE NAT LAB, 75- Concurrent Pos: Am Cancer Soc fel, 70; Helen Hay Whitney Found fel, 71. Mem: Am Phys Soc; Biophys Soc; Am Soc Microbiol; Am Soc Photobiol. Res: Study of DNA replication and the molecular mechanism of DNA repair in bacteria and bacteriophage. Mailing Add: Div of Biol Oak Ridge Nat Lab PO Box Y Oak Ridge TN 37830

MASKIT, BERNARD, b New York, NY, May 27, 35; m 57; c 3. MATHEMATICS. Educ: NY Univ, AB, 57, MS, 62, PhD(math), 64. Prof Exp: Mem, Inst Advan Study,

63-65; from asst prof to assoc prof math, Mass Inst Technol, 65-70, Sloan Found fel, 70-71; PROF MATH, STATE UNIV NY STONY BROOK, 71-, CHMN DEPT, 74- Mem: Am Math Soc. Res: Riemann surfaces; Kleinian groups and surface topology. Mailing Add: Dept of Math State Univ of NY Stony Brook NY 11790

MASKORNICK, MICHAEL J, b Feb 7, 44. PHYSICAL ORGANIC CHEMISTRY. Educ: Lehigh Univ, BS, 65; Univ Calif, Berkeley, PhD(phys org chem), 69. Prof Exp: Res chemist, 69-73, ASST SUPT RES & DEVELOP, E I DU PONT DE NEMOURS & CO, INC, 73- Mem: Am Chem Soc. Res: Polymer processing and development. Mailing Add: E I du Pont de Nemours & Co PO Box 2000 La Place LA 70068

MASLAND, RICHARD LAMBERT, b Philadelphia, Pa, Mar 24, 10; m 40; c 4. PSYCHIATRY, NEUROLOGY. Educ: Haverford Col, BA, 31; Univ Pa, MD, 35; Am Bd Neurol & Psychiat, dipl, 43, cert psychiat, 48. Hon Degrees: LLD, Haverford Col, 75. Prof Exp: Intern, Hosp Univ Pa, 35-37, fel neurol, 37-39, asst neurologist, 39-47, assoc, Univ, 40-46; from asst prof to prof psychiat & neurol, Bowman Gray Sch Med, 47-57, from asst prof to assoc prof physiol, 48-57; from asst dir to dir, Nat Inst Neurol Dis & Blindness, Md, 57-68; prof, 68-70, chmn dept, 68-73, Moses prof neurol, 70-73, H Houston Merritt prof, 73-76, EMER PROF NEUROL, COL PHYSICIANS & SURGEONS, COLUMBIA UNIV, 76- Concurrent Pos: Fel, Univ Pa, 40-46; res dir sci adv bd, Nat Asn Retarded Children, 55-56; fel psychiat, Pa Inst Ment Hyg, 56-57; mem med adv bd, Myasthenia Gravis Found, Inc, 57-; trustee, Nat Easter Seal Res Found, 59-71; mem res adv comt, United Cerebral Palsy Res & Educ Found, 61-72; mem adv bd, Muscular Dystrophy Asn Am, 65-73; dir neurol serv, Neurol Inst, Presby Hosp, 68-73; mem policy comt, World Fedn Neurol, mem res group develop dyslexia & world illiteracy, 68, vpres, 69; mem adv comt epilepsies, Dept Health, Educ & Welfare, 69-73, chmn, 70; regional chmn, Epilepsy Found Am, 71-; mem NY State Develop Disabilities Coun, 73- Mem: Fel Am Acad Neurol; hon mem Am Acad Cerebral Palsy; assoc Am Asn Neurol Surg; Am Epilepsy Soc (pres, 54); Am Neurol Asn (vpres, 64). Res: Neurophysiology; clinical neurology. Mailing Add: Neurol Inst Presby Hosp 710 W 168th St New York NY 10032

MASLEN, STEPHEN HAROLD, b Cleveland, Ohio, Jan 28, 26; m 51; c 5. APPLIED MATHEMATICS. Educ: Rensselaer Polytech Inst, BAeroEng, 45, MAeroEng, 47; Brown Univ, PhD(appl math), 52. Prof Exp: Aeronaut res scientist, Nat Adv Comt Aeronaut, 47-58, chief plasma physics br, Advan Propulsion Div, Lewis Res Ctr, NASA, 58-60; prin res scientist, Martin Co, 60-67, ASSOC DIR, MARTIN MARIETTA LABS, 67- Mem: Am Phys Soc; Am Inst Aeronaut & Astronaut. Res: Fluid dynamics. Mailing Add: Martin Marietta Labs 1450 S Rolling Rd Baltimore MD 21227

MASLIN, THOMAS PAUL, b Hankow, China, Oct 27, 09; US citizen; m 34; c 3. HERPETOLOGY. Educ: Univ Calif, BA, 33, MA, 39; Stanford Univ, PhD(zool), 45. Prof Exp: Asst prof zool, Colo Agr & Mech Col, 45-47; from asst prof to prof biol, 47-74, EMER PROF BIOL, UNIV COLO, BOULDER, 75-, CUR ZOOL, UNIV MUS, 66- Concurrent Pos: Off Coord Fisheries, Calif, 43-45. Mem: Fel Am Soc Ichthyol & Herpet; Am Asn Mus; Soc Study Amphibians & Reptiles. Res: Anatomy, phylogeny and taxonomy of reptiles. Mailing Add: Univ of Colo Museum Boulder CO 80302

MASLOW, DAVID E, b Brooklyn, NY, July 6, 43. EMBRYOLOGY, CELL BIOLOGY. Educ: Brooklyn Col, BS, 63; Univ Pa, PhD(zool), 68. Prof Exp: Cancer res scientist, 68-73, CANCER RES SCIENTIST II, ROSWELL PARK MEM INST, 73- Mem: Soc Develop Biol; Am Soc Zoologists. Res: Cell specificity in developing neoplastic systems. Mailing Add: Dept of Exp Path Roswell Park Mem Inst Buffalo NY 14263

MASLOW, PHILIP HERMAN, b New York, NY, July 27, 18; m 43; c 3. ORGANIC CHEMISTRY. Educ: City Col, New York, BS, 38, MS, 41. Prof Exp: Control chemist paint lab, Monroe Sander Corp, 38-42; assoc chemist, Norfolk Naval Shipyard, Va, 42-45; chief chemist, Red Hand Marine Paint Co, NJ, 45-56; supvr coatings tech serv lab, Ciba Prods Corp, 56-59; tech dir, Permagile Corp Am, 59-61; tech dir & tech serv mgr, Dewey & Almy Chem Div, W R Grace & Co, 61-69; tech serv mgr, Preco Chem Corp, 70-73; VPRES, MASTER MASTICS CO INC, 73- Concurrent Pos: Mem, Concrete Indust Bd; lectr, Univ Mo-Rolla, Harvard Univ, New York City Community Col, Pratt Inst & Univ Wis-Madison; consult pvt pract, currently. Mem: Am Concrete Inst; Am Chem Soc; Am Soc Testing & Mat; Fedn Soc Paint Technol. Res: Paint technology, especially marine paints; epoxy technology, especially coatings and plastics applications; sealant, concrete, adhesive and building materials technology. Mailing Add: 739 E 49th St Brooklyn NY 11203

MASNYK, IHOR JAREMA, organic chemistry, see 12th edition

MASO, HENRY FRANK, b Perth Amboy, NJ, Nov 20, 19; m 44; c 3. COSMETIC CHEMISTRY. Educ: City Col New York, BS, 40. Prof Exp: Asst bur biol res, Rutgers Univ, 41; jr chemist, Philadelphia Navy Yd, 41-44; sr res chemist, Johnson & Johnson, 44-57; dir tech serv, Am Cholesterol Prod, Inc, 57-70, VPRES, TECH SERV-MKT, AMERCHOL, CPC INT INC, 70- Honors & Awards: Medalist, Soc Cosmetic Chemists, 71. Mem: Am Chem Soc; Soc Cosmetic Chemists (pres-elect, 66, pres, 67). Res: Manufacture of raw materials for cosmetics, dermatologicals and pharmaceuticals. Mailing Add: Tech Serv Dept Amerchol Park Amerchol CPC Int Inc Edison NJ 08817

MASON, AARON S, b Russia, Mar 3, 11; US citizen; m 42; c 3. PSYCHIATRY, HEALTH ADMINISTRATION. Educ: Univ Ill Col Med, BS, 33, MD, 35; Am Bd Psychiat & Neurol, dipl, 46. Prof Exp: Staff physician, Lincoln St Sch & Colony, Ill, 37-38; staff psychiatrist, Vet Admin Hosp, North Little Rock, Ark, 38-39; staff psychiatrist, Vet Admin Hosp, Knoxville, Iowa, 39-40; staff physician, chief reconstruct serv, Admitting & Outpatient Serv & Acute Intensive Treatment Serv & asst chief prof serv, Vet Admin Hosp, Downey, Ill, 40-53; chief staff, Vet Admin Hosp, Brockton, Mass, 53-62; dir, Vet Admin Hosp, Tomah, Wis, 62-63; dir, Vet Admin Hosp, Lexington, Ky, 63-72; dir, Vet Admin Hosp, Downey, 72-74; MGR MENT HEALTH BR, BUR HEALTH SERV KY, 74- Concurrent Pos: Assoc prof psychiat, Med Ctr, Univ Ky, 62-70, prof clin psychiat, 70-72 & 74-; assoc prof psychiat, Med Sch, Northwestern Univ, Chicago, 72-74. Mem: Fel Am Psychiat Asn; Asn Mil Surg US; Am Asn Med Supt Ment Hosp; AMA; fel Am Geriat Soc. Res: Rehabilitation in psychiatry; social and clinical psychiatry; psychopharmacological agents; clinical psychopharmacology. Mailing Add: 507 Lake Tower Dr Lexington KY 40502

MASON, ALLEN SMITH, b Tulsa, Okla, Dec 9, 32; m 55; c 1. ATMOSPHERIC CHEMISTRY. Educ: Kans State Univ, BS, 54; Fla Inst Technol, MS, 67; Univ Miami, PhD(marine sci), 74. Prof Exp: Chemist inorg chem, FMC Labs, 60; mem tech staff, RCA Labs, 61-63; supt tech eval, Pan Am World Airways, 63-69; RES ASST PROF CHEM OCEANOG, UNIV MIAMI, 74- Mem: Am Chem Soc; Am Geophys Union. Res: Atmospheric tracer studies using radioactive and stable chemical species for estimation of large-scale mixing and transport processes. Mailing Add: 4600 Rickenbacker Causeway Miami FL 33149

MASON, ARTHUR ALLEN, b St Louis, Mo, May 6, 25; m 64. MOLECULAR SPECTROSCOPY. Educ: Univ Okla, BS, 51; Univ Tenn, PhD(physics), 63. Prof Exp: From asst prof to assoc prof physics, 64-74, PROF PHYSICS, SPACE INST, UNIV TENN, 74- Mem: Am Phys Soc; Optical Soc Am. Res: Intensity spectroscopy of gases, planetary atmospheres and combustion phenomena. Mailing Add: Space Inst Univ of Tenn Tullahoma TN 37388

MASON, BERYL TROXELL, b Victoria, BC, Jan 21, 07; US citizen; m 35; c 2. NEUROLOGY, PSYCHIATRY. Educ: Univ Wash, BS, 29; Univ Chicago, MS, 32, MD, 36. Prof Exp: Resident, Univ Chicago, 36; intern, St Margaret's Hosp, Pittsburgh, 36-37; instr bact, Col Med, Univ Ill, 37-38; instr phys diag, Univ NC, 42-44; pvt pract, NC, 42-44 & Ill, 45-53; res assoc neurol & neurosurg, Col Med, Univ Ill, 53-58, consult, 58-61; actg chief neurol, Vet Admin Hosp, Topeka, Kans, 61-64; clin dir, Ark Rehab Serv, 64-69; study of Mid-East dis, Istanbul, Turkey, 69-71; CONSULT IN RES, MATROX LABS, 71- Concurrent Pos: Consult to planning coun, Found of Viet-Nam ·Inst Technol, 74 & 75. Res: Central nervous system; epilepsy; brain x-radiation; public health; radiation; neurological disorders. Mailing Add: 5059 N Jones Rd Oak Harbor WA 98277

MASON, BRIAN HAROLD, b Port Chalmers, NZ, Apr 18, 17; m 43. GEOCHEMISTRY. Educ: Univ NZ, MSc, 38; Univ Stockholm, PhD(mineral), 43. Prof Exp: Res officer, NZ Govt, 43-44; sr lectr geol, Univ NZ, 44-47; assoc prof mineral, Ind Univ, 47-53; cur phys geol & mineral, Am Mus Natural Hist, 53-65; RES CUR, DIV METEORITES, US NAT MUS, 65- Mem: Fel Mineral Soc Am (pres, 65-66); Geochem Soc (pres, 64-65); Royal Soc NZ; Swedish & Norweg Geol Soc. Res: Geochemistry; petrology; regional geology; meteorites. Mailing Add: Smithsonian Inst Washington DC 20560

MASON, CAROLINE FAITH VIBERT, b Harrogate, Eng, Feb 24, 42; US citizen; m 69; c 1. INORGANIC CHEMISTRY. Educ: Univ London, BSc, 64, PhD(chem), 67. Prof Exp: Fel, State Univ NY Buffalo, 67-68; chemist, Howmet Corp, Dover, NJ, 69-70; assoc scientist chem, Ortho Res Found, Raritan, NJ, 70-71; biochemist, Los Alamos Med Ctr, 72-74, STAFF MEM CHEM, LOS ALAMOS SCI LAB, 75- Concurrent Pos: Consult, Particle Technol Inc, Coulter Electronics, 73-75. Mem: The Chem Soc; Am Chem Soc. Res: Thermochemical cycles for the decomposition of water to hydrogen and oxygen and related problems such as catalysis, separation of gases, kinetics and materials problems. Mailing Add: CMB 3 Los Alamos Sci Lab Los Alamos NM 87545

MASON, CHARLES EUGENE, b Brighton, Colo, Aug 28, 43; m 70; c 3. INSECT ECOLOGY, APICULTURE. Educ: Colo State Univ, BS, 68; Univ Mo, MS, 71; Kans State Univ, PhD(entom), 73. Prof Exp: Res asst entom, Univ Mo, 68-71 & Kans State Univ, 71, asst instr, 72-73; asst entomologist, Univ Ariz, 73-75; ASST PROF ENTOM, UNIV DEL, 75- Mem: Entom Soc Am; Ecol Soc Am. Res: Economic threshold of pest insects in vegetable crops; ecology and diversity of leafhoppers; pollination of crops by honey bees. Mailing Add: Dept of Entom Univ of Del Newark DE 19711

MASON, CHARLES MORGAN, b Kenora, Ont, July 7, 06; US citizen; m 29; c 3. PHYSICAL CHEMISTRY. Educ: Univ Ariz, BS, 28, MS, 29; Yale Univ, PhD(phys chem), 32. Prof Exp: Asst, Yale Univ, 29-31; from asst prof to assoc prof, Univ NH, 32-41; phys chemist, US Bur Mines, 41-43; res chemist, Tenn Valley Authority, 43-48; phys chemist, US Bur Mines, 48-50, chief explosives res sect, 50-56, phys res sect, 56-60, proj coordr explosive res ctr, 60-69, SUPVRY RES CHEMIST, PITTSBURGH MINING & SAFETY RES CTR, 69- Concurrent Pos: Explosives consult, Aluminum Co Am, 72- Honors & Awards: Meritorious Silver Medal, Dept Interior, 72. Mem: Am Chem Soc. Res: Thermodynamic properties of phosphates, rare earths and barium salts; water adsorption of glue; metallurgy of aluminum and lithium; magneto chemistry; non-metallic minerals; ignition of fire-damp; explosives and explosion phenomena; hazardous chemicals; ammonium nitrate. Mailing Add: 3429 Sycamore Dr Bethel Park PA 15102

MASON, CHARLES PERRY, b Newport, RI, Aug 12, 32; m 58; c 2. BOTANY. Educ: Univ RI, BS, 54; Univ Wis, MS, 58; Cornell Univ, PhD(bot), 61. Prof Exp: Instr bot, Univ Wis, Milwaukee, 57-58; from asst prof to assoc prof biol, Hamline Univ, 61-67; ASSOC PROF BIOL, GUSTAVUS ADOLPHUS COL, 67- Mem: AAAS; Phycol Soc Am; Bot Soc Am; Int Phycol Soc. Res: Life cycle and development of Cladophora gracilis; ecology, especially productivity of marine benthic algae; ecology of Cladophora in farm ponds; effect of temperature shock on DNA content of beta-chromosome containing nuclei in maize. Mailing Add: Dept of Biol Gustavus Adolphus Col St Peter MN 56082

MASON, CHARLES THOMAS, JR, b Joliet, Ill, Mar 26, 18; m 43; c 1. BOTANY. Educ: Univ Chicago, BS, 40; Univ Calif, MA, 42, PhD(bot), 49. Prof Exp: Instr bot, Univ Wis, 49-53; from asst prof to assoc prof, 53-62, PROF BOT, UNIV ARIZ, 62-, BOTANIST & CUR, HERBARIUM, 53- Mem: Fel AAAS; Bot Soc Am; Am Soc Plant Taxon; Int Asn Plant Taxon. Res: Cytotaxonomy of angiosperms, Limnanthaceae and Gentianaceae. Mailing Add: Dept of Bot Univ of Ariz Tucson AZ 85721

MASON, CONRAD JEROME, b Detroit, Mich, Jan 12, 32. MICROMETEOROLOGY. Educ: Univ Mich, BS, 53; Univ Calif, Berkeley, MA, 55, PhD(physics), 60. Prof Exp: Assoc res physicist, Radiation Lab, 60-63, ASSOC RES PHYSICIST, HIGH ALTITUDE ENG LAB, UNIV MICH, ANN ARBOR, 63-, LECTR DEPT METEOROL & OCEANOG, 70-, RES SCIENTIST & LECTR, DEPT ATMOSPHERIC & OCEANIC SCI, 74- Concurrent Pos: Pres, Aeromatrix Inc, Ann Arbor, 76- Mem: Am Phys Soc; Am Geophys Union; Am Meteorol Soc; Air Pollution Control Asn; Int Asn Aerobiol. Res: Atmospheric science; air pollution. Mailing Add: 3640 E Huron River Dr Ann Arbor MI 48104

MASON, CURTIS LEONEL, b Daingerfield, Tex, Oct 9, 19; m 42; c 2. MICROBIOLOGY. Educ: Tex Agr & Mech Col, BS, 40, MS, 42; Univ Ill, PhD(plant path), 47. Prof Exp: Agent bur plant indust, soils & agr eng, USDA, Tex, 39-42; asst, Univ Wis, 42-43; spec asst, Univ Ill, 46-47; assoc pathologist, Univ Tex, 47-48; asst prof plant path, Univ Ark, 48-52; plant pathologist, Niagara Chem Div, Food Mach & Chem Corp, 52-54; asst sales mgr, 54-57, mgr tech serv, 57-59, regional mgr, 59-62; microbiologist, Buckman Labs, Inc, 62-65, area mgr, 65-70; EXTEN PLANT PATHOLOGIST, AGR EXTEN SERV, UNIV ARK, LITTLE ROCK, 71- Mem: Am Phytopath Soc. Res: Diseases of cotton, orchard crops and peaches; testing of fungicides; fungicidal action of 8-quinolinol and some of its derivatives; industrial microorganism control. Mailing Add: 4712 Hampton Rd North Little Rock AR 72116

MASON, DAVID DICKENSON, b Adingdon, Va, Jan 22, 17; m 44; c 2. APPLIED STATISTICS. Educ: King Col, BA, 36; Va Polytech Inst, MS, 38; NC State Col, PhD(agron), 48. Prof Exp: Asst agronomist, Exp Sta, Va Polytech Inst, 38-39 & Miss State Col, 41; asst, NC State Col, 41 & 45-47; asst prof agron, Ohio State Univ, 47-49; biometrician, Bur Plant Indust, USDA, 49-53; prof statist, 53-63, PROF STATIST

& HEAD DEPT & HEAD INST STATIST, NC STATE UNIV, 63- Concurrent Pos: Statist consult, Res Div, United Fruit Co, Mass, 57-; chmn, Southern Regional Ed Bd Comt on Statist, 73- Mem: Soil Sci Soc Am; fel Am Soc Agron; fel Am Statist Asn; Biomet Soc. Res: Applied statistics; soil and plant science. Mailing Add: Dept of Statist NC State Univ Box 5457 Raleigh NC 27607

MASON, DAVID LAMONT, b Warren, Pa, Dec 24, 34; m 63; c 1. BOTANY. Educ: Edinboro State Col, BS, 63; Univ Wis, MS, 67, PhD(bot), 70. Prof Exp: Teaching asst gen bot, Univ Wis, 63-64, teaching assoc, 64-65, res asst mycol, 65-69; ASST PROF BIOL, WITTENBURG UNIV, 69- Mem: Am Phytopath Soc; Mycol Soc Am. Res: Fungal parasitism. Mailing Add: Dept of Biol Wittenburg Univ Springfield OH 45501

MASON, DAVID THOMAS, b Berkeley, Calif, Jan 7, 37. LIMNOLOGY. Educ: Reed Col, BA, 58; Univ Calif, Davis, MA, 61, PhD(zool), 66. Prof Exp: Lectr zool, Univ Calif, Davis, 65; asst prof biol, 66-70, ASSOC PROF BIOL, FAIRHAVEN COL, WESTERN WASH STATE COL, 71- Concurrent Pos: Asst prof, Univ Calif, Berkeley, 69-71. Mem: Am Soc Limnol & Oceanog; Int Soc Limnol. Res: Physical and biological limnology; saline lakes. Mailing Add: Fairhaven Col Western Wash State Col Bellingham WA 98225

MASON, DEAN TOWLE, b Berkeley, Calif, Sept 20, 32; m 57; c 2. CARDIOVASCULAR DISEASES. Educ: Duke Univ, BA, 54, MD, 58; Am Bd Internal Med, dipl, 65; Am Bd Cardiovasc Dis, dipl, 66. Prof Exp: From intern to asst resident, Osler Med Serv, Johns Hopkins Hosp, 58-61; asst resident med, Med Ctr, Duke Univ, 59-60; clin assoc, Cardiol Br, Nat Heart Inst, 61-63, asst sect chief cardiovasc diag, sr investr & attend physician, 63-68; PROF MED & PHYSIOL & CHIEF SECT CARDIOVASC MED, SCH MED, UNIV CALIF, DAVIS, 68- Concurrent Pos: Consult, Surg Br, Nat Heart Inst & Clin Ctr, NIH, 61-68; from clin asst prof to clin assoc prof med, Sch Med, Georgetown Univ, 65-68; fel, Coun on Circulation, Am Heart Asn, 66-, fel, Coun Clin Cardiol, 67-; consult, US Naval Med Ctr, Bethesda, Md, 67-68, Letterman Army Gen Hosp, San Francisco, Calif, 68- & David Grant Med Ctr, Travis AFB, Calif; mem, Am Bd Internal Med; Mem adv comt, US Pharmacopeia, 70, NIH Lipid Metab, 71 & NASA Life Sci, 73- Honors & Awards: Am Therapeut Soc Award, 65 & 73. Mem: Am Soc Clin Invest; Am Physiol Soc; fel Royal Soc Med; Am Soc Pharmacol & Exp Therapeut; Am Col Cardiol (pres-elect). Res: Adult and pediatric clinical cardiology; cardiac catheterization and diagnosis; cardiovascular medicine, physiology, biochemistry and pharmacology. Mailing Add: Dept of Internal Med Univ of Calif Sch Med Davis CA 95616

MASON, DONALD FRANK, b Chicago, Ill, Mar 17, 26; m; c 4. PHYSICAL CHEMISTRY. Educ: Univ Ill, BS, 49; Univ Wis, PhD(chem), 53. Prof Exp: Asst, Naval Res Lab, Univ Wis, 49-52; res assoc chem eng, Northwestern Univ, 52-55, asst prof, 55-58; assoc chemist, Argonne Nat Lab, 58-62; assoc prof chem, Natural Sci Div, Ill Teachers Col, Chicago-North, 62-68; PROF CHEM, NORTHEASTERN ILL UNIV, 68- Concurrent Pos: Consult, Vern Alden Co, 54-55. Mem: AAAS; Am Chem Soc; Sigma Xi. Res: Heterogeneous reaction kinetics; mass spectrometry; instrumentation. Mailing Add: Dept of Chem Northeastern Ill Univ Chicago IL 60625

MASON, DONALD JOSEPH, b Kokomo, Ind, July 24, 31; m 53; c 4. MICROBIOLOGY. Educ: Purdue Univ, BS, 53, MS, 55, PhD, 58. Prof Exp: Res assoc microbiol, 58-66, sect head anal microbiol, 66-68, MGR FED DRUG ADMIN, UPJOHN CO, 68- Mem: Am Soc Microbiol; AAAS; Am Fedn Clin Res. Res: Microbial cytology and biochemistry; antibiotic production. Mailing Add: Upjohn Co Kalamazoo MI 49001

MASON, DOROTHY STAFFORD, bGreensboro, NC, Aug 15, 36; m 60; c 2. GEOGRAPHY. Educ: Univ NC, Greensboro, AB, 57; Univ Ga, MA, 60; Univ NC, Chapel Hill, PhD(geog), 66. Prof Exp: Teacher math & sci, Goldsboro Jr High Sch, NC, 57-58; instr geog, Univ NC, Greensboro, 60-61; from asst prof to assoc prof, Elon Col, 61-69; assoc prof, 69-74, PROF GEOG, NC A&T STATE UNIV, 74- Mem: Asn Am Geog; Am Geog Soc. Res: Geography of Anglo-America. Mailing Add: 2707 Hill-N-Dale Dr Greensboro NC 27408

MASON, EARL JAMES, b Marion, Ind, Aug 26, 23; m 46; c 2. PATHOLOGY, MICROBIOLOGY. Educ: Ind Univ, BS, 44, AB & MA, 47; Ohio State Univ, PhD(bact), 50; Western Reserve Univ, MD, 54. Prof Exp: Damon Runyon Cancer fel, 54-56; fel path, Postgrad Sch Med, Univ Tex, 58-59; asst prof path, Col Med, Baylor Univ, 59-60; asst pathologist, Michael Reese Hosp, Chicago, 60-61; assoc pathologist, Mercy Hosp, Chicago, 61-65, chmn dept biol sci, 62-65; DIR LABS, ST MARY MERCY HOSP, 65- Concurrent Pos: From intern to resident, Case Western Reserve Univ, 54-56; assoc prof, Dept Path, Chicago Med Sch, 65- Mem: Am Asn Path & Bact; Am Asn Cancer Res; Am Soc Exp Path; Am Soc Hemat. Res: Mechanism of action of viruses on cells; cellular production of antibodies; thrombocytopathic action of viruses. Mailing Add: St Mary Mercy Hosp 540 Tyler St Gary IN 46402

MASON, EDWARD ALLEN, b Atlantic City, NJ, Sept 2, 26; m 52; c 4. CHEMICAL PHYSICS. Educ: Va Polytech Inst, BS, 47; Mass Inst Technol, PhD(phys chem), 51. Prof Exp: Res assoc chem, Mass Inst Technol, 50-52; Nat Res Coun fel, Univ Wis, 52-53; asst prof chem, Pa State Univ, 53-55; from assoc prof to prof molecular physics, Inst Molecular Physics, Univ Md, 55-67, dir, 66-67; PROF CHEM & ENG, BROWN UNIV, 67- Honors & Awards: Sci achievement Award, Wash Acad Sci, 62. Mem: AAAS; Am Asn Physics Teachers; fel Am Phys Soc. Res: Diffusion and thermal diffusion in gases; molecular and ionic scattering; equation of state of gases; theory of transport phenomena; intermolecular forces; statistical mechanics. Mailing Add: Dept of Chem Brown Univ Providence RI 02912

MASON, EDWARD EATON, b Boise, Idaho, Oct 16, 20; m 44; c 4. SURGERY. Educ: Univ Iowa, BA, 43, MD, 45; Univ Minn, PhD(surg), 53. Prof Exp: Intern surg, Univ Minn Hosps, 45-46, fel surg, 48-52; from asst prof to assoc prof, 53-60, PROF SURG, COL MED & UNIV HOSPS, UNIV IOWA, 60- Mem: AAAS; Soc Univ Surgeons; Soc Exp Biol & Med; AMA; Am Col Surgeons. Res: Diseases of thyroid, parathyroid and gastrointestinal tract; pneumoperit- oneum in giant hernia repair; side-to-side spenorenal shunt; fluid, electrolyte and nutritional balance; gastric bypass for obesity; fatty acid toxicity. Mailing Add: Dept of Surg Univ of Iowa Hosps Iowa City IA 52240

MASON, ELLIOTT BERNARD, b Detroit, Mich, July 29, 43; m 71; c 2. PHYSIOLOGY. Educ: Loyola Univ, Chicago, BS, 65; Wayne State Univ, MS, 69, PhD(biol), 72. Prof Exp: Asst prof biol, George Mason Col, Univ Va, 71-73; asst prof, 73-75, ASSOC PROF BIOL, STATE UNIV NY COL CORTLAND, 75- Mem: AAAS; Am Inst Biol Sci; Am Soc Mammalogists; Am Soc Zool; Ecol Soc Am. Res: Environmental physiology of vertebrates; endocrinology of nonmammalian vertebrates; stress responses of vertebrates. Mailing Add: Dept of Biol Sci State Univ of NY Col Cortland NY 13045

MASON, GEORGE ROBERT, b Rochester, NY, June 10, 32; m 56; c 3. SURGERY, PHYSIOLOGY. Educ: Oberlin Col, BA, 55; Univ Chicago, MD, 57; Stanford Univ, PhD(physiol), 68. Prof Exp: Teaching asst path, Univ Chicago, 54-56, teaching asst physiol, Stanford Univ, 60-62, actg instr surg, 65-66, from instr to assoc prof, 66-71; PROF SURG & PHYSIOL & CHMN DEPT SURG, UNIV MD, BALTIMORE CITY, 71- Concurrent Pos: Markle scholar acad med, Univ Md, 69-74; mem Gov adv comt emergency med serv & adv comt chronic dis hosps, Dept Health & Ment Hyg, Md; mem bd dirs regional planning, Coun Emergency Med Serv Develop Corp; consult, Mercy Hosp, Baltimore, Md Gen Hosp, SBaltimore Gen Hosp, Baltimore Vet Admin Hosp & Mem Hosp, Easton. Mem: Am Asn Thoracic Surg; Am Col Chest Physicians; Am Col Surgeons; Am Surg Asn; Asn Acad Surg. Res: Gastrointestinal physiology, particularly autonomic control of visceral function; thoracic surgery. Mailing Add: Dept of Surg Univ of Md Sch of Med Baltimore MD 21201

MASON, GRANT WILLIAM, b Waialua, Hawaii, Aug 8, 40; m 64; c 4. COSMIC RAY PHYSICS. Educ: Brigham Young Univ, BA, 61; Univ Utah, PhD(physics), 69. Prof Exp: Res assoc & assoc instr physics, Univ Utah, 68-69; sci co-worker, Physics Inst, Aachen Tech Univ, 69-70; asst prof, 70-74, ASSOC PROF PHYSICS, BRIGHAM YOUNG UNIV, 74- Mem: Am Phys Soc. Res: High energy cosmic ray studies. Mailing Add: Dept of Physics & Astron Brigham Young Univ Provo UT 84602

MASON, GRENVILLE R, b Rush Lake, Sask, Aug 8, 34; m 56; c 4. NUCLEAR PHYSICS. Educ: Univ BC, BASc, 56; McMaster Univ, MEng, 59; Univ Alberta, PhD(physics), 64. Prof Exp: Engr, Can Westinghouse Co, Ltd, 56-58; lectr physics, Univ Victoria, BC, 62-64, from instr to asst prof, 64-68, ASSOC PROF PHYSICS, UNIV VICTORIA, BC, 68- Mem: Can Asn Physicists. Res: Mesonic atoms; pion production cross-sections. Mailing Add: Dept of Physics Univ of Victoria Victoria BC Can

MASON, HAROLD FREDERICK, b Porterville, Calif, Feb 15, 25; m 54; c 3. PHYSICAL CHEMISTRY. Educ: Cornell Univ, BChE, 50; Univ Wis, PhD(phys chem), 55. Prof Exp: Chem engr, Rohm & Haas Co, Pa, 50-51; res chemist, Chevron Res Co Div, Standard Oil Co Calif, 54-59, group supvr, 59-64, sect supvr, 64-67, mgr petrol process develop div, 67-71, MGR PETROL PROCESS RES DIV, CHEVRON RES CO DIV, STANDARD OIL CO, CALIF, 71- Mem: Am Chem Soc; Am Inst Chem Eng. Res: Chemical reaction kinetics; catalysis; petroleum processing; hydrogenation and hydrocracking; solid state reactions. Mailing Add: Chevron Res Co 576 Standard Ave Richmond CA 94802

MASON, HARRY, b Portland, Ore, Nov 27, 21; m 53; c 3. PHYSICS. Educ: Univ of the Pac, BS, 43; Cath Univ Am, MS, 49. Prof Exp: Physicist, Appl Physics Lab, Johns Hopkins Univ, 43-49; PROF PHYSICS & HEAD DEPT, JAMESTOWN COL, 49- Concurrent Pos: Res assoc & consult, Boeing Airplane Co, 51-57. Mem: Am Phys Soc; Am Asn Physics Teachers. Res: Radio noise from the sun; earth currents and geomagnetic disturbances. Mailing Add: Dept of Physics Jamestown Col Jamestown ND 58401

MASON, HERMAN CHARLES, b Chicago, Ill, Sept 26, 10. PUBLIC HEALTH, IMMUNOLOGY. Educ: Univ Chicago, BS, 32; Univ Ill, MS, 37, PhD, 39. Prof Exp: Instr in chg lab med, dent & grad students, Dept Bact & Pub Health, Univ Ill Col Med, 35-39; fel, Johns Hopkins Univ, 39-40; bacteriologist venereal dis, Health Dept, Chicago, Ill, 40-41; asst prof bact, State Col Wash, 41-42; assoc prof bact & immunol, Sch Med, Univ NC, 42-45; in chg of bact & immunol, Schering Corp, 45; consult, Chicago, Ill, 46-49; dir labs, Navy Med Res Unit 4, Ill, 49-50; head bact & consult, State Dept Pub Welfare, Ill & Psychopath Inst, 50-57; head labs, State Dept Health, Wash, 57-59; res prof, Univ Wash, 59; sr pathologist & chief med serv group, Space Med Off, Boeing Ariplane Co, 59-61; adminr labs & res biochemist, Vet Admin Hosp, Topeka, Kans, 61-64; dir res labs, Ark State Hosp, 64-69; vis prof, Int Atomic Energy Agency, Cekmece Nuclear Reactor, Istanbul Univ, 69-71; CONSULT, 71- Concurrent Pos: Spec lectr, Cook County Sch Nursing, 38 & Grinnell Col, 41; chief adv, Korean Nat Labs, Seoul, 47-48. Mem: Fel Am Geog Soc; fel Am Pub Health Asn; fel Royal Soc Health; fel Royal Soc Trop Med & Hyg. Res: Filtrable viruses; laboratory methods; central nervous system infections; behavioral sciences; radiobiology. Mailing Add: 5059 N Jones Rd Oak Harbor WA 98277

MASON, HOWARD STANLEY, b Melrose, Mass, Aug 20, 14; m 50; c 2. BIOCHEMISTRY. Educ: Mass Inst Technol, SB, 35, SM, 36, PhD(org chem), 39. Prof Exp: Instr chem, Ruston Acad, Cuba, 36-37; Michael fel, Harvard Univ, 39-41; chemist, NIH, 41-49; NIH spec fel, Dept Chem, Cambridge Univ, 49-50; res assoc, Princeton Univ, 50-52; from asst prof to assoc prof, 52-61, PROF, DEPT BIOCHEM MED SCH, UNIV ORE HEALTH SCI CTR, 61- Concurrent Pos: NIH spec fel, Dept Chem, Cambridge Univ, 59-60; Commonwealth fel & fel, Clare Hall, Cambridge Univ, 69-70. Mem: Fel Zool Soc London; Am Chem Soc; Am Soc Biol Chemists; The Chem Soc; Brit Biochem Soc. Res: Mechanisms of biological oxidation-reduction; comparative biochemistry; fundamental aspects of disease. Mailing Add: Dept of Biochem Univ Ore Health Sci Ctr Portland OR 97201

MASON, JAMES MICHAEL, b Kingsport, Tenn, Mar 19, 43; m 69. IMMUNOLOGY, EXPERIMENTAL PATHOLOGY. Educ: Memphis State Univ, BS, 66; Univ Tenn, PhD(exp path), 71. Prof Exp: Instr, 71-74, ASST PROF PATH, CTR HEALTH SCI, UNIV TENN, MEMPHIS, 74- Concurrent Pos: Consult, Chief Med Examr, State of Tenn, 71-; lectr, Nat Inst Child Health & Human Develop, 74- Mem: AAAS; Reticuloendothelial Soc. Res: Infectious disease aspects of sudden infant death syndrome; control of cell division; immunology of carcinogenesis. Mailing Add: Dept of Path Univ Tenn Ctr Health Sci Memphis TN 38163

MASON, JAMES WILLARD, b Hollywood, Calif, Apr 5, 33; m 56; c 2. SYNTHETIC ORGANIC CHEMISTRY. Educ: Univ Calif, BS, 56, PhD(org chem), 60. Prof Exp: Chemist, Papermate Pen Co, 56; res assoc med chem, Merck Sharp & Dohme Res Labs, 60-64; sr scientist, Aeronutronic Div, Philco-Ford Corp, 64-69; scientist, Havens Int, 69; PRIN SCIENTIST & CONSULT AERONUTRONIC DIV, AERONUTRONIC-FORD CORP, 69- Concurrent Pos: Consult water & waste treat, 69- Mem: AAAS; Am Inst Chem; Am Chem Soc; NY Acad Sci. Res: Environmental sciences; water and waste treatment; polymer chemistry; adhesives; membrane processes. Mailing Add: Biosci Staff Aeronutronic-Ford Corp Newport Beach CA 92663

MASON, JESSE DAVID, US citizen. MATHEMATICS. Educ: Univ Mo, Kansas City, BS, 62; Univ Calif, Riverside, PhD(math), 68. Prof Exp: Prod designer, Vendo Co, 58-62; dynamics engr, Gen Dynamics, Pomona, 62-65; res assoc math, Univ Calif, Riverside, 65-67; asst prof, Calif State Univ, San Bernardino, 67-68; asst prof, Univ Ga, 68-71; ASSOC PROF MATH, UNIV UTAH, 71- Mem: Inst Math Statist; Am Math Soc; Math Asn Am. Res: Limit theorems in probability theory and stochastic differential equations. Mailing Add: Dept of Math Univ of Utah Salt Lake City UT 84112

MASON, JOHN, b New York, NY, May 11, 19; m 51; c 2. EPIDEMIOLOGY, PUBLIC HEALTH. Educ: Middlesex Univ, DVM, 44; Nat Vet Sch, Alfort, France, DVM, 49; Univ Minn, MPH, 53. Prof Exp: Pvt pract, Va, NC & Mass, 44-46; consult, UNRRA, DC, 46-47; area supvr US & Mex foot & mouth dis eradication comn, USDA, Mex, 49-51; pub health vet, Commun Dis Ctr, 52-56, dir div commun dis, NMex Dept Pub Health, 56-60, malaria adv, Indonesia, 60-62, Honduras, 63 & Haiti, 64-68, chief malaria adv, Malaria Eradication Prog, Philippines, 68-71, epidemiologist, Cent Am Malaria Res Sta, USPHS, San Salvador, El Salvador, 71-74, VET OFFICER, USDA FOOT & MOUTH DIS PROG, MEX, 74- Mem: Am Pub Health Asn; Am Vet Med Asn. Res: Communicable disease control; malariology; veterinary public health; tuberculosis control. Mailing Add: Mexico City-Dept of State Washington DC 20521

MASON, JOHN CHRISTOPHER, b June 4, 32; Can citizen; m 67; c 3. ECOLOGY, FISHERIES BIOLOGY. Educ: Ore State Univ, BS, 59, MS, 63, PhD(fisheries), 66. Prof Exp: RES SCIENTIST, FISHERIES RES BD CAN, 67- Mem: Am Fisheries Soc; Am Soc Limnol & Oceanog; Animal Behav Soc; Ecol Soc Am; Int Asn Astacol (vpres & pres-elect, 74-76). Res: Behavior and ecology of aquatic animals. Mailing Add: Pac Biol Sta Nanaimo BC Can

MASON, JOHN FREDERICK, b Los Angeles, Calif, Nov 25, 13; m 39; c 4. PETROLEUM GEOLOGY. Educ: Univ Southern Calif, AB, 34, AM, 35; Princeton Univ, PhD(geol), 41. Prof Exp: Field geologist, Socony-Vacuum Oil Co, Egypt, 37-40; instr earth sci, Univ Pa, 41-42; field geologist, Venezuelan Atlantic Ref Co, Barcelona & Caracas, 42-46; geologist, Foreign Prod Dept, Atlantic Ref Co, Pa, 46-51; asst to gen mgr, Foreign Opers Dept, Union Oil Co, Calif, 52-54; mgr explor, Standard Vacuum Oil Co, India, 54-56, resident mgr, Prod Div, Pakistan, 56-59; staff geologist, Foreign Dept, Continental Oil Co, 59-65, sr explor adv, 65-75; CONSULT, WEEKS NATURAL RESOURCES, LTD, 75- Concurrent Pos: Field geologist, Pa Geol & Topog Surv, 41- Mem: Fel Am Asn Petrol Geol; Geol Soc London. Res: Sedimentary basins of the world as to petroleum prospects. Mailing Add: 240 Fisher Place Princeton NJ 08540

MASON, JOHN GROVE, b Louisville, Ky, Dec 4, 29; m 56; c 2. ANALYTICAL CHEMISTRY. Educ: Univ Louisville, BS, 50; Ohio State Univ, PhD(chem), 55. Prof Exp: Instr chem, Ill Inst Technol, 56-59; assoc prof, 59-66, PROF CHEM, VA POLYTECH INST & STATE UNIV, 66- Mem: Am Chem Soc. Res: Polarography; electrode processes. Mailing Add: Dept of Chem Va Polytech Inst & State Univ Blacksburg VA 24061

MASON, JOHN HUGH, b Batavia, NY, Mar 8, 29; m 53; c 2. POLYMER CHEMISTRY. Educ: Univ Rochester, BS, 50; Carnegie Inst Technol, PhD(org chem), 55. Prof Exp: Res chemist, Union Carbide Plastics Co Div, Union Carbide Corp, 54-61; sr res assoc, 62-70, PROJS MGR, CARBORUNDUM CO, 70- Mem: Am Chem Soc. Res: Inorganic fibers and composites; analytical chemistry. Mailing Add: 5205 Brookfield Lane Clarence NY 14031

MASON, JOHN LESLIE, b Birkenhead, Eng, June 2, 13; m 45; c 1. PLANT NUTRITION. Educ: Univ BC, BSA, 48; Wash State Col, MS, 51; Ore State Col, PhD(plant physiol), 56. Prof Exp: Res scientist, 48-66, HEAD SOILS SECT, RES STA, CAN DEPT AGR, 66- Concurrent Pos: Ed Can J Plant Sci, 65-69; mem study comt, Can-BC Okanagan Basin Agreement, 72-74. Mem: Am Soc Hort Sci; Can Soc Hort Sci; Can Soc Soil Sci; Agr Inst Can; Int Soc Hort Sci. Res: Nutrition of tree fruits; effects of calcium on fruit quality. Mailing Add: Res Sta Can Dept of Agr Summerland BC Can

MASON, JOHN WAYNE, b Chicago, Ill, Feb 9, 24; m 50; c 3. NEUROENDOCRINOLOGY. Educ: Ind Univ, AB, 44, MD, 47. Prof Exp: Asst physiol, Ind Univ, 43-45; intern surg, NY Hosp-Cornell Med Ctr, 47-48, resident path, 48-50; pathologist, Ft Riley, Kans & Brooke Army Hosp, Tex, 50-53; chief neuroendocrinol dept, 53-74, SCI ADV, DIV NEUROPSYCHIAT, WALTER REED ARMY INST RES, 74- Mem: Endocrine Soc; Am Psychosom Soc (pres, 70). Res: Emotional and psychosomatic mechanisms. Mailing Add: Div of Neuropsychiat Walter Reed Army Inst of Res Washington DC 20012

MASON, KARL ERNEST, b Kingston, NS, May 30, 00; nat US; m 27; c 1. ANATOMY, NUTRITION. Educ: Acadia Univ, AB, 21; Yale Univ, PhD(zool, anat), 25. Hon Degrees: DSc, Acadia Univ, 49. Prof Exp: Asst zool, Yale Univ, 21-25; Nat Res Coun fel, 25-26; from instr to assoc prof anat, Sch Med, Vanderbilt Univ, 27-40; prof & head dept, 40-65, EMER PROF ANAT, SCH MED & DENT, UNIV ROCHESTER, 65-; CONSULT, NAT INST ARTHRITIS, METAB & DIGESTIVE DIS, 75- Concurrent Pos: Hon res assoc, Univ Col, Univ London, 39; consult, USPHS; nutrit prog dir, Nat Inst Arthritis, Metab & Digestive Dis, 65-75. Honors & Awards: Mead Johnson Award, 35. Mem: Am Asn Anat; Am Inst Nutrit; Brit Nutrit Soc. Res: Vitamin deficiency and reproduction; histopathology of nutritional deficiencies; bioassay of vitamin E; experimental leprosy; muscular dystrophy; trace elements and reproduction. Mailing Add: 8114 Jeb Stuart Rd Bethesda MD 20854

MASON, LARRY GORDON, b Wyandotte, Mich, Jan 19, 37. POPULATION BIOLOGY. Educ: Univ Mich, BS, 58, MA, 59, Univ Kans, PhD(entom), 64. Prof Exp: Res assoc biol, Stanford Univ, 64-65; asst prof, 66-72, ASSOC PROF BIOL, STATE UNIV NY ALBANY, 72- Mem: Soc Study Evolution; Am Soc Nat. Res: Population phenomena, especially quantitative aspects, in natural animal populations. Mailing Add: Dept of Biol State Univ of NY at Albany Albany NY 12203

MASON, LEO SUMNER, physical chemistry, see 12th edition

MASON, LEONARD EDWARD, b Seattle, Wash, June 26, 13; m 39; c 3. APPLIED ANTHROPOLOGY. Educ: Univ Minn, BA, 35, MA, 41; Yale Univ, PhD(anthrop), 55. Prof Exp: Res asst anthrop, Sci Mus, St Paul, Minn, 35-41; res asst, Cross-Cult Surv, Inst Human Rels, Yale Univ, 41-43, instr area studies, Army Specialized Training Prog, 43-44; res analyst, Off Strategic Servs, 44-45; res analyst, Off Res & Intel, US Dept State, 45-46; from assoc prof to prof, 47-69, EMER PROF ANTHROP, UNIV HAWAII, 69-; CONSULT PAC ISLANDS AFFAIRS, 71- Concurrent Pos: Anthropologist, US Com Co, 46; Honolulu officer, Pac Sci Bd, Nat Acad Sci-Nat Res Coun, 47; mem adv comt educ in Trust Territory & Guam, US Navy Admin, 47-51; consult civil admin, US Navy, 48; dir, Human Rels Area Files, Univ Hawaii, 49-62, NSF res grant, 65-67, organizer, Pac Islands Workshops, 74; mem, Mgt Surv Trust Territory Pac Islands, US Dept Interior, 50; chmn dept anthrop, Univ Hawaii, 50-54 & 57-65, chmn, Pac Islands Studies Prog, 50-65; NSF sr fel, Yale Univ, 56-57; coordr field trip prog, Verde Valley Sch, Sedona, Ariz, 69-70, headmaster, 70; coordr, Youth Develop Conf, East-West Ctr, Honolulu, 71; tour dir & prof, Howard Tours, Inc, S Pac, 72; Mem: Fel Am Anthrop Asn; Asn Social Anthrop Oceania; Soc Appl Anthrop. Res: Pacific Islands ethnology, with special interest in Micronesia; social and cultural change; resettled populations; applied anthropology; political education; applied research in contemporary sociocultural change in Pacific Islands, especially Micronesia. Mailing Add: 5234 Keakealani St Honolulu HI 96821

MASON, LYSLE C, b Mont Ida, Kans, Oct 24, 17; m 38; c 4. MATHEMATICS. Educ: Pittsburgh State Col, BS, 38; Univ Mich, MS, 42; Okla State Univ, EdD(math), 65. Prof Exp: Teacher math, Cherokee County Community High Sch, 38-42; PROF MATH, PHILLIPS UNIV, 42- Mem: Math Asn Am. Mailing Add: Dept of Math Phillips Univ Enid OK 73701

MASON, MARCUS M, b New York, NY, Mar 23, 11; m 32; c 3. VETERINARY PATHOLOGY. Educ: Cornell Univ, BS, 33, MS, 34; NY State Col Vet Med, DVM, 38. Prof Exp: Vet, Lederle Lab, Am Cyanamid Co, 38-39 & US Bur Animal Indust, 39-40; asst prof path, Vet Col, Middlesex Univ, 40-41; vet, Civilian Conserv Corps, 41& Animal Clin, Mass, 46-57; dir biol res, Biologics Testing Lab, 57-60 & Mason Res Inst, Inc, 60-75; DIR, WORCESTER FOUND FOR EXP BIOL, 75- Concurrent Pos: Partic cycad conf, NIH, 65; mem geriat comt, Inst Lab Animal Resources, Nat Acad Sci, 65- Mem: Am Vet Med Asn; Endocrine Soc; Am Asn Lab Animal Sci; Soc Toxicol; Am Soc Lab Animal Practrs. Res: Prostatic biology; comparative neuropathology; mammalian endocrinology bioassay and immunoassay; toxicology, viral oncology, carcinogenesis and cancer chemotherapy. Mailing Add: Worcester Found for Exp Biol 222 Maple Ave Shrewsbury MA 01545

MASON, MARION, b Toronto, Ont, Nov 29, 33; US citizen. NUTRITION. Educ: Miami Univ, BS, 55; Ohio State Univ, MS, 59; Cornell Univ, PhD(nutrit), 69. Prof Exp: Instr nutrit, Univ Rochester, 56-58; consult, Vis Nurse Asn, Chicago, 59-63; asst prof, Univ Rochester, 63-66; assoc prof med dietetics, Ohio State Univ, 69-72; prof, 73-76, RUBY WINSLOW LINN PROF NUTRIT, SIMMONS COL, 76- Concurrent Pos: Clin consult dietetics, Peter Bent Brigham Hosp, Boston, 73-; res assoc, Eastman Dent Ctr, Rochester, NY, 74-75; vis prof, Univ Rochester, 75-76. Mem: Am Dietetic Asn; Sigma Xi; Soc Nutrit Educ. Res: Health care compliance and intervention; clinical dietetic practice; behavioral aspects of human nutrition. Mailing Add: Simmons Col 300 The Fenway Boston MA 02115

MASON, MAX GARRETT, b Roanoke, Va, Jan 15, 44; m 67; c 2. SURFACE PHYSICS. Educ: Johns Hopkins Univ, BA, 65, PhD(chem), 70. Prof Exp: Sr res assoc chem, Univ Southern Calif, 70-72; SR RES CHEMIST, KODAK RES LABS, EASTMAN KODAK CO, 72- Mem: Am Vacuum Soc. Res: Ultraviolet and x-ray photoemission studies of solid surfaces; chemistry and physics of adsorbed species. Mailing Add: Kodak Res Labs Kodak Park Rochester NY 14650

MASON, MERLE, b Coldspring, Mo, Aug 9, 20; m 42; c 1. BIOCHEMISTRY. Educ: Univ Iowa, BS, 47, PhD(biochem), 50. Prof Exp: From instr to asst prof, 50-58, ASSOC PROF BIOCHEM, UNIV MICH, ANN ARBOR, 59- Mem: AAAS; Am Chem Soc; Am Soc Biol Chemists; Soc Study Reprod. Res: Amino acid metabolism; steroid metabolism. Mailing Add: Dept of Biol Chemists Univ of Mich Ann Arbor MI 48104

MASON, MICHAEL E, b Ft Lupton, Colo, Dec 12, 29; m 54; c 4. BIOCHEMISTRY, FOOD SCIENCE. Educ: Univ Colo, BA, 55; Univ Ark, MS, 61; Okla State Univ, PhD(biochem), 63. Prof Exp: Res chemist, Res Labs, Swift & Co, 55-58; res asst animal sci, Univ Ark, 58-60; instr chem, Okla State Univ, 60-61, from asst prof to assoc prof biochem, 63-68; asst dir, 68-71, vpres & dir res, Flavor Res & Develop, 72-74, VPRES & GEN MGR, FLAVOR DIV, INT FLAVORS & FRAGRANCES, 74- Concurrent Pos: Rep, Indust Res Inst, 73-; adj prof food sci, Pa State Univ, 74-; mem adv bd, Food Sci Dept, Rutgers Univ, 75- Mem: AAAS; Am Chem Soc. Res: Flavor chemistry and nutrition. Mailing Add: Int Flavors & Fragrances 521 W 57th St New York NY 10019

MASON, MORTON FREEMAN, b Pasadena, Calif, Nov 12, 02; m 29; c 2. BIOCHEMISTRY. Educ: Ore State Col, BSc, 25; Duke Univ, PhD(biochem), 34. Prof Exp: Asst chem, Exp Sta, Mich State Col, 26-30; asst biochem, Sch Med, Duke Univ, 32-34; from instr to assoc prof, Sch Med, Vanderbilt Univ, 34-44; prof path chem, 44-55, PROF FORENSIC MED & TOXICOL, UNIV TEX HEALTH SCI CTR DALLAS, 55- Concurrent Pos: Toxicologist, Dallas City-County, 44-74, dir, Criminal Invest Lab, 55-74; chemist, Parkland Mem Hosp, 44-74; sr consult, US Vet Admin, 46-74. Mem: Am Soc Biol Chem; Am Chem Soc; Soc Exp Biol & Med; Am Asn Clin Chem; Am Indust Hyg Asn. Res: Analytical toxicology. Mailing Add: 3172 Brookhollow Dr Dallas TX 75234

MASON, NORMAN RONALD, b Rochester, Minn, Nov 20, 29; m 53; c 2. BIOCHEMISTRY. Educ: Univ Chicago, AB, 50, BS, 53; Univ Utah, MA, 56, PhD(biochem), 59. Prof Exp: From res instr to res asst prof biochem, Endocrinol Lab, Sch Med, Univ Miami, 59-64; SR SCIENTIST, RES LABS, ELI LILLY & CO, 64- Concurrent Pos: Investr, Howard Hughes Med Inst, 59-64. Mem: AAAS; Endocrine Soc; Am Chem Soc. Res: Endocrinology; ovarian function; gonadotropin action; cyclic nucleotides; prostaglandins; steroid hormone synthesis and metabolism; hormone action. Mailing Add: Lilly Res Labs Eli Lilly & Co Indianapolis IN 46206

MASON, PERRY SHIPLEY, JR, b Lubbock, Tex, Oct 2, 38; m 60; c 2. ORGANIC CHEMISTRY. Educ: Harding Col, BS, 59; La State Univ, PhD(org chem), 63. Prof Exp: Asst prof sci, Okla Christian Col, 63-64; res assoc chem, Grad Inst Technol, Univ Ark, 64-66; asst prof, Ark State Col, 66-71; PROF CHEM & HEAD DEPT, LUBBOCK CHRISTIAN COL, 71- Concurrent Pos: NIH fel, 63-66. Mem: Am Chem Soc. Res: Organometallic chemistry; reaction mechanism; gas chromatography. Mailing Add: Dept of Chem Lubbock Christian Col Lubbock TX 79407

MASON, PETER F, geography, see 12th edition

MASON, REGINALD G, JR, b Washington, NC, July 9, 33; m 64; c 1. PATHOLOGY, BIOCHEMISTRY. Educ: Univ NC, BS, 57, MD, 62, PhD(exp path), 64. Prof Exp: From intern to resident path, 62-64, from instr to prof path, Sch Med, Univ NC, Chapel Hill, 64-75, PATHOLOGIST-IN-CHIEF, MEM HOSP, PAWTUCKET, RI & PROF PATH, BROWN UNIV, 75- Concurrent Pos: Mem, Path A Study Sect, NIH; fel exp path, Sch Med, Univ NC, Chapel Hill, 62-65, Markle scholar acad med, 65. Mem: AAAS. Res: Thrombosis and hemorrhage; blood coagulation; blood platelet agglutination and white thrombus formation; cellular cohesion and adhesion. Mailing Add: Mem Hosp Pawtucket RI 02860

MASON, RICHARD CANFIELD, b Indianapolis, Ind, Aug 12, 23; m 44; c 2. PHYSIOLOGY. Educ: Ind Univ, AB, 48, PhD(zool), 52. Prof Exp: Asst, Ind Univ, 48-49; res assoc, Merck Inst Therapeut Res, 52-56; asst prof physiol, Seton Hall Col Med & Dent, 56-61; asst prof, Col Physicians & Surgeons, Columbia Univ, 61-71, asst dean student affairs, 70-71; ASSOC PROF PHYSIOL & ASSOC DEAN ADMIS & STUDENT AFFAIRS, RUTGERS MED SCH-COL MED & DENT NJ, 71- Mem: AAAS; Am Soc Zool; Am Physiol Soc; Harvey Soc. Res: Renal physiology. Mailing Add: Rutgers Med Sch Col of Med & Dent NJ New Brunswick NJ 08903

MASON, RICHARD PATRICK, b Cawker City, Kans, Aug 3, 09; m 35; c 2. MEDICINE. Educ: Wash Univ, AB, 32, MD, 36; Am Bd Prev Med, dipl, 50. Prof Exp: Chief lab serv, Sta Hosp, Ft Benning, Ga, Med Corps, US Army, 39-42, Commanding officer, 4th Med Labs, Ft Sam Houston, Tex & NAfrica Opers, 42-44, dep chief prev med, Mediter & Europ Opers, 44-45, med liaison officer, Gen Staff Res & Develop Bd, War Dept, 46-47, trainee, NIH, 47-48, asst chief virus & rickettsial dis, Army Med Serv Grad Sch, 48-51, commanding officer, 406th Med Gen

Lab & dir & lab consult, Med Res Activ, Far East Command, Tokyo, Japan, 51-54, chief Res & Develop Div, Off Surg Gen, 54-56, dir, Walter Reed Army Inst Res, 56-61; from vpres res to sr vpres res, Am Cancer Soc Inc, 61-74; RETIRED. Concurrent Pos: Mem adv sci bd, Gorgas Mem Inst, 56- Mem: Am Asn Immunol; Am Soc Trop Med & Hyg; fel Am Pub Health Asn; Am Asn Cancer Res; fel NY Acad Med. Res: Preventive medicine; infectious disease control; medical research administration. Mailing Add: Am Cancer Soc Inc 219 E 42nd St New York NY 10017

MASON, RICHARD RANDOLPH, b St Louis, Mo, Oct 3, 30; m 56; c 4. FORESTRY, ENTOMOLOGY. Educ: Univ Mich, BS, 52, MF, 56, PhD(forestry), 66. Prof Exp: Forest entomologist, Bowaters Southern Paper Corp, 56-58, res forester, 58-65; RES ENTOMOLOGIST, FORESTRY SCI LAB, US FOREST SERV, 65- Mem: Soc Am Foresters; Entom Soc Am; Ecol Soc Am; Entom Soc Can. Res: Protection of commercial forests from destructive insect pests by regulating insect populations. Mailing Add: Forestry Sci Lab US Forest Serv Corvallis OR 97331

MASON, ROBERT C, b Anthony, Idaho, July 9, 20; m 46; c 3. PHARMACOLOGY, MEDICINAL CHEMISTRY. Educ: Univ Utah, BS, 50; Univ Wis, PhD(pharmaceut chem), 54. Prof Exp: PROF MED CHEM, UNIV UTAH, 54- Mem: Am Chem Soc; Am Pharmaceut Asn. Res: Isolation, characterization and synthesis of natural products and related substances. Mailing Add: Col of Pharm Univ of Utah Salt Lake City UT 84112

MASON, ROBERT EDWARD, b Thunder Bay, Ont, Jan 21, 34; m 57; c 3. STATISTICS, ECOLOGY. Educ: Univ Toronto, BSA, 57, MSA, 62; NC State Univ, PhD(statist), 71. Prof Exp: Dist biologist, Ont Dept Lands & Forests, 57-59, fish & wildlife supvr, 59-64; from asst statistician to assoc statistician, NC State Univ, 65-71; statistician, 71-73, SR STATISTICIAN, RES TRIANGLE INST, 73- Mem: Ecol Soc Am; Biomet Soc. Res: Design and analysis of probability samples; nonlinear variance estimation; statistical ecology. Mailing Add: Res Triangle Inst PO Box 12194 Research Triangle Park NC 27709

MASON, ROBERT WILLIAM, inorganic chemistry, see 12th edition

MASON, RODNEY JACKSON, b New York, NY, Feb 27, 39; m 69. PLASMA PHYSICS. Educ: Cornell Univ, BA, 60, PhD, 64. Prof Exp: Fulbright grant, Inst Plasma Physics, Garching, WGer, 64-65; asst prof aeronaut & astronaut, Mass Inst Technol, 65-67; mem tech staff, Bell Tel Labs, 67-72; STAFF MEM, LOS ALAMOS SCI LAB, 72- Mem: Am Phys Soc. Res: Kinetic theory of shock formation and structure; computer simulation of ion-acoustic and magnetosonic collisionless shocks; computational physics; laser-plasma interaction studies; implosion and thermonuclear burn physics. Mailing Add: Div T Los Alamos Sci Lab Los Alamos NM 87545

MASON, RONALD GEORGE, b Southampton, Eng, Dec 24, 16; m 46. GEOPHYSICS. Educ: Univ London, BSc, 38, MSc, 39, PhD(geophys), 51. Prof Exp: Lectr geophys, Imp Col, London, 47-63; asst res geophysicist, Scripps Inst, Univ Calif, 52-62; reader, 63-65, PROF GEOPHYS, IMP COL, UNIV LONDON, 67-; RES AFFIL, HAWAII INST GEOPHYS, UNIV HAWAII, 63- Mem: AAAS; Seismol Soc Am; Soc Explor Geophys; Am Geophys Union; Europ Asn Explor Geophys. Res: Crustal and upper mantle structure of the earth; earthquake and volcano mechanisms. Mailing Add: Dept of Geophys Imp Col London England

MASON, RONALD JAMES, b Windsor, Ont, Oct 11, 29; US citizen; m 58; c 2. ANTHROPOLOGY, ARCHAEOLOGY. Educ: Univ Pa, BA, 57; Univ Mich, MA, 57, PhD(anthrop), 64. Prof Exp: Asst dir, Neville Pub Mus, Green Bay, Wis, 58-61; from asst prof to assoc prof, 61-69, PROF ANTHROP, LAWRENCE UNIV, 69- Mem: AAAS; Soc Am Archaeol; Am Anthrop Asn. Res: Archaeology of North America; early man in America; correlations of geomorphic features and archaeological distributions, particularly in Great Lakes region of North America; historic period American Indian archaeology. Mailing Add: Dept of Anthrop Lawrence Univ Appleton WI 54911

MASON, STANLEY GEORGE, b Montreal, Que, Mar 20, 14; m 43; c 2. PHYSICAL CHEMISTRY. Educ: McGill Univ, BE, 36, PhD(phys chem), 39. Prof Exp: Instr phys chem, Trinity Col, Conn, 39-41; res engr, Suffield Exp Sta, Dept Nat Defence, Alta, 41-45; assoc res chemist, Div Atomic Energy, Nat Res Coun Can, 45-46; res assoc, 46-50, HEAD PHYS CHEM DIV, PULP & PAPER RES INST CAN, 50-; PROF CHEM, McGILL UNIV, 66- Concurrent Pos: Res assoc, McGill Univ, 45-66. Mem: Tech Asn Pulp & Paper Indust; fel Royal Soc Can; assoc Can Pulp & Paper Asn; fel Chem Inst Can. Res: Colloids; cellulose; pulp and paper. Mailing Add: Dept of Chem McGill Univ Montreal PQ Can

MASON, THOMAS JOSEPH, b St Louis, Mo, Aug 8, 42. BIOSTATISTICS, EPIDEMIOLOGY. Educ: St Bernard Col, BA, 64; Univ Ga, MS, 68, PhD(statist & comput sci), 73. Prof Exp: Aerospace engr, NASA Manned Spacecraft Ctr, 64-65; statistician epidemiol, Ctr Dis Control, 67-69; STATISTICIAN EPIDEMIOL, NAT CANCER INST, 71- Mem: Sigma Xi. Res: Assessing carcinogenic exposures among residents of areas in the United States which have a markedly different cancer experience from that of this country as a whole. Mailing Add: Rm A521 Landow Bldg NIH Nat Cancer Inst Bethesda MD 20014

MASON, TIM ROBERT, b Hereford, Tex, Apr 26, 30; m 53; c 3. ANIMAL NUTRITION, REPRODUCTIVE PHYSIOLOGY. Educ: Abilene Christian Col, BS, 53; Tex Tech Col, MS, 55; Tex A&M, PhD(animal nutrit), 63. Prof Exp: High sch teacher, Tex, 55-56; instr animal husb, Abilene Christian Col, 56-59, asst prof, 63-64; res asst, Tex A&M, 59-61 & 62-63; dir agr develop, Tex Power & Light Co, 61-62; dir livestock res, Beacon Div, Textron, Inc, 64-65; dir tech serv, 65-66; assoc prof animal husb, 66-68, PROF AGR, TARLETON STATE COL, 68- Mem: Am Soc Animal Sci. Res: Beef cattle; sheep, dairy and swine nutrition research. Mailing Add: Dept of Agr Tarleton State Univ Stephenville TX 76402

MASON, W ROY, III, b Charlottesville, Va, Feb 6, 43; m 63; c 2. INORGANIC CHEMISTRY. Educ: Emory Univ, BS, 63, MS & PhD(chem), 66. Prof Exp: Instr chem, Emory at Oxford, summer 64; res fel, Calif Inst Technol, 66-67; asst prof, 67-70, ASSOC PROF CHEM, NORTHERN ILL UNIV, 70- Concurrent Pos: Vis res fel, H C Ørsted Inst, Univ Copenhagen, Denmark. Mem: Am Chem Soc. Res: Heavy metal coordination compounds; electronic structure and reactivity; molecular orbital and ligand field theory. Mailing Add: Dept of Chem Northern Ill Univ DeKalb IL 60115

MASON, WALTER HARRY, b Dover, NH, Feb 8, 35; m 62. PHYSIOLOGY, PHARMACOLOGY. Educ: Univ WVa, AB, 61, MS, 63, PhD(pharmacol), 64. Prof Exp: Fel, Tulane Univ, 64-66; asst dir environ toxicol sect, Inst Agr Med Sch Med, Univ Iowa, 66-69; ASST PROF BIOL, THIEL COL, 69- Mem: AAAS; Am Inst Biol Sci. Res: Intracellular pharmacology by microinjection into amoebae; induced liver enzymes. Mailing Add: Dept of Biol Thiel Col Greenville PA 16125

MASON, WARREN PERRY, b Colorado Springs, Colo, Sept 28, 00; m 29, 56; c 1. ACOUSTICS. Educ: Univ Kans, BSEE, 21; Columbia Univ, MA, 24, PhD(physics), 28. Prof Exp: Mem tech staff, Bell Tel Labs, 21-31, head piezoelec res, 31-48, head mech res, 48-65; VIS PROF CIVIL ENG & ENG MECH, COLUMBIA UNIV, 65-SR RES ASSOC, 69- Concurrent Pos: Res prof, George Washington Univ, 69- Honors & Awards: Arnold O Beckman Award, Instrument Soc Am; Benjamin Lamme Award, Inst Elec & Electronics Eng, 67; Gold Medal, Acoust Soc Am, 71; First Hon Mem, Brit Inst Acoust. Mem: AAAS; fel Acoust Soc Am (pres, 55-56); fel Am Phys Soc; fel Inst Elec & Electronics Eng; Instrument Soc Am. Res: Physical acoustics and the properties of materials; piezoelectricity and ferroelectricity; internal friction, acoustic emission and fatigue in metals. Mailing Add: 50 Gilbert Pl West Orange NJ 07052

MASON, WILLIAM BURKETT, b Warren, Ohio, Aug 20, 20; m 47. CLINICAL CHEMISTRY. Educ: Univ Rochester, BS, 42, MD, 50; Princeton Univ, MA, 44, PhD(chem), 46; Am Bd Clin Chem, Dipl, 56. Prof Exp: Asst, Princeton Univ, 42-46, instr, 46; asst, Atomic Energy Proj, Univ Rochester, 46-47, assoc, 47-51, instr biochem, Sch Med & Dent, 51-57, from asst prof to assoc prof biochem & med, 57-70, path, 61-70; DIR, AFFILIATED LABS, BIO-SCI ENTERPRISES, 70- Concurrent Pos: Anal chemist, Manhattan Dist, Princeton Univ, 44-46; intern, Strong Mem Hosp, Rochester, 50-51, asst resident, 51-52; fel clin path, Clin Ctr, NIH, 61-62; chief med scientist, Med Diag Opers, Xerox Corp, 68-70. Mem: Am Chem Soc; Am Asn Clin Chem (pres, 67); Acad Clin Lab Physicians & Scientists. Res: Applications of analytical chemistry to medicine; quantitative analytical procedures; infrared microspectrophotometry. Mailing Add: Bio-Sci Enterprises 7600 Tyrone Ave Van Nuys CA 91405

MASON, WILLIAM HICKMON, b Bradford, Ark, June 16, 36; m 55; c 3. ZOOLOGY. Educ: Ark Polytech Col, BS, 58; Univ Ga, MEd, 64, DEd(sci educ), 66. Prof Exp: Asst prof, 66-72, ASSOC PROF ZOOL & ENTOM, AUBURN UNIV, 72-, COORDR GEN BIOL, 68- Mem: Am Inst Biol Sci; AAAS; Entom Soc Am; Ecol Soc Am. Res: Ecosystem analysis through the use of radio nuclide cycling and improvement of undergraduate teaching through the use of audio, tutorial and modular concepts. Mailing Add: Dept Gen Biol Auburn Univ Auburn AL 36830

MASON, WILLIAM RICHARDSON MILES, b Lucknow, India, Nov 29, 21; Can citizen; m 51; c 2. SYSTEMATIC ENTOMOLOGY. Educ: Univ Alberta, BSc, 42; Cornell Univ, PhD(entomol), 53. Prof Exp: Seasonal agr res officer, 46-49, AGR RES OFFICER, CAN DEPT AGR, 49- Concurrent Pos: Ed, Can Entomologist, 61-64. Mem: Entom Soc Can. Res: Taxonomy of Hymenoptera, Ichneumonidae and Braconidae; systematics of Braconidae, especially Microgasterinae. Mailing Add: Biosyst Res Inst Can Agr Ottawa ON Can

MASON, WILLIAM VAN HORN, b Pittsburgh, Pa, Jan 8, 30; m 65; c 2. AEROSPACE MEDICINE. Educ: Harvard Univ, AB, 51; Baylor Univ, MD, 61. Prof Exp: Jr geophysicist, Humble Oil & Refining Co, 54-56; physician, Hood River Med Group, 62-65; RES PHYSICIAN, LOVELACE FOUND, 65- Concurrent Pos: NIH grants, 65- Mem: Aerospace Med Asn; Am Inst Aeronaut & Astronaut. Res: Advanced diagnostic instrumentation; physiology of unusual environments. Mailing Add: Lovelace Found 5200 Gibson Blvd SE Albuquerque NM 87108

MASORO, EDWARD JOSEPH, b Oakland, Calif, Dec 28, 24; m 47. PHYSIOLOGY. Educ: Univ Calif, AB, 47, PhD(physiol), 50. Prof Exp: Asst physiol, Univ Calif, 47-48; asst prof, Queens Univ, 50-52; from asst prof to assoc prof, Med Sch, Tufts Univ, 52-62; from res assoc prof to res prof, Univ Wash, 62-64; prof & chmn dept, Med Col Pa, 64-73; PROF PHYSIOL & CHMN DEPT, UNIV TEX HEALTH SCI CTR, SAN ANTONIO, 73- Mem: Am Physiol Soc; Am Chem Soc; Can Biochem Soc; Can Physiol Soc; Am Soc Biol Chem. Res: Intermediary metabolism; environmental physiology; muscle physiology; membrane transport. Mailing Add: Dept of Physiol Univ of Tex Health Sci Ctr San Antonio TX 78284

MASOUREDIS, SERAFEIM PANOGIOTIS, b Detroit, Mich, Nov 14, 22; m 43; c 2. HEMATOLOGY, IMMUNOHEMATOLOGY. Educ: Univ Mich, AB, 44, MD, 48; Univ Calif, Berkeley, PhD(med physics), 52. Prof Exp: Clin instr med, Univ Calif, San Francisco, 50-52, res assoc med physics, Donner Lab, Berkeley, 54; from asst prof to assoc prof path, Sch Med, Univ Pittsburgh, 55-59, asst dir cent blood bank, 55-59; assoc prof prev med, Sch Med, Univ Calif, San Francisco, 59-62, assoc prof med, 62-66, assoc prof clin path & lab med, 66-67; prof med & microbiol, Sch Med, Marquette Univ, 67-69; PROF PATH & DIR UNIV HOSP BLOOD BANK, SCH MED, UNIV CALIF, SAN DIEGO, 69- Concurrent Pos: Res assoc, Cancer Res Inst, 59-67; chief H C Moffitt Blood Bank, 62-67; spec fel, Univ Lausanne, 65-66; exec dir, Milwaukee Blood Ctr, 67-69. Mem: Am Asn Cancer Res; Am Asn Immunol; Am Soc Hemat; Soc Exp Biol & Med; Int Soc Hemat. Res: Blood group antigens; red cell membranes; membrane ultrastructure; immunological reactions involving red cell, premolytic anemias. Mailing Add: Dept of Path Univ of Calif Sch of Med La Jolla CA 92037

MASRI, MERLE SID, b Jerusalem, Palestine, Sept 12, 27; nat US; m 52; c 4. AGRICULTURAL CHEMISTRY, MAMMALIAN PHYSIOLOGY. Educ: Univ Calif, AB, 50, PhD(physiol), 53. Prof Exp: Res assoc hemat, Michael Reese Hosp, Chicago, Ill, 54-56; res chemist pharmacol, 56-71, RES CHEMIST FIBER SCI, WESTERN REGIONAL RES LAB, USDA, 71- Honors & Awards: Spec Serv Merit Award, USDA, 66. Mem: AAAS; Am Chem Soc; Am Asn Cereal Chem; NY Acad Sci; fel Am Inst Chem. Res: Chemistry metabolism and pharmacology of mycotoxins; toxicology; fiber science, especially wool; protein chemistry; metallic ion interactions with proteins and bio polymers; polymers and enzyme immobilization; pollution abatement. Mailing Add: Western Regional Res Lab US Dept of Agr Berkeley CA 94710

MASSA, DENNIS JON, b Myrtle Beach, SC, Sept 29, 45; m 66; c 2. PHYSICAL CHEMISTRY, POLYMER PHYSICS. Educ: Bradley Univ, BA, 66; Univ Wis-Madison, PhD(phys chem), 70. Prof Exp: NSF fel phys biochem, Univ Calif, San Diego, 70-71; SR RES CHEMIST, RES LABS, EASTMAN KODAK CO, 71- Mem: AAAS; Am Chem Soc; Am Phys Soc; Soc Rheol. Res: Physico-chemical investigation of biological macromolecules; hydrodynamics of polymers in solution; molecular mobility in the solid state; polymer physics; physical chemistry of polymers and biopolymers; molecular motion in the solid state; polymer rheology. Mailing Add: Res Labs Eastman Kodak Co Rochester NY 14650

MASSA, LOUIS, b Aug 4, 40; US citizen. CHEMICAL PHYSICS. Educ: LeMoyne Col, BS, 61; Clarkson Col, MS, 62; Georgetown Univ, PhD(physics), 66. Prof Exp: Res fel chem, Brookhaven Nat Lab, 66-69; ASSOC PROF CHEM, HUNTER COL, 69- Concurrent Pos: Petrol Res Fund grant, Hunter Col, 70-, City Univ New York Res Found grant, 71- Mem: AAAS; Am Phys Soc; Am Chem Soc. Res: Theoretical chemical physics; quantum mechanics. Mailing Add: Dept of Chem Hunter Col 695 Park Ave New York NY 10021

MASSAR, ANN ROLLER, b New Brunswick, NJ, Nov 2, 32; m 69; c 2. MOLECULAR BIOLOGY. Educ: Sarah Lawrence Col, BA, 52; Georgetown Univ, MS, 57; Calif Inst Technol, PhD, 61. Prof Exp: Physiol chemist, Nat Heart Inst, 53-

54, biochemist, Nat Inst Neurol Dis & Blindness, 54-57; NSF fel, Free Univ Brussels, 62-63; Damon Runyon Cancer Res Found fel, Pasteur Inst, Paris, 63-64; res assoc biochem, Col Physicians & Surgeons, Columbia Univ, 65-69; SCI WRITER, 69- Res: Mechanism of action of DNA. Mailing Add: c/o AIDR BP 100 Butare Rwanda

MASSARO, DONALD JOHN, b Jamaica, NY, Aug 7, 32; m 57; c 2. MEDICINE. Educ: Hofstra Col, BA, 53; Georgetown Univ, MD, 57. Prof Exp: Am Thoracic Soc fel, 60-62; from isntr to asst prof med, Georgetown Univ, 62-67; assoc prof, Duke Univ, 67-68; assoc prof, 68-72, PROF MED, GEORGE WASHINGTON UNIV, 72-; CHIEF CHEST SECT, VET ADMIN HOSP, 68- Concurrent Pos: fel physiol chem, Johns Hopkins Univ, 64-65. Mem: Am Physiol Soc; Am Soc Clin Invest; Soc Exp Biol & Med; Am Fedn Clin Res; Am Thoracic Soc. Res: Pulmonary diseases; lung biochemistry; phagocytosis; pulmonary physiology. Mailing Add: Chest Sect Vet Admin Hosp Washington DC 20422

MASSARO, EDWARD JOSEPH, b Passaic, NJ, June 7, 33; m 53; c 4. TOXICOLOGY, BIOCHEMISTRY. Educ: Rutgers Univ, AB, 55; Univ Tex, MA, 58, PhD(biochem), 62. Prof Exp: USPHS fel, Univ Tex, 62-63; fel, Med Sch, Johns Hopkins Univ, 63-64 & univ, 64-65, res assoc biol, 65; res assoc, Yale Univ, 65-68; from asst prof to assoc prof, 68-75, PROF BIOCHEM, STATE UNIV NY BUFFALO, 75- Mem: AAAS; Am Soc Biol Chemists; Am Soc Cell Biol; Am Soc Pharmacol & Exp Therapeut; Teratology Soc. Res: Environmental, developmental and comparative toxicology, biochemistry and physiology. Mailing Add: Dept of Biochem State Univ of NY Buffalo NY 14214

MASSE, NORMAN G, chemistry, see 12th edition

MASSEE, TRUMAN WINFIELD, b Joseph, Ore, May 5, 30; m 51; c 3. SOIL FERTILITY. Educ: Ore State Univ, BS, 52, AgM, 53; Mont State Univ, PhD, 73. Prof Exp: Soil scientist, Northern Mont Br Exp Sta, Agr Res Serv, 55-57, res soil scientist, Tetonia Br Exp Sta, Univ Idaho, 58-64, RES SOIL SCIENTIST, SNAKE RIVER RES CTR, AGR RES SERV, USDA, 65- Mem: Am Soc Agron; Soil Conserv Soc Am; Soil Sci Soc Am; Sigma Xi. Res: Dryland soil moisture-fertility-plant growth relationships. Mailing Add: Rte 2 Jerome ID 83338

MASSEL, GARY ALAN, b Trenton, NJ, May 5, 39; m 59; c 2. PLASMA PHYSICS, OPERATIONS RESEARCH. Educ: NC State Univ, BS, 61, PhD(physics), 67. Prof Exp: Asst physics, NC State Univ, 61 & 63; Aerospace engr, Langley Res Ctr, NASA, Va, 65-67; res staff mem, Inst Defense Anal, 67-70; dir land force progs, 70-72 & naval force progs, Off Asst Secy Defense Systs Anal, 72-73; assoc adminr, Social & Rehabilitation Serv, Off Health, Educ & Welfare, 73-75; VPRES, JRB ASSOCS, 75- Concurrent Pos: Consult, Defense Atomic Support Agency, 68- Mem: Am Phys Soc; Opers Res Soc Am; Am Pub Health Asn. Res: Computer modeling of many-body systems; high-temperature hydrodynamics; weapon systems analysis. Mailing Add: 8020 Birnam Wood Dr McLean VA 22101

MASSELL, PAUL BARRY, b Boston, Mass, June 26, 48. NUMBER THEORY. Educ: Univ Chicago, AB, 70; City Univ New York, PhD(math), 75. Prof Exp: Sr programmer, Nat Bur Econ Res, 74-75; RES ANALYST, JWK INT CORP, 75- Concurrent Pos: Lectr math, Brooklyn Col, 71-74. Mem: Am Math Soc; Math Asn Am. Res: Class field theory, especially examination of class groups of real quadratic number fields; relationship between class field theory and theorems of elementary number theory. Mailing Add: 2209 N Van Dorn St Apt T1 Alexandria VA 22304

MASSENGALE, MARTIN ANDREW, b Monticello, Ky, Oct 25, 33; m 59; c 2. AGRONOMY, CROP PHYSIOLOGY. Educ: Western Ky Univ, BS, 52; Univ Wis, MS, 54, PhD(agron), 56. Prof Exp: Asst agron, Univ Wis, 52-56; from asst prof & asst agronomist to assoc prof & assoc agronomist, 58-65, head dept agron & plant genetics, 66-74, PROF & AGRONOMIST, UNIV ARIZ, 65-, ASSOC DEAN COL AGR & COOP EXT SERV & ASSOC DIR ARIZ AGR EXP STA, 74- Concurrent Pos: Assoc ed, Agron J & Crop Sci, 69-72; consult to ministry agr & water, Saudi Arabia, 74; mem nat coord comt for cotton res, 74. Mem: AAAS; fel Am Soc Agron; Crop Sci Soc Am (pres, 72-73); Am Soc Plant Physiol. Res: Forage crops physiology, production and management; water-use efficiency, photosynthesis, respiration and dry-matter production. Mailing Add: Col of Agr Univ of Ariz Tucson AZ 85721

MASSENGILL, RAYMOND, b Bristol, Va, Dec 8, 37; m 59; c 3. SPEECH PATHOLOGY, AUDIOLOGY. Educ: Univ Tenn, BS, 58, MS, 59; Univ Va, EdD(speech path & audiol), 68. Prof Exp: Dir speech path, audiol & speech sci, Palmer Rehab Ctr, Tenn, 60-62; DIR SPEECH PATH & SPEECH SCI & DIR SPEECH SCI LAB, MED CTR, DUKE UNIV, 64- Concurrent Pos: NIH grant, 67-; United Med Res Found grant, 67-68; Nat Inst Dent Res grant; consult speech path, audiol & speech sci, Univ Tenn, 67- & Nat Inst Dent Res, 68- Mem: AAAS; Am Speech & Hearing Asn; Am Inst Physics; Int Asn Rehab Facil; Int Asn Logopedics & Phoniatrics. Res: Speech physiology as it relates to oral and pharyngeal mechanisms and how this mechanism is altered due to certain plastic surgery procedures. Mailing Add: Speech Sci Lab Duke Univ Med Ctr Durham NC 27706

MASSERMAN, JULES HOMAN, b Chudnov, Poland, Mar 10, 05; nat US; m 43. NEUROPHYSIOLOGY, PSYCHOANALYSIS. Educ: Wayne State Univ, MB, 30, MD, 31. Prof Exp: Resident neurol, Stanford Univ, 31-32; asst psychiatrist, Johns Hopkins Univ, 32-35; resident psychiat, Univ Chicago, 35-36, from instr to asst prof, 36-46; assoc prof, 46-50, PROF NEUROL & PSYCHIAT, NORTHWESTERN UNIV, CHICAGO, 50-, CO-CHMN DEPT, 64- Concurrent Pos: Chief consult, Downey Vet Hosp, 46-; sci dir, Nat Found Psychiat Res, 46-; consult, Great Lakes Naval Hosp, 47- & WHO, 50-; H M Camp lectr, 64; Karen Horney lectr, 65; dir ed, Ill State Psychiat Inst; vis prof psychiat, Univ Louis, Univ Zagreb. Honors & Awards: Lasker Award, Am Pub Health Asn, 47; Taylor Manor Award, 73; Sigmund Freud Award, 74. Mem: Soc Biol Psychiat (pres, 57-58); Int Asn Social Psychiat (pres, 69-); Am Asn Social Psychiat (pres-elect); fel Am Psychiat Asn (vpres, 74-75, secy, 75-77); Acad Psychoanal (pres, 57-58). Res: Experimental neuroses; physiology of emotion; music; occultisms; dynamics of language; dynamics of phantasy; dynamics of political action. Mailing Add: 8 S Mich Ave Chicago IL 60603

MASSEY, CALVIN LEROY, forest entomology, see 12th edition

MASSEY, DOUGLAS GORDON, b Clinton, Ont, Oct 14, 26; m 66; c 3. MEDICINE. Educ: Univ Toronto, MD, 51; MRCP(E), 61; McGill Univ, MSc, 63; FRCP(C), 63. Prof Exp: Consult, Estab Pulmonary Labs, Repatriation Dept, Australia, 55-57; dir pulmonary lab, Hosp St Luke, Montreal, Que, 64-66; from asst prof to assoc prof med, Univ Sherbrooke, 66-73, dir serv pneumology, Univ Hosp, 66-72; PROF MED, SCH MED, UNIV HAWAII, MANOA, 73- Concurrent Pos: Sir Edward Beatty fel, McGill Univ, 62-63; grants, Med Res Coun Can, 66-68, Nat Cancer Inst, 66-68, Inst Occup & Environ Health, 67-68 & Minister of Educ, 67-69. Mem: Fel Am Col Chest Physicians. Res: Medical education, curriculum development, programmed texts; pulmonary practice; industrial medicine. Mailing Add: Univ of Hawaii Sch Med 3675 Kilauea Ave Honolulu HI 96816

MASSEY, EDDIE H, b Canadian, Tex, July 14, 39; m 57; c 2. PHARMACEUTICAL CHEMISTRY. Educ: McMurry Col, BA, 61; Vanderbilt Univ, PhD(org chem), 66. Prof Exp: SR PHARMACEUT CHEMIST, DEPT PHARMACEUT RES, ELI LILLY & CO, 66- Mem: Am Chem Soc. Res: chemistry and chemical modification of macrolide antibiotics. Mailing Add: Dept of Pharmaceut Res Eli Lilly & Co 740 S Alabama St Indianapolis IN 46225

MASSEY, FRANK JONES, JR, b Portsmouth, NH, Nov 22, 19; m 43; c 2. MATHEMATICAL STATISTICS. Educ: Univ Calif, AB, 41, MA, 44, PhD(math statist), 47. Prof Exp: Asst prof math, Univ Md, 47-48; from asst prof to assoc prof, Univ Ore, 48-59; PROF BIOSTATIST, PREV MED & PUB HEALTH, UNIV CALIF, LOS ANGELES, 59-, BIOMATH, 70- Concurrent Pos: Ford fel, 53-54. Mem: Am Statist Asn; Inst Math Statist. Res: Non-parametric statistical analysis. Mailing Add: Dept of Pub Health Univ of Calif Los Angeles CA 90024

MASSEY, FREDRICK ALAN, b Birmingham, Ala, Dec 20, 38; m 67; c 1. APPLIED MATHEMATICS. Educ: Samford Univ, BS, 61; Auburn Univ, 63, PhD(math), 66. Prof Exp: Asst prof math, Auburn Univ, 66-67; from asst prof to assoc prof, 67-74, PROF MATH, GA STATE UNIV, 74- Mem: Am Math Soc; Math Asn Am; Soc Indust & Appl Math. Res: Theoretical physics; economic theory; nonlinear programming. Mailing Add: Dept of Math Ga State Univ Atlanta GA 30303

MASSEY, GAIL AUSTIN, b El Paso, Tex, Dec 2, 36; m 60. LASERS. Educ: Calif Inst Technol, BS, 59; Stanford Univ, MS, 67, PhS(elec eng), 70. Prof Exp: Engr, Raytheon Co, Santa Barbara, 59-63; sr eng specialist, Electro-optics orgn, GTE Sylvania, 63-72; PROF APPL PHYSICS, ORE GRAD CTR, 72- Concurrent Pos: Consult laser fusion group, Lawrence Livermore Lab, Univ Calif, 75- Mem: Optical Soc Am; Acoust Soc Am; Inst Elec & Electronics Engrs. Res: Nonlinear optical devices; ultraviolet and wavelength-tunable lasers; ultrafast optical pulse techniques. Mailing Add: Ore Grad Ctr 19600 NW Walker Rd Beaverton OR 97005

MASSEY, HERBERT FANE, JR, b Kerrville, Tenn, Jan 23, 26; m 51. AGRONOMY, SOIL FERTILITY. Educ: Univ Tenn, BS, 49; Univ Wis, MS, 50, PhD(soils), 52. Prof Exp: Asst, Univ Wis, 49-52; res agronomist, Int Minerals & Chem Corp, 52-53; from asst prof to assoc prof, 53-63, PROF AGRON & DIR REGULATORY SERV, UNIV KY, 63- Concurrent Pos: Vis prof, San Carlos Univ, Guatemala, 59-60, Univ Indonesia, 61-64 & Thailand, 67-70. Mem: Soil Sci Soc Am; Am Soc Agron; Int Soc Soil Sci. Res: Soil chemistry and fertility; micro element studies; tropical agriculture. Mailing Add: Div Regulatory Serv Col of Agr Univ of Ky Lexington KY 40506

MASSEY, JIMMY R, b Mart, Tex, July 9, 40; m 62. BOTANY. Educ: NTex State Univ, BSEd, 62; Tex A&M Univ, MS, 65; Univ Okla, PhD(bot), 71. Prof Exp: Instr bot, Tex A&M Univ, 64-65; vis scholar bot & genetics, Okla Col Lib Arts, 70-71; CUR HERBARIUM, UNIV NC, CHAPEL HILL, 71- Mem: Int Asn Plant Taxon; Am Soc Plant Taxonomists; Sigma Xi. Res: Vascular flora of southeastern United States; taxonomy of aquatic plants of the United States; pollination-reproductive biology. Mailing Add: Cur of Herbarium Univ of NC Dept of Bot Chapel Hill NC 27514

MASSEY, JOE THOMAS, b Raleigh, NC, Apr 22, 17; m 41; c 2. BIOMEDICAL ENGINEERING. Educ: NC State Col, BS, 38; Johns Hopkins Univ, PhD(physics), 53. Prof Exp: Instr elec eng, Clemson Col, 38-39; instr eng mech, NC State Col, 39-41; asst to dir, 72-74, DIR BIOMED PROGS, JOHNS HOPKINS UNIV, 74-, PRIN STAFF MEM, APPL PHYSICS LAB, 46- Concurrent Pos: Lectr, Johns Hopkins Univ, 57. Mem: NY Acad Sci; Inst Elec & Electronics Engrs. Res: Microwave plasma and physics; laser physics. Mailing Add: Appl Physics Lab Johns Hopkins Univ 8621 Georgia Ave Silver Spring MD 20910

MASSEY, JOHN HUBERT, b Homer, Ga, Sept 7, 16; m 56; c 1. AGRONOMY. Prof Exp: Univ Ga, BS, 48, MS, 52; La State Univ, PhD(agron), 61. Prof Exp: ASST AGRONOMIST, GA AGR EXP STA, 51-58, 60- Mem: Am Soc Agron; Crop Sci Soc Am. Res: Feed grain crop management. Mailing Add: Ga Agr Exp Sta Experiment GA 30212

MASSEY, LINDA KATHLEEN LOCKE, b Oklahoma City, Okla, Aug 27, 45. CELL BIOLOGY. Educ: Univ Okla, BS, 66, PhD(microbiol), 71. Prof Exp: Res assoc microbiol, Health Sci Ctr, Univ Okla, 71-72, instr, 72-73; NIH fel, Cancer Sect, Okla Med Res Found, 73-74; ASST PROF BIOCHEM, OKLA COL OSTEOPATH MED & SURG, 74- Concurrent Pos: NIH fel, 71-73. Mem: Tissue Cult Asn; AAAS. Res: Effects of drugs on glycoprotein composition of cellular membranes and growth dynamics. Mailing Add: Okla Col Osteopath Med & Surg 1111 W 17th St Tulsa OK 74107

MASSEY, LOUIS MELVILLE, JR, b Ithaca, NY, Apr 28, 23; m 49; c 2. PLANT BIOCHEMISTRY. Educ: Oberlin Col, AB, 47; Cornell Univ, PhD(biochem), 51. Prof Exp: Asst biochem, Cornell Univ, 47-51; res plant biochemist, US Army Biol Labs, Md, 51-57; from asst prof to assoc prof, 57-70, PROF BIOCHEM, NY STATE COL AGR & LIFE SCI, CORNELL UNIV, 70- Mem: AAAS; Am Chem Soc; Am Soc Plant Physiol; Am Soc Hort Sci; Inst Food Technol. Res: Post-harvest physiology of fruits and vegetables; irradiation effects on the physiology of plant tissues; transportation of fruits and vegetables for processing; biochemistry of ripening; fungus physiology. Mailing Add: Dept Food Sci NY State Agr Exp Sta Cornell Univ Geneva NY 14456

MASSEY, PEYTON HOWARD, JR, b Zebulon, NC, Oct 4, 22; m 42; c 3. OLERICULTURE, ACADEMIC ADMINISTRATION. Educ: NC State Col, BS, 47, MS, 51; Cornell Univ, PhD(veg crops), 52. Prof Exp: Res asst, Cornell Univ, 49-52; from assoc prof to prof hort, Va Polytech Inst & State Univ, 52-65, assoc dean grad sch, 64-65, assoc dir res, Agr Exp Sta, 65-66 & res div, 66-68, assoc dean res & grad studies, 68-69, ASSOC DEAN, COL AGR & LIFE SCI, VA POLYTECH INST & STATE UNIV, 69- Honors & Awards: Medal of City, Paris, France, 68. Mem: AAAS; Am Soc Hort Sci; Int Soc Hort Sci; Am Inst Biol Sci. Res: Vegetable production and breeding; agricultural uses of plastics; research administration. Mailing Add: Va Polytech Inst & State Univ Blacksburg VA 24061

MASSEY, ROBERT UNRUH, b Detroit, Mich, Feb 23, 22; m 43; c 2. INTERNAL MEDICINE. Educ: Wayne Univ, MD, 46. Prof Exp: From intern to resident med, Henry Ford Hosp, 46-50; assoc, Lovelace Clin, 50-68, chmn dept, 58-68; clin assoc, Sch Med & Res, Univ NMex, 62-68; assoc dean, 68-71, PROF MED, MED SCH, UNIV CONN, 68-, DEAN, 71-, ACTG EXEC DIR & ACTG VPRES HEALTH AFFAIRS, 75- Concurrent Pos: Consult, West Interstate Comn Higher Educ, 58-60; dir educ, Lovelace Found Med Educ, 60-68; consult, NMex Regional Med Prog, 65-68; mem accreditation comn, Am Asn Med Clins, 66- Mem: AAAS; Sigma Xi; Am Col Physicians; Am Diabetes Asn; Asn Am Med Cols. Res: Clinical endocrinology and diabetes; medical education; medical care and medical administration. Mailing Add: Sch of Med Univ of Conn Farmington CT 06032

MASSEY, VINCENT, b Berkeley, Australia, Nov 28, 26; m 50; c 3. BIOCHEMISTRY.

Educ: Univ Sydney, BSc, 47; Cambridge Univ, PhD(biochem), 53. Prof Exp: Res officer biochem, Commonwealth Sci & Indust Res Orgn, Australia, 47-50; Imp Chem Industs Res fel, 53-55; mem res staff, Edsel B Ford Inst, Mich, 55-57; from lectr to sr lectr, Univ Sheffield, 57-63; PROF BIOL CHEM, SCH MED, UNIV MICH, ANN ARBOR, 63- Mem: Am Soc Biol Chemists. Res: Basic enzymology; mechanisms of enzyme reactions, especially of flavoproteins and metalloflavoproteins; role of sulfide in biological oxidations. Mailing Add: Dept of Biol chem Univ of Mich Sch Med Ann Arbor MI 48104

MASSEY, WALTER EUGENE, b Hattiesburg, Miss, Apr 5, 38. THEORETICAL SOLID STATE PHYSICS. Educ: Morehouse Col, BS, 58; Wash Univ, MA & PhD(physics), 66. Prof Exp: Instr physics, Morehouse Col, 58-59; from fel to physicist, Argonne Nat Lab, 66-68; asst prof, Univ Ill, Urbana, 69-70; assoc prof, 70-75, PROF PHYSICS, BROWN UNIV, 75- Concurrent Pos: Fel, Wash Univ, 66; consult, Argonne Nat Lab, 68- Mem: Am Phys Soc; Am Asn Physics Teachers. Res: Many-body problem; quantum liquids and solids; theory of classical liquids; solid state theory. Mailing Add: Dept of Physics Brown Univ Providence RI 02912

MASSEY, WILLIAM S, b Granville, Ill, Aug 23, 20; m 53; c 3. MATHEMATICS. Educ: Univ Chicago, BS, 41, MS, 42; Princeton Univ, PhD(math), 48. Prof Exp: Off Naval Res fel, Princeton Univ, 48-50; from asst prof to assoc prof math, Brown Univ, 50-54; vis assoc prof, Princeton Univ, 54-55; prof, Brown Univ, 55-60; chmn dept, 68-71, PROF MATH, YALE UNIV, 60- Concurrent Pos: Assoc ed, Ind Univ Math J, 75-78. Mem: Am Acad Arts & Sci; Am Math Soc; Math Asn Am. Res: Algebraic topology. Mailing Add: Dept of Math Yale Univ New Haven CT 06520

MASSEY, WINSTON LOUIS, b Chattanooga, Tenn, Dec 6, 04; m 39; c 1. MATHEMATICS. Educ: Chattanooga Univ, BA, 28; Duke Univ, MA, 34. Prof Exp: High sch instr, Tenn, 28-31; from instr to prof math, 33-65, Guerry prof, 65-74, EMER GUERRY PROF MATH, UNIV TENN, CHATTANOOGA, 74- Mem: Am Phys Soc; Math Asn Am. Mailing Add: Dept of Math Univ of Tenn Chattanooga TN 37401

MASSIAH, THOMAS FREDERICK, b Montreal, Que, Aug 26, 26; m 51; c 1. ORGANIC CHEMISTRY. Educ: Sir George Williams Univ, BSc, 47; McGill Univ, MSc, 56; Univ Montreal, PhD(org chem), 62. Prof Exp: Chief control chemist, Dewey & Almy Chem Co, Que, 47-53; demonstr chem, McGill Univ, 53-56; res chemist, Merck & Co Ltd, Que, 56-59; sr demonstr chem, Univ Montreal, 59-62; chemist, Ayerst, McKenna & Harrison Ltd, 62-66; GROUP LEADER CHEM DEVELOP, CAN PACKERS LTD, 66- Concurrent Pos: Lectr, Sir George Williams Univ, 49-64. Mem: Chem Inst Can. Res: Organic, biochemical and medicinal chemistry; antibiotics, bile acids, enzymes; pharmaceuticals; steroids. Mailing Add: Res & Develop Labs Can Packers Ltd 2211 St Clair Ave Toronto ON Can

MASSIE, EDWARD, b St Louis, Mo, Nov 21, 10; m 40; c 2. CARDIOLOGY. Educ: Wash Univ, AB, 31, MD, 35. Prof Exp: Assoc prof, 53-68, PROF CLIN MED, SCH MED, WASH UNIV, 68- Concurrent Pos: Consult, Heart Sta, Barnes Hosp, 41 & Heart Sta, Jewish Hosp, 70- Mem: Am Heart Asn; Am Fedn Clin Res; fel Am Col Physicians; fel Am Col Cardiol. Res: Cardiology and cardiovascular diseases. Mailing Add: Queeny Tower Suite 4104 4989 Barnes Hosp Plaza St Louis MO 63110

MASSIE, HAROLD RAYMOND, b Brisbane, Australia, Jan 31, 43; US citizen; m 70; c 3. MOLECULAR BIOLOGY. Educ: San Diego State Col, AB, 64; Univ Calif, San Diego, PhD(chem), 67. Prof Exp: NIH res fel chem & tutor biochem, Harvard Univ, 67-70; RES SCIENTIST, MASONIC MED RES LAB, 70- Mem: AAAS; Biophys Soc; Tissue Cult Asn; Geront Soc; Am Soc Microbiol. Res: DNA replication and structural changes; cell synchrony; animal cell culture; aging. Mailing Add: Masonic Med Res Lab 2150 Bleecker St Utica NY 13501

MASSIE, SAMUEL PROCTOR, b North Little Rock, Ark, July 3, 19; m 47; c 3. CHEMISTRY. Educ: Agr Mech & Normal Col, Ark, BS, 38; Fisk Univ, MA, 40; Iowa State Univ, PhD(org chem), 46. Hon Degrees: LLD, Univ Ark, 70. Prof Exp: Lab asst chem, Fisk Univ, 39-40; assoc prof math, Agr Mech & Normal Col, Ark, 40-41; res assoc chem, Iowa State Univ, 43-46; instr, Fish Univ, 46-47; prof & head dept, Langston Univ, 47-53, Fisk Univ, 53-60 & Howard Univ, 62-63; assoc prog dir, NSF, 60-63; pres, NC Col Durham, 63-66; PROF CHEM, US NAVAL ACAD, 66- Concurrent Pos: Sigma Xi lectr, Swarthmore Col, 57. Honors & Awards: Mfg Chem Asn Award, 61. Mem: Mfg Chem Asn. Mailing Add: Dept of Chem US Naval Acad Annapolis MD 21402

MASSINGILL, JOHN LEE, JR, b Lufkin, Tex, Aug 18, 41; m 63; c 2. INDUSTRIAL ORGANIC CHEMISTRY. Educ: Tex Christian Univ, BA, 63, MS, 65, PhD(chem), 68. Prof Exp: Sr res chemist, Basic Res Dept, 68 & Hydrocarbon Process Res Dept, 70-73, RES SPECIALIST, HYDROCARBON PROCESS RES DEPT, TEX DIV, DOW CHEM USA, FREEPORT, 73- Concurrent Pos: Consult, Ionics Res, Inc, 71-72. Mem: AAAS; Am Chem Soc; Sigma Xi; The Chem Soc. Res: Hydrocarbon utilization; new product research and development. Mailing Add: Freeport TX

MASSION, WALTER HERBERT, b Eitorf, Ger, June 4, 23; nat US; m 56; c 3. ANESTHESIOLOGY, PHYSIOLOGY. Educ: Univ Cologne, BS, 47; Univ Heidelberg, MD, 51. Prof Exp: Intern med, Med Ctr, Univ Zurich, 51-52; trainee anesthesiol, Anesthesiol Ctr, WHO, Denmark, 52-53; asst prof physiol, Med Sch, Univ Basel, 53-54; asst resident anesthesiol, Med Sch, Univ Rochester, 54-56; from asst prof to assoc prof anesthesiol, 57-67, assoc prof physiol & res surg, 66-71, PROF ANESTHESIOL, COL MED, UNIV OKLA, 67-, PROF PHYSIOL, BIOPHYS & RES SURG, 71-, ADJ PROF CARDIORESPIRATORY SCI, 71- Concurrent Pos: Fel physiol, Med Sch, Univ Rochester, 54-56; fel, Cardiovasc Res Inst, Sch Med, Univ Calif, 59-60; NIH res career develop award, 61-71; John A Hartford Found res grant, 67-71; Humboldt sr US sci award, Tech Univ Munich, Ger, 74-75. Mem: AAAS; Am Physiol Soc; Am Soc Anesthesiol; Int Anesthesia Res Soc. Res: Respiration; circulation; shock; vasoactive polypeptides. Mailing Add: Dept of Anesthesiol Univ of Okla Hlth Sci Ctr Oklahoma City OK 73190

MASSLER, MAURY, b New York, NY, Mar 24, 12; m 47; c 3. DENTISTRY. Educ: NY Univ, BS, 32; Univ Ill, DDS, 39, MS, 41. Prof Exp: Instr dent histol, Col Dent, Univ Ill Med Ctr, 39-41, dir child res clin, 41-73, from asst prof to assoc prof histol, 43-46, prof pedodont, 46-73, supvr hosp dent clin, 43-53, asst dean postgrad & teacher educ, 65-73, assoc dean col dent, 69-73; CHMN DEPT RESTORATIVE DENT, COL DENT MED, TUFTS UNIV, 73- Mem: Am Soc Dent for Children; Am Pub Health Asn; Am Dent Asn; Int Asn Dent Res; Am Acad Pedodontics. Res: Pedodontics; oral medicine; gerodontics; dental education. Mailing Add: Tufts Univ Col of Dent Med One Kneeland St Boston MA 02111

MASSON, CHARLES ROBB, b Aberdeen, Scotland, Sept 8, 22; m 48; c 2. PHYSICAL CHEMISTRY. Educ: Aberdeen Univ, BS, 43, PhD(chem), 48. Prof Exp: Asst lectr, Aberdeen Univ, 43-45; fel, Nat Res Coun Can, 48-50; fel, Univ Rochester, 50-51; from asst res officer to sr res officer, 51-65, PRIN RES OFFICER, NAT RES COUN CAN, 65-, HEAD HIGH TEMPERATURE CHEM SECT, 54- Concurrent Pos: Vis scientist, Brit Iron & Steel Res Asn, 54; hon prof fac grad studies, Dalhousie Univ, 63-; vis prof, Imp Col, Univ London, 63, Univ Strathclyde, 64 & Aberdeen Univ, 71; hon res assoc, Fac Grad Studies, Univ NB, 73- Mem: Am Chem Soc; fel Chem Inst Can; fel Chem Soc London; Can Inst Mining & Metall. Res: Kinetics and equilibria of chemical reactions at high temperatures; polymer, silicate and metallurgical chemistry. Mailing Add: Atlantic Regional Lab Nat Res Coun Can 1411 Oxford St Halifax NS Can

MASSON, GEORGES MARIE CHARLES, b Dieulouard, France, Nov 7, 11; nat US; m 38; c 2. ENDOCRINOLOGY. Educ: Vet Sch France, DVM, 34; Univ Montreal, lic es sc, 37; McGill Univ, PhD(endocrinol), 42. Prof Exp: Prof physiol, Vet Sch, Univ Montreal, 34-42; res assoc, McGill Univ, 42-43, res fel, 43-46; assoc prof endocrinol, Univ Montreal, 46-48; MEM STAFF, CLEVELAND CLIN, 48- Mem: AAAS; Am Physiol Soc; Endocrine Soc; Am Heart Asn. Res: Renal-adrenal relationships in hypertension. Mailing Add: Cleveland Clin Cleveland OH 44106

MASSOPUST, LEO CARL, JR, b Milwaukee, Wis, Nov 12, 20; m 43; c 2. ANATOMY. Educ: Marquette Univ, BS, 43, MS, 47; Univ Colo, PhD, 53. Prof Exp: Asst prof biol, Westminster Col (Mo), 47-48; instr anat, Univ Colo, 48-54; neuroanatomist, NIH, 54-58; sr res physiologist, Southeast La Hosp, 58-60; dir div neurophysiol, Cleveland Psychiat Inst & Hosp, 60-73; ASSOC PROF ANAT, MED SCH, ST LOUIS UNIV, 74- Mem: Am Asn Anat; Am Physiol Soc; Soc Exp Biol & Med; Am Acad Neurol. Res: Physiology of ergot alkaloid; hypothermia; electrophysiology of vision; psychophysiology of audition; neurophysiology of brain function. Mailing Add: 1402 S Grand Blvd St Louis MO 63104

MASSOTH, FRANKLIN E, physical chemistry, see 12th edition

MASSOVER, WILLIAM H, b Chicago, Ill, 41; m 69. CELL BIOLOGY, BIOPHYSICS. Educ: Univ Chicago, AB, 63, MD, 67, PhD(cell biol), 70. Prof Exp: NATO fel electron micros, Lab Electronic Optics, CNRS, Toulouse, France, 71-72; res assoc dept physics, Ariz State Univ, 72-73; ASST PROF BIOL, BROWN UNIV, 73- Concurrent Pos: Mem spec study sect, NIH, 75. Mem: Am Soc Cell Biol; Biophys Soc; Electron Micros Soc Am. Res: Subcellular cytology of oogenesis; atomic level ultrastructure of biological macromolecules and macromolecular systems; experimental applications of high voltage electron microscopy to the study of biological structure. Mailing Add: Div of Biol & Med Sci Brown Univ Providence RI 02912

MAST, CECIL B, b Chicago, Ill, Feb 21, 27; m 59; c 2. GEOMETRY, THEORETICAL PHYSICS. Educ: De Paul Univ, BS, 50; Univ Notre Dame, PhD(physics), 56. Prof Exp: Instr physics, 56-57, from instr to asst prof math, 59-63, ASSOC PROF MATH, UNIV NOTRE DAME, 63- Concurrent Pos: Vis lectr, St Andrews, 65-66. Mem: Am Phys Soc; Am Math Soc; Math Asn Am. Res: Nuclear physics and group representations as used in nuclear models; relativity theory; differential geometry and lie groups; foundations of physics. Mailing Add: Dept of Math Univ of Notre Dame Notre Dame IN 46556

MAST, GEORGE WINFIELD, medicine, see 12th edition

MAST, MORRIS GLEN, b Kalona, Iowa, Dec 8, 40; m 64; c 2. FOOD SCIENCE. Educ: Goshen Col, BS, 62; Ohio State Univ, MS, 69, PhD(food sci), 71. Prof Exp: Diag parasitologist, Evanston Hosp, Evanston, Ill, 62-65; mgr qual control, V F Weaver, Inc, New Holland, Pa, 65-67; res assoc poultry sci, Ohio State Univ, 67-71; ASST PROF FOOD SCI, PA STATE UNIV, UNIVERSITY PARK, 71- Mem: Sigma Xi; Poultry Sci Asn; Inst Food Technologists. Res: Microbiological, biochemical, and organoleptic changes occurring in poultry and egg products during processing and storage. Mailing Add: Dept of Food Sci Pa State Univ University Park PA 16802

MAST, ROY CLARK, b Wheeling, WVa, Nov 28, 24; m 48; c 3. PHYSICAL CHEMISTRY, INORGANIC CHEMISTRY. Educ: Univ Cincinnati, BS, 49, MS, 51, PhD(inorg chem), 53. Prof Exp: Student asst instr, Univ Cincinnati, 49-51; RES CHEMIST PHYS & INORG CHEM, MIAMI VALLEY LABS, PROCTER & GAMBLE CO, 52- Mem: Am Chem Soc. Res: Surfactant solutions; adsorption of surfactants; emulsion formation; surface chemistry; phase studies; fundamentals of detergency; micellar solubilization; diffusion studies. Mailing Add: Procter & Gamble Co Miami Valley Labs PO Box 39175 Cincinnati OH 45247

MAST, TERRY STEVEN, b Los Angeles, Calif, Jan 2, 43; m 69; c 2. ELEMENTARY PARTICLE PHYSICS. Educ: Calif Inst Technol, BS, 64; Univ Calif, Berkeley, PhD(physics), 71. Prof Exp: PHYSICIST, LAWRENCE BERKELEY LAB, 71- Res: Experimental elementary particle physics; astrophysics. Mailing Add: Lawrence Berkeley Lab Univ of Calif Berkeley CA 94720

MAST, WILLIAM CARLTON, b Beaverton, Mich, Feb 7, 16; m 41; c 2. POLYMER CHEMISTRY. Educ: Univ Mich, BS, 38, MS, 39. Prof Exp: Analyst asst, Dow Chem Co, Mich, 35-36, chemist, 39-42; asst org chem, Univ Mich, 37-39; chemist, Eastern Regional Res Lab, Bur Agr & Indust Chem, USDA, 42-48; res chemist, 48-72, HEAD THERMOPLASTICS & LATICES RES, RES DIV, GOODYEAR TIRE & RUBBER CO, 72- Mem: Am Chem Soc. Res: High styrene paint; reinforcing injection molding resins; latices; paper coating resins. Mailing Add: Goodyear Tire & Rubber Co Res Div 142 Goodyear Blvd Akron OH 44316

MASTALERZ, JOHN W, b Mass, Mar 16, 26; m 54; c 3. FLORICULTURE, HORTICULTURE. Educ: Univ Mass, BS, 48; Purdue Univ, MS, 50; Cornell Univ, PhD(floricult), 53. Prof Exp: Asst prof res floricult, Waltham Field Sta, Univ Mass, 52-56; PROF FLORICULT, PA STATE UNIV, 56- Mem: AAAS; Am Soc Agron; Crop Sci Soc Am; Am Hort Soc. Res: Post-harvest life of cut flowers; photoperiodic, temperature, soil mixture and fertilization requirements of flower crops; growth regulators. Mailing Add: Dept of Hort Pa State Univ University Park PA 16802

MASTELLER, EDWIN C, b Independence, Iowa, Aug 11, 34; m 57; c 3. BIOLOGY, ENTOMOLOGY. Educ: Northern Iowa Univ, BA, 58; Univ SDak, MA, 61; Iowa State Univ, PhD(entom), 67. Prof Exp: Pub sch instr, Minn, 58-64; ASST PROF BIOL, PA STATE UNIV, BEHREND CAMPUS, 67- Mem: Nat Asn Biol Teachers; Am Inst Biol Sci; Entom Soc Am; Am Micros Soc; Nat Sci Teachers Asn. Res: Effects of temperature stress on larval insect diapause, physiological and histological changes; insect embryology; insect-plant association, ecological considerations; insect-arthropod survey in vineyards. Mailing Add: Dept of Biol Pa State Univ Behrend Campus Erie PA 16510

MASTERS, BETTIE SUE SILER, b Lexington, Va, June 13, 37; m 60; c 2. BIOCHEMISTRY. Educ: Roanoke Col, BS, 59; Duke Univ, PhD(biochem), 63. Prof Exp: Res assoc biochem, Duke Univ, 65-67, assoc, 67-68; asst prof, 68-72, ASSOC PROF BIOCHEM, UNIV TEX HEALTH SCI CTR DALLAS, 72- Concurrent Pos: Am Heart Asn estab investr, 68-73; ed bd, J Biol Chem, 76-81; mem pharmacol-toxicol prog res rev comt, Nat Inst Gen Med Sci, NIH, 75-79; Am Cancer Soc fel biochem, Duke Univ, 63-65, Am Heart Asn advan res fel, 66-68. Mem: AAAS; Am Soc Biol Chem; Am Chem Soc; Am Soc Pharmacol & Exp Therapeut. Res:

Microsomal electron transport in various tissues with specific reference to nicotinamide adenine dinucleotide phosphate-cytochrome c (P-450) reductase. Mailing Add: Dept of Biochem Univ of Tex Health Sci Ctr Dallas TX 75235

MASTERS, BRUCE ALLEN, b Terre Haute, Ind, Nov 3, 36; m 63. MICROPALEONTOLOGY. Educ: Univ Valparaiso, BS, 59; Univ Calif, Berkeley, MA, 62; Univ Ill, Urbana, PhD(geol), 70. Prof Exp: Jr geologist, Humble Oil & Ref Co, 62-63, from asst prof to assoc prof geol, Hartwick Col, 69-74; SR RES SCIENTIST, AMOCO PROD CO, 74- Mem: Am Asn Petrol Geol; Paleont Res Inst; Paleont Soc; Soc Econ Paleont & Mineral; Swiss Geol Soc. Res: Morphology, taxonomy, phylogeny, paleoecology and biostratigraphy of Mesozoic and Cenozoic planktonic foraminifers. Mailing Add: Amoco Prod Co Res Ctr PO Box 591 Tulsa OK 74102

MASTERS, BURTON JOSEPH, b Casper, Wyo, Sept 8, 29; m 53; c 5. SOLID STATE SCIENCE. Educ: Univ Calif, Los Angeles, BS, 50; Ore State Col, PhD(chem), 54. Prof Exp: Staff mem, Los Alamos Sci Lab, 54-63; SR CHEMIST, IBM CORP, 63- Mem: Am Chem Soc; Am Nuclear Soc; Am Phys Soc. Res: Radiochemistry; radiation chemistry; reaction kinetics; physics and chemistry of semiconductors; diffusion and ion implantation of semiconductor devices and circuits. Mailing Add: 23 Timberline Dr Poughkeepsie NY 12603

MASTERS, CHARLES DAY, b Pawhuska, Okla, Aug 4, 29; m 53; c 3. GEOLOGY. Educ: Yale Univ, BS, 51, PhD(geol), 65; Univ Colo, MS, 57. Prof Exp: Hydrographic officer, US Navy, 52-54; explor geologist, Pan Am Petrol Corp, 57-68, res geologist, 68-70; chmn div sci & math, WGa Col, 70-73; CHIEF OFF ENERGY RESOURCES & ACTG CHIEF OFF MARINE GEOL, US GEOL SURV, 73- Mem: AAAS; Sigma Xi; Geol Soc Am; Am Asn Petrol Geologists. Mailing Add: US Geol Surv Nat Ctr MS 915 Reston VA 22092

MASTERS, CHRISTOPHER FANSTONE, b Ashridge, Eng, Dec 26, 42; US citizen; m 66; c 2. MATHEMATICS. Educ: Doane Col, AB, 64; Fla State Univ, MS, 66; Univ Northern Colo, DA, 74. Prof Exp: Instr math, Fla Southern Col, 66-68; ASST PROF MATH, DOANE COL, 68- Mem: Math Asn Am; Nat Coun Teachers Math. Mailing Add: Dept of Math Doane Col Crete NE 68333

MASTERS, EDWARD JOSEPH, organic chemistry, see 12th edition

MASTERS, EDWIN M, b Everette, Mass, Nov 21, 31; m 64; c 4. ANATOMY. Educ: Harvard Univ, AB, 52; Ind Univ, AM, 55; Univ Minn, PhD(anat), 65. Prof Exp: Instr anat, Univ Pittsburgh, 58-64; instr, 64-65, asst prof, 65-75, ASSOC PROF ANAT, JEFFERSON MED COL, 75- Concurrent Pos: NIH fel, 51-64. Mem: AAAS; Am Asn Anat; NY Acad Sci. Res: Physiology of fat cells in tissue cultrue; histogenesis of elastic tissue; survival of homologous grafts in the brain. Mailing Add: Dept of Anat Jefferson Med Col Philadelphia PA 19107

MASTERS, FRANK WYNNE, b Pittsburgh, Pa, Nov 1, 20; m; c 3. PLASTIC SURGERY. Educ: Hamilton Col, AB, 43; Univ Rochester, MD, 45; Am Bd Plastic Surg, dipl, 55; Am Bd Surg, dipl, 56. Prof Exp: Intern, Strong Mem Hosp, Rochester, NY, 45-46; trainee gen surg, 48-51; trainee plastic surg, Med Ctr, Duke Univ, 51-53, assoc, 53-54; chief plastic surg, Charleston Mem Hosp, WVa, 54-58; from asst prof to assoc prof, 58-67, PROF PLASTIC SURG, UNIV KANS MED CTR, KANSAS CITY, 67-, CHIEF SECT PLASTIC SURG & VCHMN DEPT SURG, 72-, ASSOC DEAN CLIN AFFAIRS, 73- Concurrent Pos: Consult, Vet Admin Hosps, Kansas City, Mo & Wadsworth, Kans; mem, Am Bd Plastic Surg, 68-74, co-chmn exam comt, 68-69, chmn, 69-73, rep, Am Bd Med Spec, 71-74, mem exec comt, 72-74, chmn, 73-74. Mem: Am Burn Asn; Am Asn Surg Trauma; Am Cleft Palate Asn; Am Soc Plastic & Reconstruct Surg; fel Am Col Surg. Mailing Add: Sect of Plastic Surg Univ of Kans Med Ctr Kansas City KS 66103

MASTERS, JOHN ALAN, b Shenandoah, Iowa, Sept 20, 27; m 51; c 3. PETROLEUM GEOLOGY. Educ: Yale Univ, BA, 48; Univ Colo, MS, 51. Prof Exp: Dist geologist, AEC, 51-53; chief geologist, Kerr-McGee Oil Industs, Inc, 53-66, mgr Can explor, Kerr-McGee Corp, 66-69, pres, Kerr-McGee Can Ltd, 69-73; PRES, CAN HUNTER EXPLOR, 73- Honors & Awards: Mattson Award, Asn Petrol Geol, 57. Mem: Geol Soc Am; Am Asn Petrol Geol. Res: Stratigraphy; oil exploration by means of subsurface and surface geology. Mailing Add: Can Hunter Explor 705-603 Seventh Ave SW Calgary AB Can

MASTERS, JOHN EDWARD, b Greeneville, Tenn, June 20, 13; m 38; c 2. ORGANIC CHEMISTRY. Educ: Tusculum Col, AB, 36; Univ Tenn, MS, 38. Prof Exp: Resin chemist high polymers, Jones-Dabney Co, 39-46, chief chemist resin div, 46-49; res dir, Devoe & Raynolds Co, 49-65; mgr, Trade Sales Labs, 65-69, SR RES ASSOC, CELANESE COATINGS CO, JEFFERSONTOWN, 69- Mem: AAAS; Am Chem Soc; Am Oil Chem Soc; Fedn Socs Paint Technol. Res: Exploratory research in the field of high polymers and protective coatings. Mailing Add: Celanese Coatings Co 9800 Bluegrass Pkwy Jeffersontown KY 40299

MASTERS, WILLIAM HOWELL, b Cleveland, Ohio, Dec 27, 15; m 71; c 2. OBSTETRICS & GYNECOLOGY. Educ: Hamilton Col, BS, 38, ScD, 73; Univ Rochester, MD, 43; Am Bd Obstet & Gynec, dipl, 51. Prof Exp: Intern path, 44, asst obstet & gynec, 44-47, from instr to assoc prof, 47-63, assoc prof clin obstet & gynec, 64-69, PROF CLIN OBSTET & GYNEC, SCH MED, WASH UNIV, 69-, DIR CYTOL SERV, 67-; DIR REPROD BIOL RES FOUND, 64- Concurrent Pos: Intern, Barnes Hosp, 43-45, asst resident, 44, resident, 46-47, assoc obstetrician & gynecologist, 47-; intern, St Louis Maternity Hosp, 43, asst resident, 44, resident, 45-46, assoc obstetrician & gynecologist, 47; dir div reproductive biol, Sch Med, Wash Univ, 60-63, assoc obstetrician & gynecologist, Univ Clins; assoc gynecologist, St Louis Children's Hosp; consult gynecologist, St Louis City Infirmary & Salem Mem Hosp, Ill; dir, Family & Children's Serv & Health & Welfare Coun. Honors & Awards: Paul H Hoch Award, Am Psychopath Asn, 71; Sex Info & Educ Coun US award, 72. Mem: AAAS; Am Fertil Soc; Endocrine Soc; NY Acad Sci; Am Geriat Soc. Res: Infertility and sterility; geriatric endocrinology; sexual inadequacy. Mailing Add: 4910 Forest Park Blvd St Louis MO 63108

MASTERSON, JAMES EDWARD, organic chemistry, see 12th edition

MASTERSON, JOHN G, b Brooklyn, NY, July 6, 20; m 45; c 3. OBSTETRICS & GYNECOLOGY. Educ: Georgetown Univ, BS, 41; Columbia Univ, MD, 44; Am Bd Obstet & Gynec, dipl, 53. Prof Exp: From instr to assoc prof obstet & gynec, State Univ NY Downstate Med Ctr, 51-64; prof & chmn dept, 64-69, actg dean, 68-69, V PRES FOR MED CTR, STRITCH SCH MED, LOYOLA UNIV CHICAGO, 69- Concurrent Pos: Fel gynec, Mem Hosp, New York, NY, 51-52; dir, Gynec Tumor Serv, King's County Hosp, 51-64; consult, Brooklyn Vet Hosp, 53-64 & St Alban's Naval, Maimonides, Mercy, Methodist, Meadowbrook, Glen Cove & St Joseph's Hosps, 55-; dir, Gynec Tumor Serv, St Johns Hosp, 58-64; chmn adj chemother study, NIH, 58-64; mem exec comt, Clin Panel, Nat Chemother Serv Ctr, 60-64; consult, Bur State Serv, USPHS, 60-; dir, Gynec Tumor Serv, Loyola Univ Hosp, 64-; attend, Cook County Hosp, 64-, co-dir, Gynec Tumor Serv, 65-; consult, Hines Vet Admin Hosp, 65-; examr, Am Bd Obstet & Gynec, 65-; mem bd dir, Ill Regional Med Prog, 68-70 & Human Life Found, DC, 69-; consult, Great Lakes Naval Hosp, 69-; chmn bd dir, Loretto Hosp, Chicago, Ill, 70- Mem: Am Col Surg; Am Col Obstet & Gynec; Am Fertil Soc; Soc Gynec Invest; Am Soc Cytol. Res: Cancer of the female genital tract; cancer chemotherapy; endocrinology. Mailing Add: Loyola Univ Med Ctr 2160 S First Ave Maywood IL 60153

MASTERSON, KLEBER S, JR, theoretical physics, systems analysis, see 12th edition

MASTERTON, WILLIAM LEWIS, b Conway, NH, July 24, 27; m 53; c 2. PHYSICAL CHEMISTRY. Educ: Univ NH, BS, 49, MS, 50; Univ Ill, PhD(chem), 53. Prof Exp: Instr chem, Univ Ill, 53-55; from instr to assoc prof, 55-66, PROF CHEM, UNIV CONN, 66- Mem: Am Chem Soc; Sigma Xi; Am Assoc Univ Prof. Res: Thermodynamics of solutions; activity coefficients of electrolytes; solubility of gases in salt solutions. Mailing Add: Dept of Chem Univ of Conn Storrs CT 06268

MASTIN, CHARLES WAYNE, b Salinas, Calif, Apr 23, 43; m 71; c 2. MATHEMATICAL ANALYSIS. Educ: Austin Peay State Col, BS, 64; Miami Univ, MS, 66; Tex Christian Univ, PhD(math), 60. Prof Exp: Asst prof math, Miss State Univ, 69-75; vis scientist, Inst Comput Appln Sci & Eng, NASA Langley Res Ctr, 75-76; ASSOC PROF MATH, MISS STATE UNIV, 76- Mem: Am Math Soc; Soc Indust & Appl Math; Asn Comput Mach. Res: Practical application of transformation methods to the solution of fluid dynamics problems. Mailing Add: Drawer MA Mississippi State MS 39762

MASTIN, STEPHEN HOWARD, inorganic chemistry, photochemistry, see 12th edition

MASTRANGELO, SEBASTIAN VITO ROCCO, b New York, NY, July 1, 25; m 49; c 2. PHYSICAL CHEMISTRY. Educ: Queens Col, NY, BS, 47; Pa State Univ, MS, 48, PhD(chem), 51. Prof Exp: Chemist, Barrett Div, Allied Chem Corp, 51-52; phys chemist, Dextran Corp, 52-53, dept supvr, 53-56; res chemist, Jackson Lab, E I du Pont de Nemours & Co, Inc, 56-62, res assoc, Exp Sta, 62-64, res supvr, 64-69, RES FEL, EXP STA, E I DU PONT DE NEMOURS & CO, INC, 69- Mem: Am Chem Soc; Am Phys Soc; Am Inst Chem. Res: Low temperature purification; third law thermodynamics; adsorption thermodynamics; adsorption thermodynamics at liquid helium temperatures; high temperature adiabatic calorimetry; molecular weight distribution of high polymers; raman and infrared spectroscopy; free radical chemistry; elctrochemistry. Mailing Add: 8 Yorkridge Trail Hockessin DE 19707

MASTROIANNI, LUIGI, JR, b New Haven, Conn, Nov 8, 25; m 57; c c 3. OBSTETRICS & GYNECOLOGY. Educ: Yale Univ, AB, 46; Boston Univ, MD, 50; Am Bd Obstet & Gynec, dipl, 59. Prof Exp: From instr to asst prof obstet & gynec, Sch Med, Yale Univ, 55-61; prof, Univ Calif, Los Angeles, 61-65; PROF OBSTET & GYNEC & CHMN DEPT, UNIV PA, 65- Concurrent Pos: Ed, J Fertil & Steril; res fel infertility & endocrinol, Harvard Med Sch, 54-55. Mem: Endocrine Soc; Am Fertil Soc (pres, 76-77); Am Gynec Soc; fel Am Col Obstet & Gynec; fel Am Col Surg. Res: Human infertility; reproductive physiology. Mailing Add: Hosp of the Univ of Pa 3400 Spruce St Philadelphia PA 19104

MASTROMARINO, ANTHONY JOHN, b Brooklyn, NY, June 13, 40; m 73; c 1. MEDICAL MICROBIOLOGY, CANCER. Educ: Iona Col, BS, 61; Syracuse Univ, MS, 71; Baylor Col Med, PhD(exp biol, microbiol), 75. Prof Exp: Res assoc microbiol, Naylor Dana Inst Dis Prev, Am Health Found, 74-76; ASST DIR SCI OPERS, NAT LARGE BOWEL CONCER PROJ, UNIV TEX SYST CANCER CTR, M D ANDERSON HOSP & TUMOR INST, 76- Mem: Am Soc Microbiol; Asn Gnotobiotics; AAAS. Res: Analysis of bacterial enzyme systems potentially useful in prognosis of risk for colon carcinogenesis; interaction of diet, neutral and acid sterols, and intestinal anaerobes in metabolic epidemiology of colon cancer. Mailing Add: Univ of Tex Syst Cancer Ctr M D Anderson Hosp & Tumor Inst Prudential 1801 6723 Bertner Ave Houston TX 77030

MASTROMATTEO, ERNEST, b Toronto, Ont, Dec 16, 23; m 49; c 7. MEDICINE. Educ: Univ Toronto, MD, 47, dipl pub health, 50, dipl indust health, 58; Am Bd Prev Med, dipl & cert occup med, 58. Prof Exp: Jr intern med, St Michael's Hosp, Toronto, 47-48; sr intern, Ottawa Gen Hosp, 48-49; med dir pub health, Govt of Man, 49-52; physician, Govt of Ont Health Dept, 52-66; chief, Occup Health Serv, Ont, 66-68; DIR ENVIRON HEALTH, ONT HEALTH DEPT, 68-; ASSOC PROF OCCUP HEALTH, SCH HYG, FAC MED, UNIV TORONTO, 58- Concurrent Pos: Consult, Workmen's Compensation Bd, 55-; trustee, Am Bd Prev Med, 68- Honors & Awards: Can Centennial Medal, 67. Mem: Am Col Prev Med; Am Conf Govt Indust Hygienists; Am Indust Hyg Asn; Can Pub Health Asn; Can Med Asn. Res: Occupational, environmental and public health. Mailing Add: 19 Carey Rd Toronto ON Can

MASUDA, MINORU, b Seattle, Wash, Apr 10, 15; m 39; c 2. PSYCHOPHYSIOLOGY. Educ: Univ Wash, BS, 36, MS, 38, PhD(physiol), 56. Prof Exp: From res instr to res asst prof, 56-69, assoc prof, 69-72, PROF PSYCHIAT, UNIV WASH, 72- Mem: Am Psychosom Soc; Soc Psychophysiol Res; NY Acad Sci; Int Col Psychosom Med. Res: Biochemistry and physiology of mental health and illness; psychophysiology of illness; psychosocial factors in illness; ethnic identification: biochemistry of aging with special emphasis on learning. Mailing Add: Dept of Psychiat Univ of Wash Seattle WA 98195

MASUELLI, FRANK JOHN, b Masio, Italy, Apr 16, 21; nat US; m 63; c 2. ORGANIC CHEMISTRY. Educ: Manhattan Col, BS, 42; Va Polytech Inst, MS, 48, PhD(chem), 53. Prof Exp: Instr chem, Va Polytech Inst, 46-53; chemist, 53-55, supvr chem, 55-60, CHIEF PROPELLANTS RES BR, PICATINNY ARSENAL, 60- Mem: Am Chem Soc; Tech Asn Pulp & Paper Indust; Am Inst Chemists. Res: Nitrocellulose chemistry; artillery and rocket propellants. Mailing Add: Propellants Res Lab Picatinny Arsenal Dover NH 07801

MASUI, YOSHIO, b Kyoto, Japan, Oct 6, 31; m 59; c 2. DEVELOPMENTAL BIOLOGY. Educ: Kyoto Univ, BSc, 53, MS, 55, PhD(zool), 61. Prof Exp: Lectr biol, Konan Univ, Japan, 58-65, asst prof, 65-68; lectr, Yale Univ, 69; ASSOC PROF ZOOL, UNIV TORONTO, 69- Mem: Int Soc Develop Biol; Japanese Soc Develop Biol. Res: Developmental biology relating to nucleocytoplasmic interactions in early development and gametogenesis. Mailing Add: Dept of Zool Univ of Toronto Toronto ON Can

MASUOKA, DAVID TAKASHI, b Los Angeles, Calif, Oct 13, 21; m 44; c 2. PHARMACOLOGY. Educ: Univ Southern Calif, BS, 48, PhD(pharmacol), 51. Prof Exp: Asst pharmacol, Univ Southern Calif, 49-51, res assoc, 51-52; asst res pharmacologist, Sch Med, Univ Calif, Los Angeles, 53-60; pharmacologist, 60-69, CHIEF NEUROPHARMACOL RES, VET ADMIN HOSP, 69- Concurrent Pos: Giannini Found fel, Univ Southern Calif, 52-53; Commonwealth Fund fel, Stockholm, 64-65. Mem: Am Soc Pharmacol & Exp Therapeut; Am Soc Neurochem. Res: Central and peripheral monoamines; amphetamine. Mailing Add: Neuropharmacol Res Lab Vet Admin Hosp Sepulveda CA 91343

MASUREKAR, PRAKASH SHARATCHANDRA, b Bombay, India, Jan 23, 41; m 68; c 1. INDUSTRIAL MICROBIOLOGY, BIOCHEMICAL ENGINEERING. Educ: Univ Bombay, BSc, Hons, 62, BSc, 64, MSc, 66; Mass Inst Technol, SM, 68, PhD(biochem eng), 73. Prof Exp: SR RES CHEMIST BIOCHEM ENG, EASTMAN KODAK CO, 73- Res: Microbiology and engineering of industrial fermentations; microbial physiology and genetics; biological conversions and enzyme technology. Mailing Add: Res Labs Eastman Kodak Co Rochester NY 14650

MASURSKY, HAROLD, b Ft Wayne, Ind, Dec 23, 22; m 52; c 4. ASTROGEOLOGY. Educ: Yale Univ, BS, 43, MS, 51. Prof Exp: Geologist, US Geol Surv, 51-67, chief br astrogeol studies, 67-71, CHIEF SCIENTIST, CTR ASTROGEOL, US GEOL SURV, 71- Concurrent Pos: Team leader & prin investr, TV exp, Mariner Mars, 71; co-investr, Appolo Field Geol Team, Apollo 16 & 17, mem Apollo Orbital Sci Photog Team, Apollo Site Selection Group, leader, Viking Landing Site Staff, dep team leader, Orbiter Visual Imaging System, Viking Mars 75; comt space res rep Inter-Union Comn for Studies of the Moon; Int Astron Union Task Group on Nomenclature; Moon, Mars, Venus del, USA-USSR planetary data exchange; radar team mem & chmn, Surface & Interiors Group, Venus Pioneer, 78. Honors & Awards: Medal for Except Sci Achievement, NASA, 72 & 73. Mem: Geol Soc Am; Am Geophys Union; AAAS; Geochem Soc; Meteoritical Soc. Res: Geology of Owl Creek Mountains, Wyoming; uranium bearing coal in the Red Desert, Wyoming; structure stratigraphy and volcanic rocks in central Nevada; stratigraphy and structure of the moon; geology of Mars; crustal formation, eolian deposits, volcanic history, channel formation; geology of Mercury;impact history. Mailing Add: US Geol Surv 601 E Cedar Flagstaff AZ 86001

MATA, LEONARDO J, b Dota, Costa Rica, Dec 6, 33; m 56; c 4. PUBLIC HEALTH, NUTRITION. Educ: Univ Costa Rica, BS, 56; Univ PR, dipl, 58; Harvard Univ, MS, 60, DSc(trop pub health), 62. Prof Exp: Chief bact & parasitol, San Juan de Dios Hosp, Costa Rica, 56-59; lab instr microbiol, Sch Nursing, Univ Costa Rica, 57-58; chief enteric bact, Inst Nutrit Cent Am & Panama, 59; lab instr parasitol, Harvard Med Sch, 61-62; chief microbiol, Inst Nutrit Cent Am & Panama, Guatemala, 62-75 & prof microbiol, Sch Nutrit, 66-75; PROF NUTRIT, FAC MED, UNIV COSTA RICA, 75- Concurrent Pos: Nat Inst Allergy & Infectious Dis grant, Inst Nutrit Cent Am & Panama, Guatemala, 62-71; mem, Pan Am Health Orgn-WHO-Inst Nutrit Cent Am & Panama internal coun, 65-74; mem, US-Japan coop proj, NIH grant, Guatemala, 68-71; US Armed Forces Res & Develop Command grant, Cent Am, 68-71; vis prof, San Carlos Univ Guatemala, 65-70 & sch med, Univ El Salvador, 70-; mem informal study group, WHO, 71-; dir, Inst Res Health, 75- Mem: AAAS; Am Soc Trop Med & Hyg; Latin Am Nutrit Soc; NY Acad Sci; Am Soc Microbiol. Res: Tissue culture and virology; enteric bacteriology; nutrition and human growth; tropical public health; research on public health interventions. Mailing Add: Univ of Costa Rica San Pedro Costa Rica

MATACIC, SLAVICA SMIT, b Lipik, Yugoslavia, Oct 12, 33; m 64; c 2. BIOCHEMISTRY. Educ: Univ Zagreb, Yugoslavia, BS, 59, PhD(biochem), 62. Prof Exp: Lab instr chem, Sch Pharm, Univ Zagreb, 56-57; res asst radiobiol, R Boskovic Inst, Zagreb, 57-59, fel, 62-64; RES ASSOC & LECTR BIOL, HAVERFORD COL, 64- Mem: AAAS; Am Soc Microbiol. Res: Distribution and function of isopeptide bonds in proteins from different sources. Mailing Add: Dept of Biol Haverford Col Haverford PA 19041

MATALON, REUBEN, b Bagdad, Iraq, Aug 10, 35; m 58; c 3. PEDIATRICS, BIOCHEMICAL GENETICS. Educ: Hebrew Univ, Jerusalem, MD, 59. Prof Exp: Intern med, Kaplan Hosp, Rehovoth, Israel, 58-59; resident, 62-64, instr, 67-69, asst prof, 69-75, ASSOC PROF PEDIAT, UNIV CHICAGO, 75- Concurrent Pos: La Rabida fel & grant, La Rabida Inst, Univ Chicago, 64-68, Jr Philanthropic Soc grant, 64- Res: Biochemical and genetic aspects of mucopolysaccharide and lipid storage diseases; enzymatic aspects of such diseases investigated at the cellular level. Mailing Add: Dept of Pediat Univ of Chicago Med Sch Chicago IL 60637

MATANOSKI, GENEVIEVE M b Salem, Mass, Aug 26, 30; m 56; c 5. PEDIATRICS, EPIDEMIOLOGY. Educ: Radcliffe Col, AB, 51; Johns Hopkins Univ, MD, 55, MPH, 62, DPH, 64. Prof Exp: Intern & asst resident pediat, Johns Hopkins Hosp, 55-57, res assoc epidemiol, 57-60, from instr to asst prof, 60-69, ASSOC PROF EPIDEMIOL, JOHNS HOPKINS UNIV, 69- Concurrent Pos: Assoc prof, Schs Med & Dent, Univ Md; NIH grants, 65-66 & 70- Mem: AAAS; NY Acad Sci; Soc Epidemiol Res; Int Epidemiol Asn; AMA. Res: Epidemiology of streptococcal infections; etiology of rheumatic fever; risks of occupational exposure to radiation; etiology of oral cancer; evaluation of medical care programs. Mailing Add: Sch of Hyg & Pub Health Johns Hopkins Univ Baltimore MD 21205

MATCHA, ROBERT LOUIS, b Omaha, Nebr, Oct 22, 38; m 60; c 3. THEORETICAL CHEMISTRY. Educ: Univ Omaha, BA, 60; Univ Wis-Madison, PhD(theoret chem), 65. Prof Exp: Fel, IBM Corp, Calif, 65-66 & Battelle Inst, 66-67; asst prof, 67-72, ASSOC PROF THEORET CHEM, UNIV HOUSTON, 72- Concurrent Pos: Consult, Battelle Inst, 70-74. Mem: AAAS; Am Phys Soc; Am Chem Soc. Res: Theoretical study of electromagnetic interactions; effects of nuclear motion on expectation values, molecular Hartree Fock calculations. Mailing Add: Dept of Chem Univ of Houston Cullen Blvd Houston TX 77004

MATCHES, ARTHUR GERALD, b Portland, Ore, Jan 28, 29; m 52; c 3. AGRONOMY. Educ: Ore State Univ, BS, 52, MS, 54; Purdue Univ, PhD(crop physiol & ecol), 60. Prof Exp: Asst farm crops, Ore State Univ, 52-54; instr agron, Purdue Univ, 56-60; asst prof, Southeastern Substa, NMex State Univ, 60-61; RES AGRONOMIST, AGR RES SERV, USDA, 61-; PROF AGRON, UNIV MO-COLUMBIA, 61- Concurrent Pos: Mem coun agr sci & technol, Am Soc Animal Sci. Honors & Awards: Merit Cert, Am Forage & Grassland Coun, 72. Mem: Am Forage & Grassland Coun (vpres, 65); Soc Range Mgt; Am Soc Agron; Crop Sci Soc Am; Am Soc Animal Sci. Res: Pasture systems for nearly year long grazing; use of multiple assignment tester animals in grazing trials; pasture research methods. Mailing Add: Dept of Agron Univ of Mo 210 Waters Hall Columbia MO 65201

MATCHES, JACK RONALD, b Portland, Ore, May 20, 30; m 54; c 2. FOOD SCIENCE, MICROBIOLOGY. Educ: Ore State Univ, BS, 57, MS, 58; Iowa State Univ, PhD(microbiol), 63. Prof Exp: Res assoc microbiol & food sci, Iowa State Univ, 58-63; sr microbiologist, 63-65, asst prof microbiol & food sci, 65-68, ASSOC PROF MICROBIOL & FOOD SCI, COL FISHERIES, UNIV WASH, 68- Mem: Inst Food Technologists; Am Soc Microbiol. Res: Food microbiology; low temperature microbiology; anaerobic microbiology. Mailing Add: Inst for Food Sci & Technol Univ of Wash Col of Fisheries Seattle WA 98195

MATCHETT, WILLIAM H, b Pinehurst, NC, Jan 4, 32; m 54; c 2. MICROBIAL PHYSIOLOGY. Educ: Univ Ill, BS, 53, MS, 58, PhD(plant physiol), 60. Prof Exp: Asst bot, Univ Ill, 57-60; NSF fel microbiol, Sch Med, Yale Univ, 60-61; res fel, Univ Calif, San Diego, 61-63; res scientist biol, Hanford Labs, Gen Elec Co, 63-65; mgr cell biol, Pac Northwest Lab, Battelle Mem Inst, 65-69, coordr life sci, Battelle Seattle Res Ctr, 69-71; assoc dean grad sch, 71-74, CHMN DEPT BOT, WASH STATE UNIV, 71- Concurrent Pos: Vis lectr, Wash State Univ, 65, adj assoc prof chem, 67-71. Mem: AAAS; Am Soc Microbiol; Am Soc Biol Chemists. Res: Biochemical genetics of Neurospora crassa; metabolism of tryptophan in Neurospora; enzymology of tryptophan biosynthetic enzymes. Mailing Add: Dept of Bot Wash State Univ Pullman WA 99163

MATEER, FRANK MARION, b Pittsburgh, Pa, June 21, 21; m 44; c 3. MEDICINE. Educ: Univ Pittsburgh, BS, 41, MD, 44; Am Bd Internal Med, dipl, 52. Prof Exp: From instr to asst prof res med, Univ Pittsburgh, 50-62; dir, 65-70, MEM SR STAFF & CHIEF DIV MED EDUC & RES, WESTERN PA HOSP, 70- Concurrent Pos: Am Heart Asn estab investr, Univ Pittsburgh, 54-59; clin asst prof med, Univ Pittsburgh, 62-72; sr teaching fel physiol, Univ Pittsburgh, 47-48, res fel med, 48-50, Am Heart Asn res fel, 52-54. Mem: AAAS; Am Diabetes Asn; Am Fedn Clin Res; fel Am Col Physicians; Am Heart Asn. Res: Renal disease; endocrinology. Mailing Add: Western Pa Hosp 4800 Friendship Ave Pittsburgh PA 15224

MATEER, RICHARD AUSTIN, b Ashland, Ky, July 30, 40; m 62; c 2. PHYSICAL ORGANIC CHEMISTRY. Educ: Centre Col, BA, 62; Tulane Univ, PhD, 66. Prof Exp: ASSOC PROF CHEM, UNIV RICHMOND, 66-, DEAN, 75- Mem: Am Chem Soc; Sigma Xi. Res: Synthetic photochemistry; organic reaction mechanisms; nuclear magnetic resonance and infrared spectroscopy; organometallics. Mailing Add: Ryland Hall Room 307 Univ of Richmond Richmond VA 23173

MATEKER, EMIL JOSEPH, JR, b St Louis, Mo, Apr 25, 31; m 54; c 3. GEOPHYSICS. Educ: St Louis Univ, BS, 56, MS, 59, PhD(geophys), 64. Prof Exp: Geophysicist, Standard Oil Co Calif, 58-60; instr geophys, St Louis Univ, 60-63; from asst prof to assoc prof, Wash Univ, 63-69; mgr geophys res, Western Geophys Co, 69-70, VPRES RES & DEVELOP, WESTERN GEOPHYS CO, 70-, PRES, AERO SERV DIV & WESTREX DIV, 77- Concurrent Pos: Res assoc, Proj Vela Uniform, Dept Defense, 60-63 & Pan Am Petrol Corp, 64-69. Mem: AAAS; Am Geophys Union; Seismol Soc Am; Soc Explor Geophys; Europ Asn Explor Geophys. Res: Solid earth geophysics; exploration geophysics; seismic energy sources; tectonics; lithology from seismic reflections; satellite navigation; digital seismic data processing. Mailing Add: Res & Develop Western Geophys Co PO Box 2469 Houston TX 77001

MATEO, JOSE, geophysics, civil engineering, see 12th edition

MATESE, JOHN J, b Chicago, Ill, May 1, 38; m 66; c 3. THEORETICAL PHYSICS. Educ: DePaul Univ, BS, 60; Univ Notre Dame, PhD(physics), 66. Prof Exp: Lectr physics, Univ Notre Dame, 65-66; asst prof, La State Univ, Baton Rouge, 66-74; ASST PROF PHYSICS, UNIV SOUTHWESTERN LA, 74- Mem: Am Phys Soc. Res: Atomic physics. Mailing Add: Dept of Physics Univ of Southwestern La Lafayette LA 70501

MATESICH, MARY ANDREW, b Zanesville, Ohio, May 5, 39. PHYSICAL CHEMISTRY. Educ: Ohio Dominican Col, BA, 62; Univ Calif, Berkeley, MS, 63, PhD(chem), 66. Prof Exp: Asst prof chem, 65-70, ASSOC PROF CHEM, OHIO DOMINICAN COL, 70-, CHMN DEPT, 65- Concurrent Pos: Petrol Res Fund grant, Ohio Dominican Col, 65-68; NSF grant, Case Western Reserve Univ & Ohio Dominican Col, 69-72. Mem: AAAS; Am Chem Soc. Res: Ion pumps; ion transport in membranes; models for active transport; transport processes in solution; solution thermodynamics. Mailing Add: Dept of Chem Ohio Dominican Col Columbus OH 43219

MATHAI, ARAKAPARAMPIL M, b Palai, India, Apr 28, 35; m 64; c 2. MATHEMATICAL STATISTICS. Educ: Univ Kerala, BSc, 57, MSc, 59; Univ Toronto, MA, 62, PhD(math statist), 64. Prof Exp: Lectr math, St Thomas Col, Univ Kerala, 59-61; Commonwealth scholar statist, Univ Toronto, 61-64; asst prof, 64-68, ASSOC PROF MATH, McGILL UNIV, 68- Concurrent Pos: Ed, Can J Statist, 74- Honors & Awards: Gold Medal, Univ Kerala, 59. Mem: Exec mem Statist Sci Asn Can (secy, 72-74); Inst Math Statist; Can Math Cong; Am Math Soc. Res: Statistical distributions; multivariate analysis; axiomatic foundations of statistical concepts; special functions and complex analysis; functional equations. Mailing Add: Dept of Math McGill Univ PO Box 6070 Montreal PQ Can

MATHAY, WILLIAM LEWIS, b Greenville, Pa, Dec 1, 24; m 48; c 1. APPLIED CHEMISTRY, CORROSION. Educ: Thiel Col, BS, 47. Prof Exp: Res chemist, Calgon Inc, 47-54; head, Chem & Process Corrosion Sect, Appl Res Lab, US Steel Corp, 54-58, res engr, 58-64, MGR PROCESS INDUST MKT, US STEEL CORP, 64- Mem: Nat Asn Corrosion Eng; fel Am Inst Chem; Am Inst Chem Eng; Tech Asn Pulp & Paper Indust. Res: Steels for process industries. Mailing Add: US Steel Corp 600 Grant St Pittsburgh PA 15230

MATHE, CLARENCE EUGENE, JR, b Carlstadt, NJ, July 16, 15; m 40; c 3. CHEMISTRY. Educ: Ohio Univ, BS, 39. Prof Exp: Develop chemist, Nat Oil Prod Co, NJ, 39-42; plant mgr, Metal Organics, Inc, 46-47; plant mgr, Metallic Stearates Div, Witco Chem Co, NY, 47-50; pres, Mathe Chem Co, NJ, 50-65; GEN MGR, BERKELEY CHEM DEPT, MILLMASTER CHEM CO, 65- Concurrent Pos: Admin facil, Bergen Community Col, 72. Mem: Am Chem Soc. Res: Derivatives of fatty acids including esters, amides, amines and metallic soaps; lubricant additives; plasticizers and stabilizers. Mailing Add: 567 Windsor Rd Woodridge NJ 07075

MATHENY, JAMES LAFAYETTE, b Vicksburg, Miss, Aug 28, 43; m 65. PHARMACOLOGY. Educ: Delta State Col, BS, 67; Univ Miss, PhD(pharmacol), 71. Prof Exp: ASST PROF PHARMACOL, MED COL GA, 71- Concurrent Pos: NIH, Ga Heart Assoc, Fight for Sight Inc grant, 71-72. Mem: AAAS; NY Acad Sci; fel Am Col Clin Pharmacol; Am Heart Asn; Am Soc Pharmacol & Exp Therapeut. Res: Autonomic-cardiovascular pharmacology. Mailing Add: Dept of Pharmacol Med Col of Ga Augusta GA 30902

MATHENY, RAY T, anthropology, archaeology, see 12th edition

MATHER, ADALINE NICOLES, b Joliet, Ill, Aug 10, 19. BIOCHEMISTRY. Educ: Univ Chicago, BS, 40; Univ Wis, PhD(bact, biochem), 49. Prof Exp: Asst, Univ Chicago, 42-44; bacteriologist, Hiram Walter & Sons, 44-46; asst, Univ Wis, 46-49; asst prof bact, Southern Ill Univ, 49-51 & Univ SDak, 51-56; resident res assoc, Argonne Nat Lab, 56-58; SR CHEMIST, BAXTER LABS, INC, 58- Mem: Am Chem Soc; Am Soc Microbiol. Res: Nutrition and metabolism of bacteria; hydrolytic enzymes; chemistry of thyroxine; matricized enzymes; clinical application of enzymes. Mailing Add: Pharmaceut Res & Develop Baxter Labs Inc Morton Grove IL 60053

MATHER, ALAN, b Alton, Ill, Oct 31, 10; m 40; c 2. CLINICAL BIOCHEMISTRY. Educ: Shurtleff Col, BS, 32; St Louis Univ, PhD(biochem), 39. Prof Exp: Biochemist, Neuroendocrine Found, Worcester, Mass, 39-42; res assoc, Clark Univ, 42-45; asst prof biochem, Dartmouth Med Sch, 45-55; biochemist, Mem Hosp, Wilmington, Del,

55-65; chief coronary drug proj lab, Commun Dis Ctr, USPHS, 65-67, chief clin chem sect, Ctr Dis Control, 67-73, ASSOC DIR CLIN CHEM, CTR DIS CONTROL, USPHS, 73- Mem: AAAS; Am Asn Clin Chem. Res: Analytical methodology and instrumentation in clinical chemistry. Mailing Add: Ctr for Dis Control USPHS Atlanta GA 30333

MATHER, EDWARD CHANTRY, b Iowa City, Iowa, Apr 7, 37; m 58; c 2. VETERINARY MEDICINE. Educ: Iowa State Univ, DVM, 60, Univ Mo, MS, 68, PhD(reprod physiol), 70. Prof Exp: Pvt vet pract, 60-66; instr vet med, Univ Mo-Columbia, 66-68, res assoc, 68-70, from asst prof to assoc prof, 70-74, dir theriogenology lab, 68-73; ASSOC PROF VET MED, UNIV MINN, ST PAUL, 74-, HEAD DIV THERIOGENOLOGY, 74- Concurrent Pos: Adv prog appl res on fertil regulation, AID, 74-78. Mem: Soc Study Reproduction; Am Vet Soc Study Breeding Soundness. Res: Effect of seminal constituents on endometrial metabolism; endocrinological variations in large mammals as affected by reproductive pathology. Mailing Add: Univ Vet Hosps Univ of Minn St Paul MN 55101

MATHER, EUGENE COTTON, b West Branch, Iowa, Jan 3, 18; m 44; c 2. CULTURAL GEOGRAPHY. Educ: Univ Ill, AB, 40, MS, 41; Univ Wis, PhD(geog), 51. Prof Exp: Map ed, Army Map Serv, 41-42; res analyst geog, Off Strategic Serv, 42-45; instr geog, Univ Wis, 46-47; assoc prof, Univ Ga, 47-56; PROF GEOG, UNIV MINN, MINNEAPOLIS, 57- Concurrent Pos: Vis prof, Univ BC, 56-57 & 66-67, Univ Toronto, 68, Fla State Univ, 69, Univ Fla, 70 & Univ Ky, 72-73; Asn Am Geogr fel, 60; Ford Found fel, Univ Minn, Chile, 65-66 & Univ Minn, Nicaragua, 67; adj prof geog, Univ Ky, 74- Mem: Can Asn Geog; Asn Am Geog. Res: Great Plains ranching and culture; ethnic patterns and settlement forms in Wisconsin; cultural geography of Kentucky. Mailing Add: Pierce Co Geog Soc Prescott WI 54021

MATHER, FRANK JEWETT, III, b New York, NY, Mar 16, 11; m 70; c 1. ICHTHYOLOGY. Educ: Williams Col, BA, 33; Mass Inst Technol, BS, 37. Prof Exp: Planner, United & Bethlehem Shipyards, 37-38; instr naval archit, Mass Inst Technol, 39-41; instr, Hull Sci Dept, Gibbs & Cox, Inc, 41-45; res assoc, 45-63, ASSOC SCIENTIST, WOODS HOLE OCEANOG INST, 63- Concurrent Pos: Trustee, New Eng Aquarium, 61-; mem expert panel facilitation of tuna res, Food & Agr Orgn, UN, 62-, convenor, Working Party Tuna & Gillfish Tagging, 68-71; mem subcomt stock identification, Standing Comt Res & Statist, Int Comn Conserv Atlantic Tunas, 71-, sci adv to US Deleg, 71. Mem: Am Soc Limnol & Oceanog; Am Soc Ichthyologists & Herpetologists; Soc Naval Archit & Marine Eng; Am Fisheries Soc. Res: Biology of the larger pelagic fishes. Mailing Add: Dept of Biol Box 111 Woods Hole Oceanog Inst Woods Hole MA 02543

MATHER, GEORGE ROBINSON, JR, solid state physics, see 12th edition

MATHER, GEORGE WELLS, b Sioux Rapids, Iowa, Aug 9, 11; m 38; c 2. VETERINARY MEDICINE. Educ: Iowa State Col, DVM, 35; Univ Minn, PhD, 51. Prof Exp: Chief vet, Animal Rescue League, Boston, Mass, 35-48; from instr to assoc prof vet med, 48-54, PROF VET MED, COL VET MED, UNIV MINN, ST PAUL, 54- Res: Pulmonary temperatures; veterinary ophthalmology; renal function; leptospirosis; canine geriatrics and internal medicine. Mailing Add: Univ of Minn Col of Vet Med St Paul MN 55101

MATHER, JANE H, b Green Bay, Wis, July 16, 22. BIOCHEMISTRY. Educ: Univ Wis, BA, 44; Univ Chicago, PhD(biochem), 63. Prof Exp: Chem analyst, Western Elec Co, Ill, 44-45; res chemist, Armour Res Labs, Ill, 45-53; electron microscopist, Northwestern, 54-56, res asst, Univ Chicago, 56-62; asst prof biochem, Ill Inst Technol, 62-65; ASSOC PROF CHEM, GA STATE UNIV, 65- Concurrent Pos: Consult, Armour Res Labs, 53-57; USPHS res grants, Ill Inst Technol & Ga State Univ, 65-68. Mem: AAAS; Am Chem Soc. Res: Biochemical intermediary metabolism and enzymology; analytical chemistry. Mailing Add: Dept of Chem Ga State Univ Atlanta GA 30303

MATHER, JOHN RUSSELL, b Boston, Mass, Oct 9, 23; m 46; c 3. CLIMATOLOGY. Educ: Williams Col, BA, 46; Mass Inst Technol, BS, 47, MS, 48; Johns Hopkins Univ, PhD(climat), 51. Prof Exp: Instr, McCoy Col, 49-51; asst prof, Johns Hopkins Univ, 51-53; assoc prof, Drexel Tech, 57-60; PROF GEOG, UNIV DEL, 61- Concurrent Pos: Res assoc climatologist, Lab Climat, 48-55, prin res scientist, 55-63; pres, C W Thornthwaite Assocs, 63-72; consult, World Meteorol Orgn, Yugoslavia, 57; vis lectr, Univ Chicago, 57-61; consult atmosphere & hydrol hazards from nuclear reactors, US, Iran, Italy & Israel. Mem: Am Meteorol Soc; Am Geog Soc; Am Geophys Union; Asn Am Geog. Res: Water balance; evaporation; transpiration; applied climatology. Mailing Add: Dept of Geog Univ of Del Newark DE 19711

MATHER, JOSEPH WALTER, b Hartford, Conn, Mar 31, 20; m 41; c 3. PHYSICS. Educ: Rensselaer Polytech Inst, BS, 48; Univ Calif, PhD(physics), 52. Prof Exp: Mem staff, Los Alamos Sci Lab, 52-75; RETIRED. Mem: Am Phys Soc. Res: Plasma physics. Mailing Add: 105 El Corto Los Alamos NM 87544

MATHER, KATHARINE KNISKERN, b Ithaca, NY, Oct 21, 16; m 40. GEOLOGY. Educ: Bryn Mawr Col, AB, 37. Prof Exp: Geologist, Cent Concrete Lab, NY, 42-44, engr concrete res, 44-46; GEOLOGIST, CONCRETE DIV, WATERWAYS EXP STA, 46-, CHIEF PETROG & X-RAY BRANCH, 48- Concurrent Pos: Chmn comt basic res cement & concrete, Transp Res Bd, Nat Acad Sci-Nat Res Coun. Honors & Awards: Thompson Award, Am Soc Testing & Mat, 53; Wason Res Medal, Am Concrete Inst, 53; Except Civilian Serv Medal, Secy Army, 62; Fed Woman's Award, 63; Distinguished Civilian Serv Award, Secy Defense, 64. Mem: Mineral Soc Am; Am Ceramic Soc; Am Concrete Inst; Am Inst Mining, Metall & Petrol Eng; Clay Minerals Soc (secy, 64-67, pres, 73). Res: Constitution and microstructure of concrete, its constituents and alteration products; effects of variation in composition and exposure on properties of concrete. Mailing Add: PO Box 631 Vicksburg MS 39180

MATHER, KEITH BENSON, b Adelaide, S Australia, Jan 6, 22; m 46; c 2. GEOPHYSICS, NUCLEAR PHYSICS. Educ: Univ Adelaide, BSc, 42, MSc, 44. Hon Degrees: DSc, Univ Alaska, 68. Prof Exp: Demonstr physics, Univ Adelaide, 43-45, lectr, 46; Sci & Indust Endowment Fund stud & asst, Wash Univ, 46-48; Imp Chem Indust fel, Birmingham, 49-50; lectr, Ceylon, 50-51; res officer, Commonwealth Sci & Indust Res Org Australia, 52-54; sr res officer, Australian AEC, 54-56, physicist-in-chg Antarctic Div, 56-58; lectr physics, Melbourne, 58-61; assoc prof geophys, Geophys Inst, Univ Alaska, 61-62, asst dir, 63, PROF PHYSICS & DIR GEOPHYS INST, UNIV ALASKA, 63- Concurrent Pos: Fulbright travel grant, 61-62; mem, Nat Coun Univ Res Adminrs, 64- & Polar Res Bd, Nat Acad Sci-Nat Res Coun. Mem: AAAS; Am Geophys Union; Arctic Inst N Am; Brit Inst Physics; Australian Inst Physics. Res: Optical spectroscopy; nuclear scattering and reaction studies; cosmic radiation; geomagnetism and aurora; katabatic winds; polar geophysics; theory of road corrugation and relaxation oscillations. Mailing Add: Geophys Inst Univ of Alaska Fairbanks AK 99701

MATHER, ROBERT EUGENE, b Goigoi Mission Station, Portuguese E Africa, Nov 12, 18; US citizen; m 43; c 2. ANIMAL BREEDING, STATISTICAL ANALYSIS. Educ: Purdue Univ, BS, 39; Univ Md, MS, 41; Univ Wis, PhD(dairy husb, genetics), 46. Prof Exp: Asst dairy husb, Univ Md, 39-41; from asst dairy husbandman to assoc dairy husbandman, Exp Sta, Va Polytech Inst, 45-48; assoc prof dairy husb, Rutgers Univ, New Brunswick, 48-59, from assoc res specialist to res specialist, Dairy Res Ctr, Exp Sta, 48-70, PROF DAIRY HUSB, RUTGERS UNIV, NEW BRUNSWICK, 59-, ASST TO DIR, AGR EXP STA FOR STATIST & COMPUT CONSULT, 70- Mem: Am Soc Animal Sci; Am Dairy Sci Asn. Res: Dairy cattle genetics; use of statistics and computers in agricultural research. Mailing Add: Dept of Statist & Comput Sci Cook Col Rutgers Univ New Brunswick NJ 08903

MATHER, ROBERT LAURANCE, b Clarksville, Iowa, Oct 1, 21; m 56; c 2. PHYSICS. Educ: Iowa State Univ, BS, 42; Columbia Univ, AM, 47; Univ Calif, PhD(physics), 51. Prof Exp: Physicist, Naval Ord Lab, 42-44, Radio Corp Am, 44-46 & Nevis Cyclotron Lab, Columbia Univ, 46-47; asst physics, Univ Calif, 47-48, physicist, Lawrence Radiation Lab, 48-51; physicist, Atomic Energy Div, NAm Aviation, Inc, 51-52; physicist, Nuclear Radiation Physics Br, US Naval Radiol Defense Lab, 52-69, ENG PHYSICIST, NAVAL ELECTRONICS LAB, 69- Mem: Am Phys Soc; Inst Elec & Electronics Engrs. Res: Radar and communications; electronics; accelerators; Cerenkov radiation; nuclear weapon residual radiation; radiation physics. Mailing Add: 755 Cordova St San Diego CA 92107

MATHER, WILLIAM B, JR, b Peking, China, Feb 6, 36; US citizen; m 61; c 2. ELECTROCHEMISTRY. Educ: Princeton Univ, AB, 57; Calif Inst Technol, PhD(chem), 61. Prof Exp: From chemist to sr chemist, Texaco Res Lab, 61-66, res chemist, 66-71, sr res chemist, 71-73, group leader, 73, SR FINANCIAL ANALYST, TEXACO, INC, 73- Mem: Am Chem Soc. Res: Electrochemical processing, including work on fuel cells; electroorganic chemistry and electroanalytical chemistry; zeolite chemistry; desulfurization research. Mailing Add: Texaco Inc 135 E 42nd St New York NY 10017

MATHER, WILLIAM BARDWELL, b Weston, Ont, Can, Dec 29, 03; nat US; m. ECONOMIC GEOLOGY. Educ: McMaster Univ, BA, 26, MA, 28; Univ Chicago, PhD(geol), 36. Prof Exp: Chem & prod engr, Nat Carbon Co, Can, 28-29; engr, Peoples Light, Gas & Coke Co, Ill, 29-30; chemist, Universal Atlas Cement Co, Ind, 30-31; instr chem micros, Univ Chicago, 33-35; consult geologist, Can & Ohio, 36-40; prin geologist, State Geol Surv, Mo, 44-46; econ geologist, Midwest Res Inst, 46-49; chmn dept mineral technol, Southwest Res Inst, 49-59; CONSULT MINING GEOLOGIST, 59- Concurrent Pos: Mineral adv to Cuban Mission, Int Bank for Reconstruct & Develop, 50 & Joint Brazil-US Cmn for Econ Develop, 51-52. Mem: Soc Econ Geol; Sigma Xi; Geol Soc Am; Geochem Soc. Res: Metallic, non-metallic and solid fuel deposits of North and South America; petrography; chemical microscopy; mineralography. Mailing Add: 1201 Broadmoor Dr Apt 137 Austin TX 78723

MATHERS, ALEXANDER PICKENS, b Matherville, Miss, Sept 17, 09; m 33; c 1. ORGANIC CHEMISTRY. Educ: Univ Fla, BS, 31; Tulane Univ, MS, 46; George Washington Univ, PhD(chem), 56. Prof Exp: Self employed, 31-38; teacher high sch, Miss, 38-41; chemist, 41-55, from asst chief to chief Alcohol & Tobacco Tax Lab, US Treas Dept, 55-73; CONSULT WINE & DISTILLED SPIRITS INDUSTS, 73- Concurrent Pos: Owner & operator, Exp Winery & Vineyard, 74-75. Mem: AAAS; Am Chem Soc; Am Inst Chemist (pres elect, 74-75, pres, 75-76); Asn Official Anal Chem (vpres, 65-66, pres, 66-67). Res: Analytical chemistry; effect of skins, seed and pulp on the fermentation of juice of vitis Rotundifolia. Mailing Add: 1000 Frost Bridge Road Matherville MS 39360

MATHERS, AUBRA CLINTON, b Smithville, WVa, June 28, 23; m 53; c 4. SOIL SCIENCE. Educ: Univ Mo, BS, 52, MS, 53; NC State Univ, PhD(soil chem), 56. Prof Exp: SOIL SCIENTIST, SOUTHWESTERN GREAT PLAINS RES CTR, AGR RES SERV, USDA, 56- Mem: Am Soc Agron. Res: Soil chemistry and fertility; fertility status of soils; iron chlorosis studies and effect of fertilizers on plant nutrition; use or disposal of feedlot wastes to prevent pollution. Mailing Add: Southwestern Great Plains Res Ctr US Dept of Agr Bushland TX 79012

MATHERS, CAROL K, invertebrate zoology, aquatic biology, see 12th edition

MATHES, MARTIN CHARLES, b Amherst, Ohio, Feb 18, 35; m 57; c 3. PLANT PHYSIOLOGY. Educ: Miami Univ, BA, 57; Univ Md, MS, 59, PhD(plant physiol), 61. Prof Exp: Asst bot, Univ Md, 57-61; res aide plant physiol, Inst Paper Chem, Lawrence Univ, 61-64; asst prof, Univ Vt, 64-67; assoc prof, 67-74, PROF BIOL COL WILLIAM & MARY, 74- Mem: AAAS; Am Soc Plant Physiol. Res: Growth regulators; plant tissue cultures. Mailing Add: Dept of Biol Col of William & Mary Williamsburg VA 23185

MATHESON, ALASTAIR TAYLOR, b Vancouver, BC, Can, Oct 10, 29; m 59; c 2. BIOCHEMISTRY, CELL BIOLOGY. Educ: Univ BC, BA, 51, MSc, 53; Univ Toronto, PhD(biochem), 58. Prof Exp: Res fel enzymol, Nat Res Coun Can, 58-59; res assoc biophys, Johns Hopkins Univ, 59-60; from asst res officer to assoc res officer, 60-70, SR RES OFFICER CELL BIOCHEM, NAT RES COUN CAN, 70- Concurrent Pos: Adj prof biol, Carleton Univ, Ottawa, 74- Mem: Am Soc Biol Chem; Can Biochem Soc; Brit Biochem Soc; Can Soc Cell Biol. Res: Relationship between the molecular structure and biological activity of proteins and nucleic acids. Mailing Add: Div of Biol Sci Nat Res Coun of Can Ottawa ON Can

MATHESON, ALISTER FARQUHAR, chemistry, see 12th edition

MATHESON, ARTHUR RALPH, b Kansas City, Mo, Oct 7, 15; m 41. INORGANIC CHEMISTRY. Educ: Univ Ill, BS, 40, MS, 47, PhD(inorg chem), 48. Prof Exp: Anal chemist, Univ Ill, 38-40; control chemist synthetic paints, Cook Paint & Varnish Co, Mo, 40; asst chem, Univ Ill, 46-47, asst, Off Naval Res Contract, 47-48; res chemist, Hanford Works, Gen Elec Co, 48-51; dir tech admin div, Schenectady Opers Off, US AEC, 51-54; mgr contract admin dept & asst to pres, M & C Nuclear, Inc, Mass, 54-59; mgr reprocessing, Sylvania-Corning Nuclear Corp, 59-60; sales mgr nuclear fuels dept, Spencer Chem Co, Mo, 60-61; mgr mat appln, Gen Atomic Div, Gen Dynamics Corp, 61-70, asst mgr uranium mkt, Gulf Gen Atomic Co, 70-72, mgr indust & govt activities, Uranium Supply & Distrib, Gulf Energy & Environ Systs, 72-73; CONSULT, SCI APPLNS, INC, 73-, BATTELLE MEM INST, 74- & ENERGY, INC, 75- Mem: Fel AAAS; fel Am Inst Chemists; Am Chem Soc; Am Nuclear Soc. Res: Nuclear fuels; reprocessing; coated particle fuels. Mailing Add: Rte M3 Del Mar CA 92014

MATHESON, AUDRIA, microbiology, immunology, see 12th edition

MATHESON, BALLEM HOWARD, b NS, Can, Aug 13, 26; m 51; c 2. BACTERIOLOGY. Educ: McGill Univ, BSc, 50, PhD(bact, immunol), 57; Dalhousie Univ, MSc, 53. Prof Exp: Asst chemist & demonstr org chem, Nova Scotia Agr Col, 47-48; demonstr gen microbiol, Macdonald Col, McGill Univ, 49-50; chemist & bacteriologist, Path Inst NS, 50-52; bacteriologist, Food & Drug Labs, Can Dept Nat

Health & Welfare, 53-55; from lectr to asst prof bact & immunol, 59-63, assoc prof microbiol & immunol, 63-70, PROF MICROBIOL & IMMUNOL, FAC MED, McGILL UNIV, 70- Concurrent Pos: Demonstr, Fac Med, Dalhousie Univ, 50-53. Mem: Am Soc Microbiol; Can Soc Microbiol. Res: Glomerular nephritis; staphylococcus food poisoning. Mailing Add: Dept of Microbiol & Immunol Fac of Med McGill Univ Montreal PQ Can

MATHESON, HARRY, b Spring Valley, NY, Feb 20, 12; m 38; c 3. PHYSICAL CHEMISTRY. Educ: Pa State Col, BS, 35; Univ Md, MS, 39. Prof Exp: Chemist, Nat Bur Standards, 35-51, physicist, 51-66, physicist, Environ Sci Serv Admin, 66-68 & Nat Oceanic & Atmospheric Admin, 68-72; PVT CONSULT, 72- Mem: AAAS; Am Chem Soc; Am Geophys Union; Am Inst Aeronaut & Astronaut. Res: Telemetering; physics of upper atmosphere; seismic instrumentation. Mailing Add: 3114 Jennings Rd Kensington MD 20795

MATHESON, MAX SMITH, b McGill, Nev, May 24, 13; m 39, 67; c 2. RADIATION CHEMISTRY. Educ: Univ Utah, AB, 36; Brown Univ, MS, 38; Univ Rochester, PhD(chem), 40. Prof Exp: Res chemist, Gen Labs, US Rubber Co, 40-50; assoc chemist, 50-52, dir chem div, 65-71, SR CHEMIST, ARGONNE NAT LAB, 52- Concurrent Pos: Guggenheim fel phys chem lab, Paris, 60-61; vis prof, Hebrew Univ, Jerusalem, 72. Mem: AAAS; Am Chem Soc; Am Phys Soc; Radiation Res Soc; The Chem Soc. Res: Photochemistry; kinetics; rate constants of vinyl polymerization; radiation chemistry; flash-photolysis; pulsed radiolysis. Mailing Add: Chem Div Argonne Nat Lab 9700 S Cass Ave Argonne IL 60439

MATHESON, WILLARD EDWARD, b Penticton, BC, Mar 6, 19; nat US; m 53; c 2. PHYSICS. Educ: Univ BC, BA & BASc, 47; Purdue Univ, MS, 49, PhD(physics), 54. Prof Exp: Res engr, Powell River Paper Co, Can, 47; asst physics, Purdue Univ, 47-48; res engr, Linde Air Prod Co Div, Union Carbide Corp, 54-56; engr, Douglas Aircraft Co, Inc, 56-58, adv systs engr, 58-60, exec adv, 60-61; dir eng, Acoustica Assocs, Inc, 61, vpres & gen mgr, 62-63; chief adv systs engr nuclear proj, Douglas Aircraft Co, Inc, 63-64, chief engr future systs, 64, dir new bus develop, 64-65, DIR, DONALD W DOUGLAS LABS, DOUGLAS AIRCRAFT CO, INC, 65- Concurrent Pos: Mem, Atomic Indust Forum; mem, Wash State Adv Coun Nuclear Energy & Radiation, & 68- Mem: Am Phys Soc; Am Nuclear Soc; Sigma Xi. Res: Nuclear power systems; space systems development; electrical engineering; solid state energy conversion. Mailing Add: 119 Jackson Court Richland WA 99352

MATHEW, MATHAI, b Mavelikara, India; m 67. X-RAY CRYSTALLOGRAPHY, INORGANIC CHEMISTRY. Educ: Univ Kerala, India, BS, 53; Univ Agra, India, MS, 56; Univ Western Ont, PhD(chem), 66. Prof Exp: Lectr chem, Cath Col, India, 56-58 & 60-62; res asst, Atomic Energy Estab, India, 58-60; fel, Nat Res Coun Can, 65-67, Univ Waterloo, 67-70; res assoc, Univ Fla, 70-74; RES ASSOC CHEM, AM DENT ASN, NAT BUR STAND, 75- Mem: Am Chem Soc; Am Crystallog Asn. Res: X-ray crystallographic structural studies of dental materials and of compounds related to constituents of tooth, bone and dental calculus with a view to correlate the properties with structural variations. Mailing Add: Am Dent Asn Health Found Nat Bur of Stand Washington DC 20234

MATHEWES, DAVID A, b Gastonia, NC, Sept 22, 31; m 57; c 4. ORGANIC CHEMISTRY. Educ: Davidson Col, BS, 53; Univ Kans, MS, 55; Duke Univ, PhD(chem), 63. Prof Exp: Teaching asst chem, Univ Kans, 53-55; instr, Ga Inst Technol, 55-57; instr, Hampden-Sydney Col, 57-58; res asst, Duke Univ, 58-62; asst prof, 62-67, head dept, 67-69, PROF CHEM, WESTERN CAROLINA UNIV, 67- Concurrent Pos: Instr, Westminster Schs, 56-57. Mem: AAAS; Am Chem Soc; The Chem Soc. Res: Organophosphorus and organo-metallic chemistry; curriculum development; science education. Mailing Add: Dept of Chem Western Carolina Univ Cullowhee NC 28723

MATHEWES, ROLF WALTER, b Berleburg, WGer, Nov 11, 46; Can citizen; m 72. PALYNOLOGY. Educ: Simon Fraser Univ, B C, Can, BSc; Univ BC, PhD(bot). Prof Exp: Vis asst prof biogeog, Simon Fraser Univ, 73; Nat Res Coun fel palynology, Sch Bot, Cambridge Univ, Eng, 74; environ consult plant ecol, F F Slaney & Co Ltd, Vancouver, 74-75; vis asst prof, 74, ASST PROF BIOL, SIMON FRASER UNIV, 74- Concurrent Pos: Environ consult plant ecol, F F Slaney & Co Ltd, Vancouver, 74-75. Mem: Brit Ecol Soc; Can Bot Asn. Res: Palynology and paleoecology of postglacial vegetation; application of pollen analysis to archaeological and zoological problems. Mailing Add: Simon Fraser Univ Burnaby BC Can

MATHEWS, A L, b Whittier, NC, Mar 28, 40; m 60; c 3. CHEMISTRY. Educ: Western Carolina Univ, BS, 61; Univ Miss, PhD(phys chem), 65. Prof Exp: Asst prof phys chem, Western Carolina Univ, 65-69; ADMIN ASST BIOCHEM, MICH STATE UNIV, 69- Mem: AAAS; Am Chem Soc. Res: Research administration; communication and information exchange in solving significant problems; use of computers in designing experiments; thermodynamics of multiple phase systems. Mailing Add: Dept of Biochem Mich State Univ East Lansing MI 48823

MATHEWS, CHARLES WILLARD, mathematics, see 12th edition

MATHEWS, CHRISTOPHER KING, b New York, NY, May 5, 37; m 60; c 2. BIOCHEMISTRY. Educ: Reed Col, BA, 58; Univ Wash, PhD(biochem), 62. Prof Exp: Asst prof biol, Yale Univ, 63-67; assoc prof, 67-73, PROF BIOCHEM, COL MED, UNIV ARIZ, 73- Concurrent Pos: USPHS fel biochem, Univ Pa, 62-63; Am Cancer Soc scholar, Univ Calif, San Diego, 73-74. Mem: AAAS; Am Soc Biol Chem; Am Soc Cell Biol; Am Soc Microbiol; Am Chem Soc. Res: Microbial and viral enzymology; metabolic regulation; mechanism of action of antimetabolites; enzymatic aspects of bacteriophage structure and replication; metabolism of coenzymes, nucleotides, and nucleic acids. Mailing Add: Dept of Biochem Univ of Ariz Col of Med Tucson AZ 85724

MATHEWS, COLLIS WELDON, b Troy, Ala, July 19, 38; m 59; c 2. PHYSICAL CHEMISTRY, MOLECULAR SPECTROSCOPY. Educ: Univ Ala, BS, 60; Vanderbilt Univ, PhD(phys chem), 65. Prof Exp: Res assoc ultraviolet spectros, Vanderbilt Univ, 64-65; fel div pure physics, Nat Res Coun Can, 65-67; asst prof, 67-72, ASSOC PROF PHYS CHEM, OHIO STATE UNIV, 72- Mem: Am Chem Soc; Optical Soc Am. Res: Investigations of high-resolution visible and ultraviolet molecular spectra for the purposes of obtaining their geometric and electronic structures especially of unstable molecular species. Mailing Add: Dept of Chem Ohio State Univ 140 W 18th Ave Columbus OH 43210

MATHEWS, DANIEL MONROE, physical chemistry, see 12th edition

MATHEWS, FRANCES HOUTARI, biochemistry, see 12th edition

MATHEWS, FRANCIS SCOTT, b Albany, Ore, Mar 2, 34; m 59; c 3. BIOCHEMISTRY. Educ: Univ Calif, BS, 55; Univ Minn, Minneapolis, PhD(phys chem), 59. Educ: Corp fel chem, Harvard Univ, 59-61; USPHS res fel biol, Mass Inst Technol, 61-63, spec fel protein crystallog, Lab Molecular Biol, 63-65; ASSOC PROF PHYSIOL & BIOPHYS, SCH MED, WASH UNIV, 66- Mem: Am Crystallog Asn; Biophys Soc; Am Chem Soc. Res: X-ray crystallographic study of biological materials, especially the structure and function of proteins. Mailing Add: Dept of Physiol & Biophys Wash Univ Sch of Med St Louis MO 63110

MATHEWS, FREDERICK JOHN, b Columbus, Wis, Dec 20, 18; m 52; c 2. ORGANIC CHEMISTRY. Educ: Carroll Col, BA, 40; Univ Wis, PhD(org chem), 43. Prof Exp: Asst, Univ Wis, 41-43; res chemist, Rohm & Haas Co, Pa, 43-46; asst prof chem, Kent State Univ, 46-47; from asst prof to assoc prof, 53-59; OWNER, LAB CRAFTSMEN, 59- Mem: Am Chem Soc. Res: Benzoquinoline compounds synthesis; silicones; design and manufacture of scientific equipment. Mailing Add: 2925 Bartells Dr Beloit WI 53511

MATHEWS, GEOFFREY WILLIAM, b Urbana, Ill, July 28, 38; m 61; c 2. PETROLOGY, GEOLOGY. Educ: Lawrence Univ, BA, 60; Case Western Reserve Univ, PhD(geol), 69. Prof Exp: Teacher pub sch, Conn, 62-64; asst prof, 67-74, ASSOC PROF GEOL, IND UNIV, FT WAYNE, 74- Res: Geochemical variability in epizonal igneous plutons. Mailing Add: Dept of Geol Ind Univ Ft Wayne IN 46805

MATHEWS, HARRY T, b Atlanta, Ga, Nov 13, 31; m 59; c 3. MATHEMATICS. Educ: Ga Inst Technol, BS, 59; Tulane Univ, PhD(math), 64. Prof Exp: Asst prof math, Wayne State Univ, 63-65; from asst prof to assoc prof, 65-70, PROF MATH & HEAD DEPT, UNIV TENN, KNOXVILLE, 70- Mem: Am Math Soc; Math Asn Am. Res: Boundary behavior of functions of a complex variable. Mailing Add: Dept of Math Univ of Tenn Knoxville TN 37916

MATHEWS, HENRY MABBETT, b Thomasville, Ga, May 19, 40; m 62; c 3. MEDICAL PARASITOLOGY. Educ: Univ Ga, BS, 62; Emory Univ, MS, 65, PhD(biol), 67. Prof Exp: Resident microbiol, Nat Commun Dis Ctr, 67-69; RES MICROBIOLOGIST, CTR DIS CONTROL, 69- Mem: Am Soc Parasitologists; Am Soc Trop Med & Hyg; Sigma Xi. Res: Serology and sero-epidemiology of parasitic diseases. Mailing Add: Ctr of Dis Control Parasitol Div 1600 Clifton Rd Atlanta GA 30333

MATHEWS, J RODNEY, b Ione, Wash, Sept 3, 11; m 49; c 3. ORTHODONTICS. Educ: Univ Calif, Los Angeles, AB, 34; Univ Calif, Berkeley, MA, 36; Univ Calif, San Francisco, DDS, 49. Prof Exp: ASSOC PROF ORTHOD, MED CTR, UNIV CALIF, SAN FRANCISCO, 49- Res: Bacteriology; human growth and development of the jaws, face and dental apparatus. Mailing Add: Div of Orthod Univ of Calif Sch of Dent San Francisco CA 94122

MATHEWS, JEROLD CHASE, b Des Moines, Iowa, Sept 12, 30; m 59; c 2. MATHEMATICS Educ: Iowa State Univ, BS, 55, MS, 57, PhD(math), 59. Prof Exp: Asst prof math, Univ Okla, 60-61; mathematician, Mathematica, Inc, NJ, 61-62; from asst prof to assoc prof, 62-70, PROF MATH, IOWA STATE UNIV, 70- Mem: AAAS; Math Asn Am; Am Math Soc. Res: History of mathematics. Mailing Add: Dept of Math Iowa State Univ Ames IA 50011

MATHEWS, JON, b Los Angeles, Calif, Feb 10, 32; m 52; c 4. THEORETICAL PHYSICS. Educ: Pomona Col, BA, 52; Calif Inst Technol, PhD(physics), 57. Prof Exp: From instr to assoc prof, 57-66, PROF PHYSICS, CALIF INST TECHNOL, 66- Mem: AAAS; Am Phys Soc. Res: Elementary particle physics. Mailing Add: Dept of Physics Calif Inst of Technol Pasadena CA 91109

MATHEWS, KENNETH PINE, b Schenectady, NY, Apr 1, 21; m 52; c 3. ALLERGY. Educ: Univ Mich, AB, 41, MD, 43; Am Bd Internal Med, dipl, 55; Am Bd Allergy, dipl, 59. Prof Exp: From intern to asst resident, 43-45, from instr to assoc prof, 48-61, PROF INTERNAL MED, UNIV HOSP, UNIV MICH, ANN ARBOR, 61- Concurrent Pos: Consult, Ann Arbor Vet Admin Hosp & Wayne Co Gen Hosp; ed, J Allergy & Clin Immunol, 68-72; mem training grant comt, Nat Inst Allergy & Infectious Dis, 71-72, chmn allergy & immunol res comt, 73-75. Mem: AMA; Am Fedn Clin Res; fel Am Col Physicians; Am Acad Allergy (pres, 64-65). Res: Various aspects of allergy. Mailing Add: Dept of Intern Med Univ of Mich Med Ctr, Ann Arbor MI 48104

MATHEWS, MARTIN B, b Chicago, Ill, May 30, 12; m 42; c 2. BIOCHEMISTRY. Educ: Univ Chicago, BSc, 36, MSc, 41, PhD, 49. Prof Exp: Asst pediat, Univ Chicago, 49-52, from res assoc instr to asst prof pediat & biochem, 52-59, assoc prof biochem, univ & assoc prof pediat, La Rabida Inst, 59-67, PROF BIOCHEM, UNIV CHICAGO & PROF PEDIAT, LA RABIDA INST, 67- Concurrent Pos: Estab investr, Am Heart Asn, 54- Mem: Am Chem Soc; Am Soc Biol Chemists. Res: Biochemistry of connective tissue; physical chemistry of acid mucopolysaccharides Mailing Add: 5471 Dorchester Chicago IL 60615

MATHEWS, ROBERT THOMAS, b Indianapolis, Ind, Aug 30, 19; m 52; c 4. ASTRONOMY. Educ: Wesleyan Univ, BA, 40; Univ Calif, MA, 54. Prof Exp: From jr astronr to asst astronr, US Naval Observ, 42-44; observing asst, Lick Observ, Mt Hamilton, 47-48; instr astron, Wesleyan Univ, 48-54; from instr to asst prof astron & math, 54-67, ASSOC PROF ASTRON, CARLETON COL, 67- Mem: AAAS; Am Astron Soc. Res: Stellar parallax; photoelectric and spectroscopic study of galactic star clusters; visual double stars. Mailing Add: Dept of Physics & Astron Carleton Col Northfield MN 55057

MATHEWS, WALTER KELLY, b Columbus, Ga, Jan 16, 37; m; c 3. ORGANIC CHEMISTRY, BIOCHEMISTRY. Educ: Univ Ga, BS, 60, MS, 61; Univ Louisville, PhD(chem), 67. Prof Exp: Chemist, Sinclair Res, Inc, Ill, 61-63; instr gen chem, Wingate Col, 63-64; vis instr org chem, Univ Louisville, 65-66; ASSOC PROF ORG CHEM, GA SOUTHWESTERN COL, 67- Mem: Am Chem Soc. Res: Cationic polymerization mechanisms; selected oxidation processes; mechanisms of counterion binding to colloids and surfactants. Mailing Add: Dept of Org Chem Ga Southwestern Col Americus GA 31709

MATHEWS, WILLIAM HENRY, b Vancouver, BC, Feb 2, 19; m 48; c 3. GEOLOGY. Educ: Univ BC, BASc, 40, MASc, 41; Univ Calif, PhD(geol), 48. Prof Exp: Assoc mining engr, BC Dept Mines, Can, 42-49; asst prof geol, Univ Calif, 49-51; assoc prof, 51-59, head dept, 64-71, PROF GEOL, UNIV BC, 59- Concurrent Pos: Nat Res Coun Can sr fel, 63-64; mem, Can Nat Comt for Int Hydrologic Decade, 64-; mem, Int Comt Marine Geol, 66-; mem, Can Nat Adv Comt Res Geol Sci, 67-69; chmn standing comt solid earth sci, Pac Sci Asn, 67-71; Killam sr fel, 71-72. Mem: Fel Geol Soc Am; Am Asn Petrol Geologists; Royal Soc Can; Geol Asn Can; Glaciol Soc; Int Soc Soil Mech & Found Eng. Res: Geomorphology and glacial geology; glaciology, sedimentology and geological oceanography; sub-glacial vulcanism. Mailing Add: Dept of Geol Univ of BC Vancouver BC Can

MATHEWS, WILLIS WOODROW, b Wendling, Ore, May 27, 17; m 42; c 3. EMBRYOLOGY. Educ: Ore State Col, BA, 40; Univ Wis, PhD(zool), 45. Prof Exp: Asst zool, Univ Wis, 40-44; from instr to asst prof biol, Univ Chattanooga, 44-47; asst prof, 47-56, from actg chmn dept to chmn dept, 62-65, ASSOC PROF BIOL,

WAYNE STATE UNIV, 56- Mem: AAAS; Am Soc Zool. Res: Experimental embryology of chick; microscopy; growth factors. Mailing Add: Dept of Biol Wayne State Univ Detroit MI 48202

MATHEWSON, FRANCIS ALEXANDER LAVENS, b New Westminster, BC, Feb 1, 05; m 36; c 2. INTERNAL MEDICINE. Educ: Univ Man, MD, 31, BSc, 33; Royal Col Physicians Can, cert internal med, 46; Am Bd Prev Med, dipl & cert aviation med, 54. Prof Exp: ASSOC PROF MED, FAC MED, UNIV MAN, 45- Concurrent Pos: Physician, Winnipeg Gen Hosp, mem attend staff, 35, chmn, 51-52; consult, Can Nat Rwy; mem asn comt aviation med res, Nat Res Coun Can, 42-44; mem panel aviation med res, Defense Res Bd, Can, 50-54; chmn, Royal Can Air Force Med Adv Comt, 54; fel coun clin cardiol, Am Heart Asn; Col Physicians & Surgeons Man Gordon Bell res fel, 33-34. Mem: Fel Am Col Cardiol; Asn Life Ins Med Dirs (pres, 68-69); Can Cardiovasc Soc (pres, 57-58); Can Life Ins Med Off Asn (pres, 55-56); Defense Med Asn Can (pres, 54-55). Res: Cardiology; prospective epidemiological study of coronary heart disease. Mailing Add: 711 Med Arts Bldg 233 Kennedy St Winnipeg MB Can

MATHEWSON, JAMES H, b Norwalk, Conn, Nov 24, 29; m 58; c 3. BIO-ORGANIC CHEMISTRY, OCEANOGRAPHY. Educ: Harvard Univ, AB, 51; Johns Hopkins Univ, MA, 57, PhD(org chem), 59. Prof Exp: Res assoc chem, Johns Hopkins Univ, 59-60; guest investr, Rockefeller Inst, 60-61; res fel, Univ Calif, Berkeley, 61-63; asst prof chem, Western Wash State Col, 63-64; from asst prof to assoc prof, 64-72, PROF CHEM, SAN DIEGO STATE UNIV, 72- Concurrent Pos: USPHS fel, 60-63; Nat Inst Arthritis & Metab Dis res grant, 65-68, NSF sea grant prog res grant, 69-70; Nat Oceanic & Atmospheric Admin res grant, 70-71; actg dir, Bur Marine Sci, San Diego State Univ, 67-70; vis scientist, Lab Chem Enzyme Shell Res, Sittingbourne, Kent, UK, 74-75. Mem: AAAS; Am Chem Soc. Res: Organic and biological chemistry, especially tetrapyrroles; environmental chemistry, especially oceanic; science education, especially general education and interdisciplinary courses; chlorophyll chemistry; marine biochemistry; pollution measurement. Mailing Add: Dept of Chem San Diego State Univ San Diego CA 92182

MATHEWSON, JOHN ANGELL, b Providence, RI, Aug 7, 14. ENTOMOLOGY. Educ: Brown Univ, AB, 37; Northwestern Univ, MA, 40; Yale Univ, MSc, 45. Prof Exp: Asst genetics, Carnegie Inst Technol, 38; asst zool, Northwestern Univ, 38-41 & Yale Univ, 43-45; entomologist, RI State Dept Agr & Conserv, 49-61; asst prof entom, 61-68, asst prof zool, 68-70, ASSOC PROF ZOOL, UNIV RI, 70- Mem: Entom Soc Am; Int Union Study Soc Insects. Res: Internal morphology of insects; ectoparasitic insects. Mailing Add: Dept of Zool Univ of RI Kingston RI 02881

MATHEY, WILLIAM JOSEPH, JR, veterinary medicine, see 12th edition

MATHIAS, MELVIN MERLE, b Columbia City, Ind, Feb 22, 39; m 63; c 3. NUTRITION. Educ: Purdue Univ, BS, 61; Cornell Univ, PhD(nutrit), 67. Prof Exp: Asst nutrit, Cornell Univ, 62-66; ASST PROF NUTRIT, COLO STATE UNIV, 68- Concurrent Pos: Fac partic, AEC prog, Donner Lab, Univ Calif, Berkeley, 71; co-investr, Nat Heart & Lung Inst, 71-75; sabbatical, Dept Biochem Nutrit, Hoffmann-La Roche, 74-75. Mem: AAAS; Am Inst Nutrit; Sigma Xi. Res: Effects of diet and B vitamin deficiencies on intermediary metabolism. Mailing Add: Dept of Food Sci & Nutrit Colo State Univ Ft Collins CO 80523

MATHIAS, MILDRED ESTHER (MRS GERALD L HASSLER), b Sappington, Mo, Sept 19, 06; m 30; c 4. BOTANY. Educ: Wash Univ, AB, 26, MS, 27, PhD(syst bot), 29. Prof Exp: Asst, Mo Bot Garden, 29-30; res assoc, NY Bot Garden, 32-36 & Univ Calif, 37-42; herbarium botanist, 47-51, lectr bot, 51-55, from asst prof to prof, 55-74, dir bot garden, Exp Sta, 56-74, EMER PROF BOT, UNIV CALIF, LOS ANGELES, 74- Concurrent Pos: Asst specialist, Bot Garden, Exp Sta, Univ Calif, Los Angeles, 51-55, asst plant systematist, 55-57, vchmn bot garden, 55-62 & assoc plant systematist, 57-62; pres, Orgn Trop Studies, 68-70; Secy, Bd of Trustees, Inst Ecol, 75- Honors & Awards: Merit Award, Bot Soc Am, 73; Sci Citation, Am Hort Soc, 74. Mem: AAAS; Bot Soc Am; Am Soc Plant Taxon (pres, 64); Soc Study Evolution; Am Soc Naturalists. Res: Classification of plants of western United States; monographic studies of the Umbelliferae, especially of North and South America; subtropical ornamental plants; tropical medicinal plants. Mailing Add: Dept of Biol Univ of Calif Los Angeles CA 90024

MATHIASON, DENNIS R, b Fairmont, Minn, Feb 6, 41; m 63; c 2. INORGANIC CHEMISTRY. Educ: Mankato State Col, BS, 62; Univ SDak, PhD(chem), 66. Prof Exp: ASSOC PROF CHEM, MOORHEAD STATE COL, 66- Mem: Am Chem Soc. Res: Ylide chemistry; organophosphorus systems. Mailing Add: Dept of Chem Moorhead State Col Moorhead MN 56560

MATHIES, ALLEN WRAY, JR, b Colorado Springs, Colo, Sept 23, 30; m 56; c 2. PEDIATRICS, INFECTIOUS DISEASES. Educ: Colo Col, BA, 52; Columbia Univ, MS, 56, PhD(parasitol), 58; Univ Vt, MD, 61. Prof Exp: Res assoc path, Col Med, Univ Vt, 57-61, from intern to resident pediat, Los Angeles Co Gen Hosp, 61-64; res assoc, 63-64, from asst prof to assoc prof, 64-71, assoc dean, Sch Med, 70-74, interim dean, 74-75, DEAN SCH MED, UNIV SOUTHERN CALIF, 75-, PROF PEDIAT, 71- Concurrent Pos: Head physician commun dis, Los Angeles Co Gen Hosp, 64-75. Mem: Am Soc Parasitol; Am Soc Trop Med & Hyg; Soc Pediat Res; Infectious Dis Soc Am; Royal Soc Trop Med & Hyg. Res: Infectious diseases; central nervous system infections; tropical medicine. Mailing Add: 2025 Zonal Ave Los Angeles CA 90033

MATHIES, JAMES CROSBY, b Seattle, Wash, July 20, 19; m 42; c 2. BIOCHEMISTRY. Educ: Univ Wash, BS, 42, PhD(biochem), 48; Wayne State Univ, MS, 46. Prof Exp: Develop & res chemist, US Rubber Co, 42-45; assoc, Med Sch, Univ Wash, 46-48; biochemist, Edsel B Ford Inst Med Res, Henry Ford Hosp, 48-54; sr biochemist, Baxter Labs, Inc, 54-56; chief biochemist, Swedish Hosp, Seattle, 56-66; dir lab serv div, Enzomedic Labs, Inc, Wash, 66-67; CLIN BIOCHEMIST, PATH LAB, ST JOSEPH HOSP, 67- Concurrent Pos: Res biochemist, Pac Northwest Res Found, 56-66; consult, Philips Electronics, 61-65. Mem: AAAS; Am Chem Soc; Am Soc Biol Chem; Am Asn Clin Chem; NY Acad Sci. Res: Hormones and tissue enzymes; proteolytic enzymes; purification and characteristics of phosphatases; clinical chemistry; x-ray spectrochemical analysis. Mailing Add: Path Lab St Joseph Hosp 1845 Franklin St Denver CO 80218

MATHIES, MARGARET JEAN, b Colorado Springs, Colo, June 9, 35. MICROBIOLOGY, IMMUNOLOGY. Educ: Colo Col, BA, 57; Case Western Reserve Univ, PhD(microbiol), 63. Prof Exp: Asst prof biol, Haverford Col, 62-64; vis asst prof zool, Pomona Col, 64-65; from asst prof to assoc prof biol, 65-74, PROF BIOL, JOINT SCI DEPT, CLAREMONT MEN'S, PITZER & SCRIPPS COLS, 74-, MEM TRI-COL SCI PROG, SCRIPPS COL, 70- Mem: AAAS; Am Soc Microbiol. Res: Antibody formation and physico-chemical characterization of antibodies, using bacteriophage antigens; metabolic effects of viruses on host-cells; biochemistry; genetics. Mailing Add: Tri-Col Sci Prog Scripps Col Claremont CA 91711

MATHIESON, ALFRED HERMAN, b Union City, NJ, July 6, 17; m 41; c 5. PHYSICS. Educ: Pa State Teachers Col, BS, 38; Columbia Univ, MA, 39. Prof Exp: Instr physics, Springfield Col, 39-40; ASST PROF PHYSICS, UNIV MASS, AMHERST, 46-, ASST HEAD DEPT PHYSICS & ASTRON, 65- Concurrent Pos: NSF grant, 64. Mailing Add: Dept of Physics & Astron Univ of Mass Amherst MA 01002

MATHIESON, ARTHUR C, b Los Angeles, Calif, Dec 26, 37; m 58; c 3. BOTANY. Educ: Univ Calif, Los Angeles, BA, 60, MA, 61; Univ BC, PhD, 65. Prof Exp: From asst prof to assoc prof, 65-74, PROF BOT, UNIV NH, 74-, DIR, JACKSON ESTUARINE LAB, 72- Mem: Phycol Soc Am; Int Phycol Soc. Res: Morphology; distribution and ecology of marine plants in relation to oceanographic factors. Mailing Add: Dept of Bot Jackson Estuarine Lab Univ of NH Durham NH 03840

MATHIESON, DON ROMUALD, b Ft Pierre, SDak, June 11, 08; m 36, 59; c 2. IMMUNOLOGY. Educ: Univ Minn, BA, 31, MD, 36. Prof Exp: Res bacteriologist, Parke, Davis & Co, 37-42; from asst prof to assoc prof, 46-70, EMER PROF CLIN PATH, MAYO GRAD SCH MED, UNIV MINN, 70-, EMER MEM, MAYO CLIN, 70- Concurrent Pos: Consult, Mayo Clin, 46-61, sr consult sect clin path, 61-70. Mem: AMA; Am Asn Immunol. Res: Clinical pathology. Mailing Add: 1530 Durant Ct Rochester MN 55901

MATHIEU, JEAN, b Montreal, Que, Aug 7, 26; m 56; c 3. INTERNAL MEDICINE. Educ: Col Stanislas, Montreal, BA, 44; Univ Montreal, MD, 50; FRCPS(C), 54. Prof Exp: Assoc prof, 56-72, PROF MED, FAC MED, UNIV MONTREAL, 72-, STAFF MEM MED, HOSP, MAISONNEUVE, 56-, VDEAN, FAC MED, UNIV, 68- Mem: Fel Am Col Physicians. Mailing Add: Fac of Med Univ of Montreal Montreal PQ Can

MATHIEU, LEO GILLES, b Nicolet, Que, Jan 7, 32; m 56; c 2. BIOCHEMISTRY, MICROBIOLOGY. Educ: Univ Montreal, DVM, 56; Cornell Univ, MSc, 58, PhD(nutrit), 60. Prof Exp: Asst prof biochem, Col Vet Med, 60-65, asst prof molecular biol, Fac Med, 65-69, assoc prof microbiol, 69-72, PROF MICROBIOL, FAC MED, UNIV MONTREAL, 72- Concurrent Pos: Mem exec coun, Grad Sch, Univ Montreal, 71-73. Mem: Soc Gen Microbiol; Can Soc Microbiol; Can Vet Med Asn; Am Soc Microbiol. Res: Microbial physiology. Mailing Add: Dept of Microbiol Univ of Montreal Fac of Med Montreal PQ Can

MATHIEU, ROGER MAURICE, b Montreal, Que, Aug 4, 24; m 49; c 2. RADIOLOGY, PHYSICS. Educ: Univ Montreal, BSc, 46, MSc, 48, PhD(physics), 52. Prof Exp: Radiation physicist & biophysicist, Montreal Cancer Inst & X-Ray Dept, Hosp Notre Dame, 49-69; RADIATION PHYSICIST & BIOPHYSICIST, DEPT RADIOTHER & NUCLEAR MED, MAISONNEUVE-ROSEMONT HOSP, 69-; CLIN PROF RADIOL, FAC MED, UNIV MONTREAL, 70- Mem: Fr-Can Soc Radiol; Can Asn Physicists; Can Asn Radiol. Res: Radiological physics; biophysics; radiotherapy; nuclear medicine. Mailing Add: Dept of Radiother & Nuclear Med Maisonneuve-Rosemont Hosp 5415 Blvd L'Assomtion Montreal PQ Can

MATHIS, BILLY JOHN, b Henryetta, Okla, Sept 12, 32; m 57; c 2. LIMNOLOGY. Educ: Okla State Univ, BS, 59, MS, 63, PhD(zool), 65. Prof Exp: Sci teacher pub schs, Tex, 59-62; from asst prof to assoc prof, 65-72, PROF BIOL, BRADLEY UNIV, 72-, CHMN DEPT, 70- Mem: Am Soc Limnol & Oceanog; Ecol Soc Am. Res: Stream pollution; primary productivity; distribution of heavy metals in aquatic environments. Mailing Add: 6332 N Hamilton Rd Peoria IL 61614

MATHIS, JAMES L, b Dayton, Tenn, Jan 30, 25; m 48; c 4. PSYCHIATRY. Educ: Univ Mo, 44-45; St Louis Univ, MD, 49; Am Bd Psychiat & Neurol, dipl, 68. Prof Exp: Rotating intern, Fitzsimons Gen Hosp, 49-50; resident, Elk City Community Hosp-Clin, Okla, 50-51; gen practr, Crossett Health Ctr, Ark, 51-52; surg asst, Elk City Community Hosp, 52-55; pvt pract, Dayton, Tenn, 55-60; resident psychiat, Med Ctr, Univ Okla, 60-63; from instr to assoc prof, Med Sch, Rutgers Univ, 63-70; PROF PSYCHIAT & CHMN DEPT, MED COL VA, 70- Concurrent Pos: Asst chief psychiat serv, Vet Admin, Oklahoma City, 63-64; consult, Peace Corps, 65-69, Job Corps, 68-70 & NJ Correctional Syst, 68-70; asst examr, Am Bd Psychiat & Neurol, 71-; ed, Sexuality, 72-73 & Hosp Physician, 73- Mem: Am Psychiat Asn; Am Psychosom Soc; Am Col Psychiat; Am Asn Prof Psychiat. Res: Sexuality in medicine; death and dying; drug abuse; sleep and dreams. Mailing Add: Dept of Psychiat Med Col of Va Richmond VA 23298

MATHIS, JOHN BUELL, b Washington, DC, Mar 19, 39; m 64; c 2. SCIENCE ADMINISTRATION. Educ: Yale Univ, BS, 61; Mass Inst Technol, PhD(biochem), 69. Prof Exp: Asst prof chem, Bowdoin Col, 69-72; grants assoc, NIH, 72-73, SCI ADMINR, NIH, 73- Concurrent Pos: Res Corp grant, Bowdoin Col, 69-70; Am Chem Soc, Petrol Res Fund grant, 69-72. Mem: AAAS; Am Chem Soc; Sigma Xi. Res: Scientific and educational administration. Mailing Add: Div of Lung Dis Nat Heart & Lung Inst NIH Bethesda MD 20014

MATHIS, JOHN SAMUEL, b Dallas, Tex, Feb 7, 31; m 54; c 5. ASTROPHYSICS. Educ: Mass Inst Technol, BS, 53; Calif Inst Technol, PhD(astron), 56. Prof Exp: NSF res fel, Yerkes Observ, Chicago, 56-57; asst prof astron, Mich State Univ, 57-59; from asst prof to assoc prof, 59-68, PROF ASTRON, UNIV WIS-MADISON, 68- Concurrent Pos: Assoc ed, Astrophys J; sr sci awardee, Alexander-Von-Humboldt Found, Ger, 75-76. Mem: Int Astron Union; Am Astron Soc; Royal Astron Soc; Astron Soc Pac. Res: Inter- stellar matter. Mailing Add: Dept of Astron Univ of Wis Madison WI 53706

MATHIS, ROBERT FLETCHER, b Wheeling, WVa, Jan 22, 46; m 71. COMPUTER SCIENCE, MATHEMATICS. Educ: Ohio State Univ, BSc, 65, MSc, 66, PhD(math), 69. Prof Exp: ASST PROF COMPUT & INFO SCI, OHIO STATE UNIV, 69- Mem: Am Math Soc; Math Asn Am; Soc Indust & Appl Math; Inst Elec & Electronics Engrs; Asn Comput Mach. Res: Partial differential equations; functional and numerical analysis; computer algorithms. Mailing Add: Dept of Comput & Info Sci Ohio State Univ Columbus OH 43210

MATHISEN, MAURICE EARL, b Chico, Calif, Oct 21, 15; m 37. ANALYTICAL CHEMISTRY. Educ: Pacific Union Col, AB, 38, Stanford Univ, MS, 47, PhD(anal chem), 52. Prof Exp: Instr sci & math, Nevada-Utah Acad, 38-39, Glendale Union Acad, 39-40 & Mt View Acad, 40-42 & 44-47; chemist, Permanente Metals Corp, 42-43; from asst prof to prof chem, Pacific Union Col, 47-63, acad dean, 55-63; prof chem, 63-65, DIR PERSONNEL RELS, LOMA LINDA UNIV, 65- Mem: AAAS; Am Chem Soc; Sigma Xi. Res: Spectrophotometric investigations of complex ions. Mailing Add: Dept of Chem Loma Linda Univ Loma Linda CA 92354

MATHISEN, OLE ALFRED, b Oslo, Norway, Feb 9, 19; nat US; m 48; c 2. POPULATION STUDIES. Educ: Univ Oslo, Cand Mag, 41, Cand Real, 45; Univ Wash, PhD, 55. Prof Exp: Assoc prof, 46-48, PROF, FISHERIES RES INST, UNIV WASH, 68- Concurrent Pos: Inter-Univ Comt Travel Grants fel, Moscow, 60-61; Fulbright res fel, Oslo, 65-66; consult, Food & Agr Orgn of UN, 73- Mem: Am

Fisheries Soc; Biomet Soc; Inst Fishery Res Biol; Am Soc Limnol & Oceanog; Int Soc Limnol. Res: Population dynamics, especially of salmonoids; acoustical stock estimation. Mailing Add: Univ of Wash Fisheries Res Inst Seattle WA 98195

MATHISON, IAN WILLIAM, b Liverpool, Eng, Apr 17, 38. ORGANIC & MEDICINAL CHEMISTRY. Educ: Univ London, BPharm, 60, PhD(pharmaceut chem), 63. Prof Exp: Res assoc pharmaceut & med chem, Col Pharm, Univ Tenn, Memphis, 63-65, from asst prof to assoc prof med chem, 65-72, PROF MED CHEM, CTR FOR HEALTH SCI, UNIV TENN, 72- Concurrent Pos: Prin investr, Marion Labs grant, 65-74 & Beecham Pharmaceut Res grant, 74-; sr investr, NSF grant, 68-72. Mem: The Chem Soc; Am Acad Pharmaceut Sci; Brit Pharmaceut Soc; Am Chem Soc; Royal Inst Chem, London. Res: Design and synthesis of organic compounds with potential pharmacodynamic activity; influence of stereochemistry and physicochemical parameters on pharmacological potency. Mailing Add: Dept of Med Chem Col of Pharm Univ of Tenn Ctr for Health Sci Memphis TN 38163

MATHOT, CHRISTIAN, physical chemistry, immunochemistry, see 12th edition

MATHRE, DONALD EUGENE, b Frankfort, Kans, Jan 5, 38; m 61; c 2. PLANT PATHOLOGY. Educ: Iowa State Univ, BS, 60; Univ Calif, Davis, PhD(plant path), 64. Prof Exp: Asst prof plant path, Univ Calif, Davis, 64-67; from asst prof to assoc prof, 67-72, PROF PLANT PATH, MONT STATE UNIV, 72- Mem: Am Phytopath Soc; Am Soc Agron. Res: Soil-borne diseases of cereals and forages. Mailing Add: Dept of Plant Path Mont State Univ Bozeman MT 59715

MATHRE, OWEN BERTWELL, b Kendall Co, Ill, Nov 26, 29; m 55; c 3. ANALYTICAL CHEMISTRY. Educ: Harvard Univ, AB, 51; Univ Minn, PhD(anal chem), 58. Prof Exp: Lab helper, Minn Mining & Mfg Co, Minn, 54; res chemist, Electrochem Dept, Del, 56-58, NY, 58-63, Tenn, 63-65, Del, 65-72, STAFF CHEMIST, INDUST CHEMS DEPT, E I DU PONT DE NEMOURS & CO, DEL, 72- Mem: Am Chem Soc; Am Soc Testing & Mat. Res: Electrochemistry; instrumental and colorimetric analysis; gas phase catalysis; environmental pollution monitoring. Mailing Add: Indust Chems Dept E I du Pont de Nemours & Co Wilmington DE 19898

MATHSEN, RONALD M, b Minneapolis, Minn, Oct 6, 38; m 62; c 2. MATHEMATICS. Educ: Concordia Col, Moorhead, Minn, BA, 60; Univ Nebr, MA, 62, PhD(math), 65. Prof Exp: Asst prof math, Concordia Col, Moorhead, Minn, 65-67; fel, Univ Alta, 67-68, asst prof, 68-69; ASSOC PROF MATH, NDAK STATE UNIV, 69- Mem: Math Asn Am; Am Math Soc. Res: Boundary value problems for ordinary differential equations; generalized convex functions. Mailing Add: Dept of Math NDak State Univ Fargo ND 58102

MATHUR, DILIP, b Agra, India, Feb 11, 41; m 68; c 2. FISH BIOLOGY. Educ: Univ Delhi, India, BSc, 61, MSc, 64; Cornell Univ, MS, 68; Auburn Univ, PhD(fishery mgt), 72. Prof Exp: Sr fishery biologist, 67-69, FISHERIES SECT LEADER, ICHTHYOLOGICAL ASSOC INC, 72- Mem: Am Fisheries Soc; Am Inst Fishery Res Biologist. Res: Effects of thermal discharges and pumped storage facilities on fishes and fish food organisms; impact of impingement and entrainment of fishes and fish larvae; ecology of fishes. Mailing Add: Ichthyological Assoc Inc PO Box 12 Drumore PA 17518

MATHUR, RAJESH SWARUP, b Agra, India, Apr 1, 35; m 68; c 2. REPRODUCTIVE ENDOCRINOLOGY, STEROID CHEMISTRY. Educ: Agra Univ, BSc, 55, MSc, 57; McGill Univ, PhD(steroid chem), 67. Prof Exp: Asst prof, 70-74, ASSOC PROF REPROD ENDOCRINOL, MED UNIV SC, 74- Concurrent Pos: NIH grant, McGill Univ, 67-69; Med Res Coun Can fel, Karolinska Inst, Sweden, 69-70. Mem: Endocrine Soc. Res: Mechanisms involved in the regulation of steroidogenesis in human feto-placental unit and in normal human adrenals; steroids in human pregnancy. Mailing Add: Dept of Obstet & Gynec Med Univ of SC Charleston SC 29401

MATHUR, SURESH CHANDRA, b Fatehgarh, India, Mar 23, 30; m 63; c 1. NUCLEAR PHYSICS. Educ: Univ Lucknow, BS, 48, MS, 50; Univ Tex, PhD(physics), 65. Prof Exp: Asst physicist, Dept Atomic Energy, Govt India, 50-58; sr res scientist, Tex Nuclear Corp, 62-67; PROF PHYSICS, UNIV LOWELL, 67-, ACTG DIR COMPUT CTR, 71- Mem: Am Phys Soc. Res: Nuclear radiation detection techniques and instrumentation; nuclear scattering theory and experiments; nuclear particle accelerators; computer programming. Mailing Add: Dept of Physics & Appl Physics Univ of Lowell Lowell MA 01854

MATHUR, VISHWA NATH PRASAD, b Aligarh, India, Feb 8, 34; Can citizen; m 61; c 2. FOREST PRODUCTS. Educ: Agra Univ, BSc, 55; Aligarh Muslim Univ, MSc, 57; Small Scale Industs Inst, Govt India Ministry Indust, New Delhi, dipl, 59; Mich State Univ, PhD(wood technol), 64. Prof Exp: Managing partner, Timber Equip & Mach Co, India, 55-60; res engr, Tech Dept, Koppers Co Inc, Ohio, 60-61; res asst forest prod, Mich State Univ, 61-64; sr res wood technologist, 65-67, supvr specialty prod, 67-72, lumber & shingles, 72-74, SECT HEAD WOOD PRESERVATION & LUMBER, MacMILLAN BLOEDEL LTD, BC, 74- Mem: Soc Wood Sci & Technol; Forest Prod Res Soc; Int Microwave Power Inst; Indian Acad Wood Sci. Res: Wood seasoning, microwave drying; new drying processes and equipment; wood preservation, fire retardant and preservation techniques; composite boards, particle board, coated and laminated products. Mailing Add: MacMillan Bloedel Research Ltd 3350 E Broadway Vancouver BC Can

MATIJEVIC, EGON, b Otocac, Yugoslavia, Apr 27, 22; nat US; m 47. PHYSICAL CHEMISTRY, COLLOID CHEMISTRY. Educ: Univ Zagreb, Chem eng, 44, Dr Chem, 48, Dr habil, 52. Prof Exp: Instr chem, Fac Pharm, Univ Zagreb, 44-47, sr instr, Fac Sci, 48-52, privat-docent colloid chem, 52-54, docent phys & colloid chem, 55-56; res fel colloid sci, Cambridge Univ, 56-57; vis prof, 57-60, assoc prof chem, 60-62, PROF CHEM, CLARKSON COL TECHNOL, 62-, DIR, INST COLLOID & SURFACE SCI, 66- Concurrent Pos: Chmn, Gordon Conf Chem at Interfaces, 65, Div Colloid & Surface Chem, Am Chem Soc, 69-70 & 49th Nat Colloid Symp, NY, 75; vis prof, Unilever Res Lab, Port Sunlight, Eng, 71, Swedish Inst Surface Chem, Stockholm, 71 & Japan Soc Prom Sci, 73; ed, Surface & Colloid Sci. Honors & Awards: Kendall Award, Am Chem Soc, 72. Mem: Am Chem Soc; Colloid Soc Ger; Croatian Chem Soc. Res: Precipitation processes; coagulation; photogalvanic phenomena; complex ionic species; heteropoly compounds; ionized monolayers; light scattering; aerosols; monodispersed colloidal metal hydrous oxides. Mailing Add: Dept of Chem Clarkson Col of Technol Potsdam NY 13676

MATIN, ABDUL, b Delhi, India, May 8, 41; Pakistan citizen; m 68. MICROBIAL PHYSIOLOGY, MICROBIAL ECOLOGY. Educ: Univ Karachi, BS, 60, MS, 62; Univ Calif, Los Angeles, PhD(microbiol), 69. Prof Exp: Lectr microbiol, St Josephs Col Women, Karachi, 62-64; bacteriologist & asst, Univ Calif, Los Angeles, 64-69, assoc, 69-71; sci officer I class, State Univ Groningen, 71-75; ACTG ASST PROF MICROBIOL, STANFORD UNIV, 75- Mem: Am Soc Microbiol; Soc Gen Microbiol; Netherlands Soc Microbiol. Res: Microbial competition, physiology and survival at low nutrient concentration; effect of growth rate as regulated in chemostat on physiology of bacterial and mammalian cells; transport in bacterial membrane vesicles. Mailing Add: Dept of Med Microbiol Sch of Med Stanford Univ Stanford CA 94305

MATIN, SHAIKH BADARUL, b Agra, India, Feb 21, 44. PHARMACEUTICAL CHEMISTRY, CLINICAL PHARMACOLOGY. Educ: Univ Karachi, BSc, 63; Columbia Univ, MS, 65; Univ Calif, PhD(pharmaceut chem), 70. Prof Exp: Fel chem, Univ Calif, San Francisco, 70-74; RES SCIENTIST, SYNTEX RES, 74- Mem: Am Chem Soc; Am Pharmaceut Asn; Am Soc Mass Spectros; NY Acad Sci. Res: Investigation of the pharmacologic profile and time course of action, interaction and mechanism of action of synthetic drugs and naturally occurring compounds on animals and man. Mailing Add: Syntex Res 3401 Hillview Ave Palo Alto CA 94304

MATIN, SHAIKH MOIZUL, b Agra, India, Oct 5, 41; m 65; c 3. HIGH ENERGY PHYSICS. Educ: Univ Karachi, BSc, 59; Columbia Univ, AM, 61, PhD(nuclear physics), 66. Prof Exp: Res asst physics, Columbia Univ, 61-66; res fel, Cambridge Electron Accelerator, Harvard Univ, 66-70; asst prof, Upsala Col, 70-75; MEM STAFF, DEPT WEAPONS & SYSTS ENG, US NAVAL ACAD, 75- Concurrent Pos: Lectr, City Col New York, 61-66; res affil, Mass Inst Technol, 66- Mem: Am Phys Soc; Am Asn Physics Teachers. Res: Gamma, gamma angular correlations; particle gamma correlations on nuclei; wire spark chambers; on line computers; radiative corrections; theory and performance of stored beams and particle accelerators. Mailing Add: Dept Weapons & Systs Eng US Naval Acad Annapolis MD 21402

MATIOLI, GASTONE, b Volta, Italy, Aug 29, 31; m 59; c 2. CELL BIOLOGY. Educ: Univ Pavia, MD, 57. Prof Exp: Asst prof path, Univ Perugia, 57-60; res assoc cytophys, Karolinska Inst, Sweden, 60-61; asst prof microbiol, Dept Nuclear Med, Univ Calif, Los Angeles, 64-66; ASSOC PROF MICROBIOL, SCH MED, UNIV SOUTHERN CALIF, 66- Res: Studies on cell differentiation; cytogenics; cytochemistry. Mailing Add: Dept of Microbiol Univ Southern Calif Sch of Med Los Angeles CA 90033

MATIS, JAMES HENRY , b Chicago, Ill, Mar 3, 41; m 63; c 3. STATISTICS, MATHEMATICAL STATISTICS. Educ: Weber State Col, BS, 65; Brigham Young Univ, MS, 67; Tex A&M Univ, PhD(statist), 70. Prof Exp: Math statistician, Intermountain Forest & Range Exp Sta, US Forest Serv, 65-67; res assoc statist, 70, asst prof, 70-74, ASSOC PROF STATIST, TEX A&M UNIV, 74- Mem: Am Statist Asn; Biomet Soc. Res: Applied stochastic processes; compartmental analysis; time series models of fish behavior; econometrics. Mailing Add: Inst of Statist Tex A&M Univ College Station TX 77843

MATISHECK, PETER HENRY, microbiology, see 12th edition

MATKIN, ORIS ARTHUR, b Powell, Wyo, Jan 14, 17; m 42; c 3. HORTICULTURE. Educ: Univ Calif, Los Angeles, BA, 40. Prof Exp: OWNER & DIR, SOIL & PLANT LAB, INC, 46- Honors & Awards: Res Award, Calif Asn Nurserymen, 74. Mem: Am Soc Plant Physiol; Am Soc Hort Sci; Soil Sci Soc Am; Am Soc Agron. Res: Soil, plant, water and pathology analyses. Mailing Add: Soil & Plant Lab Inc Southern Calif Off PO Box 11744 Santa Ana CA 92711

MATKOVICH, VLADO IVAN, b Vrboska, Yugoslavia, Feb 17, 24; nat US; m 51; c 2. INORGANIC CHEMISTRY. Educ: Univ Zagreb, dipl, 51; Univ Toronto, PhD, 56. Prof Exp: Supvr, Aluminum Labs, Ltd, Can, 55-57; eng scientist, Allis-Chalmers Mfg Co, 57-61; res assoc, 61-69, proj mgr, Eng Br, 69-75, MGR TECH BR, CARBORUNDUM CO, NY, 75- Mem: Am Chem Soc. Res: Crystal chemistry; synthesis and development of high temperature materials; plastics engineering development; plant design and construction. Mailing Add: Eng Br Res & Develop Div Carborundum Co Niagara Falls NY 14302

MATLACK, ALBERT SHELTON, b Washington, DC, Aug 14, 23; m 53; c 2. ORGANIC POLYMER CHEMISTRY. Educ: Univ Va, BS, 44; Univ Minn, PhD(org chem), 50. Prof Exp: Res chemist, 50-67, SR RES CHEMIST, RES CTR, HERCULES INC, 67- Mem: AAAS; Am Chem Soc. Res: Reactions of organometallic compounds with quinolones; monomer and polymer synthesis; industrial organic chemistry; organic synthesis. Mailing Add: Res Ctr Hercules Inc Wilmington DE 19899

MATLACK, GEORGE MILLER, b Pittsburgh, Pa, June 14, 21; m 43; c 4. RADIOCHEMISTRY. Educ: Grinnell Col, AB, 43; Univ Iowa, MS, 46, PhD(chem), 49. Prof Exp: Chemist, Iowa Geol Surv, 43-46; asst chem, Univ Iowa, 46-47, res assoc, 47-49; MEM STAFF, LOS ALAMOS SCI LAB, UNIV CALIF, 49- Mem: AAAS; Am Inst Chem; Am Chem Soc; Am Nuclear Soc. Res: Radiochemistry of plutonium and fission products; radiation properties of plutonium-238 fuels and environmental effects. Mailing Add: Los Alamos Sci Lab Univ of Calif Los Alamos NM 87544

MATLACK, LOUIS ROGERS, b Mt Holly, NJ, Mar 29, 35; m 56; c 4. POLYMER CHEMISTRY, PHYSICAL CHEMISTRY. Educ: Haverford Col, BA, 57; Princeton Univ, MA, 59, PhD(phys chem), 60. Prof Exp: Develop chemist, Imp Floglaze Points Ltd, Can, 61-63; pres, Geo D Wetherill & Co, Inc, Pa, 63-68; dir prod res & develop, 69-73, mgr mfg, Form Div, 73-75, MGR BUS DEVELOP, FORM DIV, SCOTT PAPER CO, 75- Mem: Am Chem Soc; Fedn Socs Paint Technol; Tech Asn Pulp & Paper Indust. Res: Free radical polymerization kinetics; colloid and surface phenomena. Mailing Add: 55 E Maple Ave Moorestown NJ 08057

MATLEY, IAN MURRAY, b Edinburgh, Scotland, Oct 23, 21; m 54. GEOGRAPHY. Educ: Univ Edinburgh, MA, 48; Univ Mich, PhD(geog), 61. Prof Exp: Res officer econ, Brit Civil Serv, 49-58; assoc, Res Inst, Univ Mich, 58-61; asst prof geog, Columbia Univ, 61-63; assoc prof, 63-67, PROF GEOG, MICH STATE UNIV, 67- Mem: Asn Am Geogr; Am Geog Soc. Res: Human geography; geography of Europe and the Soviet Union. Mailing Add: Dept of Geog Mich State Univ East Lansing MI 48824

MATLIS, EBEN, b Pittsburgh, Pa, Aug 28, 23; m 42; c 2. MATHEMATICS. Educ: Univ Pittsburgh, BS, 48; Univ Chicago, MS, 56, PhD(algebra), 58. Prof Exp: From instr to assoc prof, 58-67, PROF MATH, COL ARTS & SCI, NORTHWESTERN UNIV, ILL, 67- Concurrent Pos: Mem, Inst Advan Study, 62-63. Mem: Am Math Soc. Res: Homological algebra; theory of rings and modules. Mailing Add: Dept of Math Northwestern Univ Col of Arts & Sci Evanston IL 60201

MATLOCK, RALPH S, b Norman, Okla, Aug 24, 21; m 46; c 4. AGRONOMY. Educ: Okla State Univ, BS, 48, MS, 49; Univ Nebr, PhD(genetics, plant breeding), 52. Prof Exp: Jr sci aid alfalfa, Univ Nebr, 49-51, instr genetics, 51-52, from asst prof to assoc prof agron, 52-57, PROF AGRON, OKLA STATE UNIV, 57-, HEAD DEPT, 68- Mem: Fel Am Soc Agron; Crop Sci Soc Am; Am Genetics Asn. Res: Plant genetics, breeding and crop ecology, dealing with peanuts, soybeans, mungbean, cowpeas, guar, sesame and new and special crops. Mailing Add: Dept of Agron Okla State Univ Stillwater OK 74074

MATLOCK, REX LEON, b Plain Dealing, La, Nov 27, 34; m 55; c 3. PHYSICS. Educ: Northwestern State Univ, BS, 60; La State Univ, Baton Rouge, MS, 65, PhD(physics), 67. Prof Exp: Asst prof, 67-70, ASSOC PROF PHYSICS, LA STATE UNIV, SHREVEPORT, 70-, CHMN DEPT, 74- Concurrent Pos: Grant, La State Univ, Shreveport, 70-71. Mem: Am Phys Soc. Res: High energy interactions and cosmic ray physics. Mailing Add: Dept of Physics La State Univ Shreveport LA 71105

MATLOW, SHELDON LEO, b Chicago, Ill, Aug 24, 28; m 58; c 3. CHEMICAL PHYSICS. Educ: Univ Chicago, PhB, 48, BS, 49, PhD(chem), 53. Prof Exp: Asst chem, Univ Chicago, 50-52; res assoc, Brookhaven Nat Lab, 53-54; dir chem res, Jefferson Elec Co, Ill, 55; sr physicist, Hoffman Electronics Corp, 57, unit supvr, 57-59, sect mgr, 59, tech coordr, 60; dir res & develop, Intellux, Inc, Calif, 61; pres, Inst Study Solid State, 62; sr scientist, Korad Corp, 63; mgr develop eng, Clevite Corp, 64; CONSULT, 65- Mem: Am Chem Soc; Am Phys Soc; Electrochem Soc; The Chem Soc; NY Acad Sci. Res: Quantum mechanics; solid state physics; materials science and technology; philosophy of science. Mailing Add: 2545 Booksin Ave San Jose CA 95125

MATNEY, THOMAS STULL, b Kansas City, Mo, Sept 21, 28; m 54; c 3. BACTERIOLOGY. Educ: Trinity Univ, BS, 48, BA, 49, MA, 51; Univ Tex, PhD(bact), 58. Prof Exp: Asst res biochemist, Southwest Res Inst, 50-52; med bacteriologist, Res & Develop Lab, US Dept Army, 52-55; assoc biologist & assoc prof biol, Univ Tex M D Anderson Hosp & Tumor Inst, 62-69; assoc prof, 63-70, PROF & ASSOC DEAN, UNIV TEX GRAD SCH BIOMED SCI HOUSTON, 70- Concurrent Pos: Instr, Trinity Univ (Tex), 50-51. Mem: Am Soc Microbiol; Genetics Soc Am; Environ Mutagen Soc; AAAS. Res: Bacterial genetics; radiobiology; biochemistry. Mailing Add: Univ Tex Grad Sch Biomed Sci 6415 Main St Houston TX 77030

MATOCHA, CHARLES K, b Hondo, Tex, Aug 13, 29; m 53; c 2. SPECTROCHEMISTRY. Educ: St Marys Univ, Tex, BS, 49. Prof Exp: From anal chemist to group leader, Aluminum Co Am, Tex & Pa, 49-73, SCI ASSOC CHEM, ALCOA TECH CTR, ALUMINUM CO AM, 73- Mem: Sigma Xi; Am Chem Soc; Soc Appl Spectros. Res: Analytical methods for x-ray fluorescence analysis with emphasis on nonmetallic samples; automation of analytical procedures; development of computer systems for mathematical correlation, data handling and automation. Mailing Add: Alcoa Tech Ctr Alcoa Center PA 15069

MATOLTSY, ALEXANDER GEDEON, b Kaposvar, Hungary, Feb 27, 20; US nat; m. DERMATOLOGY. Educ: Univ Budapest, MD, 44. Prof Exp: Asst prof histol, Med Sch, Univ Budapest, 43-45; res assoc, Hungarian Biol Res Inst, Tihany, 45-47; asst prof cytol, Inst Muscle Res, Woods Hole, 49; res assoc dermat, Harvard Med Sch & Mass Gen Hosp, 49-59; asst prof, Rockefeller Inst, 56-59; res prof dermat, Med Sch, Univ Miami, 59-61; RES PROF DERMAT & PATH, SCH MED, BOSTON UNIV, 61- Concurrent Pos: Spec res fel, Karolinska Inst, Sweden, 47-49. Mem: Electron Micros Soc Am; Am Soc Cell Biol; Soc Invest Dermat. Res: Keratin and keratinization. Mailing Add: Dept of Dermat & Anat Boston Univ Sch of Med Boston MA 02118

MATOLYAK, JOHN, b Johnstown, Pa, June 26, 39; m 63; c 2. MAGNETISM. Educ: St Francis Col, Pa, BS, 63; Univ Toledo, MS, 66; WVa Univ, PhD(physics), 75. Prof Exp: Instr math & physics, St Francis Col, Pa, 63-64; ASSOC PROF PHYSICS, IND UNIV, PA, 66- Mem: Am Phys Soc. Res: Magnetic properties of crystals particularly the magnetostriction of antiferromagnets and weak ferromagnets. Mailing Add: Dept of Physics Ind Univ of Pa Indiana PA 15701

MATOVICH, EDWIN, b New Chicago, Ind, Aug 29, 35; m 56; c 3. PHYSICAL CHEMISTRY. Educ: Ariz State Univ, BS, 56. Prof Exp: Anal chemist, Nat Lead Co, Utah, 56-57; res chemist, Motorola Semiconductor Prod, Inc, 57-59; device develop engr, Semiconductor Div, Hughes Aircraft Co, Calif, 59-62; electro-optical res chemist, Quantum Tech Lab, Calif, 62-63; res specialist, Autonetics Div, NAm Rockwell Corp, 63-72; CONSULT PHYS CHEM, 72- Mem: Sigma Xi. Res: Semiconductor devices and lasers; quantum behavior of fluorescent fluids; radiation devices; high temperature chemistry, particularly with regard to fuel processing, fuel gasification and desulfurization; petroleum and petrochemical engineering; chemical process systems, analysis and engineering; economic analyses. Mailing Add: PO Box 3904 Fullerton CA 92634

MATOVINOVIC, JOSIP, b Licko Cerje, Yugoslavia, Dec 22, 14; US citizen; m 43. MEDICINE. Educ: Univ Zagreb, MD, 39. Prof Exp: Resident internal med, State Gen Hosp, Zagreb, 40-45; asst prof, Med Sch, Univ Zagreb, 45-46, chief div endocrinol, 48-56, docent, 51-56; res assoc thyroid clin, Mass Gen Hosp, Harvard Med Sch, 56-58, res assoc diabetes clin, 59; from instr to assoc prof, 59-70, PROF INTERNAL MED, MED SCH, UNIV MICH, ANN ARBOR, 70- Concurrent Pos: Consult study group on endemic goiter, WHO, 52, Pakistan & Lebanon, 60; mem, Yugoslav Comn Prev Endemic Goiter, 53; ed bd, Yugoslav Encycl Med, 56; dir Mich study endemic goiter & iodine nutriture, Ctr Dis Control, 71; consult, Radiation Ctr, WHO, Bombay, India, 73; clin & res fel, Mass Gen Hosp, Harvard Med Sch, 47-48. Mem: Endocrine Soc; Am Thyroid Asn. Res: Transplantable thyroid tumor of the rat; proliferation and differentiation of thyroid carcinoma in cell culture. Mailing Add: Dept of Internal Med Univ Hosp Univ of Mich Med Sch Ann Arbor MI 48109

MATRAY, OTTO JACK, b The Hague, Neth, Feb 28, 30; m 57; c 5. ORGANIC CHEMISTRY. Educ: Univ Leiden, cand, 51, doctoraal, 53, PhD(org chem), 56. Prof Exp: Res chemist & sr res chemist, Textile Fibers Dept, Del, 56-64, tech supvr, 64-66, develop supvr, 66-69, RES ASSOC, E I DU PONT DE NEMOURS & CO INT, 69- Mem: Am Chem Soc; Royal Neth Chem Soc. Res: End-use research textile fibers. Mailing Add: E I du Pont de Nemours & Co Int 81 Rte de L'aire PO Box CH-1211 Geneva Switzerland

MATRICK, HOWARD, chemistry, see 12th edition

MATRONE, GENNARD, biochemistry, deceased

MATSCH, CHARLES LEO, b Hastings, Minn. GEOLOGY, GLACIAL GEOLOGY. Educ: Univ Maine, BA, 59; Univ Minn, MS, 62; Univ Wis-Madison, PhD(geol), 71. Prof Exp: Explor geologist petrol, Stand Oil Co Tex, 61-64; instr geol, Univ Minn, Minneapolis, 64-66, asst prof, 66-70; asst prof, 70-72, ASSOC PROF GEOL, UNIV MINN, DULUTH, 72- Concurrent Pos: Secy, INQUA Comn, Genesis Glacial Sediments, 73- Mem: Geol Soc Am; Am Asn Petrol Geologists; Am Quaternary Asn; Nat Asn Geol Teachers; AAAS. Res: Glacial geology of the midcontinent of North America; origin of quaternary continental sediments; environmental geology of glaciated terrains. Mailing Add: Dept of Geol Univ of Minn Duluth MN 55812

MATSCHINER, JOHN THOMAS, b Portland, Ore, Dec 2, 27; m 49; c 5. BIOCHEMISTRY. Educ: Univ Portland, BS, 50, MS, 51; St Louis Univ, PhD(biochem), 57. Prof Exp: Asst, Univ Va, 51-52; from instr to assoc prof biochem, Sch Med, St Louis Univ, 58-70; PROF BIOCHEM, SCH MED, UNIV NEBR, 70- Concurrent Pos: Jane Coffin Childs Fund res fel, Univ Calif, 57-58. Mem: AAAS; Am Chem Soc; Am Soc Biol Chem; Am Inst Nutrit. Res: Biochemistry and nutrition of vitamin K. Mailing Add: Dept of Biochem Univ of Nebr Med Ctr Omaha NE 68105

MATSEN, FREDERICK ALBERT, b Racine, Wis, July 26, 13; m 38; c 2. CHEMISTRY, PHYSICS. Educ: Univ Wis, BS, 37; Princeton Univ, PhD(chem physics), 50. Prof Exp: Instr chem, Bucknell Univ, 40-42; from instr to assoc prof, 42-51, PROF CHEM, UNIV TEX, AUSTIN, 51- Concurrent Pos: Vis asst prof, Univ Chicago, 45; Guggenheim fel, Oxford Univ & Univ London, 51-52; NSF fel, Univ Paris, 61. Mem: Fel Am Phys Soc; The Chem Soc. Res: Atomic and molecular structure; quantum chemistry; magnetism. Mailing Add: Dept of Chem Univ of Tex Austin TX 78712

MATSEN, JOHN MARTIN, b Salt Lake City, Utah, Feb 7, 33; m 59; c 8. MEDICINE, MICROBIOLOGY. Educ: Brigham Young Univ, BA, 48; Univ Calif, Los Angeles, MD, 63. Prof Exp: From intern to resident pediat, Univ Calif, Los Angeles, 63-66; from asst prof to prof lab med & path, pediat & microbiol, Univ Minn, Minneapolis, 68-74; PROF PATH & PEDIAT & DIR CLIN LAB, UNIV UTAH, 74- Concurrent Pos: Fel pediat infectious dis, Univ Minn, Minneapolis, 66-68. Mem: Soc Pediat Res; Infectious Dis Soc Am; fel Am Acad Pediat; fel Am Soc Clin Path. Res: Pediatric enteric infections; antibiotic evaluation and evaluation of procedures in diagnostic microbiology; studies on biology of tribe Klebsiellae. Mailing Add: Dept of Path & Pediat Univ of Utah Hosp Salt Lake City UT 84132

MATSON, DENNIS LUDWIG, b San Diego, Calif, Sept 29, 42. PLANETARY SCIENCES. Educ: San Diego State Univ, AB, 64; Calif Inst Technol, PhD(planetary sci), 72. Prof Exp: Res assoc planetology, 72-74, SR SCIENTIST PLANETOLOGY, JET PROPULSION LAB, CALIF INST TECHNOL, 74- Mem: Am Geophys Union; Am Astron Soc; AAAS. Res: Composition and morphology of planetary surfaces; astronomical photometry and spectroscopy; two-dimensional photometry and imaging of solar system bodies. Mailing Add: Mail Code 183-501 Jet Propulsion Lab Pasadena CA 91103

MATSON, EDWARD JOHN, organic chemistry, see 12th edition

MATSON, HOWARD JOHN, b Monmouth, Ill, June 8, 21; m 46; c 4. ORGANIC CHEMISTRY, PETROLEUM CHEMISTRY. Educ: Monmouth Col, Ill, BS, 43; Pa State Univ, MS, 47. Prof Exp: Res asst petrol refining, Pa State Univ, 43-47, res instr, 47-50; chemist, Sinclair Res Labs, 50-53, group leader, 53-55, sect leader, 55-63, res scientist, 63-69, asst mgr prod qual, Atlantic Richfield Co, 69-70, supvr, 70-74, MGR PROD SPECIALTIES, ATLANTIC RICHFIELD CO, 74- Mem: Am Chem Soc; Am Inst Chemists; Am Soc Lubrication Engrs; Am Soc Testing & Mat. Res: Petroleum product research and development. Mailing Add: Harvey Tech Ctr Atlantic Richfield Co Harvey IL 60426

MATSON, TED P, b Ponca City, Okla, Jan 5, 29; m 51; c 3. APPLIED CHEMISTRY, SURFACE CHEMISTRY. Educ: Univ Okla, BS, 49, EdM, 51; Okla State Univ, MS, 67. Prof Exp: Teacher & coach pub schs, Okla, 51-56; asst res chemist, Res & Develop Dept, Continental Oil Co, 57-59, assoc res chemist, 59-62, res chemist, 62-64, tech adv to managing dir, Condea Petrochem GmbH, Hamburg, Ger, 64-65; res chemist, Res & Develop Dept, Continental Oil Co, 65-66, prod develop coordr, Conoco Chem, 66-72, RES GROUP LEADER SURFACTANTS, CONTINENTAL OIL CO, 72- Mem: Am Oil Chem Soc; Am Soc Testing & Mat; Chem Specialties Mfrs Asn. Res: New product development; oil field chemicals; study of applications and synthesis of surfactants and research and development of evaluation techniques. Mailing Add: Continental Oil Co Drawer 1267 Ponca City OK 74602

MATSUDA, KEN, b Napa, Calif, Nov 30, 20; m 46; c 2. ORGANIC CHEMISTRY. Educ: Univ Md, BS, 44, PhD(chem), 51. Prof Exp: Res chemist, 51-57, sr res chemist, 57-62, group leader, 62-71, proj mgr, 71-74, MGR CHEM SECT, STAMFORD LABS, AM CYANAMID CO, 74- Mem: AAAS; Am Chem Soc. Res: Research and development in organic chemistry; catalysis; polymers. Mailing Add: Stamford Labs Am Cyanamid Co Stamford CT 06904

MATSUDA, YOSHIYUKI, b Manchuria, China, Dec 7, 43; Japanese citizen; m 71. PLASMA PHYSICS. Educ: Kyoto Univ, BS, 66, MS, 68; Stanford Univ, PhD(elec eng), 74. Prof Exp: RES ASSOC PLASMA PHYSICS, PLASMA PHYSICS LAB, PRINCETON UNIV, 74- Mem: Am Phys Soc; Inst Elec & Electronics Engrs; AAAS. Res: Theoretical and computational study of plasma physics and controlled thermonuclear fusion. Mailing Add: Plasma Physics Lab Princeton Univ PO Box 451 Princeton NJ 08540

MATSUDO, HITOSHI, b Los Angeles, Calif, Apr 7, 33. CELL BIOLOGY. Educ: Univ Southern Calif, AB, 58, MS, 61, PhD(biol), 67. Prof Exp: Biologist, Arctic Res Lab, Alaska, 61 & US Navy Ship Eltanin, 62; res asst biol, Univ Southern Calif, 63, res assoc biol sci, 67-68; res biologist, Scripps Inst Oceanog, Univ Calif, San Diego, 68-69; vis lectr biol sci, Univ Southern Calif, 69-70; electron microscopist, Vet Admin Ctr, Sawtelle, Calif, 70; ELECTRON MICROSCOPIST, RES FOUND, ST JOSEPH MED CTR, BURBANK, 70-, DIR, ELECTRON MICROS LAB, 71-; RES ASSOC BIOL SCI, UNIV SOUTHERN CALIF, 71- Mem: AAAS; Soc Protozool; Electron Micros Soc Am; Am Soc Cell Biol. Res: Cell ultrastructure; ultrastructure and morphogenesis of chonotrichous ciliate protozoan; marine biology; fish pathology. Mailing Add: Ultrastruct Res Lab St Joseph Med Ctr Burbank CA 91505

MATSUGUMA, HAROLD JOSEPH, b Honolulu, Hawaii, Oct 15, 28; m 63; c 1. INORGANIC CHEMISTRY. Educ: Univ Hawaii, BA, 51; Univ Ill, MS, 52, PhD(chem), 55. Prof Exp: Res assoc chem, Univ Ill, 52-55, assoc, 55; chemist, Explosives Res Sect, 55-57, chief, Synthesis Unit, 57-59, officer, Reactor Requirements & Explosives Res Sect, 59-63, actg chief, Explosives Lab, 63-66, CHIEF CHEM BR, FELTMAN RES LAB, PICATINNY ARSENAL, US DEPT ARMY, 66- Mem: Am Chem Soc; The Chem Soc; Am Defense Preparedness Asn. Res: Synthesis of hydrazine; hydroxylamine derivatives; chemistries of nitrogen, phosphorus and sulfur compounds; chemistry of explosives; detonation phenomena; relation of chemical constitution to explosive properties. Mailing Add: Explosives Lab SARPA-FR-E-C Picatinny Arsenal Dover NJ 07801

MATSUMOTO, CHARLES, b San Jose, Calif, Mar 25, 32; m 61; c 1. PHARMACOLOGY, BIOCHEMISTRY. Educ: San Jose State Col, BA, 53; Univ Idaho, MS, 55; Univ Wash, PhD(pharmacol), 63. Prof Exp: Chemist biol lab, US Fish & Wildlife Serv, 58-70; SR PHARMACOLOGIST, LILLY RES LAB, ELI LILLY & CO, 65- Concurrent Pos: Exec ed, Life Sci, 70-73; NIH fel, Lab Chem Pharmacol, Nat Heart Inst, 63-65. Mem: AAAS; Am Chem Soc; Am Soc Pharmacol & Exp Therapeut; Sigma Xi. Res: Autonomic, cardiovascular and biochemical pharmacology. Mailing Add: Eli Lilly Res Labs Indianapolis IN 46206

MATSUMOTO, HIROMU, b Honolulu, Hawaii, Mar 28, 20. AGRICULTURAL BIOCHEMISTRY. Educ: Univ Hawaii, BS, 44, MS, 45; Purdue Univ, PhD(biochem), 55. Prof Exp: Asst chem, Exp Sta, Univ Hawaii, 45-49, jr chemist, 49-51; asst biochem Purdue Univ, 51-53; from asst biochemist to assoc biochemist, 55-66, BIOCHEMIST, EXP STA, UNIV HAWAII, 66- Concurrent Pos: Fel, Japan Soc Advan Sci, 75-76. Mem: Am Chem Soc; Soc Toxicol; Am Asn Cancer Res. Res: Effect of toxic plant constituents on animal metabolism; mimosine, 3-nitropropanoic acid, methylazoxymethanol; carcinogenesis; neurochemistry; metabolic fate in animals of naturally occurring toxicants; cyasin, methylazoxymethanol-glucosiduronic acid; toxicology; chemical carcinogenesis. Mailing Add: Dept of Agr Biochem Univ of Hawaii Honolulu HI 96822

MATSUMOTO, KEN, b San Bernadino, Calif, Sept 8, 41; m 67; c 1. ORGANIC CHEMISTRY, MEDICINAL CHEMISTRY. Educ: Ariz State Univ, BS, 63; Univ Calif, Berkeley, PhD(org chem), 67. Prof Exp: Teaching asst, Univ Calif, Berkeley, 63-64; sr org chemist, 69-75, SR RES SCIENTIST, LILLY RES LABS, 75- Mem: Am Chem Soc. Mailing Add: Lilly Res Labs Eli Lilly & Co Indianapolis IN 46206

MATSUMOTO, YORIMI, b Yuba City, Calif, July 29, 26. PHYSIOLOGY. Educ: Whittier Col, AB, 50; Univ Calif, Los Angeles, PhD(zool), 64. Prof Exp: Instr biophys, Univ Ill, Urbana, 63-66, from asst prof to assoc prof, 66-69; ASSOC PROF PHYSIOL, EMORY UNIV, 69- Mem: AAAS; Biophys Soc; Am Soc Zoologists. Res: Mechanical analysis of muscular contraction; heat analysis of muscle-contraction; nerve-heat; birefrigency study of invertebrate muscle. Mailing Add: Dept of Physiol Emory Univ Sch of Med Atlanta GA 30322

MATSUMURA, PHILIP, b San Jose, Calif, Aug 15, 47. MICROBIAL PHYSIOLOGY. Educ: Univ Santa Clara, BS, 69; Univ Rochester, PhD(microbiol), 75. Prof Exp: Fel, Univ Calif, San Diego, 75- Mem: AAAS; Sigma Xi; Soc Gen Microbiol; Am Soc Microbiol. Res: Microbial barophysiology; microbial motility and chemotaxis. Mailing Add: Dept of Biol Univ of Calif at San Diego La Jolla CA 92093

MATSUO, KEIZO, b Osaka, Japan, Apr 23, 42. POLYMER CHEMISTRY. Educ: Kyoto Univ, BS, 66, MS, 68; Dartmouth Col, PhD(chem), 72. Prof Exp: Res assoc chem, 72-74, RES INSTR CHEM, DARTMOUTH COL, 74- Mem: Am Chem Soc; Japan Chem Soc. Res: Synthesis of new polymer; characterization; equilibrium and non-equilibrium study of polymer solution; kinetics; play with rotational state model. Mailing Add: Dept of Chem Dartmouth Col Hanover NH 03755

MATSUO, ROBERT R, b Duncan, BC, Feb 28, 32; m 61; c 2. BIOCHEMISTRY. Educ: Univ Man, BSc, 57; Univ Alta, PhD(plant biochem), 62. Prof Exp: Chemist I, Grain Res Lab, 57-59, chemist III, Durum Wheat Res, 62-66, RES SCIENTIST, GRAIN RES LAB, CAN DEPT AGR, 66- Mem: AAAS; Am Asn Cereal Chemists; Chem Inst Can; Prof Inst Pub Serv Can; Can Inst Food Sci & Technol. Res: Cereal chemistry; basic and applied research on durum wheat and durum wheat products. Mailing Add: Grain Res Lab 1404 303 Main St Winnipeg MB Can

MATSUOKA, TATS, b Seattle, Wash, Aug 24, 29; m 64; c 1. VIROLOGY. Educ: Univ Minn, BA, 52; State Col Wash, DVM, 59. Prof Exp: Asst bacteriologist, Mont State Col, 52-55, asst bacteriologist & virologist, Vet Res Lab, 61-63; practicing vet, Idaho, 59-60; vet diagnostician, Mont Livestock Sanit Bd, 60-61; res vet, 63-67, SR VIROLOGIST, GREENFIELD LABS, ELI LILLY & CO, 67- Mem: Am Vet Med Asn; Am Soc Microbio; US Animal Health Asn; Conf Res Workers Animal Dis. Res: Anaerobic bacteria pathogenic to animals; animal viruses, particularly viral diseases of bovine. Mailing Add: Greenfield Labs Box 708 Eli Lilly & Co Greenfield IN 46140

MATSUSAKA, TERUHISHA, b Kyoto, Japan, Apr 5, 26; m 50; c 5. GEOMETRY. Educ: Kyoto Univ, MS, 49, PhD(math), 54. Prof Exp: Instr math, Ochanomizu Univ, Japan, 52-53, asst prof, 53-54; res assoc, Univ Chicago, 54-57; from assoc prof to prof, Northwestern Univ, 57-61; PROF MATH, BRANDEIS UNIV, 61- Concurrent Pos: Guggenheim fel, 59. Mem: Am Math Soc; Am Acad Arts & Sci. Res: Algebra; algebraic geometry. Mailing Add: Dept of Math Brandeis Univ Waltham MA 02154

MATSUSHIMA, JOHN K, b Denver, Colo, Dec 24, 20; m 43; c 2. ANIMAL NUTRITION. Educ: Colo State Univ, BS, 43, MS, 45; Univ Minn, PhD, 49. Prof Exp: Asst animal husb, Colo State Univ, 43-45; from asst prof to prof, Univ Nebr, 49-61; PROF ANIMAL SCI, COLO STATE UNIV, 61- Mem: Am Soc Animal Sci; Soc Range Mgt; Am Dairy Sci Asn; Am Inst Nutrit. Res: Beef cattle nutrition, feeding and management. Mailing Add: Dept of Animal Sci Colo State Univ Ft Collins CO 80521

MATSUSHIMA, SATOSHI, b Fukui, Japan, May 6, 23; nat US; m 55; c 2. ASTRONOMY, ASTROPHYSICS. Educ: Univ Kyoto, MS, 46; Univ Utah, PhD(astrophys), 54; Univ Tokyo, DSc, 66. Prof Exp: Asst astron, Univ Kyoto, 46-50; res fel & asst, High Altitude Observ & Harvard Col Observ, 50-54; res assoc physics, Univ Pa & Strawbridge Observ, Haverford Col, 54-55; vis astronr, Astrophys Inst & Meudon Observ, Paris, France, 56-57; Humboldt fel, Inst Theoret Physics, Univ Kiel, 57-58; asst prof physics, Fla State Univ, 58-60; assoc prof astron, Univ Iowa, 60-67; PROF ASTRON, PA STATE UNIV, 67-, HEAD DEPT, 76- Concurrent Pos: Travel grants, Int Astron Union, 56 & 57, Ger Astron Soc, 57, US Res Coop, 58 & NSF, 65-66; guest astronr, Utrecht Observ, Neth, 56; sr res fel, Calif Inst Technol, 59-61; vis prof, US-Japan Coop Sci Prog, Univs Tokyo & Kyoto, 65-66; consult, Naval Res Lab, 62; mem, Int Astron Union Comns 12 & 36. Mem: Am Astron Soc; fel Royal Astron Soc; Am Geophys Union. Res: Theory of stellar atmospheres; solar and planetary physics; spectroscopy and spectrophometry; space and upper atmosphere physics. Mailing Add: 504 Davey Lab Pa State Univ University Park PA 16802

MATSUSHITA, SADAMI, b Ehime, Japan, Feb 12, 20; m; c 2. GEOPHYSICS. Educ: Kyoto Univ, MSc, 44, DrSc, 51. Prof Exp: Asst geophys, Kyoto Univ, 45, lectr, 45-54; mem res staff physics, Univ Col, London, 54-55; PROF ASTROGEOPHYS, UNIV COLO & MEM SR RES STAFF, HIGH ALTITUDE OBSERV, NAT CTR ATMOSPHERIC RES, 55-56, 57- Concurrent Pos: Guest worker, Nat Bur Standards, 55-56 & 57-; consult, Environ Sci Serv Admin; mem, Int Sci Radio Union & Int Union Geod & Geophys. Honors & Awards: Scientist Award, Sigma Xi, 63. Mem: Fel AAAS; Meteorol Soc Am; Am Geophys Union; Sigma Xi. Res: Relations among geomagnetism, ionosphere, space and the sun. Mailing Add: High Altitude Observ Univ of Colo Boulder CO 80302

MATSUSHITA, TATSUO, b Kearny, NJ, July 29, 37; m 67; c 1. BIOCHEMICAL GENETICS. Educ: Cornell Univ, AB, 60; Rutgers Univ, PhD(biochem), 70. Prof Exp: Teaching asst biochem, Rutgers Univ, 67-68; res assoc microbial genetics, Princeton Univ, 70-72; ASST GENETICIST, ARGONNE NAT LAB, 72- Concurrent Pos: NIH fel, Princeton Univ, 71-72. Mem: Genetics Soc Am; Am Soc Microbiol; Biophys Soc. Res: Biochemical genetics of bacterial and eukaryotic DNA replication; in vitro mutation and immunoglobulin synthesis with mouse myelomas; tryptophan photoproducts and DNA polymerases. Mailing Add: Div of Biol & Med Res Argonne Nat Lab Argonne IL 60439

MATSUYAMA, GEORGE, b Fresno, Calif, Nov 20, 18; m 45; c 2. ELECTROANALYTICAL CHEMISTRY. Educ: Univ Calif, BS, 40; Univ Minn, PhD(phys chem), 48. Prof Exp: Asst chem, Fresno State Col, 36-38; asst, Univ Minn, 40-43, instr, 43-48; asst prof, Wesleyan Univ, 48-52; res chemist, Union Oil Co, Calif, 52-55, sr res chemist, 55-57, res assoc, 57-59; sr chemist, 59-64, eng specialist, 65-69, RES SCIENTIST, BECKMAN INSTRUMENTS, INC, 70- Mem: AAAS; Am Chem Soc; Am Asn Clin Chemists; Electrochem Soc. Res: Electrometric and volumetric analysis; electroanalytical instrumentation; ion-selective electrodes; gas sensors; enzyme electroanalytical methods. Mailing Add: 548 N Stanford Ave Fullerton CA 92631

MATT, JOSEPH, b Minneapolis, Minn, Jan 10, 20; m 61; c 2. ORGANIC CHEMISTRY. Educ: Univ Pa, BS, 41; Pa State Univ, MS, 42; Purdue Univ, PhD(chem), 49. Prof Exp: Res chemist, Sharp & Dohme Div, Merck & Co, Inc, 43-45 & Armour & Co, 48-54; sr res chemist, Va-Carolina Chem Corp, 54-56; res chemist, Alkydol Labs, Inc, 56-58 & Velsicol Chem Corp, 59-62; SR RES CHEMIST, NALCO CHEM CO, 62- Mem: Am Chem Soc. Res: Organic polymers; coating resins; organophosphorus and organofluorine compounds; fatty acid derivatives; microbiocides. Mailing Add: Nalco Chem Co Clearing Res Ctr 6216 W 66th Pl Chicago IL 60638

MATT, MORRIS CHALFANT, pharmaceutical chemistry, see 12th edition

MATTA, JOSEPH EDWARD, b Philadelphia, Pa, July 29, 48. ENVIRONMENTAL PHYSICS. Educ: St Joseph's Col, Philadelphia, BS, 70; Lehigh Univ, MS, 72, PhD(physics), 74. Prof Exp: Fel physics, Lehigh Univ, 74-75; RES PHYSICIST, MINES SAFETY RES CTR, 75- Mem: Am Phys Soc. Res: Investigate various dust and methane control techniques for underground coal mines. Mailing Add: Mines Safety Res Ctr Pittsburgh PA 15213

MATTA, MICHAEL STANLEY, b Dayton, Ohio, Feb 22, 40; m 62; c 3. BIOLOGICAL CHEMISTRY, ORGANIC CHEMISTRY. Educ: Univ Dayton, BS, 62; Ind Univ, PhD(org chem), 66. Prof Exp: Sr res chemist, Mound Lab, Monsanto Res Corp, Ohio, 66-68; res assoc biol chem, Amherst Col, 68-69; asst prof, 69-74, ASSOC PROF CHEM, SOUTHERN ILL UNIV, EDWARDSVILLE, 74- Mem: AAAS; Am Chem Soc. Res: Kinetics and mechanism of enzyme action; transfer reactions of borazines, free radical rearrangement and participation. Mailing Add: Dept of Chem Southern Ill Univ Edwardsville IL 62025

MATTAIR, ROBERT, b Boston, Mass, July 19, 22; m 47; c 5. INORGANIC CHEMISTRY, ENVIRONMENTAL SYSTEMS & TECHNOLOGY. Educ: Harvard Univ, AB, 44; Ohio State Univ, MA, 48, PhD(inorg chem), 50. Prof Exp: Res chemist, Oldbury Electrochem Co, NY, 50, mgr mkt develop, 50-56; mkt develop specialist, Plastics Dept, 56-66, Electrochem Dept, 66-68, Org Chems Dept, 68-74, SR MKT SPECIALIST, PERMASEP PROD DIV, ORG CHEMS DEPT, E I DU PONT DE NEMOURS & CO, 74- Mem: Am Chem Soc; Sigma Xi; Am Electroplaters Soc. Res: Market development requiring consultation with and guidance of research and other activities. Mailing Add: 2616 Marhill Dr Wilmington DE 19810

MATTANO, LEONARD AUGUST, b Tampa, Fla, July 8, 17; m 41; c 5. ORGANIC CHEMISTRY. Educ: Univ Wis, BS, 41; Mich State Univ, PhD, 48. Prof Exp: Chemist, Allis-Chalmers Co, Wis, 41-42 & Dow Chem Co, Mich, 42-45; res chemist, Standard Oil Co, Ind, 48-56; sr res chemist, Dow Chem Co, 56-69; DIR CHEM RES, BISSELL, INC, 69- Mem: Am Chem Soc. Res: Motor oil additives; chelate resins; surfactants; general syntheses; chemical specialties; aerosols. Mailing Add: 2325 Ducoma Dr NW Grand Rapids MI 49504

MATTAX, CALVIN COOLIDGE, b Sallisaw, Okla, Feb 4, 25; m 49; c 4. PHYSICAL CHEMISTRY. Educ: Univ Tulsa, BChem, 50; La State Univ, MS, 52, PhD(phys chem), 54. Prof Exp: Eng supvr, Esso Prod Res Co, 55-75, DIV MGR, EXXON PROD RES CO, 76- Mem: Am Chem Soc; Am Inst Mining, Metall & Petrol Eng. Res: Electrochemical kinetics; properties of polymer solutions; fluid mechanics in porous media; reservoir engineering. Mailing Add: 306 Chapel Bell Lane Houston TX 77024

MATTEI, JANET AKYÜZ, b Bodrum, Turkey, Jan 2, 43; m 72. ASTRONOMY. Educ: Brandeis Univ, BA, 65; Ege Univ, Turkey, Yüksek Lisans, 70; Univ Va, MS, 72. Prof Exp: Teacher physics, astron & phys sci, Am Col Inst, Turkey, 67-69; teaching asst astron, Ege Univ, Turkey, 69-70; asst dir, 72-73, DIR ASTRON, AM ASN VARIABLE STAR OBSERVERS, 73- Mem: Am Astron Soc; Am Asn Variable Star Observers. Res: Visual and photometric studies of variable stars, particularly dwarf novae, T Tauri stars and long period variables. Mailing Add: Am Asn Variable Star Observers 187 Concord Ave Cambridge MA 02138

MATTEN, LAWRENCE CHARLES, b Newark, NJ, Sept 1, 38; m 59; c 4. PALEOBOTANY. Educ: Rutgers Univ, BA, 59; Cornell Univ, PhD(bot), 65. Prof Exp: Instr biol, State Univ NY Col Cortland, 64-65; asst prof, 65-70, ASSOC PROF BOT, SOUTHERN ILL UNIV, CARBONDALE, 70- Concurrent Pos: Mem, Int Orgn Paleobot, Int Union Biol Sci. Mem: AAAS; Bot Soc Am; Paleobot Soc; Torrey Bot Club. Res: Elucidation of Paleozoic flora, especially Devonian plants from eastern United States. Mailing Add: Dept of Bot Southern Ill Univ Carbondale IL 62901

MATTENHEIMER, HERMANN G W, b Berlin, Ger, Mar 29, 21; m 43; c 3. BIOCHEMISTRY. Educ: Univ Göttingen, MD, 47. Prof Exp: Asst physician, Helmstedt Dist Hosp, Ger, 45-49; res asst, Berlin, 49-51; res asst, Free Univ Berlin, 51-55, privat-docent, 55-59; from asst prof to assoc prof biochem, Univ Ill, 59-71, PROF BIOCHEM, RUSH MED COL, 71- Concurrent Pos: Dir clin chem, Presby-St Luke's Hosp, 59-; res fel, Theodor Kocher Inst, Switz, 51-53; Rusk Orsteel Found fel, Carlsberg Lab, Denmark, 55, WHO fel, 56-57. Mem: AAAS; Am Chem Soc; NY Acad Sci; Am Soc Biol Chem; Ger Soc Biol Chem. Res: Cell metabolism; ultramicrotechniques for enzyme determinations in single cells; clinical chemistry; clotting of casein; renal biochemistry. Mailing Add: Dept of Biochm Rush-Presby-St Luke's Med Ctr Chicago IL 60612

MATTEO, MARTHA R, b New York, NY, Mar 11, 42; m 71; c 2. BIOCHEMISTRY. Educ: Univ Rochester, BA, 62; Brandeis Univ, PhD(biochem), 67. Prof Exp: Asst prof, 67-74, ASSOC PROF BIOL, UNIV MASS, BOSTON, 74-; RES SCIENTIST BIOCHEM, UNION CARBIDE CORP, 76- Mem: AAAS; Am Soc Microbiol; Am Chem Soc. Res: Comparative enzymology of regulation, adaptation and evolution; allosteric enzymes, aspartate transcarbamylase; isozymes and response to environmental pressures. Mailing Add: Corp Res Lab Union Carbide Corp Tarrytown NY 10591

MATTERN, CARL FREDERICK THEODORE, b Baltimore, Md, Dec 23, 23; m 46; c 6. INFECTIOUS DISEASES. Educ: Univ Md, MD, 47. Prof Exp: MED OFFICER, LAB VIRAL DIS, NAT INST ALLERGY & INFECTIOUS DIS, 49- Concurrent Pos: Vis res assoc, Virus Lab, Univ Calif, 57-59. Mem: AAAS; Am Soc Microbiol;

Am Micros Soc; Soc Protozool. Res: Virology and cell biology. Mailing Add: Lab of Viral Dis Nat Inst Allergy & Infect Dis Bethesda MD 20014

MATTERN, JOHN ARTHUR, chemistry, see 12th edition

MATTERN, KENNETH LAWRENCE, b Vananda, Mont, Jan 25, 20. INORGANIC CHEMISTRY. Educ: Baldwin-Wallace Col, BS, 42; Univ Calif, PhD(inorg chem), 51. Prof Exp: Instr chem & Naval V5 progs, Williams Col, 42-43; asst chem, Univ Calif, 46-47, res chemist, Radiation Lab, 47-51; res chemist, Calif Res & Develop Co, 51-54; res specialist, Atomics Int Div, NAm Aviation, Inc, 54-68; mem tech staff, Fuel Recycle Br, Div Reactor Develop & Technol, AEC, 68-75, MEM TECH STAFF, DIV NUCLEAR FUEL CYCLE & PROD, ENERGY RES & DEVELOP ADMIN, 75- Mem: Am Chem Soc; Am Nuclear Soc. Res: Complex ions of lanthanum in aqueous solutions; problems in aqueous and high temperature processing of radioactive materials; technical and economic aspects of nuclear fuel reprocessing and refabrication as part of the overall fuel cycle; inorganic materials. Mailing Add: 10500 Rockville Pike Apt 1606 Rockville MD 20852

MATTERN, PAUL JOSEPH, b Winnetoon, Nebr, Jan 26, 22; m 50; c 4. ANALYTICAL CHEMISTRY. Educ: State Col Iowa, BA, 47; Univ Wis, MS, 51. Prof Exp: Instr biochem & nutrit, 53-59, from asst prof to assoc prof agron, 59-70, PROF AGRON, UNIV NEBR, LINCOLN, 71- Mem: Am Chem Soc; Am Asn Cereal Chemists. Res: Environmental and genetic effects on the chemical, physical and nutritional properties of wheat constituents. Mailing Add: Dept of Agron Univ of Nebr Lincoln NE 68583

MATTES, FREDERICK HENRY, b Sheboygan, Wis, Feb 18, 41; m 64; c 2. ANALYTICAL CHEMISTRY. Educ: Carroll Col, BS, 63; Ind Univ, PhD(chem), 68. Prof Exp: Instr, 67-68, ASST PROF CHEM, WILLAMETTE UNIV, 68- Mem: Am Chem Soc. Res: Electroanalytical chemistry, particularly polarography and other voltammetric methods. Mailing Add: Dept of Chem Willamette Univ Salem OR 97301

MATTESON, DONALD STEPHEN, b Kalispell, Mont, Nov 8, 32; m 53, 71; c 2. ORGANOMETALLIC CHEMISTRY. Educ: Univ Calif, BS, 54; Univ Ill, PhD(chem), 57. Prof Exp: Res chemist, E I du Pont de Nemours & Co, 57-58; from instr to assoc prof, 58-69, PROF CHEM, WASH STATE UNIV, 69- Concurrent Pos: Sloan Found fel, 66-68. Mem: Am Chem Soc; AAAS, Res: Neighboring-group effects in organoboron compounds; mechanisms of electrophilic displacement; tetrametallomethane chemistry; carboranes; boron-substituted carbanions as synthetic intermediates. Mailing Add: Dept of Chem Wash State Univ Pullman WA 99163

MATTESON, JOHN WARREN, b Flint, Mich, Nov 6, 32; m 58; c 3. ECONOMIC ENTOMOLOGY. Educ: Univ Ill, AB, 54, MS, 56, PhD, 59. Prof Exp: Asst entom, State Natural Hist Surv, Ill, 54-56; res assoc, Univ Ill, 56-58; entomologist, Entom Res Div, Agr Res Serv, USDA, 59-63 & Develop Dept, Monsanto Co, Mo, 63-67; ENTOMOLOGIST, AGRICHEM RES LAB, MINN MINING & MFG CO, 67- Mem: Entom Soc Am. Res: Plant resistance to insects; insecticide residues; insect physiology; toxicology; taxonomy. Mailing Add: Agrichem Res Lab Minn Mining & Mfg Co St Paul MN 55101

MATTESON, MAX RICHARD, b Bear Lake, Mich, Nov 10, 09; m 30; c 2. ZOOLOGY. Educ: Cent Mich Univ, BS, 32; Univ Mich, MS, 40, PhD(zool), 46. Prof Exp: Head dept biol, Flint Jr Col, 37-44; from instr to asst prof zool, 46-53, ASSOC PROF ZOOL, UNIV ILL, URBANA, 53- Mem: AAAS; Am Micros Soc; Ecol Soc Am; Am Soc Limnol & Oceanog; Am Soc Zoologists. Res: Life cycles, ecology and comparative anatomy of fresh-water mussels; gametogenesis of fresh-water mussels; former and present distribution of freshwater mussels; anthropology. Mailing Add: Dept of Zool Univ of Ill Urbana IL 61801

MATTHEIS, EULA BINGHAM, b Covington, Ky, July 9, 29; m 59; c 3. ENVIRONMENTAL HEALTH. Educ: Eastern Ky Univ, BS, 51; Univ Cincinnati, MS, 54, PhD(zool), 58. Prof Exp: Res asst deep mycoses, Jewish Hosp, Cincinnati, Ohio, 54-55; res assoc indust health, 57, asst prof environ health, 62-70, ASSOC PROF ENVIRON HEALTH, COL MED, UNIV CINCINNATI, 70- Res: Bioassay of chemical carcinogens in human environment, factors altering potency; role of alveolar macrophages in pulmonary defense; effect of metallic and carcinogenic particulates on lung; modification of toxicity by aging. Mailing Add: Dept of Environ Health Univ of Cincinnati Col of Med Cincinnati OH 45267

MATTHEIS, FLOYD E, b Ellendale, NDak, Dec 21, 31; m 55; c 5. SCIENCE EDUCATION. Educ: Univ NDak, BS, 52; Univ NC, MEd, 59, EdD, 62. Prof Exp: Teacher high sch, Minn, 54-58; assoc prof, 60-66, PROF SCI EDUC & CHMN DEPT, E CAROLINA UNIV, 66- Concurrent Pos: Dir NSF In-serv Inst Earth Sci for Elem Sch Teachers, 64-65 & Dist II, Nat Sci Teachers Asn, 72-74. Mem: AAAS; Nat Asn Res Sci Teaching; Nat Sci Teachers Asn (dir, 72-74). Res: Experimental studies in science teaching. Mailing Add: Dept of Sci Educ E Carolina Univ Greenville NC 27834

MATTHEISS, LEONARD FRANCIS, solid state physics, see 12th edition

MATTHES, THEODORE K, mathematical statistics, deceased

MATTHEWS, BRIAN WESLEY, b SAustralia, May 25, 38; m 63; c 2. MOLECULAR BIOLOGY, X-RAY CRYSTALLOGRAPHY. Educ: Univ Adelaide, BSc, 59, Hons, 60, PhD(physics), 64. Prof Exp: Mem staff, Med Res Coun Lab Molecular Biol, Eng, 63-66; vis assoc molecular biol, NIH, 67-68; assoc prof, 69-72, PROF & RES ASSOC PHYSICS & MOLECULAR BIOL, UNIV ORE, 72- Concurrent Pos: Sloan Res Found fel, 71. Mem: Am Crystallog Asn. Res: Protein structure and function; crystallography. Mailing Add: Inst of Molecular Biol Univ of Ore Eugene OR 97403

MATTHEWS, BURTON CLARE, b Kerwood, Ont, Dec 16, 26; m 51; c 2. SOIL CHEMISTRY. Educ: Ont Agr Col, BSA, 47; Univ Mo, AM, 48; Cornell Univ, PhD, 52. Prof Exp: From asst prof to assoc prof soil classification, Ont Agr Col, Guelph, 48-55, prof soil fertil & chem, 55-62, head dept soil sci, 62-66, acad vpres, 66-70; PRES, UNIV WATERLOO, 70- Concurrent Pos: Fel natural sci, Nuffield Found, 60-61; dir, Ont Educ Commun Authority, 72- Mem: AAAS; Can Soil Sci Soc. Res: Soil genesis, classification and fertility. Mailing Add: Univ of Waterloo Off of the Pres Waterloo ON Can

MATTHEWS, CHARLES GEORGE, b Minot, NDak, Mar 13, 30; m 55; c 1. NEUROPSHYCHOLOGY. Educ: St John's Univ (Minn), BA, 51; Univ SDak, MA, 55; Purdue Univ, PhD(psychol), 58. Prof Exp: Staff psychologist, Achievement Ctr for Children, Ind, 57-58; res psychologist, Ft Wayne State Sch, 58-62; from asst prof to assoc prof, 62-72, PROF NEUROL, MED SCH, UNIV WISMADISON, 72- Concurrent Pos: Consult, Community Rehab Ctr, Ft Wayne, Ind, 59-62, Madison & Tomah Vet Admin Hosps, Wis, 64-, Ft Wayne State Sch, 64- & Perinatal Res Br, Nat Inst Neurol Dis & Blindness, 67-69. Mem: Am Psychol Asn; Am Asn Ment Deficiency. Res: Psychological effects of neurological diseases; intelligence structure in mental deficiency; rehabilitation of neurologically handicapped children; learning and achievement problems in school children. Mailing Add: Dept of Neurol Neuropsychol Lab Univ of Wis Med Sch Madison WI 53706

MATTHEWS, CHARLES ROBERT, b Philadelphia, Pa, May 12, 46; m 68; c 2. BIOPHYSICAL CHEMISTRY. Educ: Univ Minn, BS, 68; Stanford Univ, MS, 69, PhD(chem), 74. Prof Exp: Fel biochem, Stanford Univ, 74-75; ASST PROF CHEM, PA STATE UNIV, UNIVERSITY PARK, 75- Mem: Am Chem Soc. Res: Conformational changes in biological macromolecules; mechanisms of reversible unfolding transitions in proteins; distribution of isoaccepting species of transfer ribonucleic acid in mammalian cells. Mailing Add: Dept of Chem 152 Davey Lab Pa State Univ University Park PA 16802

MATTHEWS, CHARLES SEDWICK, b Houston, Tex, Mar 27, 20; m 45; c 2. EARTH SCIENCES. Educ: Rice Inst, BS, 41, MS, 43, PhD(phys chem), 44. Prof Exp: Engr chem plant design, Shell Develop Co, 44-48, chemist, 48-56, sr res assoc, 56-66, mgr exploitation eng, Shell Oil Co, 66-67, dir prod res, Shell Develop Co, 67-72; MGR ENG, SHELL OIL CO, 72- Honors & Awards: Lester C Uren Award, Soc Petrol Engrs, 75. Mem: Soc Petrol Engrs; Am Petrol Inst. Res: New methods for recovery of petroleum; behavior of petroleum reservoirs; geothermal energy; recovery from tar sands and oil shale. Mailing Add: Shell Oil Co PO Box 2463 Houston TX 77001

MATTHEWS, CLIFFORD NORMAN, b Hong Kong, China, Dec 20, 21; nat US; m 47; c 2. ORGANIC CHEMISTRY. Educ: Univ London, BSc, 50; Yale Univ, PhD(chem), 55. Prof Exp: Lab supt, Birkbeck Col, London, 46-48; res chemist, Conn Hard Rubber Co, 50-51, Diamond Alkali Co, 55-59 & Monsanto Co, 59-69; PROF CHEM, UNIV ILL, CHICAGO CIRCLE, 69- Mem: AAAS; Am Chem Soc; The Chem Soc. Res: Organic and organometallic chemistry; chemical evolution. Mailing Add: Dept of Chem Univ of Ill at Chicago Circle Chicago IL 60680

MATTHEWS, DAVID ALLAN, b Washington, DC, Feb 5, 43; m 67; c 1. BIOPHYSICAL CHEMISTRY. Educ: Earlham Col, AB, 65; Univ Ill, PhD(chem), 71. Prof Exp: FEL CHEM, UNIV CALIF, SAN DIEGO, 71- Concurrent Pos: Jane Coffin Childs Mem Fund Med Res fel, 72-74; Nat Cancer Inst fel, 74-76. Res: X-ray crystallographic studies on the molecular structure and mechanism of action of dihydrofolate reductase. Mailing Add: Dept of Chem Univ of Calif San Diego La Jolla CA 92093

MATTHEWS, DAVID LESUEUR, b Ottawa, Ont, May 10, 28; m 56; c 3. PHYSICS. Educ: Queen's Univ, Ont, BSc, 49; Princeton Univ, PhD(physics), 59. Prof Exp: From jr to asst res officer, Nat Res Coun Can, 49-53; instr physics, Princeton Univ, 57-59; lectr, Carleton Univ, 59-60; sci officer, Defense Res Telecommun Estab, 60-66; RES ASSOC PROF, UNIV MD, COLLEGE PARK, 66- Concurrent Pos: Rocket sect leader, Defense Res Telecommun Estab, 63-65. Mem: Am Geophys Union; Am Phys Soc. Res: Space and upper atmosphere physics. Mailing Add: Inst for Fluid Dynamics & Appl Math Univ of Md College Park MD 20742

MATTHEWS, DAVID LIVINGSTONE, b New York, NY, Mar 13, 22; m 44; c 4. AGRONOMY, PLANT BREEDING. Educ: Rutgers Univ, BS, 48, MS, 50. Prof Exp: Asst farm crops, Rutgers Univ, 48-50; tech specialist radiation genetics, Brookhaven Nat Lab, 50-52; plant breeder, Eastern State Farmers Exchange, 52-54, mgr corn res, 55-64; mgr seed res, 65-66, dir farm eval & seed res, 66-68, DIR CROPS RES, AGWAY INC, 68- Concurrent Pos: Dir, Farmers Forage Res Coop, 65-68. Mem: Genetics Soc Am; Am Soc Agron; Crop Sci Soc Am; Am Soc Hort Sci; Coun Agr Sci & Technol. Res: Radiation genetics; corn breeding; managing field crops for maximum economic return; rate, time of application and placement studies with new fertilizer materials; mulch and irrigation management of vegetables; dairy and poultry manure management for optimum crop returns. Mailing Add: Agway Inc Box 1333 Syracuse NY 13201

MATTHEWS, DEMETREOS NESTOR, b Portchester, NY, June 28, 28; m 53; c 2. ORGANIC CHEMISTRY. Educ: Rutgers Univ, BS, 49; Polytech Inst Brooklyn, PhD(chem), 60. Prof Exp: Chemist, Res Labs, Air Reduction Co, Inc, 52-54; res scientist org chem, Res Ctr, US Rubber Co, 59-67, SR RES SCIENTIST ORG CHEM, UNIROYAL RES CTR, 67- Mem: Am Chem Soc. Res: Organic reaction mechanisms; Diels-Alder reactions; solution polymerization; free radical reactions; correlation of mechanism with structure. Mailing Add: Brookwood Rd Bethany CT 06525

MATTHEWS, DOYLE JENSEN, b Liberty, Idaho, Apr 13, 26; m 46; c 5. ANIMAL BREEDING. Educ: Utah State Univ, BS, 50, MS, 51; Kans State Univ, PhD, 59. Prof Exp: From instr to assoc prof animal husb, 51-65, PROF ANIMAL SCI, UTAH STATE UNIV, 66- Concurrent Pos: DEAN, COL AGR, 71-, DIR, AGR EXP STA, 74- Concurrent Pos: Assoc dean, Col Agr, Utah State Univ, 69-71, asst dean, 65-69. Mem: Am Soc Animal Sci. Res: Improvement of carcass characteristics and productivity of meat animals through application of breeding techniques and methods. Mailing Add: Col of Agr Utah State Univ Logan UT 84321

MATTHEWS, FREDERICK WHITE, b Carbonear, Nfld, Nov 27, 15; m 43; c 4. CHEMISTRY, INFORMATION SCIENCE. Educ: Mt Allison Univ, BSc, 36; McGill Univ, PhD(phys chem), 41. Prof Exp: Head, Tech Lit Ctr, Cent Res Lab, Can Industs Ltd, Que, 61-68, head info serv, 68-69; mgr, Cent Tech Info Unit, Imp Chem Industs, Ltd, Eng, 69-72; PROF INFO SCI, SCH LIBR SERV, DALHOUSIE UNIV, 72- Concurrent Pos: Chmn data comn, Int Union Crystallog. Mem: Chem Inst Can (dir, 58-60); Am Soc Info Sci. Res: Library catalogue systems; x-ray diffraction powder data; systems for data retrieval; systems for storage and retrieval of information on computers. Mailing Add: Sch Libr Serv Dalhousie Univ Halifax NS Can

MATTHEWS, GARY JOSEPH, b Denver, Colo, Aug 6, 42; m 64; c 3. ORGANIC CHEMISTRY. Educ: Colo State Univ, BS, 64; Univ Colo, Boulder, PhD(org chem), 68. Prof Exp: Syntex res grant, Inst Org Chem, Syntex Res, Palo Alto, 68-69, res chemist, Arapahoe Chems Div, Syntex Corp, 69-72, group leader, 72, MGR RES, ARAPAHOE CHEMS INC, 72- Mem: Am Chem Soc. Res: Process research and development on the production of fine organic chemicals. Mailing Add: Res Dept Arapahoe Chems PO Box 511 Boulder CO 80302

MATTHEWS, HAZEL BENTON, JR, b Hertford, NC, Feb 8, 40; m 65; c 2. BIOCHEMICAL PHARMACOLOGY. Educ: NC State Univ, BS, 63, MS, 65; Univ Wis-Madison, PhD(entom), 68. Prof Exp: NIH grant, Univ Calif, Berkeley, 68-70; staff fel chem, 70-71, sr staff fel, 71-74, RES CHEM, NAT INST ENVIRON HEALTH SCI, NIH, 74- Mem: Entom Soc Am; Soc Toxicol. Res: Mammalian metabolism storage; excretion of insecticides; food additives and the effects of these compounds and their metabolites on mammalian enzyme systems and the intact organism. Mailing Add: Pharmacol Br Nat Inst Environ Health Sci NIH Research Triangle Park NC 27709

MATTHEWS, HERBERT MAURICE, b Chicago, Ill, Mar 17, 26; m 51; c 2.

PHYSICS. Educ: Univ Chicago, BS, 50; Univ Ore, MS, 55, PhD(physics), 61. Prof Exp: Mem tech staff, Bell Tel Labs, Inc, 60-66; res staff mem, Sperry Res Ctr, 66-71, res prog mgr, 71-72; spec fel, Nat Eye Inst, 72-75; ADJ PROF PHYSICS & ASTRON, UNIV MASS, AMHERST, 74- Concurrent Pos: Vis scientist, Worcester Found Exp Biol, 73-75. Mem: Optical Soc; Inst Elec & Electronics Engrs. Res: Structure and function of proteins in biomembranes. Mailing Add: 23 Loblolly Lane Wayland MA 01778

MATTHEWS, HERMAN EXCELL, JR, solid state physics, see 12th edition

MATTHEWS, HEWITT WILLIAM, b Pensacola, Fla, Dec 1, 44; m 69; c 1. PHARMACEUTICAL CHEMISTRY. Educ: Clark Col, BS, 66; Mercer Univ, BS, 68; Univ Wis, MS, 71, PhD(pharm, biochem), 73. Prof Exp: Asst prof pharm, 73-75, ASSOC PROF PHARM & DIR RES, SCH PHARM, MERCER UNIV, 75- Mem: Sigma Xi; AAAS; Am Asn Cols Pharm; Nat Inst Sci. Res: Pharmacologically active agents from microbial origin; screening for agents from fermentation broths that have anti inflamatory properties. Mailing Add: Mercer Univ Sch Pharm 345 Boulevard Ave NE Atlanta GA 30312

MATTHEWS, JAMES FRANCIS, b Winston-Salem, NC, Sept 14, 35; m 61; c 2. CYTOLOGY, PLANT TAXONOMY. Educ: Atlantic Christian Col, BA, 57; Cornell Univ, MS, 60; Emory Univ, PhD(cytol), 62. Prof Exp: Asst prof biol, Western Ky State Col, 62-64; from asst prof to assoc prof, 64-72, PROF BIOL, UNIV NC, CHARLOTTE, 72-, ACAD RES & CONTRACTS OFFICER, 74- Mem: AAAS; Bot Soc Am; Am Inst Biol Sci; Am Asn Plant Taxon. Res: Speciation of plants endemic to the granite outcrops of the southeastern Piedmont, using evidence from cytology and biochemistry; floristics of urban areas. Mailing Add: Dept of Biol Univ of NC Charlotte NC 28213

MATTHEWS, JAMES HORACE, b Campbellton, NB, Mar 1, 30; m 54; c 5. NUCLEAR PHYSICS. Educ: Mt Allison Univ, BSc & cert eng, 51; Dalhousie Univ, MSc, 54; Univ London, PhD(physics), 57. Prof Exp: From asst prof to assoc prof, 57-69, PROF PHYSICS, MT ALLISON UNIV, 69- Concurrent Pos: Marjorie Young Bell fel, Univ Sussex, 66-67; vis prof, Univ Toronto, 74-75. Mem: Can Asn Physicists; Brit Inst Physics. Res: Theoretical nuclear physics; nuclear models; microwave gas discharge; picture enhancement. Mailing Add: Dept of Physics Mt Allison Univ Sackville NB Can

MATTHEWS, JAMES LESTER, b Denton, Tex, July 3, 26; m 50; c 3. MICROSCOPIC ANATOMY. Educ: NTex State Col, BS, 48, MS, 49; Univ Ill, PhD, 55. Prof Exp: Asst biol, NTex State Col, 47-48; instr biol & chem, Cisco Jr Col, 49-52; asst physiol, Univ Ill, 52-55; from res asst to assoc prof anat & physiol, 55-60, PROF MICROS ANAT & CHMN DEPT HISTOLMICROS ANAT, BAYLOR COL DENT, 60-, ASSOC DEAN, BAYLOR UNIV MED CTR, 74- Mem: Am Physiol Soc; assoc Soc Exp Biol & Med; Int Asn Dent Res; Am Asn Anat. Res: Physiology and fine structure of bone and connective tissues. Mailing Add: Dept of Histol-Micros Anat Baylor Col of Dent Dallas TX 75226

MATTHEWS, JAMES SWINTON, b Union, Miss, Jan 22, 11; m 41; c 1. GEOGRAPHY OF EUROPE, HISTORICAL GEOGRAPHY. Educ: Kent State Univ, BSEduc, 36, MA, 41; Univ Chicago, PhD(geog), 49. Prof Exp: Teacher geog, Copley High Sch, Ohio, 37-42; assoc prof, 49-65, PROF GEOG, MEMPHIS STATE UNIV, 65- Mem: Asn Am Geog. Res: Historical geography, especially its economic aspects; patterns of population distribution in 13th Century Western Europe. Mailing Add: 1136 Audubon Dr Memphis TN 38117

MATTHEWS, JANE, mathematics, see 12th edition

MATTHEWS, JERRY LEE, b Smithfield, Pa, Jan 14, 29; m 57. MARINE GEOLOGY. Educ: Allegheny Col, BS, 53; Kans State Univ, MS, 59; Univ Calif, Los Angeles, PhD(geol), 66. Prof Exp: Res geologist, Univ Calif, Los Angeles, 66-67; cur geol & mineral, Los Angeles County Mus Natural Hist, 67-69; RES SPECIALIST MARINE GEOL, SCRIPPS INST OCEANOG, UNIV CALIF, SAN DIEGO, 69- Concurrent Pos: NSF-US Antarctic Res Prog grant, 68-69. Mem: Soc Econ Paleont & Mineral. Res: Sedimentation of coastal sand dunes; sedimentation and stratigraphy of the Permo- Carboniferous Tillites of Antarctica; marine geology of the California Borderland; petrology of limestones of Pacific Ocean seamounts. Mailing Add: Scripps Inst of Oceanog Univ of Calif at San Diego La Jolla CA 92037

MATTHEWS, JOHN BRIAN, b Glazebrook, Eng, Feb 15, 38; m 67. OCEANOGRAPHY. Educ: Univ London, BSc & ARCS, 60, PhD(cloud physics), 63, Imp Col, London, dipl, 63. Prof Exp: Res asst cloud physics, Imp Col, London, 57-63; res assoc, Inst Atmospheric Physics, Univ Ariz, 63-66; asst prof phys oceanog, Inst Marine Sci, 66-70, ASSOC PROF MARINE SCI, UNIV ALASKA, FAIRBANKS, 70- Concurrent Pos: Eckert fel environ sci, IBM-Watson Res Ctr, 73-74; res scientist, Bedford Inst Oceanog, Dartmouth, Can, 75-76; chmn, Working Group Coastal & Estuarine Regimes, Int Asn Phys Sci of the Ocean, 75-79. Mem: Assoc Am Geophys Union; fel Royal Meteorol Soc; fel Royal Geog Soc; Am Soc Limnol & Oceanog; Estuarine & Brackish Water Res Asn. Res: Environmental physics; physical oceanography; coastal and estuarine dynamics; tidal and storm surge analysis; hydrodynamical numerical modeling; fjord estuary research. Mailing Add: Bedford Inst of Oceanog Dartmouth NS Can

MATTHEWS, JOHN WAUCHOPE, b Johannesburg, SAfrica, Mar 31, 32; m 61; c 4. PHYSICS. Educ: Univ Witwatersrand, BSc, 54, Hons, 56, PhD(physics), 63. Prof Exp: Lectr physics, Univ Witwatersrand, 61-65, sr lectr, 65-68, reader, 68-69; RES SCIENTIST, WATSON RES CTR, IBM CORP, 69- Concurrent Pos: Overseas fel, Churchill Col, Cambridge Univ, Eng, 73-74. Mem: SAfrican Inst Physics. Res: Growth, structure and properties of thin solid films; electron microscopy of defects in crystals. Mailing Add: Dept of Phys Sci Watson Res Ctr IBM Corp Yorktown Heights NY 10598

MATTHEWS, JOSEPH SEOANE, organic chemistry, analytical chemistry, see 12th edition

MATTHEWS, JUNE LORRAINE, b Cambridge, Mass, Aug 1, 39. NUCLEAR PHYSICS. Educ: Carleton Col, BA, 60; Mass Inst Technol, SM, 62, PhD(physics), 67. Prof Exp: NSF fel physics, Glasgow Univ, 68-71; res assoc, Rutgers Univ, 71-72; asst prof, 72-75, ASSOC PROF PHYSICS, MASS INST TECHNOL, 75- Concurrent Pos: Mem adv panel physics, NSF, 75- Res: Interactions of photons with nuclei; study of nucleon momentum distributions, short-range correlations and meson exchange effects; few-body problems. Mailing Add: Dept of Physics Mass Inst of Technol Cambridge MA 02139

MATTHEWS, KATHLEEN SHIVE, b Austin, Tex, Aug 30, 45; m 67. BIOCHEMISTRY. Educ: Univ Tex, Austin, BS, 66; Univ Calif, Berkeley, PhD(biochem), 70. Prof Exp: Am Asn Univ Women fel, Sch Med, Stanford Univ, 70-71, Giannini Found fel, 71-72; ASST PROF BIOCHEM, RICE UNIV, 72- Mem: AAAS; Soc Neurosci. Res: Chemistry and molecular biology of proteins; studies on the lactose repressor protein from Escherichia coli, including chemical modification, spectroscopy and other physical methods. Mailing Add: Dept of Biochem Rice Univ Houston TX 77001

MATTHEWS, LEE DREW, b Platteville, Wis, Mar 10, 43; m 69; c 1. PHYSICS. Educ: Platteville, Wis State Col, BS, 64; Univ Vt, MS, 67, PhD(physics), 70. Prof Exp: ASST PROF PHYSICS, SOUTHERN CONN STATE COL, 69- Mem: Am Inst Physics; Am Phys Soc; Am Asn Physics Teachers. Res: Surface physics; physics education. Mailing Add: Dept of Physics Southern Conn State Col New Haven CT 06515

MATTHEWS, MURRAY ALBERT, b Houston, Tex, June 16, 43; m 69; c 1. ANATOMY, NEUROPATHOLOGY. Educ: Univ St Thomas, BA, 65; Univ Tex Med Br, MA, 67, PhD(anat), 70. Prof Exp: ASST PROF ANAT, MED CTR, LA STATE UNIV, NEW ORLEANS, 72- Concurrent Pos: NIH trainee, Brain Res Inst, Med Sch, Univ Calif, Los Angeles, 70-72; Schlieder Educ Found res grant, 74. Mem: Am Asn Anat; Soc Neurosci. Res: Central nervous system trauma; spinal cord injury; reaction of neurons to mechanical or ischemic injury; reactive changes in nonneuronal, vascular elements. Mailing Add: Dept of Anat La State Univ Med Ctr New Orleans LA 70119

MATTHEWS, PETER WREN, b Woolwich, UK, June 11, 35. PHYSICS. Educ: Bristol Univ, BSc, 56, PhD(physics), 62. Prof Exp: Res asst physics, Clarendon Lab, Oxford Univ, 59-62; from instr to asst prof, 62-67, ASSOC PROF PHYSICS, UNIV BC, 67- Mem: Can Asn Physicists. Res: Cryogenics. Mailing Add: Dept of Physics Univ of BC Vancouver BC Can

MATTHEWS, RICHARD FINIS, b Cullman, Ala, June 1, 29; m 55; c 2. FOOD CHEMISTRY, BIOCHEMISTRY. Educ: Univ Fla, BSA, 52; Cornell Univ, MS, 57, PhD(food sci), 60. Prof Exp: Assoc technologist, Res Ctr, Gen Foods Corp, 60-63; group leader tea chem, T J Lipton Res Ctr, NJ, 63-65; assoc prof, 65-73, PROF FOOD TECHNOL, UNIV FLA, 73- Mem: Am Chem Soc; Inst Food Technologists. Res: Natural products chemistry; food additives; flavor chemistry. Mailing Add: Dept of Food Sci Univ of Fla Gainesville FL 32601

MATTHEWS, RICHARD JOHN, JR, b Scranton, Pa, Apr 11, 27; m 53; c 4. PHARMACOLOGY. Educ: Philadelphia Col Pharm, BS, 51; Jefferson Med Col, MS, 53, PhD(pharmacol), 55. Prof Exp: Head pharmacol res sect, Upjohn Co, Mich, 56-62; pres, Pharmakon, Inc, Pa, 62-65; dir pharmacol, Union Carbide Corp, 65-69; DIR RES, PHARMAKON LABS, 69- Mem: Am Soc Pharmacol & Exp Therapeut. Res: Action of drugs on synapse in peripheral and central nervous system, especially neurohumoral agents; neuropharmacology of psychotherapeutic drugs and effects of extracts of blood from schizophrenics on the central nervous system. Mailing Add: Pharmakon Labs 1140 Quincy Ave Scranton PA 18510

MATTHEWS, ROBERT WENDELL, b Detroit, Mich, Feb 17, 42; m 63; c 3. ENTOMOLOGY. Educ: Mich State Univ, BS, 63, MS, 65; Harvard Univ, PhD(biol), 69. Prof Exp: Asst prof, 69-74, ASSOC PROF ENTOM, UNIV GA, 74- Concurrent Pos: NSF res assoc, Commonwealth Sci & Indust Res Orgn, Canberra, Australia, 69-70 & Inst Miguel Lillo, Tucuman, Arg, 72; NSF res grant, 75. Mem: AAAS; Entom Soc Am; Royal Entom Soc London; Asn Trop Biol. Res: Behavior, systematics, ecology and evolution of Hymenoptera, especially Braconidae, Sphecidae and Vespidae. Mailing Add: Dept of Entom Univ of Ga Athens GA 30602

MATTHEWS, ROBLEY KNIGHT, b Dallas, Tex, Oct 6, 35; m 59; c 4. SEDIMENTOLOGY. Educ: Rice Univ, BA, 57, MA, 63, PhD(geol), 65. Prof Exp: Petrol geologist, Pan Am Petrol Corp, 57-58 & Am Int Oil Co, Libya, 58-60; geologist, Marine Geophys Serv, 60-63; asst prof, 64-71, PROF GEOL & CHMN DEPT, BROWN UNIV, 71- Mem: Geol Soc Am; Am Asn Petrol Geologists; Soc Econ Paleont & Mineral. Res: Sedimentary petrology; physical and chemical aspects of carbonate deposition and diagenesis; dynamics of climate change; Pleistocene sea levels. Mailing Add: Dept of Geol Sci Brown Univ Providence RI 02912

MATTHEWS, ROWENA GREEN, b Cambridge, Eng, Aug 20, 38; US citizen; m 60; c 2. BIOCHEMISTRY, PROTEIN CHEMISTRY. Educ: Radcliffe Col, BA, 60; Univ Mich, Ann Arbor, PhD(biophys), 69. Prof Exp: Instr biol, Univ SC, 63-64; fel biol chem, 71-74, res investr, 74-75, ASST PROF BIOL CHEM, UNIV MICH, ANN ARBOR, & RES CHEMIST, VET ADMIN HOSP, 75- Mem: Am Chem Soc; Sigma Xi; AAAS. Res: Catalytic mechanisms of flavoprotein dehydrogenases; kinetic and thermodynamic studies of lipoamide dehydrogenase. Mailing Add: Vet Admin Hosp Gen Med Res 2215 Fuller Rd Ann Arbor MI 48105

MATTHEWS, RUTH HASTINGS, b Cambridge, Md, Feb 12, 26; m 49; c 1. FOOD SCIENCE, NUTRITION. Educ: Univ Md, BS, 46; Columbia Univ, MA, 47. Prof Exp: Instr foods & nutrit, Juniata Col, 47-49 & Univ Md, 49-50; food specialist, Human Nutrit Res Div, 54-57, res food specialist, 57-69, NUTRIT ANALYST, CONSUMER & FOOD ECON INST, AGR RES SERV, USDA, 69- Mem: Am Asn Cereal Chem; Inst Food Technol; Am Home Econ Asn. Res: Foods, particularly cereal product quality; influences of functional properties of milk and fats and of quality of wheat grain; nutrient composition of baby foods, infant formulas, sugars, starches, cereals and baked products of all types for tables of food composition. Mailing Add: 430 Kentbury Dr Bethesda MD 20014

MATTHEWS, SAMUEL ARTHUR, b Grand Lake Stream, Maine, Aug 4, 02; m 27; c 1. ZOOLOGY. Educ: Boston Univ, BS, 23, MA, 24; Harvard Univ, MA, 25, PhD(zool), 28; Williams Col, ScD, 64. Prof Exp: Instr biol, Boston Univ, 23-28; instr anat, Univ Pa, 28-32, assoc, 32-37; from asst prof to prof, 37-70, EMER PROF BIOL, WILLIAMS COL, 70- Concurrent Pos: Instr & tutor, Harvard Univ, 27-28; instr, Marine Biol Lab, Woods Hole, 32-40, mem corp; chmn fac, Williams Col, 51-66. Mem: Am Soc Zool; Am Physiol Soc; Am Asn Anat; NY Acad Sci. Res: Molluscan nervous system; development of amphibians; teleost retina, melanophores and endocrines; sex cycles in teleosts; respiratory metabolism of teleosts; carbohydrate metabolism in lower vertebrates. Mailing Add: Dept of Biol Williams Col Williamstown MA 01267

MATTHEWS, VIRGIL EDISON, b LaFayette, Ala, Oct 5, 28; m 60; c 3. ORGANIC POLYMER CHEMISTRY. Educ: Univ Ill, BS, 51; Univ Chicago, SM, 52, PhD(chem), 55. Prof Exp: Asst org chem, Univ Chicago, 51-52; res chemist, Res & Develop Dept, Chem Div, 54-67, Chem & Plastics Div, 67, proj scientist, 67-75, DEVELOP SCIENTIST, CHEM & PLASTICS DIV, UNION CARBIDE CORP, 75- Concurrent Pos: Instr, WVa State Col, 55-60, part-time assoc prof & prof, 60-70. Mem: Fel AAAS; Am Chem Soc; The Chem Soc; fel Am Inst Chemists; Sigma Xi. Res: Synthesis and structure of polymers; free radicals; organic synthesis; elastomers; polymeric composites; fibers; synthetic hydrogels. Mailing Add: 835 Carroll Rd Charleston WV 25314

MATTHEWS, WILLIAM HENRY, III, b Henrietta, Okla, Mar 1, 19; m 40; c 2.

GEOLOGY. Educ: Tex Christian Univ, BA, 48, MA, 49. Prof Exp: Asst prof geol, Tex Christian Univ, 51-52; subsurface geologist, Tex Co, 52-55; from asst prof to assoc prof, 55-62, PROF GEOL, LAMAR UNIV, 62- Concurrent Pos: Consult, Tex Hwy Dept, 58-59, Tex Portland Cement Co, 59-, Earth Sci Curriculum Proj, 62- & Tex Ed Agency, 64-; dir educ, Am Geol Inst, 72-; chief tech adv, Encycl Britannica Earth Sci films, 73-; Regents Prof, Lamar Univ, 74; ed, ref series, Earth Sci Curriculum Proj. Honors & Awards: Neil Miner Award, Nat Asn Geol Teachers, 64. Mem: AAAS; Geol Soc Am; Soc Econ Paleont & Mineral; Paleont Soc; Am Asn Petrol Geologists; Nat Asn Geol Teachers. Res: Invertebrate paleontology; historical petroleum and subsurface geology; paleoecology; stratigraphy; earth science teaching. Mailing Add: Dept of Geol Lamar Univ Beaumont TX 77710

MATTHIAS, BERND T, b Frankfurt, Ger, June 8, 18; nat US; m 50. PHYSICS. Educ: Swiss Fed Inst Technol, PhD(physics), 43. Prof Exp: Sci collabr, Swiss Fed Inst Technol, 42-47; mem staff, Div Indust Coop, Mass Inst Technol, 47-48; MEM TECH STAFF, BELL TEL LABS, INC, 48-; PROF PHYSICS, UNIV CALIF, SAN DIEGO, 61-, DIR, INST PURE & APPL PHYS SCI, 71- Concurrent Pos: Asst prof, Univ Chicago, 49-51. Honors & Awards: Res Corp Award, 62; John Price Wetherill Medal, 63; Oliver E Buckley Solid State Physics Prize, 70. Mem: Nat Acad Sci; Am Acad Arts & Sci; fel Am Phys Soc; Swiss Phys Soc. Res: Ferroelectrics superconductivity, ferromagnetism; dielectric crystals and intermetallic compounds. Mailing Add: Dept of Physics Univ of Calif at San Diego La Jolla CA 92037

MATTHIES, KARL HEINRICH, b Richen, Switz, Aug 22, 26; m 51; c 3. MATHEMATICS. Educ: Univ Freiburg, Dr rer nat, 56. Prof Exp: Res mem, Math Inst, Univ Wuerzburg, 56-57; asst prof math, Univ Cincinnati, 57-59; ASSOC PROF MATH, UNIV SC, 59- Mem: Am Math Soc; Soc Indust & Appl Math. Res: Theory of differential equations. Mailing Add: 4225 Timberlane Columbia SC 29205

MATTHIJSSEN, CHARLES, b Amsterdam, Holland, July 26, 31; nat US; m 57; c 2. MICROBIOLOGY, BIOCHEMISTRY. Educ: Upsala Col, BS, 51; Rutgers Univ, MS, 55, PhD(microbiol), 57. Prof Exp: Res chemist, P Ballentine & Sons, 57-58; VCHMN DEPT ENDOCRINOL, SOUTHWEST FOUND RES & EDUC, 59-, ASSOC SCIENTIST, 73- Mem: AAAS; Am Chem Soc; NY Acad Sci; fel Am Inst Chemists; The Chem Soc. Res: Transformations and synthesis of steroid hormones; enzymology. Mailing Add: Southwest Found Res & Educ PO Box 28147 San Antonio TX 78228

MATTHYSSE, ANN GALE, b Chicago, Ill, Oct 25, 39; m 62; c 1. MICROBIOLOGY. Educ: Radcliffe Col, AB, 61; Harvard Univ, PhD(biol), 66. Prof Exp: Lectr biol, Harvard Univ, 70-71; ASST PROF MICROBIOL, SCH MED, IND UNIV, INDIANAPOLIS, 71- Concurrent Pos: NIH fel, Calif Inst Technol, 66-69 & Harvard Med Sch, 69-70. Mem: AAAS; Am Soc Microbiol; Am Soc Plant Physiol; Am Inst Biol Sci; Tissue Cult Asn. Res: Eukaryote-prokaryote interactions; molecular regulatory mechanisms. Mailing Add: Dept of Microbiol Ind Univ Sch of Med Indianapolis IN 46202

MATTHYSSE, JOHN GEORGE, entomology, see 12th edition

MATTHYSSE, STEVEN WILLIAM, b New York, NY, Aug 27, 39; m 62; c 1. MATHEMATICAL BIOPHYSICS, PSYCHIATRY. Educ: Yale Univ, BS, 59, BA, 60; Harvard Univ, PhD(clin psychol), 67. Prof Exp: Asst prof, Pfitzer Col, 66-69; ASST PROF PSYCHOBIOL, HARVARD MED SCH, 70- Concurrent Pos: Marks Found fel, Harvard Univ, 70-71; res dir, Schizophrenia Res Prog, Scottish Rite, 72- Mem: AAAS; Soc Neurosci; Am Soc Neurochem; Asn Res Nerv & Ment Dis. Res: Foundations of molecular structure calculations; phase transitions in membranes; mathematical genetics; biological aspects of schizophrenia. Mailing Add: Dept of Psychiat Mass Gen Hosp Boston MA 02114

MATTICE, WAYNE LEE, b Cherokee, Iowa, July 9, 40; m 65; c 1. PHYSICAL BIOCHEMISTRY. Educ: Grinnell Col, BA, 63; Duke Univ, PhD(biochem), 68. Prof Exp: USPHS fel, Fla State Univ, 68-70; asst prof, 70-74, ASSOC PROF BIOCHEM, LA STATE UNIV, BATON ROUGE, 74- Mem: AAAS; Am Chem Soc; Biophys Soc. Res: Physical chemistry of biopolymers. Mailing Add: Dept of Biochem La State Univ Baton Rouge LA 70803

MATTICK, JOSEPH FRANCIS, b Hudson, Pa, Nov 16, 18; m 52; c 1. BIOCHEMISTRY, BACTERIOLOGY. Educ: Pa State Univ, BS, 42, PhD(dairy technol), 50. Prof Exp: Asst prof dairy technol, Univ Md, 50-52, assoc prof, 53-58; tech consult, Venezuela, 58-60; assoc prof dairy technol, 60-65, PROF DAIRY SCI, UNIV MD, COLLEGE PARK, 65-, CHMN DEPT, 74- Concurrent Pos: Consult, Interam Develop Bank & World Bank. Mem: Am Dairy Sci Asn; Inst Food Technol. Res: Products development; curriculum of food science; food processing waste disposal acid whey utilization. Mailing Add: Dept of Dairy Sci Univ of Md Animal Sci Bldg College Park MD 20742

MATTICK, LEONARD ROBERT, b Hudson, Pa, Sept 16, 26; m 54; c 5. BIOCHEMISTRY. Educ: Pa State Univ, BS, 50, MS, 51; Univ Conn, PhD, 54. Prof Exp: Asst, Pa State Univ, 50-51; asst, Univ Conn, 51-54; chemist, USDA, 54-55; res assoc & fel, Pa State Univ, 55-57; from asst prof to assoc prof food sci, 57-70, PROF FOOD CHEM, EXP STA, NY STATE COL AGR & LIFE SCI, CORNELL UNIV, 70- Mem: Inst Food Technol. Res: Instrumental techniques to fields of research; flavor chemistry; compositional analysis of lipids and their effect on flavor; chemistry of processed food products; enological research; changes in composition of New York State wines during processing. Mailing Add: Dept of Food Sci NY Exp Sta Cornell Univ Geneva NY 14456

MATTICS, LEON EUGENE, b Butte, Mont, Mar 2, 40; m 67; c 2. NUMBER THEORY. Educ: Mont State Univ, BS, 63, PhD(math), 67. Prof Exp: Instr math, Mont State Univ, 66-67; ASSOC PROF MATH, UNIV S ALA, 67- Mem: Math Asn Am. Res: Problem solving; complex variables. Mailing Add: Dept of Math Univ Of S Ala Mobile AL 36688

MATTIKOW, MORRIS, b Russia, Jan 27, 03; nat US. CHEMISTRY. Educ: City Col New York, BS, 21; Columbia Univ, AM, 23, PhD(oil anal), 25. Prof Exp: Consult chemist, Borden Co, 30-32; chief chemist, Refining, Inc, 37-43, dir res, Refining, Unincorp, 43-63; CONSULT CHEMIST, 63- Honors & Awards: Achievement Award, Am Oil Chemists Soc, 74. Mem: Am Chem Soc; fel Am Inst Chemists; Am Oil Chem Soc; Sigma Xi; Inst Food Technologists. Res: Oils and soaps; colloid chemistry; oil refining; high temperature soap processes; phosphatides. Mailing Add: 98 Riverside Dr New York NY 10024

MATTIL, KARL FREDERICK, b Williamsport, Pa, Mar 4, 15; m 42; c 2. FOOD TECHNOLOGY. Educ: Pa State Col, BS, 35, PhD(biochem), 41. Prof Exp: Asst chem, Pa State Col, 35-39; sr fel, Univ Pittsburgh, 41-43; res chemist, Swift & Co, 44-50, head edible fats res div, 50-54, assoc dir res, 54-68; prof food sci, 68-74, PROF AGRON, TEX A&M UNIV, 74-, DIR FOOD PROTEIN RES & DEVELOP CTR, 71- Mem: Am Oil Chem Soc; Inst Food Technol; Am Asn Cereal Chem. Res: Oilseeds; vegetable proteins; engineered foods; antioxidants; glyceride distribution; interesterification; emulsifiers; digestibility of fats; flavor reversion; modifying crystalline properties of fats; hydrogenation. Mailing Add: Food Protein Res & Develop Ctr Tex A&M Univ Fac Mail Box 63 College Station TX 77843

MATTINA, CHARLES FREDERICK, JR, b Elizabeth, NJ, Dec 5, 44; m 69; c 2. PHYSICAL CHEMISTRY. Educ: Providence Col, BS, 66; Yale Univ, PhD(phys chem), 69. Prof Exp: Asst prof chem, Albertus Magnus Col, 69-71; res chemist, 71-74, HEAD CHEM SECT, SCHWEITZER DIV, KIMBERLY-CLARK CORP, 74- Mem: Am Chem Soc. Res: Electrolytic conductance; viscosity of ionic solutions; tobacco chemistry; condenser paper; cigarette paper. Mailing Add: Schweitzer Div Kimberly-Clark Corp Lee MA 01238

MATTINGLY, GLEN E, b Provo, Ark, Oct 31, 32; m 54; c 2. MATHEMATICS. Educ: Sam Houston State Univ, BS, 56, MS, 57; NMex State Univ, PhD(math), 65. Prof Exp: From instr to assoc prof, 56-67, PROF MATH & DIR DEPT, SAM HOUSTON STATE UNIV, 67- Mem: Am Math Soc; Math Asn Am. Res: Topological semi-groups; semi-topological groups; topological modules. Mailing Add: Box 2206 Sam Houston State Univ Sta Huntsville TX 77340

MATTINGLY, MARY ELLEN, b Louisville, Ky, Jan 31, 32. BIOLOGY. Educ: Brescia Col, Ky, AB, 58; Cath Univ Am, PhD(biol), 62. Prof Exp: Asst prof biol, Brescia Col, Ky, 62-64; cytologist, Biol Div, Oak Ridge Nat Lab, 64-69; ASSOC PROF ZOOL, UNIV GA, 69- Mem: AAAS; Am Soc Cell Biol. Res: Physiological phenomena associated with the cell cycle. Mailing Add: Dept of Zool Univ of Ga Athens GA 30601

MATTINGLY, PAUL FREDRICK, b Louisville, Ky, Aug 20, 30; m 56. URBAN GEOGRAPHY, ECONOMIC GEOGRAPHY. Educ: Western Ill Univ, BS, 54; Univ Mo, MA, 56; Pa State Univ, PhD(geog), 61. Prof Exp: Ed cartog, Aeronautical Chart & Info Serv, 56-57; asst prof geog, Towson State Col, 60-62; assoc prof, 62-74, PROF GEOG, ILL STATE UNIV, 74- Mem: Asn Am Geogr; Nat Coun Geog Educ. Mailing Add: Dept of Geog Ill State Univ Normal IL 61761

MATTINGLY, RICHARD FRANCIS, b Zanesville, Ohio, Oct 25, 25; m 48; c 7. OBSTETRICS & GYNECOLOGY. Educ: Ohio State Univ, AB, 49; Cornell Univ, MD, 53; Am Bd Obstet & Gynec, dipl, cert gynec oncol, 74. Prof Exp: Intern obstet & gynec, Univ Hosp, Johns Hopkins Univ, 53-54, asst resident obstet & gynec, Univ, 54-57, res gynecologist & sr resident gynec, 57-58, asst dir gynec endocrine clin, 57-61, from instr to asst prof gynec & obstet, 58-61, obstetrician-gynecologist in charge, outpatient dept, 59-61; PROF GYNEC & OBSTET & CHMN, MED COL WIS, 61-; DIR DEPT, MILWAUKEE CO HOSP, 61- Concurrent Pos: Consult staff, Columbia, Milwaukee, Mt Sinai, St Joseph's, St Luke's, & St Mary's Hosps, 61-; ed, Obstet & Gynec, Am Col Obstet & Gynec. Mem: Fel Am Col Obstet & Gynec; Am Gynec Soc; Am Asn Obstet & Gynec; Am Soc Cytol; Am Fertil Soc. Mailing Add: 8700 W Wisconsin Ave Milwaukee WI 53226

MATTINGLY, STEELE F, b Trinity, Ky, Aug 28, 27; m 49; c 2. ANIMAL HUSBANDRY, VETERINARY MEDICINE. Educ: Berea Col, BS, 50; Auburn Univ, DVM, 55; Am Col Lab Animal Med, dipl, 64. Prof Exp: Assoc teacher high sch, Ky, 50-51; mem staff primate test animals, Allied Labs, Pitman Moore Co Div & Dow Chem Co, 55-57, unit head test animals, 57-62; prod mgr, Lab Supply Co, 62-65; DIR DEPT LAB ANIMAL MED, COL MED, UNIV CINCINNATI, 65- Concurrent Pos: Consult, Vet Admin, Ohio, 65- Mem: Am Vet Med Asn; Am Asn Lab Animal Sci; NY Acad Sci. Res: Husbandry of laboratory animals; laboratory animal medicine; germ free life and its relationship to other animal research. Mailing Add: Dept of Lab Animal Med Univ of Cincinnati Cincinnati OH 45219

MATTINGLY, STEPHEN JOSEPH, b Evansville, Ind, Mar 3, 43; m 63; c 4. MICROBIAL PHYSIOLOGY. Educ: Univ Tex, Austin, BA, 65; Villanova Univ, MS, 68; Med Col Ga, PhD(microbiol), 72. Prof Exp: Microbiologist, US Food & Drug Admin, 65-66; microbiologist, Valley Forge Gen Hosp, 66-68; microbiologist, Naval Med Field Res Lab, Camp LeJeune, 68-69; res assoc, Sch Med, Temple Univ, 72-74; ASST PROF MICROBIOL, UNIV TEX HEALTH SCI CTR, 74- Concurrent Pos: Nat Inst Dent Res fel, 74. Mem: Am Soc Microbiol; Sigma Xi. Res: Bacterial physiology; regulation of cell wall and polysaccharide biosynthesis; physiology of cariogenic streptococci. Mailing Add: Dept of Microbiol Univ of Tex Health Sci Ctr San Antonio TX 78284

MATTINGLY, SUSAN CAROL, b Baltimore, Md, Mar 18, 46. AUDIOLOGY. Educ: Towson State Col, BA, 68; Ohio State Univ, MA, 69, PhD(speech & hearing sci), 72. Prof Exp: Lang develop therapist, Baltimore City Pub Sch Syst, 68; speech therapist, Baltimore County Bd Educ, 70-71; asst prof speech path & audiol, Loyola Col, Md, 72-74; ASST PROF AUDIOL, McGILL UNIV, 73-; DIR DEPT AUDIOL, CHILDRENS HOSP MONTREAL, 74- Concurrent Pos: Fel med audiol, Sch Med, Johns Hopkins Univ, 73; adj asst prof spec educ, Walden Univ, 73- Mem: Am Speech & Hearing Asn; Can Speech & Hearing Asn; Alexander Graham Bell Asn. Res: Audiological assessment; education of hearing impaired children; language development. Mailing Add: Childrens Hosp of Montreal Dept of Audiol 2300 Tupper St Montreal PQ Can

MATTINGLY, THOMAS WILLIAM, JR, organic chemistry, see 12th edition

MATTINSON, JAMES MEIKLE, b Maracaibo, Venezuela, Aug 28, 44; US citizen. GEOCHRONOLOGY, PETROLOGY. Educ: Univ Calif, Santa Barbara, BA, 66, PhD(geol), 70. Prof Exp: Fel geochronology, Geophys Lab, Carnegie Inst, Washington, 70-73; LECTR GEOL, UNIV CALIF, SANTA BARBARA, 73- Mem: Am Geophys Union; Geol Soc Am; AAAS; Sigma Xi. Res: Igneous rocks, especially calc-alkaline ingeous complexes and ophiolitic complexes. Mailing Add: Dept of Geol Sci Univ of Calif Santa Barbara CA 93106

MATTIS, ALLEN FRANCIS, b Spooner, Wis, May 3, 47; m 75. PETROLEUM GEOLOGY. Educ: Univ Wis-Superior, BS, 69; Univ Minn, Duluth, MS, 72; Rutgers Univ, MPhil, 74, PhD(geol), 75. Prof Exp: GEOLOGIST, TEXACO INC, 75- Mem: Geol Soc Am; Am Asn Petrol Geologists; Am Inst Mining, Metall & Petrol Engrs. Res: Sedimentation; provenance; regional tectonics. Mailing Add: Texaco Inc PO Box 2420 Tulsa OK 74102

MATTIS, DANIEL CHARLES, b Brussels, Belg, Sept 8, 32; nat US; m 58. SOLID STATE PHYSICS. Educ: Mass Inst Technol, BS, 53; Univ Ill, MS, 54, PhD(physics), 57. Prof Exp: Asst, Univ Ill, 54-57; asst, Nat Ctr Sci Res, France, 57-58; physicist, Res Ctr, Int Bus Mach Corp, 58-65; assoc prof, 65-66, PROF PHYSICS, BELFER GRAD SCH SCI, YESHIVA UNIV, 66- Mem: Fel Am Phys Soc. Res: Theoretical investigation of electronic properties, especially the theory of electrical conduction, with applications to metals and semiconductors; many-body theory of metal alloys. Mailing Add: Belfer Grad Sch of Sci Yeshiva Univ New York NY 10033

MATTIS, PAUL ALVIN, b Royersford, Pa, Mar 3, 09; m. TOXICOLOGY. Educ: Ursinus Col, BSc, 30; Philadelphia Col Pharm, DSc, 36. Prof Exp: Asst, Univ Pa, 33;

pharmacologist, Med Res Div, Sharp & Dohme, Inc, 36-43, asst dir pharmacol res, 43-45; head prof, Dept Pharmacog & Pharmacol, Sch Pharm, Univ Fla, 45-47; prof pharmacol, Sch Pharm, asst prof, Sch Med & dent & chief pharmacol sect, Atomic Energy Med Res Proj, Western Reserve Univ, 47-52; head pharmacol sect, Smith, Kline & French Labs, 52-60; vpres & dir, Pharmacol Res, Inc, 60-62; dir toxicol, Merck Inst Therapeut Res, 62-67, dir toxicol & path, 67-72, assoc dir safety assessment, Merck Sharp & Dohme Res Labs, 72-74; RETIRED. Mem: Am Soc Pharmacol & Exp Therapeut; Soc Exp Biol & Med; Soc Toxicol; Environ Mutagen Soc. Res: Pharmacognosy of animal drugs; bio-assay; pharmacology and toxicology of barbiturates; sulfonamides; sympathomimetics; p-aminohippuric acid; local anesthetics; pharmacology and toxicology of fissionable materials; tranquilizers; toxicological evaluation of drugs. Mailing Add: 176 Gwynedd Manor Rd North Wales PA 19454

MATTISON, LOUIS EMIL, b Lincoln, Nebr, Oct 3, 27; m 49; c 3. CHEMISTRY. Educ: La State Univ, BS, 49; Univ Del, MS, 50, PhD(org chem), 52. Prof Exp: Res chemist, Carothers Lab, Exp Sta, E I du Pont de Nemours & Co, 52-54; from assoc prof to prof chem, Davis & Elkins Col, 56-62; PROF CHEM & CHMN DEPT, KING COL, 63- Concurrent Pos: Cottrell res grant, 56-60; chmn dept chem, Davis & Elkins Col, 56-62; res assoc, Univ Ariz, 62-63. Mem: AAAS; Am Chem Soc; Sigma Xi; NY Acad Sci; Am Inst Chemists. Res: Organic synthesis of chelating agents; metal chelates; coordination compounds; photochemistry. Mailing Add: 323 Poplar St Bristol TN 37620

MATTISON, PHILLIP LEROY, b Springfield, Minn, Nov 25, 40; m 63; c 2. ORGANIC CHEMISTRY. Educ: Augsburg Col, BA, 62; Mich State Univ, PhD(org chem), 67. Prof Exp: Sr res chemist, 67-74, GROUP LEADER, GEN MILLS CHEMS, INC, 74- Mem: Am Chem Soc; Am Inst Mining, Metall & Petrol Engrs. Res: Synthesis and development of organic compounds which will function as highly selective ion exchange reagents. Mailing Add: 49 Oakwood Dr New Brighton MN 55112

MATTISON, ROLAND LEES, b Washington, DC, Nov 28, 43; m 75. COMPUTER SCIENCES. Educ: Ga Inst Technol, BS, 66; Northeastern Univ, MS, 69. Prof Exp: Mem tech staff comput prog, Aerospace Syst Div, RCA Corp, 65-68; design engr, Viatron, 69-70; mem tech staff, 70-76, RES MGR COMPUT SCI, GTE LABS, 76- Mem: Asn Comput Mach; Inst Elec & Electronic Engrs. Res: Development of computer aided design programs for digital logic simulation; integrated circuit layout and printed circuit board design; development of structured programming techniques for reliable software. Mailing Add: GTE Labs 40 Sylvan Rd Waltham MA 02154

MATTMAN, LIDA HOLMES, b Denver, Colo, July 31, 12; m 44; c 2. BACTERIOLOGY. Educ: Univ Kans, AB, 33, MA, 34; Yale Univ, PhD(bact), 40. Prof Exp: Bacteriologist, Med Dept, Endicott Johnson, NY, 34; asst, Iowa Hosp, 40-42; res bacteriologist, Nat Res Coun, 42-44, comn airborne infection, 45; mycologist, Santa Rosa Hosp, San Antonio, Tex, 46-47; sr bacteriologist, State Health Labs, Mass, 47-49; from asst prof to assoc prof bact, 49-74, PROF BIOL, WAYNE STATE UNIV, 74- Concurrent Pos: Nat Res Coun fel, Univ Pa, 43-44. Mem: Am Soc Microbiol. Res: Surface tension depressants in immunological systems; pathogenic anaerobes; L variants and mycoplasmae. Mailing Add: 1500 Seminole St Detroit MI 48214

MATTOCKS, ALBERT MCLEAN, b Wilmington, NC, Nov 5, 17; m 40; c 4. PHARMACEUTICAL CHEMISTRY. Educ: Univ NC, BS, 42; Univ Md, PhD(pharmaceut chem), 45. Prof Exp: Res chemist, Southern Res Inst, Ala, 45-47; prof pharmaceut chem & dir controls, Univ Hosp, Western Reserve Univ, 47-49; dir lab, Am Pharmaceut Asn, 49-50; mgr pharmaceut develop, McNeil Labs, 50-52; prof pharm, Univ Mich, 53-61; tech dir, R P Scherer Corp, 61-63; prof pharm & coordr hosp pharm educ & res, Univ Mich, 63-66; PROF PHARM, UNIV NC, CHAPEL HILL, 66- Mem: AAAS; Am Chem Soc; Am Pharmaceut Asn; fel Acad Pharmaceut Sci. Res: Drug metabolism and distribution; peritoneal cialysis. Mailing Add: Rte 4 Box 493A Mann's Chapel Rd Chapel Hill NC 27514

MATTONI, RUDOLF H T, population genetics, agriculture, see 12th edition

MATTOON, JAMES RICHARD, b Loveland, Colo, Dec 9, 30; m 53; c 2. BIOCHEMISTRY. Educ: Univ Ill, BS, 53; Univ Wis, MS, 54, PhD(biochem), 57. Prof Exp: From instr to asst prof chem, Univ Nebr, 57-62; asst prof, 64-70, ASSOC PROF PHYSIOL CHEM, SCH MED, JOHNS HOPKINS UNIV, 70- Concurrent Pos: Fel, Sch Med, Johns Hopkins Univ, 62-64. Mem: Am Chem Soc; Am Soc Biol Chem; Genetics Soc Am. Res: Genetics of mitochondria; mitochondrial biogenesis; oxidative phosphorylation; yeast respiration and mitochondria; lysine biosynthesis. Mailing Add: Dept of Physiol Chem Johns Hopkins Univ Sch of Med Baltimore MD 21205

MATTOON, RICHARD WILBUR, b Albuquerque, NMex, Feb 8, 12; m 45; c 3. CHEMICAL PHYSICS. Educ: Antioch Col, BS, 35; Univ Chicago, MS, 38, PhD(physics), 48. Prof Exp: Lectr physics, Mus Sci & Indust, Ill, 35; asst, Nat Bur Standards, DC, 36, Eastman Kodak Co, 37 & Rockefeller Found, Chicago, 38; asst instr radio & spectros, Univ Chicago, 42, instr phys sci, 42-44, res assoc, Off Reserve, War Prod Bd, 43-48; res physicist, Procter & Gamble Co, 48-50; RES PHYSICIST, ABBOTT LABS, 50-, DIR SPECIAL RES PROJS, 75- Concurrent Pos: Consult, AEC, 44; head chem phys lab, Abbott Labs, 65-75; Chicago scholar, Coffin fel. Honors & Awards: Award of Merit, Chicago Tech Socs Coun, 67. Mem: Fel AAAS; Am Phys Soc; Am Chem Soc; fel Am Inst Chemists; US Metric Asn (vpres, 75-). Res: Physics of pharmaceuticals; nuclear magnetic resonance and mass spectroscopy of organic compounds; x-ray diffraction analysis; high-energy radiation chemistry of organic and biological materials; surface and colloidal physics and chemistry; optical microscopy; electrostatic charge on fine powders and plastics. Mailing Add: Chem Physics Lab Abbott Labs North Chicago IL 60064

MATTOR, JOHN ALAN, b Oxford, Maine, Jan 15, 32; m 58; c 3. SYNTHETIC ORGANIC CHEMISTRY. Educ: Bates Col, BS, 58; Lawrence Univ, MS, 60, PhD(chem), 63. Prof Exp: Mem staff, 62-75, SR RES ASSOC, S D WARREN CO, SCOTT PAPER, WESTBROOK, 75- Mem: Am Chem Soc; AAAS; Soc Photog Scientists & Engrs. Res: Photochemistry; organic photoconductivity; dye sensitization. Mailing Add: Box 85 Bar Mills ME 04004

MATTOX, DONALD MOSS, b Richmond, Ky, Dec 27, 32; m 53; c 1. PHYSICS, MATERIALS SCIENCE. Educ: Eastern Ky State Col, BS, 53; Univ Ky, MS, 60. Prof Exp: Staff mem physics, 61-65, SUPVR SURFACE PHYSICS & CHEM DIV, SANDIA CORP, 65- Concurrent Pos: NSF fel, 59-61. Mem: Am Vacuum Soc; Am Phys Soc. Res: Interface formation and adhesion of deposited films; thin film technology. Mailing Add: 3416 La Sala Grande de Este NE Albuquerque NM 87111

MATTOX, KARL, b Cincinnati, Ohio, Aug 22, 36; m 57; c 3. PHYCOLOGY. Educ: Miami Univ, BS, 58, MA, 60; Univ Tex, PhD(bot), 62. Prof Exp: Asst prof bot, Univ Toronto, 62-66; asst prof, 66-70, ASSOC PROF BOT, MIAMI UNIV, 70- Concurrent Pos: Res assoc, Great Lakes Inst, 62- Mem: Bot Soc Am; Phycol Soc Am. Res: Morphology. Mailing Add: Dept of Bot Miami Univ Oxford OH 45056

MATTOX, RICHARD BENJAMIN, b Middletown, Ohio, May 15, 21; m 48. GEOLOGY. Educ: Miami Univ, BA, 48, MS, 49; Univ Iowa, PhD(geol), 54. Prof Exp: Instr geol, Miami Univ, 49-50; petrol geologist, Magnolia Petrol Co, 50; asst instr geol, Univ Iowa, 50-52; asst prof, Miss State Col, 52-54; assoc prof, 54-57, PROF GEOL, TEX TECH UNIV, 57- Concurrent Pos: Head dept geol, Tex Tech Univ, 64-70. Mem: AAAS; Nat Asn Geol Teachers; Geol Soc Am; Soc Econ Paleont & Mineral; Am Asn Petrol Geologists. Res: Eolian geology; stratigraphy; geology of Colorado plateau. Mailing Add: Dept of Geosci Tex Tech Univ Lubbock TX 79409

MATTOX, VERNON ROSS, b Union Hall, Va, Oct 19, 14; m 43. ORGANIC CHEMISTRY. Educ: Lynchburg Col, BA, 35; Va Polytech Inst, MS, 39; Univ Va, PhD(org chem), 43. Prof Exp: Anal chemist, Chesapeake Corp, Va, 35-37; Nat Res Coun asst, Univ Va, 42; from asst prof to assoc prof, 51-64, PROF BIOCHEM, MAYO GRAD SCH MED, UNIV MINN, 64-, RES BIOCHEMIST, MAYO CLIN, 42-, HEAD SECT BIOCHEM, 64- Mem: Am Chem Soc; Am Soc Biol Chem; Endocrine Soc. Res: Steroids; adrenal cortical hormones and metabolites; endocrinology. Mailing Add: Dept of Biochem Mayo Grad Sch Med Univ Minn Rochester MN 55901

MATTRAW, HAROLD CLAUDE, b Pulaski, NY, Oct 15, 15; m 34; c 4. MATERIALS SCIENCE, SPECTROSCOPY. Educ: Univ Ala, BS, 38; Cornell Univ, PhD(chem), 49. Prof Exp: Chemist, Southern Kraft Corp, 39-41 & Tenn Valley Authority, 41-44; asst, Cornell Univ, 46-49; res assoc, Knolls Atomic Power Lab, Gen Elec Co, 49-58, tech mil planning oper, 58-61; mgr chem dept, Sperry Rand Res Ctr, 61-64; phys scientist, Res & Eng Div, Autonetics Div, NAm Rockwell Corp, 64-71; EXEC DIR DEPT CHEM, CORNELL UNIV, 71- Mem: Am Chem Soc. Res: Mass, emission and infrared spectroscopy. Mailing Add: Dept of Chem Cornell Univ Ithaca NY 14850

MATTSON, DALE EDWARD, b Newberry, Mich, Apr 5, 34; m 57; c 2. BIOMETRICS. Educ: Colo Col, BA, 59; Univ Ill, MA, 61, PhD(educ psychol), 63. Prof Exp: Asst prof educ measurement, Univ Wash, 63-64; dir educ res, Am Asn Dent Schs, 64-66 & Asn Am Med Cols, 66-69; dir admis & rec, 69-72, PROF BIOMET, SCH PUB HEALTH, UNIV ILL, 72- Mem: Am Pub Health Asn; Am Statist Asn. Res: Indices of serial correlation with applications to measures of health statistics; epidemiology of sports injuries. Mailing Add: 5243 N Mason Chicago IL 60630

MATTSON, DON ARTHUR, mathematics, see 12th edition

MATTSON, DONALD EUGENE, b Chatsworth, Calif, May 19, 34; m 59; c 3. VETERINARY VIROLOGY. Educ: Univ Calif, Davis, BS, 57, DVM, 59; Wash State Univ, PhD(microbiol), 66. Prof Exp: Asst prof, 67-69, ASSOC PROF VET MED, ORE STATE UNIV, 69- Mem: Am Vet Med Asn. Res: Physical, chemical and serological properties of viruses; virus diseases of the newborn, especially bovine. Mailing Add: Dept of Vet Med Ore State Univ Corvallis OR 97331

MATTSON, FRED HUGH, b Spokane, Wash, Dec 16, 18; m 43; c 5. BIOCHEMISTRY. Educ: Loyola Univ, Calif, BS, 40; Univ Southern Calif, MS, 42, PhD(biochem), 48; Am Soc Clin Nutrit, cert specialist human nutrit, 71. Prof Exp: RES CHEMIST, PROCTER & GAMBLE CO, 48- Concurrent Pos: Adj prof, Univ Cincinnati, 70-; mem coun arteriosclerosis, Am Heart Asn. Honors & Awards: Am Chem Soc Award, 69. Mem: Am Chem Soc; Am Soc Biol Chem; Am Inst Nutrit. Res: Digestion and absorption of fat; nutritive value of fat; diet and cardio-vascular disease. Mailing Add: 1503 Kinney St Cincinnati OH 45231

MATTSON, GUY C, b Bloomfield, NJ, Jan 3, 27; m 50; c 4. ORGANIC CHEMISTRY. Educ: Union Col, NY, BS, 49; Univ Fla, PhD(chem), 55. Prof Exp: Chemist, Warner-Chilcott Labs, NJ, 49-52; instr chem, Univ Fla, 52-55; res chemist, Dow Chem Co, Mich, 55-60, facility mgr, Fla, 60-64, prod engr, Saginaw Bay, 64-65, proj mgr, Tex, 65-66, dept head, Ind, 66-71; PROF CHEM, FLA TECHNOL UNIV, 70- Res: Organic synthesis; process development. Mailing Add: Dept of Chem Fla Technol Univ Orlando FL 32816

MATTSON, HAROLD F, JR, b Ann Arbor, Mich, Dec 7, 30. APPLIED MATHEMATICS. Educ: Oberlin Col, AB, 51; Mass Inst Technol, PhD(math), 55. Prof Exp: Mathematician, Air Force Cambridge Res Ctr, 55-60; mathematician, Appl Res Lab, Sylvania Elec Prod, Inc, Gen Tel & Electronics Corp, 60-70, Eastern Opers, 70-71; PROF SYSTS & INFO SCI, SYRACUSE UNIV, 71- Concurrent Pos: Ed, Review, Soc Indust & Appl Math, 70- Mem: Am Math Soc; Math Asn Am; Soc Indust & Appl Math. Res: Combinatorial analysis; error-correcting codes. Mailing Add: Link 313 SIS Prog Syracuse Univ Syracuse NY 13210

MATTSON, JAMES STEWART, b Providence, RI, July 22, 45; m 64; c 2. CHEMICAL OCEANOGRAPHY, SURFACE CHEMISTRY. Educ: Univ Mich, Ann Arbor, BS, 66, MS, 69, PhD(water resources sci), 70. Prof Exp: Staff assoc environ sci, Gulf Gen Atomic Co, 70-71; dir res & develop, Ouachita Indust, subsid DHJ Indust Inc, 71-72; asst prof chem oceanog, Rosenstiel Sch Marine & Atmospheric Sci, Univ Miami, 72-76; PHYS SCIENTIST MARINE CHEM, DEPT COM, NAT OCEANIC & ATMOSPHERIC ADMIN, ENVIRON DATA SERV, CTR EXP DESIGN & DATA ANAL, 76- Mem: Am Chem Soc; Am Soc Testing & Mat; Am Soc Limnol & Oceanog; Coblentz Soc; Soc Appl Spectros. Res: Effects of outer continental shelf resource development on environment; natural weathering of oil spills; blood protein interactions at solid-liquid interfaces. Mailing Add: Nat Oceanic & Atmospheric Admin 3300 Whitehaven St NW Washington DC 20235

MATTSON, LELAND NEIL, b Bridgeport, Kans, Sept 17, 19; m 42; c 1. CHEMISTRY. Educ: Bethany Col, Kans, BS, 41. Prof Exp: Asst instr chem, Bethany Col, Kans, 40-41; asst instr, Univ Kans, 41-42; chemist, Atlas Powder Co, 42-43 & Nat Foundry, 43-44; asst to chief chemist, Gen Cable Co, 46-47; chemist, Cole Chem Co, St Louis, 47-48, dir control & res, 49-56; chief chemist, Sci Assoc, Inc, 56-59; dir labs, K-V Pharmacal Co & Victor M Hermelin Co, Mo, 59-64; DIR CHEM & MICROBIOL SERV, SCI ASSOCS, INC, 64-, VPRES, 66- Concurrent Pos: Consult pharmaceut mfg & lab mgr. Mem: Am Chem Soc; fel Am Inst Chemists. Res: Analytical methods for pharmaceuticals. Mailing Add: Sci Assoc Inc 6200 S Lindbergh Blvd St Louis MO 63123

MATTSON, PETER HUMPHREY, b Evanston, Ill, Apr 3, 32; m 54; c 3. GEOLOGY. Educ: Oblerin Col, BA, 53; Princeton Univ, PhD(geol), 57. Prof Exp: Geologist, US Geol Surv, 57-64; from asst prof to assoc prof, 64-72, PROF GEOL & GEOG, QUEENS COL, NY, 73- Concurrent Pos: Chmn dept geol & geog, Queens Col, NY, 65-68; consult, Commonwealth PR, 65-69. Mem: Geol Soc Am; Am Geophys Union. Res: Igneous petrology; volcanic rocks; structural geology; geology of Puerto Rico and the Caribbean area; island arcs; oceanic igneous rocks. Mailing Add: Dept of Earth & Environ Sci Queens Col Flushing NY 11367

MATTSON, RAYMOND HARDING, b Matchwood, Mich, Oct 10, 20; m 51; c 4. ORGANIC CHEMISTRY. Educ: Univ Mich, BS, 43; Univ Ill, PhD(chem), 51. Prof

Exp: Res chemist, Rohm & Haas Co, 43-44, Am Cyanamid Co, 50-52 & mkt develop, Jefferson Chem Co, 52-55; sr mkt res analyst, Am Cyanamid Co, 55-59, tech rep, 59-62, mgr sales develop rubber chem, 62-63; mkt res assoc, Glidden Co, 63-71, MGR GROUP MKT RES, GLIDDEN-DURKEE DIV, SCM CORP, 71- Mem: Am Chem Soc. Res: Restricted rotation in aryl amines. Mailing Add: 7396 Ober Lane Chagrin Falls OH 44022

MATTSON, VICTOR FRANK, b Glenridge, NJ, June 1, 25; m 51; c 2. CHEMISTRY. Educ: Union Col, BS, 49, MS, 51, PhD(paper chem), 54. Prof Exp: Res chemist, Mead Corp, 54-55 & Oxford Paper Co, 56-60; res dir, 60-64, tech dir, 64-72, DIR RES & ENG, GREAT NORTHERN PAPER CO, 72- Res: Pulp and paper chemistry. Mailing Add: 35 Crestmont Ave Millinocket ME 04462

MATTUCK, ARTHUR PAUL, b Brooklyn, NY, June 11, 30; m 59; c 1. GEOMETRY. Educ: Swarthmore Col, AB, 51; Princeton Univ, PhD(math), 54. Prof Exp: Res fel math, Harvard Univ, 54-55; C L E Moore instr, 55-57, lectr, 57-58, from asst prof to prof, 58-73, CLASS OF 1922, PROF MATH, MASS INST TECHNOL, 73- Mem: Am Math Soc; Math Asn Am. Res: Algebraic geometry. Mailing Add: Dept of Math Mass Inst Technol Cambridge MA 02139

MATTUCK, RICHARD DAVID, solid state physics, see 12th edition

MATUDA, EIZI, b Nagasaki, Japan, Apr 20, 94; m 29; c 1. PLANT TAXONOMY. Educ: Formosan Univ Sci, MSc, 15; Univ Tokyo, BSc, 62. Prof Exp: Dir & investr plant taxon, Inst Biol, Matuda Herbarium, 36-49; INVESTR PLANT TAXON, INST BIOL, NAT UNIV MEX, 49- Concurrent Pos: Forest botanist, Inst Forest Invests, Nat Univ Mex, 50-57, consult, Bot Garden, 54-; chief, Bot Explor Comn, 54- Honors & Awards: Dipl & Bot Medal, Bot Soc Mex, 59; Dipl Sci Merit, Mex Forestry Soc, 59; Dipl & Kun' Shito-Zuihosyo Medal, Japanese Emperor, 69. Mem: Am Soc Plant Taxon; Bot Soc Mex; Mex Cactus Soc; Mex Natural Hist Soc; Mex Nat Acad Sci. Res: Mexican subtropical and tropical areas, mostly on Monocotyledonae; Mexican Araceae, Bromeliaceae, Cyclantaceae and Dioscoreacea; Mexican Agavaceae, Furcraea and Beschorneria. Mailing Add: Biol Inst Nat Univ Mex Mexico DF Mexico

MATUKAS, VICTOR JOHN, b Freeport, Tex, Oct 20, 33; m 61; c 3. EXPERIMENTAL PATHOLOGY. Educ: Loyola Univ, La, DDS, 56; Univ Rochester, PhD(path), 66. Prof Exp: Resident oral surg, Charity Hosp, New Orleans, La, 58-61; asst prof path, Loyola Univ, La, 61-62 & Univ Pa, 66-68; prof stomatol & chmn dept, Sch Dent, Univ Colo, Denver, 71-74; INVESTR, INST DENT RES, UNIV ALA, BIRMINGHAM, 74- Concurrent Pos: Spec res fel, Nat Inst Dent Res, 68-71. Mem: AAAS; Am Dent Asn. Res: Synthesis, metabolism and ultrastructure of collagen and protein-polysaccharide; biological mineralization. Mailing Add: Inst of Dent Res Univ of Ala Sch of Dent Birmingham AL 35294

MATULIC, LJUBOMIR FRANCISCO, b Potosi, Bolivia, May 8, 23; US citizen; m 53; c 3. QUANTUM OPTICS. Educ: State Gym, Yugoslavia, BA, 42; Univ Chile, Lic Math & Physics, 49; Ind Univ, Bloomington, MS, 63; Univ Rochester, PhD(physics), 71. Prof Exp: Teacher high sch, Bolivia, 49-50; prof math, Collegio Normal Superior, Bolivia, 50-54; prof math & physics, Leguerrier Classical Inst, Montreal, 54-58; lectr math, Royal Mil Col, Que, 58-60; assoc prof, 63-68 & 70-73, PROF PHYSICS, ST JOHN FISHER COL, 73- Concurrent Pos: Instr, Univ San Simon, Bolivia, 49-50; vis scientist, Inst Ruder Boskovic, Univ Zagreb, Yugoslavia, 74. Mem: Arg Math Union; Am Asn Physics Teachers; Optical Soc Am; Am Phys Soc. Res: Theoretical investigation of distortionless propagation of electromagnetic fields through nonlinear absorbers, especially the phase modulation of this field due to the interaction with resonant atoms and to the bulk host medium. Mailing Add: Dept of Physics St John Fisher Col Rochester NY 14618

MATULIONIS, DANIEL H, b Lithuania, Oct 2, 38; US citizen; m 60; c 2. ANATOMY, EMBRYOLOGY. Educ: Wis State Univ-Whitewater, BEd, 63; Univ Ill, Urbana, MS, 65; Tulane Univ, PhD(anat), 70. Prof Exp: Instr biol, Eastern Ky Univ, 65-67; instr, 70-71, ASST PROF ANAT, COL MED, UNIV KY, LEXINGTON, 71- Concurrent Pos: Gen Res Support grant, Univ Ky, 70-71; Ky Tobacco Res Inst grant, 71-72. Mem: Am Asn Anat. Res: Ultrastructural analysis of keratin precursors; glycogen synthesis; ultrastructural analysis of cigarette smoke effects on the respiratory system. Mailing Add: Dept of Anat Col Med Univ of Ky Lexington KY 40506

MATULIS, RAYMOND M, b Broadview, Ill, Apr 20, 39; m 61; c 2. ANALYTICAL CHEMISTRY. Educ: Culver-Stockton Col, BA, 61; Univ Mo-Columbia, MA, 63, PhD(chem), 66. Prof Exp: Res chemist, Gulf Res & Develop Co, 66-67; mgr lab advan tech res, 67-71, mgr advan tech res, 71-73, DIR ADVAN TECH RES, HALLMARK CARDS, INC, 73- Mem: Am Chem Soc. Res: Development of adhesives, inks, coatings and plastic materials; development of physical and chemical test methods; development of processing technology. Mailing Add: 9818 W 100 Terr Overland Park KS 66212

MATUMOTO, TOSIMATU, b Tokyo, Japan, Aug 3, 26; m 55; c 2. GEOPHYSICS. Educ: Tokyo Univ, MS, 51, PhD(seismol), 60. Prof Exp: Res asst geophys, Earthquake Res Inst, Univ Tokyo, 51-61; res assoc, Lamont-Doherty Geol Observ, Columbia Univ, 60-65, sr res assoc, 66-74; MEM STAFF, MARINE SCI INST, GALVESTON, TEX, 74- Mem: Seismol Soc Am; Am Geophys Union. Res: Spectral analysis of seismic waves and its relation to magnitude; study of seismicity and microearthquake in Alaska and northwest United States. Mailing Add: Marine Sci Inst 700 The Strand Galveston TX 77550

MATURO, FRANK JUAN SARNO, JR, b Nashville, Tenn, Apr 28, 29; m 60; c 3. MARINE BIOLOGY. Educ: Univ Ky, BS, 51; Duke Univ, MA, 53, PhD(marine ecol), 56. Prof Exp: Instr zool, Duke Univ, 55-57; vis asst prof, Univ NC, 57-58; asst prof biol, 58-64, assoc prof zool, 64-72, dir marine lab, 70-74, PROF ZOOL, UNIV FLA, 72-, SUPVR MARINE LAB, 74- Concurrent Pos: Nat Acad Sci-Nat Res Coun sr vis res assoc, Smithsonian Inst Mus Natural Hist, 65-66. Mem: Am Soc Zool; Am Soc Limnol & Oceanog; Soc Syst Zool; Am Inst Biol Sci; Int Bryozool Asn. Res: Seasonal distribution and settling rates of marine invertebrates; zoogeography, ecology, and systematics of marine Bryozoa; larval behavior, metamorphosis, and astogeny of Bryozoa. Mailing Add: Dept of Zool Univ of Fla Gainesville FL 32601

MATURO, JOSEPH MARTIN, III, b Bridgeport, Conn, Nov 15, 42; m 66; c 1. BIOCHEMISTRY, PHYSIOLOGY. Educ: Fairfield Univ, BS, 64; Boston Col, PhD(biol), 69. Prof Exp: Asst prof, 69-73, ASSOC PROF BIOL, C W POST COL, LONG ISLAND UNIV, 73- Mem: AAAS. Res: Mechanism of action of insulin. Mailing Add: Dept of Biol CW Post Col Long Island Univ P O Greenvale NY 11548

MATUSZAK, ALFRED H, organic chemistry, see 12th edition

MATUSZAK, CHARLES A, b Pittsburgh, Pa, Jan 7, 32; m 55; c 2. PHYSICAL ORGANIC CHEMISTRY. Educ: Univ Okla, BS, 52, MS, 53; Ohio State Univ, PhD(org chem), 57. Prof Exp: Asst org chem, Ohio State Univ, 53-57; res chemist, Owens-Corning Fiberglass Corp, 57-58; fel org chem, Ohio State Univ, 58-59, Univ Wis, 59-60 & Univ Kans, 60-61 & 62-63; asst prof, Washburn Univ, 61-62; ASST PROF ORG CHEM, UNIV OF THE PAC, 63- Mem: Am Chem Soc; The Chem Soc; Sigma Xi. Res: Mechanisms; Birch reduction; imidazole compounds; biphenylenes. Mailing Add: Dept of Chem Univ of the Pac Stockton CA 95211

MATUSZAK, DAVID ROBERT, b Oct 2, 34; US citizen; m 53; c 3. GEOLOGY. Educ: Univ Okla, BS, 55, MS, 57; Northwestern Univ, PhD(geol), 61. Prof Exp: Lab asst geol, Univ Okla, 56-57; geologist, Kerr-McGee Oil Industs, Inc, 57-58; lab asst geol, Northwestern Univ, 58-60; res engr, Pan Am Petrol Corp, 61-63, sr res scientist, 63-68, res group supvr, 68-71, SR STAFF GEOLOGIST, AMOCO PROD CO, 71- Mem: Am Asn Petrol Geologists; Soc Prof Well Log Analysts. Res: Use of subsurface data and computers in oil exploration. Mailing Add: Amoco Prod Co PO Box 591 Tulsa OK 74102

MATUSZEK, JOHN MICHAEL, JR, b Worcester, Mass, Apr 16, 35; m 57; c 4. RADIOLOGICAL HEALTH, RADIOCHEMISTRY. Educ: Worcester Polytech Inst, BS, 57; Clark Univ, PhD(nuclear chem), 62. Prof Exp: Scientist, Southeastern Radiol Health Lab, USPHS, 62-64; asst mgr measurements div, Isotopes, Inc, 64-67; mgr physics dept, Teledyne Isotopes, 67-71; DIR RADIOL SCI LAB, NY STATE HEALTH DEPT, 71- Mem: Am Chem Soc; Am Nuclear Soc; Health Physics Soc. Res: Nuclear reaction mechanisms for intermediate energy interactions; nuclear spectroscopy; radiological health; radiochemical procedures; fission research. Mailing Add: NY State Health Dept New Scotland Ave Albany NY 12201

MATUSZKO, ANTHONY JOSEPH, b Hadley, Mass, Jan 31, 26; m 56; c 4. ORGANIC CHEMISTRY, INORGANIC CHEMISTRY. Educ: Amherst Col, AB, 46; Univ Mass, MS, 51; McGill Univ, PhD(org chem), 53. Prof Exp: Demonstr chem, McGill Univ, 50-52; instr, Lafayette Col, 52-53, from asst prof to assoc prof, 53-58; assoc head chem div, Res & Develop Dept, US Naval Propellant Plant, 58-59, head fundamental processes div, 59-62, polymer div, 62; chief org chem prog, 62-71, PROG MGR, CHEM SCI DIRECTORATE, AIR FORCE OFF SCI RES, 71- Concurrent Pos: Hon fel, Univ Wis, 67-68. Mem: Fel AAAS; Am Chem Soc; fel Am Inst Chemists; Am Ord Asn; Sigma Xi. Res: Organometallics; reactions with nitriles; pyridine derivatives; modifications and properties of cellulose nitrates; phosphonitrilic derivatives; high nitrogen compounds. Mailing Add: Chem Sci Directorate Air Force Off of Sci Res Bolling AFB Washington DC 20332

MATZ, JOHN J, JR, b Alliance, Nebr; m 58. NUTRITION, BIOCHEMISTRY. Educ: Mont State Univ, BS, 62, MS, 64; Univ Wyo, PhD(nutrit, animal husb), 67. Prof Exp: Res asst nutrit, Mont State Univ, 63-64; Allied Chem grant & supply instr nutrit, Univ Wyo, 65-67; asst prof animal husb, Univ Mo-Columbia, 67-69; nutritionist, Western Star Milling Co, Kans, 69-72; gen mgr & nutritionist, Cent Milling Co, 72-75; TECH ADV ANIMAL HEALTH & NUTRIT, INT MINERALS & CHEMS, INC, 75- Mem: AAAS; Am Soc Animal Sci; Am Dairy Sci Asn. Res: Investigations on the nutritional requirements of ruminant animals. Mailing Add: PO Box 420 Terre Haute IN 47808

MATZ, ROBERT, b New York, NY, Aug 5, 31; m 55; c 3. INTERNAL MEDICINE. Educ: NY Univ, BA, 52, MD, 56. Prof Exp: From intern to resident, Bronx Municipal Hosp Ctr, 56-60; consult, Obstet Serv, Lincoln Hosp, Bronx, 62-63; NIH trainee metab, 63-64, ASSOC PROF MED, ALBERT EINSTEIN COL MED, 71-; ASSOC DIR MED, MONTEFIORE-MORRISANIA AFFIL, 64- Concurrent Pos: Vis physician, Montefiore-Morrisania Affil, 64-76, head endocrinol & metab, 75-76, attend physician, Montefiore Hosp & Med Ctr, 75-76; attend physician, Bronx Munic Hosp Ctr, 71-76, co-dir, Diabetes Clin, 73-76; mem, Endocrine Dis Adv Comt, New York Dept Health, 72; consult, Health, Educ & Welfare Eval Unit, Albert Einstein Col Med, 73-76. Mem: Am Diabetes Asn; Harvey Soc; Am Fedn Clin Res; Am Col Physicians. Res: Clinical research in diabetes mellitus, diabetic coma, and metabolic acidoses; clinical investigation of methods to improve delivery of health care. Mailing Add: 32 Buena Vista Dr Hastings-on-Hudson NY 10706

MATZ, SAMUEL ADAM, b Carmi, Ill, July 1, 24; m 51; c 4. FOOD SCIENCE. Educ: Evansville Col, BA, 48; Kans State Col, MS, 50; Univ Calif, PhD(agr chem), 58. Prof Exp: Instr, Kans State Col, 50; chief chemist, Harvest Queen Mill & Elevator Co, 50-51; food technologist cereal & gen prod & chief br, Armed Forces Qm Food & Container Inst, 51-59; supvr refrig dough invests, Borden Foods Co, 59-65; vpres res & develop, Robert A Johnston Co, Wis, 65-69; vpres, Ovaltine Food Prod, Ill, 69-71, VPRES RES & DEVELOP, OVALTINE PROD DIV, SANDOZ-WANDER, INC, 71- Concurrent Pos: Dir, Avi Publ Co, 73- Mem: Am Chem Soc; Inst Food Technologists. Res: Food preservation methods and texture; cereal and flavor chemistry; permeability mechanisms; nutrition; regulatory affairs. Mailing Add: Ovaltine Prod Div Sandoz-Wander Inc Villa Park IL 60181

MATZ, WILLIAM HOWARD, b Pittsburgh, Pa, June 25, 22; m 44; c 1. PHYSICAL CHEMISTRY. Educ: Pa State Univ, BS, 47. Prof Exp: Res chemist, Pittsburgh Coke & Chem Co, 47-57, supvr qual control, 57-59, supvr tech serv, Activated Carbon Div, Pittsburgh Chem Co, 59-64, Pittsburgh Activated Carbon Co, 64-66; GROUP LEADER, RES ACTIVATED CARBON APPLN, CALGON CORP, 66- Mem: Am Chem Soc; Am Soc Testing & Mat; Am Inst Chemists. Res: Development of activated carbons for specific uses; new uses for activated carbons; improvement of production methods; waste water treatment. Mailing Add: 4 Perrysville Rd Ben Avon Heights Pittsburgh PA 15202

MATZINGER, DALE FREDERICK, b Alleman, Iowa, Apr 14, 29; m 60; c 2. QUANTITATIVE GENETICS. Educ: Iowa State Univ, BS, 50, MS, 51, PhD(plant breeding), 56. Prof Exp: Asst plant breeding, Iowa State Univ, 53-56; asst statistician, 56-57, asst prof statist, 57-58, statist & genetics, 58-60, assoc prof genetics, 60-64, PROF GENETICS, NC STATE UNIV, 64- Honors & Awards: Philip Morris Award Distinguished Achievement Tobacco Sci, 71. Mem: Fel Am Soc Agron; Crop Sci Soc Am; Genetics Soc Am; Biomet Soc; AAAS. Res: Statistical genetic theory and breeding methodology of self-pollinated plants; appled statistics; plant science. Mailing Add: Dept of Genetics NC State Univ Raleigh NC 27607

MATZKANIN, GEORGE ANDREW, b Chicago, Ill, June 30, 38; m 63; c 2. SOLID STATE PHYSICS. Educ: St Mary's Col, AB, 60; Univ Fla, MS, 62, PhD(physics), 66. Prof Exp: Res assoc metals physics, Argonne Nat Lab, 66-68; vis asst prof physics, Univ Ill, Chicago Circle, 68-69; SR RES PHYSICIST INSTRUMENTATION DIV, SOUTHWEST RES INST, 69- Mem: AAAS; Am Phys Soc; Am Soc Nondestructive Testing; Sigma Xi. Res: Nondestructive evaluation research; instrumentation research; nuclear magnetic resonance; magnetic and mechanical properties of materials; Barkhausen phenomena; residual stress; evaluation of metal fatigue. Mailing Add: Southwest Res Inst PO Drawer 28510 San Antonio TX 78284

MATZKE, HOWARD ARTHUR, b Winona, Minn, Mar 22, 20; m 42; c 3. NEUROANATOMY. Educ: St Louis Univ, MS, 43; Univ Minn, PhD(anat), 49. Prof Exp: Instr anat, Med Sch, Creighton Univ, 43-47; res assoc neuropath, Univ Minn, 47-50; asst prof anat, Col Med, State Univ NY Downstate Med Ctr, 50-53; assoc prof, 53-59, PROF ANAT, UNIV KANS, 59-, CHMN DEPT, 62- Mem: Soc Neurosci; Am Asn Anat; Am Acad Neurol. Res: Pathology of nervous system in bulbar poliomyelitis; nerve regeneration in parabiotic rats; comparative neuroanatomy; pathway of neurotropic agents to central nervous system. Mailing Add: 11002 W 72nd Terr Shawnee KS 66203

MATZNER, EDWIN ARTHUR, b Vienna, Austria, May 14, 28; m 53; c 1. ORGANIC CHEMISTRY. Educ: Calif Inst Technol, BS, 51; Yale Univ, PhD(org chem), 58. Prof Exp: Res chemist, 58-63, sr res group leader, 63-67, MGR RES & DEVELOP, MONSANTO CO, 67- Mem: Am Chem Soc; Soc Chem & Indust; Res & Eng Soc Am. Res: Synthetic organic chemistry; solvent extraction; chemistry of phosphates; electrochemistry; chemistry of detergents and surfactants. Mailing Add: Monsanto Co 800 N Lindbergh Blvd St Louis MO 63166

MATZNER, MARKUS, b Biala, Poland, Mar 19, 29; nat US; m 54; c 1. ORGANIC CHEMISTRY. Educ: Univ Brussels, MS, 50, PhD, 53. Prof Exp: Res chemist, Tirlemont Refinery, Belg, 53-54 & Probel Labs, 54-56; res & control chemist, Belg Petrol Refinery, 56-59; res chemist, Plastics Div, Union Carbide Corp, 59-63, proj scientist, 63-66, res scientist, 66-68, sr res scientist, 68-71, RES ASSOC CHEM & PLASTICS RES & DEVELOP DEPT, UNION CARBIDE CORP, 71- Mem: Am Chem Soc; Sigma Xi; Chem Soc Belg; Royal Netherlands Chem Soc. Res: Organic synthesis; mechanisms of reactions; polymer chemistry. Mailing Add: 23 Marshall Dr Edison NJ 08817

MATZNER, RICHARD ALFRED, b Ft Worth, Tex, Jan 2, 42; m 67; c 1. PHYSICS. Educ: Univ Notre Dame, BS, 63; Univ Md, College Park, PhD(physics), 67. Prof Exp: NSF fac assoc physics, 67-69, asst prof, 69-73, ASSOC PROF PHYSICS, UNIV TEX, AUSTIN, 73-, RES PHYSICIST, CTR RELATIVITY THEORY, 73- Concurrent Pos: Res fel physics, Wesleyan Univ, 69-70. Mem: AAAS. Res: General relativity; cosmology; gravitational collapse; geometrical optics; canonical formulations; statistical mechanics. Mailing Add: Dept of Physics Univ of Tex Austin TX 78712

MAUCK, HENRY PAGE, JR, b Richmond, Va, Feb 3, 26; c 2. CARDIOLOGY. Educ: Univ Va, BA, 48, MD, 52; Am Bd Internal Med, dipl, 59. Prof Exp: DIR A D WILLIAMS HYPERTENSION CLIN & DIR CARDIAC CATHETERIZATION LAB, MED COL VA, 70-, PROF MED & PEDIAT, 72- Concurrent Pos: Am Heart Asn fel, 56-57; consult pediat cardiol, Langley Air Force Hosp; ed consult, Am Heart J. Mem: AMA; fel Am Col Physicians; fel Am Col Cardiol; Am Fedn Clin Res; fel Am Heart Asn. Res: Neural control of the circulation. Mailing Add: Dept of Med Med Col of Va Richmond VA 23298

MAUDERLI, WALTER, b Aarau, Switz, Mar 8, 24; nat US; m 50; c 5. NUCLEAR PHYSICS. Educ: Swiss Fed Inst Technol, MS, 49, DSc(physics), 56. Prof Exp: Physicist & asst radiol, Univ Zurich, 50-56; physicist, asst prof & head isotope labs, Sch Med, Univ Ark, 56-60; assoc prof radiation physics, 60-64, PROF RADIATION PHYSICS, J HILLIS MILLER HEALTH CTR, COL MED, UNIV FLA, 65-, PHYSICIST, 60-, PROF ENVIRON ENG SCI, 74- Concurrent Pos: Lectr, Grad Inst Technol & Med Ctr, Univ Ark; consult, Vet Admin Hosp, Little Rock. Mem: Am Asn Physicists in Med; Simulation Coun; Soc Nuclear Med; AMA; Asn Comput Mach. Res: Radiation physics; computer applications in radiology; electronic instrumentation in radiation physics. Mailing Add: Dept of Radiol Univ of Fla Gainesville FL 32601

MAUDERLY, JOE L, b Strong City, Kans, Aug 31, 43; m 65; c 2. PULMONARY PHYSIOLOGY. Educ: Kans State Univ, BS, 65, DVM, 67. Prof Exp: PHYSIOLOGIST, LOVELACE FOUND MED EDUC & RES, 69- Mem: Am Vet Med Asn; Am Soc Vet Physiol & Pharmacol; assoc Am Physiol Soc; World Asn Vet Physiol, Pharmacol & Biochem; Am Soc Vet Anesthesiol. Res: Comparative cardiopulmonary physiology; pulmonary function measurements in unanesthetized animals; functional effects of inhaled toxicants; therapeutic and investigative use of bronchopulmonary lavage; anesthesiology of laboratory animals. Mailing Add: Inhal Toxicol Res Inst Lovelace Found Box 5890 Albuquerque NM 87115

MAUDLIN, LLOYD Z, b Miles City, Mont, Feb 20, 24; m 46; c 4. PHYSICS. Educ: Univ Calif, Los Angeles, AB, 49; Univ Southern Calif, MS, 52. Prof Exp: Electronic scientist, US Naval Ord Test Sta, 51-56, supvry electronic scientist & head simulation br, 56-59, supvry physicist & dir simulation & comput ctr, 59-67, head simulation & anal div, Naval Undersea Warfare Ctr, 67-71, SUPVRY PHYSICIST & HEAD COMPUT SCI & SIMULATION DIV, NAVAL UNDERSEA CTR, 71- Mem: Inst Elec & Electronics Engrs; NY Acad Sci. Res: Anti-submarine warfare, particularly guidance and control of underwater weapons; computer and simulation analysis of anti-submarine warfare weapons systems; analog and digital computing techniques. Mailing Add: Naval Undersea Ctr San Diego CA 92132

MAUE-DICKSON, WILMA, b Joliet, Ill, Apr 15, 43; m 71; c 1. DEVELOPMENTAL ANATOMY. Educ: Rockford Col, BA, 64; Northwestern Univ, MA, 68; Univ Pittsburgh, PhD(biocommun), 70. Prof Exp: Instr elem & sec sch educ, US Peace Corps, Haaraghe Prov, Ethiopia, 64-66; dir res head & neck sci, Div Otolaryngol Maxillofacial Surg, Mercy Hosp, Pittsburgh, 69-74; asst prof anat, gross anat & craniofacial res, Univ Pittsburgh, 74-75; ASSOC PROF CRANIOFACIAL MORPHOL, MAILMAN CTR CHILD DEVELOP, UNIV MIAMI, 76- Concurrent Pos: Instr, Carlow Col, 70-72; assoc dir basic sci res, Cleft Palate Ctr, Univ Pittsburgh, 70-75, res consult, 76-; Health Res & Serv Found res grant, 72. Honors & Awards: First Place Award Sci Presentation, Am Speech & Hearing Asn, 70, First Place Award Sci Merit, 75. Mem: Am Cleft Palate Asn; Am Speech & Hearing Asn; Asn Res Otolaryngol; Am Asn Anatomists. Res: Normal human embryology and developmental morphology; anatomical bases of craniofacial pathologies; specific interest in embryologic and fetal research on craniofacial pathologies such as cleft palate; early detection of craniofacial pathologies in the neonatal and pediatric clinical population. Mailing Add: Mailman Ctr Child Develop Univ Miami PO Box 520006 Miami FL 33152

MAUER, ALVIN MARX, b Le Mars, Iowa, Jan 10, 28; m 50; c 4. MEDICINE. Educ: Univ Iowa, BA, 50, MD, 53. Prof Exp: Intern, Cincinnati Gen Hosp, 53-54; from jr resident to chief resident pediat, Cincinnati Children's Hosp, 54-56; from asst prof to assoc prof pediat, Col Med, Univ Cincinnati, 59-69, prof, 69-73; PROF PEDIAT, UNIV TENN, MEMPHIS, 73-; MED DIR, ST JUDE CHILDREN'S RES HOSP, 73- Concurrent Pos: Dir div hemat, Children's Hosp Res Found, 59-; attend pediatrician & dir div hemat & hemat clin, Children's Hosp, Cincinnati, 59-; attend pediatrician, Cincinnati Gen Hosp, 59-; attend hematologist, Vet Admin Hosp, 60-; NIH fel, 56-58; res fel hemat, Univ Utah, 56-59; Am Cancer Soc fel, 58-59; NIH res career develop award, 62-; fel, Div Hemat, Children's Hosp Res Found, 63- Mem: Am Asn Cancer Res; Am Pediat Soc; Am Soc Clin Invest; Am Soc Hemat (treas). Res: Labeling techniques with radioactive materials to study leukocyte kinetics in patients with acute leukemia and disorders of granulopoiesis. Mailing Add: Dept of Pediat Univ of Cincinnati Col of Med Cincinnati OH 45221

MAUER, IRVING, b Montreal, Que, Feb 7, 27; nat US; m 52; c 3. GENETICS. Educ: McGill Univ, PhD(genetics), 60. Prof Exp: Asst cytol, Sci Serv, Can Dept Agr, 48-49; demonstr genetics, McGill Univ, 54-56, asst, 55-56; res assoc animal genetics, Storrs Agr Exp Sta, Univ Conn, 56-57; sr med writer, Squibb Inst Med Res Div, Olin Mathieson Chem Corp, 57-60; psychiat res fel, NY State Dept Ment Health, 61-62, sr res scientist, 62-67, lectr, 62 & 65; HEAD CYTOGENETICS GROUP, DEPT EXP PATH, HOFFMANN-LA ROCHE INC, 67- Mem: AAAS; Am Acad Ment Retardation; Genetics Soc Am; Am Soc Human Genetics; Am Genetic Asn. Res: Cytogenetics of man; teratology; experimental cytogenetics; mutagenicity testing. Mailing Add: Dept of Exp Path Hoffmann-La Roche Inc Nutley NJ 07110

MAUER, PAUL BERNARD, b Buffalo, NY, Apr 21, 26; m 48; c 6. OPTICAL PHYSICS. Educ: Rensselaer Polytech Inst, BS, 45. Prof Exp: Jr physicist, Distillation Prod, Inc, 45-46; teaching asst physics, Univ Rochester, 46-48; jr physicist, 48-54, from jr develop engr to sr develop engr, 54-58, proj engr, 58-69, res assoc, 69-71, SR RES ASSOC, EASTMAN KODAK CO, 71- Mem: Optical Soc Am. Res: Photovoltaic systems; fluorescent systems; thin-film optical filters; glass lasers; laser speckle; interferometry. Mailing Add: 2842 Ridgeway Ave Rochester NY 14626

MAUER, SIDNEY I, endocrinology, physiology, see 12th edition

MAUERSBERGER, KONRAD, b Lengefeld, Ger, Apr 28, 38; m 64; c 1. AERONOMY. Educ: Univ Bonn, Dipl, 64, PhD(physics), 68. Prof Exp: Res assoc physics, Univ Bonn, 68-69; res assoc, 69-74, ASST PROF PHYSICS, UNIV MINN, MINNEAPOLIS, 75- Concurrent Pos: Mem comt planetary & lunar explor, Nat Acad Sci, 75. Mem: Am Geophys Union. Res: Condition and dynamics of Earth's upper atmosphere using mass spectrometers carried on rockets and satellites; solar-atmospheric interactions at altitudes above one hundred kilometers. Mailing Add: 148 Physics Bldg Univ of Minn Minneapolis MN 55455

MAUGER, JOHN WILLIAM, b Scranton, Pa, Aug 10, 42; m 65; c 1. PHARMACEUTICS. Educ: Union Univ, NY, BS, 65; Univ RI, MS, 68, PhD(pharmaceut sci), 71. Prof Exp: Instr pharm, Univ RI, 69-71; ASST PROF PHARM, SCH PHARM, W VA UNIV MED CTR, 71- Mem: Am Pharmaceut Asn. Res: Nonelectrolyte solubility; aqueous solutions of pharmaceutical solutes. Mailing Add: Sch of Pharm WVa Univ Med Ctr Morgantown WV 26506

MAUGER, RICHARD L, b Fairdale, Pa, Sept 20, 36; m 63; c 1. GEOLOGY. Educ: Franklin & Marshall Col, BS, 58; Calif Inst Technol, MS, 60; Univ Ariz, PhD(geol), 66. Prof Exp: Asst prof geol, Univ Utah, 66-69; asst prof, 69-72, ASSOC PROF GEOL, E CAROLINA UNIV, 72- Mem: AAAS; Geol Soc Am; Am Geophys Union. Res: Mineral deposits; isotopic dating and stable isotopes. Mailing Add: Dept of Geol ECarolina Univ Greenville NC 27834

MAUGHAN, EDWIN KELLY, b Glendale, Calif, Oct 13, 26; m 51; c 4. GEOLOGY. Educ: Utah State Univ, BS, 50. Prof Exp: Geologist, Corps Engrs, US Dept Army, 51; GEOLOGIST, US GEOL SURV, 51- Concurrent Pos: Tech adv phosphate deposits, Inventaria Minero Nacional Colombia, 67-69 & stratig, Struct & Coal Resources Southeast Ky, 69-72. Mem: Geol Soc Am; Am Asn Petrol Geologists; Colombian Asn Advan Sci; Soc Econ Paleont & Mineral. Res: Areal geology vicinity of Great Falls, Mont, Thermopolis, Wyo, Middlesboro, Ky; stratigraphy and phosphate resources in Cretaceous of Colombia; stratigraphy, paleogeography and mineral resources (petroleum, salt, phosphate) in Permian, Pennsylvanian and Mississippian rocks of northern Rocky Mountains and Great Basin. Mailing Add: US Geol Surv Fed Ctr Box 25046 Denver CO 80225

MAUGHAN, GEORGE BURWELL, b Toronto, Ont, May 8, 10; m 67; c 6. OBSTETRICS & GYNECOLOGY. Educ: McGill Univ, MD & CM, 34, MSc, 38; FRCS(C), 52; FRCOG, 57. Prof Exp: Demonstr path & bact, McGill Univ, 34-35, demonstr anat, 39-40, from demonstr to asst prof obstet & gynec, 40-56, prof & chmn dept, 56-75; RETIRED. Concurrent Pos: Obstetrician & gynecologist-in-chief, Royal Victoria Hosp, 56-; consult obstetrician & gynecologist, Montreal Gen, Reddy Mem, Queen Elizabeth, St Mary's & Jewish Gen, Lakeshore Gen, Montreal Chinese, Catherine Booth & Queen Mary Vet Hosps. Mem: Can Med Asn; fel Am Col Surg; Soc Obstet & Gynaec Can (past pres); fel Am Asn Obstet & Gynec; Can Gynaec Soc. Res: Intensive care in high risk pregnancy. Mailing Add: Apt F-61 1321 Sherbrooke St W Montreal PQ Can

MAUGHAN, PAUL MCALPINE, b Spokane, Wash, May 22, 36; m 57; c 4. MARINE SCIENCES, REMOTE SENSING. Educ: Wash State Univ, BS, 59; Pa State Univ, BS, 60; Ore State Univ, MS, 63, PhD(oceanog), 66. Prof Exp: Engr, Douglas Aircraft Co, 58 & Gen Elec Co, 59; mgr ocean res & develop, Oceanog Res Inc, 65-67; spec asst biores, Bur Com Fisheries, Dept Interior, 67-71; dir oceanog applns, Earth Satellite Corp, 71-75; STAFF ADV, COMSAT GEN CORP, 75- Mem: Am Inst Astronaut & Aeronaut; Marine Technol Soc; Am Meteorol Soc. Res: Air-sea interaction; radiant energy exchange in the marine environment; remote sensing of marine resources; fishery forecasting; meteorology. Mailing Add: 950 L'Enfant Plaza SW Washington DC 20024

MAUL, JAMES JOSEPH, b Buffalo, NY, Nov 3, 38; m 62; c 2. ORGANIC CHEMISTRY. Educ: Canisius Col, BS, 60; Wayne State Univ, PhD(org chem), 66. Prof Exp: Sr chemist, Hooker Chem Corp, 68-75, ASSOC CHEMIST, HOOKER CHEMS & PLASTICS CORP, 75- Mem: Am Chem Soc. Res: Organo-fluorine chemistry; organo-halogen chemistry; chemistry for fire retarding polymers. Mailing Add: Hooker Res Ctr Main PO Box 8 Niagara Falls NY 14302

MAUL, STEPHEN BAILEY, b Parkersburg, WVa, Jan 11, 42; m 67. INDUSTRIAL MICROBIOLOGY. Educ: Abilene Christian Col, BS, 63; Univ Tex, Austin, PhD(biochem), 69. Prof Exp: Res assoc nutrit & food sci, Mass Inst Technol, 69-70; sr biochemist, Antibiotic Develop Dept, Eli Lilly & Co, 70-74; SR SCIENTIST, SCHERING CORP, 74- Mem: AAAS; Am Chem Soc; Am Soc Microbiol; Soc Indust Microbiol. Res: Antibiotic fermentations; new fermentation technology; enzyme applications; biological control mechanisms; mineral requirements in fermentations; computer control of fermentations; waste disposal. Mailing Add: Schering Corp Indust Microbiol Union NJ 07083

MAULDIN, RICHARD DANIEL, b Longview, Tex, Jan 17, 43; m 74; c 1. MATHEMATICS. Educ: Univ Tex, BA, 65, MA, 66, PhD(math), 69. Prof Exp: Asst prof, 69-75, ASSOC PROF MATH, UNIV FLA, 75- Mem: Am Math Soc. Res: Descriptive set theory, measure theory and point set topology and the interaction of these three areas. Mailing Add: Dept of Math Univ of Fla Gainesville FL 32611

MAULDING, DONALD ROY, b Evansville, Ind, Aug 15, 36; m 58; c 1. SYNTHETIC ORGANIC CHEMISTRY. Educ: Evansville Col, AB, 58; Univ Ind, PhD(org chem), 62. Prof Exp: Fel, Ohio State Univ, 62-64; RES CHEMIST, AM CYANAMID CO, 64- Mem: Am Chem Soc. Res: Organic mechanisms; photochemistry and chemiluminescence; fluorescence; rubber chemicals and polyurethanes. Mailing Add: Org Chems Div Am Cyanamid Co Bound Brook NJ 08805

MAULDING, HAWKINS VALLIANT, JR, b Foreman, Ark, Dec 21, 35; m 59; c 3. PHYSICAL CHEMISTRY, PHYSICAL PHARMACY. Educ: Univ Ark, Little Rock, BS, 58; Univ Minn, Minneapolis, PhD(med chem), 64. Prof Exp: Asst prof pharm, Univ Houston, 64-66; SR SCIENTIST & GROUP LEADER PHYS CHEM & PHYS PHARM, SANDOZ PHARMACEUT, EAST HANOVER, 66- Mem: Am Pharmaceut Asn. Res: Theoretical and applied kinetics; complexation; reaction mechanisms; stability of solid products; dosage form design. Mailing Add: Corey Lane Mendham NJ 07945

MAULDON, JAMES GRENFELL, b London, Eng, Feb 9, 20; m 53; c 4. PURE MATHEMATICS. Educ: Oxford Univ, BA & MA, 47. Hon Degrees: MA, Amherst Col, 70. Prof Exp: Lectr math, Oxford Univ, 47-68; PROF MATH, AMHERST COL, 68- Concurrent Pos: Lectr, St John's Col, Oxford Univ, 50-59, fel, Corpus Christi Col, 50-68; vis prof, Univ Calif, Berkeley, 60-61; chmn fac, Oxford Univ, 66-68; consult, IBM Corp, 74-75. Mem: Royal Statist Soc; Inst Math Statist; Am Math Soc. Res: Mathematics, including probability, algebra, analysis, geometry and combinatorics; computer languages. Mailing Add: Dept of Math Amherst Col Amherst MA 01002

MAUMENEE, ALFRED EDWARD, b Mobile, Ala, Sept 19, 13; m 49, 72; c 2. OPHTHALMOLOGY. Educ: Univ Ala, AB, 34; Cornell Univ, MD, 38; Am Bd Ophthal, dipl, 43. Hon Degrees: FRCS(E), 71. Prof Exp: Intern & asst resident ophthal, Johns Hopkins Univ, 38-42, asst, 39-42, res ophthalmologist, 42-43, from instr to assoc prof, 42-48, assoc ophthalmologist, 46-48; prof surg in ophthal, Sch Med & chief div, Hosp, Stanford Univ, 48-55; PROF OPHTHAL, UNIV & OPHTHALMOLOGIST-IN-CHIEF, HOSP, JOHNS HOPKINS UNIV, 55- Concurrent Pos: Consult ophthalmologist, San Francisco & Laguna Honda Hosps, 48-55; civilian consult, Letterman Army & US Naval Hosps, Oakland, Calif, 48-55; clin staff, Children's & Mt Zion Hosps, San Francisco, 49-55; civilian consult, Walter Reed Army Hosp, Clin Ctr, NIH & US Naval Hosp, 55-; consult, Surgeon Gen, US Navy, 63-; mem adv comt ophthalmologist, Calif Dept Pub Health, 49-55, Med Adv Bd, Bur Voc Rehab, Calif, 49-55, Sci Adv Comt, Nat Coun Combat Blindness, 50 & Adv Comt Ophthalmologists, Calif Dept Soc Welfare, 50-55; consult & mem grad training comt, Nat Inst Neurol Dis & Blindness, 53-55; mem adv comt, Nat Acad Sci, 55-58; mem adv coun, Coun Res Glaucoma & Allied Dis, 56-; mem adv comt, Ophthalmic Found, 56- & NY Eye & Ear Infirmary, 56-; mem vision res training comt, NIH, 64-; mem med adv bd, Int Eye Bank, 65; Albert C Snell & Sanford R Gifford Mem lectr, 65. Mem: Am Acad Ophthal & Otolaryngol (1st vpres, 64, pres, 71); Am Ophthal Soc; Asn Res Vision & Ophthal; Pan-Am Asn Ophthal (secy-treas, 52, pres, 71); Int Asn Prev Blindness. Res: Ophthalmic pathology; glaucoma and allied diseases; toxoplasmic uveitis; transplantation. Mailing Add: Dept of Ophthal Johns Hopkins Univ Sch of Med Baltimore MD 21205

MAUN, EUGENE KINGERY, chemistry, see 12th edition

MAUNDER, A BRUCE, b Holdrege, Nebr, May 13, 34; m 57; c 1. GENETICS, PLANT BREEDING. Educ: Univ Nebr, BS, 56; Purdue Univ, MS, 58, PhD(genetics), 60. Prof Exp: Plant breeder, 59-61, SORGHUM RES DIR, DEKALB AgRESEARCH INC, 61- Mem: Am Soc Agron. Res: Inheritance of male sterility; heterosis as regards sorghum; disease and insect resistance; genetic advances; evolution. Mailing Add: DeKalb AgResearch Inc Route 2 Lubbock TX 79415

MAUNDER, DUANE THAYER, b Joliet, Ill, July 27, 22; m 46; c 2. MICROBIOLOGY. Educ: Univ Ill, BS, 47, MS, 48, PhD(bact), 50. Prof Exp: Asst prof bact, Miami Univ, 50-57; assoc prof biol sci & head dept, Carnegie Inst Technol, 57-64; supvr microbiol sect, 64-66, ADV MICROBIOLOGIST, CONTINENTAL CAN CO, INC, 66- Mem: Am Soc Microbiol; Inst Food Technologists; Int Microwave Power Inst. Res: Canned food preservation, spoilage and public health related problems. Mailing Add: Microbiol Sect Met Res & Dev Continental Can Co Inc Chicago IL 60620

MAUNE, DAVID FRANCIS, b Washington, Mo, July 12, 39; m 61; c 2. PHOTOGRAMMETRY, RESEARCH ADMINISTRATION. Educ: Univ Mo-Rolla, BSc, 61; Ohio State Univ, MSc, 70, PhD(geod, photogram), 73. Prof Exp: Mech engr, Union Elec Co, St Louis, 61; co comdr & opers officer, 656 Eng Topog Battalion, US Army, Ger, 63-66, mapping officer, Hq, Vietnam, 66-67, opers officer, 36 Engr Group, Korea, 70-71, officer-in-charge prod, Mapping & Charting Estab, Royal Engrs, Eng, 73-74, STAFF OFFICER DIRECTORATE ARMY RES, US DEPT ARMY, HQ, WASHINGTON, DC, 74- Mem: Am Soc Photogram; Int Soc Photogram. Res: Photogrammetric calibration of scanning electron microscopes; analytical photogrammetry; satellite geodesy; operations research systems analysis; military research and development management; technological forecasting. Mailing Add: 801 Fieldcrest Dr Washingto MO 63090

MAUNEY, CHARLES URAL, microbiology, see 12th edition

MAUNSELL, CHARLES DUDLEY, b Victoria, BC, Mar 29, 24. PHYSICS. Educ: Univ BC, BA, 45, MA, 47; Univ Calif, PhD(physics), 55. Prof Exp: Sci officer, Pac Naval Lab, 55-63; SR SCI OFFICER, BEDFORD INST OCEANOG, CAN DEPT ENERGY, MINES & RESOURCES, 63- Mem: Am Phys Soc; assoc Acoust Soc Am; Can Asn Physicists. Res: Physical oceanography; underwater acoustics; scientific computing. Mailing Add: Atlantic Oceanog Lab Bedford Inst Oceanog PO Box 1006 Dartmouth NS Can

MAURER, BRUCE ANTHONY, b Springfield, Mass, Oct 22, 36; c 3. IMMUNOBIOLOGY. Educ: St Michael's Col, BA, 58; Univ Mass, MS, 60; Univ Ariz, PhD(microbiol), 66. Prof Exp: Asst prof, Miami Univ, 66-68; sr cancer res scientist, Roswell Park Mem Inst, 68-71; asst res prof virol, Roswell Park Div, State Univ NY Buffalo, 69-71; consult, Assoc Biomedic Systs, Inc, 69-71, dir biol, 71-73; IMMUNOLOGIST-VIROLOGIST, LITTON BIONETICS RES LAB, MD, 73- Concurrent Pos: United Health Fund grant, Roswell Park Mem Inst, 69-70. Mem: AAAS; Am Soc Microbiol; Brit Soc Gen Microbiol. Res: Biology and biochemistry of host-parasite relationships; virus-cell interactions; malignant transformation of cells; latent virus infections; virus carrier-states; cellular immunology of human and animal neoplasia; cellular immunology of primate transplantation rejection phenomena. Mailing Add: Litton Bionetics Res Lab 5510 Nicholson Ln Kensington MD 20795

MAURER, DONALD LEO, b Chicago, Ill, Sept 3, 34; m 67; c 4. MARINE ECOLOGY, POLLUTION BIOLOGY. Educ: Univ Ill, BS, 56; Univ Wash, MS, 58; Univ Chicago, PhD(paleozool), 64. Prof Exp: Res assoc marine ecol, Pac Marine Sta, Calif, 64-65; assoc prof biol, Old Dom Col, 65-67; asst prof, 67-73, ASSOC PROF MARINE BIOL, UNIV DEL, 73- Mem: Soc Econ Paleont & Mineral; Soc Limnol & Oceanog; Atlantic Estuarine Res Soc; Sigma Xi. Res: Ecology of marine invertebrates; paleoecology; description of macroscopic Benthic invertebrate communities in the Delaware bay and adjacent coastal zones; determination of community and specific response to pollutants in the aforementioned areas. Mailing Add: Field Sta Col of Marine Stud Univ of Del Lewes DE 19958

MAURER, EDWARD ROBERT, b San Francisco, Calif, Jan 3, 21; m 55; c 2. PHYSICAL SCIENCE, HISTORY OF SCIENCE. Educ: Stanford Univ, BS, 48, MS, 50, PhD(phys sci), 64. Prof Exp: Instr phys sci, Chico State Col, 52-55, asst prof, 55-56; vis asst prof, Stanford Univ, 56-58; assoc prof, 58-65, PROF PHYS SCI, CALIF STATE UNIV, CHICO, 65- Concurrent Pos: Head dept phys sci, Calif State Univ, Chico, 58-67, dean, Sch Prof Studies, 70-74. Mem: Fel AAAS; Hist Sci Soc; Soc Hist Technol; Sigma Xi; Nat Sci Teachers Asn. Res: The history of science and technology, with special interest in the development of chemistry during the British industrial revolution; the role of the physical sciences in general education. Mailing Add: Dept of Geol & Phys Sci Calif State Univ Chico CA 95929

MAURER, FRANK W, JR, vertebrate zoology, see 12th edition

MAURER, FRED DRY, b Moscow, Idaho, May 4, 09; m 35; c 2. VETERINARY PATHOLOGY. Educ: Univ Idaho, BS, 34; State Col Wash, BS & DVM, 37; Cornell Univ, PhD(path bact), 48. Prof Exp: Jr veterinarian dis control, Bur Animal Indust, USDA, 37; asst prof, Univ Idaho & asst bacteriologist, Exp Sta, 37-38; instr path & bact, Vet Col, Cornell Univ, 38-41; staff mem, Vet Res Lab, Vet Corps, US Army, Va, 41-43, lab officer, War Dis Control Sta, Can & Africa, 43-46, Res & Grad Sch, Army Med Ctr, 47-51, chief, Vet Path Div, Armed Forces Inst Path, 54-61, dir, Div Med, Army Med Res Lab, Ft Knox, Ky, 61-64; DISTINGUISHED PROF PATH, COL VET MED, TEX A&M UNIV, 64-, DIR INST TROP VET MED, 74- Concurrent Pos: Assoc dean, Col Vet Med, Tex A&M Univ, 64-74. Honors & Awards: 12th Int Vet Cong Prize, 68. Mem: Am Vet Med Asn; US Animal Health Asn; Am Col Vet Path (pres, 64); Conf Res Workers Animal Dis; Am Asn Lab Animal Sci. Res: Virology and pathology of infectious diseases of animals. Mailing Add: Col of Vet Med Tex A&M Univ College Station TX 77843

MAURER, HANS ANDREAS, b Frankfurt, Ger, May 7, 13; US citizen; m 40; c 1. PHYSICS. Educ: Univ Munich, BSc, 33; Univ Frankfurt, PhD(appl physics), 37. Prof Exp: Sr proj engr, Missile Systs Div, Raytheon Co, 57-61, mgr advan weapons systs, 61-63, mgr advan syst ctr, 63-64, chief engr, 64-65; TECH DIR ASST DIV MGR, MISSILE DIV, AEROSPACE GROUP, HUGHES AIRCRAFT CO, 66- Mem: AAAS; Ger Oberth Soc; fel Inst Elec & Electronics Engrs. Res: Missile guidance analysis; radar and infrared sensor analysis; synthetic array radar guidance synthesis; phased array radar design; missile electronics and inertials technology; flight test analysis; reciprocal ferrite phase shifter; electrostatic space antenna stabilization; dual passband infrared seeker. Mailing Add: 4447 Conchita Way Tarzana CA 91356

MAURER, JOHN EDWARD, b Matherville, Ill, Apr 3, 23; m 46; c 4. ORGANIC CHEMISTRY. Educ: Augustana Col, AB, 47; Univ Iowa, MS, 48, PhD(org chem), 50. Prof Exp: Res assoc, Northwestern Univ, 50-52; res chemist, Rock Island Arsenal, 52-53; from asst prof to assoc prof, 53-66, PROF CHEM, UNIV WYO, 66-, ASST HEAD DEPT, 68- Mem: AAAS; Am Chem Soc. Res: Halogenations; Van Slyke reactions; naturalproducts. Mailing Add: Dept of Chem Univ of Wyo Box 3838 Laramie WY 82070

MAURER, JOHN JOSEPH, physical chemistry, see 12th edition

MAURER, PAUL HERBERT, b New York, NY, June 29, 23; m 48; c 3. IMMUNOLOGY. Educ: City Col New York, BS, 44; Columbia Univ, PhD(immunochem), 50. Prof Exp: Res biochemist, Gen Foods Corp, 44 & 46; instr, City Col New York, 46-51; res assoc, Col Physicians & Surgeons, Columbia Univ, 50-51; asst res prof, Sch Med, Univ Pittsburgh, 51-54, assoc prof immunochem, 54-60; prof microbiol, NJ Col Med & Dent, 60-66; PROF BIOCHEM & HEAD DEPT, JEFFERSON MED COL, 66- Concurrent Pos: NIH res career award, 62- Mem: Am Chem Soc; Am Asn Immunol; NY Acad Sci; Brit Biochem Soc. Res: Immunochemistry; biochemistry; protein chemistry. Mailing Add: Dept of Biochem Jefferson Med Col Philadelphia PA 19107

MAURER, RALPH RUDOLF, b Monroe, Wis, Feb 28, 41; m 63; c 3. REPRODUCTIVE PHYSIOLOGY, REPRODUCTIVE ENDOCRINOLOGY. Educ: Univ Wis, BS, 63; Cornell Univ, MS, 66, PhD(physiol), 69. Prof Exp: SR STAFF FEL ENVIRON, NAT INST ENVIRON HEALTH SCI, 71- Concurrent Pos: Alexander von Humboldt fel, Vet Inst, Göttingen, Ger, 69-71. Honors & Awards: Lalor Found Award, 69. Mem: Soc Study reprod; Brit Soc Study Fertil; Am Soc Animal Sci; Am Inst Biol Sci; Teretology Soc. Res: Environmental factors affecting embryonic development and the reproductive process; maternal aging and embryonic mortality; storage of gametes and embryos. Mailing Add: Nat Inst of Environ Health Sci Environ Toxicol Br Box 12233 Research Triangle Park NC 27709

MAURER, RICHARD L, biological chemistry, see 12th edition

MAURER, ROBERT DISTLER, b St Louis, Mo, July 20, 24; m 51; c 3. APPLIED PHYSICS. Educ: Univ Ark, BS, 48; Mass Inst Technol, PhD(physics), 51. Prof Exp: Mem physics staff, Mass Inst Technol, 51-52; physicist, 52-62, sr res assoc, 62-63, mgr fundamental physics res, 63-70, MGR APPL PHYSICS RES, CORNING GLASS WORKS, 70- Concurrent Pos: Mem, Solid State Adv Panel, Nat Res Coun, 75-79. Mem: Am Phys Soc; fel Am Ceramic Soc; sr mem Inst Elec & Electronics Engrs. Res: Physical behavior of glasses; optical communications. Mailing Add: Corning NY

MAURER, ROBERT EUGENE, b Uhrichsville, Ohio, 25; m 53; c 1. GEOLOGY. Educ: Ohio State Univ, BS, 56; Univ Utah, PhD, 70. Prof Exp: Geologist, Texaco, Inc, 52-56; instr geol, Westminster Col, Utah, 57-59, asst prof, 59-66; asst prof, 66-70, ASSOC PROF GEOL, STATE UNIV NY COL OSWEGO, 70- Concurrent Pos: Chmn sci div, Westminster Col, Utah, 62-66; chmn dept earth sci, State Univ NY Col Oswego, 67-72. Mem: AAAS; Geol Soc Am. Res: Surface and subsurface geologic mapping; surface geologic mapping of Oswego County, New York. Mailing Add: Dept of Earth Sci State Univ of NY Col Oswego NY 13126

MAURER, ROBERT JOSEPH, b Rochester, NY, Mar 26, 13; m 40. SOLID STATE PHYSICS, RESEARCH ADMINISTRATION. Educ: Univ Rochester, BS, 34, PhD(physics), 39. Prof Exp: Res assoc, Mass Inst Technol, 39-42; instr physics, Univ Pa, 42-43; from asst prof to assoc prof, Carnegie Inst Technol, 43-49; assoc prof, 49-51, PROF PHYSICS, UNIV ILL, URBANA, 51-, DIR MAT RES LAB, 63- Concurrent Pos: Physicist, Metall Lab, Univ Chicago, 44-45; head physics br, Off Naval Res, 48. Mem: Am Phys Soc. Res: Self diffusion in solids; electrical prerties of solids; photoelectric properties of metals; optical properties of solids. Mailing Add: Mat Res Lab Univ of Ill Urbana IL 61801

MAURICE, CHARLES GEORGE, b Murphysboro, Ill, Dec 21, 11; m 53; c 3. DENTISTRY. Educ: Univ Ill, BS, 38, DDS, 40, MS, 53; Am Bd Endodont, dipl. Prof Exp: From instr to assoc prof appl mat med & therapeut, 40-60, actg head dept, 58-60, actg head dept endodont, 65-67, PROF APPL MAT MED & THERAPEUT, COL DENT, UNIV ILL MED CTR, 60-, HEAD DEPT ENDODONT, 67- Mem: Am Dent Asn; Am Asn Endodont; fel Am Col Dent. Res: Endodontics. Mailing Add: Dept of Endodont Col of Dent Univ of Ill Med Ctr Chicago IL 60612

MAURICE, DAVID MYER, b London, Eng, Apr 3, 22; m 54; c 3. PHYSIOLOGY. Educ: Univ Reading, BSc, 41; Univ London, PhD(physiol), 51. Prof Exp: Jr sci officer, Telecommun Res Estab, Ministry Aircraft Prod, 41-46; staff mem ophthal res unit, Med Res Coun, 46-63; reader physiol, Inst Ophthal, Univ London, 63-68; SR SCIENTIST OPHTHAL, MED SCH, STANFORD UNIV, 68-, ADJ PROF SURG, 74- Concurrent Pos: Ital Govt fel, Univ Rome, 51-52; Fulbright fel, Univ Calif, San Francisco, 57-58. Honors & Awards: Friedenwald Medal, Asn Res Vision & Ophthal, 67. Mem: AAAS; Asn Eye Res; Am Physiol Soc; Biophys Soc; Asn Res Vision & Ophthal. Res: Vegetative physiology of the eye; physiology and biochemistry of cornea; transport mechanisms. Mailing Add: Div of Ophthal Stanford Univ Med Ctr Stanford CA 94305

MAURIELLO, DAVID ANTHONY, b New York, NY, May 29, 41; m 68. ECOLOGY. Educ: NY Univ, BS, 63, MS, 64; Rutgers Univ, PhD(ecol), 75. Prof Exp: Mem tech staff, Bell Tel Labs, 64-65; engr, Hewlett-Packard Co, 65-69 & Burroughs Corp, 69-70; ASST PROF ECOL, SAN DIEGO STATE UNIV, 74- Mem: Ecol Soc Am; Am Soc Limnol & Oceanog. Res: Modeling and simulation of ecosystems; theoretical approaches to plankton community structure and dynamics. Mailing Add: Dept of Biol San Diego State Univ San Diego CA 92182

MAURMEYER, ROBERT, chemistry, see 12th edition

MAURO, ALEXANDER, b New Haven, Conn, Aug 14, 21; m 55. BIOPHYSICS. Educ: Yale Univ, BE, 42, PhD(biophys), 50. Prof Exp: Instr physiol, Sch Med, Yale Univ, 51-52, asst prof, 52-59; from asst prof to assoc prof, 59-71, PROF BIOPHYS, ROCKEFELLER UNIV, 71- Mem: Am Physiol Soc; Biophys Soc; Inst Elec & Electronics Eng. Res: Nerve and muscle physiology; fundamental mechanisms in electrophysiological systems; physico-chemical studies of ionic membranes and their relationship to physiological membranes; experimental and theoretical study of semipermeability and membrane transport. Mailing Add: 392 Central Park West New York NY 10025

MAURO, JACK ANTHONY, b Brooklyn, NY, Feb 21, 16; m 37; c 3. OPTICAL PHYSICS, PHYSIOLOGICAL OPTICS. Educ: Columbia Univ, BS & cert, 47; Phila Optical Col, OD, 51. Prof Exp: Mgr optics, Equitable Optical Co, 34-43; chief instr theoret optics & math, NY Inst Optics, 47-53; dir eng, Saratoga Div, Espey Mfg Co, 50-55; engr, Gen Eng Lab, Gen Elec Co, 55-60, consult optics engr, Ord Dept, Defense Electronics Div, 60-65, consult engr, Missile & Space Vehicle Div, 65-70; dir res, Shuron Continental Div, Textron Inc, 70-74; CONSULT HIGH ENERGY LASER OPTICAL SYSTS, 74- Concurrent Pos: Consult, NY Inst Optics, 47-55 & Navigational Inst Am, 47-52; mem, Bd Dirs, Columbia Univ, 48-51; consult, US Army & US Navy Ord Off, 50-51; chmn, Man-Mach Symp, US Army-Gen Elec Co, 58; mem, High Energy Laser Weapons Ad Hoc Comt, US Army Missile Command. Mem: Am Phys Soc; Optical Soc Am; fel Am Acad Optom; Soc Photo-Optical Instrument Eng. Res: Optics and electronics; physical and geometrical optics; optical design. Mailing Add: 4949 San Pedro Dr NE Albuquerque NM 87109

MAURY, LUCIEN GARNETT, b Hoisington, Kans, Aug 14, 23; m 47; c 3. PHYSICAL CHEMISTRY, ORGANIC CHEMISTRY. Educ: Ill Inst Technol, BS, 48; Northwestern Univ, PhD(chem), 52. Prof Exp: Res chemist, Hercules Inc, Del, 51-56, supvr, 56-58, mgr explosives res & high pressure lab, 58-61, mgr synthetics res, 61-64, mgr cent res, 64-67, proj mgr, 67-68, dir develop fibers & film, 68-71, dir fibers, 71-73; gen mgr, Hercules Int Dept, 73-74; PRES, HERCULES EUROPE, 74- Res: Homogeneous and heterogeneous catalysis; nitrogen chemistry. Mailing Add: Hercules Inc 910 Market St Wilmington DE 19899

MAUSEL, PAUL WARNER, b Minneapolis, Minn, Jan 2, 36; m 66; c 2. PHYSICAL GEOGRAPHY. Educ: Univ Minn, BA(chem) & BA(geog), 58, MA, 61; Univ NC, PhD(geog), 66. Prof Exp: Instr geog, Mankato State Col, 61-62; from asst prof to assoc prof, Eastern Ill Univ, 65-71; ASSOC PROF GEOG, IND STATE UNIV, TERRE HAUTE, 72- Concurrent Pos: Res grants, Eastern Ill Univ, 67-68 & 69-70; researcher, Lab Appln Remote Sensing, Purdue Univ, 72- Mem: Asn Am Geog; Am Geog Soc; Soil Sci Soc Am. Res: Remote sensing of the environment using automatically data processed multispectral sensor data, stressing land use; soils geography. Mailing Add: Dept of Geog & Geol Ind State Univ Terre Haute IN 47809

MAUSNER, JUDITH S, b New York, NY, Oct 11, 24; m 44; c 2. EPIDEMIOLOGY, PUBLIC HEALTH. Educ: Queens Col (NY), BA, 44; NY Med Col, MD, 48; Univ Pittsburgh, MPH, 61. Prof Exp: Asst prof epidemiol, Grad Sch Pub Health, Univ Pittsburgh, 61-62; asst prof, 63-70, ASSOC PROF EPIDEMIOL, MED COL PA, 70- Mem: Am Pub Health Asn; Asn Teachers Prev Med. Res: Chronic disease epidemiology. Mailing Add: Dept of Community & Prev Med Med Col of Pa Philadelphia PA 19129

MAUSNER, LEONARD FRANKLIN, b New York, NY, Mar 6, 47; m 69; c 1. NUCLEAR CHEMISTRY. Educ: Mass Inst Technol, BS, 68; Princeton Univ, MA, 72, PhD(chem), 75. Prof Exp: Instr chem, Princeton Univ, 74-75; FEL CHEM, LOS ALAMOS SCI LAB, 75- Concurrent Pos: Consult, Princeton Gamma Tech, 73. Mem: Am Phys Soc; Am Chem Soc; Sigma Xi. Res: Influence of chemical bonding on the molecular capture of negative muons; measuring type, amount, and distribution of radioactivity produced by negative pions stopped in tissue. Mailing Add: CNC-11 MS 824 Los Alamos Sci Lab Los Alamos NM 87545

MAUSS, EVELYN ABRAMS, physiology, see 12th edition

MAUSTELLER, JOHN WILSON, b West Milton, Pa, May 14, 22; m 44; c 4. PHYSICAL CHEMISTRY. Educ: Bucknell Univ, BS, 44; Pa State Univ, MS, 49, PhD(phys chem), 51. Prof Exp: Chemist, E I du Pont de Nemours & Co, 44-46; asst, Fluorine Labs, Pa State Univ, 47-50; res chemist, Callery Chem Co, 50-51; engr, Mine Safety Appliances Co, 51-52, proj engr, 52-57; res mgr, 57-60, assoc dir res, 60-74, GEN MGR, MSA RES CORP, 74- Mem: Am Chem Soc; Am Nuclear Soc. Res: Chemical oxygen; air and water pollution; safety and handling hazardous materials; life support systems; liquid metals technology; reactor coolants; chemical warfare protective technology. Mailing Add: MSA Res Corp Evans City PA 16033

MAUSTON, GLENN WARREN, b St Paul, Minn, Oct 22, 35; m 57; c 3. ENTOMOLOGY. Educ: Gustavus Adolphus Col, BS, 57; Iowa State Univ, MS, 59; NDak State Univ, PhD(entom), 69. Prof Exp: INSTR BIOL, MESABI COMMUNITY COL, 59- Res: Taxonomy and biology of the subfamily Crambinae. Mailing Add: Dept of Biol Mesabi Community Col Virginia MN 55792

MAUTE, ROBERT EDGAR, b Colorado Springs, Colo, July 1, 47; m 75; c 2. EXPERIMENTAL NUCLEAR PHYSICS. Educ: Lamar Univ, BS, 69; Ohio State Univ, MS, 72, PhD(physics), 73. Prof Exp: SR PHYSICIST, INDUST NUCLEONICS CORP, 73- Mem: Am Phys Soc. Res: Development of sensors for use in industrial control systems; sensors utilizing infrared absorption. Mailing Add: 848 Kevin Dr Columbus OH 43224

MAUTE, ROBERT LEWIS, b Springfield, Ohio, July 1, 24; m 46; c 2. ANALYTICAL CHEMISTRY. Educ: Colo Col, BS, 49; Univ Houston, MS, 50. Prof Exp: Chemist, Phillips Petrol Co, 50-51; res chemist, Monsanto Co, 51-55, asst group leader, Anal Group, 55-58, group leader, 58-64, mgr anal sect, Tex, 64-71; MGR ANAL SECT, MONSANTO POLYMERS & PETROCHEM CO, 71- Mem: Am Chem Soc; The Chem Soc; Am Inst Chemists; Sigma Xi. Res: Instrumental and chemical analyses; physical chemistry; industrial applications of radioisotopes; characterization of industrial heterogeneous catalysts. Mailing Add: PT Dept Monsanto Polymers & Petrochem Co Texas City TX 77590

MAUTNER, HENRY GEORGE, b Prague, Czech, Mar 30, 25; nat US; m 67; c 2. BIOCHEMISTRY, PHARMACOLOGY. Educ: Univ Calif, Los Angeles, BS, 46; Univ Southern Calif, MS, 49; Univ Calif, PhD(chem), 55. Hon Degrees: MS, Yale Univ, 67. Prof Exp: Lab asst, Univ Southern Calif, 47-49; res chemist, Productol Co, 50; sr res technician, Univ Calif, 51-53, asst, 53-55; from instr to assoc prof pharmacol, Sch Med, Yale Univ, 56-67, prof pharmacol & head sect med chem, 67-70; PROF BIOCHEM & PHARMACOL & CHMN DEPT, SCH MED, TUFTS UNIV, 70- Concurrent Pos: Squibb fel pharmacol, Sch Med, Yale Univ, 55-56. Mem: Am Chem Soc; Am Asn Cancer Res; Am Soc Pharmacol & Exp Therapeut; The Chem Soc; Am Soc Biol Chem. Res: Heterocyclic chemistry; purines; pyrimidines; pteridines; chemistry of selenium compounds; coenzyme analogs; choline acetyltransferase; excitable membranes; chemotherapy of cancer; antimetabolites; comparative kinetics of reactions of oxygen, sulfur and selenium isologs; molecular basis of nerve conduction. Mailing Add: Dept of Biochem & Pharmacol Tufts Univ Sch of Med Boston MA 02111

MAUTZ, CHARLES WILLIAM, b St Elmo, Ill, Apr 27, 17; m 45; c 1. PHYSICS. Educ: Univ Ill, BS, 41, MS, 43; Univ Mich, PhD(physics), 49. Prof Exp: Mem staff, Radiation Lab, Mass Inst Technol, 44-45 & Los Alamos Sci Lab, 49-60; MEM STAFF, GULF GEN ATOMIC CO, 60- Concurrent Pos: Vis assoc prof, Univ Mich, 61-62. Mem: Am Phys Soc. Res: Gas dynamics; explosives; lasers. Mailing Add: Gulf Gen Atomic Co Box 81608 San Diego CA 92138

MAUTZ, WILLIAM WARD, b Eau Claire, Wis, Apr 13, 43; m 65; c 3. WILDLIFE RESEARCH, WILDLIFE ECOLOGY. Educ: Wis State Univ-Eau Claire, BS, 65; Mich State Univ, MS, 67, PhD(wildlife ecol & physiol), 69. Prof Exp: From asst prof to assoc prof wildlife ecol, Inst Natural & Environ Resources, Univ NH, 69-75; ASST UNIT LEADER, COLO COOP WILDLIFE RES UNIT, COLO STATE UNIV, 75- Res: Ecological energetics; energy flow work with mammalian species involving efficiency of food energy utilization as related to energy requirements; development of procedures for the determination of energy utilization and requirements in wildlife species. Mailing Add: Colo Coop Wildlife Res Unit Colo State Univ Ft Collins CO 80521

MAUZERALL, DAVID CHARLES, b Sanford, Maine, July 22, 29; m 59; c 2. BIOPHYSICS. Educ: St Michael's Col, BS, 51; Univ Chicago, PhD(chem), 54. Prof Exp: Res assoc, 54-59, from asst prof to assoc prof, 59-69, PROF BIOPHYS, ROCKEFELLER UNIV, 69- Concurrent Pos: Vis assoc prof, Univ Calif, San Diego, 65-68, adj prof, 68-; Guggenheim fel, 66. Mem: AAAS; Am Chem Soc; Am Soc Biol Chemists. Res: Mechanism of photochemical and photobiological reactions; porphyrin biochemistry; photosynthesis. Mailing Add: Rockefeller Univ New York NY 10021

MAVCO, GEORGE EDWARD, b Cleveland, Ohio, Feb 22, 47; m 70. ASTRONOMY, OPTICS. Educ: Cprnell Univ, BS, 69; Rensselaer Polytech Inst, MS, 71, PhD(astron), 74. Prof Exp: PROJ SCIENTIST, US AIR FORCE, 74- Res: Photometry and spectrophotometry of faint light sources using both photomultipliers and solid state detectors; real-time computer control. Mailing Add: Air Force Avionics Lab RWI Wright-Patterson AFB OH 45433

MAVIS, JAMES OSBERT, b Mansfield, Ohio, Aug 6, 25; m 53; c 2. FOOD TECHNOLOGY. Educ: Ohio State Univ, BSc, 50, MSc, 53, PhD(food technol), 55. Prof Exp: Asst, Food Technol, Agr Exp Sta, Univ Ohio, 48-50; area rep, Topco Assocs, Inc, Ill, 50-52; chief customer res, Heekin Can Co, Ohio, 56-57; sect chief, Non-milk Frozen Foods, Pet Milk Co, 58-61, group mgr, Bakery & Hort Prods, 61-65, assoc dir res, 65-68, tech dir, Frozen Foods Div, Pet Inc, 68-69; dir res & develop, Fairmont Foods, Co, Nebr, 69-70, vpres, 70-72; DIR RES & DEVELOP, INTERSTATE BRANDS CORP, 72- Mem: Am Hort Soc; Am Soc Qual Control; Inst Food Technol. Res: New bakery and horticultural food products development and engineering and packaging research. Mailing Add: Interstate Brands Corp PO Box 1627 Kansas City MO 64141

MAVIS, RICHARD DAVID, b Fergus Falls, Minn, Aug 7, 43; m 66; c 2. BIOCHEMISTRY. Educ: St Olaf Col, BA, 65; Univ Iowa, PhD(biochem), 70. Prof Exp: Asst prof biochem, Dent Med Sch, Northwestern Univ, Chicago, 72-75; ASST PROF, DEPT RADIATION BIOL & BIOPHYS, SCH MED & DENT, UNIV ROCHESTER, 75- Concurrent Pos: USPHS fel, Wash Univ, 69-72; NIH res grant, 73. Mem: AAAS. Res: Membrane structure and function; phospholipid metabolism. Mailing Add: Dept of Radiation Biol & Biophys Univ of Rochester Sch Med & Dent Rochester NY 14642

MAVITY, JULIAN MARIS, b Paoli, Ind, Sept 27, 08; m 33, 70; c 1. ORGANIC CHEMISTRY. Educ: Earlham Col, AB, 28; Ohio State Univ, MA, 30, PhD(org chem), 31. Prof Exp: Res chemist, Universal Oil Prod Co, Ill, 31-54, mgr serv labs, 54-64, head info & methods dept, 64-73; RETIRED. Mem: Am Chem Soc. Res: Synthesis and reactions of hydrocarbons; organoaluminum compounds; catalysis. Mailing Add: 284 E Briarwood Lane Palatine IL 60067

MAVOR, HUNTINGTON, b Schenectady, NY, Mar 26, 27; m 56; c 4. NEUROLOGY. Educ: Harvard Col, AB, 48; Univ Rochester, MD, 55; Am Bd Psychiat & Neurol, dipl, 62. Prof Exp: From instr to assoc prof neurol, Univ Utah, 61-69, instr med, 61-69; assoc prof, Univ Vt, 69-74; ASSOC PROF NEUROL, UNIV UTAH, 74- Concurrent Pos: Fel EEG & clin neurophys, Montreal Neurol Inst, 59-61; asst chief neurol, Salt Lake City Vet Admin Hosp, 63-69 & 74- Mem: Am Acad Neurol; Am Electroencephalog Soc. Res: Clinical electroencephalography and neurophysiology. Mailing Add: Dept of Neurol Univ of Utah Salt Lake City UT 84114

MAVRIDES, CHARALAMPOS, b Greece, June 25, 26; Can citizen. BIOCHEMISTRY. Educ: Nat Univ Athens, BSc, 53; Univ Ottawa, PhD(biochem), 64. Prof Exp: Fel pharmacol, NY Univ, 64-66; asst prof, 66-71, ASSOC PROF BIOCHEM, UNIV OTTAWA, 71- Mem: Can Biochem Soc. Res: Regulation of enzymes in higher cells; enzymology of transamination in procaryotes. Mailing Add: Dept of Biochem Univ of Ottawa Ottawa ON Can

MAVRODINEANU, RADU, b Romania, Oct 13, 10; nat US; m 46. PHYSICAL CHEMISTRY. Educ: Univ Bucharest, Romania, BS, 32, PhD(chem), 36. Prof Exp: Asst, Dept Soil Chem, Agron Inst, Romania, 33-40, lab head, 40-47; chemist, Nat Inst Agron, France, 47-48; res chemist, J J Carnaud et Forges de Basse Indre, 48-50; head lab spectros, Sci Res Off Outre-Mer, 50-52; chemist, Boyce Thompson Inst, 52-59; sect chief, Philips Labs Div, NAm Philips Co, Inc, 59-69; RES CHEMIST, ANAL CHEM DIV, NAT BUR STANDARDS, 69- Honors & Awards: Prize, Romanian Acad; Silver Medal Award, Dept of Com, 73. Mem: Am Chem Soc; Soc Appl Spectros; Fr Phys Soc. Res: Analytical spectroscopy and chemistry; soil chemistry; air pollution; high-accuracy spectrophotometry; optical emission spectrometry using electrical discharges and flames. Mailing Add: Anal Chem Div Nat Bur of Standards Washington DC 20234

MAVROIDES, JOHN GEORGE, b Ipswich, Mass, Dec 29, 22; m 52; c 2. PHYSICS. Educ: Tufts Col, BS, 44; Brown Univ, MS, 51, PhD(physics), 53. Prof Exp: Proj engr, US Naval Underwater Sound Lab, 46-49; asst, Brown Univ, 50-51; mem staff, 52-60, group leader, 60-74, SR STAFF, LINCOLN LAB, MASS INST TECHNOL, 74- Mem: Fel Am Phys Soc. Res: Solid state physics, especially galvanometric effects; magneto-optical studies; magneto-piezo-optics; magneto-acoustic effects; cyclotron resonance; electronic bandstructure; Fermi surfaces; infrared; lasers; electrochemistry; energy conversion. Mailing Add: Lincoln Lab Mass Inst of Technol Lexington MA 02173

MAVROYANNIS, CONSTANTINE, b Athens, Greece, Nov 13, 27; Can citizen; m 61; c 2. THEORETICAL SOLID STATE PHYSICS. Educ: Athens Tech Univ, BS, 57; McGill Univ, PhD(phys chem), 61; Oxford Univ, DPhil(math), 63. Prof Exp: Nat Res Coun Can NATO sci overseas fel, 61-63; Nat Res Coun Can fel, 63-64; asst res officer, 64-65, assoc res officer, 65-74, SR RES OFFICER, NAT RES COUN CAN, 74- Mem: Am Phys Soc; Can Asn Physicists; Chem Inst Can. Res: Optical properties and many-body interactions in solids; modern quantum chemistry; spin wave theory; electromagnetic interactions in solids. Mailing Add: Nat Res Coun of Can Sussex Dr Ottawa ON Can

MAWARDI, OSMAN KAMEL, b Cairo, Egypt, Dec 12, 17; nat US; m 50. PLASMA PHYSICS, ACOUSTICS. Educ: Fuad I Univ, Egypt, BSc, 40, MSc, 46; Harvard Univ, AM, 47, PhD(acoust), 48. Prof Exp: Transmission engr, Egyptian State Tel & Tel Co, 40-41; lectr physics, Fuad I Univ, 41-46; asst prof elec eng, Mass Inst Technol, 51-56, assoc prof mech & elec eng & mem res lab electronics, 56-60; chg plasma dynamics & nuclear eng, 60-66, prof eng, 60-75, dir plasma res prog, 66-75, ENERGY COORDR, CASE WESTERN RESERVE UNIV, 75-; PRES, COLLAB PLANNERS, 73- Concurrent Pos: Consult, Bolt, Beranek & Newman, Inc, 50-54, Nat Prod Corp, 51-52, Res Found, Lowell Tech Inst, 53-54, Philco Corp, 53-64, Gen Ultrasonics Corp, 55-57, Boeing Airplane Co, 55-57, Pratt & Whitney Div, United Aircraft Corp, 58-64, Conesco, 58-61, Los Alamos Sci Lab, 59-61, 66- & Amoco Res Lab, 69-71; Guggenheim fel, 54-55; mem, Inst Advan Study, 69-70; vpres, Auctor Assocs Inc, 70-; mem, Adv Energy Task Force to Gov, Ohio, 73-74. Honors & Awards: Biennial Award, Acoust Soc Am, 52; Res Award, Sigma Xi, 64. Mem: Fel AAAS; fel Acoust Soc Am; fel Am Phys Soc; fel Inst Elec & Electronics Engrs; NY Acad Sci. Res: Controlled fusion research; acoustic holography; electric power systems. Mailing Add: Case Western Reserve Univ Cleveland OH 44106

MAWBY, JOHN EVANS, b Dayton, Ohio, Dec 7, 35; m 67. VERTEBRATE PALEONTOLOGY. Educ: Cornell Univ, BA, 58; Univ Calif, Berkeley, MA, 60, PhD(paleont), 65. Prof Exp: Instr biol & geol, Deep Springs Col, 64-67; asst prof biol, Calif State Col, Long Beach, 67-69; actg dir/dean, 75-76, ASST DEAN, DEEP SPRINGS COL, 69- Concurrent Pos: Res assoc, Los Angeles County Mus Natural Hist, 68- Mem: Soc Vert Paleont; Am Soc Mammal; Geol Soc Am; Paleont Soc; Am Soc Zoologists; Soc Study Evolution. Res: Evolution of later Cenozoic mammals; fossil mammals of the Great Basin area; paleontology of early man sites. Mailing Add: Deep Springs Col Deep Springs Calif via Dyer NV 89010

MAWE, RICHARD C, b Apr 17, 29; US citizen. CELL PHYSIOLOGY. Educ: Fordham Univ, BS, 50, MS, 51; Princeton Univ, PhD(cell physiol), 54. Prof Exp: Asst biol, Fordham Univ, 50-51; asst, Princeton Univ, 51-54, res assoc physiol, 54-55; lectr zool, Columbia Univ, 60-61; from instr to assoc prof biol, 61-71, PROF BIOL, HUNTER COL, 71-, CHMN DEPT BIOL SCI, 65- Concurrent Pos: NSF grants, 62-70. Mem: AAAS; Soc Gen Physiol. Res: Transport; membrane structure. Mailing Add: Dept of Biol Sci Hunter Col 695 Park Ave New York NY 10021

MAWHINNEY, MICHAEL G, b Honolulu, Hawaii, Aug 29, 45; m 69; c 2. PHARMACOLOGY. Educ: Grove City Col, BS, 67; WVa Univ, MS, 69, PhD(pharmacol), 70. Prof Exp: Asst prof, 71-75, ASSOC PROF PHARMACOL & UROL, MED CTR, WVA UNIV, 75-, FAC MEM REPRODUCTIVE PHYSIOL, 71- Concurrent Pos: Consult, Albert Gallatin Sch Dist, 73- & J Urol, 75- Honors & Awards: Award, Pharmaceut Mfg Asn Found, 74 & 76. Mem: Am Soc Pharmacol & Exp Therapeut; Endocrine Soc. Res: Hormonal regulation of the epithelial and stromal elements of normal, aged and neoplastic male accessory sex organs. Mailing Add: Dept of Pharmacol Med Ctr WVa Univ Morgantown WV 25606

MAX, CLAIRE ELLEN, b Boston, Mass, Sept 29, 46; m 74. PLASMA PHYSICS, ASTROPHYSICS. Educ: Radcliffe Col, AB, 68; Princeton Univ, PhD(astrophys sci), 72. Prof Exp: Res assoc physics, Univ Calif, Berkeley, 72-74; PHYSICIST, LAWRENCE LIVERMORE LAB, UNIV CALIF, 74- Mem: Am Phys Soc; Am Astron Soc; AAAS. Res: Laser-plasma interactions; applications of plasma physics to astronomical problems. Mailing Add: L-545 Lawrence Livermore Lab Livermore CA 94550

MAX, STEPHEN RICHARD, b Providence, RI, Dec 25, 40. BIOCHEMISTRY. Educ: Univ RI, BS, 62, PhD(biochem), 66. Prof Exp: Asst prof, Col Med, Howard Univ, 67-70; ASSOC PROF NEUROL, SCH MED, UNIV MD, BALTIMORE, 70- Concurrent Pos: Nat Inst Neurol Dis & Stroke fel, 68-70; Dysautonomia Found & Frank G Bressler Reserve Fund res grants, 71-72; NIH & Muscular Dystrophy Asn res grants, 75-76; guestworker neurochem, Nat Inst Neurol Dis & Stroke, 68-70; lectr fac grad sch, NIH, 72- Mem: Am Chem Soc; Soc Neurosci; Am Soc Neurochem; Int Soc Neurochem; Am Acad Neurol. Res: Neurochemistry; muscle metabolism; neuromuscular diseases; ganglioside metabolism; mitochondria; lipids; lysosomes. Mailing Add: Dept of Neurol Univ of Md Sch of Med Baltimore MD 21201

MAXCY, RUTHFORD BURT, b Fulton, Miss, Sept 24, 21; m 55; c 1. FOOD SCIENCE, MICROBIOLOGY. Educ: Miss State Univ, BS, 43; Univ Wis, MS, 47, PhD(dairy indust), 50. Prof Exp: Asst prof dairying, Kans State Univ, 50-52; tech consult, George J Meyer Mfg Co, 52-54; mgr tech sales, Diversey Corp, 54-58; PROF FOOD SCI & TECHNOL, UNIV NEBR, LINCOLN, 58- Concurrent Pos: Consult, Dairy Prod Improv Comt, Dairy Prod Inst, 48-49. Mem: Am Dairy Sci Asn; Inst Food Technol; Am Soc Microbiol; Int Asn Milk, Food & Environ Sanit. Res: Microenvironment of food processing equipment, irradiation for public health protection and microbial survival under adverse conditions of food systems. Mailing Add: Dept Food Sci & Technol Univ of Nebr Lincoln NE 68503

MAXEY, BRIAN WILLIAM, b Michigan City, Ind, Sept 13, 39. VETERINARY MEDICINE. Educ: Purdue Univ, West Lafayette, BS, 61; Mich State Univ, PhD(anal chem), 68. Prof Exp: Chemist, Dow Chem Co, 61-65; from sr scientist to sr res scientist, 68-73, proj leader, 73-75, RES HEAD VET THERAPEUT, UPJOHN CO, 75- Mem: AAAS; Am Chem Soc; Sigma Xi. Res: Agricultural science; veterinary therapeutics; research and development of veterinary pharmaceuticals. Mailing Add: Upjohn Co 9690-190-1 Kalamazoo MI 49001

MAXEY, GEORGE BURKE, b Bozeman, Mont, Apr 3, 17; m 41; c 5. HYDROLOGY. Educ: Univ Mont, AB, 39; Utah State Agr Col, MS, 41; Princeton Univ, AM, 49, PhD(geol), 51. Prof Exp: Asst geol, Utah State Agr Col, 39-41; field asst, Ground-Water Div, US Geol Surv, Utah, 41-42, jr geologist, Ky, 42-43, asst geologist, Ky & Nev, 43-45, assoc geologist, 45-46, geologist chg off, 46-48; from instr to assoc prof geol, Univ Conn, 49-55; res assoc prof, Univ Ill, 55-57, prof, 57-62; res prof, 62-69, head ctr water resources res, 67-69, PROF HYDROL & GEOL & DIR NEV CTR WATER RESOURCES RES, MACKAY SCH MINES, UNIV NEV SYST, 69- Concurrent Pos: Consult, State of Nev, 51-61, AEC, 57-, State of Mont, 68- & UN, Poland & Kenya, 68-; tech adv, Tech Coop Admin & Foreign Opers Admin, Libya, 52-54; geologist & head sect ground-water geol & geophys explor, State Geol Surv, Ill, 55-62. Honors & Awards: O E Meinzer Award Hydrogeol, Geol Soc Am, 71. Mem: Fel AAAS; fel Geol Soc Am; Soc Econ Geologists; Am Asn Petrol Geologists; Am Geophys Union. Res: Groundwater geology; stratigraphy and paleontology; Cambrian of Utah and Idaho; areal and ground-water geology of Utah, Nevada and Illinois. Mailing Add: Desert Res Inst Univ of Nev Syst Reno NV 89502

MAXFIELD, BRUCE WRIGHT, b Coronation, Alta, July 15, 39. SOLID STATE PHYSICS, BIOPHYSICS. Educ: Univ Alta, BSc, 61; Rutgers Univ, PhD(physics), 64. Prof Exp: Res assoc, 64-66, actg asst prof, 66-67, asst prof, 67-71, SR RES ASSOC PHYSICS, CORNELL UNIV, 71- Concurrent Pos: Sloan Found res fel. Mem: Am Phys Soc; Inst Elec & Electronics Engrs. Res: Transport properties in pure metals and alloys, ultrasonic studies in metals and dynamics of conformational changes in proteins and phospholipids. Mailing Add: Dept of Physics Cornell Univ Ithaca NY 14853

MAXFIELD, GALEN HARRY, b Marshall, Minn, July 9, 16; m 45. FISH BIOLOGY. Educ: Univ Minn, BS, 42; Univ Wash, MS, 52. Prof Exp: Fish biologist, State Dept Res & Educ, Md, 47-50; res fisheries biologist, Bur Commercial Fisheries, US Fish & Wildlife Serv, 50-64, lab ed, 64-66, tech ed, Biol Lab, 66-70; tech ed, Northwest Fisheries Ctr, Nat Marine Fisheries Serv, Nat Oceanic & Atmospheric Admin, 70-74; RETIRED. Mem: Am Inst Fishery Res Biol; Am Fisheries Soc. Res: Early life history of Pacific salmon and Atlantic shad; storage and retrieval of information on fisheries. Mailing Add: 7704 57th NE Seattle WA 98105

MAXFIELD, JOHN EDWARD, b Los Angeles, Calif, Mar 17, 27; m 48; c 2. ALGEBRA. Educ: Mass Inst Technol, BS, 47; Univ Wis, MS, 49; Univ Ore, PhD(math), 51. Prof Exp: Instr math, Univ Ore, 50-51; mathematician, Naval Ord Test Sta, 51-58, head math div, 58-60; prof math & head dept, Univ Fla, 60-67; PROF MATH & HEAD DEPT, KANS STATE UNIV, 67- Mem: Am Math Soc; Sigma Xi; Soc Indust & Appl Math; Math Asn Am. Res: Number theory; analog and digital computing techniques; numerical analysis. Mailing Add: Dept of Math Kans State Univ Manhattan KS 66502

MAXFIELD, MARGARET WAUGH, b Conn, Feb 23, 26; m 48; c 4. MATHEMATICS. Educ: Oberlin Col, BA, 47; Univ Wis, MS, 48; Univ Ore, PhD(algebra), 51. Prof Exp: Mathematician, Naval Ord Test Sta, Calif, 49-53 & 55-60, consult, 60-65; PVT RES & WRITING, 65- Prof Exp: Instr & lectr, Univ Calif, Los Angeles, 57-60; vis assoc prof, Univ NB, 74-75. Honors & Awards: Lester R Ford Award, 67. Res: Number theory; statistics. Mailing Add: 417 N 17th Manhattan KS 66502

MAXFIELD, MARY EVANS, b Trenton, NJ, July 2, 08. PHYSIOLOGY. Educ: Mt Holyoke Col, AB, 31; Univ Pa, MA, 35, PhD(physiol), 41. Prof Exp: Asst physics, Mt Holyoke Col, 31-32, asst prof physiol, 45-47; asst biol, Univ Del, 34-36; asst instr physiol, Univ Pa, 36-41, instr pharmacol, 42-43; instr pharmacol, Syracuse Univ, 43-45; res assoc, Col Med, Wayne State Univ, 47-48; asst prof, Col Med, Univ Tenn, 48-49; assoc prof, Med Col Pa, 49-57; res physiologist, Haskell Lab, E I du Pont de Nemours & Co, Inc, 57-73; RETIRED. Concurrent Pos: Res fel neurophysiol, Inst Pa Hosp Ment & Nerv Dis, 41-42. Mem: Am Physiol Soc. Res: Blood volume; action potentials of isolated nerves; Q-T interval of electrocardiogram and exercise; effect of muscular activity and of environment on man. Mailing Add: 116 Hoiland Dr Wilmington DE 19803

MAXFIELD, MYLES, b Portland, Maine, Aug 5, 21; m 45; c 4. EXPERIMENTAL MEDICINE, BIOPHYSICS. Educ: Harvard Univ, AB, 42, MD, 45; Mass Inst Technol, PhD(biophys), 50. Prof Exp: Med house officer, Beth Israel Hosp, Boston, 45-46; head electron micros facility, Naval Med Res Inst, Md, 46-47; res assoc biol, Mass Inst Technol, 50-52, asst prof, 52-56; assoc scientist & asst physician, Brookhaven Nat Lab, 56-59; chief phys sci div, US Army Biol Labs, 59-61; prof biophys & dir biophys prog, Univ Southern Calif, 61-68; WITH US GOVT, WASHINGTON, DC, 68- Concurrent Pos: Mem staff, Med Dept, Mass Inst Technol, 54-56, vis comt, 57-61; mem biophys & biophys chem study sect, NIH, 62-68; consult, Ford Aeronutronics, 65-66 & US Army, 62-71. Mem: AAAS; Am Physiol Soc; Electron Micros Soc Am; NY Acad Sci. Mailing Add: 259 Congressional Lane Apt 616 Rockville MD 20852

MAXFIELD, OLLIE ORLAND, b Sumner, Ill, Feb 4, 25. GEOGRAPHY. Educ: George Peabody Col, BA, 45; Ohio State Univ, MA, 46, PhD(geog), 63. Prof Exp: Instr geog, 46-50 & 51-63, from asst prof to assoc prof, 63-72, chmn dept Western civilization, 63-65, PROF GEOG, UNIV ARK, FAYETTEVILLE, 72-, CHMN DEPT, 66- Concurrent Pos: NDEA Title XI consult, 65; chmn, Gov Comt Land Resource Mgt, 73-75. Mem: AAAS; Asn Am Geogr; Nat Coun Geog Educ. Res: Geography of the South; philosophy of geography; geographical education; urban and regional planning; conservation; historical geography of the United States. Mailing Add: Dept of Geog Univ of Ark Fayetteville AR 72701

MAXFIELD, PERRY LESTER, organic chemistry, inorganic chemistry, see 12th edition

MAXIE, EDWARD CHESTER, pomology, see 12th edition

MAXIM, LESLIE DANIEL, b New York, NY, Feb 27, 41; m 62; c 2. OPERATIONS RESEARCH. Educ: Manhattan Col, BChE, 61; State Univ NY, MSc, 63; Stevens Inst Technol, MS, 66. Prof Exp: Jr chemist, Nat Starch & Chem Corp, Plainfield, 60-61, res chemist, 61-65, proj supvr phys chem res, 65-68; staff consult, 68-70, DIR OPERS RES, MATHEMATICA, 70- Concurrent Pos: Adj prof, Newark Col Eng. Mem: Am Chem Soc; Am Inst Chem Eng; Opers Res Soc Am; Am Statist Asn; NY Acad Sci. Res: Physical chemistry of polymers; statistics; statistical systems analysis. Mailing Add: Mathematica PO Box 2392 Princeton NJ 08540

MAXIMON, LEONARD CHARLES, theoretical physics, see 12th edition

MAXON, MARSHALL STEPHEN, b Syracuse, NY, June 21, 37; m 58; c 2. PLASMA PHYSICS. Educ: Syracuse Univ, BS, 58; Ind Univ, MS, 60, PhD(physics), 64. Prof Exp: Physicist, 63-69, group leader, 69-71, SR PHYSICIST, LAWRENCE LIVERMORE LAB, UNIV CALIF, 71- Mem: AAAS; Am Phys Soc. Res: X-ray emission from plasmas; solvable models in quantum field theory; thermonuclear physics. Mailing Add: Lawrence Livermore Lab L-71 Univ of Calif PO Box 808 Livermore CA 94550

MAXON, WILLIAM DENSMORE, b Detroit, Mich, Dec 8, 26; m 50; c 4. BIOCHEMISTRY. Educ: Yale Univ, BE, 48; Univ Wis, MS, 51, PhD(biochem), 53. Prof Exp: Asst biochem, Univ Wis, 49-53; res scientist antibiotics, 53-56, sect head, 56-67, GROUP MGR, FERMENTATION RES & DEVELOP, UPJOHN CO, 67- Mem: AAAS; Am Chem Soc; Am Soc Microbiol. Res: Fermentation technology and kinetics; continuous fermentation; aeration-agitation in fermentations. Mailing Add: Upjohn Co Ferment Res & Dev 7000 Portage St Kalamazoo MI 49001

MAXSON, CARLTON J, b Cortland, NY, Apr 19, 36; m 57; c 2. MATHEMATICS. Educ: State Univ NY Albany, BS, 58; Univ Ill, MA, 61; State Univ NY Buffalo, PhD(math), 67. Prof Exp: Math teacher, Hammondsport Cent Sch, 58-61; asst prof math, State Univ NY Col Fredonia, 61-66, assoc prof, 66-69; assoc prof, 69-74, PROF MATH, TEX A&M UNIV, 74- Concurrent Pos: Fac res awards, State Univ NY Col Fredonia, 67 & 68. Mem: Am Math Soc; Math Asn Am. Res: Algebra; semigroups; rings; near-rings; applications of algebraic structures to study of discrete structures. Mailing Add: Dept of Math Tex A&M Univ College Station TX 77840

MAXSON, DONALD ROBERT, b Claremont, NH, Jan 19, 24; m 57; c 2. PHYSICS. Educ: Bowdoin Col, BS, 44; Univ Ill, MS, 48, PhD(physics), 54. Prof Exp: Radio engr, US Naval Res Lab, DC, 44-47; res assoc physics, Univ Ill, 54-55; instr, Princeton Univ, 55-58, res assoc, 58-59; from asst prof to assoc prof, 59-67, PROF PHYSICS, BROWN UNIV, 67- Mem: Am Phys Soc. Res: Neutrino recoil experiments for identification of beta decay interaction; experimental studies of nuclear reactions induced by charged particles and fast neutrons; reaction mechanics and nuclear structure. Mailing Add: Dept of Physics Brown Univ Providence RI 02912

MAXSON, STEPHEN C, b Newport, RI, Apr 13, 38; m 65. PSYCHOBIOLOGY, BEHAVIOR GENETICS. Educ: Univ Chicago, SB, 60, PhD(biopsychol), 66. Prof Exp: Instr biol & res assoc behav genetics, Univ Chicago, 66-69; asst prof, 69-74, ASSOC PROF BIO-BEHAV SCI, UNIV CONN, 74- Mem: AAAS; Am Genetic Asn; Animal Behav Soc; Soc Neurosci; Am Psychol Asn. Res: Genetics and physiology of audiogenic seizures; brain mechanisms involved in motivation and emotion; genotype-environment interactions in the development and expression of behavior; sensory regulation of gene expression in nervous system. Mailing Add: Dept of Bio-Behav Sci Univ of Conn Storrs CT 06268

MAXUM, BERNARD J, b Bremerton, Wash, Nov 4, 31; m 59; c 5. ELECTROMAGNETISM, ENERGY SYSTEMS. Educ: Univ Wash, BS, 55; Univ Southern Calif, MS, 57; Univ Calif, Berkeley, PhD(cyclotron wave instabilities), 63. Prof Exp: Tech staff mem, Hughes Aircraft Co, 55-58; teaching res & mgt, 58-64; head geophys phenomenology sect, GTE Sylvania, 64-68; mgr electrosci div, MB Assocs, 68-70; mem staff, Ocean Sci Prog Develop, Cornell Aeronaut Lab, 70-73; MGR TECHNOL PLANNING & ENERGY PROG COORDR, ROCKWELL INT, 73- Mem: Am Phys Soc; Inst Elec & Electronics Engrs; Nat Energy Resources Orgn; Am Mgt Asn. Res: Electromagnetic, acoustic electronic and energy systems; business modeling. Mailing Add: 26552 Montebello Pl Mission Viejo CA 92675

MAXWELL, ARTHUR EUGENE, b Maywood, Calif, Apr 11, 25; m 46, 64; c 5. OCEANOGRAPHY. Educ: NMex State Univ, BS, 49; Univ Calif, MS, 52, PhD(oceanog), 59. Prof Exp: Asst, Scripps Inst, Univ Calif, 49-50, asst oceanog, 50-52, jr res geophysicist, 52-55; head oceanogr, Geophys Br, Off Naval Res, 55-59, head, 59-65; assoc dir, 65-69, dir res, 69-71, PROVOST, WOODS HOLE OCEANOG INST, 71- Concurrent Pos: Mem, Nat Adv Comt Oceans & Atmosphere, 72-75. Honors & Awards: Civilian Meritorious Serv Award, US Navy, 58, Superior Civilian Serv Award, 63 & Distinguished Civilian Serv Award, 64. Mem: AAAS; Marine Technol Soc (vpres, 64-65); Sigma Xi; fel Am Geophys Union (pres elect, 74-76). Res: Physical oceanography and geophysics, particularly the measurement and interpretation of heat flow through the ocean floor. Mailing Add: Woods Hole Oceanog Inst Woods Hole MA 02543

MAXWELL, CHARLES HENRY, b Las Palomas, NMex, July 9, 23; m 52; c 3. GEOLOGY. Educ: Univ NMex, BS, 50, MS, 52. Prof Exp: Geologist, Shell Oil Co, 51-52; geologist, Br Mineral Deposits, 52-56, Br Foreign Geol, Brazil, 56-61, Br Regional Geol, Ky, 61-63, Br Mil Geol, 63-66 & Br Radioactive Mat, 66-69, GEOLOGIST, BR CENT MINERAL RESOURCES, US GEOL SURV, 69- Mem: AAAS; Geol Soc Am; Mineral Soc Am. Res: Geologic mapping; field interpretive and engineering geology. Mailing Add: US Geol Surv Fed Ctr Denver CO 80225

MAXWELL, CHARLES NEVILLE, b Tuscaloosa, Ala, Oct 27, 27; m 52; c 4. MATHEMATICS. Educ: Univ Chicago, BS, 49, MS, 51; Univ Ill, PhD(math), 55. Prof Exp: Instr math, Univ Mich, 55-58; assoc prof, Univ Ala, 58-63; PROF MATH, SOUTHERN ILL UNIV, CARBONDALE, 63- Mem: Am Math Soc. Res: Topology; topological transformation groups; algebraic topology. Mailing Add: Dept of Math Southern Ill Univ Carbondale IL 62901

MAXWELL, CHARLES RICHARD, physical chemistry, see 12th edition

MAXWELL, DAVID SAMUEL, b Bremerton, Wash, Feb 13, 31; m 57; c 3. ANATOMY. Educ: Westminster Col (Mo), AB, 54; Oxford Univ, BA, 57; Univ Calif, Los Angeles, PhD, 60. Prof Exp: From instr to assoc prof, 59-68, PROF ANAT, SCH MED, UNIV CALIF, LOS ANGELES, 68-, VCHMN DEPT, 73-, PROF SURG/ANAT, CHARLES DREW POSTGRAD MED SCH, 74- Mem: AAAS; Am Asn Anat; Electron Micros Soc Am; Am Soc Cell Biol; Soc Neurosci; Asn Am Med Cols. Res: Electron microscopy; histochemistry and cytochemistry of the nervous system and eye. Mailing Add: Dept of Anat Univ of Calif Sch of Med Los Angeles CA 90024

MAXWELL, DONALD ROBERT, b Paris, France, Mar 30, 29; US citizen; m 56; c 7. PHARMACOLOGY. Educ: Cambridge Univ, BA, 52, MA, 56, PhD(pharmacol), 55. Prof Exp: Res attache, Pasteur Inst, Paris, 55-56; pharmacologist, May & Baker Ltd, Dagenham, Eng, 56-69, mgr pharmacol res, 69-74; DIR PRECLIN RES, WARNER-LAMBERT RES INST, 74- Mem: Fel Royal Soc Med; fel, Brit Inst Biol; Brit Pharmacol Soc; Brit Physiol Soc; Int Col Neuropsychopharmacol. Res: Psychopharmacology and neuropharmacology in relation to development of new drugs; cardiovascular drugs; anti-allergic drugs. Mailing Add: Warner-Lambert Res Inst Morris Plains NJ 07950

MAXWELL, DOUGLAS PAUL, b Norfolk, Nebr, Feb 12, 41; m 64; c 2. PLANT PATHOLOGY. Educ: Nebr Wesleyan Univ, BA, 63; Cornell Univ, PhD(plant path), 68. Prof Exp: Asst prof, 68-71, ASSOC PROF PLANT PATH, UNIV WIS-MADISON, 71- Mem: Am Phytopath Soc. Res: Ultrastructure of fungi; function of fungal microbodies; breeding for disease resistance in forages. Mailing Add: Dept of Plant Path Univ of Wis Madison WI 53706

MAXWELL, DWIGHT THOMAS, b Manhattan, Kans, Aug 25, 37; m 64. MINERALOGY. Educ: Univ Kans City, BS, 59; Mont State Univ, PhD(geol), 65. Prof Exp: Asst prof geol, Univ Mo-Kansas City, 64-67 & Northeast La State Col, 67-70; ASST PROF EARTH SCI, NORTHWEST MO STATE COL, 70- Mem: Clay Minerals Soc; Mineral Soc Am. Res: Clay mineralogy. Mailing Add: Dept of Earth Sci Northwest Mo State Col Maryville MO 64468

MAXWELL, ELIZABETH STARBUCK, b Birmingham, Ala, June 6, 18; m 42; c 2. BIOCHEMISTRY, MOLECULAR BIOLOGY. Educ: Univ Ala, BS, 38; Johns Hopkins Univ, DSc(microbiol), 48. Prof Exp: Technician med, State Dept Health, Ala, 39-41; biochemist, Los Alamos Sci Labs, Univ Calif, 43-46; BIOCHEMIST, NIH, 50- Mem: Am Chem Soc; Am Soc Biol Chemists. Res: Enzymology; mechanism and control of protein biosynthesis. Mailing Add: Lab of Molecular Biol Nat Inst of Arthritis Metab & Digestive Dis Bethesda MD 20014

MAXWELL, EMANUEL, b Brooklyn, NY, Dec 16, 12; m; c 3. PHYSICS. Educ: Columbia Univ, BS, 34, EE, 35; Mass Inst Technol, PhD(physics), 48. Prof Exp: Patent examr, US Patent Off, 37; geophysicist, Shell Oil Co, Inc, Tex, 37-41; staff mem, Radiation Lab, Mass Inst Technol, 41-45, res assoc physics, 45-48; physicist, Nat Bur Standards, 48-53; mem staff, Lincoln Lab, 53-63, vis assoc prof physics, 58-63, SR SCIENTIST & PROJ LEADER, FRANCIS BITTER NAT MAGNET LAB, MASS INST TECHNOL, 63- Mem: Fel Am Phys Soc. Res: Low temperature physics; superconductivity; microwave physics; applied magnetism; magnetic filtration. Mailing Add: Francis Bitter Nat Magnet Lab Mass Inst of Technol Cambridge MA 02139

MAXWELL, FOWDEN GENE, b Brownwood, Tex, Sept 29, 31; m 55; c 3. ENTOMOLOGY. Educ: Tex Tech Col, BS, 57; Kans State Univ, MS, 58, PhD(entom), 61. Prof Exp: Instr entom, Kans State Univ, 60-61; entomologist, Boll Weevil Res Lab, Entom Res Div, Agr Res Serv, USDA, 61-68; prof entom & head dept, Miss State Univ, 68-74; coordr environ activities, USDA, 74-75; PROF ENTOM & CHMN DEPT, UNIV FLA, 76- Concurrent Pos: Adj assoc prof, Miss State Univ, 62-68; lectr many US & foreign univs. Honors & Awards: J Everett Bussart Mem Award, Entom Soc Am, 72. Mem: Entom Soc Am; Am Inst Biol Sci; Agron Soc Am; Nat Soc Solar Energy; Sigma Xi. Res: Basic breeding lines resistant to boll weevil and other cotton insects; basic mechanisms of resistance and detection and isolation of biologically active substances in plants which affect or modify insect behavior; host plant resistance to cotton. Mailing Add: 9015 Greylock Dr Alexandria VA 22308

MAXWELL, GEORGE RALPH, II, b Morgantown, WVa, Mar 27, 35; m 59; c 3. ECOLOGY, ORNITHOLOGY. Educ: WVa Univ, AB, 57, MS, 61; Ohio State Univ, PhD(zool), 65. Prof Exp: Asst prof biol, The Citadel, 65-66; PROF ZOOL & DIR RICE CREEK BIOL FIELD STA, STATE UNIV NY COL OSWEGO, 66- Concurrent Pos: NSF instrnl sci equipment prog res grant, 67-69; State Univ NY Res Found grant-in-aid, 68-70; fel, Univ NC, Chapel Hill, 70; vis scientist, Fla Med Entom Lab, 73. Mem: Ecol Soc Am, Am Ornithologists Union; Wilson Ornith Soc. Res: Growth of stream mayflies; maintenance behavior of herons; breeding biology of the grackle; thermal characteristics of incubating Passerines; Heron/mosquito ecology. Mailing Add: Rice Creek Biol Field Sta State Univ of NY Oswego NY 13126

MAXWELL, GLENN, b Kent, Ohio, May 20, 31; m 59; c 3. MATHEMATICS. Educ: Kent State Univ, BS, 53, MA, 54; Ohio State Univ, PhD(math), 64. Prof Exp: Teacher high sch, Ohio, 54-56; instr, 63-64, ASST PROF MATH, KENT STATE UNIV, 64- Mem: Math Asn Am; Am Math Soc. Res: Mathematical foundations of set theory and logic. Mailing Add: Dept of Math Kent State Univ Kent OH 44242

MAXWELL, HOWARD NICHOLAS, b Zanesville, Ohio, Dec 5, 09; m 35; c 2. PHYSICS. Educ: Carnegie Inst Technol, BS, 32; Harvard Univ, AM, 33; Ohio State Univ, PhD(physics), 37. Prof Exp: Asst physics, Ohio State Univ, 33-35; instr physics & math, Hood Col, 36-40: from asst prof to assoc prof physics, Kalamazoo Col, 40-49; prof & chmn dept, 49-75, EMER PROF PHYSICS, OHIO WESLEYAN UNIV, 75- Concurrent Pos: Asst prof, Western Mich Col, 44 & 45; consult, Battelle Mem Inst, 53-55; assoc prog dir, Sec Insts, NSF, summers 61-65, 67 & 69-73, consult, Inst Sect, 62-73; AID educ prog specialist physics, Univ Mysore, 66. Mem: Am Phys Soc; Am Asn Physics Teachers. Res: Atomic spectra; Zeeman effect of gold and single ionized nitrogen. Mailing Add: Dept of Physics Ohio Wesleyan Univ Delaware OH 43015

MAXWELL, IAN DAVID, b Vancouver, BC, Mar 7, 15; m 40; c 4. PATHOLOGY. Educ: Bristol Univ, BSc, 36; Univ Edinburgh, MB, ChB, 42; Royal Col Physicians Can, cert specialist path, 51. Prof Exp: Asst pathology, Vancouver Gen Hosp, BC, 50-52; dir labs path, Royal Columbia Hosp, New Westminster, 52-56; ASSOC PROF PATH, DALHOUSIE UNIV, 56- Concurrent Pos: Assoc dir labs, Halifax Infirmary. Mem: Fel Am Soc Clin Path; Can Asn Path. Res: Clinical chemistry; hematology; immunoglobulins. Mailing Add: Dept of Path Halifax Infirmary Halifax NS Can

MAXWELL, JAMES CHRISTIE, b Boston, Mass, May 19, 26; m 51; c 4. GEOLOGY, GEOPHYSICS. Educ: Harvard Univ, AB, 51; Boston Univ, MA, 53; Columbia Univ, PhD, 61. Prof Exp: Photogeologist, Phys Res Lab, Boston Univ, 51-52, geophysicist, 52-54; lectr geol, Columbia Univ, 54-56; asst prof, Mo Sch Mines, 56-62, ASSOC PROF GEOL, UNIV MO-ROLLA, 62- Concurrent Pos: Treas & co-dir, Andean Exped, Harvard Univ, 49-54; instr, Boston Univ, 51-53; asst seismologist, Juneau Icefield Res Proj, Alaska, 55; consult, Autometric Div, Raytheon Corp, 62-64; proj dir water resources res ctr, Univ Mo-Rolla, 62-69. Mem: Geol Soc Am; Am Soc Photogram; Am Geophys Union. Res: Hydrogeology; analytical geomorphology; seismic and electrical geophysics; glaciology; photogeology; hydrogeomorphology. Mailing Add: Dept of Geol Univ of Mo Rolla MO 65401

MAXWELL, JAMES DONALD, b Mississippi Co, Ark, June 2, 40; m 63; c 2. PLANT BREEDING. Educ: Miss State Univ, BS, 62; Cornell Univ, MS, 65; NC State Univ, PhD(crop sci), 68. Prof Exp: Asst prof, 68-72, ASSOC PROF AGRON, CLEMSON UNIV, 72- Mem: Am Soc Agron; Crop Sci Soc Am. Res: Soybean breeding with emphasis on insect and disease resistance in southern region of United States. Mailing Add: Dept of Agron Clemson Univ Clemson SC 29631

MAXWELL, JAMES R, remote sensing, see 12th edition

MAXWELL, JOHN ALFRED, b Hamilton, Ont, Aug 28, 21; m 53. GEOCHEMISTRY. Educ: McMaster Univ, BSc, 49, MSc, 50; Univ Minn, PhD(geol, mineral & anal chem), 53. Prof Exp: Metall chemist, Burlington Steel Co, 39-45; asst chem, McMaster Univ, 48-50; asst petrol, Univ Minn, 51; analyst, Rock Anal Lab, 51-53; geochemist, 53-74, DIR CENT LABS & ADMIN SERV, GEOL SURV OF CAN, 74- Mem: Geol Asn Can; fel Royal Soc Can; Geochem Soc; Chem Inst Can;

Mineral Asn Can. Res: Methods of rock and mineral analysis; compilation of geochemical data; meteorites. Mailing Add: Geol Surv of Can 601 Booth St Ottawa ON Can

MAXWELL, JOHN CRAWFORD, b Xenia, Ohio, Dec 28, 14; m 39; c 2. GEOLOGY, TECTONICS. Educ: DePauw Univ, BA, 36; Univ Minn, MA, 37; Princeton Univ, PhD(geol), 46. Prof Exp: Reflections seismograph comput, Tex Co, 37; subsurface geologist, Sun Oil Co, 37-40; from instr to assoc prof geol, Princeton Univ, 46-55, prof geol eng, 55-70, chmn dept, 55-66, chmn dept geol, 66-70, chmn interdept prog water resources, 64-70; WILLIAM STAMPS FARISH PROF GEOL SCI, UNIV TEX, AUSTIN, 70- Concurrent Pos: Fulbright scholar, Italy, 52-53, NSF fel, 61-62; chmn earth sci div, Nat Res Coun, 70-72; consult, Adv Comt Reactor Safeguards, Nuclear Regulatory Comn, 74- Mem: Geol Soc Am (pres, 72-73); Am Asn Petrol Geologists; Am Geophys Union; Am Geol Inst (pres, 71-72); Ital Geol Soc. Res: Geology of Caribbean area and Montana-Wyoming; gravity tectonics in Italian Apennines and California coast ranges; high temperature high pressure on limestone, quartz sand, and sandstone; origin of rock cleavage. Mailing Add: Dept of Geol Sci Univ of Tex Austin TX 78712

MAXWELL, JOHN GARY, b Salt Lake City, Utah, Oct 5, 33; m 53; c 5. SURGERY. Educ: Univ Utah, BS, 54, MD, 58. Prof Exp: From instr to asst prof, 66-73, ASSOC PROF SURG, COL MED, UNIV UTAH, 73-, ASST DEAN ADMIS, 70-; ASST CHIEF, VET ADMIN HOSP, 66- Mem: Am Col Surg; Asn Acad Surg. Res: Gastrointestinal surgery; transplantation; vascular surgery. Mailing Add: Dept of Surg Univ of Utah Col of Med Salt Lake City UT 84112

MAXWELL, JOYCE BENNETT, b Merced, Calif, June 18, 41; m 68; c 2. GENETICS. Educ: Univ Calif, Los Angeles, AB, 63; Calif Inst Technol, PhD(genetics, biochem), 70. Prof Exp: Res asst neurohistochem, Camarillo State Hosp, 69-70; ASST PROF BIOL, CALIF STATE UNIV, NORTHRIDGE, 70- Mem: AAAS; Am Women in Sci. Res: Biochemical genetics; synthesis of serine and glycine by Neurospora crassa; multiple electrophoretic forms of tyrosinase in Neurospora crassa; high mutable serine-dependent strain of Neurospora. Mailing Add: Dept of Biol Calif State Univ Northridge CA 91324

MAXWELL, KEITH L, b Brookston, Ind, Feb 20, 18; m 50. SPEECH PATHOLOGY, SPEECH SCIENCE. Educ: Purdue Univ, BS, 40; Univ Mich, MA, 48, PhD(muscle coord), 53. Prof Exp: Teacher pub sch, Ind, 41-43, Mich, 43-44, Ind, 44-45 & Ill, 45-46; audiologist, Purdue Univ, 48-49; sr clinician, Univ Mich, 49-53; audiologist, West Side Vet Hosp, Chicago, Ill, 54-55; from asst prof to assoc prof, 55-66, PROF AUDIOL & SPEECH PATH, CENT MICH UNIV, 66-, AREA COORDR COMMUN DISORDERS, 72- Concurrent Pos: Abstractor, Deafness, Speech & Hearing Abstr. Mem: Am Speech & Hearing Asn; Speech Commun Asn; Int Asn Logopedics & Phoniatrics. Res: Cleft palate rehabilitation; articulation, causes and therapy; rehabilitation of the deaf and hard of hearing; speech and language rehabilitation of Chicanos, including reading, listening and speaking. Mailing Add: Dept of Speech & Dramatic Arts Cent Mich Univ Mt Pleasant MI 48858

MAXWELL, KENNETH EUGENE, b Huntington Beach, Calif, Sept 27, 08; m 41; c 3. ENTOMOLOGY. Educ: Univ Calif, BS, 33; Cornell Univ, PhD(entom), 37. Prof Exp: Jr entomologist, Univ Calif, Riverside, 37-39; technologist, Shell Oil Co, 39-42; mgr agr div, Chemurgic Corp, 45-47; consult, Maxwell Labs, 47-49; entomologist, E I du Pont de Nemours & Co, 49-50; mgr, Insecticide Dept, Agriform Co, 50-53; entomologist, Monsanto Chem Co, 53-59; tech dir, Moyer Chem Co, 59-63; from assoc prof to prof, 63-74, EMER PROF ENTOM, CALIF STATE UNIV, LONG BEACH, 74- Mem: AAAS; Entom Soc Am; Am Chem Soc; Weed Sci Soc Am; Am Mosquito Control Asn. Res: Toxicology of pesticides; environmental toxicology. Mailing Add: 16751 Greenview Lane Huntington Beach CA 92649

MAXWELL, MOREAU SANFORD, b Schenectady, NY, July 7, 18; m 43; c 4. ANTHROPOLOGY. Educ: Univ Chicago, AB, 39, MA, 46, PhD(anthrop), 49. Prof Exp: Dist supvr mus exten projs, Works Progress Admin, Ill, 39-42; asst prof anthrop, Beloit Col, 46-52; assoc prof, Air Univ, 52-57; cur mus, 57-71, chmn dept anthrop, 64-71, PROF ANTHROP, MICH STATE UNIV, 57-, RES ASSOC, MUS, 71- Concurrent Pos: Chief arctic sect, Arctic, Desert, Tropic Info Ctr, US Dept Air Force, 52-57. Mem: Am Anthrop Asn; Soc Am Archaeol. Res: Archaeology of lower Ohio, upper Mississippi drainage and Great Lakes; ethnology of Melanesia and Southeast Asia; archaeology of Canadian arctic. Mailing Add: Dept of Anthrop Mich State Univ East Lansing MI 48823

MAXWELL, RICHARD ELMORE, b Dallas, Tex, Aug 8, 21; m 42; c 2. BIOCHEMISTRY. Educ: Southern Methodist Univ, BS, 43; Univ Ill, PhD(biochem), 47. Prof Exp: Asst, Magnolia Petrol Co, Tex, 43-44; asst chem, Univ Ill, 44-47; asst prof, Iowa State Univ, 47-51; sr res chemist, 51-57, res leader, 57-64, dir lab biochem, 64-70, DIR BIOCHEM SECT, PHARMACOL DEPT, PARKE, DAVIS & CO, 70- Mem: Fel AAAS; Am Soc Biol Chem; Am Chem Soc; fel Am Inst Chem. Res: Actions of drugs and antibiotics on biological systems. Mailing Add: Biochem Sect Pharmacol Dept Parke Davis & Co Ann Arbor MI 48106

MAXWELL, RICHARD HOWARD, biology, botany, see 12th edition

MAXWELL, ROBERT ARTHUR, b Union City, NJ, Oct 6, 27; m 56; c 3. PHARMACOLOGY. Educ: Princeton Univ, PhD(biol), 54. Prof Exp: Assoc pharmacologist, Ciba Pharmaceut Co, 54-60, assoc dir pharmacol, 60-62; assoc prof, Col Med, Univ Vt, 62-65; HEAD PHARMACOL, WELLCOME RES LABS, 66- Concurrent Pos: Vis prof, Col Med, Univ Vt, 66-; adj prof pharmacol & exp med, Med Ctr, Duke Univ, 70-; adj prof pharmacol, Sch Med, Univ NC, Chapel Hill. Mem: AAAS; Am Soc Pharmacol & Exp Therapeut; Pharmacol Soc Can; NY Acad Sci. Res: Cardiovascular and autonomic pharmacology. Mailing Add: Wellcome Res Labs 3030 Cornwallis Rd Research Triangle Park NC 27709

MAXWELL, THOMAS JAMES, b Bryantville, Mass, June 22, 24; m 50; c 3. ANTHROPOLOGY, ARCHAEOLOGY. Educ: Col Wooster, BA, 47; Univ NMex, 47-48; Univ Mo, MA, 53; Ind Univ, PhD(anthrop), 62. Prof Exp: Instr Eng, Inst Cult Peruano-Norteamericano, 49; psychiat social worker, Massillon State Hosp, 50-51; jr high sch teacher, Bolivar, Ohio, 53-54; res asst, Indian Land Claims, Ind Univ, 54-56; chmn sociol & anthrop, Inter-Am Univ PR, 56-64; chmn div social sci, 65-66, anthrop, chmn dept sociol & 66-71, PROF ANTHROP, CALIF LUTHERAN COL, 65- Concurrent Pos: Catedratico visitante, Inst Coop, Univ PR, 62-63; acad dean lib arts, Inter-Am Univ PR, San German, 63-65; naturalist, Sequoia & Kings Canyon Nat Parks, summers 72-; consult archaeol, City of Thousand Oaks, Calif, 74- Mem: Am Anthrop Asn; Am Ethnol Soc; Inst Caribbean Studies; Am Soc Ethnohist. Res: Community studies in Puerto Rico and Central Andes; archaeological excavations in Missouri, Puerto Rico and southern California; land use and occupancy of North West Territory, especially Michigan; archaeological interpretation of land use and settlement patterns in the Conejo Valley, Ventura County, California. Mailing Add: Box 2736 Calif Lutheran Col Thousand Oaks CA 91360

MAXWELL, TRACY FRANCIS, b Rochester, NY, June 2, 48. PHYCOLOGY. Educ: State Univ NY Col, Geneseo, BA, 70, MA, 72; Ohio State Univ, PhD(bot), 76. Prof Exp: Mgr aquatic ecol, 71-73, STAFF ASSOC AQUATIC ECOL, ENVIRON RESOURCE CTR, 73-; SCI ASST PHYCOL, ACAD NATURAL SCI, 74- Mem: Phycol Soc Am; Int Phycol Soc; Brit Phycol Soc; Sigma Xi. Res: Taxonomic revision of the Cyanophyta, especially planktonic Nostocaceae. Mailing Add: Acad of Natural Sci 19th & Ben Franklin Pkwy Philadelphia PA 19103

MAXWELL, WILLIAM ANDREW, microbiology, cell biology, see 12th edition

MAXWELL, WILLIAM L, b Philadelphia, Pa, July 11, 34; m 69; c 4. OPERATIONS RESEARCH, INDUSTRIAL ENGINEERING. Educ: Cornell Univ, BME, 57, PhD(opers res), 61. Prof Exp: Asst prof indust eng, 61-64, assoc prof indust eng & opers res, 64-69, PROF OPERS RES, CORNELL UNIV, 69- Mem: Asn Comput Mach; Opers Res Soc Am; Inst Mgt Sci. Res: Scheduling theory; digital simulation; production control and data processing systems; education computing languages. Mailing Add: Upson Hall Cornell Univ Ithaca NY 14850

MAY, ARTHUR WILLIAM, b St John's, Nfld, June 29, 37; m 58; c 4. FISH BIOLOGY. Educ: Mem Univ, BSc, 58, MSc, 64; McGill Univ, PhD(marine sci), 66. Prof Exp: Scientist, 58-59, prog head, Northern Cod Res, 59-68, PROG HEAD ANADROMOUS FISH RES, FISHERIES RES BD CAN, 68- Concurrent Pos: Adv to Can Deleg, Int Comn Northwest Atlantic Fisheries, 65-, chmn subcomt, 68-69; spec lectr, Mem Univ, Nlfd, 67-68; mem, Joint Working Party, Int Coun Explor Sea-Int Comn Northwest Atlantic Fisheries, 68-; chmn coord working party, Atlantic Fishery Statist, 71; spec biol adv, Can Fisheries Serv, Ottawa, 71-72. Honors & Awards: Gov-Gen's Medal, Mem Univ Nfld, 58. Mem: Am Fisheries Soc; Can Soc Zoologists; Can Soc Wildlife & Fishery Biologists. Res: Ecology and population dynamics of fishes, especially haddock, cod and Atlantic salmon. Mailing Add: 5 Yellow Knife Plaza St John's NF Can

MAY, CHARLES EDWARD, b Hamilton, Ohio, Dec 16, 25; m 66; c 2. Prof Exp: Xavier Univ, BS, 47, MS, 49; Purdue Univ, PhD(chem), 53. Prof Exp: HEAD CHEM SECT, NASA, 53- Res: High temperature chemistry; x-ray and electron diffraction; Raman and infrared spectroscopy. Mailing Add: Lewis Res Ctr NASA 21000 Brookpark Rd Cleveland OH 44135

MAY, DANIEL STEPHEN, biochemistry, developmental biology, see 12th edition

MAY, DONALD CURTIS, JR, b Ann Arbor, Mich, May 31, 17; m 42. OPERATIONS RESEARCH. Educ: Univ Mich, AB, 38; Princeton Univ, AM, 40, PhD(math), 41. Prof Exp: Instr math, Princeton Univ, 39-40; mathematician, Bur Naval Weapons, 41-63, MATHEMATICIAN OPERS RES, SURFACE MISSILE SYSTS PROJ, US DEPT NAVY, 63- Res: Evaluation of Navy weapon systems. Mailing Add: 5931 Oakdale Rd McLean VA 22101

MAY, EDWIN ANTHONY, b Tuckahoe, NY, Nov 22, 23; m 49; c 5. BIOMEDICAL ENGINEERING. Educ: Stevens Inst Technol, ME, 47. Prof Exp: Proj engr, Prod Develop Div, Becton, Dickinson & Co, 47-57, asst to dir, 58-59, asst to vpres, Cent Res Div, 59-60, asst dir res & develop, Cardiovasc & Spec Instrument Div, 61-65, mgr bioeng, Corp Res Ctr, 66-68; dir res & mkt, Bio-Med Syst, Inc, 68-70; GROUP DIR BIOMED ENG, CORP RES & DEVELOP DIV, INT PAPER CO, 70- Mem: Am Soc Artificial Internal Organs; Am Soc Mech Engrs; Inst Elec & Electronic Engrs; AAAS. Res: Diagnostic, surgical, laboratory instrumentation; hospital systems; surgical research; artificial organs; cardiovascular instrumentation; biomedical research; biomaterials research; health care systems; advanced technology applications. Mailing Add: Corp Res & Develop Div Int Paper Co PO Box 797 Tuxedo Park NY 10987

MAY, ERNEST MAX, b Newark, NJ, July 24, 13; m 40; c 3. ORGANIC CHEMISTRY. Educ: Princeton Univ, AB, 34, AM, 35; Univ Chicago, PhD(org chem), 38. Prof Exp: Asst, Otto B May, Inc, 34-35, res dir & gen mgr, 38-52, pres, 52-73; RETIRED. Concurrent Pos: Vpres, Christ Hosp, Jersey City, NJ, 74-; trustee, Montclair State Col, NJ, 75- Mem: Am Chem Soc; fel Am Inst Chemists; Swiss Chem Soc; hon mem Synthetic Org Chem Mfrs Asn; Sigma Xi. Res: Free radical reactions in solution; azo and vat dyes; disperse dyes for polypropylene and other synthetic fibers; meaningful interdisciplinary curricula for graduate degrees. Mailing Add: 57 Colt Rd Summit NJ 07901

MAY, EVERETTE LEE, b Timberville, Va, Aug 1, 14; m 40, 65; c 4. MEDICINAL CHEMISTRY. Educ: Bridgewater Col, AB, 35; Univ Va, PhD(org chem), 39. Prof Exp: Res chemist, Nat Oil Prods Co, 39-41; from assoc chemist to sr chemist, NIH, 41-53, from scientist to sr scientist, Commissioned Corps, 53-58, scientist dir, 59, CHIEF SECT MED CHEM, NAT INSTS ARTHRITIS & METABOLIC DIS, USPHS, 60-; ADJ PROF PHARMACOL, MED COL VA, 74- Concurrent Pos: Mem expert adv panel drugs liable to cause addiction & comt probs drug dependence, 58-; chem adv panel mem, Walter Reed Army Inst, 65- Mem: Am Chem Soc. Res: Surface active agents; vitamins of the B complex; antimalarial agents; analgesic drugs; antitubercular compounds; carcinolytic agents; chemical and pharmacological investigations on central nervous system and anti-inflammatory agents. Mailing Add: Chem Lab Nat Inst Arthritis & Metabolic Dis Bethesda MD 20014

MAY, HUBERT EUGENE, b Dewey, Okla, Mar 17, 33; m 56; c 2. BIOCHEMISTRY. Educ: Bethany Nazarene Col, BS, 56; Univ Okla, MS, 60, PhD(biochem), 66. Prof Exp: Res assoc biochem, Okla Med Res Found, 60-63, fel, 66-67; from asst prof to assoc prof chem, 67-71, PROF CHEM, ORAL ROBERTS UNIV, 72- Mem: Am Chem Soc. Res: Microsomal metabolism in the liver. Mailing Add: Dept of Natural Sci Oral Roberts Univ Tulsa OK 74136

MAY, INGO W, physical chemistry, see 12th edition

MAY, IRVING, b New York, NY, Feb 16, 18; m 40; c 1. ANALYTICAL CHEMISTRY, GEOCHEMISTRY. Educ: City Col New York, BS, 38; George Washington Univ, MS, 48. Prof Exp: Testing technician, Panama Canal, 39-41; anal chemist, USPHS, NIH, 41-48; anal chemist, 48-71, CHIEF BR ANAL LABS, US GEOL SURV, 71- Mem: Am Chem Soc; Geochem Soc; Sigma Xi; AAAS. Res: Geochemical analysis, particularly determination of trace elements. Mailing Add: US Geol Surv Nat Ctr Stop 923 Reston VA 22092

MAY, JACK TRUETT, b Pike Co, Miss, Dec 24, 09; m 42; c 8. FORESTRY. Educ: La State Univ, BSF, 32; Univ Ga, MSF, 37; Mich State Univ, PhD, 57. Prof Exp: From field asst to forester, Forest Serv, USDA, 28-49; prof forestry & silvicult, Auburn Univ, 49-58; PROF SILVICULT & FOREST SOILS, UNIV GA, 58- Concurrent Pos: Tech adv & forestry consult to govt & indust orgns. Mem: Fel AAAS; Soc Am Foresters; Soil Sci Soc Am; Am Soc Agron. Res: Forest silviculture with special emphasis on soils, hardwoods and regeneration. Mailing Add: Sch of Forest Resources Univ of Ga Athens GA 30601

MAY, JACQUES M, epidemiology, deceased

MAY, JAMES AUBREY, JR, b Houston, Tex, July 15, 42; m 68. POLYMER CHEMISTRY, ORGANOMETALLIC CHEMISTRY. Educ: Tex Christian Univ, BS, 64, PhD(org chem), 68. Prof Exp: Sr res chemist, 68-72, RES SPECIALIST, DOW CHEM CO, USA, 72- Mem: Am Chem Soc; Sigma Xi; NY Acad Sci. Res: Polymer characterization; gel permeation chromatography; polyolefins; polyesters; Ziegler catalysis; free radical catalysis kinetics. Mailing Add: B-3827 Bldg Dow Chem Co USA Freeport TX 77541

MAY, JAMES DAVID, b Blue Mountain, Miss, Aug 21, 40; m 61; c 3. POULTRY PHYSIOLOGY. Educ: Miss State Univ, BS, 61, MS, 63; NC State Univ, PhD(physiol), 70. Prof Exp: RES PHYSIOLOGIST, S CENT POULTRY RES LAB, AGR RES SERV, USDA, 69- Concurrent Pos: Adj asst prof, Miss State Univ, 70. Mem: Poultry Sci Asn; World Poultry Sci Asn; AAAS; Am Physiol Soc. Res: Poultry environmental physiology; thyroid metabolism; amino acid metabolism. Mailing Add: PO Box 5367 Mississippi State MS 39762

MAY, JOHN ELLIOTT, JR, b Meriden, Conn, June 4, 21; m 45; c 2. PHYSICS. Educ: Wesleyan Univ, BA, 43; Tufts Col, MS, 49; Yale Univ, PhD(physics), 53. Prof Exp: Electronics physicist, US Naval Res Lab Field Sta, Mass, 46-49; mem tech staff, Bell Tel Labs, Inc, 52-59, supvr explor develop delay devices, 59-62, ultrasonic amplifier & evaporated film transducer develop, 62-65, head ultrasonic device dept, 65-71, head process capability dept, 71-74, physicist, 52-74; MEM STAFF, EASTMAN KODAK CO RES LABS, 74- Mem: Am Phys Soc: fel Acoust Soc Am; Inst Elec & Electronics Engrs. Res: Research and development in ultrasonic devices including delay line geometry, ceramic transducers; elastic wave guide effects; piezoelectric materials; ultrasonic amplification; mechanical filters; microphones; thin film circuits; conductors, resistors and capacitors. Mailing Add: Eastman Kodak Co Res Labs 1669 Lake Ave Rochester NY 14650

MAY, JOHN THOMAS, b Philadelphia, Pa, Jan 27, 39; m 63; c 2. EXPERIMENTAL NUCLEAR PHYSICS. Educ: US Air Force Acad, BS, 61; NC State Univ, MS, 70, PhD(physics), 74. Prof Exp: Instr, 70-72, ASSOC PROF PHYSICS, US AIR FORCE ACAD, 74- Mem: Am Phys Soc; Am Asn Physics Teachers; AAAS. Res: Internal ionization phenomena. Mailing Add: Dept of Physics US Air Force Academy CO 80840

MAY, JOHN WALTER, b London, Eng, June 25, 36; Can citizen; m 66; c 2. SURFACE PHYSICS, SURFACE CHEMISTRY. Educ: Univ BC, BA, 57, MS, 60; Oxford Univ, PhD(chem), 63. Prof Exp: Res assoc, Dept Appl Physics, Cornell Univ, 64-68; physicist, Bartol Res Found, Franklin Inst, 68-72; SR RES CHEMIST, EASTMAN KODAK RES LABS, 72- Mem: Am Vacuum Soc; Sigma Xi. Res: Surface science, including adsorption, catalysis, surface structure, surface electrostatics, triboelectrification, low energy electron diffraction, auger and x-ray photoelectron spectroscopies. Mailing Add: Eastman Kodak Res Labs Kodak Park Rochester NY 14650

MAY, KENNETH NATHANIEL, b Livingston, La, Dec 24, 30; m 53; c 2. FOOD TECHNOLOGY. Educ: La State Univ, BS, 52, MS, 55; Purdue Univ, PhD(food technol), 59. Prof Exp: Asst poultry sci, La State Univ, 52-54, res assoc, 54-56; asst state poultry supvr, State Livestock Sanit Bd, La, 54; asst poultry husb, Purdue Univ, 56-58; from asst prof to prof, Univ Ga, 58-68; prof, Miss State Univ, 68-70; dir res & qual assurance, 70-73, VPRES RES & QUAL ASSURANCE, HOLLY FARMS POULTRY INDUSTS, INC, 73- Concurrent Pos: Mem salmonella adv comt, Secy Agr, 75-; adj prof, NC State Univ, Raleigh, 75- Honors & Awards: Res Award, Inst Am Poultry Industs, 63; Res Award, Ga Egg Comn, 64; Indust Serv Award, Poultry & Egg Inst Am, 71. Mem: Am Poultry Sci Asn; Inst Food Technol; World Poultry Sci Asn. Res: Meat yields and processing losses of poultry; nutritive value and bacteriology of poultry products; biochemistry of bruised tissue. Mailing Add: Res Dept Holly Farms Poultry Industs Inc Wilkesboro NC 28697

MAY, KENNETH OWNSWORTH, b Portland, Ore, July 8, 15; m 63. MATHEMATICS, HISTORY OF SCIENCE. Educ: Univ Calif, AB, 36, MA, 37, PhD(math), 46. Prof Exp: Asst math, Univ Calif, 36-37 & 39-41; instr, Univ Study Ctr, Italy, 45; from asst prof to prof & chmn dept, Carleton Col, 46-66; dir inst hist & philos of sci & technol, 73-75, PROF MATH, UNIV TORONTO, 66- Concurrent Pos: Lectr, Univ Minn, 47 & 61, Okla Agr & Mech Col, 48, Univ Mich, 56, Univ Calif, 65 & Dominican Col, 65-66. Fund Advan Educ fel, 53-54; grants, Am Philos Soc, 49 & 65, NSF, 56-60, 61-66 & 62, Sigma Xi, 65, Off Educ, 65, Am Coun Learned Socs, 66, Can Coun, 66-71 & Killam Award, 71-73; NSF sci fac fel, 62-64; vis scholar, Univ Calif, Berkeley, 64-66. Consult, Cowles Comn Res Econ, Univ Chicago, 46-48 & Int Milling Co, 50-54; consult, Comt Sch Math, Univ Ill, 57, mem adv comt, Sch Math Study Group, 59-61, consult, Interuniv Comt Superior Student, 63; gov, Inst Current World Affairs, 57-; chmn comn hist of math, Div Hist of Sci, Int Union Hist & Philos Sci, 69-; ed, Historia Mathematica, 74- Mem: Fel AAAS; Math Asn Am; Hist Sci Soc; corresp mem Int Acad Hist Sci; Can Soc Hist & Philos Math. Res: History of mathematics; mathematics education; information retrieval; indexing. Mailing Add: Dept of Math Univ of Toronto Toronto ON Can

MAY, LEONARD, chemistry, chemical engineering, see 12th edition

MAY, LEOPOLD, b Brooklyn, NY, Nov 26, 23; m 47; c 2. PHYSICAL BIOCHEMISTRY. Educ: City Col New York, BChE, 44; Polytech Inst Brooklyn, MS, 48, PhD, 51. Prof Exp: Instr, Polytech Inst Brooklyn, 49-50; res chemist, Columbia Univ, 50-54; res assoc, Univ Md, 54-56, instr, 56-59; asst prof, 59-61, ASSOC PROF CHEM, CATH UNIV AM, 61- Concurrent Pos: Lectr, Brooklyn Col, 53; instr, Johns Hopkins Univ, 54-57; ed-in-chief, Appl Spectros, 61-64; vis assoc prof, Tel-Aviv Univ, 72-73; vis scientist, Soreg Nuclear Physics Ctr, Israel, 72-73. Mem: AAAS; Am Chem Soc; Soc Appl Spectros (pres, 71). Res: Infrared and Mössbauer spectroscopy of biological materials; effect of electric fields on brain lipids and lipoproteins. Mailing Add: Dept of Chem Cath Univ of Am Washington DC 20064

MAY, MICHAEL MELVILLE, b Marseilles, France, Dec 23, 25; nat US; m 52; c 4. PHYSICS. Educ: Whitman Col, BA, 44; Univ Calif, PhD(physics), 52. Prof Exp: Res physicist, Radiation Lab, Univ Calif, 52-57; vpres, E H Plesset Assocs, 57-60; res physicist, 60-61, div leader, 61-62, assoc dir, 62-64, lectr appl sci, 64-65, dir, 65-71, RES PHYSICIST & ASSOC DIR-AT-LG, LAWRENCE LIVERMORE LAB, UNIV CALIF, 72- Concurrent Pos: Vis physicist, Princeton Univ, 71-72; sr personal adv to Secy of Defense for Strategic Arms Limitation Talks & mem US deleg, 74- Mem: Am Phys Soc. Res: Nuclear explosions; heat and radiation; relativity. Mailing Add: 728 E Angela St Pleasanton CA 94566

MAY, MORTON, b Rock Springs, Wyo, Nov 11, 28. RANGE CONSERVATION. Educ: Univ Wyo, BS, 50, MS, 54; Agr & Mech Col, Tex, PhD(range, forestry), 58. Prof Exp: Range conserv aide, Soil Conserv Serv, USDA, 53-54; mem staff range mgt, Univ Wyo, 54-55; asst, Agr & Mech Col, Tex, 55-58; range conservationist, Rocky Mountain Forest & Range Exp Sta, US Forest Serv, 58-63; assoc prof range mgt, 63-70, PROF RANGE MGT & DIR RES, UNIV WYO, 70- Res: Systems of grazing management as related to soils; wildlife habitat and microclimate; range management. Mailing Add: Dept of Range Mgt Univ of Wyo Laramie WY 82071

MAY, PAUL DAVID, b Frankfurt, Ger, Mar 27, 14; US citizen; m 45; c 3. POLYMER CHEMISTRY, BIO-ORGANIC CHEMISTRY. Educ: Sorbonne, MS, 36. Prof Exp: Asst, High Pressure Lab, Sorbonne, 36-37; analyst, Southport Petrol Co, Tex, 37-42; res chemist, Am Oil Co, 45-60; sr res chemist, Monsanto Co, 60-64; sr res chemist, Southwest Res Inst, 64-69; SR ORG CHEMIST, GULF SOUTH RES INST, 69- Mem: Am Chem Soc. Res: Polymeric compounds for maxillofacial prostheses; synthetic polypeptides from amino acids for biomedical application; characterization and utilization of polymeric compounds; monomer and polymer synthesis and evaluation; high pressure hydrogenation; phosgenation. Mailing Add: Gulf South Res Inst PO Box 26500 New Orleans LA 70126

MAY, PAUL S, b Brooklyn, NY, July 12, 31; m 56; c 3. MICROBIOLOGY. Educ: City Col New York, BS, 51; Syracuse Univ, MS, 52; Phila Col Pharm, DSc(indust microbiol), 55; Columbia Univ, MPH, 70. Prof Exp: Instr bact, Phila Col Pharm, 52-53, instr zool, 54-55; sr res microbiologist, S B Penick & Co, 55-58; asst microbiologist, Beth Israel Hosp, NY, 58-62; sr scientist microbiol, Life Sci Lab, Melpar, Inc, 62-64; lectr, Sch Pub Health & Admin Med, Columbia Univ, 64; asst dir bur Labs, 64-71, DEP GEN DIR BUR LABS, NEW YORK DEPT HEALTH, 71-; LECTR, SCH PUB HEALTH & ADMIN MED, COLUMBIA UNIV, 70- Mem: Soc Indust Microbiol; Am Soc Microbiol; Am Pub Health Asn. Res: Antibiotics; fermentations; medical bacteriology, parasitology and mycology; public health microbiology; laboratory and public health administration. Mailing Add: 23 Fairview Lane Orangeburg NY 10962

MAY, PHILIP REGINALD ALDRIDGE, b Weymouth, Eng, May 30, 20; nat US; m 59. PSYCHIATRY. Educ: Cambridge Univ, BA, 41, MB, BCh, 44, MA, 46; Stanford Univ, MD, 44; Royal Col Physicians & Surgeons, dipl psychol med, 47; Am Bd Psychiat & Neurol, dipl, 51. Prof Exp: Resident med & neurol, Guy's Hosp, London, 45, resident psychiat, 45-46; resident, Bexley Hosp, 46-47 & Sch Med, Univ Colo, 49-50; from instr to asst prof, 50-53, from asst clin prof to assoc clin prof, 56-68, clin dir neuropsychiat inst, 62-73, PROF PSYCHIAT, UNIV CALIF, LOS ANGELES, 69-; CHIEF STAFF PROG EVAL RES & EDUC, BRENTWOOD VET ADMIN HOSP, LOS ANGELES, 70- Concurrent Pos: Chief male inpatient serv, Colo Psychopathic Hosp, 50-51, asst dir, Hosp, 51-53; consult, Fitzsimmons Army Hosp, Denver & US Armed Forces Epidemiol Bd, 51-53, Vet Admin Hosps, Denver, 51-53 & Los Angeles, 66-70; clin dir, Camarillo State Hosp, Calif, 55-59, chief res, 59-62; consult, Superior Ct, Santa Barbara & Ventura Counties & Probation Dept, Ventura, Calif, 58-66. Honors & Awards: Bronze Award, Am Psychiat Asn, 63; Paul Hoch Award, Am Psychopath Asn, 74. Mem: Fel Am Psychiat Asn; AMA; fel Am Col Neuropsychopharmacol (pres, 75); Int Col Psychopharmacol; Royal Col Physicians. Res: Outcome and treatment of schizophrenia; development and evaluation of treatment programs. Mailing Add: Dept of Psychiat Univ of Calif Los Angeles CA 90024

MAY, RALPH FORREST, b Idaho, Ohio, Oct 1, 41; m 63. AGRICULTURAL CHEMISTRY. Educ: Wilmington Col, AB, 63; Ind Univ, Bloomington, MA, 66, PhD(org chem), 67. Prof Exp: RES CHEMIST AGR CHEM, BIOCHEM DEPT, E I DU PONT DE NEMOURS & CO, INC, 67- Mem: Am Chem Soc. Res: Synthesis of fungicides, insecticides and herbicides. Mailing Add: Biochem Dept Du Pont Exp Sta Wilmington DE 19898

MAY, ROBERT CARLYLE, b San Francisco, Calif, Feb, 4, 43. FISH BIOLOGY. Educ: Univ Calif, Berkeley, BA, 64; Univ Hawaii, MS, 67; Univ Calif, San Diego, PhD(marine biol), 72. Prof Exp: ASST MARINE BIOLOGIST, UNIV HAWAII, 72- Mem: AAAS; Am Fisheries Soc; Am Soc Ichthyologists & Herpetologists. Res: Cultivation of marine fishes; factors influencing larval survival in marine fishes. Mailing Add: Hawaii Inst of Marine Biol PO Box 1346 Kaneohe HI 96744

MAY, ROBERT MCCREDIE, b Sydney, Australia, Jan 8, 36; m 62; c 1. ECOLOGY. Educ: Univ Sydney, BSc, 57, PhD(physics), 60. Prof Exp: Gordon MacKay lectr appl math, Harvard Univ, 59-61; sr lectr & reader physics, Univ Sydney, 62-69, prof, 70-73; prof biol, 73-75, CLASS OF 1877 PROF ZOOL, PRINCETON UNIV, 75- Concurrent Pos: Vis prof physics, Calif Inst Technol, 67 & Magdalene Col, Oxford, 71; vis mem, Inst Advan Study, 71-72 & King's Col, Cambridge, 76; mem comt ecosyst anal, Nat Acad Sci, 73-75; assoc ed, Theoret Pop Biol & Math Biosci, 74- & SIAM J Appl Math & Appl Ecol Abstr, 75- Honors & Awards: Pawsey Medal, Australian Acad Sci, 67. Mem: Brit Ecol Soc; Am Soc Naturalists. Res: Theoretical models which give insights into the dynamics of single populations, pairs of populations, or of entire communities of interacting populations. Mailing Add: Dept of Biol Princeton Univ Princeton NJ 08540

MAY, SHELDON WILLIAM, b Minneapolis, Minn, June 27, 46; m 68; c 1. BIOCHEMISTRY. Educ: Roosevelt Univ, BS, 66; Univ Chicago, PhD(chem), 70. Prof Exp: Sr res chemist, Corp Res Lab, Exxon Res & Eng Co, 70-73; CHEMIST, SCH CHEM, GA INST TECHNOL, 73- Concurrent Pos: NIH postdoctoral fel, 70. Mem: AAAS; Am Chem Soc. Res: Enzyme chemistry; mechanisms of biochemical reactions; biochemical oxidations; immobilized enzymes. Mailing Add: Sch of Chem Ga Inst Technol Atlanta GA 30332

MAY, SHERRY JAN, Can citizen. MATHEMATICAL ANALYSIS. Educ: Univ Sask, BA, 68, dipl math, 69; Univ Waterloo, MM, 70, PhD(appl math), 74. Prof Exp: Nat Res Coun Can fel, Univ Sask, 74- Concurrent Pos: Res grant, Can Coun, 75. Mem: Am Math Soc. Res: Rational belief change in philosophy, such as probability kinematics, as constrained optimization problems. Mailing Add: 300 Bate Crescent Saskatoon SK Can

MAY, THOMAS PENN, chemistry, see 12th edition

MAY, WALTER RUCH, b Senath, Mo, Aug 4, 37; m 57; c 3. PHYSICAL INORGANIC CHEMISTRY. Educ: Memphis State Univ, BS, 59; Vanderbilt Univ, PhD(chem), 62. Prof Exp: Instr chem, Vanderbilt Univ, 59-62; res chemist, Monsanto Co, 62-65, sr res chemist, 65-66; res chemist, 66-67, res group leader, Corp Lab, 67-73, indust chem group leader, Tretolite Div, 73-75, MGR INDUST & WATER RES, TRETOLITE DIV, PETROLITE CORP, 75- Mem: Am Chem Soc; Nat Asn Corrosion Engrs; Sigma Xi; Am Soc Mech Engrs. Res: High temperature corrosion; hydrocarbon oxidation; thermal analysis; coordination compounds; heavy petroleum fuel additives. Mailing Add: 11152 Crickett Hill Dr St Louis MO 63141

MAYA, LEON, b Mexico City, Mex, Mar 23, 38; US citizen; m 60; c 2. INORGANIC CHEMISTRY. Educ: Nat Univ Mex, BS, 60; Univ Southern Calif, PhD(inorg chem), 73. Prof Exp: Supvr qual control lab, Monsanto Mexicana SA, 60-62; sr chemist, Israel Mining Industs Res Inst, 62-68; chemist, Rainbow Beauty Supply, 68-69; fel inorg chem, Univ Southern Calif, 73-74; MEM RES STAFF, OAK RIDGE NAT LAB, 74- Mem: Am Chem Soc; Sigma Xi. Res: Synthetic inorganic chemistry; use of physical methods for structural determination; chemistry of main group elements, particularly boron, silicon, phosphorus and fluorine. Mailing Add: Oak Ridge Nat Lab PO Box X Oak Ridge TN 37830

MAYA, WALTER, b New York, NY, Oct 25, 29; m 65; c 4. ORGANIC CHEMISTRY, INORGANIC CHEMISTRY. Educ: Univ Calif, Los Angeles, BS, 54, PhD(org chem), 58. Prof Exp: Res chemist, E I du Pont de Nemours & Co, 58-59; specialist fluorine chem, Rocketdyne Div, NAm Aviation, Inc, 59-70; lectr, 71-75, ASSOC PROF CHEM, CALIF STATE POLYTECH UNIV, POMONA, 75- Concurrent Pos: Pfizer fel, Univ Ill, 58-59. Mem: AAAS; Am Chem Soc. Res: Synthesis of fluorine compounds; physical-organic chemistry. Mailing Add: Dept of Chem Calif State Polytech Univ Pomona CA 91768

MAYALL, BRIAN HOLDEN, b Nelson, Eng, Nov 14, 32; US citizen; m 55; c 4. CELL BIOLOGY, BIOPHYSICS. Educ: Cambridge Univ, BA, 54, MA, 58; Univ Western Ont, MD, 61. Prof Exp: Res assoc, Wistar Inst, 62-64; from instr radiol to asst prof radiol sci, Med Sch, Univ Pa, 64-71, assoc prof radiol, 71-72; GROUP LEADER CYTOGENETICS & CYTOMORPHOMETRY, BIOMED DIV, LAWRENCE LIVERMORE LAB, UNIV CALIF, 72- Concurrent Pos: Pa Plan scholar, Wistar Inst & Univ Pa, 63-65; consult, Med Res Coun, UK, 68 & Nat Cancer Inst, 70-; mem ed bd, J Histochem & Cytochem, 71-; adj assoc prof radiol, Univ Calif, Davis, 74- Mem: AAAS; Histochem Soc; Am Soc Cell Biol; Am Soc Human Genetics; Sigma Xi. Res: Quantitative cytochemistry; scanning cytophotometry; automated cytology; image analysis of cells and chromosomes. Mailing Add: Biomed Div Lawrence Livermore Lab Livermore CA 94550

MAYALL, MARGARET WALTON, b Iron Hill, Md, Jan 27, 02; m 27. ASTRONOMY. Educ: Swarthmore Col, AB, 25; Radcliffe Col, AM, 28. Prof Exp: Asst, Harvard Observ, 24-43; mem res staff, Heat Res Lab, Spec Weapons Group, Mass Inst Technol, 43-46; asst, Harvard Observ, 46-49; dir, Am Asn Variable Star Observers, 49-73; RETIRED. Concurrent Pos: Pickering mem astronr, Harvard Observ, 49-54; consult, Am Asn Variable Star Observers, 73-75. Honors & Awards: Cannon Prize, Am Astron Soc, 58. Mem: Fel AAAS; Am Astron Soc; Am Asn Variable Star Observers; Royal Astron Soc Can; Int Astron Union. Res: Spectroscopy; variable stars; photometry; classification of spectra of faint stars; light curves of visual observations of variable stars. Mailing Add: 5 Sparks St Cambridge MA 02138

MAYALL, NICHOLAS ULRICH, b Moline, Ill, May 9, 06; m 34; c 2. ASTRONOMY. Educ: Univ Calif, AB, 28, PhD(astron), 34. Prof Exp: Asst, Univ Calif, 28-29; asst comput, Mt Wilson Observ, 29-31; observing asst, Lick Observ, 33-35, asst astronr, 35-42; mem staff radiation lab, Mass Inst Technol, 42-43; res assoc, Calif Inst Technol, 43-45; from assoc astronr to astronr, Lick Observ, 45-60; dir, Kitt Peak Nat Observ, 60-71; RETIRED. Mem: Nat Acad Sci; Am Philos Soc; Am Astron Soc; Am Acad Arts & Sci; Int Astron Union. Res: Nebular spectroscopy; photography; radial velocities of galactic nebulae, globular star clusters; red shifts and internal motions of extragalactic nebulae. Mailing Add: 5945 Mina Vista Tucson AZ 85718

MAYBANK, JOHN, b Winnipeg, Man, Jan 23, 30; m 52; c 2. ATMOSPHERIC PHYSICS. Educ: Univ Man, BSc, 52; Univ BC, MSc, 54; Univ London, PhD(meteorol), 59. Prof Exp: Sci officer, Physics & Meteorol Sect, Defence Res Bd, 54-61; res officer, Physics Div, Sask Res Coun, 61-70; climatologist, Caribbean Meteorol Inst, Barbados, 70-71; HEAD PHYSICS DIV, SASK RES COUN, 72- Concurrent Pos: Res assoc, Univ Sask, 62-66, adj prof, 68- Mem: Can Asn Physicists; Royal Meteorol Soc; Can Meteorol Soc. Res: Cloud physics; ice nucleation phenomena; atmospheric pollution; agrometeorology. Mailing Add: Sask Res Coun Saskatoon SK Can

MAYBEE, JOHN STANLEY, b Washington, DC, Mar 23, 28; m 55; c 6. MATHEMATICS. Educ: Univ Md, BS, 50; Univ Minn, PhD, 56. Prof Exp: Mathematician, David Taylor Model Basin, US Dept Navy, 50-52; asst math, Univ Minn, 52-56; from instr to asst prof, Univ Southern Calif, 56-59; asst prof, Univ Ore, 59-61; from asst prof to assoc prof, Purdue Univ, 61-67; PROF MATH & COMPUT SCI, UNIV COLO, BOULDER, 67- Concurrent Pos: Mem, Inst Math Sci, NY Univ, 58-59. Mem: Am Math Soc; Soc Indust & Appl Math. Res: Differential equations; applied mathematics; matrix theory; numerical analysis. Mailing Add: Dept of Math Univ of Colo Boulder CO 80302

MAYBERGER, HAROLD WOODROW, b New York, NY, Aug 28, 19; m 51; c 3. OBSTETRICS & GYNECOLOGY. Educ: Univ Ala, BA, 41; Long Island Col Med, MD, 44; Am Bd Legal Med, dipl, 56; Am Bd Obstet & Gynec, dipl, 61. Prof Exp: Intern, St John's Episcopal Hosp, 44-45 & 47-48, resident obstet & gynec, 48-51; mem courtesy staff, 53, clin asst, 53-55, from asst attend obstetrician & gynecologist to assoc attend obstetrician & gynecologist, 55-58, attend obstetrician & gynecologist & asst attend pathologist, 58-64, CHIEF DIV OBSTET & GYNEC, COMMUNITY HOSP, GLEN COVE, NY, 64- Concurrent Pos: Res fel neonatal path, Beth El Hosp, 51-53; assoc prof clin obstet & gynec, State Univ NY Stony Brook; consult, St John's Episcopal Hosp, Brooklyn, 65. Mem: AAAS; fel Am Col Legal Med; fel Am Col Surg; fel Am Col Obstet & Gynec; NY Acad Sci. Res: Neonatal pathology; forensic obstetrics. Mailing Add: 4 Bear Lane Locust Valley NY 11560

MAYBERRY, JOHN PATTERSON, b New Haven, Conn, July 17, 29; m 54; c 3. OPERATIONS RESEARCH. Educ: Univ Toronto, BA, 50; Princeton Univ, MA, 54, PhD(math), 55. Prof Exp: Asst econ, Princeton Univ, 50-52, asst appl math, Anal Res Group, 53-55; engr, Defense Electronic Prod Dept, Radio Corp Am, 55-58; opers analyst, Hq Fifth Air Force, Japan, 58-61, opers analyst, Hq, US Air Force, Washington, DC, 61-64, chief res group mil opers res, 64-67; mathematician, Mathematica Inc, 67, dir math res serv, 67-69; mathematician, Lambda Corp, Va, 69-71; chmn dept, 72-75, PROF MATH, BROCK UNIV, 71-; CONSULT, JOHN P MAYBERRY ASSOCS, 71- Mem: Asn Comput Mach; Am Math Soc; Math Asn Am;Soc Indust & Appl Math; Opers Res Soc Am. Res: Topology; graph theory; decision theory; systems analysis; game theory. Mailing Add: Dept of Math Brock Univ St Catharines ON Can

MAYBERRY, LILLIAN FAYE, b Portland, Ore, May 19, 43; m 75. CELL BIOLOGY, PARASITOLOGY. Educ: Calif State Univ, San Jose, BA, 67; Univ Nev, Reno, MS, 70; Colo State Univ, PhD(zool), 73. Prof Exp: Res assoc cell biol, Colo State Univ, 73-74 & Univ Colo, Boulder, 74-76; RES AFFIL BIOL SCI, UNIV TEX, EL PASO, 76- Concurrent Pos: Protozoologist, Yugoslavian Int Biol Prog, 75-76. Mem: Am Soc Parasitologists; Am Soc Zoologists; Soc Protozoologists; AAAS; Am Inst Biol Sci. Res: Physiology and ecology of host-parasite relationships; intracellular movements of small nuclear RNA and its role in gene regulation. Mailing Add: Dept of Biol Sci Univ of Tex El Paso TX 79999

MAYBERRY, MILO GLENN, organic chemistry, see 12th edition

MAYBERRY, THOMAS CARLYLE, chemistry, see 12th edition

MAYBERRY, WILLIAM EUGENE, b Cookeville, Tenn, Aug 22, 29; m 53; c 2. ENDOCRINOLOGY. Educ: Univ Tenn, MD, 53, Univ Minn, MS, 53; Am Bd Internal Med, dipl. Prof Exp: First asst & asst to staff internal med, 56-59, from instr to assoc prof med, Mayo Grad Sch Med, Univ Minn, 60-74, chmn dept lab med, Mayo Clin, 71-75, PROF LAB MED, MAYO MED SCH, 74- Concurrent Pos: Fel internal med, Mayo Grad Sch Med, Univ Minn, 56-59; Nat Inst Arthritis & Metab Dis trainee & res fel endocrinol, New Eng Ctr Hosp, 59-60; Am Cancer Soc fel, Nat Inst Arthritis & Metab Dis, 62-64; asst, Sch Med, Tufts Univ, 59-60; consult, Mayo Clin, 60-62, mem bd gov, 71-, vchmn, 73-75, chmn, 76-; consult, Mayo Clin, 64-, mem bd trustees, Mayo Found, 71-, vchmn, 75- Mem: Endocrine Soc; Am Thyroid Asn; Am Chem Soc; Am Fedn Clin Res; fel Am Col Physicians. Res: Biochemistry and physiology of the thyroid gland; biosynthesis of thyroxine. Mailing Add: Dept of Lab Med Mayo Clin Rochester MD 55901

MAYBERRY, WILLIAM ROY, b Grand Junction, Colo, Nov 30, 38; m 67. MICROBIOLOGY, ANALYTICAL BIOCHEMISTRY. Educ: Univ Colo, BA, 61; Western State Col Colo, MA, 64; Univ Ga, PhD(microbiol & biochem), 66. Prof Exp: Chemist, AEC, Lucius Pitkin, Inc, Colo, 60-61; asst instr chem, Mesa Col, 62; res assoc microbiol, Univ Ga, 66-67; res assoc, 67-68, asst prof, 68-75, ASSOC PROF MICROBIOL, SCH MED, UNIV SDAK, 75- Mem: AAAS; Am Chem Soc; NY Acad Sci; Am Soc Microbiol; Am Inst Biol Sci. Res: Gas chromatographic analysis of biological materials; growth yields and energy relationships of bacteria; membrane structure and function. Mailing Add: Dept of Microbiol Univ of SDak Sch of Med Vermillion SD 57069

MAYBURG, SUMNER, b Boston, Mass, Feb 21, 26. SOLID STATE PHYSICS. Educ: Harvard Univ, BS, 46; Univ Chicago, MS, 48, PhD(physics), 50. Prof Exp: Asst, Univ Chicago, 49-50; sr scientist radiation damage to solids, Atomic Power Div, Westinghouse Elec Corp, 50-52; sr engr, Res Labs, Sylvania Elec Prod, Inc, 52-55, engr mgr & chief engr, Semiconductor Div, 55-59; sr eng specialist, Gen Tel & Electronics Lab Div, 59-64; dir radiation effects div, Controls for Radiation, 64-67; CO-FOUNDER, TREAS & MEM TECH STAFF, SEMICONDUCTOR PROCESSING CO, INC, 67- Mem: Am Phys Soc; Electrochem Soc; sr mem Inst Elec & Electronics Engrs; NY Acad Sci; Am Inst Physics. Res: Dielectric constants; photoconductivity in insulators and semiconductors; lattice defects in semiconductors; semiconductor devices; intermetallic semiconductors; semiconductor lasers; radiation effects in semiconductor materials and devices; surface preparation of crystalline materials. Mailing Add: Semiconductor Processing Co Inc 10 Industrial Park Rd Hingham MA 02043

MAYBURY, PAUL CALVIN, b Rio Grande, NJ, July 20, 24; m 49; c 5. PHYSICAL CHEMISTRY. Educ: Eastern Nazarene Col, BS, 47; Johns Hopkins Univ, PhD(chem), 52. Prof Exp: Sr staff chemist missiles, Appl Physics Lab, Johns Hopkins Univ, 51-52, res assoc chem, Univ, 52-54; from asst prof to assoc prof, Eastern Nazarene Col, 54-61, chmn dept, 56-61; assoc prof, 61-63, PROF CHEM, UNIV S FLA, 64-, CHMN DEPT, 62- Concurrent Pos: Res assoc, Tufts Univ, 54, vis prof, 66. Mem: AAAS; Am Chem Soc; fel Am Inst Chemists. Res: Boron hydride chemistry, including isotopic exchange studies; reactions of metal borohydrides. Mailing Add: Dept of Chem Univ of SFla Tampa FL 33620

MAYBURY, ROBERT HARRIS, physical chemistry, see 12th edition

MAYCOCK, JERRY RAY, b Sioux City, Iowa, Nov 11, 37; m 59; c 2. ORGANIC CHEMISTRY. Educ: Morningside Col, BS, 59; Washington Univ, PhD(org chem), 67. Prof Exp: Res chemist, E I du Pont de Nemours & Co, Inc, 67-71; TEACHER, NORTH CROSS SCH, 71- Mem: Am Chem Soc. Mailing Add: 3891 Hyde Park Dr SW Roanoke VA 24018

MAYCOCK, JOHN NORMAN, b Ripley, Eng, Dec 27, 37; m 62; c 2. SOLID STATE PHYSICS, CHEMICAL PHYSICS. Educ: Univ London, BS, 59, PhD(solid state chem), 62. Prof Exp: Scientist, Rias Div, Martin Co, 62-67, sr scientist, 67-69, head chem physics dept, 69-74, assoc dir, 71-74; HEAD ENERGY TECHNOL CTR, MARTIN MARIETTA LABS & CORP DIR ENERGY AFFAIRS, MARTIN MARIETTA CORP, 74- Concurrent Pos: Lectr, Univ Md, 63-; mem, Energy Adv Comn, 74- Mem: Am Phys Soc; The Chem Soc. Res: Charge transport in alkali halides; physics of explosives and oxidizers; energy conservation as related to industry. Mailing Add: Martin Marietta Corp 11300 Rockville Pike Rockville MD 20852

MAYCOCK, PAUL DEAN, b Sioux City, Iowa, Sept 2, 35; m 59; c 5. SOLID STATE PHYSICS, SCIENCE ADMINISTRATION. Educ: Iowa State Univ, BS, 57, MS, 62. Prof Exp: Res asst physics, Ames Lab, AEC, 60-62; mem tech staff, Tex Instruments Inc, 62-67, mgr new prod develop, 67-69, mgr bus develop, 69-71, sr bus analyst mat & elec prod group, 71-75; BR CHIEF ECON ANAL, SOLAR ENERGY, ENERGY RES & DEVELOP ADMIN, 75- Mem: Am Phys Soc; Inst Elec & Electronics Engrs. Res: Thermal properties of solids; energy economics. Mailing Add: Energy Res & Develop Admin Washington DC 20545

MAYCOCK, PAUL FREDERICK, b Hamilton, Ont, Aug 13, 30; m 53; c 3. PLANT ECOLOGY. Educ: Queen's Univ, Ont, BA, 54; Univ Wis, MSc, 55, PhD(bot), 57. Prof Exp: Demonstr bot & zool, Queen's Univ, Ont, 52-54; lectr bot, McGill Univ, 57-58, from asst prof to assoc prof, 58-69; PROF BOT, ERINDALE COL, UNIV TORONTO, 69- Concurrent Pos: Mem staff, Polish Acad Sci, Cracow, 64-65. Mem: Ecol Soc Am; Can Bot Asn. Res: Phytosociology; boreal forests of North America and world; vegetation of central Canada; synecology and autecology of forest species; nature reserves and conservation research. Mailing Add: Ecol Lab Erindale Col Univ Toronto 3359 Mississauga Rd Clarkson ON Can

MAYDAN, DAN, b Tel Aviv, Israel, Dec 20, 35; m 60; c 3. APPLIED PHYSICS, ELECTROOPTICS. Educ: Israel Inst Technol, BSc, 57, MSc, 62; Univ Edinburgh, PhD(physics), 65. Prof Exp: Supvr instrumentation, Soreq Res Estab, Israel AEC, 57-62, group leader devices, 65-67; mem tech staff, 67-71, supvr optical scanning & modulation, 71-72, SUPVR NEW EXPOSURE SYST GROUP, BELL LABS, 72- Mem: Inst Elec & Electronics Engrs. Res: X-ray lithography; acoustooptical devices; high resolution laser recording; display devices. Mailing Add: Bell Labs Dept 2272 Rm 2A-220 600 Mountain Ave Murray Hill NJ 07974

MAYEDA, KAZUTOSHI, b Santa Monica, Calif, June 17, 28; m 49; c 3. GENETICS. Educ: Univ Utah, BS, 57, MS, 58, PhD(genetics), 61. Prof Exp: From asst prof to assoc prof, 61-73, PROF BIOL, WAYNE STATE UNIV, 73- Concurrent Pos: Res assoc, Nat Inst Genetics Japan, Mishima, Shizuoka-Ken, 70-71. Mem: AAAS; Am Soc Human Genetics; Am Genetics Soc; Soc Study Evolution; Brit Soc Study Human Biol. Res: Immunogenetics of Drosophila and human; blood groups of lower primates and other mammals; genetics of human serum proteins; biochemical studies of amniotic fluids and cells. Mailing Add: Dept of Biol Wayne State Univ Detroit MI 48202

MAYER, ALEX, b Arad, Rumania, June 8, 22; US citizen; m 50; c 3. PHYSICS. Educ: City Col New York, BS, 43; NY Univ, MS, 50, PhD(physics), 54. Prof Exp: Instr physics, City Col New York, 46-50; res asst & instr, NY Univ, 50-54; leader antennas, Raytheon Mfg Co, Mass, 54-56; staff physicist, Lincoln Lab, Mass Inst Technol, 56-

59; head electronics lab, Convair Div, Gen Dynamics Corp, 59-60; sr eng specialist, Sylvania Electronic Defense Lab, 60-62; assoc dir appl physics lab, Western Develop Labs, Philco Corp, 62-64; sr staff scientist, Nat Eng Sci Co, 64-67; PRES, A MAYER & ASSOC, 67- Mem: Sr mem Inst Elec & Electronics Engrs; Am Inst Aeronaut & Astronaut; Sigma Xi. Res: Microwave antennas and solid state devices; electromagnetic and acoustic wave propagation; noise control. Mailing Add: A Mayer & Assocs 1202 Sesame Dr Sunnyvale CA 94087

MAYER, BROMLEY MORGAN, b Los Angeles, Calif, Sept 28, 18; c 3. MICROBIOLOGY. Educ: Univ Southern Calif, BA, 40, MS, 42. Prof Exp: Microbiologist, 46-67, DIR RES, KNUDSEN CORP, 67- Mem: Inst Food Technol; Am Soc Microbiol; Am Dairy Sci Asn. Res: Fermented dairy products; whey utilization; yeast fermentation. Mailing Add: Knudsen Corp PO Box 2335 Terminal Annex Los Angeles CA 90054

MAYER, CLAUDIUS FRANCIS, b Eger, Hungary, July 6, 99; nat US; m 27. HISTORY OF MEDICINE. Educ: Innsbruck Univ, AB, 18; Univ Budapest, MD, 25. Prof Exp: Asst librn, Med Fac, Univ Budapest, 23-25; demonstr, Path Inst, 25-26, attend physician, Urol Clin, 28-31; med dir, Lindsay Labs, NY, 31-32; ed index catalogue, Army Med Libr, 32-54; civilian med officer, US Dept Defense, 55-74; RETIRED. Concurrent Pos: Klebelsberg fel, Inst Hist Med, 28; physician, Nat Social Security Inst, Hungary, 26-28; consult, Mus Hyg, Budapest, 28-29; clin pathologist, Hosp Uzsoki St, 28-31; prosector, Tuberc Hosp, Pestujhely, 29-31; attend physician, Venerology Clin, DC, 33; mem, Int Cong Human Genetics, 71- Mem: AAAS; Asn Mil Surg US; NY Acad Sci. Res: Semantics of medical nomenclature; medieval and Arab medicine; sixteenth century medicine; bibliography; medical documentation; epidemic hemorrhagic fever; endemic panmyelotoxicosis; human genetics; occupational medicine; geopathology; human population history. Mailing Add: 5513 39th St NW Washington DC 20015

MAYER, CORNELL HENRY, b Ossian, Iowa, Dec 10, 21; m 46; c 2. ASTRONOMY. Educ: Univ Iowa, BS, 43; Univ Md, MS, 51. Prof Exp: Electronic engr, 43-49, physicist, 49-68, HEAD RADIO ASTRON, NAVAL RES LAB, 68- Concurrent Pos: Mem vis comt, Nat Radio Astron Observ, 69-72; mem nat adv comt, Owens Valley Radio Observ, Calif Inst Technol, 70-75; mem, Arecibo Adv Bd, Nat Astron & Ionosphere Ctr, 75-78. Mem: Int Astron Union; Am Astron Soc; Royal Astron Soc; Int Sci Radio Union; Inst Elec & Electronic Engrs. Res: Physical studies of space molecule regions and of the planets and satellites. Mailing Add: Space Sci Div Naval Res Lab Washington DC 20375

MAYER, DAVID JONATHAN, b Mt Vernon, NY, July 18, 42; m 72. NEUROPHYSIOLOGY. Educ: City Univ New York, BA, 66; Univ Calif, Los Angeles, PhD(psychol), 71. Prof Exp: Asst prof, 72-75, ASSOC PROF PHYSIOL, MED COL VA, 75- Concurrent Pos: NIH fel, Brain Res Inst, Univ Calif, Los Angeles, 71-72. Mem: Soc Neurosci; Am Physiol Soc; Int Asn Study Pain. Res: Neurophysiology of pain and pain inhibitory systems; neuropharmacology of narcotic analgesics. Mailing Add: Dept of Physiol Med Col of Va Richmond VA 23298

MAYER, ERNEST, physics, mathematics, see 12th edition

MAYER, EUGENE STEPHEN, b Norwalk, Conn, June 5, 38; m 63; c 1. MEDICAL EDUCATION. Educ: Tufts Univ, BS, 60; Columbia Univ, MD, 64; Yale Univ, MPH, 71. Prof Exp: Physician, USPHS & US Peace Corps, Ankara, Turkey, 65-67 & Washington, DC, 67-68; DEP DIR & ASSOC PROF FAMILY MED & INTERNAL MED, AREA HEALTH EDUC CTR PROG, SCH MED, UNIV NC, CHAPEL HILL, 71- Concurrent Pos: Consult, Bur Health Manpower, 74- Mem: Asn Am Med Cols; Asn Teachers Prev Med; AMA. Res: Distribution of health manpower and the effect of medical education on this distribution. Mailing Add: 618 Wells Ct Chapel Hill NC 27514

MAYER, FLORENCE E, b Karuizawa, Japan, Sept 26, 23; US citizen. MEDICINE, PEDIATRICS. Educ: NCent Col, BA, 45; Northwestern Univ, Chicago, MD, 50. Prof Exp: Rotating intern, Cincinnati Gen Hosp, 50-51; resident pediat, Cincinnati Children's Hosp, 51-53; resident, Children's Med Ctr, Boston, 53-54; instr pediat, Med Ctr, NY Univ, 59-60, asst clin prof, 60-61; med officer, Nat Inst Child Health & Human Develop, 63-72; SR STAFF SCIENTIST, NAT HEART & LUNG INST, 72- Concurrent Pos: Fel cardiol, Children's Med Ctr, Boston, 53-54; NIH res fel, Hosp Sick Children, London, 57-58; Med Res Fund Australia res fel, Royal Alexandra Hosp Children, Sydney, 61-62. Mem: AAAS; Am Heart Asn; Am Acad Pediat. Res: Physiology of growth and development. Mailing Add: Nat Heart & Lung Inst Bethesda MD 20014

MAYER, FOSTER LEE, JR, b Fletcher, Okla, Nov 17, 42; m 62; c 2. TOXICOLOGY, AQUATIC ECOLOGY. Educ: Southwestern State Col, BS, 65; Utah State Univ, MS, 67, PhD(toxicol), 70. Prof Exp: Leader res sect, 70-74, CHIEF BIOLOGIST, FISH-PESTICIDE RES LAB, US FISH & WILDLIFE SERV, 74- Concurrent Pos: Res assoc sch forestry, fisheries & wildlife, Univ Mo-Columbia, 71- Honors & Awards: Spec Achievement Award, US Fish & Wildlife Serv, 73. Mem: Sigma Xi; Am Chem Soc; Am Fisheries Soc; Am Soc Testing & Mat; Soc Toxicol. Res: Toxicology of chemical contaminants in aquatic organisms, including biochemical and physiological aspects; formulation of mathematical models appropriate for prediction of contaminant effects in natural aquatic ecosystems. Mailing Add: Fish-Pesticide Res Lab Rte 1 Columbia MO 65201

MAYER, GEORGE PAT, b McAlester, Okla, Sept 26, 30; m 63; c 3. ENDOCRINOLOGY, PHYSIOLOGY. Educ: Okla State Univ, DVM, 54; Univ Pa, MSc, 68. Prof Exp: Pvt pract, Okla, 54-56; instr med, Okla State Univ, 56-58; pvt pract, Okla, 58-61; from asst instr to assoc prof med, Univ Pa, 62-73; PROF PHYSIOL, COL VET MED, OKLA STATE UNIV, 73- Concurrent Pos: USPHS fel, 62-64, res career develop award, 68-72. Mem: Am Vet Med Asn; NY Acad Sci; Endocrine Soc; Am Physiol Soc. Res: Calcium homeostasis; parathyroid physiology; calcitonin physiology; parturient paresis of cows. Mailing Add: Dept of Physiol Sci Okla State Univ Col of Vet Med Stillwater OK 74074

MAYER, GERALD DOUGLAS, b Crowley, La, Jan 2, 33; m 57; c 4. MICROBIOLOGY. Educ: Southwestern La Univ, BS, 58, MS, 60; Iowa State Univ, PhD(bact), 64. Prof Exp: Res assoc virol, Charles Pfizer & Co, 64-66; sect head, Dept Infectious Dis, 66-74, DEPT HEAD INFECTIOUS DIS RES, MERRELL-NAT LABS, RICHARDSON-MERRELL INC, 74- Mem: Am Soc Microbiol. Res: Interferon and interferon inducers; antiviral chemotherapy; virology. Mailing Add: Dept of Infectious Dis Merrell-Nat Labs 110 E Amity Rd Cincinnati OH 45215

MAYER, HAROLD M, b New York, NY, Mar 27, 16; m 52; c 2. URBAN GEOGRAPHY, URBAN PLANNING. Educ: Northwestern Univ, BS, 36; Wash Univ, MS, 37; Univ Chicago, PhD(geog), 43. Prof Exp: Zoning specialist, Chicago Land Use Surv, 40-41; res planner, Chicago Plan Comn, 41-43; geogr, US Off Strategic Serv, 43-45; res planner, Chicago Plan Comn, 45; chief div planning anal, Philadelphia City Planning Comn, 45-48; dir res, Chicago Plan Comn, 48-50; from asst prof to prof geog, Univ Chicago, 50-68; univ prof, Kent State Univ, 68-74; PROF GEOG & ASSOC DIR CTR GREAT LAKES STUDIES, UNIV WIS-MILWAUKEE, 74- Concurrent Pos: Consult, Chicago, Cleveland & Baltimore Depts City Planning, Man Dept Com & Indust, Ont Dept Treas & Econ, Wis Dept Resource Develop, Chicago Asn Com & Indust, United Transp Union, Bethlehem Steel Co & others, 50-; mem, Chicago Regional Port Dist Bd, 51-53; Fulbright prof, US Fulbright Comn, Univ Auckland, 61; comnr, Northeastern Ill Planning Comn, 66-68; mem maritime transp res bd, Nat Acad Sci-Nat Res Coun, 68-74, chmn ports & cargo syst comt, 69-; mem, Transp Res Forum. Honors & Awards: Content Award, Nat Coun Geog Educ, 70; Publ Award, Geog Soc Chicago, 70. Mem: AAAS; Asn Am Geogr; fel Am Geog Soc; Nat Coun Geog Educ; Am Soc Planning Officers. Res: Urban, metropolitan and regional geography; transportation geography, city and metropolitan planning; application of geographical concepts to planning; industrial and commercial location; economic development. Mailing Add: Dept of Geog Univ of Wis Milwaukee WI 53201

MAYER, HARRIS LOUIS, b New York, NY, Feb 15, 21; m 46; c 3. THEORETICAL PHYSICS. Educ: NY Univ, BA, 40; Columbia Univ, MS, 41; Univ Chicago, PhD(physics), 47. Prof Exp: With Div War Res, Columbia Univ, 41-46; group leader theoret physics, Los Alamos Sci Lab, 47-56; dept head, Aeronutronic Systs, Inc, 56-58; vpres, E H Plesset Assoc, Inc, 58-64; spec asst to vpres res, Inst Defense Anal, 64-68; group dir survivability, 68-71, MEM TECHNOL PLANNING STAFF, AEROSPACE CORP, 71- Concurrent Pos: Consult, Avco Mfg Co, 55; consult, Los Alamos Sci Lab, 56-; mem nuclear panel, Sci Adv Bd, US Air Force, 58-62 & Lawrence Radiation Lab, Livermore, 60; mem weapons effects bd, Defense Atomic Support Agency, 60; dir nuclear technol seminar study, Advan Res Projs Agency, 68; mem space task group, Aerospace Corp, 69; study utilization space transp syst, NASA, 71. Mem: Fel Am Phys Soc. Res: Atomic physics; statistical mechanics; optical properties of ionized materials; nuclear weapons; ballistic missile systems; strategic support systems and space systems; energy management and reactor waste disposal; historical projections to year 2000; future space applications. Mailing Add: Aerospace Corp 2350 El Segundo Blvd El Segundo CA 90245

MAYER, JEAN, b Paris, France, Feb 19, 20; nat US; m 42; c 5. PHYSIOLOGY, NUTRITION. Educ: Univ Paris, BLitt, 37, MSc, 39 & 40; Yale Univ, PhD(physiol chem), 48; Sorbonne, DSc(physiol), 50. Hon Degrees: AM, Harvard Univ, 65. Prof Exp: Demonstr physiol chem, Yale Univ, 46-48; mem nutrit div, Food & Agr Orgn, UN, 48-49; res assoc pharmacol, George Washington Univ, 49; from asst prof to assoc prof, 50-65, PROF NUTRIT, HARVARD UNIV, 65-, LECTR HIST PUB HEALTH, 68- Concurrent Pos: Consult, Spec Div, UN, 48; tech secy, Int Comt Calorie Requirements, Food & Agr Off & WHO, 50 & 57, tech secy, Comt Protein Requirements, 57; assoc ed, Nutrit Revs, 51-54; consult, Children's Hosp, Boston, 57-, Ghana Govt, 58 & Ivory Coast Govt, 59; Severinghouse lectr, Med Sch, Univ Ga, 58; nutrit ed, Postgrad Med, 59-; Phi Beta Kappa scholar, 68-69; mem, Ctr Pop Studies, 68-; spec consult to the President, 69-70; chmn, White House Conf Food, Nutrit & Health, 69; mem, President's Consumer Adv Coun, 70-; chmn nutrit div, White House Conf on Aging, 71; W O Atwater Mem lectr, Agr Res Serv, USDA, 71. Honors & Awards: Silver Medal, Int Physiol Cong, 56; Alvarenga Prize, Col Physicians Philadelphia, 68; Presidential Honor Citation, Am Asn Health, Phys Educ & Recreation, 72; Bradford Washburn Award, Boston Mus Sci, 75; Sarah L Poiley Mem Award, NY Acad Sci, 75. Mem: AAAS; Am Physiol Soc; Am Inst Nutrit; Am Fedn Clin Res; fel Am Acad Arts & Sci. Res: Regulation of food and water intake; obesity; general nutrition. Mailing Add: Harvard Univ Sch Pub Health 665 Huntington Ave Boston MA 02115

MAYER, JOERG WERNER PETER, b Munich, Ger, Aug 4, 29; nat US; m 55, 65; c 5. MATHEMATICS. Educ: Univ Giessen, dipl, 53, Dr rer nat, 54. Prof Exp: Lectr math, Univ Malaya, 54-57; from asst prof to assoc prof, Univ NMex, 57-68; chmn dept, George Mason Col, 68-70; CHMN DEPT MATH, LEBANON VALLEY COL, 70-, DIR COMPUT CTR, 74- Mem: Soc Indust & Appl Math; Asn Comput Mach; Am Math Soc; Math Asn Am. Res: Order theory; topology; summability theory. Mailing Add: Dept of Math Lebanon Valley Col Annville PA 17003

MAYER, JOSEPH, organic chemistry, chemical engineering, see 12th edition

MAYER, JOSEPH EDWARD, b New York, NY, Feb 5, 04; m 30; c 2. CHEMICAL PHYSICS. Educ: Calif Inst Technol, BS, 24; Univ Calif, PhD(phys chem), 27. Hon Degrees: ScD, Univ Brussels, 62. Prof Exp: Asst, Univ Calif, 27-28; Int fel, Univ Göttingen, 29-30; assoc chem, Johns Hopkins Univ, 30-37, assoc prof, 37-39; assoc prof, Columbia Univ, 39-45; prof, Univ Chicago, 45-56, Eisendrath prof, 56-60; prof, 60-72, chmn dept, 63-66, EMER PROF CHEM, UNIV CALIF, SAN DIEGO, 72- Concurrent Pos: Gibbs lectr, Am Math Soc, 55; Kennedy lectr, Washington Univ, 67. Consult & mem adv comt, Ballistics Res Lab, 42-60, mem adv comt, 73-; consult, Los Alamos Sci Lab, 46-49, US Air Force & Midway Lab, Korea, 51 & Rand Corp, 63- Chmn div phys chem, Nat Res Coun, 51-56; pres comn thermodyn & statist mech, Int Union Pure & Appl Physics, 52-56; vpres tables of constants, Int Union Pure & Appl Chem, 55-; mem sci comn, Solvay Int Inst, 60- Honors & Awards: G N Lewis Medal, Am Chem Soc, 58, Peter Debye Award, 67; Chandler Medal, Columbia Univ, 66; J G Kirkwood Medal, Yale Univ, 67. Mem: Nat Acad Sci (pres, Class I, 65); Am Chem Soc; fel Am Phys Soc (vpres, 72, pres, 74); Am Acad Arts & Sci; NY Acad Sci. Res: Statistical and quantum mechanics. Mailing Add: 2345 via Siena La Jolla CA 92037

MAYER, JULIAN RICHARD, b New York, NY, Feb 12, 29; m 49; c 4. ENVIRONMENTAL MANAGEMENT, ENVIRONMENTAL CHEMISTRY. Educ: Union Univ, NY, BS, 50; Columbia Univ, MA, 51; Yale Univ, PhD(chem), 55. Prof Exp: Res assoc, Sterling-Winthrop Res Inst, Sterling Drug Co, 54-60, assoc mem, 60-61, group leader, 61-62; asst prog dir, NSF, 62-63; asst dir, Atmospheric Sci Res Ctr, State Univ NY, 63-64; staff assoc, NSF, 64-70; DIR LAKE ERIE ENVIRON STUDIES PROG, STATE UNIV NY COL FREDONIA, 70- Concurrent Pos: Spec consult, NSF, 63; consult, Environ Protection Agency, 72- & Union Carbide Corp, 75- Mem: AAAS; Am Chem Soc. Res: Science administration; science policy planning; environmental problems; water quality research and management; chemical and physical limnology. Mailing Add: State Univ of NY Fredonia NY 14063

MAYER, KLAUS, b May 21, 24; US citizen; m 50; c 2. INTERNAL MEDICINE, HEMATOLOGY. Educ: Queens Col, BS, 45; Univ Zurich & Groningen, MD, 50; Am Bd Internal Med, cert, 64. Prof Exp: Intern, Hosp St Raphael, New Haven, Conn, 50-51; staff mem, Dept Med, Brookhaven Nat Lab, 51-52; resident, Mem Hosp Cancer & Allied Dis, 52-55; res assoc cancer anemia, Sloan-Kettering Inst, 58-59, asst, 59-60; from instr to asst prof, 58-68, CLIN ASSOC PROF MED, MED COL, CORNELL UNIV, 68-; ASSOC, SLOANKETTERING INST, 60- Concurrent Pos: Spec fel med, Mem Hosp, 55-56; Damon Runyon fel, Sloan-Kettering Inst, 55-58; clin asst med, Mem Hosp Cancer & Allied Dis, 56-60, from asst attend physician to assoc attend physician, 60-72, dir Blood Bank & Serol Lab, 66-, dir, Hemat Lab, 71-, attend physician, 72-; attend hematologist, Hosp Spec Surg, 57-, res hematologist, 58-62, dir blood bank, 58-, assoc scientist, 62-63, sr scientist, 63-; physician to outpatients, New York Hosp, 58-68, assoc attend physician, 68-; from asst vis physician to assoc vis physician, James Ewing Hosp, 59-68; asst vis physician, Bellevue Hosp, 62-68; res collabr, Brookhaven Nat Lab, 65-66; pres, Am Asn Blood Bank, 73-74; prin, ad hoc comt to form Am Blood Comn, 74-75, secy-treas, 75- Mem: Am Soc Nuclear Med; Am Soc Hematol; Harvey Soc; fel Am Col Physicians; Int Soc Hemat. Res: Application of radioisotopic technique to hematology and transfusion therapy; quantitation of reticuloendothelial function. Mailing Add: Mem Sloan-Kettering Cancer Ctr Box 45 1275 York Ave New York NY 10021

MAYER, KURT LUDWIG, physics, see 12th edition

MAYER, MANFRED MARTIN, b Frankfurt, Ger, June 15, 16; US citizen; m 42; c 4. IMMUNOLOGY, BIOCHEMISTRY. Educ: City Col New York, BS, 38; Columbia Univ, PhD(biochem), 46. Hon Degrees: MD, Univ Mainz, 69. Prof Exp: From asst prof to assoc prof bacteriol, Sch Hyg & Pub Health, 46-59, assoc prof microbiol, Sch Med, 59-60, PROF MICROBIOL, SCH MED, JOHNS HOPKINS UNIV, 60- Concurrent Pos: Consult, USPHS, NSF, Off Naval Res & Plum Island Animal Dis Lab, USDA; assoc ed, Biol Abstracts, J Immunol & Anal Biochem; adv ed, Immunochem. Honors & Awards: Kimble Award Methodology, 53; Selman Waksman Lectr Award, 57; Karl Landsteiner Award, Am Asn Blood Banks, 74. Mem: Fel AAAS; Am Soc Biol Chem; Am Asn Immunol; Soc Exp Biol & Med; Biochem Soc. Res: Complement, an immunopathologic mediator system; properdin system; lymphokines. Mailing Add: Dept of Microbiol Johns Hopkins Univ Sch of Med Baltimore MD 21205

MAYER, MARION SIDNEY, b New Orleans, La, July 25, 35. ENTOMOLOGY, BIOCHEMISTRY. Educ: La State Univ, BS, 57; Tex A&M Univ, MS, 61, PhD(entom), 63. Prof Exp: RES ENTOMOLOGIST, AGR RES SERV, USDA, 63- Mem: AAAS; Entom Soc Am. Res: Insect attractants, isolation and behavioral characteristics leading to host or mate location; electro-physiological studies to demonstrate details of nervous activity leading to host or mate locations and thresholds. Mailing Add: USDA Agr Res Serv PO Box 14565 Gainesville FL 32604

MAYER, MEINHARD EDWIN, b Seletin, USSR, Mar 18, 29; m 54; c 2. MATHEMATICAL PHYSICS. Educ: Bucharest Polytech Inst, Dipl Ing, 51; Parhon Univ, PhD, 57. Prof Exp: From instr to assoc prof math physics, Parhon Univ, 49-61; sr res worker theoret physics, Joint Inst Nuclear Res, USSR, 57-58; vis prof theoret physics, Univ Vienna, 61-62 & Imp Col, Univ London, 62; vis physicist, Europ Orgn Nuclear Res, Switz, 62; vis assoc prof physics, Brandeis Univ, 62-64; assoc prof theoret physics, Ind Univ, 64-66; PROF MATH & PHYSICS, UNIV CALIF, IRVINE, 66- Concurrent Pos: Asst prof, Bucharest Polytech Inst, 50-52; sr res worker, Inst Atomic Physics, Acad Rumania, 51-58; vis assoc physicist, Brookhaven Nat Lab, NY, 63-; vis prof, Inst Advan Sci Studies, Bures-sur-Yvette, France, 70-71 & Tel-Aviv Univ, 71. Mem: Am Math Soc; fel Am Phys Soc. Res: Quantum field theory and statistical mechanics; differential-geometric approach to gauge theory; relativistic statistical mechanics; relativistic electron beams. Mailing Add: Dept of Math Univ of Calif Irvine CA 92664

MAYER, RAYMOND PARM, b Alma, Mich, July 25, 33; m 55; c 3. ORGANIC CHEMISTRY. Educ: Alma Col, BA, 55; Univ Mich, PhD(chem), 58. Prof Exp: Res chemist, Mich Chem Corp, 58-60 & Ludington Div, Dow Chem Co, 60-69; PROF SCI & MATH, WEST SHORE COMMUNITY COL, 69- Honors & Awards: Azbe Award, 64. Mem: Am Chem Soc. Res: Reaction mechanisms; fine organic chemicals; brine chemistry; sintering; lime. Mailing Add: 930 N Gaylord Ludington MI 49431

MAYER, RICHARD F, b Olean, NY, June 2, 29; m 59; c 5. NEUROLOGY. Educ: St Bonaventure Col, BS, 50; Univ Buffalo, MD, 54. Prof Exp: Intern med, Boston City Hosp, 54-55; resident neurol, Mass Gen Hosp, 56-57, resident neuropath, 58; res asst, Inst Neurol, Univ London, 57-58; instr neurol, Harvard Med Sch, 61-65, assoc, 65-66; assoc prof, 66-68, PROF NEUROL, SCH MED, UNIV MD, BALTIMORE, 68-, DIR NEUROMUSCULAR CLIN & EMG LAB, UNIV HOSP, 69- Concurrent Pos: Fel neurol, Mayo Found, Univ Minn, 55-56; NIH res fel, Harvard Med Sch, 60-61; NIH res grant, Boston City Hosp, 60-64; Nat Multiple Sclerosis Soc res grant, 66- Mem: AAAS; Am Neurol Asn; Am Electroencephalog Soc; Am Acad Neurol; Soc Neurosci. Res: Clinical neurophysiology; clinical and experimental animal studies of motor dysfunction; nerve and reflex activity in man; myasthenia gravis-neuromuscular transmission and ultra structure. Mailing Add: Dept of Neurol Univ of Md Med Sch Baltimore MD 21201

MAYER, RICHARD THOMAS, b Pensacola, Fla, May 11, 45; m 66; c 2. TOXICOLOGY. Educ: Univ Ga, BS, 67, PhD(entom), 70. Prof Exp: Fel entom, Univ Ga, 70-71; RES ENTOMOLOGIST, VET TOXICOL ENTOM RES LAB, AGR RES SERV, USDA, 71- Mem: Sigma Xi; Am Chem Soc; Am Entom Soc. Res: Insecticide and hormone metabolism by insects; isolation and characterization of insect metabolic systems; enzyme assay development. Mailing Add: Vet Toxicol & Entom Res Lab USDA Agr Res Serv PO Drawer GE College Station TX 77840

MAYER, STANLEY WALLACE, b New York, NY, Mar 29, 16; m 45; c 2. PHYSICAL CHEMISTRY. Educ: City Col New York, BS, 38; Univ Calif, Los Angeles, PhD(chem), 53. Prof Exp: Res scientist, NY Water Dept, 39-41 & US War Dept, 41-43; sr sci staff, Columbia Univ, 43-46, Oak Ridge Nat Lab, 46-48, US Naval Radio Lab, 48-53, US Radioisotope Res Univ, 53-56 & NAm Aviation Inc, 56-61; SR SCIENTIST, PHYS CHEM DEPT, CHEM & PHYSICS LAB, AEROSPACE CORP, 61- Mem: Am Chem Soc; Am Nuclear Soc; Am Inst Aeronaut & Astronaut; Am Phys Soc; Combustion Inst. Res: High temperature reactions and propulsion; nuclear power; properties of propellants; high-temperature materials; biomedical physics; research with electronic computers; lasers. Mailing Add: Aerospace Corp PO Box 95085 Los Angeles CA 90045

MAYER, STEVEN EDWARD, b Frankfurt am Main, Ger, Feb 11, 29; nat US; m 51; c 2. PHARMACOLOGY, BIOCHEMISTRY. Educ: Univ Chicago, BA, 47, BS, 49; Univ Ill, MS, 52, PhD(pharmacol), 54. Prof Exp: Sr asst scientist, Lab Chem Pharmacol, Nat Heart Inst, 54-56; from asst prof to prof pharmacol, Emory Univ, 57-69; PROF PHARMACOL & CHIEF DIV, UNIV CALIF, SAN DIEGO, 69- Concurrent Pos: Fel pharmacol, Wash Univ, 56-57; vis scholar, Univ Wash, 65; mem pharmacol study sect, NIH, 65-69, mem neurosci res training B comt; ed, Molecular Pharmacol, 71-74; chmn, Gordon Res Conf Heart Muscle, 72; mem res comt, Am Heart Asn, 73-; mem, USUSSR Working Group Myocardial Metab, 73-; A J Carlson Lectr, Univ Chicago, 75. Honors & Awards: John J Abel Award, Am Soc Pharmacol & Exp Therapeut, 63. Mem: AAAS; Am Soc Pharmacol & Exp Therapeut (asst ed, Jour, 61-64); Am Physiol Soc. Res: Mechanisms of drug and hormone action in metabolic control. Mailing Add: Div of Pharmacol Dept of Med M-013 Univ of Calif at San Diego La Jolla CA 92093

MAYER, THEODORE JACK, b Bridgewater, SDak, Feb 13, 33; m 59; c 4. PETROLEUM CHEMISTRY. Educ: Univ SDak, BA, 55; Pa State Univ, MS, 61; Carnegie Inst Technol, PhD(phys chem), 63. Prof Exp: Res asst petrol chem, Petrol Ref Lab, Pa State Univ, 55-59; RES CHEMIST, RES & DEVELOP DIV, SUN OIL CO, 63- Mem: Am Chem Soc; Soc Appl Spectros. Res: Composition of petroleum; analysis of petroleum fractions by instrumental methods; laboratory automation. Mailing Add: 410 Bickmore Dr Wallingford PA 19086

MAYER, THOMAS C, b Pittsburgh, Pa, Nov 30, 31; m 58. DEVELOPMENTAL BIOLOGY. Educ: Univ Tenn, AB, 53; Johns Hopkins Univ, MA, 57; La State Univ, PhD(embryol), 62. Prof Exp: Asst prof biol, Greensboro Col, 57-60; from asst prof to assoc prof, 62-67, PROF BIOL, RIDER COL, 67- Concurrent Pos: NSF res grants, 63-72. Mem: AAAS; Soc Develop Biol; Am Soc Zool. Res: Embryogenesis of spotting patterns in mice. Mailing Add: Dept of Biol Rider Col Trenton NJ 08602

MAYER, VERNON WILLIAM, JR, b Newark, NJ, Mar 29, 39; m 65; c 2. MICROBIOLOGY, GENETICS. Educ: Univ Md, BS, 63, MS, 65, PhD(microbiol), 67. Prof Exp: Nat Res Coun-Nat Acad Sci res assoc, 67-68, RES MICROBIOLOGIST, FOOD & DRUG ADMIN, 68- Mem: AAAS; Am Soc Microbiol; Genetics Soc Am. Res: Yeast genetics; chemical mutagenesis. Mailing Add: BF 156 Genetic Toxicol Br Div Toxicol Food & Drug Admin Washington DC 20204

MAYER, VICTOR JAMES, b Mayville, Wis, Mar 25, 33; m 65; c 2. EARTH SCIENCES, SCIENCE EDUCATION. Educ: Univ Wis, BS, 56; Univ Colo, MS, 60, PhD(sci educ), 66. Prof Exp: Pub sch teacher, Colo, 60-62; asst prof earth sci, State Univ NY Col Oneonta, 65-67; from asst prof to assoc prof, 67-75, PROF GEOL & SCI EDUC, OHIO STATE UNIV, 75- Concurrent Pos: Consult, NY State Dept Educ, 66-67, Pedag Inst Caracas, 71 & UNESCO, 75- Mem: Fel AAAS; Nat Sci Teachers Asn; Am Educ Res Asn; Nat Asn Res Sci Teaching; Nat Asn Geol Teachers. Res: Teacher preparation and behavior; curriculum evaluation. Mailing Add: 111 W Dominion Columbus OH 43214

MAYER, WALTER GEORG, b Silberbach, Czech, Mar 13, 27; nat US; m 59. PHYSICS. Educ: Hope Col, AB, 53; Mich State Univ, MS, 55, PhD(physics), 58. Prof Exp: Physicist high temperature res, Siemens Res Lab, Ger, 58-59; res asst prof ultrasonics, Dept Physics & Astron, Mich State Univ, 59-65; from asst prof to assoc prof, 65-72, PROF PHYSICS, GEORGETOWN UNIV, 72- Concurrent Pos: Assoc ed, Inst Elec & Electronics Engrs Trans Sonics & Ultrasonics, 72- & J Acoust Soc Am, 74- Mem: Fel Acoust Soc Am. Res: Ultrasonics, particularly measurements of wave characteristics by optical methods; application of ultrasonics to solid and liquid state. Mailing Add: Dept of Physics Georgetown Univ Washington DC 20057

MAYER, WARREN CLIFFORD, b Staten Island, NY, Dec 13, 33; m 59; c 2. FOOD SCIENCE, ENGINEERING. Educ: State Univ NY Col Forestry, Syracuse Univ, BS, 56, MS, 58. Prof Exp: Paper chemist, Cel-Fibe Div, 57-62, group leader res & develop, 62-66, res mgr, 66-71, mgr nonwoven res, Chicopee Mfg Co, 71-75, ASSOC DIR RES, DEVRO INC DIV, JOHNSON & JOHNSON, 75- Mem: Tech Asn Pulp & Paper Indust. Res: Equipment engineering in food processing. Mailing Add: 569 Morningside Dr Bridgewater NJ 08807

MAYER, WILLIAM DIXON, b Beaver Falls, Pa, Oct 5, 28; m 53; c 3. MEDICINE. Educ: Colgate Univ, AB, 51; Univ Rochester, MD, 57. Prof Exp: Intern path, Sch Med, Univ Rochester, 57-58, resident, 58-59, instr, 58-61; from asst prof to assoc prof path, 61-67, from asst dean to assoc dean, Sch Med, 61-67, dean, Sch Med & dir Med Ctr, 67-74, PROF PATH, UNIV MO-COLUMBIA, 67-, DIR, HEALTH SERV RES CTR, 75- Concurrent Pos: Buswell fel, Univ Rochester, 59-61; Markle scholar, 62-67; assoc dir div regional med progs, NIH, 66-67. Mem: AAAS; Am Soc Exp Path; Col Am Path; Asn Am Med Cols; AMA. Res: Medical education; surgical pathology; cellular aspects of antibody production; health services and health care technology. Mailing Add: Health Serv Res Ctr Univ of Mo 206 Clark Hall Columbia MO 65201

MAYER, WILLIAM JOHN, b Detroit, Mich, Mar 29, 21; m 51; c 6. PHYSICAL CHEMISTRY. Educ: Wayne State Univ, BS, 44, PhD(chem), 50. Prof Exp: Anal control chemist, Gelatin Prod Corp, 44-45; lab asst phys chem, Wayne State Univ, 48-50; res chemist, Argonne Nat Lab, 50-56; SR RES SCIENTIST, RES LABS, GEN MOTORS CORP, 56- Honors & Awards: Arch T Colwell Award, Soc Automotive Engrs, 67. Mem: Am Chem Soc; Soc Automotive Engrs. Res: Radiochemistry and isotopes; chemistry of surfaces; combustion chemistry; radiometric methods applied to automotive engines. Mailing Add: Chem Dept Res Labs Gen Motors Corp Warren MI 48090

MAYER, WILLIAM JOSEPH, b Springfield, Ohio, Sept 30, 39; m 71. INFORMATION SCIENCE. Educ: Xavier Univ, Ohio, BS, 61; Univ Mich, MS, 63, PhD(pharmaceut chem), 65. Prof Exp: Patent chemist, Res Labs, Parke, Davis & Co, 65-71; res info assoc, Olin Corp, 71-74; MGR TECH INFO SERV, JAMES FORD BELL TECH CTR, GEN MILLS, INC, 74- Mem: Am Chem Soc. Res: Synthesis of organic medicinals; patent development; research information; technology assessment; synthetic use of the Wittig and Mannich reactions. Mailing Add: 2024 Crosby Rd Wayzata MN 55391

MAYER, WILLIAM VERNON, b Vancouver, BC, Mar 25, 20; nat US; m 41; c 2. ZOOLOGY. Educ: Univ Calif, AB, 41; Stanford Univ, PhD(biol), 49. Prof Exp: Res observer, Francis Simes Hastings Natural Hist Reserve, 41; asst zool, anat & physiol, Mont State Col, 41-42; asst comp anat, Stanford Univ, 46-48, instr, 48; from instr to assoc prof anat & zool, Univ Southern Calif, 48-57, actg head dept biol, 56-57; prof biol & chmn dept, Wayne State Univ, 57-67, actg assoc dean col liberal arts, 60-61, assoc dean, 62-65; PROF BIOL, UNIV COLO, BOULDER, 67- Concurrent Pos: Writer biol sci curric study, Am Inst Biol Sci, 60-, assoc dir, 63-64, dir, 65, chmn test construct comt, 60-66; mem bd dirs, Kresge Libr Assocs; mem res adv coun & res adv comt, Mich Cancer Found; assoc mem region 8 selection comt, Woodrow Wilson Fel Found, 61-63; assoc dir biol sci curric study, Univ Colo, Boulder, 63-65, dir, 65-; consult, Nat Sci Develop Bd Philippines, 64 & DC Health & Co; mem exec comt, Mich Comn Col Accreditation, 64-65; mem coun, Assoc Midwestern Univs, 65-; pres, Educ Progs Improv Corp, 70; mem panel eval & testing, Comn Undergrad Educ Biol Sci, NSF, mem tundra biome adv panel, 71. Mem: Fel AAAS; Am Soc Mammal; Am Soc Zool; Soc Syst Zool; Nat Asn Biol Teachers (pres elect, 65, pres, 66-67). Res: Comparative vertebrate anatomy; mammalian ecology; arctic biology; hibernation; temperature phenomena. Mailing Add: Biol Sci Curric Study Univ of Colo PO Box 930 Boulder CO 80302

MAYERI, EARL MELCHIOR, b Berkeley, Calif, Dec 10, 40; m 68; c 1. NEUROPHYSIOLOGY. Educ: Univ Calif, Berkeley, BA, 63, PhD(biophys), 69. Prof Exp: ASST PROF PHYSIOL, UNIV CALIF, SAN FRANCISCO, 71- Concurrent Pos: USPHS fel neurophysiol, Med Sch, NY Univ, 59-71; fel, Pub Health Res Inst, New York, 69-71. Mem: AAAS; Am Physiol Soc; Soc Neurosci. Res: Invertebrate neurophysiology and behavior. Mailing Add: Dept of Physiol Univ of Calif San Francisco CA 94143

MAYERNIK, JOHN JOSEPH, b Manville, NJ, July 6, 16; m 46; c 1. MICROBIOLOGY. Educ: Rutgers Univ, BSc, 39, PhD(soil chem), 44; Univ Vt, MS, 41. Prof Exp: Chief microbiol & sterile prod control lab, 46-75, MGR MICROBIOL SERV, MERCK & CO, INC, 75- Concurrent Pos: Mem, US Pharmacopoeia Adv Panel on Biol Indicators. Mem: Am Soc Microbiol; Am Chem Soc; Asn Off Anat; fel

Asn Off Anal Chemists. Res: Microbiological assays of antibiotics, vitamins and amino acids; quality control of pharmaceutical products; evaluation of methods of sterilization and product sterility; evaluation of preservatives and disinfectants; bacterial monitoring of electron irradiation. Mailing Add: Qual Control Merck & Co Inc Rahway NJ 07065

MAYER-OAKES, WILLIAM JAMES, b Oskaloosa, Iowa, Oct 15, 23; m 47; c 3. ANTHROPOLOGY, ARCHAEOLOGY. Educ: Univ Chicago, MA, 49, PhD(anthrop), 54. Prof Exp: Field archaeologist, Carnegie Mus, 50-56; assoc prof anthrop, Univ Toronto, 56-59; prof anthrop & dir mus, Univ Okla, 59-62; prof & head dept, Univ Man, 62-71; PROF ANTHROP & CHMN DEPT, TEX TECH UNIV, 71- Concurrent Pos: Consult, Nat Mus Can, 58-68 & Can Coun, 62-71; mem quaternary comt, Nat Res Coun Can, 66-70; Can Coun sr fel, Ft Burgwin, NMex, 68-69; lectr, Archaeol Inst Am. Mem: Fel Am Anthrop Asn; Soc Am Archaeol (treas, 54-57); Am Quaternary Asn. Res: Archaeology of eastern and northern North America, Valley of Mexico and highland Ecuador; archaeological method and theory. Mailing Add: Dept of Anthrop Tex Tech Univ Lubbock TX 79409

MAYERS, GEORGE LOUIS, b New York, NY, Feb 22, 38; m 66; c 2. BIO-ORGANIC CHEMISTRY, IMMUNOCHEMISTRY. Educ: City Col New York, BS, 60, MA, 64; City Univ New York, PhD(org chem), 67. Prof Exp: Fel peptide chem, St John's Univ, NY, 67-70; SR CANCER RES SCIENTIST, ROSWELL PARK MEM INST, 70- Concurrent Pos: Asst res prof, Roswell Park Div, State Univ NY Buffalo, 75- Mem: Am Chem Soc; The Chem Soc. Res: Structure of the antibody site and its relationship to antigen or hapten; synthesis and physical structure of peptides and polypeptides and the relationship to proteins. Mailing Add: Biochem Res Roswell Park Mem Inst 666 Elm St Buffalo NY 14203

MAYERS, MARVIN KEENE, b Canton, Ohio, Oct 25, 27; m 52; c 2. APPLIED ANTHROPOLOGY. Educ: Wheaton Col, BA, 49; Fuller Theol Sem, BD, 52; Univ Chicago, MA, 58, PhD(social anthrop), 60. Prof Exp: Field researcher ling, Summer Inst Ling, 52-65; from assoc prof to prof anthrop, Wheaton Col, 73-74; ADJ PROF, UNIV TEX, ARLINGTON, 74-; PROF, SUMMER INST LING, 74- Concurrent Pos: Vis assoc prof ling, Univ Wash, summers 58-67. Mem: Am Anthrop Asn; Ling Soc Am. Res: Latin America; sociolinguistics; social anthropology; sociolinguistic research in United States and Latin America. Mailing Add: 7500 W Camp Wisdom Rd Dallas TX 75211

MAYERS, RICHARD RALPH, b West Brownsville, Pa, July 6, 25; m 49; c 2. NUCLEAR PHYSICS. Educ: Dartmouth Col, AB, 47; Wesleyan Univ, MA, 57. Prof Exp: Instr physics, Hood Col, 47-49; physicist, Nat Bur Standards, 49-50 & Glenn L Martin Co, 55-56; asst prof physics, Colby Col, 56-61, actg chmn dept, 57-59; from assoc prof to prof & chmn dept, Defiance Col, 61-74; RES, SURFACE COMBUSTION CO, 74- Concurrent Pos: Vis prof, Univ Surrey, 67-68; vis lectr, Tex A&M Univ, 68. Mem: AAAS; Am Phys Soc; Am Phys Soc. Res: Neutron activation analysis and gamma ray spectroscopy; industrial combustion systems. Mailing Add: 415 Monroe St Delta OH 43515

MAYERSON, HYMEN SAMUEL, b Providence, RI, Sept 10, 00; m 30; c 2. PHYSIOLOGY. Educ: Brown Univ, AB, 22; Yale Univ, PhD(physiol), 25. Hon Degrees: DSc, Brown Univ, 62. Prof Exp: Asst biol, Brown Univ, 21-22; asst physiol, Yale Univ, 22-25, instr, Sch Med, 25-26; from instr to prof & chmn dept, 26-65, EMER PROF PHYSIOL, SCH MED, TULANE UNIV, 65- Concurrent Pos: Consult, Touro Infirmary, 59-65, assoc dir prof serv & educ, 65-75; consult, Vet Hosp, 53-65; mem comt shock, Nat Res Coun, 54-61, rep biol & agr rev comt, Int Exchange Persons, 55-57; mem, Nat Bd Med Examrs, 56-60, chmn physiol test comt, 58-60; mem US nat comt, Int Union Physiol Sci, 61-65; fel panel, NSF Grad Fel Prog, Nat Acad Sci, 63-65; pres, Fedn Am Socs Exp Biol, 63. Mem: Am Physiol Soc (pres, 62); Soc Exp Biol & Med; Am Heart Asn; hon mem Int Soc Lymphology. Res: Cardiovascular effects of posture; blood volume changes in health and disease; capillary permeability to large molecules; lymph and lymphatics. Mailing Add: 1140 Seventh St New Orleans LA 70115

MAYES, BILLY WOODS, II, b Port Arthur, Tex, Feb 6, 41; m 60; c 1. PHYSICS. Educ: Univ Houston, BS, 63, MS, 65; Mass Inst Technol, PhD(physics), 69. Prof Exp: Fel, 68-69, asst prof, 69-73, ASSOC PROF PHYSICS, UNIV HOUSTON, 73- Mem: Am Phys Soc. Res: Experimental pion nucleus cross sections. Mailing Add: Dept of Physics Univ of Houston Houston TX 77004

MAYES, JARY S, b Walters, Okla, Oct 19, 38; m 58; c 3. BIOCHEMISTRY, HUMAN GENETICS. Educ: Okla State Univ, BS, 60; Mich State Univ, PhD(biochem), 65. Prof Exp: Asst prof biochem, 67-70, ASST PROF RES PEDIAT & ASSOC PROF BIOCHEM, HEALTH SCI CTR, UNIV OKLA, 70- Concurrent Pos: Fel, State Univ NY Buffalo, 65-67. Mem: AAAS; Am Chem Soc; Am Soc Human Genetics. Res: Inborn errors of metabolism; galactose metabolism; regulation of enzyme activity. Mailing Add: Dept of Biochem Univ Okla Health Sci Ctr Oklahoma City OK 73190

MAYES, MCKINLEY, b Oxford, NC, Oct 7, 30; m 59; c 1. AGRONOMY. Educ: NC Agr & Tech Col, BS, 53, MS, 56; Rutgers Univ, PhD(agron), 59. Prof Exp: PROF AGRON, SOUTHERN UNIV, BATON ROUGE, 59-, DEAN & COORDR CRS RES PROGS, 74- Mem: Am Soc Agron; Soil Conserv Soc Am. Res: Plant breeding, especially field corn and sweet corn improvement. Mailing Add: Box 9270 Southern Univ Baton Rouge LA 70813

MAYES, TERRILL W, b Evansville, Ind, Sept 4, 41; m 59; c 3. PLASMA PHYSICS, ATOMIC SPECTROSCOPY. Educ: Western Ky Univ, BS, 63; Vanderbilt Univ, MA, 65, PhD(physics), 67. Prof Exp: Asst prof, 67-74, ASSOC PROF PHYSICS, UNIV NC, CHARLOTTE, 74- Mem: Am Phys Soc. Res: Radiation and atomic physics; electricity and magnetism; classical mechanics. Mailing Add: Dept of Physics Univ of NC Charlotte NC 28205

MAYES, WILLIAM GLENN, chemistry, see 12th edition

MAYEUX, JERRY VINCENT, b Mamou, La, Apr 22, 37; div. MICROBIOLOGY, RESEARCH ADMINISTRATION. Educ: La State Univ, Baton Rouge, BS, 60, MS, 61; Ore State Univ, PhD(microbiol), 64. Prof Exp: Nat Acad Sci-Nat Res Coun res assoc exobiol, NASA Ames Res Ctr, 65-66; asst prof microbiol, Colo State Univ, 66-70; sr res scientist, Manned Exp & Life Sci Dept, Martin Marietta Corp, 70-72, chief, Life Sci, 72-74; dir, Res & Develop, Ferma Gro Corp, 74-75; LECTR MICROBIOL, BUENA VISTA COL, STORM LAKE, IOWA, 75- Concurrent Pos: Asst prof range sci, Colo State Univ, 69-70, affil prof microbiol, 70-76, col eng, 71; consult, NASA Life Sci Shuttle Planning Panel, 74-; consult, Martin Marietta Aerospace, 74-75. Mem: Soc Indust Microbiol; AAAS; Am Soc Microbiol; Am Inst Biol Sci. Res: Microbial ecology; soil and water pollution; microbial interaction; aerospace biology; waste reutilization; biological control; plant growth/disease and animal growth/disease. Mailing Add: 601 Division Alta IA 51002

MAYEWSKI, PAUL ANDREW, b Edinburgh, Scotland, July 5, 46; US citizen; m 69. GLACIAL GEOLOGY, GEOMORPHOLOGY. Educ: State Univ NY, Buffalo, BA, 68; Ohio State Univ, PhD(geol), 73. Prof Exp: Res assoc geol, Inst Polar Studies, Ohio State Univ, 68-73; fel, Inst Quaternary Studies, Univ Maine, Orono, 73-75; ASST PROF GEOL, UNIV NH, 75- Concurrent Pos: Mem res team, Climate-Long Range Invest Mapping & Prediction, 73-; panel mem, W Antarctic Ice Sheet Proj, 75- Honors & Awards: Antarctic Serv Medal, US Govt, 74. Mem: Glaciol Soc. Res: Reconstruction of former glacial events and glaciologic conditions in the Transantarctic Mountains, Antarctica; reconstruction and synthesis of the late Wisconsin glacial history of North America; former glacier flowline study in the Cocheco River drainage, New Hampshire; study of Antarctic weathering and mass wasting phenomena. Mailing Add: Dept of Earth Sci James Hall Univ of NH Durham NH 03824

MAYFIELD, DARWIN LYELL, b Somerset, Ky, Feb 22, 20; m 45; c 2. ORGANIC CHEMISTRY. Educ: Bowling Green State Univ, AB & BS, 41; Univ Chicago, MS, 44; Univ Wis, PhD(org chem), 50. Prof Exp: Res chemist, Nat Defense Res Comt, 42-43, Off Sci Res & Develop, 43-45 & Rubber Res Bd, 45-47; asst, Univ Wis, 47-50; from asst prof to assoc prof chem, Univ Idaho, 50-56; from asst prof to assoc prof, 56-62, chmn dept, 64-66, PROF CHEM, CALIF STATE UNIV, LONG BEACH, 62-, DIR RES, 67- Concurrent Pos: Fulbright lectr, Kasetsart Univ, Bangkok, 55-56 & Ain Shams Univ, Cairo, 66-67; NIH res fel, Nat Sci Res Ctr, France, 62-63. Mem: AAAS; Am Chem Soc. Res: Chemistry of plant hormones responsible for floral initiation; research administration and federal relations. Mailing Add: Off of Grad Studies & Res Calif State Univ Long Beach CA 90840

MAYFIELD, EARLE BYRON, b Oklahoma City, Okla, Jan 31, 23; m 52; c 7. SPACE PHYSICS. Educ: Univ Calif, Los Angeles, BA, 50; Univ Utah, MA, 54, PhD(physics), 59. Prof Exp: Physicist, Res Dept, US Naval Ord Test Sta, Calif, 50-59; mem tech staff, Phys Res Lab, Space Tech Labs, Inc, Thompson-Ramo-Wooldridge, Inc, 59-60; MEM TECH STAFF, AEROSPACE CORP, 60- Concurrent Pos: Asst, Univ Utah, 54-55. Mem: Am Phys Soc; Optical Soc Am; Am Geophys Union; Am Astron Soc; Int Astron Union. Res: Plasma and solar physics. Mailing Add: 5536 Michelle Dr Torrance CA 90503

MAYFIELD, ERNEST DURWARD, JR, b Medina, Tex, July 10, 39; m 61; c 2. CLINICAL BIOCHEMISTRY, PATHOLOGY. Educ: Southwest Tex State Univ, BS, 61; Univ Ill, Urbana, MS, 64, PhD(nutrit biochem), 66. Prof Exp: Res assoc pharm, 65-67, res instr pharm & path, 67-69, asst prof, 69-70, ASST PROF BIOCHEM & PATH, BAYLOR COL MED, 70-; CLIN BIOCHEMIST, ST LUKE'S EPISCOPAL HOSP, 70- Concurrent Pos: Clin biochem consult, Jack Abbott Clin Labs, 71- Mem: Electron Micros Soc Am; Am Asn Cancer Res; Am Soc Cell Biol; Am Heart Asn. Res: Nucleic acid metaboli in tissue hypertrophy. Mailing Add: St Luke's Episcopal Hosp Tex Med Ctr Houston TX 77025

MAYFIELD, HAROLD FORD, b Minneapolis, Minn, Mar 25, 11; m 36; c 4. ORNITHOLOGY. Educ: Shurtleff Col, BS, 33; Univ Ill, MA, 34. Hon Degrees: DSc, Occidental Col, 68 & Bowling Green State Univ, 75. Prof Exp: Secy, Wilson Ornith Soc, 48-52, vpres, 53-54 & 58-59, pres, 60-61; secy, Am Ornith Union, 53-58, vpres, 64-66, pres, 66-68; vpres, Cooper Ornith Soc, 73, pres, 74-76. Mem: Fel AAAS; Cooper Ornith Soc; Wilson Ornith Soc; fel Am Ornithologists Union. Honors & Awards: Brewster Mem Award for work on birds of Western Hemisphere, Am Ornithologists Union, 61. Res: Bird reproduction and mortality, social parasitism and ecology. Mailing Add: 9235 River Rd Waterville OH 43566

MAYFIELD, HAROLD GORDON, b Birmingham, Ala, Mar 31, 43; m 66; c 1. INORGANIC CHEMISTRY. Educ: Birmingham-Southern Col, BS, 66; Univ Tenn, PhD(inorg chem), 70. Prof Exp: ASST PROF CHEM, GA SOUTHERN COL, 70- Mem: Sigma Xi; Am Chem Soc. Res: Role of weakly basic anions in coordination complexes. Mailing Add: Dept of Chem Ga Southern Col Statesboro GA 30458

MAYFIELD, JOHN EMORY, b Thomasville, NC, Aug 30, 37; m 60; c 2. MYCOLOGY. Educ: Livingstone Col, BS, 59; State Univ NY Buffalo, MA, 71, PhD(biol), 72. Prof Exp: Teacher sci, Southside Sch, 59-63; teacher driver educ, Rowan Co Schs, 63-64; teacher sci, J F Kennedy Jr High Sch, 64-67; res botanist mycol, Agr Res Serv, USDA, 72-73; ASST PROF BIOL, ALA STATE UNIV, 73- Mem: AAAS; Bot Soc Am; Electron Micros Soc Am. Res: Fungal development and differentiation at the ultrastructural level. Mailing Add: Div of Biol Sci Ala State Univ Montgomery AL 36101

MAYFIELD, JOHN ERIC, b Toledo, Ohio, Dec 21, 41; m 68. MOLECULAR BIOLOGY. Educ: Col Wooster, BA, 63; Univ Pittsburgh, MA, 65, PhD(biophys), 68. Prof Exp: USPHS fel, Calif Inst Technol, 68-71, instr develop biol, 71; ASST PROF BIOL, CARNEGIE-MELLON UNIV, 71- Mem: Biophys Soc; Am Soc Cell Biol. Res: Gene control in eucaryotes, nucleic acids, chromosome structure. Mailing Add: Dept of Biol Sci Mellon Inst Sci Carnegie-Mellon Univ Pittsburgh PA 15213

MAYFIELD, MELBURN ROSS, b Island, Ky, Aug 24, 21; m 50; c 1. PHYSICS, SCIENCE EDUCATION. Educ: Western Ky State Col, AB & BS, 48; Univ Fla, MS, 50. Prof Exp: From instr to asst prof physics, Mercer Univ, 50-55, asst prof math & physics, 55-57; assoc prof physics, 57-61, chmn dept, 58-70, dir prog teachers, 68, PROF PHYSICS, AUSTIN PEAY STATE UNIV, 61-, DIR CTR FOR TEACHERS, 70-, VPRES DEVELOP & FIELD SERV, 74- Concurrent Pos: Consult under NSF grant, Acad Yr Inst Jr Col Teachers, Univ Fla, 66-67, consult, 67-68; consult, Proj Reachigh, 68. Mem: Fel AAAS; Am Phys Soc; Am Asn Physics Teachers. Res: Radioactive fallout measurement and identification; physics education at high school and college level. Mailing Add: 113 Morgan Ct Clarksville TN 37040

MAYFIELD, ROBERT CHARLES, b Oct 15, 28; US citizen; m 52; c 4. CULTURAL GEOGRAPHY, GEOGRAPHY OF SOUTH ASIA. Educ: Tex Christian Univ, BA, 52; Ind Univ, MA, 53; Univ Wash, PhD(geog), 61. Prof Exp: Lectr geog, Ind Univ, 54-56; asst prof, Southeastern State Col, 58-60; assoc prof, Tex Christian Univ, 61-64; from assoc prof to prof, Univ Tex, Austin, 64-71; PROF GEOG, BOSTON UNIV, 71-, CHMN DEPT, 72- Concurrent Pos: Partic, Symp Quant Methods Geog, Nat Acad Sci-Nat Res Coun, 60 & Int Geog Union Symp Urban Geog, Sweden, 60; off deleg, Nat Acad Sci Regional Conf Southeastern Asian Geographers, Kuala Lumpur, Malasia, 62; Fulbright-Hays fel, Mysore & Bangalore, 66-67. Mem: Asn Am Geogr; Am Geog Soc; Nat Coun Geog Educ; Regional Sci Asn; Asn Asian Studies. Res: Interpersonal communication structures in south India and the diffusion of agricultural innovation there; space-searching by a rural population and cross-caste communications. Mailing Add: Dept of Geog Boston Univ Boston MA 02215

MAYHALL, JOHN TARKINGTON, b Greencastle, Ind, Apr 7, 37; m 60. DENTAL ANTHROPOLOGY. Educ: DePauw Univ, BA, 59; Ind Univ, Indianapolis, DDS, 63; Univ Chicago, MA, 68. Prof Exp: Res assoc, 71-72, ASST PROF DENT ANAT, FAC DENT, UNIV TORONTO, 72- RES ASSOC ANTHROP, 71- Concurrent Pos: Res fel dent anthrop, Fac Dent, Univ Toronto, 71-73; abstractor, Oral Res Abstr, 71- Mem: AAAS; Int Asn Dent Res; Int Soc Craniofacial Biol; Can Asn Phys Anthrop; Am Asn Phys Anthrop. Res: Dental morphology, genetics and craniofacial growth and development of North American Eskimos and Indians; dental anatomy; forensic odontology; osteology. Mailing Add: Dept of Anthrop Univ of Toronto Toronto ON Can

MAYHEW, DENNIS ED, b Los Angeles, Calif, Apr 25, 45; m 66; c 2. PLANT PATHOLOGY, PLANT VIROLOGY. Educ: Calif State Univ, Long Beach, BS, 68; Iowa State Univ, MS, 69, PhD(plant path), 73. Prof Exp: Res assoc plant path, Mont State Univ, 73-74; plant pathologist, Calif Dept Food & Agr, 74-75; PVT BUS, 75- Mem: Am Inst Biol Sci; Am Phytopath Soc. Res: Cytological aspects of the seed transmission of barley stripe mosaic virus. Mailing Add: 1246 Meredith Way Carmichael CA 95608

MAYHEW, ERIC GEORGE, b London, Eng, June 22, 38. CELL BIOLOGY. Educ: Univ London, BSc, 60, MSc, 63, PhD(zool), 67. Prof Exp: Res asst cell biol, Chester Beatty Res Inst, London, Eng, 60-64; cancer res scientist, 64-68, sr cancer res scientist, 68-72, ASSOC CANCER RES SCIENTIST, ROSWELL PARK MEM INST, 72- Mem: AAAS; Soc Photog Scientists & Engrs. Res: Role of the cell periphery in cellular interactions; possible differences between normal and cancer cells and possible exploitation in chemotherapy. Mailing Add: Dept of Exp Path Roswell Park Mem Inst Buffalo NY 14263

MAYHEW, WILBUR WALDO, b Yoder, Colo, Mar 17, 20; m 48; c 3. VERTEBRATE BIOLOGY, DESERT ECOLOGY. Educ: Univ Calif, AB, 48, MA, 51, PhD(zool), 53. Prof Exp: Assoc zool, Univ Calif, Davis, 48-50 & 51-53; jr res biologist, Atomic Energy Proj, Univ Calif, Los Angeles, 53-54; instr biol, 54-56, from asst prof to assoc prof zool, 56-69, PROF ZOOL, UNIV CALIF, RIVERSIDE, 69- Concurrent Pos: Fulbright lectr, UAR, 65-66; Am consult, All-Indian Inst Ecol, Saurashtra Univ, India, 70; mem US deleg, Binational Conf Educ & Res Life Sci, India, 71. Mem: AAAS; Am Soc Ichthyologists & Herpetologists; Cooper Ornithologists Union; Herpetologists' League; Soc Study Amphibians & Reptiles. Res: Ecology and physiology of avian and reptilian reproduction; cliff swallow nesting and migration; ecology of deserts. Mailing Add: Dept of Biol Univ of Calif Riverside CA 92502

MAYKUT, MADELAINE OLGA, b Toronto, Ont, July 8, 25. CLINICAL PHARMACOLOGY. Educ: Univ Toronto, BA, 48, MA, 50, PhD(pharmacol), 57, MD, 64. Prof Exp: Asst cancer res, Univ Toronto, 48-49; asst biochemist, Henry Ford Hosp, Detroit, Mich, 50-51; pharmacologist, Univ Toronto, 51-59; sr pharmacologist, Res Ctr, Pitman-Moore Co Div, Dow Chem Co, Ind, 59-60; rotating intern, Univ Toronto, 64-65, Locum asst surg, Shouldice Surg, 65-66; asst dir clin res, Wm S Merrell Co, 66-67; asst dir med res, Bristol Labs, 67-72; adv, Bur Drugs, 72-73, CHIEF BIOMED RES, NONMED USE DRUGS DIRECTORATE, HEALTH PROTECTION BR, HEALTH & WELFARE, 73- Mem: Pharmacol Soc Can; NY Acad Sci; Drug Info Asn. Res: Local anesthetics; biometrics; drug combinations; anti-arrythmics; anti-inflammatory agents; narcotic antagonists; analgesics; anti-anginal agents; anti-hypertensives. Mailing Add: Non-Med Use Drugs Directorate Health Protect Br Health & Welfare Ottawa ON Can

MAYLAND, HENRY FREDERICK, b Greybull, Wyo, Dec 31, 35; m 57; c 2. SOIL SCIENCE. Educ: Univ Wyo, BS, 60, MS, 61; Univ Ariz, PhD(agr chem & soils), 65. Prof Exp: Soil scientist, 64-73, RES LEADER, AGR RES SERV, USDA, 73- Concurrent Pos: Fed collabr, Utah State Univ, 67-; affil prof, Univ Idaho, 68-; vis fel, Plant, Soil & Nutrit Lab, USDA & Cornell Univ, 73-74. Mem: Am Soc Agron; Soil Sci Soc Am; Soc Range Mgt. Res: Soil-water-plant-animal relations on rangelands. Mailing Add: Agr Res Serv USDA Kimberly ID 83341

MAYNARD, CARL WESLEY, JR, b Eveleth, Minn, June 18, 13; m 37. ORGANIC CHEMISTRY. Educ: Colo Col, AB, 34, PhD(org chem), Mass Inst Technol, 38. Prof Exp: Anal chemist, Dow Chem Co, Mich, 35-36; res chemist, Jackson Lab, 38-52, SR SUPVR INTEL, E I DU PONT DE NEMOURS & CO, INC, 52- Concurrent Pos: Civilian with Manhattan Proj, 42-44. Mem: Am Chem Soc; Am Asn Textile Chemists & Colorists. Res: Synthetic dyes; chemical literature. Mailing Add: 114 Cambridge Dr Wilmington DE 19803

MAYNARD, CHARLES DOUGLAS, b Atlantic City, NJ, Sept 11, 34; m 58; c 3. NUCLEAR MEDICINE. Educ: Wake Forest Univ, BS, 55; Bowman Gray Sch Med, MD, 59. Prof Exp: Dir nuclear med, NC Baptist Hosp, Winston-Salem, 66; from instr to assoc prof radiol, 66-73, assoc dean admis, 66-71, assoc dean student affairs, 71-75, PROF RADIOL, BOWMAN GRAY SCH MED, WAKE FOREST UNIV, 73- Concurrent Pos: Am Cancer Soc fel, 64-66; James Picker Found scholar radiol res, 66-68; consult ed, J Nuclear Med & Technol, 74; consult nuclear med, Am Registry Radiologic Technologists, 74; guest examr, Am Bd Radiol, 75. Mem: Soc Nuclear Med (vpres elect, 75); Asn Univ Radiologists; Radiol Soc NAm; Am Col Nuclear physicians; Am Col Radiol. Res: Clinical applications of radionuclides in the diagnoses of disease. Mailing Add: Dept of Radiol Bowman Gray Sch of Med Winston-Salem NC 27103

MAYNARD, DONALD EARLE, b Toronto, Ont, Nov 2, 33; US citizen; m 59; c 2. BIOCHEMISTRY. Educ: Univ Buffalo, BA, 55; Univ Ky, MSc, 58; Ohio State Univ, PhD(biochem), 61. Prof Exp: Res assoc med, asst prof physiol chem & dir clin endocrinol labs, Med Col, Ohio State Univ, 62-65; sr biochemist, Dept Clin Pharmacol, 65-67 & Chem Res Dept, 67-71, SR BIOCHEMIST, DEPT BIOCHEM & DRUG METAB, HOFFMANN-LA ROCHE, INC, 72- Mem: Am Chem Soc; Am Asn Clin Chemists; Endocrine Soc; Am Inst Chemists. Res: Biochemistry of steroid hormones; metabolism of tetrahydrocannabinols and alkaloids; pharmacokinetics. Mailing Add: Dept of Biochem & Drug Metab Hoffmann-La Roche Inc Nutley NJ 07110

MAYNARD, DONALD MORE, physiology, neurobiology, deceased

MAYNARD, DONALD NELSON, b Hartford, Conn, June 22, 32; m 74; c 1. PLANT PHYSIOLOGY. Educ: Univ Conn, BS, 54; NC State Univ, MS, 56; Univ Mass, PhD(bot), 63. Prof Exp: Instr hort, 56-62, from asst prof to assoc prof plant physiol, 62-72, PROF PLANT PHYSIOL, UNIV MASS, AMHERST, 72-, ASST DEAN COL FOOD & NATURAL RESOURCES, 74- Concurrent Pos: Consult, Greenleaf, Inc, 75- Honors & Awards: Environ Qual Award, Am Soc Hort Sci, 75. Mem: Am Soc Hort Sci; Am Soc Agron. Res: Plant nutrition, especially mechanism of absorption and mode of action of nutrient elements. Mailing Add: Dept of Plant & Soil Sci Bowditch Hall Univ of Mass Amherst MA 01002

MAYNARD, EDITH ADELE, b Flushing, NY, Jan 11, 31; m 54; c 3. NEUROBIOLOGY. Educ: Mt Holyoke Col, AB, 52; Univ Calif, Los Angeles, PhD(anat), 58. Prof Exp: Res assoc histochem of nerv tissue, Ment Health Res Inst, Univ Mich, Ann Arbor, 57-60, from instr to assoc prof anat, Med Sch, 60-70; adj prof, 70-74, PROF BIOL, UNIV ORE, 74- Mem: Am Asn Anat; Am Soc Zoologists; Am Soc Cell Biologists; Histochem Soc; Soc Neurosci. Res: Cytochemistry and ultrastructure of neural tissues; degeneration and regeneration in invertebrate nervous systems. Mailing Add: Dept of Biol Univ of Ore Eugene OR 97403

MAYNARD, FRANCIS LOUIS, b Taunton, Mass, Sept 27, 08; m 39. PHYSIOLOGY. Educ: Boston Col, AB, 31; Brown Univ, AM, 35; Boston Univ, PhD, 61. Prof Exp: Assoc prof biol, Boston Col, 35-74. Mem: AAAS; Microcirc Soc. Res: Mast cells; endocrinology; mammalian physiology. Mailing Add: Dept of Biol Boston Col Chestnut Hill MA 02167

MAYNARD, HUGH BARDEEN, b Hollywood, Calif, Dec 12, 43; m 66. MATHEMATICS. Educ: Calif Inst Technol, BS, 65; Univ Colo, MS, 67, PhD(math), 70. Prof Exp: ASST PROF MATH, UNIV UTAH, 70- Res: Functional analysis; vector-valued measures; Banach space theory. Mailing Add: Dept of Math Univ of Utah Salt Lake City UT 84112

MAYNARD, JOHN THOMAS, b Pueblo, Colo, Sept 3, 19; m 42; c 5. ORGANIC CHEMISTRY. Educ: Yale Univ, BS, 41, MS, 44, PhD(org chem), 46. Prof Exp: Instr chem, Yale Univ, 45-46; chemist, Chem Dept, 46-56, res supvr, Fibers Dept, 56-59, res chemist, Elastomer Chem Dept, 59-63, res supvr, 63-70, DIV HEAD PATENTS, ELASTOMER CHEM DEPT, E I DU PONT DE NEMOURS & CO, INC, 70- Mem: AAAS; Am Chem Soc; Sigma Xi. Res: Synthetic organic chemicals, including fluorocarbons and polymers; neoprene structure; patent information. Mailing Add: 108 Rockingham Dr Windsor Hills Wilmington DE 19803

MAYNARD, MARVIN MICHAEL, organic chemistry, see 12th edition

MAYNARD, NANCY GRAY, b Middleboro, Mass, Apr 18, 41; m 69; c 1. MARINE ECOLOGY, POLLUTION BIOLOGY. Educ: Mary Washington Col, Univ Va, BS, 63; Univ Miami, MS, 67, PhD(biol, living resources), 74. Prof Exp: Res asst malaria, Pharmaceut Lab, US Army, 63-64 & Marine biol, Rosenstiel Sch Marine & Atmospheric Sci, Univ Miami, 65-69; res assoc paleoclimat, Lamont-Doherty Geol Observ, Columbia Univ, 72-75; res fel chem, Harvard Univ, 75-76. Concurrent Pos: Res assoc oil pollution, Bermuda Biol Sta, St George, 72-76 & Am Petrol Inst, 74-76. Res: Effects of petroleum hydrocarbons on intertidal fauna and flora; general intertidal ecology; ecology of diatoms—polluted habitats, rocky intertidal zone, reefs, phytoplankton, Bermuda, whale and dolphin skins; paleoecology of diatoms. Mailing Add: 266A Harvard St Apt 5 Cambridge MA 02139

MAYNARD, ROBERT G, b Memphis, Tenn, Feb 3, 19; m 49; c 4. GEOLOGY. Educ: Univ Calif, Los Angeles, AB, 41, MA, 48. Prof Exp: Petrol geologist, Richfield Oil Corp, 45-47; petrol geologist, Sunray Oil Corp, 47-65, mgr hard mineral explor, D-X Sunray Oil Co, 65-68, ALASKA AREA GEOLOGIST, SUN OIL CO, 68- Mem: Geol Soc Am; Am Asn Petrol Geol. Res: Structural geology; economic mineral deposits; marine geology. Mailing Add: Sun Oil Co 9819 Elmcrest Dallas TX 75238

MAYNARD, RUSSELL HATTON, b Somerville, Mass, Aug 1, 12; m 36; c 3. RADIOBIOLOGY. Educ: US Naval Acad, BS, 34; Univ Calif, MS, 50. Prof Exp: In charge weather ctr, Pearl Harbor, US Navy, 42-43, aerolog officer, Aleutians Fleet, Air Wing Four, 43-44, air training bases, Tex, 44, aerol res & develop, Off Naval Oper, 45-47, sci & mil adv to chief Armed Forces Spec Weapons Proj & consult, US AEC, 50-55, off asst to Secy Navy, 55-57; staff engr, Polaris Prog, Lockheed Missiles & Space Co, 57-59, mgr plans & req, 59-62, sr staff scientist, Nuclear Space Prog Div, 62-65, sr staff engr, Polaris/Poseidon Prog, 65-70, nuclear vulnerability test & anal, Missile Systs Div, 68-70; teaching asst ecol, Cabrillo Col, 70-71; RETIRED. Mem: AAAS; assoc fel Am Inst Aeronaut & Astronaut; Am Nuclear Soc; Nat Mgt Asn. Res: Bioradiology and aerology as applied to radiological safety and radiation effects engineering. Mailing Add: 813 Vista Del Mar Aptos CA 95003

MAYNARD, RUSSELL MILTON, b Monmouth, Ill, June 6, 16; m 41; c 7. PATHOLOGY. Educ: Monmouth Col, BS, 38; Univ Ill, MD, 42; Am Bd Path, dipl, 50. Prof Exp: Instr path, Univ Colo, 54-57; asst prof, Sch Med, Marquette Univ, 57-62; pathologist, Vet Admin Hosp, Phoenix, Ariz, 62-64; clin pathologist, Good Samaritan Hosp, 64-67; CHIEF LAB SERV, VET ADMIN HOSP, VANCOUVER, 67- Concurrent Pos: Asst chief lab serv, Vet Admin Hosp, Denver, Colo, 54-57 & chief, Wood, Wis, 57-62. Mem: Am Soc Clin Path; Col Am Path; Am Col Physicians. Res: Anatomic and clinical pathology. Mailing Add: Vet Admin Hosp Vancouver WA 98661

MAYNARD, WILLIAM ROSE, JR, b Portsmouth, Va, May 31, 19; m 42. ANALYTICAL CHEMISTRY, BIOCHEMISTRY. Educ: Univ Richmond, BA, 41; Med Col Va, BS, 53. Prof Exp: Supvr drug lab, Va Dept Agr, 46-62; anal res chemist, 63-65, SR RES CHEMIST, A H ROBINS CO, INC, 65- Concurrent Pos: Mem revision comt, US Pharmacopoeia, 60-; mem, Va Bd Pharm, 67- Mem: Am Pharmaceut Asn; Am Chem Soc; fel Am Inst Chemists. Res: Metabolite study of clinical drugs under investigation; instrumentation and development of analytical methods for pharmaceuticals. Mailing Add: 7711 Brentford Dr Richmond VA 23225

MAYNE, BERGER C, b Towner, Colo, July 10, 20; m 56; c 2. PLANT PHYSIOLOGY. Educ: Western State Col Colo, AB, 46; Univ Utah, PhD(physiol), 58. Prof Exp: Res assoc, Univ Minn, 58-62; staff scientist, 62-67, INVESTR, CHARLES F KETTERING RES LAB, 67- Mem: Am Soc Plant Physiologists; Biophys Soc. Res: Photosynthesis. Mailing Add: Charles F Kettering Res Lab 150 E South College St Yellow Springs OH 45387

MAYNE, JOHN WINSTON, b Emerald, PEI, May 21, 12; m 41; c 2. MATHEMATICAL STATISTICS. Educ: Acadia Univ, BSc, 34, MSc, 35; Brown Univ, ScM, 37. Prof Exp: Instr math & physics, Prince of Wales Col, 37-39; accounts clerk, Dept Finance, Ottawa, 39-42; lectr math & mech, Carleton Univ, Can, 46-49, asst prof math & statist, 49-51; sect chief statist anal & northern opers res sect, Opers Res Group, Defence Res Bd, Ottawa, 51-53, sr opers res officer, Joint Serv Opers Res Team, 53-55; dir opers res, Royal Can Navy, 55-58; chief opers res, Air Defense Tech Ctr, Supreme Hq, Allied Powers Europe, The Hague, 58-59, dep group leader, 60-61, chief opers res br, Air Defense Div, Paris, 59-60; sr opers officer, Systs Anal Group, Defense Res Bd Can, 61-63; dir, Can Army Opers Res Estab, 63-65; dir land-air opers res, Opers Res Div, Hq, Can Forces, 65-67, dir maritime opers res, 67-68; dir gen oper res, Defense Res Bd Can, 68-75; CHIEF OPERS RES DIV, SUPREME HQ, ALLIED POWERS EUROPE TECH CTR, THE HAGUE, NETH, 75- Mem: AAAS; Opers Res Soc Am; Am Statist Asn; Inst Math Statist; Can Oper Res Soc. Res: Applied statistics; military operational research. Mailing Add: Jan Muschlaan 152 The Hague Netherlands

MAYNE, WILLIAM HARRY, b Austin, Tex, Apr 29, 13; m 39; c 1. GEOPHYSICS. Educ: Univ Tex, BS & MS, 35. Prof Exp: Asst observer, Petty Geophys Eng Co, 35, observer, 36, observer-field mgr, 36-38, res engr seismic & med instrumentation, Petty Labs, Inc, 39-57, sales coordr, Petty Geophys Eng Co, 58, vpres tech serv, 58-73, vpres tech serv, 73-75; CORP DIR NEW TECHNOL DEVELOP, GEOSOURCE INC, 75- Concurrent Pos: Mem adv bd, Exp Comt, Int Oil & Gas Educ Ctr, 69. Mem: Am Geophys Union; Am Asn Petrol Geol; Am Geol Inst; Mex Soc Explor Geophys; Soc Explor Geophys (1st vpres, 66-67, pres, 68-69, past pres, 69-70). Res: Geophysics and geophysical instrumentation; exploration seismology; underwater sound; optimum utilization of seismographic instruments and techniques. Mailing Add: PO Box 36306 Houston TX 77036

MAYNERT, EVERETT WILLIAM, b Providence, RI, Mar 18, 20. PHARMACOLOGY, CHEMISTRY. Educ: Brown Univ, ScB, 41; Univ Ill, PhD(org chem), 45; Johns Hopkins Univ, MD, 57. Prof Exp: Res chemist, Interchem Corp, NY, 45-47; res assoc pharmacol, Columbia Univ, 47-51, assoc, 51-52, asst prof, 52; assoc prof pharmacol & exp therapeut, Johns Hopkins Univ, 52-65; PROF PHARMACOL, UNIV ILL COL MED, 65- Concurrent Pos: Am Cyanamid fel, Johns Hopkins Univ, 52-57. Mem: Fel AAAS; Am Chem Soc; Harvey Soc; Am Soc Pharmacol. Res: Neuropharmacology; drug metabolism; toxicology. Mailing Add: Dept of Pharmacol Univ of Ill Col of Med Chicago IL 60680

MAYNES, ALBION DONALD, b Buffalo, NY, Jan 21, 29; Can citizen; m 52; c 2. ANALYTICAL CHEMISTRY. Educ: Univ Toronto, BA, 52, MA, 53, PhD(inorg & anal chem), 56. Prof Exp: Res assoc, Dept Physics, Univ Toronto, 56-58; anal chemist, Div Geol Sci, Calif Inst Technol, 58-65; asst prof, 65-66, ASSOC PROF CHEM, UNIV WATERLOO, 66- Concurrent Pos: Eldorado Mining & Refining Co res grant, 56-58. Mem: Meteoritical Soc. Res: Trace analysis; analysis of silicate rocks and minerals; analysis of meteorites. Mailing Add: Dept of Chem Univ of Waterloo Waterloo ON Can

MAYO, DANA WALKER, b Bethlehem, Pa, July 20, 28; m 62; c 3. ORGANIC CHEMISTRY. Educ: Mass Inst Technol, BS, 52; Ind Univ, PhD(chem), 59. Prof Exp: Res chemist, Polychem Dept, Exp Sta, E I du Pont de Nemours & Co, Del, 52; asst, Univ Pa, 52-53 & Ind Univ, 53-55 & 56-57; res assoc, Mass Inst Technol, 59-60, NIH fel, 60-62, fel, Sch Advan Study, 60-62; from asst prof to prof, 62-70, CHARLES WESTON PICKARD PROF CHEM, BOWDOIN COL, 70-, CHMN DEPT, 69- Concurrent Pos: Vis lectr, Mass Inst Technol, 62-72; NIH spec fel chem, Univ Md, 69-70; consult, Sadtler Res Labs, Inc, Philadelphia & Keyes Fibre Co. Mem: Am Chem Soc; Soc Appl Spectros; Coblentz Soc; The Chem Soc; Soc Chem Indust. Res: Natural products; organometallic compounds; application of Raman spectroscopy to organic chemistry; animal and plant chemical communication. Mailing Add: Dept of Chem Bowdoin Col Brunswick ME 04011

MAYO, EVANS BLAKEMORE, b DeKalb, Ill, Nov 16, 02; m 38; c 2. GEOLOGY. Educ: Univ SDak, BA, 27; Stanford Univ, MA, 29; Cornell Univ, PhD(petrog), 32. Prof Exp: Lab asst, Cornell Univ, 29-32, instr optical mineral & petrog, 32-36; field geologist, Augustus Locke, Calif, 37-40 & Kelowna Explor Co, BC, 40-41 & 47-50; res geologist, Cotopaxi Explor Co, NY, 42-46; chief geologist, Kelowna Mines Hedley, Ltd, 50-52; prof, 52-75, EMER PROF GEOL, UNIV ARIZ, 75- Mem: Fel Geol Soc Am; fel AAAS. Res: Structure plane of the Southern Sierra Nevada, California; lineament tectonics; ore districts of the Southwest; volcanic orogeny in the Tucson Mountains, Arizona. Mailing Add: 2702 E Seneca St Tucson AZ 85716

MAYO, FRANK REA, b Chicago, Ill, June 23, 08; m 33; c 2. PHYSICAL ORGANIC CHEMISTRY, POLYMER CHEMISTRY. Educ: Univ Chicago, BS, 29, PhD(chem), 31. Prof Exp: Lilly fel, Univ Chicago, 31-32; res chemist, E I du Pont de Nemours & Co, 33-35; instr org chem, Univ Chicago, 36-42; res chemist, US Rubber Co, 42-50; res assoc, Res Lab, Gen Elec Co, 50-56; SCI FEL, STANFORD RES INST, 56- Concurrent Pos: Lectr, Stanford Univ, 57-65. Honors & Awards: Award in Polymer Chem, Am Chem Soc, 67. Mem: AAAS; Am Chem Soc; The Chem Soc. Res: Oxidation of hydrocarbons; aging of polymers; coal chemistry; free radical reactions. Mailing Add: 89 Larch Dr Atherton CA 94025

MAYO, JAMES WELLINGTON, b Atlanta, Ga, Mar 2, 30; m 63; c 3. SOLID STATE PHYSICS. Educ: Morehouse Col, BSc, 51; Howard Univ, ScM, 53; Mass Inst Technol, MS, 61, PhD(magnetic resonance), 64. Prof Exp: Physicist, Nat Bur Stand, 52-53; instr physics, Howard Univ, 55-57; assoc prof & head dept, Morehouse Col, 64-67, prof & chmn dept, 67-71; dir col sci improv prog, 71-75, DEP DIR DIV SCI EDUC RESOURCE IMPROV, NSF, 75- Concurrent Pos: Consult resources group, Inst for Serv to Educ & sr prog assoc, Thirteen Col Curriculum Prog, 67-69; dir sci res inst, Atlanta Univ Ctr, 68-71; comnr, Comn on Col Physics, 68-72; mem adv comt sci educ, NSF, 69-71; mem bd trustees, Morehouse Col, 69-71. Res: Magnetism and magnetic measurements; electron paramagnetic spin resonance. Mailing Add: Div of Sci Educ Resource Improv Nat Sci Found Washington DC 20550

MAYO, JOSEPH WILLIAM, b Greenfield, Mass, Sept 22, 41; m 69; c 1. BIOCHEMISTRY. Educ: Univ Mass, Amherst, BS, 63; Mich State Univ, PhD(biochem), 68. Prof Exp: ASST PROF BIOCHEM & PEDIAT, CASE WESTERN RESERVE UNIV, 70-; ASST PROF BIOCHEM & PEDIAT, RAINBOW BABIES & CHILDREN'S HOSP, 70- Concurrent Pos: Nat Cystic Fibrosis Res Found fel, Sch Med, Case Western Reserve Univ, 68-70. Mem: AAAS; Am Chem Soc. Res: Carbohydrate metabolism; glycoproteins; composition and regulation of human exocrine secretions. Mailing Add: Cystic Fibrosis Lab Rm 566 Rainbow Babies & Children's Hosp Cleveland OH 44106

MAYO, MARIE JOINER, b Hammond, La, Sept 7, 26; m 46; c 2. EMBRYOLOGY, ENDOCRINOLOGY. Educ: Anderson Col, BA, 47; Tulane Univ, MS, 50. Prof Exp: From instr to asst prof, 50-63, ASSOC PROF BIOL, ANDERSON COL, 63- Concurrent Pos: Partic neurochem res, Sch Med, Ind Univ. Mem: AAAS; Am Soc Cell Biol; Soc Develop Biol. Res: Role of the superior vestibular nuclei and reticular formation in the production of decerebrate rigidity; formation of the nephric duct and kidney in the chick. Mailing Add: Dept of Biol Anderson Col Anderson IN 46012

MAYO, RALPH ELLIOTT, b Greenville, NC, May 9, 40; m 64; c 2. PHYSICAL CHEMISTRY. Educ: Emory Univ, BS, 63, PhD(phys chem), 66. Prof Exp: Sr res chemist, Perkin Elmer Corp, 66-68; SUPVR METHOD DEVELOP LAB, AIR PROD & CHEM, INC, 68- Mem: AAAS; Catalysis Soc; Am Chem Soc; Soc Appl Spectros. Res: Nuclear magnetic resonance spectroscopy; analytical instrumentation design; digital processing of scientific data; analytical methods development; mass spectrometry; spectrophotometric analyses. Mailing Add: Houdry Labs Air Prod & Chem Inc PO Box 427 Marcus Hook PA 19061

MAYO, SANTOS, b Buenos Aires, Arg, June 10, 28; m 59; c 3. MICROELECTRONICS. Educ: La Plata Univ, PhD(physics), 54. Prof Exp: Assoc res nuclear spectros, Arg AEC, 53-55, head synchrocyclotron lab, 55-68; mgr res & develop, Fate, 68-71; res mem, Cyclotron Lab, Arg AEC, 71-72; head planning, Arg Inst Indust Technol, 72-73; SOLID STATE PHYSICIST, NAT BUR STANDARDS, 74- Concurrent Pos: Instr, La Plata Univ, 50-60, assoc prof, 60-61; guest physicist, Brookhaven Nat Lab, 57; assoc res, Radiation Lab, Univ Pittsburgh, 58-59; head nuclear physics dept, Arg AEC, 62-63; Arg rep, Latin Am Physics Ctr, Brazil, 63-; consult physicist, Nat Res Coun, Arg, 64-; res fel, UN Develop Orgn, Nat Bur Standards, 73-74. Mem: Arg Physics Asn (gen secy, 68-70, pres, 70-72). Res: Beta and gamma nuclear spectroscopy; low and medium energy nuclear reactions; charged particle spectroscopy; accelerator techniques; microelectronic technology; surface probe analysis; physics of microelectronic devices. Mailing Add: Nat Bur of Standards Washington DC 20234

MAYO, THOMAS TABB, IV b Radford, Va, June 15, 32; m 57; c 3. PHYSICS. Educ: Va Mil Inst, BS, 54; Univ Va, MS, 57, PhD(physics), 60. Prof Exp: Asst instr physics, Va Mil Inst, 54-55; sr scientist res lab eng sci, Univ Va, 60-61; from asst prof physics to assoc prof physics & math, 62-67, asst acad dean, 71-73, assoc acad dean, 73-75, PROF PHYSICS, HAMPDEN-SYDNEY COL, 67-, ACTG ACAD DEAN, 75- Concurrent Pos: Res assoc quantum theory proj, Univ Fla, 69-70. Mem: Am Phys Soc; Am Asn Physics Teachers. Res: Teaching physical theory and applied mathematics at undergraduate level; classical and quantum mechanics; thermal physics; quantum theory of matter, especially the calculation of the angular distribution of photons coming from positron annihilation in solids. Mailing Add: Dept of Physics Hampden-Sydney Col Hampden-Sydney VA 23943

MAYO, Z B, b Lubbock, Tex, Mar 29, 43; m 71. ENTOMOLOGY. Educ: Tex Tech Univ, BS, 67; Okla State Univ, MS, 69, PhD(entom), 71. Prof Exp: Res assoc entom, Okla State Univ, 71-72; ASST PROF ENTOM, UNIV NEBR-LINCOLN, 72- Mem: Entom Soc Am; Am Registry Prof Entomologists; Sigma Xi. Res: Development of pest management procedures for insect pest associated with corn production. Mailing Add: Dept of Entom 202 P I Univ of Nebr Lincoln NE 68583

MAYOL, PERPETUO S, b Escalante, Philippines, Aug 12, 36; m 66; c 3. PLANT PATHOLOGY, MICROBIOLOGY. Educ: Univ Philippines, BSAgr, 57; Okla State Univ, MS, 65; Purdue Univ, West Lafayette, PhD(plant path, nematol), 68. Prof Exp: Asst plant pathologist, Bur Plant Indust, Manila, Philippines, 57; lab technician forest path, Forest Prod Res Inst, Philippines, 57-58; instr agron, Univ Philippines, 58-63; asst prof, 68-73, ASSOC PROF BIOL SCI, CALIF STATE COL, STANISLAUS, 73- Mem: AAAS; Am Soc Microbiol; Am Soc Nematol; Philippine Phytopath Soc; Inst Food Technologists. Res: Pathogenicity of some plant parasitic nematodes; microbial interrelationships in plant nematode infections. Mailing Add: Stanislaus CA

MAYOL, ROBERT FRANCIS, b Springfield, Ill, Nov 11, 41; m 62; c 4. BIOCHEMISTRY, IMMUNOCHEMISTRY. Educ: Southern Ill Univ, Carbondale, BA, 64; St Louis Univ, PhD(biochem), 68. Prof Exp: USPHS fel, Calif Inst Technol, 68-70; sr scientist biochem endocrinol, 70-75, SR INVESTR DRUG METAB, MEAD JOHNSON RES CTR, 75- Concurrent Pos: Mem assoc fac, Sch Med, Ind Univ, Evansville Ctr, 71- Mem: Sigma Xi; Endocrine Soc; AAAS. Res: Protein chemistry; development of radioimmunoassays; drug metabolism. Mailing Add: Mead Johnson Res Ctr Evansville IN 47721

MAYOR, HEATHER DONALD, b Melbourne, Australia, July 6, 30; m 56; c 2. VIROLOGY, MOLECULAR BIOLOGY. Educ: Univ Melbourne, BS, 48, MSc, 50, DSc, 70; Univ London, PhD(biophys), 54. Prof Exp: Res officer crystal physics, Defense Res Labs, Melbourne, Australia, 50-51; electron microscopist, Nat Inst Med Res, London, 52-55; res assoc virol, Walter & Eliza Hall Inst Med Res, Melbourne, 55-56; res assoc bacteriol & immunol, Harvard Med Sch, 56-59; from asst prof to assoc prof virol, 60-71, assoc prof microbiol, 71-74, PROF MICROBIOL, BAYLOR COL MED, 74- Concurrent Pos: Consult, Res Resources Br, NIH, 70-; consult, AEC, 71-, Univ Tex M D Anderson Hosp & Tumor Inst, Houston, 73- Honors & Awards: Award, Ctr Interaction, Man, Sci & Cult, 73. Mem: Am Asn Immunol; Am Soc Microbiol; Am Asn Cancer Res; Am Soc Cell Biol (treas); Sigma Xi (secy-treas). Res: Molecular biology of the growth and development of animal viruses with particular emphasis on viruses which cause cancer and on extremely small DNA-containing viruses. Mailing Add: Dept of Microbiol & Immunol Baylor Col of Med Houston TX 77025

MAYOR, JOHN ROBERTS, b La Harpe, Ill, July 9, 06; m 34; c 2. MATHEMATICS. Educ: Knox Col, BS, 28; Univ Ill, AM, 29; Univ Wis, PhD(math), 33. Hon Degrees: LLD, Knox Col, 59. Prof Exp: Instr math, Univ Wis, 29-31, 32-35 & Milwaukee Exten Div, 35; prof & chmn dept, Southern Ill Univ, 35-47; assoc prof math & educ, Univ Wis, 47-51, prof math & educ & chmn dept educ, 51-54, actg dean sch educ, 54-55; dir educ, Am Asn Advan Sci, 55-74; ASST PROVOST RES, DIV HUMAN & COMMUNITY RESOURCES, UNIV MD, COLLEGE PARK, 74- Concurrent Pos: Dir math proj & prof, Univ Md, 57-67; dir study accreditation in teacher educ, Nat Comn Accrediting, 63-65; mem adv comt, Sch Math Study Group; consult, Knox Col. Mem: AAAS; Am Math Soc; Math Asn Am; Nat Asn Res Sci Teaching; Conf Bd Math Sci (secy, 60-71, treas, 61-71). Res: Mapping rational varieties; multiple correspondences in space and hyperspace. Mailing Add: 411 Windsor St Silver Spring MD 20910

MAYOR, ROWLAND HERBERT, b Eng, Nov 5, 20; US citizen; m 48; c 3. RUBBER CHEMISTRY. Educ: Univ NH, BS, 42, MS, 44; Univ Conn, PhD(org chem), 49. Prof Exp: Instr org chem, Univ RI, 48-51; res chemist, 51-55, res sect head, 55-63, mgr stereorubber res, 64-75, ASST MGR SYNTHETIC RUBBER RES, GOODYEAR TIRE & RUBBER CO, 75- Mem: Am Chem Soc. Res: Molcular rearrangements; synthesis of diamines; condensation polymerization; synthetic rubber. Mailing Add: Res Div Goodyear Tire & Rubber Co Akron OH 44316

MAYOR, STEPHEN JOSEPH, b Detroit, Mich, Sept 9, 37; m 62; c 1. NEUROPHYSIOLOGY. Educ: Univ Mich, BSE(math) & BSE(physics), 61; Univ Ky, PhD(physiol), 69. Prof Exp: ASST PROF PHYSIOL, MED COL OHIO, 68- Concurrent Pos: Adj asst prof elec eng, Univ Toledo, 73-, adj asst prof psychol, 76- Mem: AAAS; Soc Neurosci. Res: Neurochemical correlates of behavior. Mailing Add: Dept of Physiol Med Col of Ohio Toledo OH 43614

MAYPER, STUART ALLAN, b New York, NY, July 11, 16; m 47; c 2. INORGANIC CHEMISTRY. Educ: City Col New York, BS, 38; Ohio State Univ, MSc, 39, PhD(chem), 48. Prof Exp: Asst chem, Ohio State Univ, 40-44; res chemist, Interchem Corp, NY, 44-45 & Phelps Dodge Corp, 45-46; tutor chem, Queens Col, NY, 46-48; from instr to asst prof, Brandeis Univ, 48-56; from asst prof to assoc prof, 56-73, PROF CHEM & CHMN DEPT, UNIV BRIDGEPORT, 73- Mem: Fel AAAS; Int Soc Gen Semantics. Res: Coordination complexes; logical and semantic bases of science. Mailing Add: Dept of Chem Univ of Bridgeport Bridgeport CT 06602

MAYR, ERNST, b Kempten, Ger, July 5, 04; m 35; c 2. EVOLUTIONARY BIOLOGY, HISTORY OF SCIENCE. Educ: Univ Berlin, PhD(zool), 26. Hon Degrees: DPhil, Univ Uppsala, 57 & Univ Paris, 75; DSc, Yale Univ, 59, Univ Melbourne, 59, Oxford Univ, 66 & Univ Munich, 68. Prof Exp: Asst cur zool mus, Univ Berlin, 26-32; from assoc cur to cur, Whitney-Rothschild Collection, Am Mus Natural Hist, 32-53; Agassiz prof zool, 53-75, dir mus comp zool, 61-70, EMER PROF ZOOL, HARVARD UNIV, 75- Concurrent Pos: Mem expeds, Dutch New Guinea, 28, Mandate Territory, New Guinea, 29 & Solomon Islands, 29-30; Jesup lectr, Columbia Univ, 41; ed, Soc Study Evolution, 47-49; vpres, 11th Int Zool Cong; pres, 13th Int Ornith Cong. Honors & Awards: Leidy Medal, 46; Darwin-Wallace Medal, 58; Brewster Medal, 65; Verrill Medal, 66; Daniel Giraud Eliot Medal, 67; Nat Medal of Sci, 70. Mem: Nat Acad Sci; Am Soc Naturalists; Am Soc Zool; Soc Syst Zool (pres, 66); Soc Study Evolution (secy, 46, pres, 50). Res: Ornithology; evolution; systematics; history and philosophy of biology. Mailing Add: Mus of Comp Zool Harvard Univ Cambridge MA 02138

MAYRON, LEWIS WALTER, b Chicago, Ill, Sept 20, 32; m 58; c 2. BIOLOGICAL CHEMISTRY, NUCLEAR MEDICINE. Educ: Univ Ill, MS, 56, PhD(biol chem), 59. Prof Exp: Chemist, Qm Food & Container Inst, 54; asst biochem, Univ Ill, 54-59; res assoc, Univ Southern Calif, 59-61; asst biochemist, Presby-St Lukes Hosp, 61-62, Tardanbek Labs, 62-63 & Abbott Labs, 63; res assoc, Michael Reese Hosp & Med Ctr, 64-66; consult, 66-68; RES CHEMIST, HINES VET ADMIN HOSP, 68- Concurrent Pos: Guest investr, Argonne Nat Lab, 73- Honors & Awards: Laureat, Genia Czerniak Prize Nuclear Med & Radiopharmacol, Ahavot Zion Found of Israel, 74. Mem: Soc Exp Biol & Med; Soc Nuclear Med; AAAS; Brit Biochem Soc; Am Asn Clin Chemists. Res: Biochemistry of immune mechanisms; nuclear biochemistry for medical diagnosis. Mailing Add: 5437 Suffield Terr Skokie IL 60076

MAYS, CHARLES EDWIN, b Lincoln, Nebr, May 6, 38; m 63; c 1. PHYSIOLOGY, HERPETOLOGY. Educ: Univ Nebr, BS, 63, MS, 65; Ariz State Univ, PhD(zool), 68. Prof Exp: Asst prof, 68-74, ASSOC PROF ZOOL, DePAUW UNIV, 74- Concurrent Pos: Du Pont & Nat Sci Found grants, 69-70; Ind Acad Sci grants, 69 & 71. Mem: AAAS; Am Soc Ichthyologists & Herpetologists; Soc Study Amphibians & Reptiles. Res: Natural history and physiological studies of the hellbender salamander and map turtle; growth studies on fish. Mailing Add: Dept of Zool DePauw Univ Greencastle IN 46135

MAYS, CHARLES W, radiological physics, radiobiology, see 12th edition

MAYS, DAVID ARTHUR, agronomy, see 12th edition

MAYS, DAVID LEE, b Lafayette, Ind, Sept 30, 42; m 64. ANALYTICAL CHEMISTRY. Educ: Baylor Univ, BA, 64; Purdue Univ, MS, 66, PhD(bionucleonics), 68. Prof Exp: Res scientist anal chem, 68-72, SR RES SCIENTIST ANAL CHEM, BRISTOL LABS DIV, BRISTOL MYERS CO, 72- Mem: Am Chem Soc. Res: Analytical separations and instrumental analyses of pharmaceuticals and related chemicals. Mailing Add: 7226 Mercer Circle East Syracuse NY 13057

MAYS, JOHN MOLTENO, chemical physics, see 12th edition

MAYS, ROLLAND LEE, b Buffalo, NY, Feb 21, 20; m 44; c 4. ANALYTICAL CHEMISTRY. Educ: Univ Buffalo, BA, 52. Prof Exp: Chemist, Bliss & Laughlin Co, Inc, 42-48, supvr & chief chemist, 48-52; chemist, Linde Div, 52-58, develop supvr, 58-63, develop mgr, 63-71, mgr technol, Molecular Sieve Dept, Mat Systs Div, 71-75, DIR TECHNOL, MOLECULAR SIEVE DEPT, LINDE DIV, UNION CARBIDE CORP, 75- Mem: AAAS; Am Chem Soc; Sigma Xi. Res: Sorption on solid sorbents and heterogeneous catalysis, particularly in zeolites; sorption, catalytic and ion exchange products; process development and process design. Mailing Add: Union Carbide Tech Ctr Tarrytown NY 10591

MAYSILLES, JAMES HOWARD, b Grafton, WVa, Aug 23, 21. TAXONOMY, PLANT ECOLOGY. Educ: Univ Mich, BS, 47, MS, 48, PhD, 59. Prof Exp: Asst, Herbarium, Univ Mich, 47-48; asst prof bot, 49-56, ASSOC PROF BIOL, HANOVER COL, 56- Mem: Am Soc Plant Taxon. Res: Taxonomy and ecology of vascular plants of Durango, Mexico and floral relationships of the pine forests of western Durango. Mailing Add: Box 163 Hanover Col Dept Bot Hanover IN 47243

MAYYASI, SAMI ALI, b Haifa, Palestine, Sept 20, 26; US citizen; m 53; c 2. VIROLOGY. Educ: Univ Tenn, Knoxville, BS, 49, MS, 50; Ohio State Univ, PhD(microbiol), 53. Prof Exp: Virologist, US Army Biol Labs, Ft Detrick, Md, 55-57; virologist, Ind, 57-63, ASST DIR CANCER RES, PFIZER, INC, 63- Concurrent Pos: Eli Lilly res fel, Ohio State Univ, 54-55. Mem: AAAS; Am Soc Microbiol; Am Asn Cancer Res. Res: Viral oncology; immunology. Mailing Add: Pfizer Inc Maywood NJ 07607

MAZADE, NOEL ANDRE, b Detroit, Mich, June 30, 44; m 67; c 2. PUBLIC HEALTH ADMINISTRATION. Educ: Wayne State Univ, BA, 66; Univ Mich, MSW, 68; Univ Pittsburgh, SM, 71, PhD(pub health), 72. Prof Exp: Consult ment health, Oakland County, Pontiac, Mich, 68-70; ASST PROF COMMUNITY PSYCHIAT, SCH MED, UNIV NC, CHAPEL HILL, 72-; DIR MODEL AREA PROG, NC STATE DEPT MENT HEALTH, 74- Mem: AAAS; Am Soc Pub Admin; Am Pub Health Asn. Res: Interorganizational aspects of health services delivery; assessment of local health needs. Mailing Add: 708 William Circle Chapel Hill NC 27514

MAZALESKI, STANLEY C, cellular biology, environment science, see 12th edition

MAZE, JACK REISER, b San Jose, Calif, Sept 28, 37; m 61. BOTANY. Educ: Humboldt State Col, BA, 60; Univ Wash, MS, 63; Univ Calif, Davis, PhD(bot), 65. Prof Exp: Lectr bot, Univ Calif, Davis, 65-66; asst prof, Univ Toronto, 66-68; asst prof, 68-73, ASSOC PROF BOT, UNIV BC, 73- Mem: AAAS; Am Inst Biol Sci; Am Soc Plant Taxon; Bot Soc Am; Ecol Soc Am. Res: Plant evolution and taxonomy; embryology and floret development in grasses; evolution of higher taxa; ecological morphogenesis. Mailing Add: Dept of Bot Univ of BC Vancouver BC Can

MAZEIKA, PAUL A, b Lithuania, Jan 9, 15; US citizen; m 44; c 2. PHYSICAL OCEANOGRAPHY. Educ: First Univ Navale, Italy, PhD, 43. Prof Exp: Oceanogr, US Naval Oceanog Off, 56-63 & Bur Com Fisheries, US Fish & Wildlife Serv, 63-66; OCEANOGR, US NAVAL OCEANOG OFF, 66- Mem: Am Geophys Union. Mailing Add: 7729 Brookville Rd Washington DC 20015

MAZEL, PAUL, b Norfolk, Va, Nov 27, 25; m 55; c 3. PHARMACOLOGY, BIOCHEMISTRY. Educ: Med Col Va, BS, 46; Trinity Univ, MS, 55; Vanderbilt Univ, PhD(pharmacol), 60. Prof Exp: Res asst biol, Southwest Found Res & Educ, 54-55; res asst pharmacol, Yale Univ, 55-56; res asst, Vanderbilt Univ, 56-60, instr, 60-61; from asst prof to assoc prof, 61-71, PROF PHARMACOL, GEORGE WASHINGTON UNIV, 71-, PROF ANESTHESIOL, 74- Concurrent Pos: USPHS fel, 60-61; lectr, US Naval Dent Sch, 61-62; consult, Datatrol Corp & Mediphone, Inc, 62-63 & Wallace Labs, 65-; vis prof, Fed City Col, 72-74. Mem: Soc Toxicol; Am Soc Pharmacol & Exp Therapeut. Res: Pharmacology of central nervous system acting drugs; physiological disposition of drugs; barbiturate metabolism; adaptive enzyme formation; membrane permeability; microsomal enzymes; blood-brain barrier; immunochemistry. Mailing Add: Dept of Pharmacol George Washington Univ Washington DC 20037

MAZELIS, MENDEL, b Chicago, Ill, Aug 31, 22; m 69. PLANT BIOCHEMISTRY. Educ: Univ Calif, BS, 43, PhD(plant physiol), 54. Prof Exp: Jr res biochemist, Univ Calif, 54-55; res assoc & instr, Univ Chicago, 55-57; assoc chemist, Western Regional Res Lab, USDA, 57-61; lectr, 61-65, asst biochemist, 61-64, assoc prof food sci & technol, 65-73, PROF FOOD SCI & TECHNOL, UNIV CALIF, DAVIS, 73-, ASSOC BIOCHEMIST, 64- Mem: Am Soc Biol Chem; Am Soc Plant Physiol; Brit Biochem Soc; Phytochem Soc NAm. Res: Intermediary metabolism; enzymology. Mailing Add: Dept of Food Sci & Technol Univ of Calif Davis CA 95616

MAZELSKY, ROBERT, b Middletown, NY, Mar 11, 33; m 56; c 4. SOLID STATE CHEMISTRY. Educ: Hofstra Univ, BS, 54; Univ Conn, PhD(chem), 58. Prof Exp: Res scientist solid state, 58-64, MGR CRYSTAL & VACUUM TECHNOL, WESTINGHOUSE RES, 64- Mem: Am Asn Crystal Growth. Res: Direct research programs on inorganic materials including synthesis, crystal growth and characterization; primary emphasis has been on optical, acoustic and semiconducting materials. Mailing Add: Westinghouse Res Beulah Rd Pittsburgh PA 15235

MAZENKO, GENE FRANCIS, b Coalport, Pa, July 5, 45; m 69. STATISTICAL MECHANICS. Educ: Stanford Univ, BS, 67; Mass Inst Technol, PhD(physics), 71. Prof Exp: Fel physics, Brandeis Univ, 71-72; fel, Harvard Univ & Mass Inst Technol, 72-73; fel, Stanford Univ, 73-75; ASST PROF PHYSICS, UNIV CHICAGO, 75- Mem: Am Phys Soc. Res: Interested in dynamic critical phenomenon; transport theory of classical and quantum fluids and statistical mechanics in general. Mailing Add: Dept of Physics Univ of Chicago 5640 Ellis Ave Chicago IL 60637

MAZER, MILTON, b New York, NY, Apr 5, 11; m 49; c 2. PSYCHIATRY. Educ: Univ Pa, BA, 32, MD, 35; William A White Inst NY, cert, 51. Prof Exp: Intern, Mt Sinai Hosp, Philadelphia, 35-36; resident internal med, Montefiore Hosp, New York, 36-37; clin asst, Mt Sinai Hosp, Philadelphia, 37-39; from internist to chief cardiac res unit, US Vet Admin, 39-43, resident psychiat, Vet Admin Hosp, 46-49; attend psychiatrist, Presby Hosp, 53-56; DIR PSYCHIAT, MARTHA'S VINEYARD MENT HEALTH CTR, 61-; ASST PSYCHIATRIST, MASS GEN HOSP, 63- Concurrent Pos: Pvt pract, 46-61; consult psychiat, Vet Admin, 49-55; fel, William White Inst, 52-; NIMH res grant, Martha's Vineyard Ment Health Ctr, 64-69; clin instr, Harvard Med Sch, 66-75, asst clin prof psychiat, 75-; NIMH res grant, 73-75. Mem: AAAS; Am Psychiat Asn; Am Acad Psychoanal. Res: Social factors in mental disorder; community mental health practice; epidemiology of psychiatric disorders. Mailing Add: Martha's Vineyard Ment Health Ctr Edgartown MA 02539

MAZER, RONALD STEVEN, b Lynn, Mass, Oct 21, 42; m 66; c 3. ENDOCRINOLOGY, PHYSIOLOGY. Educ: Bowdoin Col, AB, 64; Univ NH, MS, 66, PhD(zool), 68. Prof Exp: Fel, Cornell Univ, 68-69; ASST PROF BIOL, UNIV MAINE, PORTLAND-GORHAM, 69- Concurrent Pos: NSF res grant, Univ Maine, Portland-Gorham, 70-; vis prof, Westbrook Col, 74-75, Univ Maine, Augusta, 74- & Univ Maine, Lewiston, 75. Mem: Endocrine Soc; Am Fertil Soc. Res: Isolation of uterine-uteolytic factors which regulate the ovary. Mailing Add: Univ of Maine at Portland-Gorham Falmouth St Portland ME 04102

MAZERES, REGINALD MERLE, b Metairie, La, Feb 15, 34; m 57; c 3. ALGEBRA. Educ: Univ Southwestern La, BS, 59; Auburn Univ, MS, 60, PhD(math), 69. Prof Exp: Instr math, Auburn Univ, 62-63; from asst prof to assoc prof, 63-71, PROF MATH, TENN TECHNOL UNIV, 71- Res: Inflations and enlargements of semigroups. Mailing Add: Box 5054 Tenn Technol Univ Cookeville TN 38501

MAZESS, RICHARD B, b Philadelphia, Pa, June 10, 39. BIOLOGICAL ANTHROPOLOGY. Educ: Pa State Univ, BA, 61, MA, 63; Univ Wis-Madison, PhD(anthrop), 67. Prof Exp: NIH fel, 67-68, asst prof anthrop, 67-69, ASST PROF RADIOL, UNIV WIS-MADISON, 69- Mem: AAAS; Int Soc Biometeorol; Brit Soc Study Human Biol; Am Asn Phys Anthrop; Int Asn Human Biol. Res: Radionuclide measurements of skeleton and body composition; environmental physiology; adaptation of human populations. Mailing Add: Dept of Radiol Univ of Wis Hosp Madison WI 53706

MAZIA, DANIEL, b Scranton, Pa, Dec 18, 12; m 38; c 2. ZOOLOGY. Educ: Univ Pa, AB, 33, PhD(zool), 37. Prof Exp: Instr zool, Univ Pa, 35-36; Nat Res Coun fel, Princeton Univ & Marine Biol Labs, Woods Hole, 37-38; from asst prof to prof, Univ Mo, 38-50; assoc prof, 51-53, PROF ZOOL, UNIV CALIF, BERKELEY, 53- Concurrent Pos: Trustee, Marine Biol Lab, Univ Calif, Berkeley, 50-58, head physiol, 52-56. Mem: Nat Acad Sci; Am Soc Zool; Soc Gen Physiol (pres, 57-58); Am Acad Arts & Sci. Res: Ionic changes in stimulation; ion accumulation and exchange; chemistry of chromosomes; nuclear and cellular physiology; surface chemistry of enzymes; biochemistry of mitosis. Mailing Add: Dept of Zool Univ of Calif Berkeley CA 94720

MAZO, JAMES EMERY, b Bernardsville, NJ, Jan 15, 37; m 58; c 2. APPLIED MATHEMATICS. Educ: Mass Inst Technol, BS, 58; Syracuse Univ, MS, 60, PhD(physics), 63. Prof Exp: Res assoc physics, Ind Univ, 63-64; MEM TECH STAFF APPL MATH, BELL LABS, 64- Mem: Am Phys Soc; Inst Elec & Electronics Engrs. Res: Communication theory, noise theory and information theory. Mailing Add: Bell Labs Mountain Ave Murray Hill NJ 07974

MAZO, ROBERT MARC, b Brooklyn, NY, Oct 3, 30; m 54; c 3. THEORETICAL CHEMISTRY. Educ: Harvard Univ, AB, 52; Yale Univ, MS, 53, PhD(chem), 55. Prof Exp: NSF res fel, Univ Amsterdam, 55-56; res assoc, Univ Chicago, 56-58; asst prof chem, Calif Inst Technol, 58-62; assoc prof chem, 62-65, dir inst theoret sci, 64-67, assoc dean grad sch, 67-71, PROF CHEM, UNIV ORE, 65- Concurrent Pos: NSF sr fel & vis prof, Free Univ Brussels, 68-69. Mem: AAAS; Am Phys Soc. Res: Statistical mechanics; kinetic theory; irreversible thermodynamics; intermolecular forces. Mailing Add: Dept of Chem Univ of Ore Eugene OR 97403

MAZUMDAR, MAINAK, b Calcutta, India, Nov 19, 35; m 60; c 3. OPERATIONS RESEARCH, STATISTICS. Educ: Univ Calcutta, BS, 54, MS, 56; Cornell Univ, PhD(appl probability & statist), 66. Prof Exp: Res asst & res assoc opers res, Cornell Univ, 62-66; sr mathematician, 66-70, fel mathematician, 70-74, ADV MATHEMATICIAN, WESTINGHOUSE RES LABS, 74- Concurrent Pos: Lectr, Univ Pittsburgh, 67- Mem: Am Statist Asn; Inst Math Statist. Res: Mathematical theory of reliability; application of reliability methods to nuclear engineering. Mailing Add: Dept of Math Westinghouse Res Labs Pittsburgh PA 15235

MAZUMDAR, PURABI, b Feb 1, 44; Indian citizen; m 70; c 1. EXPERIMENTAL SOLID STATE PHYSICS. Educ: Univ Calcutta, BSc, 64, MSc, 67; Polytech Inst New York, PhD(physics), 76. Prof Exp: Lectr physics, J D Col, Calcutta, India, 67-70; RES FEL PHYSICS, POLYTECH INST NEW YORK, 73- Mem: Am Phys Soc; Sigma Xi. Res: Experimental solid state physics related to thin films, surface physics, magnetic materials and others. Mailing Add: Apt 10A 115 Ashland Pl Brooklyn NY 11201

MAZUMDER, BIBHUTI R, b July 1, 24; Indian citizen; m 51; c 3. SURFACE CHEMISTRY, SOLID STATE CHEMISTRY. Educ: Univ Calcutta, BS, 44; Univ Dacca, MS, 47; Howard Univ, PhD(phys chem), 58; FRIC. Prof Exp: Chemist, Standard Pharmaceut, India, 48-54; res assoc chem, Cornell Univ, 58-59; chemist, Unilever, Eng, 59-60; head phys chem, Lever Bros, India, 60-67; PROF CHEM, MORGAN STATE COL, 67- Mem: Sr mem Am Chem Soc. Res: Research and development of soaps, detergents and cosmetics. Mailing Add: Dept of Chem Morgan State Col Baltimore MD 21212

MAZUMDER, RAJARSHI, b Dacca, Bangladesh. BIOCHEMISTRY. Educ: Univ

Calcutta, BSc, 51, MSc, 53; Univ Calif, Berkeley, PhD(biochem), 59. Prof Exp: Pool officer biochem, All-India Inst Med Sci, New Delhi, 64-65, asst prof, 65-67; asst prof, 67-73, ASSOC PROF BIOCHEM, MED SCH, NY UNIV, 73- Concurrent Pos: Fel biochem, Med Sch, NY Univ, 60-63. Mem: Am Soc Biol Chem; Harvey Soc. Res: Mechanism of protein synthesis. Mailing Add: Dept of Biochem NY Univ Med Sch New York NY 10016

MAZUR, ABRAHAM, b New York, NY, Oct 8, 11; m 40; c 2. BIOCHEMISTRY. Educ: City Col New York, BS, 32, AM, 34; Columbia Univ, PhD(biochem), 38. Prof Exp: Tutor, 36-38, from instr to assoc prof, 38-59, PROF CHEM, CITY COL NEW YORK, 59-, CHMN DEPT, 69- Concurrent Pos: Carnegie Corp fel, Col Physicians & Surgeons, Columbia Univ, 38-39; Guggenheim fel, 49-50; res assoc biochem, Med Col, Cornell Univ, 41-49, asst prof, 49-66; guest investr, New York Blood Bank. Mem: AAAS; fel Soc Exp Biol & Med; Am Chem Soc; Am Soc Biol Chemists; Harvey Soc. Res: Acetylation mechanism; fat metabolism hormone; stilbestrol; components of autotrophic organisms; anticholinesterases; chemical factor in shock; ferritin; iron metabolism. Mailing Add: City Col of New York Dept Chem 140th St & Convent Ave New York NY 10021

MAZUR, BARBARA JEAN, b Los Angeles, Calif, Jan 6, 49; m 69. MOLECULAR BIOLOGY. Educ: Univ Calif, Los Angeles, AB, 70; Rockefeller Univ, PhD(genetics), 75. Prof Exp: NSF fel molecular biol, Univ Chicago, 75-76. Res: Molecular genetics of cell differentiation and nitrogen fixation in blue-green algae; DNA replication in bacterial viruses. Mailing Add: 5447 S Woodlawn Ave Chicago IL 60615

MAZUR, BOLESLAW, b Struga, Poland, Oct 19, 18; US citizen. DENTISTRY. Educ: Univ Ill, DDS, 56, MS, 61. Prof Exp: Instr crown & bridge, Col Dent, Univ Ill, 56-59; from asst prof to assoc prof ceramics & crown & bridge, Sch Dent, WVa Univ, 61-64; assoc prof fixed partial prosthodont, Col Dent, Univ Ill, 64-70; PROF FIXED PARTIAL PROSTHODONT, SCH DENT, LOYOLA UNIV CHICAGO, 70- Mem: Am Dent Asn; Am Equilibration Soc; Am Acad Crown & Bridge Prosthodont. Res: Ceramic and metallurgical investigation; biomechanical research in prosthodontics; histological investigation of vital tissues supporting prosthodontic restorations; prolem of occlusion and articulators. Mailing Add: 55 E Washington St Chicago IL 60602

MAZUR, JACOB, b Lodz, Poland, Dec 17, 21; nat US; m 51; c 2. POLYMER PHYSICS. Educ: Hebrew Univ, MSc, 45, PhD(phys chem), 48. Prof Exp: Res fel, Calif Inst Technol, 48-50; vis fel, Univ Chicago, 50-51; res scientist, Weizmann Inst Sci, Israel, 51-55; res assoc, Univ Ill, 55-57; res chemist, Dow Chem Co, 57-60; PHYS CHEMIST, NAT BUR STAND, 60- Concurrent Pos: Rockefeller Found fel, 49-50. Honors & Awards: Morrison Award, NY Acad Sci, 59. Mem: Fel Am Phys Soc. Res: Theoretical physical chemistry; high polymer physics; statistical mechanics. Mailing Add: Nat Bur of Stand Washington DC 20234

MAZUR, PETER, b New York, NY, Mar 3, 28; m 53; c 1. CELL PHYSIOLOGY, CRYOBIOLOGY. Educ: Harvard Univ, AB, 49, PhD(biol), 53. Prof Exp: Mem staff, Hq, Air Res & Develop Command, US Air Force, 53-57; NSF fel, Princeton Univ, 57-59; BIOLOGIST, OAK RIDGE NAT LAB, 59- Concurrent Pos: Mem, Am Inst Biol Sci Adv Comt, Biol & Med Br, Off Naval Res, 63-66 & Environ Biol Br, NASA, 66-; mem adv bd, Am Type Cult Collection, 66-70; vis lectr, Duke Univ, 67; chmn long-range planning off, Oak Ridge Nat Lab, 70, sci dir biophys & cell physiol, 74-75; prof, Univ Tenn-Oak Ridge Grad Sch Biomed Sci, 70-; mem, Harvard Bd Overseers Vis Comt Biol, 70- ; mem space sci bd, Nat Acad Sci, 75- Mem: Fel AAAS; Soc Gen Physiol; Biophys Soc; Bot Soc Am; Soc Cryobiol (pres, 73-74). Res: Low temperature biology; freezing and drying; cell water, membranes and permeability. Mailing Add: Biol Div Oak Ridge Nat Lab PO Box Y Oak Ridge TN 37830

MAZUR, ROBERT HENRY, b Indianapolis, Ind, June 15, 24; m 54; c 3. ORGANIC CHEMISTRY. Educ: Mass Inst Technol, BS, 48, PhD(org chem), 51. Prof Exp: NIH fel, Swiss Fed Inst Technol, 51-52; RES CHEMIST, G D SEARLE & CO, 52- Concurrent Pos: Nat Cancer Inst fel, Cambridge, 56-57. Mem: Am Chem Soc. Res: Molecular rearrangements; reaction mechanisms; alkaloids; steroids; peptides. Mailing Add: 1250 Stratford Rd Deerfield IL 60015

MAZUR, STEPHEN, b Baltimore, Md, Apr 9, 45; m 69. ORGANIC CHEMISTRY. Educ: Yale Univ, BS, 67; Univ Calif, Los Angeles, MS, 69, PhD(chem), 71. Prof Exp: Assoc chem, Univ Calif, Los Angeles, 69-70; NSF fel & res assoc, Columbia Univ, 71-72, lectr, 72-73; ASST PROF CHEM, UNIV CHICAGO, 73- Mem: Am Chem Soc; Sigma Xi. Res: Mechanistic organic chemistry, particularly chemistry of reactive intermediates, electrochemical reaction mechanisms, and the surface chemistry of graphite. Mailing Add: Univ of Chicago Dept of Chem 5747 S Ellis Ave Chicago IL 60637

MAZURAK, ANDREW PETER, b Denver, Colo, July 17, 11; m 41; c 4. SOIL PHYSICS. Educ: Mich State Col, BS, 33; Yale Univ, MF, 36; Univ Calif, PhD(soil sci), 48. Prof Exp: Tech foreman, US Forest Serv, 33-35, jr forester, 36-37; tech asst, Univ Calif, 37-40, asst, 41-42, assoc, 42-48; from asst prof to assoc prof, 48-57, PROF SOIL PHYSICS, UNIV NEBR, LINCOLN, 57- Mem: Am Soc Agron; Soil Sci Soc Am; Am Geophys Union. Res: Soil structure and clay minerals. Mailing Add: Dept of Agron Univ of Nebr Lincoln NE 65803

MAZURKIEWICZ-KWILECKI, IRENA MARIA, b Tarnow, Poland, May 14, 24; nat Can; m; c 1. PHARMACOLOGY. Educ: Jagellonian Univ, MPharm, 47; McGill Univ, MSc, 55, PhD(pharmacol), 57. Prof Exp: Res asst pharmacol, Sch Med, Jagellonian Univ, 47-48; res asst, McGill Univ, 53-57; res pharmacologist, Food & Drug Labs, Dept Nat Health & Welfare, Ont, 57-59; res pharmacologist, Prov Labs, Ministry of Health, Que, 59-60; asst prof, 60, assoc prof, 64, actg head dept, 64-65, PROF PHARMACOL, FAC MED, UNIV OTTAWA, 72- Concurrent Pos: Am Med Life Ins Fund Med Res Found fel, 57. Mem: AAAS; Fr-Can Asn Advan Sci; NY Acad Sci; Am Soc Pharmacol & Exp Therapeut; Pharmacol Soc Can. Res: Pharmacology of autonomic nervous system; neuropharmacology; catecholamines; histamine; psychoactive drugs; drugs of abuse; cardiovascular pharmacology. Mailing Add: Dept of Pharmacol Univ of Ottawa Fac of Med Ottawa ON Can

MAZZENO, LAURENCE WILLIAM, b New Orleans, La, Sept 4, 21; m 44; c 4. ORGANIC CHEMISTRY. Educ: Loyola Univ, La, BS, 42; Univ Detroit, MS, 44. Prof Exp: Asst chem, Univ Detroit, 42-44; chemist, Southern Regional Res Lab, Bur Agr & Indust Chem, 44-53 & Southern Utilization Res Br, 53-58, head new prod invests, Southern Utilization Res Div, 58-59, head chem modification invests, 59-61, asst to dir indust develop, 61-69, res chemist, Cotton Finishes Lab, Southern Mkt & Nutrit Res Div, 69-74, HEAD TECH & ECON ANAL RES, SOUTHERN REGIONAL RES CTR, USDA, 74- Mem: Am Chem Soc; Sigma Xi; Asn Textile Chem & Colorists. Res: Textile finishing, including chemical modification and resin finishing of cotton; flame and weather resistant finishes for cotton textiles. Mailing Add: Southern Regional Res Ctr USDA PO Box 19687 New Orleans LA 70179

MAZZIA, VALENTINO DON BOSCO, b New York, NY, Feb 17, 22; m; c 3. ANESTHESIOLOGY. Educ: City Col New York, BS, 43; NY Univ, MD, 50. Prof Exp: Asst prof anesthesiol, Med Col, Cornell Univ, 52-61; prof anesthesiol in oral surg, Col Dent, NY Univ, 61-73, chmn anesthesiol, Col Med & Grad Med Sch, 61-73; PROF ANESTHESIOL, UNIV COLO MED CTR, DENVER, 76- Concurrent Pos: Asst med examr & consult forensic anesthesiol, Off Chief Med Examr, New York, 62-73; prof anesthesiol & chmn dept, Charles A Drew Med Sch, 71-73; vis prof, Univ Calif, Los Angeles, 72; dep coroner & asst med examr, County Los Angeles, 72- Mem: Am Soc Pharmacol & Exp Therapeut; Soc Exp Biol & Med; Harvey Soc; fel Am Col Chest Physicians; fel NY Acad Med. Res: Correlation of eye position with amnesia for on-going events; development of monitoring systems for the operating room. Mailing Add: Univ Colo Med Ctr B113 4200 E Ninth Ave Denver CO 80220

MAZZIOTTI, ALEXANDER R, theoretical chemistry, see 12th edition

MAZZOCCHI, PAUL HENRY, b New York, NY, May 6, 39; m 61. ORGANIC CHEMISTRY. Educ: Queens Col, NY, BS, 61; Fordham Univ, PhD(org chem), 66. Prof Exp: NIH fel org chem, Cornell Univ, 65-67; asst prof, 67-71, ASSOC PROF ORG CHEM, UNIV MD, COLLEGE PARK, 71- Mem: Am Chem Soc. Res: Organic photochemistry; synthetic chemistry. Mailing Add: Dept of Chem Univ of Md College Park MD 20740

MAZZOLENI, ALBERTO, b Milan, Italy, Sept 12, 27; US citizen; m; c 2. CARDIOLOGY. Educ: Univ Milan, MD, 52; Am Bd Internal Med, dipl, 63; Am Bd Cardiovasc Dis, dipl, 68. Prof Exp: Intern med, Miriam Hosp, Providence, RI, 57-58; res asst, Lemuel Shattuck Hosp, Boston, 58-59; res asst, Boston City Hosp, 59-60; from asst prof med to assoc prof clin med, Sch Med, Univ Ky, 61-72; asst chief internal med, 64-73, chief cardiol, 67-72, DIR CORONARY CARE UNIT, VET ADMIN HOSP, LEXINGTON, 73-; ASSOC PROF MED, SCH MED, UNIV KY, 72- Concurrent Pos: Res fel cardiol, Beth Israel Hosp, Boston, 55-57 & 60-61. Mem: Fel Am Col Cardiol; fel Am Col Physicians. Res: Electrocardiogram diagnosis of cardiac hypertrophy; component heart weights. Mailing Add: 3772 Gloucester Dr Lexington KY 40511

MAZZONE, HORACE M, b Franklin, Mass, May 19, 30; m 62; c 1. BIOCHEMISTRY. Educ: Boston Col, BS, 51, MS, 53; Univ Wis, PhD(biochem), 59. Prof Exp: Res assoc biochem, Long Island Biol Asn & dept genetics, Carnegie Inst, 59; fel pharmacol, Harvard Med Sch, 59-61; res assoc path, Children's Hosp Med Ctr & Children's Cancer Res Found, Boston, 61-63; res assoc biol, Mass Inst Technol, 63-65; BIOCHEMIST, FOREST INSECT & DIS LAB, USDA, 65- Concurrent Pos: Res assoc physics, Mass Gen Hosp, 64-65; lectr, Yale Univ, 72. Mem: AAAS; Am Chem Soc; Am Soc Cell Biol; Tissue Cult Asn. Res: Viruses; properties of infectious agents; tissue culture. Mailing Add: Forest Insect & Dis Lab USDA 151 Sanford St Hamden CT 06514

MAZZUR, SCOTT RUIGH, b Princeton, NJ, m 58; c 2. MICROBIOLOGY, MEDICAL ANTHROPOLOGY. Educ: Rutgers Univ, BA, 57; Univ Pa, PhD(microbiol), 66. Prof Exp: Res scientist, NJ State Dept Health, 66-69; res assoc, Inst Cancer Res, 69-73; RES ASSOC, UNIV MUS, UNIV PA, 73-; SR RES SCIENTIST, AM NAT RED CROSS, 74- Mem: AAAS; Am Soc Microbiol; NY Acad Sci. Res: Subtyping and epidemiology in Pacific populations; tissue culture; viral serology; Hepatitis B virus. Mailing Add: Am Nat Red Cross Blood Res Lab 9312 Old Georgetown Rd Bethesda MD 20014

MCADAM, TERRY DONALD, b Arkansas City, Kans, Sept 7, 24; m 47; c 1. MATHEMATICS. Educ: Washburn Univ, AB, 47. Prof Exp: Instr math, 48-49 & 54-59, asst to pres, 49-54, asst prof math, 59-63, ASSOC PROF MATH, WASHBURN UNIV, 63- Mem: Math Asn Am; Nat Asn Teachers Math. Res: Statistics; programming. Mailing Add: Dept of Math Washburn Univ Topeka KS 66621

MCADAMS, ARTHUR JAMES, b Santa Barbara, Calif, July 16, 23; m 46; c 4. PEDIATRICS, PATHOLOGY. Educ: Johns Hopkins Univ, MD, 48. Prof Exp: Instr pediat path, Col Med, Univ Cincinnati, 56-58; dir labs, Children's Hosp East Bay, Oakland, Calif, 58-63; assoc prof, 63-69, PROF PEDIAT & PATH, COL MED, UNIV CINCINNATI, 69-; DIRECTING PATHOLOGIST, CHILDREN'S HOSP & RES FOUND, 63- Concurrent Pos: Asst pathologist, Cincinnati Children's Hosp, 56-58; clin asst prof, Sch Med, Univ Calif, 62-63. Mem: AAAS; Am Asn Path & Bact; Soc Pediat Res; Int Acad Path. Res: Role of complement in nephritis; pulmonary disease in the newborn; anoxic brain damage. Mailing Add: Dept of Path Children's Hosp Res Found Cincinnati OH 45229

MCADAMS, LOUIS VINCENT, organic chemistry, see 12th edition

MCADAMS, ROBERT ELI, b Hudson, Colo, Jan 2, 29; m 50. NUCLEAR PHYSICS. Educ: Colo State Univ, BS, 57; Iowa State Univ, PhD(physics), 64. Prof Exp: Fel physics, Iowa State Univ, 64-65; asst prof, 65-73, ASSOC PROF PHYSICS, UTAH STATE UNIV, 73- Mem: Am Phys Soc. Res: Low energy nuclear physics, especially gamma and beta ray spectroscopy, transition probabilities and internal electron conversion coefficients. Mailing Add: Dept of Physics Utah State Univ Logan UT 84322

MCADIE, HENRY GEORGE, b Montreal, Que, May 12, 30; m 56; c 4. ENVIRONMENTAL CHEMISTRY, ANALYTICAL CHEMISTRY. Educ: McGill Univ, BSc, 51; Queen's Univ, Ont, MA, 53, PhD(phys chem), 56. Prof Exp: Res fel, 56-63, sr res scientist, 64-67, prin res scientist, 67-70, from asst dir to actg dir dept phys chem, 70-71, DIR DEPT ENVIRON CHEM, ONT RES FOUND, 72- Concurrent Pos: Chmn, Can Adv Comt, Int Stand Orgn/Tech Comt-146 Air Qual, 74-77. Mem: Fel Chem Inst Can; Int Conf Thermal Anal (vpres, 74-77); Am Chem Soc; Air Pollution Control Asn; fel The Chem Soc. Res: Thermoanalytical methods and applications; heterogeneous processes; instrumentation; air pollution instrumentation; ambient and work-room monitoring; emissions testing; emission control processes. Mailing Add: Dept of Environ Chem Ont Res Found Sheridan Park Mississauga ON Can

MCADOO, DAVID JOHN, b Washington, Pa, Aug 11, 41; m 67. ANALYTICAL CHEMISTRY, PHYSICAL CHEMISTRY. Educ: Lafayette Col, AB, 63; Cornell Univ, PhD(chem), 71. Prof Exp: Chemist, Eastern Regional Res & Develop Div, Agr Res Serv, USDA, 63-64; chemist, Union Carbide Res Inst, 66-68; mem staff, Jet Propulsion Lab, 71-73; MEM, MARINE BIOMED INST, UNIV TEX MED BR GALVESTON, 73- Mem: Soc Neurosci; Am Chem Soc; Am Soc Mass Spectrometry. Res: Applications of mass spectrometry to neurochemistry; structure and fragmentation mechanisms of organic ions in the mass spectrometer; natural products. Mailing Add: Marine Biomed Inst Univ of Tex Med Br Galveston TX 77550

MCAFEE, JOHN GILMOUR, b Toronto, Ont, June 11, 26; nat US; m 52; c 3. RADIOLOGY. Educ: Univ Toronto, MD, 48. Prof Exp: Jr intern, Victoria Hosp, London, Ont, 48-49, asst resident radiol, 50-51; sr intern, Westminster Hosp, 49-50; resident radiol, Johns Hopkins Hosp, 51-52; from instr to assoc prof, Johns Hopkins Univ, 53-65; PROF RADIOL, STATE UNIV NY UPSTATE MED CTR, 65-

Concurrent Pos: Fel radiol, Johns Hopkins Univ, 52-53. Mem: Soc Nuclear Med; Radiol Soc NAm; AMA; Royal Col Physicians & Surgeons Can. Res: Use of radioactive tracers in clinical diagnosis; nuclear instrumentation and radiochemistry. Mailing Add: Dept of Radiol State Univ of NY Upstate Med Ctr Syracuse NY 13210

MCAFEE, KENNETH BAILEY, JR, b Chicago, Ill, June 22, 24; m 59; c 2. CHEMICAL PHYSICS, ATMOSPHERIC CHEMISTRY. Educ: Harvard Univ, BS, 46, MA, 47, PhD(chem physics), 50. Prof Exp: Mem tech staff, 50-66, HEAD ATMOSPHERIC RES DEPT, BELL LABS, 66- Concurrent Pos: Mem exec comt, Gaseous Electronics Conf, 65; vis fel, Joint Inst Lab Astrophys, Univ Colo, 65-66; mem, Defense Sci Bd, 66-; mem, New York Health Res Coun, 71-; mem & consult, US Environ Protection Agency Delphi Panel on Sulfur Oxides Control Technol Forecasting, 72-74; chmn panel on status of sulfur oxides technol, Nat Res Coun Comn on Sociotech Systs, 75- Mem: Fel Am Phys Soc; Am Chem Soc; Sigma Xi. Res: Semiconductors; gaseous diffusion and separation; atomic collision processes; reentry physics; upper atmosphere; atmospheric reactions and dispersion of contaminants; environmental physics and chemistry; halogen photochemistry and excited state reactions. Mailing Add: Bell Labs Murray Hill NJ 07974

MCAFEE, ROBERT DIXON, b Zamboanga City, Philippines, Sept 9, 25; m 53. PHYSIOLOGY, BIOPHYSICS. Educ: Cent Col, AB, 48; Univ Tenn, MS, 51; Tulane Univ, PhD(physiol), 53. Prof Exp: Res physiologist, 54-59, ASSOC PHYSIOL, SCH MED, TULANE UNIV, 59-; SR SCIENTIST, VET ADMIN HOSP, NEW ORLEANS, 59- Concurrent Pos: Consult prof, Sch Eng, Univ New Orleans. Res: Ion transport in membranes; physiological effects of microwave radiation on eye and central nervous system; behavioral biomedical engineering. Mailing Add: Vet Admin Hosp 1601 Perdido St New Orleans LA 70146

MCAFEE, WALTER SAMUEL, b Ore City, Tex, Sept 2, 14; m 41; c 2. THEORETICAL PHYSICS. Educ: Wiley Col, BS, 34; Ohio State Univ, MS, 37; Cornell Univ, PhD(physics), 49. Prof Exp: Teacher jr high sch, 37-42; physicist theoret studies unit, Eng Labs, 42-45, physicist & supvr, 45-46, physicist radiation physics, 48-53, chief sect electro-magnetic wave propagation, 53-57, consult physicist, Appl Physics Div, 58-65, tech dir, Passive Sensing Tech Area, 65-71, SCI ADV TO DIR RES, DEVELOP & ENG, ENG LABS, US ARMY ELECTRONICS COMMAND, FT MONMOUTH, 71- Concurrent Pos: Secy of Army fel, Harvard Univ, 57-58; lectr, West Long Br, Monmouth Col, NJ, 58- Mem: AAAS; Am Astron Soc; Am Phys Soc; Am Asn Physics Teachers; sr mem Inst Elec & Electronics Engrs. Res: Theoretical nuclear physics; electromagnetic theory. Mailing Add: 723 17th Ave South Belmar NJ 07719

MCALACK, ROBERT FRANCIS, b Camden, NJ, June 1, 40. MICROBIOLOGY, IMMUNOLOGY. Educ: Drexel Univ, BS, 64; Thomas Jefferson Univ, MS, 66, PhD(microbiol), 68. Prof Exp: Assoc immunol, 68-71, TRANSPLANT IMMUNOLOGIST, DEPT SURG, ALBERT EINSTEIN MED CTR, PHILADELPHIA, 71- Mem: AAAS; Am Soc Microbiol; Reticuloendothelial Soc. Res: Transplant immunology; autoimmune diseases; immune cell interactions; leukemia immune expressions. Mailing Add: Transplant Immunol Lab Albert Einstein Med Ctr Philadelphia PA 19141

MCALDUFF, EDWARD J, b Alberton, PEI, Dec 3, 39. SPECTROCHEMISTRY. Educ: St Francis Xavier Univ, BSc, 61; Univ Toronto, PhD(phys chem), 65. Prof Exp: Demonstr, Univ Toronto, 61-65; res fel, Univ Wash, 66-67; asst prof, 67-72, ASSOC PROF CHEM, ST FRANCIS XAVIER UNIV, 72- Concurrent Pos: Nat Res Coun grant, 67-70; res assoc, La State Univ, 75-76. Mem: Am Chem Soc; Chem Inst Can. Res: Gas phase kinetics; electronically excited states of simple molecules; photochemical aspects of air pollution; photoelectron spectroscopy of photochemically significant molecules. Mailing Add: Dept of Chem St Francis Xavier Univ Antigonish NS Can

MCALEAR, JAMES HARVEY, biophysics, see 12th edition

MCALEER, WILLIAM JOSEPH, b Philadelphia, Pa, Oct 11, 22; m 55; c 3. ORGANIC CHEMISTRY, BIOCHEMISTRY. Educ: Pa State Univ, BS, 46, MS, 47; Yale Univ, PhD(org chem), 53. Prof Exp: Sr chemist, Merck & Co, Inc, 50-57, sr chemist & mgr elctronic chem res, Merck Sharp & Dohme Res Labs, 57-63, res fel virus & cell biol, Merck Inst Therapeut Res, 64-73, SR DIR BIOL DEVELOP, VIROL & CELL BIOL, MERCK INST THERAPEUT RES, 73- Mem: Am Inst Chemists; Am Chem Soc. Res: Chemistry of natural products; electronic chemicals; vaccine production, delivery and assay. Mailing Add: Merck Inst for Therapeut Res West Point PA 19486

MCALESTER, ARCIE LEE, JR, b Dallas, Tex, Feb 3, 33; m 68; c 2. PALEOBIOLOGY. Educ: Southern Methodist Univ, BA, 54, BBA, 54; Yale Univ, MS, 57, PhD(geol), 60. Prof Exp: From instr to prof geol, Yale Univ, 59-74, from asst cur to cur, Peabody Mus, 59-74; PROF GEOL SCI & DEAN SCH HUMANITIES & SCI, SOUTHERN METHODIST UNIV, 74- Concurrent Pos: Guggenheim fel, Glasgow Univ, 64-65; res assoc, Univ Rochester, 65- Mem: AAAS; Geol Soc Am; Paleont Soc; Soc Syst Zool. Res: Invertebrate paleobiology; marine ecology and paleoecology. Mailing Add: Sch of Humanities & Sci Southern Methodist Univ Dallas TX 75205

MCALICE, BERNARD JOHN, b Providence, RI, Apr 20, 30; m 55; c 3. OCEANOGRAPHY. Educ: Univ RI, BS, 62, PhD(biol oceanog), 69. Prof Exp: Asst prof, 67-75, ASSOC PROF OCEANOG & ZOOL, UNIV MAINE, 75- Mem: Am Soc Limnol & Oceanog; Estuarine Res Fedn. Res: Ecology, distribution and succession of estuarine plankton. Mailing Add: Ira C Darling Ctr Univ of Maine Walpole ME 04573

MCALISTER, ARCHIE JOSEPH, b Birmingham, Ala, Aug 13, 32; m 64. SOLID STATE PHYSICS. Educ: Cath Univ Am, AB, 54; Univ Md, MS, 64, PhD(physics), 66. Prof Exp: Physicist, Naval Med Res Inst, 55-58 & Naval Res Lab, 59-60; Nat Acad Sci-Nat Res Coun res assoc, 65-67, PHYSICIST, NAT BUR STAND, 67- Mem: Am Phys Soc. Res: Optical properties, soft x-ray and x-ray photoemission spectroscopy of metals and alloys. Mailing Add: Nat Bur of Stand Gaithersburg MD 20760

MCALISTER, DEAN FERDINAND, b Logan, Utah, Sept 8, 10; m 32; c 3. AGRONOMY. Educ: Utah State Col, BS, 31, MS, 32; Univ Wis, PhD(plant physiol), 36. Prof Exp: Field asst agron, Exp Sta, Utah State Col, 30-32; asst plant physiol, Univ Wis, 32-36; asst physiologist, Bur Plant Indust, 36-42, from assoc plant physiologist to physiologist, Regional Soybean Lab, 46-52, AGRONOMIST, AGR EXP STA, UNIV ARIZ, USDA, 52-, ASST DIR STA, 58- Concurrent Pos: Prof agron & head dept, Univ Ariz, 52-66, chief of party, Univ Ariz-Univ Ceara, Brazil Proj, US AID, 66-68. Mem: Fel AAAS; Am Soc Plant Physiol; fel Am Soc Agron; Sigma Xi. Res: Crop physiology and production; agricultural administration. Mailing Add: Agr Exp Sta Univ of Ariz Tucson AZ 85721

MCALISTER, DONALD BEATON, b Belfast, Northern Ireland, July 15, 40; m 64; c 2. MATHEMATICS. Educ: Queen's Univ Belfast, BS, 62, MS, 63, PhD(math), 66. Prof Exp: From asst lectr to lectr math, Queen's Univ Belfast, 64-70; ASSOC PROF MATH, NORTHERN ILL UNIV, 70- Concurrent Pos: Vis asst prof, Tulane Univ, 66-67; NSF grant, Northern Ill Univ, 71-73. Mem: London Math Soc; Am Math Soc. Res: Algebraic theory of semigroups; partially ordered groups. Mailing Add: Dept of Math Northern Ill Univ De Kalb IL 60115

MCALISTER, HAROLD ALISTER, b Chattanooga, Tenn, July 1, 49; m 72. ASTRONOMY. Educ: Univ Tenn, Chattanooga, BA, 71; Univ Va, MA, 74, PhD(astron), 75. Prof Exp: Res asst, Dept Astron, Univ Va, 71-75; RES ASSOC ASTRON, KITT PEAK NAT OBSERV, 75- Mem: Am Astron Soc. Res: Astrometry; speckle interferometry of binary stars. Mailing Add: Kitt Peak Nat Observ 950 N Cherry Ave PO Box 26732 Tucson AZ 85726

MCALISTER, ROBERT HARDY, b Quinter, Kans, Sept 17, 31; m 56; c 4. FOREST PRODUCTS. Educ: Univ Idaho, BS, 54, MS, 56; Univ Ga, PhD(forest resources), 72. Prof Exp: Forest prod technologist, US Forest Prod Lab, Madison, Wis, 54-62; wood technologist, Roundwood Corp Am, 62-63; wood technologist, Moore Dry Kiln Co, 63-66; WOOD SCIENTIST, SOUTHEAST FOREST EXP STA, ASHEVILLE, NC, 66- Honors & Awards: Wood Award, Woodworking Dig, 54. Mem: Forest Prod Res Soc. Mailing Add: Forestry Sci Lab Carlton St Athens GA 30601

MCALISTER, WILLIAM BRUCE, b Seattle, Wash, Aug 11, 29. MARINE ECOLOGY. Educ: Univ Wash, BS, 49, MS, 58; Ore State Univ, PhD(oceanog), 62. Prof Exp: From instr to asst prof oceanog, Ore State Univ, 58-64; proj dir, Bur Com Fisheries, Wash, 64-70; prog dir phys oceanog, NSF, 70-72; dep dir marine fish, Northwest Fisheries Ctr, 72-74, DEP DIR MARINE MAMMAL DIV, NAT MARINE FISHERIES SERV, 74- Concurrent Pos: Fulbright res fel, Water Res Inst, Oslo, Norway, 63-64; asst prof, Univ Wash, 65-70. Mem: Am Geophys Union; Am Soc Limnol & Oceanog; AAAS; Am Inst Fish Res Biologists; Sigma Xi. Res: Descriptive oceanography, ecosystem dynamics, marine mammals and fisheries. Mailing Add: Marine Mammal Div Nav Sup Act Nat Marine Fish Serv Bldg 32 Seattle WA 98115

MCALISTER, WILLIAM H, b Highland Park, Mich, Jan 19, 30. MEDICINE, RADIOLOGY. Educ: Wayne State Univ, BS, 50, MD, 54. Prof Exp: Intern, Detroit Receiving Hosp, Mich, 54-55; resident, Cincinnati Gen Hosp, Ohio, 57-60; from instr to assoc prof radiol, 60-68, PROF RADIOL & PEDIAT IN RADIOL, EDWARD MALLINCKRODT INST RADIOL, SCH MED, WASH UNIV, 68- Concurrent Pos: Fel, Cincinnati Gen Hosp, Ohio, 57-60. Mem: Radiol Soc NAm; Asn Univ Radiol; AMA. Res: Pediatric roentgenology. Mailing Add: Mallinckrodt Inst of Radiol Wash Univ Sch of Med St Louis MO 63110

MCALLESTER, DAVID PARK, b Everett, Mass, Aug 6, 16; m 40; c 2. ANTHROPOLOGY. Educ: Harvard Univ, AB, 38; Columbia Univ, PhD(anthrop), 49. Prof Exp: Instr, Brooklyn Col, 45-46; from instr to assoc prof, 47-67, PROF ANTHROP, WESLEYAN UNIV, 67- Concurrent Pos: Asst, Columbia Univ, 45-46; res assoc, Mus Navajo Ceremonial Art, NMex, 51-; consult, State Hosp, 51-; ed, J Soc Ethnomusicol, 58-61. Mem: Fel Am Anthrop Asn; Soc Ethnomusicol (secy-treas, 55-58, pres, 62-64). Res: Personality and culture; culture and linguistics; ethnomusicology. Mailing Add: Dept of Anthrop Wesleyan Univ Middletown CT 06457

MCALLISTER, ALAN JACKSON, b Shelbyville, Ky, Aug 19, 45; m 65; c 2. ANIMAL BREEDING. Educ: Univ Ky, BS, 67; Ohio State Univ, MS, 70, PhD(animal breeding), 75. Prof Exp: Biomet geneticist poultry, DeKalb Agr Res Inc, 72-75; FEL POULTRY GENETICS, ANIMAL RES INST, AGR CAN, 75- Mem: Am Soc Animal Sci; Int Biomet Soc. Res: Genetic improvement of animal populations through application of optimal genetic evaluation procedures; where performance includes both directly observable economic traits and basic physiological measurements. Mailing Add: Animal Res Inst Agr Can Genetics Bldg Ottawa ON Can

MCALLISTER, ARNOLD LLOYD, b Petitcodiac, NB, Dec 24, 21. GEOLOGY. Educ: Univ NB, BSc, 43; McGill Univ, MSc, 48, PhD, 50. Prof Exp: Res geologist, Int Nickel Co, 50-52; assoc prof, 52-62, head dept, 62-74, PROF GEOL, UNIV NB, 62- Mem: Royal Soc Can; Geol Asn Can; Can Inst Mining & Metall. Res: Economic and structural geology. Mailing Add: Dept of Geol Univ of NB Fredericton NB Can

MCALLISTER, BYRON LEON, b Midvale, Utah, Apr 29, 29; m 57; c 3. MATHEMATICS. Educ: Univ Utah, BA, 51, MA, 55; Univ Wis, PhD(math), 66. Prof Exp: From asst prof to assoc prof math, SDak Sch Mines & Technol, 58-67; assoc prof, 67-71, PROF MATH, MONT STATE UNIV, 71- Concurrent Pos: Instr, Fox Valley Ctr, Univ Wis, 61-62. Mem: Am Math Soc; Math Asn Am. Res: General topology, particularly Whyburn cyclic element theory and extensions, multifunctions and hyperspaces. Mailing Add: Dept of Math Mont State Univ Bozeman MT 59715

MCALLISTER, CYRUS RAY, b Portland, Ore, Apr 22, 22; m 53; c 3. MATHEMATICS. Educ: Univ Minn, BA, 48; Univ Ore, MA, 51. Prof Exp: Instr math, Univ Idaho, 48-50; analyst, US Dept Defense, 52; mathematician, Sandia Corp, 52-57, consult, 57; sr res scientist & maj proj supvr nuclear div, Kaman Aircraft Corp, 57-59, consult, 59; pres & tech dir, McAllister & Assocs, Inc, 59-63; res dir, Booz, Allen Appl Res, Inc, 63-66; dir sci & consult div, Genge Industs, Inc, 66-68; staff engr, Advan Concepts Off, Aerospace Corp, Calif, 68-72; CONSULT, 72- Concurrent Pos: Consult, Hamilton Watch Co, 60. Mem: Am Meteorol Soc. Res: Applied mathematics; mathematical statistics; dynamic climatology; weapons systems analysis. Mailing Add: 4729 Libbit Ave Encino CA 91436

MCALLISTER, DEVERE RICHARD, b Provo, Utah, Nov 11, 17; m 38; c 6. AGRONOMY. Educ: Utah State Agr Col, BS, 39, MS, 48; Iowa State Col, PhD(crop breeding), 50. Prof Exp: Asst soil scientist, Soil Conserv Serv, USDA, 41-44, asst soil conservationist, 46-47; asst agron, Iowa State Col, 48-50; from asst prof to assoc prof, 50-59, PROF AGRON, UTAH STATE UNIV, 59-, EXTEN AGRONOMIST, 71- Mem: Am Soc Agron; Crop Sci Soc Am. Res: Plant breeding; forage crop improvement; seed improvement; scouting. Mailing Add: Dept of Plant Sci Utah State Univ Logan UT 84321

MCALLISTER, DONALD EVAN, b Victoria, BC, Aug 23, 34; m 56; c 5. SYSTEMATIC ICHTHYOLOGY. Educ: Univ BC, BA, 55, MA, 57, PhD(ichthyol), 64. Prof Exp: CUR FISHES, NAT MUS NATURAL SCI, NAT MUS CAN, 58- Concurrent Pos: Lectr, Univ Ottawa. Mem: Am Soc Ichthyol & Herpet; Can Soc Zool; Can Soc Wildlife & Fishery Biol; Japanese Soc Ichthyol. Res: Fish systematics, evolution, arctic, Canada and world. Mailing Add: Nat Mus of Natural Sci Ottawa ON Can

MCALLISTER, GREGORY THOMAS, JR, b Boston, Mass, July 6, 34; m 61; c 3. MATHEMATICS. Educ: St Peters Col, NJ, BS, 56; Univ Calif, Berkeley, PhD(math), 62. Prof Exp: Res mathematician, US Army Ballistic Res Labs, 63-65; PROF MATH, CTR APPLN MATH, LEHIGH UNIV, 65- Mem: Am Math Soc; Soc Indust & Appl

Math. Res: Calculus of variations; partial differential equations; numerical methods. Mailing Add: Ctr for Appln of Math Lehigh Univ Bethlehem PA 18015

MCALLISTER, HARMON CARLYLE, JR, b Durham, NC, Apr 5, 36; m 63; c 1. BIOCHEMISTRY. Educ: Univ NC, BS, 58, PhD(biochem), 63. Prof Exp: Asst biochem, Univ NC, 58-61; Muscular Dystrophy Asn fel, Univ Mich, 63-65; NIH fel cell biol, Univ Ky, 65-67; ASST PROF CHEM, WAYNE STATE UNIV, 67-, COORDR CURRICULUM STUDIES, 73- Mem: Am Chem Soc. Res: Mechanism of peptide bond formation in protein synthesis. Mailing Add: Off for Acad Progs & Planning Wayne State Univ Detroit MI 48202

MCALLISTER, HOWARD CONLEE, b Cheyenne, Wyo, Mar 14, 24; m 44; c 4. PHYSICS. Educ: Univ Wyo, BS, 48, MS, 50; Univ Colo, PhD(physics), 59. Prof Exp: From asst prof to assoc prof, 59-70, PROF PHYSICS, UNIV HAWAII, 70-; PRIN INVESTR, INST ASTRON, 69- Concurrent Pos: Consult, Univ Colo, 61 & Lincoln Lab, Mass Inst Technol, 63-64; Nat Acad Sci-Nat Res Coun sr resident res assoc, Goddard Space Flight Ctr, 65-66. Mem: Am Phys Soc; Optical Soc Am. Res: Solar ultraviolet spectroscopy and upper atmospheric physics; atomic and molecular spectroscopy; spectroscopic instrumentation for space research. Mailing Add: Dept of Physics & Astron Univ of Hawaii Honolulu HI 96822

MCALLISTER, JAMES FRANKLIN, b Mayagüez, PR, Nov 11, 11; m 46; c 3. GEOLOGY. Educ: Col Wooster, AB, 33; Stanford Univ, MA, 36, PhD, 51. Prof Exp: Asst geol, Stanford Univ, 35-39; GEOLOGIST, US GEOL SURV, 39- Mem: Geol Soc Am; Soc Econ Geologists; Mineral Soc Am. Res: Tungsten, mercury, lead and borate deposits; petrology; structure. Mailing Add: 1843 Channing Ave Palo Alto CA 94303

MCALLISTER, MARIALUISA N, b Milan, Italy, Aug 22, 33; US citizen; m 61; c 3. MATHEMATICS, COMPUTER SCIENCES. Educ: Univ Rome, PhD(math), 57. Prof Exp: NSF fel, 60-61; asst prof math, Univ Dela, 62-64; assoc prof, Towson State Col, 64-65; asst prof, 65-73, ASSOC PROF MATH, MORAVIAN COL, 73- Mem: Math Asn Am; Asn Comput Mach. Res: Algebraic geometry; numerical analysis. Mailing Add: Dept of Math Moravian Col Bethlehem PA 18018

MCALLISTER, RAYMOND FRANCIS, b Ithaca, NY, June 26, 23; m 51; c 3. OCEANOGRAPHY, OCEAN ENGINEERING. Educ: Cornell Univ, BS, 50; Univ Ill, MS, 51; Agr & Mech Col, Tex, PhD(geol oceanog), 58. Prof Exp: Instr geol, Univ Ill, 50-51; res oceanogr, Scripps Inst, Univ Calif, 51-54; marine res geologist & instr geol, Agr & Mech Col, Tex, 54-58; sr oceanogr, Bermuda Sound Fixing & Ranging Sta, Columbia Univ, 58-63; asst dir marine technol group, NAm Aviation Inc, Ohio, 64-65; PROF OCEANOG, FLA ATLANTIC UNIV, 65- Concurrent Pos: mem man in the sea panel, Nat Acad Eng, 71- Res: Marine geology and surveys; bottom photography and sampling; scuba diving; science promotion via television; ocean engineering development; artificial reef construction; ocean outfall studies. Mailing Add: Dept of Ocean Eng Fla Atlantic Univ Boca Raton FL 33432

MCALLISTER, ROBERT MILTON, b Philadelphia, Pa, June 10, 22; m 49; c 6. PEDIATRICS. Educ: Ursinus Col, BS, 42; Univ Pa, MD, 45, MS, 55; Am Bd Pediat, dipl, 53. Prof Exp: From instr to asst prof pediat, Sch Med, Univ Pa, 51-59; assoc prof, 59-64, PROF PEDIAT, CHILDREN'S HOSP, LOS ANGELES, SCH MED, UNIV SOUTHERN CALIF, 64- Mem: AAAS; Soc Pediat Res; Am Asn Cancer Res; fel Am Acad Pediat. Res: Microbiology and oncology in the pediatric age group; tissue culture. Mailing Add: Children's Hosp of Los Angeles 4614 Sunset Blvd Los Angeles CA 90027

MCALLISTER, ROBERT WALLACE, b Hermosa Beach, Calif, Feb 16, 29; m 52; c 4. PHYSICS, INSTRUMENTATION. Educ: Occidental Col, BA, 51; Stanford Univ, PhD(physics), 60. Prof Exp: Asst physics, Univ Zurich, 56-60; instr, Cornell Univ, 60-64; asst prof, 64-66, ASSOC PROF PHYSICS, COLO SCH MINES, 66- Mailing Add: Dept of Physics Colo Sch of Mines Golden CO 80401

MCALLISTER, RONALD ERIC, b Halifax, NS, Apr 15, 42; m 67; c 2. CLINICAL MEDICINE, ELECTROPHYSIOLOGY. Educ: Dalhousie Univ, BSc, 62, MSc, 63, MD, 76; Oxford Univ, DPhil(physiol), 67. Prof Exp: Asst prof physiol, Dalhousie Univ, 68-76. Mem: Can Med Asn. Res: Cardiac electrophysiology; computer simulation of excitation and propagation in cardiac cells, using experimental data. Mailing Add: c/o Fac of Med Dalhousie Univ Halifax NS Can

MCALLISTER, STUART ALLAN, b Atlanta, Ga, July 20, 23; m 50; c 4. PHYSICAL CHEMISTRY. Educ: Washington & Lee Univ, BS, 47; Pa State Univ, PhD(phys chem), 52. Prof Exp: Res chemist photo prod dept, E I du Pont de Nemours & Co, 51-58, from patent chemist to sr patent chemist, 58-67; prof chem, Gordon Col, WPakistan, 67-70; ASSOC PROF CHEM, STERLING COL, 70- Mem: Am Chem Soc; NY Acad Sci. Res: Scattering of positive ions by gases at low pressures; photographic emulsions. Mailing Add: 204 N Sixth St Sterling KS 67579

MCALLISTER, WARREN ALEXANDER, b Augusta, Ga, Mar 12, 41; m 62. INORGANIC CHEMISTRY. Educ: Mercer Univ, BA, 63; Univ SC, PhD(inorg chem), 67. Prof Exp: Res assoc, Vanderbilt Univ, 66-67; asst prof, 67-70, ASSOC PROF CHEM, E CAROLINA UNIV, 70- Concurrent Pos: Grants, USPHS, Environ Protection Agency & NC Bd Sci & Technol. Mem: Am Chem Soc. Res: Infrared and Raman studies of inorganic compounds. Mailing Add: Dept of Chem ECarolina Univ PO Box 2787 Greenville NC 27834

MCALLISTER, WILLIAM ALBERT, b Youngstown, Ohio, Oct 22, 23; m 47; c 6. PHYSICAL CHEMISTRY. Educ: Bowling Green State Univ, BS, 49; Mich State Univ, PhD(phys chem), 54. Prof Exp: Res chemist, Celanese Corp Am, 53-55; from scientist to sr scientist, Atomic Power Div, 55-59, sr scientist, Lamp Div, 59-61, phosphor sect mgr, 61-65, RES CONSULT PHOSPHORS, ADVAN DEVELOP DEPT, LAMP DIV, WESTINGHOUSE ELEC CORP, 65- Mem: Am Chem Soc; Am Phys Soc; Electrochem Soc. Res: Luminescent inorganic materials, structure and spectra, especially rare earths; metal halide vapor discharges. Mailing Add: Advan Develop Dept Lamp Div Westinghouse Elec Corp Bloomfield NJ 07003

MCALLISTER, WILLIAM BARRISS, JR, b Cleveland, Ohio, May 6, 14; m 44; c 2. MEDICINE. Educ: Yale Univ, BA, 35; Johns Hopkins Univ, MD, 39. Prof Exp: Intern med, Johns Hopkins Hosp, 39-40, asst resident pathologist, 40-41; fel, 46-49, from asst prof to assoc prof path, 51-73, PROF CLIN PATH, SCH MED, YALE UNIV, 73- Concurrent Pos: Chief path, Gen Serv, Yale-New Haven Hosp, 54- Res: Infectious diseases; chronic arterial vascular diseases. Mailing Add: Dept of Path Yale Univ Sch of Med New Haven CT 06504

MCALLISTER, WILLIAM TURNER, b Philadelphia, Pa, Apr 25, 44; m 68; c 1. MOLECULAR BIOLOGY. Educ: Lehigh Univ, BA, 66; Univ NH, PhD(biochem), 71. Prof Exp: NIH fel molecular genetics, Inst Molecular Genetics, Univ Heidelberg, 70-72, res assoc, 72-73; ASST PROF MICROBIOL, RUTGERS MED SCH, COL MED & DENT NJ, 73- Mem: Am Soc Microbiol. Res: Regulation of gene expression in virus-infected cells; control of transcription in E coli; processing of RNA in eukaryotic cells. Mailing Add: Dept of Microbiol Col Med & Dent Rutgers Med Sch Piscataway NJ 08854

MCALPIN, CESARIA EUGENIO, b Rosales, Philippines, Feb 25, 40; m 72; c 2. BOTANY. Educ: Univ Philippines, BS, 61; Univ Minn, MS, 65, PhD(phytopath), 68. Prof Exp: Res asst plant path, Univ Minn, 63-68, res fel plant physiol & insect microbiol, 68-70, teaching specialist biol, 70-71; ASSOC PROF BIOL, CLAFLIN COL, 71- Mem: AAAS; Sigma Xi; Am Inst Biol Sci; Bot Soc Am. Res: Toxic metabolites produced by fungi and algae. Mailing Add: 1110 Fairfield St Orangeburg SC 29115

MCALPIN, JOHN HARRIS, b Natchez, Miss, Dec 7, 33; m 63; c 4. MATHEMATICS. Educ: Columbia Univ, AB, 58, PhD(math), 65; NY Univ, MS, 60. Prof Exp: Teacher math high schs, NY, 58-60; instr, Brown Univ, 64-66; asst prof, Univ Colo, Boulder, 66-69; PROF MATH, SC STATE COL, 69- Mem: Am Math Soc; Math Asn Am. Res: Operations research. Mailing Add: Dept of Math SC State Col Box 2011 Orangeburg SC 29117

MCALPINE, JAMES BRUCE, b Ingham, Queensland, Australia, Dec 26, 39; m 68; c 2. ORGANIC CHEMISTRY. Educ: Univ New Eng, Australia, BSc, 62, MSc, 64, PhD(org chem), 69. Prof Exp: Fel biochem, Med Sch, Northwestern Univ, 69-71, asst prof biochem, 71-72; sr res chemist, 72-75, PROJ LEADER ANTIBIOTICS MODIFICATION, ABBOTT LABS, 75- Mem: The Chem Soc; Am Chem Soc; Sigma Xi. Res: The chemistry, mode of action and toxicity of antimicrobial agents. Mailing Add: 211 W Rockland Rd Libertyville IL 60048

MCALPINE, JAMES FRANCIS, b Maynooth, Ont, Sept 25, 22; m 50. ENTOMOLOGY. Educ: Univ Toronto, BSA, 50; Univ Ill, MSc, 54, PhD(entom), 62. Prof Exp: Tech officer, 50-53, res officer, 53-63, RES SCIENTIST, BIOSYSTS RES INST, AGR CAN, 63- Mem: Entom Soc Can. Res: Systematics of two-winged flies. Mailing Add: Biosysts Res Inst Agr Can Ottawa ON Can

MCALPINE, PHYLLIS JEAN, b Petrolia, Ont, Aug 29, 41. HUMAN GENETICS, MEDICAL GENETICS. Educ: Univ Western Ont, BSc, 63; Univ Toronto, MA, 66; Univ London, PhD(human genetics), 70. Prof Exp: Med Res Coun Can fel, Queen's Univ, 70-72; res assoc human genetics, Health Sci Children's Ctr, 72-74; ASST PROF PEDIAT, UNIV MAN, 74-; SCI STAFF, HEALTH SCI CHILDREN'S CTR, 75- Mem: Genetics Soc Can; Am Soc Human Genetics. Res: Mapping the human genome by use of somatic cell hybrids and family studies; expression of genes in human tissues. Mailing Add: Dept of Genetics Health Sci Children's Ctr 685 Bannatyne Ave Winnipeg MB Can

MCALPINE, ROBERT GOODING, forestry, silviculture, see 12th edition

MCANALLY, JOHN SACKETT, b Indianapolis, Ind, Apr 15, 18; m 43; c 2. ANALYTICAL BIOCHEMISTRY. Educ: Ind Univ, BS, 38, AM, 40, PhD(chem), 50. Prof Exp: Res asst prof biochem, Med Res Unit, Univ Miami, 50-52, asst prof, Med Sch, 52-57; from asst prof to assoc prof, 57-72, dean students, 65-68, PROF BIOCHEM, OCCIDENTAL COL, 72- Mem: AAAS; Am Chem Soc; NY Acad Sci. Res: Vitamin A; urinary estrogens and androgens; trace element analysis; seawater and marine organisms. Mailing Add: Dept of Chem Occidental Col Los Angeles CA 90041

MCANDREWS, HARRY, b Brownsville, Pa, Oct 5, 26; m 50; c 3. GEOLOGY. Educ: WVa Univ, BS, 55, MS, 56. Prof Exp: Geologist, Wyo, 56-64, dist geologist, Utah & Wyo, 64-68, regional geologist, DC, 68-71, staff geologist, 71-74, OIL & GAS SUPVR, CONSERV DIV, US GEOL SURV, 74- Mem: Geol Soc Am; Am Asn Petrol Geol; Am Inst Prof Geol. Res: Evaluation and analysis of the mineral resources of the Gulf of Mexico Outer Continental Shelf area. Mailing Add: US Geol Surv PO Box 7944 Metairie LA 70011

MCANDREWS, JOHN HENRY, b Minneapolis, Minn, Jan 16, 33; m 58; c 4. BOTANY, PLANT ECOLOGY. Educ: Col St Thomas, BS, 57; Univ Minn, MS, 59, PhD(bot), 64. Prof Exp: Res assoc paleoecol, Inst Bio-Archeol, Groningen, Neth, 63-64; asst prof biol, Jamestown Col, 64-66 & Cornell Col, 66-67; CUR GEOL, ROYAL ONT MUS, 67-, ASSOC PROF BOT, UNIV TORONTO, 68- Concurrent Pos: Vis prof ecol, Univ Minn, 74; vis lectr anthrop, Univ Man, 75. Res: Vegetation history; climatic change; pollen analysis; pollen morphology. Mailing Add: Royal Ont Mus Dept of Geol 100 Queens Park Toronto ON Can

MCANELLY, CHARLES WILLIAM, b Greeley, Colo, Sept 24, 13; m 55; c 2. PLANT PATHOLOGY. Educ: Colo Agr & Mech Col, BS, 35, MS, 53; Univ Wyo, PhD(agron), 58. Prof Exp: Field survr, Colo Agr & Mech Col, 36; supt pvt irrig co, 37-42 & 46-48; water comnr, State of Colo, 48-51; asst bot & plant path, Colo Agr & Mech Col, 51-53, temp asst horticulturist, 53, asst horticulturist, 54; instr & asst agron, 55-58, asst prof plant path & asst agronomist, 58-64, assoc plant pathologist & horticulturist, 64-68, PROF PLANT PATH & HORT, UNIV WYO, 68- Concurrent Pos: Res adv & horticulturist, Afghanistan Prog, Univ Wyo, 59-61; agr consult, Agr Specialties, Inc. Mem: Am Phytopath Soc; Potato Asn Am; Am Soc Hort Sci. Res: Diseases of potatoes and ornamental plants; virus diseases of plants; horticultural problems of vegetable crops and ornamentals. Mailing Add: Plant Sci Sect Univ of Wyo Laramie WY 82070

MCANELLY, JOHN KITCHEL, b Logansport, Ind, June 22, 31; m 53; c 2. FOOD MICROBIOLOGY. Educ: Iowa State Univ, BS, 53; NC State Univ, MS, 56; Univ Wis, PhD(bact), 60. Prof Exp: Res biochemist, Res & Develop Ctr, Swift & Co, Ill, 59-61, head biochem div, 61-66; dir res & develop, Rival Pet Foods, 66-68, vpres tech develop, 68-75; DIR CORP QUAL ASSURANCE, NABISCO, INC, 75- Mem: Am Chem Soc; Am Soc Microbiol; Inst Food Technol; fel Am Inst Chemists. Res: Biochemistry and microbiology of food products and processes; thermal processing of foods; utilization of proteins. Mailing Add: Nabisco Inc Res Ctr Fairlawn NJ 07410

MCANENY, LAURENCE RAYMOND, b Seattle, Wash, Apr 12, 26; m 46; c 3. PHYSICS. Educ: Univ Kans, BS, 46, PhD(physics), 57; Univ Calif, MA, 48. Prof Exp: Asst prof physics, Park Col, 51-57; from asst prof to assoc prof physics, 57-67, asst dean acad affairs, 63-67, dean div sci & technol, 67-73, PROF PHYSICS, SOUTHERN ILL UNIV, EDWARDSVILLE, 67- Concurrent Pos: Pres bd dirs, Cent States Univs, Inc, 75. Mem: Am Phys Soc; Am Asn Physics Teachers. Res: Theoretical physics; statistical mechanics. Mailing Add: Sch of Sci & Technol Southern Ill Univ Edwardsville IL 62026

MCANINCH, LLOYD NEALSON, b Guelph, Ont, June 6, 20; m 44; c 3. UROLOGY. Educ: Univ Western Ont, MD, 45; FRCPS(C), 53. Prof Exp: Intern, Victoria Hosp, London, Ont, 45-46; asst to Dr E D Busby, 46-49; instr anat, Fac Med, Univ Western Ont, 49-50; asst resident surg, Westminster Vet Hosp, Ont, 50-51; resident urol, Toronto Gen Hosp, 51-52; resident, Sunnybrook Hosp, 52-53; asst resident path, Westminster Hosp, 53; from instr to asst prof surg, 53-63, from clin assoc prof to clin prof urol, 63-70, chief urol, 70-73, PROF SURG, FAC MED, UNIV WESTERN ONT, 70-, CHIEF UROL, UNIV HOSP, 73- Concurrent Pos: Consult, Westminster Vet Hosp, 54-; chief urol, Victoria Hosp, Ont, 56-73, consult; attend urologist & consult, St Joseph's Hosp, Ont. Mem: Fel Am Col Surg; Can Med Asn; Can Urol Asn (pres, 74-75); fel Can Asn Clin Surg; Can Acad Urol Surg (treas, 60-65, pres, 66). Res: Vesico-ureteral reflux; renal trauma; retroperitoneal tumors; urological emergencies; chemotherapy in urology; ureteral substitutions; external meatotomy. Mailing Add: Dept of Surg Fac of Med Univ of Western Ont London ON Can

MCANULTY, WILLIAM NOEL, b Howe, Okla, Nov 26, 13; m 38; c 3. GEOLOGY. Educ: Univ Okla, BS, 38, MS, 48; Univ Tex, PhD, 53. Prof Exp: Assoc prof sci, Sul Ross State Col, 47-51; consult geologist, 51-53; chief geologist, Dow Chem Co, 53-64; head dept, 64-70, PROF GEOL, UNIV TEX, EL PASO, 64- Mem: Geol Soc Am; Am Asn Petrol Geologists; Soc Econ Geol; Am Inst Mining, Metall & Petrol Engr; Am Inst Prof Geologists. Res: General and economic geology; industrial minerals; surface mapping. Mailing Add: 220 Stratus Rd El Paso TX 79912

MCARDLE, EUGENE W, b Chicago, Ill, Oct 26, 31; m 61; c 6. ZOOLOGY. Educ: St Mary's Col, Minn, BS, 53; Marquette Univ, MS, 56; Univ Ill, PhD(zool), 60. Prof Exp: Asst prof biol, St Benedict's Col, Kans, 60-61; assoc prof, St Mary's Col, Minn, 61-69; ASSOC PROF BIOL, NORTHEASTERN ILL UNIV, 69- Concurrent Pos: Lectr, NSF In-Serv-Insts, Winona State Col, 65; visitor, Argonne Nat Lab, 68- Mem: Soc Protozoologists. Res: Cytology and behavior of ciliate protozoan Tetrahymena rostrata. Mailing Add: Dept of Biol Northeastern Ill Univ Chicago IL 60625

MCARDLE, JOSEPH JOHN, b Wilmington, Del, July 21, 45. NEUROPHYSIOLOGY, NEUROPHARMACOLOGY. Educ: Univ Del, BA, 67; State Univ NY Buffalo, PhD(pharmacol), 71. Prof Exp: Vis asst prof pharmacol, Sch Med & Dent, State Univ NY Buffalo, 71-72; ASST PROF PHARMACOL, COL MED & DENT NJ, 72- Concurrent Pos: Nat Inst Neurol Dis & Stroke fel, Col Med & Dent NJ, 73-76. Mem: Soc Neurosci. Res: Development of mechanical, electrical and pharmacological properties of skeletal muscle and the mechanisms involved in the maintenance of these properties; reinnervation of muscle; synaptic transmission. Mailing Add: Dept of Pharmacol Col Med & Dent of NJ Newark NJ 07103

MCARDLE, RICHARD EDWIN, b Lexington, Ky, Feb 25, 99; m 27; c 3. FORESTRY. Educ: Univ Mich, BS, 23, MS, 24, PhD, 30. Hon Degrees: ScD, Univ Mich, 53 & Univ Maine, 62; LLD, Syracuse Univ, 61. Prof Exp: Jr forester, US Forest Serv, 24-27, from asst silviculturist to assoc silviculturist, 26-34; dean sch forestry, Univ Idaho, 34-35; dir, Rocky Mountain Forest & Range Exp Sta, US Forest Serv, 35-38 & Appalachian Forest Exp Sta, 38-44, asst chief, 44-52, chief serv, 52-62; exec dir, Nat Inst Pub Affairs, 62-64; consult, 65-66; MEM BD DIRS, OLINKRAFT, INC, 67- Concurrent Pos: Mem, Royal Comn Forestry, Nfld & Labrador, 67-71; resources consult, Nat Wildlife Fedn, 67-; pres, Fifth World Forestry Cong. Honors & Awards: Distinguished Serv Award, USDA, 57; Awards, Nat Civil Serv League, 58, Am Forestry Asn, 58 & Pub Personnel Asn, 59; Distinguished Serv Award, NY State Col Forestry, Syracuse Univ; Rockefeller Pub Serv Award; President's Gold Medal & Order Merit Forestry of Miguel Angel de Quevedo, Mex, 61; Knight Comdr, Order Merit, Ger, 62; Sir William Schlich Mem Medal, Soc Am Foresters, 62; Rockefeller Forester-in-Residence, Univ Maine, 65. Mem: Fel Soil Conserv Soc Am; fel Soc Am Foresters; Royal Swedish Acad Agr & Forestry. Res: Executive development; forest administration. Mailing Add: 5110 River Hill Rd Washington DC 20016

MCARTHUR, CHARLES STEWART, b Stratford, Ont, Apr 11, 08; m 35; c 3. BIOCHEMISTRY. Educ: Univ Western Ont, BA, 35, MSc, 38; Univ Toronto, PhD(path chem), 43. Prof Exp: Res assoc med res, Univ Toronto, 39-41, demonstr path chem, 40-46, asst prof med res, 47-49; PROF BIOCHEM, UNIV SASK, 49- Res: Biochemistry of lipids. Mailing Add: Dept of Biochem Univ of Sask Saskatoon SK Can

MCARTHUR, CHARLES WILSON, b New Orleans, La, Nov 4, 21; m 43; c 6. MATHEMATICS. Educ: La State Univ, BS, 47; Brown Univ, MS, 50; Tulane Univ, PhD(math), 54. Prof Exp: Instr math, Univ Md, 52-53; asst prof, Ala Polytech Inst, 53-56; assoc prof, 56-64, PROF MATH, FLA STATE UNIV, 64-, CHMN DEPT, 74- Mem: AAAS; Am Math Soc; Math Asn Am. Res: Functional analysis, particularly biorthogonal systems and Schauder bases; ordered topological vector spaces. Mailing Add: Dept of Math Fla State Univ Tallahassee FL 32306

MCARTHUR, COLIN RICHARD, b Beamsville, Ont, July 18, 35; m 59; c 2. ORGANIC CHEMISTRY. Educ: Univ Western Ont, BSc, 57, MSc, 58; Univ Ill, PhD(org chem), 61. Prof Exp: Sr res chemist, Allied Chem Corp, NY, 61-67; asst prof natural sci, 67-70 & chem, 70-71, ASSOC PROF CHEM, YORK UNIV, 71- Mem: Am Chem Soc; Chem Inst Can. Res: Organic halogen compounds; organometallic reagents; carbenes; azomethines; heterocycles. Mailing Add: Dept of Chem York Univ 4700 Keele St Downsview ON Can

MCARTHUR, DAVID SAMUEL, b Nelson, NZ, May 3, 41. PHYSICAL GEOGRAPHY. Educ: Univ NZ, BSc, 62; Univ Canterbury, MSc, 64; Christchurch Teachers' Col, dipl, 64; La State Univ, PhD(geog), 69. Prof Exp: Res instr coastal geomorphol, Coastal Studies Inst, La State Univ, 68; vis asst prof geol, Mich State Univ, 69; vis asst prof geog, La State Univ, 69-70; asst prof, Univ Calif, Davis, 70-73; ASST PROF GEOG, SAN DIEGO STATE UNIV, 73- Mem: Asn Am Geog; NZ Geog Soc; Am Quaternary Asn; Coastal Soc. Res: Coastal geomorphology, especially beach sedimentation and morphology. Mailing Add: Dept of Geog San Diego State Univ San Diego CA 92182

MCARTHUR, ELDON DURANT, b Hurricane, Utah, Mar 12, 41; m 63; c 3. PLANT GENETICS. Educ: Univ Utah, BS, 65, MS, 67, PhD(biol), 70. Prof Exp: Teaching asst biol, Univ Utah, 66-70; Agr Res Coun Gt Brit fel, Sigma Xi grant & demonstr, Univ Leeds, 70-71; teaching fel biol, Univ Utah, 71; res geneticist, Great Basin Exp Area, 72-75, RES GENETICIST, SHRUB SCI LAB, INTERMOUNTAIN FOREST & RANGE EXP STA, FOREST SERV, USDA, 75- Concurrent Pos: Adj fac mem bot & range sci, Brigham Young Univ, 75- Mem: Soc Range Mgt; Bot Soc Am; Soc Study Evolution. Res: Genetics and cytology of Mimulus; cytology of Brassiceae; genetics, cytology, breeding and selection of intermountain shrubs. Mailing Add: Shrub Sci Lab USDA 735 North 500 East Provo UT 84601

MCARTHUR, JANET W, b Bellingham, Wash, June 25, 14. ENDOCRINOLOGY. Educ: Univ Wash, AB, 35, MS, 37; Northwestern Univ, MD, 42; Am Bd Internal Med, dipl, 48. Hon Degrees: DSc, Mt Holyoke Col, 62. Prof Exp: Res fel pediat, 48-50, instr, 50-51, instr gynec, 51-57, clin assoc med, 57-60, from asst clin prof to assoc clin prof, 60-71, assoc prof, 71-73, PROF OBSTET & GYNEC, HARVARD MED SCH, 73- Concurrent Pos: Asst, Children's Med Serv, Mass Gen Hosp, 48-50, asst med, 50-51, asst physician, 51-60, assoc physician, 60-; consult physician, Mass Eye & Ear Infirmary, 52; mem, Ctr Pop Studies, Sch Pub Health, Harvard Univ. Mem: AAAS; Am Fertil Soc; Endocrine Soc; AMA; Am Col Physicians. Res: Bioassay of pituitary hormones; identification of pituitary hormones in human plasma. Mailing Add: Mass Gen Hosp 32 Fruit St Boston MA 02114

MCARTHUR, NEIL M, b London, Ont, Jan 1, 21; m 46; c 2. GEOGRAPHY. Educ: Univ Western Ont, BA, 48, BA, 49, MA, 50; Univ Mich, PhD(geog), 55. Prof Exp: Asst prof geog, Mich State Norm Col, 51-57 & Univ Md, 57-61; assoc prof, Royal Mil Col, Ont, 61-68; ASSOC PROF GEOG, ATKINSON COL, YORK UNIV, 68- Concurrent Pos: Consult, Atlantic Develop Bd, 66-67 & St Clair Region Develop Coun, 68. Mem: Asn Am Geogr; Can Asn Geogr. Res: Land use geography, especially airport location; recreational geography. Mailing Add: Dept of Geog Atkinson Col York Univ Downsview ON Can

MCARTHUR, RICHARD EDWARD, b Bradford, Pa, Dec 29, 15; m 42; c 3. ORGANIC CHEMISTRY. Educ: Temple Univ, AB, 37; Pa State Col, MS, 39, PhD(org chem), 41. Prof Exp: Instr chem, Exten Sch, Temple Univ, 36-37; chem operator, Pa State Univ, 39; res chemist, Niagara Alkali Co, NY, 40; res chemist, Sherwood Refining Co, Pa, 41-45, chief chemist & operating supt, 45-46; res chemist, Olin Corp, Conn, 46-52, sect head chem res, 52-56, mgr, 56-60, assoc dir, 60-64, planning scientist, 65-69, , sect mgr customer serv-urethanes, 69-74; CHEM CONSULT, 74- Mem: Am Chem Soc; Sigma Xi. Res: Organic fluorine compounds; laboratory process development; high pressure reactions; economic analyses and chemical feasibility studies; flexible and rigid urethane foams; polyether and isocyanate synthesis and process development. Mailing Add: 104 Walter Lane Hamden CT 06514

MCARTHUR, WILLIAM GEORGE, b Kearney, NJ, July 1, 40; m 66; c 2. MATHEMATICS. Educ: Villanova Univ, BS, 66; Pa State Univ, PhD(math), 69. Prof Exp: Asst prof, 69-71, ASSOC PROF MATH, SHIPPENSBURG STATE COL, 71- Mem: Am Math Soc; Math Asn Am. Res: Realcompact topological spaces and realcompactifications. Mailing Add: Dept of Math Shippensburg State Col Shippensburg PA 17257

MCARTHUR, WILLIAM HENRY, b Selma, Ala, Oct 15, 22; m 57; c 1. ZOOLOGY. Educ: Morehouse Col, BS, 47; Atlanta Univ, MS, 48; Iowa Univ, PhD, 55. Prof Exp: Instr zool, Morehouse Col, 48-51; asst prof, 52-55, PROF ZOOL & CHMN DIV NAT SCI & MATH, KNOXVILLE COL, 55- Mem: Am Physiol Soc; Nat Inst Sci. Res: Parasitology; protozoology. Mailing Add: Dept of Biol Knoxville Col Box 173 Knoxville TN 37921

MCATEE, JAMES LEE, JR, b Waco, Tex, Aug 29, 24; m 47; c 4. COLLOID CHEMISTRY. Educ: Tex A&M Univ, BS, 47; Rice Univ, MS, 49, PhD, 51. Prof Exp: Supvr tech serv labs, Baroid Div, Nat Lead Co, 51-59; from asst prof to assoc prof, 59-71, PROF CHEM, BAYLOR UNIV, 71- Concurrent Pos: Consult, NL Indust & Mobil Oil Co. Mem: Am Crystallog Asn; Am Chem Soc; Mineral Soc Am; NAm Thermal Anal Soc; Clay Minerals Soc. Res: Clay minerals, especially montmorillonite; crystal structure of montmorillonite and organic and metal-ligand-montmorillonite complexes by means of x-ray diffraction, differential thermal techniques and electron microscopy. Mailing Add: Dept of Chem Baylor Univ Waco TX 76703

MCATEE, LLOYD THOMAS, b Lexington, Ky, July 4, 39; m 66; c 1. CELL BIOLOGY, ZOOLOGY. Educ: Hanover Col, BA, 61; Drake Univ, MA, 63; Univ Md, PhD, 69. Prof Exp: Assoc prof, 68-75, PROF MICROBIOL, PRINCE GEORGE'S COMMUNITY COL, 75-, CHMN DEPT BIOL SCI, 74- Mem: AAAS; Am Inst Biol Sci; Am Soc Microbiol. Res: Cytogenetic and kinetic effects of sublethal heat shocks on cell suspension cultures. Mailing Add: Dept of Biol Sci Prince George's Community Col Upper Marlboro MD 20870

MCATEE, PATRICIA ROONEY, b Denver, Colo, Apr 20, 31; m 54; c 1. COMMUNITY HEALTH. Educ: Loretto Heights Col, BS, 53; Univ Colo, MS, 62. Prof Exp: Dir sch health prog pub schs, Littleton, Colo, 58-60; asst prof community health, Med Ctr, Univ Colo, 62-67, spec consult curric develop proj, 67-68, acad adminr postgrad educ, 69-71; proj dir continuing educ, Western Interstate Comn Higher Educ, 72-74; PROJ CO-DIR PRIMARY HEALTH CARE, SCH MED, UNIV COLO, 74- Concurrent Pos: Secy-treas, Found Urban & Neighborhood Develop, Inc, 65-73, v chmn, 73, consult, Great Lakes Ctr, 72- & Great Plains Ctr, 73-; consult, Western Interstate Comn Higher Educ, 72-, Am Occup Ther Asn, 74- & Univ Without Walls, Union Grad Sch, 75; health consult, Town of Paonia, Colo, 74; ed, Pediat Nursing, 75; mem comt develop manpower policy, Nat Inst Med, 75- Mem: Nat Inst Med; Acad Pediat. Res: Co-developer of primary care medical practitioner; designer of curriculum in schools of medicine and medical centers and of roles for primary care providers in communities with analyses of social impact. Mailing Add: 877 E Panama Dr Littleton CO 80121

MCATEER, JAMES HOWARD, physical chemistry, see 12th edition

MCAULEY, AULEY ANDERSON, b Muddy Creek Forks, Pa, Nov 12, 12; m 38; c 4. ZOOLOGY. Educ: DePauw Univ, AB, 34; Univ Calif, Berkeley, PhD(zool), 41. Prof Exp: Teaching asst zool, Univ Calif, Berkeley, 34-37; actg head dept biol, Monmouth Col, Ill, 40-41; asst prof, Hamline Univ, 41-45; prof & head dept, Wesleyan Col, Ga, 45-46; asst prof zool, Miami Univ, 46-48; asst prof biol sci, Mich State Univ, 48-52, from asst prof to assoc prof natural sci, 52-68; PROF BIOL, EISENHOWER COL, 68- Concurrent Pos: Ford Found Fund Advan Educ fac fel, Harvard Univ, 53-54; consult, Comn Undergrad Educ in Biol Sci, 65-66 & Intermediate Sci Curric Study, Fla State Univ, 66. Mem: Hist Sci Soc; Nat Sci Teachers Asn; Am Inst Biol Sci. Res: Histology of termites; history of biological theories; undergraduate curricula in science for the non-science student; pheromones and the behavior of social insects. Mailing Add: Div of Sci & Math Eisenhower Col Seneca Falls NY 13148

MCAULEY, LOUIS FLOYD, b Travelers Rest, SC, Aug 21, 24; m 65; c 2. MATHEMATICS. Educ: Okla State Univ, BS, 49, MS, 50; Univ NC, PhD(math), 54. Prof Exp: Asst math, Okla State Univ, 49-50, instr, 50; instr Univ NC, 51-54 & Univ Md, 54-56; from instr to assoc prof, Univ Wis, 56-63; prof, Rutgers Univ, New Brunswick, 63-69; PROF MATH & CHMN DEPT MATH SCI, STATE UNIV NY BINGHAMTON, 69- Concurrent Pos: Vis assoc prof, La State Univ, 59-60; Off Naval Res fel, Univ Va, 62-63; mem, Inst Advan Study, 66-67. Mem: Am Math Soc; Math Asn Am. Res: Topology; point sets; structure of continua, upper semicontinuous collections; abstract spaces; fiber spaces; light open mappings; manifolds; regular mappings and generalizations. Mailing Add: Dept of Math Sci State Univ of NY Binghamton NY 13901

MCAULEY, PATRICIA TULLEY, b Middlebury, Vt, May 23, 35; m 65; c 3. TOPOLOGY. Educ: Vassar Col, AB, 55; Univ Wis-Madison, MS, 58, PhD(math), 62. Prof Exp: Asst prof math, Univ Md, 62-65; asst prof, Rutgers Univ, 65-68, assoc prof & chmn dept, Douglass Col, 68-69; ASSOC PROF MATH, STATE UNIV NY BINGHAMTON, 69- Concurrent Pos: Grant-dir undergrad res partic proj, NSF, 73. Mem: Am Math Soc. Res: Fiber spaces; shape theory; open maps; fixed point problems. Mailing Add: Dept of Math State Univ of NY Binghamton NY 13903

MCAULIFFE, CLAYTON DOYLE, b Chappell, Nebr, Aug 18, 18; m 43; c 4. SOIL SCIENCE. Educ: Nebr Wesleyan Univ, AB, 41; Univ Minn, MS, 42; Cornell Univ, PhD(soil sci), 48. Prof Exp: Res chemist, Div War Res, Columbia Univ, 43-44 &

Carbide & Carbon Chem Corp, 44-46; asst soil scientist, Bur Plant Indust, USDA, 47-48; res assoc agron, Cornell Univ, 48-50; res assoc prof, Stable Isotopes Lab, NC State Col, 50-56; sr res chemist, 56-68, SR RES ASSOC, CHEVRON OIL FIELD RES CO, 68- Mem: Fel AAAS; Am Chem Soc; Soil Sci Soc Am; Am Soc Agron; Soc Petrol Eng. Res: Environmental studies; solubility of hydrocarbons in water; multiphase fluid flow; geochemistry in petroleum exploration; soil chemistry; radio isotopes and stable isotopes in soil-plant investigations; stable isotope in surface area measurements; isotopic analysis of uranium. Mailing Add: Chevron Oil Field Res Co Box 446 La Habra CA 90631

MCAVOY, BRUCE RONALD, b Jamestown, NY, Jan 30, 33. PHYSICS. Educ: Univ Rochester, BS, 56. Prof Exp: Jr engr, Air Arm Div, 56, assoc engr, 57, res engr, Res Lab, 57-66, SR RES ENGR, RES LAB, WESTINGHOUSE ELEC CORP, 67- Concurrent Pos: Lectr, Dept Elec Eng, Carnegie-Mellon Univ, 68-; mem, Nat Patent Coun. Mem: Am Phys Soc; sr mem Inst Elec & Electronics Engrs; Int Microwave Power Inst; NY Acad Sci. Res: Optical physics; masers and lasers; microwave bulk effects in solids; microwave acoustics. Mailing Add: Res Lab Westinghouse Elec Corp Pittsburgh PA 15235

MCBAIN, JOHN KEITH, b San Francisco, Calif, Jan 3, 33; m 53; c 3. NUCLEAR MEDICINE, INTERNAL MEDICINE. Educ: Stanford Univ, BA, 53, MD, 56; Univ Rochester, MS, 66; Am Bd Nuclear Med, cert specialist, 72. Prof Exp: Intern med, Stanford Univ Hosps, 56-57, resident intern med, 57-59; resident, Ft Miley Vet Hosp, San Francisco, 59-60; chief resident & instr med, Med Ctr, Stanford Univ, 60-61; internist & chief gen med, US Air Force Hosp, Andrews AFB, Md, 61-65, chmn dept med & dir nuclear med, US Air Force Hosp, Wright-Patterson AFB, Ohio, 66-67; co-dir nuclear med, Miami Valley Hosp, Dayton, 68-69; ASSOC PROF RADIOL, ASSOC MED & DIR NUCLEAR MED, SCH MED, UNIV LOUISVILLE, 69- Concurrent Pos: Consult, Ft Knox Army Hosp, Ky, 69- & US Vet Admin Hosp, Louisville, 69-; mem consult staff, Louisville Gen Hosp, Jewish Hosp & Children's Hosp, Louisville, 69- Mem: AAAS; Soc Nuclear Med; fel Am Col Physicians; NY Acad Sci; Health Physics Soc. Res: Clinical nuclear medicine; use of tumor localizing radionuclides, especially gallium 67 by various human tumors. Mailing Add: Dept of Radiol Univ of Louisville Health Sci Ctr Louisville KY 40202

MCBANE, BRUCE NEWTON, b Nampa, Idaho, Oct 25, 17; m 39; c 3. INDUSTRIAL CHEMISTRY. Educ: Univ Idaho, BS, 39; Univ Wis, MS, 41. Prof Exp: Chemist, Pittsburgh Plate Glass Co, 41-48, asst tech dir, Ditzler Color Div, 48-59, proj leader, 59-63, mgr automotive coating develop, 63-69, dir automotive coating develop, PPG Industs Inc, 69-73, DIR INDUST COATING DEVELOP, PPG INDUSTS INC, 73- Mem: Am Chem Soc; Fedn Socs Coating Technol. Res: Protective coatings; unique industrial coatings for metal, wood and plastic, development and evaluation. Mailing Add: 4213 E Ewalt Rd Gibsonia PA 15044

MCBAY, ARTHUR JOHN, b Medford, Mass, Jan 6, 19; m 46; c 2. TOXICOLOGY, CHEMISTRY. Educ: Mass Col Pharm, BS, 40, MS, 42; Purdue Univ, PhD(chem), 48. Prof Exp: From asst prof to assoc prof chem, Mass Col Pharm, 48-55; supvr chem lab, Mass State Police, 55-63; supvr lab, Mass Dept Pub Safety, 63-69; assoc prof path, Med Sch & assoc prof toxicol, Sch Pharm, 69-74, PROF PATH, MED SCH & PROF PHARM, SCH PHARM, UNIV NC, CHAPEL HILL, 74-, CHIEF TOXICOLOGIST, OFF CHIEF MED EXAMR NC, 69- Concurrent Pos: Asst, Harvard Med Sch, 52-63; consult, Mass Col Pharm, 55-69; assoc prof, Law-Med Inst & Med Sch, Boston Univ, 63. Mem: Am Pharmaceut Asn; Am Acad Forensic Sci. Res: Organic pharmaceutical chemistry; spectrophotometric assays; toxicology-barbiturates and carbon monoxide; analytical chemistry. Mailing Add: Off of Chief Med Examr for NC PO Box 2488 Chapel Hill NC 27514

MCBAY, HENRY CECIL, b Mexia, Tex, May 29, 14; m 54; c 2. CHEMISTRY. Educ: Wiley Col, BS, 34; Atlanta Univ, MS, 36; Univ Chicago, PhD(chem), 45. Prof Exp: Instr chem, Wiley Col, 36-38 & Western Univ Kansas City, 38-39; from instr to assoc prof, 45-71, PROF CHEM, MOREHOUSE COL, 71-, CHMN DEPT, 60- Concurrent Pos: Tech expert, UNESCO, 51. Mem: Am Chem Soc. Res: Organic and inorganic chemistry; free radicals. Mailing Add: Dept of Chem Morehouse Col Atlanta GA 30314

MCBEATH, DOUGLAS KAY, b Raymond, Alta, Sept 19, 32; m 58; c 3. SOIL SCIENCE, PLANT NUTRITION. Educ: Univ Sask, BSA, 57; Univ Alta, MSc, 59; Cornell Univ, PhD(agron), 62. Prof Exp: RES SCIENTIST PLANT NUTRIT, RES BR, CAN DEPT AGR, 62- Concurrent Pos: Tech adv, All-India Coord Dryland Res Proj, 71-73. Mem: Am Soc Agron; Soil Sci Soc Am; Can Soc Soil Sci; Agr Inst Can. Res: Soil fertility; economics of fertilizer use; plant competition. Mailing Add: Can Dept of Agr Lacombe AB Can

MCBEE, EARL THURSTON, chemistry, see 12th edition

MCBEE, GEORGE GILBERT, b Eastland, Tex, Aug 15, 29; m 54; c 2. PLANT PHYSIOLOGY. Educ: Tex A&M Univ, BS, 51, MS, 56, PhD(plant physiol), 65. Prof Exp: Asst county agr agent, Agr Exten Serv, Tex A&M Univ, 53-54, res asst soil chem, 54-56, area agron specialist, Agr Exten Serv, 56-60, state agron specialist, 60-62, res asst plant physiol & biochem, 62-64, asst prof turf physiol, 64-69, resident dir res, Tex Agr Exp Sta, 69-75, PROF SOIL & CROP SCI, TEX A&M UNIV, 75- Mem: Am Soc Plant Physiol; Am Soc Agron; Weed Sci Soc Am. Res: Metabolic, nutritional and photobiological functions in turfgrasses; adaptive research for fungicides, fertilizers, herbicides and management systems in turfgrasses. Mailing Add: Dept of Soil & Crop Sci Tex A&M Univ College Station TX 77801

MCBEE, JAMES LEONARD, JR, b Philippi, WVa, May 4, 31; m 53; c 2. ACADEMIC ADMINISTRATION. Educ: WVa Univ, BS, 52, MS, 56; Univ Mo, PhD(meat sci), 59. Prof Exp: Prof food sci, WVa Univ, 59-69, actg chmn dept animal sci, 64; prof agr & chmn dept, 70-74, EXEC OFFICER, OFF OF THE PRES, ILL STATE UNIV, 74- Mem: Am Meat Sci Asn; Am Soc Animal Sci. Res: Carcass evaluation, and relationship of production practices and nutrition to carcass characteristics; administration of higher education. Mailing Add: 410 Hovey Hall Ill State Univ Normal IL 61761

MCBEE, RICHARD HARDING, b Eugene, Ore, May 15, 16; m 40; c 4. MICROBIOLOGY. Educ: Ore State Col, BS, 38, MS, 40; State Col Wash, PhD(bact), 48; Am Bd Med Microbiol, dipl. Prof Exp: Asst, Univ Md, 40-41; sr asst bacteriologist, Md State Dept Health, 41-42, assoc bacteriologist, 42-43; jr bacteriologist, State Col Wash, 47-48, asst bacteriologist, 48; AEC fel, Univ Calif, 48-49; from asst prof to assoc prof bact, 49-55, head dept bot & microbiol, 64-68, dean col lett & sci, 68-74, PROF BACT, MONT STATE UNIV, 55-, DIR McBEE LAB, 52- Mem: Fel AAAS; Am Soc Microbiol; Am Chem Soc; fel Am Acad Microbiol. Res: Metabolism of coliform bacteria; cellulose fermentations; microbiology of animal digestive systems; clinical anaerobic microbiology. Mailing Add: Dept of Microbiol Mont State Univ Bozeman MT 59715

MCBEE, WILLIAM, JR, b Dewar, Okla, Sept 17, 21; m 42; c 3. GEOLOGY. Educ: Univ Tulsa, BS, 42; Univ Kans, MS, 48. Prof Exp: Asst instr micropaleont & photog & cur geol mus, Univ Kans, 46-48; explor geologist, Calif Co, La, 48-50, Okla, 50-53; div stratigr, Stand Oil Co Tex, 53-56, div staff geologist, 56-61; gulf coast regional geologist, Monsanto Co, 61-64, western region staff geologist, 64-67; northern Can explor supvr, Sinclair Oil Co Can, 67-69; sr area geologist, 69-74, CAN DIST DEVELOP GEOLOGIST, ATLANTIC-RICHFIELD CAN, LTD, 75- Mem: Geol Soc Am; Am Asn Petrol Geol. Res: Stratigraphic geology, particularly clastic depositional patterns and environmental determinations. Mailing Add: Atlantic-Richfield Can Ltd 650 Guinness House Calgary AB Can

MCBIRNEY, ALEXANDER ROBERT, b Sacramento, Calif, July 18, 24; m 47; c 4. GEOLOGY, PETROLOGY. Educ: US Mil Acad, BS, 46; Univ Calif, Berkeley, PhD(geol), 61. Prof Exp: Asst prof geol, Univ Calif, San Diego & staff mem, Scripps Inst Oceanog, 62-65; dir ctr volcanology, 65-68, assoc prof geol, 65-70, chmn dept, 68-71, PROF GEOL, UNIV ORE, 70- Concurrent Pos: NSF sr fel, Eng, Japan & NZ, 71-72. Mem: AAAS; Geol Soc Am; Am Geophys Union. Res: Geology of Central America and Circum-Pacific orogenic regions; igneous petrology; volcanology. Mailing Add: Ctr for Volcanology Univ of Ore Eugene OR 97403

MCBLAIR, WILLIAM, b San Diego, Calif, Apr 19, 17; m 57; c 3. PHYSIOLOGY, OCEANOGRAPHY. Educ: San Diego State Univ, BA, 47; Univ Calif, Los Angeles, PhD(zool), 56. Prof Exp: Asst instr chem, San Diego State Univ, 47; asst zool, Univ Calif, Berkeley, 47-48; instr biol, 48-51, asst prof zool, 51-62, assoc prof biol, 62-68, PROF BIOL, SAN DIEGO STATE UNIV, 68- Concurrent Pos: Asst, Scripps Inst Oceanog, Univ Calif, San Diego, 48-52. Mem: AAAS; Am Soc Zool; NY Acad Sci; fel Int Oceanog Found; Nat Audubon Soc. Res: Active uptake; marine and environmental physiology; biology of spiders, especially prey-predator relationships. Mailing Add: Dept of Biol San Diego State Univ San Diego CA 92182

MCBOYLE, GEOFFREY REID, b Huntly, Scotland, Dec 16, 42; m 66; c 1. PHYSICAL GEOGRAPHY. Educ: Aberdeen Univ, BSc, 64, PhD(geog), 69. Prof Exp: Lectr geog, Univ Waikato, NZ, 66-69; asst prof, 69-72, ASSOC PROF GEOG, UNIV WATERLOO, 72-, CHMN DEPT, 74- Mem: AAAS; Asn Am Geogr; fel Royal Meteorol Soc. Res: Applied climatology; history of geography; air pollution. Mailing Add: Dept of Geog Univ of Waterloo Waterloo ON Can

MCBRADY, JOHN J, b St Paul, Minn, Feb 1, 16; m 44. PHYSICAL CHEMISTRY. Educ: Univ Minn, Minneapolis, BChem, 38, PhD(phys chem), 44. Prof Exp: Res chemist, Donnelley & Sons Co, 46 & Celanese Corp, 47-52; sr chemist, 52-65, res specialist, 65-70, SR RES SPECIALIST, MINN MINING & MFG CO, 70- Mem: Am Chem Soc; Coblentz Soc. Res: Molecular spectroscopy; infrared and nuclear magnetic resonance; photochemistry. Mailing Add: Minn Mining & Mfg Co PO Box 33221 St Paul MN 55133

MCBRAYER, JAMES FRANKLIN, b Rowan Co, Ky, July 12, 41; m 67; c 1. ECOLOGY. Educ: Miami Univ, BS, 63; Purdue Univ, West Lafayette, MS, 70; Univ Tenn, Knoxville, PhD(ecol), 73. Prof Exp: Teacher biol pub schs, Ohio, 63-67; asst res biologist, Lab Nuclear Med, Univ Calif, 73-74; ASST PROF ECOL, UNIV MINN, ST PAUL, 74- Concurrent Pos: Consult, Desert Biome, US Int Biol Prog, 74-75; mem, Interbiome Specialist Comt Elemental Cycling, 74- Mem: AAAS; Am Inst Biol Sci; Ecol Soc Am; Sigma Xi. Res: Ecosystem analysis; decompostition and elemental cycling in terrestrial ecosystems; synergistic relationships between microflora and decomposer invertebrates; role of fossorial animals in community development. Mailing Add: Dept of Ecol & Behav Biol 310 Biol Sci Ctr Univ of Minn St Paul MN 55108

MCBREEN, JAMES, b Cavan, Ireland, Sept 5, 38; US citizen; m 66; c 2. ELECTROCHEMISTRY. Educ: Nat Univ Ireland, BSc, 61; Univ Pa, PhD(phys chem), 65. Prof Exp: Res chemist, Yardney Elec Corp, 65-68; SR RES CHEMIST, RES LABS, GEN MOTORS CORP, 68- Honors & Awards: Battery Div Res Award, Electrochem Soc, 74. Mem: Electrochem Soc. Res: Ambient-temperature aqueous batteries, including work on the zinc, nickel oxide and manganese dioxide electrodes. Mailing Add: Dept of Electrochem Gen Motors Res Labs Warren MI 48090

MCBRIDE, CLIFFORD HOYT, b Massena, Iowa, June 14, 26; m 50; c 6. CHEMISTRY. Educ: Iowa State Univ, BS, 48; St Louis Univ, MS, 57. Prof Exp: Chemist, Mallinckrodt Chem Works, 49-55, supvr, 55-62; sr res chemist, Armour Agr Chem Co, 62-63, anal res mgr, 63-68; sect head, 68-72, MGR ANAL SERV, USS AGRI-CHEM DIV, US STEEL CORP, 72- Mem: Am Chem Soc. Res: Analytical chemistry. Mailing Add: US Steel Corp USS Agri-Chem Div 685 DeKalb Indust Way Decatur GA 30033

MCBRIDE, DUNCAN ELDRIDGE, b Chicago, Ill, Oct 26, 45; m 68. EXPERIMENTAL SOLID STATE PHYSICS, METAL PHYSICS. Educ: Carleton Col, BA, 67; Univ Calif, Berkeley, MA, 69, PhD(physics), 73. Prof Exp: US-France exchange fel physics, Lab Solid State Physics, Univ Paris VII, 73-74; ASST PROF PHYSICS, SWARTHMORE COL, 74- Mem: Am Phys Soc; AAAS. Res: Electron tunneling; inelastic electron tunneling spectroscopy; metal-insulator transitions; noise in transport processes. Mailing Add: Dept of Physics Swarthmore Col Swarthmore PA 19081

MCBRIDE, EARLE FRANCIS, b Moline, Ill, May 25, 32; m 56; c 2. SEDIMENTARY PETROGRAPHY. Educ: Augustana Col, AB, 54; Univ Mo, MA, 56; Johns Hopkins Univ, PhD(geol), 60. Prof Exp: From instr to assoc prof, 59-68, PROF GEOL, UNIV TEX, AUSTIN, 68- Mem: Geol Soc Am; Am Asn Petrol Geol; Soc Econ Paleont & Mineral; Int Asn Sedimentol. Res: Sedimentary petrology; primary sedimentary structures; Paleozoic rocks of Marathon Region, Texas; origin of sedimentary rocks, chiefly clastic rocks; origin of bedding; sandstone diagenesis. Mailing Add: Dept of Geol Sci Univ of Tex Austin TX 78712

MCBRIDE, EDWARD FRANCIS, b Rochester, Minn, Nov 4, 39. ORGANIC CHEMISTRY. Educ: St John's Univ, Minn, BA, 61; Univ Wis-Madison, PhD(org chem), 66. Prof Exp: RES CHEMIST, ORG CHEM DEPT, EXP STA, E I DU PONT DE NEMOURS & CO, INC, 66- Mem: Am Chem Soc. Res: Organic photochemistry; novel photoimaging systems and photopolymerization; flash photolysis of photochromic and photoimaging systems; cholesteric liquid crystals. Mailing Add: 2729 Skylark Rd Brookmeade Wilmington DE 19808

MCBRIDE, ELNA BROWNING, b Halls, Tenn, Jan 22, 11; m 35. APPLIED MATHEMATICS. Educ: Univ Tenn, BS, 30, MS, 31; Univ Mich, EdD(math), 66. Prof Exp: Teacher high sch, Tenn, 31-34; teacher, Harding Col, 34-35 & David Lipscomb Col, 35-42; teacher comput, Surv Sect, US Eng, 42-43; teacher elem sch, Tenn, 43-46; teacher math, 46-49, assoc prof, 50-67, PROF MATH, MEMPHIS STATE UNIV, 67- Mem: Math Asn Am; Am Math Soc. Res: Methods of obtaining generating functions for the special functions of chemistry and physics; study of applications of generating functions. Mailing Add: Dept of Math Memphis State Univ Memphis TN 38152

MCBRIDE, GORDON E, b Ft Bragg, Calif, June 6, 36; m 56; c 2. PHYCOLOGY. Educ: Humboldt State Col, BS, 59, MA, 64; Univ Calif, Berkeley, PhD(bot), 68. Prof Exp: ASST PROF BOT, UNIV MICH, ANN ARBOR, 68- Mem: Bot Soc Am; Phycol Soc Am; Int Phycol Soc. Res: Light and electron microscope studies of cytokinesis, cytodifferentiation and general cell ultrastructure in the algae. Mailing Add: Dept Bot 4004 Natural Sci Bldg Univ of Mich Ann Arbor MI 48104

MCBRIDE, JAMES MICHAEL, b Lima, Ohio, Feb 25, 40; m 64; c 2. PHYSICAL ORGANIC CHEMISTRY. Educ: Harvard Univ, BA, 62, PhD(chem), 67. Prof Exp: Asst prof, 66-72, ASSOC PROF CHEM, YALE UNIV, 72- Mem: Am Chem Soc; The Chem Soc. Res: Free radical reactions; effect of viscous and rigid media on the course of organic reactions; solid state chemistry; electron paramagnetic resonance. Mailing Add: Dept of Chem Yale Univ New Haven CT 06520

MCBRIDE, JOHN ALEXANDER, b Altoona, Pa, Mar 29, 18; m 42; c 3. CHEMISTRY. Educ: Miami Univ, AB, 40; Ohio State Univ, MSc, 41; Univ Ill, PhD(org chem), 44. Prof Exp: With Phillips Petrol Co, 44-51, asst supt, Philtex Exp Sta, 51-52, mgr develop, Rocket Fuels Div, 52-56, tech planning, 57-58, chief appln eng, Astrodyne, Inc, 58-59, tech dir, Idaho Chem Processing Plant, 59-62, dir chem tech, Atomic Energy Div, 62-65; dir div mat licensing, AEC, 65-70; VPRES, E R JOHNSON ASSOCS, INC, 70- Mem: Am Chem Soc; Am Inst Chem Engrs; Am Nuclear Soc; AAAS. Res: Irradiated nuclear fuel reprocessing; nuclear waste disposal; organic syntheses; applications and synthesis of organic sulfur compounds; petroleum derivatives; solid propellants. Mailing Add: Box 192 Merrifield VA 22116

MCBRIDE, JOHN BARTON, b Philadelphia, Pa, July 6, 43. PHYSICS. Educ: St Joseph's Col, Pa, BS, 65; Dartmouth Col, MA, 67, PhD(physics), 69. Prof Exp: Res instr physics, Dartmouth Col, 69; res assoc, Princeton Univ, 69-70; mem staff, US Naval Res Lab, 70-74; MEM STAFF, LAB APPL PLASMA SCI, SCI APPLN INC, 74- Mem: Am Phys Soc. Res: Theoretical plasma physics. Mailing Add: Lab of Appl Plasma Sci Sci Appl Inc 1200 Prospect St Box 2351 La Jolla CA 92037

MCBRIDE, JOHN JOSEPH, b New York, NY, Apr 10, 30; m 53; c 2. PHYSICAL CHEMISTRY. Educ: NY Univ, BA, 51; Univ Mich, MS, 53, PhD(phys chem), 56. Prof Exp: Chemist, Oak Ridge Nat Lab, 55-63; RES CHEMIST, TEXTILE FIBERS DEPT, E I DU PONT DE NEMOURS & CO, INC, 63- Mem: Am Chem Soc; Sigma Xi. Res: Thermodynamics; polymer physical chemistry; materials science; surface chemistry. Mailing Add: 2626 Boxwood Dr The Timbers Wilmington DE 19810

MCBRIDE, JOSEPH JAMES, JR, b Philadelphia, Pa, Dec 10, 22; m 48; c 7. ORGANIC CHEMISTRY. Educ: St Joseph's Col, BS, 43; Univ Del, MS, 47, PhD(chem), 50. Prof Exp: Res chemist, Tidewater Assoc Oil Co, 49-54 & citrus exp sta, Univ Fla, 54-59; sect head org res, Armour Indust Chem Co, 59-61; DIR DEVELOP, ARIZ CHEM CO, 61- Mem: Am Chem Soc. Res: Organosilicon compounds; organic chemistry of nitrogen, fatty acids; rosin; terpenes. Mailing Add: Ariz Chem Co Develop Lab PO Box 2447 Panama City FL 32401

MCBRIDE, LANDY JAMES, b Durant, Okla, Nov 23, 31; m 56; c 2. PLANT PHYSIOLOGY. Educ: Univ Calif, Berkeley, BA, 54, MA, 56; Univ Wis, PhD(bot), 61. Prof Exp: Res assoc biol, Argonne Nat Lab, 61-63; asst prof, Tex Tech Col, 63-64; res plant physiologist, Int Minerals & Chem Corp, Ill, 64-71; RES PLANT PHYSIOLOGIST, AM CAN CO, 71- Mem: Am Soc Plant Physiol; Bot Soc Am. Res: Aquatic biology; water pollution control. Mailing Add: American Can Co 1915 Marathon Ave Neenah WI 54956

MCBRIDE, MOLLIE ELIZABETH, b Montreal, Que, May 7, 29; m 51; c 5. MEDICAL MICROBIOLOGY. Educ: Dalhousie Univ, BSc, 55; Bryn Mawr Col, MA, 57; McGill Univ, PhD(bact, immunol), 59. Prof Exp: Res fel, McGill Univ, 59-60; bacteriologist, Halifax Children's Hosp, 60-63; asst prof, 64-75, ASSOC PROF MICROBIOL, BAYLOR COL MED, 75- Mem: AAAS; Brit Soc Gen Microbiol; Can Soc Microbiol; Am Soc Microbiol; Soc Invest Dermat. Res: Microbial ecology of the skin; comparison in health and disease, factors influencing staphylococcal and streptococcal skin infections; gram negative colonization; microbial degradation of keratin; taxonomy of Corynebacteria. Mailing Add: Dept of Dermat & Microbiol Baylor Col of Med Houston TX 77025

MCBRIDE, RALPH BOOK, b Slippery Rock, Pa, Feb 1, 28; m 54; c 3. MATHEMATICS EDUCATION. Educ: Slippery Rock State Col, BS, 51; Indiana Univ Pa, ME, 65; Univ Mich, PhD(math educ), 70. Prof Exp: Teacher math, North Butler County Schs, 51-52, Apollo Area Schs, 54-56 & Kiski Area Schs, 56-65; ASSOC PROF MATH, MANCHESTER COL, 65-, CHMN DEPT, 71-, ISAAC & ETTA H OPPENHEIM PROF, 74- Res: Algebra; number systems. Mailing Add: Dept of Math Manchester Col Box 103 North Manchester IN 46962

MCBRIDE, RAYMOND ANDREW, b Houston, Tex, Dec 27, 27; m 58; c 4. IMMUNOBIOLOGY, PATHOBIOLOGY. Educ: Tulane Univ, BS, 52, MD, 56; Am Bd Path, dipl, 61. Prof Exp: Intern surg, Baylor Col Med, 56-57; asst resident pathologist, Peter Bent Brigham Hosp, 57-58, asst in path, 58-60, sr resident pathologist, 60-61; asst prof path, Col Physicians & Surgeons, Columbia Univ, 63-65; assoc prof surg & immunogenetics, Mt Sinai Sch Med, 65-68; prof path, New York Med Col, Flower & Fifth Ave Hosps, 68-73; exec dean, 73-75, PROF PATH, NEW YORK MED COL, VALHALLA, 68-; EXEC DIR, WESTCHESTER MED CTR DEVELOP BD, INC, 74- Concurrent Pos: Teaching fel path, Harvard Med Sch, 58-61; Nat Cancer Inst spec fel, McIndoe Mem Res Unit, East Grinstead, Eng, 61-63; resident pathologist, Free Hosp for Women, 59; asst resident pathologist, Children's Hosp Med Ctr, 60; asst attend pathologist, Presby Hosp, New York, 63-65; career scientist, NY Health Res Coun, 67-73; attend pathologist, New York Med Col, Flower & Fifth Ave Hosps, 68- & Metrop Hosp, 68- Mem: AAAS; Am Soc Exp Path; Am Asn Immunol; Am Asn Path & Bact. Res: Antigen and antibody interactions in isoimmune systems; population dynamics of antibody forming cells in isoimmune systems; factors involved in the recognition of immunogenicity; DNA and RNA tumor viruses; tumor immunology. Mailing Add: Dept of Path New York Med Col Valhalla NY 10595

MCBRIDE, RICHARD PHILLIPS, b Nelson, BC, July 3, 42; m 65; c 2. MICROBIOLOGY, ECOLOGY. Educ: Univ BC, BSc, 64, MSc, 65; Univ Edinburgh, PhD(ecol), 70. Prof Exp: ASST PROF BIOL, DALHOUSIE UNIV, 70- Res: Microbial ecology; interactions of microorganisms with macrophytes; biological control of plant pathogens; role of microorganisms in algal life cycles, especially nutrition and disease. Mailing Add: Dept of Biol Dalhousie Univ Halifax NS Can

MCBRIDE, TOM JOSEPH, b Coffeyville, Kans, Sept 2, 24; m 47; c 3. BACTERIOLOGY. Educ: Univ Kans, AB, 49, MS, 51; Northwestern Univ, PhD, 53. Prof Exp: Bacteriologist, 53-60, mem staff cancer chemother, 61-64, MGR CHEMOTHER SCREENING, PFIZER THERAPEUT INST, 64- Mem: AAAS; Am Soc Microbiol; NY Acad Sci. Res: Chemotherapy transplantable tumors; antiviral chemotherapy; experimental tuberculosis and bacterial infections; new antibiotics and synthetic and anti-infectives. Mailing Add: Pfizer Therapeut Inst 199 Maywood Ave Maywood NJ 07450

MCBRIDE, WILLIAM JOSEPH, b Philadelphia, Pa, Dec 24, 38; m 62; c 4. NEUROCHEMISTRY, NEUROBIOLOGY. Educ: Rutgers Univ, BA, 64; State Univ NY Buffalo, PhD(biochem), 68. Prof Exp: ASSOC PROF BIOCHEM & PSYCHIAT MED CTR, IND UNIV, INDIANAPOLIS, 71- Concurrent Pos: NIH fel neurobiol, Ind Univ, Bloomington, 68-71. Mem: Int Soc Neurochem; Soc Neurosci; Am Soc Neurochem. Res: Study of the mechanisms of how nerve cells communicate with one another and how alterations in this process may have an effect on the behavior of animals and man. Mailing Add: Inst of Psychiat Res Ind Univ Med Ctr Indianapolis IN 46202

MCBRIDE, WILLIAM ROBERT, b Topeka, Kans, May 9, 28; m 52; c 3. INORGANIC CHEMISTRY. Educ: Univ Calif, Los Angeles, BS, 50; Univ Tex, PhD(inorg chem), 55. Prof Exp: Res chemist, US Naval Ord Test Sta, 50-51 & 55-59, head inorg chem br, 59-67, HEAD INORG CHEM BR, NAVAL WEAPONS CTR, 67- Mem: Am Chem Soc; Sigma Xi; Optical Soc Am. Res: Chemistry of hydronitrogens and derivatives; nonaqueous solutions and solvents; propellants; absorption spectroscopy of optical materials and filters; crystal growth of inorganic compounds; reaction kinetics. Mailing Add: Code 6054 Chem Div Res Dept Naval Weapons Ctr PO Box 5565 China Lake CA 93555

MCBRIDE, WOODROW H, b Milton, NDak, May 23, 18; m 44; c 1. MATHEMATICS. Educ: Jamestown Col, BA, 40; Univ NDak, MS, 47. Prof Exp: Teacher & prin, High Schs, NDak, 40-44; prin, High Schs, NDak & Minn, 44-46; from instr to assoc prof, 47-69, PROF MATH, UNIV NDAK, 69- Mem: Math Asn Am. Mailing Add: Dept of Math Univ of NDak Grand Forks ND 58202

MCBRIEN, VINCENT OWEN, b Attleboro, Mass, Apr 21, 16; m 48; c 4. MATHEMATICS. Educ: Providence Col, BS, 37; Cath Univ Am, MS, 40, PhD(math), 42. Prof Exp: Physicist, David Taylor Model Basin, US Dept Navy, 42 & Off Sci Res & Develop, 42-43; instr math, Hamilton Col, 43-44; from asst prof to assoc prof, 43-60, chmn dept, 60-70, PROF MATH, COL HOLY CROSS, 60- Concurrent Pos: Ford Found fel, Harvard Univ, 52-53; NSF fac fel, Univ Calif, Berkeley, 60-61; vis prof, Trinity Col, Dublin, 71-72. Mem: Am Math Soc; Math Asn Am; Soc Indust & Appl Math. Res: Algebraic geometry and topology; abstract algebraic geometry. Mailing Add: Dept of Math Col of the Holy Cross Worcester MA 01610

MCBROOM, MARVIN JACK, b Cherokee, Okla, Apr 13, 41; c 1. PHYSIOLOGY. Educ: Northwestern State Col, BS, 63; Okla State Univ, MS, 64; Univ Okla, PhD(physiol), 68. Prof Exp: Asst prof, 68-75, ASSOC PROF PHYSIOL, SCH MED, UNIV SDAK, VERMILLION, 75- Mem: AAAS; fel Geront Soc; Soc Exp Biol & Med. Res: Physiology of aging of bone and soft tissues of mammals as related to fluid and electrolyte metabolism. Mailing Add: Dept of Physiol & Pharmacol Univ of SDak Sch of Med Vermillion SD 57069

MCBRYDE, ANGUS MURDOCH, b Red Springs, NC, May 25, 02; m; c 3. PEDIATRICS. Educ: Davidson Col, BS, 24; Univ Pa, MD, 28. Prof Exp: Intern, Hosp Univ Pa, 28-29, resident pediat, 29-30; asst resident, Johns Hopkins Hosp, 30-31; from instr to prof, 31-70, EMER PROF PEDIAT, SCH MED, DUKE UNIV, 70- Concurrent Pos: Dir newborn & premature nurseries, Duke Hosp. Mem: Soc Pediat Res; Am Pediat Soc; Am Acad Pediat; fel AMA. Res: Newborn and premature infants. Mailing Add: Box 3967 Duke Med Ctr Durham NC 27706

MCBRYDE, FELIX WEBSTER, b Lynchburg, Va, Apr 23, 08; m 34; c 3. GEOGRAPHY, ANTHROPOLOGY. Educ: Tulane Univ, BA, 30; Univ Calif, Berkeley, PhD(geog), 40. Hon Degrees: LLD, Tulane Univ, 67. Prof Exp: Field asst archaeol, Mid Am Res Dept, Tulane Univ, Mex & Guatemala, 27-28; field asst, Univ Utah & Smithsonian Inst, 31; res asst Guatemala geog & archaeol, Clark Univ, 32; res asst geog, Mid Am Res Inst, Tulane Univ, 32-33; teaching asst, Univ Calif, Berkeley, 37; instr, Ohio State Univ, 37-42; chief, Latin Am Sect, Topog Br, Mil Intel Serv, Dept War, 42-45; dir inst social anthrop, Smithsonian Inst, Peru, 45-47; spec rep Lima archaeol, Inst Andean Res, Peru, 47-48; geogr consult, Off Coordr Int Statist, US Bur Census, 48-56; dir regional planning & surv, Gordon A Friesen, Assocs, Inc, DC & Costa Rica, 56-58; pres, F W McBryde Assocs, Inc, DC & Guatemala, 58-64; chief phys & cult geog br, Inter-Am Geog Surv, US Army, Ft Clayton, CZ, 64-65; field dir bioenviron prog, Atlantic-Pac Interoceanic Canal Studies, Battelle Mem Inst, Columbus Labs, Panama-Colombia Field Off, 65-70; DIR, McBRYDE CTR HUMAN ECOL, 70-; DIR RECRUITMENT & INT BUS INTEL SERV, TRANSEMANTICS, INC, WASHINGTON, DC, 75- Concurrent Pos: Collabr Mex & Guatemala, USDA, 40-41; Nat Res Coun fel, Guatemala, Mex & Univ Calif, Berkeley, 40-41, Pan Am Airways travel fel, 40-41 & Ohio State Univ grant, 40-41; lectr & consult prof, Univ Md, 48-63; chief mission & US census adv, Statist & Census Off, First Nat Census Ecuador, 49-51; spec consult, Tech Coop Admin, US Dept State, WI, 52-53; lectr Latin Am geog, Foreign Serv Inst, US Dept State, 53-57; consult Latin Am geog, Inst Mod Lang, 63-70; pres, Inter-Am Inst Mod Lang, Guatemala, 63-66; field dir, Andean Ecol Proj, Battelle Mem Inst, 67-68, dir, Cent Am Proj Develop Prog, 68-69 & consult ecol, 70-; expert consult ecologist, UN Develop Prog, Jamaica, 71 & Arg, 72; census geog adv, Int Statist Prog, US Bur Census, Honduras, 72; consult ecologist, World Bank, Bayano Hydroelec Proj, Panama, 73; consult ecologist, Battelle Mem Inst-US Dept Transp, Darien Gap Hwy Proj, Panama-Colombia, 73; expert consult-ecologist, Engr Agency Resources Inventories, US Army Corps Engrs, Washington, DC, 74. Mem: Am Anthrop Asn; Am Geophys Union; Am Inst Biol Sci; Asn Am Geogr (pres, Am Soc Prof Geogr, 43-45); Ecuadorian Inst Anthrop & Geog. Res: Regional ethnoecology; geodemography; natural and human resources; health; environmental impact studies; cartography; graphic statistical representation; world map projections; Latin American populations, food supply, native material culture, origin, dissemination and acculturation. Mailing Add: 10100 Falls Rd Potomac MD 20854

MCBRYDE, WILLIAM ARTHUR EVELYN, b Ottawa, Ont, Oct 20, 17; m 49; c 2. CHEMISTRY. Educ: Univ Toronto, BA, 39, MA, 40; Univ Va, PhD(chem), 4?. Prof Exp: Asst chem, Univ Toronto, 39-41; chemist, Welland Chem Works, 42-44; asst prof chem, Univ Va, 47-48; from asst prof to assoc prof, Univ Toronto, 48-60; prof & chmn dept, Univ Waterloo, 60-64, dean fac sci, 61-69; vis fel, Australian Nat Univ, 69-70; PROF CHEM & CHMN DEPT, UNIV WATERLOO, 71- Mem: Am Chem Soc; Chem Inst Can. Res: Chemistry of precious metals; colorimetric analysis; coordination chemistry. Mailing Add: Dept of Chem Univ of Waterloo Waterloo ON Can

MCBURNEY, LANE FORDYCE, b Pittsburgh, Pa, July 26, 13; m 40; c 1. ORGANIC CHEMISTRY. Educ: Drexel Inst, BS, 36; Univ Pa, MS, 40, PhD(org chem), 42. Prof Exp: Asst mgr insulin mfg dept, Sharp & Dohme, Inc, 36-37, res chemist, 37-41, mgr synthetic chem mfg dept, 41-42; instr org chem, Univ Pa, 42-43; res chemist, Hercules Inc, 43-46, mgr basic res div, 46-51, mgr cellulosic pioneering res div, 51-54, Va cellulose res div, 54-58 & assoc res div, 58-64, DIR, HERCULES RES CTR, 64-

Mem: Fel AAAS; Sigma Xi; Am Chem Soc. Res: Organic medicinals; organic sulphur compounds; cellulose and cellulose derivatives; resin acids; antiseptics. Mailing Add: Hercules Res Ctr Wilmington DE 19899

MCBURNEY, WENDELL FARIS, b Spring Valley, NY, Feb 2, 33; m 56; c 3. SCIENCE EDUCATION. Educ: Geneva Col, BS, 55; Ind Univ, MAT, 66, EdD(sci educ), 67. Prof Exp: Teacher high sch, Pa, 55-64; asst prof sci educ & coordr sch sci, Ind Univ, Bloomington, 67-73; ASSOC PROF SCI EDUC & ASST DEAN RES & SPONSORED PROGS, IND UNIV-PURDUE UNIV, INDIANAPOLIS, 73- Concurrent Pos: Acad analyst, Indianapolis Ctr Advan Res, 73- Mem: Nat Sci Teachers Asn; Nat Asn Biol Teachers; Nat Coun Univ Res Adminrs; Sigma Xi. Res: Instructional designs on quantification of subjective judgment in student investigation in the biological sciences. Mailing Add: 355 Lansing Ind Univ-Purdue Univ Indianapolis IN 46202

MCCAA, CONNIE SMITH, b Lexington, Miss, Dec 6, 37; m 57; c 4. BIOCHEMISTRY. Educ: Miss Col, BS, 58; Univ Miss, PhD(biochem), 63. Prof Exp: Asst, 59-63. from instr to assoc prof, 63-73, PROF BIOCHEM, MED CTR, UNIV MISS, 73-, ASSOC PROF PHYSIOL & BIOPHYS, 70- Concurrent Pos: Nat Heart & Lung Inst res grant, 63-74, spec res fel, 67-70; Miss Heart Asn grant, 63-66; Miss Cancer Soc grant, 65-66; mem cardiovasc & renal study sect, NIH, 74- Mem: Am Chem Soc; Am Physiol Soc; Am Heart Asn; Endocrine Soc; Am Soc Nephrology. Res: Role of aldosterone in the production of hypertension and congestive heart failure and the mechanism whereby normal animals escape from its sodium-retaining effect. Mailing Add: Dept of Biochem Univ of Miss Med Ctr Jackson MS 39216

MCCABE, BRIAN FRANCIS b Detroit, Mich, June 16, 26; m 51; c 2. OTOLARYNGOLOGY. Educ: Univ Detroit, BS, 50; Univ Mich, MD, 54. Prof Exp: From instr to assoc prof otolaryngol, Med Sch, Univ Mich, 59-64; PROF OTOLARYNGOL & CHMN DEPT, COL MED, UNIV IOWA, 64- Concurrent Pos: Consult to Surgeon Gen, USPHS, 66; mem bd dirs, Am Bd Otolaryngol, 67. Honors & Awards: Mosher Award, Am Laryngol, Rhinol & Otol Soc, 65. Mem: Am Laryngol Asn; Am Otol Soc; Am Acad Ophthal & Otolaryngol; Am Laryngol, Rhinol & Otol Soc; Col Oto-Rhino-Laryngol Amicitiae Sacrum. Res: Maxillofacial surgery; neurphysiology of the vestibular apparatus; mechanism of the quick component of nystagmus; surgery of the major salivary glands, particularly cancer. Mailing Add: Dept of Otolaryngol Univ of Iowa Iowa City IA 52240

MCCABE, CHESTER CHARLES, b Shoreham, Vt, May 9, 24; m 58. POLYMER SCIENCE. Educ: Univ Chicago, BS, 44; Columbia Univ, MS, 48; Purdue Univ, PhD(physics), 54. Prof Exp: Instr physics, Purdue Univ, 54; res physicist, Elastomers Lab, 55-59, head new polymers div, 59-72, HEAD TIRE RES & DEVELOP DIV, ELASTOMERS RES & DEVELOP LAB, E I DU PONT DE NEMOURS & CO, INC, 72- Mem: Am Phys Soc; Am Chem Soc. Res: Basic characterizations and rheological behaviors of high polymers; rheology of elastomers and plastics. Mailing Add: Elastomers Res & Develop Lab E I du Pont de Nemours & Co Inc Wilmington DE 19898

MCCABE, EDWARD MATHEW, biochemistry, see 12th edition

MCCABE, GEORGE PAUL, JR, b Brooklyn, NY, Apr 2, 45; m 67; c 3. STATISTICS. Educ: Providence Col, BS. 66; Columbia Univ, PhD(math statist), 70. Prof Exp: Asst prof, 70-75, ASSOC PROF STATIST, PURDUE UNIV WEST LAFAYETTE, 75-, HEAD STATIST CONSULT, 70- Mem: Sigma Xi; Inst Math Statist; Am Statist Asn. Res: Mathematical statistics; applied statistics; statistical computing; regression analysis. Mailing Add: Dept of Statist Purdue Univ West Lafayette IN 47907

MCCABE, JOHN PATRICK, b New York, NY, Aug 17, 35. MATHEMATICS. Educ: Manhattan Col, BS, 57; Harvard Univ, AM, 58, PhD(math), 68. Prof Exp: Systs analyst, Gen Elec Co, 62-66; from instr to asst prof, 66-74, ASSOC PROF MATH & CHMN DEPT, MANHATTAN COL, 74- Mem: Am Math Soc; Soc Indust & Appl Math. Res: Abelian varieties over local fields. Mailing Add: Dept of Math Manhattan Col New York NY 10471

MCCABE, LEO JAMES, organic chemistry, air pollution, see 12th edition

MCCABE, LOUIS CORDELL, b Graphic, Ark, Feb 5, 04; m 36; c 4. GEOLOGY. Educ: Univ Ill, BS, 31, MS, 33, PhD(geol), 37. Prof Exp: From asst geologist to geologist, Ill State Geol Surv, 30-41; chief off air & stream pollution, US Bur Mines, 49-51, chief fuels & explosives div, 51-55; sci dir, USPHS, 55; pres, Resources Res, Inc, 56-68; pres & chmn, Environ Develop Inc, 68-70; RETIRED. Concurrent Pos: Chief utilities, Off Chief Engrs, US War Dept, 41-44, dep power procurement off, 43-44; chief, Belg & Ger Solid Fuels Sect, SHAEF, 44-45; US Dept State rep coal mining comt, Int Labor Orgn, Geneva, 47, 51; chmn bd, Hazleton Labs, Inc; consult, WHO, 57- Honors & Awards: Order of Brit Empire; Order of Crown, Belg; Frank A Chambers Award, Air Pollution Control Asn, 66. Mem: AAAS; Am Chem Soc; Air Pollution Control Asn (vpres, 63-66); fel Geol Soc Am; Am Inst Mining, Metall & Petrol Engrs. Res: Mineral resources; industrial wastes; air and water pollution research and development. Mailing Add: 7102 Pomander Lane Chevy Chase MD 20015

MCCABE, MICHAEL S, b Lawrence, Kans, June 21, 42; m 66; c 3. PSYCHIATRY. Educ: Univ Kans, AB, 63, MD, 67. Prof Exp: Resident psychiat, Sch Med, Wash Univ, 68-71; res fel, Univ Aarhus, Denmark, 71-72; chief consult serv & staff psychiatrist, US Air Force Med Ctr, Lackland AFB, 72-74; ASST PROF PSYCHIAT, HOSP & CLIN, UNIV IOWA, 74- Concurrent Pos: Asst clin prof, Univ Tex, 72-74; attend & consult psychiatrist, Iowa City Vet Admin Hosp, 74- Mem: Am Psychiat Asn; Royal Col Psychiatrists; Behav Genetics Asn; AAAS. Res: Epidemiology and genetics of functional psychoses; psychiatric illness and chronic medical disease. Mailing Add: Dept of Psychiat Univ of Iowa Hosp & Clin Iowa City IA 52242

MCCABE, ROBERT ALBERT, b Milwaukee, Wis, Jan 11, 14; m 41; c 4. WILDLIFE MANAGEMENT. Educ: Carroll Col, BA, 39; Univ Wis, MS, 43, PhD(wildlife mgt, zool), 49. Prof Exp: Biologist, Arboretum, 43-46, from instr to assoc prof, 46-56, PROF WILDLIFE MGT, UNIV WIS-MADISON, 56-, CHMN DEPT, 52- Concurrent Pos: Mem res adv coun, Wis Dept Natural Resources, 54-; secy adv comt, Wis Dept Resource Develop, 60-64; mem bd dirs, Wis Expos Dept, 60-68; chmn subcomt vertebrates as pests, Nat Acad Sci, 64-; Fulbright prof, Univ Col, Univ Dublin, 69-70; adv to Irish Nat Parks, 70- Mem: Wildlife Soc; Am Soc Mammal; Wilson Ornith Soc; Cooper Ornith Soc; fel Am Ornith Union. Res: Wildlife ecology; techniques in wildlife management; land use in wildlife relationship. Mailing Add: Dept of Wildlife Ecol Col of Agr & Life Sci Univ Wis Madison WI 53706

MCCABE, ROBERT LYDEN, b Tarrytown, NY, Mar 5, 36; m 59; c 3. MATHEMATICS. Educ: Union Col, NY, BS, 57; San Diego State Col, MA, 60; Boston Univ, PhD(math), 71. Prof Exp: Asst prof, 64-68, ASSOC PROF MATH, SOUTHEAST MASS UNIV, 68- Concurrent Pos: Instr, Upperward Bound, 68. Res: Ergodic and dilation theories; Markov processes. Mailing Add: Dept of Math Southeast Mass Univ North Dartmouth MA 02747

MCCABE, WILLIAM R, b Hugo, Okla, Sept 13, 28; m 51; c 3. INFECTIOUS DISEASES, MICROBIOLOGY. Educ: Univ Okla, BS, 49, MD, 53. Prof Exp: Intern med, Univ Okla Hosps, 53-54, resident, 56-58; res asst & chief med resident infectious dis, Col Med, Univ Ill, 58-60, asst prof med, 62-63; from asst prof to assoc prof, Sch Med, 63-71, asst prof microbiol, 63-71, PROF MED & MICROBIOL, MED CTR, BOSTON UNIV, 71-, HEAD MED BACT & ASSOC VIS PHYSICIAN, UNIV HOSP, 63- Concurrent Pos: NIH trainee cardiovasc dis, Univ Okla Hosps, 57-58; res fel infectious dis, Univ Ill Res & Educ Hosp, 58-60; clin investr, West Side Vet Admin Hosp, Chicago, 60-63; asst physician, Boston City Hosp, 68-; consult physician, Vet Admin Hosp, Boston, 68-; mem drug eval panel, Nat Acad Sci-Nat Res Coun, 69-71; mem bact & micol study sect, Nat Inst Allergy & Infectious Dis, 69-73; assoc ed, J Infectious Dis, 69-; mem ed bd, Infection & Immunity. Mem: AAAS; Am Soc Clin Invest; Infectious Dis Soc Am; Am Soc Microbiol; Am Soc Clin Pharmacol & Therapeut. Res: Host defense mechanisms and bacterial virulence factors in clinical infections. Mailing Add: Infectious Dis Div Univ Hosp 75 E Newton St Boston MA 02118

MCCAFFERTY, EDWARD, b Swoyerville, Pa, Nov 28, 37; m 66; c 2. CHEMISTRY. Educ: Wilkes Col, BS, 59; Lehigh Univ, MS, 64, PhD(chem), 68. Prof Exp: Res engr, Bethlehem Steel Corp, Pa, 59-64; Robert A Welch fel chem, Univ Tex, Austin, 68-70; RES CHEMIST, ENG MAT DIV, NAVAL RES LAB, 70- Honors & Awards: Victor K LaMer Award, Am Chem Soc, 71. Mem: Electrochem Soc; Am Chem Soc. Res: Surface chemistry; corrosion science; electrochemistry of corrosion processes, kinetics and inhibition; adsorption on metals and oxides. Mailing Add: Eng Mat Div Naval Res Lab 4555 Overlook Ave Washington DC 20390

MCCAFFERTY, ROBERT EUGENE, b Butler, Pa, Apr 30, 20; m 49; c 1. ANATOMY. Educ: Grove City Col, BS, 43; Univ Pittsburgh, MS, 48, PhD(zool), 52. Prof Exp: Asst comp anat, Univ Pittsburgh, 46-49, instr comp anat & endocrinol, 49-51; instr gross anat, Sch Med, Univ Md, 51-59, assoc, Univ Hosp, 57-59; asst prof anat, Col Med, Univ Ky, 59-69; ASSOC PROF ANAT & RES ASSOC OBSTET & GYNEC, MED CTR, W VA UNIV, 69- Concurrent Pos: Lectr, St Agnes Hosp, Baltimore, Md, 57-59. Mem: AAAS; Am Asn Anat; Asn Am Med Cols; Microcirc Soc; Biol Stain Comn. Res: Radioactive isotopes; amniotic fluid; malformations; porphyrins; fetal tumor tissues; placental steroids. Mailing Add: Dept of Anat Univ Med Ctr Morgantown WV 26506

MCCAFFREY, FRANCIS, b Taunton, Mass, Dec 30, 20; m 46; c 11. SOLID STATE PHYSICS. Educ: Providence Col, BS, 43; Univ Notre Dame, PhD(physics), 52. Prof Exp: Instr physics, Providence Col, 46-48; asst, Univ Notre Dame, 48-52; electronic scientist, US Naval Ord Test Sta, 52-56; ASSOC PROF PHYSICS, BOSTON COL, 56- Mem: AAAS; Am Phys Soc; Am Asn Physics Teachers. Res: Diffusion and ionic conductivity in solids. Mailing Add: Dept of Physics Boston Col Chestnut Hill MA 02167

MCCAFFREY, JOSEPH CLIFFORD, b Marinette, Wis, Jan 10, 09; m 37; c 5. MICROBIOLOGY. Educ: St Norbert Col, BS, 32, MA, 34; Univ Wis, MS, 34; Johns Hopkins Univ, MSPH, 43. Prof Exp: Instr biol & bact, St Norbert Col, 32-34; head sci dept, Cherokee Jr Col, 34-37; prof biol & bact, St Ambrose Col, 37-41; dean, Springfield Jr Col, 41-42; chief, Bur Sanit Bact, State Dept Pub Health, Ill, 43-67, asst chief div labs, 67-70, chief div lab serv, Ill Environ Protection Agency, 70-72; RETIRED. Concurrent Pos: Prof, Loyola Univ, Ill, 43-49; exec secy, Nat Conf Interstate Milk Shipments, 73- Mem: Am Soc Microbiologists; Am Pub Health Asn. Res: Dairy bacteriology and chemistry; water biology and bacteriology. Mailing Add: 3306 Glouster St Sarasota FL 33580

MCCAIN, ARTHUR HAMILTON, b San Francisco, Calif, Aug 31, 25; m 59; c 1. PLANT PATHOLOGY. Educ: Univ Calif, BS, 49, PhD, 59. Prof Exp: LECTR PLANT PATH, UNIV CALIF, BERKELEY & AGRICULTURIST & EXTEN PLANT PATHOLOGIST, AGR EXP STA, 59- Concurrent Pos: Consult, US Forest Serv. Mem: AAAS; Am Phytopath Soc. Res: Forest tree diseases; control of plant dieases; ornamental diseases. Mailing Add: Dept of Plant Path Univ of Calif Berkeley CA 94720

MCCAIN, FRANCIS SAXON, b Ashland, Ala, Aug 13, 21; m 50; c 1. PLANT BREEDING. Educ: Auburn Univ, BS, 42, MS, 48; Purdue Univ, PhD, 50. Prof Exp: Assoc prof agron & soils & assoc plant breeder, Auburn Univ, 50-59, prof agron & soils & plant breeder, 59-66; CHMN DIV AGR, HOME ECON & FORESTRY, ABRAHAM BALDWIN AGR COL, 66-, ASST DIR RURAL DEVELOP CTR, 69- Mem: Am Soc Agron. Res: Corn breeding. Mailing Add: 2205 Murray Ave Tifton GA 31794

MCCAIN, GEORGE HOWARD, b Flora, Ind, Nov 15, 24; m 48; c 1. ORGANIC CHEMISTRY. Educ: Franklin Col, AB, 49; Univ Ill, MS, 50, PhD(chem), 53. Prof Exp: Res chemist, Electrochem Dept, E I du Pont de Nemours & Co, Inc, 52-55; sr chemist, 55-59, RES ASSOC, DIAMOND SHAMROCK CORP, 59- Mem: Am Chem Soc; fel Am Inst Chemists. Res: Vinyl polymerization; ionogenic polymers; membrane chemistry. Mailing Add: T R Evans Res Ctr Diamond Shamrock Corp Box 348 Painesville OH 44077

MCCAIN, JAMES HERNDON, b Little Rock, Ark, Sept 18, 41; m 63; c 2. ORGANIC CHEMISTRY. Educ: Southwestern at Memphis, BS, 63; Northwestern Univ, PhD(org chem), 67. Prof Exp: CHEMIST, UNION CARBIDE CORP, 67- Mem: Am Chem Soc. Res: Applied research; development; plant problems. Mailing Add: 818 Beaumont Rd Charleston WV 25314

MCCAIN, JOHN CHARLES, b Ft Worth, Tex, Aug 11, 39; m 56; c 2. MARINE BIOLOGY. Educ: Tex Christian Univ, BA, 62; Col William & Mary, MA, 64; George Washington Univ, PhD(zool), 67; Univ Hawaii, MPH, 75. Prof Exp: Res asst, Va Inst Marine Sci, 62-64; mus technologist syst zool, Smithsonian Inst, 64-65, mus specialist, 65, asst cur, 65-67; res assoc oceanog, NSF Antarctic Prog res grant to Dr Joel W Hedgpeth, Ore State Univ, 67-69; in-chg benthic invert div, Oceanog Sorting Ctr, Smithsonian Inst, Washington, DC, 69-70; aquatic ecologist, TRW/Hazleton Labs, Inc, 70-71; SR MARINE BIOLOGIST, ENVIRON DEPT, HAWAIIAN ELEC CO, 71- Concurrent Pos: Res assoc, Bernice P Bishop Mus, 73- Mem: Soc Syst Zool; Am Fisheries Soc; Am Soc Limnol & Oceanog; Biomet Soc; Nat Asn Underwater Instr. Res: Taxonomy and ecology of invertebrates, particularly Amphipoda; marine ecology; pollution biology; biostatistics. Mailing Add: Environ Dept Hawaiian Elec Co PO Box 2750 Honolulu HI 96840

MCCALEB, JOHN EARL, agronomy, see 12th edition

MCCALEB, KIRTLAND EDWARD, b Brighton, Mass, Sept 10, 26; m 49; c 2. ORGANIC CHEMISTRY. Educ: Dartmouth Col, AB, 46; Univ Wis, PhD(org chem), 49. Prof Exp: Jr chemist, Eastman Kodak Co, 46; fel, Univ Wis, 49-50; sr org chemist,

Res Labs, Gen Mills, Inc, Minn, 50-53, leader nitrogen prod sect, 53-60; mgr tech sales serv, Foremost Chem Prod Co, Calif, 60-65; indust economist, Chem Econ Handbook, 65-68, ed, 68-72, DIR CHEM-ENVIRON PROG, STANFORD RES INST, 72- Mem: Am Chem Soc; Am Oil Chem Soc. Res: Surface-active agents; chemical market research; environmental chemicals. Mailing Add: 121 Durazno Way Menlo Park CA 94025

MCCALEB, STANLEY B, b Santa Barbara, Calif, Oct 29, 19; m 52; c 3. CLAY MINERALOGY. Educ: Univ Calif, BS, 42; Cornell Univ, MS, 48, PhD(soils), 50. Prof Exp: From asst prof to assoc prof soil genesis, NC State Col, 50-56; sr soil correlator, Soil Conserv Serv, USDA, Calif, 56-58; sr res geologist & head clay mineral res sect, 58-70, supvr chem eng, Prod Serv Lab, 70-74, MGR PROD SERV LAB, SUN OIL CO, 74- Mem: Clay Minerals Soc (pres, 75-76); Soil Sci Soc Am; Am Soc Agron; Int Soc Soil Sci. Res: Soil morphology, genesis and classification; mineral weathering; land use and management; analysis labs; pollution evaluation; petroleum exploration and production. Mailing Add: 200 W Shore Dr Richardson TX 75080

MCCALL, CHARLES B, b Memphis, Tenn, Nov 2, 28; m 51; c 4. MEDICINE. Educ: Vanderbilt Univ, BA, 50, MD, 53. Prof Exp: Nat Acad Sci-Nat Res Coun pulmonary fel, 57-58; instr med, Med Col Ala, 58-59; from asst prof to assoc prof, Univ Tenn, 59-69; prof, Univ Tex Med Br, Galveston, 69-72, asst vchancellor & coordr regional med prog, Univ Tex Syst, 69-72; assoc dean clin affairs, Univ Tex Southwestern Med Sch, 72-75; DEAN COL MED, CTR HEALTH SCI, UNIV TENN, MEMPHIS, 75- Mem: Am Fedn Clin Res; fel Am Col Physicians; fel Am Col Chest Physicians; Am Thoracic Soc. Res: Pulmonary physiology and mechanics. Mailing Add: Col of Med Univ of Tenn Ctr for Health Sci Memphis TX 38163

MCCALL, CHARLES EMORY, b Lenoir, NC, Jan 30, 35; m 57; c 3. INFECTIOUS DISEASES. Educ: Wake Forest Col, BS, 57; Bowman Gray Sch Med, MD, 61. Prof Exp: From intern to resident, Harvard Med Sch, 61-66, teaching asst, 66-68; from asst prof to assoc prof, 68-75, PROF MED, BOWMAN GRAY SCH MED, 75-, DIR INFECTIOUS DIS, 71- Concurrent Pos: Bowman Gray Sch Med fac award, 61; NIH fel, Thorndike Res Lab, Harvard Med Sch, 66-68; NIH res career develop award, 74. Mem: Fel Royal Soc Med; Infectious Dis Soc Am; Am Fedn Clin Res; Am Asn Immunologists; Am Soc Exp Biol. Res: Neutrophil biology and host defense. Mailing Add: Dept of Med Bowman Gray Sch Med Winston-Salem NC 27103

MCCALL, CHESTER HAYDEN, JR, b Vandergrift, Pa, Aug 6, 27; m 52, 72; c 2. APPLIED STATISTICS. Educ: George Washington Univ, AB, 50, AM, 52, PhD, 57. Prof Exp: Asst prof statist, George Washington Univ, 52-56; res dir, Booz-Allen Appl Res, Inc, Md, 59-63, vpres & dir western opers, 63-69, managing vpres, Booz-Allen Appl Res, Inc, 69-71; pres, Int Careers Inst, Inc, 71-74; MGR SYSTS EVAL DEPT, CACI, INC, 74- Concurrent Pos: Consult, Corn Industs Res Found, 53-59; eng res & develop lab & logistics res, US Dept Army, 55-59. Mem: Am Statist Asn; Am Math Soc; Opers Res Soc Am; Inst Mgt Sci; fel Am Soc Qual Control. Res: Evaluation planning; survey sampling; experimental design; systems analysis; operations research; audio-visual and audio-manual education, training, and information transfer. Mailing Add: 7823 W 80th St Playa Del Rey CA 90201

MCCALL, DANIEL FRANCIS, b Westfield, Mass, Mar 3, 18; m 46; c 2. ANTHROPOLOGY. Educ: Boston Univ, AB, 49; Columbia Univ, PhD(anthrop), 55. Prof Exp: Prof social sci, Univ Liberia, 53; PROF ANTHROP, BOSTON UNIV, 54- Concurrent Pos: Vis lectr, Univ Col of Ghana, 60-61; Rockefeller Found travel & res grant, 63-64. Mem: Am Anthrop Asn; Am Soc Ethnohist; African Studies Asn; Int African Inst. Res: Urbanization; ethnology; historical reconstruction. Mailing Add: Dept of Anthrop & Hist Boston Univ Col of Liberal Arts Boston MA 02215

MCCALL, DAVID WARREN, b Omaha, Nebr, Dec 1, 28; m 55; c 2. PHYSICAL CHEMISTRY. Educ: Univ Wichita, BS, 50; Univ Ill, MS, 51, PhD(chem), 53. Prof Exp: Asst chem dir, 69-72, MEM TECH STAFF, BELL TEL LABS, INC, 53-, HEAD PHYS CHEM, 61-, CHEM DIR, 72- Mem: AAAS; Am Chem Soc; fel Am Phys Soc; The Chem Soc. Res: Nuclear magnetic resonance; diffusion in liquids; polymer relaxation; dielectric properties; materials for communications systems. Mailing Add: Bell Tel Labs Inc 600 Mountain Ave Murray Hill NJ 07974

MCCALL, ELIZABETH REGINA, b Columbus, Ohio, May 16, 22. SPECTROSCOPY. Educ: La State Univ, BS, 43. Prof Exp: Asst sci aide anal chem, 43-44, chemist, 44-61, anal chemist, 61-65, res chemist, Southern Mkt & Nutrit Res Div, 65-70, RES CHEMIST, SOUTHERN REGIONAL RES CTR, USDA, 70- Honors & Awards: Superior Serv Award, USDA, 52. Mem: Am Chem Soc; Soc Appl Spectros; Am Asn Textile Chem & Colorists; Sigma Xi; Coblentz Soc. Res: Application of spectroscopy, primarily infrared, to the structural characterization of cotton and chemically modified cotton cellulose; development of analytical methods for textile materials and various agricultural products. Mailing Add: Southern Regional Res Ctr USDA Box 19687 New Orleans LA 70179

MCCALL, GEORGE LELAND, biology, see 12th edition

MCCALL, JERRY C, applied mathematics, see 12th edition

MCCALL, JOHN TEMPLE, b Davenport, Iowa, May 1, 21; m 50; c 3. BIOCHEMISTRY. Educ: Rollins Col, BS, 48; Univ Fla, MS, 56, PhD(animal nutrit), 58. Prof Exp: Lab asst animal husb & nutrit, Univ Fla, 50-51, asst, 51-58, asst chemist, Agr Exp Sta, 58-60; asst prof, Iowa State Univ, 60-64; ASSOC PROF BIOCHEM, MAYO GRAD SCH MED, UNIV MINN & BIOCHEMIST, MAYO CLIN, 65- Mem: Am Chem Soc; Am Asn Clin Chemists; Am Inst Nutrit. Res: Trace mineral metabolism, especially interactions among trace mineral elements; effects of protein-mineral chelates on mineral metabolism and mechanism of mineral transport. Mailing Add: Dept of Lab Med Mayo Clin Rochester MN 55901

MCCALL, KEITH BRADLEY, b Oak Park, Ill, Dec 29, 19; m 42; c 5. BIOCHEMISTRY. Educ: Univ Nebr, BSc, 41, MA, 43; Univ Wis, PhD(biochem), 47. Prof Exp: Asst biochem, Univ Wis, 43-46; instr chem, Mich State Univ, 47-48, asst prof, 48-51; chief blood derivatives unit, Div Labs, State Dept Health, Mich, 51-62; head biol prods develop, Squibb Inst Med Res, NJ, 62-65; ASST CHIEF BIOL PRODS DIV, BUR LABS, MICH DEPT PUB HEALTH, 65- Mem: AAAS; Am Chem Soc. Res: Blood plasma protein fractionation; biologic products development. Mailing Add: Mich Dept of Public Health 3500 N Logan St Lansing MI 48914

MCCALL, MARVIN ANTHONY, b Pitts, Ga, Feb 7, 18; m 45; c 2. ORGANIC CHEMISTRY. Educ: Univ Ga, BS, 42, MS, 44; Univ Rochester, PhD(org chem), 51. Prof Exp: Instr chem, Univ Ga, 43-44; res chemist, Eastman Kodak Co, 44-51, sr res chemist, Tenn Eastman Co, 51-69, RES ASSOC, TENN EASTMAN CO, 69- Mem; Am Chem Soc. Res: Synthetic organic, polymer and organophosphorus chemistry; organometallic compounds; synthesis of organic polymer intermediates, additives and flame retardants. Mailing Add: Res Lab Bldg 150B Tenn Eastman Co Kingsport TN 37662

MCCALL, MYRON THOMAS, chemistry, photochemistry, see 12th edition

MCCALL, RICHARD C, b Inavale, Nebr, Sept 13, 29; m 56; c 3. RADIOLOGICAL PHYSICS. Educ: Mass Inst Technol, BS, 52, PhD, 57. Prof Exp: Physicist, Hanford Atomic Prod Oper, Gen Elec Co, 56-59; NIH fel, Univ Lund, 59-61; sr physicist, Controls Radiation, Inc, Mass, 61-64; HEAD HEALTH PHYSICS GROUP, STANFORD LINEAR ACCELERATOR CTR, 64- Concurrent Pos: Consult, Varian Assocs. Mem: Am Asn Physicists in Med; Health Physics Soc. Res: Dosimetry; radiation shielding; electron accelerators; health physics and medical physics instrumentation. Mailing Add: Stanford Linear Accelerator Ctr Stanford Univ PO Box 4349 Stanford CA 94305

MCCALL, WADE WILEY, b Day, Fla, Aug 13, 20; m 41; c 3. SOIL FERTILITY. Educ: Univ Fla, BS, 42, MA, 47; Mich State Univ, PhD(soil sci), 53. Prof Exp: Soil chemist, Agr Exp Sta, Fla, 46-47, asst prof soils, Univ Fla, 47-51; asst prof soil sci, Mich State Univ, 53-62; assoc specialist soil mgt, 62-68, PROF SOIL SCI & SPECIALIST SOIL MGT, UNIV HAWAII, 68- Concurrent Pos: Agron & soils adv, Kasetsart Univ, Bangkok & dept agr, Thailand, 64-65; consult, Nat Sch Agr, El Salvador, 70. Mem: Soil Sci Soc Am; Int Soc Soil Sci. Res: Soil fertility research in floricultural greenhouse crops; fruit, vegetable and ornamental crops; basic soils; fertilizer technology. Mailing Add: Dept of Agron & Soil Sci Univ of Hawaii Honolulu HI 96822

MCCALLA, ARTHUR GILBERT, b St Catharines, Ont, Mar 22, 06; m 31; c 3. PLANT BIOCHEMISTRY, PLANT PHYSIOLOGY. Educ: Univ Alta, BSc, 29, MSc, 31; Univ Calif, PhD, 33. Prof Exp: Asst plant biochem, 29-31, assoc comt on grain res, 32-41, prof field crops, 41-44, prof plant sci, 44-71, chmn dept, 44-51, dean agr fac, 51-59, dean fac grad studies, 57-71, EMER PROF PLANT BIOCHEM, UNIV ALTA, 71- Concurrent Pos: Researcher, Inst Phys Chem, Uppsala, Sweden, 39-40; mem, Nat Res Coun Can, 50-56; mem, Can Commonwealth Scholar & Fel Comt, 59-72; pres, Can Asn Grad Schs, 62-63. Mem: Fel Royal Soc Can; Can Biochem Soc; fel Agr Inst Can. Res: Chemistry of plant proteins; biochemistry and quality of cereals. Mailing Add: 11455 University Ave Edmonton AB Can

MCCALLA, DENNIS ROBERT, b Edmonton, Alta, July 30, 34; m 57; c 2. BIOCHEMISTRY, CELL BIOLOGY. Educ: Univ Alta, BSc, 57; Univ Sask, MSc, 58; Calif Inst Technol, PhD(biochem, genetics), 61. Prof Exp: From asst prof to assoc prof biochem, 61-71, dean fac sci, 67-75, PROF BIOCHEM, McMASTER UNIV, 71- Mem: AAAS; Am Soc Plant Physiol; Can Soc Plant Physiol; fel Chem Inst Can; Am Soc Biol Chem. Res: Biochemical genetics; chloroplast mutagenesis; mechanism of action of mutagens. Mailing Add: Dept of Biochem McMaster Univ Hamilton ON Can

MCCALLA, THOMAS MARK, b Corinth, Miss, Nov 29, 09; m 32; c 5. SOIL MICROBIOLOGY. Educ: Miss State Univ, BS, 34; Univ Mo, MS, 35, PhD(soil microbiol), 37. Prof Exp: Instr bact, Kans State Univ, 37-41; from assoc bacteriologist to bacteriologist, 41-59, microbiologist, 59-69, CHIEF MICROBIOLOGIST, USDA, 69-; PROF AGRON, UNIV NEBR, LINCOLN, 42- Honors & Awards: Outstanding Performance Award, Univ Nebr, Lincoln, 61, Award of Merit. Mem: Fel AAAS; fel Am Soc Agron; fel Soil Conserv Soc Am. Res: Ion adsorption by bacteria; variation of bacteria; soil structure; moisture conservation methods; pollution of surface and ground water from beef cattle feedlots; conservation production system. Mailing Add: Dept of Agron Univ of Nebr Lincoln NE 68583

MCCALLAN, SAMUEL EUGENE ALAN, b St George's, Bermuda, Apr 22, 02; nat US; m 29; c 2. PLANT PATHOLOGY. Educ: Univ Toronto, BSA, 23; Cornell Univ, PhD(plant path), 29. Prof Exp: Asst plant path, Cornell Univ, 23-27, instr, 28-29; plant pathologist, Frasch Fungicide & Insecticide Proj, Boyce Thompson Inst, 29-39, plant pathologist, 40-43, head plant pathologist fungicide proj, 43-59, secy, 59-73, consult, 73-75, mem bd dirs, 67-75; RETIRED. Concurrent Pos: Consult, Union Carbide Corp, 39-59; governor, Crop Protection Inst, 39-65. Honors & Awards: Northeastern Div Merit Award, Am Phytopath Soc, 63. Mem: Fel AAAS; fel Am Phytopath Soc (secy, 51-53, vpres, 59, pres, 61); fel NY Acad Sci. Res: Fungicidal action of copper, sulfur and organic compounds; laboratory and greenhouse bioassay of fungicides; standardization of fungicidal tests; measurement of spores; uptake of fungicides; history of plant pathology. Mailing Add: Boyce Thompson Inst 1086 N Broadway Yonkers NY 10701

MCCALLEY, RODERICK CANFIELD, b Portland, Ore, Aug 2, 43. MAGNETIC RESONANCE. Educ: Calif Inst Technol, BS, 64; Harvard Univ, PhD(phys chem), 71. Prof Exp: NSF fel phys chem, Stanford Univ, 71-72; ASST PROF CHEM, DARTMOUTH COL, 72- Mem: Sigma Xi. Res: Electron spin resonance; radiation damage in solids; spin-label dynamics in solution; magnetic relaxation; electron structure of free radicals. Mailing Add: Dept of Chem Dartmouth Col Hanover NH 03755

MCCALLION, DAVID JOHN, b Toronto, Ont, Sept 25, 16; m 44; c 6. EMBRYOLOGY, TERATOLOGY. Educ: McMaster Univ, BA, 42, MA, 47; Brown Univ, PhD(cytol, embryol), 50. Prof Exp: Lectr zool, McMaster Univ, 45-47; asst prof, Acadia Univ, 49-51, prof, 51-55; from asst prof to prof, Univ Toronto, 55-68; PROF ANAT, FAC HEALTH SCI, SCH MED, MCMASTER UNIV, 68- Mem: Am Asn Anat; Can Asn Anat; Int Soc Develop Biol; Teratology Soc; Soc Develop Biol. Res: Origin and establishment of organization patterns in vertebrate embryos with special reference to nervous system in chick embryo; experimental teratology in rat and mouse. Mailing Add: Dept of Anat Sch of Med Fac Health Sci McMaster Univ Hamilton ON Can

MCCALLION, WILLIAM JAMES, b Toronto, Ont, Aug 20, 18; m 44; c 4. MATHEMATICS, ASTRONOMY. Educ: McMaster Univ, BA, 43, MA, 46. Prof Exp: Lectr math, 43-52, from asst prof to assoc prof, 53-70, asst dir exten, 52-53, dir, 55-70, dir educ servs, 62-70, PROF MATH & DEAN SCH ADULT EDUC, McMASTER UNIV, 70- Mem: Am Math Soc; Math Asn Am; Royal Astron Soc Can; Can Math Cong. Res: Modern algebra; mathematics of business. Mailing Add: Sch of Adult Educ McMaster Univ Hamilton ON Can

MCCALLISTER, LAWRENCE P, b Chicago, Ill, Mar 27, 43; m 73. MICROSCOPIC ANATOMY. Educ: Univ Ill, Champaign, BS, 66; Loyola Univ Chicago, PhD(anat), 71. Prof Exp: USPHS fel med & physiol, Pritzker Sch Med, Univ Chicago, 71-72; ASST PROF ANAT, MILTON S HERSHEY MED SCH, PA STATE UNIV, 72- Concurrent Pos: Nat Heart & Lung Inst grant, Milton S Hershey Med Sch, Pa State Univ, 72-75 & 76-79, Nat Heart & Lung Inst res career develop award, 76- Mem: Am Asn Anatomists; Sigma Xi. Res: Ultrastructure and function of muscle and nerve. Mailing Add: Dept of Anat Milton S Hershey Med Ctr Pa State Univ Hershey PA 17033

MCCALLUM, CHARLES ALEXANDER, JR, b North Adams, Mass, Nov 1, 25; m 55; c 4. ORAL SURGERY. Educ: Tufts Univ, DMD, 51; Med Col Ala, MD, 57; Am Bd Oral Surg, dipl. Hon Degrees: DSc, Univ Ala, 75. Prof Exp: Intern oral surg, Univ

Hosp, 51-52, resident, 52-54, from instr to assoc prof, Sch Dent, 56-59, intern med, Univ Hosp, 57-58, chmn dept oral surg, Sch Dent, 58-65, chief sect oral surg, Med Col, 65-70, PROF ORAL SURG, SCH DENT, UNIV ALA, BIRMINGHAM, 59-, DEAN, 62- Concurrent Pos: Consult, Vet Admin Hosps, Birmingham, Tuscaloosa & Tuskegee, 58-, mem med adv comt, 62-; consult, Ariz Med Study, 60 & Martin Army Hosp, Ft Benning, Ga, 61-; mem oral surg test construct comt, Nat Bd Dent Exam, 62-70; mem adv comt hosps & clins, USPHS, 64-66, mem prev med & dent rev panel, 65; mem bd dirs, Am Bd Oral Surg, 64-71, pres, 71; cent off consult, Dept Med & Surg, Vet Admin, 65-70; mem nat adv dent res, coun, Nat Inst Dent Res, 68-; pres, Am Asn Dent Schs, 70. Mem: Am Dent Asn; AMA; Am Soc Oral Surg (pres, 75-76); fel Am Col Dent. Res: Dental infection and effects of various systemic drugs on control of hemorrhage. Mailing Add: Sch of Dent Univ of Ala Birmingham AL 35233

MCCALLUM, CHARLES JOHN, JR, b Tacoma, Wash, Apr 26, 43; m 67; c 2. OPERATIONS RESEARCH. Educ: Mass Inst Technol, SB, 65; Stanford Univ, MS, 67, PhD(opers res), 70. Prof Exp: Mem tech staff opers res, 70-75, SUPVR OPERS RES APPL GROUP, BELL TEL LABS, 75- Mem: Opers Res Soc Am; Inst Mgt Sci; Math Prog Soc. Res: Mathematical programming; network flow theory; the linear complementarity problem. Mailing Add: Bell Labs Holmdel NJ 07733

MCCALLUM, GEORGE ALEXANDER, b Oceanbeach, Calif, Feb 4, 13; m 36; c 2. GENETICS. Educ: Stanford Univ, AB, 35, PhD(cytogenetics), 40. Prof Exp: Lab asst, Stanford Univ, 35-39, asst, 36-40; instr biol, 40-42, from asst prof to assoc prof, 42-50, prof, 50-76, chmn dept biol sci, 54-61 & 63-66, EMER PROF BIOL, SAN JOSE STATE UNIV, 76- Concurrent Pos: Res assoc, Off Chancellor, Calif State Cols, 61-63. Res: Plant cytogenetics and physiology; meiosis in Aloestriata, especially chromosome structure. Mailing Add: Dept of Biol Sci San Jose State Univ San Jose CA 95114

MCCALLUM, JOHN, physical chemistry, see 12th edition

MCCALLUM, KEITH STUART, b Fond du Lac, Wis, June 21, 19; m 42; c 2. ANALYTICAL CHEMISTRY. Educ: Univ Wis, BS, 42, PhD(chem), 50. Prof Exp: Group leader anal chem, Redstone Arsenal Res Div, 50-59, PROJ LEADER, ROHM AND HAAS CO, 59- Mem: AAAS; Am Chem Soc. Res: Trace analysis of air and water pollutants, and impurities from chemical and biodegradation. Mailing Add: Rohm and Haas Co Res Labs Spring House PA 19477

MCCALLUM, KENNETH JAMES, b Scott, Sask, Apr 25, 18; m 50; c 2. PHYSICAL CHEMISTRY. Educ: Univ Sask, BSc, 36, MSc, 39; Columbia Univ, PhD(chem), 42. Prof Exp: From asst prof to assoc prof chem, 43-52, head dept chem, 59-70, PROF CHEM, UNIV SASK, 52-, DEAN COL GRAD STUDIES & RES, 70- Concurrent Pos: Nuffield traveling fel natural sci, Cambridge Univ, 50-51. Mem: Fel AAAS; fel Royal Soc Can; Chem Inst Can. Res: Chemical effects of nuclear reactions; radiation chemistry; isotopic exchange reactions. Mailing Add: Col of Grad Studies & Res Univ of Sask Saskatoon SK Can

MCCALLUM, MALCOLM E, b Springfield, Mass, July 28, 34; m 63; c 2. GEOLOGY. Educ: Middlebury Col, AB, 56; Univ Tenn, MS, 58; Univ Wyo, PhD(geol), 64. Prof Exp: From asst prof to assoc prof geol, 62-71, PROF GEOL, COLO STATE UNIV, 71-; GEOLOGIST, US GEOL SURV, 56- Prof Exp: Res grants, Wyo Geol Surv, 59-61 & Geol Soc Am, 60 & 70. Mem: AAAS; Geol Soc Am; Mineral Soc Am; Geochem Soc. Res: Petrology, structure, and mineral resources of Precambrian crystalline rocks, northern Colorado and southern Wyoming; Precambrian-Tertiary structure and stratigraphy of disturbed belt in northern Big Belt Mountains, Montana; petrology and chemistry of kimberlitic diatremes of northern Colorado and southern Wyoming. Mailing Add: Dept of Earth Resources Colo State Univ Ft Collins CO 80523

MCCALLUM, RODERICK EUGENE, b Denver, Colo, Aug 14, 44; m 67; c 2. MICROBIOLOGY, BIOCHEMISTRY. Educ: Univ Kans, BA, 67, PhD(microbiol), 70. Prof Exp: Instr microbiol, Univ Kans, 70; res assoc, Univ Tex, 70-72, instr, 71; asst prof, 72-75, ASSOC PROF MICROBIOL, UNIV OKLA, 75- Concurrent Pos: USPHS trainee, Univ Tex, Austin, 70-72. Mem: AAAS; Am Soc Microbiol; Reticuloendothelial Soc. Res: Metabolism in disease; endotoxin. Mailing Add: Dept of Microbiol & Immunol Univ of Okla Health Sci Ctr Oklahoma City OK 73190

MCCALLY, MICHAEL, b Cleveland, Ohio, Jan 5, 35; m 58; c 2. PHYSIOLOGY, AEROSPACE MEDICINE. Educ: Princeton Univ, AB, 56; Case Western Reserve Univ, MD, 60; Ohio State Univ, PhD(physiol), 70. Prof Exp: Intern med, Univ Hosps, Cleveland, 60-61; resident med officer environ physiol, Aerospace Med Res Lab, Wright-Patterson AFB, 61-66, chief environ physiol br, 66-67, chief environ med div, 67-71; ASSOC DEAN, SCH MED & HEALTH SCI, GEORGE WASHINGTON UNIV, 74-, ASSOC PROF MED & PHYSIOL, 72- Concurrent Pos: Asst resident obstet & gynec, Grace-New Haven Community Hosp, Conn, 63-64; clin instr prev med, Ohio State Univ, 66-71; fel, Aschoff Div, Max Planck Inst Physiol of Behav, 70-71; mem comn hearing bioacoust biomech, Nat Acad Sci-Nat Res Coun Air Force Systs. Honors & Awards: Borden Award. Mem: Am Fedn Clin Res; Am Heart Asn; Soc Biol Rhythm; Biomed Eng Soc; Am Physiol Soc. Res: Physiological mechanisms of body fluid volume regulation; physiological effects of weightlessness; sleep and biological rhythms. Mailing Add: 2300 Eye St Washington DC 20037

MCCAMAN, MARILYN WALES, b Oak Park, Ill, Oct 10, 28; m 53; c 5. NEUROSCIENCES. Educ: Grinnell Col, AB, 50; Wash Univ, PhD(pharmacol), 57. Prof Exp: Asst pharmacol, Wash Univ, 56-57; neurochemist, Neuromuscular Res Lab, Sch Med, Ind Univ, Indianapolis, 59-60, res assoc neurol, 60-61, from instr to asst prof, 61-70, res assoc, Inst Psychiat Res, 60-68; ASSOC RES SCIENTIST, CITY OF HOPE MED CTR, 70- Res: Biochemistry of muscular dystrophy; metabolism of neurotransmitters in the invertebrate central nervous system. Mailing Add: Dept of Neurosci City of Hope Med Ctr Duarte CA 91010

MCCAMAN, RICHARD EUGENE, b Barberton, Ohio, Mar 28, 30; m 53; c 5. BIOCHEMISTRY, PHARMACOLOGY. Educ: Wabash Col, AB, 52; Wash Univ, PhD(pharmacol), 57. Prof Exp: Res assoc & instr biochem & pharmacol, Inst Psychiat Res, Sch Med, Ind Univ, Indianapolis, 57-62, from asst prof to assoc prof pharmacol, 62-68, prin investr neurochem, 62-68; CHIEF MOLECULAR NEUROBIOL, CITY OF HOPE MED CTR, 68- Mem: AAAS; Asn Res Nerv & Ment Dis; Am Soc Neurochem; Int Soc Neurochem; Am Soc Biol Chem. Res: Biochemistry, physiology and chemical pathology of nervous system; chemistry, pharmacology and physiology of neurotransmitters; invertebrate neurobiology; intracellular electrophysiology. Mailing Add: Div of Neurosci City of Hope Med Ctr Duarte CA 91010

MCCAMEY, BENJAMIN FRANKLIN, JR, biology, see 12th edition

MCCAMMON, HELEN MARY, paleobiology, see 12th edition

MCCAMMON, MARY, b Eng, Aug 23, 27; m 60. MATHEMATICS. Educ: Univ London, BSc, 49, MSc, 50, PhD(math), 53. Prof Exp: Res assoc math, Mass Inst Technol, 53-54; asst prof, 54-60, ASSOC PROF MATH, PA STATE UNIV, UNIVERSITY PARK, 60- Mem: Am Math Soc; Asn Comput Mach. Res: Numerical analysis; uses of digital computers. Mailing Add: Dept of Math McAllister Bldg Pa State Univ University Park PA 16802

MCCAMMON, RICHARD B, b Indianapolis, Ind, Dec 3, 32; m 56; c 2. GEOLOGY. Educ: Mass Inst Technol, BSc, 55; Univ Mich, MSc, 56; Ind Univ, PhD(geol), 59. Prof Exp: Rockefeller fel statist, Univ Chicago, 59-60; asst prof geol, Univ NDak, 60-61; res geologist, Gulf Res & Develop Co, 61-68; assoc prof, 68-70, PROF GEOL, UNIV ILL, CHICAGO CIRCLE, 70- Mem: Geol Soc Am; Am Statist Asn; Soc Econ Paleont & Mineral. Res: Statistical application to problems in petroleum exploration; multivariate methods in biostratigraphy and paleoecology; computer applications in geology. Mailing Add: Dept of Geol Sci Univ of Ill at Chicago Circle Chicago IL 60680

MCCAMMON, ROBERT DESMOND, b Northern Ireland, July 23, 32; m 60. SOLID STATE PHYSICS. Educ: Queen's Univ, Ireland, BSc, 53; Oxford Univ, DPhil(physics), 57. Prof Exp: Vis res assoc, 57-59, asst prof physics, 59-69, ASSOC PROF PHYSICS, PA STATE UNIV, 69- Concurrent Pos: Vis fel nat standards lab, Commonwealth Sci & Indust Res Orgn, Australia, 62-63. Res: Plastic deformation of metals at low temperatures; internal friction and dielectric behavior of high polymers at low temperatures; thermal expansion of solids; thermal properties of disordered solids at low temperatures. Mailing Add: Dept of Physics Pa State Univ University Park PA 16802

MCCAMPBELL, JOHN CALDWELL, b Morgantown, NC, Nov 9, 11; m 42; c 1. GEOLOGY. Educ: Univ NC, BS, 34, PhD(geol), 44; Vanderbilt Univ, MS, 36. Prof Exp: Asst geol, Univ NC, 38-40, instr, 40-42, lectr, 42-44, asst prof, 44; asst prof, Rutgers Univ, 44-48; assoc prof, Tulane Univ, 48-54; prof & head dept, Univ Southwestern La, 54-75. Concurrent Pos: Res partic, Oak Ridge Inst Nuclear Studies, 59. Mem: Geol Soc Am; Am Asn Petrol Geologists; Am Inst Prof Geologists. Res: Geophysical studies; further geophysical evidences to the origin of the Carolina Bays. Mailing Add: 1600 Diana Blvd Merritt Island FL 32952

MCCAMY, KEITH, geophysics, see 12th edition

MCCANDLESS, BYRON HOWARD, b Florence, Colo, June 30, 24; m 49; c 2. MATHEMATICS. Educ: Colo State Univ, BS, 48; Ind Univ, AM & PhD(math), 53. Prof Exp: Instr math, Rutgers Univ, 52-54, from asst prof to assoc prof, 54-65; prof, Western Wash State Col, 65-66; PROF MATH, KENT STATE UNIV, 66- Mem: Am Math Soc; Math Asn Am. Res: Dimension theory; topology. Mailing Add: 1880 Carlton Dr Kent OH 44240

MCCANDLESS, ESTHER LEIB, b Brooklyn, NY, Mar 18, 23. PHYSIOLOGY. Educ: Bethany Col, BS, 44; Cornell Univ, MS, 47, PhD(animal physiol), 48. Prof Exp: Asst biol, Bethany Col, 43-44; teacher high sch, Pa, 44-45; lab technician, Harrisburg Hosp, Pa, 45; asst human physiol, Cornell Univ, 45-48, physiol chem, NY State Vet Col, 48-49; assoc physiol, Woman's Med Col Pa, 49-53; res assoc med, Jefferson Med Col, 53; asst physiol, Univ Tenn, 54-57, asst prof, 57-59; res physiologist, Chronic Dis Res Inst, Univ Buffalo, 59-64; assoc prof biol, 64-69, PROF BIOL, McMASTER UNIV, 69- Mem: AAAS; Am Physiol Soc; Can Physiol Soc; Can Soc Cell Biol; Can Soc Immunologists. Res: Polysaccharide immunology; algal polysaccharides; connective tissue metabolism; experimental diabetes; ruminant and phospholipid metabolism. Mailing Add: Dept of Biol McMaster Univ Hamilton ON Can

MCCANDLESS, ROBERT WILLIAM, b Stockton, Calif, July 3, 23; div; c 2. INSTRUMENTATION, MATERIALS SCIENCE. Educ: Stanford Univ, AB, 45, AM, 47, PhD(psychol), 55; Univ Calif, Berkeley, PhD(chem), 54. Prof Exp: Sr chemist, Nat Motor Bearing Co, 53-56, lab dir, 56-58, res dir, 58-60; asst dir res, Org Polymers Lab, Chem Process Co, 60-63; chief chemist res lab, Farah Mfg Co, 63-65; dir res, Baw Mfg Co, 65-67, mgr, 68-72; VPRES, DON PARKER MFG & PURITAN CAN, 72- Mem: Am Chem Soc (secy, 59-61); Soc Aerospace Mat & Process Eng; fel Am Inst Chemists; NY Acad Sci. Res: Polymer chemistry; fiber and plastics technology; instrumentation analysis; electron microscopy; spectrophotometry; colloids; cosmetic chemistry. Mailing Add: 11710 100th Ave Edmonton AB Can

MCCANE, DONALD IRWIN, b Covington, Ky, Sept 2, 26; m 50; c 4. ORGANIC CHEMISTRY. Educ: Yale Univ, BS, 48; Univ Mich, MS, 50, PhD(chem), 54. Prof Exp: Instr, Univ Mich, 51-53; res chemist exp sta, 53-60, Wash Works Lab, 60-62, tech rep, Chestnut Run Lab, 62-66, supvr tech serv lab, Plastics Dept, 66-71, consult, 71-73, SR CONSULT, TECH SERV LAB, PLASTICS DEPT, E I DU PONT DE NEMOURS & CO, INC, 73- Mem: Am Chem Soc. Res: C14 and N15 tracer techniques in organic chemistry; fluorocarbon chemistry; processing fluorocarbon polymers. Mailing Add: Chestnut Run Tech Serv Lab E I du Pont de Nemours & Co Inc Wilmington DE 19898

MCCANN, DAISY S, b Hamburg, Ger, Mar 8, 27; Can citizen; m 50; c 1. BIOCHEMISTRY, IMMUNOLOGY. Educ: Univ Toronto, BA, 50; Wayne State Univ, MS & PhD(biochem), 58. Prof Exp: Teacher, Riverside High Sch, Ont, 50-51; tech asst, John Wyeth & Bros Co Ltd, Ont, 51-54; res fel, Wayne State Univ, 55-58, res assoc, 59-67, adj assoc prof, 67-70; RES BIOCHEMIST, WAYNE COUNTY GEN HOSP, 70-; ASST PROF BIOL CHEM, DEPT MED, UNIV MICH, ANN ARBOR, 70- Concurrent Pos: A J Boyle fel, Wayne State Univ, 58-59. Mem: AAAS; Am Chem Soc; NY Acad Sci; Am Asn Clin Chem. Res: Hormonal control of connective tissue, particularly that of parathyroid hormone immunology as involved in radioimmunoassay techniques; immunology and its relationship to oncology. Mailing Add: Wayne County Gen Hosp Dept of Med PO Box 124 Eloise MI 48132

MCCANN, FRANCES VERONICA, b Manchester, Conn, Jan 15, 27; m 62. PHYSIOLOGY, ELECTROPHYSIOLOGY. Educ: Univ Conn, AB, 52, PhD(physiol), 59; Univ Ill, MS, 54. Prof Exp: Res assoc, Marine Biol Lab, Woods Hole, 52-58; from instr to assoc prof, 59-73, ADJ PROF BIOL SCI, DARTMOUTH MED SCH, 74-, PROF PHYSIOL, 73- Concurrent Pos: Estab investr, Am Heart Asn, 65-70, mem coun basic sci, 72- & New Eng res rev comt, 74-; mem physiol study sect, NIH, 73-75. Mem: Soc Gen Physiol; Am Physiol Soc; Biophys Soc. Res: Electrophysiology of neuro-muscular systems; cardiac electrophysiology, insect flight mechanisms. Mailing Add: Dept of Physiol Dartmouth Med Sch Hanover NH 03755

MCCANN, GILBERT DONALD, JR, b Glendale, Calif, Jan 12, 12; m 36; c 2. INFORMATION SCIENCE. Educ: Calif Inst Technol, BS, 34, MS, 35, PhD(elec eng), 39. Prof Exp: Engr cent sta, Westinghouse Elec Corp, Pa, 38-41, consult transmission engr, 41-46; prof elec eng, 47-65, dir, Willis H Booth Comput Ctr, 61-71, PROF APPL SCI, CALIF INST TECHNOL, 65- Concurrent Pos: Westinghouse lectr, Univ Pittsburgh, 40-46; mem adv comt lightning protection, US Army Ord; mem, Int Cong High Tension Transmission, 47. Mem: Am Soc Mech Eng; Am Soc Eng Educ; fel Inst Elec & Electronics Eng. Res: Electrical transmission and lightning research; large-scale computing devices; fulchronograph; Westinghouse transients analyzer; electric analog computer; bioengineering. Mailing Add: Dept of Info Sci 286-80 Calif Inst Technol Pasadena CA 91125

MCCANN, JAMES ALWYN, b Boston, Mass, May 20, 34; m 57; c 3. FISHERIES, BIOMETRICS. Educ: Univ Mass, BS, 56; Iowa State Univ, MS, 58, PhD(fishery mgt, biomet), 60. Prof Exp: Asst, Iowa State Univ, 56-60; marine biologist, US Bur Commercial Fisheries, Mass, 60-63, leader fisheries res, Mass Coop Fishery Unit, US Bur Sport Fisheries & Wildlife, 63-72, chief, Br Fish Ecosyst Res, 72-73, ACTG CHIEF, DIV FISHERY RES, US BUR SPORT FISHERIES & WILDLIFE, 73- Concurrent Pos: Assoc prof, Univ Mass, 64- Mem: Am Fisheries Soc; Am Inst Fishery Res Biol. Res: Fishery biometrics; Great Lakes fishery research; reservoir research; Atlantic salmon; stream channelization; fishery ecosystems. Mailing Add: Bur Sport Fisheries & Wildlife Matomic Bldg 1717 H St Washington DC 22040

MCCANN, LESTER J, b Minneapolis, Minn, Sept 16, 15; m 44; c 7. VERTEBRATE ZOOLOGY. Educ: Univ Minn, BS, 37, MS, 38; Univ Utah, PhD(vert zool), 55. Prof Exp: Instr biol, Boise Col, 49-55; PROF BIOL, COL ST THOMAS, 55- Concurrent Pos: NSF res grant, 60-61; head dept natural sci, St Mary's Jr Col, 61-69, assoc dean tech educ, 68-69. Mem: AAAS; Am Soc Zool; Sigma Xi. Res: Comparative vertebrate anatomy and embryology; human anatomy and physiology; role of predation in wildlife ecology. Mailing Add: Dept of Biol Col of St Thomas St Paul MN 55105

MCCANN, LEWIS PAUL, plant breeding, cytogenetics, see 12th edition

MCCANN, ROGER C, b Lakewood, Ohio, Aug 30, 42. MATHEMATICS. Educ: Pomona Col, BA, 64; Case Inst Technol, MS, 66; Case Western Reserve Univ, PhD(math), 68. Prof Exp: Asst prof math, Calif State Col, Los Angeles, 68-69; vis asst prof, 69-70, ASST PROF MATH, CASE WESTERN RESERVE UNIV, 70- Mem: Am Math Soc. Res: Dynamical systems; ordinary differential equations. Mailing Add: Dept of Math Case Western Reserve Univ Cleveland OH 44106

MCCANN, SAMUEL MCDONALD, b Houston, Tex, Sept 8, 25; m 50; c 3. NEUROENDOCRINOLOGY, PHYSIOLOGY. Educ: Rice Inst, 42-44; Univ Pa, MD, 48. Prof Exp: Intern internal med, Mass Gen Hosp, 48-49, asst resident, 49-50; instr physiol, Sch Med, Univ Pa, 48 & 52-53, assoc, 53-54, from asst prof to prof, 54-64; PROF PHYSIOL & CHMN DEPT, UNIV TEX HEALTH SCI CTR DALLAS, 65- Concurrent Pos: Rockefeller Found traveling grant, Royal Vet Col, Sweden, 54-55; sr asst, Royal Vet Col, Sweden, 54-55; consult, Schering Corp, NJ, 58; mem gen med B study sect, NIH, 65-67, mem endocrinol study sect, 67-69, mem corpus luteum panel, 71-74; mem pop res comn, Nat Inst Child Health & Human Develop, 74-; mem sci adv bd, Wis Regional Primate Ctr, 74- Honors & Awards: Oppenheimer Award, Endocrine Soc, 66. Mem: Am Physiol Soc; NY Acad Sci; Am Soc Clin Invest; Int Brain Res Orgn; Int Soc Res Reproduction. Res: Experimental neurogenic hypertension; hypothalamic regulation of thirst and pituitary function. Mailing Add: Dept of Physiol Univ of Tex Health Sci Ctr Dallas TX 75235

MCCANN, WILLIAM PETER, b Baltimore, Md, Dec 12, 24; m 54; c 2. PHARMACOLOGY, INTERNAL MEDICINE. Educ: Princeton Univ, AB, 49; Cornell Univ, MD, 49. Prof Exp: Intern internal med, Barnes Hosp, St Louis, Mo, 49-50; instr clin pharmacol & med, Johns Hopkins Univ, 57-58; asst prof pharmacol, Sch Med, Univ Colo, 58-63, asst prof med, 59-63; assoc prof pharmacol, 63-67, asst prof med, 63-71, PROF PHARMACOL, UNIV ALA, BIRMINGHAM, 67-, ASSOC PROF MED, 71- Concurrent Pos: Res fel pharmacol, Sch Med, Wash Univ, 52-54; res fel physiol chem, Sch Med, Johns Hopkins Univ, 54-55, res fel clin pharmacol & med, 55-57. Mem: AAAS; Am Soc Pharmacol & Exp Therapeut; Am Soc Clin Pharmacol & Therapeut. Res: Human pharmacology. Mailing Add: Dept of Pharmacol Univ of Ala Univ Sta Birmingham AL 35294

MCCANTS, CHARLES BERNARD, b Andrews, SC, Sept 14, 24; m 47; c 2. SOIL SCIENCE. Educ: NC State Col, BS, 49, MS, 50; Iowa State Univ, PhD(soils), 55. Prof Exp: Instr soils, NC State Col, 50-52; assoc agronomist, Clemson Col, 55-56; from asst prof to assoc prof, 56-63, prof, 63-71, HEAD DEPT SOIL SCI, NC STATE UNIV, 71- Mem: Am Soc Agron; Soil Sci Soc Am. Res: Soil fertility; ion absorption; plant nutrition; tobacco production. Mailing Add: Dept of Soil Sci NC State Univ Raleigh NC 27607

MCCANTS, DAVID, JR, organic chemistry, see 12th edition

MCCARGO, MATTHEW, materials science, physical chemistry, see 12th edition

MCCARL, RICHARD LAWRENCE, b Grove City, Pa, July 6, 27; m 51; c 3. BIOCHEMISTRY. Educ: Grove City Col, BS, 50; Pa State Univ, MS, 58, PhD(biochem), 61. Prof Exp: Teacher high sch, Pa, 50-55; asst biochem, 55-57, instr, 57-61, from asst prof to assoc prof, 61-74, PROF BIOCHEM, PA STATE UNIV, 74- Concurrent Pos: NSF grant, 64-65; NIH grants, 66-68 & 70-73. Mem: AAAS; Am Chem Soc; Am Soc Biol Chem; Tissue Cult Asn. Res: Lipid biosynthesis; animal tissue culture; physiology of cultured heart cells; lipid, RNA, DNA and protein patterns during differentiation and ageing. Mailing Add: 303 Althouse Lab Pa State Univ University Park PA 16802

MCCARLEY, ROBERT EUGENE, b Denison, Tex, Aug 17, 31; m 52; c 4. INORGANIC CHEMISTRY. Educ: Univ Tex, BS, 53, PhD(chem), 56. Prof Exp: From instr to assoc prof chem, Univ, 56-70, res assoc, Ames Lab, 56-57, assoc chemist, 57-63, chemist, 63-74, PROF CHEM, IOWA STATE UNIV, 70-, SR CHEMIST, AMES LAB, 74- Mem: Am Chem Soc. Res: Chemistry of transition elements; metal halide vaporization equilibria; metal carbonyl compounds; metal-metal bonding. Mailing Add: Dept of Chem Iowa State Univ Ames IA 50010

MCCARLEY, WARDLOW HOWARD, b Pauls Valley, Okla, May 2, 26; m 50; c 3. VERTEBRATE ZOOLOGY. Educ: Austin Col, AB, 48; Univ Tex, MA, 50, PhD(ecol), 53. Prof Exp: Instr biol, Austin Col, 48; instr zool, Stephen F Austin State Col, 50-52, from asst prof to assoc prof, 53-39; prof biol & head dept, Southeastern State Col, 59-61; chmn dept, 63-71, PROF BIOL, AUSTIN COL, 61- Concurrent Pos: AEC res grant, 54-58; vis prof biol sta, Univ Okla, 58-; NSF res grants, 59-62 & 64-71. Mem: AAAS; Am Soc Mammal; Ecol Soc Am; Wildlife Soc. Res: Population ecology; speciation processes and behavior of vertebrates; population biology, behavior and vocalization of wild canids. Mailing Add: Dept of Biol Austin Col Sherman TX 75090

MCCARRELL, JANE DINSMORE, b Washington, Pa, Oct 22, 09. PHYSIOLOGY. Educ: Mt Holyoke Col, BA, 32; Vassar Col, MA, 34; Radcliffe Col, PhD(physiol), 40. Prof Exp: Asst physiol, Vassar Col, 32-35, instr, 35-37; asst sch pub health, Harvard Univ, 40-41; res fel anesthesia, Mass Gen Hosp, 41-45; asst prof, Vassar Col, 45-46; prof biol & chmn dept, 46-71, EMER PROF BIOL, HOOD COL, 71- Concurrent Pos: Instr, Wellesley Col, 44. Mem: Am Physiol Soc. Res: Lymphatic physiology; circulation and respiration; anesthesia; flow and protein content of cervical lymph. Mailing Add: 1111 Fairview Ave Frederick MD 21701

MCCARROLL, BRUCE, b Chicago, Ill, July 28, 33; m 59; c 1. PHYSICAL CHEMISTRY. Educ: Univ Calif, Los Angeles, BS, 55; Univ Chicago, PhD(phys chem), 58. Prof Exp: Asst, AEC, Univ Chicago, 55-56, res assoc, Dept Chem & Enrico Fermi Inst Nuclear Studies, 58-60; phys chemist, Metall & Ceramics Lab, Gen Elec Res & Develop Ctr, NY, 60-71, sr develop engr, 71-74, MGR MAT ANAL & TEST, GEN ELEC CO, 74- Mem: Am Phys Soc; Am Chem Soc. Res: Advanced development; electro-mechanical systems; food preparation systems; head transfer; technological forecasting. Mailing Add: Gen Elec Co 1285 Boston Ave Bridgeport CT 06602

MCCARROLL, JAMES RENWICK, b New York, NY, Nov 14, 21; m. MEDICAL ADMINISTRATION. Educ: Colby Col, AB, 42; Cornell Univ, MD, 46. Prof Exp: Asst prof pub health & dir div epidemiol res, Med Col, Cornell Univ, 57-66; prof prev med & dir environ health div, Sch Med, Univ Wash, 66-70, prof environ health & chmn dept, Sch Pub Health & Community Med, 70-73; CLIN PROF, COMMUNITY ENVIRON MED, UNIV CALIF, IRVINE, 75-; ASST MED DIR, CITY OF LOS ANGELES, 73- Concurrent Pos: Dir personnel health serv, New York Hosp, 53-66. Mem: Am Pub Health Asn; Am Occup Med Asn; AMA; Am Heart Asn; Int Epidemiol Asn. Res: Health effects of air pollution. Mailing Add: Med Serv Div 1401 W Sixth St Los Angeles CA 90017

MCCARROLL, WILLIAM HENRY, b Brooklyn, NY, Mar 19, 30; m 58; c 4. INORGANIC CHEMISTRY. Educ: Univ Conn, BA, 53, MS, 55, PhD(chem), 57. Prof Exp: Asst chem, Univ Conn, 54-56; mem tech staff, RCA Labs, 56-67; asst prof, 67-73, ASSOC PROF CHEM, RIDER COL, 73-, CHMN DEPT, 72- Mem: Am Chem Soc. Res: Inorganic solid state; crystal structures; photoelectronic materials; transition metal oxides in low valance states; oxides with metal atom clusters. Mailing Add: Dept of Chem Rider Col Lawrenceville NJ 08648

MCCARRON, EDWARD M, organic chemistry, see 12th edition

MCCARRON, MARGARET MARY, b Chicago, Ill. INTERNAL MEDICINE, PHARMACY. Educ: St Xavier Col, Ill, BA, 50; Loyola Univ Chicago, MD, 54. Prof Exp: Asst prof, 60-68, ASSOC PROF MED & PHARM, UNIV SOUTHERN CALIF, 60-; ASST MED DIR, LOS ANGELES COUNTY-UNIV SOUTHERN CALIF MED CTR, 62- Concurrent Pos: NIH fel diabetes, Univ Southern Calif, 58-59; consult, Vet Admin & Indian Health Serv; consult task force on prescription drugs, Dept Health, Educ & Welfare, consult rev comt task force on drugs; consult comt rev, US Pharmacopeia; consult task force on role of pharmacist, Nat Ctr Health Serv Res & Develop. Honors & Awards: Dart Award, Univ Southern Calif, 70. Mem: AMA; Am Col Physicians; Am Soc Clin Pharmacol & Chemother; Am Asn Cols Pharm; Am Pharmaceut Asn. Mailing Add: Dept of Med & Pharmacol Univ of Southern Calif Sch of Med Los Angeles CA 90007

MCCART, BRUCE RONALD, b Omaha, Nebr, Dec 21, 38; m 61; c 2. PHYSICS. Educ: Carleton Col, BA, 60; Iowa State Univ, PhD(physics), 65. Prof Exp: Asst prof, 65-71, ASSOC PROF PHYSICS, AUGUSTANA COL, ILL, 71-, ASSOC DEAN & DIR INSTNL RES, 70- Mem: Am Phys Soc; Am Asn Physics Teachers. Res: Nuclear magnetic resonance in solids; optimal structural design. Mailing Add: Dept of Physics Augustana Col Rock Island IL 61201

MCCARTER, JOHN ALEXANDER, b Eng, Jan 25, 18; nat Can; m 41; c 4. VIROLOGY. Educ: Univ BC, BA, 39, MA, 41; Univ Toronto, PhD(biochem), 45. Prof Exp: Asst res officer, Nat Res Coun Can, 45-48; from assoc prof to prof biochem, Dalhousie Univ, 48-65; PROF BIOCHEM & DIR CANCER RES, UNIV WESTERN ONT, 65- Concurrent Pos: Brit Empire Cancer Campaign fel, 59-60. Mem: Can Biochem Soc; Royal Soc Can; Brit Biochem Soc. Res: Carcinogenesis. Mailing Add: Cancer Res Lab Univ of Western Ont London ON Can

MCCARTER, STATES MARION, b Clover, SC, Sept 30, 37; m 63; c 2. PLANT PATHOLOGY, PLANT BREEDING. Educ: Clemson Univ, BS, 59, MS, 61, PhD(plant path), 65. Prof Exp: Exten plant pathologist & nematologist, Auburn Univ, 65-66; res plant pathologist, Ga Coastal Plain Exp Sta, Agr Res Serv, USDA, 66-68; asst prof, 68-72, ASSOC PROF PLANT PATH, UNIV GA, 72- Mem: Am Phytopath Soc. Res: Studies on soil-borne plant pathogens; influence of soil environmental and ecological factors on disease development; bacterial diseases of plants. Mailing Add: Dept Plant Path & Plant Genetics Univ of Ga Col of Agr Athens GA 30602

MCCARTHY, ALBERT JOSEPH PATRICK, b Oswego, NY, July 28, 17. HISTORICAL GEOGRAPHY, GEOGRAPHY OF EUROPE. Educ: Niagara Univ, AB, 40; Syracuse Univ, MA, 48; St Louis Univ, PhD(Am studies), 65. Prof Exp: Asst prof air sci, Univ Mass, Amherst, 51-52; asst prof geog, Colgate Univ, 53-54; lectr, Ind Univ, 56-61; ASST PROF GEOG, ST LOUIS UNIV, 62- Mem: Asn Am Geogr; Am Geog Soc; Nat Coun Geog Educ; Am Studies Asn; Am Name Soc. Res: Historical geography of eastern United States and Western Europe. Mailing Add: Dept of Geog St Louis Univ St Louis MO 63103

MCCARTHY, BRIAN JOHN, b London, Eng, Mar 27, 34; m 58; c 4. MICROBIOLOGY, MOLECULAR BIOLOGY. Educ: Oxford Univ, BA, 55, MA & DPhil(chem), 58. Prof Exp: Fel microbiol, Carnegie Inst Dept Terrestrial Magnetism, 58-60, staff mem, 60-64; assoc prof microbiol & genetics, Univ Wash, 64-69, prof, 69-72; PROF BIOCHEM, UNIV CALIF, SAN FRANCISCO, 72- Honors & Awards: Eli Lilly Award Microbiol & Immunol, 68. Mem: Biophys Soc; Am Soc Microbiol. Res: Chemistry of biology of nucleic acids; evolution and development at the molecular level. Mailing Add: Dept of Biochem Univ of Calif San Francisco CA 94122

MCCARTHY, CHARLES ALAN, b Rochester, NY, Aug 7, 36; m 57; c 2. MATHEMATICAL ANALYSIS. Educ: Univ Rochester, BA, 56; Yale Univ, PhD(math), 59. Prof Exp: CLE Moore instr math, Mass Inst Technol, 59-61; from asst prof to assoc prof, 61-67, PROF MATH, UNIV MINN, MINNEAPOLIS, 67- Concurrent Pos: Alfred P Sloan Res Found fel, 62-64. Mem: Am Math Soc; Math Asn Am. Res: Functional analysis, especially the theory of operators. Mailing Add: Sch of Math Univ of Minn Minneapolis MN 55455

MCCARTHY, DENNIS DEAN, b Oil City, Pa, Sept 22, 42; m 66; c 2. ASTRONOMY. Educ: Case Inst Technol, BS, 64; Univ Va, MA, 70, PhD(astron), 72. Prof Exp: ASTRONR, US NAVAL OBSERV, WASHINGTON, DC, 65- Mem: Am Astron Soc; Am Geophys Union. Res: Astronomical research into rotational speed of the Earth; variation of astronomical latitude; precise star positions and proper motions; fundamental astronomical constants. Mailing Add: US Naval Observ 34th & Massachusetts Aves NW Washington DC 20390

MCCARTHY, DONALD JOHN, b New York, NY, Dec 22, 38; m 63; c 4. ALGEBRA. Educ: Manhattan Col, BS, 61; NY Univ, MS, 63, PhD(math), 65. Prof Exp: Asst prof math, Fordham Univ, 65-70; asst prof, 70-72, ASSOC PROF MATH, ST JOHN'S UNIV, NY, 72- Mem: Am Math Soc; Math Asn Am. Res: Group theory;

general algebra; combinatorial mathematics. Mailing Add: Dept of Math & Comput Sci St John's Univ Jamaica NY 11439

MCCARTHY, DOUGLAS ROBERT, b New York, NY, July 10, 43; m 64; c 1. MATHEMATICS. Educ: Rensselaer Polytech Inst, BS, 64; Rutgers Univ, MS, 66, PhD(math), 70. Prof Exp: Asst prof math, 69-75, ASSOC PROF MATH, PURDUE UNIV, FT WAYNE, 75- Mem: Am Math Soc; Math Asn Am; Soc Indust & Appl Math. Res: Nonlinear existence theory in differential equations; topological degree; optimization; control theory; differential games. Mailing Add: Dept of Math Purdue Univ 2101 E US 30 Ft Wayne IN 46805

MCCARTHY, DUNCAN ARTHUR, JR, b Kalispell, Mont, Sept 6, 25; m 52; c 2. PHARMACOLOGY. Educ: Univ Mont, BA, 49; Northwestern Univ, MS, 52; Univ Mich, PhD(pharmacol), 60. Prof Exp: Chemist, Mont Veg Oil & Feed Co, 49-51; parasitologist, 52-55, pharmacologist, 59-68, sect dir neuropharmacol, 68-71, DIR PHARMACOL RES, PARKE, DAVIS & CO, 71- Concurrent Pos: Lectr, Univ Mich, 64-74; vis prof, Mich State Univ, 66- Mem: Am Soc Pharmacol & Exp Therapeut; Soc Exp Biol & Med; Soc Neurosci. Res: Chemotherapy of parasitic diseases; drug abuse and addiction; general anesthetics; neuropharmacology; behavioral pharmacology; pharmacodynamics of centrally acting drugs. Mailing Add: Parke Davis & Co Res Div 2800 Plymouth Rd Ann Arbor MI 48106

MCCARTHY, EUGENE GREGORY, b Boston, Mass, Nov 8, 34; m 58; c 3. PUBLIC HEALTH. Educ: Boston Col, AB, 56; Yale Univ, MD, 60; Johns Hopkins Univ, MPH, 62. Prof Exp: Asst prof pub health, Med Sch, Columbia Univ, 65-69; ASSOC PROF PUB HEALTH, MED SCH, CORNELL UNIV, 70- Concurrent Pos: Consult, Int Dept, US Dept State, 64-67, Sisters of Charity, St Vincent's Hosp, 65-67, health coun to Off of Senator Robert Kennedy, 65-68, US Social Security Agency, 66-68, Cath Hosp Asn, 66-69 & Med Sch, Tufts Univ, 69-70; consult health affairs, Roman Cath Dioceses of Brooklyn, 66-69; consult & co-dir, Mercy Cath Med Ctr, 66-75; mem health comt, Greater New York Coun & Med Adv Comt, Piel Comn. Mem: Am Fedn Clin Investrs; NY Acad Med; Am Pub Health Asn. Res: Delivery of health services; surgical health care; established second opinion elective surgical consultations as a medical care instrument for improvement in quality of care and reduction of in-hospital elective surgery. Mailing Add: Dept of Pub Health Cornell Univ Med Sch New York NY 10021

MCCARTHY, FRANCIS DAVEY, b Sioux City, Iowa, Apr 30, 43; m 68; c 1. BIOLOGICAL OCEANOGRAPHY. Educ: Marquette Univ, BS, 68; Tex A&M Univ, PhD(biol oceanog), 73. Prof Exp: ASST PROF BIOL SCI, CALIF STATE COL, DOMINGUEZ HILLS, 73- Mailing Add: Dept Biol Sci Calif State Col 1000 E Victoria St Dominguez Hills CA 90747

MCCARTHY, FRANCIS WADSWORTH, b Boston, Mass, May 10, 13; m 40; c 3. PHYSICS. Educ: Boston Col, PhB, 36, MS, 38; Harvard Univ, EdD, 51. Prof Exp: Jr physicist, Tech Lab, Fed Bur Invest, 40-41; instr physics, Univ Conn, 41-42; sci teacher, High Sch, Mass, 42-43 & 46-47, jr master, 47-49; asst prof, 49-52, PROF PHYSICS & CHMN DEPT, BOSTON STATE COL, 52- Concurrent Pos: Lectr, Northwestern Univ, 46-49; consult, Mass Dept Educ, 54-62. Mem: Am Asn Physics Teachers. Res: Analytical mechanics; electronics. Mailing Add: 229 Jefferson St Dedham MA 02026

MCCARTHY, FRANK JOHN, b Mt Carmel, Pa, Mar 7, 28; m 51; c 3. MICROBIOLOGY. Educ: Mt St Mary's Col, Md, BS, 49; Fordham Univ, MS, 51; Lehigh Univ, PhD(biol), 60. Prof Exp: Asst biol, Fordham Univ, 49-51; res assoc, Merck Inst Therapeut Res, 53-63; mgr virus prods, 63-75, ASSOC DIR BIOL PRODS, WYETH LABS, INC, 75- Mem: Am Soc Microbiol; Tissue Cult Asn; NY Acad Sci. Res: Viral and bacterial vaccines; tissue culture; cytogenetics; anti-viral and anti-tumor agents; purification of virus and bacterial products. Mailing Add: Wyeth Labs Inc Marietta PA 17547

MCCARTHY, FRANK MARTIN, b Olean, NY, Aug 27, 24; m 49; c 2. DENTISTRY, ANESTHESIOLOGY. Educ: Univ Pittsburgh, BS, 44, DDS, 45, MD, 49; Georgetown Univ, MS, 54. Hon Degrees: ScD, St Bonaventure Univ, 56. Prof Exp: Instr oral surg, 57-66, clin prof, 66-75, PROF PAIN/ANXIETY CONTROL, SCH DENT, UNIV SOUTHERN CALIF, 75-, ASSOC DEAN EXTRAMURAL AFFAIRS, 76- Concurrent Pos: Mem adv panel dent, US Pharmacopeia, 70- & Am Fire Protection Asn, 72-; mem ad hoc comt res in pain, Nat Inst Dent Res, 71; active mem sci rev panel, Am Pharmaceut Asn, 71-; ed anesthesia sect, J Oral Surg, Am Dent Asn, 72-; consult anesthesiol, Coun Dent Therapeut, 73- & Coun Dent Mat & Devices, 74-; consult, Calif Select Comt Med Malpract, 73-75; mem comt MD-156, Am Nat Standards Inst, 74- Honors & Awards: Heidbrink Award, Am Dent Soc Anesthesiol, 77. Mem: Am Soc Oral Surgeons; Am Dent Soc Anesthesiol; fel Int Asn Oral Surgeons; Am Soc Advan Anesthesia Dent; Int Anesthesia Res Soc. Res: Pain/anxiety control, physical evaluation and emergency medicine in dentistry. Mailing Add: Univ of Southern Calif Sch of Dent 925 W34th St Los Angeles CA 90007

MCCARTHY, JAMES FRANCIS, b Sharon, Pa, Apr 29, 14; m 47; c 3. ORGANIC CHEMISTRY. Educ: Grove City Col, BS, 35. Prof Exp: Chemist, Sharon Steel Corp, 35-37; res fel biochem, Mellon Inst, 38-42; dir res spec textile finishes, Treesdale Labs, Inc & Pittsburgh Metals Purifying Co, 42-70, DIR RES & DEVELOP, TREESDALE, INC, PITTSBURGH METALS PURIFYING CO, 70- Mem: Am Chem Soc; fel Am Inst Chem; Am Soc Testing & Mat. Res: Durable flame proofing; special finishes for textiles; fluxes and exothermic compounds for metals industries; exothermic reactions; high temperature insulation. Mailing Add: Treesdale Inc Box 337 Saxonburg PA 16056

MCCARTHY, JAMES JOSEPH, b Ashland, Ore, Jan 25, 44; m 69; c 2. BIOLOGICAL OCEANOGRAPHY. Educ: Gonzaga Univ, BS, 66; Scripps Inst Oceanog, Univ Calif, San Diego, PhD(biol oceanog), 71. Prof Exp: Res assoc biol oceanog, Chesapeake Bay Inst, Johns Hopkins Univ, 71-72, assoc res scientist, 72-74; ASST PROF BIOL OCEANOG, HARVARD UNIV, 74- Mem: AAAS; Physiol Soc Am; Sigma Xi; Am Soc Limnol & Oceanog. Res: Investigation of the recycling of aquatic plant nutrients employing isotopic techniques to quantitate the rate of flux of elemental material via various processes within planktonic ecosystems. Mailing Add: Dept of Biol Harvard Univ Cambridge MA 03138

MCCARTHY, JAMES RAY, JR, b Bronx, NY, Dec 16, 43; m 67; c 2. ORGANIC CHEMISTRY. Educ: Ariz State Univ, BS, 65; Univ Utah, PhD(org chem), 69. Prof Exp: RES CHEMIST, CHEM BIOL LAB, DOW CHEM USA, 68- Mem: Am Chem Soc. Res: Organic synthesis in the area of natural products and biologically active compounds. Mailing Add: Chem Biol Lab Dow Chem USA Midland MI 48640

MCCARTHY, JOHN, b Boston, Mass, Sept 4, 27; m 54; c 2. COMPUTER SCIENCE. Educ: Calif Inst Technol, BS, 48; Princeton Univ, PhD(math), 51. Prof Exp: Res instr math, Princeton Univ, 51-53; actg asst prof, Stanford Univ, 53-55; asst prof, Dartmouth Col, 55-58; from asst prof to assoc prof commun sci, Mass Inst Technol, 58-61; PROF COMPUT SCI, STANFORD UNIV, 62-, DIR ARTIFICIAL INTEL LAB, 66- Honors & Awards: A M Turing Award, Asn Comput Mach, 71. Mem: Am Math Soc; Asn Comput Mach; Asn Symbolic Logic. Res: Programming languages; time-sharing; mathematical theory of computation; artificial intelligence. Mailing Add: Dept of Comput Sci Stanford Univ Stanford CA 94305

MCCARTHY, JOHN, b New Orleans, La, July 3, 42. METEOROLOGY. Educ: Grinnell Col, BA, 64; Univ Okla, MS, 67; Univ Chicago, PhD(geophys sci), 73. Prof Exp: Res meteorologist, Weather Sci, Inc, 66-67 & Cloud Physics Lab, Univ Chicago, 67-68; ASST PROF METEOROL, UNIV OKLA, 73- Concurrent Pos: Mem, Univ Rels Comt, Univ Corp Atmospheric Res, 74- Mem: Am Meteorol Soc; Sigma Xi. Res: Radar and aircraft studies of severe thunderstorms and tornadoes; source of thunderstorm rotation; applied aircraft hazards near thunderstorms; weather modification. Mailing Add: Dept of Meteorol Univ of Okla Norman OK 73069

MCCARTHY, JOHN A, physics, see 12th edition

MCCARTHY, JOHN F, pharmaceutical chemistry, organic chemistry, see 12th edition

MCCARTHY, JOHN LAWRENCE, JR, b New Haven, Conn, Aug 13, 29; m 55; c 2. PHYSIOLOGICAL CHEMISTRY. Educ: Univ Miami, Fla, BS, 51; Purdue Univ, MA, 53, PhD(endocrinol), 58. Prof Exp: From asst prof to assoc prof biol, 58-69, PROF BIOL, SOUTHERN METHODIST UNIV, 69- Concurrent Pos: NIH fel, 64-65. Mem: AAAS; Am Soc Zool; Am Physiol Soc; Endocrine Soc. Res: Physiological and chemical relationship between the thyroid and adrenal glands; adrenal steroid biosynthesis; steroid metabolism. Mailing Add: Dept of Biol Southern Methodist Univ Dallas TX 75275

MCCARTHY, JOHN RANDOLPH, b Huntington, WVa, Mar 1, 15; m 49. ORGANIC CHEMISTRY. Educ: Columbia Univ, PhD(chem), 49. Prof Exp: Anal chemist, Stand Ultramarine Co, 39-41; res chemist, Del, 49-74, SUPVR, MED LAB, CHAMBERS WORKS, E I DU PONT DE NEMOURS & CO, DEEPWATER, NJ, 74- Mem: Am Chem Soc. Res: Process control; elastomers; petroleum; dyes and intermediates; urine and blood analyses for biological monitoring of workplace exposure. Mailing Add: 230 Meredith St Kennett Square PA 19348

MCCARTHY, KATHRYN AGNES, b Lawrence, Mass, Aug 7, 24. SOLID STATE PHYSICS. Educ: Tufts Univ, AB, 44, MS, 46; Radcliffe Col, PhD(appl physics), 57. Prof Exp: Asst physics, 45-46, instr, 46-53, from asst prof to assoc prof, 57-62, actg chmn dept, 61-62, dean grad sch arts & sci, 69-74, PROF PHYSICS, TUFTS UNIV, 62-, PROVOST/SR VPRES, 73- Concurrent Pos: Res physicist, Baird Assocs, Inc, 47-49 & 51; res fel phys metall, Harvard Univ, 57-59; assoc res engr & res consult, Univ Mich, 57-59; dir, Doble Eng Co, 67-70; dir, Mass Elec Co, 73-; mem exec bd, New Eng Conf Grad Educ, 70-74; mem comn inst higher educ, New Eng Asn Cols & Sec Schs, Inc, 72-, vchmn, 73-74, chmn, 74-; mem comt disadvantaged students, Coun Grad Schs, 72-74, mem exec comt, 73-75; mem comn inst affairs, Asn Am Cols, 73-; mem grad adv acad comt, Bd High Educ Mass, 72-74; mem comn admin affairs & educ statist, Am Coun Educ, 73-; trustee, Southern Mass Univ, 72-74 & Merrimack Col, 74-; mem, Corp of Lawrence Mem Hosp, Medford, 75- Mem: Fel Am Phys Soc; fel Optical Soc Am. Res: Low-temperature thermal conductivity; color centers in alkali halides; ultrasonic attenuation in solids. Mailing Add: Ballou Hall Tufts Univ Medford MA 02155

MCCARTHY, LAURENCE RAY, microbial physiology, see 12th edition

MCCARTHY, MARTIN, b Lowell, Mass, July 10, 23. ASTRONOMY. Educ: Boston Col, AB, 46, MA, 47; Georgetown Univ, PhD(astron), 51; Weston Col, STL, 55. Prof Exp: STAFF MEM ASTRON, VATICAN OBSERV, 56- Concurrent Pos: Vis prof, Georgetown Univ, 62-63; guest investr, Carnegie Inst, 65-66, 71, 73 & 74; observ consult, Cerro Tololo Interam Observ, Asn Univs Res in Astron, 72 & consult adv, 74. Mem: AAAS; Am Astron Soc; Royal Astron Soc; Sigma Xi; Int Astron Union. Res: Spectral classification of stars; galactic structure; photoelectric and photographic photometry. Mailing Add: Vatican Observ Castel Gandolfo 00040 Rome Italy

MCCARTHY, MARY ANNE, b Tarentum, Pa, May 26, 39; m 69. NUMERICAL ANALYSIS, SYSTEM ANALYSIS. Educ: Wellesley Col, BA, 61; Rice Univ, MA, 66, PhD(math sci), 73. Prof Exp: Comput programmer, Bell Tel Labs Inc, 61-63; physicist, Shell Develop Co, Shell Oil Co, 67-69; assoc syst analyst, Lulejian & Assoc Inc, 73-75; RESEARCHER, R & D ASSOC INC, 75- Mem: Am Math Soc; Am Phys Soc; Soc Indust & Appl Math. Res: Development of a mathematical method which yields closed form solutions to various types of probabilistic problems which previously have required Monte Carlo simulation. Mailing Add: 3031 Stoner Ave Los Angeles CA 90066

MCCARTHY, MILES DUFFIELD, b Camden, NJ, Oct 12, 14; m 59. REPRODUCTIVE BIOLOGY. Educ: Pa State Teachers Col, BS, 36; Univ Pa, PhD(zool), 43. Prof Exp: Asst, Inst Med Res, Pa, 37-39; instr zool, Univ Pa, 39-42, tech assoc, Med Sch, 42, instr vert anat, Univ, 44-45, genetics, Vet Sch, 45-46; asst prof zool, Pomona Col, 46-49, from assoc prof to prof, 49-59; vpres acad affairs, 70-74, chmn div sci & math, 59-64, dean sch letters, arts & sci, 64-70, PROF BIOL, CALIF STATE UNIV, FULLERTON, 59- Concurrent Pos: Harrison res fel, Med Sch, Univ Pa, 44-46, vis asst prof, 51-52; responsible investr, US Dept Army contract, Pomona Col, 54-56; NIH fel, 57-61. Mem: AAAS; Soc Exp Biol & Med; Am Physiol Soc. Res: Growth and development; burn shock and infusion fluids; hemolytic and hematopoietic changes in relation to survival following severe thermal injury; reproduction and human sexuality. Mailing Add: Dept of Biol Calif State Univ Fullerton CA 92634

MCCARTHY, NEIL JUSTIN, JR, b Chicago, Ill, Oct 25, 39; m 63; c 2. ORGANIC POLYMER CHEMISTRY. Educ: Georgetown Univ, BS, 62; Cornell Univ, PhD(org chem), 69. Prof Exp: Res chemist, 66-71, PROJ SCIENTIST ORG POLYMER CHEM, UNION CARBIDE CORP, 71- Mem: Am Chem Soc. Res: Physical-organic and physical polymer chemistry and composite science. Mailing Add: Union Carbide Corp Res & Develop Lab PO Box 670 Bound Brook NJ 08805

MCCARTHY, PAUL JAMES, b Rochester, NY, June 15, 24. PHYSICAL INORGANIC CHEMISTRY. Educ: Spring Hill Col, AB, 50; Col of Holy Cross, MS, 52; Clark Univ, PhD(chem), 55. Prof Exp: Instr high sch, NY, 50-51; instr chem, St Peters Col, 59-60; instr, 60-62, from asst prof to assoc prof, 62-72, PROF CHEM, CANISIUS COL, 72- Concurrent Pos: NATO fel, Copenhagen Univ, 66-67. Mem: AAAS; Am Chem Soc; The Chem Soc. Res: Coordination complexes, especially their preparation, structure and absorption spectra. Mailing Add: Dept of Chem Canisius Col 2001 Main St Buffalo NY 14208

MCCARTHY, PAUL JOSEPH, b Chicago, Ill, Oct 23, 28; m 52; c 2. MATHEMATICS. Educ: Univ Notre Dame, BS, 50, MS, 52, PhD(math), 55. Prof Exp: Instr math, Univ Notre Dame, 54-55; instr, Col Holy Cross, 55-56; from asst prof to assoc prof, Fla State Univ, 56-61; assoc prof, 61-65, PROF MATH, UNIV

KANS, 65- Mem: Am Math Soc; Math Asn Am. Res: Rings; graph theory; number theory. Mailing Add: Dept of Math Univ of Kans Lawrence KS 66045

MCCARTHY, PHILIP JOHN, b Friendship, NY, Feb 9, 18; m 42; c 4. STATISTICS. Educ: Cornell Univ, AB, 39; Princeton Univ, MA, 41, PhD(math), 47. Prof Exp: Res assoc, Columbia Univ, 41-42; opers res, US Dept Navy, 42-43; Princeton Univ, 43-46, Soc Sci Res Coun fel, 46-48; PROF APPL STATIST, STATE UNIV NY SCH INDUST & LABOR RELS, CORNELL UNIV, 47- Concurrent Pos: Consult, Nat Ctr Health Statist, 64- & Comt Nat Statist, 74- Mem: Fel Am Statist Asn; Inst Math Statist; Int Asn Surv Statisticians. Res: Sampling methods; application of statistics in the social sciences. Mailing Add: NY State Sch Indust & Labor Rels Cornell Univ Ithaca NY 14850

MCCARTHY, RAYMOND LAWRENCE, b Jersey City, NJ, Sept 6, 20; m 43; c 4. PHYSICS. Educ: Fordham Univ, AB, 41, MA, 43; Yale Univ, MS, 47, PhD(physics), 48. Prof Exp: Physicist, Naval Ord Lab, 42-44; asst, Yale Univ, 46-48; res physicist, 48-55, res mgr, Radiation Physics Lab, 55-59, asst dir, 59-61, process supt, Org Chem Dept, 61-64, DIR FREON PROD LAB, E I DU PONT DE NEMOURS & CO, INC, 64- Mem: Sigma Xi. Res: Chemical and radiation physics; instrumentation; fluorocarbon compounds. Mailing Add: Freon Prod Lab E I du Pont de Nemours & Co Inc Wilmington DE 19898

MCCARTHY, ROBERT DAVID, b Manoa, Pa, Sept 25, 32; m 56; c 6. DAIRY SCIENCE. Educ: Pa State Univ, BS, 54, MS, 56; Univ Md, PhD(dairy sci), 58. Prof Exp: Asst dairy sci, Pa State Univ, 54-56; asst, Univ Md, 56-58, res assoc, 58; res assoc, 58-61, from asst prof to assoc prof, 61-70, PROF DAIRY SCI, PA STATE UNIV, UNIVERSITY PARK, 70- Mem: Am Dairy Sci Asn; Am Chem Soc. Res: Ruminant metabolism and physiology; lipid metabolism. Mailing Add: Dept Dairy Sci 116 Borland Lab Pa State Univ Col of Agr University Park PA 16802

MCCARTHY, ROBERT ELMER, b Washington, DC, May 19, 26; m 49; c 1. IMMUNOLOGY, MICROBIOLOGY. Educ: Univ Md, BS, 51, MS, 53; Brown Univ, PhD(biol), 56. Prof Exp: Asst poultry, Univ Md, 51-53; asst biol, Brown Univ, 53-55; asst path, Harvard Med Sch, 57-62, res assoc, 62-69, assoc, 69-70; ASSOC PROF MED MICROBIOL & IMMUNOL, UNIV NEBR MED CTR, OMAHA, 70- Concurrent Pos: USPHS fel, Inst Cell Res & Genetics, Karolinska Inst, Sweden, 62-63; asst path, Children's Hosp, Boston, 56-59, res assoc, 59-70; asst path, Children's Cancer Res Found, 56-59, res assoc, 59-70, chief lab of immunol, 68-69; assoc oral path, Sch Dent Med, Harvard Univ, 69-70. Mem: Am Soc Exp Path; NY Acad Sci; Am Asn Immunol; Am Soc Cancer Res; Environ Mutagen Soc. Res: Radiation biology, antigen structure, cytochemistry and cytogenetics of mammalian cells in culture; study of inhibition of allograft rejection by ascites tumors; immunochemistry of synthetic polypeptides; effect of pesticides on immune response. Mailing Add: Dept of Med Microbiol Univ of Nebr Med Ctr Omaha NE 68105

MCCARTHY, TIMOTHY EDWARD, b Boston, Mass, July 12, 13; m 44; c 1. BIOCHEMISTRY. Educ: Boston Col, BS, 35, MS, 37; Georgetown Univ, PhD(biochem), 40. Prof Exp: Lab instr chem, Boston Col, 35-37 & Georgetown Univ, 37-40; instr, St Francis Col, Pa, 40-42; biochemist, Armour Res Lab, Ill, 46; asst prof chem, Univ Mass, 46-47; asst prof, 47-58, ASSOC PROF CHEM, BOSTON COL, 58- Mem: Am Chem Soc. Res: Biochemistry and gas chromatography of amino acids. Mailing Add: Dept of Chem Boston Col Chestnut Hill MA 02167

MCCARTHY, VINCENT CORMAC, b Yonkers, NY, May 18, 32. PARASITOLOGY, MICROBIOLOGY. Educ: Univ Toronto, BA, 53; Univ Md, MS, 61, PhD(zool), 67. Prof Exp: Asst zool, Univ Md, 57-61; asst parasitol, Sch Med, NY Univ, 61-62; res assoc, 63-67, ASST PROF INT MED, SCH MED, UNIV MD, BALTIMORE CITY, 67- Mem: Entom Soc Am. Res: Arthropods of medical importance; pathogenic Protozoa and other microbes; tropical medicine. Mailing Add: Dept of Int Med Univ of Md Sch of Med Baltimore MD 21228

MCCARTHY, WALTER CHARLES, b Brooklyn, NY, Oct 25, 22; m 45; c 3. MEDICINAL CHEMISTRY. Educ: Mass Inst Technol, BS, 43; Univ Ind, PhD(org chem), 49. Prof Exp: From asst prof to assoc prof, 49-65, PROF PHARMACEUT CHEM, UNIV WASH, 65- Mem: Am Chem Soc; Am Pharmaceut Asn; The Chem Soc; Swiss Chem Soc. Res: Sympathomimetic amines; organic sulfur compounds; thiophene and furan derivatives. Mailing Add: Univ of Wash Sch of Pharm Seattle WA 98195

MCCARTHY, WILLIAM JOHN, b Mamaroneck, NY, June 27, 41. VIROLOGY, BIOCHEMISTRY. Educ: Univ Del, BA, 63; NY Univ, MS, 68, PhD(biol), 70. Prof Exp: Fel virol, Boyce Thompson Inst Plant Res, 70-73; ASSOC RES SCIENTIST, PESTICIDE RES LAB, PA STATE UNIV, 74- Res: Cellular and viral protein and nucleic acid synthesis in lepidoptera; isolation and characterization of viruses; proteins and nucleic acids; characterization of nucleic acid and proteins of baculoviruses, entomopoxviruses, and cytoplasmic polyhedrosis viruses. Mailing Add: Pesticide Res Lab Pa State Univ University Park PA 16802

MCCARTNEY, ALLEN PAPIN, b Fayetteville, Ark, Aug 8, 40. ARCHAEOLOGY, ANTHROPOLOGY. Educ: Univ Ark, BA, 62; Univ Wis-Madison, MA, 67, PhD(anthrop), 71. Prof Exp: Instr/lectr anthrop, Univ Wis Ctr Systs, 66-70; asst prof, 70-74, ASSOC PROF ANTHROP, UNIV ARK, FAYETTEVILLE, 74- Concurrent Pos: Ad hoc archaeol consult, US Fish & Wildlife Serv, 72-; proj dir, Thule Archaeol Conserv Proj, Archaeol Surv Can, Nat Mus of Man, 75- Mem: Fel AAAS; Soc Am Archaeol; Am Anthrop Asn; Can Archaeol Asn; Arctic Inst NAm. Res: Archaeological field research in the Aleutian Islands; museum study of Aleutian prehistoric collections; archaeological field research in arctic Canada; Alaska Peninsula prehistory; paleoclimatology. Mailing Add: Dept of Anthrop Univ of Ark Fayetteville AR 72701

MCCARTNEY, JOHN RICHARD, b Canton, NY, Jan 14, 20; m 43; c 2. PHYSICAL CHEMISTRY. Educ: Cornell Univ, AB, 41, PhD(phys chem), 46. Prof Exp: Asst, Ithaca Lab, USDA, 41-43 & Rubber Reserve Corp, 44-46; res chemist, 46-50, res assoc, 50-53, res supvr, 53-57, res mgr pioneering res div, 57-65, dacron res div, 65-69, RES MGR NYLON RES DIV, E I DU PONT DE NEMOURS & CO, INC, 69- Concurrent Pos: Asst, Off Sci Res & Develop, Cornell Univ, 42-43. Mem: Am Chem Soc. Res: Fibers; physical chemistry of high polymers. Mailing Add: Carothers Res Lab E I du Pont de Nemours & Co Inc Wilmington DE 19898

MCCARTNEY, MICHAEL SCOTT, b Buffalo, NY, Oct 21, 47; m; c 2. PHYSICAL OCEANOGRAPHY, FLUID DYNAMICS. Educ: Case Inst Technol, BS, 70; Case Western Reserve Univ, MS, 72, PhD(fluid dynamics), 73. Prof Exp: Investr, 74-75, ASST SCIENTIST PHYS OCEANOG, WOODS HOLE OCEANOG INST, 75- Concurrent Pos: Mem exec comt, Int Southern Ocean Studies, 75- Mem: Am Geophys Union. Res: Effects of bottom topography on ocean currents, laboratory ocean modeling, winter formation of water masses and their spreading mechanisms; southern ocean dynamics; general circulation of the world oceans. Mailing Add: Dept of Phys Oceanog Woods Hole Oceanog Inst Woods Hole MA 02543

MCCARTNEY, MORLEY GORDON, b Picton, Ont, Mar 4, 17; nat US; m 46; c 5. GENETICS. Educ: Ont Agr Col, BSA, 40; Univ Md, MS, 47, PhD, 49. Prof Exp: Poultry specialist, Ont Agr Col, 40-41; asst poultry husb, Univ Md, 46-49; poultry physiologist, USDA, Md, 49; from asst prof to assoc prof poultry, Pa State Univ, 49-55; prof & assoc chmn dept, Agr Exp Sta, Ohio State Univ, 55-64; PROF POULTRY SCI, HEAD DEPT & CHMN DIV POULTRY, UNIV GA, 64- Honors & Awards: Award, Nat Turkey Fedn, 59. Mem: AAAS; Am Genetic Asn; Poultry Sci Asn. Res: Population genetics and physiology of reproduction of poultry. Mailing Add: Dept of Poultry Sci Univ of Ga Athens GA 30601

MCCARTNEY, WILLIAM DOUGLAS, b Edmonton, Alta, Feb 9, 22; m 48; c 2. ECONOMIC GEOLOGY. Educ: Univ BC, BASc, 50; Harvard Univ, AM, 52, PhD(geol), 59. Prof Exp: Geologist, Dept Mines & Tech Surv, Geol Surv Can, 52-67; assoc prof geol sci, Queen's Univ, Ont, 66-71; chmn, We Healdath Consults Ltd, 72-75; GEOLOGIST, DEPT MINES & PETROL RESOURCES, BC, 75- Mem: Soc Econ Geol; fel Geol Asn Can; Can Inst Mining & Metall. Res: Structural and stratigraphic studies; metallogeny related to geosynclinal development and application to mineral exploration; regional mineral capability appraisal and maps related to land-use; socio-economic planning; mineral exploration. Mailing Add: Dept of Mines & Petrol Resources Victoria BC Can

MCCARTY, BILLY DEAN, b Wymore, Nebr, Oct 31, 23; m 46; c 2. ANALYTICAL CHEMISTRY. Educ: Nebr Wesleyan Univ, AB, 48. Prof Exp: Anal chemist oil shale demonstration plant, US Bur Mines, Colo, 50-53; spectrographer, Los Alamos Sci Lab, 53-58 & Union Carbide Nuclear Co, Colo, 58-62; spectrographer, 62-66, anal unit supvr, 66-70, ADVAN SCIENTIST, MARATHON OIL CO RES CTR, 70- Honors & Awards: Outstanding Serv Award, Rocky Mt Sect, Soc Appl Spectros, 72. Mem: Soc Appl Spectros. Res: Optical emission spectroscopy; x-ray diffraction and spectroscopy; isotopic ratio mass spectroscopy. Mailing Add: Marathon Oil Co PO Box 269 Littleton CO 80120

MCCARTY, CHARLES NORMAN, b Albion, Mich, June 3, 09; m 32; c 2. CHEMISTRY. Educ: Albion Col, AB, 31; Univ Ill, MS, 33, PhD(chem), 35. Prof Exp: From asst prof to prof, 35-74, EMER PROF CHEM, MICH STATE UNIV, 74- Mem: Am Chem Soc. Res: Quantitative estimation of the rare earths by means of their arc spectra; general chemistry. Mailing Add: Dept of Chem Mich State Univ East Lansing MI 48823

MCCARTY, CLARK WILLIAM, b Kansas City, Mo, Feb 27, 16; m 40; c 5. PHYSICAL CHEMISTRY. Educ: Univ Kans City, AB, 37; Cent Mo State Col, BSE, 40; Univ Nebr, MS, 39; Univ Mo, AM, 47, PhD(chem), 53. Prof Exp: Asst chem, Univ Nebr, 37-39; chemist & supvr bact, Cerophyl Labs, Inc, 40-42; instr chem & math, Kemper Mil Sch, 42-43; instr, Jefferson City Jr Col, 46; instr chem, Univ Mo, 46-47; instr chem, Southwest Mo State Col, 47-48; instr math, Univ Mo, 48-49; assoc prof chem & math, 50 & 52-58, PROF CHEM & PHYSICS & CHMN DEPT PHYSICS, OUACHITA BAPTIST UNIV, 58- Mem: Am Chem Soc; Am Asn Physics Teachers; Am Physical Soc; Optical Soc Am; Am Meteorol Soc. Res: Partial vapor pressures of ternary systems. Mailing Add: PO Box 517 Arkadelphia AR 71923

MCCARTY, DANIEL J, JR, b Philadelphia, Pa, Oct 31, 28; m 54; c 4. INTERNAL MEDICINE. Educ: Villanova Univ, BS, 50; Univ Pa, MD, 54. Prof Exp: Intern med, Fitzgerald-Mercy Hosp, Darby, Pa, 54-55; resident internal med, Hosp, Univ Pa, 57-58; resident, Philadelphia Vet Admin Hosp, 58-59; from asst prof to assoc prof internal med, Hahnemann Med Col, 60-67, head sect rheumatol, 60-63; prof internal med, Univ Chicago, 67-74, head sect arthritis & metab dis, 71-74; PROF MED & CHMN DEPT, MED COL WIS, 74-; CHIEF MED, MILWAUKEE COUNTY GEN HOSP, 74- Concurrent Pos: Nat Inst Arthritis & Metab Dis fel rheumatol, Hosp, Univ Pa, 59-60; Markle scholar, 62-; attend physician & chief arthritis clin, Div B, Philadelphia Gen Hosp, 60-; consult, Vet Admin Hosp, Philadelphia, 60-; consult arthritis training progs, Surgeon Gen, USPHS, 67-71; mem fel subcomt, Arthritis Found, 67-71, res comt chmn nat reference ctr rheumatol, 69; mem subcomt rheumatol, Am Bd Internal Med, 70-76; chief ed, Arthritis & Rheumatism, Am Rheumatism Asn, 65. Honors & Awards: Hektoen Silver Medal, AMA, 63; Russell S Cecil Award, Arthritis Found & Gairdner Found Award, 65. Mem: AAAS; Am Fedn Clin Res; Am Rheumatism Asn; Am Soc Clin Pharmacol & Therapeut; NY Acad Sci. Res: Rheumatic diseases; crystallography, mechanisms of inflammation; measurement of inflammation, clinical, physiological and biochemical parameters. Mailing Add: Dept of Med Med Col Wis Milwaukee WI 53226

MCCARTY, FREDERICK JOSEPH, b Chillicothe, Ohio, May 2, 27; m 61; c 4. ORGANIC CHEMISTRY, MEDICINAL CHEMISTRY. Educ: Univ Ohio, BS, 51; Mich State Univ, MS, 53; Univ Mich, PhD(med chem), 59. Prof Exp: Asst res chemist, Wm S Merrell Co Div, Richardson-Merrell, Inc, 53-56, proj leader med chem res, 59-64; sect head org res, Aldrich Chem Co, 64-65; group leader med chem res, Nat Drug Co Div, 65-70, SECT HEAD CHEM DEVELOP, MERRELL-NAT LABS DIV, RICHARDSON-MERRELL, INC, 70- Mem: Am Chem Soc. Res: Mannich reaction. Mailing Add: Chem Develop Merrell-Nat Labs 110 E Amity Rd Cincinnati OH 45215

MCCARTY, HAROLD HULL, b Hiteman, Iowa, Sept 10, 01; m 24; c 2. ECONOMIC GEOGRAPHY, GEOGRAPHY OF THE UNITED STATES. Educ: Univ Iowa, BS, 23, MA, 25, PhD(econ), 29. Prof Exp: Instr geog, 23-27, prof, 27-69, EMER PROF GEOG, UNIV IOWA, 69- Mem: Asn Am Geog; Regional Sci Asn; Am Geog Soc. Res: Theoretical and regional economic geography; location of industry. Mailing Add: 2292 D Via Puerta Laguna Hills CA 92653

MCCARTY, JOHN EDWARD, b Iowa City, Iowa, Aug 27, 28; m 55; c 2. ORGANIC CHEMISTRY. Educ: Univ Iowa, BS, 50; Univ Calif, Los Angeles, PhD, 57. Prof Exp: Res fel, Univ Kans, 56-59; from asst prof to assoc prof chem, 59-67, prof, Environ Inst, 72-75, PROF CHEM, MANKATO STATE UNIV, 67- Mem: Am Chem Soc. Res: Mechanism of organic reactions; heterocyclic compounds; environmental chemistry. Mailing Add: Dept of Chem Mankato State Univ Mankato MN 56001

MCCARTY, KENNETH SCOTT, b Dallas, Tex, June 20, 22; m 44; c 2. BIOCHEMISTRY. Educ: Georgetown Univ, BS, 44; Columbia Univ, PhD, 58. Prof Exp: Res biochemist, US Food & Drug Admin, 44-45; asst to dir cancer res, Hyde Found, Interchem Corp, NY, 45-52; mem staff tissue cult course, NY & Colo, 54-58; instr biochem, bol Physicians & Surgeons, Columbia Univ, 57-59; assoc prof, 59-69, PROF BIOCHEM, SCH MED, DUKE UNIV, 69- Concurrent Pos: Electron micros consult, Vet Admin Hosp, NY, 49-54; consult, Organon Corp, NJ, 59- & Parke, Davis & Co, Mich, 59- Mem: Am Chem Soc; Am Soc Biol Chemists; Electron Micros Soc Am; Harvey Soc; Am Asn Cancer Res. Res: Molecular mechanisms in the control of RNA transcription, RNA transport and RNA translation; development of procedures to maintain homeostasis in tissue cultures to study malignant transformation, nucleic acid synthesis, amino acid metabolism and hormone induction in tissue culture. Mailing Add: Dept of Biochem Duke Med Ctr Durham NC 27710

MCCARTY, LESLIE PAUL, b Detroit, Mich, May 30, 25; m 48; c 4.

PHARMACOLOGY. Educ: Salem Col, BS, 47; Ohio State Univ, MSc, 49; Univ Wis, PhD(pharmacol), 60. Prof Exp: Res chemist, Upjohn Co, 49-55; res pharmacologist, 60-64, SR RES PHARMACOLOGIST, DOW CHEM CO, 64- Mem: AAAS; Am Soc Pharmacol & Exp Therapeut; Sigma Xi. Res: Cardiovascular, autonomic and neuromuscular pharmacology. Mailing Add: Chem Biol Res Lab Dow Chem Co Midland MI 48640

MCCARTY, LEWIS VERNON, b Buffalo, NY, July 7, 19; m 46; c 1. CHEMISTRY. Educ: Oberlin Col, AB, 41; Univ Rochester, PhD(phys chem), 45. Prof Exp: Res chemist, 45-64, mgr feasibility invests, 64-66, RES ADV INORG CHEM, GEN ELEC CO, 66- Mem: Am Chem Soc; Electrochem Soc. Res: Reaction kinetics and mechanisms; inorganic syntheses; high temperature chemistry. Mailing Add: Lighting Res Lab Gen Elec Co Nela Park Cleveland OH 44112

MCCARTY, MACLYN, b South Bend, Ind, June 9, 11; m 34; c 4. MEDICAL BACTERIOLOGY. Educ: Stanford Univ, AB, 33; Johns Hopkins Univ, MD, 37. Prof Exp: Asst pediat, Sch Med, Johns Hopkins Univ, 39-40; fel sulfonamide drugs, NY Univ, 40-41; Nat Res Coun fel, 41-42, assoc med, 46-48, assoc mem, 48-50, mem & physician, 50-60, physician-in-chief, Hosp, 60-74, PROF MED, ROCKEFELLER UNIV, 60-, VPRES, 65- Concurrent Pos: Vpres & chmn sci adv comt, Helen Hay Whitney Found; chmn, Health Res Coun New York, 72-75; chmn res coun & mem bd trustees, Pub Health Res Inst New York. Honors & Awards: Eli Lilly Award, 46. Mem: Nat Acad Sci; Nat Inst Med; Am Soc Microbiol; Am Soc Clin Invest; Soc Exp Biol & Med (pres, 74-75); Harvey Soc (secy, 47-50, pres, 71-72). Res: Transformation of pneumococcal types; biology of hemolytic streptococci; rheumatic fever. Mailing Add: Rockefeller Univ York Ave & E 66th St New York NY 10021

MCCARTY, MACLYN, JR, b Kenosha, Wis, Sept 8, 35; m 56; c 4. PHYSICAL CHEMISTRY. Educ: Johns Hopkins Univ, AB, 57, MA, 59, PhD(chem), 60. Prof Exp: Asst prof chem eng, Johns Hopkins Univ, 60-67; res scientist, Res Inst Advan Studies, 67-72, SR RES SCIENTIST, ENERGY TECHNOL CTR, MARTIN MARIETTA CORP, 72- Mem: AAAS; Am Phys Soc; Am Chem Soc; Am Inst Aeronaut & Astronaut. Res: Photochemistry; combustion; solid state chemistry; electronic spectroscopy. Mailing Add: Martin Marietta Corp 1450 S Rolling Rd Baltimore MD 21227

MCCARTY, MELVIN KNIGHT, b Farmington, NMex, Sept 1, 17; m 43; c 4. WEED SCIENCE. Educ: Colo Agr & Mech Col, BSF, 39; Univ Nebr, MS, 51, PhD, 55. Prof Exp: Range exam, US Indian Forest Serv, 39-46; work unit conservationist, Soil Conserv Serv, 46-49, assoc agronomist, Weed Invests Sect, Agr Res Serv, 53-58, AGRONOMIST, WEED INVESTS SECT, AGR RES SERV, USDA, 58- Mem: Fel AAAS; Weed Sci Soc Am; Soc Range Mgt. Res: Pasture weed control; life cycles pasture and rangeland weeds; biological control of weeds. Mailing Add: 316 Keim Hall Univ of Nebr Lincoln NE 68583

MCCARTY, RICHARD EARL, b Baltimore, Md, May 3, 38; m 61; c 3. BIOCHEMISTRY. Educ: Johns Hopkins Univ, BA, 60, PhD(biochem), 64. Prof Exp: NIH fel biochem, Pub Health Res Inst, New York, 64-66; asst prof, 66-71, ASSOC PROF BIOCHEM, CORNELL UNIV, 71- Concurrent Pos: NIH career develop award, 69. Mem: Am Soc Biol Chem; Am Soc Plant Physiol. Res: Photophosphorylation and electron flow in chloroplasts. Mailing Add: Sect of Biochem & Molecular Biol Wing Hall Cornell Univ Ithaca NY 14853

MCCARTY, ROBERT CLARKE, mathematical statistics, see 12th edition

MCCARTY, STUART WAYNE, inorganic chemistry, see 12th edition

MCCARVILLE, MICHAEL EDWARD, b Moorland, Iowa, Aug 27, 36; m 62; c 3. ENZYMOLOGY. Educ: Loras Col, BS, 58; Iowa State Univ, PhD(biophys), 67. Prof Exp: Fel chem, La State Univ, Baton Rouge, 67-68; ASST PROF CHEM, WESTERN MICH UNIV, 68- Concurrent Pos: Sabbatical leave, Upjohn Co, 75-76. Mem: AAAS; Am Chem Soc. Res: Purification and characterization of microbial enzymes involved in waste treatment and drug biotransformation. Mailing Add: Dept of Chem Western Mich Univ Kalamazoo MI 49001

MCCARVILLE, WILLIAM JOHN, organic chemistry, see 12th edition

MCCASHLAND, BENJAMIN WILLIAM, b Geneva, Nebr, May 7, 21; m 43; c 4. CELL PHYSIOLOGY. Educ: Univ Nebr, Lincoln, AB, 47, MS, 48, PhD(zool, physiol), 55. Prof Exp: Instr physiol, Univ Nebr, Lincoln, 47-55, from asst prof to assoc prof, 55-62, prof, 62-70, asst dean grad col, 65-70; PROF BIOL & DEAN GRAD STUDIES, MOORHEAD STATE UNIV, 70- Mem: Fel AAAS; Am Soc Cell Biol; Soc Exp Biol & Med; Sigma Xi. Res: Protozoan growth and respiration. Mailing Add: Moorhead State Univ Moorhead MN 56560

MCCASKEY, THOMAS ANDREW, b Jacksonville, Ohio, Mar 23, 28; m 60; c 1. MICROBIOLOGY. Educ: Ohio Univ, BS, 60; Purdue Univ, MS, 63, PhD(dairy microbiol), 66. Prof Exp: Teaching asst microbiol, Purdue Univ, 60-62, teaching assoc, 62-63, res asst dairy microbiol, 63-66, res associate pesticide anal, 66-67; asst prof, 67-74, ASSOC PROF MICROBIOL, AUBURN UNIV, 74- Mem: Inst Food Technologists; Am Dairy Sci Asn. Res: Pesticide residues in poultry and processed milk; microbial spoilage of milk; prevalence of salmonellae in food; water and soil pollution by dairy farm wastes. Mailing Add: Dept of Dairy Sci Auburn Univ Auburn AL 36830

MCCASLAND, GIFFORD EWING, b Springfield, Mo, Oct 31, 13; m 40; c 2. ORGANIC CHEMISTRY, COMPUTER SCIENCE. Educ: Univ Calif, Los Angeles, BA, 35; Columbia Univ, AM, 38; Calif Inst Technol, PhD(org chem), 44. Prof Exp: Chemist, Consumers Union of US, Inc, 36-37 & 39-40; asst org chem, Calif Inst Technol, 41-43; res assoc, Squibb Inst Med Res, 44; res assoc, Univ Rochester, 44-46; spec asst, Univ Ill, 46-48; asst prof chem, Univ Toronto, 49-56; assoc res prof, 58-67, RES PROF CHEM, UNIV SAN FRANCISCO, 67- Concurrent Pos: Vis fel & asst prof, Ohio State Univ, 55-57; res assoc sch med, Stanford Univ, 57-66, vis scholar, 70. Mem: Am Chem Soc; Asn Comput Mach; fel Chem Inst Can. Res: Stereochemistry and conformational analysis; carbohydrates, especially cyclitols; alicyclics; application of nuclear magnetic resonance to configurational problems; role of molecular configuration in biology; nomenclature of stereoisomers. Mailing Add: Dept of Chem Univ of San Francisco San Francisco CA 94117

MCCASLIN, BOBBY DUANE, b Wichita, Kans, July 26, 43; m 64; c 2. SOIL FERTILITY. Educ: Friends Univ, BS, 66; Colo State Univ, MS, 69; Univ Minn, PhD(soil sci), 74. Prof Exp: ASST PROF AGRON, N MEX STATE UNIV, 74- Mem: Am Soc Agron; Soil Sci Soc Am; Crop Sci Soc Am; Sigma Xi; Int Soc Soil Sci. Res: Fertilizers; fertility; plant nutrition; usage of sewage sludge as a fertilizer material. Mailing Add: Dept of Agron Box 3Q NMex State Univ Las Cruces NM 98003

MCCASLIN, JOHN GARFIELD, b Broken Bow, Nebr, Jan 1, 22; m 50; c 5. PHYSICS. Educ: NMex Inst Mining & Technol, BS, 50, MS, 52. Prof Exp: Assoc res engr, Boeing Airplane Co, Kans, 52-55; from assoc prof to prof physics, Dept Physics, 55-72, PROF PHYSICS & GEOPHYS ENG, DEPT GEOPHYS ENG & DIR, COMPUT CTR, MONT COL MINERAL SCI & TECHNOL, 72-, HEAD, DEPT PHYSICS, 55- Mem: Am Soc Eng Educ; Instrument Soc Am; Am Asn Physics Teachers; Soc Explor Geophys; Nat Soc Prof Eng. Res: Instrumentation; geophysics; underground water research; rock mechanics and seismology. Mailing Add: Dept of Geophys Eng Mont Col of Mineral Sci & Technol Butte MT 59701

MCCASLIN, JOHN WEAVER, b Kansas City, Mo, Dec 22, 17; m 43; c 3. HEALTH PHYSICS, SAFETY ENGINEERING. Educ: Univ Kans, BS, 40; Am BD Health Physics, dipl, 60; Bd Cert Safety Prof, cert, 71. Prof Exp: Assoc res chemist, Phillips Petrol Co, 40-41, 46-47, res & develop safety engr, 47-51, supvr radiation safety, Atomic Energy Div, 51-53, health & safety bd mgr, 53-66; SAFETY STANDARDS BR MGR, AEROJET NUCLEAR CORP, 66- Concurrent Pos: Mem, Am Bd Health Physics, 61-66, secy-treas, 62-66; affil prof, Univ Idaho, 72- Mem: Health Physics Soc (secy, 59-61); Am Indust Hyg Asn; Am Soc Safety Eng. Res: Radiation protection associated with test and power reactors, critical facilities and chemical separations; industrial hygiene and toxicology; safety engineering and fire protection. Mailing Add: 1610 S Higbee Idaho Falls ID 83401

MCCASLIN, MURRAY FREW, b New Castle, Pa, Feb 21, 04; m 36; c 3. OPHTHALMOLOGY. Educ: Univ Pittsburgh, BS, 26, MD, 29. Prof Exp: Prof ophthal & chmn dept, 53-71, EMER PROF OPHTHAL & EMER CHMN, SCH MED, UNIV PITTSBURGH, 71- Concurrent Pos: Consult to various hosps, 48-; chief eye serv, Eye & Ear Hosp, 53; mem sr staff, Children's Hosp, 54- Mem: Am Ophthal Soc; Am Med Asn; Am Col Surg; Am Acad Ophthal & Otol; Chilean Soc Ophthal. Res: Magnetism; localization technique as applied to intraocular foreign bodies. Mailing Add: 550 Grant St Pittsburgh PA 15219

MCCAUGHAN, DANIEL VINCENT, physics, mathematics, see 12th edition

MCCAUGHEY, MARGIE, biochemistry, see 12th edition

MCCAUGHEY, MICHAEL PAUL, atomic physics, see 12th edition

MCCAUGHEY, T J, b Tyrone Co, Ireland, Dec 3, 25; m 52; c 5. ANESTHESIOLOGY. Educ: Univ Col, Dublin, MB, BCh, 50. Prof Exp: Chief dept anesthesia, Children's Hosp, Winnipeg, 57-70; assoc prof anesthetics, Univ Man, 68-70; ASSOC PROF ANESTHESIA, McGILL UNIV, 70-; DIR DEPT ANESTHESIA, MONTREAL GEN HOSP, 70- Concurrent Pos: Chmn comt standardization, Paediat Anesthetic Equip, Can, 63- Mem: Am Acad Pediat; Am Soc Anesthesiol; Can Med Asn; Can Anaesthetists Soc; Brit Med Asn. Res: Muscle relaxants; antinauseants; premedicants; burns; respiratory distress of newborn; intensive care; resuscitation. Mailing Add: Dept of Anesthesia Montreal Gen Hosp Montreal PQ Can

MCCAUGHEY, WILLIAM FRANK, b Chicago, Ill, Oct 1, 21; m 43; c 2. NUTRITIONAL BIOCHEMISTRY. Educ: Purdue Univ, BS, 42; Northwestern Univ, MS, 48; Univ Ariz, PhD(agr chem), 51. Prof Exp: Asst biochem & nutrit, 51-55, asst agr biochemist, 55-57, from asst prof to assoc prof biochem & nutrit, 57-64, PROF BIOCHEM & NUTRITION, UNIV ARIZ, 64- Mem: AAAS; Sigma Xi. Res: Honey bee nutrition; intermediary metabolism of amino acids. Mailing Add: Agr Sci 409 Univ of Ariz Tucson AZ 85721

MCCAUGHRAN, DONALD ALISTAIR, b Vancouver, BC, July 9, 32; m 52; c 3. BIOMETRICS, STATISTICS. Educ: Univ BC, BSc, 59, MSc, 62; Cornell Univ, PhD(biomet, statist), 70. Prof Exp: Regional wildlife biologist, BC Fish & Game Br, 61-65; ASST PROF STATIST & BIOMET, UNIV WASH, 69- Mem: Am Statist Asn; Biomet Soc. Res: Experimental design; mathematical models in biology. Mailing Add: Dept of Forestry Univ of Wash Seattle WA 98105

MCCAULEY, CHARLES EDWARD, b Bayonne, NJ, July 24, 13. PHYSIOLOGY. Educ: Woodstock Col, BA, 35, Georgetown Univ, MS, 40, MD, 62; NY Univ, PhD, 52. Prof Exp: Develop engr, P R Mallory & Co, 44-46; instr chem, Jersey City Jr Col, 46-49; asst prof, St Peter's Col NJ, 49-52; fel, Notre Dame Univ, 52-53; asst prof, Fordham Univ, 53-54; assoc prof, St Peter's Col NJ, 54-58; intern med, 62-63, ASST PROF PHYSIOL, SCH MED, GEORGETOWN UNIV, 62- Concurrent Pos: Fel, Georgetown Univ, 63-64. Mem: Am Chem Soc. Res: Medical statistics. Mailing Add: Depts Physiol & Community Med Georgetown Univ Sch of Med Washington DC 20007

MCCAULEY, GARRY NATHAN, b Duncan, Okla, Apr 1, 47; m 65; c 2. SOIL PHYSICS. Educ: Okla State Univ, BS, 69, MS, 71, PhD(soil sci), 74. Prof Exp: ASST PROF SOIL SCI, TEX AGR EXP STA, TEX A&M UNIV, 75- Mem: Am Soc Agron; Soil Sci Soc Am; Sigma Xi. Res: Water management for rice aimed at both maximum economic production and water conservation. Mailing Add: Tex A&M Univ Agr Res & Exten Ctr Rte 5 Box 784 Beaumont TX 77706

MCCAULEY, GERALD BRADY, b Missoula, Mont, Apr 18, 35; m 58; c 1. ANALYTICAL CHEMISTRY. Educ: Univ Mont, BA, 57; Univ of the Pac, PhD(inorg chem), 67. Prof Exp: Physicist, US Naval Ord Lab, Calif, 57-63; res scientist anal chem, Lockheed Aircraft Corp, 67-70, RES SCIENTIST ANAL CHEM, LOCKHEED RES LAB, LOCKHEED MISSILES & SPACE CO, 70- Mem: Am Chem Soc. Res: Macrocyclic complexes of group eight metals; trace contamination analysis; vacuum degassing analysis using thermogravimetric techniques. Mailing Add: Mat Sci Lockheed Res Lab Lockheed Missiles & Space Co Palo Alto CA 94304

MCCAULEY, JAMES A, b New York, NY, Jan 15, 41; m 68; c 2. PHYSICAL CHEMISTRY. Educ: St Vincent Col, BS, 62; Fordham Univ, PhD(chem), 68. Prof Exp: SR RES CHEMIST, MERCK SHARP & DOHME RES LABS, MERCK & CO, 67- Mem: Am Chem Soc. Res: The application of thermodynamics to the evaluation of the purity of biologically active compounds; isolation and identification of impurities in these compounds; polymorphism. Mailing Add: Merck Sharp & Dohme Res Labs Merck & Co Rahway NJ 07065

MCCAULEY, JAMES ELIAS, b Fowler, Ind, Mar 9, 20; m 49; c 3. BIOLOGICAL OCEANOGRAPHY. Educ: Univ Wash, Seattle, BS, 46, MS, 49; Ore State Univ, PhD, 54. Prof Exp: Asst zool, Univ Wash, Seattle, 47-49; instr, Olympic Col, 49-51; instr, Orange Coast Col, 54-56; res assoc parasitol, Ore State Univ, 56-69; asst prof vet parasitol, Mont State Col, 60-61; ASSOC PROF BIOL OCEANOG, ORE STATE UNIV, 61- Mem: AAAS; Am Inst Biol Sci; Am Micros Soc; Wildlife Dis Asn. Res: Parasitology, trematodes of deepsea fishes and trematode life histories; ecology and taxonomy of bottom invertebrates of the deep-sea; coastal pollution ecology; marine biomedicinals; estuarine dredging ecology. Mailing Add: Dept of Oceanog Ore State Univ Corvallis OR 97331

MCCAULEY, JAMES WEYMANN, b Philadelphia, Pa, Mar 21, 40; m 64; c 3. MATERIALS SCIENCE, CRYSTALLOGRAPHY. Educ: St Joseph's Col, Ind, BS, 61; Pa State Univ, MS, 65, PhD(solid state sci), 68. Prof Exp: Res asst solid state sci, Pa State Univ, 66-68; res chemist, 68-76, SUPVRY MAT RES ENGR, ARMY MAT

& MECH RES CTR, 76- Mem: Am Crystallog Asn; Am Ceramic Soc; Mineral Soc Am; Am Asn Crystal Growth. Res: Microstructural control of material properties; fracture mechanics; oxides; oxynitrides; micas; carbonates. Mailing Add: Ceramics Div Army Mat & Mech Res Ctr Watertown MA 02172

MCCAULEY, JOHN CORRAN, JR, b Rochester, Pa, Feb 25, 01; m 39; c 2. ORTHOPEDIC SURGERY. Educ: Univ Colo, MD, 27. Prof Exp: From instr to asst prof orthop surg, 30-44, assoc prof clin orthop surg, 44-47, assoc prof orthop surg, 47-56, PROF ORTHOP SURG, MED SCH, NY UNIV, 56- Concurrent Pos: Clin asst vis orthopedist, Bellevue Hosp, 30-33, asst vis surgeon, 33-35, assoc vis surgeon, 35-41, vis surgeon, 41-; attend surgeon, Univ Clin, NY Univ, 30-; consult, Flushing Hosp, 36-; assoc attend surgeon, Neponsit Beach Hosp, 37-43; attend surgeon in charge serv, Seaside Hosp, 38-40; consult, USPHS, 39; asst attend surgeon, NY State Rehab Hosp, 41-42, attend surgeon, 42-43, sr surgeon, 43-47, surgeon-in-chief, 47-; asst attend surgeon in charge club foot clin, NY Orthop Dispensary & Hosp, 41-; attend surgeon, Univ Hosp; consult orthop surgeon, Nyack, Lawrence & St Agnes Hosps. Mem: Fel AMA; Am Orthop Asn; fel Am Col Surgeons; fel Am Acad Orthop Surg; fel NY Acad Sci. Res: Problems of orthopedic surgery, particularly congenital deformities of the feet; surgical treatment of neurotrophic joints; evaluation of the crippled for proper rehabilitation. Mailing Add: Dept of Orthop Surg NY Univ Med Sch New York NY 10016

MCCAULEY, JOHN FRANCIS, b New York, NY, Apr 2, 32; m 56; c 3. ASTROGEOLOGY. Educ: Fordham Univ, BS, 53; Columbia Univ, MA, 57, PhD(geol), 59. Prof Exp: Coop geologist, Pa Geol Surv, 56-58; asst prof, Univ SC, 58-62, assoc prof, 62-63; geologist, 63-70, BR CHIEF, BR ASTROGEOL STUDIES, US GEOL SURV, 70- Concurrent Pos: Lectr, Columbia Univ, 57-58; proj geologist, Div Geol, SC Develop Bd, 58-63; consult geologist, 58-63; co-investr, Mars Mariner TV Team, 71. Mem: Fel Geol Soc Am; Am Astron Soc. Res: Uranium occurrences in Pennsylvania; metamorphism, structure and mineral deposits of the Southern Appalachian Piedmont; lunar structure and stratigraphy; use of television systems for planetary exploration; geology of Mars. Mailing Add: 60 Wilson Lane Flagstaff AZ 86001

MCCAULEY, ROBERT HENRY, JR, b Hagerstown, Md, June 13, 13; m 38; c 3. VERTEBRATE ZOOLOGY. Educ: Washington & Lee Univ, AB, 35; Cornell Univ, PhD(vert zool), 40. Prof Exp: Instr biol & wildlife mgt, Goddard Col, 40-42; biologist, Vt State Fish & Game Serv, 42-44; entomologist, Malaria Control in War Areas, USPHS, Ark, 44-46, entomologist, Tech Develop Labs, Commun Dis Ctr, Ga, 46-50, entomologist, Malaria Invest Field Sta, SC, 50-51, commun dis ctr rep, Pac Northwest Drainage Basins, Ore, 51-54, toxicologist, Tech Develop Labs Commun Dis Ctr, Ga, 54-56, entomologist, Div Int Health, Pan-Am Sanit Bur, Mex, 56-58, exec secy anat sci & genetics training comts, Div Gen Med Sci, NIH, 58-63, dep chief res training grants br, Nat Inst Gen Med Sci, 63-74; RETIRED. Mem: AAAS; Am Soc Ichthyol & Herpet. Res: Natural history; revision of genus Microrhopala; reptiles of Maryland; status of deer in Vermont; use of insecticides for public health; hookworm, housefly and mosquito larvicides; arthropods of public health importance in Pacific Northwest; toxicology of insecticides; insecticide residues on adobe surfaces; graduate research training grants. Mailing Add: 6405 Orchid Dr Bethesda MD 20034

MCCAULEY, ROBERT WILLIAM, b Toronto, Ont, July 8, 26; m 56; c 3. FISH BIOLOGY. Educ: Univ Toronto, BA, 50, MA, 57; Univ Western Ont, PhD(zool), 62. Prof Exp: Res scientist, Fisheries Res Bd, Can, 55-62; biologist in charge fish cultural prog, Ont Dept Lands & Forests, 62-63, res scientist, 63-65; ASSOC PROF BIOL, WILFRID LAURIER UNIV, 65- Concurrent Pos: Consult, Ont Hydro-Prov of Ont, 74-76. Mem: Am Fisheries Soc; Am Inst Biol Sci; Can Soc Zool. Res: Fish physiology; thermal ecology; effects of warm effluents on fish. Mailing Add: 118 Forest Hill Dr Kitchener ON Can

MCCAULEY, WILLIAM JOHN, b Ray, Ariz, Aug 11, 20; m 47; c 2. PHYSIOLOGY, ANATOMY. Educ: Univ Ariz, BS, 47; Univ Southern Calif, PhD(zool), 55. Prof Exp: Res assoc pharmacol, Lederle Labs, Am Cyanamid Co, NY, 47-49; lab assoc zool, Univ Southern Calif, 49-51, instr anat, Sch Dent, 52-54, Sch Med, 54-55; instr zool, 55-56, from asst prof to assoc prof, 56-64, PROF BIOL SCI, UNIV ARIZ, 64-, ASSOC DEAN GRAD COL, 74- Res: Vertebrate physiology; biological education. Mailing Add: Grad Col Univ of Ariz Tucson AZ 85721

MCCAULLY, RONALD JAMES, b West Reading, Pa, Dec 25, 36; m 60; c 2. ORGANIC CHEMISTRY, MEDICINAL CHEMISTRY. Educ: Mass Inst Technol, SB, 58; Harvard Univ, PhD(org chem), 65. Prof Exp: Chemist, Arthur D Little, Inc, 60-61; sr res chemist, 63-68, GROUP LEADER, WYETH LABS, AM HOME PRODS CORP, 68- Mem: Am Chem Soc. Res: Synthesis of new organic compounds as potential medicinal products; investigation of novel chemical reactions; syntheses of benzodiazepines antimicrobials and respiratory drugs. Mailing Add: Wyeth Labs Inc PO Box 8299 Philadelphia PA 19101

MCCAWLEY, ELTON LEEMAN, b Long Beach, Calif, Nov 1, 15; m 40; c 2. PHARMACOLOGY. Educ: Univ Calif, AB, 38, MS, 39, PhD(pharmacol), 42. Prof Exp: Asst pharmacol, Sch Med, Univ Calif, 39-42, Int Cancer Res Found fel, lectr & res assoc, 42-43; from instr to asst prof pharmacol, Sch Med, Yale Univ, 43-49; assoc prof, 49-60, PROF PHARMACOL, MED SCH, UNIV ORE, 60- Concurrent Pos: Dir, Ore Poison Control Ctr, 57-64; consult, Providence Hosp; trustee, Ore Mus Sci & Indust, 60-66; mem bd dirs, Portland Ctr Speech & Hearing, 63-66; chmn, Gov Adv Comt Methadone Treatment Heroin Addicts, 69-72; inst rev comt drug studies in humans, Ment Health Div, 71-72. Mem: Am Soc Pharmacol & Exp Therapeut; AMA; Am Fedn Clin Res. Res: Cardiovascular pharmacology; morphine derivatives; drugs in cardiac arrhythmias; anesthetics. Mailing Add: Dept of Pharmacol Univ of Ore Med Sch Portland OR 97201

MCCAY, MYRON STANLEY, b Pocataligo, Ga, Nov 5, 11; m 35; c 2. PHYSICS. Educ: Univ Ga, AB, 32; Univ NC, AM, 34; Ohio State Univ, PhD(physics), 37. Prof Exp: Asst physics, Univ Ga, 30-32; asst, Univ NC, 32-34; asst, Ohio State Univ, 34-37; instr, Va Polytech Inst, 37-39, asst prof, 39-41, from assoc prof to prof, 41-48; prof, 48-65, GUERRY PROF PHYSICS, UNIV TENN, CHATTANOOGA, 65-, CHMN DEPT, 48- Concurrent Pos: Physicist appl physics lab, Johns Hopkins Univ, 44; res partic, Oak Ridge Nat Lab, 57-58; coord, Univ Chattanooga-Univ Tenn Resident Grad Prog in Chattanooga, 67-69. Mem: AAAS; Am Phys Soc; Am Asn Physics Teachers; Optical Soc Am. Res: Band spectroscopy; rotation temperatures of gases from band-line intensities; reversal of subordinate series; Zeeman spectra. Mailing Add: Dept of Physics & Astron Univ of Tenn Chattanooga TN 37401

MCCAY, PAUL BAKER, b Tulsa, Okla, June 5, 24. BIOCHEMISTRY. Educ: Univ Okla, BS, 48, MS, 50, PhD(physiol, biochem), 55. Prof Exp: USPHS fel biochem, Okla Med Res Inst, 56-58, mem biochem sect, 58-71; from asst prof to assoc prof, 59-68, PROF BIOCHEM, SCH MED, UNIV OKLA, 68-; HEAD BIOMEMBRANE RES LAB, OKLA MED RES FOUND, 72- Concurrent Pos: NIH grants, 56-; mem nutrit study sect, NIH, 74-78. Mem: Am Inst Nutrit; Am Soc Biol Chemists; Brit Biochem Soc; Am Oil Chemists Soc. Res: Role of membrane-bound electron transport systems in promoting lipid peroxidation in biological membranes; alterations of membrane function caused by free radicals generated by the activity of oxidoreductases; role of dietary antioxidants and fat in chemical carcinogenesis; influence of dietary states on carcinogen metabolism. Mailing Add: Biomembrane Res Lab Okla Med Res Found Oklahoma City OK 73104

MCCHESNEY, EVAN WILLIAM, b Oconto Falls, Wis, Dec 11, 04; m 33; c 3. BIOCHEMISTRY. Educ: Univ Chicago, SB, 26, MS, 28; Northwestern Univ, PhD(biochem), 31. Prof Exp: Prof math, Westmar Col, 26-27; asst chem, Univ Chicago, 27-28; asst biochem, Med Sch, Northwestern Univ, 28-31; assoc prof, Sch Med, Univ NC, 31-37; assoc prof, Col Med, Baylor Univ, 37-38; sr biochemist, Sterling-Winthrop Res Inst, 36-68; res prof toxicol, Albany Med Col, 68-75; CONSULT TOXICOL, 75- Mem: Am Physiol Soc; Soc Toxicol. Res: Biochemical analysis; metabolism of drugs. Mailing Add: Inst of Comp & Human Toxicol Albany Med Col Albany NY 12208

MCCHESNEY, JAMES DEWEY, b Hatfield, Mo, Aug 27, 39; m 59; c 6. PLANT SCIENCE, BIO-ORGANIC CHEMISTRY. Educ: Iowa State Univ, BS, 61; Ind Univ, MA, 64, PhD(org chem), 65. Prof Exp: Chemist, Battelle Mem Inst, summer, 61; from asst prof to assoc prof bot & med chem, 65-73, PROF BOT & MED CHEM, UNIV KANS, 73- Mem: AAAS; Am Chem Soc; Am Soc Plant Physiol; NY Acad Sci; Am Asn Cols Pharm. Res: Chemistry and biochemistry of biologically active secondary plant substances. Mailing Add: RR 1 Lawrence KS 66044

MCCLAIN, ERNEST PAUL, b Columbus, Ohio, Jan 11, 26; m 51; c 4. METEOROLOGY. Educ: Univ Chicago, SM, 50; Fla State Univ, PhD(meteorol), 58. Prof Exp: Instr meteorol, Univ Wash, Seattle, 50-54; instr, Fla State Univ, 56-57; instr, Univ Chicago, 57-58, asst prof, 59-62; res meteorologist, US Weather Bur, 62-67, DIR ENVIRON SCI GROUP, NAT ENVIRON SATELLITE SERV, NAT OCEANIC & ATMOSPHERIC ADMIN, 67- Concurrent Pos: NOAA Mem, USA/USSR Joint Working Group Nat Environ, 71- Mem: AAAS; Am Meteorol Soc; Am Geophys Union. Res: Use of aircraft and satellite data in synoptic meteorology; develop applications of earth satellite data to oceanography and hydrology. Mailing Add: Environ Sci Group Nat Oceanic & Atmos Admin FB 4 (Stop D) Suitland MD 20233

MCCLAIN, JOHN A, b Ridgway, Pa, July 15, 08; m 40; c 2. ZOOLOGY. Educ: Villanova Col, BS, 30; Univ Pa, MS, 34, PhD(zool), 39. Hon Degrees: DSc, Villanova Univ, 75. Prof Exp: Instr biol, 30-35, from asst prof to assoc prof, 35-48, chmn dept, 54-67, PROF BIOL, VILLANOVA UNIV, 48- Concurrent Pos: Instr, Rittenhouse Col, 47-49. Mem: AAAS; Am Soc Zool; Am Inst Biol Sci; Am Micros Soc. Res: Vertebrate embryology and anatomy. Mailing Add: Dept of Biol Villanova Univ Villanova PA 19085

MCCLAIN, JOHN WILLIAM, b Dayton, Ohio, Nov 3, 28. PHYSICS. Educ: Antioch Col, BS, 52; Princeton Univ, MA, 53, PhD(physics), 57. Prof Exp: Instr physics, Princeton Univ, 55-56; asst lectr, Univ Manchester, 56-57; from asst prof to assoc prof, 57-73, PROF PHYSICS, AM UNIV BEIRUT, 73- Concurrent Pos: Res assoc, Harvard Univ, 67. Mem: AAAS; Am Phys Soc; Am Asn Physics Teachers. Res: Science education; history of science. Mailing Add: Dept of Physics Am Univ of Beirut Beirut Lebanon

MCCLAIN, PHILIP EDWIN, b Independence, Mo, Feb 13, 35; m 59; c 2. NUTRITIONAL BIOCHEMISTRY. Educ: Univ Mo-Columbia, BS, 61; La State Univ, MS, 65; Mich State Univ, PhD(food sci, human nutrit), 68. Prof Exp: Res scientist food res div, Armour & Co, 61-63; res assoc animal sci, La State Univ, 63-65; res asst food sci & human nutrit, Mich State Univ, 65-68; RES CHEMIST, NUTRIT INST, AGR RES SERV, USDA, 68- Concurrent Pos: Vis scientist, Dept Nutrit Sci, Univ Calif, Berkeley, 70-71. Mem: AAAS; Am Chem Soc. Res: Protein isolation and characterization; nutritional influence on protein synthesis and metabolism. Mailing Add: Nutrit Inst Agr Res Serv USDA Beltsville MD 20705

MCCLAIN, WILLIAM MARTIN, molecular spectroscopy, see 12th edition

MCCLAINE, LESLIE ANDREW, physical chemistry, see 12th edition

MCCLAMROCH, N HARRIS, b Houston, Tex, Oct 7, 42; m 63. APPLIED MATHEMATICS. Educ: Univ Tex, Austin, BS, 63, MS, 65, PhD(eng mech), 67. Prof Exp: Asst prof aerospace, 67-71, ASSOC PROF COMPUT INFO & CONTROL ENG, UNIV MICH, ANN ARBOR, 71- Mem: Math Asn Am; Inst Elec & Electronics Eng; Soc Indust & Appl Math. Res: Optimal control theory; theory of differential equations; stability theory. Mailing Add: Dept of Aerospace Univ of Mich Ann Arbor MI 48104

MCCLANAHAN, BEATRICE J, b Buffalo, NY, July 30, 25. RADIOBIOLOGY. Educ: Ind Univ, BS, 47, MS, 49; Wash State Univ, PhD(nutrit), 65. Prof Exp: Teaching assoc chem, Ind Univ, 47-48; res chemist, Gen Elec Co, Wash, 48-54 & 55-60, biol scientist, 62-65; asst nutrit, Wash State Univ, 60-62; SR BIOL SCIENTIST, PAC NORTHWEST LAB, BATTELLE MEM INST, 65- Mem: AAAS; Am Chem Soc; Soc Exp Biol & Med. Res: Radiochemistry; mammalian mineral metabolism. Mailing Add: Biol Dept Battelle Mem Inst Richland WA 99352

MCCLANAHAN, JAMES LEE, organic chemistry, physical chemistry, see 12th edition

MCCLANAHAN, ROBERT JOSEPH, b Rainy River, Ont, Jan 22, 29; m 56; c 3. ENTOMOLOGY. Educ: McMaster Univ, BA, 51; Univ Western Ont, MSc, 54; Mich State Univ, PhD(entom), 62. Prof Exp: Asst entomologist, Entom Lab, 54-65, RES SCIENTIST, RES STA, CAN DEPT AGR, 65- Mem: Entom Soc Can; Entom Soc Am; Int Orgn Biol Control. Res: Integrated control of greenhouse insects. Mailing Add: Res Sta Can Dept of Agr Harrow ON Can

MCCLAREN, MILTON, JR, b Vancouver, BC, Sept 3, 40; m 63. BOTANY, MYCOLOGY. Educ: Univ BC, BEd, 63, PhD(bot), 67. Prof Exp: Asst prof biol sci, 67-74, ASSOC PROF BIOL SCI & ASSOC PROF EDUC, SIMON FRASER UNIV, 67- Concurrent Pos: Nat Res Coun Can res grants, 67-69. Mem: AAAS; Mycol Soc Am; Can Bot Asn. Res: Species genetics in Basidiomycetes; fungal ecology and cytology. Mailing Add: Dept of Biol Sci Simon Fraser Univ Burnaby BC Can

MCCLARY, ANDREW, b Chicago, Ill, Apr 15, 27; m 54; c 2. ZOOLOGY, ANTHROPOLOGY. Educ: Dartmouth Col, AB, 50; Univ Mich, MA, 54, PhD(zool), 60. Prof Exp: Res asst city planning, Chicago Plan Comn, 50-51; shrimp fishery, 51-52; from asst prof to assoc prof zool, Univ Wis-Milwaukee, 59-64; from asst prof to assoc prof natural sci, 64-69, PROF NATURAL SCI, MICH STATE UNIV, 69- Mem: AAAS; Am Inst Biol Sci. Res: Socio-cultural implications of biology, especially history of hygiene, sanitation; views of man's nature, man's place in nature; biology of man's near environment. Mailing Add: Dept of Natural Sci Mich State Univ East Lansing MI 48823

MCCLARY, CECIL FAY, b Grayson, La, June 22, 13; m 40; c 2. GENETICS. Educ: La State Univ, BS, 36, MS, 38; Univ Calif, PhD(genetics), 50. Prof Exp: Asst poultryman, Western Wash Exp Sta, Wash State Univ, 38-52, assoc poultry scientist, 52-56; geneticist, Heisdorf & Nelson Farms, Inc, 56-71; DIR GENETICS RES, H&N INC, 72- Concurrent Pos: Asst prof, Wash State Univ, 46; asst, Univ Calif, 46-49. Mem: Biomet Soc; Poultry Sci Asn. Res: Disease resistance and heredity; genetics of egg quality. Mailing Add: H&N Inc 15305 NE 40th St Redmond WA 98052

MCCLARY, DANIEL OTHO, b Ft Towson, Okla, Apr 24, 18; m 42; c 2. MICROBIOLOGY. Educ: Southeastern State Col, BS, 40; Wash Univ, PhD, 51. Prof Exp: From asst prof to assoc prof, 51-70, PROF MICROBIOL, SOUTHERN ILL UNIV, CARBONDALE, 70- Mem: AAAS; Am Soc Microbiol; Am Inst Biol Sci; Am Acad Microbiol. Res: Physiology and cytology of yeasts. Mailing Add: 203 S Tower Rd Carbondale IL 62901

MCCLARY, JOSEPH EDWARD, b Durant, Okla, Sept 8, 09; m 35; c 3. BIOCHEMISTRY. Educ: Southeastern State Col, BS, 32; Univ Mo, MA, 37, PhD(plant physiol), 41. Prof Exp: Teacher, High Sch, Okla, 32-35 & Mo, 35-38; asst, Univ Mo, 38-41, instr, 41-42; res chemist, Anheuser-Busch, Inc, 43, asst dir res, 43-47, dir res, 47-58; supvr biochem res, Clinton Corn Processing Co, Iowa, 58-65; from asst prof to assoc prof biol, 65-70, PROF BIOL, WESTMINSTER COL, MO, 70-, CHMN DEPT, 74- Mem: Am Chem Soc; Am Soc Brewing Chemists; Soc Indust Microbiol. Res: Production, evaluation and industrial use of amylases from fungi, especially Aspergillus niger; utilization of the products from corn in industrial fermentations. Mailing Add: Dept of Biol Westminster Col Fulton MO 65251

MCCLATCHEY, ROBERT ALAN, b Rockville, Conn, July 26, 38; m 61; c 2. ATMOSPHERIC PHYSICS. Educ: Mass Inst Technol, BS, 60, MS, 61; Univ Calif, Los Angeles, PhD(meteorol), 66. Prof Exp: Sr scientist, Jet Propulsion Lab, Calif Inst Technol, 63-67; sr scientist, AVCO Space Systs Div, 67-68; res scientist, 68-75, BR CHIEF, AIR FORCE CAMBRIDGE RES LABS, 75- Mem: AAAS; Am Meteorol Soc. Res: Atmospheric radiation, the transmittance and emission of radiation in the atmosphere; molecular spectroscopy involved in the problem of atmospheric transmittance; radiative transfer in the atmosphere and its relationship to meteorology. Mailing Add: Air Force Cambridge Res Lab (OPI) L G Hanscom Field Bedford MA 01730

MCCLATCHY, JOSEPH KENNETH, b Brownwood, Tex, July 5, 39; m 60; c 4. MEDICAL MICROBIOLOGY. Educ: Tex Tech Col, BS, 61; Univ Tex, MA, 63, PhD(microbiol), 66. Prof Exp: Fel, Ind Univ, 65-66; fel, 66-67, CHIEF CLIN LABS, NAT JEWISH HOSP & RES CTR, 67- Mem: AAAS; Am Soc Microbiol; Conf Pub Health Lab Dirs. Res: Modes of action of antituberculosis drugs; diagnostic microbiology; resistance to microbial infections; immunotherapy of cancer; mechanisms of drug resistance. Mailing Add: Nat Jewish Hosp & Res Ctr Dept of Clin Labs 3800 E Colfax Denver CO 80206

MCCLEARY, CHARLES DAVID, b Huntsville, Ohio, Nov 19, 14; m 42; c 4. CHEMISTRY. Educ: Wittenberg Col, BA, 36; Ohio State Univ, PhD(org chem), 40. Prof Exp: Asst chemist, Ohio State Univ, 36-40; res chemist, Chem Div, US Rubber Co, 40-44, develop chemist synthetic rubber develop, 44-46, mgr plastics process develop, Chem Div, 46-49, mgr process develop, 49-52, mgr basic res, 52-54, mgr, Marvinol Develop, 54-56, asst dir res & develop, 56-61, DIR RES & DEVELOP, UNIROYAL CHEM, DIV UNIROYAL, INC, 61- Mem: Fel AAAS; Am Chem Soc. Res: Organic synthesis; emulsion polymerization of rubber and plastics; shortstop for synthetic rubber; isomeric esters of the benzoyl benzoic acids. Mailing Add: Uniroyal Chem Div of Uniroyal Inc Naugatuck CT 06770

MCCLEARY, HAROLD RUSSELL, b Huntsville, Ohio, Oct 11, 13; m 41; c 3. PHYSICAL CHEMISTRY. Educ: Monmouth Col, BS, 37; Columbia Univ, PhD(chem), 41. Prof Exp: Asst, Columbia Univ, 37-40; res chemist, 41-45, asst chief chemist, 45-52, asst mgr pigments res, 52-54, mgr, 54-57, dir res serv, 57-64, mgr dyes res & develop, 64-71, mgr dyes & textiles chem res & develop, 71-75, MGR BOUND BROOK RES LAB, AM CYANAMID CO, 75- Mem: Am Chem Soc. Res: Kinetics and mechanism of organic reactions; spectrophotometric methods; dyeing and textile chemistry; analytical methods; pigments. Mailing Add: Am Cyanamid Co Res Lab Bound Brook NJ 08805

MCCLEARY, JAMES A, b Bridgeport, Ohio, Mar 26, 17; m 38; c 4. BOTANY. Educ: Asbury Col, AB, 38; Univ Ohio, MS, 46; Univ Mich, PhD, 52. Prof Exp: Asst bot, Univ Ohio, 39-40; asst prof, Ariz State Univ, 47-51, from assoc prof to prof, 52-60; prof, Calif State Univ, Fullerton, 60-69; PROF BOT & CHMN DEPT BIOL SCI, NORTHERN ILL UNIV, 69- Concurrent Pos: Instr, Univ Mich, 51; NSF fac fel, 59-60; assoc prog dir, NSF Undergrad Res Participation Prog, Washington, DC. Mem: AAAS; Am Bryol & Lichenological Soc; Ecol Soc Am; Int Soc Plant Taxon. Res: Bryophytes; desert plants. Mailing Add: Dept of Biol Sci Northern Ill Univ DeKalb IL 60115

MCCLEARY, STEPHEN HILL, b Houston, Tex, Feb 6, 41; div. ALGEBRA. Educ: Rice Univ, BA, 63; Univ Wis, MS, 64, PhD(math), 67. Prof Exp: Instr math, Univ Wis, 67; vis asst prof, Tulane Univ, 67-68; ASSOC PROF MATH, UNIV GA, 68- Mem: Am Math Soc. Res: Ordered algebraic structures, especially lattice-ordered groups; infinite permutation groups. Mailing Add: Dept of Math Univ of Ga Athens GA 30601

MCCLEAVE, JAMES DAVID, b Atchison, Kans, Dec 17, 39; m 62; c 1. ZOOLOGY. Educ: Carleton Col, AB, 61; Mont State Univ, MS, 63, PhD(zool), 67. Prof Exp: Instr zool, Mont State Univ, 63-64; asst prof, Western Ill Univ, 67-68; ASSOC PROF ZOOL, UNIV MAINE, ORONO, 68- Concurrent Pos: NSF grants, Mont State Univ, 68-70 & Univ Maine, 70-71; Off Naval Res grant, Univ Maine, 71-72. Mem: AAAS; Am Soc Zool; Am Fisheries Soc; Animal Behav Soc. Res: Orientation of migratory fishes; effects of thermal pollution on ecology of fishes. Mailing Add: Dept of Zool Univ of Maine Orono ME 04473

MCCLELLAN, AUBREY LESTER, b Oklahoma City, Okla, Feb 5, 23; m 46; c 2. PHYSICAL CHEMISTRY. Educ: Centenary Col, BS, 43; Univ Tex, MS, 46, PhD(phys chem), 49. Prof Exp: Res fel, Univ Calif, 48-50; res fel, Mass Inst Technol, 50-51; res chemist, 51-54, staff asst to vpres res, 54-56, mem pres's staff, 56-57, sr res chemist, 57-65, SR RES ASSOC, CHEVRON RES CO, STANDARD OIL CO CALIF, 65- Mem: Am Chem Soc. Res: Hydrogen bonding; chemical education in secondary schools; dipole moments; spectroscopic analysis, FT—nuclear magnetic resonance. Mailing Add: Chevron Res Co 576 Standard Ave Richmond CA 94802

MCCLELLAN, BETTY JANE, b Little Rock, Ark, Oct 26, 32. PATHOLOGY. Educ: Southern Methodist Univ, BS, 55; Univ Ark, MD, 57; Am Bd Path, dipl, 67. Prof Exp: From intern to asst resident path, Mass Gen Hosp, 57-60; from instr to asst prof, Johns Hopkins Univ, 61-64; assoc prof, 64-69, PROF PATH, SCH MED, UNIV OKLA, 69-, PROF CYTOTECHNOL, SCH HEALTH RELATED PROFESSIONS, 70-, DIR SURG PATH, MED CTR, 64- Concurrent Pos: Pathologist, Johns Hopkins Hosp, 62-64, assoc dir sch cytotechnol, 63-64. Mem: Am Soc Clin Pathologists; Col Am Pathologists; Am Soc Cytol; Int Acad Path. Res: Anatomic, surgical and forensic pathology; cytopathology; cancer detection. Mailing Add: Dept of Surg Path Univ of Okla Med Ctr Oklahoma City OK 73190

MCCLELLAN, BOBBY EWING, b Cayce, Ky, July 5, 37; m 59. ANALYTICAL CHEMISTRY. Educ: Murray State Univ, BA, 59; Univ Miss, PhD(anal chem), 63. Prof Exp: Res fel, Univ Ariz, 63-64; group leader anal chem, PPG Industs, 64-65; Comt Instnl Studies & Res Grant, 66-67, PROF CHEM, MURRAY STATE UNIV, 65- Mem: Am Chem Soc. Res: Solvent extraction separations; extraction kinetics; atomic absorption spectrometry; nuclear chemistry as applied to analytical chemistry. Mailing Add: Dept of Chem Murray State Univ Murray KY 42072

MCCLELLAN, CATHARINE, b York, Pa, Mar 1, 21; m 74. ANTHROPOLOGY. Educ: Bryn Mawr Col, AB, 42; Univ Calif, Berkeley, PhD(anthrop), 50. Prof Exp: Adv ethnog, Nat Mus Can, 50-51; vis asst prof anthrop, Univ Mo, 52; asst prof, Univ Wash, 52-56; asst prof & chmn dept, Barnard Col, Columbia Univ, 56-61; assoc prof, 61-65, prof, 65-73, BASCOM PROF ANTHROP, UNIV WIS-MADISON, 73- Concurrent Pos: Margaret Snell fel, Yukon Territory, Can, 50-51; vis lectr, Bryn Mawr Col, 54; consult, Arctic Health Res Ctr, 56; vis scholar, Nat Mus Can, 70; field res, Alaska & Can, 48-68; ed, Arctic Anthrop, 74- Mem: Fel AAAS; Am Anthrop Asn; Arctic Inst NAm; Royal Anthrop Inst; Am Ethnol Soc (secy-treas, 58, pres, 65). Res: Culture history of northwestern North America. Mailing Add: Dept of Anthrop Univ of Wis Madison WI 53706

MCCLELLAN, GENE ELVIN, b Guthrie Center, Iowa, Feb 11, 43; m 69; c 1. ELEMENTARY PARTICLE PHYSICS. Educ: Iowa State Univ, BS, 65; Cornell Univ, MS, 68, PhD(physics), 71. Prof Exp: Res assoc, 70-73, ASST PROF EXP HIGH ENERGY PHYSICS, UNIV MD, COLLEGE PARK, 73- Mem: Am Phys Soc; Am Asn Physics Teachers. Res: Interactions of positive K mesons with protons at low energies; interactions of protons and pions with protons at Fermilab energies; interactions of protons and electrons. Mailing Add: Dept of Physics & Astron Univ of Md College Park MD 20742

MCCLELLAN, GUERRY HAMRICK, b Gainesville, Fla, Nov 1, 39; m 62; c 2. MINERALOGY, GEOLOGY. Educ: Univ Fla, BS, 61, MS, 62; Univ Ill, PhD(geol, clay mineral, chem), 64. Prof Exp: Res chemist, Nat Fertilizer Res Ctr, 65-66, proj leader, Nat Fertilizer Develop Ctr, 66-73, SR PROJ LEADER, NAT FERTILIZER DEVELOP CTR, TENN VALLEY AUTH, 73- Concurrent Pos: Fel, Nat Petrol Co of Aquitaine & Univ Bordeaux, France, 64-65. Mem: Mineral Soc Am; Clay Mineral Soc; Mineral Soc Gt Brit & Ireland. Res: Evaluation of phosphate rocks for industrial and agricultural uses; modification of elemental sulfur melts; sludge from stack gas removal processes; crystallography; geochemistry, inorganic and organic; phosphate rock beneficiation. Mailing Add: Fundamental Res Br Tenn Valley Auth Muscle Shoals AL 35660

MCCLELLAN, JAMES T, b Claremore, Okla, July 3, 16; m 44; c 4. PATHOLOGY. Educ: Univ Okla, BS, 37, AB & BSM, 40, MD, 42; Univ Minn, MS, 50; Am Bd Path, dipl & cert anat path & clin path, 51. Prof Exp: Intern, Emory Univ Hosp, 42-43; pres med staff, Hosp, 56-57, CHIEF DEPT PATH & DIR LABS & SCH MED TECHNOL, ST JOSEPH HOSP, 56- Concurrent Pos: Clin prof, Univ Ky, 51-; consult, Eastern State Hosp & Fed Correctional Inst, Lexington; dir, Physicians Serv Labs. Mem: Am Soc Clin Pathologists; Sigma Xi; AMA; fel Col Am Pathologists. Mailing Add: 1400 Harrodsburg Rd Lexington KY 40504

MCCLELLAN, JOHN FORBES, b Pembrook, Ont, Aug 30, 17; nat US; m 50; c 3. ZOOLOGY, PHYSIOLOGY. Educ: Univ Ill, Urbana, BS, 47, MS, 48, PhD(zool), 54. Prof Exp: Asst instr zool, Univ Ill, Urbana, 47, asst instr protozool, 47-49; from instr to assoc prof, Univ Detroit, 51-63; ASSOC PROF ZOOL, COLO STATE UNIV, 63-, ASSOC PROF ENTOM, 74- Concurrent Pos: Grants, Nat Cancer Inst, Univ Detroit, 57-60 & AEC, 62; consult, Wayne County Rd Comnr, Mich, 62; grant consult, NSF Undergrad Sci Equip Prog, 64-65, dir, NSF Summer Insts in Field Biol for Col Teachers, Colo State Univ, Mountain Campus, 68-72, proposal consult, NSF Col Teachers Progs, 69-71; consult, Colo Utility Power Co, 70-71. Mem: Fel AAAS; Soc Protozool; Am Soc Zool. Res: Cell biology; cytology; electron microscopy of protozoa; ecological role of aquatic and soil protozoa. Mailing Add: Dept of Zool Colo State Univ Ft Collins CO 80521

MCCLELLAN, ROGER ORVILLE, b Tracy, Minn, Jan 5, 37; m 62; c 3. INHALATION TOXICOLOGY, VETERINARY TOXICOLOGY. Educ: Wash State Univ, DVM, 60; Am Bd Vet Toxicol, dipl, 67. Prof Exp: Biol scientist radiobiol, Hanford Labs, Gen Elec Co, Wash, 60-63, sr scientist, 63-65; sr scientist, Pac Northwest Labs, Battelle Mem Inst, 65; scientist div biol & med, US AEC, 65-66; dir fission prod inhalation prog & asst dir res, 66-73, VPRES & DIR RES ADMIN & DIR INHALATION TOXICOL RES INST, LOVELACE FOUND MED EDUC & RES, 73- Concurrent Pos: Lectr sch vet med, Wash State Univ, 63; mem div biol & med, Adv Comt Space Nuclear Systs Radiol Safety, 67-73; adv, Lab Animal Biol & Med Training Prog, Univ Calif, Davis, 68-70; consult, Nat Inst Environ Health Sci, NIH, 68-71, mem toxicol study sect, NIH, 69-; chmn sci comt 30, Nat Coun Radiation Protection & Measurements, 69-; adj prof sch pharm, Univ Ark, 70-; pres, Am Bd Vet Toxicol, 70-73; clin assoc sch med, Univ NMex, 71-, adj prof biol, 73-; mem, Transuranium Tech Group, Adv to US AEC Div Biomed & Environ Res, 72-; mem environ radiation exposure adv comt, Environ Protection Agency, 72-, chmn, 75-, mem sci adv bd, 74-. Honors & Awards: Elda Anderson Award, Health Physics Soc, 74. Mem: Fel AAAS; Am Col Vet Toxicol; Am Vet Med Asn; Health Physics Soc; Radiation Res Soc. Res: Metabolism and toxicity of radionuclides, especially effects of inhaled radionuclides and late radiation effects of bone-seeking radionuclides on bone and hematopoietic tissue; comparative medicine. Mailing Add: Inhalation Toxicol Res Inst Lovelace Found PO Box 5890 Albuquerque NM 87115

MCCLELLAN, WILBUR DWIGHT, b LaVerne, Calif, Jan 24, 14; m 37; c 3. PLANT PATHOLOGY, RESEARCH ADMINISTRATION. Educ: Univ Calif, BS, 36; Cornell Univ, PhD(plant path), 41. Prof Exp: Instr plant path, Univ Md, 40-41; pathologist, Bur Plant Indust, USDA, 41-51; dir res, Mid-State Chem Supply Co, Calif, 51-53; pathologist, Bur Plant Indust, USDA, 53-56; prin plant pathologist, State Exp Sta Div, 56-60, asst chief crops protection res br, Crops Res Div, 60-67, asst dir div, 68-71, asst to dep adminr plant sci & entom, 71-72, AREA DIR, WESTERN REGION, AGR RES SERV, USDA, 72- Mem: AAAS; Am Phytopath Soc; Am Inst Biol Sci. Res: Administration of agricultural and market quality research. Mailing Add: 2021 S Peach Ave PO Box 8143 Fresno CA 93727

MCCLELLAN, WILLIAM ALAN, b Seattle, Wash, Feb 14, 40. ENVIRONMENTAL GEOLOGY. Educ: Univ Ariz, BS, 63; Univ Cincinnati, MS, 65; Univ Wash, Seattle, PhD(geol), 69. Prof Exp: Jr geologist, Atlantic Richfield Co, summer, 64, geologist, 65-66; asst prof geol, 69-73, ASSOC PROF GEOL, UNIV NEV, LAS VEGAS, 73- Concurrent Pos: Consult paleontologist, Nev Archeol Soc, 74-75. Mem: Paleont Soc; Paleont Res Inst; Am Asn Petrol Geol; Int Paleontol Soc; Am Forestry Asn. Res:

Invertebrate paleontology; micropaleontology; biostratigraphy of Paleozoic microfossils. Mailing Add: Dept of Geosci Univ of Nev Las Vegas NV 89109

MCCLELLAN, WILLIAM ROBERT, b Toulon, Ill, Dec 23, 13; m 39; c 4. ORGANOMETALLIC CHEMISTRY. Educ: Univ Ill, BS, 36, PhD(org chem), 40. Prof Exp: RES CHEMIST, EXP STA, E I DU PONT DE NEMOUS & CO, INC, 39- Mem: Am Chem Soc. Res: Heterogeneous catalysis in petroleum chemistry. Mailing Add: RD 1 Kennett Square PA 19348

MCCLELLAND, ALAN LINDSEY, b Galesburg, Ill, Sept 19, 25; m 47; c 4. INORGANIC CHEMISTRY, RESEARCH ADMINISTRATION. Educ: Northwestern Univ, BS, 45; Univ Ill, PhD(chem), 50. Prof Exp: Asst, Univ Ill, 47-49; Nat Res Coun fel, Univ Birmingham, 50-51; instr chem, Univ Conn, 51-54; res chemist, Cent Res Dept, E I du Pont de Nemours & Co, Inc, 54-60, col rels rep, Employee Rels Dept, 60-64; vpres eng, Cherry-Burrell Corp, Iowa, 64-67; staff asst personnel div, Employee Rels Dept, 67-69, PERSONNEL ADMINR, CENT RES DEPT, E I DU PONT DE NEMOURS & CO, INC, 69- Mem: AAAS; Am Chem Soc. Res: Technical administration. Mailing Add: Cent Res Dept E I du Pont de Nemours & Co Inc Wilmington DE 19898

MCCLELLAND, CHARLES PAUL, b Hartsdale, NY, Oct 19, 13; m 37; c 3. ORGANIC CHEMISTRY. Educ: Wesleyan Univ, BA, 34, MA, 36. Hon Degrees: DSc, MacMurray Col, Jacksonville, Ill, 66. Prof Exp: Asst, Wesleyan Univ, 34-36; res fel, Mellon Inst, 36-44; prod mgr, Carbide & Carbon Chems Corp, 44-55, asst mgr tech serv, Union Carbide Chem Co, 55-59, assoc dir tech serv lab, 59-66, prod dir chem, Belg, 66-71, MGR INT CHEM MKT, UNION CARBIDE CORP, 71- Mem: Am Chem Soc; Commercial Develop Asn. Res: Technical service research. Mailing Add: Int Chem Mkt Union Carbide Corp 270 Park Ave New York NY 10017

MCCLELLAND, CLYDE LLOYD, b Middletown, Ohio, July 21, 23; m 51; c 2. NUCLEAR PHYSICS. Educ: Ohio State Univ, BSc, 49; Mass Inst Technol, MS, 52, PhD(nuclear physics), 54. Prof Exp: Asst exp nuclear physics, Knolls Atomic Power Lab, AEC, 49-51; asst nuclear physics, Mass Inst Technol, 51-54; physicist, US Dept Defense, 54-57; sci adv, US Mission to Int Atomic Energy Agency, 57-60; physicist, Brookhaven Nat Lab, 60-64; phys sci consult, US Arms Control & Disarmament Agency, 62-66; sci attache, 66-73, SCI COUNR, EMBASSY SCI & TECHNOL AFFAIRS, US DEPT STATE, 73- Mem: AAAS; Am Phys Soc. Res: Nuclear spectroscopy; low-energy nuclear reactions; particle accelerators; nuclear reactors. Mailing Add: Am Embassy Bonn Germany

MCCLELLAND, GEORGE ANDERSON HUGH, b Bushey, Eng, May 12, 31; m 58, 75; c 2. MEDICAL ENTOMOLOGY, GENETICS. Educ: Univ Cambridge, BA, 55; Univ London, PhD(zool, entom), 62. Prof Exp: Sci officer, EAfrican Virus Res Inst, Entebbe, Uganda, 55-59; res assoc entom, London Sch Hyg & Trop Med, 59-62; res assoc, Univ Notre Dame, 62-63; from asst prof to assoc prof, 63-75, PROF ENTOM, UNIV CALIF, DAVIS, 75- Concurrent Pos: Proj leader, EAfrican Aedes Res Univ, WHO, Tanzania, 69-70. Mem: Entom Soc Am; Am Soc Naturalists; Am Soc Trop Med & Hyg; Royal Entom Soc London; Zool Soc London. Res: Biology and ecology of mosquitoes, particularly genetics; yellow fever mosquito Aëdes aegypti. Mailing Add: Dept of Entom Univ of Calif Davis CA 95616

MCCLELLAND, JOHN EDWARD, b Herbert, Sask, Oct 4, 17; nat US; m 41; c 3. SOILS. Educ: Univ Sask, BSA, 39, MSc, 46; Iowa State Col, PhD(soils), 49. Prof Exp: Soil surveyor cent exp farm div, Can Dept Agr, 39-41; soil scientist, 49-56, sr soil correlator, Soil Conserv Serv, 56-60, asst prin soil correlator, Western States, 60-68, prin soil correlator, Midwestern States, 68-72, DIR SOIL SURV OPERS DIV, SOIL CONSERV SERV, USDA, 72- Concurrent Pos: Asst prof, Iowa State Col, 49-51; assoc prof, NDak Agr Col, 51-56. Mem: Fel AAAS; Am Soc Agron; Soil Sci Soc Am; Soil Conserv Soc Am; Int Soc Soil Sci. Res: Rates of weathering of soil forming minerals; mode of deposition; properties and classification of soils. Mailing Add: Soil Conserv Serv USDA Room 5204 South Bldg Washington DC 20250

MCCLELLAND, LAURELLA, b Moose Jaw, Sask, Dec 15, 12. VIROLOGY. Educ: Univ Sask, BSc, 32; Univ Toronto, MD, 38. Prof Exp: Fel, Banting Inst, 39-40; res assoc, Connaught Med Res Lab, Univ Toronto, 40-49; res assoc, Merck Inst Therapeut Res, Merck & Co, Inc, 49-60, res assoc med affairs, 60-73, DIR DRUG EXPERIENCE, MED AFFAIRS, MERCK SHARP & DOHME MED RES LABS, 73- Res: Virus and rickettsial diseases. Mailing Add: 320 Edgewood Dr Ambler PA 19002

MCCLELLAND, NINA IRENE, b Columbus, Ohio, Aug 21, 29. ENVIRONMENTAL CHEMISTRY. Educ: Univ Toledo, BS, 51, MS, 63; Univ Mich, MPH, 64, PhD(environ chem), 68. Prof Exp: Chemist-bacteriologist, City of Toledo, 51-56, from chemist to chief chemist, 56-63; prof dir water, 68-74, VPRES TECH SERV, NAT SANIT FOUND, 74- Concurrent Pos: Consult, Ann Arbor Sci Publ, 70-; resident lectr, Univ Mich, 70- Mem: Am Chem Soc; Am Pub Health Asn; Am Water Works Asn; Nat Environ Health Asn; Water Pollution Control Fedn. Res: Development of electrochemical instrumentation for potable water quality characterization, continuous monitoring in distribution systems and in-plant treatment process control; chemical leaching from plastics pipe and piping systems for potable water applications. Mailing Add: Nat Sanit Found PO Box 1468 Ann Arbor MI 48106

MCCLELLAND, ROBERT NELSON, b Gilmer, Tex, Nov 20, 29; m 58; c 3. SURGERY. Educ: Univ Tex, BA, 52; Univ Tex Med Br Galveston, MD, 54. Prof Exp: Intern, Med Ctr, Univ Kans, 54-55; resident gen surg, Parkland Mem Hosp, Dallas, 57-59; gen practitioner, Neblett Hosp, Canyon, Tex, 59-60; resident gen surg, Parkland Mem Hosp, Dallas, 60-62; from instr to assoc prof, 62-71, PROF GEN SURG, UNIV TEX HEALTH SCI CTR DALLAS, 71- Concurrent Pos: Nat Inst Gen Med Sci res grant, Univ Tex Health Sci Ctr Dallas, 65-67, NIH res grant, 67-; surg consult, 4th Army, US Darnall Gen Hosp, 68-; ed, Audio-Jour Rev-Gen Surg, 71-; chmn ad hoc comt study liver injuries in Viet Nam, Vet Admin, 71- Mem: Soc Surg Alimentary Tract; Am Gastroenterol Asn; Am Surg Asn; fel Am Col Surg. Res: Gastroenterology, especially splanchnic blood flow and gastroduodenal stress ulceration. Mailing Add: 3601 Potomac Dallas TX 75205

MCCLELLAND, WILSON MELVILLE, JR, b San Francisco, Calif, Aug 22, 32; m 58; c 2. NUCLEAR PHYSICS. Educ: Univ Calif, Berkeley, 54; Cornell Univ, PhD(exp physics), 60. Prof Exp: PHYSICIST, LAWRENCE LIVERMORE LAB, UNIV CALIF, 60- Mem: AAAS; Am Phys Soc. Res: Physics of nuclear weapons and their effects. Mailing Add: Lawrence Livermore Lab PO Box 808 Livermore CA 94550

MCCLEMENS, DAVID JEFFERSON, analytical chemistry, see 12th edition

MCCLEMENT, JOHN HENRY, b Watertown, NY, May 6, 18. PULMONARY DISEASES. Educ: Univ Rochester, MD, 43. Prof Exp: Asst med, Columbia Univ, 47-48, assoc, 49-52; instr, Cornell Univ, 48-49; asst prof, Univ Utah, 52-55; assoc prof, Col Physicians & Surgeons, Columbia Univ, 55-68; PROF MED, NY UNIV, 68- Concurrent Pos: Chief tuberc serv, Vet Admin Hosp, Ft Douglas, Salt Lake City, Utah, 52-55; vis physician in chg chest serv, Bellevue Univ Hosp, 68-; adj assoc prof, Col Physicians & Surgeons, Columbia Univ, 68-72, lectr med, 72-; consult, Vet Admin. Mem: Am Thoracic Soc; NY Acad Med. Res: Clinical pulmonary physiology; acute and chronic pulmonary disease. Mailing Add: Chest Serv Bellevue Hosp New York NY 10016

MCCLENACHAN, ELLSWORTH, b Chicago, Ill, Mar 13, 34; m 58; c 1. ORGANIC CHEMISTRY. Educ: Univ Chicago, BA, 54, SM, 55; Univ Mich, PhD(org chem), 59. Prof Exp: Res chemist, Am Cyanamid Co, 59-61; res dir plastics, Miller-Stephenson Chem Co, 61-64; vpres, 64-67, PRES, R H CARLSON CO, 67- Mem: Am Chem Soc. Res: Organic reaction mechanisms; thermosetting resins to use in electronic and aerospace applications; epoxy resins, polyesters and silicones. Mailing Add: 55 North St Greenwich CT 06830

MCCLENAHAN, JAMES BRICE, b Des Moines, Iowa, Feb 13, 31; m 55; c 3. MEDICINE. Educ: DePauw Univ, BA, 53; Washington Univ, MD, 57. Prof Exp: USPHS fel cardiopulmonary physiol, 61-63; mem staff, 63-67, DIR COWELL HEALTH CTR, STANFORD UNIV, 67-, ASST PROF MED, SCH MED, 68- Concurrent Pos: Sr res assoc, Palo Alto Med Res Found, 63- Mem: Am Heart Asn; fel Am Col Physicians; Am Fedn Clin Res. Res: Internal medicine; cardiology. Mailing Add: Cowell Health Ctr Stanford Univ Stanford CA 94305

MCCLENAHAN, WILLIAM ST CLAIR, b Brainerd, Minn, Sept 24, 12; m 41; c 4. ORGANIC CHEMISTRY. Educ: Carleton Col, BA, 33; Mass Inst Technol, PhD(org chem), 38. Prof Exp: Chemist, Northwest Paper Co, Minn, 33-35; res assoc div chem, NIH, 38-40; indust fel, Mellon Inst, 40-43, sr fel, 43-55; coordr chem & phys res div, Standard Oil Co (Ohio), 55-60, chief spec projs mkt res & prod develop div, Chem Dept, 60-63, sr mkt res analyst, Corp Planning & Develop Dept, 63; assoc, Skeist Labs, 64-65; chief chem resources group, 65-69, DIR DIV INFO SERV, INST PAPER CHEM, 69- Concurrent Pos: Corn Industs Res Found fel, NIH, 38-40; lectr, Univ Pittsburgh, 47-48. Mem: Tech Asn Pulp & Paper Indust; Am Chem Soc; Am Soc Test & Mat; Chem Mkt Res Asn. Res: Oxidation of glycosides with lead tetraacetate; chemistry of starch and dextrins; documentation; patents; market research; pulp and paper; information storage and retrieval. Mailing Add: Div of Info Serv Inst of Paper Chem Box 1039 Appleton WI 54911

MCCLENAHEN, JAMES RICHARD, b Lewistown, Pa, July 20, 41; m 63; c 1. ENVIRONMENTAL BIOLOGY, FOREST ECOLOGY. Educ: Pa State Univ, BS, 63, MS, 64, PhD(forest ecol), 74. Prof Exp: Fel air pollution, 72-74, ASST PROF FORESTRY, OHIO AGR RES & DEVELOP CTR, 74- Mem: Soc Am Foresters. Res: Interaction and effects of contaminants and environmental factors on plant ecosystems. Mailing Add: Lab for Environ Studies Ohio Agr Res & Develop Ctr Wooster OH 44691

MCCLENDON, JAMES FRED, b Alexandria, La, Jan 1, 38; m 59; c 2. MATHEMATICS. Educ: Tulane Univ, BA, 58; Univ Calif, Berkeley, MS, 63, PhD(math), 66. Prof Exp: Instr math, Yale Univ, 66-68; asst prof, 68-72, ASSOC PROF MATH, UNIV KANS, 71- Mem: Am Math Soc. Res: Algebriac topology. Mailing Add: Dept of Math Univ of Kans Lawrence KS 66044

MCCLENDON, JOHN HADDAWAY, b Minneapolis, Minn, Jan 17, 21; m 47; c 3. PLANT PHYSIOLOGY. Educ: Univ Minn, BA, 42; Univ Pa, PhD(bot), 51. Prof Exp: Res assoc, Hopkins Marine Sta, Stanford Univ, 51-52; res assoc bot, Univ Minn, 52-53; asst res prof agr biochem & food tech, Univ Del, 53-64, actg chmn dept, 59-64; ASSOC PROF BOT, UNIV NEBR, LINCOLN, 65- Concurrent Pos: Civilian with sci & tech div, Supreme Comdr Allied Powers, Tokyo, 46-47. Mem: AAAS; Bot Soc Am; Am Soc Plant Physiol. Res: Intracellular physiology and biochemistry of plant tissues; plant cell wall enzymology; plant metabolism. Mailing Add: Sch of Life Sci Univ of Nebr Lincoln NE 68588

MCCLENNY, WILLIAM ARTHUR, plasma physics, see 12th edition

MCCLENON, JOHN R, b Grinnell, Iowa, May 1, 37; m 59; c 3. ORGANIC CHEMISTRY. Educ: Grinnell Col, BA, 59; Univ Calif, Los Angeles, PhD(org chem), 64. Prof Exp: Asst prof chem, Milton Col, 63-65; asst prof, 65-71, ASSOC PROF CHEM, SWEET BRIAR COL, 71- Mem: Am Chem Soc. Res: Chemistry of allenes; use of differential thermal analysis for analysis of organic compounds; construction of inexpensive instruments for teaching and research. Mailing Add: Dept of Chem Sweet Briar Col Sweet Briar VA 24595

MCCLINTIC, JOSEPH ROBERT, b Fayette, Mo, July 13, 28. PHYSIOLOGY. Educ: San Diego State Col, BA, 49; Univ Calif, PhD(physiol), 55. Prof Exp: Asst instr biol, San Diego State Col, 49-50; res physiologist, Univ Calif, 53-54; from instr to assoc prof biol, 54-70, PROF BIOL, CALIF STATE UNIV, FRESNO, 70- Res: Sodium metabolism; endocrinology. Mailing Add: Dept of Biol Calif State Univ Fresno CA 93726

MCCLINTOCK, BARBARA, b Hartford, Conn, June 16, 02. GENETICS, CYTOLOGY. Educ: Cornell Univ, BS, 23, MA, 25, PhD(bot), 27. Hon Degrees: ScD, Univ Rochester, 47, Western Col, 49, Smith Col, 58, Univ Mo, 68. Prof Exp: Asst bot, Cornell Univ, 24-27, instr, 27-31; Nat Res Coun fel, Calif Inst Technol, 31-33; Guggenheim Mem Found fel, Bot Inst, Univ Freiburg, 33-34; asst plant breeding, Cornell Univ, 34-36; asst prof bot, Univ Mo, 36-41; staff mem, 41-67, DISTINGUISHED SERV MEM, GENETICS RES UNIT, CARNEGIE INST, 67-, RESIDENT INVESTR, DEPT GENETICS, 73- Concurrent Pos: Andrew D White prof-at-large, Cornell Univ, 65. Honors & Awards: Kimber Genetics Award, 67; Nat Medal Sci, 70. Mem: Nat Acad Sci; Am Soc Naturalists; Am Philos Soc; Bot Soc Am; Genetics Soc Am (vpres, 39, pres, 45). Res: Cytogenetics of maize. Mailing Add: Genetics Res Unit Carnegie Inst Cold Spring Harbor NY 11724

MCCLINTOCK, DAVID K, b Springfield, Ohio, May 1, 38; m 61; c 2. BIOCHEMISTRY. Educ: State Univ NY Buffalo, BA, 64, PhD(biochem), 69. Prof Exp: Biochemist fibrinolysis, 70-74, group leader atherosclerosis, 74, head dept cardiovasc-renal pharmacol, 74-75, SECT DIR MED PROD, RES & DEVELOP, LEDERLE LABS DIV, AM CYANAMID CO, 75- Mem: AAAS. Res: Research and development of medical products and devices; clinical diagnostic materials, sutures and hospital products. Mailing Add: Lederle Labs Pearl River NY 10954

MCCLINTOCK, ELIZABETH, b Los Angeles, Calif, July 7, 12. SYSTEMATIC BOTANY. Educ: Univ Calif, Los Angeles, BA, 37, MA, 39; Univ Mich, PhD, 56. Prof Exp: Herbarium botanist, Univ Calif, Los Angeles, 41-47; assoc cur, 48-69, CUR BOT, CALIF ACAD SCI, 69- Mem: AAAS; Bot Soc Am; Am Soc Plant Taxon; Soc Study Evolution. Res: Monographic studies in Labiatae and Hydrangeaceae; taxonomy of woody ornamentals. Mailing Add: Bot Dept Calif Acad of Sci San Francisco CA 94118

MCCLINTOCK, MICHAEL, b Ft Pierce, Fla, Jan 3, 27; m 59; c 3. QUANTUM ELECTRONICS. Educ: Univ Ariz, BS, 49; Univ Colo, MS, 56, PhD, 67. Prof Exp:

Mech engr, Reynolds Metals Co, 49-50; mech engr, Cryogenic Eng Lab, Nat Bur Stand, 53-61, proj leader, 58-61, physicist, 62-70; sr scientist & dir global air pollution res prog, Space Sci Ctr, Univ Wis-Madison, 70-73; VIS PROF PHYSICS, CLARK UNIV, 73- Concurrent Pos: Consult, Denver Res Inst, 58-60 & Convair Astronaut Div, Gen Dynamics Corp, 59. Honors & Awards: Outstanding Serv Award, US Dept Com, 57. Mem: AAAS; Am Phys Soc; Am Geophys Union. Res: Lasers; light scattering; cryogenics; mechanical properties of structural solids at low temperatures. Mailing Add: Dept of Physics Clark Univ Worcester MA 01610

MCCLINTOCK, WILLIAM JOHN, pharmacy, see 12th edition

MCCLOSKEY, ALLEN LYLE, b Granville, NDak, Aug 25, 22; m 47; c 3. INDUSTRIAL CHEMISTRY. Educ: Whittier Col, AB, 46; Univ Wis, PhD(org chem), 51. Prof Exp: Instr chem, Univ Calif, Los Angeles, 51-52; instr, Univ Pa, 52-54; res chemist, US Borax & Chem Corp, 55-57, group leader, US Borax Res Corp, 57-59, from assoc dir chem res to dir chem res, 59-69, VPRES & DIR RES, US BORAX RES CORP, 69- Mem: Am Chem Soc. Res: Steroid synthesis; glyicodnitriles; acyloin reaction in ammonia; mechanism of Stobbe reaction; addition of carboxylic acids to olefins; boron and organoboron chemistry. Mailing Add: US Borax Res Corp 412 Crescent Way Anaheim CA 92801

MCCLOSKEY, CHESTER MARTIN, b Fresno, Calif, July 21, 18; m 44; c 2. ORGANIC CHEMISTRY. Educ: Whittier Col, BA, 40; Univ Iowa, MS, 42, PhD(org chem), 44. Prof Exp: With Nat Defense Res Comt Proj, Iowa, 44-45 & comt med res proj, Calif Inst Technol, 45-46; chief chemist, Alexander H Kerr & Co, Inc, 46-48; phys scientist, Off Naval Res, 48-54, chief scientist, Calif, 55-57; exec dir & sr res fel, Off Indust Assocs, Calif Inst Technol, 57-62; PRES & TECH DIR, NORAC CO, INC, 62- Honors & Awards: Bartow Award, Am Chem Soc. Mem: Am Chem Soc; Soc Plastics Indust; Nat Fire Protection Asn. Res: Organic peroxides; vinyl polymerization; photopolymerization; propellants; carbohydrates. Mailing Add: Norac Co Inc 405 S Motor Ave Azusa CA 91702

MCCLOSKEY, JAMES, b Gastonia, NC, Dec 12, 21; m 71; c 4. PHYSICS. Educ: Catawba Col, 50, BA, 50; Univ NC, MA, 58, PhD(sci ed, physics), 63. Prof Exp: Naval architect, US Naval Shipyard, Pa, 40-42 & 46-47; instr high schs, NC, 50-59; chmn physics & math dept, Fla Southern Col, 62-70; PROF PHYS SCI, EMBRY-RIDDLE AERONAUT UNIV, 70- Mem: Math Asn Am; Am Phys Soc. Res: Biostatistical study of population of fruit fly; malaise of space-man and its possible relation to serotonin. Mailing Add: Div of Phys Sci & Math Embry-Riddle Aeronaut Univ Daytona Beach FL 32015

MCCLOSKEY, JAMES AUGUSTUS, JR, b San Antonio, Tex, June 25, 36; m 60; c 4. CHEMISTRY. Educ: Trinity Univ, Tex, BS, 57; Mass Inst Technol, PhD(chem), 63. Prof Exp: Asst prof chem, Baylor Col Med, 64-67, from assoc prof to prof, 67-74; PROF CHEM, UNIV UTAH, 74- Concurrent Pos: NIH fel, Nat Ctr Sci Res, Ministry Educ, France, 63-64; sabbatical, Nat Cancer Ctr Res Inst, Tokyo, 71 & vis prof, Univ Utah, 72. Mem: Am Chem Soc; Am Soc Biol Chem; Am Soc Mass Spectrometry. Res: Mass spectrometry and its applications to structural problems in organic chemistry and biochemistry. Mailing Add: Dept Biopharmaceut Sci Univ of Utah Salt Lake City UT 84112

MCCLOSKEY, JOHN W, b Dayton, Ohio, Mar 2, 38; m 60; c 3. MATHEMATICAL STATISTICS. Educ: Univ Dayton, BS, 60; Mich State Univ, MS, 62, PhD(statist), 65. Prof Exp: Asst prof math, 65-69, ASSOC PROF MATH, UNIV DAYTON, 69- Mem: Am Statist Asn. Res: Computer simulation; error analysis; mathematical modeling; regression techniques. Mailing Add: Dept of Math Univ of Dayton Dayton OH 45469

MCCLOSKEY, JOSEPH FRANCIS, b Scranton, Pa, Aug 26, 17; m 44; c 2. MEDICINE. Educ: Univ Scranton, BS, 39; Jefferson Med Col, MD, 43; Am Bd Path, dipl. Prof Exp: Patterson fel path, 46-47, instr, 47-49, assoc, 49-52, from asst prof to assoc prof, 52-66, PROF PATH, JEFFERSON MED COL, 66- Mem: Fel Am Soc Clin Pathologists; Am Asn Path & Bact; fel Col Am Pathologists; Am Asn Blood Bank; Am Cytol Soc. Res: Study of carbon tetrachloride and chlorophyll on liver of rats; study of substitute material for insulin; liver disease; chemical poisoning. Mailing Add: Jefferson Med Col 1025 Walnut St Philadelphia PA 19107

MCCLOSKEY, KENNETH EMORY, b St Louis, Mo, July 9, 11; m 39; c 2. PHYSICAL CHEMISTRY, FOOD SCIENCE. Educ: Oberlin Col, AB, 33; Ohio State Univ, MSc, 34, PhD(phys chem), 37. Prof Exp: Unit engr, Servel, Inc, Ind, 38-39; from res chemist to chemist in charge res, Peter Cailler Kohler Swiss Chocolates Co, Inc, NY, 39-50; res chemist northern regional res lab, USDA, 50-51; res chemist, Diamond Alkali Co, Inc, 51-52; chemist, Guittard Chocolate Co, Calif, 52-55; res chemist, Bowey's Inc, 55-57; res chemist, Wilbur Chocolate Co, Inc, 57-70, tech dir, 70-73; RETIRED. Mem: Am Chem Soc. Res: Chocolate and cocoa, products and manufacturing machinery. Mailing Add: 318 Canyon Dr N Lehigh Acres FL 33936

MCCLOSKEY, LAWRENCE RICHARD, b Philadelphia, Pa, May 5, 39; m. ZOOLOGY. Educ: Atlantic Union Col, AB, 61; Duke Univ, MA, 65, PhD(zool), 68. Prof Exp: Res assoc, Friday Harbor Labs, Univ Wash, 67-68; res fel, Systs-Ecol Prog, Marine Biol Lab, Woods Hole, 69-71; ASST PROF BIOL, WALLA WALLA COL, 70- Mem: AAAS; Am Soc Limnol & Oceanog; Ecol Soc Am; Soc Syst Zool; Am Inst Biol Sci. Res: Marine and community ecology; coral biology; effects of various pollutants on marine animals; coral respiration. Mailing Add: Dept of Biol Walla Walla Col College Place WA 99324

MCCLOSKEY, RICHARD VENSEL, b Wilkinsburg, Pa, Dec 27, 33; m 56; c 3. INFECTIOUS DISEASES. Educ: Washington & Jefferson Col, AB, 55; Univ Rochester, MS & MD, 60; Am BD Internal Med, dipl, 67; Am Bd Infectious Dis, dipl, 72. Prof Exp: Intern med, Med Ctr, Duke Univ, 60-61; res assoc, Lab Trop Virol, Nat Inst Allergy & Infectious Dis, 61-63; jr & sr asst resident med, Med Ctr, Duke Univ, 63-65, assoc med & immunol, 65-66, asst prof med & assoc immunol, 66-67; from assoc prof to prof physiol & med, Med Sch, Univ Tex San Antonio, 67-72, head sect infectious dis, 67-72; PROF MED, JEFFERSON MED COL, THOMAS JEFFERSON UNIV, 72-; HEAD SECT INFECTIOUS DIS, DAROFF DIV, ALBERT EINSTEIN MED CTR, PHILADELPHIA, 73-, VCHMN DEPT MED, 75- Concurrent Pos: Am Col Physicians-Mead Johnson scholar, 64-65; head sect infectious dis & immunol, Durham Vet Admin Hosp, NC, 66-67. Mem: Am Soc Microbiol; Am Fedn Clin Res; Infectious Dis Soc Am. Res: Antibiotic therapy of infectious diseases; immunology of infectious diseases as one of the determinants of infection. Mailing Add: Sect Infectious Dis Daroff Div Albert Einstein Med Ctr Philadelphia PA 19147

MCCLOSKEY, TERESEMARIE, b Toledo, Ohio, July 1, 26. MATHEMATICAL LOGIC. Educ: Notre Dame Col, Ohio, BS, 49; John Carroll Univ, MS, 53; Case Western Reserve Univ, PhD(math), 72. Prof Exp: Teacher math & sci, St Mary High Sch, Warren, Ohio, 49-50, Notre Dame Acad, Cleveland, 50-53 & Elyria Dist Cath High Sch, 53-58, 60-63; instr math, Notre Dame Col, 63-65; teacher math & chmn dept, Elyria Dist Cath High Sch, 67-68; from instr to asst prof, 68-74, ASSOC PROF MATH & CHMN DEPT, NOTRE DAME COL, 74- Mem: Math Asn Am; Asn Comput Mach; Nat Coun Teachers Math. Res: Abstract families of languages, leading to more complete treatment of context sensitive languages; computer simulation techniques for small machines. Mailing Add: Notre Dame Col 4545 College Rd Cleveland OH 44121

MCCLOUD, DARELL EDISON, b Cass Co, Ind, Mar 7, 20; m 40; c 3. AGRONOMY, AGRICULTURE. Educ: Purdue Univ, BSA, 45, MS, 47, PhD(agron), 49. Prof Exp: Asst prof agron, Univ Fla, 48-54, assoc agronomist, 54-57; head humid pasture & range res sect, Agr Res Serv, USDA, 57-65; PROF AGRON & CHMN DEPT, INST FOOD & AGR SCI, UNIV FLA, 65- Concurrent Pos: Consult, Lab Climat, Johns Hopkins Univ, 53; consult bioclimat sect, US Weather Bur, 55. Mem: Crop Sci Soc Am (pres); fel Am Soc Agron (pres, 73-74). Res: Agricultural administration; agroclimatology; forage and pasture production and management research. Mailing Add: 304 Newell Hall Dept of Agron Univ of Fla Gainesville FL 32601

MCCLOUD, HAL EMERSON, JR, b Kansas City, Mo, Feb 15, 38; m 66. SOLID STATE PHYSICS, ELECTRICAL ENGINEERING. Educ: Univ Kans, BS, 61; Univ Mo, Rolla, MS, 64, PhD(eng physics), 67. Prof Exp: Asst prof, 66-71, ASSOC PROF PHYSICS, ARK STATE UNIV, 71- Mem: Am Phys Soc. Res: Electromagnetic theory. Mailing Add: Dept of Physics Ark State Univ State University AR 72467

MCCLUER, ROBERT HAMPTON, b San Angelo, Tex, Apr 13, 28; m 49; c 4. BIOCHEMISTRY. Educ: Rice Inst, BA, 49; Vanderbilt Univ, PhD(biochem), 55. Prof Exp: Res fel biochem, Univ Ill, 55-57; from instr to assoc prof physiol chem & psychiat, Ohio State Univ, 57-68, prof physiol chem, 68-69; asst dir biochem res, 70-75, DIR BIOCHEM RES, EUNICE KENNEDY SHRIVER CTR MENT RETARDATION, 75-; PROF BIOCHEM, BOSTON UNIV, 71- Mem: Am Chem Soc; NY Acad Sci; Am Soc Neurochem; Int Soc Neurochem; Soc Neurosci. Res: Chemistry and metabolism of gangliosides; brain RNA chemistry and metabolism; neurochemistry; high performance liquid chromatography of lipids. Mailing Add: Eunice Kennedy Shriver Ctr for Ment Retardation Waltham MA 02154

MCCLUGAGE, SAMUEL GARDNER, JR, b Peoria, Ill, Feb 28, 43. ANATOMY. Educ: Millikin Univ, BA, 66; Univ Cincinnati, PhD(anat), 70. Prof Exp: Asst instr zool, Millikin Univ, 66; res assoc anat, Col Med, Univ Cincinnati, 70-71; ASST PROF ANAT, LA STATE UNIV MED CTR, NEW ORLEANS, 71- Concurrent Pos: Consult, Procter & Gamble Co, 72- Mem: AAAS; Microcirc Soc; Soc Exp Hemat. Res: Vital microscopy; hematology. Mailing Add: Dept of Anat La State Univ Med Ctr New Orleans LA 70119

MCCLUNG, ANDREW COLIN, b Morgantown, WVa, Oct 15, 23; m 47; c 2. SOIL FERTILITY. Educ: Univ WVa, BS, 47; Cornell Univ, MS, 49, PhD(soils), 50. Prof Exp: Res assoc prof agron, NC State Col, 50-56; agronomist, IBEC Res Inst, Sao Paulo, Brazil, 56-60; soil scientist, Rockefeller Found, 60-64; assoc dir, Int Rice Res Inst, Philippines, 64-71; dep dir gen, Int Ctr Trop Agr, 72-73; DEP DIR AGR SCI, ROCKEFELLER FOUND, 73- Mem: Soil Sci Soc Am; Am Soc Agron. Res: Mineral nutrition of tree crops; methods of assessing the fertility status of soils; tropical agriculture; rural development. Mailing Add: Rockefeller Found 1133 Ave of the Americas New York NY 10036

MCCLUNG, FRED J, JR, electronics, optics, see 12th edition

MCCLUNG, LELAND SWINT, b Atlanta, Tex, Aug 4, 10; m 44. MICROBIOLOGY. Educ: Univ Tex, AB, 31, AM, 32; Univ Wis, PhD(bact), 34. Prof Exp: Res bacteriologist, Res Div, Am Can Co, 34-36; instr fruit prod & jr bacteriologist, Col Agr, Univ Calif, 36-37, instr res med, Hooper Found, Med Sch, 37-39; Guggenheim fel, Harvard Med Sch, 39-40; asst prof in chg dept, 40-44, assoc prof bact, 44-48, chmn dept, 47-66, asst dir div biol sci, 65-68, PROF BACT, IND UNIV, BLOOMINGTON, 48- Concurrent Pos: Secy, Pub Health & Nutrit Sect, Pac Sci Cong, 39; mem comt educ, Am Inst Biol Sci, 59-65, bd gov, 67-74. Mem: AAAS; Soc Indust Microbiol (vpres, 58); Soc Exp Biol & Med; Nat Asn Biol Teachers (pres, 65); fel Am Acad Microbiol. Res: Applied microbiology; history of microbiology; bacteriophage; Clostridium; science education. Mailing Add: Dept of Microbiol Jordan Hall 336 Ind Univ Bloomington IN 47401

MCCLUNG, MARVIN RICHARD, b Bamboo, WVa, Apr 3, 17; m 43; c 2. ANIMAL BREEDING. Educ: Univ WVa, BS, 41; Univ Md, MS, 42; Iowa State Univ, PhD(animal breeding, statist), 53. Prof Exp: County agent, Exten Serv, WVa, 42-45; agr counr, Monongahela Power Co, 45-47; asst prof animal indust & vet sci, Univ Ark, 47-52; geneticist & dir poultry div, Southern Farms Asn, 52-57; chmn, Dept Poultry Sci, Univ RI, 57-64; chmn, Dept Animal Indust & Vet Sci, 64-70, PROF ANIMAL SCI & ANIMAL SCIENTIST, WVA UNIV, 70- Mem: Fel AAAS; Poultry Sci Asn; Am Soc Animal Sci; Genetics Soc Am; Am Genetic Asn. Res: Population genetics; selection experiments in poultry and beef cattle and correlated responses. Mailing Add: Div of Animal Sci WVa Univ Col Agr & Forestry Morgantown WV 26506

MCCLUNG, NORVEL MALCOLM, b Bingham, WVa, June 9, 16; m 45; c 4. MICROBIOLOGY. Educ: Glenville State Col, AB, 36; Univ Mich, MS, 40, PhD(bot), 49. Prof Exp: Teacher high sch, WVa, 36-41; asst prof bot, Univ Kans, 48-57; assoc prof bact, Univ Ga, 57-66; PROF BIOL, UNIV S FLA, 66- Concurrent Pos: Fulbright res scholar, Japan, 62-63. Mem: AAAS; fel Am Acad Microbiol; Am Soc Microbiol; Mycol Soc Am; Bot Soc Am. Res: Biology of Actinomycetes, especially the genus Nocardia; ultrastructure of microorganisms; medical mycology. Mailing Add: Dept of Biol Univ of S Fla Tampa FL 33620

MCCLUNG, RONALD EDWIN DAWSON, b Victoria, BC, Oct 27, 41; m 64; c 2. CHEMISTRY, PHYSICS. Educ: Univ Alta, BSc, 63; Univ Calif, Los Angeles, PhD(chem), 67. Prof Exp: Nat Res Coun Can fel, Univ Leeds, 68-69; ASST PROF CHEM, UNIV ALTA, 69- Mem: Am Inst Physics. Res: Molecular motion in fluids; spectroscopy; magnetic resonance; transition metal chemistry. Mailing Add: Dept of Chem Univ of Alta Edmonton AB Can

MCCLURE, BENJAMIN THOMPSON, b Rochester, Minn, Oct 14, 25; m 51; c 3. ELECTRONIC PHYSICS. Educ: Univ Minn, BA, 45, MA, 47; Harvard Univ, PhD(physics), 51. Prof Exp: Physicist, Tracerlab Inc, 48-49; res asst, Harvard Univ, 51; mem tech staff, Bell Tel Labs, Inc, 51-59; staff physicist, Int Bus Mach Corp, 59-60; prin res scientist, 60-64, staff scientist, 64-65, sect head, 65-69, dept mgr, 69-71, SR STAFF SCIENTIST, CORP RES CTR, HONEYWELL INC, 71- Concurrent Pos: Instr, Augsburg Col, 62-63 & Macalester Col, 64-65. Mem: Am Phys Soc; Am Meteorol Soc. Res: Charge transport in dielectrics; conductivity-type atmospheric impurity detection. Mailing Add: 140 Meadowbrook Rd Hopkins MN 55343

MCCLURE, CHARLES FREDERICK, b Madison, Wis, July 9, 40. STATISTICAL MECHANICS, OPERATIONS RESEARCH. Educ: Johns Hopkins Univ, BA, 62;

Univ Md, PhD(physics), 72. Prof Exp: PHYSICIST OPERS RES, US NAVAL SURFACE WEAPONS CTR, 63- Mem: Am Phys Soc; AAAS. Res: Application of non-equilibrium kinetic theory methods to problems involving the flow of dense gases and fluids in the presence of boundaries. Mailing Add: US Naval Surface Weapons Ctr White Oak Silver Spring MD 20910

MCCLURE, CLAIR WYLIE, b Greenville, Pa, Oct 17, 27; m 48; c 3. MATHEMATICS. Educ: Thiel Col, BS, 50; Ohio State Univ, MA, 62, PhD(math educ), 71. Prof Exp: Teacher math, Mercer Area Sch Dist, Pa, 51-63; assoc prof, 63-72, chmn dept, 72-75, PROF MATH, SLIPPERY ROCK STATE COL, 72- Mem: Math Asn Am. Res: Mathematics laboratories for junior high school students; non-Euclidean geometries for high school students. Mailing Add: Dept of Math Slippery Rock State Col Slippery Rock PA 16057

MCCLURE, CLAUDE, b Dungannon, Va, Nov 21, 23; m 50; c 3. BIOCHEMISTRY, NEUROSURGERY. Educ: Wake Forest Univ, BS, 47, MD, 50, MS, 58; Univ NC, PhD(biochem), 60. Prof Exp: Instr neurol, Duke Univ, 53-54 & 55-56; asst, Univ Tenn, 56-57; asst prof biochem, Univ NC, 61-62; Med Corps, US Army, 62-, chief neurosurg, William Beaumont Army Med Ctr, 70-74, DIR BIOMED LAB, EDGEWOOD ARSENAL, 74- Concurrent Pos: Consult chem warfare to personal physician to the President, 62-64. Mem: AMA; Am Chem Soc; Am Asn Neurol Surgeons; NY Acad Sci. Res: Fractionation of deoxypentose nucleoproteins; physical chemistry of proteins. Mailing Add: Biomed Lab Edgewood Arsenal MD 21010

MCCLURE, COYE WILLARD, b Hollis, Okla, Nov 5, 14; m 41; c 2. OPHTHALMOLOGY. Educ: Univ Okla, BS, 38, MD, 40; Am Bd Ophthal, dipl, 50. Prof Exp: From intern to resident, Univ Okla Hosps, 40-44; from instr to asst prof, 46-59, ASSOC PROF OPHTHAL, MED CTR, UNIV OKLA, 59- Concurrent Pos: Mem, Am Bd Ophthal, 54- Mem: AMA. Mailing Add: 415 NW 11th St Oklahoma City OK 73103

MCCLURE, DAVID ALBERT, b Pittsburgh, Pa, Feb 22, 30; m 56; c 3. PHARMACOLOGY. Educ: Ore State Univ, BS, 58, MS, 59; Univ Ore, PhD(pharmacol), 62. Prof Exp: Pharmacologist, Riker Labs, Inc Div, Rexall Drug & Chem Corp, 62-64, sr pharmacologist & head drug eval & pharmacodynamic sect, 64-71; head pharmacol dept, 71-73, DIR CLIN RES DEPT, ALLERGAN PHARMACEUT, 73- Mem: Am Pharmaceut Asn. Res: Autonomic and central nervous systems; drug evaluation. Mailing Add: Allergan Pharmaceut 2525 DuPont Dr Irvine CA 92664

MCCLURE, DAVID WARREN, b Yakima, Wash, Sept 12, 36; m 61; c 2. PHYSICAL CHEMISTRY. Educ: Wash State Univ, BS, 58; Univ Wash, PhD(phys chem), 63. Prof Exp: Gatty fel, Cambridge Univ, 63-64, Oppenheimer res grant, 64-65; chemist, Shell Develop Co, 65-66; ASSOC PROF CHEM, PORTLAND STATE UNIV, 66- Concurrent Pos: Asst prof, Med Sch, Univ Ore. Mem: Am Phys Soc. Res: Thermodynamics of solutions; theory of phase transitions. Mailing Add: Dept of Chem Box 751 Portland State Univ Portland OR 97207

MCCLURE, DONALD ALLAN, b Omaha, Nebr, Apr 18, 42; c 1. NUCLEAR PHYSICS. Educ: Nebr Wesleyan Univ, BA, 64; Univ Mo-Rolla, MS, 66, PhD(physics), 70. Prof Exp: NSF res asst cryogenics, Univ Mo-Rolla, 64; fel nuclear physics, 69-72, ASST PROF PHYSICS, GA INST TECHNOL, 72- Mem: Am Phys Soc. Res: Investigation of nuclei far from stability using the density on live isotope separator at Oak Ridge plus thermal and resonance reaction capture reactive studies. Mailing Add: Sch of Physics Ga Inst of Technol Atlanta GA 30332

MCCLURE, DONALD ERNEST, b Portland, Ore, Oct 22, 44; m 71. APPLIED MATHEMATICS. Educ: Univ Calif, Berkeley, AB, 66; Brown Univ, PhD(appl math), 70. Prof Exp: From instr to asst prof, 69-75, ASSOC PROF APPL MATH, BROWN UNIV, 75- Mem: Am Math Soc; Inst Math Statist; Soc Indust & Appl Math: Math Asn Am. Res: Pattern analysis; approximation theory; mathematical statistics. Mailing Add: Div of Appl Math Brown Univ Providence RI 02912

MCCLURE, DONALD STUART, b Yonkers, NY, Aug 27, 20; m 49; c 3. PHYSICAL CHEMISTRY. Prof Exp: Univ Minn, BCh, 42; Univ Calif, Berkeley, PhD(phys chem), 48. Prof Exp: Res scientist, War Res Div, Columbia Univ, 42-44, SAM Labs & Carbide & Carbon Chem Corp, 44-46; asst chem, Univ Calif, Berkeley, 46-47, from instr to asst prof, 48-55; mem tech staff, RCA Labs, Inc, 55-62; prof chem, Univ Chicago, 62-67; PROF CHEM, PRINCETON UNIV, 67- Concurrent Pos: Guggenheim fel, 72. Mem: Am Chem Soc; fel Am Phys Soc. Res: Ultraviolet spectra and triplet states of organic molecules; spectra and electronic processes in molecular crystals; crystal field theory and spectra of inorganic ions and crystals; photochemistry. Mailing Add: Dept of Chem Princeton Univ Princeton NJ 08540

MCCLURE, FRANK JAMES, b Lafayette, Ind, Dec 30, 96. NUTRITION. Educ: Purdue Univ, BS, 19, MS, 24; Univ Ill, PhD(animal nutrit), 30. Prof Exp: Asst prof chem, NMex Col, 20-22; asst, Purdue Univ, 22-26; assoc nutrit, Univ Ill, 26-30; assoc animal nutrit, Pa State Univ, 30-35; assoc chemist, Agr Exp Sta, Univ Tex, 35-36; biochemist, NIH, 36-49, chief lab biochem, Nat Inst Dent Res, 49-66; RETIRED. Honors & Awards: Superior Serv Award, US Dept Health, Educ & Welfare, 55; H Trendley Dean Mem Award, Int Asn Dent Res, 75. Mem: AAAS; Am Inst Nutrit; hon mem Am Dent Asn; Int Asn Dent Res. Res: Dental caries; physiological effects of fluorine; cariostatic effects of phosphate. Mailing Add: 4545 Connecticut Ave Washington DC 20008

MCCLURE, FRANK TRELFORD, physical chemistry, deceased

MCCLURE, GEORGE RICHARD, b Chattanooga, Tenn, Dec 25, 29; m 58; c 3. ORGANIC CHEMISTRY, PHARMACEUTICAL CHEMISTRY. Educ: Univ Ga, BS, 51, MS, 56; Emory Univ, PhD(org chem), 60. Prof Exp: RES CHEMIST, E I DU PONT DE NEMOURS & CO, INC, 59- Res: Natural products; biologically active compounds; structure and property relation of polymers. Mailing Add: Fabric & Finish Dept Exp Sta E I du Pont de Nemours & Co Inc Wilmington DE 19898

MCCLURE, GEORGE W, JR, b Pittsburgh, Pa, Aug 19, 36; m 67. SOIL SCIENCE, PLANT PHYSIOLOGY. Educ: Col Wooster, BA, 58; Pa State Univ, MS, 61; NC State Univ, PhD(soils), 64. Prof Exp: Instr biol, Nassau Community Col, 64-65; asst prof, Adelphi Suffolk Col, 65-66; asst plant physiologist, Boyce Thompson Inst Plant Res, 66-70; lectr quant anal, Univ Puget Sound, 71-72; instr chem, S Seattle Community Col, 72-73; INSTR BIOL, BOT & ENVIRON, SHORELINE COMMUNITY COL, 73- Res: Mineral nutrient distribution in root zones; microbiological degradation of pesticides. Mailing Add: Dept of Biol Shoreline Community Col Seattle WA 98133

MCCLURE, GORDON WALLACE, b Mt Pleasant, Iowa, Mar 22, 23; m 44; c 3. PHYSICS. Educ: Univ Ill, BS, 44; Univ Chicago, PhD(physics), 50. Prof Exp: Mem staff, Radiation Lab, Mass Inst Technol, 44-45; physicist, Bartol Res Found, Franklin Inst, 49-55; PHYSICIST, SANDIA LABS, 55- Mem: Am Phys Soc. Res: Gaseous electronics; atomic collisions; physical electronics. Mailing Add: 2324 Cutler Ave NE Albuquerque NM 87106

MCCLURE, HAROLD MONROE, b Hayesville, NC, Oct 2, 37; m 58; c 3. VETERINARY PATHOLOGY. Educ: NC State Col, BS, 63; Univ Ga, DVM, 63. Prof Exp: NIH fel exp path, Univ Wis-Madison, 63-66; CLIN INSTR PATH, EMORY UNIV & VET PATHOLOGIST, YERKES PRIMATE CTR, 66- Concurrent Pos: Grants, USDA, Emory Univ, 69-72 & Nat Cancer Inst, 71-76 & Nat Inst Dent Res, 75-78. Mem: AAAS; Am Vet Med Asn; Am Asn Lab Animal Sci; Int Acad Path; Am Soc Vet Clin Pathologists. Res: Comparative and clinical pathology; electron microscopy; cytogenetics; cancer research. Mailing Add: Yerkes Primate Ctr Emory Univ Atlanta GA 30322

MCCLURE, HOWE ELLIOTT, b Chicago, Ill, Apr 29, 10; m 33; c 2. ECOLOGY, WILDLIFE MANAGEMENT. Educ: Univ Ill, BS, 33, MS, 36; Iowa State Col, PhD(econ zool), 41. Prof Exp: Asst entomologist, State Natural Hist Surv, Ill, 30-33; tree expert, Ill Tree Serv Co, 34-35 & 36-37; asst, Iowa State Col, 38-41; biologist & wildlife technician, State Game, Forestation & Parks Comn, Nebr, 41-44; biologist, USPHS & Hooper Found, Calif, 46-50; med ornithologist, 406th Med Gen Lab, Japan, 50-58; ecologist, US Army Med Res Unit, Inst Med Res, Malaya, 58-63; ornithologist, Migratory Animal Path Surv, Japan, 63-66, dir, Thailand, 66-75; RETIRED. Concurrent Pos: Mem, Brit Trust Ornith. Mem: Entom Soc Am; Wildlife Soc; Cooper Ornith Soc; Wilson Ornith Soc; Am Ornith Union. Res: Faunistic ecology; ornithology; medical ornithology. Mailing Add: Migratory Animal Path Surv SEATO Med Lab APO San Francisco CA 96346

MCCLURE, JAMES DOUGLAS, b Glen Cove, NY, May 24, 32; m 64; c 1. ORGANOMETALLIC CHEMISTRY. Educ: Polytech Inst Brooklyn, BS, 53; Univ Chicago, PhD(org chem), 57. Prof Exp: STAFF RES CHEMIST, SHELL DEVELOP CO, 57- Mem: Am Chem Soc. Res: Hydrogen peroxide chemistry; chemistry of epichlorohydrin; phosphorus ylid chemistry; catalysis with phosphines; homogeneous catalysis with transition metals; heterogeneous catalysis with supported transition metals. Mailing Add: Shell Develop Co Westhollow Res Ctr PO Box 1380 Houston TX 77001

MCCLURE, JAMES HERBERT, b Wooster, Ohio, Aug 9, 22; m 48; c 2. OBSTETRICS & GYNECOLOGY. Educ: Ohio State Univ, BA & MD, 50, MMSc, 54. Prof Exp: Instr gynec & obstet, Col Med, Ohio State Univ, 53-54; instr, Sch Med, Emory Univ, 54-56, Joseph B Whitehead res fel, 54-55; from asst prof to assoc prof, Univ Ill Col Med, 56-61; prof, Med Col, Univ Ala, 61-63; chmn dept, 63-73, PROF GYNEC & OBSTET, UNIV CALIF, IRVINE-CALIF COL MED, 63- Honors & Awards: Centennial Achievement Award, Ohio State Univ, 70. Mem: AMA; fel Am Col Obstet & Gynec; Soc Gynec Invest; Sigma Xi. Res: Physiology of pregnancy, labor, delivery and the newborn; medical education. Mailing Add: Dept of Gynec & Obstet Univ of Calif-Calif Col Med Irvine CA 92717

MCCLURE, JAMES NATHANIEL, JR, b Washington, Mo, Aug 20, 30; m 56; c 2. PSYCHIATRY. Educ: Washington Univ, AB, 51, MD, 55. Prof Exp: Intern med, Kings County Hosp, Brooklyn, NY, 55-56; asst resident psychiat, Renard Hosp, Sch Med, Washington Univ, 56-58, chief resident, 58-59; res instr med & psychiat, Univ Rochester, 62-63; instr psychiat, 63-65, ASSOC PROF PSYCHIAT & RES ASSOC, SOC SCI INST, SCH MED, WASHINGTON UNIV, 65-, DIR PSYCHIAT DIV, STUDENT HEALTH SERV, 64- Concurrent Pos: USPHS fel neuroendocrinol, Sch Med & Dent, Univ Rochester, 61-63; NIMH career teacher award, Washington Univ, 64-66; consult, Malcolm Bliss Ment Health Ctr, St Louis, 66- Mem: Am Psychiat Asn; AMA; Am Col Health Asn; Royal Col Psychiat. Res: Hypothalamic regulation of temperature and endocrine responses, premenstrual and postpartum psychiatric disorders; university student psychiatric problems; psychiatric aspects of obesity and suicide prevention. Mailing Add: Dept of Psychiat Washington Univ Sch of Med St Louis MO 63110

MCCLURE, JERRY WELDON, b Floydada, Tex, May 3, 33; m 54; c 2. PLANT CHEMISTRY. Educ: Tex Tech Col, BS, 54, MS, 61; Univ Tex, PhD(bot), 64. Prof Exp: From asst prof to assoc prof, 64-73, PROF BOT, MIAMI UNIV, 73- Concurrent Pos: NSF grants, 65 & 70; Fulbright hon res fel, WGer, 74-75; Humboldt sr scientist award, WGer Govt, 74-75; mem adv screening comt Life Sci, Coun Int Exchange Scholars, 75-78. Mem: AAAS; Bot Soc Am; Phytochem Soc NAm (secy, 71-74, pres elect, 75-76, pres, 76-77); Am Soc Plant Physiol. Res: Phytochemistry, especially biological role of polyphenolics; Lemnaceae; physiology and functions of flavonoids. Mailing Add: Dept of Bot Miami Univ Oxford OH 45056

MCCLURE, JOEL WILLIAM, JR, b Irvine, Ky, Aug 8, 27; m 53; c 2. THEORETICAL PHYSICS, SOLID STATE PHYSICS. Educ: Northwestern Univ, BS, 49, MS, 51; Univ Chicago, PhD, 54. Prof Exp: Asst prof physics, Univ Ore, 54-56; res physicist, Nat Carbon Co Div, Union Carbide Corp, 56-61; assoc prof, 61-65, PROF PHYSICS, UNIV ORE, 65- Mem: Am Phys Soc; Am Asn Physics Teachers. Res: Band theory; transport and magnetic properties. Mailing Add: Dept of Physics Univ of Ore Eugene OR 97403

MCCLURE, JOHN ARTHUR, b Belle Center, Ohio, Jan 19, 34; m 60; c 3. MATHEMATICAL ANALYSIS. Educ: Geneva Col, BS, 56; Univ Rochester, MS, 57; Va Polytechnic Inst, PhD(physics), 62. Prof Exp: Scientist reactor physics, Phillips Petrol Co, 62-69; assoc scientist comput sci, Aerojet Nuclear Corp, 69-73; sr scientist physics, 73-75, PRIN ANALYST PHYSICS, ENERGY INC, 75- Mem: Am Phys Soc; Soc Indust & Appl Math. Res: Application of numerical methods to problems in science and engineering. Mailing Add: Energy Inc Box 736 Idaho Falls ID 83401

MCCLURE, JOHN HIBBERT, b Truro, NS, Feb 7, 18; nat US; m 43; c 2. ANALYTICAL CHEMISTRY. Educ: Mt Allison Univ, BSc, 39, MSc, 41; Iowa State Col, PhD, 51. Prof Exp: Control chemist, Can Industs, Ltd, 41-44, supvr prod, 44-45, res chemist, 45-46; res chemist, Polychem Dept, 51-55, group supvr, Org Chem Dept, 56, chief supvr, 57-60, supvr anal res, 60-62, SUPVR CHEM & PHYS ANAL, ORG CHEM DEPT, E I DU PONT DE NEMOURS & CO, 62- Mem: Am Chem Soc. Res: General analytical research as applied to organic compounds; colorimetry; gas chromatography; infrared; polarography. Mailing Add: Org Chem Dept R&D Jackson Lab E I du Pont de Nemours & Co Wilmington DE 19898

MCCLURE, JOSEPH ANDREW, JR, b Sumter, SC, Jan 22, 34; m 57; c 2. THEORETICAL PHYSICS. Educ: Univ NC, BS, 56; Vanderbilt Univ, MA, 60, PhD(theoret physics), 63. Prof Exp: Res assoc & NSF fel theoret physics, Stanford Univ, 63-64; res assoc, Tufts Univ, 64-67; asst prof, 67-72, ASSOC PROF THEORET PHYSICS, GEORGETOWN UNIV, 72- Mem: Am Phys Soc. Res: High energy theoretical physics. Mailing Add: Dept of Physics Georgetown Univ Washington DC 20007

MCCLURE, JUDSON P, b Longmont, Colo, Feb 7, 34; m 61; c 3. ORGANIC CHEMISTRY. Educ: Bob Jones Univ, BS, 55; Univ Colo, PhD(chem), 61. Prof Exp: Res chemist, Esso Res & Eng Co, NJ, 61-63; from asst prof to assoc prof chem, Ohio

Northern Univ, 63-68; fel, Case Western Reserve Univ, 68-69; assoc prof, Ohio Northern Univ, 69-70; from asst prof to assoc prof, 70-75, PROF CHEM, MERCY COL NY, 75- Concurrent Pos: NSF-Acad Year Exten fel, 64-66. Mem: Am Chem Soc. Res: Reaction mechanisms; carbene or methylene; additives for lubricating oils; fluorinated organic compounds. Mailing Add: Dept of Natural Sci Mercy Col 555 Broadway Dobbs Ferry NY 10522

MCCLURE, MARK STEPHEN, b Whitinsville, Mass, Oct 27, 48; m 71. ENTOMOLOGY, ECOLOGY. Educ: Univ Mass, Boston, BA, 70; Univ Ill, Urbana, MS, 73, PhD(entom), 75. Prof Exp: Med technician, Leonard Morse Hosp, Mass, 70-71; USPHS trainee entom, Univ Ill, Urbana, 71-75, teaching asst insect ecol, 72-73; ASST ENTOMOLOGIST, CONN AGR EXP STA, 75- Mem: AAAS; Ecol Soc Am; Entom Soc Am; Entom Soc Can. Res: The biology, ecology and control of scale insects. Mailing Add: Conn Agr Exp Sta Box 1106 Dept Entom New Haven CT 06504

MCCLURE, MICHAEL ALLEN, b San Diego, Calif, Jan 19, 38; m 56; c 3. NEMATOLOGY. Educ: Univ Calif, Davis, BS, 59, PhD(nematol), 64. Prof Exp: Nematologist, Univ Calif, Davis, 64-65; asst prof nematol, Rutgers Univ, 65-68; assoc prof plant path & assoc plant pathologist, 68-74, PROF PLANT PATH, UNIV ARIZ, 74- Mem: Soc Nematol. Res: Nematode biology and physiology; culture and nutrition of plant parasitic nematodes; physiology of host-parasite relationships. Mailing Add: Dept of Plant Path Univ of Ariz Tucson AZ 85721

MCCLURE, MICHAEL EDWARD, b Goshen, Ind, Aug 15, 41; m 62. CELL BIOLOGY, BIOCHEMISTRY. Educ: Purdue Univ, BS, 63; Univ Tex Grad Sch Biomed Sci, Houston, MS, 66, PhD(cell biochem), 70. Prof Exp: R Welch Found fel biochem, Univ Tex M D Anderson Hosp & Tumor Inst Houston, 70-71, asst biochemist, 71-72, asst prof biochem, 72-73; ASST PROF CELL BIOL, BAYLOR COL MED, 73- Mem: AAAS; NY Acad Sci; Am Soc Cell Biologists; Tissue Cult Asn. Res: Biochemistry of the cell cycle, chromatin, genetic control mechanisms and development; regulatory mechanisms of cell growth and reproduction. Mailing Add: Dept of Cell Biol Baylor Col of Med Houston TX 77025

MCCLURE, ROBERT CHARLES, b Grinnell, Iowa, Feb 15, 32; m 54; c 2. VETERINARY ANATOMY. Educ: Iowa State Univ, DVM, 55; Cornell Univ, PhD(vet anat surg physiol), 64. Prof Exp: Instr vet anat, Iowa State Univ, 55-56; instr vet anat, Cornell Univ, 56-60, asst comp neurol, 60; chmn dept, 60-69, PROF VET ANAT, UNIV MO-COLUMBIA, 60- Concurrent Pos: Mem, Int Comt Vet Anat Nomenclature, 64-; Fulbright lectr, Inst Vet Med, Austria, 72-73; chmn, Nomina Embryologia Vet Comt, World Asn Vet Anatomists, 75- Mem: Am Asn Vet Anat (secy-treas, 65-66, pres elect, 66-67, pres, 67-68); Am Asn Anat; World Asn Vet Anat; Am Vet Med Asn; Am Asn Lab Animal Sci. Res: Comparative gross veterinary and developmental anatomy; laboratory animal anatomy; tooth development; comparative neuroanatomy; comparative anatomical nomenclature; veterinary medical and anatomical education and history. Mailing Add: Dept Vet Anat-Physiol Col Vet Med Univ Mo Columbia MO 65201

MCCLURE, ROBERT D, b North Bay, Ont, Sept 26, 39; m 73. ASTRONOMY. Educ: Queen's Univ Ont, BSc, 62; Univ Toronto, MA, 64, PhD(astron), 68. Prof Exp: Nat Res Coun Can fel astron, Kitt Peak Nat Observ, Ariz, 68-69; asst prof, 69-73, ASSOC PROF ASTRON, YALE UNIV, 73- Mem: Am Astron Soc; Int Astron Union. Res: Multicolor photoelectric photometry of stars, star clusters and galaxies; studies of stellar populations. Mailing Add: Yale Univ Observ Box 2023 Yale Sta New Haven CT 06520

MCCLURE, THEODORE DEAN, b New Virginia, Iowa, Oct 10, 36; m 59; c 3. NEUROANATOMY. Educ: Simpson Col, BA, 62; Univ Okla, MS, 65, PhD(med sci), 70. Prof Exp: Instr anat, Med Ctr, 66-70, ASST PROF ANAT SCI, HEALTH SCI CTR, UNIV OKLA, 70- Mem: Am Asn Anat; Am Soc Zool; Soc Exp Biol & Med. Res: Gross anatomy; histology; embryology; teratology. Mailing Add: Dept of Anat Sci Univ of Okla Health Sci Ctr Oklahoma City OK 73190

MCCLURE, THOMAS TRAQUAIR, plant pathology, deceased

MCCLURE, WILLIAM OWEN, b Yakima, Wash, Sept 29, 37; m 67; c 2. NEUROCHEMISTRY. Educ: Calif Inst Technol, BSc, 59; Univ Wash, PhD(biochem), 64. Prof Exp: Guest investr biochem, Rockefeller Univ, 64-65, res assoc, 65-66, asst prof, 66-68; asst prof biochem, Univ Ill, Urbana, 68-75; ASSOC PROF BIOL SCI, UNIV SOUTHERN CALIF, 75- Concurrent Pos: USPHS fel, 64-66; fel neurosci, Alfred P Sloan Found, 72-76; mem res coun, Nelson Res & Develop, 72- Mem: Am Chem Soc; Am Soc Neurochem; Soc Neurosci; Ny Acad Sci; Am Soc Biol Chemists. Res: Axoplasmic transport; mechanism of neurotransmitter release; mechanism of action of psychoactive drugs; mechanism of action of neurotoxins. Mailing Add: Dept of Biol Sci Univ of Southern Calif Los Angeles CA 90007

MCCLURG, JAMES EDWARD, b Bassett, Nebr, Mar 23, 45. BIOCHEMISTRY, REPRODUCTIVE BIOLOGY. Educ: Nebr Wesleyan Univ, BS, 67; Univ Nebr, PhD(biochem), 73. Prof Exp: INSTR BIOCHEM & RES INSTR OBSTET & GYNEC, UNIV NEBR MED CTR, OMAHA, 73- Mem: AAAS; Am Chem Soc. Res: Local immune response; normal and malignant cell growth. Mailing Add: Univ of Nebr Med Ctr Omaha NE 68105

MCCLURKIN, ARLAN WILBUR, b Clay Center, Kans, July 8, 17; m 54; c 4. VETERINARY MEDICINE, VIROLOGY. Educ: Kans State Univ, DVM, 43; Univ Wis, PhD(virol), 56. Prof Exp: Asst vet, Pvt Vet Hosp, 43; vet & prof animal husb, Allahabad Agr Inst, India, 47-51; res asst & instr, Univ Wis, 54-56; VET VIROLOGIST, NAT ANIMAL DIS LAB, AGR RES SERV, USDA, 56- Mem: Am Vet Med Asn; Conf Res Workers Animal Dis; NY Acad Sci; Sigma Xi. Res: Virus diseases of animals; characterization of the virus, epizootiology, pathogenesis and pathology produced in the animal. Mailing Add: 1612 Duff Ave Ames IA 50010

MCCLURKIN, DOUGLAS CHARLES, forest soils, see 12th edition

MCCLURKIN, IOLA TAYLOR, b Kinston, NC, May 22, 30; m 58; c 3. BIOLOcY. Educ: Duke Univ, BA, 52; E Carolina Col, MA, 57; Univ Miss, PhD(physiol), 65. Prof Exp: Asst zool, Duke Univ, 52-53; med technologist, Kafer Mem Hosp, New Bern, NC, 53-54; teacher high sch, NC, 54-57; asst zool, Duke Univ, 57-58; from instr to assoc prof biol, 58-74, PROF BIOL, UNIV MISS, 74- Concurrent Pos: Fac res grants, Univ Miss, 64-67. Mem: AAAS; Am Soc Microbiol; Histochem Soc; Electron Micros Soc Am, Sigma Xi. Res: Cytochemical studies involving transport enzyme locations in cellular membrane systems; histology of avian kidney; histology; histochemistry. Mailing Add: Dept of Biol Univ of Miss University MS 38677

MCCLURKIN, JOHN IRVING, JR, b Conway, Ark, Dec 20, 13; m 38; c 2. BIOLOGY. Educ: Univ Ark, BS, 34; Univ Colo, MS, 35; Stanford Univ, PhD, 53. Prof Exp: Asst biol, Univ Colo, 34-35; asst, Stanford Univ, 35-37; plant quarantine inspector, Bur Entom & Plant Quarantine, USDA, 43-44, entomologist, Div Control Invests, 44-51; prof biol, Lambuth Col, 53-56; assoc prof, Memphis State Univ, 56-59; head dept, 59-71, PROF BIOL, RANDOLPH-MACON COL, 59- Res: Control methods, especially vacuum, atmospheric and soil fumigation; insecticidal dips; plant tolerance; camellias. Mailing Add: Dept of Biol Randolph-Macon Col Ashland VA 23005

MCCLUSKEY, ELWOOD STURGES, b Loma Linda, Calif, June 1, 25; m 48; c 4. BIOLOGY. Educ: Walla Walla Col, BA, 50, MA, 52; Stanford Univ, PhD(biol), 59. Prof Exp: Asst bot, Walla Walla Col, 50-51; asst embryol, Stanford Univ, 51; asst prof biol, Atlantic Union Col, 52-54; asst instr physiol, 54-59, from instr to asst prof, 59-75, ASSOC PROF PHYSIOL & BIOL, LOMA LINDA UNIV, 75- Concurrent Pos: NSF fel, Harvard Univ, 59-60. Mem: AAAS; Am Soc Zoologists; Ecol Soc Am. Res: Comparative physiology, periodicity and ant biology; entomology. Mailing Add: 11899 Myrtlewood Colton CA 92324

MCCLUSKEY, GEORGE E, JR, b Hammonton, NJ, Aug 28, 38; m 64. ASTROPHYSICS. Educ: Univ Pa, AB, 60, MS, 63, PhD(astron), 65. Prof Exp: Asst prof, 65-68, ASSOC PROF ASTRON, LEHIGH UNIV, 68- Concurrent Pos: NASA grant, Lehigh Univ, 71-75; guest investr, Copernicus Satellite. Mem: AAAS; Am Astron Soc; Int Astron Union. Res: Eclipsing binary stars; double stars; x-ray and gamma-ray generation in close binaries; general relativistic effects in close binary systems containing neutron stars. Mailing Add: Dept of Math & Astron Lehigh Univ Bethlehem PA 18015

MCCLUSKEY, ROBERT TIMMONS, b New Haven, Conn, Jan 16, 23; m 57; c 2. PATHOLOGY. Educ: Yale Univ, AB, 44; NY Univ, MD, 47. Prof Exp: Intern, Kings County Hosp, Brooklyn, NY, 47-49, asst resident path, 49-50; asst, Sch Med, NY Univ, 50-52, from instr to prof & dir univ hosp labs, 52-68; prof path & chmn dept, Sch Med & Dent, State Univ NY Buffalo, 68-71; S Burt Wolbach Prof, 71-74, MALLINCKRODT PROF PATH, HARVARD MED SCH, 75-; CHIEF PATH, MASS GEN HOSP, 74- Concurrent Pos: Resident, Bellevue Hosp, NY, 50-52, asst pathologist, 52-53, 55-60, assoc pathologist, 60; consult, Manhattan Vet Admin Hosp, NY, 59-68; pathologist-in-chief, Children's Hosp Med Ctr, 71-74. Mem: Am Soc Exp Path; AMA; Asn Path & Bact; Col Am Path. Res: Pathogenesis of glomerulonephritis; pathogenesis of delayed sensitivity. Mailing Add: Mass Gen Hosp Boston MA 02114

MCCLYMONT, JOHN WILBUR, b Geneseo, NY, Feb 13, 25; m 52; c 1. BOTANY. Educ: Syracuse Univ, AB, 49, MS, 50; Univ Mich, PhD(bot), 55. Prof Exp: Asst prof biol, Milwaukee-Downer Col, 54-59; assoc prof, Morris Harvey Col, 59-61; prof, Findlay Col, 61-65; PROF BIOL, SOUTHERN CONN STATE COL, 65- Mem: Am Bryol & Lichenological Soc; Bot Soc Am. Res: Spores of the Bryophyta. Mailing Add: Dept of Biol Southern Conn State Col New Haven CT 06515

MCCOART, RICHARD F, JR, b Providence, RI, Apr 17, 28. MATHEMATICS. Educ: Yale Univ, BA, 48; Univ Ala, MA, 53; Univ NC, PhD(math), 61. Prof Exp: Aerophys engr, Convair Aircraft Corp, Tex, 53-54; instr math, Univ Ala, 54-56; asst prof, Georgetown Univ, 61-66; assoc prof, 66-71, PROF MATH, LOYOLA COL, MD, 71-CHMN DEPT, 70- Mem: Am Math Soc; Math Asn Am. Res: Algebra. Mailing Add: Dept of Math Loyola Col Baltimore MD 21210

MCCOLL, DANIEL CLYDE, b Arkansas City, Kans, Apr 10, 40. THEORETICAL PHYSICS, ELEMENTARY PARTICLE PHYSICS. Educ: Univ Kans, BS, 62; Univ Wis, MS, 63, PhD(physics), 69. Prof Exp: ASST PROF PHYSICS, UNIV WIS-STEVENS POINT, 69- Mem: Am Phys Soc. Res: Elementary particle physics, particularly N/D calculations with a field theory input Hamiltonian. Mailing Add: Dept of Physics Univ of Wis Stevens Point WI 54481

MCCOLL, JAMES RENFREW, b Albany, NY, Oct 30, 40; m 62; c 2. EXPERIMENTAL SOLID STATE PHYSICS. Educ: Univ Rochester, BS, 62; Univ Calif, Berkeley, PhD(physics), 67. Prof Exp: NATO fel, Clarendon Lab, Oxford, Eng, 67-68; asst prof, 68-73, ASSOC PROF PHYSICS, YALE UNIV, 73- Res: Liquid crystals; amorphous solids. Mailing Add: Dept of Physics Yale Univ New Haven CT 06520

MCCOLL, JOHN DUNCAN, b London, Ont, Nov 11, 25; m 54; c 3. PHARMACOLOGY. Educ: Univ Western Ont, BA, 46, MSc, 50; Univ Toronto, PhD(pharmacol), 53. Prof Exp: Asst res chemist, Parke, Davis & Co, Mich, 50-51; dir pharmacol res, Frank W Horner, Ltd, Can, 53-62, asst dir res, 62-69; vpres biol sci, Mead Johnson Res Ctr, 70-75; VPRES & DIR RES & DEVELOP, CHATTEM DRUG & CHEM CO, 75- Concurrent Pos: Mem toxicol panel, Defence Res Bd Can, 69-72. Mem: Am Soc Pharmacol & Exp Therapeut; Am Soc Clin Pharmacol & Therapeut; NY Acad Sci; Pharmacol Soc Can (secy, 61-63); Can Biochem Soc. Res: Toxicology; neuropharmacology; central nervous system agents; biochemistry. Mailing Add: Chattem Drug & Chem Co Chattanooga TN 37409

MCCOLL, JOHN GRAHAM, ecology, soil science, see 12th edition

MCCOLL, ROBERT WILLIAM, b San Diego, Calif, Sept 2, 38; m 59; c 2. GEOGRAPHY OF ASIA, POLITICAL GEOGRAPHY. Educ: Pomona Col, BA, 60; Univ Wash, PhD(geog), 64. Prof Exp: Asst prof geog, Univ Calif, Santa Barbara, 63-66; asst prof geog & EAsian studies, 66-68, ASSOC PROF GEOG & E ASIAN STUDIES, UNIV KANS, 68- Mem: Asn Am Geog; Asn Asian Studies. Res: Political geography of insurgency at national, local and urban scales; role of proxemics or geo-space in cultural behavior patterns, especially violence. Mailing Add: Dept of Geog Univ of Kans Lawrence KS 66045

MCCOLLESTER, DUNCAN L, b Tyringham, Mass, July 2, 25; m 57; c 1. CANCER. Educ: Harvard Univ, BA, 48; Tufts Univ, MD, 53; Univ Cambridge, PhD(biochem), 64. Prof Exp: From intern to resident med, Bellevue Hosp, New York, 53-56; clin assoc, Nat Heart Inst, 56-58; RES FEL, COLUMBIAPRESBY MED CTR, 65- Mem:- AAAS; Brit Biochem Soc; Biophys Soc; Am Chem Soc. Res: Separation and properties of cell surface membranes; cancer immunology. Mailing Add: VC12-220 Columbia-Presby Med Ctr 630 W 16th St New York NY 10032

MCCOLLISTER, ROBERT JOHN, b Iowa City, Iowa, July 27, 28; m 58; c 2. INTERNAL MEDICINE. Educ: State Univ Iowa, BA, 49, MD, 52. Prof Exp: Instr med, Sch Med, Univ Minn, Minneapolis, 59-61; instr, Duke Univ, 61-62; from instr to asst prof, 62-69, ASSOC PROF MED, SCH MED, UNIV MINN, MINNEAPOLIS, 69-, ASST DEAN SCH, 64- Mem: Am Fedn Clin Res; Am Soc Hemat. Res: Hematology; medical education. Mailing Add: Box 33 Mayo Mem Bldg Univ of Minn Med Ctr Minneapolis MN 55455

MCCOLLOCH, ROBERT JAMES, b Manhattan, Kans, July 26, 20; m 48; c 2. PHYSICAL CHEMISTRY, BIOCHEMISTRY. Educ: Kans State Col, BS, 41, PhD(phys biochem), 48. Prof Exp: Asst & asst chemist agr biochem, Purdue Univ, 42-44; from investr to res assoc, Plant Biochem, Cornell Univ, 44-48; chemist, Fruit & Veg Chem Lab, USDA, 48-52, unit supvr agr res serv, 52-56; assoc prof agr res chem, 56-58, PROF BIOCHEM & HEAD DIV, UNIV WYO, 58-, DEAN GRAD SCH, 72-

Concurrent Pos: Instr Kans State Col, 46-47. Mem: AAAS; Am Chem Soc; Am Soc Plant Physiol; Am Soc Animal Sci; NY Acad Sci. Res: Plant biochemistry; citrus products biochemistry and technology; pectic substances and pectic enzymes; instrumentation. Mailing Add: Grad Sch Univ Wyo PO Box 3108 Univ Sta Laramie WY 82070

MCCOLLOM, RICHARD EUGENE, physiology, genetics, see 12th edition

MCCOLLOUGH, EDWARD HERON, b Greenville, SC, Dec 16, 02; m 28; c 2. GEOLOGY. Educ: Univ Okla, BS, 24, MS, 25. Prof Exp: Geologist, 25-28, mgr, Calif Opers, 28-33, vpres, 33-48, vpres corp, 48-52, exec vpres, 52-55, pres, 55-67, chief exec off, chmn bd, 67-69, DIR, AMERADA PETROL CORP, 48- Concurrent Pos: Dir, Am Petrol Inst, 52-; mem, Nat Petrol Coun, 65-68. Mem: Am Asn Petrol Geol; fel Geol Soc Am. Res: Petroleum geology and seismology as applied to exploration for petroleum. Mailing Add: Amerada Petrol Corp Box 2040 Tulsa OK 74102

MCCOLLOUGH, FRED, JR, b Crawfordsville, Ind, July 19, 28; m 52; c 2. INORGANIC CHEMISTRY. Educ: Wabash Col, AB, 50; Univ Ill, MS, 52, PhD(chem), 55. Prof Exp: Sr res chemist, Victor Chem Works, 55-57, supvr inorg res, 57-64; assoc prof, 64-71, PROF CHEM, MacMURRAY COL, 71-, CHMN DEPT, 69- Mem: Am Chem Soc. Res; Coordination and phosphorus chemistry; computer application to undergraduate teaching. Mailing Add: 407 Sandusky St Jacksonville IL 62650

MCCOLLUM, ANTHONY WAYNE, b Macomb, Ill, Sept 3, 44; c 3. ORGANIC CHEMISTRY. Educ: Western Ill Univ, BS, 65; Univ Ark, Fayetteville, PhD(org chem), 69. Prof Exp: Chemist, 69-74, GROUP LEADER APPL CHEM RES & SR CHEMIST, TEX EASTMAN CO, DIV EASTMAN KODAK CO, 74- Mem: Am Chem Soc. Res: Organic synthesis; new monomer synthesis; natural products synthesis; aldol chemistry. Mailing Add: Box 7444 Longview TX 75601

MCCOLLUM, CLIFFORD GLENN, b Macon Co, Mo, May 12, 19; m 40; c 2. ZOOLOGY. Educ: Univ Mo, BS, 39, AM, 47, EdD(zool), 49. Prof Exp: Instr high schs, Mo, 38-43; asst prof, State Col Iowa, 49-55; prof State Univ NY Col Oneonta, 55-56; assoc prof sci, 56-59, head dept, 57-68, PROF BIOL, UNIV NORTHERN IOWA, 59-, DEAN COL NATURAL SCI, 68- Concurrent Pos: Consult, Coronet Instructional Films. Mem: AAAS; Nat Sci Teachers Asn; Nat Asn Res Sci Teaching; Nat Asn Biol Teachers. Res: Small mammal populations; science curricula in elementary and secondary schools. Mailing Add: Col of Nat Sci Univ of Northern Iowa Cedar Falls IA 50613

MCCOLLUM, DONALD CARRUTH, JR, b Baltimore, Md, Nov 6, 30; m 52; c 1. PHYSICS. Educ: Univ Calif, BA, 53, MA, 54, PhD(physics), 60. Prof Exp: From instr to assoc prof, 58-72, PROF PHYSICS, UNIV CALIF, RIVERSIDE, 72- Concurrent Pos: NSF fac fel, 65. Mem: Am Phys Soc; Am Asn Physics Teachers. Res: Low temperature physics. Mailing Add: Dept of Physics Univ of Calif Riverside CA 92502

MCCOLLUM, DONALD E, b Winston Salem, NC, Dec 7, 27; m 53; c 1. ORTHOPEDIC SURGERY. Educ: Wake Forest Col, BS, 49; Bowman Gray Sch Med, MD, 53. Prof Exp: Resident orthop surg, 58-62, from asst prof to assoc prof, 62-72, PROF ORTHOP SURG, MED CTR, DUKE UNIV, 72- Concurrent Pos: NIH trainee rheumatology, Med Ctr, Duke Univ, 57-58; consult, Vet Admin Hosp, 62-; orthop consult, US Air Force & Voc Rehab, 64- Mem: Am Acad Orthop Surg; Am Rheumatism Asn. Res: Rheumatology. Mailing Add: Dept of Orthop Surg Duke Univ Med Ctr Durham NC 27710

MCCOLLUM, GILBERT DEWEY, JR, b Bellingham, Wash, Aug 25, 29. PLANT CYTOGENETICS. Educ: Wash State Univ, BS, 51, MS, 53; Univ Calif, PhD(genetics), 58. Prof Exp: Geneticist, Idaho Agr Exp Sta, 58-64, RES GENETICIST, AGR RES SERV, USDA, 64- Mem: Bot Soc Am; Am Genetic Asn. Res: Botany; cytology; genetics; species relationships. Mailing Add: Plant Genetics & Germplasm Inst Veg Lab Agr Res Ctr USDA Beltsville MD 20705

MCCOLLUM, JOHN DAVID, b Evanston, Ill, Jan 8, 29; m 50; c 2. CHEMISTRY. Educ: Univ Ill, Urbana, BSc, 49; Harvard Univ, AM, 51, PhD(chem), 57. Prof Exp: Sr res scientist, Res & Develop Dept, Am Oil Co, 53-75, RES ASSOC, AMOCO OIL CO, 75- Mem: Am Chem Soc. Res: Homogeneous and heterogeneous catalysis of hydrocarbon reactions; organometallic chemistry; photo and radiation chemistry of hydrocarbons; high pressure hydrothermal reactions. Mailing Add: Amoco Oil Co Res & Develop Dept PO Box 400 Naperville IL 60540

MCCOLLUM, JOHN PASCHAL, b Emerson, Ark, Jan 11, 05; m 34; c 3. HORTICULTURE. Educ: Okla Agr & Mech Col, BS, 27; Cornell Univ, PhD(veg crops physiol), 32. Prof Exp: Asst veg crops, Cornell Univ, 27-29, in chg, Long Island Veg Res Farm, 29-30; from asst prof olericult to prof veg crops, Univ Ill, Urbana, 31-72; PROF HORT FOOD SCI, UNIV ARK, 72- Mem: Am Soc Hort Sci; Am Soc Plant Physiol. Res: Factors affecting the quality constituents in vegetables. Mailing Add: Dept of Hort Food Sci Univ of Ark Fayetteville AR 72701

MCCOLLUM, ROBERT EDMUND, b Reidsville, NC, Jan 3, 22; m 46; c 6. SOILS, PLANT BIOCHEMISTRY. Educ: NC State Col, BS, 52, MS, 53; Univ Ill, PhD, 57. Prof Exp: Asst, NC State Col, 52-53 & Univ Ill, 53-57; ASSOC PROF SOIL FERTIL, NC STATE UNIV, 57- Concurrent Pos: NC State Univ-US Agency Int Develop Proj, Peru, 63-66. Mem: Am Soc Agron. Res: Soil chemistry and fertility. Mailing Add: Dept of Soil Sci NC State Univ Raleigh NC 27607

MCCOLLUM, ROBERT WAYNE, b Waco, Tex, Jan 29, 25; m 54; c 2. EPIDEMIOLOGY. Educ: Baylor Univ, AB, 45; Johns Hopkins Univ, MD, 48; Univ London, DPH, 58. Prof Exp: Asst prev med, 51-52, from asst prof to assoc prof epidemiol & prev med, 54-65, PROF EPIDEMIOL, SCH MED, YALE UNIV, 65-, CHMN DEPT EPIDEMIOL & PUB HEALTH, 69- Concurrent Pos: Assoc mem comn viral infections, Armed Forces Epidemiol Bd, 59-61, mem, 61-72; consult, Surgeon Gen, 61- & WHO, 62-63, 70, 72 & 74. Mem: Soc Epidemiol Res; Int Epidemiol Asn; Am Epidemiol Soc; Infectious Dis Soc Am; Am Pub Health Asn. Res: Infectious diseases; viral hepatitis. Mailing Add: Dept of Epidemiol & Pub Health Yale Univ Sch of Med New Haven CT 06510

MCCOLLUM, WILLIAM HOWARD, b Traveller's Rest, Ky, Aug 17, 23. VIROLOGY. Educ: Univ Ky, BS, 47, MS, 49; Univ Wis, PhD(bact, biochem), 54. Prof Exp: Bacteriologist virol, 54-60, prof animal path, 60, PROF VET SCI, UNIV KY, 63- Mem: Am Soc Microbiol; Poultry Sci Asn; Tissue Cult Asn. Res: Virus diseases of animals, especially the horse; immunology. Mailing Add: Dept of Vet Sci Univ of KY Lexington KY 40506

MCCOLM, DOUGLAS WOODRUFF, b Kisaran, Sumatra, Sept 26, 33; US citizen; m 55; c 3. PHYSICS. Educ: Oberlin Col, BA, 55; Yale Univ, PhD(physics), 61. Prof Exp: Staff physicist, Lawrence Radiation Lab, Univ Calif, Berkeley, 61-66; ASSOC PROF PHYSICS, UNIV CALIF, DAVIS, 66- Mem: Am Phys Soc. Res: Atomic physics. Mailing Add: Dept of Physics Univ Calif 1000 Plum Lane Davis CA 95616

MCCOMAS, MURRAY RATCLIFFE, b Columbia, SC, Mar 18, 33; m 58; c 3. GROUNDWATER GEOLOGY. Educ: Univ Colo, BA, 56; Colo State Univ, MS, 66; Univ Ill, PhD(hydrogeol), 69. Prof Exp: Asst engr, Core Labs, Inc, Tex, 59-60, engr-in-chg, NMex, 60-64; asst geologist, Ill State Geol Surv, Urbana, 66-70; assoc prof, 70-75, PROF GEOL, KENT STATE UNIV, 75- Honors & Awards: E B Burwell, Jr, Award, Geol Soc Am, 74. Mem: Geol Soc Am; Am Inst Prof Geol; Asn Eng Geologists; Sigma Xi. Res: Environmental geology; engineering geological study of natural and man made slopes and hydrology of lakes. Mailing Add: Dept of Geol Kent State Univ Kent OH 44242

MCCOMAS, WILBUR HARRISON, JR, b York, Pa, Mar 1, 16; m 42; c 2. ANALYTICAL CHEMISTRY. Educ: Oberlin Col, AB, 37; Rutgers Univ, MS, 39, PhD(anal chem), 42. Prof Exp: Asst anal chem, Rutgers Univ, 37-42; res chemist, Calco Chem Div, 42-50, anal develop chemist, Org Chem Div, 50-56, group leader, Res & Develop Dept, 56-68, RES ASSOC, ORG CHEM DIV, RES & DEVELOP DEPT, AM CYANAMID CO, 68- Mem: Am Chem Soc. Res: Chromatography, especially ion-exchange and liquid chromatography. Mailing Add: R&D Dept Org Chem Div Am Cyanamid Co Bound Brook NJ 08805

MCCOMB, ANDREW LOGAN, b Vandergrift, Pa, Jan 1, 08; m 35; c 2. FORESTRY, WATERSHED MANAGEMENT. Educ: Pa State Col, BS, 32; Iowa State Col, MS, 33, PhD(plant physiol), 41. Prof Exp: Jr forester, Allegheny Forest Exp Sta, US Forest Serv, 33-34; actg nursery dir, Exp Sta, Iowa State Col & soil conserv serv, USDA, 34-35, from instr to asst prof forestry, 35-43; forester, Forest Econ Admin, Colombia, 43-45; assoc prof forestry, Iowa State Col, 46, prof & res prof, 46-59; prof watershed mgt & head dept, Univ Ariz, 59-65; proj mgr, Savanna Forestry Res Sta, Food & Agr Orgn, UN, Zaria, Nigeria, 65-70, forestry off, Indonesia, 70-72, proj mgr, Upper Solo Watershed Mgt Proj, Solo Indonesia, 72-74; RETIRED. Concurrent Pos: Fulbright res scholar, Austria, 53-54. Mem: Soc Am Foresters; Am Soc Plant Physiol; AAAS. Res: Physiology and ecology of forest tree growth; water relations and forest hydrology; watershed management. Mailing Add: 1529 N Alamo Place Tucson AZ 85712

MCCOMBIE, ALEN MILNE, b Toronto, Ont, May 26, 21. LIMNOLOGY, FISHERIES. Educ: Univ Toronto, MA, 51, PhD(limnol), 57. Prof Exp: Asst & demonstr zool, Univ Toronto, 47-57; res scientist, Southern Res Sta, Ministry Natural Resources, 57-73; MEM STAFF, ONT DEPT LANDS & FORESTS, 73- Mem: Am Soc Limnol & Oceanog; Am Fisheries Soc. Res: Ecology of planktonic, freshwater algae; productivity of freshwaters; ecology of freshwater fishes. Mailing Add: Ont Dept of Lands & Forests Maple ON Can

MCCOMBS, CLARENCE LESLIE, b Westerville, Ohio, Nov 5, 20; m 42; c 3. PLANT PHYSIOLOGY. Educ: Ohio State Univ, BSc, 47, MSc, 48, PhD(hort), 55. Prof Exp: Agent horticulturist, Wenatchee Field Sta, USDA, 49-50; from instr to prof hort, NC State Univ, 50-71; PROF HORT & HEAD DEPT, VA POLYTECH INST & STATE UNIV, 71- Mem: Am Soc Hort Sci. Res: Biochemical changes in vegetables during the post-harvest period; mechanism of disease resistance in vegetable varieties. Mailing Add: Dept of Hort Va Polytech Inst & State Univ Blacksburg VA 24061

MCCOMBS, FREDA SILER, b Asheville, NC, Feb 26, 33; m 62; c 3. SCIENCE EDUCATION. Educ: Salem Col, BS, 55; Univ NC, Chapel Hill, MEd, 59, EdD(sci educ), 63. Prof Exp: Teacher gen sci & biol city schs, Va, 55-56; teacher biol & chem pub schs, NC, 57-58; instr biol & phys sci, Longwood Col, 61-63; assoc prof biol, Cent Va Community Col, 68-70; asst prof, 70-71, ASSOC PROF PHYS SCI & SCI EDUC, LONGWOOD COL, 71- Mem: AAAS. Res: Curriculum materials; teaching aids and student resource materials for elementary science, particularly kindergarten and primary grades. Mailing Add: Dept of Natural Sci Longwood Col Farmville VA 23901

MCCOMBS, ROBERT PRATT, b Philadelphia, Pa, Dec 13, 09; m 41; c 2. INTERNAL MEDICINE. Educ: Yale Univ, BS, 31; Univ Pa, MD, 35; Am Bd Internal Med, dipl, 42. Prof Exp: Intern, Ind Univ Hosps, 35-36; resident med, Abington Mem Hosp, Pa, 36-37; New Eng Med Ctr fel, Sch Med, Tufts Univ, 37-38; asst demonstr, Jefferson Med Col, 39-46; dir postgrad teaching, 46-66, from asst prof to assoc prof med, 46-52, prof grad med, 52-66, pres postgrad med inst, 61-73, PROF MED, SCH MED, TUFTS UNIV, 66- Concurrent Pos: Sr physician, New Eng Ctr Hosp, Boston, 46-75, chief allergy serv, 59- Mem: Fel AMA; fel Am Col Physicians; fel Am Acad Allergy. Res: Allergy; clinical immunology; collagen diseases; pulmonary diseases. Mailing Add: Dept of Med Tufts Univ Sch of Med Boston MA 02111

MCCONAGHY, JOHN STEAD, JR, b Philadelphia, Pa, Feb 3, 42; m 65; c 2. PHYSICAL ORGANIC CHEMISTRY. Educ: Haverford Col, BA, 63; Yale Univ, MS, 64, PhD(org chem), 66. Prof Exp: SR RES CHEMIST, POLYMERS & PETROCHEM RES DEPT, MONSANTO CO, 66- Mem: Am Chem Soc; Sigma Xi. Res: Mechanism; catalysis; vapor phase reactions; organometallic chemistry. Mailing Add: Monsanto Co 800 N Lindbergh Blvd St Louis MO 63166

MCCONAHEY, WILLIAM MCCONNELL, JR, b Pittsburgh, Pa, May 7, 16; m 40; c 3. INTERNAL MEDICINE, ENDOCRINOLOGY. Educ: Washington & Jefferson Col, AB, 38; Harvard Med Sch, MD, 42; Univ Minn, MS, 48. Prof Exp: From instr to prof, Mayo Grad Sch Med, 50-72, chmn dept endocrinol, 67-74, PROF MED, MAYO MED SCH, 72- Concurrent Pos: Sr consult, Mayo Clin & Hosps, 49- Honors & Awards: Am Thyroid Asn Distinguished Serv Award, 73. Mem: Endocrine Soc; Am Diabetes Asn; Am Thyroid Asn (treas, 61-65, secy, 66-70, pres, 75-76); Am Fedn Clin Res; fel Am Col Physicians. Res: Diseases of the thyroid gland. Mailing Add: Div of Endocrinol Mayo Clin Rochester MN 55901

MCCONATHY, WALTER JAMES, b McAlester, Okla, Nov 3, 41; m 66; c 1. LIPID CHEMISTRY, PROTEIN CHEMISTRY. Educ: Univ Okla, Norman, BA, 64, BS, 66; Univ Okla, Oklahoma City, PhD(biochem), 71. Prof Exp: Fel, 71-74, staff scientist, 74-75, ASST MEM LIPOPROTEINS, OKLA MED RES FOUND, 75- Mem: Am Heart Asn; Sigma Xi; AAAS. Res: Investigating the structure, function and interrelations of human plasma lipoproteins in both normal and pathological conditions for potential application as diagnostic or prognostic tools in human disease states. Mailing Add: Lipoprotein Lab Okla Med Res Found 825 NE 13th St Oklahoma City OK 73104

MCCONAUGHY, DAVID LESTER, b Pittsburgh, Pa, Nov 7, 21; m 43; c 4. CHEMICAL PHYSICS. Educ: Muskingum Col, BS, 43. Prof Exp: Sr fel protective coatings, Mellon Inst, 46-60; res chemist, 60-65, prod res chemist, 65-67, DIR RES & DEVELOP, ELEC MAT CO, 67- Mem: Am Foundrymen's Soc; Am Soc Metals; Am Inst Mining, Metall & Petrol Eng. Res: Development of processes and materials in producing and fabrication of copper alloys into mill products, forgings, castings and special machinery; development of new materials, alloys and circuits for electric commutators; copper heating and melting equipment. Mailing Add: 182 Eastwood Dr North East PA 16428

MCCONE, ROBERT CLYDE, b Redfield, SDak, Sept 30, 15; m 43; c 2. CULTURAL ANTHROPOLOGY, ANTHROPOLOGICAL LINGUISTICS. Educ: Wessington Springs Col, BA, 46; SDak State Col, MS, 56; Mich State Univ, PhD(anthrop), 61. Prof Exp: Instr sociol, SDak State Col, 56-57; instr soc sci, Mich State Univ, 60-61; PROF ANTHROP, CALIF STATE UNIV, LONG BEACH, 61- Mem: AAAS; fel Am Anthrop Asn; fel Soc Appl Anthrop; fel Royal Anthrop Inst of Gt Brit & Ireland. Res: Acculturation phenomena among the Dakota Indian; world views of primitive cultures and of the anthropologists who study them. Mailing Add: Dept of Anthrop Calif State Univ Long Beach CA 90840

MCCONKEY, JOHN WILLIAM, b Portadown, North Ireland. PHYSICS. Educ: Queen's Univ Belfast, BSc, 58, PhD(physics), 62. Prof Exp: Lectr physics, Queen's Univ Belfast, 63-70; PROF PHYSICS, UNIV WINDSOR, 70- Mem: Fel Brit Inst Physics; Am Phys Soc; Can Asn Physicists. Res: Electron collisions; atmospheric processes; spectroscopy. Mailing Add: Dept of Physics Univ of Windsor Windsor ON Can

MCCONN, JAMES D, biochemistry, see 12th edition

MCCONN, RITA, b Liverpool, Eng, Jan 14, 40. BIOCHEMISTRY, PHYSIOLOGY. Educ: Univ Liverpool, BSc, 62, PhD(clin pharmacol), 66. Prof Exp: Asst prof, 66-73, ASSOC PROF SURG, CLIN RES CTRACUTE, ALBERT EINSTEIN COL MED, 73-, DIR CORE LAB, 66-, STAFF MEM BURN & SURG ICU, 66- Mem: NY Acad Sci; Am Col Angiology; Brit Med Res Soc; Am Inst Chemists; Critical Care Soc. Res: Biochemical and physiological changes in acutely ill patients; role of the scientist and laboratory in the multidisciplinary approach to acute illness and intensive patient care; blood storage and transfusion; role of humoral factors in development of pulmonary failure and/or shock in the acutely ill patient. Mailing Add: Dept of Surg Albert Einstein Col of Med Bronx NY 10461

MCCONNACHIE, PETER ROSS, b Montreal, Que, Mar 31, 40; m 64; c 2. IMMUNOBIOLOGY. Educ: Univ BC, BSc, 62, MSc, 65; Univ Alta, PhD(immunol), 70. Prof Exp: Sr scientist tissue typing, Dept Clin Path, Univ Hosp, Edmonton, Alta, 70-74, sr scientist, Transplant Lab, 74-75; DIR TRANSPLANT LAB, MEM MED CTR, 75- Concurrent Pos: Immunologist II, Dept Surg, Univ Calif, Los Angeles, 72-73; sessional lectr path, Med Lab Sci, Univ Alta, 73-, prof asst, Med Res Coun Can Transplant Immunol Unit, 74-75. Mem: AAAS; Can Soc Immunol; Can Soc Clin Invest. Res: Transplantation immunology; histoincompatibility. Mailing Add: Transplant Lab Mem Med Ctr Springfield IL 62704

MCCONNAUGHEY, BAYARD HARLOW, b Pittsburgh, Pa, Apr 21, 16; m 49; c 5. MARINE BIOLOGY. Educ: Pomona Col, BA, 38; Univ Hawaii, MA, 41; Univ Calif, PhD(zool), 48. Prof Exp: Asst zool, Univ Calif, 38-40 & Univ Hawaii, 40-41; asst, Scripps Inst, Univ Calif, 47-48, res assoc, 48; from asst prof to assoc prof, 48-72, PROF BIOL, UNIV ORE, 72- Concurrent Pos: Ky Contract Team assoc prof, Inst Agr Sci, Fac Fisheries, Univ Indonesia, 63-66; guest scientist, Coun Sci, Indonesia, 66. Mem: AAAS; Soc Syst Zool; Am Soc Parasitol; Am Soc Limnol & Oceanog. Res: Taxonomy and life history of the Mesozoa; microbiology; association analysis; plankton communities; delimiting and characterizing marine biotic communities. Mailing Add: Dept of Biol Univ of Ore Eugene OR 97403

MCCONNEL, ROBERT MERRIMAN, b Rochester, Pa, Mar 19, 36; m 62. MATHEMATICS. Educ: Washington & Jefferson Col, BA, 58; Duke Univ, PhD(math), 62. Prof Exp: Asst prof math, Univ Ariz, 62-64; asst prof, 64-66, ASSOC PROF MATH, UNIV TENN, KNOXVILLE, 66- Mem: Am Math Soc; Math Asn Am. Res: Number theory; finite field theory; polynomials over finite fields. Mailing Add: Dept of Math Univ of Tenn Knoxville TN 37916

MCCONNELL, ALBERT LAWRENCE, b Philadelphia, Pa, Nov 29, 17; m 46; c 1. CHEMISTRY. Educ: Pa Mil Col, BS, 49. Prof Exp: Res mgr plastics & chem dept, 49-67, res mgr paper & chem dept, 67-69, res mgr corp growth res & develop, 69-71, sr sci specialist, res & develop new prod & processes, 71-74, PROJ ENGR, RES & DEVELOP, SCOTT PAPER CO, 74- Mem: Am Chem Soc; Tech Asn Pulp & Paper Indust. Res: Development of plastics and chemical based new products and processes. Mailing Add: Res & Develop Scott Paper Co 50 W Powhattan Ave Essington PA 19029

MCCONNELL, ANDREW POLLOCK, JR, b Seattle, Wash, Oct 15, 19; m 44; c 3. PETROLEUM GEOLOGY. Educ: Univ Wash, BS, 42. Prof Exp: Geologist, 45-56, dist geologist, 56-71, SPEC PROJ GEOLOGIST, 71- Mem: Am Asn Petrol Geol. Res: Petroleum exploration. Mailing Add: Texaco Inc 5151 W Fremont Dr Littleton CO 80120

MCCONNELL, BRUCE, b Pittsburgh, Pa, Sept 12, 32; m 64; c 1. BIOCHEMISTRY. Educ: Grove City Col, BS, 54; Univ Vt, PhD(biochem), 66. Prof Exp: Tech rep, Atlas Powder Co, Del, 54-56; pilot plant supvr, Koppers Co, Inc, Pa, 56-59; NIH fel molecular biol, Dartmouth Med Sch, 66 & Univ Ore, 66-69; asst prof, 69-74, ASSOC PROF BIOPHYS, UNIV HAWAII, HONOLULU, 74- Concurrent Pos: NSF res grant. Mem: Am Chem Soc. Res: Physical chemistry and biosynthesis of macromolecules; protein chemistry and isolation; molecular configurations of DNA by hydrogen exchange and physical-chemical; nuclear magnetic resonance-proton exchange. Mailing Add: Dept of Biochem & Biophys Univ of Hawaii Honolulu HI 96822

MCCONNELL, DAVID GRAHAM, b Bronxville, NY, Dec 3, 26; m 48; c 4. BIOPHYSICS, BIOCHEMISTRY. Educ: Columbia Univ, AB & AM, 49; Ind Univ, PhD(exp psychol), 57. Prof Exp: Res assoc psychol, Ohio State Univ, 56-61, res assoc chem, 60-61; proj dir med prog, Britannica Ctr, Calif, 61-62; assoc prof physiol optics, Ohio State Univ, 62-65, assoc prof biophys, 65-71, assoc prof biochem, 67-71, prof biochem & biophys, 71-73; PROF BIOCHEM & BIOMECH, MICH STATE UNIV, 73- Concurrent Pos: Vis assoc prof, Enzyme Inst, Univ Wis, 64-65. Mem: Psychonomic Soc; Biophys Soc; Am Soc Cell Biol; Am Soc Biol Chemists; Am Soc Neurochem. Res: Neurochemistry; photochemistry; photobiology; learning; electron microscopy of biological membranes; biochemistry of the retina. Mailing Add: 436 E Fee Hall Mich State Univ East Lansing MI 48824

MCCONNELL, DENNIS BROOKS, b Waupun, Wis, Aug 18, 38; m 64; c 2. ORNAMENTAL HORTICULTURE. Educ: Wis State Univ, River Falls, 66; Univ Wis-Madison, MS, 67, PhD(bot, hort), 70. Prof Exp: Asst prof ornamental hort, Agr Res Ctr, 70-75, ASSOC PROF ORNAMENTAL HORT, UNIV FLA, 75- Mem: Am Soc Hort Sci; Am Hort Soc; Bot Soc Am. Res: Morphology and anatomy of normal and experimentally modified subtropical and tropical plants. Mailing Add: Dept of Ornamental Hort Univ of Fla Gainesville FL 32601

MCCONNELL, DUNCAN, b Chicago, Ill, Jan 30, 09; m 34; c 3. DENTAL RESEARCH, MATERIALS SCIENCE. Educ: Washington & Lee Univ, BS, 31; Cornell Univ, MS, 32; Univ Minn, PhD(mineral), 37. Prof Exp: Instr mineral, Univ Tex, Austin, 37-41; chemist-petrographer, US Bur Reclamation, Denver, 41-47; actg asst div chief geol, Gulf Res & Develop Co, 47-50; prof mineral, 50-56, PROF DENT RES, OHIO STATE UNIV, 57-, PROF MINERAL, 64- Concurrent Pos: USPHS spec res fel, Ohio State Univ, 57-61, USPHS grants, 57-68, US Army Med Res & Develop Command grant, 66-67; consult & site visitor, Nat Inst Dent Res, 73; examr, Dept Orthop Surg, Yale Univ, 73; consult, NSF, 73, Nat Inst Dent Res, 74 & Prev Dent Res Inst, Ind Univ, Ft Wayne, 74. Mem: Fel AAAS; Int Asn Dent Res; fel Mineral Soc Am; Brit Mineral Soc; Soc Econ Geologists. Res: Crystal chemistry of inorganic component of teeth and bones, restorative materials and mineralization process. Mailing Add: Dept of Dent Res Ohio State Univ Col of Dent Columbus OH 43210

MCCONNELL, ELLICOTT, b Providence, RI, Mar 4, 24; m 50; c 2. MEDICAL PARASITOLOGY. Educ: Univ Conn, BS, 49; Univ Minn, MS, 53, PhD(entom), 57. Prof Exp: Asst, Univ Minn, 50-55; entomologist, State Dept Agr, Minn, 56-57; microbiologist, Gorgas Mem Lab, Panama, 57-62; parasitologist, US Naval Med Res Unit 3, Cairo, 62-67, head dept parasitol, 63-67; entomologist, 67-72, HEAD MED ZOOL, US NAVAL MED RES UNIT 5, ADDIS ABABA, ETHIOPIA, 72- Mem: AAAS; Entom Soc Am; Am Soc Trop Med & Hyg; Am Soc Parasitol; Royal Soc Trop Med & Hyg. Res: African trypanosomiasis; leishmaniasis transmission; schistosomiasis. Mailing Add: US Naval Med Res Unit 5 APO New York NY 09319

MCCONNELL, FREEMAN ERTON, b West Point, Ind, Apr 20, 14; m 41; c 2. AUDIOLOGY. Educ: Univ Ill, BS, 39, MA, 46; Northwestern Univ, PhD(audiol), 50. Prof Exp: Instr, Ill High Sch, 39-43; high sch & jr col instr, 46-48; lectr speech correction & audiol, Northwestern Univ, 48-50; asst prof audiol, Inst Logopedics Wichita & Wichita Univ, 50-51; from asst prof to prof, Vanderbilt Univ, 51-60; prof & head dept, Univ Tenn, 60-63; PROF AUDIOL & HEAD DEPT, VANDERBILT UNIV, 63- Concurrent Pos: Consult, Arnold Eng Develop Ctr, US Air Force, 53-71 & Bur Res, US Off Educ, 65-73. Mem: AAAS; Acoust Soc Am; fel Am Speech & Hearing Asn; Acad Rehab Audiol (pres, 71). Res: Clinical audiology; language impaired children; hereditary deafness. Mailing Add: Dept of Hearing & Speech Sci Vanderbilt Univ Med Ctr Nashville TN 37203

MCCONNELL, HARDEN MARSDEN, b Richmond, Va, July 18, 27; m 56; c 3. BIOPHYSICAL CHEMISTRY. Educ: George Washington Univ, BS, 47; Calif Inst Technol, PhD(chem), 51. Prof Exp: Nat Res Coun fel physics, Univ Chicago, 50-52; res chemist, Shell Develop Co, 52-56; from asst prof to prof chem, Calif Inst Technol, 56-65; PROF CHEM, STANFORD UNIV, 65- Concurrent Pos: Consult, Calif Res Corp, 59-65, Varian Assocs, 64-66 & Syva, 69-; chem ed, W H Freeman Co; Harkins lectr, Univ Chicago, 67; Falk-Plaut lectr, Columbia Univ; Phillips lectr, Haverford Col, 68; Renaud lectr, Mich State Univ, 69; Debeye lectr, Cornell Univ, 71. Honors & Awards: Pure Chem Award, 62, Harrison Howe Award, 68 & Irving Langmuir Award, Am Chem Soc, 72. Mem: Nat Acad Sci; fel Am Phys Soc; Am Soc Biol Chemists; Biophys Soc; Am Chem Soc. Res: Spin distributions and hyperfine interactions in organic radicals, electron and nuclear magnetic resonance spectroscopy; excitons in molecular crystals; spin labels; biological membranes. Mailing Add: Dept of Chem Stanford Univ Stanford CA 94305

MCCONNELL, HENRY ELDEN, b Star, Idaho, June 20, 14; m 38; c 6. MICROBIOLOGY. Educ: Univ Idaho, BS, 37; Univ Mich, Ann Arbor, MPH, 48; Univ NC, Chapel Hill, DrPH(lab pract), 67. Prof Exp: Assayer, Bunker Hill Smelter, Kellogg, Idaho, 37; bacteriologist, Coeur d'Alene br lab, Idaho State Dept Pub Health, 37-41, serologist, Boise Lab, 41-44, microbiologist, 48-52, chief chemist, 52-60; DIR LABS, NEBR STATE DEPT HEALTH, 60- Concurrent Pos: Lectr, Boise Col, 52. Mem: Fel Am Pub Health Asn; Conf Pub Health Lab Dirs. Res: Medical microbiology; public health; sanitary chemistry. Mailing Add: State Health Lab Box 2755 Lincoln NE 68502

MCCONNELL, JACK BAYLOR, b Crumpler, WVa, Feb 1, 25; m 58. MEDICINE. Educ: Univ Tenn, MD, 49. Prof Exp: Resident, Baylor Univ, 50-51; assoc dir prof serv, Lederle Labs, Am Cyanamid Co, NY, 54-57, dir clin invest, 57-59, dir clin develop, 59-61; exec dir new prod div, McNeil Labs, Inc, 61-67, vpres new prod, 67-69, CORP DIR COMMERCIAL DEVELOP, JOHNSON & JOHNSON, 69- Mem: Am Thoracic Soc; AMA; Am Med Writers' Asn. Res: Biomedical instruments and materials in the field of respiratory care, laboratory instruments and patient monitoring. Mailing Add: Johnson & Johnson Com Develop 501 George St New Brunswick NJ 08903

MCCONNELL, JAMES E, b Grove City, Pa, June 27, 37; m 62; c 2. ECONOMIC GEOGRAPHY. Educ: Slippery Rock State Col, BS, 60; Miami Univ, MA, 61; Ohio State Univ, PhD(geog), 69. Prof Exp: Instr geog, Indiana Univ Pa, 62-65; from instr to asst prof, 68-72, ASSOC PROF GEOG, STATE UNIV NY, BUFFALO, 72- Concurrent Pos: Joint Awards Coun State Univ NY-Res Found Univ Awards Comt res fel projs US-Can trade, 69-71 & res & travel in Cent Am, 71-73. Mem: Asn Am Geog. Res: International trade and development; economic integration; transportation. Mailing Add: Dept of Geog State Univ of NY Amherst NY 14226

MCCONNELL, JAMES FRANCIS, b Syracuse, NY, Dec 15, 36; m 59; c 4. ORGANIC CHEMISTRY. Educ: Le Moyne Col, NY, BS, 58; Syracuse Univ, PhD(org chem), 59. Prof Exp: Asst prof org chem, Mansfield State Col, 62-64; asst prof, 64-74, ASSOC PROF ORG CHEM, STATE UNIV NY COL CORTLAND, 74- Mem: Am Chem Soc. Res: Stereochemical approach to the mechanism of the Mannich reaction; rate of loss of optical activity of nitroalkanes compared to the rate in which they form Mannich base products in Mannich reaction. Mailing Add: Dept of Chem State Univ of NY Cortland NY 13045

MCCONNELL, JANE ROYLE, cytology, see 12th edition

MCCONNELL, JOHN EARL WILLARD, b St John, NB, May 25, 15; nat US; m 38; c 1. FOOD TECHNOLOGY. Educ: Queen's Univ, Ont, BS, 41; Univ Mass, MS, 42, PhD(food technol), 45. Prof Exp: Asst food technol, Exp Sta, Univ Mass, 42-46, asst res prof, 46-48; process engr, Western Br Labs, Nat Canners Asn, 48-57, head processing & bact dept, 57-58; HEAD DEPT CHEM, VENTURA COL, 58- Concurrent Pos: With USDA, 44; bacteriologist, Hooper Found, Univ Calif, 55-58. Mem: Am Chem Soc; Inst Food Technol. Res: Rancidity in fats; thermal processing of foods; chemical sterilization. Mailing Add: Dept of Chem Ventura Col 4620 Loma Vista Rd Ventura CA 93003

MCCONNELL, KENNETH PAUL, b Rochester, NY, June 19, 11; m 50. BIOCHEMISTRY. Educ: Univ Rochester, AB, 35, MS, 37, PhD(biochem, nutrit), 42. Prof Exp: Asst, Sch Med & Dent, Univ Rochester, 35-37, instr, Dept Radiation Biol & assoc, Atomic Energy Proj, 48-49; asst, Med Sch, Ohio State Univ, 37-38; asst, Univ Iowa, 38-39; biochemist, USPHS, 46; biochemist, Med Dept, Field Res Lab, Ft Knox, Ky, 46-48; asst prof biochem & nutrit & dir med physics lab, Univ Tex Med

Br, 50-52; res assoc, 52-60, assoc prof, 60-69, PROF BIOCHEM, SCH MED, UNIV LOUISVILLE, 69- Concurrent Pos: Asst dir radioisotope serv, Vet Admin Hosp, 52-60, biochemist & prin scientist, 52-; consult, Oak Ridge Inst Nuclear Studies, 52-59. Mem: Fel AAAS; Am Chem Soc; Am Inst Nutrit; Am Soc Biol Chemists; Soc Exp Biol & Med. Res: Metabolism of trace elements in the mammalian organism in health and disease; use of isotopes in research and medicine. Mailing Add: Dept of Biochem Univ of Louisville Sch of Med Louisville KY 40202

MCCONNELL, RICHARD LEON, organic chemistry, polymer chemistry, see 12th edition

MCCONNELL, ROBERT A, b McKeesport, Pa; m. BIOPHYSICS, PARAPSYCHOLOGY. Educ: Carnegie Inst Technol, BS, 35; Univ Pittsburgh, PhD(physics), 47. Prof Exp: Asst geophysicist, Gulf Res Develop Co, 37-39; asst physicist, US Naval Aircraft Factory, 39-41; mem staff, Radiation Lab, Mass Inst Technol, 41-44, group leader, 44-46; asst prof physics, 47-53, from asst res prof to assoc res prof, 53-63, RES PROF BIOPHYS, UNIV PITTSBURGH, 64- Mem: Inst Elec & Electronics Eng; Parapsychol Asn (pres, 57-58). Res: Radar moving target indication; theory of the iconoscope; ultrasonic microwaves; extrasensory perception; psychokinesis. Mailing Add: Dept of Life Sci Univ of Pittsburgh Pittsburgh PA 15260

MCCONNELL, ROBERT KENDALL, b Toronto, Ont, Mar 1, 37; m 58; c 2. GEOPHYSICS, GEOLOGY. Educ: Princeton Univ, BSc E, 58; Univ Toronto, MASc, 60, PhD(physics), 63. Prof Exp: Geologist, Invex Corp, Ltd, 57-58; geophysicist, Hunting Surv Corp, 59-60; geophysicist, Arthur D Little, Inc, 63-69; PRES, EARTH SCI RES, INC, 70- Concurrent Pos: Vis prof, Univ RI, 70. Mem: Am Geophys Union; Soc Explor Geophys; Europ Asn Explor Geophys; Seismol Soc Am. Res: Tectonophysics; planetary evolution; geological and geophysical interpretation. Mailing Add: Earth Sci Res Inc 133 Mt Auburn St Cambridge MA 02138

MCCONNELL, STEWART, b El Paso, Tex, Apr 25, 23; m 49; c 3. VIROLOGY, IMMUNOLOGY. Educ: Ohio State Univ, MS, 60; Tex A&M Univ, DVM, 50. Prof Exp: Vet virol & res, Walter Reed Army Inst Res, Vet Corps, US Army, 60-63, adv, US Agency Int Develop, 63-65, vet malaria res, Walter Reed Army Inst Res, 65-66, vet infectious dis res, US Army Med Res Inst Infectious Dis, 66-68; ASSOC PROF VET MICROBIOL, TEX A&M UNIV, 68- Concurrent Pos: Mem subcomt fish standards, comt standards, Inst Lab Animal Resources, Nat Res Coun-Nat Acad Sci, foreign animal dis comt, US Animal Health Asn, Western Hemisphere Comt Animal Virus Characterization & bd comp virol, WHO-Food & Agr Orgn. Honors & Awards: Medal, Bolivian Ministry Agr, 64; Spec Commendation, Bolivian Hemorrhagic Fever Comn, 65. Mem: Am Asn Lab Animal Sci; Am Soc Lab Animal Practitioners; Am Soc Microbiol; fel NY Acad Sci. Res: Virus that play a role in respiratory diseases of feedlot cattle; poxvirus of animals; virus diseases of marine fish and shellfish. Mailing Add: Dept of Vet Microbiol Tex A&M Univ College Station TX 77843

MCCONNELL, VIRGINIA FENNER, b New Orleans, La, July 22, 05; m 28; c 3. ORGANIC CHEMISTRY. Educ: Tulane Univ, BA, 26, MS, 27. Prof Exp: Lab instr biochem, Sch Med, Tulane Univ, 27-29, lab instr biochem, Newcomb Col, 42-44, from instr to assoc prof chem, 44-71, head dept, 64-70, EMER ASSOC PROF CHEM, NEWCOMB COL, TULANE UNIV, 71- Mem: AAAS; Am Chem Soc; Hist Sci Soc. Res: History of organic chemistry; biographical catalogue of chemists; life of Emil Fischer. Mailing Add: 1731 Calhoun St New Orleans LA 70118

MCCONNELL, WALLACE BEVERLY, b Metiskow, Alta, Sept 24, 16; m 40; c 2. BIOCHEMISTRY. Educ: Univ Alta, BSc, 46, MSc, 47; McGill Univ, PhD(phys chem), 49. Prof Exp: Jr res officer protein chem, Prairie Regional Lab, Nat Res Coun Can, 49-51, from asst res officer to sr res officer, 51-66; prof chem, Univ Sask, Regina, 66-73, chmn dept, 67-73, assoc dean, Div Natural Sci, 73-74; DEAN FAC SCI, UNIV REGINA, 75- Mem: Fel Chem Inst Can. Res: Plant biochemistry; amino acids and proteins; biosynthesis of selenium and sulphur containing compounds in plants. Mailing Add: Fac of Sci Univ of Regina Regina SK Can

MCCONNELL, WILLIAM J, limnology, fisheries, see 12th edition

MCCONNON, MYLES, mathematics, see 12th edition

MCCONVILLE, BRIAN JOHN, b Queenstown, NZ, May 11, 33; m 59; c 3. PSYCHIATRY. Educ: Univ Otago, NZ, MB, ChB, 57; Royal Col Physicians Can, cert psychiat, 66, fel psychiat, 72. Prof Exp: Lectr psychiat, 65-67, lectr pediat, 65-69, asst prof psychiat, 67-71, ASST PROF PEDIAT, QUEEN'S UNIV, ONT, 69- DIR CHILD PSYCHIAT TRAINING, 70-, ASSOC PROF PSYCHIAT, 71- Concurrent Pos: Dir, Regional Children's Centre, Kingston Psychiat Hosp, 67-; liaison off, Can Psychiat Asn-Can Pediat Soc, 70-; chmn child-adolescent comt, Child, Adolescent & Retardate Sect, Can Psychiat Asn, 70-, mem sci coun, 72- Mem: Can Psychiat Asn; Can Med Asn; Soc Psychophysiol Res; Am Psychiat Asn; Am Acad Child Psychiat. Res: Child psychiatry; measurement of behavioral change in children's inpatient units; phenomena of childhood depression-aggression; family therapy; illness after depression. Mailing Add: Dept of Psychiat Queen's Univ Kingston ON Can

MCCONVILLE, DAVID RAYMOND, b Benson, Minn, Feb 9, 46. LIMNOLOGY, FISHERIES. Educ: St Cloud State Univ, BS, 68, MA, 69; Univ Minn, St Paul, PhD(fisheries), 72. Prof Exp: Asst, St Cloud State Univ, 68-70, res assoc, 70-71; instr biol, Univ Minn Tech Col-Waseca, 71-73; ASSOC PROF BIOL & DIR AQUATIC STUDIES, ST MARY'S COL, MINN, 73- Concurrent Pos: Consult ecol studies, Northern States Power Co, Minneapolis, 73-75; prin investr Off Water Resources res grant fisheries studies, St Mary's Col, 73-75, dir NSF environ equip grant, 75-77, dir US Fish & Wildlife Serv biol res contract, 75-78, dir US Army Corps Engrs biol res contract, 75-76. Mem: Am Fisheries Soc; Ecol Soc Am; Sigma Xi. Res: Current flows and fisheries; use of artificial substrates; emergence studies of insects; relative abundance, age and growth, food habits and selected movement studies of Mississippi River fish. Mailing Add: Dept Aquatic Studies Biol St Mary's Col Terrace Heights Winona MN 55987

MCCONVILLE, GEORGE T, b St Paul, Minn, Sept 23, 34; m 59. PHYSICS. Educ: Carleton Col, AB, 56; Purdue Univ, MS, 59; Rutgers Univ, PhD(physics), 64. Prof Exp: Tech staff assoc low temperature physics, RCA Labs, 59-61; sr res physics, 64-69, res specialist, 69-75, SR RES SPECIALIST, MONSANTO RES CORP, 75- Mem: Am Phys Soc. Res: Low temperature physics; superconductivity; calorimetry; gas transport. Mailing Add: Mound Lab Monsanto Res Corp Miamisburg OH 45342

MCCONVILLE, JOHN THEODORE, b Centerville, Iowa, Nov 15, 27; m 54; c 3. PHYSICAL ANTHROPOLOGY. Educ: Univ NMex, BA, 53; Univ Ariz, MA, 59; Univ Minn, PhD, 69. Prof Exp: Res analyst, US Army Electronic Proving Grounds, 56-59; assoc dir anthrop res proj, Antioch Col, 59-69; ASSOC DIR ANTHROP RES PROJ, WEBB ASSOCS, INC, 70- Mem: Am Asn Phys Anthrop. Res: Human biology, particularly man as a component of complex systems; research in body size variability and human physical characteristics such as body strength and composition. Mailing Add: Anthrop Res Proj Webb Assocs Inc Box 308 Yellow Springs OH 45387

MCCOOK, GEORGE PATRICK, b July 23, 37; US citizen; m 61; c 3. ASTRONOMY. Educ: Villanova Univ, BS & MA, 61; Univ Pa, MS, 65, PhD(astron), 68. Prof Exp: Asst prof, 61-72, ASSOC PROF ASTRON, VILLANOVA UNIV, 72-, CHMN DEPT, 74- Mem: Am Astron Soc. Res: Analysis of light variations of eclipsing binaries and astronomical instrumentation. Mailing Add: Dept of Astron Villanova Univ Villanova PA 19085

MCCOOK, ROBERT DEVON, b Columbus, Ga, Jan 23, 29; m 57. PHYSIOLOGY. Educ: La State Univ, BS, 50, MS, 52; Loyola Univ, Ill, PhD(physiol), 64. Prof Exp: Instr chem, Concord Col, 53-54; asst biochem, Stritch Sch Med, Loyola Univ Chicago, 54-57, asst physiol, 57-62, from instr to asst prof physiol, 62-74; CHIEF BIOMED ENG, VET ADMIN HOSP, LEXINGTON, 74- Concurrent Pos: Res physiologist, Vet Admin Hosp, Hines, Ill, 71-74. Mem: AAAS; Am Physiol Soc; Inst Elec & Electronics Engrs; Sigma Xi. Res: Central control of temperature regulation; physiological instrumentation; computer applications to medicine and biology. Mailing Add: Lexington Vet Admin Hosp Leestown Rd Lexington KY 40507

MCCOOL, JOHN CUMMINGS, organic chemistry, see 12th edition

MCCOOL, STEPHEN FORD, forest recreation, see 12th edition

MCCORD, CAREY PRATT, b Bibb Co, Ala, Sept 25, 86; m 16; c 3. OCCUPATIONAL HEALTH. Educ: Howard Col, AB, 06, LLD, 33; Univ Mich, MD, 12; Am Bd Prev Med, dipl, 49, cert, Occup Med, 55. Prof Exp: Dir res, Mich Home & Training Sch, 14-16; med dir indust health conserv labs, Univ Detroit, 19-51; lectr, Sch Pub Health, 41-51, res lectr, 51-64, CONSULT, INST INDUST HEALTH, UNIV MICH, ANN ARBOR, 51- Concurrent Pos: Assoc prof, Col Med, Univ Cincinnati, 19-35, lectr, Col Eng, 21-32; consult, Crysler Corp, 34-51; dir, Bur Indust Hyg, Dept Pub Health, Detroit, Mich, 36-39 & State Dept Health, 38-39; prof, Wayne Univ, 37-40; ed, Indust Med & Surg, 51-64 & J Occup Med, 64-70. Honors & Awards: Knudsen Award, 47. Mem: Acoust Soc Am; fel Am Pub Health Asn; AMA; Indust Med Asn. Res: Occupational diseases. Mailing Add: 2522 Sch of Pub Health I Univ of Mich Ann Arbor MI 48109

MCCORD, COLIN WALLACE, b Chicago, Ill, May 15, 28; m 54; c 3. PUBLIC HEALTH, SURGERY. Educ: Williams Col, BA, 49; Columbia Univ, MD, 53. Prof Exp: Instr & asst prof surg, Sch Med, Univ Ore, 61-65; asst prof & assoc prof, Columbia Univ, 65-71; ASSOC PROF INT HEALTH, SCH HYG, JOHNS HOPKINS UNIV, 71- Concurrent Pos: Attend clin surgeon, St Lukes Hosp, NY, 65-71; resident dir, Rural Health Res Ctr, Narangwal, Punjab, India, 71-72; dir, Companiganj Health Proj, Noakhali, Bangladesh, 72-75. Mem: Am Col Surgeons; Am Asn Thoracic Surgeons. Res: Cost and effectiveness of health programs in developing countries; cost and effectiveness of nutrition interventions. Mailing Add: 615 N Wolfe St Baltimore MD 21205

MCCORD, MICHAEL CAMPBELL, b Knoxville, Tenn, Apr 28, 36; m 58; c 3. INFORMATION SCIENCE. Educ: Univ Tenn, BA, 58; Univ Tenn, MA, 60; Yale Univ, PhD(math), 63. Prof Exp: Actg instr math, Yale Univ, 62-63; asst prof, Univ Wis, 63-64; from asst prof to assoc prof, Univ Ga, 64-69; assoc prof math, 69-75, ASSOC PROF COMPUT SCI, UNIV KY, 75- Concurrent Pos: Mem, Inst Advan Study, NJ, 67-68. Mem: Am Math Soc; Asn Computational Ling; Ling Soc Am. Res: Computational linguistics; syntax and semantics; logic. Mailing Add: Dept of Math Univ of Ky Lexington KY 40506

MCCORD, THOMAS BARD, b Elverson, Pa, Jan 18, 39; m 62. ASTRONOMY. Educ: Pa State Univ, BS, 64; Calif Inst Technol, MS, 66, PhD(planetary sci & astron), 68. Prof Exp: Res fel planetary sci, Calif Inst Technol, 68; asst prof, 68-71, ASSOC PROF PLANETARY PHYSICS, MASS INST TECHNOL & DIR GEORGE R WALLACE ASTROPHYS OBSERV, 71- Concurrent Pos: Vis assoc, Calif Inst Technol, 69-72. Mem: Fel AAAS; fel Am Geophys Union; Am Astron Soc. Res: Study of solar system objects using remote sensing techniques; development of instrumentation with which to make such studies. Mailing Add: Dept of Earth & Planetary Sci Mass Inst of Technol Cambridge MA 02139

MCCORD, TOMMY JOE, b Lubbock, Tex, Feb 19, 32; m 51; c 4. BIOCHEMISTRY. Educ: Abilene Christian Col, BS, 54; Univ Tex, MA, 58, PhD(biochem, org chem), 59. Prof Exp: Asst, Abilene Christian Col, 53-54; res scientist, Biochem Inst, Univ Tex, 56-58; from asst prof to assoc prof chem, 58-64, asst head dept, 65-67, PROF CHEM, ABILENE CHRISTIAN COL, 64-, HEAD DEPT, 67-, DIR CHEM RES INST, 60- Concurrent Pos: Res grants, USPHS, NSF & Welch Found. Mem: AAAS; Am Chem Soc. Res: Preparation and biological testing of structural analogs of important biological compounds as potential metabolic antagonists. Mailing Add: Dept of Chem Abilene Christian Col Abilene TX 79601

MCCORD, WILLIAM MELLEN, b Durban, SAfrica, Jan 24, 07; m 30; c 2. BIOCHEMISTRY. Educ: Oberlin Col, AB, 28; Yale Univ, PhD(org chem), 31; La State Univ, MB, 39, MD, 40. Prof Exp: From instr to assoc prof biochem, Sch Med, La State Univ, 31-45; PROF BIOCHEM, MED UNIV SC, 45-, PRES, 64- Concurrent Pos: Intern, Charity Hosp, New Orleans, 39-40, vis biochemist, 40- Mem: AAAS; Am Chem Soc; Am Fedn Clin Res. Res: Aldehyde condensations; blood chemistry; pathological chemistry; toxicology; clinical biochemistry; porphyrin determinations; sickle cell anemia. Mailing Add: Med Univ of SC Charleston SC 29401

MCCORKLE, HORACE J, b Center Point, Tex, Feb 5, 05; m; c 5. SURGERY. Educ: Univ Calif, AB, 30, MD, 34; Am Bd Surg, dipl, 51. Prof Exp: Intern surg, Univ Calif Serv, San Francisco Hosp, 33-34, asst resident, 34-35, house officer, 35-36; house officer, Cincinnati Gen Hosp, 36-37; asst resident, San Francisco Hosp, 37-38; from instr to assoc prof, 38-53, PROF SURG, SCH MED, UNIV CALIF, SAN FRANCISCO, 53- Concurrent Pos: Resident surgeon, Univ Calif Hosp, 38-39, exec officer, 39-42. Mem: Soc Univ Surgeons; Am Gastroenterol Asn; Int Soc Surgeons; Am Surg Asn; Am Col Surgeons. Mailing Add: 35 San Fernando Way San Francisco CA 94127

MCCORKLE, JOHN EARL, organic chemistry, polymer chemistry, see 12th edition

MCCORKLE, LOIS PAKE, b Akron, Ohio, Feb 8, 27; m 49; c 3. EPIDEMIOLOGY, BIOSTATISTICS. Educ: Oberlin Col, BA, 47; Western Reserve Univ, MD, 51. Prof Exp: Demonstr biostatist, Case Western Reserve Univ, 53-54, instr, 54-55, from instr to sr instr, Dept Prev Med, 59-63, from sr instr to assoc prof, Div Biometry & Dept Prev Med, 63-70; BIOSTATISTICIAN, UNIV HOSPS CLEVELAND, 70-, ASST DIR BLOOD BANK, 59- Concurrent Pos: Fel biostatist, Case Western Reserve Univ, 52-53, USPHS fel immunohematology, Sch Med, 55-57, univ fel, 58-59; res statistician, Univ Hosps Cleveland, 62-70; USPHS grant study utilization of facil of univ hosp, Univ Hosps Cleveland, 64-70. Mem: Am Pub Health Asn; Soc Epidemiol Res; Am Med Women's Asn; Am Asn Blood Banks. Mailing Add: Univ Hosps Univ Circle Cleveland OH 44106

MCCORKLE, RICHARD ANTHONY, b Gastonia, NC, Aug 6, 40; m 64; c 1. PLASMA PHYSICS, LASERS. Educ: NC State Univ, BS, 62, PhD(physics), 70. Prof Exp: Asst prof physics, E Carolina Univ, 68-73; RES SCIENTIST, IBM, THOMAS J WATSON RES CTR, 73- Concurrent Pos: Res dir, NSF grant, 69-70. Honors & Awards: Bisplinghoff Award, 72. Mem: Am Phys Soc. Res: Laser physics; short wavelength lasers, x-ray; electron streams. Mailing Add: IBM Watson Res Ctr PO Box 218 Yorktown Heights NY 10598

MCCORKLE, THOMAS, b Tex, July 30, 14; m 38; c 3. ANTHROPOLOGY. Educ: Univ Calif, Berkeley, AB, 38, PhD(anthrop), 54. Prof Exp: Instr anthrop, Univ Calif, Santa Barbara, 54-55; asst prof, Col Med, Univ Iowa, 56-60; dir, Div Behav Sci, Pa State Health Dept, 60-63; supvry soc sci analyst, USPHS, 63-65; assoc prof, 66-72, PROF ANTHROP, CALIF STATE UNIV, LONG BEACH, 72- Mem: Fel AAAS; fel Am Anthrop Asn; fel Soc Appl Anthrop. Res: Modern peoples of the United States and Latin America; cultural change; applications of anthropology; urban anthropology in Latin America, especially Mexico. Mailing Add: Dept of Anthrop Calif State Univ Long Beach CA 90840

MCCORKLE, WILLARD HOMER, b Palo, Iowa, Apr 21, 02; m 28; c 3. PHYSICS. Educ: Univ Iowa, BA, 24, MS, 28, PhD(physics), 35. Prof Exp: From instr to prof physics, Agr & Mech Col Tex, 24-44; physicist, Metall Lab, Chicago, 44-45; physicist, Phillips Petrol Co, Okla, 45-46; sr physicist, Argonne Nat Lab, 46-51, dir, Reactor Opers Div, 51-61; CHIEF REACTOR DIV, AMES LAB, IOWA STATE UNIV, 61- Mem: AAAS; Acoustical Soc Am; Am Nuclear Soc; Am Asn Physics Teachers. Res: Reactor design, operation and research. Mailing Add: Reactor Div Ames Lab Iowa State Univ Ames IA 50010

MCCORKLE, WILLIAM C, JR, b Johnson City, Tenn, Sept 9, 28; m 61. PHYSICS, AEROSPACE SCIENCES. Educ: Univ Richmond, BS, 50; Univ Tenn, PhD(physics), 56. Prof Exp: Asst tech dir missile syst develop, 60, sci adv to dir physics, Res & Develop Directorate, US Army Ballistic Missile Agency, 60-62, assoc dir aeroballistics, guidance & control lab, 62, dir adv systs lab, 62-71, DIR AEROBALLISTICS RES & DEVELOP, ENG & MISSILE SYSTS LAB, RES & DEVELOP DIRECTORATE, US ARMY MISSILE COMMAND, 71- Concurrent Pos: Mem adv comt space vehicle aerodyn, NASA, 62-68; sci adv weapon systs directorate, Off Dep Chief Staff, Res & Develop & Acquisition, Hq, Dept of Army, 74-75. Honors & Awards: Res & DevelopAchievement Award, US Dept Army, 63, Decoration Meritorious Civilian Serv Award, 64. Mem: Am Inst Aeronaut & Astronaut. Res: Rocket ballistics; missile guidance, control and aerodynamics. Mailing Add: Cmndg Gen US Army Missile Comd ATTN: AMSMI-RD Redstone Arsenal AL 35809

MCCORMAC, BILLY MURRAY, b Zanesville, Ohio, Sept 8, 20; m 48 & 69; c 5. ASTROPHYSICS. Educ: Ohio State Univ, BS, 43; Univ Va, MS, 56, PhD(nuclear physics), 57. Prof Exp: Physicist, Off Spec Weapons Develop, US Army, 57-60, scientist, Off Chief Staff, 60-61, physicist, Defense Atomic Support Agency, 61-62, chief electromagnetic br, 62-63; sci adv, IIT Res Inst, 63, dir geophys div, 63-68; sr consult scientist, 68-69, mgr, Radiation Physics Lab, 69-74, MGR, ELECTRO-OPTICS LAB, LOCKHEED PALO ALTO RES LAB, 74- Concurrent Pos: Ed, J Water, Air & Soil Pollution; chmn adv study inst radiation trapped in earth's magnetic field, Norway, 65, aurora & airglow, Eng, 66, Norway, 68, Can, 70, France, 72 & Belg, 74, earth's particles & fields, Ger, 67, Calif, 69, Italy, 71, Eng, 73 & Austria, 75. Mem: AAAS; sr mem Am Astronaut Soc; Am Geophys Union; Marine Technol Soc; assoc fel Am Inst Aeronaut & Astronaut. Res: Nuclear weapons effects, especially high altitude effect; multidisciplinary research in aeronomy and the earth's magnetosphere. Mailing Add: Lockheed Palo Alto Res Lab Dept 52-54 B202 3251 Hanover St Palo Alto CA 94304

MCCORMACK, CHARLES ELWIN, b Gurnee, Ill, Jan 3, 38; m 61; c 2. PHYSIOLOGY. Educ: Carroll Col, Wis, BS, 59; Univ Wis-Madison, PhD(endocrinol), 63. Prof Exp: Fel zool, Univ Wis, 63-64; from instr to asst prof physiol, 64-70, ASSOC PROF PHYSIOL, CHICAGO MED SCH, 70- Mem: Am Soc Zoologists; Endocrine Soc; Am Physiol Soc; Int Soc Neuroendocrinol; Soc Study Reproduction. Res: Influence of the brain and gonadal steroids on ovulation. Mailing Add: Dept of Physiol Chicago Med Sch Chicago IL 60612

MCCORMACK, DONALD EUGENE, b Chicago, Ill, Apr 22, 30; m 49; c 4. SOIL SCIENCE. Educ: Univ Ill, BS, 53, MS, 60; Ohio State Univ, PhD(soil sci), 73. Prof Exp: Soil scientist field, 53-60, asst state soil scientist, Mich, 60-63, state soil scientist, Ohio, 63-72, asst dir soil surv interpretations div, 72-76, LEADER RESOURCE & MGT INFO SYST, US SOIL CONSERV SERV, 76- Concurrent Pos: Mem Transportation Res Bd, Nat Acad Sci, 72- Honors & Awards: Outstanding Performance Award, US Soil Conserv Serv, 74. Mem: Soil Sci Soc Am; Am Soc Agron; Int Soil Sci Soc; Soil Conserv Soc Am. Res: The relationship of soil properties used in identifying, characterizing and classifying soils by pedologists to the engineering behavior of soils. Mailing Add: Soil Conserv Serv S Agr Bldg Rm 5210 Washington DC 20250

MCCORMACK, FRANCIS JOSEPH, b Mobile, Ala, Dec 6, 38; m 67. PLASMA PHYSICS. Educ: Spring Hill Col, BS, 60; Fla State Univ, PhD(physics), 64. Prof Exp: Asst prof, 67-71, ASSOC PROF PHYSICS, UNIV NC, GREENSBORO, 71- Mem: Am Phys Soc; Am Asn Physics Teachers. Res: Kinetic theory of ordinary and ionized gases. Mailing Add: Dept of Physics Univ of NC Greensboro NC 27412

MCCORMACK, GRACE, b Rochester, NY, Feb 16, 08. MICROBIOLOGY. Educ: Univ Rochester, AB, 41; Univ Md, MS, 51; Nat Registry Microbiologists, regist, 75. Prof Exp: Technician endocrinol, Sch Med & Dent, Univ Rochester, 42-48; bacteriologist fish & wildlife serv, US Dept Interior, 48-53; assoc bacteriologist, Md State Dept Health, 53-55; bacteriologist-technologist, Vet Admin Hosp, Canandaigua, NY, 55-66; from asst prof to assoc prof, 66-75, PROF BIOL SCI, MONROE COMMUNITY COL, 75- Concurrent Pos: Lectr, Univ Md, 50. Honors & Awards: Sustained Superior Serv Award, Vet Admin Hosp, 63. Mem: AAAS; fel Am Inst Chem; Am Soc Microbiol; Am Chem Soc; Am Pub Health Asn. Res: Assays of gonadotropic hormone; enterococci; antibiotic study of preservation of crabmeat; pollution studies of clam and oyster beds; comparative study of various media; cause of discoloration of oysters; collaborating on research problem; testing of viability of brewery yeast. Mailing Add: Dept of Biol Sci Monroe Community Col Rochester NY 14607

MCCORMACK, HAROLD ROBERT, b St Louis, Mo, Nov 25, 22; m 46; c 4. GEOPHYSICS. Educ: St Louis Univ, BS, 46. Prof Exp: Geophysicist, 43-53, dir geophys, 54-67, mgr geophys res, 68-75, MGR GEOPHYS DATA PROCESSING, SUN OIL CO, 75- Mem: Am Geophys Union; Soc Explor Geophys (secy-treas, 75-76); Seismol Soc Am. Res: Geophysical research applied to petroleum exploration. Mailing Add: Sun Oil Co PO Box 2880 Dallas TX 75221

MCCORMACK, JOHN JOSEPH, JR, b Boston, Mass, Jan 20, 38; m 62; c 4. ORGANIC CHEMISTRY, PHARMACOLOGY. Educ: Boston Col, BS, 59; Yale Univ, PhD(pharmacol), 64. Prof Exp: Nat Cancer Inst res fel med chem, Australian Nat Univ, 64-66; asst prof pharmacol, 66-69, ASSOC PROF PHARMACOL, UNIV VT, 69- Mem: AAAS; Am Chem Soc; The Chem Soc; Am Soc Pharmacol & Exp Therapeut. Res: Heterocyclic and medicinal chemistry; chemotherapy of neoplastic and protozoal diseases. Mailing Add: Dept of Pharmacol Univ of Vt Given Med Bldg Burlington VT 05401

MCCORMACK, KENT, solid state physics, molecular spectroscopy, see 12th edition

MCCORMACK, MAXWELL LELAND, JR, b Atlanta, Ga, July 18, 34; m 60; c 3. FORESTRY. Educ: Univ Maine, BS, 56; Duke Univ, MF, 59, DF(silvics), 63. Prof Exp: Forester-photogrammetrist, Rayonier Inc, Wash, 56-57; asst, Duke Univ, 60-61; from instr to asst prof, Southern Ill Univ, 61-64; asst prof, 64-69, ASSOC PROF FORESTRY, UNIV VT, 69- Mem: Soc Am Foresters. Res: Forest biology and growth; root development of conifers; silviculture of New England Christmas tree species. Mailing Add: Dept of Forestry Univ of Vt Burlington VT 05401

MCCORMACK, MICHAEL KEVIN, b South Amboy, NJ, Nov 3, 49. HUMAN GENETICS. Educ: Seton Hall Univ, NJ, BA, 71; Univ Minn, Minneapolis, PhD(genetics), 75. Prof Exp: ASST PROF GENETICS, RUTGERS UNIV, NEW BRUNSWICK, 75-; RES SCIENTIST GROWTH & DEVELOP, W M KROGMAN CTR RES CHILD GROWTH & DEVELOP, CHILDREN'S HOSP OF PHILADELPHIA, 75- Concurrent Pos: Fel pediat hemat, Dept Pediat, Univ Minn Med Ctr, Minneapolis, 75; NIH fel med genetics, 75. Mem: Am Soc Human Genetics; Behav Genetics Asn; Am Soc Hemat; Am Soc Cell Biol. Res: Biochemical genetics of hemoglobin structure and synthesis; structure-function relationships in mutant human hemoglobins; chemical modification of human hemoglobins; growth and development of children with various hemoglobinopathies. Mailing Add: Div Human Genetics Rutgers Univ 32 Bishop St New Brunswick NJ 08903

MCCORMACK, ROBERT MORRIS, b Sheboygan, Wis, June 11, 18; m 44; c 3. PLASTIC SURGERY. Educ: Swarthmore Col, BA, 40; Univ Chicago, MD, 43; Am Bd Plastic Surg, dipl. Prof Exp: From instr to assoc prof, 50-57, PROF PLASTIC SURG, SCH MED, UNIV ROCHESTER, 57-, VCHMN DEPT SURG, 68- Concurrent Pos: Am Soc Plastic & Reconstruct Surg scholar, 53. Mem: Am Soc Surg of Hand (pres, 64-65); Am Soc Plastic & Reconstruct Surg; Am Col Surg; Am Asn Plastic Surg (secy, 62-64, pres, 69); Am Burn Asn (pres, 72). Res: Transplantation of tissues. Mailing Add: Dept of Med Univ of Rochester Sch of Med Rochester NY 14642

MCCORMACK, SHIRLEY ANN, endocrinology, biochemistry, see 12th edition

MCCORMACK, WILLIAM BREWSTER, b Mirror, Alta, Nov 20, 23; US citizen; m 45; c 2. ORGANIC CHEMISTRY. Educ: Univ Alta, BSc, 44; Univ Wis, PhD(chem), 48. Prof Exp: Res chemist, 48-53, res supvr, 53-55, RES ASSOC, JACKSON LAB, E I DU PONT DE NEMOURS & CO, INC, 55-, INTERNAL TECH CONSULT, 70- Mem: Am Chem Soc; Combustion Inst; The Chem Soc. Res: Phosphorus; fluorine; organometallics; dyes and textile chemicals; nucleophilic and free radical processes; scientific philosophy; combustion; permeation separation. Mailing Add: Org Chem Dept Jackson Lab E I du Pont de Nemours & Co Inc Wilmington DE 19898

MCCORMACK, WILLIAM C, b Louisville, Ky, May 7, 25; m 52; c 4. PEDIATRICS, PHYSIOLOGY. Educ: Bowdoin Col, AB, 48; Univ Rochester, MD, 52. Prof Exp: Resident pediat, Strong Mem Hosp, 52-56; PEDIATRICIAN, McFARLAND CLIN, 56-; RES PROF PHYSIOL & PHARMACOL, SCH VET MED, IOWA STATE UNIV, 65-, PROF BIOMED ENG, 74- Concurrent Pos: Grants, Iowa State Univ, 65-68 & Iowa Thoracic Soc, 68-69. Mem: Am Acad Pediat. Res: Pediatric-respiratory distress of infancy. Mailing Add: Dept of Vet Anat Pharmacol & Physiol Iowa State Univ Sch Vet Med Ames IA 50010

MCCORMACK, WILLIAM CHARLES, b Sutherland, Iowa, Mar 31, 29; m 62. ANTHROPOLOGY, ANTHROPOLOGICAL LINGUISTICS. Educ: Univ Chicago, BA, 48; Stanford Univ, BA, 49, MA, 50; Univ Chicago, PhD(anthrop), 56. Prof Exp: Assoc res anthropologist, Modern India Proj, Univ Calif, Berkeley, 58-59; vis lectr anthrop of India, Sch Oriental & African Studies, Univ London, 59-60; asst prof anthrop, ling & Indian studies, Univ Wis-Madison, 60-64; assoc prof anthrop, Duke Univ, 64-69; head dept ling, Univ Calgary, 72-73; PROF ANTHROP & LING, UNIV CALGARY, 69- Concurrent Pos: Rockefeller Found Jr Ling fel, Deccan Col, India, 56-58; US Off Educ grant, Univ Wis-Madison & Mysore, India, 62-64; Am Coun Learned Soc grant in aid, Mysore, India, 62-63; Res Coun res grants, Duke Univ, 64-66, Endowment Found grants, 65-66 & 66-67; Endowment Found grants, Neth & Scotland, 67-68 & Duke Univ Comt Comp Study Social Systs & Insts, 67-68; res grant, Univ Calgary, 69-71; Ad hoc mem, Nat Anthrop Selection Comt, Nat Inst Ment Health, 66-67; mem fel eval comt, Shastri Indo-Can Inst, 70 & 71; linguist, Prog Comt, Int Cong Anthrop & Ethnol Sci, Ill, 73; ad hoc consult, Peace Corps, Govt Agencies, Jour & Univ Presses. Mem: Fel Am Anthrop Asn; fel Royal Anthrop Inst Gt Brit & Ireland; Ling Soc Am; Can Sociol & Anthrop Asn; Am Acad Relig. Res: Interactional strategies in language evolution and language ecology; comparative popular religion; linguistic symbols and structure of religious discourse in complex societies. Mailing Add: Dept of Anthrop Univ of Calgary Calgary AB Can

MCCORMICK, BAILIE JACK, b Amarillo, Tex, Aug 20, 37; m 58; c 2. INORGANIC CHEMISTRY. Educ: West Tex State Univ, BS, 59; Okla State Univ, PhD(chem), 62. Prof Exp: Robert A Welch Found fel, Univ Tex, 62-64; from asst prof to assoc prof, 64-73, PROF CHEM, W VA UNIV, 73- Mem: Am Chem Soc. Res: Structure, spectra and reactions of coordination compounds; unusual oxidation states; non-aqueous solvents; donor properties of sulfur; sulfur-nitrogen compounds; biological aspects of metal binding. Mailing Add: Dept of Chem WVa Univ Morgantown WV 26506

MCCORMICK, CARROLL GENE, solid state physics, see 12th edition

MCCORMICK, CLYDE TRUMAN, b Oblong, Ill, Mar 6, 08; m 37; c 3. APPLIED MATHEMATICS. Educ: Univ Ill, AB, 30, AM, 32; Ind Univ, PhD(math physics), 37. Prof Exp: Prof math & physics, Greenville Col, 37-38; from asst prof to prof math, 38-44, head dept, 54-71, EMER PROF MATH, ILL STATE UNIV, 71-; VIS PROF, PALM BEACH ATLANTIC COL, 71- Mem: Am Math Soc; Math Asn Am. Res: Applied mathematics in mechanics and sound. Mailing Add: Dept of Math Palm Beach Atlantic Col West Palm Beach FL 33401

MCCORMICK, DONALD BRUCE, b Front Royal, Va, July 15, 32; m 55; c 3. BIOCHEMISTRY. Educ: Vanderbilt Univ, BA, 53, PhD(biochem), 58. Prof Exp: Asst biochem, Vanderbilt Univ, 53-58; consult biochemist, Interdepartment Comt Nutrit for Nat Defense, Madrid, 58; USPHS res fel, Univ Calif, 58-60; asst prof nutrit, 60-65, assoc prof biochem & biol sci, 65-69, PROF NUTRIT, BIOCHEM & MOLECULAR BIOL, CORNELL UNIV, 69- Concurrent Pos: Guggenheim Mem Found fel, 66-67. Honors & Awards: Mead Johnson Award, Am Inst Nutrit, 70. Mem: AAAS; Am Soc Biol Chemists; Am Inst Nutrit; Am Chem Soc; Am Soc Microbiol. Res: Chemistry and biochemistry of cofactors; enzymology. Mailing Add: Div of Nutrit Sci Cornell Univ Ithaca NY 14850

MCCORMICK, FRANCIS B, b New Concord, Ohio, Apr 2, 16; m 40, 70. AGRICULTURAL ECONOMICS. Educ: Ohio State Univ, BS, 39, MS, 47, PhD(agr econ), 53. Prof Exp: County supvr, Farm Security Admin, Jackson, Vinton & Perry Counties, Ohio, USDA, 40-43; asst, 46-49, from instr to assoc prof agr econ, 49-64, staff mem, Dept Agr Econ & Rural Soc, 61-66, actg chmn, Dept Agr Econ, 66-67, PROF AGR ECON, OHIO STATE UNIV, 64-, ASSOC CHMN DEPT, 68- Mem: Int Conf Agr Econ; Am Agr Econ Asn; Am Financial Mgt Asn. Res: Agricultural policy and agricultural marketing. Mailing Add: 760 Beauty View Ct Columbus OH 43214

MCCORMICK, GEORGE M, II, b Memphis, Tenn, Apr 3, 39; m 64; c 3. PATHOLOGY, ENDOCRINOLOGY. Educ: Southwestern at Memphis, BS, 61, PhD(physiol, biophys), 65; Univ Tenn, MD, 69. Prof Exp: Prin investr, Jane Coffin Childs Mem Fund grant, 67-68, asst resident path, 69-70, chief resident, 70-71; ASST PROF PATH, SCH MED, LA STATE UNIV, SHREVEPORT, 71-; PATHOLOGIST, VET ADMIN HOSP, 71- Concurrent Pos: USPHS fel oncol, Med Units, Univ Tenn, Memphis, 65-66, Am Cancer Soc instnl grants, 65-66 & 68-71, clin cancer fel, 69-71; asst resident, City of Memphis Hosps, 69-70, chief resident, 70-71. Mem: AAAS. Res: Elucidation of hormonal states which affect the production and growth of chemically induced mammary tumors in experimental animals. Mailing Add: Dept of Path La State Univ Sch of Med Shreveport LA 71103

MCCORMICK, GEORGE R, b Columbus, Ohio, Apr 12, 36; m 62; c 2. MINERALOGY, PETROLOGY. Educ: Ohio Wesleyan Univ, BA, 58; Ohio State Univ, MSc, 60, PhD(mineral), 64. Prof Exp: Geologist, Ohio Div, Geol Surv, 61-62 & Norris Lab, US Bur Mines, 62-65; asst prof geol, Boston Univ, 65-68; ASSOC PROF GEOL, UNIV IOWA, 68- Concurrent Pos: Consult, Kennecott Copper Co, 67- & Humble Minerals, 71-; geologist, US Geol Surv, 72- Mem: Am Ceramic Soc; Mineral Soc Am; Mineral Soc Gt Brit & Ireland; Geol Soc Am. Res: Phase equilibrium studies of silicate and germanate systems; microchemical study of igneous and metamorphic complexes; chemical and mineralogic study of Iowa clays; geochemical studies of copper, lead, zinc and silver deposits of Utah, Nevada and Arizona. Mailing Add: Dept of Geol Univ of Iowa Iowa City IA 52240

MCCORMICK, GERALD, biochemistry, see 12th edition

MCCORMICK, J FRANK, b Indianapolis, Ind, Oct 25, 35; m 59; c 3. ECOLOGY. Educ: Butler Univ, BS, 58; Emory Univ, MS, 60, PhD(biol), 61. Prof Exp: Asst prof biol, Vanderbilt Univ, 61-63; asst prof zool, Univ Ga, 63-64; asst prof bot, 64-71, PROF BOT & ECOL, UNIV NC, CHAPEL HILL, 71- Concurrent Pos: Grants, NSF, Natural Sci Comt, Vanderbilt Univ, 62; AEC, 64-, radiation ecol training, 66-; Univ NC fac res coun, 65 & 66; USPHS ecol training prog, 66-; consult, Oak Ridge Nat Lab, 62-; PR Nuclear Ctr, 64- Honors & Awards: Sigma Xi Res Award, Emory Univ, 61. Mem: AAAS; Ecol Soc Am. Res: Environmental and radiation biology; evolution and natural selection; environmental analysis; population; pollution studies. Mailing Add: Dept of Bot Univ of NC Chapel Hill NC 27514

MCCORMICK, J JUSTIN, b Detroit, Mich, Sept 26, 33. CANCER, MOLECULAR BIOLOGY. Educ: St Paul's Col, Washington, DC, BA, 57, MA, 59; Cath Univ Am, MS, 61, PhD(biol), 67. Prof Exp: Fel, McArdle Lab Cancer Res, Univ Wis, 67-70, res assoc, 70-71; res scientist, 71-73, CHIEF, MOLECULAR BIOL LAB, MICH CANCER FOUND, 73- Mem: Am Asn Cancer Res; Am Soc Cell Biol; Biophys Soc; Tissue Cult Asn; Sigma Xi. Res: Mutagenic and transforming effect of chemical and physical carcinogens on human cells in culture; DNA repair; nucleic acids of RNA tumor viruses; biophysical characterization of RNA tumor viruses. Mailing Add: Mich Cancer Found 110 E Warren Detroit MI 48201

MCCORMICK, J ROBERT D, b St Albans, WVa, Feb 24, 21; m 53; c 1. BIOCHEMISTRY. Educ: Rensselaer Polytech Inst, BS, 46; Univ Calif, Los Angeles, PhD(chem), 54. Prof Exp: Res chemist, Winthrop Chem Co, 43-46; develop chemist, 49-54, biochem group leader, 54-58, head biochem dept, 58-60, res assoc fermentation biochem, 60-65, RES FEL FERMENTATION BIOCHEM, LEDERLE LABS, AM CYANAMID CO, 65- Concurrent Pos: Am Cyanamid award advan study sr res fel, Univ Leicester, 70-71. Mem: Fel AAAS; fel Am Inst Chem; Am Chem Soc; NY Acad Sci. Res: Natural products isolation and structure. Mailing Add: McNamara Rd Spring Valley NY 10977

MCCORMICK, JACK SOVERN, b Indianapolis, Ind, Jan 19, 29; m 70; c 2. ENVIRONMENTAL MANAGEMENT, ECOLOGY. Educ: Butler Univ, BS, 51; Rutgers Univ, PhD(bot), 55. Prof Exp: Counr bot, Indianapolis Children's Mus, 48-49; naturalist, Ind Dept Conserv, 49-51; consult ecol, Am Mus Natural Hist, 54-55, adv to dir, 55-61; res assoc ecol, Inst Polar Studies & asst prof bot, Ohio State Univ, 61-63; cur & chmn ecol, Acad Natural Sci Philadelphia, 63-71; PRES, JACK McCORMICK & ASSOCS, 71- Concurrent Pos: Res assoc, Am Mus Natural Hist, Kalbfleisch Field Res Sta, 61-74; mem res adv comt, Northeastern Forest Exp Sta, US Forest Serv, 62-73; lectr, Univ Pa, 63-72, landscape architect, 65-74; Barnes Found lectr, 64; mem hwy res bd, Nat Res Coun, 70- Mem: AAAS; Ecol Soc Am; Torrey Bot Club; Int Soc Biometeorol; Int Soc Trop Ecol. Res: Mechanisms of vegetation change; vegetation geography of North America; vegetation mapping and management; methodological techniques in vegetation sampling; bibliography of vegetation literature; natural area preservation; environmental impact assessment. Mailing Add: Jack McCormick & Assocs 860 Waterloo Rd Devon PA 19333

MCCORMICK, JAMES E, b Providence, RI, Nov 5, 27; m 52; c 6. GEOLOGY. Educ: Boston Univ, AB, 53. Prof Exp: Jr engr, 53-54, jr geologist, 54-57, geologist, 57-58, asst regional geologist, 58-64, div geologist, 64-67, div explor mgr, 67-70, key areas regional geologist, 71, EXPLOR PROGS MGR, SUN OIL CO, 72- Mem: Am Asn Petrol Geologists; Am Petrol Inst. Mailing Add: Sun Oil Co Box 1501 Houston TX 77001

MCCORMICK, JOHN PAULING, b Lansing, Mich, May 26, 43; m 68; c 2. BIO-ORGANIC CHEMISTRY. Educ: DePauw Univ, BA, 65; Stanford Univ, PhD(chem), 71. Prof Exp: Assoc terpenoid biosynthesis, Swiss Fed Inst Technol, 70-71; vis prof org chem, State Univ Groningen, 71-72; ASST PROF ORG CHEM, UNIV MO-COLUMBIA, 72- Mem: Am Chem Soc; AAAS. Res: Terpene synthesis and biosynthesis; mechanism of biochemical reactions; structure of physiologically active natural products; development of novel, stereospecific reactions. Mailing Add: 415 Maplewood Dr Columbia MO 65201

MCCORMICK, JON MICHAEL, b Portland, Ore, Feb 8, 41; m 71. MARINE ECOLOGY. Educ: Portland State Univ, BS, 63; Ore State Univ, MS, 65, PhD(biol sci), 69. Prof Exp: Res asst oceanog, Ore State Univ, 63-65; asst prof biol, Millersville State Col, 68-70; ASST PROF BIOL, MONTCLAIR STATE COL, 70- Concurrent Pos: NSF fel, Marine Biol Lab, Woods Hole, 69, partic, NSF instr Sci Equip Prog, Millersville State Col, 69-70; instr, Pa Marine Sci Consortium, 69-70; adj instr, Jersey City State Col, 70-71; instr, NJ Marine Sci Consortium, 71. Mem: AAAS; Am Inst Biol Sci; Am Soc Limnol & Oceanog; Int Phycol Soc. Res: Estuarine ecology; marine zooplankton; marine benthos, hydrozoa, population dynamics and trophic ecology. Mailing Add: Dept of Biol Montclair State Col Upper Montclair NJ 07043

MCCORMICK, KENNETH JAMES, b Toledo, Ohio, Sept 11, 37; m 63; c 3. CANCER. Educ: Univ Toledo, BS, 59; Univ Mich, Ann Arbor, MS, 62, PhD(microbiol), 65. Prof Exp: From instr to assoc prof exp biol, Dept Surg, Baylor Col Med, 65-75; HEAD SECT TUMOR IMMUNOL, LAB CANCER RES, STEHLIN FOUND CANCER RES, ST JOSEPH HOSP, HOUSTON, 75- Concurrent Pos: Am Cancer Soc instnl grants, Baylor Col Med, 67-68 & 70-71, vis assoc prof exp biol, 75- Mem: AAAS; Am Asn Cancer Res; Am Soc Microbiol. Res: Tumor immunology and virology. Mailing Add: St Joseph Hosp Lab Cancer Res 1717 Pierce Houston TX 77003

MCCORMICK, MACK HAROLD, biochemistry, see 12th edition

MCCORMICK, MICHAEL PATRICK, b Canonsburg, Pa, Nov 23, 40; m 62; c 2. ATMOSPHERIC PHYSICS. Educ: Washington & Jefferson Col, BA, 62; Col William & Mary, MA, 64, PhD(physics), 67. Prof Exp: Head photoelectronic instrumentation sect, 67-75, CHIEF, AEROSOL MEASUREMENTS RES BR, LANGLEY RES CTR, NASA, 75- Mem: Optical Soc Am. Res: Design, development and application of advanced sensors to atmospheric research; light scattering and atmospheric optics; electrooptics; laser radars. Mailing Add: Langley Res Ctr M/S 475 NASA Hampton VA 23665

MCCORMICK, NEIL GLENN, b Everett, Wash, Aug 17, 27; m 57; c 3. MICROBIOLOGY. Educ: Univ Wash, BS, 51, MS, 57, PhD(microbiol), 60. Prof Exp: Anal chemist, Wash State Dept Agr, 51-53; res assoc microbiol, Univ Wash, 60-61; USPHS fel, Univ Wis, 61-63; asst prof, Univ Va, 63-68; RES MICROBIOLOGIST, FOOD SCI LAB, US ARMY NATICK DEVELOP CTR, 68- Concurrent Pos: Nat Inst Arthritis & Infectious Dis grants, 64-66. Mem: AAAS; Am Soc Microbiol. Res: Biodegradation and biotransformation of organic nitro-compounds; biotransformation efficiency improvement by selection of radiation-induced mutants. Mailing Add: Food Sci Lab US Army Natick Develop Ctr Natick MA 01760

MCCORMICK, PATRICK GARY, b Milwaukee, Wis, Dec 20, 39; m 65. ANALYTICAL CHEMISTRY. Educ: Mass Inst Technol, BS, 62; Univ Minn, PhD(anal chem), 67. Prof Exp: Fel biochem, Univ Minn, 66-67; asst prof chem, Marquette Univ, 67-74, PROJ LEADER, QUAL CONTROL DIV, PFIZER INC, 74- Mem: AAAS; Am Chem Soc. Res: Pharmaceutical analysis and methods development; automation; dissolution testing. Mailing Add: Qual Control Div Pfizer Inc 630 Flushing Ave Brooklyn NY 11206

MCCORMICK, PHILIP THOMAS, b Oak Park, Ill, Nov 28, 26; m 57; c 5. THEORETICAL PHYSICS, PLANETARY SCIENCE. Educ: Univ Notre Dame, BS, 48, PhD(physics), 54. Prof Exp: Instr physics, DePaul Univ, 52-54; head sidewinder simulation sect, Naval Ord Test Sta, 54-57, physicist weapons planning group, 57-58; from asst prof to assoc prof, 58-69, PROF PHYSICS, UNIV SANTA CLARA, 70- Concurrent Pos: Sr eng specialist advan progs sect, Philco Corp, 59-61; Fulbright lectr, Univ Sind, Pakistan, 64-65; consult weapons planning group, Naval Weapons Ctr, 67- Mem: AAAS; Am Phys Soc; Am Asn Physics Teachers. Res: Space science; electromagnetic interactions with nuclei; weapons and communications systems analysis; planetary environments; radio wave investigations of planetary atmospheres and ionospheres; aeronomy; mathematical modeling of planetary atmospheres. Mailing Add: Dept of Physics Univ of Santa Clara Santa Clara CA 95053

MCCORMICK, ROY L, b Hillsboro, Ind, Feb 8, 29; m 49; c 2. MATHEMATICS. Educ: Wabash Col, AB, 50; Wash Univ, MA, 60; Purdue Univ, PhD(math ed), 65. Prof Exp: Teacher high schs, Ind, 50-51 & 56-60; instr math ed, Purdue Univ, 60-62; from asst prof to assoc prof, 62-71, PROF MATH, BALL STATE UNIV, 71- Concurrent Pos: Dir, NSF Inst, 67-72; chmn state adv coun, Elem & Sec Educ Act Title III, 70-72. Mem: Math Asn Am. Res: Training of secondary and elementary teachers of mathematics. Mailing Add: Dept of Math Sci Ball State Univ Muncie IN 47306

MCCORMICK, STEPHEN FAHRNEY, b Glendale, Calif, Sept 16, 44; m 68. NUMERICAL ANALYSIS. Educ: San Diego State Col, BA, 66; Univ Southern Calif, PhD(math), 71. Prof Exp: Mem tech staff data anal, Hughes Aircraft Co, 66-68; asst prof math, Pomona Col, 70-72; asst dir, Inst Educ Comput, Claremont Cols, 70-73; ASST PROF NUMERICAL ANAL, COLO STATE UNIV, 73- Concurrent Pos: Partner & dir comput, Solar Environ Eng Co, 74- Mem: Soc Indust & Appl Math; Am Math Soc; Asn Comput Mach. Res: Matrix linear algebra, optimization, approximation theory, and algebraic iterative methods in general. Mailing Add: Dept of Math Colo State Univ Ft Collins CO 80523

MCCORMICK, WILLIAM CONNER, b Ft White, Fla, July 24, 20; m 55; c 3. ANIMAL BREEDING. Educ: Univ Fla, BS, 42; Kans State Col, MS, 47; Agr & Mech Col, Tex, PhD, 54. Prof Exp: Asst animal husbandman, NFla Agr Exp Sta, 42-44; from asst animal husbandman to animal husbandman, 45-67, PROF & HEAD DEPT ANIMAL SCI, COASTAL PLAIN EXP STA, UNIV GA, 67- Res: Animal breeding and feeding of cattle and swine. Mailing Add: Coastal Plain Exp Sta Univ of Ga Tifton GA 31794

MCCORMICK, WILLIAM DEVLIN, b Tacoma, Wash, May 9, 31; m 70. PHYSICS. Educ: Calif Inst Technol, BS, 53; Duke Univ, PhD(physics), 59. Prof Exp: Fulbright travel grant, Univ Padua, 59-60; asst prof physics, Univ Wash, 63-68; ASSOC PROF PHYSICS, UNIV TEX, AUSTIN, 68- Concurrent Pos: Sloan fel, Univ Wash, 63-65; Richland fac fel, Battelle-Northwest, 67-68. Mem: Am Inst Physics. Res: Low temperature and solid state physics; liquid helium; phase transitions; nuclear magnetic resonance and calorimetry of molecular solids; surface physics. Mailing Add: Dept of Physics Univ of Tex Austin TX 78712

MCCORMICK, WILLIAM F, b Riverton, Va, Sept 9, 33; m 54; c 2. PATHOLOGY, NEUROPATHOLOGY. Educ: Univ Chattanooga, BS, 53; Univ Tenn, MD, 55, MS, 57; Am Bd Path, dipl, 60. Prof Exp: Asst path, Univ Tenn, 57-60, from instr to asst prof, 60-64; from assoc prof neuropath to prof path, Univ Iowa, 64-71, prof neurol, 71-73; PROF NEUROL & PATH, UNIV TEX MED BR GALVESTON, 73- Concurrent Pos: NIH spec fel & instr neuropath, Col Physicians & Surgeons, Columbia Univ, 60-61; scientist, Armed Forces Inst Path, 65; consult, Vet Admin Hosp, Iowa City, 64-73. Mem: Am Asn Path & Bact; Am Soc Exp Pathologists; Am Soc Human Genetics; Am Asn Neuropath; fel Am Col Pathologists. Res: Aneurysms, angiomas and infections of the central nervous system; diseases of muscle, brain death. Mailing Add: Div of Neuropath Univ of Tex Med Br Galveston TX 77550

MCCORMICK, WILLIAM WALLACE, b Marissa, Ill, Aug 30, 06; m 38. ATOMIC SPECTROSCOPY. Educ: Geneva Col, BS, 27, ScD, 69; Univ Mich, MS, 31, PhD(physics), 37. Prof Exp: Prin high sch, Pa, 27-29; from instr to prof physics,

Geneva Col, 29-43; res physicist, Univ Mich, 43-45; prof physics & head dept, Col William & Mary, 45-47; from asst prof to assoc prof, 47-59, PROF PHYSICS, UNIV MICH, ANN ARBOR, 59- Concurrent Pos: Regional counr, Mich State Univ, 64-65. Mem: Am Phys Soc; Optical Soc Am; Am Asn Physics Teachers. Res: Extreme ultraviolet spectroscopy; conduction of electricity through gases. Mailing Add: 749 Physics-Astron Bldg Univ of Mich Ann Arbor MI 48104

MCCORNACK, ANDREW ADAMS, b Eugene, Ore, Nov 12, 14; m 55; c 2. HORTICULTURE. Educ: Ore State Col, BS, 37; Mass State Col, MS, 39. Prof Exp: Farm adv, Agr Exten Serv, Univ Calif, 47-51; FOOD TECHNOLOGIST, FLA DEPT CITRUS, 52- Res: Control of decay of citrus fruits for sale on the fresh fruit market. Mailing Add: Fla Dept of Citrus Agr Res & Educ Ctr Box 1088 Lake Alfred FL 33850

MCCORQUODALE, DONALD JAMES, b Winnipeg, Man, Aug 27, 27; nat US; m 51; c 3. BIOCHEMISTRY. Educ: Univ BC, BA, 50; Univ Wis, PhD(cytol, biochem), 55. Prof Exp: Am Cancer Soc fels, Dept Oncol, McArdle Mem Inst, Univ Wis, 55-57 & Max Planck Inst Biochem, Munich, 57-58; from asst prof to assoc prof biochem, Emory Univ, 58-67; assoc prof, Southwest Ctr Advan Studies, 68-69, from assoc prof to prof, Univ Tex, Dallas, 69-75; PROF BIOCHEM, MED COL OHIO, 75- Concurrent Pos: USPHS res career develop award, 60-67; vis assoc prof, Mass Inst Technol, 65-66. Mem: AAAS; Am Chem Soc; Am Soc Biol Chemists; Am Soc Microbiol. Res: Biochemical virology; macromolecular biosynthesis; cellular control mechanisms. Mailing Add: Dept of Biochem Med Col of Ohio Toledo OH 43614

MCCORRISTON, JAMES ROLAND, b Sask, July 27, 19; wid; c 3. SURGERY. Educ: Univ Sask, BA, 39; Queen's Univ, Ont, MD & CM, 43; McGill Univ, MSc, 48; FRCPS(C), 50; Am Bd Surg, dipl, 52. Prof Exp: Demonstr, McGill Univ, 51-53, lectr, 53-56, Markle scholar med sci, 53-58, from asst prof to assoc prof surg, 56-63; head dept, 63-73, PROF SURG, QUEEN'S UNIV, ONT, 63-; SURGEON, KINGSTON GEN HOSP, 73- Concurrent Pos: Clin asst surg, Royal Victoria Hosp, Montreal, 51-53, asst surgeon, 53-59, assoc surgeon, 59-63, hon consult, 63-; surgeon-in-chief, Kingston Gen Hosp, 63-73; sr consult, Can Forces Hosp, Kingston; consult, Hotel Dieu Hosp, Kingston. Mem: Fel Am Col Surgeons; Whipple Surg Soc; Can Med Asn; Can Asn Clin Surg; Can Asn Gastroenterol. Res: Clinical, general and experimental surgery; peptic ulcer; blood volume; electrolyte balance; homotransplantation; esophageal physiology. Mailing Add: Dept of Surg Queen's Univ Kingston ON Can

MCCOSKER, JOHN E, b Los Angeles, Calif, Nov 17, 45. AQUATIC BIOLOGY, ICHTHYOLOGY. Educ: Occidental Col, BA, 67; Scripps Inst Oceanog, PhD(marine biol), 73. Prof Exp: Res fel ichthyol, Smithsonian Trop Res Inst, 70-71; lectr marine biol, Univ Calif, San Diego, 73; SUPT, STEINHART AQUARIUM, CALIF ACAD SCI, 73- Concurrent Pos: Adj prof, San Francisco State Univ, 75- Mem: AAAS; Am Soc Ichthyologists & Herpetologists; Soc Protection Old Fishes; Soc Syst Zool; Am Asn Zool Parks & Aquariums. Res: Systematics of marine fishes; aquarium maintenance of mesopelagic animals; behavior of sea snakes of the family Hydrophiidae; biology of primitive fishes, particularly the coelacanth. Mailing Add: Steinhart Aquarium Cal Acad Sci Golden Gate Park San Francisco CA 94118

MCCOUBREY, ARTHUR ORLAND, b Regina, Sask, Mar 11, 20; nat US; m 42; c 3. PHYSICS, SCIENCE ADMINISTRATION. Educ: Calif Inst Technol, BS, 43; Univ Pittsburgh, PhD(physics), 53. Prof Exp: Res engr, Res Labs, Westinghouse Elec Corp, 43-47, res physicist, 47-52, adv physicist, 52-57; mgr physics dept, Nat Co, Inc, 57-60; mgr res & adv develop, Bomac Labs, Inc, 60-63; mgr atomic frequency stand, Varian Assocs, Mass, 64-67, mgr res & develop, Quantum Electronics Div, Calif, 67-68, dir cent res labs, 68-72; vpres, Frequency & Time Systs, Inc, Danvers, Mass, 72-74; DIR, INST BASIC STAND, NAT BUR STANDARDS, 74- Mem: Am Phys Soc; fel Inst Elec & Electronics Eng; NY Acad Sci. Res: Microwaves; resonance physics; atomic frequency standards; quantum devices. Mailing Add: Washington DC

MCCOUBREY, JAMES A, b Shaunavon, Sask, June 27, 14; m 42; c 2. ENVIRONMENTAL CHEMISTRY. Educ: Univ Alta, BSc, 34; McGill Univ, PhD, 40. Prof Exp: Res chemist, Shawinigan Chem, Ltd, 41-50, develop chemist, 50-52; chem engr, Res & Develop Dept, Can Nat Rwys, 52-55; mkt res mgr, NAm Cyanamid Ltd, 55-58, mgr, Com Develop Dept, Cyanamid of Can, Ltd, 58-63; head mkt, Res Cent Develop Dept, Domtar Ltd, Can, 63-67; vpres & gen mgr, Laurentian Labs, Ltd, 67-69; consult, Samson Belair Riddell Stead, 70-71; MGR, LAMBTON INDUST SOC, 71- Mem: AAAS; Am Chem Soc; Chem Mkt Res Asn; Chem Inst Can; Soc Chem Indust. Res: High polymer and organometallic chemistry; industrial development and economics of the chemical industry; environmental control. Mailing Add: Lambton Indust Soc Suite 201N 242A Indian Rd S Sarnia ON Can

MCCOURT, FREDERICK RICHARD WAYNE, b New Westminster, BC, Jan 16, 40; m 71. MOLECULAR PHYSICS, CHEMICAL PHYSICS. Educ: Univ BC, BSc, 63, PhD(chem), 66. Prof Exp: Nat Res Coun Can fel, Swiss Fed Inst Technol, 66-67; Nat Res Coun Can fel, Kamerlingh Onnes Lab, Leiden, 67-68, Found Fundamental Res Matter sci co-worker, 68-69; vis asst prof, Inst Physics, Univ Genoa, 69; asst prof, 70-72, ASSOC PROF CHEM, UNIV WATERLOO, 72-, ASSOC PROF APPL MATH, 73- Concurrent Pos: Alfred P Sloan Found fel, 73-75. Mem: Am Phys Soc; Can Asn Physicists. Res: Nuclear magnetic relaxation; light scattering; spectral line broadening; molecule-surface interactions; kinetic theory of polyatomic gases; molecular collision theory. Mailing Add: Dept of Chem Univ of Waterloo Waterloo ON Can

MCCOURT, ROBERT PERRY, b Highland Park, Mich, Apr 27, 29; m 50; c 7. VERTEBRATE ZOOLOGY, RADIATION BIOLOGY. Educ: Univ Mich, AB, 50; Ohio State Univ, MS, 52, PhD, 54. Prof Exp: Asst instr zool, Ohio State Univ, 52-54; from instr to asst prof, 54-62, ASSOC PROF ZOOL, HOFSTRA UNIV, 62- Mem: AAAS. Res: Radiation biology of artemia; techniques of liquid scintillation counting. Mailing Add: Dept of Biol Hofstra Univ Hempstead NY 11550

MCCOWAN, JAMES ROBERT, b Nampa, Idaho, Feb 24, 23; m 46; c 3. PHARMACY. Educ: Univ Colo, BS, 48, MS, 49; Univ Fla, PhD(pharm), 54. Prof Exp: From instr to asst prof pharm, Loyola Univ La, 49-55; from asst prof to prof, St Louis Col Pharm, 55-68; PROF PHARMACEUT, COL PHARM, UNIV ARK FOR MED SCI, 68-, ASST DEAN COL PHARM, 70- Mem: Am Pharmaceut Asn. Res: Ophthalmic solutions; aerosols. Mailing Add: Col Pharm Univ Ark for Med Sci Little Rock AR 72201

MCCOWAN, OTIS BLAKELY, b Monterey, Tenn, June 17, 34. MATHEMATICS. Educ: Tenn Technol Univ, BS, 59; La State Univ, MA, 66; George Peabody Col, PhD, 75. Prof Exp: Mathematician, Air Force Missile Develop Ctr, Holloman AFB, NMex, 62-63; math teacher, Rhea Cent High Sch, Dayton, Tenn, 63-65; instr math, Kilgore Col, Tex, 66-67; from asst prof to assoc prof math, 67-75, PROF MATH, BELMONT COL, NASHVILLE, 75- Mem: Math Asn Am. Res: Mathematics in education; mathematics teacher education. Mailing Add: Dept of Math & Physics Belmont Col Nashville TN 37203

MCCOWEN, MAX CREAGER, b Sullivan, Ind, July 4, 15; m 46. PARASITOLOGY. Educ: Ind State Univ, BS, 37, MS, 38. Prof Exp: Instr high sch, Ind, 38-42; res assoc, Res Labs, Eli Lilly & Co, 46-47, asst chief parasitol res, 47-48, head dept, 48-58, sr parasitologist, 58-65, chief parasitologist, Lilly Res Ctr, Surrey, Eng, 65-68, head parasitol res, 68-71, RES SCIENTIST AQUATIC BIOCIDES, GREENFIELD LABS, ELI LILLY & CO, 71- Concurrent Pos: Lectr, Sch Med, Ind Univ, 60. Mem: AAAS; Am Soc Parasitol; Am Soc Microbiol; Am Soc Trop Med & Hyg; Sigma Xi. Res: Chemotherapy of parasitic diseases; immunology of parasitic infections; diagnosis of protozoan and helminthic diseases; aquatic biocides; aquatic herbicides, algicides, molluscicides and aquatic larvacides; environmental research. Mailing Add: Greenfield Labs Eli Lilly & Co Greenfield IN 46140

MCCOWEN, SARA MOSS, b Washington, NC, Jan 7, 44; m 69. MICROBIAL PHYSIOLOGY. Educ: Duke Univ, BA, 66; Univ NC, MAT, 68; Va Commonwealth Univ, MS, 73, PhD(microbiol), 75. Prof Exp: ASST PROF BIOL, VA COMMONWEALTH UNIV, 75- Mem: Am Soc Microbiol. Res: Carbohydrate metabolism in Pseudomonas species. Mailing Add: 4227 Kingcrest Pkwy Richmond VA 23221

MCCOWN, BRENT HOWARD, b Chicago, Ill, Feb 21, 43; m 68. PHYSIOLOGICAL ECOLOGY, HORTICULTURE. Educ: Univ Wis-Madison, BS, 65, MS, 67, PhD(bot, hort), 69. Prof Exp: Res & develop coordr, US Army Cold Regions, Res & Eng Lab, Hanover, NH, 70-72; ASST PROF HORT, INST ENVIRON STUDIES, UNIV WIS-MADISON, 72- Concurrent Pos: Adj asst prof, Inst Arctic Biol, Univ Alaska, 70- Mem: AAAS; Am Soc Plant Physiol; Am Soc Hort Sci; Nat Parks & Conserv Org; Am Inst Biol Sci. Res: Adaption of plants to environment; plant juvenility; resource analysis and carrying capacity; influence of man's activities on plant growth and survival; petroleum toxicity to terrestrial ecosystems; eco-systems analysis. Mailing Add: Dept of Hort Univ of Wis Madison WI 53706

MCCOWN, DEAN AUGUSTUS, b Anderson, SC, Dec 25, 15; m 45; c 2. PHYSICS. Educ: The Citadel, BS, 37; Univ Ky, MS, 40; Ohio State Univ, PhD(physics), 49. Prof Exp: Instr physics, Univ Louisville, 40-41; instr radio, US Air Force, Ill, 41-42, radar officer, 42-44, electronics officer, 44-46, res & develop officer, Field Off Atomic Energy, Kirtland Air Force Base, NMex, 49-50, Hqs, Air Force Spec Weapons Ctr, 50-54 & mil applns atomic energy, 54-58, dir develop div, Hqs, Field Command, Defense Atomic Support Agency, 58-63; PROF PHYSICS, UNIV SOUTHERN COLO, 63- Mem: Am Phys Soc; AAAS; Sigma Xi; Am Asn Physics Teachers. Res: Nuclear physics; optics; mechanics; general astronomy. Mailing Add: Dept of Physics Univ Southern Colo Pueblo CO 81001

MCCOWN, JOHN JOSEPH, b Cleveland, Ohio, Mar 14, 29; m 48, 66; c 5. ANALYTICAL CHEMISTRY, RADIO CHEMISTRY. Educ: Drury Col, BS, 51; Univ Tenn, MS, 55. Prof Exp: Jr chemist, USDA, 51-52; from jr chemist to assoc chemist anal chem div, Oak Ridge Nat Lab, 52-56; from asst chemist to assoc chemist chem eng div, Argonne Nat Lab, 56-60, assoc chemist, Idaho Div, 60-63; sr scientist, Nev Test Opers, Westinghouse Astronuclear Lab, 63-65, fel scientist, 65-68; sr res scientist, Pac Northwest Labs, Battelle Mem Inst, 68-70; res assoc, 70-74, MGR SODIUM SYSTS ANAL, HANFORD ENG DEVELOP LAB, WESTINGHOUSE HANFORD CO, 74- Mem: Am Chem Soc; Am Nuclear Soc. Res: Remote analyses of highly radioactive materials; remote analytical techniques and equipment; analytical radiochemistry; reactor core analysis by gamma spectrometry methods; sodium technology; sodium and gas characterization and fuel failure monitoring. Mailing Add: Westinghouse Hanford Co Mat Tech Dept Box 1970 Richland WA 99352

MCCOWN, JOSEPH DANA, b Moscow, Idaho, Aug 31, 40. INDUSTRIAL ORGANIC CHEMISTRY, RESEARCH ADMINISTRATION. Educ: Univ Idaho, BS, 62, MS, 64; Univ Iowa, PhD(org chem), 68. Prof Exp: Sr res chemist, 68-71, info scientist, 71-73, RES SUPVR, MINN MINING & MFG CO, 73- Mem: Am Chem Soc. Res: Aryl sulfonyl isocyanates; hydroxamic acid esters; beta-diketones; N-nitrosoenamines; thermal rearrangement reactions; fluorocarbon chemistry. Mailing Add: 2443 Southcrest Ave St Paul MN 55119

MCCOWN, MALCOLM G, b Prairie Lea, Tex, Aug 14, 19; m 51; c 1. ALGEBRA. Educ: Trinity Univ, BS, 41, MEd, 51. Prof Exp: Instr, 55-60, ASST PROF MATH, TRINITY UNIV, 60- Mem: Am Math Soc. Res: Geometry; theory of equations. Mailing Add: Dept of Math Trinity Univ San Antonio TX 78284

MCCOY, BARRY, b Trenton, NJ, Dec 14, 40; m 70. THEORETICAL PHYSICS. Educ: Calif Inst Technol, BS, 63; Harvard Univ, PhD(physics), 67. Prof Exp: Res assoc, 67-69, asst prof, 69-74, ASSOC PROF PHYSICS, STATE UNIV NY STONY BROOK, 74- Res: Statistical mechanics. Mailing Add: Dept of Physics State Univ of NY Stony Brook NY 11790

MCCOY, CHARLES RALPH, b Freeport, Ill, June 14, 27; m 55; c 1. PHYSICAL CHEMISTRY. Educ: Roosevelt Univ, BS, 56; Northwestern Univ, PhD(chem), 60. Prof Exp: From instr to asst prof chem, Loyola Univ Chicago, 59-63; assoc prof, 63-66, PROF CHEM, UNIV WIS-WHITEWATER, 66- Mem: Am Chem Soc; Am Phys Soc. Res: Heterogeneous reactions; kinetics. Mailing Add: Dept of Chem Univ of Wis Whitewater WI 53190

MCCOY, CLARENCE JOHN, b Lubbock, Tex, July 25, 35; m 57; c 2. HERPETOLOGY. Educ: Okla State Univ, BS, 57, MS, 60; Univ Colo, PhD(zool), 65. Prof Exp: From asst cur to assoc cur, 64-72, CUR AMPHIBIANS & REPTILES, CARNEGIE MUS NATURAL HIST, 72- Concurrent Pos: Vis prof, Ariz State Univ, 69-70; adj assoc prof, Univ Pittsburgh, 72-; chmn comt syst resources herpet, 75- Mem: AAAS; Am Soc Ichthyologists & Herpetologists; Am Soc Mammal; Soc Syst Zoologists; Soc Study Amphibians & Reptiles. Res: Natural history and systematics of reptiles and amphibians; reptilian reproductive cycles; biogeography of Mexico; general vertebrate zoology. Mailing Add: Carnegie Mus Natural Hist 4400 Forbes Ave Pittsburgh PA 15213

MCCOY, CLAYTON WILLIAM, b Rochester, Minn, June 22, 38; m 63; c 1. ENTOMOLOGY. Educ: Gustavus Adolphus Col, BS, 60; Univ Nebr, MSc, 63; Univ Calif, PhD(entom), 67. Prof Exp: Res entomologist, USDA, 67-72; ASSOC ENTOMOLOGIST, AGR RES & EDUC CTR, UNIV FLA, 72- Mem: Entom Soc Am; Int Orgn Biol Control; Int Soc Invert Path. Res: Biological control of economic pests of citrus. Mailing Add: Agr Res & Educ Ctr Univ Fla PO Box 1088 Lake Alfred FL 33850

MCCOY, DAVID ROSS, b Culver City, Calif, July 21, 42; m 66; c 2. ORGANIC CHEMISTRY. Educ: Univ Calif, Los Angeles, BS, 63; Univ Ore, PhD(org chem), 67. Prof Exp: From chemist to sr chemist, 67-72, RES CHEMIST, TEXACO, INC, 72- Mem: Am Chem Soc. Res: Chemistry of heterocyclic compounds; organophosphorus chemistry; mechanisms of heterogeneous catalysis; petrochemical synthesis; enhanced oil recovery. Mailing Add: Amherst Lane Wappingers Falls NY 12590

MCCOY, DONALD W, b Kansas City, Kans, Aug 4, 35; m 56; c 2. MICROBIOLOGY, BIOCHEMISTRY. Educ: Univ Md, BS, 59, MS, 61, PhD(microbiol), 66. Prof Exp:

Microbiologist, Meat Inspection Div, USDA, 62-65; MICROBIOLOGIST, LEDERLE LABS, AM CYANAMID CO, 65- Mem: AAAS; Am Soc Microbiol. Res: Antimicrobial activity of sulfonamides; food microbiology; staphylococcal enterotoxins; immunology of mycobacteria and diphasic fungi. Mailing Add: 240 Center St Pearl River NY 10965

MCCOY, DOROTHY, b Waukomis, Okla, Aug 9, 03. MATHEMATICS. Educ: Baylor Univ, BA, 25; Univ Iowa, MS, 27, PhD(math), 29. Prof Exp: Teacher high sch, Iowa, 25-26; asst, Univ Iowa, 28-29; prof math & head dept, Belhaven Col, 29-40; chmn div phys & biol sci, 49-72, PROF MATH & HEAD DEPT, WAYLAND BAPTIST COL, 49- Concurrent Pos: Fulbright prof, Col Sci, Univ Baghdad, 53-54. Mem: Am Math Soc; Math Asn Am. Res: Geometry; complete existential theory of eight fundamental properties of topological spaces. Mailing Add: Dept of Math Wayland Baptist Col Plainview TX 79072

MCCOY, ELBERT JULIUS, b Savannah, Mo, Nov 26, 24; m 62. PHYSIOLOGY. Educ: Mich State Univ, MS, 60; Univ Tex, PhD(physiol), 66. Prof Exp: Res assoc physiol, Parke, Davis & Co, 60-62; asst prof, Med Ctr, Univ Ark, 65-67; ASST PROF PHYSIOL, SCH MED, TEMPLE UNIV, 67- Mem: AAAS; Soc Exp Biol & Med; Am Physiol Soc. Res: Motility and electrical activity of gastrointestinal tract. Mailing Add: Dept of Physiol Temple Univ Sch of Med Philadelphia PA 19140

MCCOY, ELIZABETH (FLORENCE), bacteriology, see 12th edition

MCCOY, ERNEST E, b Victoria, BC, Sept 3, 23; m 50; c 4. MEDICINE. Educ: Univ Alta, BSc, 47, MD, 49. Prof Exp: Jr intern med, Royal Alexandria Hosp, Edmonton, Alta, 49-50; sr intern, Col Belcher Hosp, Calgary, 50-51; jr resident pediat, St Louis Children's Hosp, Mo, 51-52, asst resident, 52-53; clin instr, Univ BC, 54-57; res assoc microbiol, Vanderbilt Univ, 57-59, asst prof pediat, 59-61; assoc prof, Univ Mo-Columbia, 61-66; prof, Univ Va, 66-69; PROF PEDIAT & CHMN DEPT, UNIV ALTA, 70- Concurrent Pos: Res fel pediat, Washington Univ, 53; Markle scholar, 59. Res: Enzymatic abnormalities in congenital metabolic derangements. Mailing Add: Dept of Pediat Univ of Alta Edmonton AB Can

MCCOY, GEORGE, b Philadelphia, Pa, Nov 25, 14; m 43; c 2. ORGANIC CHEMISTRY. Educ: Univ Pa, BS, 38, MS, 41, PhD(chem), 43. Prof Exp: Res chemist, Sun Oil Co, 43; instr org chem, Univ Pa, 43-44; res chemist, Pa Salt Mfg Co, 44-46, group leader, 46-52, tech asst res dir, 52-55, dir org res, 55-56, mgr res & develop, Pennsalt Chem Corp, 56-63, gen mgr, 63-71, group vpres health, Pennwalt Corp, 71-75, PRES PHARMACEUT DIV, PENNWALT CORP, PHILADELPHIA, 71-, EXEC VPRES OPER, 75- Mem: Am Chem Soc. Res: Chemistry of fluorine compounds; agricultural, rubber and high energy chemicals; halogen compounds; organic and inorganic polymers; aliphatic sulfur chemistry. Mailing Add: Pennwalt Corp Three Parkway Philadelphia PA 19102

MCCOY, JAMES ERNEST, b Glendale, Calif, May 4, 41; m 62; c 2. MAGNETOSPHERIC PHYSICS. Educ: Calif Inst Technol, BS, 63; Rice Univ, PhD(space sci), 69. Prof Exp: PHYSICIST, MANNED SPACECRAFT CTR, NASA, 63- Mem: Am Geophys Union. Res: Geomagnetically trapped radiation; access of solar and cosmic ray radiation to the polar caps; magneto tail electric fields; low energy solar cosmic ray protons; solar cosmic ray physics. Mailing Add: TN2 NASA Manned Apacecraft Ctr Houston TX 77058

MCCOY, JAMES LLOYD, biochemistry, immunology, see 12th edition

MCCOY, JEROME DEAN, b Liberty, Mo, Feb 28, 31; m 54; c 2. PHYSICS. Educ: William Jewell Col, AB, 52; Univ Mo, MA, 57; Univ Helsinki, Phlic & PhD, 64. Prof Exp: Instr physics, Univ Mo, 47-58, asst prof, Univ Tulsa, 58-67; mgr adv systs, Western Div, Sylvania Electronic Systs, 67-71; PROF PHYSICS & HEAD DEPT, UNIV TULSA, 71- Concurrent Pos: NSF sci fac fel; consult, NAm Aviation, Inc & Sylvania Electronic Systs. Mem: AAAS; Am Asn Physics Teachers; Am Soc Eng Educ; Inst Elec & Electronics Eng; Am Inst Aeronaut & Astronaut. Res: Nuclear spectroscopy. Mailing Add: 4310 E 74th St Tulsa OK 74136

MCCOY, JIMMY JEWELL, b Corsicana, Tex, Mar 18,43; m 65. MATHEMATICAL PHYSICS, SOLID STATE PHYSICS. Educ: Baylor Univ, BS, 65, PhD(physics), 69. Prof Exp: AEC grant, Baylor Univ, 69-70; ASST PROF PHYSICS, TARLETON STATE COL, 70- Res: Particle size effects in x-ray scattering. Mailing Add: Dept of Physics Tarleton State State Col Stephenville TX 76401

MCCOY, JOHN FRANK, biostatistics, computer science, see 12th edition

MCCOY, JOHN HAROLD, b Conroe, Tex, May 16, 35; m 61; c 3. COMPUTER SCIENCE. Educ: Sam Houston State Univ, BS, 63, MA, 64; Ohio State Univ, PhD(elec eng), 68. Prof Exp: Exp physicist, United Aircraft Res Labs, 68-70; assoc prof math, 70-72, PROF COMPUT SCI, SAM HOUSTON STATE UNIV, 72-, DIR COMPUT FACIL, 71- Mem: Asn Comput Mach; Inst Elec & Electronics Eng. Mailing Add: Dept of Comput Sci Sam Houston State Univ Box 2206 Huntsville TX 77340

MCCOY, JOHN J, b Coalton, WVa, July 15, 27; m 49; c 2. ORNITHOLOGY. Educ: WVa Wesleyan Col, BS, 50; Univ WVa, MS, 52; Univ Fla, PhD, 60. Prof Exp: Asst prof biol, Tenn Wesleyan Col, 56-57; head dept, Jacksonville Univ, 57-63; HEAD DEPT BIOL, UNIV NC, ASHEVILLE, 63- Mem: Cooper Ornith Soc; Am Ornith Union. Res: Paleo-ornithology. Mailing Add: Dept of Biol Univ of NC Asheville NC 28800

MCCOY, JOHN ROGER, b Trenton, NJ, June 11, 16; m 45; c 1. CHEMOTHERAPY, COMPARTIVE PATHOLOGY. Educ: Univ Pa, VMD, 40. Prof Exp: Res specialist, 50-69, ADJ RES PROF, BUR BIOL RES, RUTGERS UNIV, 70-; PROF COMP PATH & DIR VIVARIUM, RUTGERS MED SCH, COL MED & DENT NJ, 70- Concurrent Pos: Consult to govt & var pharmaceut concerns; mem grant adv coun, Seeing Eye Found; mem vet adv coun, Vet Cancer Unit; mem subcomt org contaminants, Safe Drinking Water Comt, Nat Res Coun-Nat Acad Sci. Mem: AAAS; Am Vet Med Asn (pres, 71-72); Fedn Am Soc Exp Biol; fel NY Acad Sci; Am Asn Lab Animal Sci. Res: Spontaneous canine cancer; pathology of leishmaniasis; food additives; steroids; cancer chemotherapy. Mailing Add: 1007 River Rd Piscataway NJ 08854

MCCOY, JOHN SPALDING, organic chemistry, see 12th edition

MCCOY, JOSEPH HAMILTON, b Johnson City, Tenn, May 7, 34. PHYSICAL CHEMISTRY. Educ: E Tenn State Univ, BS, 56; Mich State Univ, PhD(chem), 63. Prof Exp: Res chemist, E I du Pont de Nemours & Co, 62-64; asst prof, 64-67, PROF CHEM, EMORY & HENRY COL, 67- Mem: AAAS; Am Chem Soc. Res: Polymer solution studies; gas chromatography. Mailing Add: PO Box EE Emory VA 24327

MCCOY, JOSEPH WESLEY, b Eugene, Ore, Jan 20, 30; m 51; c 6. ORGANIC CHEMISTRY. Educ: Univ Portland, BS, 52; Univ Notre Dame, PhD(org chem), 59. Prof Exp: Res chemist, Calif Res Corp Div, Standard Oil Co, Calif, 55-60; instr chem, 60-65, head div sci & math, 67-69, ASSOC PROF CHEM, UNIV PORTLAND, 65-, DIR GRAD PROG SCI, 69- Mem: Am Chem Soc. Res: Surface active agents; petrochemicals. Mailing Add: Dept of Chem Univ of Portland Portland OR 97203

MCCOY, LAYTON LESLIE, b Seattle, Wash, Mar 11, 27. ORGANIC CHEMISTRY. Educ: Univ Wash, BS, 47, PhD(chem), 51. Prof Exp: Instr chem, Columbia Univ, 51-53; NSF fel, Univ Wash, 57-58; from instr to asst prof chem, Columbia Univ, 58-62; from asst prof to assoc prof, 62-67, asst chmn dept, 73-75, PROF CHEM, UNIV MO-KANSAS CITY, 67- Mem: AAAS; Am Chem Soc; The Chem Soc. Res: Synthesis and reactions of cyclopropanes; stereospecific and stereoselective syntheses and reactions; correlation of structure and relative acidity of acids. Mailing Add: Dept of Chem Univ of Mo Kansas City MO 64110

MCCOY, LOWELL EUGENE, b Hillsboro, Ohio, June 1, 37; m 59; c 4. PHYSIOLOGY, BIOCHEMISTRY. Educ: Miami Univ, BA, 60, MA, 61; Wayne State Univ, PhD(physiol), 66. Prof Exp: Res asst cancer chemother & toxicol, Christ Hosp Inst Med Res, 61-63; res assoc blood coagulation biochem, 66-67, asst prof, 68-73, ASSOC PROF PHYSIOL, SCH MED, WAYNE STATE UNIV, 73- Concurrent Pos: Mem coun thrombosis, Am Heart Asn. Mem: AAAS; Am Chem Soc; Int Soc Thrombosis & Haemostasis; NY Acad Sci. Res: Blood coagulation protein biochemistry, including physical and enzymologic properties; hematology; erythropoiesis; peptide synthesis; amino acid sequence. Mailing Add: 540 E Canfield Detroit MI 48201

MCCOY, OLIVER RUFUS, b St Louis, Mo, Aug 1, 05; m 37; c 3. PARASITOLOGY. Educ: Wash Univ, AB, 26, MS, 27; Johns Hopkins Univ, ScD, 30; Univ Rochester, MD, 42. Hon Degrees: LLD, Keio Univ, Japan, 56. Prof Exp: Asst zool, Wash Univ, 26-27; helminth, Sch Hyg & Pub Health, Johns Hopkins Univ, 28-29, instr, 29-30; asst prof parasitol, Sch Med & Dent, Univ Rochester, 30-42; epidemiologist, Douglas Aircraft Co, E Africa, 42-43; field staff, Mem Div, Med & Pub Health, Rockefeller Found, 46-56; from assoc dir to dir, 56-69, PRES CHINA MED BD NEW YORK, 69- Concurrent Pos: Vis prof, Nat Med Col, Shanghai, 36; consult, Gorgas Mem Lab, 34; mem, State Trichinosis Comn, NY, 40-42; mem bd, Coord Malaria Studies, Nat Res Coun, 43-46. Mem: AAAS; Am Soc Parasitol; Am Soc Trop Med & Hyg; Am Pub Health Asn. Res: Life histories of trematodes; immunity of helminth parasites; hookworm; trichinosis; filariasis; malaria and tropical disease control. Mailing Add: China Med Bd of New York 420 Lexington Ave New York NY 10017

MCCOY, RALPH HINES, b Crowell, Tex, Nov 22, 40; m 74. WILDLIFE DISEASES, MEDICAL BACTERIOLOGY. Educ: Baylor Univ, BS, 63, MS, 65; Ore State Univ, PhD(microbiol), 73. Prof Exp: ASST PROF BIOL, TEX A&I UNIV, 74- Concurrent Pos: Consult, Garcia & Rangel, Attys, 75- Mem: Wildlife Dis Asn; Am Soc Microbiol; Am Soc Mammalogists; Wildlife Soc. Res: Zoonoses; infectious diseases of wild and domestic animals. Mailing Add: Dept of Biol Tex A&I Univ Kingsville TX 78363

MCCOY, RAYMOND DUNCAN, b Newberg, Ore, Dec 30, 24; m 51; c 3. PHYSICAL CHEMISTRY, INSTRUMENTATION. Educ: Willamette Univ, BS, 50; Univ Ore, PhD(phys chem), 56. Prof Exp: Instrument engr, Phillips Petrol Co, Okla, 55-71; MEM STAFF, APPL AUTOMATION, INC, 71- Mem: Am Chem Soc. Res: Kinetics of gaseous reactions; process instrumentation and automation. Mailing Add: Appl Automation Inc Pawhuska Rd RB 2 Bartlesville OK 74004

MCCOY, RICHARD HUGH, b Wilmington, Ohio, Nov 26, 08; m 52; c 1. BIOCHEMISTRY. Educ: Earlham Col, AB, 29; Univ Ill, MS, 31, PhD(chem), 35. Prof Exp: Instr chem, Earlham Col, 29-30; Rockefeller Found fel, Univ Chicago, 35-36 & Wistar Inst, 36-40; sr res fel biochem, 40-42, from asst res prof to assoc res prof, 43-52, assoc prof, 52-59, from asst dean to assoc dean div natural sci, 57-67, PROF BIOCHEM, UNIV PITTSBURGH, 59-, ASSOC DEAN FAC ARTS & SCI & DIR GRAD PROGS, 67- Mem: Fel AAAS; Am Chem Soc; Am Soc Biol Chemists; Am Inst Nutrit; NY Acad Sci. Res: Amino acids and vitamins in nutrition; factors affecting porphyrin formation; protein requirements; calcification. Mailing Add: 1028 Cathedral of Learning Univ of Pittsburgh Pittsburgh PA 15213

MCCOY, ROBERT A, b Springfield, Ill, Oct 1, 42; m 64; c 1. TOPOLOGY. Educ: Southern Ill Univ, BA, 64; Iowa State Univ, PhD(math), 68. Prof Exp: Asst prof, 68-75, ASSOC PROF MATH, VA POLYTECH INST & STATE UNIV, 75- Mem: Am Math Soc; Math Asn Am. Res: Topology and infinite-dimensional manifolds. Mailing Add: Dept of Math Va Polytech Inst & State Univ Blacksburg VA 24061

MCCOY, ROBERT EDWARD, inorganic chemistry, analytical chemistry, see 12th edition

MCCOY, ROGER MICHAEL, b Dewey, Okla, Feb 3, 33; m 55; c 2. PHYSICAL GEOGRAPHY. Educ: Univ Okla, BS, 57; Univ Colo, MA, 64; Univ Kans, PhD(geog), 67. Prof Exp: Asst prof geog, Univ Ill, Chicago, 67-69; asst prof, Univ Ky, 69-72; ASSOC PROF GEOG, UNIV UTAH, 72- Mem: Am Soc Photogram; Asn Am Geog. Res: Fluvial geomorphology; interpretation of air photos and remote sensing. Mailing Add: Dept of Geog Univ of Utah Salt Lake City UT 84112

MCCOY, RONALD EUGENE, b Hershey, Pa, May 3, 34; m 52; c 6. MATHEMATICS. Educ: Ind Univ Pa, BS, 56; Univ Pittsburgh, MEd, 57; Pa State Univ, DEd(math educ), 71. Prof Exp: Teacher pub sch, Pa, 56-57, teacher & dept chmn high sch, Pa, 57-67; ASSOC PROF MATH, IND UNIV PA, 67- Mem: Math Asn Am. Res: Mathematics education; theory of construction of deductive proofs in mathematics as applied to teaching. Mailing Add: Dept of Math Ind Univ of Pa Indiana PA 15701

MCCOY, SUE, b Charlottesville, Va, Nov 14, 35. BIOCHEMISTRY. Educ: Radcliffe Col, AB, 57; Johns Hopkins Univ, PhD(physiol chem), 64. Prof Exp: Fel physiol chem, Sch Med, Johns Hopkins Univ, 64-67; asst prof chem, Univ South Fla, 67-69; asst prof orthop, 69-73, ASST PROF SURG, UNIV VA, 73- Mem: AAAS; Am Chem Soc; The Chem Soc; Orthop Res Soc; NY Acad Sci. Res: Hemes and heme proteins; biochemistry of hemorrhagic shock; polymers for surgical implant materials. Mailing Add: Dept of Surg Univ of Va Sch of Med Charlottesville VA 22901

MCCOY, THEO CONNER, chemistry, see 12th edition

MCCOY, THOMAS AYLESBURY, b Bartlesville, Okla, Mar 22, 21; m 42; c 2. BIOCHEMISTRY. Educ: Univ Okla, BS, 42, MS, 47, PhD(chem), 52. Prof Exp: Asst zool, Univ Okla, 45-46; dir biomed div, Samuel Roberts Noble Found, Inc, Okla, 47-66; res assoc path, Children's Cancer Res Found, Children's Hosp Med Ctr, Boston, Mass, 66-68; dir res, Vivonex & Co, 68-70; DIR RES, LEE-MAC, 70- Mem: Fel AAAS; Am Soc Biol Chem; Soc Exp Biol & Med; Am Chem Soc; Am Asn Cancer Res. Res: Cancer metabolism. Mailing Add: Santa Clara CA

MCCOY, THOMAS LARUE, b Seville, Ohio, Jan 16, 33; m 57; c 1. MATHEMATICS.

Educ: Oberlin Col, BA, 54; Univ Wis, MS, 56, PhD(math), 61. Prof Exp: Instr math, Ill Inst Technol, 60, asst prof, 64; asst prof, 64-67, ASSOC PROF MATH, MICH STATE UNIV, 67- Mem: Am Math Soc. Res: Functions of a complex variable; differential equation; calculus of variation. Mailing Add: Dept of Math Mich State Univ East Lansing MI 48823

MCCOY, V EUGENE, JR, b Chicago, Ill, July 7, 33; m 57; c 3. PHYSICAL ORGANIC CHEMISTRY. Educ: Princeton Univ, AB, 55; Harvard Univ, PhD(chem), 65. Prof Exp: Res chemist, Pioneering Res Lab, Textile Fibers Dept, 64-67, sr res chemist, 67-68, SR RES CHEMIST, TEXTILE RES LAB, E I DU PONT DE NEMOURS & CO, INC, 68- Mem: Am Chem Soc; Sigma Xi. Res: Electron exchange reactions; chemistry of organophosphorus compounds; polymers and chemical reactions for textile fibers. Mailing Add: Textile Res Lab E I du Pont de Nemours & Co Inc Wilmington DE 19898

MCCOY, WILLIAM HARRISON, b Nelsonville, Ohio, May 28, 25; m 59. PHYSICAL CHEMISTRY. Educ: Youngstown Univ, BS, 50; Univ Pittsburgh, PhD(phys chem), 55. Prof Exp: Asst prof chem, Washington & Jefferson Col, 55-56 & Tex A&M Univ, 56-57; assoc prof, Youngstown Univ, 57-63; phys chemist, Off Saline Water, 63-64, chief div chem physics, 64-74, PHYS CHEMIST, OFF WATER RES & TECHNOL, US DEPT INTERIOR, 74- Honors & Awards: Achievement Award, US Dept of Interior, 71. Mem: Am Chem Soc. Res: Thermodynamic properties of liquids and electrolyte solutions; theory of liquids; properties of solutions, with special reference to contaminated water. Mailing Add: Div of Phys Chem Off of Water Res US Dept of Interior Washington DC 20240

MCCOYD, GERARD, physics, see 12th edition

MCCRACKEN, ALEXANDER WALKER, b Motherwell, Scotland, Nov 24, 31; m 60; c 2. MEDICAL MICROBIOLOGY. Educ: Univ Glasgow, MB, ChB, 55; Univ London, DCP, 62; Univ Liverpool, DTM, 66. Prof Exp: Pathologist, Royal Air Force Hosp, Akrotiri, Cyprus, 58-61; chief clin microbiol, Royal Air Force Inst Path & Trop Med, Halton, Eng, 62-68; chief microbial path, 68-71, assoc prof path & microbiol, 68-73, actg chmn dept, 71-73, ADJ PROF MICROBIOL, UNIV TEX MED SCH SAN ANTONIO, 73-; ASSOC DIR LABS, BEXAR COUNTY TEACHING HOSP, 71- Mem: Brit Asn Clin Path; Brit Soc Gen Microbiol; Col Path Eng; Int Acad Path; Royal Micros Soc. Res: Rapid diagnostic techniques in clinical microbiology; fluorescent antibody methodology; respiratory virus isolation methods and epidemiology; pathogenesis of viral infections. Mailing Add: Dept of Path Univ of Tex Med Sch San Antonio TX 78212

MCCRACKEN, CURTIS W, b Vici, Okla, June 7, 34. ASTRONOMY, PHYSICS. Educ: Panhandle Agr & Mech Col, BS, 56; Okla State Univ, MS, 59. Prof Exp: PHYSICIST, GODDARD SPACE FLIGHT CTR, NASA, 60- Mem: AAAS; Am Astron Soc; Am Geophys Union. Res: Interplanetary dust particles using rockets and spacecraft. Mailing Add: Code 672 NASA Goddard Space Flight Ctr Greenbelt MD 20771

MCCRACKEN, DEREK ALBERT, b Eng, Feb 11, 43; m 63; c 3. PLANT PHYSIOLOGY. Educ: McMaster Univ, BA, 63; Univ Toronto, MA, 66, PhD(bot), 69. Prof Exp: ASST PROF BIOL SCI, ILL STATE UNIV, 69- Mem: AAAS; Phycol Soc Am; Bot Soc Am; Am Soc Plant Physiol. Res: Starch biosynthesis; interaction between nuclear and cytoplasmic genes. Mailing Add: Dept of Biol Sci Ill State Univ Normal IL 61761

MCCRACKEN, FRANCIS IRVIN, b Bicknell, Ind, Aug 19, 35; m 59; c 2. FOREST PATHOLOGY. Educ: Ariz State Univ, BS, 57; Okla State Univ, MS, 59; Wash State Univ, PhD(plant path), 72. Prof Exp: Instr bot, Univ Idaho, 62-67; RES PLANT PATH, US FOREST SERV, USDA, 67- Mem: Am Phytopath Soc; Sigma Xi. Res: Etiology and control of diseases effecting southern bottomland hardwood tree species. Mailing Add: US Forest Serv PO Box 227 Stoneville MS 38776

MCCRACKEN, FRANK DERWOOD, fisheries, see 12th edition

MCCRACKEN, JOHN AITKEN, b Glasgow, Scotland, Aug 16, 34. ENDOCRINOLOGY. Educ: Glasgow Univ, BVMS, 58, PhD(endocrinol), 63. Prof Exp: House surgeon, Glasgow Univ, 58-59; vis res fel, Cambridge Univ, 60; Agr Res Coun res fel endocrine physiol, Univ Birmingham, 63-64; staff scientist, 64-70, SR SCIENTIST, WORCESTER FOUND EXP BIOL, 70- Concurrent Pos: Vis prof, Cornell Univ, 69-70. Mem: AAAS; Soc Study Reproduction; NY Acad Sci; Royal Col Vet Surg. Res: Steroid biochemistry; physiology and biochemistry of prostaglandins; transplantations of ovary, uterus and adrenal glands in the sheep and their various interrelationships; interrelationships between steroids, prostaglandins and polypeptide hormones. Mailing Add: Worcester Found for Exp Biol 222 Maple Ave Shrewsbury MA 01545

MCCRACKEN, MICHAEL DWAYNE, b Terrell, Tex, Feb 15, 41. BOTANY. Educ: Tex Tech Univ, BS, 63, MS, 65; Ind Univ, Bloomington, PhD(bot), 69. Prof Exp: Asst prof bot, Univ Wis-Madison, 69-71; ASST PROF BIOL, TEX CHRISTIAN UNIV, 71- Concurrent Pos: NSF fel, Univ Wis-Madison, 70-72. Mem: Phycol Soc Am; Int Phycol Soc; Am Soc Limnol & Oceanog; Int Soc Theoret & Appl Limnol. Res: Algology, differentiation and development in Volvox; algal primary productivity. Mailing Add: Dept of Biol Tex Christian Univ Ft Worth TX 76129

MCCRACKEN, RALPH JOSEPH, b Guantanamo, Cuba, July 3, 21; nat US, m 49; c 2. SOIL SCIENCE. Educ: Earlham Col, AB, 42; Cornell Univ, MS, 51; Iowa State Univ, PhD(soils), 56. Prof Exp: Soil scientist, USDA, 47-49, 51-55; assoc agronomist, Univ Tenn, 55-56; assoc prof, Soil Sci, N C State Univ, 56-62, prof, 62-73, asst dir res, Sch Agr & Life Sci, 70-73; ASSOC ADMIN, USDA AGR RES SERV, 73- Mem: Fel AAAS; Soil Sci Soc Am; Am Soc Agron; Clay Minerals Soc; Soil Conserv Soc Am. Res: Genesis, classification and mineralogy of soils. Mailing Add: US Dept of Agr Agr Res Serv Washington DC 20250

MCCRACKEN, ROBERT DALE, b Fairplay, Colo, Aug 8, 39; m 67; c 1. ANTHROPOLOGY, PSYCHOLOGY. Educ: Univ Colo, BA, 62, MA, 65, PhD(anthrop), 67. Prof Exp: Lectr anthrop, Univ Colo, 65; instr, Metrop State Col, 65-66; instr, Colo Woman's Col, 66-67; asst prof, Long Beach State Col, 68-69; asst prof anthrop & pub health, Univ Calif, Los Angeles, 69-71; indust consult, researcher & writer, 72-74; DIR RES, COLO MIGRANT COUN, 74- Concurrent Pos: Mem, Coun Anthrop & Educ, 68-; fel psychol, Wash Univ, 71-72; fac affil, Dept Anthrop, Colo State Univ, 74- Mem: Fel Am Anthrop Asn; fel Soc Appl Anthrop; Int Acad Metab; fel Int Col Appl Nutrit. Res: Culture change, interaction of culture and human behavior and human biology, nutrition; culture and health of migrant farmworkers. Mailing Add: 1414 Marion St Denver CO 80218

MCCRADY, EDWARD, b Canton, Miss, Sept 19, 06; m 30; c 4. EMBRYOLOGY, BIOPHYSICS. Educ: Col Charleston, AB, 27, LLD, 50; Univ Pittsburgh, MS, 30; Univ Pa, PhD(zool), 33; Southwestern at Memphis, ScD, 59; Concord Col, LHD, 62. Prof Exp: Asst cur pub instr, Charleston Mus, SC, 27-28; asst zool, Univ Pittsburgh, 28-30; res fel, Wistar Inst, 30-36, assoc, 36-37; prof biol & chmn dept, Univ of the South, 37-49; sr biologist div biol & med, Atomic Energy Comn, Tenn, 48-49, chief biol div, 49-51; prof biol & pres & vchancellor, Univ of the South, 51-71, SPEC LECTR, COL CHARLESTON & UNIV OF THE SOUTH, 71- Concurrent Pos: Danforth vis prof, 59-60; mem, Nat Selection Comt Fulbright Grants, 52-53; adv comt fac fel prog, NSF, 57; chmn, Rhodes Scholarship Selection Comt, 57-60; Brown tutorial fel, Univ of the South, 74- Mem: AAAS; Am Soc Zool; Acoust Soc Am; Nat Speleol Soc; Am Asn Anat. Res: Mammalian embryology; mechanism of hearing; great fossil cats; cave salamanders; determination of age of the earth by isotopic ratios; philosophy of science. Mailing Add: The Univ of the South Sewanee TN 37375

MCCRADY, EDWARD, III, b Trenton, NJ, Sept 24, 33; m 55; c 3. DEVELOPMENTAL BIOLOGY. Educ: Univ of the South, BS, 55; Univ Va, MA, 61, PhD(biol), 64. Prof Exp: Asst biol, Univ Va, 60-63; asst prof, 64-70, ASSOC PROF BIOL, UNIV N C, GREENSBORO, 70- Mem: AAAS; Am Soc Zool; Develop Biol Soc. Res: Insect tissue culture; development of imaginal discs. Mailing Add: Dept of Biol Univ of N C Greensboro NC 27412

MCCRADY, JAMES DAVID, b Beaumont, Tex, June 26, 30; m 51; c 3. VETERINARY PHYSIOLOGY. Educ: Tex A&M Univ, BS, 52, DVM, 58; Baylor Univ, PhD(physiol), 65. Prof Exp: From instr to asst prof physiol, Tex A&M Univ, 58-62; fel, instr & dir animal resources, Col Med, Baylor Univ, 62-64; assoc prof physiol, 64-65, PROF VET PHYSIOL & PHARMACOL & HEAD DEPT, TEX A&M UNIV, 65- Mem: Am Soc Vet Physiol & Pharmacol; Am Physiol Soc; Am Col Clin Pharmacol. Res: Pulmonary hypertension; atrial fibrillation; electrocardiography. Mailing Add: Dept of Vet Physiol & Pharmacol Tex A&M Univ College Station TX 77843

MCCRADY, WILLIAM B, b Forreston, Tex, July 25, 33; m 54; c 3. GENETICS. Educ: ETex State Col, BS, 54, MS, 58; Univ Nebr, PhD(zool), 61. Prof Exp: Asst prof biol, Ark State Col, 61-62; from asst prof to assoc prof, 62-74, PROF BIOL, UNIV TEX, ARLINGTON, 74- Mem: Genetics Soc Am. Res: Carbon dioxide sensitivity in Drosophila. Mailing Add: Dept of Biol Univ of Tex Arlington TX 76010

MCCRANIE, ERASMUS JAMES, b Milan, Ga, July 24, 15; m 45; c 3. PSYCHIATRY. Educ: Emory Univ, AB, 37, MS, 38; Univ Mich, PhD(bot), 42, MD, 45. Prof Exp: Instr biol, SGa Col, 41-42; from instr to assoc prof psychiat, Univ Tex Southwestern Med Sch, 51-56; PROF PSYCHIAT, MED COL GA, 56-, CHMN DEPT, 57- Concurrent Pos: Consult, Vet Admin Hosp, Augusta, 56-, Cent State Hosp, 65-, Vet Admin Hosp, Dublin, 66- & Ga Regional Hosp, 70- Mem: AMA; Am Psychiat Asn. Res: Psychosomatic medicine; medical applications of hypnosis. Mailing Add: Dept of Psychiat Med Col of Ga Augusta GA 30902

MCCRARY, ANNE BOWDEN, b Wilmington, NC, Oct 25, 26; m 44; c 2. INVERTEBRATE ZOOLOGY, MARINE BIOLOGY. Educ: Univ NC, Chapel Hill, AB, 61, MA, 65, PhD(zool), 69. Prof Exp: Asst prof, 70-75, ASSOC PROF BIOL, UNIV N C, WILMINGTON, 75- Mem: AAAS; Am Soc Limnol & Oceanog; Estuarine Res Fedn. Res: Ecology of marine invertebrates; plankton. Mailing Add: Dept of Biol Univ of N C Wilmington NC 28401

MCCRARY, JAMES HARVEY, physics, see 12th edition

MCCRAW, BRUCE MAXWELL, zoology, see 12th edition

MCCRAY, RICHARD ALAN, b Los Angeles, Calif, Nov 24, 37; m 61; c 2. ASTROPHYSICS. Educ: Stanford Univ, BS, 59; Univ Calif, Los Angeles, MA, 62, PhD(physics), 67. Prof Exp: Res fel physics, Calif Inst Technol, 67-68; asst prof astron, Harvard Col Observ, 68-71; assoc prof, 71-75, PROF PHYSICS & ASTROPHYS, UNIV COLO, BOULDER, 75-, FEL, JOINT INST LAB ASTROPHYS, UNIV COLO, 72- Concurrent Pos: Fel, John Simon Guggenheim Mem Found, 75. Mem: Am Astron Soc; Int Astron Union. Res: Theoretical astrophysics; interstellar gas dynamics; theory of x-ray stars. Mailing Add: Joint Inst Lab Astrophys Univ of Colo Boulder CO 80309

MCCREA, PETER FREDERICK, b Pawtucket, RI, Oct 6, 42; m 66; c 2. INSTRUMENTATION. Educ: Providence Col, BS, 64, MS, 66; Univ Wales, PhD(chem), 69. Prof Exp: Res scientist, 69-72, SR RES SCIENTIST, RES CTR, THE FOXBORO CO, 72- Mem: Am Chem Soc. Res: Gas chromatography; liquid chromatography; thermoanalytical techniques and ultrasonics; instrumentation for physical and chemical measurements in industrial environments. Mailing Add: Foxboro Co Res Ctr 38 Neponset Ave Foxboro MA 02035

MCCREADY, NEWTON WILLOUGHBY, physical chemistry, see 12th edition

MCCREADY, ROBERT RICHARD, mathematics, see 12th edition

MCCREADY, ROLLAND MARTIN, organic chemistry, biological chemistry, see 12th edition

MCCREADY, THOMAS ARTHUR, b Pueblo, Colo, Sept 1, 40; m 65; c 2. MATHEMATICS. Educ: Univ Calif, Berkeley, AB, 62; Stanford Univ, PhD(math), 68. Prof Exp: Sr assoc programmer, Eng & Sci Comput Lab, Int Bus Mach Corp, Calif, 66-68; assoc prof, 68-73, PROF MATH, CALIF STATE UNIV, CHICO, 73- Res: Fluid dynamics; numerical analysis; elliptic partial differential equations. Mailing Add: 4 Rosemary Circle Chico CA 95926

MCCREARY, JOHN FERGUSON, b Eganville, Ont, Jan 22, 10; m 36, 70; c 1. MEDICINE. Educ: Univ Toronto, MD, 34. Hon Degrees: FRCP(C), 57; DSc, Mem Univ, 68; LLD, Univ Toronto, 70; DSc, Univ BC. Prof Exp: Milbank fel, Harvard Med Sch, 39-41; asst prof pediat, Univ Toronto, 46-51; prof pediat & head dept, Univ BC, 51-59, dean fac med, 59-72, coordr health sci, 71-75. Concurrent Pos: Consult, Colombo Plan Med Educ, India, 57; chmn bd trustees, Queen Elizabeth II Fund Res Dis Children, 59-; mem Defence Res Bd Can, 60-71. Mem: Am Pediat Soc; Soc Pediat Res; Can Pediat Soc; Can Med Asn. Res: Nutrition in children. Mailing Add: 115 4775 Valley Dr Vancouver BC Can

MCCREARY, RALPH LEROY, b Morning Sun, Ohio, Dec 23, 16; m 43; c 3. PHYSICS. Educ: Miami Univ, Ohio, BS, 38; Univ Rochester, PhD(physics), 42. Prof Exp: Asst, Univ Rochester, 38-42, res assoc physics, 46-47, asst prof, 47-49; res assoc elec eng, Radiation Lab, Mass Inst Technol, 42-45; dir physics res, Nat Res Corp, 45-46; physicist, Collins Radio Co, Iowa, 49-58, dir res, 59-62; mgr systs res lab, Motorola, Inc, Calif, 62-63; dir adv tech dept, Autonetics Div, NAm Aviation, Inc, 63-64; vpres & mgr systs div, Watkins-Johnson Co, 64-68; vpres & gen mgr, Hoffman Info Systs, Inc, 68-74, VPRES, HOFFMAN ELECTRONICS CORP, 74- Mem: AAAS; fel Inst Elec & Electronics Engrs. Res: Nuclear physics; internal conversion; beta decay; radar; high vacuum components and processes; evaporation; coating; high energy accelerators; electronic systems; wave propagation; microwave and ultra-high frequency reconnaissance; space communications and telemetry; countermeasures and radio frequency interference instrumentation. Mailing Add: Hoffman Electronics Corp 4323 Arden Dr El Monte CA 91731

MCCREDIE, JOHN A, b Anahilt, Northern Ireland, Sept 8, 23; m 54; c 5. CANCER, SURGERY. Educ: Queen's Univ Belfast, MB, 46, MCh, 57; FRCS(E) & FRCS, 51; FRCS(C), 60. Prof Exp: Registr surg, Royal Victoria Hosp, Belfast, 49-56; instr, Univ Ill, Chicago, 57-58; lectr, 59-68, ASSOC PROF SURG, UNIV WESTERN ONT, 68-; RES ASSOC, ONT CANCER FOUND, 59- Honors & Awards: Jackson Prize, Royal Col Surgeons, Eng, 56. Mem: Am Asn Cancer Res; fel Am Col Surgeons. Res: Immunology of tumors. Mailing Add: Dept of Surg Victoria Hosp London ON Can

MCCREDIE, ROBERT JOSEPH, organic chemistry, see 12th edition

MCCREERY, ROBERT ATKESON, b Columbia, Mo, May 30, 17; m 40; c 7. AGRONOMY, SOILS. Educ: Univ Ga, BS, 46, MS, 47; State Col Wash, PhD(soils), 54. Prof Exp: Jr soil scientist & instr, State Col Wash, 47-54; soil scientist, Cotton Br, Agr Res Serv, USDA, Costa Rica, 54-57; asst prof, 57-68, ASSOC PROF AGRON, UNIV GA, 68- Mem: Am Soc Agron; Soil Sci Soc Am; Am Chem Soc; Mineral Soc Am; fel Am Inst Chem. Res: Soil chemistry and mineralogy; forest soils; reclamation. Mailing Add: Dept of Agron Univ of Ga Athens GA 30602

MCCREESH, ARTHUR HUGH, b New York, NY, May 18, 31; m 63; c 3. PHARMACOLOGY, VETERINARY TOXICOLOGY. Educ: St John's Univ, NY, BS, 53; Temple Univ, MS, 57, PhD(pharmacol), 60. Prof Exp: Instr physiol, Temple Univ, 57-60; CHIEF TOXICOL DIV, US ARMY ENVIRON HYG AGENCY, 60- Mem: AAAS; Soc Toxicol; NY Acad Sci. Res: Testing of chemical compounds for toxic hazard; development of new screening procedures; cutaneous toxicity; inhalation toxicology. Mailing Add: Toxicol Div US Army Environ Hyg Agency Aberdeen Proving Ground MD 21010

MCCREIGHT, CHARLES EDWARD, b Camden, SC, Mar 17, 13. ANATOMY. Educ: George Washington Univ, BS, 48, MS, 50, PhD(anat), 54. Prof Exp: Lab instr zool, George Washington Univ, 48-50, asst anat, 50-52; assoc, 52-53, from instr to asst prof, 53-65; ASSOC PROF ANAT, BOWMAN GRAY SCH MED, 65- Mem: Fel Geront Soc; Am Asn Anatomists; Am Asn Univ Prof; Sigma Xi. Res: Experimental studies in growth, development and regeneration of the kidney; epidermal cytology; cytology of lung. Mailing Add: Dept of Anat Bowman Gray Sch of Med Winston-Salem NC 27103

MCCREIGHT, EDWARD M, US citizen. COMPUTER SCIENCES. Educ: Col Wooster, AB, 66; Carnegie-Mellon Univ, PhD(comput sci), 70. Prof Exp: Res scientist comput sci, Boeing Sci Res Labs, Wash, 69-71; MEM RES STAFF COMPUT SCI, PALO ALTO RES CTR, XEROX CORP, 71- Mem: Asn Comput Mach; Inst Elec & Electronics Engrs. Res: Data structures; analysis of algorithms; theory of computing; computer architecture. Mailing Add: Xerox Palo Alto Res Ctr 3333 Coyote Hill Rd Palo Alto CA 94304

MCCREIGHT, ROBERT WILLIS, b Aledo, Ill, Sept 15, 15; m 49. ORGANIC CHEMISTRY. Educ: Monmouth Col, BS, 47; Univ Iowa, MS, 49, PhD(chem), 53. Prof Exp: Instr chem, Cent Col, Iowa, 49-51; PROF CHEM, WESTMINSTER COL, MO, 53- Mem: AAAS; Am Chem Soc. Res: Reduction and condensation reactions of carbonyl compounds. Mailing Add: Dept of Chem Westminster Col Fulton MO 65251

MCCRIMMON, HUGH ROSS, fisheries, see 12th edition

MCCROAN, JOHN EDGAR, JR, b Statesboro, Ga, Sept 14, 09; m 42; c 2. EPIDEMIOLOGY. Educ: Emory Univ, BS, 31, MS, 32; Univ Iowa, PhD(cytol), 38. Prof Exp: Asst cytol, Emory Univ, 31-32; asst, Univ Iowa, 32-38; from instr to asst prof, Emory Col, 38-42; with epidemiol units 14-305, Nat Naval Med Ctr, Bethesda, 42-46; sr biologist & dir epidemiol res, 46-50, asst dir epidemiol serv, 50-62, chief epidemiologist, Epidemiol Invests Br, 62-68, chief epidemiol unit, Dept Human Resources, 68-73, DIR EPIDEMIOL SECT, GA DIV PHYS HEALTH, 73- Concurrent Pos: Supvr field crew, USDA, 29; entomologist, USPHS, Ga, 46. Mem: Am Pub Health Asn. Res: Chemical control of ticks; epidemiology and control of hookworm, Rocky Mountain spotted fever, typhus and leptospirosis; animal related viruses and staphylococcal infections; microbiology of wrapped sandwiches; microbiology of carpets. Mailing Add: 1054 Mason Woods Dr NE Atlanta GA 30329

MCCRONE, ALISTAIR WILLIAM, b Regina, Sask, Oct 7, 31; US citizen; m 58; c 3. GEOLOGY, PALEOECOLOGY. Educ: Univ Sask, BA, 53; Univ Nebr, MSc, 55; Univ Kans, PhD(geol), 61. Prof Exp: From instr to prof geol, NY Univ, 59-70, chmn dept, 66-69, assoc dean, Grad Sch Arts & Sci, 69-70; prof, Univ of the Pac, 70-74, acad vpres, 70-71, 71-74; PROF GEOL, HUMBOLDT STATE UNIV, 74-, PRES, 74- UNIV, 74- Concurrent Pos: Wellsite geologist, Brit Am Oil Co, Sask, 53; field party chief, Shell Oil Co, Can, 56-68; mem fac comt educ policy, NY Univ, 67-69. Mem: Fel AAAS; fel Geol Soc Am; Am Asn Petrol Geol; NY Acad Sci; Sigma Xi. Res: Paleozoic and subsurface stratigraphy; marine ecology and paleoecology; marine and estuarine sedimentation; aquatic geochemistry; sedimentary facies analysis. Mailing Add: Humboldt State Univ Arcata CA 95521

MCCRONE, JOHN DAVID, b Somerville, Mass, Nov 9, 34; m 57; c 2. ZOOLOGY, ADMINISTRATIVE SCIENCES. Educ: Univ Fla, BS, 56, PhD(biol), 61. Prof Exp: Asst prof biol, Fairleigh Dickinson Univ, 61-62; from asst prof to assoc prof, Fla Presby Col, 62-66; assoc prof zool, Univ Fla, 66-69; assoc dean grad sch, Univ of the Pac, 69-71; dir res, Grad Sch, 71-73, ASSOC VPRES EDUC DEVELOP & RES, UNIV IOWA, 73- Mem: Fel AAAS; Soc Syst Zool; Soc Study Evolution; Int Soc Toxinol. Res: Arachnology; systematics; ecology; toxinology. Mailing Add: Univ of Iowa Iowa City IA 52240

MCCRONE, WALTER C, b Wilmington, Del, June 9, 16; m 56. ANALYTICAL CHEMISTRY. Educ: Cornell Univ, BChem, 38, PhD(chem micros), 42. Prof Exp: Chem microscopist, Off Sci Res & Develop Proj, Cornell Univ, 42-44; chem microscopist, Armour Res Found, Ill Inst Technol, 44-45, supvr anal chem, 45-46, asst chmn chem & chem eng, 46-52, sr scientist, 52-56; with McCrone Assocs, Inc, 56-, DIR & CHMN BD, McCRONE ASSOCS, INC, 67-, PRES, McCRONE RES INST, 66- Concurrent Pos: Publ & ed, Microscope Publications Ltd, 65-; mem adv bd, J Anal Chem, Am Chem Soc, 75-76. Honors & Awards: Benedetti-Pichler Award, 70 & 71. Mem: Am Chem Soc; Am Phys Soc; Am Soc Test & Mat; Am Acad Forensic Sci; Am Microchem Soc. Res: Crystallography; chemical microscopy; polymorphism; crystal growth; correlation of solid state properties and performance; ultramicroanalysis; physical methods of analysis. Mailing Add: McCrone Assocs Inc 2820 S Michigan Ave Chicago IL 60616

MCCROREY, HENRY LAWRENCE, b Philadelphia, Pa, Mar 13, 27; m 57; c 4. PHYSIOLOGY. Educ: Univ Mich, BS, 49, MS, 50; Univ Ill, MS, 58, PhD(physiol), 63. Prof Exp: Res assoc pharmacol, Sharp & Dohme, Pa, 51-55; asst physiol, Col Med, Univ Ill, 56-61, from instr to asst prof, 61-66; from asst prof to assoc prof, 66-73, PROF PHYSIOL & BIOPHYS, COL MED, UNIV VT, 73- Mem: Biophys Soc. Res: Muscle contraction; experimental hypertension; hypertensive heart disease. Mailing Add: Dept of Physiol & Biophys Univ of Vt Col of Med Burlington VT 05401

MCCRORY, RAYMOND W, chemistry, see 12th edition

MCCRORY, ROBERT LEE, JR, b Lawton, Okla, Apr 30, 46; m 69. PLASMA PHYSICS, HYDRODYNAMICS. Educ: Mass Inst Technol, ScB, 68, PhD(plasma physics), 73. Prof Exp: Scientist geophys, Pan Am Petrol Res, Standard Oil Ind, 68-73; STAFF SCIENTIST THEORET PHYSICS, THEORET DIV, LOS ALAMOS SCI LAB, UNIV CALIF, 73- Concurrent Pos: Sr sci programmer, Res Lab Electronics, Mass Inst Technol, 68-73, res assoc, Dept Nuclear Eng/Dept Aerodyn & Astrophys, 72-73; US deleg, Conf Plasma Physics & Controlled Fusion Res, Int Atomic Energy Agency, Vienna, Austria, 74. Mem: Am Phys Soc; Sigma Xi. Res: Symmetry-stability studies of laser induced implosions; plasma physics/hydrodynamics of laser-initiated fusion; microinstabilities of magnetically confined plasma; laser-plasma interaction studies; radiation hydrodynamics. Mailing Add: Theoret Div MS 210 Los Alamos Sci Lab Los Alamos NM 87545

MCCRORY, WALLACE WILLARD, b Racine, Wis, Jan 19, 20; m 43; c 3. PEDIATRICS. Educ: Univ Wis, BS, 41, MD, 44. Prof Exp: Instr path, Med Sch, Univ Wis, 42-43; from asst prof to assoc prof, Sch Med, Univ Pa, 53-58; prof & head dept, Col Med, Univ Iowa, 58-61; PROF PEDIAT & CHMN DEPT, MED CTR, CORNELL UNIV, 61-; PEDIATRICIAN-IN-CHIEF, NEW YORK HOSP, 61- Concurrent Pos: Ledyard fel pediat, Cornell Univ, 49-50; dir, Clin Chem Lab, Children's Hosp, Philadelphia, 53-58. Mem: AAAS; Am Pediat Soc; NY Acad Sci; fel Royal Soc Med. Res: Renal disease and clinical biochemistry. Mailing Add: Dept of Pediat Cornell Univ Med Ctr New York NY 10021

MCCROSKEY, JACK E, b St Louis, Okla, Oct 8, 30; m 51; c 2. ANIMAL NUTRITION. Educ: Okla State Univ, BSc, 53, MSc, 59, PhD(animal nutrit), 61. Prof Exp: From asst prof to assoc prof animal sci & industs, Okla State Univ, 61-73; PROF ANIMAL INDUSTS & HEAD DEPT, UNIV IDAHO, 73- Mem: Am Soc Animal Sci; Am Dairy Sci Asn. Mailing Add: Dept of Animal Industs Univ of Idaho Moscow ID 83843

MCCROSKEY, ROBERT LEE, b Richwood, WVa, Feb 22, 24; m 50; c 4. AUDIOLOGY, SPEECH PATHOLOGY. Educ: Ohio State Univ, BS, 48, MA, 52, PhD(speech sci), 56. Prof Exp: Res assoc speech sci, Ohio State Res Found, Pensacola, Fla, 55-56; prof speech path, Miss Southern Col, 56-59; prof speech & hearing, Emory Univ, 59-67; PROF LOGOPEDICS, WICHITA STATE UNIV, 67- Concurrent Pos: Dir, Atlanta Speech Sch, Ga, 59-67; Am Speech & Hearing Asn int travel grant, 63 & 66; consult, Am Hearing Soc, 64-67, Div Voc Rehab, Ga, 66, Bur Educ of Handicapped, US Off Educ, 71- & Glenrose Hosp, Edmonton, Ont, 73-74; dir prof servs, Inst Logopedics, 68-70; reviewer div int activities, Dept Health, Educ & Welfare, 71- Mem: Speech Commun Asn; fel Am Speech & Hearing Asn. Res: Early diagnosis of hearing impairments; home training programs for deaf; auditory temporal processing for children with learning disabilities; effects of speech expansion upon comprehension; auditory localization. Mailing Add: Dept of Logopedics Wichita State Univ Wichita KS 67219

MCCROSKY, RICHARD EUGENE, b Akron, Ohio, Apr 28, 24; m 52; c 4. ASTRONOMY. Educ: Harvard Univ, BS, 52, PhD(astron), 56. Prof Exp: Sr res asst, 56, LECTR ASTRON, HARVARD UNIV, 56-, RES ASSOC, SMITHSONIAN ASTRO OBSERV, 57- Concurrent Pos: Astronr, Smithsonian Astrophys Observ, 57-58 & 60-, consult, 58-59; scientist chg, Meteorite Photog & Recovery Proj, 62-; mem, Int Astron Union. Mem: Fel AAAS; Meteoritical Soc; Am Astron Soc; Am Geophys Union. Res: Physics of meteors; meteor statistics and orbits; optical instrumentation; meteor photography. Mailing Add: Smithsonian Astro Observ 60 Garden St Cambridge MA 02138

MCCROSSAN, ROBERT GEORGE, b Vancouver, BC, Mar 27, 24; m 51; c 1. PETROLEUM GEOLOGY. Educ: Univ BC, BA, 48; Univ Chicago, SM, 52, PhD(geol), 56. Prof Exp: Subsurface geologist, Seaboard Oil Co, 49-58; res geologist, Imperial Oil Ltd, 58-65; HEAD ENERGY SUBDIV, GEOL SURV CAN, 65- Mem: Am Asn Petrol Geol; Can Inst Mining & Metall; Can Soc Petrol Geol (pres, 65). Res: Petroleum occurrence and potential; organic geochemistry; regional stratigraphic studies; probability resource occurence. Mailing Add: Geol Surv of Can 3303 33rd St NW Calgary AB Can

MCCROSSEN, GARNER, b New York, NY, Feb 22, 23; m 49; c 2. MATHEMATICS. Educ: Univ Wyo, BA, 48, MA, 49; Univ Colo, PhD(math), 54. Prof Exp: Instr math, Univ Nev, 49-50; programmer digital comput, Holloman AFB, 55-58, chief prog & operating sect, 58-60; programmer, 60-62, applns mgr, 62-68, VPRES, DIGITAL COMPUT, CONTROL DATA CORP, 68-, SR STAFF CONSULT, 75- Res: Philosophy; development of digital computer operating systems; development of assemblers for digital computers. Mailing Add: Control Data Corp 8100 34th Ave S Minneapolis MN 55440

MCCROSSON, F JOSEPH, b Brooklyn, NY, June 27, 40; m 65; c 3. REACTOR PHYSICS. Educ: Fairfield Univ, BS, 62; Va Inst Technol & State Univ, MS, 65, PhD(physics), 68. Prof Exp: Physicist, 65-67, res physicist, 68-72, staff physicist, 72-74, RES SUPVR REACTOR PHYSICS, SAVANNAH RIVER LAB, E I DU PONT DE NEMOURS & CO, 74- Concurrent Pos: Chmn thermal data testing subcomt, Cross Section Eval Working Group, 70- Mem: Am Nuclear Soc. Res: Neutron cross-sections, including calculation, evaluation and testing; production reactor support, especially charge design and accident analysis development. Mailing Add: 1017 Brookhaven Dr Aiken SC 29801

MCCRUM, RICHARD CASWELL, b Portland, Maine, Dec 12, 22; m 46. PLANT PATHOLOGY. Educ: Univ Ariz, BS, 51; Univ Maine, MS, 53; Univ N H, PhD, 64. Prof Exp: Asst, Univ Maine, 51-53 & Cornell Univ, 54-56; teacher high sch, NY, 56-57; from asst prof to assoc prof, 57-72, PROF BOT & PLANT PATH, UNIV MAINE, 72- Mem: Am Phytopath Soc. Res: Virus diseases of deciduous tree fruits; diseases of ornamental plants. Mailing Add: Dept of Bot & Plant Pathol Univ of Maine Orono ME 04473

MCCRUM, WILBUR ROSS, b Whittemore, Mich, Feb 28, 21; m 58; c 6. ANATOMY. Educ: Alma Col, AB, 43; Univ Iowa, MS, 49, PhD, 51. Prof Exp: Instr biol, Flint Jr Col, 46-47; instr anat, Univ Colo, 51-54; res assoc neurosurg, Henry Ford Hosp, 55-57; dir biosci, Sherman Labs, 57-59; chief electroencephalog lab, Henry Ford Hosp, 59-64, dir biomed, Comput Lab, Edsel B Ford Inst Med Res, 64-69; vpres, Ralph L Sherman, Inc, 69-73; CLIN NEUROPHYSIOLOGIST, HENRY FORD HOSP, 73- Mem: Am Asn Anat; Am Acad Neurol; Instrument Soc Am; Eastern EEG Asn. Res: Application of mathematics and computer technology to the solution of biological problems; developing reliability methodology for use in biostatistics; developing technique in electromyography. Mailing Add: Dept of Neurosurg & Neurol Henry Ford Hosp Detroit MI 48202

MCCRUMB, FRED RODGERS, b Havre de Grace, Md, Dec 23, 25; m 50; c 3. MEDICINE. Educ: Univ Md, MD, 48. Prof Exp: Assoc bact, Jefferson Med Col, 52-53; assoc med, 56-58, from asst prof to assoc prof, 58-62, dir, Inst Int Med, 62-70, PROF MED, SCH MED, UNIV MD, BALTIMORE, 62- Concurrent Pos: Assoc mem comn epidemiol surv, Armed Forces Epidemiol Bd, 59- Honors & Awards: Medal of Honor, France, 52. Mem: Am Soc Microbiol; Am Soc Trop Med & Hyg; Am Acad Microbiol; Am Fedn Clin Res; Am Epidemiol Soc. Res: Antibiotic therapy of acute infectious diseases; pathogenesis of viral and bacterial diseases of man; mechanisms of immunity and immunoprophylaxis of infectious diseases; international health. Mailing Add: Dept of Med Univ of Md Sch of Med Baltimore MD 21201

MCCUBBIN, DONALD GENE, b Glencoe, Okla, Oct 24, 30; m 62; c 2. PETROLEUM GEOLOGY, SEDIMENTOLOGY. Educ: Okla State Univ, BS, 52; Harvard Univ, MA, 55, PhD(geol), 61. Prof Exp: Res geologist, 57-65, adv res geologist, 65-70, SR RES GEOLOGIST, DENVER RES CTR, MARATHON OIL CO, 70- Mem: Geol Soc Am; Am Asn Petrol Geol; Soc Econ Paleont & Mineral. Res: Sandstone petrology and facies; sedimentary structures; nearshore sediments and processes; sandstone petroleum reservoirs; diagenesis of clay sediments; hydrocarbon generation and migration. Mailing Add: Denver Res Ctr Marathon Oil Co Box 269 Littleton CO 80122

MCCUBBIN, THOMAS KING, JR, b Baltimore, Md, June 1, 25; m 51; c 2. PHYSICS. Educ: Univ Louisville, BEE, 46; Johns Hopkins Univ, PhD(physics), 51. Prof Exp: Fel physics, Johns Hopkins Univ, 51-53; mem res staff phys chem, Mass Inst Technol, 53-57; from asst prof to assoc prof, 57-64, PROF PHYSICS, PA STATE UNIV, 64- Mem: Fel Optical Soc Am; fel Am Phys Soc. Res: Optics and spectroscopy; molecular spectra. Mailing Add: Dept of Physics Davey Lab Pa State Univ University Park PA 16802

MCCUE, CAROLYN M, b Richmond, Va, June 26, 16; m 41; c 2. PEDIATRICS. Educ: Stanford Univ, BA, 37; Med Col Va, MD, 41; Am Bd Pediat, dipl, 48, cert pediat cardiol, 59. Prof Exp: Instr, 46-49, assoc, 49-52, from asst prof to assoc prof, 52-63, interim chmn dept, 58-61, PROF PEDIAT, MED COL VA, 63- Concurrent Pos: Clin dir, Richmond Rheumatic Fever & Congenital Heart Clins, 47- Mem: Am Heart Asn; fel Am Col Physicians; fel Am Acad Pediat; fel Am Col Cardiol. Res: Pediatric cardiology. Mailing Add: Dept of Pediat Med Col of Va Richmond VA 23298

MCCUE, EDMUND BRADLEY, b Worcester, Mass, Mar 8, 29. MATHEMATICS. Educ: Union Col, AB, 50; Univ Mich, MS, 51; Carnegie Inst Technol, PhD(math), 60. Prof Exp: Mathematician, US Dept Defense, 51-54; asst prof math, Ohio Univ, 58-63; prof statist, Inter-Am Statist Inst, 63-64; ASSOC PROF MATH & STATIST, AM UNIV, 64- Mem: Inst Math Statist; Am Statist Asn; Math Asn Am. Res: Mathematical statistics. Mailing Add: Dept of Math & Statist American Univ Washington DC 20016

MCCUE, JOHN FRANCIS, b Milford, NH, Nov 22, 33; m 62; c 4. PARASITOLOGY. Educ: St John's Univ, Minn, BS, 60; Univ Notre Dame, MS, 62, PhD(biol), 64. Prof Exp: NIH fel, Sch Med, Nat Univ Mex, 64-65; res assoc parasitol, Sch Vet Med, Auburn Univ, 65-67; asst prof, 67-74, ASSOC PROF BIOL, ST CLOUD STATE COL, 74- Mem: Am Soc Parasitologists. Res: Parasitic immunology; parasite life cycles; host-parasite relationships. Mailing Add: Dept of Biol St Cloud State Col St Cloud MN 56301

MCCUE, JOHN JOSEPH GERALD, b South Orange, NJ, Dec 25, 13; m 49; c 1. MICROWAVE PHYSICS. Educ: Harvard Univ, AB, 36; Cornell Univ, PhD(physics), 42. Prof Exp: Tech asst, Bell Labs, 30-32; asst physics, Cornell Univ, 36-41; from instr to asst prof, Hamilton Col, 41-44; mem staff, Radiation Lab, Mass Inst Technol, 44-45; assoc prof physics, Smith Col, 46-49; mem staff, Lab Nuclear Sci, 49-51, MEM STAFF LINCOLN LAB, MASS INST TECH, 51- Concurrent Pos: Mem fac, Summer Sch, Harvard Univ, 59-60. Honors & Awards: President's Cert Merit, 46. Mem: Am Phys Soc; Am Asn Physics Teachers; fel Inst Elec & Electronics Eng. Res: X-ray physics; nuclear energy levels; radar; echolocation by bats. Mailing Add: 20 N Hancock St Lexington MA 02173

MCCUISTION, WILLIS LLOYD, b Fowler, Colo, May 28, 37; m 59; c 2. GENETICS, PLANT BREEDING. Educ: Colo State Univ, BS, 59; State Univ-USDA, 59-62; asst plant breeding & genetics, Okla State Univ, 62-67; plant breeder, Rockefeller Found, India, 67-68; dir & plant breeder, Int Maize & Wheat Improv Ctr, Tunisia, 68-71 & Algeria, 71-75; RES ASSOC, DEPT AGRONOMIC CROP SCI, ORE STATE UNIV, 75- Concurrent Pos: Rockefeller Found fel, All-India Mex Wheat Prog & Tunisia Wheat Prog, 67-69. Mem: Am Soc Agron; Crop Sci Soc Am; Am Inst Biol Sci. Res: Increasing agricultural production in general, especially in developing countries; training local scientists in applied scientific methods. Mailing Add: Dept of Agronomic Crop Sci Ore State Univ Corvallis OR 97331

MCCULLEN, JOHN DOUGLAS, b Sioux Falls, SDak, Oct 4, 32; m 54; c 3. NUCLEAR PHYSICS. Educ: Univ Colo, BA, 54, MS, 58, PhD(nuclear physics & spectros), 60. Prof Exp: Asst physics, Univ Colo, 56-60; res assoc, Princeton Univ, 60-62, asst prof, 62-65; assoc prof, 65-67, PROF PHYSICS, UNIV ARIZ, 67- Concurrent Pos: Sloane Found fel, 66-68. Mem: Am Phys Soc. Res: Nuclear spectroscopy and structure theory; atomic and polarized beams. Mailing Add: Dept of Physics Univ of Ariz Tucson AZ 85721

MCCULLEY, WILLIAM STRAIGHT, b Omaha, Nebr, Feb 26, 09; m 37; c 3. APPLIED MATHEMATICS. Educ: Univ Iowa, BA, 32; Agr & Mech Col Tex, MSc, 36; Univ Tex, PhD, 56. Prof Exp: Prof math, Wesley Col, 36-37; instr, Tex A&M Univ, 37-46, from asst prof to assoc prof, 46-74; RETIRED. Mem: Math Asn Am. Res: Algebra; partial differential equations. Mailing Add: 205 Hensel Dr Bryan TX 77801

MCCULLOCH, CLAY YOUNG, JR, b Durango, Colo, Mar 14, 24; m 57; c 1. ECOLOGY. Educ: Colo Agr & Mech Col, BS, 48; Univ Idaho, MS, 53; Okla State Univ, PhD(zool), 59. Prof Exp: Range aid, US Forest Serv, 49; biologist chaparral deer studies, 52-55, BIOLOGIST WATERSHED MGT EFFECTS BIG GAME, STATE GAME & FISH DEPT, ARIZ, 59- Mem: Ecol Soc Am; Am Soc Mammal; Wildlife Soc; Soc Range Mgt. Res: Mammal populations and individual spatial behavior associated with climate and vegetation. Mailing Add: PO Box 157 Fredonia AZ 86022

MCCULLOCH, DAVID SEARS, b New York, NY, Jan 8, 29; m 56; c 2. GEOLOGY, GEOMORPHOLOGY. Educ: Tufts Univ, BS, 54; Univ Mich, MS, 58, PhD(geol), 63. Prof Exp: Eng geologist, Hayden, Harding & Buchanan, Consult Engrs, Mass, 54-55; tech asst geol, Univ Mich, 59; GEOLOGIST, ALASKAN GEOL BR, US GEOL SURV, 61- Concurrent Pos: Mem emergency mission study earthquakes, UNESCO, 65- Mem: AAAS; NY Acad Sci. Res: Pleistocene and Holocene geology of arctic and near arctic regions; geomorphologic processes; earthquake effects of unconsolidated sediments; landslide induced water waves. Mailing Add: Alaskan Geol Br US Geol Surv 345 Middlefield Rd Menlo Park CA 94025

MCCULLOCH, ERNEST ARMSTRONG, b Toronto, Ont, Apr 27, 26; m 53; c 5. MEDICINE. Educ: Univ Toronto, MD, 48; FRCP(C), 54. Prof Exp: From jr intern to sr intern, Toronto Gen Hosp, 49-52; asst resident, Sunnybrook Hosp, 52-53; clin teacher med, 54-60, assoc, 60-67, from asst prof to assoc prof med, 68-70, from asst prof to assoc prof med biophys, 59-66, PROF MED BIOPHYS, UNIV TORONTO, 66-, PROF MED, 70- Concurrent Pos: Ellen Mickle fel, Lister Inst, Eng, 48-49; Nat Res Coun Can res fel path, Univ Toronto, 50-51; Nat Cancer Inst Can fel, 54-57; asst physician, Toronto Gen Hosp, 54-60, physician, 60-67; head subdiv hemat, Div Biol Res, Int Cancer Inst, 57-67; mem grants comt microbiol & path, Med Res Coun Can, 66-67, immunol & transplantation, 67-69; mem panel B, Grants Comt, Nat Cancer Inst Can, 70-74; ed, J Cell Physiol, 69-; sr physician, Princess Margaret Hosp, 70- Honors & Awards: Starr Medal, Univ Toronto, 57, Goldie Prize in Med, 64; Gairdner Award, 69. Mem: Am Asn Cancer Res; Am Soc Exp Path; Soc Exp Biol & Med; Can Soc Cell Biol; Royal Soc Can. Res: Hematology; stem cell functions, especially physiologic and genetic control mechanisms; leukemia research. Mailing Add: Dept of Med Univ of Toronto Toronto ON Can

MCCULLOCH, KENNETH EVERETT, physical chemistry, see 12th edition

MCCULLOCH, PETER BLAIR, b Hamilton, Ont, Mar 7, 39; m 69; c 3. CANCER. Educ: Univ Toronto, MD, 64; FRCP(C), 69. Prof Exp: Lectr med, McGill Univ, 70-72; ASST PROF MED, McMASTER UNIV, 72- Concurrent Pos: Internist oncol, Ont Cancer Found, 72- Mem: Can Med Asn; Can Soc Clin Oncol. Res: Cytogenetics of human malignancy; chemotherapy and immunotherapy of human cancer. Mailing Add: Hamilton Cancer Clin Henderson Hosp Hamilton ON Can

MCCULLOCH, WILLIAM JOHN GILBERT, physical chemistry, see 12th edition

MCCULLOH, THANE H, b Glendale, Calif, July 25, 26; m 49, 63; c 5. PETROLEUM GEOLOGY. Educ: Pomona Col, AB, 49; Univ Calif, Los Angeles, PhD(geol), 52. Prof Exp: Res assoc geol, Univ Calif, Los Angeles, 52-53; asst prof, Calif Inst Technol, 53-55; from assoc prof to prof, Univ Calif, Riverside, 55-64; geologist, DC, 64-72, chief, Off Energy Res, 72-73, RES GEOLOGIST, US GEOL SURV, 73- Concurrent Pos: NSF fel, 52-53; Guggenheim fel, 60; affil prof oceanog, Univ Wash, 73- Honors & Awards: PL 313 Position Award, US Geol Surv, 74. Mem: Geol Soc Am; Soc Explor Geophys; Am Asn Petrol Geol; Norweg Geol Asn. Res: Areal geology of the central Mojave Desert; igneous and metamorphic petrology; gravimetry; structural geology of the Los Angeles Basin; petrophysics; petroleum exploration research and sedimentary basin resource studies. Mailing Add: US Geol Surv 1107 NE 45th St Seattle WA 98105

MCCULLOUGH, DALE RICHARD, b Sioux Falls, SDak, Dec 5, 33; m 58; c 3. WILDLIFE MANAGEMENT, ECOLOGY. Educ: SDak State Univ, BS, 57; Ore State Univ, MS, 60; Univ Calif, Berkeley, PhD(zool), 66. Prof Exp: Instr zool, Univ Calif, Berkeley, 65-66; asst prof wildlife mgt, 66-69, assoc prof wildlife & fisheries, 69-74, PROF NATURAL RESOURCES, UNIV MICH-ANN ARBOR, 74-, CHMN RESOURCE ECOL PROG, 69- Mem: Wildlife Soc; Ecol Soc Am. Res: Ecology and population dynamics of large herbivores; ecosystem processes. Mailing Add: Univ of Mich Ann Arbor MI 48104

MCCULLOUGH, EDGAR JOSEPH, JR, b Charleston, WVa, Nov 29, 31; m 54; c 2. STRUCTURAL GEOLOGY. Educ: Univ WVa, AB, 53, MS, 55; Univ Ariz, PhD(geol), 63. Prof Exp: From instr to assoc prof geol, 58-69, PROF GEOSCI & HEAD DEPT, UNIV ARIZ, 69- Mem: AAAS; Nat Asn Geol Teachers; Geol Soc Am. Res: Rheology; development of fluid structural features in solid rocks; science education. Mailing Add: Dept of Geosci Univ of Ariz Tucson AZ 85721

MCCULLOUGH, EDWIN CHARLES, b New York, NY, June 2, 42; m 70; c 1. MEDICAL PHYSICS. Educ: State Univ NY Stony Brook, BS, 64; Univ Md, MS, 67; Univ Wis-Madison, PhD(radiol physics), 71. Prof Exp: Scientist neutron physics, Med Res Coun Cyclotron Unit, Hammersmith Hosp, 71; res scientist radiation physics, Univ Wis-Madison, 71-73; STAFF PHYSICIST, MAYO CLIN, 73- Mem: Am Asn Physicists Med; Radiol Soc NAm; Sigma Xi. Res: Radiation therapy dosimetry and treatment planning; diagnostic radiology imaging; computer assisted tomography performance evaluation and quality assurance. Mailing Add: Mayo Clin 200 First St SW Rochester MN 55901

MCCULLOUGH, HERBERT ALFRED, b Pittsburgh, Pa, Dec 19, 14; m 43; c 2. BIOLOGY. Educ: Univ Pittsburgh, BS, 35, MS, 37, PhD(biol), 39. Prof Exp: Asst biol, Univ Pittsburgh, 36-39; prof & head dept, Bessie Tift Col, 39-47; assoc prof, 47-52, PROF BIOL, SAMFORD UNIV, 52-, HEAD DEPT, 57- Concurrent Pos: Ala Acad Sci res grant, 50, 54. Mem: Fel AAAS; Am Bryol & Lichenological Soc; Am Inst Biol Sci; Brit Lichen Soc. Res: Lichenology; plant ecology; conservation of natural areas. Mailing Add: Dept of Biol Samford Univ Birmingham AL 35209

MCCULLOUGH, JACK DENNIS, b San Antonio, Tex, Aug 8, 31; m 53; c 2. LIMNOLOGY. Educ: Univ Tex, BS, 55; Stephen F Austin State Univ, MA, 62; Tex A&M Univ, PhD(biol), 70. Prof Exp: Teacher pub schs, Tex, 55-63; instr biol, Tex A&M Univ, 63-64; ASSOC PROF BIOL, STEPHEN F AUSTIN STATE UNIV, 64- Concurrent Pos: Researcher, US Army Corps Engrs, 71-73 & 75. Mem: Sigma Xi; Am Soc Limnol & Oceanog. Res: Primary productivity in aquatic environments. Mailing Add: Dept of Biol Stephen F Austin State Univ Nacogdoches TX 75961

MCCULLOUGH, JAMES DOUGLAS, b Oskaloosa, Iowa, May 17, 05; m 31, 49; c 3. STRUCTURAL CHEMISTRY, INORGANIC CHEMISTRY. Prof Exp: From assoc to prof, 32-71, EMER PROF CHEM, UNIV CALIF, LOS ANGELES, 71- Mem: Am Chem Soc; Am Crystallog Asn. Res: Studies on the synthesis and on the crystal and molecular structure of organoselenium and organotellurium compounds. Mailing Add: Dept of Chem Univ of Calif Los Angeles CA 90024

MCCULLOUGH, JAMES DOUGLAS, JR, b Long Beach, Calif, Nov 28, 38; m 71, 75. ORGANIC CHEMISTRY, PHYSICAL CHEMISTRY. Educ: Univ Calif, Los Angeles, BS, 63; San Diego State Col, MS, 65; Univ Ill, Urbana, PhD(chem), 70. Prof Exp: RES CHEMIST, WESTHOLLOW RES CTR, SHELL DEVELOP CO, 76- Am Chem Soc; Sigma Xi. Res: Plastics technology; polymer science; petrochemical additives. Mailing Add: Shell Develop Co Westhollow Res Ctr PO Box 1380 Houston TX 77001

MCCULLOUGH, JAMES MATTHEW, b Pittsburgh, Pa, July 6, 25; m 50; c 2. BOTANY, ENVIRONMENTAL BIOLOGY. Educ: Pa State Univ, BS, 51, MS, 52; George Washington Univ, PhD(bot), 68. Prof Exp: Res technologist mat develop, Bur Ships, Navy Dept, 52-56; chmn dept biol & chem, Wakefield High Sch, Va, 56-63; head biosci div, Navy Sci & Tech Ctr, Navy Dept, 63-68, head dept phys sci, 68-69;

SPECIALIST LIFE SCI, SCI POLICY RES DIV, CONG RES SERV, LIBR OF CONG, 69- Concurrent Pos: Sr consult engr, Packaging Consults, Inc, 59-62; instr, Sch Gen Studies, Univ Va, 60-63 & 70-; adj prof, Va Polytech Inst & State Univ, 73-74. Mem: AAAS; Am Inst Biol Sci. Res: Effects of physical and biological factors on growth and development. Mailing Add: Sci Policy Res Div Cong Res Serv Library of Congress Washington DC 20540

MCCULLOUGH, JOHN EDMUND, physical organic chemistry, see 12th edition

MCCULLOUGH, JOHN FRANKLIN, b Jamestown, Ala, Feb 26, 25; m 56; c 3. INORGANIC CHEMISTRY. Educ: Auburn Univ, BS, 49, MS, 51. Prof Exp: Res chemist, Monsanto Chem Co, 51-57; proj engr, Cramet Inc, 57-58; res chemist, Thiokol Chem Corp, 58-60; res chemist, Grace Chem Co, 60-63, res supvr, 63; RES CHEMIST, TENN VALLEY AUTH, 63- Mem: Am Chem Soc. Res: Chemistry of phosphorus; fertilizer technology. Mailing Add: 2909 Alexander St Florence AL 35632

MCCULLOUGH, JOHN JAMES, b Belfast, Northern Ireland, Sept 27, 37; m 63. ORGANIC CHEMISTRY. Educ: Queens Univ, Belfast, BSc, 59, PhD(chem), 62. Prof Exp: Dept Sci & Indust Res fel, 62-63; res assoc, Univ Wis, 63-65; from asst prof to assoc prof, 65-73, PROF CHEM, McMASTER UNIV, 73- Mem: Am Chem Soc; Chem Inst Can; The Chem Soc. Res: Stereochemistry; photochemistry; physical organic chemistry. Mailing Add: Dept of Chem McMaster Univ Hamilton ON Can

MCCULLOUGH, JOHN MARTIN, b Chicago, Ill, Mar 9, 40; m 71; c 2. PHYSICAL ANTHROPOLOGY. Educ: Pa State Univ, BA, 62; Univ Ill, PhD(anthrop), 72. Prof Exp: ASST PROF ANTHROP, UNIV UTAH, 69- Concurrent Pos: Forensic anthropologist, Off Med Examr, Utah State Bd Health, 69- Mem: AAAS; Am Asn Phys Anthrop; Am Anthrop Asn; Brit Soc Study Human Biol; Int Asn Human Biol. Res: Human biology; adaptability; ecological genetics; ecological demography; Latin America, especially Mexico and Yucatan. Mailing Add: Dept of Anthrop Univ of Utah Salt Lake City UT 84112

MCCULLOUGH, JOHN PRICE, b Dallas, Tex, May 10, 25; m 46; c 3. PHYSICAL CHEMISTRY. Educ: Univ Okla, BS, 45; Ore State Col, MS, 48, PhD(chem), 49. Prof Exp: Phys chemist, Thermodyn Lab, Bartlesville Petrol Res Ctr, US Bur Mines, Okla, 49-56, chief, 56-63; mgr cent res div, Mobil Oil Corp, 63-69, appl res & develop, Mobil Res & Develop Corp, 69-71, GEN MGR, RES & DEVELOP, MOBIL CHEM CO, 71- Concurrent Pos: Adj Prof, Okla State Univ, 61-63. Honors & Awards: Distinguished Serv Award, US Dept Interior, 62. Mem: AAAS; Am Chem Soc. Res: Thermodynamics and molecular structure of hydrocarbons and related substances; research administration. Mailing Add: Res & Develop Mobil Chem Co PO Box 240 Edison NJ 08817

MCCULLOUGH, MARSHALL EDWARD, b Wick, WVa, July 24, 24. DAIRY NUTRITION. Educ: Berea Col, BS, 49; Univ Ky, MS, 51. Prof Exp: Asst dairy, Univ Ky, 50-51; asst dairy nutritionist, 51-59, ASSOC NUTRITIONIST, AGR EXP STA, UNIV GA, 59-, PROF ANIMAL NUTRIT, 68-, HEAD DEPT ANIMAL SCI, 71- Mem: AAAS; Soc Range Mgt; Am Soc Animal Sci; Am Dairy Sci Asn. Res: Plant and animal complex in grassland utilization; calf nutrition; systems for forage evaluation. Mailing Add: Dept of Animal Sci Agr Exp Sta Experiment GA 30212

MCCULLOUGH, NORMAN B, b Milford, Mich, Feb 23, 09; m 39; c 2. MICROBIOLOGY, INFECTIOUS DISEASES. Educ: Mich State Col, BS, 32, MS, 33; Univ Chicago, PhD(bact), 37, MD, 44; Am Bd Microbiol, dipl & cert pub health & med lab microbiol. Prof Exp: Asst, Mich State Col, 32-34; asst physiologist, Parke, Davis & Co, 34-36, res bacteriologist, 38-40, res bacteriologist, Tex, 40-42; from intern to asst resident, Univ Chicago Clins, 44-46, asst clin prof med, 47-50; chief brucellosis unit, Lab Infectious Dis, Nat Inst Allergy & Infectious Dis, 51-56, chief lab clin invests & clin dir, 52-58, chief lab bact dis, 58-68; PROF MED, MICROBIOL & PUB HEALTH, MICH STATE UNIV, 68- Concurrent Pos: Eve instr, Detroit Inst Technol, 38-40; mem comt brucellosis, Nat Res Coun, 51-58; mem expert panel brucellosis, Food & Agr Orgn, WHO, 52-; med dir, USPHS, 52-68; spec lectr, Georgetown Univ, 52-66; instr, USDA Grad Sch, 54-61; mem adv coun, Med Int Coop, 58-60; expert taxon subcomt brucella, Int Comt Bact Nomenclature, 58-; mem, Nat Brucellosis Comt, Inc, 61-69, mem bd dirs, 63-69; instr, Found Adv Educ in Sci, Inc, 61-68, fac chmn microbiol & immunol, 63-68. Mem: AAAS; Am Soc Microbiol; Am Acad Microbiol; NY Acad Sci. Res: Medical microbiology; infectious and parasitic diseases. Mailing Add: Dept of Microbiol & Pub Health Mich State Univ East Lansing MI 48824

MCCULLOUGH, ROGER STEWART, mathematical statistics, see 12th edition

MCCULLOUGH, ROY LYNN, b Hillsboro, Tex, Mar 20, 34; m 58. PHYSICAL CHEMISTRY. Educ: Baylor Univ, BS, 55; Univ NMex, PhD(chem), 60. Prof Exp: Mem staff, Los Alamos Sci Lab, 58-59; group leader, Chemstrand Corp, Monsanto Co, NC, 59-69; mem staff, Polymer Sci Lab, Boeing Sci Res Lab, Wash, 69-71; PROF CHEM ENG, UNIV DEL, 71- Mem: Am Chem Soc; Am Phys Soc; AAAS; fel Am Inst Chem. Res: Influence of molecular structure on solid state properties. Mailing Add: Dept of Chem Eng Univ of Del Newark DE 19711

MCCULLOUGH, THOMAS F, b Los Angeles, Calif, Nov 12, 22. ORGANIC CHEMISTRY. Educ: Univ Notre Dame, BS, 48, MS, 49; Univ Utah, PhD(chem), 55. Prof Exp: Anal chemist & chem engr, Union Oil Co, 50-51; instr chem, Long Beach City Col, 55-56; INSTR CHEM, ST EDWARD'S UNIV, 57- Mem: Am Chem Soc; Entom Soc Am. Res: Analytical organic chemistry in natural products. Mailing Add: Dept of Chem St Edward's Univ Austin TX 78704

MCCULLY, JOSEPH C, b Kalamazoo, Mich, Feb 6, 24; m 47; c 4. MATHEMATICS. Educ: Western Mich Univ, BA, 47; Univ Mich, MA, 49, PhD(math), 57. Prof Exp: Asst prof math, Univ SDak, 53-55; asst prof, Univ RI, 55-56; assoc prof, 56-66, PROF MATH, WESTERN MICH UNIV, 66- Mem: Am Math Soc; Math Asn Am. Res: Integral transforms; operational calculus. Mailing Add: Dept of Math Western Mich Univ Kalamazoo MI 49001

MCCULLY, KEITH ALLAN, biochemistry, see 12th edition

MCCULLY, KILMER SERJUS, b Daykin, Nebr, Dec 23, 33; m 55; c 2. EXPERIMENTAL PATHOLOGY. Educ: Harvard Univ, AB, 55; Harvard Med Sch, MD, 59. Prof Exp: Intern, Mass Gen Hosp, 59-60; biochemist, NIH, 60-62; USPHS res fel med, Harvard Med Sch, 62-63, res fel biol, Harvard Univ, 63-65, Am Cancer Soc fac res assoc, 63-68, asst path, Harvard Med Sch, 65-68, instr, 68-70; asst pathologist, 68-73, ASSOC PATHOLOGIST, MASS GEN HOSP, 74-; ASST PROF PATH, HARVARD MED SCH, 70- Concurrent Pos: Res assoc, Glasgow Univ, 63-64; clin & res fel path, Mass Gen Hosp, 65-68; NIH career develop award, 71- Mem: AAAS; Am Soc Exp Pathologists; Am Soc Clin Pathologists; Am Asn Path & Bact; Int Acad Path. Res: Nucleic acid structure; protein biosynthesis; microbial genetics; amino acid metabolism; arteriosclerosis; growth hormone; ascorbate; pyridoxine; somatomedin; cancer. Mailing Add: Dept of Path Mass Gen Hosp Boston MA 02114

MCCULLY, MARGARET E, b St Marys, Ont, July 25, 34. BIOLOGY. Educ: Univ Toronto, BSA, 56, MSA, 60; Harvard Univ, PhD(biol), 66. Prof Exp: Res assoc microbiol, Parke Davis Co, 56-57; teache high sch, Ont, 57-58 & St Felix Sch, Eng, 60-62; asst prof, 66-71, ASSOC PROF BIOL, CARLETON UNIV, 71- Mem: Soc Develop Biol; Histochem Soc; Can Soc Plant Physiol; Can Soc Cell Biol. Res: Cytology, histology and histochemistry in relation to plant development. Mailing Add: Dept of Biol Carleton Univ Ottawa ON Can

MCCULLY, WAYNE GUNTER, b New Cambria, Mo, Jan 3, 22; m 44. RANGE SCIENCE. Educ: Colo State Univ, BS, 47; Tex A&M Univ, 50, PhD, 58. Prof Exp: From asst prof to prof range mgt, 47-73, actg head dept, 71-72, PROF & RESIDENT DIR RES, TEX A&M UNIV, 74- Mem: Soc Range Mgt; Ecol Soc Am; Am Soc Plant Physiol; Weed Sci Soc Am. Res: Noxious plant control; range ecology; grassland management; roadside vegetation management. Mailing Add: Tex A&M Univ Res & Ext Ctr Box 1658 Vernon TX 76384

MCCUMBER, DEAN EVERETT, b Rochester, NY, Nov 25, 30; m 57. PHYSICS. Educ: Yale Univ, BE, 52, ME, 55; Harvard Univ, AM, 56, PhD(physics), 60. Prof Exp: Head crystal electronics res dept, 65-69, PHYSICIST, BELL LABS, INC, 61-, DIR, INTERCONNECTION TECHNOL LAB, 69- Mem: Sr mem Inst Elec & Electronics Eng; fel Am Phys Soc. Res: Lasers; vibrational structure in optical spectra of solids; electrical instabilities in solid-state materials; Gunn effect; electronic interconnection systems. Mailing Add: Bell Labs Inc Whippany NJ 07981

MCCUNE, DELBERT CHARLES, b Los Angeles, Calif, Aug 21, 34. PLANT PHYSIOLOGY. Educ: Calif Inst Technol, BS, 56; Yale Univ, MS, 57, PhD(bot), 60. Prof Exp: From asst plant physiologist to assoc plant physiologist, 60-68, PLANT PHYSIOLOGIST, BOYCE THOMPSON INST PLANT RES, 68- Mem: AAAS; Bot Soc Am; Am Soc Plant Physiol. Res: Plant growth regulators; air pollution. Mailing Add: Boyce Thompson Inst for Plant Res 1086 N Broadway Yonkers NY 10701

MCCUNE, DONALD LLOYD, plant physiology, nutrition, see 12th edition

MCCUNE, DUNCAN CHALMERS, b Chicago, Ill, Mar 16, 25; m 47; c 2. STATISTICAL ANALYSIS. Educ: Col of Wooster, AB, 48; Purdue Univ, MS, 50. Prof Exp: Res assoc, Purdue Univ, 50-52; statistician, Tubular Prods Div, Babcock & Wilcox Co, 52-55; sr statistician, 55-64, supvr appl math, 64-68, asst to dir tech serv, 69-71, MGR QUAL ASSURANCE, JONES & LAUGHLIN STEEL CORP, 71- Mem: Am Soc Qual Control; Am Soc Test & Mat; Royal Statist Soc. Mailing Add: Jones & Laughlin Steel Corp 3 Gateway Ctr Pittsburgh PA 15230

MCCUNE, EMMETT L, b Cuba, Mo, Jan 2, 27; m 52; c 3. VETERINARY MICROBIOLOGY, AVIAN PATHOLOGY. Educ: Univ Mo, BS & DVM, 56, MS, 61; Univ Minn, PhD(microbiol), 67. Prof Exp: Instr vet bact, 56-61, asst prof, 61-68, ASSOC PROF VET MICROBIOL, SCH VET MED, UNIV MO-COLUMBIA, 68- Mem: Am Col Vet Microbiol; Am Vet Med Asn; Am Asn Avian Path; Am Soc Microbiol. Res: Microbiology, expecially mycoplasma, Pasteurella and avian leucosis viruses as related to avian pathology and metabolism; resistance factor in Salmonella and Escherichia coli. Mailing Add: Dept of Vet Microbiol Sch of Vet Med Univ of Mo-Columbia Columbia MO 65201

MCCUNE, HOMER WALLACE, b Grove City, Pa, Sept 5, 23; m 49; c 3. PHYSICAL CHEMISTRY. Educ: Grove City Col, BS, 44; Cornell Univ, PhD(inorg chem), 49. Prof Exp: Mem staff, Coal Res Lab, Carnegie Inst Technol, 44-45; res chemist, Chem Div, 49-54, sect head, 54-56, dept head, 56-58, patent div, 58-59, assoc dir, Res Div, 59-71, ASSOC DIR PAPER & TOILET GOODS TECHNOL DIV, RES & DEVELOP DEPT, PROCTER & GAMBLE, 71- Mem: AAAS; Am Chem Soc. Res: Phosphates; corrosion; phosphorus compounds; surface chemistry; oral biology; administration. Mailing Add: 580 Larchmont Dr Cincinnati OH 45215

MCCUNE, MARY JOAN HUXLEY, b Lewistown, Mont, Jan 14, 32; m 65; c 2. MICROBIOLOGY. Educ: Montana State Univ, BS, 53; Wash State Univ, MS, 55; Purdue Univ, PhD(microbiol), 65. Prof Exp: Res technician, Vet Admin Hosp, Oakland, Calif, 56-59; bacteriologist, US Naval Radiol Defense Lab, Calif, 59-61; teaching assoc microbiol, Purdue Univ, 61-65, vis asst prof biol, 65-66; asst prof microbiol, Occidental Col, 66-69; fel, Univ Calif, Los Angeles, 69-70; AFFILIATE ASST PROF MICROBIOL, IDAHO STATE UNIV, 70- Mem: AAAS; Am Soc Microbiol; NY Acad Sci. Res: Microbial physiology, genetics and taxonomy; microbial regulation of protein and ribonucleic acid synthesis. Mailing Add: Dept of Microbiol & Biochem Idaho State Univ Pocatello ID 83201

MCCUNE, ROBERT FRANKLIN, b Muncie, Ind, Aug 23, 15; m 54; c 3. PHYSICS. Educ: Manchester Col, BA, 37; Univ Ill, MS, 39, PhD(physics), 41. Prof Exp: Asst physics, Univ Ill, 37-41; anal engr, Hamilton Standard Propellers, Conn, 41-46; from instr to asst prof physics, Trinity Col, Conn, 46-52; from asst prof to assoc prof, Bucknell Univ, 52-57; engr, Pratt & Whitney Aircraft Co, 57-68; assoc prof, 68-72, PROF PHYSICS, CENT CONN STATE COL, 68- Mem: Am Asn Physics Teachers. Res: Theoretical mechanics; applied mathematics; philosophy of science. Mailing Add: 115 Shipman Dr Glastonbury CT 06033

MCCUNE, RONALD WILLIAM, b Glade, Kans, Sept 23, 38; m 65; c 1. BIOCHEMISTRY. Educ: Kans State Univ, BS, 61; Purdue Univ, MS, 64, PhD(biochem), 66. Prof Exp: Trainee, Univ Calif, Los Angeles, 66-67, res biol chemist, 67-70; asst prof, 70-73, ASSOC PROF BIOCHEM, IDAHO STATE UNIV, 73-, CHMN DEPT MICROBIOL & BIOCHEM, 71- Mem: AAAS; Am Chem Soc; NY Acad Sci. Res: Effect of antibiotics on short-chain fatty acid metabolism by rumen microorganisms; urea-N metabolism in the ruminant animal; metabolic control of adrenal steroid biosynthesis; mechanisms of hormonal action; mitochondrial metabolism. Mailing Add: Dept of Microbiol & Biochem Idaho State Univ Pocatello ID 83209

MCCUNE, SHANNON, b Sonchon, Korea, Apr 6, 13; US citizen; m 36; c 3. GEOGRAPHY. Educ: Col Wooster, BA, 35; Syracuse Univ, MA, 37; Clark Univ, PhD(geog), 39. Hon Degrees: LLD, Clark Univ, 60, Univ Mass, 62 & Eastern Nazarene Col, 66. Prof Exp: Instr geog, Ohio State Univ, 39-46; from asst prof to prof, Colgate Univ, 47-55, head dept, 47-55; provost, Univ Mass, 55-61; dir dept educ, UNESCO, Paris, 61-62; civil adminr, US Civil Admin, Ryukyu Islands, 62-64; staff assoc, Off Pres, Univ Ill, 64-65; pres, Univ Vt, 65-66, res prof, 66-67; dir, Am Geog Soc, 67-69; chmn dept, 69-75, PROF GEOG, UNIV FLA, 69- Concurrent Pos: From econ analyst to head intel officer, Bd Econ Welfare, For Econ Admin, 42-45; dept dir, Far East Prog Div, Econ Coop Admin, 50-51; Fulbright vis prof, Tokyo Univ, 53-54 & Soon-Sun Univ, Seoul, Korea, 75; adj prof, Columbia Univ, 67-69. Mem: Am Geog Soc; Asn Am Geog; Am Oriental Soc; Asn Asian Studies; Geog Soc India. Res: Geography of monsoon Asia, particularly Korea and Ryukyu Islands. Mailing Add: 1617 N W Seventh Pl Gainesville FL 32601

MCCURDY, DAVID HAROLD, b East Orange, NJ, Apr 28, 30; m 56; c 1. PHARMACOLOGY. Educ: Dalhousie Univ, BSc, 53, MSc, 55; Univ Toronto, PhD(pharmacol), 58. Prof Exp: Asst pharmacol, Dalhousie Univ, 53-55; asst, Univ Toronto, 55-58; pharmacologist, Chem Res & Develop Labs, US Army Chem Ctr, Md, 58-62; sr res pharmacologist, 62-65, mgr pharmacol sect, 65-75, ASST DIR DRUG DESIGN & EVAL, ICI AM INC, 73- Mem: AAAS; Sigma Xi; NY Acad Sci; Pharmacol Soc Can; Am Soc Pharmacol & Exp Therapeut. Res: Basic pharmacology, including the molecular basis of drug action; drug screening and development; central nervous system pharmacology research and management. Mailing Add: 729 Burnley Rd Tavistock Wilmington DE 19803

MCCURDY, DAVID WHITWELL, b New York, NY, Sept 30, 35; m 57; c 4. CULTURAL ANTHROPOLOGY. Educ: Cornell Univ, BA, 57; Stanford Univ, MA, 59; Cornell Univ, PhD(anthrop), 64. Prof Exp: Asst prof anthrop, Colo State Univ, 64-66; asst prof, 66-70, ASSOC PROF ANTHROP, MACALESTER COL, 70-, CHMN DEPT SOCIOL & ANTHROP, 73- Concurrent Pos: Consult, Peace Corps Training Prog, India, 65; consult, Head Start Teacher's Training Prog, 66; mem prog comn, Am Anthrop Asn, 65. Mem: AAAS; Am Anthrop Asn; Am Ethnol Soc; Soc Appl Anthrop. Res: Ethnographic semantics; ethnography of complex societies; South Asia; United States; comparative religion; economic anthropology. Mailing Add: Dept of Sociol & Anthrop Macalester Col 1600 Grand Ave St Paul MN 55101

MCCURDY, DENNIS, b Charlotte, NC, Aug 7, 39; m 67; c 2. VETERINARY PHARMACOLOGY. Educ: Gettysburg Col, AB, 61; Univ Ga, DVM, 66. Prof Exp: Clinician small animal med & surg, Fleming Animal Hosp, 66-67; chief vet div, US Army, 67-69; clinician small animal med & surg, Fleming Animal Hosp & Martinez Animal Clin, 69-70 & Ford Animal Hosp, 70-72; HEAD CLIN PHARMACOL, PITMAN-MOORE, INC, DIV JOHNSON & JOHNSON, 73- Mem: Am Vet Med Asn; Am Soc Vet Anesthesiol. Res: Veterinary anesthesiology and anthelmintic research. Mailing Add: Pitman-Moore Inc Box 344 Washington Crossing NJ 08560

MCCURDY, HARRIET MACE, b Toronto, Ont, Sept 4, 03; wid; c 5. EMBRYOLOGY, ENDOCRINOLOGY. Educ: Univ Calif, Berkeley, BA, 27; Univ Iowa, MSc, 29, PhD(embryol, endocrinol), 30. Prof Exp: Instr zool, Univ Iowa, 27-29; librn med res, Rockefeller Univ, 57-62; instr zool, 62-64, RES ASSOC EMBRYOL & ENDOCRINOL, UNIV VICTORIA, BC, 64- Mem: AAAS; Am Soc Zool; Can Soc Zool; NY Acad Sci. Res: Developmental anatomy; pigment cell development; irradiation effects on pigment cell development with special reference to melanocytes; enzyme localization in melanogenesis; possible de-differentiation of cancered melanocyte. Mailing Add: Dept of Biol Univ of Victoria Victoria BC Can

MCCURDY, HOWARD DOUGLAS, JR, b London, Ont, Can, Dec 10, 32; m 56; c 4. MICROBIOLOGY. Educ: Western Ontario Univ, BA, 53; Assumption Univ, BSc, 54; Mich State Univ, MSc, 55, PhD(microbiol), 59. Prof Exp: Asst, Mich State Univ, 55-59; assoc prof, 59-70, PROF BIOL, UNIV WINDSOR, 70-, DEPT HEAD, 74- Mem: AAAS; Am Soc Microbiol; Can Soc Microbiol. Res: Ecology of prokaryotes; biology of myxobacteria in pure culture. Mailing Add: Dept of Biol Univ of Windsor Windsor ON Can

MCCURDY, JON ALAN, b Lacona, Iowa, Oct 18, 12; m 42; c 3. VETERINARY ANATOMY. Educ: Iowa State Univ, BS & DVM, 38. Prof Exp: Instr vet anat, Agr & Mech Col, Tex, 38-39; from asst prof to assoc prof, 46-56, PROF VET ANAT & CHMN DEPT, COL VET MED, WASH STATE UNIV, 56- Mem: Am Asn Vet Anat (secy, 53, pres, 55); Am Vet Med Asn. Res: Microscopic anatomy; embryology. Mailing Add: NE 725 Illionis St Pullman WA 99163

MCCURDY, KEITH G, b Lampman, Sask, Can, Dec 4, 37; m 60; c 3. PHYSICAL CHEMISTRY. Educ: Univ Sask, BA, 59, MA, 61; Univ Ottawa, PhD(chem), 64. Prof Exp: Asst prof chem, Univ Guelph, 63-67; asst prof, 67-69, ASSOC PROF CHEM, UNIV LETHBRIDGE, 69- Res: Chemical kinetics. Mailing Add: Dept of Chem Univ of Lethbridge Lethbridge AB Can

MCCURDY, LAYTON, b Florence, SC, Aug 20, 35; m 58; c 2. PSYCHIATRY. Educ: Univ NC, MD, 60. Prof Exp: Rotating intern, Med Col SC, 60-61; resident psychiat, NC Mem Hosp, 61-64; asst prof psychiat, Sch Med, Emory Univ, 66-68; PROF PSYCHIAT & CHMN DEPT, MED UNIV SC, 68- Concurrent Pos: Consult, Community Psychiat Clin, Bethesda, Md, 65-66, NIMH, 66-67, Clayton County Ment Health Clin, Ga, 66-68, VISTA, Washington, DC, 67-68, Charleston Vet Admin Hosp, Va & SC Dept Ment Health, 68- & Med Res Rev Bd Behav Sci, NIMH, 74-77; psychoanal trainee, Columbia Univ Psychoanal Clin, Atlanta, Ga, 67-68; mem res rev comt, Appl Res Br, NIMH, 70-74; NIMH res fel, Maudsley Hosp, London, Eng, 74-75. Mem: Am Col Psychiat; Am Psychiat Asn; Am Psychosom Soc; Asn Am Med Cols; Asn Acad Psychiat (pres, 70-72). Res: Behavioral sciences; medical education. Mailing Add: Dept of Psychiat Med Univ of SC Charleston SC 29401

MCCURDY, PAUL RANNEY, b Middletown, Conn, Sept 26, 25; m 52; c 5. INTERNAL MEDICINE, HEMATOLOGY. Educ: Wesleyan Univ, AB, 46; Harvard Med Sch, MD, 49. Prof Exp: Fel med, Sch Med, Georgetown Univ, 54-55; instr, 55-60, assoc prof, 60-72, PROF MED, SCH MED, GEORGETOWN UNIV, 72- Concurrent Pos: Consult, NIH & US Naval Hosp, Bethesda, Md, 70-; dir, Washington Regional Blood Prog, Am Red Cross, 75- Mem: Am Soc Hemat; Int Soc Hemat; Am Col Physicians; sr mem Am Fedn Clin Res. Res: Red blood cell, including enzymes G-6-PD and abnormal hemoglobins. Mailing Add: Am Red Cross Regional Blood Prog 2025 E St NW Washington DC 20003

MCCURDY, WALLACE HUTCHINSON, JR, b Pittsburgh, Pa, Nov 16, 26; m 54; c 3. ANALYTICAL CHEMISTRY. Educ: Pa State Univ, BS, 47; Univ Ill, MS, 47, PhD(chem), 51. Prof Exp: From instr to asst prof chem, Princeton Univ, 51-59; asst prof, 59-64, ASSOC PROF CHEM, UNIV DEL, 64- Concurrent Pos: Res chemist, Bell Tel Labs, Inc, NJ, 53; Textile Res Inst, 56; res fel, Nat Bur Standards, 70-71. Mem: Am Chem Soc. Res: inorganic reaction mechanisms; coordination complexes; nonaqueous solvent systems; electroanalytical instrumentation. Mailing Add: Dept of Chem Univ of Del Newark DE 19711

MCCURRY, PATRICK MATTHEW, JR, b Homestead, Pa, Nov 2, 44; m 67. CHEMISTRY. Educ: Univ Pittsburgh, BS, 65; Columbia Univ, PhD, 70. Prof Exp: Res chemist, US Bur Mines, 65 & 66; NIH fel, Stanford Univ, 70-71; ASST PROF CHEM, MELLON INST SCI, CARNEGIE-MELLON UNIV, 71- Concurrent Pos: Res fel, Alfred P Sloan Found, 75-77. Mem: Am Chem Soc; The Chem Soc. Res: Natural product synthesis; novel synthetic methods. Mailing Add: Mellon Inst of Sci Carnegie Mellon Univ 4400 Fifth Ave Pittsburgh PA 15213

MCCUSKER, JANE, b London, Eng, Aug 27, 43; m 67; c 3. EPIDEMIOLOGY. Educ: McGill Univ, MD, CM, 67; Columbia Univ, MPH, 69, DrPH(epidemiol), 74. Prof Exp: Lectr epidemiol statist, Fac Med, Univ Dar es Salaam, 73-74; ASST PROF PREV MED, SCH MED & DENT, UNIV ROCHESTER, 75- Mem: Am Pub Health Asn; Soc Epidemiol Res. Res: Cancer epidemiology; evaluation of cancer intervention programs; educational methods in epidemiology. Mailing Add: Sch of Med & Dent Univ of Rochester Rochester NY 14642

MCCUSKEY, ROBERT SCOTT, b Cleveland, Ohio, Sept 8, 38; m 58; c 3. ANATOMY. Educ: Western Reserve Univ, AB, 60, PhD(anat), 65. Prof Exp: From instr to assoc prof, 65-75, PROF ANAT, COL MED, UNIV CINCINNATI, 75- Concurrent Pos: Consult, Procter & Gamble Co, 66-68, 71-75 & Hoffmann-La Roche, 72-75; NIH res grants, Univ Cincinnati, 66-, SW Ohio Heart Asn res grant, 68-69 & 73-75, Akron Heart Asn res grant, 70-71; NIH res career develop award, 69-74; assoc ed, Microvascular Res, 74- Mem: Microcirc Soc; Am Asn Anatomists; Am Soc Hemat; Int Soc Exp Hemat; Sigma Xi. Res: In vivo microscopic anatomy of living organs; erythropoiesis; microcirculation. Mailing Add: Dept of Anat Univ of Cincinnati Col of Med Cincinnati OH 45267

MCCUSKEY, SIDNEY WILCOX, b Cuyahoga Falls, Ohio, Feb 28, 07; m 32; c 2. ASTRONOMY. Educ: Case Western Reserve Univ, BS, 29; Mass Inst Technol, MS, 30; Harvard Univ, PhD(astron), 36. Prof Exp: Instr math & astron, Case Western Reserve Univ, 30-34; asst astron, Harvard Univ, 34-36; from instr to asst prof math & astron, 36-42, assoc prof astron & physics, 42-45, chmn dept astron & dir, Warner & Swasey Observ, 59-70, chmn dept math, 45-59, PROF MATH & ASTRON, CASE WESTERN RES UNIV, 45- 45- Concurrent Pos: Tech aide, Nat Defense Res Comt, 44-45; Warner prof astron, Case Western Res Univ, 72-75. Mem: AAAS; Am Astron Soc; Math Asn Am; Royal Astron Soc. Res: Analysis of spectra; magnitudes; colors and counts of stars for studies of galactic structure. Mailing Add: Warner & Swasey Observ 1975 Taylor Rd East Cleveland OH 44112

MCCUTCHEN, CHARLES WALTER, b Princeton, NJ, Mar 9, 29. PHYSICS, BIOLOGY. Educ: Princeton Univ, BA, 50; Brown Univ, MSc, 52; Cambridge Univ, 57. Prof Exp: Res physicist, Cambridge Univ, 57-62; RES PHYSICIST, LAB EXP PATH, NAT INST ARTHRITIS & METAB DIS, 62- Mem: Am Phys Soc; Optical Soc Am. Res: Lubrication of animal joints; optical diffraction theory; applied optics; hydrodynamics. Mailing Add: Lab of Exp Path Nat Inst Arthritis & Metab Dis Bethesda MD 20014

MCCUTCHEON, ERNEST P, b Durham, NC, May 7, 33; m 56; c 3. MEDICAL PHYSIOLOGY, BIOMEDICAL ENGINEERING. Educ: Davidson Col, BS, 55; Duke Univ, MD, 59. Prof Exp: Intern med, Grady Mem Hosp, Atlanta, Ga, 59-60; res med, Duke Univ, 60-61; head bioinstrumentation & physiol data sect, Launch Site Med Opers, NASA, 61-63; NIH sr res fel physiol & biophys, Univ Wash, 63-66; from asst prof to assoc prof, Univ Ky, 66-74; sr res assoc, Nat Acad Sci-Nat Res Coun, NASA, 73-74, RES MED OFFICER, BIOMED RES DIV, AMES RES CTR, NASA, 74- Concurrent Pos: Vis assoc prof, Stanford Univ Med Ctr, 73-74. Mem: Am Physiol Soc; AAAS; Inst Elec & Electronics Engrs; Am Heart Asn; Biomed Eng Soc. Res: Cardiovascular dynamics; instrumentation; indirect blood pressure determination; ultrasonics. Mailing Add: Biomed Res Div 236-1 NASA Ames Res Ctr Moffett Field CA 94035

MCCUTCHEON, FREDERICK HAROLD, b Abernethy, Sask, Dec 31, 07; US citizen; m 30; c 2. COMPARATIVE PHYSIOLOGY. Educ: NDak State Univ, BS, 32, MS, 33; Duke Univ, PhD(physiol), 36. Prof Exp: Asst zool, NDak State Univ, 32-33; asst zool, Duke Univ, 33-35, instr, 35-36; from asst prof to prof zool & physiol, N C State Univ, 36-47; prof 47-67, chmn dept physiol & pharmacol, 47-57, animal biol, 57-59, PROF EMER COMP PHYSIOL, SCH VET MED, UNIV PA, 68- Concurrent Pos: Vis assoc prof, Sch Med, Duke, 44, independent investr, Marine Lab, 62-67. Mem: Fel AAAS; Am Soc Zool; Am Physiol Soc. Res: Comparative and developmental physiology of respiration; biology of hemoglobin; functions of swimbladder; phylogeny of homeostatic processes in animals; behavior. Mailing Add: PO Box 485 Beaufort NC 28516

MCCUTCHEON, ROB STEWART, b Idaho Falls, Idaho, May 10, 08; m 29; c 3. PHARMACOLOGY. Educ: Univ Idaho, BS, 33; Univ Wash, MS, 46, PhD(pharmaceut chem), 48. Prof Exp: Prof pharmacol, Ore State Univ, 48-64; scientist admin pharmacol, 64-65, EXEC SECY, TOXICOL STUDY SECT, DIV RES GRANTS, NIH, 65- Concurrent Pos: Res fel cardiovasc pharmacol, Med Col Ga, 55-56; NIH spec fel, Sch Med, WVa Univ, 62-63. Mem: Fel AAAS; Am Pharmaceut Asn; Am Soc Pharmacol & Exp Therapeut; Soc Toxicol. Res: Cardiovascular pharmacology; toxicology. Mailing Add: Div of Res Grants NIH Bethesda MD 20014

MCCUTCHEON, WILLIAM HENRY, b Toronto, Ont, Aug 26, 40; m 65. RADIO ASTRONOMY. Educ: Queen's Univ, Ont, BSc, 62, MSc, 64; Univ Manchester, PhD(radio astron), 69. Prof Exp: Res asst radio astron, Nuffield Radio Astron Labs, Jodrell Bank, Eng, 67-69; fel, 69-71, res assoc, 71-72, vis asst prof, 72-73, ASST PROF, DEPT PHYSICS, UNIV B C, 73- Mem: Royal Astron Soc; Can Astron Soc. Res: Spectral line studies in radio astronomy; neutral hydrogen studies of external galaxies. Mailing Add: Dept of Physics Univ of BC Vancouver BC Can

MCDANALD, EUGENE CHESTER, JR, b Gainsville, Tex, Oct 15, 14; m 46; c 4. PSYCHIATRY. Educ: Austin Col, BA, 35; Univ Tex, MD, 41. Prof Exp: Psychiatrist, 171st Sta Hosp, US Army, 42-44, commanding officer, Rehab & Convalescent Unit, 9th Gen Hosp, 44, chief neuropsychiat serv, 120th Gen Hosp, 45; psychiatrist, Regional Off, Ment Hyg Clin, Vet Admin, DC, 46-50; vis lectr neuropsychiat, Med Sch, Univ Va, 50-51; from asst prof to assoc prof, 51-65, CLIN PROF NEUROL & PSYCHIAT, UNIV TEX MED BR GALVESTON, 65-, CHIEF DIV UROL SURG, 67- Concurrent Pos: Partic psychotherapist, Group Psychother Res Proj, Vet Admin, 46-50, chief ment hyg clin, Regional Off, DC, 50, Va, 51. Mem: AMA; fel Am Psychiat Asn; Am Acad Psychoanal; Acad Relig & Ment Health. Res: Value of brief intensive psychotherapy as outpatient treatment of disturbed adolescents and their families. Mailing Add: Dept of Neurol & Psychiat Univ of Tex Med Br Galveston TX 77550

MCDANIEL, BENJAMIN THOMAS, b Pickens, SC, Feb 15, 35. ANIMAL GENETICS, ANIMAL BREEDING. Educ: Clemson Col, BS, 57; Univ Md, MS, 60; NC State Univ, PhD(animal breeding), 64. Prof Exp: Dairy husb, Animal Husb Res Div, USDA, 57-60; fel animal breeding, NC State Univ, 63-64; res dairy scientist, Animal Sci Res Div, USDA, 64-72; PROF ANIMAL GENETICS, NC STATE UNIV, 72- Mem: Am Dairy Sci Asn. Res: Genetics and breeding of domestic animals, with emphasis on dairy cattle. Mailing Add: Dept of Animal Sci NC State Univ Raleigh NC 27607

MCDANIEL, BOYCE DAWKINS, b Brevard, NC, June 11, 17; m 41; c 2. EXPERIMENTAL HIGH ENERGY PHYSICS. Educ: Ohio Wesleyan Univ, BS, 38; Case Inst Technol, MS, 40; Cornell Univ, PhD(physics), 43. Prof Exp: Asst physics, Case Inst Technol, 38-40; asst & res assoc, Cornell Univ, 40-42; mem staff, Radiation Lab, Mass Inst Technol, 43; physicist, Los Alamos Sci Lab, NMex, 43-45; from asst prof to assoc prof, 45-56, assoc dir, 60-67, PROF PHYSICS, CORNELL UNIV, 56-, DIR LAB NUCLEAR STUDIES, 67- Concurrent Pos: Fulbright res award, Australian Nat Univ, 53; Guggenheim & Fulbright award, Univ Rome & Synchrontron Lab, Frascati, Italy, 59; trustee, Assoc Univs, Inc, 63-75; res collabr, Brookhaven Nat Lab,

66; trustee, Univ Res Asn Inc, 71- Mem: Fel Am Phys Soc. Res: Electron diffraction; slow neutron and gamma ray spectroscopy; high energy nuclear physics. Mailing Add: Lab of Nuclear Studies Cornell Univ Ithaca NY 14050

MCDANIEL, BURRUSS, JR, b Ft Smith, Ark, Apr 18, 27; m 58; c 2. ENTOMOLOGY. Educ: Univ Alaska, BA, 53; Tex A&M Univ, MS, 61, PhD(entom), 65. Prof Exp: Consult entomologist, Insect Control & Res, Inc, Md, 57-59; asst prof biol, Tex Col Arts & Indust, 61-66; assoc prof entom, 66-71, PROF ENTOM, SDAK STATE UNIV, 71- Concurrent Pos: Consult, Rockefeller Found, Mex, 57-59. Mem: Soc Study Evolution; Entom Soc Am; Am Ornith Union. Res: Systematic studies on parasites of plants and animals, namely, superfamilies Coccoidea, Analgesoidea and Listrophoroidea. Mailing Add: Dept of Entom-Zool SDak State Univ Brookings SD 57006

MCDANIEL, CARL VANCE, b Grafton, WVa, Dec 4, 29; m 52; c 2. PHYSICAL CHEMISTRY. Educ: Univ Pittsburgh, BS, 57; Mass Inst Technol, PhD(phys chem), 62. Prof Exp: Res technologist, US Steel Corp, 52-58; asst phys chem, Mass Inst Technol, 58-62; res chemist, 62-65, RES SUPVR PHYS CHEM, W R GRACE & CO, COLUMBIA, 65- Mem: Am Chem Soc; AAAS; Catalysis Soc. Res: Chemical metallurgy; physicochemical properties of ion exchange systems; physicochemical properties of ion exchange systems; physicochemical properties of molecular sieves; hydrocarbon catalysis and catalysts; coal gasification and liquefaction; synthetic fuels. Mailing Add: 10725 Crestview Lane Laurel MD 20810

MCDANIEL, DARL HAMILTON, inorganic chemistry, see 12th edition

MCDANIEL, EARL WADSWORTH, b Macon, Ga, Apr 15, 26; m 48; c 2. PHYSICS. Educ: Ga Inst Technol, BA, 48; Univ Mich, MS, 50, PhD, 54. Prof Exp: Asst physics, Univ Mich, 48-54; res physicist, 54-55, from asst prof to prof, 55-70, REGENTS' PROF PHYSICS, GA INST TECHNOL, 70- Concurrent Pos: Instr, Ga Power Co, 57-; consult, Union Carbide Nuclear Co, 59-; Guggenheim fel & Fulbright sr res scholar, Ga Inst Technol, 66-67. Mem: Am Phys Soc. Res: Atomic collisions; gaseous electronics; plasma physics. Mailing Add: Dept of Physics Ga Inst of Technol Atlanta GA 30332

MCDANIEL, EDGAR LAMAR, JR, b Augusta, Ga, Dec 28, 31; m 54; c 2. INDUSTRIAL ORGANIC CHEMISTRY. Educ: Univ Tenn, BS, 53, MS, 54, PhD(chem), 56. Prof Exp: Asst chem, Univ Tenn, 52-56; res chemist, 56-59, sr res chemist, 60-68, res assoc, 68-74, ACTG DIV HEAD, ENG RES DIV, TENN EASTMAN CO, 74- Mem: Am Chem Soc; Am Inst Chem; NY Acad Sci. Res: Catalytic reactions; petrochemistry; chemical kinetics; chemical reaction engineering. Mailing Add: Res Labs Tenn Eastman Co Kingsport TN 37664

MCDANIEL, IVAN NOEL, b Martinsville, Ill, Feb 13, 28; m 57; c 3. MEDICAL ENTOMOLOGY. Educ: Eastern Ill Univ, BS, 51; Univ Ill, MS, 52, PhD(entom), 58. Prof Exp: Asst biol, Univ Ill, 53-57; asst prof, 57-63, ASSOC PROF ENTOM, UNIV MAINE, ORONO, 63- Concurrent Pos: Mem sci adv panel, World Health Orgn, 75-80. Mem: AAAS; Ecol Soc Am; Entom Soc Am; Entom Soc Can; Am Mosquito Control Asn. Res: Identification of attractants and stimulants that influence oviposition of mosquitoes; biological control of mosquitoes and black flies. Mailing Add: Dept of Entom 303 Deering Hall Univ of Maine Orono ME 04473

MCDANIEL, JAMES SCOTT, b Pittsburg, Kans, May 21, 33; m 66. PARASITOLOGY. Educ: Kans State Col, BS, 57; Univ Okla, MS, 61, PhD(zool), 65. Prof Exp: NSF fel biol, Rice Univ, 65-67; from asst prof to assoc prof, 67-73, dir grad studies, 70-73, prof & actg chmn, 73-75, PROF & CHMN BIOL, EAST CAROLINA UNIV, 75- Mem: Fel AAAS; Am Inst Biol Sci; Am Micros Soc; Am Soc Zool; Am Soc Parasitol. Res: Parasite physiology, intermediary metabolism of larval and adult helminths; parasites of wildlife. Mailing Add: Dept of Biol East Carolina Univ Greenville NC 27834

MCDANIEL, JOHN CLIFTON, photochemistry, polymer chemistry, see 12th edition

MCDANIEL, JOHN L, chemistry, physics, see 12th edition

MCDANIEL, LLOYD EVERETT, b Michigan Valley, Kans, Apr 4, 14; m 44; c 3. MICROBIOLOGY. Educ: Kans State Univ, BS, 35, MS, 37; Univ Wis, PhD(bact), 41. Prof Exp: Res microbiologist, Merck Sharp & Dohme Res Labs Div, Merck & Co, Inc, 40-46, head develop sect, 46-51, mgr microbiol develop, 51-56, asst dir microbiol res, 56-61; ASSOC PROF MICROBIOL, INST MICROBIOL, RUTGERS UNIV, 61- Mem: Am Soc Microbiol; Am Chem Soc; Soc Indust Microbiol. Res: Fermentation research and technology; polyene macrolide antibiotics; production of chemicals by microorganisms. Mailing Add: Waksman Inst of Microbiol Rutgers Univ New Brunswick NJ 08903

MCDANIEL, MAX PAUL, b Ft Worth, Tex, May 11, 47; m 69. SURFACE CHEMISTRY. Educ: Southern Ill Univ, BA, 69; Northwestern Univ, MS, 70, PhD(phys chem), 74. Prof Exp: Res assoc catalysis, Nat Ctr Sci Res, France, 74; RES CHEMIST, PHILLIPS PETROL CO, 75- Mem: Am Chem Soc; Soc Plastics Engrs. Res: Modifying and testing the Phillips polymerization catalyst, seeking to improve activity and melt index capability and a better understanding of the mechanism of polymerization. Mailing Add: 81-G Phillips Res Ctr Bartlesville OK 74004

MCDANIEL, MILTON EDWARD, b Elk City, Okla, Jan 17, 38; m 60; c 1. PLANT BREEDING, GENETICS. Educ: Okla State Univ, BS, 60; Va Polytech Inst, PhD(plant breeding), 65. ASSOC PROF SMALL GRAINS BREEDING, TEX A&M UNIV, 65- Mem: Am Soc Agron. Res: Oat and barley breeding and genetic research. Mailing Add: Dept of Soil & Crop Sci Tex A&M Univ College Station TX 77843

MCDANIEL, PAUL WILLIAM, b Robards, Ky, Jan 1, 16; m 75. PARTICLE PHYSICS. Educ: Western Ky Univ, BS, 36; Ind Univ, MS, 38, PhD(physics), 41. Prof Exp: Asst physics, Ind Univ, 37-41; instr, Auburn Univ, 41-42, prof, 46-53; chief, Info & Mat Br, Div Res, US Atomic Energy Comn, 47-48, exec off, 48-50, dep dir, 50-59, actg dir, 59-60, dir, 60-72; pres, Argonne Univs Asn, 72-75; RETIRED. Concurrent Pos: Chief, Tech Br, Res Div, Manhattan Eng Dist, 45-47. Mem: AAAS; Am Phys Soc; Am Nuclear Soc; Am Asn Physics Teachers. Res: Nuclear energy levels; nuclear disintegration schemes. Mailing Add: Argonne Univs Asn PO Box 307 Argonne IL 60439

MCDANIEL, REUBEN ROOSEVELT, mathematics, deceased

MCDANIEL, ROBERT GENE, b Wash, DC, Aug 2, 41; m 65; c 2. GENETICS, PLANT PHYSIOLOGY. Educ: Univ WVa, AB, 63, PhD(genetics), 67. Prof Exp: Asst agronomist, 67-71, assoc prof agron & plant genetics, 71-75, PROF PLANT SCI, UNIV ARIZ, 75- Concurrent Pos: Mem comt seed storage, Int Seed Test Asn. Mem: AAAS; Am Soc Agron; Crop Sci Soc Am; Am Soc Plant Physiol; NY Acad Sci. Res: Biochemical and physiological studies of heterosis in eukaryotes; biochemistry of plant hormones; genetics of mitochondria; biochemical studies of aging; genetics and biochemistry of hybrids; mitochondrial DNA; histones and nuclear proteins. Mailing Add: Dept of Plant Sci Univ of Ariz Tucson AZ 85721

MCDANIEL, ROBERT STEWART, b June 12, 40; Can citizen; m 65; c 2. CHEMICAL KINETICS, DATA PROCESSING. Educ: Univ BC, BSc, 64; Simon Fraser Univ, PhD(chem), 71. Prof Exp: Teacher math & sci, Alberni Sch Dist, 64-66; fel photochem, Univ Alta, 71-73, instr chem, 73-75; RES OFFICER, ALTA RES COUN, 75- Res: Aspects of catalysis in coal conversion to methane. Mailing Add: Alta Res Coun 11315 87th Ave Edmonton AB Can

MCDANIEL, SUSAN GRIFFITH, b Kansas City, Kans, Dec 3, 38; m 66. ZOOLOGY. Educ: Kans State Teachers Col, BS, 59, MS, 62; Univ Okla, PhD(zool), 66. Prof Exp: ASST PROF BIOL, EAST CAROLINA UNIV, 67-, ASST PROVOST, 73- Mem: Am Mus Natural Hist; Sigma Xi. Res: Animal ecology and behavior. Mailing Add: Off of the Provost East Carolina Univ Greenville NC 27834

MCDANIEL, TERRY WAYNE, b Decatur, Ind, Apr 12, 46; m 68; c 1. SOLID STATE PHYSICS. Educ: Wittenberg Univ, Ohio, BA, 68; Mich State Univ, MS, 70, PhD(physics), 73. Prof Exp: Asst physics, Mich State Univ, 68-73, res assoc, 73-74; ASST PROF PHYSICS, VA COMMONWEALTH UNIV, 74- Mem: AAAS; Am Asn Physics Teachers. Res: Dilute magnetic alloys; influence of atomic order on magnetic interactions; transport properties in metals; thin films. Mailing Add: Dept of Physics Va Commonwealth Univ Richmond VA 23284

MCDANIEL, THOMAS LEE, b Portland, Ind, Mar 1, 45; m 67. NUCLEAR CHEMISTRY, RADIOCHEMISTRY. Educ: Earlham Col, BA, 67; Purdue Univ, PhD(chem), 72. Prof Exp: Sr res chemist, 72-75, SUPVR RADIOCHEM GROUP, RES & DEVELOP DIV, BABCOCK & WILCOX CO, 75- Mem: Am Nuclear Soc; Am Chem Soc; Am Inst Physics. Res: Gamma spectroscopy; use of computers in nuclear and radiochemistry; radiological analyses; reactor corrosion products and nuclear fuel evaluation; nondestructive assay techniques; nuclear burnup and gamma scanning techniques. Mailing Add: Babcock & Wilcox Res Ctr PO Box 1260 Lynchburg VA 24505

MCDANIEL, VAN RICK, b San Antonio, Tex, Oct 29, 45; m 66; c 2. MAMMALOGY, HERPETOLOGY. Educ: Tex A&I Univ, BS, 67, MS, 69; Tex Tech Univ, PhD(zool), 73. Prof Exp: ASST PROF ZOOL, ARK STATE UNIV, 72- Mem: Am Soc Mammalogists; Am Soc Ichthyologists & Herpetologists; Soc Study Amphibians & Reptiles; Sigma Xi; Nat Speleol Soc. Res: Taxonomy and natural history of mammals and herptiles; natural history of Ozark cave communities. Mailing Add: Div of Biol Sci Ark State Univ State University AR 72467

MCDANIEL, WILBUR CHARLES, b Kans, July 19, 10; m 36; c 2. MATHEMATICS. Educ: Kans State Col, BS, 32; Univ Wis, MPh, 38, PhD(math), 39. Prof Exp: From asst prof to prof math & chmn dept, 39-74, EMER PROF MATH, SOUTHERN ILL UNIV, CARBONDALE, 74- Mem: Am Math Soc; Math Asn Am. Res: Mathematics education. Mailing Add: RR 4 Union Hill Carbondale IL 62901

MCDANIEL, WILLARD RICH, b Hammond, Ind, June 9, 34; m 57; c 2. METEOROLOGY, GEOLOGY. Educ: Miami Univ, AB, 56, MS, 57; Tex A&M Univ, PhD(meteorol), 67. Prof Exp: Geologist, Lion Oil Co, 57-58; assoc prof, 59-74, PROF & HEAD DEPT EARTH SCI, E TEX STATE UNIV, 74- Mem: Am Meteorol Soc. Res: Applied climate; air pollution. Mailing Add: Dept of Earth Sci East Tex State Univ Commerce TX 73428

MCDANIELS, DAVID K, b Hoquiam, Wash, May 21, 29; m 53, 66; c 3. NUCLEAR PHYSICS, SOLAR ENERGY. Educ: Wash State Univ, BS, 51; Univ Wash, Seattle, MS, 58, PhD(physics), 60. Prof Exp: Physicist, Hanford Atomic Prod Oper, Gen Elec Co, 51-54; res instr physics, Univ Wash, Seattle, 60-61; NSF fel, Nuclear Res Ctr, Saclay, France, 62-63; from asst prof to assoc prof, 63-69, PROF PHYSICS, UNIV ORE, 69- Concurrent Pos: Vis staff mem, Los Alamos Sci Lab, 70-71. Mem: Am Phys Soc; Am Asn Physics Teachers. Res: Fast neutron radiative capture; inelastic proton scattering at low and intermediate energies and microscopic reaction calculations; solar energy applications to flat-plate collector systems. Mailing Add: Dept of Physics Univ of Ore Eugene OR 97403

MCDAVID, JAMES ETHERIDGE, pharmaceutical chemistry, see 12th edition

MCDEARMAN, SARA, b Durham, NC, Apr 24, 13. MICROBIOLOGY, IMMUNOLOGY. Educ: Univ NC, Greensboro, AB, 34; Univ Tenn, MS, 50, PhD(microbiol), 55. Prof Exp: Res asst exp path, Sch Med, Duke Univ, 35-41; serologist, Army Hosp, SC, 41-43; serologist, Walter Reed Army Inst Res, 43-46; from instr to assoc prof serol immunol & microbiologist, Univ Tenn Med Units, Memphis, 50-73; MICROBIOLOGIST, KENNEDY VET ADMIN HOSP, 46- Mem: Am Soc Microbiol. Res: Immunology of histoplasmosis and blastomycosis; complement fixation. Mailing Add: Kennedy Vet Admin Hosp Memphis TN 38104

MCDERMID, ROBERT WESSON, b Charleston, SC, Aug 8, 09; m 38; c 2. FOREST ENGINEERING, FOREST ECONOMICS. Educ: Davidson Col, BS, 30; Yale Univ, MF, 37. Prof Exp: Party chief, US Forest Serv, SC, 33-35; forester, Crossett Lumber Co, Ark, 37-39, chief forest mgt, 39-43; proj forester timber prod war proj, US Forest Serv, Ala, 43-44, asst area forester, Tex, 44-45; forester, Tex Div, Champion Int, 45-51; assoc prof forest mgt & econ, 51-68, PROF FORESTRY, 68- Concurrent Pos: Exec Com La State Univ-Miss State Univ Logging & Forestry Oper Ctr, Nat Space Technol Labs Bay St Louis, Miss, 74- Mem: Soc Am Foresters. Res: Ergonomics; systems analysis. Mailing Add: Sch of Forestry & Wildlife Mgt La State Univ Baton Rouge LA 70803

MCDERMOTT, JOHN F, JR, b Hartford, Conn, Dec 12, 29; m 58; c 2. PSYCHIATRY, CHILD PSYCHIATRY. Educ: Cornell Univ, AB, 51; NY Med Col, MD, 55. Prof Exp: From instr to assoc prof, 62-69, PROF PSYCHIAT & CHMN DEPT, SCH MED, UNIV HAWAII, 69- Concurrent Pos: Mem comt cert in child psychiat, Am Bd Psychiat & Neurol, 70-, chmn, 73-, mem residency rev comt, 74-; consult, NIMH, 74-75. Mem: Fel Am Psychiat Asn; fel Am Orthopsychiat Asn; fel Am Acad Child Psychiat; fel Am Col Psychiat. Res: Child psychiatric training and practice; medical student selection; child abuse and neglect; child custody. Mailing Add: 3675 Kilauea Ave Honolulu HI 96816

MCDERMOTT, JOHN JOSEPH, b Newark, NJ, May 31, 27; m 54; c 2. PARASITOLOGY, MARINE BIOLOGY. Educ: Seton Hall, BS, 49; Rutgers Univ, MS, 51, PhD(zool), 54. Prof Exp: Asst biol, genetics & parasitol, Rutgers Univ, 51-53, parasitol, Bur Biol Res, 53-54; res assoc oyster invests, Agr Exp Sta, Oyster Res Lab, 54-56, asst res specialist, 56-58; from asst prof to assoc prof, 58-69, chmn dept, 63-73, PROF BIOL, FRANKLIN & MARSHALL COL, 69- Concurrent Pos: Vis prof marine sci, Va Inst Marine Sci, 68- Mem: AAAS; Am Soc Parasitol; Ecol Soc Am; Am Soc Zool. Res: Larval trematode studies; marine biology and ecology; biology of marine symbiotic relationships. Mailing Add: Dept of Biol Franklin & Marshall Col Lancaster PA 17604

MCDERMOTT, JOHN PATRICK, b Billings, Mont, Mar 28, 18; m 43. ORGANIC CHEMISTRY. Educ: Univ Portland, BS, 40; Univ Notre Dame, MS, 41, PhD(org chem), 44. Prof Exp: Chemist, Standard Oil Develop Co, 43-60, Esso Res & Eng Co, 60-62; chmn dept & div natural sci & math, 62-71, PROF CHEM, UNIV SAN DIEGO, 62- Mem: Am Chem Soc. Res: Reduction of aromatic and unsaturated aliphatic hydrocarbons; organic analysis; chromatography; absorption spectroscopy. Mailing Add: 1653 A Santa Anita Dr San Diego CA 92111

MCDERMOTT, LAWRENCE ALFRED, b Blind River, Ont, Aug 8, 13; m 41; c 3. MICROBIOLOGY. Educ: Ont Agr Col, BSA, 39; Univ Toronto, MSA, 46. Prof Exp: Lectr, 39-42, from asst prof to assoc prof, 44-56, PROF MICROBIOL, ONT AGR COL, UNIV GUELPH, 56- Mem: Am Fisheries Soc; Can Soc Microbiol. Res: Water, insect, systematic and pathogenic bacteriology; bacteriological methods. Mailing Add: Dept of Microbiol Col Biol Sci Univ of Guelph Guelph ON Can

MCDERMOTT, LEON ANSON, b Petoskey, Mich, Mar 15, 06; m 39. CHEMISTRY. Educ: East Mich Col, AB, 28; Univ Mich, MS, 37; Mich State Univ, EdD, 56. Prof Exp: Instr high sch, Mich, 28-42; anal chemist, Dow Chem Co, 42-47; from asst prof to prof chem, 47-65, actg chmn, 65-66, PROF PHYSICS, CENT MICH UNIV, 65-CHMN DEPT PHYSICS & PHYS SCI, 66- Concurrent Pos: Consult chemist, Howell Red Band Motors, 30-36; pollution consult, Ferro Mfg Corp, Mich, 68- Mem: Am Chem Soc. Res: Analytical methods for determining fungicides present in treated materials; science education. Mailing Add: Dept of Physics Cent Mich Univ Mount Pleasant MI 48858

MCDERMOTT, LILLIAN CHRISTIE, b New York, NY, Feb 9, 31; m 54; c 3. PHYSICS, SCIENCE EDUCATION. Educ: Vassar Col, BA, 52; Columbia Univ, MA, 56, PhD(physics), 59. Prof Exp: Instr physics, City Univ New York, 61-62; lectr physics, Seattle Univ, 65-69; lectr, 67-73, asst prof, 73-76, ASSOC PROF PHYSICS, UNIV WASH, 76- Mem: Sigma Xi; Am Phys Soc; Am Asn Physics Teachers; Nat Sci Teachers Asn. Res: Physics education; preparing teachers to teach physics and physical science and investigating conceptual difficulties and opportunities for intellectual development presented by subject matter. Mailing Add: Dept of Physics Univ of Wash Seattle WA 98195

MCDERMOTT, MARK NORDMAN, b Yakima, Wash, Feb 6, 30; m 54; c 3. ATOMIC PHYSICS. Educ: Whitman Col, BA, 52; Columbia Univ, MA, 56, PhD(physics), 59. Prof Exp: Asst physics, Columbia Univ, 53-59; res assoc, Univ Ill, 59-60; instr, Columbia Univ, 60-62; from asst prof to assoc prof, 62-74, PROF PHYSICS, UNIV WASH, 74- Mem: Fel Am Phys Soc. Res: Radiofrequency spectroscopy; atomic beams magnetic resonance; optical pumping. Mailing Add: 6253 54th NE Seattle WA 98115

MCDERMOTT, RICHARD P, b Weston, WVa, Jan 2, 28; m 54. SPEECH PATHOLOGY, AUDIOLOGY. Educ: Univ Mo, BA, 52; Western Mich Univ, BS, 53; Ohio State Univ, MA, 55; Univ Iowa, PhD(speech path, audiol), 62. Prof Exp: Instr speech path, Univ Iowa, 58-60; asst prof, St Cloud State Col, 60-62, assoc prof speech path & dir speech & hearing serv, 62-64; asst prof speech path, 64-67, ASSOC PROF SPEECH PATH & AUDIOL, UNIV MINN, MINNEAPOLIS, 67-, ASST DIR SPEECH & HEARING CLIN, 64- Mem: Am Speech & Hearing Asn. Res: Speech physiology; articulation skills. Mailing Add: 1368 Colonial Dr St Paul MN 55113

MCDERMOTT, ROBERT EMMET, b Maywood, Ill, Oct 5, 20; m 43; c 2. FORESTRY. Educ: Iowa State Col, BS, 43, MS, 47; Duke Univ, PhD(bot, forestry), 52. Prof Exp: Jr forester, Cook County Forest, Ill, 46; instr bot, Iowa State Col, 46-47; forestry, Univ Mo, 48-49; asst bot, Duke Univ, 50-52; from asst prof to assoc prof forestry, Univ Mo, 52-59; prof forestry & head dept forestry & wildlife, Pa State Univ, 59-69, assoc dean grad sch, 66-69, asst dir sch forest resources, 59-65; dean grad sch, Univ Ark, Fayetteville, 69-72; PROVOST, CAPITOL CAMPUS, PA STATE UNIV, 72- Mem: AAAS; Ecol Soc Am; Soc Am Foresters. Res: Forest ecology and physiology; rural land use. Mailing Add: Capitol Campus Pa State Univ Middletown PA 17057

MCDERMOTT, WALSH, b New Haven, Conn, Oct 24, 09; m 42. MEDICINE. Educ: Princeton Univ, BA, 30; Columbia Univ, MD, 34. Hon Degrees: DSc, Princeton Univ, 74. Prof Exp: From intern to asst resident, New York Hosp, 34-37; from instr to assoc prof med, 37-55, Livingston Farrand prof pub health, 55-72, prof pub affairs in med, 72-75, EMER PROF PUB AFFAIRS IN MED, MED COL, CORNELL UNIV, 75- Concurrent Pos: Spec adv to pres, Robert Wood Johnson Found. Honors & Awards: Lasker Award, Am Pub Health Asn, 55; Dyer Award, 59; NIH First Int Lectureship Award, 63; Woodrow Wilson Award, Princeton Univ, 69. Mem: Nat Acad Sci; Nat Inst Med; hon fel Royal Col Physicians; Am Soc Clin Invest; Asn Am Physicians. Res: Public health. Mailing Add: Robert Wood Johnson Found PO Box 2316 Princeton NJ 08540

MCDERMOTT, WILLIAM VINCENT, JR, b Salem, Mass, Mar 7, 17; m; c 3. SURGERY. Educ: Harvard Univ, AB, 38, MD, 42. Prof Exp: Intern surg, Mass Gen Hosp, Boston, 42-43, from asst resident surgeon to resident surgeon, 46-49; instr, 51-54, clin assoc, 54-57, asst clin prof, 57-63, DAVID W & DAVID CHEEVER PROF SURG, HARVARD MED SCH, 63-; DIR, HARVARD SURG SERV & SCI DIR, CANCER RES INST, NEW ENG DEACONESS HOSP, 73- Concurrent Pos: NIH res fel, 49-50; fel surg, Mass Gen Hosp, 50-51; consult staff & asst surg, Mass Gen Hosp, 51-54, asst surgeon, 54-57, assoc vis, 57-63, vis, 63-; dir, Fifth Surg Serv & Sears Surg Lab, Boston City Hosp, 63-73. Mem: AAAS; AMA; Soc Univ Surg; Am Col Surg; Am Fedn Clin Res. Res: Diseases of the liver and portal circulation; endocrine system; gastrointestinal physiology. Mailing Add: Harvard Surg Serv New Eng Deaconess Hosp Boston MA 02215

MCDEVIT, WILLIAM FERRIS, b Staten Island, NY, July 28, 20; m 44; c 1. PHYSICAL CHEMISTRY. Educ: Haverford Col, AB, 40; Rutgers Univ, MS, 42; Cornell Univ, PhD(chem), 50. Prof Exp: Du Pont fel, 50-51; res chemist chem dept, 51-54, res supvr nylon res div, 54-62, res mgr orlon res div, 62-64, tech supt lycra, 64-65, nylon, Tenn, 65-66, dir dacron res lab, NC, 66-68, res dir dacron tech div, Del, 68-70, prod mgr, 70-71, MFG DIR, TEXTILE FIBERS DEPT, ORLON-ACETATE-LYCRA DIV, E I DU PONT DE NEMOURS & CO, INC, 71- Mem: Am Chem Soc; Sigma Xi. Res: Theory and properties of aqueous solutions; reaction kinetics; electrochemistry; fiber and polymer physics. Mailing Add: Orlon-Acetate-Lycra Div E I du Pont de Nemours & Co Inc Wilmington DE 19898

MCDEVITT, DAVID STEPHEN, b Philadelphia, Pa, Nov 15, 40; m 63; c 2. DEVELOPMENTAL BIOLOGY. Educ: Villanova Univ, BS, 62; Bryn Mawr Col, MA, 63, PhD(biol), 66. Prof Exp: Investr, Biol Div, Oak Ridge Nat Lab, 66-68; asst prof, 68-73, ASSOC PROF ANIMAL BIOL, SCH VET MED, UNIV PA, 73- Concurrent Pos: Consult, Biol Div, Oak Ridge Nat Lab, 69-; res fel, Cancer Res Campaign, UK, 74-75. Mem: AAAS; Am Soc Cell Biol; Am Soc Zool. Res: Lens-specific proteins in lens development as studied by biochemical and immunological techniques. Mailing Add: Dept of Animal Biol Univ of Pa Sch of Vet Med Philadelphia PA 19104

MCDIARMID, DONALD RALPH, b Vancouver, BC, Apr 6, 37; m 71. IONOSPHERIC PHYSICS. Educ: Univ BC, BASc, 60, MASc, 61, PhD(elec eng), 65. Prof Exp: Asst res officer, 65-71, ASSOC RES OFFICER, IONOSPHERIC PHYSICS, UPPER ATMOSPHERE RES SECT, RADIO & ELEC ENG DIV, NAT RES COUN CAN, 71- Res: Study of radio aurora; mechanisms and relationships to other auroral phenomena. Mailing Add: Planetary Sci Sect Herzberg Inst Nat Res Coun Ottawa ON Can

MCDIARMID, IAN BERTRAND, b Carleton Place, Ont, Oct 1, 28; m 51; c 2. SPACE PHYSICS. Educ: Queen's Univ, Can, BA, 50, MA, 51; Univ Manchester, PhD(physics), 54. Prof Exp: Head cosmic ray sect, 55-69, asst dir div physics, 69-75, ASST DIR, HERZBERG INST ASTROPHYSICS, NAT RES COUN, CAN, 75- Concurrent Pos: Sessional lectr, Univ Ottawa, 56- Mem: Royal Soc Can; Am Geophys Union; Can Asn Physicists. Res: Cosmic rays; high energy particle physics; space research; rocket borne particle detectors. Mailing Add: Div Physics Herzberg Inst Nat Res Coun of Can Ottawa ON Can

MCDIARMID, ROY WALLACE, b Santa Monica, Calif, Feb 18, 40; m 67. BIOLOGY, VERTEBRATE ECOLOGY. Educ: Univ Southern Calif, AB, 61, MS, 66, PhD(biol), 69. Prof Exp: Teaching asst gen biol, Univ Southern Calif, 61-65, teaching assoc, 65-66; instr biol, Univ Chicago, 68-69; ASST PROF BIOL, UNIV S FLA, 69- Concurrent Pos: Orgn Trop Studies res grant, 64; worker, Los Angeles County Mus Natural Hist, 64-69; coordr advan biol course, Orgn Trop Studies, 71. Mem: Am Soc Naturalists; Soc Study Amphibians & Reptiles; Am Soc Ichthyologists & Herpetologists; Asn Trop Biol; Ecol Soc Am. Res: Major patterns of evolution in anuran amphibians; systematics of neotropical amphibians and reptiles; biogeography and ecology of neotropical vertebrates; biosystematics of neotropical frogs of families Bufonidae and Centrolenidae. Mailing Add: Dept of Biol Univ of S Fla Tampa FL 33620

MCDIFFETT, WAYNE FRANCIS, b Uniontown, Pa, Sept 2, 39; m 65; c 1. ECOLOGY. Educ: WVa Univ, AB, 61, MS, 64; Univ Ga, PhD(zool), 70. Prof Exp: Instr biol, Frostburg State Col, 63-65; asst prof, 69-74, ASSOC PROF BIOL, BUCKNELL UNIV, 74- Mem: Ecol Soc Am; Am Soc Limnol & Oceanog. Res: Fresh-water ecology; productivity and energy relationships. Mailing Add: Dept of Biol Bucknell Univ Lewisburg PA 17837

MCDIVITT, JAMES FREDERICK, b North Bay, Ont, July 31, 21; m 49, 64; c 3. MINERAL ECONOMICS. Educ: Western Ontario, BSc, 49; Univ Ill, MS, 51, PhD(geol), 54. Prof Exp: Instr geol, Univ Idaho, 51-52, asst prof, 52-54; mineral economist, Dept Mines & Tech Survs, Ottawa, Ont, 54-55, prof econ geol & head dept, Univ Indonesia, 55-59; consult, OEEC, Paris, France, 59-60; from asst prof to assoc prof, Pa State Univ, 60-65; prog specialist, France, 65-67, DIR, REGIONAL CTR SCI & TECHNOL, UNESCO, S E ASIA, 67- Mem: Geol Soc Am; Am Inst Mining, Metall & Petrol Eng. Res: Mineral economics and resource policy; mineral resources of developing areas; international development. Mailing Add: UNESCO c/o UNDP Indonesia Grand Cent Sta Box 20 New York NY 10017

MCDIVITT, MAXINE ESTELLE, b Rosedale, Ind, Aug 16, 12. Educ: Univ Ill, BS, 40, MS, 41; Univ Wis, PhD(foods & nutrit), 52. Prof Exp: Instr home econ, Univ Mo, 41-44, asst prof, 45-47; instr, Univ Iowa, 47-49; assoc prof, 52-59, PROF HOME ECON, Univ WIS-MILWAUKEE, 59- Concurrent Pos: On leave, UN Food & Agr Orgn, India, 64-66; USDA grant, Univ Wis, Milwaukee, 67- Mem: Am Home Econ Asn; Am Dietetics Asn. Res: Bacteriological aspects of food handling procedures; heat resistance of food poisoning organisms; heat transfer in foods; food habits in urban communities; consumer acceptance of dairy products; studies of milk flavor. Mailing Add: Sch of Social Welfare-Home Econ Univ of Wis Milwaukee WI 53201

MCDOLE, CLAYTON JUNIOR, nuclear physics, oceanography, see 12th edition

MCDOLE, ROBERT E, b Eugene, Ore, Oct 7, 30; m 65; c 3. SOIL FERTILITY, SOIL GENESIS. Educ: Ore State Univ, BS, 52; Univ Idaho, MS, 68, PhD(soils), 69. Prof Exp: Soil scientist, Bur Indian Affairs, US Dept Interior, Idaho, Ore & Wash, 52-65; asst res prof, 69-73, ASSOC RES PROF SOILS, ABERDEEN BR STA, UNIV IDAHO, 73- Concurrent Pos: Vis prof, Wash State Univ, 74-75. Mem: Soc Agron; Soil Sci Soc Am; Potato Asn Am. Res: Soil surveys of Northwest Indian reservations; study of loess soil materials in Southern Idaho; fertility of potato production. Mailing Add: Aberdeen Br Sta Univ of Idaho Aberdeen ID 83210

MCDONAGH, JAN M, b Wilmington, NC, Nov 9, 42; m 68; c 1. BIOCHEMISTRY, PATHOLOGY. Educ: Wake Forest Univ, BS, 64; Univ NC, PhD(biochem), 68. Prof Exp: NIH fels, Univ NC, Chapel Hill, 68-69, Med Univ Klinik, Basel, Switz, 69-70 & Karolinska Inst, 70-71; ASST PROF PATH, MED SCH, UNIV NC, CHAPEL HILL, 71-, ASST PROF BIOCHEM, 74- Concurrent Pos: Mem coun thrombosis, Am Heart Asn, 72- Mem: Int Soc Thrombosis & Haemostasis. Res: Biochemistry of blood coagulation and fibrinolysis; pathogenesis of thrombosis. Mailing Add: Dept of Path Univ of NC Sch of Med Chapel Hill NC 27514

MCDONAGH, RICHARD PATRICK, JR, b New York, NY, Mar 8, 44; m 68; c 1. PHYSIOLOGY, EXPERIMENTAL PATHOLOGY. Educ: Univ Calif, Los Angeles, AB, 64; Calif State Col, Los Angeles, MA, 66; Univ NC, Chapel Hill, PhD(physiol), 69. Prof Exp: Lectr zool, Calif State Col, Los Angeles, 65-66; Emil Barrel fel, Univ Basel Med Clin, 69-70; NIH res fel, Karolinska Inst, Sweden, 70-71; asst prof epidemiol & physiol, 71-73, ASST PROF PATH & PHYSIOL, UNIV NC, CHAPEL HILL, 73- Concurrent Pos: Specialized Ctrs of Res in Thrombosis res grant, Univ NC, Chapel Hill, 71-; mem coun thrombosis, Am Heart Asn; consult, Int Comt Thrombosis & Hemostasis, 72- Mem: AAAS; Int Soc Thrombosis & Haemostasis. Res: Pathophysiology of thrombosis and hemorrhage; structure and function of fibrin and fibrinstabilizing factor. Mailing Add: Dept of Path Univ of NC Sch of Med Chapel Hill NC 27514

MCDONALD, ALISON DUNSTAN, b St Albans, Eng, Aug 5, 17; m 42; c 4. EPIDEMIOLOGY. Educ: Univ London, MB & BS, 41, DCH, 47, DPH & MD, 50. Prof Exp: Lectr prev & social med, London Sch Hyg & Trop Med & Royal Free Hosp, 52-56; res assoc epidemiol, Pediat Res Unit, Guy's Hosp, London, Eng, 58-59, sr lectr, 60-64; assoc prof, 64-74, PROF EPIDEMIOL, McGILL UNIV, 74- Mem: Teratology Soc; Can Pub Health Asn; Int Epidemiol Asn. Res: Epidemiology of developmental disorders, particularly mental deficiency, congenital malformation, cerebral palsy and very low birth weight; health care research; cancer, especially associated with asbestos and with herpesviruses. Mailing Add: Dept of Epidemiol & Health McGill Univ Montreal PQ Can

MCDONALD, ALLAN W, b Oakland, Calif, Feb 23, 19; m 50; c 4. PHYSICS. Educ: Loyola Univ, BS, 42; Georgetown Univ, MS, 47; La State Univ, PhD(physics), 52. Prof Exp: Assoc prof physics, Loyola Univ, 52-56; res physicist, Aeronaut Div, Robertshaw-Fulton Controls Co, 56-57; asst proj mgr missile eval, Automation Electronics, Inc, 57-58; sr res engr, Gen Dynamic/Convair, 58-59 & Tasker Instrument Corp, 59-62; PRIN SCIENTIST, SPACE & INFO SYSTS DIV, N AM ROCKWELL CORP, 62- Res: Construction simulation models in fields of physics, mathematics and management systems; mechanics; electric and magnetic fields. Mailing Add: 2623 E Vermont Ave Anaheim CA 92806

MCDONALD, ALVIS EDWARD, mathematics, numerical analysis, see 12th edition

MCDONALD, ARTHUR BRUCE, b Sydney, NS, Aug 29, 43; m 66; c 3. PHYSICS. Educ: Dalhousie Univ, BSc, 64, MSc, 65; Calif Inst Technol, PhD(physics), 70. Prof Exp: Nat Res Coun Can & Rutherford Mem fels, 70-71, RES PHYSICIST, CHALK RIVER NUCLEAR LABS, ATOMIC ENERGY CAN, LTD, 71- Mem: Am Phys Soc; Can Asn Physicists. Res: Nuclear physics, especially investigations of structure of nuclei using particle accelerators. Mailing Add: Chalk River Nuclear Labs Nuclear Physics Br Atomic Energy of Can Chalk River ON Can

MCDONALD, BARBARA BROWN, b Ray, Ariz, Feb 14, 24; m 52. CYTOLOGY. Educ: Simmons Col, BS, 48; Columbia Univ, MA, 55, PhD, 57. Prof Exp: From instr to assoc prof, 56-73, PROF BIOL, DICKINSON COL, 73- Mem: AAAS; Am Soc Cell Biol; Histochem Soc; Genetics Soc Am; Soc Protozool. Res: Cytochemistry; protozoology; nucleic acids in Tetrahymena. Mailing Add: Dept of Biol Dickinson Col Carlisle PA 17013

MCDONALD, BARRIE C, geology, see 12th edition

MCDONALD, BERNARD ROBERT, b Kansas City, Kans, Nov 17, 40; m 63; c 2. MATHEMATICS. Educ: Park Col, BA, 62; Kans State Univ, MA, 64; Mich State Univ, PhD(math), 68. Prof Exp: Asst prof, 68-72, ASSOC PROF MATH, UNIV OKLA, 72- Mem: Am Math Soc; Math Asn Am. Res: Algebra; matrix theory; commutative rings; finite rings; combinatorial theory. Mailing Add: Dept of Math Univ of Okla Norman OK 73069

MCDONALD, BRUCE EUGENE, b Mannville, Alta, Apr 30, 33; m 62; c 5. NUTRITION. Educ: Univ Alta, BSc, 58, MSc, 60; Univ Wis, PhD(biochem, poultry nutrit), 63. Prof Exp: Res assoc, Univ Ill, 63-64; asst prof animal sci, Macdonald Col, McGill Univ, 64-68; assoc prof, 68-71, PROF NUTRIT, UNIV MAN, 71- Mem: Nutrit Soc Can; Am Inst Nutrit; Can Inst Food Sci & Technol; Am Oil Chemists Soc; Can Dietetic Asn. Res: Evaluation of protein quality in cereals; effect of diet on lipid metabolism in the human; plant proteins for human consumption. Mailing Add: Dept of Foods & Nutrit Univ of Man Winnipeg MB Can

MCDONALD, BRUCE JERALD, b Ashland, Wis, June 23, 28. MATHEMATICAL STATISTICS. Educ: Northland Col, BA, 50; Fla State Univ, MS, 62 & 63, PhD(math statist), 66. Prof Exp: Physicist, Mine Defense Lab, US Navy, Fla, 51-57, res mathematician, 57-59; mathematician, Off Adv Tech, Eglin Air Force Base, 59-62; asst prof statist & asst dir comput ctr, Fla State Univ, 66-67; DIR CONTRACT RES PROG STATIST & PROBABILITY, US OFF NAVAL RES, 67- Concurrent Pos: Lectr, Fla State Univ, 53-59; prof lectr, George Washington Univ, 69-; joint serv adv group res in appl math Navy rep, Dept of Defense, 69-, chmn, 71-; mem Inst Math Statist comt for liaison with Int Asn Statist in Phys Sci, 70; mem, Am Statist Asn Conf Comt for 38th Session of Int Statist Inst, 71; secy, Conf Bd Math Sci, 71- Mem: AAAS; Am Math Soc; Am Phys Soc; Math Asn Am; Asn Comput Mach. Res: Statistical signal processing; modern statistical and probabilistic theory; statistical signal analysis; numerical analysis; ranking and selection methods; computing and data processing systems and instrumentation, both analog and digital; foundations of mathematics and statistics. Mailing Add: Code 436 Off of Naval Res Arlington VA 22217

MCDONALD, CHARLES CAMERON, b Wyoming, Ont, Aug 5, 26; m 49; c 4. PHYSICAL CHEMISTRY. Prof Exp: Western Ontario, BSc, 49, MSc, 50; Ill Inst Technol, PhD(chem), 54. Prof Exp: Res fel chem, Ill Inst Technol, 54-55; res chemist, 55-65, RES SUPVR, CENT RES DEPT, E I DU PONT DE NEMOURS & CO, INC, 65- Res: Gas phase reaction kinetics; photochemistry; magnetic resonance; biophysics; structure of proteins; plant science; single cell protein; fermentation. Mailing Add: Cent Res Dept E I du Pont de Nemours & Co Wilmington DE 19898

MCDONALD, CLARENCE EUGENE, b McPherson, Kans, Oct 10, 1926; m 55; c 3. CEREAL CHEMISTRY. Educ: McPherson Col, AB, 50; Kans State Univ, MS, 53; Purdue Univ, PhD(biochem), 57. Prof Exp: Asst, Kans State Univ, 52-53; asst, Purdue Univ, 53-56, res fel, 57; chemist, Western Utilization Res & Develop Div, Agr Res Serv, USDA, 57-64; ASSOC PROF CEREAL TECHNOL, N DAK STATE UNIV, 64- Mem: Am Chem Soc; Am Asn Cereal Chem; Inst of Food Technol. Res: Isolation and characterization of wheat proteins and enzymes; nutrition of cereal proteins. Mailing Add: Dept of Cereal Chem & Technol N Dak State Univ Fargo ND 58102

MCDONALD, CURTIS W, analytical chemistry, physical chemistry, see 12th edition

MCDONALD, DANIEL FRANCIS, physics, see 12th edition

MCDONALD, DANIEL JAMES, b New York, NY, Jan 27, 25; m 52. GENETICS. Educ: Siena Col, BS, 50; Columbia Univ, MA, 52, PhD(zool), 55. Prof Exp: Lectr zool, Columbia Univ, 53-55, instr, 55-56; from asst prof to assoc prof biol, 56-70, PROF BIOL, DICKINSON COL, 70- Concurrent Pos: Instr, Long Island Univ, 55-56. Mem: AAAS; Am Soc Nat; Ecol Soc Am; Genetics Soc Am; Soc Study Evolution. Res: Population genetics; ecology of the flour beetles Tribolium confusum and Tribolium castaneum. Mailing Add: Dept of Biol Dickinson Col Carlisle PA 17013

MCDONALD, DAVID WILLIAM, b Shreveport, La, Aug 4, 23; m 48; c 4. ORGANIC CHEMISTRY. Educ: Southwestern La Univ, BS, 43; Univ Tex, PhD(chem), 51. Prof Exp: Jr res chemist, Humble Oil Co, 46-47; res chemist, Monsanto Chem Co, 51-54, asst res group leader, 54-56, res group leader, 56-59, res sect leader, 59-64, mgr polyolefins res, Monsanto Co, Tex, 64-67, prod adminr polyolefins, Mo, 67-69, dir res & develop, Hydrocarbons & Polymers Div, 69-71, dir plastics res, 71-74, DIR TECHNOL, MONSANTO POLYMERS & PETROCHEM CO, 74- Concurrent Pos: Mem Prod Res Comt, 74- Mem: Am Chem Soc; Soc Plastics Engrs. Res: Petrochemicals: monomers; polymer synthesis, properties and applications; polymer combustion properties; safe uses of plastics. Mailing Add: Monsanto Polymers & Petrochem Co 800 N Lindberg Blvd St Louis MO 63166

MCDONALD, DONALD ARTHUR, physiology, biophysics, deceased

MCDONALD, DONALD FIEDLER, b Chicago Heights, Ill, Aug 13, 19; m 42; c 3. UROLOGY. Educ: Univ Chicago, MD, 42. Prof Exp: Instr surg, Grad Sch Med, Univ Chicago, 46-49; assoc prof urol, Sch Med, Univ Wash, 49-58; prof, Sch Med & Dent, Univ Rochester, 58-69, head dept, 58-67; chief div urol, 69-75, PROF UROL, UNIV TEX MED BR GALVESTON, 69- Concurrent Pos: Markle scholar, 52-57. Mem: AAAS; Endocrine Soc; AMA; Am Asn Cancer Res; Am Asn Genito-Urinary Surg. Res: Cancer of genito-urinary organs; urinary calculi; renal causes of hypertension. Mailing Add: Dept of Surg Div of Urol Univ of Tex Med Br Galveston TX 77550

MCDONALD, FRANCIS GUY, b Henderson, Ky, Aug 9, 02; m 25; c 2. BIOCHEMISTRY. Educ: Purdue Univ, BS, 24; Rutgers Univ, MS, 28, PhD(biochem), 29. Prof Exp: Asst biochemist, Mead Johnson & Co, 24-27 & biochemist, 29-53, dir, Control Lab, 53-67; assoc prof, 67-75, EMER PROF PHARM & PHARMACEUT CHEM, SCH PHARM, UNIV MISS, 76- Mem: Am Chem Soc; Am Soc Biol Chem; Am Inst Nutrit. Res: Calcium metabolism; nutrition; sterols; analytical chemistry; pyrazines; industrial pharmacy and sterile injectable products. Mailing Add: 316 Thomas Oxford MS 38655

MCDONALD, FRANCIS RAYMOND, b Phila, Pa, Sept 28, 24; m 55; c 1. SPECTROSCOPY, ORGANIC CHEMISTRY. Educ: Phila Col Textiles & Sci, BS, 54. Prof Exp: Asst chief chemist, Niagara Falls Div, Int Minerals & Chem Corp, NY, 55-57; chemist, 57-61, res chemist, 61-75, PROJ LEADER, LARAMIE ENERGY RES CTR, US ENERGY RES & DEVELOP ADMIN, 75- Mem: AAAS; Soc Appl Spectros; Am Chem Soc; Coblentz Soc. Res: Ultraviolet, infrared and nuclear magnetic resonance spectroscopy; molecular structure of petrochemical compounds found in oil shale and shale oil using nuclear magnetic resonance, infrared and ultraviolet spectroscopy; mass spectroscopy. Mailing Add: Laramie Energy Res Ctr PO Box 3395 Laramie WY 82071

MCDONALD, FRANK ALAN, b Dallas, Tex, Jan 11, 37; m 64; c 2. THEORETICAL & NUCLEAR PHYSICS. Educ: Southern Methodist Univ, BS & BA, 58; Yale Univ, MS, 59, PhD(physics), 65. Prof Exp: Asst prof physics, Tex A&M Univ, 64-69; ASSOC PROF, SOUTHERN METHODIST UNIV, 69- Mem: Am Phys Soc; Am Asn Physics Teachers; Sigma Xi. Res: Scattering theory; few-nucleon problem. Mailing Add: Dept of Physics Southern Methodist Univ Dallas TX 75222

MCDONALD, FRANK BETHUNE, b Columbus, Ga, May 28, 25; m 51; c 3. PHYSICS. Educ: Duke Univ, BS, 48; Univ Minn, MS, 51, PhD, 55. Prof Exp: Asst physics, Duke Univ, 47-48; asst, Univ Minn, 48-51; asst prof, Univ Iowa, 53-59; head fields & particles br, 59-70, CHIEF LAB HIGH ENERGY ASTROPHYS, 70-, PROJ SCIENTIST, EXPLORER SATELLITES & HIGH ENERGY ASTRON OBSERV, NASA GODDARD SPACE FLIGHT CTR, 64- Concurrent Pos: Prof, Univ Med, 64- Mem: Am Phys Soc; Am Geophys Union. Res: Study of primary cosmic radiation at high altitudes by means of rockets, balloons and satellites. Mailing Add: Code 660 Lab High Energy Astrophys NASA Goddard Space Flight Ctr Greenbelt MD 20771

MCDONALD, GEORGE GORDON, b Chicago, Ill, Feb 20, 44. BIOCHEMISTRY. Educ: Loyola Univ Chicago, BS, 66; Johns Hopkins Univ, PhD(biochem), 70. Prof Exp: Fel biochem, 70-74, ASST PROF, DEPT BIOCHEM & BIOPHYSICS, UNIV PA, 74- Concurrent Pos: Career investr fel, Am Heart Asn, 71-73. Mem: AAAS; Am ChemSoc; NY Acad Sci. Res: Relationship between the structure and function of macromolecules; nuclear magnetic resonance. Mailing Add: Johnson Res Found Univ of Pa Dept of Biochem & Biophysics Philadelphia PA 19174

MCDONALD, GERAL IRVING, b Wallowa, Ore, Dec 31, 35; m 62; c 3. PLANT PATHOLOGY. Educ: Wash State Univ, BS, 63, PhD(plant path), 69. Prof Exp: Res plant geneticist, 66-68, RES PLANT PATHOLOGIST, INTERMT FOREST & RANGE EXP STA, US FOREST SERV, 68- Concurrent Pos: Affiliate prof, Univ Idaho, 68- Mem: Am Phytopath Soc; Soc Am Foresters. Res: Genetics, computer simulation and evolution of host-pest interaction in forest trees. Mailing Add: Intermt Forest & Range Exp Sta Forestry Sci Lab US Forest Serv Moscow ID 83843

MCDONALD, HARRY SAWYER, b New Orleans, La, Sept 24, 30; m 56; c 2. COMPARATIVE PHYSIOLOGY, HERPETOLOGY. Educ: Loyola Univ, La, BS, 54, Univ Notre Dame, PhD(zool), 58. Prof Exp: Nat Heart Inst fel zool, Univ Calif, Los Angeles, 58-59, cardiovasc trainee, 59-60; from asst prof to assoc prof biol, St John's Univ, NY, 60-65; from asst prof to assoc prof, 65-68, PROF BIOL, STEPHEN F AUSTIN STATE UNIV, 68- Prof Exp: NSF res grant, 67-70. Mem: Am Soc Zool; Am Soc Ichthyol & Herpet; Soc Study Amphibians & Reptiles. Res: Reptile electrocardiography; breathing in snakes; thermal acclimation; snake anatomy and organ topography. Mailing Add: Dept of Biol Stephen F Austin State Univ Nacogdoches TX 75961

MCDONALD, HECTOR O, b Casper, Wyo, Sept 4, 30; m 52; c 2. PHYSICAL INORGANIC CHEMISTRY. Educ: Cent Methodist Col, AB, 52; Ala Polytech Inst, MS, 54; Univ Ark, PhD(chem), 60. Prof Exp: Control chemist, Southern Cotton Oil Co, Ill, 54-55; res asst chem, Univ Ark, 55-59; assoc prof, West Tex State Univ, 59-63; ASSOC PROF CHEM, UNIV MO-ROLLA, 63- Mem: Am Chem Soc; The Chem Soc. Res: Chemistry of inorganic complexes; chemical kinetics in aqueous media, non-aqueous solvent systems and mass spectrometry. Mailing Add: Dept of Chem Univ of Mo Rolla MO 65401

MCDONALD, HUGH JOSEPH, b Glen Nevis, Ont, July 27, 13; nat US; m; c 3. BIOCHEMISTRY, PHYSICAL CHEMISTRY. Educ: McGill Univ, BSc, 35; Carnegie-Mellon Univ, MS, 36, DSc(phys chem), 39; Am Bd Clin Chem, dipl, 52. Hon Degrees: Grad, Univ Rio de Janeiro, 62. Prof Exp: Asst chem, Carnegie-Mellon Univ, 36-38, instr, 38-39; from instr to prof, Ill Inst Technol, 39-48; PROF BIOCHEM & BIOPHYS & CHMN DEPT, STRITCH SCH MED, LOYOLA UNIV CHICAGO, 48- Concurrent Pos: Res scientist, Manhattan Proj, Columbia Univ, 43; dir corrosion res lab, Ill Inst Technol, 44-48; ed theoret sect, Corrosion & Mat Protection, 45-48; consult, Argonne Nat Lab, 46- Mem: Am Chem Soc; Soc Exp Biol & Med; Am Soc Biol Chemists; Biophys Soc; Am Asn Clin Chemists (vpres, 52, pres, 53). Res: Biochemical aspects of diabetes and atherosclerosis; separation processes; electrophoresis; lipoproteins; oral hypoglycemic agents. Mailing Add: Dept of Biochem & Biophys Loyola Univ Stritch Sch of Med Maywood IL 60153

MCDONALD, IAN CAMERON CRAWFORD, b Flint, Mich, Feb 20, 39; m 64; c 2. ENTOMOLOGY, GENETICS. Educ: Southern Methodist Univ, BS, 62, MS, 65; Va Polytech Inst, PhD(entom), 68. Prof Exp: Asst, Va Polytech Inst, 64-67; ENTOMOLOGIST, METAB & RADIATION RES LAB, ENTOM RES DIV, AGR RES SERV, USDA, 68- Mem: Genetics Soc Am; Am Genetic Asn. Res: Insect and house fly genetics; genetic control measures. Mailing Add: Metab & Radiation Res Lab Agr Res Serv USDA State Univ Sta Fargo ND 58102

MCDONALD, IAN JOHNSON, b Bangor, NIreland, Feb 23, 21; Can citizen; m 58; c 2. MICROBIAL PHYSIOLOGY. Educ: Univ BC, BSA, 43; Univ Wis, MSc, 48, PhD(agr bact), 50. Prof Exp: From asst res officer to assoc res officer, Div Biosci, 50-68, SR RES OFFICER, DIV BIOL SCI, NAT RES COUN CAN, 68- Concurrent Pos: Vis worker, Nat Inst Res in Dairying, Reading, Eng, 67-68. Mem: Am Soc Microbiol; Can Soc Microbiol (2nd vpres, 65-66); Can Inst Food Technol; Brit Soc Appl Bacteriol; Brit Soc Gen Microbiol. Res: Casein utilization by lactobacilli and lactic streptococci; location in cells and liberation of extracellular bacterial proteinases; chemostat cultivation of streptoccocci and Neisseria; effect of environment on composition and physiology of cells. Mailing Add: Div of Biol Sci Nat Res Coun 100 Sussex Dr Ottawa ON Can

MCDONALD, IAN MACLAREN, b Regina, Sask, May 20, 28; m 53; c 5. PSYCHIATRY. Educ: Univ Man, MD, 53; Royal Col Physicians & Surgeons Can, dipl psychiat, 58. Prof Exp: Psychiatrist, Crease Clin Psychol Med, 53-54 & Psychiat Serv, Prov Sask, 54-56; fel neurol, Col Med, Univ Sask, 56; fel psychiat, Med Ctr, Univ Colo, 56-58; lectr, 58, from asst prof to assoc prof, 59-67, PROF PSYCHIAT, UNIV SASK, 67-, HEAD DEPT PSYCHIAT, UNIV & HEAD PSYCHIAT, UNIV HOSP, 71- Concurrent Pos: Chmn, Govt Sask Comn Alcoholism. Mem: Am Psychiat Asn. Res: Social psychiatry. Mailing Add: Dept of Psychiat Univ of Sask Saskatoon SK Can

MCDONALD, JAMES CLIFTON, b Oklahoma City, Okla, Apr 2, 30; m 54; c 3. MYCOLOGY. Educ: Wash Univ, St Louis, AB, 52; Univ Mo, MA, 57, PhD(bot), 60. Prof Exp: Asst bot, Univ Mo, 59-60; asst prof, 60-65, chmn dept, 71-75, ASSOC PROF BOT, WAKE FOREST UNIV, 65- Mem: AAAS; Am Inst Biol Sci; Mycol Soc Am. Res: Myxobacteria, nutrition, taxonomy and ecology. Mailing Add: Dept of Biol Wake Forest Univ Winston-Salem NC 27109

MCDONALD, JAMES E, b Chicago, Ill, Nov 9, 22; m 46; c 3. OPHTHALMOLOGY. Educ: Loyola Univ Chicago, MD, 45. Prof Exp: From asst prof to assoc prof ophthal, Col Med, Univ Ill, 53-70; CLIN PROF OPHTHAL & CHMN DEPT, STRITCH SCH MED, LOYOLA UNIV CHICAGO, 70- Mem: AMA; Am Acad Ophthal & Otolaryngol. Res: Effects of radiation on the eye; corneal wound healing. Mailing Add: 1046 Chicago Ave Oak Park IL 60301

MCDONALD, JAMES FREDERICK, b Detroit, Mich, Sept 26, 39; m 64. MATHEMATICAL PHYSICS. Educ: Wayne State Univ, BS, 61, PhD(physics), 67. Prof Exp: Asst prof, 67-71, dept univ affairs grant, 69, ASSOC PROF MATH, UNIV WINDSOR, 71- Concurrent Pos: Nat Res Coun Can grant, 67-76. Mem: Am Phys Soc. Res: Statistical theory of energy levels of complex quantum mechanical systems. Mailing Add: Dept of Math Univ of Windsor Windsor ON Can

MCDONALD, JAMES HOGUE, b Asheville, NC, May 16, 15; m 49; c 1. UROLOGY. Educ: NCent Col, Ill, AB, 38; Univ Ill, MD, 42; Am Bd Urol, dipl, 52. Prof Exp: From assoc prof to prof urol, Univ Ill Col Med, 58-68, head div, 58-68; secy, Am Bd Urol, 68-74; CHIEF UROL SECT, MARICOPA COUNTY GEN HOSP, 74- Concurrent Pos: Trustee, Am Bd Urol, 63- Mem: AMA; Am Urol Asn; Am Col Surg; Am Asn Genito-Urinary Surg. Res: Carcinogenesis of the urinary tract utilizing 2-acetylamino fluorene, indole and lead; regrowth of the ureter in normal and abnormal state; study of male infertility. Mailing Add: Dept of Surg Sect of Urol Maricopa County Gen Hosp Phoenix AZ 85008

MCDONALD, JAMES LEE, JR, b LaGrange, Ky, May 21, 39; m 63; c 2. ORAL BIOLOGY, NUTRITION. Educ: Ind Univ, AB, 62, PhD(dent sci), 68. Prof Exp: ASST PROF PREV DENT, SCH DENT, IND UNIV, INDIANAPOLIS, 68- Mem: AAAS; assoc Am Dent Asn; Int Asn Dent Res. Res: Nutritional control of dental caries. Mailing Add: Prev Dent Res Inst 410 Beauty Ave Indianapolis IN 46204

MCDONALD, JAMES ROBERT, b San Francisco, Calif, Jan 28, 34; m 60; c 2. GEOGRAPHY OF FRANCE, POPULATION GEOGRAPHY. Educ: Antioch Col, AB, 55; Univ Ill, MA, 56, PhD(geog), 64. Prof Exp: Asst prof geog, Univ Calif, Los Angeles, 63-65; assoc prof, 65-68, PROF GEOG, EASTERN MICH UNIV, 68- Concurrent Pos: NSF fel, Paris, France, 66-67; Soc Sci Res Coun fac res grant, Midwestern United States, 67-68; NSF sci fac fel, Southampton, Eng & Paris, France, 70-71. Mem: AAAS; Asn Am Geog. Res: Geography of human migrations, especially rural-to-urban flows and labor migrations, with particular reference to France; regional theory and the history of regional thought. Mailing Add: Dept of Geog Eastern Mich Univ Ypsilanti MI 48197

MCDONALD, JANET, b Wesson, Miss, Sept 3, 05. MATHEMATICS. Educ: Belhaven Col, BA, 25; Tulane Univ, MA, 29; Univ Chicago, PhD(math), 43. Prof Exp: Teacher high sch, Miss, 25-28; head math dept, Miss Synodical Col, 29-32; head math dept & registr, Hinds Jr Col, 32-41; instr math, Univ Chicago, 43-44; from instr to prof, 44-71, EMER PROF MATH, 71- Concurrent Pos: NSF fac fel, 59-60. Mem: Am Math Soc; Math Asn Am. Res: Differential and projective geometry. Mailing Add: 5050 Wayneland Dr Jackson MS 39211

MCDONALD, JIMMIE REED, b Austin, Tex, Aug 20, 42; m 63; c 2. PHYSICAL CHEMISTRY, MOLECULAR SPECTROSCOPY. Educ: Southwestern Univ, Tex, BS, 64; La State Univ, Baton Rouge, PhD(phys chem), 68. Prof Exp: Vis asst prof, La State Univ, Baton Rouge, 68-70; RES CHEMIST, NAVAL RES LAB, 70- Mem: Am Chem Soc. Res: Molecular electronic spectroscopy; photochemistry; dye laser technology. Mailing Add: Code 6110 Naval Res Lab Washington DC 20390

MCDONALD, JOHN C, b Baldwin, Miss; c 3. SURGERY, IMMUNOLOGY. Educ: Miss Col, BS, 51; Tulane Univ La, MD, 55. Prof Exp: Intern, Confederate Mem Med Ctr, Shreveport, La, 55-56; resident, E J Meyer Mem Hosp, Buffalo, NY, 58-63; Buswell res fel & instr surg, State Univ NY Buffalo, 63-65; assoc dir surg res lab, E J Meyer Mem Hosp, Buffalo, 65-68, head sect transplantation, 66-68; assoc prof surg, 68-72, PROF SURG & ASSOC PROF MICROBIOL, SCH MED, TULANE UNIV LA, 72-, DIR SURG LABS, 68- Concurrent Pos: Head sect transplantation, State Univ NY Buffalo, 66; attend surgeon, E J Meyer Mem Hosp, Buffalo, 67; attend surgeon & head sect transplantation, Deaconess Hosp, Buffalo, 68; consult, Roswell Park Mem Hosp, 68 & Masten Park Rehab Ctr, 68; dir transplantation, Tulane Univ La & Charity Hosp, New Orleans; consult surgeon, Lallie Kemp Charity Hosp, Keesler AFB, Biloxi, Miss, Pineville Vet Admin Hosp & Huey P Long Charity Hosp. Mem: Am Col Surgeons; Am Soc Artificial Internal Organs; Soc Univ Surgeons; Transplantation Soc; Am Asn Clin Histocompatibility Testing. Res: Transplantation. Mailing Add: Dept of Surg Tulane Univ Sch of Med New Orleans LA 70112

MCDONALD, JOHN CORBETT, b Belfast, Northern Ireland, Apr 20, 18; m 42; c 4. EPIDEMIOLOGY. Educ: Univ London, MB, BS, 47, MD & dipl, 49, dipl, 50; Harvard Univ, SM, 51. Prof Exp: Epidemiologist, Epidemiol Res Lab, Pub Health Lab Serv, Med Res Coun, Eng, 51-59, consult epidemiologist, 59-60, dir, 60-64; chmn dept, 64-73, PROF EPIDEMIOL & HEALTH, McGILL UNIV, 64- Mem: Am Epidemiol Soc; Brit Soc Social Med; Int Epidemiol Asn. Res: Environmental Soc; Brit Soc Social Med; Int Epidemiol Asn. Res: Environmental respiratory diseases; family planning; health care research. Mailing Add: Dept of Epidemiol & Health McGill Univ Sch of Med Montreal PQ Can

MCDONALD, JOHN E, b Sheridan, Ind, Feb 6, 31; m 54; c 2. PHYSICAL CHEMISTRY, INORGANIC CHEMISTRY. Educ: Purdue Univ, BS, 57, PhD(phys chem), 61. Prof Exp: Chemist, Sandia Corp, 61-64, div supvr surface phenomena, 64-65, dept mgr mat & process, 65-69, dir mat sci, 69-73; ASSOC GEN MGR, PROTOTYPE DEVELOP ASSOCS, INC, 73- Mem: AAAS; Am Inst Aeronautics & Astronautics; Am Chem Soc. Res: Surface chemistry and physics; thermochemistry; materials research and development. Mailing Add: 3259 Alta Laguna Blvd Laguna Beach CA 92705

MCDONALD, JOHN KENNELY, b Vancouver, BC, Oct 4, 30; US citizen; m 53; c 3. BIOCHEMISTRY, ENDOCRINOLOGY. Educ: Univ BC, BSA, 53; Purdue Univ, MS, 56; Ore State Univ, PhD(chem), 60. Prof Exp: Bacteriologist, Defence Res Bd, Can 53; protein chemist, Cutter Labs, Calif, 60-62; res scientist, Biomed Res Div, NASA-Ames Res Ctr, 62-75; vis prof biochem, 74-75, PROF BIOCHEM, MED UNIV S C, 75- Mem: AAAS; Am Soc Biol Chem; Endocrine Soc. Res: Protein chemistry; enzymology; structure and function of tissue peptidases and proteases. Mailing Add: Dept of Biochem Med Univ of S C Charleston SC 29401

MCDONALD, JOHN ROLAND, b Roland, Man, Jan 23, 10; nat US; m 34; c 3. PATHOLOGY. Educ: Univ Man, MD, 33; Univ Minn, MS, 36. Prof Exp: Instr, Mayo Found, Univ Minn, 37-41, assoc, Mayo Clin, 37-45, from asst prof to assoc prof, Med Sch, 41-49, prof path, Grad Sch Med, 49-58, head sect surg path, Mayo Clin, 45-58; PROF PATH, WAYNE STATE UNIV, 60-; PATHOLOGIST & DIR LABS, HARPER HOSP, 58- Concurrent Pos: Mem adv comt cancer control prog, USPHS, 64-68; chief staff, Harper Hosp, 66-69. Honors & Awards: Am Soc Clin Pathologists Gold Medal, 48. Mem: Fel Am Soc Clin Pathologists; fel AMA; Am Asn Thoracic Surg; Am Asn Path & Bact; fel Col Am Pathologists. Res: Surgical and experimental pathology; obstructive pneumonitis of neoplastic origin and its correlation according to presence or absence of sputum; cytology of sputum and bronchial secretions; studies on patients with miscellaneous pulmonary lesions; cytological diagnosis of bronchogenic carcinoma. Mailing Add: Harper Hosp 3825 Brush St Detroit MI 48201

MCDONALD, JOHN STONER, b Mt Hope, Wash, Sept 20, 32; m 63; c 1. VETERINARY MEDICINE, MICROBIOLOGY. Educ: Wash State Univ, BA, 54, DVM, 56; Univ Idaho, MS, 58; Iowa State Univ, PhD(vet microbiol), 67. Prof Exp: Sta vet, Univ Idaho, 56-58; pvt pract, 58-61; RES VET, NAT ANIMAL DIS LAB, AGR RES SERV, USDA, 61- Mem: Am Vet Med Asn; Am Col Vet Microbiol. Res: Pathogenesis of udder infection and mastitis. Mailing Add: Nat Animal Dis Lab Box 70 Ames IA 50010

MCDONALD, JOHN WILLIAM DAVID, b Chatham, Ont, Jan 7, 38; m 70; c 2. HEMATOLOGY, INTERNAL MEDICINE. Educ: Univ Western Ont, MD, 61, PhD(biochem), 66; FRCP(C), 69. Prof Exp: Intern med, Victoria Hosp, London, Ont, 61-62; resident, Montreal Gen Hosp, 66-67; res & clin fel med & hemat, Royal Victoria Hosp, Montreal, 67-70; asst prof path chem & med, Victoria Hosp, Univ Western Ont, 70-72, asst prof med, Univ Hosp, 72-73, ASST PROF, DEPT BIOCHEM, UNIV WESTERN ONT, 71-, ASSOC PROF, DEPT MED, 73-, ASST DEAN RES, FAC MED, 75- Concurrent Pos: Dir biochem-radioisotope-hemat lab & physician, Dept Med, Victoria Hosp, 70-72; physician & dir biochem-radioisotope-hemat, Hemat Serv, Hosp, Univ Western Ont, 72- Mem: Royal Col Physicians & Surgeons Can; Can Soc Hemat; Can Soc Clin Invest; Am Soc Hemat; Am Fedn Clin Res. Res: Effect of drugs on platelet prostaglandin synthesis and platelet function. Mailing Add: Dept Med Univ Western Ont Hosp 339 Windermere Rd London ON Can

MCDONALD, JOSEPH KYLE, b Athens, Ala, Oct 31, 42; m 69; c 2. PHYSICAL CHEMISTRY, MOLECULAR SPECTROSCOPY. Educ: Athens Col, BS, 64; Vanderbilt Univ, PhD(phys chem), 68. Prof Exp: ASSOC PROF MATH & CHEM, ATHENS STATE COL, 70- Mem: Am Chem Soc; Optical Soc Am. Res: Lifetimes, vibrational, and rotational analysis of electronic states of molecular radicals produced in various discharge systems. Mailing Add: Box 225 Athens State Col Athens AL 35611

MCDONALD, KEITH LEON, b Murray City, Utah, Apr 20, 23; m 56. THEORETICAL PHYSICS. Educ: Univ Utah, BS, 50, MS, 51, PhD(physics), 56. Prof Exp: Mem staff, Los Alamos Sci Lab, Calif, 56-57; physicist, Dugway Proving Ground, 57-60; physics fac mem, Brigham Young Univ, 60-62 & Univ Utah, 62-63; res physicist, Nat Bur Standards, 63-64; physics fac mem, Idaho State Univ, 65; theoret geophysicist, Inst Earth Sci, Environ Sci Serv Admin-Inst Environ Res, 66-69; consult, Environ Res Labs, Environ Sci Serv Admin, Colo, 6970 & Nat Oceanic & Atmospheric Admin, 71; CONSULT, 71- Mem: Optical Soc Am; Am Phys Soc; Am Astron Soc; Am Geophys Union; Am Asn Physics Teachers. Res: Electromagnetic theory and hydromagnetism; theoretical optics; kinetic theory; statistical mechanics. Mailing Add: PO Box 2433 Salt Lake City UT 84110

MCDONALD, LARRY WILLIAM, b Louisville, Nebr, May 25, 28; m 55; c 3. PATHOLOGY, NEUROPATHOLOGY. Educ: Univ Calif, Berkeley, AB, 50; Northwestern Univ, MD, 55. Prof Exp: Res pathologist, Pondville State Hosp, Norfolk, Mass, 59-61; res assoc, Univ Calif, Berkeley, 62-68; assoc prof path, Sch Med, Univ Calif, Davis, 68-75; PROF PATH, SCH MED, WRIGHT STATE UNIV, 75- Concurrent Pos: Damon Runyon grant, 60-62; USPHS trainee radiobiol, Cancer Res Inst, New Eng Deaconess Hosp, Boston, 61-62; consult, Vet Admin Hosp, Martinez, Calif, 66-73, chief electron micros in path, 67; consult, Lawrence Berkeley Lab, Univ Calif, 68-75 & Sacramento County Coroners Off, 71-75; chief lab serv, Vet Admin Ctr, Dayton, Ohio, 75- Mem: AAAS; Am Soc Clin Pathologists; Am Asn Neuropath; Radiation Res Soc Am; Electron Micros Soc Am. Res: Effects of radiation on the nervous system, particularly the mechanism of production of the delayed effects; mechanisms of induction of tumors in the central nervous system. Mailing Add: Wright State Univ Sch of Med 4100 W Third St Dayton OH 45428

MCDONALD, LESLIE ERNEST, b Middletown, Mo, Oct 14, 23; m 46; c 3. PHYSIOLOGY. Educ: Mich State Univ, BS, 48, DVM, 49; Univ Wis, MS, 51, PhD(physiol), 52. Prof Exp: Practicing vet, 49-50; instr physiol, Univ Ill, 52-53, assoc prof, 53-54; prof & head dept, Okla State Univ, 54-69; assoc dean, Col Vet Med, Univ Ga, 69-71; dean, Col Vet Med, Ohio State Univ, 71-72; PROF PHYSIOL & DIR DEVELOP, COL VET MED, UNIV GA, 72- Concurrent Pos: Consult, USPHS, 64-; mem, Comt Vet Drug Efficacy, Nat Res Coun, 66- Mem: Soc Exp Biol & Med; Am Physiol Soc; Am Vet Med Asn; Brit Soc Study Fertil. Res: Endocrinology; reproductive physiology. Mailing Add: Dept of Physiol Col of Vet Med Univ of Ga Athens GA 30601

MCDONALD, MALCOLM EDWIN, b Ann Arbor, Mich, May 29, 15; m 41; c 2. PARASITOLOGY, ECOLOGY. Educ: Parsons Col, BS, 37; Univ Iowa, MS, 39; Univ Mich, PhD(wildlife mgt), 51. Prof Exp: Asst prof biol, Parsons Col, 40-42; instr,

Beloit Col, 50-51; asst prof, Union Col, NY, 51-56 & Univ Nev, 56-57; WILDLIFE RES BIOLOGIST, US FISH & WILDLIFE SERV, US DEPT INTERIOR, 57- Concurrent Pos: Parasitologist, Malaria Surv Unit, Sanit Corps, 42-46. Mem: Ecol Soc Am; Am Soc Parasitol; Wildlife Soc; Wilson Ornith Soc; Wildlife Dis Asn. Res: Animal parasitology and ecology; parasites of waterfowl; wildlife conservation. Mailing Add. Nat Fish & Wildlife Lab Dept Vet Sci Univ Wis 1655 Linden Dr Madison WI 53706

MCDONALD, MARGARET RITCHIE, b Govan, Scotland, May 24, 10; nat US; m 42; c 1. BIOCHEMISTRY. Educ: Rutgers Univ, BSc, 30, PhD(biochem), 40. Prof Exp: Tech asst gen physiol, Rockefeller Inst, 30-40, fel, 40-41, asst, 41-43; res assoc, Dept Genetics, Carnegie Inst, 43-45, investr, 45-63; BIOCHEMIST, WALDEMAR MED RES FOUND, INC, 63- Concurrent Pos: Mem, Marine Biol Lab, Woods Hole. Res: Purification of enzymes; plasma proteins; nucleoproteins; chemical nature of cellular components; oncology. Mailing Add: Waldemar Med Res Found Inc Sunnyside Blvd Woodbury NY 11797

MCDONALD, PERRY FRANK, b Garner, Tex, July 26, 33; m 59; c 2. SOLID STATE PHYSICS. Educ: Tex Christian Univ, BA, 54; NTex State Univ, MA, 60; Univ Ala, Tuscaloosa, PhD(physics), 68. Prof Exp: Res asst, Socony Mobil Field Res Lab, Tex, 60-61; res physicist, Brown Eng CoInc, Ala, 61-66 & US Army Missile Command, Redstone Arsenal, 66-68; ASSOC PROF PHYSICS, SAM HOUSTON STATE UNIV, 68- Mem: Am Phys Soc; Sigma Xi. Res: Acoustic paramagnetic resonance; electron paramagnetic resonance; spin lattice interactions. Mailing Add: Dept of Physics Sam Houston State Univ Huntsville TX 77340

MCDONALD, PHILLIP ROBB, b Kong Moon, China, Apr 2, 09; nat US; m 59; c 4. OPHTHALMOLOGY. Educ: McGill Univ, BS, 30, MD, 34. Prof Exp: Clin prof ophthal, Div Grad Med, Univ Pa, 58-66, prof, 66-67; PROF OPHTHAL, JEFFERSON MED COL, THOMAS JEFFERSON UNIV, 67- Concurrent Pos: Chief ophthal serv, Lankenau Hosp, 48-74, sr consult, 74-; consult to Surgeon Gen, US Air Force, 58-; consult surgeon, Wills Eye Hosp, 71- Mem: Am Ophthal Soc; Asn Res Vision & Ophthal; Am Acad Ophthal & Otolaryngol. Res: Dark adaptation; means of improving surgical results of retinal detachment; glaucoma. Mailing Add: Lankenau Med Bldg Philadelphia PA 19151

MCDONALD, RALPH EARL, b Indianapolis, Ind, May 12, 20; m 42; c 3. DENTISTRY. Educ: Ind Univ, BS, 42, DDS, 44, MS, 51; Am Bd Pedodontics, dipl, 57. Prof Exp: From instr to assoc prof pedodontics, 46-57, asst dean & secy grad dent educ, 64-68, PROF PEDODONTICS, SCH DENT, IND UNIV, INDIANAPOLIS, 57-, CHMN DEPT, 53-, DEAN DENT EDUC, 69- Concurrent Pos: Consult, USPHS, 50-; mem exam bd, Am Bd Pedodontics, 55-62, chmn bd, 60-62. Honors & Awards: Distinguished Serv Award, Am Soc Dent for Children. Mem: Am Soc Dent for Children (secy-treas, 57-59, vpres, 59, pres, 62); fel Am Col Dent; Am Acad Pedodont (pres, 67; hon mem Brazilian Acad Dent & Am Dent Soc Ireland. Res: Clinical dental caries control; pathology of the dental pulp. Mailing Add: Dept of Pedodontics Ind Univ Sch of Dent Indianapolis IN 46202

MCDONALD, RAY LOCKE, b Chula Vista, Calif, Oct 5, 31; m 60; c 2. PHYSICAL CHEMISTRY. Educ: San Diego State Col, AB, 55; Ore State Univ, PhD(chem), 60. Prof Exp: Res assoc chem, Mass Inst Technol, 60-61; asst prof, NDak State Univ, 61-65; from asst prof to assoc prof, 65-74, assoc chmn dept, 71-74, PROF CHEM & DEPT CHMN, UNIV HAWAII, 74- Mem: Am Chem Soc. Res: Ionic interactions in nonaqueous solvents; solvent extraction of ions. Mailing Add: Dept of Chem Univ of Hawaii Honolulu HI 96822

MCDONALD, RICHARD NORMAN, b Detroit, Mich, Feb 26, 31; m 56; c 2. ORGANIC CHEMISTRY. Educ: Wayne State Univ, BS, 54, MS, 55; Univ Wash, PhD(org chem), 57. Prof Exp: Res chemist, Pioneering Res Div, Textile Fibers Dept, E I du Pont de Nemours & Co, 57-60; from asst prof to assoc prof, 60-68, PROF CHEM, KANS STATE UNIV, 68- Mem: Am Chem Soc; The Chem Soc. Res: Synthesis and chemistry of non-benzenoid aromatic compounds, small ring compounds and strained polycyclic structures; epoxidecarbonyl molecular rearrangement; synthesis by photochemical reactions and their mechanism. Mailing Add: 1820 Alabama Lane Manhattan KS 66502

MCDONALD, ROBERT H, JR, b Cincinnati, Ohio, Nov 15, 33; m 58; c 2. CLINICAL PHARMACOLOGY. Educ: Xavier Univ, BS, 55; Stritch Sch Med, MD, 59. Prof Exp: Intern, Health Ctr Hosps, Univ Pittsburgh, 59-60; jr resident internal med, Univ Pittsburgh, 60-61; NIH fels clin pharmacol, Emory Univ, 61-63; assoc surgeon, Nat Heart Inst, 63-65; USPHS hon res asst physiol, Univ Col, London, 65-66; asst prof, 66-70, ASSOC PROF MED & PHARMACOL, SCH MED, UNIV PITTSBURGH, 70-, CHIEF SECT CLIN PHARM, 71- Mem: Am Fedn Clin Res; Am Soc Pharmacol & Exp Therapeut; Royal Soc Med; Am Physiol Soc. Res: Human pharmacokinetics; hypertension; adrenergic mechanisms; drug abuse; inotropic modification of myocardial energetics. Mailing Add: Dept of Med Univ of Pittsburgh Sch of Med Pittsburgh PA 15261

MCDONALD, ROBERT SKILLINGS, b Pittsburgh, Pa, July 19, 18; m 43; c 3. PHYSICAL CHEMISTRY, SPECTROSCOPY. Educ: Univ Maine, BS, 41; Mass Inst Technol, PhD(phys chem), 52. Prof Exp: Jr engr, Hygrade Sylvania Co, Mass, 51; res physicist, Am Cyanamid Co, Conn, 42-46; res assoc, Mass Inst Technol, 46-50; RES ASSOC, RES & DEVELOP CTR, GEN ELEC CO, 51- Mem: Am Chem Soc; Am Phys Soc; Coblentz Soc. Res: Infrared spectroscopy; surface chemistry; computer applications in analytical chemistry; mass spectroscopy. Mailing Add: Gen Elec Res & Develop Ctr PO Box 8 Knolls 1 Schenectady NY 12301

MCDONALD, ROGER KOEFOD, b Ulen, Minn, Jan 30, 21; m 43; c 5. CLINICAL MEDICINE. Educ: Univ Minn, BS, 43, MB, 45, MD, 46. Prof Exp: Sr asst surgeon, Nat Heart Inst, 48-51, resident internal med, USPHS Hosp, Baltimore, 51-54, sr surgeon, NIMH, 54-55, chief sect med, Lab Clin Sci, 55-62, asst to dir labs & clins, 62-64; assoc prof psychiat, Sch Med, Yale Univ, 64-69; CHIEF LAB CLIN PSYCHOPHARMACOL, DIV SPEC MENT HEALTH RES, NIMH, 69- Concurrent Pos: Teaching fel path, Med Sch, Univ Minn, 48; dir res facil, Conn Ment Health Ctr. Mem: Am Physiol Soc; Soc Exp Biol & Med; Psychiat Res Soc; Am Col Neuropsychopharmacol; Int Col Neuropsychopharmacol. Res: Biology of mental disease. Mailing Add: NIMH Div of Spec Ment Health Res White Bldg St Elizabeth's Hosp Washington DC 20032

MCDONALD, RUSSELL F, agricultural economics, applied statistics, see 12th edition

MCDONALD, STUART, b Saskatoon, Sask, Nov 25, 19; m 44; c 4. BIOLOGY. Educ: Univ Western Ontario, BSc, 50, MSc, 52. Prof Exp: Res off spruce budworm, Forest Biol Lab, 48-53, RES OFF INSECT TOXICOL, RES STA, CAN DEPT AGR, 53- Mem: AAAS; Entom Soc Can. Res: Insect toxicology; qualitative and comparative laboratory evaluation of insecticides for the control of field crop pests; developmental stages and geographical distribution of a species as they affect control by chemicals; physiological effects of sublethal doses; development of bioassay techniques; insecticide resistance. Mailing Add: Res Sta Can Dept of Agr Lethbridge AB Can

MCDONALD, TED PAINTER, b Loudon, Tenn, Nov 23, 30; m 55; c 3. BIOCHEMISTRY. Educ: Univ Tenn, BS, 55, MS, 58, PhD(animal sci, biochem), 65. Prof Exp: Radiobiologist, Oak Ridge Nat Lab, 58-64; res asst blood platelet & mucopolysaccharide chem, Res Lab, AEC, 64-65, res assoc exp hemat, Mem Res Ctr, 65, res instr, 65-68, asst prof, 68-73, ASSOC PROF EXP HEMAT, MEM RES CTR, UNIV TENN, KNOXVILLE, 73- Concurrent Pos: Mem coun thrombosis, Am Heart Asn. Mem: Am Soc Hemat; Int Soc Hemat; Soc Exp Biol & Med; Soc Exp Hemat; Radiation Res Soc. Res: Blood platelet physiology; mucopolysaccharide chemistry; experimental hematology. Mailing Add: Univ of Tenn Mem Res Ctr 1924 Alcoa Hwy Knoxville TN 37920

MCDONALD, THEODORE JONATHAN, neurobiology, entomology, see 12th edition

MCDONALD, W JOHN, b Lethbridge, Alta, Sept 29, 36; m 61;c 61; c 3. NUCLEAR PHYSICS. Educ: Univ Sask, BSc, 59, MSc, 61; Univ Ottawa, Ont, PhD(physics), 64. Prof Exp: Asst prof, 65-70, ASSOC PROF PHYSICS, UNIV ALTA, 70- Mem: Am Asn Physics Teachers; Can Asn Physicists. Res: Low energy nuclear physics, especially neutron time of flight studies in the 0 to 20 mev energy range and intermediate energy physics on 500 mev cyclotron. Mailing Add: Dept of Physics Univ of Alta Edmonton AB Can

MCDONALD, WILLIAM CHARLES, b Chicago, Ill, Feb 14, 33; m 57; c 4. MICROBIOLOGY. Educ: Univ Okla, BS, 55; Univ Tex, PhD(bact), 59. Prof Exp: Res assoc biochem genetics, Children's Hosp, Buffalo, NY, 59-60; asst prof bact, Wash State Univ, 62-65; from assoc prof to prof biol, Tulane Univ, 65-73; PROF BIOL & CHMN DEPT, UNIV TEX, ARLINGTON, 73- Concurrent Pos: State of Wash initiative measure grants, 63-66; NSF res grants, 63-69. Mem: AAAS; fel Am Acad Microbiol; Am Soc Microbiol. Res: Effects of radiation on bacteria; microbial genetics; temperature sensitivity and resistance in microorganisms; biology of bacteriophages. Mailing Add: Dept of Biol Univ of Tex Arlington TX 76019

MCDONALD, WILLIAM CRAIK, b Graysville, Man, May 8, 21; m 49; c 2. PLANT PATHOLOGY. Educ: Univ Man, BSA, 50; Univ Wis, MSc, 51, PhD, 54. Prof Exp: Plant pathologist, 50-64, head cereal dis sect, 64-71, DIR CEREAL DIS SECT, RES STA, CAN DEPT AGR, 71- Mem: Am Phytopath Soc; Can Phytopath Soc; Agr Inst Can. Res: Diseases of cereal crops. Mailing Add: Res Br Res Sta Can Dept Agr 25 Dafoe Rd Winnipeg MB Can

MCDONALD, WILLIAM JOHN, b Nunda, SDak, Feb 19, 24; m 49; c 4. PHYSICS, MATHEMATICS. Educ: Iowa State Col, BS, 49. Prof Exp: Jr physicist, Res Lab, Bendix Aviation Corp, 49-51; physicist, Cent Res Lab, 52-59, physicist abrasive lab, 59-61, PHYS STUDIES SUPVR, ABRASIVE LAB, MINN MINING & MFG CO, 61- Res: Metal cutting; wear of hard minerals; instrumentation; geometric optics; electrostatic forces. Mailing Add: 3M Ctr Minn Mining & Mfg Co Bldg 2-2 St Paul MN 55101

MCDONELL, WILLIAM ROBERT, b New Rockford, NDak, Mar 8, 25; m 55; c 2. RADIATION CHEMISTRY. Educ: Univ Mich, BS, 47, MS, 48; Univ Calif, PhD(chem), 51. Prof Exp: Chemist, Radiation Lab, Univ Calif, 48-51 & Argonne Nat Lab, 51-53; chemist, 53-60, res supvr, 60-70, sr res scientist, 69-70, RES ASSOC, SAVANNAH RIVER LAB, E I DU PONT DE NEMOURS & CO, INC, 70- Mem: Am Chem Soc; Am Nuclear Soc. Res: Nuclear and radiation chemistry; effects of radiation on materials; reactor fuel behavior; development of radioisotopic heat and radiation sources. Mailing Add: Savannah River Lab E I du Pont de Nemours & Co Inc Aiken SC 29801

MCDONNEL, GERALD M, b Salt Lake City, Utah, Feb 13, 19; m 41; c 7. RADIOLOGY, RADIOBIOLOGY. Educ: Univ Utah, BA, 40; Temple Univ, MD, 43. Prof Exp: Intern, Temple Univ Hosp, 43, resident internal med, 44; ASSOC PROF RADIOL, CTR HEALTH SCI, UNIV CALIF, LOS ANGELES, 59- Concurrent Pos: Mem sci adv bd, US Air Force, 60- & Defense Sci Bd, 64-68; mem staff dept radiol, Hosp of the Good Samaritan, Los Angeles, 68- Mem: AMA; Am Col Radiol; Radiol Soc NAm. Res: Nuclear weapons effects; radiation effects and space radiation effects in humans; megavoltage diagnostic radiology. Mailing Add: Hosp of the Good Samaritan 1212 Shatto St Los Angeles CA 90017

MCDONNELL, CHILTON HORNER, organic chemistry, see 12th edition

MCDONNELL, EDMOND JOSEPH, b Singac, NJ, Sept 28, 16; m; c 3. ORTHOPEDIC SURGERY. Educ: St Peter's Col, BS, 38; Long Island Col Med, MD, 42; Am Bd Orthop Surg, dipl, 50. Prof Exp: Resident, 46-48, ATTEND ORTHOP SURGEON, JOHNS HOPKINS HOSP, 48-; ASSOC PROF ORTHOP SURG, SCH MED, JOHNS HOPKINS UNIV, 60- Concurrent Pos: Pvt pract, 48-; attend orthop surgeon, Children's Hosp, Inc, 48-, asst med dir, 58-; asst med dir, Union Mem Hosp, 50-; consult, Crippled Children's Serv, Md, 50-, WVa, 52- & King's Daughters Hosp, Martinsburg, 52- Mem: AMA; Am Acad Orthop Surg. Res: General surgery. Mailing Add: 4 E Madison St Baltimore MD 21202

MCDONNELL, JAMES MICHAEL, zoology, entomology, deceased

MCDONNELL, JOSEPH FRANCIS, JR, b Jenkintown, Pa, July 23, 05; m 38; c 2. PHARMACEUTICAL CHEMISTRY, NUTRITION. Educ: Philadelphia Col Pharm, BSc, 26, MSc, 27. Prof Exp: Sr chemist, Bur Foods & Chem, Pa Dept Agr, 35-37; chief chemist & dir lab, State Bd Pharm, Pa, 37-42; sr bus specialist, US Off Price Admin, 42-45; dir control, Schenley Labs, Inc, 45-50; asst mgr dept clin res, 50-68, asst to dir tech serv, 68-70, asst to tech dir, Borden, Inc, 70-72; RETIRED. Mem: AAAS; Am Chem Soc; Am Med Writers' Asn; Am Pharmaceut Asn. Res: Pharmaceutical development. Mailing Add: 274 Harvard Ave Rockville Centre NY 11570

MCDONNELL, WILLIAM VINCENT, b Carbondale, Pa, Oct 28, 22; m 49. PATHOLOGY. Educ: Univ Scranton, AB, 43; Jefferson Med Col, MD, 47; Am Bd Path, dipl. Prof Exp: From asst prof to assoc prof, 52-69, PROF PATH, JEFFERSON MED COL, 69-; MED DIR, WEST JERSEY HOSP, 69-, VPRES MED AFFAIRS, 72- Concurrent Pos: Asst dir clin lab, Methodist Hosp & Jefferson Med Col, 52-61; coordr cancer teaching, Jefferson Med Col, 55-60; dir clin lab, West Jersey Hosp, 61- Mem: Am Soc Clin Pathologists; AMA; Col Am Pathologists. Res: Human pathology. Mailing Add: Clin Lab West Jersey Hosp Voorhees NJ 08043

MCDONOUGH, EUGENE STOWELL, b Abingdon, Ill, Apr 16, 05; m 30, 75. MYCOPATHOLOGY, GENETICS. Educ: Marquette Univ, BS, 28, MS, 31; Iowa State Col, PhD(bot), 36. Prof Exp: Asst bot, Marquette Univ, 28-29; instr, Mich State Col, 29-30; from instr to prof, 30-70, EMER PROF BOT, MARQUETTE UNIV, 70- Concurrent Pos: Co-dir, Holton & Hunkel Res Award, 44-50; researcher, Mycol Unit, USPHS grant, Commun Dis Ctr, Ga, 58-59. Mem: AAAS; Mycol Soc Am; Am

Phytopath Soc; Am Soc Microbiol; Med Mycol Soc Americas. Res: Cytology and genetics of fungi; polyploidy; human heredity; host-parasite relations of plant diseases; diseases of flowering plants; chromosome structure; antimycotics; phytopathology; medical mycology; soil mycolysis. Mailing Add: Dept of Biol Marquette Univ Milwaukee WI 53233

MCDONOUGH, EVERETT GOODRICH, b Birmingham, Ala, Feb 22, 05; m 32; c 3. ORGANIC CHEMISTRY. Educ: Howard Col, AB, 26; Columbia Univ, PhD(chem), 32. Prof Exp: Mem fac, Howard Col, 26-27; chief chemist, Marinello Co, 28-33; asst tech dir, Inecto, Inc, 33-40; from vpres to exec vpres, 40-66, PRES, EVANS CHEMETICS, INC, 66-, DIR, 40- Concurrent Pos: Tech dir, Sales Affiliates, Inc, 41-66, vpres, 66-69, dir, 66-, pres, 69-; mem chem indust adv comt, Nat Prod Authority, 51-54; trustee, St John's Riverside Hosp, Yonkers, NY, 52-, vpres, 67-; exec vpres, Evans Res & Develop Corp, 56-66, dir, 56-69, pres, 66-69; vpres, Zotos Int, Inc, 69-71, dir, 69-, sr vpres, 71- Honors & Awards: Medal, Soc Cosmetic Chem, 52. Mem: Fel AAAS; Am Chem Soc; Soc Chem Indust; Soc Cosmetic Chem (vpres, 48 & 50, pres, 51); Cosmetic, Toiletry & Fragrance Asn. Res: Keratin chemistry; cosmetic depilatories and permanent waving containing substituted mercaptans; organic sulfur chemistry; dermatology; medicine. Mailing Add: Evans Chemetics Inc 90 Tokeneke Rd Darien CT 06820

MCDONOUGH, GEORGE FRANCIS, JR, applied mechanics, see 12th edition

MCDONOUGH, LESLIE MARVIN, organic chemistry, see 12th edition

MCDONOUGH, ROBERT JAKES, microbiology, immunochemistry, see 12th edition

MCDONOUGH, WALTER THOMAS, b New York, NY, Nov 14, 22; m 62. PHYSIOLOGICAL ECOLOGY. Educ: City Col New York, BS, 48; Rutgers Univ, MS, 55; Univ Md, PhD(bot), 58. Prof Exp: Food chemist, Nat Biscuit Co, 48-51 & A & P Tea Co, 52; asst, Rutgers Univ, 53-54 & Univ Md, 56-58; instr bot, Cornell Univ, 58-60; asst prof, Univ Cincinnati, 60-65; PLANT PHYSIOLOGIST, INTERMT FOREST & RANGE EXP STA, FORESTRY SCI LAB, 65-; ASSOC PROF BOT, UTAH STATE UNIV, 74- Mem: Ecol Soc Am. Mailing Add: Intermt Forest & Range Exp Sta Forestry Sci Lab 860 N 12th E St Logan UT 84321

MCDOUGAL, DAVID BLEAN, JR, b Chicago, Ill, Jan 31, 23; m 47; c 3. PHARMACOLOGY. Educ: Princeton Univ, BA, 45; Univ Chicago, MD, 47. Prof Exp: Instr anat, Johns Hopkins Univ, 50-53; from instr to assoc prof, 57-71, PROF PHARMACOL, SCH MED, WASH UNIV, 71- Res: Neurochemistry. Mailing Add: Dept of Pharmacol Wash Univ Sch of Med St Louis MO 63110

MCDOUGALD, LARRY ROBERT, b Broken Bow, Okla, Dec 31, 41; m 62; c 1. PARASITOLOGY. Educ: Southeastern State Col, BS, 62; Kans State Univ, MS, 66, PhD(parasitol), 69. Prof Exp: Teacher pub schs, Kans, 62-64; instr biol, Mt St Scholastica Col, 67-68; instr zool, Kans State Univ, 68-69; fel poultry sci, Univ Ga, 69-71; SR PARASITOLOGIST, PARASITOL RES DEPT, ELI LILLY & CO, 71- Mem: Am Soc Parasitol; Am Micros Soc; Am Asn Avian Pathologists. Res: Physiological and chemical aspects of host-parasite relationships; chemotherapy of protozoan parasites. Mailing Add: Parasitol Res Dept Eli Lilly & Co Greenfield IN 46140

MCDOUGALL, KENNETH J, b Ashland, Wis, Aug 31, 35; m 65; c 2. MICROBIAL GENETICS. Educ: Northland Col, BS, 57; Marquette Univ, MS, 59; Kans State Univ, PhD(genetics), 64. Prof Exp: Instr zool, Northland Col, 59-60; res assoc genetics, Rice Univ, 63-64, NIH fel, 64-65, res assoc, 65-66; asst prof, 66-69, NSF res grant, 67-69, ASSOC PROF BIOL, UNIV DAYTON, 69- Mem: AAAS; Genetics Soc Am; Am Soc Microbiol. Res: Ornithine metabolism in Neurospora crassa. Mailing Add: Dept of Biol Univ of Dayton Dayton OH 45469

MCDOUGALL, LEE ALLEN, physical chemistry, see 12th edition

MCDOUGALL, ROBERT I, physical chemistry, polymer chemistry, see 12th edition

MCDOUGALL, WALTER BYRON, b Ypsilanti, Mich, Dec 10, 83; m 08; c 3. BOTANY. Educ: Univ Mich, AB, 11, PhD(bot), 13. Prof Exp: Instr bot, Univ Ill, 13-17, assoc, 17-18, from asst prof to assoc prof, 18-29; prof, Univ Southern Calif, 30-31; regional wildlife technician, Nat Park Serv, 35-40; regional biologist, Nat Park Wildlife Sect, US Fish & Wildlife Serv, 40-42; asst park naturalist, Grand Canyon Nat Park, 42-43; naturalist, Nat Park Serv, 43-53; CUR BOT, MUS NORTHERN ARIZ, 55- Mem: AAAS; Bot Soc Am; Ecol Soc Am; Am Inst Biol Sci. Res: Plant ecology; floras of Yellowstone, Big Bend and Grand Canyon National Parks, Natchez Trace Parkway and northern Arizona; plant succession in abandoned formerly cultivated fields. Mailing Add: Mus of Northern Ariz Flagstaff AZ 86001

MCDOUGLE, PAUL E, b Peru, Ind, May 25, 28. MATHEMATICS. Educ: Purdue Univ, MS, 49; Univ Va, PhD(math), 58. Prof Exp: Asst prof, 58-67, ASSOC PROF MATH, UNIV MIAMI, 67-, ASSOC CHMN DEPT, 70- Mem: Math Asn Am. Res: Topology; study of decomposition spaces. Mailing Add: Dept of Math Univ of Miami Coral Gables FL 33124

MCDOWELL, CHARLES ALEXANDER, b Belfast, Ireland, Aug 29, 18; m 45; c 3. PHYSICAL CHEMISTRY. Educ: Queen's Univ, Belfast, BSc, 41-42, DSc, 55. Prof Exp: Asst lectr chem, Queen's Univ, Belfast, 41-42; lectr inorg & phys chem, Univ Liverpool, 45-55; PROF CHEM & HEAD DEPT, UNIV BC, 55- Concurrent Pos: Spec lectr & spec sci medal, Univ Liege, 55; Nat Res Coun Can sr res fel, Cambridge Univ, 63-64; vis prof, Kyoto Univ, 65 & 69; Killiam sr fel, 69-70; distinguished vis prof, Univ Fla, Gainesville, 74. Honors & Awards: Centennial Medal, Govt Can, 67; Chem Inst Can Award, 69. Mem: Am Phys Soc; Am Chem Soc; fel Royal Soc Can; fel Chem Inst Can; fel Am Inst Physics. Res: Mass spectrometry; chemical kinetics; electron and nuclear spin resonance spectroscopy; molecular structure; electronic structures of molecules; photoelectron spectrometry. Mailing Add: Dept of Chem Univ of BC Vancouver BC Can

MCDOWELL, CURTIS SCRANTON, organic chemistry, sanitary engineering, see 12th edition

MCDOWELL, DAWSON CLAYBORN, b Chicago, Ill, July 19, 13; m 36; c 1. METEOROLOGY. Educ: Univ Chicago, BS, 35, MS, 42. Prof Exp: Instr, Univ Chicago, 42-45; PROF METEOROL & DIR INST TROP METEOROL, UNIV PR, 45- Concurrent Pos: Mem, Nat Adv Comt Educ. Mem: Am Meteorol Soc; Am Geophys Union. Res: Tropical meteorology. Mailing Add: Dept of Meteorol Univ of PR PO Box 22931 Rio Piedros PR 00931

MCDOWELL, ELIZABETH MARY, b Kew Gardens, Eng, Mar 30, 40. PATHOLOGY. Educ: Univ London, BVetMed, 64; Univ Cambridge, PhD(path), 71, MA, 72. Prof Exp: Vet gen pract, 64-66; instr, 71-73, ASST PROF PATH, UNIV MD, BALTIMORE CITY, 73- Mem: Royal Col Vet Surg; Histochem Soc. Res: Cellular and subcellular pathology; kidney pathophysiology; pulmonary neoplasia. Mailing Add: Dept of Path Univ of Md Sch of Med Baltimore MD 21201

MCDOWELL, FLETCHER HUGHES, b Denver, Colo, Aug 5, 23; m 58; c 3. NEUROLOGY. Educ: Dartmouth Col, AB, 43; Cornell Univ, MD, 47. Prof Exp: From instr to assoc prof med, 52-68, PROF NEUROL, MED COL, CORNELL UNIV, 68-, ASSOC DEAN MED COL, 70-; MED DIR, BURKE REHAB HOSP, WHITE PLAINS, 73- Mem: Am Acad Neurol; Am Neurol Asn; Am Fedn Clin Res. Res: Cerebrovascular disease. Mailing Add: Cornell Univ Med Col 1300 York Ave New York NY 10021

MCDOWELL, FRANK, b Marshfield, Mo, Jan 30, 11; m 34; c 3. SURGERY. Educ: Drury Col, AB, 32; Washington Univ, MD, 36; Am Bd Surg, dipl; Am Bd Plastic Surg, dipl. Hon Degrees: ScD, Drury Col, 73. Prof Exp: Assoc prof maxillofacial surg, Sch Dent, Washington Univ, 42-70, assoc prof clin surg, Univ, 47-70; prof surg, Sch Med, Univ Hawaii, 68-74; PROF SURG, SCH MED, STANFORD UNIV, 74-; EDINCHIEF, J PLASTIC & RECONSTRUCT SURG, 67- Concurrent Pos: Lectr, Hadassah & Athens, 54; secy, Am Bd Plastic Surg, 55-60, chmn, 61-62; mem joint adv bd, Med Specialties, 55-62, 65-71, mem exec comt, 66-71; lectr, Nat Inst Burns, Arg, 56; trustee, Drury Col, 56-, chmn bd, 64-66; mem exec comt, Dept Surg, Univ Hawaii, 67-74; consult plastic surg, Tripler Army Hosp & chief plastic surg, Queen's Hosp, Honolulu; vis prof, Univ Calif, Los Angeles, 69, Univ Utah, 73 & Univ Mo, 74; consult ed, J Current Contents, 74-; attend staff & consult surgeon var hosps. Honors & Awards: First Dow-Corning Int Award Plastic Surg, 71. Mem: Am Soc Plastic & Reconstruct Surgeons; fel AMA; Am Asn Surg of Trauma; Am Asn Plastic Surgeons (pres, 62-63); fel Am Col Surgeons. Res: Burns; congenital deformities; radiation injuries, methods of repairing face injuries; skin grafting. Mailing Add: Stanford Univ Med Sch Stanford CA 94305

MCDOWELL, FRED WALLACE, b Abington, Pa, Sept 30, 39; m 64; c 3. GEOCHEMISTRY. Educ: Lafayette Col, AB, 61; Columbia Univ, PhD(geochem), 66. Prof Exp: Researcher geochronology, Swiss Fed Inst Technol, 66-69; RES SCIENTIST, UNIV TEX, AUSTIN, 69- Mem: Fel Geol Soc Am. Res: Application of isotopic age determination methods to geologic problems, including volcanism, orogenesis and metallization. Mailing Add: Dept of Geol Sci Univ of Tex Austin TX 78712

MCDOWELL, HARDING KEITH, b High Point, NC, Feb 5, 44; m 75. CHEMICAL PHYSICS. Educ: Wake Forest Univ, BS, 66; Harvard Univ, PhD(chem physics), 72. Prof Exp: Res assoc chem, State Univ NY Stony Brook, 72-74; ASST PROF CHEM, CLEMSON UNIV, 74- Mem: Am Phys Soc. Res: Study of many-body effects in chemical systems. Mailing Add: Dept of Chem & Geol Clemson Univ Clemson SC 29631

MCDOWELL, HERSHEL, b Gaffney, SC, Aug 9, 30; m 56; c 2. PHYSICAL CHEMISTRY. Educ: Morgan State Col, BS, 57; Howard Univ, PhD(chem), 67. Prof Exp: Res asst immunol, Johns Hopkins Univ, 57-58; res asst, Charlotte Mem Hosp, NC, 58-62; res assoc chem, Am Dent Asn, Nat Bur Standards, DC, 65-69; PROF CHEM & CHMN DEPT, FED CITY COL, DC, 69- Mem: Am Chem Soc. Res: Solubility and thermodynamic properties of the calcium orthophosphates; stability of complexes between calcium and phosphate ions and organic ligands. Mailing Add: Dept of Chem Fed City Col 1420 NY Ave NW Washington DC 20005

MCDOWELL, HORACE GREELEY, b Canton, Ohio, June 25, 22; m 46; c 3. GEOGRAPHY. Educ: Miami Univ, AB, 49; Univ Nebr, MA, 50; Univ Tenn, EdD, 71. Prof Exp: Instr geog, Univ NH, 50-56; asst prof, Univ Tenn, Martin, 56-60; asst prof, 60-73, ASSOC PROF GEOG & HEAD GEOSCI DEPT, UNIV TENN, CHATTANOOGA, 73- Mem: Asn Am Geog; Nat Coun Geog Educ. Res: Agriculture in Mexico; geographic education in Tennessee. Mailing Add: Dept of Geog Univ of Tenn Chattanooga TN 37401

MCDOWELL, JOHN PARMELEE, b Ridgewood, NJ, Feb 13, 31; m 55; c 2. SEDIMENTARY PETROLOGY. Educ: Yale Univ, BS, 53; Dartmouth Col, MA, 55; Johns Hopkins Univ, PhD, 63. Prof Exp: Geologist, US Geol Surv, 54-58; instr, 58-63, asst dean col arts & sci, 64-67, ASST PROF GEOL, TULANE UNIV LA, 63-, ASSOC DEAN COL ARTS & SCI, 67- Concurrent Pos: Spec geologist, Ont Dept Mines, Can, 56-57; fel acad admin, Am Coun Educ. Mem: Am Asn Petrol Geol; fel Geol Soc Am; Soc Econ Paleont & Mineral; Mineral Asn Can. Res: Sedimentary petrology. Mailing Add: Off of Dean Col of Arts & Sci Tulane Univ New Orleans LA 70118

MCDOWELL, JOHN ROBERT, b Chatham, Ont, June 17, 32; m 55; c 1. PHYSICAL CHEMISTRY. Educ: Univ Western Ont, BS, 55; Univ Alta, PhD(phys chem), 59. Prof Exp: Nat Res Coun Can fel, 59-60; staff mem, Union Carbide Res Inst, NY, 60-63; res chemist, Res Ctr, Philip Morris, Inc, 63-67; sr chemist, 67-68, mgr chem sci, 68-71, SR RES ASSOC, CENT RES, LORD CORP, 71- Mem: Am Chem Soc. Res: Free radical kinetics; atomic spectroscopy; high temperature kinetics; gas-solid reactions; solid state physics; dielectrics; surface chemistry. Mailing Add: Lord Corp 2000 W Grandview Blvd Erie PA 16512

MCDOWELL, JOHN WILLIS, b Honolulu, Hawaii, Dec 12, 21; m 50; c 2. PARASITOLOGY. Educ: Colo State Univ, BS, 47, MS, 48; Univ NC, MPH, 50; Okla State Univ, PhD(zool), 53. Prof Exp: Adv parasitol, USPHS, Int Coop Admin, Cambodia, Laos & Vietnam, 51-53, co-dir malaria control, Govt Iran, 54-56, tech adv malaria eradication, US Opers Mission, Philippines, 56-60, regional adv, Western Pac Region, AID, 60-61, spec asst to chief training br, Commun Dis Ctr, 61-62, asst chief vector borne dis sect, 62-64, chief eval unit, Aedes Aegypti Eradication Br, 64-66, asst chief eval sect, Malaria Eradication Br, 66-69; assoc prof, 69-72, PROF BIOL, BERRY COL, 72- Concurrent Pos: Mem independent malaria assessment team, US-WHO, Thailand & Iran, 63 & Vietnam, 64; consult, 11th Exper Comt Malaria, WHO, 64; US del, Int Cong Trop Med & Malaria, Lisbon, 58, Asian Malaria Conf, New Delhi, 58, Anti-Malaria Coord Bd, Southeast Asia, 59 & Borneo Conf Malaria Semarang, 60; Lilly Found fel environ educ, Ohio State Univ, 74-75. Mem: Am Soc Parasitol; Am Soc Trop Med & Hyg; Royal Soc Trop Med & Hyg; Am Mosquito Control Asn. Res: Field investigation and control of parasitic diseases, especially those which are arthropod borne; eradication and control of vector-borne disease or disease vectors, particularly malaria, filariasis and yellow fever; technical, organizational and administrative evaluation of such programs. Mailing Add: 5 Beaver Run Rome GA 30161

MCDOWELL, LARRY LEON, b South Norfolk, Va, Aug 5, 33; m 57; c 2. PLANT PATHOLOGY. Educ: Utah State Univ, BS, 58; Ore State Univ, PhD(plant path), 64. Prof Exp: ASST PROF BOT & PATH, STATE UNIV NY COL ENVIRON SCI & FORESTRY, 64- Mem: AAAS; Am Phytopath Soc. Res: Physiology and biochemistry of sporulation in fungi; physiology of parasitism. Mailing Add: Dept of Forest Bot SUNY Col of Environ Sci & Forest Syracuse NY 13210

MCDOWELL, LEE RUSSELL, b Sodus, NY, Apr 11, 41; m 71; c 3. ANIMAL

NUTRITION. Educ: Univ Ga, BS, 64, MS, 65; Wash State Univ, PhD(animal nutrit), 71. Prof Exp: Res asst animal nutrit, Univ Ga, 64-65; agr vol, Peace Corps, Bolivia, 65-67; res asst animal nutrit & teaching asst animal prod, Wash State Univ, 68-71; ASST PROF ANIMAL NUTRIT, UNIV FLA, 71- Concurrent Pos: Mineral res consult, Brazilian Enterprise Agr Res, 74. Mem: Am Soc Animal Sci; Am Dairy Sci Asn; Latin Am Asn Animal Prod. Res: Problems dealing with international animal nutrition and production; chemical composition and feeding value of Latin American feeds; determining the location of mineral deficiencies or toxicities that are inhibiting livestock production in Latin America. Mailing Add: Dept of Animal Sci Livestock Pavilion Univ of Fla Gainesville FL 32611

MCDOWELL, LESLIE L, soil science, geochemistry, see 12th edition

MCDOWELL, MARGARET ANN, b Coshocton, Ohio, Oct 22, 12. BACTERIOLOGY. Educ: St Mary of the Springs Col, BA, 35; Ohio State Univ, MA, 43; Inst Divi Thomae, MS, 45, PhD(bact), 54. Prof Exp: Instr biol, Ohio Dom Col, 45-50, prof & chmn dept, 50-69; asst to pres, Albertus Magnus Col, 69-74; STAFF MEM CANCER DEVELOP CHEMOTHER, ST MARY OF THE SPRINGS COL, 74- Concurrent Pos: Runyon grant, 52 & 58; AEC grant, 60; Ashland County Health grants, 62-66; exec vpres & acad dean, Ohio Dom Col, 62-69; consult, Mercenene Cancer Res Prog, Ohio Dom Col; mem bd trustees, Albertus Magnus Col, 70- Mem: Am Soc Microbiol; Am Soc Med Technol; NY Acad Sci. Res: Effect of growth promoting substances on certain green algae; preparation and effect of growth promoting substances on certain green algaoe; preparation and effect of bacterial products on bacteria and cancer. Mailing Add: Staff Cancer Develop Chemother St Mary of the Springs Col Columbus OH 43219

MCDOWELL, MARION EDWARD, b Torrington, Wyo, Nov 4, 21; m 44; c 2. MEDICINE, NUTRITION. Educ: Univ Wyo, BS, 42; Univ Rochester, MD, 45; Am Bd Internal Med, dipl, 53. Prof Exp: Instr med, Sch Med, Univ Rochester, 45-46 & 48-50, from intern to assoc resident, Sch Med, Univ Rochester & Strong Mem Hosp, 45-50; assoc resident, Walter Reed Army Med Ctr, 50-54, dir med div, Walter Reed Army Inst Res, 54-56, from asst chief to chief dept med, Tokyo Army Hosp, 56-58, chief outpatient dept, US Army Med Hosp, Camp Zama, Japan, 58-59, dep comdr & chief metab div, US Army Med Res & Nutrit Lab, 59-60, cmndg officer, 60-64; DIR MED EDUC, ST JOSEPH HOSP, DENVER, 64-; ASST CLIN PROF MED, SCH MED, UNIV COLO, 64- Concurrent Pos: Fel renal dis, Peter Bent Brigham Hosp, 50; Army liaison rep, Food & Nutrit Bd, Nat Acad Sci-Nat Res Coun & Nutrit Study Sect, NIH, 61-64; mem subcomt feeding & nutrit space, Nat Res Coun, 62-64. Mem: Fel Am Col Physicians; Am Soc Internal Med; Am Inst Nutrit; Am Soc Clin Nutrit; Am Fedn Clin Res. Res: Internal medicine and clinical research; renal disease; fluid and electrolyte metabolism; hemodialysis in renal failure; clinical nutrition; military medicine; medical education. Mailing Add: 7865 E Mississippi Ave Denver CO 80231

MCDOWELL, MAURICE JAMES, b Dannevirke, NZ, Dec 6, 22; nat US; m 57; c 4. CHEMISTRY. Educ: Univ Otago, NZ, BSc, 43, MSc, 44; Brown Univ, PhD(chem), 50. Prof Exp: Lectr, Univ Otago, NZ, 44-47; chemist, E I du Pont de Nemours & Co, Inc, 50-52; res assoc, Univ Montreal, 52; chem assoc, 53-60, res supvr, 60-63, RES ASSOC, FABRICS & FINISHES DIV, E I DU PONT DE NEMOURS & CO, INC, 64- Mem: Am Chem Soc; Fedn Soc Coating Technol. Res: Polymer research; colloid chemistry; paint research. Mailing Add: Fabrics & Finishes Div E I du Pont de Nemours & Co Inc Philadelphia PA 19146

MCDOWELL, ROBERT CARTER, b Glens Falls, NY, Feb 15, 35; m 63; c 3. GEOLOGY. Educ: Va Polytech Inst, BS, 56, MS, 64, PhD(geol), 68. Prof Exp: Instr geol, Va Polytech Inst, 65-66; asst prof, Wis State Univ, River Falls, 66-67; GEOLOGIST, US GEOL SURV, 67- Mem: Geol Soc Am. Res: Structural geology and petrography of the Arvonia slate quarries, Virginia; structural geology and stratigraphy of lower Cambrian quartzites, Pulaski County, Virginia; field geology of northeastern Kentucky. Mailing Add: 3517 Berwin Court Lexington KY 40503

MCDOWELL, ROBERT DUANE, wildlife management, see 12th edition

MCDOWELL, ROBERT E, JR, b Charlotte, NC, June 27, 21; m 45; c 3. ANIMAL SCIENCE, ANIMAL PHYSIOLOGY. Educ: NC State Col, BS, 42; Univ Md, MS, 49, PhD(dairy husb), 54. Prof Exp: Dist supvr, Vet Training Prog, NC, 46; dairy husbandman, Dairy Husb Res Br, Agr Res Serv, USDA, 46-58, sr res scientist, Dairy Cattle Res Br, Animal Husb Res Div, 58-67; PROF ANIMAL SCI, CORNELL UNIV, 67- Concurrent Pos: Instr, Johns Hopkins Univ, 53-57; consult, Food & Agr Orgn, UN, 56, US Marine Corps, 67-71, USDA, 67-, US Peace Corps, 69-, Govt Egypt, 71 & Govt Iran, 71; vis prof, Cornell Univ, 66; prof, Univ PR, 69- & Dept & Inst Agron, Cent Univ Venezuela. Mem: Am Soc Animal Sci; Am Dairy Sci Asn; Int Soc Biometeorol. Res: Determination of factors affecting cattle production in tropical areas. Mailing Add: Dept of Animal Sci Cornell Univ Ithaca NY 14850

MCDOWELL, ROBERT HULL, b Chicago, Ill, Dec 1, 27; m 56; c 3. MATHEMATICS. Educ: Univ Chicago, PhD, 47, BS & MS, 50; Purdue Univ, PhD, 60. Prof Exp: Asst mathematician, Argonne Nat Lab, 53-54; asst math, Purdue Univ, 54-55; instr & asst prof, Rutgers Univ, 59; asst prof, Purdue Univ, 59-60; from asst prof to assoc prof, 60-74, PROF MATH, WASH UNIV, 74- Concurrent Pos: Dir comt undergrad prog math, NSF, 64-66, mem panel teacher training, mem adv group commun. Mem: Am Math Soc; Math Asn Am. Res: Set theoretic topology and algebra and their interrelationships; mathematical structures; category theory. Mailing Add: Dept of Math Wash Univ St Louis MO 63130

MCDOWELL, ROBIN SCOTT, b Greenwich, Conn, Nov 14, 34; m 63; c 2. MOLECULAR SPECTROSCOPY. Educ: Haverford Col, BA, 56; Mass Inst Technol, PhD(phys chem), 60. Prof Exp: Asst chem, Mass Inst Technol, 56-60; STAFF MEM PHYS CHEM, LOS ALAMOS SCI LAB, 60- Mem: AAAS; Optical Soc Am; Coblentz Soc; Soc Appl Spectros. Res: Vibrational spectra of inorganic and isotopically-substituted molecules, especially fluorides; infrared laser spectroscopy; molecular dynamics and force fields; calculated thermodynamic properties of ideal gases; infrared analytical methods; atmospheric optics. Mailing Add: 885 Camino Encantado Los Alamos NM 87544

MCDOWELL, SAM BOOKER, b New York, NY, Sept 13, 28; m 52; c 2. HERPETOLOGY. Educ: Columbia Univ, AB, 47, PhD(zool), 57. Prof Exp: From asst instr to assoc prof biol, 56-68, PROF ZOOL, RUTGERS UNIV, NEWARK, 68- Mem: Am Soc Ichthyol & Herpet; Soc Study Evolution; Zool Soc London; Soc Study Amphibians & Reptiles; Linnean Soc London. Res: Vertebrate paleontology; teleost and reptilian anatomy; origin of higher taxonomic categories; New Guinea snakes; primitive teleosts. Mailing Add: Dept of Zool Rutgers Univ Newark NJ 07102

MCDOWELL, SAMUEL CHARLES THOMAS, physical chemistry, see 12th edition

MCDOWELL, TERENCE LEE, organic chemistry, photochemistry, see 12th edition

MCDOWELL, THEODORE C, b Ashland, Ohio, Mar 21, 41; m 68; c 2. FLORICULTURE, HORTICULTURE. Educ: Ohio State Univ, BS, 63, PhD, 67; NC State Univ, MS, 65. Prof Exp: ASST PROF FLORICULT, OHIO AGR RES & DEVELOP CTR, 68-Mem: AAAS; Am Soc Hort Sci; Am Inst Biol Sci; Scand Soc Plant Physiol. Res: Growth retardants and their interaction with nutrition; foliar analysis; anatomical and histochemical studies of flower initiation and flower bud development. Mailing Add: Dept of Hort Ohio Res & Develop Ctr Wooster OH 44691

MCDOWELL, WILBUR BENEDICT, b Omaha, Nebr, Feb 27, 20; m 47; c 3. ORGANIC CHEMISTRY, PHARMACEUTICAL CHEMISTRY. Educ: Ohio State Univ, BSc, 41, MSc, 42, PhD(chem), 44. Prof Exp: Asst chem, Ohio State Univ, 42-43; res assoc, Squibb Inst Med Res, 44-52; head synthetic org develop sect, Olin Mathieson Chem Corp, 53-59, sr res chemist, 59-66; assoc mgr, New York, 66-69, SCI MGR PROF SERV, E R SQUIBB & SONS, INC, PRINCETON, 69- Mem: AAAS; Am Chem Soc; fel Am Inst Chem; Soc Nuclear Med; NY Acad Sci. Res: Synthesis and isolation of antibiotics; synthetic organic process development. Mailing Add: 4 Fairview Ave East Brunswick NJ 08816

MCDUFF, JAMES MILTON, b Ft Worth, Tex, Sept 26, 24; m 51; c 3. INDUSTRIAL ORGANIC CHEMISTRY. Educ: Tex Christian Univ, BA, 49. Prof Exp: Res chemist, Tex Div, 49-53, lab supvr polymer res, 53-55, lab group leader, 55-66, sect mgr, 66-68, asst dir lab, 68-69, RES MGR, TEX DIV, DOW CHEM USA, 69- Mem: Am Chem Soc. Res: Research and development in industrial organic chemistry. Mailing Add: Bldg B-1214 Dow Chem USA Freeport TX 77541

MCDUFFIE, BRUCE, b Atlanta, Ga, Aug 25, 21; m 50; c 3. CHEMISTRY. Educ: Princeton Univ, AB, 42, MA, 46, PhD(chem), 47. Prof Exp: Asst chem, Cornell Univ, 42; asst, Princeton Univ, 43-47; instr, Emory Univ, 47-51; assoc prof, Washington & Jefferson Col, 51-58; assoc prof, 58-61, PROF CHEM, HARPUR COL, STATE UNIV NY BINGHAMTON, 61- Mem: AAAS; Am Chem Soc. Res: Electroanalytical chemistry; trace methods and environmental analysis of mercury, lead and other toxic metals; river pollution studies. Mailing Add: Dept of Chem Harpur Col State Univ NY Binghamton NY 13901

MCDUFFIE, FREDERIC CLEMENT, b Lawrence, Mass, Apr 27, 24; m 52; c 4. MEDICINE. Educ: Harvard Med Sch, MD, 51. Prof Exp: From intern to jr asst resident, Peter Bent Brigham Hosp, Boston, 51-53; fel phys chem, Harvard Univ, 53-54; fel microbiol, Col Physicians & Surgeons, Columbia Univ, 54-56; sr asst resident, Peter Bent Brigham Hosp, 56-57; from asst prof to assoc prof med & microbiol, Sch Med, Univ Miss, 57-65; asst prof, 65-69, assoc prof med & microbiol, Mayo Grad Sch Med, 69-74, PROF MED, IMMUNOL & MICROBIOL, MAYO MED SCH, UNIV MINN, 74-, CONSULT, MAYO CLIN, 65- Mem: Am Asn Immunologists; Am Rheumatism Asn; Am Fedn Clin Res: Am Col Physicians; Soc Exp Biol & Med. Res: Complement system in connective tissue diseases; antibody production in chickens. Mailing Add: Rheumatology Res Lab Mayo Clin Rochester MN 55901

MCDUFFIE, HAROLD FRITZ, JR, chemistry, see 12th edition

MCDUGLE, WOODROW GORDON, JR, b Baton Rouge, La, May 13, 42; m 67. INORGANIC CHEMISTRY, PHYSICAL CHEMISTRY. Educ: Baylor Univ, BS, 64; Univ Ill, Urbana, MS, 66, PhD(chem), 68. Prof Exp: Res assoc phys chem, Mich State Univ, 68-70; SR RES CHEMIST, RES LABS, EASTMAN KODAK CO, 70- Res: Physical properties of inorganic materials as studied by various spectroscopic methods, including electron spin resonance. Mailing Add: Eastman Kodak Co Res Labs 343 State St Rochester NY 14650

MCEACHRAN, ROBERT PAUL, theoretical physics, see 12th edition

MCELANEY, JAMES H, b Boston, Mass, Aug 28, 23. PHYSICS. Educ: Boston Col, BA, 47, MA, 48; Weston Col, PhL, 48, STL, 55; Johns Hopkins Univ, PhD(physics), 64. Prof Exp: Instr physics & math, Boston Col, 48-50 & Col Holy Cross, 50-51; from asst prof to assoc prof, 63-75, PROF PHYSICS, FAIRFIELD UNIV, 75-, CHMN DEPT, 64- Mem: Optical Soc Am; Am Asn Physics Teachers; Am Phys Soc. Res: Free-ion spectroscopy; spectroscopy of the middle and far ultraviolet. Mailing Add: Dept of Physics Fairfield Univ Fairfield CT 06430

MCELCHERAN, DONALD ELMO, b Hamilton, Ont, Apr 14, 25; m 53; c 4. PHYSICAL CHEMISTRY. Educ: McMaster Univ, BSc, 49, MSc, 50; Univ Leeds, PhD(chem), 54. Prof Exp: Fel, Nat Res Coun Can, 54-56; res assoc physics, Laval Univ, 56-57; asst prof, 57-67, ASSOC PROF PHYS CHEM, LOYOLA FAC ARTS & SCI, CONCORDIA UNIV, 67- Mem: The Chem Soc; Chem Inst Can. Res: Gas phase kinetics; mass spectrometry. Mailing Add: Dept of Chem Loyola Fac Arts & Sci Concordia Univ Montreal PQ Can

MCELFRESH, ARTHUR EDWARD, JR, b Louisville, Ky, Feb 27, 23; m 48; c 4. MEDICINE. Educ: Cornell Univ, MD, 47. Prof Exp: Asst prof pediat, Med Col Ala, 53; instr, Sch Med, Temple Univ, 54-55, assoc, 55-56, from asst prof to assoc prof, 56-65; PROF PEDIAT & CHMN DEPT, SCH MED, ST LOUIS UNIV, 65- Mem: AAAS; Soc Pediat Res; Am Soc Hemat; Am Acad Pediat; Am Pediat Soc. Res: Pediatric hematology. Mailing Add: 1465 S Grand Blvd St Louis MO 63104

MCELHANEY, RONALD NELSON, b Youngstown, Ohio, Jan 5, 42; m 65; c 2. BIOCHEMISTRY, BIOPHYSICS. Educ: Wash & Jefferson Col, BA, 64; Univ Conn, PhD(biochem), 69. Prof Exp: NIH fel biochem, State Univ Utrecht, 69-70; asst prof, 70-74, ASSOC PROF BIOCHEM, UNIV ALTA, 74- Concurrent Pos: Med Res Coun Can res grant, Univ Alta, 70- Mem: Am Inst Biol Sci; Can Biochem Soc; Am Soc Microbiol; Am Soc Biol Chemists; Sigma Xi. Res: Biological membrane structure and function; role of membrane polar lipids and cholesterol in passive and mediated membrane transport and in membrane enzyme activity; biosynthesis of fatty acids and polar lipids. Mailing Add: Dept of Biochem Univ of Alta Edmonton AB Can

MCELHENY, GEORGE CLARK, b Tiffin, Ohio, Apr 21, 20; m 45; c 3. ORGANIC CHEMISTRY. Educ: Harvard Univ, BS, 42; Wash Univ, PhD(org chem), 50. Prof Exp: Res chemist, Mallinckrodt Chem Works, 42-46 & 50-56, group leader org synthesis, 56-67, com develop specialist, 67-70, RES ASSOC, MALLINCKRODT, INC, 71- Mem: Am Chem Soc. Res: Organic synthesis; medicinal chemicals; alkaloids of opium. Mailing Add: Mallinckrodt Inc 3600 N Second St St Louis MO 63147

MCELHINNEY, JOHN, b Philadelphia, Pa, Mar 25, 21; m 42; c 2. PHYSICS. Educ: Ursinus Col, BS, 42; Univ Ill, MS, 43, PhD(physics), 47. Prof Exp: Asst physics, Univ Ill, 42-44 & Manhattan Dist, 44-45, spec res assoc, 47-48; assoc scientist, Los Alamos Sci Lab, 48-49; physicist, Nat Bur Standards, 49-55; head, Nuclear Interactions Br, 55-62, assoc supt, Nucleonics Div, 57-66, head, Linac Br, 62-66, supt, Nuclear Sci Div, 66-74, SUPT, RADIATION TECH DIV, NAVAL RES LAB, 74- Concurrent Pos: Partic sabbatical study & res prog, Stanford Univ-Lawrence Livermore Lab, 69-70. Mem: AAAS; Am Phys Soc; Sigma Xi; Am Nuclear Soc; Inst Elec & Electronics Engrs. Res: Nuclear physics with betatron, synchrotron, electron linac, Van de Graaff, cyclotron; nuclear and x-ray instrumentation; application of nuclear technology; radiation dosimetry. Mailing Add: 11601 Stephen Rd Silver Spring MD 20904

MCELHINNEY, MARGARET M (COCKLIN), b Grandview, Iowa; m 45; c 2. SCIENCE EDUCATION. Educ: Iowa Wesleyan Col, BS, 45; Univ Northern Colo, MS, 60; Ball State Univ, EdD(sci educ), 66. Prof Exp: Teacher pub schs, Iowa, seven years; assoc prof, 61-74, PROF BIOL, BALL STATE UNIV, 74- Concurrent Pos: Nat Sci Teachers Asn; Nat Asn Biol Teachers. Res: Museum technology; developing instructional materials and preparing exhibits for direct experience learning. Mailing Add: Dept of Biol Ball State Univ Muncie IN 47306

MCELHOE, FORREST LESTER, JR, b Carnation, Wash, Feb 15, 23; m 53; c 2. PHYSICAL GEOGRAPHY, ECONOMIC GEOGRAPHY. Educ: Univ Wash, BA, 48, MA, 50; Ohio State Univ, PhD(geog), 55. Prof Exp: Instr geog, Ohio State Univ, 55-56; asst prof, Univ Ky, 56-68; asst prof, 68-72, ASSOC PROF GEOG, CALIF STATE POLYTECH UNIV, POMONA, 72- Mem: Asn Am Geog; Am Geog Soc; Nat Coun Geog Educ. Res: Climate. Mailing Add: Dept of Soc Sci Calif State Polytech Univ Pomona CA 91766

MCELIN, THOMAS (WELSH), b Janesville, Wis, Aug 27, 20; m 45; c 2. MEDICINE. Educ: Dartmouth Col, AB, 42; Harvard Med Sch, MD, 44; Univ Minn, MS, 48; Am Bd Obstet & Gynec, dipl, 53. Prof Exp: Intern, Passavant Mem Hosp, 44-45; resident, Mayo Found, Univ Minn, 45-50, assoc, Mayo Clin, 50; clin asst, 50-53, instr, 53-54, assoc, 54-56, from asst prof to assoc prof, 56-67, PROF OBSTET & GYNEC, NORTHWESTERN UNIV MED SCH, CHICAGO, 67-, ASST CHMN DEPT, 74- Concurrent Pos: Mem attend staff, St Mary's Hosp, Rochester, Minn, 50; mem courtesy staff & jr attend physician, Evanston Hosp, 50-52, mem attend staff, 52-, chmn dept obstet & gynec, 65- Mem: AMA; Am Asn Obstet & Gynec; fel Am Col Obstetricians & Gynecologists; fel Am Col Surgeons. Res: Obstetrics and gynecology. Mailing Add: Evanston Hosp 2650 Ridge Ave Evanston IL 60201

MCELLIGOTT, JAMES GEORGE, b New York, NY, June 20, 38; m 67; c 2. NEUROSCIENCE, NEUROPHARMACOLOGY. Educ: Fordham Univ, BS, 60; Columbia Univ, MA, 63; McGill Univ, PhD(psychol), 66. Prof Exp: NIH fel neuroanat, Univ Calif, Los Angeles, 67-68, asst res anatomist, Sch Med, 68-71; ASST PROF PHARMACOL, SCH MED, TEMPLE UNIV, 71- Mem: AAAS; Soc Neurosci. Res: Neurophysiology; bioengineering; computer analysis of neuroscientific data. Mailing Add: Dept of Pharmacol Temple Univ Sch of Med Philadelphia PA 19140

MCELLIGOTT, PETER EDWARD, b New York, NY, Apr 4, 35; m 62; c 2. SURFACE CHEMISTRY, TRIBOLOGY. Educ: NC State Col, BS, 57; Rensselaer Polytech Inst, MS, 61, PhD(metall), 66. Prof Exp: Physicist, Alco Prod, Inc, NY, 57-59; physicist, Knolls Atomic Power Lab, 59-62, PHYS CHEMIST, RES & DEVELOP CTR, GEN ELEC CO, 62- Mem: Fel Am Inst Chem; Am Soc Lubrication Eng. Res: Friction, lubrication and wear of materials; electrical contact phenomena; surface chemistry of metals and semiconductors. Mailing Add: Res & Develop Ctr Gen Elec Co PO Box 8 Schenectady NY 12301

MCELLISTREM, MARCUS THOMAS, b St Paul, Minn, Apr 19, 26; m 57; c 4. NUCLEAR PHYSICS. Educ: Col St Thomas, BA, 50; Univ Wis, MS, 52, PhD, 56. Prof Exp: Asst physics, Univ Wis, 50-55; res assoc, Ind Univ, 55-57; from asst prof to assoc prof, 57-65, PROF PHYSICS, UNIV KY, 65- Concurrent Pos: Pres, Adena Corp, 71- Mem: AAAS; fel Am Phys Soc. Res: Analysis of nuclear reactions; measurements of differential cross sections of nuclear reactions; fast neutron physics, neutron induced reaction cross sections. Mailing Add: Dept of Physics & Astron Univ of Ky Lexington KY 40506

MCELREE, HELEN, b Waxahachie, Tex, Nov 22, 25. IMMUNOBIOLOGY. Educ: Col Ozarks, BS, 47; Univ Okla, MS, 54; Univ Kans, PhD, 59. Prof Exp: Asst prof bact, Univ Kans, 59-62; assoc prof, 62-68, PROF BIOL, EMPORIA KANS STATE COL, 68- Mem: AAAS; Am Soc Microbiol. Res: Role of reticulo-endothelial system in viral infections. Mailing Add: Dept of Biol Emporia Kans State Col Emporia KS 66801

MCELROY, ALBERT D, inorganic chemistry, see 12th edition

MCELROY, DONALD L, b Chicago, Ill, Oct 26, 08; m 34; c 2. DENTISTRY. Educ: Univ Ill, DDS, 32, MS, 45. Prof Exp: Instr oper dent, 33-35, instr appl mat med, 40-55, from asst prof to prof oral diag, 55-67, actg head dept, 65-67, assoc dean oral diag, 67-74, EMER PROF ENDODONT, COL DENT, UNIV ILL MED CTR, 74- Mem: Fel AAAS; Sigma Xi; Int Asn Dent Res. Res: Dental root canal filling materials; clinical endodontics; oral pathology and diagnosis. Mailing Add: 108 Carriage Way Dr Hinsdale IL 60521

MCELROY, FRED DEE, b Seattle, Wash, Oct 3, 34; m 55; c 3. PLANT PATHOLOGY, NEMATOLOGY. Educ: Wash State Univ, BS, 60; Univ Calif, Riverside, PhD(plant path), 67. Prof Exp: Aide nematol, Wash State Univ, 60-63; RES SCIENTIST, CAN DEPT AGR, 67- Concurrent Pos: Hon lectr, Univ BC, 72- Mem: Am Phytopath Soc; Am Sci Affiliation; Soc Nematol; Can Phytopath Soc; Soc Europ Nematol. Res: Host-parasite relations of phytoparasitic nematodes; nematode problems of small fruits; nematodes as vectors of plant viruses. Mailing Add: 6660 N W Marine Dr Vancouver BC Can

MCELROY, MARY KIERAN, b Philadelphia, Pa, Feb 24, 18. INORGANIC CHEMISTRY, ANALYTICAL CHEMISTRY. Educ: Chestnut Hill Col, AB, 56; Univ Pa, PhD(chem), 64. Prof Exp: Teacher parochial schs, Pa, 37-54, dept head math & chem, 54-62; lectr, 62-64, from instr to asst prof, 64-70, ASSOC PROF CHEM, CHESTNUT HILL COL, 70-, CHAIRWOMAN DEPT, 75- Mem: Am Chem Soc; Sigma Xi; AAAS. Res: Electrochemical and spectral studies of the iron-polyphosphate systems. Mailing Add: Dept of Chem Chestnut Hill Col Philadelphia PA 19118

MCELROY, MICHAEL BRENDAN, b Shercock Co Cavan, Ireland, May 18, 39; m 63. APPLIED MATHEMATICS, PHYSICS. Educ: Queen's Univ, Belfast, BA, 60, PhD(math), 62. Prof Exp: Proj assoc, Theoret Chem Inst, Univ Wis, 62-63; from asst physicist to physicist, Kitt Peak Nat Observ, 63-71; PHYSICIST, HARVARD CTR FOR EARTH & PLANETARY PHYSICS, 71- Concurrent Pos: Mem Mars panel, Lunar & Planetary Missions Bd, NASA, 68-69. Honors & Awards: James B Macelwane Award, Am Geophys Union, 68. Mem: AAAS; Am Astron Soc; Am Geophys Union. Res: Physics and chemistry of planetary atmospheres. Mailing Add: Ctr for Earth & Planetary Physics Harvard Univ Cambridge MA 02138

MCELROY, PAUL TUCKER, b Boston, Mass, Sept 25, 31; m 63; c 4. UNDERWATER ACOUSTICS. Educ: Harvard Univ, AB, 53, MA, 60, PhD(solid state physics), 68. Prof Exp: Asst scientist underwater acoustics, Woods Hole Oceanog Inst, 68-75; SR SCIENTIST, BOLT BERANEK & NEWMAN, INC, 75- Mem: AAAS; Marine Technol Soc; Sigma Xi. Res: Scattering of sound from marine organisms; statistical analysis of relationship of acoustic scattering and biological targets; signal processing; acoustic arrays; medical ultrasonics; computer control of real-time processes. Mailing Add: Bolt Beranek & Newman Inc 50 Moulton St Cambridge MA 02138

MCELROY, ROBERT C, b Wheeling, WVa, Dec 12, 11; m 45; c 2. MEDICINE. Educ: Ohio State Univ, AB, 33; Jefferson Med Col, MD, 37; Am Bd Obstet & Gynec, dipl, 45. Prof Exp: Instr, 39-45, assoc, 45-55, asst prof, 55-60, ASSOC PROF OBSTET & GYNEC, SCH MED, UNIV PA, 60- Mem: AMA; Am Col Obstetricians & Gynecologists; Am Col Surgeons. Res: Gynecology and obstetrics. Mailing Add: Dept of Clin Obstet & Gynec Univ of Pa Sch of Med Philadelphia PA 19104

MCELROY, WILBUR RENFREW, b Fayetteville, Pa, Aug 20, 14; m 39; c 1. POLYMER CHEMISTRY, POLYMER ENGINEERING. Educ: Gettysburg Col, AB, 36; Pa State Univ, MS, 38; Purdue Univ, PhD(chem), 43. Prof Exp: Res chemist, Tex Co, NY, 39-40 & 43-45; proj leader, Jefferson Chem Co, 45-49; CHEM ENG CONSULT & OWNER, WAYNE LABS, PA, 49-; PRES, ACTION PROD, INC, 69- Concurrent Pos: Dir chem technol, Cent Res Labs, Westinghouse Air Brake Co, Va, 52-55; sr group leader, Mobay Chem Co, 55-64; exec vpres, Conap, Inc, 64-69. Mem: Am Chem Soc; fel Am Inst Chem; Am Soc Test & Mat; NY Acad Sci; Nat Soc Prof Eng. Res: Process and product development in chemical intermediates and urethanes; orthopedic pads and other health care products. Mailing Add: Admin Action Prods Inc 514 W Sullivan St Olean NY 14760

MCELROY, WILLIAM DAVID, b Rogers, Tex, Jan 22, 17; m 40, 67; c 5. BIOLOGY, BIOCHEMISTRY. Educ: Stanford Univ, BA, 39; Reed Col, MA, 41; Princeton Univ, PhD(biol), 43. Hon Degrees: DSc, Univ Buffalo, 62, Mich State Univ & Loyola Univ, Chicago, 70, Del State Col, 71 & Notre Dame Univ, 75; DPS, Providence Col, 70; LLD, Univ Pittsburgh, 71, Univ Calif, San Diego, 72 & Fla State Univ, 73. Prof Exp: Asst, Reed Col, 39-41; asst, Princeton Univ, 43-44, res assoc, 44-45; Nat Res Coun fel, Stanford Univ, 45; from instr to prof biol, Johns Hopkins Univ, 45-69, chmn dept, 56-69, dir, McCollum-Pratt Inst, 49-69; dir, NSF, 69-72; CHANCELLOR, UNIV CALIF, SAN DIEGO, 72- Concurrent Pos: Trustee, Brookhaven Nat Lab, 54; ex officio mem, Nat Sci Bd, 69-72. Honors & Awards: Barnett Cohen Award, Am Soc Microbiol, 58; Andrew White Medal, Loyola Col, Md, 71; Howard N Potts Medal, Franklin Inst, 71; Rumford Award, Am Acad Arts & Sci. Mem: Nat Acad Sci; AAAS (pres-elect, 75-); Am Chem Soc; Am Acad Arts & Sci; Am Soc Microbiol. Res: Bioluminescence; bacterial mutations; biochemical genetics; mechanism of inhibitor action; bacterial and mold metabolism. Mailing Add: Univ of Calif San Diego La Jolla CA 92093

MCELROY, WILLIAM NORDELL, b Minneapolis, Minn, Nov 28, 26; m 50; c 4. REACTOR PHYSICS. Educ: Univ Southern Calif, BA, 51; Ill Inst Technol, MA, 60, PhD(physics), 65. Prof Exp: Sr engr, Atomics Inst, 51-57, sr tech specialist, 65-67; mgr reactor oper, Ill Inst Technol, 57-65; res assoc, Battelle-Northwest, 67-70; MGR & FEL SCIENTIST IRRADIATION CHARACTERIZATION & ANAL, HANFORD ENGR DEVELOP LAB, WESTINGHOUSE HANFORD CO, 70- Concurrent Pos: Mem, Nat Acad Sci-Nat Acad Engr-Nat Res Coun Panel For Nat Bur Standards Ctr For Radiation Res, 73- Mem: Am Nuclear Soc; AAAS. Res: Reactor environmental characterization and fuels and materials data correlation and damage analysis for light water, fast reactor and comtrolled thermonuclear reactors. Mailing Add: Westinghouse Hanford Co PO Box 1970 Richland WA 99352

MCELROY, WILLIAM TYNDELL, JR, b Shreveport, La, Sept 29, 24. PHYSIOLOGY. Educ: La State Univ, BS, 49; Univ Minn, MS, 54; Stanford Univ, PhD, 56. Prof Exp: Asst prof physiol, SDak State Col, 56-58; res assoc, Hahnemann Med Col, 59, asst prof, 60-67; PROF PHYSIOL & BIOPHYS & ASST DEAN STUD AFFAIRS, SCH MED, LA STATE UNIV, SHREVEPORT, 67- Mem: Am Physiol Soc; Am Fedn Clin Res. Res: Cardiovascular lipid metabolism; immunology. Mailing Add: Dept of Physiol & Biophys La State Univ Sch of Med Shreveport LA 71101

MCELVAIN, SAMUEL MARION, organic chemistry, see 12th edition

MCELWEE, EDGAR WARREN, b Centreville, Miss, Nov 11, 07; m 34; c 2. ORNAMENTAL HORTICULTURE. Educ: Miss State Col, BS, 30; Ala Polytech Inst, MS, 32; Ohio State Univ, PhD, 51. Prof Exp: Asst, Ala Polytech Inst, 30-32, lab technician, 32-33, from asst prof to assoc prof hort, 35-47; asst, Ohio State Univ, 33-35; prof floricult, Miss State Col & Agr Exp Sta, 48-52; assoc & prof res, NC Agr Exp Sta, 52-53; prof ornamental hort, 56-73, ornamental horticulturist, Coop Exten Serv, 53-73, chmn dept ornamental hort, 56-67, EMER ORNAMENTAL HORTICULTURIST, UNIV FLA, 73- Mem: Am Soc Hort Sci. Res: Photoperiodism in flowering plants; nutrition and physiology of greenhouse plants. Mailing Add: 1403 NW 14th Ave Gainesville FL 32605

MCELWEE, ROBERT L, b Elkins, WVa, Oct 4, 27; m 51; c 3. FOREST MANAGEMENT, GENETICS. Educ: WVa Univ, BSF, 51; NC State Col, MS, 60, NC State Univ, PhD, 70. Prof Exp: Mgt forester, Gaylord Container Corp, 51-56; liaison geneticist, NC State Univ, 56-63, dir hardwood res prog, 63-70; PROJ LEADER, EXTEN FORESTRY, WILDLIFE & OUTDOOR RECREATION, VA POLYTECH INST & STATE UNIV, 70- Concurrent Pos: Assoc prof, Univ Maine, 70-71. Mem: Soc Am Foresters. Mailing Add: Ext Forest Wildlife & Outdoor Rec Va Polytech Inst & State Univ Blacksburg VA 24060

MCENALLY, TERENCE ERNEST, JR, b Richmond, Va, Apr 21, 27; m 64; c 2. PHYSICS, ELECTRONIC ENGINEERING. Educ: Va Polytech Inst, BS, 50, MS, 55; Mass Inst Technol, PhD(physics), 66. Prof Exp: Instr physics, Va Polytech Inst, 52-55; instr elec eng, NC State Univ, 61-63; instr physics, Mass Inst Technol, 66-67; ASST PROF PHYSICS, E CAROLINA UNIV, 67- Mem: Am Phys Soc. Res: Electron spin resonance of transition metal ions in crystals; exchange coupling of paramagnetic ions in crystals; theory of nuclear magnetic resonance of fluid molecules. Mailing Add: Dept of Physics E Carolina Univ Greenville NC 27834

MCENTEE, KENNETH, b Oakfield, NY, Mar 30, 21; m 52; c 2. VETERINARY PATHOLOGY. Educ: State Univ NY, DVM, 44. Hon Degrees: Dr, Royal Vet Col, Sweden, 75. Prof Exp: Asst path, 47-48, from asst prof to assoc prof vet path, 48-55, assoc dean clin studies, 69-73, PROF VET PATH, NY STATE VET COL, CORNELL UNIV, 55-, CHMN DEPT LARGE ANIMAL MED, OBSTET & SURG, 65- Concurrent Pos: Commonwealth Sci & Indust Res Orgn sr res fel, Univ Melbourne, 65-66; vis lectr & researcher, Royal Vet Col, Sweden, 73; vis prof, Vet Col, Belo Horizonte, Brazil, 73. Honors & Awards: Borden Award, Am Vet Med Asn, 71. Mem: Am Vet Med Asn; Am Col Vet Path; Int Fertil Asn; Int Acad Path. Res: Pathology of reproduction in dairy cattle; vibriosis of cattle. Mailing Add: Vet Col Cornell Univ Ithaca NY 14850

MCENTEE, THOMAS EDWIN, b San Mateo, Calif, Feb 27, 43; m 63, 74; c 2. SYNTHETIC ORGANIC CHEMISTRY. Educ: Univ Vt, BA, 66; Univ Colo, PhD(chem), 72. Prof Exp: RES CHEMIST, ARAPAHOE CHEM, INC, DIV, SYNTEX CORP, 73- Mem: Am Chem Soc. Res: Synthetic organic chemistry;

alkaloid synthesis; organosulfur chemistry; boride catalysts; industrial methods for organic synthesis. Mailing Add: Arapahoe Chem Inc PO Box 511 Boulder CO 80302

MCENTIRE, RICHARD WILLIAN, b Miami, Fla, Sept 10, 42; m 66; c 1. SPACE PHYSICS. Educ: Mass Inst Technol, BS, 64; Univ Minn, PhD(physics), 72. Prof Exp: Res asst physics, Univ Minn, 65-72; SR STAFF PHYSICIST, APPL PHYSICS LAB, JOHNS HOPKINS UNIV, 72- Mem: Am Phys Soc; Am Geophys Union; AAAS. Res: Planetary magnetospheric physics; interplanetary phenomena; solar and galactic cosmic rays. Mailing Add: Appl Physics Lab Johns Hopkins Univ Laurel MD 20810

MCEUEN, ROBERT BLAIR, b Riverside, Calif, May 22, 29; m 55. GEOPHYSICS. Educ: Univ Calif, MA, 54; Univ Utah, PhD(geophys), 59. Prof Exp: Geophysicist, Stanolind Oil & Gas Co, 54-56; sr res scientist, Pure Oil Co, Ill, 58-66, res assoc, Union Oil Co Calif, 66-67; staff geoscientist, Rucker Co, 67-69; assoc prof geol, 69-72, PROF GEOL, SAN DIEGO STATE UNIV, 72- Mem: Soc Explor Geophys. Res: Geophysical exploration; physical properties of rock; electro-viscous fluids; geothermics. Mailing Add: Dept of Geol San Diego State Univ San Diego CA 92115

MCEVILLY, THOMAS V, b East St Louis, Ill, Sept 2, 34; m 55; c 6. GEOPHYSICS, SEISMOLOGY. Educ: St Louis Univ, BS, 56, PhD(geophys), 64. Prof Exp: Geophysicist, Calif Co, 57-60; res asst, St Louis Univ, 60-64; asst prof seismol, 64-68, assoc prof geol & geophys, 68-74, PROF GEOL & GEOPHYS, UNIV CALIF, BERKELEY, 74-, ASST DIR SEISMOL STA, 68- Concurrent Pos: Vpres eng, W F Sprengnether Instrument Co, 63-69; mem, Earthquake Eng Res Inst, 71- Mem: AAAS; Seismol Soc Am; Am Geophys Union; Soc Explor Geophys; Royal Astron Soc. Res: Structure of the earth as revealed by seismic surface and body wave propagation; nature of earthquake sequences; seismic instrumentation. Mailing Add: Dept of Geol & Geophys Univ of Calif Berkeley CA 94720

MCEVOY, DONALD, b Birkenhead, Eng, Apr 5, 30; US citizen; m 63; c 2. BIOCHEMISTRY, ENDOCRINOLOGY. Educ: Univ Liverpool, BSc, 56, PhD(biochem), 59; Nat Registry Clin Chem, cert, 68. Prof Exp: Demonstr biochem, Univ Liverpool, 56-59; NIH res fel lipid chem, Univ Tenn, Memphis, 60-61; instr biochem, Univ Pa, 61-66, assoc path, 66-70, asst prof path, 70-74; MEM STAFF, ENDOCRINE CTR, CLEVELAND, OHIO, 74- Concurrent Pos: Med res scientist I, Eastern Pa Psychiat Inst, 61-63, med res scientist II, 63-66; lectr biochem, Philadelphia Col Pharm & Sci, 64-70, lectr endocrinol, 69-74; dir endocrine div, Pepper Lab, Hosp Univ Pa, 66-74. Mem: Am Asn Clin Chemists. Res: Biochemical assessment of fetoplacental function; timing of ovulation; effect of psychiatric states on adrenocortical function. Mailing Add: Endocrine Ctr 200 Univ Res Ctr 11001 Cedar Ave Cleveland OH 44106

MCEVOY, FRANCIS JOSEPH, b New York, NY, Jan 18, 23; m 57; c 4. CHEMISTRY. Educ: Seton Hall Univ, BS, 48; Fordham Univ, MS, 50. Prof Exp: SR CHEMIST RES, LEDERLE LABS, DIV AM CYANAMID CO, 50- Mem: Am Chem Soc. Res: Medicinal chemical research. Mailing Add: Lederle Labs Pearl River NY 10965

MCEVOY, JAMES EDWARD, b London, Eng, Aug 5, 20; US citizen; m 41; c 2. PETROLEUM CHEMISTRY. Educ: Temple Univ, BA, 55. Prof Exp: Res chemist, Houdry Process & Chem Co, 55-64, proj dir catalyst res, 64-66, sect head explor process res, 66-69, asst dir chem res, Houdry Labs, 69-71, asst dir res & develop, Houdry Div, 71-72, DIR CONTRACT RES & DEVELOP, AIR PROD & CHEM, INC, 72- Concurrent Pos: Mem, Inst Cong Catalysis. Mem: Am Chem Soc. Res: Applied catalysis; catalytic cracking; hydrogenation; desulfurization; fuel cell catalysts; catalytic oxidation of environmental pollutants; auto emissions control catalysts; synthetic fuels from coal. Mailing Add: Air Prods & Chem Inc PO Box 538 Allentown PA 18105

MCEWAN, IAN HUGH, b Cardiff, Wales, Aug 13, 33, Can citizen; m 57; c 3. POLYMER CHEMISTRY. Educ: Univ Wales, BSc & ARIC, 54, PhD(phys org chem), 58. Prof Exp: Tech off oil cracking, Heavy Org Chem Div, Imperial Chem Industs, 57-58; chemist polymers & paints, 58-66, GROUP LEADER PAINTS DIV, CAN INDUSTS, LTD, 66- Mem: Chem Inst Can; The Chem Soc; Royal Inst Chem. Res: Free radical polymerization; design of polymers for use in paints and formulation of paints, especially water-borne systems. Mailing Add: Paint Res Lab Paints Div Can Industs Ltd 1330 Castlefield Ave Toronto ON Can

MCEWAN, WILLIAM SHELLEY, b Long Beach, Calif, Mar 25, 15; m 41; c 4. PHYSICAL CHEMISTRY. Educ: Utah State Univ, BS, 38; Harvard Univ, PhD(phys chem), 42. Prof Exp: Chemist asst, Agr Res Lab, Am Smelting & Refining Co, Utah, 40, res chemist, 45-46; res asst, Div VIII, Nat Defense Res Coun, 40-42; sr chemist, US Naval Ord Test Sta, 46-51, head phys chem br, 51-54, actg asst tech dir & head res dept, Naval Weapons Ctr, 59-60, HEAD CHEM DIV, RES DEPT, NAVAL WEAPONS CTR, 54- Concurrent Pos: Consult, Proj Squid, Off Naval Res, Foreign Projs Br, US Dept Defense & Naval Propellant Plant; consult to write nat prog for weather modification for Asst Secy Com. Honors & Awards: Tech Dir Award Outstand Tech Accomplishment, Naval Weapons Ctr, 73. Mem: Am Chem Soc; Am Phys Soc. Res: Thermochemistry; thermodynamics of propellants; explosives; weather modification; nucleation agents; crystal growth; ferrite materials; chemical luminescence. Mailing Add: Chem Div Code 605 Michelson Lab Naval Weapons Ctr China Lake CA 93555

MCEWEN, BRUCE SHERMAN, b Ft Collins, Colo, Jan 17, 38; m 60; c 2. NEUROBIOLOGY. Educ: Oberlin Col, AB, 59; Rockefeller Univ, PhD(biol), 64. Prof Exp: USPHS fel, Inst Neurobiol, Gothenburg Univ, Sweden, 64-65; asst prof zool, Univ Minn, Minneapolis, 66; asst prof, 66-71, ASSOC PROF ZOOL, ROCKEFELLER UNIV, 71- Mem: Soc Gen Physiol; Soc Neurosci; Am Soc Neurochem; Endocrine Soc. Res: Gene activity in nervous tissue, focusing on steroid hormone action. Mailing Add: Dept of Zool Rockefeller Univ New York NY 10021

MCEWEN, CHARLES NEHEMIAH, b Matoaca, Va, Feb 9, 42; m 67; c 2. MASS SPECTROMETRY. Educ: Col William & Mary, BS, 65; Atlanta Univ, MS, 70; Univ Va, PhD(chem), 73. Prof Exp: Teacher chem, Hermitage High Sch, Henrico County, Va, 65-68; RES CHEMIST MASS SPECTROMETRY, CENT RES & DEVELOP DEPT, E I DU PONT DE NEMOURS & CO, INC, 73- Mem: Am Chem Soc; Am Soc Mass Spectrometry. Res: Mass spectrometry, especially field desorption and chemical ionization; emphasis on instrumentation, techniques and new applications. Mailing Add: Exp Sta E228/113 E I du Pont de Nemours & Co Wilmington DE 19898

MCEWEN, CURRIER, b Newark, NJ, Apr 1, 02; m 30; c 4. MEDICINE. Educ: Wesleyan Univ, BS, 23; NY Univ, MD, 26. Hon Degrees: DSc, Wesleyan Univ, 50; DSc, Marietta Col, 52. Prof Exp: Intern, Bellevue Hosp, 26-27, asst res, 27-28; asst med & asst resident physiciatn, Rockefeller Inst, 28-30, assoc med, 30-32; from asst prof to prof, 33-70, asst dean & secy, 32-37, dean, 37-55, EMER PROF MED, SCH MED, NY UNIV, 70- Concurrent Pos: Vis physician, Bellevue Hosp Ctr, 32-70; consult physician, Goldwater Mem Hosp, 46-70; vis physician, Univ Hosp, 49-70; consult, Vet Admin Hosps, New York, 55-70 & Vet Admin Cent Off, 59-; consult physician, Maine Med Ctr, Regional Mem Hosp, Cent Maine Gen & Togus Vet Admin Hosps. Mem: Am Soc Clin Invest; Am Rheumatism Asn (pres, 52-53); Am Heart Asn; Asn Am Physicians; master Am Col Physicians. Res: Rheumatic and collagen diseases. Mailing Add: South Harpswell ME 04079

MCEWEN, DAVID JOHN, b Winnipeg, Man, July 23, 30; m 54; c 4. ANALYTICAL CHEMISTRY. Educ: Acadia Univ, BSc, 52; Purdue Univ, MS, 55, PhD(chem), 57. Prof Exp: Res chemist, Cent Res Lab, Can Industs, Ltd, Que, 57-60; SR RES CHEMIST, RES LABS, GEN MOTORS CORP, WARREN, 60- Mem: Am Soc Mass Spectrometry. Res: Gas chromatography; mass spectrometry; mass spectrometry of vehicle emissions; polymers; solvents; synthesized chemicals. Mailing Add: 3477 Newgate Rd Troy MI 48084

MCEWEN, FREEMAN LESTER, b Bristol, PEI, Nov 11, 26; nat US; m 47; c 3. ENTOMOLOGY. Educ: McGill Univ, BS, 50; Univ Wis, MS, 52, PhD(entom), 54. Prof Exp: Asst, Univ Wis, 50-54; agr res off, Sci Serv Can, 54; from asst prof to prof entom & head dept, NY State Agr Exp Sta, Cornell Univ, 54-69; prof zool, 69-71, PROF ENVIRON BIOL & CHMN DEPT, UNIV GUELPH, 71- Mem: Entom Soc Am. Res: Vegetable insect control; transmission of plant virus diseases by insects; microbial control of insects. Mailing Add: Dept of Environ Biol Graham Hall Univ of Guelph Guelph ON Can

MCEWEN, KATHLEEN LENORE, b Vancouver, BC, July 7, 29. THEORETICAL CHEMISTRY. Educ: Univ BC, BA, 51, MSc, 53; Cambridge Univ, PhD(theoret chem), 58. Prof Exp: Nat Res Coun Can fel, Univ Montreal, 57-59; asst prof, 59-65, ASSOC PROF CHEM, UNIV SASK, 65- Res: Molecular quantum mechanics; molecular spectroscopy. Mailing Add: Dept of Chem Univ of Sask Saskatoon SK Can

MCEWEN, MILDRED MORSE, b Charlotte, NC, Aug 6, 01; m 29. CHEMISTRY. Educ: Queens Col, NC, AB, 22; Univ NC, MA, 24, PhD(biochem), 45. Prof Exp: Asst prof chem, Queens Col, NC, 24-42; asst biochem, Univ NC, 42-45, fels asst biochem, Med Sch, 43-45; prof, 45-71, EMER PROF CHEM, QUEENS COL, NC, 71- Mailing Add: 2316 Westfield Rd Charlotte NC 28207

MCEWEN, WILLIAM EDWIN, b Oaxaca, Mex, Jan 13, 22; m 45; c 3. ORGANIC CHEMISTRY. Educ: Columbia Univ, AB, 43, MA, 45, PhD(chem), 47. Prof Exp: Asst chem, Columbia Univ, 43-45; tech engr, Carbide & Carbon Chem Corp, Oak Ridge, 45-46; from asst prof to prof chem, Univ Kans, 47-62; PROF CHEM & HEAD DEPT, UNIV MASS, AMHERST, 62- Concurrent Pos: Res collab, Brookhaven Nat Lab. Mem: Fel NY Acad Sci; fel Am Chem Soc. Res: Mechanisms of organic reactions; stereochemistry of organophosphorus compounds; heterocyclic, organosulfur and organoantimony chemistry; organometallic compounds. Mailing Add: Dept of Chem Univ of Mass Amherst MA 01002

MCEWEN, WILLIAM JOHN, b Hermosa Beach, Calif, Mar 27, 29; m 51; c 2. ANTHROPOLOGY. Educ: Univ Calif, Los Angeles, AB, 50, MA, 51; Cornell Univ, PhD(anthrop), 54. Prof Exp: Asst prof anthrop & admin, Cornell Univ, 55-57; asst prof anthrop & community health, State Univ NY Downstate Med Ctr, 57-64; CONSULT ANTHROP & PUB HEALTH, CALIF STATE DEPT PUB HEALTH, 64- Concurrent Pos: Consult, Sch Social Work, Columbia Univ, 59-60; consult, NY State Health Dept, 60-61; Russell Sage Found fel, Mex, 61-62; USPHS fel, Mex, 62-63; sr res assoc, Res Inst Study Man, 64-69; lectr, Sch Pub Health, Univ Calif, Berkeley, 72- Mem: Am Anthrop Asn; Soc Appl Anthrop; Am Pub Health Asn; Am Sociol Asn; Royal Anthrop Inst. Res: Anthropology of development; medical anthropology; community social and political organization; rural health services; research methodology; Latin American studies. Mailing Add: Res Coord Off Calif State Dept of Health 2151 Berkeley Way Berkeley CA 94704

MCEWEN, WILLIAM KIRK, b Chicago, Ill, May 29, 10; m 47; c 4. BIOCHEMISTRY. Educ: Harvard Univ, AB, 31, MA, 33, PhD(org chem), 34. Prof Exp: Res worker, Chicago, Ill, 36-39; group leader propellant res, Carnegie Inst Technol, 40-44; group leader, Monsanto Chem Co, 44-45; res assoc ophthal, Med Ctr, 47-53, asst res radiochemist & asst prof chem, 51-53, from asst prof to assoc prof biochem, 54-72, EMER PROF BIOCHEM, SCH MED, UNIV CALIF, SAN FRANCISCO, 72-, RES BIOCHEMIST, PROCTOR FOUND, 62- Mem: AAAS; Am Chem Soc; Soc Exp Biol & Med; Asn Res Vision & Ophthal; NY Acad Sci. Res: Ophthalmological biochemistry and rheology. Mailing Add: Univ of Calif Med Ctr San Francisco CA 94143

MCEWEN, WILLIAM ROBERT, b Duluth, Minn, Aug 21, 11; m 40; c 6. MATHEMATICS. Educ: Minn State Teachers Col, Duluth, BE, 35; Univ Minn, MA, 39, PhD(math), 47. Prof Exp: Teacher high schs, Minn, 35-37; asst math, Univ Minn, Minneapolis, 37-39, from instr to asst prof, 39-47, assoc prof, Univ Minn, Duluth, 47-49, head dept, 47-54, PROF MATH, UNIV MINN, DULUTH, 49-, CHMN DIV SCI & MATH, 55- Mem: Am Math Soc; Math Asn Am. Res: Orthogonal polynomials; probability; statistics. Mailing Add: Dept of Math Univ of Minn Duluth MN 55812

MCFADDEN, BRUCE ALDEN, b La Grande, Ore, Sept 23, 30; m 58; c 3. BIOCHEMISTRY, MICROBIOLOGY. Educ: Whitman Col, AB, 52; Univ Calif, Los Angeles, PhD(chem), 56. Prof Exp: Asst chem, Univ Calif, Los Angeles, 52-54; from instr to assoc prof, 56-66, PROF CHEM, WASH STATE UNIV, 66-, DIR DEVELOP, DIV SCI, 74- Concurrent Pos: NIH res career develop award, 63; vis prof, Univ Ill, Urbana, 66-67; Guggenheim fel & vis prof, Univ Leicester, 72-73; NIH spec fel, 73. Mem: AAAS; Am Soc Biol Chem; Am Soc Microbiol; Am Chem Soc; Am Soc Plant Physiol. Res: Microbial assimilation of one-carbon and two-carbon compounds; bacterial cell wall structure, synthesis and function; nematode biochemistry. Mailing Add: Dept of Chem Wash State Univ Pullman WA 99163

MCFADDEN, DAVID LEE, b Orange, Calif, Oct 18, 45; m 70. CHEMICAL PHYSICS. Educ: Occidental Col, AB, 67; Mass Inst Technol, PhD(phys chem), 72. Prof Exp: Res fel chem, Harvard Univ, 72-73; ASST PROF CHEM, BOSTON COL, 73- Mem: Am Phys Soc; Am Chem Soc; AAAS. Res: Gas phase reaction dynamics and kinetics. Mailing Add: Dept of Chem Boston Col Chestnut Hill MA 02167

MCFADDEN, ERNEST B, b Omaha, Nebr, Oct 8, 23; m 50; c 2. PHYSIOLOGY. Educ: Ohio State Univ, MS, 57. Prof Exp: Chief clin lab, Regional Off, Vet Admin, Okla, 47-49; from asst physiologist to physiologist, Civil Aeronaut Med Res Lab 49-59, res assoc, 59-60, actg chief protective equip sect, 60-61; CHIEF SURVIVAL RES UNIT, CIVIL AEROMED INST, FED AVIATION ADMIN, 61- Concurrent Pos: Consult, Univ Mich, 70-71. Honors & Awards: Harry G Moseley Award, Aerospace Med Asn, 70. Mem: Aerospace Med Asn; Survival & Flight Equip Asn; Soc Air Safety Investrs. Res: Respiration; acceleration; biomechanics; stress; survival; high altitude, aviation and space physiology. Mailing Add: Fed Aviation Admin Aeroanaut Ctr Civil Aeromed Inst Oklahoma City OK 73125

MCFADDEN, HARRY WEBBER, JR, b Greenwood, Nebr, Dec 9, 19; m 45; c 2. MEDICAL MICROBIOLOGY, INFECTIOUS DISEASES. Educ: Univ Nebr, AB,

41, MD, 43; Am Bd Path, dipl, 51 & 65. Prof Exp: From intern to resident path, 43-45, resident clin path & bact, 47-49, fel & instr path & bact, 49-52, asst prof path & microbiol, 52-55, interim chancellor, Med Ctr, 72, PROF MED MICROBIOL & CHMN DEPT, COL MED, UNIV NEBR, OMAHA, 55-, PROF PATH, 68-, INTERIM ASSOC DEAN GRAD STUDIES, MED CTR, 72- Concurrent Pos: Trustee, Am Bd Path, 70- Mem: Col Am Pathologists; Am Soc Microbiol; Am Soc Clin Pathologists; NY Acad Sci; Am Fedn Clin Res. Res: Immunology, microbiology and pathology. Mailing Add: Dept of Med Microbiol Univ of Nebr Col of Med Omaha NE 68105

MCFADDEN, JAMES DOUGLAS, b Winchester, Va, Feb 19, 34; m 61; c 4. METEOROLOGY, OCEANOGRAPHY. Educ: Va Polytech Inst, BS, 56; Univ Wis-Madison, PhD(meteorol), 65. Prof Exp: Res meteorologist, 65-73, DIR RES FLIGHT FACILITY, NAT OCEANIC & ATMOSPHERIC ADMIN, 73- Mem: Am Meteorol Sco; Am Geophys Union; Am Geog Soc. Res: Airborne meteorological and oceanographic research; cloud physics; weather modification; remote sensing of the environment. Mailing Add: Nat Oceanic & Atmospheric Admin Res Flight Facility Box 480197 Miami FL 33148

MCFADDEN, JAMES THOMPSON, b Ebensburg, Pa, Apr 22, 31; m 54; c 6. POPULATION ECOLOGY. Educ: Univ Pittsburgh, BS, 53; Ohio State Univ, MS, 54; Pa State Univ, PhD, 61. Prof Exp: Asst bot, Ohio State Univ, 53-54; proj leader fishery res, State Conserv / Dept, Wis, 54-57; asst zool, Pa State Univ, 57-58; NSF fel, Univ Chicago, 60-61; biometrician, Inst Fisheries Res, State Conserv Dept, Mich, 61-64, chief fish sect, 64; res assoc fisheries, Univ Mich, 63-64; asst prof, Dept Zool & Inst Fisheries, Univ BC, 64-66; assoc prof wildlife & fisheries, 66-69, chmn dept, 69-74, dean sch natural resources, 70-74, PROF WILDLIFE & FISHERIES, UNIV MICH, ANN ARBOR, 69- Concurrent Pos: Dir sea grant prog, Univ Mich, 69-70. Mem: Ecol Soc Am; Am Fisheries Soc. Res: Biometrics; natural resource management. Mailing Add: Dept of Wildlife & Fisheries Univ of Mich Ann Arbor MI 48104

MCFADDEN, LEONARD, b Wigan, Eng, Aug 2, 13; nat US; m 44; c 4. MATHEMATICS. Educ: Queen's Univ, Can, BA, 36, MA, 37; Brown Univ, PhD(math), 41. Prof Exp: Instr, Brown Univ, 39-41; from instr to assoc prof, 41-52, PROF MATH, VA POLYTECH INST & STATE UNIV, 52- Mem: Am Math Soc; Math Asn Am. Res: Absolute Norlund summability. Mailing Add: Dept of Math Va Polytech Inst & State Univ Blacksburg VA 24061

MCFADDEN, LORNE AUSTIN, b Lower Stewiacke, NS, Oct 12, 26; nat US; m 53; c 3. PLANT PATHOLOGY. Educ: McGill Univ, BSc, 49; Cornell Univ, MS, 54, PhD(plant path), 56. Prof Exp: Plant pathologist, NS Agr Col, 49-50; asst, Cornell Univ, 50-56; asst plant pathologist, Subtrop Exp Sta, Univ Fla, 56-63; assoc prof plant sci, Univ NH, 63-69, prof plant path, 69-72; PROF BIOL & HEAD DEPT, NS AGR COL, 72- Concurrent Pos: Exten horticulturist, Agr Exp Sta, Univ NH, 63-69. Mem: Am Phytopath Soc; Can Phytopath Soc; Agr Inst Can. Res: Bacterial plant pathogens; diseases of ornamental and other crop plants. Mailing Add: Dept of Biol NS Agr Col Truro NS Can

MCFADDEN, MAX WULFSOHN, b Mt Vernon, NY, Apr 11, 36; m 58; c 2. ENTOMOLOGY, RESEARCH ADMINISTRATION. Educ: Univ Md, BS, 59; Univ Alta, PhD(entom), 62. Prof Exp: Head biol & agr sci, Sci Info Exchange, Smithsonian Inst, 62-63; res entomologist, Entom Res Div, Agr Res Serv, USDA, 63-68; res assoc entom, Wash State Univ, 68-72; insect ecologist, Food & Agr Orgn, UN, 72-74; RES COORDR, EXPANDED DOUGLAS-FIR TUSSOCK MOTH PROG, USDA, 74- Mem: Entom Soc Am; Soc Syst Zool; AAAS; Sigma Xi. Res: Systematics and biology of Stratiomyidae; population suppression techniques; interests in integrated pest management; insect ecology. Mailing Add: Douglas-Fir Tussock Moth Prog PO Box 3141 USDA Portland OR 97208

MCFADDEN, ROBERT, b Belfast, Northern Ireland, Oct 7, 34; m 57; c 2. MATHEMATICS. Educ: Queen's Univ Belfast, BA, 55, MA, 59, PhD(math), 62. Prof Exp: Lectr math, Queen's Univ Belfast, 62-68; PROF, NORTHERN ILL UNIV, 68- Concurrent Pos: Asst prof, Ind Univ, 64-65 & La State Univ, 65-66; NSF res grants, Northern Ill Univ, 69-73. Mem: London Math Soc; Am Math Soc; Math Asn Am. Res: Theory of partially ordered semigroups; algebraic theory of semigroups. Mailing Add: Dept of Math Northern Ill Univ De Kalb IL 60115

MCFADDEN, WILLIAM HAMILTON, physical chemistry, see 12th edition

MCFALL, ELIZABETH, b San Diego, Calif, Oct 28, 28. BIOCHEMISTRY. Educ: San Diego State Col, BS, 50; Univ Calif, MA, 54, PhD(biochem), 57. Prof Exp: Asst resident biochemist, Univ Calif, 57; res assoc microbiol, Mass Inst Technol, 60-61, 62-63; from asst prof to assoc prof, 63-72, PROF MICROBIOL, SCH MED, NY UNIV, 72- Concurrent Pos: Res fel bact, Harvard Med Sch, 57-60; Nat Inst Med Res fel, London, 61-62; USPHS career develop award, 63- Mem: Am Soc Biol Chemists; Am Soc Microbiol. Res: Microbial genetics; regulatory mechanisms in microorganisms. Mailing Add: Dept of Microbiol NY Univ Sch of Med New York NY 10016

MCFARLAN, EDWARD, JR, b Brooklyn, NY, Mar 24, 21; m 49; c 2. PETROLEUM GEOLOGY, SEDIMENTOLOGY. Educ: Williams Col, BA, 43; Univ Tex, MA, 48. Prof Exp: Res geologist, Humble Oil & Ref Co, 48-59, area stratigrapher, La, 59-64, area explor geologist, 64-65; div stratigrapher, STex, 65-66, Mgr basin geol div, Esso Prod Res Co, 66-69, mgr stratig geol div, Explor Dept, 69-73, EXPLORATION ADV, EXPLORATION DEPT, HQ, EXXON CO USA, 73- Mem: Soc Econ Paleont & Mineral; Geol Soc Am; Am Inst Prof Geol; Am Asn Petrol Geol. Res: Recent and Pleistocene geology; sedimentation; photogeology; geomorphology; Cenozoic and Mesozoic structure and stratigraphy of the Gulf Coast and adjacent offshore areas; Regional petroleum geology of the Mesozoic sequence in the US Gulf Coast from Florida to Mexico. Mailing Add: Exxon Co USA PO Box 2180 Rm 3977-Exxon Bldg Houston TX 77001

MCFARLAND, CHARLES ELWOOD, b Kirkwood, Mo, June 20, 27; m 50; c 2. PHYSICS. Educ: Mo Sch Mines, BS, 49; Wash Univ, PhD(physics), 55. Prof Exp: Sr scientist, Bettis Atomic Power Div, Westinghouse Elec Corp, 55-56; nuclear physicist, Internuclear Co, 56-60; ASSOC PROF PHYSICS, UNIV MO-ROLLA, 60- Mem: Am Phys Soc. Res: Solid state physics; nuclear physics. Mailing Add: Dept of Physics Univ of Mo Rolla MO 65401

MCFARLAND, CHARLES R, b Columbus, Ohio, Aug 26, 27; m 50. MICROBIOLOGY. Educ: Otterbein Col, BS, 49; Ohio State Univ, MS, 50; WVa Univ, PhD(microbiol), 68. Prof Exp: Clin microbiologist, Clifton Springs Sanitarium & Clin, NY, 55-57, Aultman Hosp, Canton, Ohio, 57-60 & Good Samaritan Hosp, Dayton, 60-65; asst prof, 67-74, ASSOC PROF MICROBIOL, WRIGHT STATE UNIV, 74- Concurrent Pos: Consult, Barney's Children's Med Ctr, Dayton, Ohio, 69- Mem: AAAS; Am Soc Microbiol. Res: Chemistry, metabolism and biological significance of microbial lipids; bacterial endogenous carbon and energy reserves. Mailing Add: Dept of Biol Sci Wright State Univ Dayton OH 45431

MCFARLAND, JAMES THOMAS, b Junction City, Kans, Dec 7, 42; m 64; c 2. BIOCHEMISTRY. Educ: Col Wooster, BA, 64; Calif Inst Technol, PhD(chem), 68. Prof Exp: USPHS fel, Inst Molecular Biol, Univ Ore, 68-70; ASST PROF CHEM, UNIV WIS-MILWAUKEE, 70- Mem: Am Chem Soc. Res: Elucidation of the mechanism of action of enzymes; the use of rapid kinetic measurements, magnetic resonance and raman spectroscopy to investigate biological systems. Mailing Add: Dept of Chem Univ of Wis-Milwaukee Milwaukee WI 53201

MCFARLAND, JAMES WILLIAM, b Sacramento, Calif, Nov 16, 31; m 61; c 3. MEDICINAL CHEMISTRY. Educ: Chico State Col, BA, 54; Univ Calif, Berkeley, PhD(org chem), 57. Prof Exp: Fulbright fel, Univ Munich, 57-58; res fel org chem, Univ Calif, Berkeley, 58-59; res chemist, 60-68, PROJ LEADER, PFIZER, INC, 68- Mem: Am Chem Soc. Res: Heterocyclic syntheses; quantitative structure-activity correlations; pesticides; anthelmintic and antibacterial agents. Mailing Add: Dept of Chem Res Pfizer Inc Groton CT 06340

MCFARLAND, JAMES WILLIS, b Wyndmere, NDak, Feb 10, 17; m 46; c 6. BOTANY. Educ: Asbury Col, AB, 39; Univ Ky, MS, 46. Prof Exp: Instr sci, Cascade Col, 40-41, assoc prof, 46, chmn dept, 48-51; teaching asst, Univ Calif, 46-48; with off atomic energy, US Air Force Pentagon, 53-57, chief prog officer, Aerospace Intel Ctr, Ohio, 57-59, asst prof & dir human physiol, US Air Force Acad, 59-63, chief space sci div, Air Force Space Systs, Calif, 64 & spacexExpxDivspace exp div, 65-68; ASSOC PROF SCI & CHMN DIV SCI & MATH, AZUSA PAC COL, 68- Concurrent Pos: Consult, Vis Lectr Prog, NSF for Colo & Wyo, 60-63. Mem: AAAS; Nat Asn Biol Teachers; Am Sci Affiliation. Res: Cytogenetics; field biology; radiobiology; ecology. Mailing Add: Dept of Biol Azusa Pac Col Azusa CA 91702

MCFARLAND, JOHN WILLIAM, b Elkton, Tenn, Aug 16, 23; m 47; c 4. ORGANIC CHEMISTRY. Educ: DePauw Univ, AB, 49; Vanderbilt Univ, PhD(chem), 53. Prof Exp: Fel, Mass Inst Technol, 53-55; res chemist, E I du Pont de Nemours & Co, Inc, Del, 55-61; assoc prof, 61-67, PROF CHEM, DEPAUW UNIV, 67- Concurrent Pos: Fel, Univ Groningen, 71. Mem: Am Chem Soc. Res: Organic syntheses and reaction mechanisms; polymerization. Mailing Add: Dept of Chem DePauw Univ Greencastle IN 46135

MCFARLAND, KAY FLOWERS, b Daytona Beach, Fla, Jan 27, 42; m 63; c 3. INTERNAL MEDICINE, ENDOCRINOLOGY. Educ: Wake Forest Col, BS, 63; Bowman Gray Sch Med, MD, 66. Prof Exp: From instr to asst prof internal med & endocrinol, 71-75, ASST PROF INTERNAL MED, HYPERTENSION & CLIN PHARMACOL, MED COL GA, 75- Concurrent Pos: Endocrinologist, pvt pract, 72-75. Mem: Am Diabetes Asn; fel Am Col Physicians; Am Soc Internal Med. Res: Clinical research on diabetes and hypertension. Mailing Add: 3203 Candace Dr Augusta GA 30904

MCFARLAND, MACK, b Houston, Tex, Sept 9, 47; m 68; c 2. AIR POLLUTION. Educ: Univ Tex, Austin, BS, 70; Univ Colo, PhD(chem physics), 73. Prof Exp: Chemist, Aeronomy Lab, Nat Oceanic & Atmospheric Admin, US Dept Com, 73-74; proj scientist air pollution, York Univ, 74-75; CHEMIST, AERONOMY LAB, NAT OCEANIC & ATMOSPHERIC ADMIN, US DEPT COM, 75- Res: Monitoring trace gas constituents of the troposphere and stratosphere. Mailing Add: US Dept Com Nat Oceanic & Atmospheric Admin Boulder CO 80302

MCFARLAND, ROBERT HAROLD, b Severy, Kans, Jan 10, 18; m 40; c 2. PHYSICS. Educ: Kans State Teachers Col, BS & BA, 40; Univ Wis, PhM, 43, PhD(physics), 47. Prof Exp: Instr high sch, Kans, 40-41 & Radio Sch, Univ Wis, 42-43; sr engr, Sylvania Elec Prod, Inc, 44-46; from asst prof to prof physics, Kans State Univ, 47-60, dir nuclear lab, 58-60; physicist, Lawrence Livermore Lab, Univ Calif, 60-69; actg vpres acad affairs, 74-75, PROF PHYSICS & DEAN GRAD SCH, UNIV MO-ROLLA, 69- Concurrent Pos: Consult, Well Survs, Inc, Okla, 53-54; mem, Grad Rec Exam Bd, 71- Mem: Fel Am Phys Soc; Am Asn Physics Teachers. Res: Atomic spectra; gaseous electronics; nuclear physics; effect of humidity on low pressure discharges; fluorescent and discharge spectra of mercury with zinc, thallium and indium metals; use of tracers; electron impact ionization of alkali metals; threshold polarization of helium radiation. Mailing Add: 204 Parker Hall Univ of Mo-Rolla Rolla MO 65401

MCFARLAND, ROSS ARMSTRONG, b Denver, Colo, July 18, 01; m 50. PHYSIOLOGICAL PSYCHOLOGY, AEROSPACE MEDICINE. Educ: Univ Mich, BA, 23; Harvard Univ, PhD(physiol psychol), 28. Hon Degrees: ScD, Park Col, 42, Rutgers Univ, 64, Trinity Col, 65, Univ Denver, 71. Prof Exp: Instr psychol, Columbia Univ, 28-37; asst prof indust res fatigue lab, 37-39, asst prof, Grad Sch Bus Admin, 39-48, from asst prof to assoc prof indust hyg, Sch Pub Health, 48-58, dir, Guggenheim Ctr Aerospace Health & Safety, 57-72, prof environ health & safety, Sch Pub Health, 58-62, Guggenheim prof aerospace health & safety, 62-72, EMER GUGGENHEIM PROF AEROSPACE HEALTH & SAFETY, SCH PUB HEALTH, HARVARD UNIV, 73- Concurrent Pos: Mem int high altitude exped, Andes, 35; med coordr, Pan-Am World Airways, 36-52; mem exec comt selection & training aircraft pilots, Nat Res Coun, 39-41, mem comt emergency med serv, 67-69, mem sci comt probs alcohol; mem US nat comn, Permanent Aeronaut Comn, 41-44; mem comt operating probs, Nat Adv Comt Aeronaut, 47-53, mem comt flight safety, 54-59; mem, President's Conf Indust Safety, 48-; dir comn accidental trauma, Armed Forces Epidemiol Bd, US Dept Defense, 50-69, mem adv panel med sci, res & develop, 55-60; consult, Surgeon Gen, US Dept Army, 51-69; mem panel social & occup health, WHO, 51-; consult, Surgeon Gen, Us Air Force, 54-56, consult sci adv bd panel human factors, 56-57; consult accident prev adv comt, USPHS, 57-60, consult adv comt US Nat Health Surv, 57-63, consult spec comt environ health probs & facil, 59-60, consult div accident prev, 62-65; mem adv panel accident prev div res grants, NIH, 58-61; mem med adv coun, Fed Aviation Agency, 60-66, mem tech adv bd, 65-66; mem res adv comt biotechnol & human res, NASA, 63-65, mem space med adv group, 64-66; mem nat adv environ control coun, Consumer Protection & Environ Health Serv, 67-70; mem, White House Conf Health, 65, Presidential Task Force Hwy Safety, 69-70 & White House Conf Aging, 71. Honors & Awards: Longacre Award, Aerospace Med Asn, 47, Boothby Award, 62; Award, Flight Safety Found, 53, Williams Mem Safety Award, 65; Jeffries Award, Am Inst Aeronaut & Astronaut, 56; Taylor Award Eng Psychol, Am Psychol Asn, 68; Except Serv Award, US Air Force, 69; Award, Airlines Med Dir Asn, 70; Distinguished Civilian Serv Award, Dept Army, 71; Godfrey L Cabot Award, Aero Club New Eng, 71; Laura Taber Barbour Air Safety Award, 74. Mem: Fel Am Psychol Asn; fel Aerospace Med Asn (vpres, 70-71); assoc fel Am Inst Aeronaut & Astronaut; fel Am Acad Arts & Sci; fel Human Factors Soc (pres, 70-71). Res: Industrial health and safety; effects of high altitude; fatigue; human factors in air transport design and operation; highway safety; human engineering problems. Mailing Add: Harvard Univ Sch of Pub Health 665 Huntington Ave Boston MA 02115

MCFARLAND, WILLIAM, b Wilkinsburg, Pa, Nov 4, 19; m 45; c 2. HEMATOLOGY. Educ: Washington & Jefferson Col, BS, 41; Univ Pittsburgh, MD, 44; Am Bd Internal Med, dipl, 55. Prof Exp: Intern, St Francis Hosp, Pittsburgh, Pa, 44-45; resident internal med, Henry Ford Hosp, Detroit, 45-48; pvt pract, Pa, 48-54; instr med, Univ Louisville, 54-55; res fel hemat, New Eng Ctr Hosp, Boston, 56-58;

hematologist, Nat Naval Med Ctr, Bethesda, Md, 58-61; from asst prof to assoc prof med, Georgetown Univ, 61-73; PROF MED, UNIV CALIF, DAVIS, 73- Concurrent Pos: Consult bone dent res lab, Pa State Univ, 52-54; instr, Sch Med, Tufts Univ, 56-58; mem hemat study sect, NIH, 60-65; consult, Nat Naval Med Ctr, 61-; mem erythropoietin subcomt, Nat Heart Inst, 63-67; chief hemat res sect, Vet Admin Hosp, Washington, DC, 61-73; mem staff, Vet Admin Hosp, Martinez, Calif. Mem: AAAS; Am Fedn Clin Res; Am Soc Hemat. Res: Relative roles and interactions of blood cells in immunologic reactions. Mailing Add: Vet Admin Hosp Martinez CA 94553

MCFARLAND, WILLIAM NORMAN, b Toronto, Ont, Sept 11, 25; nat US; m 50; c 2. ZOOLOGY. Educ: Univ Calif, Los Angeles, BA, 51, MA, 53, PhD(zool), 59. Prof Exp: Chemist & biologist, Marineland of Pac, Calif, 54-57; marine biologist, State Game & Fish Comn, Univ Tex, 58; res scientist, Inst Marine Sci & lectr zool, Univ Tex, 58-61; from asst prof to assoc prof, 61-71, PROF ZOOL, CORNELL UNIV, 71- Mem: AAAS; Am Soc Ichthyol & Herpet; Am Soc Zool; Am Fisheries Soc; Ecol Soc Am. Res: Fish physiology; comparative physiology and ecology. Mailing Add: Sect Ecol & Syst Div Biol Sci Cornell Univ Ithaca NY 14850

MCFARLANE, ELLEN SANDRA, b Halifax, NS, May 19, 38. ONCOLOGY, BIOCHEMISTRY. Educ: DAlhousie Univ, BSc, 59, MSc, 61, PhD(biochem), 63. Prof Exp: ASST PROF MICROBIOL & LECTR BIOCHEM, DALHOUSIE UNIV, 71- Concurrent Pos: Nat Cancer Inst Can fel, Univ Man, 63-64; Med Res Coun Can fel, Harvard Univ, 64 & Dalhousie Univ, 65; Med Res Coun Can scholar microbiol, Dalhousie Univ, 66-71. Mem: Can Biochem Soc; Brit Biochem Soc. Res: The study of viral oncology with particular reference to human tumors. Mailing Add: Dept of Microbiol Dalhousie Univ Halifax NS Can

MCFARLANE, FINLEY EUGENE, b Atkins, Va, Nov 24, 40; m 64; c 2. POLYMER CHEMISTRY. Educ: King Col, AB, 63; Univ NC, PhD(chem), 68. Prof Exp: Sr res chemist, 67-75, RES ASSOC, TENN EASTMAN CO, 75- Mem: Am Chem Soc. Res: Polyester chemistry; catalysis, kinetics and mechanisms of polymerization and degradation reactions; polymer synthesis; preparation and properties of liquid-crystalline polyesters. Mailing Add: Bldg 150B Tenn Eastman Co Kingsport TN 37662

MCFARLANE, HUGH MURRAY, b Winnipeg, Man, Jan 7, 19; m 47; c 3. CHEMISTRY. Educ: Univ Man, BSc, 41, MSc, 42; McGill Univ, PhD(chem), 50. Prof Exp: Res chemist, Abitibi Power & Paper Co, 50-51; sr res chemist, Bathurst Power & Paper Co, 51-56; supvr, Wastes & By-prod Sect, Cent Res Div, Abitibi Power & Paper Co, 56-64; tech dir, Sandwell & Co, Ltd, BC, 64-71, mgr, PR Sandwell & Co, UK, Ltd, London, 71-74; FORESTRY OFFICER, IBRD CORP PROG, FOOD & AGR ORGN, 74- Mem: Tech Asn Pulp & Paper Indust; Chem Inst Can; Can Pulp & Paper Asn. Res: Cellulose chemistry; pulp and paper processes. Mailing Add: IBRD Corp Prog Food & Agr Orgn Via Delle Terme Di Caracolla 00100 Rome Italy

MCFARLANE, JOHN ELWOOD, b Tisdale, Sask, Aug 1, 29; m 60; c 3. INSECT PHYSIOLOGY. Educ: Univ Sask, BA, 49, MA, 51; Univ Ill, PhD(entom), 55. Prof Exp: From asst prof to assoc prof, 55-72, PROF ENTOM, MACDONALD COL, MCGILL UNIV, 73- Mem: Entom Soc Can; Can Soc Zoologists. Res: Physiology of development and aging in insects. Mailing Add: Fac of Agr Macdonald Campus McGill Univ Ste Anne de Bellevue PQ Can

MCFARLANE, JOHN SPENCER, b Lothair, Mont, Aug 26, 15; m 47. GENETICS. Educ: Mont State Col, BS, 38; Univ Wis, PhD(genetics), 43. Prof Exp: Asst, Univ Wis, 38-43, instr genetics & entom, 43-44; asst olericulturist, Univ Hawaii, 44-46; assoc geneticist, Curly Top Resistance Breeding Comt, Utah, 46-47; geneticist, Bur Plant Indust Soils & Agr Eng, Calif, 47-53, supt agr res sta, 64-72, leader sugar beet invests, 69-72, GENETICIST, AGR RES SERV, USDA, 53-, LOCATION LEADER, AGR RES STA & RES LEADER SUGARBEET PROD, 72- Mem: AAAS; Am Phytopath Soc; Am Genetic Asn; Am Soc Sugar Beet Technol; Int Inst Sugar Beet Res. Res: Genetics of sugar beet; breeding for disease resistance; breeding hybrid sugar beet varieties. Mailing Add: Agr Res Sta USDA Box 5098 Salinas CA 93901

MCFARLANE, ROSS ALEXANDER, b Toronto, Ont, Can, June 10, 31; m 60. PHYSICS. Educ: McMaster Univ, BSc, 53; McGill Univ, MSc, 55, PhD, 59. Prof Exp: Asst physics, Eaton Electronics Lab, McGill Univ, 59; mem staff, Res Lab Electronics, Mass Inst Technol, 59-61; mem staff, Bell Tel Labs, Inc, 61-69; assoc prof, 69-74, PROF ELEC ENG, CORNELL UNIV, 74- Mem: Optical Soc Am; Inst Elec & Electronics Eng; Can Asn Physicists. Res: Microwave physics and frequency standards; noise and signal studies of electron beam type microwave devices; paramagnetic chemical lasers; reaction kinetics. Mailing Add: Sch of Elec Eng Cornell Univ Ithaca NY 14850

MCFARLANE, WALTER KENNETH, b Glasgow, Scotland, Mar 7, 37; m 61; c 2. PHYSICS. Educ: Glasgow Univ, BSc, 58; Univ Birmingham, PhD(high energy physics), 64. Prof Exp: Res investr high energy physics, Univ Pa, 63-66, asst prof, 66-69; assoc prof, 69-75, chmn dept, 70-75, PROF PHYSICS, TEMPLE UNIV, 75- Mem: Am Phys Soc. Res: Proton-nucleon scattering and meson production; K meson decays; pi meson decays. Mailing Add: Dept of Physics Barton Hall Temple Univ Philadelphia PA 19122

MCFARLIN, RICHARD FRANCIS, b Oklahoma City, Okla, Oct 12, 29; m 53; c 4. INORGANIC CHEMISTRY. Educ: Va Mil Inst, BS, 51; Purdue Univ, MS, 53, PhD, 56. Prof Exp: Sr res chemist, Monsanto Chem Co, Mo, 56-60; supvr inorg chem, Int Minerals & Chem Corp, 60-62; mgr Atlanta Res Ctr, Armour Agr Chem Co, 62-64, tech dir, 64-65; vpres & tech dir, Armour Agr Chem Co, 65-68; vpres com develop, 68-69 & develop & admin, 69-74, VPRES OPERS, USS AGRI-CHEM, INC, US STEEL CORP, 74- Concurrent Pos: Mem adv mgt prog, Columbia Univ, 67. Mem: AAAS; Am Chem Soc. Res: Nitrogen and industrial chemicals; industrial explosives; rocket oxidants; fertilizer materials; metal hydrides; inorganic phosphates. Mailing Add: 455 Forest Dale Dr NE Atlanta GA 30342

MCFARREN, EARL FRANCIS, b Akron, Ohio, June 30, 19; m 45; c 3. ANALYTICAL CHEMISTRY. Educ: Bowling Green State Univ, BA, 41. Prof Exp: Asst chem, Bowling Green State Univ, 41-42; chemist, Nat Dairy Res Labs, Inc, Div Nat Dairy Prod Corp, 46-52; chemist, Robert A Taft Sanit Eng Ctr, USPHS, 52-64, chief training prog, Environ Protection Agency, 64-69, chief anal qual control, Water Supply Prog Div, 69-72, CHIEF WATER SUPPLY PROG SUPPORT ACTIV, ENVIRON PROTECTION AGENCY, 72- Concurrent Pos: Consult, Dept Interior, Marshall Islands, 57. Honors & Awards: USPHS Awards, 58 & 63. Mem: Am Chem Soc; Am Water Works Asn; Am Soc Testing & Mat; Sigma Xi. Res: Paper chromatography of amino acids and sugars; enzymatic digestion of casein; reactivation of normal alkaline milk phosphatase; bioassay and chemical assay of paralytic shellfish poison and poisonous fishes; gas chromatography of pesticides; chemical analysis of water; statistics. Mailing Add: 4676 Columbia Pkwy Cincinnati OH 45268

MCFEAT, TOM FARRAR SCOTT, b Montreal, Que, Feb 5, 19; m 47; c 2. ETHNOLOGY. Educ: McGill Univ, BA, 50; Harvard Univ, AM, 54, PhD, 57. Prof Exp: Assoc prof anthrop, Univ NB, 54-59; sr ethnologist, Nat Mus Can, 59-63; sr ethnologist, Carleton Univ, 63-64; chmn dept, 64-74, PROF ANTHROP, UNIV TORONTO, 64- Mem: Fel Am Anthrop Asn; Can Polit Sci Asn. Res: Culture process, particularly concepts of growth, evolution and pattern in diachronic analysis; small group culture, especially influence of information on structure of n-generation groups; Canadian Indian and other ethnic communities; certain aspects of mass media analysis. Mailing Add: Dept of Anthrop Univ of Toronto 1000 St George St Toronto ON Can

MCFEE, ALFRED FRANK, b Knoxville, Tenn, Aug 19, 31; m 51; c 2. CYTOGENETICS, RADIOBIOLOGY. Educ: Univ Tenn, BS, 53, MS, 59; Cornell Univ, PhD(animal breeding), 63. Prof Exp: From asst prof to assoc prof, Agr Res Lab, Univ Tenn-AEC, 63-75, PROF, COMP ANIMAL RES LAB, UNIV TENN-ENERGY RES & DEVELOP ADMIN, 75- Mem: Radiation Res Soc Am. Res: Adverse effects of energy-related pollutants on chromosome structure and cell behavior as they relate to embryonic survival. Mailing Add: Univ Tenn-Energy Res & Develop Adm 1299 Bethel Valley Rd Oak Ridge TN 37830

MCFEE, ARTHUR STORER, b Portland, Maine, May 1, 32; m 67. SURGERY. Educ: Harvard Univ, BA, 53, MD, 57; Univ Minn, MS, 66, PhD(surg), 67; Am Bd Surg, dipl, 67. Prof Exp: Intern surg, Univ Minn Hosp, 57-58, spec fel surg, 58-65; from asst prof to assoc prof, 67-73, PROF SURG, UNIV TEX HEALTH SCI CTR SAN ANTONIO, 73- Mem: Asn Acad Surgeons; Asn Hist Med; AMA; Am Col Surgeons; Soc Surg Alimentary Tract. Res: Local gastric hypothermial gastric physiology; prevention of hepatic metastatic diseases. Mailing Add: Dept of Surg Univ of Tex Health Sci Ctr San Antonio TX 78284

MCFEE, DONALD RAY, b Union Co, Ind, July 4, 29; m 54; c 3. INDUSTRIAL HEALTH, MECHANICAL ENGINEERING. Educ: Purdue Univ, BSc, 51; Univ Cincinnati, MSc, 60, DrSc(indust health), 62; Am Bd Indust Hyg, cert comprehensive pract indust hyg, 64; Bd Cert Safety Prof, cert, 71. Prof Exp: Tool engr, Boeing Airplane Co, Wash, 51-53, facil engr, 53-54, sr facil engr, Pilotless Aircraft Div, 58; assoc indust hygienist, Indust Hyg & Safety Div, Argonne Nat Lab, 61-71, supvr indust hyg group, 71-72; VPRES, OCCUSAFE, INC, WILMETTE, 72- Concurrent Pos: Instr, Nat Safety Coun, 63-71. Mem: Am Indust Hyg Asn; Am Acad Indust Hyg. Res: Environmental health, including industrial hygiene; air pollution; plutonium fume characteristics; sand filtration of fumes; toxicology of organic solvents; noise, safety and air sampling; air cleaning media; combustible gas detection systems. Mailing Add: 25 W 210 Highview Dr Naperville IL 60540

MCFEE, JAMES H, physics, see 12th edition

MCFEE, MALCOLM, b Seattle, Wash, Feb 13, 17; m 41; c 1. ANTHROPOLOGY. Educ: San Jose State Col, BA, 56; Stanford Univ, MA, 58, PhD(anthrop), 62. Prof Exp: Asst prof anthrop, Univ Ariz, 62-65; asst prof, 65-73, ASSOC PROF ANTHROP, UNIV ORE, 73- Mem: Am Anthrop Asn; Am Ethnol Soc. Res: Ethnology and problems of culture change among the Blackfeet Indians of Montana. Mailing Add: Dept of Anthrop Univ of Ore Eugene OR 97403

MCFEE, RAYMOND HERBERT, b Somerville, Mass, Mar 1, 16; m 38; c 2. PHYSICS. Educ: Mass Inst Technol, SB, 37, SM, 38, PhD(physics), 43. Prof Exp: Tech asst, Geophys Res Corp, Okla, 38-40; asst physics, Mass Inst Technol, 41-42, physicist, Div Indust Coop, 42-43; chief physicist, White Res Assocs, Boston, 43-45; physicist, Cambridge Thermionic Corp, 45-46; sr elec engr, Submarine Signal Co, 46; res physicist, Electronics Corp of Am, 46-53, dir eng, 53-54, res, 54-56; dir avionics div, Aerojet-Gen Corp, 56-60 & Azusa Plant, 60-64; sect mgr, Jet Propulsion Lab, Calif Inst Technol, 64-67; assoc dir, Douglas Advan Res Labs, McDonnell Douglas Corp, Calif, 67-70, sr staff engr, McDonnell Douglas Astronaut Co, 70-73, PRIN ENGR/SCIENTIST, McDONNELL DOUGLAS ASTRONAUT CO, 73- Mem: AAAS; Am Phys Soc; fel Optical Soc Am; assoc fel Am Inst Aeronaut & Astronaut. Res: Solid state physics; semiconductors; spectroscopy; electronics; optical system design; purification of materials by recrystallization from the melt; infrared systems and techniques; space science; optics of solar power systems Mailing Add: 5163 Belmez Laguna Hills CA 92653

MCFEE, WILLIAM WARREN, b Concord, Tenn, Jan 8, 35; m 57; c 3. FOREST SOILS, SOIL FERTILITY. Educ: Univ Tenn, BS, 57; Cornell Univ, MS, 63, PhD(soils), 66. Prof Exp: Asst soils, Cornell Univ, 61-65; from asst prof to assoc prof, 65-73, PROF AGRON, PURDUE UNIV, WEST LAFAYETTE, 73-, DIR NATURAL RESOURCE & ENVIRON SCI PROG, 75- Res: Relationship of soils to plant nutrition; mechanisms of ion uptake; forest tree-site relationships; forest tree species; common agricultural plants. Mailing Add: Dept of Agron Purdue Univ West Lafayette IN 47906

MCFEELEY, JAMES CALVIN, b Altoona, Pa, Aug 6, 40; m 63; c 2. BOTANY. Educ: Otterbein Col, BS, 65; Ohio Univ, MS, 68; Ohio State Univ, PhD(plant path), 71. Prof Exp: Teacher, High Sch, Ohio, 65-66; ASST PROF BIOL, E TEX STATE UNIV, 72- Concurrent Pos: Res assoc, Purdue Univ, 71-72; consult, Tulsa Dist, US Corps Engrs, 73 & Tex Hwy Dept, 73- Mem: Am Phytopath Soc; Am Bot Soc; Mycol Soc Am; Sigma Xi. Res: Study of the affect of strong 60- hertz electric fields on biological systems. Mailing Add: Dept of Biol ETex State Univ Commerce TX 75428

MCFEELY, RICHARD AUBREY, b Trenton, NJ, Dec 3, 33; m 56; c 3. VETERINARY MEDICINE, CYTOGENETICS. Educ: Pa State Univ, BS, 55; Univ Pa, VMD, 61, MS, 67; Am Col Theriogenologists, dipl, 73. Prof Exp: Pvt pract, 61-62; res asst comp cardiol, Sch Vet Med, 62, USPHS fels reprod physiol, 62-63, grad div, Sch Med, 63-65 & cytogenetics, Sch Vet Med, 65-66, from asst prof to assoc prof clin reprod, 66-75, chief, Sect Reprod, 68-73, PROF CLIN REPROD, SCH VET MED, UNIV PA, 75-, CHIEF OF STAFF, 73- Concurrent Pos: NIH res grant, 64-; Lalor fel award, 71-72. Mem: Am Vet Med Asn; Am Asn Vet Clinicians; Am Asn Equine Practitioners; Am Asn Bovine Practitioners; Soc Theriogenol. Res: Chromosome abnormalities in domestic mammals; especially sex determination and altered reproductive function. Mailing Add: Sect of Reprod New Bolton Ctr Kennett Square PA 19348

MCFERRAN, JOE, b Charleston, Ark, Mar 22, 17; m 41; c 2. VEGETABLE CROPS. Educ: Univ Ark, BSA, 41, MS, 50, Cornell Univ, PhD, 55. Prof Exp: Rural supvr, Farm Security Admin, 41; asst, 46, from instr to assoc prof, 47-62, PROF HORT & FORESTRY, UNIV ARK, FAYETTEVILLE, 62- Mem: Am Soc Hort Sci. Res: Plant breeding; irrigation of horticultural crops; variety and strain. Mailing Add: Dept of Hort Univ of Ark Fayetteville AR 72701

MCFETERS, GORDON ALWYN, b Ayer, Mar 28, 39; m 63; c 2. MICROBIAL PHYSIOLOGY, BIOCHEMISTRY. Educ: Andrews Univ, BA, 61; Loma Linda Univ, MS, 63; Ore State Univ, PhD(microbiol), 67. Prof Exp: Asst prof, 67-72, ASSOC PROF MICROBIOL, MONT STATE UNIV, 72- Mem: Am Soc Microbiol; fel Am

Acad Microbiol. Res: Microbial control mechanisms, enzymology and electron transport mechanisms; water microbiology. Mailing Add: Dept of Bot & Microbiol Mont State Univ Bozeman MT 59715

MCGAHA, YOUNG JOHN, b Cosby, Tenn, Sept 13, 10; m 47; c 1. ZOOLOGY. Educ: Lincoln Mem Univ, AB, 38; Univ Tenn, MS, 40; Univ Mich, MA, 50, PhD(zool), 51. Prof Exp: Asst zool, Univ Mich, 40-42; asst prof biol, Bradley Univ, 46-48; from asst prof to assoc prof, 50-60, PROF BIOL, UNIV MISS, 60- Mem: Am Soc Limnol & Oceanog; Micros Soc Am; Am Fisheries Soc. Res: Limnology; aquatic entomology; food relationships of aquatic animals; fishery biology. Mailing Add: Dept of Biol Univ of Miss University MS 38677

MCGAHAN, MERRITT WILSON, b Los Angeles, Calif, Mar 8, 20; m 46; c 2. PLANT MORPHOLOGY. Educ: Univ Calif, Berkeley, AB, 48; Univ Calif, Davis, PhD(bot), 52; Union Theol Sem, NY, BD, 66. Prof Exp: Teaching asst bot, Univ Calif, Davis, 49-51; res assoc chem debarking, State Univ NY Col Forestry, Syracuse Univ, 51-52; from instr to asst prof bot, Univ RI, 52-59; anatomist-morphologist, United Fruit Co, Mass, 59-61 & Honduras, 61-63; prof bot, dean grad sch & dir, Inst Sci Res, 66-70, dean acad affairs, 70-74, PROF BIOL & CHMN DEPT, NMEX HIGHLANDS UNIV, 74- Concurrent Pos: Vis assoc prof, Brandeis Univ, 60-61. Mem: AAAS; Bot Soc Am. Res: Effects of trace metals on growth of Southwestern range grasses. Mailing Add: NMex Highlands Univ Las Vegas NM 87701

MCGAHEN, JOE WINFIELD, b Haileyville, Okla, July 27, 22; m 46. MICROBIOLOGY. Educ: Okla Agr & Mech Col, BS, 47; La State Univ, MS, 49, PhD(bot), 51. Prof Exp: Microbiologist, 51-74, RES ASSOC, BIOCHEM DEPT, E I DU PONT DE NEMOURS & CO, INC, 74- Mem: Am Inst Biol Sci. Res: Chemical action; virology. Mailing Add: 1108 N Hilton Rd Oak Lane Manor Wilmington DE 19803

MCGAHREN, WILLIAM JAMES, b Ballyshannon, Ireland, Feb 16, 24; US citizen; m 58; c 4. ORGANIC CHEMISTRY. Educ: Chelsea Polytech Col, BSc, 53; Brooklyn Col, MA, 57; Brooklyn Polytech Inst, PhD(org chem), 66. Prof Exp: Chemist, Charles Pfizer Co, 53-63; res fel chem, Brooklyn Polytech Inst, 64-66; SR RES SCIENTIST NATURAL PROD CHEM, LEDERLE LABS, AM CYANAMID CO, PEARL RIVER, NY, 66- Mem: Am Chem Soc; Sigma Xi. Res: Isolation from microbial sources of novel products that exhibit a specific biological activity; elucidation of structure and sterochemistry of these materials; microbial enzyme transformation of substrates of commercial interest. Mailing Add:

MCGANDY, EDWARD LEWIS, b Washington, DC, Apr 22, 30; m 55; c 2. PHYSICAL CHEMISTRY, PUBLIC HEALTH. Educ: George Washington Univ, BS, 51; Boston Univ, PhD(chem), 60. Prof Exp: Lab technician, Geophys Lab, Carnegie Inst, 47-51; anal chemist, Nat Bur Stand, Washington, DC, 51-54, phys chemist, 54-55; asst prof biol, Purdue Univ, West Lafayette, 64-67; ASST PROF BIOCHEM, GRAD SCH PUB HEALTH, UNIV PITTSBURGH, 67- Concurrent Pos: Am Cancer Soc exchange fel, Med Res Coun Lab Molecular Biol, Cambridge Univ, 60-63. Mem: AAAS; Am Chem Soc; Am Crystallog Asn. Res: Molecular structure of biological molecules; x-ray crystallography; automatic x-ray diffractometry; amino acid and peptide structures and conformations. Mailing Add: Grad Sch of Pub Health Univ of Pittsburgh Pittsburgh PA 15213

MCGANITY, WILLIAM JAMES, b Kitchener, Ont, Sept 21, 23; nat US; m 48; c 3. OBSTETRICS & GYNECOLOGY. Educ: Univ Toronto, MD, 46; FRCS(C), 53; Am Bd Nutrit, dipl. Prof Exp: Intern, Toronto Gen Hosp, Ont, Can, 46-47; resident obstet & gynec, Univ Toronto & Toronto Gen Hosp, 49-52; from instr to assoc prof, Sch Med, Vanderbilt Univ, 52-59; PROF OBSTET & GYNEC & CHMN DEPT, UNIV TEX MED BR, GALVESTON, 60- Concurrent Pos: Res fel nutrit, Univ Toronto, 47-48; res fel, Vanderbilt Univ, 48-49; Lowell M Palmer sr fel, 54-56; part time lectr, Univ Toronto, 49-52; consult, Interdept Comt Nutrit Nat Defense, DC, 56-73, Dept Air Force, 60-, Dept Army, 60-67 & Comt Consider Folic Acid, Food & Drug Admin, 60-63; mem, Comts Dietary Allowances & Int Nutrit, Food & Nutrit Bd, Nat Res Coun, 59-63; dean fac, Univ Tex Med Br, Galveston, 64-67; mem comt maternal nutrit, 66-70; chmn, Comt Maternal Health & Sci Activ, Tex Med Asn, 61-67, Coun Foods & Nutrit, AMA, 63-66, Nutrit Study Sect, NIH, 63-67 & Res Panel Maternal & Child Health, Children's Bur, 64-66; panel mem, White House Conf Nutrit & Health, 69; co-chmn panel nutrit & health, US Senate Select Comt Nutrit & Human Needs, 74. Honors & Awards: Hendry Prize, Univ Toronto, 47. Mem: Am Col Obstet & Gynec; Am Inst Nutrit; AMA; Am Soc Clin Nutrit (past pres); Asn Prof Gynec & Obstet. Res: Nutrition in reproduction; nutrition among underdeveloped populations; physiology of human reproduction. Mailing Add: Dept of Obstet & Gynec Univ of Tex Med Br Galveston TX 77550

MCGANN, LOCKSLEY EARL, b Kingston, Jamaica, Aug 11, 46; Can citizen; m 69; c 2. CRYOBIOLOGY. Educ: Univ Waterloo, BSc, 69, MSc, 70, PhD(physics), 73. Prof Exp: Med Res Coun Can fel, Div Cryobiol, Clin Res Ctr, Harrow, Eng, 73-75; ASST PROF BIOMED ENG, UNIV ALTA, 75- Mem: Soc Cryobiol; Soc Low Temperature Biol; Biophys Soc. Res: Interactions of living systems with the environment during cooling to and warming from low temperatures; development of methods for the frozen preservation of cells, tissues and organs. Mailing Add: Div of Biomed Eng & Appl Sci Univ of Alta Edmonton AB Can

MCGANNON, DONALD E, JR, b Minneapolis, Minn, Oct 12, 29; m 55; c 3. GEOLOGY, GEOCHEMISTRY. Educ: Syracuse Univ, BA, 52; Univ Minn, MS, 57, PhD(geol), 60. Prof Exp: Assoc prof, 59-71, PROF GEOL, TRINITY UNIV, TEX, 71-, CHMN DEPT, 59- Mem: AAAS; Geol Soc Am; Nat Asn Geol Teachers. Res: Petrology; mineralogy. Mailing Add: Dept of Geol Trinity Univ 715 Stadium Dr San Antonio TX 78282

MCGARITY, WILLIAM CECIL, b Jersey, Ga, Oct 5, 21; m 50; c 3. SURGERY. Educ: Emory Univ, BA, 42, MD, 45. Prof Exp: Instr, 51-54, assoc, 54-59, asst prof, 59-64, ASSOC PROF SURG, SCH MED, EMORY UNIV, 64- Concurrent Pos: Mem, Am Bd Surg. Mem: AMA; Am Heart Asn; fel Am Col Surg. Res: Burns treated with steroids; transminase values in biliary tract and liver pathology; prognosis and course of dogs following litigation of common duct; arterial emboli. Mailing Add: Dept of Surg Emory Univ Med Sch Atlanta GA 30322

MCGARR, ARTHUR, b San Francisco, Calif, May 24, 40; m 71. SEISMOLOGY. Educ: Calif Inst Technol, BS, 62, MS, 63; Columbia Univ, PhD(geophys), 68. Prof Exp: Res asst, Lamont Geol Observ, Columbia Univ, 63-68; sr res fel, 68-70, SR RES OFFICER GEOPHYS, BERNARD PRICE INST GEOPHYS RES, UNIV WITWATERSRAND, 70- Mem: AAAS; Am Geophys Union; Seismol Soc Am. Res: Modes of rock deformation in high-stress environments; earthquakes associated with mining and fluid injection; magnitude statistics of earthquakes; seismic near-field measurements of tilt, strain and acceleration. Mailing Add: Bernard Price Inst Geophys Res Univ of Witwatersrand Johannesburg South Africa

MCGARRITY, GERARD JOHN, b Brooklyn, NY, Oct 10, 40; m 64; c 2. MICROBIOLOGY, INFECTIOUS DISEASES. Educ: St Joseph Col, Pa, BS, 62; Jefferson Med Col, MS, 65, PhD(microbiol), 70. Prof Exp: Asst prof biol, Glassboro State Col, 64-65; res assoc cell biol, 65-71, HEAD DEPT MICROBIOL, INST MED RES, 71- Concurrent Pos: J A Hartford Found grant, 69-72; instr, Univ Pa, 71-; instr, Univ Pa, 71-74; exchange visitor, Czechoslovak Acad Sci, 75. Mem: AAAS; Am Soc Microbiol; Tissue Culture Asn. Res: Infection in hospitals research and animal laboratories; mycoplasma; murine viruses; equipment design; biohazard research. Mailing Add: Dept of Microbiol Inst for Med Res Copewood St Camden NJ 08103

MCGARRY, ELEANOR E, b Montreal, Que, Dec 17, 15. MEDICINE, ENDOCRINOLOGY. Educ: McGill Univ, BSc, 37, MD & CM, 47, MSc, 51; FRCPS(C). Prof Exp: Med Res Coun Can traveling fel, Univ Col Hosp Med Sch, London, 51-52; lectr med & clin med, McGill Univ, 52-54; asst med, Sch Med, Johns Hopkins Univ, 54-55; lectr med & clin med, 56-58, from asst prof to assoc prof, 58-74, PROF MED, MCGILL UNIV, 74- Concurrent Pos: Sr fel, Nat Res Coun Can, 52-54; med res assoc, Med Res Coun Can, 57; sr physician, Royal Victoria Hosp, 75- Mem: Am Acad Allergy; Am Col Physicians; Am Diabetes Asn; Endocrine Soc; Can Soc Clin Invest. Res: Metabolism; protein hormone, particularly those of pituitary and immunochemistry of protein and polypeptide hormones. Mailing Add: Royal Victoria Hosp 687 Pine Ave W Montreal PQ Can

MCGARRY, JOHN DENIS, b Widnes, Eng, Dec 6, 40; m 67; c 3. BIOCHEMISTRY. Educ: Victoria Univ, Manchester, BSc, 62, PhD(biochem), 66. Prof Exp: Asst prof, 69-73, ASSOC PROF INTERNAL MED & BIOCHEM, UNIV TEX HEALTH SCI CTR DALLAS, 73- Concurrent Pos: Fels, Univs Liverpool & Wales, 65-67; fel, Univ Tex Southwestern Med Sch Dallas, 68-69. Mem: Am Diabetes Asn; Brit Biochem Soc; Am Soc Biol Chem. Res: Regulation of carbohydrate and lipid metabolism with particular emphasis on the control of ketogenesis in starvation and diabetes. Mailing Add: Dept of Internal Med Univ of Tex Health Sci Ctr Dallas TX 75235

MCGARRY, MARGARET, b Boston, Mass, Apr 11, 28. INORGANIC CHEMISTRY, ANALYTICAL CHEMISTRY. Educ: Regis Col, Mass, AB, 57; Univ Pa, PhD(chem), 64. Prof Exp: From instr to asst prof, 64-70, ASSOC PROF CHEM, REGIS COL, MASS, 70- Mem: Am Chem Soc. Res: Liability of aqueous solutions of alkali silicates; spectral properties of dyes in colloidal systems; flocculation of colloids; electron microscopy of colloidal substances; ion exchange separations; gas chromatography. Mailing Add: Dept of Chem Regis Col Weston MA 02193

MCGARRY, PAUL ANTHONY, b Warren, Pa, Nov 13, 28; m 55; c 3. PATHOLOGY, NEUROPATHOLOGY. Educ: Pa State Univ, BS, 50; Temple Univ, MD, 54; Am Bd Path, dipl & cert anat & clin path, 63, cert neuropath, 68. Prof Exp: From instr to assoc prof, 63-72, PROF PATH, SCH MED, LA STATE UNIV, 72- Concurrent Pos: Asst vis pathologist, Charity Hosp, New Orleans, La, 63-65, vis pathologist, 65-; vis pathologist, New Orleans Vet Admin, La, 68- Mem: Am Asn Neuropath. Res: Cerebrovascular disease; cervical spondylotic myelopathy; cerebrospinal fluid cytology. Mailing Add: Dept of Path La State Univ New Orleans LA 70112

MCGARVEY, BRUCE RITCHIE, b Springfield, Mo, Mar 10, 28; m 54; c 3. PHYSICAL CHEMISTRY. Educ: Carleton Col, BA, 50; Univ Ill, MA, 51, PhD(chem), 53. Prof Exp: From instr to asst prof chem, Univ Calif, 53-57; from asst prof to assoc prof, Kalamazoo Col, 57-62; from assoc prof to prof, Polytech Inst Brooklyn, 62-72, actg head dept, 71-72; PROF CHEM, UNIV WINDSOR, 72- Concurrent Pos: Guggenheim fel, Imp Col, Univ London, 67-68. Mem: Am Chem Soc; Am Phys Soc; Chem Inst Can. Res: Nuclear magnetic and electron spin resonance. Mailing Add: Dept of Chem Univ of Windsor Windsor ON Can

MCGARVEY, DAVID CARTER, mathematics, see 12th edition

MCGARY, CHARLES WESLEY, JR, b New Castle, Pa, Dec 12, 29; m 49; c 5. POLYMER CHEMISTRY. Educ: Westminster Col, Pa, BS, 51; Purdue Univ, PhD(phys org chem), 55. Prof Exp: Res chemist, Chem Div, 54-61, group leader, 61-66, asst dir chem & plastics div, 66-73, PROD MGR, UNION CARBIDE CORP, 73- Mem: Am Chem Soc. Res: Plastics intermediates, epoxy resins, polyesters, plasticizers, vinyl resins, latex paints, coatings and water soluble polymers. Mailing Add: Chem & Plastics Div Union Carbide Corp 270 Park Ave New York NY 10017

MCGAUGH, JAMES L, b Long Beach, Calif, Dec 17, 31; m 52; c 3. NEUROBIOLOGY, PSYCHOPHARMACOLOGY. Educ: San Jose State Univ, BA, 53; Univ Calif, Berkeley, PhD(psychol), 59. Prof Exp: Nat Res Coun sr fel physiol psychol, Adv Inst Health, Italy, 61-62; assoc prof psychol, Univ Ore, 62-64; prof psychobiol & chmn dept, 64-67, dean, Sch Biol Sci, 67-70, chmn psychobiol dept, 71-75, PROF PSYCHOBIOL, UNIV CALIF, IRVINE, 71-, VICE CHANCELLOR ACAD AFFAIRS, 75- Concurrent Pos: Mem biol sci training rev comt, Nat Inst Ment Health, chmn, 71-72, mem preclin psychopharmacol res rev comt, 75-; consult, Vet Admin; ed, Behav Biol. Mem: Fel AAAS; Int Brain Res Orgn; Soc Neurosci; fel Am Psychol Asn. Res: Biological bases of behavior; central nervous system functions in memory. Mailing Add: Dept of Psychobiol Univ of Calif Irvine CA 92664

MCGAUGH, MAURICE EDRON, b Rayville, Mo, June 9, 06; m 29. GEOGRAPHY. Educ: Univ Kans, BS, 39, AM, 41; Univ Chicago, PhD, 50. Prof Exp: Instr geog, Army Air Corps Col Training Detachment, Southwest Mo State Col, 43-44; instr, Army Specialized Training Prog, Univ Kans, 44; from asst prof to prof, 49-74, head dept, 56-71, EMER PROF GEOG, CENT MICH UNIV, 74- Mem: Asn Am Geog; Nat Coun Geog Educ. Res: Cartography; regional geography of Great Plains and Michigan. Mailing Add: Dept of Geog Cent Mich Univ Mt Pleasant MI 48858

MCGAUGHEY, ALBERT WAYNE, b Russellville, Ind, July 16, 14; m 41; c 4. MATHEMATICS. Educ: Wabash Col, AB, 35; Univ Iowa, MS, 37; Univ Cincinnati, PhD(math), 40. Prof Exp: Instr math, Purdue Univ, 40-41; from instr to asst prof, US Naval Acad, 41-46; prof & chmn dept, Westminster Col, 46-48; assoc prof, 48-53, PROF MATH, BRADLEY UNIV, 53-, CHMN DEPT, 71- Concurrent Pos: Vis lectr, Eureka Col, 56-58. Mem: Math Asn Am. Res: Lacunary double Fourier series. Mailing Add: Dept of Math Bradley Univ Peoria IL 61606

MCGAUGHEY, CHARLES GILBERT, b San Diego, Calif, Sept 8, 25. BIOCHEMISTRY, DENTAL RESEARCH. Educ: Univ Calif, BA, 50; Univ Southern Calif, MA, 52. Prof Exp: Asst biochem, Univ Southern Calif, 51-52; radiol biochemist, US Naval Radiol Defense Lab, 52; res biochemist med, 52-63, RES BIOCHEMIST, ORAL DIS RES LAB, VET ADMIN HOSP, LONG BEACH, CALIF, 63- Mem: AAAS; Am Chem Soc. Res: Mechanism of formation of dental plaque and calculus; biochemistry and physiology of cancer cell; biochemistry of saliva; relation of cyclic nucleotides to oral squamous cell carcinoma; effects of polyphosphates on parameters related to dental caries. Mailing Add: Oral Dis Res Lab R-4 Vet Admin Hosp Long Beach CA 90801

MCGAUGHEY, WILLIAM HORTON, b Jack Co, Tex, Apr 6, 41; m 61; c 2. ENTOMOLOGY. Educ: Tex Tech Col, BS, 63; Iowa State Univ, MS, 65, PhD(entom), 67. Prof Exp: Instr entom, Iowa State Univ, 66-67; RES SCIENTIST,

AGR RES SERV, US GRAIN MKT RES CTR, USDA, 67- Concurrent Pos: Adj entomologist, Dept of Entomology, Kans State Univ, 73- Mem: Entom Soc Am; Entom Soc Can; Am Mosquito Control Asn; Soc Invert Path. Res: Medical and stored-product entomology. Mailing Add: US Grain Mkt Res Ctr 1515 College Ave Manhattan KS 66502

MCGAUGHY, ROBERT EARL, biophysics, see 12th edition

MCGAVIC, JOHN SAMUEL, b St Louis, Mo, Feb 18, 11; m 33; c 3. OPHTHALMOLOGY. Educ: Iowa Wesleyan Col, BS, 34; Univ Iowa, MD, 34; Univ Cincinnati, MS, 37; Am Bd Ophthal, dipl. Prof Exp: Instr ophthal, Univ Cincinnati, 37-38; instr, Col Physicians & Surgeons, Columbia Univ, 39-44; prof clin ophthal, Grad Sch Med, Univ Pa, 55-73; PROF OPHTHAL, JEFFERSON MED COL, 73- Concurrent Pos: Fel Columbia Univ, 38-39; asst pathologist, Eye Inst, Presby Hosp, New York, 40-44; asst ophthalmologist, Mem Hosp Cancer & Allied Dis, 41-44; attend ophthalmologist, Bryn Mawr Hosp, 47-; consult, Valley Forge Army Hosp, 47-74; prof ophthal, Sch Med, Temple Univ, 56-73; attend ophthalmologist, Wills Eye Hosp, 73- Mem: Am Ophthal Soc; AMA; Asn Res Vision & Ophthal; Am Acad Ophthal & Otolaryngol. Mailing Add: Dept of Ophthal Jefferson Med Col Philadelphia PA 19107

MCGAVIN, MATTHEW DONALD, b Goondiwindi, Australia, July 25, 30; m 61; c 3. VETERINARY PATHOLOGY. Educ: Univ Queensland, BVSc, 52, MVSc, 62; Mich State Univ, PhD(path), 64; Am Col Vet Path, dipl, 63. Prof Exp: Vet off, Animal Res Inst, Yeerongpilly, Australia, 52-59, sr histopathologist, 59-61 & 64-68; ASSOC PROF PATH, KANS STATE UNIV, 68-, RES PATHOLOGIST, 74- Concurrent Pos: Muscular Dystrophy Asn Am grant, Kans State Univ, 72. Mem: Am Vet Med Asn; Am Col Vet Path; Australian Vet Asn. Res: Ovine muscular dystrophy; comparative neuropathology. Mailing Add: Dept of Path Kans State Univ Manhattan KS 66506

MCGAVIN, RAYMOND E, b Youngstown, Ohio, Jan 13, 21; m 50; c 6. PHYSICS. Educ: Niagara Univ, BA, 46; Univ Colo, BA, 50. Prof Exp: Physicist, Nat Bur Standards, 50-52, electronics scientist, 52-59, supvr elec sci, 59-61, physicist, 61-64, supvry physicist, Environ Sci Serv Admin, 64-69, SUPVRY PHYSICIST, NAT OCEANIC & ATMOSPHERIC ADMIN, 69- Concurrent Pos: Mem, Inter-Range Instrumentation Group, 65- Mem: Am Geophys Union; Sigma Xi; Inst Elec & Electronics Engrs. Res: Radio meteorology; atmospheric turbulence; tropospheric radio wave propagation. Mailing Add: Wave Propagation Lab Nat Oceanic & Atmospheric Admin Boulder CO 80302

MCGAVOCK, WALTER DONALD, b Nashville, Tenn, Apr 18, 33. INVERTEBRATE ZOOLOGY, PARASITOLOGY. Educ: Mid Tenn State Univ, BS, 56, MA, 58; Univ Tenn, PhD(zool), 67. Prof Exp: Chmn dept biol, Limestone Col, 58-62; asst prof, Wofford Col, 62-64; instr & res asst zool, Univ Tenn, 66-67; assoc prof biol, ETenn State Univ, 67-70; assoc prof, 70-74, PROF BIOL, TUSCULUM COL, 74-, DIR DIV NATURAL SCI & MATH, 70- Concurrent Pos: Lectr, Spartanburg Gen Hosp, SC, 62-64. Mem: Am Soc Parasitol. Res: Chemotherapy of cestodes; taxonomy of helminths; host-parasitic relationships. Mailing Add: Div of Natural Sci & Math Tusculum Col Box 4 Greeneville TN 37743

MCGAVOCK, WILLIAM CREWS, b Springfield, Ill, Dec 20, 06; m 38. INORGANIC CHEMISTRY. Educ: Univ Mo, AB, 28, MA, 29; Univ Calif, PhD(chem), 36. Prof Exp: Chemist, Welsbach Mfg Co, 28-29; asst, Univ Mo, 29-30 & Univ Calif, 30-33; chemist, Shell Oil Co, 33-35, Great Western Electro-Chem Co, 35-37 & Great Western Div, Dow Chem Co, 37-39; prof chem, Trinity Univ, Tex, 39-72; RETIRED. Concurrent Pos: Chemist, Tex Veg Oil Co, 43-45; Welch grant, 64-68; consult. Honors & Awards: Piper Prof Award. Mem: Am Chem Soc; The Chem Soc; Am Inst Chem; The Chem Soc. Res: Iodine; layer crystals; reaction kinetics; gel permeation chromatography. Mailing Add: 1702 Waverly Ave San Antonio TX 78201

MCGAVOCK, WILLIAM GILLESPIE, b Franklin, Tenn, Dec 31, 09; m 31; c 2. MATHEMATICS. Educ: Davidson Col, BS, 30; Duke Univ, MA, 33, PhD(math), 39. Prof Exp: Teacher, Branham-Hughes Mil Acad, Tenn, 30-32; teacher, Morgan Sch, Tenn, 33-34; from asst prof to prof math, Davidson Col, 34-44; vis assoc prof, Duke Univ, 44-45; vis prof, Emory Univ, 45-46; prof, 46-70, DANA PROF, DAVIDSON COL, 70- Concurrent Pos: Ford fel, Univ Wis, 54-55. Mem: Math Asn Am. Res: Invariant of Pfaffian systems; annihilators of quadratic forms with applications to Pfaffian systems. Mailing Add: Dept of Math Davidson Col Davidson NC 28036

MCGAVRAN, EDWARD GRAFTON, public health, deceased

MCGAVRAN, MALCOLM HOWARD, b Harda, India, Oct 18, 29; US citizen; c 5. PATHOLOGY. Educ: Bethany Col, WVa, AB, 51; Wash Univ, MD, 54; Am Bd Path, dipl, 59. Prof Exp: From instr to assoc prof path, Med Sch, Wash Univ, 57-70; prof path, Hershey Col Med, Pa State Univ & dir anat path, Hershey Med Ctr Hosp, 70-75; PROF PATH & DIR DIV ANAT PATH, BAYLOR COL MED, 75-; CHIEF ANAT PATH, THE METHODIST HOSP, 75- Concurrent Pos: Consult, Vet Admin Hosp, Houston. Mem: AAAS; Am Asn Path & Bact; Int Acad Path. Res: Surgical pathology; oncology; electron microscopy. Mailing Add: The Methodist Hosp Tex Med Ctr Houston TX 77025

MCGEACHIN, ROBERT LORIMER, b Pasadena, Calif, May 13, 17; m 47; c 2. BIOCHEMISTRY. Educ: Univ Nebr, BS, 39, MS, 40; Wash Univ, PhD(org chem), 42. Prof Exp: Asst chem, Univ Nebr, 39-40; asst, Univ Ill, 46-47; from asst prof to assoc prof, 47-66, PROF BIOCHEM, SCH MED, UNIV LOUISVILLE, 66- Concurrent Pos: USPHS fel, Univ Ill, 46-47; chmn dept biochem, Sch Med, Univ Louisville, 72-75. Mem: Am Chem Soc; Soc Exp Biol & Med; Am Soc Biol Chem. Res: Chemistry and physiology of mammalian amylases. Mailing Add: Univ of Louisville Sch of Med Louisville KY 40201

MCGEARY, DAVID F R, b Bellefonte, Pa, Dec 23, 40; m 67; c 1. GEOLOGY, OCEANOGRAPHY. Educ: Williams Col, BA, 62; Univ Ill, Urbana, 64; Scripps Inst Oceanog, PhD(oceanog), 69. Prof Exp: Asst prof, 69-74, ASSOC PROF GEOL, CALIF STATE UNIV, SACRAMENTO, 74- Mem: AAAS; Geol Soc Am; Am Geophys Union; Am Asn Petrol Geologists. Res: Marine geology. Mailing Add: Dept of Geol Calif State Univ Sacramento CA 95819

MCGEE, CHARLES E, b Baylor Co, Tex, July 29, 35; m 60; c 1. ANALYTICAL CHEMISTRY, RADIOLOGICAL HEALTH. Educ: ETex State Univ, BS, 62; Purdue Univ, MS, 66, PhD(bionucleonics), 68. Prof Exp: Res asst physiol, Southwestern Med Sch, Univ Tex, 60-62; asst prof, 67-71, ASSOC PROF RADIOCHEM, SCH PHARM, TEMPLE UNIV, 71-, PROJ DIR, RADIOL HEALTH SPECIALIST TRAINING PROJ, 68- Mem: Am Chem Soc; Am Pub Health Asn. Res: Environmental health; automation; enzymes as analytical reagents; clinical biochemistry, preclinical sensors. Mailing Add: Dept of Pharmaceut Chem Temple Univ Sch of Pharm Philadelphia PA 19140

MCGEE, EDWARD ARTHUR, b Glen Falls, NY, Feb 11, 38; m 65; c 5. OPERATIONS RESEARCH. Educ: Univ Notre Dame, BSChE, 59, MSChE, 64; Univ Chicago, MBA, 69. Prof Exp: Engr, Tech Serv Dept, Standard Oil Co Ind, Whiting, 61-65, analyst opers res, 65-67, supvr tech comput prog, Chicago, 67-69; CONSULT OPERS RES, W R GRACE & CO, NEW YORK, 69- Concurrent Pos: Teaching asst chem eng, Univ Notre Dame, 59-61. Mem: Inst Mgt Sci; Am Inst Chem Eng. Res: Linear programming and mixed integer programming applications in industrial and business fields. Mailing Add: 21 Nassau Rd Upper Montclair NJ 07043

MCGEE, JAMES FRANCIS, b Philadelphia, Pa, Sept 23, 15; m 42. ATOMIC PHYSICS. Educ: St Joseph's Col, BS, 38; Univ Notre Dame, MS, 40; Stanford Univ, PhD(physics), 56. Prof Exp: Res engr aerophysics div, NAm Aviation, Inc, 46-49; res engr, Atomics Int, 49-52; asst, Stanford Univ, 52-55, res assoc, 55-56; from asst prof to prof, 56-69, PROF PHYSICS, ST LOUIS UNIV, 69- Concurrent Pos: Res Corp grants, 57-59; NSF grants, 58-60; Nat Acad Sci-Nat Res Coun grant, 61; mem, Governor's Sci Adv Comt, Mo, 62-64; Fulbright lectr, Finland, 64-65; Am Cancer Soc grant, 67-69; NIH grant, 67-73. Mem: AAAS; Optical Soc Am; Am Phys Soc; Biophys Soc; Am Asn Physics Teachers. Res: Optics; optics of x-ray microscopes and x-ray telescopes; radiation biology in the low energy x-ray and thermal neutron regions; solid-state radiation damage and x-ray astronomy. Mailing Add: Dept of Physics St Louis Univ 221 N Grand Blvd St Louis MO 63103

MCGEE, LLOYD R, organic chemistry, polymer chemistry, see 12th edition

MCGEE, THOMAS HOWARD, b New York, NY, Dec 19, 41; m 66; c 3. CHEMICAL KINETICS, PHOTOCHEMISTRY. Educ: St John's Univ, NY, BS, 62; Univ Conn, PhD(phys chem), 66. Prof Exp: Fel chem, Univ Tex, Austin, 66-67; asst prof, 67-74, ASSOC PROF CHEM, YORK COL, NY, 74- Concurrent Pos: NSF acad year award, 68-69; res collabr, Chem Dept, Brookhaven Nat Lab, 75-76. Mem: Am Chem Soc. Res: Chemical kinetics; photochemistry. Mailing Add: Dept of Chem York Col 150-14 Jamaica Ave Jamaica NY 11432

MCGEE, VICTOR ERROL, applied statistics, management sciences, see 12th edition

MCGEE, WILLIAM WALTER, b Toledo, Ohio, June 27, 39; m 63; c 1. ANALYTICAL CHEMISTRY. Educ: Univ Toledo, BS, 61, MS, 63; Univ Fla, PhD(chem), 66. Prof Exp: Group leader gas chromatography lab, B F Goodrich Co, 66-68; asst prof, 68-69, ASSOC PROF ANAL CHEM, FLA TECHNOL UNIV, 69- Mem: Am Chem Soc. Res: Fundamental processes which occur in the flame, as used in flame spectroscopy. Mailing Add: Dept of Chem Fla Technol Univ PO Box 25000 Orlando FL 32816

MCGEER, EDITH GRAEF, b New York, NY, Nov 18, 23; m 54; c 3. ORGANIC CHEMISTRY. Educ: Swarthmore Col, BA, 44; Univ Va, PhD(org chem), 46. Prof Exp: Technician, Squibb Inst Med Res, E R Squibb & Sons, 43; res chemist, Exp Sta, E I du Pont de Nemours & Co, Inc, 46-54; res assoc, 54-74, ASSOC PROF, KINSMEN LAB NEUROL RES, UNIV BC, 74- Mem: NY Acad Sci; Can Biochem Soc; Soc Neurosci. Res: Synaptic transmission; neurochemistry and behavior. Mailing Add: Kinsmen Lab Neurol Res Univ of BC Vancouver BC Can

MCGEER, JAMES PETER, b Vancouver, BC, May 14, 22; m 48; c 4. PHYSICAL CHEMISTRY. Educ: Univ BC, BA, 44, MA, 46; Princeton Univ, MA, 48, PhD(chem), 49. Prof Exp: Res phys chemist & group leader, Aluminum Labs, Ltd, 49-68; tech supt, Aluminum Co Can Ltd, 68-70, asst mgr reduction div, Arvida Works & mgr reduction technol, Quebec Smelters, Alcan, 70-75, MEM STAFF, ALCAN SMELTERS SERV, 75- Mem: Am Inst Mining, Metall & Petrol Engrs; Chem Inst Can. Res: Electrometallurgy of aluminum; carbon; smelting furnace operation. Mailing Add: Alcan Smelter Serv PO Box 6090 Montreal PQ Can

MCGEER, PATRICK L, b Vancouver, BC, Can, June 29, 27; m 54; c 3. BIOCHEMISTRY. Educ: Univ BC, BA, 48, MD, 58; Princeton Univ, PhD(phys chem), 51. Prof Exp: Res chemist, Polychem Dept, E I du Pont de Nemours & Co, 51-54; res assoc, 56-58, from asst prof to assoc prof, 59-74, PROF, KINSMEN LAB NEUROL RES, UNIV BC, 74- DIR, 64- Concurrent Pos: Intern Vancouver Gen Hosp, 58-59. Mem: Fel AAAS; Soc Neurosci; Can Biochem Soc; Am Neurochem Soc; Int Brain Res Orgn. Res: Medical biochemistry; neurochemistry; neurophysiology; biogenic amine metabolism. Mailing Add: Kinsmen Lab of Neurol Res Univ of BC Vancouver BC Can

MCGEE-RUSSELL, SAMUEL M, b Sutton, Eng, Aug 24, 27; m 55; c 3. BIOLOGY, ELECTRON MICROSCOPY. Educ: Oxford Univ, BA, 51, MA, 54, DPhil(zool), 55. Prof Exp: Asst lectr zool, Birkbeck Col, Univ London, 54-56; vis scientist fel, NIH, 56-57; lectr zool, Birkbeck Col, Univ London, 57-61; lab dir & staff scientist, Virus Res Unit, Med Res Coun, 62-68; sr res assoc & lectr cell biol, 68-72, PROF CELL BIOL & ELECTRON MICROS, STATE UNIV NY ALBANY, 72- Concurrent Pos: Dir, Co of Biologists, 67- Mem: Fel Zool Soc London; fel Royal Micros Soc; Brit Soc Exp Biol (secy, 62-66); Am Soc Cell Biol; Electron Micros Soc Am. Res: Light microscopy; microtechnique and microtomy; cell biology; histochemistry; protozoology; invertebrate zoology especially of mollusca; virology. Mailing Add: Dept of Biol Sci State Univ of NY Albany NY 12203

MCGEHEE, CHARLES LEROY, b Joplin, Mo, Jan 24, 24; m 60; c 2. ANALYTICAL CHEMISTRY. Educ: Mo Sch Mines, BS, 48. Prof Exp: Analyst, Cities Serv Refining Corp, 49-53, chemist, 54-57, res chemist, Cities Serv Res & Develop Co, 58-61, sect supvr, 62-63; assoc prof scientist, Columbian Carbon Co, NJ, 64-71; SR RES SCIENTIST, CITIES SERV CO, 71- Mem: AAAS; Am Chem Soc; Soc Appl Spectros. Res: Mass spectrometry; nuclear magnetic resonance; differential thermal analysis. Mailing Add: Petrochem Res Drawer 4 Cities Serv Co Cranbury NJ 08512

MCGEHEE, FIELDING M, JR, operations research, physics, see 12th edition

MCGEHEE, OSCAR CARRUTH, b Baton Rouge, La, Nov 29, 39. MATHEMATICS. Educ: Rice Univ, BA, 61; Yale Univ, MA, 63, PhD(math), 66. Prof Exp: From instr to asst prof math, Univ Calif, Berkeley, 65-71; ASSOC PROF, LA STATE UNIV, BATON ROUGE, 71- Concurrent Pos: NATO fel, Fac Sci, Orsay, France, 67-68. Mem: Am Math Soc; Math Asn Am; Math Soc France; London Math Soc. Res: Commutative harmonic analysis; functional analysis. Mailing Add: Dept of Math La State Univ Baton Rouge LA 70803

MCGEHEE, RALPH MARSHALL, b Magnolia, Miss, Apr 23, 21; m 45; c 1. APPLIED MATHEMATICS, NUMERICAL ANALYSIS. Educ: La Col, BA, 42; NC State Col, BEE, 49, MS, 50, PhD(elec eng), 53. Prof Exp: Staff mem, Sandia Lab, US AEC, 53-61; mem tech staff, Tucson Eng Lab, Hughes Aircraft Co, 61-62; assoc prof math, 62-66, head dept comput sci, 66-67, ASSOC PROF COMPUT SCI, NMEX INST MINING & TECHNOL, 66- Mem: Inst Elec & Electronics Eng; Asn Comput Mach; Soc Indust & Appl Math; Opers Res Soc Am; Am Meteorol Soc. Res: Applied analysis; modeling of numerical analysis; atmospheric dynamics on thunderstorn scales. Mailing Add: Dept of Comput Sci NMex Inst of Mining & Technol Socorro NM 87801

MCGEHEE, RICHARD PAUL, b San Diego, Calif, Sept 20, 43; m 67. MATHEMATICS. Educ: Calif Inst Technol, BS, 64; Univ Wis-Madison, MS, 65, PhD(math), 69. Prof Exp: Vis mem, Courant Inst Math Sci, NY Univ, 69-70; asst prof, 70-75, ASSOC PROF MATH, UNIV MINN, MINNEAPOLIS, 75- Mem: Am Math Soc. Res: Qualitative theory of ordinary differential equations. Mailing Add: Sch of Math Univ of Minn Minneapolis MN 55455

MCGEHEE, RICHARD VERNON, b Tyler, Tex, Aug 1, 34; m 58; c 2. GEOLOGY. Educ: Univ Tex, BS, 55, PhD(geol), 63; Yale Univ, MS, 56. Prof Exp: Petrol geologist, Phillips Petrol Co, 56-57; instr geol, Univ Kans, 63; asst prof, Western Mich Univ, 63-66 & SDak Sch Mines & Technol, 66-67; assoc prof, Western Mich Univ, 67-72; vis prof geol, Inst Geol, Nat Univ Mex, 72-74; ASSOC PROF GEOL, DIV EARTH & PHYS SCI, UNIV TEX AT SAN ANTONIO, 74- Mem: Geol Soc Am; Nat Asn Geol Teachers; Am Asn Petrol Geologists. Res: Igneous and metamorphic petrology; structural and economic geology. Mailing Add: Div of Earth & Phys Sci Univ of Tex San Antonio TX 78285

MCGERRIGLE, HAROLD WILLIAM, geology, deceased

MCGERVEY, JOHN DONALD, b Pittsburgh, Pa, Aug 9, 31; m 57; c 2. EXPERIMENTAL SOLID STATE PHYSICS. Educ: Univ Pittsburgh, BS, 52; Carnegie Inst Technol, MS, 55, PhD(physics), 61. Prof Exp: Instr math, Carnegie Inst Technol, 57-60; asst prof, 60-65, ASSOC PROF PHYSICS, CASE WESTERN RESERVE UNIV, 65- Concurrent Pos: Vis scientist, Inst Solid State Physics & Nuclear Res, W Ger, 72-73. Mem: AAAS; Am Phys Soc; Am Asn Physics Teachers; Am Asn Univ Prof; Sigma Xi. Res: Study of solids and liquids by positron annihilation. Mailing Add: Dept of Physics Case Western Reserve Univ Cleveland OH 44106

MCGETCHIN, THOMAS R, b La Jolla, Calif, July 31, 36; m 61; c 2. GEOLOGY. Educ: Occidental Col, BA, 59; Brown Univ, ScM, 61; Calif Inst Technol, PhD(geol), 68. Prof Exp: Teaching asst geol, Brown Univ, 60-61 & Calif Inst Technol, 61-67; from instr to asst prof geophys, Air Force Inst Technol, 67-69; asst prof geol, Mass Inst Technol, 69-74; res assoc, Smithsonian Inst, 74; GROUP LEADER, GEOSCI GROUP, LOS ALAMOS SCI LAB, UNIV CALIF, LOS ALAMOS, NM, 74- Mem: AAAS; Am Geophys Union; Geol Soc Am. Res: Igneous petrology; volcanism; planetary science. Mailing Add: Univ of Calif Los Alamos Sci Lab PO Box 1663 Q-21 MS 978 Los Alamos NM 87545

MCGHEE, CHARLES ROBERT, b Chattanooga, Tenn, July 17, 34; m 64; c 1. SYSTEMATIC ZOOLOGY. Educ: Mid Tenn State Univ, BS, 61, MA, 62; Va Polytech Inst, PhD(zool), 70. Prof Exp: Asst prof, 68-73, ASSOC PROF BIOL, MID TENN STATE UNIV, 73- Mem: Am Arachnol Soc; Sigma Xi. Res: Systematics of phalangid genus Leiobunum. Mailing Add: Dept of Biol Mid Tenn State Univ Box 292 Murfreesboro TN 37130

MCGHEE, JERRY ROGER, b Knoxville, Tenn, June 25, 41; m 61; c 2. MICROBIOLOGY, IMMUNOLOGY. Educ: Univ Tenn, Knoxville, BS, 64; Univ Tenn, Memphis, PhD(microbiol), 69. Prof Exp: Res asst prof, Univ Tenn, Memphis, 69; instr microbiol & dent res, 69-71, asst prof microbiol, 72-75, ASSOC PROF MICROBIOL & SCIENTIST INST DENT RES, UNIV ALA, BIRMINGHAM, 75- Concurrent Pos: Teaching fel microbiol, Med Units, Univ Tenn, Memphis, 64-66; NIH fel microbiol, Univ Chicago, 71-72. Mem: Am Soc Microbiol; Am Asn Immunol; Soc Exp Biol & Med; Reticuloendothelial Soc; Tissue Cult Asn. Res: Host-parasite interrelationships; cellular rates of synthesis; effective cholera immunity; biological activity of cholera toxin; local immunity to infectious diseases; dental caries immunity. Mailing Add: Dept of Microbiol Univ of Ala Med Ctr Birmingham AL 35233

MCGHEE, ROBERT BARCLAY, b Cleveland, Tenn, Feb 22, 18; m 46; c 3. PARASITOLOGY. Educ: Berea Col, AB, 40; Univ Ga, MS, 42; Univ Chicago, PhD(bact, parasitol), 48. Prof Exp: Asst animal path, Rockefeller Inst, 48-51, assoc, 51-54; assoc prof biol, 54-55, head dept zool, 55-64, prof, 59-64, ALUMNI FOUND DISTINGUISHED PROF ZOOL, UNIV GA, 64- Concurrent Pos: Vis lectr, Univ Mex, 64-65; mem study sect, Trop Med & Parasitol, NIH, 66-71, chmn, 70-71. Mem: AAAS; Am Soc Nat; Am Soc Zool; Am Soc Parasitol; Am Soc Protozool. Res: Relationship of host to parasite, as expressed in various species of plasmodia and trypanosomids; innate and acquired immunity; evolution of parasitic organisms; medical bacteriology and parasitology; insects affecting man and animals; invertebrate zoology. Mailing Add: Dept of Zool Univ of Ga Athens GA 30601

MCGIBBON, WILLIAM HENRY, b Moore's Mills, NB, Mar 29, 11; nat US; m 36; c 7. GENETICS. Educ: McGill Univ, BSA, 32; Univ Wis, PhD(poultry genetics), 42. Prof Exp: Asst poultry supt, Dept Agr, NB, 34-38; asst genetics, 38-41, instr poultry husb, 41-42, from asst prof to assoc prof, 42-50, PROF POULTRY HUSB, UNIV WIS-MADISON, 50- Mem: AAAS; Genetics Soc Am; Poultry Sci Asn; Am Genetic Asn. Res: Poultry genetics; inheritance of antigenic characters in the red blood cells of domesticated birds; inbreeding of chickens. Mailing Add: Dept of Poultry Sci Univ of Wis Madison WI 53706

MCGIFF, JOHN C, b New York, NY, Aug 6, 27; m 58; c 5. PHARMACOLOGY, INTERNAL MEDICINE. Educ: Georgetown Univ, BS, 47; Columbia Univ, MD, 51; Am Bd Internal Med, dipl, 64. Prof Exp: Res assoc, Sch Med, St Louis Univ, 58-59; from asst instr to instr, Univ Pa, 60-62; assoc med & pharmacol, 62-64, asst prof, 64-65; assoc prof internal med & chief cardiovasc sect, Sch Med, St Louis Univ, 65-70, prof med, 70-71; prof pharmacol & med & dir clin pharm sect, Med Col Wis, 71-75; PROF & CHMN DEPT PHARMACOL, UNIV TENN, CTR HEALTH SCI, 75- Concurrent Pos: Fel med, Col Physicians & Surgeons, Columbia Univ, 57-58; Am Heart Asn res fel, 57-59; fel & co-recipient inst grant, USPHS, 63-65; Burroughs Wellcome Fund scholar; estab investr, Am Heart Asn, 64-69; vis prof, Tulane Univ, 69-; mem, Cardiovasc B Study Sect, NIH; mem med adv bd, Coun High Blood Pressure Res. Mem: Am Fedn Clin Res; Am Physiol Soc; NY Acad Sci; Am Soc Pharmacol; Am Soc Clin Invest. Res: Cardiorenal pharmacology and physiology; clinical pharmacology. Mailing Add: Dept of Pharmacol Univ of Tenn Ctr Health Sci Memphis TN 38163

MCGILL, DAVID A, b Albany, NY, Sept 18, 30. OCEANOGRAPHY. Educ: Bucknell Univ, BSc, 52; Columbia Univ, MA, 56; Yale Univ, PhD(zool), 63. Prof Exp: Technician chem oceanog, Woods Hole Oceanog Inst, 56-59, fel, 59-62, asst scientist, 62-66; asst prof, Southeastern Mass Tech Inst, 66-68; PROF OCEAN SCI, US COAST GUARD ACAD, 68- Concurrent Pos: Mem, Int Geophys Year Surv Atlantic Ocean, 57-58; Sci Comt Oceanic Res-UNESCO grants, Hawaii, 61 & Australia, 62; mem, Int Indian Ocean Expeds, 62-65. Mem: AAAS; Soc Study Evolution; Am Soc Limnol & Oceanog; Marine Biol Asn UK. Res: Distribution of nutrient elements in sea water, chiefly inorganic and total phosphorus. Mailing Add: Dept of Ocean Sci US Coast Guard Acad New London CT 06320

MCGILL, DAVID PARK, b Waverly, Nebr, Sept 3, 19; m 44; c 1. AGRONOMY. Educ: Univ Nebr, BS, 41, MS, 49; Iowa State Col, PhD(agron), 54. Prof Exp: Asst agron, 46-51, from asst agronomist to assoc agronomist, 51-62, PROF AGRON, UNIV NEBR, LINCOLN, 62- Mem: Am Soc Agron. Res: Genetics teaching. Mailing Add: Dept of Agron Univ of Nebr Lincoln NE 68583

MCGILL, DOUGLAS B, b New York, NY, Aug 11, 29; m 53; c 4. GASTROENTEROLOGY. Educ: Yale Univ, BA, 51; Tufts Univ, MS, 55; Univ Minn, MS, 61. Prof Exp: Intern med, Boston City Hosp, 55-56; resident, Mayo Grad Sch Med, Univ Minn, 58-61; CONSULT MED & GASTROENTEROL, MAYO CLIN, 61-; ASSOC PROF MED, MAYO MED SCH, 73- Mem: AMA; fel Am Col Physicians; Am Gastroenterol Asn; Am Asn Study Liver Dis; Am Fedn Clin Res. Res: Liver disease. Mailing Add: Div of Gastroenterol Mayo Clin 200 First St SW Rochester MN 55901

MCGILL, GEORGE EMMERT, b Des Moines, Iowa, June 10, 31; m 55; c 3. GEOLOGY. Educ: Carleton Col, BA, 53; Univ Minn, MS, 55; Princeton Univ, PhD(geol), 58. Prof Exp: Asst prof, 58-65, ASSOC PROF GEOL, UNIV MASS, AMHERST, 65- Mem: Am Geophys Union; Geol Soc Am; Am Asn Petrol Geol. Res: Structural and regional geology; astrogeology. Mailing Add: Dept of Geol Univ of Mass Amherst MA 01002

MCGILL, HENRY COLEMAN, JR, b Nashville, Tenn, Oct 1, 21; m 45; c 3. PATHOLOGY. Educ: Vanderbilt Univ, BA, 43, MD, 46. Prof Exp: Intern, Vanderbilt Univ, 46-47; asst, Sch Med, La State Univ, 47-48, from asst prof to prof path, 50-66, head dept, 66-72, PROF PATH, UNIV TEX HEALTH SCI CTR, SAN ANTONIO, 72- Concurrent Pos: Mem coun arteriosclerosis, Am Heart Asn; chmn dept path, Univ Tex Health Sci Ctr, 60-66. Mem: Am Soc Exp Path; Am Asn Path & Bact; Int Acad Path. Res: Arteriosclerotic heart disease. Mailing Add: Dept of Path Univ of Tex Health Sci Ctr San Antonio TX 78284

MCGILL, JOHN JOSEPH, b Wilkes-Barre, Pa, July 18, 08; m 37. ENTOMOLOGY. Educ: Fordham Univ, BS, 30, MS, 34, PhD(biol), Concurrent Pos: Chmn physiol panel, Rumen Function Conf, 69- Honors & Awards: Am Feed Mfrs Award, 72. Mem: Am Dairy Sci Asn; Am Soc Animal Sci. Res: Nutritional physiology and biochemistry; experimental surgery of digestive tract; cardiovascular and lymphatic systems. Mailing Add: Dept of Animal Sci Iowa State Univ Ames IA 50010

MCGILL, JOHN THOMAS, b Memphis, Tenn, June 19, 21; m 43; c 2. GEOLOGY. Educ: Univ Calif, Los Angeles, AB, 43, MA, 48, PhD(geol), 51. Prof Exp: Lectr geol, Univ Calif, Los Angeles, 48-50 & 51-52, asst res geologist, 52-53; geologist, 53-64, admin geologist & chief, 64-74, GEOLOGIST, ENG GEOL BR, US GEOL SURV, 74- Mem: Geol Soc Am; Asn Eng Geol (vpres, 59-60). Res: Urban engineering geology and environmental geology; landslides; coastal geomorphology. Mailing Add: US Geol Surv Fed Ctr Denver CO 80225

MCGILL, JULIAN EDWARD, b Blacksburg, SC, Oct 22, 32; m 59; c 2. ANALYTICAL CHEMISTRY. Educ: Erskine Col, BA, 55; Clemson Univ, MS, 68, PhD(chem), 71. Prof Exp: Qual control engr, Celanese Fibers Co, 59-61; ASST PROF PHARM CHEM, COL PHARM, MED UNIV SC, 70- Mem: Am Chem Soc. Res: Spectrophotometric analysis; pharmaceutical analysis with emphasis on drugs of abuse. Mailing Add: Col of Pharm Med Univ of SC 80 Barre St Charleston SC 29401

MCGILL, LAWRENCE DAVID, b Lincoln, Nebr, Mar 24, 44; m 66. VETERINARY PATHOLOGY. Educ: Okla State Univ, BS, 66, DVM, 68; Tex A&M Univ, PhD(vet path), 72. Prof Exp: NIH fel, Tex A&M Univ, 68-71, teaching asst vet path, 71; asst prof, Univ Minn, St Paul, 71-72; ASST PROF VET PATH, UNIV NEBR, LINCOLN, 72- Mem: Am Vet Med Asn. Res: Immunopathology; pathogenesis of viral infections; ultrastructural pathology; host-virus relationships. Mailing Add: Dept of Vet Sci Univ of Nebr Lincoln NE 68583

MCGILL, LOIS SATHER, b Wilsonville, Ore, July 5, 23; m 69; c 6. FOOD SCIENCE. Educ: Ore State Col, BS, 45. Prof Exp: Instr food technol, 45-58, from instr to assoc prof food sci, 52-72, PROF FOOD SCI, ORE STATE UNIV, 72- Mem: Inst Food Technol; Am Dairy Sci Asn; Am Home Econ Asn. Res: Sensory evaluation of foods and product development. Mailing Add: Dept of Food Sci & Technol Ore State Univ Corvallis OR 97331

MCGILL, ROBERT MAYO, b Marianna, Ark, Nov 29, 25. PHYSICAL INORGANIC CHEMISTRY. Educ: Univ Ark, BS, 45, MA, 48; Univ Tenn, PhD(phys chem), 55. Prof Exp: CHEMIST, RES LAB, K-25 PLANT, UNION CARBIDE CORP, 48- Mem: Am Chem Soc; AAAS. Res: Molecular weight determination; adsorption of fatty acids on metals; chemistry of uranium and fluorine; kinetics of precipitation; isotope separations. Mailing Add: Union Carbide Corp Nuclear Div Oak Ridge TN 37830

MCGILL, SUZANNE, b Port Arthur, Tex, Mar 10, 44. MATHEMATICS. Educ: Univ St Thomas, BA, 67; Tex Christian Univ, MS, 70, PhD(math), 72. Prof Exp: Instr, Pub Sch, Tex, 67-68; instr math, Tex Christian Univ, 72; PROF MATH, UNIV S ALA, 72- Mem: Math Asn Am. Res: Research activities are of general nature in the field of abstract algebra. Mailing Add: Dept of Math Univ of SAla Mobile AL 36688

MCGILL, THOMAS CONLEY, JR, b Port Arthur, Tex, Mar 20, 42; m 66. PHYSICS, ELECTRICAL ENGINEERING. Educ: Lamar State Col, BS, 63 & 64; Calif Inst Technol, MS, 65, PhD(elec eng, physics), 69. Prof Exp: Asst prof, 71-74, ASSOC PROF APPL PHYSICS, CALIF INST TECHNOL, 74- Concurrent Pos: Mem tech staff, Hughes Res Labs, 67-73, consult, 73-; NATO fel theoret physics, Bristol Univ, 69-70; Air Force Nat Res Coun fel, Princeton Univ, 70-71; Alfred P Sloan Found fel, 74-76. Mem: AAAS; Am Inst Physics; Inst Elec & Electronics Eng. Res: Solid state physics, particularly semiconductors and insulators. Mailing Add: Dept of Appl Physics Calif Inst of Technol Pasadena CA 91125

MCGILLIARD, A DARE, b Stillwater, Okla, Oct 15, 26; m 51; c 3. ANIMAL NUTRITION. Educ: Okla State Univ, BS, 51, MS, 52; Mich State Univ, PhD(animal nutrit), 61. Prof Exp: From asst prof to assoc prof, 57-72, PROF ANIMAL SCI, IOWA STATE UNIV, 72- Concurrent Pos: Chmn physiol panel, Rumen Function Conf, 69- Honors & Awards: Am Feed Mfrs Award, 72. Mem: Am Dairy Sci Asn; Am Soc Animal Sci. Res: Nutritional physiology and biochemistry; experimental surgery of digestive tract; cardiovascular and lymphatic systems. Mailing Add: Dept of Animal Sci Iowa State Univ Ames IA 50010

MCGILLIARD, LON DEE, b Manhattan, Kans, Aug 9, 21; c 4. ANIMAL BREEDING, DAIRY SCIENCE. Educ: Okla State Univ, BS, 42; Mich State Univ, MS, 47; Iowa State Univ, PhD(animal breeding, dairy husb), 52. Prof Exp: Asst dairy prod, Mich State Univ, 46-47; asst animal breeding, Iowa State Univ, 48-49, from res assoc to assoc prof animal breeding & dairy husb, 49-55; assoc prof, 55-62, PROF DAIRY CATTLE BREEDING, MICH STATE UNIV, 62- Concurrent Pos: Assoc ed, J Dairy Sci, 69-72, ed, 73- Mem: Am Dairy Sci Asn. Res: Improving dairy cattle through breeding; population genetics. Mailing Add: Dept of Dairy Sci Mich State Univ East Lansing MI 48824

MCGILLIARD, MICHAEL LON, b Lansing, Mich, Sept 29, 47; m 70. DAIRY SCIENCE, ANIMAL GENETICS. Educ: Mich State Univ, BS, 69; Iowa State Univ, MS, 70, PhD(animal breeding), 74. Prof Exp: EXTEN SPECIALIST DAIRY SCI, VA POLYTECH INST & STATE UNIV, 75- Mem: Am Dairy Sci Asn. Res: Determining the influence of quantitative measures on dairy herd management. Mailing Add: Dept of Dairy Sci Va Polytech Inst & State Univ Blacksburg VA 24061

MCGILVERY, ROBERT WARREN, b Coquille, Ore, Aug 25, 20; m 43; c 4. BIOCHEMISTRY. Educ: Ore State Col, BS, 41; Univ Wis, PhD(physiol chem), 47. Prof Exp: From asst prof to assoc prof physiol chem, Univ Wis, 48-57; assoc prof, 57-62, PROF BIOCHEM, UNIV VA, 62- Concurrent Pos: USPHS sr res fel, Univ Wis, 47-48; vis prof biochem, State Univ NY Buffalo, 71. Mem: AAAS; Am Soc Biol Chem; Am Chem Soc. Res: Metabolic economy. Mailing Add: PO Box 3852 Charlottesville VA 22903

MCGINLEY, PATTON HOPKINS, radiological physics, health physics, see 12th edition

MCGINN, CLIFFORD, b New York, NY, Aug 22, 22; m 54; c 4. ORGANIC CHEMISTRY. Educ: Fordham Univ, BS, 43, MS, 46; Syracuse Univ, PhD(chem), 57. Prof Exp: From instr to assoc prof, 49-70, PROF CHEM, LEMOYNE COL, NY, 70- Mem: Am Chem Soc. Res: Theoretical organic chemistry; chemical bonding. Mailing Add: Dept of Chem LeMoyne Col Syracuse NY 13214

MCGINN, GEORGE A, theoretical atomic physics, molecular physics, see 12th edition

MCGINN, LAURENCE CODORI, mathematics, see 12th edition

MCGINNES, BURD SHELDON, b Pittsburgh, Pa, Aug 10, 21; m 46; c 1. WILDLIFE RESEARCH. Educ: Pa State Univ, BS, 48, MS, 49; Va Polytech Inst & State Univ, PhD(biol), 58. Prof Exp: Proj leader, Del Game & Fish Comn, 49-51 & 52-55; leader wild turkey invest, Pa Game Comn, 58; LEADER COOP WILDLIFE RES UNIT, VA POLYTECH INST & STATE UNIV, 58- Mem: Wildlife Soc. Res: Animal ecology; wildlife diseases and management techniques. Mailing Add: Coop Wildlife Res Unit Va Polytech Inst & State Univ Blacksburg VA 24061

MCGINNES, EDGAR ALLAN, b Chestertown, Md, Feb 15, 26; m 51; c 3. WOOD CHEMISTRY, WOOD TECHNOLOGY. Educ: Pa State Univ, BS, 50, MF, 51; State Univ NY Col Forestry, Syracuse Univ, PhD(wood technol), 55. Prof Exp: Jr res chemist, Am Viscose Corp, 51-52, res chemist, 55-58, res specialist, 59-60; assoc prof, 60-63, PROF FORESTRY, UNIV MO-COLUMBIA, 63- Mem: Fel AAAS; fel Am Inst Chemists; Am Chem Soc; Soc Am Foresters; Forest Prod Res Soc. Res: Wood anatomy; influence of environmental stress on wood structure. Mailing Add: Sch of Forestry Fisheries & Wildlife Univ of Mo Columbia MO 65201

MCGINNESS, JAMES DONALD, b Evansville, Ind, June 23, 30; m 50; c 5. ANALYTICAL CHEMISTRY, SPECTROSCOPY. Educ: Univ Evansville, AB, 52. Prof Exp: Chemist, 52-55, chemist & group leader spectros, 55-61, dir anal res dept, Ill, 61-70, mgr reliability, Ohio, 70-74, GROUP MGR AUTOMOTIVE CHEM COATINGS, SHERWIN-WILLIAMS CO, 74- Mem: Am Chem Soc; Soc Appl Spectros; Fedn Socs Paint Technol; Am Soc Qual Control. Res: Infrared spectroscopy; gas chromatography; x-ray emission; flame photometry; coatings technology; color science; polymer characterization; nuclear magnetic resonance. Mailing Add: Sherwin-Williams Co 8250 St Aubin Detroit MI 48212

MCGINNESS, JOHN ED, quantum biochemistry, see 12th edition

MCGINNIES, WILLIAM GROVENOR, b Steamboat Springs, Colo, Aug 14, 99; m 25; c 1. PLANT ECOLOGY. Educ: Univ Ariz, BSA, 22; Univ Chicago, PhD(plant ecol), 32. Prof Exp: Asst range examr, US Forest Serv, 24-26; asst prof animal husb, univ & range specialist, Exp Sta, Univ Ariz, 27-29, assoc prof & range ecologist, 30-32, actg head dept bot, 32-35; mgr & dir land mgt, Navajo Dist, Soil Conserv Serv, USDA, 35-38, chief div range res, Southwest Forest & Range Exp Sta, US Forest Serv, 38-42; chief div surv & invest, Guayule Emergency Rubber Proj, 42-44; dir, Rocky Mountain Forest & Range Exp Sta, US Forest Serv, 45-53 & Cent States Forest Exp Sta, 54-60; dir lab tree-ring res & coordr arid lands res prog, 60-65, prof dendrochronol & proj mgr, Off Arid Lands Res, 65-70, EMER PROF DENDROCHRONOL & EMER DIR ARID LANDS STUDIES, UNIV ARIZ, 70- Mem: Ecol Soc Am; Soc Range Mgt; Tree Ring Soc. Res: Dendrochronology; ecology of desert plants; frequency and abundance of plants; water requirement of Xerophytes. Mailing Add: 530 E Cambridge Dr Tucson AZ 85704

MCGINNIES, WILLIAM JOSEPH, b Tucson, Ariz, Jan 2, 27; m 49; c 3. RANGE SCIENCE, PLANT ECOLOGY. Educ: Colo State Univ, BS, 48, PhD, 67; Univ Wis, MS, 52. Prof Exp: Forester watershed mgt, Forest Serv, Idaho, 48, range conservationist grazing mgt & range reseeding, Utah, 49-53, range conservationist range reseeding, Agr Res Serv, Utah, 54-56 & Colo, 56-66, RANGE SCIENTIST RANGE RESEEDING, AGR RES SERV, COLO, USDA, 66- Concurrent Pos: Fac affil & mem grad fac, Colo State Univ, 66- Mem: AAAS; Soc Range Mgt; Crop Sci Soc Am; Am Soc Agron; Soil Sci Soc Am. Res: Adaptability of species for range seeding; methods of establishing seeded grasses on rangeland; techniques for measuring range herbage and environmental requirements of range plants; rehabilitation of disturbed areas. Mailing Add: Crops Res Lab Agr Res Serv Colo State Univ Ft Collins CO 80523

MCGINNIS, ARTHUR JAMES, b Edmonton, Alta, Dec 28, 21; m 48; c 4. BIOCHEMISTRY. Educ: Univ Alta, BSc, 46; Mont State Univ, MS, 50; Ore State Univ, PhD(biochem), 55. Prof Exp: Entomologist, Alta, 57-66, head cereal crop protection sect, 66-72, DIR RES STA, CAN DEPT AGR, 72- Mem: Am Chem Soc; Entom Soc Can; Agr Inst Can. Res: Nutrition and biochemistry of insects. Mailing Add: Res Sta Can Dept of Agr PO Box 185 Vineland Station ON Can

MCGINNIS, CARL LEONARDT, b Los Angeles, Calif, Apr 24, 20; m 62. PHYSICS. Educ: Mass Inst Technol, BS, 42; Univ Calif, PhD(physics), 52. Prof Exp: Physicist, Nuclear Data Group, Nat Bur Standards, 52-54 & Nat Res Coun, 54-62; corp secy, Southwestern Portland Cement Co, 63-70; vpres, Leonardt Found, 70-75; RETIRED. Mem: Am Phys Soc. Res: Beta decay. Mailing Add: Suite 500 1801 Ave of the Stars Los Angeles CA 90067

MCGINNIS, CHARLES HENRY, JR, b Newark, NJ, Nov 24, 34; m 56; c 2. AVIAN PHYSIOLOGY. Educ: Rutgers Univ, New Brunswick, BS, 56; Purdue Univ, MS, 61; Mich State Univ, PhD(poultry physiol), 65. Prof Exp: Asst prof physiol, Sch Vet Med, Univ Minn, St Paul, 65-67; avian physiologist, 67-69, HEAD POULTRY RES SECT, HESS & CLARK DIV, RHODIA INC, 69- Mem: Poultry Sci Asn; World Poultry Sci Asn; Am Physiol Soc; Am Inst Biol Sci. Res: Physiology, nutrition and toxicology in poultry. Mailing Add: Hess & Clark Div Rhodia Inc Seventh & Orange Ashland OH 44805

MCGINNIS, EDGAR LEE, b La Ceiba, Honduras, Jan 4, 38; US citizen. PETROLEUM CHEMISTRY. Educ: Col William & Mary, BSc, 59; Oxford Univ, DPhil(org chem), 63. Prof Exp: SR RES CHEMIST, GULF RES & DEVELOP CO, 63- Mem: Am Chem Soc. Res: Organic intermediates and reaction kinetics; spectroscopic and stereochemical determinations of steroidal and triterpene compounds; petroleum and separations research; coal science; catalysis. Mailing Add: Phys Sci Div Gulf Res & Develop Co Pittsburgh PA 15230

MCGINNIS, EUGENE A, b Jessup, Pa, May 15, 21; m 41; c 5. PHYSICS. Educ: Univ Scranton, BS, 48; NY Univ, MS, 53; Fordham Univ, PhD(physics), 60. Prof Exp: Instr, Clarks Summit Sch, 46-48; from instr to assoc prof, 48-63, PROF PHYSICS, UNIV SCRANTON, 63-, CHMN DEPT, 67- Mem: Am Asn Physics Teachers; Soc Appl Spectros. Res: Emission and molecular spectroscopy, particularly Raman spectorscopy of gases. Mailing Add: Dept of Physics Univ of Scranton Scranton PA 18510

MCGINNIS, GARY DAVID, b Everett, Wash, Oct 1, 40; m 64; c 3. CARBOHYDRATE CHEMISTRY. Educ: Pac Lutheran Univ, BS, 62; Univ Wash, MS, 68; Univ Mont, PhD(org chem), 70. Prof Exp: Prod chemist, Am Cyanamid Co, 64-67; fel, Univ Mont, 70-71; ASST PROF WOOD CHEM, FOREST PROD LAB, MISS STATE UNIV, 71- Mem: Am Chem Soc (secy carbohydrate div, 75); Forest Prod Res Soc; Sigma Xi. Res: Thermal decomposition of wood; high pressure liquid chromatography of natural products; chemical composition of wood barks. Mailing Add: Miss State Univ PO Drawer FP Mississippi State MS 39762

MCGINNIS, JAMES, b Cliffside, NC, Apr 10, 18; m 52; c 1. NUTRITION. Educ: NC State Col, BS, 40; Cornell Univ, PhD(nutrit), 44. Prof Exp: Asst nutrit, Cornell Univ, 40-43, res assoc, 43-44; from asst prof to assoc prof poultry sci, 44-50, PROF POULTRY SCI, POULTRY SCIENTIST & EXTEN POULTRY SCI SPECIALIST, WASH STATE UNIV, 50- Concurrent Pos: Asst dir agr develop dept, Chas Pfizer & Co, 52-53. Mem: AAAS; Am Chem Soc; Soc Exp Biol & Med; Poultry Sci Asn; Am Inst Nutrit. Res: Nutritional requirements of poultry; biochemistry of proteins; nutritional and endocrinological interrelationships; Browning reaction in foods; antibiotics and nutrition. Mailing Add: Dept of Animal Sci Wash State Univ Pullman WA 99163

MCGINNIS, JOHN THURLOW, b Mattoon, Ill, July 9, 34; m 56; c 6. ECOLOGY, BIOSTATISTICS. Educ: Eastern Ill Univ, BS, 56; Univ Tenn, MS, 58; Emory Univ, PhD(biol), 64. Prof Exp: Res assoc biol, Emory Univ, 60-64; asst statistician, Inst Ecol, Univ Ga, 64-69, asst prof bot, 66-69, sr ecologist, 69-70; assoc chief ecol & ecosysts anal sect, 70-72, mgr ecol & ecosysts anal sect, 73-76; RES LEADER, ECOL & ECOSYSTS ANAL SECT, BATTELLE COLUMBUS LABS, 76- Concurrent Pos: Consult, Off Technol Assessment, 76. Mem: Fel AAAS; Ecol Soc Am; Am Trop Biol; Sigma Xi. Res: Environmental effects, assessment and systems ecology for government and industrial projects. Mailing Add: Battelle Columbus Labs 505 King Ave Columbus OH 43201

MCGINNIS, LYLE DAVID, b Appleton, Wis, Mar 5, 31; m 59; c 4. GEOPHYSICS. Educ: St Norbert Col, BSc, 54; St Louis Univ, MSc, 60; Univ Ill, PhD(geol), 65. Prof Exp: Geophys trainee, Carter Oil Co, Okla, 54-55; asst geophysicist, Arctic Inst NAm, 57-59; assoc geophysicist, Ill State Geol Surv, 59-66; tech expert, UN Develop Prog, Afghanistan, 66-67; assoc prof, 67-72, PROF GEOL, NORTHERN ILL UNIV, 72- Concurrent Pos: Consult geophysicist, Desert Res Inst, Univ Nev, 61-62; consult, Minn Geol Surv, 70-; NSF coordr, Antarctic Drilling Prog, 71-; geophysicist, US Geol Surv, Woods Hole, Mass, 75-76. Mem: AAAS; fel Geol Soc Am; Soc Explor Geophys; Am Geophys Union; Glaciol Soc. Res: Polar geology and geophysics gravity field related to crustal densities and tectonics; geophysical studies applied to ground water. Mailing Add: Dept of Geol Northern Ill Univ De Kalb IL 60115

MCGINNIS, MICHAEL RANDY, b Hayward, Calif, Oct 22,. MYCOLOGY. Educ: Calif State Polytech Col, BS, 61; Iowa State Univ, PhD(mycol), 69. Prof Exp: Fel, Ctr Dis Control, USPHS, 71-73; supvr, Mycol Lab, SC Dept Health & Environ Control, 74-75; ASSOC DIR, CLIN MICROBIOL LABS, NC MEM HOSP, UNIV NC, 75- Concurrent Pos: Consult & instr, Analytab Prod Inc, 75-76. Mem: Mycol Soc Am; Am Soc Microbiol; Int Soc Human & Animal Mycol; Paleont Soc; Am Phytopath Soc. Res: Taxonomic relations of the human pathogenic dematiaceous hyphomycetes. Mailing Add: Clin Microbiol Labs NC Mem Hosp Univ of NC Chapel Hill NC 27514

MCGINNIS, ROBERT CAMERON, b Edmonton, Alta, Aug 18, 25; m 51; c 3. AGRONOMY. Educ: Univ Alta, BSc, 49, MSc, 51; Univ Man, PhD(cytogenetics), 54. Prof Exp: Asst, Univ Alta, 49-51; cytologist, Can Dept Agr, 51-60; assoc prof plant sci, Univ Man, 60-65, prof & Head dept, 65-75; ASSOC DIR, INT CROPS RES INST FOR SEMI-ARID TROPICS, 75- Concurrent Pos: Dir, Plant Breeding Sta, Njoro, Kenya, 73-75. Honors & Awards: Centennial Medal, Can Govt, 67. Mem: Sigma Xi; NY Acad Sci; Genetics Soc Can; Agr Inst Can. Res: Improving crops and farming systems for the semi-arid tropics. Mailing Add: ICRISAT 1-11-256 Begumpet Hyderabad AP 500016 India

MCGINNIS, SAMUEL M, b Milwaukee, Wis, Apr 18, 36; m 60. HERPETOLOGY. Educ: Univ Wis, BS, 58; Univ Calif, Berkeley, PhD(physiol ecol), 65. Prof Exp: From asst prof to assoc prof, 64-74, PROF BIOL, CALIF STATE COL, HAYWARD, 74- Res: Physiological ecology of reptiles, amphibians and fish, especially temperature physiology and the use of thermal telementary techniques. Mailing Add: Dept of Biol Sci Calif State Col Hayward CA 94542

MCGINNIS, WILLIAM JOSEPH, b St Louis, Mo, Apr 30, 23; m 51; c 4. INORGANIC CHEMISTRY, PHYSICAL CHEMISTRY. Educ: Colo Col, BA, 48; Iowa State Col, MS, 51. Prof Exp: Chemist, Los Alamos Sci Lab, Univ Calif, 48-49; res asst, Inst Atomic Res, Iowa State Col, 49-51; chemist, Int Minerals & Chem Corp, 51-56; chemist, Pigments Dept, E I du Pont de Nemours & Co, Inc, 56-63, sr res chemist, 63, tech supvr, Tenn, 63-69, prod specialist, Del, 69-70, tech serv supvr, 70-74, TECH SERV CONSULT, PIGMENTS DEPT, E I DU PONT DE NEMOURS & CO, INC, 74- Mem: Am Chem Soc; Tech Asn Pulp & Paper Indust. Res: Rare earth metals; potassium and magnesium compounds; phosphate chemistry; titanium chemistry and compounds, particularly titanium dioxide pigments. Mailing Add: Pigments Dept Tech Serv Labs E I du Pont de Nemours & Co Inc Wilmington DE 19899

MCGINNISS, VINCENT DANIEL, b Philadelphia, Pa, Feb 9, 42; m 68; c 1. POLYMER CHEMISTRY. Educ: Univ Fla, BS, 63; Univ Ariz, PhD(phys org chem), 70. Prof Exp: Anal chemist agr, Agr Dept, Univ Fla, 62-63; res chemist polymers, Peninsular Chem Res, 64-65; teaching asst org chem, Univ Ariz, 66-68; sr res chemist, 70-72, SCIENTIST POLYMERS, GLIDDEN-DURKEE, 72- Honors & Awards: Roon Award, Fedn Socs Paint Technol, 73. Mem: Am Chem Soc. Res: Kinetics and mechanisms of photopolymerization reactions; photoinitiation and network formation

MCGINNISS

as applied to coatings and films. Mailing Add: Glidden-Durkee 16651 Sprague Rd Strongsville OH 44136

MCGIRK, RICHARD HEATH, b Boonville, Mo, Jan 25, 45; m 66; c 2. ORGANIC POLYMER CHEMISTRY. Educ: Tex Christian Univ, BS, 67; Univ Colo, PhD(chem), 71. Prof Exp: Fel, Univ Col, Univ London, 71-72 & Johns Hopkins Univ, 72-73; RES CHEMIST ORG POLYMER CHEM, ELASTOMER CHEM DEPT, E I DU PONT DE NEMOURS & CO, 73- Mem: Sigma Xi; Am Chem Soc. Res: Polymer, vulcanization of elastomers, isocyanate and polyurethane chemistry. Mailing Add: Elastomers Chem Dept Exp Sta E I du Pont de Nemours & Co Wilmington DE 19898

MCGIRT, ALBERT FRANKLIN, nuclear science, see 12th edition

MCGLAMERY, MARSHAL DEAN, b Mooreland, Okla, July 29, 32; m 57; c 2. AGRONOMY, WEED SCIENCE. Educ: Okla Agr & Mech Col, BS, 56; Okla State Univ, MS, 58; Univ Ill, Urbana, PhD, 65. Prof Exp: Instr soils, Panhandle Agr & Mech Col, 58-60; agronomist, Agr Bus Co, 60-61; instr soils, 61-63, from res assoc to assoc prof weed control, 64-75, PROF WEED CONTROL, UNIV ILL, URBANA, 75- Concurrent Pos: AID weed control consult, India, 67-68; vis assoc prof, Univ Minn, 71-72. Mem: Weed Sci Soc Am; Coun Agr Sci & Technol. Res: Herbicide residue; weed taxonomy; herbicide selectivity. Mailing Add: Weed Control Dept of Agron Univ of Ill Urbana IL 61871

MCGLASSON, ALVIN GARNETT, b Boone Co, Ky, July 27, 25; m 49; c 3. MATHEMATICS. Educ: Eastern Ky State Col, BS, 49; Univ Ky, MS, 52. Prof Exp: From instr to assoc prof, 50-74, PROF MATH, EASTERN KY UNIV, 74- Concurrent Pos: NSF fel, Univ Kans, 59. Mem: Math Asn Am; Nat Coun Teachers of Math. Res: Geometry and preparation of high school teachers of mathematics; computer science. Mailing Add: Dept of Math Eastern Ky Univ Richmond KY 40475

MCGLINN, EDWARD JAMES, b Leavenworth, Kans, May 13, 28; m 64; c 2. TECHNOLOGICAL FORECASTING. Educ: Kans Univ, BS, 50. Prof Exp: Analyst math, Continental Motor Corp, 50-51; analyst syst, 51-54, mgr comput facil, 54-58, syst anal acoust, 58-60, syst anal weapon syst, 60-68, syst anal arms control, 61-63, mgr prog develop, civil syst, 68-73, MGR TECHNOL FORECASTING, RES LAB, BENDIX CORP, 73- Res: Analysis of trends in the technology development for management awareness and corporate technology planning, including automotive, energy systems, electronics and materials; technology assessment. Mailing Add: 29933 Barwell Farmington Hills MI 48024

MCGLINN, WILLIAM DAVID, b Leavenworth, Kans, Feb 15, 30; m 53; c 5. THEORETICAL PHYSICS. Educ: Univ Kans, BS, 52, PhD(physics), 59. Prof Exp: From instr to asst prof physics, Northwestern Univ, 58-63; physicist, Argonne Nat Lab, 63-65; assoc prof, 65-68, PROF PHYSICS, UNIV NOTRE DAME, 68- Mem: Am Phys Soc. Res: Particle physics; strong and weak interaction theory. Mailing Add: Dept of Physics Univ of Notre Dame Notre Dame IN 46556

MCGLOHON, NORMAN EDWARD, plant pathology, plant nematology, see 12th edition

MCGLOIN, PAUL ARTHUR, b Woburn, Mass, Jan 15, 23; m 53; c 3. MATHEMATICS, COMPUTER SCIENCE. Educ: Boston Univ, AB & AM, 49; Rensselaer Polytech Inst, PhD(math), 68. Prof Exp: Instr math, Univ Conn, 49-51 & 53-55; asst prof, 55-65, ASSOC PROF MATH, RENSSELAER POLYTECH INST, 65- Mem: Asn Comput Mach. Res: Graph theory, non-Hamiltonian graphs. Mailing Add: Dept of Math Rensselaer Polytech Inst Troy NY 12181

MCGLONE, ROBERT ERNEST, b Grayson, Ky, Sept 12, 35; m 57; c 2. SPEECH SCIENCES, HEARING SCIENCES. Educ: Miami Univ, BS, 57; Wichita State Univ, MA, 61; Univ Iowa, PhD(speech), 63. Prof Exp: Speech therapist, Inst Logopedics, Wichita State Univ, 57-60; specialist speech, Nat Inst Dent Res, 63-65; from asst prof to assoc prof speech, State Univ NY Buffalo, 65-74; PROF & CHMN DEPT SPEECH & DRAMATIC ARTS & VCHMN & DIR SPEECH PATH & AUDIOL, UNIV NEBR-LINCOLN, 74- Concurrent Pos: Res assoc, Dept Otolaryngol, Med Ctr, Univ Calif, 68-70; res assoc orthodont, Col Dent, Univ Ky, 68-70, vis assoc prof, 71; trainee, Vet Admin Hosp, San Francisco. Mem: AAAS; fel Am Speech & Hearing Asn; Acoust Soc Am; Int Asn Dent Res; NY Acad Sci. Res: Physiology of speech and phonation in children and adults as related to the acoustic signal. Mailing Add: Dept of Speech & Dramatic Arts 204 Temple Bldg Lincoln NE 28588

MCGLYNN, SEAN PATRICK, b Ireland, Mar 8, 31; nat US; m 55; c 5. PHYSICAL CHEMISTRY. Educ: Nat Univ Ireland, BSc, 51, MSc, 52; Fla State Univ, PhD(phys chem), 56. Prof Exp: Fel chem, Fla State Univ, 56 & Univ Wash, 56-57; from asst prof to prof, 57-68, BOYD PROF CHEM, LA STATE UNIV, BATON ROUGE, 68- Concurrent Pos: Assoc prof, Yale Univ, 61; NSF lectr, 63 & 64; consult biophys prog, Mich State Univ, 63-65; consult, Am Optical Co & Am Instrument Co, 63- & Bell Tel Labs, 65; Alfred E Sloan Found fel, 64- Honors & Awards: Southwestern US Award, Am Chem Soc, 67 & Fla Award, 70. Mem: AAAS; Am Chem Soc; Am Phys Soc. Res: Biophysics; excitons; spectroscopy; fluorescence and phosphorescence; energy transfer; electronic structure of inorganic anions; charge transfer; conductivity of aromatics; spin-orbital perturbation; rotation barriers; vacuum ultraviolet. Mailing Add: Dept of Chem La State Univ Baton Rouge LA 70803

MCGONIGAL, PAUL J, b Philadelphia, Pa, Feb 11, 36; m 57; c 3. PHYSICAL CHEMISTRY, CHEMICAL ENGINEERING. Educ: Univ Pa, AB, 56; Temple Univ, PhD(phys chem), 64. Prof Exp: Chemist, Cent Res Lab, Borden Chem Co, 60-61; res chemist, Res Inst, Temple Univ, 61-64; staff chemist, Marshall Lab, E I du Pont de Nemours & Co, Inc, 64-70; res dir, Polysci, Inc, Warrington, 70-71, vpres, 71-72; asst dir res & develop, Certain-Teed Prod Corp, 72-75; DIR DEVELOP, INDUST CHEM DEPT, AIR PROD & CHEM INC, 75- Mem: Am Chem Soc; Am Inst Chem Engrs. Res: Chemical economics; plant and process design; industrial organic syntheses; polymer research; biomedical materials; adhesion; high temperature condensed phase research; physical properties of matter. Mailing Add: 659 Lindley Rd Glenside PA 19038

MCGONIGAL, WILLIAM E, b Decatur, Ga, Aug 20, 39; m 59; c 1. ORGANIC CHEMISTRY. Educ: Ga Inst Technol, BS, 61, PhD(org chem), 65. Prof Exp: M A Ferst Sigma Xi res award, 65; res chemist, Indust & Biochem Dept, Del, 65-67, res supvr, 68-69, admin asst, 69-73, tech area supvr, Biochem Dept, Tex, 73-75, BIOCHEM PROD SUPT, BIOCHEM DEPT, E I DU PONT DE NEMOURS & CO, INC, 75- Mem: Am Chem Soc; The Chem Soc. Res: Synthetic organic, carbohydrate, heterocyclic, agricultural and pharmaceutical chemistry; nuclear magnetic resonance. Mailing Add: Biochem Dept E I du Pont de Nemours & Co Inc Belle WV 25015

MCGONIGLE, EUGENE JOSEPH, b Philadelphia, Pa, Apr 6, 42; m 65; c 2. ANALYTICAL CHEMISTRY. Educ: La Salle Col, BA, 64; Villanova Univ, MS, 71, PhD(anal chem), 73. Prof Exp: From chem to res chemist, 64-73, SR RES CHEMIST, MERCK SHARP & DOHME RES LABS, 73- Mem: Am Chem Soc: AAAS. Res: Analytical method development of pharmaceutical dosage forms; chromatography; flurorescence analysis; polarography. Mailing Add: Pharmaceut Res & Develop Merck Sharp & Dohme Res Labs West Point PA 19486

MCGONNAGLE, WARREN JAMES, b Newcastle, Nebr, Jan 3, 19; m 43; c 2. PHYSICS. Educ: Univ Nebr, AB, 40; Univ Okla, MS, 42, PhD(physics), 47. Prof Exp: Asst physics, Univ Okla, 41-42, instr, 42-44; jr physicist, Manhattan Proj, Tenn, 44-45; instr physics, Univ Okla, 46-47; assoc prof, Western Mich Univ, 47-50; from assoc physicist to group leader nondestructive testing, Argonne Nat Lab, 50-61; asst dir electronics & elec eng, Southwestern Res Inst, 61-64; sci adv nondestructive testing, Ill Inst Technol Res Inst, 64-66; ASSOC PROF RADIOL, MED SCH, NORTHWESTERN UNIV, 69-; RADIATION PHYSICIST, CHICAGO WESLEY MEM HOSP, 69- Mem: Am Phys Soc; Am Soc Testing & Mat; Health Physics Soc; Am Asn Physicists in Med; Am Soc Nondestructive Testing. Res: Raman and mass spectroscopy; ultrasonics; eddy currents; nondestructive testing techniques. Mailing Add: Dept of Radiation Ther Chicago Wesley Mem Hosp Chicago IL 60611

MCGOOKEY, DONALD PAUL, b Sandusky, Ohio, Sept 19, 28; m 51; c 4. PETROLEUM GEOLOGY. Educ: Bowling Green State Univ, BS, 51; Univ Wyo, MA, 52; Ohio State Univ, PhD(geol), 58. Prof Exp: Div stratigr, 52-69, exec producing comnr, 69-71, CHIEF GEOLOGIST, TEXACO, INC, 71- Mem: AAAS; Geol Soc Am; Am Asn Petrol Geologists. Res: Investigation of regional tectonic relations by detailed study of the chronology of structural relations and related sedimentary deposits. Mailing Add: Texaco Inc Box 52332 Houston TX 77052

MCGOUGH, WILLIAM EDWARD, b Union City, NJ, Nov 12, 28. PSYCHIATRY, PHYSIOLOGICAL PSYCHOLOGY. Educ: St Peter's Col, BS, 50; Duke Univ, MD, 56; Am Bd Psychiat & Neurol, dipl, 63. Prof Exp: Instr psychiat & med, Sch Med, Univ Rochester, 60; from instr to asst prof psychiat, Sch Med, Duke Univ, 61-64; asst prof, Rutgers Med Sch, 64-66, assoc prof, Col Med & Dent NJ, 66-71, PROF PSYCHIAT, RUTGERS MED SCH, COL MED & DENT NJ, 71-, ASSOC DEAN, 72- Concurrent Pos: NIMH res fel, Sch Med, Duke Univ, 60-62; staff psychiatrist, Duke Hosp, 61-64; consult, Vet Admin Hosp, Durham, NC, 63-64, Fayetteville, NC, 64 & Lyons, NJ, 67-; consult, NJ Rehab Comn, 64-, Family Life & Ment Health Proj, Cornell Univ, 65-66 & Family Life & Ment Health Proj, Harvard Sch Pub Health, Harvard Med Sch, 66-71; dir, Fifth Channel in NJ, Col Med & Dent NJ, 72- Mem: Am Psychiat Asn; Am Psychosom Asn. Res: Culture and psychiatric illness; depression and culture; economic forms of speech therapy for stuttering. Mailing Add: Laird Rd Colts Neck NJ 07722

MCGOVERN, JOHN JOSEPH, b Pittsburgh, Pa, June 21, 20; m 47; c 6. CHEMISTRY. Educ: Carnegie Inst Technol, BS, 42, MS, 44, ScD(chem), 46. Prof Exp: Asst, Carnegie Inst Technol, 43-44, instr, 44-45; fel spectros, Mellon Inst, 45-48, sr fel spectros & chem anal groups, 48-50; asst chief chemist, Kobuta Plant Chem Div, Koppers Co, 50-56, chief chemist, 56-58; head res serv, Mellon Inst, 58-71, dir educ & res serv, Carnegie-Mellon Univ, 71-73; ASST DIR, CARNEGIE-MELLON INST RES, 73- Mem: Am Chem Soc; Am Mass Spectros Soc. Res: Administration of technical activities. Mailing Add: Carnegie-Mellon Inst of Res 4400 Fifth Ave Pittsburgh PA 15213

MCGOVERN, JOHN PHILLIP, b Washington, DC, June 2, 21. ALLERGY. Educ: Duke Univ, MD, 45; Am Bd Pediat, dipl, 52. Prof Exp: Intern pediat, New Haven Gen Hosp, 45-46; asst resident, Sch Med, Duke Univ, 48-49; asst chief resident, Children's Hosp, Washington, DC, 49, chief resident, 50; assoc, Sch Med, George Washington Univ, 50-51, asst prof, 51-54; assoc prof, Sch Med, Tulane Univ, La, 54-56; from assoc clin prof to clin prof allergy, 56-69, PROF HIST MED, UNIV TEX GRAD SCH BIOMED SCI, 69-; CLIN PROF PEDIAT & MICROBIOL, BAYLOR COL MED, 69-, CHIEF ALLERGY SECT, DEPT PEDIAT, 58- Concurrent Pos: Markle scholar, 50-55; asst, Sch Med, Yale Univ, 45-46; chief outpatient dept, Children's Hosp, Washington, DC, 50-51, asst to pres res found, 51-52, mem assoc attend staff, 51-54, mem attend staff, Allergy Clin, 53-54; assoc, Univ Hosp, George Washington Univ, 50-54; attend physician & chief, George Washington Univ Pediat Div, DC Gen Hosp, 51-54; attend physician, Doctors Hosp, 52-54; chief Tulane pediat allergy clin & vis physician, Charity Hosp, New Orleans, La, 54-55; clin assoc prof, Baylor Col Med, 56-69, clin assoc prof microbiol, 57-69; mem staff, Tex Children's Hosp, Houston, 56-59, chief allergy serv & dir allergy clin, 57-; dir, McGovern Allergy Clin, Houston; consult, Episcopal Eye, Ear & Throat Hosp, Washington, DC, 53-54, Huey Long Charity Hosp, La, 54-56 & Lackland AFB, Tex, 59-; regional consult, Children's Asthma Res Inst & Hosp, Denver, 60- & Nat Found Asthmatic Children, Sahuaro Sch, Tucson, Ariz, 60. Mem: Am Geriat Soc; AMA; Asn Res Nerve & Ment Dis; Am Asn Hist Med; Int Asn Allergol. Res: Clinical allergy; immunology of hypersensitivity. Mailing Add: Dept of Pediat Baylor Col of Med Houston TX 77025

MCGOVERN, TERRENCE PHILLIP, b Kenosha, Wis, July 30, 30; m 55; c 3. ORGANIC CHEMISTRY. Educ: Univ Md, BS, 59, PhD(org chem), 62. Prof Exp: RES ORG CHEMIST, USDA, 62- Mem: AAAS; Entom Soc Am; Am Chem Soc. Res: Organic synthesis directed toward development of insect attractants, repellents and insect growth regulating chemicals as well as other insect behavior modifying chemicals. Mailing Add: Agr Environ Qual Inst Rm 309 Beltsville Agr Res Ctr E Bldg 306 Beltsville MD 20705

MCGOVERN, WAYNE ERNEST, b Orange, NJ, June 3, 37; m 63; c 2. METEOROLOGY. Educ: Newark Col Eng, BSME, 59, MS, 64; NY Univ, PhD(meteorol), 67. Prof Exp: Aerospace scientist, NASA, 67-68; meteorologist, Nat Meteorol Ctr, 68-69; assoc prof meteorol, Univ Ariz, 69-72; METEOROLOGIST, ENVIRON MONITORING & PREDICTION, NAT OCEANIC & ATMOSPHERIC ADMIN, 72- Concurrent Pos: Dept of Com sci fel & prog leader, Global Environ Monitoring Syst Prog, Dept of State, 75-76. Mem: AAAS; Am Meteorol Soc. Res: Program development and research within the areas of environmental assessment and mesoscale meteorology. Mailing Add: Environ Monitoring & Prediction Nat Oceanic & Atmospheric Admin Rockville MD 20852

MCGOWAN, BLAINE, JR, b San Francisco, Calif, Oct 3, 21; m 49; c 2. VETERINARY MEDICINE. Educ: Univ Calif, Davis, BS, 50, DVM, 52. Prof Exp: From instr to prof vet med, 52-69, assoc dean sch vet med, 64-69, prof clin sci, 69-74, PROF VET MED, SCH VET MED, UNIV CALIF, DAVIS, 74- Concurrent Pos: Vet consult & mem bd dirs, Calif Wool Growers Asn, 53-; chmn, Nat Comt Foot Rot Sheep, 64- Mem: Am Vet Med Asn; Am Asn Lab Animal Sci. Res: Transmissable diseases of sheep and beef cattle; preventive veterinary medicine; veterinary medical education. Mailing Add: Sch of Vet Med Univ of Calif Davis CA 95616

MCGOWAN, CLEMENT LEO, III, b Orange, NJ, Nov 13, 42; m 66; c 2. COMPUTER SCIENCE. Educ: Col of the Holy Cross, BS, 65; Cornell Univ, MA, 68, PhD(comput sci), 71. Prof Exp: Instr, 69-70, ASST PROF APPL MATH, BROWN UNIV, 70- Concurrent Pos: NSF res grant, Brown Univ, 70-; adv programmer, IBM, 73-75; consult, SofTech, Inc, 75- Mem: Asn Comput Mach; Sigma Xi; Inst Elec & Electronics Engr; Comput Soc. Res: Software engineering; programming languages; programming methodology; structured programming. Mailing Add: Div of Appl Math Brown Univ Providence RI 02912

MCGOWAN, FRANCIS KEITH, b Baileyville, Kans, May 2, 21; m 44. PHYSICS. Educ: Kans State Teachers Col, Emporia, AB, 42; Univ Wis, PhM, 44; Univ Tenn, PhD(physics), 51. Prof Exp: Asst physics, Univ Wis, 42-44; res engr, Sylvania Elec Prod, Inc, NY, 44-46; PHYSICIST, OAK RIDGE NAT LAB, 46- Mem: Fel Am Phys Soc. Res: Short-lived nuclear states; short-lived isomers of nuclei; gamma-ray spectroscopy; coulomb excitation of nuclei; cross sections for charged-particle induced reactions. Mailing Add: Oak Ridge Nat Lab PO Box X Oak Ridge TN 37830

MCGOWAN, H CHRISTOPHER, b Elizabeth, NJ, Feb 29, 40; m 65; c 3. CERAMICS, CERAMIC ENGINEERING. Educ: Rutgers Univ, BS & AB, 63, PhD(ceramic eng), 73. Prof Exp: Sales engr, Paper Makers Importing Co, 63-67; sr res engr, Thermokinetic Fibers, Inc, 67-68; res assoc, Rutgers Univ, 68-73; sr res scientist, 73-74, GROUP LEADER, ENGELHARD MINERALS & CHEM CORP, 74- Mem: Am Ceramic Soc; Nat Inst Ceramic Engr. Res: Preparation, fabrication and characterization of ceramic materials and products. Mailing Add: Engelhard Minerals & Chem Corp Menlo Park Edison NJ 08817

MCGOWAN, JAMES WILLIAM, b Pittsburgh, Pa, July 5, 31; m 58; c 6. ATOMIC PHYSICS, MOLECULAR PHYSICS. Educ: St Francis Xavier Univ, BSc, 53; Carnegie-Mellon Univ, MS, 58; Laval Univ, DSc(physics), 61. Prof Exp: Instr physics, St Francis Xavier Univ, 55-56; res asst, Westinghouse Res Labs, 57-58; instr physics, St Lawrence Col, Laval Univ, 58-59; staff mem, Gen Atomic Div, Gen Dynamics Corp, 62-69; chmn dept physics, 69-72, chmn, Ctr Interdisciplinary Studies Chem Physics, 73-76, PROF PHYSICS, UNIV WESTERN ONT, 69- Concurrent Pos: Fel, Joint Inst Lab Astrophys, Univ Colo, Boulder, 66; adv, NSF, Nat Bur Standards, AEC, Advan Res Proj Agency, US Dept Defense, Defense Atomic Support Agency & Defense Res Bd Can; consult, Vacuum Electronics Corp, 59-61 & Bach-Simpson Ltd, 71- Mem: AAAS; Asn Sci, Eng & Technol Community Can; fel Am Phys Soc; Can Asn Physicists; Am Vacuum Soc. Res: Experimental atomic and molecular collision physics; electron, ion, positron, photon and atom collisions with atoms, ions and molecules; energy deposition studies; positrons in solids; charge transfer lasers; laser interaction with eyes; x-ray studies of biologically significant molecules. Mailing Add: Dept of Physics Univ of Western Ont London ON Can

MCGOWAN, JOHN ARTHUR, b Oshkosh, Wis, Aug 22, 24. BIOLOGICAL OCEANOGRAPHY. Educ: Ore State Univ, BS, 50, MS, 51; Univ Calif, San Diego, PhD(oceanog), 60. Prof Exp: Marine biologist, Trust Territory of Pac Islands, 56-58; asst res biologist, 60-62, from asst prof to assoc prof, 62-72, PROF OCEANOG, SCRIPPS INST OCEANOG, UNIV CALIF, SAN DIEGO, 72- Concurrent Pos: Mem UNESCO consult comt, Indian Ocean Biol Ctr, 63-66. Mem: AAAS; Am Soc Limnol & Oceanog. Res: Ecology and zoogeography of oceanic plankton; taxonomy of Thecosomata and Heteropoda; biology of squid. Mailing Add: Div of Oceanic Res Scripps Inst of Oceanog La Jolla CA 92037

MCGOWAN, ROBERT WILLIAM, b Paris, Tenn, June 5, 22; m 44; c 2. ORNITHOLOGY. Educ: Lambuth Col, AB, 46; George Peabody Col, MA, 47. Prof Exp: Naturalist, State Dept Conserv, Tenn, 48-49; assoc prof, 49-69, consult, NSF Inst Biol, 60, PROF BIOL, MEMPHIS STATE UNIV, 69- Concurrent Pos: Consult, Memphis & Shelby County Schs. Mem: AAAS. Res: Plant taxonomy and ecology; flowering plants of west Tennessee; general botany; natural history; breeding census of pileated woodpecker in Shelby County, Tennessee. Mailing Add: Dept of Biol Memphis State Univ Memphis TN 38111

MCGOWAN, WILLIAM COURTNEY, JR, b Mobile, Ala, Aug 8, 37; m 65; c 2. SOLID STATE PHYSICS. Educ: Spring Hill Col, BS, 59; Univ NC, PhD(physics), 65. Prof Exp: Electronic engr, White Sands Missile Range, 65-66, asst chief anti-tank div & res engr, 66; ASSOC PROF PHYSICS & CHMN DEPT, WESTERN CAROLINA UNIV, 66- Mem: Am Phys Soc. Res: Imperfections in ionic crystals; motion and charge on dislocations in silver chloride; range and velocity measurement accuracy as affected by radar signal design. Mailing Add: Dept of Physics Western Carolina Univ Cullowhee NC 28723

MCGOWEN, JOSEPH HOBBS, b Baird, Tex, Jan 15, 32. GEOLOGY. Educ: Hardin-Simmons Univ, BA, 60; Baylor Univ, MA, 64; Univ Tex, Austin, PhD(geol), 69. Prof Exp: GEOLOGIST, BUR ECON GEOL, UNIV TEX, AUSTIN, 66- Concurrent Pos: Geologist, Gen Land Off, State of Tex, 71-72. Res: Coastal geology; sedimentary processes and facies of coastal and fluvial environments; application of geology in solving environmental problems. Mailing Add: Bur of Econ Geol Univ of Tex Austin TX 78712

MCGRADY, ANGELE VIAL, b New Rochelle, NY, Dec 31, 41; m 68; c 1. PHYSIOLOGY. Educ: Chestnut Hill Col, BS, 63; Mich State Univ, MS, 66; Univ Toledo, PhD(physiol), 72. Prof Exp: Instr vet physiol, Wash State Univ, 65-66; res assoc physiol Kresge Eye Inst, Wayne State Univ, 66-68; instr, 68-72, ASST PROF PHYSIOL, MED COL OHIO, 72- Mem: AAAS; Soc Neurosci; Biofeedback Res Soc. Res: Electrophysiology of spermatozoa; neurosciences, biofeedback for control of hypertension and migraine headache. Mailing Add: Dept of Physiol Med Col of Ohio Toledo OH 43614

MCGRADY, M MERCY, inorganic chemistry, physical chemistry, see 12th edition

MCGRANAGHAN, FRANCIS RICHARD, physical chemistry, organic chemistry, see 12th edition

MCGRATH, CHARLES MORRIS, b Seattle, Wash, Sept 17, 43. ONCOLOGY. Educ: Univ Portland, BS, 65; Univ Calif, Berkeley, PhD(virol), 70. Prof Exp: Cell biologist, Univ Calif, Berkeley, 70-71; viral oncologist, Inst Jules Bordet, Brussels, Belg, 71-72; res scientist viral oncol, 72-83, CHIEF TUMOR BIOL LAB VIRAL ONCOL, MICH CANCER FOUND, 73- Concurrent Pos: NIH res grant, Nat Cancer Inst, 75. Mem: Tissue Cult Asn; Am Asn Cancer Res. Res: Viral and hormonal involvement in malignant conversion of breast epithelial cells. Mailing Add: Mich Cancer Found Detroit MI 48214

MCGRATH, HAROLD BENTON, veterinary science, deceased

MCGRATH, JAMES EDWARD, b Easton, NY, July 11, 34; m 59; c 6. ORGANIC POLYMER CHEMISTRY, POLYMER SCIENCE. Educ: Siena Col, BS, 56; Univ Akron, MS, 64, PhD(polymer sci), 67. Prof Exp: Res chemist, ITT-Rayonier, 56-59 & Goodyear Tire & Rubber Co, 59-65; fel polymer sci, Univ Akron, 67; res scientist, Union Carbide Corp, 67-75; ASSOC PROF POLYMER SCI, VA POLYTECH INST & STATE UNIV, 75- Concurrent Pos: Polymer consult, Union Carbide Corp, 75- Mem: Am Chem Soc; NY Acad Sci. Res: Anionic polymerization; block and graft copolymers; reactions of polymers. Mailing Add: Dept of Chem Va Polytech Inst & State Univ Blacksburg VA 24061

MCGRATH, JAMES J, b Brooklyn, NY, Oct 30, 31. PLANT ANATOMY. Educ: Univ Notre Dame, AB, 55; Univ Calif, Davis, MA, 64, PhD(bot), 66. Prof Exp: Asst prof biol & rector, 65-71, ASSOC PROF BIOL, UNIV NOTRE DAME, 71-, ASST CHMN DEPT, 74- Mem: AAAS; Am Forestry Asn; Bot Soc Am. Res: Developmental plant anatomy; seasonal changes in secondary phloem of angiosperms. Mailing Add: Box 369 Univ of Notre Dame Notre Dame IN 46556

MCGRATH, JAMES JOSEPH, b Hempstead, NY, Apr 23, 36; m 59; c 4. INHALATION TOXICOLOGY. Educ: Hofstra Univ, BA, 58; Ind Univ, PhD(physiol), 68. Prof Exp: Fel environ physiol, Ind Univ, Bloomington, 68-69; from asst prof to assoc prof, Rutgers Univ, 69-73; assoc prof physiol, Col Med, Univ Ill, 73-76; SR TOXICOLOGIST, GEN MOTORS RES LABS, 76- Concurrent Pos: Nat Acad Sci exchange scientist, Inst Physiol, Czech Acad Sci, 71. Mem: Am Physiol Soc; Soc Exp Biol & Med; Int Soc Res Can Metab; Int Soc Biometeorol. Res: Acute and chronic effects of environmental alterations on the respiratory and cardiovascular systems; respiratory and cardiovascular physiology. Mailing Add: Gen Motors Res Labs Warren MI 48090

MCGRATH, JAMES RUSSELL, b Chicago, Ill, Dec 2, 32; m 59; c 4. PHYSICS, AERONAUTICAL ENGINEERING. Educ: Cath Univ, BAE, 57; MS, 65, PhD, 71; George Washington Univ, BA, 63. Prof Exp: Aeronaut res engr, Res Div, Bur Aeronaut, 57-60, opers res analyst, Bur Naval Weapons, 60-61, res physicist, Naval Res Lab, 62-67 & Maury Ctr Ocean Sci, 67-68, RES PHYSICIST, ACOUST DIV, NAVAL RES LAB, DEPT NAVY, 68- Mem: Acoust Soc Am; Sigma Xi; Brit Acoust Soc; Brit Inst Physics; Eng & Sci Asn Ireland. Res: Preliminary aircraft design and performance; exploding wire phenomena; underwater explosions and acoustics; air-sea interactions. Mailing Add: 5900 Madawaska Rd Washington DC 20016

MCGRATH, JAMES THOMAS, metal physics, see 12th edition

MCGRATH, JAMES WILLIAMSON, b Kirkland, Ill, May 17, 12; m 34; c 1. PHYSICS. Educ: Ft Hays Kans State Col, BA, 33; Univ Kans, MA, 34; Univ Iowa, PhD(physics), 39. Prof Exp: Instr sci, Univ High Sch, Iowa, 34-37; from instr to assoc prof physics, Mich State Univ, 39-46; assoc prof, 46-49, PROF PHYSICS, KENT STATE UNIV, 49-, DEAN GRAD SCH & RES, 68- Mem: Fel Am Phys Soc. Res: X-ray absorption; ultrasonics; dielectric constants of gases; nuclear magnetic resonance; chemical physics; missile guidance. Mailing Add: Grad Sch Kent State Univ Kent OH 44242

MCGRATH, JOHN F, b New York, NY, Dec 30, 27; m 50; c 3. PHYSICAL CHEMISTRY, INORGANIC CHEMISTRY. Educ: Siena Col, NY, BS, 49; State Univ NY, MS, 51; Rensselaer Polytech Inst, MS, 60. Prof Exp: Teacher, Colonie Cent High Sch, 53-60; from asst prof to assoc prof, 61-71, PROF PHYS SCI, COL ST ROSE, 71-, CHMN DIV NATURAL SCI, 73- Concurrent Pos: Partic, NSF Summer Inst, Rensselaer Polytech Inst, 55-59; consult geol Lake Superior region res, Mich Technol Univ, 65; partic, Inst Sci, Ithaca Col, 68 & Eastern Regional Inst, 68-71; ed, Bull Sci Teachers Asn NY State, 71-74. Mem: Am Chem Soc. Res: Instrumental analysis; determination of stability constants of series of esters of some complexes of tripositive cobalt. Mailing Add: Dept of Chem Col of St Rose 432 Western Ave Albany NY 12203

MCGRATH, JOHN THOMAS, b Philadelphia, Pa, Sept 27, 18; m 47; c 3. PATHOLOGY. Educ: Univ Pa, VMD, 43. Prof Exp: Instr path, Vet Sch, 47-48, assoc, 48-50, from asst prof to assoc prof, 50-58, PROF PATH, GRAD SCH MED, UNIV PA, 58- Concurrent Pos: Attend veterinarian, Philadelphia Gen Hosp. Mem: Am Vet Med Asn; Am Asn Neuropath; Am Col Vet Path; Am Acad Neurol. Res: Nervous disorders of animals; pathology of central nervous system diseases of animals. Mailing Add: Dept of Vet Path Univ of Pa Philadelphia PA 19104

MCGRATH, JOSEPH JOHN, organic chemistry, see 12th edition

MCGRATH, MICHAEL GLENNON, b St Louis, Mo, Oct 12, 41; m 67. ORGANIC CHEMISTRY. Educ: Col Holy Cross, BS, 63; Mass Inst Technol, PhD(chem), 67. Prof Exp: Asst prof, 67-72, ASSOC PROF CHEM, COL OF THE HOLY CROSS, 72- Mem: Am Chem Soc. Res: Synthesis and reactions of small ring carbon compounds; carbene chemistry; synthesis of non-benzenoid aromatic compounds. Mailing Add: Dept of Chem Col of the Holy Cross Worcester MA 01610

MCGRATH, ROBERT L, b Iowa City, Iowa, Nov 2, 38; m 60; c 2. NUCLEAR PHYSICS. Educ: Oberlin Col, BA, 60; Univ Iowa, MS, 62, PhD(physics), 65. Prof Exp: Fel physics, Univ Iowa, 65; res assoc, Lawrence Radiation Lab, Univ Calif, 66-68; asst prof, 68-73, ASSOC PROF PHYSICS, STATE UNIV NY STONY BROOK, 73- Concurrent Pos: Alexander von Humboldt Found sr US scientist award, 74-75. Mem: Am Phys Soc. Res: Low energy nuclear reaction research on the structure of nuclei; heavy ion reactions; properties of analog states of light nuclei; high-isospin multiplets. Mailing Add: Dept of Physics State Univ of NY Stony Brook NY 11794

MCGRATH, THOMAS FREDERICK, b Braddock, Pa, Jan 4, 29; m 52; c 4. ORGANIC CHEMISTRY. Educ: Franklin & Marshall Col, BS, 50; Univ Pittsburgh, PhD(org chem), 55. Prof Exp: Res chemist, Reilly Tar & Chem Corp, 54-55 & Am Cyanamid Co, 57-63; assoc prof, 63-69, PROF CHEM, SUSQUEHANNA UNIV, 70- Mem: Am Acad Arts & Sci; Am Chem Soc; Sigma Xi. Res: Organometallic and organofluorine chemistry. Mailing Add: Dept of Chem Susquehanna Univ Selinsgrove PA 17870

MCGRATH, WILLIAM PATRICK, b Pittsburgh, Pa, Jan 28, 47; m 69. CLINICAL CHEMISTRY. Educ: Georgetown Univ, BS, 68, PhD(biochem), 73. Prof Exp: Chief clin chem, 10th US Army Med Lab, 73-76; CLIN CHEMIST, BIO-SCI LAB, 76- Mailing Add: 7600 Tyrone Ave Van Nuys CA 91405

MCGRATH, WILLIAM ROBERT, b Weymouth, Mass, Sept 25, 33; m 57; c 3. PHARMACOLOGY. Educ: Mass Col Pharm, BS, 55, MS, 57; Univ Wash, PhD(pharmacol), 60. Prof Exp: Asst prof pharmacol, Univ Wash, 57-60; head sect cent nerv syst pharmacol, Wm S Merrell Co Div, Richardson-Merrell, Inc, 60-62, head dept pharmacol, 62-67; dir labs, Enzomedic Labs, Inc, 67-69; dir div pharmacol & toxicol, 69-70, dir biol res, 70, dir, Res & Support, 70-74, DIR RES & DEVELOP, USV PHARMACEUT CORP, 74- Concurrent Pos: Lectr, Univ Tex, Med Br, 64-67; affil assoc prof, Sch Med, Univ Wash, 67-69. Mem: Am Soc Pharmacol & Exp Therapeut. Res: Biochemical pharmacology; central nervous system stimulants; pharmacology of smooth muscle. Mailing Add: USV Pharmaceut Corp 1 Scarsdale Rd Tuckahoe NY 10707

MCGRATH, WILLIAM THOMAS, b Lincoln, Nebr, Feb 27, 33. FOREST PATHOLOGY. Educ: Mont State Univ, BSF, 60; Univ Wis, PhD(plant path), 67. Prof Exp: Tech asst, Australian Forestry & Timber Bur, 61-62; ASSOC PROF

FORESTRY, STEPHEN F AUSTIN STATE UNIV, 68- Mem: Soc Am Foresters; Am Phytopath Soc; Am Inst Biol Sci; Sigma Xi. Res: Pine rusts; effectiveness of Peniophora gigantea as a control for annosus root rot in east Texas; American mistletoe host specificity. Mailing Add: Sch of Forestry Stephen F Austin State Univ Nacogdoches TX 75961

MCGRAW, CHARLES PATRICK, b Sherman, Tex, Feb 17, 42; m 67; c 3. BIOMEDICAL ENGINEERING, NEUROPHYSIOLOGY. Educ: Belmont Abbey Col, BCh, 64; ETex State Univ, MS, 67; Baylor Univ, cert biomed eng, 68; Tex A&M Univ, PhD(physiol), 69. Prof Exp: Mem fac, ETex State Univ, 65-67; mem fac, Tex A&M Univ, 67-69; neurophysiologist, Univ Tex Med Br, Galveston, 69-71; NEUROPHYSIOLOGIST, BOWMAN GRAY SCH MED, 71- Concurrent Pos: Dept Health, Educ & Welfare clin head injury grant; mem, Brain Res Found. Mem: AAAS; Am Inst Biol Sci; Am Heart Asn; Am Physiol Soc; Am Soc Clin Hypnosis. Res: Tritium labeled testosterone locating centers in the rat brain with radioaudiography; neural mechanisms involved in the immobility reflex from phylogenetic ontogenetic, pharmacological and surgical approaches; head injury and stroke; neurological and neurosurgical research; cerebral blood flow; autoregualtion; patient monitoring; monitoring of intracranial pressure. Mailing Add: Dept of Neurol Bowman Gray Sch Med Winston-Salem NC 27106

MCGRAW, DELFORD ARMSTRONG, b Keyrock, WVa, May 13, 17; m 41; c 2. PHYSICS. Educ: Concord Col, AB, 37; WVa Univ, MS, 39. Prof Exp: Supvr, Ballistics Lab, Hercules Powder Co, 41-45; res physicist, Owens-Ill Glass Co, 46-55, sr physicist, Owens-Ill, Inc, 55-66, mgr res instrumentation, 66-69, assoc dir res, 69-70, DIR ENG RES, OWENS-ILL, INC, 70- Honors & Awards: Forrest Award, Am Ceramic Soc, 52 & 59. Mem: Fel Am Ceramic Soc; fel Brit Soc Glass Technol. Res: Velocity of sound in gases; ballistics of small arms and rocket powders; mechanical and thermal properties of glass and plastics. Mailing Add: 3410 Chapel Dr Toledo OH 43615

MCGRAW, GARY EARL, b Wellsville, NY, Sept 26, 40; m 63; c 3. PHYSICAL CHEMISTRY. Educ: Univ Mich, BS, 62; Pa State Univ, PhD(phys chem), 65. Prof Exp: US Dept Health, Educ & Welfare fel, Div Air Pollution, Univ Calif, Berkeley, 66-67; res chemist, Tenn Eastman Co, 67-68, sr res chemist, 69-74, res assoc, 74-75, actg div head, Phys & Anal Chem Div, 75-76, ACTG DIV DIR, ANAL SCI DIV, EASTMAN KODAK CO, 76- Mem: AAAS; Am Chem Soc; Am Phys Soc; NY Acad Sci. Res: Spectroscopy; polymer morphology. Mailing Add: 301 Lanewood Circle Kingsport TN 37663

MCGRAW, GERALD WAYNE, b Tampa, Fla, Apr 30, 43; m 67; c 1. BIOCHEMISTRY, ORGANIC CHEMISTRY. Educ: Ouachita Baptist Univ, BS, 65; Fla State Univ, PhD(org chem), 71. Prof Exp: Asst prof, 70-74, ASSOC PROF CHEM, LA COL, 74- Mem: Am Chem Soc. Res: Synthesis and enzymolysis of polypeptides related to collagen. Mailing Add: La Col Dept of Chem College Station Pineville LA 71360

MCGRAW, HUGO RICHARD, b Parkersburg, WVa, Oct 2, 12; m 44; c 2. ORGANIC CHEMISTRY. Educ: Ohio Univ, BS, 34; WVa Univ, MS, 36, PhD(org chem), 41. Prof Exp: Asst, Tech Dept, Libbey-Owens-Ford Glass Co, WVa, 36-38; fel, Mellon Inst, 41-47; lab supvr, Chem Div, Koppers Co, Inc, 47-52, tech coordr, 52-55, sr chemist, Res Dept, 55-58; res chemist, 58-60, res specialist, 60-62, group leader, 62-70, AUTOMATION SPECIALIST II, MOUND LAB, MONSANTO RES CORP, 70- Mem: Am Chem Soc. Res: Syntheses of organic compounds; ebulliometry; separation of organic compounds; explosives. Mailing Add: 642 Larriwood Ave Kettering OH 45429

MCGRAW, JAMES CARMICHAEL, b Martins Ferry, Ohio, Mar 20, 28; m 51; c 3. ZOOLOGY, PARASITOLOGY. Educ: Oberlin Col, AB, 51; Ohio State Univ, MS, 57, PhD(zool, parasitol), 68. Prof Exp: ASST PROF BIOL, STATE UNIV NY COL PLATTSBURGH, 64- Mem: AAAS; Am Soc Parasitol; Am Inst Biol Sci. Res: Nematodes parasitic in amphibians, especially the genus Cosmocercoides. Mailing Add: Dept of Biol State Univ of NY Col Plattsburgh NY 12901

MCGRAW, JAMES LORENZ, b Syracuse, NY, Sept 7, 17. OPHTHALMOLOGY. Educ: Syracuse Univ, AB, 38, MD, 41; Columbia Univ, DrMedSci, 47. Prof Exp: Intern med, Univ Grad Hosp, Univ Pa, 41-42; resident ophthal, Inst Ophthal, Presby Hosp, New York, 42-46; from clin instr to clin asst prof, 47-57, PROF OPHTHAL & CHMN DEPT, STATE UNIV NY UPSTATE MED CTR, 57- Concurrent Pos: Teaching & res fel, Inst Ophthal, Presby Hosp, 45-46; instr, Col Physicians & Surgeons, Columbia Univ, 45-46. Mem: AMA; Am Acad Ophthal & Otolaryngol; Pan-Am Asn Ophthal; Asn Res Vision & Ophthal; Nat Soc Prev Blindness. Res: Prevention of blindness. Mailing Add: Dept of Ophthal State Univ of NY Upstate Med Ctr Syracuse NY 13210

MCGRAW, JOHN LEON, JR, b Port Arthur, Tex, June 16, 40; m 57; c 3. CELL BIOLOGY, PARASITOLOGY. Educ: Lamar State Col, BS, 62; Tex A&M Univ, MS, 64, PhD(biol), 68. Prof Exp: Res grants, Res Ctr, 67-74, ASST PROF BIOL, LAMAR UNIV, 67- Mem: Am Soc Parasitologists; Soc Protozool; Sigma Xi; Bot Soc Am. Res: Parasites of fishes; morphology and ecological distribution of Myxomycetes; Tardigrada-taxonomy. Mailing Add: Dept of Biol Lamar Univ Box 10037 Beaumont TX 77710

MCGRAW, LESLIE DANIEL, b Hutchinson, Minn, Dec 19, 20; m 47; c 2. CHEMISTRY. Educ: Col St Thomas, BS, 42; Carnegie Inst Technol, MS, 43, DSc(phys chem), 46. Prof Exp: Instr chem, Pa Col Women, 45-46; asst prof, Webster Col, 46-48; from assoc consult chemist to asst div chief, Battelle Mem Inst, 48-63; dir res phys chem, North Star Res & Develop Inst, 63-65; DIV CHIEF CHEM PROCESSES & RES SERV, NAT STEEL CORP, 65- Mem: Am Chem Soc; Electrochem Soc; Am Electroplaters Soc. Res: Analytical chemistry; electro-mechanical research; operations research; industrial hygiene; corrosion; electrochemical processes. Mailing Add: 336 County Club Blvd Weirton WV 26062

MCGRAY, ROBERT JAMES, b Fond du Lac, Wis, June 4, 25; m 47; c 4. MICROBIOLOGY. Educ: Marquette Univ, PhB, 50, MS, 52. Prof Exp: Bacteriologist, Microbiol Labs, Pabst Brewing Co, 52-53; assoc microbiologist, Froedtert Malt Corp, 53-55; instr biol, Marquette Univ, 56; asst biologist, 57-60, microbiologist, 60-65, sr microbiologist, 65-71, mkt develop supvr, 71-75, MKT MGR, SC JOHNSON & SON, INC, 75- Mem: Soc Indust Microbiol; Am Pub Health Asn; Am Soc Microbiol. Res: Industrial microbiology; preservation of raw mater ls an materials and finished products; anti-microbial products; public health microbiology; antiseptics and disinfectants; environmental aspects of epidemiology of infectious diseases. Mailing Add: SC Johnson & Son Inc Racine WI 53403

MCGREAL, DOUGLAS ANTHONY, b Hull, Eng, May 19, 23; m 56; c 2. MEDICINE. Educ: Univ St Andrews, MB, BCh, 46, MD, 57; Royal Col Physicians, dipl, 50; Royal Col Physicians & Surgeons Can, cert, 59. Prof Exp: Asst neurol, Hosp Sick Children, Toronto, 59; clin teacher, 59-71, ASST PROF PEDIAT, UNIV TORONTO, 71-; CONSULT NEUROL, CRIPPLED CHILDRENS CTR, 62- Mem: Can Neurol Soc; Can Soc Electroencephalog. Res: Pediatric neurology. Mailing Add: Med Arts Bldg 170 St George St Toronto ON Can

MCGREER, DONALD EDWARD, b Calgary, Alta, Mar 25, 32; m 56; c 4. ORGANIC CHEMISTRY. Educ: Univ Alta, BSc, 54, MSc, 56; Univ Ill, PhD(org chem), 59. Prof Exp: From instr to assoc prof, 59-74, PROF CHEM, UNIV BC, 74- Mem: Chem Inst Can; Am Chem Soc. Res: Pyrolysis or photolysis of compounds giving rearrangements or products by loss of a molecule. Mailing Add: Dept of Chem Univ of BC Vancouver BC Can

MCGREGOR, DONALD NEIL, b Cameron, Tex, Mar 27, 36; m 57; c 1. ORGANIC CHEMISTRY. Educ: Rice Univ, BA, 57; Mass Inst Technol, PhD(org chem), 61. Prof Exp: Res asst chem, Lever Bros Co, 56 & 57; teaching asst org chem, Mass Inst Technol, 57-58; sr res scientist, 61-69, asst dir org chem res, 69-73, DIR ANTIBIOTIC CHEM RES, BRISTOL LABS, 73- Mem: Am Chem Soc. Res: Chemistry of medicinal agents and antibiotics; amino acid chemistry. Mailing Add: Bristol Labs PO Box 657 Syracuse NY 13201

MCGREGOR, DOUGLAS D, b Hamilton, Ont, Mar 5, 32; m 63; c 1. EXPERIMENTAL PATHOLOGY. Educ: Univ Western Ont, BA, 54, MD, 56; Oxford Univ, DPhil(path), 63. Prof Exp: Asst prof path, Case Western Reserve Univ, 63-68; assoc prof, Univ Conn, 68-69; MEM, TRUDEAU INST, 69- Res: Development and immunological activity of lymphocytes. Mailing Add: Trudeau Inst Saranac Lake NY 12983

MCGREGOR, DUNCAN COLIN, b St Catharines, Ont, Dec 27, 29; m 59; c 3. PALYNOLOGY. Educ: McMaster Univ, BA, 51, MSc, 53, PhD(paleobot), 57. Prof Exp: Geologist, 57-65, RES SCIENTIST, GEOL SURV CAN, 65- Mem: Am Paleont Soc; Can Bot Asn; Am Asn Stratig Palynologists; Brit Palaeont Asn; Int Asn Plant Taxon. Res: Plant microfossils, particularly Devonian spores and their botanical and geological implications. Mailing Add: Geol Surv of Can 601 Booth St Ottawa ON Can

MCGREGOR, DUNCAN J, b St Joseph, Mo, Jan 3, 21; m 46; c 3. GEOLOGY. Educ: Univ Kans, BS, 43, MS, 48; Univ Mich, PhD(geol), 53. Prof Exp: Geologist indust minerals, State Geol Surv, Kans, 47; asst dist geologist petrol geol, Sinclair Oil & Gas Co, 48; asst prof geol & head indust minerals sect, Geol Surv, Ind Univ, 53-63; PROF GEOL & STATE GEOLOGIST, STATE GEOL SURV, UNIV SDAK, 63- Mem: AAAS; Asn Am State Geologists; Geol Soc Am; Soc Econ Geologists; Soc Econ Paleontologists & Mineralogists; Am Asn Petrol Geologists. Res: Industrial minerals; economic geology. Mailing Add: SDak State Geol Surv Univ of SDak Sci Ctr Vermillion SD 57069

MCGREGOR, JOHN ROBERT, b Flagstaff, Ariz, Feb 4, 36; m 63; c 2. ECONOMIC GEOGRAPHY. Educ: Univ Ill, Urbana, BS, 58, MA, 61, PhD(geog), 65. Prof Exp: Asst prof geog, Univ Ga, 63-70; assoc prof, 70-75, PROF GEOG, IND STATE UNIV, TERRE HAUTE, 75- Concurrent Pos: Off Water Resources res grant, Univ Ga, 67-68; assoc prof & consult, Ctr Econ Develop, Ind State Univ, 73; consult, Amax, 74-75 & 75-76. Mem: Asn Am Geogr; Regional Sci Asn. Res: Manufacturing geography, especially the functional nature of industrial areas; applied geography; mineral resources. Mailing Add: Dept of Geog Ind State Univ Terre Haute IN 47809

MCGREGOR, ROBERT FINLEY, b Chicago, Ill, Sept 16, 28; m 57; c 2. BIOCHEMISTRY, ENDOCRINOLOGY. Educ: Univ Ill, BS, 48; Univ Wyo, MS, 55; Baylor Univ, PhD(biochem), 61. Prof Exp: Chemist, Armour Labs, Ill, 48-52; state food & drug chemist, Wyo State Dept Agr, 55-57; instr biochem, Sch Med, Univ NDak, 61-63; asst biochemist, 63-67, ASSOC BIOCHEMIST, UNIV TEX M D ANDERSON HOSP & TUMOR INST, 67- Mem: AAAS; Am Chem Soc. Res: Biochemistry of cancer; hormone dependent tumors; chromatographic methodology; steroid hormones; lipid metabolism. Mailing Add: Univ of Tex M D Anderson Hosp & Tumor Inst Houston TX 77025

MCGREGOR, RONALD LEIGHTON, b Manhattan, Kans, Apr 4, 19; m 42. BOTANY. Educ: Univ Kans, AB, 41, MA, 47, PhD, 54. Prof Exp: Asst instr, 41-42 & 46, from instr to assoc prof, 47-60, chmn dept, 61-69, PROF BOT, UNIV KANS, 61-, CHMN DIV BIOL SCI, 69-, DIR HERBARIUM & DIR, STATE BIOL SURV, 73- Mem: Am Bryol Soc; Bot Soc Am; Am Fern Soc; Am Soc Plant Taxonomists; Brit Bryol Soc. Res: Systematic botany; ecology; flora of Kansas and the southwest United States. Mailing Add: Dept of Bot Univ of Kans Lawrence KS 66045

MCGREGOR, SANDY, b Corvallis, Ore, Aug 16, 38; m 60; c 1. VIROLOGY. Educ: Ore State Univ, BS, 61, MS, 63; Baylor Col Med, PhD(virol), 69. Prof Exp: Res asst virol, Baylor Col Med, 64-65; RES ASSOC VIROL, UNIV WIS-MADISON, 72- Concurrent Pos: NIH training grant, Baylor Col Med, 69-70; Nat Cancer Inst fel, Univ Wis-Madison, 70-72. Mem: Am Soc Microbiol. Res: Virus structure and morphogenesis; virus-host cell interactions. Mailing Add: Biophys Lab Univ of Wis Madison WI 53706

MCGREGOR, STANLEY DANE, b Endicott, NY, Nov 11, 38; m 60; c 2. ORGANIC CHEMISTRY. Educ: WVa Wesleyan Col, BS, 60; Univ Wis, MS, 61, PhD(org chem), 66. Prof Exp: Instr chem, Davis & Elkins Col, 61-63; fel, Univ Fla, 66-67; res chemist, E C Britton Res Lab, 68-71, RES CHEMIST, AGR SYNTHESIS LAB, DOW CHEM USA, 71- Mem: Am Chem Soc; AAAS. Res: Synthesis of biologically active organic compounds; heterocyclic synthesis. Mailing Add: 3903 Chestnut Hill Dr Midland MI 48640

MCGREGOR, WHEELER KESEY, JR, physics, see 12th edition

MCGREGOR, WILLIAM HENRY DAVIS, b SC, Mar 25, 27; m 50; c 2. PLANT PHYSIOLOGY, FORESTRY. Educ: Clemson Col, BS, 51; Univ Mich, BSF & MF, 53; Duke Univ, PhD(physiol), 58. Prof Exp: Res forester, Forest Serv, USDA, 53-57, plant physiologist, 57-60; from assoc prof to prof forestry, 60-69, head dept, 69-70, DEAN COL FOREST & RECREATION RESOURCES, CLEMSON UNIV, 70- Concurrent Pos: Asst, Duke Univ, 55-57. Mem: Sigma Xi; Soc Am Foresters; Nat Parks Asn. Res: Forest ecology and tree physiology, particularly metabolism, water relationships and growth processes. Mailing Add: Col of Forest & Recreation Resources Clemson Univ Clemson SC 29631

MCGREW, ELIZABETH ANNE, b Faribault, Minn, Aug 30, 16. PATHOLOGY. Educ: Carleton Col, AB, 38; Univ Minn, MB, 44, MD, 45. Prof Exp: From instr to asst prof path, Col Med, 46-47, from asst pathologist to assoc pathologist, Res & Educ Hosps, 47-62, PROF PATH, UNIV ILL COL MED, 62-, PATHOLOGIST, UNIV HOSPS, 62- Mem: Am Soc Clin Path; Col Am Path; Am Asn Path & Bact; Am Asn Cancer Res; Am Soc Cytol. Res: Cytology and cancer. Mailing Add: Dept of Path Univ of Ill Col of Med Chicago IL 60612

MCGREW, FRANK CLIFTON, organic chemistry, see 12th edition

MCGREW, GEORGE THOMAS, b Silverton, WVa, Sept 27, 25; m 46; c 2. INDUSTRIAL CHEMISTRY. Educ: Western Md Col, BS, 51; Johns Hopkins Univ, MA, 55, PhD(chem), 67. Prof Exp: VP tech dir, Miller Chem & Fertilizer Corp, Div Alco Stand, 51-71, VP TECH DIR, ALCO CHEM CORP, DIV ALCO STAND, 71- Mem: Am Chem Soc; Tech Asn Pulp & Paper Indust; Sigma Xi. Res: Development from inception to final marketing of chemicals for the water treatment industry. Mailing Add: Alco Chem Corp Div Alco Stand Corp Trenton & Williams Sta Philadelphia PA 19134

MCGREW, JOHN ROBERTS, b Washington, DC, Jan 17, 23; m 52. PHYTOPATHOLOGY, VITICULTURE. Educ: Harvard Univ, AB, 47; Univ Md, MS, 51, PhD, 54. Prof Exp: Asst, Univ Md, 50-53; PLANT PATHOLOGIST, PLANT INDUST STA, USDA, 53- Mem: AAAS; Am Inst Biol Sci; Am Soc Enol; Am Phytopath Soc. Res: Virus diseases of strawberry; grape diseases and breeding for resistance. Mailing Add: Plant Indust Sta USDA Beltsville MD 20705

MCGREW, PAUL ORMAN, b Ottumwa, Iowa, Aug 27, 09; m 34; c 3. GEOLOGY, PALEONTOLOGY. Educ: Univ Nebr, AB, 33; Univ Calif, MA, 35; Univ Chicago, PhD(geol paleont), 42. Prof Exp: Asst, Mus, Univ Nebr, 29-33; asst paleont, Field Mus Natural Hist, 38-39, asst cur, 40-46; from asst prof to assoc prof, 46-58, PROF GEOL, UNIV WYO, 58- Concurrent Pos: Mem expeds, Great Plains, 29-37 & 39-40, Honduras, 37-38 & 41-42; ed, J Soc Vert Paleont, 45-47. Mem: Soc Study Evolution; Soc Vert Paleont; fel Geol Soc Am. Res: Mammalian paleontology; Tertiary stratigraphy. Mailing Add: Dept of Geol Univ of Wyo Laramie WY 82070

MCGRIFF, RICHARD BERNARD, organic chemistry, biochemistry, see 12th edition

MCGROARTY, JOSEPH A, b Wilkes-Barre, Pa, May 26, 16; m 47; c 3. INDUSTRIAL CHEMISTRY. Educ: Rensselaer Polytech Inst, BS, 41. Prof Exp: Supvr process develop, Merck & Co, Inc, 41-49, supvr indust eng, 49-50, supvr commercial develop, 50-56, supvr indust sales, 56-64, sales mgr, Indust & Agr Chem, 64-74; CHEM & MKT CONSULT, 74- Concurrent Pos: Chief facilities br, Chem Div, Nat Prod Authority, 51-52; consult, Merck & Co, Inc, 74- & C P Chem, Inc, 75- Honors & Awards: Cert, Secy of Com, 52. Mem: Am Chem Soc; Commercial Develop Asn; assoc Consult Chemists & Engrs. Res: Commercial development; chemical marketing; organic and inorganic chemistry; Environmental Protection Agency pesticide registration; chemical toxicity. Mailing Add: 925 Lawrence Ave Westfield NJ 07090

MCGRODDY, JAMES CLEARY, b New York, NY, Apr 6, 37; m 62; c 3. SOLID STATE PHYSICS. Educ: St Joseph's Col, Pa, BS, 58; Univ Md, PhD(physics), 64. Prof Exp: Res assoc physics, Univ Md, 64-65; RES SCIENTIST, IBM CORP, 65- Concurrent Pos: Vis prof, Tech Univ Denmark, 70-71. Mem: Fel Am Phys Soc; sr mem Inst Elec & Electronics Engrs. Res: Solar energy; experimental semiconductor and surface physics. Mailing Add: IBM Thomas J Watson Res Ctr Yorktown Heights NY 10598

MCGUCKIN, WARREN FRANCIS, b Mankato, Minn, Dec 6, 13; m 41; c 1. CLINICAL CHEMISTRY, INSTRUMENTATION. Educ: Iowa State Col, BS, 48, MS, 50; Univ Minn, PhD(chem), 53. Prof Exp: Res chemist, Mayo Clin, 32-44; res assoc, Iowa State Col, 48-50; asst, Mayo Clin, 50-54, med staff consult, 54-69; med researcher, Res & Develop Div, Xerox Corp, 69-70; res consult, Baxter Travenol Labs, 70-73; OWNER-DIR CHELMO RES LAB, 72- Concurrent Pos: Assoc prof, Mayo Grad Sch Med, Univ Minn, 60-69. Mem: Fel AAAS; Am Asn Clin Chem; Am Chem Soc. Res: Steroid and protein chemistry; biomedical research; research and development of new instruments for medical diagnostics. Mailing Add: 9172 Annik Dr Huntington Beach CA 92646

MCGUGAN, ALAN, b Belfast, Northern Ireland, Apr 17, 24. GEOLOGY, PALEONTOLOGY. Educ: Queen's Univ, Ireland, BSc, 49, MSc, 55, DSc, 66. Prof Exp: Geologist, Brit Petrols, 49-51; micropaleontologist, Shell Condor SAm, 54-55; geologist, Imperial Oil Co, BC, 56-57; lectr geol & paleont, Queen's Univ, Can, 57-59; asst prof geol, Univ Alta, 64-66; assoc prof, 66-74, PROF GEOL, UNIV CALGARY, 74- Concurrent Pos: Consult, 59- Mem: Geol Soc Am; Paleont Soc; Sigma Xi; Can Soc Petrol Geologists. Res: Cretaceous Foraminifera; general micropaleontology; late Paleozoic biostratigraphy; conodonts; Cambrian microfossils. Mailing Add: Dept of Geol Univ of Calgary Calgary AB Can

MCGUGAN, WESLEY ALEXANDER, b Weyburn, Sask, Oct 31, 20; m 58. FOOD CHEMISTRY. Educ: Univ Sask, BSA, 49; Univ Wis, MS, 53, PhD(dairy & food indust, biochem), 54. Prof Exp: Agr res officer, Dairy Technol Res Inst, 54-62, RES SCIENTIST, FOOD RES INST, AGR CAN, 62- Mem: Am Dairy Sci Asn; Agr Inst Can; Can Inst Food Sci & Technol. Res: Dairy chemistry; flavors. Mailing Add: Food Res Inst Agr Can Ottawa ON Can

MCGUIGAN, JAMES E, b Paterson, NJ, Aug 20, 31; m 56; c 3. GASTROENTEROLOGY, IMMUNOLOGY. Educ: Seattle Univ, BS, 52; St Louis Univ, MD, 56; Am Bd Gastroenterol, dipl. Prof Exp: Intern med, Pa Hosp, Philadelphia, 56-57; resident internal med, Sch Med, Univ Wash, 60-62; asst prof med, Wash Univ, 66-69; PROF MED & CHIEF DIV GASTROENTEROL, COL MED, UNIV FLA, 69- Concurrent Pos: NIH fel gastroenterol, 62-64; Nat Inst Allergy & Infectious Dis spec fel immunol, Sch Med, Wash Univ, 64-66; NIH res career develop award, 66-69. Mem: AAAS; Am Soc Clin Invest. Res: Immunological phenomena associated with gastrointestinal diseases. Mailing Add: Col of Med Univ of Fla Gainesville FL 32601

MCGUIGAN, ROBERT ALISTER, JR, b Evanston, Ill, July 21, 42; m 65. MATHEMATICS. Educ: Carleton Col, BA, 64; Univ Md, College Park, PhD(math), 68. Prof Exp: Asst prof math, Univ Mass, Amherst, 68-74; ASST PROF MATH, WESTFIELD STATE COL, 74-, CHMN DEPT, 75- Mem: Am Math Soc; Math Asn Am. Res: Banach spaces; spaces of continuous functions; non-standard analysis; history of mathematics; theoretical linguistics. Mailing Add: Dept of Math Westfield State Col Westfield MA 01085

MCGUINE, THOMAS HARRY, b Oshkosh, Wis, Feb 2, 17; m 42; c 2. CHEMISTRY. Educ: Marquette Univ, BS, 39; DePaul Univ, MS, 45. Prof Exp: Res chemist, Chicago Printed String Co, 39-43 & Wilson & Co, Inc, 43-53; res dir, Wilson-Martin Div, 53-61; develop supvr indust chem, Archer Daniels Midland Co, Minn, 62-68; vpres corp planning & develop, Glyco Chem Co, NY, 68-71; PRES, T M INNOVATIONS, 71- Concurrent Pos: Pres, T M Abstr Co & Turtle Lake Realty Inc, Wis, 71-; assoc dir, Oasis-2000, 74- Honors & Awards: First Spec Serv Award, Fatty Acid Producers' Coun, 61. Mem: Am Oil Chemists Soc; Commercial Develop Asn. Res: Electrochemistry; edible fats and oils; fatty acids. Mailing Add: T M Innovations Box 356 Turtle Lake WI 54889

MCGUINNESS, AIMS CHAMBERLAIN, b Chester, NY, Mar 10, 05; m 33; c 3. PEDIATRICS. Educ: Princeton Univ, AB, 27; Columbia Univ, MD, 31. Prof Exp: From asst instr to instr pediat, Sch Med, Univ Pa, 34-39, assoc, 40-45, from asst prof to assoc prof, 46-59; exec secy med educ, NY Acad Med, 59-66; pres, Educ Coun Foreign Med Grads, 63-66, assoc dir, 67-74; RETIRED. Concurrent Pos: Intern, Hosp Univ Pa, 31-33; resident, Children's Hosp, Philadelphia, 33-34, mem vis staff, 34-35, sr vis physician, 47-54, dir, 51-54, consult, 54-; dir, Philadelphia Serum Exchange, 34-54; pvt pract, 34-47; consult, Secy War, 41-43, assoc mem, Comn Immunization, Armed Forces Epidemiol Bd, 47-58; examr, Am Bd Pediat, 50-65; dean, Grad Sch Med, Univ Pa, 51-54; clin dir, Miners Mem Hosp Asn; 54-56; spec asst for health & med affairs, Secy Health, Educ & Welfare & mem interagency adv group, President's Coun Youth Fitness, 57-59; US del, World Health Assemblies, 57-59, Fed Coun Sci & Tech, 59 & Health Res Coun, New York, 62-68; trustee, Keene Valley Hosp, NY & Princeton Hosp, 63-68; adj prof pediat, Univ Pa, 67- Mem: Am Pediat Soc (secy, 51-60); Soc Pediat Res; Asn Med Cols; Am Acad Pediat. Res: Immunology and infectious diseases; development of pertussis hyperimmune serum; medical administration. Mailing Add: 721 Prospect Ave Princeton NJ 08540

MCGUINNESS, EUGENE T, b Newark, NJ, Feb 2, 27; m 59; c 2. BIOCHEMISTRY. Educ: St Peter's Col, BS, 49; Fordham Univ, MS, 54; Rutgers Univ, PhD(biochem), 61. Prof Exp: Anal chemist, Wallace & Tiernan, Inc, 49-50; res asst, Fordham Univ, 51-52; asst lab control supvr, P Ballantine & Sons, 52-55; from instr to asst prof, 55-63, ASSOC PROF CHEM, SETON HALL UNIV, 63- Mem: AAAS; Am Chem Soc. Res: Mechanisms of enzyme action; biological oxidation; metalloproteins; radioisotope chemistry. Mailing Add: Dept of Chem Seton Hall Univ South Orange NJ 07079

MCGUINNESS, JAMES ANTHONY, b Staten Island, NY, Nov 4, 41. ORGANIC CHEMISTRY. Educ: St Peter's Col, NJ, BS, 63; Columbia Univ, MA, 64, PhD(chem), 68. Prof Exp: RES CHEMIST RES & DEVELOP, UNIROYAL INC, 67- Mem: Am Chem Soc; The Chem Soc. Res: Polymer degradation, especially photochemical; polymer modification by organic reactions; synthesis of organic chemicals as agricultural products. Mailing Add: Res & Develop Uniroyal Inc Middlebury CT 06749

MCGUINNESS, JAMES L, b Philadelphia, Pa, Nov 22, 17; m 41; c 3. HYDROLOGY. Prof Exp: Anal statistician, 55-64, RES STATISTICIAN, SOIL & WATER CONSERV RES DIV, USDA, 64- Mem: AAAS; Am Geophys Union. Res: Hydrologic research, particularly on small areas. Mailing Add: Soil & Water Conserv Res Div USDA PO Box 478 Coshocton OH 43812

MCGUINNESS, MICHAEL JOSEPH, JR, b Santa Fe, NMex, Apr 15, 13; m 42; c 2. PHYSICAL CHEMISTRY. Educ: Univ NMex, BS, 34; Syracuse Univ, MS, 38, PhD(phys chem), 41. Prof Exp: Asst chem, Univ NMex, 33-34; asst, Syracuse Univ, 34-41; res assoc, Battelle Mem Inst, 41-42, res engr, 42-44; res chemist, Distillation Prod, Inc, NY, 44-48 & E I du Pont de Nemours & Co, Va, 48-50; asst prof chem, Fordham Univ, 50-53; assoc prof, 53-71, PROF CHEM, MARYMOUNT COL, 71-, HEAD DEPT, 53- Concurrent Pos: Consult, US Air Force, 50-51; adj assoc prof, Fordham Univ, 53-; consult, Burroughs, Wellcome & Co, Inc, 57- Mem: Am Chem Soc; NY Acad Sci. Res: Thermodynamics; kinetics; olefin extraction, molecular distillation, cellulose acetate fiber studies; electrochemistry of non-aqueous solutions. Mailing Add: Dept of Chem Marymount Col Tarrytown NY 10591

MCGUIRE, AUSTIN DOLE, b Malden, Mass, Oct 13, 24; m 50; c 1. PARTICLE PHYSICS. Educ: Mass Inst Technol, BS, 49; Univ Rochester, PhD(physics), 54. Prof Exp: Jr physicist, Univ Rochester, 49; staff mem, Los Alamos Sci Lab, 54-65; prog mgr, Electro-Optical Systs, Inc, 65-66; sr staff mem, TRW Systs, Calif, 66-67; mem staff, EG&G Inc, 67-70; SPEC PROJ OFFICER, LOS ALAMOS SCI LAB, 70- Mem: Fel Am Phys Soc. Res: Fundamental particles; interaction of mesons with nucleii; detection of free neutrino; space radiation; particle accelerator design; nuclear weapon testing. Mailing Add: Los Alamos Sci Lab PO Box 1663 Los Alamos NM 87545

MCGUIRE, CHARLES FRANCIS, b Heber City, Utah, May 13, 29; m 52; c 5. AGRONOMY, CEREAL CHEMISTRY. Educ: Brigham Young Univ, BS, 54; Utah State Univ, MS, 65; NDak State Univ, PhD(agron), 68. Prof Exp: Lab technician & mgr qual control, Pillsbury Co, 55-63; asst prof, 68-71, ASSOC PROF AGRON & CEREAL TECHNOL, MONT STATE UNIV, 71- Mem: Am Asn Cereal Chemists; Am Soc Agron. Res: Improvement in wheat and barley quality through improved breeding and cultural practices. Mailing Add: Dept of Plant & Soil Sci Mont State Univ Bozeman MT 59715

MCGUIRE, DAVID KELTY, b Pittsburgh, Pa, Dec 18, 34; m 65. SCIENCE POLICY. Educ: St Vincent Col, BS, 57; Univ Pittsburgh, PhD(anal chem), 64. Prof Exp: Res assoc, Brookhaven Nat Lab, 64-65; asst prof chem, Rider Col, 65-67; asst prof, 67-70, ASSOC PROF CHEM, UPSALA COL, 70-, CHMN DEPT, 71- Mem: AAAS; Am Chem Soc. Res: Science, technology, and public policy relationships, with special attention to the energy situation. Mailing Add: 94 Christopher St Montclair NJ 07042

MCGUIRE, DONALD CHARLES, b Clearwater, Kans, Sept 20, 15; m 42; c 4. GENETICS. Educ: Univ Wash, BS, 46; Univ Calif, PhD(genetics), 50. Prof Exp: From asst prof to prof agr, Univ Hawaii, 50-62, asst olericulturist, Agr Exp Sta, 50-54, chmn dept agr, Univ, 56-59, chmn dept trop crop sci, 57-62; assoc prog dir, 62-66, dir undergrad student prog, 66-69, DIR PRE-SERV TEACHER EDUC PROG, NSF, 69- Concurrent Pos: Mem environ adv comt, Washington Tech Inst, 73- Mem: Fel AAAS; Am Inst Biol Sci; Am Educ Res Asn; Asn Educ Teachers of Sci. Res: Self-incompatibility of wild tomato; inheritance of disease resistance in cultivars; science education, particularly at the undergraduate level; development of programs for enhancement of science education. Mailing Add: 4206 N 35th St Arlington VA 22207

MCGUIRE, EDWARD JOHN, chemistry, see 12th edition

MCGUIRE, EUGENE J, b New York, NY, May 15, 38; m 65; c 3. ATOMIC PHYSICS. Educ: Manhattan Col, BEE, 59; Cornell Univ, PhD, 65. Prof Exp: Mem tech staff physics res, 65-74, SUPVR LASER THEORY DIV, SANDIA LABS, 74- Mem: Am Phys Soc. Res: High power gas laser theory; x-ray lasers; auger transitions and electron spectroscopy; gas discharges; inelastic atomic processes. Mailing Add: Sandia Labs Box 5800 Albuquerque NM 87115

MCGUIRE, FRANCIS JOSEPH, b Baltimore, Md, Aug 29, 32; m 63. ORGANIC CHEMISTRY, ANALYTICAL CHEMISTRY. Educ: Loyola Col, Md, BS, 54; Johns Hopkins Univ, MA, 56, PhD(org chem), 61. Prof Exp: Res chemist, E I du Pont de Nemours & Co, 61-63; from instr to asst prof, 63-67, chmn dept, 65-67, ASSOC PROF CHEM & DEAN STUDIES, LOYOLA COL, MD, 67- Concurrent Pos: Acad internship prog fel, Am Coun on Educ, 69-70. Mem: Am Chem Soc. Res: Structure determination and mechanism of cyclization reactions of natural products; synthesis of polymers; problems in chemical education. Mailing Add: Off of Dean Studies Day Div Loyola Col 4501 N Charles St Baltimore MD 21210

MCGUIRE, JAMES B, b Newcastle, Ind, May 7, 34; m 56; c 4. THEORETICAL PHYSICS. Educ: Purdue Univ, BSEE, 56; Univ Calif, Los Angeles, MS, 60, PhD(physics), 63. Prof Exp: Mem tech staff, Hughes Aircraft Co, 56-57 & Space

Technol Labs, 57-63; asst res physicist, Univ Calif, Los Angeles, 63-64; asst prof, 64-70, ASSOC PROF PHYSICS, FLA ATLANTIC UNIV, 70-, CHMN DEPT, 74- Concurrent Pos: Vis asst physicist, Brookhaven Nat Lab, 65. Mem: Am Phys Soc. Res: Scattering theory; many particle systems; one dimensional models. Mailing Add: Dept of Physics Fla Atlantic Univ Boca Raton FL 33432

MCGUIRE, JAMES MARCUS, b Gassville, Ark, July 30, 35; m 56; c 4. PLANT PATHOLOGY. Educ: Univ Ark, BS, 56, MS, 57; NC State Univ, PhD(plant path), 61. Prof Exp: Asst prof plant path, SDak State Univ, 61-63; res assoc, 63-65, assoc prof, 65-70, PROF PLANT PATH, UNIV ARK, FAYETTEVILLE, 70- Mem: Am Phytopath Soc; Soc Nematol. Res: Nematode transmission of viruses; interactions between nematodes and other plant pathogens; diseases of ornamental plants. Mailing Add: Dept of Plant Path Univ of Ark Fayetteville AR 72701

MCGUIRE, JOHN ALBERT, b Banner, Miss, July 31, 31; m 55. ANIMAL BREEDING, ANIMAL GENETICS. Educ: Miss State Univ, BS, 51, MS, 57; Auburn Univ, PhD(animal breeding), 69. Prof Exp: Instr animal sci, Miss State Univ, 55-56, supt, Natchez Br Exp Sta, 56-65; ASSOC PROF BIOSTATIST, AUBURN UNIV, 68- Mem: Am Soc Animal Sci; Biomet Soc. Res: Animal physiology; improvement of important production traits in beef cattle and sheep by selection for genetically superior animals. Mailing Add: Res Data Anal Auburn Univ Auburn AL 36830

MCGUIRE, JOHN J, b Mar 19, 30; US citizen; m 54; c 6. ORNAMENTAL HORTICULTURE. Educ: Rutgers Univ, BS, 58; Univ RI, MS, 61, PhD(biol sci), 68. Prof Exp: Instr hort, Va Polytech Inst & State Univ, 61-62; from instr to asst prof, 62-72, ASSOC PROF HORT, UNIV RI, 72- Mem: Am Soc Hort Sci; Int Plant Propagators Soc. Res: Winter hardiness of woody ornamental plants; marketing technology of container grown ornamental plants; propagation of woody plants. Mailing Add: Dept of Plant & Soil Sci Univ of RI Kingston RI 02881

MCGUIRE, JOHN L, b Kittanning, Pa, Nov 3, 43; m 69. PHARMACOLOGY. Educ: Butler Univ, BS, 65; Marquette Sch Med, 65-66; Princeton Univ, MS, 68, PhD(physiol, biochem), 69. Prof Exp: From assoc scientist pharmacol to scientist pharmacol, 69-72, sect head biochem res, 72-75, EXEC DIR RES, ORTHO PHARMACEUT CORP, 75- Mem: Am Soc Pharmacol & Exp Therapeut; Endocrine Soc; Am Chem Soc; Soc Study Reproduction; Biochem Soc Gt Brit. Res: Mechanism of steroid hormone & drug action; radioimmunoassays; drug metabolism; endocrinology. Mailing Add: Ortho Pharmaceut Corp Raritan NJ 08869

MCGUIRE, JOHN MURRAY, b New Bedford, Mass, May 15, 29; m 54; c 5. ANALYTICAL CHEMISTRY. Educ: Univ Miami, BS, 48, MS, 51; Univ Fla, PhD(chem), 55. Prof Exp: Prod chemist, Gen Elec Co, 55-57; sr res chemist, Wash Res Ctr, W R Grace & Co, 57-60; sr chemist, Cathode Ray Tube Dept, Gen Elec Co, 60-63, supvr fluid develop eng, TV Receiver Dept, 63-68, supvr fluids eval, Advan Eng Proj Oper, Visual Commun Dept, 68-70, advan mat engr, Audio Prod Dept, 70-71; res chemist, Contaminants Characterization Prog, Southeast Water Lab, 71-73, CHIEF ORG ANAL SECT, ANAL CHEM BR, ENVIRON RES LAB, ENVIRON PROTECTION AGENCY, 73- Concurrent Pos: Ed adv, Biomed Mass Spectrometry, 74- Mem: Am Chem Soc; Am Soc Mass Spectrometry; Am Soc Testing & Mat. Res: Mass spectrometric and gas chromatographic analysis of water contaminants; computer control of analytical instrumentation; computerized spectrum matching. Mailing Add: Environ Res Lab College Station Rd Athens GA 30601

MCGUIRE, JOHN R, b Milwaukee, Wis, Apr 20, 16; m 45; c 1. FOREST ECONOMICS. Educ: Univ Minn, BS, 39; Yale Univ, MF, 41; Univ Pa, MA, 54. Prof Exp: Res ctr leader, Northeastern Forest Exp Sta, US Forest Serv, Pa, 49-51, res analyst & actg div chief, 51-55, chief div forest econ res, 55-57, chief economist, Southwestern Forest & Range Exp Sta, Calif, 57-62, staff asst res, Hq, Washington, DC, 62-63, dir, Southwestern Forest & Range Exp Sta, Calif, 63-67, dep chief prog planning & legis, Hq, Washington, DC, 67-71, assoc chief, US Forest Serv, 71-72, CHIEF, US FOREST SERV, 72- Mem: Soc Am Foresters. Res: Timber and range management; wildlife habitat; forest protection; recreation; management sciences. Mailing Add: US Forest Serv Washington DC 20250

MCGUIRE, JOSEPH SMITH, JR, b Logan, WVa, Apr 19, 31; m 54; div; c 4. MEDICINE. Educ: WVa Univ, AB, 52; Yale Univ, MD, 55. Prof Exp: Intern med, Yale-New Haven Med Ctr, 55-56; clin assoc, Nat Inst Arthritis & Metab Dis, 56-58, investr, 58-59; from asst prof to assoc prof med, 61-72, PROF DERMAT, SCH MED, YALE UNIV, 72- Concurrent Pos: USPHS spec fel med, Yale Univ, 59-61, USPHS award, Sch Med, 61-63, dir clin res training prog, Yale Univ, 64-69. Mem: AAAS; Soc Invest Dermat; Am Dermat Asn; Am Soc Cell Biol; Am Acad Dermat. Res: Control of pigmentation in vertebrate melanocytes; control of cell division in cultivated malignant and non-malignant cells. Mailing Add: Dept of Dermat Yale Univ Sch of Med New Haven CT 06510

MCGUIRE, JUDSON ULERY, JR, b Maricao, PR, July 4, 17; m 43; c 2. BIOMETRICS. Educ: Iowa State Col, BS, 41, MS, 47, PhD(entom), 54. Prof Exp: Sr asst sanitarian, USPHS, 47-51; biometrician, Biomet Serv, Plant Indust Sta, Md, 54-57, chg Hoja Blanca Res Lab, Camagüay, Cuba, 57-59, entomologist, Entom Res Div, 57-62, invest leader rice insects, La, 60, res entomologist, Ecol Studies Face Fly, Paris, 61-62, biometrician, Biomet Serv, 62-74, STAFF SPECIALIST, DATA SYSTS APPLN DIV, AGR RES SERV, USDA, 74- Mem: AAAS; Entom Soc Am; Sigma Xi; Biomet Soc. Res: Application of statistical procedures to entomology; population studies of the European corn borer; discrete distribution theory; statistical services. Mailing Add: Nat Agr Libr Bldg Beltsville MD 20705

MCGUIRE, ODELL, b Knoxville, Tenn, Apr 19, 27; m 57; c 3. GEOLOGY. Educ: Univ Tulsa, BS, 56; Columbia Univ, MA, 58; Univ Ill, PhD(geol), 62. Prof Exp: Assoc prof, 62-71, PROF GEOL, WASHINGTON & LEE UNIV, 71- Concurrent Pos: Geologist, Va Div Mining Resources. Res: Fossil population studies; geology of Appalachian region. Mailing Add: Dept of Geol Washington & Lee Univ Lexington VA 24450

MCGUIRE, ROBERT FRANCIS, b Rochester, NY, Mar 3, 38; m 67; c 4. BIOPHYSICS, BIOCHEMISTRY. Educ: St John Fischer Col, BA, 60; Univ Ky, MS, 62; Univ Rochester, PhD(biophys), 68. Prof Exp: Instr, Univ Ky, 62; fel, Cornell Univ, 68-71; scientist biophys, Worcester Found Exp Biol, 71-74; ASST PROF BIOCHEM, SCH MED, UNIV MASS, WORCESTER, 75- Mem: Biophys Soc. Res: Physical chemical studies on structure-function relationships in cell membranes; hormone-membrane receptor interactions; hormone binding; activation of membrane enzymes; membrane isolation techniques and characterization; spectroscopy in membranes. Mailing Add: Dept of Biochem Univ of Mass Sch of Med Worcester MA 01605

MCGUIRE, ROBERT FRANK, b Greeneville, Tenn, Oct 8, 37; m 62. PHYCOLOGY. Educ: Union Col, Ky, BA, 60; Univ Tenn, Knoxville, MS, 64, PhD(bot), 71. Prof Exp: Teacher, Pub Schs, Ky, 60-62; instr math & sci, Eastern Ky Univ, 64-67; from teaching asst to instr bot, Univ Tenn, Knoxville, 67-71; asst prof, 71-74, ASSOC PROF BIOL, UNIV MONTEVALLO, 74- Concurrent Pos: NDEA fel, Univ Tenn, Knoxville, 70. Mem: Am Inst Biol Sci; Int Phycol Soc; Phycol Soc Am. Res: Application of the principles of numerical taxonomy to the algal divisions, especially the Chlorophyta; these procedures have been used with selected species of Chlorococcum and clones of Chara. Mailing Add: Dept of Biol Univ of Montevallo Montevallo AL 35115

MCGUIRE, ROBERT LEE, animal science, see 12th edition

MCGUIRE, STEPHEN EDWARD, b Excelsior Springs, Mo, Mar 25, 42; m 63; c 2. INDUSTRIAL CHEMISTRY. Educ: Lamar State Col Technol, BS, 63; Univ Tex, Austin, PhD(org chem), 67. Prof Exp: Res chemist, 67-70, sr res chemist, 70-73, RES GROUP LEADER, CONTINENTAL OIL CO, 73- Mem: Am Chem Soc. Res: Friedel-Crafts alkylations; carbonium ions; detergent intermediates. Mailing Add: Res & Develop Dept Res Div Continental Oil Co Ponca City OK 74601

MCGUIRE, WILLIAM SAXON, b Prescott, Ark, Oct 9, 22; m 52; c 3. AGRONOMY. Educ: Univ Ark, BS, 47; Univ NZ, MAgSc, 51; Wash State Univ, PhD(agron), 52. Prof Exp: Asst prof agron & asst agronomist, Miss State Univ, 52-55; assoc prof agron & assoc agronomist, NMex State Univ, 56; PROF AGRON & AGRONOMIST, ORE STATE UNIV, 57- Mem: Am Soc Agron; Crop Sci Soc Am; Am Forage & Grassland Coun. Res: Pasture and forage production and management. Mailing Add: Dept of Crop Sci Ore State Univ Corvallis OR 97331

MCGUIRE, WILLIS CLARKE, parasitology, protozoology, see 12th edition

MCGURK, DONALD J, b Wichita, Kans, June 2, 40. ORGANIC CHEMISTRY. Educ: Univ Nebr, BS, 62; Okla State Univ, PhD(org chem), 68. Prof Exp: Asst prof, 68-72, ASSOC PROF CHEM, SOUTHWESTERN STATE COL, OKLA, 72- Mem: AAAS; Am Chem Soc. Res: Chemistry of natural products; structure of molecules in oil of catnip; identity of volatile compounds produced by ants. Mailing Add: Dept of Chem Southwestern State Col Weatherford OK 73096

MCHAFFEY, DAVID GEORGE, b Pueblo, Colo, Nov 8, 32; m 58; c 2. ENTOMOLOGY, PLANT PATHOLOGY. Educ: Colo Agr & Mech Col, BS, 54; Univ Ariz, MS, 58; Wash State Univ, PhD(entom, plant path), 69. Prof Exp: Entomologist, Algodonera Escandone, SA, Mexicali, Mex, 53; exp aide, Exp Sta, Univ Ariz, Mesa, 57; control aide Khapra beetle inspection, USDA, Ariz, 58; res technician, Exp Sta, Univ Calif, Riverside, 58-59; entomologist malaria eradication projs, US AID, Indonesia & SVietnam, 59-65; area supvr, Aedes Aegypti Eradication Proj, Commun Dis Ctr, USPHS, Ga, 65; asst, Wash State Univ, 66-69; res entomologist, Southern Res Inst, 69-74, MEM STAFF, AGR RES SERV, USDA, 74- Mem: Entom Soc Am; Am Mosquito Control Asn. Res: Medical entomology; insect chemosterilants for the boll weevil. Mailing Add: ARS USDA Agr Res Ctr East Bldg 306 Room 112 Beltsville MD 20705

MCHALE, EDWARD THOMAS, b Hazleton, Pa, Dec 10, 32; m 57; c 3. PHYSICAL CHEMISTRY. Educ: King's Col, Pa, BS, 54; Pa State Univ, PhD(fuel sci), 64. Prof Exp: Res chemist, Gen Chem Div, Allied Chem Corp, 54-58; chemist, Reaction Motors Div, Thiokol Chem Corp, 58-60; PHYS CHEMIST, ATLANTIC RES CORP, 64- Mem: Am Chem Soc; Combustion Inst. Res: Chemical kinetics; reaction mechanisms; fire research; gaseous explosions; combustion of gases and solids; thermal decompositions; phase equilibrium studies; analytical methods and instrumental techniques. Mailing Add: Atlantic Res Corp Alexandria VA 22314

MCHALE, JOHN LAWRENCE, JR, physics, see 12th edition

MCHALE, JOHN T, b New York, NY, Nov 2, 33; m 62; c 1. BOTANY, CYTOLOGY. Educ: Iona Col, BS, 55; Univ Tex, PhD(bot), 65. Prof Exp: Asst prof, 65-69, ASSOC PROF BIOL, LOYOLA UNIV, LA, 69- Concurrent Pos: Sigma Xi grant-in-aid-of res, 67; consult cytol & cytogenetics, Gulf South Res Inst, La, 68-70 & 72-74; NIH training grant, Univ Calif, Berkeley, 70-71, res assoc, 71-72; res physiologist consult, Vet Admin Hosp, Martinez, Calif, 70-72. Mem: Am Physiol Soc. Res: Mechanisms of aging on the cellular level; bioenergetics of subcellular organelles; ultrastructure of plant cells; plant morphogenesis; cytology and cytogenetics. Mailing Add: Dept of Biol Sci Loyola Univ New Orleans LA 70118

MCHARD, JAMES ALVAH, analytical chemistry, see 12th edition

MCHARDY, GEORGE GORDON, b New Orleans, La, Mar 7, 10; m; c 3. MEDICINE. Educ: Spring Hill Col, BA, 32; Tulane Univ, MD, 36; Am Bd Internal Med, dipl & cert gastroenterol, 43. Prof Exp: Asst prof med, Tulane Univ La, 38-51; chief div gastroenterol, 60-71, assoc prof, 51-58, PROF MED, SCH MED, LA STATE UNIV MED SCH, NEW ORLEANS, 59- Concurrent Pos: Mem, World Cong Gastroenterol, 58-62; lectr, Univ Brazil, 60; sr vis physician, Charity Hosp. Mem: AMA; Am Gastroenterol Asn (treas, 53-59, vpres, 59-62, pres, 62); Am Soc Gastrointestinal Endoscopy (pres, 65); Am Col Physicians. Res: Gastroenterology. Mailing Add: Med Ctr of New Orleans 3638 St Charles Ave New Orleans LA 70115

MCHARRIS, WILLIAM CHARLES, b Knoxville, Tenn, Sept 12, 37; m 60; c 1. NUCLEAR CHEMISTRY. Educ: Oberlin Col, BA, 59; Univ Calif, Berkeley, PhD(nuclear chem), 65. Prof Exp: From asst prof to assoc prof, 65-71, PROF CHEM & PHYSICS, COL NATURAL SCI, MICH STATE UNIV, 71- Concurrent Pos: Consult, Heavy Elements Group, Argonne Nat Lab, 66-; sabbatical, Lawrence Berkeley Lab, Univ Calif, 71-72; Sloan fel, 72-76. Mem: AAAS; Am Phys Soc; Am Chem Soc; Sigma Xi. Res: Nuclear spectroscopy and reactions in actinides, lead region and deformed rare earths; on-line spectroscopy with cyclotron; beta-decay theory. Mailing Add: Dept of Chem Mich State Univ East Lansing MI 48823

MCHENRY, HENRY MALCOLM, b Los Angeles, Calif, May 19, 44; m 66. BIOLOGICAL ANTHROPOLOGY. Educ: Univ Calif, Davis, BA, 66, MA, 67; Harvard Univ, PhD(anthrop), 72. Prof Exp: ASST PROF ANTHROP, UNIV CALIF, DAVIS, 71- Mem: AAAS; Sigma Xi; Am Asn Phys Anthrop; Am Anthrop Asn; Brit Soc Study Human Biol. Res: Paleoanthropology; australopithecine postcranial anatomy. Mailing Add: Dept of Anthrop Univ of Calif Davis CA 95616

MCHENRY, HUGH LANSDEN, b Baxter, Tenn, Aug 19, 37; m 63; c 2. BIOSTATISTICS, MATHEMATICAL STATISTICS. Educ: Tenn Technol Univ, BS, 60; George Peabody Col, MA, 61, PhD(math), 70. Prof Exp: Instr math, Okla Christian Col, 61-63, asst prof, 65-68; from asst prof to assoc prof, Memphis State Univ, 68-73; ASST PROF BIOSTATIST, UNIV TENN, 73- Mem: Am Statist Asn; Math Asn Am. Res: Mathematics and statistics education. Mailing Add: Dept of Community Med Univ of Tenn 800 Madison Ave Memphis TN 38163

MCHENRY, JOHN ROGER, b Gering, Nebr, Nov 3, 16; m 42; c 4. SOIL SCIENCE. Educ: Nebr State Teachers Col, BSc, 39; Univ Nebr, MSc, 41; Iowa State Col, PhD(soil physics), 44. Prof Exp: Res assoc soils, Iowa State Col, 42-44; asst prof agron, Univ Nebr, 44-47; soil physics, State Col Wash, 47-49; soil scientist, Div Soil

Surv, USDA, NDak, 49-52; chemist, Gen Elec Co, Wash, 52-57; soil scientist, Soil & Water Conserv Res Div, 57-69, chemist, Sedimentation Lab, 69-74, SOIL SCIENTIST, SEDIMENTATION LAB, AGR RES SERV, USDA, 74-; PROF CHEM & CHEM ENG, UNIV MISS, 59- Concurrent Pos: Hydrologist & consult, Int Atomic Energy Agency-Indonesian Atomic Energy Agency, Jakarta, Indonesia, 70. Mem: Am Chem Soc; Soil Sci Soc Am; Am Geophys Union; Am Soc Agron; Am Water Resources Asn. Res: Mechanics of soil aggregation; soil structure, moisture, irrigation; ionic exchange properties of soils; clay minerals as related to soils; radiochemistry; watershed hydrology; reservoir sedimentation; remote sensing appled to sedimentation; fallout 137-cesium in erosion and sedimentation. Mailing Add: 114 Leighton Rd Oxford MS 38655

MCHUGH, JAMES ANTHONY, JR, b Stockton, Calif, Oct 7, 37; m 57; c 4. PHYSICAL CHEMISTRY, NUCLEAR CHEMISTRY. Educ: Univ Pac, BS, 59; Univ Calif, Berkeley, PhD(chem), 63. Prof Exp: Phys chemist, 63-74, mgr, Mass Spectrometry Res & Develop, 74-75, MGR, CHEM LAB, KNOLLS ATOMIC POWER LAB, GEN ELEC CO, 75- Mem: Am Soc Mass Spectrometry. Res: Mass spectrometry; systematics of nuclear fission; medium energy ion-surface interactions and secondary positive ion emission; ionization phenomena. Mailing Add: Chem Lab Knolls Atomic Power Lab Gen Elec Co PO Box 1072 Schenectady NY 12301

MCHUGH, JAMES PAUL, physical chemistry, see 12th edition

MCHUGH, JOHN LAURENCE, b Vancouver, BC, Nov 24, 11; nat US; m 41; c 3. FISHERIES MANAGEMENT. Educ: Univ BC, BA, 36, MA, 38; Univ Calif, Los Angeles, PhD(zool), 50. Prof Exp: Asst, Univ BC, 36-38; fishery biologist, Pac Biol Sta, BC, 38-41; asst ichthyol, Scripps Inst, Univ Calif, 47-48, res assoc, 48-51; dir, Va Fisheries Lab, 51-59; chief div biol res, Bur Commercial Fisheries, US Fish & Wildlife Serv, 59-63, asst dir biol res, 63-66, dept dir bur, 66-68; actg dir, Off Marine Resources, US Dept Interior, 68-70; head, Off Int Decade Ocean Explor, NSF, 70; PROF MARINE RESOURCES, MARINE SCI RES CTR, STATE UNIV NY STONY BROOK, 70- Concurrent Pos: Mem, Nat Res Coun, 65-70; adv comn marine resources to dir gen, Food & Agr Orgn, 66-70; mem US nat sect, Int Biol Prog, 67-70; comt int marine sci affairs policy, Nat Acad Sci, 70-; US comnr, Inter-Am Trop Tuna Comn, 60-70; vchmn, Int Whaling Comn, 68-71, chmn, 71-72; mem hard clam adv comt, Nassau-Suffolk Regional Marine Resources Coun, 73-; consult, Town of Islip, NY & Islip Town Shellfish Mgt Comn, 74- Mem: AAAS; Am Inst Biol Sci; Am Fisheries Soc; Inst Fishery Res Biol. Res: Oceanography; fishery biology and management; resolution of social-political impediments to application of scientific knowledge in fishery utilization and management. Mailing Add: Marine Sci Res Ctr State Univ of NY Stony Brook NY 11794

MCHUGH, KENNETH LAURENCE, b Brooklyn, NY, Mar 22, 27; m 53; c 5. ORGANIC CHEMISTRY. Educ: Hofstra Col, BA, 51; Univ Conn, PhD(chem), 59. Prof Exp: Jr res chemist, Evans Res & Develop Corp, 51-53; res chemist, Conn Hard Rubber Co, 53-56; asst instr chem, Univ Conn, 56-58; sr res chemist, 58-60, res group leader, Spec Proj Dept, 60-61 & Org Chem Div, 61-64, proj mgr org develop dept, 64-66, sr proj mgr, 66-68, proj mgr commercial develop, Chemstrand Res Ctr, New Enterprises Div, 68-71, MGR COMMERCIAL DEVELOP, NEW ENTERPRISES DIV, MONSANTO CO, 71- Mem: Am Chem Soc; Am Soc Lubrication Engrs; Coord Res Coun. Res: High temperature stable fluids, principally those derived from organometallic, organophosphorus and polyaromatic ether chemistry for use as turbine lubricants, lubricant additives, power transmission fluids and thermodynamic fluids for Rankine and Brayton power cycles; air pollution control by catalytic processes. Mailing Add: A3NF New Enterprises Div Monsanto Co St Louis MO 63166

MCHUGH, PAUL RODNEY, b Lawrence, Mass, May 21, 31; m 59; c 3. NEUROLOGY, PSYCHIATRY. Educ: Harvard Univ, AB, 52, MD, 56. Prof Exp: Intern med, Peter Bent Brigham Hosp, 56-57; res neurol, Mass Gen Hosp, 57-60; clin asst psychiat, Maudsley Hosp, London, Eng, 60-61; res asst neuroendocrinol, Walter Reed Army Inst Res, 61-64; asst prof neurol & psychiat, Med Col, Cornell Univ, 64-71, prof psychiat, 71-73; prof psychiat & chmn dept, Med Sch, Univ Ore, 73-75; CHMN & PSYCHIATRIST-IN-CHIEF DEPT PSYCHIAT & BEHAV SCI, SCH MED, JOHNS HOPKINS UNIV, 75- Concurrent Pos: Clin dir & supvr psychiat educ, Westchester Div, New York Hosp-Cornell Med Ctr, 69-73. Mem: Am Psychiat Asn; Am Neurol Asn; Am Physiol Soc; Harvey Soc; Am Psychopath Asn. Res: Neural mechanisms of visceral, endocrine and behavioral control. Mailing Add: Dept of Psychiat & Behav Sci Johns Hopkins Univ Hosp Baltimore MD 21205

MCHUGH, RICHARD B, b Villard, Minn, Oct 25, 23; m 51; c 5. BIOMETRICS, BIOSTATISTICS. Educ: Univ Minn, PhD, 55. Prof Exp: Assoc prof psychol statist, Iowa State Univ, 50-56; assoc prof, 56-62, PROF BIOMET, UNIV MINN, MINNEAPOLIS, 62-, DIR BIOMET, 68- Concurrent Pos: Nat Inst Gen Med Sci fel, 64-65; vis scholar, Univ London, 71-72; consult, Dept Med & Surg, Vet Admin; mem pancreatic cancer study sect, Nat Cancer Inst, 74- Mem: Hon fel AAAS; Pop Asn Am; hon fel Am Statist Asn; hon fel Am Pub Health Asn; Biomet Soc. Res: Research design in the health and life sciences; mathematical demography; epidemetrics; biomathematical models in assay; biostatistics in health services research; cancer clinical trials and surveys. Mailing Add: Sch of Pub Health Univ of Minn Mayo Box 197 Minneapolis MN 55455

MCHUGH, WILLIAM PAUL, b Highland Park, Ill. ANTHROPOLOGY, ARCHAEOLOGY. Educ: Univ Wis, PhD(anthrop), 71. Prof Exp: Asst prof anthrop, Univ Wis-Milwaukee, 67-74; ASST PROF ANTHROP, MURRAY STATE UNIV, 74- Concurrent Pos: Mem, Am Res Ctr in Egypt. Mem: Soc Am Archaeol; Soc Field Archaeol; Am Anthrop Asn. Res: General Old World, post-Pleistocene, Near East, Neolithic and urban origins; archaeology methods and theory; teaching of archaeology; role of archaeology in education and society; teaching and training in archaeology; Northeastern African prehistory; Mid-South prehistory; preservation of cultural resources. Mailing Add: Dept of Sociol & Anthrop Murray State Univ Murray KY 42071

MCILHENNY, HUGH M, b Gettysburg, Pa, Sept 25, 38; m 62; c 2. DRUG METABOLISM. Educ: Pa State Univ, BS, 60; Univ Mich, MS, 64, PhD(pharm, chem), 66. Prof Exp: Res asst antibiotics, Parke-Davis & Co, Inc, 60-62; anal chemist, 66-75, RES INVESTR, PFIZER, INC, 75- Mem: Am Chem Soc; affil mem Am Soc Pharmacol & Exp Therapeut. Res: Fate for foreign substances in biological systems; factors influencing drug metabolism; drug bioavailability; development of assay methods for the measurement of drugs and their metabolites. Mailing Add: Cent Res Dept of Drug Metab Pfizer Inc Groton CT 06340

MCILRATH, THOMAS JAMES, astrophysics, see 12th edition

MCILRATH, WAYNE JACKSON, b Laurel, Iowa, Oct 18, 21; m 42; c 3. PLANT PHYSIOLOGY. Educ: Iowa State Teachers Col, BA, 43; Univ Iowa, MS, 47, PhD(plant physiol), 49. Prof Exp: From asst to instr bot, Univ Iowa, 46-49; asst prof plant physiol, Agr & Mech Col Tex & Agr Exp Sta, 49-51; from asst prof to prof bot, Univ Chicago, 51-64; dean grad sch, 64-73, PROF BIOL SCI, NORTHERN ILL UNIV, 64- Concurrent Pos: Consult, Argonne Nat Lab, 56-; OEEC sr vis fel, Sorbonne, 61; res assoc, Univ Chicago, 64-66; chmn coun, Cent States Univs, Inc, 67-68, bd dirs, 69-, treas, 71-; mem comn scholars, Ill Bd Educ, 68-71; mem coun & gov bd, Quad-Cities Grad Study Ctr, 70-73. Mem: AAAS; Am Soc Plant Physiologists; Bot Soc Am; Am Inst Biol Sci; Soc Res Adminr. Res: Mineral nutrition of plants; physiology of growth and development. Mailing Add: Dept of Biol Sci Northern Ill Univ DeKalb IL 60115

MCILRATH, WILLIAM OLIVER, b Coulterville, Ill, Aug 30, 36; m 63; c 2. PLANT BREEDING, GENETICS. Educ: Univ Ill, BS, 59, MS, 64; Okla State Univ, PhD(agron), 67. Prof Exp: From asst prof to assoc prof agron, La State Univ, 67-73; RES AGRONOMIST, AGR RES SERV, USDA, 73- Mem: Am Soc Agron; Crop Sci Soc Am; Am Genetic Asn. Res: Effects of artificial shading on development and morphology of corn; heterosis; combining ability and quantitative genetics of hexaploid wheat; breeding, genetics of rice. Mailing Add: Box 1429 Rice Exp Sta Crowley LA 70526

MCILREATH, FRED J, b Amsterdam, NY, Apr 1, 29; m 52; c 5. PHYSIOLOGY, PHARMACOLOGY. Educ: Siena Col, BS, 51; Univ Ky, MS, 55; McGill Univ, PhD(physiol), 59. Prof Exp: Sr pharmacologist, Strasenburgh Labs, 59-62; from asst dir to assoc dir clin invest, Riker Labs, 62-71; asst dir, 71-73, DIR REGULATORY AFFAIRS, G D SEARLE & CO, 73- Concurrent Pos: Head pulmonary dis sect, Riker Labs, 68-71. Mem: AAAS; Am Physiol Soc; Am Soc Clin Pharmacol & Therapeut; NY Acad Sci. Res: Clinical pharmacology and drug development; effects of drugs on air flow dynamics. Mailing Add: Regulatory Affairs G D Searle & Co Skokie IL 60076

MCILROY, MALCOLM B, b Stone, Eng, Mar 10, 21; m 45; c 5. INTERNAL MEDICINE. Educ: Cambridge Univ, BA, 41, MB, BCh, 44, MD, 51; FRCP, 68. Prof Exp: Casualty physician, St Bartholomews Hosp, London, Eng, 52-55; first asst, Inst Cardiol, Univ London, 55-58; assoc prof, 58-64, PROF MED, SCH MED, UNIV CALIF, SAN FRANCISCO, 64- Concurrent Pos: Res fel, Harvard Med Sch, 53-54. Mem: Am Physiol Soc; Am Fedn Clin Res; Am Soc Clin Invest. Res: Physiology and diseases of the heart and lungs. Mailing Add: Univ of Calif Sch of Med San Francisco CA 94122

MCILROY, MALCOLM DOUGLAS, computer science, see 12th edition

MCILROY, WILLIAM, aeronautical engineering, mathematics, see 12th edition

MCILWAIN, CARL EDWIN, b Houston, Tex, Mar 26, 31; m 52; c 2. PHYSICS. Educ: NTex State Col, BME, 53; Univ Iowa, MS, 56, PhD(physics), 60. Prof Exp: Asst prof physics, Univ Iowa, 60-62; assoc prof, 62-66, PROF PHYSICS, UNIV CALIF, SAN DIEGO, 66- Concurrent Pos: Guggenheim fels, Eng, 67-68 & Eng, Ger & Sweden, 72. Honors & Awards: Space Sci Award, Am Inst Aeronaut & Astronaut, 70. Mem: Am Phys Soc; Am Geophys Union; Am Astron Soc. Res: Energetic particles in solar system. Mailing Add: Dept of Physics Univ of Calif at San Diego La Jolla CA 92037

MCILWAIN, DAVID LEE, b Memphis, Tenn, Jan 7, 38. NEUROCHEMISTRY. Educ: Vanderbilt Univ, BA, 60; Wash Univ, MD, 64. Prof Exp: Fel biochem, Univ Calif, Berkeley, 64-66; fel neurochem, Col Physicians & Surgeons, Columbia Univ, 68-72; ASST PROF NEUROCHEM, UNIV NC, CHAPEL HILL, 72- Mem: Soc Neurosci; Am Soc Neurochem; NY Acad Sci. Res: Chemistry of electrically excitable membranes. Mailing Add: Dept of Physiol Sch of Med Univ of NC Chapel Hill NC 27514

MCILWAIN, ROBERT LESLIE, JR, b Lancaster, SC, Apr 16, 29; m 54; c 3. HIGH ENERGY PHYSICS. Educ: Carnegie Inst Technol, BS, 53, PhD, 60. Prof Exp: Instr physics, Princeton Univ, 59-62; asst prof, 62-65, ASSOC PROF PHYSICS, PURDUE UNIV, WEST LAFAYETTE, 65- Mem: Am Phys Soc. Res: High energy nuclear physics and elementary particles. Mailing Add: Dept of Physics Purdue Univ West Lafayette IN 47906

MCINERNEY, EUGENE FRANCIS, organic chemistry, see 12th edition

MCINERNEY, JOHN EDWARD, b Halifax, NS, June 24, 35; m 59; c 3. ICHTHYOLOGY. Educ: Univ Ottawa, BSc, 59; Univ BC, MSc, 61, PhD(zool), 63. Prof Exp: Asst prof biol, Dalhousie Univ, 63-65; asst prof, 65-70, ASSOC PROF BIOL, UNIV VICTORIA, BC, 70- Concurrent Pos: Dir, Bamfield Marine Sta, 75- Mem: Can Soc Zoologists. Res: Physiology; behavior of fishes. Mailing Add: Dept of Biol Univ of Victoria Victoria BC Can

MCINNIS, BERTRAM CAMPBELL, theoretical physics, high energy physics, see 12th edition

MCINROY, ELMER EASTWOOD, b Ont, Can, Nov 2, 21; m 46; c 3. AGRICULTURAL CHEMISTRY. Educ: Ont Agr Col, BSA, 44. Prof Exp: Res chemist, Defense Industs, Ltd, Can, 43; metall chemist, Deloro Smelting & Refining Co, 44-46; res biochemist, Cent Res Labs, Gen Foods Corp, NJ, 46-50, lab supvr, Gaines Div, 50-53; dir animal nutrit, Foxbilt, Inc, Iowa, 53-56; dir animal nutrit, 56-65, vpres nutrit & tech serv, 65, DIR, ARBIE MINERAL FEED CO, INC, 65- Mem: AAAS; Am Chem Soc; Am Soc Animal Sci; Am Feed Mfrs Asn. Res: Animal, human and dog nutrition; food technology; audio-visual training; technical sales; technical service; public speaking. Mailing Add: Arbie Mineral Feed Co Inc 404 S Center St Marshalltown IA 50158

MCINTEER, BERTHUS BOSTON, JR, physics, see 12th edition

MCINTIRE, CHARLES DAVID, b St Louis, Mo, Sept 20, 32; m 65; c 3. AQUATIC ECOLOGY. Educ: Southern Methodist Univ, BBA, 54; Ore State Univ, BS, 58, MS, 60, PhD(bot), 64. Prof Exp: Res asst fisheries, 58-60, res asst bot, 61-63, asst prof, 64-69, ASSOC PROF BOT, ORE STATE UNIV, 69- Concurrent Pos: Acad Natural Sci Philadelphia McHenry Fund grant, 67; NSF sci fac fel, Ctr Quant Sci, Univ Wash, 70-71. Mem: Ecol Soc Am; Am Soc Limnol & Oceanog; Phycol Soc Am. Res: Physiological ecology of marine and freshwater algae; trophic ecology of aquatic ecosystems; mathematical ecology and systems analysis; diatom systematics. Mailing Add: Dept of Bot & Plant Path Ore State Univ Corvallis OR 97331

MCINTIRE, ELLIOT GREGOR, b Inglewood, Calif, Aug 30, 40; m 63; c 2. CULTURAL GEOGRAPHY, HISTORICAL GEOGRAPHY. Educ: Univ Calif, Riverside, BA, 63; Johns Hopkins Univ, MA, 65; Univ Ore, PhD(geog), 68. Prof Exp: Lectr geog, Univ Calif, Riverside, 67-68; asst prof, 68-71, ASSOC PROF GEOG, CALIF STATE UNIV, NORTHRIDGE, 71-, CHMN DEPT, 74- Mem: AAAS; Asn Am Geogr. Res: Cultural and spatial changes among contemporary American Indian populations; cultural and historical geography of the American Southwest. Mailing Add: Dept of Geog Calif State Univ Northridge CA 91324

MCINTIRE, FLOYD COTTAM, b Price, Utah, Sept 10, 14; m 43; c 3.

BIOCHEMISTRY. Educ: Brigham Young Univ, AB, 36, MA, 37; Univ Wis, PhD(biochem), 40. Prof Exp: Asst, Brigham Young Univ, 37-72; asst, Univ Wis, 37-40; asst agr chemist, Exp Sta, NDak State Col, 41-42; res biochemist, Abbott Labs, Ill, 42-57, sect head, 57-58, head biochem res, 58-66, head molecular biol res, 66-70, dir div exp biol, 70-74; PROF ORAL BIOL, SCH DENT, UNIV COLO MED CTR, 74- Concurrent Pos: Indust fel biochem, Univ Wis, 40-41; assoc prof biochem, Med Sch, Northwestern Univ, 67-74. Mem: AAAS; Am Chem Soc; Am Soc Biol Chem; Am Soc Cell Biol; Am Soc Microbiol. Res: Molecular interactions responsible for adherence phenomena in dental plaque; molecular requirements for immunogenicity and adjuvant effects; regulation of the immune response. Mailing Add: Dept of Oral Biol Sch of Dent Univ of Colo Med Ctr Denver CO 80220

MCINTIRE, JOHN MICHAEL, organic chemistry, polymer chemistry, see 12th edition

MCINTIRE, JUNIUS MERLIN, b Price, Utah, Jan 27, 18; m 47; c 2. BIOCHEMISTRY. Educ: Brigham Young Univ, AB, 40; Univ Wis, MS, 42, PhD(biochem), 44. Prof Exp: Res biochemist, Western Condensing Co, Wis, 44-50; chief dairy oil & fat div, Qm Food & Container Inst, 50-55; asst dir res, 55-67, GEN MGR RES, CARNATION CO, 67- Mem: AAAS; Am Chem Soc. Res: Animal nutrition; vitamin assays; protein chemistry of milk; food processing of milk products; distribution and nutritional significance of certain members of vitamin B complex. Mailing Add: Carnation Labs 8015 Van Nuys Blvd Van Nuys CA 91412

MCINTIRE, KENNETH ROBERT, b Portland, Ore, Mar 31, 33; m 54; c 3. BIOLOGY, IMMUNOLOGY. Educ: Univ VA, BA, 55, MD, 59. Prof Exp: From intern to resident internal med, Univ Hosps of Cleveland, Western Reserve Univ, 59-61; res assoc, 61-63, staff scientist, Lab Biol, 63-70, STAFF SCIENTIST, LAB CELL BIOL, NAT CANCER INST, 70- Mem: AAAS; Am Asn Cancer Res. Res: Study of pathogenesis and pathophysiology of reticular neoplasms, particularly synthesis of abnormal immunoglobulins by the tumors; study of tumor associated antigens used for diagnosis and monitoring of therapy. Mailing Add: Lab of Cell Biol Bldg 8 Rm 204 Nat Cancer Inst Bethesda MD 20014

MCINTIRE, MATILDA S, b Brooklyn, NY, July 15, 20; c 1. PEDIATRICS. Educ: Mt Holyoke Col, BA, 42; Albany Med Col, MD, 46; Am Bd Pediat, dipl. Prof Exp: Intern, Flower & Fifth Ave Hosps, New York, 46-47; pediatrician, 8th Army Hq, Japan, 48-49; resident pediat, St Louis Univ Med Sch, 52-53 & Univ Nebr Hosp, 53-54; pvt pract, 54-56; instr pediat, Sch Med, Creighton Univ, 55-61; pediat consult & asst res prof, Col Med, Univ Nebr, 61-66; from assoc clin prof to clin prof pediat, 66-73, CLIN PROF PUB HEALTH & PREV MED, SCH MED, CREIGHTON UNIV, 68-, PROF PEDIAT & DIR COMMUNITY PEDIAT, 73- Concurrent Pos: Children's Mem Hosp trainee, 73; dir div prev dis control, Omaha-Douglas County Health Dept, 56-61, dir div maternal & child health, 66-; asst prof pediat, Col Med, Univ Nebr, 61-72, assoc prof, 72-, asst prof food & nutrit, 70-; consult, co-investr & prin investr, USPHS grants, 66-74; consult toxicol info prog, Nat Libr Med; consult fac, Inst Clin Toxicol, Houston, Tex; mem infant stand comt, Nebr Dept Pub Welfare; mem comt accident prev, Am Acad Pediat, 74-77. Mem: AAAS; Am Col Prev Med; AMA; fel Am Pub Health Asn; fel Am Acad Pediat. Mailing Add: Dept of Pediat Creighton Univ Sch of Med Omaha NE 68108

MCINTIRE, SUMNER HARMON, b Essex, Mass, July 7, 12; m 36; c 2. PHYSICS, ACADEMIC ADMINISTRATION. Educ: Bowdoin Col, AB, 33; NY Univ, MA, 37. Hon Degrees: DSc, Norwich Univ, 64. Prof Exp: Instr, Westbrook Jr Col, 34-36 & Thornton Acad, 37-38; instr chem, 36-37, from instr to assoc prof physics, 38-57, chmn dept, 62-72, PROF PHYSICS, NORWICH UNIV, 57-, DEAN UNIV, 72- Concurrent Pos: State exam consult, Vt, 44-46; res worker, Bur Indust Res, Northfield, 47; consult, Vt Dept Pub Safety, 54-; sci adv to Gov, 71- Mem: Nat Asn Res Sci Teaching; Am Asn Physics Teachers; Am Asn Higher Educ; Am Conf Acad Deans. Res: Wood-machining practices in Vermont. Mailing Add: Dean's Off Norwich Univ Northfield VT 05663

MCINTIRE, WILLIAM GRANT, b Price, Utah, June 28, 18; m 48; c 3. GEOGRAPHY. Educ: Brigham Young Univ, BS, 50; La State Univ, PhD, 54. Prof Exp: Asst, Univ, 51-54, from instr to assoc prof, 53-64, asst dir, Ctr Wetland Resources, 70-74, PROF GEOG, LA STATE UNIV, BATON ROUGE, 64-, DIR COASTAL STUDIES INST, 66-, ASSOC DIR, CTR WETLAND RESOURCES, 74- Mem: AAAS; Asn Am Geogr; Am Antiq Soc; Soc Am Archaeol. Res: Coastal morphology and environments, delta formation; comparisons of coastal environments on a worldwide basis; aerosols and coastal information management. Mailing Add: Coastal Studies Inst Dept Geomorphol La State Univ Baton Rouge LA 70803

MCINTOSH, A VERN, JR, organic chemistry, see 12th edition

MCINTOSH, ALEXANDER OMAR, b Acton, Ont, Oct 27, 13. INDUSTRIAL CHEMISTRY. Educ: Univ Toronto, BA, 39; Univ Minn, MS, 40; Univ Glasgow, PhD(phys chem), 51. Prof Exp: Asst geol, Univ Minn, 39-40; chemist, Defence Industs, Ltd, 41-44; res chemist, Can Indust, Ltd, 44-48; demonstr, Univ Glasgow, 48-50; res chemist, 51-56, group leader, Cent Res Lab, 56-62, patent asst, Legal Dept, 62-65, PATENT AGT, LEGAL DEPT, CAN INDUST, LTD, 65- Mem: Fel Chem Inst Can. Res: Protection of chemical inventions. Mailing Add: Can Industs Ltd PO Box 10 Montreal PQ Can

MCINTOSH, ARTHUR HERBERT, b St Thomas, VI, Apr 2, 34; m 64; c 3. MICROBIOLOGY. Educ: McMaster Univ, BA, 59; Univ Guelph, MS, 62; Harvard Univ, DSc(microbiol), 69. Prof Exp: NIH fel, Stanford Res Inst, 69-71; staff researcher microbiol, Boyce Thompson Inst, NY, 71-74; ASST RES PROF, WAKSMAN INST MICROBIOL, RUTGERS UNIV, NEW BRUNSWICK, 74- Concurrent Pos: NSF grant, 75. Mem: Soc Invert Path; Tissue Culture Asn. Res: Invertebrate and vertebrate tissue culture and virology; in vitro safety testing of viral insecticides; plant and animal mycoplasmas. Mailing Add: Waksman Inst of Microbiol Rutgers Univ New Brunswick NJ 08903

MCINTOSH, BRUCE ANDREW, b Walkerton, Ont, Oct 30, 29; m 54; c 4. PHYSICS, ELECTRONICS. Educ: Western Ont Univ, BSc, 52, MSc, 53; McGill Univ, PhD(electron beams), 58. Prof Exp: Asst res officer, 53-55, assoc res officer, 58-70, secy assoc comt meteorites, 60, SR RES OFFICER, NAT RES COUN CAN, 70- Mem: Can Asn Physicists; Meteoritical Soc; Int Astron Union; Can Astron Soc. Res: Meteoritics; upper atmosphere physics. Mailing Add: Nat Res Coun Can Herzberg Inst Astrophys Ottawa ON Can

MCINTOSH, CHARLES BARRON, b Edgemont, SDak, Sept 5, 16; div; c 4. GEOGRAPHY. Educ: Huron Col, AB, 38; Univ Nebr, BSc, 39, MA, 51, PhD, 55. Prof Exp: Lectr geog, Univ Canterbury, 53; instr, Univ Tex, 55-56; assoc prof, Eastern Ill Univ, 56-58; PROF GEOG, UNIV NEBR-LINCOLN, 58- Mem: Asn Am Geogr; Am Geog Soc. Res: Climatology; geography of soils; historical geography. Mailing Add: Dept of Geog Univ of Nebr Lincoln NE 68508

MCINTOSH, DAVID LIVINGSTON, b Victoria, BC, Dec 26, 19; m 51; c 3. PLANT PATHOLOGY. Educ: Univ BC, BSA, 48; Univ Toronto, PhD(plant path, cytol), 51. Prof Exp: RES SCIENTIST, PLANT PATH SECT, AGR CAN, 47- Mem: Can Phytopath Soc; Agr Inst Can. Res: Tree fruit parasitic diseases; Phytophthora root diseases. Mailing Add: Plant Path Sect Agr Can Summerland BC Can

MCINTOSH, DOUGLAS CARL, b Ottawa, Ont, Aug 30, 21; m 45; c 4. BOTANY, WOOD TECHNOLOGY. Educ: Queen's Univ, Ont, BA, 44, MA, 45; State Univ NY Col Forestry, Syracuse Univ, PhD(wood technol), 54. Prof Exp: Forest prod engr, Can Forest Prod Lab, Univ Ottawa, 45-54; wood & fiber technologist, Int Cellulose Res, Ltd, Can Int Paper Co, 54-57; sr res engr, Cent Res Lab, Mead Corp, 57-67, res fel wood & fiber technol, 67-69; consult, Arthur D Little, Inc, Mass, 69-71; res scientist, Corp Res Ctr, 72, SUPVR, MICROSTRUCT LAB, CORP RES & DEVELOP DIV, INT PAPER CO, 72- Res: Wood, fiber and paper structure and properties; microscopy of fibers and paper coatings and pigments. Mailing Add: Microstruct Lab Int Paper Co PO Box 797 Tuxedo Park NY 10987

MCINTOSH, ELAINE NELSON, b Webster, SDak, Jan 30, 24; m 55; c 3. PHYSIOLOGICAL BACTERIOLOGY, NUTRITION. Educ: Augustana Col, SDak, AB, 45; Univ SDak, Vermillion, MA, 49; Iowa State Univ, PhD(phys bact & biochem), 54. Prof Exp: From instr to asst prof chem, Sioux Falls Col, SDak, 45-48; res fel biochem, Univ SDak, Vermillion, 48-49; instr bact, Iowa State Univ, 49-54; res assoc dairy sci, Univ Ill, Urbana, 54-55; res assoc home econ res, Iowa State Univ, 55-62; from asst prof to assoc prof nutrit sci, 68-75, asst to vchancellor, 74-75, ASSOC PROF NUTRIT SCI, UNIV WIS-GREEN BAY, 72-, SPEC ASST TO CHANCELLOR, 75- Concurrent Pos: Sigma Xi res grant, 66; pres, Wis Nutrit Coun, 75. Honors & Awards: Chancellor's Res Award, Univ Wis-Green Bay, 69. Mem: Sigma Xi; Am Soc Microbiol; Inst Food Technologists; Am Dietetic Asn; Am Pub Health Asn. Res: Food additives and their relationship to human nutrition and food safety. Mailing Add: Libr Learning Ctr 730E Univ of Wis Green Bay WI 54302

MCINTOSH, HAMISH WILLIAM, b Vancouver, BC, Jan 13, 20; m; c 2. MEDICINE. Educ: Cambridge Univ, MB, 38; McGill Univ, MD, CM, 43, dipl & MSc, 50; FRCPS(C), 49. Prof Exp: Clin instr, 52-60, from asst prof to assoc prof, 60-66, PROF MED, UNIV BC, 66- Concurrent Pos: Dir res, Shaughnessy Hosp, 50-60; assoc, Vancouver Gen Hosp, 52- Mem: Endocrine Soc; Can Soc Clin Invest (past pres). Res: Endocrinology and metabolism. Mailing Add: Dept of Med Univ of BC Vancouver BC Can

MCINTOSH, HAROLD LEROY, b Fairfax, Mo, Dec 25, 31; m 53; c 3. PHYSICS. Educ: Tarkio Col, BA, 53; Univ Colo, MS, 61; Ore State Univ, PhD(physics, chem), 68. Prof Exp: Pub sch teacher, Mo, 56-61; PROF PHYSICS, TARKIO COL, 62-, DEAN EDUC PROGS, 74- Mem: Am Asn Physics Teachers. Res: Chemistry; mathematics. Mailing Add: Dept of Physics Tarkio Col Tarkio MO 64491

MCINTOSH, HENRY DEANE, b Gainesville, Fla, July 19, 21; m 45; c 3. INTERNAL MEDICINE. Educ: Davidson Col, BS, 43; Univ Pa, MD, 50; Am Bd Internal Med, dipl, 57; Am Bd Cardiovasc Dis, dipl, 64. Prof Exp: Intern med, Duke Univ Hosp, 50-51; asst res, Lawson Vet Admin Hosp, 51-52; from instr to assoc med, Sch Med, Duke Univ, 54-58, from asst prof to prof, 58-70; PROF MED & CHMN DEPT, BAYLOR COL MED, 70- Concurrent Pos: Am Heart Asn fel cardiol, Duke Univ Hosp, 52-54; consult, Vet Admin Hosps, Durham, NC, 56-70 & Fayetteville, 57-70, Womack Army Hosp, Ft Bragg, 57-70, Watts Hosp, Durham, 57-70 & Portsmouth Naval Hosp, Va, 57-70; dir cardiovasc lab, Med Ctr, Duke Univ, 56-70, dir cardiovasc div, 66-70; asst ed, Mod Concepts Cardiovasc Dis, 65-68; mem cardiovasc study sect, Nat Heart Inst, 65-69; mem, Subspecialty Bd Cardiovasc Dis, Am Bd Internal Med, 68-70, mem, Am Bd Internal Med, 70-; chief med serv, Methodist Hosp, Houston, 70-, Ben Taub Gen Hosp, Jefferson Davis Hosp & Houston Vet Admin Hosp; consult, St Luke's Hosp; ed consult, Am J Cardiol & Circulation, 70-; chmn coun clin cardiol, Am Heart Asn, 74-76. Mem: Am Soc Internal Med; Am Heart Asn; Am Col Physicians; Am Fedn Clin Res; Am Col Cardiol (pres, 73-74). Res: Cardiovascular hemodynamics, especially factors controlling cardiac output. Mailing Add: Dept of Med Baylor Col of Med Houston TX 77025

MCINTOSH, JERRY LEON, b St James, Mo, Apr 11, 33; m 62. SOIL CHEMISTRY, AGRONOMY. Educ: Univ Mo, BS, 58, MS, 59, PhD(soil chem), 62. Prof Exp: Asst soil scientist & soil plant relationship specialist, Univ Vt, 62-64, from asst prof to assoc prof plant & soil sci, 64-73; MULTIPLE CROPPING AGRONOMIST, INT RICE RES INST, INT PROG INDONESIA, 73- Concurrent Pos: Rockefeller Found vis scientist, Multiple Cropping Proj, Int Rice Res Inst, Los Banos, Philippines, 69-70; consult, US Fed Exten Serv, S Vietnam, 71. Mem: Am Soc Agron; Int Soc Soil Sci. Res: Plant nutrition, improvement of soil testing methods for different soils; understanding factors affecting mineral availability to plants; evaluation and implementation of cropping systems for different agro-climatic zones. Mailing Add: IRRI-CRIA Coop Prog in Indonesia PO Box 107 Bogor Indonesia

MCINTOSH, JOHN MCLENNAN, b Galt, Ont, Apr 16, 40; m 65; c 2. SYNTHETIC ORGANIC CHEMISTRY. Educ: Queen's Univ, Ont, BSc, 62; Mass Inst Technol, PhD(org chem), 66. Prof Exp: Fel org chem, Nat Res Coun Can, 66-68; teaching fel, Univ Waterloo, 68; asst prof, 68-73, ASSOC PROF ORG CHEM, UNIV WINDSOR, 73- Mem: Am Chem Soc; Chem Inst Can. Res: Synthesis of complex molecules; new methods of organic synthesis, especially those utilizing organo-sulfur and organo-phosphorus compounds. Mailing Add: Dept of Chem Univ of Windsor Windsor ON Can

MCINTOSH, JOHN RICHARD, b New York, NY, Sept 25, 39; m 61; c 3. CELL BIOLOGY. Educ: Harvard Univ, BA, 61, PhD(biophys), 67. Prof Exp: Sch teacher, Mass, 61-63; from instr to asst prof biol, Harvard Univ, 67-70; asst prof, 71-73, ASSOC PROF BIOL, UNIV COLO, BOULDER, 73- Concurrent Pos: Consult, Educ Serv Inc, 62-63; NIH fel, Biol Labs, Harvard Univ, 68. Mem: Am Soc Cell Biologists. Res: Mitosis and cell motion; control of cell form. Mailing Add: Dept of Molecular, Cellular & Develop Biol Univ of Colo Boulder CO 80302

MCINTOSH, JOHN STANTON, b Ford City, Pa, Jan 6, 23. THEORETICAL PHYSICS. Educ: Yale Univ, BS, 48, MS, 49, PhD(physics), 52. Prof Exp: Res assoc physics, Proj Matterhorn, Princeton Univ, 52-53, instr, Univ, 53-56; from instr to asst prof, Yale Univ, 56-63; assoc prof, 63-65, PROF PHYSICS, WESLEYAN UNIV, 65- Concurrent Pos: Consult, Los Alamos Sci Lab & Rand Corp, 53; res assoc, Peabody Mus Natural Hist, Yale Univ, 65-71. Mem: Am Phys Soc; Soc Vert Paleont. Res: Low energy nuclear and nuclear scattering theories; heavy ion scattering and reactions. Mailing Add: Dept of Physics Wesleyan Univ Middletown CT 06457

MCINTOSH, ROBERT LLOYD, b Montreal, Que, Feb 16, 15; m 42; c 3. PHYSICAL CHEMISTRY. Educ: Dalhousie Univ, BA, 35, MSc, 36; McGill Univ, PhD, 39. Hon Degrees: DSc, McGill Univ, 72. Prof Exp: Res chemist, Shawinigan Chem, Ltd, 39-40; asst dir extramural res chem, Nat Res Coun Can, 41-45, sr res chemist, 47-48; from assoc prof to prof chem, Univ Toronto, 48-64; prof & head dept, 65-69, assoc dean arts & sci, 69-71, DEAN SCH GRAD STUDIES & RES, QUEEN'S UNIV,

ONT, 71- Honors & Awards: Mem, Order Brit Empire, 42. Mem: Fel Royal Soc Can. Res: Properties of physically absorbed gas films in solids such as dielectric constant chemisorption of gases on metal films by surface potential measurements. Mailing Add: Sch of Grad Studies & Res Queen's Univ 131 Union St Kingston ON Can

MCINTOSH, ROBERT PATRICK, b Milwaukee, Wis, Sept 24, 20; m 47; c 2. PLANT ECOLOGY. Educ: Lawrence Col, BS, 42; Univ Wis, MS, 48, PhD, 50. Prof Exp: Asst bot, Univ Wis, 46-50; from instr to asst prof, Middlebury Col, 50-53; asst prof, Vassar Col, 53-58; asst prof, 58-67, PROF BIOL, UNIV NOTRE DAME, 67- Concurrent Pos: Ed, Am Midland Naturalist, 70- Mem: AAAS; Ecol Soc Am; Brit Ecol Soc; Soc Am Naturalists. Res: Forest ecology; history of ecology. Mailing Add: Dept of Biol Univ of Notre Dame Notre Dame IN 46556

MCINTOSH, THOMAS HENRY, b Ames, Iowa, July 3, 30; m 55; c 3. ENVIRONMENTAL SCIENCES. Educ: Iowa State Univ, BS, 56, MS, 58, PhD(soil microbiol), 62. Prof Exp: Instr soils, Iowa State Univ, 60-62; from asst prof to assoc prof, Univ Ariz, 62-68; asst dean col environ sci, 68, assoc prof sci & environ change, 68-71, asst dean cols, 70-73, assoc dean, 73-75, PROF SCI & ENVIRION CHANGE, UNIV WIS-GREEN BAY, 71-, ASST CHANCELLOR STUDENT & ADMIN SERV, 75- Concurrent Pos: Vis prof, Iowa State Univ, 67. Mem: Am Soc Agron; Soil Sci Soc Am; Am Soc Microbiol; Am Chem Soc; Soil Conserv Soc Am. Res: Soil organic matter chemistry; nitrogen; transformations in soil and aquatic systems; biogeochemistry. Mailing Add: Asst Chancellor for Student & Admin Serv LLC-835 Univ of Wis Green Bay WI 54302

MCINTOSH, WILLIAM DAVID, b Pryor, Okla, Oct 14, 36. MATHEMATICS. Educ: Southwestern Col, Kans, BA, 58; Univ Kans, MA, 60, PhD(math), 65. Prof Exp: Asst instr math, Univ Kans, 58-63, asst, 63-64, asst instr, 64-65; asst prof, Univ Mo-Columbia, 65-70; PROF MATH & CHMN DEPT, CENTRAL METHODIST COL, 70- Mem: Am Math Soc; Math Asn Am. Res: Theory of retracts. Mailing Add: Dept of Math Cent Methodist Col Fayette MO 65248

MCINTURFF, ALFRED D, b Clinton, Okla, Feb 24, 37; m 60, 64; c 2. HIGH ENERGY PHYSICS, CRYOGENICS. Educ: Okla State Univ, BS, 59; Vanderbilt Univ, MS, 60, PhD(physics), 64. Prof Exp: Res assoc, Vanderbilt Univ, 64-66, sr res engr, Atomics Int, 66-67; mem staff, Advan Accelerator Design Div, 67-71, PHYSICIST, ACCELERATOR DEPT, BROOKHAVEN NAT LAB, 71- Mem: Am Phys Soc. Res: High field super conductors; photostar research; photosigma production; sigma magnetic moment studies. Mailing Add: T-923 Brookhaven Nat Lab Upton NY 11973

MCINTYRE, ALAN DAVID, b Vancouver, BC, Sept 18, 29; nat US; m 54; c 3. POLYMER CHEMISTRY. Educ: Univ BC, BA, 50, MA, 52; Cornell Univ, PhD(chem), 55. Prof Exp: Asst, Univ BC, 49-51 & Cornell Univ, 52-53; res chemist, Torrance Res Lab, Shell Chem Co Div, Shell Oil Co, 56-57, sr res chemist, 57-59, group leader, Plastics & Resins Div, 59-60, sect leader, 60-62, mgr polymer chem, 62-63, mgr olefin plastics, 64-65, mgr plastics develop, 65-69; HEAD APPL CHEM DIV, BC RES COUN, 69- Honors & Awards: Award, Los Angeles Sch Syst, 59 & 60. Mem: Am Chem Soc; Am Soc Testing & Mat. Res: Polymer molecular weights and distributions; dynamic mechanical properties of polymers; impact phenomena; testing of polymers; melt rheology of plastics; economic evaluation of laboratory programs; air pollution. Mailing Add: Appl Chem Div BC Res Coun 3650 Wesbrook Crescent Vancouver BC Can

MCINTYRE, ANDREW, b Weehawken, NJ, Sept 17, 31; m 67. MARINE GEOLOGY, MARINE BIOLOGY. Educ: Columbia Univ, BA, 55, PhD(marine biol), 67. Prof Exp: Lectr geol, Barnard Col, Columbia Univ, 61-62; instr, City Col New York, 63-64; res assoc, 62-67, SR VIS RES ASSOC OCEANOG, LAMONT GEOL OBSERV, COLUMBIA UNIV, 67-; ASSOC PROF GEOL, QUEENS COL, NY, 69- Concurrent Pos: NSF res grants, 62-; asst prof geol, Queens Col, NY, 67-69. Mem: AAAS; Geol Soc Am; NY Acad Sci; Soc Econ Paleontologists & Mineralogists; Paleont Soc Japan. Res: Biogeography and ecology of Coccolithophorida, Pleistocene paleoclimatology and paleooceanography; nanoplankton of the oceans; skeletal ultramicrostructure of nanoplankton and microplankton. Mailing Add: Dept of Earth & Environ Sci Queens Col Flushing NY 11367

MCINTYRE, DAVID H, b Yakima, Wash, July 13, 36; m 67. GEOLOGY. Educ: Univ Wash, BS, 59, MS, 60; Wash State Univ, PhD(geol), 66. Prof Exp: GEOLOGIST, US GEOL SURV, 66- Mem: Geol Soc Am; assoc Am Geophys Union. Res: Cenozoic volcanic rocks of Southwestern and Central Idaho; geology of Western Puerto Rico. Mailing Add: US Geol Surv Bldg 25 Denver Fed Ctr Denver CO 80225

MCINTYRE, DONALD, b Detroit, Mich, Sept 8, 28; m 57; c 4. POLYMER CHEMISTRY. Educ: Lafayette Col, BA, 49; Cornell Univ, PhD(chem), 54. Prof Exp: Chemist, Monsanto Chem Co, Mass, 53-54; chemist, Nat Bur Standards, 56-62, sect chief macromolecules sect, 62-66; PROF CHEM, UNIV AKRON, 66- Mem: Am Chem Soc; Am Phys Soc. Res: Physical chemistry of polymers; solution properties; kinetics; phase equilibria of polymers; characterization of polymers; cycloparaffins; structure of polymers. Mailing Add: Inst of Polymer Sci Univ of Akron Akron OH 44325

MCINTYRE, DONALD B, b Edinburgh, Scotland, Aug 15, 23; m 57; c 1. GEOLOGY. Educ: Univ Edinburgh, BSc, 45, PhD, 47. Hon Degrees: DSc, Univ Edinburgh, 51. Prof Exp: Lectr econ geol, Univ Edinburgh, 48-52, lectr petrol, 52-54; assoc prof, 54-57, PROF GEOL, POMONA COL, 57-, CHMN DEPT, 55- Concurrent Pos: Res assoc, Univ Calif, 52; Guggenheim Found fel, 69-70; consult, IBM Sci Ctr. Honors & Awards: Pigeon Award, Geol Soc London, 52. Mem: Geol Soc Am; Geol Soc London; Royal Soc Edinburgh. Res: Structural geology of deformed rocks; petrogenesis; sampling methods; computer applications; James Hutton and the foundation of geology. Mailing Add: Dept of Geol Pomona Col Claremont CA 91711

MCINTYRE, DONALD PATRICK, b Toronto, Ont, Mar 17, 16; m 41; c 3. METEOROLOGY, ATMOSPHERIC SCIENCES. Educ: Univ Toronto, BA, 38, MA, 39; Univ Chicago, PhD(meteorol), 49. Prof Exp: Meteorologist, Meteorol Serv Can, 39-40, forecasting, BC, 40-43, meteorologist in chg, Hq, Royal Can Air Force Group, 43-44, Forecasting Off, BC, 44-45, meteorologist forecasting, Que, 46-47, researcher, 49-50, chief res & training div, 50-72; DIR GEN, ATMOSPHERIC RES DIRECTORATE, ATMOSPHERIC ENVIRON SERV, 72- Concurrent Pos: Hon lectr, Univ Toronto, 49-63; mem, Arrangements Comt, Joint Royal Meteorol Soc-Am Meteorol Soc Conf, 52-53; Can rep, Comn Atmospheric Sci, World Meteorol Orgn, 53-, chmn, Working Group Int Projs Meteorol, 61-65, Working Group Rev Tech Regulations, Comn Atmospheric Sci, 70-; partic, Scand-Am Meteorol Conf, Bergen, 57-58; ed, J Appl Meteorol, 61-66; mem, Assoc Comt Geod & Geophys, Nat Res Coun Can, 61-67, Subcomt Meteorol & Atmospheric Sci, 61-67, adv, 67-, mem, Assoc Comt Space Res, 61-; mem & reporter, Can Comt Int Quiet Sun Yrs, 62-68; mem, Off Can Deleg, Pac Sci Cong, Tokyo, 66; mem, Exec & Organizing Comts, Int Conf Cloud Physics, Int Asn Meteorol & Atmospheric Physics, 67-68; mem, Interdept Appraisal Comt Res Scientists & Res Mgt, 68-72; mem, Ad Hoc Adv Comt Interim Can Space Coun, Can Sci Secretariat, 69; transport contact, Prog Planning Off & alt, Interdept Comt Resource Satellites & Remote Atmospheric Sensing, 69-71, alt, Coord Comt, Can Global Atmospheric Res Prog, 69-; transp rep, Ad Hoc Subcomt Environ Qual, Interdept Comt Resources, 70-71; mem, Space Subcomt, Ministry Transp, 70-71; mem, Shaping Comt, Environ Sci & Eng Prog, Univ Toronto, 70-71, UN Subcomt, Interdept Comt Int Environ Activities, 70-72 & Interdept Comt Air Pollution, 70-; meteorol consult, 71- Honors & Awards: Darton Prize, Royal Meteorol Soc, 53 & 55. Mem: Can Meteorol Soc (pres, 53-55); fel Royal Meteorol Soc (vpres, 55-57); fel Am Meteorol Soc; fel NY Acad Sci. Res: From the management and direction viewpoint, all fields of atmospheric sciences and applications. Mailing Add: 55 Queen St E Suite 1300 Toronto ON Can

MCINTYRE, GARY A, b Portland, Ore, July 16, 38; m 63. PLANT PATHOLOGY. Educ: Ore State Univ, BS, 60, PhD(plant path), 64. Prof Exp: From asst prof to prof plant path, Univ Maine, Orono, 64-75, actg head dept bot & plant path, 68-69, chmn dept, 69-75; CHMN DEPT BOT & PLANT PATH, COLO STATE UNIV, 75- Mem: Am Phytopath Soc; Potato Asn Am; Sigma Xi. Res: Potato diseases; physiology of parasitism. Mailing Add: Dept of Bot & Plant Path Colo State Univ Ft Collins CO 80521

MCINTYRE, GEORGE FRANCIS, b Chicago, Ill, Mar 12, 43; m 65; c 2. APPLIED CHEMISTRY. Educ: Univ Dayton, BS, 65; Ill Inst Technol, PhD(phys chem), 70. Prof Exp: Applns chemist, 69-72, prod specialist, 72-74, MINING INDUST MGR, NALCO CHEM CO, 74- Mem: Am Chem Soc; Soc Mining Engrs; Am Mining Cong; Am Inst Metall Engrs. Res: Coagulation and flocculation of mineral process slurries; new water and process chemical development for mining and mineral processing industry. Mailing Add: Nalco Chem Co 2901 Butterfield Rd Oak Brook IL 60521

MCINTYRE, JAMES DOUGLASS EDMONSON, b Toronto, Ont, Feb 16, 34; m 60; c 3. PHYSICAL CHEMISTRY. Educ: Univ Toronto, BA, 56, MA, 58; Rensselaer Polytech Inst, PhD(phys chem), 61. Prof Exp: Res assoc chem, Princeton Univ, 60-62; MEM TECH STAFF CHEM RES, BELL LABS, 62- Mem: Am Chem Soc; Electrochem Soc. Res: Electrode kinetics and adsorption; electrodeposition; chemistry and physics of surfaces; catalysis; optical properties of thin films; modulation spectroscopy. Mailing Add: Chem Physics Res Dept Bell Labs Murray Hill NJ 07974

MCINTYRE, JOHN ARMIN, b Seattle, Wash, June 2, 20; m 47; c 1. NUCLEAR PHYSICS. Educ: Univ Wash, BS, 43; Princeton Univ, MA, 48, PhD(physics), 50. Prof Exp: Instr elec eng, Carnegie Inst Technol, 43-44; radio engr, Westinghouse Elec Corp, 44-45; res assoc, Stanford Univ, 50-57; from asst prof to assoc prof physics, Yale Univ, 57-63; assoc dir cyclotron inst, 65-70, PROF PHYSICS, TEX A&M UNIV, 63- Concurrent Pos: Consult, Varian Assocs, 55; vis scientist, NSF, 60-65; Oak Ridge Inst Nuclear Studies res partic grant, 63-67; Welch Found grants, 64-66 & 67-70; councilor, Oak Ridge Assoc Univs, 65-71. Mem: AAAS; fel Am Phys Soc; Soc Nuclear Med; Am Sci Affil. Res: Scintillation counters; Compton effect; high energy electron scattering; heavy ion scattering and transfer reactions; gamma ray spectroscopy; three-nucleon scattering; nuclear instrumentation for medicine and technology. Mailing Add: Dept of Physics Tex A&M Univ College Station TX 77843

MCINTYRE, JOHN BOWIE, b Long Beach, Calif, July 13, 24; m 51; c 4. GEOLOGY. Educ: Ore State Univ, BS, 48; Am Univ, MA, 55. Prof Exp: Geologist, US Geol Surv, 48-51; res analyst, Army Coastal Eng Res Ctr, 51-62, Army Topographic Command, 62-63 & Dept Defense, 64-67; opers res analyst, Army Combat Develop Command, 67-72; OPERS RES ANALYST, ARMY CONCEPTS ANAL AGENCY, 72- Concurrent Pos: Consult to asst chief force develop, Dept Army, 68-69. Mem: Soc Econ Paleontologists & Mineralogists. Res: Neritic zone coastal hydrography; operations research and systems analysis; coastal oceanography; coastal processes; force development; human behavior; military cost analysis. Mailing Add: 6907 Baylor Dr Alexandria VA 22307

MCINTYRE, JOHN LEE, b Portland, Ore, Jan 31, 47; m 71; c 1. PLANT PATHOLOGY. Educ: Ore State Univ, BS, 69, MS, 71; Purdue Univ, PhD(plant path & biochem), 74. Prof Exp: ASST PLANT PATHOLOGIST, CONN AGR EXP STA, 74- Mem: Am Phytopath Soc. Res: Host-parasite interactions between plants and plant pathogenic microorganisms. Mailing Add: Conn Agr Exp Sta PO Box 1106 New Haven CT 06504

MCINTYRE, JOHN MARK, physical chemistry, see 12th edition

MCINTYRE, LAURENCE COOK, JR, b Knoxville, Tenn, July 9, 34; m 57; c 4. NUCLEAR PHYSICS. Educ: Stanford Univ, BS, 57; Univ Wis, MS, 61, PhD(physics), 65. Prof Exp: Instr physics, Princeton Univ, 65-66; asst prof, 66-70, ASSOC PROF PHYSICS, UNIV ARIZ, 70- Mem: Am Phys Soc. Res: Low energy nuclear physics; polarization and charged particle reactions. Mailing Add: Dept of Physics Univ of Ariz Tucson AZ 85721

MCINTYRE, MICHAEL PERRY, b Seattle, Wash, May 16, 21; m 45; c 3. PHYSICAL GEOGRAPHY, GEOGRAPHY OF THE PACIFIC BASIN. Educ: Univ Wash, BA, 43, MA, 47; Ohio State Univ, PhD(geog), 51. Prof Exp: Asst prof geog, Kent State Univ, 49-52; asst prof, Wayne State Univ, 52-56; from asst prof to assoc prof, 56-60, PROF GEOG & CHMN DEPT, CALIF STATE UNIV, SAN JOSE, 60- Mem: AAAS; Asn Am Geogr. Res: Post World War II evaluation of the national economy of New Zealand and Australia; Philippine agriculture. Mailing Add: Dept of Geog Calif State Univ San Jose CA 95192

MCINTYRE, OSWALD ROSS, b Chicago, Ill, Feb 13, 32; m 57; c 3. HEMATOLOGY. Educ: Dartmouth Col, AB, 53; Harvard Med Sch, MD, 57. Prof Exp: Intern, Hosp Univ Pa, 57-58; resident med, Dartmouth Affiliated Hosps, 58-60; trainee hemat, Dartmouth Med Sch, 60-61; asst chief clin res sect, Pakistan SEATO Cholera Res Lab, 61-62; mem bact prof, Div Biol Standards, NIH, 62-63; trainee heamt, 63-64, from asst prof med to assoc prof med, 66-76, PROF MED, DARMOUTH MED SCH, 76- Concurrent Pos: Markle scholar acad med, 65; counsult, Hitchcock Clin, 64; attend physician, Vet Admin Hosp, White River Junction, Vt, 64. Mem: AAAS; Am Fedn Clin Res; Am Soc Hemat; NY Acad Sci. Res: Cellular immunology; in vitro lymphocyte culture and production of effecoeffectors; chemotherapy of hematologic malignancies; genetic aspects of hematologic illness. Mailing Add: Norris Cotton Cancer Ctr Hinman Box 7000 Hanover NH 03755

MCINTYRE, PATRICIA ANN, b Christopher, Ill, Sept 1, 26. MEDICINE. Educ: Kalamazoo Col, AB, 48; Johns Hopkins Univ, MD, 52; Am Bd Internal Med, dipl, 62; Am Bd Nuclear Med, dipl, 72. Prof Exp: Intern, Mass Gen Hosp, 52-53; asst resident internal med, Johns Hopkins Hosp, 53-55, instr med, 57-58, clin dir dept med, Mem Med Ctr, 58-64; res assoc hemat div, Sch Med, 64-65, instr med, 65-67, instr, Univ, 66-67, asst prof med & radiol sci, 67-71, assoc prof radiol sci, Sch Hyg, 71-73, ASSOC PROF MED & RADIOL, SCH MED, JOHNS HOPKINS UNIV, 71-, ASSOC PROF RADIOL SCI, SCH MED & ASSOC PROF ENVIRON HEALTH, SCH HYG, 73- Concurrent Pos: Fels hemat, Sch Med, Johns Hopkins Univ, 55-57;

dir med educ, Hosp for Women, 57-58; mem panel int comt standardization hemat. Mem: Am Fedn Clin Res; Am Soc Hemat; Soc Nuclear Med; fel Am Col Physicians. Res: Hematology; nuclear medicine. Mailing Add: Rm 2001 Johns Hopkins Med Inst 615 N Wolfe St Baltimore MD 21205

MCINTYRE, ROBERT ALLEN, JR, b Lumberton, NC, Feb 25, 26. ZOOLOGY. Educ: Wake Forest Col, BS, 46; Univ NC, MA, 48. Prof Exp: Asst prof biol, Coker Col, 48-51; assoc prof, Presby Col, SC, 55-62; ASSOC PROF BIOL, CAMPBELL COL, 62- Mailing Add: Dept of Biol Campbell Col Buies Creek NC 27506

MCINTYRE, ROBERT GERALD, b Cleveland, Okla, Mar 26, 24; m 59; c 3. APPLIED MATHEMATICS, THEORETICAL PHYSICS. Educ: US Naval Acad, BS, 45; Univ Okla, PhD(physics), 59. Prof Exp: Res mathematician, Sohio Petrol Co, 56-59; asst prof math & physics, Grad Inst Technol, Univ Ark, 57-60; mem tech staff & theoret physicist, Tex Instruments, Inc, 60-63; asst prof math, Okla State Univ, 63-65; from assoc prof to prof physics, 65-69, prof math, 69-74, PROF PHYSICS, UNIV TEX, EL PASO, 74- Mem: Am Math Soc; Am Phys Soc. Mailing Add: Dept of Physics Univ of Tex El Paso TX 79902

MCINTYRE, ROBERT JOHN, b Bathurst, NB, Dec 19, 28; m 55; c 2. SOLID STATE PHYSICS. Educ: St Francis Xavier Univ, Can, BSc, 50; Dalhousie Univ, MSc, 53; Univ Va, PhD(physics), 56. Prof Exp: Sr mem sci staff, RCA Victor Co, Ltd, 56-67; DIR SEMICONDUCTOR ELECTRONICS LAB, RCA LTD, 67- Mem: Sr mem Inst Elec & Electronics Engrs; Can Asn Physicists. Res: Solid state surface, semiconductor and device physics. Mailing Add: RCA Ltd Ste Anne de Bellevue PQ Can

MCINTYRE, RUSSELL THEODORE, b Alexis, Ill, Mar 20, 25; m 48; c 3. BIOCHEMISTRY. Educ: Monmouth Col, BS, 49; Kans State Col, MS, 50, PhD(chem), 52. Prof Exp: Asst prof chem, La State Univ, 52-53; asst biochemist agr chem & biochem, 53-56; biochemist, Haynie Prod, Inc, NJ, 56-61; dir qual control, 61-65, DIR SPEC PROD DEVELOP, CAPITAL CITY PROD CO, INC, COLUMBUS, 65- Mem: AAAS; Am Chem Soc; Inst Food Technologists; Am Oil Chemists Soc. Res: Electrophoresis of plant proteins; biochemistry of fishes; lipid chemistry; polyglycerols and their derivatives. Mailing Add: 3780 Smiley Rd Hilliard OH 43026

MCINTYRE, THOMAS WILLIAM, b Chicago, Ill, Apr 22, 41; m 60; c 3. ORGANIC CHEMISTRY, RESEARCH ADMINISTRATION. Educ: DePauw Univ, BA, 62; Univ Iowa, PhD(org chem), 67. Prof Exp: Sr org chemist, 67-71, res scientist, 71-73, MGR PROD INTROD, ELI LILLY & CO, 73- Mem: AAAS; Am Chem Soc. Res: Electrochemical reactions of organic compounds; synthesis, modification and purification of cephalosporin antibiotics; process research on plant fungicides and herbicides. Mailing Add: Eli Lilly & Co 307 E McCarty St Indianapolis IN 46206

MCINTYRE, THOMAS WOODFORD, b Los Angeles, Calif, July 11, 36; m 58; c 3. BIOPHYSICS, COMPUTER SCIENCE. Educ: Park Col, BA(biol) & BA(chem), 61; Univ Calif, Los Angeles, PhD(physiol), 65. Prof Exp: Asst prof physiol, Univ Calif, Los Angeles, 65-67, asst prof biophys, 66-67; lectr biosci, Brock Univ, 67-69; asst prof, 69-73, ASSOC PROF PHYSIOL & BIOPHYS, WVA UNIV, 73- Concurrent Pos: Muscular Dystrophy Asn Can grant, Brock Univ, 67-69. Mem: AAAS; Biophys Soc. Res: Mechanical aspects of muscle contraction; on line laboratory data analysis; cardiovascular mechanics. Mailing Add: Dept of Physiol & Biophys WVa Univ Med Ctr Morgantown WV 26506

MCINTYRE, WALLACE EDWARD, b Streator, Ill, Nov 18, 18; m 43; c 1. ECONOMIC GEOGRAPHY, ADMINISTRATIVE SCIENCES. Educ: Ill State Normal Univ, BEd, 40; Clark Univ, MA, 47, PhD(geog), 51. Prof Exp: Mem staff, Univ NH, 47-49; prof geog & head dept, W Tex State Col, 50-51; assoc prof, Ill State Normal Univ, 51-58; ADMINR, CENT INTEL AGENCY, WASHINGTON, DC, 58- Concurrent Pos: Fulbright lectr, Univ Philippines, 52-53; mem overseas staff, Univ Md, 54-55; univ fel, Harvard Univ, 65-66. Mem: Am Geog Soc; Asn Am Geogr. Res: Geography of Africa and the Near East; physical geography. Mailing Add: 1618 Craig Lane McLean VA 22101

MCINTYRE, WILLIAM ERNEST, JR, b Alvy, WVa, Dec 19, 25; m 50; c 1. CHEMISTRY. Educ: Salem Col, WVa, BS, 48; Carnegie Inst Technol, MS, 51, PhD(org chem), 53. Prof Exp: Lab instr chem, Carnegie Inst Technol, 48-51; res chemist, 52-69, STAFF SCIENTIST POLYMER CHEM, FILM DEPT, E I DU PONT DE NEMOURS & CO, INC, 69- Mem: AAAS; Am Chem Soc; NY Acad Sci. Res: Polymer synthesis, evaluation and chemistry; plastic film manufacture. Mailing Add: 610 Plaza Dr Circleville OH 43113

MCIRVIN, RONALD RAY, b Flagler, Colo, Apr 10, 41; m 64; c 2. CULTURAL ANTHROPOLOGY. Educ: Univ Colo, BA, 62; Univ Kans, PhD(anthrop), 70. Prof Exp: Instr, 68-70, ASST PROF ANTHROP, UNIV NC, GREENSBORO, 70-, DIR LATIN AM STUDIES, 72- Concurrent Pos: Mem, Southeastern Conf Latin-Am Studies. Mem: Latin Am Studies Asn; Soc Appl Anthrop. Res: Latin American culture; socioeconomic development in Latin America; culture change. Mailing Add: Dept of Anthrop Univ of NC Greensboro NC 27412

MCIRVINE, EDWARD CHARLES, b Winnipeg, Man, Dec 19, 33; nat US; m 54; c 2. THEORETICAL PHYSICS, INFORMATION SCIENCE. Educ: Univ Minn, BS, 54; Cornell Univ, PhD(theoret physics), 59. Prof Exp: Asst physics, Cornell Univ, 54-58; mem res staff, Theoret Physics, Gen Atomic Div, Gen Dynamics Corp, 58-60; res scientist, Sci Lab, Ford Motor Co, 60-66; mgr mfg systs planning, Gen Parts Div, 66-68, mgr comput applns proj new bus develop off, 68-69, mgr tech anal, Xerox Corp, 69-61, mgr physics res lab, Xerox Webster Res Ctr, 71-73, DIR, FORWARD TECH PLANNING, XEROX CORP, 74- Mem: Am Phys Soc; Am Asn Physics Teachers; Asn Comput Mach; Sigma Xi. Res: Theory of solids; quantum theory of transport phenomena; real-time computer applications; research management. Mailing Add: Xerox Corp Stamford CT 06904

MCISAAC, ROBERT JAMES, b Brooklyn, NY, Jan 9, 23; m 48; c 2. PHARMACOLOGY. Educ: Univ Buffalo, BS, 49, PhD, 54. Prof Exp: Assoc pharmacol, Sch Med, Univ Buffalo, 54-55; from asst prof to assoc prof, 58-66, PROF PHARMACOL, SCH MED, STATE UNIV NY BUFFALO, 66- Concurrent Pos: Res fel, Grad Sch Med, Univ Pa, 56-58. Mem: AAAS; Am Soc Pharmacol & Exp Therapeut; NY Acad Sci; Soc Neurosci. Res: Synaptic transmission; neuroeffector release; distribution and transport of drugs. Mailing Add: Dept of Pharmacol State Univ of NY Sch of Med Buffalo NY 14214

MCIVER, JAMES W, JR, b Portland, Ore, Dec 19, 40; m 63; c 1. QUANTUM CHEMISTRY. Educ: Johns Hopkins Univ, BA, 62; Univ Pa, PhD(chem), 66. Prof Exp: Res assoc chem, Carnegie Inst Technol, 66-67, fel chem & dir comput lab, Mellon Inst, 67-69; asst prof, 69-73, ASSOC PROF CHEM, STATE UNIV NY BUFFALO, 73- Mem: Am Chem Soc; Am Phys Soc. Res: Molecular quantum mechanics; magnetic resonance. Mailing Add: Dept of Chem State Univ of NY Buffalo NY 14214

MCIVER, NORMAN L, b Abbey, Sask, Sept 7, 31; m 58; c 3. GEOLOGY. Educ: Univ Sask, BA, 56; Johns Hopkins Univ, PhD(geol), 61. Prof Exp: Res geologist, Shell Develop Co, Tex, 60-68, staff geologist, Shell Oil Co, La, 68-71, staff geologist, Shell Can Ltd, 71, SR STAFF GEOLOGIST, SHELL OIL CO, 71- Mem: Soc Econ Paleontologists & Mineralogists; Am Asn Petrol Geologists. Res: Stratigraphy and sedimentation of clastic sediments. Mailing Add: Shell Oil Co Box 991 Houston TX 77001

MCIVER, RICHARD DONALD, b South Haven, Mich, Dec 1, 29; m 51; c 3. GEOCHEMISTRY. Educ: John Brown Univ, BA, 51; Ohio State Univ, PhD(chem), 54. Prof Exp: Res chemist, Carter Oil Co, 54-56; res chemist, Jersey Prod Res Co, 56-61, sr res chemist, 61-64, sect head org geochem, 64-65, res assoc & supvr org geochem res, Esso Prod Res Co, 65-67; planning specialist, Explor Dept, Humble Oil & Refining Co, 68-69; res assoc, Offshore Geol Div, Esso Prod Res Co, 70-73, SR RES ASSOC, EXXON PROD RES CO, 73- Concurrent Pos: Chmn, Org Geochem Div, Geochem Soc, 72-73; mem, Adv Panel Geochem, Deep Sea Drilling Proj, Joint Oceanog Inst Deep Earth Sampling Prog. Mem: Am Chem Soc; Geochem Soc; Am Asn Petrol Geologists; Int Asn Geochem & Cosmochem. Res: Origin, migration and diagenesis of petroleum; chemistry of oils, natural gases, kerogens, asphalts, coal, sulfur compounds; oceanography; economics. Mailing Add: Reservoir Evaluation Div Exxon Prod Res Co PO Box 2189 Houston TX 77001

MCIVER, SUSAN BERTHA, b Hutchinson, Kans, Nov 6, 40. ENTOMOLOGY. Educ: Univ Calif, Riverside, BA, 62; Wash State Univ, MS, 64, PhD(entom), 67. Prof Exp: Asst prof, 67-71, ASSOC PROF PARASITOL, UNIV TORONTO, 71- Mem: Entom Soc Am; Can Soc Zoologists; Entom Soc Can. Res: Medical entomology; behavior and sensory physiology and morphology of blood feeding arthropods. Mailing Add: Dept of Microb & Parasitol Univ of Toronto Toronto ON Can

MCKAGUE, ALLAN BRUCE, b Weston, Ont, Oct 21, 40; m 63; c 2. ORGANIC CHEMISTRY. Educ: McMaster Univ, BSc, 62; Univ BC, PhD(chem), 67. Prof Exp: Chemist I, Toronto Food & Drug Directorate, 62-63; Nat Res Coun fel, Australian Nat Univ, 67-69; res chemist & res dir indust org chem, Orchem Res Co, 69-71; RES OFF POLLUTION CONTROL BC RES COUN, 71- Mem: Am Chem Soc; Chem Inst Can; Entom Soc Am; Entom Soc Can. Res: Applied organic synthesis; organic chemical aspects of pollution; non-poisonous insect control through insect hormones and attractants. Mailing Add: Div of Appl Biol BC Res Coun Vancouver BC Can

MCKAGUE, HERBERT LAWRENCE, b Altoona, Pa, June 24, 35; m 63; c 2. GEOLOGY, GEOCHEMISTRY. Educ: Franklin & Marshall Col, BS, 57; Wash State Univ, MS, 60; Pa State Univ, PhD(mineral, petrol), 64. Prof Exp: Fel gem minerals, Gemol Inst Am, 64-66; asst prof geol, Rutgers Univ, New Brunswick, 66-72; PHYSICIST, LAWRENCE LIVERMORE LAB, UNIV CALIF, 72- Mem: Geol Soc Am. Res: Genesis, mineralogy, petrology and geochemistry of alpine ultramafic rocks; mineralogy of gem minerals; genesis of stratiform type ore deposits; geochemistry of chromite; geologic and geophysical criteria for containment of explosions, geologic interpretation of boizehole gravimetry. Mailing Add: Lawrence Livermore Lab Univ of Calif Livermore CA 94550

MCKAVENEY, JAMES P, b Pittsburgh, Pa, Sept 26, 25; m 52; c 2. ANALYTICAL CHEMISTRY. Educ: Univ Pittsburgh, BS, 49, MS, 51, PhD(anal chem), 57. Prof Exp: Asst technologist chem, Res Div, US Steel Corp, 51-56; assoc chemist, Res Div, Crucible Steel Co Am, 56-58, res engr, 58-59, supvr anal chem, 59-64, mgr, 64-67, assoc dir chem & melting, 67-69; MGR ANAL SERV, OCCIDENTAL RES CORP, 69- Concurrent Pos: Pres, Pittsburgh Conf Anal Chem & Appl Spectros, 65-66. Mem: Am Chem Soc; Am Inst Chemists; Am Soc Testing & Mat; NY Acad Sci; Electrochem Soc. Res: Spectroscopy of steel and refractory metals; analysis of industrial acids; semiconductor electrodes as analytical sensors; separation of mercury and other heavy metals from water; analysis of coal and non-ferrous minerals. Mailing Add: 940 Fenn Ct Claremont CA 91711

MCKAY, ALEXANDER SCOTT, b Alvinston, Ont, June 4, 18; nat US; m 53; c 3. PHYSICS. Educ: Univ Sask, MA, 42; Cornell Univ, PhD(physics), 50. Prof Exp: Res physicist, 49-55, asst to dir res, 55-57, dir explor res, 57-60, sr res technologist, 60-64, planning dir, explor & prod, 64-69, arctic res mgr, 69-72, MGR ARCTIC RES, TEXACO EXPLOR CAN, LTD, 72- Mem: Soc Explor Geophysicists; Am Inst Mining, Metall & Petrol Engrs; Can Asn Physicists. Res: Cosmic rays; radioactive well logging. Mailing Add: Texaco Explor Can Ltd Box 3333 Calgary AB Can

MCKAY, ARTHUR FERGUSON, organic chemistry, see 12th edition

MCKAY, DAVID STEWART, b Titusville, Pa, Sept 25, 36; m 71. VOLCANOLOGY. Educ: Rice Univ, BA, 58, PhD(geol), 64; Univ Calif, Berkeley, MA, 60. Prof Exp: Geophysicist, Humble Oil Co, 60-61; NSF fel, Univ Calif, Berkeley, 64-65; GEOLOGIST, JOHNSON SPACE STR, NASA, 65- Res: Lunar geology; lunar soils and breccias; size and particle characterization; lunar surface processes which form soil and breccias; lunar and terrestial volcanic geology; volcanic ashes; scanning electron microscope studies. Mailing Add: Geol Br TNG NASA Johnson Space Ctr Houston TX 77058

MCKAY, DONALD EDWARD, b Washington, DC, Aug 21, 38; m 63; c 4. ORGANIC CHEMISTRY. Educ: Univ Calif, BS, 60; Univ Ill, PhD(org chem), 66. Prof Exp: Asst, Univ Ill, 60-66; res chemist, 66-74, SR RES CHEMIST, ORG CHEM DIV, AM CYANAMID CO, 74- Mem: Am Chem Soc. Res: Synthetic organic chemistry; small ring compounds; dyestuffs. Mailing Add: Org Chem Div Am Cyanamid Co Bound Brook NJ 08805

MCKAY, DONALD GEORGE, b Sacramento, Calif, July 24, 21; m 45; c 2. PATHOLOGY. Educ: Univ Calif, BS, 43, MD, 45. Prof Exp: Instr, Sch Med, Boston Univ, 47-49; asst prof path, Harvard Med Sch, 57-60; Delafield prof & chmn dept, Col Physicians & Surgeons, Columbia Univ, 60-67; PROF PATH, SCH MED, UNIV CALIF, SAN FRANCISCO, 67-, VCHMN DEPT, 69- Concurrent Pos: Instr, Sch Med, Tufts Univ, 47-48; mem staff, Dept Path, San Francisco Gen Hosp, 67- Mem: AMA; Am Col Path; Am Gynec Soc; Am Col Obstet & Gynec; Am Soc Exp Path. Res: Maternal pathology; disseminated intravascular coagulation; toxemia of pregnancy. Mailing Add: Dept of Path Univ of Calif Sch of Med San Francisco CA 94122

MCKAY, DOUGLAS WILLIAM, b Howland, Maine, Mar 10, 27; c 3. ORTHOPEDIC SURGERY. Educ: Univ Maine, BA, 51; Tufts Univ, MD, 55. Prof Exp: Intern, Maine Gen Hosp, Bangor, 55-56; resident adult orthop surg & trauma, Vet Admin Prog, McKinney & Dallas, Tex, 56-57; children's orthop surg, Newington Hosp Crippled Children, Conn, 59-60; pvt pract orthop surg, Covington, Ky, 60-61; chief surgeon, Carrie Tingley Hosp Crippled Children Truth or Consequences, NMex, 61-67; chief surgeon, Shriner's Hosp Crippled Children, 67-72; PROF ORTHOP SURG, MED

SCH, GEORGE WASHINGTON UNIV, 72-; CHIEF PEDIAT ORTHOP SURG, CHILDREN'S HOSP NAT MED CTR, 72- Concurrent Pos: Clin instr, Univ Colo, 61-67; consult, William Beaumont Gen Hosp, US Army, El Paso, Tex, 64-67; adj prof, Univ NMex, 66-67; consult, US Air Force Hosp, Lackland AFB, 67- & Vet Admin Hosp, Shreveport; dir orthop residency prog, Confederate Mem Med Ctr, 68-74; prof orthop & head dept, Sch Med, La State Univ, Shreveport, 69-74. Mem: Pediat Orthop Soc (secy, 71-72). Res: Orthopedic pathology. Mailing Add: Childrens Hosp Nat Med Ctr 2125 Thirteen St NW Washington DC 20009

MCKAY, JAMES BRIAN, b Uniontown, Pa, Sept 2, 40; m 65; c 3. ANALYTICAL CHEMISTRY. Educ: Phila Col Pharm, BSc, 62; Wayne State Univ, MS, 64, PhD(anal chem), 66. Prof Exp: Asst prof chem, Univ Conn, 66-72; CHMN DEPT CHEM, EDINBORO STATE COL, 72- Mem: Am Chem Soc. Res: Solvent extraction; column chromatography; organic reagents; radiochemistry; ion exchange. Mailing Add: Dept of Chem Edinboro State Col Edinboro PA 16444

MCKAY, JAMES HAROLD, b Seattle, Wash, July 23, 28; m 47; c 4. MATHEMATICS. Educ: Univ Seattle, BS, 48; Univ Wash, MS, 50, PhD(math), 53. Prof Exp: Assoc math, Univ Wash, 51-53, instr, 53-54; instr, Mich State Univ, 54-56, asst prof, 56-57; asst prof, Univ Seattle, 57-59; assoc prof, 59-63, PROF MATH, OAKLAND UNIV, 63- Concurrent Pos: Assoc dean sci & eng, Oakland Univ, 61-65, chmn dept math, 63-65; NSF fac fel, Univ Calif, Berkeley, 66-67; mem, Nat Coun, Am Asn Univ Profs, 75-78. Mem: Am Math Soc; Math Asn Am. Res: Algebra. Mailing Add: Dept of Math Oakland Univ Rochester MI 48063

MCKAY, JERRY BRUCE, b Wyandotte, Mich, Dec 20, 35; m 58; c 2. ORGANIC CHEMISTRY, POLYMER CHEMISTRY. Educ: Mich State Univ, BS, 58; Stanford Univ, MS, 60; Univ Ohio, PhD(org chem), 66. Prof Exp: Chemist, Stanford Res Inst, 59-62; res chemist, Dacron Res Lab, 65-68, res supvr, Textile Fibers Dept, 68-71, res & develop supvr, Kinston Plant, 71-73, PROCESS SUPVR, KINSTON PLANT, E I DU PONT DE NEMOURS & CO, 73- Mem: Am Chem Soc; Sigma Xi. Res: Structure of synthetic fibers, polymer structure and property analysis; monomer and polymer synthesis; organic synthesis; structural analysis of alkaloids; mechanism of organic reactions; polymer process development. Mailing Add: E I du Pont de Nemours & Co Kinston NC 28501

MCKAY, JOHN FRANCIS, organic chemistry, petroleum chemistry, see 12th edition

MCKAY, KENNETH ALEXANDER, b Toronto, Ont, Sept 10, 14; m 42; c 4. BACTERIOLOGY. Educ: Univ Toronto, BA, 46, MA, 50, MSc, 54; Ont Vet Col, DVM, 58. Prof Exp: Assoc prof bact, Ont Vet Col, Guelph, 47-64; res bacteriologist, 64-66, HEAD DEPT BACT, ANIMAL PATH DIV, ANIMAL DIS RES INST, CAN DEPT AGR, 66- Concurrent Pos: Med Res Coun Can grant, 59-64; mem, Int Asn Microbiol Socs, 62- Mem: Can Soc Microbiol. Res: Pleuropneumonia-like organisms and L forms of bacteria. Mailing Add: Animal Dis Res Inst Can Dept of Agr Hull PQ Can

MCKAY, KENNETH GARDINER, b Montreal, Que, Apr 8, 17; m 42; c 2. SOLID STATE ELECTRONICS, TELECOMMUNICATIONS. Educ: McGill Univ, BSc, 38, MSc, 39; Mass Inst Technol, ScD(physics), 41. Prof Exp: Demonstr physics, McGill Univ, 38-39; jr radio res engr, Nat Res Coun Can, 41-46; res physicist, Bell Tel Labs, 46-52, head phys electronics res, 52-54, physics of solids res, 54-57, dir solid state device develop, 57-59, vpres systs eng, 59-62, exec vpres, 62-66, vpres eng, Am Tel & Tel Co, 66-73; EXEC VPRES, BEL TEL LABS, INC, 73- Concurrent Pos: Mem tech adv bd, Dept Com, 70-72; counr, Nat Acad Eng, 71- Mem: Nat Acad Sci; Nat Acad Eng; fel Am Phys Soc; fel Inst Elec & Electronics Engrs. Res: Secondary electron emission; electron bombardment conductivity; electrical breakdown; light emitting diodes; semiconducting nuclear detectors. Mailing Add: Bell Labs 600 Mountain Ave Murray Hill NJ 07974

MCKAY, LARRY LEE, b Oregon City, Ore, June 3, 43; m 64; c 2. MICROBIOLOGY. Educ: Univ Mont, BA, 65; Ore State Univ, PhD(microbiol), 69. Prof Exp: Res assoc, Mich State Univ, 69-70; ASST PROF MICROBIOL, UNIV MINN, ST PAUL, 70- Mem: Am Dairy Sci Asn; Am Soc Microbiol; Int Asn Milk, Food & Environ Sanit. Res: Improvement of production strains of microorganisms useful to the dairy and food industry; bacteriophages in food fermentation; physiological and genetic control of acid and flavor production by lactic streptococci. Mailing Add: Dept of Food Sci & Indust Univ of Minn St Paul MN 55101

MCKAY, MICHAEL DARRELL, b Temple, Tex, Jan 5, 44; m 66; c 3. STATISTICS. Educ: Univ Tex, BA, 66; Tex A&M Univ, PhD(statist), 72. Prof Exp: Res mathematician, Southwest Res Inst, 66-68; asst prof statist, Tex A&M Univ, 71-73; STAFF MEM, LOS ALAMOS SCI LAB, 73- Mem: Am Statist Asn; Biomet Soc; Inst Mgt Sci. Res: Optimization; mathematical modeling; computer applications in statistics. Mailing Add: Los Alamos Sci Lab MS254 Los Alamos NM 87545

MCKAY, ROBERT HARVEY, b Cordova, Alaska, June 12, 27; m 58; c 3. BIOCHEMISTRY. Educ: Univ Wash, BS, 53; Univ Calif, PhD(biochem), 59. Prof Exp: Asst biochem, Univ Calif, 53-58; fel, Brandeis Univ, 58-61; Am Cancer Soc fel, Harvard Med Sch, 61-63; asst prof, 63-68, ASSOC PROF BIOCHEM, UNIV HAWAII, 68- Concurrent Pos: NIH fel, Univ Hawaii, 64-73. Mem: AAAS; Am Chem Soc; Am Soc Biol Chem. Res: Enzymology; protein chemistry; optical and fluorescent properties of proteins; enzyme-coenzyme interactions; iron metabolism; iron storage proteins. Mailing Add: Dept of Biochem & Biophys Univ of Hawaii Honolulu HI 96822

MCKAY, ROBERT JAMES, JR, b New York, NY, Oct 8, 17; m 43; c 4. PEDIATRICS. Educ: Princeton Univ, AB, 39; Harvard Univ, MD, 43. Prof Exp: From asst prof to assoc prof, 50-55, PROF PEDIAT, UNIV VT, 55-, CHMN DEPT, 50- Concurrent Pos: Markle scholar med sci, 50-55; Fulbright lectr, State Univ Groningen, 60. Mem: Soc Pediat Res; Am Pediat Soc; Am Soc Human Genetics; Can Pediat Soc; fel Am Acad Pediat. Res: Rheumatoid arthritis; chromosomes; genetics. Mailing Add: Dept of Pediat Univ of Vt Burlington VT 05401

MCKAY, RUTH BLUMENFELD, b Brooklyn, NY, Oct 19, 32; m 65; c 1. ANTHROPOLOGY. Educ: Brooklyn Col, AB, 54; Univ Pa, MA, 61, PhD(anthrop), 65. Prof Exp: Res asst psychol, Univ Pa, 58-59; intern anthrop, Vet Admin Hosp, Perry Point, Md, 59-60; res instr, Univ Md, 60-62; instr sociol & anthrop, Howard Univ, 62-63; asst prof pediat psychiat & anthrop, 65-71, prof lectr anthrop, 71-75, RES PROF DEPT ANTHROP, GEORGE WASHINGTON UNIV, 75-; TEACHING ASSOC PSYCHIAT, CHILDREN'S HOSP, WASHINGTON, DC, 63- Concurrent Pos: Consult anthrop, Med Ctr, Georgetown Univ, 63-64. Res: Sexual egalitarianism, family size and sex ratios in elite Black populations. Mailing Add: Dept of Anthrop George Washington Univ Washington DC 20052

MCKAY, WILLIAM NEIL, b Calgary, Alta, June 30, 27; m 55; c 4. PHYSICAL CHEMISTRY, ANALYTICAL CHEMISTRY. Educ: Univ Alta, BSc, 47, MSc, 54; Univ Colo, PhD(chem), 57. Prof Exp: Chemist, Consol Mining & Smelting Co, 47-52; CHEMIST, IMP OIL, LTD, 57- Mem: Am Chem Soc; fel Chem Inst Can. Res: Mass transfer; sulphur chemistry; electrochemical phenomena. Mailing Add: Imperial Oil Ltd 339-50th Ave SE Calgary AB Can

MCKAYE, KENNETH ROBERT, b Camp Lejeune, NC, July 12, 47. ECOLOGY, BEHAVIORAL BIOLOGY. Educ: Univ Calif, Berkeley, AB, 70, MA, 72, PhD(zool), 75. Prof Exp: ASST PROF BIOL, YALE UNIV, 75- Mem: Ecol Soc Am; Soc Study Evolution; Animal Behav Soc; Am Soc Ichthyologists & Herpetologists. Res: Evolution of social behavior and the role behavior plays in determining the community structure of animals; study of the behavioral ecology of cichlid fishes. Mailing Add: Dept of Biol Osborn Mem Labs Yale Univ New Haven CT 06520

MCKEAGUE, JUSTIN ALEXANDER, b Sibbald, Alta, Nov 7, 24; m 52; c 5. SOIL GENESIS, SOIL CLASSIFICATION. Educ: Univ BC, BA, 47, BSA, 55; Univ Alta, MSc, 58; Cornell Univ, PhD(soils), 61. Prof Exp: Res officer, Alta Soil Surv, Can Agr, Edmonton, 55-59, RES SCIENTIST SOIL GENESIS, SOIL RES INST, CAN AGR, OTTAWA, 62-71, 72- Concurrent Pos: Res scientist, Res Sta, Ste Foy, Que, 71-72; mem, Can Soil Surv Comt, 65-75; actg dir, Soil Res Inst, Can Agr, Ottawa, 74-75. Mem: Can Soc Soil Sci (pres, 71-72); Soil Sci Soc Am; Agr Inst Can; Int Soc Soil Sci. Res: Diagnostic criteria of soil classification; genesis of soils; soil micromorphology. Mailing Add: Soil Res Inst Ottawa ON Can

MCKEAN, HARLLEY ELLSWORTH, b Chicago, Ill, June 23, 31; m 67; c 5. MATHEMATICS, STATISTICS. Educ: Cornell Col, BA, 52; Purdue Univ, MS, 54, PhD(math statist), 58. Prof Exp: Instr math & statist, Purdue Univ, 56-58, from asst prof to assoc prof, 58-67; assoc prof math, NMex State Univ, 67-71, dir statist lab, 67-70; PROF STATIST, UNIV KY & CHIEF STATISTICIAN, TOBACCO HEALTH RES INST, 71- Concurrent Pos: Consult, Univ Ill Health Surv Proj, Western Elec Corp, Ill, 61-; assoc, Bayer & McElrath, Inc, Mgt Consults, 63-67; vis assoc prof, Univ Calif, Berkeley, 65-66. Mem: Biomet Soc; Am Statist Asn; Inst Math Statist. Res: Statistical genetics; design of experiments; probability. Mailing Add: Dept of Statist Univ of Ky Lexington KY 40506

MCKEAN, HERBERT BALDWIN, b Williamsport, Pa, Nov 28, 11; m 35; c 3. WOOD TECHNOLOGY. Educ: Syracuse Univ, BS, 33, MF, 34; Univ Mich, PhD(wood technol), 42. Prof Exp: Tie & log inspector, T J Moss Tie Co, 34-35; instr wood technol, La State Univ, 35-37, asst prof, 37-42; wood technologist, Forest Prod Lab, US Forest Serv, Wis, 42-45; asst dir res, Timber Eng Co, 45-50, assoc dir res, 50-54; dir res, 54-67, corp dir res & develop, 67-69, VPRES RES & DEVELOP, POTLATCH FORESTS, INC, 69- Mem: Soc Am Foresters; Soc Wood Sci & Technol (vpres, 59). Res: Timber mechanics; wood preservation; mechanical properties and design procedures for glued laminated beams of two wood species. Mailing Add: Potlatch Forests Inc PO Box 600 Lewiston ID 83501

MCKEAN, JOEL MERCER, mathematics, see 12th edition

MCKEAN, JOSEPH WALTER, JR, b Sewickly, Pa, June 11, 44; m 64; c 2. STATISTICS. Educ: Geneva Col, Pa, BS, 66; Univ Ariz, MS, 68; Pa State Univ, PhD(statist), 75. Prof Exp: Instr math, Waynesburg Col, Pa, 68-72; ASST PROF MATH SCI, UNIV TEX, DALLAS, 75- Mem: Inst Math Statist; Am Statist Asn; Math Asn Am. Res: Non-parametric statistics, particularly robust analysis of linear models based on ranks. Mailing Add: Univ of Tex Dallas Box 688 Richardson TX 75080

MCKEAN, THOMAS ARTHUR, b Boise, Idaho, Jan 27, 41. PHYSIOLOGY. Educ: Whitman Col, AB, 63; Univ Ore, PhD(physiol), 68. Prof Exp: Fel physiol, Sch Med, Univ Minn, 68-69; asst dir sci mkt, Hoechst Pharmaceut Co, 69-70; asst prof zool & physiol, Univ Wyo, 70-74; ASSOC PROF ZOOL, UNIV IDAHO, 74- Mem: AAAS; Soc Neurosci; Am Physiol Soc. Res: Oxygen transport to tissues; temperature regulation in wild ruminants. Mailing Add: Dept of Biol Sci Univ of Idaho Moscow ID 83843

MCKEE, ALBERT PRESTON, b Brandon, Iowa, Aug 31, 13; m; c 4. BACTERIOLOGY. Educ: Drake Univ, BA, 36; Univ Iowa, MD, 41. Prof Exp: From instr to prof bact, Univ Iowa, 41-70; DIR DIV VIROL & MICROBIOL DIAG RES, ORTHO RES FOUND, 70- Concurrent Pos: Dir regional lab, WHO; coordr, Sect Res Microbiol, USPHS, 50-56, assoc, Comn Influenza, Armed Forces Epidemiol Bd, 49- Mem: Royal Soc Health; Am Soc Microbiol; Soc Exp Biol & Med; Am Asn Immunologists; fel Am Pub Health Asn. Res: Medical bacteriology; immunology; virology; mycology. Mailing Add: Diag Res Div of Virol & Microbiol Ortho Res Found Raritan NJ 08869

MCKEE, BATES, b Mt Kisco, NY, Jan 10, 34; m 54, 67; c 6. REGIONAL GEOLOGY. Educ: Yale Univ, BS, 55; Stanford Univ, PhD(geol), 59. Prof Exp: Eng geologist, Calif State Dept Water Resources, 57; from asst prof to assoc prof geol, Univ Wash, 58-72; PRES, MCKEE & MOONEY, INC, 72- Concurrent Pos: Affil prof, Dept Geol Sci, Univ Wash, 72- Mem: Geol Soc Am; Geol Asn Can; Asn Eng Geologists. Res: Engineering geology consulting; petrology and structure; regional tectonics. Mailing Add: McKee & Mooney Inc 301 W Kinnezr Place Seattle WA 98119

MCKEE, CHRISTOPHER FULTON, b Washington, DC, Sept 6, 42; m 65; c 1. ASTROPHYSICS. Educ: Harvard Univ, AB, 63; Univ Calif, Berkeley, PhD(physics), 70. Prof Exp: Physicist, Lawrence Livermore Lab, Univ Calif, 69-70; res fel astrophys, Calif Inst Technol, 70-71; asst prof astron, Harvard Univ, 71-73; ASST PROF PHYSICS, UNIV CALIF, BERKELEY, 73- Mem: Am Astron Soc; Am Phys Soc. Res: High energy and plasma astrophysics. Mailing Add: Dept of Physics Univ of Calif Berkeley CA 94720

MCKEE, CLAUDE GIBBONS, b Md, June 30, 30; m 58; c 2. AGRONOMY. Educ: Univ Md, BS, 51, MS, 55, PhD(agron), 59. Prof Exp: Asst, Univ Md, 51-54, exten instr, 54-56; exec secy, Md Tobacco Improv Found, Inc, 57-62; EXTEN TOBACCO SPECIALIST, UNIV MD, 62- Res: Production of Maryland tobacco, especially quality constituents. Mailing Add: Box 2005 Largo Rd Upper Marlboro MD 20870

MCKEE, DOUGLAS WILLIAM, b Toronto, Ont, Can, Oct 6, 30; m 59; c 4. PHYSICAL CHEMISTRY. Educ: Univ London, BSc, 51, PhD(chem), 54. Prof Exp: Res fel, Nat Res Coun Can, 54-55; Welch fel, Rice Univ, 55-56; res chemist, Linde Co div, Union Carbide Corp, 56-60; RES CHEMIST, RES LABS, GEN ELEC CO, 60- Concurrent Pos: Lectr, Canisius Col, 58-60. Mem: The Chem Soc; assoc Royal Inst Chem; Catalysis Soc. Res: Surface chemistry; adsorption; diffusion; ion exchange; catalysis; surface chemistry of carbon; carbon fibers; colloid chemistry and surface activity; fuel cells; corrosion; high temperature materials. Mailing Add: Gen Elec Co PO Box 1088 Schenectady NY 12305

MCKEE, EDWIN DINWIDDIE, b Washington, DC, Sept 24, 06; m 29; c 3. GEOLOGY. Educ: Cornell Univ, AB, 28. Hon Degrees: ScD, Ariz State Univ, 57. Prof Exp: Park naturalist, Grand Canyon Nat Park, Nat Park Serv, 29-40; asst dir,

Mus Northern Ariz, 41-53; from asst prof to prof geol & head dept, Univ Ariz, 42-53; head, Paleotectonic Mapping Proj, US Geol Surv, 53-63, RES GEOLOGIST, US GEOL SURV, 63- Concurrent Pos: Res assoc, Carnegie Inst; collabr, Nat Park Serv, 41-; trustee, Mus Northern Ariz, 54-; prin investr, ERTS-A proj, NASA, 72-74, discipline expert deserts, Sci Support Team, SKYLAB IV, NASA, 73-74; mem, Team Earth Observs, Apollo-Soyuz Test Proj, 74-75. Honors & Awards: Achievement Award, NASA, 74; Twenhofel Medal, Soc Econ Paleont & Mineral, 74. Mem: Geol Soc Am; Soc Econ Paleont & Mineral; Paleont Soc; Am Asn Petrol Geologists; AAAS. Res: Sedimentation; stratigraphy; paleogeography; global sand seas; the Supai of Grand Canyon. Mailing Add: US Geol Surv Fed Ctr Denver CO 80225

MCKEE, EDWIN H, b Phoenix, Ariz, Nov 30, 35; m 65; c 2. GEOLOGY. Educ: Yale Univ, BS, 57; Univ Calif, Berkeley, PhD(geol), 62. Prof Exp: GEOLOGIST, REGIONAL GEOL BR, US GEOL SURV, 64- Mem: Geol Soc Am. Res: General geology state of Nevada; potassium-argon radiometric dating. Mailing Add: US Geol Surv 345 Middlefield Rd Menlo Park CA 94025

MCKEE, ELMER S, physical chemistry, see 12th edition

MCKEE, FRANK WRAY, b Beaver, Pa, Feb 18, 15; m 43; c 2. PATHOLOGY. Educ: Hamilton Col, AB, 36; Univ Rochester, MD, 43. Prof Exp: Instr path, Univ Rochester, 46-51; dir labs, Genesee Hosp, 51-54; prof path & dir clin labs, Univ Calif, Los Angeles, 54-52; prof path, Sch Med & Dent, Univ Rochester, 62-67; dir div physician manpower, Bur Health Manpower, 67-70; prof path & dean sch med, WVa Univ, 70-73; ASSOC DIR, EDUC COMN FOREIGN MED GRAD, 73- Concurrent Pos: Med dir, Rochester Regional Hosp Coun, NY, 66-67; consult path, Genesee Hosp, Rochester, 66-67. Mem: Am Soc Exp Path; Am Asn Path & Bact. Res: Phlorhizin glucosuria; Ascites; liver injury; chloroform. Mailing Add: Educ Comn Foreign Med Grad 3624 Market St Philadelphia PA 19104

MCKEE, GUY WILLIAM, b Renovo, Pa, Apr 14, 19; m 45; c 5. AGRONOMY. Educ: Pa State Univ, BS, 52, MS, 54, PhD(agron), 59. Prof Exp: Soil conservationist, Soil Conserv Serv, USDA, 49-51 & 54; from instr to assoc prof, 54-70, PROF AGRON, PA STATE UNIV, 70- Concurrent Pos: Chmn fac comt environ resource mgt, Pa State Univ, 71- Honors & Awards: Pugh Medal, 52. Mem: Fel AAAS; Am Soc Plant Physiol; Am Meteorol Soc; Ecol Soc Am; Am Soc Agron. Res: Crop ecology and physiology; microclimatology; seed technology and physiology; improvement of quality of the environment; intra-specific taxonomy of crop species; revegetation of disturbed sites. Mailing Add: 219 Tyson Bldg Pa State Univ University Park PA 16802

MCKEE, JAMES JOSEPH, b New York, NY, May 11, 32; m 58; c 3. MICROBIOLOGY. Educ: Fordham Univ, BS, 54; Brown Univ, MS, 58; Rutgers Univ, PhD(microbiol), 61. Prof Exp: ASST DIR, SMITH KLINE & FRENCH LABS, 61- Mem: AAAS; Am Soc Microbiol; NY Acad Sci. Res: Chemotherapy and immunology of infectious diseases in man and animals. Mailing Add: Smith Kline & French Labs 1500 Spring Garden St Philadelphia PA 19101

MCKEE, JAMES ROBERT, b Denver, Colo, June 22, 46; m 70; c 1. ORGANIC CHEMISTRY. Educ: Univ Md, BS, 70, PhD(chem), 74. Prof Exp: Fel ultrasonics, 74-75, FEL SYNTHESIS, JOHNS HOPKINS UNIV, 75- Mem: Am Chem Soc. Res: Synthesis of possible anti-cancer agents and the use of nitrosoamides in the labeling of enzyme active sites. Mailing Add: 3424 Tulane Dr Apt 34 Hyattsville MD 20783

MCKEE, JAMES STANLEY COLTON, b Belfast, Northern Ireland, June 6, 30; m 62; c 2. PHYSICS. Educ: Queens Univ Belfast, BSc, 52, PhD(theoret physics), 56; Univ Birmingham, DSc, 68. Prof Exp: Asst lectr physics, Queens Univ Belfast, 54-56; lectr, Univ Birmingham, 56-64, sr lectr, 64-74; vis prof, Lawrence Radiation Lab, 66-67 & 72; PROF PHYSICS & DIR CYCLOTRON LAB, UNIV MAN, 74- Mem: Fel Inst Physics London. Res: Few body problems, nuclear reaction mechanisms and polarisation phenomena. Mailing Add: Cyclotron Lab Dept Physics Univ of Man Winnipeg MB Can

MCKEE, JAMES W, b Lawrenceburg, Ky, Sept 25, 32; m 50; c 3. PALEONTOLOGY, SEDIMENTOLOGY. Educ: La State Univ, BS, 60, MS, 64, PhD(geol), 67. Prof Exp: Sea sampler, US Fish & Wildlife Serv, 53-54; laborer, Nat Distillers Corp, 56-58; instr, 65-66, asst prof, 66-69, ASSOC PROF GEOL, UNIV WIS-OSHKOSH, 69- Concurrent Pos: Off Water Resources Res Grant, 68-69; chmn dept geol, Univ Wis-Oshkosh, 73-75. Mem: AAAS; Paleont Soc. Res: Sediment-fossil relationships; limnology, especially lake sediments; Cretaceous stratigraphy of Central Coahuila, Mexico. Mailing Add: Geol Dept Univ of Wis-Oshkosh Oshkosh WI 54901

MCKEE, JOHN W, b Muscatine, Iowa, Sept 25, 26; m 47; c 3. PHYSICS, BIOLOGY. Educ: Iowa State Univ, BS, 49; Calif Inst Technol, PhD(physics, biol), 55. Prof Exp: Asst chief nuclear engr, Douglas Aircraft Co, Inc, Calif, 54-62; MGR TECH PROGS, TEMPO, CTR ADVAN STUDIES, GEN ELEC CO, 62- Mem: AAAS; Sigma Xi; Am Inst Aeronaut & Astronaut. Res: Military technology, especially plasma physics and nuclear weapons effects; space technology, especially fluid mechanics and exobiology; bacteriophage. Mailing Add: Ctr for Advan Studies Gen Elec Co Santa Barbara CA 93102

MCKEE, KELLY TILSON, b Bristol, Tenn, Mar 9, 16; m 48; c 2. INTERNAL MEDICINE, ALLERGY. Educ: Emory & Henry Col, BA, 38; Univ Va, MD, 41. Prof Exp: From assoc to assoc prof, 49-67, PROF MED, MED UNIV SC, 67- Mem: AMA; Am Col Physicians; Am Acad Allergy; Am Thoracic Soc. Res: Allergy and chest disease. Mailing Add: Dept of Med Med Univ of SC Charleston SC 29403

MCKEE, PATRICK ALLEN, b Tulsa, Okla, Apr 30, 37; m 63; c 5. MEDICAL RESEARCH, HEMATOLOGY. Educ: Univ Okla, MD, 62. Prof Exp: From intern to resident internal med, Med Ctr, Duke Univ, 62-65; clin assoc epidemiol heart dis, Nat Heart Inst, 65-67; chief resident & fel internal med & hemat, Med Ctr, Univ Okla, 67-69; from assoc med to assoc prof internal med & hemat, 69-75, ASST PROF BIOCHEM, MED CTR, DUKE UNIV, 71-, PROF INTERNAL MED & HEMAT, 75- Concurrent Pos: Assoc ed, Circulation, 73-; mem hemat study sect, NIH, 73-; fel coun thrombosis, Am Heart Asn. Mem: Am Fedn Clin Res; Am Heart Asn; Am Soc Hemat; Am Soc Biol Chemists; Am Soc Clin Invest. Res: Biochemistry of structure-function relationships of human blood coagulation proteins, particularly with respect to mechanisms of thromboses in human disease. Mailing Add: Box 3390 Duke Univ Med Ctr Durham NC 27710

MCKEE, RALPH WENDELL, b Boynton, Okla, Nov 13, 12; m 38; c 3. BIOLOGICAL CHEMISTRY. Educ: Kalamazoo Col, AB, 34, MS, 35; St Louis Univ, PhD(biochem), 40. Prof Exp: Asst, St Louis Univ, 35-39, sr asst, 39-40; instr indust hyg, Sch Pub Health, Harvard Univ, 40-45, assoc biochem, Harvard Med Sch, 45-47, asst prof, 47-52; PROF BIOL CHEM, UNIV CALIF, LOS ANGELES, 53-, ASST DEAN, SCH MED, 58- Concurrent Pos: Head biochem, Cancer Res Inst, New Eng Deaconess Hosp, 50-52. Mem: Fel AAAS; Am Chem Soc; Am Soc Biol Chemists; Soc Exp Biol & Med. Res: Isolation and chemistry of vitamin K; physiology and toxicology of carbon disulfide; biochemical studies of lewisite and mustard gas; growth, metabolism and nutrition of malarial parasites; interrelationship of ascorbic acid and cortical hormones; cytochemistry; chemistry and metabolism of cancer cells. Mailing Add: Dept of Biol Chem Univ of Calif Sch of Med Los Angeles CA 90024

MCKEE, ROBERT LAMBERT, b Wilkes-Barre, Pa, June 8, 16; m 44. CHEMISTRY. Educ: Rice Inst, AB, 38; Univ Tex, MA, 40, PhD(org chem), 43. Prof Exp: Instr anal chem, Univ Tex, 38-42 & Naval Flight Prep Sch, 43; Merrell fel, 43-45, from asst prof to assoc prof, 46-55, PROF CHEM, UNIV NC, CHAPEL HILL, 55- Mem: Am Chem Soc. Res: Synthesis of heterocyclic organic compounds to be tested for possible therapeutic use; preparation of N-substituted hydantoins; synthesis of fused pyrimidine systems. Mailing Add: Dept of Chem Univ of NC Chapel Hill NC 27514

MCKEE, RODNEY ALLEN, b Freeport, Tex, Oct 1, 47; m 69; c 1. SOLID STATE KINETICS. Educ: Lamar Univ, BS, 70; Univ Tex, PhD(mat sci), 75. Prof Exp: Metallurgist, Nat Bur Standards, 74-75; MEM STAFF, METALS & CERAMICS DIV, OAK RIDGE NAT LAB, 75- Concurrent Pos: Nat Acad Sci-Nat Res Coun assoc, Nat Bur Standards, 74-75. Mem: Am Phys Soc; Am Crystallog Asn. Res: Kinetics of diffusion processes in solids; thermomigration; oxidation of metals and alloys. Mailing Add: Metals & Ceramics Div Oak Ridge Nat Lab PO Box X Oak Ridge TN 37830

MCKEE, RUTH STAUFFER, b Harrisburg, Pa, July 16, 10; m 37; c 2. ALGEBRA. Educ: Swarthmore Col, AB, 31; Bryn Mawr Col, MA, 33, PhD(math), 35. Prof Exp: Teacher pvt sch, Md, 35-36 & NJ, 36-37; instr algebra, Bryn Mawr Col, 38-39; ANALYST MATH STATIST, JOINT STATE GOVT COMN, 53- Mem: Am Math Soc. Res: Abstract algebra and statistics. Mailing Add: Joint State Govt Comn Main Capital Bldg Harrisburg PA 17120

MCKEE, WILLIAM HENRY, JR, b Murphysboro, Ill, Sept 17, 34; m 62; c 1. AGRONOMY. Educ: Southern Ill Univ, BS, 57, MS, 61; Va Polytech Inst, PhD(agron), 65. Prof Exp: Instr agron, Va Polytech Inst, 63-66; RES SOIL SCIENTIST, SOUTHERN FOREST EXP STA, FOREST SERV, USDA, 66- Mem: Am Soc Agron. Res: Development of intensive management practices of southern pine, including soil fertility, chemistry of forest soils, and physiology of trees as related to mineral nutrition. Mailing Add: Southern Forest Exp Sta US Dept of Agr Pineville LA 71360

MCKEEHAN, CHARLES WAYNE, b Greencastle, Ind, Nov 16, 29; m 53; c 4. PHARMACEUTICAL CHEMISTRY, RESEARCH ADMINISTRATION. Educ: Purdue Univ, BS, 51, MS, 53, PhD(pharmaceut chem), 57. Prof Exp: Sr pharmaceut chemist, Eli Lilly & Co, 57-62, corp trainee, 62-64, proj coordr new prod, 64-68, head liquid-ointment parenteral prod pilot plants, 68-69, head pharmaceut res, 69-71, HEAD LIQUID-OINTMENT PROD DEVELOP, ELI LILLY & CO, 72- Mem: Am Chem Soc; Am Pharmaceut Asn; NY Acad Sci. Res: Planning and coordination of new product development; physical pharmacy; research and development administration in the area of new drug dosage forms. Mailing Add: 304 Daffon Dr Indianapolis IN 46227

MCKEEHAN, WALLACE LEE, b Texarkana, Tex, Jan 22, 44. BIOCHEMISTRY. Educ: Univ Fla, BS, 65; Univ Tex, Austin, PhD(biochem), 69. Prof Exp: Res assoc protein synthesis, Univ Tex, Austin, 69-71; res scientist, Basel Inst Immunol, 71-73; RES ASSOC MOLECULAR, CELLULAR & DEVELOP BIOL, UNIV COLO, BOULDER, 74- Mem: Am Chem Soc. Res: Regulation of hemoglobin syntheisi; control of polypeptide initiation, elongation and termination; isolation and characterization of growth factors for primary cells in culture. Mailing Add: MCD-Biol Univ of Colo Boulder CO 80302

MCKEEN, COLIN DOUGLAS, b Strathroy, Ont, June 23, 16; m 42; c 2. PLANT PATHOLOGY. Educ: Univ Western Ont, BA, 38; Univ Toronto, MA, 40, PhD(plant path), 42. Prof Exp: Class asst bot, Univ Toronto, 38-46; mem, Harrow Res Sta, 46-62, head plant path, 62-73, res coordr plant path, 73, HEAD PLANT PATH SECT, RES STA, CAN DEPT AGR, 74- Mem: AAAS; Am Phytopath Soc; Can Phytopath Soc; Can Microbiol Soc. Res: Vegetable diseases in greenhouse and field crops; virology. Mailing Add: Plant Path Sect Res Sta Can Dept Agr PO Box 370 Harrow ON Can

MCKEEN, WILBERT EZEKIEL, b Strathroy, Ont, Can, Feb 20, 22; m 50; c 5. PLANT PATHOLOGY. Educ: Univ Western Ont, BSc, 45, MSc, 46; Univ Toronto, PhD, 49. Prof Exp: Plant pathologist, Sci Serv Ont, 49-51, BC, 51-57; assoc prof, 57-60, PROF BOT, UNIV WESTERN ONT, 60- Honors & Awards: Gold Medal, Univ Western Ont, 45; Wintercorbyn Award, Univ Toronto. Mem: AAAS; Am Phytopath Soc; Can Soc Phytopath; Can Soc Microbiol. Mailing Add: Dept of Plant Sci Univ of Western Ont London ON Can

MCKEEVER, CHARLES H, b Gibson City, Ill, June 6, 12. ORGANIC CHEMISTRY. Educ: Ill Wesleyan Univ, BS, 36; Univ Ill, PhD(chem), 40. Prof Exp: Org res chemist, 40-60, res supvr, 60-72, PROCESS & EMPLOYEE HEALTH ADV, ROHM & HAAS CO, 72- Mem: Am Chem Soc. Res: Process research. Mailing Add: 1406 Holcomb Rd Meadowbrook PA 19046

MCKEEVER, L DENNIS, b Pittsburgh, Pa, Jan 23, 41; m 63; c 3. POLYMER CHEMISTRY, PHYSICAL CHEMISTRY. Educ: Univ Pittsburgh, BS, 62; Univ Calif, Irvine, PhD(phys chem), 66. Prof Exp: Lab dir, Dow Chem Co, 66-73, DIR CENT RES PLASTICS LAB, DOW CHEM CO, 73- Mem: Am Chem Soc. Res: Organo-electrochemistry; organo-alkali metal chemistry; anionic and free radical polymerization; thermoplastics; styrene based polymers; plastic foams. Mailing Add: Dow Chem Co 1712 Bldg Midland MI 48640

MCKEEVER, STURGIS, b Renick, WVa, Sept 6, 21; m 46. ECOLOGY, MAMMALOGY. Educ: NC State Univ, BS, 48, MS, 49, PhD(animal ecol), 55. Prof Exp: Proj leader biol res, WVa Conserv Comn, 49-51; biologist, Commun Dis Ctr, USPHS, 55-57; asst zoologist, Agr Exp Sta, Univ Calif, Davis, 57-63; from asst prof to assoc prof, 63-67, PROF BIOL, GA SOUTHERN COL, 67- Mem: Ecol Soc Am; Wildlife Soc; Wildlife Dis Asn. Res: Animal ecology; mammalian reproduction; wildlife diseases and pathology; parasitology of mammals. Mailing Add: Dept of Biol Ga Southern Col Statesboro GA 30458

MCKEEVER, WILLIAM PAUL, b Dublin, Ireland, June 27, 23; nat US; m 53; c 3. PHYSIOLOGY. Educ: Nat Univ Ireland, MB & BCh, 47; Am Bd Internal Med, dipl, 57; Am Bd Cardiovasc Dis, dipl, 61. Prof Exp: Intern, Mater Hosp, Dublin, Ireland, 47; demonstr physiol, Univ Col, Nat Univ Ireland, 48; asst res med, Johns Hopkins Univ, 49-52; instr, 52-57, assoc, 57-60, asst prof, 60-65, ASSOC PROF MED, MED SCH, NORTHWESTERN UNIV, EVANSTON, 65- Concurrent Pos: Traveling fel, Nat Univ Ireland, 49; fel physiol, Johns Hopkins Univ, 49-52; USPHS fel, 50; pvt pract, 57-; mem coun clin cardiol, Am Heart Asn. Mem: Cent Soc Clin Res; Am Fedn Clin Res; Am Physiol Soc. Res: Myocardial metabolism; renal function. Mailing Add: 2500 Ridge Ave Evanston IL 60201

MCKELL, CYRUS MILO, b Payson, Utah, Mar 19, 26; m 47; c 3. RANGE MANAGEMENT, ENVIRONMENTAL SCIENCES. Educ: Univ Utah, BS, 49, MS, 50; Ore State Univ, PhD(bot), 56. Prof Exp: Asst plant ecol, Univ Utah, 49-50; prin high sch, Utah, 52-53; asst bot, Ore State Univ, 53-55, instr, 55-56; agr res scientist plant physiol, Agr Res Serv, USDA, Univ Calif, Davis, 56-61, assoc prof agron & vchmn dept, Univ Calif, Davis & Univ Calif, Riverside, 61-66, chmn dept, Univ Calif, Riverside, 66-69; head dept, 69-71, dir environ & man prog, 69-76, PROF RANGE SCI, UTAH STATE UNIV, 69-, DIR INST LAND REHAB, 76- Concurrent Pos: Consult, Ford Found Mex, 65 & US AID, Bolivia & Ford Found, Agr, 71; Fulbright res fel, Spain, 67-68; panelist, Nat Acad Sci, Brazil, 74 & 76; consult, Food & Agr Orgn, UN, 75. Mem: Am Soc Agron; Soc Range Mgt; Soil Conserv Soc Am; Sigma Xi. Res: Environmental physiology of range plants; rehabilitation of distrubed arid lands; land use planning; range improvement. Mailing Add: UMC 48 Utah State Univ Logan UT 84322

MCKELLIPS, TERRAL LANE, b Terlton, Okla, Dec 2, 38; m 58; c 2. MATHEMATICS. Educ: Southwestern State Col, BSEd, 61; Okla State Univ, MS, 63, EdD(math), 68. Prof Exp: Asst prof math, Southwestern State Col, 62-66; instr math & educ, Okla State Univ, 67-68; PROF MATH & CHMN DEPT, CAMERON UNIV, 68- Concurrent Pos: Vis assoc prof, Okla State Univ, 72-; consult, Consult Bur, Math Asn Am, 73- Mem: Math Asn Am. Res: Point set topology. Mailing Add: Dept of Math Cameron Univ Lawton OK 73501

MCKELVEY, DONALD RICHARD, b Indiana, Pa, June 19, 38; m 60. PHYSICAL ORGANIC CHEMISTRY. Educ: NMex Inst Mining & Technol, BS, 60; Carnegie Inst Technol, PhD(chem), 64. Prof Exp: Res assoc & instr chem, Univ Pittsburgh, 64-66; asst prof, Cornell Univ, 66-69; assoc prof, 69-73, PROF CHEM, INDIANA UNIV PA, 73- Concurrent Pos: Mem staff, Mellon Inst, 64-66. Mem: Am Chem Soc; The Chem Soc. Res: Reaction mechanisms and solvent effects. Mailing Add: Dept of Chem Indiana Univ of Pa Indiana PA 15701

MCKELVEY, EUGENE MOWRY, b Greensburg, Pa, June 13, 34; m 60; c 2. INTERNAL MEDICINE, HEMATOLOGY. Educ: Yale Univ, BS, 56; Johns Hopkins Univ, MD, 60. Prof Exp: From Intern to resident, Boston City Hosp, 60-62; clin assoc, Med Br, Nat Cancer Inst, 62-64; assoc staff physician, Cleveland Clin, 66-67; asst prof med, Northwestern Univ, Chicago, 67-73; ASSOC PROF MED & ASSOC INTERNIST, M D ANDERSON HOSP & TUMOR INST, 73- Concurrent Pos: Fel med, Cleveland Clin, 64-66; hematologist, Vet Admin Res Hosp, Chicago, 67-73. Mem: Am Fedn Clin Res:; Am Hemat Soc; Am Soc Clin Oncol; Am Asn Cancer Res. Res: Oncology, investigational chemotherapy, combined modality treatment; antibody and antigen detection in malignancy. Mailing Add: Dept of Develop Therapeut M D Anderson Hosp & Tumor Inst Houston TX 77025

MCKELVEY, JOHN BERNARD, organic chemistry, see 12th edition

MCKELVEY, JOHN JAY, JR, b Albany, NY, July 16, 17; m 41; c 4. ENTOMOLOGY. Educ: Oberlin Col, AB, 39; Va Polytech Inst, MS, 41; Cornell Univ, PhD(entom), 45. Prof Exp: Entomologist, Mex Agr Prog, Rockefeller Found, 45-52, asst dir, Div Natural Sci & Agr, 52-55, Agr Div, 55-60, assoc dir agr sci, 60-66, dep dir, 66-68, ASSOC DIR AGR SCI, ROCKEFELLER FOUND, 68- Concurrent Pos: Consult agr res, Thailand, UN Spec Fund, 64; mem, Surv Team Natural Resources, northern region, Nigeria, Food & Agr Orgn-Int Coop Admin, 60, Study Comt Manpower Needs & Educ Capabilities in Africa, Educ & World Affairs, 64-65, Rev Comn, Univ Basutoland, Bechuanaland Protectorate & Swaziland, 66, Overseas Liason Comt Study Team Higher Educ, Sierra Leone, 68 & Agr Educ Mission, Univ SPac, Overseas Ministry, UK, 70; mem, Adv Comt on Africa, Nat Acad Sci, 61-68, vchmn, 64-68, mem, Sci Orgn Develop Bd, 66-68, chmn, Africa Sci Panel, 68-, chmn Comt African Agr Res Capabilities, 71- & Agr Bd Adv Panel, Arid Lands of Sub Saharan Africa; trustee, Int Inst Trop Agr, Ibadan, Nigeria & mem, Bd Govs, Inst Agr Res, Ahmadu Bello Univ, 71- Mem: AAAS; Am Phytopath Soc; Entom Soc Am; Asn Advan Agr Sci in Africa. Res: Man against tsetse in Africa. Mailing Add: 1133 Ave of the Americas New York NY 10036

MCKELVEY, JOHN LEYLAND, b Kingston, Ont, Can, Apr 11, 01; nat US; m 30; c 3. OBSTETRICS, GYNECOLOGY. Educ: Queen's Univ, Ont, BA, 23, MD, 26. Hon Degrees: LLD, Queen's Univ, Ont, 54. Prof Exp: Instr obstet, Johns Hopkins Hosp, 33-34; from assoc prof to prof & head dept obstet & gynec, Peiping Union Med Col, 34-38; prof, 38-67, EMER PROF OBSTET & GYNEC, MED SCH, UNIV MINN, MINNEAPOLIS, 67- Mem: Fel Am Gynec Soc. Res: Hypertensive pregnancy toxemias; gynecological-obstetrical histopathology; gynecological malignant diseases. Mailing Add: 1916 Summit Ave St Paul MN 55105

MCKELVEY, JOHN MURRAY, b Stanley, NC, Nov 1, 37; m 73; c 1. THEORETICAL CHEMISTRY, ORGANIC POLYMER CHEMISTRY. Educ: Mercer Univ, AB, 61; Univ of Ga, MS, 65; Ga Inst Technol, PhD(phys org chem), 71. Prof Exp: Fel theoret chem, Chem Lab, Advan Normal Sch for Women, Paris, France, 71-73 & Dept of Chem, Univ Calif, Berkeley, 73-75; RES SCIENTIST, EASTMAN KODAK CO, 75- Mem: Am Chem Soc. Res: Applied quantum mechanics in polymer and organic chemistry. Mailing Add: Color Photog Chem Lab Eastman Kodak Co Res Div Rochester NY 14650

MCKELVEY, JOHN PHILIP, b Ellwood City, Pa, Nov 9, 26; m 50; c 2. SOLID STATE PHYSICS. Educ: Pa State Univ, BS, 49, MS, 50; Univ Pittsburgh, PhD(physics), 57. Prof Exp: Asst physics, Pa State Univ, 49-50, instr math, 50-51; res physicist, Res Labs, Westinghouse Elec Corp, 51-59, supvry physicist, 59-62; from assoc prof to prof physics, Pa State Univ, 62-74; PROF PHYSICS & HEAD DEPT PHYSICS & ASTRON, CLEMSON UNIV, 74- Concurrent Pos: Asst dean col sci, Pa State Univ, 69-73. Mem: Am Asn Physics Teachers; fel Am Phys Soc. Res: Semiconductor physics; solid state theory; statistical physics. Mailing Add: Dept of Physics & Astron Clemson Univ Clemson SC 29631

MCKELVEY, KATHERINE K, mathematics, see 12th edition

MCKELVEY, ROBERT WILLIAM, b Ligonier, Pa, Apr 27, 29; m 52; c 2. MATHEMATICS. Educ: Carnegie Inst Technol, BS, 50; Univ Wis, MS, 52, PhD(math), 54. Prof Exp: Instr math, Purdue Univ, 54; res fel, Inst Fluid Dynamics & Appl Math, Univ Md, 54-56; from asst prof to prof, Univ Colo, 56-70; PROF MATH, UNIV MONT, 70- Concurrent Pos: Fac fel, Univ Colo, 60, chmn dept math, 65-66; mem, Math Res Ctr, Madison, Wis, 64-65; vis prof, Univ Utah, 66-67; exec dir, Rocky Mountain Math Consortium, 67-75. Mem: Am Math Soc; Math Asn Am; Soc Indust & Appl Math. Res: Asymptotic theory of differential equations; differential boundary value problems; linear operations in Hilbert space; functional analysis; mathematical models. Mailing Add: Dept of Math Univ of Mont Missoula MT 59801

MCKELVEY, RONALD DEANE, b Battle Creek, Mich, June 24, 44; m 72; c 1. PHYSICAL ORGANIC CHEMISTRY, PHOTOCHEMISTRY. Educ: Western Mich Univ, BS, 66; Univ Wis-Madison, PhD(org chem), 71. Prof Exp: Fel org chem, Univ Calif, Berkeley, 72; ASSOC PROF ORG CHEM, INST PAPER CHEM, 72- Mem: Am Chem Soc. Res: Free radical reactions of carbohydrates; oxidation reactions lignin; synthesis of insect hormones and naturally occurring mimics. Mailing Add: Inst of Paper Chem Appleton WI 54911

MCKELVEY, VINCENT ELLIS, b Huntington, Pa, Apr 6, 16; m 37; c 1. GEOLOGY. Educ: Syracuse Univ, BA, 37; Univ Wis, MA, 39, PhD(geol), 47. Prof Exp: Asst geol, Univ Wis, 37-40; geologist, 41-60, asst chief geologist, 60-65, res geologist, 65-71, chief geologist, 71, DIR, US GEOL SURVEY, 71- Honors & Awards: Distinguished Serv Award, US Geol Surv, 61; Career Serv Award, Nat Civil Serv League, 72. Mem: AAAS; Geol Soc Am; Geochem Soc; Soc Econ Geol; Am Geophys Union. Res: Economic geology; physical stratigraphy; mineral economics; seabed resources. Mailing Add: US Geol Surv 12201 Sunrise Valley Dr Reston VA 22092

MCKELVIE, DOUGLAS H, b Collbran, Colo, Mar 15, 27; m 50; c 3. VETERINARY PATHOLOGY. Educ: Colo Agr & Mech Col, BS, 50, DVM, 52; Univ Calif, Davis, PhD(comp path), 68; Am Col Lab Animal Med, dipl. Prof Exp: Pvt pract, 52-60; res vet, Radiobiol Lab, Univ Calif, 60-68; clin prof sci & dir lab animal med, Col Vet Med & Biomed Sci, Colo State Univ, 68-74; DIR DIV ANIMAL RESOURCES & ASSOC PROF PATH, COL MED, UNIV ARIZ, 74- Concurrent Pos: Consult, Hill Found dog breeding grant, Med Sch, Univ Ore, 66-; consult & mem bd trustees, Am Asn Accreditation Lab Animal Care. Mem: Am Vet Med Asn; Am Asn Lab Animal Sci; Radiation Res Soc; NY Acad Sci; Am Soc Vet Clin Path. Res: Production and care of laboratory dogs; serum chemistry of the dog; effects of internal and external irradiation on maturing and adult bone; laboratory animal medicine and biology. Mailing Add: Div Animal Resources Col Med Univ of Ariz Tucson AZ 85724

MCKELVIE, NEIL, b Welwyn Garden City, Eng, Dec 9, 30; US citizen; m 59; c 1. ORGANIC CHEMISTRY. Educ: Cambridge Univ, BA, 53; Columbia Univ, PhD(chem), 61. Prof Exp: Chemist, Fairey Aviation Co, Eng, 53-54; res chemist, Am Cyanamid Co, Conn, 59-61; res asst, Yale Univ, 61-62; asst prof, 62-70, ASSOC PROF CHEM, CITY COL NEW YORK, 71- Mem: Am Chem Soc. Res: Organophosphorus chemistry; chemistry of other elements in groups IV, V and VI. Mailing Add: Dept of Chem City Col of New York New York NY 10031

MCKELVY, JEFFREY FORRESTER, b Akron, Ohio, Aug 25, 38; m 63; c 2. NEUROBIOLOGY. Educ: Univ Akron, BSc, 63; Johns Hopkins Univ, PhD(biochem), 68. Prof Exp: Asst prof anat, Health Ctr, Univ Conn, 71-76; ASSOC PROF BIOCHEM, HEALTH SCI CTR, UNIV TEX, DALLAS, 76- Concurrent Pos: Jane Coffin Childs Found fel, Weizman Inst Sci, Rehovot, Israel, 68-69 & Roche Inst Molecular Biol, 69-71; panel mem adv panel neurobiol, NSF, 74-77. Mem: Am Soc Neurochem; Am Soc Biol Chemists; Soc Neurosci; Int Soc Neuroendocrinol. Res: Biochemical aspects of neuroendocrinology; biosynthesis of hypothalmic releasing hormones; regulation of hypothalamic hormone metabolism; structure and role of central nervous system peptides. Mailing Add: Dept of Biochem Univ of Tex Health Sci Ctr Dallas TX 75235

MCKENNA, CHARLES EDWARD, b Long Beach, Calif, May 9, 44; m 74. BIO-ORGANIC CHEMISTRY. Educ: Oakland Univ, BA, 66; Univ Calif, San Diego, PhD(chem), 71. Prof Exp: Res assoc chem, Univ Calif, San Diego, 71-72; NIH fel, Harvard Univ, 72-73; Nat Acad Sci exchange scholar, Bakh Inst, Moscow, 73; ASST PROF CHEM, UNIV SOUTHERN CALIF, 73- Concurrent Pos: Consult, Jessen & Co, Inc, 75- Mem: Am Chem Soc. Res: Mechanism of biological dinitrogen fixation; molybdoenzymes; biologically important organophosphorous compounds. Mailing Add: Dept of Chem Univ of Southern Calif Los Angeles CA 90007

MCKENNA, FRANCIS EUGENE, b Globe, Ariz, July 29, 21. INFORMATION SCIENCE. Educ: Univ Calif, BS, 41; Univ Wash, PhD(phys chem), 44. Prof Exp: Res chemist, SAM Labs, Columbia, 44-45; res supvr, Carbide & Carbon Chem Corp, NY, 45-46; fel vapor pressures of metals, Inst Nuclear Studies, Univ Chicago, 46-47; res chemist, Air Reduction Sales Co, 47-49, sr chemist, Air Reduction Co, Inc, 50-53, sr info specialist, 53-59, supvr info ctr, 59-67; ed, Spec Libr, 67-70, EXEC DIR, SPEC LIBR ASN, 70- Concurrent Pos: Dana lectr, Pratt Inst, 61; mem, US/USSR tech exchange sci info, 66; vis prof, Univ Tokyo, Univ Osaka & Nat Col Librarianship, Tokyo, 72. Mem: Fel AAAS; Am Soc Info Sci; Am Chem Soc; Spec Libr Asn (pres, 66-67); Inst Info Sci. Res: Chemical reactions in the gas phase; physical properties of gases; fluorine and fluorocarbons; analytical fluorine and gases; cryoscopy; hemiacetal formation in alcohol-aldehyde systems; organization and dissemination of information. Mailing Add: Spec Libr Asn 235 Park Ave S New York NY 10003

MCKENNA, GEORGE FINLEY, b Brooklyn, NY, Apr 2, 13; m 41; c 4. BOTANY. Educ: Mich State Univ, BS, 37; Univ Conn, MS, 49. Prof Exp: Asst entom, Boyce Thompson Inst Plant Res, 38-40; res scientist, Clayton Found Biochem Inst, Univ Tex, Austin, 59-64, res assoc, Col Pharm, 65-69; INDEPENDENT CONSULT MKT RES, 69- Concurrent Pos: Instr, Univ Conn, 48-50. Mem: AAAS. Res: Inhibitors in plants that will check growth of tumor tissue and application to tumor growth in humans. Mailing Add: 2933 Pecan Springs Rd Austin TX 78723

MCKENNA, JAMES, b Canton, Ohio, June 18, 29; m 71. MATHEMATICS, PHYSICS. Educ: Mass Inst Technol, BSc, 51, MSc, 54; Princeton Univ, PhD(math), 61. Prof Exp: Mem tech staff, 60-70, HEAD DEPT MATH PHYSICS & NETWORKS, BELL LABS, NJ, 70- Mem: Am Phys Soc; Soc Indust & Appl Math. Res: Electromagnetic theory; stochastic differential equations; numerical analysis; optics. Mailing Add: Bell Labs Mountain Ave Murray Hill NJ 07974

MCKENNA, JAMES FRANCIS, b Portland, Ore, Apr 24, 13; m 40; c 2. APPLIED CHEMISTRY. Educ: Univ Notre Dame, BS, 36, MS, 37, PhD(org chem), 39. Prof Exp: Res chemist, Sharples Chem Inc, Mich, 39-42 & Magnolia Petrol Co, Tex, 42-44; chemist, Pittsburgh Plate Glass Co, 44-52; chemist, 52-55, chief chemist, 55-70, GEN MGR, J G MILLIGAN CO, 70- Mem: Am Chem Soc. Res: Paints and paint raw materials; plastics; adhesives; latex compounding; polymerization reactions. Mailing Add: 4430 N Marlborough Dr Milwaukee WI 53211

MCKENNA, JOHN MORGAN, b Providence, RI, Oct 11, 27; m 54; c 5. IMMUNOLOGY. Educ: Providence Col, BS, 50, MS, 52; Lehigh Univ, PhD(biol), 59; Am Bd Microbiol, dipl. Prof Exp: Res assoc, Sharp & Dohme Res Labs div, Merck & Co, Inc, 55-59; assoc, Harrison Dept Surg Res, Sch Med, Univ Pa, 59-66; from assoc prof to prof microbiol, Sch Med, Univ Mo, 66-72, res assoc, Space Sci Res Ctr, 66-72; PROF MICROBIOL & CHMN DEPT, SCH MED, TEX TECH UNIV, 72- Mem: AAAS; fel Am Acad Microbiol; Am Soc Microbiol; Am Asn Immunologists; Soc Exp Biol & Med. Res: Antibody formation; virology; tumor immunity. Mailing Add: Dept of Microbiol Tex Tech Univ Sch of Med Lubbock TX 79409

MCKENNA, MALCOLM CARNEGIE, b Pomona, Calif, July 21, 30; m 52; c 4. VERTEBRATE PALEONTOLOGY. Educ: Univ Calif, AB, 54, PhD(paleont), 58. Prof Exp: Instr paleont, Univ Calif, 58-59; asst prof, 60-64, ASSOC PROF GEOL, COLUMBIA UNIV, 64-; FRICK CUR, DEPT VERT PALEONT, AM MUS NATURAL HIST, 68- Concurrent Pos: From asst cur to assoc cur, Dept Vert

Paleont, Am Mus Natural Hist, 60-65, Frick assoc cur, 65-68. Mem: AAAS; Soc Vert Paleont; Am Soc Mammalogists; Soc Study Evolution; Geol Soc Am. Res: Evolution of the Mammalia during the late Mesozoic and Cenozoic eras; stratigraphic paleontology of Mesozoic and Cenozoic continental sediments; mammalian order Insectivora and its close allies; plate tectonics; biogeography. Mailing Add: Dept of Geol Columbia Univ New York NY 10027

MCKENNA, ROBERT WILSON, b Graceville, Minn, Oct 30, 40; m 64; c 3. PATHOLOGY. Educ: Col St Thomas, Minn, BS, 62; Univ Minn, MD, 66. Prof Exp: Med intern, Univ Calif, San Diego, 66-67; fel, 69-73, instr, 73-74, ASST PROF PATH, UNIV MINN, MINNEAPOLIS, 74- Mem: Am Soc Hemat; Am Asn Cancer Res; Am Soc Clin Path. Res: Morphology, cytochemistry and ultrastructure of blood cells and hematologic malignancies. Mailing Add: 6939 Morgan Ave S Minneapolis MN 55423

MCKENNAN, ROBERT ADDISON, b Helena, Mont, Dec 20, 03; m 28. ANTHROPOLOGY. Educ: Dartmouth Col, AB, 25; Harvard Univ, PhD(anthrop), 33. Prof Exp: From instr to prof sociol, 30-54, prof anthrop, 54-69, EMER PROF ANTHROP, DARTMOUTH COL, 69- Concurrent Pos: Chmn dept sociol, Dartmouth Col, 37-41, chmn div soc sci, 47-51, dir, Northern Studies Prog, 53-58, actg chmn dept sociol & anthrop, 57; mem, Arctic Archaeol Comt, Arctic Inst NAm, 70-; NSF fel prehist, Univ Alaska, 69-72; mem, Ad Hoc Comt, Nat Acad Sci, 71; fel, Wenner-Gren publ subvention. Mem: Fel Am Anthrop Asn; fel Arctic Inst NAm; Soc Am Archaeologists; Can Archaeol Asn. Res: Ethnography; Alaskan Athapaskans; prehistory; sub-Arctic and arctic prehistroy. Mailing Add: Elm St Norwich VT 05055

MCKENNEY, DEAN BRINTON, b Newton, Mass, Mar 1, 40; m 62; c 1. OPTICS. Educ: Bowdoin Col, AB, 62; Univ Rochester, MS, 65; Univ Ariz, PhD(optics), 69. Prof Exp: Res assoc optics, Univ Ariz, 67-69, asst prof, 69-71; PRES, HELIO ASSOCS, INC, 71- Mem: Optical Soc Am. Res: Physical and thin film optics; optical properties of solids. Mailing Add: 8230 E Broadway Tucson AZ 85710

MCKENNEY, DONALD JOSEPH, b Eganville, Ont, May 3, 33; m 59; c 2. PHYSICAL CHEMISTRY. Educ: Univ Western Ont, BSc, 57, MSc, 58; Univ Ottawa, Can, PhD(phys chem), 63. Prof Exp: Lectr gen chem, Royal Mil Col Can, 58-60 & Exten Div, Univ Ottawa, Can, 62-63; Nat Res Coun Can fel, Cambridge Univ, 63-64; asst prof, 64-70, ASSOC PROF GEN & PHYS CHEM, UNIV WINDSOR, 70- Mem: Chem Inst Can; The Chem Soc. Res: Gas phase chemical kinetics; kinetics of atom and free radical reactions. Mailing Add: Dept of Chem Univ of Windsor Windsor ON Can

MCKENNEY, JOEL R, b Ponca City, Okla, Feb 21, 34; m 56; c 4. PSYCHOPHYSIOLOGY. Educ: Univ Kans, BS, 56, MS, 58; Univ Wash, PhD(physiol, biophys), 65; Med Col Ga, MD, 75. Prof Exp: Biol scientist, Hanford Labs, Gen Elec Co, 58-62; sr scientist physiol, Pac Northwest Labs, Battelle Mem Inst, 64-67; asst prof physiol, 67-74, RESIDENT PSYCHIAT, MED COL GA, 74- Mem: Assoc Am Physiol Soc; Radiation Res Soc. Res: Gastrointestinal transport of water and electrolytes; calcium and phosphate transport across the isolated intestine of the rat; gastrointestinal absorption and radiation effects. Mailing Add: Dept of Psychiat Med Col of Ga Augusta GA 30902

MCKENNEY, LEO FOSTER, organic chemistry, see 12th edition

MCKENNEY, THOMAS WILLIAM, b Livermore, Maine, Dec 2, 28; m 60 ICHTHYOLOGY, MARINE BIOLOGY. Educ: Tufts Univ, BS, 51; Univ Miami, MS, 58, PhD(ichthyol), 65. Prof Exp: Res aide marine biol, Inst Marine Sci, Univ Miami, 56-58, res instr, 58-63; asst prof biol, Mem Univ Nfld, 64-67; fishery biologist, Bur Com Fisheries, Ga, 67-69, NC, 69-71 & Trop Atlantic Biol Lab, Nat Marine Fisheries Serv, 71-74; MEM STAFF, NARRAGANSETT LAB, NAT MARINE FISHERIES SERV, 74- Mem: AAAS; Am Soc Ichthyol & Herpet; Am Soc Zool; Can Soc Zool; Marine Biol Asn UK. Res: Systematic ichthyology; taxonomy of larval fishes. Mailing Add: Nat Marine Fisheries Serv Narragansett Lab RR 7A Box 522A Narragansett RI 02882

MCKENNIS, HERBERT, JR, b New York, NY, Jan 29, 16; m 38; c 3. BIOCHEMISTRY. Educ: Harvard Univ, SB, 38; Cornell Univ, PhD(biochem), 45. Prof Exp: Chemist, Nuodex Prods Co, NJ, 38-39; chemist, Ciba Pharmaceut Prods, Inc, 39-42; asst biochem, Med Col, Cornell Univ, 42-45; asst prof chem, Med Col Va, 45-46; instr physiol chem, Sch Med, Johns Hopkins Univ, 46-48; assoc prof biochem, Med Col Univ Va, 48-49; head basic sci res dept & dir chem div, Naval-Civil Eng Res & Eval Lab, 49-53; assoc prof, 53-55, PROF PHARMACOL, MED COL VA, 55-, HEAD DIV BIOCHEM PHARMACOL, 72- Concurrent Pos: Vis prof, Inst Physiol, Univ Chile, 60; guest investr, Royal Vet Col Sweden, 65; ed, Va J Sci. Mem: AAAS; Soc Toxicol; Am Chem Soc; Am Soc Biol Chem; Soc Exp Biol & Med. Res: Metabolite antagonists; chemistry of natural organic compounds; production of polymers; biological effects of hydrazine compounds; biochemical pharmacology. Mailing Add: Dept of Pharmacol Med Col of Va Richmond VA 23297

MCKENZIE, BASIL EVERARD, b Jamaica, WI, Sept 14, 35; m 66; c 2. VETERINARY PATHOLOGY. Educ: Jamaica Sch Agr, dipl agr, 56; Tuskegee Inst, DVM, 66; Cornell Univ, MSc, 69; Univ Wis, PhD(vet path), 71. Prof Exp: Artificial insemination officer, Ministry of Agr & Lands, Jamaica, 56-60; res asst reprod path, Cornell Univ, 66-68; exp pathologist, Lederle Labs, Pearl River, NY, 71-73; asst prof, 73-76, ASSOC PROF PATH, TUSKEGEE INST, 76- Mem: Am Vet Med Asn; Sigma Xi; Soc Environ & Pharmacol Pathologists; Am Col Vet Pathologists; Int Acad Path. Res: Viral myocarditis; comparative cardiovascular pathology; pathology of the male and female reproductive system. Mailing Add: Dept of Vet Path Tuskegee Inst Sch of Vet Med Tuskegee Institute AL 36088

MCKENZIE, DONALD EDWARD, b Rivers, Man, Aug 9, 24; nat US; m 50; c 3. PHYSICAL CHEMISTRY, ENVIRONMENTAL CHEMISTRY. Educ: Univ Man, BSc, 45, MSc, 47; Univ Southern Calif, PhD, 50. Prof Exp: Res off, Atomic Energy Can, Ltd, Chalk River, Ont, 50-57; MGR CHEM TECHNOL, ATOMICS INT DIV, ROCKWELL INT CORP, 57- Mem: Am Chem Soc; Sigma Xi; Air Pollution Control Asn. Res: High temperature chemistry; fused salts; electrochemistry; air and water pollution; management of research and development for coal gasification and liquefaction; chemical reactions in fused salts; high temperature batteries; disposal of combustible waste; gas scrubbing. Mailing Add: Chem Technol Atomics Int PO Box 309 Canoga Park CA 91304

MCKENZIE, DOUGLAS HUGH, b Minneapolis, Minn, Feb 14, 39; m 65; c 1. ANTHROPOLOGY. Educ: Univ Minn, BA, 60; Harvard Univ, PhD(anthrop), 65. Prof Exp: Asst prof anthrop, Case Inst Technol, 64-67; assoc prof, 67-74, PROF ANTHROP, CLEVELAND STATE UNIV, 74- Concurrent Pos: Co-prin investr, NSF res grant, 65-66. Mem: Fel Am Anthrop Asn; Soc Am Archaeol. Res: Archaeology of North America, especially eastern United States. Mailing Add: Dept of Anthrop Cleveland State Univ Cleveland OH 44115

MCKENZIE, FREDERICK FRANCIS, b Lulu Island, BC, May 31, 00; nat US; m 24; c 3. ANIMAL SCIENCE. Educ: Univ BC, BSA, 21; Univ Mo, AM, 23, PhD(animal husb), 25. Hon Degrees: DAgr, Cath Univ Chile, 41. Prof Exp: Agent animal husb, USDA, 23-27, 28-41; dir agr, Int Col, Turkey, 27-28; prof animal husb & head dept, Utah State Univ, 41-43; prof, 44-60, EMER PROF ANIMAL SCI, ORE STATE UNIV, 60- Concurrent Pos: Hon mem fac vet med, Univ Chile, 45-; vis prof, Univ Indonesia, 60-63; prof & nat livestock adv, NC State Univ Mission, AID, Lima, Peru, 63-67; dean agr, Univ Ife, Nigeria, USAID-Univ Wis Contract, 67-69; consult animal sci, Ore State Univ, 69- Honors & Awards: Medal, Chilean Soc Agr, 41; Medallion & Citation, Int Cong Animal Reprod & Artificial Insemination, Milan, Ital, 64. Mem: AAAS; fel Am Soc Animal Sci (vpres, 56-57); Ital Soc Advan Zootechnol. Res: Reproduction physiology; livestock potential for production of food and fiber in developing countries in the tropics. Mailing Add: 2675 SW Fairmont Dr Corvallis OR 97330

MCKENZIE, GARRY DONALD, b Niagara Falls, Ont, June 8, 41; m 65; c 1. GEOLOGY. Educ: Univ Western Ont, BSc, 63, MSc, 64; Ohio State Univ, PhD(geol), 68. Prof Exp: Asst to dir, Inst Polar Studies, 68-69, exec officer dept, 69-72, asst prof, 69-75, ASSOC PROF GEOL, OHIO STATE UNIV, 72- Concurrent Pos: NSF grant glacial geol, Adams Inlet, Alaska, 66-69. Mem: AAAS; Geol Soc Am; Int Glaciol Soc; Geol Asn Can; Soc Environ Geochem & Health. Res: Quaternary stratigraphy and geomorphology; environmental geology. Mailing Add: Dept of Geol & Mineral Ohio State Univ Columbus OH 43210

MCKENZIE, HARVEY, b Big Bend, Wis, Jan 15, 10. MATHEMATICS. Educ: Univ Wis, BA, 30, MA, 47; Univ Colo, PhD(math), 56. Prof Exp: High sch teacher, Wis, 30-42; instr math, Univ Wis, 45-47; asst prof, Western State Col Colo, 47-49; instr, Univ Colo, 49-54; from asst prof to assoc prof, SDak State Col, 54-61; assoc prof, Northern Ill Univ, 61-68; PROF MATH, UNIV WIS-OSHKOSH, 68- Mem: Am Math Soc; Math Asn Am. Res: Entire functions; modern algebra. Mailing Add: Dept of Math Univ of Wis Oshkosh WI 54902

MCKENZIE, HUGH, b Indian Head, Sask, Nov 24, 19; m 43; c 3. PLANT BREEDING, PLANT GENETICS. Educ: Univ Sask, BSA, 49, MSc, 50; Univ Minn, PhD(plant genetics), 63. Prof Exp: Tech officer, Exp Farm, Agr Can, Sask, 50-51, res officer, Res Sta, 51-66, RES SCIENTIST, RES STA, AGR CAN, ALTA, 66- Mem: Agr Inst Can; Can Soc Agron; Prof Inst Pub Serv Can. Res: Inheritance studies of disease and insect reactions in wheat; breeding improved hard red spring wheat varieties which are resistant to the wheat stem sawfly. Mailing Add: Agr Can Res Sta Lethbridge AB Can

MCKENZIE, JESS MACK, b Woodsboro, Tex, Sept 2, 32; m 55; c 2. PHYSIOLOGY. Educ: NTex State Col, BA, 54, MA, 56; Univ Tex, PhD(physiol), 59. Prof Exp: Physiologist, Dept Space Med, US Air Force Sch Aerospace Med, 58-59; chief hemat sect, 59-67, CHIEF STRESS PHYSIOL RES, PHYSIOL LABS, CIVIL AEROMED RES INST, 67-; PROF ZOOL, UNIV OKLA, 73- Concurrent Pos: Adj asst prof zool, Univ Okla, 59-67, adj assoc prof, 67-73, asst prof, Sch Med, 59-74. Mem: Am Physiol Soc; Soc Exp Biol & Med; Sigma Xi. Res: Blood coagulation; automated and computer-aide measurement of stress in large populations; effects of stress on health. Mailing Add: AAC-115 Civil Aeromed Res Inst PO Box 25082 Oklahoma City OK 73125

MCKENZIE, JOHN MAXWELL, b Glasgow, Scotland, Nov 13, 27; m; c 3. INTERNAL MEDICINE, ENDOCRINOLOGY. Educ: Univ St Andrews, MD, ChB, 50, MD, 58. Prof Exp: House surgeon, Dundee Royal Infirmary, Scotland, 50-51; house physician therapeut unit, Maryfield Hosp, 51; asst, Dept Pharmacol & Therapeut, Univ St Andrews, 53-55; registr, Maryfield Hosp, 55-56; asst med, Sch Med, Tufts Univ, 56-57; registr, Maryfield Hosp, 57-58; instr med, Sch Med, Tufts Univ, 58-59; lectr med, 60-61, from asst prof to assoc prof med & clin med, 61-70, PROF EXP MED, McGILL UNIV, 70- Concurrent Pos: Res fel endocrinol, New Eng Ctr Hosp, 56-57; res assoc endocrinol, New Eng Ctr Hosp, 58-59; clin asst, Royal Victoria Hosp, Montreal, Que, 59-63, asst physician, 64- Honors & Awards: Ayerst Award, Endocrine Soc, 61. Mem: AAAS; Endocrine Soc; Am Fedn Clin Res; Am Soc Clin Invest; Am Thyroid Asn. Res: Pathogenesis of Graves' disease, particularly the role of the longacting thyroid stimulator in that syndrome; investigations of mode of action of thyrotropin in the thyroid gland. Mailing Add: Dept Exp Med McGill Univ Montreal PQ Can

MCKENZIE, JOHN WARD, b Dillon, SC, Sept 11, 18; m 49; c 1. HISTOLOGY, EMBRYOLOGY. Educ: The Citadel, BS, 40; Univ SC, MS, 50; Univ NC, PhD(zool), 54. Prof Exp: Asst gen biol, Univ SC, 49-50; asst gen zool, Univ NC, 51-54; from asst prof to assoc prof micros anat, 54-74, PROF ANAT, MED COL GA, 74- Concurrent Pos: Wilson scholar, Marine Biol Lab, Woods Hole, 52; mem staff, Eugene Talmadge Mem Hosp. Mem: AAAS; Electron Micros Soc Am. Res: Neurophysiology in primates. Mailing Add: Dept of Anat Med Col of Ga Augusta GA 30902

MCKENZIE, JOSEPH ADDISON, b Trinidad, WI, Nov 6, 30; Can citizen; m 57; c 2. ETHOLOGY, BIOCHEMICAL SYSTEMATICS. Educ: McMaster Univ, BA, 57; Univ Toronto, MA, 60; Univ Western Ont, PhD(ethology), 64. Prof Exp: Fisheries officer, Govt Trinidad & Tobago, 60-61; asst prof zool, Laurentian Univ, 64-67; asst prof zool, 67-69, ASSOC PROF ANIMAL BEHAV, UNIV NB, 69- Concurrent Pos: Nat Res Coun Can operating grants, 64- Mem: Can Soc Zoologists. Res: Comparative behavior of stickleback family; biochemical systematics of fish. Mailing Add: Dept of Biol Univ of NB Fredericton NB Can

MCKENZIE, LESTER, organic chemistry, see 12th edition

MCKENZIE, MALCOLM ARTHUR, b Providence, RI, Apr 21, 03. FOREST PATHOLOGY. Educ: Brown Univ, PhB & MA, 26, PhD(forest path), 35. Prof Exp: Field asst forest path, Forest Prods Lab, US Forest Serv, 26-27; instr biol, Univ NC, 27-29; agent forest path, Bur Plant Indust, USDA, 29-35; pathologist shade tree dis, 35-36, from asst res prof to res prof bot, 36-50, prof plant path & dir shade tree labs, 50-73, EMER PROF PLANT PATH, UNIV MASS, AMHERST, 73-; CONSULT SHADE TREE MGT, 73- Concurrent Pos: Lectr, Brown Univ, 29-35; actg head dept entom & plant path, Univ Mass, Amherst, 65-68. Honors & Awards: Hon life award, Int Shade Tree Conf, 75. Mem: Am Phytopath Soc; Mycol Soc Am; Soc Indust Microbiol. Res: Tree pests; Dutch Elm disease; wood decay; tree hazards in public utility work; municipal tree maintenance; continuing education programs in tree workshops and environmental pollution control. Mailing Add: PO Box 651 North Amherst MA 01059

MCKENZIE, RALPH NELSON, b Cisco, Tex, Oct 20, 41. MATHEMATICS. Educ: Univ Colo, Boulder, BA, 63, PhD(math), 66. Prof Exp: Asst prof, 67-71, ASSOC PROF MATH, UNIV CALIF, BERKELEY, 71- Concurrent Pos: NSF fel, Inst Advan Study, 71-72. Mem: Am Math Soc; Asn Symbolic Logic. Res: Direct products of relational systems; equational varieties of algebras; algorithmicity in algebra. Mailing Add: 301 Campbell Hall Dept of Math Univ of Calif Berkeley CA 94720

MCKENZIE, WALTER LAWRENCE, b Somerville, Mass, Apr 6, 38; m 59; c 4. INDUSTRIAL PHARMACY, RESEARCH ADMINISTRATION. Educ: Mass Col Pharm, BS, 60, MS, 62, PhD(med chem), 71. Prof Exp: Asst dir pharmaceut develop, Res & Develop Labs, Astra Pharmaceut Prod, Inc, 63-72; ASST DIR PHARMACEUT RES & DEVELOP, EATON LABS, NORWICH PHARMACAL CO, MORTON-NORWICH PROD, INC, 72- Mem: Acad Pharmaceut Sci; Am Chem Soc; Parenteral Drug Asn; Proprietary Asn; Soc Cosmetic Chemists. Res: Drug formulation, bioavailability, new drug entity characterization, instrumentation and analysis, specifications and controls, process development and synthetic medicinal chemistry. Mailing Add: Eaton Labs Norwich Pharmacal Co Morton-Norwich Prod PO Box 191 Norwich NY 13815

MCKENZIE, WENDELL HERBERT, b Wykoff, Minn, Nov 23, 42; m 64; c 2. HUMAN GENETICS. Educ: Westmar Col, BA, 64; NC State Univ, MS, 69, PhD(genetics), 73. Prof Exp: Teacher, Pub Schs, Iowa, 64-67; res asst genetics, NC State Univ, 68-69, univ instr human genetics, 69-71; res cytogeneticist, Univ Colo Med Ctr, Denver, 72; univ instr human genetics, 73, ASST PROF HUMAN GENETICS, NC STATE UNIV, 73- Mem: Genetics Soc Am; Am Soc Human Genetics; Am Inst Biol Sci; AAAS; Inst Soc, Ethics & Life Sci. Res: Human cytogenetics, including chromosome structure and variation, ethodologies of chromosome banding techniques and mutagenic effects of environmental agents as pollutants. Mailing Add: Dept of Genetics NC State Univ Raleigh NC 27607

MCKEON, ALFRED JEROME, b New York, NY, Oct 1, 26; m 48; c 2. MATHEMATICAL STATISTICS. Educ: George Washington Univ, BA, 49. Prof Exp: Statistician, US Vet Admin, 48-55; math statistician, US Internal Revenue Serv, 55-58; chief opers res, US Census Bur, 58-61; opers analyst mil defense, NAm Air Defense Command, 61-66; opers res analyst, US Internal Revenue Serv, 66-69, chief math statistician, 69-73; DIR STATIST PROGS & STANDARDS, US POSTAL SERV, 73- Mem: Fel AAAS; Opers Res Soc Am; Am Soc Qual Control. Mailing Add: OSP & S US Postal Serv 475 L'Enfant Plaza Washington DC 20260

MCKEON, JAMES EDWARD, b Derby, Conn, June 25, 30; m 52; c 4. ORGANIC CHEMISTRY. Educ: Wesleyan Univ, BA, 51, MA, 53; Yale Univ, PhD(org chem), 60. Prof Exp: Res chemist, Chem Div, Union Carbide Corp, 59-63, res scientist & group leader, 63-69, sr res scientist, 69-73, ASSOC DIR RES, UNION CARBIDE CHEMS & PLASTICS, 73- Mem: Am Chem Soc; Sigma Xi. Res: Reactions of molecules coordinated with metals; homogeneous catalysis; oxidation processes. Mailing Add: Chems & Plastics Union Carbide Corp Tarrytown NY 10591

MCKEON, MARY GERTRUDE, b New Haven, Conn, June 8, 26. ELECTROCHEMISTRY. Educ: Albertus Magnus Col, BA, 47; Yale Univ, MS, 52, PhD(chem), 53. Prof Exp: Lab asst, Yale Univ, 49-52; from instr to assoc prof chem, 52-71, dean sophomores, 63-69, PROF CHEM, CONN COL, 71-, MARGARET W KELLY PROF, 74- Concurrent Pos: NSF sci fac fel, Harvard Univ, 59-60; vis assoc prof, Wesleyan Univ, 70-71. Mem: Am Chem Soc. Res: Synthetic organic; organic polarography; electroanalytical chemistry. Mailing Add: Dept of Chem Conn Col New London CT 06320

MCKEOWN, JAMES JOHN, b Albert Lea, Minn, Oct 29, 30; m 58; c 3. PHYSICAL CHEMISTRY. Educ: St John's Univ, Minn, BS, 53; Iowa State Univ, PhD(chem), 58. Prof Exp: Asst chem, Iowa State Univ, 58; res chemist, Procter & Gamble Co, 58-60; sr chemist, 60-64, supvr, 64-66, mgr appl res, 66-69, tech dir dielec mat & systs lab, 69-73, TECH DIR ELECTRONIC PRODS LAB, MINN MINING & MFG CO, 73- Res: Ceramics; electronics; engineering management; materials science; microelectronics; polymers. Mailing Add: 10055 Ideal Ave N White Bear Lake MN 55115

MCKEOWN, JAMES PRESTON, b Vicksburg, Miss, Mar 2, 37; m 62; c 1. ETHOLOGY. Educ: Univ of the South, BS, 59; Univ Miss, MS, 62; Miss State Univ, PhD, 68. Prof Exp: From instr to assoc prof, 62-75, PROF BIOL & CHMN DEPT, MILLSAPS COL, 75- Concurrent Pos: Actg chmn dept biol, Millsaps Col, 67-69. Mem: AAAS; Am Soc Ichthyol & Herpet. Res: Celestial navigation by amphibians; experimental embryology. Mailing Add: Dept of Biol Millsaps Col Jackson MS 39210

MCKEOWN, JOSEPH, b N Ireland, June 26, 34; m 60; c 2. PHYSICS. Educ: Queen's Univ Belfast, BSc, 56; MSc, 59; Univ Man, PhD(physics), 70. Prof Exp: Physicist, Plessey Nucleonics, Eng, 59-61; lectr physics, Plymouth Polytech Inst, Eng, 61-66; RES OFFICER ACCELERATOR PHYSICS, CHALK RIVER NUCLEAR LABS, ATOMIC ENERGY CAN LTD, 70- Concurrent Pos: Lectr, Col Technol, Belfast, 57-59. Mem: Brit Inst Physics; Can Asn Physicists. Res: Proton-proton bremsstrahlung; wire chambers; high power nf control systems; accelerator physics. Mailing Add: Chalk River Nuclear Labs Atomic Energy of Can Ltd Chalk River ON Can

MCKERCHER, DELBERT GRANT, b Ont, Can, July 3, 14; nat US; m 43; c 3. VETERINARY VIROLOGY. Educ: Ont Vet Col, DVM, 38; Queen's Univ, Ont, MA, 42; Cornell Univ, PhD(virol), 49. Prof Exp: Agr scientist, Animal Dis Res Inst, Que, 38-40; bacteriologist, Prov Dept Pub Health, BC, 46; asst, Cornell Univ, 46-49; from asst prof to assoc prof, 49-57, PROF VET MED, SCH VET MED, UNIV CALIF, DAVIS, 57- Concurrent Pos: Vis investr, Vet Res Labs, Repub SAfrica, 56; NIH sr fel, Fed Res Inst Virus Dis Animals, Univ Tübingen, Ger, 60-61; guest investr, Univ Perugia, Italy, 68 & Nat Vet Sch, Alfort, France, 75; consult, NIH, 59-64, Vet Admin Hosp, Livermore, Calif, 64-70 & USDA to Peru, 65; mem, Adv Comt, Div Vet Biol, USDA, 68- & working team viral classification, WHO, 68- Mem: Am Vet Med Asn; US Animal Health Asn; Conf Res Workers Animal Dis; World Asn Vet Microbiologists & Immunologists; NY Acad Sci. Res: Virus diseases of domestic animals; viral and chlamydial infections (respiratory and reproductive, mainly) of domestic livestock, and the immunologic aspects of these infections. Mailing Add: 749 Anderson Rd Davis CA 95616

MCKERN, THOMAS WILTON, physical anthropology, see 12th edition

MCKERNS, KENNETH (WILSHIRE), b Hong Kong, Mar 5, 19; nat US; m 43; c 2. BIOCHEMISTRY. Educ: Univ Alta, BSc, 42, MSc, 46; McGill Univ, PhD(biochem), 51. Prof Exp: Demonstr, Univ Alta, 40-42; sr demonstr & lectr, McGill Univ, 46-51; chief biochemist, Can Packers, Ltd, Toronto, 51-55; lectr, Univ St Andrews, 55-56; sr res scientist & group leader, Lederle Labs Div, Am Cyanamid Co, NY, 56-60; assoc prof, 60-65, PROF OBSTET & GYNEC, COL MED, UNIV FLA, 65- Concurrent Pos: Fel, McGill-Montreal Gen Hosp Res Inst, 50-51; NIH spec res fel, 69-70; vis lectr, Harvard Med Sch, 69-70. Mem: AAAS; Am Soc Biol Chem; Soc Gynec Invest; Endocrine Soc. Res: Endocrine regulation of intermediate metabolism; protein hormone action and purification; effect of sex hormones on growth, metabolism and disease states. Mailing Add: Dept of Obstet & Gynec Univ of Fla Col of Med Gainesville FL 32601

MCKHANN, CHARLES FREMONT, b Boston, Mass, Jan 29, 30; m 54; c 3. SURGERY, MICROBIOLOGY. Educ: Harvard Univ, BA, 51; Univ Pa, MD, 55. Prof Exp: From instr to asst prof surg, Harvard Med Sch, 64-67; PROF SURG & MICROBIOL, MET CTR, UNIV MINN, MINNEAPOLIS, 68- Concurrent Pos: Nat Cancer Inst spec res fel tumor biol, Karolinska Inst, Sweden, 61-62; Am Cancer Soc clin fel, Mass Gen Hosp, Boston, 63-64; Andres Soriano investr oncol, Mass Gen Hosp, 64-67. Mem: Am Asn Cancer Res; Transplantation Soc; Am Surg Soc; Am Asn Immunol; Am Col Surg. Res: Tumor immunology. Mailing Add: Univ of Minn Hosp Box 85 Minneapolis MN 55455

MCKHANN, GUY MEAD, b Boston, Mass, Mar 20, 32; m 57; c 4. NEUROLOGY, NEUROCHEMISTRY. Educ: Yale Univ, MD, 55. Prof Exp: Intern med, NY Hosp, 55-56; resident pediatrics, Johns Hopkins Hosp, 56-57; res assoc neurochem, Nat Inst Neurol Dis & Blindness, 57-60; resident neurol, Mass Gen Hosp, 60-63; from asst prof to assoc prof pediat & neurol, Sch Med, Stanford Univ, 63-69; PROF NEUROL & EXEC HEAD DEPT, SCH MED, JOHNS HOPKINS UNIV, 69- Concurrent Pos: Joseph P Kennedy, Jr scholar, 63-66; John & Mary R Markle scholar acad med, 64-69. Mem: Am Acad Neurol; Am Neurol Asn; Am Neurochem Soc; Am Soc Pediat Res. Res: Lipid metabolism in the developing nervous system; metabolism of myelin; neurochemsitry of seizures. Mailing Add: Dept of Neurol Johns Hopkins Univ Sch of Med Baltimore MD 21205

MCKIBBEN, JOHN JOSEPH, b Philadelphia, Pa, Jan 4, 30; m 66; c 2. MATHEMATICS. Educ: Univ Chicago, MS, 54, PhD(math), 57. Prof Exp: Asst prof math, Wesleyan Univ, 57-61; asst prof, Univ Mass, Amherst, 66-71; ASSOC PROF MATH, SKIDMORE COL, 71-, CHMN DEPT, 73- Mem: Am Math Soc; Math Asn Am. Res: Analysis; recursive functions. Mailing Add: Dept of Math Skidmore Col Saratoga Springs NY 12866

MCKIBBEN, JOHN SCOTT, b Toledo, Ohio, Jan 25, 37; m 55; c 3. VETERINARY ANATOMY. Educ: Purdue Univ, BS, 59, DVM, 63; Iowa State Univ, MS, 66, PhD(vet anat), 69. Prof Exp: Clinician vet med, Rowley Mem Animal Hosp, Soc Prev Cruelty Animals, Mass, 63-64; from instr to asst prof vet anat, Iowa State Univ, 64-69; assoc prof, 69-75, PROF ANAT & HISTOL, AUBURN UNIV, 75- Mem: Am Vet Med Asn; Am Asn Vet Anat; World Asn Vet Anat; Am Asn Anatomists; Am Asn Vet Neurol. Res: Neurology, cardiology, arthrology and teaching methods, primarily in cardiac autonomic innervation and denervation, fetlock injuries in horses and multimedia programming of teaching materials. Mailing Add: Dept of Vet Anat Auburn Univ Auburn AL 36830

MCKIBBEN, ROBERT BRUCE, b Cincinnati, Ohio, Sept 1, 43. COSMIC RAY PHYSICS. Educ: Harvard Col, BA, 65; Univ Chicago, MS, 67, PhD(physics), 72. Prof Exp: Res assoc physics, 72-74, SR RES ASSOC PHYSICS, ENRICO FERMI INST, UNIV CHICAGO, 75- Mem: Sigma Xi. Res: Distribution of solar and galactic cosmic rays within the solar system and studies of energetic charged particles in planetary magnetospheres. Mailing Add: Enrico Fermi Inst 933 E 56th St Chicago IL 60637

MCKIBBIN, JOHN MEAD, b Tucson, Ariz, Nov 15, 15; m 44; c 4. BIOCHEMISTRY. Educ: Mich State Univ, BS, 38; Univ Wis, MS, 40, PhD(biochem), 42. Prof Exp: Instr nutrit, Harvard Med Sch & Sch Pub Health, Harvard Univ, 42-45; from asst prof to prof biochem, Col Med, Syracuse Univ & State Univ NY Med Ctr, 45-61; PROF BIOCHEM & CHMN DEPT, MED CTR, UNIV ALA, BIRMINGHAM, 61- Mem: Am Chem Soc; Soc Exp Biol & Med; Am Soc Biol Chem. Res: Nutrition, chemistry and metabolism of phospholipids and glycolipids. Mailing Add: Dept of Biochem Univ of Ala Med Ctr Birmingham AL 35294

MCKIE, JAMES EDWARD, physical chemistry, biochemistry, see 12th edition

MCKIEL, JOHN ALBERT, b Fredericton, NB, Nov 13, 18; m 43; c 3. MICROBIOLOGY. Educ: Univ NB, BSc, 49, MSc, 51; Queen's Univ, Ont, PhD(entom, bact), 55. Prof Exp: Tech officer entom, Can Dept Agr, NB, 49-50; sci officer immunol & entom, Defence Res Bd, Dept Nat Defence, Ont, 50-55, opers res, 55-56; BACTERIOLOGIST & CHIEF MICROBIOL, ZOONOSES LAB, CAN COMMUN DIS CTR, DEPT NAT HEALTH & WELFARE, 56- Mem: Can Soc Microbiol; Can Pub Health Asn. Res: Allergic reactions to insect bites; epidemiology of zoonotic infections, especially arthropod-borne encephalitides; Q fever; Rocky Mountain spotted fever; tularemia. Mailing Add: LCDC Dept Nat Health & Welfare Tunneys Pasture Ottawa ON Can

MCKIERNAN, MICHEL AMEDEE, b Chicago, Ill, Feb 17, 30; m 58. MATHEMATICS. Educ: Loyola Univ, Ill, BS, 51, MA, 52; Ill Inst Technol, PhD(math), 56. Prof Exp: Mathematician, Armour Res Found, 55-56; from instr to asst prof math, Ill Inst Technol, 56-61; mathematician, Inst Air Weapons Res, Univ Chicago, 61-62; assoc prof, 62-68, PROF MATH, UNIV WATERLOO, 68- Mem: Am Math Soc; Math Asn Am. Res: Functional equations. Mailing Add: Dept of Math Univ of Waterloo Waterloo ON Can

MCKIGNEY, JOHN IGNATIUS, b Marshall, Minn, Nov 14, 24; m 53; c 4. NUTRITION, BIOCHEMISTRY. Educ: Univ Fla, BS, 51, PhD(nutrit), 56. Prof Exp: Res adv animal nutrit, AID, 56-64; nutrit adv, Off Int Res, NIH, 64-67; dep dir, Food & Agr Orgn, UN, 67-71; SCI ADMINR, NAT INST CHILD HEALTH & HUMAN DEVELOP, 71- Mem: Am Inst Nutrit. Res: Human nutrition; food economics; human development. Mailing Add: Nat Inst Child Health & Human Develop Bldg 31 Rm 3A25 Bethesda MD 20014

MCKILLIP, WILLIAM JAMES, organic chemistry, polymer chemistry, see 12th edition

MCKILLOP, J H, b Detroit, Mich, Sept 21, 27; Can citizen; m 54; c 2. GEOLOGY. Educ: St Francis Xavier Univ, BSc, 51; Mem Univ Nfld, MSc, 61. Prof Exp: Geologist, Geol Surv Can, 51; asst govt geologist, Govt Nfld, 51-61, chief geologist, 61-62, dir mineral resources, Nlfd Dept Mines, Agr & Resources, 62-72, DEP MINISTER, DEPT MINES & ENERGY, GOVT NFLD & LABRADOR, 72- Mem: Can Inst Mining & Metall (vpres, 73-74); Geol Asn Can. Mailing Add: 17 Dublin Rd St John's NF Can

MCKILLOP, LUCILLE MARY, b Chicago, Ill, Sept 28, 24. MATHEMATICS. Educ: St Xavier Col, Ill, BS, 51; Univ Notre Dame, MS, 59; Univ Wis, PhD(math educ), 65. Prof Exp: Elem teacher, St Patrick Acad, Ill, 47-51, St Xavier Acad, 51-52, Siena High Sch, 52-57 & Marquette High Sch, 57-58; from instr to prof math, St Xavier Col, Ill, 58-73; PRES, SALVE REGINA-NEWPORT COL, 73- Concurrent Pos: Consult, Archdioceasan Sch Bd Prog Math, 65-66; chmn div liberal arts & humanities, St Xavier Col, 66-73; secy, Comt Math Prep Teachers Elem Sch Math, 67-72. Res: Evolution of concepts in mathematics, particularly the evolution of concepts in finite geometries; investigations of collineations in projective planes with coordinates in a Galois field; self-generative quality of historical studies of mathematical creativity. Mailing Add: Ochre Point Dr Newport RI 02840

MCKILLOP, WILLIAM L M, b Aberdeen, Scotland, June 3, 33; m 58; c 3. FOREST ECONOMICS. Educ: Univ Aberdeen, BSc, 54; Univ NB, MSc, 59; Univ Calif, Berkeley, MA & PhD(agr econ), 65. Prof Exp: Res officer forestry, Can Dept

Forestry, 58-59, forest economist, 59-61; asst prof forestry & forest economist, Univ Calif, Berkeley, 64-69, assoc prof, 67-75, PROF FORESTRY, UNIV CALIF, BERKELEY, 75- Mem: Am Econ Asn; Economet Soc; Soc Am Foresters. Res: Econometrics; economic theory; forest economics and statistics. Mailing Add: Dept of Forestry Univ of Calif Berkeley CA 94720

MCKIM, HARLAN L, b Gothenburg, Nebr, Sept 28, 37; m 61; c 1. GEOLOGY, SOIL SCIENCE. Educ: Univ Nebr, BS, 62, MS, 67; Iowa State Univ, PhD(soil sci), 72. Prof Exp: Soil scientist, US Army Cold Regions Res & Eng Lab, 62-64, Soil Conserv Serv, 64-66 & Soil Surv Lab, 66-68; SOIL SCIENTIST, US ARMY COLD REGIONS RES & ENG LAB, 68- Concurrent Pos: Res assoc soil sci, Iowa State Univ, 68-72. Mem: Soil Conserv Soc Am; Soil Sci Soc Am; Am Quaternary Asn; Am Soc Agron; Am Soc Photogram. Res: Feasibility of land treatment of wastewater, including denitrification/nitrification, monitoring of water quality and establishment of design criteria for land treatment facilities and mathematical modeling activities; use of remote sensing in Corps of Engineers programs. Mailing Add: US Army Col Regions Res & Eng Lab PO Box 282 Hanover NH 03755

MCKIMMY, MILFORD D, b Beaverton, Mich, Dec 22, 23; m 54; c 3. WOOD TECHNOLOGY. Educ: Mich State Univ, BS, 49; Ore State Col, MS, 51; State Univ NY, PhD(wood technol), 55. Prof Exp: From instr to assoc prof, 53-72, PROF FOREST PROD, ORE STATE UNIV, 72- Concurrent Pos: Charles Bullard forest res fel, Harvard Univ, 66-67. Mem: Soc Am Foresters; Soc Wood Sci & Technol; Forest Prod Res Soc; Tech Asn Pulp & Paper Indust. Res: Growth quality relationships of wood; application of genetics to wood quality. Mailing Add: Sch of Forestry Ore State Univ Corvallis OR 97331

MCKINLEY, DANIEL LAWSON, b Dora, Mo, Oct 25, 24; m 54; c 2. ECOLOGY, NATURAL HISTORY. Educ: Univ Mo, BA, 55, MA, 57. Prof Exp: Teaching fel biol, Bowdoin Col, 57-59; instr, Salem Col, 59-61; instr, Lake Erie Col, 61-64, asst prof, 64-66; mem fac, 66-71, ASSOC PROF BIOL, STATE UNIV NY ALBANY, 71- Mem: Soc Bibliog Natural Hist. Res: Life history and bibliography of vanishing animals in Missouri; history of Carolina parakeet; ecology of man; history of natural history in early America; vertebrate zoogeography. Mailing Add: Biol Bldg State Univ of NY Albany NY 12222

MCKINLEY, DONALD WILLIAM ROBERT, b Shanghai, China, Sept 22, 12; m 41, 50; c 2. AERONOMY. Educ: Univ Toronto, BA, 34, MA, 35, PhD(exp physics), 38. Prof Exp: Demonstr physics, Univ Toronto, 35-38; res physicist, Nat Res Coun Can, 38-41, head, Air Force Res Sect, 41-47, Radio Physics Sect, 48-53, asst dir, Radio & Elec Eng Div, 54-60, assoc dir, 60-62, dir, 63-68, vpres, 68-74; RETIRED. Concurrent Pos: Chmn & mem comts, Defence Res Bd, 44; mem, Commonwealth & Empire Radio Civil Aviation Confs, 44-47; lectr, McGill Univ, 47-48; chmn, Can Nat Comt, Int Union Radio Sci, 51- Honors & Awards: Off, Order of the Brit Empire, 46. Mem: Fel Am Phys Soc; Am Astron Soc; fel Inst Elec & Electronics Engrs; Can Asn Physicists; fel Royal Soc Can. Res: Radio and radar research and development; aids to air navigation; radio studies of meteors and upper atmosphere; quartz crystals and velocity of light. Mailing Add: 1889 Fairmeadow Cres Ottawa ON Can

MCKINLEY, HARRY R, b Gaylord, Mich, Jan 8, 32; m 57; c 3. OPTICS. Educ: Univ Rochester, BS, 59. Prof Exp: Optical designer, Barnes Eng Co, Conn, 59-62; staff engr, Kollmorgen Corp, 62-67; PRES, McKINLEY OPTICS, INC, 67- Mem: Optical Soc Am. Res: Optical design of ultraviolet and infrared instruments; spectrographic systems, especially spaceborne; applied optics. Mailing Add: McKinley Optics Inc College Hwy Southampton MA 01073

MCKINLEY, JAMES ERNEST, b Scotch Hill, Pa, Sept 22, 23; m 46; c 4. MATHEMATICS. Educ: Clarion State Col, BS, 49; Pa State Univ, EdM, 53; Univ Pittsburgh, EdD, 60. Prof Exp: Teacher math, Tyrone Area Schs, Pa, 49-53; supvr & teacher math, Indiana Univ, Pa, 53-63, chmn dept, 63-67; dean acad affairs, Edinboro State Col, 67-69, vpres acad affairs, 69-73; interim pres, West Chester State Col, 73-74; VPRES ACAD AFFAIRS, EDINBORO STATE COL, 74- Mem: Math Asn Am. Res: Experimental research in teaching statistics in public schools; elements of research; mathematical education. Mailing Add: Off of the Vpres Edinboro State Col Edinboro PA 16412

MCKINLEY, JOHN DOUGLAS, JR, physical chemistry, see 12th edition

MCKINLEY, JOHN MCKEEN, b Wichita, Kans, Feb 1, 30; m 53; c 3. THEORETICAL PHYSICS. Educ: Univ Kans, BS, 51; Univ Ill, PhD(physics), 62. Prof Exp: Asst prof physics, Kans State Univ, 60-66; assoc prof, 66-71, PROF PHYSICS, OAKLAND UNIV, 71- Mem: Am Phys Soc; AAAS; Am Asn Physics Teachers. Res: Theoretical nuclear physics. Mailing Add: Dept of Physics Oakland Univ Rochester MI 48063

MCKINLEY, RAYMOND EARL, b New Lebanon, Pa, Nov 9, 22; m 49; c 3. VETERINARY MEDICINE, ANIMAL NUTRITION. Educ: Pa State Univ, BS, 51; Univ Pa, VMD, 51. Prof Exp: Farm laborer, 36-41; supvr dairy herd improv, 41-43; instr vet sci, Pa State Univ, 51-52; pvt pract vet med, Pa, 52-62; vet, Div Vet New Drugs, Bur Vet Med, Dept Health, Educ & Welfare, 62-67; asst dir, Drug Regulatory Affairs Dept, 67-75, ASST DIR, ANIMAL HEALTH RES DEPT, HOFFMANN-LA ROCHE, INC, 75- Mem: AAAS; Am Vet Med Asn; Animal Health Inst; Am Soc Animal Sci. Res: Animal management; therapeutic and diagnostic devices; regulatory medicine. Mailing Add: Hoffmann-La Roche Inc 340 Kingsland St Nutley NJ 07110

MCKINLEY, WILLIAM ALBERT, b Dallas, Tex, Aug 23, 17; m 40; c 2. PHYSICS. Educ: Univ Tex, BA, 39; Mass Inst Technol, PhD(physics), 47. Prof Exp: Mem staff, Radiation Lab, Mass Inst Technol, 44-46, res assoc, Instrumentation Lab, 46-47; from asst prof to assoc prof, 47-54, PROF PHYSICS, RENSSELAER POLYTECH INST, 54- Mem: AAAS; Am Phys Soc; Am Asn Physics Teachers. Res: Quantum field theory; theory of atomic and nuclear collissions. Mailing Add: Dept of Physics Rensselaer Polytech Inst Troy NY 12181

MCKINNELL, ROBERT GILMORE, b Springfield, Mo, Aug 9, 26; m 64; c 3. DEVELOPMENTAL BIOLOGY. Educ: Univ Mo, AB, 48; Drury Col, BS, 49; Univ Minn, PhD(zool), 59. Prof Exp: Res assoc embryol, Inst Cancer Res, 58-61; asst prof zool, Tulane Univ, 61-65, assoc prof biol, 65-69, prof, 69-70; PROF ZOOL, UNIV MINN, MINNEAPOLIS, 70- Concurrent Pos: Instr, Univ Minn, 58; sr sci fel, NATO, St Andrews Univ, Scotland, 74; mem, Adv Coun, Inst Lab Animal Resources, 74- Mem: AAAS; Soc Develop Biol; Am Soc Zoologists; Am Asn Cancer Res; Int Soc Differentiation (secretariat, 75-). Res: Transplantation of nuclei from normal and neoplastic anuran cells; use of mutant genes as nuclear markers; viral oncogenesis and epidemiology of Lucke renal adenocarcinoma. Mailing Add: Dept of Zool Univ of Minn Minneapolis MN 55455

MCKINNEY, ALBERT WILLIAM, III, mathematics, see 12th edition

MCKINNEY, ALFRED LEE, b Houston, Tex, Aug 19, 37; m 60; c 1. MATHEMATICAL ANALYSIS, NUMERICAL ANALYSIS. Educ: La Tech Univ, BS, 59, MS, 61; Univ Okla, PhD(math), 72. Prof Exp: Res mathematician, United Gas Res Lab, 61-65; chmn math dept, Okla Col Lib Arts, 68-72, chmn math & sci div, 72-74; ASSOC PROF MATH, LA STATE UNIV, SHREVEPORT, 74- Mem: Asn Comput Mach; Data Processing Mgt Asn; Math Asn Am. Res: Numerical approximations, particularly computer-oriented approaches to solutions of calculus of variations or control problems. Mailing Add: Dept of Math La State Univ Shreveport LA 71105

MCKINNEY, AUBREY REW, physical chemistry, see 12th edition

MCKINNEY, CHARLES DANA, JR, b Chattanooga, Tenn, Mar, 30, 20; m 47; c 1. PHYSICAL CHEMISTRY. Educ: Univ Chattanooga, BS, 41; Ill Inst Technol, PhD(phys chem), 50. Prof Exp: Lab shift supvr anal chem, Vol Ord Works, 44-45; asst instr, Univ Pa, 45-47; asst instr phys chem, Ill Inst Technol, 47-50; res chemist, Hercules Powder Co, 50-52, res supvr, 52-55, dir res, Allegany Ballistics Lab, Md, 55-58; proj dir, Aeroprojs, Inc & Technidyne, Inc, 58-70; SR RES CHEMIST, HERCULES, INC, 70- Mem: Am Chem Soc; Sigma Xi. Res: Polymer chemistry; explosives; reaction kinetics; rapid reactions; solid and liquid rocket propellants. Mailing Add: 705 Halstead Rd Wilmington DE 19803

MCKINNEY, CHARLES NASH, physical chemistry, see 12th edition

MCKINNEY, CHARLES ORAN, b New Braunfels, Tex, Aug 10, 42; m 65; c 2. POPULATION BIOLOGY. Educ: Univ Tex, Austin, BA, 64; Tex Tech Col, MS, 66; Univ Mich, Ann Arbor, PhD(zool), 69. Prof Exp: NSF-Univ sci develop grant & fac assoc zool, Univ Tex, Austin, 69-71; asst prof pop biol & ecol, Univ Dayton, 71-73; ASST PROF LIFE SCI, UNIV TEX, PERMIAN BASIN, 73- Honors & Awards: Wilke's Award, Southwestern Asn Naturalists, 66. Mem: AAAS; Am Soc Ichthyologists & Herpetologists; Soc Study Evolution; Soc Syst Zool. Res: Biochemical variation of natural populations of vertebrates. Mailing Add: Fac of Life Sci Univ of Tex Permian Basin Odessa TX 79762

MCKINNEY, CHESTER MEEK, b Cooper, Tex, Jan 29, 20; m 48; c 2. PHYSICS. Educ: ETex State Teachers Col, BS, 41; Univ Tex, MA, 47, PhD(physics), 50. Prof Exp: Res physicist, 45-65, DIR APPL RES LAB, UNIV TEX, 65- Concurrent Pos: Assoc prof, Tex Tech Col, 50-53; mem, Lab Adv Bd Naval Ships, Naval Res Adv Comt, 75-77; mem, US Navy Underwater Sound Adv Group, 62-64 & 75-77, chmn, 71-73; mem, Nat Res Coun Mine Adv Comt. Mem: Fel Acoust Soc Am; Am Inst Elec & Electronics Engrs; Brit Inst Acoust. Res: Underwater acoustics; electronics; microwaves; dielectric waveguides and antennae. Mailing Add: 4305 Farhills Dr Austin TX 78731

MCKINNEY, DAVID SCROGGS, b Atwood, Pa, Aug 16, 02. PHYSICAL CHEMISTRY. Educ: Carnegie Inst Technol, BS, 23, DSc(phys chem), 38. Prof Exp: Analyst, NJ Zinc Co, 23; from analyst to chief chemist, Duquesne Light Co, Pa, 24-34; asst, 34-35, instr, 36-38, from asst prof to prof, 39-71, EMER PROF CHEM, CARNEGIE-MELLON UNIV, 71- Concurrent Pos: Sect chief, Anal Sect, Chem Div, Metall Lab, Univ Chicago, 44; assoc head dept chem, Carnegie-Mellon Univ, 63-68. Mem: Am Chem Soc; NY Acad Sci. Res: Infrared absorption spectra; thermodynamics; equilibria in water solution; conductance; corrosion, industrial water treatment. Mailing Add: 114 Ridge Rd Pittsburgh PA 15237

MCKINNEY, EARL H, b Wilkinsburgh, Pa, May 24, 29; m 52; c 3. NUMERICAL ANALYSIS. Educ: Washington & Jefferson Col, AB, 51; Univ Pittsburgh, MS, 56, PhD(math), 61. Prof Exp: Instr math, Univ Pittsburgh, 55-59; asst prof, Northern Ill Univ, 59-62; PROF MATH & HEAD DEPT, BALL STATE UNIV, 62- Concurrent Pos: NSF res grant appl math, Argonne Nat Lab, 68-69. Mem: Am Math Soc; Math Asn Am; Soc Indust & Appl Math. Res: Analysis; applied mathematics; numerical analysis—interpolation and numerical integration. Mailing Add: Dept of Math Ball State Univ Muncie IN 47306

MCKINNEY, FRANK, b Ballymena, Northern Ireland, Oct 23, 28; m 63. ZOOLOGY. Educ: Oxford Univ, BA, 49; Bristol Univ, PhD(zool), 53. Prof Exp: Nuffield Found res grant, 53-54; asst dir, Delta Waterfowl Res Sta, Man, Can, 54-62; res assoc waterfowl res, Mus Natural Hist, 63-66, asst prof animal behav, Univ, 65-66, assoc prof ethology, 66-70, ETHOLOGIST, MUS NATURAL HIST, UNIV MINN, MINNEAPOLIS, 66-, PROF ETHOLOGY, UNIV, 70- Concurrent Pos: Ed, Auk Recent Lit, Am Ornithologists Union, 55-62; Wildlife Mgt Inst grants, 58 & 63. Mem: Am Ornithologists Union; Cooper Ornith Soc; Ecol Soc Am; Animal Behav Soc. Res: Behavior of waterfowl, especially comfort movements, displays, pair formation, territory, nesting and migration; communication and the evolution and survival value of behavior patterns in birds. Mailing Add: Mus of Natural Hist Univ of Minn Minneapolis MN 55455

MCKINNEY, FRANK KENNETH, b Birmingham, Ala, Apr 13, 43; m 64; c 4. INVERTEBRATE PALEONTOLOGY, BIOSTRATIGRAPHY. Educ: Old Dom Col, BS, 64; Univ NC, Chapel Hill, MS, 67, PhD(paleont), 70. Prof Exp: ASST PROF GEOL, APPALACHIAN STATE UNIV, 68- Concurrent Pos: Fel, Smithsonian Inst, 72-73. Mem: Paleont Soc; Brit Paleont Asn; Soc Econ Paleont & Mineral; Int Bryozool Asn. Res: Mississippian and Ordovician bryozoans, particularly Trepostomata. Mailing Add: Dept of Geol Appalachian State Univ Boone NC 28607

MCKINNEY, GORDON R, b Indianapolis, Ind, Oct 14, 23; m 47; c 1. ENDOCRINOLOGY, PHARMACOLOGY. Educ: DePauw Univ, AB, 46; Univ Notre Dame, MS, 48; Duke Univ, PhD(physiol), 51. Prof Exp: Lectr pharmacol, Duke Univ, 52-53; from asst prof to assoc prof, Sch Med, WVa Univ, 53-59; res pharmacologist, 59-61, sr res fel, 61-68, dir pharmacol, 68-75, DIR BIOL RES, MEAD JOHNSON & CO, 75- Concurrent Pos: Am Cancer Soc res fel med, Duke Univ, 51-53; Lederle med fac award, WVa Univ, 55-57. Mem: AAAS; Am Soc Pharmacol & Exp Therapeut; Soc Study Reproduction; Soc Exp Biol & Med; Endocrine Soc. Res: Adrenergic, biochemical and endocrine pharmacology; reproduction. Mailing Add: Biol Res Dept Mead Johnson Res Ctr Evansville IN 47721

MCKINNEY, JAMES DAVID, b Gainesville, Ga, Dec 28, 41; m 70; c 2. ENVIRONMENTAL CHEMISTRY. Educ: Univ Ga, BS, 63, PhD(org chem), 68. Prof Exp: Pub health scientist, Pesticide Toxicol Lab, Food & Drug Admin, Ga, 67-69; res scientist environ chem, 69-74, HEAD CHEM SECT, ENVIRON BIOL & CHEM BR, NAT INST ENVIRON HEALTH SCI, 74- Mem: Am Chem Soc. Res: Interactions of diverse environmental agents in biological systems on a molecular level; degradation of these agents by nonbiological routes. Mailing Add: Environ Biol & Chem Br Nat Inst of Environ Health Sci Research Triangle Park NC 27709

MCKINNEY, JAMES T, b Detroit, Mich, May 28, 38; m 61; c 2. SURFACE PHYSICS. Educ: Univ Detroit, BS, 60; Univ Wis, MS, 62, PhD(physics), 66. Prof Exp: Scientist, Fundamental Res Lab, US Steel Res Ctr, 66-71; SR PHYSICIST, 3M CO, 72- Mem: AAAS; Am Phys Soc. Res: Ion scattering; chemistry of surfaces:

interaction of ions with solid surfaces; surface analytical instrumentation. Mailing Add: Bldg 209 BS 3M Co 3M Ctr St Paul MN 55101

MCKINNEY, JOHN EDWARD, b Altoona, Pa, Apr 6, 25; m 58. THERMODYNAMICS, RHEOLOGY. Educ: Pa State Univ, BS, 50. Prof Exp: PHYSICIST, POLYMERS DIV, NAT BUR STANDARDS, 50- Concurrent Pos: Guest worker, Nat Physics Lab, Teddington, Eng, 64. Mem: Rheology Soc; Acoust Soc Am. Res: Experimental rheology, acoustics, thermodynamics and dynamic mechanical properties of polymers and glasses; related theoretical development of liquid and glassy states; development of related instrumentation. Mailing Add: Polymers Div Nat Bur of Standards Washington DC 20234

MCKINNEY, LEONARD LAURENCE, b Siloam Springs, Ark, May 28, 08; m 41; c 3. FOOD CHEMISTRY. Educ: Univ Ark, BS, 31; Bradley Univ, MS, 50. Prof Exp: Chemist-operator, Texaco Salt Prod Co, 31-34; seafood inspector, Food & Drug Admin, 35-37; res chemist, Soybean Indust Prod Lab, USDA, 37-42, supvr, Indust Protein Unit, North Utilization Res & Develop Div, 47-57, asst dir div, Agr Res Serv, 57-68, asst dir, Richard B Russell Agr Res Ctr, 68-72, asst area dir, Athens, Ga area, 72-73, ASST AREA EMER DIR, ATHENS, GA AREA, AGR RES SERV, USDA, 73- Concurrent Pos: Lectr, Foods for Peace, Israel, 62 & India, 72; adj res assoc, Univ Ga, 70-73. Honors & Awards: Cert Merit, USDA, 57, Superior Serv Award, 59. Mem: Fel AAAS; Am Chem Soc; Am Oil Chem Soc; Inst Food Technologists. Res: Chemistry of proteins, amino acids, carbohydrates, vitamins and lipids from oilseeds, cereal grains, and their uses in foods, feeds, pharmaceuticals and industrial products. Mailing Add: 210 Ponderosa Dr Athens GA 30601

MCKINNEY, MAX TERRAL, b Esto, Fla, Sept 25, 35; m 53; c 2. MATHEMATICS. Educ: Troy State Univ, BS, 56; Auburn Univ, MEd, 62, DEd, 64. Prof Exp: Proj mathematician, Vitro Corp Am, 56-57; high sch teacher, Ga, 57-61; asst prof, 64-66, ASSOC PROF MATH & CHMN DEPT, GA SOUTHWESTERN COL, 66- Res: Statistics and algebraic fields. Mailing Add: Dept of Math Ga Southwestern Col Americus GA 31709

MCKINNEY, MYRON WILLIAM, b Winslow, Ind, Feb 19, 17; m 41; c 2. PHARMACEUTICAL CHEMISTRY. Educ: Purdue Univ, BS, 39, PhD(pharmaceut chem), 51; Western Reserve Univ, MS, 41. Prof Exp: Assoc pharmaceut chemist, Pharmaceut Develop Dept, Eli Lilly & Co, 41-51, chief, Liquid Pilot Plant, 51-62, mgr, Pharmaceut Develop Depts, 62-65, staff asst to dir, Pharmaceut Develop Div, 66-72; RETIRED. Mem: Am Chem Soc; Am Pharmaceut Asn. Res: Synthesis of medicinal chemicals; development of pharmaceutical dosage forms; pharmaceutical product development management. Mailing Add: 4425 Brown Rd Indianapolis IN 46226

MCKINNEY, PAUL CAYLOR, b Otterbein, Ind, Aug 21, 30. PHYSICAL CHEMISTRY. Educ: Wabash Col, AB, 52; Northwestern Univ, PhD, 58. Prof Exp: Asst prof, 58-64, ASSOC PROF CHEM, WABASH COL, 64- Res: Molecular mechanics. Mailing Add: Dept of Chem Wabash Col Crawfordsville IN 47933

MCKINNEY, PETER, b Baltimore, Md, Nov 2, 34; m 73; c 2. PLASTIC SURGERY. Educ: Harvard Univ, AB, 56; McGill Univ, MD, CM, 60. Prof Exp: Intern, Montreal Gen Hosp & resident, New York City Hosp, 60-61; asst, Bellevue-Jacobi Hosp & teacher gen surg, Albert Einstein Col Med, 61-64; resident plastic surg, New York Hosp, Med Ctr, Cornell Univ, 64-67, chief resident, 66-67; instr & assoc surg, 67-70, asst prof, 70-74, ASSOC PROF CLIN SURG, SCH MED, NORTHWESTERN UNIV, CHICAGO, 74- Concurrent Pos: Instr surg, Sch Med, Cornell Univ, 64-67. Mem: Plastic Surg Res Coun; Am Soc Plastic & Reconstruct Surgeons; Am Asn Plastic Surgeons; Am Col Surgeons. Res: Control of cartilage bend in rabbit ears which would have clinical application in shaping of cartilaginous structures of the face. Mailing Add: Northwestern Univ 251 E Chicago Chicago IL 60611

MCKINNEY, PETER STARKWEATHER, electrochemistry, analytical chemistry, see 12th edition

MCKINNEY, RALPH VINCENT, JR, b Columbus, Ohio, Jan 9, 33; m 55; c 4. PATHOLOGY, CELL BIOLOGY. Educ: Bowling Green State Univ, BS, 54; Ohio State Univ, DDS, 61; Univ Rochester, PhD(path), 71. Prof Exp: Asst instr dent hyg, Ohio State Univ, 60-61; clin asst prof oper dent, Case Western Reserve Univ, 61-65; PROF ORAL PATH & ORAL BIOL, GRAD FAC, MED COL GA, 70- Concurrent Pos: NIH fel path, Univ Rochester, 65-70; NIH grants, 72-74 & 72-75; Nat Inst Dent Res contract, 73-76; pvt dent pract, Ohio, 61-65; mem dent staff, Talmadge Mem Hosp, Augusta, 70-; oral path diag serv, Med Col Ga, 73-; pres dent found, 74-76. Mem: AAAS; Int Acad Path; Am Acad Oral Path; Am Soc Cell Biol; Int Asn Dent Res. Res: Wound healing; microcirculation, emphasis of the biochemical, histochemical and morphological fine structure of capillary basement membrane; peripheral interests, inflammation, connective tissue. Mailing Add: Dept of Oral Path Med Col of Ga Augusta GA 30902

MCKINNEY, RICHARD LEROY, b Altoona, Pa, May 23, 28; m 56; c 3. MATHEMATICS. Educ: Syracuse Univ, AB, 51, MA, 52; Univ Wash, PhD(math), 58. Prof Exp: Asst, Univ Wash, 53-58; from instr to asst prof math, Univ Calif, Riverside, 58-62; asst prof, 62-67, ASSOC PROF MATH, UNIV ALTA, 67- Concurrent Pos: Hon res assoc, Univ Col, Univ London, 68-69. Mem: Am Math Soc; Can Math Cong. Res: Linear spaces; convex sets; functional analysis; topology. Mailing Add: Dept of Math Univ of Alta Edmonton AB Can

MCKINNEY, ROBERT WARREN, virology, epidemiology, see 12th edition

MCKINNEY, ROBERT WESLEY, b East St Louis, Ill, Dec 11, 31; m 61; c 2. ANALYTICAL CHEMISTRY. Educ: Southern Ill Univ, BA, 53; Univ Kans, PhD(anal chem), 57. Prof Exp: Anal chemist, Celanese Corp Am, 57-60; anal chemist, 60-64, group leader anal chem, 64-72, MGR ANAL CHEM, W R GRACE & CO, 72- Mem: Am Chem Soc. Res: Instrumental and wet analytical chemistry; gas chromatography. Mailing Add: W R Grace & Co Washington Res Ctr Columbia MD 21044

MCKINNEY, ROGER MINOR, b Deerbrook, Wis, May 31, 26; m 52; c 3. ORGANIC CHEMISTRY, IMMUNOCHEMISTRY. Educ: Wis State Col, River Falls, BS, 50; St Louis Univ, MS, 56, PhD(chem), 58. Prof Exp: Chemist, Lambert Pharmacal Co, Mo, 50-52; chemist, Universal Match Corp, 52-55; res chemist, USPHS, Ga, 58-66, Aedes Aegypti Eradication Prog, 66-68, Tech Develop Labs, 68-71, RES CHEMIST, TECH DEVELOP LABS, CTR DIS CONTROL, USPHS, 71- Mem: AAAS; Am Chem Soc. Res: Synthesis of radioactive isotope labeled insecticides; technical aspects of immunofluorescent staining; basic immunochemistry studies. Mailing Add: 4872 Cambridge Dr Dunwoody GA 30358

MCKINNEY, TED MEREDITH, b Huntsville, Ala, Apr 18, 38. ANALYTICAL CHEMISTRY. Educ: Harvard Univ, AB, 60; Cornell Univ, PhD(chem), 65. Prof Exp: Res assoc chem, Cornell Univ, 64-66; asst prof, Univ Calif, Riverside, 66-71; INDEPENDENT WRITER & ED, 71- Concurrent Pos: Consult, Beckman Instruments, Inc, Calif, 69. Mem: Am Chem Soc. Res: Magnetic resonance; electroanalytical chemistry; optical spectroscopy. Mailing Add: 5156 Colina Way Riverside CA 92507

MCKINNEY, WILLIAM JAN, b York, Pa, July 18, 43. ANALYTICAL CHEMISTRY. Educ: Knox Col, Ill, BA, 65; Mich State Univ, PhD(anal chem), 69. Prof Exp: CHEMIST, BIOL SCI RES CTR, SHELL DEVELOP CO, 69- Mem: Am Chem Soc. Res: Analysis of biologically active compounds. Mailing Add: Shell Develop Co Box 4248 Modesto CA 95252

MCKINNEY, WILLIAM MARK, b Spring Valley, NY, Dec 26, 23; m 51; c 1. PLANETARY SCIENCES, PHYSICAL GEOGRAPHY. Educ: New Sch Social Res, BA, 48; Univ Fla, PhD(geog), 58. Prof Exp: Consult pub health, Ga Dept Pub Health, 53-58; from instr to asst prof geog, Southern Ore Col, 58-63; from asst prof to assoc prof, 63-68, PROF GEOG, UNIV WIS-STEVENS POINT, 68- Mem: Asn Am Geogrs; Nat Coun Geog Educ. Res: Analysis and mapping of atmospheric phenomena of Mars as photographed in various wavelengths of light; study and development of instrumentation for demonstrating principles of astronomical geography. Mailing Add: Dept of Geog & Geol Univ of Wis Stevens Point WI 54481

MCKINNEY, WILLIAM MARKLEY, b Roanoke, Va, June 6, 30; m 52; c 3. NEUROLOGY. Educ: Univ NC, Chapel Hill, BA, 51; Univ Va, MD, 59. Prof Exp: From instr to asst prof neurol, 63-70, ASSOC PROF NEUROL, BOWMAN GRAY SCH MED, 70-, RES ASSOC RADIOL, 67- Concurrent Pos: Dir sonic lab, Bowman Gray Sch Med, 63-, mem subcomt on stroke, Regional Med Prog, 67-; consult, Vet Admin Hosp, Salisbury, NC. Mem: Am Acad Neurol; Am Fedn Clin Res; Asn Res Nerv & Ment Dis; am Inst Ultrasound in Med (secy, 67, pres, 74-76). Res: Diagnostic ultrasound in medicine; the application of ultrasound to medicine; cerebrovascular disease; crystallography; urinary lithiasis. Mailing Add: Dept of Neurol Bowman Gray Sch of Med Winston-Salem NC 27103

MCKINNIS, CHARLES LESLIE, b Cape Girardeau, Mo, July 10, 23; m 44; c 1. CHEMISTRY. Educ: Southeast Mo State Col, BS, 46; Univ Mo, BS, 47, MS, 48; Ohio State Univ, PhD(glass chem), 54. Prof Exp: Res engr, Res Lab, Pittsburgh Plate Glass Co, 48-50; res assoc, Res Found, Ohio State Univ, 50-54; res chemist, Midwest Res Inst, 54-55; sr res scientist, Owens Corning Fiberglas Corp, 55-75, RES ASSOC SUPPORT SCI & TECHNOL, GLASS RES & DEVELOP, TECH CTR, OWENS CORNING FIBERGLAS CORP, 75- Concurrent Pos: Chmn glass div, Am Ceramic Soc, 74. Mem: Am Ceramic Soc; Nat Inst Ceramic Engrs; Sigma Xi. Res: Glass chemistry; glass structure and physical properties; heat transfer. Mailing Add: Owens-Corning Fiberglas Corp Tech Ctr Box 415 Granville OH 43023

MCKINNIS, RONALD BISHOP, b Summerville, Ore, July 3, 03; m 39; c 2. CHEMISTRY, FOOD TECHNOLOGY. Educ: Willamette Univ, BA, 26; State Col Wash, MS, 28; Univ Pittsburgh, PhD(chem), 30. Prof Exp: Asst, Univ Pittsburgh, 27-29; indust fel, Mellon Inst Technol, 29-37; consult, 37-38; mgr & secy, Sunshine Foods, Inc, Fla, 38-42; owner, McKinnis Foods, 42-45; mgr res & develop, Am Mach Corp, 45-47; mgr tech serv, Votator Div, Girdler Corp, 47-50 & Brown Citrus Mach Corp, 50-71; VPRES TECH SERV, AUTOMATIC MACH CORP, 71- Mem: AAAS; Am Chem Soc; Instrument Soc Am; Inst Food Technologists. Res: Technical service and research citrus juice extraction, preparation and processing. Mailing Add: Box 1888 Winter Haven FL 33880

MCKINNON, DAVID M, b Scotland, Aug 11, 38; m 63; c 2. ORGANIC CHEMISTRY. Educ: Univ Edinburgh, BSc, 60, PhD(chem), 63. Prof Exp: Fel chem, Dalhousie Univ, 63-65; asst prof, 65-69, ASSOC PROF CHEM, UNIV MAN, 69- Mem: The Chem Soc; Can Inst Chemists. Res: Chemistry of heterocyclic sulphur and nitrogen compounds. Mailing Add: Dept of Chem Univ of Man Winnipeg MB Can

MCKINNON, ELIZABETH L, physiology, biology, see 12th edition

MCKINNON, JOE WALLACE, science education, see 12th edition

MCKINNON, WILLIAM MITCHELL PATRICK, b Houston, Tex, Mar 17, 24; m 53; c 6. SURGERY. Educ: St Edward's Univ, BS, 49; Baylor Univ, MD, 52. Prof Exp: From asst prof to assoc prof surg, NY Med Col, 61-67; SURGEON IN CHARGE SURG RES LAB, OCHSNER FOUND & STAFF SURGEON, OCHSNER CLIN, 68- Concurrent Pos: Nat Heart Inst fel surg res, Maimonides Hosp Brooklyn, NY, 58-59; consult, Vet Admin Hosp, Lyons, NJ, 62-; clin assoc prof, Tulane Univ & vis surgeon, Tulane Surg Div, Charity Hosp, 68- Mem: Fel Am Col Surg; Soc Surg Alimentary Tract; AMA. Res: Clinical abdominal surgery, especially acute bleeding from gastrointestinal tract and abdominal trauma. Mailing Add: Dept of Surg Ochsner Clin 1514 Jefferson Highway New Orleans LA 70121

MCKINSEY, RICHARD DAVIS, b New York, NY, May 20, 21; m 44; c 3. BOTANY. Educ: Ill Inst Technol, BS, 48; Stanford Univ, MA, 53, PhD(biol), 58. Prof Exp: From instr to asst prof, 57-65, ASSOC PROF BIOL, FAC ARTS & SCI, UNIV VA, 65- Res: Intermediate metabolism of fungi. Mailing Add: Dept of Biol Fac of Arts & Sci Univ of Va Charlottesville VA 22903

MCKINSTRY, DONALD MICHAEL, b Lancaster, Pa, June 10, 39; m 66; c 2. ANIMAL PHYSIOLOGY, HERPETOLOGY. Educ: Univ Md, BS, 64, MS, 65, PhD(dairy sci), 71. Prof Exp: Lab technician dairy sci, Univ Md, 67-70; asst prof biol, 70-75, ASSOC PROF BIOL, BEHREND COL, PA STATE UNIV, 75- Res: Bacteria of reptiles; reptile distribution surveys; drug action in mammals as influenced by diet. Mailing Add: Behrend Col Pa State Univ Station Rd Erie PA 16510

MCKINSTRY, DONALD W, b Middletown, Pa, Dec 20, 10; m 34; c 4. MICROBIOLOGY, CHEMOTHERAPY. Educ: Franklin & Marshall Col, BS, 32; Pa State Univ, MS, 37, PhD(bact), 39. Prof Exp: Microbiologist, Bloch Bros Tobacco Co, 39-40; instr biochem, WVa Univ, 41-42; res assoc chemother, Biochem Res Found, 42-44; microbiol, Lederle Labs, Am Cyanamid Co, 44-45; assoc prof, Pa State Univ, 45-48; scientist virol, NIH, 48-63; asst prof microbiol, Col Med, Howard Univ, 69-73; STAFF SCIENTIST, SCI DIV, TRACOR JITCO, INC, 73- Concurrent Pos: Res fel chemother, Upjohn Co, 40-41. Res: Chemotherapy and immunology of experimental virus infections and cancer; interferon. Mailing Add: Sci Div Tracor Jitco Inc Rockville MD 20850

MCKINSTRY, DORIS NAOMI, b McVeytown, Pa, Sept 8, 36. CLINICAL PHARMACOLOGY. Educ: Pa State Univ, BS, 58; Univ Pa, PhD(pharmacol), 65. Prof Exp: Res assoc pharmacol, Merck Sharp & Dohme Res Labs, 58-61; sr scientist, McNeil Labs, Inc, 66-69; sr res toxicologist, Merck Sharp & Dohme Res Labs, 69-70; sr res investr preclin res admin, 71-75, ASST CLIN PHARMACOL DIR, SQUIBB INST MED RES, 75- Concurrent Pos: Fel pharmacol, Univ Pa, 65-66. Mem: Am Soc Pharmacol & Exp Therapeut; Am Soc Clin Pharmacol & Therapeut; AAAS. Res: Autonomic-cardiovascular pharmacology and physiology; radiocontrast agents; anti-inflammatory agents. Mailing Add: Squibb Inst for Med Res PO Box 4000 Princeton NJ 08540

MCKINSTRY, HERBERT ALDEN, b Rochester, NY, Apr 22, 25; m 45; c 4. SOLID STATE PHYSICS. Educ: Alfred Univ, BS, 47; Pa State Univ, MS, 50, PhD, 60. Prof Exp: Res asst, 47-60, res assoc, 60-64, asst prof, 64-69, ASSOC PROF SOLID STATE SCI, PA STATE UNIV, UNIVERSITY PARK, 69- Res: X-ray diffraction; x-ray fluorescence analysis; ceramics. Mailing Add: 144 Mat Res Lab Pa State Univ University Park PA 16802

MCKINZIE, HOWARD LEE, b Olustee, Okla, Apr 16, 41; m 60; c 3. SURFACE CHEMISTRY, SOLID STATE CHEMISTRY. Educ: Cent State Col, Okla, BS, 63; Ariz State Univ, PhD(phys chem), 68. Prof Exp: Res assoc chem, 67-69, asst prof div eng, 69-73, ADJ ASSOC PROF CHEM, BROWN UNIV, 73-; MEM TECH STAFF, GTE LABS, 73- Honors & Awards: Ralph Tec tor Award, Soc Automotive Engrs, 71. Mem: Soc Automotive Engrs; AAAS; Am Chem Soc. Res: Photo-assisted catalysis: photocatalysis; semiconductor and insulating catalysts; organometallic catalysts; catalytic mechanisms. Mailing Add: GTE Labs 40 Sylvan Rd Waltham MA 02154

MCKIRAHAN, RICHARD DUNCAN, b San Francisco, Calif, May 2, 15; m 44; c 3. ORGANIC CHEMISTRY, ENVIRONMENTAL MANAGEMENT. Educ: Univ Calif, AB, 36. Prof Exp: Chemist, Canco Div, Am Can Co, 36-49, supvr tech serv, 50-60, mgr, 60-61, asst to vpres res & develop, Am Can Co, 61-62, supvr tech serv, 62-66, mgr nonmetallic mat sect, 66-68, supvr polymers group, 68-74, MGR AIR QUAL, ENVIRON AFFAIRS, AM CAN CO, 74- Mem: Inst Food Technologists; Chem Coaters Asn. Mailing Add: American Lane Greenwich CT 06830

MCKNIGHT, JAMES DAWSON, JR, b Birmingham, Ala, Aug 3, 28; m 47; c 4. MATHEMATICS. Educ: Birmingham-Southern Col, AB, 47; Purdue Univ, MS, 50, PhD(math), 53. Prof Exp: Instr math, Purdue Univ, 48-50, asst, Statist Lab, 50-52, res assoc, 52-53; sr aerophys engr, Gen Dynamics/Convair, 53-55, proj aerophys engr, 55-56; assoc prof, Univ SC, 56-57; res scientist, Missiles & Space Div, Lockheed Aircraft Corp, 57-60; sr mathematician, Comput Lab, Gen Elec Co, 60-64; assoc prof, 64-67, PROF MATH, UNIV MIAMI, 67- Mem: Am Math Soc; Math Asn Am. Res: Topology; algebra; functional analysis. Mailing Add: Dept of Math Univ of Miami Coral Gables FL 33124

MCKNIGHT, JAMES POPE, b Arlington, Tenn, Sept 19, 21; m 49; c 4. DENTISTRY. Educ: Memphis State Univ, BS, 48; Univ Tenn, DDS, 51, cert, 52; Ind Univ, MSD, 64. Prof Exp: Pvt pract, 52-56; from instr to assoc prof, 56-69, PROF PEDODONT & CHMN DEPT, COL DENT, UNIV TENN, MEMPHIS, 69- Mem: Am Dent Asn; Am Soc Dent for Children. Res: Treatment of the dental pulp; dental care of handicapped children. Mailing Add: Dept of Pedodontics Univ of Tenn Col of Dent Memphis TN 38103

MCKNIGHT, JOHN LACY, b Monroe, Mich, Sept 13, 31; m 64; c 1. THEORETICAL PHYSICS, HISTORY OF SCIENCE. Educ: Univ Mich, AB, 53; Yale Univ, MS, 54, PhD(physics), 57. Prof Exp: Asst prof, 57-59, assoc prof, 59-68, PROF PHYSICS, COL WILLIAM & MARY, 68- Mem: AAAS; Am Phys Soc; Am Asn Physics Teachers; Philos Sci Asn; Hist Sci Soc. Res: Logical foundations of quantum mechanics, particularly problems related to measuring interactions; history of scientific ideas. Mailing Add: Dept of Physics Col of William & Mary Williamsburg VA 23185

MCKNIGHT, JOSEPH SIDNEY, forestry, see 12th edition

MCKNIGHT, KENT HOWELL, mycology, see 12th edition

MCKNIGHT, LEE GRAVES, b Washington, DC, Sept 7, 33; m 55; c 2. CHEMICAL PHYSICS. Educ: Va Mil Inst, BS, 55; Univ Mich, MS & PhD(chem), 61. Prof Exp: NATO fel, Univ Col, Univ London, 61-62; lectr chem, Univ Mich, 63; mem tech staff physics, Bell Labs, Whippany, NJ, 63-73, MEM TECH STAFF, BELL LABS, MURRAY HILL, NJ, 73- Mem: Am Phys Soc. Res: Ion-molecule interactions; carbon arcs. Mailing Add: Bell Labs Murray Hill NJ 07974

MCKNIGHT, MELVIN EDWARD, b Barre, Vt, Dec 10, 35; m 56; c 3. ENTOMOLOGY. Educ: Univ Vt, BS, 56; Univ Nebr, MS, 58; Colo State Univ, PhD(zool), 67. Prof Exp: Entomologist, Entom Res Div, Agr Res Serv, USDA, 57-59, Rocky Mountain Forest & Range Exp Sta, US Forest Serv, Colo, 59-68, Shelterbelt Lab, 68-73, STAFF ENTOMOLOGIST, US FOREST SERV, USDA, 73- Mem: Entom Soc Am. Res: Biology, ecology, and control of forest insects; population dynamics and biological evaluations to evaluate the potential for damage and the need for control. Mailing Add: US Forest Serv US Dept of Agr Washington DC 20250

MCKNIGHT, RANDY SHERWOOD, b Los Angeles, Calif, June 18, 43. APPLIED MATHEMATICS. Educ: Univ Calif, BS, 66; Rice Univ, MS, 69, PhD(math sci), 72. Prof Exp: Sci programmer, IBM Corp, 66; RES SCIENTIST GEOPHYS & RESERVOIR MODELING, DENVER RES CTR, MARATHON OIL CO, 71- Mem: Soc Indust & Appl Math; Inst Elec & Electronics Engrs. Res: Application of optimization theory and numerical analysis to direct and inverse problems in exploration geophysics and petroleum engineering; system theory to processing and interpretation of seismic data. Mailing Add: Denver Res Ctr Marathon Oil Co PO Box 269 Littleton CO 80120

MCKNIGHT, ROBERT KELLOGG, b Sendai, Japan, Jan 10, 24; US citizen; m 48; c 2. APPLIED ANTHROPOLOGY. Educ: Miami Univ, BA, 50; Ohio State Univ, MA, 54, PhD(anthrop), 60. Prof Exp: Dist anthropologist, Palau Dist, Trust Territory Pac Islands, 58-63; community develop officer, Off High Comnr, Micronesia, 63-65; assoc prof anthrop, Univ Wis-Milwaukee, 65-66; PROF ANTHROP, CALIF STATE UNIV, HAYWARD, 66- Concurrent Pos: Chmn dept anthrop, Calif State Univ, Hayward, 70-74; mem, Int House, Japan. Mem: Am Anthrop Asn; Soc Appl Anthrop; Am Ethnol Soc; Soc Social Anthrop in Oceania. Res: Pacific area anthropology, especially Micronesia and Japan; modernization, especially non-western ideologies of culture change; history of thought and ethics in anthropology; anthropology of alternative life styles. Mailing Add: Dept of Anthrop Calif State Univ Hayward CA 94542

MCKNIGHT, THEODORE SAMUEL, forest products, see 12th edition

MCKNIGHT, THOMAS JOHN, b Marietta, Ohio, Nov 5, 06; m; c 3. PARASITOLOGY. Educ: Okla Agr & Mech Col, BS, 25; Univ Okla, MS, 47, PhD, 59. Prof Exp: Instr high sch, 29-30; supt schs, 31-41; instr zool, Univ Okla, 46; PROF BIOL, E CENT STATE COL, 47-, CHMN DEPT BIOL, 68- Concurrent Pos: Consult, Environ Survs, Kerr Magee Corp, Okla, 74-75 & US Corps Engrs, 73-74. Mem: AAAS; Am Soc Parasitol; Nat Sci Teachers Asn; fel Royal Soc Health. Res: Bacteriology; animal parasitology and microbiology; taxonomy and physiology of parasites; taxonomy of parasites of reptiles, soil bacteria and Actinomycetes. Mailing Add: 1020 S Stockton Ada OK 74821

MCKNIGHT, TOM, b Dallas, Tex, Oct 8, 28; m 53; c 2. GEOGRAPHY. Educ: Southern Methodist Univ, BA, 49; Univ Colo, MA, 51; Univ Wis, PhD(geog), 55. Prof Exp: Instr exten div, Univ Colo, 50; instr geog, Southern Methodist Univ, 53-55 & Univ Tex, 55-56; from asst prof to assoc prof, 56-67, PROF GEOG, UNIV CALIF, LOS ANGELES, 67- Concurrent Pos: Vis lectr, Univ Mich, 60, Univ Adelaide, 62 & 66 & New England Univ, Australia, 66 & 70. Mem: Asn Am Geogrs; Am Geog Soc; Can Asn Geogrs; Royal Can Geog Soc; Inst Australian Geogrs. Res: Geography of manufacturing; feral livestock; conservation. Mailing Add: Dept of Geog Univ of Calif Los Angeles CA 90024

MCKNIGHT, WILLIAM BALDWIN, b Macon, Ga, July 4, 23; m 55; c 2. LASERS. Educ: Purdue Univ, BS, 50; Oxford Univ, PhD(physics), 68. Prof Exp: Physicist, Navy Underwater Sound Reference Lab, 52-53; test engr, Ord Missile Labs, Redstone Arsenal, 53-56, chief, Infrared Br, 56-58, Electro-Optical Br, 58-62, chief appl physics br, Res & Develop Directorate, US Missile Command, 62-74; RES PROF, UNIV ALA, HUNTSVILLE, 74- Honors & Awards: Army Res & Develop Achievement Awards, 61 & 64, Missile Command Sci & Eng Awards, 64 & 66. Mem: Am Phys Soc; Optical Soc Am; Inst Elec & Electronics Engrs. Res: Infrared radiation and detection; rocketry; missile guidance systems; solid state, molecular and x-ray lasers. Mailing Add: 7702 Treeline Dr Hunstville AL 35802

MCKNIGHT, WILLIAM FRAULENE, b Rayville, La, Oct 23, 40; m 62; c 2. NUTRITION. Educ: Northeastern La State Col, BS, 63; La State Univ, Baton Rouge, MS, 65, PhD(poultry nutrit), 68. Prof Exp: Asst prof animal sci, Northeast La State Col, 68-69; asst prof, 69-74, ASSOC PROF NUTRIT, LA STATE UNIV, BATON ROUGE, 74- Mem: Poultry Sci Asn. Res: Human nutrition; factors associated with processing and nutritional quality of rice. Mailing Add: Sch of Home Econ La State Univ Baton Rouge LA 70803

MCKOWN, GARY LEONARD, physical chemistry, see 12th edition

MCKOY, BASIL VINCENT, b Trinidad, BWI, Mar 25, 38. THEORETICAL CHEMISTRY. Educ: NS Tech Col, BE, 60; Yale Univ, PhD(chem), 64. Prof Exp: From instr to asst prof, 64-69, ASSOC PROF CHEM, CALIF INST TECHNOL, 69- Concurrent Pos: Sloan Found fel, 69-71; Guggenheim fel, 73; consult, Lawrence Livermore Lab, Univ Calif, 75- Mem: Am Phys Soc. Res: Electron scattering and photoconization processes in molecules. Mailing Add: Dept of Chem Calif Inst of Technol Pasadena CA 91109

MCKUSICK, BLAINE CHASE, organic chemistry, see 12th edition

MCKUSICK, MARSHALL BASSFORD, b Minneapolis, Minn, Jan 13, 30; m 54; c 3. ANTHROPOLOGY, ARCHAEOLOGY. Educ: Univ Minn, BA, 52, MA, 54; Yale Univ, PhD(anthrop), 60. Prof Exp: Asst prof, 60-64, ASSOC PROF ANTHROP, UNIV IOWA, 64-, STATE ARCHAEOLOGIST OF IOWA, 60- Concurrent Pos: Nat Endowment Humanities fel, Univ Iowa, 71-73. Res: Archaeology of Upper Midwest, both prehistoric and historic. Mailing Add: Archaeol Lab Univ of Iowa Iowa City IA 52240

MCKUSICK, VICTOR ALMON, b Parkman, Maine, Oct 21, 21; m 49; c 3. MEDICINE. Educ: Johns Hopkins Univ, MD, 46; Am Bd Internal Med, dipl, 54. Prof Exp: Intern med, Johns Hopkins Hosp, 46-47, asst resident, 47-48; researcher, USPHS, 48-50; intern, Johns Hopkins Hosp, 50-51, resident, 51-52; from instr to assoc prof, Sch Med, 52-60, assoc epidemiol, Sch Hyg & Pub Health, 66-70, PROF MED, SCH MED, JOHNS HOPKINS UNIV, 60-, PROF EPIDEMIOL, SCH HYG & PUB HEALTH, 70-, CHMN DEPT MED, 73- Concurrent Pos: Exec chief cardiovasc unit, Baltimore Marine Hosp, 48-50; resident, Osler Med Clin, 51-52, physician, 53-; physician, Johns Hopkins Hosp, 53-73, chief div med genetics, 57-73, physician-in-chief. Mem: Nat Acad Sci; hon fel Am Acad Orthop Surg; Am Soc Clin Invest; Am Soc Human Genetics; fel Am Col Physicians. Res: Medical genetics; cardiology. Mailing Add: Dept of Med Johns Hopkins Hosp Baltimore MD 21205

MCLACHLAN, DAN, JR, b Arcola, Sask, Can, Dec 5, 05; nat US; m 34; c 3. CHEMICAL PHYSICS. Educ: Kans State Col, BS, 30; Pa State Col, MS, 33, PhD(phys chem), 36. Prof Exp: Asst, Sinclair Refining Co, 30-32 & Pa State Col, 33-36; phys chemist, Corning Glass Works, 36-41; physicist, Am Cyanamid Co, 41-47; prof metall, mineral & physics, Univ Utah, 47-53; asst chmn, Poulter Labs, Stanford Res Inst, 53-54, fundamental res scientist, 54-61; coordr dept physics, Univ Denver, 61-62, prof metall, 62-63; Battelle prof, 63-64, PROF MINERAL, OHIO STATE UNIV, 64- Concurrent Pos: Del, Int Cong Crystallog, Univ London, 46, Harvard Univ, 49, Univ Stockholm, 51 & Univ Toronto, 57; mem, Nat Comt Crystallog, 54, Arctic Exped, Off Naval Res, 58 & Nat Res Coun, 65-68. Mem: AAAS; fel Am Phys Soc; fel Mineral Soc Am; Am Crystallog Asn (pres, 58); fel NY Acad Sci. Res: Information theory; crystallography; mechanics of metals. Mailing Add: 1934 Langham Rd Columbus OH 43221

MCLACHLAN, EUGENE KAY, b Houston, Tex, Jan 28, 24; m 51; c 2. MATHEMATICS. Educ: Baylor Univ, BA, 44; Univ Tex, MA, 46; Rice Inst, MA, 51; Univ Kans, PhD(math), 55. Prof Exp: From instr to assoc prof math, Baylor Univ, 44-58; assoc prof, 58-63, PROF MATH, OKLA STATE UNIV, 63- Concurrent Pos: Asst, Univ Kans, 53-54. Mem: Am Math Soc; Math Asn Am. Res: Linear topological spaces; convexity; mathematical analysis; function theory; convex cones. Mailing Add: Dept of Math Okla State Univ Stillwater OK 74074

MCLACHLAN, JACK (LAMONT), b Huron, SDak, Apr 1, 30; m 51; c 2. PHYCOLOGY. Educ: Ore State Univ, BSc, 53, MA, 54, PhD, 57. Prof Exp: Asst, Ore State Univ, 55-57; NIH res fel, Woods Hole Oceanog Inst, 57-59; NIH res fel appl biol, 60-61, from asst res officer to assoc res officer, 61-74, SR RES OFFICER, DEPT MARINE PLANTS, ATLANTIC REGIONAL LAB, NAT RES COUN CAN, 74- Mem: Phycol Soc Am; Am Soc Limnol & Oceanog; Brit Phycol Soc; Int Phycol Soc. Res: Marine algae. Mailing Add: Dept of Marine Plants Atlantic Regional Lab Nat Res Coun Halifax NS Can

MCLAEN, DONALD FRANCIS, b Butte, Mont, Sept 22, 42; m 64; c 3. ORGANIC CHEMISTRY, PHOTOGRAPHIC CHEMISTRY. Educ: Carroll Col, Mont, BA, 64; Univ Nebr, Lincoln, PhD(org chem), 68. Prof Exp: NIH fel org chem, Univ Ill, Urbana, 68-69; SR CHEMIST, EASTMAN KODAK CO, 69- Mem: Am Chem Soc; Soc Photog Sci & Eng. Res: Organic chemical research related to photographic chemistry and processing. Mailing Add: Res Labs Eastman Kodak Co 343 State St Rochester NY 14650

MCLAFFERTY, FRED WARREN, b Evanston, Ill, May 11, 23; m 48; c 5. ANALYTICAL CHEMISTRY. Educ: Univ Nebr, BS, 43, MS, 47; Cornell Univ, PhD(org chem), 50. Prof Exp: Fel, Univ Iowa, 50; chemist, Dow Chem Co, 50-52, div leader, Mass Spectrometry Sect, Spectros Lab, 52-56, dir, Eastern Res Lab, Framingham, Mass, 56-64; prof chem, Purdue Univ, 64-68; PROF CHEM, CORNELL UNIV, 68- Honors & Awards: Am Chem Soc Award, 71; Spectros Soc Pittsburgh Award, 75. Mem: AAAS; Am Chem Soc; The Chem Soc. Res: Mass spectrometry; molecular structure determination; on-line computers. Mailing Add: Dept of Chem Cornell Univ Ithaca NY 14850

MCLAFFERTY, JOHN J, b Carbondale, Ill, July 6, 29; m 60; c 1. ANALYTICAL CHEMISTRY. Educ: Southern Ill Univ, BA, 51; Loyola Univ, Ill, MS, 64, PhD(anal chem), 66. Prof Exp: Lab supvr, Fansteel Metall Corp, 51-62; staff chemist, Union Carbide Corp, 66-70; CHIEF CHEMIST, STELLITE DIV, CABOT CORP, 70- Mem: Am Soc Testing & Mat; Soc Appl Spectros (pres, 75-); Am Chem Soc. Res: X-ray spectroscopy; emission spectroscopy. Mailing Add: Stellite Div Cabot Box 746 Kokomo IN 46901

MCLAIN, ALBERTSON LAMSON, b Stockton Springs, Maine, June 2, 21; m 43; c 2. FISH BIOLOGY. Educ: Univ Maine, BS, 49. Prof Exp: Mem staff, Outdoor Maine, Inc, 48-49, Maine Dept Inland Fisheries & Game, 49-50; mem staff, Great Lakes Fishery Invests, US Fish & Wildlife Serv, 50 & Hammond Bay Lab, 50-52, mem staff, Field Sta, 52-56, Wis, 56 & Mich, 56-66, supvry res fishery biologist & chief cold-water fishery prog, Ann Arbor Biol Lab, Bur Commercial Fisheries, 66-70; GREAT LAKES COORDR, SEA LAMPREY CONTROL PROG, BUR SPORT FISHERIES & WILDLIFE, 70- Concurrent Pos: Consult to var state agencies & foreign govts. Honors & Awards: Unit Award, Dept of Interior, 59. Mem: Am Fisheries Soc; Am Inst Fishery Res Biol. Res: Fishery research, especially sea lamprey life history and its control; electro-fishing and electro-physiology; Great Lakes fishery biology. Mailing Add: Bur of Sport Fisheries & Wildlife Fed Bldg Ft Snelling Twin Cities MN 55111

MCLAIN, DAVID KENNETH, b Marietta, Ga, Aug 23, 37; m 64; c 2. MATHEMATICAL ANALYSIS. Educ: Ga Inst Technol, BS, 59, MS, 61; Carnegie Inst Technol, MS, 64, PhD(math), 67. Prof Exp: Assoc scientist, Bettis Atomic Power Lab, 60-62, SR MATHEMATICIAN, RES LABS, WESTINGHOUSE ELEC CORP, 66- Mem: AAAS; Am Math Soc; Soc Indust & Appl Math. Res: Calculus of variations; differential and integral equations; linear algebra; electromagnetic fields. Mailing Add: Dept of Math Westinghouse Res Labs Pittsburgh PA 15235

MCLAIN, DONALD DAVIS, JR, b Denver, Colo, July 7, 30; m 56; c 2. MYCOLOGY, MORPHOLOGY. Educ: Univ Ill, BS, 52, MS, 56, PhD(bot), 60. Prof Exp: Res mycologist, Plant Res Inst, Res Br, Can Dept Agr, 60-62; assoc prof bot, Tex A&M Univ, 62-71; MOODY FOUND PROF BIOL, INCARNATE WORD COL, 71- Mem: Mycol Soc Am; Brit Mycol Soc; Int Asn Plant Taxon. Res: Mycological taxonomy; tropical mycology. Mailing Add: Dept of Natural Sci Incarnate Word Col San Antonio TX 78209

MCLAIN, JOSEPH HOWARD, b Weirton, WVa, July 11, 16; m 41; c 2. CHEMISTRY. Educ: Washington Col, BS, 37; Johns Hopkins Univ, PhD(phys chem), 46. Prof Exp: Chief screening smokes sect, Army Chem Ctr, Md, 42-45; from asst prof to assoc prof, 46-55, PROF CHEM & CHMN DEPT, WASHINGTON COL, 55-, PRES, 75- Concurrent Pos: Tech dir, Kent Mfg Corp, 46-54; mem staff, Olin Mathieson Chem Corp, 55-64; lectr, Franklin Inst, 66-; chmn, Water Pollution Control Comn, Md. Mem: Am Chem Soc; The Chem Soc; Combustion Inst. Res: Pyrotechnic reactions; solid propellants and explosives; solid state chemistry. Mailing Add: Dept of Chem Washington Col Chestertown MD 21620

MCLAIN, PAUL LARIMER, b Pittsburgh, Pa, July 18, 08; m 32; c 2. PHYSIOLOGY, PHARMACOLOGY. Educ: Univ Pittsburgh, BS, 29, MD, 32. Prof Exp: Asst physiol & pharmacol, 29-32, demonstr obstet, 33-34, from instr to prof physiol & pharmacol, 34-62, actg chmn dept, 57-65, PROF PHARMACOL, SCH MED, UNIV PITTSBURGH, 62- Concurrent Pos: Trustee, US Pharmacopoeia Conv, 60-75. Mem: AAAS; Am Soc Pharmacol & Exp Therapeut; Am Physiol Soc; Soc Exp Biol & Med. Res: Autonomic pharmacology; cardiac glycosides. Mailing Add: Dept of Pharmacol Univ of Pittsburgh Sch of Med Pittsburgh PA 15261

MCLAMORE, WILLIAM MERRILL, b Shreveport, La, Mar 15, 21; m 49; c 2. MEDICAL RESEARCH. Educ: Rice Inst, BA, 41, MA, 43; Harvard Univ, PhD(org chem), 49. Prof Exp: Jr chemist, Shell Develop Co, 43-45; res chemist, 50-58, res assoc, 58-61, sect mgr, 61-68, RES ADMINR, CHAS PFIZER & CO, INC, 68- Mem: Am Chem Soc. Res: Constituents of bone oil; polymers and resins; synthesis of substituted benzoquinones; synthesis of alkaloids; total synthesis of steroids; structure and synthesis of antibiotics; synthesis of organic medicinals. Mailing Add: Res Admin Pfizer Inc Eastern Point Rd Groton CT 063040

MCLANE, CHARLES KEITH, physics, see 12th edition

MCLANE, S BROOKS, JR, physics, see 12th edition

MCLANE, STANLEY REX, JR, b Kansas City, Mo, Aug 11, 22; m 45; c 3. HORTICULTURE, PLANT PHYSIOLOGY. Educ: Univ Mo, BS, 46, PhD(hort), 51. Prof Exp: Plant physiologist, Physicochem Corps Div, US Civil Serv, Ft Detrick, Md, 51-57; PLANT PHYSIOLOGIST & WEED CONTROL SPECIALIST, AGR CHEM DIV, AMCHEM PROD, INC, 57- Mem: Am Soc Plant Physiologists; Weed Sci Soc Am. Res: Supervising development of new herbicides and growth regulators on agronomic and horticultural crops. Mailing Add: Agr Chem Div Amchem Prod Inc Ambler PA 19002

MCLARDY, TURNER, b Glasgow, Scotland, Apr 16, 14; m 40; c 3. NEUROBIOLOGY, NEUROPSYCHIATRY. Educ: Glasgow Univ, BSc, 34, MB, ChB, 37, MD, 49; Univ London, DPM, 47. Prof Exp: Lectr neuropath, Inst Psychiat, Univ London, 47-52, recognized teacher path, Fac Med, 51-52; dir path & res, St Andrews Psychiat Hosp, Northampton, Eng, 53-64; DIR PATH & HEAD MYERSON RES LAB, BOSTON STATE HOSP, 64- Concurrent Pos: Assoc prof, Sch Med, Tufts Univ, 64-; consult psychiatrist, Cape Cod Ment Health Asn & Falmouth Hosp, 69- Honors & Awards: Gold Medal, Glasgow Univ, 49. Mem: AAAS; fel Royal Col Path; fel Royal Col Psychiat. Res: Experimental neurology; pathology of temporal lobe epilepsy; experimental histological and behavioral testing of tentative interpretations of especially mammalian diencephalic and limbic structures in terms of functional control circuitry. Mailing Add: Boston State Hosp 591 Morton St Boston MA 02124

MCLAREN, ARTHUR DOUGLAS, b Ipava, Ill, Sept 27, 17; m 65; c 2. BIOPHYSICS, BIOLOGY. Educ: Park Col, BA, 39; Univ Mo, PhD(org chem), 43. Prof Exp: Res chemist, E I du Pont de Nemours & Co, NY, 43-46; asst prof polymer chem, Polytech Inst Brooklyn, 46-51; assoc prof soil chem, 51-57, PROF SOIL BIOCHEM, UNIV CALIF, BERKELEY, 57-, BIOL, 70- Concurrent Pos: Instr, Univ Buffalo, 44-45; Rockefeller Found fel, Carlsberg Lab, Copenhagen & Inst Phys Chem, Uppsala, 49-50; Fulbright scholar phys chem, Univ Sydney, 59; Guggenheim fel, Rothamsted Exp Sta, Eng, 66-67; chmn dept biol, Univ Calif, Berkeley, 69-74; ed, Photochem & Photobiol. Mem: Fel AAAS; Biophys Soc; Am Soc Biol Chem; Am Soc Photobiol. Res: Photochemistry of proteins and viruses; adhesion of high polymers to cellulose; enzyme kinetics on surfaces; soil biology and biochemistry. Mailing Add: Col of Natural Resources Univ of Calif Berkeley CA 94720

MCLAREN, DIGBY JOHNS, b Carrickfergus, Northern Ireland, Dec 11, 19; m 42; c 3. GEOLOGY. Educ: Univ Cambridge, BA, 41, MA, 47; Univ Mich, PhD(geol), 51. Prof Exp: Mem staff, 48-59, chief paleontologist, 59-67, dir inst sedimentary & petrol geol, 67-73, DIR, GEOL SURV CAN, 73- Mem: Am Paleont Soc (pres, 69); fel Am Geol Soc; fel Royal Soc Can; Can Soc Petrol Geologists (pres, 71). Res: Devonian paleontology and stratigraphy of Western Canada. Mailing Add: Geol Surv of Can 601 Booth St Rm 213 Ottawa ON Can

MCLAREN, EUGENE HERBERT, b Troy, NY, Aug 3, 24; m 57; c 3. ATMOSPHERIC CHEMISTRY. Educ: NY State Col Teachers Albany, BA, 48, MA, 49; Washington Univ, PhD(phys chem), 55. Prof Exp: Instr chem, State Univ NY Col Teachers Albany, 50-52, assoc prof, 55-57; sr res chemist, Pan-Am Petrol Corp, 57-60; prof sci & math, 60-69, PROF CHEM, STATE UNIV NY ALBANY, 69- Concurrent Pos: Assoc dean, State Univ NY Albany, 68-69, chmn div sci & math, 61-68; univ fel & res scientist, Max Planck Inst Chem, 69-70. Mem: AAAS; Am Chem Soc; Geochem Soc; Am Meteorol Soc; Air Pollution Control Asn. Res: Physical chemistry of atmospheric particulates and gases; geochemistry of carbonates; x-ray diffraction and spectrometry. Mailing Add: Dept of Chem State Univ of NY Albany NY 12222

MCLAREN, GEORGE AIKEN, b New York, NY, Oct 2, 12; m 36; c 2. AGRICULTURAL BIOCHEMISTRY. Educ: Univ Va, BS, 35; Fordham Univ, MS, 40; Okla State Univ, PhD(nutrit), 55. Prof Exp: Asst biochemist, Schering Corp, 40-41; res biochemist, Am Molasses Co, 41-46, group leader biochem, 46-53; from asst prof to assoc prof agr biochem, 55-59, PROF AGR BIOCHEM & NUTRIT, WVA UNIV, 59- Concurrent Pos: Consult, Am Molasses Co, 53-54. Mem: AAAS; Am Soc Animal Sci; Am Chem Soc; Am Inst Nutrit. Res: Food and nutritional biochemistry; carbohydrates; metabolic effects of natural products; N-propyl nitrate utilization by ruminants fed low-quality roughages. Mailing Add: Div of Animal & Vet Sci WVa Univ Morgantown WV 26506

MCLAREN, IAN ALEXANDER, b Montreal, Que, Jan 11, 31; m 56; c 3. BIOLOGY. Educ: McGill Univ, BSc, 52, MSc, 55; Yale Univ, PhD(zool), 61. Prof Exp: Asst scientist, Fisheries Res Bd Can, 55-63; asst prof biol, Marine Sci Ctr, McGill Univ, 63-66; assoc prof, 66-69, PROF BIOL, DALHOUSIE UNIV, 69- Concurrent Pos: Can Coun fel, McGill Univ, 64-66. Mem: Fel Arctic Inst NAm; Can Soc Zoologists; Am Soc Naturalists; Ecol Soc Am; Am Soc Limnol & Oceanog. Res: Population; evolutionary ecology; birds; sea mammals; zooplankton. Mailing Add: Dept of Biol Dalhousie Univ Halifax NS Can

MCLAREN, JAMES BLACKBURN, animal breeding, see 12th edition

MCLAREN, LEROY CLARENCE, b Bishop, Calif, Jan 18, 24; m 45; c 3. MICROBIOLOGY, VIROLOGY. Educ: San Jose State Col, AB, 49; Univ Calif, Los Angeles, MA, 51, PhD(microbiol), 53; Am Bd Med Microbiol, dipl. Prof Exp: Asst microbiol, Univ Calif, Los Angeles, 49-50, asst bact, 50-52, instr infectious dis, Sch Med, 53-55; from instr to assoc prof bact, Sch Med, Univ Minn, 55-64; PROF MICROBIOL & CHMN DEPT, SCH MED, UNIV NMEX, 64- Concurrent Pos: USPHS career res award, 62-64; mem microbiol fel rev panel, NIH, 64-68; consult, Lilly Res Labs, 64-68; mem comt personnel for res, Am Cancer Soc, 73-; mem bd educ & training, Am Soc Microbiol, 72- Mem: Fel AAAS; fel Am Acad Microbiol; Am Soc Microbiol; Am Asn Immunol; Tissue Cult Asn. Res: Tissue culture of animal viruses; stability of viruses; animal virus multiplication; mechanisms of viral susceptibility and resistance. Mailing Add: Dept of Microbiol Univ of NMex Sch of Med Albuquerque NM 87131

MCLARTY, DUNCAN ARCHIBALD, b St Thomas, Ont, Feb 13, 13; m 41; c 6. BOTANY. Educ: Univ Western Ont, BA, 34; Columbia Univ, PhD, 40. Prof Exp: Lab asst, Univ Western Ont, 34-35 & Columbia Univ, 35-39; instr bot, Dartmouth Col, 39-42; from instr to prof bot, 43-70, PROF PLANT SCI, UNIV WESTERN ONT, 70- Res: Cytology of the lower fungi; algology; algal responses to pollutants in the Great Lakes. Mailing Add: Dept of Plant Sci Col of Sci Univ of Western Ont London ON Can

MCLAUGHLIN, BARBARA JEAN, b Miami, Fla, Nov 3, 41. NEUROBIOLOGY. Educ: Univ Fla, BS, 63; Stanford Univ, PhD(anat), 71. Prof Exp: ASST PROF ANAT, CTR HEALTH SCI, UNIV TENN, MEMPHIS, 74- Concurrent Pos: Agr Res Coun, Eng Underwood Fund grant zool, Univ Cambridge, 71-73; fel neurosci, City of Hope Med Ctr, Duarte, Calif, 73-74. Mem: Soc Neurosci; Am Asn Anat; Am Soc Cell Biol; Sigma Xi; Am Soc Zoologist. Res: Developmental neurobiology; cytochemical and immunocytochemical localization of various molecular components in adult and developing nervous tissue; cytochemistry of developing synapses. Mailing Add: Dept of Anat Univ of Tenn Ctr Health Sci Memphis TN 38163

MCLAUGHLIN, CALVIN STURGIS, b St Joseph, Mo, May 29, 36; m 60; c 3. BIOCHEMISTRY, GENETICS. Educ: King Col, BS, 58; Mass Inst Technol, PhD(biochem), 64. Prof Exp: From asst prof to assoc prof, 66-72, PROF BIOCHEM, UNIV CALIF, IRVINE, 72- Concurrent Pos: Am Cancer Soc fel, Inst Phys Chem Biol, Paris, 64-66. Mem: Genetics Soc Am; Am Soc Microbiol; Am Soc Biol Chem. Res: Biochemistry and biochemical genetics of protein and RNA synthesis; mechanism of action of antibiotics; regulation of protein and RNA synthesis; mycology. Mailing Add: Dept of Molecular Biol & Biochem Univ of Calif Irvine CA 92717

MCLAUGHLIN, CHARLES ALBERT, b Chatham, Ill, Nov 12, 26; m 49; c 2. ZOOLOGY. Educ: Univ Ill, BS, 49, MS, 51, PhD(zool), 58. Prof Exp: Asst, Mus Natural Hist, Univ Ill, 50-51, exhibit preparator, 56, sci artist zool, Univ 51-55, asst, 55-56; assoc curator ornith & mammal, Los Angeles County Mus Natural Hist, 57-62, curator, 62-65, sr curator mammal, 65-67; assoc prof zool & curator mammal, Univ Wyo, 67-71; DIR EDUC, SAN DIEGO ZOO, 71- Mem: Am Soc Mammal; Cooper Ornith Soc; Soc Syst Zool; Am Ornith Union; Am Soc Zool. Res: Mammalogy, especially taxonomy, ecology and distribution, particularly of rodents and bats. Mailing Add: Dept of Educ San Diego Zoo PO Box 551 San Diego CA 92112

MCLAUGHLIN, CHARLES WILLIAM JR, b Washington, Iowa, Feb 3, 06; m 39; c 2. SURGERY. Educ: Univ Iowa, BS, 27; Wash Univ, MD, 29; Am Bd Surg, dipl, 40. Prof Exp: From instr to assoc prof, 35-55, PROF SURG, COL MED, UNIV NEBR, OMAHA, 55- Concurrent Pos: Res fel surg, Univ Pa, 31-34; nat consult, Gen Surgeon, US Air Force. Mem: Fel Am Col Surg; AMA; Asn Mil Surg US. Res: Abdominal and pediatric surgery. Mailing Add: 409 Doctors Bldg 44th & Farnum Sts Omaha NE 68131

MCLAUGHLIN, DAVID, b Sumter, SC, Nov 1, 34; m 63. ZOOLOGY. Educ: Clark Col, BS, 56; Howard Univ, MS, 62, PhD(zool), 65. Prof Exp: Asst biochem & physiol protozoa, 57-62, asst prof, 65-71, ASSOC PROF ZOOL, HOWARD UNIV, 71- Concurrent Pos: Partic, Biospace Training Prog, NASA, Va, 65 & Gemini Summary Conf, Manned Spacecraft Ctr, 67. Mem: AAAS; Am Soc Protozool; Am Inst Biol Sci;

Am Micros Soc. Res: Biochemistry and physiology of cells. Mailing Add: Dept of Zool Howard Univ Washington DC 20001

MCLAUGHLIN, DONALD REED, b Los Angeles, Calif, Oct 6, 38; m 64; c 5. CHEMICAL PHYSICS. Educ: Univ Calif, Los Angeles, BS, 60; Univ Utah, PhD(chem), 65. Prof Exp: Asst prof, 60-65, ASSOC PROF CHEM, UNIV N MEX, 65- Concurrent Pos: Assoc, Rocky Mt Univs, Inc, Fac Orientation & Training Summers Fels, 65-69. Res: Theoretical chemistry, especially quantum mechanics, statistical mechanics, chemical kinetics and thermodynamics. Mailing Add: Dept of Chem Univ of NMex Albuquerque NM 87106

MCLAUGHLIN, ELLEN WINNIE, b Roosevelt, NY, Aug 17, 37. EXPERIMENTAL EMBRYOLOGY. Educ: State Univ NY Albany, BS, 58; Univ NC, Chapel Hill, MA, 62; Emory Univ, PhD(biol), 67. Prof Exp: Instr biol, Converse Col, 60-63; from asst prof to assoc prof, 67-75, PROF BIOL, SAMFORD UNIV, 75- Concurrent Pos: Res grants, Samford Univ, 68-69. Mem: AAAS; Am Soc Zool; Am Sci Affiliation; Am Inst Biol Sci. Res: Effects of heavy metals and pesticides on aquatic vertebrate and invertebrate embryos, including amphibians, fish and snails. Mailing Add: Dept of Biol Samford Univ Birmingham AL 35209

MCLAUGHLIN, FOIL WILLIAM, b NC, Dec 9, 23; m 48; c 4. AGRONOMY. Educ: NC State Univ, BS, 49, MS, 53. Prof Exp: Res asst prof field crops, 53-63, exten assoc prof, 63-68, DIR, NC CROP IMPROV ASN, NC STATE UNIV, 48-, EXTEN PROF FIELD CROPS, 63- Res: Crop and seed improvement, especially breeding and quality control in seed development and production; seed certification. Mailing Add: Dept of Crop Sci NC State Univ Sch Agr & Life Sci Raleigh NC 27607

MCLAUGHLIN, GERALD WAYNE, b Nashville, Tenn, Aug 16, 42; m 65; c 2. RESEARCH MANAGEMENT, APPLIED STATISTICS. Educ: Univ Tenn, BS, 64, MS, 65, PhD(orgn psychol), 69. Prof Exp: Res asst indust mgt, Univ Tenn, 64-65, res asst econ, 65-66; asst dir instnl res, US Mil Acad, 69-71; ASST PROF INSTNL RES, VA POLYTECH INST & STATE UNIV, 71- Mem: Asn Instnl Res; Psychomet Soc. Res: Administration and conducting applied research; testing and measurement; role analysis in education, industry and other fields. Mailing Add: Va Polytech Inst & State Univ Blacksburg VA 24061

MCLAUGHLIN, JACK ENLOE, b St Maries, Idaho, Aug 17, 23; m 49. MATHEMATICS. Educ: Univ Idaho, BS, 44; Calif Inst Technol, PhD(math), 50. Prof Exp: From instr to assoc prof, 50-70, PROF MATH, UNIV MICH, ANN ARBOR, 70- Concurrent Pos: Res fel, Harvard Univ, 60. Mem: Am Math Soc; Math Asn Am. Res: Lattice theory and ring theory. Mailing Add: Dept of Math Univ of Mich Ann Arbor MI 48104

MCLAUGHLIN, JAMES E, b Providence, RI, Apr 17, 25; m 48; c 7. PHYSICS. Educ: Boston Col, BS, 51; Am Bd Health Physics, cert, 60. Prof Exp: Physicist, 52-61, DIR RADIATION PHYSICS, HEALTH & SAFETY LAB, US ENERGY RES & DEVELOP ADMIN, 61- Concurrent Pos: Chmn, Nat Comt Radiation Protection & Measurements, 35. Mem: Am Nuclear Soc; Health Physics Soc. Res: Studies in radiation dosimetry; radiation safety and environmental radiation. Mailing Add: Health & Safety Lab 376 Hudson St US Energy Res & Develop Admin New York NY 10014

MCLAUGHLIN, JAMES JOSEPH, b North Adams, Mich, Apr 6, 11; m 32; c 2. MATHEMATICS. Educ: Eastern Mich Univ, AB, 33; Univ Mich, AM, 36, EdD, 53. Prof Exp: High sch teacher, Mich, 34-39; dir union & res hall, Cent Mich Col Educ, 39-41; teacher math, Ford Motor Co, 42-44; teacher, Supply Div, Mech Sch, Brazilian Air Ministry, 44-47; chmn dept math, 47-52, dean col arts & sci, 52-70, PROF MATH, UNIV WIS-RIVER FALLS, 70- Mem: Math Asn Am. Res: Mathematics in agriculture. Mailing Add: Dept of Math Univ of Wis River Falls WI 54022

MCLAUGHLIN, JAMES L, b Detroit, Mich, July 16, 42; m 64; c 1. PLANT PATHOLOGY. Educ: Eastern Mich Univ, BA, 64; Univ Ill, MS, 66, PhD(plant path), 69. Prof Exp: Asst prof biol, Univ Wis-Superior, 69-72; asst prof, 72-73, ASSOC PROF BIOL, COL ST SCHOLASTICA, 73-, CHMN DEPT, 73- Mem: Am Inst Biol Sci; Am Phytopath Soc. Res: Mycology; plant physiology; bacteriology. Mailing Add: Dept of Biol Col of St Scholastica Duluth MN 55811

MCLAUGHLIN, JAMES RICHARD, mathematics, see 12th edition

MCLAUGHLIN, JERRY LOREN, b Coldwater, Mich, Oct 14, 39; m 60; c 2. PHARMACOGNOSY. Educ: Univ Mich, BS, 63, PhD(pharmacog), 65. Prof Exp: Asst prof pharmacog, Univ Mich, 65-66 & Univ Mo-Kansas City, 66-67; mem fac, Col Pharm, Univ Wash, 67, from asst prof to assoc prof pharmacog, 67-72; assoc prof, 71-75, PROF PHARMACOG & EXEC ASST TO DEAN, SCH OF PHARM & PHARMACAL SCI, PURDUE UNIV, 75- Concurrent Pos: Nat Inst Ment Health res grants, 66-67, 69-72; NIH res grant, 74-77; Nat Sci Found res grants, 74, 75. Mem: AAAS; Am Pharmaceut Asn; Am Soc Pharmacog; Soc Econ Bot; Acad Pharmaceut Sci. Res: Cactus alkaloids, their isolation and biosynthesis; active constituents of psychotropic and poisonous plants. Mailing Add: Dept of Med Chem & Pharmacog Sch of Pharm & Pharmacal Sci Purdue Univ Lafayette IN 47907

MCLAUGHLIN, JOHN J (ANTHONY), b Long Island, NY, Aug 26, 24; m 51; c 5. MICROBIOLOGY. Educ: St Francis Col, NY, BS, 50; St John's Univ, NY, MS, 52; NY Univ, PhD, 56. Prof Exp: Asst microbiol, Haskins Labs, 52-56, mem staff, 56-69; assoc prof biol, St Francis Col, NY, 56-69; chmn dept, 69-74, PROF BIOL SCI, FORDHAM UNIV, BRONX, 69- Res: Anexic culturing of microscopic and macroscopic organisms; phytoplanktonic and symbiotic organisms from the marine environment. Mailing Add: Dept of Biol Sci Fordham Univ Bronx NY 10458

MCLAUGHLIN, JOSEPH, JR, b Marietta, Ohio, Oct 7, 15; m 52; c 4. ORGANIC CHEMISTRY. Educ: Univ Fla, BS, 39, MS, 47, PhD(chem), 49. Prof Exp: Res chemist, Biochem Dept, Army Med Serv Grad Sch, Walter Reed Army Inst Res, 50-58; biochemist, Lab Invests Br, Toxicol Eval Div, 58-70, CHIEF REPRODUCTIVE PHYSIOL BR, DIV TOXICOL, FOOD & DRUG ADMIN, 70- Mem: Soc Exp Biol & Med; Soc Toxicol; Am Indust Hyg Asn; Pan Am Med Asn. Res: Toxicity of food packaging materials; chick embryo toxicity studies. Mailing Add: Reproductive Physiol Br Div of Toxicol Bur of Food Washington DC 20204

MCLAUGHLIN, KENNETH PHELPS, b Blackfoot, Idaho, Apr 16, 17; m 41; c 3. GEOLOGY. Educ: Univ Mo, AB, 39, MA, 41; La State Univ, PhD(geol), 47. Prof Exp: Instr geol, Centenary Col, 42-43; geologist, Stanolind Oil & Gas Co, Tex, 43-45; instr geol, La State Univ, 45-47; asst prof, State Col Wash, 47-50; prof & chmn dept, Univ Mont, 50-55; staff geologist, Tex Industs Inc, 55-62, consult, 62-65; PROF GEOL & GEOL ENG & CHMN DEPT, UNIV MISS, 65- Concurrent Pos: Geologist, US Geol Surv, 48-50 & Shell Oil Co, 51-53. Mem: Am Asn Petrol Geologists; Asn Eng Geologists. Res: Industrial minerals. Mailing Add: Dept of Geol & Geol Eng Univ of Miss Sch of Eng University MS 38677

MCLAUGHLIN, PAUL JOHN, b Schenectady, NY, Apr 2, 15; m 41; c 3. CHEMISTRY. Educ: Univ Pa, BS, 37, MS, 41, PhD(chem), 43. Prof Exp: Engr, Process Eng Lab, Nat Carbon Co, NY, 37-39; instr chem, Moravian Col, 40-41; asst instr, Univ Pa, 41-43; res chemist, Exp Sta, Hercules Powder Co, 43-53; RES CHEMIST, EXP STA, ROHMN AND HAAS CO, PHILADELPHIA, 53- Mem: Am Chem Soc. Res: Cellulose; heterocyclic nitrogen compounds; organic analytical methods; acrylic polymers in paper coating; anidex fibers; technical writing. Mailing Add: Rohm and Haas Co Independence Mall W Philadelphia PA 19105

MCLAUGHLIN, RENATE, b Chemnitz, Ger, Jan 3, 41; m 64. MATHEMATICAL ANALYSIS. Educ: Univ Münster, AB, 63; Univ Mich, Ann Arbor, AM, 64, PhD(math), 68. Prof Exp: From asst prof to assoc prof, 68-75, PROF MATH, UNIV MICH-FLINT, 75- Concurrent Pos: Managing ed, Mich Math J, 68-76. Mem: Am Math Soc; Math Asn Am. Res: Complex analysis; univalent functions; special classes of functions, variational methods. Mailing Add: Dept of Math Univ of Mich Flint MI 48503

MCLAUGHLIN, ROBERT EVERETT, b Aurora, Ind, Nov 30, 19; m 45; c 3. PALEONTOLOGY. Educ: Tulane Univ, BS, 51, MS, 52; Univ Tenn, Knoxville, PhD(paleont), 57. Prof Exp: Asst bot, Tulane Univ, 50-51, asst geol, 52; from asst to instr bot, 52-54, from instr to assoc prof geol, 54-71, PROF GEOL, UNIV TENN, KNOXVILLE, 71- Concurrent Pos: Southern Fels Fund award, 55-56. Mem: Bot Soc Am; Paleont Soc; Geol Soc Am; Am Asn Stratig Palynologists. Res: Palynology of tertiary and older deposits; Paleozoic paleontology. Mailing Add: Dept of Geol Univ of Tenn Knoxville TN 37916

MCLAUGHLIN, ROBERT LAWRENCE, b Beaver, Pa, Sept 17, 23; m 46; c 2. RESEARCH ADMINISTRATION, ORGANIC CHEMISTRY. Educ: Pa State Univ, BS, 44, MS, 46, PhD(org chem), 49. Prof Exp: Mem, Am Petrol Inst Proj, Pa State Univ, 44-46, asst, 44-49; res scientist, NASA, Ohio, 49-50, group leader, 50-51; fel, Mellon Inst, 51-55, sr fel, 55-57; sr res chemist, Res & Develop Lab, Socony-Mobil Oil Co, 57-60, group leader, 60-62, sect leader, 62-64; coordr res, Velsicol Chem Co, 64-65, dir res, Resin Prod Div, 65-69; res mgr, Armour Dial Inc, 69-73; TECH DIR, DESOTO, INC, 73- Mem: Am Chem Soc; Am Oil Chemists Soc. Res: Hydrocarbon analysis, synthesis, purifications, separations, adducts, processing and properties; organic and petroleum chemicals; monomers and polymeric materials, synthesis, processing and evaluations; soaps and detergents; personal care products; cleaning products. Mailing Add: 2333 Schiller Ave Wilmette IL 60091

MCLAUGHLIN, ROY EARL, b Peoria, Ill, Sept 28, 30; m 60; c 2. ENTOMOLOGY, PROTOZOOLOGY. Educ: Univ Ill, BS, 54, MS, 59, PhD(entom), 61. Prof Exp: ENTOMOLOGIST, BOLL WEEVIL RES LAB, AGR RES SERV, USDA, 61- Mem: Soc Protozool; Soc Invert Path. Res: Insect pathology; investigation of basic relationships and development of applied methods of control of the boll weevil; development of the principle of use of pathogens with response-eliciting baits for biological control of insects. Mailing Add: Boll Weevil Res Lab USDA PO Box 5367 Mississippi State MS 39762

MCLAUGHLIN, THAD GERALD, b Valley Center, Kans, Oct 16, 13; m 42. GEOLOGY. Educ: Univ Wichita, AB, 35; Univ Kans, PhD(geol), 39. Prof Exp: Geologist, State Geol Surv, Kans, 39-42; geologist, 42-51, dist geologist, 51-59, br area chief, Rocky Mt Area, 59-64, staff hydrologist, 64-67, regional hydrologist, 67-72, ASST DIR, US GEOL SURV, 72- Mem: Geol Soc Am; Am Asn Petrol Geol; Am Water Resources Asn (vpres, 75). Res: Hydrogeology. Mailing Add: US Geol Surv Box 25046 Fed Ctr Denver CO 80225

MCLAUGHLIN, THOMAS G, b McIntosh, SDak, Nov 26, 33; m 69; c 1. MATHEMATICS. Educ: Univ Calif, Los Angeles, BA, 59, MA, 62, PhD(math), 63. Prof Exp: From instr to asst prof math, Univ Ill, Urbana, 63-66; vis asst prof, Cornell Univ, 66-67; ASSOC PROF MATH, UNIV ILL, URBANA, 67- Mem: Am Math Soc; Math Asn Am; Asn Symbolic Logic. Res: Recursive function theory. Mailing Add: Dept of Math Univ of Ill Urbana IL 61801

MCLAUGHLIN, WILLIAM IRVING, b Oak Park, Ill, Mar 6, 35; m 60; c 4. CELESTIAL MECHANICS. Educ: Univ Calif, Berkeley, BS, 63, MA, 66, PhD(math), 68. Prof Exp: Mem tech staff celestial mech, Bellcomm, Inc, 68-71; MEM TECH STAFF CELESTIAL MECH, JET PROPULSION LAB, CALIF INST TECHNOL, 71- Honors & Awards: Apollo Achievement Award, Nat Aeronautics & Space Admin, 69. Mem: Fel Brit Interplanetary Soc; Am Asn Variable Star Observers. Res: Satellite orbits; rings of Saturn; interstellar communication. Mailing Add: Jet Propulsion Lab Calif Inst Technol 4800 Oak Grove Dr Pasadena CA 91103

MCLAUGHLIN, WILLIAM LOWNDES, b Stony Point, Tenn, Mar 30, 28; m 51; c 2. RADIATION PHYSICS. Educ: Hampden-Sydney Col, BS, 49; George Washington Univ, MS, 63. Prof Exp: Physicist, Radiation Phys Div, 51-54, 56-64, proj leader, 64-69, PROJ LEADER, CTR RADIATION RES, NAT BUR STANDARDS, 69- Concurrent Pos: Ed, Int J Appl Radiation & Isotopes, 73- Honors & Awards: Silver Medal, Dept Commerce, 69. Mem: AAAS; Am Phys Soc, Optical Soc Am; Health Physics Soc; Radiation Res Soc. Res: Measurement of ionizing radiation; x-ray, gamma-ray and electron spectrometry, absorption and scattering measurement and computation; radiation chemistry; photographic prosesses. Mailing Add: Ctr for Radiation Res Nat Bur Standards Washington DC 20234

MCLAURIN, JAMES WALTER, b Natchez, Miss, July 1, 10; m 35. MEDICINE, SURGERY. Educ: Univ Ark, MD, 34. Prof Exp: Intern, Mary's Help Hosp, San Francisco, Calif, 34-35; resident, Los Angeles Children's Hosp, 35-36; instr, Eye, Ear, Nose & Throat Clins, Sch Med, Univ Ark, 36; instr otolaryngol, Sch Med, La State Univ, 46-49; chmn dept otolaryngol, 49-58, med dir, Speech & Hearing Ctr, 52-63, OTTO JOACHIEM PROF OTOLARYNGOL, SCH MED, TULANE UNIV, 49- Concurrent Pos: Mem, Nat Comt Deafness Res Found. Mem: Am Laryngol, Rhinol & Otol Soc; Am Otol Soc; AMA; fel Am Col Surg; Am Acad Opthal & Otolaryngol. Res: Otitis externa and otitis media; speech and hearing problems. Mailing Add: 3888 Government St Baton Rouge LA 70806

MCLAURIN, ROBERT L, b Dallas, Tex, Jan 5, 22; m 46; c 5. NEUROSURGERY. Educ: Rice Inst, BA, 44; Harvard Med Sch, MD, 44. Prof Exp: Asst surg, Harvard Med Sch, 51-53; from instr to assoc prof surg, 53-60, actg dir div neurosurg, 54-55, PROF SURG, COL MED, UNIV CINCINNATI, 53-, DIR DIV NEUROSURG, 55- Concurrent Pos: Res fel neurosurg, Children's Hosp & Peter Bent Brigham Hosp, Boston, Mass, 51-53. Mem: Cong Neurol Surg; Am Asn Neurol Surg; Am Acad Neurol Surg (secy-treas, 58-63, pres, 71-72); Soc Brit Neurol Surg; Soc Neurol Surg (treas, 70-). Res: Clinical and experimental aspects of intracranial trauma and hemorrhage. Mailing Add: Dept of Surg Univ of Cincinnati Col of Med Cincinnati OH 45229

MCLAY, DAVID BOYD, b Toronto, Ont, Feb 29, 28; m 53. MOLECULAR PHYSICS. Educ: McMaster Univ, BSc, 50, MSc, 51; Univ BC, PhD(physics), 56. Prof Exp: Lectr physics, Victoria Col, 54-56; asst prof, Univ New Brunswick, 56-62; asst prof, 62-64,

ASSOC PROF PHYSICS, QUEEN'S UNIV, ONT, 64- Mem: Can Asn Physicist; Am Asn Physics Teachers; Can Astron Soc. Res: Microwave spectroscopy and paramagnetic spectroscopy of gases; dielectric and nuclear paramagnetic relaxation; paramagnetic resonance. Mailing Add: Dept of Physics Queen's Univ Kingston ON Can

MCLEAN, ALASTAIR, b Edinburgh, Scotland, Jan 26, 21; Can citizen; m 49; c 3. RANGE ECOLOGY. Educ: Univ BC, BSA, 44; Utah State Univ, MS, 53; Wash State Univ, PhD, 69. Prof Exp: Agronomist, Exp Farm, Can Dept Agr, Sask, 44-48; RANGE ECOLOGIST, RES STA, BC, 48- Mem: Ecol Soc Am; Soc Range Mgt (sec pres, 69); Agr Inst Can; Brit Ecol Soc; Can Bot Asn. Res: Range productivity; plant succession; soil-plant relations. Mailing Add: Agr Can Res Sta Box 940 Kamloops BC Can

MCLEAN, DARRELL MARSHALL, b Cedarville, Ohio, Mar 27, 12; m 39; c 2. PLANT PATHOLOGY. Educ: Okla State Univ, BS, 38; Ohio State Univ, MS, 39, PhD(plant path), 43. Prof Exp: Asst prof plant path, Mich State Univ, 43-46; PLANT PATHOLOGIST, VEG CROPS INVESTS, CROPS RES DIV, AGR RES SERV, USDA, 46- Mem: AAAS; Am Phytopath Soc. Res: Vegetable pathology. Mailing Add: Regional Veg Breeding Lab PO Box 3348 Charleston SC 29407

MCLEAN, DONALD LEWIS, b Norwood, Mass, Oct 2, 28; m 52; c 3. ENTOMOLOGY. Educ: Tufts Univ, BS, 53; Univ Mass, MS, 55; Univ Calif, PhD(entom), 58. Prof Exp: PROF ENTOM, UNIV CALIF, DAVIS, 58-, CHMN DEPT, 74- Mem: Entom Soc Am; AAAS. Res: Insect transmission of plant viruses, especially aphid vectors; biological studies dealing with aphid feeding; culturing and chemical studies with aphid cells and symbiotes. Mailing Add: Dept of Entom Univ of Calif Davis CA 95616

MCLEAN, DONALD MILLIS, b Melbourne, Australia, July 26, 26; Can citizen. VIROLOGY, MEDICAL MICROBIOLOGY. Educ: Univ Melbourne, BSc, 47, MD, 50; FRCP(C), 67. Prof Exp: Dir virol, Hosp Sick Children, Toronto, 58-67; PROF MED MICROBIOL & HEAD DEP, UNIV BC, 67- Concurrent Pos: Consult microbiol, Children's Hosp, Vancouver Gen Hosp, Shaughnessy Hosp & St Paul's Hosp, Vancouver, 68- Mem: Am Soc Trop Med & Hyg; Am Soc Microbiol; Am Acad Pediat; Am Epidemiol Soc; Can Med Asn. Res: Arbovirus vectors and reservoirs and enteroviruses in water. Mailing Add: Div of Med Microbiol Univ of BC 2075 Wesbrook Pl Vancouver BC Can

MCLEAN, EDGAR ALEXANDER, b Gastonia, NC, July 25, 27; m 51; c 5. PLASMA PHYSICS. Educ: Univ NC, Chapel Hill, BS, 49; Univ Del, MS, 51. Prof Exp: RES PHYSICIST, NAVAL RES LAB, 51- Concurrent Pos: Res consult, Space Sci Dept, Cath Univ Am, 64-69 & Univ Western Ont, 73- Honors & Awards: Res Publ Award, Naval Res Lab, 71 & 74. Mem: AAAS; Am Phys Soc; Sigma Xi. Res: Optical diagnostics in the field of plasma physics, including spectroscopic, interferometric, laser-scattering, and high-speed photographic measurements on shock tubes, laser-produced plasmas, and various controlled fusion devices. Mailing Add: Naval Res Lab Washington DC 20375

MCLEAN, EDWARD BRUCE, b Washington Court House, Ohio, Jan 10, 37; m 68; c 3. ORNITHOLOGY, ECOLOGY. Educ: Ohio State Univ, BSc, 58, MSc, 63, PhD(zool), 68. Prof Exp: Asst prof biol, Southern Univ, Baton Rouge, 68-70; asst prof, 70-73, ASSOC PROF BIOL, JOHN CARROLL UNIV, 73- Concurrent Pos: Southern Univ Found grant, Southern Univ, Baton Rouge, 69-70. Mem: Am Ornith Union; Wilson Ornith Soc; Am Inst Biol Sci. Res: Bioacoustics; vertebrate ecology; ethnology. Mailing Add: Dept of Biol John Carroll Univ Cleveland OH 44118

MCLEAN, EUGENE OTIS, b Nixa, Mo; m 43; c 2. SOIL CHEMISTRY. Educ: Univ Mo, BS, 42, MA, 43, PhD(soil chem), 48. Prof Exp: Instr math, Univ Mo, 42-43, soils, 43-48, asst prof, 48-50; from asst prof to assoc prof agron, Univ Ark, 50-56; assoc prof, 56-58, PROF AGRON, OHIO STATE UNIV, 58- Mem: AAAS; Am Chem Soc; Am Soc Agron; Soil Sci Soc Am; Int Soil Sci Soc. Res: Activities and bonding energies of cations in relation to plant availability from colloidal systems; chemistry of phosphate and potassium in soils with reference to forms available to plants; strontium-calcium relationships in soils; exchangeable aluminum as a factor in soil acidity and lime requirement of acid soils. Mailing Add: Dept of Agron 108 Townshend Hall Ohio State Univ Columbus OH 43210

MCLEAN, FLYNN BREVARD, b Ft Bragg, NC, Nov 15, 41; m 66; c 2. SOLID STATE PHYSICS. Educ: Rensselaer Polytech Inst, ScB, 63; Brown Univ, ScM, 66, PhD(physics), 68. Prof Exp: Physicist, Harry Diamond Labs, 68-69; res assoc solid state physics, Brookhaven Nat Lab, 70-71; PHYSICIST, HARRY DIAMOND LABS, 71- Mem: Am Phys Soc; Am Asn Physics Reachers. Res: Radiation effects in solids; electrical transport in amorphous insulators; physics of metal-oxide-semiconductor systems. Mailing Add: Radiation Effects Physics Br Harry Diamond Labs Adelphi MD 20783

MCLEAN, GORDON WILLIAM, soil chemistry, see 12th edition

MCLEAN, I WILLIAM, JR, medicine, see 12th edition

MCLEAN, IAN WILLIAM, b Durham, NC, Sept 21, 43; m 66; c 2. PATHOLOGY, OPHTHALMOLOGY. Educ: Univ Mich, BS, 65, MD, 69; Univ Colo Med Ctr, cert path, 73; Am Bd Path, cert anat path, 74. Prof Exp: Fel ophthal path, Armed Forces Inst Path, 71-72; vis scientist, Inst Biol Sci, Oakland Univ, 72; STAFF PATHOLOGIST, OPHTHAL DIV, ARMED FORCES INST PATH, 73- Mem: Asn Res Vision & Ophthalmol. Res: Evaluation of factors relating to mortality in patients with intraocular malignant melanomas and development of a mathematical model describing the transport of ions in the ocular lens. Mailing Add: Ophthal Div Armed Forces Inst of Path Washington DC 20305

MCLEAN, JAMES AMOS, b Flint, Mich, Dec 2, 21; m 54; c 2. INTERNAL MEDICINE, ALLERGY. Educ: Univ Mich, BS & MD, 46; Baylor Univ, MS, 52. Prof Exp: From asst prof to assoc prof, 56-67, PROF INTERNAL MED, MED CTR, UNIV MICH, ANN ARBOR, 67-, ASST ALLERGY, HEALTH SERV, 65- Mem: AMA; Am Acad Allergy; Am Fedn Clin Res. Res: Clinical allergy. Mailing Add: Univ of Mich Hosp Ann Arbor MI 48104

MCLEAN, JAMES DENNIS, b Bay City, Mich, Nov 23, 40; m 68. ANALYTICAL CHEMISTRY, POLAROGRAPHY. Educ: Univ Mich, Ann Arbor, BS, 62; Mich State Univ, PhD(anal chem), 67. Prof Exp: ANAL SPECIALIST, DOW CHEM CO, 67- Honors & Awards: A D Beckmann Award, Instrument Soc Am, 75. Mem: Am Chem Soc; Instrument Soc Am; Sigma Xi. Res: Polarographic analysis; anodic stripping voltammetry; electrochemical flo-thru detectors and liquid chromatography detectors. Mailing Add: Dow Chemical Co 574 Bldg Midland MI 48640

MCLEAN, JAMES DOUGLAS, b Regina, Sask, Feb 12, 20; m 50; c 4. DENTISTRY. Educ: Univ Toronto, DDS, 42. Prof Exp: From lectr to asst prof dent, Univ Alta, 47-53; PROF DENT, DALHOUSIE UNIV, 53-, DEAN FAC DENT, 54- Mem: Fel Int Col Dent; fel Am Col Dent. Mailing Add: Fac of Dent Dalhousie Univ Halifax NS Can

MCLEAN, JAMES H, b Detroit, Mich, June 17, 36. INVERTEBRATE ZOOLOGY. Educ: Wesleyan Univ, BA, 58; Stanford Univ, PhD(biol), 66. Prof Exp: CURATOR INVERT ZOOL, LOS ANGELES COUNTY MUS NATURAL HIST, 64- Mem: Am Malac Union; Soc Syst Zool. Res: Systematics of marine mollusks, especially prosobranch gastropods and chitons of the eastern Pacific. Mailing Add: Invert Zool 900 Exposition Blvd Los Angeles CA 90007

MCLEAN, JOHN A, JR, b Chapel Hill, Tenn, Nov 8, 26; m 58; c 3. INORGANIC CHEMISTRY. Educ: Tenn State Univ, BS, 48; Univ Ill, MS, 56, PhD(chem), 59. Prof Exp: High sch teacher, Ill, 48-53; PROF CHEM, UNIV DETROIT, 59-, ASST CHMN UNDERGRAD PROGS, 74- Mem: Am Chem Soc. Res: Synthesis, structure and reaction kinetics and mechanisms of transition metal coordination compounds. Mailing Add: Dept of Chem Univ of Detroit Detroit MI 48221

MCLEAN, JOHN ROBERT, b St Thomas, Ont, Apr 15, 26; m 51; c 2. BIOCHEMISTRY. Educ: Queen's Univ, Ont, BSc, 50, PhD(biochem), 54. Prof Exp: Instr biochem, Yale Univ, 54-56; assoc res chemist, 56-60, sr res biochemist, 60-72, SECT DIR NEUROPHARMACOL, PARKE DAVIS & CO, 72- Mem: Am Soc Pharmacol & Exp Therapeut; Soc Neurosci. Res: Drugs acting on the central nervous system. Mailing Add: Res Div Parke Davis & Co Ann Arbor MI 48106

MCLEAN, KATHARINE WEIDMAN, b Camden, NJ, Oct 18, 27; m 51; c 3. BIOCHEMISTRY. Educ: Cornell Univ, AB, 48; Univ Ill, MS, 49, PhD(biochem), 51. Prof Exp: Org res chemist, Rohm and Haas Co, 51-52; anal chemist, Ariz Testing Labs, 52-57; TEACHER CHEM, PHOENIX COL, 57- Mem: Am Chem Soc; Am Inst Chemists. Res: Vitamin BT; explosives. Mailing Add: Dept of Chem Phoenix Col Phoenix AZ 85013

MCLEAN, MAX C, b Ogden, Utah, Apr 20, 23; m 48; c 6. PHYSICAL OCEANOGRAPHY. Educ: US Merchant Marine Acad, BS, 52. Prof Exp: Oceanog technician, Woods Hole Oceanog Inst, 48; oceanog technician, US Navy Hydrographic Off, 48-50, oceanographer, 50-51, 54-59; oceanographer, Off Naval Res, 59-60, scientific liason officer, Boston Br Off, Woods Hole Oceanog Inst, 60-64; asst to dir, Mohole Proj, Nat Sci Found, 64-67; spec asst oceanog to assoc dir, US Coast & Geod Surv, 67-69; oceanogr, Environ Sci Serv Admin, 69-70; chief, Div Marine Plans, Maritime Admin, 70-72; prog assoc sci youth progs, Sci Serv, 72; CONSULT, 72- Concurrent Pos: Mem sci adv comt, US Comn, Int Comn Northwest Atlantic Fisheries, 61-64; lectr, Vis Scientist Prog, IBM Corp, 64-; exec officer, Nat Coun Marine Sci & Eng Develop, Off of the President, 67-69; consult, Appl Data Res & Planning Res Corp, 74- Mem: AAAS; Am Geog Soc; Am Soc Limnol & Oceanog; Am Geophys Union; Am Polar Soc. Res: Physical and dynamic processes in the ocean; chemical oceanography; oceanographic instrumentation and general physical oceanography; research administration in marine sciences; ocean engineering; maritime affairs. Mailing Add: 9232 Chapel Hill Terr Fairfax VA 22030

MCLEAN, NORMAN, JR, b San Diego, Calif, May 8, 26; m 63; c 1. INVERTEBRATE ZOOLOGY. Educ: Univ Calif, Berkeley, BS, 51, PhD(zool), 65. Prof Exp: Asst prof to assoc prof, 65-74, PROF ZOOL, SAN DIEGO STATE UNIV, 74- Res: Functional morphology of the molluscan digestive tract. Mailing Add: Dept of Zool San Diego State Univ San Diego CA 92115

MCLEAN, RICHARD ALFRED, pharmacology, see 12th edition

MCLEAN, RICHARD BEA, b Raleigh, NC, Aug 27, 46; m 68; c 1. MARINE BIOLOGY, ANIMAL BEHAVIOR. Educ: Fla State Univ, BA, 68, PhD(marine biol), 75. Prof Exp: Res asst spiny lobsters, Fla State Univ, 69-70, instr biol, 70-73; RES ASSOC MARINE ENVIRON IMPACT, OAK RIDGE NAT LAB, 74- Concurrent Pos: NASA fel, 72. Mem: Am Inst Biol Sci; Am Soc Zool; Animal Behav Soc; Sigma Xi. Res: Behavioral ecology of marine benthic communities, emphasizing interactions such as predation, symbiosis and competition. Mailing Add: Environ Sci Div Oak Ridge Nat Lab Oak Ridge TN 37830

MCLEAN, ROBERT A, statistics, mathematical statistics, see 12th edition

MCLEAN, ROBERT GEORGE, b Warren, Ohio, Jan 10, 38; m 60; c 2. EPIZOOTIOLOGY, VERTEBRATE ECOLOGY. Educ: Bowling Green State Univ, BSE & BS, 61, MA, 63; Pa State Univ, PhD(zool), 66. Prof Exp: Chief parasitol br, Third US Army Med Lab, Ft McPherson, Ga, 66-68; chief rabies ecol subunit, 68-69, mem rabies control unit, Viral Zoonoses Sect, Viral Dis Br, Epidemiol Prog, 69-73, MEM VERTEBRATE ECOL BR & VECTOR-BORNE DIS BR, CTR DIS CONTROL, BUR LABS, 73- Concurrent Pos: Part-time instr biol, Ga State Univ, 67-68; fac affil zool, Colo State Univ, 73- Mem: Wildlife Dis Asn; Am Soc Trop Med & Hygiene. Res: Homing ability and courtship behavior of pigeons; population control of vertebrate pest animals with chemosterilants; ecological studies of zoonotic diseases in birds and mammals; ecology of Colorado tick fever in northern Colorado. Mailing Add: Vector-Borne Dis Br Bur of Labs Ctr for Dis Control PO Box 2087 Ft Collins CO 80522

MCLEAN, ROBERT J, b New Haven, Conn, Aug 15, 40; m 63; c 4. PHYCOLOGY. Educ: Univ Conn, BA, 62, MS, 64, PhD(phycol), 67. Prof Exp: NIH fel bot, Univ Tex, 67-68; asst prof, 68-71, ASSOC PROF BIOL SCI, STATE UNIV NY COL BROCKPORT, 71- Mem: AAAS; Am Inst Biol Sci; Phycol Soc Am; Bot Soc Am; Int Phycol Soc. Res: Algal physiology and ultrastructure. Mailing Add: Dept of Biol Sci State Univ of NY Brockport NY 14420

MCLEAN, ROBERT L, biochemistry, organic chemistry, see 12th edition

MCLEAN, ROBERT T, b Westerville, Ohio, July 18, 22; m 53. ALGEBRA. Educ: Otterbein Col, BS, 46; Bowling Green State Univ, MA, 50; Univ Pittsburgh, PhD(group theory), 61. Prof Exp: Instr high sch, Ohio, 45-49, 50-51; asst chem, Bowling Green State Univ, 49-50; from asst prof to prof math, 52-67, head dept, 57-67, chmn div natural sci, 65-67; CHMN DEPT MATH, COL ARTS & SCI, LOYOLA UNIV, LA, 67- Mem: Am Math Soc; Math Asn Am. Res: Theory of groups; higher education program design; institutional research. Mailing Add: Dept of Math Col of Arts & Sci Loyola Univ St Charles Ave New Orleans LA 70118

MCLEAN, STEWART, b Moascar, Egypt, Nov 19, 31; m 57; c 3. ORGANIC CHEMISTRY. Educ: Glasgow Univ, BSc, 54; Cornell Univ, PhD(org chem), 58. Prof Exp: Fel org chem, Univ Wis, 57-58; fel, Nat Res Coun Can, 58-60; from asst prof to assoc prof, PROF ORG CHEM, UNIV TORONTO, 70- Mem: Am Chem Soc; fel Chem Inst Can; The Chem Soc. Res: Structural and synthetic organic chemistry; mechanistic studies. Mailing Add: Dept of Chem Univ of Toronto 80 St George St Toronto ON Can

MCLEAN, WILLIAM BURDETTE, b Portland, Ore, May 21, 14; m 39; c 3. PHYSICS. Educ: Calif Inst Technol, BS, 35, MS, 37, PhD(physics), 39. Prof Exp: Instr physics, Calif Inst Technol, 35-39; res assoc, Univ Iowa, 39-41; res physicist, Nat Bur Standards, 41-45; res physicist, Naval Ord Test Sta, 45-50, head aviation ord dept, 50-54, tech dir, 54-67; tech dir, Naval Undersea Res & Develop Ctr, 67-74. Concurrent Pos: Fel, Univ Iowa, 39-41; mem mil aircraft panel, President's Sci Adv Comt, 64-68. Honors & Awards: Fed Govt Award, 56; President's Award Distinguished Fed Civilian Serv, 58; Rockefeller Pub Serv Award Sci, Technol & Eng, 65; Blandy Gold Medal, Am Ord Asn, 60; Harry Diamond Award, Inst Elec & Electronics Eng, 72. Mem: AAAS; Am Phys Soc; Am 11 Inst Aeronaut & Astronaut; NY Acad Sci, Nat Acad Sci. Res: Aviation ordnance equipment; naval ordnance projects; ocean engineering. Mailing Add: Naval Undersea Res & Develop Ctr San Diego CA 92132

MCLEAN, WILLIAM L, b Dunedin, NZ, May 22, 33. PHYSICS. Educ: Univ Otago, NZ, BSc, 54, MSc, 56; Cambridge Univ, PhD(physics), 60. Prof Exp: Asst lectr physics, Univ Otago, NZ, 56-57; instr, Univ BC, 60-61; asst prof, 61-66, ASSOC PROF PHYSICS, RUTGERS UNIV, NEW BRUNSWICK, 66- Mem: Am Phys Soc. Res: Radio-frequency techniques on superconductors; propagation of helicon waves through metals and the relation to the band structure; superconductivity; low temperature, solid state and statistical physics. Mailing Add: Dept of Physics Rutgers Univ New Brunswick NJ 08903

MCLEAN, WILLIS JOHN, crystallography, mineralogy, see 12th edition

MCLEES, BYRON D, b Little Rock, Ark, Oct 5, 37; m 61. BIOCHEMISTRY, MEDICINE. Educ: Univ Ark, BS, 54, MS, 59; Johns Hopkins Univ, PhD(biochem), 64; Duke Univ, MD, 68. Prof Exp: Assoc med, Med Ctr, Duke Univ, 66-69; chief lab non-ionizing radiation, Naval Med Res Inst, Md, 69-72; CHIEF RESIDENT & ASST PROF MED, MED CTR, DUKE UNIV, 72- Concurrent Pos: Fel rheumatol, Med Ctr, Duke Univ, 65-66. Mem: AAAS; Am Chem Soc. Res: Chemistry of metalloorganic compounds; magnetic and spectral properties of porphyrins and metallo porphyrins; mechanism of electron transport in biological systems; role of fibrous proteins in athero genesis. Mailing Add: Dept of Med Duke Univ Med Ctr Durham NC 27710

MCLEESE, DONALD WILSON, fisheries biology, see 12th edition

MCLELLAN, ALDEN, IV, b Meridian, Miss, Mar 7, 36; m 72. ENVIRONMENTAL MANAGEMENT, RESOURCE MANAGEMENT. Educ: Univ Calif, Berkeley, BA, 60, MA, 64; Univ Nev, PhD(physics), 67. Prof Exp: Theoret Naval Air Missile Test Ctr, 57; res physicist, Lawrence Radiation Lab, Univ Calif, 59; res physicist, Univ Calif, Berkeley, 64; fel physics, Univ Nev, 64-65; res assoc, Desert Res Inst, 67-69, dep dir lab space res, 69; res scientist, Atmospheric Pollution Prog, Space Sci & Eng Ctr, Univ Wis-Madison, 69-72, sr scientist, Environ Res Group, Inst Environ Studies, 73-75; PRES, IMPACT ENVIRON RES, 73- Concurrent Pos: Europ Space Res Orgn res asst, Europ Space Res Inst, Italy, 67; vis res prof, Int Ctr Theoret Physics, Int Atomic Energy Agency, 67-68; res lectr, Nat Ctr Atmospheric Res, Colo, 68. Mem: AAAS; Am Phys Soc; Am Inst Aeronaut & Astronaut; Am Geophys Union; Am Nuclear Soc. Res: Theoretical optics; radar analysis; global atmospheric pollution; neutron-proton nuclear evaporation theory; theoretical plasms and space physics; astrophysics; solar and atmospheric physics; computational mathematics; air and water pollution; environmental impact from land use; telecommunications. Mailing Add: Impact Environ Res 2020 Kendall Ave Madison WI 53705

MCLELLAN, CRAWFORD REID, b Lexington, Miss, Oct 18, 06; m 42; c 6. ORGANIC CHEMISTRY. Educ: Miss Col, BA, 29; Univ NC, MA, 30; La State Univ, PhD, 41. Prof Exp: Prof chem, Ark Agr & Mech Col, 30-38 & La Col, 41-42; civilian chief chemist, Cent Lab, Ark Arsenal, 42-45; prof, 45-74, EMER PROF CHEM, LA STATE UNIV, BATON ROUGE, 74- Mem: Am Chem Soc. Res: Preparation, properties and dehydration study of 1-alkylcyclopentanols, 2-methyl cyclopenthanols; secondary cyclopentyl alcohols. Mailing Add: Dept of Chem La State Univ Baton Rouge LA 70803

MCLELLAN, HUGH JOHN, b Sydney, NS, Mar 16, 21; nat US; m 51; c 4. OCEANOGRAPHY. Educ: Dalhousie Univ, BSc, 41, MSc, 47; Uni Calif, Los Angeles, PhD(phys oceanog), 56. Prof Exp: Jr physicist, Nat Res Coun Can, 41-42; assoc oceanogr, Atlantic Oceanog Group, Fisheries Res Bd Can, 47-56, sr scientist, 56-57; res scientist, Tex A&M Univ, 57-58, from asst prof to prof, 58-64; oceanogr, Off Naval Res, Washington, DC, 64-67; head oceanog sect, Div Environ Sci, NSF, 67-69; assoc sci & tech dir, US Naval Oceano Off, 69-71; assoc prog dir, Nat Sea Grant Prog, 71-75, DIR, GRANTS MGT, NAT OCEANIC & ATMOSPHERIC ADMIN, DEPT OF COM, 75- Mem: Am Soc Limnol & Oceanog; Am Geophys Union. Res: Physical oceanography. Mailing Add: Nat Oceanic & Atmospheric Admin Washington DC 20235

MCLELLAN, WILLIAM L, JR, b Boston, Mass, Feb 12, 30; m 57; c 6. BIOCHEMISTRY, MICROBIOLOGY. Educ: Boston Col, AB, 51; Boston Univ, MA, 54, PhD(biochem), 57. Prof Exp: Instr biochem, Sch Med, Boston Univ, 57-58; res assoc, Inst Microbiol, Rutgers Univ, 62-65, asst prof, 65-67, assoc res prof, 67-69; asst prof path, Col Physicians & Surgeons, Columbia Univ, 69-74; vis investr molecular biol, Albert Einstein Col Med, 73-75; SCIENTIST BASIC RES, NAT CANCER INST, FREDERICK CANCER RES CTR, MD, 75- Concurrent Pos: USPHS fel microbiol, Sch Med, NY Univ, 58-61. Mem: Am Soc Microbiol. Res: Tumor cell surface antigens; membrane glycoproteins of C-type virus; virally transfromed cells and turmors; regulation of viral expression. Mailing Add: Frederick Cancer Res Ctr PO Box B Frederick MD 21701

MCLEMORE, BENJAMIN HENRY, JR, b Memphis, Tenn, Oct 4, 24; m; c 2. MATHEMATICAL STATISTICS. Educ: Dillard Univ, AB, 44; Univ Ill, MS, 52, AM, 55, PhD(statist), 59. Prof Exp: High sch instr, Tex, 44-45; instr math, Dillard Univ, 45-47; instr, Jackson State Col, 47-53, head dept, 56-64; ASSOC PROF MATH, CLEVELAND STATE UNIV, 64- Mem: Am Statist Asn; Math Asn Am; Inst Math Statist. Mailing Add: Dept of Math Cleveland State Univ Cleveland OH 44115

MCLEMORE, BOBBIE FRANK, b Jasper, Tex, May 22, 32; m 50; c 2. PLANT PHYSIOLOGY. Educ: Tex A&M Univ, BS, 53; La State Univ, MS, 57, PhD(forestry), 67. Prof Exp: Lab asst bot, Tex A&M Univ, 50-53; res asst agron, Tex Agr Exp Sta, 55; SILVICULTURIST, SOUTHERN FOREST EXP STA, US FOREST SERV, 57- Mem: Soc Am Foresters. Res: Storage, processing, testing and dormancy of southern pine seed; pine seed, cone and conelet physiology. Mailing Add: Southern Forest Exp Sta US Forest Serv Pineville LA 71360

MCLEMORE, WILLIE O, b McGregor, Ga, July 16, 38; m 63; c 2. BIOCHEMISTRY. Educ: Central State Univ, BS, 60; Iowa State Univ, PhD(biochem), 64. Prof Exp: Chief clin chem, Army Med Lab, Ft Sam Houston, Tex, 65-67; ASST PROF BIOCHEM & DIR CLIN BIOCHEM, COL MED, HOWARD UNIV, 67- Mem: AAAS: Am Chem Soc. Res: Study of the purification of enzymology; kinetics and mechanism of action. Mailing Add: Dept of Biochem Howard Univ Col of Med Washington DC 20001

MCLENDON, DAVID MARK, b Lufkin, Tex, Feb 18, 47; m 69. PSYCHOPHARMACOLOGY. Educ: NTex State Univ, MS, 70; Univ Houston, BS, 69, PhD(psychol), 74. Prof Exp: Res asst behav pharmacol, Tex Res Inst Ment Sci, 70-72; res asst, 72-74, NAT INST DRUG ABUSE FEL PSYCHOPHYSIOL, BAYLOR COL MED, 74- Concurrent Pos: Staff psychologist, Fabre Clin, Houston, 74- Mem: Am Psychol Asn; Soc Neurosci. Res: Abuse liability of marihuana studied in rhesus monkeys; aggression and marihuana in rats; relationship of stress and drug use in monkeys; efficacy of benzodiazepines in humans. Mailing Add: Dept of Physiol Baylor Col of Med Tex Med Ctr Houston TX 77030

MCLENDON, WILLIAM WOODARD, b Durham, NC, Oct 29, 30; m 52; c 3. PATHOLOGY, LABORATORY MEDICINE. Educ: Univ NC, BA, 53, MD, 56. Prof Exp: Intern & resident path, Columbia-Presby Med Ctr, New York, 56-58; resident, Univ NC, Chapel Hill, 58-61; asst chief path serv, US Army Hosp, Landstuhl, Ger, 61-63; assoc pathologist, Moses Cone Hosp, Greensboro, NC, 63-69, dir labs, 69-73; PROF PATH, SCH MED, UNIV NC, CHAPEL HILL, 73-, CHMN DEPT HOSP LABS, NC MEM HOSP, 73- Concurrent Pos: Asst chief ed, Arch Path, 74- Mem: AAAS; Soc Comput Med; Acad Clin Lab Physicians & Sci; Col Am Path; Am Soc Clin Path. Res: Endocrine pathology; automation and computerization of the clinical laboratory. Mailing Add: Dept of Hosp Labs NC Mem Hosp Chapel Hill NC 27514

MCLENNAN, BARRY DEAN, bBracken, Sask, Apr 15, 40; m 63; c 2. BIOCHEMISTRY. Educ: Brandon Univ, BSc, 60; Univ Sask, MSc, 63; Univ Alta, PhD(biochem), 66. Prof Exp: Instr chem, Brandon Univ, 60-61; asst prof biochem, 69-73, ASSOC PROF BIOCHEM, UNIV SASK, 73- Concurrent Pos: Inst fel biochem, Roswell Park Mem Inst, 66-67, Nat Cancer Inst Can fel, 66-68; fel McMaster Univ, 67-69; Med Res Coun Can grant, Univ Sask, 69- Mem: AAAS; Can Biochem Soc. Res: Structure and chemistry of nucleic acids; metabolism of modified nucleosides in transfer RNA. Mailing Add: Dept of Biochem Univ of Sask Col of Med Saskatoon SK Can

MCLENNAN, CHARLES EWART, b Duluth, Minn, Dec 26, 09; m 37; c 4. OBSTETRICS & GYNECOLOGY. Educ: Univ Minn, BA, 30, MA, 32, MD, 34, PhD(obstet & gynec), 42; Am Bd Obstet & Gynec, dipl, 42. Prof Exp: From instr to assoc prof obstet & gynec, Univ Minn, 38-44; prof, Univ Utah, 44-47; prof & head dept, 47-75, EMER PROF OBSTET & GYNEC, SCH MED, STANFORD UNIV, 75-; CONSULT GYNEC, MENLO MED CLIN, MENLO PARK, CALIF, 75- Concurrent Pos: Dir, Am Bd Obstet & Gynec, 66-72. Mem: AAAS; Am Gynec Soc (pres, 73); Soc Gynec Invest (pres, 62); AMA; Am Fedn Clin Res. Res: Gynecological histopathology and cancer; exfoliative cytology. Mailing Add: Dept of Gynec & Obstet Stanford Univ Med Ctr Stanford CA 94305

MCLENNAN, DONALD ELMORE, b London, Ont, Dec 5, 19; m 43; c 5. ELECTRODYNAMICS. Educ: Univ Western Ont, BA, 41; Univ Toronto, PhD(physics), 50. Prof Exp: Res scientist ballistics, Can Armament Res & Develop Estab, 50-59; prof physics, Col William & Mary, 59-67; PROF PHYSICS & ASTRON, YOUNGSTOWN STATE UNIV, 67- Res: Unified field theory. Mailing Add: Dept of Physics & Astron Youngstown State Univ Youngstown OH 44555

MCLENNAN, HUGH, b Montreal, Que, Oct 22, 27; m 49; c 2. PHYSIOLOGY. Educ: McGill Univ, BSc, 47, MSc, 49, PhD(biochem), 51. Prof Exp: Asst lectr biophys, Univ Col, Univ London, 52-53; res fel, Montreal Neurol Inst, 53-55; asst prof physiol, Dalhousie Univ, 55-57; from asst prof to assoc prof, 57-65, PROF PHYSIOL, UNIV BC, 65- Mem: Am Physiol Soc; Can Physiol Soc (secy, 65-); Brit Biochem Soc; Brit Physiol Soc. Res: Neurophysiology. Mailing Add: Dept of Physiol Univ of BC Vancouver BC Can

MCLENNAN, JAMES ALAN, JR, b Atlanta, Ga, Nov 24, 24; m 52; c 2. STATISTICAL MECHANICS. Educ: Harvard Univ, AB, 48; Lehigh Univ, MS, 50, PhD(physics), 52. Prof Exp: Tech engr, Gen Elec Co, 52-53; from instr to assoc prof, 53-62, PROF PHYSICS, LEHIGH UNIV, 62-, CHMN DEPT, 68- Concurrent Pos: Nat Sci Found fel, Lehigh Univ, 60-61; consult, Los Alamos Sci Lab. Mem: Fel Am Phys Soc; Sigma Xi. Res: Quantum theory of elementary particles. Mailing Add: Dept of Physics Lehigh Univ Bethlehem PA 18015

MCLEOD, DONALD WINGROVE, b Rochester, NY, Feb 15, 35; m 58; c 3. HIGH ENERGY PHYSICS. Educ: Univ Rochester, BS, 56; Cornell Univ, PhD(exp physics), 62. Prof Exp: Res assoc, Argonne Nat Lab, 62-64, asst scientist, 64-66; ASST PROF EXP HIGH ENERGY PHYSICS, UNIV ILL, CHICAGO CIRCLE, 66- Concurrent Pos: Part-time instr, Univ Ill, Chicago, 64-66. Mem: Am Phys Soc. Res: Experimental high energy physics, using counter and spark chamber techniques, emphasis on strong interactions. Mailing Add: Dept of Physics Univ of Ill Box 4348 Chicago IL 60680

MCLEOD, EDWARD BLAKE, b Los Angeles, Calif, July 25, 24. MATHEMATICS. Educ: Occidental Col, BA, 47, MS, 49; Stanford Univ, PhD(math), 54. Prof Exp: Mathematician, NAm Aviation, Inc, 47-48; asst, Stanford Univ, 51-53; instr math, Univ Colo, 53-55; asst prof, Ore State Col, 55-63; sr mathematician, Dynamics Sci Corp, 63-64; assoc prof, 64-72, PROF MATH, CALIF STATE UNIV, LONG BEACH, 72- Mem: Am Math Soc; Math Asn Am; Soc Indust & Appl Math; Am Inst Aeronaut & Astronaut. Res: Complex variables; fluid dynamics; time series forecasting and applications of mathematics to biology. Mailing Add: Dept of Math Calif State Univ Long Beach CA 90804

MCLEOD, GERALD LOUIS, organic chemistry, see 12th edition

MCLEOD, GUY COLLINGWOOD, b Brockton, Mass, Feb 18, 28; m 50; c 3. PLANT PHYSIOLOGY. Educ: Harvard Univ, BA, 51; Trinity Col, MS, 54. Prof Exp: Asst geochem, Boston Univ, 51-52; sr fel, Sias Inst Res, Brooks Hosp, Brookline, Mass, 58-59, mem, 61-63; res fel, Dept Plant Biol, Carnegie Inst, 59-61; group leader photobiol, Air Force Cambridge Res Labs, Mass, 63-64; sr scientist, Tyco Labs, 64-66, asst dept head phys-chem, 66-69; DIR RES, NEW ENG AQUARIUM, 69- Concurrent Pos: Res assoc, Med Sch, Tufts Univ, 64-; lectr, Dept Environ Sci, Univ Mass, 67-; adj prof, Boston Univ, 69- Mem: Am Soc Plant Physiol; Biophys Soc; NY Acad Sci; Scandinavian Soc Plant Physiol. Res: Photosynthesis; cellular physiology. Mailing Add: New Eng Aquarium Central Wharf Boston MA 02110

MCLEOD, HENRY GEORGE, b Scranton, Pa, Jan 28, 21; m 51; c 2. ELECTROCHEMISTRY. Educ: Univ Toronto, BA, 43, MA, 44, PhD(chem), 46. Prof Exp: Res fel polymers, Ont Res Found Can, 46-48; res chemist, Electrochem Dept, E I du Pont de Nemours & Co, 48-57, sr res chemist, 57-63; assoc prof, 63-66, chmn dept, 64-70, PROF CHEM, UNIV WATERLOO, 66- Res: Electrodeposition and corrosion mechanisms. Mailing Add: Dept of Chem Univ of Waterloo Waterloo ON Can

MCLEOD, JAMES ARCHIE, b Rosedale, Man, Sept 1, 04; m 38; c 2. ZOOLOGY. Educ: Univ Man, BSc, 31, MSc, 33, PhD(zool), 39. Prof Exp: Instr zool, Univ Man, 32-33, Univ Western Ont, 33-34 & Univ Minn, 36-37; lectr, Univ Man, 37-44, from asst prof to prof, 44-66, head dept, 56-65; prof & head dept, 66-74, EMER PROF ZOOL, BRANDON UNIV, 74- Mem: Can Soc Zoologists. Res: Parasitology; ecology; parasitological survey of the genus Citellus in Manitoba; cercarial dermatitis; animal ecology. Mailing Add: Dept of Zool Brandon Univ Brandon MB Can

MCLEOD, JOHN MALCOLM, b Montreal, Que, Oct 8, 28; m 54; c 3. ENTOMOLOGY. Educ: Univ NB, BSc, 53; State Univ NY Col Forestry, Syracuse Univ, MSc, 55, PhD(insect ecol), 61. Prof Exp: RES OFFICER POPULATION ECOL FOREST INSECTS, CAN DEPT FORESTRY, CAN DEPT ENVIRON, 55- Mem: Entom Soc Can; Can Inst Forestry; Entom Soc Am; Can Pulp & Paper Asn. Res: Taxonomy and bionomics of microlepidoptera; population ecology; Swaine jack pine sawfly; mammalian and avian predators of Neodiprion swainei. Mailing Add: Laurentian Forest Res Centre CP 3800 1080 DuVallon Ste Foy PQ Can

MCLEOD, KENNETH NEIL, b Butte, Mont, Mar 5, 11; m 37; c 2. PHYSICAL CHEMISTRY. Educ: Mont State Col, BS, 32; Ore State Col, MS, 35, PhD(phys chem), 39. Prof Exp: Asst, Ore State Col, 34-36, chem analyst, 36-37; from instr to assoc prof chem, Willamette Univ, 40-46; assoc prof, 46-56, PROF CHEM, MONT COL MINERAL SCI & TECHNOL, 56-, HEAD DEPT, 57-, DEAN ACAD AFFAIRS, 66- Mem: Am Chem Soc. Res: Methods of chemical analysis; mineral dressing problems and metal recovery; surface active agents as applied to flotation; alkaline sulfide leach of cinnabar ore and electrolytic deposition of mercury from the leach solution. Mailing Add: Dept of Chem Mont Col Mineral Sci & Technol Butte MT 59701

MCLEOD, KENNETH WILLIAM, b Miami, Okla, Oct 14, 47. PLANT ECOLOGY. Educ: Okla State Univ, BS, 69, MS, 71; Mich State Univ, PhD(plant ecol), 74. Prof Exp: RES ASSOC PLANT ECOL, SAVANNAH RIVER ECOL LAB, UNIV GA, 74- Mem: Ecol Soc Am; Am Inst Biol Sci; AAAS. Res: Factors that govern the establishment and distribution of plant populations, especially seed germination and subsequent seedling growth. Mailing Add: Savannah River Ecol Lab Drawer E Aiken SC 29801

MCLEOD, LIONEL EVERETT, b Wainwright, Alta, Aug 9, 27; m 52; c 4. MEDICAL EDUCATION, MEDICAL ADMINISTRATION. Educ: Univ Alta, BSc, 49, MD, 51; McGill Univ, MSc, 56; FRCP, 57; FRCPS(C). Prof Exp: Attend physician, Univ Alta Hosp, 58-69, assoc prof med, Univ Alta, 59-69; PROF MED & HEAD DEPT, UNIV CALGARY, 69-, DEAN FAC MED, 73- Mem: Am Soc Artificial Internal Organs; Can Soc Clin Invest; Can Fedn Biol Soc. Res: Endocrinology and metabolism; application of intermittent hemodialysis in chronic renal failure. Mailing Add: Fac of Med Univ of Calgary Calgary AB Can

MCLEOD, LLOYD ALEXANDER, b Calgary, Alta, July 27, 18; m 44; c 4. PHYSICAL CHEMISTRY, RESEARCH ADMINISTRATION. Educ: Univ Alta, BSc, 41, MSc, 43; McGill Univ, PhD(phys chem), 44. Prof Exp: Lectr, Univ Alta, 41-42; res chemist, Nat Res Coun, Can, 44-50; res chemist, Polymer Corp, Ltd, 50-51, group leader res & develop, 51-59, proj supvr, 59-62, mgr, Employee Rels Div, 62-64, res & develop div, 64-71; COORDR RES GRANTS & CONTRACTS, ONT MINISTRY COLS & UNIVS, 72- Mem: Am Chem Soc; fel Chem Inst Can. Res: Polymerization and properties of elastomers; production of fluorine; osomotic pressure techniques; physical properties of synthetic rubbers; contracting research relating to policy formulation and performance evaluation of Ontario's universities and community colleges. Mailing Add: 117 Gloucester Ave Oakville ON Can

MCLEOD, RICHARD KENNETH, b Shreveport, La, Sept 30, 33; m 62; c 2. PHYSICAL ORGANIC CHEMISTRY. Educ: Rice Univ, BA, 55; Univ Va, MS, 60, PhD(phys org chem), 63. Prof Exp: Instr chem, Univ Va, 61-62; asst prof phys chem, Tex Christian Univ, 63-66; sr chemist, 66-70, PROCESS SPECIALIST, PROCESS TECHNOL DEPT, MONSANTO CO, 70- Mem: AAAS; Am Chem Soc. Res: Kinetics and mechanisms of organic reactions; chemistry of imines; Friedel-Crafts reactions; environmental chemistry. Mailing Add: Process Technol Dept Monsanto Co PO Box 1311 Texas City TX 77590

MCLEOD, ROBERT MELVIN, b Newco, Miss, June 19, 29; m 65; c 2. MATHEMATICAL ANALYSIS. Educ: Miss State Univ, BS, 50; Rice Univ, MA, 53, PhD(math), 55. Prof Exp: Instr math, Duke Univ, 55-58, asst prof, 58-61; assoc prof, Am Univ Beirut, 61-65; assoc p prof, Univ Tenn, 65-66; ASSOC PROF MATH, KENYON COL, 66- Mem: Am Math Soc; Math Asn Am. Res: Function theory. Mailing Add: Box 187 Gambier OH 43022

MCLEOD, WILLIAM D, b Toronto, Ont, Nov 16, 30; m 56; c 1. BIOMEDICAL ENGINEERING. Educ: Univ Toronto, BASc, 58, MAS, 61; Case Western Reserve Univ, PhD(mech eng), 65. Prof Exp: Asst dir cybernetic systs group, Case Western Reserve Univ, 64-66; asst prof mech eng & res assoc bioeng inst, Univ NB, 66-68; dir bioeng res, Insts Achievement Human Potential, Pa, 68-69; ASST PROF PHYS MED, EMORY UNIV, 69- Mem: Inst Elec & Electronics Eng; Int Soc Electromyographic Kinesiology. Res: Information processing from bio-electric signals, primarily myo-electric signals; electroencephalographic signals; human operator modelling with handicapped people using electromyographic and electroencephalographic signals. Mailing Add: Ga Ment Health Inst 1256 Briarcliff Rd NE Atlanta GA 30306

MCLEOD, WILLIAM STIRLING, b Calgary, Alta, Dec 15, 12; m 43; c 2. ENTOMOLOGY. Educ: Univ Alta, BSc, 39; McGill Univ, MSc, 43. Prof Exp: Asst prof entom, Univ Man, 43-48; chief entomologist, Green Cross Div, Sherwin-Williams Co, 48-49; insecticide specialist, Fruit Insects Unit, Entom Div, Can Dept Agr, 49-52, officer analyst, Pesticide Lab, Plant Prod Div, 52-60, supvr pesticide registrations, 60-64; asst mgr field res sect, 64-68, mgr field res sect, 68-73, MGR RES SCHEDULING, CHEMAGRO CORP, KANSAS CITY, 73- Mem: Entom Soc Am. Res: Biological control of forest insects; chemical control of vegetables, stored products and livestock insects; bioassay of insecticides. Mailing Add: 8232 Reeds Lane Prairie Village KS 66208

MCLERAN, JAMES HERBERT, b Audubon, Iowa, Apr 9, 31; m 57; c 1. DENTISTRY, ORAL SURGERY. Educ: Simpson Col, BS & BA, 53; Univ Iowa, DDS, 57, MS, 62; Am Bd Oral Surg, dipl, 64. Prof Exp: Instr oral surg, Univ Iowa, 59-60; resident, Univ Hosps, Iowa City, Iowa, 60-62; pvt pract, Calif, 62-63; from asst prof to assoc prof oral surg, Univ Iowa, 63-69; prof & chmn dept, Sch Dent, Univ NC, Chapel Hill, 69-72; assoc dean, 72-74, PROF ORAL SURG, COL DENT, UNIV IOWA, 72-, DEAN COL, 74- Concurrent Pos: Mem dent educ rev comt, Dept Health, Educ & Welfare, 69- Mem: Int Asn Dent Res; Am Soc Oral Surg; Am Dent Asn; Am Soc Advan Gen Anesthesia Dent. Res: Pain control in dentistry; temporomandibular joint disfunction; bacteremia following oral surgical procedures. Mailing Add: Dept of Oral Surg Univ of Iowa Col of Dent Iowa City IA 52242

MCLEROY, CAROL ANN, geology, palynology, see 12th edition

MCLEROY, EDWARD GLENN, b Atlanta, Ga, June 23, 26; c 2. PHYSICS. Educ: Emory Univ, BA, 49, MS, 51. Prof Exp: Asst math, Ga Inst Technol, 49; physicist, US Navy Mine Defense Lab, 51-53, 54-56; asst physics, Brown Univ, 53-54; asst prof, Marine Lab, Univ Miami, 57; head acoustics sect, US Navy Mine Defense Lab, 57-72; PHYSICIST, NAVAL COASTAL SYSTS LAB, 72- Mem: AAAS; Acoust Soc Am; Am Asn Physics Teachers; Am Geophys Union; Soc Explor Geophys. Res: Marine physics, particularly marine acoustics. Mailing Add: Box 4647 Panama City FL 32401

MCLICK, JEROME, organic chemistry, biochemistry, see 12th edition

MCLIMANS, WILLIAM FLETCHER, b Duluth, Minn, Aug 22, 16; m 40; c 3. VIROLOGY, MICROBIOLOGY. Educ: Univ Minn, BA, 38, PhD(bact, immunol), 46. Prof Exp: Res virologist, Minn Dept Health, 41-42; from instr to asst prof bact & immunol, Sch Med, Univ Minn, 46-49; res virologist, Rocky Mt Lab, USPHS, 49-50; head bact dept, Upjohn Co, 50-54; res & assoc prof microbiol, Vet Sch, asst prof, Med Sch & assoc mem, Wistar Inst, Univ Pa, 54-59; supvry microbiologist & chief tissue cult unit, Commun Dis Ctr, USPHS, 59-61; CAREER SCIENTIST, ROSWELL PARK MEM INST, 61- Concurrent Pos: Consult, US Air Force Proj Big Ben, 54-58; mem cell cult comt, Nat Res Coun, 57-58. Honors & Awards: Presidential Award, Int Cong Poliomyelitis, 58. Mem: AAAS; Am Soc Microbiol; Sci Res Soc Am; Am Asn Immunol; fel NY Acad Sci. Res: Cell culture; host cell-virus interaction; culture of mammalian cells; chemotherapy; cell physiology. Mailing Add: Roswell Park Mem Inst 666 Elm St Buffalo NY 14203

MCLINDEN, LYNN, bUS 23 citizen. MATHEMATICS. Educ: Princeton Univ, AB, 65; Univ Wash, PhD(math), 71. Prof Exp: Vis asst prof math, Math Res Ctr, Univ Wis-Madison, 71-73; ASST PROF MATH, UNIV ILL, URBANA, 73- Concurrent Pos: NSF res grant, 75. Mem: Am Math Soc; Soc Indust & Appl Math. Res: Convex analysis and optimization theory, including conjugate duality, nonlinear programming and saddlepoint problems. Mailing Add: Dept of Math 273 Atlgeld Hall Univ of Ill at Urbana-Champaign Urbana IL 61801

MCLINTOCK, JOHN JAMES REID, b Glasgow, Scotland, Aug 18, 12; m 43; c 4. MEDICAL ENTOMOLOGY. Educ: Univ Man, BSc, 39; McGill Univ, PhD(med entom), 51. Prof Exp: Entomologist, Dept Health & Pub Welfare, Man, 42-48; Div Entom, Sci Serv, Can Dept Agr, 48-55; arthropod-borne dis control proj, WHO, Iran, 55-57; Entom Res Inst, Cent Exp Farm, 57-65, ENTOMOLOGIST, RES STA, RES BR, CAN DEPT AGR, 65- Mem: AAAS; Am Mosquito Control Asn; Entom Soc Can; Can Soc Zool; Royal Entom Soc London. Res: Ecology of western encephalitis in prairie provinces; mosquito ecology. Mailing Add: Res Sta Res Br Can Dept Agr Univ Campus Saskatoon SK Can

M'CLOSKEY, ROBERT THOMAS, b Los Angeles, Calif, Nov 10, 40; m 64; c 2. ECOLOGY. Educ: Univ Calif, Los Angeles, BA, 64; Calif State Col, Los Angeles, MA, 66; Univ Calif, Irvine, PhD(pop biol), 70. Prof Exp: Asst prof ecol, 70-75, ASSOC PROF BIOL, UNIV WINDSOR, 75- Concurrent Pos: Nat Res Coun Can fel, Univ Windsor, 70-; Dept Indian Affairs Nat Parks Br fel, 71-72. Mem: AAAS; Am Inst Biol Sci; Ecol Soc Am; Brit Ecol Soc; Am Soc Mammal. Res: Species diversity and coexistence; habitat selection. Mailing Add: Dept of Biol Univ of Windsor Windsor ON Can

MCLOUGHLIN, DONALD KEITH, b Fairbury, Ill, July 5, 23; m 46; c 2. PROTOZOOLOGY, PARASITOLOGY. Educ: George Washington Univ, BS, 48, MS, 49; Univ Ill, PhD, 55. Prof Exp: Fishery biologist, US Fish & Wildlife Serv, 55; PARASITOLOGIST, ANIMAL PARASITOL INST, AGR RES SERV, USDA, 55- Mem: AAAS; Am Soc Parasitol; Soc Protozool. Res: Chemotherapy of protozoan diseases; control and prevention of parasitic diseases; parasite physiology. Mailing Add: Animal Parasitol Inst Beltsville Agr Res Ctr E Beltsville MD 20705

MCMAHAN, CHALMERS ALEXANDER, bSeneca, SC, July 4, 14; m 36; c 2. APPLIED STATISTICS, DEMOGRAPHY. Educ: Clemson Univ, BS, 35; Univ Ga, MA, 46; Vanderbilt Univ, PhD(demog), 49. Prof Exp: From asst prof to assoc prof, Univ Ga, 48-53; chief, Manpower Res Br, Air Force Personnel Res Lab, 53-57; PROF BIOSTATIST, MED CTR, LA STATE UNIV, NEW ORLEANS, 57-, HEAD DEPT BIOMETRY & DIR CLIN TRIALS RES CTR, 62- Concurrent Pos: Res scientist, Human Resources Res Inst, 51-52; vis lectr, Tulane Univ La, 60-61; vis prof, Univ Minn, 61 & Univ NC, 63; vis lectr, Mass Inst Technol, 64; vis prof, Yale Univ, 65. Mem: Am Statist Asn; Am Pub Health Asn; Biomet Soc; Sociol Res Asn; Int Union Sci Study Pop. Res: Biometry; forces of mortality and stochastic aspects of attrition. Mailing Add: Dept of Biometry La State Univ Med Ctr New Orleans LA 70112

MCMAHAN, ELIZABETH ANNE, b Davie Co, NC, May 5, 24. ZOOLOGY. Educ: Duke Univ, AB, 46, MA, 48; Univ Hawaii, PhD(entom), 60. Prof Exp: Res asst, Parapsychol Lab, Duke Univ, 43-54; Am Asn Univ Women fel, Univ Chicago, 60-61; from asst prof to assoc prof, 61-72, PROF ZOOL, UNIV NC, CHAPEL HILL, 72- Mem: AAAS; Animal Behav Soc; Am Inst Biol Sci; Am Soc Zool; Entom Soc Am. Res: Termite colony development; termite feeding behavior and sociobiology; biology of dragonflies. Mailing Add: Dept of Zool Univ of NC Chapel Hill NC 27514

MCMAHAN, UEL JACKSON, II, b Kansas City, Mo, July 22, 38; m 60; c 3. ANATOMY. Educ: Westminster Col, BA, 60; Univ Tenn, PhD(anat), 64. Prof Exp: Instr anat, Sch Med, Yale Univ, 65-67; instr, 67-72, asst prof, 72-75, ASSOC PROF NEUROBIOL, HARVARD MED SCH, 72- Res: Structure and function of synapses. Mailing Add: Dept of Neurobiol Harvard Med Sch Boston MA 02115

MCMAHAN, WILLIAM H, b Sylacauga, Ala, Apr 19, 37; m 61. INORGANIC CHEMISTRY. Educ: Auburn Univ, BS, 59, MS, 61; Univ Kans, PhD(chem), 65. Prof Exp: ASSOC PROF CHEM & COORDR GEN CHEM, MISS STATE UNIV, 65- Mem: Am Chem Soc. Res: Solution chemistry of low dielectric nonaqueous solvents; coordination chemistry of hydroxamic acids; chemical education. Mailing Add: Dept of Chem Miss State Univ Box CH State College MS 39762

MCMAHON, BEVERLY EDITH, b Los Angeles, Calif, Apr 11, 22. PALEOMAGNETISM. Educ: Univ Colo, Boulder, BA, 44, PhD(geol), 66. Prof Exp: Geologist, Shell Oil Co, 44-61; res asst geol, Mass Inst Technol, 66-68; asst prof, Cent State Univ, Ohio, 68-71; adj asst prof, Ohio State Univ, 71-74, res assoc, 74-75; RES ASSOC GEOSCI, UNIV TEX, DALLAS, 75- Mem: AAAS; Am Geophys Union; Geol Soc Am; Am Asn Petrol Geol. Res: Rock magnetism. Mailing Add: Inst Geosci Univ Tex at Dallas Richardson TX 75080

MCMAHON, BRIAN ROBERT, b Harrow, Eng, May 27, 36. ANIMAL PHYSIOLOGY. Educ: Univ Southampton, BSc, 64; Bristol Univ, PhD(zool), 68. Prof Exp: ASSOC PROF BIOL, UNIV CALGARY, 68- Mem: Brit Soc Exp Biol; Can Soc Zool; Can Physiol Soc; Am Soc Zoologists, Div Comp Physiol & Biochem. Res: Neural control of respiration in invertebrates; evolution of respiratory mechanisms; gas exchange dynamics across gill surfaces. Mailing Add: Dept of Biol Univ of Calgary Calgary AB Can

MCMAHON, DANIEL STANTON, b Cleveland, Ohio, Aug 2, 39. BIOCHEMISTRY. Educ: Case Western Reserve Univ, AB, 61; Univ Chicago, MS, 62, PhD, 66. Prof Exp: NIH fel, 66-67; lectr math biol, Univ Chicago, 67-68; ASST PROF BIOL, CALIF INST TECHNOL, 68- Mem: AAAS; Am Chem Soc; Am Soc Plant Physiol; Am Soc Cell Biol. Res: Development and function of the chloroplast; differentiation of cellular slime molds; cell membranes. Mailing Add: Div of Biol Calif Inst of Technol Pasadena CA 91109

MCMAHON, DAVID HAROLD, b Troy, NY, Apr 27, 42; m 68. ANALYTICAL CHEMISTRY. Educ: Col of the Holy Cross, BS, 63; Univ NH, PhD(chem), 68. Prof Exp: Res chemist, Esso Res & Eng Co, NJ, 67-70; res chemist, 70-74, GROUP LEADER RES & DEVELOP DIV, UNION CAMP CORP, 74- Mem: Am Chem Soc; Am Soc Mass Spectroscopists. Res: Chromatographic analysis of natural products; polynuclear aromatic hydrocarbon pollution analyses; characterization of natural products. Mailing Add: Res & Develop Div Union Camp Corp PO Box 412 Princeton NJ 08540

MCMAHON, DONALD HOWLAND, b Buffalo, NY, Apr 18, 34; m 54. OPTICAL PHYSICS. Educ: Univ Buffalo, BA, 57; Cornell Univ, PhD(exp physics), 64. Prof Exp: PHYSICIST, SPERRY RAND RES CTR, 63- Mem: Am Phys Soc; Inst Elec & Electronics Engrs; Optical Soc Am. Res: Paramagnetic resonance; nonlinear optics; quantum electronics; holography; optical information processing. Mailing Add: Sperry Rand Res Ctr 100 North Rd Sudbury MA 01776

MCMAHON, FRANCIS GILBERT, b Kalamazoo, Mich Sept 10, 23; m 54; c 4. INTERNAL MEDICINE. Educ: Univ Notre Dame, BS, 45; Univ Mich, MS, 51, MD, 53; Am Bd Internal Med, dipl, 59. Prof Exp: Intern, Univ Wis Hosps, 56-58; vis physician, Charity Hosp & Med Sch, La State Univ, 58-60, clin asst prof med, 59-60; dir med res, Upjohn Co, Mich, 60-64; vpres-in-charge med res, Ciba Pharmaceut Co, 64-67; exec dir, Merck Sharp & Dohme, 67-68; PROF MED & HEAD THERAPEUT & DIR CLIN PHARMACOL, MED SCH, TULANE UNIV, 68- Honors & Awards: Univ Notre Dame Sci Award, 64. Mem: Int Soc Clin Pharmacol (vpres); Am Soc Clin Pharmacol & Therapeut; fel Am Col Physicians; Endocrine Soc; AMA. Res: Hypertension; diabetes; clinical pharmacology; endocrinology; bioavailability of drugs and drug metabolism. Mailing Add: Dept of Med Tulane Univ Med Ctr New Orleans LA 70112

MCMAHON, GARFIELD WALTER, b Man, Feb 25, 32. ACOUSTICS. Educ: Univ Man, BSc, 52; Univ BC, MSc, 55. Prof Exp: Sci officer, 55-70, TRANSDUCER GROUP LEADER, DEFENSE RES ESTAB ATLANTIC, 70- Mem: Fel Acoust Soc Am. Res: Underwater acoustics; transducer calibration and design; properties of piezo-electric ceramics; vibrations of solid cylinders. Mailing Add: Defence Res Estab Atlantic Dartmouth NS Can

MCMAHON, HOWARD OLDFORD, b Killam, Alta, Sept 16, 14; m 41; c 3. PHYSICAL CHEMISTRY. Educ: Univ BC, BA, 35, MA, 37; Mass Inst Technol, PhD(phys chem), 41. Prof Exp: Res assoc phys chem, Mass Inst Technol, 41-42; res assoc, 42-52, sci dir, 52-56, vpres & sci dir, 56-62, exec vpres, 62-64, DIR & MEM EXEC COMT, ARTHUR D LITTLE, INC, 63-, PRES, 64- Concurrent Pos: Dir, Cryogenic Technol, Inc, 70-; dir, Mass Sci & Technol Found, 70-; chmn alumni fund bd, Mass Inst Technol, 71- Honors & Awards: Longstreth Medal, Franklin Inst, 51; Forrest Medal, Am Ceramic Soc, 52. Mem: AAAS; assoc Am Chem Soc; assoc Am Phys Soc; fel Am Acad Arts & Sci; Am Soc Test & Mat. Res: Liquefaction of helium; infrared absorption; molecular spectroscopic computations; flow calorimetry of gases. Mailing Add: Huckleberry Hill Lincoln MA 01773

MCMAHON, JOHN MARTIN, b Buffalo, NY, Dec 24, 15; m 42; c 6. MEDICINE. Educ: Georgetown Univ, BS, 36, MD, 40; Univ Minn, MS, 50. Prof Exp: Instr psychosom med, Sch Med, Tulane Univ, 50-52; CLIN PROF MED, SCH MED, UNIV ALA, BIRMINGHAM, 52-, DIR ARTHRITIS CLIN, MED CTR, 66-; CHIEF MED & ASSOC DIR MED EDUC, BIRMINGHAM BAPTIST HOSP, 54- Concurrent Pos: Attend consult, Vet Admin Hosp, Birmingham. Honors & Awards: Benemerenti Medal by Pope Paul VI, 67. Mem: AMA; Am Rheumatism Asn; Am Heart Asn; fel Am Col Physicians; Am Col Gastroenterol (past pres). Res: Internal medicine; gastroenterology; arthritis. Mailing Add: 800 Clinic Lane Bessemer AL 35020

MCMAHON, JOHN WALTER, invertebrate zoology, limnology, see 12th edition

MCMAHON, KENNETH JAMES, b Flandreau, SDak, July 9, 22; m 47; c 2. BACTERIOLOGY. Educ: SDak State Univ, BS, 47; Okla State Univ, MS, 49; Kans State Univ, PhD(bact), 54. Prof Exp: Asst bact, Okla State Univ, 47-48, instr, 48-49; from instr to prof, Kans State Univ, 49-70, actg head dept, 61-62, 63-64; PROF BACT & CHMN DEPT, N DAK STATE UNIV, 70- Mem: Am Soc Microbiol; fel Am Acad Microbiol. Res: Bacteriology of animal diseases and insect pathogens. Mailing Add: Dept of Bacteriol NDak State Univ Fargo ND 58102

MCMAHON, PAUL E, b Burlington, Vt, July 2, 31; m 54; c 1. PHYSICAL CHEMISTRY. Educ: St Michael's Col, Vt, BS, 54; Univ Vt, MS, 56; Univ Ill, PhD(phys chem), 61. Prof Exp: Spectroscopist, Ill State Geol Surv, 56-5756-57 & Univ Ill, 57-61; res chemist, Chemstrand Div, Monsanto Co, NC, 61-68; GROUP LEADER, CELANESE RES CO, 68- Res: Molecular structure and motion in small molecules and polymers; structure-property relations of polymers and composites; characterization and evaluation of fiber reinforced composites. Mailing Add: Celanese Res Co PO Box 1000 Summit NJ 07901

MCMAHON, RITA MARY, b New York, NY, Mar 5, 22. CYTOLOGY. Educ: Fordham Univ, BS, 49, MS, 51, PhD(biol), 53. Prof Exp: From instr to assoc prof sci, Sch Educ, Fordham Univ, 54-67, chmn dept, 57-67; ASSOC PROF BIOL, WESTERN CONN STATE COL, 67- Concurrent Pos: Adj prof, Grad Sch, New Eng Inst, 69- Mem: AAAS; Bot Soc Am; Am Inst Biol Sci; Environ Mutagen Soc; Genetics Soc Am. Res: Plant tissue and cell culture; polyploidy in development. Mailing Add: Box 333 Rte 2 Pound Ridge NY 10576

MCMAHON, ROBERT EARL, organic chemistry, see 12th edition

MCMAHON, ROBERT FRANCIS, III, b Syracuse, NY, June 17, 44; m 74. INVERTEBRATE ECOLOGY. Educ: Cornell Univ Sch Arts & Sci, BA, 66; Syracuse Univ, PhD(zool), 72. Prof Exp: From teaching asst biol to res asst, Syracuse Univ, 67-72; ASST PROF BIOL, UNIV TEX, ARLINGTON, 72- Mem: Marine Biol Lab; Malacol Soc London; Marine Biol Asn UK; Sigma Xi; Am Soc Limnol & Oceanog. Res: Physiological ecology of aquatic invertebrate organisms; study of the effects of thermal discharge on freshwater animals and the physiological basis for intertidal zonation in marine animals. Mailing Add: Dept of Biol Univ of Tex Arlington TX 76019

MCMAHON, THOMAS JOSEPH, b Rahway, NJ, 1943; m 66; c 2. SEMICONDUCTORS, ELECTROOPTICS. Educ: Univ Ill, Urbana, BS, 65, MS, 66; Univ Mo-Rolla, PhD(physics), 69. Prof Exp: RES PHYSICIST, NAVAL WEAPONS CTR, CHINA LAKE, 69- Mem: Am Phys Soc. Res: Liquid phase and vapor phase epitaxial semiconductors are grown for application in the areas of infrared and visible photo detection, photo thermal and photovoltaic solar energy conversion and electrooptic devices. Mailing Add: Code 601 Michelson Lab Naval Weapons Ctr China Lake CA 93555

MCMAHON, VERN AUGUST, b Thief River Falls, Minn, July 5, 32; m 52; c 2. PLANT BIOCHEMISTRY. Educ: St Cloud State Col, BS, 57; Univ Ill, MS, 59, PhD(bot), 63. Prof Exp: Asst bot, Univ Ill, 57-60, asst plant physiol, 60-63; trainee biochem, Univ Calif, Davis, 63-65; PROF PLANT BIOCHEM, UNIV WYO, 65- Res: Biosynthesis of lipids in a thermophilic alga, Cyanidium caldarium; single cell protein of algae as a possible food supplement. Mailing Add: Div of Biochem Univ of Wyo Box 3354 Laramie WY 82070

MCMANE, DOUGLAS GLENN, chemistry, see 12th edition

MCMANIMIE, ROBERT JOHN, organic chemistry, see 12th edition

MCMANIS, DOUGLAS R, b US. HISTORICAL GEOGRAPHY. Educ: Kent State Univ, BS in Ed, 54, MA, 55; Univ Chicago, PhD(geog), 64. Prof Exp: Instr geog, Butler Univ, 60-64; asst prof, Eastern Mich Univ, 64-66; assoc prof, 66-74, PROF GEOG & CHMN DEPT SOCIAL STUDIES, TEACHERS COL, COLUMBIA UNIV, 66- Mem: Asn Am Geogr; Am Geog Soc. Res: Historical geography of the American Midwest and New England. Mailing Add: Dept of Social Studies Teachers Col Columbia Univ New York NY 10027

MCMANNIS, WILLIAM J, geology, see 12th edition

MCMANUS, DEAN ALVIS, b Dallas, Tex, July 8, 34. GEOLOGY, OCEANOGRAPHY. Educ: Southern Methodist Univ, BS, 54; Univ Kans, MS, 56, PhD(geol), 59. Prof Exp: From res assoc to res asst prof, 59-65, from asst prof to assoc prof, 65-71, PROF OCEANOG, UNIV WASH, 71-, ADJ PROF MARINE STUDIES, 73- Concurrent Pos: NSF grants, 63-69; mem, Pac Ocean Adv Panel, Joint Oceanog Insts Deep Earth Sampling, 65- Mem: AAAS; Am Asn Petrol Geologists; Soc Econ Paleontologists & Mineralogists; Geol Soc Am; Am Geophys Union. Res: Particle size analysis; distribution of marine sediments; continental shelf topography and sediments; Arctic shelf sedimentation; oceanic rise topography in Northeast Pacific; marine geology of Gulf of Alaska. Mailing Add: Dept of Oceanog Univ of Wash Seattle WA 98195

MCMANUS, EDWARD CLAYTON, b McIntosh, Minn, Aug 19, 18; m 66. PARASITOLOGY, PHARMACOLOGY. Educ: Univ Minn, BS, 40, PhD(pharmacol), 50; Iowa State Univ, DVM, 44. Prof Exp: Asst pharmacol, Univ Minn, 44-48; res assoc, Sharp & Dohme, Inc, 48-53, res assoc, Merck Sharp & Dohme Res Labs, 53-55, pathologist, 55-58, RES ASSOC & RES FEL BASIC ANIMAL SCI RES, MERCK SHARP & DOHME RES LABS, 58- Mem: Am Vet Med Asn; Am Soc Pharmacol & Exp Therapeut; Am Soc Vet Physiol & Pharmacol; Am Asn Vet Parasitol; World's Poultry Sci Asn. Res: Gastrointestinal pharmacology; pathology; toxicology; parasitologic chemotherapy. Mailing Add: Merck Inst Rahway NJ 07065

MCMANUS, ELIZABETH CATHERINE, b Albany, NY, Sept 10, 12. PHYSICS, MATHEMATICS. Educ: Col St Rose, BA, 32; Catholic Univ, MS, 43. Prof Exp: High sch teacher parochial schs, NY, 35-43; from asst prof to assoc prof, 43-70, chmn dept, 43-68, PROF PHYSICS & COMPUT SCI, COL ST ROSE, 71- Concurrent Pos: Mem conf vibrations & waves, Reed Col, 66, Pa State Univ, 69, comput undergrad physics educ, Ill Inst Technol, 71. Mem: Am Asn Physics Teachers; Am Meteorol Soc; Am Geophys Union. Res: Teacher training in meteorology. Mailing Add: Dept of Phys Sci Col of St Rose 432 Western Ave Albany NY 12203

MCMANUS, HUGH, b West Bromwich, Eng, May 10, 18; m 53; c 3. THEORETICAL PHYSICS. Educ: Univ Birmingham, BSc, 39, PhD(math, physics), 47. Prof Exp: Res officer ionosphere & commun, Brit Admiralty, London, 40-42, res officer opers res, 42-43, res officer tech intel, 43-44; res fel theoret physics, Univ Birmingham, 47-49, lectr, 49-51; assoc res officer, Theoret Physics Div, Atomic Energy Can, Ltd, 51-60; PROF PHYSICS & ASTRON, MICH STATE UNIV, 60- Concurrent Pos: Res assoc, Mass Inst Technol, 57-58 & 70; Guggenheim fel, Nordic Inst Theoret Atomic Physics, 63-64; Sci Res Coun fel, Oxford Univ, 69-70. Mem: Fel Am Phys Soc. Res: Nuclear theory; scattering of elementary particles. Mailing Add: Dept of Physics Mich State Univ East Lansing MI 48823

MCMANUS, IVY ROSABELLE, b Erie, Pa, Oct 29, 23. BIOCHEMISTRY. Educ: Villa Maria Col, BS, 45; Western Reserve Univ, MS, 47, PhD(biochem), 51. Prof Exp: USPHS res fel, Univ Chicago, 51-52; instr biochem, Yale Univ, 52-54, asst prof, 54-57; from asst res prof to assoc res prof, Grad Sch Pub Health, 57-65, assoc prof, 65-73, PROF FAC ARTS & SCI, UNIV PITTSBURGH, 73- Mem: Am Chem Soc; Am Soc Biol Chem; Brit Biochem Soc. Res: Amino acid and protein metabolism; developmental biochemistry of skeletal muscle. Mailing Add: Life Sci Dept Fac Arts & Sci Univ of Pittsburgh Pittsburgh PA 15213

MCMANUS, JAMES MICHAEL, b Brooklyn, NY, May 22, 30; m 55; c 3. ORGANIC CHEMISTRY. Educ: Col Holy Cross, BS, 52; Niagara Univ, MS, 54; Mich State Univ, PhD(chem), 58. Prof Exp: Res chemist, 58-69, PATENT CHEMIST, PFIZER, INC, 69- Mem: Am Chem Soc. Res: Chemistry of tetrazoles, indoles and benzimidazoles; preparation and pharmacology of diuretics and sulfonylureas. Mailing Add: Craig Rd Jericho Hill Old Lyme CT 06371

MCMANUS, JOHN JOSEPH, b Manhattan, NY, July 27, 43. VERTEBRATE ZOOLOGY, ECOLOGY. Educ: Rutgers Univ, BA, 65; Cornell Univ, PhD(biol), 69. Prof Exp: Asst prof, 68-72, ASSOC PROF BIOL SCI, FAIRLEIGH DICKINSON UNIV, 72- Mem: AAAS; Am Soc Mammalogists; Ecol Soc Am; Am Inst Biol Sci. Res: Behavioral ecology of vertebrates; thermoregulation and bioenergetics; water relations; growth and development. Mailing Add: Dept of Biol Fairleigh Dickinson Univ Madison WI 07940

MCMANUS, JOSEPH FORDE ANTHONY, b Blackville, NB, July 13, 11; nat US; m 41; c 2. PATHOLOGY. Educ: Fordham Univ, BSc, 33; Queen's Univ, Ont, 38. Prof Exp: Asst path, Sch Med, Johns Hopkins Univ, 38-40; resident, NY Hosp, 40; assoc prof, Univ Ala, 46-50; assoc prof, Univ Va, 50-53; prof & chmn dept, Med Ctr, Univ Ala, 53-61; prof combined degree prog, Ind Univ, 61-65; exec dir, Fedn Am Soc Exp Biol, 65-70; dean, Col Med, 70-74, PROF PATH, MED UNIV SC, 70- Concurrent Pos: Beit Mem fel, Oxford Univ, 45-46; spec consult to Surgeon Gen, USPHS, 54-58, mem path training grant comt, 58-62. Mem: Am Asn Path & Bact; Am Soc Exp Path (secy-treas, 57-61, pres, 61-62 & 62-63); Int Acad Path. Res: Kidneys; histochemistry. Mailing Add: Col of Med Med Univ of SC Charleston SC 29401

MCMANUS, LAWRENCE ROBERT, b North Bergen, NJ, Mar 21, 21; m 53; c 2. ECOLOGY. Educ: Cornell Univ, BS, 49, MEd, 51, PhD(animal ecol), 60. Prof Exp: Instr biol, Orange County Community Col, 54-57; asst zool, Cornell Univ, 57-60; from asst prof to assoc prof, 60-73, PROF BIOL, HAMILTON COL, 73- Mem: AAAS; Ecol Soc Am; Am Micros Soc; Am Inst Biol Sci. Res: Ecology of crayfish, branchiobdellid annelids and entocytherid ostracods. Mailing Add: Dept of Biol Hamilton Col Clinton NY 13323

MCMANUS, MARGARET ANN (MARY ANNUNCIATA), b De Witt, U Iowa, Aug 6, 12. BIOLOGY. Educ: Col St Theresa, BS, 38; St Louis Univ, MS, 45, PhD(biol), 58. Prof Exp: HEAD DEPT BIOL & CHMN DIV NATURAL SCI & MATH, MT MERCY COL, IOWA, 59- Mem: AAAS; Mycol Soc Am; Bot Soc Am; Torrey Bot Club; Am Soc Zool. Res: Mycology and plant physiology; myxomycete plasmodia and life history; effects of plant growth regulating substances on cell division; electron microscopy; ultrastructure of phases of the life cycle of myxomycetes. Mailing Add: Mt Mercy Col 1330 Elmhurst Dr Cedar Rapids IA 52402

MCMANUS, SAMUEL PLYLER, b Edgemoor, SC, Oct 29, 38; m 59; c 2. ORGANIC CHEMISTRY. Educ: The Citadel, BS, 60; Clemson Univ, MS, 62, PhD(chem), 64. Prof Exp: Res chemist, Marshall Lab, E I du Pont de Nemours & Co, 64-66; from asst prof to assoc prof, 66-73, PROF CHEM, UNIV ALA, HUNTSVILLE, 73- Concurrent Pos: Consult, US Army Res Off, Durham, 68-; vis prof, Univ SC, 74-75. Honors & Awards: Army Commendation Medal; Nat Defense Serv Medal. Mem: AAAS; Am Chem Soc; fel Am Inst Chem; Sigma Xi. Res: Acid catalyzed rearrangements; structure of heteronuclear substituted carbonium ions; neighboring group participation; polymer chemistry. Mailing Add: Dept of Chem Univ of Ala Huntsville AL 35807

MCMANUS, THOMAS (JOSEPH), b Omaha, Nebr, Feb 5, 25; m 51; c 5. CELL PHYSIOLOGY, HEMATOLOGY. Educ: Antioch Col, BS, 51; Boston Univ, MD, 55. Prof Exp: Asst hemat, Sloan-Kettering Inst, NY, 49-51; res fel med, Harvard Med Sch, 55-58, asst med, 58-59, res assoc, 59-61; asst prof, 61-67, ASSOC PROF PHYSIOL & PHARMACOL, SCH MED, DUKE UNIV, 67-, DIR LAB CELL PHYSIOL, 68- Concurrent Pos: Asst physician, Peter Bent Brigham Hosp, Boston, Mass, 55-58; vis prof, NC Col, 63; mem sci staff, Res Vehicle Alpha Helix Amazon Exped, Brazil, 67. Mem: Fel AAAS; Am Physiol Soc; Biophys Soc; Soc Gen Physiol. Res: Membrane transport. Mailing Add: Dept of Physiol & Pharmacol Duke Univ Med Ctr Box 3709 Durham NC 27710

MCMASTER, MARVIN CLAYTON, JR, b Gering, Nebr, June 27, 38; m 62; c 1. BIOCHEMISTRY. Educ: SDak Sch Mines & Technol, BS, 60; Univ Nebr, PhD(org chem), 66. Prof Exp: Fel polypeptides, Inst Molecular Biophys, Fla State Univ, 70-71; Nat Heart & Lung Inst spec fel biochem, Webb-Waring Lung Inat, Univ Colo Med Ctr, Denver, 71-73; SCHOLAR BIOCHEM LIPID STORAGE DIS, MENT HEALTH RES INST, UNIV MICH, ANN ARBOR, 73- Concurrent Pos: Res chemist, Indust & Biochem Dept, E I du Pont de Nemours & Co, Inc, 65-68; sr scientist I, Indust Chem Div, Kraftco Corp, 68-70. Mem: Am Chem Soc. Res: Small ring heterocyclic compounds; polypeptide synthesis; biochemistry of lung metabolism. Mailing Add: Ment Health Res Inst Univ of Mich Ann Arbor MI 48104

MCMASTER, PAUL D, b Norwich, Conn, Feb 24, 32; m 63; c 3. BIOCHEMISTRY, ORGANIC CHEMISTRY. Educ: Col Holy Cross, BS, 54; Clark Univ, PhD(chem), 61. Prof Exp: Asst prof, 61-65, ASSOC PROF CHEM, COL HOLY CROSS, 65-, CHMN DEPT, 72- Concurrent Pos: Consult drug design, Astra Pharmaceut Prod, Inc, Worcester, Mass, 73- Mem: AAAS; Am Chem Soc. Res: Conformational analysis; drug design. Mailing Add: Dept of Chem Col of the Holy Cross Worcester MA 01610

MCMASTER, PHILIP DURYEE, physiology, immunology, deceased

MCMASTER, PHILIP ROBERT BACHE, b Cambridge, Mass, Feb 19, 30; m 58; c 2. IMMUNOLOGY, EXPERIMENTAL PATHOLOGY. Educ: Princeton Univ, AB, 52; Johns Hopkins Univ, MD, 56. Prof Exp: Intern med, New York Hosp-Cornell Med Ctr, 56-57; sr asst surgeon, Lab Immunol, Nat Inst Allergy & Infectious Dis, 57-60; USPHS surgeon, Pasteur Inst Paris, 60-62; surgeon, Lab Immunol, 62-64, SR SURGEON, LAB GERMFREE ANIMAL RES, NAT INST ALLERGY & INFECTIOUS DIS, 64- Mem: Am Soc Exp Path. Res: Immunopathology related to immune process and tissue destruction; physiology of ocular pressure in normal and abnormal states. Mailing Add: Bldg 8 Rm 319 Nat Inst of Health Bethesda MD 20014

MCMASTER, ROBERT H, b Flint, Mich, Feb 26, 16; m 39; c 2. MEDICINE. Educ: Ohio Univ, AB, 38; Univ Cincinnati, MD, 50. Prof Exp: Physician, Gallipolis Clin, Ohio, 51-56; physician, Chas Pfizer & Co, 56; assoc dir med res, Wm S Merrell Co, 56-67, dir sci training, 67-74, DIR EMPLOYEE HEALTH SERV, MERRELL-NAT LABS, 74- Concurrent Pos: Asst, Col Med, Univ Cincinnati, 57-66, instr, 66-70, asst clin prof med, 70-; clinician, Cincinnati Gen Hosp, 64- Mem: Am Soc Clin Pharm & Therapeut; AMA; Am Heart Asn; NY Acad Sci; fel Royal Soc Health. Res: Employee health and safety; training of nonprofessionals in medical subjects. Mailing Add: Merrell-Nat Labs 110 E Amity Rd Cincinnati OH 45215

MCMASTER, WILLIAM H, b Ft Lewis, Wash, Apr 16, 26; m 51; c 1. HYDRODYNAMICS. Educ: US Mil Acad, BS, 46; Univ Va, MS, 52, PhD(physics), 54. Prof Exp: Sr physicist, 55-62, group leader, 62-74, STAFF PHYSICIST, LAWRENCE LIVERMORE LAB, UNIV CALIF, 74- Mem: Am Phys Soc; NY Acad Sci. Res: Polarization of radiation; nuclear test diagnostics; computational methods; x-ray physics; equation of state-solids; calculational methods in two dimensional hydrodynamics. Mailing Add: H Div Lawrence Livermore Lab PO Box 808 Livermore CA 94550

MCMASTERS, ALAN WAYNE, b Ottawa, Kans, Dec 19, 34; m 57; c 4. OPERATIONS RESEARCH. Educ: Univ Calif, Berkeley, BS, 57, MS, 62, PhD(indust eng, opers res), 66. Prof Exp: Res engr, Pac Southwest Forest & Range Exp Sta, US Forest Serv, 55-61, opers analyst, 61-65; ASSOC PROF OPERS RES, NAVAL POSTGRAD SCH, 65- Concurrent Pos: Consult, Mellonics Systs Div, Litton Indust, 69-70. Mem: Opers Res Soc Am. Res: Flows in networks; locational problems on networks; inventory models; forest fire spread modeling. Mailing Add: Dept of Oper Res & Admin Sci Naval Postgrad Sch Monterey CA 93940

MCMASTERS, DENNIS WAYNE, b Chickasha, Okla, July 9, 40; m 68; c 1. PLANT PHYSIOLOGY. Educ: Okla Christian Col, BSE, 62; Univ Okla, MNS, 68; Univ Ark, PhD(bot), 73. Prof Exp: Teacher biol & chem, McAlester Pub Schs, 62-68; teacher, Tulsa Pub Schs, 68-69; TEACHER BOT, HENDERSON STATE UNIV, 72- Mem: Scand Soc Plant Physiologists. Res: Gibberellins; algal physiology. Mailing Add: Box H-1052 Henderson State Univ Arkadelphia AR 71923

MCMASTERS, DONALD L, b Crawfordsville, Ind, July 14, 31. ANALYTICAL CHEMISTRY. Educ: Wabash Col, AB, 53; Univ NDak, MS, 55; Ind Univ, PhD(anal chem), 59. Prof Exp: Asst prof chem, Beloit Col, 59-63; lectr, Univ Ill, 63-64; LECTR CHEM, IND UNIV, BLOOMINGTON, 64- Mem: Am Chem Soc. Res: Polarography in aqueous and non-aqueous solvents. Mailing Add: Dept of Chem Ind Univ Bloomington IN 47401

MCMASTERS, ROBERT EARL, b El Paso, Tex, Jan 18, 32. NEUROSCIENCES, CLINICAL NEUROLOGY. Educ: Univ Tex, El Paso, BS, 53; Univ Tex Med Br Galveston, MA, 56, MD, 58. Prof Exp: Intern, Univ Iowa Hosp, 58-59; resident neurol, Neurol Inst NY, Columbia-Presby Med Ctr, 59-61; chief neurol, US Naval Hosp, Portsmouth, Va, 63-66; chief neurol, Vet Admin Hosp, Durham, NC, 66-68; asst prof physiol & neurol, 68-69, ASSOC PROF NEUROL, UNIV TEX HEALTH SCI CTR, SAN ANTONIO, 69-, HEAD DIV NEUROSCI, 68-, ASSOC PROF ANAT, 75- Concurrent Pos: Fel neuroanat, Columbia Univ, 61-62; vis fel neurol, Neurol Inst NY, Columbia-Presby Med Ctr, 61-63; instr, Teachers Col, Columbia Univ, 61-63; assoc, Sch Med, Duke Univ, 66-68. Mem: Am Acad Neurol; Soc Neurosci; Am Asn Anat; Asn Res Nerv & Ment Dis; Asn Univ Prof Neurol. Mailing Add: Dept of Med Univ of Tex Health Sci Ctr San Antonio TX 78284

MCMECHAN, JAMES HOWARD, organic chemistry, see 12th edition

MCMEEKIN, DOROTHY, b Boston, Mass, Feb 24, 32. PLANT PATHOLOGY. Educ: Wilson Col, AB, 53; Wellesley Col, MA, 55; Cornell Univ, PhD(plant path), 59. Prof Exp: Prof natural sci, Upsala Col, 59-64 & Bowling Green State Univ, 64-66; PROF NATURAL SCI, MICH STATE UNIV, 66- Mem: Am Phytopath Soc; Bot Soc Am. Res: Physiology of plant disease. Mailing Add: Dept of Natural Sci Kedzie Lab Mich State Univ East Lansing MI 48824

MCMENAMIN, JOHN WILLIAM, b Tacoma, Wash, Apr 1, 17; m 42; c 2. DEVELOPMENTAL BIOLOGY. Educ: Occidental Col, AB, 40; Univ Calif, Los Angeles, MA, 46, PhD(zool), 49. Prof Exp: High sch & jr col teacher, Calif, 42-45; spec appt, 46-47, from instr to assoc prof, 47-57, PROF BIOL, OCCIDENTAL COL, 57- Mem: AAAS; Am Soc Zoologists; Am Micros Soc. Res: Role of the thyroid gland in the development of the chick embryo; lipid and porphyrin metabolism in the chick embryo as influenced by hormones; in vitro development of embryonic tissues. Mailing Add: Dept of Biol Occidental Col 1600 Campus Rd Los Angeles CA 90041

MCMENAMY, RAPIER HAYDEN, b O'Fallon, Mo, Jan 12, 17; m 43; c 1. BIOCHEMISTRY, PHYSICAL CHEMISTRY. Educ: Univ Southern Calif, AB, 47, MS, 48; Harvard Univ, PhD, 58. Prof Exp: Res chemist, Turco Prod, Inc, 47-50; biol chemist, Protein Found, 53-58; res assoc, Harvard Univ, 58-60; from asst prof to assoc prof, 60-69, PROF, STATE UNIV NY BUFFALO, 69- Mem: Am Soc Biol Chem; Am Chem Soc; Biophys Soc. Res: Properties of protein and protein solutions; protein-small molecule interactions; distribution and transport into cells; metabolism and amino acid utilization in critical illness. Mailing Add: Dept of Biochem State Univ NY Buffalo NY 14214

MCMICHAEL, EDWARD VANCE, archaeology, anthropology, see 12th edition

MCMICHAEL, JOHN CALHOUN, b Imperial, Pa, Feb 12, 41. BIOPHYSICS. Educ: Univ Pittsburgh, BS, 62, PhD(biophys), 73. Prof Exp: RES ASSOC, DEPT BIOPHYS & MICROBIOL, UNIV PITTSBURGH, 74- Mem: Sigma Xi; Am Soc Microbiol. Res: Studies of bacterial pili as virulence factors for pathogenic bacteria. Mailing Add:

MCMICHAEL, KIRK DUGALD, b Schenectady, NY, July 13, 35; m 58; c 2. ORGANIC CHEMISTRY. Educ: Shimer Col, AB, 53; Univ Chicago, MS, 56, PhD(chem), 60. Prof Exp: Res assoc chem, Univ Wis, 60-62; asst prof, 62-68, asst to chmn dept, 68-71, ASSOC PROF CHEM, SH STATE UNIV, 68- Concurrent Pos: Petrol Res Fund grant, 64-66. Mem: AAAS; Am Chem Soc. Res: Characterization of allylic rearrangement pathways using deuterium isotope effects; long range allylic rearrangements in steroidal systems; thermodynamic isotope effects and their relationship to theory. Mailing Add: Dept of Chem Wash State Univ Pullman WA 99163

MCMICKLE, ROBERT HAWLEY, b Paterson, NJ, July 30, 24; m 49; c 5. PHYSICS. Educ: Oberlin Col, BA, 47; Univ Ill, MS, 48; Pa State Univ, PhD(physics), 52. Prof Exp: Physicist, Res Ctr, B F Goodrich Co, 52-59; assoc prof, Robert Col, Istanbul, 59-63, prof & head dept, 63-71; PROF & HEAD DEPT PHYSICS, BOGAZICI UNIV, TURKEY, 71- Mem: Am Phys Soc; Am Asn Physics Teachers. Res: High pressure and polymer physics; solid state phenomena. Mailing Add: Dept of Physics Bogazici Univ Bebek PK 2 Istanbul Turkey

MCMILLAN, BROCKWAY, b Minneapolis, Minn, Mar 30, 15; m 42; c 3. MATHEMATICS. Educ: Mass Inst Technol, BS, 36, PhD(math), 39. Prof Exp: Instr math, Mass Inst Technol, 36-39; Procter fel, Princeton Univ, 39-40, Fine instr math, 40-41, res assoc, 41-42; res mathematician, 46-55, asst dir systs eng, 55-59, dir mil res, 59-61, exec dir, 65-69, VPRES MIL SYSTS, BELL TEL LABS, 69- Mem: Nat Acad Eng; AAAS; Am Math Soc; Soc Indust & Appl Math (vpres, 57-58, pres, 60); Math Asn Am. Res: Statistical mechanics of interacting systems; electrical network theory; probability theory and random processes. Mailing Add: 6 Hawthorne Pl Summit NJ 07901

MCMILLAN, CALVIN, b Murray, Utah, Feb 20, 22; m 50; c 4. PLANT ECOLOGY. Educ: Univ Utah, BS, 47, MS, 48; Univ Calif, Berkeley, PhD(bot), 52. Prof Exp: From asst prof to assoc prof bot, Univ Nebr, 52-58; assoc prof, 58-65, PROF BOT, UNIV TEX, AUSTIN, 65- Concurrent Pos: Vis assoc prof & actg dir, Bot Garden, Univ Calif, Berkeley, 64-65. Mem: AAAS; Bot Soc Am; Ecol Soc Am. Res: Ecotypes and ecosystem functions; ecology of colonizing species. Mailing Add: Plant Ecol Res Lab Univ of Tex Austin TX 78712

MCMILLAN, CAMPBELL WHITE, b Soochow, China, Jan 10, 27; US citizen; m 55; c 6. PEDIATRICS, HEMATOLOGY. Educ: Wake Forest Col, BS, 48, Bowman-Gray Sch Med, MD, 52. Prof Exp: Intern, Boston City Hosp, 52-53; asst resident pediat, Children's Hosp Med Ctr, 53-55; pediat registr, St Mary's Hosp Med Sch, London, Eng, 55-56; asst med, Nemazee Hosp, Shiraz, Iran, 56-58; fel pediat hemat, Harvard Med Sch, 58-61; from asst prof to assoc prof, 63-72, PROF PEDIAT, MED SCH, UNIV NC, CHAPEL HILL, 72-, ASSOC DIR CLIN RES UNIT, 66- Mem: Soc Pediat Res; Am Pediat Soc. Res: Coagulation. Mailing Add: Dept of Pediat Univ of NC Sch of Med Chapel Hill NC 27514

MCMILLAN, CLARA ALBERTINA, b Buenos Aires, Arg, Feb 24, 19; m 46; c 2. PHYSICAL CHEMISTRY. Educ: Univ Buenos Aires, PhD(chem), 46. Prof Exp: Asst inorg chem, Univ Buenos Aires, 50-53; fel, Arg Nat Sci Found, 53-54; prof chem, Univ Cuyo, 54-59; assoc prof, 62-72, PROF CHEM, ILL BENEDICTINE COL, 72- Concurrent Pos: Investr, Arg Atomic Energy Comn, 53-59; fac fel, Argonne Nat Lab, 59-60. Res: Nuclear chemistry; magnetochemistry. Mailing Add: Dept of Chem Ill Benedictine Col Lisle IL 60532

MCMILLAN, DANIEL RUSSELL, b Bartow, Ga, June 25, 07; m 32; c 1. PHYSICS. Educ: Ga Tech Univ, BS, 31; Emory Univ, MS, 35; Univ NC, PhD(physics), 40. Prof

Exp: From instr to prof physics, Emory Univ, 33-60, chmn dept, 41-43; prof physics & math & chmn dept, 60-74, EMER PROF PHYSICS & MATH, UNIV MONTEVALLO, 74-, DEAN GRAD STUDIES, 71- Mem: Am Phys Soc; Am Asn Physics Teachers. Res: Infrared spectroscopy; ultrasonics. Mailing Add: Dept of Physics & Math Univ of Montevallo Montevallo AL 35115

MCMILLAN, DANIEL RUSSELL, JR, b Atlanta, Ga, Feb 26; 35; m 65; c 1. TOPOLOGY. Educ: Emory Univ, BA, 56; Univ Wis, MA, 57, PhD(math), 60. Prof Exp: Res instr math, La State Univ, 60-61; asst prof, Fla State Univ, 61-62; vis mem, Inst Advan Study, 62-64; actg assoc prof, Univ Va, 64-65, assoc prof, 65-66; assoc prof, 66-69, PROF MATH, UNIV WIS-MADISON, 69- Concurrent Pos: Nat Sci Found fel, 62-63; Inst Adv Study fel, 63-64, 69-70; Sloan fel, 65-67. Mem: Math Asn Am; Am Math Asn. Res: Topology of combinatorial manifolds; local homotopy properties of topological embeddings; cellular sets in combinatorial manifolds; geometric topology of mappings, three manifolds. Mailing Add: Dept of Math Univ of Wis Madison WI 53706

MCMILLAN, DAVID JON, nuclear physics, see 12th edition

MCMILLAN, DONALD BURLEY, b Toronto, Ont, Apr 25, 29. ZOOLOGY, COMPARATIVE HISTOLOGY. Educ: Univ Western Ont, BSc, 51, MSc, 53; Univ Toronto, PhD(zool), 58. Prof Exp: Demonstr zool, Univ Toronto, 53-56; instr, 56-59, lectr, 59-60, asst prof, 60-66, ASSOC PROF ZOOL, UNIV WESTERN ONT, 66- Mem: Am Soc Zool; Can Asn Anat; Can Soc Zool; Can Soc Cell Biol. Res: Haemopoietic organs of vertebrates; comparative histology of the vertebrate kidney and gonad; ultrastructure. Mailing Add: Dept of Zool Univ of Western Ont London ON Can

MCMILLAN, DONALD EDGAR, b Butler, Pa, Sept 23, 37; m 61; c 2. PHARMACOLOGY, PSYCHOLOGY. Educ: Grove City Col, BS, 59; Univ Pittsburgh, MS, 62, PhD(psychol), 65. Prof Exp: From instr to asst prof pharmacol, State Univ NY Downstate Med Ctr, 67-69; asst prof pharmacol, 69-72, asst prof psychol, 70-72, ASSOC PROF PHARMACOL & CLIN ASSOC PROF PSYCHOL, UNIV NC, CHAPEL HILL, 72- Concurrent Pos: NIH training grant, Harvard Med Sch, 65-66. Mem: AAAS; Behav Pharmacol Soc; Am Soc Pharmacol & Exp Therapeut. Res: Behavioral pharmacology; mechanisms of drug tolerance and drug dependence; operant conditions. Mailing Add: Dept of Pharmacol Univ of NC Sch of Med Chapel Hill NC 27514

MCMILLAN, EDWIN MATTISON, b Redondo Beach, Calif, Sept 18, 07; m 41; c 3. PHYSICS. Educ: Calif Inst Technol, BS, 28, MS, 29; Princeton Univ, PhD(physics), 32. Hon Degrees: DSc, Rensselaer Polytech Inst, 61 & Gustavus Adolphus Col, 63. Prof Exp: Nat Res Coun fel, Univ Calif, Berkeley, 32-34, res assoc, 34-35, from instr to prof physics, 35-73, mem staff, Lawrence Radiation Lab, 34-54, assoc dir, 54-58, dir, 58-71, dir, Lawrence Berkeley Lab, 71-73. Concurrent Pos: Researcher, Radiation Lab, Mass Inst Technol, 40-41, Radio & Sound Lab, US Navy, 41-42 & Los Alamos Sci Lab, 42-45; mem gen adv comt, USAEC, 54-58; trustee, Rand Corp, 59-69; mem comn high energy physics, Int Union Pure & Appl Physics, 60-66; sci policy comt mem, Stanford Linear Accelerator Ctr, 62-66; physics adv comt, Nat Accelerator Lab, 67-69; chmn class I, Nat Acad Sci, 68-71; trustee, Univs Res Asn, Washington, DC, 69-74. Honors & Awards: Co-recipient, Nobel Prize in Chem, 51; Sci Award, Res Corp, 51; Atoms for Peace Award, 63. Mem: Nat Acad Sci; fel Am Phys Soc; fel Am Acad Arts & Sci; Am Philos Soc. Res: Nuclear physics; design and construction of particle accelerators. Mailing Add: Lawrence Berkeley Lab Univ of Calif Berkeley CA 94720

MCMILLAN, GARNETT RAMSAY, b Madison, SC, June 11, 32; m 58. PHOTOCHEMISTRY. Educ: Univ Ga, BS, 53; Univ Rochester, PhD(chem), 58. Prof Exp: Res chemist, Celanese Corp Am, 57-62; res assoc, Ohio State Univ, 62-64; from asst prof to assoc prof, 64-73, PROF CHEM, CASE WESTERN RESERVE UNIV, 73- Res: Reaction kinetics. Mailing Add: Dept of Chem Case Western Reserve Univ Cleveland OH 44106

MCMILLAN, GRAHAM WATSON, b Monmouth, Ill, Sept 25, 16; m 43; c 2. CHEMISTRY. Educ: Monmouth Col, BS, 37; Univ Ill, PhD(phys chem), 41. Prof Exp: Res chemist, 40-45, asst to vpres in chg prod, 46-51, mgr control div, 51-54, mgr develop dept, 54-57, VPRES RES & DEVELOP, COMMERCIAL SOLVENTS CORP, 57- Mem: AAAS; Am Chem Soc. Res: Infrared spectroscopy; electroorganic reductions; vapor phase syntheses; chemistry of microbiological products. Mailing Add: Commercial Solvents Corp 1331 S First St Terre Haute IN 47801

MCMILLAN, HARLAN L, b Cabot, Ark, Sept 28, 26; m 48; c 4. BIOLOGY, ACADEMIC ADMIN. Educ: Col Ozarks, BS, 50; Univ Ark, MS, 55; Purdue Univ, PhD(entom), 60. Prof Exp: Biol aide, Ark State Bd Health, 50-52, malaria control supvr, 52-53; med entomologist, Tech Develop Labs, Commun Dis Ctr, USPHS, Ga, 60-61; assoc prof biol, Col Ozarks, 61-64, prof & dean of men, 64-68, head dept biol, 61-69, chmn div natural sci & math, 68-69; assoc prof biol, 69-72, chmn dept, 72-73, PROF BIOL & DEAN, SCH ARTS & SCI, ARK POLYTECH COL, 72- Mem: AAAS; Am Soc Allied Health Prof. Res: Culture of insect cells in artificial media; electrophoresis studies of proteins in aquatic organisms. Mailing Add: Sch of Arts & Sci Ark Polytech Col Russellville AR 72801

MCMILLAN, JAMES ALEXANDER, b Atascadero, Calif, Dec 18, 41; m 65. NEUROPHYSIOLOGY. Educ: Univ Calif, Davis, BS(vet sci), 63, BS(animal husb), 65, MS, 70, PhD(physiol), 72. Prof Exp: ASST PROF PHYSIOL, MONT STATE UNIV, 73- Concurrent Pos: NIH training grant, Sch Med, Univ Wash, 72-73. Mem: AAAS; Sigma Xi; Soc Neurosci; Am Physiol Soc. Res: Spinal cord and brain stem integration of sensory information. Mailing Add: Dept of Biol Mont State Univ Bozeman MT 59715

MCMILLAN, JAMES MALCOLM, b Victoria, BC, Aug 10, 36; m 57; c 3. THEORETICAL PHYSICS. Educ: Univ BC, BSc, 58, MSc, 59; McGill Univ, PhD(theoret nuclear physics), 61. Prof Exp: Nat Res Coun Can fel physics, Univ Turin, 61-62 & Cambridge Univ, 62-63; from asst prof to assoc prof, 63-74, asst dean sci, 71, PROF PHYSICS, UNIV BC, 74- Mem: Am Phys Soc; Can Asn Physicists. Res: Intermediate energy nuclear physics; unification of the classifications of triton state components; construction and application of approximate triton wave functions; Regge poles in potential scattering; nuclear many-body problem; temperature dependence of magnetic resonance lines. Mailing Add: Dept of Physics Univ of BC Vancouver BC Can

MCMILLAN, JOHN FRANK, b Buford, NDak, Mar 16, 10. VERTEBRATE ECOLOGY. Educ: Southwestern Mo State Col, BSc, 33; Univ Mich, MSc, 39, PhD(zool), 50. Prof Exp: High sch teacher, Mo & Iowa, 32-41; instr biol, Lincoln Mem Univ, 41-42; chmn dept zool, Col St Thomas, 49-52; prof & chmn dept, Wichita Univ, 53-59; assoc prof, 59-62, PROF BIOL, COL ST THOMAS, 62- Mem: Ecol Soc Am; Am Soc Mammalogists; Wildlife Soc; Nat Audubon Soc. Res: Natural history of vertebrates; physiology. Mailing Add: Dept of Biol Col of St Thomas St Paul MN 55105

MCMILLAN, JOSEPH PATRICK, b San Diego, Calif, Jan 17, 45. ZOOLOGY, COMPARATIVE PHYSIOLOGY. Educ: St Norbert Col, BSc, 67; Univ Ga, PhD(zool), 71. Prof Exp: Teaching asst anat & physiol, Univ Ga, 68-70, instr anat, summer 70; NIH fel zool, Univ Tex, Austin, 71-72, asst prof, 73-75; ASST PROF, COL OF VIRGIN ISLANDS, 75- Mem: AAAS; Am Soc Zool; Am Inst Biol Sci. Res: Ecological physiology; biological clocks; extraretinal photoreception; animal migration, orientation, navigation and homing; photoperiodism. Mailing Add: Div of Sci Col of the Virgin Islands St Thomas VI 00801

MCMILLAN, JUAN ALEJANDRO, b Buenos Aires, Arg, July 15, 18; m 46; c 2. PHYSICAL CHEMISTRY. Educ: Univ Buenos Aires, PhD(chem), 46. Prof Exp: Asst physics, Univ Buenos Aires, 46-53; scientist, Arg AEC, 53-55; prof chem & chmn dept, Cuyo, 55-59; assoc chemist, 59-74, CHEMIST, ARGONNE NAT LAB, 74- Concurrent Pos: Guggenheim fel, 58. Mem: AAAS; Am Chem Soc. Res: Crystallography; electron paramagnetic resonance; irreversible phase transformations; radiation of solids. Mailing Add: 1440 N Lake Shore Dr Chicago IL 60610

MCMILLAN, LOUIS KELLY, JR, b Little Rock, Ark, Feb 8, 29; m 59; c 2. OPERATIONS RESEARCH, MATHEMATICS. Educ: US Naval Acad, BS, 52, US Naval Postgrad Sch, MS, 62. Prof Exp: Opers res analyst, Off Asst Secy Defense, 65-66; dir opers anal group, Tracor, Inc, 66-68; dir naval & com systs div, Vertex Corp, 68-70, secy, 70-72, exec vpres, 72-74; MEM PROF STAFF, CTR NAVA ANAL, 74- Concurrent Pos: Consult, US Navy, 69- & Vertex Corp, 74- Mem: Opers Res Soc Am. Res: Application of scientific analysis to decision problems of military, commercial and investigative institutions. Mailing Add: 6103 Sherborn Lane Springfield VA 22152

MCMILLAN, NEIL JOHN, b Souris, Man, Nov 11, 25; m 52; c 3. GEOLOGY, SOIL SCIENCE. Educ: Univ Man, BSc, 48; Univ Sask, MSc, 51; Univ Kans, PhD(geol), 55. Prof Exp: Geologist, Int Nickel Co, Can, summers, 48-50 & Chevron Standard, 51-52; field geologist, Gulf Oil Co, 53, Geol Surv Kans, 54 & Geol Surv Can, 55-56; sr geologist, Bur Mining Resource, Australia, 56-58; res geologist, Tenneco Oil Co, Houston, 59-71; SR GEOLOGIST, AQUITANE CO CAN LTD, 71- Honors & Awards: Medal of Merit, Can Soc Petrol Geologists, 75. Mem: Geol Soc Australia; Geol Soc Am; Geol Soc Can; Can Asn Petrol Geologists. Res: Continental break-up as it relates to basin development and oil exploration; comparison of sediments on each side of the North Atlantic; provenance of clastics in the Labrador Sea. Mailing Add: 211 Scarboro Ave SW Calgary AB Can

MCMILLAN, PAUL JUNIOR, b Atlanta, Ga, Sept 13, 30; m 55; c 4. BIOCHEMISTRY, HISTOCHEMISTRY. Educ: Southern Missionary Col, BA, 51; Loma Linda Univ, MS, 57, PhD(biochem), 60. Prof Exp: Instr, 60-61, asst prof, 63-69, ASSOC PROF ANAT, SCH MED, LOMA LINDA UNIV, 69- Concurrent Pos: USPHS fel histochem, NIH, 61-63. Mem: AAAS; Am Asn Anat; Int Soc Stereology; Histochem Soc. Res: Interactions of form and function; metabolic and endocrine interrelations at the cellular level, with reference to the morphological design of tissues. Mailing Add: Dept of Anat Sch of Med Loma Linda Univ Loma Linda CA 92354

MCMILLAN, R BRUCE, b Springfield, Mo, Dec 3, 37; m 61; c 3. ARCHAEOLOGY, BIOGEOGRAPHY. Educ: Southwest Mo State Col, BS, 60; Univ Mo-Columbia, MA, 63; Univ Colo, PhD(anthrop), 71. Prof Exp: Res assoc archaeol, Univ Mo, 64-66; assoc curator anthrop, Ill State Mus, 69-72, CUR ANTHROP, ILL STATE MUS, 72-, ASST MUS DIR, 73- Concurrent Pos: NSF fels, Ozark Pleistocene Springs, Ill State Mus, 71-72 & 72-73; lectr, Northwestern Univ, 72-73; consult, Midwest Res Inst, 72- Mem: Fel AAAS; fel Am Anthrop Asn; Soc Am Archaeol. Res: Late Pleistocene and post-Pleistocene environments, especially eastern North America; man's adaptation to environmental change during the early Holocene in the eastern North American Praire Peninsula. Mailing Add: Sect for Anthrop Ill State Mus Spring & Edwards Springfield IL 65201

MCMILLAN, ROBERT, b Pittsburgh, Pa, Nov 19, 34; m 66; c 2. IMMUNOHEMATOLOGY. Educ: Pa State Univ, BS, 56; Univ Pa, MD, 60. Prof Exp: Assoc hemat, 68-74, ASSOC MEM HEMAT & ONCOL, SCRIPPS CLIN RES FOUND, 74- Concurrent Pos: Adj asst prof hemat, Univ Calif, San Diego, 68-75, assoc adj prof, 75-; consult, Vet Admin Hosp, San Diego, 74- Mem: Am Soc Hemat. Res: Evaluation of antiplatelet antibodies in human disease, their synthesis, site of origin and antigens; study of platelet surface proteins; synthesis of human immunoglobulins. Mailing Add: Scripps Clin Res Found 476 Prospect St La Jolla CA 92037

MCMILLAN, ROBERT THOMAS, JR, b Miami, Fla, June 1, 34; m 54; c 2. PLANT PATHOLOGY. Educ: Univ Miami, BS, 61, MS, 64; Wash State Univ, PhD(plant path), 68. Prof Exp: Asst plant pathologist, 67-74, ASSOC PROF PLANT PATH & ASSOC PLANT PATHOLOGIST, AGR RES & EDUC CTR, UNIV FLA, 67- Mem: Am Phytopath Soc; Am Soc Agron; Crop Sci Soc Am. Res: Physiology, ecology and control of vegetable diseases. Mailing Add: Agr Res & Educ Ctr Univ Fla 18905 SW 280th St Homestead FL 33030

MCMILLAN, WILLIAM GEORGE, b Montebello, Calif, Oct 19, 19; m 46; c 3. CHEMICAL PHYSICS. Educ: Univ Calif, Los Angeles, BA, 41; Columbia Univ, MA, 43, PhD(chem), 45. Prof Exp: Teaching asst chem, Columbia Univ, 41-44; Guggenheim fel, Inst Nuclear Studies, Chicago, 46-47; from asst prof to assoc prof, 47-58, chmn dept, 59-65, PROF CHEM, UNIV CALIF, LOS ANGELES, 58- Concurrent Pos: Vis lectr, Harvard Univ, 51-52; mem physics dept, Rand Corp, 54-71; Sloan res fel, 57-61. Consult, Lawrence Radiation Lab, Univ Calif, 52-57, Brookhaven Nat Lab, 52-63 & President's Sci Adv Comt, 60-; consult, Open Ear Panel, Air Force Sci Adv Bd, 60-65, mem nuclear panel, 61-, mem weapons & munitions panel & writing group, Tactical Study, 64-65, chmn armament develop & testing ctr adv group, 69-72; consult, Air Force Range Tech Adv Group, 62 & Air Force Ballistic Systs Div Adv Group, 62-64; chmn sci adv group effects, Dir, Defense Res & Eng & Defense Atomic Support Agency, 61-66, consult, 69-71. Mem, Defense Sci Bd, 62-66 & 69-70, mem exec comt, 64-66; chmn ad hoc group radiation effects, US Air Force, US Navy & Dir, Defense Res & Eng, 63-66, mem, 66-; chmn advan res projs agency defense sci sem, Univ Calif, Los Angeles, 64-66; chmn, Joint Chiefs-of-Staff Panel on Nuclear Test Ban, 65-66; vchmn sci adv comt, Defense Intel Agency, 65-71, mem, 71-; sci adv to Comdr, US Mil Assistance Command, Vietnam, 66-68; mem, Army Sci Adv Panel, 69- & Oak Ridge Nat Lab Adv Group on Civil Defense, 71-; chmn, Nat Acad Army Countermine Study Group, 71-72. Honors & Awards: Distinguished Civilian Serv Award, Dept of Army, 68; Distinguished Pub Serv Award, Dept of Defense, 69; Knight, Nat Order of Vietnam, 69. Mem: AAAS; Am Phys Soc; Am Chem Soc. Res: Statistical and quantum mechanics of small molecules; adsorption; equation of state; spectroscopy at high pressure; electrolytes. Mailing Add: Dept of Chem Univ of Calif Los Angeles CA 90024

MCMILLAN, WILLIAM MARCUS, medicine, deceased

MCMILLEN, JANIS KAY, b El Dorado, Kans, Oct 21, 37. VIROLOGY. Educ: Trinity Univ, Tex, BS, 59; Univ Kans, PhD(microbiol), 71. Prof Exp: Res technician virol, Pitman-Moore Div, Dow Chem Co, 59-60 & Dept Pediat, Univ Kans Med Ctr, 60-66; Nat Inst Allergy & Infectious Dis fel, Div Biol, Kans State Univ, 71-74, res assoc, 74-76; SR RES VIROLOGIST, JENSEN-SALSBERY LABS DIV, RICHARDSON-MERRELL, INC, 76- Mem: Am Soc Microbiol; Sigma Xi; NY Acad Sci; Tissue Cult Asn. Res: Viral pathogenesis; molecular mechanisms of viral infections. Mailing Add: Virus Res Div Jensen-Salsbery Labs 2000 S 11th St Kansas City KS 66103

MCMILLEN, LARRY BYRON, b Defiance, Ohio, Apr 18, 42; m 64; c 2. PROSTHODONTICS. Educ: Ohio State Univ, DDS, 66, MS, 70. Prof Exp: Instr prosthodont, Col Dent, Ohio State Univ, 69-70; ASST PROF, SCH DENT, LA STATE UNIV MED CTR, NEW ORLEANS, 70- Mem: Int Asn Dent Res. Res: Border movements of human mandible; dental crown contours in relation to periodontal health. Mailing Add: Dept of Prosthodont La State Univ Sch of Dent New Orleans LA 70119

MCMILLEN, WARREN NEWTON, b Calera, Ala, Oct 16, 13; m 37; c 2. ANIMAL NUTRITION. Educ: Mich State Univ, PhD, 50. Prof Exp: Prof animal husb, Panhandle Agr & Mech Col, 37-44; assoc prof, Mich State Univ, 44-50; asst dir res, Allied Mills, Inc, 50-52; dir feed nutrit, A E Staley Co, 52-60; feed consult, 60-65; livestock specialist, Fla Dept Agr, 65-67; INSTR ANIMAL NUTRIT, CENT FLA JR COL, 67- Mem: AAAS; Am Soc Animal Sci; Am Genetic Asn; Poultry Sci Asn. Res: Vitamin and other nutrient requirements of farm animals. Mailing Add: 3521 SW 25th St Ocala FL 32670

MCMILLIAN, FRANK LEBARRON, b Mobile, Ala, June 9, 35; m 58; c 2. ORGANIC CHEMISTRY. Educ: Dillard Univ, BA, 54; Tuskegee Inst, MS, 56; Univ Kans, PhD(org chem), 65. Prof Exp: Instr phys sci, Ft Valley State Col, 56-59; instr chem, NC Col Durham, 59-60; asst prof, Va State Col, Norfolk, 60-62; res chemist, Explosives Dept, 65-70, plastics dept, 70-72, mkt rep, Photo Prod Dept, 72-74, TECH SERV CHEMIST, ELASTOMER CHEM DEPT, E I DU PONT DE NEMOURS & CO, 74- Concurrent Pos: Assoc prof, Dillard Univ, 68-69. Res: Synthetic and physical organic chemistry; free radicals in solution; polymer intermediates for synthesis of polyamides and polyesters. Mailing Add: Elastomer Lab Chestnut Run E I du Pont de Nemours & Co Inc Wilmington DE 19898

MCMILLIAN, WILLIAM WALLARD, b Florence, SC, Se t 20, 33; m 59; c 2. ENTOMOLOGY. Educ: Clemson Univ, BS, 57, MS, 59; Dipl, Am Registry Prof Entomologists. Prof Exp: Asst entomologist, Clemson Univ, 57-59; ENTOMOLOGIST, SOUTHERN GRAINS INSECTS RES LAB, AGR RES SERV, USDA, 59- Mem: Entom Soc Am. Res: Host plant resistance; insect vectors of plant diseases; insect biology including feeding stimulants, mating activities, food utilization. Mailing Add: Southern Grains Insects Res Lab Coastal Plain Exp Sta USDA Tifton GA 31794

MCMILLIN, CARL KENNETH, organic chemistry, see 12th edition

MCMILLIN, CARL RICHARD, b Warren, Ohio, Aug 4, 46; m 70. MEDICAL RESEARCH. Educ: Gen Motors Inst, BME, 69; Case Western Reserve Univ, MS, 71, PhD(macromolecular sci), 74. Prof Exp: Prod engr, Packard Elec Div, Gen Motors Corp, 68-69; res assoc polymer stress cracking,Queen Mary Col, Univ London, 71-72; NIH res fel, Case Western Reserve Univ & vis scientist, Artificial Organs Div, Cleveland Clin, 74-75; SR RES CHEMIST, MONSANTO RES CORP, 75- Mem: Biophys Soc; Am Inst Biol Sci. Res: Development and application of novel biomedical and biophysical techniques and instrumentation. Mailing Add: Monsanto Res Corp 1515 Nicholas Rd Dayton OH 04507

MCMILLIN, CHARLES W, b Indianapolis, Ind, Aug 22, 32. WOOD SCIENCE & TECHNOL. Educ: Purdue Univ, BS, 54; Univ Mich, MWT, 57, PhD(wood sci), 69. Prof Exp: Assoc mech engr, Am Mach & Foundry Co, 57-64; res coordr, Southern Pine Asn, 64-65; PRIN WOOD SCIENTIST, SOUTHERN FOREST EXP STA, FOREST SERV, USDA, 65- Concurrent Pos: Chmn, Pulp & Paper Tech Comt, Forest Prod Res Soc, 71, 72, chmn-elect, Div C Processes, 71, chmn, 72, chmn-elect, Mid-South Sect, 73, chmn, 73. Honors & Awards: Wood Award, Wood & Wood Prod Mag, 57. Mem: Forest Prod Res Soc; Sigma Xi; Soc Wood Sci & Technol; Int Asn Wood Anatomists; Tech Asn Pulp & Paper Indust. Res: Wood machining; mechanical pulping; modification of properties; characterization; applications of scanning electron microscopy to wood science. Mailing Add: 531 Highpoint Dr Alexandria LA 71301

MCMILLIN, KENNETH M, b Britt, Iowa, Jan 7, 19; m 47; c 1. MATHEMATICS. Educ: Cent Col, Iowa, BA, 44; Univ Minn, MS, 47; Univ Cincinnati, PhD(math), 53. Prof Exp: Instr math & mech, Univ Minn, 48-51; asst prof math, US Air Force Inst Technol, Wright Field, Ohio, 51-53 & Col of St Thomas, 53-55; assoc prof, 55-63, PROF MATH & DIR SIMULATION LAB, MICH TECHNOL UNIV, 63- Res: Computers; control devices; Fourier series; hybrid computation. Mailing Add: 3 Isle Royale Mill Houghton MI 49931

MCMILLION, C ROBERT, b Avon, WVa, Aug 5, 27; m 49; c 2. PHARMACEUTICAL CHEMISTRY, ORGANIC CHEMISTRY. Educ: Glenville State Col, AB, 51; Johns Hopkins Univ, MA, 59; Univ Md, PhD(pharmaceut chem), 64. Prof Exp: Res chemist, 51-64, CHEM PROD MGR, HYNSON WESTCOTT & DUNNING, INC, 64- Mem: Am Chem Soc. Res: Chromatography of triphenylmethane dyes; fluorescent isothiocyanates for antibody labeling; synthesis of catechol amines; spiroindolenines; micron and submicron sized particles and crystals associated with seriological reactions. Mailing Add: Hynson Westcott & Dunning Inc Charles & Chase St Baltimore MD 21201

MCMILLION, LESLIE GLEN, b Nallen, WVa, June 24, 30; m 54; c 2. HYDROLOGY. Educ: Marshall Univ, BA, 52; Mich State Univ, MS, 57. Prof Exp: Geologist, US Geol Surv, 54-56; geologist, Tex Water Comn, 56-59, div dir, Tex Water Develop Bd, 59-65; RES HYDROLOGIST, US ENVIRON PROTECTION AGENCY, 66- Concurrent Pos: Prin partic, Tex-US Study Comn, 59-60; mem, Tex Govr's Water Pollution Adv Coun, 60-61, Working Group Preparation Regulations Safe Drinking Water Act, 75-76 & chmn, Nat Aquifer Protection Comn, 67-70. Mem: Nat Water Well Asn; Am Water Resources Asn; Geol Soc Am; Am Water Works Asn. Mailing Add: US Environ Protection Agency PO Box 15027 Las Vegas NV 89114

MCMILLION, OVID MILLER, b Friars Hill, WVa, Mar 26, 05; m 50; c 4. ECONOMIC GEOGRAPHY, GEOGRAPHY OF NORTH AMERICA. Educ: Ohio Univ, AB, 30; George Peabody Col, MA, 34; Cambridge Univ, cert, 37; Univ Chicago, cert, 38; Univ Md, College Park, PhD(geog & econ), 61. Prof Exp: Teacher Bd Educ, Pocahontas County, WVa, 25-26, Greenbrier County, 26-27, Pocahontas County, 29-30 & Hayden, Ariz, 30-31; teacher soc sci, High Sch, Bd Educ, Fairbanks, Alaska, 31-33; head dept geog, Concord State Teachers Col, 34-39; from instr to prof geog & econ, DC Teachers Col, 39-66; PROF GEOG, MID TENN STATE UNIV, 66- Mem: Asn Am Geog; Nat Coun Geog Educ; Conf Latin Am Geog. Res: Economic geography; regional geography of North and South America; conservation; economics and history. Mailing Add: 207 Ridgecrest Dr Murfreesboro TN 37130

MCMILLION, THEODORE MILLER, b Friars Hill, WVa, Oct 5, 03; m 29, 49; c 2. ZOOLOGY. Educ: WVa Univ, BA, 27, MA, 29; Univ Pittsburgh, PhD(zool), 41. Prof Exp: From instr to prof, 26-74, EMER PROF BIOL, GENEVA COL, 74- Concurrent Pos: Instr, Beaver Valley Gen Hosp, Pa. Mem: AAAS; Am Inst Biol Sci. Res: Mammalian chromosomes; stain technology; spermatogenesis and chromosome complexes in the rabbit. Mailing Add: Dept of Biol Geneva Col Beaver Falls PA 15010

MCMINN, ROBERT GORDON, forest ecology, see 12th edition

MCMINN, TREVOR JAMES, b Salt Lake City, Utah, Jan 23, 21. MATHEMATICS. Educ: Univ Utah, BA, 42; Univ Calif, Berkeley, PhD(math), 55. Prof Exp: Instr math, Univ Rochester, 50, Univ Calif, Riverside, 54-55 & Univ Calif, Berkeley, 55-56; asst prof, Univ Wash, 56-63; assoc prof, 63-69, PROF MATH, UNIV NEV, RENO, 69- Mem: Am Math Soc. Res: Measure theory; real analysis. Mailing Add: Dept of Math Univ of Nev Reno NV 89507

MCMORDIE, WARREN C, JR, b Austin, Tex, May 14, 29; m 52; c 4. INORGANIC CHEMISTRY. Educ: Tex Christian Univ, BA, 50, MA, 60; Univ Tex, PhD(chem), 63. Prof Exp: Chemist, Gen Dynamics-Ft Worth, 51-57, asst supvr, 57-59, res chemist, 59-61; res chemist, Jackson Lab, E I du Pont de Nemours & Co, Inc, 63-67; mgr chem labs, 67-72, DIR RES & DEVELOP, OIL BASE, INC, 72- Mem: Am Chem Soc; The Chem Soc; Am Petrol Inst. Res: Rheology; oil well drilling fluids; asphalt. Mailing Add: 3625 Southwest Freeway Houston TX 77027

MCMORRIS, FRED RAYMOND, b Gary, Ind, Aug 28, 43; m 65; c 2. MATHEMATICS, BIOMATHEMATICS. Educ: Beloit Col, BS, 65; Univ Calif, Riverside, MA, 66; Univ Wis-Milwaukee, PhD(math), 69. Prof Exp: Asst prof, 69-74, ASSOC PROF MATH, BOWLING GREEN STATE UNIV, 74- Concurrent Pos: Nat Inst Gen Med Sci fel, Biomath Prog, NC State Univ, 71-73. Mem: Am Math Soc; Math Asn Am; Soc Math Biol. Res: Abstract algebra; mathematical biology; numerical taxonomy. Mailing Add: Dept of Math Bowling Green State Univ Bowling Green OH 43403

MCMORRIS, FREDERICK ARTHUR, b Lawton, Okla, Sept 17, 44; m 68. BIOLOGY, BIOCHEMISTRY. Educ: Brown Univ, BA, 66; Yale Univ, PhD(biol), 72. Prof Exp: Teaching asst, Brown Univ, 66 & Yale Univ, 67-70; res assoc biol, Mass Inst Technol, 72-74; RES ASSOC, WISTAR INST, 74-; ASST PROF HUMAN GENETICS, UNIV PA, PA SCH OF MED, 75- Mem: AAAS. Res: Genetic regulation of development and cell differentiation; molecular genetics; cell biology; neurobiology; somatic cell genetics. Mailing Add: The Wistar Inst 36th St at Spruce Philadelphia PA 19104

MCMORRIS, REX O, b Nevada, Iowa, Feb 15, 12; m 37; c 2. MEDICINE. Educ: Univ Nebr, BS, 48, MD, 49; Univ Minn, MS, 53. Prof Exp: Phys ther trainee, Mayo Clin, 39; staff phys therapist, US Vet Admin, 40-41; head dept phys ther, Plant Health Indust Ctr, Merck & Co, 41-45; asst prof phys med, Ohio State Univ & assoc dir dept phys med & rehab, Univ & Children's Hosps, 53-54; chmn dept phys med & rehab, Sch Med, Univ Louisville, 54-60; DIR, INST PHYS MED & REHAB, 60- Concurrent Pos: Med dir, Rehab Ctr, Inc, 54-60. Mem: Am Cong Rehab Med. Res: Industrial physical medicine; muscular hypertrophy and strength; electromyography. Mailing Add: Inst of Phys Med & Rehab 619 NE Glen Oak Ave Peoria IL 61603

MCMORRIS, TREVOR CALTHORPE, organic chemistry, see 12th edition

MCMULLEN, CARMAN CALVERT, nuclear physics, see 12th edition

MCMULLEN, CHARLES HENRY, b Lisnaskea, Northern Ireland, Dec 27, 39; m 67; c 2. ORGANIC CHEMISTRY. Educ: Queen's Univ, Belfast, BS, 62, PhD(org chem), 65. Prof Exp: Nat Sci Found grant, Univ Pittsburgh, 65-66; PROJ SCIENTIST, CHEM & PLASTIC DIV, UNION CARBIDE CORP, 67- Mem: Am Chem Soc; The Chem Soc. Res: Mixed sandwich pi— complexes of benzenoid compounds; stereochemistry of addition of nucleophiles to acetylenes; mechanism of diazonium salt decomposition; oxidation chemistry; free radical chemistry; homogeneous and heterogeneous catalysis. Mailing Add: Chem & Plastics Div Union Carbide Corp PO Box 65 Tarrytown NY 10591

MCMULLEN, EUGENE JOSEPH, b Minneapolis, Minn, Oct 15, 20; m 49; c 3. CHEMISTRY. Educ: Col St Thomas, BS, 42; Univ Tex, PhD(chem), 47. Prof Exp: Instr org chem, Univ Tex, 45-47; res chemist, Monsanto Chem Co, 47-59; RES CHEMIST, SHAMROCK SPECIALTIES CO, 59- Mem: Am Chem Soc. Res: Organic chemistry; commercial development; chemical specialties manufacture. Mailing Add: 428 22nd Ave N Texas City TX 77590

MCMULLEN, JAMES CLINTON, b Alton, Ill, July 6, 42; m 63; c 3. CHEMISTRY. Educ: Wis State Univ-Superior, BS, 65; Univ SDak, PhD(chem), 69. Prof Exp: ASST PROF CHEM, ST CLOUD STATE COL, 69- Mem: Am Chem Soc. Res: Analytical instrumentation; computer applications to chemistry. Mailing Add: Dept of Chem St Cloud State Col St Cloud MN 56301

MCMULLEN, JOHN LLOYD, b Tuscola, Ill, Apr 14, 13; m 46. BOTANY. Educ: Eastern Ill Univ, BEd, 34; Wash State Univ, MS, 48, PhD(cytogenetics), 66. Prof Exp: Instr biol sci, N Idaho Jr Col, 39-42, 46-47; instr bot, Univ Wyo, 49-51; from instr to asst prof, 51-67, asst to dean, 67-69, asst dean, Col Letters & Sci, 69-75, ASSOC PROF BOT, UNIV IDAHO, 67-, CHMN REGISTRATION, 67-, ASSOC DEAN, COL LETTERS & SCI, 75- Mem: Bot Soc Am; Phycol Soc Am. Res: Cytogenetics; algae; plant morphology; algae of north Idaho. Mailing Add: Col of Letters & Sci Univ of Idaho Moscow ID 83843

MCMULLEN, JOHN THOMAS CREIGHTON, chemistry, see 12th edition

MCMULLEN, ROBERT MICHAEL, b Toronto, Ont, Nov 20, 35; m 62; c 1. SEDIMENTOLOGY, INFORMATION SCIENCE. Educ: Univ Alta, BSc, 57, MSc, 59; Univ Reading, PhD(sedimentology), 65. Prof Exp: Info specialist in-chg explor, Imp Oil Ltd, Alta, 59-61; explor geologist & trainee, Imp Oil Explor, 64-65; explor geologist, Hudson's Bay Oil & Gas Co, 65; res sci, Bedford Inst Oceanog, NS, 66-68, head sci info serv & libr, Atlantic Oceanog Lab, 68-70; chief, Commun Data & Libr Serv, Dept Commun, 70-71, dir info retrieval serv, 71-73, WITH DEPT ENVIRON, OCEAN & AQUATIC AFFAIRS, ENVIRON CAN, 73- Mem: Am Asn Petrol Geol; fel Geol Asn Can; fel Geol Soc London; Brit Inst Info Sci; Am Soc Info Sci. Res: Sedimentary petrology and structures; sediment transport; marine geology; scientific information and data storage and retrieval; intertidal sedimentation; information technology; on-line systems; information networks; science policy. Mailing Add: Dept of the Environ Ocean & Aquatic Affairs Fontaine Bldg 13th Floor Ottawa ON Can

MCMULLEN, WARREN ANTHONY, b Faulkton, SDak, Oct 22, 07; m 29; c 3. ORGANIC CHEMISTRY. Educ: Greenville Col, BS, 28; Univ Nebr, MA, 36; NY Univ, PhD(sci ed), 60. Prof Exp: Teacher high sch, Nebr, 28-43; teacher chem & physics, Cent Col, Kans, 43-45; from asst prof to prof, 45-75, EMER PROF CHEM, GREENVILLE COL, 75-; PRIN & SCI TEACHER, OAKDALE CHRISTIAN HIGH SCH, JACKSON, KY, 75- Mem: AAAS; Am Chem Soc; Nat Sci Teachers Asn. Res: Assessment of environmental quality, especially water quality and pesticides in aqueous environment. Mailing Add: 320 N First St Greenville IL 62246

MCMUNN, JOHN CARTER, applied mechanics, see 12th edition

MCMURCHY, KENNETH ALLAN, b Woodrow, Sask, June 23, 23; m 49; c 4. ORAL PATHOLOGY. Educ: Univ Sask, BA, 45; Univ Alta, DDS, 47; Columbia Univ, cert, 51; Inst Dent Surg, London, 63. Prof Exp: From lectr to assoc prof, 49-70, PROF ORAL PATH, UNIV ALTA, 70-, CHMN DEPT ORAL BIOL, 71- Mem: Can Acad Oral Path. Res: Ultrastructure of multinucleated cells. Mailing Add: Dept of Oral Biol Univ of Alta Fac of Dent Edmonton AB Can

MCMURCHY, ROBERT CONNELL, b Arden, Man, Sept 15, 02; m 36; c 3. GEOLOGY. Educ: Univ Man, BSc, 30; Univ Minn, PhD(geol), 34. Prof Exp: Lectr geol, Univ Man, 34-35; asst geologist, Geol Surv Can, 35-39; geologist, Powell-Rouyn Gold Mines, Ltd, 39-51; explor engr, Chesterville Mines, Ltd, 51-53; CONSULT GEOLOGIST, 53- Mem: Geol Asn Can; Can Inst Mining & Metall; Asn Prof Engrs. Res: Mining exploration and development. Mailing Add: 95 Southvale Dr Toronto ON Can

MCMURDIE, HOWARD FRANCIS, b Detroit, Mich, Feb 5, 05; m 28; c 3. CHEMISTRY. Educ: Northwestern Univ, BS, 28. Prof Exp: Jr chemist, DC, 28-33, Calif, 33-35, jr petrographer, DC, 35-44, gen phys scientist, 44-48, chief sect crystallog, 48-65, RES FEL, JOINT COMT POWDER DIFFRACTION STANDARDS, NAT BUR STANDARDS, DC, 66- Mem: Fel Am Ceramic Soc; Mineral Soc Am; Am Crystallog Asn. Res: Crystal chemistry; phase equilibrium of refractory oxides; high temperature x-ray diffraction; chemical analysis by x-ray diffraction; effect of heat on crystals; data compilation. Mailing Add: Nat Bur Standards Washington DC 20234

MCMURPHY, WILFRED E, b Lamont, Okla, Aug 3, 34; m 59; c 1. AGRONOMY. Educ: Okla State Univ, BS, 56, MS, 59; Univ Kans, PhD(agron), 63. Prof Exp: Res asst agron, Kans State Univ, 59-62; asst prof, SDak State Univ, 62-64; asst prof, 64-70, ASSOC PROF AGRON, OKLA STATE UNIV, 70- Mem: Am Soc Agron; Crop Sci Soc Am; Soc Range Mgt; Ecol Soc Am. Res: Range and pasture management; pasture production and ecology. Mailing Add: Dept of Agron Okla State Univ Stillwater OK 74075

MCMURRAY, BIRCH LEE, b Polk Co, NC, Oct 18, 31; m 56; c 3. VETERINARY MEDICINE, AGRICULTURE. Educ: NC State Univ, BS, 54; Univ Ga, DVM, 57. Prof Exp: Dir lab, Fla Dept Agr, 58-60; vet, Dr Robert E Lee Vet Hosp, 60-62; res vet, 62-64, mgr vet res, 64-72, DIR INT FEED RES, CENT SOYA CO, INC, 72- Mem: Am Vet Med Asn; Am Asn Avian Pathologists; US Animal Health Asn. Res: Programmed preventive veterinary medicine; preventive programs for coccidiosis, fowl cholera, Marek's disease, respiratory diseases of cattle, enteridides or swine, mastitis-metritis complex; interrelations of disease, nutrition and management. Mailing Add: Res Dept Cent Soya Co Inc Decatur IL 46733

MCMURRAY, GERTRUDE SMIT, organic chemistry, see 12th edition

MCMURRAY, VIRGINIA M (VOLLMER), b Birmingham, Ala, Aug 29, 23. PHYSIOLOGY. Educ: Univ Cincinnati, AB, 47, MS, 49; Univ Iowa, PhD, 53. Prof Exp: Res asst, Rockefeller Inst, 54-56; asst prof physiol, Mt Holyoke Col, 56-57; instr, Sch Med, Univ Mich, 57-59; proj analyst, Collab Res Study, Nat Inst Neurol Dis & Blindness, 59-60; prof anat, Univ Tenn, 60-62; assoc prof, Tex Woman's Univ, 65-67; health sci consult, 67-71; ASST PROF INSTRNL COMMUN, UNIV TEX HEALTH SCI CTR, DALLAS, 71- Concurrent Pos: Fulbright fel, Hubrecht Lab, Netherlands, 53-54; lectr, Nat Drug Educ Ctr, Med Ctr, Univ Okla, training consult, Methodist Hosp, Dallas. Mem: Am Physiol Soc; Am Soc Zool. Res: Experimental embryology; human physiology. Mailing Add: Dept of Instrnl Commun Univ of Tex Health Sci Ctr Dallas TX 75235

MCMURRAY, WALTER JOSEPH, b Montague, Mass, Aug 22, 35; m 60; c 4. BIOCHEMISTRY. Educ: Amherst Col, AB, 58; Univ Ill, PhD, 62. Prof Exp: Res assoc chem, Mass Inst Technol, 63-64; instr med, 65-66, asst prof pharmacol, 66-67, asst prof health sci resources, 67-70, ASSOC PROF HEALTH SCI RESOURCES, SCH MED, YALE UNIV, 70- Mem: Am Chem Soc; The Chem Soc. Res: Organic biochemistry; application of mass spectrometry to problems in organic chemistry; use of computers in chemistry. Mailing Add: Div of Health Sci Resources Yale Univ Sch of Med New Haven CT 06520

MCMURRAY, WILLIAM COLIN CAMPBELL, b Bangor, Northern Ireland, Mar 16, 31; nat Can; m 53; c 2. BIOCHEMISTRY. Educ: Univ Western Ont, BSc, 53, PhD(biochem), 56. Prof Exp: Proj assoc, Inst Enzyme Res, Univ Wis, 56-58; asst prof cancer res & lectr biochem, Univ Sask, 58-59; asst prof biochem & Jr Red Cross Res Prof ment retardation, 59-65, assoc prof biochem, 65-70, PROF BIOCHEM, UNIV WESTERN ONT, 70- Concurrent Pos: Vis scientist, Inst Animal Physiol, Agr Res Coun, Cambridge, Eng, 67-68. Mem: Am Soc Biol Chem; Can Biochem Soc; Brit Biochem Soc. Res: Biochemistry of the nervous system; metabolism and enzymology of differentiation and development; lipid metabolism; respiratory enzymes and oxidative phosphorylation; membrane synthesis; mitochondrial biogenesis. Mailing Add: Dept of Biochem Univ of Western Ont Fac of Med London ON Can

MCMURRY, EARL WILLIAM, b Arnolds Park, Iowa, Jan 4, 31; m 63; c 1. PHYSICS, GEOPHYSICS. Educ: Iowa State Univ, BSc, 53, MSc, 57; Univ London, PhD(physics) & Imp Col, dipl, 68. Prof Exp: Assoc physics, Inst Atmospheric Physics, Univ Ariz, 58-61; instr, Southern Colo State Col, 61-63; res asst, Imp Col, Univ London, 63-69; asst prof, Univ Alta, 69-75; ASST PROF PHYSICS, NORTHEAST MO STATE UNIV, 75- Mem: Am Inst Physics; Am Geophys Union. Res: The use of paleomagnetism to study the ancient geomagnetic field and continental drift; magnetic properties of rocks; computer applications. Mailing Add: Sci Div Northeast Mo State Univ Kirksville MO 63501

MCMURRY, HENRY LEWIS, physics, see 12th edition

MCMURRY, JOHN EDWARD, b New York, NY, July 27, 42; m 64; c 2. SYNTHETIC ORGANIC CHEMISTRY. Educ: Harvard Univ, BA, 64; Columbia Univ, MA & PhD(chem), 67. Prof Exp: From asst prof to assoc prof, 67-75, PROF CHEM, UNIV CALIF, SANTA CRUZ, 75- Concurrent Pos: Alfred P Sloan fel, 69-71. Honors & Awards: Career Develop Award, NIH, 75-80. Mem: AAAS; Am Chem Soc; The Chem Soc. Res: Natural product synthesis; new synthetic reactions. Mailing Add: Dept of Chem Natural Sci Div Univ of Calif Santa Cruz CA 95064

MCMURTREY, MARION JOHN, b Ririe, Idaho, Dec 5, 26; m 49; c 5. SURGERY, MOLECULAR BIOLOGY. Educ: Idaho State Col, BS, 51; Johns Hopkins Univ, MD, 59; Am Bd Surg, dipl. Prof Exp: Intern surg, Emory Univ Hosp, 59-60, resident, 60-61; resident, Univ Utah Hosp, 64-65; ASSOC PROF SURG, UNIV TEX M D ANDERSON HOSP & TUMOR INST HOUSTON, 66- Concurrent Pos: NIH grant cardiac metab, 69-71. Mem: AAAS; AMA; fel Am col Surg. Res: Electron microscopy; carcinoembryonic antigen; tumor antigen localization. Mailing Add: Univ of Tex M D Anderson Hosp & Tumor Inst Houston TX 77025

MCMURTRY, JAMES A, b Lodi, Calif, Sept 21, 32; m 54; c 2. ENTOMOLOGY. Educ: San Jose State Col, AB, 54; Univ Calif, Davis, PhD(entom), 60. Prof Exp: Res assoc entom, Univ Calif, Davis, 56-60; asst entomologist, 60-66, ASSOC ENTOMOLOGIST & LECTR BIOL CONTROL, UNIV CALIF, RIVERSIDE, 66- Concurrent Pos: Guggenheim fel, 68. Mem: Entom Soc Am; Entom Soc Can. Res: Biological control and population ecology of phytophagous mites. Mailing Add: Div of Biol Control Univ of Calif Riverside CA 92502

MCNAB, IAN RODERICK, b Cleckheaton, Eng, Feb 28, 39; m 64; c 4. MAGNETOHYDRODYNAMICS, CURRENT COLLECTION. Educ: Univ Leeds, BSc, 60; Univ Reading, PhD(appl sci), 74. Prof Exp: Physicist, Int Res & Develop Co, Newcastle, Eng, 60-66, group leader direct energy conversion, 66-68, group leader current collectors, 69-73, group leader laser systs, 73-75; SR SCIENTIST MAGNETOHYDRODYN & CURRENT COLLECTION, WESTINGHOUSE RES & DEVELOP CTR, 75- Concurrent Pos: UK mem, Joint Europ Nuclear Energy Agency/Int Atomic Energy Agency Int Liaison Group Magnetohydrodyn, 68-75; mem comt elec & magnetic phenomena ionized gases, UK Inst Elec Engrs, 71-74; mem plasma physics comt, UK Inst Physics. Mem: Fel Brit Inst Physics; corp mem Brit Inst Elec Engrs; assoc fel Brit Interplanetary Soc. Res: Magnetohydrodynamics of plasma and liquid metal flows; fundamental aspects of current transfer through solid/solid and liquid/solid interfaces. Mailing Add: Liquid Metal Technol Group Westinghouse Res & Develop Ctr 1310 Beulah Rd Pittsburgh PA 15235

MCNABB, CLARENCE DUNCAN, JR, b Beloit, Wis, July 7, 28; m 53; c 7. LIMNOLOGY. Educ: Loras Col, BA, 51; Univ Wis, MS, 57, PhD(algal ecol), 60. Prof Exp: Pub health biologist, Wis State Bd Health, 57-59; from asst prof to assoc prof biol, Wis State Univ, Whitewater, 59-63; assoc prof biol & assoc dir hydrobiol sta, St Mary's Col, Minn, 63-68; assoc prof, 68-72, PROF LIMNOL, MICH STATE UNIV, 72- Concurrent Pos: Dir field col progs, Wis State Pigeon Lake Field Sta, 62-63; mem, Smithsonian Assoc. Mem: AAAS; Am Soc Limnol & Oceanog; Ecol Soc Am; Weed Sci Soc Am; Int Asn Ecol. Res: Primary production of aquatic angiosperms in hypereutrophic environments; movement of minerals in aquatic angiosperm communities; ecological implications of aquatic weed control. Mailing Add: Dept of Fisheries & Wildlife Mich State Univ East Lansing MI 48823

MCNABB, COY GAYLORD, agricultural economics, see 12th edition

MCNABB, F M ANNE, b Edmonton, Alta, Jan 17, 39; m 63; c 1. COMPARATIVE PHYSIOLOGY. Educ: Univ Alta, BEd, 60, BSc, 61; Univ Calif, Los Angeles, MA, 65, PhD(zool), 68. Prof Exp: Asst prof biol & allied health, Quinnipiac Col, 68-69; adj asst prof, 69-72, ASST PROF ZOOL, VA POLYTECH INST & STATE UNIV, 72- Concurrent Pos: Fel zool, Yale Univ, 68-69. Mem: AAAS; Am Soc Zoologists. Res: Avian kidney function, particularly nitrogen excretion, including urate solubility problems, electrolyte excretion, water balance in different environments; role of the thyroid gland in development of temperature regulation in precocial and altricial birds. Mailing Add: Dept of Biol Va Polytech Inst & State Univ Blacksburg VA 24061

MCNABB, HAROLD SANDERSON, JR, b Lincoln, Nebr, Nov 20, 27; m 49; c 2. FOREST PATHOLOGY. Educ: Univ Nebr, BSc, 49; Yale Univ, MS, 51, PhD(plant sci), 54. Prof Exp: Asst bot, Univ Nebr, 46-49; asst plant sci, Yale Univ, 49-52; asst prof bot, 53-54, asst prof bot & forestry, 54-56, assoc prof, 56-64, PROF FORESTRY & PLANT PATH, IOWA STATE UNIV, 64-, ASSOC, INST ATOMIC RES, 54- Concurrent Pos: Sci adv, Tree Res Inst; coordr forestry res, Iowa State Conserv Comn, 53-59; mem, Int Bot Cong, France, 54, Can, 59 & Gt Brit, 64; ed, Iowa Acad Sci, 64- Mem: AAAS; Soc Am Foresters; Am Phytopath Soc; Mycol Soc Am. Res: Diseases of forest and shade trees; deterioration of wood. Mailing Add: 223 B Charles E Bessey Hall Iowa State Univ Ames IA 50010

MCNABB, ROGER ALLEN, b Moose Jaw, Sask, May 21, 38; m 63. COMPARATIVE PHYSIOLOGY. Educ: Univ Alta, BSc, 61, MSc, 63; Univ Calif, Los Angeles, PhD(zool), 68. Prof Exp: Res staff biologist, Yale Univ, 68-69; asst prof, 69-75, ASSOC PROF BIOL, VA POLYTECH INST & STATE UNIV, 75- Mem: AAAS; Am Soc Zool; Am Inst Biol Sci. Res: Physiology of environmental adaptations in lower vertebrate animals. Mailing Add: Dept of Biol Va Polytech Inst & State Univ Blacksburg VA 24061

MCNABNEY, RALPH, inorganic chemistry, see 12th edition

MCNAIR, ANDREW HAMILTON, b Victor, Mont, May 29, 09; m 39; c 3. GEOLOGY. Educ: Univ Mont, AB, 31, MA, 33; Univ Mich, PhD(geol), 35. Hon Degrees: MA, Dartmouth Col, 44. Prof Exp: Asst geol, Univ Mont, 31-33; asst Ky field camp, Univ Mich, 34-35; from instr to prof, 35-74, EMER PROF GEOL, DARTMOUTH COL, 74- Concurrent Pos: Chmn dept geol, Dartmouth Col, 39-42 & 50-54; geologist, NH State Planning Comn, 41 & 46, US Geol Surv, 42-45, Phillips Petrol Co, 48-50 & Gulf Oil Co, 51-52; consult geologist, Mica & Feldspar Mining Co, 47-; mem, Int Geol Field Inst, Gt Brit, 61; leader Dartmouth Col Geol Exped, Victoria Island, Can, 64-66; vis scientist, Am Geol Inst, 65. Mem: Fel Geol Soc Am; fel Paleont Soc; Am Asn Petrol Geol. Res: Micropaleontology; Paleozoic Bryozoa; stratigraphy; diatomaceous earth; mica, feldspar and beryl bearing pegmatites; petroleum and Arctic geology. Mailing Add: Dept of Earth Sci Dartmouth Col Hanover NH 03755

MCNAIR, DOUGLAS MCINTOSH, b Rockingham, NC, July 19, 27; m 51; c 1. PSYCHOPHARMACOLOGY, PSYCHOLOGY. Educ: Univ NC, Chapel Hill, AB, 48, PhD(clin psychol), 54. Prof Exp: Sr clin psychologist, Guildford County Ment Health Clin, 52-55; psychologist, Col Infirmary & asst prof psychol, Univ NC, Greensboro, 55-56; resident psychologist & asst chief outpatient psychiat res lab, Vet Admin, Washington, DC, 56-64; asst prof, Psychopharmacol Lab, 64-65, assoc prof psychiat, 65-70, PROF PSYCHIAT, SCH MED, BOSTON UNIV, 70-, HEAD CLIN STUDIES UNIT, PSYCHOPHARMACOL LAB, 65- Concurrent Pos: Adv ed, J Consult & Clin Psychol, 72. Mem: Psychomet Soc; Am Col Neuropsychopharmacol; Am Psychol Asn. Res: Drugs and behavior; psychotherapy; experimental design. Mailing Add: Psychopharmacol Lab Boston Univ Sch of Med Boston MA 02181

MCNAIR, HAROLD MONROE, b Miami, Ariz, May 31, 33; m 60; c 3.

ANALYTICAL CHEMISTRY. Educ: Univ Ariz, BS, 55; Purdue Univ, MS, 57, PhD(anal chem), 59. Prof Exp: Fulbright & univ fels, Eindhoven Technol Univ, 59-60, Perkin Elmer res fel, 60; res chemist, Esso Res & Eng, NJ, 60-61; tech dir, Europe Div, F&M Sci Corp, Amsterdam, 61-63, gen mgr, 63-64; dir int opers, Varian Aerograph, Switz, 64-66, dir mkt, Calif, 66-68; assoc prof, 68-71, PROF ANAL CHEM, VA POLYTECH INST & STATE UNIV, 71- Concurrent Pos: Consult, Instrument Group, Varian Assocs, 68-; adj prof, Nat Univ Mex; Air Pollution Control Off, Res Triangle, NC. Mem: Am Chem Soc; Am Soc Test & Mat. Res: Gas chromatography; quantitative analysis of ionization detectors and temperature programming; trace gas analysis by ionization detectors; selection of selective liquid phases; theory of chromatography; liquid chromatography; computer analysis of chromatographic peaks. Mailing Add: Dept of Chem Va Polytech Inst & State Univ Blacksburg VA 24061

MCNAIR, RUTH DAVIS, b Flint, Mich, Mar 18, 21; m 50; c 2. CLINICAL CHEMISTRY. Educ: Univ Mich, BS, 41; Univ Cincinnati, BS, 42; Wayne State Univ, PhD(biochem), 48. Prof Exp: DIR BIOCHEM, PROVIDENCE HOSP, 48- Mem: AAAS; Am Chem Soc; fel Am Asn Clin Chemists. Res: Protein, nutrition and metabolism. Mailing Add: Lab Providence Hosp 16001 W Nine Mile Rd Southfield MI 48075

MCNAIRN, ROBERT BLACKWOOD, b San Francisco, Calif, Mar 13, 40; m 62; c 2. PLANT PHYSIOLOGY. Educ: Univ Calif, Berkeley, BS, 62; Univ Calif, Davis, MS, 64, PhD(plant physiol), 67. Prof Exp: Asst prof , 67-71, ASSOC PROF BOT, CALIF STATE UNIV, CHICO, 71- Concurrent Pos: Plant physiologist, Thornton Wholesale Florist & Flower Grower, 74-75. Mem: Am Soc Plant Physiol. Res: Phloem translocation; heat injury in plants. Mailing Add: Dept of Biol Calif State Univ Chico CA 95926

MCNAIRY, SIDNEY A, JR, b Memphis, Tenn, Oct 16, 37; m 65. BIOCHEMISTRY, ORGANIC CHEMISTRY. Educ: LeMoyne Col, Tenn, BS, 59; Purdue Univ, MS, 62, PhD(biochem), 65. Prof Exp: Lab asst chem, LeMoyne Col, Tenn, 58-59; asst biochem, Purdue Univ, 59-65, instr physiol chem, 62-65; PROF BIOCHEM, SOUTHERN UNIV, BATON ROUGE, 65-, DIR, HEALTH RES CTR, 71- Concurrent Pos: Instr, In-serv Inst, Southern Univ, Baton Rouge, 65-66, coordr black exec exchange prog, 70-71, exten of thirteen col concept chem, 71-; res chemist, Chevron Res Co, Calif. Mem: Am Chem Soc. Res: Metabolic disorders; isolation characterization and studies of triterpenoid glycosides from alfalfa tops. Mailing Add: Dept of Chem Southern Univ Baton Rouge LA 70813

MCNALL, EARL GEORGE, b Guernsey, Wyo, Sept 18, 24; m 48; c 2. BIOCHEMISTRY. Educ: Univ Southern Calif, AB, 52; Univ Calif, Los Angeles, PhD(chem), 57. Prof Exp: Asst res biochemist, Sch Med, Univ Calif, Los Angeles, 58-61; asst prof biochem, Sch Med, Univ Southern Calif, 61-68; assoc dir artificial heart prog, Nat Inst Sci Res, Calif, 68-70; VPRES RES & DEVELOP, CONSYNE CORP, 70- Mem: AAAS; Am Chem Soc. Res: Protein and polymer chemistry; intermediary metabolism; enzymology; immunochemistry. Mailing Add: Consyne Corp Glendora CA 91740

MCNALL, JOHN FAIRMAN, electrical engineering, see 12th edition

MCNALL, JOHN WILLIAM, b Cleveland, Ohio, Oct 19, 14; m 42; c 1. APPLIED PHYSICS. Educ: Case Western Reserve Univ, BS, 36; Mass Inst Technol, PhD(physics), 42. Prof Exp: Res engr, Westinghouse Elec Corp, Pa, 36-38, lamp div, NJ, 39-40; mem staff, Radiation Lab, Mass Inst Technol, 41-43; res engr, Lamp Div, 43-47, engr res div, 50-59, dir res lamp div, 59-70, adv engr, Res Labs, 70-72, MGR LAMP RES, RES LABS, WESTINGHOUSE ELEC CORP, 72- Concurrent Pos: Lamme scholar, Westinghouse Elec Corp. Mem: Am Phys Soc; Inst Elec & Electronics Engrs; Illum Eng Soc Res: Electron emission; gas discharges; magnetrons; getters; oxide coated cathodes; secondary emission cathode for magnetron; lamps and light generation. Mailing Add: Appl Sci Div Res Labs Westinghouse Elec Labs Beulah Rd Pittsburgh PA 15235

MCNALL, LESTER R, b Gaylord, Kans, Oct 28, 27. PLANT NUTRITION. Educ: Univ Wis, BS, 50; Univ Calif, Los Angeles, PhD(org chem), 55. Prof Exp: Res chemist, Esso Res & Eng Co, 55-56; res chemist, Paper Mate Mfg Co, Gillette Co, 56-58, head chem res, 58-62, res div, 62-64; TECH DIR, LEFFINGWELL CHEM CO, 66- Mem: AAAS; Am Chem Soc; Am Soc Hort Sci. Res: Chemistry of natural products; agricultural chemistry; minor elements in plant nutrition; foliar feeding. Mailing Add: Leffingwell Chem Co 11 S Berry St Brea CA 92621

MCNALLY, JAMES GREEN, JR, b Rochester, NY, Jan 27, 34; m 52; c 3. ORGANIC CHEMISTRY. Educ: Duke Univ, BS, 55; Univ Rochester, PhD(org chem), 58. Prof Exp: Res chemist, Fabrics & Finishes Div, E I du Pont de Nemours & Co, 58-64; res assoc, 70, SR RES CHEMIST, EASTMAN KODAK CO, 64- Mem: Am Inst Chem; Soc Photog Sci & Eng. Res: Organic photochemistry; non-silver imaging processes. Mailing Add: Res Labs Eastman Kodak Co 343 State St Rochester NY 14650

MCNALLY, JAMES HENRY, b Orange, NJ, Dec 18, 36. NUCLEAR PHYSICS, LASERS. Educ: Cornell Univ, BEng Phys, 59; Calif Inst Technol, PhD(physics), 66. Prof Exp: Physicist & prog mgr, Los Alamos Sci Lab, 65-74; ASST DIR LASERS & ISOTOPE SEPARATION, ENERGY RES & DEVELOP ADMIN, 74- Concurrent Pos: Mem subpanel impact on nat security of nuclear physics, Nat Acad Sci-Nat Res Coun, 70; mem US delegation, Conf on Disarmament, Geneva, 69, 73, 74 & threshold test ban treaty talks, Moscow, 74. Mem: Am Phys Soc. Res: Low energy nuclear spectroscopy; neutron time-of-flight; nuclear weapons design; laser fusion; laser isotope separation. Mailing Add: Energy Res & Develop Admin Washington DC 20545

MCNALLY, JAMES RAND, JR, b Boston, Mass, Nov 10, 17; m 42; c 7. PLASMA PHYSICS. Educ: Boston Col, BS, 39; Mass Inst Technol, SM, 41; PhD(physics), 43. Prof Exp: Asst spectros, Mass Inst Technol, 39-41, spectros & physics, 41-44, instr physics, 44-48; physicist, Stable Isotopes Div, 48-55, assoc dir div, 55-57, sr physicist, Physics Div, 57-60, SR PHYSICIST, THERMONUCLEAR EXP DIV, OAK RIDGE NAT LAB, 60- Mem: Fel AAAS; Am Phys Soc; fel Optical Soc Am; Soc Appl Spectros; Am Asn Physics Teachers. Res: Atomic physics; atomic spectroscopy; fusion physics; advanced fusion fuels; fusion chain reactions. Mailing Add: Thermonuclear Div Oak Ridge Nat Lab PO Box Y Oak Ridge TN 37830

MCNALLY, JOHN G, b Brooklyn, NY, Mar 5, 32; m 64. ANALYTICAL CHEMISTRY. Educ: Polytech Inst Brooklyn, BS, 55, MS, 63. Prof Exp: Technician, Chas Pfizer & Co, 49-55, chemist, 55-57; chemist, Escambia Chem Corp, 57-61, head anal dept, 61-68; proj leader, 68-70, HEAD ANAL GROUP, AM CYANAMID CO, STAMFORD, 70- Mem: Am Chem Soc. Res: Quantitative gas chromatography; mass spectrometry; electroanalytical chemistry; high performance liquid chromatography; gel permeation chromatography; polymer characterization. Mailing Add: Am Cyanamid Co 1937 W Main St Stamford CT 06904

MCNALLY, RICHARD THOMAS, analytical chemistry, see 12th edition

MCNAMARA, ALLEN GARNET, b Regina, Sask, Feb 28, 26; m 52; c 1. AERONOMY. Educ: Univ Sask, BE, 47, MSc, 49, PhD(physics), 54; Univ Mich, MA, 51. Prof Exp: From res officer to assoc res officer, 51-63, SR RES OFFICER, NAT RES COUN CAN, 63-, HEAD PLANETARY SCI SECT, HERZBERG INST ASTROPHYS, 75- Concurrent Pos: Mem, Can Nat Comt Radio Sci, Int Sci Radio Union. Mem: Can Asn Physicists; Inst Elec & Electronics Eng; Am Geophys Union. Res: Physics of upper atmosphere; scattering of radio waves by meteor ionization and aurora; rocket investigators of the aurora and upper atmosphere. Mailing Add: Nat Res Coun Ottawa ON Can

MCNAMARA, BERNARD PATRICK, b Baltimore, Md, Feb 17, 15; m 47; c 5. PHARMACOLOGY. Educ: Univ Md, BS, 36, MS, 39, PhD(pharmacol), 42. Prof Exp: Pharmacist, Univ Md Hosp, 36-37; asst pharmacol, Univ Md, 37-41; asst toxicologist, Med Res Div, Edgewood Arsenal, Md, 42-43, pharmacologist, 46-47, asst chief pharmacol br, Med Div, US Army Edgewood Arsenal, 47-54, chief aerosol br, 54-58, CHIEF TOXICOL DIV, BIOMED LAB, US ARMY EDGEWOOD ARSENAL, 58- Res: Effects of mustard on muscle tissue; stability of peroxides and solid oxidizing agents; chronic toxicity of dichlorodiphenyl-trichloro-ethane in dogs; pharmacology of chemical warfare agents, benzene hexachloride, dichloro-diphenyl-trichloro-ethane and antichlorlinesterase compounds. Mailing Add: Toxicol Div Biomed Lab US Army Edgewood Arsenal Aberdeen Proving Ground MD 21010

MCNAMARA, DAN GOODRICH, b Waco, Tex, Oct 19, 22; m 49; c 5. PEDIATRICS. Educ: Baylor Univ, BS, 43, MD, 46. Prof Exp: Resident, Hermann Hosp, Houston, Tex, 49-50; from asst prof to assoc prof, 53-69, PROF PEDIAT, BAYLOR COL MED, 69- Concurrent Pos: Fel pediat cardiol, Cardiac Clin, Harriet Lane Home, Johns Hopkins Hosp, 51-53; dir cardiac clin, Tex Children's Hosp, Houston, Tex, 53- Mem: Am Acad Pediat. Res: Pediatric cardiology, especially secondary pulmonary hypertension and malfunctions of heart in infancy. Mailing Add: 6621 Fannin St Houston TX 77025

MCNAMARA, DELBERT HAROLD, b Salt Lake City, Utah, June 28, 23; m 45; c 3. ASTROPHYSICS, ASTRONOMY. Educ: Univ Calif, BA, 47, PhD(astron), 50. Prof Exp: Asst astronomer, Univ Calif, Berkeley, 50-52, assoc res astronomer, 52-55; asst prof astron, 55-57, assoc prof, 57-62, PROF PHYSICS & ASTRON, BRIGHAM YOUNG UNIV, 62- Concurrent Pos: Prin scientist space sci lab, NAm Aviation, Inc. Mem: Am Astron Soc; Int Astron Union. Res: Stellar spectroscopy and photometry; variable stars; eclipsing binaries. Mailing Add: Dept of Physics & Astron Brigham Young Univ Provo UT 84602

MCNAMARA, FRANK L, optics, solid state electronics, see 12th edition

MCNAMARA, JAMES ALYN, JR, b San Francisco, Calif, June 11, 43; m 70. ORTHODONTICS, ANATOMY. Educ: Univ Calif, Berkeley, AB, 64; Univ Calif, San Francisco, BS, DDS & cert orthod spec, 68; Univ Mich, Ann Arbor, MS, 69, PhD(anat), 72. Prof Exp: Res assoc, Ctr Human Growth & Develop, 70-72, ASSOC RES SCIENTIST & PROG DIR EXP CRANIOFACIAL RES, CTR HUMAN GROWTH & DEVELOP & ASST PROF ANAT, UNIV MICH, ANN ARBOR, 72- Honors & Awards: Milo Hellman Res Award, Am Asn Orthodontists, 73. Mem: Int Asn Dent Res; Am Asn Orthod; Am Asn Anat; Am Asn Phys Anthrop; Am Dent Asn. Res: Experimental studies of musculoskeletal interaction, craniofacial growth in man and non-human primates; embryology and congenital anomalies. Mailing Add: Dept of Anat Univ of Mich Ann Arbor MI 48109

MCNAMARA, JAMES HENRY, b Pittsfield, Mass, Aug 21, 29; m 55; c 5. ORGANIC CHEMISTRY. Educ: St Michael's Col, BS, 50; Univ Detroit, MS, 53; Pa State Univ, PhD, 56. Prof Exp: Instr, Univ Detroit, 50-52 & Pa State Univ, 52-56; sect head org sect, Process Metall Div, Alcoa Res Labs, Aluminum Co Am, 56-61; asst dir res, 61-67, dir res, 67-68, VPRES RES & DEVELOP, ROME CABLE DIV, CYPRUS MINES, 68- Mem: Am Chem Soc; Soc Plastics Eng. Res: Adaption of organic chemistry to aluminum industry; carbon products; purification; coloring; lubricants; compounding and processing thermoplastic and thermosetting materials for electrical use as insulations; solid dielectrics, especially compatibility with metals such as aluminum and copper. Mailing Add: Res Lab Rome Cable Div Cyprus Wire & Cable Co PO Box 71 Rome NY 13440

MCNAMARA, JOHN EDWARD, b Galesburg, Ill, June 14, 26; m 47; c 2. PHYSICAL INORGANIC CHEMISTRY. Educ: Knox Col, Ill, AB, 47; Univ Ill, MA, 49, PhD(chem), 52. Prof Exp: Asst chem, Univ Ill, 47-52; res chemist, Calif Res Corp, Stand Oil Co Calif, 52-54, Bell Tel Labs, Inc, 54-55, Semiconductor Prod Div, Motorola, Inc, 55-67 & Integrated Circuits Eng Corp, 67-71; CONSULT, SOLAR ENERGY ASSOCS, 71- Mem: Am Chem Soc. Res: Semiconductor materials and processes. Mailing Add: 4508 E Earll Dr Phoenix AZ 85018

MCNAMARA, JOSEPH JUDSON, b Oakland, Calif, Sept 12, 36; m 58; c 4. SURGERY. Educ: Wash Univ, MD, 61; Am Bd Surg, dipl, 67, cert thoracic surg, 68. Prof Exp: Surgeon, Peter Bent Brigham & Mass Gen Hosps, 61-66; thoracic & cardiovasc surgeon, Baylor Univ Hosp, Dallas, 67-68; actg dir div surg, Walter Reed Army Inst Res, 69-70; PROF SURG, SCH MED, UNIV HAWAII, MANOA, 70- Concurrent Pos: Dir surg educ & cardiovasc res lab, Queen's Med Ctr, 70- Honors & Awards: Sheard-Sanford Award, Am Soc Clin Path, 61. Mem: Soc Univ Surg; Soc Thoracic Surg; Am Heart Asn. Res: Myocardial viability after coronary occulsion; defective function of blood elements with storage; platelet function.

MCNAMARA, MARY COLLEEN, b Albuquerque, NMex, Apr 5, 47. NEUROBIOLOGY. Educ: Univ NMex, BS, 71, MS, 72; Univ NC, Chapel Hill, PhD(neurobiol), 75. Prof Exp: Teaching asst, Univ NMex, 71-72; instr introd psychol, 74-76, FEL PHYSIOL, SCH MED, UNIV NC, 75- Mem: Sigma Xi; Soc Neurosci; Geront Soc. Res: Delineating age related changes in central neurotransmitters in response to stress. Mailing Add: Dept of Physiol Sch of Med Univ of NC Chapel Hill NC 27514

MCNAMARA, MICHAEL JOSEPH, b New York, NY, May 16, 29; m 57; c 3. COMMUNITY HEALTH, PREVENTIVE MEDICINE. Educ: Fordham Univ, AB, 51; NY Univ, MD, 55; Am Bd Prev Med, dipl. Prof Exp: Intern med, Bellevue Hosp, NY Univ, 55-56, asst resident internal med, 56-58; from asst prof to assoc prof community med, Col Med, Univ Ky, 61-70; PROF COMMUNITY MED & CHMN DEPT, MED COL OHIO, 70- Concurrent Pos: Nat Found fel virus res, Col Med, Cornell Univ, 58-59. Mem: Am Soc Microbiol; Soc epidemiol Res; Asn Teachers Prev Med; Am Pub Health Asn; fel Am Col Prev Med. Res: Epidemiology of viral agents in human diseases; epidemiology of hospital acquired infection; teaching of community medicine to medical students and residents in general preventive medicine. Mailing Add: Dept of Community Med Med Col of Ohio PO Box 6190 Toledo OH 43614

MCNAMARA, THOMAS FRANCIS, bacteriology, therapeutics, see 12th edition

MCNANEY, JOHN A, b Seneca Falls, NY, Feb 8, 23; m 43. INORGANIC CHEMISTRY. Educ: Mass State Col Fitchburg, BSE, 49, MEd, 53; Univ of the Pac,

PhD(chem), 68. Prof Exp: Chmn dept sci & math, High Sch, Mass, 49-53, dept sci, 53-58; instr chem & physics, 58-61, from asst prof to assoc prof, 61-69, chmn dept chem & physics, 65-68, PROF CHEM, FITCHBURG STATE COL, 69-, CHMN DEPT, 68- Concurrent Pos: NSF fel, Univ of the Pac, 62-63, fac fel, 66-67. Mem: Am Chem Soc; fel Am Inst Chem. Res: Physical chemistry; phase diagram studies on dioxane-water-potassium salts and tetrahydrofuran-water and potassium salts. Mailing Add: Dept of Chem Fitchburg State Col Fitchburg MA 01420

MCNARY, ROBERT REED, b Dayton, Ohio, Oct 9, 03; m 48. BIOCHEMISTRY. Educ: Univ Cincinnati, Chem E, 26, PhD(biochem), 36; Antioch Col, AM, 33. Prof Exp: Res chemist, Thomas & Hochwalt Labs, Ohio, 26-28, Frigidaire Corp, 28-32, Kettering Lab, Col Med, Univ Cincinnati, 36-43 & Citrus Exp Sta, Fla Citrus Comn, 46-59; CONSULT CHEMIST, 59- Res: By-product development of citrus fruits; methods of treating citrus cannery waste water; industrial hygiene chemistry; chlorophyll decompositions; freon refrigerants. Mailing Add: 31 Cunningham Dr New Smyrna Beach FL 32069

MCNARY, WILLIAM FRANCIS, JR, b Attleboro, Mass, Nov 17, 26; m 49; c 2. ANATOMY. Educ: Tufts Col, BS, 49; Boston Univ, PhD(anat), 54. Prof Exp: From instr to asst prof, 54-66, dir student labs, 67-72, dean student affairs, 70-75, ASSOC PROF ANAT, BOSTON UNIV, 66-, ASSOC DEAN STUDENT AFFAIRS, 75- Mem: AAAS; Am Asn Anat. Res: Histochemistry; histology; electron microscopy. Mailing Add: 80 E Concord Boston MA 02118

MCNATT, EUGENE MELTON, b Aurora, Mo, Aug 3, 13; m 41; c 2. BIOPHYSICS. Educ: Univ Colo, AB, 35; Wash Univ, MS, 37, PhD(physics), 39. Prof Exp: Asst physics, Wash Univ, 35-39; jr res geophysicist, Carter Oil Co, 39-46, head geophys res, 46-49, asst chief res, 49-53, asst mgr, Geophys Div, 53-56, GEOPHYS ADV, EXXON CORP, NJ, 56- Mem: Am Phys Soc; Soc Explor Geophys; Am Inst Mining, Metall & Petrol Eng. Res: Honls theory of the dispersion of x-rays in crystals; petroleum geophysics. Mailing Add: 23 Ridge Rd Bronxville NY 10708

MCNAUGHT, DONALD CURTIS, b Detroit, Mich, May 1, 34; m 62; c 4. LIMNOLOGY. Educ: Univ Mich, BS, 56, MS, 57; Univ Wis, PhD(limnol), 65. Prof Exp: Res asst zool, Univ Wis, 57-60, proj asst, 60-64, proj assoc, 65; asst prof, Mich State Univ, 65-68; ASSOC PROF ZOOL, STATE UNIV NY, ALBANY, 68- Concurrent Pos: Dir, Cranberry Lake Biol Sta, 68-; consult, Commonwealth Edison, 71- Mem: Am Soc Limnol & Oceanog; Ecol Soc Am; Am Soc Zool; Am Fisheries Soc; Int Asn Theoret & Appl Limnol. Res: Physiological ecology of the zooplankton; migratory behavior and photochemistry of the visual pigments of the Cladocera; thermal pollution. Mailing Add: Dept of Biol Sci State Univ NY Albany NY 12203

MCNAUGHTON, DUNCAN ANDERSON, b Cornwall, Ont, Dec 7, 10; m 46; c 3. GEOLOGY. Educ: Univ Southern Calif, BA, 33, PhD(geol), 51; Calif Inst Technol, MS, 34. Prof Exp: Jr geologist, Can Geol Surv, 35-39; geologist, Tex Petrol Co, Colombia, 39-41; lectr geol, Univ Southern Calif, 45-47, instr, 47-48, asst prof, 48-51; geologist, Gulf Oil Co, 51-54; consult, Meyer & Achtschin, 54-58; consult, Bednar & McNaughton, 58-60; CONSULT FOR EXPLOR, 60- Mem: AAAS; Am Asn Petrol Geol (vpres, 74-75); Geol Soc Am; Soc Econ Geol; Soc Explor Geophys. Res: Exploration and development of oil and gas reserves in central Australia salt tectonics; fracture patterns; relationship between porosity in reservoir rocks and time; static loading; tectonism. Mailing Add: 1325 Fidelity Union Life Bldg Dallas TX 75201

MCNAUGHTON, MICHAEL WALFORD, b Durban, SAfrica, Mar 2, 43; Eng citizen; m 69; c 1. EXPERIMENTAL NUCLEAR PHYSICS. Educ: Univ London, BSc, 62, PhD(physics), 72; Oxford Univ, MA, 66. Prof Exp: Physicist, Crocker Nuclear Lab, Univ Calif, Davis, 72-75; SR RES ASSOC NUCLEAR PHYSICS, CASE WESTERN RESERVE UNIV, 75- Res: Few nucleon nuclear physics, neutron physics, polarised proton targets. Mailing Add: Los Alamos Sci Lab Los Alamos NM 87545

MCNAUGHTON, ROBERT, b Brooklyn, NY, Mar 13, 24; div; c 2. MATHEMATICS, COMPUTER SCIENCES. Educ: Columbia Univ, BA, 48; Harvard Univ, PhD(philos), 51. Prof Exp: Asst prof philos, Stanford Univ, 54-57; from asst prof to assoc prof elec eng, Univ Pa, 57-64; vis assoc prof elec eng & mem staff, Proj MAC, Mass Inst Technol, 64-66; PROF MATH, RENSSELAER POLYTECH INST, 66- Honors & Awards: Levy Medal, 56. Mem: Asn Comput Mach; Asn Symbolic Logic. Res: Theory of automata and applications of symbolic logic. Mailing Add: Dept of Math Sci Rensselaer Polytech Inst Troy NY 12181

MCNAUGHTON, SAMUEL J, b Takoma Park, Md, Aug 10, 39; m 59; c 2. PLANT ECOLOGY. Educ: Northwest Mo State Col, BS, 61; Univ Tex, PhD(bot), 64. Prof Exp: Asst prof biol, Portland State Col, 64-65; USPHS trainee, Stanford Univ, 65-66; from asst prof to assoc prof bot, 66-73, PROF BOT, SYRACUSE UNIV, 73- Concurrent Pos: Res scientist, Serengeti Res Inst, Tanzania, 74-; adj prof, Univ Dar es Salaam, 74-75. Mem: Am Soc Plant Physiol; Bot Soc Am; Ecol Soc Am; Brit Ecol Soc. Res: Ecotype physiology, community organization, and grazing ecology. Mailing Add: Biol Res Labs Syracuse Univ Syracuse NY 13210

MCNEAL, BRIAN LESTER, b Cascade, Idaho, Jan 27, 38, m 58; c 4. SOIL CHEMISTRY. Educ: Ore State Univ, BS, 60, MS, 62; Univ Calif, Riverside, PhD(soil chem), 65. Prof Exp: Lab asst soil chem, Ore State Univ, 58-59; student trainee, Agr Res Serv, USDA, 59-60, soil scientist, 60-61, res soil scientist, US Salinity Lab, Calif, 61-70; ASSOC PROF SOILS, WASH STATE UNIV, 70- Mem: Soil Sci Soc Am; Am Soc Agron. Res: Pollution chemistry; chemistry of salt-affected soils; soil physical chemistry; modeling of soil chemical processes. Mailing Add: Dept of Agron & Soils Wash State Univ Pullman WA 99163

MCNEAL, DALE WILLIAM, JR, b Kansas City, Kans, Nov 23, 39; m 66. BOTANY. Educ: Colo Col, AB, 62; State Univ NY Col Forestry, Syracuse Univ, MS, 65; Wash State Univ, PhD(bot), 69. Prof Exp: Asst prof biol, 69-74, ASSOC PROF BIOL, UNIV OF THE PAC, 74- Mem: Am Bot Soc; Am Soc Plant Taxon; Int Soc Plant Taxon. Res: Biosystematics of Allium; Alpine floristics. Mailing Add: Dept of Biol Sci Univ of the Pac Stockton CA 95211

MCNEAL, FRANCIS H, b Bartlett, Ore, Dec 9, 20; m 47; c 2. PLANT BREEDING, PLANT GENETICS. Educ: Ore State Col, BS, 43, MS, 48; Univ Minn, PhD(genetics), 53. Prof Exp: RES AGRONOMIST, AGR RES SERV, USDA, 47- Concurrent Pos: Mem, Hard Red Spring Wheat Regional Comt, 49-; secy & tech adv, Western Wheat Improv Comt, 56-; mem, Nat Wheat Improv Comt, 59- Mem: Fel Am Soc Agron. Res: Development of improved wheat varieties for western states; genetic and related studies of the wheat plant. Mailing Add: Dept of Plant & Soil Sci Mont State Univ Bozeman MT 59715

MCNEAL, ROBERT JOSEPH, b Knoxville, Tenn, Dec 23, 37; m 62; c 2. CHEMICAL PHYSICS, SPACE PHYSICS. Educ: Univ Calif, Berkeley, BS, 59; Columbia Univ, MA, 61, PhD(chem), 62. Prof Exp: NSF fel, Harvard Univ, 63-64; head lab aeronomy dept, Space Physics Lab, 64-70, ASST DIR, CHEM & PHYSICS LAB, AEROSPACE CORP, LOS ANGELES, 70- Mem: Fel Am Phys Soc; Am Geophys Union. Res: Atomic and molecular physics; chemical kinetics; aeronomy. Mailing Add: 404 21st St Santa Monica CA 90402

MCNEAL, ROY WILSON, b Dallas Co, Mo, June 23, 91; m 13; c 1. GEOGRAPHY. Educ: Univ Ariz, BS, 17; Univ Calif, MS, 41; Nat Univ Mexico, DLett, 50. Prof Exp: Prof chem & geol, Albany Col, 17-23; dir athletics, Col Puget Sound, 23-25; prof, 27-74, EMER PROF GEOG, SOUTHERN ORE COL, 74- Mem: Nat Coun Geog Educ; Am Chem Soc. Res: Field studies in geographic analysis for planning and land utilization; methods and materials in teaching of geography. Mailing Add: Dept of Geog Southern Ore Col Ashland OR 97520

MCNEARY, SAMUEL STUART, b Philadelphia, Pa, Sept 16, 13; m 40; c 3. APPLIED MATHEMATICS. Educ: Haverford Col, BS, 36; Univ Pa, MA, 43. Prof Exp: Instr math, Drexel Inst Technol, 39-43; design engr, Westinghouse Elec Corp, 43-46; from instr to assoc prof, 46-57, PROF MATH, DREXEL UNIV, 58- Mem: Am Math Soc; Math Asn Am; Soc Indust & Appl Math. Res: Partial differential equations. Mailing Add: Dept of Math Drexel Univ Philadelphia PA 19104

MCNEE, ROBERT BRUCE, b Big Timber, Mont, Aug 20, 22; m 49, 54; c 5. ECONOMIC GEOGRAPHY, URBAN GEOGRAPHY. Educ: Wayne State Univ, BA, 49; Syracuse Univ, MA, 50, PhD(geog), 53. Prof Exp: Lectr geog, City Col New York, 51-52, from asst prof to assoc prof, 52-63; chmn dept, Univ Cincinnati, 63-69 & 73-74, PROF GEOG, UNIV CINCINNATI, 63- Concurrent Pos: Mem & chmn, Comn Int Fel, Italy & Greece, 61-63; Found Econ Educ fels, Socony-Mobil Oil Corp, 56 & Sun Oil Co, 61; consult, Ford Found, 62-63; City Col Fund fel, City Col New York, 62-63; Comn Col Geog fel, Univ Cincinnati, 65-68; mem steering comt rapid excavation, Nat Res Coun, 67-69; dir, Am Geog Soc, 73-76. Mem: Asn Am Geog; Am Geog Soc; Nat Coun Geog Educ. Res: Geography of formal organizations; history and philosophy of geography; geographic education; energy. Mailing Add: Dept of Geog Univ of Cincinnati Cincinnati OH 45221

MCNEEL, BURDETT HARRISON, b Toronto, Ont, May 10, 09; m 38; c 2. PSYCHIATRY. Educ: Univ Toronto, BA, 33, MD, 36, dipl psychiat, 40. Prof Exp: Med officer, Ont Dept Health, 37-41, area consult psychiat, 46-51; supt, Ont Hosp, St Thomas, 51-55; dir community ment health, Ont Dept Health, 55-57, chief ment health br, 57-66, dir prof serv br, Ment Health Div, 66-69; assoc prof psychiat, Univ Toronto, 60-68, asst prof, Sch Hyg, 69-72; SR CONSULT PSYCHIAT, ONT WORKMEN'S COMPENSATION BD, 69-; CHIEF OF PSYCHIAT, NORTH YORK GEN HOSP, 71- Concurrent Pos: Dir res schizophrenia-Scottish rite proj, Toronto Psychiat Hosp, 40-41; instr & sr assoc, Univ Western Ont, 46-55; mem bd dirs, Can Coun Hosp Accreditation, 57-64; mem adv comt ment health, Dept Nat Health & Welfare, Ont, 57-64; mem adv comt, Ont Ment Health Found, 64-70; mem bd trustees, Clarke Inst Psychiat, 64-71. Mem: Fel Am Psychiat Asn; Can Psychiat Asn; Can Ment Health Asn. Res: Development and operation of community mental health services. Mailing Add: North York Gen Hosp 4001 Leslie St Willowdale ON Can

MCNEELY, ROBERT LEWIS, b Morganton, NC, June 5, 38. ANALYTICAL CHEMISTRY. Educ: Duke Univ, BS, 60; Univ NC, Chapel Hill, PhD(chem), 70. Prof Exp: Teaching asst chem, Calif Inst Technol, 60-62; Peace Corps teacher chem, Govt Col Nigeria, 63-65; res asst, Univ NC, Chapel Hill, 65-69; asst prof, 69-73, ASSOC PROF CHEM, UNIV TENN, CHATTANOOGA, 73- Mem: AAAS; Am Chem Soc. Res: Instrumental analysis, particularly in elecro- chemistry and absorption spectroscopy; analysis of air and water pollutants. Mailing Add: Dept of Chem Univ of Tenn Chattanooga TN 37401

MCNEELY, WILLIAM HAROLD, b Hillsboro, Mo, Sept 15, 15; m 40; c 4. CHEMISTRY, MICROBIOLOGY. Educ: Univ Calif, BS, 40; Calif Inst Technol, PhD(chem), 43. Prof Exp: Hoffmann-La Roche fel, Ohio State Univ, 43; res chemist, Kelco Co, 44-52, sect head org chem, 52-60, sect head microbial develop, 60-67, mgr org & biochem develop, 67-70, tech dir, 70-74, dir tech opers, 74-75, VPRES RES, KELCO, DIV OF MERCK & CO, INC, 75- Honors & Awards: Indust Achievement Award, Inst Food Technologists, 74. Mem: AAAS; Am Chem Soc; Am Soc Microbiol. Res: Water soluble colloids; chromatography; organic derivatives of polysaccharides; animal feeding; analytical chemistry; microbial metabolism; microbial production of polysaccharides; biochemical engineering; industrial fermentations. Mailing Add: Kelco Div of Merck & Co Inc 8355 Aero Dr San Diego CA 92123

MCNEER, REMBERT DURBIN, JR, chemistry, see 12th edition

MCNEIGHT, SAMUEL ARNOLD, physical chemistry, see 12th edition

MCNEIL, ARTHUR LOUIS, b Waubaushene, Ont, Nov 6, 05. ANALYTICAL CHEMISTRY. Educ: Gonzaga Univ, AB, 31, AM, 32; Cath Univ, PhD, 36. Prof Exp: Prof chem & chmn dept, Gonzaga Univ, 40-70; PROF CHEM, SEATTLE UNIV, 70- Mem: Am Chem Soc. Res: Rapid growth of plants using high carbon dioxide concentrate and high intensity lights. Mailing Add: Dept of Chem Seattle Univ Seattle WA 98122

MCNEIL, CHARLES WINSLOW, b Tecumseh, Mich, July 19, 16; m 40; c 3. ZOOLOGY. Educ: Univ Mich, BS, 38, MS, 40, PhD(parasitol), 42. Prof Exp: Instr zool, Univ Mich, 46; from instr to assoc prof, 46-62, PROF ZOOL, WASH STATE UNIV, 62- Concurrent Pos: Actg chmn dept zool, 64-65 & 68-69, actg chmn biol prog, 66-67. Mem: Am Soc Parasitol; Am Micros Soc; Ecol Soc Am. Res: Parasitology; snail ecology. Mailing Add: Dept of Zool Wash State Univ Pullman WA 99163

MCNEIL, CRICHTON, b Evanston, Ill, Jan 4, 13; m 41; c 3. CLINICAL PATHOLOGY. Educ: Univ Buffalo, MD, 38. Prof Exp: Instr path, Univ Buffalo, 41; instr, Univ Utah, 46-47; pathologist & dir labs, 47-69, SR STAFF DIR LABS, HOLY CROSS HOSP & DIR CLIN PATH, ST MARK'S HOSP LAB, 69- Concurrent Pos: Assoc clin prof, Sch Med, Univ Utah, 47-71; pathologist, Shriners Hosp Crippled Children, 47-; consult, Am Red Cross, 54-; res malnutrition, Mt Silinda Hosp, Rhodesia, Africa, 73-74. Mem: Am Soc Clin Path; Am Asn Path & Bact; Am Asn Immunol; Int Acad Path; Int Soc Hemat. Res: Immunopathology, especially blood groups; cancer research; malnutrition. Mailing Add: St Mark's Hosp Lab 1200 E 39th S Salt Lake City UT 84117

MCNEIL, HARRY DANIEL, JR, b Bangor, Maine, Oct 2, 24; m 51; c 3. ORGANIC CHEMISTRY, POLYMER CHEMISTRY. Educ: Bowdoin Col, BS, 45; Univ Maine, MS, 47; Purdue Univ, PhD(chem), 51. Prof Exp: Assistantship, Univ Maine, 45-47; assistantship, Purdue Univ, 47-49; res chemist, Hercules Inc, 51-60; prin engr, Fibers Co, Celanese Corp Am, 60-62; instr chem, Charlotte Col, 63-65; lectr, Pfeiffer Col, 65-66; tech dir, Rutland Plastics, Inc, 66-75; RETIRED. Mem: Fel AAAS; Am Chem Soc; Am Inst Chem Eng; fel Am Inst Chem; Sigma Xi. Res: Organic synthesis and development; synthetic polymers and fibers. Mailing Add: 101 Edgewood Ave Longmeadow MA 01106

MCNEIL, PHILLIP EUGENE, b Cincinnati, Ohio, May 13, 41; m 66; c 3. ALGEBRA. Educ: Ohio Univ, BS, 63; Pa State Univ, MA, 65, PhD(math), 68. Prof Exp: Asst prof math, Xavier Univ, 68- 70 & Univ Cincinnati, 70-73; prog assoc math, Inst Serv Educ, 71-73; ASSOC PROF MATH, NORFOLK STATE COL, 73- Concurrent Pos: Consult, Educ Develop Prog, Univ Cincinnati, 70-71, Minorities Comt, Nat Res Coun, 72, Norfolk Pub Schs, Va, 73 & Inst Serv Educ, 73-; consult & lectr, SEEK Proj, Hunter Col, 72-73 & Racine Pub Schs, Wis, 73; dir, Minority Inst Sci Improv Prog, 74-; nat pres, Men's & Women's Develop Asn, 75. Honors & Awards: Inst Serv Educ Plaque, 73. Mem: Am Math Soc; Math Asn Am; Nat Asn Mathematicians; Asn Educ Data Systs. Res: Development of structure theorems in the area of algebraic semigroups; curriculum development in undergraduate mathematics. Mailing Add: Dept of Math Norfolk State Col 2401 Corprew Ave Norfolk VA 23504

MCNEIL, RAYMOND, b St Fabien de Panet, Que, Nov 30, 36; m 63; c 1. ORNITHOLOGY. Educ: Laval Univ, BA, 59; Univ Montreal, BSc, 62, MSc, 64, PhD(ornith), 68. Prof Exp: Teacher ecol & ornith, Oriente, Venezuela, 65-67; ASST PROF BIOL SCI, ECOL RES CTR MONTREAL, UNIV MONTREAL, 67- Mem: Am Ornith Union; Cooper Ornith Soc; Wilson Ornith Soc; Brit Ornith Union. Res: Population ecology of birds; natural history of birds; fat deposition in migratory birds and its relationships with flyways and the phenomenon of summering in southern latitudes. Mailing Add: Dept of Biol Sci Ecol Res Ctr Univ of Montreal Montreal PQ Can

MCNEIL, RICHARD JEROME, b Marquette, Mich, Dec 22, 32; m 60; c 3. ECOLOGY. Educ: Mich State Univ, BS, 54, MS, 57; Univ Mich, PhD(wildlife mgt), 63. Prof Exp: Biologist, Mich Dept Conserv, 57-59, res biologist deer, 60-64; asst prof, 64-69, ASSOC PROF CONSERV, CORNELL UNIV, 70- Concurrent Pos: Fulbright scholar, NZ, 62-63. Mem: Ecol Soc Am; Wildlife Soc. Res: International natural resource problems; ecology of ungulates; man and environment; conservation education; social surveys in environmental affairs. Mailing Add: Fernow Hall Col of Agr Cornell Univ Ithaca NY 14850

MCNEIL, WILLIAM J, b Portland, Ore, Sept 13, 30; m 60; c 1. FISHERIES. Educ: Ore State Univ, BS, 52, MS, 56; Univ Wash, PhD(fisheries), 62. Prof Exp: Fishery res biologist, Bur Com Fisheries, US Fish & Wildlife Serv, 62-66; assoc prof fisheries, Ore State Univ, 66-72; INVESTS CHIEF, NAT MARINE FISHERIES SERV, NAT OCEANIC & ATMOSPHERIC ADMIN, 72- Concurrent Pos: Assoc prof, Ore State Univ, 72- Mem: Fel Am Inst Fishery Res Biol; Am Fisheries Soc. Res: Population ecology, physiology and husbandry of salmanid fishes; natural populations and environments; aquaculture. Mailing Add: Auke Bay Fisheries Lab Nat Marine Fisheries Serv Box 155 Auke Bay AK 99821

MCNEILL, JOHN, b Edinburgh, Scotland, Sept 15, 33; m 61; c 2. BOTANY. Educ: Univ Edinburgh, BS, 55, PhD(plant taxon), 60. Prof Exp: From asst lectr to lectr agr bot, Univ Reading, 57-61; lectr bot, Univ Liverpool, 61-69; actg assoc prof pop & environ biol, Univ Calif, Irvine, 69; sect chief taxon & econ bot, Plant Res Inst, Can Dept Agr, 69-73, RES SCIENTIST, BIOSYST RES INST, AGR CAN, 73- Concurrent Pos: Vis assoc prof, Univ Wash, 69; adj prof, Carleton Univ, Ottawa, 74- Mem: Am Soc Plant Taxon; Bot Soc Brit Isles; Can Bot Asn; Int Asn Plant Taxon; Soc Syst Zool. Res: Taxonomy and biosystematics of vascular plants, especially weeds; applications of numerical taxonomy to plant classification. Mailing Add: Biosyst Res Inst Cent Exp Farm Agr Can Ottawa ON Can

MCNEILL, JOHN HUGH, b Chicago, Ill, Dec 5, 38; Can citizen; m 63; c 2. PHARMACOLOGY. Educ: Univ Alta, BSc, 60, MSc, 62; Univ Mich, PhD(pharmacol), 67. Prof Exp: Lab asst pharm, Univ Alta, 59-62, lectr, 63; lectr, Dalhousie Univ, 62-63; asst instr pharmacol, Mich State Univ, 66-65, asst prof, 67-71; assoc prof pharmacol, 71-75, PROF & CHMN, DIV PHARMACOL & TOXICOL, UNIV BC, 75- Concurrent Pos: Teaching fel pharmacol, Univ Mich, 63-66. Mem: AAAS; Am Fedn Clin Res; Pharmacol Soc Can; Am Soc Pharmacol & Exp Therapeut; NY Acad Sci. Res: Drug interactions with the adrenergic amines on cardiac cyclic AMP; role of cyclic AMP in the cardiac actions of drugs. Mailing Add: Fac of Pharmaceut Sci Univ of BC Vancouver BC Can

MCNEILL, JOHN J, b Washington, DC, Dec 4, 22; m 65; c 2. MICROBIOLOGY. Educ: Univ Md, BS, 51, MS, 53, PhD(bact), 57. Prof Exp: Asst bact, Univ Md, 51-55; asst prof animal sci & microbiol, 56-66, ASSOC PROF ANIMAL SCI & MICROBIOL, NC STATE UNIV, 66- Mem: AAAS; Am Soc Microbiol; fel Am Acad Microbiol; Brit Soc Gen Microbiol. Res: Bacterial lipid metabolism; rumen microbiology. Mailing Add: 310 Polk Hall NC State Univ Raleigh NC 27607

MCNEILL, KENNETH GORDON, b Appleton, Eng, Dec 21, 26; m 59; c 1. NUCLEAR PHYSICS, NUCLEAR MEDICINE. Educ: Oxford Univ, BA, 47, MA & DPhil(physics), 50. Prof Exp: Fel nuclear physics, Yale Univ, 50-51; fel, Glasgow Univ, 51-52, lectr physics, 52-57; assoc prof, 57-63, PROF PHYSICS, UNIV TORONTO, 63-, PROF MED, 69- Concurrent Pos: Spec staff mem, Toronto Gen Hosp, 74- Mem: Am Phys Soc; Can Asn Physicists; fel Inst Nuclear Eng; Can Soc Clin Invest; Soc Nuclear Med. Res: Low energy nuclear physics; photodisintegration; applications of nuclear physics to medicine. Mailing Add: Dept of Physics Univ of Toronto Toronto ON Can

MCNEILL, MICHAEL JOHN, b Algona, Iowa, Sept 12, 42; m 67; c 1. GENETICS, PLANT BREEDING. Educ: Iowa State Univ, BS, 64, MS, 67, PhD(plant breeding), 69. Prof Exp: Res plant pathologist plant path div, US Biol Res Lab, 69-71; RES GENETICIST, FUNK SEEDS INT, INC, 71- Mem: Am Phytopath Soc. Res: Plant breeding and pathology dealing mainly with cereal crops. Mailing Add: Funk Seeds Int Inc 1300 W Washington St Bloomington IL 61701

MCNEILL, ROBERT BRADLEY, b Martinsburg, WVa, June 20, 41; m; c 2. MATHEMATICS. Educ: Univ WVa, AB, 63; Pa State Univ, MA, 65, PhD(math), 68. Prof Exp: Asst prof math, 68-72, ASSOC PROF MATH, NORTHERN MICH UNIV, 72- Mem: AAAS; Am Math Soc; Math Asn Am. Res: Qualitative behavior of solutions of differential equations and differential systems. Mailing Add: Dept of Math Northern Mich Univ Marquette MI 49855

MCNELIS, EDWARD JOSEPH, b Philadelphia, Pa, Aug 17, 30; m 56; c 2. ORGANIC CHEMISTRY. Educ: Villanova Univ, BS, 53; Columbia Univ, PhD(chem), 60. Prof Exp: Res chemist, Sun Oil Co, Pa, 60-67; chmn dept chem, 70-73, ASSOC PROF CHEM, WASH SQ COL NY UNIV, 67- Concurrent Pos: Vis assoc prof, Haverford Col, 66-67. Mem: Am Chem Soc. Res: Phenolic oxidation; olefin metathesis. Mailing Add: Dept of Chem Wash Sq Col NY Univ New York NY 10003

MCNERNEY, JAMES MURTHA, b Pittsburgh, Pa, Apr 3, 27; m 56; c 3. TOXICOLOGY, ENVIRONMENTAL HEALTH. Educ: Univ Pittsburgh, BS, 51, ML, 56, MPH, 57. Prof Exp: Res asst, Indust Hyg Found, Mellon Inst, 51-53, res assoc, 53-55, res toxicologist, 55-57, chief toxicologist, 57-64; assoc dir toxic hazard res unit, Aerojet-Gen Corp, 64-66; chief animal toxicol, Environ Health Lab, Am Cyanamid Co, 66-68, toxicol group leader, 68-69; dir inhalation toxicol dpet, TRW Hazleton Labs, 69-70; STAFF TOXICOLOGIST, AM PETROL INST, 70- Mem: Air Pollution Control Asn; Am Indust Hyg Asn; Soc Toxicol. Res: Occupational and environmental toxicology; industrial hygiene. Mailing Add: Am Petrol Inst 1801 K St NW Washington DC 20006

MCNERNEY, WILLIAM NORTON, analytical chemistry, physical chemistry, see 12th edition

MCNESBY, JAMES ROBERT, b Bayonne, NJ, Apr 16, 22; m 49; c 3. PHYSICAL CHEMISTRY. Educ: Univ Ohio, Athens, BS, 43; NY Univ, PhD(chem), 52. Prof Exp: Res chemist, Interchem Corp, 45-49; phys chemist, US Naval Ord Test Sta, Calif, 51-56; phys chemist, 57-62, chief photochem sect, 62-67, CHIEF PHYS CHEM DIV, NAT BUR STANDARDS, 67-, MGR MEASURES FOR AIR QUAL, 70- Concurrent Pos: Rockefeller pub serv fel, Univ Leeds, 58-59. Mem: Am Chem Soc; The Chem Soc. Res: Kinetics of free radical reactions; photochemistry. Mailing Add: Nat Bur of Standards Washington DC 20234

MCNETT, CHARLES WILLIAM, JR, b Alexandria, Va, Nov 6, 36; m 64; c 2. ANTHROPOLOGY, ARCHAEOLOGY. Educ: Tulane Univ, BA, 59, PhD(anthrop), 67. Prof Exp: Instr anthrop, Bethany Col, 62-63; asst prof, East Stroudsburg State Col, 63-64; asst prof, Baylor Univ, 64-67; from asst prof to assoc prof, 67-73, PROF ANTHROP, AM UNIV, 73- Concurrent Pos: NSF fel, Archeol Surv of Potomac River, Am Univ, 69-71; adv ed, Behav Sci Res, Yale Univ, 75-; Early Man Proj, Nat Geog Soc, 74- & NSF, 75- Mem: Fel Am Anthrop Asn; Soc Am Archeol; Soc Cross-Cult Res. Res: Multivariate analysis of anthropological data; data storage and retrieval; Paleo-Indian archeology. Mailing Add: Dept of Anthrop Am Univ Washington DC 20016

MCNEW, GEORGE LEE, b Alamogordo, NMex, Aug 22, 08; m 32; c 1. PLANT PATHOLOGY, MICROBIOLOGY. Educ: Univ NMex, BS, 30; Iowa State Univ, MS, 31, PhD(plant path), 35. Hon Degrees: DSc, Univ NMex, 54. Prof Exp: Asst, Iowa Agr Exp Sta, 31-35; fel, Rockefeller Inst, 35-39; assoc res, NY Exp Sta, Geneva, 39-43; mgr res & develop agr chem, US Rubber Co, Conn, 43-47; prof bot & head dept, Iowa State Univ, 47-49; managing dir, 49-74, DISTINGUISHED SCIENTIST, BOYCE THOMPSON INST PLANT RES, 74- Concurrent Pos: Mem res adv comt crops res & entom, USDA, 63-70; mem adv coun, State Univ NY Col Agr, Cornell Univ, 64-74; mem adv comt air qual criteria, US Environ Protection Admin, 75- Mem: AAAS; Am Chem Soc; Am Phytopath Soc (pres, 51-52); Torrey Bot Club (pres, 59); Am Inst Biol Sci (pres, 70). Res: Diseases of nursery and vegetable crops and their control; fungicides; herbicides; microbial antagonists; soil conditioning. Mailing Add: Boyce Thompson Inst for Plant Res 1086 N Broadway Yonkers NY 10701

MCNICHOLS, JOSEPH L, JR, physics, see 12th edition

MCNICKLE, D'ARCY, cultural anthropology, applied anthropology, see 12th edition

MCNIEL, NORBERT ARTHUR, b Moody, Tex, Dec 22, 14; m 39; c 4. GENETICS. Educ: Agr & Mech Col, Tex, BS, 35, MEd, 52, PhD(genetics), 55. Prof Exp: Teacher high sch, Tex, 35-41; teacher & supvr, McLeunan County Voc Sch, 46-49; consult for prog, Pakistan, Agr & Mech Col Syst, Tex, 55-56, asst prof genetics, 57-64; assoc prof, 64-71, PROF GENETICS, TEX A&M UNIV, 72- Concurrent Pos: Partic genetics conf, NSF, 60. Mem: Am Genetic Asn. Res: Human heredity. Mailing Add: Genetics Sect Dept of Plant Sci Tex A&M Univ College Station TX 77843

MCNIFF, EDWARD J, JR, b Danvers, Mass, Sept 26, 35; m 58; c 4. EXPERIMENTAL SOLID STATE PHYSICS. Educ: Boston Col, BS, 57; Northeastern Univ, MS, 61. Prof Exp: Electronics engr missile systs div, Sylvania Elec Prods, Inc, 57-61; staff physicist, Arthur D Little Co, 61-64; STAFF PHYSICIST, NAT MAGNET LAB, MASS INST TECHNOL, 64- Mem: Am Phys Soc. Res: Nuclear magnetic resonance; electron paramagnetic resonance; antiferromagnetic resonance; tunneling effects in thin films; magnetic effects in metals and dilute alloys; high temperature superconductors. Mailing Add: Nat Magnet Lab Bldg NW 14 Mass Inst of Technol Cambridge MA 02139

MCNINCH, JOSEPH HAMILTON, b Indianapolis, Ind, Oct 5, 04; m; c 1. PREVENTIVE MEDICINE. Educ: Ohio State Univ, AB, 27, MD, 30; Johns Hopkins Univ, MPH, 51. Prof Exp: Med Corps, US Army, 30-62, intern, Ft Sam Houston Hosp, 30-31, dir biol prods lab & instr biomet, Army Med Sch, 37-39, staff med officer, 40-45, dir army med libr & ed, Hist Med Dept, US Army, World War II, 46-50, prev med officer, US Forces, Far East, 51-54, cmndg officer, Army Environ Health Lab, 54-55, chief personnel div, Off Surgeon Gen, 55-57, surgeon, Army Forces Hq, Far East, 57-58, cmndg officer, Army Med Res & Develop Command, Off Surgeon Gen, 58-60, chief surgeon, Europe Hq, 60-62; dir res, Am Hosp Asn, 62-63; chief med dir, US Vet Admin, Washington, DC, 63-66; dir Atlanta off, Am Hosp Asn, 66-71; STAFF CONSULT, HEALTH SYSTS RES CTR, GA INST TECHNOL, 71- Concurrent Pos: Trustee, Comn Prof Hosp Activities, 63-74 & 68-71; comnr, Nat Comn Community Health Serv, 64- Mem: Asn Mil Surg US (pres, 64-65); Am Col Physicians. Mailing Add: 4565 Angelo Dr NE Atlanta GA 30319

MCNITT, JAMES R, b Chicago, Ill, May 10, 32; m 63; c 4. GEOLOGY. Educ: Univ Notre Dame, BS, 53; Univ Ill, MS, 54; Univ Calif, PhD(geol), 61. Prof Exp: Geologist, Calif Div Mines & Geol, 58-65; inter-regional adv, 65-70, proj mgr, Geothermal Explor Proj, Kenya, 70-74, SR TECH ADV GEOTHERMAL ENERGY, UN, 74- Mem: Geol Soc Am; Geophys Union. Res: Exploration and development of geothermal energy. Mailing Add: UN Room TH1383 PO Box 20 Grand Cent Post Off New York NY 10017

MCNITT, RAND EDWIN, b Coldwater, Mich, Sept 16, 45; m 67. MYCOLOGY. Educ: Albion Col, BA, 67; Univ NC, Chapel Hill, PhD(bot), 73. Prof Exp: Appointee, Div Biol & Med Res, Argonne Nat Lab, 73-75; vis lectr bot, 75-76, RES ASSOC BIOL CONTROL MOSQUITOS, UNIV NC, CHAPEL HILL, 75- Mem: Mycol Soc Am. Res: Biology of host/parasite interactions concerning mosquitos and copepods parasitized by fungi, especially by species of Coelomomyces of the class Phycomycetes; taxonomy, host range and world distribution of species Coelomomyces. Mailing Add: Dept of Bot Univ of NC Chapel Hill NC 27514

MCNIVEN, NEAL LINDSAY, b Kingston, Ont, Sept 15, 14; m 45; c 2. ORGANIC CHEMISTRY. Educ: McGill Univ, BSc, 36, MSc, 39; St Andrews Univ, PhD(chem), 50. Prof Exp: Chemist, Shawinigan Works, Can Industs, Ltd, 37-39, res chemist, Cent Res Lab, 39-45; res chemist, Ont Res Found, 45-47; res chemist, Worcester Found Exp Biol, 51-67, sr scientist, 67-69; CONSULT, ORGANOMATION ASSOCS, WORCESTER, 69- Mem: AAAS; Am Soc Test & Mat; Am Chem Soc; NY Acad Sci; The Chem Soc. Res: Terpenes; steroids; manufactured gas; infrared, ultraviolet and nuclear magnetic resonance spectroscopy; gas chromatography; radioisotopes; health physics; alkaloids. Mailing Add: 17 Arcturus Dr Shrewsbury MA 01545

MCNULTY, CHARLES LEE, JR, b Dallas, Tex, Feb 4, 18; m 42; c 3. PALEONTOLOGY. Educ: Southern Methodist Univ, BS, 40; Syracuse Univ, MS, 48; Univ Okla, PhD, 55. Prof Exp: Asst geol, Syracuse Univ, 40-42; asst prof Arlington State Col, 46-48; instr, Univ Okla, 48-49; assoc prof, Arlington State Col, 50-51; geologist, Concho Petrol Co, 51-53 & Continental Oil Co, 53-57; PROF GEOL, UNIV TEX, ARLINGTON, 57- Mem: Geol Soc Am; Soc Vert Paleont; Soc Econ Paleont & Mineral; Am Asn Petrol Geol; Swiss Geol Soc. Res: Micropaleontology, mainly small foraminifera; stratigraphy of Texas. Mailing Add: Dept of Geol Univ of Tex Arlington TX 76010

MCNULTY, GEORGE FRANK, b Palo Alto, Calif, June 18, 45. MATHEMATICAL LOGIC. Educ: Harvey Mudd Col, BS, 67; Univ Calif, Berkeley, Cand Phil, 69, PhD(math), 72. Prof Exp: Nat Res Coun fel, Univ Man, 72-73; res instr, Dartmouth Col, 73-75; ASST PROF MATH, UNIV SC, 75- Mem: Am Math Soc; Asn Symbolic Logic. Res: Foundations of mathematics and general theory of algebraic structures, especially on the connections between model theory, set theory and the theory of equational classes. Mailing Add: Dept of Math & Comput Sci Univ of SC Columbia SC 29208

MCNULTY, IRVING BAZIL, b Salt Lake City, Utah, Jan 6, 18; m 43; c 3. PLANT PHYSIOLOGY. Educ: Univ Utah, BS, 42, MS, 47; Ohio State Univ, PhD(plant physiol), 52. Prof Exp: Instr biol, bot & plant physiol, 47-53, from asst prof to assoc prof, 53-65, PROF BIOL, UNIV UTAH, 65- Concurrent Pos: Head dept biol, Univ Utah, 60-69. Mem: AAAS; Bot Soc Am; Am Soc Plant Physiol. Res: Physiology of sodium-potassium nutrition of plants; fluoride effects on plant metabolism. Mailing Add: Dept of Biol Univ of Utah Salt Lake City UT 84112

MCNULTY, JOHN KNEELAND, marine biology, see 12th edition

MCNULTY, MICHAEL LEIGH, b Charleroi, Pa, Apr 7, 41; m 62; c 2. GEOGRAPHY. Educ: California State Col, Pa, BS, 62; Northwestern Univ, MA, 64, PhD(geog), 66. Prof Exp: Asst prof geog, Ind Univ, Bloomington, 66-69; ASSOC PROF GEOG, UNIV IOWA, 69- Mem: AAAS; Asn Am Geog. Res: Urban and economic geography, especially the geographical aspects of the process of economic development. Mailing Add: Dept of Geog Univ of Iowa Iowa City IA 52240

MCNULTY, PETER J, b New York, NY, Aug 2, 41; m 66. BIOPHYSICS, HIGH ENERGY PHYSICS. Educ: Fordham Univ, BS, 62; State Univ NY Buffalo, PhD(physics), 65. Prof Exp: Asst physics, State Univ NY Buffalo, 62-65, fel, 65-66; from asst prof to assoc prof, 66-75, PROF PHYSICS, CLARKSON COL TECHNOL, 75- Concurrent Pos: Nat Acad Sci-Nat Res Coun sr resident res assoc, 70-71; vis assoc scientist, Brookhaven Nat Lab, 72-73; res collabr, Med Dept, 71- Mem: AAAS; Am Phys Soc; NY Acad Sci; Radiation Res Soc. Res: Biological effects of radiation; theory of vision; particle physics. Mailing Add: Dept of Physics Clarkson Col of Technol Potsdam NY 13676

MCNUTT, CLARENCE WALLACE, b Ozan, Ark, Aug 5, 13; m 39; c 4. GENETICS. Educ: Henderson State Col, AB, 35; La State Univ, MS, 38; Brown Univ, PhD(biol, genetics), 41. Prof Exp: Instr gross anat, Univ Wis, 46-50; assoc prof anat, Univ Tex Med Br Galveston, 50-67; PROF ANAT, UNIV TEX HEALTH SCI CTR, SAN ANTONIO, 67- Concurrent Pos: Muellhaupt fel, Ohio State Univ, 41-42; vis staff dept surg, Brooke Gen Hosp, Ft Sam Houston, Tex. Mem: Fel AAAS; Soc Exp Biol & Med; Genetics Soc Am; Am Soc Human Genetics; Am Asn Anat. Res: Mammalian developmental genetics; human genetics; neurological conditions in mice and man. Mailing Add: Dept of Anat Univ of Tex Health Sci Ctr San Antonio TX 78284

MCNUTT, DOUGLAS P, b Rome, Ga, Apr 24, 35; m 59; c 4. PHYSICS. Educ: Wesleyan Univ, BA, 56; Univ Wis, MS, 57, PhD(physics), 62. Prof Exp: Proj assoc interference spectros, Univ Wis, 62-63; RES PHYSICIST, US NAVAL RES LAB, 63- Mem: AAAS; Optical Soc Am; Am Geophys Union; Am Astron Soc. Res: Spectroscopic determination of atmospheric sodium; infrared and microwave rocket astronomy and aeronomy. Mailing Add: Code 7122 US Naval Res Lab Washington DC 20375

MCNUTT, JOHN DEWIGHT, b Detroit, Mich, Apr 29, 38; m 67; c 1. ATOMIC PHYSICS, ELECTRON PHYSICS. Educ: Univ Mich, BS, 60; Wayne State Univ, MS, 62, PhD(physics), 66. Prof Exp: Asst physics, Wayne State Univ, 60-66, res assoc, 66-67; from asst prof to assoc prof, 67-75, PROF PHYSICS, UNIV TEX, ARLINGTON, 75- Concurrent Pos: Adj asst prof, Univ Tex Health Sci Ctr, Dallas, 69-75, adj prof, 75- Mem: Am Phys Soc; Am Asn Physics Teachers. Res: Annihilation mechanisms of positrons and positronium atoms in materials; atomic and molecular structure of materials through positron annihilations; positron and positronium chemistry; liquid crystals. Mailing Add: Dept of Physics Univ of Tex Arlington TX 76019

MCNUTT, ROBERT HAROLD, b Moncton, NB, July 4, 37; m 64; c 3. PETROLOGY, GEOCHEMISTRY. Educ: Univ NB, BSc, 59; Mass Inst Technol, PhD(geol), 65. Prof Exp: Asst prof geol, 65-70, ASSOC PROF GEOL, McMASTER UNIV, 70 Mem: Am Geophys Union; Geol Asn Can. Res: Genesis of anorthosites-field and rubidium-strontium isotopic studies; petrology and geochemistry of Archean greenstone belts; rubidium-strontium isotopic studies in the Sudbury area. Mailing Add: Dept of Geol McMaster Univ Hamilton ON Can

MCNUTT, RONALD CLAY, b Birmingham, Ala, Oct 29, 29; m 54; c 1. ANALYTICAL CHEMISTRY, INORGANIC CHEMISTRY. Educ: Athens Col, BS, 59; Vanderbilt Univ, MS, 61, PhD(chem), 66. Prof Exp: Res technician chem, Chemstrand Corp, Ala, 53-59, res chemist, Chemstrand Res Ctr, NC, 61-62; assoc prof, 66-68, PROF CHEM & CHMN DEPT, ATHENS COL, 68- Mem: Am Chem Soc. Res: Analysis of polymers related to textile and tire industry; reactivity of coordinated ligands; analysis of liquid rocket fuels; analysis of contaminants on spacecraft surfaces. Mailing Add: Dept of Chem Athens Col Athens AL 35611

MCNUTT, WALTER SCOTT, b Ozan, Ark, July 21, 18; m 54; c 2. BIOCHEMISTRY, PHARMACOLOGY. Educ: Henderson State Teachers Col, AB, 40; Brown Univ, MSc, 43; Univ Wis, PhD(biochem), 49. Prof Exp: Asst prof biochem, Vanderbilt Univ, 51-53; res fel & sr res fel plant biochem, Calif Inst Technol, 53-58; assoc scientist, Conn Exp Sta, 58-59; assoc prof, 59-65, PROF PHARMACOL, SCH MED, TUFTS UNIV, 65-, PROF BIOCHEM, 68- Concurrent Pos: Jane Coffin Childs res fel biochem, Inst Cytophysiol, Copenhagen Univ, 49-50; Jane Coffin Childs res fel chem, Cambridge Univ, 50-51. Mem: Am Soc Biol Chem. Res: Metabolism of nucleosides; biogenesis of riboflavin and other pteridines. Mailing Add: Dept of Pharmacol Tufts Univ Sch of Med Boston MA 02111

MCPARTLAND, RICHARD PAUL, b Jamaica, NY, Jan 3, 48; m 69; c 1. BIOCHEMISTRY. Educ: St John's Univ, NY, BA, 69; State Univ NY Buffalo, PhD(biochem), 75. Prof Exp: Asst enzymol, 70-74, NIH FEL ENDOCRINOL, ROSWELL PARK MEM INST, 74- Mem: AAAS. Res: Mechanism of glucocorticoid resistance is under investigation in hepatomas and lymphomas to determine if alterations in specific receptors or the absence of receptors is a cause for resistance. Mailing Add: Roswell Park Mem Inst 666 Elm St Buffalo NY 14263

MCPETERS, ARNOLD LAWRENCE, b Sept 13, 25; m 51; c 4. POLYMER SCIENCE. Educ: Univ NC, BS, 50, PhD(chem), 54. Prof Exp: Res chemist, Am Enka Corp, 53-58, head develop sect, Rayon Res Dept, 58-60; res chemist, Chemstrand Corp, 60-61, group leader, 62-67, SR GROUP LEADER, MONSANTO TEXTILES CO, MONSANTO CO, 68- Mem: Am Chem Soc; Fiber Soc. Res: Improved acrylic and modacrylic fibers polymer structure; fiber morphology; fiber production processes; textile performance. Mailing Add: 412 Dartmouth Rd Raleigh NC 27609

MCPETERS, RICHARD DOUGLAS, b Florence, Ala, July 3, 47. ATMOSPHERIC PHYSICS. Educ: Mass Inst Technol, BS, 69; Univ Fla, PhD(physics), 75. Prof Exp: ASSOC PHYSICS, UNIV FLA, 75- Mem: Sigma Xi. Res: Atmospheric optics involving the transmission of sunlight through the atmosphere as affected by aerosols, ozone, and air pollutants. Mailing Add: Dept of Physics Univ of Fla 221 SSRB Gainesville FL 32611

MCPHAIL, ANDREW TENNENT, b Glasgow, Scotland, Sept 23, 37; m 61; c 2. CHEMISTRY. Educ: Glasgow Univ, BSc, 59, PhD(chem), 63. Prof Exp: Asst lectr chem, Glasgow Univ, 61-64; res assoc, Univ Ill, Urbana, 64-66; lectr, Univ Sussex, 66-68; assoc prof, 68-73, PROF CHEM, DUKE UNIV, 73- Mem: The Chem Soc; Am Crystallog Asn. Res: X-ray crystal structure analysis of organic molecules, particularly biologically active compounds; molecular conformations; studies of structure and bonding in transition metal complexes and in organometallic compounds. Mailing Add: Paul M Gross Chem Lab Duke Univ Durham NC 27706

MCPHAIL, CLARENCE WILMER BERNARD, b Rockwood, Ont, May 31, 16; m 46; c 4. PUBLIC HEALTH, DENTISTRY. Educ: Univ Alta, BSc & DDS, 43; Northwestern Univ, MSD, 54; Univ Toronto, MSD, 67; FRCD(C). Prof Exp: Regional dent consult, Dept Pub Health, Govt BC, Can, 52-59; assoc prof prev & commun dent, Fac Dent, Univ Alta, 59-63, assoc prof, head dept & asst dean, 63-66; PROF SOCIAL & PREV DENT & HEAD DEPT, COL DENT & LECTR SOCIAL & PREV MED, UNIV SASK, 67-, DEAN, 74- Concurrent Pos: Actg dir dent health, Dept Pub Health, Govt BC, Can, 56-57; prov dir prev & commun dent, Dept Pub Health, Govt Alta, Can, 59-63; actg dean col dent, Univ Sask, 73-74. Mem: Can Dent Asn; Can Pub Health Asn. Res: Epidemiology; preventive dentistry; use of dental auxiliary services. Mailing Add: Col of Dent Univ of Sask Saskatoon SK Can

MCPHAIL, MURCHIE KILBURN, b Kilburn, NB, Jan 26, 07; m 37; c 3. PHYSIOLOGY, BIOCHEMISTRY. Educ: Univ BC, BA, 29; McGill Univ, PhD(biochem physiol), 32; FRS(C), 68. Prof Exp: Asst prof physiol & pharmacol, Univ Alta, 34-38; assoc prof pharmacol, Dalhousie Univ, 38-40, prof & head dept, 40-45; chief pharmacologist, Vick Chem Soc, NY, 45-48; head physiol sect, Suffield Exp Sta, Defence Res Bd, Alta, 48-63, dir biosci res, Nat Defence Hq, 63-67, dir chem & microbiol res, Defense Res Estab, Shirley Bay, Ottawa, 67-69, with Nat Defence Hq, 69-72; RETIRED. Concurrent Pos: Mem, Nat Defence Col, 59-60. Mem: Royal Soc Can; Am Soc Pharmacol; Can Physiol Soc (vpres, 58-59, pres, 59-60); Pharmacol Soc Can; Brit Biochem Soc. Res: Endocrinology, especially progestin and the effects of hypophysectomy; toxicity and mode of action of anti-cholinesterase agents. Mailing Add: 2 Galt St Ottawa ON Can

MCPHAUL, JOHN J, JR, b Washington, DC, May 11, 30; m 54; c 2. INTERNAL MEDICINE, IMMUNOLOGY. Educ: Georgetown Univ, BS, 52, MD, 56. Prof Exp: US Air Force, 56-, from intern to asst resident internal med, DC Gen Hosp, 56-58, from jr to sr asst resident, Med Ctr, Duke Univ, 58-60, chief renal serv, Wilford Hall Air Force Med Ctr, 61-67, investr, Clin Res Lab, 69-71, DIR RES, WILFORD HALL AIR FORCE MED CTR, 71- Concurrent Pos: Fel nephrol, Med Ctr, Georgetown Univ, 61; fel exp path, Scripps Clin & Res Found, 67-69. Mem: Fel Am Col Physicians; AMA; Am Soc Nephrol; Int Soc Nephrol; Transplantation Soc. Res: Clinical nephrology, immunopathogenesis of human glomerulonephritis, transplantation biology. Mailing Add: Aerospace Med Lab Wilford Hall Air Force Med Ctr Lackland AFB TX 78236

MCPHERRON, ALAN, b Chicago, Ill, Apr 3, 29. ANTHROPOLOGY, ARCHAEOLOGY. Educ: Univ Chicago, BA, 57; Univ Mich, MA, 62, PhD(anthrop), 65. Prof Exp: Instr anthrop, 63-65, asst prof, 66-70, ASSOC PROF ANTHROP, UNIV PITTSBURGH, 71- Mem: AAAS; Soc Am Archaeol; Archaeol Inst Am. Res: Neolithic of Eastern Europe; agricultural origins and spread; theory of culture change; archaeological theory and methods. Mailing Add: Dept of Anthrop Univ of Pittsburgh Pittsburgh PA 15260

MCPHERRON, ROBERT LLOYD, b Chelan, Wash, Jan 14, 37; m 58; c 2. GEOPHYSICS, SPACE PHYSICS. Educ: Univ Wash, BS, 59; Univ Southern Calif, MS, 61; Univ Calif, Berkeley, PhD(physics), 68. Prof Exp: Res physicist space sci lab, Univ Calif, Berkeley, 66-68; res geophysicist, Inst Geophys & Planetary Physics, 68-69, asst prof space physics, 69-73, ASSOC PROF PLANETARY PHYSICS & GEOPHYS, UNIV CALIF, LOS ANGELES, 73- Mem: AAAS; Am Geophys Union. Res: Magnetic field variations within the magnetosphere, including both macroscopic currents and wave phenomena and the part they play in magnetic storms and substorms; particles and fields; auroral phenomena. Mailing Add: Dept of Geophys & Space Sci Univ of Calif Los Angeles CA 90024

MCPHERSON, ALEXANDER, b Columbus, Ohio, Feb 28, 44. BIOLOGICAL STRUCTURE. Educ: Duke Univ, BS, 66; Purdue Univ, West Lafayette, PhD(biol), 70. Prof Exp: Damon Runyon res fel biol, Mass Inst Technol, 70-71, Am Cancer Soc res fel, 71-73, res assoc, 74-75; ASST PROF BIOL CHEM, HERSHEY MED CTR, PA STATE UNIV, 75- Mem: Am Crystallog Asn; Am Soc Biol Chemists. Res: Analysis and determination of the atomic structures of biological macromolecules by x-ray diffraction techniques and their correlation with mechanistic properties. Mailing Add: Dept of Biol Chem Milton S Hershey Med Ctr Hershey PA 17033

MCPHERSON, CHARLES ALLEN, b Chattanooga, Tenn, Jan 7, 44; m 66; c 1. ORGANIC CHEMISTRY. Educ: Univ Kans, BS, 66; Univ Calif, San Diego, PhD(phys org chem), 69. Prof Exp: Res assoc org chem, Ind Univ, Bloomington, 69-71; lectr, 71-72; MEM RES STAFF ORG CHEM, WESTERN ELEC RES CTR, 72- Concurrent Pos: NIH fel, Ind Univ, Bloomington, 69-71. Mem: Am Chem Soc. Res: Organic synthesis; mechanisms of organic and organometallic reactions; polymer chemistry; surface chemistry. Mailing Add: Western Elec Res Ctr Box 900 Princeton NJ 08540

MCPHERSON, CHARLES WILLIAM, b Rugby, NDak, Feb 24, 32; m 56; c 2. LABORATORY ANIMAL MEDICINE. Educ: Univ Minn, BS, 54, DVM, 56; Univ Calif, Berkeley, MPH, 64; Am Col Lab Animal Med, dipl. Prof Exp: Vet animal hosp sect, 56-57, head primate unit, 57-58, vet microbiologist comp path sect, 58-60, chief animal prod sect, Lab Aids Br, 60-64, head pathogen free unit & asst to chief, 64-66, chief lab animal med & vivarium sci sect, Animal Resources Br, Div Res Resources,

66-70, CHIEF BR, ANIMAL RESOURCES BR, DIV RES RESOURCES, NIH, 71- Honors & Awards: Co-recipient Res Award, Am Asn Lab Animal Sci, 63. Mem: Am Vet Med Asn; Am Asn Lab Animal Sci. Res: Diseases of laboratory animals; production of microbiologically defined laboratory animals. Mailing Add: Animal Resources Br Room 5B31 Dir of Res Resources Bldg 31 NIH Bethesda MD 20014

MCPHERSON, CLARA, b Roscoe, Tex, Mar 10, 22; m 43; c 3. NUTRITION, FOODS. Educ: Tex Tech Col, BS, 43, MS, 47. Prof Exp: Instr food & nutrit, 47-48 & 55-60, asst prof, 61-68, ASSOC PROF FOOD & NUTRIT, TEX TECH UNIV, 68- Mem: Am Dietetic Asn; Am Home Econ Asn; Inst Food Technol; Soc Nutrit Educ. Res: Dietary studies of college students; frozen foods; development of high protein foods using cottonseed and soy protein; determination of quality of pork fed various rations. Mailing Add: Dept of Food & Nutrit Tex Tech Univ Lubbock TX 79409

MCPHERSON, CLINTON MARSUD, b Gainesville, Tex, Oct 6, 18; m 43; c 3. CHEMISTRY. Educ: Tex Tech Col, BS, 47, MEd, 52, DEd(psychol), 59. Prof Exp: Teacher pub schs, Tex, 50-56; instr chem, 56-59, asst prof, 60-74, ASSOC PROF FOOD & NUTRIT, TEX TECH UNIV, 74- Res: Inorganic chemistry; use of audio-visual materials. Mailing Add: Dept of Chem Tex Tech Univ Lubbock TX 79409

MCPHERSON, DONALD ATTRIDGE, b Cleveland, Ohio, Jan 1, 34; m 57; c 3. SPACE PHYSICS. Educ: Ohio State Univ, BA, 57; Univ Calif, Berkeley, MA, 58, PhD(physics), 62. Prof Exp: Engr, Battelle Mem Inst, 56-57; staff mem high energy nuclear physics, Lawrence Radiation Lab, 61-62; mem tech staff, Aerospace Corp, 62-73, dir survivability, 73-75; DEP MGR SYSTS RES DIV, SCI APPLN, INC, 75- Mem: Am Phys Soc; Am Geophys Union. Res: Nuclear reactor technology; high energy, plasma and magnetospheric physics; atmospheric reentry. Mailing Add: Sci Appln Inc 101 Continental Bldg Suite 310 El Segundo CA 90245

MCPHERSON, DONALD CARMAN, b Toronto, Ont, Nov 26, 10; m 53; c 1. PLANT ECOLOGY. Educ: Univ Toronto, BA, 33, MA, 35, PhD(plant ecol), 38. Prof Exp: Instr sci, Western Tech Sch, 39-43; instr, Lawrence Park Col, 43-57; INSTR SCI, N TORONTO COLLEGIATE, 57- Res: Cortical air spaces in the roots of Zea mays. Mailing Add: 36 McRae Dr Toronto ON Can

MCPHERSON, DONALD FRANK, b Kittanning, Pa, Dec 7, 31; m 58; c 1. AUDIOLOGY, SPEECH PATHOLOGY. Educ: Indiana Univ, Pa, BS, 61; Univ Hawaii, MA, 63; Purdue Univ, West Lafayette, PhD(audiol, speech sci), 66. Prof Exp: Asst prof audiol & speech sci, Ohio Univ, 66-68; assoc prof audiol, Univ Northern Colo, 68-70; ASSOC PROF SPEECH PATH & AUDIOL, UNIV HAWAII, 70- Mem: Am Speech & Hearing Asn; Acoust Soc Am. Res: Psychoacoustics; speech science; language development. Mailing Add: Div of Speech Path & Audiol Univ of Hawaii Honolulu HI 96822

MCPHERSON, HAROLD JAMES, b Newry, NIreland, May 28, 39; Can citizen. PHYSICAL GEOGRAPHY, RESOURCE GEOGRAPHY. Educ: Queen's Univ, Ont, BA, 61; Univ Alta, MSc, 63; McGill Univ, PhD(geog), 67. Prof Exp: Asst prof geog, Queen's Univ, Ont, 66-70; asst prof, 70-71, ASSOC PROF GEOG, UNIV ALTA, 71- Concurrent Pos: Vis scholar, Univ Ariz, 74-75; Can Coun leave fel, Can Coun, 74. Mem: Asn Am Geog. Res: Sediment yields; natural hazards; environmental quality; perception of environment; mountain geomorphology. Mailing Add: Dept of Geog Univ of Alta Edmonton AB Can

MCPHERSON, JAMES BEVERLEY, JR, b Cleveland, Ohio, May 9, 20; m 51; c 2. ORGANIC CHEMISTRY. Educ: Kenyon Col, AB, 41; Univ Ill, MS, 43, PhD(org chem), 48. Prof Exp: Res chemist, Standard Oil Develop Co, 48-52; from res chemist to sr res chemist, 52-56, group leader, 56-66, mgr, 66-70, DEPT HEAD, AM CYANAMID CO, 70- Mem: Am Chem Soc; Am Asn Textile Technol. Res: Surgical sutures; synthetic fibers; agricultural chemicals. Mailing Add: 33 Mimosa Dr Cos Cob CT 06807

MCPHERSON, JAMES C, JR, b Hamilton, Tex, Dec 27, 26; m 45; c 4. BIOCHEMISTRY, MEDICINE. Educ: NTex State Col, BS, 46; Univ Tex, MA, 55, MD, 60. Prof Exp: Res scientist, Univ Tex Southwestern Med Sch Dallas, 60-61, from instr to asst prof biochem, 61-63; asst res prof biochem, 63-70, ASSOC PROF SURG, CELL & MOLECULAR BIOL, MED COL GA, 70- Mem: AAAS; Am Chem Soc; Am Oil Chem Soc; Am Asn Clin Chem; Soc Exp Biol & Med. Res: Lipid absorption and metabolism. Mailing Add: Dept of Biochem Med Col of Ga Augusta GA 30902

MCPHERSON, JAMES KING, b Tucson, Ariz, Nov 11, 37; m 62. PLANT ECOLOGY. Educ: Univ Idaho, BS, 59; Univ Calif, Santa Barbara, MA, 66, PhD(bot), 68. Prof Exp: Asst prof bot, 68-74, ASSOC PROF BOT, OKLA STATE UNIV, 74- Mem: AAAS; Bot Soc Am; Ecol Soc Am. Res: Ecological aspects of forest tree water relations; allelopathy and competition among plants. Mailing Add: Sch of Biol Sci Okla State Univ Stillwater OK 74074

MCPHERSON, JAMES LOUIS, b Chattanooga, Tenn, June 25, 22; m 48; c 3. POLYMER CHEMISTRY. Educ: Ga Inst Technol, BS, 44; Univ Tex, MA, 49; Ohio State Univ, PhD(org chem), 53. Prof Exp: Asst lab, Univ Tex, 47-48; res chemist explor sect, Plastics Dept, Exp Sta, E I du Pont de Nemours & Co, 53-58; proj leader, Cent Res Lab, Gen Aniline & Film Corp, 59-61; lab dir basic polymer res, Cent Res & Eng Div, Continental Can Co, 61-64; sr fel, Mellon Inst, 64-67; mgr chem activ div, DeBell & Richardson, Inc, Mass, 67-69; ASSOC PROF CHEM, LEE COL, TENN, 69- Mem: Am Chem Soc; The Chem Soc. Res: Synthesis, properties and applications of polymers. Mailing Add: Shorttail Springs Rd Route 2 Harrison TN 37341

MCPHERSON, JOE WAYNE, b Greensboro, NC, Jan 16, 47; m 69. SOLID STATE PHYSICS. Educ: Guilford Col, NC, BS, 69; Fla State Univ, Tallahassee, MS, 71, PhD(physics), 75. Prof Exp: ASST PROF PHYSICS, UNIV NC, GREENSBORO, 75- Mem: Am Phys Soc. Res: Theoretical solid state physics, primarily in magnetism, effective-field theories, and in cooperative Jahn-Teller phase transitions, pseudospin Hamiltonian development. Mailing Add: Dept of Physics Univ of NC Greensboro NC 27412

MCPHERSON, JOHN EDWIN, b San Diego, Calif, June 8, 41; m 66; c 2. ENTOMOLOGY. Educ: San Diego State Univ, BS, 63, MS, 64; Mich State Univ, PhD(entom), 68. Prof Exp: Asst prof, 69-74, ASSOC PROF ZOOL, SOUTHERN ILL UNIV, CARBONDALE, 74- Mem: Entom Soc Am; Entom Soc Can; Sigma Xi. Res: Bionomics and taxonomy of North American Scutellerodea of the group Insecta; effects of photoperiod on morphology and pigmentation of Pentatomidae of the order Hemiptera. Mailing Add: Dept of Zool Southern Ill Univ Carbondale IL 62901

MCPHERSON, ROSS, b Buffalo, NY, May 30, 34; m 57; c 3. PHYSICS. Educ: Queen's Univ, Ont, BSc, 59, MSc, 61; McGill Univ, PhD(physics), 64. Prof Exp: Res assoc, Brookhaven Nat Lab, 64-66; asst prof physics, Cornell Univ, 66-72; ASSOC PROF PHYSICS, UNIV GUELPH, 72- Mem: AAAS; Am Phys Soc; Can Asn Physicists. Res: Nuclear physics; nuclear and digital instrumentation; nuclear engineering. Mailing Add: Dept of Physics Univ of Guelph Guelph ON Can

MCPHERSON, THOMAS ALEXANDER, b Calgary, Alta, Mar 1, 39; m 67; c 4. IMMUNOLOGY. Educ: Univ Alta, MD, 62; Univ Melbourne, PhD(med, immunol), 69. Prof Exp: Sr resident med officer, Royal Adelaide Hosps, SAustralia, 63-64, med registr, 64-65, sr med registr, Renal Unit, 65; asst physician, Clin Res Univ, Walter & Eliza Hall Inst Med Res, Melbourne, 66-68; asst prof med, 69-70, ASSOC PROF MED, UNIV ALTA, 70-, DIR DEPT MED, CROSS INST, 73- Concurrent Pos: R S McLaughlin Res Found traveling fel, Southeast Asia & Europe, 68-69. Mem: Australasian Soc Med Res; Can Soc Immunol; Can Soc Clin Invest. Res: Induction and inhibition of experimental allergic encephalomyelitis using human encephalitogenic basic protein and synthetic polypeptides; the carcinoembryonic antigen in the human colon; trial of anti-thymocyte globulin in acute relapses of multiple sclerosis. Mailing Add: Dept of Med Cross Inst 11560 Univ Ave Edmonton AB Can

MCPHERSON, WILLIAM HAKES, b Poplar Bluff, Mo, Oct 10, 20; m 48. PAPER CHEMISTRY. Educ: Syracuse Univ, BS, 42; Inst Paper Chem, MS, 44, PhD, 48. Prof Exp: Chemist, Tenn Eastman Corp, 44-46; res chem engr, Int Paper Co, 48-50; res chem engr, Mead Corp, 50-52; res mgr, Minn & Ont Paper Co Div, Boise Cascade Corp, 52-65, dir pulp & paper res lab, 65-73; PVT CONSULT, WILLIAM H McPHERSON, CONSULT, 73- Concurrent Pos: Consult, Forest Prod Dept, Univ Minn, 74. Mem: Am Chem Soc; Tech Asn Pulp & Paper Indust; Can Pulp & Paper Asn. Res: Pulping and bleaching of wood fibers; development of paper specialties; paper coatings; mechanism of softening paper; insulation board research; waste water treatment; papermaking; properties of Tropical Hardwoods. Mailing Add: Island View Route PO Box 218 International Falls MN 56649

MCPHILLIPS, JOSEPH JOHN, b Philadelphia, Pa, Oct 10, 34; m 59; c 4. PHARMACOLOGY. Educ: St Joseph's Col, Pa, BS, 56; Jefferson Med Col, MS, 59, PhD(pharmacol), 63. Prof Exp: From instr to asst prof pharmacol, Med Col Va, 62-67; from asst prof to assoc prof pharmacol, WVa Univ, 70-73; ASSOC MED DIR, ASTRA PHARMACEUT PROD, INC, 73- Concurrent Pos: Nat Inst Child Growth & Human Develop grant, 65-68. Res: Pharmacology and toxicology of cholinesterase inhibitors; autonomic pharmacology; drug tolerance; pharmacology of bronchodilators. Mailing Add: Astra Pharmaceut Prod Inc Framingham MA 01701

MCQUADE, HENRY ALONZO, b St Louis, Mo, Nov 1, 15. CYTOLOGY, CYTOGENETICS. Educ: Wash Univ, AB, 38, PhD, 49; Univ Mo, MA, 40. Prof Exp: Asst prof biol, Harris Teachers Col, 49-54; res assoc, Mallinckrodt Inst Radiol, Sch Med, Wash Univ, 54-56; res assoc, Radiation Res Lab, Col Med, Univ Iowa, 56-57; assoc prof radiobiol, Sch Med, 57-64, dir radioisotope lab, Med Ctr, 57-70, PROF RADIOBIOL, SCH MED, UNIV MO-COLUMBIA, 64-, CHIEF RADIOL SCI SECT, MED CTR, 70- Mem: AAAS; Genetics Soc Am; Bot Soc Am. Res: Effects of radioisotopes in corporated in cell structures; radiation cytology; electron microscopy of cells. Mailing Add: Dept of Radiol Univ of Mo Med Ctr Columbia MO 65201

MCQUAID, RICHARD WILLIAM, b Woodland, Calif, Jan 6, 23; m 44; c 3. FUEL SCIENCE, FUEL TECHNOLOGY. Educ: Univ Calif, AB, 43; Johns Hopkins Univ, AM, 50, PhD(chem), 51. Prof Exp: Jr instr phys chem, Johns Hopkins Univ, 50-51; res chemist, Mutual Chem Co, 51-55; res chemist, Catalyst Res Corp, 55-57, mgr res & develop, 57-61; prin staff scientist, Aircraft Armaments Inc, 61-65; sr proj eng, 65-74, HEAD FUELS & LUBRICANTS BR, ANNAPOLIS LAB, DAVID TAYLOR NAVAL SHIP RES & DEVELOP CTR, 74- Mem: AAAS; Am Chem Soc; Sigma Xi. Res: Thermodynamics and structural inorganic chemistry; properties of highly desiccated silica and alumina gels; chemistry of chromium compounds; crystal optics; electrochemistry; corrosion; pyrotechnics; fuels and lubricants chemistry and technology. Mailing Add: Fuels Lubricant Br Annapolis Lab Naval Ship Res & Develop Ctr Annapolis MD 21402

MCQUARRIE, BRUCE CALE, b Easton, Pa, June 6, 29; m 48; c 3. ALGEBRA. Educ: Lafayette Col, AB, 51; Univ NH, MA, 56; Boston Univ, PhD(math), 71. Prof Exp: Instr math, 60-63, asst prof, 63-71, ASSOC PROF MATH, WORCESTER POLYTECH INST, 71- Concurrent Pos: Vis instr, Tex A&M Univ, 69-70. Mem: Am Math Soc; Can Math Cong; Math Asn Am. Res: Near rings; endomorphisms of nonabelian groups. Mailing Add: 125 Hampton St Auburn MA 01501

MCQUARRIE, DONALD ALLAN, b Lowell, Mass, May 20, 37; m 59; c 2. THEORETICAL CHEMISTRY. Educ: Lowell Technol Inst, BS, 58; Johns Hopkins Univ, MA, 60; Unive Ore, PhD(chem), 62. Prof Exp: Asst prof chem, Mich State Univ, 62-64; mem tech staff, NAm Aviation Sci Ctr, Calif, 64-68; PROF CHEM, IND UNIV, BLOOMINGTON, 68- Mem: Am Chem Soc; Biophys Soc. Res: Statistical thermodynamics; stochastic processes; biophysics. Mailing Add: Dept of Chem Ind Univ Bloomington IN 47401

MCQUARRIE, DONALD G, b Richfield, Utah, Apr 17, 31; m 56; c 2. SURGERY, COMPUTER SCIENCES. Educ: Univ Utah, BS, 53, MD, 56; Univ Minn, Minneapolis, PhD(surg), 65. Prof Exp: From instr to assoc prof, 65-72, PROF SURG, MED SCH, UNIV MINN, MINNEAPOLIS, 72-; STAFF SURGEON & DIR SURG RES LAB, MINNEAPOLIS VET ADMIN HOSP, 64- Concurrent Pos: USPHS fel, Univ Minn, Minneapolis, 62-65; Navy liaison mem, Div Med Sci, Nat Res Coun, 59-61; chmn surg partic comt, Vet Admin Ctr Off; dir, Biophys Comput Proj. Mem: AAAS; Asn Acad Surg; Soc Exp Biol & Med; Soc Univ Surg. Res: Tissue immunology; respiratory physiology; computer sciences. Mailing Add: Minneapolis Vet Admin Hosp 54th & 48th Ave S Minneapolis MN 55417

MCQUARRIE, IRVINE GRAY, b Ogden, Utah, June 27, 39; m 67; c 2. EXPERIMENTAL NEUROLOGY. Educ: Univ Utah, BS, 61; Cornell Univ, MD, 65. Prof Exp: Chief neurosurg, Naval Hosp, Boston, 73-74; intern & asst surgeon gen surg, Cornell-New York Hosp Med Ctr, 65-68, asst surgeon neurosurg, 68-71, surgeon, 72-73; res fel physiol, Col Med, Cornell Univ, 71-72, res assoc, 74, res fel, 74-76; CONSULT, NAVAL MED RES INST, BETHESDA, MD, 76- Concurrent Pos: Extramural res fel, Nat Inst Neurol & Commun Disorders & Strokes, 74-76. Mem: AAAS; Soc Neurosci. Res: Nerve regeneration in rat peripheral nerves and goldfish optic nerves, studied by means of histochemistry, axonal transport of radioactive proteins, axolemmal uptake of radioactive neurotransmitters, and electron microscopy. Mailing Add: 1300 York Ave New York NY 10021

MCQUATE, JOHN TRUMAN, b Upper Sandusky, Ohio, Aug 28, 21; m 46; c 3. GENETICS. Educ: Heidelberg Col, BS, 43; Ind Univ, PhD(zool), 51. Prof Exp: From asst prof to assoc prof, 51-70, PROF ZOOL, OHIO UNIV, 70- Concurrent Pos: Am Cancer Soc res fel, Case Inst Technol, 56-57, USPHS res fel, 57-59. Mem: AAAS; Genetics Soc Am; Am Soc Zool; Environ Mutagen Soc. Res: Radiation and chemical mutagenesis in Drosophila; human genetics and cytogenetics. Mailing Add: Dept of Zool & Microbiol Ohio Univ Athens OH 45701

MCQUATE, ROBERT SAMUEL, b Lebanon, Pa, Set 4, 47; m 70; c 1. BIO-

INORGANIC CHEMISTRY. Educ: Lebanon Valley Col, BS, 69; Ohio State Univ, PhD(chem), 73. Prof Exp: Res fel, NMex State Univ, 73-74; ASST PROF CHEM, WILLAMETTE UNIV, 74- Concurrent Pos: Sigma Xi grant-in-aid, 75; Petrol Res Fund grant-in-aid, 75. Mem: Am Chem Soc. Res: Study of the involvement of metal ions in metalloenzyme systems and metal ion catalysis of interligand reactions serving as models for biological systems. Mailing Add: Dept of Chem Willamette Univ Salem OR 97301

MCQUAY, RUSSELL MICHAEL, JR, b Olean, NY, Sept 6, 21; m 50; c 2. MEDICAL PARASITOLOGY. Educ: Bucknell Univ, AB, 43; Tulane Univ, MS, 49, PhD(med parasitol), 51. Prof Exp: Asst med parasitol, Sch Med, Tulane Univ, 46-49, instr, 49-51; asst, Chicago Med Sch, 53-59, instr, 59-64, asst prof microbiol & pub health, 64-74; DIR MICROBIOL, DIV MED PARASITOL, MT SINAI HOSP MED CTR, 53-; ASSOC PROF MICROBIOL & PATH, RUSH MED COL, 75- Concurrent Pos: Instr, schs Med Technol, Mt Sinai Hosp, Chicago, 53-, WSuburban Hosp, Oak Park, Ill, 61-, Med Sch, Northwestern Univ, Chicago, 68-, Presby St Luke's-Rush Med Ctr, 70- & Northwestern Mem Hosp, 71-; consult, WSuburban Hosp & Northwestern Mem Hosp; lectr, Rush Med Col. Mem: Am Sci Affiliation; Am Soc Trop Med & Hyg; assoc mem Am Soc Clin Path. Res: Schistosomiasis and amebiasis; filariasis. Mailing Add: Mt Sinai Hosp Med Ctr 15th & California Ave Chicago IL 60608

MCQUEEN, DONALD JAMES, b Vancouver, BC, Sept 8, 43; m 65; c 1. ECOLOGY. Educ: Univ BC, BSc, 66, MSc, 68, PhD(ecol), 70. Prof Exp: Asst prof biol, 70-74, ASSOC PROF BIOL, YORK UNIV, 74- Concurrent Pos: Nat Res Coun Can fel, 70-75; res fund, Nat Res Coun, 70-76. Mem: Can Soc Zool; Int Asn Theoret & Appl Limnol; Ecol Soc Am. Res: Components studies of competition in cellular slime mold and isopod species; competitive interactions in isopod and spider populations. Mailing Add: Dept of Biol York Univ 4700 Keele St Toronto ON Can

MCQUEEN, GEORGE ROBERT, agronomy, see 12th edition

MCQUEEN, JAMES LEE, b Jefferson City, Mo, Dec 28, 32; m 58; c 3. CANCER, EPIDEMIOLOGY. Educ: Univ Mo, BS & DVM, 57; Univ Mich, MPH, 61, DrPH(virol), 64. Prof Exp: Vet dir virol & epidemiol, Ctr Dis Control, USPHS, Ga, 57-74; chief virol labs, Lunar Receiving Lab, Manned Spacecraft Ctr, NASA, 67-69, chief microbiol, Prev Med Div, 69-70, dep chief microbiol & environ biol, 70-72, chief microbial ecol invests off, 72-73, spec asst advan mission planning, Health Maintenance Br, 73-74; CHIEF DETECTION, DIAG & PRETREAT EVAL BR, DIV CANCER CONTROL & REHAB, NAT CANCER INST, NIH, 74- Mem: Am Pub Health Asn; Soc Epidemiol Res. Res: Rabies diagnosis; fluorescent microscopy; arbovirus ecology; epizootiology of respiratory disease; relationships of influenza in man and animals; microbial ecology of manned space flight; epidemiology; cancer control; detection/diagnosis of cancer. Mailing Add: Div of Cancer Control & Rehab Nat Cancer Inst NIH Blair Bldg Bethesda MD 20014

MCQUEEN, JOHN DONALD, b Bently, Alta, Oct 17, 23; m 55; c 3. MEDICINE. Educ: Univ Toronto, MD, 46, MA, 50. Prof Exp: Neurosurgeon, Johns Hopkins Hosp, 56; from instr to asst prof neurol surg, 56-65, ASSOC PROF NEUROSURG, SCH MED, JOHNS HOPKINS UNIV, 65- Concurrent Pos: Asst chief surg in charge neurosurg, Baltimore City Hosps, 56-61, assoc chief, 61-; consult, Perry Point Vet Hosp, 62; consult, Loch Raven Vet Admin Hosp, 66, chief neurosurg, 70- Mem: AAAS; Am Soc Cell Biol; AMA; Am Asn Neurol Surg; fel Am Col Surg. Res: Experimental head injury; intracranial hypertension; acute ethanolism. Mailing Add: Div of Neurol Surg Baltimore City Hosp Baltimore MD 21224

MCQUEEN, RALPH DAVID, reproductive physiology, veterinary medicine, see 12th edition

MCQUIGG, JAMES DONALD, atmospheric science, see 12th edition

MCQUIGG, ROBERT DUNCAN, b Wooster, Ohio, Apr 7, 36; m 59; c 4. PHYSICAL CHEMISTRY. Educ: Muskingum Col, BS, 58; Ohio State Univ, PhD(phys chem), 64. Prof Exp: Asst prof chem, Univ Toledo, 64-65; asst prof, 65-70, ASSOC PROF CHEM, OHIO WESLEYAN UNIV, 70- Concurrent Pos: Vis scientist, Nat Ctr Atmospheric Res, Colo, 69-70. Mem: Am Chem Soc. Res: Photochemistry of air pollutants; photoreduction of inorganic coordination compounds. Mailing Add: Dept of Chem Ohio Wesleyan Univ Delaware OH 43015

MCQUISTAN, RICHMOND BECKETT, b West New York, NJ, June 12, 27; m 51; c 2. PHYSICS, STATISTICAL MECHANICS. Educ: Purdue Univ, BS, 50, MS, 52, PhD(physics, heat transfer), 54. Prof Exp: Instr, Purdue Univ, 52-54; head solid state sect, Farnsworth Electronics Co, 54-56; res supvr phys electronics, Honeywell Res Ctr, 56-61; assoc prof elec eng, Univ Minn, Minneapolis, 61-63; staff physicist, Honeywell Res Ctr, 63-66; chmn dept physics, 68-69, assoc dean col lett & sci, 69-71, asst to vchancellor, 71-72, dean grad sch, 72-75, PROF PHYSICS, UNIV WIS, MILWAUKEE, 66- Concurrent Pos: Vis assoc prof, Univ Minn, Minneapolis, 63-64; vis prof, Univ Liverpool, 73-76; sr res fel, Brit Sci Res Coun, 75-76. Mem: Am Optical Soc; Am Phys Soc; Am Vacuum Soc. Res: Infrared detectors and detection theory; electrical and optical properties of vacuum deposited thin films; radiative heat transfer and thermodynamics; infrared gonimetric reflectance measurements; resistance-temperature characteristics of semiconductors; infrared modulation by free carrier absorption; statistical mechanics of lattice spaces. Mailing Add: Dept of Physics Univ of Wis 1900 E Kenwood Ave Milwaukee WI 53201

MCQUISTION, WILLIAM EDGAR, organic chemistry, see 12th edition

MCRAE, D HAROLD, plant physiology, organic chemistry, see 12th edition

MCRAE, DANIEL GEORGE, b Jordan, Mont, Nov 25, 38; m 63; c 2. MATHEMATICS. Educ: Univ Mont, BA, 60, MA, 61; Univ Wash, PhD(math), 67. Prof Exp: Instr math, Univ Mont, 61-62; asst prof, Univ Ill, Urbana, 67-70; asst prof, 70-74, ASSOC PROF MATH, UNIV MONT, 74- Mem: Am Math Soc; Math Asn Am; Opers Res Soc Am. Res: Ring theory; homological algebra. Mailing Add: Dept of Math Univ of Mont Missoula MT 59801

MCRAE, DONALD LANE, b Toronto, Ont, July 23, 12; m 40; c 2. RADIOLOGY. Educ: Univ Western Ont, MD, 38. Prof Exp: From asst prof to assoc prof radiol, McGill Univ, 46-67; PROF RADIOL, UNIV TORONTO, 67-; DIR DEPT RADIOL, SUNNYBROOK HOSP, 67- Concurrent Pos: Radiologist-in-chief, Montreal Children's Hosp, 45-50 & Montreal Neurol Inst, 45-67. Mem: AAAS; Am Roentgen Ray Soc; fel Am Col Radiol; Am Soc Neuroradiol (pres, 65-66); Can Asn Radiol (pres, 61-62). Res: Neurological radiology. Mailing Add: Sunnybrook Hosp Dept of Radiol 2075 Bayview Ave Toronto ON Can

MCRAE, EION GRANT, b Tambellup, Australia, Dec 25, 30; m 54; c 2. PHYSICAL CHEMISTRY, PHYSICS. Educ: Univ Western Australia, BSc, 52, MSc, 54; Fla State Univ, PhD(phys chem), 57. Prof Exp: Fel chem, Ind Univ, 57-58; sr res officer, Commonwealth Sci & Indust Res Orgn, Australia, 58-63; MEM TECH STAFF, BELL LABS, 63- Mem: Fel Am Phys Soc. Res: Theoretical and experimental work on the development of the technique of low-energy electron diffraction as a means of studying the surface structures of crystals. Mailing Add: Room 1D463 Bell Labs Murray Hill NJ 07974

MCRAE, LARRY GENE, inorganic chemistry, see 12th edition

MCRAE, LORIN POST, b Tucson, Ariz, Feb 20, 36; m 56; c 7. BIOMEDICAL ENGINEERING, ELECTRONICS. Educ: Univ Ariz, BS, 61, PhD(elec eng), 68; NY Univ, MEE, 63. Prof Exp: Mem tech staff, Bell Tel Labs, 61-63; instr elec eng, Univ Ariz, 63-68; asst prof, Univ Wyo, 68-70; DIR MED ELECTRONICS, TUCSON MED CTR, 70- Concurrent Pos: Chmn elec safety comt, Southern Ariz Hosp Coun, 71-72. Mem: Asn Advan Med Instrumentation. Res: Signal processing and pattern recognition in diagnostic medicine; evaluation of cerebrovascular insufficiency for the prevention of strokes; information theory and tactile communications; neurologic evaluation and treatment of pain. Mailing Add: Tucson Med Ctr 5301 E Grant Tucson AZ 85712

MCRAE, ROBERT JAMES, b Jordan, Mont, Aug 16, 31; m 56; c 7. SCIENCE EDUCATION. Educ: Univ Mont, BA, 54, MA, 57; Univ Wis, Madison, PhD(hist sci), 69. Prof Exp: Asst prof physics, 58-60, assoc prof, 60-62 & 66-69, PROF PHYSICS, EASTERN MONT COL, 69- Mem: Hist Sci Soc; Am Asn Physics Teachers; Nat Sci Teachers Asn. Res: History of nineteenth and early twentieth century physics. Mailing Add: 2606 Raymond Place Billings MT 59102

MCRAE, VINCENT VERNON, b Columbia, SC, Sept 2, 18; m 41; c 1. SCIENCE POLICY, INFORMATION SCIENCE. Educ: Miner Teachers Col, BS, 40; Cath Univ, MS, 44, PhD(math), 55. Prof Exp: Teacher high sch, DC, 41-52; analyst opers res off, Johns Hopkins Univ, 52-60, chmn Stratspiel group & chief strategic div, 60-61; mem sr staff, Res Anal Corp, 61-64; tech asst, Off Sci & Technol, White House, 64-74; ASST BUS PLANNING, FED SYSTS DIV, IBM CORP, 74- Concurrent Pos: Mem tech staff, Gaither Comt, 57; adv to del & mem US del, Surprise Attack Conf, Geneva, 58; res assoc, Coolidge Comt, 59; consult, Dept of State & Nat Security Coun, Dept of Defense, 73- Mem: Fel AAAS; Am Math Soc; Opers Res Soc Am. Res: Operations research; military operations research. Mailing Add: 1501 Emerson St NW Washington DC 20011

MCRAE, WALTER BRUCE, theoretical chemistry, see 12th edition

MCRAE, WAYNE ALAN, b Chicago, Ill, Aug 8, 25; m 49; c 2. PHYSICAL CHEMISTRY. Educ: Harvard Univ, AB, 47, AM, 48. Prof Exp: Chemist, 49-50, patents supvr, 50-52, asst tech dir, 52-57, dir org div, 57-59, VPRES & DIR RES, IONICS, INC, 59- Res: Membrane transport processes; membrane permation; reversed osmosis; electrodialysis; electrochemistry; fuel cells; batteries; environment engineering; saline water conversion. Mailing Add: Ionics Inc 65 Grove St Watertown MA 02172

MCREYNOLDS, MARY MAUREEN (SIMS), genetics, human ecology, see 12th edition

MCRIPLEY, RONALD JAMES, b Detroit, Mich, Nov 17, 32; m 55; c 3. MEDICAL MICROBIOLOGY. Educ: Mich State Univ, BS, 54; Univ Mich, MS, 61, PhD(microbiol), 64. Prof Exp: Bacteriologist, Mich State Dept Health, 58-59; res asst microbiol, St Margaret's Hosp, 64-67; sr res scientist, Norwich Pharmacal Co, 67-69; sr res bacteriologist, 69-74, RES FEL, SQUIBB INST MED RES, 74- Concurrent Pos: Fel sch med, Tufts Univ, 64- Mem: AAAS; Am Burn Asn; NY Acad Sci; Am Soc Microbiol. Res: Host defense mechanisms against bacterial infection; chemotherapy in infectious diseases; host-parasite interactions. Mailing Add: Squibb Inst for Med Res Box 4000 Princeton NJ 08540

MCROBERTS, J WILLIAM, b Rochester, Minn, Dec 6, 32; m 59; c 2. UROLOGY. Educ: Princeton Univ, BA, 55; Cornell Univ, MD, 59. Prof Exp: Intern surg, New York Hosp & Cornell Univ Med Col, 59-60; resident surg, Mayo Clin, 60-61, resident urol, 61-64; from instr to assoc prof urol, Sch Med, Univ Wash, 67-72; assoc prof, 72-74, PROF SURG, COL MED, UNIV KY, 74-, CHIEF DIV UROL, 72- Mem: Am Col Surg; Am Urol Asn; Am Acad Pediat; Soc Pediat Urol; Soc Univ Urol. Res: Renal transplantation; pediatric urology, intersex. Mailing Add: Div of Urol Univ of Ky Col of Med Lexington KY 40506

MCROBERTS, MILTON R, b Hazleton, Ind, Dec 11, 24. BIOCHEMISTRY. Educ: Purdue Univ, BS, 49, MS, 58; Aberdeen Col, PhD(biochem, nutrit), 61. Prof Exp: Teacher high schs, Ind, 49-51; res chemist, Purdue Univ, 55-58; res chemist, Rowett Res Inst, 58-61, Univ Reading, 61 & Hy-Line Poultry Farms & Pioneer Hi-Bred Corn Co, 62-65; asst dir res life sci, Res Div, W R Grace & Co, 65-67, NUTRIT OFFICER, FOOD & AGR ORGN, UN, 67- Mem: AAAS; Am Inst Nutrit; Am Soc Animal Sci; Poultry Sci Asn; Brit Nutrit Soc. Res: Genetic-nutrition interactions; mineral and vitamin metabolism in tooth and bone formation; thyroid activity; nutrient biological availability. Mailing Add: Food & Agr Orgn of UN Bangkok Thailand

MCRORIE, ROBERT ANDERSON, b Statesville, NC, May 7, 24; m 46; c 3. BIOCHEMISTRY. Educ: NC State Col, BS, 49, MS, 51; Univ Tex, PhD(biochem), 54. Prof Exp: From asst prof to assoc prof, 53-63, PROF BIOCHEM, UNIV GA, 63- Concurrent Pos: Res partic, Oak Ridge Inst Nuclear Study, 57 & 58; assoc dean grad sch & dir gen res, Univ Ga, 59-71, asst vpres res, 68-71. Mem: AAAS; Am Chem Soc; Am Soc Microbiol; Am Soc Biol Chem. Res: Microbial nutrition; intermediary metabolism; enzymology; research administration. Mailing Add: Dept of Biochem Univ of Ga Athens GA 30601

MCROWE, ARTHUR WATKINS, b San Francisco, Calif, Aug 16, 37; m 65; c 3. ORGANIC CHEMISTRY. Educ: Univ Calif, Berkeley, BS, 60; Univ Wis, Madison, PhD(org chem), 66. Prof Exp: Res chemist, 66-71, SR RES CHEMIST, B F GOODRICH RES CTR, 71- Mem: Am Chem Soc. Res: Physical organic chemistry; carbonium ion rearrangements; monomer processes; metallo-organic chemistry; polymer flammability. Mailing Add: B F Goodrich Res Ctr Brecksville OH 44141

MCROY, C PETER, b East Chicago, Ind, Jan 17, 41. BIOLOGICAL OCEANOGRAPHY. Educ: Mich State Univ, BS, 63; Univ Wash, MS, 66; Univ Alaska, College, PhD(marine sci), 70. Prof Exp: Asst prof marine sci, 69-74, ASSOC PROF MARINE SCI, INST MARINE SCI, UNIV ALASKA, COLLEGE, 74- Concurrent Pos: Fel, Univ Ga, 71-72; magus, Int Seagrass Ecosyst Study, 74- Mem: AAAS; Am Soc Limnol & Oceanog; Ecol Soc Am; Arctic Inst NAm; Int Asn Aquatic Vascular Plant Biologists. Res: Ecology of seagrass communities; productivity of ice-covered seas. Mailing Add: Inst of Marine Sci Univ of Alaska College AK 99701

MCSHAN, WILLIAM HARTFORD, b Lonoke, Ark, May 4, 03; m 54; c 1.

ENDOCRINOLOGY, BIOCHEMISTRY. Educ: Ark State Teachers Col, BS, 28; Univ Mo, AM, 33, PhD(org chem), 36. Prof Exp: Asst, Univ Mo, 33-36; res assoc, 36-41, from asst prof to assoc prof, 41-51, PROF ENDOCRINOL, DEPT ZOOL, UNIV WIS-MADISON, 51- Mem: AAAS; Am Chem Soc; Am Soc Biol Chem; Am Soc Zool; Endocrine Soc. Res: Preparation and biochemistry of gonadotropic hormones; isolation and study of pituitary cytoplasmic organelles. Mailing Add: Zool Res Bldg Univ of Wis Madison WI 58706

MCSHANE, EDWARD JAMES, b New Orleans, La, May 10, 04; m; c 3. MATHEMATICS. Educ: Tulane Univ, BE & BS, 25, MS, 27; Univ Chicago, PhD(math), 30. Hon Degrees: ScD, Tulane Univ, 47. Prof Exp: Instr math, Tulane Univ, 25-27; asst prof, Univ Wichita, 28-29; Nat Res Coun fel, Univ Chicago, Princeton Univ, Ohio State Univ & Harvard Univ, 30-32; instr math, Princeton Univ, 33-34, asst prof, 34-35; prof, 35-57, alumni prof, 57-74, ALUMNI EMER PROF MATH, UNIV VA, 74- Concurrent Pos: Head mathematician, Ballistic Res Labs, Aberdeen Proving Grounds, 42-45. Mem: Nat Acad Sci; Am Math Soc (ed, Trans, 44-46, pres, 58-59); Am Philos Soc; Math Asn Am (pres, 53-54); Soc Indust & Appl Math. Res: Existence theorems in calculus of variations; integration; exterior ballistics; problems of Bolza in calculus of variations; stochastic processes. Mailing Add: Dept of Math Univ of Va Charlottesville VA 22903

MCSHARRY, JAMES JOHN, b Newark, NJ, May 28, 42; m 67; c 2. MICROBIOLOGY, VIROLOGY. Educ: Manhattan Col, BS, 65; Univ Va, MS, 67, PhD(microbiol), 70. Prof Exp: ASST PROF MICROBIOL, ALBANY MED COL, 73- Concurrent Pos: Nat Inst Allergy & Infectious Dis fel, Rockefeller Univ, 70-73. Mem: AAAS; Am Soc Microbiol; Harvey Soc; Sigma Xi. Res: Studies on the structure and function of the membrane proteins and surface glycoproteins of enveloped RNA viruses. Mailing Add: Dept of Microbiol Albany Med Col Albany NY 12208

MCSHARRY, WILLIAM OWEN, b Hammels, NY, Oct 14, 39; m 70; c 1. ANALYTICAL CHEMISTRY. Educ: Fordham Univ, BS, 61, MS, 67, PhD(chem), 69. Prof Exp: Sr scientist anal chem, 69-75, GROUP LEADER ANAL CHEM, HOFFMANN-LaROCHE INC, 75- Mem: Am Chem Soc; Sigma Xi. Res: Determination of vitamin D in pharmaceutical and food preparations by liquid chromatography; measurement of color in the pharmaceutical industry. Mailing Add: Hoffmann-LaRoche Inc Nutley NJ 07110

MCSHEFFERTY, JOHN, b Akron, Ohio, Mar 14, 29; m 59; c 2. PHARMACEUTICAL CHEMISTRY. Educ: Univ Glasgow, BSc, 53, PhD(medicinal chem), 57. Prof Exp: Asst lectr pharm, Univ Strathclyde, 53-54; res assoc, Sterling-Winthrop Res Inst, NY, 57-62; sr scientist, 62-63, DIR PHARMACEUT DEVELOP, ORTHO PHARMACEUT CORP, RARITAN, 63- Mem: AAAS; Am Chem Soc; Am Pharmaceut Asn; NY Acad Sci; Am Acad Pharmaceut Sci. Res: Pharmaceutical research and development. Mailing Add: 13 Nottingham Way Somerville NJ 08876

MCSHERRY, CHARLES K, b New York, NY, Nov 22, 31; m 57; c 1. SURGERY. Educ: Fordham Univ, BS, 53; Cornell Univ, MD, 57; Am Bd Surg, dipl, 65; Bd Thoracic Surg, dipl, 66. Prof Exp: Asst, 58-63, instr, 63-64, from clin instr to clin asst prof, 65-71, ASSOC PROF SURG, MED COL, CORNELL UNIV, 71- Concurrent Pos: Asst attend surgeon, New York Hosp, 65-72, assoc attend surgeon, 72- Mem: Am Col Surg; Am Gastroenterol Asn; Asn Study Liver Dis; Soc Univ Surg; Am Surg Asn. Res: Gastrointestinal malignancy; gallstone formation. Mailing Add: 525 E 68th St New York NY 10021

MCSPADEN, JAY BYRON, b Malden, Mass, Mar 6, 39; m 61; c 3. AUDIOLOGY. Educ: Mt Angel Col, BA, 67; Ore Col Educ, MSEd, 68; Univ Wash, PhD(audiol), 71. Prof Exp: Trainee, Vet Admin Outpatient Clin, Seattle, Wash, 68-71; audiologist, 71-74, CHIEF AUDIOL & SPEECH PATH CLIN, VET ADMIN HOSP, 74-; ASST PROF AUDIOL, BAYLOR COL MED, 71- Concurrent Pos: Mem, Coun Educ of the Deaf. Mem: Am Speech & Hearing Asn; Acad Rehab Audiol. Res: Auditory research into speech intelligibility; speech processing in central auditory disorders. Mailing Add: Vet Admin Hosp 2002 Holcombe Blvd Houston TX 77211

MCSWAIN, BARTON, b Paris, Tenn, Jan 19, 06; m 44; c 2. SURGERY. Educ: Vanderbilt Univ, BA, 27, MD, 30. Prof Exp: Instr surg, Cornell Univ, 44; from asst prof to assoc prof, 46-58, PROF SURG, SCH MED, VANDERBILT UNIV, 58-, PROF SURG PATH, 63- Concurrent Pos: Consult, Vet Admin Hosp, 53- Mem: Soc Univ Surg; AMA; Am Asn Cancer Res; fel Am Col Surg; Am Surg Asn. Res: Cancer. Mailing Add: Dept of Surg Vanderbilt Univ Hosp Nashville TN 37232

MCSWAIN, BERAH DAVIS, b Albany, NY, Feb 6, 35; m 65; c 1. BIOPHYSICS. Educ: Univ Rochester, BS, 56, MS, 62; Univ Calif, Berkeley, PhD(biophys), 68. Prof Exp: Optical engr, Northrop Aircraft, Inc, Calif, 56-58; optical physicist, US Naval Ord Lab, 58-60; res assoc chem eng, Univ Rochester, 60-62; res asst cell physiol, 62-68, ASST BIOPHYSICIST & LECTR CELL PHYSIOL, UNIV CALIF, BERKELEY, 68- Mem: AAAS; Am Chem Soc; Optical Soc Am; Am Soc Plant Physiol; Soc Photo-Optical Instrument Eng. Res: Photosynthesis; studies of the light reactions of photosynthesis, including electron transport and photophosphorylation. Mailing Add: Dept of Cell Physiol 251 Hilgard Hall Univ of Calif Berkeley CA 94720

MCSWEENEY, ELLSWORTH EDWARD, b Jersey City, NJ, Mar 19, 14; m 38; c 4. CHEMISTRY. Educ: Oberlin Col, AB, 34; Univ Rochester, PhD(chem), 38. Prof Exp: Asst, Univ Rochester, 34-38; res assoc, Columbus Labs, Battelle Mem Inst, 38-39, res engr, 39-44, asst supvr, 44-47, supvr, 47-52, mgr chem res, 53-63, mgr dept chem & chem eng, 63-66; asst dir chem res & develop, 66-70, TECH DIR CHEM DIV, UNION CAMP CORP, NJ, 70- Mem: Am Soc Testing & Mat; Fedn Socs Paint Technol; Chem Mkt Res Asn; Am Chem Soc; Brit Soc Chem Indust. Res: Organotin compounds; kinetics of reaction of solution; organic coatings; plastics; petroleum resins; drying oils; hydrolysis of sulfonic acid esters; tall oils; rosin and rosin derivatives; fatty acids; research management. Mailing Add: Union Camp Corp 1600 Valley Rd Wayne NJ 07470

MCSWEENEY, JEAN, b Rockford, Ill, Apr 3, 30. ORGANIC CHEMISTRY. Educ: Rosary Col, BA, 51; Univ Notre Dame, PhD(org chem), 65. Prof Exp: Teacher high sch, Ill, 53-57; teacher high sch, Wis, 57-60; assoc prof chem, Edgewood Col, 64-73, chmn div chem & physics, 64-68 & natural sci & math, 68-73; PROF CHEM, ROSARY COL, 73- Mem: Am Chem Soc. Res: Ylide methylation of aromatic nitro compounds. Mailing Add: Dept of Chem Rosary Col River Forest IL 60305

MCTAGGART, KENNETH C, b Vancouver, BC, Can, Aug 10, 19; m 51; c 2. PETROLOGY. Educ: Univ BC, BASc, 43; Queen's Univ, Ont, MSc, 46; Yale Univ, PhD(geol), 48. Prof Exp: Geologist, Geol Surv Can, 48-50; PROF GEOL, UNIV BC, 50- Mem: Geol Asn Can; fel Geol Soc. Res: Petrology and structure. Mailing Add: Dept of Geol Univ of BC Vancouver BC Can

MCTAGGART, WILLIAM DONALD, b Edinburgh, Scotland, Aug 16, 33; m 55; c 5. GEOGRAPHY OF SOUTHEAST ASIA, POPULATION GEOGRAPHY. Educ: Univ St Andrews, MA, 56; Australian Nat Univ, PhD(geog), 63. Prof Exp: Lectr geog, Univ Malaya, 63-71; ASSOC PROF GEOG, ARIZ STATE UNIV, 71- Mem: Asn Asian Studies. Res: Social, demographic and economic problems of developing countries in Southeast Asia. Mailing Add: Dept of Geog Ariz State Univ Tempe AZ 85281

MCTAGGART-COWAN, PATRICK DUNCAN, b Edinburgh, Scotland, May 31, 12; m 39; c 2. METEOROLOGY, SCIENCE POLICY. Educ: Univ BC, BA, 33; Oxford Univ, BA, 36. Hon Degrees: LLD, St Francis Xavier Univ, 70; DSc, Univ BC, 60, McGill Univ, 74, Lakehead Univ, 74. Prof Exp: Instr physics, Univ BC, 33-34; meteorologist in training, British Meteorol Off, 36; officer in charge, Meteorol Off, Botwood & Newf, 37-42; chief meteorol officer, Royal Air Force Ferry Command, Que, 42-45; secy air navig, Provisional Int Civil Aviation Orgn, 45; asst dir & chief forecast div, Meteorol Br, Can Dept Transport, 46-57, from assoc dir to dir, 58-64; pres, Simon Fraser Univ, 63-68; exec dir, Sci Coun Can, 68-75; RETIRED. Educ: Mem exec comt, World Meteorol Orgn, UN, 60-63; head task force, Oper Oil, 70-72. Honors & Awards: Mem, Order of Brit Empire, 44; Losey Award, Inst Aeronaut Sci, 59; Paterson Medal, Meteorol Serv Can, 65; Charles F Brookes Award, Am Meteorol Soc, 65. Mem: Am Meteorol Soc (vpres, 60); Arctic Inst NAm; Can Asn Physicists; Royal Meteorol Soc (pres, Can Br, 52). Res: Environment; climatic change. Mailing Add: High Falls Rd RR2 Bracebridge ON Can

MCTAGUE, JOHN PAUL, b Jersey City, NJ, Nov 28, 38; m 61; c 4. CHEMICAL PHYSICS. Educ: Georgetown Univ, BS, 60; Brown Univ, PhD(phys chem), 65. Prof Exp: Mem tech staff sci ctr, NAm Aviation Inc, 64-70; from asst prof to assoc prof chem, 70-74, PROF CHEM, UNIV CALIF, LOS ANGELES, 74- Concurrent Pos: Alfred P Sloan res fel, 71-74; John Simon Guggenheim Mem fel, 75-76. Mem: Am Phys Soc; Am Chem Soc. Res: Spectroscopic studies of molecular interactions; Raman scattering; neutron scattering; dynamics of condensed phases; surface physics; quantum solids and liquids. Mailing Add: Dept of Chem Univ of Calif Los Angeles CA 90024

MCTERNAN, EDMUND J, b Hollis, NY, June 15, 30; m 52; c 6. PUBLIC HEALTH, HEALTH ADMINISTRATION. Educ: New Eng Col, BS, 56; Columbia Univ, MS, 58; Univ NC, Chapel Hill, MPH, 63; Boston Univ, EdD, 74. Prof Exp: Asst adminr, Emerson Hosp, Concord, Mass, 58-60; adminr, Mem Hosp, North Conway, NH, 60-62; adminr, Boston Dispensary & Rehab Inst, Mass, 63-65; assoc prof & dean div health sci, Northeastern Univ, 65-69; PROF HEALTH SCI & DEAN, SCH ALLIED HEALTH PROFESSIONS, HEALTH SCI CTR, STATE UNIV NY, STONY BROOK, 69- Concurrent Pos: USPHS fel core curric allied health educ, Northeastern Univ, 65-67; Bruner Found fel physician assoc educ, State Univ NY Stony Brook, 70-; mem nat adv allied health professions coun, NIH, 67-70; consult col proficiency exam prof health educ, State Dept Educ, NY, 71- Mem: Asn Schs Allied Health Professions. Res: Education patterns and role relationships of allied health profession; continuing education needs in allied health professions. Mailing Add: Sch Allied Health Professions State Univ of NY Health Sci Ctr Stony Brook NY 11790

MCTIGUE, FRANK HENRY, b Holyoke, Mass, Dec 19, 19; m 43; c 2. PLASTICS CHEMISTRY. Educ: Williams Col, BA, 41; Yale Univ, PhD(org chem), 49. Prof Exp: Chemist, Monsanto Chem Co, 41-42; lab asst org chem, Yale Univ, 46-49; res chemist, 49-69, res supvr, 69-75, MGR PROD TECHNOL, POLYMERS DEPT, HERCULES INC, 75- Mem: Am Chem Soc; Soc Plastics Eng; Am Soc Testing & Mat. Res: Polymers, plastics and fibers; plastics formulation, stabilization, degradation. Mailing Add: 2551 Deepwood Dr Wilmington DE 19810

MCURDY, ORVILLE L, b Miles City, Mont, Sept 5, 29; m 59; c 2. POLYMER CHEMISTRY, ORGANIC CHEMISTRY. Educ: Mont State Col, BS, 51; Univ Mich, MS, 52, PhD(org chem), 56. Prof Exp: Res chemist, Fabric & Finishes Div, E I du Pont de Nemours & Co, 56-58; res chemist, Aerojet Gen Corp, 58-62, sr res chemist, 62-64; mgr coatings develop, Memorex Corp, Calif, 64-70; TECH DIR, MAGNETIC TAPE DIV, AMPEX CORP, REDWOOD CITY, 70- Mem: Am Chem Soc. Res: Synthetic organic chemistry; alkaloids; protective coatings; polymer preparation and characterization; polymerization kinetics; magnetic tape. Mailing Add: 12543 Palmtag Dr Saratoga CA 95070

MCVAUGH, ROGERS, b Brooklyn, NY, May 30, 09; m 37; c 2. BOTANY. Educ: Swarthmore Col, AB, 31; Univ Pa, PhD(bot), 35. Prof Exp: Asst inst bot, Univ Pa, 33-35; instr, Univ Ga, 35-36, asst prof, 36-38; assoc botanist, Div Plant Explor & Introd, Bur Plant Indust, USDA, 38-43, Div Soils & Agr Eng, 43-46; assoc prof bot, 46-51, prof, 51-74, dir, Univ Herbarium, 72-75, HARLEY HARRIS BARTLETT PROF BOT, UNIV MICH, ANN ARBOR, 74-, CUR PHANEROGAMS, HERBARIUM, 46-, VASCULAR PLANTS, 56- Concurrent Pos: Prog dir syst biol, NSF, 55-56. Mem: Am Soc Plant Taxon (pres, 56); Int Asn Plant Taxon (vpres, 69-72, pres, 72-75). Res: Taxonomy of flowering plants; history of botany; flora of Mexico. Mailing Add: Univ Herbarium Univ of Mich Ann Arbor MI 48104

MCVAY, CHESTER BIDWELL, b Yankton, SDak, Aug 1, 11; m 38; c 3. SURGERY. Educ: Yankton Col, AB, 33; Northwestern Univ, MS, 37, MD & PhD(gross anat), 39; Am Bd Surg, dipl, 46. Prof Exp: Instr anat, Med Sch, Northwestern Univ, 34-38; resident surg, Med Sch, Univ Mich, 39-41, instr path, 41-42, instr surg, 42-43; CLIN PROF SURG, SCH MED, UNIV S DAK, VERMILLION, 46-, ASSOC PROF ANAT, 53- Concurrent Pos: Chief surg, Yankton Clin & Sacred Heart Hosp. Mem: Soc Univ Surg; Am Asn Anat; Am Surg Asn; Am Col Surg. Res: Gross anatomy; anatomy of the abdominal wall, especially the inguinal and hypogastric regions as they relate to the occurrence and surgical repair of the inguinal and femoral herniae. Mailing Add: Yankton Clin Dept of Surg Fourth & Park St Yankton SD 47078

MCVAY, FRANCIS EDWARD, b Peace Dale, RI, Sept 27, 17; m 49. APPLIED STATISTICS. Educ: Univ RI, BS, 42; NC State Univ, MS, 44; Univ NC, Chapel Hill, PhD(econ), 46. Prof Exp: Agr statistician, USDA, 45-47; asst prof econ & statist, NC State Univ, 47-52; statist consult, US Naval Ord Test Sta, Calif, 52-54, head test design & eval br, 54-57; assoc prof statist, NC State Univ, 57-62, PROF STATIST, NC STATE UNIV, 62- Mem: AAAS; Am Statist Asn; Biomet Soc. Mailing Add: Dept of Statist 612 Cox Hall NC State Univ Raleigh NC 27607

MCVEAN, DUNCAN EDWARD, b Pontiac, Mich, June 8, 36; m 60; c 2. PHARMACEUTICAL CHEMISTRY. Educ: Univ Mich, BS, 58, MS, 60, PhD(pharmaceut chem), 63. Prof Exp: Pharmaceut res chemist, William S Merrell Co, 65-71, head methods develop sect, Qual Control Dept, 71-74, PROJ ASST TO VPRES QUAL OPERS, MERRELL-NAT LABS, 74- Mem: AAAS; Am Pharmaceut Asn; Am Acad Pharmaceut Sci; affiliate Am Med Asn. Res: Analytical chemistry; physical pharmacy; biopharmaceutics; pharmaceutical product development. Mailing Add: Qual Opers Div Merrell-Nat Lab 110 E Amity Rd Cincinnati OH 45215

MCVEIGH, ILDA, b Fulton, Mo, Feb 12, 05. BACTERIOLOGY, BOTANY. Educ: Univ Mo, BS, 31, MA, 33, PhD(bot), 37. Prof Exp: Asst bot, Univ Mo, 31-37, instr, 37-39, asst, 39-40; asst prof, Northwestern State Col, 40-41; res asst, Yale Univ, 41-42; instr, Conn Col, 42-43; res asst, Yale Univ, 43-46; tech asst, New York Bot

Garden, 45-48; from asst prof to prof biol, 48-71, EMER PROF BIOL, VANDERBILT UNIV, 71- Mem: Fel AAAS; Bot Soc Am; Am Soc Microbiol; Mycol Soc Am. Res: Vegetative reproduction; nutrition; vitamins; growth factors; antibiotics. Mailing Add: 406 Douglas Ave Fulton MO 65251

MCVEY, WILLIAM HENRY, b Falls City, Nebr, Mar 13, 21; m 48; c 2. CHEMISTRY. Educ: Univ Calif, BS, 43, PhD(chem), 48. Prof Exp: Asst, Manhattan Proj & US AEC, Univ Calif, 43-48; res chemist, Hanford Eng Works, Gen Elec Co, 48-50; res chemist, Calif Res & Develop Co, 50-54; asst sect chief, Nat Reactor Test Sta, Phillips Petrol Co, 54-57, proj leader, Combat Develop Exp Ctr, Calif, 57-58, chief plans & forecasts br, Div Opers Anal & Forecasting, US AEC, 58-66, Chem & Chem Separations Br, Div Reactor Develop & Technol, 66-68, chief, Fuel Recycle Br, 68-75; CHIEF, FUEL REPROCESSING BR, DIV NUCLEAR FUEL CYCLE & PROD, US ENERGY RES & DEVELOP ADMIN, 75- Res: Chemistry of plutonium; redox potentials; compounds and complexes; chemistry of zirconium; nuclear fuel processing; operations analysis; nuclear forecasts. Mailing Add: Div of Nuclear Fuel Cycle & Prod US Energy Res & Develop Admin Washington DC 20545

MCVICAR, GEORGE ARCHIBALD, biochemistry, see 12th edition

MCVICAR, JOHN WEST, b Rochester, NY, Oct 15, 28; m 56; c 4. VETERINARY MEDICINE, VIROLOGY. Educ: State Univ NY Vet Col, Cornell Univ, DVM, 52. Prof Exp: Partner, Loudoun Animal Hosp, 56-66; RES VET, AGR RES SERV, USDA, 66- Mem: US Animal Health Asn; Wildlife Dis Asn. Res: Transmission, early growth and pathogenesis of foot-and-mouth disease virus in cattle, sheep and goats; methods of virus detection and detection of infection. Mailing Add: Plum Island Animal Dis Lab USDA Vet Sci Res Div Agr Res Serv Greenport NY 11944

MCVICKAR, DAVID LANGSTON, b Lansdowne, Pa, Feb 24, 13; m 35; c 3. MEDICAL MICROBIOLOGY. Educ: Harvard Univ, AB, 35, MA, 37, PhD(biol), 40; Vanderbilt Univ, MD, 48. Prof Exp: Res fel surg mycol, Sch Med, Yale Univ, 40-41; instr med mycol, Sch Med & Dent, Univ Rochester, 42-46; from instr to asst prof bact, Sch Med, Vanderbilt Univ, 46-51; ASSOC PROF MED MICROBIOL & IMMUNOL, SCH MED, UNIV CALIF, LOS ANGELES, 51- Concurrent Pos: Attend physician, US Vet Admin Hosp, Nashville, Tenn, 48-51; consult, US Vet Admin Hosps, San Fernando, Calif, 51-56, Long Beach, 65- Mem: AAAS; Am Soc Microbiol; fel Am Acad Microbiol. Res: Physiology, epidemiology, immunology, pathology and treatment of mycotic infections of man and animals; immunology and allergy in general. Mailing Add: Ctr for the Health Sci Univ of Calif Sch of Med Los Angeles CA 90024

MCVICKAR, JOHN S, b May 12, 09; m 42; c 1. AGRONOMY. Educ: Univ Ill, AB, 35, MS, 39, PhD(agron), 42. Prof Exp: Instr & prin high schs, Ill, 31-38; asst soil surv, Ill, 39-42; prof, 46-74, EMER PROF AGR & CONSERV, WESTERN ILL UNIV, 74- Concurrent Pos: Head dept agr, Western Ill Univ, 46-69. Mem: Am Soc Agron. Res: Soil fertility; pasture and plant composition. Mailing Add: Dept of Agr Western Ill Univ Macomb IL 61455

MCVOY, KIRK WARREN, b Minneapolis, Minn, Feb 22, 28; m 53; c 3. THEORETICAL NUCLEAR PHYSICS. Educ: Carleton Col, BA, 50; Oxford Univ, BA, 52; Univ Göttingen, dipl, 53; Cornell Univ, PhD(physics), 56. Prof Exp: Res assoc physics, Brookhaven Nat Lab, 56-58; asst prof, Brandeis Univ, 58-62; from asst prof to assoc prof, 62-67, PROF PHYSICS, UNIV WIS-MADISON, 67- Concurrent Pos: Fulbright res grant, Univ Utrecht, 60-61; vis distinguished prof, Brooklyn Col, 70-71; vis prof, Ind Univ, 71-72. Mem: Fel Am Phys Soc. Res: Nuclear structure; scattering theory; elementary particles; nuclear reaction theory. Mailing Add: Dept of Physics Univ of Wis Madison WI 53706

MCWHAN, DENIS B, b New York, NY, Dec 10, 35; m 59; c 3. SOLID STATE PHYSICS. Educ: Yale Univ, BS, 57; Univ Calif, Berkeley, PhD(phys chem), 61. Prof Exp: Am-Scand Found Berquist fel, Royal Inst Technol, Sweden, 61-62; MEM TECH STAFF, PHYS RES LAB, BELL TEL LABS, 62- Mem: AAAS; Am Phys Soc; Am Crystallog Asn. Res: Metal-insulator transitions; magnetism and crystal structure at high pressure and low temperature. Mailing Add: Phys Res Lab Bell Tel Labs Murray Hill NJ 07974

MCWHINNEY, IAN RENWICK, b Burnley, Eng, Oct 11, 26; m 55; c 2. MEDICINE. Educ: Cambridge Univ, BA, 46, MB, ChB, 49, MD, 59. Prof Exp: Pvt pract, Eng, 55-68; PROF FAMILY MED, UNIV WESTERN ONT, 68- Concurrent Pos: Nuffield travelling fel, US, 64-65. Mem: Fel Royal Col Gen Practitioners; Col Family Physicians Can; Soc Teachers Family Med. Res: Diagnostic process; social and behavioral aspects of medical practice. Mailing Add: Dept of Family Med Univ of Western Ont London ON Can

MCWHINNIE, DOLORES J, b Elmhurst, Ill, Sept 13, 33. DEVELOPMENTAL BIOLOGY, COMPARATIVE ENDOCRINOLOGY. Educ: DePaul Univ, BS, 55, MS, 58; Marquette Univ, PhD(develop biol), 65. Prof Exp: Res assoc radiation biol, Argonne Nat Lab, 59-60; ASSOC PROF DEVELOP BIOL & ENDOCRINOL, DePAUL UNIV, 65- Mem: AAAS; Am Soc Zool; Am Inst Biol Sci. Res: Metabolic aspects of bone growth and differentiation; metabolism of amphibian bone; effects of hormones on mineral metabolism. Mailing Add: Dept of Biol Sci DePaul Univ 1036 W Belden Ave Chicago IL 60614

MCWHINNIE, MARY ALICE, b Chicago, Ill, Aug 10, 22. POLAR BIOLOGY, COMPARATIVE ANIMAL PHYSIOLOGY. Educ: DePaul Univ, BS, 44, MS, 46; Northwestern Univ, PhD, 52. Prof Exp: Asst biol sci, 44-50, instr, 50-52, from asst prof to assoc prof, 52-60, chmn dept, 64-68, PROF BIOL SCI, DePAUL UNIV, 60- Concurrent Pos: Lilly fel, Woods Hole Marine Biol Lab, 52; NSF grants, 58-66, Antarctic grants, 62-63, 65, 67, 70-73 & 74-. Mem: AAAS; fel Am Soc Physiol; Biophys Soc. Res: Crustacean metabolism, with special reference to carbohydrates during the molt cycle; metabolism of low temperature adaptation. Mailing Add: Dept of Biol Sci DePaul Univ 1036 W Belden Ave Chicago IL 60614

MCWHIRTER, NOLAN, b Texhoma, Tex, Apr 21, 14; m 39; c 2. GEOLOGY. Educ: Panhandle Agr & Mech Col, BS, 38 & 39; Colo State Univ, MA, 53, EdD(sci educ), 61. Prof Exp: Field geologist, Stovall Mus, Univ Okla, 37-38, mus asst, 38-39; field geologist, Statewide Paleont Surv, Bur Econ Geol, Univ Tex, 39-41; prof sci geol & sci educ, Panhandle Agr & Mech Col, 46-65, head dept sci, 64-65; asst prof, 65-72, head dept earth sci, 66-72, PROF EARTH SCI, E TEX STATE UNIV, 72- Concurrent Pos: Consult, Frontiers of Sci of Okal, 61-65 & Okla Acad Sci, 63-65. Mem: AAAS; Soc Vert Paleont; Nat Sci Teachers Asn; Nat Asn Geol Teachers. Res: General geology; vertebrate paleontology; science education. Mailing Add: Dept of Earth Sci ETex State Univ Commerce TX 75428

MCWHORTER, CHESTER GRAY, b Brandon, Miss, May 3, 27; m 52; c 2. PLANT PHYSIOLOGY, WEED SCIENCE. Educ: Miss State Univ, BS, 50, MS, 52; La State Univ, PhD(bot), 58. Prof Exp: Agronomist, Agr Exp Sta, Miss State Univ, 52-56; plant physiologist, 58-75, LAB CHIEF, SOUTHERN WEED SCI LAB, AGR RES SERV, USDA, 75- Concurrent Pos: Plant physiologist, Agr Exp Sta, Miss State Univ, 58-, adj assoc prof weed control, 73- Mem: AAAS; Am Soc Plant Physiol; Weed Sci Soc Am; Sigma Xi; NY Acad Sci. Res: General weed control; Johnson grass control; weed control in soybeans. Mailing Add: Southern Weed Sci Lab PO Box 225 Stoneville MS 38776

MCWHORTER, CLARENCE AUSTIN, b Wheatland, Wyo, July 19, 18; m 47; c 3. MEDICINE. Educ: Univ Nebr, BS & MD, 44. Prof Exp: From instr to asst prof, 49-59, PROF PATH, UNIV NEBR MED CTR, OMAHA, 59-, CHMN DEPT, 64- Mem: Col Am Path (pres, 69-71); Am Soc Clin Path; Int Acad Path. Res: Neoplasms. Mailing Add: Dept of Path Sch of Med Univ Nebr Med Ctr Omaha NE 68105

MCWHORTER, EARL JAMES, b Argyle, NY, Sept 12, 29. ORGANIC CHEMISTRY. Educ: Rensselaer Polytech, BS, 50; Cornell Univ, PhD(org chem), 55. Prof Exp: Instr chem, 54-56, asst prof, 56-59, ASSOC PROF CHEM, UNIV MASS, AMHERST, 69- Concurrent Pos: Res assoc, New York Bot Garden, 63-64. Mem: Am Chem Soc. Res: Polynuclear aromatic hydrocarbons; fungal polyacetylenes. Mailing Add: Dept of Chem Univ of Mass Amherst MA 01002

MCWILLIAMS, DONALD A, b Kansas City, Mo, Nov 10, 29; m 53; c 2. PHYSICS. Educ: Iowa State Univ, BS, 56, PhD(physics), 62. Prof Exp: Instr physics, Iowa State Univ, 61-62; sr tech specialist, Autonetics, NAm Aviation, 62-65; mem tech staff & asst sect head, TRW Systs Group, 65-68; mgr microelectronics, Lockheed Missiles & Space Co, 68-69; dir res & govt affairs, Calif State Col, Los Angeles, 69-71; PRES, PLYCON INDUSTS, 71- Concurrent Pos: Guest lectr, Aachen Tech Univ, 64. Mem: Am Phys Soc; Am Asn Physics Teachers. Res: Microelectronics research and development; radiation hardened solid state circuit development; optical properties of semiconductors; surface state phenomena. Mailing Add: Plycon Industs 1916 N Gilbert Fullerton CA 92633

MCWILLIAMS, EDWARD LACAZE, b Shreveport, La, May 22, 41; m 65; c 2. ORNAMENTAL HORTICULTURE. Educ: Univ Southwestern La, BS, 63; Iowa State Univ, MS, 65, PhD(hort plant ecol), 66. Prof Exp: Hort botanist, Bot Gardens & asst prof bot, Univ Mich, Ann Arbor, 67-72; ASSOC PROF HORT, TEX A&M UNIV, 72- Concurrent Pos: Horace H Rackham Grad Sch grant, Univ Mich, Ann Arbor, 68. Mem: AAAS; Am Soc Plant Taxon; Ecol Soc Am; Am Asn Bot Gardens & Arboretums; Am Soc Hort Sci. Res: Horticultural taxonomy; introduction and propagation of native ornamental plants; ecology and evolution of Amaranthus and Billbergia. Mailing Add: Dept of Hort Sci Tex A&M Univ College Station TX 77843

MCWILLIAMS, GERALD VERNON, b Ft Worth, Tex, Dec 13, 37; m 59; c 1. NUMERICAL ANALYSIS. Educ: Tex Tech Univ, BSEE, 60, MS, 65, PhD(math), 69. Prof Exp: Mem tech staff, Bell Tel Labs, 65-66; sr engr, LTV Electrosysts, Inc, 66-67; dept head, United Technol Labs, 69-70; asst prof math, Angelo State Univ, 70-74; SYSTS ANALYST, ELECTROSPACE SYSTS, INC, 74- Mem: Math Asn Am. Mailing Add: Electrospace Systs Inc 320 Richardson Ave Dallas TX 75215

MCWILLIAMS, KENNETH LEROY, b Grand Junction, Colo, Sept 13, 39; m 62. ENTOMOLOGY. Educ: Colo State Univ, BS, 62; Ind Univ, PhD(entom), 68. Prof Exp: ASSOC PROF BIOL, CALIF STATE UNIV, FULLERTON, 68- Mem: Soc Syst Zool; Entom Soc Am; Audio-Tutorial Cong. Res: Systematics, ecology, evolution and biogeography of aquatic beetles in the western hemisphere; audio-tutorial and innovative instruction. Mailing Add: Dept of Biol Calif State Univ Fullerton CA 92634

MCWILLIAMS, KENNETH RICHARD, b Oklahoma City, Okla, Mar 2, 38; m 61; c 1. PHYSICAL ANTHROPOLOGY. Educ: Univ Okla, BA, 64, MA, 69; Ariz State Univ, PhD(anthrop), 74. Prof Exp: Phys anthropologist, Fed Aviation Admin, 67-69; cur phys anthrop, Pueblo Grande Mus, 70-71; ASST PROF ANTHROP, WAKE FOREST UNIV, 74- Concurrent Pos: Consult, Okla State Med Examr, 67-69, Maricopa County, Ariz, Med Examr, 71-72 & NC State Med Examr, 73-75. Mem: Am Asn Phys Anthropologists; fel Am Acad Forensic Sci; Am Anthrop Asn; Soc Am Archeol; Am Asn Mus. Res: Improved or alternate methods for determining age, race, sex and individuation from the human skeleton; sources and movement of early American Indian populations. Mailing Add: Dept of Anthrop Wake Forest Univ Winston-Salem NC 27109

MCWILLIAMS, MARGARET ANN, b Osage, Iowa, May 26, 29; m 53; c 2. NUTRITION, FOOD. Educ: Iowa State Univ, BS, 51, MS, 53; Ore State Univ, PhD(food, nutrit), 68. Prof Exp: From asst prof to assoc prof food & nutrit, 61-68, PROF HOME ECON & CHMN DEPT, CALIF STATE UNIV, LOS ANGELES, 68- Concurrent Pos: Mem, Nat Coun Home Econ Adminr, 68-; mem nat coord comt, Col Teachers Food & Nutrit, 69-; mem nat steering comt plan local prof involvement, White House Conf Food, Nutrit & Health, 69- Mem: Inst Food Technol; Am Dietetic Asn; Am Home Econ Asn; Soc Nutrit Educ; Nutrit Today Soc. Res: Experimental foods; nutrition education; organic chemistry; psychology; consumerism and relation to food and nutrition. Mailing Add: Dept Home Econ Calif State Univ 5151 State University Dr Los Angeles CA 90032

MCWILLIAMS, RALPH DAVID, b Ft Meyers, Fla, Nov 5, 30; m 59; c 2. MATHEMATICAL ANALYSIS. Educ: Fla State Univ, BS, 51, MS, 53; Univ Tenn, PhD(math), 57. Prof Exp: Instr math, Univ Tenn, 56-57; instr, Princeton Univ, 57-59; from asst prof to assoc prof, 59-69, PROF MATH & ASSOC CHMN DEPT, FLA STATE UNIV, 69- Mem: AAAS; Am Math Soc; Math Asn Am. Res: Functional analysis; weak topologies in Banach spaces. Mailing Add: Dept of Math Fla State Univ Tallahassee FL 32306

MCWILLIAMS, ROBERT GENE, b Junction City, Ore, Dec 26, 39; m 66; c 2. STRATIGRAPHY. Educ: Stanford Univ, BS, 62; Univ Wash, MS, 65, PhD(geol), 68. Prof Exp: Consult micropaleontologist, Phillips Petrol Co, 66; asst prof geol, 68-73, ASSOC PROF GEOL, MIAMI UNIV, 73-; COORDR SCI & MATH, MIAMI UNIV-HAMILTON, 74- Concurrent Pos: Penrose grant, Geol Soc Am, 72. Mem: AAAS; Geol Soc Am; Am Asn Univ Prof. Res: Biostratigraphy and paleoecology of West Coast Tertiary Foraminifera and Mollusca; plate tectonics of Pacific Northwest United States. Mailing Add: Dept of Geol Miami Univ Oxford OH 45056

MCWRIGHT, CORNELIUS GLEN, b Sebree, Ky, Aug 3, 29; m 57; c 3. IMMUNOLOGY, MICROBIOLOGY. Educ: Univ Evansville, BA, 52; George Washington Univ, MS, 65, PhD(microbiol), 70. Prof Exp: Supvry spec agent forensic biol, 56-73, CHIEF BIOL SCI RES, FED BUR INVEST LAB, 73- Concurrent Pos: Assoc prof, 70-75, adj prof biol & forensic sci, George Washington Univ, 75-, res consult immunochem & forensic biol, Grad Sch, 72-73, immunochemist, Lab Virus & Cancer Res, Sch Med, 72-73; res consult physiol fluids, Law Enforcement Assistance Admin, 75- Mem: Fel Am Acad Forensic Sci; AAAS; Am Soc Microbiol. Res: Molecular biology and genetics of cell membranes and plasma proteins; immunochemistry; immunohematology; immunogenetics. Mailing Add: Fed Bur Invest Lab Washington DC 20535

MEACHAM, ROBERT COLEGROVE, b Moultrie, Ga, May 1, 20; m 43; c 4. MATHEMATICS. Educ: Southwestern Univ, Memphis, AB, 42; Brown Univ, ScM, 48, PhD(appl math), 49. Prof Exp: Instr math, Carnegie Inst Technol, 49-50, asst prof, 50-54; assoc prof, Univ Fla, 54-60; PROF MATH, ECKERD COL, 60- Concurrent Pos: Consult, RCA Serv Co, 58-64; NSF fel comput sci, Stanford Univ, 65-66; mem panel on res, Sch Math Study Group, 68-72. Mem: Am Math Soc; Math Asn Am; Soc Indust & Appl Math; Nat Coun Teachers Math. Res: Mechanics; numerical analysis; differential equations. Mailing Add: Dept of Math Eckerd Col PO Box 12560 St Petersburg FL 33733

MEACHAM, ROGER HENING, JR, b Richmond, Va, Sept 10, 42; m 65; c 2. PHARMACOLOGY. Educ: Univ Richmond, BS, 65, MS, 67; Med Col Va, PhD(pharmacol), 71. Prof Exp: SR SCIENTIST METAB CHEM, WYETH LABS, INC, 71- Mem: Sigma Xi. Res: Drug biotransformation. Mailing Add: Wyeth Labs Inc Metab Chem Sect PO Box 8299 Philadelphia PA 19101

MEACHAM, WILLIAM FELAND, b Washington, DC, Dec 12, 13; m 44; c 4. SURGERY. Educ: Western Ky State Col, BS, 36; Vanderbilt Univ, MD, 40; Am Bd Surg, dipl, 47; Am Bd Neurol Surg, dipl, 48. Prof Exp: Intern surg, Univ Hosp, 40-41, asst surg, Sch Med, 41-43, instr, 43-44, from asst prof to assoc prof clin surg, 47-53, assoc prof neurol surg, 53, CLIN PROF NEUROL SURG & HEAD DIV, SCH MED, VANDERBILT UNIV, 54- Concurrent Pos: Howe fel neurosurg, Sch Med, Vanderbilt Univ, 45-47; asst resident, Vanderbilt Univ Hosp, 41-43, resident surgeon, 43-44, asst vis surgeon, 44-, assoc vis surgeon, Outpatient Serv, 44-, neurosurgeon-in-chief; vol asst, Montreal Neurol Inst, 47-; asst prof, Meharry Med Col, 50; attend neurosurgeon & consult, hosps; chmn, Dept Neurol Surg. Mem: Neurosurg Soc Am (pres, 52); Am Asn Neurol Surg; Soc Univ Surg; AMA; Am Col Surg. Res: Intracranial tumors and aneurysms; steroatactic surgery. Mailing Add: Dept of Neurol Surg Vanderbilt Univ Hosp Nashville TN 37232

MEACHAM, WILLIAM ROSS, b Ft Worth, Tex, Jan 12, 23; m 50; c 3. VERTEBRATE ZOOLOGY. Educ: Agr & Mech Col, Tex, BS, 48; NTex State Col, MS, 50; Univ Tex, PhD(vert zool), 58. Prof Exp: From asst prof to assoc prof, 50-63, PROF BIOL & HEAD DEPT, 63- Res: Animal ecology; evolution; genetics; vertebrate population dynamics. Mailing Add: Dept of Biol Univ of Tex Arlington TX 76010

MEACHUM, ZACKEY DONNELL, JR, microbiology, see 12th edition

MEAD, ALBERT RAYMOND, b San Jose, Calif, July 17, 15; m; c 2. MALACOLOGY. Educ: Univ Calif, BS, 38; Cornell Univ, PhD(zool), 42. Prof Exp: With Marine Biol Lab, 41-42; instr, US Army Col, Brit WAfrica, 44-45; from instr to assoc prof, 46-52, head dept zool, 56-67, coordr marine sci prog, 67-70, cur invert, 67-71, PROF ZOOL, UNIV ARIZ, 52-, COORDR UNDERGRAD PROG, BIOL SCI, 70- Concurrent Pos: Res fel, Univ Calif, 46; res assoc, Pac Sci Bd, Nat Res Coun, 48 & 49 & NSF, 54; Pac Sci Coun observer, UNESCO Adv Comt Humid Tropics Res, 61; mem, Invert Consults Comt Pac, Pac Sci Bd, Nat Res Coun-Nat Acad Sci, 63-; guest scientist, Royal Mus Cent Africa, Tervuren, Belg, 74-75. Mem: Soc Invert Path; AAAS; Am Soc Zool; Am Malacol Union (pres, 63). Res: Giant African snail ecology, control; comparative genital anatomy and physiology of Gastropoda; speciation and taxonomy of Gastropoda; economic malacology; molluscan pathology. Mailing Add: Dept of Gen Biol Univ of Ariz Tucson AZ 85721

MEAD, CHARLES G, biochemistry, genetics, see 12th edition

MEAD, CHESTER ALDEN, b St Louis, Mo, Dec 9, 32. THEORETICAL CHEMISTRY. Educ: Carleton Col, BA, 54; Wash Univ, PhD, 57. Prof Exp: Res assoc chem, Brookhaven Nat Lab, 57-58; from asst prof to assoc prof phys chem, 58-66, PROF PHYS CHEM, UNIV MINN, MINNEAPOLIS, 66- Concurrent Pos: Consult, Brookhaven Nat Lab, 59-63. Mem: AAAS; Am Phys Soc; Am Chem Soc. Res: Quantum mechanics; quantum theory of dispersion and absorption line shapes; theory of excitons in solids; gravitation and quantum theory; algebraic techniques in theoretical chemistry. Mailing Add: Dept of Chem Univ of Minn Minneapolis MN 55455

MEAD, DARWIN JAMES, b Dowagiac, Mich, June 27, 10; m 36; c 3. CHEMISTRY. Educ: Kalamazoo Col, AB, 32; Brown Univ, ScM, 33, PhD(chem), 36. Prof Exp: Instr chem, Colby Col, 36-38; res chemist, Gen Elec Co, 38-46; ASSOC PROF PHYSICS, UNIV NOTRE DAME, 46- Mem: Am Phys Soc. Res: Conductance of electrolytes; properties of high polymers; dielectric properties of polymers. Mailing Add: 1101 Cleveland Ave South Bend IN 46628

MEAD, EDWARD JAIRUS, b Cleveland, Ohio, Oct 3, 28; m 59; c 2. INORGANIC CHEMISTRY. Educ: Va Mil Inst, BS, 49; Purdue Univ, MS, 52, PhD(inorg chem), 55. Prof Exp: Instr chem, Va Mil Inst, 49-50; res chemist, 54-63, res supvr, 64-65, prod develop mgr, 65-73, DIR LAB, PIGMENTS DEPT, EXP STA, E I DU PONT DE NEMOURS & CO, INC, 74- Mem: Am Chem Soc. Res: Substituted borohydrides; textile fibers; synthesis and properties of organic and inorganic pigments. Mailing Add: Pigments Dept Exp Sta E I du Pont de Nemours & Co Inc Wilmington DE 19898

MEAD, FRANK WALDRETH, b Columbus, Ohio, June 11, 22; m 45; c 2. ENTOMOLOGY. Educ: Ohio State Univ, BS, 47, MS, 49; NC State Univ, PhD(entom), 68. Prof Exp: Asst entom, Ohio State Univ, 48-49; scout Japanese beetle control proj, Bur Entom & Plant Quarantine, USDA, 48, biol aid, Div Forest Insect Invest, 50-53; entomologist, 53-71, TAXONOMIC ENTOMOLOGIST, DIV PLANT INDUST, FLA DEPT AGR & CONSUMER SERV, 71- Concurrent Pos: Res asst insect mus, NC State Univ, 58-60. Honors & Awards: Cert Appreciation for Serv Rendered in Field of Entom, Fla Entom Soc, 75. Mem: Entom Soc Am; Ecol Soc Am; Soc Syst Zool; Am Mosquito Control Asn. Res: Fulgoroidea, especially Oliarus and other Cixiidae; Auchenorhynchus Homoptera; Culicidae. Mailing Add: Div of Plant Indust PO Box 1269 Fla Dept of Agr & Consumer Serv Gainesville FL 32602

MEAD, GILBERT DUNBAR, b Madison, Wis, May 31, 30; m 51, 68; c 4. GEOPHYSICS. Educ: Yale Univ, BS, 52, MA, 53; Univ Calif, Berkeley, PhD(physics), 62. Prof Exp: Instr sci high sch, Calif, 53-55; res asst, Lawrence Radiation Lab, Univ Calif, 57-62; physicist lab theoret studies, 62-68, physicist lab space physics, 68-73, HEAD GEOPHYS BR, GODDARD SPACE FLIGHT CTR, NASA, 73- Concurrent Pos: Lectr dept space sci & appl physics, Cath Univ Am, 64-67. Mem: Fel Am Phys Soc; Am Geophys Union; Geol Soc Am; Seismol Soc Am; Soc Explor Geophysicists. Res: Experimental high energy physics; space and magnetospheric physics; planetary sciences; Jupiter's magnetosphere; geomagnetism; magnetospheric models; seismology; plate tectonics; earthquake prediction; geodesy. Mailing Add: Geophys Br Code 922 NASA Goddard Space Flighr Ctr Greenbelt MD 20771

MEAD, GILES WILLIS, b New York, NY, Feb 5, 28. ICHTHYOLOGY. Educ: Stanford Univ, AB, 49, AM, 52, PhD(biol), 53. Prof Exp: Fishery res biologist, US Fish & Wildlife Serv, 49-51, syst zoologist, 51-54, dir ichthyol lab, 56-60; cur fishes, Mus Comp Zool, Harvard Univ, 60-70; DIR, LOS ANGELES COUNTY MUS NATURAL HIST, 70- Concurrent Pos: Chmn, Calif Natural Areas Coord Coun, 70- Res: Systematics, distribution and ecology of oceanic fishes. Mailing Add: Los Angeles County Mus of Natural Hist 900 Exposition Blvd Los Angeles CA 90007

MEAD, JAMES FRANKLYN, b Evanston, Ill, Oct 24, 16; m 42; c 2. BIOCHEMISTRY. Educ: Princeton Univ, AB, 38; Calif Inst Technol, PhD(org chem), 42. Prof Exp: Asst, Calif Inst Technol, 42; from instr to asst prof, Occidental Col, 45-48; res coordr, Off Naval Res, Calif, 48; head synthetic br, Biochem Dept, Atomic Energy Proj, 48-50, res biochemist & chief biochem div, 50-69, assoc clin prof physiol chem, 51-56, prof, 56-63, prof biol chem & biophys, Med Sch, 63-69, PROF BIOL CHEM, MED SCH & ASSOC DIR LABS NUCLEAR MED & RADIATION BIOL, UNIV CALIF, LOS ANGELES, 69-, PROF PUB HEALTH, 73- Concurrent Pos: NIH career res award. Honors & Awards: E A Bailey Award, 71. Mem: Am Chem Soc; Am Soc Biol Chem; Am Oil Chem Soc. Res: Lipid, brain lipid and fatty acid metabolism; essential fatty acids. Mailing Add: Dept of Biol Chem Univ of Calif Sch of Med Los Angeles CA 90024

MEAD, JAYLEE MONTAGUE, b Clayton, NC, June 14, 29; m 68. ASTRONOMY. Educ: Univ NC, BA, 51; Stanford Univ, MA, 54; Georgetown Univ, PhD(astron), 70. Prof Exp: Eng asst math, Knolls Atomic Power Lab, Gen Elec Co, 51-52; teacher, Van Antwerp Sch, NY, 52-53; counsr & instr, Univ NC, 54-56; mathematician opers res off, Johns Hopkins Univ, 57-59; mathematician, 59-68, astronomerlab theoret studies, 68-71, ASTRONOMER LAB OPTICAL ASTRON, GODDARD SPACE FLIGHT CTR, NASA, 71- Mem: Am Astron Soc; Am Geophys Union. Res: Statistical astronomy; stellar dynamics; planetary atmospheres, planet Mars; star catalogues. Mailing Add: Lab for Optical Astron Code 671 NASA Goddard Space Flight Ctr Greenbelt MD 20771

MEAD, JOHN MARCUS, b Western Springs, Ill, Feb 13, 14; m 40; c 2. ORGANIC CHEMISTRY. Educ: Univ Ill, BS, 36; Ohio State Univ, MS, 39, PhD(chem), 40. Prof Exp: Asst plant supt, Columbia Powder Co, Wash, 40-42; explosives supt, US Rubber Co, NC, 42-45; res & develop chemist, Naugatuck Chem Co, Conn, 45-54, chief pilot plant engr, 54-64; prod supvr, Baton Rouge Plants, US Rubber Co, 64-67; asst tech supt, 67-68, mgr qual control, 68-72, SAFETY DIR, JOLIET ARMY AMMUNITION PLANT, UNIROYAL, INC, 72- Mem: Am Chem Soc. Res: Organic chemicals; diazonium salts; incendiaries and pyrotechnics; photochemical decomposition of diazonium salt solutions; aromatics; plastics; synthetic rubber. Mailing Add: 1304 Mason Ave Joliet IL 60435

MEAD, JUDSON, b Madison, Wis, Sept 16, 17; m 44; c 3. GEOPHYSICS. Educ: Mass Inst Technol, BS, 40, PhD(geophys), 49. Prof Exp: Proj supvr, Airborne Instruments Lab, 41-45; from asst prof to assoc prof, 49-60, dir geol field sta, 65-70, PROF GEOPHYS, IND UNIV, BLOOMINGTON, 60-, DIR GEOL FIELD STA, 74- Mem: Geol Soc Am; Am Geophys Union; Am Inst Mining, Metall & Petrol Eng. Res: Structure of the crust; exploration geophysics. Mailing Add: Dept of Geol Ind Univ Bloomington IN 47405

MEAD, MARGARET, b Philadelphia, Pa, Dec 16, 01; m 36; c 1. CULTURAL ANTHROPOLOGY. Educ: Columbia Univ, BA, 23, MA, 24, PhD(anthrop), 29. Hon Degrees: DSc, Wilson Col, 40, Elmira Col, 47, Western Col Women, 55, Univ Leeds, 57, Women's Med Col Pa, 63, Univ Cincinnati, 65, Harvard Univ, 73 & Univ Delhi, 73; LLD, Rutgers Univ, 41; LHD, Kalamazoo Col, 57 & Skidmore Col, 58; Dr, Univ Calif, Berkeley, 67. Prof Exp: From asst cur ethnol to cur ethnol, 26-69, dir studies in contemp cult, 51-53, EMER CUR, AM MUS NATURAL HIST, 69- Concurrent Pos: Soc Sci Res Coun fel, Admiralty Islands, 28-29; exec secy, Comt on Food Habits, Nat Res Coun, 42-45; dir, Wellesley Sch Community Affairs, 44; guest prof, Sch Gen Studies, Columbia Univ, 52-53, dir res contemp cult, 48-50, adj prof anthrop, 54-; vis prof anthrop, Dept Psychiat, Col Med, Univ Cincinnati, 57-; Sloan prof, Menninger Found, Topeka, Kans, 59; vis prof, Emory Univ, 64; vis prof anthrop, NY Univ, 65-67 & Yale Univ, 66-67; prof soc sci & chmn dept, Col Lib Arts, Fordham Univ, Lincoln Ctr, 68-70; Fogerty scholar in residence, NIH, Md, 73. Vis lectr child study, Vassar Col, 39-41; lectr child psychol, NY Univ, 40; Gimbel lectr psychol of sex, Stanford Univ & Univ Calif, 46; lectr, Harvard Sem Am Civilization, Salzburg, Austria, 47; lectr anthrop, Teachers Col, Columbia Univ, 47-51; Mason lectr, Birmingham, Eng, 49; Inglis lectr, Grad Sch Educ, Harvard Univ, 50; lectr, World Fedn Ment Health Int Sem Ment Health & Infant Develop, Chichester, Eng, 52; Terry lectr, Yale Univ, 57; Jones lectr, Brit Psychoanal Soc, 57; Alumni distinguished lectr, Univ RI, 70-71. Mem expeds, Samoa, 25-26; Manus tribe, Admiralty Islands, 29, 53, 64, 65, 67, 71, 75, Am Indian Tribe, 30, New Guinea, 31-33, 38, 53, 64, 65, 67, 71, 73 & 75, Bali, 36-38 & 57-58, Iatmul, New Guinea, 38, 67 & 71. Honors & Awards: Nat Achievement Award, 40; Women Geogr Medal, 42; Viking Medal Gen Anthrop, Wenner-Gren Found Anthrop Res, 57; William Proctor Prize for Sci Achievement, Sci Res Soc Am, 69; Arches of Sci Award, Pac Sci Ctr, 71; Kalinga Prize, UNESCO & Govt of India, 71; Wilder Penfield Award, Vanier Inst of the Family, 72; Woman of Conscience, Nat Coun Women, 75; Ceres Medal, Food & Agr Orgn UN, 75. Mem: Nat Acad Sci; AAAS (pres, 75); Am Acad Arts & Sci; Am Anthrop Asn (pres, 60); Soc Appl Anthrop (pres, 49). Res: Personality and culture; child development; application of psychoanalytic theory, learning theory, ethology, cybernetics in studies of seven oceanic cultures; native languages and applications to the fields of national character, mental health, education. Mailing Add: Am Mus of Natural Hist 15 W 77th St New York NY 10024

MEAD, MARSHALL WALTER, b Franklin, Ind, Jan 15, 21; m 47; c 3. ANALYTICAL CHEMISTRY. Educ: Franklin Col, AB, 42. Prof Exp: Anal chemist, Ala Ord Works, E I du Pont de Nemours & Co, 42-43; chief anal chemist, Oldbury Electrochem Co, 48-54, prod supt, Miss Works, 54-55; admin asst res & develop, Nat Aniline Div, Allied Chem Corp, 55-62; mgr local sect activ off, 62-69, asst dir mem activ div, 69-71, HEAD, MEM ACTIV DEPT, AM CHEM SOC, 71- Mem: Am Chem Soc (ed, Double Bond, 51-52). Res: Analytical chemistry of chlorates; phosphorus compounds; perchlorates; instrumental methods of analysis; research management. Mailing Add: 1507 Milestone Dr Silver Spring MD 20904

MEAD, RICHARD OLIVER, nuclear physics, see 12th edition

MEAD, ROBERT WARREN, b Yonkers, NY, Mar 3, 40; m 61; c 3. ANIMAL PARASITOLOGY, INVERTEBRATE ZOOLOGY. Educ: Colo State Univ, BS, 62, MS, 63, PhD(zool), 68. Prof Exp: Asst prof biol, Davis & Elkins Col, 65-67; NIH fel, Univ Mass, Amherst, 68-70; asst prof, 70-74, ASSOC PROF BIOL, SCH MED, UNIV NEV, RENO, 74-, CHMN DEPT, 76- Mem: AAAS; Am Soc Zool; Am Soc Parasitol; NY Acad Sci. Res: Physiology and nutrition of parasitic and free living invertebrates; host-parasite interactions; vertebrate difestive physiology. Mailing Add: Dept of Biol Univ of Nev Reno NV 89507

MEAD, RODNEY A, b Moline, Ill, Apr 28, 38; m 61; c 2. REPRODUCTIVE PHYSIOLOGY, ENDOCRINOLOGY. Educ: Univ Calif, Davis, AB, 60, MA, 62; Univ Mont, PhD(zool), 66. Prof Exp: USPHS fel steroid biochem, Col Med, Univ

Utah, 66-68; ASSOC PROF ZOOL, UNIV IDAHO, 68- Concurrent Pos: Mem biol reprod study sect, Inst Child Health & Human Develop, 75-79. Honors & Awards: A Brazier Howell Award, Am Soc Mammalogists, 66. Mem: AAAS; Soc Study Reproduction; Am Soc Mammal; Am Soc Zool. Res: Hormonal control of delayed implantation in mustelids. Mailing Add: Dept of Biol Sci Univ of Idaho Moscow ID 83843

MEAD, SYLVESTER WARREN, III, b New Brunswick, NJ, Jan 26, 23; m 54; c 3. PHYSICS. Educ: Univ Calif, BA, 48, PhD(physics), 57. Prof Exp: Assoc math, 49-50, PHYSICIST, LAWRENCE LIVERMORE LAB, UNIV CALIF, 57- Mem: Am Phys Soc. Res: Lasers; laser-produced plasmas; environmental sciences. Mailing Add: Lawrence Livermore Lab Univ of Calif PO Box 808 Livermore CA 94550

MEAD, THOMAS EDWARD, b Norwalk, Conn, Jan 9, 33; m 60; c 1. ORGANIC CHEMISTRY, MASS SPECTROMETRY. Educ: Ohio Wesleyan Univ, AB, 55; Brown Univ, MS, 58. Prof Exp: Chemist, 57-65, group leader mass spectrometry, 65-70, SR RES SCIENTIST, STAMFORD RES LABS, AM CYANAMID CO, 70- Mem: Am Chem Soc; Am Soc Mass Spectrometry. Res: Acid-base reactions in nonaqueous solvents; countercurrent distribution; gas-liquid chromatography. Mailing Add: Stamford Res Labs Am Cyanamid Co 1937 W Main St Stamford CT 06904

MEADE, ALSTON BANCROFT, b Jamaica, WI, June 28, 30; m 57; c 5. ENTOMOLOGY. Educ: Fisk Univ, BA, 56; Univ Minn, MS, 59, PhD(entom), 62. Prof Exp: Res biologist, Del, 64-71, SR RES BIOLOGIST, BIOCHEM DEPT, E I DU PONT DE NEMOURS & CO, INC, 71- Mem: Entom Soc Am; Royal Entom Soc London; Sigma Xi. Res: Resistance of plants to insect attack; ecology of Empoasca fabae and Macrosteles fascifrons; insecticidal controls of vegetable pests; development of new insecticides; insect attractants. Mailing Add: Box 166A RD 1 Glen Mills PA 19342

MEADE, DALE M, b Portage, Wis, Aug 7, 39; m 58; c 2. PLASMA PHYSICS. Educ: Univ Wis, BA, 61, MS, 62, PhD(physics), 65. Prof Exp: Res assoc physics, Univ Wis, 65-66; res assoc, Princeton Univ, 66-67; from asst prof to assoc prof, 67-72, PROF PHYSICS, UNIV WIS-MADISON, 72- Mem: Am Phys Soc. Res: Experimental studies of equilibrium and stability of plasma confined by magnetic fields with emphasis on applications in controlled thermonuclear fusion. Mailing Add: Dept of Physics Sterling Hall Univ of Wis Madison WI 58706

MEADE, GRAYSON EICHELBERGER, b Palacios, Tex, Apr 8, 12; m 37; c 4. GEOLOGY. Educ: Univ Nebr, AB, 35, MA, 37; Univ Chicago, PhD(geol), 46. Prof Exp: From instr to assoc prof geol, Tex Tech Col, 41-46; asst geologist, Bur Econ Geol, Tex, 44-45; geologist, Tex Mem Mus, 46-49; assoc prof, Tex Tech Col, 49-52; geologist, Union Oil Co, Calif, 52-58, chief geologist, Can Div, 58-61, STAFF GEOLOGIST, UNION OIL CO CAN, 61- Concurrent Pos: Sessional instr, Dept Archaeol, Univ Calgary, 68- Mem: Fel Geol Soc Am; Soc Vert Paleont; Am Asn Petrol Geol. Res: Cenozoic and petroleum geology; vertebrate paleontology; Devonian stratigraphy. Mailing Add: Agate Springs Ranch Harrison NE 69346

MEADE, JAMES HORACE, JR, b Vicksburg, Miss, Nov 1, 32; m 58; c 3. BIOMETRICS. Educ: Miss State Univ, BS, 54, MS, 59; Univ Fla, PhD(animal genetics), 61. Prof Exp: Fel biomath, NC State Col, 61-62, asst statistician, 62-63; from asst prof to assoc prof biostatist, Med Ctr, Univ Ala, 63-65, sr biostatistician, 63-65; assoc prof biomet, 65-69, PROF BIOMET & HEAD DIV, MED SCH, UNIV ARK, LITTLE ROCK, 69- Concurrent Pos: Consult, Vet Admin, 66- Mem: Biomet Soc; Am Statist Asn. Res: Applications of mathematics and statistics in medical research; use of computers in medicine; teaching statistics to biologists. Mailing Add: Div Biomet Univ of Ark Med Ctr 4301 W Markham Little Rock AR 72201

MEADE, JAMES MONTGOMERY, b Honolulu, Hawaii, Sept 23, 38; m 60; c 3. RESOURCE GEOGRAPHY, POPULATION GEOGRAPHY. Educ: Ore State Univ, BS, 65, MS, 67; Univ NC, Chapel Hill, PhD(geog), 71. Prof Exp: Planner, Linn Planning Comn, 67; res assoc, Mountain States Regional Med Prog, 67-68; asst prof geog, NC Cent Univ, 69-71; asst prof, Sir George Williams Univ, 71-72; dir res, Health Systs, Inc, 72-75; CHIEF HEALTH DATA SECT, DEPT HEALTH & WELFARE, 75- Concurrent Pos: Consult, Nat Ctr Health Statist, 75-76; corresp, Georgetown Univ Study, 75-76. Mem: Asn Am Geog; Am Geog Soc. Res: Provision of medical services to specific patient populations; health service area planning. Mailing Add: Dept of Health & Welfare 1520 W State Boise ID 83702

MEADE, JOHN ARTHUR, b Coldwater, Mich, Aug 29, 28; m 49; c 2. AGRONOMY. Educ: Univ Md, BS, 54, MS, 55; Iowa State Col, PhD(plant physiol), 58. Prof Exp: From asst prof to assoc prof agron, Univ Md, 58-66; EXTEN SPECIALIST, RUTGERS UNIV, NEW BRUNSWICK, 66- Mem: Weed Sci Soc Am. Res: Herbicides. Mailing Add: Dept of Soils & Crops Rutgers Univ New Brunswick NJ 08903

MEADE, LINDA CELIDA, b London, WVa, Aug 16, 44. BIOCHEMICAL GENETICS. Educ: WVa State Col, BS, 64; City Univ New York, MA, 69, PhD(biochem), 72. Prof Exp: ASST PROF BIOL SCI, STATE UNIV NY COL OLD WESTBURY, 72- Concurrent Pos: Vis asst prof, Rockefeller Univ, 73-74; consult, Dept Nutrit, Sch Pub Health, Harvard Univ, 73-; NIH res fel, Col Med, Univ Ariz, 75-; mem exec bd, Nat Orgn Prof Advan Black Chemists & Chem Engrs, 75- Mem: AAAS; assoc Sigma Xi; Asn Women in Sci; Am Chem Soc; Biophys Soc. Res: Biochemical mechanisms of regulation of gene expression, specifically cloning of bacteriophage genes in Eschericha coli and origin and direction of replication in T4 bacteriophage. Mailing Add: Dept of Biol Sci Col at Old Westbury Box 210 Old Westbury NY 11568

MEADE, MELINDA SUE, b New York, NY, Nov 2, 45. GEOGRAPHY. Educ: Hofstra Univ, BA, 66; Mich State Univ, MA, 70; Univ Hawaii, PhD(geog), 74. Prof Exp: Res officer geog, Univ Calif Int Ctr Med Res, Kuala Lumpur, Malaysia, 72-74; ASST PROF GEOG, UNIV CALIF, LOS ANGELES, 74- Mem: AAAS; Asn Am Geogrs; Am Pub Health Asn; Asian Studies Asn; Pop Asn Am. Res: The human ecology of diseases approached through population exposure to microenvironments resulting from spatial patterns of land use and mobility, and the alterations effected by developmental change. Mailing Add: Dept of Geog 1255 Bunche Hall Univ of Calif Los Angeles CA 90024

MEADE, REGINALD ESON, b Great Bend, Kans, Mar 1, 11; m 52; c 2. FOOD SCIENCE, FOOD TECHNOLOGY. Prof Exp: Engr process develop, Western Condensing Co, 33-42, vpres res & develop process & prod develop, 43-48, exec vpres & gen mgr, 49-52; independent eng consult process & prod develop, 53-62; engr & res assoc process eng & food process res, 63-72, SR RES ASSOC FOOD RES, PILLSBURY CO, 73- Concurrent Pos: Exec consult, Int Exec Serv Corps, Manila, Philippines, 71-72. Honors & Awards: Cert Appreciation, Nat Inst Sci & Technol, Repub Philippines, 72. Mem: Am Dairy Sci Asn; NZ Soc Dairy Sci & Technol; Inst Food Technol; Am Asn Cereal Chemists. Res: Thermal processes and water activity in food dehydration and evaporation processes; prevention and recovery of liquid and solid wastes in food processing; industrial fermentations; plant nutrition. Mailing Add: 1421 N First St Stillwater MN 55082

MEADE, ROBERT HEBER, JR, b Brooklyn, NY, Dec 27, 30; m 56; c 3. GEOLOGY. Educ: Univ Okla, BS, 52; Stanford Univ, MS, 57, PhD(geol), 60. Prof Exp: Geologist, Calif Co, 52 & 55-56; GEOLOGIST, US GEOL SURV, 57- Concurrent Pos: Assoc ed, J Geophys Res, 74-76; adj prof, State Univ NY, Stony Brook, 75- Mem: Soc Econ Paleont & Mineral; Int Asn Sedimentol; Am Geophys Union; Am Soc Civil Eng; fel Geol Soc Am. Res: Sedimentology; erosion, transport, deposition and compaction of sediments; river morphology; coastal hydrology and oceanography. Mailing Add: US Geol Surv Denver CO 80225

MEADE, ROBERT J, b Tecumseh, Nebr, Oct 8, 18; m 49; c 1. ANIMAL SCIENCE, NUTRITION. Educ: Univ Nebr, BSc, 49; Univ Ill, MSc, 52, PhD(nutrit), 55. Prof Exp: Asst animal sci, Univ Ill, 49-52; from asst prof to assoc prof animal sci, Univ Nebr, 51-56; assoc prof, 56-59, PROF ANIMAL SCI, UNIV MINN, ST PAUL, 59- Concurrent Pos: Travel fel animal nutrit, Nat Feed Ingredients Asn, 72. Honors & Awards: Nutrit Res Award, Am Feed Mfrs Asn, 72. Mem: AAAS; Am Soc Animal Sci; Am Inst Nutrit. Res: Influence of dietary protein and of supplemental amino acids on development of young swine, concentrations of plasma-free amino acids, carcass characteristics and composition; amino acid requirements of swine. Mailing Add: Dept of Animal Sci Inst of Agr Univ of Minn St Paul MN 55108

MEADE, THOMAS GERALD, b Pound, Va, Sept 3, 37. PARASITOLOGY, INVERTEBRATE ZOOLOGY. Educ: Whitman Col, BA, 59; Purdue Univ, MS, 62; Ore State Univ, PhD(zool), 64. Prof Exp: Grad coun fel, Ore State Univ, 64-65; from asst prof to assoc prof parasitol, 65-75, PROF PARASITOL, SAM HOUSTON STATE UNIV, 75- Mem: Am Soc Parasitol; Helminthol Soc Washington. Res: Helminth parasites of fishes; immuno-parasitology; larval trematode snail interaction. Mailing Add: Dept of Life Sci Sam Houston State Univ Huntsville TX 77340

MEADE, THOMAS LEROY, b Center Junction, Iowa, July 4, 20; m 42; c 2. ANIMAL NUTRITION. Educ: Univ Fla, BS, 50, MS, 51, PhD, 53. Prof Exp: Asst animal nutritionist, Univ Fla, 53-54; animal nutritionist, Chas Pfizer & Co, 54-55; vpres & dir res, Hayne Prod Inc, 55-63; dir res, J Howard Smith, Inc, 63-68; assoc prof, 68-74, PROF FISHERIES & MARINE TECHNOL, UNIV RI, 74- Res: Marine resource utilization, primary efforts in process and product development for industrial fisheries; aquaculture systems development, including nutrition and physiology of salmonoids. Mailing Add: Dept of Animal Sci Univ of RI Kingston RI 02881

MEADER, ARTHUR LLOYD, JR, b Clarksville, Tenn, Dec 13, 20; m 43; c 2. PETROLEUM CHEMISTRY. Educ: Univ Ky, BS, 41; Univ Wis, MS, 44, PhD(chem), 47. Prof Exp: From assoc res chemist to res chemist, Calif Res Corp Div, Standard Oil Co, Calif, 47-62, sr res chemist, Chevron Res Co, 62-67, SR RES ASSOC, CHEVRON RES CO, 67- Mem: Am Chem Soc. Res: Surface-active chemicals; plastics; fibers; surface coatings; elastomers; asphalt specialties. Mailing Add: Fuels & Asphalts Div Chevron Res Co 576 Standard Ave Richmond CA 94802

MEADER, RALPH GIBSON, b Eaton Rapids, Mich, Sept 6, 04; m 28; c 1. ANATOMY. Educ: Ohio Wesleyan Univ, AB, 25; Hamilton Col, AM, 27; Yale Univ, PhD(comp anat), 32. Prof Exp: From instr to asst prof biol, Hamilton Col, 25-28; instr, Wesleyan Univ, 28-29; from instr to assoc prof anat, Sch Med, Yale Univ, 31-48; chief cancer res grants br, Nat Cancer Inst, 48-53, chief res grants & fels br, 53-60, assoc dir grants & training, 60-70; DEP DIR RES ADMIN, MASS GEN HOSP, 70- Concurrent Pos: Biologist, State Biol Surv, NY, 29-30; mem corp, Bermuda Biol Sta, 32-; asst to dir bd sci advs, Jane Coffin Childs Mem Fund Med Res, 42-43, asst dir, 43-48; exec secy, Nat Adv Cancer Coun, 47- Mem: AAAS; Soc Develop Biol; Am Asn Anat; Am Asn Cancer Res. Res: Comparative anatomy of nervous system; neuroanatomy of teleosts; electrical characteristics of living organisms; history of medicine; sequence of nerve degeneration. Mailing Add: Res Admin Mass Gen Hosp Boston MA 02114

MEADER, ROLAND DARRELL, b Cambridge, Nebr, Feb 27, 19; m 48; c 1. ANATOMY, PATHOLOGY. Educ: Nebr State Teachers Col, BS, 41; Univ Colo, MA, 48; Univ Minn, PhD(anat, path), 56. Prof Exp: Asst zool, Univ Colo, 46-48; asst instr anat, Med Sch, Univ Minn, 52-54, instr, 54-56; instr, Sch Med, Tulane Univ, 56-58; asst prof, Sch Med, La State Univ, 58-63; PROF ANAT, COL MED, UNIV NEBR, OMAHA, 63- Mem: Electron Micros Soc Am; Am Asn Anat; Asn Am Med Cols. Res: Electron and light microscopic observations on relationships between lymphocytes and intestinal epithelium; studies of lymphocytes in hibernating animals during homothermic and heterothermic stages. Mailing Add: Dept of Anat Univ of Nebr Col of Med Omaha NE 68105

MEADOR, CLIFTON KIRKPATRICK, b Selma, Ala, Sept 7, 31; m 55; c 4. MEDICINE, ENDOCRINOLOGY. Educ: Vanderbilt Univ, BA, 52, MD, 55. Prof Exp: Instr med, Sch Med, Vanderbilt Univ, 60-61; from asst prof to prof, Sch Med, Univ Ala, Birmingham, 62-73, dean, 68-73, prog dir clin res ctr, Med Ctr, 63-73; PROF MED, SCH MED, VANDERBILT UNIV, 73-; CHIEF MED, ST THOMAS HOSP, 73- Concurrent Pos: USPHS fel, Sch Med, Vanderbilt Univ, 60-61; Markle scholar, 63- Mem: Endocrine Soc; Am Diabetes Asn; AMA. Res: Clinical patient care. Mailing Add: Dept of Med Vanderbilt Univ Sch of Med Nashville TN 37203

MEADOR, WILLIAM RALPH, organic chemistry, see 12th edition

MEADORS, VICTOR GERALD, b Princeton, Ky, Oct 24, 12; m 46. PHYSICAL CHEMISTRY. Educ: Univ Ill, BS, 35, PhD(phys chem), 39. Prof Exp: Res chemist, US Rubber Co, NJ, 39-40, Naugatuck Chem Co, Conn, 40-46, US Rubber Co, NJ, 46-47, Carter Oil Co, 47-58 & Jersey Prod Res Co, 58-65; asst prof, Muskingum Col, 65; res chemist, 65-69, assoc tech dir, 69-71, DIR ANAL SERV, OZARK MAHONING CO, 71- Mem: AAAS; Am Chem Soc. Res: Resin and rubber technology; surface chemistry; petroleum production; analytical methods; industrial waste disposal. Mailing Add: 3325 W 74th St S Tulsa OK 74132

MEADOWS, BRIAN T, b London, Eng, May 20, 40; m 63. HIGH ENERGY PHYSICS. Educ: Oxford Univ, BA, 62, MA, 67, PhD(physics), 67. Prof Exp: Sci Res Coun Gr Brit res fel high energy physics, Oxford Univ, 66-67; res assoc, Syracuse Univ, 67-68, vis asst prof, 68-69, asst prof, 69-72; ASSOC PROF HIGH ENERGY PHYSICS, UNIV CINCINNATI, 72- Res: Experimental high energy physics. Mailing Add: Dept of Physics Univ of Cincinnati Cincinnati OH 54221

MEADOWS, CHARLES MILTON, b Merryville, La, Nov 8, 12; m 41; c 2. ENTOMOLOGY. Educ: La State Norm Col, AB, 36; La State Univ, MS, 38; Ohio State Univ, PhD, 42. Prof Exp: Asst entom, Ohio State Univ, 38-42; in-charge cotton insect invest, 42-44; tech rep, Sherwin-Williams Co, 46-50; PRES & GEN MGR, SOUTHWEST SPRAYER & CHEM CO, 50- Concurrent Pos: Consult, Tenneco Oil Co, Tex, 64- Mem: AAAS; Nat Agr Chem Asn; Entom Soc Am. Res: Toxicity of insecticides; development and application of selective weed killers; spray machinery. Mailing Add: Southwest Sprayer & Chem Co 2632 Cedar Ridge Waco TX 76708

MEADOWS, CLINTON ELWOOD, b Ark, Jan 16, 13; m 42; c 2. ANIMAL BREEDING. Educ: Henderson State Col, BA, 35; Univ Ark, BS, 52; Iowa State Univ, MS, 55, PhD, 59. Prof Exp: PROF DAIRYING, MICH STATE UNIV, 57- Mem: Am Soc Animal Sci; Am Dairy Sci Asn. Res: Dairy cattle breeding. Mailing Add: Dept of Dairying Mich State Univ East Lansing MI 48823

MEADOWS, GEOFFREY WALSH, b Bury, Eng, Jan 16, 21; m 45; c 3. INDUSTRIAL CHEMISTRY. Educ: Univ Manchester, BSc, 42, MSc, 43, PhD(chem), 48. Prof Exp: Res chemist, Shell Co, Eng, 43-45; asst lectr, Univ Manchester, 45-49; Nat Res Coun Can fel, 49-51; res chemist, 51-65, res assoc, 65-66, RES SUPVR, E I DU PONT DE NEMOURS & CO, INC, 66- Mem: Am Chem Soc; Sigma Xi. Res: Ionic catalysed polymerization; physical properties of polymers; reaction kinetics; high temperature synthesis. Mailing Add: 312 Marshall St Kennett Square PA 19348

MEADOWS, JAMES LAWSON, organic chemistry, see 12th edition

MEADOWS, JAMES WALLACE, JR, b Meridian, La, Aug 16, 23; m 50; c 2. PHYSICAL CHEMISTRY. Educ: La Polytech Inst, BS, 44; La State Univ, MS, 48, PhD(chem), 50. Prof Exp: Anal chemist, Cities Serv Refining Corp, La, 44-46; asst, La State Univ, 46-50; tech asst, Harvard Univ, 50-58; assoc chemist, 58-70, CHEMIST, ARGONNE NAT LAB, 70- Mem: Am Chem Soc; Am Phys Soc; Am Nuclear Soc. Res: Nuclear reactions; neutron diffusion; fission; neutron physics. Mailing Add: Appl Physics Div Argonne Nat Lab Bldg 316 9700 S Cass Ave Argonne IL 60439

MEADOWS, JERRIANE KUJIE STAFFORD, b Staten Island, NY, Oct 9, 43; m 66. NUTRITION. Educ: Mich State Univ, BS, 65; Univ Ga, MS, 68; Univ NC, Greensboro, PhD(nutrit), 74. Prof Exp: Jr high teacher sci, Telfair County High Sch, McRae, Ga, 68-69; instr, 69-71, ASST PROF FOODS & NUTRIT, DIV HOME ECON, GA SOUTHERN COL, 74- Mem: Am Dietetic Asn; Am Home Econ Asn; Soc Nutrit Educ; Nutrit Today Soc; Inst Food Technologists. Res: Determination of palatability, tenderness, and vitamin retention of meat cooked in a selected oven film; food and nutrition education methods developed for single concept and self-study packets to be used in introductory foods and nutrition courses with laboratory experiences. Mailing Add: Box 8034 Div of Home Econ Ga Southern Col Statesboro GA 30458

MEADOWS, WILLIAM R, b Chicago, Ill, Feb 3, 19. INTERNAL MEDICINE, CARDIOLOGY. Educ: Northwestern Univ, BS, 41, MD, 44. Prof Exp: Intern, Cook County Hosp, 44, resident med, 47-48, 49-50; ward physician med, Livermore, Calif, 53-55, ward physician, Palo Alto, 55-60, asst chief cardiopulmonary lab, Hines, Ill, 60-72, chief cardiac catheterization lab, 72-74, CHIEF GRAPHICS SECT CARDIOL, VET ADMIN HOSP, HINES, 74- Concurrent Pos: Fel hemat, Cook County Hosp, Chicago, Ill, 50-51; fel cardiol, New Eng Deaconess Hosp, Boston, Mass, 59-60; asst clin prof, Stritch Sch Med, Loyola Univ Chicago, 61-62, asst prof, 62-65, assoc prof, 65- Mem: Fel Am Col Physicians; fel Am Col Cardiol. Mailing Add: Cardiol Sect Vet Admin Hosp Hines IL 60141

MEADS, MANSON, b Oakland, Calif, Mar 25, 18; m 45; c 1. MEDICINE. Educ: Univ Calif, AB, 39; Temple Univ, MD, 43. Hon Degrees: DSc, Temple Univ, 56. Prof Exp: Asst med, Thorndike Mem Lab, Harvard Med Sch, 44-46, asst bact & immunol, 46-47; instr med, 47-50, asst prof internal med, 51-56, from assoc prof to prof prev med, 51-57, dir dept, 51-57, assoc dean, 55-58, acad dean, 58-59, exec dean, 59-63, dean, 63-71, PROF INTERNAL MED, BOWMAN GRAY SCH MED, 57-, VPRES HEALTH AFFAIRS, 67-, DIR MED CTR, 74- Concurrent Pos: Ernst fel, Thorndike Mem Lab, Harvard Med Sch, 46-47; Markle scholar, 48-53; med officer, vis prof & adv, USPHS, Thailand, 53-55. Mem: Am Soc Clin Invest; AMA; fel Am Col Physicians. Res: Medical school administration. Mailing Add: Off of VPres Bowman Gray Sch of Med Winston-Salem NC 27103

MEADS, PHILIP F, b Oakland, Calif, Dec 4, 07; m 36; c 3. PHYSICAL CHEMISTRY. Educ: Univ Calif, Berkeley, BS, 28, PhD(chem), 32. Prof Exp: Chief chemist, Calif & Hawaiian Sugar Co, 53-59; tech dir, 59-72, ASST TO REFINERY MGR ENVIRON AFFAIRS, CALIF & HAWAIIAN SUGAR CO, CROCKETT, 72- Concurrent Pos: Partic cane sugar refiners res proj, US Nat Comt Sugar Anal. Honors & Awards: Hon Award, Sugar Indust Technicians, 69. Mem: Am Chem Soc; Inst Food Technol; Sugar Indust Technicians (pres, 71-72). Res: Refining of cane sugar; sugar analysis and products; sugar refinery wastewater treatment. Mailing Add: 133 Farallon Dr Vallejo CA 94590

MEADS, PHILIP FRANCIS, JR, b Oakland, Calif, May 19, 37; m 66; c 2. THEORETICAL PHYSICS, COMPUTER SCIENCE. Educ: Univ Calif, Berkeley, AB, 58, PhD(physics), 63. Prof Exp: Asst physics, Lawrence Radiation Lab, Univ Calif, 59-63; physicist, Midwest Univ Res Asn, Wis, 63-65; physicist, 65-69, SR PROJ ENGR, WILLIAM M BROBECK & ASSOCS, 69- Concurrent Pos: Consult, William M Brobeck & Assocs, 60, Argonne Nat Lab, Lawrence Radiation Lab, Univ Calif, 65- & Los Alamos Sci Lab, 70- Mem: AAAS; Health Physics Soc; Am Phys Soc; Audio Eng Soc. Res: Aberrations of quadrupole focusing magnets; accelerator design, particularly injection and extraction studies; optical design of beam transport systems; digital computer systems and subsystems, particularly microprocessors. Mailing Add: 7053 Shirly Dr Oakland CA 94611

MEAGHER, JAMES FRANCIS, b Sydney, NS, Oct 23, 46; m 73. PHYSICAL CHEMISTRY. Educ: St Francis Xavier Univ, BSc, 67; Cath Univ Am, PhD(phys chem), 71. Prof Exp: Res assoc chem kinetics, Univ Wash, 70-73; res asst atmospheric chem, Pa State Univ, 73-75; RES ANALYST ATMOSPHERIC CHEM, TENN VALLEY AUTHORITY, 76- Mem: Chem Inst Can; Sigma Xi. Res: Chemical kinetics and photochemistry; including the measurement and study of elementary reactions and photochemical processes important in the chemistry of polluted atmospheres. Mailing Add: Air Qual Studies Sect Tenn Valley Authority Muscle Shoals AL 35660

MEAGHER, RALPH ERNEST, b Chicago, Ill, Sept 22, 17; m 41; c 1. APPLIED PHYSICS, COMPUTER SCIENCE. Educ: Univ Chicago, BS, 38; Mass Inst Technol, MS, 39; Univ Ill, PhD(physics), 49. Prof Exp: Mem staff radiation lab, Mass Inst Technol, 41-45; res asst prof physics, Univ Ill, 48-50, res assoc prof, 50-51, res prof, 51-57, res prof physics & elec eng, 57-58; CONSULT, 58- Concurrent Pos: Head digital comput lab, Univ Ill, 57-58. Honors & Awards: President's Cert of Merit, 47. Mem: Fel Am Phys Soc; Asn Comput Mach; fel Inst Elec & Electronics Eng. Res: Radar and radar indicators; nuclear physics; electronics; proton-proton scattering; electronic computers. Mailing Add: PO Box 356 South Bend IN 46624

MEAGHER, RICHARD BRIAN, b Chicago, Ill, Sept 30, 47; m 68; c 1. MOLECULAR GENETICS, ENZYMOLOGY. Educ: Univ Ill, BS, 69; Yale Univ, MPhil, 71, PhD(biol), 73. Prof Exp: Am Cancer Soc fel biochem, Univ Calif, Berkeley, 73-74, lectr, 73-74; NIH RES FEL BIOCHEM & MICROBIOL, UNIV CALIF, SAN FRANCISCO, 74- Mem: Am Chem Soc. Res: Evolution of biochemical pathways and their regulation; the molecular cloning and expression of Cauliflower Mosaic Virus in E coli; techniques of genetic engineering applied to the developing of new plant phenotypes. Mailing Add: Dept of Biochem & Biophys Univ of Calif San Francisco CA 94143

MEAKIN, JAMES WILLIAM, b Smith Falls, Ont, May 28, 29; m 53; c 2. INTERNAL MEDICINE. Educ: Queen's Univ, Ont, MD, CM, 53; Univ Toronto, MA, 57; FRCP(C), 60. Prof Exp: Fel med, Harvard Univ & asst med, Peter Bent Brigham Hosp, 57-59; clin teacher, 60-65, assoc, 65-68, ASSOC PROF MED, FAC MED, UNIV TORONTO, 68-; PHYSICIAN, ONT CANCER INST & PRINCESS MARGARET HOSP, 60- Concurrent Pos: Am Col Physicians fel, Harvard Univ, 57-58, Life Ins Med Res Fund fel, 58-59. Honors & Awards: Starr Medal, 57. Mem: Can Med Asn; Can Biochem Soc; Can Soc Clin Invest; Endocrine Soc. Res: Effect of steroid hormones on cells, particularly tumour cell growth in animals and man. Mailing Add: Ont Cancer Inst 500 Sherbourne St Toronto ON Can

MEAKIN, PAUL, b Burton-on-Trent Staffs, Eng, Mar 29, 44. PHYSICAL CHEMISTRY. Educ: Manchester Univ, BSc, 65; Univ Calif, Santa Barbara, PhD(chem), 69. Prof Exp: CHEMIST, E I DU PONT DE NEMOURS & CO, INC, 69- Mem: Am Chem Soc. Res: Magnetic resonance, nuclear magnetic resonance and electron spin resonance; chemical dynamics of transition metal complexes; solid state chemistry and physics, solid electrolytes; atmospheric chemistry. Mailing Add: Cent Res & Develop Dept Exp Sta E I du Pont de Nemours & Co Inc Wilmington DE 19898

MEAL, HARLAN C, b Rush Co, Ind, Jan 31, 25; m 53; c 3. PHYSICAL CHEMISTRY. Educ: Harvard Univ, AB, 50, MA, 53, PhD(phys chem), 54. Prof Exp: Opers analyst oper res off, Johns Hopkins Univ, 53-57; opers analyst, Dunlap & Assocs, Inc, 57-59; opers analyst, 59-72, HEAD LOGISTICS UNIT, ARTHUR D LITTLE, INC, 72- Mem: Opers Res Soc Am. Res: Industrial operations research; industrial logistics; management systems for industrial operations; decision rules, control systems, organizational structures. Mailing Add: Arthur D Little Inc 35 Acorn Park Cambridge MA 02140

MEAL, LARIE L, b Cincinnati, Ohio, June 15, 39. PHYSICAL CHEMISTRY, ANALYTICAL CHEMISTRY. Educ: Univ Cincinnati, BS, 61, PhD(phys chem), 66. Prof Exp: Res chemist, US Indust Chem Co, Nat Distillers & Chem Corp, 66-67; instr chem, Ohio Col Appl Sci, 68-69; asst prof chem technol, 69-75, ASSOC PROF CHEM TECHNOL, UNIV CINCINNATI, 75- Concurrent Pos: Consult & chem analyst, Cincinnati Fire Div, 74- Mem: AAAS; Am Chem Soc; NY Acad Sci. Res: Ethylenediaminetetraacetic acid titrations; chemical analysis of arson debris. Mailing Add: 2231 Slane Ave Norwood OH 45212

MEALEY, EDWARD H, b Boston, Mass, July 28, 25; m 51; c 3. BIOCHEMISTRY. Educ: Tufts Col, BS, 48; Univ Kans, PhD(biochem), 60. Prof Exp: Supvr chem test unit lab blood & blood prod, Div Biologics Standards, NIH, 60-63, chief blood & blood derivatives sect, 64-67; dir qual assurance, Hyland Labs, 67-70; VPRES & TECH DIR, INT CLIN LAB INDUSTS, 70- Mem: AAAS; NY Acad Sci. Res: Physical and chemical studies on whole blood; plasma and plasma protein solutions. Mailing Add: 17646 Fremont St Fountain Valley CA 92708

MEALEY, JOHN, JR, b Providence, RI, Aug 30, 28; m 52; c 3. NEUROSURGERY. Educ: Brown Univ, BA, 49; Johns Hopkins Univ, MD, 52; Am Bd Neurol Surg, dipl, 62. Prof Exp: Intern surg, Johns Hopkins Hosp, 52-53; clin & res fel neurosurg, Harvard Sch Med, 55-56; from asst resident to resident, Mass Gen Hosp, 56-60; from instr to assoc prof surg, 60-68, PROF SURG, SCH MED, IND UNIV, INDIANAPOLIS, 69- Mem: AMA; Am Asn Neurol Surg; Cong Neurol Surg; Am Col Surg. Res: Brain tumors; radioactive methods; chemotherapy; head injuries. Mailing Add: 1100 W Michigan St Indianapolis IN 46202

MEANS, ANTHONY R, b Bartlesville, Okla, May 7, 41; m 61. ENDOCRINOLOGY. Educ: Okla State Univ, BA, 63, MS, 64; Univ Tex, Austin, PhD(physiol), 67. Prof Exp: Res assoc molecular biol, Southwest Found Res & Educ, 68-69; asst prof obstet & gynec, Med Sch, Vanderbilt Univ, 69-72, asst prof physiol & asst dir, Ctr Pop Res, 71-72; assoc prof, 73-75; PROF CELL BIOL & ASSOC DIR, CTR POP RES, BAYLOR COL MED, 75- Concurrent Pos: Res fel, Australian Res Coun, Russell Grimwade Sch Biochem, Univ Melbourne, 67-68. Mem: Soc Study Reproduction; Am Soc Cell Biol; Endocrine Soc; Am Soc Biol Chem. Res: Biochemical mechanism of action of estrogen upon oviduct and liver and of follicle-stimulating hormone upon testis. Mailing Add: Dept of Cell Biol Baylor Col of Med Houston TX 77025

MEANS, BRUCE, b Los Angeles, Calif, Mar 9, 41; m 60; c 2. ECOLOGY, HERPETOLOGY. Educ: Fla State Univ, BS, 68, MS, 72, PhD(ecol), 75. Prof Exp: Teaching asst, Fla State Univ, 68-70; Gerald Beadel res scholar, 70-75; ASST DIR, TALL TIMBERS RES STA, 75- Concurrent Pos: Adj asst prof, Fla State Univ, 76- Mem: AAAS; Ecol Soc Am; Soc Study Evolution; Am Soc Naturalists; Am Soc Ichthyologists & Herpetologists. Res: Population biology; ecology of reproduction and life history phenomena; evolution at the species and population level; geographical ecology. Mailing Add: Tall Timbers Res Sta Rte 1 Box 160 Tallahasse FL 32303

MEANS, CRAIG RAY, b Shreveport, La, Aug 16, 22; div. PROSTHODONTICS. Educ: Southern Univ, BS, 50; Howard Univ, DDS, 54; Ohio State Univ, MSc, 63. Prof Exp: Asst prof prosthodont, 61-62 & 64-66, supvr dent technicians, 64-67, assoc prof & chief div removable partial & complete dentures, 66-68, actg chmn dept removable prosthodont, 68-69, assoc prof, 69-70, PROF REMOVABLE PROSTHODONT, COL DENT, HOWARD UNIV, 70-, CHMN DEPT, 69-, ASSOC DEAN UNDERGRAD AFFAIRS, 70- Mem: Am Dent Asn; Nat Dent Asn; Am Prosthodont Soc. Res: Damage to the oral tissues resulting from the use of home reline denture materials; temporomandibular joint function in complete denture patients. Mailing Add: Off Assoc Dean Undergrad Affairs Howard Univ Col of Dent Washington DC 20059

MEANS, GARY EDWARD, b Wykoff, Minn, Aug 31, 40. BIOCHEMISTRY. Educ: San Jose State Col, BS, 64; Univ Calif, Davis, PhD(biochem), 68. Prof Exp: USPHS fel, Virus Lab, Univ Calif, Berkeley, 68-70; instr biochem, 71-73, ASST PROF BIOCHEM, OHIO STATE UNIV, 74- Mem: Am Chem Soc. Res: Protein chemistry; structure-function relationships of proteins, enzyme mechanisms. Mailing Add: Dept of Biochem Ohio State Univ 484 W 12th Ave Columbus OH 43210

MEANS, JOHN AUGUST, analytical chemistry, see 12th edition

MEANS, LYNN L, b Kansas City, Mo, Jan 26, 14; m 36; c 4. METEOROLOGY, PHYSICAL SCIENCE. Educ: Univ Chicago, BS, 42, MS, 44. Prof Exp: Instr & res assoc, Univ Chicago, 42-45; forecaster, Chicago Forecast Ctr, US Weather Bur, 45-47, res forecaster, 47-55, leading analyst, Nat Weather Anal Ctr, 55-59, chief pub & agr forecast sect, 59-65, dep to dir user affairs, Environ Sci Serv Admin, 65-68, sr prog analyst, 68-74, CONSULT & GEN PHYS SCIENTIST, NAT OCEANIC & ATMOSPHERIC ADMIN, 75- Concurrent Pos: Consult air weather serv, US Army Air Force, 44-46. Mem: Am Meteorol Soc. Res: Applied meteorology; forecast improvement; economic benefits of science services. Mailing Add: 4901 Stan Haven Rd Camp Springs MD 20031

MEANS, THOMAS MARION, animal physiology, see 12th edition

MEANY, JOHN EAGLETON, b Brooklyn, NY, Sept 28, 37; m 59; c 2. CHEMISTRY. Educ: Seattle Univ, BS, 62; Univ Wash, PhD, 66. Prof Exp: Anal chemist, Wash State Horse Racing Comn, 60-62; res assoc acid-base catalysis, Univ Wash, 66-67; asst prof chem, Loyola Univ, 67-68; asst prof, 68-70, ASSOC PROF CHEM, CENT WASH STATE COL, 70- Concurrent Pos: NIH fel, 66-67. Mem: Am Chem Soc. Res: Catalysis of organic reactions. Mailing Add: Dept of Chem Cent Wash State Col Ellensburg WA 98926

MEARNS, ALAN JOHN, b Los Angeles, Calif, Oct 4, 43; m. FISHERIES, POLLUTION BIOLOGY. Educ: Calif State Univ, Long Beach, BS, 65, MA, 68; Univ Wash, PhD(fisheries), 71. Prof Exp: Biologist, Allan Hancock Found, Univ Southern Calif, Los Angeles/Arctic Res Lab, Barrow, Alaska, 65-66; res assoc, Fisheries Res Inst, Univ Wash, 68-70; consult physiol, Auke Bay Lab, Nat Marine Fisheries Serv, 70; sr environ scientist pollution biol, 71-73, DIR BIOL DIV, SOUTHERN CALIF COASTAL WATER RES PROJ, 73- Concurrent Pos: Environ Protection Agency grant, Corvallis, Ore, 72-75; consult, Calif State Water Resources Conserv Bd, 75- Mem: Am Inst Fishery Res Biologists; Am Fisheries Soc; Sigma Xi. Res: Directing research on coastal and marine pollution problems, including virus and bacteria survival, fish pathology, fish physiology, trace metal and hydrocarbon metabolism, fish and invertebrate systematics, benthic ecology and biological oceanography. Mailing Add: Southern Calif Coastal Water Res 1500 E Imperial Hwy El Segundo CA 90245

MEARS, BRAINERD, JR, b Williamstown, Mass, June 24, 21; m 48; c 4. GEOMORPHOLOGY. Educ: Williams Col, AB, 43; Columbia Univ, PhD(geol), 50. Prof Exp: Lectr geomorphol, Columbia Univ, 47-49; from asst prof to assoc prof, 49-63, PROF GEOL, UNIV WYO, 63- Mem: AAAS; Geol Soc Am; Am Geophys Union. Res: Pleistocene geology. Mailing Add: Dept of Geol Univ of Wyo Laramie WY 82070

MEARS, DANA CHRISTOPHER, b Pittsburgh, Pa, Sept 4, 40; m 64; c 2. CHEMICAL METALLURGY, MEDICINE. Educ: Cornell Univ, BA, 62, Cambridge Univ, PhD(metall), 65; Oxford Univ, MD, 69; MRCP, UK, 72. Prof Exp: Res asst metall, Cent Res Lab, Broken Hill Proprietary Co, Ltd, 62; res worker, Cambridge Univ, 62-66; res fel physiol, Nuffield Dept Orthop Surg, Oxford Univ, 66-70, house physician, Radcliffe Infirmary, 70; intern surg, Univ Pittsburgh, 70-71; sr house officer rheumatol, Nuffield Orthop Ctr, Oxford Univ, 71-72, registr in metab med, Nuffield Dept Orthop Surg, 72; resident orthop surg, Children's Hosp Pittsburgh, 72-75; ASST PROF ORTHOP SURG, UNIV PITTSBURGH, 75- Concurrent Pos: Mem, Brit Standards Comt Surg Implants, 64- Mem: Am Soc Metals; Brit Corrosion & Protection Asn; Brit Corrosion Sci Soc. Res: Corrosion and passivity; selection of materials for surgical implants; bone physiology, the relationship between mechanical stress on bone and bone cell metabolism. Mailing Add: Children's Hosp of Pittsburgh De Sota St Pittsburgh PA 15213

MEARS, FLORENCE MARIE, b Baltimore, Md, May 18, 96. MATHEMATICS. Educ: Goucher Col, AB, 17; Cornell Univ, AM, 24, PhD(math), 27. Prof Exp: Head dept math, Ala Col, 27-28; actg asst prof, Pa State Col, 28-29; asst prof to prof, 29-66, EMER PROF, GEORGE WASHINGTON UNIV, 66-; PROF LECTR, HOWARD UNIV, 66- Mem: Am Math Soc; Math Asn Am. Res: Transformations in infinite series.

MEARS, JAMES AUSTIN, b Baytown, Tex, May 18, 44; m 66; c 2. BIOCHEMISTRY, SYSTEMATICS. Educ: Univ Tex, Austin, BA, 66, PhD(biol), 70. Prof Exp: ASST CUR DEPT BOT, ACAD NATURAL SCI PHILADELPHIA, 70- Concurrent Pos: Adj asst prof, Univ Pa, 70-; NSF grants, Smithsonian Inst, 71 & Acad Natural Sci, 71-72; res assoc, Morris Arboretum, Philadelphia, 71-; Am Philos Soc grants, 75. Mem: Am Soc Plant Taxon; Phytochem Soc NAm; Am Inst Biol Sci; Sigma Xi. Res: Studies of the processes of organism and molecule evolution, primarily through analyses of morphological and biochemical characteristics. Mailing Add: Dept of Bot Acad Natural Sci 19th & Pkwy Philadelphia PA 19103

MEARS, ROBERT BRUCE, b Scranton, Pa, Jan 28, 07; m 29; c 2. METALLURGICAL CHEMISTRY. Educ: Pa State Univ, BS, 28; Cambridge Univ, PhD(metall), 35. Prof Exp: Sect leader chem div, Bell Tel Labs, 28-32; instr, Cambridge Univ, 32-33, res grant, Brit Non-Ferrous Metals Res Inst, 34; from res chemist in-chg to chief chem metall div, Res Labs, Aluminum Co, Am, 35-46; dir appl res lab, US Steel Corp, 46-60, asst vpres appl res, 60-64, vpres new prod develop, 64-72. Concurrent Pos: Instr, Carnegie Inst Technol, 37-40 & Univ Pittsburgh, 40-43; consult, Manhattan Dist, 43-45; vol exec, Cent Found Technol, Minas Gerais, Belo Horizonte, MG, Brazil, 74. Honors & Awards: Whitney Award, Nat Asn Corrosion Eng, 47. Mem: Electrochem Soc (vpres, 39-42); Am Soc Metals; Nat Asn Corrosion Eng (pres, 46); Am Inst Mining, Metall & Petrol Engrs; Brit Iron & Steel Inst. Res: Electroplating; electrochemistry of corrosion; mechanisms of stress corrosion; cathodic protection. Mailing Add: 628 California Ave Oakmont PA 15139

MEARS, THOMAS WOOD, b Washington, DC, May 6, 18; m 44; c 2. ORGANIC CHEMISTRY. Educ: Univ Md, BS, 39. Prof Exp: Asst, 39-40, jr chemist, 40-42, asst chemist, 42-44, chemist, 44-64, CHIEF ORG STAND, NAT BUR STAND, 64- Honors & Awards: Silver Medal, Dept Com, 63. Mem: Am Soc Testing & Mat; Am Chem Soc. Res: Synthesis of hydrocarbons and organometallic compounds; analysis of hydrocarbon mixtures; determination of hydrocarbon structures; compounds of light elements; new standard reference materials. Mailing Add: Nat Bur of Stand Washington DC 20234

MEARS, WHITNEY HARRIS, b Williamstown, Mass, June 8, 12; m; c 2. CHEMISTRY. Educ: Williams Col, BA, 33; Harvard Univ, AM, 35, PhD(phys chem), 37. Prof Exp: Rockefeller grant, Mt Sinai Hosp, New York, 37-38; Am Philos Soc grant, Univ Pa, 38-39; res chemist, Interchem Co, NY, 40; res chemist, Specialty Chem Div, 41-43 & 46-70, RES ASSOC, ALLIED CHEM CORP, 70- Mem: Am Chem Soc; Am Soc Heat, Refrig & Air Conditioning Eng. Res: Thermodynamic properties of fluorine compounds; gaseous dielectrics, mixtures. Mailing Add: 139 Darwin Dr Amherst NY 14226

MEASURES, RAYMOND MASSEY, b London, Eng, Feb 17, 38; m 62; c 2. LASER PHYSICS. Educ: Univ London, BSc, 60, Imp Col, dipl, PhD(physics), 64. Prof Exp: Asst prof, 64-69, ASSOC PROF PHYSICS, INST AEROSPACE STUDIES, UNIV TORONTO, 69- Res: Laser environmental sensing; application of laser mapping of air and water pollution; laser studies of plasmas; development of special facility for evaluation of new form of environmental signature, fluorescence decay spectrum. Mailing Add: Inst Aerospace Studies Univ Toronto 4925 Dufferin St Downsview ON Can

MEATH, WILLIAM JOHN, b Toronto, Ont, Apr 8, 36; m 60; c 2. THEORETICAL CHEMISTRY. Educ: Carleton Univ, BSc, 60; Univ Wis, PhD(chem), 65. Prof Exp: Proj assoc theoret chem inst, Univ Wis, 65; from asst prof to assoc prof chem, 65-71, PROF CHEM, UNIV WESTERN ONT, 71- Concurrent Pos: Vis asst prof, Theoret Chem Inst, Univ Wis, 66. Mem: Am Phys Soc; fel Chem Inst Can. Res: Atomic and molecular quantum mechanics; intermolecular forces; stationary state and time dependent atomic and molecular properties. Mailing Add: Dept of Chem Univ of WEstern Ont London ON Can

MEBS, RUSSELL WILLIAM, alloy physics, see 12th edition

MEBUS, CHARLES ALBERT, b Paterson, NJ, Sept 10, 32; m 55; c 3. VETERINARY PATHOLOGY. Educ: Cornell Univ, DVM, 56; Kans State Univ, MS, 62, PhD(vet path), 63. Prof Exp: Pvt pract, Del, 58-60; assoc prof vet path, Kans State Univ, 63-65; PROF VET SCI, UNIV NEBR, LINCOLN, 65- Mem: Am Vet Med Asn; Am Col Vet Path. Res: Viral animal diseases. Mailing Add: Dept of Vet Sci Univ of Nebr Lincoln NE 68503

MECAY, WILLIAM LLOYD, b Scotia, Nebr, Jan 1, 16; m 45; c 1. INORGANIC CHEMISTRY. Educ: Case Western Reserve Univ, BS, 51, MS, 53, PhD(inorg chem), 54. Prof Exp: Asst prof, 54-64, ASSOC PROF CHEM & PHYSICS, TEX WOMAN'S UNIV, 64- Concurrent Pos: Res partic, Oak Ridge Inst Nuclear Studies, 61. Mem: AAAS; Am Chem Soc; fel Am Inst Chem. Res: Boron hydrides, alkyls, esters, diamines; reactions of titanium tetrachloride; radioisotopes; neutron activation analysis. Mailing Add: Dept of Chem Tex Women's Univ Denton TX 76204

MECCA, CHRISTYNA EMMA, b Brooklyn, NY, Oct 23, 36. BIOLOGY, BIOCHEMISTRY. Educ: George Washington Univ, BS, 60, MS, 63, PhD(biol), 69. Prof Exp: Med biol technician, NIH, 58-60, biologist, 60-62, chemist, 62-65; instr gen biol, Montgomery Col, 68-69; res biologist, Bur Radiol Health, USPHS, 69-70; staff biologist, Coastal Plains Ctr Marine Develop Serv, Washington, DC, 70-72; prog analyst, NIH, 72-74; ANALYST-BIOLOGIST, SMITHSONIAN INST SCI INFO EXCHANGE, 75- Mem: AAAS; Am Inst Biol Sci. Res: Qualitative biochemical characteristics of cellular and organismic systems. Mailing Add: 297 John's Circle Deale MD 20751

MECCA, STEPHEN JOSEPH, b New York, NY, Jan 15, 43; m 64; c 3. NUCLEAR PHYSICS. Educ: Providence Col, BS, 64, MS, 66; Rensselaer Polytech Inst, PhD(physics), 69. Prof Exp: ASSOC PROF PHYSICS, PROVIDENCE COL, 69- Mem: Am Phys Soc; Am Inst Physics; Am Asn Physics Teachers. Res: Nuclear physics; especially nuclear spectroscopy and photonuclear physics; systems approach to complex problem solving; systems analysis, systems science and engineering. Mailing Add: Dept of Physics Providence Col Providence RI 02918

MECH, LUCYAN DAVID, b Auburn, NY, Jan 18, 37; m 58; c 4. WILDLIFE ECOLOGY. Educ: Cornell Univ, BS, 58; Purdue Univ, PhD(vert ecol), 62. Prof Exp: NIH fel animal movements & telemetry, Univ Minn, Minneapolis, 63-64, res assoc, 64-66; res assoc biol, Macalester Col, 66-69; WILDLIFE RES BIOLOGIST, US FISH & WILDLIFE SERV, 69- Honors & Awards: Award, Wildlife Soc, 72. Mem: Am Soc Mammal; Ecol Soc Am; Animal Behav Soc; Wildlife Soc. Res: Predator-prey relations; mammal behavior and natural history; animal movements and factors affecting them; ecology, behavior and sociology of wolves; spatial organization of mammals; telemetry and radio-tracking. Mailing Add: US Fish & Wildlife Serv NCent Forest Exp Sta Folwell Ave St Paul MN 55101

MECH, WILLIAM PAUL, b La Crosse, Wis, Mar 10, 42; m 64; c 3. MATHEMATICS. Educ: Wash State Univ, BA, 64; Univ Ill, MS, 65, PhD(math), 70. Prof Exp: Asst prof math, 70-73, ASSOC PROF MATH, BOISE STATE UNIV, 73-, CHMN DEPT, 75- Concurrent Pos: Dir honors prog, Boise State Univ, 70- Mem: AAAS; Math Asn Am; Am Math Soc. Res: Analysis and functional analysis; extension of positive operators; graphs of groups. Mailing Add: Dept of Math Boise State Univ Boise ID 83725

MECHAM, JOHN STEPHEN, b Austin, Tex, Feb 29, 28; m 50; c 2. HERPETOLOGY, EVOLUTIONARY BIOLOGY. Educ: Univ Tex, BA, 50, PhD(zool), 55; Univ Fla, MS, 52. Prof Exp: Asst prof zool, Univ Tulsa, 55-56; from asst prof to assoc prof, Auburn Univ, 56-65; assoc prof, 65-69, PROF ZOOL, TEX TECH UNIV, 69- Concurrent Pos: NSF res grants, 58-61, 62-65, 67-68 & 69-71. Mem: Am Soc Ichthyol & Herpet; Soc Study Evolution; Soc Syst Zool. Res: Systematics and evolutionary mechanisms of anuran amphibians. Mailing Add: Dept of Biol Tex Tech Univ Lubbock TX 79409

MECHAM, LORRIE ELMER, physical chemistry, see 12th edition

MECHAM, MERLIN J, b Neola, Utah, Jan 31, 23; m 47; c 2. SPEECH PATHOLOGY, AUDIOLOGY. Educ: Brigham Young Univ, BA, 48; Utah State Univ, MS, 49; Ohio State Univ, PhD(speech path, audiol), 54. Prof Exp: Instr speech, Utah State Univ, 49-50; instr, Ohio State Univ, 52-54; assoc prof speech path, Brigham Young Univ, 54-61; actg chmn dept speech, 69-70, PROF SPEECH & DIR SPEECH PATH & AUDIOL, UNIV UTAH, 61- Concurrent Pos: Book abstractor, DSH Abstracts, 60-; consult, Utah State Training Sch, American Fork, 68-; Am Speech & Hearing Asn accrediting site visitor, Am Bd Exam, 70-; consult ed, J Speech & Hearing Disorders, 74- Mem: Fel Am Speech & Hearing Asn; Am Asn Mental Retardation. Res: Developmental aspects of normal and disordered audiolinguistic skills in children; exploratory model of audiolinguistic dysfunction. Mailing Add: Div of Speech Path & Audiol Univ of Utah Salt Lake City UT 84112

MECHANIC, GERALD, b New York, NY,Jan 7, 27; m 52; c 2. ORGANIC CHEMISTRY, BIOCHEMISTRY. Educ: City Col New York, BS, 51; NY Univ, MS, 53, PhD(chem), 58. Prof Exp: Asst acetylene chem, NY Univ, 52-53, res chemist biochem, 57-58; head biochem res lab, Manhattan State Hosp, 58-59; res assoc, Inst Med Res & Studies, NY, 59-60; res fel orthop surg, Mass Gen Hosp, 60-69, asst biochem, 66-69; res assoc biol chem, Harvard Med Sch, 63-69; assoc prof, 69-72, PROF ORAL BIOL, SCH DENT & PROF BIOCHEM, SCH MED, UNIV NC, CHAPEL HILL, 72- Mem: AAAS; Am Chem Soc; NY Acad Sci; The Chem Soc. Res: Chemistry of amino acids, peptides and proteins, with special reference to connective tissue. Mailing Add: Dent Res Ctr Univ of NC Chapel Hill NC 27514

MECHLER, MARK VINCENT, b Fredericksburg, Tex, Feb 5, 25; m 57; c 3. PHYSICS. Educ: Univ Tex, BA, 51, MA, 57, PhD(physics), 67. Prof Exp: Res scientist, Defense Res Lab, Univ Tex, 51-57; res engr, Collins Radio Co, 57-58; from res scientist to head underwater missile div, Defense Res Lab, Univ Tex, Austin, 58-69; SR SCIENTIST RES DIV, UNITECH, INC, 69- Mem: Acoust Soc Am. Res: Underwater sound; electro-acoustic transducers; sound propagation and scattering; electronics. Mailing Add: 1003 Floradale Dr Austin TX 78753

MECHLIN, GEORGE FRANCIS, JR, b Pittsburgh, Pa, July 23, 23; m 49. PHYSICS. Educ: Univ Pittsburgh, BS, 44, MS, 47, PhD(physics), 50. Prof Exp: Sr scientist, Bettis Atomic Power Div, 49-57, dir adv systs eng, Sunnyvale Div, Calif, 57-64, mgr missile launching & handling, 64-68, gen mgr, Underseas Div, Md, 68-71, gen mgr astronuclear & oceanic div, Md, 71-73, V PRES, RES & DEVELOP, WESTINGHOUSE ELEC CORP, 73- Concurrent Pos: Mem, Res Adv Comt, US

Coast Guard, 73-; vchmn, Marine Bd, Nat Res Coun, 75-; mem naval res adv comt, Lab Adv Bd for Naval Ships, 75- Honors & Awards: Meritorious Pub Serv Award, US Navy, 61; Order of Merit, Westinghouse Elec Corp, 61. Mem: Nat Acad Eng; Am Phys Soc; Am Inst Aeronaut & Astronaut; Marine Technol Soc; Nat Soc Aerospace Prof. Mailing Add: Westinghouse Elec Corp Res Labs Beulah Rd Pittsburgh PA 15235

MECHLINSKI, WITOLD, b Grudziadz, Poland, Sept 6, 35. MEDICINAL CHEMISTRY, ANALYTICAL CHEMISTRY. Educ: Gdansk Tech Univ, MSEng, 58, PhD(chem), 65. Prof Exp: Teaching asst org chem Gdansk Tech Univ, 58-61; res assoc med chem, Polish Acad Sci, 62-65; Nat Res Coun Can fel org chem, 66-68; res assoc, 69-71, RES ASST PROF MED CHEM, INST MICROBIOL, RUTGERS UNIV, NEW BRUNSWICK, 71- Mem: Am Chem Soc; Am Crystallog Asn; Am Soc Microbiol; AAAS. Res: Antibiotics isolation, structure, derivatives, natural and synthetic antifungal, antibacterial, antitumor compounds; organic synthesis; chromatographic techniques; x-ray crystal structure analysis. Mailing Add: Waksman Inst of Microbiol Rutgers Univ New Brunswick NJ 08903

MECHTLY, EUGENE A, b Red Lion, Pa, Feb 14, 31; m 65; c 3. RADIOPHYSICS. Educ: Western Md Col, BS, 52; Pa State Univ, MS, 58, PhD(physics), 62. Prof Exp: Physicist, Army Missile Labs & Marshall Space Flight Ctr, NASA, 54-65; res assoc elec eng, 65-67, asst prof, 67-69, asst dean col eng, 75-76, ASSOC PROF ELEC ENG, UNIV ILL, URBANA, 69- Concurrent Pos: Mem comn 3, US Nat Comt, Int Union Radio Sci. Mem: Am Geophys Union; Int Union Radio Sci. Res: Aeronomy; propagation of radio waves in the ionosphere; physics of the upper atmosphere; metrology. Mailing Add: 804 Mumford Dr Urbana IL 61801

MECK, ROBERT ALLEN, b Fredonia, Kans, Aug 6, 41; div; c 2. CANCER, HEMATOLOGY. Educ: Univ Ore, BS, 64; Ore State Univ, MS, 68; Univ Calif, Berkeley, PhD(biophys), 73. Prof Exp: Health physicist, Atomic Energy Div, Phillips Petrol Co, 64-66; radiol health officer, Ore State Univ, 66-67; biophys trainee, NIH, 71-73; res assoc cancer, Brookhaven Nat Lab, 73-75; RES ASSOC CANCER, UNIV KY, 76- Concurrent Pos: Res collabr, Brookhaven Nat Lab, 75- Mem: Int Soc Exp Hemat; AAAS. Res: Experimental quantitative growth of human tumors in vivo; quantitation of effects of chemotherapy and radiotherapy biophysics of stem cell differentiation. Mailing Add: Dept of Cell Biol MS 409 Univ of Ky Med Ctr Lexington KY 40506

MECKEL, ALFRED HANS, b Munich, Ger, Mar 12, 24; US citizen; m 51; c 1. DENTISTRY. Educ: Univ Munich, Dr med dent, 49; Northwestern Univ, DDS, 57. Prof Exp: Dentist, US Army, Ger, 49-54; RES DENTIST, PROCTER & GAMBLE CO, 54- Mem: Am Dent Asn. Res: Organic films on teeth; reactions of tin and fluorides; enamel structure; electron microscopy. Mailing Add: WHTCTIN 48 6110 Center Hill Rd Cincinnati OH 45224

MECKEL, LAWRENCE DANIEL, JR, geology, see 12th edition

MECKLENBORG, KENNETH THOMAS, b Cincinnati, Ohio, Oct 15, 27; m 56; c 4. ORGANIC CHEMISTRY. Educ: Xavier Univ, Ohio, BS, 48; Univ Cincinnati, MS, 54, PhD(org chem), 57. Prof Exp: Heart Found fel, Univ Cincinnati, 57-58; proj chemist, Standard Oil Co, Ind, 58-59; sr chemist, Emery Industs, 59-71; MGR APPLN, HUMKO SHEFFIELD CHEM, 71- Mem: Am Chem Soc; Am Oil Chemists Soc. Res: Structure definition of dimer acid; synthesis of esters; process definition; fatty acid and derivative application. Mailing Add: 6441 Keswick Dr Memphis TN 38138

MECKLENBURG, ROY ALBERT, b Elmhurst, Ill, Feb 10, 33; m 60; c 3. PLANT PHYSIOLOGY, MICROCLIMATOLOGY. Educ: Mich State Univ, BS, 58; Cornell Univ, MS, 61, PhD(agr), 63. Prof Exp: Asst ornamental hort, Cornell Univ, 58-63; asst prof Landscape hort, 63-70, ASSOC PROF HORT, MICH STATE UNIV, 70- Mem: Am Soc Hort Sci. Res: Physiology of low temperature hardiness in higher plants; the effect of plants on urban noise, dust and microclimate. Mailing Add: Dept of Hort Mich State Univ East Lansing MI 48823

MECKLER, ALVIN, b New York, NY, Apr 20, 26; m 47; c 3. THEORETICAL PHYSICS. Educ: City Col New York, BS, 47; Mass Inst Technol, PhD(physics), 52. Prof Exp: Mem staff solid state physics, Lincoln Lab, Mass Inst Technol, 52-55; chief div phys sci, Nat Security Agency, Md, 55-67; ASSOC PROF PHYSICS, UNIV MD, BALTIMORE COUNTY, 67- Mem: Am Phys Soc. Mailing Add: Dept of Physics Univ of Md Baltimore County Baltimore MD 21228

MECKLOSKY, MORTON, b New York, NY, Apr 13, 32; m 52; c 2. MATHEMATICS. Educ: Hunter Col, AB, 59; Columbia Univ, MA, 61, prof dipl math, 65; Rutgers Univ, MA, 63. Prof Exp: PROF MATH, SUFFOLK COUNTY COMMUNITY COL, 63- Concurrent Pos: Lectr, State Univ NY Stony Brook, 68- Mem: Math Asn Am. Res: Logic. Mailing Add: 35 Shelbourne Lane Stony Brook NY 11790

MECKSTROTH, GEORGE R, b Cincinnati, Ohio, Aug 26, 35; m 57; c 2. RADIOLOGICAL PHYSICS. Educ: Univ Cincinnati, BS, 58, MS, 60; PhD(radiol physics), 63. Prof Exp: PROF RADIOL, SCH MED, TULANE UNIV, 64- Concurrent Pos: Consult, Charity Hosp La, New Orleans, 65-, USPHS Hosp, 65-, Vet Admin Hosp, 65-, West Jefferson Gen Hosp, 68-, Hotel Dieu Hosp, 68-, East Jefferson Gen Hosp, 70- & St Charles Gen Hosp, 73- Mem: Am Col Radiol; Am Asn Univ Prof; Am Asn Physicists in Med; Health Physics Soc; Soc Nuclear Med. Mailing Add: Dept of Radiol Tulane Univ Sch of Med New Orleans LA 70112

MECOM, JOHN ODEN, b Winnsboro, La, Oct 9, 39; m 61; c 2. ZOOLOGY, LIMNOLOGY. Educ: La Polytech Inst, BS, 61; Northwestern Univ, MS, 63; Univ Colo, PhD(zool), 69. Prof Exp: Instr biol, Hartnell Col, 65-68; ASST PROF BIOL, SOUTHERN METHODIST UNIV, 69- Mem: AAAS; Am Soc Limnol & Oceanog; Ecol Soc Am. Res: Substrate ecology and trophic relationships of aquatic insects. Mailing Add: Dept of Biol Southern Methodist Univ Dallas TX 75222

MEDAK, HERMAN, b Vienna, Austria, Apr 26, 14; nat US; m 45; c 4. ORAL PATHOLOGY. Educ: Univ Toledo, BS, 43; Northwestern Univ, MS & DDS, 46; Univ Ill, PhD(anat), 59; Am Bd Oral Path, dipl, 64. Hon Degrees: MD, Univ Vienna, 73. Prof Exp: Med technician, Flower Hosp, Toledo, Ohio, 39-43; med technician, Chicago Wesley Mem Hosp, 43-47; res asst histol, 48-51, from instr to prof oral path, 53-67, actg head dept, 64-67, PROF PREV MED & COMMUNITY HEALTH, COL MED, UNIV ILL MED CTR, 67-, CHIEF CLIN ORAL PATH, DEPT ORAL DIAG, COL DENT, 67- Concurrent Pos: Dent consult, Ill Res Hosp & Tumor Clin, 48-53. Mem: AAAS; Am Dent Asn; Am Soc Clin Path; Am Acad Dent Med; Am Acad Oral Path. Res: Effect of irradiation on teeth and oral structures; epithelium of the oral mucosa; oral cytology. Mailing Add: Dept of Oral Diag Col of Dent Univ of Ill Med Ctr Chicago IL 60680

MEDALIA, AVROM IZAK, b Boston, Mass, Feb 3, 23; m 43, 56; c 4. COLLOID CHEMISTRY, RUBBER CHEMISTRY. Educ: Harvard Univ, AB, 42; Univ Minn, PhD(anal chem), 48. Prof Exp: Asst, Cornell Univ, 42-43; chemist, Brookhaven Nat Lab, 49-52; asst dir polymer res, Boston Univ, 52-55; sr res chemist, Godfrey L Cabot, Inc, 56-58, head fundamental res sect, 59-62, assoc dir res carbon black div, 63-70, GROUP LEADER, RES & DEVELOP DIV, CABOT CORP, 70- Concurrent Pos: Mem ed bd, Rubber Reviews. Mem: Am Chem Soc; Soc Rheology. Res: Colloids; polymers. Mailing Add: Cabot Corp Concord Rd Billerica MA 01821

MEDARIS, L GORDON, JR, b Memphis, Tenn, July 14, 36; m 58; c 3. PETROLOGY. Educ: Stanford Univ, BS, 58; Univ Calif, Los Angeles, PhD(geol), 66. Prof Exp: ASSOC PROF GEOL, UNIV WIS-MADISON, 71- Mem: AAAS; Geol Soc Am; Mineral Soc Am; Geochem Soc; Am Geophys Union. Res: Experimental petrology; partitioning of elements between coexisting phases; petrology of alpine peridotites; areal geology of northwest California and southwest Oregon; precambrian geology of Wisconsin. Mailing Add: Dept of Geol & Geophys Univ of Wis Madison WI 53706

MEDCALF, DARRELL GERALD, b Tillamook, Ore, Feb 10, 37; m 60; c 3. CARBOHYDRATE CHEMISTRY. Educ: Lewis & Clark Col, BA, 59; Purdue Univ, MS, 62, PhD(biochem), 64. Prof Exp: Asst prof cereal technol, NDak State Univ, 63-67; assoc prof chem, 67-73, PROF CHEM, UNIV PUGET SOUND, 73- Mem: Am Chem Soc; Am Asn Cereal Chemists. Res: Organic chemistry of carbohydrates, particularly polysaccharides; plant biochemistry, particularly structure and biosynthesis of algal polysaccharides. Mailing Add: Dept of Chem Univ of Puget Sound Tacoma WA 98416

MEDEARIS, DONALD N, JR, b Kansas City, Kans, Aug 22, 27; m 56; c 3. PEDIATRICS, MICROBIOLOGY. Educ: Univ Kans, AB, 49; Harvard Med Sch, MD, 53. Prof Exp: Intern internal med, Barnes Hosp, St Louis, Mo, 53-54; resident pediat, Children's Hosp, Cincinnati, Ohio, 54-56; res fel, Harvard Med Sch & Res Div Infectious Dis, Children's Med Ctr, Boston, 56-58; asst prof pediat, Sch Med, Johns Hopkins Univ, 58-63, asst prof microbiol, 59-63, assoc prof pediat & microbiol, 63-65; chmn dept, 65-69, PROF PEDIAT SCH MED, UNIV PITTSBURGH, 65-, DEAN, 69- Concurrent Pos: Med dir, Children's Hosp Pittsburgh, 65-69. Mem: AAAS; Am Asn Immunol; Soc Pediat Res; Infectious Dis Soc Am; Soc Exp Biol & Med. Res: Pathogenesis of infection in immature animals. Mailing Add: Off of the Dean Univ of Pittsburgh Sch of Med Pittsburgh PA 15213

MEDEIROS, ROBERT WHIPPEN, b Newburgh, NY, Mar 10, 31; m 59; c 2. ORGANIC CHEMISTRY. Educ: Univ Maine, BS, 52, Univ Del, MS, 57, PhD(org chem), 60. Prof Exp: Res chemist, Scott Paper Co, Pa, 52-55; res chemist, Newburgh Res Lab, E I du Pont de Nemours & Co, NY, 59-60 & Armstrong Cork Co, Pa, 60-63; asst prof chem, PMC Cols, 63-68; assoc prof, 68-74, PROF CHEM, WEST CHESTER STATE COL, 74- Concurrent Pos: Petrol Res Fund grant, 67-69. Mem: Am Chem Soc. Res: Chemistry of heterocyclic nitrogen compounds and vinyl polymers. Mailing Add: West Chester State Col West Chester PA 19380

MEDERSKI, HENRY JOHN, b Chicago, Ill, Jan 24, 22; m 48; c 3. PLANT PHYSIOLOGY. Educ: Mich State Col, BS, 47; Ohio State Univ, PhD(soils, agron), 50. Prof Exp: From asst prof to assoc prof agron, Agr Res & Develop Ctr, 50-59, PROF AGRON, AGR RES & DEVELOP CTR, UNIV OHIO, 59- Concurrent Pos: Consult, Farm Bur Coop Asn, 57- & Int Atomic Energy Agency, 63 & 66. Mem: AAAS; Am Soc Plant Physiol; Soil Sci Soc Am; Am Soc Agron. Res: Soil-plant-water relations; internal plant water relations; photosynthesis; plant-climate relations; plant nutrition. Mailing Add: Ohio Agr Res & Develop Ctr Wooster OH 44691

MEDICI, JOHN COE, biochemistry, nutrition, see 12th edition

MEDICI, PAUL T, b New York, NY, May 10, 19; m 43; c 3. HEMATOLOGY, ENDOCRINOLOGY. Educ: St John's Univ, NY, BS, 42 & 48, MS, 51; NY Univ, PhD, 56. Prof Exp: Instr biol sci, Col Pharm, 48-52, asst prof bact & pub health, 52-56, from assoc prof to prof, Grad Sch, 56-65, PROF HEMAT & ENDOCRINOL, GRAD SCH, ST JOHN'S UNIV, NY, 65-, CHMN DEPT BIOL, 65-, DEAN GRAD SCH ARTS & SCI, 69- Concurrent Pos: Lectr, Guggenheim Dent Clin, New York, 51-53. Mem: Fel AAAS; fel NY Acad Sci; Soc Study Blood; Am Soc Hemat. Res: Endocrinology of blood. Mailing Add: Grad Sch of Art & Sci St John's Univ Jamaica NY 11432

MEDICUS, GUSTAV KONRAD, b Immenstadt, Ger, Apr 28, 06; nat US; m 41; c 5. ELECTRON PHYSICS. Educ: Munich Tech Univ, Dipl Ing, 33, Dr Ing(tech physics), 35, Dozent Phys, 47. Prof Exp: Asst electrophys, Munich Tech Univ, 31-35; physicist, R Bosch AG, Stuttgart, 35-36; asst prof physics, Munich Tech Univ, 36-47, assoc prof, 47-49; res physicist, Wright Air Develop Command, 49-62, SR SCIENTIST, AIR FORCE AVIONICS LAB, WRIGHT-PATTERSON AFB, 62- Concurrent Pos: Adj prof, US Air Force Inst Technol, 69. Honors & Awards: Air Force Sci Achievement Award, 62. Res: Corona discharge and voltage control tubes; geiger counter; polarized light by metallic reflection; secondary electron multipliers; low voltage arcs; cathode sputtering; second derivative of probe curves; energy spectra of plasma electrons; Langmuir-type and related probes; improved thermionic electron emitters. Mailing Add: 7521 W Hyland Dr Dayton OH 45424

MEDICUS, HEINRICH ADOLF, b Zurich, Switz, Dec 24, 18; m 61. NUCLEAR PHYSICS. Educ: Swiss Fed Inst Technol, DrScNat(physics), 49. Prof Exp: Res assoc physics, Swiss Fed Inst Technol, 43-50; visitor radiation lab, Univ Calif, 50-51; guest, Mass Inst Technol, 51-52, instr physics, 52-54, vis asst prof, 54-55; assoc prof, 55-72, PROF PHYSICS, RENSSELAER POLYTECH INST, 72- Concurrent Pos: Swiss Nat Scholarship, 50-52; vis scientist, Atomic Energy Res Estab, Harwell, Eng, 67-68 & Swiss Inst Nuclear Res, Villigen, 75-76. Mem: Am Phys Soc; Swiss Phys Soc. Res: Radioactivity; meson physics; photonuclear reactions; nuclear structure. Mailing Add: Dept of Physics Rensselaer Polytech Inst Troy NY 12181

MEDINA, ANTONIO SAMUEL, b Nov 12, 20; US citizen; m; c 1. MEDICINE, PUBLIC HEALTH. Educ: George Washington Univ, MD, 43; Harvard Univ, MPH, 60. Prof Exp: Med officer, Pub Health Unit, PR Health Dept, 48-52; dir maternity serv, Rio Piedras Munic Hosp, 52-53; consult & dep dir bur maternal & child health, PR Health Dept, 53-58, from actg dir to dir, 58-60; from instr to assoc prof maternal & child health, 60-74, PROF MATERNAL & CHILD HEALTH, SCH MED, UNIV PR, SAN JUAN, 74-, DIR DEPT HUMAN DEVELOP, SCH PUB HEALTH, 70- Concurrent Pos: Lectr, Sch Pub Health, Harvard Univ, 65-75; vis prof, Univ Calif, Berkeley, 66-72; dir, San Juan City Family Planning Proj, 69-; consult, Nat Ctr Family Planning Serv, 70-73; dir, San Juan Children & Youth Comprehensive Health Care Proj, 71-; consult maternal & child health & family planning, Ecuador & Nicaragua, 73 & Dominican Repub & Paraguay, 74, Honduras, 75; WHO consult, Southeast Asia Regional Off, 73-74; mem, Nat Family Planning Forum; mem comt int health, Am Pub Health Asn, 74-75. Mem: Am Pub Health Asn; Asn Teachers of Maternal & Child Health. Mailing Add: Univ of PR Sch of Pub Health PO Box 13577 Santurce PR 00908

MEDINA, DANIEL, b New York, NY, Mar 6, 41; m 63; c 3. ONCOLOGY. Educ: Univ Calif, Berkeley, BA, 63, MA, 66, PhD(zool), 69. Prof Exp: ASSOC PROF CELL BIOL, BAYLOR COL MED, 69- Concurrent Pos: USPHS res grant chem carcinogenesis, Baylor Col Med, 71-77. Mem: AAAS; Am Asn Cancer Res. Res: Tumor biology; chemical carcinogenesis of mouse mammary glands; chemical-virus interactions; biology of preneoplastic lesions. Mailing Add: Dept of Cell Biol Baylor Col of Med Houston TX 77025

MEDINA, JOSE ENRIQUE, b Santurce, PR, May 1, 26; m 48; c 3. DENTISTRY. Educ: Univ Md, DDS, 48. Prof Exp: From instr to prof oper dent, Baltimore Col Dent Surg, Sch Dent, Univ Md, 48-67, from actg head dept to head dept, 57-67, asst dean col, 63-67; from assoc dean to dean, Col Dent, 67-74, PROF CLIN DENT, UNIV FLA, 67-, DIR HEALTH CTR SPACE PLANNING & UTILIZATION, 74- Concurrent Pos: Spec lectr, Walter Reed Army Med Ctr, Washington, DC, 58-64; consult, Univ Md Hosp, 58-67, USPHS Hosp, 60- & US Naval Dent Sch, 64-; hon prof, San Carlos Univ Guatemala, 60. Mem: Fel AAAS; Am Dent Asn; fel Am Col Dent; Int Asn Dent Res; hon Dent Soc Guatemala. Res: Restorative procedures; new materials for dental use. Mailing Add: J Hillis Miller Health Ctr Univ of Fla Col of Dent Gainesville FL 32601

MEDINA, MIGUEL ANGEL, b Laredo, Tex, July 5, 32; m 63; c 3. PHARMACOLOGY, BIOCHEMISTRY. Educ: St Mary's Univ, Tex, BS, 57, MS, 63; Univ Tex, Dallas, PhD(pharmacol), 68. Prof Exp: Jr chemist, Res & Develop Div, Am Oil Co, Tex, 57-59; res biochemist, Sch Aerospace Med, Brooks AFB, 59-64, res pharmacologist, 67-70; ASSOC PROF PHARMACOL, UNIV TEX MED SCH SAN ANTONIO, 70- Concurrent Pos: Lectr, St Mary's Univ, Tex, 67- Mem: Am Chem Soc; Am Soc Pharmacol & Exp Therapeut; Int Soc Biochem Pharmacol. Res: Brain and drug metabolism; histamine. Mailing Add: 120 West Summit San Antonio TX 78212

MEDIOLI, FRANCO, b Parma, Italy, Apr 1, 35; m 65; c 1. MICROPALEONTOLOGY. Educ: M Luigia Col, BA, 53; Univ Parma, PhD(geol), 59. Prof Exp: Fel, French Inst Petrol, Paris, 60-61; asst prof micropaleont, Univ Parma, 61-65; fel, Inst Oceanog, 65-66, asst prof, 66-69, ASSOC PROF MICROPALEON, DALHOUSIE UNIV, 69- Concurrent Pos: Micropaleontologist, Italian Geol Surv, 61-62, field geologist, 62-65. Mem: Italian Geol Soc; Int Paleont Union. Res: Recent foraminifera-living and dead; recent nannoplankton. Mailing Add: Dept of Geol Life Sci Bldg Dalhousie Univ Halifax NS Can

MEDLEN, AMMON BROWN, b Lockhart, Tex, Sept 12, 08; m 36; c 2. HISTOLOGY, ENDOCRINOLOGY. Educ: Baylor Univ, AB, 30, AM, 32; Agr & Mech Col Tex, PhD, 52. Prof Exp: Instr zool, Baylor Univ, 32-34; teacher & prin pub schs, 35-40; asst prof biol, Ouachita Col, 40-43; instr biol, Univ Houston, 46; instr biol & dir dept at annex, Tex A&M Univ, 46-51, from asst prof to assoc prof, 51-65; prof biol sci & head dept, 65-74, EMER PROF BIOL, TARLETON STATE UNIV, 74- Mem: AAAS; Am Fisheries Soc; Am Soc Zool; NY Acad Sci. Res: Histological effects of low chronic irradiation with cobalt, especially upon the reproduction and embryonic abnormalities of rats. Mailing Add: Dept of Biol Sci Tarleton State Univ Stephenville TX 76401

MEDLER, JOHN THOMAS, b Las Cruces, NMex, May 28, 14; m 64; c 4. ENTOMOLOGY. Educ: NMex State Col, BS, 36, MS, 37; Univ Minn, PhD(entom), 40. Prof Exp: Asst entom, Univ Minn, 37-40; res fel exp sta, NMex State Col coop with Tex Gulf Sulfur Co, 41; Guggenheim Mem Found fel, Univ Calif, 42; asst entomologist, USPHS, 42-43; from asst prof to assoc prof, 46-62, PROF ENTOM, UNIV WIS-MADISON, 62- Concurrent Pos: Head dept entom, Univ Ife, Nigeria, 69-73. Mem: Entom Soc Am. Res: Chemotropism of codling moth; taxonomy and biology of Cicadellidae; physiology of Homoptera; control of cotton and legume insects; malaria control; insects and legume seed production; ecology of bumblebees and native wild bees. Mailing Add: Dept of Entom Univ of Wis Madison WI 58706

MEDLIN, GENE WOODARD, b Greensboro, NC, Oct 5, 25; m 45; c 5. MATHEMATICS. Educ: Wake Forest Col, BS, 48; Univ NC, MA, 50, PhD(math), 53. Prof Exp: Assoc prof math, Wake Forest Col, 52-56; mathematician, Oak Ridge Nat Lab, 56-57; NSF grant, Swiss Fed Inst Tech, 57-58; assoc prof, 58-65, PROF MATH, STETSON UNIV, 65-, CHMN DEPT, 58- Concurrent Pos: Vis lectr grad sch, Univ Tenn, 56-57. Mem: Am Math Soc; Math Asn Am; Asn Comput Mach. Res: Matrix theory. Mailing Add: 600 N McDonald Ave Deland FL 32720

MEDLIN, WILLIAM LOUIS, b Harlingen, Tex, Aug 25, 28; m 58. SOLID STATE PHYSICS. Educ: Univ Tex, BS, 51, MS, 54, PhD(physics), 56. Prof Exp: Res assoc, Mobil Oil Corp, 56-57, RES ASSOC, FIELD RES LAB, MOBIL RES & DEVELOP CORP, 67- Res: Rock mechanics; luminescence; color centers. Mailing Add: Field Res Lab Mobil Res & Develop Corp Dallas TX 75221

MEDLIN, WILLIAM VIRGIL, b Stockton, Calif, Feb 14, 11; m 36, 59; c 2. CHEMISTRY. Educ: Univ Calif, BS, 32; Calif Inst Technol, PhD(chem), 35. Prof Exp: Technologist, Shell Oil Co, 35-46, engr, Shell Develop Co, NY, 46-55, supvr, 55-59, asst to pres, 59-67, licensing rep, 67-69, mgr oil licensing, 69-72, MGR OIL PROD, PATENTS & LICENSING, SHELL DEVELOP CO, 72- Mailing Add: 2323 Shakespeare Rd Houston TX 77025

MEDNICK, MORTON L, b Tamaqua, Pa, Sept 15, 21; m 50. ORGANIC CHEMISTRY. Educ: Univ Chicago, PhB, 48, MS, 52; Boston Univ, PhD(org chem), 57. Prof Exp: Res asst org chem, Ben May Lab Cancer Res, Univ Chicago, 51-52 & Worcester Found Exp Biol, 52-55; res assoc, Boston Univ, 55-56; res chemist, Atlas Powder Co, Pa, 56-58; sr res assoc, Horizons, Inc, Ohio, 58-59; res chemist, Northern Regional Labs, USDA, Ill, 59-62; ORG RES CHEMIST, US ARMY CHEM RES & DEVELOP LABS, EDGEWOOD ARSENAL, 62- Mem: AAAS; Am Chem Soc. Res: Organic synthesis and research; medicinal chemistry; chemical structure; biological activity relationships. Mailing Add: 343 B-1 Trimble Rd Joppa MD 21085

MEDOFF, GERALD, b New York, NY, Nov 9, 36; m 60; c 2. MICROBIOLOGY. Educ: Columbia Univ, AB, 58, Wash Univ, MD, 62; dipl, Am Bd Internal Med. Prof Exp: Fel infectious dis, Mass Gen Hosp, Boston, 65-68, instr med & pediat, Harvard Med Sch, 68-70; ASST PROF MED, SCH MED, WASH UNIV, 70- Res: Mycology; infectious diseases; medicine. Mailing Add: Dept of Med Box 8051 Wash Univ Sch of Med St Louis MO 63110

MEDORA, RUSTEM SOHRAB, b Deolali, India, May 4, 34; m 64; c 2. PHARMACOGNOSY, BIOLOGY. Educ: Gujarat Univ, India, BPharm, 58, MPharm, 60; Univ RI, PhD(pharmaceut sci), 65. Prof Exp: Tutor pharmacog, L M Col Pharm, Gujarat Univ, 58-61; asst, Univ RI, 61-65; asst prof, Idaho State Univ, 65-66; Nat Res Coun Can fel bot, McGill Univ, 66-67; asst prof pharmacog, 67-72, ASSOC PROF PHARM, SCH PHARM, UNIV MONT, 72- Concurrent Pos: Smith Kline Found & Mont Heart Asn res grants. Mem: Int Tissue Cult Asn; Am Soc Pharmacog; Soc Econ Bot; Tissue Cult Asn. Res: Pharmacognosy and tissue culture of plants of medicinal interest. Mailing Add: Sch of Pharm Univ of Mont Missoula MT 59801

MEDOVY, HARRY, b Winnipeg, Man, Oct 22, 04; m 34; c 2. PEDIATRICS. Educ: Univ Man, BA, 23, MD, 28. Prof Exp: Head dept, 54-70, prof pediat, 54-75, EMER PROF PEDIAT, FAC MED, UNIV MAN, 75- Concurrent Pos: Consult, City Health Dept, Winnipeg, 50-; physician-in-chief, Children's Hosp, 54- Res: Am Pediat Soc; Am Acad Pediat; Can Pediat Soc (pres, 57). Res: Newborn; nutrition and diabetes. Mailing Add: Dept of Pediat Univ of Man Fac of Med Winnipeg MB Can

MEDRUD, RONALD CURTIS, b Tracy, Minn, July 9, 34; m 59; c 2. X-RAY CRYSTALLOGRAPHY, CERAMICS. Educ: Augustana Col, SDak, BA, 56; Univ Iowa, PhD(phys chem), 63. Prof Exp: Nat Acad Sci-Nat Res Coun res assoc, US Naval Ord Lab, 63-64; SR RES CHEMIST, CORNING GLASS WORKS, 64- Mem: Am Crystallog Asn; Am Chem Soc; Am Ceramic Soc. Res: X-ray crystallography for materials evaluation; laboratory automation. Mailing Add: Corning Glass Works Sullivan Park Corning NY 14830

MEDSGER, GERALD WILLIAM, b Los Angeles, Calif, Aug 18, 27; m 48; c 2. PHYSICS, MATHEMATICS. Educ: US Mil Acad, BS, 48; Calif Inst Technol, MS, 57; NY Univ, MS, 68. Prof Exp: Corps Engrs, US Army, 44-, airport engr, Air Force, Europe, 49-52, instr, Army Engr Sch, 52-53, cmndg officer, Airborne Engr Co, Ft Bragg, NC, 53-54, resident engr, Army Eastern Ocean Engr Dist, 55-56, asst prof mil sci, hist & eng, Mo Sch Mines, 57-59, engr construct group, Army, Europe, 59-60, opers officer construct battalion, 60-62, chief electronics br, Army Airborne, Electronics & Spec Warfare Bd, 63-65, assoc prof math, US Mil Acad, 65-69, DIR RES, US MIL ACAD, 69- Mem: Fel AAAS; fel Am Soc Civil Engrs; Math Asn Am; Nat Soc Prof Engrs; Am Asn Physics Teachers. Res: Psychometrics; civil engineering; psychology. Mailing Add: Off of Instnl Res US Mil Acad West Point NY 10996

MEDSKER, LARRY ROBERT, b Bloomington, Ind, Apr 25, 44; m 66. EXPERIMENTAL NUCLEAR PHYSICS. Educ: Ind Univ, BA, 65, MS, 67, PhD(physics), 71. Prof Exp: Res assoc physics, Argonne Nat Lab, 71-73 & Tandem Accelerator Lab, Univ Pa, 73-75; ASST PROF PHYSICS, FLA STATE UNIV, 75- Concurrent Pos: Nuclear info res assoc, Nat Res Coun, 71-73. Mem: Am Phys Soc. Res: Nuclear spectroscopy by means of (helium-3, d) and (helium-3, p), reactions; studies of heavy ion reactions using beams of lithium, boron, carbon and oxygen; studies of models for nuclei with masses eighty to one hundred. Mailing Add: Dept of Physics Fla State Univ Tallahassee FL 32306

MEDVE, RICHARD J, b California, Pa, Jan 28, 36; m 58; c 5. PLANT ECOLOGY. Educ: California State Col, Pa, BS, 57; Kent State Univ, MA, 59; Ohio State Univ, PhD(bot), 68. Prof Exp: Counr, Kent State Univ, 57-58; teacher pub schs, 58-66; TEACHER BIOL, SLIPPERY ROCK STATE COL, 66- Mem: Torrey Bot Club; Nat Sci Teachers Asn; Nat Asn Biol Teachers. Res: Mycorrhizae; science education; stripmine revegetation. Mailing Add: Dept of Biol Vincent Sci Hall Slippery Rock State Col Slippery Rock PA 16057

MEDVED, DAVID BERNARD, b Philadelphia, Pa, Feb 21, 26; m 47; c 4. ELECTROOPTICS. Educ: Univ Pa, BA, 49, MSc, 51, PhD(physics), 55. Prof Exp: Res physicist, Philco Corp, 49-51; sr res engr, Gen Dynamics/Convair, 54-57, head solid state physics group, 57-63; mgr advan concept, Electro-Optical Systs, Inc, Calif, 63-67, chief scientist, Measurement Systs Div, 67-68, chief scientist, Advan Systs & Requirements, 68-69; pres & tech dir, Meret Co, 69-72, PRES & TECH DIR, MERET INC, 72- Concurrent Pos: Consult, Remington Rand & Univac Divs, Sperry Rand Corp, 52; vis asst prof & lectr, Univ Calif, Los Angeles, 56-; assoc prof, San Diego State Col, 58- Mem: Am Phys Soc; Inst Elec & Electronics Eng. Res: Experimental solid state and plasma physics, especially particle-surface interactions, physical electronics, electronic properties of semiconductors and interaction of electromagnetic radiation with matter. Mailing Add: Meret Inc 1815 24th St Santa Monica CA 90404

MEDWAY, WILLIAM, b Man, Can, Feb 23, 27; m 71. VETERINARY MEDICINE, CLINICAL MEDICINE. Educ: Univ Man, BS, 47; Ont Vet Col, DVM, 54; Cornell Univ, PhD(physiol), 58. Hon Degrees: MA, Univ Pa, 71. Prof Exp: Instr biochem, Ont Agr Col, Univ Toronto, 48-49; asst physiol chem, Cornell Univ, 54-58; assoc med, Univ Pa, 58-60; asst prof physiol & res assoc, Ont Vet Col, 60-62; from asst prof to assoc prof, 62-68, PROF CLIN LAB MED, UNIV PA, 68- Mem: Am Soc Vet Clin Path (pres, 68-69); Int Asn Aquatic Animal Med (secy-treas, 69-); Am Soc Vet Physiol & Pharmacol; Am Physiol Soc; NY Acad Sci. Res: Clinical chemistry; clinical pathology as applied to aquatic animals; veterinary diagnostics; veterinary physiology. Mailing Add: Dept of Clin Studies Univ of Pa Philadelphia PA 19174

MEDWICK, THOMAS, b Jersey City, NJ, Oct 15, 29. ANALYTICAL CHEMISTRY. Educ: Rutgers Univ, BS, 52, MS, 54; Univ Wis, PhD(pharmaceut chem), 58. Prof Exp: Res analyst, Merck & Co, Inc, NJ, 58-60; asst prof pharmaceut chem, Col Pharm, Rutgers Univ, Newark, 60-63, from assoc prof to prof, 63-71, PROF PHARMACEUT CHEM, COL PHARM, RUTGERS UNIV, NEW BRUNSWICK, 71- Concurrent Pos: Sci adv, NY Dist, Food & Drug Admin, 70- Mem: Am Pharmaceut Asn; Am Chem Soc; fel Am Inst Chemists. Res: Theoretical analytical and hydrazine chemistry; acid-base reactions in nonaqueous solvents; analysis of pharmaceuticals. Mailing Add: Col of Pharm Rutgers Univ Busch Campus New Brunswick NJ 08903

MEDWIN, HERMAN, b Springfield, Mass, Apr 9, 20; m 45. UNDERWATER ACOUSTICS. Educ: Worcester Polytech Inst, BS, 41; Univ Calif, Los Angeles, MS, 48, PhD(physics), 54. Prof Exp: Asst physics, Univ Calif, Los Angeles, 46-53, res assoc, 53-54; consult acoustics, Bolt, Beranek & Newman, Inc, 54-55; assoc prof physics, 55-60, PROF PHYSICS, US NAVAL POSTGRAD SCH, 60- Concurrent Pos: Instr, Los Angeles City Col, 48-54; liaison scientist, Off Naval Res, London, 61-62, ed, Europ Sci Notes, 62; consult, Hudson Labs, Columbia Univ, 64-68; vis prof, Imp Col, Univ London, 65-66; vchmn acoustics panel, Nat Acad Sci-Nat Res Coun Physics Surv Comt, 70-71; vis scientist, Royal Australian Naval Res Lab, 72-73; chmn tech comt underwater acoustics, Acoust Soc Am, 75. Honors & Awards: Res Award, Naval Postgrad Sch Chap, Sigma Xi, 72. Mem: Fel Acoust Soc Am; Am Geophys Union; Inst Noise Control Eng. Res: Correlations between ocean microstructure and acoustical fluctuations; surface and volume scattering of sound in the sea; effects of high intensity sounds. Mailing Add: Dept of Physics US Naval Postgrad Sch Monterey CA 93940

MEDZ, ROBERT B, b Los Angeles, Calif, Nov 7, 19; m 66. ENVIRONMENTAL MANAGEMENT. Educ: Univ Calif, BA, 42, MA, 47; Univ Wash, PhD, 64. Prof Exp: Fel, Univ Wash, 64, instr chem, 64-65; res chemist, Southeast Water Lab, Fed Water Pollution Control Admin, 65-67; phys sci adminr, Bur Dis Prev & Environ Controls, USPHS, 68-69, phys sci adminr, Environ Health Serv, 70, assoc dir, Lab Opers Div, 70-72, chief standardization br, Qual Assurance Div, Off Res & Develop, 72-75, SR PROG ADV MONITORING QUAL ASSURANCE, MONITORING TECHNOL DIV, OFF RES & DEVELOP, ENVIRON PROTECTION AGENCY, 75- Concurrent Pos: Tech adv, Pilot Secretariat Pollution Measurement, Int Orgn Legal Metrol, 74- Mem: AAAS; Am Chem Soc; NY Acad Sci. Res: Impact of chemical pollutants on the quality of the environment and upon human health and the well being of all life forms. Mailing Add: Off Res & Develop Environ Protection Agency Washington DC 20460

MEDZIHRADSKY, FEDOR, b Kikinda, Yugoslavia, Feb 4, 32; m 67; c 2. BIOCHEMISTRY. Educ: Munich Tech Univ, MS, 61, PhD(biochem), 65. Prof Exp: Instr biochem, Univ Munich, 65-66; asst prof, 69-73, ASSOC PROF BIOCHEM, MED SCH, UNIV MICH, ANN ARBOR, 73-, ASSOC PROF PHARMACOL, 75- Concurrent Pos: NIH fel, Univ Wis, 66-67; Nat Inst Neurol Dis & Blindness trainee, Wash Univ, 67-69; vis assoc prof pharmacol, Stanford Univ Med Ctr, 75-76; Nat Res Serv Award, USPHS-Alcohol, Drug Abuse & Mental Health Admin, 75-76. Mem: Ger Soc Biol Chem; Am Soc Neurochem; Am Chem Soc; Am Soc Biol Chem; Am Soc Pharmacol & Exp Therapeut. Res: Neurochemistry; biochemical pharmacology; biological transport. Mailing Add: Dept of Biol Chem Univ of Mich Med Sch Ann Arbor MI 48109

MEDZON, EDWARD LIONEL, b Winnipeg, Man, May 26, 36; m 61; c 3. VIROLOGY. Educ: Univ Man, BSc, 57, MSc, 60; McGill Univ, PhD(virol, immunol), 64. Prof Exp: Instr microbiol, Univ Mich, 63-65; asst prof, 65-69, ASSOC PROF MICROBIOL, UNIV WESTERN ONT, 69- Concurrent Pos: Vis lectr, Eastern Mich Univ, 64-65; ed-in-chief, Dict Microbiol, Am Soc Microbiol, 73- Mem: AAAS; Am Soc Microbiol; Can Soc Microbiol; Can Soc Cell Biol; Sigma Xi. Res: Interaction with viruses and the changes in surface properties of animal cells through the cell cycle; early detection of virus infection in vitro. Mailing Add: Dept of Bact & Immunol Health Sci Ctr Univ Western Ont London ON Can

MEE, JACK EVERETT, b Brainerd, Minn, July 6, 30; m 52; c 3. INORGANIC CHEMISTRY, SOLID STATE ELECTRONICS. Educ: Dakota Wesleyan Univ, BA, 52; Iowa State Univ, PhD(inorg chem), 62. Prof Exp: Jr chemist, Gen Elec Co, Wash, 52-57; res asst inorg chem, Inst Atomic Res, Iowa State Univ, 57-62; sr res engr, Autonetics Div, NAm Rockwell Corp, 62-64, specialist res, 64-67, supvr, 67-74, MGR SOLID STATE MAT RES BR, AUTONETICS GROUP, ROCKWELL INT, 74- Mem: Am Chem Soc; Inst Elec & Electronics Engrs; Sigma Xi. Res: Chemical vapor deposition; liquid phase epitaxy; epitaxial garnets; bubble domain materials; epitaxial III-V's. Mailing Add: Rockwell Int Electronics Res Div 3370 Miraloma Ave Anaheim CA 92803

MEECHAM, WILLIAM CORYELL, b Detroit, Mich, June 17, 28; m 48; c 2. MATHEMATICAL PHYSICS, CLASSICAL PHYSICS. Educ: Univ Mich, BS & MS, 48, PhD(physics), 54. Prof Exp: Asst physics, Univ Mich, 48-53 & Brown Univ, 53-54; res assoc, Univ Mich, 54-55, assoc res physicist, 55-56, instr, 56-57, asst prof, 57-60, res physicist & head fluid & solid mech lab, 59-60; prof fluid mech, Univ Minn, Minneapolis, 60-66; sr scientist, Lockheed Palo Alto Res Labs, 66-67; head div appl mech, Col Eng, 68-69, PROF FLUID MECH & HEAD DIV, UNIV CALIF, LOS ANGELES, 67- Concurrent Pos: Res assoc, Univ Calif, San Diego, 63; consult, TRW, Inc, 59-65, Rand Corp, Calif, 64-72, Inst Sci & Technol, Univ Mich, 60-67 & Bolt, Beranek & Newman, Inc, 68-74; consult, Aerospace Corp, 74-, res & develop assoc, 74- Mem: Am Phys Soc; fel Acoust Soc Am; assoc fel Am Inst Aeronaut & Astronaut. Mem: Fluid dynamics; acoustics; diffraction theory; stochastic processes; wave propagation problems. Mailing Add: Sch Eng & Appl Sci Univ of Calif Los Angeles CA 90024

MEECHAN, CHARLES JAMES, b Usk, Wash, Aug 7, 28; m 51; c 6. SOLID STATE PHYSICS. Educ: Ore State Col, BS, 51. Prof Exp: Res physicist, Atomics Int Div, NAm Aviation, Inc, 51-61, staff physicist, Sci Ctr, 61-63, res adv, Corp Off, 63-67, exec dir res & eng, NAm Rockwell Corp, 67-69, vpres indust systs, 69-71, VPRES RES & ENG & DIR SCI CTR, ROCKWELL INT CORP, 71- Mem: Am Phys Soc; Sigma Xi. Res: Experimental research in study of lattice imperfections; radiaton damage and diffusion phenomena in solids; Mossbauer spectroscopy. Mailing Add: Rockwell Int Corp 600 Grant St Pittsburgh PA 15219

MEECHAN, ROBERT JOHN, b Newport, Wash, Aug 25, 26; m 53; c 3. PEDIATRICS. Educ: Ore State Col, BA, 51; Univ Ore, MS & MD, 53. Prof Exp: From instr to assoc prof, 57-68, PROF PEDIAT, MED SCH, UNIV ORE, 68- Mailing Add: Dept of Pediat Univ of Ore Sch of Med Portland OR 97201

MEEHAN, EDWARD JOSEPH, b Oakland, Calif, July 21, 12; m 45; c 3. CHEMISTRY. Educ: Univ Calif, BS, 33, PhD(phys chem), 36. Prof Exp: Instr chem, Univ Calif, 36-39; from instr to assoc prof, 39-52, PROF CHEM, UNIV MINN, MINNEAPOLIS, 52- Concurrent Pos: With Off Rubber Reserve, 44. Mem: Am Chem Soc; Optical Soc Am. Res: Absorption spectra of solids; spectrophotometry; physical and chemical properties of high polymers; light scattering; reaction mechanisms. Mailing Add: Dept of Chem Univ of Minn Minneapolis MN 55455

MEEHAN, JOHN PATRICK, b San Francisco, Calif, May 22, 23; m 49; c 4. PHYSIOLOGY. Educ: Univ Southern Calif, MD, 48. Prof Exp: Instr, 47-49, asst prof, 49-51 & 54-55, assoc prof, 55-62, PROF PHYSIOL, SCH MED, UNIV SOUTHERN CALIF, 62-, CHMN DEPT, 66- Mem: AAAS; Aerospace Med Asn. Res: Aviation physiology; central nervous system control of the vascular system; cardiovascular and respiratory physiology; aerospace medicine. Mailing Add: Dept of Physiol Sch of Med Univ of Southern Calif Los Angeles CA 90033

MEEHAN, WILLIAM ROBERT, b Buffalo, NY, Apr 9, 31. FISH BIOLOGY. Educ: Univ Buffalo, BA, 52; Univ Ore, MA, 55; Mich State Univ, PhD(fisheries, wildlife), 58. Prof Exp: Res biologist, Alaska Dept Fish & Game, 58-66; FISHERY RES BIOLOGIST, FORESTRY SCI LAB, US FOREST SERV, 66- Mem: Am Fisheries Soc; Am Soc Limnol & Oceanog; Wildlife Soc; Am Inst Fishery Res Biologists. Res: Wildlife biology; salmon investigations; aquatic entomology. Mailing Add: Forestry Sci Lab US Forest Serv 3200 Jefferson Way Corvallis OR 97331

MEEK, ALEXANDER MILLAR, animal husbandry, deceased

MEEK, BURL DEAN, b Canton, Okla, Aug 13, 36; m 64; c 1. SOIL CHEMISTRY. Educ: Colo State Univ, BS, 58; Univ Calif, Riverside, MS, 67, PhD(soil sci), 70. Prof Exp: Soil scientist, Agr Res Serv, 60-75, DIR SOIL SCI, IMP VALLEY CONSERV RES CTR, AGR RES SERV, USDA, 75- Mem: Am Soc Agron. Res: Effects of soil aeration on plant growth, denitrification and reducing conditions in soils. Mailing Add: Imp Valley Conserv Res Serv 4151 Hwy 86 Brawley CA 92227

MEEK, DEVON WALTER, b River, Ky, Feb 24, 36; m 65. INORGANIC CHEMISTRY. Educ: Berea Col, BA, 58; Univ Ill, MS, 60, PhD(chem), 61. Prof Exp: Asst chem, Univ Ill, 58-60, res fel, 60-61; from asst prof to assoc prof inorg chem, 61-69, PROF INORG CHEM, OHIO STATE UNIV, 69- Concurrent Pos: Vis assoc prof, Northwestern Univ, 67; sr res fel, Univ of Sussex, England, 74. Mem: Am Chem Soc; The Chem Soc; Sigma Xi. Res: Studies of the syntheses, electronic and magnetic properties and structures of transition metal complexes with unusual coordination number; homogeneous catalysis and activition and/or stabilization of small molecules. Mailing Add: Dept of Chem Ohio State Univ Columbus OH 43210

MEEK, JACK HENRY, b Toronto, Ont, July 30, 18; m 41; c 7. PHYSICS. Educ: Univ Toronto, BA, 40; Univ Sask, MA, 53, PhD, 55. Prof Exp: Forecaster, Meteorol Serv Can, 40-42; with radio physics lab, Defense Res Bd Can, 47-51, with physics dept, Saskatchewan, 51-56, dep dir phys res, 56-59, supt commun lab, 60-67, mem plans staff, 68-72, DIR SCI & TECHNOL INFO ANAL, DEPT NAT DEFENCE, 73- Mem: Am Geophys Union; Inst Elec & Electronics Eng; Can Asn Physicists. Res: Upper atmospheric physics and meteorology; geomagnetism; oceanography; radio communications; cybernetics; technological forecasting. Mailing Add: 2365 Ridgecrest Pl Ottawa ON Can

MEEK, JAMES LATHAM, b San Antonio, Tex, Apr 10, 37; m 56; c 3. MATHEMATICS. Educ: Univ Tex, BA, 62, MA, 63, PhD(math), 67. Prof Exp: Instr math, San Antonio Col, 63-64; res assoc acoust & math, Defense Res Lab, Univ Tex, Austin, 67; asst prof, 67-74, ASSOC PROF MATH, UNIV ARK, FAYETTEVILLE, 74- Concurrent Pos: Consult, Defense Res Lab, Univ Tex, Austin, 67-68. Mem: Am Math Soc; Math Asn Am. Res: Underwater acoustics; boundary behavior of analytic, harmonic, and subharmonic functions; harmonic analysis. Mailing Add: Dept of Math Univ of Ark Fayetteville AR 72701

MEEK, JOHN SAWYERS, b Madison, Wis, Aug 12, 18; m 45; c 2. ORGANIC CHEMISTRY. Educ: Univ Wis, BA, 41; Univ Ill, MS, 44, PhD(org chem), 45. Prof Exp: Asst inorg chem, Univ Ill, 41-44, Allied Chem & Dye fel, 45; from instr to assoc prof, 45-60, PROF ORG CHEM, UNIV COLO, BOULDER, 60- Concurrent Pos: Asst, Univ Wis, 43. Mem: Am Chem Soc. Res: Diels-Alder reactions; Bridgehead compounds. Mailing Add: Dept of Chem Univ of Colo Boulder CO 80302

MEEK, JOSEPH CHESTER, JR, b Sabetha, Kans, July 16, 31; m 54; c 3. INTERNAL MEDICINE, ENDOCRINOLOGY. Educ: Univ Kans, AB, 54, MD, 57. Prof Exp: Intern, San Diego County Gen Hosp, Calif, 57-58; resident, Univ Kans Med Ctr, Kansas City, 58-60; res asst space med, US Naval Sch Aviation Med, 60-62; from instr to asst prof, 64-69, assoc prof med, 69-75, PROF MED, UNIV KANS MED CTR, KANSAS CITY, 75- Concurrent Pos: Am Col Physicians Mead Johnson scholar, Univ Kans, 59-60; fel endocrinol, Scripps Clin & Res Found, La Jolla, Calif, 62-63, trainee, 63-64; attending physician, Vet Admin Hosp, 64- Mem: Am Fedn Clin Res: Am Thyroid Asn; Am Diabetes Asn; fel Am Col Physicians; Endocrine Soc. Res: Metabolism; long acting throid stimulator; insulin A and B chains. Mailing Add: Dept of Med Univ of Kans Med Ctr Kansas City KS 66103

MEEK, VIOLET IMHOF (MRS DEVEON W), b Geneva, Ill, June 12, 39; m 65; c 1. INORGANIC CHEMISTRY. Educ: St Olaf Col, BA, 60; Univ Ill, MS, 62, PhD(inorg chem), 64. Prof Exp: Instr chem, Mt Holyoke Col, 64-65; asst prof, 65-70, ASSOC PROF CHEM, OHIO WESLEYAN UNIV, 70-, CHMN DEPT, 75- Mem: AAAS; Am Chem Soc; The Chem Soc. Res: Preparation of coordination compounds of transition metals and studies of their structures and properties using physical methods. Mailing Add: Dept of Chem Ohio Wesleyan Univ Delaware OH 43015

MEEKER, C IRVING, b Schenectady, NY, Jan 8, 29; m 52; c 4. MEDICINE. Educ: Middlebury Col, BA, 50; Cornell Univ, MD, 54; Am Bd Obstet & Gynec, dipl, 65. Prof Exp: Intern, Mary Fletcher Hosp, Burlington, Vt, 54-55; asst resident obstet & gynec, Boston Lying-In Hosp, 58-59; resident affiliated hosps, Col Med, Univ Vt, 59-60; from instr to asst prof, 62-70, ASSOC PROF OBSTET & GYNEC, COL MED, UNIV VT, 71- Concurrent Pos: NIH trainee, Univ Vt, 60-61 & McGill Univ, 61-62. Mem: AAAS; Am Col Obstet & Gynec; AMA. Res: Endocrinology of reproduction. Mailing Add: Dept of Obstet & Gynec Univ of Vt Burlington VT 05401

MEEKER, MICHAEL ELLIOTT, b Ft Worth, Tex, Oct 2, 35. CULTURAL ANTHROPOLOGY. Educ: Mass Inst Technol, SB, 58; Univ Chicago, AM, 66, PhD(anthrop), 70. Prof Exp: Res physicist, Thermo-Electron Eng Corp, 61-63; asst prof anthrop, Cornell Univ, 71-75; ACTG ASSOC PROF ANTHROP, UNIV CALIF, SAN DIEGO, 75- Concurrent Pos: Soc for Humanities fel, 73-74. Mem: Mid East Studies Asn. Res: Cultural interpretation of kinship, religion and politics; popular literary traditions of Turkish and Arabic speaking peoples. Mailing Add: Dept of Anthrop Univ of Calif San Diego La Jolla CA 92037

MEEKER, RALPH DENNIS, b Chicago, Ill, Nov 15, 45; m 68; c 1. PHYSICS. Educ: St Procopious Col, BS, 67; Iowa State Univ, PhD(physics), 70. Prof Exp: Asst prof physics, 70-73, ASSOC PROF PHYSICS, ILL BENEDICTINE COL, 73- Concurrent Pos: Resident assoc, Argonne Nat Lab, consult, 71- Mem: Am Phys Soc; Am Asn Physics Teachers. Res: Mössbauer effect of actinide elements and compounds. Mailing Add: Dept of Physics Ill Benedictine Col Lisle IL 60532

MEEKER, ROBERT ELDON, physical chemistry, see 12th edition

MEEKER, ROBERT LESLEY, analytical chemistry, see 12th edition

MEEKER, THRYGVE RICHARD, b Pottstown, Pa, Mar 9, 29; m 54; c 3. PHYSICAL CHEMISTRY. Educ: Ursinus Col, BS, 51; Univ Del, MS, 54, PhD(phys chem), 56. Prof Exp: MEM TECH STAFF, BELL LABS, 55- Mem: Am Chem Soc; Am Phys Soc; Am Inst Chemists; Inst Elec & Electronics Eng; NY Acad Sci. Res: Chemical kinetics; spectroscopy; ultrasonics and elastic properties; piezoelectric, dielectric and ferroelectric phenomena; applied mathematics; quantum mechanics; thermodynamics; wave phenomena. Mailing Add: 2956 Lindberg Ave Allentown PA 18103

MEEKER, WILLIAM QUACKENBUSH, JR, b New York, NY, Nov 28, 49; m 75. STATISTICS. Educ: Clarkson Col Technol, BS, 72; Union Col, MS, 73, PhD(admin & eng systs), 75. Prof Exp: Res fel statist, Inst Admin & Mgt, Union Col, 73-75; ASST PROF STATIST, IOWA STATE UNIV, 75- Concurrent Pos: Statistician, Corp Res & Develop, Gen Elec Co, 73-75. Mem: Am Statist Asn; Inst Math Statist; Opers Res Soc Am. Res: Applied areas of statistics, including life data analysis, sequential analysis and experimental design. Mailing Add: Dept of Statist Iowa State Univ Ames IA 50011

MEEKINS, JOHN FRED, b Boston, Mass, Oct 4, 37; m 61; c 2. SOLAR PHYSICS, X-RAY ASTRONOMY. Educ: Bowdoin Col, BA, 59; Cath Univ Am, PhD(physics), 73. Prof Exp: RES PHYSICIST, NAVAL RES LAB, 59- Mem: Am Geophys Union; Sigma Xi. Res: Astrophysics, especially concerning high temperature astrophysical plasmas. Mailing Add: Naval Res Lab Code 7125 Washington DC 20375

MEEKS, BENJAMIN SPENCER, JR, b Florence, SC, Nov 17, 24; m 49; c 2. ORGANIC CHEMISTRY. Educ: Univ SC, BS, 44; Cornell Univ, PhD(org chem), 51. Prof Exp: Res chemist, US Rubber Co, 51-52 & Tenn Eastman Co, 52-56; assoc prof chem, Mercer Univ, 56-58; Univ Ky Contract Team, Bandung Tech Inst, 58-62; assoc prof, 62-65, PROF CHEM, MOORHEAD STATE UNIV, 65-, CHMN DEPT, 74- Concurrent Pos: Res grant, Univ Col, London, 69-70. Mem: AAAS; Am Chem Soc; The Chem Soc. Res: Synthesis of pteridines; general organic synthesis; condensation polymerization; ionic reaction mechanisms; diazo ring enlargements. Mailing Add: Dept of Chem Moorhead State Univ Moorhead MN 56560

MEEKS, FRANK ROBERT, b Ft Worth, Tex, Dec 5, 28. PHYSICAL CHEMISTRY. Educ: Tex Christian Univ, BA, 49; Polytech Inst of NY, PhD, 56. Prof Exp: Asst prof phys chem, 57-60, ASSOC PROF PHYS CHEM, UNIV CINCINNATI, 60- Concurrent Pos: Res scholar, Univ Montpellier, France, 63-64. Mem: Am Chem Soc; Sigma Xi. Res: Periodic precipitation and condensation kinetics; critical phenomena in binary liquid systems; statistical mechanics; thermodynamics of irreversible processes. Mailing Add: Dept of Chem Univ of Cincinnati Cincinnati OH 45221

MEEKS, HAROLD AUSTIN, b New Haven, Conn, Aug 22, 30; m 54; c 2. GEOGRAPHY. Educ: Middlebury Col, AB, 56; Univ Minn, MA, 58, PhD(geog), 64. Prof Exp: Instr geog, Exten Div, Univ Minn, 57-58; instr, Mich Technol Univ, 59-63, asst prof, 63-64; asst prof, Suomi Col, 60-64; asst prof, 64-69, ASSOC PROF GEOG, UNIV VT, 69- Concurrent Pos: Asst prof geog, Suomi Col, 60-64; lectr, Univ Durham, 70-71; Fulbright travel grant, 70; fac travel grant, Univ Vt, 71. Mem: Asn Am Geog; Am Geog Soc. Res: Resources of the United States and Soviet Union; settlement geography and environmental problems in New England; geography and land use in New England, particularly Vermont. Mailing Add: Dept of Geog Univ of Vt Burlington VT 05401

MEEKS, MARION LITTLETON, b Gainesville, Ga, Oct 1, 23; m 44, 70; c 3. RADIO ASTRONOMY. Educ: Ga Inst Technol, BS, 43, MS, 48; Duke Univ, PhD(physics), 51. Prof Exp: Instr physics, Ga Inst Technol, 46-47 & Duke Univ, 48-49; asst prof, Clemson Col, 50-51; from asst prof to assoc prof, Ga Inst Technol, 51-61; staff mem, Lincoln Lab, 61-72, STAFF MEM, HAYSTACK OBSERV, MASS INST TECHNOL, 72- Concurrent Pos: Physicist, Harvard Observ, 59-60; vis prof, Univ Mass, Amherst, 71; mem comn 5, Int Union Radio Sci; mem comn 40, Int Astron Union; adj prof, Five Col Astron Dept, 74-76. Mem: Am Astron Soc. Res: Microwave spectral lines; computer control and data processing in radio astronomy; development of computer animation for teaching astronomy. Mailing Add: Haystack Observ Mass Inst Technol Northeast Radio Observ Corp Westford MA 01886

MEEKS, WILKISON (WINFIELD), b Pittsburgh, Pa, Apr 4, 15; m 46; c 2. ACOUSTICS. Educ: Maryville Col, AB, 37; Northwestern Univ, MS, 39, PhD(physics), 41. Prof Exp: Contract employee, US Naval Ord Lab, 41-44; staff physicist, Haskins Labs, Inc, NY, 44-46; assoc prof physics, Western Md Col, 46-47 & Southern Ill Univ, 47-48; asst prof, Western Reserve Univ, 48-55; physicist res ctr, B F Goodrich Co, Ohio, 55-58; assoc prof physics, 58-60, chmn dept, 60-68, PROF PHYSICS, ROSE-HULMAN INST TECHNOL, 60- Mem: Am Phys Soc; Acoust Soc Am; Am Asn Physics Teachers. Res: Magneto-mechanical effects; underwater sound. Mailing Add: Dept of Physics Rose-Hulman Inst of Technol Terre Haute IN 47803

MEELHEIM, RICHARD YOUNG, b Cape Charles, Va, Aug 30, 25; m 51; c 3. PHYSICAL CHEMISTRY. Educ: Univ Va, BS, 50, PhD(chem), 58. Prof Exp: Anal chemist, Monsanto Chem Co, 50-54; res chemist fiber surface res, 58-62, chemist, Dacron Res Lab, 62-67, SR RES CHEMIST, DACRON RES LAB, E I DU PONT DE NEMOURS & CO, INC, 67- Mem: Am Chem Soc. Res: Research and development fibers. Mailing Add: 1218 Stockton Rd Kinston NC 28501

MEEN, RONALD HUGH, b Can, Nov 25, 25; m 68. SYNTHETIC ORGANIC CHEMISTRY. Educ: Univ Toronto, BA, 47, MA, 49, PhD(chem), 53. Prof Exp: Fel & res assoc org chem, Iowa State Col, 53-54; RES CHEMIST, TENN EASTMAN CO, 54- Mem: Am Chem Soc. Res: Chemistry; organic chemical development. Mailing Add: 2121 Cypress St Kingsport TN 37664

MEENTEMEYER, VERNON GEORGE, b Centralia, Ill, Nov 7, 42; m 66; c 2. CLIMATOLOGY. Educ: Southern Ill Univ, BA, 65, MA, 68, PhD(climat), 71. Prof Exp: Asst prof phys geog, Southern Ill Univ, 73; ASST PROF PHYS GEOG, UNIV GA, 73- Mem: Asn Am Geogrs; AAAS; Soc Gen Systs Res. Res: Climatic influences on decomposer food chains; atmospheric hazards; impact on natural ecosystems, and probability mapping; radiant energy flows in forests; biogeography of poison ivy; forest floor hydrology. Mailing Add: Dept of Geog Univ of Ga Athens GA 30602

MEERBAUM, SAMUEL, b Brno, Czech, May 1, 19; m 46; c 3. BIOENGINEERING. Educ: Mass Inst Technol, BS & MS, 46; Univ Calif, Los Angeles, PhD(bioeng), 71. Prof Exp: Resident mech engr, M W Kellogg Co, 46-51, sect head, 51-53, develop staff engr, 53-58, sect head, 58-62; asst chief engr, Arde Inc, 62-65; prin scientist, Rocketdyne, NAm Aviation, 65-68; SR RES SCIENTIST, CEDARS-SINAI MED CTR, 71- Honors & Awards: Award for Exhibit on Circulatory Assistance, AMA, Am Col Cardiol & Am Col Chest Physicians. Mem: Fel Am Col Cardiol; Instrument Soc Am. Res: Cardiology; physiology; bioengineering; mechanisms of cardiovascular system hemodynamics; treatment of heart disease; mechanical assistance to the failing heart; fluid dynamics; heat and mass transfer. Mailing Add: 5741 El Canon Ave Woodland Hills CA 91364

MEERBOTT, WILLIAM KEDDIE, b Jersey City, NJ, Aug 31, 18; m 43; c 2. PETROLEUM CHEMISTRY. Educ: St Peters Col, BS, 40; Lehigh Univ, MS, 42. Prof Exp: Sr res chemist, Shell Oil Co, 49, group leader, 51-66, staff res chemist, 66-72; STAFF RES CHEMIST, SHELL DEVELOP CO, 72- Mem: Am Chem Soc. Res: Catalysis in the field of petroleum chemistry and related to hydro processing, desulfurization and catalytic reforming. Mailing Add: 3006 Winslow Houston TX 77025

MEERKAMPER, BEAT, physical chemistry, see 12th edition

MEEROVITCH, EUGENE, b Vladivostok, Russia, July 11, 19; Can citizen; m 61; c 2. PARASITOLOGY. Educ: St John's Univ, China, BSc, 47; McGill Univ, MSc, 53, PhD(parasitol), 57. Prof Exp: Res asst parasitol, Hebrew Univ, Israel, 48-53; res asst, 53-57, from asst prof to assoc prof, 57-71, PROF PARASITOL, MACDONALD COL, McGILL UNIV, 71- Concurrent Pos: US Acad Sci Donner fel, Nat Inst Med Res, Eng, 58-59; consult, WHO, 69-; vis sr scientist, Wellcome Res Labs, Eng, 75-76. Mem: AAAS; Am Soc Parasitol; Am Soc Trop Med & Hyg; Royal Soc Trop Med & Hyg; Can Soc Zoologists. Res: Amoebiasis, especially immunology, serology and host-parasite relations; trichinosis, especially bionomics and strain variation; in vitro cultivation of parasites; experimental concomitant helminth-protozoal infections. Mailing Add: Inst of Parasitol MacDonald Campus of McGill Univ MacDonald College PQ Can

MEERS, JOSEPH TINSLEY, b Hardin Co, Ky, June 20, 23; m 47; c 3. RESEARCH ADMINISTRATION. Educ: Western Ky State Teachers Col, BS, 44; Univ Ky, MS, 47; Univ NC, PhD(physics), 54. Prof Exp: Staff scientist, Nat Carbon Co, 54-60, group leader physics, 60-67, asst dir process eng, 67-70, dir electrode technol, 70-74, DIR RES, CARBON PROD DIV, UNION CARBIDE CORP, 74- Mem: Am Carbon Soc. Res: Formulation and administration of research programs on carbon and its precursors, with special emphasis on the growth and motivation of scientists. Mailing Add: Parma Tech Ctr Union Carbide 12900 Snow Rd Parma OH 44130

MEESE, JON MICHAEL, b Indianapolis, Ind, Aug 5, 38; m 63; c 1. SOLID STATE PHYSICS. Educ: Univ Cincinnati, BS, 61; Purdue Univ, Lafayette, MS, 64, PhD(physics), 70. Prof Exp: Jr physicist, Wabash Magnetics, 60-61; asst instr solid state physics, Purdue Univ, Lafayette, 61-62, assoc instr, 62-65, res asst, 65-70; RES PHYSICIST, UNIV DAYTON, 70- Concurrent Pos: In-house contractor, Aerospace Res Labs, Wright-Patterson AFB, 70-72; adj prof physics, Wright State Univ, 75- Mem: Am Phys Soc. Res: Radiation damage and ion implantation in semiconductors; luminescence; electro-optical devices; solar cells. Mailing Add: Dept of Physics Univ of Dayton Dayton OH 45440

MEETZ, GERALD DAVID, b Aurora, Ill, Aug 22, 37; m 74. IMMUNOLOGY, CELL BIOLOGY. Educ: NCent Col, Ill, BA, 59; Univ Ill, MS, 67, PhD(anat), 69. Prof Exp: USPHS fel cellular, molecular & develop biol, Univ Colo, 69-70; fel pediat, Univ Minn, Minneapolis, 70-71, fel path, 71-72; asst prof anat, Med Col Va, 72-75; ASST PROF ANAT, MARQUETTE UNIV SCH DENT, 75- Mem: Am Soc Cell Biol; Reticuloendothelial Soc. Res: Role of histones in development of avian erythrocyte; gene activation in lymphoid cells. Mailing Add: Div of Anat Marquette Univ Sch of Dent Milwaukee WI 53233

MEEUSE, BASTIAAN JACOB DIRK, b Sukabumi, Indonesia, May 9, 16; m 42; c 2. PLANT BIOCHEMISTRY, PLANT PHYSIOLOGY. Educ: Univ Leiden, BSc, 36, Drs, 39; Univ Delft, DTechSc, 43. Prof Exp: Lab asst tech bot, Univ Delft, 43-46, chief asst, 46-49, lectr, 49-52; from asst prof to assoc prof bot, 52-60, PROF BOT, UNIV WASH, 60- Concurrent Pos: Rockefeller Found fel, Bot Lab, Univ Pa, 47-49; actg head biochem lab, Univ Delft, 50-52 & NSF sr fel, Lab Microbiol, 62-63. Mem: AAAS; Bot Soc Am; Am Soc Plant Physiol; Royal Neth Bot Soc; cor mem Royal Neth Acad Sci. Res: Plant respiration and photosynthesis; carbohydrate biochemistry; physiology and biochemistry of algae; pollination; animal behavior; entomology; photorespiration; dispersal of seeds and fruits. Mailing Add: Dept of Bot AK-10 Johnson Hall Univ of Wash Seattle WA 98195

MEEUWIG, RICHARD O'BANNON, b St Louis, Mo, Dec 8, 27; m 67; c 2. FOREST ECOLOGY. Educ: Univ Calif, Berkeley, BS, 51, MS, 60; Utah State Univ, PhD(soil physics), 64. Prof Exp: Forester timber mgt, US Forest Serv, 51-55, forester watershed mgt, 56-64, soil scientist, 64-75, RES FORESTER, INTERMOUNTAIN FOREST & RANGE EXP STA, 75- Concurrent Pos: Adj prof, Univ Nev, Reno. Mem: Soc Range Mgt. Res: Ecology and management of pinyon-juniper woodlands. Mailing Add: 1200 Monroe Reno NV 89502

MEEZAN, ELIAS, b New York, NY, Mar 5, 42; m 67. BIOCHEMISTRY, PHARMACOLOGY. Educ: City Col New York, BS, 62; Duke Univ, PhD(biochem), 66. Prof Exp: Asst prof pharmacol, Duke Univ, 69-70; asst prof pharmacol, 70-75, ASSOC PROF PHARMACOL, MED SCH, UNIV ARIZ, 75- Concurrent Pos: Helen Hay Whitney fel, 66-69. Mem: AAAS; Am Soc Pharmacol & Exp Therapeut; NY Acad Sci. Res: Biochemistry and pharmacology of carbohydrate containing compounds in health and disease. Mailing Add: Dept of Pharmacol Univ of Ariz Med Sch Tucson AZ 85721

MEFFORD, DAVID ALLEN, b Keokuk, Iowa, Dec 28, 28; m 51; c 2. ANALYTICAL CHEMISTRY. Educ: Randolph-Macon Col, BS, 51. Prof Exp: Asst biol, Randolph-Macon Col, 50; chemist, 53-73, with qual assurance, 57-62, DIR QUAL ASSURANCE, A H ROBINS CO, INC, 73- Mem: AAAS; Am Chem Soc; Am Soc Qual Control; Pharmaceut Mfrs Asn. Res: Microelemental analysis; analytical problems related to pharmaceutical products; infrared quantitative methods. Mailing Add: Qual Assurance A H Robins Co Inc 1407 Cummings Dr Richmond VA 23220

MEGARD, ROBERT O, b Garretson, SDak, Dec 4, 33; m 58. ZOOLOGY. Educ: St Olaf Col, BA, 56; Univ NMex, MS, 58; Ind Univ, PhD(zool), 62. Prof Exp: Res fel limnol, Univ Minn, Minneapolis, 61-64, res assoc, 64-67, asst prof ecol, 67-71, asst prof, Col Biol Sci, St Paul, 71-72, ASSOC PROF ECOL, COL BIOL SCI, UNIV MINN, ST PAUL, 72- Mem: AAAS; Am Soc Limnol & Oceanog; Ecol Soc Am. Res: Limnology ecology of plankton; paleolimnology; biology and ecology of plankton populations. Mailing Add: Dept of Ecol & Behav Biol Biol Sci Ctr Univ of Minn St Paul MN 55101

MEGARGLE, ROBERT G, b Flushing, NY, Oct 11, 41; m 63. ANALYTICAL CHEMISTRY. Educ: Clarkson Col, BS, 63, PhD(chem), 68. Prof Exp: NSF fel chem, Univ Minn, Minneapolis, 66-67; asst prof, Univ Mo-Columbia, 67-72; ASSOC PROF CHEM, CLEVELAND STATE UNIV, 72- Mem: AAAS; Am Chem Soc. Res: Development of instrumentation for measurement and control of experiments, including the use of the computer in laboratory experimentation; titration methods of analysis in aqueous and nonaqueous solvents. Mailing Add: Dept of Chem Cleveland State Univ Cleveland OH 44115

MEGAW, WILLIAM JAMES, b Belfast, Northern Ireland, July 8, 24; m 46; c 3. ATMOSPHERIC PHYSICS. Educ: Univ Liverpool, BSc, 51, DSc(physics), 73. Prof Exp: Sci officer, UK Atomic Energy Authority, 51-53; sr sci officer, Atomic Energy Res Estab, Eng, 53-57, prin sci officer, 57-61, sr prin sci officer, 61-71; PROF ATMOSPHERIC PHYSICS, YORK UNIV, 71-, DIR CTR RES ENVIRON QUAL, 74- Concurrent Pos: Mem, Comt Nucleation, Comn Cloud Physics, Int Asn Meteorol & Atmospheric Physics, 65-; subcomn ions, aerosols & radioactivity, Int Comn Atmospheric Elec, 67-; consult, Danish AEC, 69- Mem: Fel Brit Inst Physics. Res: Physics of particles in the atmosphere; inhibition of condensation on cloud nuclei; fog modification; environmental protection. Mailing Add: Dept of Physics York Univ 4700 Keele St Downsview ON Can

MEGEL, HERBERT, b Newark, NJ, Nov 10, 26; m 51; c 2. IMMUNOLOGY. Educ: NY Univ, BS, 48, MS, 50, PhD(exp biol), 54. Prof Exp: Res physiologist endocrine physiol, Princeton Labs, 54-59; res physiologist environ physiol, Boeing Co, 59-62; res biochemist, Nat Drug Co, 62-70; SECT HEAD IMMUNOL, MERRELLNAT LABS, 70- Mem: AAAS; Reticuloendothelial Soc; Am Soc Pharmacol & Exp Therapeut; Soc Exp Biol & Med; NY Acad Sci. Res: Immunopharmacology; biochemical pharmacology; endocrine physiology. Mailing Add: Merrell-Nat Labs Div of Richardson-Merrell Inc Cincinnati OH 45215

MEGGERS, BETTY JANE, b Washington, DC, Dec 5, 21; m 46. ANTHROPOLOGY. Educ: Univ Pa, AB, 43; Univ Mich, MA, 44; Columbia Univ, PhD(anthrop), 52. Prof Exp: Instr anthrop, Am Univ, 50-51; RES ASSOC ANTHROP, NAT MUS NATURAL HIST, SMITHSONIAN INST, 54- Concurrent Pos: Consult, Battelle Mem Inst, 65-68; chmn, Behav Sci Panel, Nat Res Coun, 68-69. Honors & Awards: Gold Medal, Int Cong Americanists, 66; Order of Merit, Govt of Ecuador, 66. Mem: AAAS; Soc Am Archaeol; Am Geog Soc; Asn Trop Biol. Res: Development of culture in the New World, particularly in terms of the influence of transpacific contact and the operation of natural selection on cultural behavior. Mailing Add: Nat Mus Natural Hist Smithsonian Inst Washington DC 20560

MEGGISON, DAVID LAURENCE, b Lynn, Mass, Dec 24, 28; m 52; c 4. FOOD TECHNOLOGY. Educ: Johns Hopkins Univ, BA, 48; Univ Mass, MS, 50, PhD(food sci), 53. Prof Exp: Assoc technologist cent labs, Gen Foods Corp, 52-55, proj leader frozen foods, Birds Eye Labs, 55-59; food prod res lab, Borden Foods Co, NY, 59-62;

sect chief, Hunt Foods & Industs, Inc, 62-64, assoc dir res, Prod Develop, Hunt-Wesson Foods, 64-74; VPRES RES & DEVELOP, UNITED VINTNERS, INC, 74- Mem: Inst Food Technol. Res: Development of new products and processes in instant foods; frozen and canned foods; wines. Mailing Add: United Vintners Inc 601 Fourth St San Francisco CA 94107

MEGGITT, WILLIAM FREDRIC, b Green Springs, Ohio, Feb 9, 28; m 48; c 1. AGRONOMY. Educ: Ohio State Univ, BS, 50, MS, 51; Rutgers Univ, PhD(weed control, farm crops), 54. Prof Exp: Res agronomist, Agr Res Serv, USDA, 57-58; asst prof farm crops & weed control, Rutgers Univ, 58-60; assoc prof crop sci & weed control, 60-66, PROF CROP SCI, MICH STATE UNIV, 66- Mem: Am Soc Agron; Crop Sci Soc Am; Weed Sci Soc Am. Res: Weed control, chemical and cultural means; weed life cycles and competition; soil residues, penetration, translocation, accumulation and sites of action of herbicides. Mailing Add: Dept of Crop & Soil Sci Mich State Univ East Lansing MI 48824

MEGIBBEN, CHARLES KIMBROUGH, b Lexington, Ky, Oct 22, 36; m 57; c 4. MATHEMATICS. Educ: Southern Methodist Univ, BS, 59; Auburn Univ, PhD(math), 63. Prof Exp: Asst prof math, Tex Tech Col, 63-64; res assoc, Off Naval Res, Univ Wash, 64-65; asst prof, Univ Houston, 65-67; ASSOC PROF MATH, VANDERBILT UNIV, 67- Concurrent Pos: NSF res grants, 66-72. Mem: Am Math Soc. Res: Theory of Abelian groups; rings and modules. Mailing Add: Box 1589 Vanderbilt Univ Nashville TN 37235

MEGILL, LAWRENCE REXFORD, b Potsdam, Ohio, July 5, 25; m 46; c 2. PHYSICS. Educ: Univ Nebr, BSc, 49, MA, 51; Univ Colo, PhD(physics), 59. Prof Exp: Mem staff, Los Alamos Sci Lab, Univ Calif, 51-53; physicist inst telecommun sci & aeronomy, Environ Sci Serv Admin, Colo, 55-69; PROF PHYSICS & ELEC ENG, UTAH STATE UNIV, 69- Concurrent Pos: Prog dir aeronomy, NSF, 74-75. Res: Atmospheric physics, particularly photometry and spectroscopy. Mailing Add: Dept of Physics Utah State Univ Logan UT 84322

MEGIRIAN, ROBERT, b New York, NY, June 18, 26; m 57; c 1. PHARMACOLOGY. Educ: Colgate Univ, AB, 51; Univ Rochester, MS, 53; Boston Univ, PhD(pharmacol), 57. Prof Exp: Res assoc pharmacol, Univ Rochester, 51-53; asst, Boston Univ, 53-56; pharmacologist, US Food & Drug Admin, 57-61; asst prof, 61-68, assoc prof, 68-74, PROF PHARMACOL, ALBANY MED COL, 75- Concurrent Pos: Res fel, Boston City Hosp, Mass, 57. Mem: Reticuloendothelial Soc; Am Soc Pharmacol & Exp Therapeut; Soc Exp Biol & Med; Am Soc Clin Pharmacol & Exp Therapeut. Res: Effect of drugs on the reticuloendothelial system. Mailing Add: Dept of Pharmacol Albany Med Col Albany NY 12208

MEGLEN, ROBERT RUDOLF, physical chemistry, analytical chemistry, see 12th edition

MEGLITSCH, PAUL ALLEN, b Harvel, Ill, Mar 3, 14; m 38; c 3. PROTOZOOLOGY. Educ: Univ Ill, BS, 35, MS, 36, PhD(zool), 38. Prof Exp: Teaching fel zool & protozool, Univ Ill, 37-38; instr biol, Wright Jr Col, 39-42, Gary Col, 43 & Herzl Jr Col, 43-49; assoc prof, 49-58, PROF BIOL, DRAKE UNIV, 58- Concurrent Pos: Fulbright fel, 58; vis fel, Hull Col, Eng, 65-66. Mem: Am Soc Zool; Soc Protozool; Soc Syst Zool; Am Micros Soc; Wildlife Dis Asn. Res: Cytology; systematics; protozoa; myxosporida systematics; bark-lichen association and air pollution. Mailing Add: Dept of Biol Drake Univ Des Moines IA 50311

MEGNA, IGNAZIO SALVATORE, organic chemistry, polymer chemistry, see 12th edition

MEGO, JOHN L, b Pukwana, SDak, Sept 29, 22; m 53; c 2. BIOCHEMISTRY. Educ: Johns Hopkins Univ, PhD(biol), 60. Prof Exp: Res assoc plant physiol, Johns Hopkins Univ, 60; res asst neurosurg, Baltimore City Hosps, 60-62, res assoc, 62-65, res chief, 65-67; assoc prof biol, 67-73, PROF BIOL, UNIV ALA, TUSCALOOSA, 73- Concurrent Pos: Co-prin investr, NIH-AEC grant, 64-; Nat Acad Sci exchange fel, Univ Bratislava, 67; Nat Inst Environ Health Sci grant award, 71-74; Nat Inst Gen Med Sci res grant, 75-78. Mem: AAAS; Am Soc Cell Biol; Am Chem Soc. Res: Biochemical properties of lysosomes and related subcellular particles. Mailing Add: Dept of Biol Box 1927 Univ of Ala in Tuscaloosa University AL 35486

MEGRAW, ROBERT ARTHUR, b Rochester, Minn, Dec 5, 39; m 69. FORESTRY, WOOD TECHNOLOGY. Educ: Univ Minn, BS, 62, PhD(forest prod eng), 66. Prof Exp: Res scientist pioneering dept, Res Div, 66-68, sr scientist, Res & Eng Div, 68-74, MGR WOOD SCI & MORPHOL SECT, RES & ENG DIV, WEYERHAEUSER CO, 74- Honors & Awards: Wood Award, 67. Mem: Forest Prod Res Soc; Soc Wood Sci & Technol; Tech Asn Pulp & Paper Indust. Res: Wood and fiber properties; tree growth-fiber property relationships; flame spread control in wood and fiber products; x-ray transmission as a wood research tool; scanning electron microscopy. Mailing Add: Weyerhaeuser Co Seattle Lab 3400-13th Ave SW Seattle WA 98134

MEGRAW, ROBERT ELLIS, b Philadelphia, Pa, Feb 10, 30; m 71; c 5. CLINICAL CHEMISTRY. Educ: Fla State Univ, BA, 56, MS, 60; Iowa State Univ, PhD(bact), 64. Prof Exp: Fel biochem, Albert Einstein Med Ctr, 64-66; scientist, Warner-Lambert Res Inst, 66-71; res biochemist, Sigma Chem Co, 71-73; MGR UNITEST CHEM, BIO-DYNAMICS/bmc, 73- Mem: AAAS; Am Chem Soc; Sigma Xi; Soc Exp Biol & Med; Am Asn Clin Chemists. Res: Research and development of clinical diagnostic reagents. Mailing Add: PO Box 50100 Indianapolis IN 46250

MEGREGIAN, STEPHEN, b Smyrna, Turkey, Jan 22, 15; US citizen; m 38; c 4. ANALYTICAL CHEMISTRY. Educ: Wayne State Univ, BS, 38; Univ Md, MS, 55. Prof Exp: Jr chemist, USPHS, Cincinnati, Ohio, 40-43; vpres, Aja & Mfg Corp, Mich, 47-50; dir labs, Metrop Sanit Dist, Chicago, Ill, 66-67; dir res, Hach Chem Co, 67-68; vpres, Environ Develop Inc, Md, 68-70; DIR WATER QUAL PROGS, WAPORA, INC, DC, 70- Concurrent Pos: Consult, Pan Am Health Orgn, 65. Honors & Awards: Commendation Medal, USPHS, 65. Mem: Am Chem Soc. Res: Water and wastewater analysis; pollution control; fluoridation and defluoridation; water resource management; control of thermal pollution. Mailing Add: 4536 Hornbeam Dr Rockville MD 20853

MEGRUE, GEORGE HENRY, b Jamaica, NY, Mar 23, 36; m 58; c 4. GEOCHEMISTRY, RESEARCH ADMINISTRATION. Educ: Amherst Col, BA, 57; Columbia Univ, MA, 59, PhD(geol), 62. Prof Exp: Res assoc chem, Brookhaven Nat Lab, 62-64, assoc chemist, 64-66; geochemist & cosmochemist, Smithsonian Inst Astrophys Observ & Harvard Col Observ, 66-74; FOUNDER & PRES, MEGRUE MICROANAL SYSTS CO, 75- Concurrent Pos: Sci leader, Nat Geog Ethiopian Rift Valley Exped, 69; prin investr, Apollo 12, 14 & 15 Manned Space Flights, NASA, 71-73. Mem: AAAS; Am Geophys Union; Am Chem Soc. Res: Thermal history of meteorites; tectonic history of Ethiopian Rift Valley; distribution and origin of helium, neon, and argon isotopes in meteorites and lunar rocks; laser probe mass spectrometry; chemistry of cosmic dust; application of laser microchemical analyses to scientific research and development. Mailing Add: Megrue Microanal Systs Co Box 523 New Canaan CT 06840

MEGSON, FREDERIC HOUGHTON, b Duluth, Minn, Feb 27, 17; m 44; c 2. ORGANIC CHEMISTRY. Educ: Columbia Univ, AB, 39; NY Univ, MS, 47, PhD(org chem), 55. Prof Exp: Analyst res labs, Tex Co, 39-41; SR RES CHEMIST, AM CYANAMID CO, 41- Mem: Am Chem Soc. Res: Synthetic organic chemistry; intermediates; dyes; textile chemicals; analytical chemistry; identification of organic structures; plant genetics. Mailing Add: Res Div Am Cyanamid Co Bound Brook NJ 08805

MEGUERIAN, GARBIS H, b Turkey, Sept 10, 22; nat US; m 51; c 2. PHYSICAL ORGANIC CHEMISTRY. Educ: Am Univ, Beirut, BS, 47; Brown Univ, PhD(chem), 50. Prof Exp: Fel & res chemist, Harvard Univ, 50-52; chemist, Standard Oil Co Ind, 52-58, group leader, 58-64, RES ASSOC AMOCO OIL CO, 64- Mem: Am Chem Soc. Res: Reactions of elementary sulfur; oxidation of mercaptans; high temperature oxidation of hydrocarbons; nitrogen-oxygen control in automotive emissions. Mailing Add: Amoco Oil Co PO Box 400 Naperville IL 60540

MEHAL, EDWARD WALTER, b Detroit, Mich, Dec 9, 31; m 59; c 2. INORGANIC CHEMISTRY. Educ: Wayne State Univ, BS, 54; Univ Mich, MS, 56. Prof Exp: Mem tech staff, Tex Instruments, Inc, 57-69; MGR, SPECTRONICS, INC, 69- Mem: Am Chem Soc; Electrochem Soc. Res: Synthesis of semiconductor materials and crystal growth. Mailing Add: Spectronics Inc 830 E Arapaho Richardson TX 75080

MEHENDALE, HARIHARA MAHADEVA, b Philya, India, Jan 12, 42; m 68; c 1. INSECT TOXICOLOGY. Educ: Karnatak Univ, India, BSc, 63; NC State Univ, MS, 66, PhD(entom, physiol), 69. Prof Exp: Fel entom, Univ Ky, 69-71; NIH vis fel, Anal & Synthetic Chem Br, Nat Inst Environ Health Sci, 71-72, staff fel, Environ Pharmacol & Toxicol Br, 72-75; ASST PROF PHARMACOL & TOXICOL, UNIV MISS MED CTR, 75- Mem: Entom Soc Am; Am Chem Soc; Entom Soc India; Indian Sci Cong Asn. Res: Use of isolated perfused organs in studies of environmental toxicology; pharmacokinetics of environmental agents; effect of toxic agents on hepato-biliary and pulmonary systems. Mailing Add: Dept of Pharmacol & Toxicol Univ of Miss Med Ctr Jackson MS 39216

MEHERIUK, MICHAEL, b Derwent, Alta, June 5, 36; m 63. PLANT BIOCHEMISTRY. Educ: Univ Alta, BSc, 57, BEd, 59, PhD(biochem), 65. Prof Exp: RES SCIENTIST, CAN DEP AGR, 65- Res: Post-harvest physiology; storage of tree fruits, especially pears and apples. Mailing Add: RR 2 Summerland BC Can

MEHL, CLARENCE ROBERT, theoretical physics, see 12th edition

MEHL, JAMES BERNARD, b Minneapolis, Minn, May 5, 39; m 61; c 2. PHYSICS. Educ: Univ Minn, BPhys, 61, MS, 64, PhD(physics), 66. Prof Exp: Res assoc physics, Univ Ore, 66-68; asst prof, 68-74, ASSOC PROF PHYSICS, UNIV DEL, 74- Mem: Am Phys Soc; Am Asn Physics Teachers. Res: Low temperature physics; superfluidity; acoustics. Mailing Add: Dept of Physics Univ of Del Newark DE 19711

MEHL, JOHN WILBUR, b Upland, Calif, July 8, 10; m 39; c 4. BIOCHEMISTRY. Educ: Calif Inst Technol, BS, 31; Univ Calif, AM, 33, PhD(biochem), 36. Prof Exp: Asst, Univ Calif, 33-36; instr & tutor biochem sci, Harvard Univ, 36-40; asst prof biochem, Sch Med, Univ Southern Calif, 40-45, from assoc prof to prof, 45-68; dep div dir biol & med sci, NSF, 68-75; RETIRED. Concurrent Pos: Res assoc, Harvard Med Sch, 38-40; head chemist, Los Angeles County Gen Hosp, 46-49; prog dir, NSF, 62-63. Mem: Am Chem Soc; Soc Gen Physiologists; Am Soc Biol Chem. Res: Chemistry of proteins and enzymes. Mailing Add: 212 Calle De Anza San Clemente CA 92672

MEHLENBACHER, LYLE E, b Dansville, NY, Mar 4, 10; m 42; c 2. MATHEMATICAL ANALYSIS. Educ: NY State Col Teachers, AB, 31; Univ Mich, AM, 34, PhD(math), 36. Prof Exp: Teacher high sch, NY, 31-33; instr math, SDak State Col, 36-37; prof & head dept, Ariz State Col, 37-45; assoc prof, Univ Detroit, 45-48, chmn dept math, 47-63, prof, 48-75, coordr sponsored educ progs, 63-75, assoc dean grad sch, 66-72, dir grants admin, 72-75; RETIRED. Mem: Am Math Soc; Math Asn Am. Res: Analysis-asymptotic developments of functions defined by Maclaurin series; interrelations of fundamental solutions of hypergeometric equation; differential equations. Mailing Add: 1957 Edgewood Blvd Berkley MI 48072

MEHLER, ALAN HASKELL, b St Louis, Mo, May 24, 22; m 43; c 4. BIOCHEMISTRY. Educ: Wash Univ, AB, 42; NY Univ, PhD(biochem), 48. Prof Exp: Res assoc, Rheumatic Fever Res Inst, Northwestern Univ, 49; asst prof, Inst Radiobiol & Biophys, Univ Chicago, 49-51; vis scientist, NIH, 51-52, chemist, 52-60, chief enzyme chem sect, Nat Inst Dent Res, 60-65; PROF BIOCHEM & CHMN DEPT, MED COL, UNIV WIS, 65- Concurrent Pos: Mem, Weizmann Inst, 48; NSF sr fel, Dept Genetics, Sorbonne, 58-59. Mem: AAAS; Am Chem Soc; Am Soc Biol Chem; Soc Gen Physiol; Soc Exp Biol & Med. Res: Enzyme chemistry; transfer ribonucleic acid; intermediary metabolism. Mailing Add: Dept of Biochem Med Col of Wis 561 N 15th St Milwaukee WI 53233

MEHLER, ERNEST LOUIS, b Amsterdam, Holland, Sept 25, 38; US citizen; m 64; c 2. THEORETICAL CHEMISTRY. Educ: Ill Inst Technol, BS, 60; Johns Hopkins Univ, MA, 64; Iowa State Univ, PhD(theoret chem), 68. Prof Exp: Instr theoret chem, Univ Groningen, 71-73; SR RES ASSOC THEORET BIOCHEM, UNIV BASEL, 74- Concurrent Pos: Fel, Univ Wash, 68-70. Mem: Int Soc Quantum Biol. Res: Electronic structures and properties of atoms and small molecules; ab initio methods for large molecules with application to biochemistry. Mailing Add: Biocenter Univ of Basel Klingelbergstr 70 Basel Switzerland

MEHLER, WILLIAM RAPHAEL, b Cleveland, Ohio, Apr 26, 26; m 53; c 2. NEUROANATOMY. Educ: John Carroll Univ, BS, 49; St Louis Univ, MS, 51; Univ Md, PhD, 59. Prof Exp: Physiologist, NIH, 54-55; neuroanatomist, Walter Reed Army Inst Res, DC, 55-62; res scientist, 62-68, CHIEF NEUROBIOL BR, AMES RES CTR, NASA, MOFFETT FIELD, 68- Concurrent Pos: Fel anat, St Louis Univ, 49-52; fel, Univ Tenn, 52-54; lectr, George Washington Univ, 59-60; lectr, Univ Calif, San Francisco, 62-72, adj assoc prof anat, 72-; ed sect structure, J Pain. Mem: Am Asn Anat; Soc Neurosci; Int Asn Study Pain. Mailing Add: NASA Ames Res Ctr Moffett Field CA 94035

MEHLHAFF, LEON CURTIS, b Lodi, Calif, Apr 28, 40; m 65; c 3. ANALYTICAL CHEMISTRY, POLYMER CHEMISTRY. Educ: Univ Calif, Berkeley, BS, 61; Univ Wash, PhD(anal chem), 65. Prof Exp: Chemist plastics dept, E I du Pont de Nemours & Co, 65-68; asst prof anal chem, 68-71, ASSOC PROF ANAL CHEM, UNIV PUGET SOUND, 71- Concurrent Pos: Regional coordr, Wash Shoreline Tech Adv Bd, 71-73. Mem: Am Chem Soc; Electrochem Soc. Res: Analytical chemistry of polymer systems; instrumental analytical chemistry; electrochemistry; environmental science program. Mailing Add: Dept of Chem Univ of Puget Sound Tacoma WA 98416

MEHLIG, JOSEPH PARKE, b Tipton, Ind, June 26, 85; m 33. ANALYTICAL CHEMISTRY. Educ: Purdue Univ, BS, 08, MS, 10, PhD(anal chem), 31. Prof Exp: Asst, Purdue Univ, 08-10; asst chemist, Bur Chem, USDA, 10-12 & US Forest Serv, Wis, 12-13; chemist, Pittsburgh Testing Lab, Pa, 13; teacher high sch, Colo, 15-17 & Utah, 17-20; from instr to prof anal chem, Ore State Univ, 20-50, emer prof chem, 50-67, chemist, Agr Exp Sta, 52-67; RETIRED. Mem: Am Chem Soc. Res: Spectrophotometric and cerate methods of analysis; methods of analysis for trace elements in animal feeds. Mailing Add: 835 S 11th St Corvallis OR 97330

MEHLQUIST, GUSTAV ARTHUR LEONARD, b Eskilstuna, Sweden, Apr 13, 06; nat US; m 29; c 1. HORTICULTURE, PLANT BREEDING. Educ: Univ Conn, BS, 36; Univ Calif, PhD(genetics), 39. Prof Exp: Asst genetics, Univ Calif, 36-38, assoc, 38-39; instr floricult, Univ Calif, Los Angeles, 39-43, asst prof, 43-45; assoc prof bot, Wash Univ, 45-49, prof, 49-52; PROF PLANT BREEDING, COL ARTS & NATURAL RESOURCES, UNIV CONN, 52- Concurrent Pos: Res horticulturist, Mo Bot Garden, 45-52; Guggenheim fel, 47. Honors & Awards: Vaughan Mem Res Award, 46; Am Carnation Soc res award, 59; Soc Am Florists res award, 60. Mem: AAAS; Am Soc Hort Sci; Bot Soc Am; Genetics Soc Am; Am Soc Plant Taxon. Res: Mutation breeding; genetic and cytological studies; interspecific hybridization in Antirrhinum, Delphinium, Dianthus, Lilium and certain orchids. Mailing Add: Dept of Plant Sci Univ of Conn Col of Arts & Natural Resources Storrs CT 06268

MEHLTRETTER, CHARLES LOUIS, organic chemistry, see 12th edition

MEHMEDBASICH, ENVER, b Detroit, Mich, Oct 29, 31; m 57; c 3. ORGANIC CHEMISTRY. Educ: Univ Mich, BS, 54; Wayne State Univ, PhD(org chem), 61. Prof Exp: Res chemist, Fuel Additives Sect, 61-65, sr res chemist, Lubricants Res Dept, 66-70, SR RES ASSOC, LUBRICANTS RES DEPT, CHEVRON RES CO, 71- Mem: Am Chem Soc; The Chem Soc. Res: Detergents, dispersants, corrosion inhibitors and anti-icers for motor and diesel fuels. Mailing Add: Lub Res Dept Chevron Res Co 576 Standard Ave Richmond CA 94802

MEHNER, JOHN FREDERICK, b Grove City, Pa, July 15, 21. ORNITHOLOGY. Educ: Grove City Col, BS, 42; Univ Pittsburgh, MS, 50; Mich State Univ, PhD(zool), 58. Prof Exp: Teacher high sch, Pa, 42-59; assoc prof biol, Edinboro State Col, 59-63; assoc prof, 63-66, PROF BIOL, MARY BALDWIN COL, 66-, CHMN DEPT, 63- Mem: Am Ornith Union; Ecol Soc Am; Wilson Ornith Soc; AAAS; Am Inst Biol Sci. Res: Effects of insecticides on passerine bird populations; ecology and ethology of the evening grosbeak; ecological distribution of birds. Mailing Add: Dept of Biol Mary Baldwin Col Staunton VA 24401

MEHR, CYRUS B, b Tehran, Iran, July 7, 27; US citizen; m 55; c 4. MATHEMATICS, ELECTRICAL ENGINEERING. Educ: La State Univ, BS, 52; Purdue Univ, MS, 53, PhD(math, elec eng), 64. Prof Exp: Develop engr, Elec Prod Co, 53-55; systs engr, Gen Elec Co, 55-57; mem develop lab, Square D Elec Co, 57-59; mem res staff, Int Bus Mach Corp, 63-65; asst prof, 65-67, ASSOC PROF MATH, OHIO UNIV, 67- Mem: Am Math Soc; Math Asn Am. Res: Functional analysis. Mailing Add: Dept of Math Ohio Univ Athens OH 45701

MEHR, LOUIS, organic chemistry, see 12th edition

MEHRA, JAGDISH, theoretical physics, see 12th edition

MEHRA, KRISHNA NANDAN, b Bareilly, India, Sept 20, 23; m 48; c 2. PARASITOLOGY. Educ: Agra Univ, BSc, 43, MSc, 45; Univ Ill, Urbana, MS, 58, PhD(parasitol), 60. Prof Exp: Res asst parasitol, Indian Vet Res Inst, 46-54; asst prof, Madhya Pradesh Vet Col, India, 60-61; asst prof, Uttar Pradesh Agr Univ, India, 61-66; asst prof biol, Ga Southern Col, 66-68; assoc prof, 68-70, PROF BIOL & CHMN DEPT, LIMESTONE COL, 70- Res: Life cycles of cestodes; the biochemical nature of parasites. Mailing Add: Dept of Biol Limestone Col Gaffney SC 29340

MEHRA, MOOL CHAND, b Lahore, Pakistan, July 31, 36; Can citizen; m 64; c 2. RADIOCHEMISTRY, INORGANIC CHEMISTRY. Educ: Univ Rajasthan, BSc, 55, MSc, 57; Laval Univ, DSc(chem), 68. Prof Exp: Asst prof chem, Univ Rasjasthan, 57-59; sci off radiochem, Atomic Energy Estab, India, 59-65; from asst prof to assoc prof, 68-74, head dept, 71-74, PROF CHEM, UNIV MONCTON, 74- Concurrent Pos: Univ Moncton rep, Atlantic Prov Inter-Univ Comt Sci, Chem & Water Resources, 70-72. Mem: Chem Inst Can; fel Indian Chem Soc. Res: Radioanalytical chemistry; analytical chemistry coordination complexes in solution. Mailing Add: Dept of Chem Univ of Moncton Moncton NB Can

MEHRAN, FARROKH, b Tehran, Iran, June 29, 36; m 63; c 2. PHYSICS. Educ: Univ Calif, Berkeley, BS, 59; Harvard Univ, PhD(physics), 64. Prof Exp: Fel, Harvard Univ, 64-65; asst prof physics, Sacramento State Col, 65-67; staff physicist, 67-70, MEM RES STAFF, IBM CORP, 70- Mem: AAAS; Am Phys Soc. Res: Molecular beams; quantum electronics; electron paramagnetic resonance. Mailing Add: Dept of Phys Sci Thomas J Watson Res Ctr IBM Corp PO Box 218 Yorktown Heights NY 10598

MEHRING, ARNON LEWIS, JR, b Washington, DC, Apr 24, 15; m 36; c 3. POULTRY NUTRITION. Educ: Univ Md, BS, 36. Prof Exp: Mkt inspector eggs, State Dept Mkt, Exten Serv, Univ Md, 36-39; farm owner, 39-43; farm mgr, 43-45; poultryman in-chg nutrit & breeding invests, Lime Crest Res Lab, 45-63; nutritionist, 63-70, QUAL SUPVR, LIMESTONE PROD CORP AM, 70- Mem: Poultry Sci Asn; Am Soc Animal Sci; Am Dairy Sci Asn; Am Soc Testing & Mat. Res: Mineral nutrition of livestock and poultry. Mailing Add: Limestone Prod Corp of Am 122 Main St Newton NJ 07860

MEHRING, JEFFREY SCOTT, b Cleveland, Ohio, July 6, 42; m 65; c 2. NUTRITION, TOXICOLOGY. Educ: Ohio State Univ, BS, 64, MS, 66, PhD(nutrit), 69. Prof Exp: Lectr nutrit, Ohio State Univ, 69; LAB MGR NUTRIT, GAINES NUTRIT CTR, GEN FOODS CORP, 69- Concurrent Pos: Mem subcomt lab animal nutrit, Nat Acad Sci-Nat Res Coun, 74- Mem: Animal Nutrit Res Coun; Am Soc Animal Sci; Inst Food Technologists. Res: Comparative animal nutrition; quantification of nutrient requirements; clinical assessment of nutritional status and applied toxicology. Mailing Add: RFD 1 Peotone IL 60468

MEHRINGER, PETER JOSEPH, JR, b Lawrence, Kans, Dec 9, 33; m 54; c 3. ECOLOGY, PALYNOLOGY. Educ: Calif State Col, Los Angeles, BA, 59, MA, 62; Univ Ariz, PhD, 68. Prof Exp: Instr biol, Glendale Jr Col, 60-61; res assoc geochronology, Univ Ariz, 64-68, asst prof earth sci, 68-69; asst prof anthrop, Univ Utah, 69-71; ASSOC PROF ANTHROP, WASH STATE UNIV, 71- Mem: AAAS; Am Quaternary Asn. Res: Quaternary biogeography of North America, particularly the desert west; quaternary chronology, geology and paleoecology; influence of climatic change on prehistoric North American population. Mailing Add: Dept of Anthrop Wash State Univ Pullman WA 99163

MEHRLE, PAUL MARTIN, JR, b Caruthersville, Mo, Dec 13, 45; m 64; c 2. BIOCHEMISTRY, PHYSIOLOGY. Educ: Southwestern at Memphis, BA, 67; Univ Mo-Columbia, MA, 69, PhD(biochem), 71. Prof Exp: PHYSIOLOGIST, FISH-PESTICIDE RES LAB, US DEPT INTERIOR, 71- Concurrent Pos: Res assoc, Univ Mo-Columbia, 73- Mem: Am Chem Soc; Sigma Xi. Res: Directing research concerned with evaluating the biochemical and physiological effects of pesticides and other xenobiotics on fish; major areas of interest are enzymatic metabolism of xenobiotics, amino acid metabolism and nutrition. Mailing Add: Fish-Pesticide Res Lab Rte 1 Columbia MO 65201

MEHRLICH, FERDINAND PAUL, b Cincinnati, Ohio, 05; m 33; c 3. FOOD SCIENCE. Educ: Butler Univ, AB, 27; Univ Wis, PhD(plant physiol), 30. Prof Exp: Asst plant physiol, Univ Wis, 28-30; assoc pathologist pineapple res, Pineapple Producers Coop Asn, Hawaii, 30-35; sci adv, Hawaiian Pineapple Co, Ltd, 35-43, dir res, 43-44, asst vpres in-charge res, 44-49; vpres, Res Inst, Int Basic Econ Corp, 50-54, trustee & dir res div, 54-58; sci dir, Qm Food & Container Inst Armed Forces, 58-63, dir, Food Labs, US Army Natick Labs, 63-75; CONSULT & TECH WRITER, 75- Concurrent Pos: Consult, Cocoa Res Inst, DC, 56-, Nat Planning Asn, 56- & Arthur D Little, Inc, Mass, 56-; consult to vpres res, Gen Foods Corp, NY, 56-57; mem survs, WAfrica, Belgian Cong, Brazil, Costa Rica, Cuba, Mex, PR, Venezuela & Peru. Honors & Awards: Meritorious Civilian Serv award, 64. Mem: Fel AAAS; Am Soc Plant Physiol; Inst Food Technol; Nutrit Today; NY Acad Sci. Res: Food technology including radiation preservation of food; product and process development; agricultural research and planning; pineapple and coffee production and processing; tropical and subtropical crop production and upgrading. Mailing Add: 96 Pilgrim Rd Wellesley MA 02181

MEHROTRA, BAM DEO, b Meerut, India, Sept 4, 33; m 65; c 1. BIOCHEMISTRY. Educ: Agra Univ, BSc, 50, MSc, 52; Ind Univ, Bloomington, PhD(biochem), 64. Prof Exp: Lectr org chem, J V Col, India, 52-58; fel biochem, Inst Enzyme Res, Univ Wis-Madison, 62-64; asst prof, All-India Inst Med Sci, New Delhi, 65-66; res assoc, Ind Univ, Bloomington, 67-69; PROF CHEM & CHMN DEPT, TOUGALOO COL, 69-, CHMN NATURAL SCI DIV, 72- Mem: Am Chem Soc. Res: Chemistry of nucleic acids, their interactions with small ions, characterization and physical, chemical and biological characteristics. Mailing Add: Dept of Chem Tougaloo Col Tougaloo MS 39174

MEHROTRA, KISHAN GOPAL, b Kashipur, India, Dec 9, 41; m 71; c 1. MATHEMATICAL STATISTICS. Educ: Univ Lucknow, BSc, 60, MSc, 62; Univ Wis-Madison, MS, 69, PhD(statist), 71. Prof Exp: Lectr statist, Banaras Hindu Univ, 62-66; asst prof, 71-72, ASSOC PROF STATIST, SYRACUSE UNIV, 72- Mem: Inst Math Statist; Am Statist Asn. Res: Nonparametric statistics; pattern recognition. Mailing Add: Dept Systs & Info Sci Syracuse Univ Syracuse NY 13210

MEHTA, AVINASH C, b Rehlu, India, Nov 1, 31; m 70; c 1. ORGANIC CHEMISTRY. Educ: Panjab Univ, India, BSc, 52, MSc, 54; Univ Delhi, PhD(org chem), 58. Prof Exp: Lectr chem, Deshbandhu Col, New Delhi, India, 57-58 & Univ Delhi, 59-62; res assoc org chem, Univ Mich, 62-64; res scientist, Uniroyal Res Labs, Ont, 64-67; sr org chemist, Arthur D Little, Inc, 68-70; scientist, 70-74, RES GROUP LEADER, POLAROID CORP, 74- Mem: Am Chem Soc; The Chem Soc. Res: Organic chemical reaction mechanisms; organic synthesis, polymer chemistry and photographic chemistry; synthesis of novel monomers and polymers. Mailing Add: 12 Brookside Ave Belmont MA 02178

MEHTA, BIPINCHANDRA MOHANLAL, b Bombay, India, July 25, 35; m 60. MICROBIAL GENETICS, MICROBIAL PHYSIOLOGY. Educ: Univ Bombay, BSc, 55, BSc, 57, PhD(microbial genetics & nutrit), 63. Prof Exp: Teaching asst microbiol & biochem, Univ Bombay, 60-61; sr res asst microbial genetics, Coun Sci & Indust Res, Govt India, 61-65; teaching asst & res assoc molecular biol, State Univ NY Downstate Med Ctr, 65-66; vis res fel molecular genetics, Sloan-Kettering Inst Chem Res, 66-69; res assoc microbiol, Univ Ottawa, 69-72; res assoc, 72-74, ASSOC, SLOAN-KETTERING INST CANCER RES, 74-; ASST PROF, SLOAN-KETTERING DIV, GRAD SCH MED SCI, CORNELL UNIV, 75- Concurrent Pos: Instr, Sloan-Kettering Div, Grad Sch Med Sci, Cornell Univ, 73-75. Mem: Am Soc Microbiol; Soc Gen Microbiol; Can Soc Cell Biol; AAAS; NY Acad Sci. Res: Study of the distribution kinetics of cancer chemotherapeutic agents in body fluids and tissues of patients and experimental animals; study of mechanism of resistance to drugs. Mailing Add: Sloan-Kettering Inst Cancer Res 145 Boston Post Rd Rye NY 10580

MEHTA, HIMATLAL R, b Bombay, India, July 2, 27; nat US; m 54. PHARMACY. Educ: Univ Bombay, BSc, 49; Univ Colo, BS, 52, MS, 53, PhD(pharmaceut chem), 60. Prof Exp: Asst chem lab, Univ Colo, 50-53; chief pharmaceut inspector, Warner-Chilcott Labs Div, Warner-Hudnut, Inc, NJ, 53-55; staff pharmacist, Denver Gen Hosp, Colo, 55-57; instr pharm, Col Pharm, Univ Colo, Boulder, 57-60, asst prof, 60-67; dir hosp serv, Fox-Vliet Drug Co, 67-72; PRES, INT HOSP SERV, 72- Concurrent Pos: Consult, Vet Admin Hosp, Ft Lyon, Colo & Montevista Hosp. Mem: Am Soc Hosp Pharmacists; Am Pharmaceut Asn; Am Hosp Asn. Res: Pharmaceutical chemistry; hospital and general pharmacy. Mailing Add: 6850 Jordan Dr Denver CO 80221

MEHTA, KISHOR KALIDAS, b Bombay, India. NUCLEAR PHYSICS, REACTOR PHYSICS. Educ: St Xavier's Col, India, BSc, 58; Univ Md, PhD(nuclear eng & physics), 66. Prof Exp: SR PHYSICIST, ATOMIC ENERGY CAN, LTD, 66- Mem: Am Nuclear Soc. Res: In-pile radiation dosimetry; computer simulation of reactor systems. Mailing Add: Atomic Energy of Can Ltd Whiteshell Nuclear Res Estab Pinawa MB Can

MEHTA, MAHESH J, b Ahmedabad, Gujarat, Nov 19, 35; m 64. PHYSICAL CHEMISTRY. Educ: Gujarat Univ, India, BS, 56, MS, 64; Univ Baroda, PhD(phys chem), 69. Prof Exp: Lectr chem, Fac Sci, Univ Baroda, 64-69; res assoc polymer chem, Stevens Inst Technol, 70-72; polymer chemist, 73-74, PROJ MGR, POLYMER RES CORP AM, 75- Mem: Am Chem Soc. Res: Developing graft copolymers to obtain special surface characteristics such as wetting, adhesion coating on plastics, metals, human hair; development of synthetic polymeric membranes and diaphragms for electrodialysis, reverse osmosis, storage batteries. Mailing Add: 3586 Kennedy Blvd Jersey City NJ 07307

MEHTA, NARIMAN BOMANSHAW, b Bombay, India, Apr 8, 20; m 54; c 3. DRUG METABOLISM, PHARMACEUTICAL CHEMISTRY. Educ: Univ Bombay, BSc, 41, BA, 42; Univ Kans, PhD, 52. Prof Exp: Lectr physics, Univ Bombay, 41-46; trainee, J E Seagram & Sons, Inc, Ky, 47-48; fel, Univ Toronto, 53-54; prof chem, Cent State Col, 54-57; SR RES SCIENTIST, RES LABS, BURROUGHS WELLCOME & CO, INC, 57- Concurrent Pos: Consult, Charles F Kettering Found, Ohio, 55-57. Mem: Am Chem Soc; The Chem Soc; fel Indian Chem Soc. Res: Organo-physical and heterocyclic chemistry; reaction mechanisms and kinetics; high vacuum techniques; medicinal chemistry; structure-activity studies in design of drugs of central nervous system effects; mechanism of reactions. Mailing Add: 3030 Cornwallis Rd Research Triangle Park NC 27709

MEHTA, TARA, b Gujarat, India, Oct 23, 42. NUTRITIONAL BIOCHEMISTRY. Educ: Univ Jabalpus, India, BSc, 61; Univ Baroda, India, MSc, 63; London Sch Hyg & Trop Med & Ibadan Univ, Nigeria, dipl, 65; Univ Ill, Urbana, PhD(nutrit sci), 72. Prof Exp: Lectr & head dept foods & nutrit, M H Col Home Sci, Univ Jabalpus, M P State Govt, India, 63-72; RES ASSOC, DEPT FOOD SCI, UNIV ILL, URBANA-CHAMPAIGN, 72- Mem: Brit Nutrit Soc; Nutrit Soc India. Res: Mode of action of neurotoxin, present in Lathyrus sativus and responsible for causing paralysis of lower limbs. Mailing Add: Dept of Food Sci Univ of Ill Urbana IL 61801

MEI, ALEXIS ITALO, b San Francisco, Calif, Aug 8, 98. SEISMOLOGY. Educ: Gonzaga Univ, AB, 22, MA, 23; Gregorian Univ, STD, 30; Univ Calif, PhD(seismol), 42. Prof Exp: Instr high sch, Calif, 24-26; asst prof physics, Univ San Francisco, 31-33, from assoc prof to prof, 42-56, head dept, 45-48, dean col sci, 47-56, dean col lib arts, 50-56; acad vpres & trustee, 56-68, DIR RICHARD OBSERV & EMER ACAD VPRES, UNIV SANTA CLARA, 68- Mem: Assoc Seismol Soc Am; assoc Am Asn Physics Teachers; assoc Am Geophys Union. Res: Ratio of the amplitudes of reflected seismic waves to those of direct waves as possible means of distinguishing between geologic structure under the Pacific Ocean and that in other regions of the earth's crust. Mailing Add: Univ of Santa Clara Santa Clara CA 95053

MEIBOHM, EDGAR PAUL HUBERT, b New Orleans, La, Dec 13, 15; m 55; c 4. POLYMER SCIENCE. Educ: Guilford Col, BS, 36; Univ NC, MS, 39; Ohio State Univ, PhD(phys chem), 47. Prof Exp: Instr chem, Kans State Univ, 41-42; group leader, Nat Defense Res Comt Div Eight, Explosive Res Lab, Pa, 42-45, res sect leader, Los Alamos Sci Lab, 45; res chemist, Cent Res Dept, 47-55, sr res chemist, Rayon Res Lab, 55-57, sr res chemist, Dacron Res Lab, 57-62, sr res chemist, Kinston Plant Tech, 62-67, STAFF CHEMIST, MARSHALL LAB, E I DU PONT DE NEMOURS & CO, INC, PHILADELPHIA, 67- Mem: Am Chem Soc; Am Crystallog Asn. Res: Small-angle scattering of x-rays; crystal structures; applications of x-ray diffraction to high polymers; physics and physical chemistry of polymers; polymer characterization. Mailing Add: 521 Shadeland Ave Drexel Hill PA 19026

MEIBOOM, SAUL, b Antwerp, Belg, Apr 7, 16; US citizen; m 46; c 2. CHEMICAL PHYSICS. Educ: Univ Delft, PhysEng, 39; Hebrew Univ, Israel, PhD(physics), 55. Prof Exp: Instr physics, Hebrew Univ, Israel, 40-48; sr scientist, Weizmann Inst, 48-58; MEM TECH STAFF, BELL LABS, 58- Mem: Am Phys Soc. Res: Nuclear magnetic resonance spectroscopy; physics of liquid crystals. Mailing Add: Bell Labs PO Box 261 1B-308 Murray Hill NJ 07974

MEIBUHR, STUART GENE, b Cleveland, Ohio, Jan 6, 34; m 60. ELECTROCHEMISTRY. Educ: Western Reserve Univ, BS, 55, MS, 58, PhD(electrochem), 60. Prof Exp: Chemist, Harshaw Chem Co, Ohio, 54-57; technologist appl res lab, US Steel Corp, Pa, 60-63; sr res chemist, Fuel Cell Corp, Mo, 63-64; assoc sr res chemist, Res Lab, 64-65, SR RES CHEMIST, RES LAB, GEN MOTORS CORP, 65- Mem: Am Chem Soc; Electrochem Soc. Res: Electrodeposition; electrode kinetics; fuel cells; batteries; organic electrolytes. Mailing Add: Gen Motors Res Lab Electrochem Dept RCEL Bldg Warren MI 48090

MEIDAV, TSVI, b Poland, June 20, 30; m; c 3. EXPLORATION GEOPHYSICS. Educ: Wash Univ, BA, 55, MA, 56, PhD(geophys), 61. Prof Exp: Geophysicist, Geophys Inst Israel, 61-65; sr geophysicist, Huntec Ltd, 65-67; assoc prof geophys, Univ Calif, Riverside, 67-71, assoc geophysicist, Inst Geophys & Planetary Physics, 69-71; tech adv geothermal energy, UN, New York, 71-73; PRES, GEONOMICS, INC, 74- Concurrent Pos: Mem site invest comt, Israel Bur Stand, 63-65; mem subcomt geod & geomagnetism, Nat Res Coun Can, 66-67; mem, NSF-Sponsored Hickel Comt Geothermal Energy; assoc ed, Geothermics; app to geothermal liaison adv group, Energy Res & Develop Agency; vis res scholar, Univ NMex, 73-74; consult on innovative technol explor, development of geothermal resources, Lawrence Berkeley Lab, 74-; partic, NATO-CCMS Small Geothermal Power Plant Workshops, 75- Mem: Soc Explor Geophysicists; Am Geophys Union; Asn Eng Geologists; Seismol Soc Am; Europ Asn Explor Geophysicists. Res: Development of methodology, electronics, innovative techniques and application of geophysics to exploration and exploitation of geothermal resources; economics of geothermal power generation. Mailing Add: Geonomics Inc 3165 Adeline St Berkeley CA 94703

MEIENHOFER, JOHANNES ARNOLD, b Dresden, Ger, Mar 3, 29; c 2. BIOCHEMISTRY, ORGANIC CHEMISTRY. Educ: Univ Heidelberg, dipl, 54, PhD(chem), 56. Prof Exp: Res assoc, Med Sch, Cornell Univ, 57-59; res assoc, Univ Calif, Berkeley, 59-60; proj chief, Ger Wool Res Inst, Aachen, 61-64; HEAD PEPTIDE & PROTEIN CHEM LAB, CHILDREN'S CANCER RES FOUND, 65- Concurrent Pos: Fulbright travel grant, 57-61; res assoc, Farbenfabriken Bayer, A G, Ger, 61-64; assoc, Harvard Med Sch, 69-71, lectr, 71- Mem: AAAS; Am Chem Soc; Am Soc Biol Chem; Ger Chem Soc; Ger Soc Biol Chem. Res: Peptide and protein chemistry; hormones; antibiotics; antitumor agents. Mailing Add: Children's Cancer Res Found Boston MA 02115

MEIER, ALBERT HENRY, b New Haven, Mo, June 29, 29; m 54; c 3. ZOOLOGY, PHYSIOLOGY. Educ: Washington Univ, AB, 56; Univ Mo, MA, 59, PhD(zool), 62. Prof Exp: NIH fel, Wash State Univ, 62-64; from asst prof to assoc prof zool, 64-72, PROF ZOOL, LA STATE UNIV, 72- Concurrent Pos: NIH res career develop award, 69-74. Honors & Awards: Distinguished Res Master, La State Univ, 73. Mem: Am Soc Zool; Am Ornith Union; Ecol Soc Am; Soc Exp Biol & Med; Int Soc Chronobiol. Res: Comparative endocrinology and physiology of vertebrates; biological rhythms in hormonal control of seasonal and developmental conditions. Mailing Add: Dept of Zool La State Univ Baton Rouge LA 70803

MEIER, DALE JOSEPH, b The Dalles, Ore, Apr 21, 22; m 48; c 2. POLYMER PHYSICS. Educ: Calif Inst Technol, BS, 47, MS, 48; Univ Calif, Los Angeles, PhD(chem), 51. Prof Exp: Chemist, Shell Develop Co, 51-55, supvr res, 55-68, exchange scientist, Shell Lab, Amsterdam, 68-69, proj leader, Shell Chem Co, 69-71, supvr res, Shell Develop Co, 71-72; SR RES SCIENTIST, MIDLAND MACROMOLECULAR INST, 72- Concurrent Pos: Consult, Alza Corp, 73- Mem: Am Phys Soc; Am Chem Soc; Soc Rheology. Res: Physics of high polymers and polymer solutions; rheology; statistical mechanics; physics of interfaces. Mailing Add: Midland Macromolecular Inst 1910 St Andrews Dr Midland MI 48640

MEIER, EUGENE PAUL, b Rosenberg, Tex, Oct 3, 42; m 63; c 2. ENVIRONMENTAL CHEMISTRY, ANALYTICAL CHEMISTRY. Educ: Tex A&M Univ, BS, 65; Univ Colo, Boulder, PhD(anal chem), 69. Prof Exp: Biochemist protein chem br, Med Res Labs, Edgewood Arsenal, 69-70, chemist, Phys Protection Br, Defense Eng & Develop, Defense Systs Div, 71-72; RES CHEMIST, ENVIRON PROTECTION RES BR, US ARMY MED BIOMECH RES & DEVELOP LAB, 72- Mem: Am Chem Soc. Res: Gas chromatography; tr trace analysis; metals and organics in water; environmental analysis. Mailing Add: USAMBRDL Ft Detrick Frederick MD 21701

MEIER, GILBERT W, b St Louis, Mo, Apr 10, 27; m 53; c 1. PHYSIOLOGICAL PSYCHOLOGY. Educ: Wash Univ, AB, 48, PhD(embryol psychol), 53; Univ Ill, MA, 49. Prof Exp: Asst prof psychol, Vanderbilt Univ, 53-58, assoc prof, 58-61; res psychologist, Neuropsychol Sect Lab Perinatal Physiol, Nat Inst Neurol Dis & Blindness, 61-65; prof psychol, George Peabody Col, 65-72; PROF PSYCHOL, UNIV & PROF PEDIAT PSYCHIAT, MED CTR, UNIV NEBR, OMAHA, 72- Concurrent Pos: Consult, clin psychol training prog, Vet Admin, 58-62; ed, Develop Psychobiol. Mem: AAAS; Int Soc Develop Psychobiol; Animal Behav Soc. Res: Behavioral development and teratology; maternal-fetal-infant relations; behavioral ecology; primatology. Mailing Add: Nebr Psychiat Inst Univ Nebr Col of Med 42nd St & Dewey Ave Omaha NE 68105

MEIER, HANS, b Ruemlang, Switz, July 21, 29; nat US; m 57, 65; c 2. EXPERIMENTAL PATHOLOGY. Educ: Univ Zurich, DVM, 54, PhD, 57. Prof Exp: Asst path, Angel Mem Hosp, Boston, Mass, 54-55, resident, 55-56, head dept, 56-57; asst path, Harvard Med Sch, 57-60; assoc staff scientist, 60-62, staff scientist, 62-70, SR STAFF SCIENTIST, JACKSON LAB, 70- Concurrent Pos: Asst, Children's Cancer Res Found & asst pathologist, Children's Hosp, 57-59, res assoc, Children's Med Ctr, 59-60; consult, Charles River Breeding Labs, 58-60, Worcester Biol Testing Labs, 59-60 & Nat Cancer Inst, 70-; consult sci adv bd, Coun Tobacco Res, 71-; mem breast cancer virus working group, Nat Cancer Inst, 72-; mem sci adv bd, Inst de la Vie, 74- Mem: Am Soc Exp Path; NY Acad Sci; Am Vet Med Asn; Am Asn Cancer Res; Am Asn Lab Animal Sci. Res: Histochemistry; metabolic disorders; diabetes; cancer; infectious diseases; immunopathology. Mailing Add: Jackson Lab Bar Harbor ME 04609

MEIER, JAMES ARCHIBALD, b New Salem, NDak, May 6, 36; m 67. POLYMER CHEMISTRY. Educ: NDak State Univ, BS, 59, PhD(phys chem), 71. Prof Exp: SUPVR RESIN DEVELOP, AUTOMOTIVE DEVELOP CTR, INMOUNT CORP, 71- Mem: Am Chem Soc; Fedn Socs Coatings Technol. Res: Development of resins for coatings. Mailing Add: Automotive Develop Ctr Inmount Corp Whitehouse OH 43571

MEIER, JOHN WARREN, b Madison, Wis, Jan 26, 16; m 45; c 3. POLYMER CHEMISTRY. Educ: Univ Wis, BS, 38; Univ Ill, PhD(org chem), 42. Prof Exp: Chemist, Nat Defense Res Comt, 40-41; STAFF SCIENTIST, FILM DEPT, E I DU PONT DE NEMOURS & CO, INC, 42- Mem: Am Chem Soc. Res: Polyamides; vinyl polymers; emulsion polymerization; cellophane; coatings for packaging films. Mailing Add: E I du Pont de Nemours & Co Inc Clinton IA 52732

MEIER, JOSEPH FRANCIS, b Sharon, Pa, Nov 7, 36. POLYMER CHEMISTRY. Educ: John Carroll Univ, BS, 58; Univ Akron, MS, 60, PhD(polymer chem), 63. Prof Exp: Res chemist, Gen Tire & Rubber Co, 62-63 & 64-66; sr res chemist, 66-72, MGR ELASTOMERS GROUP, WESTINGHOUSE ELEC CORP, 73- Honors & Awards: Mat Eng Award, Mat Eng Mag, 67-68. Res: Development of rigid-brittle foam for energy absorbing applications; molded and cast elastomers for missile launch systems, including dynamic and static compressive testing and creep measurements on missile support pads. Mailing Add: Westinghouse Elec Corp Beulah Rd Pittsburgh PA 15235

MEIER, MANFRED JOHN, b Milwaukee, Wis, July 17, 29; m 54; c 2. NEUROPSYCHOLOGY. Educ: Univ Wis, BA, 52, MS, 53, PhD(psychol), 56. Prof Exp: Instr psychol, Univ Wis, 56-57; from asst prof to assoc prof, 57-66, PROF PSYCHOL & DIR NEUROPSYCHOL LAB, UNIV MINN, MINNEAPOLIS, 66- Concurrent Pos: Staff psychologist, Clin Psychol, Vet Admin Hosp, Wood, Wis, 56-57; Nat Inst Neurol Dis & Blindness res career develop award, 62-72. Mem: Am Psychol Asn; Am Acad Neurol; Am Heart Asn. Res: Effects of brain lesions on behavior in man. Mailing Add: Box 390 Mayo Med Sch Univ of Minn Minneapolis MN 55455

MEIER, MARK FREDERICK, b Iowa City, Iowa, Dec 19, 25; m 55; c 3. GLACIOLOGY. Educ: Univ Iowa, BS, 49, MS, 51; Calif Inst Technol, PhD(geol, appl mech), 57. Prof Exp: Engr geol, US Bur Reclamation, Wash, 48-49; geologist, US Geol Surv, Alaska, 51; instr geol, Occidental Col, 52-55; Fulbright grant, Innsbruck Univ, 55-56; GEOLOGIST, US GEOL SURV, 56-; RES PROF GEOPHYS, UNIV WASH, 64- Concurrent Pos: Mem tech panel, US Nat Comt, Int Geophys Year, 57-59; mem glaciol panel, Comt Polar Res, Nat Acad Sci, 59-68; vis assoc prof, Dartmouth Col, 64; US Nat Comt Int Hydrol Decade, 64-66, chmn working group combined balances & glacial basins, 65-; pres, Int Comn Snow & Ice, 67-71; dir, World Data Ctr A, Glaciol, 70-; mem gov bd, Permanent Serv Glacier Fluctuations, Zurich. Honors & Awards: Distinguished Serv Award, Us Dept Interior, 68; Medal of 150th Anniversary of Discovery of Antarctica & Medal of Inst Geog, Acad Sci, USSR, US Antarctic Serv Medal. Mem: Fel AAAS; fel Geol Soc Am; fel Am Geophys Union; fel Arctic Inst NAm; Glaciol Soc (vpres, 66-69). Res: Seasonal snowcover, glaciers, remote sensing of snow and ice, mountain and arctic hydrology, structural glaciology, flow of ice and rock. Mailing Add: US Geol Surv 1305 Tacoma Ave S Tacoma WA 98408

MEIER, MICHAEL MCDANIEL, b Chicago, Ill, Oct 14, 40; m 70. NUCLEAR PHYSICS. Educ: St Procopius Col, BS, 62; Duke Univ, PhD(nuclear physics), 69. Prof Exp: Teaching asst physics, Duke Univ, 62-63, res asst nuclear physics, 63-69, res assoc, 69-70; PHYSICIST, NAT BUR STAND, 70- Mem: Am Phys Soc. Res: Neutron polarization; neutron standards. Mailing Add: Bldg 245 Rm B119 Nat Bur of Stand Washington DC 20234

MEIER, PAUL, b New York, NY, July 24, 24; m 48; c 3. STATISTICS. Educ: Oberlin Col, BS, 45; Princeton Univ, MA, 47, PhD(math), 51. Prof Exp: Asst prof math, Lehigh Univ, 48-49; res secy, Philadelphia Tuberc & Health Asn, 49-51; res assoc math anal, Forrestal Res Ctr, Princeton Univ, 51-52; biostatist, Sch Hyg & Pub Health, Johns Hopkins Univ, 52-53, from asst prof to assoc prof, 53-57; assoc prof statist, 57-62, chmn dept statist, 60-66, dir biol sci comput ctr, 62-69, actg chmn dept statist, 70-71, PROF PHAR & PHYSIOL SCI, UNIV CHICAGO, 74-, PROF STATIST, 62-, PROF THEORET BIOL, 68- Concurrent Pos: Mem spec study sect biomath & statist, Nat Inst Gen Med Sci, 65-70, therapeut eval comt, Nat Heart Inst, 67-71 & diet-heart feasibility study rev comt, 68; comt biol effects of atmospheric pollution, Nat Acad Sci-Nat Inst Health spec fel, Sch Hyg & Trop Med, Imp Col, Univ London, 66-67; consult, statist probs to indust & govt. Mem: Fel AAAS; fel Am Statist Asn (vpres, 65-67); Inst Math Statist; Soc Indust & Appl Math; fel Am Thoracic Soc. Res: Estimation from incomplete observations. Mailing Add: Dept of Statist Univ of Chicago 5801 S Ellis Ave Chicago IL 60637

MEIER, PETER M, b Buffalo, NY, Dec 12, 41; m 72; c 1. BIOMEDICAL ENGINEERING. Educ: Univ Pa, BS & BA, 64; Case Western Reserve Univ, PhD(chem eng), 72. Prof Exp: Spec assignments engr, Dow Chem Co, Inc, 64-66; RES ENGR, BIOMED DIV, ABCOR, INC, 73- Mem: Am Inst Chem Engrs. Res: Sustained-release drug delivery systems; microencapsulation; engineering advances in human plasma fractionation. Mailing Add: Abcor Inc Biomed Div 341 Vassar St Cambridge MA 02139

MEIER, ROBERT R, b Pittsburgh, Pa, Nov 21, 40; m 63; c 2. AERONOMY. Educ:

Duquesne Univ, BS, 62; Univ Pittsburgh, PhD(physics), 66. Prof Exp: Res asst physics, Univ Pittsburgh, 62-66, res assoc, E O Hulburt Ctr Space Res, US Naval Res Lab & Univ Pittsburgh, 66-68, RES PHYSICIST, US NAVAL RES LAB, 68- Concurrent Pos: Assoc ed, J Geophycical Res, Am Geophys Union, 73-75. Mem: Am Phys Soc; Am Geophys Union. Res: Aeronomy, especially airglow, radiative transfer theory, ionospheric physics and model atmospheres; interplanetary medium. Mailing Add: Code 7121 US Naval Res Lab Washington DC 20375

MEIER, RUDOLF H, b Heiligenstadt, Ger, Feb 27, 18; US citizen; m 52; c 1. OPTICAL PHYSICS. Educ: Univ Göttingen, Vordiplom, 46; Univ Jena, dipl physics, 49, Dr rer nat(physics), 51. Prof Exp: Assoc res scientist, Zeiss Werke, Ger, 49-53; physicist with Dr J & H Krautkramer, Cologne, 53-54; consult engr, Sperry Prod, Conn, 54-55; sr physicist & group leader, Perkin-Elmer Corp, 55-60; sect supvr, Aeronutronic Div, Philco Corp, Calif, 60-66; br chief, Electro-Optics, 66-73, PRIN SCIENTIST TECH STAFF, McDONNELL DOUGLAS ASTRONAUTICS CO, HUNTINGTON BEACH, 73- Mem: Optical Soc Am; Ger Soc Appl Optics. Res: Space optics; radiometry; infrared physics. Mailing Add: 11001 Limetree Dr Santa Ana CA 92705

MEIER, WALTER ERICH, organic chemistry, see 12th edition

MEIERE, FORREST T, b Atlanta, Ga, Oct 12, 37; m 57; c 2. THEORETICAL PHYSICS. Educ: Carnegie Inst Technol, BS(math) & BS(physics), 59; Mass Inst Technol, PhD(physics), 64. Prof Exp: Res assoc physics, Mass Inst Technol, 64; asst prof, Purdue Univ, Lafayette, 64-69; assoc prof, 69-72, PROF PHYSICS, IND UNIV-PURDUE UNIV, INDIANAPOLIS, 72-, CHMN DEPT, 69- Mem: Am Phys Soc; Am Asn Physics Teachers. Res: Theory of elementary particles; field theory; dispersion relations; biophysics. Mailing Add: 1201 E 38th St Indianapolis IN 46205

MEIERHOEFER, ALAN WILLIAM, b Humboldt, Tenn, July 24, 44; m 65; c 2. ORGANIC CHEMISTRY. Educ: Southwest at Memphis, BS, 65; Miss State Univ, PhD(chem), 70. Prof Exp: RES CHEMIST, AM ENKA CORP, 69- Mem: Am Chem Soc. Res: Fiber forming polymers; polymer additives to impart unusual properties to fibers. Mailing Add: East Res Div Am Enka Corp Enka NC 28728

MEIGGS, THEODORE ORMSBEE, physical organic chemistry, water chemistry, see 12th edition

MEIGHAN, CLEMENT WOODWARD, b San Francisco, Calif, Jan 21, 25; m 60; c 1. ANTHROPOLOGY, ARCHAEOLOGY. Educ: Univ Calif, Berkeley, BA, 49, PhD(anthrop), 53. Prof Exp: From instr to assoc prof anthrop, 52-62, PROF ANTHROP, UNIV CALIF, LOS ANGELES, 62- Mem: AAAS; Am Geog Soc; Soc Am Archaeol. Res: New World prehistory; ecological studies in archaeology; obsidian dating; archaeology of Western United States, Mexico and Chile. Mailing Add: Dept of Anthrop Univ of Calif Los Angeles CA 90024

MEIGHAN, RICHARD MERL, physical chemistry, see 12th edition

MEIGHEN, EDWARD ARTHUR, b Vancouver, BC, Dec 27, 42; m 62; c 2. BIOCHEMISTRY. Educ: Univ Alta, BSc, 64; Univ Calif, Berkeley, PhD(biochem), 69. Prof Exp: Res fel biochem, Dept Molecular Biol & Virol, Univ Calif, Berkeley, 69; res fel biol, Biol Labs, Harvard Univ, 69-71; ASST PROF BIOCHEM, McGILL UNIV, 71- Mem: Can Fedn Biol Socs. Res: Enzyme regulation and relationship to subunit structure; mechanisms and control of enzyme induction in bioluminescent bacteria. Mailing Add: Dept of Biochem McGill Univ 3655 Drummond St Montreal PQ Can

MEIGS, FREDERICK MADISON, b Quincy, Ill, Apr 26, 06; m 41; c 3. CHEMISTRY. Educ: Univ Chicago, BS, 27, PhD(org chem), 30. Prof Exp: Res chemist, E I du Pont de Nemours & Co, 30-36, res chemist mkt develop, 37-40, res chemist develop, 40-42; dir develop dept, Gen Aniline & Film Corp, 42-53, gen mgr foreign opers, 53-58; exec dir indust res, Merck Sharp & Dohme Res Labs, 58-65; consult, 65-67; PRES, OXFORD CHEM RES, INC, 67- Mem: Am Chem Soc; fel Am Inst Chem. Res: Amino acid chemistry; polypeptides; synthetic fibers, films and coatings; surfactants; acetylene chemistry; organic specialties; organic synthesis. Mailing Add: Oxford Chem Res Inc Box 127 Oxford MD 21654

MEIHOEFER, HANS-JOACHIM, geography, see 12th edition

MEIJER, PAUL HERMAN ERNST, b The Hague, Neth, Nov 14, 21; nat US; m 49; c 4. THEORETICAL PHYSICS. Educ: Univ Leiden, PhD(physics), 51. Prof Exp: Vis lectr physics, Case Univ, 53-54; res assoc, Duke Univ, 54-55; asst prof, Univ Del, 55-56; assoc prof, 56-60, PROF PHYSICS, CATH UNIV AM, 60- Concurrent Pos: Fulbright grant, 53-55; consult, Nat Bur Standards, 58-, Naval Ord Lab, Naval Res Lab, Ft Belvoir & Lawrence Radiation Lab; Guggenheim Mem Found grant, Lab Magnetic Resonance, Univ Paris, 64-65. Mem: Am Phys Soc; Europ Phys Soc; Neth Phys Soc; Sigma Xi. Res: Statistical mechanics; critical phenomena; phase transitions; irreversible thermodynamics; solid state; magnetism; paramagnetic resonance; surface phenomena; superconductivity; mathematical physics; group theory; liquid state. Mailing Add: Dept of Physics Cath Univ of Am Washington DC 20064

MEIJER, ROBERT RANDAL, b Staunton, Va, Dec 1, 16; m 42; c 3. PHYSICS. Educ: Oberlin Col, AB, 39; Cornell Univ, PhD(physics), 48. Prof Exp: Asst physics, Cornell Univ, 39-43 & 46-47; mem staff, Radiation Lab, Mass Inst Technol, 43-46; assoc prof, George Washington Univ, 47-53; tech adv, Red Bank Div, Bendix Aviation Corp, 53-60, mkt mgr, Bendix Corp, 60-66; prof physics, Parsons Col, 66-73, chmn dept, 69-73; PROF PHYSICS, STATE UNIV NY UTICA/ROME, 73- Concurrent Pos: Consult, Div Math & Phys Sci, Nat Res Coun, 48-52 & US Bur Mines, 49; res analyst, Opers Res Off, Johns Hopkins Univ, 52. Mem: Am Asn Physics Teachers; Nat Sci Teachers Asn. Res: Nuclear physics; absorption of slow neutrons in rhodium and high-speed coincidence measurements; atomic, electron and statistical physics; semiconductors. Mailing Add: Dept of Physics SUNY 811 Court St Utica NY 13502

MEIJER, WILLEM, b The Hague, Neth, June 27, 23; m 51; c 3. PLANT TAXONOMY. Educ: Univ Amsterdam, BSc, 47, MSc & PhD(philos), 51. Prof Exp: Botanist, Bot Gardens of Indonesia, Dept Agr, 51-54; lectr plant taxonomy, plant geog & ecol, Univ Cent Sumatra, 55-56, prof, 56-58; forest botanist, Dept Natural Resources, Sabah, 59-68; ASSOC PROF BOT, UNIV KY, 68- Concurrent Pos: Res assoc, Mo Bot Garden, 72- Mem: Bot Soc Am; Am Soc Plant Taxon; Soc Econ Bot. Res: Bryology; economic botany; forest botany in Southeast Asia; flora of Kentucky; flora neotropica. Mailing Add: Sch of Biol Sci Univ of Ky Lexington KY 40506

MEIKLE, MARY B, b Springfield, Mass, Aug 30, 34. AUDITORY PHYSIOLOGY. Educ: Vassar Col, AB, 54; Univ Ore, MS, 67, PhD(physiol psychol), 69. Prof Exp: NIH res fel, Kresge Hearing Res Lab, 69-71, res assoc physiol of the ear, 71-72, ASST PROF OTOLARYNGOL & MED PSYCHOL, KRESGE HEARING RES LAB, MED SCH, UNIV ORE, 72- Concurrent Pos: Vis lectr, Reed Col, 70. Mem: Acoustical Soc Am; Asn Res Otolaryngol; AAAS. Mailing Add: Kresge Lab Dept of Otolaryngol Univ of Ore Health Sci Ctr Portland OR 97201

MEIKLE, RICHARD WILLIAM, b Chicago, Ill, June 15, 22; m 50; c 4. ORGANIC CHEMISTRY, BIOCHEMISTRY. Educ: Ill Wesleyan Univ, BS, 43; Univ Ill, MS, 47, PhD(chem), 50. Prof Exp: Res assoc org chem, Univ Wash, 50; SR RES CHEMIST BIOPROD RES, DOW CHEM CO, 51- Mem: Am Chem Soc. Res: Agricultural research. Mailing Add: Agr-Organics Res Dow Chem Co Walnut Creek CA 94598

MEIKLEJOHN, GORDON, b Providence, RI, Apr 8, 11; m 40; c 3. MEDICINE. Educ: McGill Univ, MD, CM, 37; Am Bd Internal Med, dipl, 47. Prof Exp: Intern, Montreal Gen Hosp, 37-38, resident, 38-39; resident, Univ Calif Hosp, 39-40; resident, Burton Cairns Gen Hosp, Casa Grande, Ariz, 40-41; clin asst, Univ Calif, 42-43, clin instr, 43-44, lectr, 46-47, asst clin prof, 47-48, asst prof, 48-51; PROF MED & CHMN DEPT, UNIV COLO MED CTR, DENVER, 51- Concurrent Pos: Rockefeller Found res fel med, Med Sch, Univ Calif, 41-42; res assoc, State Dept Pub Health, Calif, 42-48; consult, Comn Neurotropic Virus Dis, US Army Epidemiol Bd, Japan & Guam, 47-48, 6th Army Surgeon, Calif, 48-51, Fitzsimons Army Hosp, Viral & Rickettsial Dis Lab, State Dept Pub Health, 42-48, WHO, 68-69 & Vet Admin Hosps; dir comn influenza, Armed Forces Epidemiol Bd. Mem: Soc Exp Biol & Med; Am Soc Trop Med & Hyg; Am Soc Clin Invest; AMA; Am Asn Immunol. Res: Viral diseases. Mailing Add: Dept of Med Univ of Colo Med Ctr Denver CO 80220

MEIKLEJOHN, ROBERT BAIKIE, b Harriston, Ont, Sept 26, 07; m 53; c 2. OBSTETRICS & GYNECOLOGY. Educ: Univ Toronto, MD; FRCOG, 48. Prof Exp: SR OBSTETRICIAN & GYNECOLOGIST, TORONTO WESTERN HOSP, 57-; PROF OBSTET & GYNEC, UNIV TORONTO, 63- Mem: Soc Obstet & Gynec Can (treas, 58-66). Mailing Add: Suite 329 Toronto Western Hosp Bathurst St Toronto ON Can

MEIKLEJOHN, RUDD ANDREW, analytical chemistry, see 12th edition

MEILI, JAY ERNEST, organic chemistry, see 12th edition

MEILING, RICHARD L, b Springfield, Ohio, Dec 21, 08; m 40; c 1. OBSTETRICS & GYNECOLOGY. Educ: Wittenberg Col, AB, 30; Univ Munich, MD, 37; Am Bd Obstet & Gynec, dipl, 47; Am Bd Prev Med, dipl aviation med, 53. Hon Degrees: DSc, Wittenberg Col, 50. Prof Exp: Demonstr gynec, Western Reserve Univ, 46-47; clin asst prof obstet & gynec, 47-51, clin instr path, 47-60, assoc med dir, Health Ctr, 51-60, from assoc prof to prof obstet & gynec, 51-67, from assoc dean to dean, Col Med, 51-70, dir, Univ Hosps, 61-72, vpres med affairs, 70-74, PROF ALLIED MED PROFESSIONS, OHIO STATE UNIV, 67-, EMER VPRES MED AFFAIRS, 74- Concurrent Pos: Med adv, Comn Reorgn Exec Br Govt, Nat Security Comt, 48; mem, Armed Forces Med Adv Comt, 48-49; consult, Surgeon Gen, US Dept Air Force, 48-49, 51-53; asst secy med & health affairs, US Dept Defense, 49-51, consult, 51-53; adv to US deleg world health assembly, WHO, Geneva, 50-51, mem coun, NIH, 50-51; med adv, Fed Civil Defense Adminr, 53. Honors & Awards: Commendation Award, Secy War; Cert, US Dept Defense, 51. Mem: AMA; fel Aerospace Med Asn; fel Am Col Surg; fel Col Obstet & Gynec; Int Col Surg. Mailing Add: 91 N Columbia Ave Columbus OH 43209

MEIMAN, JAMES R, b Louisville, Ky, Dec 10, 33. WATERSHED MANAGEMENT. Educ: Univ Ky, BS, 55, MS, 59; Colo State Univ, PhD(watershed mgt), 62. Prof Exp: Soil conservationist, Soil Conserv Serv, USDA, 55, 57-58 & 59; soil conservationist, Forest Serv, 62; from instr to assoc prof, 62-74, PROF WATERSHED MGT, COLO STATE UNIV, 74- Concurrent Pos: Assoc ed, Water Resources Res; hydrologist, Rocky Mountain Forest & Range Exp Sta, US Forest Serv, Ft Collins, Colo, 75- Mem: Am Geophys Union; Glaciol Soc. Res: Water yields and quality of wildland watersheds and impact of land use thereon; snow hydrology. Mailing Add: Dept of Earth Resources Colo State Univ Ft Collins CO 80521

MEINCKE, P P M, b Winnipeg, Man, Jan 21, 36; m 58; c 2. PHYSICS. Educ: Queen's Univ, Ont, BSc, 59; Univ Toronto, MA, 60, PhD(physics), 63. Prof Exp: Asst prof physics, Royal Mil Col, Ont, 62-65; mem tech staff, Bell Tel Labs, 65-67; asst prof physics, 67-69, assoc dean, Erindale Col, 70-72, ASSOC PROF PHYSICS, UNIV TORONTO, 69-, V PROVOST, 72- Mem: Am Phys Soc. Res: Low temperature solid state physics; thermal expansion of solids at low temperatures; magnetism; superconductivity. Mailing Add: Dept of Physics Univ of Toronto Toronto ON Can

MEINEKE, HOWARD ALBERT, b Cincinnati, Ohio, July 2, 21; m 46; c 5. ANATOMY. Educ: Maryville Col, BA, 47; Univ Cincinnati, MS, 49, PhD(zool), 53. Prof Exp: Asst zool, 47-49, 50-51, instr, 49-50, from instr to assoc prof anat, 51-69, PROF ANAT, COL MED, UNIV CINCINNATI, 69- Mem: AAAS; Am Asn Anat. Mailing Add: 1339 Delta Ave Cincinnati OH 45208

MEINEL, ADEN BAKER, b Pasadena, Calif, Nov 25, 22; m 44; c 7. ASTROPHYSICS, SOLAR ENERGY. Educ: Univ Calif, AB, 47, PhD(astron), 49. Prof Exp: From instr to assoc prof astrophys, Univ Chicago, 50-53, assoc dir, Yerkes & McDonald Observ, 53-56; dir, Kitt Peak Nat Observ, 56-60; chmn dept astron, 61-65, dir, Steward Observ, 62-67, dir optical sci, 67-73, PROF ASTRON & OPTICAL SCI, UNIV ARIZ, 61- Concurrent Pos: Regent, Calif Lutheran Col, 62-71; mem pres comn 9, Int Astron Union, 73-76; consult, Energy Res & Develop Admin, 75- Honors & Awards: Lomb Medal, Optical Soc Am, 52; Warner Prize, Am Astron Soc, 54. Mem: AAAS; Am Acad Arts & Sci; Optical Soc Am (pres, 72); Am Astron Soc; Royal Astron Soc. Res: Spectroscopy of night air-glow and aurora borealis; identification of new band systems of OH and N2; detection of extraterrestrial protons in aurora; reproduction auroral spectrum in laboratory by accelerated particles; optical design; volcanic aerosols; ionic impact spectroscopy; stellar spectroscopy; glass technology; space telescopes; solar power. Mailing Add: Dept of Astron & Optical Sci Univ of Ariz Tucson AZ 85721

MEINERS, ALFRED FRANCOIS, organic chemistry, see 12th edition

MEINERS, JACK PEARSON, b Walla Walla, Wash, Sept 9, 19; m 45; c 3. PHYTOPATHOLOGY. Educ: Wash State Univ, BS, 42, PhD(plant path), 49. Prof Exp: Jr pathologist, Forage Div, Bur Plant Indust, USDA, Wash, 46-49, assoc pathologist, Fruit & Veg Div, Idaho, 49-50; from asst prof to assoc prof plant path, Wash State Univ, 50-53; pathologist, Cereal Crops Res Br, Wash, 53-58, asst chief Md, 58-65, asst dir crops res div, 65-70, leader bean & pea invests, Veg & Ornamentals Res Br, Plant Sci Res Div, 70-72, chmn, Plant Protection Inst, Beltsville Agr Res Ctr, 72-74, CHIEF APPL PLANT PATH LAB, BELTSVILLE AGR RES CTR, AGR RES SERV, USDA, 74- Mem: Am Phytopath Soc. Res: Diseases, breeding and physiology of peas, beans, lentils and other edible legumes. Mailing Add: Appl Plant Path Lab Beltsville Agr Res Ctr W Beltsville MD 20705

MEINERSHAGEN, FRED HENRY, dairy husbandry, see 12th edition

MEINERT, CURTIS LYNEA, b Sleepy Eye, Minn, June 30, 34; m 57; c 3.

BIOSTATISTICS. Educ: Univ Minn, BA, 56, MS, 59, PhD(biostatist), 64. Prof Exp: Res assoc, Sch Pub Health, Univ Minn, 60-62; res assoc, Inst Int Med, 62-65, from asst prof to assoc prof, 65-72, PROF SOCIAL & PREV MED & HEAD SECT BIOMET, UNIV MD BALTIMORE CITY, 72- Concurrent Pos: Mem, Biomet Comt, Food & Drug Admin, 72-74, Epidemiol & Statist Adv Comt, Perinatal Res Br, Nat Inst Neurol Dis & Stroke, 72-, Vet Admin Coop Studies Eval Comt, 72- & Protocol Comt, Gallstone Adv Bd, NIH, 73- Mem: Am Diabetes Asn; Am Heart Asn; Am Pub Health Asn; Am Statist Asn; Biomet Soc. Res: Design and conduct of long-term cooperative clinical trials primarily in the fields of diabetes and cardiovascular diseases and development of the methods for clinical trials. Mailing Add: Dept of Social & Prev Med Univ of Md Baltimore MD 21201

MEINERT, WALTER THEODORE, b Walcott, Iowa, May 18, 22; m 46; c 3. ORGANIC CHEMISTRY. Educ: St Ambrose Col, BS, 47, MS, 49. Prof Exp: Tech serv rep, 49-51, develop rep, 51-53, asst dir develop & tech serv, 53-56, dir, 56-65, gen mgr, Western Opers, Org Chem Div, 65-67, dir int opers, 67-69, VPRES INT OPERS, EMERY INDUSTS, INC, 69- Concurrent Pos: Adv to pres, Unilever-Emery NV, Neth, 63-65. Mem: Am Chem Soc; Am Inst Chemists; Commercial Develop Asn. Res: Markets and uses for chemicals derived from fat sources; ozone oxidation and polymerization of unsaturated fatty chemicals; fatty acids. Mailing Add: Emery Industs Inc 1400 Carew Tower Cincinnati OH 45202

MEINHARD, JAMES EDGAR, b Ill, 1919; m 45; c 3. APPLIED CHEMISTRY, EXPERIMENTAL SOLID STATE PHYSICS. Educ: Univ Wis, BS, 47, PhD(chem), 50. Prof Exp: Prob leader anal separations, Hanford Atomic Prod Oper, Gen Elec Co, 50-56; with Nat Cash Register Co, 56-57 & Hughes Aircraft Co, 57-59; pres, Crystech, Inc, 59-69; PRES, J E MEINHARD ASSOCS, 69- Concurrent Pos: Mem staff, Astropower Lab, Douglas Aircraft Co, 62-64, NAm Aviation, Inc, 64-71, NAm Rockwell Corp, 71-73 & Dept Physics, Calif State Col, Fullerton, 66-69. Mem: Fel AAAS; fel Am Inst Chemists. Res: Solid state physics; chemistry; organic semiconductors; development of analytical instrument components; solid state devices. Mailing Add: 12472 Ranchwood Rd Santa Ana CA 92705

MEINHARDT, NORMAN ANTHONY, b Davenport, Iowa, Jan 19, 19; m 51; c 5. ORGANIC POLYMER CHEMISTRY. Educ: St Ambrose Col, BS, 40; Univ Iowa, PhD(org chem), 49. Prof Exp: Asst chemist ferrous anal, Rock Island Arsenal, 40-44; fel rubber res, Univ Ill, 49-50; res chemist org phosphorous compounds, 50-53, res supvr develop group, 53-56, fundamental group, 56-61 & lubricant additives sect, 61-65, RES SUPVR ASHLESS DISPERSANTS, LUBRIZOL CORP, 65- Res: Highly arylated ethylenes; initiation systems for emulsion polymerization; sulfinic acid reactions; organic phosphorous reactions; additive systems and intermediates for diesel engine lubricants; ashless inhibitor systems. Mailing Add: Lubrizol Corp Box 3057 Cleveland OH 44117

MEINHOLD, CHARLES BOYD, b Boston, Mass, Nov 1, 34; m 56; c 5. HEALTH PHYSICS. Educ: Providence Col, BS, 56; Am Bd Health Physics, cert. Prof Exp: AEC fel radiol physics, 56; from jr scientist to sr scientist health physics, 57-71; HEAD, SAFETY & ENVIRON PROTECTION DIV, ASSOC UNIVS, BROOKHAVEN NAT LAB, 72- Concurrent Pos: Chmn sci comt oper radiation safety, Nat Coun Radiation Protection & Measurements, 73-; mem comt 3 external radiation, Int Comn Radiol Protection, 73- Mem: Health Physics Soc (treas, 75-77); Am Nuclear Soc; AAAS. Res: Radiation dosimetry and radiation protection standards development. Mailing Add: Safety & Environ Protection Div Brookhaven Nat Lab Upton NY 11973

MEINIG, DONALD WILLIAM, b Palouse, Wash, Nov 1, 24; m 46; c 3. CULTURAL GEOGRAPHY, GEOGRAPHY OF THE UNITED STATES. Educ: Georgetown Univ, BS, 48; Univ Wash, MA, 50, PhD(geog), 53. Prof Exp: From instr to assoc prof geog, Univ Utah, 50-59; assoc prof, 59-62, PROF GEOG, SYRACUSE UNIV, 62- Concurrent Pos: Fulbright res scholar, Univ Adelaide, 58; Guggenheim fel, Am West, 66-67; adv ed geog, Harper & Row Publ, 66- Honors & Awards: Meritorious Contrib Award, Asn Am Geog, 65; Award Merit, Am Asn State & Local Hist, 69. Mem: Asn Am Geog; Am Geog Soc; Inst Australian Geog. Res: Historical geography; cultural geography; American regionalism; landscape interpretation. Mailing Add: Dept of Geog Syracuse Univ Syracuse NY 13210

MEINKE, GERALDINE CHCIUK, b Detroit, Mich, Jan 21, 44; m 69; c 1. IMMUNOLOGY. Educ: Madonna Col, BS, 65; Wayne State Univ, PhD(microbiol), 70. Prof Exp: Res fel immunol, 70-74, RES ASSOC IMMUNOL, DEPT IMMUNOPATH, SCRIPPS CLIN & RES FOUND, 75- Concurrent Pos: Leukemia Soc Am spec fel, Scripps Clin & Res Found, 73-75. Mem: Am Asn Immunologists. Res: Inhibition of plasma cell tumors by immunization with myeloma proteins; structural studies of SV40 and papilloma viral proteins; N-terminal sequence studies of Bence Jones cryoglobulins. Mailing Add: Dept of Immunopath Scripps Clin & Res Found La Jolla CA 92037

MEINKE, WILLIAM JOHN, b Troy, Mich, May 16, 42; m 69. MICROBIOLOGY, VIROLOGY. Educ: Albion Col, BA, 64; Wayne State Univ, MS, 67, PhD(microbiol), 69. Prof Exp: Nat Cancer Inst fel, 69-72, ASSOC MICROBIOL, SCRIPPS CLIN & RES FOUND, 72- Mem: AAAS; Am Soc Microbiol; Biophys Soc; Tissue Cult Asn; NY Acad Sci. Res: Cell regulation in normal and neoplastic cells; mechanisms of virus replication; etiology of multiple sclerosis. Mailing Add: Scripps Clin & Res Found 476 Prospect St La Jolla CA 92037

MEINKE, WILLIAM WAYNE, b Elyria, Ohio, June 27, 24; m 47; c 2. ANALYTICAL CHEMISTRY. Educ: Oberlin Col, AB, 47; Univ Calif, Berkeley, PhD(chem), 50. Prof Exp: From instr to prof chem, Univ Mich, 50-63; chief anal chem div, Nat Bur Standards, 63-73; DIR RADIATION APPLN, KMS FUSION, INC, 73- Concurrent Pos: Consult to govt & indust, 53-63; chmn subcomt radiochem, Nat Res Coun, 58-62; mem adv comt, Anal Div, Oak Ridge Nat Lab, 59-63; mem adv comt, Anal & Inorg Chem Div, Nat Bur Standards, 60-63, chief off stand ref mat, 64-69; mem adv comt isotopes & radiation develop, AEC, 61-65; mem subcomt physiochem standards, Nat Acad Sci-Nat Res Coun, 63-69 & geochem panel, Apollo Adv Comt, NASA, 63-65; mem comt anal radiochem & nuclear mat & comt data & stand, Int Union Pure & Appl Chem, 65-71. Honors & Awards: Spec Award Distinguished Serv, Am Nuclear Soc, 68; Hevesy Medal, 68; Rosa Award, Nat Bur Standards, 68; Gold Medal, US Dept Com, 71; Fisher Award, Am Chem Soc, 72. Mem: Fel AAAS; Am Chem Soc; Am Phys Soc; Am Nuclear Soc; hon mem Brit Soc Anal Chem. Res: Standard reference materials; application of physical techniques to chemical characterization; trace analysis; activation analysis; radiochemical separations; applications of nuclear radiations, especially as applied to fusion; energy sources. Mailing Add: KMS Fusion Inc 3941 Research Park Dr Ann Arbor MI 48106

MEINKE, WILMON WILLIAM, b Yoakum, Tex, Nov 8, 14; m 39. ORGANIC CHEMISTRY, BIOCHEMISTRY. Educ: Agr & Mech Col, Tex, BS, 36, PhD(org & biol chem), 49. Prof Exp: Asst chemist, Tex State Dept, 36-43; asst res chemist, 43-50, assoc res chemist, 50-65, PROF CHEM ENG, ENG EXP STA, TEX A&M UNIV, 65- Mem: Am Chem Soc. Res: Chemurgic research: upgrading of agricultural products and residues through chemistry. Mailing Add: Eng Exp Sta Tex A&M Univ College Station TX 77843

MEINKOTH, NORMAN AUGUST, b New Baden, Ill, Jan 29, 13; m 38; c 1. ZOOLOGY. Educ: Southern Ill Normal Univ, BEd, 38; Univ Ill, MS, 44, PhD(zool), 47. Prof Exp: Pub sch instr, Ill, 38-41; asst zool, Univ Ill, 41-47; from instr to assoc prof, 47-66, PROF ZOOL & CHMN DEPT BIOL, SWARTHMORE COL, 66- Concurrent Pos: With Marine Biol Lab Corp; Fulbright lectr, Chulalongkorn Univ, Bangkok, 57-58; mem staff, Univ NH, summers 59, 60, 63-65 & 67-75. Mem: AAAS; Am Soc Zoologists; Am Soc Parasitologists; Am Micros Soc. Res: Parasitology; cestodes; bird blood protozoa; invertebrate zoology; free-living flatworms; nemerteans; Stauromedusae. Mailing Add: Dept of Biol Swarthmore Col Swarthmore PA 19081

MEINS, FREDERICK, JR, b New York, NY, May 31, 42; m 70. DEVELOPMENTAL BIOLOGY, BIOCHEMISTRY. Educ: Univ Chicago, BS, 64; Rockefeller Univ, PhD(life sci), 69. Prof Exp: Asst prof biol, Princeton Univ, 69-76; ASSOC PROF BOT & GENETICS & DEVELOP, SCH LIFE SCI, UNIV ILL, URBANA, 76- Mem: NY Acad Sci; Soc Develop Biol; Ger Soc Biol Chem; Am Soc Plant Physiol. Res: Studies of cell heredity and the stability of the differentiated state using chemical approaches and crown gall tumors as test objects. Mailing Add: Sch of Life Sci Univ of Ill Urbana IL 61801

MEINSCHEIN, WARREN G, b Slaughters, Ky, Nov 12, 20; m 44; c 3. ORGANIC CHEMISTRY. Educ: Univ Mich, BS, 48; Univ Tex, PhD, 51. Prof Exp: Res assoc geochem, Field Res Lab, Magnolia Petrol Co Div, Socony Mobil Oil Co, Inc, 51-58; sr chemist, Esso Res & Eng Co, NJ, 58-61, res assoc, 61-66; PROF GEOCHEM, IND UNIV, BLOOMINGTON, 66-, ASSOC DEAN ACAD PROGS, SCH PUB & ENVIRON AFFAIRS, 75- Mem: AAAS; Am Chem Soc; Am Geophys Union; NY Acad Sci; Geochem Soc. Res: Geochemistry; origin of petroleum composition of naturally occurring hydrocarbons; paleobiochemistry; evidence for life in Precambrian rocks and meteorites; intramolecular distribution of stable carbon isotopes in organic compounds. Mailing Add: Dept of Geol Ind Univ Bloomington IN 47401

MEINTS, CLIFFORD LEROY, b Kansas City, Mo, May 23, 30; m 54; c 4. BIOCHEMISTRY. Educ: Purdue Univ, BS, 53; Ohio Univ, MS, 54; Univ Okla, PhD(biochem), 57. Prof Exp: Asst prof chem, 57-58, Kresge-Carver assoc prof natural sci, 58-64, KRESGE-CARVER PROF NATURAL SCI, SIMPSON COL, 64-, CHMN DIV, 59- Concurrent Pos: Consult, Armstrong Rubber Mfg Co, 57-58; Nat Cancer Inst fel, Sci Res Inst, Ore State Univ, 64-65. Mem: Am Chem Soc. Res: Acetobacter suboxydans intermediary metabolism; Wiswesser line notation use in laboratory courses; computer-assisted instruction applications in chemistry. Mailing Add: Dept of Chem Simpson Col Indianola IA 50125

MEINTS, RUSSEL H, b Clara City, Minn, Apr 13, 39; m 59; c 3. DEVELOPMENTAL BIOLOGY. Educ: Macalester Col, AB, 60; Kent State Univ, MA, 62, PhD(cell biol), 65. Prof Exp: Instr biol, Kent State Univ, 62-63; from asst prof to assoc prof, 65-74, PROF ZOOL, UNIV NEBR-LINCOLN, 74-, DIR SCH LIFE SCI, 75- Concurrent Pos: Res assoc, Dept Biochem, Univ Chicago & Argonne Cancer Res Hosp, 70-71; NIH fel, 71. Mem: AAAS; Am Soc Cell Biologists; Sigma Xi. Res: Developmental biology of hemopoietic stem cells; colony forming units; hemoglobin synthesis. Mailing Add: Sch of Life Sci Univ of Nebr Lincoln NE 62508

MEINTZER, ROBER BRUCE, b Fargo, NDak, July 5, 27; m 54; c 6. BIOCHEMISTRY. Educ: NDak Agr Col, BS, 50, MS, 52; Univ Wis, PhD(biochem), 54. Prof Exp: Res assoc biochem, Univ Wis, 54-55; instr med sch, Northwestern Univ, 55-57; assoc prof, NDak State Univ, 57-67; assoc prof, 67-70, PROF CHEM & CHMN DEPT, UNIV LETHBRIDGE, 70-, COORDR CONTINUING EDUC, 74- Mem: AAAS; Am Chem Soc; NY Acad Sci. Res: Mechanism of action of vitamin D and parathyroid hormone; chemistry of citric acid and other organic acids; their assay and metabolism as related to vitamins and endocrine functions. Mailing Add: Dept of Chem Univ of Lethbridge Lethbridge AB Can

MEINWALD, JERROLD, b New York, NY, Jan 16, 27; m 55; c 2. ORGANIC CHEMISTRY. Educ: Univ Chicago, PhB, 47, BS, 48; Harvard Univ, PhD(chem), 52. Prof Exp: Du Pont fel, Cornell Univ, 52, from instr to prof chem, 52-72; prof, Univ Calif, San Diego, 72-73; PROF CHEM, CORNELL UNIV, 73- Concurrent Pos: Sloan Found fel, 58; Guggenheim fel, 60-61; NIH spec fel, 67-68; vis prof, Rockefeller Univ & Univ Calif, San Diego, 70; mem med chem study sect A, NIH, 64-68; chmn vis comt, Dept Chem, Brookhaven Nat Lab & res dir, Int Ctr Insect Physiol & Ecol, Nairobi, 70-; consult, Schering Corp & Norwich Pharmacal Co; adv bd mem, Petrol Res Fund, 70-73; Louderman lectr, Wash Univ, 64; four-col lectr, Mt Holyoke Col, Smith Col, Amherst Col & Univ Mass, 65; Sigma Xi-Sci Res Soc Am nat lectr, 65 & 75; A Burger lectr, Univ Va, 66; Rennebohm lectr, Univ Wis, 67; F B Dains lectr, Univ Kans, 68; F P Venable lectr, Univ NC, 70; 70; Frontiers of Chem lectr, Case-Western Reserve Univ, 71; Julius Stieglitz lectr, Univ Chicago, 72; distinguished lectr, Howard Univ, 72; Reilly lectr, Univ Notre Dame, 73 73; Priestly lectr, Pa State Univ, 73. Mem: Nat Acad Sci; AAAS; fel Am Acad Arts & Sci; Am Chem Soc; The Chem Soc. Res: Problems of structure, synthesis and reaction mechanism from the field of natural products; synthesis and reactions of highly strained systems; molecular rearrangements; photochemistry; chemical defense mechanisms of arthropods; chemistry of pheromones. Mailing Add: Dept of Chem Cornell Univ Ithaca NY 14853

MEINWALD, YVONNE CHU, b Shanghai, China, Feb 24, 29; nat US; m 55; c 2. ORGANIC CHEMISTRY. Educ: Bryn Mawr Col, BA, 52; Cornell Univ, PhD(chem), 55. Prof Exp: Res assoc, 55-67, LECTR CHEM, CORNELL UNIV, 71- Concurrent Pos: Lectr chem, Univ Calif, San Diego, 72-73; mem comt scholarly relationships with People's Repub of China, Nat Acad Sci, 74- Mem: Am Chem Soc. Res: Synthesis, solvolysis and rearrangement reaction of medium-sized rings; synthesis of bisketenes; synthesis and reactions of highly unsaturated cyclobutane derivatives; 1, 2-cycloaddition reactions of tetracyanoethylene; nitrosyl chloride addition reactions; chemistry of arthropod defensive secretions and pheromones. Mailing Add: Dept of Chem Cornell Univ Ithaca NY 14853

MEISBURGER, EDWARD PAUL, b Kansas City, Mo, June 30, 25; m 52; c 5. MARINE GEOLOGY. Educ: Univ Mo, BA, 50. Prof Exp: MARINE GEOLOGIST, COASTAL ENG RES CTR, WASHINGTON, DC, 65- Mem: Am Geophys Union. Res: Military geology; application of marine geology to problems of engineering research and development; geology of the Continental Shelf. Mailing Add: 1810 Warren Dr Woodbridge VA 22191

MEISEL, DAN, b Tel-Aviv, Israel, July 4, 43; m 65; c 2. PHYSICAL CHEMISTRY, RADIATION CHEMISTRY. Educ: Hebrew Univ, Jerusalem, BSc, 67, MSc, 69, PhD(phys chem), 74. Prof Exp: Asst phys chem, Hebrew Univ, Jerusalem, 70-73, instr, 73-74; res fel radiation chem, Carnegie-Mellon Univ, 74-76; RES FEL CHEM, ARGONNE NAT LAB, 76- Res: Radiation chemistry of organic and inorganic systems; fast kinetics in solutions; electron spin resonance and radical reactions;

electron-transfer reactions. Mailing Add: Argonne Nat Lab Chem Div 9700 S Cass Ave Argonne IL 60439

MEISEL, DAVID DERING, b Fairmont, WVa, Mar 28, 40; m 62; c 2. ASTRONOMY, ASTROPHYSICS. Educ: WVa Univ, BS, 61; Ohio State Univ, MS, 63, PhD(astron), 67. Prof Exp: From instr to asst prof astron, Univ Va, 65-70; ASST PROF ASTRON & PLANETARIUM DIR, COL ARTS & SCI, STATE UNIV NY COL GENESEO, 70- Concurrent Pos: Mem US nat comt & mem comn radio astron, Int Union Radio Sci; res assoc, C E K Mees Observ, Univ Rochester, 73-; consult, NASA, 74-; res assoc, Kellogg Observ, Buffalo Mus of Sci, 74-; guest investr, Copernicus Space Telescope, Princeton Univ, 74- Mem: Fel AAAS; Am Astron Soc; Am Meteor Soc; fel Royal Astron Soc. Res: Astrophysical studies of multiple star systems; solar system physics; comets and meteors; studies of solar-terrestrial relations, especially during solar eclipses. Mailing Add: Dept of Physics & Astron State Univ of NY Geneseo NY 14454

MEISEL, LAWRENCE VICTOR, solid state physics, see 12th edition

MEISEL, SEYMOUR LIONEL, b Albany, NY, Aug 14, 22; m 46; c 3. ORGANIC CHEMISTRY, RESEARCH ADMINISTRATION. Educ: Union Col, NY, BS, 44; Univ Ill, MS, 46, PhD(org chem), 47. Prof Exp: Res chemist, Socony-Vacuum Oil Co, 47-50; sr res chemist, 50-56, asst supvr, 56-58, supvr, 58-61, tech dir, 61-64, mgr appl res & develop div, 64-68, VPRES RES, MOBIL OIL CORP, 68- Mem: AAAS (secy, Sect C, 60-64); Am Chem Soc; Am Petrol Inst; Indust Res Inst; Asn Res Dirs. Res: Citronellal; azo dyes containing boron; thiophene chemistry; petrochemicals; hydrocarbon catalysis; petroleum composition; shale oil; combustion; lubrication; additives for fuels and lubricants; petroleum and chemical processes. Mailing Add: 150 E 42nd St New York NY 10017

MEISELAS, LEONARD E, b Brooklyn, NY, Oct 17, 18; c 2. INTERNAL MEDICINE. Educ: Rutgers Univ, BA, 39; NY Univ, MD, 43; Univ Va, MSc, 48. Prof Exp: Instr path, State Univ NY Downstate Med Ctr, 47-48, instr med, 52-54, from asst prof to prof, 54-70; PROF MED & ASSOC DEAN SCH MED, STATE UNIV NY STONY BROOK, 70- Concurrent Pos: Dir rheumatic dis res lab & chief rheumatic dis clin, Maimonides Hosp, Brooklyn, NY, 56-66; consult physician, Vet Admin Hosp, Brooklyn, 62-70, Nassau Hosp, Mineola, 66-, South Nassau Communities Hosp, Oceanside, 66- & Vet Admin Hosp, Northport, 70-; mem med & sci comt & chmn med adv comt, NY Chap Arthritis Found, 67-; chmn dept med, Nassau County Med Ctr, East Meadow, NY, 69-70; co-chmn dept med, Vet Admin Hosp, Northport. Mem: Am Rheumatism Asn; Am Fedn Clin Res; Am Heart Asn; AMA; Am Soc Nephrology. Res: Clinical effectiveness of immunosuppressive agents in rheumatic diseases. Mailing Add: Sch of Med Health Scis Ctr State Univ of NY at Stony Brook Stony Brook NY 11790

MEISELMAN, NEWTON, b Mineola, NY, Apr 5, 30; m 61; c 2. BOTANY, BIOLOGY. Educ: Syracuse Univ, AB, 51; Hofstra Univ, MS, 52; Rutgers Univ, PhD(bot), 56. Prof Exp: Asst, Rutgers Univ, 52-54; res assoc, Brookhaven Nat Lab, 55-56; from asst prof to assoc prof, 56-65, chmn dept, 68-74, PROF BIOL, C W POST COL, LONG ISLAND UNIV, 65- Concurrent Pos: Res collabr, Brookhaven Nat Lab, 59-62; researcher, Sect Biol Ultrastruct, Weizmann Inst, Israel, 65-66 & Plant Breeding Inst, Cambridge Univ, Eng, 75. Mem: AAAS; Bot Soc Am; Torrey Bot Club. Res: Morphogenesis; radiation morphology and cytology; electron microscopy; plant cytology.

MEISELS, ALEXANDER, b Berlin, Ger, Feb 18, 26; Can citizen; m 53; c 5. CYTOLOGY. Educ: Nat Univ Mex, MD, 51. Prof Exp: Asst dir lab cytol, Nat Cancer Inst, Mex, 56-60; dir lab clin cytol, 60-70, from asst prof to assoc prof path, 61-68, PROF PATH, UNIV LAVAL, 68-; DIR REGIONAL CYTODIAGNOSTIC CTR, ST SACREMENT HOSP, QUEBEC, 70- Concurrent Pos: Consult cytopathologist, Hotel-Dieu Hosp, Levis, 64-; mem, Can Cytol Coun, vchmn, 65, chmn, 66. Mem: Am Soc Cytol; Can Asn Path; Can Asn Anat; Int Acad Cytol (secy-treas, 71-). Res: Clinical cytology, particularly hormone, urinary and vaginal cytology; carcinogenesis, particularly effect of sex hormones and flora on evolution of cancer of the cervix. Mailing Add: Regional Cytodiag Ctr St Sacrement Hosp Quebec PQ Can

MEISELS, GERHARD GEORGE, b Vienna, Austria, May 11, 31; nat US; m 58; c 1. PHYSICAL CHEMISTRY, ANALYTICAL CHEMISTRY. Educ: Univ Notre Dame, MS, 52, PhD(phys chem), 56. Prof Exp: Res assoc, AEC, Univ Notre Dame, 53-56; radiation chemist, Gulf Res & Develop Co, 56-59; chemist, Union Carbide Nuclear Co, 59-63, asst group leader, 63-65; assoc from assoc prof to prof chem, Univ Houston, 65-75, assoc chmn dept, 69-72, chmn dept, 72-75; PROF CHEM & CHMN DEPT, UNIV NEBR-LINCOLN, 75- Concurrent Pos: Consult, Tech Div, Union Carbide Corp, 65. Mem: AAAS; Am Chem Soc; Am Phys Soc; fel Am Inst Chemists; Am Soc Mass Spectrometry. Res: Radiation chemistry; mass spectrometry; photochemistry; ion molecule reactions and collision dynamics. Mailing Add: Dept of Chem Univ of Nebr Lincoln NE 68588

MEISENHEIMER, JOHN LONG, b Olney, Ill, June 21, 33; m 56; c 2. ORGANIC CHEMISTRY. Educ: Evansville Col, BA, 54; Ind Univ, PhD(org chem), 63. Prof Exp: From asst prof to assoc prof, 63-68, PROF CHEM, EASTERN KY UNIV, 68- Mem: Am Chem Soc. Res: Medicinal and heterocyclic chemistry. Mailing Add: Dept of Chem Eastern Ky Univ Richmond KY 40475

MEISENHELDER, WILLIAM CROSBY, organic chemistry, see 12th edition

MEISER, JOHN H, b Cincinnati, Ohio, Nov 21, 38; m 67; c 2. CHEMISTRY. Educ: Xavier Univ, BS, 61; Univ Cincinnati, PhD(chem), 66. Prof Exp: Asst prof chem, Univ Dayton, 66-69; asst prof, 69-74, ASSOC PROF CHEM, BALL STATE UNIV, 74- Mem: Am Chem Soc; Am Phys Soc. Res: Diffusion of metals; counterdiffusion of ions; enzyme kinetics. Mailing Add: Dept of Chem Ball State Univ Muncie IN 47306

MEISINGER, MELVIN A, organic chemistry, see 12th edition

MEISKE, JAY C, b Hartley, Iowa, June 22, 30; m 56; c 4. ANIMAL HUSBANDRY. Educ: Iowa State Col, BS, 52; Okla State Univ, MS, 53; Mich State Univ, PhD, 57. Prof Exp: Asst animal husb, Okla State Univ, 52-53 & Mich State Univ, 53-57; from instr to assoc prof, 57-70, PROF ANIMAL HUSB, UNIV MINN, ST PAUL, 70- Mem: Am Soc Animal Sci; Am Inst Nutrit. Res: Ruminant nutrition; rumen microbiology and biochemistry. Mailing Add: Dept of Animal Sci Univ of Minn Inst of Agr St Paul MN 55108

MEISLER, ARNOLD IRWIN, b New York, NY, Mar 28, 32; m 59; c 3. ONCOLOGY, MOLECULAR BIOLOGY. Educ: Columbia Univ, AB, 52; NY Univ, MD, 56. Prof Exp: ASSOC PROF MED & MICROBIOL, SCH MED & DENT, UNIV ROCHESTER, 68- Concurrent Pos: Helen Hay Whitney Found fel, Royal Postgrad Med Sch, Univ London & Harvard Med Sch, 60-63; Am Cancer Soc fel, Harvard Med Sch, 65-68; Leukemia Soc Am scholar, 67- Mem: AAAS; Am Col Physicians. Res: Control of cell growth in tissue culture; contact inhibition of cell growth. Mailing Add: Dept of Med Sch of Med & Dent Univ of Rochester Rochester NY 14642

MEISLER, HAROLD, b New York, NY, Feb 7, 31; m 54; c 2. HYDROGEOLOGY. Educ: City Col New York, BS, 52; Univ Mich, MS, 53. Prof Exp: Geophysicist, Carter Oil Co, 55-56; geologist, Ground Water Br, 56-75, CHIEF NJ DIST, US GEOL SURV, 75- Mem: Geol Soc Am. Res: Hydrogeology of carbonate rocks, sandstones and shales; geomorphology of limestone terrain; computer simulation modeling of aquifer systems. Mailing Add: US Geol Surv Fed Bldg PO Box 1238 Trenton NJ 08607

MEISLER, MIRIAM HOROWITZ, b New York, NY, Mar 28, 43; m 63; c 2. BIOCHEMISTRY. Educ: Queens Col, BA, 64; Ohio State Univ, PhD(biochem), 68. Prof Exp: Cancer res scientist, Roswell Park Mem Inst, 71-73; ASST PROF BIOCHEM, SCH MED, STATE UNIV NY BUFFALO, 73- Concurrent Pos: NIH fel biochem, Roswell Park Mem Inst, 69-70; Nat Found March of Dimes res grant, 74-76; NIH proj grant, 74-77. Mem: AAAS; Am Soc Human Genetics; NY Acad Sci. Res: Regulation of gene expression in mammals; developmental biochemistry. Mailing Add: Dept of Biochem G-56 Capen Hall State Univ of NY Buffalo NY 14214

MEISLICH, HERBERT, b Brooklyn, NY, Mar 26, 20; m 55; c 3. ORGANIC CHEMISTRY. Educ: Brooklyn Col, AB, 40; Columbia Univ, AM, 47, PhD(chem), 51. Prof Exp: Chemist control & res, Edgewood Arsenal, 41-43; from teacher to assoc prof 46-68, PROF CHEM, CITY COL NEW YORK, 68- Concurrent Pos: Chemist, Med Sch, Columbia Univ, 50-52; fel, Med Sch, NY Univ, 52-53; lectr, Brooklyn Col, 52-58; Sloan-Kettering fel, 55-57. Mem: Am Chem Soc. Res: Mechanisms of organic reactions; heterocyclic chemistry; structure and reactivity of organic compounds. Mailing Add: 338 Lacey Dr New Milford NJ 07646

MEISNER, DONALD F, b Sabetha, Kans, Mar 26, 16; m 44; c 2. CEREAL CHEMISTRY. Educ: Wichita Univ, BA, 41; Univ Ill, MS, 43. Prof Exp: Chemist org res, Wm S Merrill Co, 44-46; chemist cereal res, C J Patterson Co, 46-48; dir lab baking technol, Am Inst Baking, 48-53; baking prod supt, Omar, Inc, 53-55; sect leader cereal res, Gen Mills, Inc, 55-66; GROUP LEADER INDUST FOODS DEPT, RES & DEVELOP DIV, KRAFTCO CORP, 66- Mem: Am Asn Cereal Chemists; Am Chem Soc; Am Soc Bakery Engrs. Res: Baking technology; production of baked cereal foods; product and process development. Mailing Add: Kraftco Corp Res & Develop Div 801 Waukegan Rd Glenview IL 60025

MEISNER, GERALD WARREN, b Mt Kisco, NY, Aug 16, 38; m 62; c 2. ELEMENTARY PARTICLE PHYSICS. Educ: Hamilton Col, AB, 60; Univ Calif, Berkeley, PhD(physics), 66. Prof Exp: Res assoc & guest lectr physics, Univ Mass, 66-70; ASSOC PROF PHYSICS, UNIV NC, GREENSBORO, 70- Concurrent Pos: Res Corp grant, 72. Mem: Am Phys Soc. Res: Experimental high-energy physics. Mailing Add: Dept of Physics Univ of NC Greensboro NC 27412

MEISNER, HERMAN M, cell physiology, biochemistry, see 12th edition

MEISS, ALFRED NELSON, b Philadelphia, Pa, Mar 27, 18; m 42; c 3. BIOLOGY. Educ: Rutgers Univ, BSc, 41, MSc, 43; Yale Univ, PhD(plant sci), 50. Prof Exp: Asst veg crops, Rutgers Univ, 43-44 & 46; asst bot, Yale Univ, 46-47; asst biochemist, Conn Agr Exp Sta, 49-52; asst res specialist soils, Rutgers Univ, 52-54, assoc prof, 54-57; sci adv, Ted Bates & Co, 57-68; sr assoc, 68-71, VPRES, SIDNEY M CANTOR ASSOCS, INC, 71- Mem: AAAS; Am Soc Plant Physiologists; Am Chem Soc. Res: Foods and nutrition; conservation; planning and management of natural resources; international development; analysis of national and regional food systems. Mailing Add: 14 N Main St Cranbury NJ 08512

MEISS, RICHARD ALAN, b Philadelphia, Pa, Aug 25, 43; m 65; c 1. PHYSIOLOGY. Educ: Univ Del, BA, 65; Univ Ill, Urbana, PhD(physiol), 69. Prof Exp: ASST PROF PHYSIOL, OBSTET & GYNEC, ASST PROF MED BIOPHYS, MED CTR, IND UNIV, INDIANAPOLIS, 71- Concurrent Pos: Nat Heart Inst fel, Harvard Med Sch, 69-71. Mem: AAAS; Am Physiol Soc. Res: Physiology and mechanical properties of muscle, in particular, cardiac muscle and the smooth muscle of the female reproductive system. Mailing Add: Dept of Obstet & Gynec Ind Univ Med Ctr Indianapolis IN 46202

MEISSNER, CHARLES ROEBLING, JR, b Joliet, Ill, May 4, 23; m 44; c 5. GEOLOGY. Educ: Lehigh Univ, BA, 48. Prof Exp: Explor geologist, US Geol Surv, Colo, 48-49; field geologist, Gulf Oil Corp, Okla, 49-52, field coordr, 52, well-site geologist, 52-53; Foreign Opers Admin mutual security mission consult & petrol geologist, Chinese Petrol Corp, Taiwan, 53-55; asst area geologist, Stanolind Oil & Gas Co, Tex, 55-56; sr geologist, Standard-Vacuum Oil Co, Sumatra & India, 56-61; geologist adv, Pakistan, 61-66, econ geologist, Saudi Arabia, 66-70, PROF GEOLOGIST, SOUTHWEST VA, US GEOL SURV, 70- Mem: Am Inst Prof Geologists; Am Asn Petrol Geologists. Res: Regional and detailed geologic mapping; mineral investigations; stratigraphy; petroleum geology; coal geology. Mailing Add: US Geol Surv 956 Nat Ctr Reston VA 22092

MEISSNER, HANS WALTER, b Berlin, Ger, Mar 19, 22; nat US; m 47; c 3. EXPERIMENTAL PHYSICS. Educ: Univ Munich, BS, 46, MS & PhD, 48. Hon Degrees: MS, Stevens Inst Technol, 62. Prof Exp: Asst physics, Low Temperature Inst, Acad Sci, Bavaria, 48-52; res engr, Heat Transfer Lab, Ill Inst Technol, 52-53; asst prof physics, Johns Hopkins Univ, 53-59; assoc prof, 59-62, PROF PHYSICS, STEVENS INST TECHNOL, 62- Honors & Awards: Jess H Davis Award, Stevens Inst Technol, 74. Mem: Fel Am Phys Soc. Res: Low temperature physics; superconductivity; metal physics; time dependent phenomena in superconductivity. Mailing Add: Dept of Physics Stevens Inst of Technol Castle Point Hoboken NJ 07030

MEISSNER, LOREN PHILLIP, b Los Angeles, Calif, Nov 24, 28; m 49; c 3. COMPUTER SCIENCE, APPLIED MATHEMATICS. Educ: Univ Calif, Berkeley, BA, 49, MA, 63, PhD(appl math), 65. MATHEMATICIAN, LAWRENCE BERKELEY LAB, UNIV CALIF, 59-, LECTR COMPUT SCI, UNIV, 68- Mem: Asn Comput Mach; Soc Indust & Appl Math. Res: Non-numeric applications of computers. Mailing Add: Lawren Berkeley Lab Univ of Calif Berkeley CA 94720

MEISSNER, WILLIAM AVISON, b Oregon City, Ore, May 20, 13; m 36; c 2. PATHOLOGY. Educ: Univ Ore, AB, 35, MD, 38. Prof Exp: From asst clin prof to clin prof, 52-71, PROF PATH, NEW ENG DEACONESS HOSP, HARVARD MED SCH, 71- Concurrent Pos: Pathologist, New Eng Baptist Hosp, 42-, pathologist-in-chief, 63-71; pathologist, New Eng Deaconess Hosp, 42-, pathologist-in-chief, 63-72, consult, Children's Hosp, Peter Bent Brigham Hosp, Vet Admin Hosp, West Roxbury & Armed Forces Inst Path, 67. Mem: Am Soc Clin Path; Am Cancer Soc; Col Am Path; Am Asn Cancer Res; Am Asn Path & Bact. Res: Thyroid disease; cancer. Mailing Add: 185 Pilgrim Rd Boston MA 02215

MEISTER, ALTON, b New York, NY, June 1, 22; m 43; c 2. BIOCHEMISTRY. Educ:

Harvard Univ, BS, 42; Cornell Univ, MD, 45. Prof Exp: From intern to asst resident, New York Hosp, 45-46; res worker, NIH, 46-55; prof biochem & chmn dept, Sch Med, Tufts Univ, 56-67; PROF BIOCHEM & CHMN DEPT, MED COL, CORNELL UNIV, 67-; BIOCHEMISTINCHIEF, NEW YORK HOSP, 71- Concurrent Pos: Consult comt growth, Nat Res Coun, 54; consult biochem study sect, USPHS, 55-60, consult biochem training comt, 60-63; consult, Am Cancer Soc, 58-61, 71-73; vis prof, Univ Wash, 59 & Univ Calif, Berkeley, 62; mem, US Nat Comt Biochem, 62-65; mem sci adv comn, New Eng Enzyme Ctr, 63-, chmn, 63-67; chmn physiol chem study sect, USPHS, 64-67; mem bd sci counr, Nat Cancer Inst, chmn, 72; ed, Advan Enzymol. Honors & Awards: Paul-Lewis Award, Am Chem Soc, 54. Mem: Nat Acad Sci; AAAS; fel Am Acad Arts & Sci; Am Chem Soc; Am Soc Biol Chem. Res: Biochemistry of the amino acids and proteins; mechanism of enzyme action. Mailing Add: Med Col Dept of Biochem Cornell Univ 1300 York Ave New York NY 10021

MEISTER, ARNOLD GEORGE, b Chicago, Ill, May 30, 12; m 39; c 1. MOLECULAR SPECTROSCOPY. Educ: Cent YMCA Col, BS, 36; Ill Inst Technol, PhD(physics), 47. Prof Exp: Cost analyst, Commonwealth Edison Co, Ill, 30-42; from instr to assoc prof physics, Ill Inst Technol, 42-57; PROF PHYSICS, ARIZ STATE UNIV, 57- Mem: AAAS; fel Am Phys Soc; Optical Soc Am; Can Asn Physicists; fel Brit Inst Physics & Phys Soc. Res: Analysis of rotation-vibration spectra of polyatomic molecules. Mailing Add: Dept of Physics Ariz State Univ Tempe AZ 85281

MEISTER, CHARLES WILLIAM, b Hackensack, NJ, Oct 5, 40. PLANT PATHOLOGY, PLANT PHYSIOLOGY. Educ: Rutgers Univ, New Brunswick, BS, 63; Univ Nebr, Lincoln, MS, 66; Univ Ariz, PhD(plant path), 72. Prof Exp: Res asst, Boyce Thompson Inst Plant Res, 65-66; asst prof biol, Catawba Col, 67-69; res asst agr biochem, Univ Ariz, 72; citrus virologist, Peace Corps, USDA, Fiji, 72-74; PLANT VIROLOGIST, USDA, SUVA, FIJI, 74- Concurrent Pos: Consult citrus virol, UNDP/Food Agr Org Surv Plant Pests & Dis in SPac, 75 & 76. Mem: Am Phytopath Soc; Am Inst Biol Sci. Res: Plant virology, cytology; control of plant pathogens; cell physiology; enzymes and other biochemicals found in diseased plants; established citrus-virus-indexing programs in Fiji and the Cook Islands. Mailing Add: 666 Kent Ave Teaneck NJ 07666

MEISTER, PETER DIETRICH, b Schaffhausen, Switz, May 24, 20; m 51; c 2. ORGANIC CHEMISTRY. Educ: Swiss Fed Inst Technol, MSc, 44, PhD(chem), 47. Prof Exp: Fel, Swiss Fed Inst Technol, 47-49 & Nat Res Coun Can, 49-50; res chemist, 50-55, sect head, 56-63, mgr, 63-66, asst dir, 66-68, DIR SUPPORTIVE RES, UPJOHN CO, 68- Mem: Am Chem Soc; Am Pharmaceut Asn. Res: Synthesis of steroids; degradation of lycopodium alkaloids; microbiological transformations of steroids; dosage forms. Mailing Add: 1001 Wilshire Blvd Kalamazoo MI 49008

MEISTERS, GARY HOSLER, b Ottumwa, Iowa, Feb 17, 32; m 52; c 2. MATHEMATICS. Educ: Iowa State Univ, BS, 54, PhD(math), 58. Prof Exp: Instr math, Iowa State Univ, 57-58; res instr, Duke Univ, 58-59; from asst prof to assoc prof, Univ Nebr, 59-63; from asst prof to prof, Univ Colo, Boulder, 63-72; PROF MATH, UNIV NEBR, LINCOLN, 72- Concurrent Pos: Res fel, Res Inst Advan Study, Md, 60-62. Mem: Am Math Soc; Math Asn Am. Res: Almost periodic functions; ordinary differential equations; abstract and functional analysis. Mailing Add: Dept of Math Univ of Nebr Lincoln NE 68508

MEISTERS, MARTS, analytical chemistry, see 12th edition

MEISTRICH, MARVIN LAWRENCE, b Brooklyn, NY, Oct 10, 41. BIOPHYSICS, CELL BIOLOGY. Educ: Rensselaer Polytech Inst BS, 62; Cornell Univ, PhD(physics), 67. Prof Exp: Mem tech staff, Bell Tel Labs, 67-69; res assoc biophys, Ont Cancer Inst, 69-72; ASST PROF BIOPHYS, UNIV TEX M D ANDERSON HOSP & TUMOR INST, 72- Mem: Biophys Soc; Genetics Soc Am; Am Soc Cell Biol; Soc Study Reprod; Radiation Res Soc. Res: Biophysical methods for cell separation; biochemical mechanisms in cell differnedifferentiation; cytotoxic and mutagenic effects of radiation and chemicals on spermatogenic cells; spermatogenesis; nucleoproteins. Mailing Add: Sect Exp Radiother Univ Tex M D Anderson Hosp Tex Med Ctr Houston TX 77025

MEITES, JOSEPH, b Kishinev, Russia, Dec 22, 13; nat US; m 43. PHYSIOLOGY. Educ: Univ Mo, BS, 38, MA, 40, PhD(exp endocrinol), 47. Prof Exp: Asst, Exp Sta, Univ Mo, 40-42, 46-47; from asst prof to assoc prof, 47-53, PROF PHYSIOL, MICH STATE UNIV, 53-, MEM STAFF, AGR EXP STA, 47- Concurrent Pos: Weizmann fel, Weizmann Inst Sci, Israel, 55-56; mem subcomt, use hormones in domestic animals, Nat Acad Sci-Nat Res Coun, 60-; mem endocrinol study sect, NIH, 66-70, comn neuroendocrinol, Int Union Physiol Sci, 71- Mem: AAAS; Am Asn Cancer Res; Am Physiol Soc; Endocrine Soc; Soc Exp Biol & Med. Res: Endocrinology as related to nutrition, lactation, reproduction and tumors; brain-pituitary relationships. Mailing Add: Dept of Physiol Mich State Univ East Lansing MI 48823

MEITES, LOUIS, b Baltimore, Md, Dec 6, 26; m 47; c 3. PHYSICAL CHEMISTRY, ANALYTICAL CHEMISTRY. Educ: Middlebury Col, BA, 45; Harvard Univ, MA, 46, PhD(chem), 47. Prof Exp: Instr chem, Princeton Univ, 47-48; from instr to asst prof, Yale Univ, 48-55; from assoc prof to prof, Polytech Inst Brooklyn, 55-68; PROF CHEM & CHMN DEPT, CLARKSON COL TECHNOL, 68- Concurrent Pos: Founding ed, Critical Rev in Anal Chem, 69-74; mem comn electroanal chem, Int Union Pure & Appl Chem. Mem: Assoc mem Int Union Pure & Appl Chem. Res: Thermochemical and electrochemical investigations of reaction kinetics and equilibria; differential thermometry and ebulliometry, controlled-potential electrolysis, polarography; titrimetric theory; chemical applications of non-linear regression; machine decisions. Mailing Add: Dept of Chem Clarkson Col of Technol Potsdam NY 13676

MEITES, SAMUEL, b St Joseph, Mo, Jan 3, 21; m 45; c 1. CLINICAL CHEMISTRY. Educ: Univ Mo, AB, 42; Ohio State Univ, PhD(biochem), 50; Am Bd Clin Chem, dipl. Prof Exp: Biochemist, Vet Admin Hosp, Poplar Bluff, Mo, 50-52 & Toledo Hosp, Ohio, 53; res asst prof, 54-66, assoc prof, 66-72, PROF PEDIAT, OHIO STATE UNIV, COL MED, 72-, PROF PATH, 74- Concurrent Pos: Biochemist, Children's Hosp, Columbus, Ohio, 54- Honors & Awards: Bernard J Katchman Award, Am Asn Clin Chemists, 72. Mem: Fel AAAS; Am Chem Soc; fel Am Asn Clin Chemists (secy, 75-77). Res: Amylase isoenzymes; micro methods; measurement of jaundice; normal values in pediatric clinical chemistry. Mailing Add: Clin Chem Lab Children's Hosp 561 S 17th St Columbus OH 43205

MEITZLER, ALLEN HENRY, b Allentown, Pa, Dec 16, 28; m 53; c 3. PHYSICS. Educ: Muhlenberg Col, BS, 51; Lehigh Univ, MS, 53, PhD, 55. Prof Exp: Asst physics, Lehigh Univ, 51-54; mem tech staff, Bell Tel Labs, Inc, NJ, 55-72; PRIN RES SCIENTIST, RES STAFF, FORD MOTOR CO, 72- Mem: Am Phys Soc; fel Acoust Soc Am; Inst Elec & Electronics Engrs. Res: Solid state physics; ultrasonic devices, ferroelectric ceramic and liquid crystal display devices; automotive emission control systems. Mailing Add: Ford Motor Co PO Box 2053 Dearborn MI 48121

MEIZEL, STANLEY, b New York, NY, May 1, 38; m 68; c 2. REPRODUCTIVE BIOLOGY, DEVELOPMENTAL BIOLOGY. Educ: Queens Col, NY, BS, 56; Univ Rochester, PhD(biochem), 66. Prof Exp: Asst prof, 67-74, ASSOC PROF HUMAN ANAT, SCH MED, UNIV CALIF, DAVIS, 74- Concurrent Pos: NIH fel develop biol, Yale Univ, 65-67. Mem: Am Asn Anat; Soc Study Reproduction; Soc Develop Biol; Am Soc Cell Biol; Histochem Soc. Res: Biochemistry of fertilization and sperm capacitation. Mailing Add: Dept of Human Anat Univ of Calif Sch of Med Davis CA 95616

MEJIA, GASTON RENE, b La Paz, Bolivia, Oct 11, 36; m 65; c 2. ATMOSPHERIC PHYSICS. Educ: Univ La Paz, Ing, 61. Prof Exp: Lectr fluid mech, Fac Indust Eng, Univ La Paz, 59-61, lectr electronics, 60-61; res asst physics, Mass Inst Technol, 61-62; res asst ionospheric physics, Lab Cosmic Physics, 63-64, from jr researcher to sr researcher, 64-67, dep dir, 67-68, assoc prof physics, Inst Basic Sci, 67-68, DIR LAB COSMIC PHYSICS & PROF PHYSICS, INST BASIC SCI, UNIV SAN ANDRES, 68- Concurrent Pos: Mem, Bolivian Comn for Solar Eclipse, 65-66 & Eclipse Sci Meeting Organizing Comt, 68; pres, Nat Comn Geophys Res, Bolivia, 69-70; secy-gen, Latin Am Coun Cosmic Rays & Space Physics, 70-; mem organizing comt, Interam Sem Cosmic Rays, La Paz, 70. Mem: AAAS; Am Phys Soc; Am Inst Phys; Am Meteorol Soc; Am Geophys Union. Res: Electronics; ionspheric physics and cosmic rays. Mailing Add: Lab Cosmic Physics Univ of San Andres La Paz Bolivia

MEKJIAN, ARAM ZAREH, b New York, NY, Sept 26, 41; m 69; c 2. NUCLEAR PHYSICS. Educ: Calif Inst Technol, BS, 63; Univ Md, College Park, PhD(physics), 68. Prof Exp: Fel, Rutgers Univ, 68-69; Alexander von Humboldt fel, Univ Heidelberg, 69-71; asst prof, 71-74, ASSOC PROF PHYSICS, RUTGERS UNIV, 74- Concurrent Pos: NSF grant, 71-72. Res: Nuclear structure; nuclear reactions; heavy ion physics; intermediate energy physics. Mailing Add: Dept of Physics Rutger Univ New Brunswick NJ 08903

MEKLER, ARLEN B, b New York, NY, May 4, 32; m 61; c 6. ORGANIC CHEMISTRY. Educ: San Jose State Col, BS, 53; Iowa State Univ, MS, 55; Ohio State Univ, PhD(phys org chem), 58; Temple Univ, JD, 72. Prof Exp: E I du Pont de Nemours & Co res fel, Ohio State Univ, 58-59; res chemist, Polychem Dept, E I du Pont de Nemours & Co, 59-61, res chemist, Explosives Dept, 61-63; sr res chemist, Arco Chem Res Div, Atlantic Richfield Co, 63-69; RES ASSOC, RCEO INC, 69- Mem: AAAS; Am Chem Soc; Sigma Xi; The Chem Soc. Res: Benzil-ammonia reaction; synthesis of steroid intermediates; steric effects in addition reaction; stereospecific polymerizations; selective oxidation and oxidative coupling reactions; transition metal-olefin complexes; synthesis and polymerization of small ring compounds; patent law; forensic science. Mailing Add: 1108 N Rodney St Wilmington DE 19806

MEL, HOWARD CHARLES, b Oakland, Calif, Jan 14, 26; m 49; c 3. BIOPHYSICS. Educ: Univ Calif, Berkeley, BS, 48, PhD(phys chem), 53. Prof Exp: Mem staff, Calo Dog Food Co, Calif, 48-50; asst chem, Univ Calif, Berkeley, 50-51, chemist radiation lab, 51-53; Fulbright fel, Free Univ Brussels, 53-54; instr chem, 55, USPHS fel, spec fel & lectr med physics & biophys, 55-59, from asst prof to assoc prof, 60-73, PROF BIOPHYS, UNIV CALIF, BERKELEY, 73-, STAFF MEM, LAWRENCE RADIATION LAB, 60- Concurrent Pos: Am Inst Physics vis scientist, 63-; NSF sr fel, France, 65-66. Mem: AAAS; Biophys Soc; Am Chem Soc; Am Asn Physics Teachers. Res: Cellular and subcellular biophysics; biological separations including electrophoresis; cellular development and differentiation, hematopoiesis; thermodynamics of open and closed systems. Mailing Add: Div of Med Physics Donner Lab Univ of Calif Berkeley CA 94720

MELA, LEENA MARJA, b Viiala, Finland, Apr 5, 35. BIOCHEMISTRY, PHYSIOLOGY. Educ: Univ Helsinki, BSc, 60; Turku Univ, MD, 64. Prof Exp: USPHS grant, Johnson Res Found, Univ Pa, 66-69, res assoc, Harrison Dept Surg Res Hosp, 69-70, asst prof, 70-75, ASSOC PROF PHYS BIOCHEM IN SURG, HARRISON DEPT SURG RES, UNIV PA, 75- Concurrent Pos: NIH career develop award, 71-76; mem prog proj adv comt, Nat Inst Neurol & Commun Disorders & Stroke, 74- Mem: AAAS; Am Soc Biol Chemists; Am Chem Soc; Am Physiol Soc; Soc Gen Physiologists. Res: Mitochondrial metabolism during shock and trauma; structural and functional damage of the cell after injury; mitochondrial membrane transport and energy-linked functions. Mailing Add: Harrison Dept of Surg Res Univ of Pa Hosp Philadelphia PA 19174

MELACHOURIS, NICHOLAS, b Piraeus, Greece, May 24, 34; US citizen; m 67. FOOD SCIENCE, BIOCHEMISTRY. Educ: Col Agr Athens, Greece, BS, 58; Univ Ill, Urbana, MS, 63, PhD(food sci), 66. Prof Exp: Res chemist, Miles Labs, 67-69; SUPVR RES, STAUFFER CHEM CO, 69- Mem: Am Chem Soc; NY Acad Sci; Inst Food Technologists; Am Dairy Sci Asn. Res: Protein chemistry, yeasts, phosphates and biochemistry. Mailing Add: Eastern Res Ctr Stauffer Chem Co Dobbs Ferry NY 10522

MELAMED, MYRON ROY, b Cleveland, Ohio, Aug 9, 27; m 57; c 2. MEDICINE, PATHOLOGY. Educ: Western Reserve Univ, BS, 47; Univ Cincinnati, MD, 50. Prof Exp: From asst attend pathologist to assoc attend pathologist, 58-69, ATTEND PATHOLOGIST, MEM HOSP CANCER & ALLIED DIS, 69-, CHIEF CYTOL SERV, 73-; ASSOC PROF PATH, MED SCH, CORNELL UNIV, 73- Concurrent Pos: Assoc mem, Sloan-Kettering Inst Cancer Res, 58-; consult, USPHS Hosp, Staten Island, NY, 61-64; med res consult, Int Bus Mach Corp, 64-68, NY State Dept Health, 65 & Col Dent, NY Univ, 65; asst vis pathologist, James Ewing Hosp, 65; consult, Hosp Spec Surg, 73- Honors & Awards: Papanicolaou Award, Am Soc Cytol, 75. Mem: AAAS; Am Asn Path & Bact; Am Soc Cytol; James Ewing Soc; Am Soc Clin Path. Res: Pathology and cytology of cancer. Mailing Add: Mem Hosp Cancer & Allied Dis 1275 York Ave New York NY 10021

MELAMED, NATHAN T, b Poland, May 1, 23; nat US; m 67; c 2. CHEMISTRY. Educ: City Col New York, BS, 43; Polytech Inst Brooklyn, PhD(chem), 49. Prof Exp: Res chemist, Manhattan Dist Proj, SAM Labs, Columbia Univ, 43-45; instr chem, Polytech Inst Brooklyn, 48; res physicist solid state physics, Horizons, Inc, 49-50; mgr optical electronics, 72-74, ADV SCIENTIST, WESTINGHOUSE RES LABS, 50- Concurrent Pos: Lectr, Univ Pittsburgh, 55-56; prof, Fed Univ Rio de Janeiro, 73 & 74. Honors & Awards: IR-100 Award, 65. Mem: Am Phys Soc; NY Acad Sci. Res: Luminescence of inorganic solids-fundamental theory; laser physics; semiconductors; spectroscopy. Mailing Add: 6636 Dalzell Pl Pittsburgh PA 15217

MELAMED, SIDNEY, b Philadelphia, Pa, Oct 5, 20; m 43; c 2. POLYMER CHEMISTRY. Educ: Philadelphia Col Pharm, BSc, 41; Univ Ill, PhD(org chem), 44. Prof Exp: Res chemist, Univ Ill, 43-44; res chemist, Univ Md, 44-45, res chemist, Comt Med Res, 45-46; res chemist, US Navy, 46-47; res chemist, 47-58, lab head, 58-71, supvr new prod develop, Fibers Div, 71-73, SR RES ASSOC LIFE SCI, ROHM & HAAS CO, 73- Mem: AAAS; Am Chem Soc. Res: Nylon, polyesters, elastic and antistatic fibers; vinyl and condensation polymers; monomers; polymers for paper, textiles, leather and coatings; ion-exchange resins and fibers; insecticides; fiber and fabric technology; biomedical research; agricultural chemicals. Mailing Add: 8270 Thomson Rd Elkins Park PA 19117

MELAMID, ALEXANDER, b Freiburg, Ger, Mar 28, 14; US citizen; wid; c 2. ECONOMIC GEOGRAPHY, REGIONAL ECONOMICS. Educ: London Sch Econ, BSc, 38; New Sch Social Res, PhD(econ), 53. Prof Exp: From instr to assoc prof econ geog, New Sch Social Res, 49-57; assoc prof, NY Univ, 57-68, PROF ECON GEOG, NY UNIV, 68- Concurrent Pos: Fel, New Sch Social Res, 54-55; NY Univ fel, 65-68 & 69-72; mem steering comt, High Sch Geog Proj, NSF, 63-66; vis prof & res coordr, Univ Lagos, 64; vis prof, Univ Pa, 66; consult, Tidelands Comn, State of La, 66-71; vis prof, Monash Univ, 68; vis prof, London Sch Econ, 69. Mem: Asn Am Geog; hon fel, Am Geog Soc; Am Econ Asn; Royal Geog Soc; Asn Australian Geog. Res: Geography and economic development; petroleum economics and pricing; industrial location. Mailing Add: Dept of Econ NY Univ Tisch Hall New York NY 10003

MELAMPY, ROBERT MAURICE, b Lebanon, Ohio, Apr 1, 09; m 37; c 1. PHYSIOLOGY. Educ: Wilmington Col, BS, 30; Haverford Col, MA, 31; Cornell Univ, PhD(animal nutrit), 35. Prof Exp: Asst animal nutrit, Cornell Univ, 31-35; asst physiologist, USDA, 36-41; asst prof physiol, La State Univ, 41-42, assoc prof, 42-46 & 48-49; assoc prof, Univ Ill, 46-48; assoc prof, 49-50, PROF ANIMAL SCI, IOWA STATE UNIV, 50- Concurrent Pos: Res fel, Cambridge Univ, 66-67; consult, NIH, 66-71. Honors & Awards: Award, Am Soc Animal Sci, 66. Mem: AAAS; Am Soc Zool; Soc Exp Biol & Med; Am Physiol Soc; Soc Study Reproduction (pres, 67-68). Res: Physiology of reproduction. Mailing Add: Dept of Animal Sci Iowa State Univ Ames IA 50010

MELANCON, MARK J, JR, biochemistry, see 12th edition

MELAVEN, ARTHUR DAVID, b Rapid City, SDak, Feb 10, 07. INORGANIC CHEMISTRY. Educ: SDak Sch Mines & Technol, BS, 29; Pa State Univ, MS, 32, PhD(chem), 35. Prof Exp: Anal chemist, Res Labs, Aluminum Co Am, Pa, 29-31; from instr to prof, 35-72, EMER PROF CHEM, UNIV TENN, KNOXVILLE, 72- Mem: AAAS; Am Chem Soc. Res: Recovery of rhenium from molybdenite ores; chemistry of rhenium. Mailing Add: 6619 E El Paso St Mesa AZ 85205

MELBY, EDWARD C, JR, b Burlington, Vt, Aug 10, 29; m 53; c 4. VETERINARY MEDICINE. Educ: Cornell Univ, DVM, 54; Am Col Lab Animal Med, dipl, 67. Prof Exp: Pvt pract, Vt, 54-62; from instr to prof & dir dept, Lab Animal Med, Sch Med, Johns Hopkins Univ, 62-74; DEAN & PROF MED, NY STATE COL VET MED, CORNELL UNIV, 74- Concurrent Pos: Consult, Vet Admin, 64-74; mem, White House Conf on Health, 65; mem coun accreditation, Am Asn Accreditation Lab Animal Care, 66-73; chmn, Inst Lab Animal Resources, Nat Res Coun-Nat Acad Sci, 75-; mem adv comt, Div Res Resources, NIH; consult, Nat Inst Child Health & Human Develop, 71-; pres, Am Col Lab Animal Med, 74-75. Mem: AAAS; Am Asn Lab Animal Sci; Am Vet Med Asn; NY Acad Sci. Res: Laboratory animal medicine and comparative pathology, especially lymphoproliferative diseases and transplantation. Mailing Add: NY State Col of Vet Med Cornell Univ Ithaca NY 14850

MELBY, JAMES CHRISTIAN, b Duluth, Minn, Feb 14, 28; m 55; c 2. MEDICINE. Educ: Univ Minn, BS, 51, MD, 53. Prof Exp: Lectr endocrine biochem, Med Sch, Univ Minn, 58, instr med & dir clin chem, 58-59; asst prof med & biochem, Sch Med, Univ Ark, 59-62; assoc prof, 62-69, PROF MED, SCH MED, BOSTON UNIV, 69-, PROF PHYSIOL, 71-, MEM FAC, DIV MED SCI, GRAD SCH, 71- Concurrent Pos: Consult, Merck & Co, 55-65; head sect endocrinol, Evans Mem Hosp, 62- Mem: AAAS; Am Soc Clin Invest; Asn Am Physicians; Am Chem Soc; Endocrine Soc. Res: Metabolism of steroid hormones, physical interaction of steroid hormones and macromolecules; endocrinology; internal medicine. Mailing Add: Univ Hosp Boston MA 02118

MELBY, LESTER RUSSELL, b Calgary, Alta, Oct 9, 27; nat US; div; c 3. ORGANIC CHEMISTRY. Educ: Univ Alta, BS, 48, MS, 50; Univ Ill, PhD(chem), 53. Prof Exp: RES ORG CHEMIST, CENT RES & DEVELOP DEPT, E I DU PONT DE NEMOURS & CO, INC, 63- Mem: Am Chem Soc. Res: Synthetic oligonucleotides; coordination compounds, organic ligands and organic reactions of polymers. Mailing Add: Cent Res & Develop Dept E I du Pont de Nemours & Co Inc Wilmington DE 19898

MELBYE, SUSANNE WARNER, b Pittsfield, Mass, Nov 8, 41; m 69. BIOCHEMISTRY, DERMATOLOGY. Educ: Harvard Univ, BA, 63; Univ Calif, Berkeley, PhD(biochem), 70. Prof Exp: Res assoc dermat, Sch Med, Stanford Univ, 70-71; RES FEL DERMAT, BETH ISRAEL HOSP-HARVARD MED SCH, 71- Mem: AAAS; Am Chem Soc. Res: Protein structure and synthesis; control of macromolecular synthesis in mammalian epidermis. Mailing Add: Beth Israel Hosp Dept of Dermat 330 Brookline Ave Boston MA 02215

MELCER, IRVING, b Havana, Cuba, Nov 15, 31; nat US; m 54; c 5. BIOCHEMISTRY, FOOD SCIENCE. Educ: Wayne State Univ, BS, 53, PhD(biochem), 58. Prof Exp: USPHS res fel pharmacol, Yale Univ, 58-59; res chemist, Wilson Labs, 59-64; res chemist, 65-73, LAB MGR CHEM PROCESSING RES, GRIFFITH LABS, 73- Mem: Am Chem Soc; Inst Food Technol. Res: Soy proteins and protein hydrolysates; porphyrin and serotonin biosynthesis; isolation of enzymes, hormones, proteins and other natural products; chemistry of meat and meat processing. Mailing Add: Griffith Labs 1415 W 37th St Chicago IL 60609

MELCHER, ANTONY HENRY, b Johannesburg, SAfrica, July 1, 27; m 53; c 2. HISTOLOGY. Educ: Univ Witwatersrand, BDS, 49, HDD, 58, MDS, 60; Univ London, PhD(morphol), 64. Prof Exp: PROF HISTOL, FAC DENT, UNIV TORONTO, 69-, DIR MED RES COUN GROUP PERIODONT PHYSIOL, 74- Concurrent Pos: Leverhulme Found res fel, Royal Col Surgeons Eng, 64-67, col res fel morphol, 64-69; ed, Oral Sci Rev, 71- Mem: Int Asn Dent Res; Orthop Res Soc; Tissue Culture Asn; Brit Bone & Tooth Soc. Res: Repair of bone; structure and function of periodontium. Mailing Add: Fac of Dent Univ of Toronto Toronto ON Can

MELCHER, ROBERT LEE, b Marshalltown, Iowa, Jan 27, 40; m 66; c 2. SOLID STATE PHYSICS. Educ: Southern Methodist Univ, BS, 62; Wash Univ, MA, 65, PhD(physics), 68. Prof Exp: Res assoc physics, Cornell Univ, 68-70; MEM RES STAFF, WATSON RES CTR, IBM CORP, 70- Mem: Am Phys Soc. Res: Elastic and magnetoelastic properties of materials; nuclear spin-phonon interactions; elastic properties of materials undergoing phase transitions; polarization echoes in piezoelectric and magnetoelastic materials, echo holography in piezoelectric semiconductors. Mailing Add: IBM Watson Res Ctr PO Box 218 Yorktown Heights NY 10598

MELCHER, ULRICH KARL, b London, Eng, July 7, 45; US citizen; m 68; c 2. MOLECULAR BIOLOGY, IMMUNOCHEMISTRY. Educ: Univ Chicago, BS, 65; Mich State Univ, PhD(biochem), 70. Prof Exp: Fel molecular biol, Univ Aarhus, Denmark, 70-71; res scientist immunol, Med Ctr, NY Univ, 72; res asst microbiol, Univ Tex Health Sci Ctr, Dallas, 72-74, asst prof, 74-75; ASST PROF BIOCHEM, OKLA STATE UNIV, 75- Concurrent Pos: NATO fel, 70-71 & NIH fel, 73-74. Mem: Am Soc Plant Physiologists. Res: Attachment of immunoglobulin to membranes; control of protein synthesis in seed development. Mailing Add: Dept of Biochem PS II Okla State Univ Stillwater OK 74074

MELCHING, J STANLEY, b New York, NY, Mar 4, 23; m 41; c 2. PLANT PATHOLOGY. Educ: Univ Maine, BS, 54, MS, 56; Cornell Univ, PhD(plant path), 61. Prof Exp: Res plant pathologist, Field Crops & Animal Prods Br, USDA, Watseka, Ill, 61-63 & Biol Br, Corps Div, US Army Biol Labs, Ft Detrick, Frederick, Md, 63-70; RES PLANT PATHOLOGIST, CROP RESPONSE FOREIGN DIS RES UNIT, PLANT DIS RES LAB, AGR RES SERV, USDA, 70- Mem: Am Phytopath Soc; Am Soc Plant Physiologists; Am Inst Biol Sci; Sigma Xi. Res: Disease dynamics of the rusts of corn and the soybean rust; development of quantitative techniques for measurement of the infection process, disease spread and special instrumentation for assessment of environmental factors. Mailing Add: Plant Dis Res Lab Agr Res Serv USDA Box 1209 Frederick MD 21701

MELCHIOR, JACKLYN BUTLER, b Sacramento, Calif, May 19, 18; m 39; c 2. BIOCHEMISTRY. Educ: Univ Calif, BS, 40, PhD(biochem), 46. Prof Exp: Res assoc chem, Northwestern Univ, 46-49; from instr to asst prof biochem, Sch Med, Loyola Univ, Ill, 49-57, assoc prof pharmacol, 57-59; PROF BIOCHEM, CHMN DEPT & ASST DEAN, BASIC SCI, CHICAGO COL OSTEOP MED, 59- Concurrent Pos: Lederle Med Fac Award, 57. Mem: Am Chem Soc; Am Soc Biol Chem. Res: Enzyme chemistry; protein conformation. Mailing Add: Dept of Biochem Chicago Col of Osteop Med Chicago IL 60615

MELCHIOR, NORTEN CASS, b Trinidad, Colo, Feb 15, 13; m 39; c 2. ORGANIC CHEMISTRY, BIOLOGICAL CHEMISTRY. Educ: Univ Calif, BS, 44, PhD(chem), 46. Prof Exp: Instr chem, Northwestern Univ, 46-49; asst prof, 49-54, ASSOC PROF BIOCHEM, STRITCH SCH MED, LOYOLA UNIV CHICAGO, 54-, DIR MULTIDISCIPLINE LABS, 68- Mem: Am Chem Soc. Res: Absorption spectra; metal-organic compounds; biological transport mechanisms. Mailing Add: Dept of Biochem & Biophys Loyola Univ Stritch Sch of Med Maywood IL 60153

MELCHIORE, JOHN J, b Philadelphia, Pa, Jan 8, 32; m 53; c 3. PETROLEUM CHEMISTRY. Educ: La Salle Col, BA, 53; Purdue Univ, PhD(phys chem), 57. Prof Exp: Res chemist, 57-68, appl res proj leader, 68-75, MEM PLANNING & ECON STAFF, SUN OIL CO, 75- Mem: Am Chem Soc. Res: Planning and economic studies; industrial product development; electrical and refrigeration oils; aromatic and oxidation chemistry. Mailing Add: 302 Cooper Dr Wallingford PA 19086

MELCHOR, JACK L, b Mooresville, NC, July 6, 25; m 46; c 4. PHYSICS. Educ: Univ NC, BS, 48, MS, 50; Univ Notre Dame, PhD(physics), 53. Prof Exp: Group leader, Sylvania Elec Co, 53-56; pres & treas, Melabs, Inc, 56-61; pres, HP Assocs, 61-67; gen mgr, Palo Alto Div, Hewlett Packard Co, 67-68; CHMN BD, PALO ALTO INVEST CO, 69- Concurrent Pos: Consult, Triad Am Capital Mgt, 74- Honors & Awards: Notre Dame Centennial of Sci Award, 67. Mem: Fel Inst Elec & Electronics Engrs; Sigma Xi. Res: Microwave ferrites; semiconductor devices; photo conductors; microwave semiconductors; small company development in the Middle East. Mailing Add: 26000 Westwind Way Los Altos Hills CA 94022

MELECA, COSMO BENJAMIN, b Batavia, NY, Nov 8, 37; m 58; c 4. MEDICAL EDUCATION. Educ: State Univ NY Col Brockport, BS, 63; Syracuse Univ, MS, 66, PhD(sci educ), 68. Prof Exp: High sch teacher, NY, 63-65; instr biol, Syracuse Univ, 67-68; asst dir, Biol Core Prog, 68-70, asst prof biochem, 68-72, dir introductory biol prog, 70-74, ASSOC PROF BIOCHEM, OHIO STATE UNIV, 72-, ASSOC PROF PREV MED & DIR DIV RES & EVAL IN MED EDUC, COL MED, 74- Concurrent Pos: Pres, Int Audio-Tutorial Cong, 71-72; chmn, Audio-Tutorial Conf, 71-, chmn conf comt, 71- Mem: AAAS; Am Chem Soc; Nat Asn Biol Teachers; Am Asn Med Cols. Res: Science education; biological education; computer assisted instruction; pre and post MD education; research, evaluation, and development in continuum of medical education; instructional development. Mailing Add: 3190 Graves Hall Ohio State Univ Col of Med Columbus OH 43210

MELECHEN, NORMAN EDWARD, b New York, NY, Jan 26, 24; m 53; c 3. GENETICS. Educ: Columbia Univ, AB, 44; Univ Pa, PhD(zool), 54. Prof Exp: Spec investr genetics, Carnegie Inst Wash, 54-56; instr, 56-57, sr instr, 57-58, from asst prof to assoc prof, 58-64, PROF MICROBIOL, SCH MED, ST LOUIS UNIV, 64- Concurrent Pos: Commonwealth Fund fel, Stanford Univ, 66-67. Mem; Genetics Soc Am; Am Soc Microbiol; Brit Soc Gen Microbiol. Res: Structure and function of bacteriophage chromosomes; genetics and chemistry of bacteriophage infection and induction. Mailing Add: Dept of Microbiol St Louis Univ Sch of Med St Louis MO 63104

MELENDEZ, LUIS VARGAS, b Santiago, Chile, Feb 14, 27; US citizen; m 53; c 3. VIROLOGY. Educ: Univ Chile, DVM, 50. Prof Exp: Virologist, Dept Microbiol, Vet Sect, Inst Bact Chile, 50-54, sr virologist, 57-62; res asst virol, Univ Wis, 62-64; res assoc bact & immunol, 64-66, chmn div microbiol, Primate Res Ctr, 66-74, mem fac med & prin assoc microbiol & molecular genetics, 67-74, assoc prof microbiol, 70-74, LECTR MICROBIOL & MOLECULAR GENETICS, HARVARD MED SCH, 74-; REGIONAL ADV VET MED, PAN AM HEALTH ORGN, 74- Concurrent Pos: Rockefeller Found fel infectious dis, Children's Med Ctr, Harvard Univ, 55-56; Rockefeller Found travel grant, 60; res consult virologist, Shields Warren Radiation Lab & New Eng Deaconess Hosp, Boston, Mass, 66-67. Mem: AAAS; Am Soc Microbiol; Am Asn Immunol; Am Asn Lab Animal Sci; Am Cancer Soc. Res: Herpesviruses; lymphoma viruses of monkeys and viral diseases of South American monkeys. Mailing Add: Pan Am Health Orgn 525 23rd St NW Washington DC 20037

MELENDEZ, PEDRO LUIS, phytopathology, nematology, see 12th edition

MELERA, ATTILIO, b Bellinzona, Switz, Sept 19, 29; m 59; c 2. ORGANIC CHEMISTRY, SPECTROSCOPY. Educ: Swiss Fed Inst Technol, PhD(chem), 57. Prof Exp: NIH res grant, Stanford Univ, 57-60; res chemist, Varian Res Labs, Zurich, Switz, 60-65, mgr nuclear magnetic resonance, 66-67; sr res chemist, Cent Res Lab, 68-70, SR RES CHEMIST, SCI INSTRUMENTS DIV, HEWLETT PACKARD CO, PALO ALTO, 70- Res: Organic structural analysis; nuclear magnetic resonance; electro spectroscopic chemical analysis. Mailing Add: 1071 Suffolk Way Los Altos CA 94304

MELERA, PETER WILLIAM, b Union City, NJ, Feb 19, 42; m 63; c 2. CELL BIOLOGY, BIOCHEMISTRY. Educ: Univ Ga, BS, 65, PhD(bot), 69. Prof Exp: Res assoc, Walker Lab, 72-75, ASSOC, WALKER LAB, SLOANKETTERING INST CANCER RES, 75-; ASST PROF BIOCHEM, CORNELL GRAD SCH MED SCI, 76- Concurrent Pos: NIH fel, McArdle Lab Cancer Res, Univ Wis-Madison, 69-72. Mem: Am Chem Soc; Am Soc Plant Physiol; Can Soc Plant Physiol; Am Soc Cell

Biol. Res: Metabolism of ribonucleic acids, their catalytic role in protein synthesis and the involvement of these molecules in the control of mitotic division. Mailing Add: Walker Lab Sloan-Kettering Inst Cancer Res Rye NY 10580

MELEZIN, ABRAHAM, b Odessa, Russia, June 29, 10; US citizen; m 48; c 1. GEOGRAPHY. Educ: Stephen Batory Univ, Poland DrSc(geog), 39. Prof Exp: Asst prof geog, Trenton State Col, 60-64; from asst prof to assoc prof, 64-74, PROF GEOG, CITY COL NEW YORK, 74- Mem: Am Asn Geog; Nat Coun Geog Educ; Am Geog Soc; Can Asn Geog. Res: Human geography; geography of the Union Soviet Socialist Republics, especially its socio-economic aspects. Mailing Add: Dept of Econ City Col of New York New York NY 10031

MELGAARD, KENNETT GILBERT, b Lawton, Iowa, Aug 5, 13; m 42; c 3. ANALYTICAL CHEMISTRY. Educ: Univ SDak, AB, 35; Columbia Univ, MA, 39; Univ Colo, PhD(chem), 57. Prof Exp: Mech engr, Carrier Corp, 36-38; high sch teacher, SDak, 40-42; chemist, E I du Pont de Nemours & Co, 42-44; chem engr, Houdaille Hershey Corp, 44-45; chemist, Grigoleit Co, 45-48; assoc prof, 48-64, PROF CHEM, NMEX STATE UNIV, 64- Mem: Fel AAAS; Am Chem Soc; Sigma Xi. Res: Electrochemistry; chelated compounds. Mailing Add: Dept of Chem NMex State Univ Box 3C Las Cruces NM 88001

MELGARD, RODNEY, b Carrington, NDak, Feb 24, 36; m 60; c 3. RADIOCHEMISTRY, NUCLEAR PHYSICS. Educ: Jamestown Col, BS, 57. Concurrent Pos: Actinide element group leader, Tracerlab, Inc, 58-61; radioactivity measurement supvr, 61-73, LAB OPERS MGR, ENVIRON DIV, LFE CORP, RICHMOND, 74- Mem: Am Chem Soc. Res: High sensitivity studies of natural and man-made radiation related to environmental surveillance programs; geothermal radiochemistry; reactor coolant chemistry; nuclear rocket ablation studies; snap devices; alpha and gamma spectroscopy. Mailing Add: 1000 Alberdan Circle Pinole CA 94564

MELGES, FREDERICK TOWNE, b Battle Creek, Mich, Dec 2, 35; m 58; c 2. PSYCHIATRY, EXPERIMENTAL PSYCHOLOGY. Educ: Princeton Univ, AB, 57; Columbia Univ, MD, 61. Prof Exp: Med intern, Univ Mich, 61-62; res psychiat, Strong Mem Hosp, Univ Rochester, 62-64, chief res & instr, 64-65; NIMH spec fel, Sch Med, Stanford Univ, 65-67, ASST PROF PSYCHIAT & NIMH RES CAREER DEVELOP AWARD, STANFORD UNIV, 67- Concurrent Pos: Dir psychiat educ & res, Stanford Univ Prog at Santa Clara Valley Med Ctr, San Jose, Calif. Res: Time sense, identification and emotion in mental illness; hormones and behavior. Mailing Add: Dept of Psychiat Med Ctr Stanford Univ Palo Alto CA 94304

MELHORN, WILTON NEWTON, b Sistersville, WVa, July 8, 20; m 61. GEOMORPHOLOGY, QUATERNARY GEOLOGY. Educ: Mich State Univ, BS, 42, MS, 51; NY Univ, MS, 43; Univ Mich, PhD(geol), 55. Prof Exp: Hydrogeologist, Geol Surv, Mich, 46-49; meteorologist, US Weather Bur, 49-50; from asst prof to assoc prof eng geol, 55-67, head dept geosci, 67-71, PROF ENG GEOL, PURDUE UNIV, WEST LAFAYETTE, 67- Concurrent Pos: Vis prof, Univ Ill, 60-61; vis prof, Univ Nev, Reno, 71-72, adj prof, Mackay Sch Mines, 73- Mem: Fel AAAS; Geol Soc Am; Am Meteorol Soc; Soc Econ Paleontologists & Mineralogists; Am Asn Petrol Geologists. Res: Geomorphology and Pleistocene geology; remote sensing; geomorphology of arid lands. Mailing Add: Dept of Geosci Purdue Univ West Lafayette IN 47906

MELI, ALBERTO L G, b Florence, Italy, Sept 19, 21; US citizen; m 55; c 1. PHYSIOLOGY. Educ: Univ Pisa, DVM, 47; Univ Milan, PhD(physiol), 58, PhD(pharmacol), 69. Prof Exp: Lectr physiol, Univ Pisa, 47-49; group leader pharmacol, C Erba Res Inst, Italy, 50-57; group leader endocrinol, Vister Labs, 58-59; group leader pharmacol, Sterling-Winthrop Res Inst, NY, 60-61; sr res assoc physiol, Warner-Lambert Res Inst, NJ, 61-67; dir pharmacol dept, Vister Labs, Italy, 67-68, dir biol res, 68-69; DIR RES, MANARINI LABS, 69- Mem: AAAS; Soc Study Reproduction; Endocrine Soc; Am Physiol Soc; Soc Exp Biol & Med. Res: General endocrinology and pharmacology; reproductive physiology; metabolism. Mailing Add: Lab Chimico Farmaceutico A Menarini Via Sette Santi 1 Florence 50131 Italy

MELICK, WILLIAM F, b St Louis, Mo, Nov 9, 14; m 39; c 4. UROLOGY. Educ: Wash Univ, MD, 39; St Louis Univ, MS, 43. Prof Exp: From asst prof to assoc prof, 53-60, dir dept, 59-66, PROF CLIN UROL, ST LOUIS UNIV, 60- Mem: Am Urol Asn; Am Col Surg. Res: Renal arteriography; urographic media; cancer of bladder; ureteral physiology. Mailing Add: 1035 Bellevue St Louis MO 63117

MELISSINOS, ADRIAN CONSTANTIN, b Thessaloniki, Greece, July 28, 29; m 60; c 2. PARTICLE PHYSICS. Educ: Mass Inst Technol, MS, 56, PhD, 58. Prof Exp: Asst, Univ Athens, Greece, 54-55 & Mass Inst Technol, 55-58; from instr to assoc prof, 58-67, PROF PHYSICS, UNIV ROCHESTER, 67-, CHMN DEPT PHYSICS & ASTRON, 74- Concurrent Pos: Guest physicist, Brookhaven Nat Lab, 63-; vis scientist, Europ Orgn Nuclear Res, 68-69. Mem: Fel Am Phys Soc. Mailing Add: Dept of Physics & Astron Univ of Rochester Rochester NY 14627

MELIUS, MELVIN EUGENE, JR, zoology, reproductive physiology, see 12th edition

MELIUS, PAUL, b Livingston, Ill, Nov 21, 27; m 53; c 4. BIOCHEMISTRY. Educ: Bradley Univ, BS, 50; Univ Chicago, MS, 52; Loyola Univ, Ill, PhD(chem), 56. Prof Exp: Chemist, Nat Aluminate Corp, 52-53; biochemist, Med Sch, Northwestern Univ, 56-57; assoc prof, 57-65, PROF BIOCHEM, AUBURN UNIV, 65- Prof Exp: NIH spec fel biochem, Univ Ky, 62 & Univ Calif, Los Angeles, 68. Mem: Am Chem Soc; NY Acad Sci; Am Soc Biol Chemists. Res: Thermal polymerization of amino acids, pyridoxine analogs, platinum complexes and enzymes and protein chemistry. Mailing Add: Dept of Chem Auburn Univ Auburn AL 36830

MELKANOFF, MICHEL ALLAN, b Russia, July 3, 23; nat US; m 65. NUCLEAR PHYSICS. Educ: NY Univ, BS, 43; Univ Calif, Los Angeles, MA, 50, PhD(physics), 55. Prof Exp: Aeronaut engr, 43-44 & 46-47; vis asst prof physics, 56-58, from asst res physicist to assoc res physicist, 58-61, assoc prof eng, 62-66, PROF ENG, UNIV CALIF, LOS ANGELES, 66-, CHMN DEPT, 69- Concurrent Pos: Vis physicist, Saclay Nuclear Res Ctr, France, 61-62; consult, Thompson Ramo Wooldridge, Inc, 59-60, Opers Res Ctr, France, 62, Mass Inst Technol, 63, Univ Hawaii, 64, Univ Md, 65, Douglas Aircraft Inc, 65, Rand Corp, 65-66, Nat Bank of Mex, 71 & US Army Comput Syst Command, 71. Mem: Asn Comput Mach. Res: Digital computers; computer languages, compilers, design automated design; file organization; management; information systems. Mailing Add: Dept Comput Sci Univ of Calif Los angeles CA 90024

MELKONIAN, EDWARD, b Alexandria, Egypt, June 29, 20; US citizen; m 54; c 3. PHYSICS. Educ: Columbia Univ, AB, 40, AM, 41, PhD(physics), 49. Prof Exp: Asst physics, Columbia Univ, 41-42; tech engr, Carbide & Carbon Chem Corp, Tenn, 45-46; res scientist, Atomic Energy Comn Contract, 46-63, assoc prof, 63-68, PROF NUCLEAR SCI & ENG, COLUMBIA UNIV, 68- Concurrent Pos: Res scientist, Nat Defense Res Comt, Off Sci Res & Develop & Manhattan Dist, Columbia Univ, 41-44; res assoc, Atomic Energy Res Estab, Harwell, Eng, 58-59. Mem: AAAS; fel Am Phys Soc; Am Nuclear Soc. Res: Porous membranes; gas flow; mathematical study of diffusion cascade; neutron spectroscopy of gases; neutron resonances and physics; fission and nuclear reactor physics; application of computers to scientific problems. Mailing Add: Dept of Nuclear Sci & Eng Columbia Univ New York NY 10027

MELL, GALEN P, b Modesto, Calif, Sept 20, 34; m 61; c 3. BIOCHEMISTRY. Educ: Univ Idaho, BS, 56; Univ Wash, PhD(biochem), 61. Prof Exp: Res fel folic acid, Scripps Clin & Res Found, 64-68; asst prof biochem, 68-73, Am Cancer Soc grant, 70-72, ASSOC PROF BIOCHEM, UNIV MONT, 73- Mem: Am Chem Soc. Res: Enzymology; clinical biochemistry; role of folic acid coenzymes in intermediary metabolism; proteases from plant tissue cultures. Mailing Add: Dept of Chem Univ of Mont Missoula MT 59801

MELL, LEROY DAYTON, JR, b Carlisle, Pa, Oct 26, 45; m 69; c 1. ANALYTICAL CHEMISTRY. Educ: Juniata Col, BS, 67; WVa Univ, PhD(anal chem), 75. Prof Exp: Forensic chemist, US Army Criminal Invest Lab, Ft Gordon, Ga, 75; RES CHEMIST, NAVAL MED RES INST, NAT NAVAL MED CTR, 75- Res: New analytical methods for catecholamine analysis; analytical methods involved in studies related to heat stress, hypertension and obesity; use of electroanalytical chemistry and high pressure liquid chromatography in biomedical and forensic research. Mailing Add: Naval Med Res Inst Nat Naval Med Ctr Bethesda MD 20014

MELLEN, FREDERIC FRANCIS, b State College, Miss, Aug 21, 11; m 35; c 4. GEOLOGY. Educ: Miss State Univ, BS, 34; Univ Miss, MS, 37. Prof Exp: Geol aide, Tenn Valley Authority, 33-34; asst geol, Miss State Univ, 34-35; jr geologist, Tenn Valley Authority, 35-37; geologist & asst state geologist, Works Progress Admin, Miss Geol Surv, 37-40; consult geol, 40-44; dist geologist, Brit-Am Oil Producing Co, 44-45; consult geol, 45-62; dir & state geologist, Miss Geol Surv, 62-65; GEOLOGIST, FREDERIC F MELLEN, GEOL ASSOCS, 65- Mem: Paleont Soc; Am Asn Petrol Geologists; Geol Soc Am; Asn Prof Geol Scientists; Sigma Xi. Res: Micropaleontological and well sample work; structural and stratigraphic geology; evaluation of oil properties. Mailing Add: 5540 Queen Christina Lane Jackson MS 39209

MELLEN, GILBERT EMERY, b Akron, Iowa, Mar 16, 21; m 44; c 3. PHYSICS. Educ: Iowa State Col, BS, 43. Prof Exp: Physicist, Radiation Lab, Univ Calif, 43-45, Tenn Eastman Corp, Oak Ridge, 45-47 & Carbide & Carbon Chem Co, 47-61; PHYSICIST, CHEM & PLASTICS DIV, UNION CARBIDE CORP, 61- Mem: Am Soc Mass Spectrometry. Res: Development of calutron ion source; mass spectrometer analysis; data processing systems; analytical and preparative gas chromatography. Mailing Add: Chem & Plastics Div Bldg 106 Union Carbide Corp PO Box 8004 South Charleston WV 25303

MELLEN, ROBERT HARRISON, b New Haven, Conn, Nov 12, 19; m 42; c 2. PHYSICS. Educ: Wesleyan Univ, BA, 41; Univ Conn, MA, 53, PhD, 55. Prof Exp: Res physicist, US Naval Res Lab, DC, 41-46, Underwater Sound Lab, Conn, 46-64; marine electronics off, Avco Corp, 64-67; RES PHYSICIST, NAVAL UNDERWATER SYSTS CTR, NEW LONDON, 67- Mem: AAAS; Am Phys Soc; Acoust Soc Am. Res: Underwater sound. Mailing Add: RD 3 Old Lyme CT 06371

MELLEN, WALTER ROY, b Newark, NJ, Mar 10, 28; m 50; c 5. ELEMENTARY PARTICLE PHYSICS, THEORETICAL PHYSICS. Educ: Mass Inst Technol, SB, 48; Lowell Technol Inst, MS, 62. Prof Exp: Tech writer, Sperry Gyroscope Co, 50-53; instr physics, Adelphi Col, 53-55; tech writer, Sperry Gyroscope Co, 55-56; asst prof, Alfred Univ, 56-59; asst prof elec eng, 59-63, ASSOC PROF PHYSICS, LOWELL TECHNOL INST, 63- Mem: AAAS; Am Asn Physics Teachers; Am Phys Soc. Res: Electromagnetic field theory and its applications to the structure of elementary particles, to astronomical problems and to plasmas. Mailing Add: Dept of Physics Lowell Technol Inst Lowell MA 01854

MELLEN, WILLIAM JAMES, b Brattleboro, Vt, Sept 10, 23; m 43; c 3. AVIAN PHYSIOLOGY. Educ: Univ Mass,BS, 49; Cornell Univ, MS, 51, PhD(genetics), 53. Prof Exp: Asst animal genetics, Cornell Univ, 50-53, res assoc avian physiol, 53; asst prof animal & poultry physiol, Univ Del, 53-57, assoc prof, 57-58; assoc prof poultry physiol, 58-60, PROF POULTRY PHYSIOL, UNIV MASS, AMHERST, 60-, ASSOC DEAN RES, COL AGR, 70- Concurrent Pos: Vis prof, Hokkaido Univ, 60-61. Mem: Fel AAAS; Poultry Sci Asn. Res: Avian thyroid physiology; physiological genetics of poultry. Mailing Add: Col of Agr Univ of Mass Amherst MA 01002

MELLENTHIN, WALTER M, b Ririe, Idaho, July 10, 20; m 46; c 3. HORTICULTURE. Educ: Ore State Univ, BS, 50, MS, 51; Prof Exp: Instr hort, 51-52, from asst prof & supt to assoc prof & supt, 52-65, PROF HORT & SUPT, MID-COLUMBIA BR EXP STA, ORE STATE UNIV, 65- Mem: Am Pomol Soc; Am Soc Hort Sci. Res: Pome fruit rootstocks; effects of environment on growth and maturity of pears; effect of environments on quality; physiological disorders of pears and apples. Mailing Add: Mid-Columbia Br Exp Sta Rt 5 Box 240 Hood River OR 97031

MELLER, EMANUEL, chemistry, see 12th edition

MELLETT, JAMES SILVAN, b New York, NY, July 12, 36; m 61; c 2. VERTEBRATE PALEONTOLOGY. Educ: Iona Col, BS, 59; Columbia Univ, MA, 64, PhD(geol), 66. Prof Exp: Asst prof biol, Iona Col, 63-67, dir NSF In-Serv Inst, 66-67; asst prof, 67-71, ASSOC PROF GEOL & CHMN DEPT, NY UNIV, 71- Mem: AAAS; Soc Vert Paleont; Soc Study Evolution. Res: Taxonomy, geologic history of fossil mammals; paleoecology of Tertiary mammals; mortality rates in fossil mammals; biomechanical analyses of mammalian jaw muscles. Mailing Add: Dept of Geol NY Univ New York NY 10003

MELLETT, LAWRENCE B, b Detroit, Mich, Mar 13, 24; m 51; c 8. PHARMACOLOGY. Educ: Univ Detroit, BS, 48, MS, 51; Univ Mich, PhD(pharmacol), 57. Prof Exp: Asst, Univ Mich, 52-56, from instr to asst prof pharmacol, Med Sch, 57-65; head pharmacol sect, 65-68, HEAD BIOCHEM PHARMACOL DIV, SOUTHERN RES INST, 68-; PROF, MED CTR, UNIV ALA, BIRMINGHAM, 74- Concurrent Pos: Mem cancer chemother collab rev comt, Nat Cancer Inst, 67-69; assoc, Med Ctr, Univ Ala, 65-74. Mem: AAAS; Am Soc Pharmacol; NY Acad Sci; Am Asn Cancer Res; Am Soc Microbiol. Res: Drug metabolism; biological disposition of narcotic analgesics and antitumor agents; biochemical mechanisms of drug action. Mailing Add: Southern Res Inst 2000 Ninth Ave S Birmingham AL 35205

MELLETTE, RUSSELL RAMSEY, JR, b Orangeburg, SC, Dec 11, 27; m 51; c 5. CHILD PSYCHIATRY. Educ: Clemson Col, BS, 46; Med Col SC, MD, 50. Prof Exp: Intern, Wayne County Gen Hosp, Eloise, Mich, 50-51; resident psychiat, Edgewood Sanatorium, Orangeburg, SC, 51-52, clin dir, 52, med dir, 53; staff psychiatrist, US Army Hosp, Ft Sam Houston, Tex, 53-54; resident med neuropsychiat, Med Ctr Hosps, Charleston, SC, 55-56; from asst resident to resident & jr clin instr, Neuropsychiat Inst & Children's Psychiat Hosp, Sch Med, Univ Mich, Ann Arbor,

MELLETTE

56-58, instr neuropsychiat, 58; asst prof psychiat, 58-65, chief child psychiat sect, 63-73, actg dir youth serv div, Dept Psychiat & Behav Sci, 73-74, ASSOC PROF PSYCHIAT, MED UNIV SC, 65-, ASSOC PROF PEDIAT, 67-, ASSOC PROF BEHAV SCI, 73-, DIR YOUTH OUTPATIENT SERV, DEPT PSYCHIAT & BEHAV SCI, 74- Concurrent Pos: Dir, Charleston County Ment Health Clin, SC, 58-60; sr teaching fel & chief children's psychiat unit, Sch Med, Univ NC, Chapel Hill, 60-61; consult child psychiat, Child Develop Clin, Dept Pediat, Med Univ SC, 61-66; consult, SC Dept Ment Health, 61-69 & NIMH Sch Proj, Sumter, SC, 61-74; consult & lectr, SC State Hosp, Columbia, 61-; NIMH GP educ grant, 64-74. Mem: Fel Am Geriat Soc; fel Am Orthopsychiat Asn; fel Am Psychiat Asn; fel Am Acad Child Psychiat; AMA. Res: Inpatient child psychiatry sevices; postgraduate education of nonpsychiatric physicians; voodoo; bromism; iatrogenic illnesses; involutional depressive states; infectious mononucleosis. Mailing Add: Dept of Psychiat & Behav Sci Med Univ of SC Charleston SC 29401

MELLIERE, ALVIN L, b Praire du Rocher, Ill, Aug 9, 39; m 61; c 3. NUTRITION. Educ: Univ Ill, BS, 61, MS, 64, PhD(nutrit biochem), 65. Prof Exp: Asst prof nutrit, Univ Minn, 65-67; RES SCIENTIST, LILLY RES LABS, ELI LILLY & CO, 67- Mem: AAAS; Am Soc Animal Sci. Res: Dietary and chemical factors affecting the metabolism of nutrients in monogastric animals. Mailing Add: Dept G708 Eli Lilly & Co Greenfield IN 46140

MELLIN, GILBERT WYLIE, b Manorville, Pa, Sept 22, 25; m 55; c 2. MEDICINE, PEDIATRICS. Educ: Bethany Col, BS, 45; Johns Hopkins Univ, MD, 49; Am Bd Pediat, dipl, 54. Prof Exp: Intern med, Med Ctr, Univ Pittsburgh, 49-50; from jr resident to sr resident pediat, Bellevue Hosp, New York, 50-52; instr, Med Sch, NY Univ, 52-53; clin instr, Med Sch, Georgetown Univ, 53-55; actg chmn dept, 70-71, instr, 55-57, assoc, 57-58, asst prof, 58-67, ASSOC PROF PEDIAT, COL PHYSICIANS & SURGEONS, COLUMBIA UNIV, 67- Concurrent Pos: Dir fetal life study, Columbia-Presby Med Ctr; chief resident, Bellevue Hosp, 52-53; jr assoc, Children's Hosp, Washington, DC, 53-55; asst pediatrician, Babies Hosp & Vanderbilt Clin, Presby Hosp, New York, 55-57, asst attend pediatrician, 57-67, assoc attend pediatrician, 67-, actg dir pediat serv, 70-71; mem tech adv comt cleft palate, City New York Dept Health, 61-; proj consult, Children's Bur, US Dept Health, Educ & Welfare & Govt Pakistan, 67- Mem: AAAS; Am Acad Pediat; fel Am Pub Health Asn; Am Statist Asn; Soc Pediat Res. Res: Epidemiological approach to fetal life and pregnancy outcome by means of direct prospective observation; establishment of documented magnetic tape data set banks for rapid data tabulation and analysis by electronic computer. Mailing Add: Col of Physicians & Surgeons Columbia Univ New York NY 10032

MELLIN, THEODORE NELSON, b Paterson, NJ, Dec 24, 37; m 59; c 1. PHYSIOLOGY, BIOCHEMISTRY. Educ: Univ Vt, BS, 59; Univ Maine, MS, 61; Purdue Univ, PhD(reproductive physiol), 65. Prof Exp: Res asst animal sci, Univ Maine, 59-61; res asst reproductive physiol, Wash State Univ, 61-62; res asst, Purdue Univ, 62-63, res fel, 63-65; RES PHYSIOLOGIST, MERCK INST THERAPEUT RES, 66- Concurrent Pos: Nat Cancer Inst fel steroid biochem, Worcester Found Exp Biol, 64-66. Mem: Am Dairy Sci Asn; Am Soc Animal Sci. Res: Reproductive physiology and endocrinology of large animals; metabolism and conjugation of steroid sex hormones; development and application of steroid sex hormone assays to the bovine; rumen microbiology. Mailing Add: Dept of Reproductive Biol Merck Inst for Therapeut Res Rahway NJ 07452

MELLINGER, CLAIR, b Ephrata, Pa, Mar 29, 42; m 66; c 1. PLANT ECOLOGY. Educ: Eastern Mennonite Col, BS, 64; Univ NC, Chapel Hill, PhD(bot), 72. Prof Exp: Asst instr, 64-65, ASSOC PROF BIOL, EASTERN MENNONITE COL, 70- Mem: AAAS; Ecol Soc Am; Am Inst Biol Sci. Res: Population dynamics of plant species under environmental stress. Mailing Add: Biol Dept Eastern Mennonite Col Harrisonburg VA 22801

MELLINGER, GEORGE T, b New Orleans, La, Nov 8, 19; m 45; c 2. UROLOGY. Educ: Tulane Univ, BS, 41, MD, 43; Am Bd Urol, dipl, 52. Prof Exp: Chief urol, Cincinnati, 55-60, chief urol, Minneapolis, 60-69, chief staff, Kansas City, 69-72, MEM STAFF, VET ADMIN HOSP, WICHITA, 72- Concurrent Pos: Chmn, Vet Admin Coop Urol Res Group, 59-; assoc prof, Med Sch, Univ Minn, Minneapolis, 60-69; prof urol & asst dean sch, Univ Kans, 69-72. Mem: Am Urol Asn; Am Col Surg; Pan-Am Med Asn; Asn Mil Surg US. Res: Carcinoma of the prostate. Mailing Add: Vet Admin Hosp 5500 E Kellogg Ave Wichita KS 67218

MELLINGER, MELVIN WAYNE, b Quincy, Ill, Mar 10, 41; m 65; c 3. NEUROPSYCHOLOGY. Educ: Univ Mo-Kansas City, BA, 68, MA, 70; Univ Kans, PhD(anat), 75. Prof Exp: ASST PROF ANAT, UNIV TEX MED BR, GALVESTON, 75- Mem: Sigma Xi; Soc Neurosci. Res: Forebrain regeneration and behavioral implications in lower vertebrates. Mailing Add: Dept of Anat Univ of Tex Med Br Galveston TX 77550

MELLINGER, MICHAEL VANCE, b Harrisburg, Pa, Dec 21, 45; m 68; c 2. PLANT ECOLOGY, ENVIRONMENTAL MANAGEMENT. Educ: Bloomsburg State Col, BA, 67; Syracuse Univ, PhD(plant ecol), 72. Prof Exp: PLANT ECOLOGIST, SARGENT & LUNDY ENGRS, 72- Mem: Ecol Soc Am; Am Inst Biol Sci; AAAS; Sigma Xi. Res: Structure and function of terrestrial communities; analysis of environmental effects of electric power facilities. Mailing Add: Sargent & Lundy Engrs 55 E Monroe St Chicago IL 60603

MELLINKOFF, SHERMAN MUSSOFF, b McKeesport, Pa, Mar 23, 20; m 44; c 2. MEDICINE. Educ: Stanford Univ, BA, 41, MD, 44. Prof Exp: Intern med, Stanford Univ Hosp, 44, asst resident, 44-45; asst resident, Osler Serv, Johns Hopkins Hosp, 47-49, resident & instr, 50-51, physician in charge gastroenterol, Outpatient Dept, 51-53; from asst prof to assoc prof, 53-62, PROF MED & DEAN, SCH MED, UNIV CALIF, LOS ANGELES, 62- Concurrent Pos: Fel, Hosp Univ Pa, 49-50; instr med, Johns Hopkins Univ, 51-53; attend consult, Wadsworth Gen Hosp, Vet Admin Ctr, 53-; sr attend physician, Harbor Gen Hosp, Torrance, 53- Mem: Nat Inst Med; AAAS; Am Gastroenterol Asn; Am Fedn Clin Res; Am Col Physicians. Res: Gastroenterology. Mailing Add: Univ of Calif Sch of Med Los Angeles CA 90024

MELLINS, HARRY ZACHARY, b Brooklyn, NY, May 23, 21; m 50; c 3. RADIOLOGY. Educ: Columbia Univ, AB, 41; Long Island Col Med, MD, 44; Univ Minn, MS, 51; Am Bd Radiol, dipl. Hon Degrees: AM, Harvard Univ, 69. Prof Exp: From instr to asst prof radiol, Med Sch, Univ Minn, 50-53; clin asst prof, Col Med, Wayne State Univ, 53-56; prof & chmn dept, Col Med, State Univ NY Downstate Med Ctr, 56-69; PROF RADIOL, HARVARD MED SCH, 69-; DIR DIV DIAG RADIOL, PETER BENT BRIGHAM HOSP, BOSTON, 69- Concurrent Pos: AEC-Nat Res Coun fel, Rice Inst, 49-50; nat consult, Surgeon-Gen, US Air Force; consult, Vet Admin Hosp, West Roxbury. Mem: Roentgen Ray Soc; Am Col Radiol; Asn Univ Radiol; Radiol Soc NAm. Res: Intestinal obstruction; renal medullary function. Mailing Add: 25 Shattuck St Boston MA 02115

MELLINS, ROBERT B, b New York, NY, Mar 6, 28; m 59; c 2. PEDIATRICS, CARDIOPULMONARY PHYSIOLOGY. Educ: Columbia Univ, AB, 48; Johns Hopkins Univ, MD, 52; Am Bd Pediat, dipl. Prof Exp: Intern pediat, Johns Hopkins Hosp, 52-53; clin instr, Col Med, Univ Ill, 54-55; asst resident pediat, New York Hosp, 55-56; asst resident, Columbia-Presby Med Ctr, 56-57, asst, 57-60, instr, 60-65, assoc, 65-66, asst prof, 66-70, assoc prof pediat, 70-75, PROF PEDIAT, COL PHYSICIANS & SURGEONS, COLUMBIA UNIV, 75-, DIR PEDIAT PULMONARY DIV, 73- Concurrent Pos: NY Heart Asn trainee, 61-63, res fel, 63-66; NIH career develop award, 66-71; dir, Chicago Poison Control Ctr, 53-55; mem subcomt poisoning, Am Standards Asn, 54-55; asst pediatrician, Presby Hosp, 57-65, asst attend pediatrician, 65-70, assoc attend pediatrician, 70- Mem: Am Fedn Clin Res; Soc Pediat Res; Am Thoracic Soc; Am Physiol Soc; NY Acad Sci. Res: Cardiopulmonary research; lung mechanics; regulation of respiration and transcapillary exchange of fluid in lung, pleura and pericardium. Mailing Add: Presby Hosp Dept of Pediat 622 W 168th St New York NY 10032

MELLITS, E DAVID, b Philadelphia, Pa, Sept 5, 37; m 58; c 2. BIOSTATISTICS. Educ: Johns Hopkins Univ, BES, 59, ScD(biostatist), 65. Prof Exp: Systs analyst, Strong Mem Hosp, Rochester, NY, 59-61; asst prof, 65-71, ASSOC PROF BIOSTATIST, SCHS HYG & MED, JOHNS HOPKINS UNIV, 71- Mem: Am Statist Asn; Biomet Soc; Am Inst Math Statist; NY Acad Sci; Soc Pediat Res. Res: Application of statistics in the biological and medical sciences. Mailing Add: Dept of Pediat Johns Hopkins Hosp Baltimore MD 21218

MELLMAN, WILLIAM JULES, b Philadelphia, Pa, May 7, 28; m 64; c 2. MEDICAL GENETICS. Educ: Univ Pa, AB, 48, MD, 52. Prof Exp: Intern, Philadelphia Gen Hosp, 52-53; resident, Univ Pa Hosp, 55-57; fel endocrinol, Children's Hosp Philadelphia, 57-58; res fel metab & cytogenetics, 58-63, from instr to assoc pediat, 59-61, asst prof, 61-66, assoc prof pediat & med genetics, 66-72, PROF PEDIAT, SCH MED, UNIV PA, 72-, PROF HUMAN GENETICS & CHMN DEPT, 72-, DIR HUMAN GENETICS CTR, 74- Concurrent Pos: Am Cancer Soc scholar, Med Res Coun Human Biochem Genetics Res Unit, Univ London, 62-63; NIH res career develop award, Nat Inst Child Health & Human Develop, 63-73; dir, Genetics Div, Children's Hosp Philadelphia, 65-, sr attend physician, 71-; vis prof pharmacol, Stanford Univ, 70-71; mem genetics study sect, NIH, 74-; ed, Am J Human Genetics, 75; contrib ed, J Pediat; mem, Coun Acad Socs. Mem: Am Soc Human Genetics; Am Pediat Soc; Soc Pediat Res; Am Soc Clin Invest. Res: Identification and characterization of human gene loci involved in control of biochemical function; analysis of the control of enzyme activity in human cells; characterization of the genetics and population distribution of hereditary traits responsible for huaml human disease. Mailing Add: Dept Human Genetics Univ Pa Richards Bldg G4 Sch of Med Philadelphia PA 19174

MELLO, JAMES FRANCIS, b Providence, RI, Aug 24, 36; m 57; c 5. PALEONTOLOGY, COMPUTER SCIENCE. Educ: Brown Univ, AB, 58; Yale Univ, MSc, 60, PhD(paleont), 62. Prof Exp: Geologist, Paleont & Stratig Br, US Geol Surv, 62-70; res assoc & spec asst to dir electronic data processing, 70-73, ASST DIR, NAT MUS NATURAL HIST, SMITHSONIAN INST, 73- Concurrent Pos: Fel, Cushman Found Foraminiferal Res; mem bd dirs, Mus Computer Network. Res: Paleont Soc. Res: Cretaceous and Paleocene Foraminifera of the western interior of the United States and the Atlantic Coastal Plain. Mailing Add: Off of the Dir Rm 419 Nat Mus of Natural Hist Washington DC 20560

MELLON, DEFOREST, JR, b Cleveland, Ohio, Dec 18, 34; m. NEUROPHYSIOLOGY. Educ: Yale Univ, BS, 57; Johns Hopkins Univ, PhD(biol), 61. Prof Exp: Air Force Off Sci Res-USPHS fel neurophysiol, Stanford Univ, 61-63; asst prof, 63-68, ASSOC PROF BIOL, UNIV VA, 68- Prof Exp: Guggenheim mem fel, 66-67. Mem: Fel AAAS; Soc Gen Physiol; Am Physiol Soc; Soc Neurosci; Am Soc Zoologists. Res: Physiology of sensory cells; sensory integration in invertebrates; neural correlates of simple behavior patterns. Mailing Add: Dept of Biol Gilmer Hall Univ of Va Charlottesville VA 22903

MELLON, EDWARD FRANCIS, physical chemistry, see 12th edition

MELLON, EDWARD KNOX, JR, b Rochester, NY, Oct 8, 36; m 66. INORGANIC CHEMISTRY, SCIENCE EDUCATION. Educ: Univ Tex, BS, 59, PhD(chem), 63. Prof Exp: Instr chem, St Edward's Univ, 62-63; fel, Univ Mich, 63-65, lectr, 65-66; asst prof, 66-72, ASSOC PROF CHEM, FLA STATE UNIV, 72- Mem: Am Chem Soc; The Chem Soc. Res: Chemical education; synthesis and structure of borazines and organometallic compounds. Mailing Add: Dept of Chem Fla State Univ Tallahassee FL 32306

MELLON, GEORGE BARRY, b Edmonton, Alta, Aug 5, 31. GEOLOGY. Educ: Univ Alta, BSc, 54, MSc, 55; Pa State Univ, PhD(mineral, petrol), 59. Prof Exp: Sr res off & head geol div, Res Coun Alta, 58-74; MEM STAFF, ALTA ENERGY & NATURAL RESOURCES, DEP MINISTER ENERGY RESOURCES, 74- Mem: Geol Asn Can; Soc Econ Paleont & Mineral. Res: Sedimentary petrology and applied statistics in geology. Mailing Add: Alta Energy & Nat Resources Dep Minister Energy Resources Edmonton AB Can

MELLOR, CLIVE SIDNEY, b Derbyshire, Eng, May 12, 32; Can citizen; m 57; c 5. PSYCHIATRY. Educ: Univ Manchester, MB & ChB, 56, MD, 66, PhD(genetics of alcoholism), 69. Prof Exp: Lectr psychiat, Univ Manchester, 64-69, sr lectr, 69-70; assoc prof, 71-75, PROF PSYCHIAT, MEM UNIV, NFLD, 75- Concurrent Pos: Consult in chg, Regional Alcoholism Serv, Manchester, Eng, 66-70; consult psychiat, Manchester Royal Infirmary, 68-70; mem, Sci Rev Comt, Fed Nonmed Use of Drugs Directorate, Ottawa, Can, 75- Mem: Can Psychiat Asn; AMA. Res: Psychophysiology of alcohol dependence; alcoholism in native peoples; phenomenology of schizophrenia. Mailing Add: Dept of Psychiat Mem Univ St Johns NF Can

MELLOR, DAVID BRIDGWOOD, b Brockton, Mass, June 25, 29; m 62; c 1. POULTRY SCIENCE, FOOD TECHNOLOGY. Educ: Pa State Univ, BS, 56; Tex A&M Univ, MS, 57; Purdue Univ, PhD(food tech), 65. Prof Exp: Overseas trainee, Int Coop Admin, 57-58, livestock adv, Agency Int Develop, 58-62; poultry adv, IRI Res Inst Inc, 65-67; ASSOC PROF POULTRY MKT, TEX A&M UNIV, 67- Mem: Poultry Sci Asn; World Poultry Sci Asn; Inst Food Technol. Res: Tenderness of poultry meat; salmonella contamination of shell eggs; poultry marketing. Mailing Add: Dept of Poultry Sci Tex A&M Univ College Station TX 77843

MELLOR, JESSE LYNN, b Manti, Utah, Aug 5, 20; m 47; c 2. SOILS. Educ: Utah State Agr Col, BS, 47; Colo Agr & Mech Col, MS, 50; NC State Col, PhD(soils), 53. Prof Exp: Asst prof, Colo Agr & Mech Col, 47-50; agt & soil scientist, Bur Plant Indust, Soils & Agr Eng, USDA, 53; asst prof agron, Univ Wyo, 53-55; DIR PLANT FOODS, OLIN CORP, 55- Mem: Am Soc Agron. Res: Soil chemistry and ferility. Mailing Add: Olin Corp PO Box 991 Little Rock AR 77203

MELLOR, JOHN, b Guilford, Conn, Apr 28, 33; m 56; c 3. PHYSICAL CHEMISTRY. Educ: Univ Conn, BA, 56; Tufts Univ, MS, 58; Mass Inst Technol, PhD(phys chem), 62. Prof Exp: Chemist, Gen Elec Co, Conn, 56; res asst phys chem,

Tufts Univ, 58; asst chemist, Arthur D Little, Inc, Mass, 59, consult physics, 60-62; res assoc, Brookhaven Nat Lab, 62-64; asst prof, 64-74, ASSOC PROF PHYS CHEM, UNIV BRIDGEPORT, 74- Concurrent Pos: Vis asst physicist, Brookhaven Nat Lab, 65, res collabr, 65-66. Mem: Am Crystallog Asn. Res: Crystal structure analysis by x-ray and neutron diffraction; molecular motions and lattice dynamics by neutron inelastic scattering. Mailing Add: Dept of Chem Univ of Bridgeport Bridgeport CT 06602

MELLOR, MALCOLM, b Stalybridge, Eng, May 24, 33; m 58; c 2. APPLIED PHYSICS, ENGINEERING. Educ: Univ Nottingham, BS, 55; Univ Melbourne, MS, 59, DSc(appl sci), 69; Univ Sheffield, PhD(civil & struct eng), 70. Prof Exp: Asst engr, State Rivers & Water Supply Comn, Australia, 56; glaciologist, Australian Nat Antarctic Res Expeds, Dept External Affairs, 56-59; contract engr, Snow, Ice & Permafrost Res Estab, US Dept Army, Thayer Sch Eng, Dartmouth Col, 59-61, res engr, Cold Regions Res Lab, 61-75, RES PHYS SCIENTIST, COLD REGIONS RES & ENG LAB, US DEPT ARMY, 75- Concurrent Pos: Assoc consult, Creare Inc, 70-74; mem ed bd, Int Glaciological Soc, 72-; mem comt glaciol, Polar Res Bd, 74-; secy, Int Comn Snow & Ice, 75- Honors & Awards: Brit Polar Medal, 59; Antarctic Medal, 67; Spec Award, Nat Res Coun, Nat Acad Sci-Nat Acad Eng, 72. Mem: Int Glaciological Soc; Glaciol Soc. Res: Applied mechanics; explosives; physics and mechanics of snow, ice and frozen ground; cold regions engineering; glaciology. Mailing Add: Cold Regions Res & Eng Lab Hanover NH 03755

MELLOR, ROBERT SYDNEY, b Casper, Wyo, June 25, 31; m 60; c 3. PLANT PHYSIOLOGY. Educ: Colo State Univ, BS, 54, MS, 59, PhD(plant physiol), 62. Prof Exp: Asst prof, 62-67, ASSOC PROF BOT, UNIV ARIZ, 67- Mem: Am Soc Plant Physiologists; Sigma Xi. Res: Plant-water relations; plant biochemistry. Mailing Add: Dept of Biol Sci Univ of Ariz Tucson AZ 85721

MELLORS, ALAN, b Mansfield, Eng, Feb 26, 40. BIOCHEMISTRY. Educ: Univ Liverpool, BSc, 61, PhD(biochem), 64. Prof Exp: Res biochemist, Univ Calif, Davis, 64-67; res scientist, Food Res Inst, Can Dept Agr, 67-68; asst prof, 68-71, ASSOC PROF CHEM, UNIV GUELPH, 71- Concurrent Pos: Fulbright scholar, 64-67; Nuffield Found scholar, 75. Mem: Am Soc Biol Chemists; Brit Biochem Soc; Can Biochem Soc; Can Soc Cell Biologists. Res: Lysosomes and lysosomal enzymes; intracellular digestion by lysosomes; cathepsins and other hydrolases; steroids and enzymes; drug design; steroid structure-activity relationships; cell damage by cannabinoids. Mailing Add: Dept of Chem Univ of Guelph Guelph ON Can

MELLORS, ROBERT CHARLES, b Dayton, Ohio, June 18, 16; m 44; c 4. PATHOLOGY. Educ: Western Reserve Univ, AB, 37, MA, 38, PhD(biochem), 40; Johns Hopkins Univ, MD, 44. Prof Exp: Instr biochem, Western Reserve Univ, 40-42; asst epidemiol, Sch Hyg & Pub Health & asst poliomyelitis, Res Ctr, Johns Hopkins Univ, 42-44; assoc prof path, 53-58, assoc dir res, 58-69, PROF PATH, MED COL, CORNELL UNIV, 61-, DIR RES, HOSP SPEC SURG, 69-, PATHOLOGISTINCHIEF & DIR LABS, 58- Concurrent Pos: Spec fel med, Mem Hosp Ctr, New York, 46-47, Am Cancer Soc sr fel, 47-50, Runyan Fund sr fel, 50-53; ed, Anal Cytol & Anal Path Mem, Nat Found Infantile Paralysis Res, 43; assoc, Sloan-Kettering Inst, 50-53; assoc attend pathologist, Mem Hosp & Ewing Hosp, 53-58; mem res adv comt, USPHS, 62-66; attend pathologist, New York Hosp, 71- Mem: Am Soc Exp Path; Am Soc Biol Chem; Am Asn Immunol; Am Asn Path & Bact; fel Royal Col Path. Res: Immunopathology; glomerulonephritis; rheumatoid arthritis; autoimmune diseases; systemic lupus erythematosus; cancer; viruses in cancer; experimental pathology. Mailing Add: Hosp for Spec Surg Dept of Res 535 E 70th St New York NY 10021

MELMED, ALLAN JACK, physics, see 12th edition

MELMON, KENNETH LLOYD, b San Francisco, Calif, July 20, 34; m 58; c 2. CLINICAL PHARMACOLOGY. Educ: Stanford Univ, AB, 56; Univ Calif, MD, 59; Am Bd Internal Med, dipl, 66. Prof Exp: Intern internal med, Moffitt Hosp, Univ Calif, San Francisco, 59-60, asst resident, 60-61; clin assoc exp therapeut, Nat Heart Inst, 61-64; chief resident med, King County Hosp, Seattle, Wash, 64-65; chief sect clin pharmacol, asst prof med & pharmacol & assoc mem cardiovasc res inst, 65-68, CHIEF DIV CLIN PHARMACOL, SR STAFF CARDIOVASC RES INST & PROF MED & PHARMACOL, SCH MED, UNIV CALIF, SAN FRANCISCO, 68- Concurrent Pos: Mosby scholar, 59; consult, NIH, 65-; Burroughs Wellcome scholar clin pharmacol, 66-71; mem consult comt, Food & Drug Admin, Senate Subcomt on Health, House Ways & Means Comt; mem coun basic scis high blood pressure res & circulation, Am Heart Asn; mem bd, Am Bd Internal Med, 68; Guggenheim fel exp studies clin biochem & immunopharmacol, 71; NIH spec fel, 71. Mem: AAAS; Am Fedn Clin Res; Am Soc Pharmacol & Exp Therapeut; fel Am Col Physicians; Am Physiol Soc. Res: Mechanical action cyclic amp on cell growth and death; hormone receptors on cells; leukocyte; role of vasoactive amines in immunology; carcinoid syndrome. Mailing Add: 1089 Moffitt Hosp Univ of Calif Med Ctr San Francisco CA 94143

MELNGAILIS, JOHN, b Riga, Latvia, Feb 4, 39; US citizen; m 68; c 2. APPLIED PHYSICS. Educ: Carnegie Inst Technol, BS, 60, MS, 62, PhD(solid state theory), 65. Prof Exp: Assoc engr, Westinghouse Res Lab, 60-65; STAFF MEM, LINCOLN LAB, MASS INST TECHNOL, 67- Concurrent Pos: NSF fel, Max Planck Inst Metall Res, Stuttgart, Ger, 65; res attache, Nat Ctr Sci Res, Bellevue, France, 66-67. Mem: Am Phys Soc; Inst Elec & Electronics Engrs. Res: Surface acoustic wave devices; solid state physics; band structure of semiconductors. Mailing Add: Lincoln Lab Mass Inst of Technol Lexington MA 02173

MELNICK, DANIEL, b Boston, Mass, Jan 4, 10; m 35; c 3. BIOCHEMISTRY. Educ: Yale Univ, PhB, 31, PhD(biochem), 36; Am Bd Nutrit, dipl. Prof Exp: Asst pharmacologist, E R Squibb & Son, NJ, 31-32; asst, Yale Univ, 34-36, Coxe fel, 36-37; Upjohn fel, Univ Mich, 37-40; supvr res, Food Res Labs, Inc, NY, 40-47; chief food develop div, Qm Food & Container Inst Armed Forces, Ill, 47-49; chief technologist, Best Foods Div, CPC Int, NJ, 49-61, dir res & qual control, 61-68, vpres & mem bd dirs, 68-74; ADJ PROF CHEM, FLA ATLANTIC UNIV, 74- Concurrent Pos: Chmn comts for collab studies of methods for determining thiamine & niacin, Res Corp; chmn food & nutrit sect, Gordon Res Confs; pres, NY Inst Food Technol, Inc. Honors & Awards: Res Award, Am Oil Chemists Soc. Mem: Am Chem Soc; Am Soc Biol Chemists; Am Oil Chemists Soc; Am Asn Cereal Chemists; Inst Food Technologists. Res: Serum protein formation; protein metabolism; vitamin methodology and metabolism; nutrition; food chemistry; product development; fat chemistry; methodology and nutrition; development of teaching manual on food and nutrition for non-science majors and a modus operandi for teaching such. Mailing Add: Dept of Chem Fla Atlantic Univ Boca Raton FL 33432

MELNICK, DONALD A, b Philadelphia, Pa, June 14, 26. PHYSICS. Educ: Univ Pa, BA, 46, PhD(physics), 54; Univ Calif, Los Angeles, MA, 49. Prof Exp: Sr physicist, Opers Res, Inc, 54-58; pres, Sci Planning Assocs Corp, 58-61; dir opers res dept, C-E-I-R, Inc, 61-63; vpres, Res for Opers Mgt, 63; pvt consult opers res, 63-69; dir phys res, Raff Assocs Inc, 69-71; PVT CONSULT OPERS RES, 71- Mem: Am Phys Soc; Opers Res Soc Am; Inst Mgt Sci; NY Acad Sci. Res: Operations research. Mailing Add: 1900 Lyttonsville Rd Silver Spring MD 20910

MELNICK, EDWARD LAWRENCE, b Ann Arbor, Mich, Dec 12, 38; m 63; c 2. MATHEMATICS, STATISTICS. Educ: Lehigh Univ, BA, 60; Va Polytech Inst & State Univ, MS, 63; George Washington Univ, PhD(math statist), 70. Prof Exp: Math statistician, US Census Bur, Dept of Com, 63-69; asst prof, 69-73, ASSOC PROF STATIST, GRAD SCH BUS ADMIN, NY UNIV, 73-, FAC RES GRANT, 69- Concurrent Pos: Lectr, Grad Sch, USDA, 64-69; statist consult, Bellevue Hosp & others, 69- Mem: Am Statist Asn; fel Royal Statist Soc; Inst Math Statist. Res: Time series analysis with emphasis upon signal detection and prediction theory; collection and analysis of data. Mailing Add: 6BA Quantitative Anal Area NY Univ Grad Sch of Bus Admin New York NY 10006

MELNICK, JOSEPH LOUIS, b Boston, Mass, Oct 9, 14; m 36; c 1. VIROLOGY. Educ: Wesleyan Univ, BA, 36, Yale Univ, PhD(biochem), 39; Am Bd Med Microbiol, dipl. Hon Degrees: DSc, Wesleyan Univ, 71. Prof Exp: Asst physiol chem, Sch Med, Yale Univ, 37-39, from instr to asst prof prev med, 42-49, assoc prof microbiol, 49-54, prof epidemiol, 54-57; chief virus labs, Div Biologics Standards, NIH, 57-58; PROF VIROL & EPIDEMIOL & CHMN DEPT, BAYLOR COL MED, 58-, DEAN GRAD SCI, 68-, DISTINGUISHED SERV PROF, 74- Concurrent Pos: Finney-Howell Res Found fel, Sch Med, Yale Univ, 39-41, Nat Res Coun fel prev med & pediat, 41-42; Am-Scand Found fel viruses, Karolinska Inst, Sweden, 49; mem, Viral & Rickettsial Registry Comt, 51-, mem exec comt, 57-60; mem panel virol & immunol, Comt on Growth, Nat Res Coun, 52-56 & Comt Viral Hepatitis, 71-; chmn comt echoviruses, Nat Found, 55-57, chmn comt enteroviruses, 57-60; expert adv panel virus dis, WHO, 57-, dir int reference ctr enteroviruses, 61-74, mem comt polio vaccine, 73-, dir Collaborating Ctr Virus Reference & Res, 74-; mem comt live polio virus vaccine, USPHS, 58-61, mem nat adv cancer coun, 65-69; mem viruses & cancer bd, Nat Cancer Inst, 60-62, mem human cancer virus task force, 62-66, mem etiology prog adv comt, 70-73; chmn comt enteroviruses, NIH, 60-63; mem bd virus reference reagents, Nat Inst Allergy & Infectious Dis, 62-65, mem allergy & infectious dis training grant comt, 62-65, mem panel picornaviruses, 63-65; nat lectr, Found for Microbiol, 63-64 & 66-67; trustee, George Washington Carver Res Found, Tuskegee Inst, 64-; mem, Int Comt Nomenclature of Viruses, 66-74; chmn, Picornavirus & Papovavirus Comts, 68-; mem, Surgeon Gen Comt Hepatitis, Dept Army, 66-73; secy-gen, Int Congs Virol, Helsinki, 68 & Budapest, 71; chmn virol sect & mem exec bd, Int Asn Microbiol Socs, 70-75, mem int comt microbial ecology, 72-; mem coun res & clin investr, Am Cancer Soc, 71-74; chmn task group active immunization against viral hepatitis, NIH, 73-74; co-chmn, Duran-Reynals Int Symp Viral Oncol, Barcelona, Spain, 73; ed-in-chief, Progress Med Virol, 58-, Monographs in Virol, 60-, Intervirol, 73-; chmn int conf viruses in water, Am Pub Health Asn & WHO, Mexico City, 74; mem, Int Comt Taxon Viruses, 74- Honors & Awards: Int Gold Medal, Arg Found Against Infantile Paralysis, 49; Polio Hall of Fame, 58; Mod Med Distinguished Achievement Award, 65; Indust Res-100 Award, 71; Inventor of Year Award, 72; Freedman Found Award Res Virol, NY Acad Sci, 73. Mem: Fel AAAS; Am Epidemiol Soc; Soc Exp Biol & Med; Am Asn Immunol; Am Soc Microbiol. Res: Virology, especially infectious diseases and cancer. Mailing Add: Dept of Virol & Epidemiol Baylor Col of Med Houston TX 77025

MELNICK, LABEN MORTON, b Pittsburgh, Pa, June 10, 26; m 49; c 2. ANALYTICAL CHEMISTRY. Educ: Univ Pittsburgh, BS, 49, MS, 50, PhD(anal chem), 54. Prof Exp: Anal chemist, Res Lab, Jones & Laughlin Steel Corp, 50, x-ray diffractionist & spectroscopist, 51-52; technologist, 53-56, supv technologist, 56-58, sect supvr anal chem, 58-75, DIV CHIEF PHYSICS & ANAL CHEM, RES LAB, US STEEL CORP, 75- Concurrent Pos: Mem, Nat Acad Sci-Nat Res Coun adv panels to anal chem div, Nat Bur Standards, 69-71 & measures for air qual off, 71-73. Mem: AAAS; Israel Chem Soc; Am Chem Soc; Soc Appl Spectros. Res: Analysis of raw materials and metals; determination of gases and second phase inclusions in steel; analysis of water. Mailing Add: Res Lab Physics & Anal Chem Div US Steel Corp Mail Sta 15 Monroeville PA 15146

MELNICK, RONALD L, b New York, NY, May 19, 43; m 71; c 1. CELL PHYSIOLOGY. Educ: Rutgers Univ, BS, 65; Univ Mass, MS, 68, PhD(food sci), 71. Prof Exp: Fel cell physiol, Univ Calif, Berkeley, 71-73; ASST PROF BIOL SCI, POLYTECH INST NEW YORK, 73- Concurrent Pos: Reviewer, Arch Biochem & Biophys, 74- Mem: AAAS; Biochem Soc; Biophys Soc; NY Acad Sci. Res: Organization of protein components in normal and tumor cell membranes; regulation of activities of membrane bound enzymes; mechanisms of mitochrondrial ion transport and oxidative phosphorylation. Mailing Add: Dept of Life Sci Polytech Inst New York Brooklyn NY 11201

MELNYCHYN, PAUL, b Big Valley, Alta, Oct 9, 26; m 52. AGRICULTURAL BIOCHEMISTRY. Educ: Univ Alta, BSc, 48, MSc, 51; Univ Minn, PhD(agr biochem), 57. Prof Exp: Asst, Univ Alta, 49-51; chemist, J R Short Can Mills, 51-52; engr, Halliburton Oil Well Cementing Co, 52-53; from asst to instr agr biochem, Univ Minn, 53-56; res chemist, Puget Sound Pulp & Timber Co, 57-60; collabr, Fruit & Veg Chem Lab, USDA, Calif, 60-62; group leader proteins, Carnation Res Labs, Calif, 62-66, mgr biochem res, 66-71; vpres & dir, SNC Protein Consults Ltd, Montreal, 71-72; CONSULT PROTEIN TECHNOL & FOOD SCI, 73- Concurrent Pos: Vis prof, Trop Ctr Food Invest & Technol, Campinas, Brazil, 70. Mem: AAAS; Am Chem Soc; Am Asn Cereal Chemists; Inst Food Technologists. Res: Natural polymers chemistry; lignins; cellulose; proteins from milk, oilseeds, dry beans. Mailing Add: PO Box 509 Hudson PQ Can

MELNYK, ANDREW ROSTYSLAW, solid state physics, plasma physics, see 12th edition

MELNYK, JOHN H, b Winnipeg, Man, Oct 20, 23; m 53; c 5. CYTOGENETICS. Educ: Univ Man, BSc, 56, MSc, 57; Univ Alta, PhD(plant cytogenetics), 61. Prof Exp: Prof cytogenetics, Univ Southern Calif & Children's Hosp, Los Angeles, 64-69; PRIN INVESTR, CITY OF HOPE MED CTR, 69- Concurrent Pos: Consult, Los Angeles County Gen Hosp, 66- Mem: AAAS; Am Soc Human Genetics; Tissue Cult Asn; Genetics Soc Can. Res: Human genetics; mammalian cytogenetics; relationship of chromosome abnormalities to mental and physical retardation; effects of chromosome abnormalities on early development. Mailing Add: City of Hope Med Ctr Dept of Biol Duarte CA 91010

MELNYKOVYCH, GEORGE, b Halych, Ukraine, Oct 14, 24; nat US; m 49; c 2. CELL BIOLOGY. Educ: Univ Minn, MS, 53, PhD, 56. Prof Exp: Res chemist, Elgin State Hosp, Ill, 58 & Lederle Labs Div, Am Cyanamid Co, 58-63; asst prof microbiol, 63-70, PROF MICROBIOL, SCH MED, UNIV KANS, 70-; RES MICROBIOLOGIST, VET ADMIN HOSP, 63- Concurrent Pos: USPHS fel biochem, Univ Tex, 56 & Univ Calif, 56-58. Mem: AAAS; Am Chem Soc; Tissue Cult Asn; Am Soc Exp Path. Res: Microbial and tissue culture nutrition and metabolism; biochemistry. Mailing Add: US Vet Admin Hosp 4801 Linwood Blvd Kansas City MO 64128

MELOAN, CLIFTON E, b Bettendorf, Iowa, Aug 4, 31; m 57; c 3. ANALYTICAL

MELOAN

CHEMISTRY. Educ: Iowa State Univ, BS, 53; Purdue Univ, PhD(anal chem), 59. Prof Exp: From asst prof to assoc prof, 59-68, PROF CHEM, KANS STATE UNIV, 68- Concurrent Pos: Sci adv, Food & Drug Admin, 66- Mem: Am Chem Soc. Res: Liquid-liquid extractions; metal chelates; infrared; gas chromatography; spectrophotometry. Mailing Add: Dept of Chem Kans State Univ Manhattan KS 66502

MELOCHE, HENRY PAUL, b Detroit, Mich, Nov 15, 28; m 54; c 6. MICROBIOLOGY, PUBLIC HEALTH. Educ: Univ Detroit, BS, 51; Mich State Univ, MS, 53, PhD(microbiol), 56. Prof Exp: Food chemist, Swift & Co, 56-57; microbiologist, Fermentation Lab, Agr Res Serv, USDA, 57-60; res assoc agr chem, Mich State Univ, 60-64; res assoc, 64-69, ASST MEM, INST CANCER RES, 69- Concurrent Pos: NIH fel, 62-63. Mem: AAAS; Am Chem Soc; Am Soc Microbiol; Am Soc Biol Chem. Res: Enzyme chemistryl active site chemistry. Mailing Add: Inst for Cancer Res 7701 Burholme Ave Philadelphia PA 19111

MELONI, EDWARD GEORGE, b East Boston, Mass, Aug 2, 32; m 59; c 2. INORGANIC CHEMISTRY. Educ: Columbia Univ, AB, 53; Tufts Univ, MS, 55; Rutgers Univ, PhD(chem), 61. Prof Exp: Res chemist, Pennsalt Chem Corp, 60-63; res chemist, Esso Res & Eng Co, 63-65; chief chemist, Alfa Inorg Inc, 65-71; TECH DIR, VENTRON CORP, 71- Mem: Am Inst Chem; Am Chem Soc. Res: Synthetic inorganic chemistry; inorganic polymers. Mailing Add: Chem Div Ventron Corp Congress St Beverly MA 01915

MELOON, DANIEL THOMAS, JR, b Buffalo, NY, Aug 13, 35; m 59; c 5. ANALYTICAL CHEMISTRY. Educ: Univ Buffalo, BA, 57, MA, 60, PhD(inorg chem), 63. Prof Exp: Sr chemist, Carborundum Co, Niagara Falls, 63-66; asst prof, 66-68, ASSOC PROF CHEM, STATE UNIV NY COL BUFFALO, 68- Mem: Am Chem Soc; Sigma Xi. Res: High temperature oxidation catalysts; water pollution and inorganic complex ions. Mailing Add: 186 Calvert Blvd Tonawanda NY 14150

MELOY, CARL RIDGE, b Detroit, Mich, Sept 21, 12; m 32, 65; c 4. ORGANIC CHEMISTRY. Educ: Univ Mich, BS, 32, MS, 34; Mich State Univ, PhD(org chem), 42. Prof Exp: Instr chem, Highland Park Jr Col, Mich, 34-42; res chemist, Stamford Res Lab, Air Reduction Co, Conn, 42-43; asst prof chem & physics, Baldwin-Wallace Col, 43-45; asst prof phys sci, Mich State Univ, 45-47; from asst prof to prof phys sci & head dept, Univ Ill, 47-64; chmn div sci, 64-69, PROF CHEM, GRAND VALLEY STATE COL, 64- Concurrent Pos: Chemist, Process Chem Co, Mich, 39-40 & McGean Chem Co, Ohio, 45; res consult, Culligan, Inc, 51-53; Smith-Mundt exchange prof, Kabul Univ, Afghanistan, 61-62; fel, Imp Col, Univ London, 62; vis prof, Univ Kent, 71. Mem: AAAS; Am Chem Soc; Geochem Soc; The Chem Soc. Res: Organic synthesis; aromatic and heterocyclic compounds. Mailing Add: 1769 Greenwoods Dr Jenison MI 49428

MELROSE, JAMES C, b Spokane, Wash, Mar 27, 22; m 52; c 1. PHYSICAL CHEMISTRY. Educ: Harvard Univ, SB, 43; Stanford Univ, PhD(chem), 58. Prof Exp: Jr res chemist, Shell Oil Co, 43-46; res chemist, Mobil Oil Corp, 47-49, sr res chemist, 49-51 & 54-58, res assoc, 58-65, sr res assoc, 65-67; SR RES ASSOC, MOBIL RES & DEVELOP CORP, 67- Concurrent Pos: Vis lectr, Univ Tex, 64; vis assoc, Calif Inst Technol, 65. Mem: AAAS; Am Chem Soc; fel Am Inst Chemists; Soc Petrol Engrs. Res: Thermodynamics of interfaces; capillary phenomena in porous media; properties of petroleum reservoir fluids. Mailing Add: Field Res Lab Mobil Res & Develop Corp PO Box 900 Dallas TX 75221

MELROSE, RAYMOND JOHN, b Chicago, Ill, Nov 27, 36. ORAL PATHOLOGY. Educ: Northwestern Univ, DDS, 62. Prof Exp: Intern dent, Vet Admin Hosp, Long Beach, Calif, 64-65; spec lectr path, 66-68, asst prof, 68-72, ASSOC PROF PATH, SCH DENT, UNIV SOUTHERN CALIF, 72- Concurrent Pos: Resident oral path, Vet Admin Hosp, Long Beach, 65-68, consult & oral pathologist, 68-; consult test construct comt path & microbiol, oral path & radiol, Nat Bd Dent Examr-Am Dent Asn. Mem: Am Dent Asn; Am Acad Oral Path; Am Soc Clin Path; Am Asn Cancer Educ. Res: Oral cancer; odontogenic tumors; minor salivary gland tumors of the oral cavity. Mailing Add: Sch of Dent Dept of Path Univ of Southern Calif Los Angeles CA 90007

MELSON, GORDON ANTHONY, b Sheffield, Eng, July 6, 37; m 62; c 2. INORGANIC CHEMISTRY. Educ: Univ Sheffield, BSc, 59, PhD(chem), 62. Prof Exp: Res assoc chem, Ohio State Univ, 62-64; lectr, Univ Strathclyde, 64-69; asst prof, Mich State Univ, 69-75; ASSOC PROF CHEM, VA COMMONWEALTH UNIV, 75- Mem: Am Chem Soc; The Chem Soc. Res: Coordination chemistry of scandium; reactions of coordinated ligands; mechanisms of mactocyclic ligand formation. Mailing Add: Dept of Chem Va Commonwealth Univ Richmond VA 23284

MELSON, WILLIAM GERALD, b Washington, DC, Oct 23, 38; m 61; c 1. PETROLOGY, GEOCHEMISTRY. Educ: Johns Hopkins Univ, AB, 61; Princeton Univ, MA, 63, PhD(petrol), 64. Prof Exp: Instr geol, Johns Hopkins Univ, 60-61; geologist, US Geol Surv, 61; NSF fel, 61-64; CUR PETROL, SMITHSONIAN INST, US NAT MUS, 64- Mem: Am Geophys Union; Am Mineral Soc; fel Geol Soc Am. Res: Inorganic geochemistry; phase equilibria in metamorphic rocks; petrology of oceanic rocks, particularly those from the mid-Atlantic ridge; studies of active volcanoes. Mailing Add: US Nat Mus Smithsonian Inst Washington DC 20560

MELSTED, SIGURD WALTER, b Gardar, NDak, Nov 23, 11. AGRONOMY, SOILS. Educ: NDak State Col, BS, 38; Rutgers Univ, MS, 40; Univ Ill, PhD(soil chem), 43. Prof Exp: Asst, NDak State Col, 36-38 & Rutgers Univ, 38-40; soil surv anal, 40-43, assoc prof, 46-55, PROF SOILS, UNIV ILL, URBANA, 55- Concurrent Pos: Guggenheim fel, 56-57; guest lectr, Mid East Tech Univ Ankara, 63-64; AID adv, Njala Univ Col, Sierra Leone, 65-67, adv soils, 71; adv agr, US State Dept Atoms for Peace Exhib, Tehran, Iran, 67; AID-A T Kearndy Co consult fertilizer technol, Chile, 69- Mem: AAAS; Am Chem Soc; Soil Sci Soc Am; fel Am Soc Agron; Instrument Soc Am; fel Am Inst Chemists. Res: Soil colloids; fertilizer use; soil testing methods and calibration; spectrographic analysis of soils and plants; waste disposal on land; heavy metal toxicity in soils: phosphorus pollution of waters; international agriculture. Mailing Add: S-510 Turner Hall Univ of Ill Urbana IL 61801

MELSTROM, DONALD STANLEY, organic chemistry, deceased

MELTER, ROBERT ALAN, b New York, NY, Mar 20, 35; m 65; c 1. MATHEMATICS. Educ: Cornell Univ, AB, 56; Univ Mo, AM, 60, PhD(math), 62. Prof Exp: Instr math, Univ Mo, 58-62; asst prof, Univ RI, 62-64 & Univ Mass, 64-67; assoc prof, Univ SC, 67-71; ASSOC PROF MATH, SOUTHAMPTON COL, LONG ISLAND UNIV, 71- Concurrent Pos: Vis asst prof, Amherst Col, 67; Fulbright lectr, Univ Nismey, Niger, 74-75. Mem: Am Math Soc; Math Asn Am; Sigma Xi. Res: Algebra; abstract distance spaces and valuations; graph theory. Mailing Add: Dept of Math Southampton Col Long Island Univ Southampton NY 11968

MELTON, BILLY ALEXANDER, JR, b Wheeler, Tex, Aug 21, 32; m 51; c 2. PLANT BREEDING. Educ: NMex Col Agr & Mech Arts, BS, 54; Univ Ill, MS, 56, PhD, 58. Prof Exp: Asst plant breeding, Univ Ill, 54-58; from asst prof to assoc prof, 58-69, PROF AGRON, N MEX STATE UNIV, 69- Mem: Am Soc Agron. Res: Genetics of sorghum and forage crops, primarily alfalfa. Mailing Add: Dept of Agron NMex State Univ Las Cruces NM 88001

MELTON, CARL WESLEY, b Barnesville, Ohio, July 7, 21; m 45; c 3. ZOOLOGY, ELECTRON MICROSCOPY. Educ: Kent State Univ, AB, 47. Prof Exp: Sr microscopist, 47-74, PRIN CHEMIST, BATTELLE-COLUMBUS LAB, 74- Mem: AAAS; Electron Micros Soc Am. Res: Demonstration, by means of chemical reactions, of microscopical distributions of elements or reactive groups within materials; microscopic and electron probe analysis of air pollutants. Mailing Add: Battelle-Columbus Labs 505 King Ave Columbus OH 43201

MELTON, CARLTON EARL, JR, b Allen, Tex, June 1, 24; m 57; c 3. PHYSIOLOGY. Educ: NTex State Col, BS, 48; Univ Ill, MS, 50, PhD(physiol), 53. Prof Exp: Asst elec eng, Univ Ill, 49-51, asst physiol, 51-53; instr, Sch Med, Western Reserve Univ, 53-55; from instr to asst prof, Univ Tex Southwestern Med Sch Dallas, 55-61; assoc prof, 61-66, PROF RES PHYSIOL, SCH MED, UNIV OKLA, 66-, ADJ PROF ZOOL, 73-; CHIEF, PHYSIOL LAB, CIVIL AEROMED INST, FED AVIATION ADMIN, 61- Mem: Am Physiol Soc; Soc Exp Biol & Med; Am Asn Anat. Res: Physiology of muscle and nerve; special senses; stress. Mailing Add: Civil Aeromed Inst Fed Aviation Admin Oklahoma City OK 73125

MELTON, CHARLES ESTEL, b Fancy Gap, Va, May 18, 24; m 46; c 3. CHEMICAL PHYSICS. Educ: Emory & Henry Col, BA, 52; Vanderbilt Univ, MS, 54; Univ Notre Dame, PhD(phys chem), 64. Hon Degrees: DSc, Emory & Henry Col, 67. Prof Exp: Physicist, Oak Ridge Nat Lab, 54-67; PROF CHEM, UNIV GA, 67-, HEAD DEPT CHEM, 72- Honors & Awards: DeFriece Award Physics, Emory & Henry Col, 59. Mem: Fel AAAS; fel Am Inst Chemists; Am Chem Soc; Sigma Xi. Res: Chemical kinetics; mass spectrometry; radiation chemistry; catalysis; theoretical chemistry and geochemistry. Mailing Add: Rte 2 Box 18 Hull GA 30646

MELTON, CHARLES GILBERT, JR, developmental biology, see 12th edition

MELTON, JAMES RAY, b Paris, Tex, Aug 24, 40; m 62; c 2. ANALYTICAL CHEMISTRY, SOIL CHEMISTRY. Educ: Tex Tech Col, BS, 62; Mich State Univ, MS, 64, PhD(soil sci), 68. Prof Exp: ASSOC PROF AGR ANAL SERV, TEX A&M UNIV, 68- Mem: Sigma Xi; Asn Off Anal Chemists; Am Soc Agron; Soil Sci Soc Am. Res: Zinc fertilization of various crops; movement and availability of heavy metals in soils; development of methods for determination of heavy metals by atomic absorption and atomic emission. Mailing Add: Agr Anal Serv Tex A&M Univ College Station TX 77843

MELTON, LYNN AYRES, b Huntsville, Tex, Aug 7, 44; m 67; c 1. PHYSICAL CHEMISTRY. Educ: Calif Inst Technol, BS, 66; Harvard Univ, MA & PhD(chem), 72. Prof Exp: ASST PROF CHEM, UNIV TEX, DALLAS, 71- Mem: Am Chem Soc; Am Phys Soc; AAAS. Res: Gas phase energy transfer, detailed state-to-state rate measurements by laser excited fluorescence. Mailing Add: Dept of Chem Univ of Tex at Dallas Box 688 Richardson TX 75080

MELTON, REX EUGENE, b Ozark, Mo, Dec 4, 21; m 42; c 4. FORESTRY. Educ: Univ Mo, BS, 46; Univ Mich, BS & MF, 47. Prof Exp: Asst prof forestry, 47-58, dir exp forest, 58-74, ASSOC PROF FORESTRY, 58-, ADMIN ASST TO DIR SCH FOREST RESOURCES, 74- Mem: Soc Am Foresters. Res: Forest entomology and management; watershed management; silviculture. Mailing Add: Sch of Forest Resources 107 Ferguson Bldg Pa State Univ University Park PA 16802

MELTON, THOMAS MASON, b Rockville, Va, Jan 8, 27; m 49, 57; c 2. ORGANIC CHEMISTRY, INDUSTRIAL HYGIENE. Educ: Col William & Mary, BS, 48; Am Acad Indust Hyg, dipl. Prof Exp: Res chemist, Sterling-Winthrop Res Inst, 48-51; sr chemist, Va-Carolina Chem Corp, 51-65; group leader agr chem, Mobil Chem Co, 65-71; CHEM SPECIALIST & CONSULT CHEM & INDUST HYG & SAFETY, TRAVELERS CORP, 71- Mem: Am Chem Soc. Res: Synthesis of insecticides; organic phosphorus and organic medicinal chemicals. Mailing Add: 1914 LeSuer Rd Richmond VA 23229

MELTON, WILLIAM GROVER, JR, b Oakland, Calif, Jan 1, 23; m 56; c 5. VERTEBRATE PALEONTOLOGY. Educ: Univ Mont, BA, 53; Univ Mich, MS, 69. Prof Exp: Geologist, US Geol Surv, 53-56; preparator vert paleont, Univ Mich, 57-66; asst, 66-69, lectr, 69-75, CUR GEOL, UNIV MONT, 66-, ADJ ASSOC PROF, 75- Mem: Soc Vert Paleont; Paleont Soc; Int Paleont Asn. Res: Paleozoic ray finned fish; conodonts and conodont bearing animals; soft anatomy and classification. Mailing Add: Dept of Geol Univ of Mont Missoula MT 59801

MELTZER, ALAN SIDNEY, b New York, NY, Apr 26, 32; m 57. ASTRONOMY. Educ: Syracuse Univ, BS, 53; Princeton Univ, PhD(astron), 56. Prof Exp: Res assoc observ, Harvard Col, 56; physicist, Smithsonian Astrophys Observ, 56-57; assoc prof, 57-74, PROF ASTRON, RENSSELAER POLYTECH INST, 74- Mem: Fel AAAS; Am Astron Soc. Res: Solar and stellar spectroscopy; interstellar extinction of polarization. Mailing Add: Dept of Physics Rensselaer Polytech Inst Troy NY 12181

MELTZER, CARL MARTIN, b New York, NY, Dec 23, 34; m 56; c 3. PHYSICS. Educ: Duke Univ, BS, 56, PhD(physics), 60. Prof Exp: Res assoc physics, Duke Univ, 60-61 & Northwestern Univ, 61-62; res physicist, Carnegie-Mellon Univ, 62-63, from asst prof to assoc prof physics, 63-73; PROD MGR SCI INSTRUMENTS, ELSCINT INC, 73- Mem: Am Phys Soc. Res: Experimental studies of the structure and the interactions of elementary particles at high energy. Mailing Add: 169 Lorraine Ave Montclair NJ 07043

MELTZER, HERBERT LEWIS, b New York, NY, Apr 23, 21; m 49; c 2. BIOCHEMISTRY. Educ: Long Island Univ, BS, 42; Columbia Univ, PhD, 50. Prof Exp: Res assoc, 51-56, ASST PROF BIOCHEM, COL PHYSICIANS & SURGEONS, COLUMBIA UNIV, 56-; ASSOC RES SCIENTIST, NY STATE PSYCHIAT INST, 52- Mem: Am Chem Soc; Am Soc Biol Chem. Res: Neurochemical and behavioral effects of rubidium; membrane transport systems. Mailing Add: 722 W 168th St New York NY 10032

MELTZER, HERBERT YALE, b Brooklyn, NY, July 29, 37; m 60; c 2. PSYCHIATRY. Educ: Cornell Univ, BA, 58; Harvard Univ, MA, 59; Yale Univ, MD, 63. Prof Exp: Res assoc pharmacol & anat, Sch Med, Yale Univ, 59-63; clin assoc, Lab Clin Sci, NIH, 66-68, instr grad training prog, 67-68; from asst prof to assoc prof, 68-74, PROF PSYCHIAT, UNIV CHICAGO, 74-; RES ASSOC, ILL STATE PSYCHIAT INST, 68- Concurrent Pos: Teaching fel psychiat, Harvard Med Sch, 64-66; consult, Peter Bent Brigham Hosp & Mass Ment Health Ctr, 65-66 & VISTA, 68-; assoc ed, Schizophrenia Bull. Mem: Am Col Neuropsychopharmacol; Am Psychosom Soc. Res: Biological study of mental illness; muscle physiology and ultrastructure. Mailing Add: Pritzker Sch of Med Univ of Chicago Chicago IL 60637

MELTZER, RICHARD S, b Paterson, NJ, May 10, 21; m 43; c 2. PEDIATRICS, MEDICAL EDUCATION. Educ: Colgate Univ, AB, 42; Univ Rochester, MD, 45. Prof Exp: Chief pediat, Genesee Hosp, 64-71; med dir, Proj Hope Jamaica, 71-73, DIR GRAD MED EDUC, PROJ HOPE & DIR CARIBBEAN PROGS, PEOPLE TO PEOPLE HEALTH FOUND, 73- Concurrent Pos: From assoc prof to prof pediat, Sch Med, Univ Rochester, 65-71; prof med educ, Univ WIndies, 71-73. Mem: AMA; Am Acad Pediat; Asn Am Med Cols. Res: Respiratory distress syndrome and infectious diseases of newborns; angiocardiography as a diagnostic tool in congenital heart disease and in allergic diseases in childhood. Mailing Add: 2233 Wisconsin Ave NW Washington DC 20007

MELTZER, ROBERT ISRAEL, b New York, NY, Oct 4, 18; m 44; c 2. ORGANIC CHEMISTRY. Educ: City Col New York, BS, 39; Ohio State Univ, MS, 40; Univ Ill, PhD(org chem), 43. Prof Exp: Asst chem, Univ Ill, 41-43; res chemist, 43-62, dir med chem, 62-65, DIR CHEM RES, WARNER LAMBERT RES INST, 65- Mem: AAAS; Am Chem Soc; NY Acad Sci; The Chem Soc. Res: Synthetic organic medicinals. Mailing Add: Warner Lambert Res Inst 170 Tabor Rd Morris Plains NJ 07950

MELTZER, ROBERT JASON, optics, see 12th edition

MELTZER, THEODORE H, polymer chemistry, marketing, see 12th edition

MELVEGER, ALVIN JOSEPH, b New York, NY, July 9, 37; m 61; c 1. CHEMISTRY. Educ: Brooklyn Col, BS, 59; Northeastern Univ, MS, 64; Univ Md, College Park, PhD(chem), 68. Prof Exp: Scientist mat, Avco Corp, Mass, 60-64; grant, Ctr Mat Res, Univ Md, College Park, 68-69; res chemist, Allied Chem Corp, 69-72; SECT MGR INSTRUMENTAL ANAL, ETHICON, INC, 72- Mem: AAAS; Am Chem Soc; Am Phys Soc; Soc Appl Spectros; Am Soc Testing & Mat. Res: Molecular, analytical and high pressure spectroscopy; laser-Raman and infrared spectroscopy of polymers and inorganic materials; structure-property relationships of polymers. Mailing Add: Ethicon Inc Somerville NJ 08876

MELVILLE, DONALD BURTON, b Netherton, Eng, Jan 30, 14; nat US; m 40; c 2. BIOCHEMISTRY. Educ: Univ Ill, BS, 36, MS, 37, PhD(biochem), 39. Prof Exp: Asst biochem, Univ Ill, 36-38; res assoc, Med Col, Cornell Univ, 39-46, from asst prof to assoc prof, 46-60; PROF BIOCHEM & CHMN DEPT, COL MED, UNIV VT, 60- Mem: Am Chem Soc; Am Soc Biol Chem; NY Acad Sci. Res: Determination of structure and study of biological effects of biotin; chemistry of penicillin; syntheses with radio isotopes; biochemistry of ergothioneine. Mailing Add: Dept of Biochem Univ of Vt Col of Med Burlington VT 05401

MELVILLE, GEORGE S, JR, b Yonkers, NY, June 21, 24; m 48; c 3. ORGANIC CHEMISTRY. Educ: Johns Hopkins Univ, BA, 49, MA, 51, PhD(org chem), 53. Prof Exp: Chemist, Chem Biol & Radiol Prog, US Air Force, Ft Detrick, Md, 53-54, radiobiologist, Biol Div, Oak Ridge Nat Lab, 54-56, chemist-radiobiologist, Radiobiol Lab, Univ Tex & US Air Force, 56-64, chemist-radiobiologist, Radiobiol Br, Sch Aerospace Med, 64-68, chief pharmacol, 68-70, chief cellular physiol br, 70-71; LECTR CHEM, SAN ANTONIO COL, 72- Mem: Am Chem Soc; Radiation Res Soc. Res: Synthetic organic chemistry; small rings; proof of structure; ultraviolet and infrared spectroscopy; radiobiology; effects of radiation on primates; radioprotective chemicals and techniques. Mailing Add: 110 Chesterfield Dr San Antonio TX 78223

MELVILLE, KENNETH IVAN, pharmacology, deceased

MELVILLE, MARJORIE HARRIS, b Baltimore, Md, Aug 30, 27; m 48; c 3. ORGANIC CHEMISTRY. Educ: Agnes Scott Col, BA, 47; Johns Hopkins Univ, MA, 49, PhD(chem), 53. Prof Exp: From asst prof to assoc prof, 64-75, PROF CHEM, SAN ANTONIO COL, 75- Mem: Am Chem Soc; Sigma Xi. Mailing Add: 110 Chesterfield Dr San Antonio TX 78223

MELVILLE, ROBERT SEAMAN, b Worcester, Mass, Nov 20, 13; m 42; c 5. CLINICAL BIOCHEMISTRY. Educ: Clark Univ, AB, 37; Univ Iowa, PhD(biochem), 50; Am Bd Clin Chem, cert, 55. Prof Exp: Res chemist, Mass Gen Hosp, 39-42; clin biochemist, St Luke's Hosp, Chicago, 50-54 & Vet Admin Hosp, Iowa City, 54-63; chief biochemist, Vet Admin Cent Off, Washington, DC, 63-65; scientist adminr, 65-71, CHIEF BIOENG & AUTOMATION CLIN LABS SECT, RES GRANTS BR, NAT INST GEN MED SCI, 71-, EXEC SECY AUTOMATION IN MED LAB SCI REV COMT, 68- Concurrent Pos: Dir, Am Bd Clin Chem, 65-76; chmn adv comt to chem & chem-tech div, Nat Acad Sci-Nat Res Coun, 66-71; pres, Nat Registry Clin Chem, 73-75; clin prof path, Div Lab Med, George Washington Univ, 73- Honors & Awards: Joseph H Roe Award, Am Asn Clin Chemists, 72. Mem: AAAS; Am Chem Soc; sr mem Instrument Soc Am; fel Am Asn Clin Chemists; Asn Clin Sci. Res: Clinical chemistry methodology; instrumentation; training of clinical chemists; improve functions of the clinical laboratory. Mailing Add: Nat Inst of Gen Med Sci 5333 Westband Ave Bethesda MD 20014

MELVIN, DOROTHY MAE, b Fayetteville, NC. MEDICAL PARASITOLOGY. Educ: Univ NC, Greensboro, AB, 42; Univ NC, Chapel Hill, MS, 45; Rice Univ, PhD(parasitol), 51. Prof Exp: Training officer & med parasitologist, 45-49, 51-62, chief parasitol training unit, 62-74, CHIEF PARASITOL TRAINING SECT, CTR DIS CONTROL, 74- Concurrent Pos: La State Univ fel trop med, PR & Haiti, 58; asst prof, Sch Med, Emory Univ, 67- Honors & Awards: Super Performance Award, Ctr Dis Control, 58. Mem: Am Soc Trop Med & Hyg; Am Soc Parasitol; Am Sci Film Asn; Int Col Trop Med. Res: Methodology of technical training for health care personnel. Mailing Add: Parasitol Training Sect Ctr for Dis Control Atlanta GA 30333

MELVIN, ERNEST EUGENE, b Monmouth, Ill, May 15, 23; m 52; c 1. URBAN GEOGRAPHY. Educ: Western Ill Univ, BS, 47; Syracuse Univ, MA, 49; Northwestern Univ, PhD(geog), 52; Univ Chicago, cert, 57. Prof Exp: Geogr, Far East Command, Dept Army, 52-54; economist, Evert Kincaid & Assoc, 54-56; planning asst, City of Evanston, 56-57; prin planner, Dept of City Planning, City of Chicago, 57-60, sect head & actg dir planning, 60-61, sect head, Community Renewal Prog, 61; assoc prof geog & assoc dir, Inst Community & Area Develop, 61-68, PROF GEOG & DIR, INST COMMUNITY & AREA DEVELOP, UNIV GA, 68- Honors & Awards: Cert of Achievement, Dept of Army, 53; Geog Educ Citation, Nat Coun Geog Educ, 57. Mem: AAAS; Am Inst Planners; Am Geog Soc; Nat Coun Geog Educ; Am Soc Planning Officers. Res: Planning and citizen participation; rural depopulation. Mailing Add: Inst of Community & Area Develop Univ of Ga Athens GA 30601

MELVIN, HORACE WILLIS, b Portsmouth, Va, July 5, 13; m 45; c 1. ORGANIC CHEMISTRY. Educ: Fisk Univ, AB, 36, AM, 38; Iowa State Col, PhD(chem), 54. Prof Exp: Instr chem, Miss Indust Col, 39-41; instr, Tex Col, 41-43; instr, Houston Col, 43-44; assoc prof, Agr Mech & Norm Col, Ark, 49-54, prof, 54-55; PROF CHEM, HAMPTON INST, 55-, CHMN DEPT PHYS SCI, 74- Mem: AAAS; Am Chem Soc; Nat Inst Sci; fel Am Inst Chem. Res: Friedel-Crafts reaction; alkylation; organosilicon compounds. Mailing Add: Box 6164 Hampton Inst Hampton VA 23368

MELVIN, JOAN BARKER, biology, cytology, see 12th edition

MELVIN, JOHN HARPER, b Chillicothe, Ohio, Dec 11, 06; m 33; c 6. GEOLOGY. Educ: Ohio Wesleyan Univ, BA, 31; Ohio State Univ, MA, 33. Prof Exp: Asst geol, Ohio State Univ, 31-33 & Univ Iowa, 33-34; inspector core boring, US Engr Off, Ohio, 34; chief party, Field Eng Dept, Muskingum Watershed Conservancy Dist, 34-39; res geologist, Norfork Dam, Ark, 39-40; geologist & found engr, Pa Drilling Co, 40-42, treas & geologist, 42-46; head geol sect, Mo River Control Proj, Corps Engrs, Nebr, 46-47; state geologist & chief, Div Geol Surv, Ohio, 47-56; gen mgr, Pa Drilling Co, 56-61; consult & exec officer, Ohio Acad Sci, 61-75; CONSULT, 75- Concurrent Pos: Prof, Ohio State Univ, 47-56. Honors & Awards: Orton Award, Ohio State Univ, 71. Mem: AAAS; Am Soc Civil Engrs; Geol Soc Am; Am Inst Mining, Metall & Petrol Engrs. Res: Engineering and economic geology. Mailing Add: 8535 Winchester Rd Carroll OH 43112

MELVIN, JOHN LEWIS, b Columbus, Ohio, May 26, 35; m 57; c 4. PHYSICAL MEDICINE & REHABILITATION, ELECTROMYOGRAPHY. Educ: Ohio State Univ, BSc, 55, MD, 60, MMSc, 66. Prof Exp: From asst prof to assoc prof phys med, Col Med, Ohio State Univ, 66-73; PROF PHYS MED & REHAB & CHMN DEPT, MED COL WIS, 73- Concurrent Pos: Mem adv comt, Joint Comt Stroke Facil, 69-; med dir, Curative Workshop Milwaukee, 73-; sr attend staff, Milwaukee County Med Complex, 73-; dir phys med, Sacred Heart Rehab Hosp, 74-; mem consult staff, Milwaukee Children's Hosp, 73-, St Luke's Hosp, 73- & West Allis Mem Hosp, 74- Honors & Awards: Sci Exhib-Gold Medal Award, Am Cong Rehab Med, 71. Mem: Am Acad Phys Med & Rehab; Am Cong Rehab Med; Am Asn Electromyog & Electrodiag (first vpres, 75); Am Asn Acad Physiatrists (secy, 75). Res: Testing models for delivery of rehabilitation services; conceptualizing, classifying, and describing disability states; identifying electrophysiologic responses associated with diseases of the peripheral nervous system. Mailing Add: Med Col of Wis 10437 W Watertown Plank Rd Milwaukee WI 53226

MELVIN, JONATHAN DAVID, b New York, NY, Apr 7, 47. ATOMIC PHYSICS. Educ: Yale Univ, BA & MA, 68; Calif Inst Technol, PhD(physics), 74. Prof Exp: Instr math, Albertus Magnus Col, 68-69; res physicist, Lab Nuclear Med & Radiation Biol, Univ Calif, Los Angeles, 74-75; ROBERT A MILLIKEN RES FEL, KELLOGG LAB, CALIF INST TECHNOL, 75- Concurrent Pos: Mem res staff, Hughes Res Lab, 71-74; consult physicist, Technion, Inc, 74-75; vis scientist, Nat Univ Mex & INEN, Mexico City, 76. Res: Experimental and theoretical study of sputtering relevant to lunar sample and controlled thermonuclear reactor chamber wall erosion. Mailing Add: Kellog Lab Calif Inst of Technol Pasadena CA 91125

MELVIN, MAEL AVRAMY, b Palestine, Mar 27, 13; US citizen; m 46; c 2. PHYSICS. Educ: Univ Chicago, BS, 33, MS, 35, PhD(physics), 38. Prof Exp: Metallurgist, Carnegie Ill Steel Corp, 37-38; assoc metall, Columbia Univ, 38-40, instr, 40-42, instr physics, 42-46, asst prof metal physics, 46-47, assoc prof, 47-48; vis prof, Univ Ore, 51; Guggenheim fel, Princeton Univ, 51-52; prof physics, Fla State Univ, 52-66; PROF PHYSICS, TEMPLE UNIV, 66- Concurrent Pos: Guggenheim fel & vis prof, Univ Upsala, 57-58; Int Atomic Energy Agency exchange prog vis prof, Inst Physics, Bariloche, Arg, 59-60; Nat Res Coun sr resident res assoc, Jet Propulsion Lab, Calif Inst Technol, 71-72. Consult, Chem Warfare Serv, US Dept Army, 41, Nat Defense Res Comt, 42, Hazeltine Serv Corp, NY, 43, Heat Transfer Res Lab, 43 & Off Sci Res & Develop, 43-45. Mem: Fel Am Phys Soc. Res: Symmetry methods in physics; elementary particles and fields; relativity; astrophysics; cosmology; homogeneous and anisotropic cosmologies. Mailing Add: Dept of Physics Temple Univ Philadelphia PA 19122

MELVIN, ROBERT BURROW, b Ft Collins, Colo, May 19, 31; m 54; c 2. BIOCHEMISTRY. Educ: Colo State Univ, BS, 53, MS, 55. Prof Exp: Res biochemist, Great Western Sugar Co Agr Exp Sta, 63-68; res biologist, Univ Colo, 68-69; res physiologist, Colo State Univ, 70-71; INSTR BIOCHEM, KANSAS CITY COL OSTEOP MED, 71- Mem: AAAS; Am Chem Soc. Res: Carbohydrates; lipids; nucleic acids; metabolic pathways. Mailing Add: Dept of Biochem Kansas City Col of Osteop Med Kansas City MO 64124

MELVOLD, ROGER WAYNE, b Wadena, Minn, Mar 21, 46. GENETICS. Educ: Moorhead State Col, BS, 68; Univ Kans, PhD(genetics), 73. Prof Exp: ASSOC RADIATION BIOL, MED SCH, HARVARD UNIV, 72- Mem: Genetics Soc Am; Am Genetic Asn; Soc Develop Biol. Res: Immunogenetics—identification and analysis of mouse histocompatibility mutants with particular regard to mutation rates, gene mapping, regulation and functional and structural studies of histocompatibility loci. Mailing Add: Shields Warren Radiation Lab Harvard Med Sch 50 Binney St Boston MA 02115

MEMEGER, WESLEY, JR, b Riverdale, Fla, Sept 21, 39; m 63; c 2. ORGANIC CHEMISTRY, ANALYTICAL CHEMISTRY. Educ: Clark Col, BS, 61; Adelphi Univ, PhD(org chem), 66. Prof Exp: Res chemist, 65-71, SR RES CHEMIST, E I DU PONT DE NEMOURS & CO, INC, 71- Mem: Am Chem Soc; Sigma Xi; NY Acad Sci. Res: Field effects in nucleophilic substitution reactions; reaction kinetics; synthesis and characterization of addition and condensation polymers; high temperature fibers; basic research aimed at new high strength high modulus polymers and fibers. Mailing Add: 109 Hoiland Dr Shipley Heights Wilmington DE 19803

MEMORY, JASPER DURHAM, b Raleigh, NC, Dec 10, 36; m 61; c 2. PHYSICS. Educ: Wake Forest Col, BS, 56; Univ NC, PhD(physics), 60. Prof Exp: From asst prof to assoc prof physics, Univ SC, 60-64; assoc prof, 64-67, PROF PHYSICS, NC STATE UNIV, 67-, ASSOC DEAN SCH PHYS & MATH SCI, 68- Mem: Am Phys Soc. Res: Nuclear magnetic resonance, quantum theory of molecular electronic structure; molecular biophysics. Mailing Add: Sch of Phys Sci & Appl Math NC State Univ Raleigh NC 27607

MENA, ROBERTO ABRAHAM, b Merida, Mex, Mar 12, 46; m 69. ALGEBRA. Educ: Univ Houston, BS, 68, MS, 71, PhD(math), 73. Prof Exp: ASST PROF MATH, UNIV WYO, 73- Mem: Am Math Soc; Math Asn Am. Res: Partially-ordered algebraic structures and related areas. Mailing Add: Dept of Math Univ of Wyo Laramie WY 82071

MENAKER, MICHAEL, b Vienna, Austria, May 19, 34; US citizen; m 55; c 1. COMPARATIVE PHYSIOLOGY. Educ: Swarthmore Col, BA, 55; Princeton Univ, MA, 58, PhD(biol), 60. Prof Exp: NSF fel, Harvard Univ, 59-61, NIH fel, 61-62; from asst prof to assoc prof, 62-72, PROF ZOOL, UNIV TEX, AUSTIN, 72- Concurrent Pos: Guggenheim Mem Found fel, 71. Mem: Am Physiol Soc; Am Soc Photobiol; Soc Neurosci. Res: Biological clocks; celestial orientation in animals; time measurement in animal photoperiodism; annual cycles; physiology of mammalian hibernation; brain photoreception and pineal physiology. Mailing Add: Dept of Zool PAT 30 Univ of Tex Austin TX 78712

MENAPACE, LAURENE MARY, chemistry, see 12th edition

MENAPACE, LAWRENCE WILLIAM, b Brooklyn, NY, Apr 13, 37; m 60; c 3. ORGANIC CHEMISTRY. Educ: St Peter's Col, BS, 60; Univ NH, PhD(org chem), 64. Prof Exp: Chemist, Texaco Exp, Inc, 63-65, sr chemist, Texaco, Inc, 65-68; asst prof, 68-71, ASSOC PROF CHEM, MARIST COL, 71-; VPRES, R-2 ENVIRON CONSULTS, 73- Concurrent Pos: Mem, Environ Adv Comt, 73- Mem: Am Chem Soc. Res: Mechanism and scope of organotin hydride reductions; fundamentals of chemical vapor plating; synthesis of petroleum based chemicals. Mailing Add: Dept of Chem Marist Col Poughkeepsie NY 12601

MENARD, HENRY WILLIAM, JR, b Calif, Dec 10, 20; m 46; c 3. GEOLOGY. Educ: Calif Inst Technol, BS, 42, MS, 47; Harvard Univ, PhD(geol), 49. Prof Exp: Oceanogr, US Navy Electronics Lab, 49-55; assoc prof, 55-60, PROF GEOL, INST MARINE RESOURCES & SCRIPPS INST OCEANOG, UNIV CALIF, 61- Concurrent Pos: Dir, Geol Diving Consults, Inc, 54-58; vis prof, Calif Inst Technol, 59; Guggenheim fel, 62; tech asst, Off Sci & Technol, 65; fel, Churchill Col, Cambridge Univ, 70. Mem: Nat Acad Sci; fel Geol Soc Am; Am Geophys Union; Am Asn Petrol Geologists. Res: Marine geology; tectonics; sedimentation; sociology of science; mineral resources and environment. Mailing Add: Scripps Inst of Oceanog La Jolla CA 92037

MENARD, MARCEL R, medicinal chemistry, see 12th edition

MENARDI, PETER JOHN, analytical chemistry, inorganic chemistry, see 12th edition

MENASHE, VICTOR D, b Portland, Ore, July 13, 29; m 52; c 2. PEDIATRICS, CARDIOLOGY. Educ: Univ Ore, BS, 51, MD, 53. Prof Exp: Intern gen med, Univ Hosps & Clins, 53-54, resident pediat, 54-56, from instr to assoc prof, Med Sch, 58-71, PROF PEDIAT, MED SCH, UNIV ORE, 71-, DIR CRIPPLED CHILDREN'S DIV & ASST DEAN, HEALTH SCI CTR, 72- Concurrent Pos: Pediat consult, Shriners Hosp Crippled Children, 59- Mem: Am Acad Pediat; Am Heart Asn. Res: Epidemiology of congenital heart disease. Mailing Add: Crippled Children's Div Univ of Ore Med Sch Portland OR 97201

MENASHI, JAMEEL, b Teheran, Iran, Apr 1, 38; m 64; c 1. INORGANIC CHEMISTRY, PHYSICAL CHEMISTRY. Educ: Univ London, BS, 60, PhD(phys chem), 63. Prof Exp: Fel chem, 63-64; group leader, Harshaw Chem Co, 64-68; MEM TECH STAFF, CABOT CORP, 68- Mem: Am Chem Soc. Res: Kinetics of electron exchange reactions; inorganic pigment systems; thermodynamic constants of complexes; synthesis and kinetics of catalytic systems. Mailing Add: 68 Gleason Rd Lexington MA 02173

MENCH, JOHN WARREN, b Chicago, Ill, May 13, 16; m 41; c 2. ORGANIC CHEMISTRY. Educ: Antioch Col, BS, 39; Purdue Univ, MS, 42, PhD(org chem), 44. Prof Exp: Chemist, Eastman Kodak Co, 35-41; lab asst microanal, Purdue Univ, 41; RES CHEMIST, EASTMAN KODAK CO, 44- Mem: Am Chem Soc. Res: Oxidation of glucose and its derivatives; cellulose chemistry; polymer chemistry. Mailing Add: Polymer Technol Div Bldg 46 Eastman Kodak Co Kodak Park Rochester NY 14650

MENCHER, ALAN GEORGE, b New York, NY, May 17, 25; m 46, 64; c 3. SCIENCE POLICY. Educ: Mass Inst Technol, SB, 45; Yale Univ, ME, 47; Univ Calif, Los Angeles, PhD(physics), 52. Prof Exp: Engr, Airborne Instruments Lab, Inc, 47-48; asst physics, Univ Calif, Los Angeles, 48-52, instr, 53-54; mem tech staff, Hughes Aircraft Co, 52-53; Swiss Govt-Inst Int Educ fel, Swiss Fed Inst Technol, 54-55, Swiss-Am Found Sci Exchange fel physics, 55-56; mem tech staff, Missile Systs Div, Ramo-Wooldridge Corp, 56-57; foreign serv officer, Am Embassy, Paris, 57-60, dep sci attache, London, 61-63, Paris, 63-65, actg sci attache, 65-66, sci attache, 66-67, sci attache, London, 67-73; vis fel, London Grad Sch Bus Studies, 73-75; ANALYST POLITICO-MIL-SCI AFFAIRS, DEPT OF STATE, 75- Concurrent Pos: Consult, Ctr Policy Alternatives, Mass Inst Technol, 73-74; prin investr, NSF, 73-75. Mem: Am Phys Soc. Res: Semiconductor physics; optical properties of solids; international scientific and technological affairs; science and government; national and international science policy. Mailing Add: Dept of State Washington DC 20520

MENCHER, ELY, b New York, NY, Dec 14, 13; m 51; c 1. GEOLOGY. Educ: City Col New York, BS, 34; Mass Inst Technol, PhD(geol), 38. Prof Exp: Prof geol, Univ Cent de Venezuela, 38-42; geologist, Socony-Vacuum Oil Co, Venezuela, 43-52; assoc prof geol, Mass Inst Technol, 52-67; chmn dept, 68-72, PROF EARTH & PLANETARY SCI, CITY COL NEW YORK, 67- Concurrent Pos: Hon cur paleont & conchol, Mus Natural Hist, Caracas, 41-44. Mem: AAAS; Geol Soc Am; Paleont Soc; Mineral Soc Am; Am Asn Petrol Geologists. Res: Stratigraphy and geology of Venezuela; geology of northern Maine. Mailing Add: Dept of Earth & Planetary Sci City Col of New York New York NY 10031

MENCHER, JOAN PHYLLIS, b New York, NY, Jan 29, 30. APPLIE ANTHROPOLOGY, ECONOMIC ANTHROPOLOGY. Educ: Smith Col, BA, 50; Columbia Univ, PhD(anthrop), 58. Prof Exp: Res asst anthrop, Bank St Col Educ, 55-56; res asst anthrop, City Col New York, 57-58; instr anthrop, Hofstra Col, 60-61; vis asst prof anthrop, Cornell Univ, 64-65; res assoc anthrop, Columbia Univ, 65-67, vis assoc prof, 67-68; assoc prof, 68-74, PROF ANTHROP, LEHMAN COL & GRAD CTR, CITY UNIV NY, 74- Concurrent Pos: Am Asn Univ Women res fel, India, 58-60; Ogden Mills fel, Am Mus Natural Hist, 61-62; NSF res fels, India, 62-64 & 69-73; NIMH res fel, India, 65-67; sr res assoc, Columbia Univ, 69-74; chmn, Comt Status Women Anthrop, 69-70, co-chmn, 72-73; Guggenheim Found fel, 74 & 75-76; Wenner-Gren grant-in-aid, 75. Mem: Fel Am Anthrop Asn; Asn Asian Studies; fel Soc Appl Anthrop. Res: Social and economic development and change in South Asia. Mailing Add: c/o Dept of Anthrop Lehman Col Bedford Park Blvd W Bronx NY 10468

MENCHER, JORDAN RONALD, microbiology, see 12th edition

MENCIK, ZDENEK, b Jihlava, Czech, July 7, 27; m 50; c 1. PHYSICAL CHEMISTRY, X-RAY CRYSTALLOGRAPHY. Educ: Slovak Tech Univ Bratislava, MS, 56; Charles Univ, Prague, PhD(phys chem polymers), 61. Prof Exp: Res scientist polymer struct & properties, Res Inst Macromolecular Chem, Brno, Czech, 53-68; fel supermolecular struct polysaccharides, State Univ NY Col Forestry, Syracuse Univ, 68-69; SR RES SCIENTIST, SCI RES STAFF, FORD MOTOR CO, 69- Mem: AAAS; NY Acad Sci; Am Phys Soc; Am Crystallog Asn. Res: Physical chemistry of high polymers; use of physico-chemical methods in the study of macromolecular systems; x-ray crystallography of polymeric and non-polymeric systems. Mailing Add: Sci Res Staff Ford Motor Co PO Box 2053 Dearborn MI 48121

MENCZEL, JEHUDA H, b Vienna, Austria, Jan 29, 36; US citizen; m 71. ENVIRONMENTAL CHEMISTRY. Educ: Univ Minn, Minneapolis, BA, 61; Rutgers Univ, PhD(phys chem), 67. Prof Exp: NSF res asst, Rutgers Univ, 65-66; staff scientist, Aerospace Res Ctr, Singer-Gen Precision, Inc, 66-69 & Res Labs, Olivetti Corp, Am, 69-71; chemist, 71-73, SECT CHIEF AIR FACIL BR, US ENVIRON PROTECTION AGENCY, 73- Mem: Am Chem Soc; Air Pollution Control Asn. Res: Photochemistry and gas phase kinetics; photoconductivity; thermochromic phenomena; electrochemical processes in conjunction with non-impact printing; waste water treatment; air pollution. Mailing Add: Environ Protection Agency 26 Fed Plaza New York NY 10007

MENDALL, HOWARD LEWIS, b Augusta, Maine, Nov 21, 09; m 33. WILDLIFE BIOLOGY, ORNITHOLOGY. Educ: Univ Maine, BA, 31, MA, 34. Prof Exp: Asst zool, Univ Maine, 34-36; wildlife technician, US Resettlement Admin, 36-37; asst leader, 37-42, LEADER, MAINE COOP WILDLIFE RES UNIT, US BUR SPORT FISHERIES & WILDLIFE & PROF WILDLIFE RESOURCES, UNIV MAINE, ORONO, 42- Honors & Awards: Terrestrial Pub Award, Wildlife Soc, 59 & John Pearce Mem Award, 66. Mem: Wildlife Soc; fel Am Ornithologists Union. Res: Field ornithology and general wildlife ecology, especially food habits; habitat influences and breeding biology; fish-eating birds, woodcock and waterfowl. Mailing Add: Maine Coop Wildlife Res Unit 240 Nutting Hall Univ of Maine Orono ME 04473

MENDE, THOMAS JULIUS, b Budapest, Hungary, Oct 3, 22; m 49; c 2. BIOCHEMISTRY. Educ: Univ of Sciences, Budapest, PhD(org chem), 48. Prof Exp: Res assoc chem embryol, NY Univ, 49-50; res asst prof biochem, 54-58, asst prof, 58-60, assoc prof, 60-74, PROF BIOCHEM, SCH MED, UNIV MIAMI, 74- Concurrent Pos: Res fel enzymol, Med Sch, Univ Budapest, 48; res fel, Lobund Inst, Univ Notre Dame, 50-54. Mem: Am Chem Soc; Am Soc Pharmacol & Exp Therapeut; Int Soc Haemostasis & Thrombosis; Sigma Xi; fel Gerontological Soc. Res: Natural products chemistry; blood coagulation; thrombosis. Mailing Add: PO Box 520875 Sch of Med Univ Miami Miami FL 33152

MENDE, WILLIAM CARL, pharmaceutical chemistry, see 12th edition

MENDEL, ARTHUR, b Ger, Dec 14, 31; nat US; m 56; c 1. ORGANIC CHEMISTRY. Educ: Univ Ill, BS, 54; Univ Mo, MA, 56, PhD, 58. Prof Exp: Res chemist, Petrol chem, Inc, 58-60; RES CHEMIST, MINN MINING & MFG CO, ST PAUL, 60- Mem: Am Chem Soc. Res: Organic synthesis; medicinals; organometallic and analytical organic chemistry; chromatography; electrophoresis. Mailing Add: 4525 Oak Leaf Dr White Bear Lake MN 55110

MENDEL, CLIFFORD WILLIAM, b St Louis, Mo, Sept 16, 07; m 31, 69; c 3. MATHEMATICS. Educ: Univ Chicago, BS, 27, MS, 28, PhD(math), 30. Prof Exp: Instr math, Univ Chicago, 30-31; Nat Res Coun fel, 31-33; instr math, Univ Cincinnati, 33-37; assoc, 37-43, from asst prof to prof, 43-71, EMER PROF MATH, UNIV ILL, URBANA, 71- Mem: Am Math Soc. Res: Projective geometry; theory of numbers; analysis. Mailing Add: Sherwood Forest Cedar Mountain NC 28718

MENDEL, GERALD ALAN, b New York, NY, May 9, 29; m 69; c 2. INTERNAL MEDICINE, HEMATOLOGY. Educ: Col William & Mary, BS, 50; Wash Univ, MD, 54; Am Bd Internal Med, dipl, 62. Prof Exp: Intern, Univ Chicago, 54-55, resident med, 57-60, from instr to asst prof, 60-66; ASST PROF MED, SCH MED, NORTHWESTERN UNIV, EVANSTON, 66- Concurrent Pos: Schweppe Found grant, 62-65. Honors & Awards: Joseph A Capps Prize, 61. Mem: Am Fedn Clin Res; Am Soc Hemat. Res: Iron metabolism. Mailing Add: Dept of Med Sch of Med Northwestern Univ Evanston IL 60201

MENDEL, JOHN RICHARD, b St Paul, Minn, Nov 24, 36; m 68. COLLOID CHEMISTRY, POLYMER CHEMISTRY. Educ: Univ Wash, BS, 58; Boston Col, MS, 69. Prof Exp: Chemist adhesives, Am Marietta Co, 59-61; chemist dispersions, Hercules Inc, 68-70; SR RES CHEMIST PHOTO RES, EASTMAN KODAK CO, 70- Mem: Am Chem Soc; Soc Photog Scientists & Engrs. Res: Polymer colloids; synthesis, characterization and determination of all physical and chemical properties. Mailing Add: Eastman Kodak Co Bldg 59 Kodak Park Rochester NY 14617

MENDEL, JULIUS LOUIS, b Amarillo, Tex, Jan 13, 25. CLINICAL CHEMISTRY. Educ: Univ Tex, BA, 46; Univ Southern Calif, MS, 49, PhD(biochem), 50. Prof Exp: Biochemist, US Vet Admin, 51-63; res chemist, US Vet Admin, 51-63 & US Food & Drug Admin, Washington, DC, 63-66; ASST TO DIR PATH, VET ADMIN CENT OFF, 66- Mem: AAAS; Am Chem Soc; Am Asn Clin Chemists; NY Acad Sci; Am Inst Chemists. Res: Hematology; data processing in clinical laboratories. Mailing Add: Path & Allied Sci Vet Admin Cent Off Washington DC 20420

MENDEL, MARYANN MADELIENE, b Philadelphia, Pa; m 68. PHOTOGRAPHIC CHEMISTRY, SYNTHETIC ORGANIC CHEMISTRY. Educ: Immaculata Col, BA, 65; Boston Col, PhD(org chem), 74. Prof Exp: Develop chemist, United Merchants & Mfrs, 68-70; RES CHEMIST, EASTMAN KODAK CO, 71- Res: Precipitating and chemically sensitizing silver halide emulsions for photographic applications. Mailing Add: Eastman Kodak Co Res Labs 1669 Lake Ave Rochester NY 14650

MENDEL, VERNE EDWARD, b Lewistown, Mont, Apr 28, 23; m 46; c 3. PHYSIOLOGY. Educ: Univ Idaho, BSc, 55, MSc, 58; Univ Calif, PhD(animal physiol), 60. Prof Exp: Asst prof animal physiol, Univ Alta, 60-63; asst physiologist & lectr, 63-71, assoc prof, 71-75, PROF ANIMAL PHYSIOL, UNIV CALIF, DAVIS, 75-, CHMN DEPT, 73- Mem: Am Physiol Soc. Res: Chemical and physiological basis of food intake control. Mailing Add: Dept of Animal Physiol Univ of Calif Davis CA 95616

MENDEL, WERNER MAX, b Hamburg, Ger, June 11, 27; US citizen; c 2. PSYCHIATRY. Educ: Univ Calif, Los Angeles, BA, 48; Stanford Univ, MA, 49, MD, 53; Am Bd Psychiat & Neurol, cert psychiat, 59; Southern Calif Psychoanal Inst, dipl, 64. Prof Exp: Intern, Los Angeles County Gen Hosp, 53-54; resident psychiat, St Elizabeth's Hosp, Washington, DC, 54-55 & Winter Vet Admin Hosp, Topeka, Kans, 55-57; staff psychiatrist & dir rehab proj, Metrop State Hosp, Norwalk, Calif, 57-58; clin instr, 58-60, from asst prof to assoc prof, 60-67, PROF PSYCHIAT, SCH MED, UNIV SOUTHERN CALIF, 67- Concurrent Pos: Fel, Menninger Sch Psychiat, 55-57; dir outpatient serv, Metrop State Hosp, Norwalk, Calif, 58-60; chief sr clerkship psychiat, Sch Med, Univ Southern Calif, 60-62; consult, Calif State Dept Ment Hyg, 60-, mem spec residency training rev bd, 65 & res adv comt, 70-74; mem attend staff, Los Angeles County Gen Hosp Psychiat Unit, 60-; chief teaching serv, Psychiat Hosp, Los Angeles County-Univ Southern Calif Med Ctr, 62-65, clin dir, Adult Inpatient Serv, 65-67, dir, Div Prof & Staff Develop, 67-; consult, Ment Health Comt, Am Acad Gen Pract, 63-66; asst examr, Am Bd Neurol & Psychiat, 64-68; spec proj ed, Basic Bks, Inc, 66-69; mem exp & spec training rev comt, NIMH, 68-72; ed-in-chief, Mara Bks, Inc, 69-; chmn med adv bd, Human Resource Inst, Inc, 70-74; Calif State Dept Ment Hyg grant; Attend Staff Fund grants, Los Angeles County-Univ Southern Calif Med Ctr; Vet Admin grant; NIMH grants. Mem: AMA; Am Psychiat Asn; Asn Am Med Cols; Am Psychoanal Asn. Res: Determinants of the decision for psychiatric hospitalization; reversal of soft neurological signs in brain-damaged patients; perceptual changes in schizophrenic patients; effectiveness of outpatient treatment in chronic schizophrenia; phenomenological theory of schizophrenia; mental health care delivery systems. Mailing Add: Dept of Psychiat Univ of Southern Calif Sch of Med Los Angeles CA 90033

MENDELHALL, VON THATCHER, b Soda Springs, Idaho, Nov 1, 37; m 56; c 4. FOOD SCIENCE. Educ: Utah State Univ, BS, 62, MS, 67; Ore State Univ, PhD(food sci), 70. Prof Exp: Asst prof food sci, Univ Fla, 70-72; ASST PROF FOOD SCI, UTAH STATE UNIV, 72- Mem: Inst Food Technologists. Res: Protein degradation in shellfish; formaldehyde production in frozen fish tissue; lipid oxidation in Florida mullet and turkey products. Mailing Add: Dept of Food Sci Utah State Univ Logan UT 84322

MENDELL, DAVID, b New York, NY, May 10, 09; m 45; c 2. PSYCHIATRY. Educ: City Col New York, BS, 29; Univ Vienna, MD, 34. Prof Exp: Instr child psychiat, Univ Calif, 47-48; assoc prof, Postgrad Sch Med, 52-72, CLIN PROF PSYCHIAT, UNIV TEX MED SCH HOUSTON, 72-; ASSOC PROF PSYCHIAT, BAYLOR COL MED, 51- Concurrent Pos: Pvt pract med, 36-42, psychiat, 48-; dir child psychiat, City & County Hosp, Houston, Tex, 50-55; consult, Family Serv Bur, Vet Admin, Houston, Tex, 50-55 & Tex Children's Hosp, 55- Res: Neurotic patterns in families; group dynamics in teaching and psychotherapy. Mailing Add: 1630 Doctor's Ctr Houston TX 77025

MENDELL, JAY STANLEY, b New York, NY, Mar 13, 36; m 61; c 2. TECHNOLOGICAL FORECASTING. Educ: Rensselaer Polytech Inst, BS, 56, PhD(physics), 64; Vanderbilt Univ, MA, 58. Prof Exp: Health physicist, Oak Ridge Nat Lab, 57-58; asst physics, Rensselaer Polytech Inst, 58-60, asst elec eng, 60-63; asst proj engr, Pratt & Whitney Aircraft Div, United Aircraft Corp, 63-68, sr staff analyst, Advan Planning, 68-74; ASSOC PROF, SCH TECHNOL, FLA INT UNIV, 74- Concurrent Pos: Adj asst prof, Univ Hartford, 65-66; lectr, Univ Conn, 70 & Can Armed Forces Sch Mgt, 71; innovation ed, The Futurist, World Future Soc, 69-; contrib ed, Planning Digest, 71-; mem adv bd, Technol Forecasting & Social Change, 71- Mem: AAAS; World Future Soc; Inst Elec & Electronics Engrs; Int Soc Technol Assessment. Res: Creativity; corporate and research planning; technology assessment; futures research. Mailing Add: Sch of Technol Fla Int Univ Miami FL 33144

MENDELL, LORNE MICHAEL, b Montreal, Que, Nov 6, 41; m 67. NEUROPHYSIOLOGY. Educ: McGill Univ, BSc, 61; Mass Inst Technol, PhD(neurophysiol), 65. Prof Exp: Asst prof, 68-73, ASSOC PROF PHYSIOL, MED CTR, DUKE UNIV, 73- Concurrent Pos: USPHS fel, Harvard Med Sch, 65-68; USPHS grant, Med Ctr, Duke Univ, 69-; NIH career develop award, 71. Mem: AAAS; Am Physiol Soc; Soc Neurosci; Soc Gen Physiol. Res: Neuroembryology. Mailing Add: Dept of Physiol Duke Univ Med Ctr Durham NC 27710

MENDELL, ROSALIND B, b New York, NY, Oct 20, 20; m 41; c 2. COSMIC RAY PHYSICS. Educ: Hunter Col, BA, 40; NY Univ, PhD(physics), 63. Prof Exp: Instr physics, NY Univ, 42-43; physicist, US Bur Standards, 44-46; assoc res scientist, 63-74, adj assoc prof, 72-75, RES SCIENTIST, NY UNIV, 74-, ASSOC RES PROF, 75- Mem: Am Phys Soc; Am Geophys Union. Res: Study of neutrons in cosmic radiation; cosmic ray modulation. Mailing Add: Dept of Physics NY Univ 4 Washington Pl New York NY 10003

MENDELSOHN, LAWRENCE BARRY, b Brooklyn, NY, Apr 19, 34; m 58; c 2. PHYSICS. Educ: Brooklyn Col, BS, 55; Columbia Univ, MA, 59; NY Univ, PhD(physics), 65. Prof Exp: Physicist, Combustion Eng, Conn, 56-57, Walter Kidde Nuclear Labs, 58-59 & Tech Res Group, Inc, 59-62; instr physics, Cooper Union, 62-65; from asst prof to assoc prof, Polytech Inst Brooklyn, 65-73; PROF PHYSICS, NEW SCH LIB ARTS, BROOKLYN COL, 73- Concurrent Pos: Consult, Sandia Corp, 69- Mem: Am Phys Soc. Res: Many body problem, particularly calculation of correlation effects in atoms and molecules; industrial experience comprises; nuclear reactor and shielding calculations; x-ray scattering cross sections. Mailing Add: New Sch of Lib Arts Brooklyn Col 210 Livingston St Brooklyn NY 11201

MENDELSOHN, MORTIMER LESTER, b New York, NY, Dec 1, 25; m 48; c 3. BIOPHYSICS, CANCER. Educ: Harvard Univ, MD, 48; Cambridge Univ, PhD, 58. Prof Exp: Intern med, Mass Gen Hosp, 48-49; resident, Med Ctr, NY, 49-52; from asst prof to prof radiol, Sch Med, Univ Pa, 57-72; DIR BIOMED DIV, LAWRENCE LIVERMORE LAB, UNIV CALIF, 72- Concurrent Pos: Res fel, Sloan-Kettering Inst Cancer Res, 52-53; Am Cancer Soc Brit-Am exchange fel, 55-57; mem, comput res study sect, NIH, 67-70, chmn, 70-71. Mem: Radiation Res Soc; Histochem Soc; Am Asn Cancer Res; Environ Mutagenic Soc. Res: Experimental and clinical cancer research; radiation effects; cell division; biophysical cytology; flow cytometry; computer analysis of cell images; environmental mutagenesis. Mailing Add: Lawrence Livermore Lab Univ of Calif Livermore CA 94550

MENDELSOHN, NATHAN SAUL, b Brooklyn, NY, Apr 14, 17; m 40; c 2. MATHEMATICS. Educ: Univ Toronto, BA, 39, MA, 40, PhD, 42. Prof Exp: Supvr munitions gauge lab, Nat Res Coun Can, 42; scientist, Proof & Develop Estab, Que, 42-45; lectr math, Queen's Univ, Ont, 45-47; PROF, UNIV MAN, 47-, HEAD DEPT, 63- Mem: Soc Indust & Appl Math; Am Math Soc; Math Asn Am; fel Royal Soc Can; Can Math Cong (pres). Res: Abstract algebra and geometry; combinatory statistics; a group-theoretic characterization of the general projective collineation group; ballistics; theory of error in computing machines; graph and matroid theory. Mailing Add: Dept of Math Univ of Man Winnipeg MB Can

MENDELSON, BERT, b Brooklyn, NY, Jan 3, 26; m 52; c 3. MATHEMATICS. Educ: Columbia Univ, BA, 45, PHD(math), 59; Univ Nebr, MA, 51. Prof Exp: PROF MATH, SMITH COL, 57-, DIR COMPUT CTR, 74- Mem: Am Math Soc; Math Asn Am. Res: Algebraic topology. Mailing Add: Dept of Math Smith Col Northampton MA 01060

MENDELSON, ELLIOTT, b New York, NY, May 24, 31; m 59; c 3. MATHEMATICAL LOGIC. Educ: Columbia Univ, AB, 52; Cornell Univ, MA, 54, PhD(math), 55. Prof Exp: Instr math, Univ Chicago, 55-56; jr fel, Harvard Univ, 56-58; J F Ritt instr, Columbia Univ, 58-61; assoc prof, 61-64, PROF MATH, QUEENS COL, NY, 64- Mem: Am Math Soc; Math Asn Am; Asn Symbolic Logic. Res: Axiomatic set theory. Mailing Add: Dept of Math Queens Col Flushing NY 11367

MENDELSON, JACK H, b Baltimore, Md, Aug 30, 29; m 52; c 3. MEDICINE, PSYCHIATRY. Educ: Univ Md, MD, 55. Prof Exp: Intern med serv, Boston City Hosp, 55-56; asst, 59-71, PROF PSYCHIAT, HARVARD MED SCH, 71-; DIR, ALCOHOL & DRUG ABUSE RES CTR, McLEAN HOSP, 73- Concurrent Pos: Teaching fel psychiat, Harvard Med Sch, 56-59; res fel psychiat, Boston City & Mass Gen Hosps, 56-59; consult, Washingtonian Hosp, Boston, 58-59; consult, psychiat res labs, Sch Med, Univ Md, 59-; asst, Mass Gen Hosp, 59-; dir dept psychiat, Boston City Hosp, 71-73. Mem: Am Psychiat Asn; Asn Res Nerv & Ment Dis; Am Acad Neurol; Am Soc Pharmacol & Exp Therapeut. Res: Psychiatric research, especially in alcohol and drug abuse. Mailing Add: Alcohol & Drug Abuse Res Ctr McLean Hosp Belmont MA 02178

MENDELSON, KENNETH SAMUEL, b Chicago, Ill, Aug 24, 33; m 61; c 3. PHYSICS. Educ: Ill Inst Technol, BS, 55; Purdue Univ, MS, 57, PhD(physics), 63. Prof Exp: Assoc physicist, IIT Res Inst, 62-65; asst prof, 65-70, ASSOC PROF PHYSICS, MARQUETTE UNIV, 70- Concurrent Pos: Sr sci fel, NATO, 74. Mem: Am Phys Soc. Res: Fields in heterogeneous media. Mailing Add: Physics Dept Marquette Univ Milwaukee WI 53233

MENDELSON, MARTIN, b New York, NY, Apr 16, 37; m 58; c 2. PHYSIOLOGY. Educ: Cornell Univ, AB, 58; Calif Inst Technol, PhD(biol), 62; State Univ NY Stony Brook, MD, 76. Prof Exp: Res assoc physiol, Col Physicians & Surgeons, Columbia Univ, 61-63; from instr to assoc prof, Sch Med, NY Univ, 63-71; assoc prof physiol, Health Sci Ctr, 71-73, MEM ADJ STAFF, DEPT MED, DIV NEUROL, NASSAU COUNTY MED CTR, STATE UNIV NY STONY BROOK, 73- Concurrent Pos: Mem, Corp Marine Biol Lab, Woods Hole, Mass. Mem: Soc Neurosci; Am Soc Zool; Soc Gen Physiol. Res: Sensory mechanisms in Crustacea and mammals; neuromuscular transmission in Crustacea; central nervous mechanisms of rhythmicity and integration in Curstacea. Mailing Add: c/o Health Sci Ctr State Univ of NY Stony Brook NY 11790

MENDELSON, MYER, b Lithuania, Dec 5, 20; nat US; m 56; c 1. PSYCHIATRY. Educ: Dalhousie Univ, BA, 45, BSc, 46, MD, CM, 50. Prof Exp: Instr psychiat, Johns Hopkins Univ, 54-56; asst prof, Dalhousie Univ, 56-58; from asst prof to assoc prof, 58-71, PROF CLIN PSYCHIAT, SCH MED, UNIV PA, 71- Mem: Am Psychosom Soc; Am Psychiat Asn; Can Psychiat Asn. Res: Theoretical models in psychoanalysis; depression; manic-depressive illness; psychopharmacology; obesity. Mailing Add: 1220 Wingate Rd Wynnewood PA 19096

MENDELSON, NEIL HARLAND, b New York, NY, Nov 15, 37; m 59; c 2. GENETICS. Educ: Cornell Univ, BS, 59; Ind Univ, PhD(genetics, bact), 64. Prof Exp: Asst prof biol sci, Univ Md, Baltimore County, 66-69; assoc prof, 69-74, PROF MICROBIOL & MED TECHNOL, UNIV ARIZ, 74- Concurrent Pos: NSF fel, Med Res Coun Microbial Genetics Res Unit, Hammersmith Hosp, London, Eng, 65-66; Nat Inst Gen Med Sci res career develop award, 73-77; prin investr res grant, Div Biol & Med Sci, NSF, 67-69, 69-71 & Nat Inst Gen Med Sci, 71-. Mem: Fel AAAS; Am Soc Microbiol; Genetics Soc Am; Sigma Xi. Res: Molecular and microbial genetics; genetic control of DNA replication, growth and cell division in Bacillus subtilis. Mailing Add: Dept of Microbiol & Med Technol Univ of Ariz Tucson AZ 85721

MENDELSON, ROBERT ALEXANDER, JR, b Los Angeles, Calif, Jan 24, 41. NUCLEAR PHYSICS, MOLECULAR BIOLOGY. Educ: Occidental Col, AB, 62; Univ Iowa, MS, 64, PhD(physics), 68. Prof Exp: Res assoc physics, Lawrence Berkeley Lab, Univ Calif, 68-71; RES ASSOC BIOPHYS, UNIV CALIF, SAN FRANCISCO, 71- Mem: Am Phys Soc; Biophys Soc. Res: Nuclear structure physics. Mailing Add: HSW-831 Univ of Calif San Francisco CA 94143

MENDELSON, ROBERT ALLEN, b Cleveland, Ohio, Dec 17, 30; m 71. POLYMER SCIENCE, RHEOLOGY. Educ: Case Western Reserve Univ, BS, 52, PhD(chem), 56. Prof Exp: Res chemist, Monsanto Chem Co, 56-61, res specialist, 61-69, SCI FEL, MONSANTO CO, 69- Mem: AAAS; Am Chem Soc; Soc Rheol (secy, 74-78). Res: Polymer solution properties; polymer molecular weight and structure characterization; polymer rheology and processing. Mailing Add: Res Dept Monsanto Co 730 Worcester St Indian Orchard MA 01151

MENDELSON, WILFORD LEE, b Baltimore, Md, July 8, 37. ORGANIC CHEMISTRY. Educ: Johns Hopkins Univ, AB, 58, MA, 60, PhD(chem), 63. SR CHEMIST, SMITH KLINE & FRENCH LABS, 63- Mem: Am Chem Soc; The Chem Soc. Res: Incorporation of C-14 and H-3 into existing and potential pharmaceuticals; organic synthesis with isotopes. Mailing Add: 320-B Hermit St Philadelphia PA 19128

MENDENHALL, CHARLES L, internal medicine, biochemistry, see 12th edition

MENDENHALL, GEORGE DAVID, b Iowa City, Iowa, Feb 12, 45; m 73; c 1. PHYSICAL ORGANIC CHEMISTRY. Educ: Univ Mich, BS, 66; Harvard Univ, PhD(chem), 71. Prof Exp: Fel, Nat Res Coun Can, 71-73; fel, Stanford Res Inst, 73-74; STAFF MEM CHEM, COLUMBUS LABS, BATTELLE MEM INST, 74- Mem: Am Chem Soc. Res: Reactions of ozone, singlet molecular oxygen; electron spin resonance studies of organic radicals; kinetics of smog-producing reactions; composition of atmospheric aerosols; autoxidation processes studied by chemiluminescence. Mailing Add: Columbus Labs Battelle Mem Inst 505 King Ave Columbus OH 43201

MENDENHALL, RICHARD MASON, biochemistry, see 12th edition

MENDENHALL, ROBERT VERNON, b Geneva, Ind, Dec 27, 20; m 44; c 3. MATHEMATICAL LOGIC. Educ: Ohio State Univ, BA, 47, MA, 49, PhD(math), 52. Prof Exp: Instr math, Ohio State Univ, 47-53; mathematician, NAm Aviation, Inc, 53-55 & Vitro Labs, Inc, 55; asst prof math, Univ Miami, 55-62; assoc prof, 62-66, PROF MATH, OHIO WESLEYAN UNIV, 66- Concurrent Pos: Consult, NSF Math Insts, India, 65, 66 & 70. Mem: AAAS; Am Math Soc; Math Asn Am; Indian Math Soc. Res: Measure theory; integration. Mailing Add: Dept of Math Ohio Wesleyan Univ Delaware OH 43015

MENDENHALL, WILLIAM, III, b Pa, Apr 20, 25; m 49; c 2. STATISTICS. Educ: Bucknell Univ, BS, 45, MS, 50; NC State Col, PhD(statist), 57. Prof Exp: Asst prof statist, NC State Col, 58-59; assoc prof math, Bucknell Univ, 59-63; PROF STATIST, UNIV FLA, 63- Concurrent Pos: Consult, London Sch Econ, 57-58, Westinghouse Elec Co, Pa, 59-60, Armstrong Cork Co, Pa, 60-, Burroughs Corp, Mich, 60-61, Lewis Res Ctr, NASA, Ohio, 60-61, Merck & Co, Pa, 61-62 & WVa Pulp & Paper Co, 65. Mem: Am Statist Asn; Inst Math Statist; Royal Statist Soc. Res: Design of experiments; distribution theory. Mailing Add: Dept of Statist 522 NSB Univ of Fla Gainesville FL 32601

MENDES, ERASMO GARCIA, b Santos, Brazil, May 28, 15; m 50; c 3. COMPARATIVE PHYSIOLOGY. Educ: Univ Sao Paulo, Lic, 39, DSc, 44. Prof Exp: From instr to assoc prof gen physiol, 41-69, PROF PHYSIOL, UNIV SAO PAULO, 70- Concurrent Pos: Rockefeller fel, Yale Univ & Woods Hole, 46; Int Union Biol Sci fel, Zool Sta, Naples, Italy; Brit Coun Bursary, Cambridge & Plymouth, 63; Ger Acad Exchange Serv Bursary, Kiel, 63. Res: Amphibian respiration; color change; metabolism and enzymes in smooth muscle; comparative pharmacology. Mailing Add: Dept of Physiol Inst of Biosci Univ of Sao Paulo CP11461 Sao Paulo Brazil

MENDES, ROBERT W, b Fall River, Mass, Apr 6, 38; m 60. INDUSTRIAL PHARMACY. Educ: New Eng Col Pharm, BS, 60; Univ NC, MS, 64, PhD(pharm), 66. Prof Exp: Asst prof pharm, 65-75, ASSOC PROF & CHMN INDUST PHARM, MASS COL PHARM, 75- Concurrent Pos: Consult & grantee, several material suppliers and mfg concerns, 71- Mem: Am Pharmaceut Asn; Acad Pharmaceut Sci. Res: Formulations; product development. Mailing Add: Dept of Indust Pharm Mass Col of Pharm 179 Longwood Ave Boston MA 02115

MENDEZ, EUGENIO FERNANDEZ, b Cayey, PR, Nov 6, 24; m 46; c 1. ANTHROPOLOGY. Educ: Univ PR, BA, 47; Columbia Univ, PhD, 51. Prof Exp: From instr to assoc prof anthrop, 49-71, PROF ANTHROP, UNIV PR, SAN JUAN, 71- Mem: Fel Am Anthrop Asn. Res: Puerto Rican culture history; Caribbean societies and peoples; American Indian mythologies. Mailing Add: Dept of Sociol Univ of Puerto Rico San Juan PR 00931

MENDEZ, JOSE DE LA VEGA, b Tapachula, Mex, Aug 17, 21; m 50; c 5. PHYSIOLOGY. Educ: San Carlos Univ Guatemala, BS, 47; Univ Ill, MS, 48; Univ Minn, PhD(physiol), 57. Prof Exp: Chief spec projs, Inst Nutrit Cent Am & Panama, Guatemala, 49-53, dir training progs & co-chief physiol, 59-65; assoc prof nutrit, Mass Inst Technol, 65-66; PROF HEALTH & APPL PHYSIOL, LAB HUMAN PERFORMANCE RES, PA STATE UNIV, 66- Concurrent Pos: Adv, Sch Med, San Carlos Univ Guatemala, 59-60. Mem: AAAS; Am Inst Nutrit; NY Acad Sci; Guatemala Soc Natural Sci & Pharm; Latin Am Nutrit Soc. Res: Nutritional studies and nutrition training in developing countries; body composition and anthropology; fat metabolism and nutritional factors in atherosclerosis; physiology of work and nutrition; adaptation of man to different environmental stresses. Mailing Add: Lab of Human Performance Res Pa State Univ University Park PA 16802

MENDEZ-BAUER, CARLOS, b Montevideo, Uruguay, Feb 3, 30; m 68. OBSTETRICS & GYNECOLOGY, PHYSIOLOGY. Educ: Univ Repub, Uruguay, MD, 55. Prof Exp: Instr physiol, Sch Med, Univ Repub, Uruguay, 51, asst prof, 57-71, obstet physiol, 61-66, assoc prof physiopath & chmn serv obstet physiol, 66-69; assoc prof obstet, gynec & physiol, Sch Med, Univ Minn, Minneapolis, 69-70; SR INVESTR, MED RES LAB, MASONIC FOUND MED RES & HUMAN WELFARE, 70- Concurrent Pos: Rockefeller Found fel obstet physiol, 55; partic, World Cong Int Fedn Obstet & Gynec, Montreal, 58; lectr, Int Cong Obstet, 65; partic, Int Cong Physiol Sci, DC, 68. Honors & Awards: Louis Calzada Prize, Univ Repub, Uruguay, 66. Mem: Int Soc Res Human Reproduction; Soc Gynec Invest; Latin Am Asn Physiol Sci; Latin Am Soc Human Reproduction; Uruguayan Soc Endocrinol. Res: Physiology and paraphysiology of mother and fetus during pregnancy, labor and delivery; prevention and treatment of fetal distress. Mailing Add: Med Res Lab Masonic Found Med Res Human Welf Utica NY 13501

MENDICINO, JOSEPH FRANK, b Cleveland, Ohio, Nov 22, 30; m 52; c 4. BIOCHEMISTRY. Educ: Case Western Reserve Univ, BS, 53; PhD, 57. Prof Exp: NSF res fel, Inst Biochem Invest, Argentina, 59-60 & Case Western Reserve Univ, 60-62; asst prof agr biochem, Ohio State Univ, 62-68; ASSOC PROF BIOCHEM, UNIV GA, 68- Mem: Am Chem Soc; Am Soc Biol Chem. Res: Enzymology. Mailing Add: Dept of Biochem Univ of Ga Athens GA 30601

MENDIS, EUSTACE FRANCIS, b Colombo, Ceylon, June 22, 37; m 71. SOLID STATE PHYSICS. Educ: Univ Ceylon, BSc, 58; Univ Wis, PhD(physics), 68. Prof Exp: Fel physics, Univ NB, 69-70; lectr, Univ of Toronto, 70-75; HEAD PHYSICS DEPT, ONTARIO SCI CTR, 75- Mem: Am Phys Soc; Can Asn Physicists; Ceylon Asn Advan Sci. Res: Nuclear magnetic resonance in ferromagnets. Mailing Add: Dept of Physics Ontario Sci Ctr 770 Don Mills Rd Toronto ON Can

MENDLOWITZ, HAROLD, b New York, NY, Aug 23, 27; m 50; c 3. THEORETICAL PHYSICS. Educ: City Col New York, BS, 47; Columbia Univ, AM, 48; Univ Mich, PhD(physics), 54. Prof Exp: Asst prin & chmn dept, Beth Yehudah Schs, Mich, 48-49; asst exp physics, Columbia Univ, 49-50; physicist, Nat Bur Standards, 51-52; instr aeronaut eng, Univ Mich, 52-53, asst theoret physics, 53-54; theoret physicist, Nat Bur Standards, 54-65; PROF PHYSICS, HOWARD UNIV, 65- Concurrent Pos: Sr res fel, Hebrew Univ, Israel, 61-62; consult, Nat Bur Standards; vis prof, Hebrew Univ, Israel, 71-72. Mem: Fel Am Phys Soc; Sigma Xi. Res: Photomeson production; electron physics, scattering, interference and polarization; Dirac theory; optical properties of solids; magnetic moment of electron; characteristic electron energy losses in solids; atomic spectroscopy; transition probabilities. Mailing Add: Dept of Physics Howard Univ Washington DC 20001

MENDLOWITZ, MILTON, b New York, NY, Dec 30, 06; m 40; c 3. INTERNAL MEDICINE. Educ: City Col New York, AB, 27; Univ Mich, MD, 32. Prof Exp: Clin asst, 39-40, sr asst, 40-42, adj physician, 46-53, assoc attend physician, 53-59, ATTEND PHYSICIAN, MT SINAI HOSP, 59-, JOE LOWE & LOUIS PRICE PROF MED, MT SINAI MED SCH, 72- Concurrent Pos: Blumenthal fel, Mt Sinai Hosp, 36; Libman fel, Michael Reese Hosp, Chicago, Ill, 37 & Univ Col Hosp, London, 38; Dazian fel, Mt Sinai Hosp, 42; res fel, Goldwater Mem Hosp, 51-; Pvt pract, 39-42 & 46-; sr physician, NY Regional Off, US Vet Admin, 46-52; asst clin prof med, Columbia Univ, 56-61, assoc clin prof, 66-; assoc clin prof, Mt Sinai Med Sch, 66- Mem: AAAS; Am Soc Clin Invest; Am Physiol Soc; Soc Exp Biol & Med; AMA. Res: Physiology and pathological physiology of digital circulation; mechanism of heart failure; physiological effects of coronary occlusion and pulmonary embolism; mechanism and treatment of hypertension. Mailing Add: Dept of Med Mt Sinai Sch of Med New York NY 10029

MENDLOWSKI, BRONISLAW, b Tarnopol, Poland, June 28, 14; US citizen; m 50; c 3. PATHOLOGY, IMMUNOLOGY. Educ: Univ Lwow, Poland, DVM, 44; Univ Edinburgh, MRCVS, 47; Univ Ill, MS, 63. Prof Exp: Pathologist, West of Scotland Agr Col, 45-46, Wis State Diag Lab, 51-60 & Univ Ill, 60-63; res fel path, 63-74, SR RES FEL PATH, MERCK INST THERAPEUT RES, 74- Mem: Am Vet Med Asn; NY Acad Sci. Res: Etiologic and immunologic aspects of arthritides in animals. Mailing Add: Merck Inst for Therapeut Res West Point PA 19486

MENDOZA, CELSO ENRIQUEZ, b Bocaue, Bulacan, Philippines, Mar 28, 33; m 68; c 3. BIOCHEMISTRY, ENTOMOLOGY. Educ: Univ Philippines, BS, 59; Iowa State Univ, MS, 61, PhD(entom), 64. Prof Exp: Res assoc med entom, Cornell Univ, 64-65; Nat Res Coun Can fel pesticide residue anal, 65-67, res scientist I, 67-69, RES SCIENTIST II, HEALTH PROTECTION BR, NAT HEALTH & WELFARE DEPT, CAN, 69- Concurrent Pos: Assoc referee esterase methods, Asn Off Anal Chem, 70-; NATO spec study travel grant, Alta, Can, 71; vis scientist, Biochem Dept, Arrhenius Lab, Stockholm Univ, 73-74. Mem: AAAS; Am Chem Soc; Entom Soc Am; Philippine Entom Soc. Res: Development of analytical method for pesticide residues in foods; chromatographic-enzyme inhibition techniques for insecticides; toxicological and biochemical determination of pesticide effects on animals particularly neonates. Mailing Add: Health Protection Br Nat Health & Welfare Dept Ottawa ON Can

MENDOZA, GUILLERMO, b Mexico City, Mex, July 5, 09; nat US; m 43; c 2. ZOOLOGY. Educ: Northwestern Univ, BS, 32, MS, 34, PhD(zool), 37. Prof Exp: Instr zool, Univ Col, Northwestern Univ, 37-41, asst prof, 41-43; assoc prof, 43-46, chmn dept, 44-52, 61-63, 65-67, chmn div natural sci, 60-62, PROF BIOL, GRINNELL COL, 46- Concurrent Pos: AAAS res grant, 48; NSF grants, 54, 56, 58, 61 & 63; res assoc, Univ Calif, Berkeley, 63-64; res assoc, Argonne Nat Lab, 69-70. Mem: AAAS; Am Soc Zool; Am Inst Biol Sci. Res: Embryology and histology of some fresh-water fishes; early embryology and cytology of gonads of fresh-water fishes. Mailing Add: Dept of Biol Grinnell Col Grinnell IA 50112

MENDUKE, HYMAN, b Warsaw, Poland, Aug 20, 21; US citizen; m 46; c 1. BIOSTATISTICS. Educ: Univ Pa, BA, 43, MA, 48, PhD(econ statist), 52. Prof Exp: Instr soc & econ statist, Univ Pa, 47-53; asst prof biostatist, 53-58, assoc prof, 58-63, PROF COMMUNITY HEALTH & PREV MED & DIR SPONSORED PROGS, JEFFERSON MED COL, 63- Mem: Am Statist Asn; Biomet Soc; Franklin Inst. Res: Applied statistics in the design of surveys, clinical trials and laboratory experiments and in the analysis, interpretation and presentation of results. Mailing Add: Jefferson Med Col 1025 Walnut St Philadelphia PA 19107

MENEELY, GEORGE RODNEY, b Hempstead, NY, Sept 30, 11; m 35, 68; c 3. MEDICINE. Educ: Princeton Univ, BS, 33; Cornell Univ, MD, 37; Am Bd Internal Med, dipl, 44. Prof Exp: Intern med, Strong Mem Hosp, Rochester, NY, 37-38, asst resident, 39-40; instr, Sch Med, La State Univ, 41-43; from instr to assoc prof, Sch Med, Vanderbilt Univ, 43-62, asst to dean, 55-58, dir radioisotope ctr & div nuclear med & biophys, 55-62; assoc prof med, Med Sch, Northwestern Univ, 62-63; prof nuclear med, Grad Sch Biomed Sci, Univ Tex, 63-66; chief cardiopulmonary function lab & internist, Sect Nuclear Med, Univ Tex M D Anderson Hosp & Tumor Inst, 63-66, actg head dept biomath, 64-65; assoc dean sch med, 66-73, PROF & HEAD DEPT PHYSIOL & BIOPHYS, PROF MED, SCH MED, LA STATE UNIV, SHREVEPORT, 66- Concurrent Pos: Fel med, Sch Med & Dent, Univ Rochester, 38-39, James Gleason fel, 40-41; asst, Sch Med & Dent, Univ Rochester, 39-40; vis physician & dir lung sta, Charity Hosp, New Orleans, 41-43; dir heart sta, Vanderbilt Univ Hosp, 43-48, vis physician, 43-62; assoc prof, Meharry Med Col, 46-62; consult, Thayer Vet Admin Hosp, Nashville, 46-48, 57-62, asst chief med serv, 48-55, dir res lab & radioisotope unit, 48-55; consult, Oak Ridge Inst Nuclear Studies, 55-62, St Thomas Hosp, Nashville, 56-62, Social Security Admin, Dept Health, Educ & Welfare, 57-61, 64-, Bur Employee's Compensation, Dept Labor, 58-62, 65-, Surgeon Gen Army, 60-62, Tex Inst Rehab & Res Inst, Tex Med Ctr, 65-69 & M D Anderson Hosp & Tumor Inst, 66-; Mem, US-UK Epidemiol Bd, 60-65, Sci Adv Comt, United Health Founds, 63-70, Atomic Energy Adv Comt, State of Tex, 64-66 & Sci Adv Comt, Grayson Found, 65-; fel coun epidemiol & clin cardiol, Am Heart Asn. Mem: Fel AAAS; AMA; fel Am Col Physicians; fel Am Col Cardiol (pres, 57-58); Am Soc Exp Path. Res: Internal and nuclear medicine; clinical cardiopulmonary physiology; body composition; sodium and potassium metabolism. Mailing Add: La State Univ Sch of Med PO Box 3932 Shreveport LA 71130

MENEES, JAMES H, b Checotak, Okla, Nov 24, 29; m 53; c 1. HISTOLOGY, EMBRYOLOGY. Educ: San Jose State Col, AB, 53; Cornell Univ, MS, 57, PhD(entom), 59. Prof Exp: Asst, Cornell Univ, 56-59; PROF ENTOM, LONG BEACH STATE COL, 59- Concurrent Pos: Res grant, USDA, 56-59. Mem: Entom Soc Am; Am Soc Parasitol. Res: Insect anatomy; arthropod histology, embryology and physiology. Mailing Add: Dept of Biol Sci Calif State Col Long Beach CA 90804

MENEFEE, EMORY, b Wichita Falls, Tex, June 30, 29; m 53; c 3. PHYSICAL CHEMISTRY. Educ: Tex Tech Col, BS, 50; Mass Inst Technol, PhD(phys chem), 56. Prof Exp: Chem engr, Amarillo Helium Plant, US Bur Mines, Tex, 50-52; chemist, Plastics Dept, E I du Pont de Nemours & Co, Del, 56-60; CHEMIST WOOL LAB, WESTERN REGIONAL RES LAB, US DEPT AGR, 60- Mem: Soc Rheol. Res: Low temperature adsorption and phase equilibrium; viscoelasticity of molten polymers; compressibility of gases; rheology and physical chemistry of wool fibers. Mailing Add: Western Regional Res Lab 800 Buchanan St Albany CA 94710

MENEFEE, MAX GENE, b Perry, Mo, Mar 30, 25; m 51; c 3. PATHOLOGY. Educ: Wash Univ, PhD(anat), 56; State Univ NY, MD, 61. Prof Exp: Spec lectr surg, Col Med, State Univ NY Upstate Med Ctr, 57-61, asst prof anat, 61-66, ASSOC PROF PATH & ANAT, COL MED, UNIV CINCINNATI, 66-; CONSULT, US VET ADMIN HOSP, 71- Concurrent Pos: USPHS fel, 56-57. Mem: Am Asn Anat; Electron Micros Soc; Sigma Xi. Res: Fine structure of tissue culture cells; morphological aspects of vascular transport and disease; virus structure and replication. Mailing Add: Dept of Path Univ of Cincinnati Col of Med Cincinnati OH 45267

MENEFEE, ROBERT WILLIAM, b Akron, Ohio, Aug 8, 29; m 54; c 3. SCIENCE EDUCATION. Educ: Univ Akron, BS, 52; Kent State Univ, ME, 57; Ohio State Univ, PhD(sci ed), 65. Prof Exp: Teacher pub schs, Ohio, 54-62; instr unified sci, Ohio State Univ Sch, 63-65, asst prof zool, Ohio State Univ, 65-67, core prog dir biol, 67-68, asst dean & core dir biol sci, 68-69; assoc prof biol, Univ Md, College Park, 69-71; CHMN DIV MATH SCI, MONTGOMERY COL, TAKOMA PARK, 71- Mem: Fel AAAS; Am Inst Biol Sci; Nat Asn Res Sci Teaching; Nat Sci Teachers Asn. Res: Televised biology instruction; audio-tutorial laboratory instruction in biology; unified science; environmental education. Mailing Add: Div of Math Sci Montgomery Col Takoma Park MD 20012

MENEGHETTI, DAVID, b Chicago, Ill, May 8, 23; m 50; c 2. NUCLEAR PHYSICS. Educ: Univ Chicago, BS, 44; Ill Inst Technol, PhD(physics), 54. Prof Exp: Asst physicist, Argonne Nat Lab, 46-49 & 52-54; lab asst nuclear physics, Med Sch, Univ Ill, 51-52; assoc physicist, Armour Res Found, Ill Inst Technol, 54-55; SR PHYSICIST, ARGONNE NAT LAB, 55- Concurrent Pos: Consult, Centro di Calcolo & lectr, Bologna, 62-63. Mem: Am Phys Soc; fel Am Nuclear Soc. Res: Neutron physics; neutron diffraction; magnetics and antiferromagnetics; reactor physics. Mailing Add: Argonne Nat Lab 9700 Cass Ave Argonne IL 60439

MENELEY, DANIEL ALLISON, reactor physics, see 12th edition

MENELEY, WILLIAM ALLISON, geology, see 12th edition

MENENDEZ, ALFREDO, geology, see 12th edition

MENENDEZ, MANUEL GASPAR, b New York, NY, June 15, 35; m 58; c 3. ATOMIC PHYSICS, MOLECULAR PHYSICS. Educ: Univ Fla, BChE, 58, PhD(chem physics), 63. Prof Exp: Fel chem physics, Oak Ridge Nat Lab, 63-65; atomic physicist, Nat Bur Standards, 65-66; staff scientist, Martin Marietta Corp, 66-69; ASSOC PROF PHYSICS, UNIV GA, 69- Concurrent Pos: Consult, Martin Marietta Corp, 70-72. Mem: Am Phys Soc. Res: Ionization mechanisms at intermediate and low energies; molecular aspects of ion-atom collisions. Mailing Add: Dept of Physics & Astron Univ of Ga Athens GA 30602

MENES, MEIR, b Berlin, Ger, Oct 3, 25; nat US. PHYSICS. Educ: Cooper Union, BE, 48; NY Univ, PhD(physics), 52. Prof Exp: Physicist, Res Labs, Westinghouse Elec Corp, 52-60; ASSOC PROF PHYSICS, POLYTECH INST BROOKLYN, 60- Mem: Am Phys Soc. Res: Electrical discharges through gases; nuclear magnetic resonance; acoustic studies of solids. Mailing Add: Dept of Physics Polytech Inst Brooklyn Brooklyn NY 11201

MENEZ, ERANANI GUINGONA, b Manila, Philippines, Aug 15, 31; m 61; c 2. MARINE PHYCOLOGY. Educ: Univ Philippines, BS, 54; Univ Hawaii, MS, 62. Prof Exp: Res asst bot, Univ Philippines, 53-54; instr bot & zool, Southeastern Col, Philippines, 54-58; asst bot, Univ Hawaii, 58-61 & Univ BC, 62-64; SUPVR ALGAE,

SMITHSONIAN OCEANOG SORTING CTR, SMITHSONIAN INST, 64- Concurrent Pos: Dir, Mediterranean Marine Sorting Ctr, Tunisia, 73-75. Mem: Phycol Soc Philippines; Int Phycol Soc. Res: Taxonomy and ecology of tropical marine benthic algae; marine floristics of Tunisia. Mailing Add: Smithsonian Oceanog Sorting Ctr Smithsonian Inst Washington DC 20560

MENEZES, JOSE PIEDADE CAETANO AGNELO, b Curtorim, India, July 25, 39; Can citizen; m 70. MICROBIOLOGY, IMMUNOLOGY. Educ: Nat Col Goa, India, dipl, 58; Univ Perugia, DVM, 63; Pasteur Inst, Paris, dipl bact, 65; Univ Montreal, MS, 67; Univ Ottawa, PhD(microbiol), 71. Prof Exp: ASST PROF MICROBIOL & IMMUNOL, FAC MED, UNIV MONTREAL, 73- Concurrent Pos: Med Res Coun Can fel, Univ Montreal, 67-73, Univ Ottawa, 68-71 & Dept Tumor Biol, Karolinska Inst, Sweden, 71-73; Med Res Coun Can scholar, Univ Montreal, 73- Mem: Am Soc Microbiol; Can Soc Microbiol; Fr Soc Microbiol; NY Acad Sci; Tissue Cult Asn. Res: Virology; cellular immunology; tumor biology; cell culture; cell biology; electron microscopy. Mailing Add: Dept of Microbiol Univ of Montreal Fac of Med Montreal PQ Can

MENG, HEINZ KARL, b Baden, Ger, Feb 25, 24; nat US, m 53. BIOLOGY. Educ: Cornell Univ, BS, 47, PhD(ornith), 51. Prof Exp: Asst prof, 51-56, assoc prof, 56-61, PROF BIOL, STATE UNIV NY COL NEW PALTZ, 61- Mem: Assoc Wildlife Soc; Assoc Cooper Ornith Soc; assoc Wilson Ornith Soc; assoc Am Ornith Union. Res: Ornithology; entomology; falconry; vertebrate zoology. Mailing Add: Dept of Biol State Univ of New York Col New Paltz NY 12561

MENG, RAYMOND HSIEN CHANG, b Wu-Ching Co, China, Aug 12, 17; nat US; m 60; c 3. PHYSIOLOGY. Educ: Cheeloo Univ, MD, 41; Northwestern Univ, MS, 46, PHD, 47. Prof Exp: Asst physiol, Med Sch, Nat Cent Univ, China, 40-41; asst physiol & pharmacol, Cheeloo Univ, 41-42, instr, 42-44; asst, Med Sch, Northwestern Univ, 45-46, asst physiol, 46-47; from instr to assoc prof, 47-66, PROF PHYSIOL, MED SCH, VANDERBILT UNIV, 66- Concurrent Pos: Guggenheim fel, 60-61; vis prof, Karolinska Inst, Sweden, 67; mem adv comt nutrit, US Army Med Res & Develop Command, 56- Mem: AAAS; Am Physiol Soc; Am Heart Asn; Geront Soc; Am Inst Nutrit; Am Fedn Clin Res; Am Gastroenterol Asn. Res: Complete parenteral alimentation; clearing factor lipase; lipid transport and metabolism; gastrointestinal physiology. Mailing Add: Dept of Physiol Vanderbilt Univ Med Sch Nashville TN 37203

MENGE, ALAN C, b Marengo, Ill, Apr 8, 34; m 57; c 4. REPRODUCTIVE PHYSIOLOGY. Educ: Univ Ill, BSc, 56; Univ Wis, MSc, 58, PhD(endocrinol), 61. Prof Exp: Asst prof animal sci, Rutgers Univ, 61-65, assoc prof, 65-67; ASSOC PROF REPROD BIOL, MED CTR, UNIV MICH, ANN ARBOR, 67- Concurrent Pos: Mem, Int Coord Comt Immunol of Reprod USA, 58. Mem: Soc Study Reproduction; Am Soc Animal Sci; Brit Soc Study Fertil; Sigma Xi; Soc Exp Biol & Med. Res: Problems related to endocrine and immunologic causes of infertility. Mailing Add: Dept of Obstet & Gynec Univ of Mich Med Ctr Ann Arbor MI 48104

MENGE, BRUCE ALLAN, b Minneapolis, Minn, Oct 5, 43; m 71. ECOLOGY. Educ: Univ Minn, Minneapolis, BA, 65; Univ Wash, PhD(ecol), 70. Prof Exp: Ford Found fel, Univ Calif, Santa Barbara, 70-71; ASST PROF BIOL, UNIV MASS, BOSTON, 71- Mem: Ecol Soc Am; Am Soc Nat; Am Soc Zool; Soc Study Evolution. Res: Population and community ecology in the marine environment; effect of biological interactions; life history strategies. Mailing Add: Dept of Biol Univ of Mass Boston MA 02116

MENGE, HENRY, b Lancaster, Pa, July 29, 08; m 45; c 1. ANIMAL NUTRITION. Educ: Pa State Col, BS, 45; Univ Md, MS, 50, PhD, 52. Prof Exp: Owner poultry farm, 28-42 & Wene Hatchery, 45; mem staff, Soil Conserv Serv, USDA, 45-48; asst, Univ Md, 48-52; RES ANIMAL SCIENTIST, AGR RES CTR, US DEPT AGR, 52- Mem: Poultry Sci Asn; Am Inst Nutrit. Res: Unidentified growth factors; vitamin B12; amino acids; fatty acid metabolism; cholesterol. Mailing Add: Agr Res Ctr US Dept of Agr Beltsville MD 20705

MENGEBIER, WILLIAM LOUIS, b New York, NY, Dec 2, 21; m; c 3. PHYSIOLOGY. Educ: The Citadel, BS, 43; Oberlin Col, MA, 49; Univ Tenn, PhD(zool), 53. Prof Exp: Instr chem & biol, The Citadel, 46-49, asst prof, 49-54; prof biol, Madison Col, Va, 54-67; PROF BIOL, BRIDGEWATER COL, 67- Res: Effects of anoxia on cellular respiration; relative survival times of hibernators and mammals to anoxia; cellular physiology; effect of vertebrate hormones on invertebrates. Mailing Add: Dept of Biol Bridgewater Col Bridgewater VA 22812

MENGEL, CHARLES E, b Baltimore, Md, Nov 29, 31; m; c 5. MEDICINE, HEMATOLOGY. Educ: Lafayette Col, AB, 53; Johns Hopkins Univ, MD, 57. Prof Exp: Intern, Osler Ward Med Serv, Johns Hopkins Hosp, 57-58; asst resident med, Duke Univ Hosp, 58-59; clin assoc exp pharmacol & therapeut, Leukemia Serv, Nat Cancer Inst, 59-61; chief resident & instr med, Duke Univ Hosp, 61-62, assoc med, sr staff & fac, Med Ctr & dir med emergency room facilities, 62-65, organizer & dir med educ, Lincoln Hosp, 62-65; assoc prof med, Ohio State Univ & dir div hemat & oncol, Univ Hosp, 65-70; PROF MED & CHMN DEPT, UNIV MOCOLUMBIA, 70- Concurrent Pos: Markle scholar acad med, 63-68. Mem: Am Fedn Clin Res; Am Soc Hemat; Am Asn Cancer Res; NY Acad Sci. Res: Nonimmune hemolytic mechanisms; abnormalities of tryptophan metabolism; adult and childhood leukemia; medical education programs, particularly bridging of preclinical and clinical areas and student and fellow research training. Mailing Add: Dept of Med Univ of Mo Columbia MO 65202

MENGEL, JOHN GEIST, b Lebanon, Pa, Apr 30, 45. ASTRONOMY. Educ: Franklin & Marshall Col, AB, 67; Harvard Univ, MA, 69; Yale Univ, PhD(astron), 72. Prof Exp: Res assoc, 72-74, RES STAFF ASTRONR & LECTR ASTRON, YALE UNIV, 74- Mem: Am Astron Soc. Res: Stellar evolution. Mailing Add: Yale Univ Observ Box 2023 Yale Sta New Haven CT 06520

MENGEL, ROBERT MORROW, b Glenview, Ky, Aug 19, 21; m 63. ZOOLOGY. Educ: Cornell Univ, BS, 47; Univ Mich, MA, 50, PhD, 58. Prof Exp: Ornith bibliographer, univ libr & res assoc, 53-65, lectr zool, 65-67, assoc prof, 67-71, instr, 58-65, assoc cur birds, 67-68, PROF SYST & ECOL, UNIV KANS, 72-, CUR BIRDS, MUS NAT HIST, 69- Concurrent Pos: Ed, The Auk, Am Ornith Union, 63-67, ed, monogr, 69. Mem: Fel Am Ornith Union; Wilson Ornith Soc; Cooper Ornith Soc; Soc Systs Zool; Soc Study Evolution. Res: Distribution, ecology, evolution, systematics and paleontology of birds and mammals; bibliography of ornithology. Mailing Add: Rte 4 Lawrence KS 66044

MENGELING, WILLIAM LLOYD, b Elgin, Ill, Apr 1, 33; m 58; c 2. VETERINARY MICROBIOLOGY. Educ: Kans State Univ, BS, 58, DVM, 60; Iowa State Univ, MS, 66, PhD(microbiol, biochem), 69; Am Col Vet Microbiol; Dipl. Prof Exp: Vet, St Francis Animal Hosp, Albuquerque, NMex, 60-61; RES VET, NAT ANIMAL DIS LAB, 61- Mem: Am Vet Med Asn; Animal Health Asn. Res: Respiratory diseases of swine; virology. Mailing Add: Virol Dept Nat Animal Dis Lab Ames IA 50010

MENGENHAUSER, JAMES VERNON, b Armour, SDak, Oct 12, 33; m 60; c 1. PHYSICAL CHEMISTRY, PETROLEUM CHEMISTRY. Educ: Univ SDak, BA, 54; NMex State Univ, MS, 64; Univ Colo, PhD(chem), 69. Prof Exp: Chemist, Ames Lab, AEC, 54-55; chemist, El Paso Natural Gas Co, 59-61; CHEMIST, US ARMY MOBILITY EQUIP RES & DEVELOP COMMAND, FT BELVOIR, 68- Mem: Am Chem Soc; Sigma Xi. Res: Analysis of oily wastes in water; nuclear magnetic resonance; pyrolysis chromatography; solubility of hydrocarbon fuels in water; turbulent drag reduction. Mailing Add: 8905 Camfield Dr Alexandria VA 22308

MENGER, EVA L, b South Bend, Ind, Feb 18, 43; m 64; c 2. CHEMICAL PHYSICS. Educ: Carleton Col, BA, 64; Harvard Univ, MA, 65, PhD(chem), 68. Prof Exp: Asst ed, Accts Chem Res, 69-74; ASST PROF CHEM, UNIV CALIF, SANTA CRUZ, 74- Mem: Am Chem Soc; Sigma Xi. Res: Theoretical and experimental studies of solution phase reactions; special interest focuses on nonsteady state effects in diffusion controlled processes and excited state reactions; ultrafast processes monitored by picosecond laser techniques. Mailing Add: Natural Sci II Univ of Calif Santa Cruz CA 95064

MENGER, FRED M, b South Bend, Ind, Dec 13, 37; m 62. ORGANIC CHEMISTRY. Educ: Johns Hopkins Univ, AB, 58; Univ Wis, PhD(chem), 63. Prof Exp: NIH fel, Northwestern Univ, 64-65; asst prof, 65-69, ASSOC PROF CHEM, EMORY UNIV, 69- Honors & Awards: Camille & Henry Dreyfus teacher-scholar award, 70; NIH career develop award, 70-75. Res: Bio-organic and physical organic chemistry; interfaces and colloidal systems; reaction mechanisms; enzyme models. Mailing Add: Dept of Chem Emory Univ Atlanta GA 30322

MENGERT, WILLIAM FELIX, b Wash, DC, Nov 13, 99; m 24, 71; c 2. OBSTETRICS & GYNECOLOGY. Educ: Haverford Col, SB, 21; Johns Hopkins Univ, MD, 27. Prof Exp: Instr chem & math, Gallaudet Col, 21-23; res fel obstet & gynec, Univ Iowa, 28-32, from asst prof to assoc prof, 34-43; res fel, Gynecean Hosp Inst, Pa, 32-34; prof & chmn dept, Southwestern Med Sch, Univ Tex, 43-55; prof & head dept, 55-69, EMER PROF OBSTET & GYNEC, UNIV ILL COL MED, 69- Concurrent Pos: Chmn dept obstet & gynec, Parkland Hosp, Dallas, 43-55; obstetrician & gynecologist-in-chief, Res & Educ Hosps, 55-68; mem study sect human embryol, NIH. Mem: Fel Am Gynec Soc; Am Med Asn; fel Am Asn Obstet & Gynec (asst secy, 47-50, secy, 50-53, pres, 57-58); fel Am Col Obstet & Gynec (pres, 54-55). Res: Obstetric pelvic capacity; toxemia of pregnancy; correlation of radiographic pelvimetries with subsequent course of labor; production of human placental abruption by digital compression vena cava; proved maternal red blood cells cross human placental barrier; establishment of mechanisms uterine support of fresh human cadavers. Mailing Add: 333 S East Ave Apt 307 Oak Park IL 60302

MENGES, ROBERT M, agronomy, see 12th edition

MENGOLI, HENRY FRANCIS, b Plymouth, Mass, June 8, 28; m 54; c 2. IMMUNOLOGY. Educ: Boston Univ, AB, 50; Cath Univ, MS, 53, PhD(biol), 57. Prof Exp: Bacteriologist, Clin Ctr, NIH, 57-59, res biologist, Peripheral Blood Proj, Collab Area, Diag Res Br, Nat Cancer Inst, 59-63; instr, Cancer Res Lab, 63-65, ASST PROF PATH & MICROBIOL, IMMUNOL RES LAB, MED CTR, WVA UNIV, 65- Mem: NY Acad Sci; Am Chem Soc. Res: Biology and immunology of cancer; autoimmune mechanisms in human disease; properties of cell populations. Mailing Add: Dept of Path Immunol Res Lab WVa Univ Med Ctr Morgantown WV 26506

MENGUY, RENE, b Prague, Czech, Feb 4, 26; nat US; m; c 2. MEDICINE. Educ: Univ Paris, MD, 51, Univ Minn, PhD, 57. Prof Exp: Fulbright grant, Am Hosp Chicago, Ill, 51-52, fel, Mayo Clin, 52-57; from instr to asst prof surg, Med Ctr, Univ Okla, 57-58; from assoc prof to prof, Med Ctr, Univ Ky, 61-65, assoc prof physiol, 64-65; prof surg & chmn dept, Univ Chicago, 65-71; PROF SURG, SCH MED & DENT, UNIV ROCHESTER, 71-; SURGEON-IN-CHIEF, GENESEE HOSP, 71- Concurrent Pos: Markle scholar, 58; asst chief surgeon Vet Admin Hosp, Okla, 59-61. Mem: Soc Exp Biol & Med; Am Gastroenterol Asn; AMA; Am Fedn Clin Res; Am Asn Cancer Res. Res: Surgery; gastroenterological surgery; experimental biology. Mailing Add: Genesee Hosp 224 Alexander St Rochester NY 14607

MENHINICK, EDWARD FULTON, b Cambridge, Mass, May 18, 35; m 61; c 3. ECOLOGY, PHYSIOLOGY. Educ: Emory Univ, BA, 57; Cornell Univ, MS, 60; Univ Ga, PhD(zool), 62-63; fel, Health Physics Div, Oak Rdige Nat Lab, 63-65; asst prof zool, 65-71, ASSOC PROF BIOL, UNIV NC, CHARLOTTE, 71- Concurrent Pos: Consult, Duke Power Co. Mem: Ecol Soc Am; Entom Soc Am. Res: Water pollution; radiation ecology; environmental physiology; statistical analysis of density, diversity and energy flow. Mailing Add: Dept of Biol Univ of NC Charlotte NC 28213

MENHUSEN, BERNADETTE REMUS, b Concordia, Kans, Oct 10, 27; m 47; c 1. BOTANY, ECOLOGY. Educ: Kans State Teachers Col, BS, 57; Univ Kans, PhD(bot), 63. Prof Exp: Teacher pub sch, Kans, 45-58; instr biol, Christian Col, 63-64; assoc prof, Kans State Teachers Col, 64-73; consult-writer, Curriculum Develop, AAAS, Washington, DC, 73-74; CONSULT-WRITER, CURRICULUM DEVELOP, BIOL SCI CURRICULUM STUDY, UNIV COLO, 74- Concurrent Pos: Adj prof, Univ Colo, 75- Mem: Am Bryol & Lichenological Soc; Ecol Soc Am; Am Soc Plant Taxon; Nat Sci Teachers Asn. Res: Taxonomy and morphology of vascular plants and bryophytes; science education. Mailing Add: Biol Sci Curriculum Study Univ of Colo Box 930 Boulder CO 80302

MENIUS, ARTHUR CLAYTON, JR, b Salisbury, NC, Apr 30, 16; m 46. PHYSICS. Educ: Catawba Col, AB, 37; Univ NC, PhD(physics), 42. Hon Degrees: DSc, Catawba Col, 68. Prof Exp: From asst prof to prof physics, Clemson Col, 42-49; sr physicist, Appl Physics Lab, Johns Hopkins Univ, 44-46; prof physics, 49-56, prof & head dept, 56-60, DEAN SCH PHYS & MATH SCI, NC STATE UNIV, 60- Concurrent Pos: Chmn NC Atomic Energy Comt, 60-71; NC State Univ Coun rep, Oak Ridge Assoc Univs, 63-69, mem bd, 69-; mem exec comt, NC State Univ Courn rep, Oak Ridge Inst Nuclear Studies; mem permanent adv comt, Southern Interstate Nuclear Bd. Mem: Fel Am Phys Soc; assoc fel Am Inst Aeronaut & Astronaut; Am Nuclear Soc. Res: Nuclear and space physics; laser technology. Mailing Add: 541 Hertford Street Raleigh NC 27609

MENKART, JOHN, b Prague, Aug 20, 22; m; c 4. COSMETIC CHEMISTRY, RESEARCH ADMINISTRATION. Educ: Univ Leeds, BSc, 44, PhD(textile chem), 46. Prof Exp: Res chemist, Denham & Hargrave, Ltd, 46-48; sci liaison officer, Int Wool Secretariat, Eng, 48-50; res chemist, Patons Baldwins, Ltd, 50-53; asst dir res, Textile Res Inst, Princeton Univ, 54-58; group leader, Harris Res Labs Inc, Gillette Co, 58-65, from asst dir to assoc dir, 65-67, vpres, Gillette Res Inst, 67-68, pres, 68-71; VPRES TECHNOL, CLAIROL, INC, 71- Mem: Soc Cosmetic Chem; Am Chem Soc; Fiber Soc; fel Brit Textile Inst. Res: Chemistry and physical properties of fibers; formulation and properties of topical products. Mailing Add: Clairol Inc 2 Blachley Rd Stamford CT 06902

MENKE, ANDREW GIEDRIUS, b Erzvilkas, Lithuania, Aug 24, 44; US citizen; m 71; c 1. ORGANOMETALLIC CHEMISTRY. Educ: Wayne State Univ, BS, 67; Univ Toledo, MS, 73. Prof Exp: Asst chem, Wayne State Univ, 63-67; sr res chemist, Libbey-Owens-Ford Co, Ohio, 68-75; DIR RES & DEVELOP, ANGLASS INDUST INC, 76- Mem: Am Chem Soc; Am Vacuum Soc. Res: Selective coating for solar collectors, including work on organotin compounds used to produce conductive tin oxide by pyrolysis of same; properties of these compounds and other related organometallic compounds. Mailing Add: 11733 Monogram Ave Granada Hills CA 91344

MENKE, ARNOLD STEPHEN ERNST, b Glendale, Calif, Nov 22, 34; m 61; c 1. ENTOMOLOGY. Educ: Univ Calif, Berkeley, BS, 57; Univ Calif, Davis, MS, 59, PhD(entom), 65. Prof Exp: Asst res entomologist, Univ Calif, Davis, 65-67; RES ENTOMOLOGIST, US DEPT AGR, 68- Concurrent Pos: Res assoc, Los Angeles County Mus Natural Hist, 66- Mem: Entom Soc Can. Res: Taxonomic research in the Sphecidae and Belostomatidae. Mailing Add: Syst Entomol Lab USDA c/o US Nat Mus Washington DC 20560

MENKE, JOHN ROGER, b New York, NY, Apr 16, 19; m 45; c 2. PHYSICS. Educ: Columbia Univ, BS, 43. Prof Exp: Mem sr sci staff, Sam Labs, Columbia Univ, 42-46 & Argonne Nat Labs, 46; prin physicist, Oak Ridge Nat Labs, 46-48; pres, Nuclear Develop Assocs, Inc, 48-55; pres, Nuclear Develop Corp Am, 55-61; DIR, UNITED NUCLEAR CORP, 61- Concurrent Pos: Dir, Hudson Inst, Verde Explor & Standard Shares, Inc. Mem: Am Soc Mech Eng; Am Nuclear Soc. Res: Lubrication; gaseous and neutron diffusion; heat transfer; nuclear reactors; exploration geophysics. Mailing Add: United Nuclear Corp 101 Executive Blvd Elmsford NY 10523

MENKES, JOHN H, b Vienna, Austria, Dec 20, 28; US citizen; m 57; c 3. PEDIATRICS, NEUROLOGY. Educ: Univ Southern Calif, AB, 47, MS, 51; Johns Hopkins Univ, MD, 52. Prof Exp: Intern & asst resident pediat, Boston Children's Hosp, Mass, 52-54; asst prof neurol med & assoc prof pediat, Johns Hopkins Univ, 60-66; prof pediat & neurol, 66-70; CLIN PROF PEDIAT, NEUROL & PSYCHIAT, UNIV CALIF, LOS ANGELES, 70-; CHIEF NEUROL, NEUROCHEM LAB, BRENTWOOD VET ADMIN HOSP, 70- Concurrent Pos: Fel pediat neurol, NY, 57-60; Joseph P Kennedy, Jr scholar ment retardation, 60-66. Mem: AAAS; fel Am Acad Neurol; fel Am Neurol Asn; Am Pediat Soc; Soc Pediat Res. Res: Metabolic disorders of the nervous system; child neurology. Mailing Add: Dept of Neurol Brentwood Vet Admin Hosp West Los Angeles CA 90073

MENKES, JOSHUA, b Vienna, Austria, Aug 14, 25; US citizen; m 61; c 2. APPLIED MATHEMATICS. Educ: Polytech Inst Brooklyn, BS, 53, MS, 54; Univ Mich, Ann Arbor, PhD(appl math), 56. Prof Exp: Analyst physics, Jet Propulsion Lab, Calif Inst Technol, 57-61; sr staff mem systs anal, Inst Defense Anal, 62-70; prof eng, Univ Colo, Boulder, 70-75; DIR DIV SYSTS ANAL, NSF, 75- Res: Resource allocation and policy analysis. Mailing Add: 7510 Alaska Ave NW Washington DC 20012

MENN, JULIUS JOEL, b Free City of Danzig, Feb 20, 29; m 52; c 3. ENTOMOLOGY, ENVIRONMENTAL TOXICOLOGY. Educ: Univ Calif, BS, 53, MS, 54, PhD(toxicol), 58. Prof Exp: Res asst, Univ Calif, 53-57; head entom & insecticide biochem sect, Agr Res Ctr, 57-67; sr sect mgr, Biochem & Entom, 67-69, sr sect mgr, Insecticide Res & Biochem, 69-74, MGR BIOCHEM DEPT, STAUFFER CHEM CO, 74- Concurrent Pos: NSF lectr, Univ Calif, Davis, 65; invited speaker spec conf fate of pesticides in environ, Nat Acad Sci-Nat Res Coun, 71; mem proj, Forms & Mechanisms by which Pesticides are transported in the environ, US/USSR, 74- Mem: AAAS; Am Chem Soc; Entom Soc Am; NY Acad Sci; Soc Toxicol. Res: Metabolism and mode of action of pesticides; development of pesticides. Mailing Add: Mountain View Res Ctr Stauffer Chem Co PO Box 760 Mountain View CA 94042

MENNE, THOMAS JOSEPH, b St Louis, Mo, May 13, 34; m 56; c 1. THEORETICAL PHYSICS. Educ: St Louis Univ, BS, 56; Univ Calif, Los Angeles, MS, 58, PhD(physics), 63. Prof Exp: Mem tech staff, Hughes Aircraft Co, 56-59; lectr physics, Loyola Univ Los Angeles, 60-62; res scientist, 62-65, assoc scientist, 65-69, scientist, 69-70, MGR RES, McDONNELL DOUGLAS CORP, 70- Concurrent Pos: Lectr, Washington Univ, 65-67. Mem: Am Phys Soc. Res: Theoretical research on laser dynamics, electron spin resonance spectroscopy, crystal field theory and paramagnetic ion-lattice phonon interactions; research direction in chemical and molecular laser development. Mailing Add: McDonnell Douglas Corp PO Box 516 Dept 223 St Louis MO 63166

MENNEAR, JOHN HARTLEY, b Flint, Mich, Apr 25, 35; m 56; c 4. PHARMACOLOGY. Educ: Ferris Inst, BS, 57; Purdue Univ, MS, 60, PhD(pharmacol), 62. Prof Exp: Pharmacologist, Hazleton Labs, Inc, 62-63; Pitman-Moore Div, Dow Chem Co, 63-66; from asst prof to assoc prof, 66-72, PROF TOXICOL, PURDUE UNIV, 72- Mem: Am Soc Pharmacol & Exp Therapeut; Soc Toxicol. Res: Toxicology; applied pharmacology. Mailing Add: Dept of Pharmacol & Toxicol Purdue Univ Lafayette IN 47907

MENNEGA, AALDERT, b Assen, Netherlands, July 3, 30; US citizen; m 58; c 6. ANATOMY, PHYSIOLOGY. Educ: Calvin Col, AB, 57; Mich State Univ, MA, 60, PhD(anat, zool), 64. Prof Exp: Lab technologist, Grand Rapids Osteop Hosp, Mich, 58-59; med technologist, E W Sparrow Hosp, Lansing, 59-64; from asst to assoc prof, 64-75, PROF BIOL, DORDT COL, 75-, CHMN DEPT, 70- Mem: Am Sci Affil. Res: Respiratory system, especially of birds; histology and gross anatomy of birds, mammals and snakes. Mailing Add: Dept of Biol Dordt Col Sioux Center IA 51250

MENNELL, JOHN MCMILLAN, b London, Eng, Jan 21, 16; US citizen; m 53. PHYSICAL MEDICINE. Educ: Cambridge Univ, BA, 36; dipl med radiol & electrology, 42, MB, BCh, 53; Am Bd Phys Med & Rehab, dipl, 63. Prof Exp: Staff physician, Vet Admin Hosp, Long Beach, Calif, 61-65; chief phys med & rehab, Philadelphia Gen Hosp, 65-69; chief phys med & rehab, 70-74, CHIEF REHAB MED SERV, VET ADMIN HOSP, 74-; ASSOC CLIN PROF PHYS MED & REHAB, SCH MED, UNIV CALIF, DAVIS, 72-; CLIN PROF AMBULATORY CARE & ENVIRON MED, SAN FRANCISCO COL PODIATRIC MED, 72- Concurrent Pos: Vol attend, Los Angeles County Hosp, 65; assoc prof, Univ Pa, 65-71; asst clin prof, Univ Southern Calif, 69-; consult, Letterman Gen Army Hosp, San Francisco, Calif, 75- Mem: Am Cong Rehab Med; Am Acad Phys Med & Rehab; Brit Asn Phys Med; Royal Soc Med; Worshipful Soc Apothecaries London. Mailing Add: Vet Admin Hosp 150 Muir Rd Martinez CA 94553

MENNINGA, CLARENCE, b Otley, Iowa, Apr 6, 28; m 49; c 7. NUCLEAR CHEMISTRY, PHYSICAL CHEMISTRY. Educ: Calvin Col, BA, 49; Western Mich Univ, MA, 59; Purdue Univ, PhD(chem), 66. Prof Exp: Chemist, Maytag Co, 50-56; teacher, Grand Rapids Christian High Sch, 56-61; chemist, Lawrence Radiation Lab, 66-67; ASST PROF GEOL, CALVIN COL, 67- Mem: Geochem Soc; Am Sci Affil; Meteoritical Soc. Res: Composition of meteorites; geochemistry. Mailing Add: Dept of Geol Calvin Col Grand Rapids MI 49506

MENNINGER, FLORIAN FRANCIS, JR, b Brooklyn, NY, Aug 18, 37; m 60; c 3. IMMUNOLOGY, IMMUNOCHEMISTRY. Educ: St Michael's Col, BA, 59; St John's Univ, NY, MS, 62, PhD(microbiol), 65. Prof Exp: Res assoc bact & immunol, State Univ NY Buffalo, 64-67; RES IMMUNOLOGIST, MASON RES INST, 67- Concurrent Pos: Res assoc, Urol Res Lab & head immunol & biochem div, Millard Fillmore Hosp Res Inst, Buffalo, NY, 65-66; instr, Clark Univ, 69. Mem: AAAS; Soc Indust Microbiol; NY Acad Sci; Am Pub Health Asn. Res: Applied immunology and immunochemistry; radioimmunoassay of protein hormones; tumor immunology; cancer specific antigens; non-specific enhancement of immune response; immunosuppression by anti-cancer agents. Mailing Add: Immunochem Lab Dept Immunol Mason Res Inst Worcester MA 01608

MENNINGER, JOHN ROBERT, b Columbus, Ohio, July 29, 35; m 60; c 2. MOLECULAR BIOLOGY. Educ: Harvard Univ, AB, 57, PhD(biochem), 64. Prof Exp: Whitney Found vis res fel molecular genetics, Med Res Coun Lab Molecular Biol, Cambridge Univ, 63-66; asst prof biol, Univ Ore, 66-72, res assoc molecular biol, Inst Molecular Biol, 66-72; ASSOC PROF ZOOL, UNIV IOWA, 72- Concurrent Pos: Prog dir, Cellular & Molecular Biol Training Grant, Univ Iowa Grad Col, 75- Mem: Am Soc Biol Chemists. Res: Mechanism and control of information transfer in biological systems; protein biosynthesis; peptidyl-tRNA metabolism; mechanisms of cellular aging; ion transport across bacterial membranes. Mailing Add: Dept of Zool Univ of Iowa Iowa City IA 52242

MENNINGER, KARL AUGUSTUS, b Topeka, Kans, July 22, 93; m 16, 41; c 4. PSYCHIATRY. Educ: Univ Wis, AB, 14, MS, 15; Harvard Univ, MD, 17. Hon Degrees: DSc, Washburn Univ, 49, Univ Wis, 65; LHD, Park Col, 55, St Benedict's Col, Kans, 63; LLD, Jefferson Med Col, 56, Parsons Col, 60, Kans State Univ, 62, Baker Univ, 65. Prof Exp: Intern, Kansas City Gen Hosp, Mo, 17-18; asst neuropath, Harvard Med Sch, 18-20; prof ment hyg, criminol & abnormal psychol, Washburn Col, 23-40; dean, Menninger Sch Psychiat, 46-70, dir educ, Menninger Found, 46-70, chmn bd trustees, 54-70, mem educ comt, 67-70, chief staff, Menninger Clin, 25-46, 52-70; PROF MED, SCH MED, UNIV KANS, 70- Concurrent Pos: Asst, Med Col, Tufts Univ, 18-19; from asst to instr, Boston Psychopathic Hosp, 19; ed-in-chief, Bull, Menninger Clin, 20-; col asst, Topeka State Hosp, Kans, 20, chief consult, 48-; adv, Surg Gen, US Army, 45; consult, Fed Bur Prisons, Off Voc Rehab, Dept Health, Educ & Welfare, 48-; consult, Vet Admin Hosp, Topeka, 48-; mgr, Winter Vet Admin Hosp, 45-48, chmn, Dean's Comt & sr consult, 48-55; consult, Forbes AFB Hosp, 58- & Stone-Brandel Ctr, Chicago; dir, Topeka Inst Psychoanal, 60-; mem, Adv Comt, Int Surv Correctional Res & Pract, Calif, 60-; prof-at-large, Univ Kans; neuropsychiatrist, Stormont-Vail Hosp, Topeka; vis prof, Med Sch, Univ Cincinnati, trustee, Albert Deutsch Mem Found, 61; trustee, Aspen Inst Humanistic Studies, 61-64; consult, Inst Mgt Bd Social Welfare, State Kans; consult, Res Inst, Boston Univ, Europ Educ Ctr, Asn Migros Schs, Zurich; mem, Nat Cong Am Indian; mem, Kans Bd, John F Kennedy Mem Libr, Bd Overseers, Lemberg Ctr Study Violence, Brandeis Univ & Spec Comt Psychiat, Off Sci Res & Develop. Honors & Awards: Distinguished Serv Award, Am Psychiat Asn, 65. Mem: Fel AMA; fel Am Psychiat Asn; fel Am Psychoanal Asn (pres, 41-43); fel Am Col Physicians; Am Orthopsychiat Asn (secy, 26, pres, 27); hon fel Am Asn Suicidology. Res: Influenza and mental diseases; psychological factors in somatic disease; suicide; hypertension; industrial and military psychiatry; psychiatric education; criminology; penology; religion. Mailing Add: Sch of Med Univ of Kans Kansas City KS 66103

MENNITT, PHILIP GARY, b Battle Creek, Mich, Mar 29, 37; m 61; c 4. PHYSICAL CHEMISTRY. Educ: Providence Col, BS, 58; Mass Inst Technol, PhD(phys chem), 62. Prof Exp: From instr to asst prof, 64-73, ASSOC PROF CHEM, BROOKLYN COL, 73- Mem: Am Chem Soc. Res: Nuclear magnetic resonance spectroscopy. Mailing Add: Dept of Chem Brooklyn Col Brooklyn NY 11210

MENON, ARAVINDAKSHAN I, b Peringottukara, Kerala, India, Jan 9, 33; m; c 2. BIOCHEMISTRY. Educ: Univ Bombay, BSc, 53, MSc, 55, PhD(biochem), 60. Prof Exp: Sr fel, Coun Sci Res & Indust, 60-61; res assoc biochem, Columbia Univ, 61-64 & McGill Univ, 64-66; ASST PROF BIOCHEM, UNIV TORONTO, 67- Mem: AAAS; Am Chem Soc; Can Biochem Soc; Brit Biochem Soc; Am Soc Photobiol. Res: Nucleic acid synthesis; enzymology; pigmentation; photobiology. Mailing Add: Med Sci Bldg Univ of Toronto Toronto ON Can

MENON, MANAVAZHI VIJAYA KRISHNA, b Kerala, India, Apr 25, 28. MATHEMATICS, STATISTICS. Educ: Univ Madras, BA, 48, MSc, 50; Ohio State Univ, PhD(math statist), 59; Stanford Univ, MS, 73. Prof Exp: Res assoc statist, NC State Col, 59-60; mem res staff, Res Lab, IBM Corp, Calif, 60-65; vis mem res staff, Math Res Ctr, Univ Wis, Madison, 65-66; from assoc prof to prof statist, Univ Mo-Columbia, 66-74; MATHEMATICIAN, OFF NAVAL RES, 74- Mem: Am Math Soc; Inst Math Statist. Res: Probability theory; matrix problems. Mailing Add: Off of Naval Res 536 S Clark St Chicago IL 60605

MENON, MANCHERY PRABHAKARA, b India, Aug 31, 28; m 55; c 2. NUCLEAR CHEMISTRY, RADIOCHEMISTRY. Educ: Univ Madras, BSc, 49; Agra Univ, MSc, 53; Univ Ark, PhD(nuclear chem, radiochem), 63. Prof Exp: Teacher sci, St Mary's CGH Sch, Kerala, India, 49-50; HS Vadayar, Kerala, 50-51; lectr chem, NSS Col, 53-55; lectr & head dept, Moulmein Col, Rangoon, 55-59; res asst, Univ Ark, 59-63; res assoc nuclear sci, Mass Inst Technol, 63-64; asst prof chem, univ & asst res chemist, Activation Anal Lab, Tex A&M Univ, 64-67; assoc prof, 67-71, PROF CHEM, SAVANNAH STATE COL, 71- Concurrent Pos: Tex A&M Univ Res Coun res grant nuclear fission, 65-66. Res: Nuclear organic chemistry; mass and charge distributions in nuclear fission, energetics of nuclear fission; natural radioactivity as applied to geochronology; fall-out from nuclear detonations; activation analysis and nuclear spectroscopy. Mailing Add: 903 E 33rd St E Savannah GA 31404

MENON, MAYA DEVI, b Trichur, India. INSECT ENDOCRINOLOGY. Educ: Presidency Col, Madras, BSc, 55, MSc, 57; Univ Calif, Berkeley, PhD(zool), 69. Prof Exp: Instr cytol, Govt Arts Col, Coimbatore, India, 58-60; teaching asst introd biol, Univ Calif, Berkeley, 65-66; NIH fel insect hormones & pheromones, Simon Fraser Univ, 70-72; ASST PROF PHYSIOL & ENDOCRINOL, BARNARD COL, COLUMBIA UNIV, 72- Mem: Am Soc Zoologists. Res: Hormone pheromone relationships in insects; mechanism of action of juvenile hormone by studies of receptor sites for the hormone action of target organs; radiation effects on endocrines, pheromones and reproduction in insects. Mailing Add: Dept of Biol Barnard Col Columbia Univ New York NY 10027

MENOTTI, AMEL ROMEO, b Carrara, Italy, Oct 7, 13; nat US; m; c 4. CHEMISTRY. Educ: Antioch Col, BS, 37; Ohio State Univ, PhD, 40. Prof Exp: Microchemist, AMA, Ill, 40-43; res dir, Bristol Labs, 43-46, vpres & sci dir, 46-74; RETIRED. Concurrent Pos: Mem Gordon Res Conf & Nat Comt Cancer Chemother. Mem: AAAS; Am Chem Soc; Am Pharmaceut Asn; Asn Res Dirs; NY Acad Sci. Res: Pharmaceutical industry; research administration. Mailing Add: Meadow Dr Fayetteville NY 13066

MENSAH, PATRICIA LUCAS, b Washington, DC, Feb 7, 48. NEUROANATOMY, NEUROSCIENCES. Educ: Howard Univ, BS, 70; Univ Calif, Irvine, PhD(biol sci),

74. Prof Exp: Psychologist neuropsychol, NIMH, 70; fel psychiat, State Univ NY Stony Brook, 74-75; instr neuroanat, Univ Calif, Irvine, 75, NIMH fel psychobiol, 75-76; ASST PROF ANAT, UNIV SOUTHERN CALIF, 76- Res: Detailed neuroanatomical analyses of projection patterns within and between the caudate nuclei of the mammalian striatum; correlations between ultrastructure and function using the synaptic surface of the frog motoneuron as a model system. Mailing Add: Dept of Anat Sch of Med Univ of Southern Calif Los Angeles CA 90033

MENSER, HARRY ALVIN, JR, b Pittsburgh, Pa, Dec 30, 30; m 60; c 2. PLANT PHYSIOLOGY. Educ: Univ Delaware, BS, 54; Univ Md, MS, 59, PhD(agron, bot), 62. Prof Exp: Asst county agent, Agr Exten Serv, Univ Md, 55-57, asst agron, Univ Md, 57-59; res technician, Tobacco Lab, Plant Genetics & Germplasm Inst, 60-63, RES PLANT PHYSIOLOGIST, AGR RES SERV, US DEPT AGR, 63- Mem: Am Soc Plant Physiol; Am Inst Biol Sci; Am Soc Agron; Crop Sci Soc Am. Res: Analysis of heavy metals as environmental hazards; land recycling of municipal wastes; revegetation of lands disturbed by surface mining. Mailing Add: Soil & Water Unit Agr Res Serv US Dept of Agr Agr Sci Bldg Morgantown WV 26506

MENSING, RICHARD WALTER, b Hackensack, NJ, Sept 20, 36; m 57; c 3. STATISTICS. Educ: Valparaiso Univ, BS, 60; Iowa State Univ, MS, 65, PhD(statist), 68. Prof Exp: Engr, Gen Elec Co, 60-62; from instr to asst prof, 66-74, ASSOC PROF STATIST, IOWA STATE UNIV, 74- Mem: Inst Math Statist; Am Statist Asn. Res: Engineering and industrial statistics; applied probability. Mailing Add: Dept of Statist Iowa State Univ Ames IA 50010

MENSOIAN, MICHAEL GEORGE, JR, b Providence, RI, June 24, 29; m 69; c 1. GEOGRAPHY. Educ: Clark Univ, AB, 49; Boston Univ, EdM, 51; Worcester State Col, EdM, 56; Boston Univ, dipl soc sci, 59, MA, 60; Univ Conn, PhD(geog educ), 62; Univ Pittsburgh, dipl Latin Am studies, 70. Prof Exp: Teacher pub schs, Worcester, Mass, 50-54; asst prof educ & prin, Campus Lab Schs, Fitchburg State Col, 54-56; assoc prof geog, 56-67, actg chmn dept, 61-62, PROF GEOG, BOSTON STATE COL, 67-, CHMN DEPT, 70- Concurrent Pos: Consult & ed geog, Doubleday & Co, Inc, 61-62; dir grad progs geog & earth sci, Boston State Col, 74- & dir grad progs urban studies & planning, 74- Res: Improvement of geography textbooks on the elementary and secondary levels; population migration; economic-political problems; Latin America and Middle East and Northern Africa with emphasis on urban areas. Mailing Add: Dept of Geog Boston State Col 625 Huntington Ave Boston MA 02115

MENTE, GLEN ALLEN, b Wheatland, Iowa, Feb 25, 38; m 59; c 2. ANIMAL NUTRITION. Educ: Iowa State Univ, BS, 61, MS, 63. Prof Exp: Nutritionist, Farmers Coop Soc, 63-64; nutritionist, 64-67, mgr, 67-74, VPRES NUTRIT & PROD DEVELOP, KENT FEEDS INC, 74- Mem: Am Soc Animal Sci; Agr Res Inst; Nutrit Coun Am Feed Mgrs; Nutrit Feed Ingredients Asn. Res: Continuous evaluation of nutrient requirements for swine, beef, dairy, poultry, turkeys, lambs, horses, fish, pets and miscellaneous animals; also genetic research with swine and beef. Mailing Add: Kent Feeds Inc 1600 Oregon St Muscatine IA 52761

MENTON, DAVID NORMAN, b Mankato, Minn, July 1, 38; m 61; c 2. ANATOMY, HISTOLOGY. Educ: Mankato State Col, BS, 59; Brown Univ, PhD(biol), 66. Prof Exp: Instr anat, 66-70, ASST PROF ANAT & PATH, SCH MED, WASH UNIV, 70- Mem: AAAS; Am Asn Anat; Soc Invest Dermat. Res: Fine structure of skin and its adnexa; barrier function of the stratum corneum; effects of essential fatty acid deficiency of skin. Mailing Add: Dept of Anat Wash Univ Sch of Med St Louis MO 63110

MENTONE, PAT FRANCIS, b Chicago, Ill, May 25, 42; m 70. INORGANIC CHEMISTRY, ELECTROCHEMISTRY. Educ: Col St Thomas, BS, 64; Univ Minn, PhD(inorg chem), 69. Prof Exp: Sr chemist, 69-73, plating prod mgr, 73-74, ENG SUPVR, BUCKBEE MEARS CO, 74- Mem: Am Chem Soc; Am Electroplaters Soc. Res: New product development and manufacturing engineering for microelectronic components. Mailing Add: 1756 Eleanor St Paul MN 55116

MENTZER, LOREN WILLIS, b Yates Center, Kans, Sept 12, 13; m 40; c 3. ECOLOGY. Educ: Kans State Teachers Col, BS & AB, 38, MS, 41; Univ Nebr, PhD, 50. Prof Exp: Teacher schs, Kans, 31-33, 38-40, 41-42; instr bot, biol, zool, animal ecol & genetics, Kans State Teachers Col, 45-47; prof bot, biol, genetics, ecol & conserv, St Cloud State Col, 50-57, chmn dept biol, 54-57; assoc prof biol sci, 57-70, PROF BOT, ILL STATE UNIV, 70- Mem: AAAS; Am Inst Biol Sci; Nat Sci Teachers Asn. Res: Plant and animal ecology in relation to range management and conservation. Mailing Add: Dept of Bot Ill State Univ Normal IL 61761

MENYHERT, WILLIAM R, biochemistry, medicinal chemistry, see 12th edition

MENZ, LEO JOSEPH, b Erie, Pa, Mar 7, 27; m 54; c 6. BIOPHYSICS, CRYOBIOLOGY. Educ: Gannon Col, BS, 49; St Louis Univ, MS, 52, PhD(biophys), 57. Prof Exp: Asst prof biol & dir dept, Gannon Col, 54-57; res assoc, Am Found Biol Res, 57-68; ASSOC PROF SURG, ST LOUIS UNIV, 68- Concurrent Pos: Consult, US Vet-Cochran Hosp, St Louis. Mem: Sigma Xi; Soc Cryobiol; Electron Micros Soc Am; Biophys Soc; Soc Cell Biol. Res: Effects of low temperature on biological material; freeze-drying; correlation of ultrastructural changes and function of frozen-thawed tissues, such as mammalian nerve, heart and blood cells; toxicity of cryoprotectants; nature of freezing injury. Mailing Add: Dept of Surg St Louis Univ St Louis MO 63104

MENZ, WILLIAM WOLFGANG, b Zweibruecken, Ger, Mar 2, 17; m 41; c 2. ORGANIC CHEMISTRY, CHEMICAL LITERATURE. Educ: Univ Munich, Diplom Chemiker, 39. Prof Exp: Br chief, Intel Dept, US Army Air Force & US Air Force, 46-50; ed, USPHS, 50-51; info analyst, Gen Aniline & Film Corp, 51-52 & Ethyl Corp, 52-57; sect chief tobacco res, R J Reynolds Indust, 57-70; dir tech info, 70-75, DIR RES, DAIRY RES INC, 75- Mem: Am Dairy Sci Asn; Am Chem Soc. Res: Scientific information storage and retrieval; fats and oils; dairy chemistry. Mailing Add: Dairy Res Inc 6300 River Rd Rosemont IL 60018

MENZEL, BRUCE WILLARD, b Waukesha, Wis, Aug 23, 42; m 69; c 1. ICHTHYOLOGY. Educ: Univ Wis, BS, 64; Marquette Univ, MS, 66; Cornell Univ, PhD(vert zool), 70. Prof Exp: Asst prof, 70-74, ASSOC PROF ZOOL, IOWA STATE UNIV, 74- Mem: Am Soc Ichthyol & Herpet; Am Fisheries Soc; Genetics Soc Am; Soc Syst Zool; Sigma Xi. Res: Genetics; biochemical systematics; ecology and behavior of North American freshwater fishes, amphibians and reptiles. Mailing Add: Dept of Animal Ecol Iowa State Univ Ames IA 50011

MENZEL, DANIEL B, b Cincinnati, Ohio, Sept 27, 34; m 56; c 1. NUTRITION, PHARMACOLOGY. Educ: Univ Calif, Berkeley, BS, 56, PhD(agr chem), 61. Prof Exp: Asst specialist nutrit, Univ Calif, Berkeley, 62, 65, asst prof food sci & asst biochemist, Inst Marine Sci, 62-67, asst prof food sci, Univ & asst biochemist, Agr Exp Sta, 65-67; res assoc biol, Pac Northwest Labs, Battelle Mem Inst, Wash, 67-68, mgr nutrit & food technol sect, 68-69; dir clin res; Ross Labs Div, Abbott Labs, Ohio, 69-71; ASSOC PROF PHARMACOL & EXP MED, DUKE UNIV, 71- Concurrent Pos: Biochemist, US Bur Com Fisheries, 62-65; consult, Life Sci Div, Ames Res Ctr, NASA, 63- Mem: AAAS; Am Chem Soc; Am Oil Chem Soc; Entom Soc Am; NY Acad Sci. Res: Mechanisms of aging; lipid oxidation and vitamin E; biochemistry of fat soluble vitamins; environmental effects on the lung; chronic lung disease. Mailing Add: Dept of Physiol & Pharmacol Duke Univ Med Ctr Durham NC 27710

MENZEL, DAVID WASHINGTON, b India, Feb 22, 28; m 52. OCEANOGRAPHY. Educ: Elmhurst Col, BS, 49; Univ Ill, MS, 52; Univ Mich, PhD(fisheries), 58. Prof Exp: Res biologist, Bermuda Biol Sta, 57-63; assoc scientist, Woods Hole Oceanog Inst, 63-70; DIR, SKIDAWAY INST OCEANOG, 70- Mem: Am Chem Soc; Am Soc Limnol & Oceanog. Res: Ecology and physiology of marine plankton; marine chemistry and biology. Mailing Add: Skidaway Inst of Oceanog PO Box 13687 Savannah GA 31406

MENZEL, DONALD HOWARD, b Florence, Colo, Apr 11, 01; m 26; c 2. ASTRONOMY. Educ: Univ Denver, AB, 20, AM, 21; Princeton Univ, AM, 23, PhD(astrophys), 24. Hon Degrees: ScD, Univ Denver, 54; AM, Harvard Univ, 42. Prof Exp: Instr math, Univ Denver, 19-21 & pub schs, Colo, 21; asst, Princeton Univ, 21-22; instr astron, Univ Iowa, 24-25; asst prof, Ohio State Univ, 25-26; asst astronomer, Lick Observ, Univ Calif, 26-32; asst prof astron, 32-35, assoc prof astrophys, 35-38, prof, 38-71, Paine prof practical astron, 56-71, chmn dept astron, 46-49, assoc dir solar res, 46-54, actg dir, 52-54, dir observ, 54-66, EMER PROF ASTROPHYS & EMER PAINE PROF PRACTICAL ASTRON, HARVARD UNIV, 71- Concurrent Pos: Exchange prof, Univ Calif, 28, 31; mem, Crocker Eclipse Expeds, Calif, 30, Maine, 32; dir eclipse exped, Harvard & Mass Inst Technol, USSR, 36; ed, Telescope, 37-41; chmn radio propagation comt, Joint & Combined Chiefs of Staff, 43-45; mem, US & Gt Brit Eclipse Exped, Can 45; mem vis comt, Nat Bur Standards, 49-54, chmn, 53-54; Rushton lectr, Birmingham-Southern Col, 54; coun, Boyden Observ, 54-74; Rand scholar, 55; Arthur lectr, Smithsonian Inst, 56; Nat Sci Planning Bd & Exec Comt, Seattle World's Fair, 58-62; McMillin lectr, Ohio State Univ, 60; mem exped, Venus Harvard Observ, 59; dir, Harvard Col Observ Expeds, Italy, 61, Peru, 66, Mex, 70, PEI 72, WAfrica, 73; mem bd dir, Wash Planetarium & Space Ctr, 61-71; vis prof, Univ Chile, 63; State Dept specialist & lectr, Latin Am, 64; pres, Comn 17, Int Astron Union, 64-67. Honors & Awards: Morrison Award, NY Acad Sci, 26, 28, 47; award, Edison Found, 57; Evans Award, Univ Denver, 65. Mem: Nat Acad Sci; AAAS; Am Astron Soc (vpres, 46-48, pres, 54-56); Am Math Soc; Am Philos Soc (vpres, 65-68). Res: Astrophysics; problems of the sun and interpretation of stellar nebular spectra; planetary atmospheres; wave mechanics and atomic spectra; theory of reactions and equilibria at high temperatures; radio propagation; spectroscopy; ionosphere. Mailing Add: Ctr for Astrophys Harvard Col Observ 60 Garden St Cambridge MA 02138

MENZEL, DOROTHY, anthropology, see 12th edition

MENZEL, JOERG H, b Kassel, Ger, July 27, 39; US citizen. NUCLEAR SCIENCE. Educ: Rensselaer Polytech Inst, BME, 62, MS, 64, PhD(nuclear sci), 68. Prof Exp: Staff mem res & develop safeguards, Los Alamos Sci Lab, 68-73; first officer int safeguards, Int Atomic Energy Agency, 73-75; staff mem res & develop safeguards, Los Alamos Sci Lab, 75-76; PHYS SCI OFFICER, US ARMS CONTROL & DISARMAMENT AGENCY, 76- Mem: Inst Nuclear Mat Mgt; Am Nuclear Soc. Res: Non-destructive analysis of nuclear materials. Mailing Add: US Arms Control & Disarm Agency State Dept Bldg Washington DC 20451

MENZEL, MARGARET YOUNG, b Kerrville, Tex, June 21, 24; m 49; c 3. CYTOGENETICS. Educ: Southwestern Univ, Tex, BA, 44; Univ Va, PhD(biol), 49. Prof Exp: Instr chem & bact, Lamar Col, 44-45; instr agron, Agr & Mech Col, Tex, 49-54; res assoc, 55-63, assoc prof, 63-68, assoc chmn dept, 72-73, PROF BIOL SCI, FLA STATE UNIV, 68- Concurrent Pos: Plant geneticist, Agr Res Serv, USDA, 56-; res grants, Am Philos Soc, 55-66, Sigma Xi, 55 & Atomic Energy Comn, 65-75. Mem: Am Soc Nat; Soc Study Evolution; Bot Soc Am; Am Genetic Asn; Int Asn Plant Taxon. Res: Cytotaxonomy and cytogenetics of Hibiscus; fine structure of meiotic chromosomes. Mailing Add: Dept of Biol Sci Fla State Univ Tallahassee FL 32306

MENZEL, ROBERT WINSTON, b Toano, Va, Jan 29, 20. MARINE BIOLOGY. Educ: Col William & Mary, BS, 40, MA, 43; Tex A&M Univ, PhD, 54. Prof Exp: Asst, Va Fisheries Lab, 40-42, asst biologist, 42-46; lab instr bot, Univ Va, 46-47; biologist, Res Found, Agr & Mech Col Tex, 47-52, asst prof wildlife mgt, 53, instr biol, 53-54; from asst prof to assoc prof, 54-70, PROF OCEANOG, FLA STATE UNIV, 70- Mem: AAAS; Am Fisheries Soc; Am Soc Limnol & Oceanog; Am Soc Ichthyol & Herpet; Nat Shellfisheries Asn (vpres, 71-72, pres, 72-73). Res: Clam and oyster biology; marine ecology; commercial fish. Mailing Add: Dept of Oceanog Fla State Univ Tallahassee FL 32306

MENZEL, RONALD GEORGE, b Independence, Iowa, Jan 23, 24; m 52; c 2. SOIL CHEMISTRY. Educ: Iowa State Col, BS, 47; Univ Wis, PhD(soil chem), 50. Prof Exp: Soil scientist, Agr Res Serv, 50-69, DIR WATER QUAL MGT LAB, US DEPT AGR, 69- Mem: AAAS; Soil Sci Soc Am; fel Am Soc Agron; Am Chem Soc. Res: Reactions of copper and zinc in soils and availability to plants; uptake of nuclear fission products by plants; relation of agricultural chemicals and fertilizers to water pollution. Mailing Add: Water Qual Mgt Lab US Dept of Agr Durant OK 74701

MENZEL, WOLFGANG PAUL, b Heidenheim, Ger, Oct 5, 45; US citizen. THEORETICAL SOLID STATE PHYSICS. Educ: Univ Md, BS, 67; Univ Wis, MS, 68, PhD(physics), 74. Prof Exp: Res asst solid state physics, Univ, 72-74, proj asst, 74-75, PROJ ASSOC RADIOMETRY, SPACE SCI & ENG CTR, UNIV WIS-MADISON, 75- Mem: Sigma Xi. Res: Investigation of optical properties of solids using the method of linear combinations of atomic orbitals; real time analysis of the atmosphere with infrared radiometric satellite probing. Mailing Add: Space Sci & Eng Ctr 1225 Dayton St Madison WI 53706

MENZER, ROBERT EVERETT, b Wash, DC, Dec 21, 38; m 62; c 3. INSECT TOXICOLOGY. Educ: Univ Pa, BS, 60; Univ Md, MS, 62; Univ Wis, PhD(entom, biochem), 64. Prof Exp: Res asst entom, Univ Md, 61-62, instr, 62; instr, Univ Wis, 64; from asst prof to assoc prof, 64-73, PROF ENTOM, UNIV MD, COLLEGE PARK, 73-, ASSOC DEAN GRAD STUD, 74- Concurrent Pos: Mem toxicol study sect, NIH, 71-75, chmn, 73-75. Mem: AAAS; Am Chem Soc; Entom Soc Am; Soc Toxicol. Res: Pesticide chemistry and toxicology; metabolism of organophosphorus insecticides; insect biochemistry. Mailing Add: Dept of Entom Univ of Md College Park MD 20742

MENZIES, CARL STEPHEN, b Menard, Tex, Mar 6, 32; m 52; c 2. ANIMAL HUSBANDRY. Educ: Tex Tech Col, BS, 54; Kans State Univ, MS, 56; Univ Ky, PhD(animal nutrit), 65. Prof Exp: Asst county agent, Tex Exten Serv, 54; res asst animal husb, Kans State Univ, 54-55, from instr to assoc prof, 55-68; prof animal sci & head dept, SDak State Univ, 68-72; PROF ANIMAL SCI & RESIDENT DIR RES, TAMU AGR RES & EXTEN CTR, TEX A&M UNIV, 72- Mem: Am Soc

Animal Sci; Sigma Xi. Res: Nutrition and breeding of sheep. Mailing Add: Tex A&M Univ Res & Exten Ctr Box 950 Rte 1 San Angelo TX 76901

MENZIES, JAMES DAVID, b Vancouver, BC, Jan 19, 15; nat US; m 42; c 3. MICROBIOLOGY. Educ: Univ BC, BSA, 35, MSA, 39; State Col Wash, PhD(plant path), 42. Prof Exp: Asst timber path, Forest Prod Labs, BC, 35-36; lab asst, Univ BC, 36-38; inspector plant dis, Seed Potato Cert Serv, Can Dept Agr, 38-39; asst plant path, State Col Wash, 39-42, asst plant pathologisst, 42-45; pathologist, Div Soil Mgt & Irrig, 45-53, Soil & Water Conserv Res Div, Agr Res Serv, 53-55, MICROBIOLOGIST, SOIL & WATER CONSERV RES DIV, AGR RES SERV, USDA, 55- Concurrent Pos: Consult agr prog, Rockefeller Found, Chile, 60. Mem: AAAS; Am Phytopath Soc; Am Soc Agron; Am Soc Microbiol. Res: Soil microbiology; soil borne diseases; municipal waste utilization on land. Mailing Add: US Soils Lab Plant Indust Sta Agr Res Serv Beltsville MD 20705

MENZIES, ROBERT ALLEN, b San Francisco, Calif, Nov 13, 35; m 58; c 3. BIOCHEMISTRY. Educ: Univ Fla, BS, 60, MS, 62; Cornell Univ, PhD(phys biol), 66. Prof Exp: Res chemist, Aging Res Lab, Vet Admin Hosp, Baltimore, Md, 65-67; asst prof biochem, La State Univ Med Ctr, New Orleans, 67-73; ASST PROF BIOCHEM, LIFE SCI CTR, NOVA UNIV, 73- Mem: Am Soc Biol Chem; Am Chem Soc; Biophys Soc; NY Acad Sci; Am Soc Cell Biol. Res: Control of macromolecule biosynthesis and degradation; control of cell division; involvement of macromolecules in information storage; use of biochemical technology in marine population genetics. Mailing Add: Dept of Biochem Nova Univ Life Sci Ctr Ft Lauderdale FL 33314

MENZIES, ROBERT JAMES, b Denver, Colo, Dec 2, 23; m 47; c 4. ZOOLOGY, OCEANOGRAPHY. Educ: Col Pac, AB, 45, MA, 49; Univ Southern Calif, PhD(zool), 51. Prof Exp: In-charge crustacea sect, San Diego Natural Hist Mus, 43; res asst, Hancock Found, Univ Southern Calif, 46-47; instr taxon & cur mus, Pac Marine Sta, Col Pac, 47-49; assoc zool, Univ Calif, Davis, 51-52, res biologist, Scripps Inst, 53-55; dir biol prog, Lamont Geol Observ, Columbia Univ, 55-60; res assoc, Univ Southern Calif, 60-61, assoc prof biol, 61-62; prof zool & dir oceanog, Duke Univ, 62-66; PROF OCEANOG, FLA STATE UNIV, 66- Concurrent Pos: Lectr, Modesto Jr Col, 48; Ellsworth fel, Arctic Inst NAm, 56; dir, Margarita Marine Res Sta, La Salle Found Nautral Sci, Caracas, Venezuela, 60; chmn biol, Gulf Univ Res Corp, 70- Honors & Awards: William F Clapp Mem Prize, 57. Mem: Fel AAAS; Marine Biol Asn UK. Res: Isopods, benthic marine ecology; marine fouling and corrosion; deep-sea oceanography. Mailing Add: Dept of Oceanog Fla State Univ Tallahassee FL 32303

MENZIN, MARGARET SCHOENBERG, b New York, NY, Nov 17, 42; m 68; c 1. MATHEMATICS. Educ: Swarthmore Col, BA, 63; Brandeis Univ, MA, 67, PhD(math), 70. Prof Exp: From instr to asst prof, 69-73, ASSOC PROF MATH, SIMMONS COL, 73-, CHMN DEPT, 71- Concurrent Pos: Consult & dir, Design Technol Corp, 69- Mem: Am Math Soc; Math Asn Am; Asn Women in Math; Women in Sci & Eng; Sigma Xi. Res: Ring theory; linear programming; mathematical models. Mailing Add: Dept of Math Simmons Col 300 The Fenway Boston MA 02115

MENZINGER, MICHAEL, b Bruck-Mur, Austria, Sept 13, 38; m 66. PHYSICAL CHEMISTRY. Educ: Graz Tech Univ, dipl ing, 63; Yale Univ, MSc, 64, PhD(chem), 67. Prof Exp: Teaching asst chem, Yale Univ, 63-67; res assoc, Univ Colo, 67-69; Alexander von Humboldt grant & res assoc physics, Univ Freiburg 69-70; ASST PROF CHEM, UNIV TORONTO, 70- Res: Dynamics of nonreactive and reactive molecular collisions. Mailing Add: Dept of Chem Univ of Toronto Toronto ON Can

MEOLA, SHIRLEE MAY, b Canton, Ohio, Dec 7, 35; m 56. ENTOMOLOGY. Educ: Ohio State Univ, BSc, 58, MSc, 62, PhD(zool & entom), 70. Prof Exp: Res assoc entom, Entom Res Ctr, Vero Beach, Fla, 64-69 & Univ Ga, 69-72; RES ENTOMOLOGIST, TOXICOL & ENTOM RES LAB, AGR RES SERV, USDA, 72- Mem: Am Soc Zoologist; Am Soc Electron Microscopists. Res: Histological and ultrastructural studies of the morphology of Diptera, especially the neuroendocrine and reproductive systems. Mailing Add: Vet Toxicol & Entom Res Lab Agr Res Serv USDA PO Box GE College Station TX 77840

MERANZE, DAVID RAYMOND, b Philadelphia, Pa, Dec 25, 00; m 27; c 2. PATHOLOGY. Educ: Univ Pa, BS, 21, MA, 20; Jefferson Med Col, MD, 27. Prof Exp: Asst biochem, Jefferson Med Col, 22-23; instr path, Grad Sch Med, Univ Pa, 28-30; dir labs & pathologist, Mt Sinai Hosp, Philadelphia, 30-54; dir res & med educ, 54-55, pathologist & dir labs res, South Div, 55-66, MEM & HEAD RES PATH, KORMAN RES INST, ALBERT EINSTEIN MED CTR, 66-; PATHOLOGIST & DIR LABS, PHILADELPHIA GERIATRIC CTR, 70- Concurrent Pos: Asst, Jefferson Med Col, 31-33, assoc, 34-36; dep asst pathologist, Philadelphia Gen Hosp, 38-45; assoc mem, Med Adv Bd Philadelphia, 42-44; consult pathologist, Philadelphia Psychiat Hosp, 45-64; affiliate prof path, Hahnemann Med Col, 47-66; consult, Fels Res Inst, Med Sch, Temple Univ, 62-73 & NIH, 66- Honors & Awards: Highest Award, Int Cong Radiol, 37. Mem: AAAS; Am Soc Clin Path; Am Soc Microbiol; fel AMA; Col Am Path. Res: Toxic granulations of white blood cells; phosphatase content of the blood in jaundice; syphilis and tuberculosis in rabbits; role of liver in metabolism of estrogens; galactose tolerance test; coagulation of blood; testicular histopathology; pathogenesis of atherosclerosis; histopathology; experimental schistosomiasis; ovarian histology; experimental carcinogenesis and emphysema. Mailing Add: Pelham Park Apts 229 West Upsal St Philadelphia PA 19119

MERBS, CHARLES FRANCIS, b Neenah, Wis, Sept 3, 36; m 62; c 2. PHYSICAL ANTHROPOLOGY. Educ: Univ Wis-Madison, BS, 58, MS, 63, PhD(anthrop, genetics), 69. Prof Exp: From instr to assoc prof anthrop, Univ Chicago, 63-73; assoc prof, 73-74, PROF ANTHROP, ARIZ STATE UNIV, 74-, CHMN DEPT, 73- Mem: Fel Arctic Inst NAm; Soc Am Archaeol; Am Asn Phys Anthropologists; Soc Study Human Biol. Res: Physical anthropology, human osteology, paleopathology, medical genetics; Arctic populations of America and Siberia; southwestern United States and northeastern Africa. Mailing Add: Dept of Anthrop Ariz State Univ Tempe AZ 85281

MERCADO, EDWARD J, b Peoria, Ill, Aug 13, 34; m 58; c 2. GEOPHYSICS, SEISMOLOGY. Educ: Rensselaer Polytech Inst, BS, 56, PhD(geophys), 63; Washington Univ, MA, 58. Prof Exp: Geophysicist, 58-62, res geophysicist, 62-65, sr res geophysicist, 65-67, HEAD INFO THEORY SECT, GULF RES & DEVELOP CO, 67- Mem: Soc Explor Geophysicists. Res: Application of statistical filter theory to digital processing of seismograph data. Mailing Add: Box 211 Simonton TX 77476

MERCANDO, NEIL ALDO, b New York, NY, June 5, 43; m 65; c 1. AQUATIC ECOLOGY. Educ: Bloomsburg State Col, BS, 65; Pa State Univ, MS, 71; NC State Univ, Raleigh, PhD(zool), 75. Prof Exp: Teacher biol, Abington High Sch, 65-67; biologist, Nat Inst Environ Health Sci, 73- instr zool, 72-75, VIS ASST PROF ZOOL, NC STATE UNIV, RALEIGH, 76- Mem: Sigma Xi; Am Soc Zoologists; Am Soc Limnol & Oceanog; AAAS; Am Inst Biol Sci. Res: Invertebrate symbioses and hermit crabs; the effects of subacute levels of pollutants on the ecology and behavior of benthic invertebrates and fish. Mailing Add: Biol Sci Interdept Prog NC State Univ PO Box 5577 Raleigh NC 27607

MERCER, EDWARD EVERETT, b Buffalo, NY, Mar 5, 34; m 57; c 6. PHYSICAL INORGANIC CHEMISTRY. Educ: Canisius Col, 55; Purdue Univ, PhD(phys chem), 60. Prof Exp: Res assoc chem, Lawrence Radiation Lab, Univ Calif, 60-61; from asst prof to assoc prof, 61-73, PROF CHEM, UNIV SC, 73-, ASST HEAD DEPT, 73- Mem: AAAS; Am Chem Soc. Res: Thermodynamics and kinetics of metal complexes in solution; chemistry of ruthenium; transition metals as anti-carcinogens. Mailing Add: Dept of Chem Univ of S C Columbia SC 29208

MERCER, EDWARD KING, b Santa Barbara, Calif, July 1, 31; m 63; c 2. PLANT NEMATOLOGY, FRESH WATER BIOLOGY. Educ: Univ Calif, Santa Barbara, BA, 58, MA, 62; Auburn Univ, PhD(plant nematol), 68. Prof Exp: Res asst plant nematol, Univ Calif, Riverside, 63-65; res asst, Auburn Univ, 65-68; ASSOC PROF BIOL SCI, CALIF STATE POLYTECH UNIV, POMONA, 68- Mem: AAAS; Am Inst Biol Sci; Soc Nematol; Asn Meiobenthologists. Res: Energy requirements and feeding behavior of free-living fresh water nematodes. Mailing Add: Dept of Biol Sci Calif State Polytech Univ Pomona CA 91768

MERCER, FRANK LOUIS, pharmacognosy, see 12th edition

MERCER, GERALD DEAN, b Bushnell, Nebr, Apr 9, 26; m 48; c 2. ORGANIC CHEMISTRY. Educ: Univ Nebr, BSc, 52, MSc, 55, PhD(org chem), 56. Prof Exp: Res chemist, Dow Chem Co, 56-69; RES CHEMIST, BUCKMAN LABS, INC, 69- Mem: Am Chem Soc. Res: Synthesis and product development of new cellulose derivatives and development of new analytical techniques for cellulose derivatives. Mailing Add: Buckman Labs Inc 1256 N McLean Blvd Memphis TN 38108

MERCER, HENRY DWIGHT, b Blakely, Ga, Feb 20, 39; m 60; c 2. VETERINARY PHARMACOLOGY, VETERINARY TOXICOLOGY. Educ: Univ Ga, BS, 60, DVM, 63; Univ Fla, MS, 66; Am Bd Vet Toxicol, dipl, 73. Prof Exp: Practitioner vet med, Houston Animal Clin, Blakely, 64-65; NIH fel, Univ Fla, 65-66; br chief vet med, Div New Animal Drugs, 66-68, actg dir, Div Vet Res, 68-72, dep dir, 72-74, ACTG DIR DIV VET RES, BUR VET MED, FOOD & DRUG ADMIN, 74- Concurrent Pos: Rating bd mem, Civil Serv Comn, 70-; mem res comt, Food & Drug Admin Task Force, 71 & Nat Mastitis Coun, 73-; fel, Ohio State Univ, 75-76. Honors & Awards: Award of Merit, Food & Drug Admin, 74. Mem: Am Vet Med Asn; Sigma Xi; Am Soc Vet Physiol & Pharmacol; Am Col Vet Toxicol. Res: Clinical pharmacology in domestic animals, specifically, the metabolism kinetics and pharmacokinetics of veterinary drugs in food producing animals. Mailing Add: Food & Drug Admin Agr Res Ctr Bldg 328A Beltsville MD 20705

MERCER, JAMES WAYNE, b Panama City, Fla, Dec 23, 47; m 69. HYDROGEOLOGY. Educ: Fla State Univ, BS, 69; Univ Ill, MS, 72, PhD(geol), 73. Prof Exp: HYDROLOGIST, US GEOL SURV, 72- Mem: Am Geophys Union; Soc Petrol Engrs. Res: Development of theoretical and numerical models for simulating hydrogeologic processes, with emphasis on geothermal systems. Mailing Add: Mail Stop 431 US Geol Surv Reston VA 22092

MERCER, LEONARD PRESTON, II, b Fort Worth, Tex, Jan 16, 41; m 63; c 3. NUTRITIONAL BIOCHEMISTRY. Educ: Univ Tex, Austin, BS, 68; La State Univ, PhD(biochem), 71. Prof Exp: NIH fel, Med Sch, Univ Ala, Birmingham, 71-73; instr, 73-74, ASST PROF BIOCHEM, UNIV S ALA, 74- Mem: Am Chem Soc. Res: Mathematical analysis of biochemical responses to nutritional stimuli; prediction of nutritional responses of proteins and amino acids and comparison of biological efficacy of alternate nutrient sources. Mailing Add: Dept of Biochem Col of Med Univ of S Ala Mobile AL 36688

MERCER, MALCOLM CLARENCE, b St John's, Nfld, June 20, 44. FISHERIES MANAGEMENT, SYSTEMATIC ZOOLOGY. Educ: Mem Univ Nfld, BSc, 65, MSc, 68. Prof Exp: Scientist, Fisheries Res Bd Can, 65-69, res biologist, 69-74, sect head shellfish, 74-75, prog head marine fisheries mgt pelagic & shellfish, Nfld Biol Sta, 75-76, PROG ADV, MARINE MAMMALS RESOURCE SERV DIRECTORATE, DEPT ENVIRON FISH & MARINE SERV, OTTAWA, 76- Mem: Sigma Xi; Can Soc Zoologists; Soc Syst Zool; Soc Study Evolution; Nat Shellfisheries Asn. Res: Systematics and zoogeography of cephalopods; biology and population dynamics of commercially exploited molluscs, particularly squid and scallops, and of small cetaceans. Mailing Add: Mar Mamm Res Serv Directorate Dept Environ Fish & Marine Serv Ottawa ON Can

MERCER, PAUL FREDERICK, b Guelph, Ont, Apr 21, 36; m 62; c 1. PHYSIOLOGY. Educ: Univ Toronto, DVM, 59; Cornell Univ, PhD(phys biol), 64. Prof Exp: Med Res Coun Can fel biol chem, Copenhagen Univ, 63-64; asst prof physiol, Univ Alta, 64-67; asst prof, 67-70, ASSOC PROF PHYSIOL, UNIV WESTERN ONT, 70- Mem: Soc Nephrology; Can Physiol Soc. Res: Renal physiology. Mailing Add: Dept of Physiol Univ of Western Ont London ON Can

MERCER, ROBERT ALLEN, b Providence, RI, Aug 2, 42; m 66; c 2. PHYSICS. Educ: Carnegie-Mellon Univ, BS, 64; Johns Hopkins Univ, PhD(physics), 69. Prof Exp: Res assoc physics, Johns Hopkins Univ, 69-70; asst prof, Ind Univ, Bloomington, 70-74; MEM TECH STAFF, BELL LABS, 74- Mem: Am Phys 2f6 18a c. Res: Elementary particle physics. Mailing Add: Bell Labs Crawford Corner Rd Holmdel NJ 07733

MERCER, ROBERT J, b Gordon, Nebr, Nov 28, 29; m 52; c 2. NUMERICAL ANALYSIS. Educ: Univ Calif, Berkeley, BA, 51, Los Angeles, MA, 56. Prof Exp: Mem tech staff, Space Tech Labs, 56-61; MEM TECH STAFF, AEROSPACE CORP, 61- Concurrent Pos: Lectr, Univ Calif, Los Angeles, 62-63 & Univ Southern Calif, 63-65. Mem: Soc Indust & Appl Math; Asn Comput Mach. Res: Satellite orbit reconstruction and tracking system analysis. Mailing Add: 3270 Ellenda Ave Los Angeles CA 90034

MERCER, SHERWOOD ROCK, b Manchester, Conn, June 27, 07; m 33; c 3. HISTORY OF MEDICINE, HISTORY OF SCIENCE. Educ: Wesleyan Univ, AB, 29, AM, 30. Hon Degrees: LLD, Philadelphia Col Textiles & Sci, 57. Prof Exp: Instr hist & eng, Pub Schs, Conn, 30-42; chmn div appl sci, Elmira Col, 44-45; consult higher educ, Conn Dept Pub Instr, 45-46; dean fac, Muhlenberg Col, 46-54; dean, 54-69, PROF HIST MED & OSTEOP, PHILADELPHIA COL OSTEOP MED, 54-, VPRES EDUC AFFAIRS, 67- Concurrent Pos: Consult, Columbia Univ, 54- Mem: Am Osteop Asn; Am Asn Cols Osteop Med (secy-treas, 71-). Res: Nature, structure and teaching for a liberal education, particularly in a society heavily influenced by pure and applied science. Mailing Add: 13 Thompson Dr Havertown PA 19083

MERCER, THOMAS T, b Victoria, BC, Dec 30, 20; m 42; c 3. INDUSTRIAL HYGIENE, HEALTH PHYSICS. Educ: San Jose State Col, AB, 49; Univ Rochester, PhD(indust hyg), 57. Prof Exp: Health physicist & instr, Univ Wash, 53-55; res assoc aerosol physics, Atomic Energy Proj, Univ Rochester, 55-57, chief aerosol physics sect, 57-59; nuclear physicist, US Naval Radiol Defense Lab, San Francisco, 59-61; head dept aerosol physics, Lovelace Found Med Educ & Res, NMex, 61-65; assoc prof, 65-70, PROF RADIATION BIOL & BIOPHYS, UNIV ROCHESTER, 70- Mem: AAAS; Am Indust Hyg Asn; Health Phys Soc. Res: Production and characterization of airborne particulates; interaction of radioactive vapors with particles. Mailing Add: Dept Radiation Biol & Biophys Univ of Rochester Rochester NY 14642

MERCER, WALTER ASHBY, b Comanche, Tex, May 10, 15; m 40. MICROBIOLOGY, CHEMISTRY. Educ: Univ Ariz, BS, 48; Univ Calif, MA, 50. Prof Exp: Bacteriologist, Nat Canners Asn, 50-58; res coordr, 58-60, asst dir, 60-65, MGR, WESTERN RES LAB, 65-; ASSOC DIR, RES LABS, NAT CANNERS ASN, 65- Concurrent Pos: Consult res microbiologist, Lab Canning Indust, Univ Calif, 61-, dir lab, 65-; chmn, Nat Tech Task Comt on Indust Wastes, 65. Mem: Am soc Microbiol; Inst Food Technol; Inst Sanit Mgt. Res: Bacteriological and chemical problems affecting the canning industry. Mailing Add: Nat Canners Asn 1950 Sixth St Berkeley CA 94710

MERCEREAU, JAMES EDGAR, b Sharon, Pa, Apr 3, 30; m 50; c 3. PHYSICS. Educ: Pomona Col, BA, 53; Univ Ill, MS, 54; Calif Inst Technol, PhD, 59. Hon Degrees: DSc, Pomona Col, 68. Prof Exp: Asst, Univ Ill, 53-54; physicist, Hughes Res Labs, 54-59; asst prof, Calif Inst Technol, 59-62; prin scientist, Sci Labs, Ford Motor Co, Calif, 62-65, mgr cryogenics, 65-69; PROF PHYSICS, CALIF INST TECHNOL, 69- Concurrent Pos: Consult, Hughes Aircraft Co, 59-60 & Aerospace Corp, 60-62; vis assoc, Calif Inst Technol, 64-65, res assoc, 65-; prof, Univ Calif, Irvine, 65-69. Mem: Fel Am Phys Soc. Res: Cryogenics; ferromagnetism. Mailing Add: Dept of Physics Mail Code 63-37 Calif Inst of Technol Pasadena CA 91109

MERCHANT, DONALD JOSEPH, b Biltmore, NC, Sept 7, 21; m 43; c 3. MICROBIOLOGY. Educ: Berea Col, AB, 42; Univ Mich, MS, 47, PhD(bact), 50. Prof Exp: From instr to prof bact, Univ Mich, 48-69; dir, W Alton Jones Cell Sci Ctr, Tissue Cult Asn, 69-73; PROF MICROBIOL & CELL BIOL & CHMN DEPT, EASTERN VA MED SCH, NORFOLK, 73- Concurrent Pos: Prof, Univ Vt, 69-; mem working cadre, Nat Prostatic Cancer Proj, 72- Mem: AAAS; Am Soc Microbiol; Soc Exp Biol & Med; Tissue Cult Asn (vpres, 60-64, pres, 64-66); Am Soc Cell Biol. Res: Tissue culture techniques; cell growth and metabolism; cancer cell biology; pathogenesis of infectious disease. Mailing Add: 2433 Spindrift Rd Virginia Beach VA 23451

MERCHANT, HENRY CLIFTON, b Washington, DC, Aug 7, 42; m 65; c 4. ECOLOGY, ZOOLOGY. Educ: Univ Md, College Park, BS, 64, MS, 66; Rutgers Univ, New Brunswick, PhD(zool), 70. Prof Exp: Instr zool, Rutgers Univ, 70; asst prof, 70-75, ASSOC PROF BIOL, GEORGE WASHINGTON UNIV, 75- Mem: AAAS; Ecol Soc Am; Am Inst Biol Sci. Res: Bioenergetics of species, populations and communities. Mailing Add: Dept of Biol Sci George Washington Univ Washington DC 20006

MERCHANT, ROLAND SAMUEL, b New York, NY, Apr 18, 29; m 70; c 3. BIOSTATISTICS, ECONOMIC STATISTICS. Educ: NY Univ, BA, 57, MA, 60; Columbia Univ, MS, 63, MSHA, 74. Prof Exp: Asst statistician, New York City Dept Health, 57-60, statistician, 60-63; statistician, NY Tuberc & Health Asn, 63-65; biostatistician, Inst Surg Studies, Montefiore Hosp & Med Ctr, 65-72; admin resident, Roosevelt Hosp, 73-74; DIR HEALTH & HOSP MGT, NY CITY DEPT HEALTH, 74- Mem: AAAS; fel Am Pub Health Asn; Am Statist Asn; Biomet Soc; Inst Math Statist. Res: Application of biostatistical techniques to administrative methodology in health care delivery systems. Mailing Add: Off of Prog Anal & Planning NY Dept of Health 125 Worth St New York NY 10013

MERCIER, ERNEST, b Rosaire, Que, Mar 1, 14; m 45; c 6. AGRICULTURE. Educ: Laval Univ, BA, 39, BScAgr, 43; Cornell Univ, MSc Agr, 44, PhD(physiol, reproduction), 46. Prof Exp: Agr specialist, Livestock Br, Govt of Que, 43-50; head res sta, Exp Farm, Govt of Can, 50-60; dep minister, 60-67, ADV AGR, GOVT OF QUE, DEPT INTERGOVT AFFAIRS, 67- Concurrent Pos: Mem spec comt agr res, Sci Coun Can, 67-69; various comts, Agr Inst Can, 69-70, award comt, 71. Mem: Fel Agr Inst Can; Int Asn Agr Econ; Can Soc Agron. Res: Organization of agricultural research; agricultural policy at national, regional, and international levels; international rural development. Mailing Add: Dept of Intergovt Affairs Govt Bldg Quebec PQ Can

MERCIER, PHILIP LAURENT, b Norwich, Conn, Oct 3, 26; m 55; c 5. PHYSICAL CHEMISTRY. Educ: Univ Conn, BA, 51; Brown Univ, PhD(chem), 55. Prof Exp: Res chemist, Esso Res & Eng Co, 55-61; supvr, Rexall Chem Co, 61-68; lab head, Borden Chem Co, 68; MGR PHYS CHEM GROUP, DART INDUSTS, 68- Mem: Am Chem Soc. Res: Conductance of electrolytes in aqueous and nonaqueous solvents; physical characterization of polymers. Mailing Add: Dart Industs 115 W Century Rd Paramus NJ 07452

MERCKEL, CHARLES GEORGE, b Detroit, Mich, Aug 21, 11; m 46; c 2. MEDICINE. Educ: Wayne State Univ, AB, 33, MS, 38, MB, 40, MD, 41. Prof Exp: Asst dir student health serv, Wayne State Univ, 39-40; intern, Hosp Univ Pa, 40-42; instr med, Univ Southern Calif, 46-53; med dir, Sylvania Elec Prod, Inc Div, Gen Tel & Electronics Corp, 60-71; chief benefits exam sect, 71-74, ACTG CHIEF AMBULATORY CARE SECT, VET ADMIN HOSP, 74- Concurrent Pos: Pvt pract, Calif, 46-51; attend staff, Los Angeles County Gen Hosp, 46-60 & Hosp of Good Samaritan, Los Angeles, 51-60; mem med staff, Southern Calif Edison Co, 51-60. Mem: AAAS; Asn Mil Surg US; fel Indust Med Asn. Res: Physiology of reproduction; internal and preventive medicine. Mailing Add: Vet Admin Hosp 4150 Clement St San Francisco CA 94121

MERCURE, RUEL COE, JR, b Denver, Colo, Apr 15, 31; m 51; c 2. PHYSICS. Educ: Univ Colo, BA, 51, MA, 55, PhD(physics), 57. Prof Exp: Physicist, Dow Chem Co, 51-53 & Cambridge Corp, 53-54; asst, Univ Colo, 54-55, instr, 56, mem res staff, 57; consult, Ball Bros Res Corp, 57, staff scientist, 57-58, dir, 58-68, exec vpres, 68-69, pres, 69-70, group vpres, 70- 74, PRES, BALL CORP, 74- Mem: AAAS; Am Phys Soc; Optical Soc Am. Res: Scientific measurements from rockets and satellites; design of satellite payloads for space exploration. Mailing Add: Ball Corp PO Box 1062 Boulder CO 80302

MERCURI, ARTHUR J, b Albany, NY, Mar 25, 23; m 56; c 4. FOOD MICROBIOLOGY, FOOD SCIENCE. Educ: San Jose State Col, AB, 48; Stanford Univ, PhD(bact), 54. Prof Exp: Bacteriologist, Consumer Yeast Co, Calif, 54-55; food technologist, Biol Sci Br, Agr Mkt Serv, USDA, 55-56, bacteriologist, Biol Sci Br, 56-62, leader poultry qual invest, Mkt Qual Res Div, 62-70, CHIEF ANIMAL PROD LAB, RICHARD B RUSSELL AGR RES CTR, AGR RES SERV, USDA, GA, 70- Concurrent Pos: Mem res coun, Poultry & Egg Inst Am, 62- Mem: Am Soc Microbiol; Poultry Sci Asn; Inst Food Technol; Brit Soc Appl Bact; Int Asn Milk, Food & Environ Sanit. Res: Bacterial physiology; food microbiology and technology; sanitation; Salmonellae; staph; clostridium; food poisoning; poultry and meat products technology; pollution; public health research administration. Mailing Add: Animal Prods Lab Agr Res Serv Richard B Russell Agr Res Ctr USDA PO Box 5677 Athens GA 30604

MERCURIO, ANDREW, physical chemistry, polymer chemistry, see 12th edition

MERDINGER, EMANUEL, b Austria, Mar 29, 06; nat US; m 53. BIOCHEMISTRY. Educ: Prague Tech Univ, Master Pharmacol, 31; Univ Ferrara, Dr Pharm, 34, Dr Chem, 35. Prof Exp: Prof sch eng, Univ Ferrara, 36-38, 45-47; from asst prof to assoc prof, 47-62, PROF CHEM, ROOSEVELT UNIV, 62- Concurrent Pos: Mem res dept dermat, Univ Chicago; abstractor, Chem Abstr, 49-62; Abbott Labs annual res grants, Roosevelt Univ, 61-, Ill State Acad Sci grants, 68-71; Nat Acad Sci exchange scientist, Romanian Acad Sci, 71-72 & 75; Nat Acad Sci exchange scientist, Bulgarian Acad Sci, 74-75; res assoc, Loyola Univ Stritch Sch Med, 72-, distinguished lectr, 74- Mem: Am Chem Soc; Am Soc Microbiol. Res: Microbiological biochemistry; lipid and carbohydrate metabolism and enzymology of yeasts and fungi; fungal pigments and some of their chemotherapeutic properties; relation between Pullularia pullulans, a fungus, and arthritis. Mailing Add: 7251 Randolph Forest Park IL 60130

MEREDITH, CHARLES EYMARD, b St John, NB, Apr 1, 26; US citizen; m 52; c 8. HOSPITAL ADMINISTRATION, PSYCHIATRY. Educ: Loyola Col Montreal, BA, 47; McGill Univ, MD, CM, 51. Prof Exp: Intern, Montreal Gen Hosp, 51-52; resident psychiat, Conn Valley Hosp, Middletown, Conn, 52-57, physician & psychiatrist, 57-59, clin dir & dir residency training prog in psychiat, 59-61, asst supt, 61-63; SUPT, COLO STATE HOSP, 63- Concurrent Pos: Pvt pract, Conn, 60-63; assoc clin prof psychiat, Sch Med, Univ Colo Denver, 63-; prof med technol, Southern Colo State Col, 64-; mem & consult, Hosp Staff Develop Comt, NIMH, 67-71; mem adv bd, Measurement Div, Nat League Nursing, NY, 75-; mem, PSRO Regional Workshops Adv Comt, Asn State Mental Health Prog Dirs, 75- Mem: Fel Am Psychiat Asn; Am Col Hosp Adminr; Am Asn Med Supt Ment Hosps. Res: Mental health administration; staff development and training in mental health and related health fields; delivery of public mental health services; quality assurance programs in psychiatric settings. Mailing Add: Colo State Hosp 1600 W 24th St Pueblo CO 81003

MEREDITH, CURTIS LOCKE, physical chemistry, see 12th edition

MEREDITH, FARRIS RAY, b Denver, Colo, Mar 15, 29; m 50; c 3. BOTANY, SOILS. Educ: Colo State Univ, BS, 51; NMex Highlands Univ, MS, 58; Wash State Univ, PhD(bot, plant ecol), 65. Prof Exp: Asst prof bot, Humboldt State Col, 63-65; asst prof, NMex Highlands Univ, 65-66; from asst prof to assoc prof, 66-74, PROF BOT, HUMBOLDT STATE UNIV, 74- Mem: AAAS; Ecol Soc Am; Torrey Bot Club. Res: Plant autecology and synecology; autecology and physiology of coniferous trees. Mailing Add: Dept of Biol Humboldt State Univ Arcata CA 95521

MEREDITH, HARVEY L, b Nebraska City, Nebr, Mar 10, 31; m 69; c 2. SOIL PHYSICS. Educ: La State Univ, BS, 58, MS, 59; Purdue Univ, PhD(soil physics), 64. Prof Exp: Area dir, Tenn Valley Authority, Mich, 64-65; AGRICULTURIST, UNIV MINN, ST PAUL, 65- Res: Crop Production, particularly soil nutrition. Mailing Add: Dept of Soil Sci Univ of Minn St Paul MN 55101

MEREDITH, HOWARD VOAS, b Birmingham, Eng, Nov 5, 03; nat US; m 26; c 2. CHILD GROWTH, MORPHOLOGY. Educ: Univ Iowa, BA, 31, MA, 32, PhD(phys growth), 35. Prof Exp: Asst, Child Welfare Res Sta, Univ Iowa, 31-35, res assoc, 35-39, from asst prof to assoc prof, 39-48, prof phys growth, 48-49; prof, Sch Health & Phys Educ, Univ Ore, 49-52; prof, Child Welfare Res Sta, 52-63, consult Res Sta & Col Dent, 52-56, prof child somatol, Inst Child Behav & Develop, 63-72, actg dir, 71-72, EMER PROF CHILD SOMATOL, INST CHILD BEHAV & DEVELOP, UNIV IOWA, 72- TCPExchange fel, Harvard Univ, 35; vis lectr, Univ Southern Calif, 48, vis prof, 51; affil prof human somatology, Univ SC, 73- Mem: Soc Res Child Develop (pres, 53-55); Am Asn Phys Anthrop; Int Asn Human Biol; Sigma Xi. Res: Physical growth of the child. Mailing Add: Col of Health & Phys Educ Univ of SC Columbia SC 29208

MEREDITH, JESSE HEDGEPETH, b Fancy Gap, Va, Mar 19, 23; m; c 3. MEDICINE. Educ: Elon Col, BA, 43; Western Reserve Univ, MD, 51; Am Bd Surg, dipl, 58; Am Bd Thoracic Surg, dipl, 59. Prof Exp: Intern med, Bellevue Hosp, New York, 51-52; asst surgeon, NC Baptist Hosp, Winston-Salem, 52-56, resident gen & thoracic surg, 56-57, cardiovasc surg, 57-58; asst surgeon, 52-58, from instr to assoc prof, 58-70, PROF SURG, BOWMAN GRAY SCH MED, 70-, DIR SURG RES, 59- Concurrent Pos: NIH res fel, 56-57, spec res fel, 59-62; res fel, Bowman Gray Sch Med, 58. Mem: Fel Am Col Surgeons; Am Asn Thoracic Surgeons; AMA; Am Soc Artificial Internal Organs. Res: Cardiovascular surgery and physiology; cancer chemotherapy; biomedical engineering; cadaver blood in transfusions; kidney transplantation. Mailing Add: Dept of Surg Bowman Gray Sch of Med Winston-Salem NC 27103

MEREDITH, LESLIE HUGH, b Birmingham, Eng, Oct 23, 27; m 48; c 3. SPACE PHYSICS. Educ: Univ Iowa, BA, 50, MS, 52, PhD(physics), 54. Prof Exp: Res assoc, Univ Iowa, 51-53; asst, Proj Matterhorn, Princeton Univ, 53-54; sect head, Naval Res Lab, 54-58, br head, 58; br head, 58-59, div chief space sci, 59-70, dep dir space & earth sci directorate, 70-72, ASST DIR, GODDARD SPACE FLIGHT CTR, 72- Mem: AAAS; Am Phys Soc; Am Geophys Union. Mailing Add: Goddard Space Flight Ctr Greenbelt MD 20771

MEREDITH, ORSELL MONTGOMERY, b Jamestown, NY, Oct 19, 23; m 49; c 1. RADIOLOGICAL HEALTH, NUCLEAR MEDICINE. Educ: Univ Chicago, BS, 48; Univ Southern Calif, MS, 51, PhD(pharmacol, toxicol), 53; Am Univ, Washington, DC, MS, 74. Prof Exp: Asst pharmacol & toxicol, Sch Med, Univ Southern Calif, 49-52; pharmacologist, Carlborg Labs, Calif, 52-53; chief nuclear physiol sect & asst res pharmacologist, Lab Nuclear Med & Radiation Biol, Med Ctr, Univ Calif, Los Angeles, 53-62; res scientist, Lockheed Missiles & Space Co, Calif, 62-66; tech mgr, US Naval Radiological Defense Lab, 66-69; Biologist, Nuclear Prog Off, Adv Planning & Anal Staff, Naval Ord Lab, 69-75; EXEC SECY, SPEC PROGS BR, DIV RES GRANTS, NIH, 75- 25 Concurrent Pos: Consult, Nuclear Div, Am Electronics Inc, Calif. Mem: AAAS; Health Physics Soc; Soc Nuclear Med; Radiation Res Soc; NY Acad Sci. Res: Catecholamine action on intestinal smooth muscle; anticholinesterase action of organic phosphate insecticides; inhalation toxicity of radioactive fallout debris; radioisotope clinical diagnosis; bioastronautics; mammalian radiation biology; military operations research. Mailing Add: 5333 Westbard Ave Bethesda MD 20014

MEREDITH, ROBERT E, physics, spectroscopy, see 12th edition

MEREDITH, RUBY FRANCES, b Sedalia, Mo, Feb 6, 48. CANCER, GENETICS. Educ: Univ Mo, BA, 69; Ind Univ, AM, 71, PhD(genetics), 74. Prof Exp: Asst prof biol, Baylor Univ, 74-75; fel, 75-76, RES ASSOC VIRAL ONCOGENESIS, CANCER RES UNIT, ALLEGHENY GEN HOSP, 76- Honors & Awards: Harold C Bold Award, Phycol Soc Am, 74. Mem: Genetics Soc Am; Int Soc Exp Hemat; Phycol Soc Am; Radiation Res Soc. Res: Pathology, genetics and treatment of murine viral leukemogenesis including chemotherapy, radiotherapy, bone marrow transplantation, immunotherapy and combinations of these. Mailing Add: Cancer Res Unit Allegheny Gen Hosp Pittsburgh PA 15212

MEREDITH, WILLIAM EDWARD, b Dennison, Ohio, Nov 30, 32; m 57; c 3. MICROBIOLOGY. Educ: Ohio Univ, BSc, 59; Ohio State Univ, MSc, 61, PhD(microbiol), 64. Prof Exp: Asst microbiol, Ohio State Univ, 59-62, Ohio State Univ Res Found, 62-64; microbiologist, Hess & Clark Div, Richardson-Merrell Inc, 64-68; PROF BIOL, ASHLAND COL, 68- Concurrent Pos: Researcher, Ohio Agr Res & Develop Ctr, 69, 70, 73. Mem: AAAS; Am Soc Microbiol. Res: General microbiology; immunology; microbial physiology. Mailing Add: Dept of Biol Ashland Col Ashland OH 44805

MEREDITH, WILLIAM G, b Fairmont, WVa, May 16, 33; m 55; c 3. ECOLOGY. Educ: Fairmont State Col, AB, 55; WVa Univ, MS, 57; Univ Md, PhD(ecol), 67. Prof Exp: From instr to assoc prof, 57-71, chmn dept sci & math, 68-75, PROF BIOL, MT ST MARY'S COL, MD, 71- Mem: AAAS; Ecol Soc Am; Am Inst Biol Sci; Sigma Xi. Res: Comparative ecology and physiology of crayfishes; food habits of freshwater fishes; distribution of crayfishes. Mailing Add: Dept of Sci & Math Mt St Mary's Col Emmitsburg MD 21727

MERENDINO, K ALVIN AURELIUS, b Clarksburg, WVa, Dec 3, 14; m 43; c 5. SURGERY. Educ: Ohio Univ, BA, 36; Yale Univ, MD, 40; Univ Minn, PhD(surg), 46. Prof Exp: Intern, Cincinnati Gen Hosp, 40-41; asst surg, Med Sch, Univ Minn, 42-43, from instr to asst prof, 44-48, mem fac, Grad Sch, 48; dir, Exp Surg Lab, 49-72, chmn dept, 64-72, assoc prof, 49-55, PROF SURG, SCH MED, UNIV WASH, 55- Concurrent Pos: Chief surg res, Med Sch, Univ Minn, 44-45; dir, Kellog Found, Ancker Hosp, St Paul, 45-48; dir, Tumor Clin, King County Hosp, 49-59; mem, Surg A Study Sect, NIH, 58-62, chmn, 70-72; mem, Am Bd Surg, 58-64, chmn, 63-64, sr mem, 64-; mem, Training Comt, Nat Heart Inst, 65-69; mem, Adv Comt Hosps & Clins, USPHS, 63-66; mem, Conf Comt Grad Educ in Surg, 67-73, vchmn, 71-72; mem, Vet Admin Surg Merit Rev Bd, 72-74. Mem: Soc Univ Surgeons; fel Am Col Surgeons; Int Soc Surg. Res: Gastrointestinal, cardiovascular nnd thoracic problems. Mailing Add: Univ of Wash Sch of Med Seattle WA 98195

MERESZ, OTTO, b Rima-Sobota, Czech, Jan 16, 32; m 55; c 1. ORGANIC CHEMISTRY, ENVIRONMENTAL CHEMISTRY. Educ: Budapest Tech Univ, Dipl org chem, 56; Univ London, PhD(org chem), 65. Prof Exp: Tech officer, Imp Chem Indust Ltd, 57-58; res chemist, Res Inst, May & Baker Ltd, 58-61, sect head synthetic perfumes, 61-65, dept head, 66-67; asst prof chem, Univ Toronto, 67-73; dir res, Kemada Res Corp, 73-74; MGR, ORG CHEM SECT, ONT MINISTRY OF THE ENVIRON, 74- Concurrent Pos: Fel, Univ Toronto, 65-66; consult, Addiction Res Found, Ont, 74- Mem: The Chem Soc; Am Chem Soc; Chem Inst Can. Res: Theory of odor; correlation between chemical structure and odor; synthetic and structural organic chemistry; trace-organic analysis. Mailing Add: 8 Wallingford Rd Don Mills ON Can

MEREU, ROBERT FRANK, b Alta, Nov 1, 30; m 61; c 3. GEOPHYSICS, SEISMOLOGY. Educ: Univ Western Ont, BSc, 52, PhD(physics), 62; Univ Toronto, MA, 53. prof, Prof Exp: From asst prof to assoc prof, 63-74, PROF GEOPHYS, UNIV WESTERN ONT, 74- Mem: Seismol Soc Am; Am Geophys Union; Can Asn Physicists. Mailing Add: Dept of Geophys Univ of Western Ont London ON Can

MEREWETHER, DAVID EVAN, b Detroit, Mich, July 7, 36; m 58; c 3. ELECTROMAGNETICS. Educ: Univ NMex, BS & BA, 60, MS, 62, PhD(elec eng), 68. Prof Exp: Staff Mem, Sandia Labs, 60-74; DIR ALBUQUERQUE OPERS, MISSION RES CORP, 74- Concurrent Pos: Consult, Defense Nuclear Agency, US Air Force & USA Safeguard. Mem: Inst Elec & Electronics Engrs. Res: Design of pulse transmission antennas; shielding using non-linear ferromagnetic materials; excitation of cables by ionizing radiation; currents induced on a body of revolution by an electromagnetic pulse. Mailing Add: Albuquerque Opers Mission Res Corp 5601 Domingo Rd NE Albuquerque NM 87108

MERGEN, FRANCOIS, b Redange, Luxembourg, May 1, 25; nat US; m 47; c 1. FOREST GENETICS. Educ: Univ NB, BSF, 50; Yale Univ, MF, 51, PhD(forest genetics), 54. Prof Exp: Forest geneticist, US Forest Serv, Fla, 52-54; asst prof forest genetics, J A Hartford Mem Res Ctr, 54-60, actg dir ctr, 55-58, from assoc prof to prof, 60-66, PINCHOT PROF FORESTRY, YALE UNIV, 66-, DEAN SCH FORESTRY & ENVIRON STUDIES, 65-, FEL, SAYBROOK COL, 74- Concurrent Pos: Res collab, Brookhaven Nat Lab, 60-70. Mem: Soc Am Foresters; Can Inst Forestry. Res: Effect of storms on trees; vegetative propagation of slash pine; physiology of flow of oleo-resin; forest genetics of southern pines; cytology of conifers; effect of radiation on forest trees. Mailing Add: Sch of Forestry & Environ Yale Univ 205 Prospect St New Haven CT 06511

MERGENHAGEN, STEPHAN EDWARD, b Depew, NY, Apr 12, 30; m 55; c 3. IMMUNOLOGY, MICROBIOLOGY. Educ: Allegheny Col, BS, 52; Univ Buffalo, MA, 54; Univ Rochester, PhD(bact), 57. Prof Exp: Res microbiologist, 58-65, chief immunol sect, 65-69, CHIEF LAB MICROBIOL & IMMUNOL, NAT INST DENT RES, 69- Concurrent Pos: Fel, Univ Rochester, 57-58. Mem: Am Soc Microbiol; Soc Exp Biol & Med; Infectious Dis Soc Am; Reticuloendothelial Soc; Am Asn Immunol. Res: Natural resistance mechanisms in oral and systemic disease; immunochemistry of bacterial antigens; endotoxic lipopolysaccharides; pathogenesis of mixed infections. Mailing Add: Lab of Microbiol & Immunol Nat Inst of Dent Res Bethesda MD 20014

MERGENS, WILLIAM JOSEPH, b Queens, NY, July 26, 42; m 65; c 2. ANALYTICAL CHEMISTRY. Educ: St Johns Univ, BS, 64; Seton Hall Univ, MS, 70, PhD(chem), 76. Prof Exp: SR SCIENTIST ANAL CHEM, HOFFMANN LA ROCHE INC, 64- Mem: Am Chem Soc. Res: Formulation and analysis of product development chemistry; chemical carcinogenesis. Mailing Add: Hoffmann LaRoche Inc 340 Kingsland St Nutley NJ 07110

MERGENTIME, MAX, b Brooklyn, NY, Apr 2, 14; m 50; c 3. FOOD CHEMISTRY. Educ: Cornell Univ, BS, 35, MS, 36; Ore State Col, PhD(food tech), 41. Prof Exp: Processed foods inspector, Prod & Mkt Admin, USDA, 41-45; chief chemist, Sunshine Packing Corp, Pa, 45-50; HEAD JUICE DEPT, FRIGID FOOD PROD, INC, 50- Mem: Inst Food Technol; Am Chem Soc. Res: Low temperature studies rate; reaction proteolytic enzyme of peas. Mailing Add: 18082 Parkside Detroit MI 48221

MERGERIAN, DICKRON, solid state physics, see 12th edition

MERIANOS, JOHN JAMES, b Krokeai Sparta, Greece, Feb 12, 37; US citizen. MEDICINAL CHEMISTRY. Educ: New Eng Col Pharm, BS, 61; Univ Wis, MS, 63, PhD(med chem), 66. Prof Exp: Res chemist, Res & Develop Ctr, FMC Corp, NJ, 66-68; SR RES SCIENTIST, MILLMASTER ONYX CORP, 68- Mem: Am Pharmaceut Asn; Am Chem Soc. Res: Medicinal chemistry research and development in pharmaceutical industry synthesis of biologically active compounds; synthesis of heterocyclic compounds; synthesis of ammonium quaternary compounds; isolation structure elucidation of natural products; germicides; detergents; disinfectants; biocites. Mailing Add: 3 Towers St Jersey City NJ 07305

MERICLE, LEO WILLIS, b Weatherford, Okla, Oct 4, 15; m 53; c 1. GENETICS, RADIATION BIOLOGY. Educ: Southwestern State Col, Okla, BS, 34; Univ Okla, MS, 41; Univ Tex, PhD(bot), 49. Prof Exp: Instr high sch & jr col, Okla, 36-40; Naval Res fel, Univ Tex, 49-50, from asst prof to assoc prof, 50-60, PROF BOT & PLANT PATH, MICH STATE UNIV, 60- Concurrent Pos: Res grants, AEC, Mich State Alumni Fund & Res Fund; ed, Radiation Bot. Mem: Genetics Soc Am; Am Genetic Asn; Radiation Res Soc. Res: Developmental genetics, gene expression, and radio-sensitivity; modification by environment and fine structure changes during development. Mailing Add: Dept of Bot & Plant Path Mich State Univ East Lansing MI 48823

MERICLE, R BRUCE, b Omaha, Nebr, June 4, 38; m 63; c 3. MATHEMATICS. Educ: Iowa State Univ, BS, 60; Univ Md, College Park, MS, 64; Wash State Univ, PhD(math), 70. Prof Exp: Instr math, Univ Maine, 64-66; instr, Wash State Univ, 66-70; asst prof, Mankato State Col, 70-74; DIR ACAD COMPUT SERV, MICH TECH UNIV, 74- Mem: Am Math Soc; Math Asn Am. Res: Measure theory; measures in topological spaces. Mailing Add: Dept of Acad Comput Serv Mich Tech Univ Houghton MI 49931

MERICLE, RAE PHELPS, b Toledo, Ohio, Mar 22, 26; m 53. DEVELOPMENTAL GENETICS. Educ: Western Reserve, BA, 46; Univ Tex, MA, 48, PhD(genetics), 50. Prof Exp: Instr med genetics, Med Sch, Tulane Univ, 49-51; sr instr, Sch Nursing, E W Sparrow Hosp, 52-59; fel, 51-53, cancer consult, 52-54, instr natural sci, 60-61, RES ASSOC, MICH STATE UNIV, 61- Concurrent Pos: Bibliog ed, Am J Human Genetics, 51-55; ed sect human genetics, Biol Abstracts, 55- Mem: AAAS; Am Soc Human Genetics; Genetics Soc Am; Radiation Res Soc. Res: Developmental human and plant genetics; effects of ionizing radiation on plant embryogeny; biological effect of natural background radiations. Mailing Add: PO Box 852 East Lansing MI 48823

MERICOLA, FRANCIS CARL, b Franklin, Md, Jan 27, 11; m 37. INORGANIC CHEMISTRY. Educ: Kent State Univ, BS, 33; Western Reserve Univ, PhD(org chem), 38. Prof Exp: Asst instr qual & quant anal, Western Reserve Univ, 35-38; staff mem, Res Dept, 38-45, supvr plant res dept, 45-50, prod supt, S Works, 50-53, prod res dept, 53-55, mgr, 55-63, tech asst to vpres mfg opers, 63-66, tech asst to dir res opers, 66-68, process engr, Eng Dept, 68-73, CONSULT, BASF WYANDOTTE CORP, 73- Mem: AAAS; Am Chem Soc. Res: Chemical engineering. Mailing Add: 2071 17th St Wyandotte MI 48192

MERIDETH, CHARLES WAYMOND, b Atlanta, Ga, Nov 2, 40; m 61; c 2. PHYSICAL INORGANIC CHEMISTRY. Educ: Morehouse Col, BS, 61; Univ Calif, Berkeley, PhD(phys chem), 65. Prof Exp: Fel phys inorg chem, Univ Ill, 64-65; from asst prof to assoc prof, 65-73, PROF PHYS CHEM, MOREHOUSE COL, 73- Mem: Am Phys Soc; Am Chem Soc. Res: Ligand field theory; electronic structure of inorganic complex ions and the nature of the chemical bonds in these complexes. Mailing Add: Dept of Chem Morehouse Col Atlanta GA 30314

MERIDETH, GEORGE THOMAS, b Colorado Springs, Colo, Mar 22, 08; m 38. PHYSICS. Educ: Colo Col, AB, 30; Univ Colo, AM, 33, PhD(physics), 38. Prof Exp: Asst physics, Univ Colo, 30-31, instr, 31-39; from instr to assoc prof, 40-53, prof, 53-74, EMER PROF PHYSICS, COLO SCH MINES, 74-, CONSULT, RES FOUND, 52 Concurrent Pos: Consult, Heiland Res Corp, Denver, Colo, 44-50 & Bear Creek Mining Co, 55-57. Mem: Am Phys Soc. Res: Cosmic radiation; ionization of gases. Mailing Add: Dept of Geophys Colo Sch of Mines Golden CO 80401

MERIFIELD, PAUL M, b Santa Monica, Calif, Mar 17, 32; m 68; c 2. GEOLOGY. Educ: Univ Calif, Los Angeles, AB, 54, MA, 58; Univ Colo, PhD(geol), 63. Prof Exp: Res scientist, Lockheed-Calif Co, 62-64; DIR GEOSCI, EARTH SCI RES CORP, 64- Concurrent Pos: Lectr, Univ Calif, Los Angeles; partner, Lamar-Merifield, 64- Mem: AAAS; Am Soc Photogram; Asn Eng Geol. Res: Interpretation of satellite photography; age and origin of the earth-moon system; remote sensing; engineering and environmental geology. Mailing Add: 1318 Second St Suite 27 Santa Monica CA 90401

MERIGAN, THOMAS CHARLES, JR, b San Francisco, Calif, Jan 18, 34; m 59; c 1. INFECTIOUS DISEASES, VIROLOGY. Educ: Univ Calif, Berkeley, BA, 55; Univ Calif, San Francisco, MD, 58; Am Bd Internal Med, dipl, 65. Prof Exp: Intern med, Boston City Hosp, Mass, 58-59, asst resident, 59-60; clin assoc, Nat Heart Inst, 60-62; assoc, Nat Insts Arthritis & Metab Dis, 62-63; from asst prof to assoc prof med, 63-72, dir diag microbiol lab, 66-72, PROF MED, SCH MED, STANFORD UNIV, 72-, CHIEF DIV INFECTIOUS DIS & HOSP EPIDEMIOLOGIST, 66-, DIR DIAG VIROL LAB, 69- Concurrent Pos: Mem microbiol training grant comt, Nat Inst Gen Med Sci, 69-73, mem virol study sect, Div Res Grants, NIH, 74- Mem: AAAS; Am Asn Immunol; Am Fedn Clin Res; Am Soc Clin Invest; Am Soc Microbiol. Res: Host responses to viral infections and antiviral agents. Mailing Add: Div of Infectious Dis Stanford Univ Sch of Med Stanford CA 94305

MERIJANIAN, ARIS, b Abadan, Iran, Dec 29, 29; US citizen; m 53; c 4. ORGANIC CHEMISTRY. Educ: Kans State Univ, BS, 57; Southwest Tex State Col, MA, 59; Tex A&M Univ, PhD(chem), 63. Prof Exp: Asst prof chem, Tex A&M Univ, 59-60, teaching fel, 60-61; from asst prof to assoc prof, 62-67, PROF CHEM & CHMN DEPT, UNIV MONTEVALLO, 67- Concurrent Pos: NSF grant, 64-65. Mem: Am Chem Soc; Am Inst Chem. Res: Organophosphorous chemistry leading to medicinals; organometallic chemistry; inorganic complexes and their possible inclusion-compound formation with urea and thiourea as means of separation of their isomers. Mailing Add: Dept of Chem Univ of Montevallo Montevallo AL 35115

MERILAN, CHARLES PRESTON, b Lesterville, Mo, Jan 14, 26; m 49; c 2. DAIRY HUSBANDRY. Educ: Univ Mo, BS, 48, AM, 49, PhD(dairy husb), 52. Prof Exp: Instr dairy husb, 50-52, bact & prev med, 52-53, from asst prof to assoc prof, 53-59, chmn dept, 61-62, assoc dir agr exp sta, 62-63, PROF DAIRY HUSB, UNIV MO-COLUMBIA, 59- Mem: AAAS; Am Chem Soc; Am Soc Animal Sci; Am Dairy Sci Asn; Soc Cryobiol. Res: Cellular physiology; biophysics; reproductive physiology. Mailing Add: 201 Eckles Hall Univ of Mo Columbia MO 65201

MERILEES, PHILIP, b Chatham, Ont, Sept 3, 40; m 63; c 3. DYNAMIC METEOROLOGY. Educ: Sir George Williams Univ, BSc, 60; Carleton Univ, Ont, MSc, 62; McGill Univ, PhD(meteorol), 66. Prof Exp: Res assoc meteorol, Univ Mich, 66; asst prof, Fla State Univ, 66-67; res meteorologist, Govt of Can, 67-68; asst prof, 68-71, ASSOC PROF, METEOROL, McGILL UNIV, 71- Concurrent Pos: Vis scientist, Nat Ctr Atmospheric Res, 72-74. Mem: Am Meteorol Soc; Can Meteorol Soc. Res: Dynamics of large scale atmospheric motions. Mailing Add: Dept of Meteorol McGill Univ Montreal PQ Can

MERIN, ROBERT GILLESPIE, b Glens Falls, NY, June 16, 33; m 58; c 3. ANESTHESIOLOGY, PHARMACOLOGY. Educ: Swarthmore Col, BA, 54; Cornell

Univ, MD, 58. Prof Exp: Instr anesthesiol, Albany Med Col, 63-66, res assoc pharmacol, 65-66; asst prof, 66-70, ASSOC PROF ANESTHESIOL, SCH MED, UNIV ROCHESTER, 70- Concurrent Pos: NIH career develop award, Sch Med, Univ Rochester, 72-77; consult, Vet Admin Hosp, Albany, 63-66. Mem: Am Soc Anesthesiol; Int Anesthesia Res Soc; NY Acad Sci; Circanes. Res: Effect of anesthesia on the cardiovascular system and metabolism. Mailing Add: Dept of Anesthesiol Univ of Rochester Sch of Med Rochester NY 14642

MERINO, WILLIAM MICHAEL, pharmacology, see 12th edition

MERITS, ILMAR, biochemistry, see 12th edition

MERIWETHER, JOHN R, b Beaumont, Tex, May 22, 37; m 56; c 4. NUCLEAR PHYSICS, COMPUTER SCIENCE. Educ: Univ Southwestern La, BS, 58, MS, 59; Fla State Univ, PhD(nuclear physics), 62. Prof Exp: Assoc, Lawrence Radiation Lab, 62-65, staff physicist, 65-66; asst prof nuclear physics & comput sci, 66-71, assoc prof, 71-75, PROF PHYSICS, UNIV SOUTHWESTERN LA, 75-, CHMN DEPT, 71- Mem: Am Phys Soc; Sigma Xi. Res: Nuclear spectroscopy using inelastic scattering of particles from nuclei; application of computer based numerical analysis to physical problems. Mailing Add: Dept of Physics Univ of Southwestern La Lafayette LA 70501

MERIWETHER, JOHN WILLIAMS, JR, b Louisville, Ky, Apr 14, 42; m 73. AERONOMY. Educ: Mass Inst Technol, SB, 64; Univ Md, PhD(physics), 70. Prof Exp: Nat Acad Sci res assoc, Goddard Space Flight Ctr, 69-71; res assoc atmospheric physics, Univ Mich, 71-73; staff physicist atmospheric physics, PhotoMetrics, Inc, 73-74; RES ASSOC IONOSPHERIC PHYSICS, ARECIBO OBSERV, CORNELL UNIV, 75- Mem: Am Inst Physics; Am Geophys Union. Res: Aeronomy of the earth's atmosphere using observations of airglow and auroral emissions as supplemented with incoherent scatter radar observations of the ionosphere. Mailing Add: Arecibo Observ Box 995 Arecibo PR 00612

MERIWETHER, LEWIS SMITH, b Washington, DC, May 23, 30; m 53; c 3. PHYSICAL CHEMISTRY, ORGANIC CHEMISTRY. Educ: Harvard Univ, AB, 52; Univ Chicago, PhD(chem), 56. Prof Exp: Res chemist, 55-59, SR RES CHEMIST, AM CYANAMID CO, 59-, GROUP LEADER, 60- Concurrent Pos: Cyanamid Sr Award, 64-65. Mem: NY Acad Sci; Am Chem Soc. Res: Homogeneous catalysis; transition metal complexes; polymerization; photochemistry; enzyme model systems; membranes; surgical adhesives; artificial kidney systems; biomaterials; pharmaceuticals. Mailing Add: Chem Res Div Am Cyanamid Co 1937 W Main St Stamford CT 06904

MERKAL, RICHARD STERLING, b Lansing, Mich, Nov 12, 28; m 49; c 3. MICROBIOLOGY. Educ: Univ Md, BS, 53; Auburn Univ, MS, 61; Iowa State Univ, PhD(physiol bact), 65. Prof Exp: Bacteriologist regional animal dis res lab, Ala, 53-61, RES MICROBIOLOGIST, NAT ANIMAL DIS LAB, USDA, 61- Mem: Am Soc Microbiol. Res: Mycobacterial metabolism and nutrition; relationship of metabolic products to host; histochemistry; serology and immunology. Mailing Add: Nat Animal Dis Lab USDA Ames IA 50011

MERKEL, EDWARD PAUL, b Waterbury, Conn, Nov 27, 23; m 49; c 4. FOREST ENTOMOLOGY. Educ: State Univ NY Col Forestry, Syracuse, BS, 46. Prof Exp: Entomologist, Bur Entom & Plant Quarantine, Forest Insect Invest, Beltsville, Md, 46-49, Ft Collins, Colo, 49-51, Asheville, NC, 51-53, entomologist, Forest Serv, 53-56, res entomologist, Fla, 56-59, SUPVRY RES ENTOMOLOGIST, FOREST SERV, SOUTHEASTERN FOREST EXP STA, US DEPT AGR, 59- Mem: Soc Am Foresters; Entom Soc Am; Entom Soc Can. Res: Development of integrated control system for pine seed orchard insects; and studies on biology and control of pine bark beetles associated with naval stores trees and lightwood trees. Mailing Add: Naval Stores & Timber Prod Lab US Forest Serv PO Box 70 Olustee FL 32072

MERKEL, JOSEPH ROBERT, b Alburtis, Pa, Dec 21, 24; m 48; c 1. MICROBIAL BIOCHEMISTRY, MARINE MICROBIOLOGY. Educ: Moravian Col, BS, 48; Purdue Univ, MS, 50; Univ Md, PhD(bact), 52. Prof Exp: Waksman-Merck fel, Rutgers Univ, 52-53, res assoc, 53-54, res investr, Inst Microbiol, 54-55; dir, Ft Johnson Marine Biol Lab, Col Charleston, 55-62; assoc prof biochem, 62-65, PROF DEPT CHEM & MARINE MICROBIOLOGIST, MARINE SCI CTR, LEHIGH UNIV, 65- Mem: Am Soc Biol Chem; Am Chem Soc; Am Soc Microbiol; Brit Soc Gen Microbiol. Res: Proteolytic enzymes of marine bacteria. Mailing Add: Dept of Chem Lehigh Univ Bethlehem PA 18015

MERKEL, PAUL BARRETT, b Rochester, NY, May 14, 45. PHOTOCHEMISTRY, PHOTOGRAPHIC CHEMISTRY. Educ: St John Fisher Col, BS, 67; Univ Notre Dame, PhD(chem), 70. Prof Exp: Res assoc chem, Univ Calif, Riverside, 70-71; SR RES CHEMIST, EASTMAN KODAK CO, 71- Mem: Am Chem Soc; Sigma Xi. Res: In photochemistry and photophysics, interests include laser and flash photolysis, photo-oxidation reactions, electronic excitation and luminescence; research in photographic chemistry involves kinetics and mechanisms, thermoanalytical methods and dye imaging. Mailing Add: Res Labs Eastman Kodak Co Rochester NY 14650

MERKEL, ROBERT ANTHONY, b Marshfield, Wis, Feb 7, 26; m 54; c 4. MEAT SCIENCE. Educ: Univ Wis-Madison, BS, 51, MS, 53, PhD(meat sci, biochem), 57. Prof Exp: From asst prof to assoc prof meat sci, Kans State Univ, 57-62; assoc prof, 62-67, PROF MEAT SCI, MICH STATE UNIV, 67- Concurrent Pos: Sect ed meat sci & muscle biol, J Animal Sci, 74-76. Mem: AAAS; Am Meat Sci Asn; Am Soc Animal Sci; Inst Food Technologists. Res: Differentiation, histogenesis and growth of muscle and adipose tissues; biosynthesis of muscle proteins and subcellular and molecular study of meat tenderness. Mailing Add: Dept of Animal Husb Mich State Univ East Lansing MI 48824

MERKEL, TIMOTHY FRANKLIN, b Jersey Shore, Pa, June 24, 42; m 66; c 2. ORGANIC POLYMER CHEMISTRY. Educ: Lycoming Col, AB, 64; Pa State Univ, MS, 66; Univ Mich, PhD(org chem), 73. Prof Exp: Res chemist, Whitmoyer Labs Inc, Rohm & Haas Co, 66-68; MGR PROD DEVELOP, SARTOMER CO, 73- Mem: Am Chem Soc; AAAS; Sigma Xi. Res: Specialty monomers, cyclic azo compounds. Mailing Add: Sartomer Co Nields & Bolmar Sts West Chester PA 19380

MERKEN, HENRY, b Peabody, Mass, Oct 14, 29; m 53; c 4. POLYMER CHEMISTRY. Educ: Northeastern Univ, BS, 53. Prof Exp: Asst engr, Res & Develop Dept, Am Polymer Corp, Mass, 49-53, develop engr, 53; develop engr, Polyco Dept, Borden Co, 55-56; develop engr, Polyvinyl Chem, Inc, 56-64, dir mfg, 64-68, asst to pres, 68-71, vpres int, 71-72; DIR INT OPER, BEATRICE CHEM, 72- Concurrent Pos: Instr, Lowell Tech Inst, 59-62. Mem: Am Chem Soc; Am Inst Chem Eng. Res: Organic chemistry; emulsion polymers; plasticizers. Mailing Add: 730 Main St Wilmington MA 01887

MERKEN, MELVIN, b Peabody, Mass, Jan 19, 27; m 56; c 3. CHEMISTRY, SCIENCE EDUCATION. Educ: Tufts Univ, BS, 50, AM, 51; Boston Univ, EdD(sci ed), 67. Prof Exp: Teacher high schs, Conn, 51-58; assoc prof, 58-67, PROF CHEM WORCESTER STATE COL, 67-, CHMN DEPT, 58- Mem: Fel AAAS; fel Am Inst Chem; Am Chem Soc; Am Asn Physics Teachers. Res: Teaching science to non-scientists in general education program at college level; promotion of scientific literacy and understanding; environmental chemistry. Mailing Add: Dept of Chem Worcester State Col Worcester MA 01602

MERKER, JERRY WHEELER, b Salina, Kans, Jan 17, 41; m 61; c 2. REPRODUCTIVE PHYSIOLOGY. Educ: Kans State Univ, BS, 64, MS, 69, PhD(animal breeding), 70. Prof Exp: Fel reproduction, Fertil & Gamete Physiol Training Prog, Woods Hole, Mass, 70; fel biochem, Univ Mass, 70-72; sr scientist, Ayerst ResL abs, Res Labs, Montreal, 72-75; LECTR BIOL, TEX A&M UNIV, 75- Mem: AAAS; Sigma Xi; Soc Study Reproduction; Am Soc Zoologists. Res: Use of brain multiple unit activity to access neurological changes in response to hormones and neurotransmitters. Mailing Add: Dept of Biol Tex A&M Univ College Station TX 77843

MERKER, MILTON, b New York, NY, Sept 15, 41; m 63; c 2. COSMIC RAY PHYSICS, ASTROPHYSICS. Educ: City Col New York, BS, 63; NY Univ, MS, 65, PhD(physics), 70. Prof Exp: Res asst physics, NY Univ, 65-70; res assoc, 69-71, RES ASST PROF ASTROPHYS, UNIV PA, 71-, ASST CHMN DEPT, 73- Mem: AAAS; Am Phys Soc; Am Geophys Union. Res: Cosmic rays; astrophysical spallation and heavy-ion reactions; quasars; nuclear cascade; high-energy shielding and radiologic dosimetry; atmospheric neutrons. Mailing Add: Dept of Astron Univ of Pa Philadelphia PA 19174

MERKER, PHILIP CHARLES, b New York, NY, July 23, 22; m 52; c 2. PHARMACOLOGY. Educ: Brooklyn Col, BA, 46; Long Island Univ, BS, 51; Purdue Univ, MS, 53, PhD, 55. Prof Exp: Lab asst physiol, Brooklyn Col, 46-47; teaching asst mat med & chmn, Long Island Univ, 48-51; asst pharm, Purdue Univ, 51-53; asst, Sloan-Kettering Inst Cancer Res, 56-62, head sect, 58-62, assoc mem, 62; prof pharmaceut, Col Pharm, Univ Tenn, 62-64; chmn dept pharmacol & animal sci, Col Pharmaceut Sci, Columbia Univ, 65, prof pharmacol & chmn div biol sci & pharmacol, 65-72; PHARMACOLOGIST, ARTHUR D LITTLE, INC, 72- Concurrent Pos: Res fel, Sloan-Kettering Inst Cancer Res, 54-56; asst prof, Sloan-Kettering Div, Cornell Univ, 58-62. Mem: AAAS; Am Asn Cancer Res; Am Soc Exp Path; Am Soc Pharmacol & Exp Therapeut. Res: Experimental cancer chemotherapy; drug screening; crude drug extraction for glycosides and alkaloids; hormone metabolism; organic synthesis-brominations; biochemical pharmacology. Mailing Add: 144 Commonwealth Ave Boston MA 02116

MERKER, ROBERT LESLIE, physical chemistry, see 12th edition

MERKES, EDWARD PETER, b Chicago, Ill, Apr 14, 29; m 56. MATHEMATICS. Educ: DePaul Univ, BS, 50; Northwestern Univ, PhD, 58. Prof Exp: Lectr math, De Paul Univ, 50-54, instr, 56-58, asst prof, 58-59; asst prof, Marquette Univ, 59-63, assoc prof, 62-63; assoc prof, 63-69, PROF MATH, UNIV CINCINNATI, 69-, HEAD DEPT, 70- Concurrent Pos: Vis assoc prof, Math Res Ctr, Univ Wis, 62-63. Mem: Am Math Soc; Math Asn Am. Res: Complex variable and continued fractions. Mailing Add: 3108 Hanna Cincinnati OH 45211

MERKL, MARVIN EUGENE, b Ala, July 23, 21; m 51; c 4. ENTOMOLOGY. Educ: Ala Polytech Inst, BS, 48, MS, 50; Agr & Mech Col, Tex, PhD(entom), 53. Prof Exp: Asst entom, Ala Polytech Inst, 48-50, 50-51; asst, Agr & Mech Col Tex, 51-53; entomologist in charge field eval methods & mat, Bol Weevil Res Lab, 53-55, asst in charge, 55-63, ENTOMOLOGIST & HEAD BOLL WEEVIL RES LAB, ENTOM RES LAB, AGR RES SERV, USDA, STONEVILLE, 63- Mem: Entom Soc Am. Res: Life history and chemical control of corn ear worm; control of cotton insects with chlorinated compounds; control of boll weevil with various dust formulations.

MERKLE, F HENRY, b Newark, NJ, Aug 31, 31; m; c 3. PHARMACEUTICAL CHEMISTRY. Educ: Rutgers Univ, BS, 54, MS, 61, PhD(pharmaceut sci), 64. Prof Exp: Instr pharm, Rutgers Univ, 58-62, lectr, 62-63; res scientist, Res Ctr, FMC Corp, NJ, 64-65; sr res scientist, 65-70, dept head pharmaceut prod develop, 70-75, MGR PHARM PROD DEVELOP, BRISTOL MYERS PROD, 75- Mem: Am Pharmaceut Asn; Am Chem Soc. Res: Pharmaceutical analysis, products, and development. Mailing Add: 2217 Shawnee Path Scotch Plains NJ 07076

MERKLE, JOHN, b Norman, Okla, Feb 2, 13; m 47; c 1. ECOLOGY. Educ: Univ Okla, BA, 36; Ore State Col, PhD(bot), 48. Prof Exp: Asst bot, Univ Okla, 36-37 & Mich State Col, 37-39; asst, Ore State Col, 39-42, instr, 47-48; asst prof, Agr & Mech Col Tex, 48-55; vis prof, Purdue Univ, 55-56; INSTR BOT, C S MOTT COMMUNITY COL, 56- Mem: Ecol Soc Am; Bot Soc Am; Wildlife Soc. Res: Plant community analysis; plant taxonomy and geography; soil science; paleobotany. Mailing Add: Dept of Sci & Math C S Mott Community Col Flint MI 48503

MERKLE, OWEN GEORGE, b Meade, Kans, Nov 22, 29; m 52; c 5. PLANT BREEDING, GENETICS. Educ: Okla State Univ, BS, 51, MS, 54; Tex A&M Univ, PhD(plant breeding), 63. Prof Exp: RES AGRONOMIST, US DEPT AGR, 57- Mem: Am Soc Agron; Crop Sci Soc Am. Res: Plant breeding; breeding wheat for disease resistance; improved quality and agronomic characters and associated genetic studies. Mailing Add: USDA-ARS Agron Dept Okla State Univ Stillwater OK 74074

MERKLEY, JOSEPH HERBERT, organometallic chemistry, see 12th edition

MERKLEY, WAYNE BINGHAM, b Murray, Utah, Apr 1, 41; m 59; c 4. POLLUTION BIOLOGY, AQUATIC ECOLOGY. Educ: Univ Utah, BS, 63, MA, 66, PhD(limnol), 69. Prof Exp: Instr biol, Univ Utah, 68; asst prof, 69-73, ASSOC PROF BIOL, DRAKE UNIV, 73- Mem: Am Inst Biol Sci; Am Soc Limnol & Oceanog; Water Pollution Control Fedn; NAm Benthol Soc. Res: Ecological impact of impoundments and urban areas on aquatic environments. Mailing Add: Dept of Biol Drake Univ Des Moines IA 50311

MERLIS, JEROME K, b New York, NY, Feb 24, 14; m 41; c 3. NEUROPHYSIOLOGY. Educ: Univ Louisville, BS, 33, MD, 37, MS, 38; Am Bd Psychiat & Neurol, dipl, 48. Prof Exp: Asst physiol, Sch Med, Univ Louisville, 37-38, from instr to asst prof, 38-46; neurologist in chg, Nat Vet Epilepsy Ctr, Cushing Vet Admin Hosp, Mass, 46-52, chief, 52-56; assoc prof physiol, 59-75, PROF NEUROL & CLIN NEUROPHYSIOL, SCH MED, UNIV MD, BALTIMORE CITY, 56-, HEAD EEG DEPT, UNIV HOSP, 56- Concurrent Pos: Coman fel, Univ Chicago, 39; Commonwealth Fund fel, Yale Univ, 40-41; ed, Epilepsia, 52-56; clin assoc neurol, Mass Gen Hosp, 53-56; asst, Harvard Univ, 53-54, instr, 54-56. Mem: Am Neurol Asn; Am Acad Neurol; Am Electroencephalog Soc (secy, 56-59, pres, 60); Am Epilepsy Soc (secy-treas, 51-55, pres, 57); Int League Against Epilepsy (treas, 57-65, vpres, 65-68, pres, 68-69). Res: Electroencephalography; experimental epilepsy; neurology. Mailing Add: EEG Dept Univ of Md Hosp Baltimore MD 21201

MERLIS, SIDNEY, b New York, NY, Apr 13, 25; m 46; c 3. PSYCHIATRY,

NEUROLOGY. Educ: Creighton Univ, 46, MD, 48. Prof Exp: Sr psychiatrist, 51-53, supv psychiatrist, 53-56, DIR RES, CENT ISLIP PSYCHIAT CTR, 56- Concurrent Pos: Chief psychiat, Southside Hosp, 53, dir EEG Labs; dir EEG Labs, Smithtown Gen Hosp & Cent Gen Hosp; consult, EEG, South Oaks Hosp; vis prof, State Univ NY Stony Brook, clin prof, Sch Med; fac assoc, Hofstra Univ. Mem: AAAS; Soc Biol Psychiat; Am Electroencephalog Soc; Am Psychiat Asn; Am Col Phys. Res: Nervous and mental diseases; psychopharmacology; electroencephalography. Mailing Add: Cent Islip Psychiat Ctr Central Islip NY 11722

MERMELSTEIN, ROBERT, b Mukacevo, Czech; Can citizen. POLYMER CHEMISTRY. Educ: Sir George Williams Univ, BSc, 57; Univ Alta, PhD(phys org chem), 64. Prof Exp: Fel, Brandeis Univ, 64-65 & Childrens' Cancer Res Found, Boston, 65-66; SCIENTIST POLYMER CHEM, XEROX CORP, 66- Mem: AAAS; Am Chem Soc. Res: Synthesis, characterization of vinyl and condensation polymers; biopolymers; structure-activity relationships; rheological behaviour; kinetics and mechanism of organic reactions. Mailing Add: 345 Pelham Rd Rochester NY 14610

MERMIN, N DAVID, b New Haven, Conn, Mar 30, 35; m 57. PHYSICS. Educ: Harvard Univ, AB, 56, AM, 57, PhD(physics), 61. Prof Exp: NSF fel physics, Univ Birmingham, 61-63; res assoc, Univ Calif, San Diego, 63-64; from asst prof to assoc prof, 64-72, PROF PHYSICS, CORNELL UNIV, 72- Concurrent Pos: Alfred P Sloan Found fel, 66-70; John Simon Guggenheim Found fel, 70-71. Mem: Fel Am Phys Soc. Res: Theoretical solid state and statistical physics. Mailing Add: Dept of Physics Cornell Univ Ithaca NY 14853

MERNER, RICHARD RAYMOND, b Chicago, Ill, Sept 23, 18; m 51; c 2. INDUSTRIAL ORGANIC CHEMISTRY, SCIENCE ADMINISTRATION. Educ: Univ Ill, BS, 39; Northwestern Univ, PhD(chem), 49. Prof Exp: Asst chem, Univ Mo, 39-40; asst chem electrochem dept, Res & Develop, 40-44, org chem dept, Res, 49-53, tech supvr process develop, 53-67, SUPVR TECH EMPLOY & PERSONNEL DEVELOP, E I DU PONT DE NEMOURS & CO, INC, 67- Concurrent Pos: Adj prof, Col Bus & Econs, Univ Del, 70- Mem: AAAS; Am Chem Soc; Am Inst Chem Eng. Res: Development research; intermediates; dyes and pigments; fluorocarbons; behavior science. Mailing Add: RD 2 Box 326 Sullivan Rd Avondale PA 19311

MEROLA, A JOHN, b Freehold, NJ, July 21, 31; m 58; c 3. BIOCHEMISTRY, MICROBIOLOGY. Educ: Univ Tex, BA, 53; Rutgers Univ, MS, 59, PhD(bact), 61. Prof Exp: Fel biochem, Enzyme Inst, Univ Wis, 61-63; res biologist, Sterling-Winthrop Res Inst, 63-65; from asst prof to assoc prof, 66-73, PROF PHYSIOL CHEM, OHIO STATE UNIV, 73- Res: Energy conservation; drug and cholesterol metabolism; electron transport. Mailing Add: Dept of Physiol Chem Ohio State Univ Columbus OH 43210

MERONEY, WILLIAM HYDE, III, b Murphy, NC, Dec 27, 17; m 52. INTERNAL MEDICINE. Educ: Univ NC, BS, 43; NY Univ, MD, 45; Am Bd Internal Med, dipl. Prof Exp: Instr pharmacol, Sch Med, Univ NC, 43; Med Corps, US Army, 46-75; instr internal med, Yale Univ, 50-51, lectr, 51-52, chief renal insufficiency ctr, Korea, 53, from res clinician to chief dept metab, Walter Reed Army Inst Res, 53-57, dep dir inst, 61-64, dir trop res med lab, San Juan, PR, 57-61, chief res div, Med Res & Develop Command, Washington, DC, 64-65, dep dir personnel & training directorate, Off Surgeon Gen, 66-68, dir & commandant, Walter Reed Army Inst Res, 68-71, commanding gen, Walter Reed Gen Hosp, 71-72, commanding gen, Madigan Army Med Ctr, Med Corps, US Army, 72-75; CONSULT MED, DEPT HEALTH, PROVIDENCE, RI, 75- Concurrent Pos: Asst clin prof, Sch Med, Georgetown Univ, 57; clin assoc prof, Univ PR, 57-61; consult, Bayamon Dist Hosp, PR & Surgeon Gen, US Army. Mem: AAAS; Endocrine Soc; Soc Exp Biol & Med; AMA; fel Am Col Physicians. Res: Metabolic processes. Mailing Add: 602 Black Point Farm Portsmouth RI 02871

MEROW, WILLIAM WAYNE, b Sparta, Wis, Mar 22, 22; m 44; c 2. ORTHODONTICS. Educ: Univ Md, DDS, 51, Ohio State Univ, MS, 61. Prof Exp: Instr dent anat, Baltimore Col Dent Surg, Univ Md, 51-52; instr orthod, Col Dent, Ohio State Univ, 61-65; PROF ORTHOD & CHMN DEPT, SCH DENT, W VA UNIV, 65- Concurrent Pos: Orthod consult, Children's Hosp, Columbus, 61-65. Mem: Am Dent Asn; Am Asn Orthod. Res: Growth and development, including proportional growth changes in the facial and cranial skeleton of the growing child; force distribution accompanying rapid palatal expansion. Mailing Add: Sch of Dent WVa Univ Morgantown WV 26506

MERRELL, DAVID JOHN, b Bound Brook, NJ, Aug 20, 19; m 45; c 4. GENETICS. Educ: Rutgers Univ, BS, 41; Harvard Univ, MA, 47, PhD(zool), 48. Prof Exp: From instr to assoc prof, 48-64, PROF ZOOL, UNIV MINN, MINNEAPOLIS, 64- Mem: Soc Study Evolution; Genetics Soc Am; Am Genetics Asn; Am Soc Nat; Behav Genetics Asn. Res: Ecological and behavioral genetics. Mailing Add: Dept of Zool Univ of Minn Minneapolis MN 55455

MERRELL, THEODORE REED, JR, b Superior, Wis, June 12, 23; m 46; c 4. FISH BIOLOGY. Educ: St Olaf Col, BA, 48; Univ Mich, MS, 49. Prof Exp: Res biologist, State Fish Comn, Ore, 49-56; chief salmon ecol, 56-67, COORDR ECOL PROGS, NAT MARINE FISHERIES SERV, DEPT COMMERCE, 67- Mem: Am Fisheries Soc (pres, 75-76); Inst Fisheries Res Biol. Res: Salmon ecology; fish passage at dams; marine pollution; intertidal biological baselines for oil pollution. Mailing Add: Rte 5 Box 5679 Juneau AK 99801

MERRELL, WILLIAM JOHN, b Grand Island, Nebr, Feb 16, 43. PHYSICAL OCEANOGRAPHY. Educ: Sam Houston State Col, BS, 65, MA, 67; Tex A&M Univ, PhD(oceanog), 71. Prof Exp: Res assoc oceanog, Tex A&M Univ, 71-72; mgr mgt systs, Off Int Decade Ocean Explor, 72-73; res assoc oceanog, Tex A&M Univ, 73-74; staff assoc, 74-76, EXEC OFFICER, OFF INT DECADE OCEAN EXPLOR, NSF, 76- Res: Dynamics of internal waves; descriptive physical oceanography; marine resource management. Mailing Add: Off Int Decade Ocean Explor NSF 1800 G St NW Washington DC 20550

MERRENS, HARRY ROY, b Salford, Eng, July 21, 31; US citizen. GEOGRAPHY. Educ: Univ London, BA, 54; Univ Md, MA, 57; Univ Wis, PhD(geog), 62. Prof Exp: Actg instr geog, Rutgers Univ, 60-61; actg instr, Univ Wis, 61-62; asst prof, San Fernando Valley State Col, 62-65; vis lectr, Univ Wis, 65-66; assoc prof, San Fernando Valley State Col, 67-68; assoc prof, 68-73, PROF GEOG, YORK UNIV, 73- Concurrent Pos: Guggenheim fel, 66-67. Honors & Awards: Gerfurth Award, Univ Wis, 65. Mem: Asn Am Geog; Am Geog Soc; Can Asn Geographers. Res: Historical geography of North America, with emphasis upon the colonial Atlantic seaboard. Mailing Add: Dept of Geog York Univ Downsview ON Can

MERRIAM, ALAN PARKHURST, b Missoula, Mont, Nov 1, 23; div; c 3. ANTHROPOLOGY. Educ: Mont State Univ, BA, 47; Northwestern Univ, MM, 48, PhD(anthrop), 51. Prof Exp: Instr anthrop, Northwestern Univ, 53-54; asst prof sociol & anthrop, Univ Wis-Milwaukee, 54-56; from asst prof to assoc prof anthrop, Northwestern Univ, 56-62; chmn dept anthrop, 66-69, PROF ANTHROP, IND UNIV, BLOOMINGTON, 62- Concurrent Pos: Belg-Am Educ Found & Wenner-Gren Found Anthrop Res grant, Belg Congo & Ruanda Urundi, 51-52; ed, Ethnomusicology, Soc Ethnomusicol, 53-58; rev ed, J Am Folklore, 57-58; Northwestern Univ grant field res with Flathead Indians, Western Mont, 58-59; mem comt human resources in Cent Africa, Nat Res Coun, 58; NSF & Belg-Am Educ Found grant ethnog & ethnomusicol res with Basongye people, Repub of Congo, 59-60 & Guggenheim Mem Found fel, 69-70; mem, President's Task Force for Africa, 60; mem joint comt African studies, Soc Sci Res Coun-Am Coun Learned Socs, 60-66, chmn, 62-65; mem Africa nat screening comt, Foreign Area Fel Prog, 68-69; Soc Sci Res Coun, Am Coun Learned Socs, Joint Comt African Studies fel, Ind Univ, 73; mem bd dirs, Comt Res on Dance, 75-76. Mem: Am Folklore Soc; Soc Ethnomusicol (vpres, 60-62, pres, 62-64); African Studies Asn; Am Anthrop Asn. Res: Ethnology, ethnomusicology and general arts in Africa and among North American Indians. Mailing Add: Dept of Anthropology Ind Univ Bloomington IN 47401

MERRIAM, CHARLES N, physical chemistry, see 12th edition

MERRIAM, CHARLES WARREN, geology, paleontology, see 12th edition

MERRIAM, DANIEL FRANCIS, b Omaha, Nebr, Feb 9, 27; m 46; c 5. GEOLOGY. Educ: Univ Kans, BS, 49, MS, 53, PhD, 61; Univ Leicester, MSc, 69, DSc, 75. Prof Exp: Geologist, Union Oil Co, Calif, 49-51; asst instr geol, Univ Kans, 51-53; geologist, Kans Geol Surv, 53-58, div head basic geol, 58-63, chief geol res, 63-71; JESSIE PAGE HEROY PROF GEOL & CHMN DEPT, SYRACUSE UNIV, 71- Concurrent Pos: Res assoc, Univ Kans, 63-71, instr, 54, 71; vis res scientist, Stanford Univ, 63; Fulbright-Hays sr res fel, UK, 64-65; dir, Am Geol Inst Int Field Inst, Japan, 67; vis prof geol, Wichita State Univ, 68-70; Am Geol Inst Vis Geol Scientist, 69; mem, Sci Comt 4, 75-, chmn, 76; participant, Project COMPUTe, Dartmouth Col, 74. Honors & Awards: Erasmus Haworth Grad Award Geol, Univ Kans, 55. Mem: Fel AAAS; fel Geol Soc London; Int Asn Math Geol; fel Geol Soc Am; Am Asn Petrol Geol. Res: Carboniferous and Mesozoic stratigraphy; geologic history of the mid-continent; cyclic sedimentation; petroleum geology; computers and computer applications in the earth sciences; quantitative stratigraphic analysis. Mailing Add: Dept of Geol 204 Heroy Geol Lab Syracuse Univ Syracuse NY 13210

MERRIAM, ESTHER VIRGINIA, b Pittsburgh, Pa, Apr 9, 40; m 63; c 1. VIROLOGY, BIOCHEMISTRY. Educ: Elizabethtown Col, BS, 62; Univ Wash, PhD(biochem), 66. Prof Exp: USPHS fel biol div, Oak Ridge Nat Lab, 66-67; fel, Calif Inst Technol, 67-69; actg asst prof molecular biol in bact, Univ Calif, Los Angeles, 69-71; asst prof, Loyola Univ Los Angeles, 71-74, ASSOC PROF BIOL, LOYOLA MARYMOUNT UNIV, 74- Mem: Am Soc Microbiol; Genetics Soc Am. Res: Nucleic acid interactions, especially in connection with viral systems. Mailing Add: Dept of Biol Loyola Marymount Univ Los Angeles CA 90045

MERRIAM, FREDERIC CUTTER, organic chemistry, see 12th edition

MERRIAM, GEORGE RENNELL, JR, b Harrisburg, Pa, May 22, 13; m 36; c 4. MEDICINE. Educ: Brown Univ, AB, 34; Columbia Univ, MD, 41; Am Bd Ophthal, dipl, 49. Prof Exp: Instr ophthal, 49-56, assoc, 56-59, instr, Univ, 49-56, PROF CLIN OPHTHAL, COL PHYSICIANS & SURGEONS, COLUMBIA UNIV, 59- Concurrent Pos: Asst ophthalmologist, Presby Hosp, New York, 49-56, attend ophthalmologist, 56-; asst ophthalmologist, Mem Hosp, New York, 49-57, assoc ophthalmologist, 57-59, ophthalmologist, 59-69; assoc ophthalmologist, Francis Delafield Hosp, 51- Mem: Am Ophthal Soc; Am Radium Soc; Asn Res Ophthal; fel Am Col Surgeons; AMA. Res: Ophthalmic radiotherapy; cataracts; relative biological effectiveness of various qualities of radiation. Mailing Add: Edward S Harkness Eye Inst Columbia Univ New York NY 10032

MERRIAM, HOWARD GRAY, b Smithville, Ont, July 8, 32; m 56; c 2. ECOLOGY. Educ: Univ Toronto, BSA, 56; Cornell Univ, PhD, 60. Prof Exp: Asst gen zool, Cornell Univ, 56-58, asst animal ecol, 58-60; from asst prof to assoc prof animal ecol, Univ Tex, 60-68; ASSOC PROF BIOL, CARLETON UNIV, 68- Concurrent Pos: Consult, Can Ministry State Urban Affairs, 73-74, Can Wildlife Serv, 74-75 & Parks Can, 75- Mem: Can Soc Zoologists; Can Soc Environ Biologists; AAAS; Ecol Soc Am; Am Soc Mammal. Res: Population ecology and quantitative autecology; ecology of land isopods and marmots; ecology of decomposer ecosystems; heterogencity in natural systems. Mailing Add: Dept of Biol Carleton Univ Ottawa ON Can

MERRIAM, JOHN ROGER, b Kenosha, Wis, Jan 6, 40; m 63; c 2. GENETICS. Educ: Univ Wis, BS, 62; Univ Wash, MS, 63, PhD(genetics), 66. Prof Exp: USPHS fels biol, Oak Ridge Nat Lab, 66-67 & Calif Inst Technol, 67-69; asst prof, 69-74, ASSOC PROF GENETICS, UNIV CALIF, LOS ANGELES, 74- Concurrent Pos: Vis fel, Res Sch Biol Sci, Australian Nat Univ, 75-76. Mem: AAAS; Genetics Soc Am. Res: Neurological genetics of Drosophila; gene regulation; chromosome mechanics; somatic crossing over and mosaic analysis of development. Mailing Add: Dept of Biol Univ of Calif Los Angeles CA 90024

MERRIAM, LAWRENCE CAMPBELL, JR, b Portland, Ore, Aug 31, 23; m 47; c 5. FOREST MANAGEMENT. Educ: Univ Calif, BS, 48; Ore State Univ, MF, 58, PhD(forest mgt), 63. Prof Exp: Log scaler-compassman, Shasta Forests Co, Calif, 48; forestry aide, Ore Bur Land Mgt, 49; retail sales millworker, Willamette Nat Lumber Co, 49-50; log pond foreman bookkeeper, M&M Woodworking Co, 50; state parks historian, planner & forester state parks div, Ore State Hwy Dept, 51-59; from asst prof to assoc prof forestry, Univ Mont, 59-66; PROF FORESTRY, UNIV MINN, ST PAUL, 66- Concurrent Pos: Consult, Bur Land Mgt, DC, 65-66 & UN Food & Agr Orgn, Paraguay, 69. Mem: Soc Am Foresters; fel Soc Park & Recreation Educr; Nat Parks Asn. Res: Park wilderness management and policy. Mailing Add: Col of Forestry Univ of Minn St Paul MN 55108

MERRIAM, RICHARD HOLMES, b San Marcos, Calif, Nov 22, 12. GEOLOGY. Educ: Pomona Col, BA, 34; Univ Calif, PhD(geol), 40. Prof Exp: Field asst, Yale Univ, 41; rodman, US Geol Surv, 41; asst geologist, US Bur Reclamation, 41-42, assoc geologist, 43-46; asst prof mineral petrol, Colo Sch Mines, 46-47; asst prof, Mt Holyoke Col, 47-48; from asst prof to assoc prof, 48-56, PROF GEOL, UNIV SOUTHERN CALIF, 56- Concurrent Pos: Guggenheim fel, 56. Mem: Fel Geol Soc Am. Res: Geology of Southern California peninsular range; concrete aggregates; engineering geology; geology of southwest part of Romana Quadrangle, California. Mailing Add: Dept of Geol Univ of Southern Calif Los Angeles CA 90007

MERRIAM, ROBERT ARNOLD, b Keokuk, Iowa, Apr 30, 27; m 53; c 3. FOREST HYDROLOGY. Educ: Iowa State Univ, BS, 51; Univ Calif, Berkeley, MS, 57. Prof Exp: Range conservationist, Calif Forest & Range Exp Sta, 53-55, res forester, Pac Southwest Forest & Range Exp Sta, 55-60, Intermountain Forest & Range Exp Sta, 60-63 & Pac Southwest Forest & Range Exp Sta, 63-73, ASST MGR HAWAII WATER RESOURCES REGIONAL STUDY, PAC SOUTHWEST FOREST & RANGE EXP STA, US FOREST SERV, 73- Mem: Soc Am Foresters; Am Geophys Union; Int Asn Sci Hydrol. Res: Watershed management; soil moisture measurement techniques, including neutron probe; interception and fog drip; river basin surveys. Mailing Add: 616 Pamaele St Kailua HI 96734

MERRIAM, ROBERT WILLIAM, b Waverly, Iowa, Nov 21, 23; m 50; c 2. DEVELOPMENTAL BIOLOGY. Educ: Univ Iowa, AB, 47; Ore State Univ, MS, 49; Univ Wis, PhD(zool), 53. Prof Exp: Asst zool, Ore State Univ, 47-50 & Univ Wis, 50-53; from instr to asst prof, Univ Pa, 53-60; ASSOC PROF BIOL, STATE UNIV NY STONY BROOK, 60- Concurrent Pos: Vis fel zool, Columbia Univ, 60-61; NIH spec fel, Oxford Univ, 67-68. Honors & Awards: Lalor Found Awards, 55 & 57. Mem: Am Soc Zool; Am Inst Biol Sci; Soc Develop Biol; Am Soc Cell Biol. Res: Cellular biology; nuclear-cytoplasmic interaction; control mechanisms in oogenesis. Mailing Add: Dept of Cellular & Comp Biol State Univ of NY Stony Brook NY 11794

MERRIAM, WILLIS BUNGAY, b Spokane, Wash, May 14, 05; m 28; c 2. GEOGRAPHY, ANTHROPOLOGY. Educ: Univ Wash, BS, 31, MS, 33, PhD(geog), 45. Prof Exp: Instr geog, Eastern Wash State Col, 37-40; asst prof, Ore Col Educ, 40-42 & Univ Ore, 42-45; from asst prof to prof, 45-70, EMER PROF GEOG, WASH STATE UNIV, 70- Mem: Fel AAAS. Res: Regional, historic and political geography of the Pacific Northwest, North America and Europe, including Union of Soviet Socialist Republics. Mailing Add: SE 420 Derby Pullman WA 99163

MERRICK, ARTHUR WEST, b Great Falls, Mont, Dec 22, 17; m 45; c 5. PHYSIOLOGY. Educ: Univ Mont, AB & BS, 50; Univ Mo, MA, 52, PhD(physiol), 54. Prof Exp: Asst physiol, Univ Mo, 51-52, asst instr, 53-54; instr, Univ Kans, 54-55; from asst prof to assoc prof, Med Ctr, Univ Mo-Columbia, 55-68; prof, Ill State Univ, 68-72; HEALTH SCIENTIST ADMINR, NAT HEART & LUNG INST, 72- Concurrent Pos: Wyeth Drug Corp fel, 61-62; Nat Heart & Lung Inst grant; exec secy, Rev Br, Div Extramural Affairs, Nat Heart & Lung Inst, 73- Mem: Fel AAAS; Am Physiol Soc; Am Heart Asn. Res: Carbohydrate metabolism of cardiac and nervous tissue; intrinsic nervous system of mammalian heart. Mailing Add: Rev Br Div Extramural Affairs Nat Heart & Lung Inst Bethesda MD 20014

MERRICK, JOSEPH M, b Welland, Ont, Mar 20, 30; m 55; c 3. BIOCHEMISTRY, MICROBIOLOGY. Educ: Mich State Univ, BS, 51, MS, 53; Univ Mich, PhD(biochem), 58. Prof Exp: Arthritis & Rheumatism Found fel, Univ Calif, 59-61; assoc biochem, State Univ NY Buffalo, 61-62, asst prof, 62-65; assoc prof bact & bot, Syracuse Univ, 65-70; PROF MICROBIOL, STATE UNIV NY BUFFALO, 70- Mem: Am Soc Microbiol; Brit Biochem Soc; Am Soc Biol Chem. Res: Metabolism of poly-beta-hydroxybutyrate by bacteria; mechanism of enzyme secretion by bacteria. Mailing Add: Dept of Microbiol State Univ of NY Buffalo NY 14207

MERRIELL, DAVID MCCRAY, b Minneapolis, Minn, Oct 25, 19; m 51; c 2. MATHEMATICS. Educ: Yale Univ, BA, 41; Univ Chicago, MS, 47, PhD(math), 51. Prof Exp: Instr math, Univ Chicago, 49-51; asst prof, Robert Col, Turkey, 51-54, assoc prof & head dept, 54-57; from asst prof to assoc prof, Univ Calif, Santa Barbara, 57-68; chmn dept math, 71-74, PROF MATH, VASSAR COL, 68- Mem: Am Math Soc; Math Asn Am. Res: Non-associative algebras; graphy theory. Mailing Add: Dept of Math Vassar Col Poughkeepsie NY 12601

MERRIFIELD, D BRUCE, b Chicago, Ill, June 13, 21; m 49; c 3. PHYSICAL ORGANIC CHEMISTRY. Educ: Princeton Univ, BS, 42; Univ Chicago, MS, 48, PhD(phys org chem), 50. Prof Exp: Res chemist, Monsanto Co, 50-56; group leader res, Tex-US Chem Co, 56-60, mgr polymer res, 60-63; dir res, Petrolite Corp, 63-68; dir res, Res Ctr, 68-70, VPRES RES & DEVELOP, RES CTR, HOOKER CHEM CORP, 70- Mem: Fel AAAS; fel Am Inst Chemists; Indust Res Inst; Am Chem Soc. Res: Mechanisms of free radical reactions; oxidation mechanisms; polymer and surface chemistry; electrochemistry and electronics. Mailing Add: Res Ctr Hooker Chem Corp Box 61569 Houston TX 77208

MERRIFIELD, PAUL ELLIOTT, b Springvale, Maine, Dec 31, 22; m 44; c 4. COLLOID CHEMISTRY. Educ: Colby Col, AB, 47; Rice Inst, AM, 49, PhD(chem), 51. Prof Exp: Res chemist, 51-59, plant chief chemist, 60-67, MGR FELT MFG, ARMSTRONG CORK CO, 67- Mem: Am Chem Soc. Res: Colloidal properties of fibers; fiber products development. Mailing Add: 615 W First St Fulton NY 13069

MERRIFIELD, RICHARD EBERT, b Seattle, Wash, Feb 18, 29; m 56; c 2. CHEMICAL PHYSICS. Educ: Mass Inst Technol, PhD(phys chem), 53. Prof Exp: Res chemist, 53-59, RES SUPVR CENT RES DEPT, E I DU PONT DE NEMOURS & CO, INC, 59- Mem: Am Phys Soc. Res: Molecular spectra and structure; solid state theory; exciton physics; physics of molecular crystals. Mailing Add: 2633 Longwood Dr Wilmington DE 19810

MERRIFIELD, ROBERT BRUCE, b Ft Worth, Tex, July 15, 21; m 49; c 6. BIOCHEMISTRY. Educ: Univ Calif, Los Angeles, BA, 43, PhD(chem), 49. Hon Degrees: DSc, Univ Colo, 69 & Yale Univ, 71; PhD, Uppsala Univ, 70. Prof Exp: Chemist, Philip R Park Res Found, 43-44; asst chem, Med Sch, Univ Calif, Los Angeles, 48-49; asst biochem, 49-53, assoc, 53-57, from asst prof to assoc prof, 57-66, PROF BIOCHEM, ROCKEFELLER UNIV, 66- Concurrent Pos: Nobel guest prof, Uppsala Univ, 68; assoc ed, Int J Peptide & Protein Res, 69- Honors & Awards: Lasker Award Basic Med Res, 69; Gairdner Award, 70; Intra-Sci Award, 70; Award for Creative Work in Synthetic Org Chem, Am Chem Soc, 72, Nichols Medal, 73. Res: Development of solid phase peptide synthesis, first synthesis of an enzyme; relation of structure to function in synthetic, biologically active peptides and proteins. Mailing Add: Rockefeller Univ 66th St & York Ave New York NY 10021

MERRIFIELD, ROBERT G, b Carthage, Mo, July 26, 30; m 52; c 4. SILVICULTURE. Educ: Ark Agr & Mech Col, BS, 53; La State Univ, MF, 58; Duke Univ, DF(silvicult), 62. Prof Exp: From asst prof to assoc prof forestry, La State Univ, 58-67; PROF FORESTRY, TEX A&M UNIV, 67-, HEAD DEPT FOREST SCI, 69- Mem: Soc Am Foresters. Res: Artificial regeneration and plantation management of southern pines; intensive culture of pulping hardwood species. Mailing Add: Dept of Forest Sci Tex A&M Univ College Station TX 77843

MERRIGAN, JOSEPH A, b Maryville, Mo, Feb 8, 40; m 62; c 2. PHOTOGRAPHIC CHEMISTRY. Educ: Northwest Mo State Col, BS, 62; Univ Nebr, MS, 65, PhD(phys chem), 66; Mass Inst Technol, SM, 74. Prof Exp: Sr res chemist, Eastman Kodak Co, 67-69, res assoc & head silver halide chem lab, 69-74, head radiography lab, Kodak Res Labs, 74-75, HEAD SPEC PROCESSES LAB, KODAK RES LABS, 75- Mem: Am Chem Soc. Res: Hot atom reactions of neutron irradiated bromine with organic molecules; positronium interactions in solid systems; mechanisms of photographic recording; radiographic recording; photovoltaic cells. Mailing Add: Kodak Res Labs Kodak Park B59 Rochester NY 14650

MERRIL, CARL R, b Brooklyn, NY, Dec 6, 36; m 61; c 2. MOLECULAR BIOLOGY, MEDICINE. Educ: Col William & Mary, BS, 58; Georgetown Univ, MD, 62. Prof Exp: Intern med, USPHS Hosp, Boston, Mass, 62-63; res assoc molecular biol, Lab Neurochem Sect, Phys Chem, 63-65, mem staff, 65-69, SR STAFF SCIENTIST, LAB GEN & COMP BIOCHEM, NIMH, 69- Mem: AAAS; AMA; Biophys Soc; NY Acad Sci. Res: Bacteriophage interactions with eukaryotic systems; galactosemia; inborn metabolic diseases; gene transfer; primary structure of biopolymers. Mailing Add: Lab of Gen & Comp Biochem NIMH Bldg 36 Rm 3A15 Bethesda MD 20014

MERRILL, ANNE PATTERSON, biology, see 12th edition

MERRILL, CLAUDE IRVIN, organic chemistry, analytical chemistry, see 12th edition

MERRILL, DOROTHY, b Abington, Mass, Jan 1, 27. ZOOLOGY. Educ: Mass State Col Bridgewater, BS, 47; Univ Mich, AM, 59, PhD(zool), 64. Prof Exp: Teacher high sch, Mass, 47-60; from instr to asst prof zool, Smith Col, 64-70; assoc prof biol, Western Col, 70-74; ASSOC PROF BIOL, COL IV, GRAND VALLEY STATE COLS, 74- Mem: Am Soc Zool; Animal Behav Soc. Res: Neural mechanisms in insect behavior; biology of caddis larvae. Mailing Add: Col IV Grand Valley State Cols Allendale MI 49401

MERRILL, GLEN KENTON, b Columbus, Ohio, Aug 28, 35; m 64. GEOLOGY, PALEONTOLOGY. Educ: Ohio Univ, BS, 57; Univ Tex, Austin, MA, 64; La State Univ, PhD(geol), 68. Prof Exp: Instr geol, Northwestern La State Col, 64; asst prof, Monmouth Col, 68-71; asst prof, Univ Tex, Arlington, 71-74; ASST PROF GEOL, COL CHARLESTON, 74- Mem: Geol Soc Am; Paleont Soc; Soc Econ Paleontologists & Mineralogists; Nat Speleol Soc. Res: Biostratigraphy, paleoecology and systematics of lake Paleozoic conodonts; carbonate petrography. Mailing Add: Dept of Geol Col of Charleston Charleston SC 29401

MERRILL, GORDON CLARK, b Windsor, Ont, Sept 23, 19; m 50; c 2. GEOGRAPHY. Educ: McGill Univ, BA, 49, MA, 51; Univ Calif, PhD, 57. Prof Exp: Lectr geog, Univ Ind, 54-55; asst prof, McGill Univ, 55-57; assoc prof, Carleton Univ, 57-64, assoc dean arts, 66-69, dean, 69-71, PROF GEOG, CARLETON UNIV, 64- Concurrent Pos: Mem regional geog comt, Pan-Am Inst Geog & Hist, 66- Mem: Am Geog Soc; Asn Am Geog; Can Asn Geog. Res: Historical geography; geography of the Caribbean area. Mailing Add: Dept of Geog Carleton Univ Ottawa ON Can

MERRILL, HOWARD EMERSON, b Laconia, NH, Aug 6, 30; m 53; c 4. PETROLEUM CHEMISTRY. Educ: Stetson Univ, BS, 52; Univ Pittsburgh, PhD(chem), 57. Prof Exp: Sr res chemist, Esso Res Labs, Exxon Co, USA, 57-75, SR STAFF CHEMIST, EXXON CHEM CO USA, 75- Mem: AAAS. Res: Catalysis. Mailing Add: 5045 Sequoia Dr Baton Rouge LA 70814

MERRILL, JAMES ALLEN, b Cedar City, Utah, Oct 27, 25; m 49; c 4. OBSTETRICS & GYNECOLOGY, PATHOLOGY. Educ: Univ Calif, Berkeley, AB, 45, Univ Calif, San Francisco, MD, 48. Prof Exp: Fel path, Harvard Med Sch, 50-51; fel, Cancer Res Inst, Univ Calif, San Francisco, 58-61; from instr to asst prof obstet & gynec, Sch Med, Univ Calif, San Francisco, 57-61, asst clin prof path, 59-61, res asst, Cancer Res Inst, 58-61; PROF GYNEC & OBSTET & HEAD DEPT, SCH MED, UNIV OKLA, 61-, CONSULT PROF PATH, 61-, PROF CYTOTECHNOL, COL HEALTH REL PROFESSIONS, 70- Concurrent Pos: Markle scholar med sci, 57-62; consult, Vet Admin Hosp, Oklahoma City, 61-, US Army Hosp, Ft Sill, Okla & Tinker AFB Hosp, Midwest City, 61-; nat consult, Air Force Hosp, Lackland AFB, San Antonio, Tex, 63. Honors & Awards: Aesculapian Award, Univ Okla Student Body, 63 & 69; Regents Award Superior Teaching, Univ Okla Bd Regents, 69. Mem: Am Asn Obstetricians & Gynecologists; Soc Gynec Invest; Asn Profs Gynec & Obstet (pres, 66-67); Int Soc Advan Humanistic Studies Gynec (pres, 72-73); Am Gynec Soc (treas, 70-75). Res: Gynecologic oncology. Mailing Add: Dept of Obstet & Gynec Univ of Okla Sch of Med Norman OK 73069

MERRILL, JERALD CARL, b Las Vegas, Nev, Aug 12, 40; m 63; c 1. PHYSICAL CHEMISTRY. Educ: Univ Nev, Reno, BS, 62, PhD(phys chem), 71. Prof Exp: US AEC fel, Univ Calif, Davis, 71-72; presidential res intern, Brookhaven Nat Lab, 72-73; LECTR CHEM, UNIV UTAH, 73- Concurrent Pos: Lectr, Univ Calif, Davis, 72. Mem: Am Chem Soc. Res: Kinetics of fast reactions in solution and the gas phase; use of radiotracers in determining mechanisms and kinetics of chemical reactions; applications of computers for analyzing chemical data. Mailing Add: Dept of Chem Univ of Utah Salt Lake City UT 84112

MERRILL, JOHN ELLSWORTH, b Parsonsfield, Maine, May 10, 02; m 25; c 2. ASTRONOMY. Educ: Univ Boston, AB, 23; Case Inst Technol, MS, 27; Princeton Univ, AM, 29, PhD(astron), 31. Prof Exp: Instr math, Case Inst Technol, 24-28, Cleveland Col, 25-28 & Princeton Univ, 29-30; asst prof astron, Univ Ill, 31-32; cur, Buffalo Mus Sci, 32-36; asst, Princeton Univ, 36-37; instr, Hunter Col, 37-38, from asst prof to assoc prof, 38-50; from asst prof to assoc prof, Ohio Wesleyan Univ & Ohio State Univ, 50-51, prof, 51-59; sr staff engr, Franklin Inst, 59-61, prin scientist astron, 61-63; prof astron & math & dir, Morrison Observ, 64-67, Dearing prof astron, 67-69, EMER PROF ASTRON, CENT METHODIST COL, 69-; VIS PROF, UNIV FLA, 69- Concurrent Pos: Am Philos Soc grants, 41-42; dir pilot training prog, Princeton Univ, 42-43, vis asst prof, 43-45, res assoc, 47-63; adj prof, Univ Pa, 59-63; pres comn 42, Int Astron Union, 61-67. Mem: AAAS; Am Astron Soc. Res: Photometry of eclipsing variables; solutions for orbits of eclipsing binaries; effects of eccentricity of orbit. Mailing Add: Dept of Physics & Astron Univ of Fla Gainesville FL 32611

MERRILL, JOHN JAY, b Nampa, Idaho, Jan 24, 33; m 53; c 6. PHYSICS. Educ: Calif Inst Technol, BS, 55, MS, 56, PhD(physics), 60. Prof Exp: Instr physics, Harvey Mudd Col, 59-60; med physicist, Dee Mem Hosp, Ogden, Utah, 60-62; assoc prof physics, Utah State Univ, 62-69; pres, Tronac, Inc, 69-71; PROF PHYSICS, BRIGHAM YOUNG UNIV, 70- Concurrent Pos: Consult med physics. Mem: Am Phys Soc; Am Asn Physics Teachers. Res: Medical physics; precision x-ray spectroscopy; atomic and molecular beam interactions. Mailing Add: 763 S 600 West Orem UT 84057

MERRILL, JOHN PUTNAM, b Hartford, Conn, Mar 10, 17; m 42; c 3. MEDICINE. Educ: Dartmouth Col, AB, 38; Harvard Med Sch, MD, 42; Am Bd Internal Med, dipl, 50. Hon Degrees: DSc, Colby Col, 69, Univ Paris, 74. Prof Exp: House officer med, Peter Bent Brigham Hosp, 42-43, asst resident physician, 47-48; instr, 50-52, assoc, 52-56, from asst prof to assoc prof, 56-69, PROF MED, HARVARD MED SCH, 69- Concurrent Pos: Milton fel med, Harvard Med Sch, 48-49, res fel, 49-50; Am Heart Asn res fel, 49-50; asst, Peter Bent Brigham Hosp, 48-50, jr assoc, 50-54, sr assoc, 55-63, physician, 63-; estab investr, Am Heart Asn, 50-57; investr, Howard Hughes Med Inst, 57-; consult, Surgeon Gen, US Air Force, 64 & Nat Inst Arthritis & Metab Dis, 65; res collab, Brookhaven Nat Labs, 65-66; mem Nat Adv Coun, Regional Med Prog Servs, 71; nat adv coun, Nat Inst Arthritis & Metabolic Dis, 71-75. Honors & Awards: Alvarenga prize, 60; Mod Med Distinguished Achievement Award, 65; Gairdner Int Award, 69; Valentine Medal, NY Acad Sci, 70. Mem: Am Soc Clin Invest (pres, 63); Am Physiol Soc; Am Clin & Climat Asn; AMA; fel Am Col Physicians. Res: Internal medicine. Mailing Add: Peter Bent Brigham Hosp 721 Huntington Ave Boston MA 02115

MERRILL, JOHN RAYMOND, b Englewood, NJ, Aug 25, 39; m 60; c 4. SOLID

STATE PHYSICS. Educ: Swarthmore Col, AB, 61; Cornell Univ, PhD(solid state physics), 66. Prof Exp: Instr & res assoc physics, Cornell Univ, 66-67; asst prof, Dartmouth Col, 67-73; ASSOC PROF PHYSICS & DIR CTR EDUC DESIGN, FLA STATE UNIV, 73- Concurrent Pos: Res Corp res grant, 68; AEC res grant, 70-73. Mem: AAAS; Am Phys Soc; Am Asn Physics Teachers; NY Acad Sci. Res: Computers in teaching; instructional design and development. Mailing Add: Ctr for Educ Design Fla State Univ Tallahassee FL 32306

MERRILL, JOHN RICHARD, b Perth Amboy, NJ, June 20, 33. PHYSICAL CHEMISTRY. Educ: Rutgers Univ, BS, 55; Princeton Univ, MA, 57, PhD(phys chem), 58. Prof Exp: Asst, Princeton Univ, 55-57; RES CHEMIST, CENT RES & DEVELOP DEPT, E I DU PONT DE NEMOURS & CO, INC, 58- Mem: AAAS; Am Chem Soc. Res: Electronically excited molecules; photopolymerization; photoimaging; flame retardance. Mailing Add: Cent Res & Develop Dept E I du Pont de Nemours & Co Inc Wilmington DE 19898

MERRILL, JOSEPH MELTON, b Andalusia, Ala, Dec 8, 23. MEDICAL ADMINISTRATION. Educ: Harvard Med Sch, MD, 48; Am Bd Internal Med, dipl, 56. Prof Exp: Intern, Louisville Gen Hosp, Mo, 48-49; intern, Vanderbilt Univ Hosp, 49-50, asst resident med, 50-51; instr med & attend physician, Med Col, Univ Ala, 53-54; asst resident med res, Vet Admin Hosp, Nashville, Tenn, 54-55; res assoc, Postgrad Med Sch, Univ London, 55-56; chief clin physiol, Vet Admin Hosp, 56-64; chief Gen Clin Res Ctrs Br, Div Res Facil, NIH, 64-67; PROF MED & EXEC VPRES, BAYLOR COL MED, 67- Concurrent Pos: Clin investr, Vet Admin Hosp, Nashville, Tenn, 56-59, asst chief radioisotope serv, 57-64, asst dir, Prof Servs for Res, 60-64; instr, Med Sch, Vanderbilt Univ, 60-; Wellcome assoc, Royal Soc Med, 55-56. Mem: Am Physiol Soc; AMA; Am Heart Asn; Am Fedn Clin Res; fel Am Col Physicians. Res: Medical science and education administration. Mailing Add: Baylor Col of Med Houston TX 77025

MERRILL, LELAND (GILBERT), JR, b Danville, Ill, Oct 4, 20; m 49; c 2. ENVIRONMENTAL SCIENCE. Educ: Mich State Col, BS, 42; Rutgers Univ, MS, 48, PhD, 49. Prof Exp: Asst, Rutgers Univ, 46-49; asst prof entom, Mich State Col, 49-53; exten specialist, 53-59, res specialist, 59-61, dean col agr & environ sci, 61-71, DIR INST ENVIRON STUDIES, RUTGERS UNIV, NEW BRUNSWICK, 71- Mem: AAAS; Entom Soc Am. Res: Applied environmental studies; natural resource inventory applications to land use management. Mailing Add: Inst for Environ Studies Rutgers Univ New Brunswick NJ 08903

MERRILL, MALCOLM HENDRICKS, b Richmond, Utah, June 28, 03; m 26; c 3. PUBLIC HEALTH. Educ: Utah State Col, BS, 25; St Louis Univ, MS, 27, MD, 32; Univ Calif, MPH, 47; Am Bd Prev Med & Pub Health, dipl, 49. Prof Exp: Asst bact, St Louis Univ, 25-32 & Rockefeller Inst, Princeton, 32-35; intern med, Univ Calif, 35-36; asst dermat & resident med, 36-37; chief bur venereal dis, Calif State Dept Pub Health, 37-41, chief div labs, 41-54, dep dir pub health, 44-54, actg dir, 51-52, dir, 54-65; dir health serv, Off Tech Coop & Res, AID, 65-68; dir community health action planning serv, 68-71, DIR DIV INT HEALTH PROGS, AM PUB HEALTH ASN, 71- Concurrent Pos: Lectr sch pub health, Univ Calif, 46-65; mem exp therapeut study sect, NIH, 48-53, cancer control comt, 55-58, nat adv comt on pub health traineeships, 56-60, nat adv comt on community air pollution, 57-61 & nat adv health coun, 57-61; consult pub health for India, Tech Coop Admin, 52; mem tech adv comt, Inst Nutrit Cent Am & Panama, 55, 56, 58, 59; mem, US Pub Health Mission to Russia, 57. Honors & Awards: Bronfman Prize, Am Pub Health Asn, 64. Mem: AAAS; Am Pub Health Asn (pres, 59-60); AMA; Asn State & Territorial Health Officers (pres, 61-62); Asn State & Territorial Pub Health Lab Dirs. Res: Virus diseases; etiology and transmission of virus encephalitides; immunology of virus diseases; metabolism of the mycobacteria; epidemiology of typhoid fever and Q fever; problems in financing public health administration; serology. Mailing Add: Am Pub Health Asn 1015 18th St NW Washington DC 20036

MERRILL, REYNOLD CLUFF, b Richmond, Utah, May 5, 20; m 47; c 5. MEDICINE, BIOLOGY. Educ: Utah State Univ, BS, 39; Stanford Univ, MA, 40, PhD(chem), 42; Johns Hopkins Univ, MD, 53. Prof Exp: Res chemist, Lever Bros Co, 41; res chemist, USDA, 42-45; from res chemist to res mgr, Philadelphia Quartz Co, Pa, 45-50; res contractor, E I du Pont de Nemours & Co, 51-53; mem house staff, Johns Hopkins Hosp, 53-54; physician pvt pract, Calif, 54-56; asst med dir, Organon, Inc, 56-57, med dir, 57-58; dir clin res, Squibb Inst Med Res, 58-59, assoc dir res & develop labs, 60-64, dir biol res, 64, vpres med res, E R Squibb & Sons, Inc, NJ, 65-67; exec dir biomed sci div, Parke Davis & Co, Mich, 67-69; med dir, Winthrop Prod, 69-70, vpres, 70, sr vpres, 71-74; MED DIR, ALZA CORP, 74- Concurrent Pos: Mem, Bd Dirs, Winthrop Prod, Winthrop Labs & Sterling Winthrop Res Inst. Mem: Am Chem Soc; AMA; NY Acad Sci. Res: Application of chemistry to biology and medicine; clinical research; colloidal electrolytes. Mailing Add: Apt 101 1812 Willow Rd Palo Alto CA 94304

MERRILL, ROBERT KIMBALL, b Lima, Peru, Oct 11, 45; US citizen; m 72. EXPLORATION GEOLOGY. Educ: Colby Col, AB, 67; Ariz State Univ, MS, 70, PhD(geol), 74. Prof Exp: Stratigr, Am Stratig Co, 69-70; instr geol, Ariz State Univ, 73-74; STAFF GEOLOGIST, CITIES SERV OIL CO, 74- Mem: Sigma Xi; Geol Soc Am; Am Asn Petrol Geologists; Am Quaternary Asn. Res: Analysis of the tectonic development within the structural and stratigraphic framework of established hydrocarbon provinces and frontier areas to better understand the regional setting in the search for hydrocarbons. Mailing Add: 8207 S Florence Ave Tulsa OK 74136

MERRILL, RONALD EUGENE, b Salem, Ore, Aug 7, 47; m 74. SYNTHETIC ORGANIC CHEMISTRY. Educ: Mass Inst Technol, BS, 68; Univ Ore, PhD(chem), 73. Prof Exp: Fel chem, Syracuse Univ, 73-74; VIS ASST PROF CHEM, ROCHESTER INST TECHNOL, 74- Mem: Am Chem Soc. Res: New synthetic reactions; organoborane chemistry; heterocyclic chemistry; nuclear magnetic resonance spectroscopy. Mailing Add: Dept of Chem Rochester Inst of Technol Rochester NY 14623

MERRILL, RONALD THOMAS, b Detroit, Mich, Feb 5, 38; m 61; c 1. GEOPHYSICS. Educ: Univ Mich, BS, 59, MS, 61; Univ Calif, Berkeley, PhD(geophys), 67. Prof Exp: Asst prof, 67-72, ASSOC PROF GEOPHYS, UNIV WASH, 72- Concurrent Pos: Hon vis fel, Res Sch Earth Sci, Australia Nat Univ, 75. Mem: AAAS; Am Geophys Union; Soc Terrestrial Magnetism & Elec Japan. Res: Geomagnetism, especially paleomagnetism and rock magnetism. Mailing Add: Dept of Oceanog Univ of Wash Seattle WA 98195

MERRILL, SAMUEL, III, b New Orleans, La, Oct 27, 39; m 69; c 2. MATHEMATICS. Educ: Tulane Univ, BA, 61; Yale Univ, MA, 63, PhD(math), 65. Prof Exp: Instr math, Univ Rochester, 65-67, asst prof, 67-73; ASSOC PROF MATH, WILKES COL, 73- Mem: Am Math Soc; Math Asn Am. Res: Functional analysis, especially Banach spaces of analytic functions; mathematical applications to political science, especially voting power; harmonic analysis; complex variables. Mailing Add: Dept of Math Wilkes Col Wilkes-Barre PA 18703

MERRILL, STEWART HENRY, b Andover, Ohio, Sept 8, 26; m 50; c 4. ORGANIC POLYMER CHEMISTRY. Educ: Case Inst Technol, BS, 50; Ohio State Univ, PhD(chem), 53. Prof Exp: Res chemist, 53-64, RES ASSOC, RES LAB, EASTMAN KODAK CO, 64- Res: Synthesis and polymerization of monomers; reactions, structures and properties of polymers. Mailing Add: Eastman Kodak Co Bldg 82 Kodak Park Rochester NY 14650

MERRILL, WARNER JAY, JR, b Springfield, Ill, Jan 27, 23; m 45; c 2. STATISTICS. Educ: Univ Del, BA, 47; Ohio State Univ, PhD(statist), 56. Prof Exp: Asst psychol statist, Ohio State Univ, 49-51; staff statistician, Am Power Jet Co, 51-53; statistician, Gen Elec Co, NY, 56-61; prof mgt, Rensselaer Polytech Inst, 61-64; mgt scientist, Dunlap & Assocs, 64-65; mem tech staff, Hughes Aircraft Co, 65-66; sr assoc, Planning Res Corp, 66-68; OPERS RES ANALYST, LAW ENFORCEMENT ASSISTANCE ADMIN, DEPT JUSTICE, 68- Concurrent Pos: Prof lectr, George Washington Univ, 68-70; prof lectr, Am Univ, 69-70, adj prof, 70-72. Mem: Am Statist Asn; Inst Mgt Sci. Res: Applications of statistics to crime research; systems analysis, including human factors; statistical applications of computers; non-parametric statistics; operations research. Mailing Add: 9904 Inglemere Dr Bethesda MD 20034

MERRILL, WILLIAM, b Haverhill, NH, Sept 5, 33; m 61; c 2. FOREST PATHOLOGY, FOREST PRODUCTS. Educ: Univ NH, BS, 58; Univ Minn, MS, 61, PhD(plant path), 63. Prof Exp: Instr plant path, Univ Minn, 61-64; res staff pathologist & fel, Yale Univ, 64-65; from asst prof to assoc prof, 65-75, PROF PLANT PATH, PA STATE UNIV, UNIVERSITY PARK, 75- Mem: Am Phytopath Soc; Mycol Soc Am. Res: Etiology, epidemiology and control of forest tree pathogens; biodeterioration of wood. Mailing Add: 210 Buckhout Lab Pa State Univ University Park PA 16802

MERRILL, WILLIAM GEORGE, b Wilmington, Del, Oct 19, 31; m 58; c 1. ANIMAL NUTRITION. Educ: Univ Md, BS, 53; Univ Wis, MS, 54; Cornell Univ, PhD(animal nutrit), 59. Prof Exp: Asst prof, Exten Div, 59-64, ASSOC PROF ANIMAL HUSB, DAIRY CATTLE DIV, CORNELL UNIV, 64- Mem: Am Soc Animal Sci; Am Dairy Sci Asn. Res: Dairy husbandry; dairy cattle management systems; manure handling, manure re-feeding; milking systems; feeding systems. Mailing Add: Dept of Animal Sci Cornell Univ Ithaca NY 14850

MERRILL, WILLIAM MEREDITH, b Detroit, Mich, Dec 1, 18; m 43; c 3. GEOLOGY. Educ: Mich State Univ, BS, 46; Ohio State Univ, MA, 48, PhD(geol), 50. Prof Exp: Geologist, Ohio Div Geol Surv, 46-50; from instr to assoc prof geol, Univ Ill, 50-58; prof & chmn dept, Syracuse Univ, 58-63; chmn dept, 63-72, PROF GEOL, UNIV KANS, 63- Concurrent Pos: Asst, Ohio State Univ, 47-48, res assoc, Res Found, 55-57; consult, Ohio Div Geol Surv, 50-60; geologist, Nfld Dept Mines, 54 & Res Coun Alta, 58-62; vis scientist, Am Geol Inst, 62 & 64; team capt, Geo-Study, 62-63; chmn panel earth sci teacher prep, Coun Educ Geol Sci, 64-67, mem steering comt, Earth Sci Curriculum Proj, 65-, mem writers conf, 66. Mem: Fel Geol Soc Am; Soc Econ Paleont & Mineral; Am Asn Petrol Geol; Int Asn Sedimentol; Int Asn Math Geol. Res: Mesozoic stratigraphy of western United States and Canada; computer simulation in geology. Mailing Add: Dept of Geol Univ of Kans Lawrence KS 66045

MERRIMAN, CHARLES RICHARD, b Woodward, Okla, Oct 14, 46; m 65; c 2. BIOCHEMISTRY. Educ: Northwestern State Col, BS, 68; Okla State Univ, PhD(biochem), 72. Prof Exp: RES ASSOC BIOMED RES, S R NOBLE FOUND, 72- Mem: AAAS; Sigma Xi. Res: Investigation of the mechanisms of the acute phase response during neoplastic disease. Mailing Add: S R Noble Found Rte 1 Ardmore OK 73401

MERRIMAN, DANIEL, b Cambridge, Mass, Sept 17, 08; m 34, 71. BIOLOGICAL OCEANOGRAPHY. Educ: Univ Wash, BS, 33, MS, 34; Yale Univ, PhD(zool), 39. Prof Exp: Aquatic biologist, Conn State Bd Fish & Game, 36-38; from instr to asst prof biol, 38-46, dir, Bingham Oceanog Lab, 42-66, master, Davenport Col, 46-66, ASSOC PROF BIOL, YALE UNIV, 46-, DIR SEARS FOUND MARINE RES & OCEANOG HIST, 66- Concurrent Pos: Chmn comt food resources coastal waters, Nat Res Coun, 43-46; trustee, Bermuda Biol Sta & Woods Hole Oceanog Inst, 44-64; res assoc dept fishes & aquatic biol, Am Mus Natural Hist, 45-62; mem adv comt biol, Off Naval Res, 49-52; consult, NSF, 51-54 & President's Sci Adv Comt, 60 & 61; comnr, Conn State Bd Fish & Game, 53-56; mem adv comt, Susquehanna Fishery Study, 57-60; dir, Conn River Study, Conn Yankee Atomic Power Co, 64-74. Mem: Am Soc Limnol & Oceanog; Am Soc Zool; Hist Sci Soc; Am Soc Ichthyol & Herpet. Res: Ichthyology; fisheries biology; oceanography; morphology; history of sciences. Mailing Add: 298 Sperry Rd Bethany CT 06525

MERRIMAN, JOHN EDWARD, b Hamilton, Ont, Aug 12, 24; m 47; c 4. INTERNAL MEDICINE. Educ: Queen's Univ, Ont, MD, CM, 47; FRCP(C), 53. Prof Exp: Jr intern, Kingston Gen Hosp, Ont, Can, 47-48; sr intern med, Ottawa Civic Hosp, 48-49, sr intern path, 49-50; asst resident med, Sunnybrook Hosp, 50-51; instr & res asst, Univ Western Ont, 52-54; from asst prof to assoc prof, 54-70, prof med, Univ Sask, 70-75; PROF MED, ORAL ROBERTS UNIV, 75- Concurrent Pos: Res fel, Harvard Med Sch, 51-52; Nat Res Coun Can res fel, Univ Western Ont, 52-53. Mem: Fel Am Col Physicians; fel Am Col Cardiol; Can Soc Clin Invest; Can Physiol Soc; Can Med Asn. Res: Exercise testing; computer application in medicine; rehabilitation, including exercise training in cardiac patients. Mailing Add: Sch of Med Oral Roberts Univ Tulsa OK 74102

MERRIN, SEYMOUR, b Brooklyn, NY, Aug 13, 31; m 63; c 2. PHYSICAL CHEMISTRY, GEOCHEMISTRY. Educ: Tufts Univ, BS, 52; Univ Ariz, MS, 54; Pa State Univ, PhD(geochem), 62. Prof Exp: Geologist, US Geol Surv, 56-58; asst geochem, Pa State Univ, 59-62; sr assoc chemist, IBM Corp, 62-64; package develop dept mgr, Sperry Semiconductor Div, Sperry Rand Corp, 65-68; independent consult prod & prod develop electronics, 68-69; vpres technol, Innotech Corp, 69-74; PROJ MGR, EXXON ENTERPRISES, 74- Mem: AAAS; Am Ceramic Soc; Am Crystallog Asn; fel Geol Soc Am; Mineral Soc Am. Res: Glass; experimental petrology; microelectronic processing; phase equilibrium. Mailing Add: 235 Old Spring Rd Fairfield CT 06430

MERRINER, JOHN VENNOR, b Winchester, Va, Sept 13, 41. FISH BIOLOGY, MARINE BIOLOGY. Educ: Rutgers Univ, BA, 64; NC State Univ, MS, 67, PhD(zool), 73. Prof Exp: Assoc marine scientist, 70-74, actg head dept, 74-75, HEAD DEPT ICHTHYOL, VA INST MARINE SCI, 75- Concurrent Pos: Asst prof marine sci, Col William & Mary & Univ Va, 75- Mem: Am Fisheries Soc; Am Soc Ichthyologists & Herpetologists; Am Soc Limnol & Oceanog. Res: Ecology and life history of estuarine and marine fishes, as well as response of fishes to natural and pollution stress factors. Mailing Add: Dept of Ichthyol Va Inst of Marine Sci Gloucester Point VA 23062

MERRIS, RUSSELL LLOYD, b Calif, 1943. ALGEBRA. Educ: Harvey Mudd Col, BS, 64; Univ Calif, Santa Barbara, MA & PhD(math), 67. Prof Exp: Nat Acad Sci-Mat Res Coun assoc, Nat Bur Stand, 69-71; asst prof, 71-74, ASSOC PROF MATH,

CALIF STATE COL, HAYWARD, 74- Mem: Am Math Soc. Res: Multilinear algebra; group representation theory. Mailing Add: Dept of Math Calif State Col 25800 Hillary St Hayward CA 94542

MERRITHEW, PAUL BURTON, b Boston, Mass, May 17, 42; m 67. CHEMISTRY. Educ: Williams Col, BA, 64; Univ Mich, PhD(chem), 69. Prof Exp: ASST PROF CHEM, WORCESTER POLYTECH INST, 69- Concurrent Pos: Am Chem Soc-Petrol Res Fund grant, Worcester Polytech Inst, 69-72. Mem: Am Chem Soc; Am Phys Soc. Res: Electron paramagnetic resonance and Mössbauer spectroscopy employed to investigate the electronic structure of metal complexes. Mailing Add: Dept of Chem Worcester Polytech Inst Worcester MA 01609

MERRITT, ALFRED M, II, b Boston, Mass, Apr 10, 37; m 63; c 2. GASTROENTEROLOGY, VETERINARY MEDICINE. Educ: Bowdoin Col, AB, 59; Cornell Univ, DVM, 63, Univ Pa, MS, 69. Prof Exp: Asst instr vet med, Sch Vet Med, Univ Calif, 63-64; instr vet med, Sch Vet Med, 64-66, fel, Grad Sch Arts & Sci, 66-69, asst prof vet med, 69-72, ASSOC PROF VET MED, SCH VET MED, UNIV PA, 72- Concurrent Pos: Ed comp gastroenterol, Am J Digestive Dis, 72- Mem: Am Vet Med Asn; Comp Gastroenterol Soc (pres, 75); NY Acad Sci. Res: Neurohumoral control of gastric and pancreatic secretion in swine; pathophysiology of chronic diarrhea in horses. Mailing Add: Sch of Vet Med Univ of Pa New Bolton Ctr Kennett Square PA 19348

MERRITT, ARTHUR DONALD, b Shawnee, Okla, June 11, 25; m 53; c 2. INTERNAL MEDICINE. Educ: George Washington Univ, AB, 49, MD, 52; Am Bd Internal Med, dipl, 60. Prof Exp: Intern, Washington, DC Gen Hosp, 52-53; asst resident med, George Washington Univ Hosp, 53-54; asst resident med, Duke Univ Hosp, 54-55, chief resident & instr, 56-57; clin assoc, Nat Inst Arthritis & Metab Dis, 57-58, chief, 59-60; assoc prof med & biochem, 61-66, chmn med genetics prog, 62-66, PROF MED & CHMN DEPT MED GENETICS, MED CTR, IND UNIV INDIANAPOLIS, 66- Concurrent Pos: Fel, Duke Univ Hosp, 55-56; consult, Durham Vet Admin Hosp, NC, 57-58; lectr & clin instr, Sch Med, George Washington Univ, 59-60. Mem: Am Soc Human Genetics (treas, 64-70); Genetics Soc Am; Am Fedn Clin Res; NY Acad Sci; Am Col Physicians. Res: Human genetics; medical education; intermediary metabolism. Mailing Add: Ind Univ Med Ctr 129 Riley Res Wing Indianapolis IN 46202

MERRITT, CHARLES, JR, b Lynn, Mass, Mar 15, 19; m 42; c 3. ANALYTICAL CHEMISTRY. Educ: Dartmouth Col, AB, 41; Univ Vt, MS, 48; Mass Inst Technol, PhD(anal chem), 53. Prof Exp: Finish engr, W Lynn Works Lab, Gen Elec Co, 41-46; instr chem, Univ Vt, 46-49; asst, Mass Inst Technol, 49-53; res anal chemist, Nat Bur Stand, 53; asst prof anal chem, Polytech Inst Brooklyn, 53-56; supvry anal chemist, 56-57, HEAD ANAL CHEM LAB, US ARMY NATICK LABS, 57- Concurrent Pos: Mem staff grad sch arts & sci, Northeastern Univ, 56-; vis lectr, Mass Inst Technol, 65-; adj prof food sci, Univ Mass, 75-; mem subcomt stand ref mat, Nat Acad Sci-Nat Res Coun. Mem: AAAS; Inst Food Technol; Am Soc Mass Spectrometry; Am Chem Soc; Soc Appl Spectros. Res: Electrodeposition; spectrophotometry; gas chromatography; mass spectrometry; irradiation techniques; determination of the volatile components of foodstuffs; pollution and environmental studies. Mailing Add: US Army Natick Labs Natick MA 01760

MERRITT, CLAIR, b Quakertown, Pa, Jan 27, 22; m 43; c 3. FORESTRY. Educ: Univ Mich, BSF, 43, MF, 48, PhD(forestry), 59. Prof Exp: Asst dist forester, Md, 47; self-employed, 48; from instr to asst prof forestry, State Univ NY Col Forestry, Syracuse Univ, 49-56; assoc prof, 56-58, PROF FORESTRY, PURDUE UNIV, WEST LAFAYETTE, 68- Concurrent Pos: Vpres, Foresters, Inc; mem, Poplar Coun, Int Union Forest Res Orgn. Mem: Soc Am Foresters; Int Soc Trop Foresters. Res: Silvics and silviculture, light relationships in forest openings; establishment of hardwood regeneration. Mailing Add: Dept Forestry & Nat Resources Purdue Univ West Lafayette IN 47907

MERRITT, DORIS HONIG, b New York, NY, July 16, 23; m 53; c 2. RESEARCH ADMINISTRATION, PEDIATRICS. Educ: Hunter Col, BA, 44; George Washington Univ, MD, 52. Prof Exp: Exec secy cardiovasc & gen med dis, NIH, 57-60; from asst to assoc prof pediat, 61-73, from asst to assoc dean grants admin, 62-68, from asst to assoc dean sponsored prog, 65-70, DEAN SPONSORED PROGS, IND UNIVPURDUE UNIV, INDIANAPOLIS, 70-, PROF PEDIAT, 73- Concurrent Pos: Nat Heart Inst fel cardiovasc dis, Duke Univ, 56-57; Nat Heart & Lung Inst grant & interim dir, Indianapolis Sickle Cell Ctr, Sch Med, Ind Univ-Purdue Univ, 73-, Lilly endowment urban educ, 74-76; consult div res grants, Nat Heart & Lung Inst, 63-; mem regional adv group, Ind Regional Med Prog, 69-; mem & chmn biomed libr rev comt, Nat Libr Med, 70-73; chmn, Consortium for Urban Educ, Indianapolis, 71-75; mem, Nat Coun Univ Res Adminr. Mem: AAAS; Am Acad Pediat. Res: Genetic counseling; sickle cell disease; improvement of research administration. Mailing Add: Off of Sponsored Res Ind Univ-Purdue Univ Indianapolis IN 46202

MERRITT, EDISON S, b St John, NB, Mar 7, 23; m 49; c 2. POPULATION GENETICS, POULTRY BREEDING. Educ: McGill Univ, BSc, 49; Iowa State Univ, PhD(quant genetics), 57. Prof Exp: Geneticist, Animal Res Inst, 50-73, RES COORDR ANIMAL BREEDING, CENT EXP FARM, AGR CAN, 73- Mem: Genetics Soc Can; Can Soc Animal Prod; Poultry Sci Asn; World Poultry Sci Asn. Res: Quantitative genetics of poultry populations and direct and correlated responses to selection for quantitavie traits. Mailing Add: Rm 1087 K W Neatby Bldg Res Br Agr Can Cent Exp Farm Ottawa ON Can

MERRITT, JACK, b Sacramento, Calif, May 2, 18; m 42; c 1. PHYSICS. Educ: Pomona Col, AB, 39; Univ Calif, PhD(physics), 53. Prof Exp: Admin analyst, US Bur Budget, 46-47; admin officer res div, AEC, 47-49; physicist radiation lab, Univ Calif, 53-54; instrumentation, Shell Develop Co Div, Shell Oil Co, 55-57, Spectros, 57-66; PROF PHYSICS, CLAREMONT MEN'S COL, 66- Mem: Am Phys Soc. Res: High energy nuclear physics; instrumentation; meson production; spectroscopy; x-rays. Mailing Add: Joint Sci Dept Claremont Men's Col Claremont CA 91711

MERRITT, JAMES MANLEY, entomology. see 12th edition

MERRITT, KATHARINE, b Bridgeport, Conn, Apr 11, 38; m 70. IMMUNOLOGY, MICROBIOLOGY. Educ: Vassar Col, AB, 60; Univ Mich, MS, 62, PhD(microbiol), 64. Prof Exp: Res assoc path & microbiol, 64-66, instr, 66-68, ASST PROF MICROBIOL, DARTMOUTH MED SCH, 68- Mem: Am Soc Microbiol; Can Soc Immunol. Res: Mechanism of antibody formation and resistance to infection with study of the alteration of the responses with endotoxin, drugs and changes in host physiology. Mailing Add: Dept of Microbiol Dartmouth Med Sch Hanover NH 03755

MERRITT, LYNNE LIONEL, JR, b Alba, Pa, Sept 10, 15; m 37; c 4. ANALYTICAL CHEMISTRY. Educ: Wayne State Univ, BS, 36, MS, 37; Univ Mich, PhD(anal chem), 40. Prof Exp: Instr chem, Wayne State Univ, 36-37, 39-42; from asst prof to assoc prof chem, 42-53, assoc dean col arts & sci, 59-62, dir bur instnl res, 60-65, assoc dean faculties, 62-64, actg dean, 63-64, vpres & dean res & advan studies, 65-75, PROF CHEM, IND UNIV, BLOOMINGTON, 53-, SPEC ASST TO PRES & DEAN RES COORD & DEVELOP, 75- Concurrent Pos: Vis prof, Calif Inst Technol, 49-50; Guggenheim fel & res assoc, 55-56; pres & dir, Ind Instrument & Chem Corp, 59-; Fulbright fel, Nat Sci Res Ctr, France, 63. Mem: AAAS; Am Chem Soc; Am Crystallog Asn; Coblentz Soc; Am Inst Chemists. Res: Organic reagents; instrumental methods of analysis; x-ray diffraction crystal structure determinations. Mailing Add: Dept of Chem Ind Univ Bloomington IN 47401

MERRITT, MARGARET VIRGINIA, b Springfield, Ohio, June 30, 42. ANALYTICAL CHEMISTRY. Educ: Col Wooster, BA, 64; Cornell Univ, PhD(anal chem), 68. Prof Exp: Fel electrochem, Univ Calif, Riverside, 68-69; fel, Radiation Res Lab, Mellon Inst, 69-70; asst prof anal chem, Franklin & Marshall Col, 70-72; RES CHEMIST, UPJOHN CO, 72- Mem: Am Chem Soc; Sigma Xi. Res: High pressure liquid chromatography; trace organic analysis; electron spin resonance; spin labeling; spin trapping. Mailing Add: Phys & Anal Chem Upjohn Co Kalamazoo MI 49001

MERRITT, MELVIN LERÒY, b Juneau, Alaska, Nov 12, 21; m 49; c 4. GEOPHYSICS. Educ: Calif Inst Technol, BS, 43, PhD(physics), 50. Prof Exp: Test engr, Gen Elec Co, 43-46; mem staff, 50-56, mgr phys res dept, 56-61, DIV SUPVR, SANDIA LABS, 61- Mem: AAAS; Am Phys Soc; Am Asn Physics Teachers; Arctic Inst NAm; Health Physics Soc. Res: Cosmic rays; shock waves; effects of nuclear weapons; environmental studies and assessments. Mailing Add: Dept 1150 Sandia Labs Albuquerque NM 87115

MERRITT, PAUL EUGENE, b Watertown, NY, Oct 23, 20; m 46; c 2. ANALYTICAL CHEMISTRY. Educ: NY State Col Teachers, Albany, BA, 42, MA, 47; Rensselaer Polytech Inst, PhD, 54. Prof Exp: Line foreman, Gen Chem Defense Corp, 42-43; teacher pub sch, 47-48; asst gen chem, Rensselaer Polytech Inst, 48-51, instr anal chem, 53-54; from asst prof to assoc prof, St Lawrence Univ, 54-63; chmn dept chem, 70-73, PROF ANAL CHEM, STATE UNIV NY COL POTSDAM, 63- Mem: Fel Am Inst Chemists; Am Chem Soc; Sigma Xi; NY Acad Sci. Res: Physico-chemical methods of analysis; infrared analysis of inorganic complexes, minerals and ores; structure studies of complex-inorganic salts; trinitrotoluene. Mailing Add: Dept of Chem State Univ of NY Col Potsdam NY 13676

MERRITT, RICHARD FOSTER, b South Braintree, Mass, Nov 16, 34; m 56; c 3. ORGANIC CHEMISTRY. Educ: Bowdoin Col, AB, 56; Mass Inst Technol, PhD(org chem), 62. Prof Exp: Res chemist, Redstone Arsenal Res Div, 62-67, group leader, 67-70, LAB HEAD PROCESS RES & DEVELOP, ROHM AND HAAS CO, 70- Concurrent Pos: Guest lectr, St Bernard Col, 62-64. Mem: Am Chem Soc. Res: Fluorine and oxygen difluoride chemistry; fluorine nuclear magnetic resonance spectroscopy; strained smallring hydrocarbons. Mailing Add: Rohm and Haas Co Res Labs Spring House PA 19477

MERRITT, RICHARD HOWARD, b Jersey City, NJ, Mar 28, 33; m 55; c 4. HORTICULTURE, ACADEMIC ADMINISTRATION. Educ: Rutgers Univ, BSc, 55, MSc, 56, PhD(hort, plant physiol), 61. Prof Exp: Lectr pomol, 58-61, from asst prof ornamental hort to assoc prof hort, 61-70, dir res instr & assoc dean col agr & environ sci, 62-73, PROF HORT, RUTGERS UNIV, 70-, ASSOC DEAN COOK COL, 73-, DEAN INSTR, 74- Mem: Am Soc Hort Sci. Res: Educational research; post harvest studies on apples; physiological studies on Easter lily. Mailing Add: 823 Meadow Rd Bridgewater NJ 08807

MERRITT, RICHARD WILLIAM, b San Francisco, Calif, July 26, 45; m 67; c 2. ENTOMOLOGY. Educ: Calif State Univ, San Jose, BA, 68; Wash State Univ, MS, 70; Univ Calif, Berkeley, PhD(entom), 74. Prof Exp: ASST PROF ENTOM, MICH STATE UNIV, 74- Mem: Entom Soc Am; NAm Benthological Soc; Sigma Xi; Freshwater Biol Asn. Res: Aquatic and veterinary entomology and ecology; biosystematics of Diptera; aquatic biting flies; insects in sewage oxidation ponds; degradation and recycling of organic material by insects. Mailing Add: Dept of Entom Mich State Univ East Lansing MI 48823

MERRITT, ROBERT BUELL, b Topeka, Kans, Nov 20, 42; m 65. POPULATION GENETICS. Educ: Univ Kans, BA, 64, PhD(zool), 70. Prof Exp: Trainee genetics, Univ Rochester, 70-72; ASST PROF BIOL, SMITH COL, 72- Mem: AAAS; Soc Study Evolution; Genetics Soc Am. Mailing Add: Dept of Biol Smith Col Northampton MA 01060

MERRITT, ROBERT EDWARD, b Coudersport, Pa, Aug 1, 30. LEATHER CHEMISTRY, INFORMATION SCIENCE. Educ: Hanover Col, AB, 51; Univ Cincinnati, MS, 53. Prof Exp: Dir res, Barrentan Testing & Res Corp, Pa, 53-59; asst ed, 59-63, assoc ed, 63-68, group leader, 68-71, sr assoc indexer, 71-72, SR ASSOC ED, CHEM ABSTRACTS SERV, OHIO STATE UNIV, 72- Concurrent Pos: Ed, The Chem Record, Am Chem Soc, 69- Mem: Am Chem Soc; Am Leather Chemists Asn. Res: Leather chemistry, especially synthetic tanning materials and lignosulfonates; polymer chemistry; information science. Mailing Add: 3468 Colchester Rd Columbus OH 43221

MERRITT, THOMAS PARKER, b Chicago, Ill, June 21, 14; m 43; c 1. ATOMIC SPECTROSCOPY. Educ: NCent Col, BS, 37; Northwestern Univ, MS, 39; Boston Univ, PhD, 49. Prof Exp: From jr physicist to physicist, Fed Bur Invest, Washington, DC, 40-43; res assoc, Underwater Sound Lab, Harvard Univ, 43-45; physicist, Polaroid Corp, Mass, 45-46; asst prof physics, Simmons Col, 46-49; prof math & physics & head dept, Albright Col, 49-55; optics group engr, Lockheed Aircraft Corp, 55-56; sr mem prof staff, Systs Labs Corp, 56-57; sr mem tech staff, Int Tel & Tel Labs, 57-59; sr sci adv, Lockheed Aircraft Corp, 59-64; from sci analyst to sr sci analyst, 64-70, mgr res-Europe, 70-71, dir long range planning serv-Europe, London, 71-74, SR INDUST ECONOMIST, STANFORD RES INST, 75- Mem: Am Phys Soc; Am Asn Physics Teachers. Res: Emission spectroscopy; theoretical structure of solids; infrared theory and technology; nuclear physics. Mailing Add: Stanford Res Inst 333 Ravenswood Ave Menlo Park CA 94025

MERROW, RAYMOND THEODORE, b Allport, Pa, Aug 5, 20; m 43, 58; c 3. ORGANIC CHEMISTRY. Educ: Juniata Col, BS, 40; Univ Colo, PhD(chem), 51. Prof Exp: Teacher high sch, Pa, 40-41; analyst, Union Steel Castings Co, 41-42; chemist, Biochem Div, Eastern Regional Res Lab, USDA, 42-43, 46-47; org chemist, Res Dept, 51-58, head solid propellants br, 58-64, head systs anal br, 64-71, OPERS RES ANALYST, WEAPONS PLANNING GROUP, NAVAL WEAPONS CTR, 71- Mem: Am Chem Soc. Res: Mechanism of organic reactions; diene syntheses; chemistry of nitrate esters; solid propellants; military operations research. Mailing Add: 312 Howell St Ridgecrest CA 93555

MERRY, WILLIAM JAMES, b Pontiac, Mich, June 6, 14; m 41; c 3. BOTANY. Educ: Univ Mich, BS, 36, MS, 37, PhD(bot), 40. Prof Exp: Instr biol, Denison Univ, 40-42; asst Arnold Arboretum, Harvard Univ, 42-43; instr bot, Barnard Col, Columbia Univ, 43-46; asst prof biol, Champlain Col, 46-53; prof & head dept, Norwich Univ, 53-58; PROF BIOL, NORTHERN MICH UNIV, 58- Res: Experimental embryology;

plant taxonomy. Mailing Add: Dept of Biol Northern Mich Univ Marquette MI 49885

MERRYMAN, EARL L, physical chemistry, radiochemistry, see 12th edition

MERSEREAU, J MARK, organic chemistry, see 12th edition

MERSHEIMER, WALTER LYON, b New York, NY, Mar 25, 11; m 41; c 4. SURGERY. Educ: Norwich Univ, BS, 33; NY Med Col, MD, 37, MS, 42; Am Bd Surg, dipl. Hon Degrees: DSc, Norwich Univ, 62. Prof Exp: From instr to assoc prof, 42-60, clin prof, 60-62, PROF SURG & CHMN DEPT, NY MED COL, 62-, COORDR CANCER EDUC, 48- Concurrent Pos: Consult surgeon, US Naval Hosp, St Albans, NY, 47-60; consult, Vet Admin Hosp, Lyons, 48-62 & Bronx, 55-; mem, End Results Comt, Nat Cancer Inst, 50- Mem: AMA; Asn Am Med Cols; Am Col Surgeons; Am Col Gastroenterol; Soc Surg Alimentary Tract. Res: Teaching and training of medical students and surgical residents; administration. Mailing Add: NY Med Col Flower & Fifth Ave Hosps New York NY 10029

MERSKEY, CLARENCE, b Alberton, SAfrica, July 20, 14; US citizen; m 39; c 3. PHYSIOLOGY, MEDICINE. Educ: Univ Cape Town, MB, ChB, 37, MD, 47. Prof Exp: Res asst, Radcliffe Infirmary, Oxford, Eng, 49-50; sr lectr med, Univ Cape Town, 50-59; from asst prof to assoc prof, 59-72, PROF MED, ALBERT EINSTEIN COL MED, 72-, PROF LAB MED, 74- Concurrent Pos: Cecil John Adams travelling fel, SAfrica, 49-50; Rockefeller travelling fel, 57-; fel biol chem, Harvard Univ, 57; fel, Col Physicians & Surgeons, SAfrica, 58; fel, Univ Cape Town, 58. Mem: Am Physiol Soc; Soc Exp Biol & Med; Am Soc Hemat; Int Soc Hemat. Res: Internal medicine and hematology, especially blood coagulation, thrombosis and atheroma; congenital hemorrhagic disorders. Mailing Add: Dept of Med Albert Einstein Col of Med New York NY 10461

MERSMANN, HARRY JOHN, b St Louis, Mo, Nov 13, 36; m 59; c 2. BIOCHEMISTRY. Educ: St Louis Univ, BS, 58, PhD(biol), 63. Prof Exp: Res assoc biochem, Auburn Univ, 63-65, Univ Calif, San Francisco, 65-66 & State Univ NY Buffalo, 66-68; asst prof life sci, Ind State Univ, 68-69; BIOCHEMIST, SHELL DEVELOP CO, 69- Mem: Am Soc Biol Chemists. Res: Carbohydrate and lipid metabolism; regulation of enzymes; neonatal biochemistry and physiology. Mailing Add: Dept Animal Physiol & Growth Shell Develop Co Modesto CA 95352

MERTEL, HOLLY EDGAR, b Springfield, Mo, Sept 26, 20; c 1. ORGANIC CHEMISTRY. Educ: Drury Col, BS, 41; Univ Nev, MS, 43; Univ Southern Calif, PhD(org chem), 50. Prof Exp: Res assoc, Columbia Univ, 50-51; RES CHEMIST, MERCK SHARP & DOHME RES LABS DIV, MERCK & CO, INC, 51-, SR RES FEL, 70- Mem: AAAS; Am Chem Soc; The Chem Soc; Health Physics Soc. Res: Radiochemical preparations; cortical steroid hormones. Mailing Add: 721 Harding St Westfield NJ 07090

MERTEN, HELMUT LUDWIG, b Vienna, Austria, Apr 2, 22; nat US; m 57. ORGANIC CHEMISTRY. Educ: Univ Vienna, PhD(chem), 51. Prof Exp: Fel, Iowa State Univ, 52-54 & Univ Toronto, 54-55; res chemist, 55-58, group leader, 58-60, res adv, 60-66, sr res specialist, 66-71, SR RES GROUP LEADER, MONSANTO CO, 71- Mem: Am Chem Soc; Soc Chem Indust. Res: Dielectrics; lubricants; oil additives; functional fluids; reactor coolants; food ingredients; pharmaceutical intermediates; flavor and fragrance components; low calorie foods; rubber chemistry. Mailing Add: 1660 Mayflower Lane Hudson OH 44236

MERTEN, ULRICH, b Houston, Tex, Feb 27, 30; m 53; c 2. PHYSICAL CHEMISTRY. Educ: Calif Inst Technol, BS, 51; Wash Univ, PhD(chem), 55. Prof Exp: Mem staff chem, Knolls Atomic Power Lab, Gen Elec Co, 55-56; mem staff chem, Gen Atomic Div, Gen Dynamics Corp, 56-67, Gulf Gen Atomic, 67-71; VPRES PROCESS SCI DEPT, GULF RES & DEVELOP CO, 71- Mem: AAAS; Am Chem Soc; Am Nuclear Soc. Res: High temperature chemistry; membrane phenomena; synthetic fuels. Mailing Add: Gulf Res Ctr PO Drawer 2038 Pittsburgh PA 15230

MERTENS, DAVID ROY, b Jefferson City, Mo, Sept 11, 47; m 72; c 2. RUMINANT NUTRITION, DAIRY SCIENCE. Educ: Univ Mo-Columbia, BS, 69, MS, 70; Cornell Univ, PhD(nutrit), 73. Prof Exp: Asst prof animal sci, Iowa State Univ, 73-75; ASST PROF DAIRY SCI, UNIV GA, 75- Mem: Sigma Xi; Am Dairy Sci Asn; Am Soc Animal Sci; AAAS. Res: Mathematical and chemical study of ruminal metabolism of forages and fibrous carbohydrates; development of optimal nutrition and management of dairy animals. Mailing Add: 322 Livestock-Poultry Bldg Univ of Ga Athens GA 30602

MERTENS, EDWARD WILLIAM, b White Sulfur Springs, Mont, Mar 12, 22; m 45; c 3. ORGANIC CHEMISTRY, MARINE BIOLOGY. Educ: Univ Calif, BS, 43. Prof Exp: Res chemist, Calif Res Corp Div, Stand Oil Co Calif, 45-59, group supvr, 59-63, sect supvr, Chevron Res Co, 63-67, mgr asphalts div, 67-69, sr staff engr, 69-73, SR RES ASSOC, CHEVRON RES CO, STAND OIL CO CALIF, 73- Mem: Am Chem Soc. Res: Asphalts and asphalt products; fundamental properties of asphalts; oil spill prevention, control and retrieval technology; fate and behavior of oil in marine waters; effect of oil on marine organisms. Mailing Add: Chevron Res Co Process Eng Dept PO Box 1627 Richmond CA 94802

MERTENS, FREDERICK PAUL, b Danbury, Conn, June 10, 35; m 62; c 1. PHYSICAL CHEMISTRY, CORROSION. Educ: Worcester Polytech Inst, BS, 57, PhD(chem), 65. Prof Exp: Res chemist, Columbia-Southern Chem Corp, 57-60; res asst, Worcester Polytech Inst, 60-64; chemist, 64-65, sr chemist, 65-69, res chemist, 69-72, PROJ CHEMIST, TEXACO INC, PORT ARTHUR, 72- Mem: Am Chem Soc; Nat Asn Corrosion Engrs; Sigma Xi. Res: Infrared reflectance study of metallic corrosion and adsorbed molecules; failure analysis; evaluation and recommendation of corrosion inhibitors, antifoulants, coatings and materials of construction in petroleum industry. Mailing Add: 3216 Lawrence Ave Nederland TX 77627

MERTENS, THOMAS ROBERT, b Ft Wayne, Ind, May 22, 30; m 53; c 2. GENETICS, SCIENCE EDUCATION. Educ: Ball State Univ, BS, 52; Purdue Univ, MS, 54, PhD(genetics), 56. Prof Exp: Res assoc genetics, Univ Wis, 56-57; from asst prof sci to assoc prof biol, 57-66, PROF BIOL, BALL STATE UNIV, 66- Concurrent Pos: NSF fac fel, Stanford Univ, 63-64. Mem: AAAS; Nat Asn Biol Teachers; Genetics Soc Am; Am Genetic Asn. Res: Plant genetics and taxonomy; allotetraploidy in Tragopogon; cytotaxonomy of genus Polygonum in North America; cygotegentics of Rhoeo; programmed instruction in biology education. Mailing Add: Dept of Biol Ball State Univ Muncie IN 47306

MERTES, DAVID H, b Pittsburg, Calif, Nov 22, 29; m 62. CHEMICAL EMBRYOLOGY. Educ: San Francisco State Col, BA, 52; Univ Calif, Berkeley, MA, 59, PhD(zool), 66. Prof Exp: Mem fac zool, San Joaquin Delta Col, 59-62; Am Cancer Soc Dernham res fel, Univ Calif, Berkeley, 66-68; chmn div sci, 68-69, acad dean, 69-71, PRES, COL SAN MATEO, 71- Mem: AAAS; Am Soc Zool; Am Inst Biol Sci; Soc Develop Biol. Res: Biochemical embryology; genetic read-off and protein synthesis during the early embryogenesis of invertebrate embryos, differentiation. Mailing Add: Off of the Pres Col of San Mateo San Mateo CA 94402

MERTES, MATHIAS PETER, b Chicago, Ill, Apr 22, 32; div; c 4. MEDICINAL CHEMISTRY. Educ: Univ Ill, BS, 54; Univ Tex, MS, 56; Univ Minn, PhD(med chem), 60. Prof Exp: From asst prof to assoc prof, 60-68, PROF MED CHEM, SCH PHARM, UNIV KANS, 68- Concurrent Pos: NIH career develop award, 67-72. Mem: Am Chem Soc. Res: Organic reaction mechanisms and drug-enzyme interactions. Mailing Add: Malott Hall Univ of Kans Sch of Pharm Lawrence KS 66044

MERTS, ATHEL LAVELLE, b Paragould, Ark, Oct 14, 25; m 47; c 1. ATOMIC PHYSICS, ASTROPHYSICS. Educ: Univ Mo-Rolla, BS, 50, MS, 51; Univ Kans, PhD(physics), 57. Prof Exp: Instr physics, Univ Tulsa, 51-52; staff mem continuum physics, Los Alamos Sci Lab, 57-64; sr scientist, Gulf-Gen Atomics, 64-65; STAFF MEM ATOMIC PHYSICS OPACITIES, LOS ALAMOS SCI LAB, 65- Mem: Am Phys Soc; Am Astrophys Soc. Res: Study of the effects of collisional excitation and dielectronic recombination processes on the radiative power loss from optically thin plasmas. Mailing Add: 125 Aztec Ave Los Alamos NM 87544

MERTZ, DAN, b Allen Co, Ohio, Sept 19, 28; m 56; c 1. PLANT PHYSIOLOGY. Educ: Ohio Univ, BA, 54; Univ Tex, PhD(plant physiol), 60. Prof Exp: Asst bot & biol, Univ Tex, 54-57, res scientist, 57-60; res assoc bot, 60-61, assoc prof, 61-71, PROF BIOL SCI, UNIV MO-COLUMBIA, 71- Concurrent Pos: Visitor, Univ Glasgow, 67-68. Mem: Am Soc Plant Physiol; Bot Soc Am; Scand Soc Plant Physiol. Res: Physiological and biochemical changes associated with growth and development. Mailing Add: Div of Biol Sci Univ of Mo Columbia MO 65202

MERTZ, DAVID B, b Sandusky, Ohio, July 10, 34; m 61. ANIMAL ECOLOGY. Educ: Univ Chicago, BS, 60, PhD(zool), 65. Prof Exp: NSF vis scholar statist, Univ Calif, Berkeley, 65-66; asst prof biol, Univ Calif, Santa Barbara, 66-69; assoc prof, 69-74, PROF BIOL SCI, UNIV ILL, CHICAGO CIRCLE, 74- Mem: AAAS; Am Soc Naturalists; Ecol Soc Am; Soc Study Evolution; Am Soc Zool. Res: Population ecology and ecological genetics of flour beetles. Mailing Add: Dept of Biol Sci Univ of Ill at Chicago Circle Chicago IL 60680

MERTZ, EDWIN THEODORE, b Missoula, Mont, Dec 6, 09; m 36; c 2. BIOCHEMISTRY. Educ: Univ Mont, AB, 31; Univ Ill, MS, 33, PhD(biochem), 35. Prof Exp: Res chemist, Armour & Co, 35-37; instr biochem, Univ Ill, 37-38; res assoc path, Univ Iowa, 38-40; instr agr chem, Univ Mo, 40-43; res chemist, Exp Sta, Hercules Powder Co, Del, 43-46; from asst prof agr chem to assoc prof biochem, 46-57, PROF BIOCHEM, PURDUE UNIV, WEST LAFAYETTE, 57- Concurrent Pos: Consult, Ind State Hosps, 57- & US-Japan Malnutrit Panel, 70-73. Honors & Awards: McCoy Award, 67; John Scott Award, 67; Hoblitzelle Award, 68; Cong Medal, Fed Land Banks, 68; Spencer Award, Am Chem Soc, 70; Osborne-Mendel Award, Am Inst Nutrit, 72; Distinguished Serv Award, Univ Mont, 73; Edward W Browning Award, Am Soc Agron, 74. Mem: Nat Acad Sci; Am Chem Soc; Am Soc Biol Chem; Am Inst Nutrit; Am Asn Cereal Chem. Res: Amino acid requirements of humans and animals; purification of plasminogens; biochemistry of mental retardation; opaque-2 and floury-2 high lysine maize. Mailing Add: Dept of Biochem Purdue Univ West Lafayette IN 47907

MERTZ, JAMES LAURENCE, biochemistry, see 12th edition

MERTZ, JANET ELAINE, b Bronx, NY, Aug 9, 49. MOLECULAR BIOLOGY. Educ: Mass Inst Technol, BS(biol) & BS(elec eng), 70; Stanford Univ, PhD(biochem), 75. Prof Exp: Jane Coffin Childs Mem Fund fel, Med Res Coun Lab Molecular Biol, Cambridge, Eng, 75-76; ASST PROF ONCOL, McARDLE LAB CANCER RES, UNIV WIS-MADISON, 76- Mem: AAAS; Am Soc Microbiol; Fedn Am Scientists. Res: Molecular biology of tumor viruses. Mailing Add: McArdle Lab Cancer Res Univ of Wis Madison WI 53706

MERTZ, ROBERT THEODORE, b Floral Park, NY, Jan 18, 28; m 53; c 2. APPLIED MATHEMATICS. Educ: Harvard Univ, AB, 48; Columbia Univ, AM, 51, PhD(appl math), 61. Prof Exp: MATHEMATICIAN & PROGRAMMER SYSTS ANAL, IBM CORP, 50- Concurrent Pos: Bus exec fel, Brookings Inst, 68-69. Mem: AAAS; Soc Indust & Appl Math; Am Math Soc; Math Asn Am; Opers Res Soc Am. Res: Large scale computer programming; applications in satellite orbits, missile trajectories; English-to-Braille translation; portfolio selection; linear, quadratic, non-linear programming and optimization; management information systems; econometric modelling. Mailing Add: 51 Mansfield Ave Darien CT 06820

MERTZ, WALTER, b Mainz, Ger, May 4, 23; m 53. NUTRITION, BIOCHEMISTRY. Educ: Univ Mainz, MD, 51. Prof Exp: Intern surg, County Hosp, Ger, 52-53; asst internal med, Univ Hosp, Univ Frankfurt, 53; res fel nutrit, NIH, 53-56, vis scientist, Exp Liver Dis Sect, 56-61; res biochemist, Walter Reed Army Inst Res, Washington, DC, 61-64, chief dept biol chem, 64-69; chief vitamin & mineral nutrit lab, Human Nutrit Res Div, 69-72, CHMN NUTRIT INST, AGR RES SERV, USDA, 72- Mem: Am Soc Biol Chem; Am Diabetes Asn; Am Inst Nutrit. Res: Biochemistry and nutrition of trace elements. Mailing Add: Nutrit Inst Agr Res Serv USDA Beltsville MD 20705

MERWIN, HENRY DENISON, analytical chemistry, see 12th edition

MERWIN, RUTH MINERVA, b Washington, DC, July 3, 16. ZOOLOGY. Educ: Mt Holyoke Col, AB, 37, AM, 39; Univ CHicago, PhD(zool), 44. Prof Exp: Asst, Mt Holyoke Col, 37-39; asst, Univ Chicago, 39-41; instr zool, Conn Col, 44-47; BIOLOGIST, NAT CANCER INST, 47- Mem: Am Soc Exp Path; Am Asn Cancer Res; Radiation Res Soc; Microcirc Soc. Res: Ecology; cancer study. Mailing Add: Viral Biol Br Nat Cancer Inst Bethesda MD 20014

MERWINE, NORMAN CHARLES, b Westerville, Ohio, Dec 27, 21; m 45; c 1. AGRONOMY, CROP SCIENCE. Educ: Ohio State Univ, BSc, 43; Miss State Univ, MS, 53, PhD(cytogenetics, plant breeding), 56. Prof Exp: Teacher pub schs, Ohio, 46-47 & Miss, 47-50; asst, 50-51, 52-54, asst agronomist, Exp Sta, 55-59, assoc agronomist & assoc prof agron, 59-62, dean col agr, 66-67, PROF AGRON & AGRONOMIST, MISS STATE UNIV, 62-66 & 67- Mem: Am Soc Agron; Sigma Xi. Res: Genetics and improvement of grain sorghum; intensive cropping systems; minimum tillage technology. Mailing Add: Dept of Agron Miss State Univ Box 5248 Mississippi State MS 39762

MERYMAN, HAROLD THAYER, b Washington, DC, Feb 5, 21; m 47; c 4. CRYOBIOLOGY, MEDICAL RESEARCH. Educ: Long Island Col Med, MD, 46. Prof Exp: Intern, US Naval Hosp, Md, 46-47; physiologist, Naval Med Res Inst, 47-54; Am Cancer Soc fel, Yale Univ, 54-56, res fel, Sch Med, 55-57; physiologist, Naval Med Res Inst, Md, 57-68; ASSOC RES DIR BLOOD PROG, AM NAT RED CROSS, 68- Mem: Am Physiol Soc; Biophys Soc; Am Soc Cell Biol; Electron Micros

Soc Am; NY Acad Sci. Res: Mechanism of freezing and drying injury in biological media; physiology of cold injury; preservation of cells and tissues by freezing; blood banking. Mailing Add: Red Cross Blood Res Lab 9312 Old Georgetown Rd Bethesda MD 20014

MERZ, EARL H, b Chicago, Ill, Nov 26, 08; m 36; c 2. OPHTHALMOLOGY. Educ: Valparaiso Univ, BS, 38; Northwestern Univ, MD, 38; Am Bd Ophthal, dipl, 43. Prof Exp: Chmn dept ophthal, 66-69, SR ATTEND OPHTHALMOLOGIST, WESLEY MEM HOSP, CHICAGO, 69-; ASSOC PROF OPHTHAL, MED SCH, NORTHWESTERN UNIV, CHICAGO, 68- Mem: Am Acad Ophthal & Otolaryngol; AMA; Asn Res Vision & Ophthal; Pan-Am Asn Ophthal. Mailing Add: Dept of Ophthal Northwestern Univ Med Sch Chicago IL 60201

MERZ, EDMUND HERMAN, b Springfield, Mass, Nov 28, 22; m 44; c 3. POLYMER CHEMISTRY. Educ: Univ Mich, BSE, 44; Polytech Inst Brooklyn, MS, 45, PhD(polymer sci), 47. Prof Exp: Prof org chem, St Francis Col, 45-46; instr phys chem, Polytech Inst Brooklyn, 46-47; res chemist, Monsanto Chem Co, 47-60; dir res, Gen Packaging Div, 60-64, DIR BASIC RES, CORP RES & DEVELOP, CONTINENTAL CAN CO, INC, 64- Mem: Am Chem Soc; assoc Soc Rheol; AAAS. Res: Kinetics of polymerization; physical and solution properties of high polymers; efficiency of fractionation using carbon 13 as a tracer. Mailing Add: Continental Can Co Inc 7622 S Racine Ave Chicago IL 60620

MERZ, JAMES L, b Jersey City, NJ, Apr 14, 36; m 62; c 3. SOLID STATE PHYSICS. Educ: Univ Notre Dame, BS, 59; Harvard Univ, MA, 61, PhD(appl physics), 67. Prof Exp: MEM STAFF PHYSICS, BELL LABS, 66- Concurrent Pos: Vis lectr, Harvard Univ, 72. Mem: Am Phys Soc. Res: Study of semiconducting compounds for optical communications and integrated optics. Mailing Add: Bell Labs Murray Hill NJ 07974

MERZ, PAUL LOUIS, b New Haven, Conn, June 1, 18; m 50; c 4. RUBBER CHEMISTRY, POLYMER CHEMISTRY. Educ: Union Col, NY, BS, 40; Yale Univ, PhD(org chem), 52. Prof Exp: Res chemist, Beech-Nut Packing Co, 40-43, head polymer lab, 45-47, consult, 47-51; sr res chemist, Naugatuck Chem Div, US Rubber Co, 51-56, group leader, 56-59, sr res specialist, 59-61, proj leader high temperature elastomers, 61-62; plastics chemist, Lawrence Radiation Lab, 62-64; STAFF SCIENTIST, GEN DYNAMICS/CONVAIR, 64- Mem: AAAS; Am Chem Soc; Soc Advan Mat & Process Eng. Res: Vinyl polymerization; antioxidant and antiozonant research; rubber reclaiming; high temperature and cryogenic seals for aerospace applications; polymeric systems for radiation shielding; advanced structural and thermoformable composites. Mailing Add: Gen Dynamics/Convair Mail Zone 643-1 PO Box 80847 San Diego CA 92138

MERZ, ROBERT WILLIAM, forestry, see 12th edition

MERZ, TIMOTHY, b Philadelphia, Pa, Jan 11, 27; m 53; c 2. CYTOGENETICS, RADIOBIOLOGY. Educ: Johns Hopkins Univ, AB, 51, PhD, 58. Prof Exp: NIH fel, Johns Hopkins Univ, 58-60, res assoc, 60-61, from asst prof to assoc prof cytogenetics, 64-75; PROF RADIOL & CHMN DIV RADIATION BIOL, MED COL VA, VA COMMONWEALTH UNIV, 75- Mem: AAAS; Radiation Res Soc; Am Soc Cell Biol; Genetics Soc Am; Am Soc Human Genetics. Res: Chromosome structure and behavior. Mailing Add: Med Col Va Div Radiation Biol Box 87 1200 E Broad St Richmond VA 23298

MERZ, WALTER JOHN, b Cheadle Hulme, Eng, Oct 10, 20; US citizen; m 49; c 2. SOLID STATE PHYSICS. Educ: Swiss Fed Inst Technol, MS, 44, PhD(physics), 48. Prof Exp: Res asst solid state physics, Mass Inst Technol, 48-51; vis prof, Pa State Univ, 51; mem tech staff, Bell Tel Labs, Inc, 51-56; mem tech staff, RCA Labs, Inc, NJ, 56-57, mgr res, 57-68, DIR RES, LABS, RCA LTD, SWITZ, 68- Mem: Am Phys Soc. Res: Ferroelectrics; dielectrics; semiconductors; photoconductors. Mailing Add: Labs RCA Ltd Badenerstrasse 569 CH-8048 Zurich Switzerland

MERZ, WILLIAM GEORGE, b Orange, NJ, Dec 20, 41. MEDICAL MYCOLOGY, IMMUNOLOGY. Educ: Drew Univ, BA, 63; WVa Univ, MS, 65, PhD(microbiol), 68. Prof Exp: NIH fel microbiol, Columbia-Presby Med Ctr, 68-70; instr dermat, Col Physicians & Surgeons, Columbia Univ, 70-73; instr dermat, 73-75, ASST PROF LAB MED, JOHNS HOPKINS UNIV, 73-, ASST PROF DERMAT & EPIDEMIOL, 75-; CLIN MICROBIOLOGIST, DEPT LAB MED, JOHNS HOPKINS HOSP, 73- Concurrent Pos: Brown-Hazen grant dermat, Columbia Univ, 70-73. Honors & Awards: Bot Award, Ciba Pharmaceut Co, 63. Mem: Am Soc Microbiol; Mycol Soc Am; Med Mycol Soc of the Americas; Int Soc Human & Animal Mycoses. Res: Rapid techniques for the identification of fungi; immune responses to mycotic infections. Mailing Add: Dept of Lab Med (Path) Johns Hopkins Hosp Baltimore MD 21205

MERZBACHER, CLAUDE F, b Philadelphia, Pa, Oct 29, 17; m 45; c 2. NATURAL HISTORY, CHEMICAL ENGINEERING. Educ: Univ Pa, BS, 39; Claremont Grad Sch, MA, 50; Univ Poitiers, cert, 51; Univ Calif, Los Angeles, EdD, 61. Prof Exp: Teacher math high sch, Fla, 45-46; instr math, physics & chem, Oceanside High Sch & Jr Col, 46-47; instr chem, 47-50, from asst prof to assoc prof phys sci, 50-65, chmn dept, 64-69, PROF PHYS SCI, SAN DIEGO STATE UNIV, 65-, PLANETARIUM LECTR, 53-, COUNR, 54- Concurrent Pos: Dir, NSF Coop Col-Sch Sci Prog, 67-69; pvt pract psychother; consult mgt & leadership creativity. Mem: AAAS; Am Chem Soc; fel Am Inst Chemists. Res: Affective interference with mathematics performance; design of integrated courses in physical science; statistical methods; noncognitive processes. Mailing Add: Dept of Phys Sci San Diego State Univ San Diego CA 92115

MERZBACHER, EUGEN, b Berlin, Ger, Apr 9, 21; nat US; m 52; c 4. THEORETICAL PHYSICS. Educ: Istanbul Univ, Licentiate, 43; Harvard Univ, AM, 48, PhD(physics), 50. Prof Exp: Mem, Inst Advan Study, 50-51; vis asst prof physics, Duke Univ, 51-52; from asst prof to prof, 52-69, KENAN PROF PHYSICS, UNIV NC, CHAPEL HILL, 69- Concurrent Pos: NSF fac fel, Inst Theoret Physics, Copenhagen, 59-60. Mem: AAAS; fel Am Phys Soc; Am Asn Physics Teachers. Res: Quantum mechanics; atomic and nuclear theory. Mailing Add: Dept of Physics & Astron Univ of NC Chapel Hill NC 27514

MERZENICH, MICHAEL MATTHIAS, b Lebanon, Ore, May 15, 42; m 66; c 2. NEUROPHYSIOLOGY, NEUROANATOMY. Educ: Univ Portland, BS, 64; Johns Hopkins Univ, PhD(physiol), 68. Prof Exp: NIH fel, Univ Wis, 68-71; asst prof, 72-75, ASSOC PROF PHYSIOL & OTOLARYNGOL, UNIV CALIF, SAN FRANCISCO, 75-; DIR, COLEMAN MEM LAB, 71- Concurrent Pos: Consult, NIH, 74. Mem: AAAS; Soc Res Otolaryngol; Acoust Soc Am. Res: Auditory neurophysiology; aids for the profoundly deaf; sensation coding; anatomy and physiology of the central auditory nervous system. Mailing Add: Dept of Physiol & Otolaryngol Univ of Calif San Francisco CA 94143

MESCHAN, ISADORE, b Cleveland, Ohio, May 30, 14; m 43; c 4. RADIOLOGY. Educ: Case Western Reserve Univ, BA, 35, MA, 37, MD, 39; Am Bd Radiol, dipl, 57. Prof Exp: Intern, Cleveland City Hosp, 39-40; resident, Univ Hosps, Case Western Reserve Univ, 40-42, instr radiol, Univ, 46-47; prof & head dept, Sch Med, Univ Ark, 47-55; PROF RADIOL & DIR, BOWMAN GRAY SCH MED, WAKE FOREST UNIV, 55- Concurrent Pos: Consult, Walter Reed Army Hosp; chmn comt radiol, Nat Acad Sci-Nat Res Coun, 74. Mem: Radiol Soc NAm; Radiation Res Soc; Am Roentgen Ray Soc; Soc Nuclear Med; fel Am Col Radiol. Res: Radioisotopes and nuclear medicine; radiation biology; diagnostic and therapeutic clinical radiology. Mailing Add: Dept of Radiol Bowman Gray Sch Med Winston-Salem NC 27103

MESCHI, DAVID JOHN, b East Chicago, Ind, May 1, 24. HIGH TEMPERATURE CHEMISTRY. Educ: Univ Chicago, BA, 49, MS, 52; Univ Calif, PhD(chem), 56. Prof Exp: Asst res chemist, Inst Eng Res, Univ Calif, 56-59; resident res assoc, Argonne Nat Lab, 59-60; assoc res chemist, 60-65, CHEMIST, INORG MAT RES DIV, LAWRENCE BERKELEY LAB, UNIV CALIF, 65- Mem: Am Chem Soc; Am Phys Soc; Am Ceramic Soc. Res: High temperature physics. Mailing Add: Inorg Mat Res Div Bldg 62 Lawrence Berkeley Lab Berkeley CA 94720

MESCHIA, GIACOMO, b Milan, Italy, Feb 7, 26; nat US; m 61; c 3. PHYSIOLOGY. Educ: Univ Milan, MD, 50. Prof Exp: Asst prof physiol, Univ Milan, 51-53; Toscanini res fel, Sch Med, Yale Univ, 53-55; asst prof physiol, Univ Milan, 55-56; res fel, Josiah Macy Found, Sch Med, Yale Univ, 56-58, res asst, 58-59, asst prof, 59-65; assoc prof, 65-69, PROF PHYSIOL, MED CTR, UNIV COLO, DENVER, 69- Mem: AAAS; Am Physiol Soc. Res: Fetal physiology. Mailing Add: Dept of Physiol Univ of Colo Med Ctr Denver CO 80220

MESCHINO, JOSEPH ALBERT, b Cranston, RI, Aug 23, 32; m 54; c 3. ORGANIC CHEMISTRY, MEDICINAL CHEMISTRY. Educ: Brown Univ, ScB, 54; Rice Univ, PhD(org chem), 58. Prof Exp: NIH fel org synthesis, Mass Inst Technol, 58-59; sr scientist, 59-61, group leader chem develop, 61-67, DIR CHEM RES, McNEIL LABS, JOHNSON & JOHNSON, 67- Mem: Am Chem Soc. Res: Design and synthesis of new biologically active agents; process development of fine organic chemicals. Mailing Add: 14 Sandy Knoll Dr Doylestown PA 18901

MESCON, HERBERT, b Toronto, Ont, Apr 3, 19; nat US; m 46; c 4. DERMATOLOGY. Educ: City Col New York, BS, 38; Boston Univ, MD, 42. Prof Exp: Resident, Mallory Inst, Boston City Hosp, 42-43; intern med, Lebanon Hosp, Bronx, NY, 46-47 & Tufts Med Serv, 47-48; asst instr dermat, Sch Med, Univ Pa, 48-52, instr, Grad Sch Med, 50-52; PROF DERMAT & CHMN DEPT, SCH MED, BOSTON UNIV, 52-; CHIEF DERMAT & GENITO-INFECTIOUS DIS, UNIV HOSPS & MEM EVANS MEM DEPT PREV MED, 52- Concurrent Pos: Res fel, Hosp Univ Pa, 48-51; Runyon clin cancer res fel, 49-51; consult, Vet Admin Hosps, Wilmington, Del, 50-52 & Boston, Mass, 54- & Lemuel Shattuck Hosp, Jamaica Plain, 55-; area consult, Vet Admin, Boston, 59-64; mem comt cutaneous med, Nat Res Coun, 62-65; dir dermat, Boston City Hosp, 74- Mem: Soc Invest Dermat (vpres, pres); Histochem Soc; Am Asn Path & Bact; Am Dermat Asn; Am Acad Dermat (vpres). Res: Dermatopathology; clinical dermatology. Mailing Add: 720 Harrison Ave Boston MA 02118

MESECAR, RODERICK SMIT, b Hot Springs, SDak, May 24, 33; m 52; c 4. PHYSICAL OCEANOGRAPHY, ELECTRICAL ENGINEERING. Educ: Ore State Univ, BS, 56, MS, 58, EE, 64, PhD(oceanog), 67. Prof Exp: Design engr res lab, Raytheon Co, 58-61; asst prof comput res, 61-64, asst prof oceanog res, 64-74, ASSOC PROF OCEANOG RES, ORE STATE UNIV, 74-, ASST CHMN DEPT OCEANOG, 71- Mem: Inst Elec & Electronics Eng; Marine Technol Soc. Res: Application of electronic circuit designs and instrumentation to computer development and oceanographic research. Mailing Add: Dept of Oceanog Ore State Univ Corvallis OR 97331

MESEL, EMMANUEL, b Recife, Brazil, June 6, 24; m 57; c 3. PEDIATRIC CARDIOLOGY, INFORMATION SCIENCE. Educ: Rensselaer Polytech Inst, BEE, 48; Univ Cincinnati, MD, 59; Am Bd Pediat, dipl, 66, cert pediat cardiol, 67. Prof Exp: Res fel pediat, Albert Einstein Col Med, 61-62, res instr, 62-64, res assoc, 64; asst prof, Sch Med, Stanford Univ, 64-69, assoc prof clin pediat, 69; ASSOC PROF INFO SCI & PEDIAT, UNIV ALA, BIRMINGHAM, 69- Concurrent Pos: Consult, Ala State Dept Pub Health, 69- Honors & Awards: Borden Award, 59. Mem: AAAS; Inst Elec & Electronics Eng; AMA. Res: Cardiovascular physiology; clinical information systems. Mailing Add: Dept of Info Sci Univ of Ala Birmingham AL 35294

MESELSON, MATTHEW STANLEY, b Denver, Colo, May 24, 30; m 69; c 2. PHYSICAL CHEMISTRY, MOLECULAR BIOLOGY. Educ: Univ Chicago, PhB, 51; Calif Inst Technol, PhD, 57. Hon Degrees: DSc, Oakland Univ, 66, Columbia Univ, 71 & Univ Chicago, 75. Prof Exp: Asst prof chem, Calif Inst Technol, 58-60; assoc prof, 60-64, PROF BIOL, HARVARD UNIV, 64- Concurrent Pos: Consult, US Arms Control & Disarmament Agency, 63- Honors & Awards: Nat Acad Sci Prize Molecular Biol, 63; Eli Lilly Award Microbiol & Immunol, 64. Mem: Nat Acad Sci; Inst of Med of Nat Acad Sci; fel Am Acad Arts & Sci. Res: Biochemistry of molecular biology of nucleic acids; molecular mechanism of genetic recombination; control of gene action in higher organisms. Mailing Add: Dept of Biol Harvard Univ Cambridge MA 02138

MESERVE, BRUCE ELWYN, b Portland, Maine, Feb 2, 17; m 61; c 3. MATHEMATICS. Educ: Bates Col, AB, 38; Duke Univ, AM, 41, PhD(math), 47. Prof Exp: Teacher, Moses Brown Sch, RI, 38-41; asst math, Duke Univ, 41-42, 45-46; instr, Univ Ill, 46-47, asst prof, 48-54; from assoc prof to prof, 54-Montclair State Col, 54-64, chmn dept, 57-63; PROF MATH, UNIV VT, 64- Concurrent Pos: Consult, Metrop Sch Study Coun. Mem: AAAS; Am Math Soc; Math Asn Am. Res: Geometry; mathematical training of prospective teachers. Mailing Add: Dept of Math Univ of Vt Burlington VT 05401

MESERVE, PETER LAMBERT, b Buffalo, NY, Sept 22, 45; m 69. ECOLOGY. Educ: Univ Calif, Davis, BA, 67; Univ Nebr-Lincoln, MS, 69; Univ Calif, Irvine, PhD(biol), 72. Prof Exp: Asst prof ecol, Cath Univ, Santiago, 73-75; ASST PROF ZOOL, UNIV IDAHO, 75- Mem: Ecol Soc Am; Am Soc Mammalogists; Am Soc Naturalists. Res: Population and community ecology of vertebrates; biogeography; behavioral ecology. Mailing Add: Dept of Biol Sci Univ of Idaho Moscow ID 83843

MESERVE, RICHARD ANDREW, solid state physics, molecular physics, see 12th edition

MESERVEY, EDWARD BLISS, plasma physics, see 12th edition

MESERVEY, ROBERT H, b Hanover, NH, Apr 1, 21; m 53; c 2. SOLID STATE PHYSICS. Educ: Dartmouth Col, BA, 43; Yale Univ, PhD(physics), 61. Prof Exp: Physicist, US Army Eng Res & Develop Lab, 51-55; consult, Perkin Elmer Corp, 55-60; physicist, Lincoln Lab, 61-63, SR SCIENTIST, FRANCIS BITTER NAT MAGNET LAB, MASS INST TECHNOL, 63- Mem: Fel Am Phys Soc. Res: Superconductivity; magnetism; low temperature physics; fluid mechanics; optics.

Mailing Add: Francis Bitter Nat Magnet Lab Mass Inst of Technol Cambridge MA 02139

MESETH, EARL HERBERT, b Chicago, Ill, Nov 29, 38; m 65; c 1. ZOOLOGY. Educ: Ill Col, BS, 61; Wash Univ, MA, 62; Southern Ill Univ, Carbondale, PhD(zool), 68. Prof Exp: Asst prof, 68-72, ASSOC PROF BIOL, ELMHURST COL, 72- Mem: Sigma Xi; Am Ornithologists Union; Cooper Ornith Soc; Soc Study Evolution. Res: Behavior, ecology and biology of albatrosses, Diomedea immutabilis; relationship of courtship rituals to nest site selection and pair bonding. Mailing Add: Dept of Biol Elmhurst Col Elmhurst IL 60126

MESHKOV, SYDNEY, b Philadelphia, Pa, June 5, 27; m 56; c 3. PHYSICS. Educ: Univ Pa, AB, 47, PhD(physics), 54; Univ Ill, MS, 49. Prof Exp: Asst physics, Univ Ill, 47-49; asst instr, Univ Pa, 49-54; asst prof, Univ Del, 54-55; lectr, Univ Pa, 55-56; asst prof, Univ Pittsburgh, 56-62; PHYSICIST, RADIATION THEORY SECT, RADIATION PHYSICS DIV, NAT BUR STAND, 62- Concurrent Pos: Instr, LaSalle Col, 51-52; res assoc, Princeton Univ, 60; res assoc, Weizmann Inst, 61-62; vis assoc, Calif Inst Technol, 73; secy, Aspen Ctr Physics, 75- Mem: Fel Am Phys Soc. Res: Elementary particle theory. Mailing Add: Radiation Theory Sect 240.01 Nat Bur of Stand Washington DC 20234

MESHRI, DAYALDAS TANUMAL, b Kaloi, WPakistan, Mar 11, 36; m 66; c 1. INORGANIC CHEMISTRY, PHYSICAL CHEMISTRY. Educ: Gujarat Univ, India, BSc, 58, MSc, 62; Univ Idaho, PhD(inorg & phys chem), 68. Prof Exp: Demonstr chem, St Xavier's Col, India, 58; demonstr, Gujarat Col, 58-61, asst lectr, 61-62; assoc fluorine chem, Cornell Univ, 67-69; res chemist, 69-70, head fluorine res dept, 70-72, DIR FLUORINE & INORG RES, SPEC CHEM DIV, OZARK-MAHONING CO, 72- Mem: Fel Am Inst Chemists; Am Chem Soc; AAAS; Electrochem Soc; Sigma Xi. Res: Neutron activation analysis; coordination and fluorine chemistry; nitrogen-fluorine, oxygen fluorine chemistry; electrophilic substitution; hydrogen fluoride chemistry. Mailing Add: Ozark-Mahoning Co Spec Chem Div 1870 S Boulder Tulsa OK 74119

MESIROV, JILL PORTNER, b Philadelphia, Pa, May 12, 50. MATHEMATICS. Educ: Univ Pa, AB, 70; Brandeis Univ, MA, 71, PhD(math), 74. Prof Exp: LECTR MATH, UNIV CALIF, BERKELEY, 74- Mem: Am Math Soc. Res: Global analysis; calculus of variations. Mailing Add: Dept of Math Univ of Calif Berkeley CA 94720

MESKILL, VICTOR PETER, b Albertson, NY, May 9, 35; m 56; c 5. MICROBIOLOGY, CYTOLOGY. Educ: Hofstra Univ, BA, 61, MA, 62; St John's Univ, NY, PhD(microbiol), 67. Prof Exp: Teacher high sch, NY, 63-64; from instr to assoc prof biol & asst dean col, 64-69, VPRES ADMIN, C W POST CTR, LONG ISLAND UNIV, 70-, PROF BIOL, 75- Mem: AAAS; Soc Protozool; Soc Indust Microbiol; Am Soc Zool; Nat Asn Biol Teachers. Res: Growth, nutrition and ecology of protozoa; sex chromatin studies with mammalian and invertebrate cells. Mailing Add: Box 324 C W Post Ctr Long Island Univ Greenvale NY 11548

MESKIN, LAWRENCE HENRY, b Detroit, Mich, July 21, 35; m 59; c 2. DENTAL EPIDEMIOLOGY. Educ: Univ Detroit, DDS, 61; Univ Minn, Minneapolis, MSD, 63, MPH, 64, PhD(epidemiol), 66. Prof Exp: Instr oral path, Sch Dent, 63-66, assoc prof prev dent & chmn div, 66-68, CHMN DIV HEALTH ECOL, UNIV MINN, MINNEAPOLIS, 68-, HILL RES PROF DELIVERY OF DENT HEALTH SERV, 70-, LECTR PEDIAT, SCH MED, 63- Concurrent Pos: USPHS fel epidemiol, 63-; consult, Cleft Palate Clin, Univ Ill, 64-; partic, Inst Advan Educ Dent Res, 64; WHO traveling fel, 68. Mem: Am Dent Asn; Am Acad Oral Path; Int Asn Dent Res; Am Pub Health Asn. Res: Preventive dentistry; dental public health; craniofacial malformations; health care delivery. Mailing Add: Div of Health Ecol Univ of Minn Sch of Dent Minneapolis MN 55455

MESLOW, E CHARLES, b Waukegan, Ill, Aug 25, 37; m 59; c 3. ECOLOGY, WILDLIFE RESEARCH. Educ: Univ Minn, BS, 59, MS, 66; Univ Wis, PhD(wildlife ecol), 70. Prof Exp: Asst prof zool & vet sci, NDak State Univ, 68-71; ASST LEADER ORE COOP WILDLIFE RES UNIT, ORE STATE UNIV, 71- Mem: Ecol Soc Am; Wildlife Soc; Am Soc Mammal; Am Ornith Union. Res: Population dynamics; predation; wildlife ecology. Mailing Add: Coop Wildlife Res Unit 104 Nash Hall Ore State Univ Corvallis OR 97331

MESMER, GUSTAV, b Bromberg, Ger, July 2, 05; nat US; m 30; c 3. APPLIED MECHANICS. Educ: Univ Göttingen, PhD(appl mech), 29. Prof Exp: Asst, Kaiser Wilhelm Inst, Univ Göttingen, 30-34; asst prof appl mech, Aachen Tech Univ, 34-39; group leader, Junkers Aircraft Corp, 39-41; prof aeronaut eng & appl mech, Darmstadt Tech Univ, 40-51; prof, 50-64, head dept, 52-64, dir, Sever Inst, 56-64, distinguished serv prof, 64-74, EMER PROF APPL MECH, WASH UNIV, 74- Mem: Soc Exp Stress Anal; Am Soc Mech Eng; Am Soc Eng Educ. Res: Photoelasticity; vibrations; stability. Mailing Add: 6641 Waterman Ave St Louis MO 63130

MESMER, ROBERT EUGENE, physical inorganic chemistry, see 12th edition

MESNIKOFF, ALVIN MURRAY, b Asbury Park, NJ, Dec 25, 25; m 52; c 4. PSYCHIATRY, PSYCHOANALYSIS. Educ: Rutgers Univ, BA, 48; Univ Chicago, MD, 54. Prof Exp: Asst chief male serv, NY State Psychiat Inst, 58-60, chief female psychiat serv, 60-65, dir, Washington Heights Community Serv, 65-68; dir, South Beach Psychiat Ctr, 68-75, NEW YORK CITY REGIONAL DIR, NY STATE DEPT MENT HYG, 75-; PROF PSYCHIAT, STATE UNIV NY DOWNSTATE MED CTR, 68- Concurrent Pos: Assoc clin prof psychiat, Columbia Univ, 60-68; assoc attend psychiatrist, Columbia Presby Hosp, 68-69; attend psychiatrist, Kings County Hosp, 69- & St Vincent's Med Ctr, 70- Mem: AAAS; fel Am Psychiat Asn; NY Acad Sci. Res: Community psychiatry; design of programs; evaluation of mental health services. Mailing Add: Dept of Psychiat State Univ NY Downstate Med Ctr Brooklyn NY 11203

MESROBIAN, ROBERT BENJAMIN, b New York, NY, July 31, 24; m 50; c 5. CHEMISTRY. Educ: Princeton Univ, BA, 44, MS, 45, PhD(phys chem), 47. Prof Exp: Res assoc & proj adminr, Polytech Inst Brooklyn, 47-49, from asst prof to assoc prof polymer chem, 49-54, prof & assoc dir, Polymer Res Inst, 55-57; assoc dir res high polymer chem, Cent Res & Eng Div, 57-58, gen mgr, Gen Packaging Res & Develop Div, 59-64, gen mgr, Cent Res & Eng Div, 64-67, gen mgr res & eng, 67-69, VPRES RES & ENG, CONTINENTAL CAN CO, INC, 69- Concurrent Pos: Co-holder, Chaire Franqui lectr, Univ Liege, 47-48; US Educ Found vis prof, State Univ Groningen, 50; consult, Nuclear Eng Div, Brookhaven Nat Lab, 51-; US State Dept adv, Atoms for Peace Conf, Geneva, 58. Mem: Am Chem Soc; Soc Plastics Eng; Am Inst Chem. Res: Synthesis and properties of polymers; oxidation of hydrocarbons; organic peroxides; effects of ionizing radiation on polymers; application of polymers for coatings, adhesives, inks and packaging. Mailing Add: 632 S Elm St Hinsdale IL 60521

MESSENGER, AUBREY STEVEN, b Palmyra, Va, Mar 9, 31; m 56; c 3. SOIL SCIENCE, FOREST ECOLOGY. Educ: NC State Univ, B8, 53; Mich State Univ, PhD(soil sci), 66. Prof Exp: Forester, Va Div Forestry, 56-58; instr soils & phys geog, 63-64, ASST PROF SOILS & FOREST ECOL, NORTHERN ILL UNIV, 64- Concurrent Pos: Consult ecologist, Argonne Nat Lab, 67-71; res assoc, Morton Arboretum, 71- Mem: Soil Sci Soc Am; Am Soc Agron; Ecol Soc Am. Res: Soil biology and genesis; tree-soil interactions. Mailing Add: Dept of Geog Northern Ill Univ DeKalb IL 60115

MESSENGER, JOHN COWAN, b Green Bay, Wis, Apr 2, 20; m 47. CULTURAL ANTHROPOLOGY. Educ: Lawrence Univ, BS, 47; Northwestern Univ, PhD(anthrop), 57. Prof Exp: From instr to assoc prof anthrop, African Studies & Soc Sci, Mich State Univ, 53-63; assoc prof anthrop, Carleton Col, 63-64; prof anthrop, Folklore & African Studies, Ind Univ, 64-71; PROF ANTHROP, OHIO STATE UNIV, 71- Concurrent Pos: Exchange scholar, Univ Col, Galway, 59-60; Ford Int Prog grant, Ind Univ, 64 & Montserrat, 65; mem exec comt, Am Comt Irish Studies, 63-; Fulbright-Hayes sr lectr, Inst Irish Studies, Queen's Univ Belfast, 68-69. Mem: Am Anthrop Asn; Royal Anthrop Inst; Int African Inst; Am Folklore Soc; African Studies Asn. Res: Ethnographic research among the Anang of southeastern Nigeria; Aran islanders of western Ireland; Montserrat islanders of the West Indies; folklore research in Dublin, Ireland and Belfast, Northern Ireland. Mailing Add: Dept of Anthrop Ohio State Univ 65 S Oval Dr Columbus OH 43210

MESSENGER, JOSEPH UMLAH, b Medicine Hat, Alta, Aug 5, 13; US citizen; m 41; c 3. INORGANIC CHEMISTRY, PHYSICAL CHEMISTRY. Educ: Univ Calif, Berkeley, AB, 35, BS, 39; Univ Southern Calif, MS, 42. Prof Exp: Res chemist, Nat Defense Res Coun, Univ Southern Calif, 41-42 & Univ Chicago, 42-43; from asst res chemist to res chemist, Field Res Lab, Socony Mobil Oil Co, Inc, 43-46, from sr chemist to sr res chemist, 46-50, sr res technologist, 50-54, drilling mud engr, Mobil Oil Can, Ltd, Alta, 54-58, sr staff engr, 58-60, chem eng sect chief, 60-62, eng specialist, Field Res Lab, Mobil Oil Corp, Tex, 62-67, ENG ASSOC, FIELD RES LAB, MOBIL RES & DEVELOP CORP, 67- Mem: Am Chem Soc; Soc Petrol Eng; Am Inst Mining, Metall & Petrol Eng; Sigma Xi; fel Am Inst Chemists. Res: Chemical and petroleum engineering; chemical well stimulation, drilling muds, cements, lost circulation, water injection and corrosion; contact catalysis; fluorine, boron and uranium chemistry. Mailing Add: Mobil Res & Develop Corp PO Box 900 Dallas TX 75221

MESSENGER, POWERS SLATER, b Redding, Calif, Aug 14, 20; m 50; c 2. ENTOMOLOGY. Educ: Univ Calif, BS, 42, PhD(agr chem), 51. Prof Exp: Asst specialist, 50, from jr entomologist to assoc entomologist & lectr insect ecol, 51-65, vchmn dept, 70-72, PROF ENTOM & ENTOMOLOGIST, UNIV CALIF, BERKELEY, 65-, CHMN DEPT, 73- Concurrent Pos: Collabr, Entom Res Div, Agr Res Serv, USDA, 51-; Guggenheim fel, 64; insect ecologist, Food & Agr Orgn, Bangkok, Thailand, 72-73. Mem: Entom Soc Am; Ecol Soc Am; Japanese Soc Pop; Entom Soc Can; Brit Ecol Soc. Res: Insect physiology and ecology; biology and ecology of tephritid fruit flies; bioclimatology and population dynamics of parasites and predators of insect pests; influence of physical environmental factors on insect growth, development and geographic distribution; design and operation of environment control chambers; population ecology. Mailing Add: Dept of Entom Sci Univ of Calif Berkeley CA 94720

MESSER, CHARLES EDWARD, b Baltimore, Md, Aug 16, 15; m 58. PHYSICAL CHEMISTRY. Educ: Johns Hopkins Univ, AB, 36, PhD(phys chem), 40. Prof Exp: Res chemist, Biochem Res Found, Del, 41 & Nat Defense Res Comt, US Bur Mines, Pa, 41-42; instr chem, Clarkson Tech, 42-44 & Dartmouth Col, 44-46; from instr to asst prof, 46-71, ASSOC PROF CHEM, TUFTS UNIV, 71- Mem: Am Chem Soc. Res: Calorimetry; phase studies. Mailing Add: Dept of Chem Tufts Univ Medford MA 02155

MESSER, HAROLD HENRY, b Maryborough, Australia, Dec 30, 42; m 68; c 1. DENTAL RESEARCH. Educ: Univ Queensland, BDSc, 64, MDSc, 67; Univ Minn, PhD(biochem), 72. Prof Exp: Sr res officer dent, Univ Queensland, 64-67; fel physiol, Univ BC, 72-74; ASST PROF ORAL BIOL, UNIV MINN, 74- Concurrent Pos: Fulbright travel award, Australian-Am Educ Found, 67; res award, Int Asn Dent Res (Australian Sect), 67. Mem: Int Asn Dent Res. Res: Calcium transport mechanisms in bone; essential status of fluoride. Mailing Add: 17-238 Health Sci Unit A Univ of Minn Minneapolis MN 55455

MESSER, WAYNE RONALD, b Cedar Rapids, Iowa, Nov 7, 42; m 64; c 2. ORGANIC CHEMISTRY, PHOTOCHEMISTRY. Educ: Iowa State Univ, BS, 64; Univ Ill, PhD(chem), 68. Prof Exp: RES CHEMIST, CENT RES DIV, HERCULES, INC, 68- Mem: Am Chem Soc. Res: Nitrogen heterocycles; cycloadditions; concerted reactions; physical organic chemistry; photopolymerization. Mailing Add: Res Ctr Hercules Inc Wilmington DE 19899

MESSERLY, GEORGE HENRY, b Beech Creek, Pa, Nov 17, 11; m 38; c 2. PHYSICAL CHEMISTRY. Educ: Pa State Col, BS, 33, PhD(phys chem), 38. Prof Exp: Asst, Pa State Col, 38-39, instr chem, 39-40; assoc chemist, US Bur Mines, 40-45; res engr, M W Kellogg Co, 45-51; res dir, Redstone Arsenal, 51-53; sr res engr, Bendix Prod Div, 58-65; CONSULT, 65- Mem: AAAS; Am Chem Soc; Am Inst Aeronaut & Astronaut. Res: Cryogenics; high explosives; combustion; rocket propellants. Mailing Add: RD 2 Port Murray NJ 07865

MESSERSMITH, DONALD HOWARD, b Toledo, Ohio, Dec 17, 28; m 57; c 4. ENTOMOLOGY, ORNITHOLOGY. Educ: Univ Toledo, BEd, 51; Univ Mich, MS, 53; Va Polytech Inst, PhD(entom), 62. Prof Exp: Prof biol, Radford Col, 57-64; PROF ENTOM, UNIV MD, COLLEGE PARK, 64- Mem: Entom Soc Am; Am Ornith Union. Res: Biology and taxonomy of Culicoides and Forcipomyia; control of bird depredations. Mailing Add: Dept of Entom Univ of Md College Park MD 20742

MESSERSMITH, JAMES DAVID, b Paintsville, Ky, Sept 14, 31; m 60; c 4. FISHERIES MANAGEMENT. Educ: Ore State Univ, BS, 53, MS, 58. Prof Exp: Res fel, Ore Coop Wildlife Res, 56-58; fishery biologist, 58-60, marine biologist, 61-63, assoc marine biologist, 63-69, sr marine biologist, 69-72, coordr state-fed fisheries mgt progs, 72-75, CONSERV PROG OFFICER & LEGIS COORDR, CALIF DEPT FISH & GAME, 75- Concurrent Pos: Proj mgr, Dungeness Crab Mgt Proj, Pac Fishery Biol. Mem: Am Fisheries Soc; Am Inst Fishery Res Biol. Res: Legislation and management with reference to marine fauna of the northeastern Pacific Ocean with emphasis on fish, mollusks and crustaceans of sport and commercial importance. Mailing Add: Calif Dept of Fish & Game Rm 1236-4, 1416 Ninth St Sacramento CA 95814

MESSERSMITH, ROBERT E, b Trenton, NJ, Mar 15, 30; m 57; c 3. VETERINARY MEDICINE. Educ: Cornell Univ, DVM, 54. Prof Exp: Vet, pvt pract, 54-61; vet, Agr Div, Am Cyanamid Co, NJ, 61-63, mgr swine prog, 63-68; clin vet, Animal Health Res Dept, 68-74, PROF SERV VET, DEPT AGR & ANIMAL HEALTH, CHEM DIV, HOFFMANN-LA ROCHE, INC, 74- Mem: Am Vet Med Asn. Res: Cause of problems in animal production and development of practical methods of control.

Mailing Add: Dept of Agr & Animal Health Chem Div Hoffman-LaRoche Inc Nutley NJ 07110

MESSIEH, SHOUKRY NASEEF, b Assiut, Egypt, Nov 27, 28; m 57; c 3. FISHERIES. Educ: Univ Cairo, BSc, 49; Univ Alexandria, dipl oceanog, 55; McGill Univ, PhD, 73. Prof Exp: Teacher, Ramleh High Sch, Egypt, 49-52 & St Marc Col, 52-55; fishery biologist, Inst Hydrobiol & Fisheries, 55-60; dir, Fisheries Res Sta, Port-Taufig, 61-66; RES SCIENTIST, MARINE & FISHERIES SERV, BIOL STA, FISHERIES RES BD CAN, 67- Concurrent Pos: Colombo Plan Sci Res grant, Tokai Fisheries Res Lab, Japan, 60. Mem: Am Fisheries Soc. Res: Fishery biology, particularly herring population studies. Mailing Add: Marine & Fisheries Serv Fisheries Res Bd Can Biol Sta St Andrews NB Can

MESSIER, BERNARD, b Montreal, Que, May 4, 26; m 53; c 4. EXPERIMENTAL PATHOLOGY. Educ: Univ Montreal, BS, 49; McGill Univ, MSc, 56, PhD(anat), 60. Prof Exp: Asst path, Col Physicians & Surgeons, Columbia Univ, 61-62; asst prof med, 64-67, from asst prof to assoc prof anat, 67-75, PROF ANAT, UNIV MONTREAL, 75- Res: Radioautography; cell renewal in normal tissues. Mailing Add: Dept of Anat Univ of Montreal Montreal PQ Can

MESSIER, ROBERT LOUIS, b Worcester, Mass, Sept 13, 17. BIOCHEMISTRY. Educ: Worcester Polytech Inst, BS, 40; Univ Mass, MS, 41; Cornell Univ, PhD(biochem), 45. Prof Exp: Dir food packing, NGrafton State Hosp, Mass, 41; assoc ed, 45-59, SR ASSOC ED, CHEM ABSTR, 59- Mem: Am Chem Soc. Res: Chemical information. Mailing Add: Chem Abstr Dept 54 Ohio State Univ Columbus OH 43210

MESSINA, EDWARD JOSEPH, b Brooklyn, NY, May 28, 37; m 60; c 1. CARDIOVASCULAR PHYSIOLOGY. Educ: St John's Univ, BSc, 60; New York Med Col, PhD(physiol), 72. Prof Exp: NIH fel, 72-73, instr, 73-74, ASST PROF PHYSIOL, NEW YORK MED COL, 74- Mem: AAAS; Microcirc Soc; NY Acad Sci; Am Physiol Soc. Res: Understanding the interrelationships between those local factors which contribute to the regional regulation of blood flow. Mailing Add: Dept of Physiol New York Med Col Valhalla NY 10595

MESSINEO, LUIGI, b Bronte, Italy, May 25, 26; US citizen; m 68. BIOCHEMISTRY, BIOPHYSICS. Educ: Univ Palermo, Lic clas, 46, PhD(natural sci), 53; Inst Philos, Messina, Italy, lic philos, 49. Prof Exp: Vis investr biochem physiol, Univ Calif, Berkeley, 58-59; vis investr biophys, Univ Pittsburgh, 59-61; res chemist, Vet Admin Hosp, Buffalo, NY, 62-67; dir biochem res lab, Vet Admin Ctr, 67-70; PROF BIOL & CHEM, CLEVELAND STATE UNIV, 70- Concurrent Pos: Vis scientist, LaStazione Zoologica, Italy, 53; vis investr, Univ Pittsburgh, 57; Damon Runyon Mem Found res fel, 58-61; Health Res & Serv Found grant, 60-61, Leukemia Found, 62, Health Res Found Western NY, 63, Nat Cancer Inst, 64-67 & Am Heart Asn, 69-70; res assoc, Nat Cancer Inst, 61-62; res asst prof, State Univ NY Buffalo, 62-67 & Xavier Univ, Ohio, 67- Mem: Biophys Soc; Am Inst Biol Sci; Am Soc Zool. Res: Physiochemical and immunological properties of deoxyribonucleoproteins from normal and abnormal sources. Mailing Add: Dept of Biol Cleveland State Univ Cleveland OH 44115

MESSING, ALAN WALLACE, organic chemistry, see 12th edition

MESSING, RALPH ALLAN, b Newark, NJ, Jan 16, 27; m 55; c 3. ENZYMOLOGY. Educ: City Col New York, BS, 49; Purdue Univ, MS, 51. Prof Exp: Res assoc neurol, Sch Med, Univ Ark, 52-54; assoc res biochemist, Ethicon, Inc, Div Johnson & Johnson, 55-61; chief enzymologist, Schwarz Labs, Inc, 61-63; SR RES ASSOC, CORNING GLASS WORKS, 63- Mem: AAAS; Am Inst Chem; Am Chem Soc; NY Acad Sci. Res: Immobilization, preparation, purification and characterization of enzymes; reactions of proteins on inorganic surfaces; metabolism; industrial and clinical microbiology. Mailing Add: Sullivan Park Corning Glass Works Corning NY 14830

MESSING, SHELDON HAROLD, b Apr 6, 47; US citizen; m 67; c 1. ORGANIC CHEMISTRY. Educ: Brooklyn Col, BS, 67; Polytech Inst Brooklyn, PhD(org chem), 72. Prof Exp: SR RES CHEMIST, DOW CHEM CO, 72- Mem: Am Chem Soc; Sigma Xi. Res: Physical organic chemistry applied to the synthesis and development of processes for compounds applicable in the agricultural field. Mailing Add: 2309 Brookfield Dr Midland MI 48640

MESSING, SIMON D, b Frankfurt-am-Main, Ger, July 13, 22; US citizen; m 67; c 1. CULTURAL ANTHROPOLOGY, MEDICAL ANTHROPOLOGY. Educ: City Col New York, BSS, 49; Univ Pa, PhD(anthrop), 57. Prof Exp: Interdisciplinary res, Behav Inst, Univ Pa, 52-53; asst prof soc sci, Paine Col, 56-58; assoc prof anthrop, Hiram Col, 58-60; assoc prof, Univ SFla, 60-64; researcher & field consult, US AID-Ethiopia, 61-67; PROF ANTHROP, SOUTHERN CONN STATE COL, 68- Mem: Fel AAAS; Am Anthrop Asn; fel Soc Appl Anthrop; fel Am Pub Health Asn; fel Am Sociol Asn. Res: Applied anthropology of Africa, especially in public health attitudes and practices. Mailing Add: Dept of Anthrop Southern Conn State Col New Haven CT 06515

MESSMER, DENNIS A, b Wessington Springs, SDak, Dec 22, 37; m 65; c 1. MICROBIOLOGY, BIOCHEMISTRY. Educ: SDak State Univ, BS, 63, MS, 64; Kans State Univ, PhD(bact), 68. Prof Exp: Asst prof, 68-72, ASSOC PROF MICROBIOL, SOUTHWESTERN STATE COL, OKLA, 72- Concurrent Pos: Danforth fel, 71. Mem: AAAS; Am Soc Microbiol. Res: Metabolic interrelationships among bacteria in regard to substrate utilization. Mailing Add: Dept of Biol Sci Southwestern State Col Weatherford OK 73096

MESSMER, RICHARD PAUL, b Pittsburgh, Pa, Nov 24, 41; m 67. CHEMICAL PHYSICS. Educ: Carnegie Inst Technol, BS, 63; Univ Alta, PhD(theoret chem), 67. Prof Exp: Res assoc theoret chem, Mass Inst Technol, 67-68; lectr chem, Univ Alta, 68-69; STAFF MEM, RES & DEVELOP CTR, GEN ELEC CO, 69- Concurrent Pos: Vis scientist, Dept of Mat Sci & Eng, Mass Inst Technol, 73- Mem: AAAS; Am Chem Soc; Am Phys Soc. Res: Quantum theory of solid state, especially chemically related problems; theoretical studies of chemisorption, surfaces, deep defect levels in semiconductors, transition metal complexes and anisotropic molecular crystals. Mailing Add: Res & Develop Ctr Gen Elec Co PO Box 8 Schenectady NY 12301

MESSMER, TRUDY OTTILIA, b Long Beach, Calif, Apr 1, 47. CELL BIOLOGY. Educ: Univ Calif, Irvine, BS, 69; Univ Calif, San Diego, PhD(biol), 73. Prof Exp: Damon Runyon fel cell biol, 73-75, RES ASSOC CELL BIOL, SALK INST BIOL SCI, 75- Res: The control of cell proliferation especially with respect to the cancerous state versus the noncancerous state, and factors controlling the commitment to and initiation of DNA synthesis and mitosis. Mailing Add: Salk Inst PO Box 1809 San Diego CA 92112

MESSNER, ROBERT LEE, b Long Beach, Calif, Apr 4, 46. EXPERIMENTAL HIGH ENERGY PHYSICS, ELEMENTARY PARTICLE PHYSICS. Educ: Univ Chicago, BA, 68; Univ Colo, MS, 70, PhD(physics), 74. Prof Exp: RES ASSOC PHYSICS, UNIV ILL, URBANA, 74- Mem: Am Phys Soc. Mailing Add: Dept of Physics Univ of Ill Urbana IL 61801

MESTECKY, FRANK JOSEPH, mathematics, see 12th edition

MESTECKY, JIRI, b Prague, Czech, June 3, 41; m; c 1. IMMUNOLOGY, IMMUNOCHEMISTRY. Educ: Charles Univ, Prague, MD, 64. Prof Exp: Asst, Inst Microbiol & Immunol, Fac Med, Charles Univ, Prague, 61-63; sr res asst immunol, Inst Microbiol, Czech Acad Sci, 63-65; instr, Inst Microbiol, Fac Med, Charles Univ, Prague, 65-66; vis res assoc, 67-68, from instr to asst prof, 68-72, ASSOC PROF MICROBIOL, UNIV ALA, BIRMINGHAM, 72-, SCIENTIST, INST DENT RES, 72-, SCIENTIST, CANCER RES INST, 73- Concurrent Pos: Vis fel microbiol, Univ Ala, Birmingham, 67; WHO travel stipend. Mem: Am Asn Immunol; Soc Exp Biol & Med. Res: Protein chemistry; secretory antibodies; lysozyme. Mailing Add: 2244 Garland Dr Birmingham AL 35216

MESZLER, RICHARD M, b Peekskill, NY, Aug 30, 42; m 65; c 1. CELL BIOLOGY, NEUROANATOMY. Educ: NY Univ, BA, 64; Univ Louisville, PhD(anat), 69. Prof Exp: NIH res fel anat, Albert Einstein Col Med, 69-71; instr, 71-72, ASST PROF ANAT, SCH DENT, UNIV MD, 72- Mem: Am Soc Cell Biol; Am Asn Anatomists. Res: Function-structure relationships in nervous system; active transport of ions across biological membranes. Mailing Add: Dept of Anat Univ of Md Dent Sch Baltimore MD 21201

MESZOELY, CHARLES ALADAR MARIA, b Szekesfehervar, Hungary, Apr 24, 33; US citizen; m 61; c 2. PALEONTOLOGY, PARASITOLOGY. Educ: Northeastern Univ, BS, 61; Boston Univ, MA, 63, PhD(biol), 67. Prof Exp: Instr biol, Northeastern Univ, 66-68; res assoc biophys, Armed Forces Inst Path, 68-70; ASSOC PROF BIOL, NORTHEASTERN UNIV, 70- Mem: Soc Vert Paleont. Res: Paleontology, especially evolution and systematics of fossil and recent anguid lizards, and other Cenozoic lower vertebrates; parasitology, especially ultrastructure and ultrastructural changes in the malarial parasite. Mailing Add: Dept of Biol Northeastern Univ Boston MA 02115

METAKIDES, GEORGE, b Thessaloniki, Greece, Sept 22, 45. MATHEMATICAL LOGIC. Educ: Cornell Univ, BSc, 67, MS, 70, PhD(math), 71. Prof Exp: C L E Moore instr math, Mass Inst Technol, 71-72; ASST PROF MATH, UNIV ROCHESTER, 72- Concurrent Pos: NSF res grant, 71- Mem: Am Math Soc; Math Asn Am; Asn Symbolic Logic. Res: The investigation of recursively enumerably presented models of appropriate theories from a recursive model theory viewpoint, in particular, the uniqueness of universal homogeneous recursively enumerably presented models. Mailing Add: Dept of Math Univ of Rochester Rochester NY 14627

METANOMSKI, WLADYSLAW VAL, b Vienna, Austria, Oct 3, 23; US citizen; m 66; c 1. POLYMER CHEMISTRY, INFORMATION SCIENCE. Educ: Univ London, BSc, 52; Univ Toronto, MASc, 60, PhD(chem eng), 64. Prof Exp: Chemist, Anal & Res Lab, Dearborn Chem Co Ltd, Ont, 52-56, chem engr, Tech Field Serv, 56-58; demonstr chem eng, Univ Toronto, 58-64; asst ed, 64-66, group leader, 66-71, asst to ed, 71-72, MGR ED PLANNING & DEVELOP, CHEM ABSTR SERV, AM CHEM SOC, OHIO STATE UNIV, 72- Mem: Am Chem Soc. Res: High polymers; electron exchangers; chemical information science; indexing chemical literature; chemical compound nomenclature; vocabulary control; development of computer-based information processing system. Mailing Add: Chem Abstr Serv Ohio State Univ Columbus OH 43210

METCALF, ARTIE LOU, b Dexter, Kans, July 5, 29. ZOOLOGY. Educ: Kans State Col, BS, 56; Univ Kans, MA, 57, PhD(zool), 64. Prof Exp: From instr to assoc prof, 62-68, PROF ZOOL, UNIV TEX, EL PASO, 69- Mem: AAAS; Am Malacol Union; Am Quaternary Asn; Conchol Soc Gt Brit & Ireland; Ger Malacozool Soc. Res: Systematics and paleoecology of terrestrial mollusks. Mailing Add: Dept of Biol Sci Univ of Tex El Paso TX 79968

METCALF, EDWARD ALBERT, organic chemistry, see 12th edition

METCALF, FREDERIC THOMAS, b Oak Park, Ill, Dec 28, 35; m 57; c 2. APPLIED MATHEMATICS. Educ: Lake Forest Col, BA, 57; Univ Md, MA, 59, PhD(appl math), 61. Prof Exp: Asst engr, Electronics Div, Westinghouse Elec Co, 57-58; mathematician, Phys Chem Div, US Naval Ord Lab, 60-62, res mathematician, Math Dept, 62-63; asst res prof appl math, Inst Fluid Dynamics & Appl Math, Univ Md, 63-66; assoc prof, 66-69, chmn dept, 68-70, PROF MATH, UNIV CALIF, RIVERSIDE, 69- Concurrent Pos: Consult, US Naval Ord Lab, 64-66. Mem: AAAS; Am Math Soc; Math Asn Am. Res: Dynamical systems with two degrees of freedom; finite difference schemes for partial differential equations; inequalities; second order ordinary differential equations; fluid flow about bodies of revolution. Mailing Add: Dept of Math Univ of Calif Riverside CA 92502

METCALF, HAROLD, b Boston, Mass, June 11, 40; m 63; c 3. PHYSICS. Educ: Mass Inst Technol, ScB, 62; Brown Univ, PhD(physics), 68. Prof Exp: Res assoc physics, Brown Univ, 67-68; res assoc, 68-70, ASST PROF PHYSICS, STATE UNIV NY STONY BROOK, 70- Mem: Am Phys Soc; Optical Soc Am. Res: Experimental atomic physics; precision measurements; experimental quantum electrodynamics; level crossing spectroscopy; astrophysics; problems of human visual perception. Mailing Add: Dept of Physics State Univ of NY Stony Brook NY 11790

METCALF, HOMER NOBLE, b Ellington, Conn, Mar 9, 17. HORTICULTURE. Educ: Univ Conn, BS, 39; Cornell Univ, MS, 43. Prof Exp: Asst, Tree Fruit Br Exp Sta, Wenatchee, Wash, 44-45 & Agr Res Div, Gen Lab, Libby, McNeill & Libby, 45-47; from asst prof to assoc prof, 47-62, PROF HORT, MONT STATE UNIV, 62- Mem: Fel AAAS; Bot Soc Am; Torrey Bot Club; Am Plant Life Soc; Int Soc Hort Sci. Res: Pomology; ornamental horticulture; horticultural biosystematics. Mailing Add: Dept of Plant & Soil Sci Mont State Univ Bozeman MT 59715

METCALF, ISAAC STEVENS HALSTEAD, b Cleveland, Ohio, Aug 17, 12; m 41; c 3. GROSS ANATOMY, COMPARATIVE ANATOMY. Educ: Oberlin Col, BA, 34; Columbia Univ, MA, 36; Case Western Reserve Univ, PhD(biol), 40. Prof Exp: From asst prof to assoc prof biol & chem, The Citadel, 37-57, prof biol, 57-66; assoc prof, 66-70, PROF ANAT, MED UNIV SC, 70- Mem: AAAS; Nat Audubon Soc. Res: Neuroanatomy; sense organs of Elasmobranchs. Mailing Add: Dept of Anat Med Univ of SC Charleston SC 29401

METCALF, ROBERT ALAN, b Riverside, Calif, Feb 1, 49. POPULATION BIOLOGY. Educ: Univ Ill, BA, 71; Harvard Univ, PhD(biol), 75. Prof Exp: ASST PROF ZOOL, UNIV CALIF, DAVIS, 75- Res: Social behavior and its evolution, particularly in insects. Mailing Add: Dept of Zool Univ of Calif Davis CA 95616

METCALF, ROBERT HARKER, b Chicago, Ill, Aug 29, 43; m 68. MICROBIOLOGY. Educ: Earlham Col, AB, 65; Univ Wis-Madison, MS, 68, PhD(bact), 70. Prof Exp: Asst prof, 70-75, ASSOC PROF BIOL SCI, CALIF STATE UNIV, SACRAMENTO,

75- Mem: Am Soc Microbiol. Res: Food and water microbiology. Mailing Add: Dept of Biol Sci Calif State Univ Sacramento CA 95819

METCALF, ROBERT LEE, b Columbus, Ohio, Nov 13, 16; m 40; c 3. ENTOMOLOGY. Educ: Univ Ill, BA, 39, MA, 40; Cornell Univ, PhD(entom), 43. Prof Exp: From asst entomologist to assoc entomologist, Tenn Valley Authority, Ala, 43-46; from asst entomologist to assoc entomologist, Citrus Exp Sta, Univ Calif, Riverside, 46-53, prof entom & entomologist, 53-68; PROF ENTOM, UNIV ILL, URBANA-CHAMPAIGN, 68-, HEAD DEPT ZOOL, 69- Concurrent Pos: Vchancellor, Univ Calif, Riverside, 62-67; consult, WHO; President's Sci Adv Comt. Mem: Nat Acad Sci; Am Chem Soc; Entom Soc Am (pres, 58). Res: Insect physiology and toxicology; mosquito control. Mailing Add: Dept of Zool 287 Morrill Hall Univ of Ill Urbana IL 61801

METCALF, THEODORE GORDON, b Rockville, Conn, Oct 9, 18; m 42; c 4. MICROBIOLOGY. Educ: Mass Col Pharm, BS, 40; Univ Kans, PhD(bact), 50. Prof Exp: Asst virus res, Parke, Davis & Co, 42-44; from instr to assoc prof bact, Univ Kans, 46-56; assoc prof, 59-61, PROF BACT, UNIV NH, 61- Concurrent Pos: NSF fel, 59. Mem: Am Soc Microbiol; Am Acad Microbiol. Res: Respiratory and enteric viruses; rickettsiae; pathogenic bacteriology; immunology; environmental virology-viruses of vertebrates occurring in polluted surface waters, shellfish and estuarine waters. Mailing Add: Dept of Microbiol Spaulding Life Sci Bldg Univ NH Durham NH 03824

METCALF, WILLIAM, b Norwood, Mass, Dec 31, 07; m 50. SURGERY. Educ: Mass Inst Technol, BSc & MSc, 31; Johns Hopkins Univ, MD, 37. Prof Exp: Res fel surg, Johns Hopkins Univ, 38-39; Cushing fel, Sch Med, Yale Univ, 39-40, asst surg, 41-43; asst chief, Vet Admin Hosp, Hines, NY, 47-48; teaching fel, St Vincents Hosp, New York, 50-52; instr, Sch Med, NY Univ, 52-55; from asst prof to assoc prof, 54-62, PROF SURG, ALBERT EINSTEIN COL MED, YESHIVA UNIV, 62- Mem: AAAS; Am Soc Surg of Hand; AMA; Am Col Surg; Am Soc Clin Nutrit. Res: General and hand surgery; surgical metabolism and shock; plasma and plasma expanders; nitrogen metabolism; anabolic steroids. Mailing Add: Albert Einstein Col of Med Yeshiva Univ New York NY 10461

METCALF, WILLIAM GERRISH, b New York, NY, Aug 2, 18; m 44; c 4. OCEANOGRAPHY. Educ: Oberlin Col, BA, 40; Amherst Col, MA, 42. Prof Exp: Asst phys oceanog, 52-58, phys oceanogr, 58-63, ASSOC SCIENTIST, WOODS HOLE OCEANOG INST, 63- Concurrent Pos: Dir oceanog prog, Dept Pub Works, Commonwealth PR, 71-72; hon vis prof, Univ PR, Mayagüez, 71-72. Mem: AAAS; Arctic Inst NAm. Res: Arctic and physical oceanography; tropical ocean circulation. Mailing Add: Woods Hole Oceanog Inst Woods Hole MA 02543

METCALF, WILLIAM KENNETH, b Whitley Bay, Eng, Apr 30, 21; m 44; c 7. HUMAN ANATOMY. Educ: Univ Durham, MB & BS, 43; Bristol Univ, MD, 60. Prof Exp: From lectr to sr lectr anat, Bristol Univ, 48-64, reader, 64-68; prof, Univ Iowa, 68-73; PROF ANAT & CHMN DEPT, UNIV NEBR MED CTR, 73- Concurrent Pos: USPHS fel, 69-73. Honors & Awards: Media Fair 1st Prize, Health Educ Media Asn-Health Sci Communications Asn, 75. Mem: Am Asn Anat; Anat Soc Gt Brit & Ireland; Brit Soc Hemat; Brit Physiol Soc. Res: Hematology; cell kinetics; physical properties of cells; cellular immunology; education, especially effect of objectives and efficiency of programmed instruction. Mailing Add: Dept of Anat Univ Nebr Med Ctr Omaha NE 68105

METCALFE, DARREL SEYMOUR, b Arkansaw, Wis, Aug 28, 13; m 42; c 2. AGRONOMY. Educ: Univ Wis, BS, 41; Kans State Univ, MS, 42; Iowa State Univ, PhD(plant physiol, crop breeding), 50. Prof Exp: Asst, Kans State Univ, 40-42; prof agron, Iowa State Univ, 46-56, asst dir stud affairs, 56-58; PROF AGRON, ASSOC DEAN OF COL AGR, DIR RESIDENT INSTR OF COL AGR & ASST DIR OF AGR EXP STA, UNIV ARIZ, 58- Concurrent Pos: Mem comt educ in agr & nat res, Nat Acad Sci-Nat Res Coun, 66-70; Agency Int Develop consult to Brazil; Orgn Econ Coop & Develop, Europe. Honors & Awards: Agron Educ Award, Am Soc Agron, 58. Mem: Fel Am Soc Agron. Res: Seed production of forage grasses and legumes. Mailing Add: Col of Agr Univ of Ariz Tucson AZ 85721

METCALFE, DAVID RICHARD, b Carroll, Man, Oct 14, 23; m 49; c 2. PLANT BREEDING, GENETICS. Educ: Univ Man, BSA, 50, PhD(genetics, plant breeding), 60; Univ Wis, MS, 53. Prof Exp: RES SCIENTIST GENETICS & PLANT BREEDING, RES BR, AGR CAN, 50- Mem: Agr Inst Can; Genetic Soc Can. Res: Breeding two-rowed barley varieites for the eastern prairie region of Canada. Mailing Add: Agr Can Res Sta Univ Man Campus 25 Dafoe Rd Winnipeg MB Can

METCALFE, GRANT E, b Albany, NY, July 21, 06; m 33; c 2. PSYCHIATRY. Educ: Hahnemann Med Col, BS, 28, MD, 30. Prof Exp: Asst prof, 53-57, ASSOC PROF PSYCHIAT, MED CTR, IND UNIV, 57-, LECTR PSYCHOANAL, UNIV EXTEN SOUTH BEND, 53- Mem: Fel Am Psychiat Asn; Am Psychosom Soc; Nat Asn Ment Health; fel Royal Soc Health; Pan-Am Med Asn. Res: Promotion of mental health, education and hospital and community services. Mailing Add: 919 E Jefferson Blvd South Bend IN 46622

METCALFE, JAMES, b New Bedford, Mass, Aug 16, 22; m 44; c 4. MEDICINE. Educ: Brown Univ, AB, 44; Harvard Univ, MD, 46; Am Bd Internal Med, dipl, 53. Prof Exp: Med house officer, Peter Bent Brigham Hosp, Boston, 46-47, asst, 50-51, sr asst, 51-52; ward med officer, US Naval Hosp, Newport, RI, 47-49; instr, Harvard Med Sch, 53-55, assoc med, 55-59, tutor, 57-58, asst prof, 59-61; assoc prof, 61-64, PROF MED, MED SCH, UNIV ORE, 64- Concurrent Pos: Res fel, Peter Bent Brigham Hosp, 50-51; res fel physiol, Harvard Med Sch, 49-50, 52-53; fel, Boston Lying-in Hosp, 52-53; assoc physician, Bosotn Lying-in Hosp, 52-59, vis physician, 59-61; Am Heart Asn estab investr, 53-59; jr assoc, Peter Bent Brigham Hosp, 53-56, assoc, 56-58, sr assoc, 58-61; chmn cardiovasc res, Ore Heart Asn, 61- Mem: Am Physiol Soc; Am Fedn Clin Res; Am Clin & Climat Asn; Am Soc Clin Invest. Res: Modifications of maternal physiology during pregnancy and their effects on the course of disease. Mailing Add: Heart Res Lab Univ of Ore Sch of Med Portland OR 97201

METCALFE, JOSEPH EDWARD, III, b Fallowfield Twp, Pa, May 27, 38; m 59; c 3 PHYSICAL CHEMISTRY, FUEL TECHNOLOGY. Educ: Pa State Univ, BS, 60, MS, 62, PhD(fuel technol), 65. Prof Exp: Res asst fuel technol, Pa State Univ, 60-62 & 63-65; sr res chemist, 65-74, res supvr electrokinetics, 68-71 & gasoline phys res, 71-74, supvr technol assessment, 74-76, SUPVR ALT ENERGY SOURCES, STAND OIL CO OHIO, 76- Mem: Am Chem Soc. Res: Fused salt batteries; carbon technology; molecular sieves; adsorption; physical properties of gasoline; coal technology; catalysts. Mailing Add: Res Dept Broadway Lab Stand Oil Co Ohio 3092 Broadway Cleveland OH 44115

METCALFE, LINCOLN DOUGLAS, b Melstone, Mont, Feb 11, 21; m 48; c 1. ANALYTICAL CHEMISTRY. Educ: Univ Chicago, BS, 47. Prof Exp: Head anal sect, Res Div, Armour & Co, 47-60, head anal res sect, Res Labs, Armour Indust Chem Co, 60-66, ASST RES DIR ANAL & PHYS CHEM & INSTRUMENTAL RES, RES LABS, ARMAK CO, 66- Concurrent Pos: Mem lipid anal comt, Nat Heart Inst, 58-59. Honors & Awards: Bond Award, Am Oil Chem Soc, 64; Nat Merit Award, Am Soc Testing & Mat, 66. Mem: Am Chem Soc; Am Soc Testing & Mat; Am Oil Chem Soc. Res: Nonaqueous titrations; gas and high pressure liquid chromatography; infrared and ultraviolet spectrophotometry as applied to lipid and protein chemistry, especially fatty acid derivatives. Mailing Add: Res Labs Armak Co 8401 W 47th St McCook IL 60525

METCOFF, JACK, b Chicago, Ill, Feb 2, 17; m 43; c 2. PHYSIOLOGY. Educ: Northwestern Univ, BS, 38, BM, 42, MD, 43, MS, 44; Harvard Univ, MPH, 44. Prof Exp: From asst to assoc pediat, Harvard Med Sch, 48-53, asst prof, 53-56; chmn dept, Michael Reese Hosp, 56-70; PROF PEDIAT BIOCHEM & MOLECULAR BIOL, UNIV OKLA HEALTH SCI CTR, 70- Concurrent Pos: Prof, Med Sch, Northwestern Univ, 56-63; prof & chmn dept, Chicago Med Sch, 63-68. Mem: Fel Am Acad Pediat; Am Physiol Soc; Am Soc Nephrology; Am Soc Clin Nutrit; Am Inst Nutrit. Res: Cell metabolism; electrolyte and renal physiology; relations between intracellular ions and intercellular metabolites during prematurity; normal growth and development of the human infant; severe chronic infantile malnutrition; Kwashiorkor; fetal malnutrition; metabolism of isolated kidneys; leukocyte metabolism; nutrition in pregnancy; renal disease in children; metabolism in uremia. Mailing Add: Dept Pediat Health Sci Ctr Univ Okla PO Box 26901 Oklahoma City OK 73190

METEER, JAMES WILLIAM, b Columbus, Ohio, Apr 7, 21; m 44; c 4. FORESTRY, COMPUTER SCIENCE. Educ: Univ Mich, BSF, 44, MF, 47. Prof Exp: Asst prof forestry, Agr Exp Sta, Ohio State Univ, 47-54, consult forester, 54-65; from asst prof to assoc prof, 64-73, PROF FOREST RES & FORESTRY, FORD FORESTRY CTR, MICH TECHNOL UNIV, 73- Concurrent Pos: Lectr, Purdue Univ, 60. Mem: Soc Am Foresters. Res: Forest management and growth investigations; continuous forest inventory control with electronic machine handling of data. Mailing Add: Sch of Forestry Ford Forestry Ctr Mich Technol Univ L'Anse MI 49946

METEVIA, LOUIS ANTHONY, b Baton Rouge, La, Oct 20, 31; m 66; c 1. CELL BIOLOGY, ELECTRON MICROSCOPY. Educ: Southern Univ, Baton Rouge, BS, 58; Pa State Univ, MS, 67. Prof Exp: From instr to asst prof, 60-70, ASSOC PROF BIOL & DIR ELECTRON MICROS LAB, SOUTHERN UNIV, BATON ROUGE, 70- Mem: Am Soc Cell Biol. Res: Ultrastructure of calcium transport, neurosecretions, unorthodox cell divisions and thick-thin filament development in rhabdocoel flatworms. Mailing Add: 6130 Crestway Ave Baton Rouge LA 70812

METEYER, THOMAS EDWARD, b Rochester, NY, Oct 18, 45. SYNTHETIC ORGANIC CHEMISTRY. Educ: St John Fisher Col, BS, 67; Syracuse Univ, PhD(chem), 72. Prof Exp: Nat Acad Sci overseas res fel & vis prof chem, Univ Sao Paulo, 72-74; SR CHEMIST ION EXCHANGE RES, ROHM AND HAAS CO, 74- Mem: Am Chem Soc. Res: Natural products synthesis and structure elucidation; polymer synthesis; novel synthetic reactions. Mailing Add: Ion Exchange Res Rohm and Haas Co Norristown Rd Spring House PA 19477

METHOD, PETER FRANCIS, b St Louis, Mo, Aug 25, 43; m 70; c 1. PHYSICAL CHEMISTRY. Educ: Marian Col, Ind, BS, 65; WVa Univ, PhD(chem), 73. Prof Exp: Instr chem, Univ Ky, 71-73, res assoc biochem, 73-75; ASST PROF CHEM, ROSE-HULMAN INST TECHNOL, 75- Mem: Am Chem Soc; Am Crystallog Asn. Res: Use of computers in the teaching of undergraduate chemistry. Mailing Add: Rose-Hulman Inst Technol Terre Haute IN 47803

METLAY, MAX, b New York, NY, Mar 18, 22; m 51; c 4. CLINICAL CHEMISTRY. Educ: City Col New York, BS, 42; Columbia Univ, AM, 44, PhD(phys chem), 48. Prof Exp: Asst, Columbia Univ, 42-43, sam labs, 43-44, Univ, 44-45 & 47; researcher, Northwestern Univ, 47-48, instr chem, 48; researcher, Pa State Col, 48-49; res assoc, Sch Mines, Columbia Univ, 50-52; asst prof chem, Harpur Col, 52-57, assoc prof, 57-58; prof chem & physics, New Eng Col Pharm, 58-60; phys chemist, Res & Develop Ctr, Gen Elec Co, 60-67; sr fel, Mellon Inst, 67-72; clin chemist, Univ Health Ctr Pittsburgh, 72-73; ASSOC PROF CLIN CHEM, UNIV PITTSBURGH, 73- Mem: AAAS; Am Chem Soc; Am Asn Clin Chemists. Res: Electron affinities; memechanism of metallographic etching; flash photolysis; photochemistry; photo-bleaching of dyes; photochromism; thermal analysis and radiation chemistry of polymers; immobilized enzymes. Mailing Add: 6553 Darlington Rd Pittsburgh PA 15217

METRAKOS, JULIUS DEMETRIUS, b Montreal, Aug 12, 15; m 50. MEDICAL GENETICS. Educ: McGill Univ, BSc, 47, MSc, 48, PhD(human genetics), 50. Prof Exp: Demonstr, 48-52, lectr, 53-56, from asst prof to assoc prof, 56-66, PROF GENETICS, MCGILL UNIV, 66- Concurrent Pos: Lectr, Loyola Col, 59; res scientist, Montreal Children's Hosp. Mem: Genetics Soc Am; Am Soc Human Genetics. Res: Hereditary factors in children's diseases; genetic counselling. Mailing Add: Dept of Biol McGill Univ Montreal PQ Can

METRAUX, RHODA, b Brooklyn, NY, Oct 18, 14; m 41; c 1. CULTURAL ANTHROPOLOGY, APPLIED ANTHROPOLOGY. Educ: Vassar Col, BA, 34; Columbia Univ, PHD(anthrop), 51. Prof Exp: Mem res staff anthrop, Res Contemp Cult, Columbia Univ, 47-48, asst dir, 49-52; res dir anthrop, Mus Natural Hist, 52-53; res fel human ecol, Cornell Univ Med Col at New York Hosp, 54-57; consult anthrop, Inst Intercult Studies, 57; dir surv anthrop, Educ Div, Riverside Church, 58; lectr, Educ Div, NY Univ, 59; assoc dir allo-physics, 60-65, proj dir perceptual commun, 65-69, RES ASSOC ANTHROP, AM MUS NATURAL HIST, 70- Concurrent Pos: Soc Sci Res Coun fel, 53-54. Mem: Fel AAAS; fel Am Anthrop Asn; fel Am Ethnol Soc; fel Soc Appl Anthrop; fel NY Acad Sci. Res: Comparative studies including peasant cultures of the Caribbean area, especially Haiti and Montserrat and complex western and non-western cultures; communication, verbal and non-verbal; changing American attitudes toward scientists and science. Mailing Add: 211 Central Park W New York NY 10024

METRIONE, ROBERT M, b Teaneck, NJ, Aug 22, 33; m 57; c 3. BIOCHEMISTRY. Educ: Bowling Green State Univ, BS, 55; Univ Nebr, MS, 60, PhD(biochem), 63. Prof Exp: Res assoc biochem, Yale Univ, 63-65, asst prof, 65-67; asst prof, 67-73, ASSOC PROF BIOCHEM, THOMAS JEFFERSON UNIV, 73- Mem: AAAS; Am Chem Soc; Am Soc Biol Chem. Res: Structure-function relationships of proteolytic enzymes; cathepsins. Mailing Add: Dept of Biochem Thomas Jefferson Univ Philadelphia PA 19107

METROPOLIS, NICHOLAS CONSTANTINE, b Chicago, Ill, June 11, 15; m 55; c 3. APPLIED MATHEMATICS, THEORETICAL PHYSICS. Educ: Univ Chicago, BS, 36, PhD(physics), 41. Prof Exp: Res assoc, Univ Chicago, 41, res assoc, Metall Lab & instr physics, 42; res assoc, Columbia Univ, 42; res assoc, Los Alamos Sci Lab, 43-46, consult, 46-48, mem staff, 48-57; prof physics, Univ Chicago & Enrico Fermi Inst Nuclear Studies, 57-65, dir inst comput res, 58-65; STAFF MEM, LOS ALAMOS SCI LAB, 65- Concurrent Pos: Asst prof, Univ Chicago & Inst Nuclear Studies, 46-48; consult, Argonne Nat Lab, Brookhaven Nat Lab & Lawrence Radiation Lab, Univ Calif; mem adv panel univ comput facilities, NSF, 59- & adv comt comput activities; chmn comput adv group, AEC, S9-; mem, UN Tech Mission to India, 61; vis prof,

Univ Colo, 64; mem adv comt res, NSF, 73-75. Mem: Fel Am Phys Soc; Am Math Soc; Soc Indust & Appl Math. Res: Theoretical nuclear physics; electronic computing; logical design of general purpose computers; pure and applied mathematical analysis of inherent error propagation; studies of non-linear differential equations; theoretical investigations of nuclear cascades. Mailing Add: 1502 44th St Los Alamos NM 87544

METTALIA, JOSEPH BERNARD, JR, organic chemistry, see 12th edition

METTE, HERBERT L, b Kassel, Ger, Oct 26, 25; US citizen; m 58; c 1. SOLID STATE PHYSICS. Educ: Univ Göttingen, dipl physics, 52. Prof Exp: Res asst solid state physics, Univ Göttingen, 52-53; scientist, Ger Nat Bur Standards, 53-56; physicist, Res & Develop Lab, 56-58, leader device physics sect, Solid State Devices Div, 58-65, chief integrated devices tech br, Integrated Electronics Div, 65-73, LEADER ADVAN IC TECHNOL TEAM, INTEGRATED ELECTRONICS TECH AREA, US ARMY ELECTRONICS COMMAND, FT MONMOUTH, 73- Mem: Fel Am Phys Soc; sr mem Inst Elec & Electronics Eng. Res: Organic semiconductors; photomagneto and magnetothermal effects in semiconductors; integrated electronics; materials, devices and processes research. Mailing Add: 650 Valley Rd Brielle NJ 08730

METTEE, HOWARD DAWSON, b Boston, Mass, Aug 6, 39; m 63; c 1. SPECTROCHEMISTRY. Educ: Middlebury Col, BA, 61; Univ Calgary, PhD(phys chem), 64. Prof Exp: Fel spectros & photochem, Nat Res Coun Can, 64-66 & Univ Tex, Austin, 66-68; asst prof, 68-74, ASSOC PROF CHEM, YOUNGSTOWN STATE UNIV, 74- Mem: The Chem Soc. Res: Energy transfer and relaxation; primary photochemical events and gas phase kinetics; chemical applications of spectroscopy. Mailing Add: Dept of Chem Youngstown State Univ Youngstown OH 44555

METTEE, MAURICE FERDINAND, b Mobile, Ala, Apr 28, 43; m 68. AQUATIC BIOLOGY. Educ: Spring Hill Col, BS, 65; Univ Ala, MA, 67, MS, 70, PhD(biol), 74. Prof Exp: Res asst biol, Univ Ala, 74-75; ENVIRON BIOLOGIST, ENVIRON DIV, GEOL SURV ALA, 75- Concurrent Pos: US Forest Serv grant ichthyol, Univ Ala, 74; aquatic biologist, US Air Force Acad, 74; res grant, US Fish & Wildlife Serv, 75, consult, Okaloosa Darter in Fla, 75- Mem: Am Ichthyol & Herpet; Am Fisheries Soc; Sigma Xi. Res: Studies on the systematics, ecology, reproductive behavior, embryology and development of freshwater and marine fishes; endangered and threatened vertebrate life in the southeastern United States. Mailing Add: Ala Geol Surv Environ Div PO Box O University AL 35486

METTENET, WILLIAM JOSEPH, b Pittsburgh, Pa, May 16, 19; m 44; c 3. GEOGRAPHY, CARTOGRAPHY. Educ: Univ Pittsburgh, BA, 42; George Washington Univ, MA, 56. Prof Exp: Cartogr, US Dept State, 45-49; geogr & cartogr, US Govt, 49-64; asst prof, 64-67, ASSOC PROF GEOG, EASTERN NMEX UNIV, 67-, DIR, ROOSEVELT COUNTY MUS, 66- Mem: Asn Am Geog; Nat Coun Geog Educ. Res: Cartographic methods of teaching geography; interdisciplinary environmental studies; geography of Roosevelt County; maps of the location of New Mexico towns that have had newspapers since 1900; maps of Roosevelt County: postal locations, former settlements. Mailing Add: Roosevelt County Mus Number 30 Eastern NMex Univ Portales NM 88130

METTER, DEAN EDWARD, b Champaign, Ill, Aug 1, 32; m 54; c 3. HERPETOLOGY. Educ: Eastern Ill Univ, BS, 57; Wash State Univ, MS, 60; Univ Idaho, PhD(zool), 63. Prof Exp: Instr zool, Univ Idaho, 63-64; asst prof, 64-69, ASSOC PROF ZOOL, UNIV MO-COLUMBIA, 69- Mem: Am Soc Ichthyol & Herpet. Res: DiDistribution and differentiation of amphibian populations. Mailing Add: Dept of Zool Univ of Mo Columbia MO 65201

METTER, GERALD EDWARD, b Los Angeles, Calif, Oct 15, 44. BIOSTATISTICS, ONCOLOGY. Educ: Univ Calif, Berkeley, AB, 66, PhD(biostatist), 72. Prof Exp: Statistician, Ctr Dis Control, USPHS, 67-69; asst prof biostatist, Tulane Univ, 72-73; ASST PROF STATIST & HUMAN ONCOL, UNIV WIS-MADISON, 73- Concurrent Pos: Consult, Forest Prod Lab, USDA, 75- Mem: Am Statist Asn; Biomet Soc; AAAS. Res: Conduct of and statistical methods in clinical trials; statistical methods in epidemiology. Mailing Add: Dept of Statist Univ of Wis 1210 W Dayton St Madison WI 53706

METTER, RAYMOND EARL, b Champaign, Ill, Aug 25, 25; m 48; c 2. GEOLOGY. Educ: Ohio State Univ, MS, 52, PhD(geol), 55. Prof Exp: Instr math, Ohio State Univ, 48-50, asst geol, 50-52, res assoc physics, Res Found, 53-54; res geologist, Res Lab, Carter Oil Co Div, Stand Oil Co (NJ), 55-58, res geologist & group head, Jersey Prod Res Co, 58-65; res geologist, 65-66, RES ASSOC, EXXON PROD RES CO, 66- Mem: Am Asn Petrol Geol; Geol Soc Am. Res: Petroleum geology; applied organic geochemistry; geology of subsurface fluids. Mailing Add: 13323 Indian Creek Dr Houston TX 77024

METTLER, FREDERICK ALBERT, b New York, NY, June 13, 07; wid; c 3. NEUROLOGY, ANATOMY. Educ: Clark Univ, AB, 29; Cornell Univ, MA, 31, PhD(anat), 33; Univ Ga, MD, 37. Hon Degrees: ScD, Clark Univ, 51. Prof Exp: Asst anat, Med Col, Cornell Univ, 30-31, instr, 31-33; instr physiol, Med Col, St Louis Univ, 33-34; from asst prof to prof anat, Sch Med, Univ Ga, 34-42; assoc prof, 41-51, PROF ANAT, COL PHYSICIANS & SURGEONS, COLUMBIA UNIV, 51- Concurrent Pos: Guest investr, Univ Ill, 31-38, Harvard Univ, 35 & Univ Rochester, 38-40; Commonwealth Fund vis prof, Long Island Med Col, 43-44; lectr, Univ Edinburgh, 45, Rutgers Univ, 48 & Univ PR, 53; mem, Div Med Sci, Nat Res Coun, 47-50 & Intersoc Comt, NSF, 47-50; res consult, NJ State Hosp, Greystone Park, 47-71; coordr res, State Dept Ment Hyg, NY, 48-49, dir, 49-50; chmn comt psychosurg, Div Ment Hyg, Nat Adv Ment Health Coun, USPHS, 49-52; mem, US Nat Conf Deleg, UNESCO, 52-54; adv, Yerkes Primate Ctr, Atlanta, Ga, 73-; res consult, Res Ctr, Rockland State Hosp, Orangeburg, NY, 74- Mem: Soc Exp Biol & Med; Am Asn Anat; Am Neurol Asn; fel NY Acad Med; assoc NY Acad Sci. Res: Gross anatomy; physiology and pathology of the nervous system. Mailing Add: Pippin Hill Exp Area Blairstown NJ 07825

METTLER, LAWRENCE EUGENE, b Harrison, Ohio, Feb 18, 29; m 55; c 2. BIOLOGY. Educ: Miami Univ, AB, 50; Univ Ky, MS, 52; Univ Tex, PhD(zool, genetics), 56. Prof Exp: Res fel genetics, Johns Hopkins Univ, 56-59, lectr, 58; from asst prof to assoc prof, 59-69, PROF GENETICS & ZOOL, NC STATE UNIV, 69- Concurrent Pos: Lectr, NC Col Durham, 59. Mem: Genetics Soc Am; Soc Study Evolution. Res: Genetics of Drosophila; experimental population genetics. Mailing Add: Dept of Genetics NC State Univ Raleigh NC 27607

METTRICK, DAVID FRANCIS, b London, Eng, 1932; Can citizen. PARASITOLOGY, PATHOLOGICAL PHYSIOLOGY. Educ: Univ Wales, BSc, 54; Univ London, PhD(parasitol), 57, DSc(parasitol), 72. Prof Exp: Lectr zool, Univ Rhodesia & Nyasaland, 57-61; sr lectr zool, Univ WI, 61-67; assoc prof, 67-71, assoc chmn dept, 73-75, PROF ZOOL, UNIV TORONTO, 71-, CHMN DEPT, 75- Concurrent Pos: Prof parasitol, Fac Med, Univ Toronto, 71-; pres, Biol Coun Can, 74-; Can rep to coun, World Fedn Parasitologists, 74-; chmn, Can Coun Animal Care, 75-; chmn, Animal Biol Grants Comt, Nat Res Coun Can, 75- Mem: Am Soc Parasitologists; Can Soc Zoologists; Brit Soc Parasitol. Res: Ecology and physiology of intestinal parasites; pathophysiology; membrane transport; metabolism of intestinal parasites; symbiology. Mailing Add: Dept of Zool Univ of Toronto Toronto ON Can

METZ, CHARLES BAKER, b New York, NY, Dec 27, 16; m 40; c 2. DEVELOPMENTAL BIOLOGY. Educ: Johns Hopkins Univ, AB, 39; Calif Inst Technol, PhD(embryol), 42. Prof Exp: Instr biol, Wesleyan Univ, 42-46; instr zool, Yale Univ, 46-47, asst prof, 47-52; asst prof, Univ Calif, 52; assoc prof, Univ NC, 52-53; from assoc prof to prof, Fla State Univ, 53-64, assoc dir oceanog inst, 58-62, prof, Inst Space Biosci, 61-64; prof biol, 64-70, PROF MOLECULAR EVOLUTION & ZOOL, INST MOLECULAR EVOLUTION, UNIV MIAMI, 70- Concurrent Pos: Nat Res Coun fel, Ind Univ, 45-46; Gosney fel, Calif Inst Technol, 50; Lillie fel, Marine Biol Lab, Woods Hole, 53, instr, 47-52, mem, Corp Trustees, 56-65. Mem: AAAS; Am Soc Nat; Soc Exp Biol & Med; Am Soc Zool (secy, 61-63); Soc Gen Physiol. Res: Echinoderm development; physiology of reproduction in protozoa and marine invertebrates; immuno-reproduction; electron microscopy of protozoa and gametes. Mailing Add: Inst of Molecular Evolution Univ of Miami 521 Anastasia Ave Coral Gables FL 33134

METZ, CHARLES EDGAR, b Bayshore, NY, Sept 11, 42; m 67; c 1. MEDICAL PHYSICS. Educ: Bowdoin Col, BA, 64; Univ Pa, MS, 66, PhD(radiol physics), 69. Prof Exp: Instr, 69-71, ASST PROF RADIOL, UNIV CHICAGO, 71- Mem: Am Asn Physicists in Med; Soc Nuclear Med. Res: Analysis of medical imaging systems; medical image enhancement by computer; evaluation of human observer performance in terms of signal detection theory. Mailing Add: Dept of Radiol Univ of Chicago Chicago IL 60637

METZ, CHARLES FRANKLIN, physical chemistry, see 12th edition

METZ, CHARLES WILLIAM, zoology, deceased

METZ, CLYDE, b Gary, Ind, May 3, 40; m 61; c 2. PHYSICAL CHEMISTRY. Educ: Rose Polytech Inst, BS, 62; Ind Univ, PhD(phys chem), 66. Prof Exp: Asst, Ind Univ, 62-66; asst prof, 66-70, ASSOC PROF CHEM, IND UNIV-PURDUE UNIV, INDIANAPOLIS, 70- Honors & Awards: Leiber Teaching Assoc Award, Ind Univ, 66. Mem: AAAS; Am Chem Soc; Electrochem Soc. Res: Fused slat electrochemistry, phase equilibria, thermodynamics and x-ray crystallography. Mailing Add: 4217 Melbourne Rd Indianapolis IN 46208

METZ, DAVID A, b Cleveland, Ohio, Sept 10, 33; m 60; c 3. AUDIOLOGY, SPEECH PATHOLOGY. Educ: Western Reserve Univ, BA, 60, MA, 65, PhD(aural harmonics), 67. Prof Exp: Res asst speech path, Western Reserve Univ, 64-67; asst prof, 67-71, ASSOC PROF AUDIOL, CLEVELAND STATE UNIV, 71-, DIR SPEECH PATH & AUDIOL PROG, 67-, CHMN DEPT SPEECH & HEARING, 74- Mem: Am Speech & Hearing Asn; Acoust Soc Am. Res: Aural harmonics; temporary threshold shift; speech discrimination. Mailing Add: Speech & Hearing Clin Cleveland State Univ Cleveland OH 44115

METZ, DONALD J, b Brooklyn, NY, May 18, 24; m 47. PHYSICAL CHEMISTRY. Educ: St Francis Col, NY, BS, 47; Polytech Inst Brooklyn, MS, 49, PhD(phys chem), 55. Prof Exp: From assoc chemist to chemist, 54-74, HEAD DIV MOLECULAR SCI, BROOKHAVEN NAT LAB, 74- Concurrent Pos: Instr chem & physics, St Francis Col, NY, 47-54, from asst prof to assoc prof, 56-63, prof, 63- Mem: AAAS; fel Am Inst Chem; Am Chem Soc. Res: Radiation polymerization and radiation chemistry of organic compounds; mechanisms of vinyl polymerizations; biomaterials. Mailing Add: Brookhaven Nat Lab Upton NY 11973

METZ, FLORENCE IRENE, b Willard, Ohio, Sept 1, 29. PHYSICAL CHEMISTRY, INORGANIC CHEMISTRY. Educ: Case Western Reserve Univ, AB, 51, MS, 56; Iowa State Univ, PhD(phys chem), 60. Prof Exp: Res chemist, Lewis Lab, Nat Adv Comt Aeronaut, Ohio, 51-55; instr & res fel, Iowa State Univ, 56-60; sr chemist, 60-63, dir germanium info ctr, 62-67, prin chemist, 63-67, sr adv chem, 67-68, head phys & anal chem sect, 68-72, ASST DIR PHYS SCI, MIDWEST RES INST, 72- Concurrent Pos: Lectr, Univ Mo, Kansas City, 61-65. Mem: AAAS; Am Chem Soc; Sigma Xi; Am Asn Corrosion Eng; Am Soc Test & Mat. Res: Biochemistry, psychology and physiology related to disease, especially cancer; chemistry of materials; evaluation of effects of radiation on materials; vacuum evaporation of thin metallic films on inorganic oxidizers; analytical chemistry; behavioral sciences. Mailing Add: Midwest Res Inst 425 Volker Blvd Kansas City MO 64110

METZ, FRED LEWIS, b McComb, Ohio, Apr 23, 35; m 69; c 2. ORGANIC CHEMISTRY. Educ: Bowling Green State Univ, BA, 57; Ind Univ, PhD(org chem), 62. Prof Exp: Summer res chemist, Monsanto Chem Co, 57 & 59; sr res chemist, 62-69, RES ASSOC, T R EVANS RES CTR, DIAMOND SHAMROCK CORP, 69- Mem: AAAS; Am Chem Soc; Am Leather Chem Asn; Am Dairy Sci Asn. Res: Organic synthesis; amino acids; organic fluorine chemicals; leather chemicals; arsenic chemicals; paper chemicals; immobilized enzymes. Mailing Add: T R Evans Res Ctr Diamond Shamrock Corp PO Box 348 Painesville OH 44077

METZ, ROBERT, b New York, NY, June 2, 38; m 61; c 2. STRATIGRAPHY. Educ: City Col New York, BS, 61; Univ Ariz, MS, 63; Rensselaer Polytech Inst, PhD(stratig), 67. Prof Exp: Asst prof geol, State Univ NY Col Potsdam, 66-67; asst prof, Newark State Col, 67-70, ASSOC PROF GEOL, KEAN COL NJ, 70- Concurrent Pos: Sigma Xi grant-in-aid res; NY State grad fel award. Mem: Geol Soc Am; Soc Econ Paleontologists & Mineralogists; Nat Asn Geol Teachers. Res: Structure and taconic stratigraphy of the Cambridge Quadrangle, New York; stratigraphy and petrography of the Raritan Formation of New Jersey. Mailing Add: Dept of Earth & Planetary Environ Kean Col of NJ Union NJ 07083

METZ, ROBERT JOHN SAMUEL, b Johannesburg, SAfrica, Jan 23, 29; m 53; c 4. INTERNAL MEDICINE, ENDOCRINOLOGY. Educ: Univ Witwatersrand, MB, BCh, 51; Northwestern Univ, MS, 59; Univ Toronto, PhD(physiol), 62. Prof Exp: Res fel med & physiol, Northwestern Univ, 57-59; res fel, Banting & Best Dept Med Res, Univ Toronto, 59-61; assoc med, Northwestern Univ, Chicago, 62-64, asst prof, 64-69; chief metab res, 69-71, PRES, VIRGINIA MASON RES CTR, 71- Concurrent Pos: Chief, Med Div, Northwestern Univ, Chicago, 62-65; chief, Diabetes & Metab Serv, Cook County Hosp, Chicago, 62-65; attend physician, Passavant Mem Hosp, Chicago, 65-69; clin asst prof med, Univ Wash, 69-; attend physician, Virginia Mason Hosp, 69- Mem: Am Diabetes Asn; Am Fedn Clin Res; fel Am Col Physicians; Endocrine Soc; NY Acad Sci. Res: Diabetes and allied diseases. Mailing Add: Mason Clin 1118 Ninth Ave Seattle WA 98101

METZ, ROBERT WINFIELD, plant pathology, see 12th edition

METZ, ROGER N, b Bedford, Ohio, Dec 15, 38; m 58; c 1. THEORETICAL PHYSICS. Educ: Oberlin Col, BA, 60; Cornell Univ, PhD(physics), 68. Prof Exp: Res asst physics, Lab Atomic & Solid State Physics, Cornell Univ, 65-67; NSF intern physics & math, Antioch Col, 67-68; ASST PROF PHYSICS, COLBY COL, 68-, CHMN DEPT, 74- Mem: Am Phys Soc; Am Asn Physics Teachers. Res: Statistical mechanics of fluids. Mailing Add: Dept of Physics Colby Col Waterville ME 04901

METZ, WILLIAM CLINTON, b Leominster, Mass, Oct 13, 44; m 73. GEOGRAPHY. Educ: Bates Col, BA, 66; Univ RI, MA, 71; Univ Pittsburgh, PhD(geog), 74. Prof Exp: SR SCIENTIST, ENVIRON SYSTS DEPT, WESTINGHOUSE ELEC CORP, 73- Mem: Asn Am Geogrs; Nat Audubon Soc. Res: Socioeconomic impact of energy development, especially mining, processing and producing, on the workers, community and the region. Mailing Add: 255 Darlan Hill Dr Pittsburgh PA 15239

METZENBERG, ROBERT LEE, JR, b Chicago, Ill, June 11, 30; m 54; c 2. BIOCHEMISTRY. Educ: Pomona Col, AB, 51; Calif Inst Technol, PhD(biochem), 56. Prof Exp: From instr to assoc prof, 55-68, PROF PHYSIOL CHEM, UNIV WIS-MADISON, 68- Concurrent Pos: Am Cancer Soc fel, 55-58; Markle investr, 58-; res fel, Univ Zurich, 59-60; USPHS career development awardee, 63-73; genetics study sect, NIH, 69-73; assoc ed, Genetics J, 75- Mem: AAAS; Am Soc Biol Chemists; Am Chem Soc; Genetics Soc Am; Am Soc Microbiol. Res: Mechanism of action of urea cycle enzymes; genetic control of metabolism in Neurospora; control of metabolism in eucaryotes. Mailing Add: Dept of Physiol Chem Univ of Wis Madison WI 53706

METZGAR, DON P, b Hastings, Nebr, June 7, 29; m 50; c 2. VIROLOGY, IMMUNOLOGY. Educ: Hastings Col, BA, 56; Purdur Univ, PhD(microbiol), 61. Prof Exp: NSF fel cell physiol, Purdue Univ, 61-62; res fel virol, Merck Inst Therapeut Res, 62-67; sr res virologist, Nat Drug Co, 67-71; SECT HEAD CELL BIOL RES, MERRELL NAT LABS, 71- Mem: AAAS; NY Acad Sci. Res: Tissue culture; radiochemistry; electron microscopy; vaccine development and relationship to antigenic potentiation. Mailing Add: Biol Mfg & Develop Merrell Nat Labs Swiftwater PA 18370

METZGAR, LEE HOLLIS, b Olean, NY, Jan 10, 41; m 61; c 1. ECOLOGY. Educ: State Univ NY Col Fredonia, AB, 62; Univ Mich, MS, 64, MA, 66, PhD(zool), 68. Prof Exp: Teacher jr high sch, NY, 62-63; asst prof, 68-74, ASSOC PROF ZOOL, UNIV MONT, 74-, CHMN DEPT, 75- Mem: Am Soc Mammal; Ecol Soc Am; Wildlife Soc. Res: Relationships between social organization and population dynamics in small mammals. Mailing Add: Dept of Zool Univ of Mont Missoula MT 59801

METZGAR, RICHARD STANLEY, b Erie, Pa, Feb 2, 30; m 52; c 2. IMMUNOLOGY. Educ: Univ Fla, BS, 51; Univ Buffalo, MA, 57, PhD(immunol), 59. Prof Exp: Sr cancer res scientist, Roswell Park Mem Inst, 59-62; assoc prof, 62-72, PROF IMMUNOL, SCH MED, DUKE UNIV, 72- Concurrent Pos: Mem staff, Yerkes Primate Res Ctr, Emory Univ. Mem: AAAS; Am Asn Immunol; Am Asn Cancer Res; NY Acad Sci. Res: Cancer immunity; transplantation biology; tumor virology. Mailing Add: Dept of Microbiol & Immunol Duke Univ Sch of Med Durham NC 27710

METZGER, ALBERT E, b New York, NY, Sept 10, 28; m 58; c 2. NUCLEAR CHEMISTRY. Educ: Cornell Univ, AB, 49; Columbia Univ, MA, 51, PhD(nuclear chem), 58. Prof Exp: Chemist, Sylvania Elec Corp, 51-53; asst, Columbia Univ, 53-54; from scientist to sr scientist, 59-61, RES GROUP SUPVR, SPACE SCI DIV, JET PROPULSION LAB, CALIF INST TECHNOL, 61- Mem: Fel AAAS; Am Astrom Soc; Am Phys Soc; Am Geophys Union. Res: Geochemistry; gamma ray astronomy; x-ray physics; radiation physics; space science instrumentation. Mailing Add: 380 Olive Tree Lane Sierra Madre CA 91024

METZGER, DANIEL SCHAFFER, b Greenville, Mich, Sept 3, 36; div; c 2. PHYSICS. Educ: Kalamazoo Col, BA, 58; Ohio State Univ, MSc, 62, PhD(physics), 65. Prof Exp: Vis asst prof physics, Ohio State Univ, 65-66; STAFF MEM EXP PHYSICS, LOS ALAMOS SCI LAB, 66-, ALT GROUP LEADER, 74- Mem: Am Phys Soc. Res: X-ray spectroscopy; nuclear magnetic resonance in solids; spectral and diagnostic measurements of radiation associated with nuclear weapons testing; airborne measurements of infrared radiation in aurorae. Mailing Add: Los Alamos Sci Lab Box 1663 Los Alamos NM 87544

METZGER, GERSHON, b New York, NY, June 25, 35; m. CHEMISTRY, RESEARCH ADMINISTRATION. Educ: Yeshiva Univ, BA, 55; Columbia Univ, MA, 56, PhD, 59. Prof Exp: Asst, Columbia Univ, 55-59, Sloan res assoc, 59-60; res chemist, Esso Res & Eng Co, NJ, 60-64; sr chemist, Chem & Phosphates Co, Ltd, Israel, 64-65; assessor of patents, 65-68, head patent exploitation div, 68-69, dir phys sci div, 69-74, DEP DIR PLANNING, PRIME MINISTER'S OFF, NAT COUN RES & DEVELOP, 74- Concurrent Pos: Instr, Yeshiva Univ, 60-62; assoc prof chem, Jerusalem Col Technol, 73- Mem: AAAS; Am Chem Soc. Mailing Add: Prime Minister's Off State Israel Nat Coun Res & Develop Bldg 3 Hakirya Jerusalem Israel

METZGER, H PETER, b New York, Ny, Feb 22, 31; m 56; c 4. BIOCHEMISTRY, SCIENCE WRITING. Educ: Brandeis Univ, BA, 53; Columbia Univ, PhD, 65. Prof Exp: Res scientist, NY State Psychiat Inst, 65-66; sr res scientist, NY State Inst Neurochem & Drug Addiction, 66; res assoc biochem, Univ Colo, 66-68; staff scientist, Ball Bros Res Corp, 68-69, mgr adv progs, Environ Instrumentation Dept, 69-70; dir & consult, Colspan Environ Systs, Inc, 70-73; SYNDICATED COLUMIST, NEWSPAPER ENTERPRISE ASN, NEW YORK, 74- Concurrent Pos: Prin investr, USPHS res grant, 67-68; sci ed, Rocky Mt News, Denver, 74- Mem: AAAS; Am Chem Soc. Res: Mechanisms of enzyme action; neurochemistry; biochemical basis of memory; protein hormone production. Mailing Add: 2595 Stanford Ave Boulder CO 80303

METZGER, HENRY, b Mainz, Ger, Mar 23, 32; US citizen; m 57; c 3. BIOCHEMISTRY. Educ: Univ Rochester, AB, 53; Columbia Univ, MD, 57. Prof Exp: Intern internal med, Presby Hosp, New York, 57-58, asst resident, 58-59; res assoc, NIH, 59-61; Helen Hay Whitney Found fel, 61-63; SR INVESTR, ARTHRITIS & RHEUMATISM BR, NAT INST ARTHRITIS, METABOLISM & DIGESTIVE DIS, 63-, CHIEF SECT CHEM IMMUNOL, 73- Mem: Am Asn Immunol (secy-treas); Am Soc Biol Chemists. Res: Immunochemistry; protein chemistry; structure of immunoglobulins. Mailing Add: Nat Inst Arthritis Metab & Digestive Dis Bethesda MD 20014

METZGER, JAMES DOUGLAS, b Allentown, Pa, Feb 10, 42; m 64. ORGANIC CHEMISTRY, AGRICULTURAL CHEMISTRY. Educ: Univ Nev, Reno, BS, 64, PhD(org chem), 69. Prof Exp: RES CHEMIST, E I DU PONT DE NEMOURS & CO, INC, 68- Mem: Am Chem Soc; AAAS. Res: Organic synthesis. Mailing Add: E I du Pont de Nemours & Co Inc Wilmington DE 19898

METZGER, ROBERT MELVILLE, b Yokohama, Japan, May 7, 40; US citizen; m 70. CHEMISTRY, CHEMICAL PHYSICS. Educ: Univ Calif, Los Angeles, BS, 62; Calif Inst Technol, PhD(chem), 68. Prof Exp: Jr res asst chem, Atomics Int Div, NAm Aviation Corp, Calif, 61; res assoc, Stanford Univ, 68-71, lectr Italian, 69-71, res assoc chem eng, 71-72. ASST PROF CHEM, UNIV MISS, 71- Concurrent Pos: Instr, Chabot Col, 71-72. Mem: AAAS; Am Chem Soc; Am Crystallog Asn; Am Phys Soc. Res: Solid state chemistry; organic crystals and conductors; crystallography; quantum mechanics; Madelung energy calculations; tight binding Hartree-Fock calculations; neutron stars; computers in chemistry; x-ray radial distribution function studies of platinum catalysts; polarizabilities of organic molecules and ions. Mailing Add: Dept of Chem Univ of Miss University MS 38677

METZGER, ROBERT P, b San Jose, Calif, Jan 28, 40; m 68; c 1. BIOCHEMISTRY. Educ: Univ Calif, Los Angeles, BS, 61; San Diego State Col, MS, 63, PhD(chem), 67; Univ Calif, San Diego, PhD(chem), 67. Prof Exp: Lectr chem, San Diego State Col, 63-68, asst prof phys sci, 68-71, ASSOC PROF PHYS SCI, SAN DIEGO STATE UNIV, 71- Mem: Fel AAAS; Am Chem Soc; The Chem Soc. Res: Enzymology; carbohydrate metabolism. Mailing Add: Dept of Phys Sci San Diego State Univ San Diego CA 92182

METZGER, SIDNEY HENRY, JR, b Atlanta, Ga, Mar 29, 29; m 52; c 3. APPLIED CHEMISTRY. Educ: Univ Ala, BS, 51; Texas A&M Univ, MS, 56; Univ Ill, PhD(org chem), 62. Prof Exp: Chemist, Monsanto Co, 51-54; res chemist, Jefferson Chem Co, 60-62; sr chemist, 62-67, group leader, 67-71, DIR, MOBAY CHEM CO, 71-, MGR ELASTOMERIC A/D, MOBAY CHEM CORP, 73- Mem: Am Chem Soc; Am Inst Chemists. Res: Organophosphorus chemistry; reactions of epoxides; synthesis and reactions of carbodiimides and isocyanates. Mailing Add: Mobay Chem Co Penn-Lincoln Pkwy W Pittsburgh PA 15205

METZGER, THOMAS ANDREW, b Paterson, NJ, July 14, 44; m 70; c 3. PURE MATHEMATICS. Educ: Seton Hall Univ, BS, 61; Creighton Univ, MS, 69; Purdue Univ, West Lafayette, PhD(math), 71. Prof Exp: Asst prof math, Tex A&M Univ, 71-73; ASST PROF MATH, UNIV PITTSBURGH, 73- Mem: Am Math Soc; Math Asn Am. Res: Automorphic forms, applications to Riemann surfaces; weighted areal approximation in the complex plane; function theory. Mailing Add: Dept of Math Univ of Pittsburgh Pittsburgh PA 15260

METZGER, WESLEY JAMES, b Gary, Ind, Apr 25, 39; m 65. PHYSICS. Educ: Wabash Col, AB, 61; Univ Rochester, PhD(physics), 67. Prof Exp: Res assoc elem particle physics, Brookhaven Nat Lab, 66-68; physicist, Polytech Sch, Paris France, 68-71; MEM STAFF, PHYSICS LAB, ROMAN CATH UNIV, NIJMEGEN, 71- Mem: Am Phys Soc. Res: Elementary particle physics; experimental high energy physics. Mailing Add: Physics Lab Roman Cath Univ of Nijmegen Toernooiveld Nijmegen Netherlands

METZGER, WILLIAM HENRY, JR, b Richmond, Va, Feb 17, 22; m 49. ELECTROCHEMISTRY, PHYSICAL CHEMISTRY. Educ: Univ Richmond, BS, 43. Prof Exp: From chemist to supv chemist, Dept Commerce, Nat Bur Standards, 43-68; SUPV CHEMIST, GEN SERV ADMIN CENT LAB, 68- Mem: Am Chem Soc; Am Electroplaters Soc; Am Soc Testing & Mat. Res: Industrial electroplating and metal finishing; development of practices and processes in electroforming; measurement of physical properties and thickness of metal coatings; calorimetry in aqueous and fused salt media; test methodology for commodity evaluation. Mailing Add: Gen Serv Admin FSS/MEDL Washington DC 20405

METZGER, WILLIAM IRWIN, b Peekskill, NY, Oct 29, 15; m 41; c 2. MICROBIOLOGY. Educ: Purdue Univ, BS, 37, MS, 39; Univ Ill, PhD(bact), 46; Am Bd Med Microbiol, Dipl. Prof Exp: Asst gen biol, Purdue Univ, 37; instr bact, SDak State Col, 39-42 & Univ Ill, 42-46; res bacteriologist, Lederle Labs, Am Cyanamid Co, 46-54; DIR MICROBIOL, HEKTOEN INST MED RES, COOK COUNTY HOSP, 54-; RES ASSOC PREV MED, COL MED, UNIV ILL, 60- Mem: AAAS; Am Soc Microbiol; Am Pub Health Asn; Am Acad Microbiol. Res: Antibiotics and chemotherapeutics; diagnostic microbiology; taxonomy; microbial ecology and antagonisms. Mailing Add: Hektoen Inst for Med Res Cook County Hosp 629 S Wood St Chicago IL 60612

METZGER, WILLIAM JOHN, b Freeport, Ill, Nov 7, 35; m 57; c 2. GEOLOGY. Educ: Beloit Col, BS, 57; Univ Ill, MS, 59, PhD(geol), 61. Prof Exp: Asst geol, Univ Ill, 57-61; from asst prof to assoc prof geol, 61-71, actg chmn dept, 63-65, PROF GEOL, STATE UNIV NY COL FREDONIA, 71- Mem: AAAS; fel Geol Soc Am; Am Asn Petrol Geol; Soc Econ Paleont & Mineral; Clay Minerals Soc. Res: Stratigraphy and sedimentation; clay mineralogy and x-ray analysis of sedimentary rocks. Mailing Add: Dept of Geol State Univ of NY at Fredonia Fredonia NY 14063

METZLER, CARL MAUST, b Masontown, Pa, Dec 13, 31; m 53; c 3. STATISTICS, BIOMATHEMATICS. Educ: Goshen Col, BS, 55; NC State Univ, PhD(biomath), 65. Prof Exp: Asst prof math, Goshen Col, 60-62; res scientist, 65-70, HEAD RES, CLIN BIOSTATIST, 70- Concurrent Pos: Adj asst prof, Western Mich Univ, 68- Mem: AAAS; Soc Indust & Appl Math; Biomet Soc; Am Statist Asn; Am Soc Clin Pharmacol & Therapeut. Res: Application of mathematical and statistical methods to chemical biological and medical research. Mailing Add: Clin Biostatist Upjohn Co 301 Henrietta St Kalamazoo MI 49001

METZLER, DAVID EVERETT, b Palo Alto, Calif, Aug 12, 24; m 48; c 5. BIOCHEMISTRY. Educ: Calif Inst Technol, BS, 48; Univ Wis, MS, 50, PhD(biochem), 52. Prof Exp: Res scientist, Univ Tex, 51-53; from asst prof to assoc prof, 53-61, PROF BIOCHEM, IOWA STATE UNIV, 61- Mem: Am Chem Soc; Am Soc Biol Chem. Res: Mechanisms of coenzyme action; electronic absorption spectra of vitamins, coenzymes and proteins. Mailing Add: Dept of Biochem & Biophys Iowa State Univ Ames IA 50011

METZLER, RICHARD CLYDE, b Cleveland, Ohio, Oct 19, 37; m 60; c 2. PURE MATHEMATICS. Educ: Univ Mich, BS, 49; Wayne State Univ, MA, 62, PhD(math), 66. Prof Exp: Asst prof, 65-71, ASSOC PROF MATH, UNIV N MEX, 71- Mem: Math Asn Am; Am Math Soc. Res: Ordered topological vector spaces. Mailing Add: Dept of Math Univ of NMex Albuquerque NM 87131

METZNER, JEROME, b New York, NY, Apr 14, 11; m 32; c 3. BOTANY, CYTOLOGY. Educ: City Col New York, BA, 32; Columbia Univ, MA, 33, PhD(bot), 44. Prof Exp: Teacher high schs, NY, 44-49; asst prof educ, City Col New York, 49-50; chmn dept biol & gen sci, Jamaica High Sch, 50-53, dept biol & introd sci, High Sch Sci, 53-60 & dept biol, Francis Lewis High Sch, 60-67; PROF BIOL, JOHN JAY COL, CITY UNIV NEW YORK, 67- Concurrent Pos: Lectr, Hunter Col, 46-48 & City Col New York, 47-; admin officer, Education Mission, US Dept Army, Korea, 48; educ dir, Nature Ctrs Young Am, Inc, 59-; mem gifted student comt biol sci curric study, Am Inst Biol Sci. Mem: AAAS; Torrey Bot Club; Am Soc Protozoologists; Nat Asn Biol Teachers. Res: Protozoology; phycology; cytology. Mailing Add: Dept of Biol John Jay Col 445 W 59th St New York NY 10019

METZNER, WENDELL PHILLIPS, b Bryant, Ind, Feb 16, 12; m 35; c 2. CHEMISTRY. Educ: Ind Univ, AB, 33; Univ Chicago, PhD(org chem), 37. Prof Exp: Res chemist, Monsanto Co, 37-40, res group leader, 40-47, assoc dir res, 47-59, admin dir, Res Ctr, 59-63, patent tech specialist, 63-69, MGR PATENT SERV,

MONSANTO CO, 69- Mem: AAAS; Am Chem Soc. Res: Organic chemistry related to rubber chemicals, detergents, intermediates and themoplastic materials; uracil derivatives. Mailing Add: Patent Dept Monsanto Co 800 Lindbergh St Louis MO 63166

MEUDT, WERNER J, b Koblenz, Ger, Nov 22, 31; US citizen; m 57; c 2. PLANT PHYSIOLOGY. Educ: Okla State Univ, BS, 59; Yale Univ, MS, 62; Rutgers Univ, PhD(plant physiol), 64. Prof Exp: RES PLANT PHYSIOLOGIST, USDA, 64- Mem: AAAS; Am Soc Plant Physiol; Bot Soc Am; Japanese Soc Plant Physiol; Scandinavian Soc Plant Physiol. Res: Metabolism and mode of action of plant growth hormones in tobacco; enzymatic oxidation on indole-3-acetic acid in plants; flowering physiology; photoperiodism; vernalization. Mailing Add: Plant Hormone & Regulator Lab USDA Agr Res Ctr West Beltsville MD 20705

MEULY, WALTER C, b Langenthal, Switz, Nov 5, 98; nat US; m; c 1. ORGANIC CHEMISTRY. Educ: Swiss Fed Inst Technol, AB, 21, PhD(org chem), 23. Prof Exp: Res chemist, Newport Co, 24-30; area supt, Fine Chem Div, New Brunswick Works, E I du Pont de Nemours & Co, 30-41, prod mgr, 41-44, chem supt, 44-53, res assoc, Jackson Lab, 53-58; dir res, 58-63, vpres res & develop, 63-75, CONSULT, RHODIA, INC, 75- Mem: AAAS; Am Chem Soc; Am Inst Chemists; NY Acad Sci. Res: Synthetic dyestuffs; perfume chemicals; pharmaceuticals; organic fine chemicals. Mailing Add: 685 River Rd Piscataway NJ 08854

MEUNIER, JEAN-LOUIS, b Iberville, Que, Dec 14, 30; m 55; c 3. EXPERIMENTAL NUCLEAR PHYSICS, EXPERIMENTAL SOLID STATE PHYSICS. Educ: Univ Montreal, BA, 51, BSc, 54, MSc, 66. Prof Exp: Lectr physics, Royal Mil Col (Que), 54-59, spec lectr, 59-66; dir gen, Fr-Can Asn Advan Sci, 66-67; exec dir, Can Asn Physicists, 68-71; assoc awards officer, 71-73, AWARDS OFFICER, OFF GRANTS & SCHOLAR, NAT RES COUN CAN, 73- Mem: AAAS; Can Asn Physicists; Am Phys Soc; Fr-Can Asn Advan Sci (treas, 68-69); Asn Sci Eng & Technol Community Can (treas, 72-73). Res: Dielectrics and nuclear emulsions. Mailing Add: Off of Grants & Scholar Nat Res Coun of Can Ottawa ON Can

MEUNIER, VINCENT C, chemistry, see 12th edition

MEURISSE, ROBERT THOMAS, soil science, see 12th edition

MEUTER, RALPH F, b Oakland, Calif, Dec 2, 41; m 64; c 3. URBAN GEOGRAPHY, ECONOMIC GEOGRAPHY. Educ: Chico State Col, AB, 63; Univ Okla, MA, 65, PhD(geog), 70. Prof Exp: Instr geog, NTex State Univ, 65-67; asst prof, Calif State Univ, Chico, 70-72, ASSOC V PRES ACAD AFFAIRS & DEAN CONTINUING EDUC, CALIF STATE UNIV, CHICO, 72- Concurrent Pos: Thesis coordr, Indust Develop Inst, Okla Ctr Continuing Educ, 69- Mem: Asn Am Geog; Nat Coun Geog Educ. Res: Rural, non-metropolitan planning and development; threshold population studies. Mailing Add: Calif State Univ Chico First & Normal St Chico CA 95926

MEUX, JOHN WESLEY, b Little Rock, Ark, Apr 25, 28; m 53; c 2. MATHEMATICS. Educ: Henderson State Teachers Col, BS, 53; Univ Ark, MS, 57; Univ Fla, PhD(math), 60. Prof Exp: Instr math, Univ Fla, 59-60; asst prof, Kans State Univ, 60-64; asst prof, 64-65, chmn dept, 64-68, ASSOC PROF MATH, MIDWEST UNIV, 65-, DEAN SCH SCI & MATH, 68- Mem: Math Asn Am. Res: Orthogonal functions; numerical analysis. Mailing Add: Sch of Sci & Math Midwestern Univ Wichita Falls TX 76308

MEWALDT, LEONARD RICHARD, b La Crosse, Wis, May 31, 17; m 41; c 2. ORNITHOLOGY. Educ: Univ Iowa, BA, 39; Univ Mont, MA, 48; Wash State Univ, PhD, 53. Prof Exp: From instr to assoc prof, 53-63, PROF ZOOL, SAN JOSE STATE UNIV, 63- Mem: Fel Am Ornith Union; Cooper Ornith Soc. Res: Biology of Nucifraga ecology and behavior. Mailing Add: Avian Biol Lab San Jose State Univ San Jose CA 95192

MEWBORN, ANCEL CLYDE, b Greene Co, NC, Sept 22, 32; m 54; c 2. ALGEBRA. Educ: Univ NC, AB, 54, MA, 57, PhD(math), 59. Prof Exp: Instr math, Yale Univ, 59-61; from asst prof to assoc prof, 61-70, PROF MATH, UNIV NC, CHAPEL HILL, 70- Mem: Am Math Soc; Math Asn Am. Res: Structure of prime and semi-prime rings. Mailing Add: Dept of Math Univ of NC Chapel Hill NC 27514

MEWISSEN, DIEUDONNE JEAN, b Ans, Belg, Oct 25, 24; m 53; c 3. RADIOLOGY, RADIOBIOLOGY. Educ: Univ Liege, MD, 50, Agrege, 61; Am Bd Radiol, cert therapeut radiol, 71. Prof Exp: Asst prof, Cancer Inst, Univ Liege, 50-53; Fulbright grant nuclear med, Oak Ridge Assoc Univs, 54-56; WHO grant radiobiol, Nuclear Ctr, Saclay, France, 56; res assoc, Nuclear Sci Inst Brussels, 57-68; PROF RADIOL, PRITZKER SCH MED, UNIV CHICAGO, 68-Concurrent Pos: Head radiobiol lab, Univ Brussels, 61-, prof radiobiol, Univ, 69-; mem, Belg Adv Coun Cancer, 63- & Hosps, 64- Honors & Awards: Prize, Cong Radiol & Electrol Latin Cult, Lisbon, 57. Mem: AAAS; Radiation Res Soc; Royal Soc Med; Belg Cancer Soc (secy-gen, 71); Belg Radiol Soc. Res: Radiation carcinogenesis; long term effects of radiation; toxicity and carcinogenicity of tritium and tritiated compounds. Mailing Add: A J Carlson Lab Pritzker Sch Med Univ of Chicago Chicago IL 60637

MEYBOOM, PETER, b Barneveld, Netherlands, Apr 26, 34; Can citizen; m 57; c 3. HYDROGEOLOGY. Educ: Univ Utrecht, BSc, 56, MSc, 58, PhD(hydrogeol), 60. Prof Exp: Res officer, Alta Res Coun, Can, 58-60; res scientist, Geol Surv Can, Can Fed Dept Energy, Mines & Resources, 60-66, sect head hydrogeol, Inland Waters Br, 66-67, head groundwater subdiv, 67-69, sci adv, Can Fed Dept Finance, 70-71; dir sci policy, Can Fed Dept Environ, 71-73, DIR GEN SCI CENTRE, CAN FED DEPT SUPPLY & SERV, 73- Concurrent Pos: Distinguished lectr, Can Inst Mining & Metall, 62-63. Honors & Awards: Can Centennial Medal, 67. Res: Hydrology; science policy. Mailing Add: 4 Cedarcrest Ave Ottawa ON Can

MEYDRECH, EDWARD F, b Oak Park, Ill, July 21, 43; m 65; c 2. BIOSTATISTICS, EPIDEMIOLOGY. Educ: Univ Fla, BS, 65, MS, 67; Univ NC, PhD(biostatist), 72. Prof Exp: Health serv officer, USPHS, 67-69; ASST PROF BIOMET, VA COMMONWEALTH UNIV, 72- Mem: Am Statist Asn; Biomet Soc; Sigma Xi. Res: Response surfaces; nonparametric statistics; epidemiological studies. Mailing Add: Med Col of Va Sta PO Box 32 Richmond VA 23298

MEYER, ALBERT WILLIAM, b Schenectady, NY, Nov 29, 06; m 31; c 4. PHYSICAL CHEMISTRY. Educ: Univ Chicago, BS, 27, PhD(phys chem), 30. Prof Exp: Res chemist, E I du Pont de Nemours & Co, 30-31; group leader, A O Smith Corp, Wis, 31-34; dept head, US Rubber Co, 34-54; dir explor res, Diamond Alkali Co, 54-57; head atomic & radiation res, US Rubber Co, 57-59, dir tech personnel & univ rel, 59-63; asst dir res, Stevens Inst Technol, 63-67; exec secy, Plastics Inst Am, 67-71; mem coun, Eng & Sci Soc Execs; CONSULT, 72- Concurrent Pos: Assoc prof eve div, NY Univ, 46-48; mem Gordon Res Conf, chmn, 51; chmn subcomt continuing educ, Coun Comt Chem Educ, Am Chem Soc, 69- Mem: Soc Plastics Eng; Am Chem Soc. Res: Liquid ammonia as a solvent; contact catalysis; polymerization, emulsion and oil phase; rubber technology; agricultural chemicals; radiation chemistry; nuclear processes; polymer science and technology. Mailing Add: 138 Alexander Ave Upper Montclair NJ 07043

MEYER, ALFRED HERMAN LUDWIG, b Venedy, Ill, Feb 27, 93; m 26; c 1. GEOGRAPHY. Educ: Univ Ill, AB, 21, AM, 23; Univ Mich, PhD(geog), 34. Prof Exp: Teacher elem schs, Linneman, Ill, 11-14; sec sch teacher, South Side Br, Englewood Luther Inst, Chicago, 25-26; instr geol & zool, Valparaiso Univ, 26-28, instr geol, 28-31, asst prof, 31-33, head dept geog & geol, 31-67, assoc prof, 33-42, prof, 42-67, distinguished serv prof, 67-74, DISTINGUISHED SERV EMER PROF GEOG, VALPARAISO UNIV, 74- Concurrent Pos: Mem, Ill State Geol Surv, 22-23. Honors & Awards: Distinguished Serv Award, Chicago Geog Soc, 58-59 & Nat Coun Geog Educ, 69. Mem: Fel AAAS; Nat Coun Geog Educ (pres, 47); Asn Am Geog. Res: Systematic geography; geography of world affairs; historical and educational geography; geography of Calumet region, Indiana-Illinois. Mailing Add: Dept of Geog 101 Heine Hall Valparaiso Univ Valparaiso IN 46383

MEYER, ALVIN F, JR, b Shreveport, La, Sept 3, 20; m 42; c 2. ENVIRONMENTAL HEALTH. Educ: Va Mil Inst, BS, 41; Am Acad Environ Eng, dipl, 56; Indust Col Armed Forces, dipl, 62. Prof Exp: Chief environ health eng, hq, Air Mat Command, US Air Force, 49-61, hq, Strategic Air Command, 61-65, chief bioenviron eng, Off Surgeon Gen, 62-65, chief biomed sci corps, 65-69; spec asst legis, Off Adminr, Consumer Protection & Environ Health Serv, Dept Health, Educ & Welfare, 69-71; DIR OFF NOISE ABATEMENT & CONTROL, ENVIRON PROTECTION AGENCY, 71- Concurrent Pos: Vis fac mem, Va Mil Inst, 55-; asst prof, Creighton Univ, 55-61; mem nat adv coun environ health, 63-69; chmn environ pollution control comt, Dept Defense, 64-69. Honors & Awards: Legion of Merit, 65; Distinguished Serv Medal, 69. Mem: Aerospace Med Asn; Am Indust Hyg Asn; Am Soc Civil Eng; Am Pub Health Asn. Res: Environmental pollution control from toxic aerospace propellants; engineering control of environmental stresses on man; bioacoustics and noise control. Mailing Add: 1600 Longfellow St McLean Va 22101

MEYER, ANDRE S, organic chemistry, biochemistry, see 12th edition

MEYER, ANDREW LEO, organic chemistry, see 12th edition

MEYER, AXEL, b Copenhagen, Denmark, Mar 3, 26; nat US; m 50; c 3. METAL PHYSICS. Educ: City Col New York, BS, 48 & 50; Ga Inst Technol, MS, 52; Ill Inst Technol, PhD(physics), 56. Prof Exp: Res fel nuclear physics, Armour Res Found, Ill Inst Technol, 52-54; from asst prof to assoc prof physics, Univ Fla, 55-59; solid state physicist, Neutron Physics Div, Oak Ridge Nat Lab, 59-62, Solid State Div, 61-67; assoc prof, 67-69, PROF PHYSICS, NORTHERN ILL UNIV, 69- Concurrent Pos: Res Corp grant, 58. Mem: Am Phys Soc; fel Brit Inst Physics. Res: Theoretical solid state and metal physics, especially electronic structure; transport properties and thermodynamics of metals and alloys in both liquid and solid state. Mailing Add: Dept of Physics Northern Ill Univ DeKalb IL 60115

MEYER, BERNARD SANDLER, b Nantucket, Mass, July 20, 01; m 31, 48. BOTANY. Educ: Ohio State Univ, BA, 21, MA, 23, PhD(bot), 26. Prof Exp: Instr bot, Ohio State Univ, 23-27; assoc forest ecologist, Cent States Forest Exp Sta, US Forest Serv, 27-28; from asst prof to prof bot, Ohio State Univ, 28-74, chmn dept, 46-67, chmn dept bot & plant path, Agr Exp Sta, 48-66, EMER PROF BOT, OHIO STATE UNIV, 74- Concurrent Pos: Ed-in-chief J Bot, Bot Soc Am, 46-51. Mem: AAAS; Bot Soc Am (vpres, 52). Res: Plant physiology; water relations; photosynthesis; mineral nutrition; photoperiodism. Mailing Add: Dept of Bot Ohio State Univ 1735 Neil Ave Columbus OH 43210

MEYER, BERNARDINE (HELEN), home economics, deceased

MEYER, BURNETT CHANDLER, b Denver, Colo, Mar 24, 21. MATHEMATICS. Educ: Pomona Col, BA, 43; Brown Univ, ScM, 45; Stanford Univ, PhD(math), 49. Prof Exp: Asst, Stanford Univ, 46-49; from asst prof to assoc prof math, Univ Ariz, 49-57; from asst prof to assoc prof, 57-68, PROF MATH, UNIV COLO, BOULDER, 68- Mem: Am Math Soc; Math Asn Am. Res: Complex analysis; real analysis; history of mathematics. Mailing Add: Dept of Math Univ of Colo Boulder CO 80302

MEYER, CARL BEAT, b Zurich, Switz, May 5, 34; m 61; c 1. PHYSICAL INORGANIC CHEMISTRY. Educ: Univ Zurich, PhD(inorg chem), 60. Prof Exp: Fel McGill Univ, 60-61; res chemist & fel Lawrence Radiation Lab, Univ Calif, 61-64; from asst prof to assoc prof, 64-75, PROF CHEM, UNIV WASH, 75- Concurrent Pos: Consult, Lawrence Berkeley Lab, Univ Calif, Berkeley, 64-; NSF grants, 64; mem subcomt handling hazardous mat, US Coast Guard-Nat Acad Sci, 64-; Environ Protection Agency grants; dir indust res, Sulphur Inst, Wash, DC, 65-69. Mem: AAAS; Am Phys Soc; Am Chem Soc; NY Acad Sci. Res: Inorganic physical chemistry of free radicals, high temperature molecules and sulphur-containing compounds; optical spectroscopy of diatomics; cryochemistry. Mailing Add: Dept of Chem Univ of Wash Seattle WA 98195

MEYER, CAROL DIANE, b Plainfield, NJ, Nov 30, 49. ORGANOMETALLIC CHEMISTRY. Educ: Bucknell Univ, BSc, 71; Brown Univ, PhD(chem), 75. Prof Exp: Res asst chem, Univ Rochester, 73-74, fel chem, 75; ASST PROF CHEM, MASS INST TECHNOL, 75- Mem: Am Chem Soc. Res: Transition metal stabilization of strained olefins; homogeneous catalysis of oxygen transfer reactions; organometallic chemistry of nitrosyl complexes; coordination chemistry of stable organic free radicals. Mailing Add: Rm 18-282 Dept of Chem Mass Inst of Technol Cambridge MA 02139

MEYER, CHARLES, b St Louis, Mo, Sept 30, 15; m 40; c 2. GEOLOGY. Educ: Wash Univ, AB, 37, MS, 39; Harvard Univ, AM, 41, PhD(geol), 50. Prof Exp: Geologist, Anaconda Copper Mining Co, 41-53; PROF GEOL, UNIV CALIF, BERKELEY, 53-, CHMN DEPT, 69- Concurrent Pos: Guggenheim fel, 60. Mem: Fel Geol Soc Am; Soc Econ Geol; Geochem Soc; fel Mineral Soc Am; Am Inst Mining, Metall & Petrol Eng. Res: Genesis of ore deposits; chemical and mineralogical studies of ore forming systems. Mailing Add: Dept of Geol Univ of Calif Berkeley CA 94720

MEYER, CHARLES FREDERICK, analytical chemistry, see 12th edition

MEYER, CHARLES FREDERICK, b Gloucester, NJ, Apr 15, 13; m 38; c 3. MATHEMATICS, STATISTICS. Educ: Pa State Col, BS, 35; George Wash Univ, MA, 37, PhD(physics), 42. Prof Exp: From instr to asst prof physics, Wayne State Univ, 38-44; physicist, Johns Hopkins Univ, 44-46, group supvr, Appl Physics Lab, 47-57, DIV SUPVR, APPL PHYSICS LAB, JOHNS HOPKINS UNIV, 57-, PRIN STAFF PHYSICIST, 51- Concurrent Pos: Consult, Joint Res & Develop Bd, Wash, DC, 48-51 & Weapons Syst Eval Group, US Dept Defense, 50-51 & 56-; asst secy, Defense Res & Develop, 54- Mem: Am Phys Soc; Opers Res Soc Am. Res: Military operations research. Mailing Add: Appl Physics Lab Johns Hopkins Univ Laurel MD 20810

MEYER, CONRAD FREDERICK, b Santa Monica, Calif, Apr 23, 20; m 54; c 4.

DEVELOPMENTAL PLANT ANATOMY. Educ: Syracuse Univ, BS, 49, MS, 50; Cornell Univ, PhD(plant anat), 54. Prof Exp: Asst bot, Syracuse Univ, 49-50; asst Cornell Univ, 51-53, instr, 53-54; asst prof biol, Tex Western Col, 54-56; from instr to asst prof, 56-60, ASSOC PROF BOT, RUTGERS UNIV, 60- Mem: AAAS; Bot Soc Am; Am Inst Biol Sci; Sigma Xi; Int Soc Plant Morphol. Res: Plant embryogeny; morphogenesis; developmental anatomy; histochemistry. Mailing Add: Dept of Bot Rutgers Univ 195 Univ Ave Newark NJ 07102

MEYER, DALLAS KREMER, b Tamalco, Ill, Dec 17, 16; m 44; c 1. PHYSIOLOGY. Educ: SDak State Teachers Col, BS, 38; Univ SDak, MA, 39; Univ Mo, PhD(physiol), 47. Prof Exp: Asst zool, Univ SDak, 38-39; asst, Univ Iowa, 39-41; aquatic biologist, US Fish & Wildlife Serv, 42-47; from asst prof to assoc prof, 47-55, PROF PHYSIOL, SCH MED, UNIV MO-COLUMBIA, 55- Mem: AAAS; Am Physiol Soc; Soc Exp Biol & Med. Res: Cardiac metabolism; myocardial function and metabolism, especially as affected by circadian rhythms. Mailing Add: Dept of Physiol Univ of Mo-Columbia Columbia MO 65201

MEYER, DANIEL L, b New York, NY, Jan 31, 32; m 53; c 2. BACTERIOLOGY. Educ: State Univ Col Albany, BA, 54; Rutgers Univ, MS, 61, PhD(bact), 65. Prof Exp: Teacher, Mex Acad & Cent Sch, 54-56; teacher, Bd Educ, Essex, Md, 56-58 & Edison, NJ, 58-59; res asst bur biol res, Rutgers Univ, 60-64; asst prof, 64-66, ASSOC PROF BIOL, STATE UNIV NY COL GENESEO, 66- Mem: Am Soc Microbiol. Res: Endogenous metabolism studies of cultured cells; Walker carcinoma in rats and in vitro. Mailing Add: Dept of Biol State Univ of New York Col Geneseo NY 14454

MEYER, DAVID BERNARD, b Rochester, NY, Jan 20, 23; m 52; c 3. ANATOMY. Educ: Wayne State Univ, BA, 48, PhD, 57; Univ Mich, MS, 50. Prof Exp: Instr biol, Wayne State Univ, 51-52, asst, 56-57, USPHS res fel, 57-58; NIH fel, 58-59; USPHS fel, 59-60, from instr to assoc prof, 60-70, PROF ANAT, SCH MED, WAYNE STATE UNIV, 70- Concurrent Pos: Vis prof, Graz Univ, 62-63; vis res anatomist, Carnegie Lab Embryol, Davis, Calif, 75-76. Mem: AAAS; Asn Res Vision & Ophthal; Am Asn Anat; Histochem Soc; Pan-Am Asn Anat. Res: Ocular development and histochemistry; prenatal ossification of the human skeleton; embryological histochemistry; origin, ultrastructure and chemistry of avian visual cells. Mailing Add: Dept of Anat Wayne State Univ Sch Med Detroit MI 48201

MEYER, DAVID LACHLAN, b East Orange, NJ, Dec 26, 43; m 70. INVERTEBRATE PALEONTOLOGY. Educ: Univ Mich, BS, 66; Yale Univ, Mphil, 69, PhD(geol), 71. Prof Exp: Fel, Smithsonian Inst, 70- 71, biologist, Smithsonian Trop Res Inst, 71-75; ASST PROF GEOL, UNIV CINCINNATI, 75- Mem: Paleont Soc; Soc Econ Paleontologists & Mineralogists; Geol Soc Am; assoc Sigma Xi. Res: Ecology and functional morphology of recent and ancient crinoids (Echinodermata); coral reef ecology. Mailing Add: Dept of Geol Univ of Cincinnati Cincinnati OH 45221

MEYER, DELBERT EUGENE, b Eau Claire, Wis, Nov 16, 27; m 55; c 2. ZOOLOGY, BIOCHEMISTRY. Educ: NCent Col, Ill, BA, 51; Univ Wis, MS, 55, PhD(zool, bot), 59. Prof Exp: Instr zool & bot, NCent Col, Ill, 58-60; asst prof zool, Univ Wis-Milwaukee, 60-66; assoc prof biol, Keene State Col, 66-67; from assoc prof to prof zool, Ctr Syst, Univ Wis-Milwaukee, 67-75; VPRES ACAD AFFAIRS, EASTERN CONN STATE COL, 74- Concurrent Pos: Assoc dean admin, Univ Wis-Milwaukee, 68-73; vpres acad affairs, Baldwin-Wallace Col, 73-74. Mem: AAAS; Fedn Am Sci; Am Inst Biol Sci; Am Soc Ichthyol & Herpet. Res: Animal behavior; behavioral and physiological adaptations to desert climates. Mailing Add: Eastern Conn State Col Willimantic CT 06226

MEYER, DELBERT HENRY, b Maynard, Iowa, Aug 28, 26; m 49; c 5. ORGANIC CHEMISTRY. Educ: Wartburg Col, BA, 49; Univ Iowa, PhD(chem), 53. Prof Exp: Chemist, Standard Oil Co, Ind, 53-61; chemist, Amoco Chem Corp, 61-67, RES SUPVR, RES & DEVELOP DEPT, AMOCO CHEM CORP, 67- Mem: Am Chem Soc. Res: Esterification and oxidation reactions; polymerization; synthetic fibers; flame retardant polymers. Mailing Add: Res & Develop Dept Amoco Chem Corp PO Box 400 Naperville IL 60540

MEYER, DIANE HUTCHINS, b Springfield, Mass, Feb 3, 37; m 67; c 4. CELL BIOLOGY. Educ: Russell Sage Col, BA, 58; Univ Vt, PhD(zool), 72. Prof Exp: Res asst pharmacol res, Sterling-Winthrop Res Inst, 58-59; res technician, Burroughs Wellcome Co, 60-62; res technician med res, Med Col, Univ Vt, 64-68; res officer, Med Col, Univ Western Australia, 73; RES ASSOC BASIC RES, MED COL, UNIV VT, 74- Res: Investigation of the role of ribonucleases, esterases and proteases in the turnover of RNA and protein in normal, developing, denervated and dystrophic muscle and in dystrophic human muscle. Mailing Add: Dept of Biochem Given Bldg Univ of Vt Med Col Burlington VT 05401

MEYER, DONALD IRWIN, b St Louis, Mo, Feb 13, 26; m 50; c 2. PHYSICS. Educ: Mo Sch Mines, BS, 46; Univ Wash, PhD(physics), 53. Prof Exp: Mem staff, Los Alamos Sci Lab, 46-48; asst prof physics, Univ Okla, 52-54; with Brookhaven Nat Lab, 54-56; asst prof, 56, PROF PHYSICS, UNIV MICH, ANN ARBOR, 56- Mem: Am Phys Soc. Res: High energy nuclear physics. Mailing Add: 2740 Parkridge Ann Arbor MI 48103

MEYER, DOUGLAS KERMIT, b Lansing, Mich, Jan 1, 39; m 60; c 1. CULTURAL GEOGRAPHY. Educ: Concordia Teachers Col, Ill, BS, 60; Wayne State Univ, MA, 68; Mich State Univ, PhD(geog), 70. Prof Exp: Elem teacher, First Lutheran Sch, Little Rock, Ark, 60-62; sec teacher, Lutheran High E, Harper Woods, Mich, 62-67; teaching asst geog, Mich State Univ, 67-68, res asst, Social Sci Teaching Inst, 68-69, instr social sci, 69-70; ASST PROF GEOG, EASTERN ILL UNIV, 70- Mem: Asn Am Geogrs; Am Geog Soc; Nat Coun Geog Educ; Pop Asn Am. Res: Reassessing the process of migration operating in nineteenth century Illinois; examining the diffusion and development of nineteenth century folk housing patterns in Illinois. Mailing Add: Dept of Geog-Geol Eastern Ill Univ Charleston IL 61920

MEYER, DWAIN WILBER, b Fremont, Nebr, Jan 11, 44; m 66. AGRONOMY. Educ: Univ Nebr, Lincoln, BS, 66; Iowa State Univ, PhD(crop prod & physiol), 70. Prof Exp: ASST PROF AGRON, N DAK STATE UNIV, 70- Mem: Am Soc Agron; Crop Sci Soc Am; Am Forage & Grassland Coun. Res: Forage management, production and physiology; irrigated forage problems and systems; Sudangrass and crested wheatgrass digestibility. Mailing Add: Dept of Agron NDak State Univ Fargo ND 58102

MEYER, EDGAR F, b El Campo, Tex, July 19, 35; m 65; c 3. Educ: NTex State Univ, BS, 59; Univ Tex, PhD(chem), 63. Prof Exp: Fel lab org chem, Swiss Fed Inst Technol, 63-65; fel dept biol, Mass Inst Technol, 65-67; asst prof, 65-74, ASSOC PROF BIOPHYS, TEX A&M UNIV, 74- Concurrent Pos: Res collabr chem, Brookhaven Nat Lab, 68-76; acad guest, Swiss Fed Inst Technol, 75-76. Mem: AAAS; Sigma Xi; Am Crystallog Asn; Am Chem Soc. Res: Computer driven display of molecular structures; crystallographic structure determinations with the assistance of digital computers; x-ray crystallography of biologically related substances; chemical information processing and the application of computational methods to chemical problems. Mailing Add: Dept of Biochem & Biophys Tex A&M Univ College Station TX 77843

MEYER, EDMOND GERALD, b Albuquerque, NMex, Nov 2, 19; m 41; c 3. PHYSICAL CHEMISTRY. Educ: Carnegie Inst Technol, BS, 40, MS, 42; Univ NMex, PhD(chem), 50. Prof Exp: Jr chemist, Harbison Walker Refractories Co, 40-41; instr, Carnegie Inst Technol, 41-42; asst phys chem, US Bur Mines, 42-44; chemist, US Naval Res Lab, 44-46; res div, NMex Inst Mining & Technol, 46-48; head dept sci, Univ Albuquerque, 50-52; prof chem, NMex Highlands, 52-63, head dept, 52-58, dir inst sci res, 58-63, grad dean, 60-63; dean col arts & sci, Univ Wyo, 63-75, PROF CHEM, UNIV WYO, 63-, V PRES-RES, 75- Concurrent Pos: Grants, Res Corp, AEC, NSF, USPHS, US Dept Interior & Am Heart Asn; Fulbright prof, Chile, 59; state sci adv, Wyo, 72-; consult, Los Alamos Sci Lab, NSF & US Dept Health, Educ & Welfare. Mem: Fel AAAS; Am Chem Soc; Biophys Soc; fel Am Inst Chem. Res: Kinetics; radiochemistry; thermodynamics; science education. Mailing Add: Off of VPres/Res Univ of Wyo Laramie WY 82071

MEYER, EDWARD RAYMOND, b Chicago, Ill, Nov 6, 43. ENVIRONMENTAL SCIENCES. Educ: Ind Univ, Bloomington, AB, 66; Univ Louisville, MS, 69; Ariz State Univ, PhD(ecol), 72. Prof Exp: Fac res assoc paleoecol, Ariz State Univ, 71-72; asst prof biol & ecol, Va Commonwealth Univ, 72-74; ecologist water pollution abatement, Nat Comn Water Qual, 74-76; ECOLOGIST DEEPWATER PORT EFFECTS, NAT OCEANIC & ATMOSPHERIC ADMIN, DEPT COM, 76- Mem: AAAS; Am Inst Biol Sci; Ecol Soc Am; Int Soc Limnol; Am Quaternary Asn. Res: Ecological effects of municipal and industrial water pollution abatement; quaternary paleoecology; freshwater and marine ecology. Mailing Add: Nat Oceanic & Atmospher Admin 2001 Wisconsin Ave NW Washington DC 20235

MEYER, EDWIN F, b Chicago, Ill, July 30, 37; m 59; c 6. PHYSICAL CHEMISTRY. Educ: DePaul Univ, BS, 59; Northwestern Univ, PhD(phys chem), 62. Prof Exp: NATO fel, Queen's Univ, Belfast, 62-63; res assoc phys adsorption, Naval Res Lab, 65-67; asst prof, 67-72, ASSOC PROF CHEM, DePAUL UNIV, 72- Mem: AAAS; Am Chem Soc. Res: Heterogeneous catalysis; intermolecular interactions; thermodynamics; gas chromatography. Mailing Add: 1022 Dobson St Evanston IL 60202

MEYER, ERNEST ALAN, b San Jose, Calif, Aug 30, 25; m 50; c 4. MICROBIOLOGY. Educ: Univ Calif, AB, 49; Purdue Univ, MS, 53; Johns Hopkins Univ, ScD, 58. Prof Exp: From instr to assoc prof bact, 61-73, PROF MICROBIOL & IMMUNOL, MED SCH, UNIV ORE, 73- Mem: AAAS; Am Soc Microbiol; Am Soc Trop Med & Hyg; Brit Soc Gen Microbiol. Res: Mechanisms of microbial pathogenicity; bacterial toxins; protozoan physiology. Mailing Add: Univ of Ore Med Sch Portland OR 97201

MEYER, EUGENE, b New York, NY, June 7, 15; m 40; c 4. PSYCHIATRY. Educ: Yale Univ, AB, 37; Johns Hopkins Univ, MD, 41. Prof Exp: From instr to assoc prof psychiat, 49-66, from instr to assoc prof med, 50-70, PROF PSYCHIAT, JOHNS HOPKINS UNIV, 66-, PROF MED, 70- Concurrent Pos: Instr, Wash Sch Psychiat, 50-51; psychiatrist in chg, Psychiat Liaison Serv, 51-; trustee, William Alanson White Psychiat Found, 52- Mem: Am Psychosom Soc; Am Psychiat Asn; AMA; Am Col Physicians. Res: Clinical psychiatry; psychosomatic medicine. Mailing Add: 809 W Lake Ave Baltimore MD 21210

MEYER, EUGENE FRANK, JR, b Michigan City, Ind, May 6, 38; m 64; c 3. PHYSICAL CHEMISTRY. Educ: Lewis Col, BS, 60; Fla State Univ, PhD(chem), 64. Prof Exp: Fel, Inst Nuclear Physics Res, Netherlands, 64-65; from asst prof to assoc prof chem, Lewis Univ, 65-69, acting chmn dept, 67-69, PROF CHEM, LEWIS UNIV, 69- Mem: Am Chem Soc; Combustion Inst. Res: Nuclear structure problems; nuclear fission; thermodynamics of lanthanide salt solutions, chemistry of hazardous materials. Mailing Add: Dept of Chem Lewis Univ Lockport IL 60441

MEYER, FERDINAND CLARK, b Dorchester, Ill, Jan 1, 19; m 40; c 5. ORGANIC CHEMISTRY. Educ: Shurtleff Col, AB, 38; Univ Ill, MS, 40; Univ Wis, PhD(org chem), 43. Prof Exp: Res chemist, Monsanto Chem Co, 43-49, group leader, 49-53, asst dir res, 54-57, assoc dir, 57-63, mgr & dir tech eval & control, 64-67, tech mgr, 68-71, dir tech opers, 72-73, DIR BIOMED TECHNOL, NEW ENTERPRISE DIV, MONSANTO CO, 74- Mem: Am Chem Soc. Res: Organic synthesis and processes; aromatic intermediates; fine chemicals and biologicals; biomedical product development. Mailing Add: Monsanto Co 800 N Lindbergh St Louis MO 63166

MEYER, FRANCIS JOHN, pharmacology, toxicology, see 12th edition

MEYER, FRANK HENRY, b Brooklyn, NY, July 11, 16; m 46; c 2. SOLID STATE PHYSICS, PHILOSOPHY OF SCIENCE. Educ: City Col New York, BS, 36; Polytech Inst Brooklyn, MS, 51; Univ Minn, MA, 68. Prof Exp: X-ray crystallogr, Textile Res Inst, NJ, 51-53; res physicist, Continental Oil Co, 54-60; res engr, Kaiser Aluminum & Chem Corp, 60-63; sr develop engr, Univac Div, Sperry Rand Corp, 63-65; teacher pub sch, Wis, 65-66; ASST PROF PHYSICS & PHILOS, UNIV WIS-SUPERIOR, 66- Concurrent Pos: ED, Reciprocity, 71- Mem: Am Phys Soc; Am Crystallog Asn; Am Asn Physics Teachers; Fedn Am Sci; Philos Sci Asn. Res: Solid state defect physics; solid surface chemistry; philosophy and history of science and education; solid and liquid cohesion theory; three-dimensional time; space time progression. Mailing Add: Depts of Physics & Philos Univ of Wis-Superior Superior WI 54881

MEYER, FRANKLIN VINCENT, b Perth Amboy, NJ, Feb 25, 48; m 68; c 2. NUMBER THEORY. Educ: Hamline Univ, BA, 69; Univ Minn, Minneapolis, MS, 71, PhD(math), 75. Prof Exp: From teaching asst to teaching assoc math, Univ Minn, Minneapolis, 69-75; ASST PROF MATH, BETHEL COL, MINN, 75- Res: Transcendence measures for transcendental numbers related to the exponential function; problems in the theory of diophantine approximation. Mailing Add: Dept of Math Bethel Col 3900 Bethel Dr St Paul MN 55112

MEYER, FRANZ, b Berlin, Ger, July 3, 23; nat US; m 60. BIOCHEMISTRY. Educ: Univ Heidelberg, MD, 53. Prof Exp: Res assoc, Univ Chicago, 54-60; res fel, Harvard Univ, 60-63; assoc prof, 63-74, PROF MICROBIOL, STATE UNIV NY UPSTATE MED CTR, 74- Concurrent Pos: NIH spec res fel, 60-62. Res: Lipid metabolism in microorganisms and lower invertebrates. Mailing Add: Dept of Microbiol State Univ NY Upstate Med Ctr Syracuse NY 13210

MEYER, FRED, applied mathematics, physical chemistry, see 12th edition

MEYER, FRED PAUL, b Holstein, Iowa, Aug 15, 31. PARASITOLOGY, FISHERIES. Educ: Univ Northern Iowa, BA, 53; Iowa State Univ, MS, 57, PhD(parasitol), 60. Prof Exp: Teacher high sch, Iowa, 53-56; asst, Iowa State Univ, 56-59; parasitologist & asst dir, Fish Farming Exp Sta, US Fish & Wildlife Serv, 60-73, DIR FISH CONTROL LAB, BUR SPORT FISHERIES & WILDLIFE, US FISH & WILDLIFE SERV, 73- Mem: Am Soc Parasitol; Am Fisheries Soc. Res: Fish parasitology and pathology with reference to diseases of fish; toxicology; entomology. Mailing Add: Fish Control Lab US Fish & Wildlife Serv LaCrosse WI 54601

MEYER, FREDERICK GUSTAV, b Olympia, Wash, Dec 7, 17; m 46. BOTANY. Educ: Wash State Univ, BSc, 39, MSc, 41; Wash Univ, PhD(bot), 49. Prof Exp: Lab asst bot, Wash State Univ, 39-51; dendrologist, Mo Bot Garden, 51-56; botanist, New Crops Res Br, Agr Res Serv, USDA, 57-63, RES BOTANIST CHG HERBARIUM, US NAT ARBORETUM, USDA, 63- Concurrent Pos: Mo Bot Garden grant, Univ Col & Univ London, 49-51; NSF fel, 55- Mem: Bot Soc Am; Am Soc Plant Taxon; Am Hort Soc; Soc Study Evolution; Int Soc Plant Taxon. Res: Taxonomic botany of the flowering plants; studies in Coffea; evolution and taxonomy of cultivated plants; Valeriana. Mailing Add: US Nat Arboretum Washington DC 20002

MEYER, FREDERICK RICHARD, b Brooklyn, NY, May 26, 38; m 62; c 2. MAMMALIAN PHYSIOLOGY. Educ: Valparaiso Univ, BS, 60; Ind Univ, MA, 62, PhD(physiol), 66. Prof Exp: Asst prof biol, Wilson Col, 65-67; ASSOC PROF BIOL, VALPARAISO UNIV, 67- Mem: Am Physiol Soc; Sigma Xi; AAAS. Res: Temperature regulation in the laboratory rat and man. Mailing Add: Niels Sci Ctr Rm 243 Valparaiso Univ Valparaiso IN 46383

MEYER, GARSON, b Rochester, NY, Nov 22, 96; m 20; c 2. INDUSTRIAL CHEMISTRY. Educ: Univ Rochester, BS, 19. Hon Degrees: DCL, Pace Col, 72; LHD, Ithaca Col, 73. Prof Exp: Chemist, Res Labs, Eastman Kodak Co, 19-20, chief chemist, Camera Works, 20-45, dir chem, Metall & Plastics Lab, Apparatus & Optical Div, 45-62, spec tech adv, 62-65; CHMN BD, GENESEE FED SAVINGS & LOAN ASN, 66- Mem: AAAS; Am Chem Soc; Am Soc Testing & Mat; Soc Plastics Indust; Soc Plastics Eng. Res: Chemical induction and photographic development; industrial chemical development. Mailing Add: 1600 East Ave Rochester NY 14610

MEYER, GEORGE G, b Frankfurt, Ger, Nov 13, 31; nat US; m 53; c 3. PSYCHIATRY, SOCIAL PSYCHIATRY. Educ: Johns Hopkins Univ, BA, 51; Univ Chicago, MD, 55; Am Bd Psychiat & Neurol, cert psychiat, 64. Prof Exp: Resident, Univ Chicago, 58-61, chief resident psychiat, 60-61, from instr to assoc prof, 61-69; assoc prof, 69-71, PROF PSYCHIAT, UNIV TEX MED SCH, SAN ANTONIO, 71- Concurrent Pos: Nat Inst Ment Health career teacher grant, 61-63; consult to many orgn, 61-; assoc chief, Psychiat Inpatient Serv, Univ Chicago, 61-65, chief, 66-69; dir, Northwest San Antonio Ment Health Ctr, 69-74; vis lectr, Univ Edinburgh, 66; mem consult staff, Santa Rosa Med Ctr, San Antonio, 71-; mem bd, Econ Develop Corp, Mex-Am Unity Coun, 71-; mem exec bd, Crisis Ctr, San Antonio, 71-74; psychiat consult, Ecumenical Ctr Bexar County, 72- Mem: Fel Am Psychiat Asn; fel Am Orthopsychiat Asn; Am Group Psychother Asn; Am Asn Med Cols; World Psychiat Asn. Mailing Add: Dept of Psychiat Univ of Tex Med Sch San Antonio TX 78284

MEYER, GERALD, b Newport News, Va, Dec 31, 22; m 47; c 2. GEOLOGY. Educ: Univ NC, BS, 48. Prof Exp: Lab instr geol & geog, Univ NC, 47-48; geologist, Md, 48-57, dist geologist, WVa, 57-64, ASST CHIEF, GROUNDWATER BR, US GEOL SURV, DC, 64- Mem: Geol Soc Am; Am Geophys Union; Am Water Resources Asn; Nat Water Well Asn; Am Water Works Asn. Res: Groundwater geology and hydrology. Mailing Add: Water Resources Div US Geol Surv Washington DC 20242

MEYER, GLEN ERNEST, b Crabtree, Ore, Feb 10, 13; m 64; c 1. Educ: Albany Col, Ore, BS, 35; Johns Hopkins Univ, PhD(phys chem), 39. Prof Exp: Fel chem, Johns Hopkins Univ, 39-40; chemist, Edgewood Arsenal, Md, 40; res chemist, Plantation Div, US Rubber Co, Sumatra, 41-42, Naugatuck Chem Div, Conn, 42-46 & Plantation Div, Malaya, 46-49; tech sales, Rubber & Latex, Inc, Md, 50-53; res chemist, 53-56, sect head synthetic rubber, 56-71, MGR EMULSION POLYMERS RES DEPT, GOODYEAR TIRE & RUBBER CO, 71- Mem: Am Chem Soc. Res: Natural and synthetic elastomers. Mailing Add: Res Div Goodyear Tire & Rubber Co Akron OH 44316

MEYER, GLENN ARTHUR, b Baraboo, Wis, Mar 8, 34; m 61; c 3. NEUROSURGERY. Educ: Univ Wis, Madison, BS, 57, MD, 60. Prof Exp: Surg extern, Univ Wis Hosps, 59-60, res asst neurophysiol, Med Sch, Univ, 63-64, instr neurosurg & staff physician, Univ Hosps, 66; neurosurg consult, St Elizabeth's Hosp, 67-68; assoc prof, Univ Tex Med Br Galveston, 69-72; ASSOC PROF NEUROSURG, MED COL WIS, 72- Concurrent Pos: Staff physician, Mendota State Psychiat Hosp, 62-65; med adv, Social Security Admin, 70-72. Mem: AAAS; Cong Neurol Surg; Soc Neurosci; Am Col Surg; Am Asn Neurol Surg. Res: Spinal autonomic mechanisms in spinal shock; reconstruction of craniofacial anomalies; pain control with electrical stimulation; chronic monitoring of intracranial pressure with a fully implantable device; developing new lesion targets in psychosurgery. Mailing Add: Dept of Neurosurg Med Col of Wis Milwaukee WI 53226

MEYER, GOLDYE W, b Wilkes Barre, Pa; c 3. SCIENCE EDUCATION. Educ: Wilkes Col, BS, 62; Temple Univ, MS, 64; Univ Conn PhD(coun psychol), 75. Prof Exp: Instr chem, Wilkes Col, 62-64; instr chem, 64-65, instr sec educ, 66-68, dir student teaching & educ placement, 68-69, assoc prof sec educ, 69-76, ASSOC PROF COUNR EDUC & HUMAN RESOURCES, UNIV BRIDGEPORT, 76- Concurrent Pos: Consult, Conn State Dept Educ, 74- Mem: Am Educ Res Asn; Am Psychol Asn; Am Personnel & Guid Asn. Res: Science teaching as a second language; humanistic education models applied to the teaching of science in secondary schools. Mailing Add: Univ of Bridgeport Bridgeport CT 06602

MEYER, GORDON BENJAMIN, nutrition, physiology, see 12th edition

MEYER, GREGORY CARL, b Willmar, Minn, Feb 10, 18; m 42, 68; c 4. ORGANIC CHEMISTRY. Educ: Southwestern Texas State Univ, AB, 38; Univ Nebr, MA, 40, PhD(chem), 43. Prof Exp: Res chemist, 42-57, tech ed, 57-63, PATENT CHEMIST, E I DU PONT DE NEMOURS & CO, INC, 63- Mem: Am Chem Soc. Mailing Add: Jackson Lab E I du Pont de Nemours & Co, Inc Wilmington DE 19898

MEYER, GUNTER HUBERT, b Stettin, Ger, Aug 19, 39; US citizen; m 66; c 2. NUMERICAL ANALYSIS, APPLIED MATHEMATICS. Educ: Univ Utah, BA, 61; Univ Md, MA, 63, PhD(math), 67. Prof Exp: ASSOC PROF MATH, GA INST TECHNOL, 67- Concurrent Pos: Res mathematician, Mobil Res & Develop Corp, 67-71; consult, UNESCO-Univ Simon Bolivar, Caracas, Venezuela, 74; sr vis fel, Brunel Univ, Uxbridge, Eng, 75. Mem: Soc Indust & Appl Math; Am Math Soc; Math Asn Am; Am Soc Lubrication Engrs. Res: Numerical solution of boundary value problems, especially of free boundary problems. Mailing Add: Sch of Math Ga Inst of Technol Atlanta GA 30332

MEYER, HAROLD DAVID, b Indianapolis, Ind, Oct 17, 39; m 68; c 2. NUMERICAL ANALYSIS. Educ: Mass Inst Technol, SB & SM, 62; Univ Chicago, MS, 68, PhD(math), 69. Prof Exp: Sci & math analyst, Foreign Sci & Technol Ctr, US Army, 69-70, br chief & opers res analyst, Mil Assistance Command, Vietnam, 70-71; ASST PROF MATH, TEX TECH UNIV, 71- Mem: Am Math Soc. Res: Numerical solution of partial differential equations, finite element approaches, ill-posed problems and representations of solutions to be used numerically in conjunction with such problems. Mailing Add: Dept of Math Tex Tech Univ PO Box 4319 Lubbock TX 79409

MEYER, HARRY MARTIN, JR, b Palestine, Tex, Nov 25, 28; m 49; c 3. VIROLOGY, PEDIATRICS. Educ: Hendrix Col, BS, 49; Univ Ark, MD, 53. Prof Exp: Chief diag sect, Dept Virus & Rickettsial Dis, Walter Reed Army Inst Res, DC, 54-57; asst resident pediat, Sch Med, Univ NC, 57-59; chief sect gen virol, 59-64, chief lab viral immunol, 64-72, DIR BUR BIOLOGICS, FOOD & DRUG ADMIN, 72- Honors & Awards: Chevalier de l'Ordre Nat, Repub of Upper Volta, Africa, 63; Lett of Commendation from Pres, 66; Meritorious Serv Medal, Dept Health, Educ & Welfare, 66, Distinguished Serv Medal, 69; Mead Johnson Award for Pediat Res, 67; Max Weinstein Award for Med Res, United Cerebral Palsy Asns, 69; Int Award for Distinguished Sci Res, Joseph P Kennedy, Jr Found, 71. Mem: Am Asn Immunol; Am Pediat Soc; Am Acad Pediat; Am Epidemiol Soc; Am Soc Microbiol. Res: Virus vaccine; virus and rickettsial diseases. Mailing Add: Bur of Biologics Food & Drug Admin Rockville MD 20852

MEYER, HARUKO, b Tokyo, Japan, Jan 17, 29; US citizen; m 60. BIOCHEMISTRY, MICROBIOLOGY. Educ: Toho Women's Col Sci, Japan, BS, 49; Tokyo Col Sci, MS, 51; State Univ NY Upstate Med Ctr, PhD(microbiol), 66. Prof Exp: RES ASSOC MICROBIOL, STATE UNIV NY UPSTATE MED CTR, 67- Mem: Am Soc Microbiol. Res: Lipid metabolism of parasitic organisms. Mailing Add: Dept of Microbiol State Univ of NY Upstate Med Ctr Syracuse NY 13210

MEYER, HARVEY JOHN, b St Paul, Minn, July 16, 35; m 62; c 2. GEOLOGY, COMPUTER SCIENCE. Educ: Univ Minn, BA, 57; Calif Inst Technol, MS, 59; Pa State Univ, PhD(sedimentary petrol), 64. Prof Exp: Res fel, Antarctica Proj, Univ Minn, 62-63; sr res scientist statist geol, Pan Am Petrol Corp, 63-66; mgr data processing, Can Stratig Serv Ltd, 66-71; lectr geol, Univ Calif, Davis , 71-72; EXPLOR SYSTS ANALYST, AMOCO PROD CO, 72- Mem: Geol Soc Am; Am Asn Petrol Geol; Int Asn Math Geol; Soc Econ Paleont & Mineral. Res: Digital computer and statistical methods applied to geological problems; hydrocarbon exploration; sedimentary petrology of clastic rocks and Antartica geology. Mailing Add: Amoco Prod Co Security Life Bldg Denver CO 80202

MEYER, HEINZ FRIEDRICH, b Suedmoslesfehn, Ger, Jan 4, 32; m 59; c 4. CHEMISTRY. Educ: Univ Frankfurt, Dr phil nat(org chem), 59. Prof Exp: Res asst, Ohio State Univ, 60-61 & C H Boehringer, Ger, 61-63; RES ASST, UPJOHN CO, 63- Mem: Am Chem Soc; Soc Ger Chem. Res: Alkaloids; antibiotics; steroids. Mailing Add: Upjohn Co Kalamazoo MI 49002

MEYER, HENRY, b Grand Rapids, Mich, June 28, 08; m; c 2. ZOOLOGY. Educ: Calvin Col, AB, 30; Univ Mich, MS, 31, PhD(embryol, cytol, ornith), 36. Prof Exp: Asst zool, Univ Mich, 30-35; instr biol, Lawrence Col, 35-37; asst prof zool, Univ Tenn, 37-45; from assoc prof to prof biol, Ripon Col, 45-55, chmn dept, 45-55; prof, 55-74, chmn dept, 72-74, EMER PROF BIOL, UNIV WIS-WHITEWATER, 74- Mem: Fel AAAS; Wilson Ornith Soc; assoc Am Ornith Union. Res: Reproductive behavior of Molliensia; nesting and migration of birds. Mailing Add: Dept of Biol Univ of Wis Whitewater WI 53190

MEYER, HENRY IRVING, b Can, Nov 22, 21; US citizen; m 48; c 3. MATHEMATICS. Educ: Wash Univ, BS, 43; Harvard Univ, MA, 51. Prof Exp: Asst engr, United Gas Pipeline Co, 47-48, mem res staff, United Gas Corp, 51-55, supvr math sect, Res Lab, 55-61, mgr comput dept, 61-69; dir mgt sci, Pennzoil United Inc, 69-74, DIR MGT SCI, PENNZOIL CO, 74- Mem: Am Math Soc; Soc Indust & Appl Math; Asn Comput Mach; Can Math Cong. Res: Corporate financial modeling; operation simulations. Mailing Add: Pennzoil Co 900 Southwest Tower Houston TX 77002

MEYER, HENRY OOSTENWALD ALBERTIJN, b Warwickshire, Eng, Jan 18, 37; m 59; c 5. MINERALOGY, GEOCHEMISTRY. Educ: Univ London, BSc, 59, PhD(geol), 62. Prof Exp: Res asst mineral, Univ Col, Univ London, 61-66; sr fel mineral & exp petrol, Geophys Lab, Carnegie Inst Technol, 66-69; sr res assoc, Nat Res Coun, Goddard Space Flight Ctr, NASA, 69-71; assoc prof, 71-74, PROF PETROL, PURDUE UNIV, LAFAYETTE, 74- Concurrent Pos: NSF & NASA grants, Purdue Univ, 71- Mem: AAAS; fel Geol Soc London; Mineral Soc Gt Brit & Ireland; Mineral Soc Am; Am Geophys Union. Res: Mineralogy and petrology of basic igneous rocks, including lunar samples; high pressure phase equilibria studies pertinent to basic igneous rocks; origin of diamond and kimberlite rock; electron microprobe and x-ray diffraction studies. Mailing Add: Dept of Geosci Purdue Univ Lafayette IN 47907

MEYER, HERBERT A, b Howells, Nebr, Apr 13, 11; m 45; c 4. INORGANIC CHEMISTRY, ANALYTICAL CHEMISTRY. Educ: Univ Nebr, BA, 38, MA, 44, PhD(chem math), 61. Prof Exp: Teacher & prin elem sch, 30-37; instr phys sci & math, Concordia Teachers Col, Nebr, 38-41, assoc prof phys sci, math & phys educ, 41-59, chmn dept chem, 69-73, PROF CHEM, CONCORDIA TEACHERS COL, NEBR, 59- Res: Vertical variations in chemical content of Greenhorn Limestone outcrops. Mailing Add: Dept of Chem Concordia Teachers Col Seward NE 68434

MEYER, HERMAN, b Chicago, Ill, May 29, 12; m 40; c 3. MATHEMATICS. Educ: Armour Inst Technol, BS, 33, MS, 35; Univ Chicago, PhD(math), 41. Prof Exp: Jr sanit engr, Bd Health, Chicago, 33-34, bur eng, 35-42; assoc pub health engr, USPHS, Wash, 42; from asst prof to assoc prof, 42-46, chmn dept, 47-58, PROF MATH, UNIV MIAMI, 47- Mem: Am Math Soc; Math Asn Am. Res: Analysis; abstract spaces; polynomial approximations to functions defined on abstract spaces. Mailing Add: Dept of Math Univ of Miami Coral Gables FL 33146

MEYER, HERMANN, b Frauenfeld, Switz, Mar 29, 27; nat US; m 52; c 3. VETERINARY ANATOMY. Educ: Univ Zurich, DVM, 50, Dr med vet(anat), 52; Cornell Univ, PhD(anat), 57. Prof Exp: Asst vet anat, Univ Zurich, 51-52; instr, Cornell Univ, 53-56, actg asst prof, 56-57; from asst prof to prof anat, Colo State Univ, 57-72; head vet anat dept, Univ Zurich, 72-73; pub rels dir, Poudre Valley Mem Hosp, 73-75; PROF VET ANAT & PHYSIOL, UNIV MO, COLUMBIA, 75- Concurrent Pos: Guest auditor, Univ Basel, 51-52; guest lectr, Univ Zurich, 63; consult, Nat Defense Educ Act Title IV grad fel prog, 66-68; vis prof, Cornell Univ, 67-68. Mem: Am Vet Med Asn; Am Asn Vet Anat; Am Asn Anat; World Asn Vet Anat; Ger Anat Soc. Res: Histology; embryology; neuromorphology; history of morphology. Mailing Add: Dept of Vet Anat & Physiol Univ of Mo Columbia MO 65201

MEYER, IRVING, b Springfield, Mass, Mar 19, 20; m 53; c 3. ORAL SURGERY, ORAL PATHOLOGY. Educ: Univ Mass, BS, 41; Tufts Univ, DMD, 44; Univ Pa, MSc, 50, DSc, 58; Am Bd Oral Surg, dipl, 54. Prof Exp: Resident oral surg, Metrop Hosp, New York, 48-49; resident, Philadelphia Gen Hosp, Pa, 49-50; instr oral surg & path, 50-60, assoc res prof, 60-65, assoc prof oral path, 65-66, RES PROF ORAL PATH, SCH DENT MED, TUFTS UNIV, 66-; DIR DEPT ORAL SURG, SPRINGFIELD HOSP MED CTR, 74- Concurrent Pos: Instr grad sch med, Univ Pa,

49-51; chief oral surgeon, Wesson Mem Hosp, Springfield, Mass, 51-, secy-treas med staff, 72-74; oral surgeon, Mercy Hosp, 55-; oral surgeon, Springfield Hosp, 56-, pathologist, 57-; ed, Ann Conf Oral Cancer, 60 & 63; lectr, Grad Sch, Boston Univ, 62- & Harvard Univ, 67-; assoc ed & sect ed oral path, J Oral Surg, 65-; consult ed oral surg, oral med & oral path, New Eng Soc Oral Surg, 67-; mem, Adv Comt & examr, Am Bd Oral Surg, 70-, mem, Bd Dirs, 73-80; oral surgeon, Cancer Div, Western Mass Hosp; consult to var hosps, Mass; physician pvt pract. Mem: Am Soc Oral Surg; fel Am Acad Oral Path; NY Acad Sci. Res: Cancer of oral cavity and adnexia, especially in clinical aspects, etiology, therapy and pathology. Mailing Add: 50 Maple St Springfield MA 01103

MEYER, JAMES HENRY, b Lewiston, Idaho, Apr 13, 22; m 47; c 5. NUTRITION. Educ: Univ Idaho, BS, 47; Univ Wis, MS, 49, PhD(nutrit), 51. Prof Exp: Asst nutrit, Univ Wis, 47-51; from asst prof to prof animal husb & chmn dept, Univ Calif, Davis, 50-63, dean col agr, 63-69, CHANCELLOR, UNIV CALIF, DAVIS, 69- Concurrent Pos: Mem comt animal nutrit, Nat Acad Sci-Nat Res Coun, 65-67 & comn undergrad educ in biol. Mem: AAAS; Am Soc Animal Sci; Am Inst Nutrit. Res: Pasture and fiber nutrition; nutrient requirements; undernutrition. Mailing Add: Chancellor's Off Univ of Calif Davis CA 95616

MEYER, JAMES HENRY, b St Marys, Pa, July 20, 28; m 60; c 3. METEOROLOGY. Educ: Pa State Univ, BS, 53, MS, 55. Prof Exp: Meteorologist, Res & Develop Ctr, Intel & Reconnaissance Lab, Griffiss Air Force Base, NY, 54-55; atmospheric physicist, Lincoln Lab, Mass Inst Technol, 55-63; proj meteorologist, Tech Oper Inc, Mass, 63-64; proj mgr, Electromagnetic Res Corp, Md, 64-67; meteorologist, 67-72, SR STAFF METEOROLOGIST, APPL PHYSICS LAB, JOHNS HOPKINS UNIV, 72- Mem: Am Meteorol Soc; Am Geophys Union; Am Geol Inst. Res: Radar and radio meteorology; meteorological instrumentation; atmospheric and cloud physics; power plant siting and atmospheric pollution meteorology. Mailing Add: 12926 Allerton Lane Silver Spring MD 20904

MEYER, JAMES MELVIN, b West Palm Beach, Fla, Jan 18, 43; m 69; c 2. POLYMER CHEMISTRY. Educ: Ind Univ, BS, 64; Northwestern Univ, PhD(inorg chem), 68. Prof Exp: Asst prof chem, Univ Ill, 67-69; res chemist, 69-74, DIV HEAD CHEM, E I DU PONT DE NEMOURS & CO, INC, 74- Res: Development of uses for elastomeric polymers. Mailing Add: Elastomers Lab E I du Pont de Nemours Wilmington DE 19898

MEYER, JAMES WAGNER, b Rhineland, Mo, May 22, 20; m 49; c 4. ENERGY CONVERSION. Educ: Univ Wis, PhB, 48, PhD(physics), 56; Dartmouth Col, MA, 50. Prof Exp: Asst physics, Dartmouth Col, 48-49, Univ Calif, 49-50 & Univ Wis, 50-52; mem staff, Lincoln Lab, Mass Inst Technol, 52-57, group leader, 57-59, assoc head radar div, 59-62, assoc head solid state div, 62-63, head radio physics div, 63-65; sr scientist, Educ Serv, Inc, 65-67; mem tech staff, Lincoln Lab, Mass Inst Technol, 67-70, prog mgr, Ctr Space Res, 70-73, PROG DIR, ENERGY LAB, MASS INST TECHNOL, 73- Mem: AAAS; Am Phys Soc; Am Asn Physics Teachers; Int Solar Energy Soc. Res: Microwave spectroscopy of solid state; cryogenics, electronics; radio physics; education; energy conservation; alternative energy sources; solar energy. Mailing Add: Rm 3-137 Energy Lab Mass Inst Technol Cambridge MA 02139

MEYER, JEAN-PIERRE, b Lyon, France, Aug 5, 29; US citizen; m 58; c 6. MATHEMATICS. Educ: Cornell Univ, BA, 50, MA, 51, PhD(math), 54. Prof Exp: Asst prof math, Syracuse Univ, 56; res assoc, Brown Univ, 56-57; from vis asst prof to asst prof, 57-65, ASSOC PROF MATH, JOHNS HOPKINS UNIV, 65- Mem: Am Math Soc. Res: Algebraic topology. Mailing Add: Dept of Math Johns Hopkins Univ Baltimore MD 21218

MEYER, JOACHIM DIETRICH, b Berlin, Ger, Dec 13, 35; US citizen; m 61; c 2. MINERALOGY, PETROLOGY. Educ: Univ Tex, Austin, BS, 58, MA, 61; Univ Giessen, PhD(geol), 63. Prof Exp: Geologist, Humble Oil & Ref Co, 64-66; Chevron Oil Co, 66; asst prof, 66-71, ASSOC PROF MINERAL & PETROL, TULANE UNIV, 71- Mem: Geol Soc Am; Soc Econ Paleont & Mineral. Res: Stratigraphy of volcanic ash deposits; environmental geology; planetology; volcanology. Mailing Add: Dept of Earth Sci Tulane Univ New Orleans LA 70118

MEYER, JOE, biochemistry, see 12th edition

MEYER, JOHANNES HORST MAX, b Berlin, Ger, Mar 1, 26; m 53; c 2. PHYSICS. Educ: Univ Geneva, Lic es sc, 49; Univ Zurich, PhD(physics), 52. Prof Exp: Swiss Soc res fel math & physics, Clarendon Lab, Oxford Univ, 53-55, Nuffield fel, 55-57; lectr appl physics, Harvard Univ, 57-59; from asst prof to assoc prof, 59-64, PROF PHYSICS, DUKE UNIV, 64- Concurrent Pos: Sloan fel, 60-63; vis prof, Munich Tech Univ, 65; vis fel, Japanese Soc Sci, 71; consult, NSF, 72-; vis sci, Inst Laue-Langevin, Grenoble, France, 74. Mem: Fel Am Phys Soc; Swiss Soc Natural Sci. Res: Phenomena at low temperatures; magnetism; properties of clathrate compounds; liquid and solid helium and hydrogen; critical phenomena. Mailing Add: 2716 Montgomery St Durham NC 27705

MEYER, JOHN AUSTIN, b St Mary's, Pa, Sept 18, 19; m 55. ANALYTICAL CHEMISTRY. Educ: Pa State Univ, BS, 49, MS, 50; State Univ NY Col Environ Sci & Forestry, PhD(org chem), 58. Prof Exp: Asst chemist, Anal Lab, Speer Carbon Co, Pa, 45-57; chemist, Gulf Res & Develop Co, 50-52; asst head anal dept, Verona Res Ctr, Koppers Co, 52-54; PROF NUCLEAR & RADIATION CHEM & RADIOL SAFETY OFF STATE UNIV NY COL ENVIRON SCI & FORESTRY, 58-, DIR, ANAL & TECH SERV, 72- Concurrent Pos: Fel, Oak Ridge Inst Nuclear Studies, 57 & 66; consult various US & foreign industs. Mem: AAAS; Am Chem Soc; Am Nuclear Soc; Forest Prod Res Soc; NY Acad Sci. Res: Development of wood-polymer materials by the heat-catalyst method; neutron activation analysis; trace analytical methods; radiation chemistry. Mailing Add: Dept of Chem State Univ of NY Col of Environ Sci & Forestry Syracuse NY 13210

MEYER, JOHN L, inorganic chemistry, see 12th edition

MEYER, JOHN ROGER, b Russell, Iowa, Dec 10, 36; m; c 3. PHYSIOLOGY, ZOOLOGY. Educ: Faith Baptist Bible Col, BA, 61; Kearney State Col, BS, 64; Univ Iowa, PhD(zool), 70. Prof Exp: NIH fel respiratory physiol, Cardiovasc Pulmonary Res Lab, Med Sch, Univ Colo, Denver, 69-73; ASST PROF PHYSIOL & BIOPHYS, HEALTH SCI CTR, UNIV LOUISVILLE, 73- Honors & Awards: Outstanding Achievement Award, Accrediting Asn Bible Cols, 72. Mem: AAAS; Soc Neurosci; Am Soc Zool; Am Heart Asn; Instrument Soc Am. Res: Control of respiration; carotid body function during hypoxic stress; comparative aspects of environmental adaptations. Mailing Add: Dept of Physiol & Biophys Univ Louisville Health Sci Ctr Louisville KY 40201

MEYER, JOHN SIGMUND, b Princeton, Ill, May 12, 37; m 63; c 3. STATISTICS. Educ: Wartburg Col, BA, 59; Northwestern Univ, MS, 61; Iowa State Univ, PhD(statist), 73. Prof Exp: Instr math, Wartburg Col, 61-67; asst prof, 67-74, ASSOC PROF MATH, CORNELL COL, 74-, CHMN DEPT, 75- Mem: Am Statist Asn; Math Asn Am. Res: Confidence intervals for quantiles of finite populations. Mailing Add: Dept of Math Cornell Col Mt Vernon IA 52314

MEYER, JOHN STIRLING, b London, Eng, Feb 24, 24; nat US; m 47; c 5. NEUROLOGY. Educ: McGill Univ, MD, CM, 48, MSc, 49; Am Bd Psychiat & Neurol, dipl. Prof Exp: Demonstr histol, McGill Univ, 45-46, demonstr clin micros, 46-47, fel, 48-49; asst med, Sch Med, Yale Univ, 49-50; demonstr neuropath & teaching fel neurol, Harvard Med Sch, 50-52; USPHS sr res fel, 52-54; instr med, Harvard Med Sch, 54-56; assoc vis physician neurol, Boston City Hosp, 56-57; consult & lectr neurol, US Naval Hosp, Chelsea, Mass, 57; prof neurol & chmn dept, Sch Med, Wayne State Univ, 57-69; PROF NEUROL & CHMN DEPT, BAYLOR COL MED, 69- Concurrent Pos: Asst, Montreal Neurol Inst, McGill Univ, 45-46; from jr intern to sr intern, New Haven Hosp, Conn, 49-50, vis neurologist & supvr EEG lab, 54-56; head dept neurol, Detroit Gen Hosp, 57-69; consult, Grace, Children's Sinai, Detroit Mem & Dearborn Vet Hosps, 57-69; chief neurol dept, Harper Hosp, Detroit, 63-69; chief neurol serv, Methodist & Ben Taub Gen Hosps, Houston, 69-; consult, Vet Admin & Hermann Hosps, 69- Mem, President's Comn Heart Dis, Cancer & Stroke, 64-65, Nat Adv Comt Neurol Dis & Blindness, 65- & Subcomt cerebrovasc dis, Nat Heart Inst-Nat Inst Neurol Dis & Blindness Joint Coun, 65-; chmn res subcomt, Nat Inst Neurol Dis & Blindness, 66. Mem: Am Neurol Asn; NY Acad Sci; Am Heart Asn; AMA. Res: Cerebral blood flow and metabolism studies in stroke patients. Mailing Add: Dept of Neurol Baylor Col of Med Houston TX 77025

MEYER, JOSEPH H, biostatistics, see 12th edition

MEYER, KARL, b Kerpen, Ger, Sept 4, 99; nat US; m 30; c 2. BIOCHEMISTRY. Educ: Univ Cologne, MD, 24; Univ Berlin, PhD(chem), 28. Prof Exp: Asst, Univ Berlin, 27-28; Int Educ Bd fel, Univ Zurich, 28-29; asst prof exp biol, Univ Calif, 31-33; from asst prof to prof, 33-67, EMER PROF BIOCHEM & SPEC LECTR & CONSULT, DEPT MED, COL PHYSICIANS & SURGEONS, COLUMBIA UNIV, 67-; PROF BIOCHEM, BELFER GRAD SCH SCI, YESHIVA UNIV, 67- Concurrent Pos: Chemist, Presby Hosp, New York, 33- Honors & Awards: Lasker Award, 56; Duckett-Jones Mem & Gairdner Award, 59. Mem: Nat Acad Sci; fel Am Acad Arts & Sci; AAAS; Soc Exp Biol & Med; Am Chem Soc. Res: Mucopolysaccharides; glycoproteins; connective tissues; mucopolysaccharides and mycolytic enzymes. Mailing Add: Belfer Grad Sch of Sci Yeshiva Univ New York NY 10033

MEYER, KARL FRIEDERICH, pathology, bacteriology, deceased

MEYER, LAWRENCE DONALD, b Concordia, Mo, Apr 14, 33; m 54; c 3. SOIL CONSERVATION. Educ: Univ Mo, BS, 54, MS, 55; Purdue Univ, PhD(agr eng), 64. Prof Exp: Asst soil & water eng, Univ Mo, 54-55; agr engr, USDA, Purdue Univ, 55-73, from asst prof to assoc prof agr eng, 65-73, AGR ENGR, SEDIMENTATION LAB, USDA, 73- Concurrent Pos: Adj prof agr & biol eng, Miss State Univ, 75- Honors & Awards: Outstanding Performance Rating & Award, Agr Res Serv, USDA, 59. Mem: Am Soc Agr Eng; Soil Conserv Soc Am; Soil Sci Soc Am. Res: Soil and water conservation engineering; mechanics of the soil erosion process; rainfall simulation; erosion research techniques; erosion control on construction sites. Mailing Add: Sedimentation Lab USDA PO Box 1157 Oxford MS 38655

MEYER, LEO FRANCIS, b Pittsburgh, Pa, July 19, 29; m 54; c 4. ORGANIC POLYMER CHEMISTRY. Educ: Duquesne Univ, BS, 56; Univ Richmond, MS, 66. Prof Exp: Res chemist, Gulf Res & Develop Ctr, 56-61; assoc chemist, 61-63, res chemist, 63-69, MGR RES CTR, PHILLIP MORRIS INC, 69- Mem: Am Chem Soc. Res: Polymers; catalysis; aromatic alkylations; aerosol filtration; gas adsorbants; tobacco processing; cigarette making; plastics extrusion; paper coating. Mailing Add: 8526 Hanford Dr Richmond VA 23229

MEYER, LEO MARTIN, b New York, NY, Jan 14, 06; m 28; c 1. HEMATOLOGY. Educ: City Col New York, BS, 26; Cornell Univ, AM, 27; Univ Md, MD, 31. Prof Exp: Assoc vis physician, Kings County Hosp, 33-51; dir labs, South Nassau Communities Hosp, 41-64; hematologist in chg, Queen's Hosp Ctr, Jamaica, NY, 64-71; chief hemat & oncol res, 71-74, COORDR SICKLE CELL PROGS, VET ADMIN CENT OFF, WASHINGTON, DC, 74- Concurrent Pos: Attend hematologist, Vet Admin Hosp, Bronx, 46-52; consult hematologist, Goldwater Mem Hosp, 47-; asst prof clin med, Sch Med, NY Univ, 50-; attend hematologist, Meadowbrook Hosp, 53-55; res collarbr, Brookhaven Nat Lab, 55-; mem med serv teams to Marshall Islands, AEC, 57-; Fulbright-Hays res scholar, Turku Univ, 67; NSF exchange scientist to India, 68-69; chief hemat sect, Vet Admin Hosp, Brooklyn, 70- Mem: Am Fedn Clin Res; Am Asn Path & Bact; Soc Exp Biol & Med; Am Asn Cancer Res; Am Soc Hemat. Res: Macrocytic anemia; leukemia; applications of radioisotopes in hematology. Mailing Add: Hemat Sect Vet Admin Hosp Brooklyn NY 11209

MEYER, LEON HERBERT, b Navasota, Tex, Sept 4, 26; m 58; c 2. PHYSICAL CHEMISTRY, CHEMICAL ENGINEERING Educ: Ga Inst Technol, BChE, 49, MS, 51; Univ Ill, PhD(chem), 53. Prof Exp: Asst, Univ Ill, 52-53; res chemist & engr, Atomic Energy Div, Explosives Dept, 53-64, res mgr, Separations Chem Div, Savannah River Lab, 64-67, dir separations chem & Eng Sect, 67-69, ASST DIR, SAVANNAH RIVER LAB, E I DU PONT DE NEMOURS & CO, INC, 69- Mem: Am Chem Soc; Am Inst Chem Eng. Res: Vapor-liquid equilibrium; nuclear magnetic resonance; fused salt electrolysis; cryogenics; radiochemical separations. Mailing Add: 2219 Dartmouth Rd Augusta GA 30904

MEYER, LESTER WILLIAM AREND, organic chemistry, physical chemistry, see 12th edition

MEYER, LOTHAR, physical chemistry, see 12th edition

MEYER, MARGARET E, microbiology, infectious diseases, see 12th edition

MEYER, MARTIN MARINUS, JR, b Wichita, Kans, Dec 24, 36; m 61; c 2. ORNAMENTAL HORTICULTURE, PLANT PHYSIOLOGY. Educ: Kans State Univ, BS, 58; Cornell Univ, MS, 61, PhD(hort), 65. Prof Exp: Asst nursery mgr, M Meyer & Son Nursery, Kans, 58-59; asst ornamental hort, Cornell Univ, 59-64; asst prof, 65-71, ASSOC PROF NURSERY MGT, UNIV ILL, URBANA, 71- Mem: AAAS; Am Soc Hort Sci; Sigma Xi. Res: Physiology of propagation; growth and development of woody plants; tissue culture propagation. Mailing Add: Dept of Hort Univ of Ill Urbana IL 61801

MEYER, MARVIN CHRIS, b Detroit, Mich, Sept 19, 41; m 66; c 2. CLINICAL PHARMACOLOGY. Educ: Wayne State Univ, BS, 63, MS, 65; State Univ NY Buffalo, PhD(pharmaceut), 69. Prof Exp: Teaching asst pharmaceut, Wayne State Univ, 64-65 & State Univ NY Buffalo, 66-68; asst prof, 69-72, ASSOC PROF PHARMACEUT, UNIV TENN CTR HEALTH SCI, 72-, DIR DIV DRUG METAB & BIOPHARMACEUT, 69- Concurrent Pos: Expert, US Food & Drug Admin, 73-; consult Cooper Labs, 75- Honors & Awards: Mead Johnson Undergrad Res Award,

Am Asn Cols Pharm, 70. Mem: Am Pharmaceut Asn; Acad Pharmaceut Sci; Am Asn Cols Pharm. Res: Study and quantitation of the time course of drugs in humans and animals, especially studies of drug absorption, metabolism, distribution and elimination. Mailing Add: Dept of Med Chem Col of Pharm Univ of Tenn Ctr Health Sci Memphis TN 38163

MEYER, MARVIN CLINTON, b Jackson, Mo, Dec 20, 07; m 46; c 3. ANIMAL PARASITOLOGY. Educ: Mo State Col, BS, 32; Ohio State Univ, AM, 36; Univ Ill, PhD(parasitol), 39. Prof Exp: Asst biol, Mo State Col, 30-32; prin high sch, Mo, 32-36; asst zool, Univ Ill, 36-39; instr, Univ Ky, 39-41; adj prof & actg head dept, Douglass Col, Rutgers Univ, 41-42, adj prof, 46; from instr to assoc prof, 46-54, PROF ZOOL, UNIV MAINE, ORONO, 54- Concurrent Pos: Fulbright res scholar, NZ, 55-56; mem adv comt, Smithsonian Oceanog Sorting Ctr, 65-; sr vis res assoc, Smithsonian Inst, US Nat Mus, 67-68; mem nat screening comt, Inst Int Educ, 67- Mem: Am Micros Soc; Am Soc Parasitol; Am Soc Zool; Soc Syst Zool. Res: Parasitology; morphology and taxonomy of Hirudinea; parasites of fish and wildlife; invertebrate morphology; faunistic zoology. Mailing Add: Dept of Zool Univ of Maine Orono ME 04473

MEYER, MARVIN WILLIAM, plant physiology, see 12th edition

MEYER, MAURICE WESLEY, b Long Prairie, Minn, Feb 13, 25; m 46; c 2. PHYSIOLOGY, DENTISTRY. Educ: Univ Minn, BS, 53, DDS, 57, MS, 59, PhD(physiol), 60. Prof Exp: Res fel, 57-60, from instr to asst prof, 60-64, lectr, 61-73, ASSOC PROF DENT, PHYSIOL & NEUROL, UNIV MINN, MINNEAPOLIS, 64- Concurrent Pos: Res career develop award, 63-69. Mem: AAAS; Am Physiol Soc; Am Dent Asn; Soc Exp Biol & Med; Int Asn Dent Res. Res: Circulation; blood flow in teeth and supporting structures; cerebral blood flow. Mailing Add: Dept of Physiol 424 Millard Hall Univ of Minn Minneapolis MN 55455

MEYER, MERLE P, b Eldridge, Iowa, Feb 11, 20; m 56; c 2. FORESTRY. Educ: Univ Minn, BS, 49, PhD(forestry), 56; Univ Calif, MF, 50. Prof Exp: Instr, 52, PROF FORESTRY & FOREST RESOURCES, UNIV MINN, ST PAUL, 52- Concurrent Pos: Consult, Okla Res Inst, 52-54, US Forest Serv, 56 & Geotechnics & Resources, Inc, 59-62; Fulbright lectr, Norway, 61-62; guest lectr, Finland & Poland, 62; consult, UN Arg, 64-67; NSF vis scientist lectr, 65; consult, Bur Land Mgt, 70- Mem: Soc Am Foresters; Am Soc Photogram; Soc Range Mgt; Wildlife Soc. Res: Application of aerial photography to forest and rangeland resource inventory and management. Mailing Add: Univ of Minn Col of Forestry St Paul MN 55101

MEYER, MORTON A, b New York, NY, Dec 18, 17; m 44; c 2. GEOGRAPHY. Prof Exp: Chief pop processing off, US Bur Census, 50-51, statist procedures sect, Pop & Housing Div, 51-54 & electronic systs, Tabulating Sect, 52-53, actg chief opers br, 54-55; adv census & statist, Int Co-op Admin, Uruguay, 55-58; chief prog br, US Bur Census, 58-59, asst chief decennial opers div, 60-61, chief, 61-62, chief demog opers div, 62-66, coordr, Civil Defense Serv, 66-68; tech adv to subcomt on census & statist, Post Office & Civil Serv Comt, US House Rep, 68-71; spec asst to assoc dir admin, US Dept Com, 71; CHIEF GEOG DIR, US BUR CENSUS, 71- Concurrent Pos: Orgn Am States adv, Nat Develop Coun Repub Arg, 62; consult, Agency Int Develop, Nat Planning Inst Repub Peru, 64 & Brazilian Inst Geog & Statist, 67. Mem: AAAS; Am Statist Asn; Acad Polit Sci. Res: Statistical administration and operations. Mailing Add: US Bureau of the Census Washington DC 20031

MEYER, NORMAN JAMES, b Wolsey, SDak, Feb 17, 26; m 52; c 2. PHYSICAL CHEMISTRY. Educ: Univ SDak, BA, 49; Univ Kans, PhD(chem), 56. Prof Exp: Res engr, Continental Oil Co, 54-56; res chemist, Monsanto Chem Co, 56-58; res assoc inorg & nuclear chem, Mass Inst Technol, 58-59; from asst prof to assoc prof, 59-74, PROF CHEM, BOWLING GREEN STATE UNIV, 74- Concurrent Pos: Vis prof, Middle East Tech, Ankara, 65-66. Mem: Am Chem Soc. Res: Hydrolysis of metal ions; ion exchange; thermodynamics of electrolytic solutions. Mailing Add: 36 Ranch Ct Bowling Green OH 43402

MEYER, NORMAN JOSEPH, b Wilkes-Barre, Pa, Aug 5, 30. ACOUSTICS. Educ: Pa State Univ, BS, 51, MS, 53; Univ Calif, Los Angeles, PhD(physics), 59. Prof Exp: Sr engr, HRB-Singer, 51-52; res engr, Lockheed Aircraft Corp, 55-56; scientist space systs, Aeronutronic Div, Philco-Ford, 59-61; staff physicist, Marshall Labs, 61-62; sr scientist, West Div, Ling-Temco-Vought Res Ctr, 62-67, dir, 68-70; pres & prin scientist, OAS-Western, 70-72; DIR, WYLE RES-WYLE LABS, 73- Concurrent Pos: Mem comt hearing & bioacoust, Nat Acad Sci-Nat Res Coun, 68- Mem: Acoust Soc Am. Res: Experimental acoustics, primarily gases, transducers and instrumentation; noise control technology. Mailing Add: Wyle Labs 128 Maryland St El Segundo CA 90245

MEYER, PAUL RICHARD, b New York, NY, Feb 2, 30; m 55; c 4. PURE MATHEMATICS, TOPOLOGY. Educ: Dartmouth Col, AB, 51, MS, 52; Columbia Univ, MA, 60, PhD(math), 64. Prof Exp: Engr, Eastman Kodak Co, NY, 54-55; from lectr to instr math, Columbia Univ, 56-61; asst prof, St John's Univ, 62-64 & Hunter Col, 64-67; assoc prof, 68-71, PROF MATH, LEHMAN COL, 72-, CHMN DEPT, 75- Concurrent Pos: NSF res grant, 66-68; vis assoc prof, Univ Tex, 68-69; sr vis fel, Westfield Col, Univ London, 71-72. Mem: Am Math Soc; Math Asn Am; London Math Soc; fel NY Acad Sci, 74. Res: Spaces of real-valued functions; general topology. Mailing Add: Dept of Math Herbert H Lehman Col Bronx NY 10468

MEYER, PETER, b Berlin, Ger, Jan 6, 20; nat US; m 46; c 2. COSMIC RAY PHYSICS. Educ: Tech Univ, Berlin, dipl, 42; Univ Göttingen, PhD(physics), 48. Prof Exp: Mem staff physics, Univ Göttingen, 46-49; fel, Cambridge Univ, 49-50; mem res staff, Max Planck Inst Physics, Ger, 50-52; res assoc, 53-56, from asst prof to assoc prof, 56-65, PROF PHYSICS, ENRICO FERMI INST, UNIV CHICAGO, 65- Concurrent Pos: Consult, NASA; mem cosmic ray comn, Int Union Pure & Appl Physics, 66-72 & space sci bd, Nat Acad Sci, 75-; chmn cosmic physics div, Am Phys Soc, 72-73; foreign mem, Max Planck Inst Physics & astrophys, 73- Mem: Fel Am Phys Soc; Am Geophys Union; Am Astron Soc. Res: Origin of cosmic radiation; astrophysics. Mailing Add: Enrico Fermi Inst Univ of Chicago 933 E 56th St Chicago IL 60637

MEYER, RALPH A, JR, b Washington, DC, July 3, 43; m 69. PHYSIOLOGY, ENDOCRINOLOGY. Educ: Univ Md, College Park, BS, 65, PhD(zool), 69. Prof Exp: Nat Inst Dent Res fel biol, Rice Univ, 69-71; asst prof physiol & pharmacol, Col Dent, NY Univ, 71-73; ASST PROF PHYSIOL & HEAD DIV, SCH DENT, MARQUETTE UNIV, 73- Mem: AAAS; Am Physiol Soc; Am Soc Zool; Endocrine Soc. Res: Actions of hormones in promoting mineral homeostasis and in regulating bone physiology. Mailing Add: Dept of Basic Sci Marquette Univ Sch of Dent Milwaukee WI 53233

MEYER, RALPH O, b Covington, Ky, May 28, 38; m 59; c 2. SOLID STATE PHYSICS, FUEL TECHNOLOGY. Educ: Univ Ky, BS, 60; Univ NC, PhD(physics), 66. Prof Exp: Res assoc physics, Univ Ariz, 65-68; asst metallurgist, Mat Sci Div, Argonne Nat Lab, 68-73; REACTOR ENGR, US NUCLEAR REGULATORY COMN, 73- Mem: Am Nuclear Soc. Res: Diffusion in solids; reactor fuel analysis. Mailing Add: US Nuclear Regulatory Comn Washington DC 20555

MEYER, RALPH ROGER, b Milwaukee, Wis, Feb 18, 40; m; c 3. CELL BIOLOGY, BIOCHEMISTRY. Educ: Univ Wis-Milwaukee, BS, 61, Madison, MS, 63 & PhD(zool), 66. Prof Exp: Res assoc Yale Univ, 66-67; NIH fel, State Univ NY Stony Brook, 67-69; asst prof, 69-75, PROF BIOL SCI, UNIV CINCINNATI, 75- Concurrent Pos: NSF grant, Univ Cincinnati, 69-71, Am Cancer Soc grants, 69-73; NIH res grant, 75- Mem: AAAS; Am Soc Cell Biol; Am Soc Microbiol; Soc Protozool. Res: Biogenesis of cellular organelles; regulation of mitochondrial and nuclear DNA replication in normal and neoplastic tissues. Mailing Add: Dept of Biol Sci Univ of Cincinnati Cincinnati OH 45221

MEYER, RAYMOND ERWIN, agronomy, soil physics, see 12th edition

MEYER, RICH BAKKE, JR, b Houston, Tex, Nov 6, 43; m 69. MEDICINAL CHEMISTRY, ORGANIC CHEMISTRY. Educ: Rice Univ, BA, 65; Univ Calif, Santa Barbara, PhD(org chem), 68. Prof Exp: Res scientist, ICN Pharmaceut, Inc, 70-73, head dept bio-org chem, 73-75; ASST PROF PHARMACEUT CHEM, UNIV CALIF, SAN FRANCISCO, 75- Mem: Am Chem Soc. Res: Design and synthesis of cancer chemotherapeutic agents, enzyme inhibitors and analogs of cyclic adenosine monophosphate; enzyme mechanisms. Mailing Add: Dept of Pharmaceut Chem Univ of Calif San Francisco CA 94143

MEYER, RICHARD ADLIN, b Norwood, Mass, Dec 12, 33; m 56; c 3. NUCLEAR CHEMISTRY & PHYSICS. Educ: Northeastern Univ, BS, 56, MS, 58; Univ Ill, Urbana, PhD, 63. Prof Exp: Asst chemist, Bird & Son, Inc, Mass, 52-56; res asst chem, Northeastern Univ, 56-58; nuclear physics, Univ Ill, Champaign-Urbana, 59-63, NATO fel nuclear chem, Danish AEC Res Estab, Roskilde, Denmark, 65-66; res scientist nuclear struct, 66-75, GROUP LEADER & PROG LEADER, LAWRENCE LIVERMORE LAB, UNIV CALIF, 75- Concurrent Pos: Lectr, Northeastern Univ, 60. Mem: Am Chem Soc; Am Phys Soc. Res: Nuclear structure; radiation chemistry; radiation preservation of foods; electron dosimetry; high energy photonuclear reactions in complex nuclei. Mailing Add: Lawrence Livermore Lab PO Box 808 Livermore CA 94550

MEYER, RICHARD CHARLES, b Cleveland, Ohio, May 2, 30; m 63; c 2. VETERINARY MICROBIOLOGY, ANIMAL VIROLOGY. Educ: Baldwin-Wallace Col, BSc, 52; Ohio State Univ, MSc, 57, PhD(cellulose digestion), 61. Prof Exp: Asst microbiol, Ohio State Univ, 56-61, res assoc virol & germ free res, Ohio State Univ Res Found, 61-62; microbiologist, Virol Res Resources Br, Nat Cancer Inst, 62-64; asst prof swine dis & germ free res, Col Vet Med, Univ Ill, Urbana-Champaign, 65-68, assoc prof, 68-73, PROF VET PATH & HYG, COL VET MED & MICROBIOL, SCH LIFE SCI, 73- Concurrent Pos: Mem spec review comt, Extramural Activ, Cancer Ther Eval Br, Nat Cancer Inst, 66. Mem: AAAS; Am Soc Microbiol; Soc Cryobiol; fel Am Acad Microbiol. Res: Procine and bovine viruses; tissue culture; enteric and respiratory tract infections; diseases of baby pigs; development and application of germ free techniques to studies on host-parasite relationships. Mailing Add: Dept Vet Path & Hyg Col Vet Med Univ of Ill at Urbana-Champaign Urbana IL 61801

MEYER, RICHARD DAVID, b Allentown, Pa, Apr 26, 43. INFECTIOUS DISEASES. Educ: Univ Pittsburgh, BS, 63, MD, 67. Prof Exp: Intern med, Bellevue Hosp, NY Univ, 67-68, jr asst resident, 68-69; sr asst resident med, Albert Einstein Col Med-Bronx Munic Hosp, 69-70; clin res trainee infectious dis, Mem Hosp, Cornell Univ Med Col, 70-72; Lt Comdr microbiol, US Naval Med Res Inst, 72-74; ASST CHIEF INFECTIOUS DIS SECT, WADSWORTH VET ADMIN HOSP, LOS ANGELES, 74- Concurrent Pos: Asst prof med, Sch Med, Univ Calif, Los Angeles, 74- Mem: Am Soc Microbiol; Am Fedn Clin Res; Am Col Physicians. Res: Clinical evaluation of antibiotics; immunology of anaerobic infections; infections in leukemia. Mailing Add: Infectious Dis Sect 691/111F Wadsworth Hosp Ctr Los Angeles CA 90073

MEYER, RICHARD ERNST, b 19. MATHEMATICS, GEOPHYSICS. Educ: Swiss Fed Inst Technol, Dipl Mech Eng, 42, Dr Sc Techn, 46; Brown Univ, MA, 62. Prof Exp: Jr sci officer math, Brit Ministry Aircraft Prod, 45-46; asst lectr, Univ Manchester, 46-47, Imp Chem Industs res fel, 47-52; sr lectr aeronaut, Univ Sydney, 53-56, reader, 56-57; assoc prof appl math, Brown Univ, 57-59, prof, 59-64; PROF MATH, UNIV WIS-MADISON, 64- Concurrent Pos: Vis mem, Courant Inst Math Sci, NY Univ, 63-64; consult, Aeronaut Res Labs, Australian Dept Supply, 55-57 & Rand Corp, 61-62; mem, NSF Postdoctoral Panel, Nat Res Coun, 68-69, chmn panel for math, 70; sr fel, Fluid Mech Res Inst, Univ Essex, 71-72. Mem: Fel Australian Acad Sci; Royal Aeronaut Soc; Am Math Soc; Am Geophys Union; Soc Indust & Appl Math. Res: Asymptotic analysis; partial differential equations; plasma physics; water waves; meteorology; gas dynamics. Mailing Add: Dept of Math Univ of Wis Madison WI 53706

MEYER, RICHARD FASTABEND, b Covington, Ky, Sept 13, 21. ECONOMIC GEOLOGY. Educ: Dartmouth Col, AB, 47; Harvard Univ, MA, 50; Univ Kans, PhD(geol), 68. Prof Exp: Sr geologist, Humble Oil & Ref Co, 51-61; geologist, US Geol Surv, 64-66; petrol specialist, Off Oil & Gas, US Dept Interior, 66-72; GEOLOGIST, US GEOL SURV, 72- Mem: AAAS; Am Asn Petrol Geol; Geol Soc Am; Soc Petrol Eng. Res: Petroleum origin and occurrence; stratigraphy. Mailing Add: US Geol Surv Nat Ctr Reston VA 22092

MEYER, RICHARD IRWIN, b Chicago, Ill, Jan 30, 23; m 49; c 3. FOOD SCIENCE, FOOD TECHNOLOGY. Educ: Univ Ill, BS, 48, MS, 49. Prof Exp: Food technologist, Armed Forces Qm Food & Container Inst, 49-56, head dairy, oil & fat prods function, Applns Eng Br, 56-58, chief, Dairy, Oil & Fat Prods Br, 58-61; SR SPECIALIST DAIRY, OIL & FAT FOODS, DIV FOOD TECHNOL, FOOD & DRUG ADMIN, 61- Mem: Am Oil Chem Soc; Am Dairy Sci Asn; Inst Food Technol. Res: Development of processes for production of specialized food products, cheeses, butterlike foods, eggs; standards of identity for food products within specialty. Mailing Add: 1707 Wilmart St Rockville MD 20852

MEYER, RICHARD LEE, b Independence, Mo, June 5, 31; m 60; c 1. PHYCOLOGY. Educ: Mo Valley Col, BS, 54; Univ Minn, PhD(bot), 65. Prof Exp: Instr bot & biol, Univ Minn, 58-59; staff scientist, NSF-Int Indian Ocean Exped, 61-62; asst prof phycol & biol, Chico State Col, 65-68; ASSOC PROF PHYCOL & BIOL, UNIV ARK, FAYETTEVILLE, 68- Concurrent Pos: Water Resources Res Off res grant, 68-71; consult, Phillips Petrol Co, 68-, Reserve Mining Co, Silver Bay, Minn, 72-, Minn Pollution Control Agency, 74- & Limnetics, Inc, 75. Mem: AAAS; Bot Soc Am; Int Phycol Soc; Phycol Soc Am; Am Inst Biol Sci. Res: Morphology, cytology, life-history and systematics of the algal class Chrysophyceae and other flagellated algae; algal ecology. Mailing Add: Dept of Bot & Bact Univ of Ark Fayetteville AR 72701

MEYER, RICHARD LEE, b Red Wing, Minn, Aug 23, 37; m 61; c 2. AGRICULTURAL ECONOMICS. Educ: Univ Minn, BS, 59; Cornell Univ, MS, 67, PhD(agr econ), 70. Prof Exp: Peace Corps vol, Chile, 62-64, vol liaison officer, Peace

Corps, Washington, DC, 64-65; chief party agr econ res, 70-72, res adv, 72-73, asst prof, 73-74, ASSOC PROF AGR ECON, OHIO STATE UNIV, 74- Mem: Am Agr Econ Asn; Am Econ Asn. Res: Agricultural development in developing countries; allocation and productivity of agricultural credit; part-time farming and off-farm income. Mailing Add: Ohio State Univ 2120 Fyffe Rd Columbus OH 43210

MEYER, RICHARD THOMAS, b St Peter, Minn, June 26, 34; c 3. PHYSICAL CHEMISTRY. Educ: Univ Wis, BS, 56; Univ Calif, PhD(chem), 61. Prof Exp: Asst phys chem, Univ Wis, 55-56; chem, Univ Calif, 57, Petrol Res Fund asst, 59; mem tech staff, Res Orgn 5000, Sandia Labs, 49-75; SCI & RES ADV, WESTERN GOV REGIONAL ENERGY POLICY OFF, 75- Mem: AAAS; Am Chem Soc; Am Soc Mass Spectros. Res: Gas phase kinetics; flash photolysis, time resolved mass spectrometry; high temperature chemistry; laser-induced vaporization and reactions of carbon; research administration and policy-making for Western States. Mailing Add: W Gov Regional Energy Policy Off 4730 Oakland St Denver CO 80239

MEYER, ROBERT BRUCE, b St Louis, Mo, Oct 13, 43; m 66. SOLID STATE PHYSICS, MATERIALS SCIENCE. Educ: Harvard Univ, BS, 65, PhD(appl physics), 70. Prof Exp: Res fel & lectr, 70-71, asst prof, 71-74, ASSOC PROF APPL PHYSICS, HARVARD UNIV, 74- Concurrent Pos: Sloan Found fel, Harvard Univ, 71-73. Mem: AAAS; Am Phys Soc. Res: Liquid crystals; amorphous materials; physics of molecular systems. Mailing Add: Pierce Hall Harvard Univ Cambridge MA 02138

MEYER, ROBERT EARL, b Chicago, Ill, May 18, 32; m 62. PLANT PHYSIOLOGY. Educ: Purdue Univ, BS, 54, MS, 56; Univ Wis, PhD(agron), 61. Prof Exp: PLANT PHYSIOLOGIST, AGR RES SERV, USDA, 61- Mem: Weed Sci Soc Am; Am Soc Plant Physiol. Res: Development of more efficient methods of brush control on rangeland of the Southwest using good conservation and management practices; evaluation of chemicals; developmental anatomy of woody plants. Mailing Add: USDA Dept of Range Sci Tex A&M Univ College Station TX 77843

MEYER, ROBERT F, b Switz, Mar 7, 25; nat US; m 52; c 4. PHARMACEUTICAL CHEMISTRY. Educ: Swiss Fed Inst Technol, PhD(chem), 50. Prof Exp: With pharmaceut chem, Univ Kans, 50-51; res chemist, 52-58, SR RES CHEMIST, PARKE, DAVIS & CO, 58- Res: Chemistry of chloromycetin and analogs; hydroxyl aminderivatives; heterocycles; cardiovascular drugs; diuretics; antianginals. Mailing Add: Res Dept Parke Davis & Co Ann Arbor MI 48106

MEYER, ROBERT PAUL, geology, geophysics, see 12th edition

MEYER, ROGER J, b Olympia, Wash, May 14, 28; m 59; c 6. PEDIATRICS. Educ: Univ Wash, BS, 51; Wash Univ, MD, 55; Harvard Univ, MPH, 59. Prof Exp: Instr pediat, Med Sch, Harvard Univ, 59-62; asst prof, Col Med, Univ Vt, 62-65; assoc prof, Sch Med, Univ Va, 65-68; assoc prof, Sch Med, Northwestern Univ, Evanston, 68-74; ASSOC PROF PUB HEALTH, SCH PUB HEALTH, UNIV ILL, 74-, ASST DEAN CONTINUING EDUC, 74- Concurrent Pos: Dir, Infant Welfare Soc, Chicago, 68-70; regional med coordr, Social & Rehab Serv, Dept Health, Educ & Welfare, Chicago, 70-74; mem exec bd, Nat Comt Prev Child Abuse; mem child safety comt, Nat Safety Coun; lectr, Northwestern Univ & Chicago Med Sch; health adv, Mayor's Comt Senior Citizens, Chicago. Mem: Am Cong Rehab Med; Am Acad Pediat; fel Am Pub Health Asn. Res: Epidemiology and control of childhood injury; diagnosis and management of child and family disorders. Mailing Add: 712 Judson Ave Evanston IL 60202

MEYER, ROLAND KENNETH, b Fond du Lac, Wis, May 4, 04; m 47; c 2. ZOOLOGY. Educ: Milton Col, AB, 26; Univ Wis, MA, 28, PhD(zool), 30. Prof Exp: Asst zool, Milton Col, 25-26 & Univ Wis, 26-30; Nat Res Coun fel biol, Sch Med & Dent, Univ Rochester, 30-32, spec fel anat, 32-33; biol res staff mem, Upjohn Co, 33-35; from asst prof to prof, 35-57, Marshall prof, 57-74, chmn dept zool, 62-64, endocrinol reproduction physiol prog, 63-70, EMER PROF ZOOL, UNIV WIS-MADISON, 74- Concurrent Pos: Consult, Endocrinol Sect, NIH, 58-61, health endocrinol pharmacol panel, 62-66, mem res career award comt, 66-, pop res comt, 67-; consult, Ford Found, India, 64. Honors & Awards: Fred Conrad Koch Award, Endocrine Soc, 68. Mem: AAAS; Am Soc Zool; Soc Exp Biol & Med; Endocrine Soc; Am Physiol Soc. Res: Endocrinology of reproduction, especially hypothalmic control of early mammalian development; endocrine regulation of ovarian function, particularly ovulation; biological action of steroids. Mailing Add: Dept of Zool Zool Res Bldg Univ of Wis 1117 W Johnson St Madison WI 53706

MEYER, RONALD HARMON, b Walsh, Ill, Dec 30, 29; m 51; c 4. ECONOMIC ENTOMOLOGY. Educ: Univ Ill, BS, 51, MS, 56, PhD(econ entom), 63. Prof Exp: Asst entomologist, 56-65, ASSOC ENTOMOLOGIST, ILL STATE NATURAL HIST SURV, 65- Mem: Entom Soc Am. Res: Integrated control of insects and mites on fruit crops. Mailing Add: Ill Natural Hist Surv 163 Natural Resources Bldg Urbana IL 61801

MEYER, RONALD LEO, b Turlock, Calif, Aug 30, 44. NEUROEMBRYOLOGY. Educ: Don Bosco Col, BA, 66; Calif Inst Technol, PhD(biol), 74. Prof Exp: RES FEL BIOL, CALIF INST TECHNOL, 74- Mem: AAAS. Res: The growth and development of patterned axonal connectivity in the nervous system, in particular in the retinotectal system of lower vertebrates. Mailing Add: Div of Biol Calif Inst of Technol Pasadena CA 91125

MEYER, RONALD WARREN, b Battle Creek, Iowa, July 10, 29; m 52. PLANT PATHOLOGY. Educ: Univ Mo, BS, 56, MA, 57; Univ Calif, Berkeley, PhD(plant path), 65. Prof Exp: Asst prof plant path, Everglades Exp Sta, Fla, 65-66; bot, 66-70, ASSOC PROF BOT, CALIF STATE UNIV, FRESNO, 70- Mem: Mycol Soc Am; Am Phytopath Soc; AAAS; Am Inst Biol Sci; Bot Soc Am. Res: Heterokaryosis; soil and root inhabiting fungi; lichenology. Mailing Add: Dept of Biol Calif State Univ Fresno CA 93710

MEYER, RUBEN, b Jan 16, 14; c 2. PEDIATRICS, PUBLIC HEALTH. Educ: Wayne State Univ, BA, 35, MB, 38, MD, 39; Univ Mich, Ann Arbor, MPH, 71. Prof Exp: From instr to clin prof pediat, Wayne State Univ, 42-73, dir community med & actg chmn dept community & family med, 70, prof community & family med & chmn dept, Sch Med, 71-73; PROF & DIR MATERNAL & CHILD HEALTH, SCH PUB HEALTH, UNIV MICH, ANN ARBOR, 73- Concurrent Pos: Asst pediatrician, Children's Hosp Mich, 42-46, assoc pediatrician, 46-52, attend pediatrician, 52-73; sr res assoc, Child Res Ctr Mich, 61-69; pediat consult, Children's Hosp Mich, Sinai Hosp Detroit, Mt Carmel Mercy Hosp & Wyandotte Gen Hosp; nephrology consult, Plymouth State Home. Mem: Am Acad Pediat; Am Pub Health Asn. Res: Kidney diseases in children with specific emphasis on congenital malformations; health care research. Mailing Add: Sch of Pub Health Univ of Mich Ann Arbor MI 48104

MEYER, SAMUEL LEWIS, b Steinmetz, Mo, Nov 9, 06; m 35; c 2. Educ: Cent Col, Mo, AB, 30, LLD, 53; Vanderbilt Univ, MS, 32; Cornell Univ, MS, 32; Univ Va, PhD(bot), 40; Ohio Wesleyan Univ, LLD, 66. Hon Degrees: PhD(sci), Dan Kook Univ, Seoul, Korea, 74. Prof Exp: Instr biol, Cent Col, Mo, 30-31; instr, Vanderbilt Univ, 32-36 & Univ Va, 37-40; from asst prof to prof bot, Univ Tenn, 40-51, head dept, 46-51; assoc prof, Emory Univ, 45-46; prof & head dept bot, Fla State Univ, 51-55; dean, Cent Col, Mo, 55-58; acad vpres, Univ of the Pac, 58-65; PROF BIOL & PRES, OHIO NORTHERN UNIV, 65- Concurrent Pos: Exec secy div biol & agr, Nat Res Coun, 52-53; chmn Ohio Found Independent Cols, 72-74; mem univ senate, United Methodist Church, 72-76. Honors & Awards: Award of Hon, Wisdom Soc for Advan Knowledge, Learning and Res in Educ, 70; Citation, Sigma Phi Epsilon, 75. Mem: Newcomen Soc NAm. Res: Educational administration; biological effects of heavy water; physiology of mosses; taxonomy and genetics of ascomycetes. Mailing Add: Off of the Pres Ohio Northern Univ Ada OH 45810

MEYER, STEPHEN FREDERICK, b Berkeley, Calif, Aug 9, 47. EXPERIMENTAL SOLID STATE PHYSICS, NUCLEAR MAGNETIC RESONANCE. Educ: Whitman Col, BA, 69; Stanford Univ, MS, 70, PhD(appl physics), 73. Prof Exp: RES ASSOC PHYSICS, UNIV ILL, URBANA-CHAMPAIGN, 73- Mem: Am Phys Soc. Res: Properties of the layered transition metal dichalcogenides. Mailing Add: Dept of Physics Univ of Ill Urbana IL 61801

MEYER, STUART LLOYD, b New York, NY, May 28, 37; m 59; c 3. PHYSICS, TELECOMMUNICATIONS. Educ: Columbia Univ, AB, 57; Princeton Univ, PhD(physics), 62. Prof Exp: Res physicist, Nevis Cyclotron Labs, Columbia Univ, 61-63; asst prof physics, Rutgers Univ, 63-67, Rutgers fac fel, Rutherford High Energy Lab, Eng, 66-67; assoc chmn dept, 68-70, ASSOC PROF PHYSICS, NORTHWESTERN UNIV, 67- Concurrent Pos: Consult, Nat Accelerator Lab, 70-71 & NSF, 74-; chmn gen fac comt, Northwestern Univ, 71-72, assoc prof decision sci, Grad Sch, 75-; vis staff mem, Los Alamos Sci Lab, 72-; prog dir intermediate energy physics, NSF, 74-75; fel, Ctr for Teaching Professions, 74-75. Mem: AAAS; Am Phys Soc; NY Acad Sci. Res: High energy interactions; muon physics; weak interactions; counter and spark-chamber techniques; data analysis; probability and statistics; radiation shielding; neutrino physics; telecommunications. Mailing Add: Dept of Physics Northwestern Univ Evanston IL 60201

MEYER, THOMAS J, b Dennison, Ohio, Dec 3, 41; m 63; c 2. INORGANIC CHEMISTRY. Educ: Ohio Univ, BS, 63; Stanford Univ, PhD(chem), 66. Prof Exp: NATO fel, Univ Col, Univ London, 66-67; asst prof, 68-75, PROF CHEM, UNIV NC, CHAPEL HILL, 75- Concurrent Pos: Alfred P Sloan Found fel, 75. Mem: Am Chem Soc; Am Soc Univ Prof. Res: Kinetics and mechanisms of inorganic and organo-metallic reactions; photochemistry; electrochemistry; catalysis. Mailing Add: Dept of Chem Univ of NC Chapel Hill NC 27514

MEYER, THOMAS S, analytical biochemistry, deceased

MEYER, VICTOR BERNARD, b New York, NY, Dec 17, 20; m 47; c 2. ORGANIC POLYMER CHEMISTRY. Educ: City Col New York, BS, 42; Columbia Univ, AM, 49, PhD(org chem), 53. Prof Exp: Chemist, Montrose Chem Co, 42 & 46-47; asst, Columbia Univ, 48-50; group leader textile chem, United Merchants Labs, Inc, 53-57; res chemist, W R Grace & Co, 57; sr chemist, Air Reduction Co, Inc, 58-60; chemist & dir, Viburnum Assocs, 60-74, pres & tech dir, Viburnum Resins, Inc, 62-74; RES ASSOC, J P STEVENS & CO, INC, 74- Concurrent Pos: Consult chemist, 62-74. Mem: Am Chem Soc; Fedn Soc Paint Technol; Am Asn Textile Chemists & Colorists. Res: Synthetic latex and resin development; emulsion polymers in paints, textiles, nonwoven fabrics and paper; aqueous coatings and adhesives; paint testing and evaluation. Mailing Add: 83 Briarwood Dr E Berkeley Heights NJ 07922

MEYER, VINCENT D, b McKees Rocks, Pa, Nov 7, 32; m 62; c 3. PHYSICAL CHEMISTRY, CHEMICAL PHYSICS. Educ: Duquesne Univ, BS, 54; Ohio State Univ, PhD(phys chem), 62. Prof Exp: Fel electron-impact spectros, Mellon Inst, 62-65; mem tech staff, Gen Tel & Electronics Labs Inc, NY, 65-72 & GTE Sylvania, Inc, Pa, 72-74, ENG SPECIALIST, GTE SYLVANIA LIGHTING CTR, 74- Mem: Am Chem Soc; Am Phys Soc. Res: Electron scattering; molecular structure; energy transfer; high vacuum technique; quantum chemistry; cathodoluminescence; electro-optic phenomena; spectroscopy; mass spectrometry. Mailing Add: E I du Pont de Nemours & Co Inc 100 Endicott St Danvers MA 01923

MEYER, WALTER CARL, b Calgary, Alta, Oct 3, 31; m 57; c 5. RESTORATIVE DENTISTRY. Educ: Univ Alta, DDS, 61, BMus, 69. Prof Exp: ASSOC PROF OPER DENT, UNIV ALTA, 57- Mem: Am Acad Gold Foil Opers; Can Acad Restorative Dent. Res: Operative dentistry. Mailing Add: Dept of Dent Univ of Alta Edmonton AB Can

MEYER, WALTER DAVIDSON, b Deshler, Ohio, Apr 9, 40; m 63; c 3. DYNAMIC METEOROLOGY. Educ: Capital Univ, BS, 62; Univ Utah, BS, 63; Univ Wash, PhD(atmospheric sci), 69. Prof Exp: Weather officer, 63-70, ADVAN WEATHER OFFICER, AIR WEATHER SERV, US AIR FORCE, 70- Concurrent Pos: Adj prof, St Louis Univ, 75; spec asst, Defense Meteorol Satellite Prog, Air Force/Army. Honors & Awards: Commendation Medal, US Air Force, 73. Mem: Am Meteorol Soc. Res: The application of meteorological satellite data of all kinds to operational weather support problems; design of future meteorological satellite systems. Mailing Add: Hq Air Weather Serv OL-F Hq SAMSO/YDA PO Box 92960 Worldway Postal Ctr Los Angeles CA 90009

MEYER, WALTER EDWARD, b Hackensack, NJ, Sept 15, 29; m 58; c 2. PHARMACEUTICAL CHEMISTRY. Educ: Rutgers Univ, BS, 51; NY Univ, MS & PhD(org chem), 64. Prof Exp: Biochemist, 51-53, RES CHEMIST, LEDERLE LABS DIV, AM CYANAMID CO, 53- Mem: Am Chem Soc. Res: Structure determination on compounds having pharmaceutical interest; synthesis of compounds related to physiologically active materials. Mailing Add: Dept 913 Bldg 6513 R312 Lederle Labs Middletown Rd Pearl River NY 10965

MEYER, WALTER H, b Cincinnati, Ohio, Aug 19, 22; m 44; c 3. NUTRITION. Educ: Mich State Univ, BS, 48. Prof Exp: Proj engr, Procter & Gamble Co, 48-51, supvr qual control, Procter & Gamble Defense Corp, 51-54, sect head toilet goods prod, Procter & Gamble Co, 54-57 & food prod, 57-66, mgr prof & regulatory rels, 66-68, ASSOC DIR FOOD PROD, PROD DEVELOP DEPT, PROCTER & GAMBLE CO, 68- Concurrent Pos: Indust Liaison Panel Food & Nutrit Bd, Nat Acad Sci; chmn, Tech Comt Indust Shortening & Edible Oils, Inc, Tech Comt Peanut Butter Mfrs & Nut Salters Asn & Sci Adv Group Nat Coffee Asn. Res: Food safety and regulations. Mailing Add: Procter & Gamble Co 6071 Center Hill Rd Cincinnati OH 45224

MEYER, WALTER JOSEPH, b New York, NY, Jan 12, 43; m 67; c 1. MATHEMATICS. Educ: Queen's Col, BA, 64; Univ Wis, MS, 66, PhD(math), 70. Prof Exp: Asst prof, 69-74, ASSOC PROF MATH, ADELPHI UNIV, 74- Res: Graph theory; convex sets. Mailing Add: Dept of Math Adelphi Univ Garden City NY 11530

MEYER, WALTER LESLIE, b Toledo, Ohio, Feb 28, 31; m 54; c 3. ORGANIC CHEMISTRY. Educ: Univ Mich, BS, 53, MS, 55, PhD(chem), 57. Prof Exp: Instr &

res assoc chem, Univ Mich, 57; NSF fel, Univ Wis, 57-58; from instr to assoc prof chem, Univ Ind, 58-65; assoc prof, 65-68, chmn dept, 67-73, PROF CHEM, UNIV ARK, FAYETTEVILLE, 68- Mem: Am Chem Soc; The Chem Soc; AAAS. Res: Chemistry of natural products; stereochemistry; nuclear magnetic resonance; organic synthesis. Mailing Add: Dept of Chem Univ of Ark Fayetteville AR 72701

MEYER, WERNER FRANZ, b New NY, Dec 10, 38. MATHEMATICS. Educ: Manhattan Col, BS, 60; NY Univ, MS, 63. Prof Exp: Analyst, Int Bus Mach Corp, 63-66; RES MATHEMATICIAN, METRIC RES LABS, UNIVAC DEFENSE SYSTS DIV, RAND CORP, 66- Concurrent Pos: Vis mathematician, Bell Tel Labs, 67-68; consult, Metric Res Labs, 66- Mem: Asn Comput Mach. Res: Exterior ballistics and orbital mechanics; nonlinear programing and continuous games; numerical analysis; scientific and engineering applications of digital computers; computer sciences. Mailing Add: 24 Fairview Rd Scarsdale NY 10583

MEYER, WILLIAM ELLIS, b Bonne Terre, Mo, July 22, 36; m 62; c 2. ANIMAL SCIENCE. Educ: Univ Mo, BS, 60, MS, 62, PhD(agr), 65. Prof Exp: ASSOC PROF AGR, SOUTHEAST MO STATE UNIV, 65-, CHMN DEPT, 70- Mem: Am Soc Animal Sci. Res: Meat technology. Mailing Add: Dept of Agr Southeast Mo State Univ Cape Girardeau MO 63701

MEYER, WILLIAM HERMAN LEWIS, b Bottineau, NDak, July 27, 15; m 51; c 2. MATHEMATICS. Educ: Westminster Col, AB, 36; Univ Chicago, SM, 37, PhD(math), 47. Prof Exp: Instr math, Bowling Green State Univ, 37-38, Ga Sch Technol, 38-39, Col Wooster, 41-42 & Univ Chicago, 42-44; res mathematician, Ballistic Res Lab, Aberdeen Proving Ground, 44-46; PROF MATH, UNIV CHICAGO, 46-, ASSOC CHMN DEPT, 68- Concurrent Pos: Vis assoc prof, Univ Calif, 53-54; NSF Fel, 65-66. Mem: Am Math Soc; Math Asn Am. Res: Analysis; foundations of mathematics. Mailing Add: Dept of Math Univ of Chicago Chicago IL 60637

MEYER, WILLIAM KEITH, physical chemistry, see 12th edition

MEYER, WILLIAM LAROS, b Keyser, WVa, May 27, 36; m 67; c 5. BIOLOGICAL CHEMISTRY. Educ: Yale Univ, BS, 56; Univ Wash, PhD(biochem), 62. Prof Exp: From instr to asst prof, 62-70, ASSOC PROF BIOCHEM, UNIV VT, 70- Concurrent Pos: Res group, World Fedn Neurol, 72-; vis assoc prof & NIH spec res fel, Univ Western Australia, 73. Mem: AAAS; Am Chem Soc; Am Soc Biol Chemists. Res: Physiological control of enzyme activity and turnover in muscle, cartilage and other tissues; proteases and ribonucleases and their relationship to neuromuscular diseases; enzyme mechanisms; fructose metabolism; calcium metabolism; insect biochemistry. Mailing Add: Dept of Biochem Univ of Vt Col of Med Burlington VT 05401

MEYER, WILLIAM PAUL, b Okauchee, Wis, May 9, 42; m 73. ORGANIC CHEMISTRY. Educ: Univ Calif, Los Angeles, BS, 65; Univ Ill, Urbana, MS, 73, PhD(org chem), 75. Prof Exp: RES CHEMIST, EXP STA, E I DU PONT DE NEMOURS & CO, 75- Mem: Am Chem Soc. Res: Textile fibers; polymers. Mailing Add: Carothers Lab Exp Sta Textile Fibers E I du Pont de Nemours & Co Wilmington DE 19898

MEYER, WILLIS GEORGE, b Bellwood, Nebr, Jan 21, 06; m 37; c 2. GEOLOGY. Educ: Univ Nebr, AB, 30; Univ Cincinnati, MA, 32, PhD(geol), 41. Prof Exp: Asst geol, Univ Cincinnati, 31-34; geologist, Amerada Petrol Corp, Tex, 34-38; De Golyer & MacNaughton, 38-42, partner, 42-47; Meyer & Achtschin, 47-58; Willis G Meyer & Assocs, 58-61; CONSULT & INDEPENDENT OIL OPERATOR, 61- Mem: AAAS; fel Geol Soc Am; Am Asn Petrol Geol (vpres, 69-70); Am Geophys Union; Soc Independent Prof Earth Sci (pres, 66-67). Res: Gulf coast stratigraphy and historical geology; structural geology of oil fields; origin of porosity in limestone oil and gas reservoirs; estimation of recoveries from oil and gas reservoirs. Mailing Add: 1400 Republic Nat Bank Bldg Dallas TX 75201

MEYERAND, RUSSELL GILBERT, JR, b St Louis, Mo, Dec 2, 33; m 56; c 1. PLASMA PHYSICS. Educ: Mass Inst Technol, SB, 55, SM, 56, ScD(plasma physics), 59. Prof Exp: Mem staff, res lab electronics, Mass Inst Technol, 56-57; prin scientist, United Aircraft Res Labs, 58-64, chief res scientist, 64-67, DIR RES, UNITED TECHNOLOGIES RES CTR, 67- Concurrent Pos: Consult, Atomic Power Equip Dept, Gen Elec Co, 55-56 & Army Sci Adv Panel, 70-74; mem eng adv comt, Hartford Grad Ctr, Rensselaer Polytech Inst, 65-, educ coun, Mass Inst Technol, 69-; adv comt on corp assocs, Am Inst Physics, 71-73; NASA Space Prog Adv Coun, 74-; chmn eval panel, Quantum Elec Div, Inst Basic Standards, Nat Bur Standards, 70-72; mem ad hoc laser adv panel, NASA Res & Tech Coun, 71-; panel on productivity enhancement, Off Sci & Technol, Exec Off of the President, 71- Mem: Am Phys Soc; assoc fel Am Inst Aeronaut & Astronaut; sr mem Inst Elec & Electronics Engrs; Sigma Xi. Res: Plasma and laser physics; electronics. Mailing Add: United Technol Res Ctr 400 Main St East Hartford CT 06108

MEYER-ARENDT, JURGEN RICHARD, b Berlin, Ger, Oct 4, 21; US citizen; m 49; c 3. OPTICS. Educ: Univ Würzburg, MD, 45; Univ Hamburg, PhD(biophys), 52. Prof Exp: Prof histol, Sao Paulo, 54-55; prof path, Ohio State Univ, 55-59; prof physics, Utah State Univ, 60-63 & Univ Colo, 63-66; PROF PHYSICS & OPTICS, PAC UNIV, 66- Concurrent Pos: Consult, Corning Glass Works. Mem: Am Phys Soc; Optical Soc Am. Res: Physical optics. Mailing Add: Dept of Physics Pac Univ Forest Grove OR 97116

MEYERHOF, WALTER ERNST, bKiel, Ger, Apr 29, 22; nat US; m 47; c 2. EXPERIMENTAL PHYSICS, NUCLEAR PHYSICS. Educ: Univ Pa, MA, 44, PhD(physics), 46. Prof Exp: Asst instr physics, Univ Pa, 43, res physicist, 44-46; asst prof physics, Univ Ill, 46-49; from asst prof to assoc prof, 49-59, PROF PHYSICS, STANFORD UNIV, 59-, CHMN DEPT, 70- Concurrent Pos: Sloan Found sr res fel, 55-59. Mem: Fel Am Phys Soc. Mailing Add: Dept of Physics Stanford Univ Stanford CA 94305

MEYERHOFER, DIETRICH, b Zurich, Switz, Sept 19, 31; nat US; m 54; c 2. SOLID STATE PHYSICS. Educ: Cornell Univ, BEng Phys, 54; Mass Inst Technol, PhD(physics), 58. Prof Exp: Asst solid state physics, Mass Inst Technol, 54-56; MEM TECH STAFF, RES LABS, RCA CORP, 58- Mem: Inst Elec & Electronics Eng; Am Phys Soc; Optical Soc Am. Res: Galvanomagnetic and optical measurements in semiconductors and insulators; molecular lasers; holography; electrical and optical properties of liquid crystals. Mailing Add: David Sarnoff Res Ctr RCA Corp Princeton NJ 08540

MEYERHOFF, ARTHUR AUGUSTUS, b Northampton, Mass, Sept 9, 28; m 51; c 3. GEOLOGY. Educ: Yale Univ, BA, 47; Stanford Univ, MS, 50, PhD(geol), 52. Prof Exp: Geologist, US Geol Surv, 48-52; geologist, Calif Explor Co, Standard Oil Co Calif, 52-56, sr geologist, Cuba Calif Oil Co, 56-59, geophysicist, Chevron Oil Co, 59-60, res geologist, Calif Co, 60-65; pub mgr, Am Asn Petrol Geol, 65-75; PROF GEOL, OKLA STATE UNIV, 75- Concurrent Pos: Dir, Tulsa Sci Found, 66-72; pres northeast div, Frontiers Sci Found Okla, Inc, 71-72, mem exec comt, 71-72. Honors & Awards: George C Mattson Award, Am Assoc Petrol Geologists. Mem: AAAS; Geol Soc Am; Soc Econ Paleont & Mineral; Am Asn Petrol Geologists; Asn Earth Sci Educ (pres, 69-70). Res: Structural geology; geotectonics; stratigraphy; carbonate rock; paleobotany; plate tectonics, continental drift; Caribbean geology; petroleum resources of USSR and Peoples Republic of China. Mailing Add: PO Box 4602 Tulsa OK 74104

MEYEROTT, ROLAND EDWARD, b Baldwin, Ill, Nov 20, 16; m 44; c 6. ASTROPHYSICS, ATOMIC PHYSICS. Educ: Univ Nebr, BA, 38, MA, 40; Yale Univ, PhD(physics), 43. Prof Exp: Asst physics, Univ Nebr, 38-40; asst, Yale Univ, 40-41, phys asst, Med Sch, 41, instr physics, 42-47, res assoc, 47-49; sr physicist, Argonne Nat Lab, 49-53 & Rand Corp, 53-56; mgr physics, Missiles & Space Div, Lockheed Aircraft Corp, 56-62, mgr phys sci lab, Lockheed Missiles & Space Co, 62-66, dir sci, res & develop div, 66-69, asst to vpres res & develop div, 69-71, dir sci, res & develop div, 71-73; SCI CONSULT, 73- Mem: Fel Am Phys Soc. Res: Theoretical molecular physics; spectroscopy; atomic wave function calculations; opacity and equation of state of matter at higher temperatures; effects of minor species on atmosphere. Mailing Add: 27100 Elena Rd Los Altos Hills CA 94022

MEYEROWITZ, SANFORD, b New York, NY, Feb 17, 27; m 54; c 4. MEDICINE, PSYCHIATRY. Educ: City Col New York, BS, 48; Univ Rochester, MD, 54. Prof Exp: Intern med, Yale-New Haven Med Ctr, 54-55; resident psychiat, 55-58, instr, 60-61, asst prof psychiat & med, 62-68, assoc prof psychiat, 68-74, ASSOC PROF MED, MED CTR, UNIV ROCHESTER, 68-, PROF PSYCHIAT, 75- Mem: Am Psychiat Asn; Am Rheumatism Asn; Am Psychosom Soc; Sigma Xi. Res: Psychosomatic medicine; psychologic aspects of rheumatoid arthritis and other diseases; psychoanalysis. Mailing Add: Strong Mem Hosp 260 Crittenden Blvd Rochester NY 14642

MEYERS, ALBERT IRVING, b New York, NY, Nov 22, 32; m 57; c 3. ORGANIC CHEMISTRY. Educ: NY Univ, AB, 54, PhD(chem), 57. Prof Exp: Res chemist, Cities Serv Res & Develop Co, 57-58; from asst prof to prof chem, La State Univ, 58-69, Boyd prof, 69-70; prof, Wayne State Univ, 70-72; PROF CHEM, COLO STATE UNIV, 72- Concurrent Pos: Res grants, NIH, 58-78, Res Corp, 58-59, New Orleans Cancer Soc, 59-60, Eli Lilly, 65-66, US Army, 66-69, NSF, 69-76, G D Searle, 75, Hoffmann-La Roche, 70-75 & Petrol Res Fund, 69-75; ed bd, J Heterocyclic Chem, 63-, J Org Chem, 73-78 & J Heterocycles, 74-; vis res fel, Harvard Univ, 65-66; adj Boyd prof, La State Univ, 70-73; consult, G D Searle, 72- & Midwest Res Inst, 75-; exec comt orgn div, Am Chem Soc, 75-77. Honors & Awards: Distinguished Fac Award, La State Univ, 64. Mem: Am Chem Soc; The Chem Soc; Sigma Xi; Int Soc Heterocyclic Chem (vpres). Res: Synthetic organic chemistry; chemistry of heterocyclis compounds; asymmetric syntheses, total synthesis of natural products. Mailing Add: Dept of Chem Colo State Univ Ft Collins CO 80521

MEYERS, ALLAN RICHARD, b Holyoke, Mass, Nov 25, 46; m 68; c 2. ANTHROPOLOGY. Educ: Dartmouth Col, AB, 68; Cornell Univ, MA, 70, PhD(anthrop), 74. Prof Exp: Lectr anthrop, Tufts Univ, 74-75; lectr relig studies, Brown Univ, 75; VIS ASST PROF ANTHROP, CONN COL, 75- Mem: Royal Anthrop Inst London; Am Anthrop Asn; MidE Studies Asn NAm. Res: History and culture of North Africa and the Middle East, especially politics and ethnicity in Morocco. Mailing Add: 7 Mt Vernon Pl Boston MA 02108

MEYERS, BERNARD LEE, analytical chemistry, inorganic chemistry, see 12th edition

MEYERS, CAL YALE, b Utica, NY, Nov 14, 27. ORGANIC CHEMISTRY. Educ: Cornell Univ, AB, 48; Univ Ill, PhD(org chem), 51. Prof Exp: Res fel, Princeton Univ, 51-53; res chemist, Union Carbide Plastics Co, 53-60; vis res prof, Bologna, 60-63; sr res assoc, Univ Southern Calif, 63; vis scholar, Univ Calif, Los Angeles, 63-64; assoc prof, 64-68, PROF CHEM, SOUTHERN ILL UNIV, CARBONDALE, 68- Concurrent Pos: Consult, Heliodyne Corp, 64 & Scripps Clin & Res Found, Calif, 64-; NSF res grant, 73. Honors & Awards: Res Award, Union Carbide Corp, 57; Int Travel Award, NSF, 61, 70 & 71; Res Award, Am Chem Soc-Petrol Res Fund, 62; Res Award, Intra-Sci Res Found, 64. Mem: Am Chem Soc; Italian Chem Soc. Res: Organosulfur bonding; reactions of carbon tetrahalides with alcohols, ketones and sulfones; isomerizations and eliminations of allylic ethers, sulfides, sulfoxides and sulfones; comparison of nucleophilic and one-electron-transfer reactivities of anions; aromatic-substitution reactions via one-electron transfers. Mailing Add: Dept of Chem Southern Ill Univ Carbondale IL 62901

MEYERS, DONALD BATES, b Cedar Rapids, Iowa, Jan 30, 22; m 50; c 4. PHARMACOLOGY. Educ: Univ Iowa, BS, 44, MS, 48, PhD(pharmaceut chem, pharmacol), 49. Prof Exp: Asst pharmaceut chem, Univ Iowa, 47-49; from asst prof to prof pharmacol, Butler Univ, 49-62, Baxter distinguished prof, 58; prof, Univ Tex, 62-63; sr pharmacologist, 63-68, res scientist, 68-72, RES ASSOC, TOXICOL DIV, ELI LILLY & CO, 72- Concurrent Pos: Lectr, Sch Nursing, Ind Methodist Hosp, 52-60; lectr, Butler Univ, 65- Mem: AAAS; Acad Pharmaceut Sci; Soc Toxicol. Res: Toxicology; neuropharmacology; drug metabolism. Mailing Add: Toxicol Div Eli Lilly & Co Box 708 Greenfield IN 46140

MEYERS, EARL LAWRENCE, b Victor, Iowa, Nov 1, 07; m 41; c 2. PHYSICAL CHEMISTRY. Educ: Coe Col, BS, 30; Univ Ill, PhD(chem), 34. Prof Exp: Res fel, 34; chem consult, 37-38; inspector, US Food & Drug Admin, 39-41, resident inspector, 46-51, new drug off, new drug br, 52-58, chief chemist div new drugs, 58-63, chief controls eval br, 63-66, DIR DIV ONCOL & RADIOPHARMACEUT, US FOOD & DRUG ADMIN, 67- Mem: Fel AAAS; Am Chem Soc; NY Acad Sci; fel Am Inst Chem; Soc Nuclear Med. Res: X-rays; spectroscopy; rare earths; radiopharmaceuticals; stability of drugs; quality control of drugs. Mailing Add: 5225 S Seventh Rd Arlington VA 22204

MEYERS, EDWARD, b New York, NY, Aug 17, 27; m 62; c 1. MICROBIOLOGY. Educ: City Col New York, BS, 49; Univ Ky, MS, 51; PhD(bact), Univ Wis, 58. Prof Exp: Microbiologist, Nepera Chem Co, NY, 52; med bacteriologist, Ft Detrick, Md, 52-54; RES GROUP LEADER, SQUIBB INST MED RES, 58- Mem: Am Soc Microbiol; Japanese Antibiotics Res Asn. Res: Area of microbial biochemistry dealing with fermentation, isolation and characterization of antibiotics and other microbial products; full or partial biosynthesis of new compounds; fermentation and analytical techniques. Mailing Add: Squibb Inst for Med Res New Brunswick NJ 08903

MEYERS, ELWOOD WILLIAM, b Dallastown, Pa, June 6, 08; m 33; c 3. CHEMISTRY. Educ: Lebanon Valley Col, BS, 30. Prof Exp: Chemist, 30-47, chief chemist & dir res, 47-66, DIR RES, HERSHEY FOODS RES LABS, HERSHEY CHOCOLATE CORP, 66- Mem: Fel Am Chem Soc; Inst Food Technol; Am Dairy Sci Asn; Am Asn Candy Technol. Res: New products and process development for the manufacture of chocolate and chocolate products. Mailing Add: Hershey Foods Res Labs PO Box 54 Hershey PA 17033

MEYERS, FREDERICK HENRY, b Ft Wayne, Ind, June 16, 18; m 47; c 3. PHARMACOLOGY. Educ: Univ Calif, MD, 49. Prof Exp: Intern, Univ Calif Hosp,

49-50; from instr to asst prof pharmacol, Univ Tenn, 50-53; from asst prof to assoc prof, 53-64, PROF PHARMACOL, SCH MED, UNIV CALIF, SAN FRANCISCO, 64- Res: Cardiovascular and autonomic physiology and pharmacology; problems of drug abuse. Mailing Add: Dept of Pharmacol Univ of Calif Sch of Med San Francisco CA 94143

MEYERS, GENE HOWARD, b Chicago, Ill, Dec 6, 42; m 71. PHYSICAL CHEMISTRY, COMPUTER SCIENCES. Educ: Univ Ill, Urbana, BS, 64; Univ Calif, Berkeley, PhD(phys chem), 69. Prof Exp: SCI PROGRAMMER, KAISER ALUMINUM & CHEM CORP, PLEASANTON, 69- Mem: Am Chem Soc; Asn Comput Mach. Res: Laboratory automation; interfacing of man, laboratory and computer for real-time data acquisition; experimental control and data reduction; microwave spectroscopy. Mailing Add: Kaiser Aluminum & Chem Corp 6177 Sunol Blvd Pleasanton CA 94566

MEYERS, H RUSSELL, b Brooklyn, NY, Feb 25, 05; m; c 8. NEUROPHYSIOLOGY, NEUROSURGERY. Educ: Brown Univ, AB, 27, ScM, 29; Cornell Univ, MD, 32. Prof Exp: Asst psychol, Brown Univ, 27-28; instr, NY Univ, 29-32, lectr, 36-39; instr neurol, Long Island Col Med, 38-46; from asst prof to prof, Univ Iowa, 46-63, chmn, Div Neurosurg, 49-63; CHIEF NEUROL & NEUROSURG SERV, HIGHLANDS CLIN & APPALACHIAN REGIONAL HOSPS, 63- Concurrent Pos: Fel, Lahey Clin, Boston, 35-36; asst, Med Col, Cornell Univ, 31-32; lectr, Columbia Univ, 39; instr, Long Island Col Med, 39-46; vis lectr, Univ Wyo, 71; sr consult, Vet Admin Hosp, Iowa City. Mem: Soc Neurol Surg; Am Asn Neurol Surg; Am Neurol Asn; fel Am Col Surgeons; Am Acad Cerebral Palsy (vpres, 61-62, pres, 62-63). Res: Neurophysiology of basal ganglia and sleep; language-thought relationships. Mailing Add: Neurol Suite Highlands Clin Bldg Williamson WV 25661

MEYERS, HARVEY I, b Chelsea, Mass, Sept 9, 24; m 50; c 3. RADIOLOGY. Educ: Harvard Univ, BA, 47; Tufts Univ, MD, 51; Am Bd Radiol, dipl. Prof Exp: Intern, Los Angeles County Hosp, 51-52; USPHS clin cancer trainee, Beth Israel Hosp, Boston, 52-53, resident radiol, 52-55; instr, 55-56, asst clin prof, 56-57, from asst prof to assoc prof, 57-66, PROF RADIOL, SCH MED, UNIV SOUTHERN CALIF, 66- Mem: AMA; Am Col Radiol. Res: Clinical research in diagnostic radiology. Mailing Add: Dept of Radiol Univ Southern Calif Sch Med Los Angeles CA 90033

MEYERS, HERBERT, b New York, NY, Nov 15, 31; m 67; c 4. GEOPHYSICS. Educ: City Col New York, BS, 58. Prof Exp: Geophysicist geomagnetism, Coast & Geod Surv, 58-66; DIR CHIEF GEOPHYS, NAT OCEANIC & ATMOSPHERIC ADMIN, 66- Concurrent Pos: Mem, Geothermal Resource Coun, 74- Mem: Am Geophys Union; Soc Explor Geophysicists; Sigma Xi. Res: Geophysical data management, including seismology, geomagnetism, marine geology. Mailing Add: Geophys & Solar-Terrest Data Ctr Solid Earth Data Serv EDS-NOAA Boulder CO 80302

MEYERS, JAMES HARLAN, b Fountain Springs, Pa, Sept 5, 45; m 66; c 2. GEOLOGY. Educ: Franklin & Marshall Col, AB, 67; Ind Univ, Bloomington, MA, 69, PhD(geol), 71. Prof Exp: Lectr geol, Ind Univ, Bloomington, 71; ASST PROF GEOL, MUSKINGUM COL, 71- Mem: Geol Soc Am; Am Asn Petrol Geologists; Soc Econ Paleontologists & Mineralogists; Clay Minerals Soc. Res: Sedimentary petrology, paleoenvironments of sedimentary rocks; clay mineralogy, relating to provenance and environment of deposition of sedimentary rocks; sedimentology and recent analogs of ancient depositional environments. Mailing Add: Dept of Geol Muskingum Col New Concord OH 43762

MEYERS, KENNETH PURCELL, b Jamaica, NY; m 57; c 2. CELL BIOLOGY, ENDOCRINOLOGY. Educ: NY Univ, AB, 53; Rutgers Univ, New Brunswick, MS, 65, PhD(endocrinol), 67. Prof Exp: Sr scientist, Worcester Found Exp Biol, Mass, 67-69; sr pharmacologist, Dept Pharmacol, 69-74, SR SCIENTIST, DEPT CELL BIOL, HOFFMAN-LA ROCHE INC, 74- Mem: Brit Soc Endocrinol; Soc Study Reprod; Sigma Xi. Res: Influence of hormones on development and growth of hormone dependent tumors. Mailing Add: Res Div Hoffmann-La Roche Inc Nutley NJ 07110

MEYERS, LEROY FREDERICK, b New York, NY, June 30, 27. MATHEMATICS. Educ: Queens Col, NY, BS, 48; Syracuse Univ, MA, 50, PhD(math), 53. Prof Exp: Actg asst prof math, Univ Va, 53-54; from instr to asst prof, 54-62, ASSOC PROF MATH, OHIO STATE UNIV, 62- Concurrent Pos: Partic, numerical anal training prog, NSF-Nat Bur Standards, 59. Mem: Am Math Soc; Asn Comput Mach; Math Asn Am; Asn Symbolic Logic. Res: Mechanical translation; Hilbert space. Mailing Add: Dept of Math Ohio State Univ 231 W 18th Ave Columbus OH 43210

MEYERS, M DOUGLAS, b Mt Sterling, Ill, Jan 30, 33; m 61; c 2. INORGANIC CHEMISTRY. Educ: Univ Ill, BS, 55; Mass Inst Technol, PhD, 59. Prof Exp: From res chemist to sr res chemist, Am Cyanamid Co, 59-71; group leader, Kennecott Copper Co, 71-75; TECH DIR, KOCIDE CHEM CORP, 75- Mem: Am Chem Soc. Res: Transition metal complexes; copper chemistry; fungicides; aquatic chemicals; thixotropic materials. Mailing Add: Kocide Chem Corp 12701 Almeda Rd Houston TX 77045

MEYERS, MARTIN BERNARD, b Newark, NJ, Sept 12, 33; m 62. ORGANIC CHEMISTRY. Educ: Polytech Inst Brooklyn, BS, 54; Yale Univ, MS, 56, PhD(chem), 58. Prof Exp: NIH fel, Queen's Univ Belfast, 58-59; proj leader steroid res, Gen Mills, Inc, 59-61; Imp Chem Industs fel, Glasgow Univ, 61-64; sr lectr org chem, Col Technol, Belfast, 61-71; SR LECTR ORG CHEM, NORTHERN IRELAND POLYTECH, 71- Mem: The Chem Soc; Royal Inst Chem. Res: Plant biochemistry and biosynthesis; fungicides and chemistry of color photography. Mailing Add: Sch of Phys Sci Northern Ireland Polytech Jordanstown Northern Ireland

MEYERS, MARVIN HAROLD, b Bridgeport, Conn, Nov 3, 18; m 46; c 2. ORTHOPEDIC SURGERY. Educ: Univ Calif, Los Angeles, AB, 40, MA, 41; Univ Calif, San Francisco, MD, 45. Prof Exp: Clin assoc prof, 54-67, ASSOC PROF ORTHOP SURG, SCH MED, UNIV SOUTHERN CALIF, 67- Mem: Am Acad Orthop Surg; Asn Acad Surg. Res: Biology of hip fractures and avascular necrosis of the femoral head; functional aspects of injury to the ligaments of the knee; osteochondral allograft as joint replacements; muscle pedicle grafts in treatment of hip fractures. Mailing Add: Dept of Orthop Surg Univ Southern Calif Sch Med Los Angeles CA 90033

MEYERS, MURIEL CHARLOTTE, b New York, NY, Mar 31, 16. MEDICINE. Educ: Hood Col, AB, 37; Duke Univ, MD, 41; Am Bd Internal Med, dipl. Hon Degrees: ScD, Hood Col, 63. Prof Exp: From intern to asst resident med, Univ Hosp, Duke Univ, 41-43; resident, Univ Hosp, 44-45, from instr to assoc prof, Sch Med, 45-62, PROF MED, MED SCH, UNIV MICH, ANN ARBOR, 62- Concurrent Pos: Res asst hemat, Simpson Mem Inst, 44-52; res assoc, 52-59, actg dir, 59-60, assoc dir, 60- Mem: Am Soc Hemat; Am Fedn Clin Res; fel Am Col Physicians; Int Soc Hemat. Res: Hematology; chemotherapy leukemia and lyphomas; autoimmune disorders; medical education. Mailing Add: Simpson Mem Inst Univ of Mich Ann Arbor MI 48109

MEYERS, NORMAN GEORGE, b Buffalo, NY, June 29, 30; m 58; c 4. MATHEMATICS. Educ: Univ Buffalo, BA, 52; Ind Univ, MA, 54, PhD(math), 57. Prof Exp: From instr to assoc prof, 57-68, PROF MATH, INST TECHNOL, UNIV MINN, MINNEAPOLIS, 68- Mem: Am Math Soc. Res: Partial differential equations; calculus of variations. Mailing Add: Dept of Math Univ of Minn Minneapolis MN 55455

MEYERS, PAUL, b Philadelphia, Pa, Sept 23, 39; m 61; c 2. MICROBIOLOGY, VIROLOGY. Educ: Temple Univ, BA, 61; State Univ NY Upstate Med Ctr, PhD(microbiol), 70. Prof Exp: Res assoc, Med Sch, Univ Miami, 70-71, from instr to asst prof microbiol, 72-73; CONSULT & ASST PROF MICROBIOL, MAYO CLIN, UNIV MINN, 73- Mem: AAAS; Transplantation Soc; Soc Gen Microbiol; NY Acad Sci; Am Soc Microbiol. Res: Viral oncology; immunology. Mailing Add: Dept of Microbiol Mayo Clin & Med Sch Rochester MN 55901

MEYERS, PHILIP ALAN, b Hackensack, NJ, Mar 3, 41; m 65; c 2. ORGANIC GEOCHEMISTRY, OCEANOGRAPHY. Educ: Carnegie-Mellon Univ, BS, 64; Univ RI, PhD(oceanog), 72. Prof Exp: Chemist, Interchem Corp, 67-68; ASST PROF OCEANOG, UNIV MICH, ANN ARBOR, 72- Mem: AAAS; Am Soc Limnologists & Oceanogrs; Geochem Soc. Res: Organic geochemistry of water and sediments; distribution of fatty acids and hydrocarbons in natural waters, sediments and organisms. Mailing Add: 2215 Space Res Bldg Univ of Mich Ann Arbor MI 48105

MEYERS, PHILIP HENRY, b Chicago, Ill, Feb 24, 33; m; c 3. RADIOLOGY. Educ: Univ Minn, BA, 52, BS, 53, MD, 55; Am Bd Radiol, dipl. Prof Exp: Intern, Kings County Hosp, Brooklyn, 55-56; resident radiol, Bellevue Hosp, New York, 56-57; asst prof, 62-64, ASSOC PROF RADIOL, SCH MED, TULANE UNIV, 64- Concurrent Pos: Fel diag radiol, New York Hosp, 59-60; Nat Cancer Inst fel radiation ther & radioisotopes, NY Univ-Bellevue Med Ctr, 60-61; vis radiologist, Charity Hosp, New Orleans, La, 62-; consult radiologist, St Barnabas Hosp Chronic Dis, New York, 62-; consult assoc scientist, Biomed Comput Ctr, Tulane Univ, 64-65. Mem: Am Col Radiol; Soc Nuclear Med; Radiol Soc NAm. Mailing Add: 401 Emerald St New Orleans LA 70124

MEYERS, PHILIP ROBERT, b Brooklyn, NY, Oct 8, 37. TOPOLOGY, BIOMATHEMATICS. Educ: Brooklyn Col, BS, 59; Univ Md, MA, 62, PhD(math), 66. Prof Exp: Physicist celestial mech, Theoret Div, Goddard Space Flight Ctr, NASA, 59-62; mathematician opers res, Appl Math Div, Nat Bur Stands, DC, 62-67; sr analyst, IBM World Trade Corp, 67-70; asst prof biomath & health sci commun, Health Sci Ctr, State Univ NY Stony Brook, 70-71; SPEC ASST TO COMNR, NEW YORK CITY POLICE DEPT, 71- Concurrent Pos: Vpres, Dynamic Access Corp, NY, 70. Mem: Am Math Soc; AAAS; Math Asn Am. Res: Point set topology combinatorics. Mailing Add: 400 Broome St Rm 708 New York City NY 10013

MEYERS, RICHARD G, solid state physics, see 12th edition

MEYERS, RICHARD KAYTON, organic chemistry, see 12th edition

MEYERS, ROBERT ALLEN, b Los Angeles, Calif, May 15, 36; m 61; c 2. AIR POLLUTION, ORGANIC CHEMISTRY. Educ: San Diego State Col, BA, 59; Univ Calif, Los Angeles, PhD(chem), 63. Prof Exp: Fel, Calif Inst Technol, 63-64; sr res chemist, Bell & Howell Res Ctr, 64-66; head org chem sect, Chem & Chem Eng Dept, 66-73, MGR COAL DESULFURIZATION, SYSTS & ENERGY DIV, TRW SYSTS, 73- Concurrent Pos: Mem, US-USSR working group on air pollution control, 74- Mem: Am Chem Soc. Res: Inorganic and organic sulfur chemistry; hydrometallury of sulfur; organic synthesis; aromatic nucleophilic substitution; oxidative mechanisms; polymer synthesis; desulfurization of fossil fuels through chemical reaction; design and construction of pilot test units for desulfurization. Mailing Add: TRW Systs One Space Park Redondo Beach CA 90278

MEYERS, ROBERTA LEE, b New York, NY, July 5, 37; m 61; c 2. IMMUNOLOGY. Educ: San Diego State Col, BS, 59; Univ Calif, Los Angeles, MS, 62, PhD(med microbiol, immunol), 64. Prof Exp: Immunologist, Sch Med, Univ Calif, Los Angeles, 60-62, immunochemist, 62-64; immunochemist, Calif Inst Technol, 64-67; asst prof ophthal, 73-75, ASSOC PROF OPHTHAL, UNIV CALIF, LOS ANGELES, 75-, IMMUNOLOGIST, JULES STEIN EYE INST, 67- Concurrent Pos: Nat Inst Allergy & Infectious Dis fel, 64-67; Nat Eye Inst res career develop award, 75- Mem: Am Acad Allergy; Am Asn Immunol; Asn Res Vision & Ophthal; Reticuloendothelial Soc; Am Soc Microbiol. Res: Viral immunology and immunopathology; ocular immunology and inflammation; cellular immunity. Mailing Add: Jules Stein Eye Inst Univ of Calif Ctr Health Sci Los Angeles CA 90024

MEYERS, SAMUEL PHILIP, b Asbury Park, NJ, Feb 21, 25; m 52; c 3. MARINE MICROBIOLOGY. Educ: Univ Fla, BS, 50; Univ Miami, MS, 52; Columbia Univ, PhD(bot), 57. Prof Exp: Res aide marine microbiol, Marine Lab, Univ Miami, 52-54, asst prof, 57-61, assoc prof, Inst Marine Sci, 61-68; PROF FOOD SCI & TECHNOL, LA STATE UNIV, BATON ROUGE, 68- Mem: AAAS; Mycol Soc; Am Soc Indust Microbiol; Am Soc Microbiol; Brit Soc Gen Microbiol. Res: Biology of marine fungi; microbial ecology; bionomics of marine yeasts; ecology of marine nematodes. Mailing Add: Dept of Food Sci & Technol La State Univ Baton Rouge LA 70803

MEYERS, THEODORE RALPH, b Cincinnati, Ohio, Dec 21, 02; m 27; c 2. GEOLOGY. Educ: Ohio State Univ, AB, 26, AM, 29. Prof Exp: From asst prof to prof, 27-74, EMER PROF GEOL, UNIV NH, 74- Concurrent Pos: State geologist, State Planning & Develop Comn, NH, 42-63. Mem: Fel AAAS; fel Geol Soc Am; Am Mineral Soc. Res: Geology and mineral resources of New Hampshire. Mailing Add: Dept of Geol Univ of NH Durham NH 03824

MEYERS, WAYNE MARVIN, b Aitch, Pa, Aug 28, 24; m 53; c 4. MEDICAL MICROBIOLOGY. Educ: Juniata Col, BS, 47; Univ Wis, MS, 53, PhD(microbiol), 55; Baylor Univ, MD, 59. Prof Exp: Res assoc, Univ Wis, 51-54; from asst to instr microbiol, Col Med, Baylor Univ, 54-59; intern med, Conemaugh Valley Mem Hosp, Johnstown, Pa, 59-60; staff physician, Berrien County Hosp, 60-61; dir, Nyankanda Leprosarium, Burundi, 61-62; staff physician, Oicha Leprosarium, Repub of Zaïre, 62-64; med dir, Kivuvu Leprosarium, Repub of Zaïre, 65-73; prof path, Univ Hawaii Sch Med, 73-75; CHIEF DIV MICROBIOL, ARMED FORCES INST PATH, 75- Mem: NY Acad Sci; Int Leprosy Asn; Int Soc Trop Dermat; Am Soc Trop Med & Hyg. Res: Leprosy; filariasis; mycobacterium ulcerans infections; tropical and parasitic diseases. Mailing Add: Div of Microbiol Dept of Infectious & Parasitic Dis Path Armed Forces Inst of Path Washington DC 20306

MEYERSBURG, HERMAN ARNOLD, b New York, NY, Dec 8, 13; m 37; c 4. PSYCHIATRY. Educ: New York Univ, ScB, 33, MD, 38. Prof Exp: Rotating intern, Greenpoint Hosp, Brooklyn, 38-40; asst res neuropsychiat serv, Goldwater Mem Hosp, NY Univ, 40-41, fel med serv, Psychiat Div, Bellevue Hosp, 41-42; asst res & res neuropsychiat, Hosp, Va, 42-43; consult neurologist & psychiatrist, US Marine Hosp, Norfolk & Vet Admin Hosp, Kecoughtan, 47-50; sr psychiatrist ment hyg clin regional off, US Vet Admin, 50-52, consult psychiatrist, 52-56; clin instr psychiat, Sch

Med, George Washington Univ, 55-60, assoc, 60-64, mem fac, Washington Psychoanal Inst, 56-59, CLIN PROF PSYCHIAT, SCH MED, GEORGE WASHINGTON UNIV, 64-, TRAINING PSYCHOANALYST, WASHINGTON PSYCHOANAL INST, 59- Concurrent Pos: Res assoc, Wash Sch Psychiat; consult psychiatrist, US Army Med Servs Grad Sch, Walter Reed Hosp, 52-58 & Hampton Inst, 53-61; consult, Hillcrest Residence, 59-66, Nat Inst Ment Health, 66-, drug treatment & res, Vet Admin Hosp, DC, 70-73 & Montgomery County Child Day Care Asn, 70- Mem: Am Psychiat Asn; Am Psychoanal Asn. Res: Dynamics of psychotherapeutic processes; psychodynamics of depressive disorders. Mailing Add: 9910 Summit Ave Kensington MD 20795

MEYER-SCHÜTZMEISTER, LUISE, b Magdeburg, Ger, Jan 20, 15; US citizen; m 46; c 2. NUCLEAR PHYSICS. Educ: Tech Univ, Berlin, Dipl Eng, 39, PhD(physics), 43. Prof Exp: Mem teaching staff physics, Tech Univ, Berlin, 43-45; assoc, Univ Göttingen, 46-48; group leader, Max Planck Inst Med Res, 48-52; res assoc, Univ Chicago, 53-55; assoc scientist, 56-73, SR SCIENTIST, ARGONNE NAT LAB, 73- Mem: Fel Am Phys Soc. Res: Low and medium energy nuclear physics. Mailing Add: Argonne Nat Lab Physics Div 9700 Cass Ave Argonne IL 60440

MEYERSON, ARTHUR LEE, b East Orange, NJ, June 30, 38; m 61; c 1. MARINE GEOLOGY, MARINE GEOCHEMISTRY. Educ: Univ Pa, BA, 59; Lehigh Univ, MS, 61, PhD(geol), 71. Prof Exp: Instr geol, Upsala Col, 61-62; from instr to assoc prof, 62-75, PROF EARTH & PLANETARY ENVIRON, KEAN COL NJ, 75-, CHMN DEPT, 73- Mem: AAAS; Geol Soc Am. Res: Holocene stratigraphy and palynology; estuarine geochemistry. Mailing Add: Dept Earth & Planetary Environ Kean Col of NJ Union NJ 07083

MEYERSON, MARK DANIEL, b Alexandria, Va, Feb 14, 49; m 69. TOPOLOGY. Educ: Univ Md, BS, 71; Stanford Univ, MS, 73, PhD(math), 75. Prof Exp: VIS LECTR MATH, UNIV ILL, URBANA, 75- Mem: Am Math Soc; Math Asn Am. Res: Geometric topology. Mailing Add: Dept of Math Univ of Ill Urbana IL 61801

MEYERSON, SEYMOUR, b Chicago, Ill, Dec 4, 16; m 43; c 2. CHEMISTRY, MASS SPECTROMETRY. Educ: Univ Chicago, SB, 38. Prof Exp: Chemist, Res Dept, 46-61, chemist, Res & Develop Dept, Am Oil Co, 61-62, res assoc, 62-72, SR RES ASSOC, RES DEPT, STANDARD OIL CO IND, 72- Mem: Am Chem Soc; Am Soc Mass Spectrometry. Res: Mass spectrometry of organic compounds and applications thereof to study of molecular structures and reaction mechanisms. Mailing Add: Standard Oil Co Ind Res Dept Box 400 Naperville IL 60540

MEYMARIS, ELIAS, physical organic chemistry, see 12th edition

MEYN, RAYMOND EVERETT, JR, b Mobile, Ala, Aug 29, 42; m 68; c 2. BIOPHYSICS. Educ: Univ Kans, BS, 65, MS, 67, PhD(radiation biophys), 69. Prof Exp: USPHS fel, 69-70, asst physicist, 71-75, ASSOC PHYSICIST & ASSOC PROF BIOPHYS, UNIV TEX, M D ANDERSON HOSP & TUMOR INST, HOUSTON, 75- Concurrent Pos: Assoc mem, Univ Tex Grad Sch Biomed Sci, Houston, 71- Mem: Biophys Soc; Radiation Res Soc; Am Soc Photobiol. Res: Replication and repair of DNA in mammalian cells; regulation of growth and division; radiobiology of fast neutrons. Mailing Add: Dept of Physics Univ Tex MD Anderson Hosp & Tumor Inst Houston TX 77025

MEZEI, CATHERINE, b Budapest, Hungary, July 27, 31; Can citizen; m 54; c 1. BIOCHEMISTRY, PHARMACY. Educ: Univ Budapest, BS, 54; Univ BC, MS, 60, PhD(biochem), 64. Prof Exp: Fel, Ore State Univ, 64-67; asst prof, 67-73, ASSOC PROF BIOCHEM, DALHOUSIE UNIV, 73- Concurrent Pos: Med Res Coun Can res scholar, Dalhousie Univ, 67-73. Mem: Can Biochem Soc; Int Soc Neurochem. Res: Biochemistry of nerve development. Mailing Add: Dept of Biochem Dalhousie Univ Halifax NS Can

MEZEI, MICHAEL, b Mezokovesd, Hungary, Oct 7, 27; Can citizen; m 54; c 1. PHARMACEUTICS. Educ: Med Univ Budapest, Dipl pharm, 54; Ore State Univ, PhD(pharm), 67. Prof Exp: Instr pharm, Univ BC, 57-64 & Ore State Univ, 65-67; asst prof, 67-71, ASSOC PROF PHARM, DALHOUSIE UNIV, 71- Mem: Am Pharmaceut Asn; Acad Pharmaceut Sci; Can Pharmaceut Asn; Ny Acad Sci; Soc Cosmetic Chem. Res: Physiological properties of nonionic surfactants; dermatitic effect of nonionic surfactants; formulation of dermatological preparations; biochemistry of skin; biopharmaceutics. Mailing Add: Dept of Pharm Dalhousie Univ Col Pharm Halifax NS Can

MEZEY, EUGENE JULIUS, b Cleveland, Ohio, Apr 9, 26; m 56; c 2. INORGANIC CHEMISTRY. Educ: Ohio Univ, BS, 50; Ohio State Univ, MS, 54, PhD(chem), 57. Prof Exp: Asst, Res Found, Ohio State Univ, 52-54, assoc, 54-55, res fel, 55-57; sr res chemist, Pittsburgh Plate Glass Co, 57-60, supvr explor inorganic group, 60-63; SR CHEMIST, BATTELLE MEM INST, 63- Concurrent Pos: Instr, Univ Akron, 59-63; lectr, Ohio State Univ, 64-65, assoc, 65- Mem: Fel AAAS; Am Chem Soc; Am Ceramic Soc; Am Inst Chem; Int Microwave Power Inst. Res: Alkylation of boron hydrides; boron-nitrogen chemistry; halogens and halogen oxides; metal and nonmetal halides and subhalides; carbides; oxides; high energy processes; reactions induced with microwave energy and the use of microwaves in chemical processing. Mailing Add: Battelle Columbus Labs 505 King Ave Columbus OH 43201

MEZEY, KALMAN C, b Nagyvarad, Hungary, Sept 18, 09; US citizen; m 35; c 3. MEDICINE, PHARMACOLOGY. Educ: Univ Basel, MD, 33. Hon Degrees: Dr, Univ Javeriana Bogota, 75. Prof Exp: Asst internal med, Univ Hosp, Univ Basel, 33-36; asst, Univ Clin, Univ Vienna, 36-37; prof pharmacol, Pontifical Univ Javeriana, Colombia, 42-58; VPRES MED SCI, MERCK SHARP & DOHME INT, 59- Prof Exp: Biol Sci Bogota award, 52-57; vis prof, Med Sch, Univ Vienna, 72; clin prof med, NJ Med Sch, 74- Honors & Awards: Cross of Boyaca Award, 57; Kalman C Mezey Chair of Pharmacol, Univ Javeriana Bogota, 75. Mem: Am Soc Pharmacol & Exp Therapeut; fel Am Col Clin Pharmacol & Chemother; Am Soc Trop Med & Hyg; AMA. Res: Pharmacology of drugs acting on the cardiovascular system; medicinal plants; arrow poisons; clinical pharmacology. Mailing Add: Merck Sharp & Dohme Int Rahway NJ 07065

MEZGER, FRITZ WALTER WILLIAM, b Bryn Mawr, Oct 19, 28. THEORETICAL PHYSICS. Educ: Harvard Col, AB, 48; Univ Cincinnati, PhD(physics), 57. Prof Exp: Reactor physicist nuclear energy propulsion aircraft proj, Fairchild Co, 48-51; leader reactor physics subunit, Aircraft Nuclear Propulsion Dept, 51-54, supvr nuclear anal unit, 54-56, mgr appl math subsect, 56-59, controls & instrumentation develop subsect, 59-60 & physics & math subsect, 60-61, consult physicist space sci lab & proj scientist, Adv Space Proj Dept, 61-64, mgr space power & propulsion res, 64-68, mgr advan studies, 68-74, MGR RESOURCE PLANNING & MGT, ADVAN STUDIES, SPACE DIV, GEN ELEC CO, 74- Mem: Am Phys Soc; Am Nuclear Soc; Asn Comput Mach; Inst Elec & Electronics Eng; Am Inst Aeronaut & Astronaut. Res: Plasma propulsion; magneto-hydrodynamic power generation; laser applications; reactor physics; nuclear engineering; electrodynamics; control engineering; digital computer applications. Mailing Add: Gen Elec Co Space Div PO Box 8555 Philadelphia PA 19101

MEZGER-FREED, LISELOTTE, b Berlin, Ger, Feb 6, 26; US citizen; m 51; c 4. GENETICS, EMBRYOLOGY. Educ: Bryn Mawr Col, AB, 46; Wash Univ, MA, 48; Columbia Univ, PhD(zool), 52. Prof Exp: Instr biol, Brooklyn Col, 52-53; USPHS fel embryol, Inst Cancer Res, 54-57; instr, Temple Univ, 59-60; res assoc, Bryn Mawr Col, 61-66; res assoc, 66-70, ASST MEM, INST CANCER RES, 70- Mem: Am Soc Cell Biol; Am Asn Cancer Res; Genetics Soc Am; Soc Develop Biol. Res: Somatic cell genetics; haploid cell lines; drug resistance; nuclear transfer. Mailing Add: Inst for Cancer Res 7701 Burholme Ave Philadelphia PA 19111

MEZICK, JAMES ANDREW, biochemistry, see 12th edition

MEZL, ZDENEK, b Prague, Czech, May 30, 07; Can citizen; m 46; c 2. DENTISTRY, ORAL PATHOLOGY. Educ: Univ Prague, MD, 32; Univ Alta, DDS, 57. Prof Exp: Asst privat dozent dent, Univ Prague, 34-49, privat dozent, 46; asst privat dozent, Univ Groningen, 49-52; PROF ORAL PATH & CHMN DEPT, UNIV MONTREAL, 57- Res: Dental pathology; anomalies of development; pulp pathology; dental caries. Mailing Add: Fac of Dent Surg Univ of Montreal Montreal PQ Can

MEZZINO, MICHAEL JOSEPH, JR, b Galveston, Tex, Sept 5, 40; m 65; c 1. MATHEMATICS. Educ: Austin Col, BA, 62; Kans State Col Pittsburg, MA, 63; Univ Tex, Austin, PhD(math), 70. Prof Exp: Res mathematician, Tracor Inc, Tex, 65-66; instr math, Univ Tex, Austin, 70; asst prof, Southwestern Univ Tex, 70-74; ASST PROF MATH, UNIV HOUSTON, CLEAR LAKE CITY, 74- Concurrent Pos: Consult, Tracor Inc, Tex, 66- & J & J Marine Diving Co, 74- Mem: AAAS; Am Math Soc. Res: Mathematical modeling; computer graphics. Mailing Add: Sch of Sci & Technol Univ Houston Clear Lake City Houston TX 77058

MGBODILE, MARCEL UME, b Abor, Eastern Nigeria, Sept 25, 41; m 73. BIOCHEMISTRY, NUTRITON. Educ: Cuttington Col, Liberia, BS, 66; Tuskegee Inst, MS, 68; Va Polytech Inst & State Univ, PhD(biochem), 72. Prof Exp: Res asst biochem, Va Polytech Inst & State Univ, 68-72; ASST PROF BIOCHEM & NUTRIT, MEHARRY MED COL, 72- Concurrent Pos: Nutritionist-toxicologist maternal & child health, Family Planning Ctr, Meharry Med Col, 72-; res assoc, Ctr Environ Toxicol, Med Sch, Vanderbilt Univ, 72-; Future Leader in Nutrit grant, Nutrit Found, 73-75. Mem: Am Chem Soc; Inst Food Technologists. Res: Mechanism of nutrient-drug interactions; toxicological and biochemical consequences of undernutrition. Mailing Add: Dept of Biochem Meharry Med Col Nashville TN 37208

M'GONIGLE, JOHN WILLIAM, geology, see 12th edition

MHATRE, NAGESHSHAMRAO, b Bombay, India, July 6, 32; m 61; c 2. BIOCHEMISTRY, ENZYMOLOGY. Educ: Univ Bombay, BSc, 54; Ore State Univ, MS, 55; Rutgers Univ, PhD(biochem), 62. Prof Exp: Res scientist, Miles Labs, Inc, 62-66, sect head & sr res scientist, Ames Res Lab, Ames Co Div, 67-71, managing dir, Miles-Yeda Ltd, Rehoust, Israel, 71-75, DIR SCI AFFAIRS, LAB-TEK PROD, DIV MILES LABS, INC, 75- Mem: Am Chem Soc; NY Acad Sci. Res: Microbiology; immunochemistry; microbial analyses; immunochemical studies of clinically significant isoenzymes; immobilization of biopolymers; carbonic anhydrase inhibition by DDT; irreversible inhibition of beta-glucuronidase. Mailing Add: Div Miles Lab Inc Lab-Tek Prod 30 W 475 N Aurora Naperville IL 60540

MI, MING-PI, b Shanghai, China, May 24, 33; m 65. GENETICS. Educ: Taiwan Univ, BS, 54; Univ Wis, MS, 59, PhD(genetics), 63. Prof Exp: NSF fel genetics, 63-64, asst geneticist, 64-65, from asst prof to assoc prof, 65-74, PROF GENETICS, UNIV HAWAII, 74- Concurrent Pos: NIH res grant, 64-67. Mem: Am Soc Human Genetics. Res: Statistical and population genetics. Mailing Add: Dept of Genetics Univ of Hawaii Honolulu HI 96822

MIA, ABDUL JABBAR, b Dacca, Bangladesh, Feb 1, 29; m 70. PLANT MORPHOGENESIS, ELECTRON MICROSCOPY. Educ: Univ Dacca, BSc, 50, MSc, 52; NC State Univ, PhD(bot, plant path), 59. Prof Exp: Res asst bot, Jute Res Inst, Dacca, Bangladesh, 54-56 & NC State Univ, 57-59; botanist, Jute Res Inst, Dacca, 59-61; US Educ Exchange assoc prof biol, Bishop Col, 61-63; Nat Res Coun Can fel electron micros, Carleton Univ, 64-66; res scientist, Can Dept Fisheries & Forestry, 66-70; PROF BIOL & BOT, BISHOP COL, 71- Mem: Bot Soc Am; Forest Prod Res Soc. Res: Cytochemical and sub-cellular organization of dormant and germinating pine embryos as affected by imbibition of water; ultrastructures and synthesis of cell walls in fibers and vessels in forest trees. Mailing Add: Dept of Biol Bishop Col Dallas TX 75241

MIALE, JOHN BUYER, b Rochester, NY, May 16, 16; m 42; c 1. PATHOLOGY. Educ: Cornell Univ, BS, 36; Univ Rochester, MD, 40. Prof Exp: Intern, Strong Mem Hosp, 40-42; resident path, Med Col, Cornell Univ, 42-43; instr, Sch Med, Univ NC, 44-46; preceptor, Med Sch, Univ Wis, 46-53; actg chmn dept, 53-72, PROF PATH, SCH MED, UNIV MIAMI, 53-, HEAD SECT EXP PATH, 72- Concurrent Pos: Pathologist, Brady Urol Inst, New York Hosp, 42-43; dir labs, Watts Hosp, Durham, NC, 44-46, St Joseph's Hosp & Marshfield Clin, Wis, 46-53; dir clin path, Jackson Mem Hosp, 53- Mem: Am Soc Clin Path; fel Soc Exp Biol & Med; fel AMA; fel Col Am Pathologists. Res: Morphologic hematology; blood coagulation; immunohematology; anatomy and pathology of coronary arteries. Mailing Add: 3764 Carmen Ct Coconut Grove FL 33133

MIALE, JOSEPH NICOLAS, b Johnston, RI, May 9, 19; m 55; c 3. PETROLEUM CHEMISTRY. Educ: Providence Col, BS, 40; Tex A&M Univ, MS, 47. Prof Exp: Instr chem, Agr & Mech Col, Tex, 47; res chemist, Res Dept, 47-62, SR RES CHEMIST, CENT RES DIV, MOBIL RES & DEVELOP CORP, 62- Mem: Am Chem Soc; Int Cong Catalysis. Res: Exploratory research on processes and catalysts for hydrocarbon conversions. Mailing Add: Mobil Res & Develop Corp PO Box 1025 Princeton NJ 08540

MIALE, JOSEPH PETER, b Acquaviva delle Fonti, Italy, Aug 10, 04; US citizen; m 30; c 2. PHARMACEUTICAL CHEMISTRY, ORGANIC CHEMISTRY. Educ: Columbia Univ, PhCh, 26; Polytech Inst Brooklyn, BS, 32; Columbia Univ, MA, 34. Prof Exp: From instr to assoc prof pharm, Col Pharm, Columbia Univ, 26-43; mem staff prod develop, Carroll Dunham Smith Pharm Co, 43-44; lab dir, Walker Vitamin Prod, 44-51, dir res & develop, Walker Labs, Inc, 51-56, plant mgr, 56-58, dir prof serv, Walker Div, Richardson-Merrell Inc, 58-67, mem staff, Dept Sci Info, Vick Div Res & Develop, 67-71; RETIRED. Mem: AAAS; Am Pharmaceut Asn; Am Chem Soc; Am Soc Pharmacog; fel Am Inst Chemists. Res: Quality control; research and development of drugs and pharmaceutical products; manufacture and control of medicinal products. Mailing Add: 867 Country Club Dr North Palm Beach FL 33408

MIANO, RALPH R, inorganic chemistry, physical chemistry, see 12th edition

MIATECH, GERALD JAMES, b Stambaugh, Mich, Dec 31, 22; m 45; c 3. GEOPHYSICS, SPACE SCIENCES. Educ: Mich Technol Univ, BS, 49; St Louis Univ, MS, 56; Univ Wis, PhD(geophys), 61. Prof Exp: Geophysicist, M A Hanna Co, Mich, 49-50; geophysicist & aerospace engr, US Air Force, 50-65 & Ames Res Ctr, NASA, Moffett Field, Calif, 66-72; geophysicist & aerospace engr, 72-74, SR MEM TECH STAFF, ELECTROMAGNETICS SYSTS LAB, SUNNYVALE, CALIF, 74- Mem: AAAS; Am Geophys Union; Geol Soc Am; Sigma Xi; Am Inst Mining, Metall & Petrol Eng. Res: Space systems; geophysical investigations for earth resources (vector aeromagnetometry). Mailing Add: 19300 Chablis Ct Saratoga CA 95070

MICALE, FORTUNATO JOSEPH, b Niagara Falls, NY, Aug 11, 32. PHYSICAL CHEMISTRY, COLLOID CHEMISTRY. Educ: St Bonaventure Univ, BA, 56; Niagara Univ, BS, 59; Purdue Univ, MS, 61; Lehigh Univ, PhD(phys chem), 65. Prof Exp: Res asst prof, 66-70, ASSOC PROF PHYS CHEM, LEHIGH UNIV, 70- Mem: Am Chem Soc. Res: Colloid and surface properties of silica and carbon; dispersion stability; solution adsorption. Mailing Add: Ctr Surface & Coatings Res Sinclair Lab Lehigh Univ Bethlehem PA 18015

MICCIOLI, BRUNO R, inorganic chemistry, see 12th edition

MICELI, ANGELO SYLVESTRO, b New York, NY, Dec 24, 13; m 38; c 6. PHYSICAL CHEMISTRY. Educ: Wayne Univ, BS, 34, MS, 36; Univ Mich, PhD(chem), 42. Prof Exp: From asst to instr chem, Wayne Univ, 34-42; sr scientist, US Rubber Co, 42-48, head chem res & develop, 48-51, from asst mgr res & develop to mgr res & develop, 51-60, sect mgr tires, 60-62, dept mgr, 62-65, dir develop int div, 65-67, DIR RES & DEVELOP, UNIROYAL INT, 67- Mem: Am Chem Soc. Res: Adhesion of rubber to metal; general industrial adhesives; electrodeposition of alloys, especially brass; resinoid and rubber-bonded grinding wheels; development of gum plastics; kinetics of isotopic exchange reaction; organizational planning; tire technology and product development. Mailing Add: Uniroyal Tire Co Uniroyal Inc 6600 E Jefferson Detroit MI 48232

MICETICH, RONALD GEORGE, b Madras, India, May 28, 31; Can citizen; m 58; c 4. ORGANIC CHEMISTRY, MEDICINAL CHEMISTRY. Educ: Loyola Col, Madras, India, BSc, 52; Univ Madras, MA, 55; Univ Sask, PhD(org chem), 62. Prof Exp: Nat Res Coun Can fel, Prairie Regional Lab, Sask, Can, 62-63; res chemist, R&L Molecular Res Ltd, Alta, 63-69; from asst res dir to assoc res dir org & med chem, 69-71, res mgr, 71-75, ACTG RES DIR ORG & MED CHEM, RAYLO CHEM LTD, 75- Mem: The Chem Soc; Chem Inst Can; Int Soc Heterocyclic Chem. Res: Heterocyclic chemistry; synthetic organic chemistry; organometallic chemistry; chemical modification of B-lactam antibiotics. Mailing Add: Raylo Chem Ltd 8045 Argyll Rd Edmonton AB Can

MICH, THOMAS FREDERICK, b Milwaukee, Wis, May 26, 39; m 66. ORGANIC CHEMISTRY. Educ: Marquette Univ, BS, 61; Northwestern Univ, MS, 64; State Univ NY Buffalo, PhD(chem), 68. Prof Exp: Res assoc, Dartmouth Col, 67-68; sr res chemist, Monsanto Co, 68-69; RES CHEMIST, CHEM DEPT, PARKE DAVIS & CO, 69- Mem: Am Chem Soc; The Chem Soc. Res: Synthetic organic and medicinal chemistry. Mailing Add: Chem Dept Ann Arbor Res Labs Parke Davis & Co 2800 Plymouth Ann Arbor MI 48106

MICHA, DAVID ALLAN, b Villa Mercedes, Argentina, Sept 12, 39; US citizen; m 65; c 2. CHEMICAL PHYSICS. Educ: Nat Univ Cuyo, Lic Physics, 62; Univ Uppsala, Fil Lic, 65, Fil Doctor, 66. Prof Exp: Res assoc, Theoret Chem Inst, Univ Wis-Madison, 66-67; asst res physicist, Inst Pure & Appl Phys Sci, Univ Calif, San Diego, 67-69; assoc prof, 69-74, PROF CHEM & PHYSICS, UNIV FLA, 74- Concurrent Pos: Res grants, Am Chem Soc-Petrol Res Fund, 69-72, NSF, 71-73, Nat Res Coun, 71 & Alfred P Sloan fel, 71-73; docent, Univ Uppsala, 68; vis Lamberg Prof, Univ Gothenburg, 70. Mem: Am Phys Soc; Am Chem Soc. Res: Molecular dynamics; electronic structure of matter; computational methods in theoretical chemistry. Mailing Add: Dept of Chem Univ of Fla Gainesville FL 32604

MICHAEL, ALFRED FREDERICK, JR, b Philadelphia, Pa, Aug 10, 28; m 52; c 3. PEDIATRICS, NEPHROLOGY. Educ: Temple Univ, MD, 53. Prof Exp: Intern, Philadelphia Gen Hosp, 53-54; resident pediat, St Christopher's Hosp Children, Sch Med, Temple Univ, 54-55; from jr resident to sr resident, Children's Hosp & Col Med, Univ Cincinnati, 57-59, chief resident & instr, 59-60; USPHS fel, 60-63, Am Heart Asn estab investr, 63-68, assoc prof, 65-68, PROF PEDIAT & PATH, MED SCH, UNIV MINN, MINNEAPOLIS, 68- Concurrent Pos: Vis investr & Guggenheim fel, Neurochem Inst, Copenhagen, Denmark, 66-67; mem rev comt, Nat Kidney Found, 70-74; mem subspecialty bd pediat nephrol, Am Bd Pediat, 73- Mem: Soc Pediat Res; Am Soc Exp Path; Am Asn Immunol; Am Soc Nephrology; Am Soc Pediat Nephrology. Res: Renal disease; biochemical pathology; immunopathology and mechanisms of glomerulonephritis. Mailing Add: Dept of Pediat Univ Of Minn Hosps Minneapolis MN 55455

MICHAEL, CHARLES REID, b Bucyrus, Ohio, June 30, 39; m 64; c 2. NEUROPHYSIOLOGY. Educ: Harvard Col, BA, 61; Harvard Univ, PhD(biol), 65. Prof Exp: Fel biophys, Johns Hopkins Univ, 65-68; asst prof, 68-71, ASSOC PROF PHYSIOL, SCH MED, YALE UNIV, 71- Mem: AAAS; Soc Gen Physiol; Am Physiol Soc. Res: Sensory physiology of the mammalian central nervous system. Mailing Add: Dept of Physiol Yale Univ Sch of Med New Haven CT 06510

MICHAEL, EDWIN DARYL, b Mannington, WVa, Jan 22, 38; m 60; c 2. WILDLIFE MANAGEMENT. Educ: Marietta Col, BS, 59; Tex A&M Univ, MS, 63, PhD(wildlife ecol), 66. Prof Exp: From asst prof to assoc prof biol, Stephen F Austin State Col, 64-70; assoc prof, 70-74, PROF WILDLIFE MGT, WVA UNIV, 74- Mem: Am Soc Mammal; Wildlife Soc; Cooper Ornith Soc. Res: Ecology and management of forest wildlife. Mailing Add: Div of Forestry WVa Univ Morgantown WV 26506

MICHAEL, ERNEST ARTHUR, b Zurich, Switz, Aug 26, 25; nat US; m 56; div; c 3; m 66; c 2. TOPOLOGY. Educ: Cornell Univ, BA, 47; Harvard Univ, MA, 48; Univ Chicago, PhD(math), 51. Prof Exp: Fel, AEC, Inst Advan Study, 51-52 & Univ Chicago, 52-53; from asst prof to assoc prof, 53-60, PROF MATH, UNIV WASH, 60- Concurrent Pos: Mem, Inst Advan Study, 56-57, 60-61 & 68; ed, Gen Topology & Proceedings, Am Math Soc; vis, Math Res Inst, Swiss Fed Inst Technol, 73-74. Mem; Am Math Soc; Math Asn Am. Res: General topology. Mailing Add: Dept of Math Univ of Wash Seattle WA 98105

MICHAEL, ERNEST DENZIL, JR, b Lewiston, Maine, Jan 18, 22; m 45; c 3. PHYSIOLOGY, ERGONOMICS. Educ: Purdue Univ, BPE, 47; Univ Ill, MS, 49, PhD, 52. Prof Exp: Dir athletics high sch, 47-48; instr phys ed, Univ Ill, 49-50 & 51-52; asst, USPHS, 50-51; from instr to assoc prof phys educ & physiol, 52-67, chmn dept ergonomics & phys educ, 73-76, PROF PHYS EDUC, UNIV CALIF, SANTA BARBARA, 67-, RES ASSOC, INST ENVIRON STRESS, 64- Concurrent Pos: Am Physiol Soc res fel, Lankenau Hosp, Philadelphia, Pa, 59; res fel, Valley Forge Heart Hosp, 59-60; consult, Water Safety Prog, YMCA, 59-; res prof, Inst Physiol, Univ Glasgow, Scotland, 67-68. Mem: Am Physiol Soc; Sigma Xi (treas, 74-76); fel Am Col Sports Med; Am Asn Health, Phys Educ & Recreation; NY Acad Sci. Res: Effects of training on the cardiovascular system; perception of levels of exertion; body composition and performance. Mailing Add: Dept Ergonomics & Phys Educ Univ of Calif Santa Barbara CA 93106

MICHAEL, HENRY N, b Pittsburgh, Pa, July 14, 13; m 43; c 3. ANTHROPOLOGY, DENDROCHRONOLOGY. Educ: Univ Pa, BA, 48, MA, 51, PhD, 54. Prof Exp: Instr, Univ Pa, 48-54; ed, Smith, Kline & French & J B Lippincott, 54-58; from asst prof to assoc prof, 59-65, chmn dept, 65-73, PROF GEOG, TEMPLE UNIV, 65- Concurrent Pos: Field work, US Southwest, 49, Bering Strait Region, Alaska, 52 & Calif & Nev, 59-60; dir & ed series, Anthrop of the North, Transl from Russian Sources, Arctic Inst NAm, 59-, mem publ comt, Arctic Inst, Montreal, 67-70; res assoc, Univ Mus, Univ Pa, 59- & Radiocarbon Lab, 60-; partic archaeol field surv, Czech, 62, Lebanon & Egypt, 63-65, Greece, Egypt & Czech, 71 & US Southwest, 74 & 75. Mem: AAAS; fel Asn Am Geog; fel Am Anthrop Asn; fel Arctic Inst NAm; Am Asn Advan Slavic Studies. Res: Research to establish correction factors for radiocarbon dates, the former being based on dendrochronological studies. Mailing Add: Philadelphia PA

MICHAEL, IRVING, b Pittsburgh, Pa, June 28, 29. SPACE PHYSICS, NUCLEAR PHYSICS. Educ: George Washington Univ, BSc, 50; Univ Wis, MSc, 51, PhD(physics), 58. Prof Exp: Res asst phys chem, Geophys Lab, Carnegie Inst Technol, 50; res asst nuclear physics, Univ Wis, 51-58, res assoc, 58-60; res fel, Univ Notre Dame, 60-62; sr scientist, Northrop Space Labs, Calif, 62-63; res physicist space physics br, Air Force Weapons Lab, NMex, 64; RES PHYSICIST SPACE PHYSICS LAB, AIR FORCE CAMBRIDGE RES LABS, 64- Concurrent Pos: Consult, Radiation Dynamics Corp, NY, 60. Mem: Am Phys Soc; Am Geophys Union; Am Vacuum Soc; Int Orgn Vacuum Sci & Technol. Res: Satellite and deep-space probe measurements of the solar wind; low-energy nuclear physics; electrostatic accelerator development; high-voltage breakdown in vacuum; techniques and measurements in extreme-high vacuum; measurements of electric fields in space. Mailing Add: Space Physics Lab Air Force Cambridge Res Labs L G Hanscom Field Bedford MA 01730

MICHAEL, JACOB GABRIEL, b Rimavska Sobota, Czech, July 2, 31; US citizen; m 58; c 4. IMMUNOLOGY. Educ: Hebrew Univ, Israel, BA, 55, MSc, 56; Rutgers Univ, PhD(microbiol), 59. Prof Exp: Res assoc & vis scientist, Nat Cancer Inst, 59-61; res assoc, Harvard Med Sch, 61-66; assoc prof, 66-73, PROF MICROBIOL & IMMUNOL, MED CTR, UNIV CINCINNATI, 73- Concurrent Pos: Am Soc Microbiol pres fel, 62; USPHS career develop award, 65- Mem: AAAS; Am Soc Microbiol; Am Asn Immunol. Res: Microbial immunity; regulation of immune response; immediate hypersensitivity. Mailing Add: Dept of Microbiol Univ of Cincinnati Med Ctr Cincinnati OH 45267

MICHAEL, JAMES RICHARD, b Peoria, Ill, Dec 6, 32; m 56; c 2. ORGANIC CHEMISTRY. Educ: Univ Ill, BS, 53; Cornell Univ, MS, 55, PhD(org chem), 57. Prof Exp: Res chemist, Esso Res & Eng Co, 57-62, tech serv coordr, Esso Chem Co, 62-63, supply coordr, 63-65, planning coordr, Gen Chem Div, NY, 65-66, mkt mgr indust chem, Esso Chem SA, Belg, 66-69; mgr int contracts, 69-70, mgr indust chem, coatings, planning & tech dept, Esso Stand Sekiyu K K, Akasaka, Japan, 70-71, dir mgr chem & opers, Esso Kagaku K K, Akasaka, Japan, 71-73, prod exec, Exxon Chem Co, 73-74, MGR, SOLVENTS TECHNOL DIV, EXXON CHEM CO, 74- Mem: AAAS; Am Chem Soc; fel Am Inst Chem; NY Acad Sci. Res: Chemical marketing in the international market; chemistry of organic nitrogen compounds; chemistry and synthesis of solid rocket propellants and ferrocene related compounds; mechanism and synthesis of new polymers; surface coating chemistry; oxgenated and hydrocarbon solvents. Mailing Add: Exxon Chem Co PO Box 536 Linden NJ 07036

MICHAEL, JOE VICTOR, b South Whitley, Ind, Oct 2, 35; m 60; c 3. PHYSICAL CHEMISTRY. Educ: Wabash Col, BA, 57; Univ Rochester, PhD(chem), 63. Prof Exp: Res assoc, Harvard Univ, 62-64 & Brookhaven Nat Lab, 64-65; asst prof, 65-70, ASSOC PROF CHEM, CARNEGIE-MELLON UNIV, 70- Mem: NY Acad Sci; Am Chem Soc. Res: Photochemistry; chemical kinetics; time-of-flight mass spectroscopy; shock tubes; flow reactors; resonance photometry. Mailing Add: Dept of Chem Carnegie-Mellon Univ 4400 5th Ave Pittsburgh PA 15213

MICHAEL, JOEL ALLEN, b Chicago, Ill, Mar 8, 40; m 65; c 2. NEUROPHYSIOLOGY, BIOENGINEERING. Educ: Calif Inst Technol, BS, 61; McGill Univ, MSc, 64; Mass Inst Technol, PhD(physiol), 65. Prof Exp: Carnegie fel neurophysiol, Nat Phys Lab, Teddington, Eng, 65-66; res fel psychiat, Mass Gen Hosp & Harvard Med Sch, Boston, 66-67; asst prof bioeng, Univ Ill, Chicago, 67-70, asst prof physiol, Col Med, 68-70; assoc prof biomed eng & neurol sci, 70-74, ASSOC PROF PHYSIOL & ACTG CHMN DEPT, RUSH MED COL, 74- Concurrent Pos: Asst attend bioengr, Presby-St Luke's Hosp, Chicago, 67-70. Mem: Am Physiol Soc; Inst Elec & Electronics Eng; Soc Neurosci. Res: Data processing in mammalian visual system; control of cerebral circulation and effects of hypoxia; neurophysiological basis of symptoms in multiple sclerosis; use of digital computers in biomedicine. Mailing Add: Dept of Physiol Rush Med Col Rush-Presby St Luke's Med Ctr Chicago IL 60612

MICHAEL, LESLIE WILLIAM, b San Francisco, Calif, Jan 26, 33; m 54. CHEMISTRY. Educ: Univ Calif, Berkeley, BA, 58; Fresno State Col, MS, 62; Univ Cincinnati, PhD(chem), 69. Prof Exp: Technician, Ortho Div, Chevron Chem Co, 58-60; AMA Educ Res Fund grant environ health, 68-69, res assoc chem, 69-70, ASST PROF ENVIRON HEALTH, COL MED, UNIV CINCINNATI, 70- Mem: Am Chem Soc; Am Crystallog Asn; Am Soc Testing & Mat. Res: Origin and chemistry of environmental materials with biological interactions; qualitative and quantitative chemical analysis; structure and physical chemical constants particularly with regard to essential and toxic metals. Mailing Add: Dept Environ Hlth Kettering Lab Univ Cincinnati Col Med Cincinnati OH 45267

MICHAEL, MAX, JR, b Athens, Ga, Feb 14, 16; m 44; c 5. INTERNAL MEDICINE. Educ: Univ Ga, BS, 35; Harvard Univ, MD, 39. Prof Exp: Asst med, Sch Med, Johns Hopkins Univ, 41-42; asst, Emory Univ, 45-47, assoc, 47-50, from asst prof to assoc prof, 52-54; prof, Col Med, State Univ NY, 54-58; exec dir, 58-67, ASST DEAN EDUC PROG, JACKSONVILLE HOSPS, UNIV FLA, 67-, PROF MED, COL MED, 58- Concurrent Pos: Chief med serv, Vet Admin Hosp, Atlanta, 47-54; dir med serv, Maimonides Hosp, Brooklyn, 54-58; mem bd regents, Nat Libr Med, 68-72. Mem: Inst of Med of Nat Acad Sci; Am Tuberc Soc; Am Soc Clin Invest; Am Clin & Climat Asn; assoc Am Col Physicians. Res: Infectious diseases. Mailing Add: 655 W Eighth St Jacksonville FL 32209

MICHAEL, NORMAN, b New York, NY, Dec 12, 31; m 63; c 1. INORGANIC CHEMISTRY. Educ: Columbia Univ, AB, 55. Prof Exp: Radiol chemist, US Navy Mat Lab, 54-56; nuclear chemist, Alco Prod Inc, 56-57; scientist, Atomic Power Div, Westinghouse Elec Corp, 57-62; sr res chemist, Astropower Lab, Douglas Aircraft Co, 62-63; chemist, Vallecitos Atomic Lab, Gen Elec Co, 63-66; SR ENGR, RES LABS, WESTINGHOUSE ELEC CORP, 66- Mem: Am Chem Soc. Res: Oxidation, corrosion and radioactive contamination of metals and alloys; fuel cell electrolytes and battery separators; inorganic ion exchangers for water purification. Mailing Add: Phys & Inorg Chem Dept R&D Ctr Westinghouse Elec Corp Pittsburgh PA 15235

MICHAEL, PAUL ANDREW, b New York, NY, July 6, 28; m 53; c 2. PHYSICS. Educ: NY Univ, AB, 49, PhD, 59; Univ Chicago, BS, 53, MS, 55. Prof Exp: Jr test engr, Curtiss-Wright Corp, 53, physicist, Res Div, 55-56; physicist, 58-72, leader meteorol group, 72-75, HEAD ATMOSPHERIC SCI DIV, BROOKHAVEN NAT LAB, 75- Concurrent Pos: Instr, NY Univ, 56-58, adj asst prof, 58-59. Mem: Am Phys Soc; Am Nuclear Soc; Am Meteorol Soc. Res: Reactor and neutron physics; fluid dynamics; atmospheric diffusion. Mailing Add: Dept of Appl Sci Brookhaven Nat Lab Upton NY 11973

MICHAEL, PAUL LEE, b Fairmont, WVa, June 7, 25; m 46; c 4. PHYSICS, ENVIRONMENTAL HEALTH. Educ: Fairmont State Col, BS, 48; Univ WVa, MS, 49; Univ Pittsburgh, PhD(occupational health), 55. Prof Exp: Instr physics, Univ WVa, 49-51; physicist, Nat Bur Stand, 51-52; sr physicist, Mine Safety Appliances Co, Pa, 52-54, supvr acoustics sect, 54-56, chief physicist, 56-59; assoc prof eng res, 59-69, PROF ENVIRON ACOUST & HEAD ENVIRON ACOUST LAB, PA STATE UNIV, 70- Concurrent Pos: Indust consult, 59-; mem bioacoust comt & writing group, Am Nat Stand Inst. Mem: Acoust Soc Am; Am Conf Govt Indust Hyg; Audio Eng Soc; Am Speech & Hearing Asn; Am Indust Hyg Asn. Res: Hearing conservation; noise measurement and control; instrumentation and calibration; development of clinical test procedures for determining. Mailing Add: Pa State Univ Environ Acoust Lab 110 Moore Bldg University Park PA 16802

MICHAEL, THOMAS HUGH GLYNN, b Toronto, Ont, May 20, 18; m 42; c 3. CHEMISTRY, SCIENCE ADMINISTRATION. Educ: Univ Toronto, BA, 40. Prof Exp: Chemist, Ont Res Found, 40-41 & Protective Coatings Lab, Nat Res Coun Can, 41-46; chief chemist, Woburn Chems, Ltd, 46-53; dir res, Howards & Sons Can, Ltd, 53-58; GEN MGR & SECY, CHEM INST CAN, 58- Concurrent Pos: Treas, Youth Sci Found, 61-71. Honors & Awards: Can Centennial Medal, 67. Mem: AAAS; Am Chem Soc; fel Chem Inst Can (treas, 53-56); Coun Eng & Sci Soc Exec (pres, 69-70); Am Soc Asn Execs (pres, 71-72). Res: Chemistry of synthetic resins, particularly alkyds and plasticizers; chemistry of protective coatings; management of scientific societies and publications. Mailing Add: The Chem Inst of Can 151 Slater St Ottawa ON Can

MICHAEL, WILLIAM ALEXANDER, b Peoria, Ill, Mar 20, 25; m 55; c 2. MATHEMATICS. Educ: Univ Ill, AB, 48, MS, 49, PhD(math), 54. Prof Exp: Asst math, Univ Ill, 51-54; instr, Northwestern Univ, 54-55; staff mathematician, Res Lab, Int Bus Mach Corp, 55-65; mem res staff, Ampex Corp, 65-67 & Systs Develop Div, IBM Corp, 67-68; asst prof, 59 & 68-73, ASSOC PROF MATH, SAN JOSE STATE UNIV, 73- Mem: Am Math Soc. Res: Numerical and functional analysis. Mailing Add: Dept of Math San Jose State Univ San Jose CA 95114

MICHAEL, WILLIAM EARL, b Mannington, WVa, June 13, 14; m 34; c 2. BIOLOGY, ZOOLOGY. Educ: Alderson-Broaddus Col, BS, 38; WVa Univ, MS, 46. Prof Exp: Teacher elem & high sch, Marion County Pub Sch Syst, 40-46; from instr to assoc prof, 46-56, PROF BIOL & CHMN DEPT, POTOMAC STATE COL, WVA UNIV, 56- Mem: AAAS; Am Inst Biol Sci. Res: Science education. Mailing Add: Dept of Biol Potomac State Col of WVa Univ Keyser WV 26726

MICHAEL, WILLIAM HERBERT, JR, b Richmond, Va, Dec 10, 26; m 52; c 2. SPACE SCIENCES. Educ: Princeton Univ, BS, 48, MS, 64, PhD(aerospace sci), 67; Univ Va, MS, 51; Col William & Mary, MA, 62. Prof Exp: Res scientist, Nat Adv Comn Aeronaut-NASA, Langley Res Ctr, 48-58, head trajectory anal group, 58-60, head mission anal sect, 60-69, head lunar & planetary sci br, 69-70, CHIEF ENVIRON & SPACE SCI DIV, LANGLEY RES CTR, NASA, 70- Concurrent Pos: Prin investr, Lunar Orbiter Selenodesy Exp, Langley Res Ctr, NASA, 65-69, team leader, Viking Mars Missions Radio Sci Team, 69-; mem work groups figure & motion of moon & laser tracking & appl, Comn 17, Int Astron Union & tracking & dynamics of satellites, Comt on Space Res. Honors & Awards: NASA Spec Serv Award, 67, Lunar Orbiter Proj Achievement Award, 68, Apollo Prog Spec Achievement Award, 69. Mem: Am Inst Aeronaut & Astronaut; Am Geophys Union; Int Astron Union. Res: Lunar and planetary exploration; space flight experiments in gravitational fields, geophysics, atmospheric properties and radio science; research administration. Mailing Add: M/S 401 NASA Langley Res Ctr ESSD Hampton VA 23365

MICHAEL, WILLIAM R, b Peoria, Ill, May 23, 30; m 53; c 4. PHARMACOLOGY, BIOCHEMISTRY. Educ: Univ Ill, BS, 52; Bradley Univ, MS. 57; St Louis Univ, PhD(biochem), 61. Prof Exp: Chemist, Northern Utilization Res Lab, USDA, 54-56; res biochemist, Miami Valley Lab, 61-74, MEM STAFF, SHARON WOODS TECH CTR, PROCTER & GAMBLE CO, 74- Mem: AAAS; Am Chem Soc; Soc Toxicol. Res: Drug metabolism; calcium and phosphate metabolism; pharmacokinetics; toxicology. Mailing Add: Sharon Woods Tech Ctr 11520 Reed Hartman Hwy Cincinnati OH 45241

MICHAELI, DOV, b Tel Aviv, Israel, May 28, 35; US citizen; m 62; c 2. IMMUNOCHEMISTRY, BIOCHEMISTRY. Educ: Hebrew Univ, Israel, BS, 60; Univ Calif, Berkeley, PhD(toxicol), 62. Prof Exp: Asst res scientist, Lab Med Entom, Kaiser Found Res Inst, 62-67, assoc res scientist, 67-71; ASSOC PROF BIOCHEM & SURG, SCH MED, UNIV CALIF, SAN FRANCISCO, 71- Concurrent Pos: NIH res career develop award. Mem: Am Asn Immunol; NY Acad Sci; Am Chem Soc. Res: Hematology; interaction of platelets with macramolecules; immunochemistry of collagen and of acetylcholinesterases from various sources; biochemistry and immunology of connective tissue. Mailing Add: Dept of Surg Univ of Calif Sch of Med San Francisco CA 94122

MICHAELIS, ARTHUR FREDERICK, b Bronx, NY, July 24, 41; m 64; c 2. PHARMACEUTICS, PHYSICAL PHARMACY. Educ: Bucknell Univ, BS, 63; Univ Wis, MS, 65, PhD(pharm), 67; Fairleigh Dickinson Univ, MBA, 75. Prof Exp: Sr chemist, Hoffmann-La Roche, Inc, Nutley, 67-70; dir qual control, Sandoz Pharmaceut, 70-71, DIR PHARM & ANAL RES, SANDOZ, INC, EAST HANOVER, 71- Concurrent Pos: Lectr, Sch Pharm, Univ Md, 69. Mem: AAAS; Am Pharmaceut Asn; Acad Pharmaceut Sci; fel Am Inst chem; NY Acad Sci. Res: Physical chemistry of the ion pair extraction of pharmaceutical amines; physical pharmacy of drugs used in the prophylaxis and treatment of nerve gas casualties; new methods of optimizing drug delivery; high speed liquid chromatography. Mailing Add: 27 Tuxford Terr Basking Ridge NJ 07920

MICHAELIS, CARL I, b Paxico, Kans, May 11, 18. ORGANIC CHEMISTRY. Educ: Univ Kans, AB, 45, AM, 47; Univ Fla, PhD(org chem), 53. Prof Exp: Asst chem, Univ Kans, 45-47; instr, Fla State Univ, 47-50; from instr to asst prof, Univ Fla, 50-57; assoc prof, 57-62, PROF CHEM, UNIV DAYTON, 62-, DIR PREMED DEPT, 60- Mem: Am Chem Soc. Res: Quaternary ammonium compounds and their derivatives; epoxides; amines. Mailing Add: Dept of Chem Univ of Dayton Dayton OH 45409

MICHAELIS, MORITZ, biochemistry, deceased

MICHAELS, ADLAI ELDON, b Alma, Wis, Nov 22, 13; m 40; c 2. PHYSICAL CHEMISTRY. Educ: Univ Wis, BS, 35; Ohio State Univ, PhD, 40. Prof Exp: Asst, Ohio State Univ, 35-39; instr chem, Univ Tenn, 40-43; res chemist, Esso Res & Eng Co, 43-59; from asst prof to assoc prof, 59-67, PROF CHEM, WASHINGTON & JEFFERSON COL, 67-, SECY FAC, 65- Concurrent Pos: Ed, Topic, 65-66. Mem: Am Chem Soc. Res: Corrosion; electrochemistry; motor fuels; lubricants; fuel and lubricant additives; air pollution. Mailing Add: 73 Crest Vue Rd Washington PA 15301

MICHAELS, DAVID D, b Cologne, Ger, July 16, 25; nat US; m 53; c 4. OPHTHALMOLOGY. Educ: Northern Ill Col Optom, OD, 47; Ill Inst Technol, BS, 51; Chicago Col Optom, MS, 52, DOS, 53; Roosevelt Univ, BS, 56; Univ Ill, MD, 57; Am Bd Ophthal, dipl, 64. Prof Exp: Instr neurol, Northern Ill Col Optom, 47-48, asst prof physiol optics, 48-49; asst physiol optics & ocular anat, Ill Col Optom, 49-51, chmn dept biol, 51-55, assoc prof path, 55-56; resident surgeon, Dept Ophthal, Cook County Hosp, Ill, 58-60; instr, 60-71, ASST PROF SURG, UNIV CALIF, LOS ANGELES, 71-; CHMN DEPT OPHTHAL, SAN PEDRO COMMUNITY HOSP, 60- Concurrent Pos: Lectr, Loyola Univ, Ill, 54-55; mem attend staff, Los Angeles County Hosp. Mem: Fel Optical Soc Am; fel Am Phys Soc; fel Am Acad Optom; fel Am Acad Ophthal & Otolarnygol; fel Int Col Surgeons. Res: Visual adaptation; chemistry; electrical responses of the eye; neuroanatomy; ocular anatomy and surgery; color vision; clinical diagnosis; physiologic optics; biochemistry. Mailing Add: 1350 W Seventh St San Pedro CA 90732

MICHAELS, ELIZABETH LOUISE, mathematics, algebra, see 12th edition

MICHAELS, IRA A L, b New York, NY. CHEMICAL PHYSICS. Educ: Brooklyn Col, BS, 70; Mass Inst Technol, PhD(phys chem), 74. Prof Exp: Res asst phys chem, Mass Inst Technol, 72-74; PROG HEALTH SCI & TECHNOL FEL, HARVARD UNIV & MASS INST TECHNOL, 74-; NIH FEL BIOPHYS, 74- Res: Hydrodynamic effects in fluids, two and three dimensions; transport in fluids; ionic transport through membranes. Mailing Add: Mass Inst of Technol 13-2130 Cambridge MA 02139

MICHAELS, JOHN EDWARD, b Boston, Mass, Feb 2, 39; m 65; c 1. CELL BIOLOGY. Educ: Harvard Univ, AB, 64; Boston Univ, PhD(biol), 70. Prof Exp: Fel anat, McGill Univ, 71-74; ASST PROF ANAT, COL MED, UNIV CINCINNATI, 74- Mem: Am Soc Cell Biol; AAAS. Res: Formation of cell coat glycoproteins and their transport from the Golgi apparatus to the cell surface studied by electron microscopy and related techniques. Mailing Add: Dept of Anat Col of Med Univ of Cincinnati Cincinnati OH 45267

MICHAELS, NICHOLAS, organic chemistry, polymer chemistry, see 12th edition

MICHAELSEN, JOHN DANIEL, physical chemistry, see 12th edition

MICHAELSON, EVALYN JACOBSON, b Los Angeles, Calif, Jan 14, 31; m 71; c 2. ANTHROPOLOGY. Educ: Univ Calif, Los Angeles, BA, 55, PhD(anthrop), 66. Prof Exp: Asst prof anthrop, Univ Ark, 66-67; asst prof, 67-72, ASSOC PROF ANTHROP, CALIF STATE UNIV, NORTHRIDGE, 72- Concurrent Pos: Actg instr, Univ Calif, Los Angeles, 70. Mem: Fel Am Anthrop Asn. Res: Family organization and values among peasants; cultural roles of women; student attitudes toward teaching and learning. Mailing Add: Dept of Anthrop Calif State Univ Northridge CA 91324

MICHAELSON, I ARTHUR, b New York, NY, Mar 15, 25; m 58; c 3. PHARMACOLOGY. Educ: NY Univ, BA, 50; George Washington Univ, PhD(pharmacol), 59. Prof Exp: Fel, Lab Chem Pharmacol, Nat Heart Inst, 59-61; USPHS fel, Agr Res Coun Inst Animal Physiol, Cambridge Univ, 61-63; fel, Lab Chem Pharmacol, Nat Heart Inst, 63-65; asst prof, 65-67, ASSOC PROF PHARMACOL, COL MED, UNIV CINCINNATI, 67- Concurrent Pos: USPHS res career develop award, 67-72; vis scientist, Toxicol Unit, Med Res Coun, Carshalton, Surrey, Eng, 71-72. Mem: Am Soc Pharmacol & Exp Therapeut. Res: Intermediary metabolism of drugs and biochemical pharmacology; subcellular localization of biogenic amines and the effect of drugs on synthesis, storage and release; toxicology; effect of metals on brain development and neurochemistry. Mailing Add: Dept of Environ Health Univ Cincinnati Col Med Cincinnati OH 45219

MICHAELSON, KAREN L, b New York, NY, Mar 11, 44. ETHNOLOGY. Educ: Univ Miami, AB, 66; Univ Wis, Madison, MA, 68, PhD(anthrop), 73. Prof Exp: Asst prof anthrop, Univ Louisville, 73-74; ASST PROF ANTHROP, STATE UNIV NY BINGHAMTON, 74- Mem: Am Anthrop Asn; Soc Appl Anthrop; Asn Asian Studies, Res: Development of social class and the political and economic anthropology of South Asia; social organization of the contemporary United States, including kinship, economic and political relations, ethnicity and class. Mailing Add: Dept of Anthrop State Univ NY Binghamton NY 13901

MICHAELSON, MERLE EDWARD, b Hudson, Wis, Feb 14, 21; m 44; c 3. PLANT PATHOLOGY. Educ: Wis State Col, River Falls, BS, 43; Colo State Univ, MS, 48; Univ Minn, PhD(plant path, bot), 53. Prof Exp: Asst prof bot, Colo State Univ, 46-49; res asst plant path, Univ Minn, St Paul, 49-52; asst prof bot, Univ Mo, 52-54; plant pathologist crops res div, Agr Res Serv, SDak Agr Exp Sta, USDA, 54-59; prof biol, St Cloud State Col, 59-67, actg dean grad sch, 66-67; PROF BIOL, UNIV WIS-RIVER FALLS, 67- Concurrent Pos: Chmn dept biol, Univ Wis-River Falls, 67-72. Mem: AAAS; Am Phytopath Soc; Mycol Soc Am; Nat Asn Biol Teachers. Res: Mycology; diseases of corn and flax. Mailing Add: Dept of Biol Univ of Wis River Falls WI 54022

MICHAELSON, NEIL ELBERT, b Minneapolis, Minn, Feb 1, 18; m 50; c 7. SOILS. Educ: Univ Minn, BS, 48; Univ Nebr, MS, 50. Prof Exp: Soil scientist, Alaska Agr Exp Sta, 48-67 & Pac Southwest Watershed Surv, US Forest Serv, 67-71; soil scientist, Northern Alaska Planning Proj, 71-75, RESOURCE SPECIALIST PLANNING & PROG COORD, BUR LAND MGT, 75- Res: Soil physical studies; watershed hydrology; soil survey and management; sprinkler irrigation. Mailing Add: Bur of Land Mgt 555 Cordova St Anchorage AK 99501

MICHAELSON, SOLOMON M, b New York, NY, Apr 23, 22; m 50; c 2. RADIATION BIOLOGY, PHYSIOLOGY. Educ: City Col New York, BS, 42; Middlesex Univ, DVM, 46; Am Col Lab Animal Med, dipl. Prof Exp: Asst prof immunol, Univ Ark, 47-48; sr pharmacologist, Eaton Labs, Norwich Pharmacal Co, 48-53; from asst prof to assoc prof radiation biol, 58-72, CHIEF RADIATION PHYSIOL & THER, ATOMIC ENERGY PROJ, UNIV ROCHESTER, 53-, ASSOC PROF MED & LAB ANIMAL MED, SCH MED & DENT, 62-, PROF RADIATION BIOL & BIOPHYS, 72- Concurrent Pos: Consult, UNRRA, 46-47, Armed Forces Radiobiol Res Inst, 63-70, Walter Reed Army Inst Res, 65-70, Vet Admin, 66- & Nat Acad Sci-Nat Res Coun, 72-; vis lectr, Am Inst Biol Sci; assoc ed, J Microwave Power, 74-; consult, Elec Power Res Inst, 75. Mem: Am Col Vet

MICHAELSON

Toxicol; Am Physiol Soc; Radiation Res Soc; Health Physics Soc; Sigma Xi. Res: Mechanisms of injury and recovery from electromagnetic radiations, especially neuroendocrine physiology. Mailing Add: Dept Radiation Biol & Biophys Univ Rochester Sch Med & Dent Rochester NY 14642

MICHAL, EDWIN KEITH, b Independence, Kans, Sept 17, 32; m 56; c 4. PHYSIOLOGY. Educ: Kans Wesleyan Univ, BA, 54; Univ Ill, MS, 62, PhD(physiol), 65. Prof Exp: Asst prof, 65-72, ASSOC PROF PHYSIOL, OHIO STATE UNIV, 72-Res: Neural mechanisms in behavior; neuroendocrinology. Mailing Add: Dept of Physiol Ohio State Univ Col Med Columbus OH 43210

MICHALEK, JOEL EDMUND, b Detroit, Mich, Aug 30, 44. STATISTICS. Educ: Wayne State Univ, BS, 66, MA, 68, PhD(statist), 73. Prof Exp: Instr math, RI Col, 70-71; ASST PROF STATIST, SYRACUSE UNIV, 73- Mem: Am Statist Asn; Inst Math Statist. Res: Non-parametric tests of hypothesis and estimation. Mailing Add: 927 Ackerman Ave Syracuse NY 13210

MICHALIK, EDMUND RICHARD, b Munhall, Pa, Aug 5, 15; m 46; c 1. APPLIED MATHEMATICS, STATISTICS. Educ: Univ Pittsburgh, BA, 37; Univ Chicago, MA, 40. Prof Exp: Instr math, Univ Pittsburgh, 40-42, asst prof math statist, 46-52; sr analyst rev planning, US Dept Army, 52-53; consult appl math & statist, Atlantic Res Co, 53-54; head dept appl math, Mellon Inst, 54-57; SR STAFF ENGR, GLASS RES CTR, PPG INDUST, INC, 57- Mem: Math Asn Am; Brit Soc Glass Technol. Res: Use of electronic digital machines in mathematic, statistics, industrial problems; glass strength and glass strengthening. Mailing Add: 3711 Spring St West Mifflin PA 15122

MICHALKA, JACK, b Summit, NJ, Nov 22, 41; m 71; c 1. MICROBIAL GENETICS. Educ: Philadelphia Col Pharm & Sci, BSc, 65; Univ Pa, PhD(microbiol), 70. Prof Exp: NIH fel, Pub Health Res Inst, NY, 70-72; LECTR GENETICS, UNIV AUCKLAND, 72- Mem: Genetics Soc Am. Mailing Add: Dept of Cell Biol Univ of Auckland Auckland New Zealand

MICHALOWICZ, JOSEPH VICTOR, b Oct 23, 41; m 63; c 2. MATHEMATICS. Educ: Catholic Univ Am, BA, 63, PhD(math), 67. Prof Exp: Mathematician, Res Anal Corp, 66-67; asst prof math, Catholic Univ Am, 67-73; MATHEMATICIAN, HARRY DIAMOND LABS, 73- Concurrent Pos: Lectr elec eng, Catholic Univ Am, 66-67; NSF fel category theory, Bowdoin Col, 69; consult, Res Anal Corp, 67-70 & Harry Diamond Labs, 70-73. Mem: Am Math Soc; Math Asn Am. Res: Systems analysis and cost effectiveness studies; statistical analyses; research in category theory. Mailing Add: 5855 Glen Forest Dr Falls Church VA 22041

MICHALOWSKI, JOSEPH THOMAS, b Newburgh, NY, Dec 17, 43. GEOCHEMISTRY. Educ: Marist Col, BA, 68; Univ Calif, Santa Barbara, PhD(chem), 72. Prof Exp: Res assoc chem, Univ Calif, Santa Barbara, 72-73, lectr, 73; res assoc biochem, St Louis Univ, 73-74; RES SCIENTIST GEOCHEM, PHILLIPS PETROL CO, 74- Mem: Am Chem Soc; AAAS; Geol Soc Am; Sigma Xi. Res: Electrochemistry and mechanics of migration of multi-phase fluids through porous rocks; problems of petroleum migration. Mailing Add: Res Bldg 3 Rm 111-B Phillips Petrol Co Bartlesville OK 74004

MICHALSKI, CHESTER JAMES, b Detroit, Mich, June 7, 42; m. MOLECULAR BIOLOGY. Educ: Mich State Univ, BS, 65, MS, 67; Univ NC, PhD(biochem), 71. Prof Exp: Fel biochem, St Judes Children's Res Hosp, Memphis, 71-72; fel, 72-75, ASST PROF MOLECULAR BIOL & DIR MULTIDISCIPLINE LABS, FAC MED, MEM UNIV NFLD, 75- Mem: Am Soc Microbiol; Can Soc Microbiol; Asn Multidiscipline Educ Health Sci. Res: Structure and function of the macromolecular components involved in the protein synthesizing system of E coli cells and the molecular events involved in hormone regulation of the fungi. Mailing Add: Fac of Med Mem Univ of Nfld St John's NF Can

MICHALSKI, RAYMOND J, b Harvey, Ill, Oct 30, 28; m 57; c 3. ENVIRONMENTAL CHEMISTRY. Educ: Univ Ill, BS, 52. Prof Exp: Microbiologist, Nat Dairy Co, 52-56; microbiologist, 56-65, group leader antifoam, 65-68, weed control-antifoam, 68-71 & surface active chem, 71-73, SR GROUP LEADER POLLUTION CONTROLS & MINERAL PROCESSING, NALCO CHEM CO, 73- Mem: Am Soc Microbiol; Soc Indust Microbiol; Sigma Xi; Tech Asn Pulp & Paper Indust. Res: Additives in paper processing areas, especially slime control, dispersants, wire life improvers, mold proofing agents and antifoams for varied industrial applications; pollution control and mineral process aids. Mailing Add: Nalco Chem Co 6216 W 65th St Chicago IL 60638

MICHALSKI, RYSZARD STANISLAW, b Kalusz, Poland, May 7, 37. COMPUTER SCIENCES. Educ: Warsaw Tech Univ, BS, 59; Leningrad Polytech Inst, MS, 61; Silesia Tech Univ, Poland, PhD(comput sci), 69. Prof Exp: Logical designer comput sci, Inst Math Mach, Polish Acad Sci, Warsaw, 61-62, res scientist, Inst Automatic Control, 62-70; vis asst prof, 70-72, ASSOC PROF COMPUT SCI, UNIV ILL, URBANA, 72- Concurrent Pos: Lectr comput sci & electronics, State Tech Col, Warsaw, Poland, 64-68; Fulbright fel, US State Dept, 70; NSF res award, 75. Mem: Asn Comput Mach; Pattern Recognition Soc; Sigma Xi; Polish Inst Arts & Sci in Am. Res: Automatic inductive inference and machine learning; multi and variable-valued logic development of inferential, intelligent computer consultants with applications in plant pathology, medicine and other areas. Mailing Add: Comput Sci Digital Comput Lab Univ of Ill Urbana IL 61801

MICHAUD, GEORGES JOSEPH, b Quebec, Que, Apr 30, 40; m 66; c 1. ASTROPHYSICS. Educ: Univ Laval, BA, 61, BSc, 65; Calif Inst Technol, PhD(astron), 69. Prof Exp: Asst prof, 69-73, ASSOC PROF PHYSICS, UNIV MONTREAL, 73- Mem: Am Astron Soc; Can Asn Physicists; Can Astron Soc; Int Astron Union. Res: Nucleosynthesis of iron peak elements; diffusion in the atmosphere of stars; mechanisms of heavy ion reactions. Mailing Add: Dept of Physics Univ of Montreal Box 6128 Montreal PQ Can

MICHAUD, HOWARD H, b Berne, Ind, Oct 12, 02; m 28; c 1. ENVIRONMENTAL MANAGEMENT. Educ: Bluffton Col, AB, 25; Ind Univ, MA, 30. Prof Exp: Teacher pub schs, Ind, 25-45; prof, 45-71, EMER PROF CONSERV, PURDUE UNIV, 71- Concurrent Pos: Del, Int Union Conserv Nature & Natural Resources, 48 & 66. Honors & Awards: Osborn Wildlife Conserv Award, 59. Mem: Conserv Educ Asn (pres, 56-57); Am Asn Biol Teachers (pres, 48); hon mem Soil Conserv Soc Am; Wildlife Soc. Res: Science and conservation education. Mailing Add: 301 E Stadium Ave West Lafayette IN 47906

MICHAUD, LAURENT, b Montreal, Que, May 25, 15; nat US; m 49; c 3. NUTRITION, VETERINARY MEDICINE. Educ: Univ Montreal, DVM, 36; Univ Wis, PhD(biochem), 48. Prof Exp: Pvt pract, Que, Can, 36-39; dir, Que Exp Fur Farm, 39-41, Biochem & Vet Res Labs, Que, 48-49 & Biochem Diag Lab, Waterbury Hosp, Conn, 49-50; MGR VET NUTRIT & DIR PROG PLANNING & CONTROL, MERCK SHARP & DOHME RES LABS, 50- Mem: Am Soc Animal Sci; Am Asn Vet Nutrit; Am Vet Med Asn; Am Dairy Sci Asn; Animal Nutrit Res Coun. Res: Animal physiology and nutrition. Mailing Add: Merck Sharp & Dohme Res Labs Rahway NJ 07065

MICHAUD, RONALD NORMAND, b Madawaska, Maine, July 7, 37; m 58; c 2. MICROBIOLOGY. Educ: Univ Maine, BS, 63; Cornell Univ, MS, 66, PhD(microbiol), 68. Prof Exp: Group leader & assoc res biologist, 68-74, RES MICROBIOLOGIST, STERLING-WINTHROP RES INST DIV, STERLING DRUG INC, 74-, SECT HEAD, 75- Mem: Am Soc Microbiol; Soc Indust Microbiol; Am Soc Testing & Mat. Res: Natural microbiological aspects of pharmaceutical product development. Mailing Add: Bact Sect Sterling Winthrop Res Inst Rensselaer NY 12144

MICHAUD, TED C, b Ft Wayne, Ind, Oct 5, 29; m 55; c 3. ZOOLOGY. Educ: Purdue Univ, BS, 51; Univ Mich, MS, 54; Univ Tex, PhD(zool), 59. Prof Exp: From asst prof to assoc prof, 59-70, PROF BIOL, CARROLL COL, WIS, 70- Concurrent Pos: Mem, Demog Inst, Cornell Univ, 71; mem steering comt, Cent States Col Asn Environ Studies Comt. Mem: AAAS; Soc Study Evolution; Am Soc Ichthyol & Herpet; Ecol Soc Am; Nat Biol Soc. Res: Amphibian behavior and evolution. Mailing Add: Dept of Biol Carroll Col 100 N East Ave Waukesha WI 53186

MICHEJDA, CHRISTOPHER JAN, b Kielce, Poland, Dec 19, 37; US citizen; m 64; c 1. PHYSICAL ORGANIC CHEMISTRY, BIO-ORGANIC CHEMISTRY. Educ: Univ Ill, BS, 59; Univ Rochester, PhD(org chem), 64. Prof Exp: NSF fel, Harvard Univ, 63-64; from asst prof to assoc prof, 64-75, PROF ORG CHEM, UNIV NEBR, LINCOLN, 75-; ASSOC PROG DIR FOR CHEM DYNAMICS, NAT SCI FOUND, DC, 75- Concurrent Pos: NIH spec fel, Swiss Fed Inst Technol, 72-73. Mem: Am Chem Soc. Res: Free radical chemistry; chemical carcinogenesis. Mailing Add: Nat Sci Found Washington DC 20550

MICHEL, ALOYS ARTHUR, b Brooklyn, NY, Sept 9, 28; m 57; c 3. GEOGRAPHY. Educ: Harvard Univ, AB, 50; Columbia Univ, MBA, 53, PhD(geog), 59. Prof Exp: Lectr geog, Columbia Univ, 56-57, vis lectr, 61; Off Naval Res grant, Afghanistan, 57-58; from asst prof to assoc prof geog, Yale Univ, 58-66; assoc prof, 66-67, chmn dept, 67-69, assoc dean grad sch, 69-71, PROF GEOG & REGIONAL PLANNING, UNIV RI, 67-, DEAN GRAD SCH, 74- Concurrent Pos: Jr fac fel, Yale Univ & Am Coun Learned Soc res grant Asia, 63-64; vis assoc prof, Columbia Univ, 66; actg dean grad sch, Univ RI, 71-74. Mem: Asn Am Geog; Am Geog Soc; Am Acad Polit & Soc Sci. Res: Regional geography of the Soviet Union; south and southwest Asia and Europe; economic geography, especially irrigation agriculture, transportation, regional planning and economic development. Mailing Add: Dept of Geog Univ of RI Kingston RI 02881

MICHEL, ALWIN EARL, solid state physics, see 12th edition

MICHEL, BEDE EUGENE, b St Paul, Minn, Mar 21, 09. CHEMISTRY. Educ: St John's Univ, Minn, BA, 31; Univ Notre Dame, PhD(org chem), 39. Prof Exp: From instr to assoc prof, 39-63, PROF CHEM, ST JOHN'S UNIV, MINN, 63- Mem: Am Chem Soc. Res: Preparation and properties of aromatic tertiary amino alcohols; dehydration of aromatic amino alcohols. Mailing Add: Dept of Chem St John's Univ Collegeville MN 56321

MICHEL, BURLYN EVERETT, b Ladoga, Ind, Mar 7, 23; m 46; c 3. PLANT PHYSIOLOGY. Educ: Univ Chicago, SB, 48, PhD(bot), 50. Prof Exp: Asst prof bot, Univ Iowa, 51-58; assoc prof, 58-64, PROF BOT, UNIV GA, 64- Concurrent Pos: Plant physiologist, Agr Res Serv, USDA, 59-62. Mem: Am Inst Biol Sci; Bot Soc Am; Am Soc Plant Physiol. Res: Plant-water relations. Mailing Add: Dept of Bot Univ of Ga Athens GA 30602

MICHEL, EDWARD L, b May 11, 26; US citizen; m; c 4. PHYSIOLOGY. Educ: St Anselm's Col, AB, 50; Univ Notre Dame, MS, 52. Prof Exp: Aviation physiologist, Air Crew Equip Lab, Dept Navy, Pa, 52-57, head respiratory physiol sect, 57-61; head respiratory thermal physiol sect, Crew Systs Div, 61-66, asst chief space physiol br, Biomed Res Off, 66-68, actg chief, 68 & 69-70, asst chief, 68-69, CHIEF BIOMED LABS DIV, LYNDON B JOHNSON SPACE CTR, NASA, 70- Concurrent Pos: Liaison mem first working group gaseous environ for manned spacecraft, Nat Acad Sci. Honors & Awards: NASA Group Achievement Award, Proj Mercury & Proj Gemini, Superior Achievement Award, 69. Mem: Aerospace Med Asn; Undersea Med Soc. Res: Development, implementation and conducting of biomedical research programs related to human tolerance and performance limitations providing requirements and reference data for the development of manned space flight equipment and operations; research support for in-flight medical experiments. Mailing Add:

MICHEL, F CURTIS, b La Crosse, Wis, June 5, 34; m 58; c 2. ASTROPHYSICS, SPACE PHYSICS. Educ: Calif Inst Technol, BS, 55, PhD(physics), 62. Prof Exp: Res fel astrophys, Calif Inst Technol, 62-63; from asst prof to assoc prof space sci, 63-70, PROF PHYSICS, SPACE PHYSICS & ASTRON, RICE UNIV, 70-, CHMN DEPT SPACE PHYSICS & ASTRON, 74- Concurrent Pos: Scientist-astronaut, NASA, 65-69; mem lunar atmosphere working group, planetary atmospheres subcomt, Space Sci Steering Comt, 66-67; mem sch natural sci, Inst Advan Study, 71-72; trustee, Univs Space Res Asn, 75- Mem: AAAS; Am Phys Soc; Am Astron Soc; Am Geophys Union. Res: Gravitational collapse; particle acceleration; pulsars; magnetospheric tail structure; solar wind interaction with moon and planets; elementary particles; weak magnetism; nuclear parity violation; symmetries; gravitationally induced electric fields. Mailing Add: Dept of Space Physics & Astron Rice Univ Houston TX 77001

MICHEL, GERD WILHELM, b Darmstadt, Ger, July 4, 30; m 59. NATURAL PRODUCTS CHEMISTRY. Educ: Darmstadt Tech Univ, Dipl chem, 56, Dr rer nat, 59. Prof Exp: Res assoc chem, Urbana, Ill, 59-62; res chemist res div, Photo Prod Dept, E I du Pont de Nemours & Co, 62-64, tech serv specialist, 64-65; asst to dir sales indust chem, E Merck A G, Darmstadt, 66-67; res fel, 67, SECT HEAD, SQUIBB INST MED RES, 67- Mem: Am Chem Soc; Soc Ger Chem; NY Acad Sci. Res: Chemistry of Mannich-bases, nitrones, alkaloids, steroids; antibiotics; photo polymerization.

MICHEL, HARDING B, b Louisville, Ky, Aug 17, 24; m 48, 70; c 1. MARINE ZOOLOGY. Educ: Duke Univ, AB, 46; Univ Miami, MS, 49; Univ Mich, PhD, 57. Prof Exp: Asst zool, 46-48, asst instr, 48-50, instr, Marine Lab, 54-57, asst prof, 57-67, assoc prof, Inst Marine Sci, 67-70, PROF BIOL OCEANOG, ROSENSTIEL SCH MARINE & ATMOSPHERIC SCI, UNIV MIAMI, 70- Concurrent Pos: Asst ed, Bull Marine Sci of the Gulf & Caribbean, 52-53. Mem: AAAS; Soc Syst Zool; Soc Study Evolution; Marine Biol Asn UK; Sigma Xi. Res: Invertebrate embryology; marine zooplankton; distribution of oceanic zooplankton in Caribbean Sea; ecology of estuaries in South Vietnam. Mailing Add: Rosenstiel Sch Marine & Atms Sci 46 Rickenbacker Causeway Miami FL 33149

MICHEL, HARRY OSCAR, b Oakland, Calif, May 15, 10; m 42; c 3.

BIOCHEMISTRY. Educ: Univ Calif, BS, 33; Duke Univ, PhD(biochem), 38. Prof Exp: Asst biochem, Duke Univ, 34-38; asst ophthal, Johns Hopkins Univ, 38-40; res biochemist, Edgewood Arsenal, Md, 40-42; biochemist, Res Directorate, Army Chem Ctr, 46-62; biochemist, Directorate Defensive Systs, 62-72, CONSULT, EDGEWOOD ARSENAL, 72- Mem: AAAS; Am Soc Biol Chemists; Sigma Xi. Res: Proteins; enzyme kinetics. Mailing Add: 6 Lombardy Pl Towson MD 21204

MICHEL, KARL HEINZ, b Marklissa, Ger, Nov 9, 29; m 59. BIO-ORGANIC CHEMISTRY. Educ: Weihenstephan Univ, Ger, BS, 56, MS, 64; Landau Univ, BS, 58. Prof Exp: Chem engr, Cent Lab, Swedish Pharmaceut Soc, 59-60; res asst org chem, Royal Inst Pharm, Stockholm, 60-65; res assoc org chem, Iowa State Univ, Ames, 65-66; res assoc, Royal Inst Pharm, Stockholm, 66-67; proj leader, Fleischmann Lab, Stamford, Conn, 67-69; sr biochemist, 69-75, RES SCIENTIST, ELI LILLY & CO, INDIANAPOLIS, IND, 75- Res: Isolation, characterization, structure determination and biological evaluation of new antibiotics and other biological active compounds. Mailing Add: Eli Lilly & Co MC-539 Indianapolis IN 46206

MICHEL, KENNETH EARL, b Chicago, Ill, May 22, 30; m 58; c 2. CYTOGENETICS. Educ: Northern Ill Univ, BS, 51, MS, 52; Univ Minn, PhD(genetics), 66. Prof Exp: Instr sci, Gavin Sch, 54-57; instr biol, Waldorf Col, 57-62; assoc prof, 66-68, PROF GENETICS, SLIPPERY ROCK STATE COL, 68- Concurrent Pos: Chmn dept genetics, Slippery Rock State Col, 66-74; mem, Maize Genetics Coop. Mem: Genetics Soc Am; Am Genetics Asn. Res: Interrelated behavior of non-homologous chromosomes in maize; chromosome pairing and disjunction. Mailing Add: Dept of Biol Slippery Rock State Col Slippery Rock PA 16057

MICHEL, LESTER ALLEN, b Mexico, Ind, Mar 5, 19; m 42; c 5. CHEMISTRY. Educ: Taylor Univ, AB, 41; Purdue Univ, MS, 44; Univ Colo, PhD(phys chem), 47. Prof Exp: Asst chem, Purdue Univ, 41-44; tech adv, Manhattan Proj, Linde Air Prod Co, NY, 44-45; asst chem, Univ Colo, 45-46; from instr to prof, 47-70, chmn dept chem, 59-70, VERNER Z REED PROF CHEM, COLO COL, 70- Concurrent Pos: Res Corp grant, 48. Mem: AAAS; Am Chem Soc. Res: Calorimetry; crystal growth; vapor pressures; isothermal flow calorimeter for vapor phase reactions; surface chemistry. Mailing Add: Dept of Chem Colo Col Colorado Springs, CO 80903

MICHEL, MARSHALL LOUIS, JR, b Biloxi, Miss, July 11, 13; m 42; c 3. SURGERY. Educ: Tulane Univ, MD, 37. Prof Exp: From asst prof to assoc prof, 49-70, PROF SURG, TULANE UNIV, 71- Concurrent Pos: Vis prof, Univ Antioquia, Colombia; chief surg, Touro Infirmary, New Orleans, 62-66; sr vis surgeon, Charity Hosp; consult, Crippled Children's Hosp. Mem: Am Cancer Soc; Am Col Surgeons; Soc Surg Alimentary Tract; assoc Int Soc Surg; Pan-Am Med Asn. Res: Clinical surgery. Mailing Add: 1430 Tulane Ave New Orleans LA 70112

MICHEL, MAYNARD C, nuclear chemistry, see 12th edition

MICHEL, RICHARD EDWIN, b Saginaw, Mich, Oct 31, 28; m 51; c 3. SOLID STATE PHYSICS. Educ: Mich State Univ, BS, 50, MS, 53, PhD(physics), 56. Prof Exp: Mem tech staff, RCA Labs, 56-62; sr res physicist, Gen Motors Res Labs, 62-73; instr, 71-73, DEAN, LAWRENCE INST TECHNOL, 73- Res: Magnetic resonance; magnetic materials; polymers; semiconductors. Mailing Add: Lawrence Inst of Technol 21000 W Ten Mile Rd Southfield MI 48075

MICHEL, ROBERT LOUIS, veterinary pathology, see 12th edition

MICHEL, RUDOLPH HENRY, b Landau, Ger, Mar 23, 25; nat US; m 49; c 2. ORGANIC CHEMISTRY. Educ: City Col New York, BS, 48; Univ Notre Dame, PhD(chem), 52. Prof Exp: Res chemist, 51-59, staff scientist, 59-63, RES ASSOC, E I DU PONT DE NEMOURS & CO, INC, 63- Mem: Am Chem Soc. Res: Mechanism of organic reactions; polymer chemistry. Mailing Add: 19 Rockland Ct Wilmington DE 19810

MICHELAKIS, ANDREW M, b Greece, Aug 12, 27; US citizen; m 64; c 2. MEDICINE. Educ: Athens Col Agr, BS, 52; Univ Kans, MS, 56; Ohio State Univ, PhD(chem), 59; Western Reserve Univ, MD, 64. Prof Exp: Asst chem, Univ Kans, 54-55; asst, Ohio State Univ, 55-59; intern, Mt Sinai Hosp, Cleveland, 64-65; resident, Vet Admin Hosp, 65-66; from instr to assoc prof med, Sch Med, Vanderbilt Univ, 66-74, from asst prof to assoc prof pharmacol, 68-74; PROF PHARMACOL & MED & DIR CLIN PHARMACOL, MICH STATE UNIV, 74- Concurrent Pos: Fel endocrinol, Vanderbilt Univ, 66-68. Mem: Am Fedn Clin Res; Endocrine Soc; Soc Exp Biol & Med; Am Soc Pharmacol & Exp Therapeut. Res: Endocrinology; hypertension and cardiovascular diseases; clinical pharmacology. Mailing Add: Dept of Pharmacol Mich State Univ Col Human Med East Lansing MI 48824

MICHELE, ARTHUR A, b Repub San Marino, Dec 22, 11; US citizen; m 38; c 3. ORTHOPEDIC SURGERY. Educ: Long Island Univ, BS, 32; NY Med Col, MD, 35; Univ Iowa, MS, 40; Am Bd Orthop Surg, dipl. Prof Exp: DIR ORTHOP SURG, USPHS HOSP, STATEN ISLAND, 40-; PROF ORTHOP SURG & CHMN DEPT, NY MED COL, 60- Concurrent Pos: Consult, city, state & fed govts, 40-; consult, St John's Episcopal & St Giles Orthop Hosps, Brooklyn, Newark Crippled Children's & St Barnabas Hosps, NJ, 42-, St Joseph's Hosp, Yonkers, NY & Highland Hosp, Beacon, NY; dir orthop surg, Metrop, Flower & Fifth Ave & Bird S Coler Mem Hosps, New York, 60 & Westchester County Med Ctr, Valhalla, NY, 60. Mem: Fel Am Col Surgeons; fel Am Rheumatism Asn. Res: Kinesiomechanics; cerebrospastics, including stroke; osteoarthritis; rheumatoid arthritis; osteochondroses; herniated nucleus pulposus; torsional syndromes of lower extremities. Mailing Add: Dept of Orthop Surg NY Med Col New York NY 10029

MICHELI, ROBERT ANGELO, b San Francisco, Calif, Dec 31, 22. ORGANIC CHEMISTRY. Educ: Univ Calif, BS, 51; Duke Univ, PhD(org chem), 54. Prof Exp: Asst, Duke Univ, 51-53; Nat Cancer Inst res fel, Harvard Univ, 54-56; res chemist, Dow Chem Co, 56-58, Western Regional Res Lab, USDA, Calif, 58-62 & Basel, 62-64; RES CHEMIST, HOFFMANN-LA ROCHE, INC, 64- Mem: Am Chem Soc. Res: Steroid and medicinal chemistry. Mailing Add: Chem Res Dept Hoffmann-La Roche Inc Nutley NJ 07110

MICHELL, ARTHUR STEPHEN, b Toronto, Ont, Sept 7, 17; m 54; c 1. FORESTRY. Educ: Univ Toronto, BScF, 40; Duke Univ, MF, 51. Prof Exp: Forester, Dept Lands & Forests, Ont, 40-42; from asst prof to assoc prof, 46-64, PROF LOGGING, UNIV TORONTO, 64- Mem: Assoc Can Pulp & Paper Asn; Royal Can Inst (pres, 63-64); Can Inst Forestry. Res: Cost analysis; logging and milling. Mailing Add: Fac of Forestry Univ of Toronto Toronto ON Can

MICHELL, JOHN HUMFREY, b Eng, July 30, 15; nat Can; m 41; c 3. ORGANIC CHEMISTRY. Educ: Univ Toronto, BA, 37, MA, 38; Mass Inst Technol, ScD(org chem), 41. Prof Exp: Res chemist, Cent Res Lab, Can Industs, Ltd, 41-44, group leader, 44-54, tech rep, London, 55-57, sect mgr develop dept, 57-59, tech mgr, Paints Div, 59-71; instr, Centennial Col Appl Arts & Technol, 72-75; VPRES, SOLCO ENERGY SYSTS LTD, 75- Mem: Fel Chem Inst Can; Solar Energy Soc Can. Res: Surface coatings; solar energy systems. Mailing Add: 61 Waterford Dr Apt 606 Weston ON Can

MICHELL, WILSON DOE, b Madison, Wis, Mar 23, 14; m 46; c 1. ECONOMIC GEOLOGY. Educ: Univ Wis, BA & MA, 35; Harvard Univ, AM, 37, PhD(econ geol), 41. Prof Exp: Geologist, State Geol Surv, Wis, 35-40, Desert Silver, Inc, Nev, 41-43, Magma Copper Co, Ariz, 44-46, Ste Nord Africaine du Plomb, French Morocco, 46-48, Resurrection Mining Co & Newmont Mining Corp, Colo, 49-50 & Reynolds Jamaica Mines, Ltd, WIndies, 51; geologist, 52-57, ASST CHIEF GEOLOGIST, REYNOLDS METALS CO, 57- Concurrent Pos: Teacher, Wis State Teachers Col, Superior, 36; mining geologist, Ramshorn Mines Co, Idaho, 38; asst, Harvard Univ, 39-40. Mem: AAAS; Soc Econ Geologists; Geol Soc Am; Am Inst Mining, Metall & Petrol Eng; Am Inst Prof Geologists. Res: Metallic ore deposits; silver ore deposits; fluorspar; bauxite. Mailing Add: Reynolds Metals Co Geol Div PO Box 27003 Richmond VA 23261

MICHELOTTI, FRANCIS WILLIAM, organic chemistry, polymer chemistry, see 12th edition

MICHELS, DONALD JOSEPH, b Brooklyn, NY, Apr 17, 32; m 61; c 6. PHYSICS. Educ: St Peter's Col NJ, BS, 54; Fordham Univ, MS, 56; Cath Univ Am, PhD(physics), 70. Prof Exp: RES PHYSICIST, NAVAL RES LAB, 61- Mem: Optical Soc Am; Am Geophys Union. Res: Extreme ultraviolet spectroscopy; solar radiation and solar-terrestrial effects; space physics and spacecraft instrumentation. Mailing Add: E O Hulburt Ctr Space Res Code 7143 M Naval Res Lab Washington DC 20375

MICHELS, HORACE HARVEY, b Philadelphia, Pa, Dec 9, 32; m 58, 75; c 2. CHEMICAL PHYSICS. Educ: Drexel Inst Technol, BSChE, 55; Univ Del, MChE, 57, PhD, 60. Prof Exp: Res engr, G & W H Corson Co, Inc, Pa, 51-53; sr anal engr, 59-62, sr res scientist, 62-68, SR THEORET PHYSICIST, UNITED TECHNOLOGIES CORP, 68- Concurrent Pos: Adj asst prof, Rensselaer Polytech, Hartford Grad Ctr, 60-65, adj assoc prof, 65-69, adj prof, 69-72; Nat Bur Stand vis fel, Joint Inst for Lab Astrophys, Univ Colo, 70; vis scholar, Quantum Inst, Univ Calif, Santa Barbara, 71; adj prof, Univ Hartford, 75- Mem: Am Chem Soc; fel Am Phys Soc; Sigma Xi. Res: Quantum mechanics of the electronic structure of atoms and molecules; thermochemistry of reacting gaseous systems at high temperatures; atomic recombination and transport processes. Mailing Add: Physics Dept United Technologies Res Ctr East Hartford CT 06108

MICHELS, JOSEPH WILLIAM, b Chicago, Ill, Nov 1, 36; m 59; c 2. ANTHROPOLOGY, ARCHAEOLOGY. Educ: Univ Calif, Los Angeles, BA, 58, MA, 63, PhD(anthrop), 65. Prof Exp: Asst prof anthrop, Pa State Univ, 65-68; asst prof, Univ Calif, San Diego, 68-69; assoc prof, 69-73, PROF ANTHROP, PA STATE UNIV, 73- Concurrent Pos: NSF inst grant, Pa State Univ, 65-66, Nat Parks Serv grant, 66-68, Pa State Mus & Hist Comn grant, 66-68, NSF res grant, 66-68 & 68-74. Mem: Fel AAAS; fel Am Anthrop Asn; Soc Am Archaeol. Res: Archaeological theory and method; African and Mesoamerican prehistory; obsidian hydration dating; computer applications in anthropology; complex society archaeology. Mailing Add: Dept of Anthrop Pa State Univ 409 Soc Sci Bldg University Park PA 16802

MICHELS, JULIAN GETZ, b Savannah, Ga, July 30, 20; m 46; c 5. ORGANIC CHEMISTRY. Educ: Univ Ga, BS, 41; Univ Tenn, MS, 43; Lehigh Univ, PhD(org chem), 49. Prof Exp: Res chemist explosives, Trojan Powder Co, 43-44 & Atlas Powder Co, 49-51; res chemist explosives, 51-59, chief sect phys & anal chem, 59-70, RES ASSOC, NORWICH PHARMACAL CO, 70- Mem: NY Acad Sci; Sigma Xi. Res: Heterocyclic chemistry. Mailing Add: Star Rte Norwich NY 13815

MICHELS, ROBERT, b Chicago, Ill, Jan 21, 36; c 2. PSYCHIATRY, PSYCHOANALYSIS. Educ: Univ Chicago, BA, 53; Northwestern Univ, MD, 58; Am Bd Psychiat & Neurol, dipl, 64; Columbia Univ, cert psychoanal med, 67. Prof Exp: Res fel, Lab Clin Sci, NIMH, 62-64; from instr to assoc prof psychiat, Col Physicians & Surgeons, Columbia Univ, 64-74; PROF PSYCHIAT & PSYCHIATRIST IN CHIEF, NY HOSP-CORNELL MED CTR, 74- Concurrent Pos: Spec lectr & instr psychiat, Columbia Univ, 60-74, attend psychiatrist, Student Health Serv, 66-74, mem fac & supv & training analyst, Psychoanal Clin Training & Res, 67-74; assoc fel, Inst Policy Studies, Washington, DC, 63-64; NIMH career teacher award, Columbia Univ, 64-66; from asst to attend psychiatrist, Vanderbilt Clin & Presby Hosp, 64-74; from asst to attend psychiatrist, St Lukes Hosp Ctr, New York, 66-; asst examr, Am Bd Psychiat & Neurol, 67-; secy, Inst Soc, Ethics & Life Sci, 71- Mem: Fel Am Psychiat Asn; Royal Medico-Psychol Asn; Asn Res Nerv & Ment Dis; Am Psychoanal Asn; Group Advan Psychiat. Res: Psychiatric education. Mailing Add: Dept of Psychiat NY Hosp-Cornell Med Ctr New York NY 10021

MICHELSOHN, MARIE-LOUISE, b New York, NY, Oct 8, 41; c 2. TOPOLOGY. Educ: Univ Chicago, BS, 62, MS, 63, PhD(math), 74. Prof Exp: Asst prof math, Univ Calif, San Diego, 74-75; LECTR MATH, UNIV CALIF, BERKELEY, 75- Mem: Am Math Soc. Res: Homotopy theory; cohomology operations; vector fields on manifolds; immersions and embeddings of manifolds. Mailing Add: Dept of Math Univ of Calif Berkeley CA 94720

MICHELSON, EDWARD HARLAN, b St Louis, Mo, June 6, 26; m 52 & 68; c 6. MALACOLOGY, PUBLIC HEALTH. Educ: Univ Fla, BS, 49, MS, 51; Harvard Univ, PhD(biol), 56. Prof Exp: Instr biol, Cambridge Jr Col, 51-53; asst, 53-5S, res assoc, 55-57, from instr to asst prof, 57-69, ASSOC PROF TROP PUB HEALTH, SCH PUB HEALTH, HARVARD UNIV, 69-, ASSOC MOLLUSKS, MUS COMP ZOOL, 57- Concurrent Pos: La State Univ-China Med Bd fel, 59; advisor Schistosomiasis, Pan Am Health Orgn, Orgn Am States, WHO, 70. Mem: Am Soc Trop Med & Hyg; Am Soc Parasitol; Am Malacol Union; NY Acad Sci; Netherlands Malacol Soc. Res: Ecology of the terrestrial mollusca of Florida; taxonomy of West Indian land and fresh water mollusca; biological control of the intermediate snails host of Schistosomiasis. Mailing Add: Harvard Sch of Pub Health Dept of Trop Pub Health 665 Huntington Ave Boston MA 02115

MICHELSON, ISRAEL DAVID, b Baltimore, Md, July 8, 97; m 36; c 1. MEDICAL BACTERIOLOGY, PATHOLOGY. Educ: Johns Hopkins Univ, AB, 18, MD, 22. Prof Exp: Intern, Mt Sinai Hosp, Baltimore, 22-23; from instr to assoc prof, 23-58, PROF BACT, PATH INST, UNIV TENN, MEMPHIS, 58- Mem: AAAS; Am Soc Microbiol; Am Asn Path & Bact. Res: Anerobic and fungus infections; bacterial mutations. Mailing Add: Dept of Microbiol Univ of Tenn Memphis TN 38103

MICHELSON, LOUIS, b Lynn, Mass, Mar 24, 19; m 41; c 1. PHYSICS. Educ: Mass Inst Technol, BS, 40. Prof Exp: Physicist, Corning Glass Works, 40-41; electronic engr, Sanborn Instruments Co, 45-46; gen mgr, Allied Cement & Chem Co, 46-47; tech dir electromagnetics, US Army Ord Submarine Mine Lab, 47-50; chief, Mine Div, US Naval Ord Lab, 50-51; tech dir torpedo hydrodynamics & acoust, US Naval Underwater Ord Sta, 51-55; mgr rocket engines, Flight Propulsion Lab, Gen Elec Co, 55-60, space environ simulator fac, Missile & Space Vehicle Dept, 60-61, Nimbus

Proj, Spacecraft Dept, 61-64, NASA progs, 64-65 & adv requirements, 65-66; pres, Spacerays, 66-67; PRES, LION PRECISION CORP, 67- Concurrent Pos: Mem acoust & ord panels, Res & Develop Bd, 47-50 & planning coun & torpedo planning adv comt, Bur Ord, 51-55. Mem: Am Mgt Asn; Am Ord Asn; Am Inst Aeronaut & Astronaut; Inst Elec & Electronics Eng. Res: Underwater sound and electric phenomena; electromagnetic fields; electronic control systems; rocket propulsion; high vacuum techniques. Mailing Add: 25 Beechcroft Rd Newton MA 02158

MICHELSON, MALVIN J, b Brooklyn, NY, Feb 3, 33; m 57; c 2. ORGANIC CHEMISTRY. Educ: City Col New York, BS, 55; Univ Ariz, MS, 57; Univ Colo, PhD(org chem), 62. Prof Exp: Res chemist, Texaco Res Ctr, 62-69; ASST PROF CHEM, MARIST COL, NY, 69- Mem: Am Chem Soc; The Chem Soc. Res: Heterocyclic compounds. Mailing Add: Dept of Chem Marist Col Poughkeepsie NY 12601

MICHENER, BRYAN PAUL, b Kenya, Sept 20, 37; US citizen; m 59; c 2. ANTHROPOLOGY. Educ: Univ Colo, BA, 62, MA, 66, PhD(anthrop), 71. Prof Exp: Community developer, Am Indian Develop & Navajo Tribe, 58-59; social worker, Steele Community Ctr, Denver, Colo, 59-60; field researcher, Inst Behav Sci, Univ Colo, 65-67; instr anthrop, Colo State Univ, 68; field dir, Nat Study Am Indians, Univ Chicago, 68-69, regional dir, 69-70; instr anthrop, Univ Colo, 70-71; ASST PROF ANTHROP, UNIV CONN, 71-, ASST TO DEAN STUDENTS, 72- Concurrent Pos: Consult, Nat Indian Youth Coun, 69-70, Nat Workshop Am Indian Educ, 69-70, Urban Systs Res & Eng, US Dept Labor, 70, Nat Study Am Indian Educ, 70-71 & Rough Rock Demonstration Sch, Navajo Reservation, 71; Dir, Native Am Prog Legis Advocacy, Friends Comt on Nat Legis, Washington, DC, 75-76. Mem: Soc Appl Anthrop; Am Anthrop Asn. Res: Cross-cultural testing; motivation; achievement; anthropological studies of education; longitudinal follow-up of Indian high school graduates; comparative educational studies; culture change; urbanization; community development; ongoing longitudinal study of motivational factors in a cross-cultural predictions of post school success, using Anglo, Navajo and individual criteria. Mailing Add: Dept of Anthrop U-158 Univ of Conn Storrs CT 06268

MICHENER, CHARLES DUNCAN, b Pasadena, Calif, Sept 22, 18; m 40; c 4. ENTOMOLOGY. Educ: Univ Calif, BS, 39, PhD(entom), 41. Prof Exp: Tech asst entom, Univ Calif, 39-42; from asst cur to assoc cur Lepidoptera & Hymenoptera, Am Mus Natural Hist, 42-48; from assoc prof to prof, 48-59, chmn dept entom, 49-61 & 72-75, ELIZABETH M WATKINS PROF ENTOM, UNIV KANS, 59-, PROF SYSTS & ECOL, 69- Concurrent Pos: State entomologist, Southern Div, Kans, 49-61; Am ed, Insectes Sociaux, 54-55, 60-; Guggenheim fel & res prof, Univ Parana, 55-56; pres, Am sect, Int Union Study Soc Insects, 57-60; Fulbright scholar, Univ Queensland, 58-59; ed, Evolution, Soc Study Evolution, 62-64; Guggenheim fel, Africa, 66-67. Mem: Nat Acad Sci; Entom Soc Am; Soc Study Evolution (pres, 67); Soc Syst Zoologists (pres, 68); hon fel Am Entom Soc. Res: Biology and taxonomy of bees; behavior of social insects; principles of systematics. Mailing Add: Dept of Entom Univ of Kans Lawrence KS 66045

MICHENER, CHARLES EDWARD, b Red Deer, Alta, Can, Jan 4, 07; m 36; c 3. GEOLOGY. Educ: Univ Toronto, BA, 31; Cornell Univ, MS, 32; Univ Toronto, PhD, 40. Prof Exp: Explor geologist, 32-35; geologist, Can Nickel Co, Ltd, 35-39, res geologist, 39-44, chief explor geologist, 44-45, vpres, 55-69; consult geologist, C E Michener & Assoc, Ltd, 69-70; CONSULT GEOLOGIST, DERRY, MICHENER & BOOTH, 70- Mem: Soc Econ Geologists; Geol Soc Am; Am Inst Mining, Metall & Petrol Eng; Can Inst Mining & Metall. Res: Mining geology. Mailing Add: Derry Michener & Booth 2302-401 Bay St Toronto ON Can

MICHENER, HAROLD DAVID, b Pasadena, Calif, Dec 21, 12; m 39; c 4. MICROBIOLOGY. Educ: Calif Inst Technol, BS, 34, PhD(plant physiol), 37. Prof Exp: Asst, Scripps Inst, Calif, 37-38; jr pomologist, Exp Sta, Univ Hawaii, 38-40; asst, Calif Inst Technol, 40-42; from jr chemist to assoc chemist, USDA, 42-55, from chemist to sr chemist, Western Mkt & Nutrit Res Ctr, 55-65, PRIN CHEMIST, WESTERN REGIONAL RES CTR, AGR RES SERV, USDA, 65- Mem: AAAS; Inst Food Technol; Am Soc Microbiol; Am Inst Biol Sci. Res: Heat resistance of bacterial spores; heat resistant fungi; food poisoning and spoilage of microbial origin; microbiological standards for foods; psychrophils; growth and survival of microorganisms at low temperatures; microbiology of frozen and chilled foods. Mailing Add: Western Regional Res Ctr USDA Berkeley CA 94710

MICHENER, JOHN WILLIAM, b Wilkinsburg, Pa, May 14, 24; m 53; c 2. PHYSICS. Educ: Carnegie Inst Technol, BS, 46, MS & PhD(physics), 53. Prof Exp: Asst physics, Carnegie Inst Technol, 42-50; physicist, Owens-Corning Fiberglas Corp, 51-59; head, Dept Physics, 59-73, MGR, TEXTILE TESTING DEPT, DEERING MILLIKEN RES CORP, 75- Mem: Am Phys Soc; AAAS; Am Soc Testing & Mat. Res: Glass structure; mechanical properties of glass; radiation damage in glass; structure and properties of textile fibers; physics of textiles; static electricity in textile materials. Mailing Add: Deering Milliken Res Corp Textile Testing Dept Box 1927 Spartanburg, SC 29304

MICHIE, DAVID DOSS, b Aniston, Ala, Feb 23, 36; m 66. CARDIOVASCULAR PHYSIOLOGY. Educ: Trinity Univ, Tex, BS, 58, MSc, 59; Univ Tex, PhD(physiol), 66. Prof Exp: Sr res physiologist, Technol Inc, 65-67; asst prof physiol, Med Sch, Creighton Univ, 67-70; asst prof surg, Sch Med, Univ Miami, 70-73; PROF PHYSIOL & BIOENG & CHMN DEPT, EASTERN VA MED SCH, 73- Mem: Fel Am Col Cardiol; Am Physiol Soc; Soc Exp Biol & Med; Am Fedn Clin Res. Res: Cardiodynamics; cardiac pacemakers; experimental surgery. Mailing Add: PO Box 1980 Norfolk VA 23507

MICHINI, LOUIS JOSEPH, microbiology, see 12th edition

MICHL, JOSEF, b Prague, Czech, Mar 12, 39; m 69. CHEMISTRY. Educ: Charles Univ, Prague, MS, 61; Czech Acad Sci, PhD(chem), 65. Prof Exp: Fel, Univ Houston, 65-66 & Univ Tex, Austin, 66-67; res chemist, Inst Phys Chem, Czech Acad Sci, 67-68; asst prof, Aarhus Univ, 68-69; fel chem, 69-70, res assoc prof, 70-71, assoc prof, 71-75, PROF CHEM, UNIV UTAH, 75- Concurrent Pos: A P Sloan Found fel, 71-75. Mem: Am Chem Soc; Czech Chem Soc; The Chem Soc; Europ Photochem Asn. Res: Quantum organic chemistry; electronic spectroscopy of organic molecules; low temperature photochemistry, especially preparation of new species and photochemical mechanisms. Mailing Add: Dept of Chem Univ of Utah Salt Lake City UT 84112

MICHLMAYR, MANFRED, b Thorn, Poland, Aug 14, 43; Austrian citizen; m 68; c 2. INORGANIC CHEMISTRY, CATALYSIS. Educ: Vienna Tech Univ, BSc, 63, MSc, 66, PhD(inorg chem), 67. Prof Exp: Asst prof inorg chem, Vienna Tech Univ, 66-67; Austrian Govt res asassoc electrochem, Czech Acad Sci, 67; NSF fel electrochem, Univ Calif, Riverside, 67-68; res chemist, Shell Develop Co, 68-72; SR RES CHEMIST, CHEVRON RES CO, STANDARD OIL CO CALIF, 72- Concurrent Pos: Session chmn, Gordon Res Conf, 75. Honors & Awards: Karoline Krafft Medal, Austrian Govt, 68. Mem: Am Chem Soc; Catalysis Soc; Electrochem Soc. Res: Catalysis in petroleum and synthetic fuel processing, including search for novel catalysts and new processes; mechanistic studies of heterogeneous catalytic systems. Mailing Add: Chevron Res Co 576 Standard Ave Richmond CA 94801

MICHNE, WILLIAM F, organic chemistry, see 12th edition

MICHNICK, MICHAEL JOSEPH, chemistry, see 12th edition

MICICH, THOMAS JOSEPH, chemistry, see 12th edition

MICKAL, ABE, b Talia, Lebanon, June 15, 13; US citizen; m 42; c 4. OBSTETRICS & GYNECOLOGY. Educ: La State Univ, BS, 36, MD, 40; Am Bd Obstet & Gynec, dipl, 51. Prof Exp: Instr anat, 45-46, from clin instr to clin assoc prof, 49-59, PROF OBSTET & GYNEC & HEAD DEPT, LA STATE UNIV MED CTR, NEW ORLEANS, 59- Mem: Fel Am Col Surgeons; AMA; Am Col Obstetricians & Gynecologists; Am Asn Obstetricians & Gynecologists; Soc Gynec Oncol. Mailing Add: Dept of Obstet & Gynec La State Univ Sch of Med New Orleans LA 70112

MICKEL, BLANCHARD LEROY, physical chemistry, see 12th edition

MICKEL, HUBERT SHELDON, b Bridgeton, NJ, Aug 27, 37; div; c 3. NEUROLOGY, NEUROCHEMISTRY. Educ: Eastern Nazarene Col, BS, 58; Harvard Med Sch, MD, 62; Am Bd Neurol & Psychiat, dipl, 71. Prof Exp: Intern, Mary Fletcher Hosp, Burlington, Vt, 62-63; resident internal med, Royal Victoria Hosp, Montreal, Que, 63-64; resident neurol, Boston City Hosp, 64-67; consult neurol, Travis State Sch, Austin, Tex, 68-70; instr, 70-71, ASST PROF NEUROL, HARVARD MED SCH, 71-; ASST NEUROL, CHILDREN'S HOSP MED CTR, 70- Concurrent Pos: Res fel neurol, Harvard Med Sch, 64-67, NIH spec fel chem, Harvard Univ, 67-68; consult, Boston State Hosp, 70-71; instr, Sch Med, Boston Univ, 70-; pre-med adv, Leverett House, Harvard Col, 71-; asst neurol, Beth Israel Hosp, Boston, 71-; mem consult staff, Emerson Hosp, Concord, 71-75; dir med & res, Wrentham State Sch, Mass, 73-, dir dept neurol, Wrentham State Sch Div, Children's Hosp Med Ctr, 74- Mem: AAAS; Am Acad Neurol; NY Acad Sci; Am Chem Soc; Am Oil Chemists Soc. Res: Lipid neurochemistry; peroxidation of unsaturated lipids; neurology and neurochemistry of mental retardation; biological effects of lipid peroxidation. Mailing Add: Dept of Neurol Children's Hosp Med Ctr Boston MA 02115

MICKEL, JOHN THOMAS, b Cleveland, Ohio, Sept 9, 34; m 59; c 4. PLANT TAXONOMY, PLANT MORPHOLOGY. Educ: Oberlin Col, BA, 56; Univ Mich, MA, 58, PhD(fern taxon), 61. Prof Exp: From asst prof to assoc prof bot, Iowa State Univ, 61-69; CUR FERNS, NY BOT GARDEN, 69- Concurrent Pos: Sigma Xi res grant, 62-63; Iowa State Alumni Asn res grant, 62-63; NSF grant, 63-66 & 69-; Nat Acad Sci-Nat Res Coun sr vis res assoc, Smithsonian Inst, 67-68; adj prof, City Univ New York, 69-; ed, Fiddlehead Forum, Am Fern Soc, 74- & Brittonia, 76. Mem: Am Fern Soc (vpres, 70-71, pres, 72-73); Bot Soc Am; Am Soc Plant Taxon; Int Asn Plant Taxon; Brit Pteridological Soc. Res: Monographic studies in the fern genus Anemia; taxonomic work on the ferns of southern Mexico; phylogeny of the ferns. Mailing Add: NY Bot Garden Bronx NY 10458

MICKELBERRY, WILLIAM CHARLES, b Seattle, Wash, May 26, 33; m 58; c 4. FOOD SCIENCE. Educ: Wash State Univ, BS, 55; Purdue Univ, MS, 60, PhD(food sci), 63. Prof Exp: From asst prof to assoc prof food sci & biochem, Clemson Univ, 62-68; mgr prod develop, Western Farmers Asn, 68-73; PROD DEVELOP MGR, ORE FREEZE DRY FOODS, INC, 73- Concurrent Pos: Mem, Res & Develop Assocs, Food & Container Inst, 63-68; mem poultry & egg inst, Am Res Coun, 68- Mem: Inst Food Technol; Poultry Sci Asn; Potato Asn Am. Res: Influence of dietary and environmental factors upon the food quality attributes of poultry meats; poultry meat tenderness; freeze dried foods, processes of freeze dried compressed foods research and development. Mailing Add: Ore Freeze Dry Foods Inc PO Box 1048 Albany OR 97321

MICKELSEN, JOHN RAYMOND, b Portland, Ore, June 1, 28; m 50; c 4. PHYSICAL CHEMISTRY. Educ: Linfield Col, BA, 50; Ore State Col, MA, 53, PhD(phys chem), 56. Prof Exp: Instr chem, Ore State Col, 54-55; from instr to asst prof, 55-61, ASSOC PROF CHEM, PORTLAND STATE UNIV, 61- Concurrent Pos: Rask-Ørsted fel, Copenhagen Univ, 66-67. Mem: Am Chem Soc; Am Electroplaters Soc. Res: Ionic equilibria in nonaqueous solvents; equilbria of complex ions. Mailing Add: Dept of Chem Portland State Univ 1620 SW Park Ave Portland OR 97201

MICKELSEN, OLAF, b Perth Amboy, NJ, July 29, 12; m 39, 53; c 2. BIOCHEMISTRY. Educ: Rutgers Univ, BS, 35; Univ Wis, MS, 37, PhD(biochem), 39; Am Bd Nutrit, dipl. Prof Exp: Chemist, Univ Hosps, Univ Minn, 39-41, assoc scientist, Physiol Hyg Lab, 42-44, from asst prof to assoc prof, 44-48; mem staff, nutrit br, USPHS, 48-51, mem lab nutrit & endocrinol, Nat Insts Arthritis & Metab Dis, 51-62; PROF NUTRIT, BIOCHEM & HUMAN GROWTH & DEVELOP, MICH STATE UNIV, 62- Concurrent Pos: Consult, Secy War, 42-43. Mem: AAAS; Am Inst Nutrit (secy, 63-66, pres, 73-74); Am Chem Soc; Am Soc Biol Chemists; Soc Exp Biol & Med. Res: Human nutrition; evaluation of nutritional status and influence of starvation on normal subjects; laboratory investigation of experimental obesity and toxic substances naturally present in foods. Mailing Add: Dept of Food Sci & Human Nutrit Mich State Univ East Lansing MI 48823

MICKELSEN, W DUANE, b Coulee City, Wash, June 27, 36; m 56; c 4. VETERINARY MEDICINE. Educ: Wash State Univ, DVM, 70. Prof Exp: Practr, Weiser Vet Clin, Idaho, 70-71 & Isenhart Vet Clin, Wenatchee, Wash, 71-72; ASST PROF LARGE ANIMAL MED, WASH STATE UNIV, 72- Concurrent Pos: Embryo transplant consult, Embryonics, Inc, Tenn, 74-75. Mem: Am Vet Med Asn; Am Asn Bovine Practrs; Soc Theriogenology. Res: Long term preservation of the bovine embryo; superovulation and embryo transplant in the mare and in the cow. Mailing Add: Dept of Vet Clin Med & Surg Wash State Univ Pullman WA 99163

MICKELSON, GRANT ALLAN, soil chemistry, see 12th edition

MICKELSON, JOHN CHESTER, b Winter, Wis, Nov 16, 20; m 47; c 4. GEOLOGY. Educ: Augustana Col, AB, 41; Univ Iowa, MS, 48, PhD(geol), 49. Prof Exp: Asst geol, Univ Iowa, 47-49; asst prof, Wash State Univ, 49-54; staff geologist, Sohio Petrol Co, 54-60; sr geologist, DX Sunray Oil Co, 61; assoc prof, 61-66, PROF GEOL & GEOL ENG, SDAK SCH MINES & TECHNOL, 66-, CHMN DEPT, 68- Mem: Geol Soc Am; Am Asn Petrol Geol. Res: Cretaceous stratigraphy and sedimentation of the Rocky Mountains; geomorphology and Pleistocene geology of Iowa and eastern Washington, particularly loesses. Mailing Add: Dept Geol & Geol Eng SDak Sch Mines & Technol Rapid City SD 57701

MICKELSON, JOHN CLAIR, b Canton, SDak, Aug 4, 29; m 52; c 2. MICROBIOLOGY. Educ: SDak State Col, BS, 51, MS, 57; Iowa State Univ, PhD(dairy bact), 60. Prof Exp: From asst prof to assoc prof, 60-71, PROF MICROBIOL, MISS STATE UNIV, 71- Mem: AAAS; Am Soc Microbiol; Am Inst

Biol Sci. Res: Dairy microbiology; microbial lipases active on butter oil; electrolytic decomposition of human wastes; electrolytic demineralization of algae. Mailing Add: Dept of Microbiol Miss State Univ State College MS 39762

MICKELSON, MICHAEL EUGENE, b Columbus, Ohio, May 3, 40; m 66; c 2. MOLECULAR SPECTROSCOPY, PLANETARY ATMOSPHERES. Educ: Ohio State Univ, BSc, 62, PhD(physics), 69. Prof Exp: ASST PROF PHYSICS, DENISON UNIV, 69- Mem: Optical Soc Am; Am Astron Soc; Am Asn Physics Teachers; Sigma Xi. Res: Spectroscopic studies under high resolution of both laboratory and telescopic spectra of molecules of astrophysical interest. Mailing Add: Dept of Physics & Astron Denison Univ Granville OH 43023

MICKELSON, MILO NORVAL, b Iowa Co, Wis, Feb 27, 11; m 41; c 4. PHYSIOLOGICAL BACTERIOLOGY. Educ: Univ Wis, BS, 35; Iowa State Univ, PhD(physiol bact), 39. Prof Exp: Res bacteriologist, Com Solvents Corp, Ind, 39-40; instr bact, Univ Mich, 40-45; sr res bacteriologist, Midwest Res Inst, 45-61; MEM STAFF, NAT ANIMAL DIS LAB, 61- Concurrent Pos: Assoc, Med Ctr, Kans, 52-61. Mem: AAAS; Am Chem Soc; Am Soc Microbiol; Am Acad Microbiol. Res: Industrial fermentations; growth requirements of microorganisms; intermediary metabolism of microorganisms. Mailing Add: 1803 Meadow Lane Ames IA 50010

MICKENS, RONALD ELBERT, b Petersburg, Va, Feb 7, 43; m 66. CHEMICAL KINETICS. Educ: Fisk Univ, BA, 64; Vanderbilt Univ, PhD(physics), 68. Prof Exp: Lab instr physics, Fisk Univ, 61-64, lectr, 66-67, lectr math, 67-68; NSF res fel physics, Mass Inst Technol, 68-70; asst prof, 70-72, ASSOC PROF PHYSICS, FISK UNIV, 72- Concurrent Pos: Vis Prof, Howard Univ, 70-71 & Mass Inst Technol, 73-74. Mem: AAAS; Am Phys Soc. Sigma Xi. Res: Complex angular momentum; Regge theory; analytic properties of collision amplitudes; asymptotic bounds on the behavior of scattering amplitudes. Mailing Add: Dept of Physics Fisk Univ Nashville TN 37203

MICKEY, DONALD LEE, b Fairfield, Iowa, Mar 28, 43; m 62; c 2. ASTROPHYSICS. Educ: Harvard Univ, AB, 64; Princeton Univ, PhD(astrophys sci), 68. Prof Exp: Res assoc astrophys, Princeton Univ, 68-69; res fel physics, Calif Inst Technol, 69-70; ASST ASTRONR ASTROPHYS, INST ASTRON, UNIV HAWAII, 70- Res: Solar physics; spectroscopy. Mailing Add: Inst for Astron Univ of Hawaii PO Box 135 Kula Maui HI 96790

MICKEY, GEORGE HENRY, b Claude, Tex, Jan 26, 10; m 32; c 2. CYTOGENETICS. Educ: Baylor Univ, AB, 31; Univ Okla, MS, 34; Univ Tex, PhD(genetics), 38. Prof Exp: Asst zool, Univ Okla, 32-34; asst genetics, Univ Tex, 34-35, instr zool, 35-38; from instr to assoc prof, La State Univ, 38-48; assoc prof, Northwestern Univ, Ill, 49-56; prof & chmn dept, La State Univ, 56-59, dean grad sch, 59-60; cytogeneticist, 60-66, prof biol, 66-69, assoc dean grad sch, 69-70, actg dean, 70-71, DEAN GRAD SCH, NEW ENG INST, 71- Concurrent Pos: Guggenheim fel, 48; res fel, Calif Inst Technol & Univ Tex, 48; prin biologist, Oak Ridge Nat Lab, 53. Mem: AAAS; Am Soc Nat; Genetics Soc Am; Soc Study Evolution; Am Soc Zool. Res: Genetics and cytology of Drosophila; cytology of Romalea; radiation genetics; mutation studies; cytogenetic effects of radio frequency waves; human cytogenetics; tissue culture. Mailing Add: New Eng Inst PO Box 308 Ridgefield CT 06877

MICKEY, MAX RAY, JR, b Pagosa Springs, Colo, Mar 24, 23; m 48; c 2. STATISTICS. Educ: Va Polytech Inst, BS, 47; Iowa State Col, PhD(statist), 52. Prof Exp: Asst prof statist, Iowa State Col, 52-55; assoc mathematician, Rand Corp, 55-58; statistician, Gen Anal Corp, 58-60 & CEIR, Inc, 60-63; RES STATISTICIAN, DEPT BIOMATH, UNIV CALIF, LOS ANGELES, 63-, LECTR BIOMATH, 74- Mem: Economet Soc; Inst Math Statist; fel Am Statist Asn. Res: Application of statistical concepts and methods to applied problems of research, particularly in medicine. Mailing Add: Dept of Biomath Univ of Calif Los Angeles CA 90024

MICKEY, WENDELL VADEN, geophysics, see 12th edition

MICKLE, EARL JOHN, b Sandusky, Ohio, Feb 5, 11; m 42; c 2. MATHEMATICS. Educ: Mt Union Col, BS, 33; Bowling Green State Univ, BS, 35; Ohio State Univ, MA, 39, PhD(math), 41. Prof Exp: Teacher high sch, Ohio, 35-37; asst math, 37-41, from instr to assoc prof, 41-51, PROF MATH, OHIO STATE UNIV, 51- Mem: AAAS; Am Math Soc; Math Asn Am. Res: Calculus of variations; analysis; surface area; topology; measure theory. Mailing Add: Dept of Math Ohio State Univ, Columbus, OH 43210

MICKLE, JAMES BURKET, food science, deceased

MICKLES, JAMES, b Rochester, NY, May 17, 23; m 46; c 2. MEDICINAL CHEMISTRY. Educ: Brigham Young Univ, BS, 44; Purdue Univ, MS, 49. Prof Exp: Chemist Qm Corps Proj, Columbia Univ, 45-47; from asst prof to assoc prof, 50-67, PROF CHEM, MASS COL PHARM, 67-, DEANS STUDENTS & DIR ADMIS, 74- Mem: Am Chem Soc; Am Pharmacuet Asn; fel Am Inst Chem. Res: Chelates of pharmacologically active compounds; synthesis of anti-radiation compounds. Mailing Add: Dept of Chem Mass Col of Pharm 179 Longwood Ave Boston MA 02115

MICKLEWRIGHT, MALCOLM A, b Winnipeg, Man, Sept 22, 20; m 42; c 1. GEOGRAPHY, ECONOMICS. Educ: Univ Wash, BA, 65, PhD(geog, econ), 70. Prof Exp: Gen mgr cartog, Windsor Plate Makers Ltd, 51-54; supvr cartog, Van Dyke & Brant, 56-63; asst prof geog, Mem Univ Nfld, 69-71; ASST PROF GEOG, UNIV VICTORIA, 71-, GRAD STUDIES ADV, DEPT GEOG, 72- Concurrent Pos: Mem, Brit Econ Res Coun. Mem: Can Asn Geog; Am Econ Asn; Can Econ Asn; Fr Lang Asn Regional Sci. Res: Regional planning and economic development; socio-economic problems of resource based communities. Mailing Add: Dept of Geog Univ of Victoria Victoria BC Can

MICKLICH, JOHN R, b Pueblo, Colo, Aug 14, 25; m 47 & 56; c 3. MATHEMATICS. Educ: Baker Univ, AB, 48; Eastern NMex Univ, MS, 56; Univ NMex, EdD(math educ), 69. Prof Exp: Storekeeper, Kans Power & Light Co, 48-52; instr math, NMex Mil Inst, 52-63; asst prof, 63-68, ASSOC PROF MATH, NORTHERN ARIZ UNIV, 68- Concurrent Pos: Dir stud sci training prog, NSF, 69- Mem: Math Asn Am. Res: Mathematics education; effect of homework on student performance. Mailing Add: 538 Charles Rd Flagstaff AZ 86001

MICKLIN, PHILIP PATRICK, b Gig Harbor, Wash, May 8, 38; m 61; c 1. RESOURCE GEOGRAPHY. Educ: Univ Wash, BA, 60, MA, 66, PhD(geog), 71. Prof Exp: Instr geog, Western Mich Univ, 66-67 & 69-71, ASST PROF GEOG, WESTERN MICH UNIV, 71- Mem: Asn Am Geog. Res: Environmental problems in the Union of Soviet Socialist Republics, particularly hydrologic in nature; environmental problems in the United States. Mailing Add: Dept of Geog Western Mich Univ Kalamazoo MI 49001

MICKO, MICHAEL M, b Trebisov, Czechoslvakia, Nov 9, 35; m 61; c 2. ENVIRONMENTAL CHEMISTRY. Educ: Slovak Tech Univ, Bratislava, BEng, 59, PhD(polymer chem), 66; Univ BC, PhD(wood sci & technol), 73. Prof Exp: Asst engr polymers, Slovak Acad Sci, 59-61; asst prof fiber technol, Slovak Tech Univ, Bratislava, 66-69; RES ASSOC BIORESOURCE ENG, UNIV BC, 73- Res: Pollution standards and control monitoring. Mailing Add: Univ of BC Dept Bioresource Eng 2075 Wesbrook Pl Vancouver BC Can

MICKS, DON WILFRED, b Mt Vernon, NY, Nov 23, 18; m 44; c 4. PREVENTIVE MEDICINE, COMMUNITY HEALTH. Educ: NTex State Univ, BS, 40, MS, 42; Johns Hopkins Univ, ScD(parasitol), 49. Prof Exp: Instr zool, NTex State Univ, 40-42; asst, Univ Mich, 42; asst med entom, Sch Hyg & Pub Health, Johns Hopkins Univ, 46-48; from asst prof to assoc prof, 49-59, PROF PREV MED & COMMUNITY HEALTH, UNIV TEX MED BR GALVESTON, 59-, CHMN DEPT, 66- Concurrent Pos: Fulbright scholar, Italy, 53-54; scientist-biologist, Div Environ Health, WHO, Switz, 58-59, consult, Pakistan, 69. Mem: Asn Teachers Prev Med; fel Am Pub Health Asn; Entom Soc Am; Am Mosquito Control Asn (pres, 63); Am Soc Trop Med & Hyg. Res: Insect transmission of disease; arthropod venoms; development of new, non-toxic vector control agents; health applications of remote sensing. Mailing Add: Dept Prev Med & Commun Health Univ of Tex Med Br Galveston TX 77550

MICUCCI, DOMINIC DONALD, organic chemistry, see 12th edition

MICZAIKA, GERHARD ROBERT, astrophysics, see 12th edition

MIDDAUGH, PAUL RICHARD, b Fargo, NDak, Feb 11, 20; m 43; c 2. BACTERIOLOGY. Educ: NDak Agr Col, BS, 42; Univ Wis, MS, 48, PhD(bact, biochem), 51. Prof Exp: Asst bact, Univ Wis, 46-48, exten div, 48-51; med bacteriologist chem corps, US Dept Army, Utah, 51-52, Ft Detrick, Md, 52-53, supvry bacteriologist, 53-59; sr microbiologist, Grain Processing Corp, Iowa, 59-64; assoc prof, 64-67, PROF BACT, SDAK STATE UNIV, 67- Mem: AAAS; Am Soc Microbiol; Sigma Xi; Brit Soc Gen Microbiol. Res: Nitrogen metabolism of lactic acid bacteria; industrial fermentation. Mailing Add: Dept of Bact SDak State Univ Brookings SD 57006

MIDDAUGH, RICHARD LOWE, b Salamanca, NY, Oct 2, 38. INORGANIC CHEMISTRY. Educ: Harvard Univ, AB, 60; Univ Ill, MS, 62, PhD(chem), 65. Prof Exp: Asst prof, 64-69, ASSOC PROF CHEM, UNIV KANS, 69- Mem: Am Chem Soc; The Chem Soc. Res: Structure, reactions and properties of polyhedral molecules. Mailing Add: Dept of Chem Univ of Kans Lawrence KS 66044

MIDDELKAMP, JOHN NEAL, b Kansas City, Mo, Sept 29, 25; m 49, 74; c 4. PEDIATRICS. Educ: Univ Mo, BS, 46; Wash Univ, MD, 48. Prof Exp: Med intern, Gallinger Munic Hosp, 48-49; asst resident pediat, St Louis Children's Hosp, 49-50 & 52-53, co-chief resident, 53; from instr to assoc prof, 54-70, PROF PEDIAT, SCH MED, WASH UNIV, 70- Concurrent Pos: Consult, Homer G Phillips Hosp, St Louis, Mo, 54-, Barnes & Allied Hosp, 54-, Crippled Children's Servs, Univ Ill, 55- & Univ Mo, 66-; fel internal med, Wash Univ, 60-61, USPHS fel anat, 61-62. Mem: AAAS; Am Acad Pediat; Am Soc Microbiol; Am Pediat Soc; Infectious Dis Soc Am. Res: Infectious diseases; virology; bacteriology; electron microscopy. Mailing Add: Dept of Pediat Wash Univ Sch of Med St Louis MO 63110

MIDDENDORF, DONALD FLOYD, b Templeton, Iowa, Feb 26, 31; m 62; c 1. POULTRY NUTRITION, BIOCHEMISTRY. Educ: Iowa State Univ, BS, 54; Univ Md, MS, 58, PhD(poultry nutrit), 59. Prof Exp: Poultry res specialist, Cent Soya Co, Inc, 59-65; res specialist, Upjohn Co, Inc, Mich, 65-67; nutrit specialist, 67-68, POULTRY FEEDS DIR, CENT SOYA CO, INC, 68- Concurrent Pos: Mem, Alfalfa Res Coun, 64-72. Mem: Poultry Sci Asn; NY Acad Sci; Am Inst Chem. Res: General poultry nutrition, especially amino acids; protein and energy areas; effect of environment upon performance of growing and laying chickens and turkeys. Mailing Add: Cent Soya Co Inc 1200 N Second St Decatur IN 46733

MIDDLEBROOK, GARDNER, b Lowell, Mass, Dec 6, 15; m 48; c 4. MEDICINE. Educ: Harvard Univ, AB, 38, MD, 44. Prof Exp: Asst path & bact, Rockefeller Inst, 45-50, assoc, 50-52; asst prof microbiol, Sch Med, Univ Colo, 53-54, prof, 54-64; PROF PATH, SCH MED, UNIV MD, 65- Concurrent Pos: Dir res & labs, Nat Jewish Hosp, Denver, Colo, 52-64; mem, Study Sect Bact & Mycol, NIH, 58-63, Training Grant Comt, 63-; mem, Comn Acute Respiratory Dis & chmn, Comt Tuberc Control, Armed Forces Epidemiol Bd, 63-70; chmn, US Tuberc Panel, US-Japan Coop Med Sci Prog, Off Sci & Technol, 65-70. Honors & Awards: Pasteur Medal, Pasteur Inst, Paris, 54. Mem: Am Soc Clin Invest; Soc Exp Biol & Med; Harvey Soc. Res: Experimental pathology and infectious diseases. Mailing Add: Dept of Path Univ of Md Sch of Med Baltimore MD 21201

MIDDLEBROOK, ROBERT E, organic chemistry, see 12th edition

MIDDLEBROOKS, BOBBY L, virology, rickettsiology, see 12th edition

MIDDLEHURST, BARBARA MARY, b Penarth, Wales, Sept 10, 15. ASTRONOMY. Educ: Cambridge Univ, BA, 36, MA, 47. Prof Exp: Observer astron, Univ Observ, St Andrews Univ, 51-54, lectr, Univ, 54-59; res assoc, Yerkes Observ, Chicago, 59-60; res assoc, Lunar & Planetary Lab, Univ Ariz, 60-68; astron ed, Encycl Britannica, 68-72; MEM STAFF, LUNAR SCI INST, NASA BAY, 72- Concurrent Pos: Goethe Link fel, Ind Univ, 53, Fulbright travel grant & res assoc, 53-54; Carnegie Trust Scottish Univs res grant, 54; prin investr, Off Naval Res Proj Grant, 63; mem comn, Int Astron Union, 64; NSF proj grant, 66; consult, Lockheed Electronics, 69; consult, Chicago Sch Dist 97, 69-70. Mem: Hon fel Am Geophys Union; Am Astron Soc; Royal Astron Soc; Brit Astron Asn. Res: Stellar, lunar and planetary research; discovery of the Middlehurst effect, that is, tidally related periodicity in reported shortlived lunar phenomena, substantiated in a similar periodicity in seismic signals recorded by instruments landed on the moon. Mailing Add: Lunar Sci Inst NASA Bay Houston TX 77058

MIDDLEKAUFF, WOODROW WILSON, b Hagerstown, Md, Feb 11, 13; m 38; c 2. ENTOMOLOGY. Educ: Juniata Col, BS, 35; Cornell Univ, MS, 36, PhD(entom), 41. Prof Exp: From asst to instr entom, Cornell Univ, 36-42; from asst prof to assoc prof, 46-58, PROF ENTOM & ENTOMOLOGIST, EXP STA, UNIV CALIF, BERKELEY, 59-, ASST DEAN COL AGR, 64- Concurrent Pos: Assoc entomologist, Exp Sta, Univ Calif, Berkeley, 51-59. Mem: Entom Soc Am. Res: Taxonomy of sawflies; ecology of Acrididae; bionomics and control of economic insect pests of California. Mailing Add: Dept of Entom Giannini Hall Univ of Calif Berkeley CA 94720

MIDDLEMISS, ROSS RAYMOND, b Marysville, Kans, July 30, 03; m 28. MATHEMATICS. Educ: Univ Colo, BS, 26, MS, 29. Prof Exp: Instr eng math, Univ Colo, 26-29; asst prof math, 29-42, assoc prof appl math, 42-45, prof math, 45-69, EMER PROF MATH, WASH UNIV, 69- Mem: Math Asn Am. Res: Differential and integral calculus; analytic geometry; college algebra. Mailing Add: 1343 Woodlawn Ave Canon City CO 81212

MIDDLETON, ALEX LEWIS AITKEN, b Banchory, Scotland, May 20, 38; Can citizen; m 62; c 3. ZOOLOGY, ECOLOGY. Educ: Univ Western Ont, BSc, 61, MSc, 62; Monsah Univ, Australia, PhD(zool), 66. Prof Exp: Asst prof, 66-70, ASSOC PROF ZOOL, UNIV GUELPH, 70- Mem: Am Ornith Union; Can Soc Zoologists; Royal Australasian Union. Res: Ecology of birds, particularly histology and physiology of breeding cycles. Mailing Add: Dept of Zool Univ of Guelph Guelph ON Can

MIDDLETON, ARTHUR EVERTS, b Erie, Pa, June 10, 19; m 41; c 2. SOLID STATE PHYSICS, ELECTRONICS. Educ: Westminster Col, BS, 40; Purdue Univ, MS, 42, PhD(physics), 44. Prof Exp: Asst physics, Purdue Univ, 40-43, instr, 43-45; res engr, Fed Tel & Radio Co, NJ, 45; res engr, Battelle Mem Inst, 45-47, asst supvr res, 47-51, supvr, 51-53; dir physics & phys chem labs, P R Mallory & Co, Inc, 53-57; tech counr & group leader, Large Lamp Eng Dept, Gen Elec Co, 58-59; mgr & dir solid state div, Harshaw Chem Co, 59-62; chief scientist, Ohio Semiconductors Div, Tecumseh Prod, Inc, 62-64; VPRES & DIR, OHIO SEMITRONICS, INC, 64-; PROF ELEC ENG & DIR ELECTRONICS MAT & DEVELOP LAB, OHIO STATE UNIV, 65- Concurrent Pos: Consult, Adv Group Electronic Parts, 55-57; mem adv panel dielectrics, Nat Adv Bd, 56; dir, N Pittsburgh Tel Co, 56-; mem adv panel passive components, Wright Air Develop Ctr, 57; dir, Ohio Semiconductors, Inc, 58-60; lectr, Univ Mich, 65. Mem: Electrochem Soc; Am Phys Soc; Inst Elec & Electronics Engrs. Res: Nuclear reactions in photographic emulsions; galvanomagnetic properties of semiconductors; thermoelectric materials and devices; Hall effect and electroluminescent devices; integrated circuit technology; solid state radiation detection; solar energy convertors; electrophotographic plates; new semiconductors and other electronic components. Mailing Add: Dept of Elec Eng Ohio State Univ Columbus OH 43210

MIDDLETON, CHARLES CHEAVENS, b Pilot Point, Tex, Apr 12, 30; m; c 2. EXPERIMENTAL PATHOLOGY. Educ: Univ Mo-Columbia, BS & DVM, 58; Mich State Univ, MS, 61; Am Col Lab Animal Med, dipl, 66. Prof Exp: Instr vet surg, Univ Pa, 58-59; instr physiol & pharmacol, Mich State Univ, 60-62; fel cardiovasc res, Bowman Gray Sch Med, 62-63, from instr to asst prof lab animal med, 63-66; asst prof community health & med practices, 66-70, assoc prof vet path, 66-75, ASSOC PROF COMMUNITY HEALTH & MED PRACTICES, UNIV MO-COLUMBIA, 70-, PROF VET PATH, 75-, DIR SINCLAIR RES FARM, 66- Concurrent Pos: Co-investr, Dept Health, Educ & WElfare grants, 67-68, 68-70 & 68-72; post doctoral thesis adv, Sch Med & Sch Vet Med, Univ Mo-Columbia, 67-; mem coun arterosclerosis Am Heart Asn. Mem: Am Inst Biol Sci; Int Primatol Soc; Am Col Lab Animal Med; Am Vet Med Asn; NY Acad Sci. Res: Atherosclerosis; pathology of laboratory animals. Mailing Add: Sinclair Res Farm R R 3 Columbia MO 65201

MIDDLETON, DAVID, b New York, NY, Apr 19, 20; m 45; c 4. APPLIED PHYSICS, APPLIED MATHEMATICS. Educ: Harvard Univ, AB, 42, AM, 45, PhD(physics), 47. Prof Exp: Res assoc, Off Sci Res & Develop proj, Harvard Univ, 42-45, res fel electronics, 47-49, asst prof appl physics, 49-54; CONSULT PHYSICIST, 54- Concurrent Pos: Consult, Govt & Industs, 49-; adj prof, Columbia Univ, 60-61, Rensselaer Polytech Inst, 61-70 & Univ RI, 66-; mem, Naval Res Adv Comt, 70-, Navy Lab Adv Bd Undersea Warfare, 70-, Navy Lab Adv Bd Res, 71- & US study group 1A, Int Radio Consult Comt, 71-; contractor, Off Naval Res, Dept Defense, Inst Telecommun Sci, Off Telecommun, Dept Commerce, NASA & Nat Oceanic & Atmospheric Admin; consult, Off Telecommun Policy-Exec Off of Pres, 74- Honors & Awards: Award, Nat Electronic Conf, 56. Mem: Fel AAAS; fel Am Phys Soc; Am Math Soc; fel Inst Elec & Electronics Eng; Inst Math Statist. Res: Communication theory in radar, radio, underwater sound, seismology, optics, mechanics, electronics, space sciences; applied mathematics; scattering theory; wave surface oceanography; mass-made and natural EM environments; electromagnetic compatability. Mailing Add: 127 E 91st St New York NY 10028

MIDDLETON, EDWARD JAMES, b London, Ont, Sept 13, 22; m 45. TOXICOLOGY. Educ: Univ Western Ont, BSc, 52, MSc, 55; Rutgers Univ, PhD(physiol, biochem), 59. Prof Exp: Chemist, Atomic Energy Can Ltd, 52-53; chemist, 55-66, sci adv, 66-70, HEAD FOOD ADDITIVES SECT, DIV TOXICOL EVAL, FOOD & DRUG DIRECTORATE, 70- Mem: Nutrit Soc Can (treas, 65-68); Prof Inst Pub Serv Can; Soc Toxicol. Res: Toxicological assessment of food additives, cosmetics, sanitizers used in food industry and food contaminants; effect of nutritional state on the toxicity of various food additives. Mailing Add: Food Addit Sect Div Toxic Eval Bur Chem Safety Hlth Prot Br Ottawa ON Can

MIDDLETON, ELLIOT, JR, b Glen Ridge, NJ, Dec 15, 25; m 48; c 4. INTERNAL MEDICINE. Educ: Columbia Univ, MD, 50; Princeton Univ, AB, 51; Am Bd Internal Med, dipl, 58, Am Bd Allergy, dipl, 62. Prof Exp: Intern, Presby Hosp, New York, 50-51, asst resident, 51-53; asst med, Col Physicians & Surgeons, Columbia Univ, 56-57, instr, 57-60, assoc, 60-69; DIR CLIN SERV & RESEARCHER, CHILDREN'S ASTHMA RES INST & HOSP, NAT ASTHMA CTR, 69- Concurrent Pos: Nat Heart Inst clin fel, Presby Hosp, New York, 51-53; clin fel, NIH, 53-54; clin & res fel, Inst Allergy, Roosevelt Hosp, 55; asst, Immunochem Lab, Col Physicians & Surgeons, Columbia Univ, 52; physician pvt pract, 56-69; asst attend physician, Mountainside Hosp, Montclair, NJ, 58-62, assoc attend, 62-; asst physician, Presby Hosp, NY, 60. Mem: AAAS; Am Fedn Clin Res; Am Asn Immunologists; Am Acad Allergy (pres, 72-73); Harvey Soc. Res: Allergy; immunology; biochemical mechanisms of human allergic reactions by in vitro techniques; chemical mediators of allergic reactions. Mailing Add: 1999 Julian St Denver CO 80204

MIDDLETON, FRANCIS MARVIN, b Wilmington, Ohio, Mar 28, 12; m 42; c 3. SANITARY CHEMISTRY. Educ: Wilmington Univ, BS, 34; Univ Mich, MPH, 49. Prof Exp: Teacher high sch, Ohio, 36-39; chemist, USPHS, 39-48, chemist in charge org contaminants unit, Water Prog, 49-60, chief advan waste treatment res prog, 60-66; dir res, Cincinnati Water Res Lab, Fed Water Pollution Control Admin, 66-68; dir, Robert A Taft Water Res Ctr, 68-72, DEP DIR, NAT ENVIRON RES CTR, ENVIRON PROTECTION AGENCY, 72- Honors & Awards: Bartow Award, Am Chem Soc, 55; Am Water Works Asn Award, 58; Meritorious Serv Award, Dept Health, Educ & Welfare, 57 & USPHS, 66. Mem: Am Water Works Asn; Water Pollution Control Fedn. Res: Organic contaminants that affect the quality of water; sampling and analytical procedures; taste and odor aspects; total role in environment; sanitary chemistry. Mailing Add: Nat Environ Res Ctr Environ Protection Agency Cincinnati OH 45268

MIDDLETON, GERARD VINER, b Capetown, SAfrica, May 13, 31; m 59; c 3. GEOLOGY. Educ: Imp Col, Univ London, BSc, 52, dipl & PhD(geol), 54. Prof Exp: Geologist, Standard Oil Co Calif, 54-55; from asst prof to assoc prof, 55-67, chmn dept, 59-62, PROF GEOL, McMASTER UNIV, 67- Concurrent Pos: Consult, Shell Oil Co, 56-57 & 59. Mem: Soc Econ Paleontologists & Mineralogists; Am Asn Petrol Geol; Geol Asn Can; fel Royal Soc Can; Int Asn Sedimentol. Res: Sedimentary petrography; sedimentology. Mailing Add: Dept of Geol McMaster Univ Hamilton ON Can

MIDDLETON, HUGH WILLIAM, b Watford, Eng. CHEMISTRY, INFORMATION SCIENCE. Educ: Univ London, BSc, 64; Univ Toronto, PhD(chem), 68. Prof Exp: Med Res Coun fel, Univ Toronto, 69-70; group mgr chem appl to microbiol, Thomson Res Assocs, 70-72; PRES, SEARCHFAST SYSTS, LTD, 72- Mem: Am Chem Soc. Res: Mechanism of action of antimicrobial compounds; design and efficiency of computerized information systems; design of artificially intelligent computer systems; development of high level computer languages. Mailing Add: 2240 Mississauga Rd N Mississauga ON Can

MIDDLETON, JOHN F M, b London, Eng, May 22, 21; m; c 2. ANTHROPOLOGY. Educ: Univ London, BA, 41; Oxford Univ, BSc, 49, DPhil(anthrop), 53. Prof Exp: Lectr anthrop, Univ London, 53-54; sr lectr, Univ Cape Town, 54-56; lectr anthrop, Univ London, 56-63; prof, Northwestern Univ, 63-66; PROF ANTHROP, NY UNIV, 66- Mem: Royal Anthrop Inst Gt Brit & Ireland. Res: Social anthropology, especially of Africa; religion and politics. Mailing Add: Dept of Anthrop New York Univ Univ Heights Bronx NY 10453

MIDDLETON, PETER JAMES, b Leeston, NZ, Sept 29, 32; m 59. MEDICAL VIROLOGY. Educ: Univ NZ, MB & ChB, 57; Univ Otago, NZ, MD, 63; FRCPath(A), 65; CRCP(C), 70. Prof Exp: Asst lectr path, Med Sch, Univ Otago, NZ, 60-62; trainee clin virol, Regional Virus Lab, Ruchill, Glasgow, Scotland, 63-64; from lectr to sr lectr microbiol, Med Sch, Univ Otago, NZ, 64-68; CHIEF VIROL, HOSP SICK CHILDREN, TORONTO, 68- Concurrent Pos: Assoc prof med microbiol, Fac Med & asst prof, Sch Hyg, Univ Toronto, 68- Mem: Can Pub Health Asn; Can Soc Microbiol; Australian Col Path; Am Soc Microbiol; Can Microscopal Soc. Res: Gastro-enteritis viral; evaluation of viral diagnostic techniques; multiple sclerosis. Mailing Add: Dept Virol Hosp Sick Children 555 Univ Ave Toronto ON Can

MIDDLETON, RICHARD B, b Rockford, Ill, Nov 24, 36; m 59; c 2. MICROBIAL GENETICS, MEDICAL LAW. Educ: Harvard Univ, AB, 58, Am, 60, PhD(biol), 63. Prof Exp: Res fel bact genetics, Brookhaven Nat Lab, 62-64; asst prof biol, Am Univ Beirut, 64-65; asst prof genetics, McGill Univ, 65-71; assoc prof, 71-75, PROF MICROBIOL & GENETICS, MEM UNIV NFLD, 75- Concurrent Pos: Res grants, Rockefeller Found, 64-65, Med Res Coun Can, 66-68 & 72-75, Res Corp, 66-68, Nat Res Coun Can, 66-, Int Cell Res Orgn, Int Lab Genetics & Biophys, Naples, 67, Food & Drug Directorate, Dept Nat Health & Welfare Can, 69-71, World Health Orgn, Ctr Immunol, State Univ NY Buffalo, 71, Tissue Cult Asn, Jones Cell Sci Ctr, Lake Placid, NY, 72, Europ Molecular Biol Orgn, Biozentrum, Univ Basel, Switz, 72 & March Dimes, Nat Found, Jackson Lab, Bar Harbor, Maine, 74. Mem: AAAS; Am Soc Microbiol; Can Soc Cell Biol; Genetics Soc Am; NY Acad Sci. Res: Genetic homology of Salmonella typhimurium and Escherichia Coli; fertility of intergeneric crosses of enteric bacteria; legal reform for donation of human tissues for scientific uses. Mailing Add: Fac of Med Mem Univ St John's NF Can

MIDDLETON, SAMUEL, b Santiago, Chile, May 3, 12; m 40; c 2. PHYSIOLOGY. Educ: Univ Chile, MD, 38. Prof Exp: Asst exp physiol, Univ Chile, 33-36, from asst prof to assoc prof, 36-46, prof extraordinary, 47-56; chief prof educ, Pan Am Health Orgn, 56-59; PROF PHYSIOL & DIR DEPT PHYSIOL & BIOPHYS, UNIV CHILE, 59-, SECY GEN, FAC MED, 68- Concurrent Pos: Rockefeller Found scholar, Western Reserve Univ, 42-43; Brit Coun fel, Eng, 49; Univ Chile-Pan Am Sanit Bur fel, Can & US, 53-54; res guest, Johns Hopkins Univ, 53 & Univ Wis, 72. Mem: AAAS; Chilean Biol Soc; Latin Am Asn Physiol Sci; Cardiol Soc Chile. Res: Release of neurohormones in cardiac sympathetic; central control of cardiac activity. Mailing Add: Dept of Physiol & Biophys Univ of Chile Santiago Chile

MIDDLETON, WILLIAM JOSEPH, b Amarillo, Tex, Apr 9, 27; m 48; c 2. ORGANIC CHEMISTRY. Educ: NTex State Col, BS, 48, MS, 49; Univ Ill, PhD(chem), 52. Prof Exp: RES CHEMIST, E I DU PONT DE NEMOURS & CO, INC, 52- Mem: Am Chem Soc; Sigma Xi. Res: Cyanocarbon, organic fluorine, heterocyclic and medicinal chemistry. Mailing Add: R D 2 Box 62A Ridge Rd Chadds Ford PA 19317

MIDER, GEORGE BURROUGHS, b Windsor, NY, Aug 9, 07; m 39; c 1. PATHOLOGY. Educ: Cornell Univ, AB, 30, MD, 33. Prof Exp: Intern, Albany Hosp, NY, 33-34, from asst resident to resident surgeon, 34-36; asst resident surg, Sch Med & Dent, Univ Rochester, 36-37; instr path, Med Col, Cornell Univ, 41-42, asst prof, 42-44; assoc prof med, Univ Va, 44-45; res assoc cancer, Sch Med & Dent, Univ Rochester, 45-48, prof cancer res, 48-52; assoc dir in chg res, Nat Cancer Inst, 52-60, dir labs & clins, NIH, 60-68; spec asst to dir med prog develop & eval, Nat Libr Med, 68-70, dep dir, 70-72; EXEC OFFICER, UNIVS ASSOC FOR RES & EDUC PATH & EXEC OFFICER, AM SOC EXP PATH, 72- Concurrent Pos: Fel, Univ Rochester, 37-38; Nat Cancer Inst res fel, 38-41. Honors & Awards: Distinguished Serv Award, Dept Health, Educ & Welfare, 60; Award, James Ewing Soc, 68. Mem: Am Soc Exp Path; Harvey Soc; Am Asn Pathologists & Bacteriologists; Am Asn Cancer Res; James Ewing Soc (lectr, 68). Res: Neoplastic and related diseases; surgical pathology. Mailing Add: Am Soc Exp Path 9650 Rockville Pike Bethesda MD 20014

MIDGLEY, A REES, JR, b Burlington, Vt, Nov 9, 33; m 55; c 3. ENDOCRINOLOGY, PATHOLOGY. Educ: Univ Vt, BS, 55, MD, 58. Prof Exp: Sarah Mellon Scaife fel path, Univ Pittsburgh, 58-61; Sarah Mellon Scaife fel, 61-62, res assoc, 62-63, from instr to assoc prof, 63-70, PROF PATH, UNIV MICH, ANN ARBOR, 70- Concurrent Pos: Nat Inst Child Health & Human Develop career develop award, 66- Honors & Awards: Parke Davis Award, Am Soc Exp Path, 70. Mem: Am Soc Exp Path; Soc Study Reproduction; Endocrine Soc; Soc Exp Biol & Med; Am Physiol Soc. Res: Reproductive endocrinology; immunoendocrinology; experimental pathology. Mailing Add: Dept of Path Med Sci Bldg 1 Univ of Mich Ann Arbor MI 48104

MIDGLEY, JAMES EARDLEY, b Kansas City, Mo, Sept 18, 34; m 61; c 3. SYSTEMS THEORY. Educ: Univ Mich, BS(eng phys), BS(eng math) & BS (eng mech), 56; Calif Inst Technol, PhD(physics), 63. Prof Exp: Res assoc magnetosphere, Univ Tex, Dallas, 63-64, asst prof gravity waves, 64-67, ASSOC PROF PHYSICS, UNIV TEX, DALLAS, 67- Concurrent Pos: NSF fel, Calif Inst Technol, 58-61. Mem: Asn Comput Mach. Res: Computer operating systems; data structures. Mailing Add: Univ of Tex at Dallas Dept Physics Box 688 Richardson TX 75080

MIDLAND, MICHAEL MARK, b Ft Dodge, Iowa, Jan 1, 46; m 72. ORGANIC CHEMISTRY, ORGANOMETALLIC CHEMISTRY. Educ: Iowa State Univ, BS, 68; Purdue Univ, PhD(org chem), 72. Prof Exp: Assoc chem, Purdue Univ, 72-75; lectr, 75-76, ASST PROF CHEM, UNIV CALIF, RIVERSIDE, 76- Mem: Am Chem Soc. Res: New chemistry of organoboranes and organolithiums; new synthetic reactions; investigation of reaction mechanisms. Mailing Add: Dept of Chem Univ of Calif Riverside CA 92502

MIDLIGE, FREDERICK HORSTMANN, JR, b Hoboken, NJ, June 13, 35; m 61; c 2. VIROLOGY, MICROBIOLOGY. Educ: Muhlenberg Col, BS, 57; Lehigh Univ, MS, 59, PhD(biol), 68. Prof Exp: From instr to asst prof, 63-72, ASSOC PROF BIOL,

FAIRLEIGH DICKINSON UNIV, 72- Mem: AAAS; Am Soc Microbiol. Res: Virus diseases of fish; morphology and maturation of lymphocystis virus. Mailing Add: Dept of Biol Fairleigh Dickinson Univ Madison NJ 07940

MIDURA, THADDEUS, b Chicopee, Mass, Dec 2, 31; m 65; c 2. FOOD TECHNOLOGY. Educ: Univ Mass, BS, 57, MS, 59; Univ Mich, MPH, 61, PhD(environ health), 64. Prof Exp: Sanitarian, Food & Milk Lab, Springfield Health Dept, Mass, 56; sanitarian, Environ Health, Philadelphia Health Dept, Pa, 57; instr food technol, Univ Mass, 58-60; fel, 64-66, RES MICROBIOLOGIST, DIV LABS, CALIF STATE DEPT PUB HEALTH, 66- Mem: Am Soc Microbiol; Am Pub Health Asn; Inst Food Technol. Res: Public health microbiology; anaerobic bacteriology and microorganisms significant in food-borne diseases; laboratory aspects of environmental associated disease outbreaks. Mailing Add: 2151 Berkeley Way Berkeley CA 94704

MIECH, RALPH PATRICK, b South Milwaukee, Wis, Aug 17, 33; m 57; c 5. BIOCHEMISTRY, PHARMACOLOGY. Educ: Marquette Univ, BS, 55, MD, 59; Univ Wis, PhD(pharmacol), 63. Prof Exp: Intern med, St Mary's Hosp, Duluth, Minn, 59-60; Nat Cancer Inst fel, Univ Wis, 61-63; asst prof, 63-69, ASSOC PROF MED SCI, BROWN UNIV, 69- Concurrent Pos: Nat Inst Neurol Dis & Blindness grant, 69-72; NSF fel, 69-72. Mem: Am Soc Pharmacol & Exp Therapeut; Am Col Emergency Physicians. Res: Enzymes of nucleotide synthesis; theophylline metabolism; nucleotide metabolism in parasitic organisms; nucleotide metabolism in the brain; purine transport via the blood; inhibitor of bronchial camp phosphodiesterase. Mailing Add: Dept of Med Sci Brown Univ Providence RI 02912

MIECH, RONALD JOSEPH, b Milwaukee, Wis, Feb 23, 35; m 60; c 2. MATHEMATICS. Educ: Univ Ill, Urbana, BS, 59, PhD(math), 63. Prof Exp: Res fel math, Nat Bur Standards, Washington, DC, 63-64; from asst prof to assoc prof, 64-74, PROF MATH, UNIV CALIF, LOS ANGELES, 74- Mem: Am Math Soc. Res: Number theory; group theory. Mailing Add: Dept of Math Univ of Calif Los Angeles CA 90024

MIECZKOWSKI, ZBIGNIEW TED, b Bydgoszcz, Poland, Jan 28, 23; US citizen; m 62. GEOGRAPHY OF THE SOVIET UNION. Educ: Sch Econ, Warsaw, MA, 47; Moscow Econ Inst, PhD(econ geog), 54; Univ Vienna, PhD(geog), 61. Prof Exp: Teaching asst econ geog, Warsaw Econ Univ, 45-48; asst prof geog, Univ Warsaw, 54-57; asst prof, 62-66, ASSOC PROF GEOG, UNIV MAN, 66- Mem: Can Asn Geographers. Res: Geography of tourism and recreation; geography of the Soviet Union, especially resources, development of the north, tourism and recreation and economic regionalization. Mailing Add: Dept of Geog Univ of Man Winnipeg MB Can

MIEDEMA, EDDY, b Broadland, SDak, Oct 22, 37; m 58; c 6. BIOCHEMISTRY, ORGANIC CHEMISTRY. Educ: Southern State Col, SDak, BS, 59; SDak State Univ, MS, 61; Ohio State Univ, PhD(biochem), 64. Prof Exp: Asst res chemist, S R Noble Found, Ardmore, Okla, 64-65, res chemist, 65-67, sr res chemist, 67-68; PROF CHEM, UNIV SDAK, SPRINGFIELD, 68-, ASST DIR PHYS & NATURAL SCI, 72- Concurrent Pos: Chmn div sci & math, Univ SDak, Springfield, 68-72. Mem: AAAS; Am Chem Soc; Soc Exp Biol & Med; Tissue Cult Asn; Am Cancer Soc. Res: Tissue culture nutrition and metabolism; oncology-control mechanisms. Mailing Add: Phys & Natural Sci Univ of SDak Springfield SD 57062

MIEHLE, WILLIAM, b Ulm, Ger, Mar 31, 15; nat US; m 48; c 2. MATHEMATICS. Educ: Mass Inst Technol, SB, 38; Univ Pa, MS, 59. Prof Exp: Engr, Radio Corp Am, 40-46; res engr, Philco Corp, 46-49; designer & comput analyst, Res Ctr, Burroughs Corp, 49-56; mem staff, Inst Coop Res, Pa, 56-58; asst prof math, Pa Mil Col, 58-60; ASST PROF MATH, VILLANOVA UNIV, 60- Concurrent Pos: Consult, Auerbach Electronics Corp, 60 & Appl Psychol Serv, 61- Mem: Soc Indust & Appl Math; Opers Res Soc Am; Math Asn Am. Res: Operations research; information retrieval; numerical analysis. Mailing Add: 1240 Steel Rd Havertown PA 19083

MIELCZAREK, EUGENIE V, b New York, NY, Apr 22, 31; m 54; c 2. SOLID STATE PHYSICS. Educ: Queens Col NY, BS, 53; Catholic Univ, MS, 57, PhD(physics), 63. Prof Exp: Physicist, Nat Bur Standards, 53-57; from res asst to res assoc physics, Catholic Univ, 57-62, asst res prof, 62-65; PROF PHYSICS, GEORGE MASON UNIV, 65- Concurrent Pos: Mem vis scientist prog, Am Inst Physics, 64- Mem: Am Phys Soc; Am Asn Physics Teachers. Res: Solid state low temperature physics; semiconductors; Fermi surfaces of metals. Mailing Add: Dept of Physics George Mason Univ Fairfax VA 22030

MIELENZ, KLAUS DIETER, b Berlin, Ger, May 8, 29; m 59; c 2. OPTICS, SPECTROSCOPY. Educ: Univ Berlin, BSc, 49; Free Univ Berlin, MSc, 52, PhD(physics), 55. Prof Exp: Asst physics, Free Univ Berlin, 49-52; physicist, R Fuess Optical Co, Ger, 52-58 & Nat Bur Standards, 58-60; tech mgr, R Fuess Optical Co, 60-63; proj leader optical masers, 63-71, PROJ LEADER SPECTROPHOTOM & LUMINESCENCE SPECTROMETRY, NAT BUR STANDARDS, 72- Concurrent Pos: Guest res worker, Inst Appl Spectros, Dortmund, Ger, 57; adj prof, George Washington Univ, 68-72; chmn comt, Int Comn Illum, 73- Honors & Awards: Silver Medal, US Dept Commerce, 66; Superior Accomplishment Award, US Dept Commerce, 72. Mem: Optical Soc Am; German Soc Appl Optics. Res: Physical optics; spectrochemistry; spectroscopic instruments; spectrophotometry, luminescence spectrometry; metrology; thin films; optical masers; vacuum techniques. Mailing Add: 6 Waycross Ct Kensington MD 20795

MIELKE, EUGENE ALBERT, b Visalia, Calif, Mar 1, 46; m 68; c 1. POMOLOGY, PLANT PHYSIOLOGY. Educ: Calif State Polytech Col, San Luis Obispo, BS, 69; Mich State Univ, MS, 70, PhD(hort, pomol), 74. Prof Exp: Lectr crop prod, Calif State Polytech Col, San Luis Obispo, 68-69; asst prof hort & pomol, Mich State Univ, 74-75; ASST PROF POMOL & CITRUS, UNIV ARIZ, 75- Concurrent Pos: Ariz rep, Western Region Coord Comt, Flowering & Fruit Set, 75-; regional ed, Pecan Quart, 75- Mem: Am Soc Hort Sci; Am Soc Plant Physiologists; Japanese Soc Plant Physiologists; Scand Soc Plant Physiol. Res: Flowering, fruit-set, and fruit growth and development of pecans and pistachios; extraction, purification, identification and measurement of plant growth substances; identification of physiological factors limiting yield; hormonal metabolism. Mailing Add: Dept of Plant Sci Bldg 36 Univ of Ariz Col of Agr Tucson AZ 85721

MIELKE, JAMES EDWARD, b Toledo, Ohio, Oct 6, 40; m 66; c 2. GEOCHEMISTRY. Educ: Mass Inst Technol, BS, 62; Univ Ariz, MS, 65; George Washington Univ, PhD(geochem), 74. Prof Exp: Geologist, Universal Eng Corp, 63-64; geologist, Radiation Biol Lab Serv, Smithsonian Inst, 64-73; SCI POLICY ANALYST, CONG RES SERV, LIBR CONG, 73- Mem: AAAS; Am Geophys Union. Res: Science policy in the earth and marine sciences with regard to matters of current and future interest to Congress. Mailing Add: Cong Res Serv Sci Policy Div Libr of Cong Washington DC 20540

MIELKE, MARVIN V, b Marshfield, Wis, May 2, 39; m 69. TOPOLOGY. Educ: Univ Wis, BS, 60, MS, 61; Ind Univ, PhD(math), 65. Prof Exp: Teaching assoc math, Ind Univ, 61-65; mem, Inst Advan Studies, 65-66; RES ASSOC MATH, UNIV MIAMI, 66- Mem: Am Math Soc; Math Asn Am. Res: Differential and algebraic topology. Mailing Add: Dept of Math Univ of Miami Coral Gables FL 33124

MIELKE, PAUL THEODORE, b Racine, Wis, Sept 28, 20; m 46; c 3. MATHEMATICS. Educ: Wabash Col, AB, 42; Brown Univ, ScM, 46; Purdue Univ, PhD(math), 51. Prof Exp: Instr math, Brown Univ, 43-44; from instr to asst prof, Wabash Col, 46-51; sr group engr digital comput, Dynamics Staff, Boeing Airplane Co, 52-57; assoc prof math, Wabash Col, 57-63, prof & chmn dept, 63-69; assoc dir, Comt Undergrad Prog Math, 69-70, exec dir, 70-71; PROF MATH & CHMN DEPT, WABASH COL, 71- Concurrent Pos: Mem bd gov, Math Asn Am, 72-75; assoc ed math educ, Am Math Monthly, 74-78. Mem: Math Asn Am; Am Math Soc. Res: Digital computing; linear algebra. Mailing Add: Dept of Math Wabash Col Crawfordsville IN 47933

MIELKE, PAUL W, JR, b St Paul, Minn, Feb 18, 31; m 60; c 1. STATISTICS, MATHEMATICS. Educ: Univ Minn, BA, 53, PhD(biostatist), 63; Univ Ariz, MA, 58. Prof Exp: Asst math, Univ Ariz, 57-58; asst biostatist, Univ Minn, 58-62, lectr, 62-63; from asst prof to assoc prof, 63-74, PROF STATIST, COLO STATE UNIV, 74- Mem: AAAS; Am Statist Asn; Biomet Soc; Math Asn Am. Res: Theoretical development and application of statistical techniques which may be appropriate to various fields involving chance phenomena. Mailing Add: Dept of Statist Colo State Univ Fort Collins CO 80521

MIENTKA, WALTER EUGENE, b Amherst, Mass, Oct 1, 25; m 54; c 4. MATHEMATICS. Educ: Univ Mass, BS, 48; Columbia Univ, MA, 49; Univ Colo, PhD, 55. Prof Exp: Instr math, Univ Mass, 49-52; instr & asst, Univ Colo, 52-55; instr, Univ Mass, 55-56; asst prof, Univ Nev, 56-57; from asst prof to assoc prof, 57-70, vchmn dept, 70-75, PROF MATH, UNIV NEBR, LINCOLN, 70- Concurrent Pos: Fac fel, Univ Nebr, Lincoln, 60 & 64-65; res scholar, Univ Calif, Berkeley, 64-65. Mem: Am Math Soc; Math Asn Am; Indian Math Soc. Res: Theory of numbers. Mailing Add: Dept Math 915 Oldfather Hall Univ of Nebr Lincoln NE 68588

MIERS, RICHARD ERNEST, physics, see 12th edition

MIES, FREDERICK HENRY, b New York, NY, Oct 3, 32; m 53; c 3. QUANTUM CHEMISTRY. Educ: City Col New York, BS, 56; Brown Univ, PhD(phys chem), 61. Prof Exp: Nat Bur Standards-Nat Res Coun fel phys chem, 61-62, PHYS CHEMIST, NAT BUR STANDARDS, 62- Mem: Am Phys Soc. Res: Scattering theory; chemical kinetics and energy transfer; pressure broadening and continuum spectroscopy; autoionization and predissociation. Mailing Add: B164 Chem Bldg Nat Bur Standards Washington DC 20234

MIESCH, ALFRED THOMAS, b Hammond, Ind, May 10, 27; m 50; c 2. GEOLOGY. Educ: St Joseph's Col Ind, BS, 50; Ind Univ, MA, 54; Northwestern Univ, PhD, 61. Prof Exp: Geol technician, NMex Bur Mines & Mineral Resources, 51-52; asst, Ind Univ, 52-53; RES GEOLOGIST, REGIONAL GEOCHEM BR, US GEOL SURV, 53- Concurrent Pos: Asst, Northwestern Univ, 56-57. Honors & Awards: Meritorious Serv Award, US Dept Interior, 73. Mem: Geol Soc Am; Geochem Soc; Int Asn Math Geol; Soc Environ Geochem & Health. Res: Distribution of minor elements in rocks and ores; Colorado Plateau uranium deposits; statistical methods in geologic and geochemical research; geochemical prospecting; environmental geochemistry. Mailing Add: US Geol Surv Fed Ctr Box 25046 Denver CO 80225

MIESCHER, GUIDO, b Zurich, Switz, Dec 13, 21; nat US; m 54; c 1. MICROBIOLOGY. Educ: Swiss Fed Inst Technol, dipl, 47, PhD(microbiol, plant path), 49. Prof Exp: RES MICROBIOLOGIST, COM SOLVENTS CORP, 49- Mem: Am Chem Soc. Res: Nutrition and metabolism of plants and microorganisms; development of industrial fermentations. Mailing Add: Com Solvents Corp Terre Haute IN 47808

MIESEL, JOHN LOUIS, b Erie, Pa, Nov 26, 41; m 64; c 3. ORGANIC CHEMISTRY. Educ: Univ Notre Dame, BS, 62; Univ Ill, PhD(org chem), 66. Prof Exp: Sr org chemist, 66-74, RES SCIENTIST, ELI LILLY & CO, 74- Mem: Am Chem Soc. Res: Synthesis of heterocyclic compounds; structure-activity relationships in insecticides; photochemistry as a synthetic tool. Mailing Add: Eli Lilly & Co Indianapolis IN 46206

MIETLOWSKI, WILLIAM LEONARD, b Buffalo, NY, Sept 25, 47. BIOSTATISTICS. Educ: Canisius Col, BS, 69; Univ Rochester, MA, 71, PhD(statist), 74. Prof Exp: Teaching asst statist, Univ Rochester, 69-73, tech assoc biostatist, Heart Res Follow-up Study, 73-74; RES ASST PROF STATIST SCI, STATIST LAB, STATE UNIV NY BUFFALO, 74- Concurrent Pos: Coord statistician, Lung Cancer Group, Vet Admin, 74- Mem: Am Statist Asn; Biomet Soc; Inst Math Statist. Res: Application of biometric methods to lung cancer clinical trials; multivariate descriptive statistics; discriminant analysis with mixed data; comparison of correlated covariance matrices. Mailing Add: State Univ NY at Buffalo Statist Lab 4230 Ridge Lea Rd Amherst NY 14226

MIEURE, JAMES PHILIP, b McLeansboro, Ill, July 5, 41; m 66; c 2. CHEMISTRY. Educ: Kenyon Col, AB, 63; Purdue Univ, MS, 66; Tex A&M Univ, PhD(chem), 68. Prof Exp: Sr res chemist, 68-73, res specialist, 73-74, RES GROUP LEADER, MONSANTO CO, 74- Mem: Am Chem Soc; Soc Appl Spectros; Am Soc Mass Spectrometry. Res: Analytical chemistry; applications of chromatography and membrane separations to environmental problems. Mailing Add: Monsanto Co Technol Dept 800 N Lindberg St Louis MO 63166

MIEYAL, JOHN JOSEPH, b Cleveland, Ohio, Feb 17, 44; m 66; c 3. BIOCHEMISTRY, PHARMACOLOGY. Educ: John Carroll Univ, BS, 65; Case Western Reserve Univ, PhD(biochem), 69. Prof Exp: NIH fel, Brandeis Univ, 69-71; asst prof pharmacol & biochem, Med Sch, Northwestern Univ, Chicago, 71-76; ASSOC PROF PHARMACOL, MED SCH, CASE WESTERN RESERVE UNIV, 76- Concurrent Pos: Grants, Res Corp Am, 71-, Chicago Heart Asn, 74-76 & Nat Inst Gen Med Sci, 74-77. Mem: Am Soc Pharmacol & Exp Therapeut; AAAS; Am Chem Soc; Sigma Xi. Res: Physicochemical studies of molecular interactions; mechanisms of enzymic reactions; drug metabolism. Mailing Add: Dept Pharmacol Med Sch Case Western Reserve Univ Cleveland OH 44106

MIFFLIN, MARTIN DAVID, b Olympia, Wash, Mar 29, 37; m 59; c 4. HYDROGEOLOGY. Educ: Univ Wash, BS, 60; Mont State Univ, MS, 63; Univ Nev, Reno, PhD(hydrogeol), 68. Prof Exp: Res assoc, Desert Res Inst Nev, 63-69; ASSOC PROF GEOL, UNIV FLA, 69- Res: Delineation of ground-water flow systems; ground-water hydrology of carbonate and arid terrane; pluvial lakes in Great Basin; mechanics of mud lump formation; isostatic rebound in Lahontan Basin. Mailing Add: Dept of Geol Univ of Fla Gainesville FL 32601

MIGAKI, GEORGE, b Troy, Mont, Apr 26, 25; m 52; c 2. VETERINARY PATHOLOGY, COMPARATIVE PATHOLOGY. Educ: Wash State Univ, BS &

MIGAKI

DVM, 52; Am Col Vet Path, dipl, 62. Prof Exp: Pvt pract, 52-54; vet meat inspector, Meat Inspection Div, USDA, 54-57, vet pathologist, 57-62, head vet pathologist, 62-68; VET PATHOLOGIST, REGISTRY COMP PATH, ARMED FORCES INST PATH, 68- Concurrent Pos: Mem Nat Conf Vet Lab Diagnosticians, 60- Mem: Am Vet Med Asn; US Animal Health Asn. Res: Comparative pathology of diseases in a wide variety of animals in which a similar disease exists in humans, including the documentation and classification of animal models of human disease. Mailing Add: Registry Compar Path Armed Forces Inst of Path Washington DC 20306

MIGDALOF, BRUCE HOWARD, b Brooklyn, NY, July 19, 41; m 67; c 3. DRUG METABOLISM. Educ: Cornell Univ, BA, 62; Purdue Univ, MS, 65; Univ Pittsburgh, PhD(org chem), 69. Prof Exp: Sr scientist drug metab, Sandoz Pharmaceuts, Sandoz-Wander Inc, 69-72; sr scientist, 72-74, GROUP LEADER DRUG DISPOSITION, McNEIL LABS INC, 74- Mem: AAAS; Am Chem Soc; NY Acad Sci. Res: Drug metabolism and disposition; absorption, distribution, metabolism and excretion studies in animals and man. Mailing Add: Dept of Biochem Res McNeil Labs Inc Camp Hill Rd Ft Washington PA 19034

MIGDALSKI, EDWARD CHARLES, b Conn, May 3, 18; m 53; c 2. ICHTHYOLOGY. Prof Exp: Ichthyologist, Bingham Oceanog Lab, 50-66, chief preparator of fishes, Peabody Mus, 58-66, DIR OUTDOOR EDUC & RECREATION, YALE UNIV, 66- Concurrent Pos: Consult, State Reservoir Invest, Conn & Ronald Press Co, NY; scientist expeds, India & Nepal, Yale-Smithsonian Inst, 46-47, NZ, Yale Univ, 48, India & Nepal, Nat Geog Soc-Yale-Smithsonian Inst, 48-49, SAm, Yale Univ, 49, Africa, 49-50, Aleutian Islands, 50, Alaska, 51, Bahama Islands, 54, Kaibab Forest, Ariz, 56, Mex, 57 & Arctic Dew Line, US Air Force-Western Elec Co, 58. Mem: Am Soc Ichthyologists & Herpetologists; Am Fisheries Soc. Res: Fresh-water and salt water sport fishes. Mailing Add: Outdoor Recreation Yale Univ New Haven CT 06520

MIGEON, BARBARA RUBEN, b Rochester, NY, July 31, 31; m 60; c 3. MEDICAL GENETICS. Educ: Smith Col, BA, 52; Univ Buffalo, MD, 56. Prof Exp: Intern pediat, Johns Hopkins Hosp, 56-57, asst resident, 57-59; fel endocrinol, Med Sch, Harvard Univ, 59-60; fel genetics, 60-62, from instr to asst prof pediat, 62-70, ASSOC PROF PEDIAT, SCH MED, JOHNS HOPKINS UNIV, 70- Concurrent Pos: Pediatrician, Johns Hopkins Hosp, 62-; mem genetics study sect, NIH, 75- Honors & Awards: Citation, Nat Bd Med Col Pa, 71. Mem: Am Soc Human Genetics; Am Soc Pediat Res. Res: Somatic cell genetics; regulation of expression of X-linked genes; X chromosome inactivation; complementation analysis of human inborn errors. Mailing Add: Dept of Pediat Johns Hopkins Hosp Baltimore MD 21205

MIGEON, CLAUDE JEAN, b Lievin, France, Dec 22, 23; m 60; c 3. PEDIATRICS, ENDOCRINOLOGY. Educ: Lycee de Reims, France, BA, 42; Univ Paris, MD, 50. Prof Exp: Asst med biochem, Univ Paris, 47-50; Am Field Serv fel, 50-51; res fel pediat, Johns Hopkins Univ, 51-52; res instr biochem, Univ Utah, 52-54; from asst prof to assoc prof, 54-71, PROF PEDIAT, JOHNS HOPKINS UNIV, 71- Concurrent Pos: Fulbright traveling fel, 50; Mayer fel, 51-52; NIH res career award, 64. Mem: Endocrine Soc; Soc Pediat Res; Am Soc Clin Invest; Am Physiol Soc; Am Pediat Soc. Res: Pediatric endocrinology, particularly steroids biochemistry; adrenal function; transplacental passage of steteroids from mother to fetus. Mailing Add: Dept of Pediat Johns Hopkins Hosp Baltimore MD 21205

MIGET, RUSSELL JOHN, b Long Beach, Calif, Oct 22, 42; m 63; c 2. MARINE MICROBIOLOGY. Educ: Univ Fla, BS, 64; Fla State Univ, PhD(oceanog), 71. Prof Exp: Res assoc marine microbiol, Inst Marine Sci, Univ Tex, 71-75; PRES, TURTLE COVE LAB, INC, 75- Res: Marine microbial ecology. Mailing Add: Turtle Cove Lab Inc No 10 Tarpon St Box 219 Port Aransas TX 78373

MIGHTON, CHARLES JOSEPH, b Saskatoon, Sask, Oct 13, 13; nat US; m 37; c 2. ORGANIC CHEMISTRY. Educ: Univ Sask, BSc, 33, MSc, 35; Univ Chicago, PhD(org chem), 37. Prof Exp: Res chemist, Inst Am Meat Packers, 36; res chemist exp sta, Del, 37-42, res supvr, 42-46, mgr rubber chem div, Akron Lab, Ohio, 46-50, asst to lab mgr, Indust & Biochem Dept, 50-51, res sect mgr, 51-60, RES ASSOC GOVT LIAISON, INDUST & BIOCHEM DEPT, E I DU PONT DE NEMOURS & CO, INC, 61- Mem: AAAS; Am Chem Soc; fel Am Inst Chem. Res: Raman spectroscopy; plastics; coating compositions and adhesives; synthetic rubbers and derivatives; inorganic colloids; pharmaceuticals. Mailing Add: Indust & Biochem Dept E I du Pont de Nemours & Co Inc Wilmington DE 19898

MIGHTON, HAROLD RUSSELL, b Saskatoon, Sask, Can, Jan 6, 19; nat US; m 43; c 2. CHEMISTRY. Educ: Univ Sask, BA, 39, MA, 41; Columbia Univ, PhD(org chem), 45. Prof Exp: Res chemist comt med res, Off Sci Res & Develop Proj, Columbia Univ & Rockefeller Inst, 44-45 & Goodyear Tire & Rubber Co, Ohio, 45-48; res chemist rayon dept, 48-50 & film dept, 50-52, res assoc, 52-53, res supvr, 53-57, res mgr, 57-68, mgr tech liaison, 68-72, PLANNING MGR CENT RES & DEVELOP DEPT, E I DU PONT DE NEMOURS & CO, 72- Mem: Am Chem Soc; Sigma Xi; Chem Inst. Res: Surface chemistry; antimalarials; high polymers; thermodynamics of crystallization in high polymers. Mailing Add: Cent Res & Develop Dept E I du Pont de Nemours & Co Wilmington DE 19898

MIGICOVSKY, BERT BARUCH, b Winnipeg, Man, Mar 14, 15; m 43; c 2. BIOCHEMISTRY. Educ: Univ Man, BSA, 35; Univ Minn, PhD, 39; Carleton Univ, DSc, 70. Prof Exp: Agr scientist, Animal Res Inst, Cent Exp Farm, 45-55, chief biochem, 55-64, asst dir-gen res insts, Res Br, 64-68, dir-gen, 68-75, ASST DEP MINISTER, RES BR, AGR CAN, 75- Concurrent Pos: Chmn, Gordon Conf, 59-60; spec lectr fac med, Univ Ottawa, 61; chmn, Nat Comt Pac Sci Cong, 75. Honors & Awards: Montreal Medal, Chem Inst Can, 75. Mem: Am Asn Biol Chemists; Can Biochem Soc; Chem Inst Can; Agr Inst Can; Can Physiol Soc. Res: Cholesterol synthesis; action of Vitamin D; radioisotopes in food. Mailing Add: Res Br Agr Can Ottawa ON Can

MIGLIARESE, JOSEPH FRANCIS, biochemistry, physiology, see 12th edition

MIGLIORE, PHILIP JOSEPH, b Pittsburgh, Pa, Dec 18, 31; m 57; c 3. MEDICINE. Educ: Univ Pittsburgh, BS, 54, MD, 56. Prof Exp: Asst pathologist, Univ Tex M D Anderson Hosp & Tumor Inst, 64-69; ASST PROF PATH, BAYLOR COL MED, 69-; ASST PATHOLOGIST, METHODIST HOSP, HOUSTON, 69- Mem: AAAS; AMA; Am Soc Clin Path. Res: Gamma globulins; myeloma proteins. Mailing Add: Dept of Path Methodist Hosp Houston TX 77025

MIGLIOZZI, JOSEPH ANDREW, b Worcester, Mass, Jan 14, 42; m 65; c 1. EXPERIMENTAL PATHOLOGY. Educ: St Anselm's Col, BA, 64; Univ Ill, Chicago, PhD(path), 73. Prof Exp: Biochemist endocrinol, Worcester Found Exp Biol, 65-68; instr anat, Sch Pharm, 71, ASST PROF PATH, PEORIA SCH MED, UNIV ILL, 73- Mem: Sigma Xi; AAAS. Res: The effect of anti-neoplastic agents on tumor stromal enzymes; histochemical and immunochemical study of the effects of various cytotoxins on the desmoplastic response in carcinomas. Mailing Add: Dept of Path Peoria Sch Med Univ Ill Peoria IL 61606

MIGNAULT, JEAN DE L, b Sherbrooke, Que, Feb 3, 24; m 53; c 3. MEDICINE, CARDIOLOGY. Educ: Univ Montreal, BA, 45, MD, 51; FRCP(C). Prof Exp: Clin monitor cardiol, Maisonneuve Hosp, 56-58; asst prof, Univ Montreal & Inst Cardiol Montreal, 58-61, from asst prof to assoc prof, Univ Montreal & Hotel-Dieu Hosp, 62-69, dir cardiac lab, Hosp, 62-69, head dept cardiol, 65-69; chmn dept med, 69-73, dean sch med, 70-73, PROF MED, MED SCH, UNIV SHERBROOKE, 69- Mem: Fel Am Col Cardiol; fel Am Col Chest Physicians; Can Med Asn; Can Cardiovasc Soc; fel Am Col Physicians. Res: Development of the new cardiac catheterization unit, the Saturn. Mailing Add: Fac of Med Univ of Sherbrooke Sherbrooke PQ Can

MIGNERY, ARNOLD LOUIS, b West Unity, Ohio, Apr 18, 18; m 42; c 4. FORESTRY. Educ: Univ Mich, BS, 40, MF, 49. Prof Exp: Res forester, 46-56, res ctr leader, 56-64, PRIN SILVICULTURIST & PROJ LEADER, SOUTHERN FOREST EXP STA, US FOREST SERV, 64- Res: Silviculture and forest management techniques; southern tree species. Mailing Add: R R Box 120 Sewanee TN 37375

MIHAILOVSKI, ALEXANDER, b Sofia, Bulgaria, Nov 8, 37; US citizen; m 69. ORGANIC CHEMISTRY. Educ: Pa State Univ, BS, 60; Univ Calif, Los Angeles, PhD(org chem), 67. Prof Exp: Asst gen & org chem, Univ Calif, Los Angeles, 63-66 & org chem, 66-67; from res chemist to sr res chemist, 67-72, SECT SUPVR, WESTERN RES CTR, STAUFFER CHEM CO, 72- Mem: AAAS; Am Chem Soc; The Chem Soc. Res Acetylene-allene chemistry; organic reaction mechanisms; synthetic organic chemistry in agricultural pest control. Mailing Add: Western Res Ctr Stauffer Chem Co 1200 S 47th St Richmond CA 94804

MIHAJLOV, VSEVOLOD S, b Kladanj, Yugoslavia, Feb 12, 25; nat US; m 51; c 3. PHYSICAL CHEMISTRY, ORGANIC CHEMISTRY. Educ: Univ Munich, BS, 49; Clark Univ, AM, 54, PhD, 56. Prof Exp: Sr scientist, 56-68, mgr process & mat develop, 68-69, mgr color technol area, 69-74, MGR COPY QUAL TECHNOL, XEROX CORP, 74- Concurrent Pos: Abstractor, Chem Abstr, 53- Mem: Am Chem Soc; Soc Photog Sci & Eng. Res: Photographic science; graphic arts; photosensitive systems; imaging; xerography; color; color vision. Mailing Add: Xerox Corp Xerox Sq PO Box 1540 Rochester NY 14603

MIHALAS, DIMITRI, b Los Angeles, Calif, Mar 20, 39; m 63; c 2. ASTROPHYSICS. Educ: Univ Calif, Los Angeles, AB, 59; Calif Inst Technol, MS, 60, PhD(astron, physics), 64. Prof Exp: Mem tech staff, TRW Space Tech Labs, 59; Higgins vis fel astron, Princeton Univ, 63-64, asst prof, 64-67; asst prof physics & astrophys & mem joint inst lab astrophys, Univ Colo, 67-68; from assoc prof to prof astron, Univ Chicago, 68-71; RES SCIENTIST, HIGH ALTITUDE OBSERV, 71- Concurrent Pos: Alfred P Sloan res fel, 69-71; mem comn 36, Int Astron Union; mem astron adv panel, NSF, 72-75. Honors & Awards: Helen B Warner Prize, Am Astron Soc, 74. Mem: Int Astron Union (vpres, 73-76); Am Astron Soc. Res: Physics of stellar atmospheres and abundances of elements in the stars; theory of radiative transfer. Mailing Add: High Altitude Observ Boulder CO 80302

MIHALISIN, TED WARREN, b Houston, Tex, Feb 11, 40; m 61; c 3. LOW TEMPERATURE PHYSICS. Educ: Cornell Univ, BA, 61; Univ Rochester, PhD(physics), 67. Prof Exp: ASSOC PROF PHYSICS, TEMPLE UNIV, 69- Concurrent Pos: Assoc scientist dept physics, Gulf Gen Atomic Inc, Calif, 67-69. Mem: Am Phys Soc. Res: Critical phenomena in magnetic systems via measurements of thermodynamic and transport properties. Mailing Add: Dept of Physics Temple Univ Philadelphia PA 19122

MIHALYI, ELEMER, b Deva, Roumania, Jan 11, 19; nat US; m 48; c 2. BIOCHEMISTRY. Educ: Univ Kolozsvar, Hungary, MD, 43; Cambridge Univ, PhD, 63. Prof Exp: Instr med chem, Univ Kolozsvar, 41-44 & biochem, Univ Budapest, 46-48; guest investr, Nobel Inst Med, Stockholm, 48-49; res assoc, Inst Muscle Res, Woods Hole, 49-51; res fel, Harrison Dept Surg Res, Pa, 51-55; CHEMIST, LAB CELL BIOL, NAT HEART & LUNG INST, 55- Mem: Am Soc Biol Chemists; Am Chem Soc; Int Soc Hemat. Res: Protein chemistry; proteins involved in blood coagulation; heavy intermediates in proteolytic breakdown of proteins; physicochemical and chemical studies of the transformation of fibrinogen into fibrin. Mailing Add: Nat Heart & Lung Inst Lab Cell Biol Bldg 3Rm B1-17 Bethesda MD 20014

MIHALYI, LOUIS JAMES, b Kismalas, Czech, Oct 4, 23; US citizen; c 2. GEOGRAPHY. Educ: San Francisco State Col, BA, 61; Stanford Univ, MA, 64; Univ Calif, Los Angeles, PhD(geog), 69. Prof Exp: Teacher soc sci, Teacher Educ for EAfrica, Columbia Univ, 61-63; ASSOC PROF GEOG, CALIF STATE UNIV, CHICO, 67- Mem: Asn Am Geogrs; African Studies Asn; Zambia Geog Asn. Res: Agricultural and rural development in tropical Africa; land-use mapping and interpretation through photo-reconnaisance. Mailing Add: Dept of Geog Calif State Univ Chico CA 95926

MIHELICH, JOHN WILLIAM, b Colorado Springs, Colo, Jan 2, 22; m 46; c 3. NUCLEAR PHYSICS. Educ: Colo Col, AB, 42; Univ Ill, PhD(physics), 50. Prof Exp: Assoc physicist, Brookhaven Nat Lab, 50-54; from asst prof to assoc prof, 54-61, PROF PHYSICS, UNIV NOTRE DAME, 61- Mem: Fel Am Phys Soc. Res: Radioactivity; decay schemes; internal conversion of gamma ray transitions; gamma ray spectra. Mailing Add: Dept of Physics Univ of Notre Dame South Bend IN 46556

MIHICH, ENRICO, b Fiume, Italy, Jan 4, 28; m 54; c 1. PHARMACOLOGY. Educ: Univ Milan, MD, 51. Prof Exp: Instr, Inst Pharmacol, Univ Milan, 51, asst prof, 52 & 54-56; from sr cancer res scientist to prin cancer res scientist, Roswell Park Mem Inst, 57-71; assoc prof, 62-68, PROF BIOCHEM PHARMACOL, ROSWELL PARK DIV GRAD SCH, STATE UNIV NY BUFFALO, 68-, CHMN PROG PHARMACOL, 69-; DIR DEPT EXP THERAPEUT, ROSWELL PARK MEM INST, 71- Concurrent Pos: Vis res fel, Sloan-Kettering Inst Cancer Res, 52-54; dir lab pharmacol, Valeas Pharmaceut Indust, Italy, 54-56; docent pharmacol, Univ Milan, 62. Mem: AAAS; Am Soc Pharmacol & Exp Therapeut; Soc Exp Biol & Med; Am Asn Cancer Res; NY Acad Sci. Res; General and pre-clinical pharmacology; cancer biology and experimental therapy. Mailing Add: Dept of Exp Therapeut Roswell Park Mem Inst Buffalo NY 14203

MIHINA, JOSEPH STEPHEN, b New York, NY, May 4, 18; m 49; c 2. ORGANIC CHEMISTRY. Educ: NY Univ, BS, 38; Mich State Col, MS, 48, PhD(chem), 50. Prof Exp: Asst foreman, Oil Tempering, Washburn Wire Co, 38-41; field inspector, Chem Warfare Serv, 41-43; fel, Northwestern Univ, 50-51; res chemist, 51-66, MGR CHEM MFG, G D SEARLE & CO, 66- Mem: Am Chem Soc. Res: Emulsion polymerization; tetrazoles steroids. Mailing Add: 8959 N Lockwood Ave Skokie IL 60076

MIHM, MARTIN C, JR, b Pittsburgh, Pa. DERMATOLOGY, PATHOLOGY. Educ: Duquesne Univ, BA, 55; Univ Pittsburgh, MD, 61; Am Bd Dermat, dipl, 69; Am Bd Path, dipl, 74, cert dermatopath, 75. Prof Exp: Clin & res fel dermat, Mass Gen Hosp, 64-67, clin fel path, 69-72, asst pathologist & asst dermatologist, 72-75; res fel dermat,

69-72, ASST PROF PATH, HARVARD MED SCH, 72-; ASSOC PATHOLOGIST, MASS GEN HOSP, 75- Concurrent Pos: Assoc staff, Peter Bent Brigham Hosp, 75-; consult path, Cambridge City Hosp & Children's Hosp Med Ctr, 75-; consult dermatopath, Addison Gilbert Hosp, 75- Mem: Am Acad Dermat; Soc Invest Dermat; fel Am Col Physicians. Res: Biology of malignant melanoma, host response to this tumor and its histology; morphology of delayed hypersensitivity reactions in man; other aspects of cutaneous inflammation. Mailing Add: Dept of Dermatopath Warren Bldg Mass Gen Hosp Fruit St Boston MA 02114

MIHRAM, GEORGE ARTHUR, b Norman, Okla, Sept 21, 39; m 65. MATHEMATICAL STATISTICS, SYSTEMIC SCIENCES. Educ: Univ Okla, BS, 60; Okla State Univ, MS & PhD(statist), 65. Prof Exp: Mathematician, Opers Res Inc, 65-66; analyst, Orgn Joint Chiefs of Staff, 66-68; asst prof statist, Univ Pa, 68-74. Concurrent Pos: Consult, Hq US Air Force, 68-69, Acad Natural Sci, 70-71, Opers Res Inc, 72-73 & IBM Corp, 73; NSF int travel grant, 75. Honors & Awards: Outstanding Serv Award, Joint Chiefs of Staff. Mem: AAAS; Soc Gen Systs Res; Am Statist Asn; Biomet Sci; Sigma Xi. Res: Theory of modelling; history and philosophy of science; simulation; telecybernetics; epistemology; computerized modelling; computer science. Mailing Add: PO Box 234 Haverford PA 19041

MIHRAN, THEODORE GREGORY, b Detroit, Mich, June 28, 24; m 53; c 3. MICROWAVE ELECTRONICS. Educ: Stanford Univ, AB, 44, MS, 47, PhD(elec eng), 50. Prof Exp: PHYSICIST, CORP RES & DEVELOP, GEN ELEC CO, 50- Concurrent Pos: Lectr, Union Col, NY, 52-53 & 60-61; vis assoc prof, Cornell Univ, 63-64; assoc ed, Inst Elec & Electronics Engrs Trans on Electron Devices, 70-73. Mem: Inst Elec & Electronics Eng; Am Phys Soc; Int Microwave Power Inst. Res: Electron physics; microwave tubes and electronics; kylstrons; space charge wave amplification; plasmas; MOSFET modeling; microwave ovens. Mailing Add: 898 Ashtree Lane Schenectady NY 12309

MIJAL, CHESTER FRANCIS, b Manchester, NH, May 3, 22; m 47; c 1. PHYSICAL CHEMISTRY, ORGANIC CHEMISTRY. Educ: St Anselm's Col, BS, 50; Georgetown Univ, MS, 54, PhD(chem), 57. Prof Exp: Res chemist, US Naval Ord Lab, Md, 51; res chemist, Nat Bur Standards, 51-54, proj leader, 54-58; proj leader, Qm Food & Container Inst, 58-59; supvr res & develop, Union Bay State Chem Co, Inc, Mass, 59-63; tech dir polymer div, 63-71, TECH DIR, K J QUINN CO, INC, 71- Honors & Awards: Superior Achievement Award, Nat Bur Standards, 56. Mem: Am Chem Soc. Res: Polymer chemistry; free radicals; instrumentation design and technique. Mailing Add: K J Quinn Co Inc 195 Canal St Malden MA 02148

MIKA, EDWARD STANLEY, b Whiting, Ind, Aug 16, 20; m 64. PHARMACOGNOSY. Educ: Univ Chicago, BS, 42, PhD(bot), 54; Wash State Univ, MS, 50. Prof Exp: Jr radiobiologist, Argonne Nat Lab, 46-49; res assoc bot & pharmacol, Univ Chicago, 53-60; from asst prof to assoc prof, 61-69, PROF PHARMACOG, COL PHARM, UNIV ILL, CHICAGO, 69- Honors & Awards: Newcomb Award Pharmacog. Mem: Soc Econ Bot; Am Soc Pharmacog; Am Asn Col Pharm; Acad Pharmaceut Sci. Res: Medicinal plants; chemical composition, growth and development. Mailing Add: Dept of Pharmacog & Pharmacol Rm 309 Univ of Ill Col of Pharm Chicago IL 60612

MIKA, LEONARD ALOYSIUS, b Bay City, Mich, Apr 17, 17; m 43; c 2. MICROBIOLOGY. Educ: Univ Mich, BS, 47, MS, 49; George Wash Univ, PhD, 55; Am Bd Microbiol, Dipl, 63. Prof Exp: Asst bact, Univ Mich, 47-49, sr investr, Ft Detrick, 49-62, staff microbiologist, 62-68, phys sci admin, 68-73, DIR, US ARMY FOR SCI & TECHNOL CTR, 73- Mem: Am Soc Microbiol; Soc Exp Biol & Med; Am Asn Immunol; fel Am Acad Microbiol; NY Acad Sci. Res: Research administration. Mailing Add: 102 Melissa Pl Charlottesville VA 22901

MIKAMI, HARRY M, b Seward, Alaska, Dec 28, 15; m 55. MINERALOGY, CERAMICS. Educ: Univ Alaska, BS, 37; Yale Univ, MS, 42, PhD(petrol), 45. Prof Exp: With US Smelting, Ref & Mining Co, Alaska, 37 & Am Creek Operating Co, Alaska, 38-39; geologist, Conn Geol Surv, 43-45; instr mineral, Yale Univ, 45; res scientist, E J Lavino & Co, 45-60, res mgr, 60-65, dir res, 65-67, dir res & develop, Lavino Div, Int Minerals & Chem, Corp, 67-74; RES MGR, BASIC REFRACTORIES, KAISER ALUMINUM & CHEM CORP, 74- Concurrent Pos: Instr, Pa State Univ, 48-49; consult, Villanova Univ, 57-58. Mem: Fel Am Ceramic Soc; fel Mineral Soc Am; Geochem Soc; fel Geol Soc Am; Am Chem Soc. Res: Ceramic mineralogy and microstructure; phase equilibria of periclase-chromite-orthosilicate systems; high temperature materials; basic oxygen furnace refractories. Mailing Add: Ctr for Technol PO Box 870 Pleasanton CA 94566

MIKAWA, YUKIO, spectroscopy, chemical physics, see 12th edition

MIKELL, WILLIAM GAILLARD, b Columbia, SC, Sept 13, 23; m 53; c 2. POLYMER PHYSICS, TEXTILES. Educ: Univ SC, BS, 44; Univ Va, MS, 47, PhD(physics), 52. Prof Exp: Engr, Westinghouse Elec Corp, 44-45; physicist, Inst Textile Tech, 47-49; instr, Univ Va, 50-51; res physicist, 52-56, res supvr, 56-57, res mgr, Spunbonded Prod Div, Textile Fibers Dept, 57-61, tech supt, May Plant, Orlon Div, 61-63, res mgr, Spunbonded Div, 63-66, asst mgr & mgr indust mkt, Indust Fibers Div, 66-71, ASST TO TECH DIR, TEXTILE FIBERS DEPT, INDUST FIBERS DIV, E I DU PONT DE NEMOURS & CO, INC, 71- Mem: Am Phys Soc. Res: Textile and synthetic fiber technology; physics of high polymers and pneumatic tires; environmental and toxicological technology. Mailing Add: Nemours Bldg E I du Pont de Nemours & Co Inc Wilmington DE 19898

MIKENBERG, GIORA, b Buenos Aires, Arg, July 15, 47; Israeli citizen; m 69; c 1. HIGH ENERGY PHYSICS. Educ: Cath Univ Chile, Licenciate, 69; Weizmann Inst, Rehovot, Israel, MSc, 71, PhD(physics), 74. Prof Exp: RES ASST PHYSICS, FERMI NAT ACCELERATOR LAB, 74- Mem: Israeli Phys Soc. Res: Hadronic physics, studying single particle inclusive spectra and its associated multiplicity. Mailing Add: Physics Dept Fermilab PO Box 500 Batavia IL 60510

MIKES, JOHN ANDREW, b Budapest, Hungary, Jan 20, 22; m 48; c 1. CHEMISTRY. Educ: Pazmany Peter Univ, Budapest, dipl, 45, PhD(org chem), 48; Eötvös Lorand Univ, Budapest, DSc, 68. Prof Exp: Asst prof polymer chem, Budapest Tech Univ, 45-48; tech mgr, Hutter & Lever, Co, Budapest, 48-49; dir res & consult serv plastics inst, Nat Polymer Res Ctr, 50-69; dir res & develop, Water Treatment Develop Co, Tata, 69-70; MGR RES & DEVELOP, LUNDY ELECTRONICS, INC, 71- Concurrent Pos: Guest lectr postgrad sch, Budapest Inst Technol, 53-57; consult, Inst Coloristics, Budapest, 68-69, Treadwell Corp, New York, 70-71 & Mocatta Metals Corp, 71; ed, Ion Exchange & Membranes, 71-; mem, Scientist's Comt Pub Info, 71. Mem: Am Chem Soc; fel Am Inst Chem. Res: Polymer synthesis; water treatment process development; polymer structure research; ion exchange; membranes; adsorption; diffusion; porosity; cross linking. Mailing Add: 71 Grace Ave Great Neck NY 11021

MIKES, PETER, b Prague, Czech, Oct 28, 38; m 69; c 3. POLYMER PHYSICS. Educ: Czech Tech Univ, ing Phys, 61; Charles Univ, Prague, CSc, 65. Prof Exp: Scientist polymers, Inst Macromolecular Chem, Czech, 65-69; fel spectroscopy, Dept Polymer Sci, Case Western Reserve Univ, 69-71; sr scientist syst sci, Res Ctr, Rockland State Hosp, NY, 71-73; SCIENTIST RHEOLOGY, XEROX CORP, 73- Concurrent Pos: Imp Chem Indust fel textiles, Univ Leeds, 66-67. Mem: Am Phys Soc; Mat Res Soc. Res: Mechanical properties of polymers; rheology of elastomers; printing physics. Mailing Add: Xerox Corp 1341 Mockingbird Lane Dallas TX 75247

MIKESELL, JAN ERWIN, b Macomb, Ill, Feb 19, 43; m 65. PLANT ANATOMY. Educ: Western Ill Univ, BSc, 65, MSc, 66; Ohio State Univ, PhD(bot), 73. Prof Exp: Teaching asst biol, Western Ill Univ, 65-66; researcher virol & immunol, Viral & Immunol Lab, Sixth US Army Med Labs, 67-69; teaching assoc bot, Ohio State Univ, 69-73; ASST PROF BOT, GETTYSBURG COL, 73- Mem: Bot Soc Am; AAAS; Sigma Xi; Can Soc Plant Physiologists. Res: Investigations of anomalous secondary thickening in vascular plants; especially patterns of development and directions of differentiation of anomalous types of cambia in dicotyledonous plants. Mailing Add: Dept of Biol Gettysburg Col Gettysburg PA 17325

MIKESELL, MARVIN WRAY, b Kansas City, Mo, June 16, 30; m 57. CULTURAL GEOGRAPHY, BIOGEOGRAPHY. Educ: Univ Calif, Los Angeles, BA, 52, MA, 53; Univ Calif, Berkeley, PhD(geog), 59. Prof Exp: From instr to assoc prof, 58-66, PROF GEOG, UNIV CHICAGO, 67-, CHMN DEPT, 69- Concurrent Pos: Soc Sci Res Coun-Am Coun Learned Socs joint comt on Near & Mid Eastern studies grant, Lebanon & Syria, 62-63; ed, Monogr, Asn Am Geog, 64-72. Mem: Asn Am Geog; Am Geog Soc; Am Anthrop Asn; Mid East Studies Asn. Res: Cultural, historical and biogeography, especially of the Mediterranean region; history of geography. Mailing Add: Dept of Geog Univ of Chicago 5928 University Ave Chicago IL 60637

MIKESELL, SHARELL LEE, b Coshocton, Ohio, Nov 24, 43; m 65. POLYMER CHEMISTRY. Educ: Olivet Nazarene Col, AB, 65; Ohio State Univ, MS, 68; Univ Akron, PhD(polymer chem), 71. Prof Exp: Prod develop engr polymer chem, 71-72, proj mgr, 72-74, mgr indust mkt develop tech mkt, 74-75, MGR INDUST PROD DEVELOP POLYMER CHEM, LAMINATED & INSULATING MAT BUS DEPT, GEN ELEC CO, 75- Mem: Am Chem Soc. Res: Development of high performance thermosetting epoxy and phenolic resins used in high pressure copper clad and multilayer laminates for printed wiring applications. Mailing Add: Laminated Prod Bus Dept Gen Elec Co Coshocton OH 43812

MIKESKA, EMORY EUGENE, b Abbott, Tex, Aug 24, 27; m 48; c 2. PHYSICS. Educ: Univ Tex, BS, 47, MA, 50. Prof Exp: Jr seismic observer, Magnolia Petrol Co, 47-49; res physicist, Defense Res Lab, Univ Tex, Austin, 50-61; sr physicist &proj dir, Tracor, Inc, 61-67; RES SCIENTIST ASSOC, APPL RES LAB, UNIV TEX, AUSTIN, 67- Concurrent Pos: Consult, Boner & Lane, 54-59; consult & partner, Lane & Mikesa, 60-61. Mem: Fel Acoust Soc Am; Audio Eng Soc. Res: Noise control; architectural acoustics; underwater sound. Mailing Add: 7613 Rustling Rd Austin TX 78731

MIKHAIL, ADEL AYAD, b Cairo, Egypt, Nov 8, 34; US citizen; m 58; c 2. MEDICINAL CHEMISTRY, BIOMEDICAL ENGINEERING. Educ: Univ Alexandria, BPharm, 55, MPharmaceut Chem, 60; Univ Minn, Minneapolis, PhD(med chem), 66. Prof Exp: NIH fel & assoc scientist, Cancer Res Lab, Vet Admin Hosp, Minneapolis, Minn, 66; asst prof, Col Pharm, Univ Alexandria, 66-70; USDA Forest Serv fel & res assoc pharm, Ohio State Univ, 70-72; RES MGR, MED, INC, 73- Mem: Am Chem Soc. Res: Drug design; synthesis; structure elucidation of natural products; biocompatibility of polymers; cardiovascular prosthesis and devices; quality control of medical devices. Mailing Add: 2332 W 111th St Bloomington MN 55431

MIKHAIL, WADIE F, b Egypt, Apr 16, 29; US citizen; m 63; c 2. STATISTICS. Educ: Cairo Univ, BSc, 49; Univ NC, PhD(math statist), 60. Prof Exp: Lectr math, Teachers Col, Cairo, 49-57; assoc statistician, 60-62, staff statistician, 62-66, adv statistician, 66-69, SR STATISTICIAN, IBM CORP, 69- Mem: Inst Math Statist; Am Statist Asn. Res: Probabilistic modeling of wiring space requirements for large scale integration technology. Mailing Add: 75 Pleasant Ridge Dr Poughkeepsie NY 12603

MIKITEN, TERRY MICHAEL, b New York, NY, June 1, 37; m 60; c 3. NEUROPHYSIOLOGY. Educ: NY Univ, BA, 60; Albert Einstein Col Med, PhD(pharmacol), 67. Prof Exp: Res asst neurosurg, Mt Sinai Hosp, NY, 59-60; asst pharmacologist, Schering Corp, NJ, 60-61; USPHS fel, Med Sch, Columbia Univ, 67-69; asst prof physiol, 69-75, ASSOC PROF PHYSIOL, UNIV TEX MED SCH SAN ANTONIO, 75- Concurrent Pos: Consult, Bexar County Hosp, 69-; regional chmn, Osteogenesis Imperfecta Found, 71- Mem: AAAS; Soc Neurosci; NY Acad Sci. Res: Neurophysiology and pharmacology of synaptic transmission; desensitization of cholinergic receptors; electrophysiology of excitable and non-excitable membranes. Mailing Add: Dept of Physiol & Med Univ of Tex Med Sch San Antonio TX 78229

MIKKELSEN, DUANE SOREN, b Payson, Utah, Nov 1, 21; m 43; c 4. SOIL FERTILITY, AGRONOMY. Educ: Brigham Young Univ, BS, 46; Rutgers Univ, PhD, 49. Prof Exp: From asst prof to assoc prof, 49-63, PROF AGRON, UNIV CALIF, DAVIS, 63- Concurrent Pos: Consult, Rockefeller Found, Colombia, 62; IRI Res Inst, Brazil, 62, 68-71, Chile-Calif Proj, 63, Int Rice Comn, Manila, Philippines, 64 & 66, Int Atomic Energy Agency, Hong Kong, 66, & Peace Corp Training Progs, 67; vis scientist & Rockefeller Found grant, Int Rice Res Inst, Manila, Philippines, 67-68; chmn, US Rice Tech Working Group, 67-68; mem rice fertilizer adv comt, Tenn Valley Auth, 68-71. Mem: Soil Sci Soc Am; Crop Sci Soc Am; fel Am Soc Agron; Am Soc Plant Physiol. Res: Plant-soil interrelations; mineral nutrition of plants; mineral nutrition of rice; chemistry of flooded soils. Mailing Add: 617 Oeste Dr Davis CA 95616

MIKKELSEN, HARRY E, b Chicago, Ill, Nov 9, 11; m 37; c 4. PHYSICS. Educ: US Mil Acad, BS, 36; Univ Colo, MSc, 62. Prof Exp: Asst prof ord eng, US Mil Acad, 40-42 & 47-50; instr physics, Pueblo Jr Col, 58-63; head dept, 63-75, ASSOC PROF PHYSICS, UNIV OF SOUTHERN COLO, 63- Mem: Am Asn Physics Teachers. Res: teaching; atomic energy. Mailing Add: Dept of Physics Univ of Southern Colo 2200 Bonforte Blvd Pueblo CO 81001

MIKKELSEN, WILLIAM MITCHELL, b Minneapolis, Minn, May 25, 23; m 48; c 6. INTERNAL MEDICINE. Educ: Univ Mich, MD, 49. Prof Exp: From asst prof to assoc prof, 57-69, asst dir health prog, 59-62, actg dir, 62-66, PROF INTERNAL MED, MED SCH, UNIV MICH, ANN ARBOR, 69-, DIR PERIODIC HEALTH APPRAISAL UNIT, 66- Concurrent Pos: Assoc physician, Rackham Arthritis Res Unit, Univ Mich, Ann Arbor, 55-; attend physician, Vet Admin Hosp, 55- Mem: Soc Advan Med Syst; Am Soc Clin Pharmacol & Therapeut; Am Col Physicians; Am Fedn Clin Res; Am Geriatrics Soc. Res: Rheumatic diseases; evaluation and application of periodic health appraisal techniques. Mailing Add: Univ of Mich Med Ctr Ann Arbor MI 48104

MIKKELSON, RAYMOND CHARLES, b Blue Earth, Minn, Mar 22, 37; m 60; c 1. SOLID STATE PHYSICS. Educ: St Olaf Col, BA, 59; Univ Ill, MS, 61, PhD(physics), 65. Prof Exp: From asst prof to assoc prof, 65-75, PROF PHYSICS,

MACALESTER COL, 75- Concurrent Pos: Assoc Cols Midwest physics fac mem, Argonne Nat Lab, 71, vis scientist, Physics Div, 71-72. Mem: Am Asn Physics Teachers; Am Phys Soc. Res: Interactions of radiation with solids; channeling; environmental radiation levels; teaching of university-college physics; digital instrumentation. Mailing Add: Dept of Physics Macalester Col St Paul MN 55105

MIKNIS, FRANCIS PAUL, b DuBois, Pa, Jan 31, 40; m 60; c 3. NUCLEAR MAGNETIC RESONANCE. Educ: Univ Wyo, BS, 61, PhD(chem), 66. Prof Exp: Sr scientist, Aeronutronic Div, Philco-Ford Corp, 66-67; RES CHEMIST, LARAMIE ENERGY RES CTR, ENERGY RES & DEVELOP ADMIN, 67-, PROJ LEADER, 75- Mem: Am Chem Soc; Am Asn Physics Teachers; Sigma Xi. Res: Nuclear magnetic resonance of solids, particularly oil shales; elemental analysis of fossil fuels. Mailing Add: 1819 W Hill Rd Laramie WY 82070

MIKOLAJCIK, EMIL MICHAEL, b Colchester, Conn, Jan 14, 26; m 53; c 2. DAIRY MICROBIOLOGY. Educ: Univ Conn, BS, 50; Ohio State Univ, MS, 51, PhD(dairy microbiol), 59. Prof Exp: Prof dairy mfg, Univ PR, 51-61; from asst prof to assoc prof, 61-74, PROF DAIRY TECHNOL, OHIO STATE UNIV, 74- Concurrent Pos: NIH grants, 61- Honors & Awards: Pfizer Award in Cheese Res, 74. Mem: Sigma Xi; Am Dairy Sci Asn; Am Soc Microbiol; Int Asn Milk, Food & Environ Sanit; Inst Food Technol. Res: Investigations on the mechanisms of bacteriophage action on lactic organisms; bacterial metabolism of organisms associated with the dairy industry, particularly on lactic streptococci and spore formers; immunoglobulins of bovine milk and colostrum. Mailing Add: Dept of Food Sci & Nutrit Ohio State Univ Columbus OH 43210

MIKOLAJCZAK, KENNETH LEE, b Elcho, Wis, Oct 23, 32; c 2. NATURAL PRODUCTS CHEMISTRY. Educ: Wis State Col, Stevens Point, BS, 58. Prof Exp: From asst chemist to chemist, 58-68, RES CHEMIST, NORTHERN REGIONAL RES LAB, USDA, 68- Mem: Am Chem Soc; Phytochem Soc NAm. Res: Gas-liquid chromatographic analysis of fatty acid methyl esters; isolation and characterization of unknown and unusual fatty acids from plant seed oils; identification and synthesis of other natural compounds found in plant extracts; antitumor compounds. Mailing Add: Northern Regional Res Ctr 1815 N University St Peoria IL 61604

MIKOLASEK, DOUGLAS GENE, b Menominee, Mich, Aug 23, 30; m 59; c 3. MEDICINAL CHEMISTRY. Educ: Univ Mich, BS, 52, PhD(med chem), 62. Prof Exp: Control chemist, Marinette Paper Co, Scott Paper Co, 54; develop chemist, Abbott Labs, 55-56, res chemist, 56-58; sr scientist, 62-68, group leader chem develop, 68-71, PRIN INVESTR, MEAD JOHNSON RES CTR, 71- Concurrent Pos: Parke-Davis fel, Univ Kans, 58-60. Mem: AAAS; Am Chem Soc. Res: Quinoline chemistry and antimalarial research. Mailing Add: Mead Johnson Res Ctr 2404 Pennsylvania Ave Evansville IN 47721

MIKOVSKY, RICHARD JOSEPH, physical chemistry, see 12th edition

MIKSAD, RICHARD WALTER, b Trenton, NJ, Aug 24, 40; m 70; c 2. FLUID DYNAMICS, METEOROLOGY. Educ: Bradley Univ, BSME, 63; Cornell Univ, MSc, 64; Mass Inst Technol, ScD(oceanog), 70. Prof Exp: Res staff pollution, Ctr Study Responsive Law, 70; res scientist fluid dynamics, Imperial Col, 70-72; asst res prof atmospheric sci, Univ Miami, 72-74; ASST PROF ATMOSPHERIC SCI, UNIV TEX, AUSTIN, 74- Concurrent Pos: Nat Res Coun fel, Imperial Col, 70-72. Mem: Am Phys Soc; Am Inst Aeronaut & Astronaut; Am Meteorol Soc. Res: Turbulent transport; non-linear hydrodynamic stability; air-sea interaction. Mailing Add: Atmospheric Sci Group Bldg ECJ 9-218 Univ Tex Austin TX 78712

MIKSCH, EDMOND STEWART, b Houston, Tex, Apr 17, 32; m 66; c 4. ENERGY CONVERSION. Educ: Reed Col, BA, 54; Harvard Univ, AM, 59, PhD(appl physics), 64. Prof Exp: Res assoc mat sci, Mass Inst Technol, 64-65; inventor aircraft syst, 65-67; scientist appl physics, Fundamental Lab, US Steel, 67-70; consult eng, 71-73; ed, Whitaker House, 73-74; SR PHYSICIST, BASIC TECHNOL, INC, 74- Res: Energy conversion technologies; electromechanical technologies; development of computer programs for finite element analysis; development of computer logic for data processing.

MIKSCHE, JEROME PHILLIP, b Breckenridge, Minn, June 11, 30; m; c 3. PLANT MORPHOLOGY. Educ: Moorhead State Col, BS, 54; Miami Univ, MS, 56; Iowa State Univ, PhD(plant morphol), 59. Prof Exp: Instr gen bot, Iowa State Univ, 58-59; fel radiobot, Brookhaven Nat Lab, 59-61, staff position, Biol Dept, 61-65; RES BOTANIST CYTOL, INST FOREST GENETICS, US FOREST SERV, 65- Concurrent Pos: Asst prof, C W Post Col, Long Island Univ, 60-61; lectr, Adelphi Suffolk Col, 62-65; radiobiol adv, AEC exhibit, Brazil & Lebanon, 61. Mem: Bot Soc Am; Radiation Res Soc; Am Soc Plant Physiol; Am Soc Cell Biol; fel Royal Micros Soc. Res: Developmental plant anatomy; plant cytology, genetics and physiology; radiation botany; nucleic acids in connection with genetic relatedness between organisms. Mailing Add: Inst of Forest Genetics US Forest Serv Box 898 Rhinelander WI 54501

MIKULA, BERNARD C, b Johnstown, Pa, Aug 29, 24; m 51; c 2. GENETICS, TAXONOMY. Educ: Col William & Mary, BS, 51; Univ Wash, St Louis, PhD(bot), 56. Prof Exp: Asst, Mo Bot Garden, 51-56; proj assoc genetics, Univ Wis, 56-60; from asst prof to assoc prof, 60-67, PROF GENETICS, DEFIANCE COL, 67- Concurrent Pos: Vis fel ctr biol of natural syst, Wash Univ, 66-67. Mem: AAAS; Genetics Soc Am; Am Inst Biol Sci. Res: Mechanisms of allelic variation. Mailing Add: Dept of Genetics Defiance Col Defiance OH 43512

MIKULA, JAMES J, b Philadelphia, Pa, July 24, 26; m 50; c 3. PHYSICAL CHEMISTRY. Educ: Univ Pa, BA, 47; Temple Univ, PhD(phys chem), 68. Prof Exp: Chemist, Samuel P Sadtler & Son Inc, 47-50 & Stanton Labs, 50; from org chemist to chemist, 51-55, phys chemist, 55-69, res chemist, 69-73, CHIEF CHEM RES DIV, PITMAN-DUNN RES LABS, FRANKFORD ARSENAL, 73- Concurrent Pos: Adj asst prof chem, Temple Univ, 68-70, adj assoc prof, 70- Mem: Am Chem Soc; Sigma Xi; Am Defense Preparedness Asn. Res: Conventional and laser spectroscopy studies to determine energy transfer mechanisms and kinetics of molecules involved in propellant ignition and combustion; laser countermeasures; lubricants; chelating compounds. Mailing Add: Pitman-Dunn Res Labs Chem Res Div Frankford Arsenal Philadelphia PA 19137

MIKULCIK, JOHN D, b Ilasco, Mo, July 30, 36; m 61; c 3. AGRONOMY. Educ: Univ Mo, BS, 58, MS, 59, PhD(soils), 64. Prof Exp: From asst prof to assoc prof, 63-73, PROF AGRON, MURRAY STATE UNIV, 73- Mem: AAAS; Am Soc Agron; Soil Sci Soc Am. Res: Soil testing; levels of nitrate and phosphate in runoff from rural watersheds; levels of nitrate in soils under barn lot conditions. Mailing Add: Dept of Agr Murray State Univ Murray KY 42072

MIKULEC, RICHARD ANDREW, b New Brighton, Pa, Feb 26, 28; m 63; c 1. ORGANIC CHEMISTRY. Educ: Wayne State Univ, BS, 51; Wash State Univ, MS, 53, PhD(chem), 56. Prof Exp: Res fel chem, Univ Minn, 56-57; RES INVESTR, G D SEARLE & CO, 57- Mem: Am Chem Soc. Res: Organic chemistry applied to pharmaceutical products; polypeptides. Mailing Add: G D Searle & Co PO Box 5110 Chicago IL 60680

MIKULECKY, DONALD CASIMIR, biophysics, physiology, see 12th edition

MIKULSKI, CHESTER MARK, b Philadelphia, Pa, Nov 26, 46; m 70; c 2. INORGANIC CHEMISTRY. Educ: Drexel Univ, BS, 69, PhD(inorg chem), 72. Prof Exp: Fel, Dept Chem, Univ Pa, 72-76; ASST PROF INORG CHEM, BEAVER COL, 76- Mem: Am Chem Soc; Am Inst Chemists. Res: The synthesis and characterization of metal complexes with organo-phosphoryl, -nitryl and -sulfuryl ligands; decomposition of phosphoryl and thiophosphoryl esters in the presence of metal salts; paramagnetic non-metal silicon and phosphorus compounds; synthesis of polymeric metallic conductors. Mailing Add: Dept of Chem & Physics Beaver Col Glenside PA 19038

MIKULSKI, FLORIAN A, organic chemistry, see 12th edition

MIKULSKI, PIOTR W, b Warsaw, Poland, July 20, 25; m 60; c 1. MATHEMATICAL STATISTICS. Educ: Sch Planning & Statist, Warsaw Tech Univ, Dipl, 50, MS, 51; Univ Calif, Berkeley, PhD(statist), 61. Prof Exp: Adj statist, Sch Planning & Statist, Warsaw Tech Univ, 50-57, Inst Math, Polish Acad Sci, 52-57; asst prof, Univ Ill, Urbana, 61-62; from asst prof to assoc prof, 62-70, PROF STATIST, UNIV MD, COLLEGE PARK, 70- Mem: Inst Math Statist. Res: Nonparametric methods in statistics; asymptotic optimal properties of statistical procedures. Mailing Add: Dept of Math Univ of Md College Park MD 20742

MIKUS, FELIX F, b Bomarton, Tex, Jan 1, 16; m 45; c 4. PHYSICAL CHEMISTRY. Educ: Southwest Tex State Teachers Col, BS, 38; CAtholic Univ, MS, 42; Iowa State Univ, PhD(phys chem), 46. Prof Exp: Asst, Nat Defense Res Comt, Catholic Univ, 42; instr chem, Iowa State Univ, 42-43, 45-46; res chemist, US Rubber Co, Iowa, 45; asst prof phys chem, Mich Col Mining & Technol, 46-48; prof, Tex Col Arts & Indust, 48-57; res chemist, Knolls Atomic Power Lab, Gen Elec Co, 57-58; phys chemist, 58-66, ENG MGR, SYLVANIA ELEC PROD INC DIV, GEN TEL & ELECTRONICS CORP, 66- Mem: Am Chem Soc; Electrochem Soc. Res: Thermodynamics; heterogeneous equilibrium; starch; ternary systems; amylose complexes of organic acids; phosphors; solid state; electrochemistry; photochemical machining. Mailing Add: 305 Bridge St Towanda PA 18848

MILAKOFSKY, LOUIS, b Philadelphia, Pa, Feb 21, 41; m 63; c 1. ORGANIC CHEMISTRY. Educ: Temple Univ, BA, 62; Univ Wash, PhD(org chem), 67. Prof Exp: Chemist, Dupont Co, 62; asst chem, Univ Wash, 62-67; fel & instr, Ind Univ, 67-68; asst prof, Pa State Univ, Scranton, 68-71, ASST PROF CHEM, PA STATE UNIV, BERKS CAMPUS, 71- Mem: Am Chem Soc. Res: Mechanistic studies of hydrolysis of iminium ions and solyolysis of brosylates. Mailing Add: Dept of Chem Pa State Univ Berks Campus 814 Hill Ave Wyomissing PA 19608

MILAM, DENVER FRANKLIN, b Charleston, WVa, Aug 23, 19; m 53; c 1. MEDICINE. Educ: WVa Univ, AB, 42, BS, 43; Univ Pa, MD, 44; Am Bd Urol, dipl, 54. Prof Exp: Asst instr urol, Univ Pa, 45-46 & 48-49, instr, 49-50; instr, Sch Med, Creighton Univ, 50-54; assoc prof, 60-63, PROF UROL, SCH MED, WVA UNIV, 63-, CHMN DEPT, 60- Mem: AAAS; Am Urol Asn; fel Am Col Surgeons; Int Soc Urol. Res: Genito-urinary surgery; urodynamics; physiology of sex; accessory organs. Mailing Add: Dept of Surg WVa Univ Med Ctr Morgantown WV 26506

MILAN, FREDERICK ARTHUR, b Waltham, Mass, Mar 10, 24; m 59; c 3. PHYSICAL ANTHROPOLOGY. Educ: Univ Alaska, BA, 52; Univ Wis, MS, 59, PhD(anthrop), 62. Prof Exp: Observer meteorol, Mt Wash Observ, NH, 43-44, 46-47; observer meteorol, Arctic Sect, US Weather Bur, 47-48; res physiologist, US Air Force Arctic Aeromed Lab, 53-54, 56-57, 59-61; res physiologist, Oper Deepfreeze, Little Am V, Antarctica, 57-58; chief environ protection br, US Air Force Arctic Aeromed Lab, 62-67; assoc scientist-lectr anthrop, Univ Wis-Madison, 67-71; chief behav sci br, Arctic Health Res Ctr, USPHS, 71-73; PROF ANTHROP, UNIV ALASKA, 73-, CHMN DEPT, 75- Concurrent Pos: Dir, Int Study of Eskimos, US Nat Comt Int Biol Prog, US Nat Acad Sci, DC, 67-; adj assoc prof, Inst Arctic Biol, Univ Alaska, 71-73. Res: Comparative physiology of aboriginal populations in polar regions; general anthropology of polar regions; human ecology of arctic populations. Mailing Add: Inst of Arctic Biol Univ of Alaska Fairbanks AK 99701

MILANI, SALVATORE, b Akron, Ohio, Feb 1, 27; m 55; c 2. NUCLEAR PHYSICS. Educ: Univ Akron, BS, 50; Ohio State Univ, PhD(physics), 56. Prof Exp: Scientist reactor physics, Gen Elec Co, 56; sr scientist reactor exp physics, 57-58, supvr, 58-60, mgr, 60-72, MGR NUCLEAR DESIGN & ANAL, BETTIS ATOMIC POWER LAB, 72- Mem: Am Phys Soc; Am Nuclear Soc. Res: Reactor experiments, design and analysis. Mailing Add: Bettis Atomic Power Lab Westinghouse Elec Corp PO Box 79 West Mifflin PA 15122

MILANI, VICTOR JOHN, b Mt Vernon, NY, Sept 26, 45; m 73. MICROBIOLOGY. Educ: City Univ New York, BS, 67; NY Univ, MS, 71, PhD(microbiol), 73. Prof Exp: Teaching assoc biol, NY Univ, 67-71; asst prof biol, Manhattan Community Col, 71-74; ASSOC PROF SCI & CHMN DEPT, BAY PATH JR COL, 74- Concurrent Pos: Res assoc, Lab Plant Morphogenesis, Manhattan Col, 73-74. Mem: AAAS; Am Inst Biol Sci; Am Soc Microbiol; NY Acad Sci; Sigma Xi. Res: The effects of concanavalin A on growth and tumor inducing ability of Agrobacterium tumefaciens. Mailing Add: Dept of Sci Bay Path Jr Col Longmeadow St Longmeadow MA 01106

MILANICH, JERALD THOMAS, b Painesville, Ohio, Oct 13, 45; m 70; c 1. ANTHROPOLOGY. Educ: Univ Fla, BA, 67, MA, 68, PhD(anthrop), 71. Prof Exp: Nat Endowment for Humanities fel, Smithsonian Inst, 71-72; asst prof anthrop, 72-75, ASST CUR ANTHROP, FLA STATE MUS, UNIV FLA, 75- Mem: Am Anthrop Asn; Soc Am Archaeol; Soc Hist Archaeol. Res: Prehistoric peoples and cultural dynamics of the eastern United States; ethnohistory and ethnography of aborigines of southeastern United States. Mailing Add: Dept of Social Sci Fla State Mus Univ of Fla Gainesville FL 32611

MILANO, MICHAEL JOHN, b Poughkeepsie, NY, July 12, 47; m 69. ANALYTICAL CHEMISTRY. Educ: Union Col, NY, BS, 69; Purdue Univ, MS, 72, PhD(chem), 74. Prof Exp: ASST PROF CHEM, STATE UNIV NY BUFFALO, 74- Mem: Am Chem Soc; Am Soc Testing & Mat; Soc Appl Spectros. Res: Development and applications of rapid scanning array spectrometers to anlytical problems in clinical chemistry and biomedical engineering. Mailing Add: Dept of Chem State Univ of NY Buffalo NY 14214

MILANOVICH, FRED PAUL, b Rochester, Pa, Nov 22, 44; m 68; c 2. BIOPHYSICS. Educ: US Air Force Acad, BS, 67; Univ Calif, Davis, MS, 68, PhD(appl sci), 74. Prof Exp: Proj officer, Air Force Weapons Lab, 68-71; BIOPHYSICIST, LAWRENCE LIVERMORE LAB, 74- Mem: Am Chem Soc. Res: Application of lasers to biological systems; Raman spectroscopic investigations of macromolecular and membrane structure. Mailing Add: PO Box 808 L-523 Livermore CA 94550

MILAZZO, FRANCIS HENRY, b Syracuse, NY, Aug 7, 28; m 56; c 2. MICROBIOLOGY, PHYSIOLOGY. Educ: WVa Wesleyan Col, BS; Syracuse Univ, MSc, 53, PhD(microbiol), 60. Prof Exp: Res asst bact, microbiol & biochem, Res Inst, Syracuse Univ, 54-60; from asst prof to assoc prof, 60-70, PROF MICROBIOL, QUEEN'S UNIV, ONT, 70- Mem: Am Soc Microbiol; Can Soc Microbiol. Res: Sulfur metabolism of wood destroying fungi and Proteus species; sulfur metabolism of Pseudomonas species and algae with particular reference to synthesis and control of aryl and alkylsulfatases. Mailing Add: Dept of Microbiol & Immunol Queen's Univ Kingston ON Can

MILBERG, MORTON EDWIN, b New York, NY, July 21, 26; m 62; c 3. SOLID STATE CHEMISTRY. Educ: Rutgers Univ, BS, 46; Cornell Univ, PhD(phys chem), 49. Prof Exp: Asst chem, Cornell Univ, 46-48; fel, Univ Minn, 49-50; instr, Univ NDak, 50-52; res chemist, 52-59, supvr phys & inorg chem sect, Chem Dept, 59-61, staff scientist, 61-69, PRIN RES SCIENTIST, SCI LAB, FORD MOTOR CO, 69- Concurrent Pos: Chmn, Gordon Res Conf Glassy State, 71; prog chmn, Glass Div, Am Ceramic Soc, 74-75. Mem: Am Chem Soc; Am Crystallog Asn; Am Ceramics Soc. Res: Structure and properties of noncrystalline solids; diffusion in glass; high temperature ceramic materials. Mailing Add: Res Staff Ford Motor Co Box 2053 Dearborn MI 48121

MILBERGER, ERNEST CARL, b Galatia, Kans, Apr 2, 21; m 45; c 2. CHEMISTRY. Educ: Univ Mo, AB, 41, MA, 43; Case Western Reserve Univ, PhD(org chem), 57. Prof Exp: Chemist, Tex Co, NY, 42-46; sr chemist, 46-60, sect leader, 60-63, SR RES ASSOC, STANDARD OIL CO, 63- Mem: Am Chem Soc. Res: Petrochemical process research; heterogeneous catalysis. Mailing Add: Standard Oil Co 4440 Warrensville Center Rd Cleveland OH 44128

MILBOCKER, DANIEL CLEMENT, b Gaylord, Mich, May 25, 31; m 59; c 3. HORTICULTURE. Educ: Mich State Univ, BS, 65, MS, 66; Pa State Univ, PhD(hort), 69. Prof Exp: Asst prof ornamental hort, Univ Ky, 69-74; PLANT PHYSIOLOGIST, VA TRUCK & ORNAMENTALS RES STA, 74- Mem: Am Soc Hort Sci; Weed Sci Soc Am. Res: Herbicide evaluation for nursery crops; propagation and container culture of ornamental plants; improvement of ornamental species through genetic and cytological research. Mailing Add: Va Truck & Ornamentals Res Sta PO Box 2160 Norfolk VA 23501

MILBRATH, GENE MCCOY, b Corvallis, Ore, Feb 15, 41; m 64; c 2. PLANT PATHOLOGY. Educ: Ore State Univ, BS, 63; Univ Ariz, MS, 66, PhD(plant path), 70. Prof Exp: Asst prof plant path, Univ Hawaii, 70-71; ASST PROF PLANT PATH, UNIV ILL, URBANA, 71- Mem: Am Phytopath Soc. Res: Epidemiology of plant viruses; characterization of plant viruses of economic plants. Mailing Add: 113 Hort Field Lab Univ of Ill Urbana IL 61801

MILBURN, BURTON BURNETT, b Los Angeles, Calif, Apr 5, 38; m 61; c 1. INDUSTRIAL HYGIENE. Educ: Univ Calif, Los Angeles, BS, 62, MS, 66, DrPH, 71. Prof Exp: Health physicist, Univ Calif, Los Angeles, 62-64; LECTR SOCIAL ECOL, UNIV CALIF, IRVINE, 71- Concurrent Pos: Consult dir environ health & safety, Heavy Metals Technol Corp, 68-70; res consult, Los Angeles County Pub Health Found, 70-71. Mem: AAAS; Am Indust Hyg Asn; Am Pub Health Asn. Res: Behavioral effects of electromagnetic fields; human behavior aspects of occupational health. Mailing Add: Prog in Social Ecol Univ of Calif Irvine CA 92664

MILBURN, NANCY STAFFORD, b Syracuse, NY, Sept 7, 27; m 51; c 2. PHYSIOLOGY, ELECTRON MICROSCOPY. Educ: Radcliff Col, AB, 49, PhD, 58; Tufts Univ, MS, 50. Prof Exp: Asst, Tufts Univ, 49-52; asst, Harvard Univ, 56-58; from instr to assoc prof, 58-71, actg chmn dept, 67-68, PROF PHYSIOL, TUFTS UNIV, 71-, RES ASSOC NEUROPHYSIOL, 58-, DEAN, JACKSON COL, 72- Concurrent Pos: Mem, Nat Res Coun; mem, Coun for Int Exchange of Scholars & Corp Woods Hole Oceanog Inst, 75- Mem: Am Soc Zool; Am Physiol Soc; Biophys Soc; Am Soc Cell Biol. Res: Neurophysiology, especially synaptic transmission, neurohormones and synaptic transmitters; electron microscopy of insect nervous system, receptor organs and effectors. Mailing Add: Dept of Biol Tufts Univ Medford MA 02155

MILBURN, RICHARD HENRY, b Newark, NJ, June 3, 28; m 51; c 2. NUCLEAR PHYSICS. Educ: Harvard Univ, AB, 48, AM, 51, PH PhD(physics), 54. Prof Exp: Instr physics, Harvard Univ, 54 & 56-57, asst prof, 57-61; assoc prof, 61-65, PROF PHYSICS, TUFTS UNIV, 65- Concurrent Pos: Guggenheim fel, Univ Geneva, 60. Mem: AAAS; Am Phys Soc; Am Asn Physics Teachers. Res: Physics of elementary particles; lasers. Mailing Add: Dept of Physics Tufts Univ Medford MA 02155

MILBURN, RONALD MCRAE, b Wellington, NZ, May 29, 28; m 55. INORGANIC CHEMISTRY. Educ: Victoria Univ, BSc, 49, MSc, 51; Duke Univ, PhD(chem), 54. Prof Exp: Demonstr chem, Victoria Univ, NZ, 51-52; asst, Duke Univ, 52-54; lectr, Victoria Univ, NZ, 55; res assoc, Univ Chicago, 56-57; from asst prof to assoc prof, 57-68, PROF CHEM, BOSTON UNIV, 68- Concurrent Pos: Fulbright grant, 52; instr & res assoc, Duke Univ, 54 & 56; NIH fel, Oxford Univ, 65-66; vis fel, Australian Nat Univ, 74-75. Mem: Am Chem Soc. Res: Reactions and stabilities of complex ions; mechanisms of inorganic reactions. Mailing Add: Dept of Chem Boston Univ Boston MA 02215

MILBY, THOMAS HUTCHINSON, b South Bend, Ind, Feb 7, 31; m 53; c 3. OCCUPATIONAL MEDICINE, TOXICOLOGY. Educ: Purdue Univ, BS, 53; Univ Cincinnati, MD, 57, MS, 65; Univ Calif, Berkeley, MPH, 66; Am Bd Prev Med, dipl & cert occup med, 66. Prof Exp: Intern med, Ohio State Univ Hosp, 58; med officer, Div Occup Health, USPHS, 59-62; med officer, Bur Occup Health, Calif State Dept Pub Health, 62-66, chief, 66-73; CONSULT OCCUP MED, TOXICOL & EPIDEMIOL, 73- Concurrent Pos: Mem comn pesticides & environ health, Secy Health, Educ & Welfare, 69; mem study sect, Nat Inst Occup Safety & Health, 69-72; spec consult, WHO, 70; assoc prof, Sch Pub Health, Univ Calif, Berkeley, 70-; chmn task group on occup exposure to pesticides, Fed Working Group on Pest Mgt, Nat Inst Occup Safety & Health, 72-74. Mem: Am Occup Med Asn. Res: Toxicology and epidemiology and chemical-related diseases. Mailing Add: 524 Woodmont Ave Berkeley CA 94708

MILCH, ALFRED EDWARD, physical chemistry, see 12th edition

MILCH, LAWRENCE JACQUES, b New York, NY, Sept 5, 18; m 42; c 4. PHARMACOLOGY, BIOCHEMISTRY. Educ: Univ Iowa, AB, 40; Rutgers Univ, PhD(physiol, biochem), 50. Prof Exp: Biophysicist, US Air Force Sch Aviation Med, 50-55, chief dept pharmacol & biochem, 55-59, dep comdr, 6102 Air Base Wing, Yakota Air Base, 59-61, head space biophys task group, Univ Calif, Berkeley, 61-62; staff res dir, Miles Labs, 62-66; asst res dir, Human Health Res & Develop Div, 66-71, DIR DEVELOP, DOW CHEM USA, ZIONSVILLE, 72- Concurrent Pos: Asst, Rutgers Univ, 47-50. Mem: Soc Exp Biol & Med; Am Soc Pharmacol & Exp Therapeut; Am Soc Clin Pharmacol & Therapeut; Soc Toxicol. Res: Coronary artery disease; traumatic injury; biophysical instrumentation; drug mechanisms; toxicology. Mailing Add: 1775A Pemberton Lane Indianapolis IN 46260

MILCH, PAUL R, b Budapest, Hungary, May 1, 34; US citizen; m 62; c 2. OPERATIONS RESEARCH, STATISTICS. Educ: Brown Univ, BS, 57; Stanford Univ, PhD(statist), 66. Prof Exp: ASSOC PROF OPERS RES DEPT, NAVAL POSTGRAD SCH, 63- Concurrent Pos: Statistician, Data Dynamics Inc, Calif, 65-66; opers analyst, Mellonics Inc, Litton Industs, 68-70 & BDM, Calif, 71- Mem: Am Statist Asn; Opers Res Soc Am. Res: Probability theory; queueing theory; stochastic processes; reliability; birth and death processes. Mailing Add: Dept of Opers Res & Admin Sci Naval Postgrad Sch Monterey CA 93950

MILCZAREK, CHESTER J, b Chicago, Ill, Mar 16, 18; m 43; c 1. INORGANIC CHEMISTRY, ANALYTICAL CHEMISTRY. Educ: Northwestern Univ, Ill, BS, 49. Prof Exp: Waste control supvr radioactive mat, Argonne Nat Lab, 51-54; chemist & group leader alkalies & chlorine, Columbia-Southern Chem Corp Div, Pittsburgh Plate Glas Co, Tex, 54-57, chief chemist chromium chem, NJ, 57-61, supvr exp lab, 61-70, ANAL SUPVR, CORPUS CHRISTI TECH CTR, PPG INDUSTS, INC, 70- Mem: Fel Am Inst Chem. Res: Inorganic chemicals; inorganic analytical chemistry. Mailing Add: 529 Fairfield Dr Corpus Christi TX 78412

MILDER, JACK WALTER, b Chicago, Ill, May, 21, 25; m 52; c 2. INTERNAL MEDICINE. Educ: Univ Notre Dame, BS, 48; Northwestern Univ, BS, 50, MD, 51. Prof Exp: Resident internal med, Hines Vet Admin Hosp, Ill, 51-54; pvt pract, Ill, 54-57; exec secy surg study sect, Div Res Grants, NIH, 57-58; RES ADMINR, AM CANCER SOC, 58- Concurrent Pos: Consult, St Mary's Hosp, Kankakee, Ill, 54-57. Mem: AAAS. Res: Therapy and diagnosis of cancer; biochemical approach to mechanisms of action and metabolism of chemotherapeutic agents; immunologic approach to cancer. Mailing Add: 43 Clinton St Mt Vernon NY 10552

MILDVAN, ALBERT S, b Philadelphia, Pa, Mar 3, 32; m 57; c 3. BIOPHYSICS, ENZYMOLOGY. Educ: Univ Pa, AB, 53; Johns Hopkins Univ, MD, 57. Prof Exp: Intern med, Baltimore City Hosps, Md, 57-58; res assoc cell physiol, Geront Br, NIH, 58-60; NIH res fel biochem, Inst Animal Physiol, Cambridge, Eng, 60-62; NIH res fel biophys, 62-64, assoc, 64-65, from asst prof to assoc prof phys biochem, 65-74, assoc mem, Inst, 68-73, PROF PHYS BIOCHEM, JOHNSON FOUND, SCH MED, UNIV PA, 74-, MEM, INST CANCER RES, 73- Concurrent Pos: Advan fel, Am Heart Asn, 63-65, estab investr, 65-70, mem coun basic sci, 71-; NIH res grant, 65-75; mem adv panel molecular biol, NSF, 71- Mem: Am Soc Biol Chemists; Brit Biochem Soc. Res: Biology of aging; mechanisms of enzymes action and metal activation of enzymes. Mailing Add: Inst for Cancer Res 7701 Burholme Ave Philadelphia PA 19111

MILES, CHARLES BURKE, b Salt Lake City, Utah, Jan 2, 15; m 36; c 2. CHEMISTRY. Educ: Univ Utah, BA, 35; Purdue Univ, PhD(phys chem), 40. Prof Exp: Anal chemist, Kalunite, Inc, Utah, 35-36; asst, Purdue Univ, 37-40; res chemist, US Rubber Co, Mich, 40-41; group leader, Pa Salt Mfg Co, 41-43; group leader, Magnolia Petrol Co, Tex, 43-46; dir res & develop, Newark Sect, Westvaco Chem Div, Food Mach & Chem Corp, 46-54, res dir, Westvaco Mineral Prods Div, 54-58, asst res dir, Inorg Chem Dept, 58-60; tech dir, Chem Res Div, Socony Mobil Oil Co, 60, mgr res & develop dept, Mobil Chem Co Div, 60-63; tech adv to pres, 63-64, corp tech dir, 64-68, tech adv, Patent Dept, 68-71, SR RES ASSOC, KAISER ALUMINUM & CHEM CORP, 71- Mem: AAAS; Am Chem Soc. Res: Inorganic chemicals; fluorides; phosphorus; magnesia; barium; thermochemistry; research and development administration and planning. Mailing Add: Kaiser Aluminum & Chem Corp Ctr for Technol Pleasanton CA 94566

MILES, CHARLES DAVID, b Kansas City, Mo, Aug 11, 26; m 53; c 2. INVERTEBRATE ZOOLOGY. Educ: Univ Kans, AB, 50, MA, 56; Univ Ariz, PhD(zool), 61. Prof Exp: Asst prof biol, Eureka Col, 61-64; from asst prof to assoc prof, 64-72, PROF ZOOL, UNIV MO-KANSAS CITY, 72- Mem: Am Malacol union; Am Soc Parasitologists; Sigma Xi. Res: Land snail taxonomy and distribution; anatomy; physiology. Mailing Add: Dept of Biol Univ of Mo 5100 Rockhill Rd Kansas City MO 64110

MILES, CHARLES DONALD, b Franklin, Ind, Dec 17, 38; m 66; c 1. PLANT PHYSIOLOGY. Educ: Franklin Col, AB, 63; Indiana Univ, PhD(bot), 67. Prof Exp: NIH fel plant biochem, Cornell Univ, 67-69; asst prof, 69-75, ASSOC PROF BOT, UNIV MO-COLUMBIA, 75- Mem: AAAS; Am Soc Plant Physiol; Bot Soc Am; Am Inst Biol Sci; Phytochem Soc NAm. Res: Development and control of pigmentation in higher plants; mechanism of photosynthetic phosphorylation and electron transport; chloroplast fluorescence and luminescence. Mailing Add: Div of Biol Univ of Mo Columbia MO 65201

MILES, CHARLES P, b Chicago, Ill, June 19, 22; m 54; c 3. PATHOLOGY, CYTOLOGY. Educ: Univ Calif, Berkeley, BA, 47; Univ Calif, San Francisco, MD, 53; Am Bd Path, dipl, 59. Prof Exp: Bank Am-Giannini Found fel, 55-56; Nat Cancer Inst fel, 57-58; asst prof in residence nuclear med & radiation biol, Univ Calif, Los Angeles, 58-59; from instr to asst prof path, Stanford Univ, 59-62; assoc, Sloan-Kettering Inst, 62-66; assoc prof, Univ Utah, 66-69; assoc prof, Univ Calif, San Francisco, 69-70; PROF PATH, UNIV UTAH, 70- Concurrent Pos: Asst attend pathologist, Mem Hosp, NY, 62-66. Mem: Am Soc Exp Path; Am Asn Path & Bact; Am Asn Cancer Res; Soc Human Genetics. Res: Cytology analysis in cancer and in tissue culture strains. Mailing Add: Dept of Path Univ of Utah Salt Lake City UT 84132

MILES, DAVID HARRY, organic chemistry, see 12th edition

MILES, DELBERT HOWARD, b Warrior, Ala, Jan 4, 43; m 63; c 2. CHEMISTRY. Educ: Birmingham-Southern Col, BS, 65; Ga Inst Technol, PhD(org chem), 70. Prof Exp: NIH res fel org chem, Stanford Univ, 69-70; asst prof, 70-74, ASSOC PROF CHEM, MISS STATE UNIV, 74- Honors & Awards: Sigma Xi Award, 74. Mem: Am Chem Soc. Res: Isolation, structure elucidation and synthesis of natural products which exhibit biological activity of some type or which possess some biosynthetic significance; conformation analysis. Mailing Add: Dept of Chem Miss State Univ State College MS 39762

MILES, DONALD ORVAL, b Callaway, Nebr, May 29, 39; m 60; c 1. MEDICAL MICROBIOLOGY. Educ: Hastings Col, BA, 64; Univ Nebr, Lincoln, MS, 67, PhD(microbiol), 72. Prof Exp: Asst lectr biol, Univ Nebr, Lincoln, 70-71, instr med bact, 72; ASST PROF MICROBIOL, SCH HEALTH SCI, GRAND VALLEY STATE COLS, 73- Concurrent Pos: Res assoc, Dept Microbiol, Univ Nebr, Lincoln, 71 & Dept Oral Biol, Col Dent, 73; clin microbiologist & microbiol consult, Microbiol Lab, Dept Path, St Mary's Hosp, Grand Rapids, 74-, mem infection control comt, 75-, assoc mem med & dent staff, 75- Mem: Am Soc Microbiol; Am Inst Biol Sci; Sigma Xi; Soc Indust Microbiol. Res: Amino acid metabolism of gram negative non-spore forming anaerobes; periplasmic enzymes of gram negative bacilli; microbiology of infection control in hospitals. Mailing Add: Sch of Health Sci Grand Valley State Cols Allendale MI 49401

MILES, EDWARD JERVIS, b London, Ont, Mar 6, 26; US citizen; m 55; c 2. GEOGRAPHY. Educ: Univ Western Ont, BA, 48; Syracuse Univ, MA, 50, PhD(geog), 58. Prof Exp: Instr geog, Concord Col, 50-51; instr geog & phys sci, Univ Fla, 51-52; instr geog & econ, Univ Md, 52-55; asst prof geog & geol, Valparaiso Univ, 58-62; assoc prof geog, Univ Vt, 62-66, chmn dept, 66-73, PROF GEOG, UNIV VT, 66-, DIR, CAN STUDIES PROG, 63- Concurrent Pos: Vis fel, Inst Can Studies, Carleton Univ, 69-70; regional counr, Asn Am Geographers, 74- Mem: Am Geog Soc; Asn Am Geog; Nat Coun Geog Educ; fel African Studies Asn; Asn Can Studies in US (vpres, 73-75, pres, 75-). Res: Political geography; regional geography of Canada; regional geography of Africa; historical geography of North America. Mailing Add: Dept of Geog 112 Old Mill Bldg Univ of Vt Burlington VT 05401

MILES, ERNEST PERCY, JR, b Birmingham, Ala, Mar 16, 19; m 45; c 2. MATHEMATICS. Educ: Birmingham-Southern Col, BA, 37; Duke Univ, MA, 39, PhD(math), 49. Prof Exp: Teacher high sch, Ala, 38-39; instr math, NC State Col, 40-41; assoc prof, Ala Polytech Inst, 49-58; mem staff, Nat Sci Found, 58; assoc prof math, 58-61, dir comput ctr, 61-71, PROF MATH, FLA STATE UNIV, 61- Concurrent Pos: Vis assoc prof & Air FOrce Off Sci Res contract, Inst Fluid Dynamics & Appl Math, Univ Md, 57-58; participant, Nat Sci Found Training Prog Numerical Anal, Nat Bur Standards, 59; res grants, Air Force Off Sci Res, 60-61, Nat Sci Found, 62-72; consult, Nat Sci Found, 59-63, 65-70 & US Off Educ, 65-71; coun mem, Conf Bd Math Socs, 74- Mem: Fel AAAS; Am Math Soc; Math Asn Am; Asn Comput Mach; Soc Indust & Appl Math. Res: Partial differential equations; numerical methods; information retrieval; computer uses in education. Mailing Add: Dept of Math Fla State Univ Tallahassee FL 32306

MILES, FRANCIS TURQUAND, b Princeton, NJ, July 10, 09; m 36; c 4. INORGANIC CHEMISTRY, NUCLEAR ENGINEERING. Educ: Princeton Univ, AB, 31, AM, 32, PhD(inorg chem), 36. Prof Exp: Res chemist, Monsanto Chem Co, Ala, 36-43; SAM labs, Columbia Univ, 43-45, dir div, 45-46; Clinton Labs, Univ Tenn, 46-48; Oak Ridge Nat Lab, 48; asst to chmn dept appl sci, Brookhaven Nat Lab, 48-65, dep chmn, 65-73; RETIRED. Concurrent Pos: Dir div nuclear power & reactors, Int Atomic Energy Agency, 62-65. Mem: Am Chem Soc; Am Nuclear Soc. Res: Fluid reactors, especially liquid metal; separation processes in atomic energy. Mailing Add: Dept of Appl Sci Brookhaven Nat Lab Upton NY 11973

MILES, FRANK BELSLEY, b Champaign, Ill, May 15, 30; m 66; c 1. MATHEMATICS. Educ: Univ Ill, BS, 61; Univ Calif, Berkeley, PhD(chem), 65; Univ Wash, MS, 70. Prof Exp: NIH fel, Univ Calif, Los Angeles, 64-65; asst prof chem, Univ Calif, Santa Barbara, 65-68; NSF fel, Univ Wash, 69-71, instr math, 71, res asst, 71-72; ASST PROF MATH, CALIF STATE COL, DOMINGUEZ HILLS, 72- Mem: Am Math Soc. Res: Harmonic analysis. Mailing Add: Dept of Math Calif State Col Dominguez Hills CA 90747

MILES, GEORGE BENJAMIN, b Erin, Tenn, May 14, 26; m 56; c 1. ORGANIC CHEMISTRY. Educ: Univ Tenn, BS, 50, PhD(chem), 58. Prof Exp: Chemist, US Naval Ord Lab, 53-54; instr chem, Univ Tenn, 57-58; res chemist, Dacron Res Lab, Textile Fibers Dept, E I du Pont de Nemours & Co, 58-61; from asst prof to assoc prof, 61-69, PROF & CHMN DEPT CHEM, APPALACHIAN STATE UNIV, 69- Mem: Am Chem Soc. Res: Polymer and steroid chemistry; organic mechanisms. Mailing Add: Dept of Chem Appalachian State Univ Boone NC 28607

MILES, HARRY MCCAULEY, b San Antonio, Tex, Apr 17, 42; m 62; c 1. ECOLOGY, PHYSIOLOGY. Educ: Tex A&M Univ, BS, 66; Univ Wash, MS, 67, PhD(fisheries), 69. Prof Exp: ASST PROF BIOL, MARQUETTE UNIV, 69- Concurrent Pos: Wis State Dept Natural Resources grant, Marquette Univ, 69-72, Fed Water Resources Res grant, 71-72. Mem: Ecol Soc Am; Am Fisheries Soc. Res: Sublethal physiological effects of heavy metal water pollutants on fish; radiotelemetry of physiological parameters from free-swimming fish. Mailing Add: Dept of Biol Marquette Univ Milwaukee WI 53233

MILES, HENRY HARCOURT WATERS, b Burnside, La, Sept 18, 15; m 39; c 2. PSYCHIATRY, PSYCHOANALYSIS. Educ: Tulane Univ, BS, 36, MD, 39. Prof Exp: Res fel psychiat, Harvard Med Sch, 46-48; asst, Harvard Univ & Mass Gen Hosp, 49-52; from asst prof to assoc prof clin psychiat, 52-66, PROF PSYCHIAT, SCH MED, TULANE UNIV, 66- Concurrent Pos: Training & supv analyst, New Orleans Psychoanal Inst, 56-; consult, Family Serv Soc New Orleans, 57-64. Mem: Am Psychosom Soc; AMA; fel Am Psychiat Asn; Am Psychoanal Asn. Res: Evaluation of psychotherapy; personality factors in cardiovascular diseases. Mailing Add: 1446 Arabella St New Orleans LA 70115

MILES, HENRY JARVIS, b St George, Utah, Jan 1, 00; m 31; c 2. MATHEMATICS. Educ: Univ Calif, AB, 25, MA, 26, PhD(math), 29. Prof Exp: From instr to prof, 29-68, EMER PROF MATH, UNIV ILL, URBANA, 68- Mem: Am Math Soc; Math Asn Am. Res: Generalization of Plucker's surface. Mailing Add: 507 W High St Urbana IL 61801

MILES, JAMES FRANKLIN, agricultural economics, see 12th edition

MILES, JAMES LOWELL, b Buckhannon, WVa, Aug 15, 37; m 68. CHEMISTRY, BIOCHEMISTRY. Educ: WVa Univ, BS, 59, MS, 61, PhD(biochem), 64. Prof Exp: Trainee clin chem, Hosp Univ Pa, 64-65; RES BIOCHEMIST, E I DU PONT DE NEMOURS & CO, INC, WILMINGTON, 66- Mem: Am Chem Soc; Am Asn Clin Chemists. Res: Alpha-chymotrypsin; clinical and analytical chemistry; enzyme assay systems; lipoprotein electrophoresis. Mailing Add: 44 Quartz Mill Rd Newark DE 19711

MILES, JAMES S, b Baltimore, Md, Apr 16, 21; m 44; c 4. MEDICINE. Educ: Grinnell Col, AB, 42; Univ Chicago, MD, 45; Am Bd Orthop Surg, dipl, 54. Prof Exp: Intern, Univ Clins, Sch Med, Univ Chicago, 45-46, resident, 48-51, instr orthop, 51-52; from instr to assoc prof orthop surg, 52-65, chmn div, 58-73, actg chmn dept orthop, 73-74, PROF ORTHOP SURG, SCH MED, UNIV COLO, DENVER, 65-, CHMN DEPT ORTHOP, 74- Concurrent Pos: Am Orthop Asn traveling fel, 59; consult, Vet Admin Hosp, Grand Junction, Colo & Fitzsimons Army Hosp, Denver. Mem: Orthop Res Soc; fel Am Col Surgeons; Am Acad Orthop Surg; Am Orthop Asn; Clin Orthop Soc. Res: Histochemistry of articular cartilage; vascular supply of femoral head. Mailing Add: Dept of Orthop Univ of Colo Sch of Med Denver CO 80220

MILES, JAMES THOMAS, dairy husbandry, see 12th edition

MILES, JAMES WILLIAM, b Henderson, Ky, Sept 19, 18; m 51; c 1. PESTICIDE CHEMISTRY. Educ: Western Ky State Col, BS, 40; Univ Ill, MS, 47, PhD(anal chem), 53. Prof Exp: Instr chem, Louisville Col Pharm, 41-42; from asst prof to assoc prof, Univ Ky, 47-55, prof pharmaceut chem & head dept, 55-58; asst chief, Chem Sect, Tech Develop Lab, 58-64, chief chem sect, Tech Develop labs, Commun Dis Ctr, 64-74, CHIEF PESTICIDES BR, BUR OF TROP DIS, CTR FOR DIS CONTROL, USPHS, 74- Concurrent Pos: Mem expert adv panel on insecticides, WHO, 71-, mem sci & tech adv comt, WHO Onchocerciasis Control Prog, 74- Mem: Am Chem Soc; Sigma Xi; Res: Colorimetric determination of vanillin and levulose; spectrophotometric determination of xanthine derivatives, salicylic and benzoic acids; spectrophotometric titrations of chromium and vanadium; development of methods of analysis of pesticide residues in the environment; research on pesticide formulations and analysis of formulations. Mailing Add: Bur of Trop Dis Ctr for Dis Control USPHS Atlanta GA 30333

MILES, JOHN BLANCHARD, physics, see 12th edition

MILES, JOHN LEONARD, physics, see 12th edition

MILES, JOHN WILDER, b Cincinnati, Ohio, Dec 1, 20; m 43; c 3. GEOPHYSICS. Educ: Calif Inst Technol, BS, 42, MS, 43, PhD(elec eng), 44. Prof Exp: Staff mem radiation lab, Mass Inst Technol, 44; res eng, Lockheed Aircraft Corp, Calif, 44-45; from asst prof to assoc prof eng, Univ Calif, Los Angeles, 45-55, prof eng & geophys, 55-61; prof appl math, Inst Adv Studies, Australian Nat Univ, 62-64; chmn appl mech & eng sci, 68-74, PROF APPL MECH & GEOPHYS, UNIV CALIF, SAN DIEGO, 65- Concurrent Pos: Fulbright lectr, Univ NZ, 51; vis lectr, Univ London, 52; Guggenheim fel, 58-59 & 68-69; Fulbright res fel, Cambridge Univ, 69. Mem: AAAS; fel Am Inst Aeronaut & Astronaut; Fedn Am Sci; fel Am Acad Arts & Sci; fel NY Acad Sci. Res: Wave propagation and generation; hydrodynamic stability; geophysical fluid mechanics. Mailing Add: Inst Geophys & Planetary Physics Univ of Calif at San Diego La Jolla CA 92037

MILES, JOSEPH BELSLEY, b Champaign, Ill, June 17, 42; m 70. MATHEMATICS. Educ: Univ Ill, BS, 63; Univ Wis, MS, 64, PhD(math), 68. Prof Exp: Res assoc, Cornell Univ, 68-69; asst prof, 69-74, ASSOC PROF MATH, UNIV ILL, URBANA, 74- Concurrent Pos: Off Naval Res fel, Cornell Univ, 68-69; res assoc, Univ Md, 75-76. Mem: Am Math Soc. Res: Functions of a complex variable. Mailing Add: Dept of Math Univ of Ill Urbana IL 61801

MILES, KELLY GEORGE, b Miles, NC, Nov 5, 10; m 36; c 2. PHYSICS. Educ: Appalachian State Teachers Col, BS, 33; Univ NC, MS, 36. Prof Exp: Asst physics, Appalachian State Teachers Col, 30-33, head sci dept, 36-41; instr in charge theory, Capitol Radio Eng Inst, DC, 41-44; physicist, Appl Physics Lab, Johns Hopkins Univ, 44-48; physicist & electronic scientist, US Naval Res Lab, DC, 48-55; assoc prof, 55-65, PROF PHYSICS, UNIV EVANSVILLE, 65-, HEAD DEPT, 63- Mem: AAAS; Inst Elec & Electronics Eng; Am Asn Physics Teachers. Res: Electrical measurements; astronomy. Mailing Add: 737 S Norman Ave Evansville IN 47714

MILES, MARION LAWRENCE, b Columbus, Ga, Sept 5, 29; m 56; c 3. ORGANIC CHEMISTRY. Educ: Univ Ga, BS, 57, MS, 59; Univ Fla, PhD(org chem), 63. Prof Exp: Fel, Duke Univ, 63-64; asst prof, 65-69, ASSOC PROF ORG CHEM, NC STATE UNIV, 69- Mem: Am Chem Soc. Res: Physical properties of multiple carbanions; mechanisms of condensation reactions. Mailing Add: Dept of Chem NC State Univ PO Box 5247 Raleigh NC 27607

MILES, MAURICE HOWARD, b St George, Utah, Nov 20, 33; m 60; c 2. SOLID STATE PHYSICS. Educ: Univ Utah, BS, 55, PhD(physics), 63. Prof Exp: Res assoc metall, Univ Ill, 63-65; ASST PROF PHYSICS, WASH STATE UNIV, 65- Mem: Am Phys Soc. Res: Internal friction in metals; electronic properties of dislocations in semiconductors. Mailing Add: Dept of Physics Wash State Univ Pullman WA 99163

MILES, MAURICE JARVIS, b St George, Utah, Nov 24, 07; m 31; c 11. ANALYTICAL CHEMISTRY, ENVIRONMENTAL CHEMISTRY. Educ: Brigham Young Univ, AB, 30; Univ Utah, MA, 33. Prof Exp: Dir phys sci, Dixie Jr Col, 33-53; chief chemist, Titanium Metals Corp Am, 53-70; RES ASSOC, DESERT RES INST, UNIV NEV SYST, 70- Concurrent Pos: Consult, US Bur Mines, Boulder City, 74. Mem: AAAS; Am Chem Soc; Am Soc Test & Mat. Res: Chemical analysis of titanium; determination of beryllium, magnesium, zirconium, oxygen, hydrogen and nitrogen in titanium; rapid x-ray ion-exchange analyses of titanium alloys; water pollution analyses related to Lake Mead; trace metals in geothermal waters; trace metals in soils; atmospheric dusts. Mailing Add: PO Box 168 Henderson NV 89015

MILES, MELVIN HENRY, b St George, Utah, Jan 18, 37; m 62; c 2. ELECTROCHEMISTRY, PHYSICAL CHEMISTRY. Educ: Brigham Young Univ, BA, 62; Univ Utah, PhD(phys chem), 66. Prof Exp: NATO res fel electrochem, Munich Tech, 65-66; res chemist, Naval Weapons Ctr, 67-69; asst prof, 69-72, ASSOC PROF CHEM, MID TENN STATE UNIV, 72- Mem: Electrochem Soc; Am Inst Chem; Sigma Xi. Res: Fast reaction kinetics; electrode kinetics; electrochemical energy conversion; electrode catalysis; fuel cells; viscosity; specific conductivity; water electrolysis; properties of mixed solvents; hydrogen production; oxygen electrode reaction. Mailing Add: Mid Tenn State Univ Box 323 Murfreesboro TN 37130

MILES, NEIL WAYNE, b River Falls, Wis, June 22, 37; m 59; c 2. HORTICULTURE, PLANT PHYSIOLOGY. Educ: Univ Minn, BS, 59, MS, 64, PhD(hort), 65. Prof Exp: Exten horticulturist, Univ Minn, St Paul, 65-66; POMOLOGIST, KANS STATE UNIV, 66- Mem: Am Soc Hort Sci; Am Pomol Soc. Res: Physiological studies on fruit crops. Mailing Add: Dept of Hort & Forestry Kans State Univ Waters Hall Manhattan KS 66506

MILES, PHILIP GILTNER, b Olean, NY, Aug 10, 22; m 49; c 3. BOTANY. Educ: Yale Univ, BA, 48; Indiana Univ, PhD(bot), 53. Prof Exp: Res assoc bot, Univ Chicago, 53-54; res fel, Harvard Univ, 54-56; from asst prof to assoc prof, 56-70, PROF BIOL, STATE UNIV NY BUFFALO, 70- Concurrent Pos: Fulbright res scholar, Japan, 63-64; vis scientist, US-China Coop Sci Prog, 70-71; vis prof, Nat Taiwan Univ, 70-71. Mem: AAAS; Bot Soc Am; Genetics Soc Am; Mycol Soc Am; Soc Study Evolution. Res: Genetics and physiology of sexual mechanisms in fungi. Mailing Add: Dept of Biol State Univ NY Buffalo NY 14214

MILES, RALPH FRALEY, JR, b Philadelphia, Pa, May 15, 33. OPERATIONS RESEARCH. Educ: Calif Inst Technol, BS, 55, MS, 60, PhD(physics), 63. Prof Exp: Sr engr, Jet Propulsion Lab, 63-65, supvr syst eng, 65-69; vis fel econ systs, Stanford Univ, 69-70; vis asst prof aeronaut & environ eng sci, Calif Inst Technol, 70-71; mgr, Mission Anal & Eng, Outer Planets Missions, 71-75, SUPVR SYST ANAL, JET PROPULSION LAB, 75- Mem: AAAS; Am Phys Soc; Opers Res Soc Am; Am Inst Aeronaut & Astronaut. Res: Density of cosmic ray neutrons in the atmosphere; systems analysis; design analy analysis. Mailing Add: 3608 Canon Blvd Altadena CA 91001

MILES, WALTER RICHARD, b Silverleaf, NDak, Mar 29, 85; m 08, 27; c 4. PHYSIOLOGICAL PSYCHOLOGY. Educ: Pac Col, BS, 06; Earlham Col, AB, 08; Iowa State Univ, MA, 10, PhD(psychol), 13. Hon Degrees: AM, Yale Univ, 31; ScD,

Earlham Col, 52. Prof Exp: Asst psychol, Iowa State Univ, 10-13; assoc prof, Wesleyan Univ, 13-14; res psychologist, Nutrit Lab, Carnegie Inst, 14-22; prof exp psychol, Stanford Univ, 22-32; prof, 31-53, EMER PROF PSYCHOL, YALE UNIV, 53- Concurrent Pos: Res assoc, Inst Human Rels, 30-31; assoc psychologist, New Haven Hosp & Dispensary, 32-54; fel, Jonathan Edwards Col, 35-53, assoc fel, 53-59, emer prof & fel, 59-; pres, Psychol Corp, NY, 39-44, chmn bd dirs, 44-54; mem comt selection & training aircraft pilots, Nat Res Coun, 39-46, comt aviation med, 40-46, chmn div anthrop & psychol, 45-46; consult, Royal Air Force, 45-46; mem, Conn Bd Exam Psychol, 45-54; chmn bd dirs, Am Inst Res, 47-54; ord prof, Istanbul Univ, 54-57; sci dir, Med Res Lab, US Naval Submarine Base, Conn, 57-65, consult, 65-; consult, Nat Coun Naval Sr Scientists. Honors & Awards: Presidential Cert of Merit, 48; Warren Medal, Soc Exp Psychol, 49; Gold Medal, Am Psychol Found, 62. Mem: Nat Acad Sci; AAAS; Soc Exp Psychol (pres, 32); Am Physiol Soc; Optical Soc Am. Res: Vision; later maturity and old age; habitability for submarines and space capsules; human retina; space perception; olfaction. Mailing Add: RFD 3 Harvard Terr Gales Ferry CT 06335

MILES, WILLIAM RAYMOND, b Boise, Idaho, Nov 22, 22; m 49; c 7. FORESTRY. Educ: Univ Minn, BS, 51, MF, 59, PhD(forestry), 71. Prof Exp: Land agent, Weyerhauser Co, 50-56, forester, 56-57, logging foreman, 57-59; from instr to assoc prof, 59-74, PROF FORESTRY, UNIV MINN, ST PAUL, 74-, EXTEN FORESTER, 61- Honors & Awards: Outstanding Serv, Environ Protection Agency, 74. Mem: Soc Am Foresters. Res: Forest management; environmental studies in natural resources; forestry and environmental education. Mailing Add: Col of Forestry 110 Green Hall Univ of Minn St Paul MN 55108

MILES, WYNDHAM DAVIES, b Wilkes-Barre, Pa, Nov 21, 16; m 52; c 4. HISTORY OF CHEMISTRY. Educ: Philadelphia Col Pharm, BS, 42; Pa State Univ, MS, 44; Harvard Univ, PhD(hist of sci), 55. Prof Exp: Instr chem, Pa State Univ, 44-50, asst prof, 52-53; historian, US Army Chem Corps, 53-60; specialist in sci, Nat Arch, 60-61; historian, Polaris Proj, 61-62; HISTORIAN, NIH, 62- Honors & Awards: Dexter Award in Hist of Chem, 71. Mem: Am Chem Soc. Mailing Add: 24 Walker Ave Gaithersburg MD 20760

MILEWICH, LEON, b Buenos Aires, Arg, Mar 26, 27; US citizen; m 59; c 3. ORGANIC CHEMISTRY. Educ: Univ Buenos Aires, BS, 56, MS, 58, PhD(org chem), 59. Prof Exp: Chemist, Res Inst Armed Forces, Arg, 55-58, res chemist, 60-61; fel, Sch Pharm, Univ Md, 61-64; fel, Sch Med, Johns Hopkins Univ, 64-66, instr gynec & obstet, 66-67; res assoc, Southwest Found Res & Educ, 67-72; ASST PROF, DEPT OBSTET & GYNEC, UNIV TEX SOUTHWESTERN MED SCH DALLAS, 72- Concurrent Pos: NIH fel, 63-64. Mem: AAAS; Am Chem Soc; The Chem Soc; Arg Chem Asn; NY Acad Sci. Res: Steroids. Mailing Add: Dept of Obstet & Gynec Univ of Tex Southwestern Med Sch Dallas TX 75235

MILFORD, ALAN HACKNEY, b Johnstown, NY, Mar 5, 29; m 54; c 4. INDUSTRIAL CHEMISTRY. Educ: Hamilton Col, BA, 51. Prof Exp: Res chemist, 51-63, supvr res group liquid propellants, 63-66, res assoc ballistics, 66-71, supvr res group propellants, 71-73, res assoc, 73-75, SUPVR RES GROUP URETHANE FOAM, OLIN CORP, 75- Mem: Am Chem Soc. Res: Polymer chemistry, reaction rates, formulation in urethane foam technology. Mailing Add: 120 Wakefield St Hamden CT 06517

MILFORD, FREDERICK JOHN, b Cleveland, Ohio, July 1, 26; m 51; c 1. PHYSICS. Educ: Case Inst Technol, BS, 49; Mass Inst Technol, PhD(physics), 52. Prof Exp: Asst physics, Mass Inst Technol, 49-51; instr, Case Western Reserve Univ, 49-51, 52-56, asst prof, 56-59; div consult, 59-64, sr fel & chief theoret physics div, 64-66, inst scientist, 73, mgr physics & electronics dept, 73-74, DIR INST RES PHYS SCI, BATTELLE MEM INST, 65-, MGR PHYSICS ELECTRONICS & NUCLEAR TECHNOL, 74- Mem: Fel Am Phys Soc; Am Math Soc; Am Nuclear Soc; AAAS; Int Glaciological Soc. Res: Meson field theory; theoretical nuclear physics; cosmic ray primaries; nuclear magnetic resonance; nuclear moments; electronic structure of solids; magnetism; helium films at low temperature. Mailing Add: Battelle Mem Inst 505 King Ave Columbus OH 43201

MILFORD, GEORGE NOEL, b Victoria, PEI, Can, May 4, 24; nat US; m 48; c 3. POLYMER CHEMISTRY. Educ: Mt Allison Univ, BSc, 44; Dalhousie Univ, MSc, 48; McGill Univ, PhD(chem), 53. Prof Exp: Asst chemist, Best Yeast Co, 45-46; res chemist, Dom Steel & Coal Corp, 48-50; res chemist, 53-60, SR RES CHEMIST, E I DU PONT DE NEMOURS & CO, 60- Mem: Am Chem Soc. Res: Preparation of monomers and polymers; synthetic fibers, Orlon, Lycra, hollow fibers, Nomex. Mailing Add: Textile Fibers Dept Benger Lab E I du Pont de Nemours & Co Waynesboro VA 22980

MILFORD, MURRAY HUDSON, b Honey Grove, Tex, Sept 29, 34; m 61; c 2. SOIL SCIENCE, SOIL MINERALOGY. Educ: Tex A&M Univ, BS, 55, MS, 59; Univ Wis, PhD(soil sci), 62. Prof Exp: Fel soil chem & res specialist, Cornell Univ, 62-63, from asst prof to assoc prof soil sci, 63-68; assoc prof, 68-74, PROF SOIL SCI, TEX A&M UNIV, 74- Mem: Fel AAAS; Am Soc Agron; Soil Sci Soc Am; Clay Minerals Soc; Soil Conserv Soc Am. Res: Compacted layers in soils; potassium and magnesium chemistry of soils; movement and degradation of clay minerals in soils in relation to drainage; clay-organic interactions. Mailing Add: Dept of Soil & Crop Sci Tex A&M Univ College Station TX 77840

MILFORD, SIDNEY NEVIL, b Melbourne, Australia, Feb 11, 25; nat US; wid; c 2. ENVIRONMENTAL PHYSICS, PHYSICAL OCEANOGRAPHY. Educ: Univ Melbourne, BSc, 44, BA, 45; Univ London, PhD(physics), 48. Prof Exp: Exchange vis astrophysicist, Astrophys Inst, Univ Paris, 49-50 & Univ Chicago, 50-51; assoc prof physics, Evansville Col, 51-52; prof & chmn dept, St John's Univ, NY, 52-62; head geo-astrophys res, Grumman Aircraft Eng Corp, 62-71; READER DEPT PHYSICS, UNIV QUEENSLAND, 71- Concurrent Pos: Vis prof, State Univ NY, Stony Brook, 67-70; res assoc, Lamont Geol Observ, Columbia Univ, 75; consult, Allies Res Assoc & Repub Aviation Corp. Mem: Am Geophys Union; Am Phys Soc; Am Meteorol Soc; Australian Water & Waste Water Asn. Res: Field measurements and mathematical models of circulation and mixing in estuaries, rivers and coastal waters; tides, current, turbulence, seicles, surges; salinity, temperature, water quality, diffusion and dispersion theories. Mailing Add: Dept of Physics Univ of Queensland Brisbane Queensland 4067 Australia

MILFRED, CLARENCE JAMES, b Cazenovia, Wis, Aug 18, 29; m 66; c 3. GEOGRAPHY, SOIL SCIENCE. Educ: Univ Wis, BS, 60, MS, 62, PhD(soils), 66. Prof Exp: Soil scientist soil surv div, Geol & Natural Hist Surv, Univ Wis-Madison, 66-70; asst prof, 70-73, ASSOC PROF GEOG, UNIV WIS-STEVENS POINT, 73- Mem: AAAS; Soil Sci Soc Am; Soc Econ Paleont & Mineral; Am Soc Photogram. Res: Soil genesis, classification and survey; remote sensing; air photo interpretation. Mailing Add: Dept of Geog Univ of Wis Stevens Point WI 54481

MILGRAM, RICHARD JAMES, b South Bend, Ind, Dec 5, 39; m 64; c 2. MATHEMATICS. Educ: Univ Chicago, BSc & MSc, 61; Univ Minn, PhD(math), 64. Prof Exp: Instr math, Univ Minn, 63-64; instr, Princeton Univ, 64-66; from asst prof to assoc prof, Univ Ill, Chicago, 66-69; PROF MATH, STANFORD UNIV, 69- Concurrent Pos: Assoc mem inst advan study, Univ Ill, 67-68; vis prof, Princeton Univ, 69-70; ed, Duke J of Math & Pac J of Math. Mem: Am Math Soc. Res: Algebraic and differential topology; theory of H-spaces; construction of classifying spaces; structure and classification of manifolds and Poincare duality spaces; structure of the Steenrod algebras. Mailing Add: Dept of Math Stanford Univ Stanford CA 94305

MILGROM, FELIX, b Rohatyn, Poland, Oct 12, 19; nat; m 41; c 2. MEDICAL BACTERIOLOGY, IMMUNOLOGY. Educ: Wroclaw Univ, MD, 47. Prof Exp: From asst prof to assoc prof microbiol, Sch Med, Wroclaw Univ, 46-53, prof & dir in charge, 54; dir in charge, Inst Immunol & Exp Ther, Polish Acad Sci, 54; prof microbiol & head dept, Silesian Med Sch, 54-57; from res assoc to res assoc prof bact & immunol, 58-62, assoc prof, 62-67, PROF BACT, IMMUNOL & MICROBIOL & CHMN DEPT MICROBIOL, SCH MED, STATE UNIV NY BUFFALO, 67- Mem: Am Asn Immunol; Soc Exp Biol & Med; Am Acad Microbiol; Int Soc Blood Transfusion; Transplantation Soc. Res: Serology of syphilis and rheumatoid arthritis; natural antibodies; autoimmune processes; transplantation; tissue antigens; tumor immunology. Mailing Add: Dept of Microbiol Sch of Med State Univ of NY Buffalo NY 14214

MILGROM, HARRY, b New York, NY, Feb 29, 12; m 37; c 2. SCIENCE EDUCATION, SCIENCE WRITING. Educ: City Col New York, BS, 32; Columbia Univ, MA, 33. Prof Exp: Teacher physics, New York City Bd Educ, 35-53, supvr elem sci, 53-61, asst dir sci, 61-67; dir educ serv, Hall of Sci of the City of New York, 67-69; DIR SCI, NEW YORK CITY BD EDUC, 69- Concurrent Pos: Res assoc, Electrophys Labs, 42-44; consult, NY State Educ Dept, 56-60; Coun Chief State Sch Off, 65; Mfg Chem Asn, 58-60. Mem: Fel AAAS; Nat Sci Teachers Asn; NY Acad Sci; Nat Sci Supvrs Asn. Res: Design and development of new approaches, techniques and materials for the improvement and enrichment of science learning. Mailing Add: 185 E 85th New York NY 10028

MILGROM, JACK, b Chicago, Ill, May 21, 27; m 48; c 3. POLYMER CHEMISTRY. Educ: Univ Chicago, AB, 50, MS, 51, PhD(org chem), 59. Prof Exp: Sr chemist, Ninol Labs, Chicago, Ill, 51-56; proj chemist, Standard Oil Co Ind, 56-60; group leader polymerization catalysis, Gen Tire & Rubber Co, 60-66; mgr polymerization & process res, Foster Grant Co, Inc, 66-68; SR STAFF MEM, ARTHUR D LITTLE INC, 68- Concurrent Pos: Adv, Acad Sci, 74-75. Mem: Am Chem Soc; NY Acad Sci; Am Soc Test & Mat. Res: Free-radical and coordination chemistry; organometallics; catalysis; environmental studies; radiation chemistry; polymer technology; impact of technology on society; packaging; solid waste management. Mailing Add: Arthur D Little Inc 35 Acorn Park Cambridge MA 02140

MILHAM, ROBERT CARR, b Grand Haven, Mich, June 20, 22; m 56. ENVIRONMENTAL CHEMISTRY. Educ: Alma Col, BSc, 44; Univ Wis, PhD(inorg chem), 51. Prof Exp: Tester, Petrol Lab, Leonard Refining, Mich, 40-42; calculator, Sugar Lab, Hawaiian Sugar Planters Asn, 45-46; asst radiochem, Univ Wis, 46-52; chemist, 52-64, engr, Reactor Eng Div, 64-70, chemist, Radiol Sci Div, 70-73, CHEMIST, ENVIRON EFFECTS DIV, SAVANNAH RIVER LAB, E I DU PONT DE NEMOURS & CO, INC, 73- Res: Radiochemistry; radiological physics; trace element analysis in environmental samples; activation analysis; determination of radionuclides in environmental water and air samples. Mailing Add: Savannah River Lab Bldg 735-A E I du Pont de Nemours & Co Inc Aiken SC 29801

MILHAM, SAMUEL, JR, b Albany, NY, Mar 12, 32; m 56; c 3. HUMAN GENETICS, EPIDEMIOLOGY. Educ: Union Col, NY, BS, 54; Albany Med Col, MD, 58; Johns Hopkins Univ, MPH, 61. Prof Exp: Develop consult, NY State Dept Health, 62-67; assoc prof pub health, Univ Hawaii, 67-68; CHIEF CHRONIC DIS EPIDEMIOL, WASH STATE DEPT HEALTH, 68- Mem: Am Soc Human Genetics; Am Pub Health Asn. Res: Chronic disease epidemiology; population genetics; twin studies. Mailing Add: Wash State Dept of Health Olympia WA 98501

MILIAN, ALWIN S, JR, b Tampa, Fla, May 29, 32. FLUORINE CHEMISTRY. Educ: Mass Inst Technol, BS, 54; Univ Calif, PhD(chem), 58. Prof Exp: Org chemist, 58-67, SR RES CHEMIST, PLASTICS DEPT, DU PONT EXP STA, 67- Mem: Am Chem Soc. Res: Fluorocarbon chemistry; synthetic organic chemistry; analytical chemistry; industrial hygiene. Mailing Add: Plastics Dept Du Pont Exp Sta Wilmington DE 19898

MILIC-EMILI, JOSEPH, b Sesana, Yugoslavia, May 27, 31; m 57; c 4. PHYSIOLOGY. Educ: Univ Milan, MD, 55. Prof Exp: Asst prof physiol, Univ Milan, 55-58; asst prof, Univ Liege, 59-60; NIH res fel, Sch Pub Health, Harvard Univ, 60-63; from asst prof to assoc prof, 64-70, PROF PHYSIOL, McGILL UNIV, 70-, CHMN DEPT, 73- Concurrent Pos: Med Res Coun Can fel, McGill Univ, 63-; prof, Univ Clin, Royal Victoria Hosp, Montreal, 64- Mem: Am Physiol Soc; Can Physiol Soc; Can Soc Clin Invest; Can Thoracic Soc. Res: Physiology of respiration. Mailing Add: Dept of Physiol McGill Univ Montreal PQ Can

MILICI, ROBERT CALVIN, b New Haven, Conn, Aug 8, 31; m 58; c 2. GEOLOGY. Educ: Cornell Univ, AB, 54; Univ Tenn, MS, 55, PhD(geol), 60. Prof Exp: Instr geol, Univ Tenn, 55-58; geologist, Tenn Div Geol, 58-62; geologist, Va Div Mineral Resources, 62-63; GEOLOGIST, TENN DIV GEOL, 63- Concurrent Pos: US Geol Surv contract, 75-76. Mem: AAAS; Geol Soc Am; Am Asn Petrol Geol; Soc Econ Paleont & Mineral. Res: Geologic mapping; stratigraphy, structural geology and mineral resources studies in Tennessee; evaluation of oil and gas resources in southern Appalachians. Mailing Add: 8101 Normandy Dr Knoxville TN 37919

MILIONIS, JERRY PETER, b New York, NY, Mar 6, 26; m 48; c 3. ORGANIC CHEMISTRY. Educ: Brooklyn Col, BS, 47; Purdue Univ, PhD(chem), 51. Prof Exp: Asst, Purdue Univ, 47-49 & 50-51; res chemist, 51-54 & new prod develop dept, 54-57, group leader res, 57-63, dir org pigments res, 63-70, mgr agr res & develop, 70-74, SR RES CHEMIST AGR RES & DEVELOP, AM CYANAMID CO, 74- Mem: Am Chem Soc. Res: Sulfur and heterocyclic chemistry; polymer degradation; pigments. Mailing Add: 58 Marcy St Somerset NJ 08873

MILITZER, WALTER ERNEST, b Aug 20, 06; m 39; c 2. BIOCHEMISTRY. Educ: Univ Wis, BS, 33, PhD(biochem), 36. Prof Exp: From instr to prof chem, 36-52, dean arts & sci, 52-67, prof, 68-74; RETIRED. Mem: AAAS; Am Soc Biol Chem. Res: Enzymes: nature of bacterial thermophily with emphasis on enzymes and temperature adaptations. Mailing Add: 648 Hamilton Hall Dept of Chem Univ of Nebr Lincoln NE 68508

MILKEY, ROBERT WILLIAM, b Washington, DC, Jan 21, 44; m 65. ASTRONOMY. Educ: Amherst Col, BA, 65; Ind Univ, Bloomington, MA, 67, PhD(astrophys), 70. Prof Exp: Res assoc, Los Alamos Sci Lab, 70-71; asst astronr, 71-75, MGR COMPUT SERV, KITT PEAK NAT OBSERV, 75- Mem: Am Astron Soc; Royal Astron Soc. Res: Solar physics; structure of the solar chromosphere; hydromagnetics of the solar atmosphere; radiative transfer and spectral line formation. Mailing Add: Solar Div Kitt Peak Nat Observ PO Box 26732 Tucson AZ 85726

MILKMAN, JOSEPH, b Brooklyn, NY, Nov 25, 12; m 40; c 3. MATHEMATICS. Educ: Brooklyn Col, BS, 34, MS, 37; NY Univ, PhD(math), 51. Prof Exp: Eve instr, Brooklyn Col, 35-41; tutor physics, City Col New York, 42-44; res physicist, Remington Rand, 44-46; PROF MATH, US NAVAL ACAD, 46-; PROF MATH, McCOY COL, JOHNS HOPKINS UNIV, 51- Concurrent Pos: Instr, Polytech Inst Brooklyn, 44-46; mem res staff, Univ Md, 55 & 56; mem appl physics lab, Johns Hopkins Univ, 57. Mem: Am Math Soc; Math Asn Am. Res: Functional analysis; applied mathematics. Mailing Add: 110 Spa View Ave Annapolis MD 21402

MILKMAN, ROGER DAWSON, b New York, NY, Oct 15, 30; m 58; c 4. GENETICS. Educ: Harvard Univ, AB, 51, AM, 54, PhD(biol), 56. Prof Exp: Asst marine embryol, Marine Biol Lab, 54-55; Nat Sci Found res fel genetics, lab genetics & physiol, Nat Ctr Sci Res, France, 56-57; instr zool, Univ Mich, 57-59, asst prof, 59-60; assoc prof, Syracuse Univ, 60-67, prof, 67-68; PROF ZOOL, UNIV IOWA, 68- Concurrent Pos: Instr, Marine Biol Lab, 62-64, investr, 61, 65-72; USPHS res fel, Biol Labs, Harvard Univ, 66-67; assoc ed, Evolution, 74- Mem: Fel AAAS; Genetics Soc Am; Soc Study Evolution; Am Soc Zool; Soc Develop Biol. Res: Genetic structure of species electrophoretic analysis of populations; Botryllus; Drosophila; E coli; Mytilus; development; polygenes; temperature effects. Mailing Add: Dept of Zool Univ of Iowa Iowa City IA 52242

MILKOVICH, RALPH, b Clairton, Pa, Apr 7, 29; m 53; c 5. POLYMER CHEMISTRY. Educ: Duquesne Univ, BS, 51; State Univ NY, MS, 57; Akron Univ, PhD, 59. Prof Exp: Cadet chemist, Koppers Co, Inc, 51-55; res chemist, Shell Chem Co, Calif, 59-63; group leader plastics polymerization res, Gen Tire & Rubber Co, 63-66, sect head explor polymers, 66-69; asst dir org & polymer res, Moffett Tech Ctr, CPC Int, Inc, 69-74; MGR EXPLOR POLYMERS, ARCO POLYMERS, INC, 74- Concurrent Pos: Chmn, Gordon Res Conf Polymers, 63, discussion leader, 71; chmn meeting arrangements, Polymer Div, Joint Polymer-Rubber Div Meeting, 73. Honors & Awards: Orr Award, Soc Plastics Engrs, 58. Mem: Am Chem Soc. Res: Anionic mechanisms; block polymerization; chemically joined phase separated systems; polymer alloys. Mailing Add: 440 College Park Dr Monroeville PA 15146

MILKOWSKI, JOHN DAVID, b Baltimore, Md, Apr 1, 40; m 62; c 2. MEDICINAL CHEMISTRY. Educ: Loyola Col, Md, BS, 62; Univ Md, PhD(med chem), 66. Prof Exp: SR RES CHEMIST, MERCK & CO, INC, 66- Mem: Am Chem Soc. Res: Synthetic organic chemistry in the field of peptides and medicinal chemistry. Mailing Add: 16 Celler Rd Edison NJ 08817

MILKS, JOHN EDWARD, organic chemistry, see 12th edition

MILL, THEODORE, b Hamilton, Ont, Apr 17, 31; nat US; m 57; c 2. ORGANIC CHEMISTRY. Educ: Wayne State Univ, BS, 53; Univ Wash, PhD(chem), 57. Prof Exp: Res fel chem, Hickrill Res Found, NY, 56-57; res chemist, Org Chem Dept, E I du Pont de Nemours & Co, 57-60; SR ORGANIC CHEMIST, STANFORD RES INST, 60-, CHMN PHYS ORG CHEM DEPT, 64- Mem: Am Chem Soc; Sigma Xi. Res: Physical organic chemistry; photochemistry; oxidation and free radical chemistry; environmental chemistry. Mailing Add: Dept of Phys Org Chem Stanford Res Inst Menlo Park CA 94025

MILLAR, CHARLES HOWARD, b Waterloo, Que, Mar 4, 20; m 50; c 3. REACTOR PHYSICS. Educ: Bishop's Univ, Can, BSc, 40; Univ Western Ont, MA, 42; McGill Univ, PhD(physics), 47. Prof Exp: Jr res physicist, Nat Res Labs Can, 42-45; asst nuclear physics, McGill Univ, 46-47; asst res officer, Atomic Energy Proj, Nat Res Coun Can, 47-52; assoc res officer, 52-59, sr res officer, 59-63, head reactor physics I br, 63-68, asst dir in-chg div advan projs & reactor physics, 68-69, DIR ADVAN PROJS & REACTOR PHYSICS DIV, ATOMIC ENERGY CAN, LTD, 69- Concurrent Pos: Head res, NORA Proj, Norweg Atomic Energy Inst, 61-63. Mem: Can Asn Physicists; Am Nuclear Soc. Res: Research administration; nuclear engineering. Mailing Add: 1 Cartier Circle Deep River Ont Can

MILLAR, DAVID BOSIE-SEURS, III, b New York, NY, Sept 19, 31; m 54. PHYSICAL BIOCHEMISTRY. Educ: City Col New York, BS, 54; Duke Univ, PhD(biochem), 61. Prof Exp: Res assoc biochem, Med Ctr, Univ Ky, 59-60, res fel, 60-62; Nat Cancer Inst fel, 62-63; Nat Acad Sci-Nat Res Coun res assoc, 63-65; res biochemist, 65-70, CHIEF, LAB PHYS BIOCHEM, NAVAL MED RES INST, 70- Honors & Awards: Outstanding Young Scientist Award, Wash Acad Sci, 66. Mem: AAAS; Am Soc Biol Chem; fel NY Acad Sci. Res: Physical chemistry of transfer RNA and synthetic polynucleotides; structure-function relationships of lactate dehydrogenase isoenzymes; fluorescent conjugates of biopolymers; protein self association; structure-function relationship of acetylcholinesterase. Mailing Add: Naval Med Res Inst Nat Naval Med Ctr Bethesda MD 20014

MILLAR, HARVEY CLIFFORD, agricultural chemistry, organic chemistry, see 12th edition

MILLAR, JACK WILLIAM, b Ogden, Utah, July 11, 22; m 46; c 4. MEDICINE. Educ: Stanford Univ, AB, 45; George Washington Univ, MD, 47; Harvard Univ, MPH, 51, MS, 52; Am Bd Prev Med, dipl, 56. Prof Exp: Intern, Naval Hosp, Bethesda, Md, 47-48; med officer in chg, Tinian Leprosarium, Tinian Island, 48-50; med dir & epidemiologist, Am Leprosy Found, Far East, 50-53; instr epidemiol, Naval Med Sch, Md, 54-55, cmndg officer, Naval Med Res Unit 1, Berkeley, Calif, 55-60, dir prev med div, Bur Med & Surg, Navy Dept, 60-67; VIVIAN GILL PROF EPIDEMIOL & ENVIRON HEALTH, SCH MED & HEALTH SCI, GEORGE WASHINGTON UNIV, 67- Concurrent Pos: Pres, Gorgas Mem Inst Trop Med; consult epidemiol & trop med, Vet Admin Hosp, Wilmington, Del. Mem: Am Soc Trop Med & Hyg; Am Pub Health Asn; AMA; fel Am Col Prev Med. Res: Epidemiology; infectious diseases; leprosy; tropical diseases. Mailing Add: Dept of Allied Health George Washington Univ Washington DC 20037

MILLAR, JOHN DAVID, b Dallas Co, Tex, May 24, 21; m 47; c 1. ANALYTICAL CHEMISTRY. Educ: Trinity Univ, BA, 47. Prof Exp: Chemist, Found Appl Res, 47-49; chemist, Southwest Res Inst, 49-52; org chemist, Celanese Corp Am, 52-53; assoc chemist, 53-62, SR RES CHEMIST, SOUTHWEST RES INST, 62- Concurrent Pos: Lab instr, Evening Div, San Antonio Col, 57-59. Mem: Am Chem Soc; Sigma Xi. Res: Process development; gas chromatography and trace analysis; technical literature and reports. Mailing Add: Southwest Res Inst 8500 Culebra Rd San Antonio TX 78284

MILLAR, JOHN DONALD, b Newport News, Va, Feb 27, 34; m 57; c 3. PREVENTIVE MEDICINE, EPIDEMIOLOGY. Educ: Univ Richmond, BS, 56; Med Col Va, MD, 59; London Sch Hyg & Trop Med, dipl trop pub health, 66. Prof Exp: Intern, Univ Utah Hosps, 59-60, asst resident med, 60-61; asst chief EIS, Epidemiol Br, 61-62, chief, 62-63, dep chief surveillance sect, 63-64, chief smallpox unit, 63-65, chief invests unit, Smallpox Eradication Prog, 66, chief prog, 66-70, DIR BUR STATE SERV, CTR DIS CONTROL, USPHS, 70- Concurrent Pos: Mem sci group smallpox eradication, WHO, 67-, mem comn eval smallpox eradication in SAm, 73, consult expanded immunization prog, 74; assoc mem comn immunization, Armed forces Epidemiol Bd, 68-71; mem rural health coord comt, USPHS, 75-, mem comt maternal & child health, 75- Honors & Awards: Surgeon General's Commendation Medal, USPHS, 65. Mem: AMA; Am Pub Health Asn; Royal Soc Trop Med & Hyg. Res: Epidemiology of acute viral illness; mass immunization against infectious disease; infectious disease eradication and control techniques. Mailing Add: Bur of State Serv Ctr for Dis Control Atlanta GA 30333

MILLAR, JOHN ROBERT, b Edinburgh, Scotland, June 3, 27; m 55; c 3. ORGANIC POLYMER CHEMISTRY. Educ: Univ Cambridge, BA, 48, ARIC, 49, MA, 52; FRIC, 59. Prof Exp: Res chemist, Howards of Ilford Ltd, 48-50; sr res chemist, Permutit Co Ltd, UK, 50-73; GROUP LEADER RES & DEVELOP, DIAMOND SHAMROCK CORP, 73- Concurrent Pos: Vchmn, Gordon Res Conf Ion Exchange, 75-77. Mem: Soc Chem Indust; The Chem Soc; Royal Inst Chem; Am Chem Soc. Res: Preparation, characterization and properties of functional polymers as a function of their chemical and physical structures, including the effect of macroporosity. Mailing Add: 1743 Edgewood Rd Redwood City CA 94062

MILLAR, KAY, b Syracuse, NY, Oct 28, 34. INTERNAL MEDICINE, CARDIOLOGY. Educ: Syracuse Univ, AB, 56; State Univ NY Upstate Med Ctr, MD, 60. Prof Exp: Resident internal med, State Univ NY Upstate Med Ctr, 61-64; NIH trainee cardiol, 64-67; asst prof internal med, State Univ NY Upstate Med Ctr, 67-68; asst prof, 68-73, ASSOC PROF INTERNAL MED, MED CTR, UNIV UTAH, 73- Mem: Am Heart Asn. Res: Electrocardiography; relation of electrical to mechanical activity in the intact heart; body surface potentials of cardiac origin. Mailing Add: Cardiol Div Bldg 100 Univ of Utah Med Ctr Salt Lake City UT 84112

MILLAR, ROBERT FYFE, b Guelph, Ont, Jan 23, 28. APPLIED MATHEMATICS. Educ: Univ Toronto, BA, 51, MA, 52; Cambridge Univ, PhD, 57. Prof Exp: Jr res officer, Microwave Sect, Radio & Elec Div, Nat Res Coun Can, 52-53, asst res officer, 57-60; visitor, Courant Inst Math Sci, NY Univ, 60-61; assoc prof math, Royal Mil Col, Ont, 61-63; sci asst electromagnetic theory lab, Tech Univ Denmark, 63-66; sr res officer, Antenna Eng Sect, Radio & Elec Eng Div, Nat Res Coun Can, 66-74; PROF MATH, UNIV ALTA, 74- Mem: Am Math Soc; Soc Indust & Appl Math; Can Math Cong. Res: Diffraction and scattering of waves; scattering by periodic structures. Mailing Add: Dept of Math Univ of Alta Edmonton AB Can

MILLAR, WAYNE NORVAL, b Beverly, Mass, Oct 10, 42; m 65; c 2. MICROBIOLOGY. Educ: Bucknell Univ, BS, 64; Pa State Univ, MS, 66, PhD(microbiol), 69. Prof Exp: From asst prof to assoc prof bact, WVa Univ, 69-73; SR SCIENTIST, ELI LILLY & CO, 73- Concurrent Pos: NSF grant, 71-73. Mem: Am Soc Microbiol; Sigma Xi. Res: Soil and water microbiology; isolation of antibiotic producing microorganisms; microbial ecology. Mailing Add: Microbiol & Fermentation Prod Res Eli Lilly & Co Indianapolis IN 46206

MILLARD, BEN, b Painswick, Eng, Dec 5, 20; m 43; c 6. PHYSICAL CHEMISTRY. Educ: Bristol Univ, BSc, 42, PhD(chem), 47. Prof Exp: Res assoc surface chem, Amherst Col, 49-50; asst prof chem, Univ NH, 50-55, assoc prof, 55-57; asst dir res, 57-67, TECH DIR RES, RES DEPT, S D WARREN CO DIV, SCOTT PAPER CO, 67- Concurrent Pos: Exp officer, Brit Ministry of Supply, 42-45. Res: Mass spectrometry; physical adsorption; surface chemistry; photochemistry. Mailing Add: Res Dept S D Warren Co Westbrook ME 04092

MILLARD, FREDERICK WILLIAM, b Johnson City, NY, Feb 10, 31; m 53; c 2. PHOTOGRAPHIC CHEMISTRY. Educ: Pa State Univ, BS, 53; Mich State Univ, PhD(org chem), 58. Prof Exp: Sr chemist, Tex US Chem Co, 57-60; res specialist, Gen Aniline & Film Corp, 60-63, tech assoc silver halide photochemistry, 63-74, MGR GRAPHIC & BLACK & WHITE FILMS & PAPERS RES & DEVELOPMENT, GAF CORP, 74- Mem: Am Chem Soc; Soc Photog Sci & Eng; Tech Asn Graphic Arts. Res: Photopolymerization theory and adaptation to a photographic system; elucidation of free radical reactions in solution; silver halide technology, particularly in graphic arts; non silver imaging systems. Mailing Add: Hinds St Box 151 Montrose PA 18801

MILLARD, GEORGE BUENTE, b Kansas City, Kans, Feb 13, 17; m 43; c 4. INORGANIC CHEMISTRY. Educ: Wash State Univ, BS, 42, MS, 55. Prof Exp: Tech sales agr chem, Sherwin-Williams Co, 46-47, asst to vpres & gen mgr, Calif, 47-49; vpres, Mid-State Chem Co, 49-54; PROF CHEM, YAKIMA VALLEY COL, 51- Concurrent Pos: Secy-treas & mem bd, Northwest Col & Univ Asn Sci, 71- Mem: Sigma Xi. Res: Investigation of some complex ions of zinc by polarographic methods. Mailing Add: 201 N 27th Ave Yakima WA 98902

MILLARD, HERBERT DEAN, b Grayling, Mich, May 22, 24; m 48; c 4. DENTISTRY. Educ: Univ Mich, DDS, 52, MS, 56. Prof Exp: From instr to assoc prof, 52-64, PROF ORAL DIAG, SCH DENT, UNIV MICH, ANN ARBOR, 64-, CHMN DEPT, 56- Mem: Am Col Dent; Am Dent Asn. Res: Relationships of oral disease to systemic disease; procedures in oral diagnosis. Mailing Add: Dept of Oral Diag & Radiol Univ of Mich Sch of Dent Ann Arbor MI 48109

MILLARD, HUGH THOMPSON, JR, b Cleveland, Ohio, Dec 23, 35; m 57; c 4. RADIOCHEMISTRY, GEOCHEMISTRY. Educ: Coe Col, BA, 57; Calif Inst Technol, PhD(chem), 62. Prof Exp: Res chemist geochem, Calif Inst Technol, 62; res chemist, Univ Calif, San Diego, 63-65; RADIOCHEMIST, US GEOL SURV, 65- Mem: AAAS; Geochem Soc. Res: Determination of chemical composition of geologic samples by radiochemical and activation analysis methods. Mailing Add: US Geol Surv Bldg 15 Denver Fed Ctr Denver CO 80225

MILLARD, KENNETH YOUNG, b Takoma Park, Md, July 10, 41; m 61; c 2. STATISTICAL MECHANICS. Educ: Case Inst Technol, BS, 63; Case Western Reserve Univ, PhD(physics), 71. Prof Exp: Res assoc physics, Case Western Reserve Univ, 71; fel, Univ Mo-Rolla, 71-72; FEL PHYSICS, SIMON FRASER UNIV, 72- Res: Statistical mechanical treatment of thermodynamic metastable states; anisotropic Heisenberg model; classical fluids in an external field; distribution functions for classical fluids; zeros of the partition function. Mailing Add: Dept of Physics Simon Fraser Univ Burnaby BC Can

MILLARD, RICHARD JAMES, b Peabody, Mass, Oct 3, 18; m 49; c 1. ELECTROCHEMISTRY. Educ: Boston Col, BS, 49, MS, 50. Prof Exp: Res engr, 50-65, MGR ENG DEPT, SPRAGUE ELEC CO, 65- Mem: Am Chem Soc; Electrochem Soc; Inst Elec & Electronic Engrs. Res: Thin films on metals; dielectric breakdown; electrolytic oxidation; identification of phenols; electrolytic capacitor development and engineering. Mailing Add: Sprague Elec Co North Adams MA 01247

MILLBURN, GEORGE P, b Cleveland, Ohio, Aug 10, 25; m 46; c 2. NUCLEAR PHYSICS. Educ: Case Western Reserve Univ, BS, 50; Univ Calif, Berkeley,

PhD(physics), 56. Prof Exp: Res physicist, Lawrence Radiation Lab, Univ Calif, 50-56, nuclear weapons, 56-59; dept mgr re-entry systs, Aeronutronic Div, Ford Motor Co, 59-62; assoc gen mgr & gen mgr reentry systs div, 62-68, corp dir technol, 68, gen mgr off develop planning, 68-71, GEN MGR TECHNOL DIV, AEROSPACE CORP, 71- Res: Reentry physics; weapon system analysis; nuclear weapon design; high energy particle physics; neutron production; military space technology. Mailing Add: 1 Sunnyfield Dr Rolling Hills Estates CA 90274

MILLEMANN, RAYMOND EAGAN, b New York, NY, Jan 18, 28; m 55; c 3. ZOOLOGY. Educ: Dartmouth Col, AB, 48; Univ Calif, Los Angeles, MA, 51, PhD(zool), 54. Prof Exp: Teaching asst zool, Univ Calif, Los Angeles, 49-53; from instr to asst prof bact & parasitol, Sch Med, Univ Rochester, 55-63; assoc prof, 63-69, PROF FISHERIES, ORE STATE UNIV, 69- Mem: Am Soc Parasitol; Soc Protozool; Am Soc Trop Med & Hyg; Am Micros Soc; Wildlife Dis Asn. Res: Taxonomy and life cycles of helminths; fish diseases and parasites; marine biology. Mailing Add: Dept of Fisheries & Wildlife Ore State Univ Corvallis OR 97331

MILLENER, DAVID JOHN, b Auckland, NZ, May 2, 44. THEORETICAL NUCLEAR PHYSICS. Educ: Univ Auckland, NZ, BSc, 66, MSc, 68; Oxford Univ, Eng, DPhil(nuclear physics), 72. Prof Exp: Int Bus Mach res fel nuclear physics, Oxford Univ, Eng, 72-74, res fel, 74-75; ASST PHYSICIST NUCLEAR PHYSICS, BROOKHAVEN NAT LAB, 76- Res: Calculations of the structure and properties of light nuclei. Mailing Add: Physics Dept Brookhaven Nat Lab Upton NY 11973

MILLER, A EUGENE, b Philadelphia, Pa, Apr 27, 29; m 57; c 2. MATHEMATICS, RESEARCH ADMINISTRATION. Educ: Univ Pa, BA, 51, MA, 53. Prof Exp: Asst res engr, Burroughs Corp, 51-55, develop engr, 55-58; mem tech staff, Auerbach Corp, Va, 58-62, prog mgr info systs eng, 62-70, dir prog develop, Auerbach Assocs, 70-72; VPRES OPERS, INS INST HWY SAFETY, 72- Mem: AAAS; Inst Elec & Electronics Engrs; Soc Indust & Appl Math; Sigma Xi; Asn Comput Mach. Res: Reducing losses, human and economic, resulting from or associated with the highway transportation system. Mailing Add: Ins Inst of Hwy Safety 600 New Hampshire Ave Washington DC 20037

MILLER, AARON, b Brooklyn, NY, Jan 17, 22; m 49. MEDICINE. Educ: City Col New York, BS, 43; Univ Rochester, MD, 50. Prof Exp: Intern & asst resident med, Strong Mem Hosp, Rochester, NY, 50-52; asst resident, Vet Admin Hosp, Boston, 52-53; ASST DIR RADIOISOTOPE UNIT, VET ADMIN HOSP, BOSTON, 55-; ASSOC RES PROF MED, SCH MED, BOSTON UNIV, 70- Concurrent Pos: USPHS res fel hemat, Mass Mem Hosp, 53-55; asst prof, Boston Univ, 59-70. Mem: Am Fedn Clin Res; Res: Absorption transport and function of vitamin B-12; synthesis and turnover of erythrocyte gluthatione. Mailing Add: Dept of Med Boston Univ Sch of Med Boston MA 02118

MILLER, AGNES CHAMBLESS, b Ruston, La, Apr 8, 18; m 50 & 65; c 2. NUTRITION, FOODS. Educ: La Tech Univ, BS, 38; La State Univ, Baton Rouge, MS, 44; Fla State Univ, PhD(food, nutrit), 64. Prof Exp: Teacher high sch, La, 38-40; dir food serv, Standard Oil Restaurant Co, 41-42; teacher high sch, La, 42-44; instr & supvr nutrit educ, La Tech Univ, 44-46, asst prof home econ, 44-49; supvr sch lunch, Natchitoches Parish Sch Bd, 50; asst prof home econ, La Tech Univ, 55-59; supvr sch lunch, Calcasieu Parish Sch Bd, 61-62; assoc prof, 64-65, PROF FOOD NUTRIT, LA TECH UNIV, 65-, DEAN COL HOME ECON, 70- Mem: Am Home Econ Asn; Inst Food Technol. Res: Antioxidant activity of aqueous vegetable extracts. Mailing Add: Col of Home Econ La Tech Univ Ruston LA 71270

MILLER, AKELEY, b Phoenix, Ariz, Mar 12, 26; m 49; c 2. PARTICLE PHYSICS. Educ: Univ SDak, BA, 50, MA, 52; Univ Mo, Columbia, PhD, 60. Prof Exp: Instr physics, Univ SDak, 52-55; asst prof, 60-65, ASSOC PROF PHYSICS, UTAH STATE UNIV, 65- Mem: AAAS; Am Phys Soc; Am Asn Physics Teachers; Sigma Xi. Res: Small angle x-ray scattering; mathematical and numerical analysis of x-ray scattering from small particles, 10 to 10,000 angstrom representative diameter, to determine particle size and shape. Mailing Add: Dept of Physics Utah State Univ Logan UT 84322

MILLER, ALBERT, b Brooklyn, NY, Mar 18, 11; m 50; c 1. MEDICAL ENTOMOLOGY. Educ: Cornell Univ, BS, 33, MS, 34, PhD(insect embryol), 38. Prof Exp: Asst entom & parasitol, Cornell Univ, 34-35, instr, 35-38; instr entom & plant path, Univ Ark, 38-40; guest genetics, Carnegie Inst Technol, 40-41; from instr to asst prof, 41-48, ASSOC PROF MED ENTOM, SCH MED, TULANE UNIV, 48- Mem: AAAS; Entom Soc Am; Am Soc Trop Med & Hyg; Soc Syst Zool; Am Mosquito Control Asn. Res: Insect morphology, embryology and taxonomy; biology of medically important arthropods; coprophilic fauna. Mailing Add: Dept of Parasitol Tulane Univ Sch of Med New Orleans LA 70112

MILLER, ALBERT, b Philadelphia, Pa, June 4, 23; m 45; c 2. METEOROLOGY. Educ: Pa State Univ, BS, 43, MS, 53, PhD(meteorol), 56. Prof Exp: Meteorologist, Pan Am Grace Airways, 43-48; analyst, US Weather Bur, 49-50, res meteorologist, 50-51; instr meteorol, Pa State Univ, 51-56; consult, US Opers Mission, 56-59; proj dir, Res Off, US Weather Bur, 59-61; PROF METEOROL, SAN JOSE STATE UNIV, 61-, CHMN DEPT, 66- Concurrent Pos: NSF res grant proj dir, 63-67, 71-; consult, AEC, 61-67; consult, US Army, 61-64; consult meteorologist, 61- Mem: AAAS; Am Meteorol Soc; Am Geophys Union. Res: Small-scale circulations; atmospheric diffusion, weather forecasting; applied meteorology; internal gravity waves. Mailing Add: Dept of Meteorol San Jose State Univ San Jose CA 95192

MILLER, ALEXANDER, immunology, immunochemistry, see 12th edition

MILLER, ALEXANDER ANDREW, b South Windsor, Conn, Dec 17, 18; m 44; c 2. PHYSICAL CHEMISTRY. Educ: Mass State Col, BS, 39; Trinity Col, Conn, MS, 41; Univ Wis, PhD, 49. Prof Exp: Asst chem, Trinity Col, Conn, 39-41; res chemist, Am Cyanamid Co, Conn, 41-46; asst, Alumni Res Found, Univ Wis, 46-49; res assoc, Knolls Res Lab, 49-67, PHYS CHEMIST, RES & DEVELOP CTR, GEN ELEC CO, 67- Mem: AAAS; Am Chem Soc; fel Am Inst Chemists. Res: Kinetics of halogen exchange reactions; autoxidation; radiation and polymer chemistry. Mailing Add: Gen Elec Res & Dev Ctr PO Box 8 Bldg K-1 Schenectady NY 12301

MILLER, ALFRED DAVID, medicinal chemistry, organic chemistry, see 12th edition

MILLER, ALLAN STEPHEN, b Arlington Heights, Ill, Feb 21, 28; m 50; c 3. SOLID STATE PHYSICS. Educ: Univ Notre Dame, BS, 49, MS, 50; Univ Ill, PhD(physics), 57. Prof Exp: Assoc physicist, Res Lab, Int Bus Mach Corp, 57-59, staff physicist, 59-61, develop physicist, Prod Develop Lab, 61-64; res assoc, Nat Res Corp, 64-66; asst dir res, Norton Res Corp, 66-70; consult solid state physics, Light Emitting Diodes, 71-72; MGR PROCESS DEVELOP, MULTIGRAPHICS DEVELOP CTR, ADDRESSOGRAPH MULTIGRAPH CORP, 72- Mem: Am Phys Soc. Res: Photoconductivity in semiconductors and insulators; light emitting diodes; transistors. Mailing Add: 86 S Hayden Pkwy Hudson OH 44236

MILLER, ALLEN H, b Brooklyn, NY, June 23, 32. PHYSICS. Educ: Brooklyn Col, AB, 53; Rutgers Univ, MS, 55, PhD(physics), 60. Prof Exp: Physicist, Electronics Corp Am, 55-56; res assoc physics, Univ Ill, 60-62; asst prof, 62-65, ASSOC PROF PHYSICS, SYRACUSE UNIV, 65- Mem: Am Phys Soc. Res: Solid state theory; many-particle problem. Mailing Add: Dept of Physics Syracuse Univ Syracuse NY 13210

MILLER, ALVIN LEON, b Bowling Green, Ky, Mar 8, 38. OCCUPATIONAL HEALTH. Educ: Western Ky Univ, BS, 60; Univ Cincinnati, MS, 62; Northwestern Univ, PhD(environ health eng), 72. Prof Exp: Res asst cancer res, Kettering Lab, Univ Cincinnati, 61-62; health serv officer, R A Taft Sanit Eng Ctr, USPHS, 62-65; health physicist, Sargent & Lundy Engrs, 72; asst prof indust hyg & occup health, 72-76, ASSOC PROF INDUST HYG & OCCUP HEALTH, SCH PUB HEALTH, UNIV ILL MED CTR, 76- Concurrent Pos: Consult, Occup Safety & Health Admin, 74-, Environ Health Resources Ctr, Ill Inst Environ Qual, 74 & State of Ill, 74-75. Mem: Am Indust Hyg Asn; Health Physics Soc; Soc Occup & Environ Health; Am Pub Health Asn. Res: Occupational and environmental health, with particular emphasis on industrial hygiene and health physics; voluntary accreditation of occupational health programs in industry; bio-effects of physical and chemical agents; radiation dosimetry. Mailing Add: Univ of Ill Med Ctr Sch Pub Health PO Box 6998 Chicago IL 60680

MILLER, ANNA KATHRINE, b East Orange, NJ, June 8, 13. BACTERIOLOGY. Educ: Moravian Col Women, BA, 34; Columbia Univ, MS, 36; Cornell Univ, PhD(bact), 42; Am Bd Med Microbiol, dipl. Prof Exp: Asst dir phys educ, Moravian Col Women, 34-35, asst prof biol, 36-41; substitute, Wells Col, 42-43; res fel, 43-74, SR RES FEL, MERCK INST, MERCK & CO, INC, 74- Mem: Fel AAAS; Am Soc Microbiol; fel Am Acad Microbiol; NY Acad Sci; Brit Soc Gen Microbiol. Res: Experimental chemotherapy. Mailing Add: Merck Inst Merck & Co Inc Rahway NJ 07065

MILLER, ANTHONY BERNARD, b Woodford, Eng, Apr 17, 31; m 52; c 5. EPIDEMIOLOGY. Educ: Univ Cambridge, BA, 52, MB & BChir, 55; FRCP(C), 72. Prof Exp: House officer, Oldchurch Hosp, Romford, Eng, 55-57; med officer, Royal Air Force, Netheravon, Eng, 57-59; med registr, Luton & Dunstable Hosp, Eng, 59-61; mem sci staff, Med Res Coun Tuberc & Chest Dis Unit, London, 61-71; DIR EPIDEMIOL UNIT, NAT CANCER INST CAN, 71-; ASSOC PROF PREV MED & STATIST, UNIV TORONTO, 72- Concurrent Pos: Mem working cadre, Bladder Cancer Proj, US, 73-75; mem epidemiol comt, Breast Cancer Task Force, US, 73-, chmn, 75-; mem, Fed Task Force Cervical Cytol Screening, Can, 74-76. Mem: Can Oncol Soc (secy-treas, 75-); Soc Epidemiol Res; Int Study Group Detection & Prev Cancer. Res: Epidemiology of breast, bladder and colo-rectal cancer; radiation, occupation and cancer; monitoring for environmental carcinogenesis; evaluation of screening for cervix and breast cancer; controlled clinical trials in cancer. Mailing Add: Nat Cancer Inst of Can 25 Adelaide St E Toronto ON Can

MILLER, ARCHIE PAUL, b Lampman, Sask, Nov 13, 40; m 63; c 2. SOLID STATE PHYSICS. Educ: Univ Sask, BSc, 62, MSc, 64; McMaster Univ, PhD(solid state physics), 69. Prof Exp: Lectr, 68-69, asst prof, 69-74, ASSOC PROF PHYSICS, BRANDON UNIV, 74- Concurrent Pos: Nat Res Coun operating grant, Brandon Univ, 70- Mem: Can Asn Physicists. Res: Lattice dynamics of metals; electronic specific heats of palladium and copper; Kohn effect in palladium and platinum; bulk specific heats of metals at high temperatures. Mailing Add: Dept of Physics Brandon Univ Brandon MB Can

MILLER, ARILD JUSTESEN, b Pine City, Minn, May 16, 18; m 43; c 3. PHYSCIAL CHEMISTRY. Educ: Carleton Col, BA, 39; Purdue Univ, PhD(phys chem), 43. Prof Exp: Asst, Carleton Col, 39-41, prof chem & chmn dept, 49-60; asst, Purdue Univ, 41-42; res chemist, Metall Lab, Univ Chicago, 43-45; chemist, Clinton Labs, Tenn, 45-46; asst prof chem, Antioch Col, 46-49; DIR ADMISSIONS & ASSOC DEAN, INST PAPER CHEM, LAWRENCE UNIV, 60- Concurrent Pos: Res chemist, Kettering Found, 46-49. Mem: Am Chem Soc. Res: Photosynthesis; photochemistry; radiation chemistry; radiochemistry; heats of combustion; solubilities of organic compounds; heats of combustion of some polynitroparaffins. Mailing Add: Inst of Paper Chem PO Box 1039 Appleton WI 54911

MILLER, ARNOLD, b New York, NY, May 8, 28; m 50; c 3. PHYSICAL CHEMISTRY. Educ: Univ Calif, Los Angeles, BS, 48, PhD(chem), 51. Prof Exp: Asst, Univ Calif, Los Angeles, 48-49; res phys chemist, William Wrigley Res Lab, 51; res phys chemist, Armour Res Found, Ill Inst Technol, 52-54, supvr phys chem, 55-56; mgr chem, Borg Warner Res Ctr, 56-59; chief mat res, Autonetics Div, NAm Aviation, Inc, 59-62, dir phys res, 62-66, dir cent microelectronics, NAm Rockwell Corp, 66-68; gen mgr res & develop div, Whittaker Corp, 68-69, gen mgr, 69-71; pres, Theta Sensors, Inc, 71-73; DIR ENG SCI, XEROX CORP, 73- Concurrent Pos: Pres, Space Sci, Inc, 68-71. Honors & Awards: Armour Res Found Award, 53; Award, Bur Ord, US Navy, 53; Indust Res 100 Awards, 64, 69. Mem: Am Chem Soc; Am Phys Soc; Am Vacuum Soc. Res: Surface physics and chemistry; photochemistry; epitaxial growth; chemical vapor deposition. Mailing Add: 92 Rolling Ridge Rd Stamford CT 06903

MILLER, ARTHUR, b New York, NY, Apr 3, 30; m 61; c 3. PHYSICAL CHEMISTRY, INORGANIC CHEMISTRY. Educ: Polytech Inst Brooklyn, BS, 51; Calif Inst Technol, PhD(chem), 57. Prof Exp: Asst chem, Brookhaven Nat Lab, 50-51; asst, Los Alamos Sci Lab, 52; MEM TECH STAFF, RCA LABS, 56- Mem: Am Crystallog Asn; Am Phys Soc. Res: Magnetic materials; crystal chemistry of spinels; x-ray and neutron diffraction crystallography; crystal optics; radiochemistry; ferroelectrics; nonlinear optics; electro-optic materials; integrated optics. Mailing Add: David Sarnoff Res Ctr RCA Labs Princeton NJ 08540

MILLER, ARTHUR FRANCIS, organic chemistry, see 12th edition

MILLER, ARTHUR I, b New York, NY, Feb 6, 40; m 62; c 2. HISTORY OF SCIENCE, THEORETICAL PHYSICS. Educ: City Col New York, BS, 61; Mass Inst Technol, PhD(physics), 65. Prof Exp: Asst prof, 65-70, ASSOC PROF PHYSICS, UNIV LOWELL, 70- Concurrent Pos: Nat Endowment for Humanities fel, Harvard Univ, 72-73. Mem: Am Phys Soc; Hist Sci Soc Am; Am Asn Physics Teachers; AAAS. Res: Interdisciplinary research in the history of 19th and 20th century science. Mailing Add: Dept of Physics & Appl Physics Univ of Lowell North Campus Lowell MA 01854

MILLER, ARTHUR JOEL, b Malden, Mass, Dec 24, 12; m 45; c 1. CHEMISTRY. Educ: Mass Inst Technol, BS, 34, MS, 35, PhD(phys chem), 39. Prof Exp: Asst, Harvard Univ, 37-40; chemist, Portland Cement Asn, Nat Bur Stand, DC, 40; chemist, Div Indust Coop, Mass Inst Technol, 40-41; chemist, Fed Tel & Radio Corp, NJ, 41-43; sr chemist, Clinton Eng Works, Tenn Eastman Corp, 43-47; actg head dept, Nuclear Energy Propulsion Aircraft Div, Fairchild Engine & Airplane Corp, 47-51; proj chemist, Patchen & Zimmerman Engrs, 51; asst dir, Aircraft Nuclear Propulsion Proj, 51-58, aircraft nuclear propulsion coordr & asst dir, Reactor Projs Div, 58-62, space progs coordr, 62-67, HOUSING & URBAN DEVELOP PROG

COORDR, REACTOR DIV, OAK RIDGE NAT LAB, 67- Mem: AAAS; Am Chem Soc; Sigma Xi; Am Nuclear Soc. Res: Atomic weights; selenium rectifiers; uranium chemistry; metal corrosion; refractory materials; nuclear power systems; limiting density of carbon dioxide and the atomic weight of carbon; space power supplies; urban area utilities. Mailing Add: Oak Ridge Nat Lab PO Box Y Bldg 9102 Oak Ridge TN 37830

MILLER, ARTHUR JOSEPH, b San Francisco, Calif, Jan 18, 43; m 65; c 2. NEUROPHYSIOLOGY, PHYSIOLOGY. Educ: Univ Calif, Los Angeles, PhD(physiol), 70. Prof Exp: Trainee, Brain Res Inst, Univ Calif, Los Angeles, 70; asst prof physiol, Univ Ill Med Ctr, 70-75, adj asst prof otolaryngol, 74-75; ASST PROF OROFACIAL ANOMALIES & DENT & LECTR PHYSIOL, UNIV CALIF, SAN FRANCISCO, 75- Concurrent Pos: NIH grants, Univ Ill Med Ctr, 71-75. Mem: Neurosci Soc; Am Physiol Soc; Fedn Am Scientists. Res: Cranial reflexes and respiration in developing and adult animals; sudden infant death syndrome; motor control of cranial musculature; swallowing. Mailing Add: Sect Orofacial Anomalies Univ Calif San Francisco CA 94143

MILLER, ARTHUR R, b Boston, Mass, Aug 6, 15; m 41; c 2. PHYSICAL OCEANOGRAPHY. Prof Exp: ASSOC SCIENTIST, WOODS HOLE OCEANOG INST, 46- Concurrent Pos: Grant from Woods Hole Oceanog Inst, Scripps Inst, Univ Calif, 50; consult, US Weather Bur, 55; mem, Int Comn Bibliog Phys Oceanog, 60-; mem working panel, Int Indian Ocean Exped, 61-63, lectr, US Biol Prog, Bermuda Biol Sta, 62. Mem: AAAS; Am Geophys Union; Am Soc Limnol & Oceanog. Res: Tide and storm surge research; cooperative investigations of the Mediterranean. Mailing Add: Woods Hole Oceanog Inst Woods Hole MA 02543

MILLER, ARTHUR SIMARD, b Sidney, Mont, Mar 4, 35; m 66. ORAL PATHOLOGY. Educ: Mont State Univ, BS, 57; Wash Univ, DDS, 59; Ind Univ, MSD, 63. Prof Exp: Instr oral path, Sch Dent, Ind Univ, 63-66; from asst prof to assoc prof, 66-72, PROF PATH, SCH DENT, TEMPLE UNIV, 72-, CHMN DEPT, 68- Concurrent Pos: Consult, Cent Am Registry Oral Path, 66-; consult dent aptitude testing, Am Dent Asn, 71- Mem: Am Dent Asn; Am Acad Oral Path; Int Asn Dent Res. Res: Oral diseases and neoplasms; use of the computer in oral pathology; teaching improvements and innovations. Mailing Add: Dept of Path Temple Univ Sch of Dent Philadelphia PA 19140

MILLER, AUGUST, b Isola, Miss, Nov 28, 33; m 63. ATMOSPHERIC PHYSICS. Educ: NMex State Univ, BS, 55, PhD(physics), 61; Univ Md, MS, 58. Prof Exp: Eng specialist, Ariz Div, Goodyear Aircraft Corp, 61-62; mem res staff, Northrop Space Labs, 62-64; PROF PHYSICS, N MEX STATE UNIV, 64- Concurrent Pos: Consult, Battelle Mem Inst, Columbus Labs, 75-76. Mem: Am Phys Soc; Am Asn Physics Teachers. Res: Scanning electro-optical image sensors; very low pressure radio frequency gaseous discharges; radio frequency plasmoids; atmospheric optics; satellite meteorology. Mailing Add: Dept of Physics NMex State Univ Las Cruces NM 88001

MILLER, AUGUSTUS TAYLOR, JR, b Arlington, Tex, Apr 14, 10; m 38; c 1. PHYSIOLOGY. Educ: Emory Univ, BS, 31, MS, 33; Univ Mich, PhD(physiol), 39; Duke Univ, MD, 53. Prof Exp: Res assoc, W H Maybury Sanatorium, Mich, 36-39; from instr to assoc prof, 39-50, PROF PHYSIOL, SCH MED, UNIV NC, CHAPEL HILL, 50- Mem: Am Physiol Soc. Res: Cell physiology; respiration. Mailing Add: Dept of Physiol Univ of NC Sch of Med Chapel Hill NC 27514

MILLER, BANNER ISOM, meteorology, see 12th edition

MILLER, BARRY, b Passaic, NJ, Jan 22, 33; m 65; c 2. ELECTROCHEMISTRY. Educ: Princeton Univ, AB, 55; Mass Inst Technol, PhD(chem), 59. Prof Exp: Instr chem, Harvard Univ, 59-62; MEM TECH STAFF, CHEM RES DEPT, BELL TEL LABS, 62- Mem: Am Chem Soc; Electrochem Soc. Res: Electrochemical kinetics; electrodeposition; corrosion. Mailing Add: Bell Labs Murray Hill NJ 07974

MILLER, BEATRICE DIAMOND, b New York, NY, May 29, 19; m 43; c 3. ANTHROPOLOGY. Educ: Univ Mich, AB, 48; Univ Wash, PhD(anthrop), 58. Prof Exp: Ed & res anthropologist, Northwest China Handbook Proj, Human Rels Area Files, Inc, 55-56; actg asst prof anthrop, Wash Univ, 57-58, eve col, 57-59; vis lectr, Beloit Col, 61-62; lectr anthrop, Univ Wis, Madison, 67-68; vis lectr, Beloit Col, 69-70. Concurrent Pos: Vis lectr, Am Anthrop Asn, 65-66, 69-70; co-prin investr, NSF grant, Univ Wis Proj; lectr anthrop, Univ Wis, 73-74. Mem: Fel Am Anthrop Asn; Asn Asian Studies; fel Soc Appl Anthrop; Tibet Soc; India Anthrop Asn. Res: Factors of cultural resistance and stability in contact situations; role of Buddhist social institutions in Inner Asia and the Himalayan borderlands; biocultural factors in high altitude populations; culture and behavior; a general systems approach to anthropological problems. Mailing Add: 1227 Sweetbriar Rd Madison WI 53705

MILLER, BENNETT, b New York, NY, Jan 18, 38; m 61; c 2. PLASMA PHYSICS, SCIENCE ADMINISTRATION. Educ: Columbia Univ, AB, 59, MA, 61, PhD(physics), 65. Prof Exp: Fel high temperature plasma physics, Columbia Univ, 65-66, res assoc plasma physics, 66-69; asst prof nuclear eng, Ohio State Univ, 69-70; physicist, Atomic Energy Comn, 70-74, dep asst dir res, 74-75, CHIEF EXP PLASMA RES BR, DIV CONTROLLED THERMONUCLEAR RES, US ENERGY RES & DEVELOP ADMIN, 74-, ASST DIR RES, 75- Concurrent Pos: Instr, Fairleigh Dickinson Univ, 66-68, adj assoc prof, 68-69; adj asst prof, Columbia Univ, 68-69; consult, Breed Corp, NJ, 68-; consult, Battelle Mem Inst, 69-70. Mem: Am Phys Soc. Res: Controlled thermonuclear research; instrumentation; shock wave research; synchrotron radiation research. Mailing Add: Div Controlled Thermonuclear Res US Energy Res & Develop Admin Washington DC 20545

MILLER, BERNARD, b Monticello, NY, Sept 1, 30; m 65. ORGANIC CHEMISTRY. Educ: City Col New York, BS, 51; Columbia Univ, MA, 53, PhD(org chem), 55. Prof Exp: NIH fel, Univ Wis, 55-56, NSF, 56-57; res chemist, Am Cyanamid Co, 57-60, sr res chemist, 60-67; assoc prof, 67-72, PROF CHEM, UNIV MASS, AMHERST, 72- Mem: Am Chem Soc. Res: Organic reaction mechanisms; molecular rearrangements; organic phosphorus chemistry. Mailing Add: Dept of Chem Univ of Mass Amherst MA 01002

MILLER, BERNARD, b New York, NY, Apr 9, 27; m 58; c 2. POLYMER SCIENCE, TEXTILES. Educ: Va Polytech Inst, BS, 48, MS, 49; McGill Univ, PhD(chem), 55. Prof Exp: Jr chemist, Hoffman-La Roche Inc, 49-52; res chemist, Celanese Corp Am, 55-56; asst prof polymer chem, Lowell Technol Inst, 56-58; Du Pont fel, Textile Res Inst, 58-59; res chemist, E I du Pont de Nemours & Co, 59; asst prof phys chem, Am Univ, 59-61, assoc prof, 61-66; sr scientist, 66-67, assoc dir chem res, 67-69, ASSOC DIR RES, TEXTILE RES INST, 69- Concurrent Pos: NASA res grant, 62-64; USPHS res grant, 62-66; consult, NIH, 63-64; Fiber Soc nat lectr, 73-74. Mem: AAAS; Am Chem Soc; Fiber Soc; Info Coun on Fabric Flammability; NAm Thermal Anal Soc. Res: Fiber science; flammability; fiber surface properties; thermal analysis; calorimetry; cellulose chemistry; thermal properties of polymers. Mailing Add: Textile Res Inst Princeton NJ 08540

MILLER, BERTRAND JOHN, b Cedar Rapids, Iowa, Sept 19, 11; m 38; c 3. PHYSICS. Educ: St Ambrose Col, BS, 33; Univ Iowa, MS, 36, PhD(physics), 39. Prof Exp: Instr physics, St Ambrose Col, 38-40; instr math, St Louis Univ, 40-43; physicist, Nat Bur Standards, 43-48; physicist, Zenith Radio Corp, 48-59, asst dir res, 59-65; PROF PHYSICS, ST AMBROSE COL, 65-; CHMN DEPT, 64- Mem: Am Phys Soc; sr mem Inst Elec & Electronics Eng; Am Asn Physics Teachers. Res: Effect of hyperfine structure on magnetic rotation of plane of polarization of resonance radiation. Mailing Add: Dept of Physics St Ambrose Col Davenport IA 52803

MILLER, BETTY M (TINKLEPAUGH), b Corunna, Mich, Apr 23, 30; m 64. PETROLEUM GEOLOGY. Educ: Cent Mich Univ, AB, 52; Mich State Univ, MS, 55, PhD(geol), 57. Prof Exp: Cartographer & geologist, McClure Oil Co, 54-55; asst natural sci, Mich State Univ, 56-57, instr, 57-58; res geologist, Pure Oil Co, 58-65; sr res geologist & geostatistician, Sun Oil Co, 66-73; geologist, 73-74, PROG CHIEF RESOURCE APPRAISAL GROUP, OIL & GAS RESOURCES BR, US GEOL SURV, 74- Concurrent Pos: Asst, Mich State Univ, 54-55. Mem: AAAS; Geol Soc Am; Am Asn Petrol Geol. Res: Geostatistical and computer applications as related to petroleum geology, including information theory and operations research; carbonate petrography and geochemistry; oil and gas resource appraisal methodology and appraisals and assessments of the national and worldwide resources. Mailing Add: US Geol Surv Oil & Gas Resources Br Denver Fed Ctr Denver CO 80225

MILLER, BILLIE LYNN, b Harrodsburg, Ky, Nov 19, 34; m 59; c 2. MEDICINE. Educ: Eastern Ky Univ, BS, 54; Univ Chicago, MD, 57; Am Bd Pediat, dipl, 68, cert cardiol, 70. Prof Exp: Intern, Hosp, Univ Mich, Ann Arbor, 57-58; resident cardiol, Hammersmith Hosp, London, Eng, 58-59; fel, St Thomas Hosp, 59-61; fel, Nat Heart Hosp, 61-62; res assoc, London Hosp, 62; asst med investr, Nat Inst Cardiol Mex, 62-65; resident pediat, 66-68, spec clin fel pediat cardiol, 68-69, instr pediat, 69-70, ASST PROF PEDIAT, COL MED, UNIV FLA, 70- Concurrent Pos: Consult, Fla Bur Crippled Children, 68- Mem: Fel Am Acad Pediat; assoc fel Am Col Cardiol; Am Heart Asn. Res: Analysis of disorders of cardiac rhythm in infants and children; application of computer techniques to the interpretation of the body surface manifestations of the electrical activity of the heart in infants and children. Mailing Add: Dept of Pediat Univ of Fla Col of Med Gainesville FL 32610

MILLER, BRADFORD, b Seattle, Wash, Nov 11, 24; m 59; c 3. CLINICAL BIOCHEMISTRY. Educ: Stetson Univ, BS, 48, MS, 50; Univ NC, PhD(biochem), 57. Prof Exp: Res assoc biochem, Univ NC, Chapel Hill, 57-65, environ sci & eng, 65-67; CLIN CHEMIST, CHEM LAB, CHARLOTTE MEM HOSP, 67- Res: Lipid metabolism. Mailing Add: Chem Lab Charlotte Mem Hosp Charlotte NC 28201

MILLER, BRINTON MARSHALL, b Delaware Co, Pa, Dec 30, 26; m 48; c 3. MICROBIOLOGY, PROTOZOOLOGY. Educ: Univ Va, BA, 50, MS, 51; Purdue Univ, PhD(plant sci), 56. Prof Exp: Teacher high sch, Va, 51-53; sr microbiologist, 56-66, sect head, 66-73, asst dir basic animal sci, 73-75, DIR ANIMAL INJECTIONS, BASIC ANIMAL SCI, MERCK SHARP & DOHME RES LABS, 75- Concurrent Pos: Asst ed, Am Biol Teacher, Nat Asn Biol Teachers, 53-56; vis biologist, Am Inst Biol Sci, 70-73; chmn coun, Proj Biotech, Am Inst Biol Sci-Nat Sci Found, 71-75. Honors & Awards: Merit Award, Soc Indust Microbiol, 70. Mem: AAAS; Soc Indust Microbiol (secy, 58-61, pres, 63); Wildlife Soc; Am Soc Microbiol (treas, 75-); fel Am Acad Microbiol. Res: Pathology of mycoplasma; parasitic Protozoa of man and animals, especially trypanosomes and Coccidia; water microbiology; biotechnology training. Mailing Add: Merck Sharp & Dohme Res Labs Rahway NJ 07065

MILLER, BRUCE JONES, chemistry, deceased

MILLER, BRUCE LINN, b Grove City, Pa, Sept 8, 23; m 48; c 2. PHYSICS, MATHEMATICS. Educ: SDak State Univ, BS, 47; Univ Kans, MS, 51, PhD(physics), 53. Prof Exp: Jr physicist, Univ Iowa, 43-44; res physicist, Sandia Corp, NMex, 53-55; from asst prof to assoc prof, 55-70, PROF PHYSICS, GRAD FAC, S DAK STATE UNIV, 70- Mem: Am Inst Physics; Am Asn Physics Teachers. Res: Geometrical and physical optics; biological effects of nuclear radiation. Mailing Add: Dept of Physics SDak State Univ Brookings SD 57006

MILLER, BRUCE NEIL, b New York, NY, Dec 12, 41; m 66. THEORETICAL PHYSICS. Educ: Columbia Univ, BA, 63; Univ Chicago, MSc, 65; Rice Univ, PhD(physics), 69. Prof Exp: Asst physicist, IIT Res Inst, 65; NIH grant light scattering, Rice Univ, 69-70; NSF grant transport theory, State Univ NY Albany, 70-71; ASST PROF PHYSICS, TEX CHRISTIAN UNIV, 71- Res: Statistical physics; transport theory; electromagnetic theory; light scattering; critical phenomena; information theory. Mailing Add: Dept of Physics Tex Christian Univ Ft Worth TX 76129

MILLER, BYRON F, b Robinson, Kans, Aug 20, 31; m 52; c 5. POULTRY NUTRITION. Educ: Kans State Univ, BS, 53, MS, 60, PhD(animal nutrit), 60. Prof Exp: From instr to asst prof, 60-66, ASSOC PROF POULTRY PRODS, COLO STATE UNIV, 66- Mem: Poultry Sci Asn; Inst Food Technol; Entom Soc Am; Soc Nutrit Educ. Res: Poultry products and food technology. Mailing Add: Dept of Animal Sci Colo State Univ Ft Collins CO 80521

MILLER, BYRON SLOANE, b Sheridan Co, Nebr, Dec 25, 17; m 43; c 2. CEREAL CHEMISTRY. Educ: Univ Nebr, BSc, 39; Purdue Univ, MS, 42; Kans State Univ, PhD(org-biochem), 48. Prof Exp: Asst chemist, Agr Res Serv, USDA, 46-48, assoc chemist, Hard Winter Wheat Qual Lab, 48-52, chemist, 52-55, sr chemist, 56-60, prin chem chemist, 60-61; prin scientist, 61-64, head spec proj dept, 64-68, HEAD MIX DEVELOP DEPT, JAMES FORD BELL TECH CTR, GEN MILLS, INC, 68- Concurrent Pos: NSF sr fel, Rothamsted Exp Sta, Eng, 58-59. Mem: Am Asn Cereal Chem (secy, 62-65, pres, 69-70). Res: Food technology; cereal product development; food chemistry; quality of wheat and flour. Mailing Add: James Ford Bell Tech Ctr Gen Mills Inc 9000 Plymouth Ave N Minneapolis MN 55427

MILLER, C ARDEN, b Shelby, Ohio, Sept 19, 24; m 48; c 4. PEDIATRICS, PUBLIC HEALTH. Educ: Yale Univ, MD, 48. Prof Exp: From instr to assoc prof pediat, Med Ctr, Univ Kans, 51-57, from asst dean to dean, Sch Med, 57-66, provost, 65-66, dir med ctr, 60-65; vchancellor health sci, 66-71, PROF MATERNAL & CHILD HEALTH, SCH PUB HEALTH, UNIV NC, CHAPEL HILL, 66- Concurrent Pos: Markle scholar med sci, 55-60; chmn exec comt, Citizens Bd Inquiry, Health Serv for Americans, 68- Mem: Am Soc Pediat Res; Am Pub Health Asn (pres, 74-). Res: Health policy; handicapped children; child health. Mailing Add: Rosenau Hall Univ of NC Chapel Hill NC 27514

MILLER, C DAVID, b Baltimore, Md, Apr 7, 31; m 61; c 2. ANALYTICAL CHEMISTRY. Educ: Columbia Univ, AB, 52; Univ Md, MS, 59; Univ Fla, PhD(anal chem), 64. Prof Exp: Anal res chemist, E I du Pont de Nemours & Co, 52-53; DIR, RES & DEVELOP & ENG, AM INSTRUMENT CO, 60-61 & 64- Mem: Am Chem Soc; Soc Appl Spectroscopy; Electron Micros Soc Am. Res: Development of clinical and analytical instrumentation; atomic and molecular spectroscopy, luminescence, electrochemical, neutron activation and enthalpometric analysis. Mailing Add: Am Instrument Co 8030 Georgia Ave Silver Spring MD 20910

MILLER, CARL ELMER, b Flint, Mich, Apr 28, 37; m 65. MOLECULAR PHYSICS. Educ: Univ Mich, BA, 59, MS, 61, PhD(physics), 67. Prof Exp: Instr physics, Flint Jr Community Col, 61-63; asst prof, 67-70, ASSOC PROF PHYSICS, MANKATO STATE COL, 70- Mem: Am Phys Soc. Res: Measurement of the rotational vibrational dependence of the hyper-fine structure constants of the alkali halides with data obtained from a molecular beam electric resonance spectrometer. Mailing Add: Dept of Physics Mankato State Col Mankato MN 56001

MILLER, CARL HENRY, JR, b Cleveland, Ohio, Sept 18, 20; m 51. PHYSIOLOGICAL CHEMISTRY, ELECTRON MICROSCOPY. Educ: Ohio State Univ, BSc, 42, PhD(physiol chem), 67. Prof Exp: Asst physics, Case Inst Technol, 45-46, sr technician biochem, 48-56; res asst, Ohio State Univ, 59-67; Nat Inst Ment Health traineeship & assoc res scientist, Med Sch, NY Univ, 67-69; ASSOC PROF BIOL, JERSEY CITY STATE COL, 69- Mem: AAAS; Electron Micros Soc Am. Res: Biochemistry of mental illness; metabolism of catecholamines. Mailing Add: Dept of Biol Jersey City State Col 2039 Kennedy Blvd Jersey City NJ 07305

MILLER, CARL STINSON, b Edmonton, Alta, July 23, 12; nat US; m 44; c 2. PHYSICAL CHEMISTRY. Educ: Univ Alta, BSc, 35, MSc, 36; Univ Minn, PhD, 40. Prof Exp: RES CHEMIST GRAPHIC ARTS, MINN MINING & MFG CO, 40- Mem: Am Chem Soc. Res: Photographic thermal and electric processes of graphic reproduction. Mailing Add: Minn Mining & Mfg Co 900 Fauquier St St Paul MN 55119

MILLER, CARLOS OAKLEY, b Jackson, Ohio, Feb 19, 23. PLANT PHYSIOLOGY. Educ: Ohio State Univ, BSc, 48, MA, 49, PhD, 51. Prof Exp: Proj assoc bot, Univ Wis, 51-55, asst prof, 55-57; from asst prof to assoc prof, 57-63, PROF BOT, IND UNIV, BLOOMINGTON, 63- Mem: Bot Soc Am; Am Soc Plant Physiol (secy, 60-61, vpres, 62). Res: Plant growth substances, particularly those of kinetin type; chemical control of plant development; control of plant growth by light. Mailing Add: Dept of Plant Sci Indiana Univ Bloomington IN 47401

MILLER, CAROL RAYMOND, b Asheville, NC, Sept 10, 38; m 59; c 2. PLANT BREEDING, PLANT PATHOLOGY. Educ: West Carolina Col, BS, 60; Clemson Univ, MS, 62, PhD(plant path), 65. Prof Exp: Res asst, Clemson Univ, 63-64; asst prof plant path, Univ Fla, 64-71; asst dir tobacco res & prod, 71-72, DIR TOBACCO RES & PROD, COKER'S PEDIGREED SEED CO, 72- Mem: Am Phytopath Soc. Res: Tobacco diseases and breeding. Mailing Add: Tobacco Res Coker's Pedigreed Seed Co Hartsville SC 29550

MILLER, CAROLYN THATCHER, biochemistry, see 12th edition

MILLER, CECIL R, b Morrow Co, Ohio, Oct 17, 33; m 54; c 4. VETERINARY MEDICINE. Educ: Ohio State Univ, DVM, 58. Prof Exp: Large animal pract, Leipsic Vet Serv, 58-67; res vet, Dow Chem Co, 67-68; FIELD DEVELOP INVESTR, SMITH KLINE LABS, 68- Mem: Indust Vet Asn; Am Asn Bovine Practitioners; Am Asn Swine Practitioners; Am Asn Avian Pathologists; Am Vet Med Asn. Res: Swine diseases, poultry, bovine nutrition; applied aspects of disease and management conditions. Mailing Add: Smith Kline Animal Health Prods 1600 Paoli Pike West Chester PA 19380

MILLER, CHARLES ALDEN, biology, biochemistry, see 12th edition

MILLER, CHARLES ALEXIS, b New Orleans, La, Jan 18, 20; m 47; c 7. ORGANIC CHEMISTRY, RESEARCH ADMINISTRATION. Educ: Loyola Univ, BS, 43; Univ Detroit, MS, 47. Prof Exp: Instr chem, Univ Detroit & Wayne State Univ, 47; sr res org chemist, 47-59, head patent liaison, 59-64, mgr res admin, 64-65, DIR RES PLANNING & CONTRACTS, PARKE, DAVIS & CO, 66- Mem: Am Chem Soc; Am Mgt Asn. Res: Medicinal research. Mailing Add: Joseph Campau at the River Detroit MI 48232

MILLER, CHARLES BENEDICT, b Minneapolis, Minn, Apr 28, 40; m 63; c 2. BIOLOGICAL OCEANOGRAPHY. Educ: Carleton Col, BA, 63; Scripps Inst Oceanog, PhD(biol oceanog), 69. Prof Exp: NSF fel, Univ Auckland, 69-70; asst prof, 70-75, ASSOC PROF BIOL, SCH OCEANOG, ORE STATE UNIV, 75- Mem: Am Soc Limnol & Oceanog. Res: Zooplankton ecology. Mailing Add: Sch of Oceanog Ore State Univ Corvallis OR 97331

MILLER, CHARLES DOUGLAS F, b Alabama, NY, Mar 20, 25; Can citizen; m 49; c 5. ENTOMOLOGY. Educ: Univ Guelph, BSA, 48; Univ BC, MSA, 51; McGill Univ, PhD, 67. Prof Exp: Scientist, Entom Res Sta, Can Dept Agr, 51-68, head entom sect, 68-73, RES COORDR BIOSYST, RES BR, AGR CAN, 73- Mem: Entom Soc Can. Res: Environmental management and taxonomy of Hymenoptera, especially Vespoidea and Chalcioloidea. Mailing Add: 2754 Draper Pl Ottawa ON Can

MILLER, CHARLES EDWARD, b Baltimore, Md, June 19, 18; m 42; c 2. ORGANIC CHEMISTRY. Educ: Wash Col, Md, BS, 40. Prof Exp: Chemist, Madison Glue Corp, Ind, 40-41; org chemist, Tech Command, Army Chem Ctr, 45-50, asst chief, Test Div, Chem & Radiol Labs, 50-56, chief, Test Div, Chem Res & Develop Labs, 56-62, dep dir, Develop Support Directorate, Edgewood Arsenal, 62-63; chief tech adv for Chem, Biol & Radiol, Hq, US Army Test & Eval Command, Aberdeen Proving Ground, 63-73; RETIRED. Honors & Awards: Meritorious Civilian Serv Award, Dept of Army, 73. Mem: Am Ord Asn. Res: Chemical warfare; materials testing; esterification; process laboratory design, erection and operation. Mailing Add: 27 Lake Dr Bel Air MD 21014

MILLER, CHARLES EDWARD, b Philadelphia, Pa, Feb 16, 25; m 44; c 1. MYCOLOGY. Educ: Furman Univ, BS, 51; Univ NC, MA, 54, PhD(bot), 57. Prof Exp: Asst bot, Univ NC, Chapel Hill, 51-57, Am Bact Soc fel, 57-58; instr biol, Emory Univ, 58-59; asst prof bot, Tex A&M Univ, 59-62; assoc prof, Univ Maine, 62-65; assoc prof, 65-70, PROF & CHMN DEPT BOT, OHIO UNIV, 70- Mem: Bot Soc Am; Mycol Soc Am; Brit Mycol Soc. Res: Taxonomy; morphology; ecology; physiology; ultrastructure. Mailing Add: Dept of Bot Ohio Univ Athens OH 45701

MILLER, CHARLES ELLSWORTH, b Tefft, Ind, Oct 13, 24; m 47; c 2. BIONUCLEONICS. Educ: Purdue Univ, BSEE, 45, MS, 46, PhD, 50. Prof Exp: Instr elec eng, Purdue Univ, 44-50; assoc scientist, Argonne Nat Lab, 50-69; prof med, biochem & biophys, Med Sch, Loyola Univ Chicago, 69-71; asst prof, 71-75, ASSOC PROF ELEC ENG TECHNOL & HEAD DEPT, PURDUE UNIV, CALUMET CAMPUS, 75- Mem: AAAS; Radiation Res Soc; Health Physics Soc. Res: Radium toxicity of humans; retention of radioisotopes in humans; whole-body counting; gamma-ray spectroscopy; nuclear medicine; advanced circuit analysis; digital logic. Mailing Add: Box 38 Tefft IN 46380

MILLER, CHARLES EVERETT, JR, b Waukesha, Wis, May 17, 29; m 52; c 8. RADIOCHEMISTRY, ANALYTICAL CHEMISTRY. Educ: Northwestern Univ, BSc, 51; Fla State Univ, MSc, 57, PhD(anal chem), 59. Prof Exp: Instr, chem, Fla State Univ, 58; assoc physicist, Armour Res Found, 59-60; chemist, Oak Ridge Nat Lab, Union Carbide Nuclear Co, 60-67; assoc chemist, Liquid Metal Fast Breeder Reactor Prog Off, 67-70, ASSOC CHEM, ARGONNE NAT LAB, REACTOR ANAL & SAFETY DIV, 70- Mem: Am Chem Soc; Am Nuclear Soc. Res: Activation analysis; nuclear safety, release and behavior of fission products from molten nuclear reactor fuels; fission product fractionation; transmutation effects in reactor materials; in-pile experiments in reactor fuel and coolant interactions under accident conditions. Mailing Add: Argonne Nat Lab 9700 S Cass Ave Argonne IL 60439

MILLER, CHARLES FREDERICK, III, b Springfield, Ill, Feb 12, 41; m 66; c 2. MATHEMATICS. Educ: Lehigh Univ, BA, 62; NY Univ, MS, 64; Univ Ill, PhD(math), 69. Prof Exp: Instr math, Univ Ill, 68-69; mem, Inst Advan Study, Princeton, NJ, 69-70; NSF fel, Oxford Univ, 70-71; ASST PROF MATH, PRINCETON UNIV, 71- Mem: Am Math Soc; Asn Symbolic Logic. Res: Mathematical logic; combinatorial group theory; decision problems in algebra; geometric methods to group theory. Mailing Add: Dept of Math Princeton Univ Princeton NJ 08540

MILLER, CHARLES G, b Greensburg, Ind, Feb 9, 40; m 65; c 3. MICROBIOLOGY, BIOCHEMISTRY. Educ: Ind Univ, Bloomington, AB, 63; Northwestern Univ, PhD(biochem), 68. Prof Exp: USPHS fel, Univ Calif, Berkeley, 68-70; ASST PROF MICROBIOL, SCH MED, CASE WESTERN RESERVE UNIV, 70- Mem: Am Chem Soc; Am Soc Microbiol. Res: Biochemical genetics. Mailing Add: Dept of Microbiol Sch of Med Case Western Reserve Univ Cleveland OH 44106

MILLER, CHARLES GARDNER, physics, see 12th edition

MILLER, CHARLES PHILLIP, b Oak Park, Ill, Aug 29, 1894; m 31; c 2. CLINICAL MEDICINE. Educ: Univ Chicago, BS, 16; Rush Univ, Chicago, MD, 19; Univ Mich, MS, 20. Prof Exp: Intern, Presby Hosp, Chicago, Ill, 18-19; asst path, Univ Mich, 19-20, asst prof, 20; asst res, Rockefeller Inst Hosp, New York City, 20-24, asst path & bact, 24-25; from asst prof to prof med, 25-60, EMER PROF MED, UNIV CHICAGO, 60- Concurrent Pos: Vol asst, Prussian Inst Infectious Dis, Berlin, 26; consult, Argonne Nat Lab; consult Secy War, 41-49; sr sci officer, Off Sci & Technol, Am Embassy, London, Eng, 48; mem div, Comt Biol & Med Sci, NSF, 55-60; vchmn, Farm Found, 58- Mem: Nat Acad Sci; AAAS; Am Soc Clin Invest (vpres, 38); Am Soc Exp Path (secy-treas, 30-34, vpres, 36, pres, 37); Radiation Res Soc. Res: Experimental meningococcal and gonococcal infection; biology of meningoccus and gonococcus; action of antibiotics on bacteria; bacterial antagonism; effect of ionizing radiation on susceptibility to infection. Mailing Add: 5757 Kimbark Ave Chicago IL 60637

MILLER, CHARLES STANDISH, b Independence Kans, Jan 24, 27; m 52; c 4. PLANT PHYSIOLOGY. Educ: Tex A&M Univ, PhD(plant physiol), 59. Prof Exp: Sci aide agr res, Bur Plant Indust, Soils & Agr Eng, USDA, 51-52; asst agronomist, Tex Agr Exp Sta, 52-53; asst prof, 58-65, ASSOC PROF PLANT PHYSIOL, TEX A&M UNIV, 65- Mem: Am Soc Plant Physiol; Weed Sci Soc Am. Res: Abscission physiology of cotton. Mailing Add: Dept of Plant Sci Tex A&M Univ College Station TX 77843

MILLER, CHARLES WESLEY, plant pathology, nematology, see 12th edition

MILLER, CHARLES WILLIAM, b Quantico, Va, June 7, 42; m 66. BIOLOGY. Educ: Purdue Univ, BS, 64; Colo State Univ, MS, 66, PhD(physiol), 69. Prof Exp: Res fel, Univ Wash, 68-70; ASST PROF RADIOL & RADIATION BIOL, COLO STATE UNIV, 71- Mem: AAAS; Am Inst Biol Sci. Res: Blood flow characteristics at bends and branch points; arterial wall mechanical properties; pulse wave velocity variations with age and blood pressure; noninvasive studies of blood flow in man; investigations of renal function as affected by age and low doses of ionizing radiation. Mailing Add: Dept of Biophys Colo State Univ Ft Collins CO 80523

MILLER, CHRIS H, b Indianapolis, Ind, Feb 28, 42; m 63; c 3. ORAL MICROBIOLOGY. Educ: Butler Univ, BA, 64; Univ NDak, MS, 66, PhD(microbiol), 69. Prof Exp: Nat Inst Gen Med Sci res fel, Purdue Univ, 69-70; asst prof med & dent microbiol, Med Ctr, Ind Univ-Purdue Univ, Indianapolis, 70-75; ASSOC PROF MICROBIOL, IND UNIV, SCH DENT, INDIANAPOLIS, 75- Mem: Am Soc Microbiol; Int Asn Dent Res; Am Asn Dent Sch. Res: Ecology of oral bacteria; mechanisms of bacterial dental-plaque formation; bacterial extra-cellular polysaccharides; pathogenicity of actinomycetes. Mailing Add: Dept of Oral Microbiol Ind Univ Sch of Dent Indianapolis IN 46202

MILLER, CLARENCE ALLAN, b Harlan, Iowa, May 15, 43; m 65; c 3. MAMMALIAN ECOLOGY, ANIMAL BEHAVIOR. Educ: Buena Vista Col, BS, 65; Mankato State Univ, MA, 68; NDak State Univ, PhD(zool), 75. Prof Exp: Instr biol, Pub Schs, Iowa, 68-70; RES ASST ZOOL, NDAK STATE UNIV, 75- Mem: Am Soc Mammalogists; Animal Behav Soc; Soc Study Evolution; Am Soc Naturalists; Sigma Xi. Res: Behavioral ecology, habitat diversity and species diversity of rodents in their natural communities; sensory modalities involved in habitat selection in rodents. Mailing Add: Dept of Zool NDak State Univ Fargo ND 58102

MILLER, CLIFFORD DANIEL, b Salem, Ore, Nov 30, 41; m 75. QUATERNARY GEOLOGY, VOLCANOLOGY. Educ: Univ Wash, BS, 65, MS, 67; Univ Colo, PhD(geol), 71. Prof Exp: Teaching asst geol, Univ Wash, 65-67 & Univ Colo, 67-70; explor geologist, Standard Oil Co Calif, 70-71; prof geol, Colgate Univ, 71-74; RES SCIENTIST GEOL, US GEOL SURV, 74- Concurrent Pos: Penrose grant, Geol Soc Am, 72; NSF grant, 72-73; consult, Cossitt Concrete Co, NY, 73-74. Mem: Geol Soc Am; Glaciol Soc; Am Quaternary Asn. Res: Stratigraphic study of Quaternary glacial and eruptive products from North American volcanoes; appraisal of future volcanic hazards at Mount Shasta, California. Mailing Add: Eng Geol Br MS 903 KAE Box 25046 Denver CO 80225

MILLER, CLIFFORD H, b Chicago, Ill, Dec 9, 32. DENTISTRY. Educ: Northwestern Univ, DDS, 57. Prof Exp: Instr oper dent, 59-61, from asst prof to assoc prof, 61-66, chmn dept oper dent, 69-72, PROF DENT HISTOL & OPER DENT, NORTHWESTERN UNIV, CHICAGO, 66-, ASSOC DEAN, 72- Mem: Fel Am Col Dent; Am Dent Asn. Res: Bio-mechanics relating to restorative dentistry; operative dentistry; dental histology. Mailing Add: Northwestern Univ Dent Sch 311 E Chicago Ave Chicago IL 60611

MILLER, CONRAD ERVE, physical chemistry, see 12th edition

MILLER, CONRAD HENRY, b Lowell, WVa, July 28, 26; m 47; c 3. PLANT PHYSIOLOGY. Educ: Va Polytech Inst, BS, 54, MS, 55; Mich State Univ, PhD(hort), 57. Prof Exp: Asst prof bot & plant path, Mich State Univ, 57; from asst prof to assoc prof, 57-69, PROF HORT, NC STATE UNIV, 69- Mem: Am Soc Hort

Sci. Res: Vegetable production research, especially plant nutrition and growth regulators. Mailing Add: Dept of Hort Sci NC State Univ Raleigh NC 27607

MILLER, CURTIS C, b Shamokin, Pa, Nov 26, 35; m 58; c 2. GENETICS, ANIMAL BREEDING. Educ: Iowa State Univ, BS, 61; Mich State Univ, MS, 65, PhD(dairy sci), 68. Prof Exp: Exten dairyman, Mich State Univ, 61-64; statistician, 68-71, mgr animal regulatory affairs & statist serv, 71-73, RES MGR ANIMAL REGULATORY AFFAIRS, STATIST SERV & THERAPEUT, ANIMAL HEALTH RES & DEVELOP, 73- Mem: Am Dairy Sci Asn; Am Genetic Asn; Crop Sci Soc Am; Am Statist Asn; Biomet Soc. Res: Animal breeding, especially response to selection; population genetics, especially effects of selection, linkage, and dominance on genetic parameters. Mailing Add: Upjohn Co Kalamazoo MI 49001

MILLER, DANIEL NEWTON, JR, bSt Louis, Mo, Aug 22, 24; m 50; c 2. GEOLOGY. Educ: Mo Sch Mines, BS, 49, MS, 51; Univ Tex, PhD(geol), 55. Prof Exp: Jr geologist, Stanolind Oil & Gas Co, 51-52, geologist, 55-57; sr geologist, Pan Am Petrol Corp, 57-59 & Lion Oil Div, Monsanto Chem Co, 59-60; consult geologist, Barlow & Haun, Inc, 60-63; prof geol & chmn dept, Southern Ill Univ, 63-69; STATE GEOLOGIST & EXEC DIR, WYO GEOL SURV, 69- Concurrent Pos: Wyo rep, Dept Interior—Oil Shale Environ Adv Comt, 75-76; Wyo rep, Fed Power Comt—Supply-Tech Adv Task Force—Prospective Expor and Develop. Mem: Soc Econ Paleont & Mineral; Am Asn Petrol Geol; Am Inst Prof Geologists; Asn Am States Geologists (secy-treas, 75-76). Res: Sedimentation; sedimentary petrology and stratigraphic interpretation of sedimentary rocks and the diagenetic alteration that they have undergone; effect of tectonism on late diagenetic alteration in sedimentary rocks related to petroleum accumulation. Mailing Add: Geol Surv of Wyo PO Box 3008 Univ Sta Laramie WY 82071

MILLER, DANIEL ROBERT, b Milwaukee, Wis, May 14, 17; m 44; c 4. CHEMISTRY, RESEARCH ADMINISTRATION. Educ: Univ Wis, BS, 41, MS, 43; Univ Calif, PhD(chem), 48. Prof Exp: Chemist, Metal Lab, Univ Chicago, 43; group leader & asst sect chief, Clinton Labs, Tenn, 43-44, 45-46; sr supvr, Hanford Eng Works, 44-45; chemist, Radiation Lab, Univ Calif, 46-48; asst prof chem, Cornell Univ, 48-51; chemist, Div Res, 51-53, asst chief chem br, 53-60, chief, 60-61, asst dir res chem prog & dep dir, AEC, 61-75; ACTG DIR, DIV PHYS RES, ENERGY RES & DEVELOP ADMIN, 75- Mem: AAAS; Am Chem Soc Res Admin. Res: Nuclear chemistry; chemistry of actinide elements. Mailing Add: Energy Res & Develop Admin Washington DC 20545

MILLER, DANIEL WEBER, b Omaha, Nebr, Jan 24, 26; m 47; c 2. EXPERIMENTAL NUCLEAR PHYSICS. Educ: Univ Mo, BS, 47; Univ Wis, PhD(physics), 51. Prof Exp: Res assoc, 51-52, from asst prof to assoc prof, 52-62, assoc dean, Col Arts & Sci, 62-64, actg chmn dept physics, 64-65, assoc dean res & advan studies, 72-73, PROF PHYSICS, IND UNIV, BLOOMINGTON, 62- Concurrent Pos: Consult, Los Alamos Sci Lab, 59-64; co-prin investr, Nuclear Reactions Res & 200 MeV Cyclotron Proj, Ind Univ, 63-; mem bd dirs, Midwest Univs Res Asn, 64-72; del, Argonne Univ Asn, 66-; chmn, Publ Comt Div Nuclear Physics, Am Phys Soc, 75- Mem: Fel Am Phys Soc; mem Sigma Xi; mem AAAS. Res: Nuclear reaction mechanism and nuclear structure investigations utilizing light- and heavy-ion charges particle beams at intermediate energies; polarization in nuclear reactions; total cross sections of nuclei for fast neutrons. Mailing Add: Physics Dept Ind Univ Bloomington IN 47401

MILLER, DARRELL ALVIN, b Lincoln, Ill, Sept 27; 32; m 53; c 4. GENETICS, PLANT BREEDING. Educ: Univ Ill, BS, 58, MS, 60; Purdue Univ, PhD(genetics, plant breeding), 62. Prof Exp: Asst prof plant breeding & genetics, NC State Univ, 62-64, asst prof in charge crop sci teaching, 64-65, asst dir of instr, Sch Agr & Life Sci, 65-67; assoc prof, 67-71, PROF PLANT BREEDING & GENETICS IN ALFALFA, UNIV ILL, URBANA, 71- AGRON TEACHING COORD, 69- Concurrent Pos: Mem, Crop Sci Writing Conf, 65-74; agron coord for jr cols, 68- Mem: Am Soc Agron; Crop Sci Soc Am; Nat Sci Teacher Asn; Sigma Xi. Res: Inheritance of cytoplasmic male sterility and its interaction in sorghum and alfalfa; alfalfa protein investigations, weevil resistance and physical genetics; allelophathic investigations with alfalfa. Mailing Add: Dept of Agron Turner Hall Univ of Ill Urbana IL 61801

MILLER, DAVID ARTHUR, b Marion, Ohio, Jan 7, 42; m 72; c 1. MEDICAL PHYSIOLOGY. Educ: Ohio Northern Univ, BSEE, 64; Ohio Univ, MSEE, 66; Ohio State Univ, PhD(physiol), 72. Prof Exp: Lectr physiol, Ohio State Univ, 72-73; ASST PROF PHYSIOL, MED COL GA, 73- Res: Respiratory mechanics and control. Mailing Add: Dept of Physiol Med Col of Ga Augusta GA 30902

MILLER, DAVID CHARLES, b Aurora, Ill, May 25, 18; m 42; c 3. PHYSICS. Educ: Ripon Col, BA, 39; Washington Univ, MS, 41, PhD(physics), 47. Prof Exp: Asst physics, Washington Univ, 39-42, asst & instr night sch, 46-47; asst prof, Univ Mont, 47-48; consult, US Res & Develop Bd, 48, Sci Warfare Adv, 48-51; asst dir res, Philips Labs, Inc, NAm Aviation Philips Co, Inc, 51-53, tech dir, Philips Electronics Instruments Div, 53-61; gen mgr, Alexandria Div, Am Mach & Foundry Co, 61-62, dir res & develop lab, Conn, 62-64, vpres & gen mgr, Alexandria Div, 64-67; sr staff assoc, Comn Eng Educ, 67-68; sr staff assoc, Comn Educ, Nat Acad Eng, 68-70; prof systs eng, Polytech Inst Brooklyn, 70-75; MEM FAC PHYSICS, ALPENA COMMUNITY COL, 75- Concurrent Pos: Vis prof, Georgetown Univ, 48-49; consult, Eng Concepts Curric. Mem: Am Phys Soc; Am Soc Testing & Mat; Electron Micros Soc; Instrument Soc Am; NY Acad Sci. Res: X-ray diffuse reflections; German electronic and infrared research in World War II; x-ray diffraction and spectrographic instrumentation; electron microscopes and microprobe instrumentation; educational systems engineering. Mailing Add: Dept of Physics Alpena Community Col Alpena MI 49707

MILLER, DAVID CLAIR, b Worcester, Mass, Apr 17, 34. SYSTEMATIC ENTOMOLOGY, INVERTEBRATE ZOOLOGY. Educ: Univ Calif, Riverside, BA, 57; Univ Wash, PhD(zool), 62. Prof Exp: Asst zool, Univ Wash, 57-61; lectr biol, 61-62, instr, 62-64, asst prof, 64-71, ASSOC PROF BIOL, CITY COL NEW YORK, 72- Mem: Am Soc Zool; Emtom Soc Am; Soc Syst Zool; Soc Study Evolution; Coleopterists' Soc. Res: Systematics; biology and evolution of the beetle family Hydrophilidae. Mailing Add: Dept of Biol City Col of New York New York NY 10031

MILLER, DAVID EUGENE, soil physics, see 12th edition

MILLER, DAVID HARRY, b Callington Cornwall, Eng, Mar 3, 39; m 61; c 2. EXPERIMENTAL HIGH ENERGY PHYSICS. Educ: Univ London, BSc, 60, PhD(high energy physics), 63. Prof Exp: Res assoc, 63-65, asst prof, 65-68, ASSOC PROF HIGH ENERGY PHYSICS, PURDUE UNIV, 68- Concurrent Pos: Guggenheim fel, 72; vis scientist, Europ Orgn Nuclear Res, Geneva, 72-73. Mem: Fel Am Phys Soc. Res: Study of elementary particles using visual techniques. Mailing Add: Dept of Physics Purdue Univ West Lafayette IN 47907

MILLER, DAVID HEWITT, b Russell, Kans, 1918; m 62; c 1. CLIMATOLOGY. Educ: Univ Calif, Los Angeles, AB, 39, MA, 44, PhD(geog), Berkeley, 53. Prof Exp: Res librn, Lockheed Aircraft Corp, 40-41; meteorologist, Corps Engrs, US Army, 41-43, forecaster, Transcontinental & Western Air Lines, 43-44; climatologist, Off Qm Gen, 44-46; meteorologist-hydrologist, snow invests, 46-50, asst dir, 50-53; chief environ anal br, Qm Res & Eng Lab, Mass, 53-58; meteorologist, US Forest Serv, Calif, 59-64; prof geog, 64-75, PROF ATMOSPHERIC SCI, DEPT GEOL SCI, UNIV WIS-MILWAUKEE, 75- Concurrent Pos: NSF res fel, 52-53; vis lectr, Clark Univ, 57-58; Univ Ga, 58; Univ Calif, Berkeley, 61 & 63; Univ Wis, 62; Fulbright lectr, Univ Newcastle, Australia, 66; Nat Acad Sci exchange scientist, USSR Acad Sci, 69; Fulbright sr scholar, Univs Newcastle & Macquarie, 71; vis scientist, Commonwealth Sci Res & Indust Orgn, 71. Mem: Am Meteorol Soc; Asn Am Geog; Ecol Soc Am; Soc Am Foresters; Am Geophys Union. Res: Heat and water budget; snow hydrology; climatology; radiation budget at the earth's surface; energy/mass analysis of environmental impact; energy and mass budget at earth's surface; especially snow cover, forests, and cities; energy-mass in environmental impact; hydrology; radiation. Mailing Add: Dept of Geol Sci Univ of Wis-Milwaukee Milwaukee WI 43201

MILLER, DAVID JACOB, b St Louis, Mo, Sept 20, 10; m 36; c 3. ANALYTICAL CHEMISTRY. Educ: Wash Univ, BS, 31. Prof Exp: Sea food inspector, US Food & Drug Admin, La, 34-37, chemist, Md & NY, 37-58, asst to dir bur field admin, Washington, DC, 58-61, chemist, Off Comnr, 61-64, asst to dir div color cert & eval, 64-65, dep dir, 65-68, asst dir div colors & cosmetics, 68-69; CONSULT, FOOD, DRUG & COSMETICS INDUSTS, 69- Mem: AAAS; Am Chem Soc; Inst Food Technol; Soc Cosmetic Chem; fel Am Inst Chem. Res: Food additives; drugs; color additives; cosmetics. Mailing Add: Suite 407 1010 Vermont Ave NW Washington DC 20005

MILLER, DAVID LEE, b Knoxville, Tenn, Aug 15, 38; m 61; c 2. PHYSICAL ORGANIC CHEMISTRY. Educ: Oberlin Col, BA, 60; Harvard Univ, PhD(chem), 66. Prof Exp: NIH fel, 65-66; asst prof chem, Oberlin Col, 66-68; staff fel, Nat Heart Inst, 68-69; res assoc, 69-71, ASST MEM, ROCHE INST MOLECULAR BIOL, 71- Mem: Am Chem Soc; Fedn Am Soc of Exp Biol. Res: Reactions of organic polyphosphates; mechanism of protein biosynthesis; control of plasma protein synthesis. Mailing Add: Dept of Biochem Roche Inst for Molecular Biol 340 Kingsland St Nutley NJ 07110

MILLER, DAVID MILROY, b Calgary, Alta, Oct 28, 22; m 49; c 3. PHYSICAL CHEMISTRY. Educ: Univ Alta, BSc, 45, MSc, 46; McGill Univ, PhD(chem), 50. Prof Exp: Res fel photochem lab, Nat Res Coun Can, Ont, 50-52; SR CHEMIST, AGR RES INST, CAN DEPT AGR, 52- Concurrent Pos: Hon lectr, Western Ont Univ, 56. Mem: Chem Inst Can. Res: Adsorption; photochemistry; polarography; physico-chemical nature and function of cytoplasmic membrane; active transport and nerve conduction. Mailing Add: Agr Res Inst Univ Substation London ON Can

MILLER, DAVID W, b Boston, Mass, Apr 26, 29; m 55; c 3. GROUNDWATER GEOLOGY. Educ: Colby Col, BA, 51; Columbia Univ, MA, 53. Prof Exp: Hydrologist, US Geol Surv, 51-53; geologist, Leggette, Brashears & Graham, 53-57; PARTNER, GERAGHTY & MILLER, CONSULT GROUNDWATER GEOLOGISTS, 57-; PRES, WATER INFO CTR, INC, 58- Mem: Am Water Works Asn; Am Inst Prof Geologists; Soc Econ Geologists; Nat Water Well Asn. Res: Development of water resources; disposal of municipal and industrial wastes and conservation; investigation of ground-water pollution problems. Mailing Add: Geraghty & Miller Water Res Bldg Port Washington NY 11743

MILLER, DEREK HARRY, b Hull, Eng, Jan 18, 24; m 47; c 3. PSYCHIATRY, PSYCHOANALYSIS. Educ: Univ Leeds, MB, ChB, 47, MD, 55. Prof Exp: Psychiatrist, Menninger Found, Topeka, Kans, 55-59; dir adolescent unit, Tavistock Clin, London, 59-69; dir adolescent psychiat prog, 69-75, PROF PSYCHIAT & ASSOC CHMN, MED SCH, UNIV MICH, ANN ARBOR, 75- Concurrent Pos: Assoc mem, Inst Psychoanal, London, 64-68; lectr, Inst Sociol, Bedford Col, 65-69, Inst Archit, Univ Cambridge, 65-69 & WHO, 67-69; training grant adolescent psychiat, Univ Mich, 72-75; consult, W S Hall Psychiat Inst, Columbia, SC, 72- Mem: Brit Psychoanal Soc; Am Psychiat Asn; AMA; Am Soc Adolescent Psychiat; Royal Col Psychol. Res: Adolescent psychiatry, drug abuse, prediction of homicidal behavior and relationship between physical maturation and psychological development. Mailing Add: Dept of Psychiat Univ Hosp Ann Arbor MI 48104

MILLER, DON CURTIS, marine ecology, invertebrate zoology, see 12th edition

MILLER, DON DALZELL, b Menomonie, Wis, Apr 8, 13. ALGEBRA. Educ: Wayne State Univ, AB, 34, MA, 36; Univ Mich, PhD(math), 41. Prof Exp: Instr math, Lawrence Inst Technol, 35-36 & Univ Ohio, 38-42; assoc prof, 46-55, PROF MATH, UNIV TENN, KNOXVILLE, 56- Concurrent Pos: Fulbright lectr, Univ Besancon, 64-65. Mem: Am Math Soc; Math Asn Am; Asn Symbolic Logic; Math Soc France. Res: Semigroups, semirings, binary relations. Mailing Add: Dept of Math Univ of Tenn Knoxville TN 37916

MILLER, DONALD ELBERT, b Germano, Ohio, Apr 6, 06; m 31; c 1. BIOLOGY. Educ: Thiel Col, AB, 25 & 28; Univ Mich, MS, 29, PhD(zool), 35. Prof Exp: Teacher high schs, Pa, 25-28, 30-31; asst zool, Univ Mich, 29-30; instr biol Gustavus Adolphus Col, 31-33; zool, Univ Idaho, 35-36; asst prof sci, 36-39, from assoc prof to prof, 39-72, EMER PROF BIOL, BALL STATE UNIV, 72- - Res: Limnology. Mailing Add: 6045 Shagway Rd RR 2 Ludington MI 16125

MILLER, DONALD FLETCHER, b Laurel, Md, Aug 9, 24; m 62; c 2. NUTRITION. Educ: Univ Md, BS, 50; State Col Wash, MS, 52. Prof Exp: Asst, State Col Wash, 50-52; tech secy comt feed composition, Nat Acad Sci-Nat Res Coun, Washington, DC, 52-58; nutrit analyst, Consumer & Food Econ Res Div, Agr Res Serv, USDA, 59-66; ACTG CHIEF NUTRIT & FOOD COMPOS SECT, DIV NUTRIT, BUR FOODS, USDA, 66- Mem: Am Inst Nutrit; Am Asn Cereal Chem; Inst Food Technol: Animal Nutrit Res Coun; Sigma Xi. Res: Food and feed compostion; food regulations; food labeling; data processing; food and nutrition resources. Mailing Add: HFF-240 Off Nutrit & Consumer Sci Food & Drug Admin 200 C St SW Washington DC 20204

MILLER, DONALD GABRIEL, b Oakland, Calif, Oct 29, 27; m 49; c 2. PHYSICAL CHEMISTRY. Educ: Univ Calif, BS, 49; Univ Ill, PhD(phys chem), 53. Prof Exp: Asst chem, Univ Ill, 49-52; asst prof, Univ Louisville, 52-54; res assoc, Brookhaven Nat Lab, NY, 54-56; CHEMIST, LAWRENCE LIVERMORE LAB, UNIV CALIF, 56- Concurrent Pos: Fulbright prof, Univ Lille & Cath Univ Lille, 60-61. Mem: Am Chem Soc; Math Asn Am; Asn Symbolic Logic. Res: Thermodynamics of irreversible processes; electrolyte solutions; history of science. Mailing Add: Lawrence Livermore Lab Univ of Calif Livermore CA 94550

MILLER, DONALD MORTON, b Chicago, Ill, July 24, 30; m 63; c 1. COMPARATIVE PHYSIOLOGY. Educ: Univ Ill, AB, 60, MA, 62, PhD(physiol), 65. Prof Exp: Sci asst org chem, Polymer Res Lab, Univ Ill, 60, asst physiol & biophys, 60-63, 63-64, comp physiol trainee, 63, 65; asst prof, 66-72, ASSOC PROF

PHYSIOL, SOUTHERN ILL UNIV, CARBONDALE, 72- Concurrent Pos: USPHS fel protozool & parasitol, Univ Calif, Los Angeles, 55-66. Mem: AAAS; Am Soc Zool; Am Micros Soc; Bot Soc Am; Am Physiol Soc. Res: Primitive motile systems and insect nervous systems. Mailing Add: Dept of Physiol Southern Ill Univ Carbondale IL 62903

MILLER, DONALD PIGUET, b New Orleans, La, Oct 11, 27; m 51; c 3. CHEMICAL PHYSICS, CRYSTALLOGRAPHY. Educ: Agr & Mech Col, Tex, BS, 48; Tulane Univ, MS, 52; Polytech Inst Brooklyn, PhD, 62. Prof Exp: Asst biophys, Tulane Univ, 51; fel physics, Polytech Inst Brooklyn, 53-54, instr, 54-57; mem tech staff, Cent Res Lab, Tex Instruments, Inc, 57-63; ASSOC PROF CHEM PHYSICS, CLEMSON UNIV, 63- Mem: Am Crystallog Asn; Am Inst Chemists. Res: X-ray crystallography; electron diffraction; solid state physical chemistry and chemical physics. Mailing Add: Dept of Physics Clemson Univ Clemson SC 29631

MILLER, DONALD RICHARD, b Hamilton, Ont, July 4, 36; m 69. MATHEMATICAL BIOLOGY. Educ: Univ Toronto, BA, 60, MA, 61, PhD(appl math), 64. Prof Exp: Instr math, Univ Toronto, 61-63; asst prof, Univ Western Ont, 63-65, assoc prof appl math, 65-68, res assoc cancer res lab, 68-69; assoc prof indust & systs eng, Univ Fla, Cape Canaveral, 69-71; assoc prof & grad coord, Univ Fla, Gainesville, 71-73; GROUP LEADER, BIOMATH & ECOL KINETICS, DIV OF BIOL SCI, NAT RES COUN CAN 73- Concurrent Pos: Dir, Ottawa River Proj, Div Biol Sci, Nat Res Coun Can, Ottawa, 73-; mem, Can Comt Man & the Biosphere, MAB Secretariat, Environ Can, Ottawa. Mem: Fel AAAS; Opers Res Soc Am; Soc Math Biol. Res: Simulation of large systems; environmental studies; pollutant transport; distribution and transport of pollutants in agnatic systems. Mailing Add: Div of Biol Sci Nat Res Coun KIA Orgn Ottawa ON Can

MILLER, DONALD SIDNEY, b Maywood, Ill, Dec 9, 08; m 36; c 3. ORTHOPEDICS. Educ: Univ Ill, BS, 30, MS & MD, 32, PhD(orthop), 42; Am Bd Orthop Surg, dipl. Prof Exp: Intern, Cook County Hosp, 32-33, resident, 33-36; PROF ORTHOP & TRAUMATIC SURG, CHICAGO MED SCH, 46-, CHMN ORTHOP SURG, 69- Concurrent Pos: Attend surgeon, Cook County Hosp & Mt Sinai Hosps, Chicago. Mem: AMA; fel Am Col Surgeons; Am Acad Orthop Surg. Res: Vascular orthopedics; Sudecks bone atrophy. Mailing Add: Dept of Orthop & Traumatic Surg Chicago Med Sch Chicago IL 60612

MILLER, DONALD SMITH, b Toronto, Ont, Mar 22, 14; nat US; m 44; c 1. MATHEMATICS. Educ: McMaster Univ, BA, 37; Cornell Univ, AM, 38, PhD, 41. Prof Exp: Instr math, Cornell Univ, 37-41 & Acadia Univ, 41-42; instr math, Yale Univ, 42-46, Nat Res Coun fel, 46-47; asst prof, Univ Rochester, 47-53; MATHEMATICIAN INDUST COMPUT, EASTMAN KODAK CO, 53- Mem: Am Math Soc; Soc Indust & Appl Math. Res: Measure theory; non-absolutely convergent integrals; linear operators. Mailing Add: Bus & Tech Personnel Eastman Kodak Co 12 Birch Crescent Rochester NY 14607

MILLER, DONALD SPENCER, b Ventura, Calif, June 12, 32; m 54; c 3. GEOCHEMISTRY. Educ: Occidental Col, AB, 54; Columbia Univ, AM, 56, PhD, 60. Prof Exp: Asst, Lamont Geol Observ, Columbia Univ, 54-59; res scientist, 59-60; from asst prof to assoc prof, 60-69, PROF GEOCHEM & CHMN DEPT, RENSSELAER POLYTECH INST, 69- Concurrent Pos: NSF sci fac fel & guest prof, Univ Berne, 66-67. Mem: Geol Soc Am; Nat Asn Geol Teachers; Geochem Soc. Res: Geochronology; Colorado Plateau uranium-lead ages; fission track techniques in geology. Mailing Add: Dept of Geol Rensselaer Polytech Inst Troy NY 12181

MILLER, DONALD WRIGHT, b Columbus, Wis, May 28, 27; m 52; c 6. MATHEMATICS. Educ: Univ Wis, BS, 50, MS, 51, PhD(math), 57. Prof Exp: From instr to assoc prof, 55-69, PROF MATH, UNIV NEBR, LINCOLN, 69- Mem: Am Math Soc; Math Asn Am; London Math Soc. Res: Algebra, especially the structure of semigroups. Mailing Add: 930 Oldfather Hall Univ of Nebr Lincoln NE 68508

MILLER, DOROTHEA STARBUCK, b Iowa City, Iowa, Nov 12, 08; m 31. ZOOLOGY. Educ: Univ Iowa, BA, 28, MS, 35, PhD(zool), 38. Prof Exp: Asst zool, Univ Iowa, 33-38; asst prof, Conn Col, 39-43; instr, Univ Wis, 44-45; res asst, Toxicity Lab, Univ Chicago, 45-46, from instr to asst prof biol sci, 46-63, res assoc zool & asst dean students, Biol Sci Div, 54-63; asst prog dir, NSF, Washington, DC, 63-64; PROG ADMINSTR, GENETICS TRAINING & ANAT SCI TRAINING COMTS, NAT INST GEN MED SCI, 64- Concurrent Pos: USPHS res grants, 48-50, 54-63. Mem: AAAS; Am Soc Zool; Radiation Res Soc; Genetics Soc Am. Res: Audiogenic seizures in mice; mammalian genetics; influence of low-level radiation on audiogenic seizures. Mailing Add: Nat Inst of Gen Med Sci Bethesda MD 20014

MILLER, DOROTHY ANNE SMITH, b New York, NY; m 54; c 3. GENETICS, CYTOGENETICS. Educ: Wilson Col, BA, 52; Yale Univ, PhD(biochem), 57. Prof Exp: From res asst to res assoc, 63-73, ASST PROF GENETICS, COL PHYSICIANS & SURGEONS, COLUMBIA UNIV, 73- Res: Chromosome analysis of human, mouse and interspecific somatic cell hybrids. Mailing Add: Dept Human Genetics & Develop Columbia Univ Col of Phys & Surg New York NY 10032

MILLER, DOUGLAS GORDON, b Cortland, NY, July 13, 29; m 57. PHYSICS. Educ: Yale Univ, AB, 51; Univ Rochester, PhD, 58. Prof Exp: From instr to asst prof physics, Harvard Univ, 57-65; ASSOC PROF PHYSICS, HAVERFORD COL, 65- Res: Scattering theory and experimental nuclear physics, particularly strong interactions. Mailing Add: Dept of Physics Haverford Col Haverford PA 19041

MILLER, DOUGLAS L, mathematics, physics, see 12th edition

MILLER, DOUGLASS ROSS, b Monterey Park, Calif, Feb 15, 42; m 64; c 2. ENTOMOLOGY. Educ: Univ Calif, Davis, BS, 64, MS, 65, PhD(entom), 69. Prof Exp: RES ENTOMOLOGIST, SYST ENTOM LAB, AGR RES SERV, USDA, 69- Concurrent Pos: Vis asst prof entom, Univ Md, 73- Mem: Entom Soc Am; Soc Syst Zool. Res: Systematics of scale insects with emphasis on the families Pseudococcidae and Eriococcidae. Mailing Add: Rm 6 Bldg 003 Syst Entom Lab Agr Res Serv USDA Beltsville MD 20705

MILLER, DUANE DOUGLAS, b Great Bend, Kans, July 15, 43; m 62; c 2. MEDICINAL CHEMISTRY. Educ: Univ Kans, BS Pharm, 66; Univ Wash, PhD(med chem), 69. Prof Exp: ASST PROF MED CHEM, COL PHARM, OHIO STATE UNIV, 69- Mem: Am Chem Soc. Res: Mechanism of adrenergic and psychotomimetic drugs. Mailing Add: Dept of Med Chem Col of Pharm Ohio State Univ Columbus OH 43210

MILLER, DUDLEY GRANT, b Santa Barbara, Calif, Sept 17, 23; m 47; c 3. NUCLEAR CHEMISTRY, CORROSION CHEMISTRY. Educ: Univ Calif, Los Angeles, BS, 48. Prof Exp: Res chemist radiochem, Gen Elec Co, 48-56; sr engr, Aerophys Develop Corp, 56-58; radio chemist, 58-67, CONSULT NUCLEAR CHEMIST, KNOLLS ATOMIC POWER LAB, GEN ELEC CO, 67- Mem: AAAS. Res: Radiochemistry; nuclear reactor technology; radiation chemistry; radiation induced corrosion effects; hydrogen diffusion. Mailing Add: Knolls Atomic Power Lab Gen Elec Co Schenectady NY 12301

MILLER, DWANE GENE, b Cheyenne, Wyo, May 15, 34; m 59; c 2. AGRONOMY. Educ: Univ Wyo, BS, 60, MS, 64, PhD(crop sci), 66. Iowa State Univ, PhD, 65. Prof Exp: Instr pub sch, Wyo, 60-62; asst prof biol, Southern Ore Col, 66-67; ASST PROF AGRON, WASH STATE UNIV, 67- Mem: Am Soc Agron. Res: Hybrid wheat, especially artificial induction of male sterility; histological and physiological studies on gametocides and their action; simulated hail studies in wheat and peas; minimum tillage practices in peas; teaching of agronomy. Mailing Add: Dept of Agron & Soils Wash State Univ Pullman WA 99163

MILLER, DWIGHT DANA, b Cedar Rapids, Iowa, Apr 27, 14; m 46; c 2. GENETICS. Educ: Whittier Col, AB, 37; Calif Inst Technol, PhD(genetics), 40. Univ Calif, PhD, 46. Prof Exp: Asst, Univ Rochester, 40-41; res assoc, Carnegie Inst Technol, 42; instr bot, Univ Wash, Seattle, 46; from instr to assoc prof, 46-55, PROF ZOOL, UNIV NEBR, LINCOLN, 55- Concurrent Pos: Prog dir genetic biol, NSF, 59-60. Mem: Soc Study Evolution; Genetics Soc Am; Am Genetic Asn. Res: Cytogenetics and population study of Drosophila. Mailing Add: Sch of Life Sci Univ of Nebr Lincoln NE 68508

MILLER, E WILLARD, b Turkey City, Pa, May 17, 15; m 41. GEOGRAPHY. Educ: Clarion Col, BS, 37; Univ Nebr, AM, 39; Ohio State Univ, PhD(geog, geol), 42. Prof Exp: Instr geog, Ohio State Univ, 42-43; asst prof geol & geog, Western Reserve Univ, 43-44; geogr, Off Strategic Serv, 44-45; assoc prof geog, 45-49, head dept geog, 45-63, asst dean resident instr, 64-72, ASSOC DEAN RESIDENT INSTR, PA STATE UNIV, 72-, PROF GEOG, 49- Concurrent Pos: Assoc ed, Producers Monthly, 46-69; geogr, Qm Gen, US Army, 47-50; geogr ed, Thomas Y Crowell, 56-; assoc ed, J Geog, 60-65; NSF travel grants, Int Geog Union, Stockholm 60, Pam Am Inst, Caracas, 63, Int Geog Union, London, 64 & New Delhi, 68; assoc ed, Pa Geogr, 65-; gen ed, Earth & Mineral Sci Bull, 67-69. Honors & Awards: Whitbeck Award, Nat Coun Geog Educ, 50. Mem: Fel AAAS; Am Soc Prof Geog (secy, 45-48, pres, 48-49); fel Am Geog Soc; fel Nat Coun Geog Educ; Am Inst Mining, Metall & Petrol Eng. Res: Industrial location; manufactural systems; resource utilization. Mailing Add: 101 Mineral Sci University Park PA 16802

MILLER, EARL FRED, II, physiological optics, see 12th edition

MILLER, EARL ROY, b Milwaukee, Wis, Oct 18, 07; m 33; c 2. RADIOLOGY. Educ: Univ Wis, AB, 29, MA, 30, MD, 36. Prof Exp: Asst biophys, Rockefeller Inst, 30-31; asst radiol, Univ Wis, 31-36; intern, Res Hosp, Kansas City, Kans, 36-37; from asst resident to resident, Med Sch, Stanford Univ, 37-39; instr, Sch Med, Yale Univ, 39-40; from instr to prof, 40-74, chmn dept, 43-46, vchmn, 52-62, EMER PROF RADIOL, SCH MED, UNIV CALIF, SAN FRANCISCO, 74- Concurrent Pos: Dir health physics, Manhattan Proj, 43-47; radiologist, Moffitt Hosp, 46- Mem: Radiol Soc NAm; AMA; fel Am Col Radiol. Res: Image quality and reading methods vs x-ray reading error. Mailing Add: 55 Ashbury Terr San Francisco CA 94117

MILLER, EDITH JOAN WILSON, b Eng, Mar 2, 23; US citizen; m 58. CULTURAL GEOGRAPHY. Educ: Cambridge Univ, BA, 44, MA, 47; Univ NC, PhD(geog), 65. Prof Exp: Head dept geog, Girls' High Sch, Batley, Yorks, 44-47; head dept geog, Girls' High Sch, Girls' Pub Day Sch Trust, London, 47-57; lectr geog, Ind Univ, 57-62; from asst prof to assoc prof, 62-74, PROF GEOG, ILL STATE UNIV, 74- Mem: AAAS; Asn Am Geog; Am Geog Soc; Nat Coun Geog Educ; Am Folklore Soc. Res: Place names and folk life especially house types, United Kingdom and United States. Mailing Add: Dept of Geog & Geol Ill State Univ Normal IL 61761

MILLER, EDSEL LEO, b Overbrook, Kans, Oct 29, 21; m 47; c 2. ORGANIC CHEMISTRY. Educ: Kans State Col, PhD(org chem), 52. Res chemist, 51-56, res chemist mkt & develop, 56-65, tech asst to res dir, 65-68, GROUP LEADER PETROL PROD DIV, CONTINENTAL OIL CO, 68- Mem: Am Chem Soc. Res: Petroleum products. Mailing Add: 2401 Hummingbird Lane Ponca City OK 74601

MILLER, EDWARD C, biochemistry, nutrition, see 12th edition

MILLER, EDWARD ERNST, b Omaha, Nebr, Oct 22, 15; m 39; c 9. APPLIED PHYSICS. Educ: Univ Mo, BS, 37; Univ Wis, PhD(physics), 41. Prof Exp: Mem staff, radiation lab, Mass Inst Technol, 41-45; from asst prof to assoc prof, 45-59, PROF PHYSICS & AGR, UNIV WIS-MADISON, 59- Mem: Am Phys Soc; Soil Sci Soc Am. Res: Fluid flow in unsaturated porous media; similitude; hysteresis; optics of plant canopies. Mailing Add: Dept of Physics Sterling Hall Univ of Wis Madison WI 53706

MILLER, EDWARD FREDERICK, b Pittsburgh, Pa, May 10, 38; m 70; c 1. PETROLEUM CHEMISTRY. Educ: Duquesne Univ, BS, 60; Univ Wash, BS, 61; Univ Mass, MS, 67, PhD(org chem), 69. Prof Exp: Sr chemist, 69-74; RES CHEMIST, TEXACO RES CTR, TEXACO INC, 74- Res: Lubricating oil additive synthesis; effects and interactions of lubricant additives in transmission and hydraulic fluids. Mailing Add: Texaco Res Ctr Texaco Inc PO Box 509 Beacon NY 12508

MILLER, EDWARD GEORGE, b Columbiana, Ohio, Mar 29, 34; m 57; c 2. ORGANIC CHEMISTRY. Educ: Manchester Col, AB, 56; Cornell Univ, PhD(org chem), 61. Prof Exp: From instr to assoc prof, 60-74, PROF CHEM & CHMN DEPT, MANCHESTER COL, 74- Concurrent Pos: USPHS fel, Princeton Univ, 65-66. Mem: AAAS; Am Chem Soc. Res: Reaction mechanisms and stereochemistry. Mailing Add: Dept of Chem Manchester Col North Manchester IN 46962

MILLER, EDWARD GODFREY, JR, b Pittsburgh, Pa, Feb 16, 41; m 64; c 1. BIOCHEMISTRY, ONCOLOGY. Educ: Univ Tex, BS, 63, PhD(chem), 69. Prof Exp: Fel, McArdle Labs, Univ Wis, 69-72; asst prof, 72-75, ASSOC PROF BIOCHEM, BAYLOR COL DENT, 75-, ASST PROF, BAYLOR UNIV, 73- Res: Ribosomal RNA synthesis; modification of nuclear proteins; synthesis of poly ADP-ribose; DNA repair. Mailing Add: Dept of Biochem Baylor Col of Dent Dallas TX 75226

MILLER, EDWARD JOSEPH, b Akron, Ohio, Oct 27, 35; m 64; c 3. BIOCHEMISTRY, RADIOBIOLOGY. Educ: Spring Hill Col, BS, 60; Univ Rochester, PhD(radiation biol), 64. Prof Exp: Res assoc biochem, Nat Inst Dent Res, 63-71; PROF BIOCHEM & SR INVESTR, INST DENT RES, UNIV ALA, BIRMINGHAM, 71- Honors & Awards: Award for Basic Res Oral Sci, Int Asn Dent Res, 71. Mem: AAAS; Am Chem Soc; Am Soc Biol Chemists. Res: Biosynthesis of connective tissue protein, collagen and elastin and their cross-linking mechanisms; alterations induced by aging and pathological processes. Mailing Add: Univ of Ala Med Ctr Univ Sta Birmingham AL 35294

MILLER, EDWARD TITUS, b Englewood, NJ, Apr 15, 27; m 50; c 3. GEOPHYSICS. Educ: Mass Inst Technol, BS, 49; Columbia Univ, MA, 52, PhD(geophys), 55. Prof Exp: Res assoc geophys, Lamont Geol Observ, Columbia Univ, 49-56; sr res geophysicist, Humble Oil & Ref Co, 56-65; dir data processing, Alpine Geophys

Assocs, Inc, 65-68; SR EXPLOR GEOPHYSICIST, EXXON CO, USA, 68- Mem: AAAS; Soc Explor Geophys; Seismol Soc Am; Acoust Soc Am; Inst Elec & Electronics Engrs. Eng. Res: Seismology; applied mathematics; oceanography; data processing; underwater acoustics; computer science. Mailing Add: Explor Dept Exxon Co USA PO Box 2180 Houston TX 77001

MILLER, EDWIN LYNN, b North English, Iowa, June 22, 06; m 34; c 2. ZOOLOGY. Educ: Iowa Wesleyan Col, BS, 28; Univ Ill, MS, 30, PhD(parasitol), 33. Prof Exp: From asst to instr zool, Univ Ill, 28-31, fel parasitol, 31-32; from instr to assoc prof zool & parasitol, La State Univ, 32-42; assoc prof biol, Lawrence Col, 42-46; supvr pre-med educ, US Mil Govt, Korea, 46-47, chief adv higher educ, 47-48; prof biol & head dept, 48-65, dean sch sci & math & coord grants, 65-74, DIR PREPROFESSIONAL PROGS, STEPHEN F AUSTIN UNIV, 67- Concurrent Pos: Res zoologist, Inst Paper Chem, Lawrence Col, 45. Mem: AAAS; fel Am Soc Zool; fel Nat Asn Biol Teachers; fel Am Soc Ichthyol & Herpet; fel Am Soc Parasitol. Res: Parasitology; faunistic zoology. Mailing Add: Off of the Dean Sch Sci & Math Stephen F Austin State Univ Nacogdoches TX 75961

MILLER, EILIF VERNER, agriculture, see 12th edition

MILLER, ELBERT ERNEST, b Ephrata, Wash, July 2, 16; m 43; c 3. GEOGRAPHY. Educ: Cent Wash Col Educ, BA, 40; Univ Wash, MA, 47, PhD(geog), 51. Prof Exp: Asst prof geog, Northwestern State Col, La, 46-47; instr, Univ Nebr, 47-48; asst prof, Univ Utah, 48-50; instr, Univ Wash, Seattle, 50-51; from asst prof to assoc prof, Univ Utah, 51-57; assoc prof, Western Wash State Col, 57-68; PROF GEOG & CHMN DEPT, UNIV LETHBRIDGE, 68- Concurrent Pos: Consult, Inter Am Inst Agr Sci, 65-67. Mem: Asn Am Geog. Res: Agricultural geography of Central America. Mailing Add: Dept of Geog Univ of Lethbridge Lethbridge AB Can

MILLER, ELDON STILES, geography, see 12th edition

MILLER, ELEANOR MARIE, b Philadelphia, Pa, Aug 5, 00. MEDICAL RESEARCH. Educ: Univ Penn, BS, 32; Catholic Univ, MS, 38. Prof Exp: Teacher high sch, Pa, 26-35; from instr to prof chem, 35-72, chmn div natural sci, 54-72, EMER PROF CHEM, CHESTNUT HILL COL, 72-; DIR, ST JOSEPH VILLA-MED TECHNOL LAB, 72- Mem: Am Chem Soc; Nat Sci Teachers Asn. Res: Column chromatographic analysis of metal ions; spectrophotometric studies of complex salts. Mailing Add: Dept of Chem Chestnut Hill Col Philadelphia PA 19118

MILLER, ELIZABETH CAVERT, b Minneapolis, Minn, May 2, 20; m 42; c 2. ONCOLOGY. Educ: Univ Minn, BS, 41; Univ Wis, MS, 43, PhD(biochem), 45. Prof Exp: Finney-Howell Found med res fel oncol, 45-47, from instr to assoc prof, 47-69, PROF ONCOL, MED CTR, UNIV WIS-MADISON, 69- Honors & Awards: Co-recipient, Teplitz-Langer Award, Ann Langer Cancer Res Found, Ill, 63, Papanicolaou Res Award, Papanicolaou Cancer Res Inst, 75 & Lewis S Rosenstiel Award, Brandeis Univ, 76; Lucy Wortham James Award, James Ewing Soc, 65; Bertner Award Cancer Res, Univ Tex M D Anderson Hosp & Tumor Inst, 71; Wis Nat Div Award, Am Cancer Soc, 73. Mem: Am Soc Biol Chemists; Am Asn Cancer Res. Res: Experimental chemical carcinogenesis. Mailing Add: McArdle Lab Univ of Wis Med Ctr Madison WI 53706

MILLER, ELIZABETH ESHELMAN, b Waxahachie, Tex, Aug 6, 19; c 44. BIOCHEMISTRY, IMMUNOLOGY. Educ: Univ Colo, BS, 43; Univ Pa, MS, 54, PhD(med microbiol), 55. Prof Exp: Technician plant path, Rockefeller Inst, 43-45; asst chem, Biochem Res Found, 45-46; asst gen biochem, Inst Cancer Res, 46-52; asst med microbiol, Henry Phipps Inst, Univ Pa, 52-55; res assoc, Cancer Res Unit & Dept Biochem, Sch Med, Tufts Univ, 56-57; res assoc protein chem & immunol, Bio-Res Inst, 57-61, res assoc, Univ Pittsburgh, 61-62; RES ASSOC CANCER, UNIV PA, 62-, ASST PROF SURG RES, 73- Res: Chemistry and immunology of cancer. Mailing Add: 1333 Prospect Hill Rd Villanova PA 19085

MILLER, ELMER S, b Elizabethtown, Pa, Apr 26, 31; m 53; c 2. ANTHROPOLOGY. Educ: Eastern Mennonite Col, BA, 54, ThB, 56; Hartford Sem Found, MA, 64; Univ Pittsburgh, PhD(anthrop), 67. Prof Exp: Fieldworker Argentine Toba, Mennonite Bd Missions & Charities, 57-63; asst prof, 66-71, ASSOC PROF ANTHROP, TEMPLE UNIV, 72-, CHMN DEPT, 70- Concurrent Pos: Temple Univ, grant-in-aid, 70 & 72, study leave, 72; fel, Wenner-Gren Found, 72; mem, Sci & Technol Adv Coun, Philadelphia, Pa, 72- Mem: Fel Am Anthrop Asn; Am Ethnol Soc. Res: Cognitive anthropology; Indian tribes of the Gran Chaco, South America; social and culture change. Mailing Add: Dept of Anthrop Temple Univ Philadelphia PA 19122

MILLER, ELWOOD MORTON, b Barnesville, Ohio, July 15, 07; m 31; c 1. ZOOLOGY. Educ: Bethany Col, BS, 29, DSc, 62; Univ Chicago, MS, 30, PhD(zool), 41. Prof Exp: From instr to prof, 30-74, EMER PROF ZOOL, UNIV MIAMI, 74- Concurrent Pos: Asst to dir biol Century Prog Expos, Chicago, 33; chmn dept zool, Univ Miami, 46-53, dean arts & sci, 53-66; mem, State Bd Exam Basic Sci, Fla, 46-67. Mem: Am Soc Zool; Entom Soc Am. Res: Termite biology; social insects. Mailing Add: Dept of Biol Col of Arts & Sci Univ of Miami Coral Gables FL 33124

MILLER, ELWYN RITTER, b Edon, Ohio, Dec 10, 23; m 51; c 5. ANIMAL NUTRITION. Educ: Mich State Univ, BS, 48, PhD(animal nutrit), 56. Prof Exp: Instr educ, 51-52, from asst to assoc prof, 52-66, PROF NUTRIT, MICH STATE UNIV, 66- Concurrent Pos: Sigma Xi res award, 64. Honors & Awards: Award, Am Soc Animal Sci, 65. Mem: AAAS; Am Soc Animal Sci; Soc Exp Biol & Med; Am Inst Nutrit. Res: Hematology, immunology and nutritional requirements of baby pig; normal growth and physiological development of swine fetus. Mailing Add: Dept of Animal Husb Mich State Univ East Lansing MI 48823

MILLER, EMERY B, b Bloomington, Ill, Aug 11, 25; m 48; c 4. ORGANIC CHEMISTRY. Educ: Univ Ill, BS, 47; Rice Univ, MS, 49, PhD(org & phys chem), 51. Prof Exp: Res chemist, Org Div, Monsanto Chem Co, Mo, 51-53; chem res dir, Maumee Chem Co, Ohio, 53-64; pres, Peninsular Chem Res, Fla, 64-65; mgr res & develop, Houston Res Inst, 65-67; res dir, Tenneco Hydrocarbon Chem Div, 67-69; PRES, EMCHEM CORP, 69- Mem: Am Chem Soc; The Chemical Soc. Mailing Add: Emchem Corp Box 876 Pearland TX 77581

MILLER, EMERY PARKER, b Duncansville, Pa, Mar 23, 08; m 35. PHYSICS. Educ: Lafayette Col, AB, 29, MA, 30; Purdue Univ, PhD(physics), 36. Prof Exp: Asst physics, Purdue Univ, 30-35, from instr to asst prof, 35-43;dir res, 43-62, vpres res & develop, Ransburg Corp, 62-73; RETIRED. Concurrent Pos: With Nat Defense Res Comt, 44. Mem: Am Phys Soc. Res: Structure of matter; electrodeposition. Mailing Add: 641 E 80th St Indianapolis IN 46240

MILLER, EMIL C, b Litchfield, Nebr, Nov 21, 07; m 40; c 1. PHYSICS. Educ: St Olaf Col, AB, 31; Univ Iowa, MS, 35, PhD, 51. Prof Exp: Asst math, St Olaf Col, 30-31; teacher high sch, Minn, 31-36, 37-39; mathematician, Corps Engrs, US Army, 36-37; prof physics & chmn dept, 39-42, 47-73, EMER PROF PHYSICS & CHMN DEPT, LUTHER COL, IOWA, 73- Mem: Am Asn Physics Teachers; Sigma Xi. Res: Space physics; astronomy; science education. Mailing Add: 107 Western Ave Decorah IA 52101

MILLER, ERNEST L, b Howard Lake, Minn, Aug 31, 13; m 35; c 4. PROSTHODONTICS. Educ: Univ Detroit, DDS, 40; Ohio State Univ, MS, 52. Prof Exp: PROF DENT, SCH DENT, UNIV ALA, BIRMINGHAM, 66- Mem: Am Dent Asn; Am Prosthodont Soc; Am Col Dent. Res: Removable partial prosthodontics. Mailing Add: Univ of Ala Sch of Dent 1919 Seventh Ave S Birmingham AL 35233

MILLER, EUGENE D, b Wilkes Barre, Pa, June 18, 31; m; c 3. PHYSICAL CHEMISTRY. Educ: King's Col, Pa, BS, 55; Catholic Univ, PhD(phys chem), 61. Prof Exp: Res chemist, Atlantic Ref Co, 60-64; assoc prof phys & anal chem, Cheyney State Col, 64-67; PROF CHEM, LUZERNE COUNTY COMMUNITY COL, 67- Mem: Am Chem Soc. Res: Free radical reactions and heterogeneous catalysis. Mailing Add: Luzerne County Community Col Nanticoke PA 18634

MILLER, EUGENE JAMES, JR, organic chemistry, see 12th edition

MILLER, FLETCHER A, b Iowa City, Iowa, May 16, 22; m 48; c 6. SURGERY. Educ: Univ Iowa, BA, 43, MD, 46; Univ Minn, PhD(surg), 55; Am Bd Surg, dipl, 55; Bd Thoracic Surg, dipl, 57. Prof Exp: Intern, Hartford Hosp, Conn, 46-47; assoc path, Col Physicians & Surgeons, Columbia Univ, 47-48; from asst prof to assoc prof, Med Sch, Univ Minn, Minneapolis, 55-63; prof & chmn dept, Sch Med, Creighton Univ, 63-71; DIR SURG EDUC & CHIEF SURG, C T MILLER HOSP, 71- Concurrent Pos: Am Col Surg Finley fel, 55-58; dir teaching & res surg, Mt Sinai Hosp, Minneapolis, 59-63. Mem: AMA; Am Asn Thoracic Surg; Am Col Surgeons; Soc Univ Surgeons; Am Surg Asn. Res: Cardiac irritability and surgery; pediatric surgery. Mailing Add: C T Miller Hosp 125 College Ave W St Paul MN 55102

MILLER, FLOYD LAVERNE, b Arbela, Mich, Apr 12, 05; m 26; c 3. PHYSICAL CHEMISTRY. Educ: Eastern Mich Univ, BA, 26; Univ Mich, MS, 27, PhD(phys chem), 29. Prof Exp: Instr chem, Eastern Mich Univ, 26-27; res chemist, Esso Res & Eng Co, 30-31, head, Chem Res Group, 31-33, lubricating sect, 33-37, asst dir, Res Div, 37-45, from assoc dir to dir, 45-53, mgr, Contract, Legal & Patent Dept, 53-55, from dep coordr to coordr, Legal, Patent & Info Dept, 55-65; CONSULT, 65- Concurrent Pos: Chem Found res grant, 29-30; chmn cmt chem, Elec Insulation Conf, Nat Res Coun, 38-40; vchmn, Res & Develop Bd, US Dept Defense, 52. Mem: Am Chem Soc; Soc Automotive Eng. Res: Surface and colloid chemistry; petroleum fuels and lubricants, petroleum processes. Mailing Add: Van Beuren Rd Morristown NJ 07960

MILLER, FOIL ALLAN, b Aurora, Ill, Jan 18, 16; m 41; c 2. PHYSICAL CHEMISTRY. Educ: Hamline Univ, BS, 37; Johns Hopkins Univ, PhD(chem), 42. Prof Exp: Nat Res Coun fel, Univ Minn, 42-44; asst prof chem, Univ Ill, 44-48; head spectros div, Mellon Inst, 48-58, sr fel fundamental res, 58-67; PROF CHEM & DIR SPECTROS LAB, UNIV PITTSBURGH, 67- Concurrent Pos: Lectr, Univ Pittsburgh, 52-63, adj prof, 63-67; Guggenheim fel, 57-58; ed, Spectrochimica Acta, 57-63; Reilly lectr, Univ Notre Dame, 67; adj sr fel, Mellon Inst, 67-74, mem comn molecular spectros, Int Union Pure & Appl Chem, 67-75, secy, 69-75. Honors & Awards: Pittsburgh Spectros Award, 64; Pittsburgh Award, Am Chem Soc, 65; Hasler Award, Soc for Appl Spectros, 73. Mem: Am Chem Soc; Optical Soc Am; Coblentz Soc (pres, 59-60); Soc Appl Spectros. Res: Infrared, Raman and electronic spectra. Mailing Add: Dept of Chem Univ of Pittsburgh Pittsburgh PA 15260

MILLER, FOREST LEONARD, JR, b Cincinnati, Ohio, June 18, 36; m 61; c 3. EXPERIMENTAL STATISTICS. Educ: Purdue Univ, BS, 58, MS, 59; NC State Univ, PhD(statist), 75. Prof Exp: CONSULT STATISTICIAN, NUCLEAR DIV, UNION CARBIDE CORP, 59- Mem: Am Statist Asn; Biomet Soc; Sigma Xi; Nat Speleol Soc. Res: Development of conditional probability integral transformations and their application; development of tests for extreme value distributions. Mailing Add: Union Carbide Corp Nuclear Div Bldg 9704-1 PO Box Y Oak Ridge TN 37830

MILLER, FRANCIS JOSEPH, b Montgomery, Ala, Aug 6, 17; m 45; c 2. ANALYTICAL CHEMISTRY. Educ: Univ Ala, AB, 39. Prof Exp: Anal chemist, Oak Ridge Nat Lab, 46-67, tech ed & writer, Isotopes Div, 67-68; ASST PROF RES & ADMIN ASST TO DIR RES, UNIV TENN, MEM RES CTR & HOSP, KNOXVILLE, 68- Mem: Am Chem Soc; Sci Res Soc Am. Res: Instrumental methods of analysis; research administration. Mailing Add: Univ of Tenn Mem Res Ctr & Hosp 1924 Alcoa Hwy Knoxville TN 37920

MILLER, FRANCIS MARION, b Central City, Ky, Dec 28, 25; m 47; c 3. ORGANIC CHEMISTRY. Educ: Western Ky State Col, BS, 46; Northwestern Univ, PhD(chem), 49. Prof Exp: Res assoc org chem, Harvard Univ, 48-49; asst prof sch pharm, Univ Md, 49-51, assoc prof, 51-61, prof & chmn dept, 61-68; PROF CHEM & HEAD DEPT, NORTHERN ILL UNIV, 68- Concurrent Pos: Consult, Chem Corps, US Army, 55-; guest prof, Univ Heidelberg, 58-59. Mem: Am Chem Soc. Res: Alkaloids; heterocyclics; mechanisms of organic reactions; medicinal chemistry. Mailing Add: Dept of Chem Northern Ill Univ De Kalb IL 60115

MILLER, FRANCIS MICHAEL, b Dubuque, Iowa, Aug 30, 34; m 56; c 6. ASTRONOMY. Educ: Loras Col, BS, 55; Univ Colo, Boulder, MBS, 59. Prof Exp: Pub sch teacher sci & math, Iowa, 55-58; teacher sci & math, Marshalltown Community Col, 59-64; ASSOC PROF PHYSICS & DIR, PLANETARIUM, LORAS COL, 64- Mem: Am Inst Physics; Am Asn Physics Teachers. Res: Applied solar energy research. Mailing Add: Dept of Physics Loras Col Dubuque IA 52001

MILLER, FRANK, b Brooklyn, NY, Mar 25, 32; m 54; c 2. ORGANIC CHEMISTRY. Educ: City Col New York, BS, 53; Univ Kans, MS, 57; Univ Del, PhD(chem), 70. Prof Exp: Chemist, Res Div, Dept Chem Eng, NY Univ, 58-59 & Houdry Labs, Air Prod Co, 59-67; sr chemist, Morehead Patterson Labs, Am Mach Foundry, 67-68; ASSOC PROF CHEM, UNIV WIS CTR-MEDFORD, 69- Mem: Am Chem Soc. Res: Synthetic organic chemistry; heterocyclic compounds; structural chemistry. Mailing Add: Dept of Chem Univ of Wis Ctr-Medford Medford WI 54451

MILLER, FRANK CHARLES, b Quincy, Ill, Feb 11, 32. ANTHROPOLOGY. Educ: Carleton Col, BA, 54; Harvard Univ, PhD(social anthrop), 60. Prof Exp: Instr anthrop, Carleton Col, 58-61, asst prof, 61-63; assoc prof, 64-68, asst dean int progs, 67-69, chmn dept anthrop, 69-72, PROF ANTHROP, UNIV MINN, MINNEAPOLIS, 68- Concurrent Pos: NSF res grant, Ojibwa Indians, 60-61; USPHS res grant, Ojibwa Indians, 61-63; NSF grant, Cambridge Univ, 63-64, Nat Endowment Humanities sr fel, New Towns, Mex & US, 72-73; consult, NIMH, 64-66, USPHS, 65-67 & Am Indian Oral Hist Proj, 67-69; mem, Coun Anthrop & Educ, 70-; pop policy fel, Ford & Rockefeller Found, 74-75. Mem: Am Anthrop Asn; Am Ethnol Soc; Soc Appl Anthrop; Latin Am Studies Asn. Res: Modernization; urbanization; educational innovation; methodology; developing societies; contemporary American Indians. Mailing Add: Dept of Anthrop Univ of Minn Minneapolis MN 55455

MILLER, FRANK L, b Kansas City, Mo, Aug 31, 30; m 58; c 3. ATOMIC PHYSICS.

COSMIC RAY PHYSICS. Educ: Univ Okla, BS, 51, MS, 57, PhD(physics), 64. Prof Exp: Asst prof physics, Ft Lewis Col, 63-66; assoc prof, 66-73; PROF PHYSICS, US NAVAL ACAD, 73- Mem: Am Phys Soc; Am Asn Physics Teachers; AAAS; Sigma Xi. Res: Atomic excitation by electron bombardment. Mailing Add: Dept of Physics US Naval Acad Annapolis MD 21402

MILLER, FRANK NELSON, JR, b Alexandria, Va, Apr 15, 19; m 54; c 2. PATHOLOGY. Educ: George Washington Univ, BS, 43, MD, 48. Prof Exp: Asst scientist chem, Allegheny Ballistics Lab, 43-44; asst resident path, Univ Hosp, 49-50, from asst prof to assoc prof, 51-63, assoc dean, 66-73, PROF PATH, GEORGE WASHINGTON UNIV, 63- Concurrent Pos: Teaching fel, George Washington Univ, 50-51; assoc path, DC Gen Hosp, 52-; attend, Mont Alto Vet Admin Hosp, 56-58; consult, US Air Force Hosp, Washington, DC, 55-58, Baker Vet Admin Hosp, Martinsburg, WVa, 60-68 & Vet Admin Hosp, Washington, DC, 68- Mem: AAAS; Am Chem Soc; AMA; Asn Am Med Cols; fel Col Am Path. Res: Neoplasms of the breast and female genital tract. Mailing Add: 2300 I St NW Washington DC 20037

MILLER, FRANKLIN, JR, b St Louis, Mo, Sept 8, 12; m 37; c 1. PHYSICS, SCIENCE EDUCATION. Educ: Swarthmore Col, AB, 33; Univ Chicago, PhD(physics), 39. Prof Exp: Asst physics, Univ Chicago, 35-37; from instr to asst prof, Rutgers Univ, 37-48; assoc prof, 48-58, chmn dept, 55-66, 69-73, PROF PHYSICS, KENYON COL, 59- Honors & Awards: Millikan Lectr Award, Am Asn Physics Teachers, 70. Mem: AAAS; Am Phys Soc; Am Asn Physics Teachers; Fedn Am Sci; Soc for Social Responsibility in Sci (pres, 53-55). Res: Musical acoustics; educational science films. Mailing Add: Dept of Physics Kenyon Col Gambier OH 43022

MILLER, FRANKLIN STUART, b Columbus, Ohio, June 3, 06; m 39; c 1. ECONOMIC GEOLOGY. Educ: Williams Col, AB, 28; Harvard Univ, AM, 31, PhD(geol), 34. Prof Exp: Geologist, Western Mining Corp, Australia, 34-36; instr geol, Univ Ill, 36-37; consult geologist, Can & Calif, 37-38; instr mineral, Univ Calif, 38-39; field supt, Roseville & Roaring River Gold Dredging Cos, 39-40; consult geologist, Univ Toronto, 40-41; asst dir, Mining Div, War Prod Bd, 41-45; consult geologist, Univ Ohio, 46-47; asst mgr, Pac Tin Consol Corp, Malaya, 48-50, vpres, NY, 50-66, pres, 66-72, chmn, 72-75; vpres, 55-66, pres, 66-72, DIR, FELDSPAR CORP, 55-; DIR, PAC TIN CONSOL CORP, 50- Concurrent Pos: Partner, Guggenheim Bros, 73-; dir, Co-Co Del, Inc, 73- Mem: Fel AAAS; fel Geol Soc Am; Soc Econ Geol; Am Inst Mining, Metall & Petrol Eng; Can Inst Mining & Metall. Res: Petrology of the intrusive rocks; geologic structure of ore deposits; graphs for geologic calculations; production and resources of tin. Mailing Add: PO Box 266 Salisbury CT 06068

MILLER, FRANKLYN DAVID, b Hazleton, Pa, Feb 6, 21; m 44; c 1. PHYSICAL CHEMISTRY, ORGANIC CHEMISTRY. Educ: Ursinus Col, BS, 42; Univ Pa, MS, 44, PhD(chem), 50. Prof Exp: Asst, Kind & Knox Gelatin Co, 42-43; asst instr chem, Univ Pa, 42-44, 46-49; MGR ANAL RES, US INDUST CHEM CO DIV, NAT DISTILLERS & CHEM CORP, 49-, RES SCIENTIST, 74- Mem: AAAS; Am Chem Soc; Am Inst Chem; Am Mgt Asn. Res: Instrumental analysis; process development; analytical chemistry; technical planning and evaluation; long range research planning. Mailing Add: 8871 Falmouth Dr Cincinnati OH 45231

MILLER, FRED DOUGLAS, protozoology, physiology, see 12th edition

MILLER, FRED KEY, b Alhambra, Calif, Dec 1, 37; m 61. GEOLOGY. Educ: Univ Calif, Riverside, BA, 61; Stanford Univ, PhD(geol), 66. Prof Exp: GEOLOGIST, US GEOL SURV, 65- Mem: Geol Soc Am. Res: Regional geologic problems in southwest Arizona and southeastern California; regional geologic problems in northeastern Washington, especially Precambrian structure and stratigraphy. Mailing Add: US Geol Surv 345 Middlefield Rd Menlo Park CA 94025

MILLER, FREDERICK ARNOLD, b La Crosse, Wis, Dec 1, 26; m 51; c 4. PHYSICAL CHEMISTRY. Educ: Luther Col, Iowa, BA, 49; Iowa State Univ, PhD(chem), 53. Prof Exp: ASSOC SCIENTIST, DOW CHEM USA, 53- Mem: Am Chem Soc. Res: Synthetic latex; colloid chemistry; organic coatings; emulsion polymerization technology; latex based formulations. Mailing Add: 1113 Glendale St Midland MI 48640

MILLER, FREDERICK POWELL, b Springfield, Ohio, Oct 17, 36; m 65. SOIL SCIENCE, AGRONOMY. Educ: Ohio State Univ, BSc, 58, MSc, 61, PhD(agron, soil classification), 65. Prof Exp: Res asst soil chem, Ohio State Univ, 59-62, soil classification, 62-63; soil scientist, Ohio Dept Natural Resources, 63-64; res asst soil classification, Ohio State Univ, 64-65; soil & water resource specialist, 65-69, from asst prof to assoc prof, 65-74, PROF SOILS, UNIV MD, COLLEGE PARK, 74- Mem: Am Soc Agron; Soil Sci Soc Am; Soil Conserv Soc Am; Int Soc Soil Sci. Res: Physical, chemical and mineralogical characterization of soil Fragipans; electrophoretic separation of soil clay minerals; soil survey interpretation. Mailing Add: Dept of Agron Univ of Md College Park MD 20742

MILLER, FREDRIC N, b Chicago, Ill, May 26, 41; m 62; c 2. PAPER CHEMISTRY. Educ: Ill Inst Technol, BSc, 62, PhD(org chem), 67. Prof Exp: Res chemist, WVa Pulp & Paper Co, 66-69; supvr converting group, 72-74, SR SCIENTIST, AM CAN CO, 69-, MGR PROD DEVELOP, TISSUE & TOWEL, 74- Mem: Am Chem Soc; The Chem Soc; Tech Asn Pulp & Paper Indust. Res: Reprography, especially electrophotography and electrofax; novel imaging systems; specialty papers; consumer paper products; non-wovens; flame retardancy of cellulosics and cellulosic blends; dry forming of paper; embossing; synthetic and natural binders; injection molding and thermoforming. Mailing Add: Am Can Co Pilot Plant 1915 Marathon Ave Neenah WI 54956

MILLER, FREEMAN DEVOLD, b Somerville, Mass, Jan 4, 09; m 33. ASTRONOMY. Educ: Harvard Univ, SB, 30, MA, 32, PhD(astron), 34. Prof Exp: Dir Swasey Observ, Denison Univ, 34-40; assoc prof astron, Univ Mich, Ann Arbor, 46-55; assoc dean Horace H Rackham Sch Grad Studies, 59-66, actg chmn dept astron, 60-61; PROF ASTRON, UNIV MICH, ANN ARBOR, 55- Mem: Am Astron Soc. Res: Comets. Mailing Add: Dept of Astron Physics-Astron Bldg Univ of Mich Ann Arbor MI 48104

MILLER, GABRIEL LORIMER, b New York, NY, Jan 18, 28. PHYSICS. Educ: Univ London, BSc, 49, MSc, 52, PhD(physics), 57. Prof Exp: Physicist, Instrumentation Div, Brookhaven Nat Lab, 57-63; MEM TECH STAFF, BELL TEL LABS, 63- Mem: Am Phys Soc; Inst Elec & Electronics Eng. Res: Nuclear instrumentation; satellite experiments; solid state; electronics. Mailing Add: Bell Tel Labs Murray Hill NJ 07974

MILLER, GAIL LORENZ, biochemistry, see 12th edition

MILLER, GARTH EDWARD, pharmacology, physiology, see 12th edition

MILLER, GARY A, b Newark, NJ, Dec 5, 45. NUTRITION, FOOD SCIENCE. Educ: Rutgers Univ, BS, 68, PhD(food sci), 74. Prof Exp: ASST PROF FOOD & NUTRIT, UNIV NEBR, LINCOLN, 74- Mem: Sigma Xi; Inst Food Technologists; NY Acad Sci; AAAS; Am Soc Testing & Mat. Res: Development and evaluation of biological assays and rapid chemical indices to measure pro- tein nutritive value. Mailing Add: 208 Home Econ Bldg Univ of Nebr Col of Home Econ Lincoln NE 68583

MILLER, GARY GLENN, b Kansas City, Mo, June 11, 41; m 68. TOPOLOGY, THEORETICAL PHYSICS. Educ: Univ Mo-Kansas City, BS, 62, MS, 64, PhD(math), 68. Prof Exp: Instr math, Univ Mo-Kansas City, 64-68; ASST PROF UNIV VICTORIA, BC, 68- Concurrent Pos: Nat Res Coun Can res grant, 70-77; vis fel physics, Princeton Univ, 75-76. Mem: Am Math Soc; Math Asn Am; Royal Astron Soc Can. Res: Topological and lattice theoretic continua; topological structures in relativity theory; foundations of mathematics, Souslin's problem; foundations of physics, philosophy of science. Mailing Add: Dept of Math Univ of Victoria Victoria BC Can

MILLER, GARY L, b Columbia, Pa, July 26, 40; m 62; c 2. ECOLOGY, BOTANY. Educ: Millersville State Col, BS, 62; Univ NC, Chapel Hill, MA, 65, PhD(bot), 68. Prof Exp: Instr bot, Univ NC, Chapel Hill, 68-69; asst prof, 69-72, ASSOC PROF BIOL, EISENHOWER COL, 72- Concurrent Pos: Vis instr, 71-73 & instr ecol, Mt Lake Biol Sta, Univ Va, 71-75. Mem: Am Inst Biol Sci; AAAS. Res: Radiation effects on plant communities; terrestrial and aquatic succession and pollution. Mailing Add: Dept of Biol Eisenhower Col Seneca Falls NY 13148

MILLER, GARY WILLIAM, b Chicago, Ill, Sept 20, 44; m 68. IMMUNOBIOLOGY. Educ: Univ Ill, Champaign-Urbana, BS, 67, MS, 68, PhD(immunol), 72. Prof Exp: Fel immunol, NY Univ Med Ctr, 71-74; INSTR PATH, LA RABIDA INST, UNIV CHICAGO, 74- Res: Involvement of the complement system and of complement receptor-bearing lymphocytes in the immune response and in immune complex diseases. Mailing Add: La Rabida Inst Univ Chicago E 65th St Chicago IL 60649

MILLER, GENE WALKER, b Utah, Dec 21, 25; m 53; c 5. PLANT BIOCHEMISTRY. Educ: Utah State Univ, BS, 50, MS, 54; NC State Univ, PhD(bot), 57. Prof Exp: Plant biochemist, Utah State Univ, 57-69, actg dean col sci, 67; dean, Col Environ Sci, Huxley Col, Western Wash State Col, 69-74; HEAD DEPT BIOL, UTAH STATE UNIV, 74- Concurrent Pos: USPHS spec fel, Univ Münster, 66, NSF fel. Mem: AAAS; Biochem Soc; Am Soc Plant Physiol; Japanese Soc Plant Physiol. Res: Mineral nutrition of plants, especially the role of metals and nutrients, electron transport system, oxidative phosphorylation and plant metabolism; effects of air pollutants on biochemical reactions; environmental study programs. Mailing Add: Dept of Biol Utah State Univ Logan UT 84322

MILLER, GEORGE ALFORD, b Madison, Wis, Dec 24, 25; m 63. PHYSICAL CHEMISTRY. Educ: Univ Wis, BS, 50; Univ Mich, PhD(chem), 55. Prof Exp: Res assoc, Univ Mich, 55-56; instr, Am Cols Istanbul, 56-57; res assoc, Univ Mich, 57-58; from asst prof to assoc prof, 58-70, PROF CHEM, GA INST TECHNOL, 70- Mem: Am Chem Soc. Res: Thermodynamics and kinetic theory of gases. Mailing Add: Sch of Chem Georgia Inst of Technol Atlanta GA 30332

MILLER, GEORGE ALLEN, organic chemistry, see 12th edition

MILLER, GEORGE C, b Portland, Ore, June 8, 25; m 59; c 2. ICHTHYOLOGY. Educ: Univ Wash, BS, 51; Ore State Col, BS, 56, MS, 60. Prof Exp: Aquatic biologist, Ore Fish Comn, 51-52, 56, 57-58; fisheries adv to Liberia, US For Opers Admin, 52-54; aquatic biologist, Wash State Dept Fisheries, 54-55 & Ore State Col, 58-59; fisheries biologist, Biol Lab, Bur Com Fisheries, US Fish & Wildlife Serv, Ga, 60-65, FISHERIES BIOLOGIST, MIAMI LAB, SOUTHEAST FISHERIES CTR, NAT MARINE FISHERIES SERV, FLA, 65- Mem: Am Soc Ichthyol & Herpet; Marine Biol Asn UK. Res: Invertebrate biology; marine ecology; systematic zoology; and zoogeography; biology of marine organisms in relation to technological development. Mailing Add: Southeast Fish Nat Marine Fish 75 Virginia Beach Dr Miami FL 33149

MILLER, GEORGE E, b Banbury, Eng, May 12, 37; m 64; c 2. PHYSICAL CHEMISTRY, RADIOCHEMISTRY. Educ: Oxford Univ, BA, 59, DPhil (chem), 63. Prof Exp: Res assoc chem, Univ Kans, 63-65; res assoc, 65-68, LECTR CHEM & REACTOR SUPVR, UNIV CALIF, IRVINE, 68- Mem: AAAS; Am Chem Soc; Am Nuclear Soc. Res: Reactor utilization in chemistry; activation analysis applications in geochemistry, archeology and medicine; synthesis of labelled molecules and particles. Mailing Add: Dept of Chem Univ of Calif Irvine CA 92664

MILLER, GEORGE EARL, b Buffalo, NY, Dec 5, 28; m 54; c 4. UNDERWATER ACOUSTICS. Educ: St Lawrence Univ, BS, 51, MS, 53. Prof Exp: Assoc res staff mem, Res Div, Raytheon Co, 53-58, sr engr, Submarine Signal Div, 58-60; proj engr, Harris Anti-Submarine Warfare Div, Gen Instrument Corp, 60-61; PHYSICIST, ARTHUR D LITTLE, INC, 61- Mem: Acoust Soc Am. Res: Industrial and medical applications of ultrasonics; ultrasonic and sonar transducer design; anti-submarine warfare systems analysis. Mailing Add: Arthur D Little Inc 15 Acorn Park Cambridge MA 02140

MILLER, GEORGE EDWARD, b St Louis, Mo, Oct 26, 28; m 49; c 2. BIOMEDICAL ENGINEERING. Educ: George Washington Univ, BSME, 65, MEA, 76. Prof Exp: Design engr mech eng, Eng Serv, 59-65, supvr nuclear eng, Cyclotron Br, 65-70, SUPVR NUCLEAR ENG, RADIOBIOL BR, NAVAL RES LAB, 70- Concurrent Pos: Consult, Dept Radiol, George Washington Univ, 74-76. Honors & Awards: Outstanding Performance Award, Naval Res Lab, 69 & 71. Mem: Am Soc Mech Eng; Am Nuclear Soc. Res: Radiation damage in materials and life cells. Mailing Add: Naval Res Lab Code 6649 Washington DC 20375

MILLER, GEORGE EDWARD, b Swarthmore, Pa, June 15, 19; m 49; c 3. MEDICINE. Educ: Univ Penn, AB, 40, MD, 43. Prof Exp: Intern, asst res & res med, Buffalo Gen Hosp, 43-45; mem staff, Niagara Sanatorium, Lockport, NY, 47-48; asst prof med, Univ Buffalo, 52-56, assoc prof, 56-59; prof med, 59-71, DIR RES IN MED EDUC, 59-, PROF MED EDUC, UNIV ILL, COL MED, 71- Concurrent Pos: Res fel, Buffalo Gen Hosp, 48-50, mem res div, 50-59, dir house staff educ, 54-59. Mem: Asn Am Med Cols. Res: Medical education. Mailing Add: Ctr Educ Develop Univ Ill Col Med 835 S Wolcott St Chicago IL 60612

MILLER, GEORGE H, JR, b Iowa City, Iowa, Dec 16, 24; m 45; c 4. UROLOGY. Educ: Princeton Univ, AB, 47; Univ Pa, MD, 48. Prof Exp: Intern, Univ Chicago, 49, resident, 54, from instr to assoc prof urol, 54-58; assoc prof, 58-62, chief div urol, 58-70, PROF UROL, COL MED, UNIV FLA, 62-, ASST DEAN VET ADMIN HOSP RELS, 70- Concurrent Pos: Chief staff & urol sect, Vet Admin Hosp, Gainesville, 70- Mem: Am Urol Asn; AMA; Am Col Surgeons. Res: Urinary tract infection and urinary stone disease. Mailing Add: Dept of Urol Univ of Fla Col of Med Gainesville FL 32610

MILLER, GEORGE HENRY, medicinal chemistry, physical pharmacy, see 12th edition

MILLER, GEORGE TYLER, JR, b Winchester, Va, Oct 18, 31; m 53; c 3. CHEMISTRY, HUMAN ECOLOGY. Educ: Va Mil Inst, BS, 53; Univ Va, PhD(phys chem), 58. Prof Exp: From asst prof to prof chem, Hampden-Sydney Col, 58-66, chmn dept, 61-66; prof chem, 66-70, PROF CHEM & HUMAN ECOL, ST ANDREWS PRESBY COL, 70- Prof Exp: Asst dean, St Andrews Presby Col, 66-70; sci educ & bldg design consult; consult, Educ Facil Labs, Educ Facil Corp, Adv Coun Col Chem, Engelhardt & Englehardt, Stanton Leggett & Assoc & var US & Can educ insts. Honors & Awards: Horsley Res Award, Va Acad Sci, 58. Mem: AAAS; Am Inst Biol Sci; Am Chem Soc; Electrochem Soc. Res: Thermal pollution; social, economic and technological implications of population and pollution; thermodynamics and society; electrochemistry; corrosion of metals. Mailing Add: Dept of Chem St Andrews Presby Col Laurinburg NC 28352

MILLER, GERALD, b Newburgh, NY, Feb 18, 17; m 47; c 2. MEDICINE. Educ: Oberlin Col, AB, 40; Amherst Col, MA, 42; Univ Rochester, MD, 45. Prof Exp: Intern, 45-46, from instr to assoc prof, 50-66, PROF PEDIAT, SCH MED & DENT, UNIV ROCHESTER, 66- Concurrent Pos: Resident res fel & Buswell fac fel, Univ Rochester, 49-52; consult, Genesee & Highland Hosps. Mem: AAAS; Am Pediat Soc; Soc Pediat Res; Am Soc Hemat; Am Fedn Clin Res. Res: Hematology; pediatrics. Mailing Add: Dept of Pediat Rochester Gen Hosp Rochester NY 14621

MILLER, GERALD R, b McClure, Ill, Dec 4, 34; m 58; c 2. WEED SCIENCE. Educ: Univ Ill, BS, 56, MS, 57; Mich State Univ, PhD(weed control, physiol), 63. Prof Exp: Exten agronomist, Purdue Univ, 63-64; EXTEN AGRONOMIST, UNIV MINN, ST PAUL, 64- Mem: Weed Sci Soc Am (secy, 75-76); Crop Sci Soc Am; Am Soc Agron. Res: Weed control; crop-weed competition; herbicide development. Mailing Add: 1428 Arden Pl St Paul MN 55112

MILLER, GERALD R, b Wellsville, NY, Dec 6, 39; m 61; c 2. MATERIALS SCIENCE, PHYSICS. Educ: Cornell Univ, BMetE, 62, MS, 63, PhD, 65. Prof Exp: ASSOC PROF MAT SCI, UNIV UTAH, 65-, ADJ ASSOC PROF PHYSICS, 68- Concurrent Pos: Sci Res Coun fel, Univ Edinburgh, 71-72. Mem: Am Phys Soc. Res: Amorphous semiconducting devices; point defects in solids; transport theory. Mailing Add: Dept of Physics Univ of Utah Salt Lake City UT 84112

MILLER, GERALD RAY, b Milwaukee, Wis, Nov 13, 36; m 58; c 2. PHYSICAL CHEMISTRY. Educ: Univ Wis, BS, 58; Univ Ill, MS, 60, PhD(chem), 62. Prof Exp: NSF fel phys chem, Oxford Univ, 61-63; asst prof, 65-71, ASSOC PROF PHYS CHEM, UNIV MD, COLLEGE PARK, 71- Concurrent Pos: Assoc dir fels, Nat Acad Sci-Nat Res Coun, 74-76. Mem: Am Chem Soc; The Chem Soc; Am Phys Soc; AAAS. Res: Nuclear magnetic resonance and electron spin resonance spectroscopy; molecular and solid state structure. Mailing Add: Dept of Chem Univ of Md College Park MD 20742

MILLER, GERALD WILLIAM, physical chemistry, see 12th edition

MILLER, GERSON HARRY, b Philadelphia, Pa, Mar 2, 24; m 61; c 2. MATHEMATICS, MEDICAL STATISTICS. Educ: Pomona Col, BA, 49; Temple Univ, MEd, 51; Univ Southern Calif, PhD(ed psychol, math), 57. Prof Exp: Instr math, Los Angeles Sch Dist, 53-57; assoc prof math & ed, Western Ill Univ, 57-60; prof, Towson State Col, 60-61; prof math, Parsons Col, 61-65; assoc prof, Wis State Univ, Whitewater, 65-66; prof & dir math educ res, Tenn Technol Univ, 66-68; PROF MATH & SYSTS ANALYST, COMPUT CTR, EDINBORO STATE COL, 68-, ASST DIR, OFF INSTNL RES, 73- Concurrent Pos: Dir, Nat Study Math Requirements for Scientists & Engrs, 65-; dir, Northwestern Pa Study on Smoking & Health. Mem: AAAS; Am Math Soc; Math Asn Am; Am Chem Soc; Am Soc Eng Educ. Res: History of mathematics; comparative mathematics education; retention and deficencies in mathematics; analysis; set theory; curricular improvements. Mailing Add: Comput Ctr Edinboro State Col Edinboro PA 16421

MILLER, GERTRUDE NEVADA, b Dover, Del, July 6, 19. PLANT TAXONOMY. Educ: Univ WVa, BS, 41, MA, 43; Cornell Univ, PhD(plant taxon), 53. Prof Exp: Asst bot & plant taxon, Cornell Univ, 44-47; instr biol, Wells Col, 47-50 & Hollins Col, 53-54; from asst prof to assoc prof, 54-57, PROF BIOL & CHMN DEPT MATH, NATURAL SCI & HEALTH PROFESSIONS, NORTHERN STATE COL, 60- Mem: Nat Asn Biol Teachers; Nat Sci Teachers Asn; Am Nature Study Soc; Am Soc Plant Taxon; Am Inst Biol Sci. Res: Taxonomic revision of the genus Fraximus. Mailing Add: Dept of Biol Northern State Col Aberdeen SD 57401

MILLER, GLEN RUSSEL, b Wellman, Iowa, May 6, 02; c 2. ORGANIC CHEMISTRY. Educ: Hesston Col, BA, 24; Univ Iowa, MS, 25, PhD(org chem), 30. Prof Exp: Instr chem, 25-28 & 30-34, prof, 34-73, EMER PROF CHEM, GOSHEN COL, 73- Mem: AAAS; Am Chem Soc. Res: Halogenation of phenols; nitration by Zincke method; rearrangement; air oxidation of ferrous sulfate; synthesis of anticancer compounds. Mailing Add: 607 College Ave Goshen IN 46526

MILLER, GLENDON RICHARD, b Columbus, Ohio, Oct 28, 38; m 66; c 2. MICROBIOLOGY. Educ: Southern Ill Univ, BA, 60, MA, 62; Univ Mo-Columbia, PhD(microbiol), 66. Prof Exp: Res microbiologist, Colgate Palmolive Res Ctr, 66-68; asst prof bact & physiol, 68-73, ASSOC PROF BIOL, WICHITA STATE UNIV, 73-, GRAD COORDR DEPT, 75- Concurrent Pos: Consult, Koch Eng, 71- Mem: AAAS; Am Soc Microbiol. Res: Radiation repair enzymes of yeasts; antibiotics; antibiotic combination; resistance and cross-resistance to antibiotics; effects of antibiotics on the ribosome. Mailing Add: Dept of Biol Wichita State Univ Wichita KS 67208

MILLER, GLENN HARRY, b Pittsburgh, Pa, Feb 10, 22; m 51; c 3. CHEMISTRY. Educ: Geneva Col, BS, 43; Brown Univ, PhD(chem), 48. Prof Exp: Chemist, Oak Ridge Nat Lab, 44; sr chemist, Butadiene Div, Koppers Co, Inc, 44-45; chemist, Tex Co, 48-49; from instr to assoc prof, 49-63, chmn dept, 60-64, PROF CHEM, UNIV CALIF, SANTA BARBARA, 63- Concurrent Pos: Nat Res Coun Can fel, 56-57; Fulbright-Hays lectr, Univ Malaya, 67-68 & Univ Liberia, 74-75. Mem: Am Chem Soc. Res: Popcorn polymerization; polymers; photolysis. Mailing Add: Dept of Chem Univ of Calif Santa Barbara CA 93106

MILLER, GLENN HOUSTON, b Washington, DC, June 15, 20; m 45; c 3. PHYSICS. Educ: Wake Forest Col, BS, 42; Cornell Univ, PhD(appl physics), 47. Prof Exp: Asst eng physics, Cornell Univ, 42-44; res engr, Stromberg-Carlson Co, NY, 44-46; asst eng physics, Cornell Univ, 46-47; asst prof physics, Iowa State Univ, 47-55; prof elec eng & asst dir ord res lab, Univ Va, 55-59; sr physicist & assoc head physics div, Denver Res Inst, 59-61; staff mem, Sandia Corp, 61-64, supvr atomic physics, 65-71, SUPVR RADIATION SOURCE DIAGNOSTICS, SANDIA LABS, 71- Mem: Fel Am Phys Soc. Res: Ionization yields for atomic particles in gases; production of high velocity molecular beams; condensation of atoms on surfaces; interaction of energetic ions with gases; electron physics. Mailing Add: Div 5224 Organ 5226 Sandia Labs PO Box 5800 Albuquerque NM 87115

MILLER, GLENN JOSEPH, b Crete, Nebr, June 28, 25; m 50; c 3. BIOCHEMISTRY. Educ: Doane Col, BA, 51; Purdue Univ, MS, 53, PhD(biochem), 56. Prof Exp: Asst, Purdue Univ, 51-56; from asst prof to assoc prof, 56-63, PROF BIOCHEM, UNIV WYO, 63-, HEAD DIV, 72- Mem: AAAS; Am Oil Chem Soc; NY Acad Sci. Res: Compostion and characteristics of storage, structural and protective lipids. Mailing Add: Div of Biochem Univ of Wyo Univ Sta Box 3944 Laramie WY 82071

MILLER, GOERGE BOEHM, physical chemistry, see 12th edition

MILLER, GORDON LEE, b Milwaukee, Wis, Sept 27, 38. MATHEMATICS. Educ: Moorhead State Col, BS, 64; NDak State Univ, MS, 65; Univ Northern Colo, EdD(math), 70. Prof Exp: From instr to asst prof, 65-72, ASSOC PROF MATH, UNIV WIS-STEVENS POINT, 72- Mem: Math Asn Am. Res: Analysis. Mailing Add: Dept of Math Univ of Wis-Stevens Point Stevens Point WI 54481

MILLER, GORDON SMITH, b Parkin, Ark, Aug 23, 17; m 40; c 1. INORGANIC CHEMISTRY, PHYSICAL CHEMISTRY. Educ: Southeastern La Col, BA, 40; La State Univ, MA, 47. Prof Exp: Head sci dept high sch, La, 40-42, 43-47; chemist, Gaylord Container Corp, 42-43; assoc prof, Southeastern La Col, 47-51 & Mercer Univ, 51-52; head dept phys sci, 52-60, dean, 60-73, PROF CHEM, TIFT COL, 52-, ASST TO PRES, 73- Mem: Am Chem Soc. Res: Coordination compounds and complex ions; chemical education. Mailing Add: Dept of Chem Tift Col Forsyth GA 31029

MILLER, GROVER CLEVELAND, b Jackson, Ky, Jan 23, 27; m 51; c 4. ZOOLOGY. Educ: Berea Col, AB, 50; Univ Ky, MS, 52; La State Univ, PhD(zool), 57. Prof Exp: Asst parasitologist, La State Univ, 56-57; from instr to assoc prof, 57-69, PROF ZOOL, NC STATE UNIV, 69- Mem: Am Soc Parasitol. Res: Invertebrate zoology; parasitology; trematodes of freshwater fishes; helminth parasites in wild animals. Mailing Add: Dept of Zool NC State Univ Raleigh NC 27607

MILLER, HALSEY WILKINSON, JR, b Camden, NJ, July 1, 30; m 66; c 4. GEOLOGY, PALEONTOLOGY. Educ: Temple Univ, AB, 54; Yale Univ, MS, 54; Univ Kans, PhD(geol), 58. Prof Exp: Geologist, State Geol Surv, Kans, 55-57; asst prof invert paleont, Univ Ariz, 57-63; assoc prof geol, High Point Col, 63-67; asst prof, Ft Hays Kans State Col, 67-69; PROF GEOL, SOUTHERN ILL UNIV, 69- Honors & Awards: Res Award, Southern Ill Univ, 71. Mem: Paleont Soc; Soc Vert Paleont; Am Asn Petrol Geol. Res: Cretaceous stratigraphy and paleontology; Cretaceous-Tertiary boundary line; fossil molluscs; lower vertebrates paleoecology. Mailing Add: Fac of Earth Sci Southern Ill Univ Edwardsville IL 62025

MILLER, HAROLD A, b St Paul, Minn, Mar 14, 21; m 46. PHYSICAL CHEMISTRY. Educ: Univ Minn, BS, 42; NY Univ, PhD(phys chem), 51. Prof Exp: Asst agr biochem, Univ Minn, 41-42; asst chem, NY Univ, 49-50; res chemist, 50-62, ASST DIR RES, MEARL CORP, 62- Mem: Am Phys Soc; Electron Micros Soc Am; NY Acad Sci. Res: Crystal growth; surface chemistry; nacreous pigments; structure of crystals and molecules; infrared spectroscopy. Mailing Add: Mearl Corp 217 N Highland Ave Ossining NY 10562

MILLER, HAROLD CHARLES, b Canton, Ohio, Nov 28, 41; m 63; c 2. MICROBIOLOGY, IMMUNOLOGY. Educ: Hiram Col, AB, 64; Mich State Univ, MS, 66, PhD(microbiol), 68. Prof Exp: Instr microbiol, Mich State Univ, 68; res assoc immunol, Roswell Park Mem Inst, 68-70; res assoc, State Univ NY Buffalo, 70-71, asst prof path & immunol, 71; ASST PROF MICROBIOL, MICH STATE UNIV, 71- Concurrent Pos: Am Cancer Soc fel, Roswell Park Mem Inst, 68-71; Nat Cancer Inst fel, Univ Buffalo, 68-71. Mem: Am Asn Immunologists; Am Soc Microbiol. Res: Cellular immunology including differentiation of primitive marrow stem cells into B and T lymphocytes and development of immunological memory; lymphocyte surface or membrane changes. Mailing Add: Dept of Microbiol & Pub Health Mich State Univ East Lansing MI 48823

MILLER, HAROLD EUGENE, food science, microbiology, see 12th edition

MILLER, HAROLD JAMES, b Rudolph, Ohio, Apr 17, 11; m 39; c 2. PLANT PATHOLOGY. Educ: Ohio State Univ, AB, 33; Pa State Univ, PhD(plant path), 42. Prof Exp: Jr pathologist, USDA, 33-34; asst, Ohio Spray Serv, 34 & Cornell Univ, 34-37; instr plant path, Pa State Univ, 37-43, asst prof, 43-48; sr plant pathologist, Pa Salt Mfg Co, 48-59, SUPVR PESTICIDE SCREENING, PENNWALT CORP, 59-, SUPVR BIOL SECT, 74- Mem: Am Phytopath Soc. Res: Laboratory and field development of new fungicides, nematicides, herbicides, and other pesticides including development of more effective and effecient screening techniques. Mailing Add: Pennwalt Technol Ctr 900 1st Ave King of Prussia PA 19406

MILLER, HAROLD NELSON, organic chemistry, see 12th edition

MILLER, HAROLD VINCENT, geography, see 12th edition

MILLER, HARRY BROWN, b Cumberland, Md, May 25, 13; m 41. ORGANIC CHEMISTRY. Educ: Univ NC, BS, 36, PhD(chem), 46. Prof Exp: With Standard Oil Co (NJ), 36-42; instr chem, Armstrong Jr Col, Ga, 45-47; from asst prof to assoc prof, 47-61, PROF CHEM, WAKE FOREST UNIV, 61- Mem: AAAS; Am Chem Soc. Res: Reactions of organic halogen compounds; fluorine chemistry. Mailing Add: Box 7241 Reynolda Sta Winston-Salem NC 27109

MILLER, HARRY GALEN, b Annapolis, Md, May 17, 37; m 57; c 2. PHYSICS. Educ: Defiance Col, BS, 59; Ohio State Univ, PhD(nuclear physics), 63. Prof Exp: Teacher high sch, 58-59; admissions counsr, 59-60, from asst prof to assoc prof, 63-73, PROF PHYSICS, DEFIANCE COL, 73- Concurrent Pos: Resident assoc, Argonne Nat Lab, Ill, 69-70, consult, 70-72. Res: Nuclear energy levels; gamma-ray spectroscopy. Mailing Add: Dept of Physics Defiance Col Defiance OH 43512

MILLER, HARVEY ALFRED, b Sturgis, Mich, Oct 19, 28; m 52; c 2. BRYOLOGY. Educ: Univ Mich, BS, 50; Univ Hawaii, MS, 52; Stanford Univ, PhD(biol), 57. Prof Exp: Asst bot, Univ Hawaii, 50-53; asst herbarium, Stanford Univ, 53-55; instr bot, Univ Mass, 55-56; from instr to asst prof bot, Univ Miami, Ohio, 56-61, assoc prof & curator herbarium, 61-67; prof biol & bot, chmn div biol sci & chmn prog gen biol, Wash State Univ, 67-69; vis prof bot, Univ Ill, Urbana, 69-70; PROF BOT & CHMN DEPT BIOL SCI, FLA TECHNOL UNIV, 70- Concurrent Pos: Asst, herbarium, Univ Mich, 51; grant-in-aid, Sigma Xi, 54; Guggenheim fel, 58-59; prin investr bryophytes, NSF res grant, Pac Islands; hon mem staff, Hattori Bot Lab, Japan; prin investr bryophytes of Micronesia; NSF US-Japan Coop Sci prog grant; vis lectr, Col Guam, 65; prin investr, Miami Univ-NSF exped, Micronesia & Philippines, 65; ed, Fla Scientist, 73-; res assoc, John Young Museum, Orlando's Museum of Sci & Technol, 75-. Mem: Am Bryol & Lichenological Soc (vpres Am Bryol Soc, 62-63, pres, 64-65); Asn Trop Biol; Am Soc Plant Taxon; Brit Bryol Soc; Int Asn Plant Taxon. Res: Biochemical taxonomy of bryophytes; taxonomy and distribution of Pacific Island bryophytes; world phytogeography; effects of growth regulators on bryophytes in culture. Mailing Add: Dept of Biol Sci Fla Technol Univ Orlando FL 32816

MILLER, HARVEY I, b Brooklyn, NY, May 25, 32; m 54; c 1. PHYSIOLOGY. Educ: City Col New York, BS, 55; Hahnemann Med Col, MS, 58, PhD(physiol), 61. Prof Exp: Biochemist, Lab & Res Div, State Dept Health, NY, 55-57; asst lipid physiol, Hahnemann Med Col, 58-61; res assoc, Lankenau Hosp, Philadelphia, 61-73; assoc prof, Cardiol Div, Hahnemann Med Col, 72-73; ASSOC PROF PHYSIOL, LA STATE UNIV MED CTR, NEW ORLEANS, 73- Concurrent Pos: Assoc prof, Jefferson Med Col, 66-74; estab investr, Am Heart Asn, 69-74. Mem: Am Physiol Soc; Am Heart Asn; Am Col Sports Med. Res: Myocardial metabolism; shock; exercise. Mailing Add: Dept of Physiol La State Univ Med Ctr New Orleans LA 70112

MILLER, HELEN CARTER, b Indianapolis, Ind, Dec 7, 25; m 57; c 3. VERTEBRATE ZOOLOGY, ETHOLOGY. Educ: Butler Univ, AB, 48; Cornell Univ, MA, 52, PhD(vert zool), 62. Prof Exp: Instr biol, Miami Univ, Ohio, 53-56; instr, 63-74, ASST PROF BIOL, OKLA STATE UNIV, 74- Res: Ethological research on fishes and birds. Mailing Add: 2616 Black Oak Dr Stillwater OK 74074

MILLER, HELENA AGNES, b Rudolph, Ohio, Apr 25, 13. BOTANY. Educ: Ohio State Univ, BA & BSc, 35, MS, 38; Radcliffe Col, PhD(biol), 45. Prof Exp: Teacher high sch, Ohio, 35-37; asst bot, Ohio State Univ, 38-39; lectr biol, Hiram Col, 39; teacher, Milton Acad, 39-41; instr bot, Conn Col, 44-45 & Wellesley Col, 45-48; from assoc prof to prof, 48-66, ASST TO DEAN ARTS & SCI, DUQUESNE UNIV, 66-, PROF BIOL, 75- Mem: AAAS; Bot Soc Am; Am Soc Plant Taxon; Soc Study Evolution; Soc Develop Biol. Res: Developmental anatomy of certain angiosperms; study of growth by culturing embryos of certain angiosperms in vitro. Mailing Add: Off of the Dean of Arts & Sci Duquesne Univ Pittsburgh PA 15219

MILLER, HENRY CHARLES, organic chemistry, see 12th edition

MILLER, HENRY CHRISTIAN, chemistry, see 12th edition

MILLER, HENRY KEITH, physiology, ecology, see 12th edition

MILLER, HERBERT CHAUNCEY, b East Orange, NJ, Nov 2, 07; m 34; c 3. PEDIATRICS. Educ: Yale Univ, AB, 30, MD, 34. Prof Exp: Asst pediat, Sch Med, Yale Univ, 34-37, from instr to assoc prof, 37-45; chmn dept, 45-72, PROF PEDIAT, MED CTR, UNIV KANS, 45- Mem: Am Pediat Soc; Soc Pediat Res; fel Am Acad Pediat. Res: Diseases of children; fetology; neonatology. Mailing Add: Dept of Pediat 130C Univ of Kans Med Ctr Kansas City KS 66103

MILLER, HERBERT CRAWFORD, b Lenoir, NC, Mar 30, 44; m 67; c 1. ANALYTICAL CHEMISTRY. Educ: Univ Ala, BS, 66, PhD(anal chem), 73. Prof Exp: RES CHEMIST, SOUTHERN RES INST, 72- Mem: Am Chem Soc. Res: Environmental chemistry; pollution abatement; chemical analyses for trace constituents. Mailing Add: Southern Res Inst 2000 Ninth Ave S Birmingham AL 35205

MILLER, HERBERT KENNETH, b New York, NY, Apr 5, 21; m 43; c 2. BIOCHEMISTRY. Educ: City Col New York, BS, 40; Univ Ill, MS, 47; Columbia Univ, PhD(biochem), 51. Prof Exp: Res assoc infectious diseases, Pub Health Res Inst City New York, Inc, 51-53, assoc, 53-56; biochemist, VA Hosp, Bronx, NY, 56-62; from adj asst prof to assoc prof, 62-72, PROF CHEM, MANHATTAN COL, 72- Concurrent Pos: Res assoc, Div Nucleoprotein Chem, Sloan-Kettering Inst Cancer Res, NY, 62-74. Mem: Am Chem Soc; Am Soc Microbiol. Res: Glutamine peptides; influenza virus nucleic acids; adenovirus infected cells; control of DNA synthesis in chick fibroblasts; purine metabolism. Mailing Add: Dept of Chem Manhattan Col Bronx NY 10471

MILLER, HERMAN LUNDEN, b Detroit, Mich, Apr 23, 24; m 51. NUCLEONICS, OPTICS. Educ: Univ Mich, BS, 48, MS, 51. Prof Exp: Physicist, Res Labs, Ethyl Corp, 48-49; prof physicist, Rocky Flats Plant, Dow Chem Co, 50-55; mem res staff, Proj Matterhorn, Princeton Univ, 55-65; staff engr, Aerospace Systs Div, Bendix Corp, 65-72; SR NUCLEAR ENGR, COMMONWEALTH ASSOCS, INC, 73- Mem: Am Phys Soc; Inst Elec & Electronics Eng; Am Nuclear Soc. Res: Nuclear radiation instrumentation; optical instrumentation. Mailing Add: 1924 Dunmore Rd Ann Arbor MI 48103

MILLER, HERMAN T, b Syracuse, Mo, Feb 28, 31; m 56. BIOCHEMISTRY, IMMUNOCHEMISTRY. Educ: Lincoln Univ, Mo, BS, 53; Kans State Univ, MS, 58; Univ Mo, PhD(biochem), 62. Prof Exp: Asst biochem, Kans State Univ, 55-58; from asst to instr, Univ Mo, 58-62; NIH fels, Univ Calif, Davis, 62-64, asst protein biochem, 64-65, res biochemist, 65-66; PROF CHEM, LINCOLN UNIV, MO, 66- Res: Immunochemistry as a tool in studying structure and function relationships of protein molecules. Mailing Add: Dept of Chem Lincoln Univ Jefferson City MO 65102

MILLER, HILLARD CRAIG, b Northampton, Pa, Dec 15, 32; m 56; c 4. APPLIED PHYSICS. Educ: Lehigh Univ, BA, 54, MS, 55; Pa State Univ, PhD(physics), 60. Prof Exp: Physicist, Gen Elec Res Labs, 60-67, PHYSICIST, GEN ELEC CO, 67- Mem: AAAS; Am Phys Soc; Inst Elec & Electronics Eng; Am Vacuum Soc; Royal Astron Soc Can. Res: Electrical discharges in gases and vacua. Mailing Add: Gen Elec Co PO Box 11508 St Petersburg FL 33733

MILLER, HOWARD ANTHONY, b Rochester, NY, Feb 1, 11; m 36; c 6. PHARMACY, ORGANIC CHEMISTRY. Educ: Univ Buffalo, BS, 35; Univ Md, MS, 37. Prof Exp: Chemist, Cosmex Labs, NY, 32-35; photog chemist, 37-48, staff asst appl photog, 48-50, staff asst to dir res, 50-65, res assoc photomat, 65-71, SR RES ASSOC, EASTMAN KODAK RES LABS, 72- Res: Chemistry of nonsilver photographic materials and processes. Mailing Add: 163 Hoover Rd Rochester NY 14617

MILLER, HOWARD CHARLES, b Syracuse, NY, Feb 6, 17; m 53; c 2. ENTOMOLOGY. Educ: State Univ NY, BS, 41; Cornell Univ, PhD(entom), 51. Prof Exp: Entomologist, Forest Insect Div, USDA, 46; asst, Cornell Univ, 47-50; from asst prof to assoc prof biol sci, 54-69, exten entomologist & pathologist, 50-71, PROF BIOL SCI, STATE UNIV NY COL ENVIRON SCI & FORESTRY, PUB SERV & CONTINUING EDUC, 73-, EXTEN SPECIALIST BIOL SCI & ASSOC PUB SERV OFFICER, TREE PEST SERV, 71- Mem: AAAS; Entom Soc Am; Soc Am Foresters; Lepidop Soc. Res: Insect ecology; forest insect and disease problems; wildlife parasites; medical entomology; biology; urban forestry. Mailing Add: Off Pub Serv & Continuing Educ State Univ NY Col Environ Sci & For Syracuse NY 13210

MILLER, HOWARD NILE, b Niota, Tenn, Apr 19, 13; m 46; c 2. PLANT PATHOLOGY. Educ: Bridgewater Col, BA, 39; Mich State Univ, MS, 41; Univ Calif, PhD(plant path), 48. Prof Exp: Asst plant path, Mich State Univ, 41-42; assoc plant pathologist, Exp Sta, 48-55, PROF PLANT PATH UNIV & PLANT PATHOLOGIST, EXP STA, UNIV FLA, 55- Mem: Am Phytopath Soc; Soc Nematol. Res: Diseases of ornamental plants; diseases caused by soil-borne organisms and their control; control of nematode diseases of plants; fungus parasites of nematodes. Mailing Add: Inst of Food & Agr Sci Univ of Fla Gainesville FL 32603

MILLER, IAN MCKENZIE, b Vancouver, BC, July 31, 24; nat US; m 51; c 4. MICROBIOLOGY. Educ: Univ BC, BSA, 45; Univ NH, MS, 48; Univ Wis, PhD(bact), 52. Prof Exp: Res assoc, Univ Wis, 51-52; MICROBIOLOGIST, MERCK & CO, INC, 52- Mem: Am Soc Microbiol; NY Acad Sci. Res: Tissue culture; lactic acid producing bacteria; antibiotics; vitamin B12. Mailing Add: Merck & Co Inc Rahway NJ 07065

MILLER, INGLIS, JR, b Columbus, Ohio, Mar 17, 43; m 63; c 2. PHYSIOLOGY, ANATOMY. Educ: Ohio State Univ, BS, 65; Fla State Univ, PhD(sensory physiol), 68. Prof Exp: USPHS trainee, Univ Pa, 68-71; ASST PROF ANAT, BOWMAN GRAY SCH MED, WAKE FOREST UNIV, 71- Mem: Soc Neurosci. Res: Gustatory neurophysiology; neuroanatomy of peripheral gustatory system. Mailing Add: Dept of Anat Bowman Gray Sch of Med Winston-Salem NC 27103

MILLER, IRVIN ALEXANDER, b Schellsburg, Pa, Nov 29, 32; m 56; c 3. THEORETICAL PHYSICS. Educ: Drexel Inst Technol, BS, 55; Univ Pa, MS, 59; Temple Univ, PhD, 68. Prof Exp: From instr to asst prof, 55-70, ASSOC PROF PHYSICS & ATMOS SCI & ASST VPRES ACAD AFFAIRS, DREXEL UNIV, 70- Concurrent Pos: Actg dean sci, Drexel Univ, 69-70. Mem: Am Soc Eng Educ; Am Asn Physics Teachers. Res: Electron paramagnetic resonance. Mailing Add: 415 W Allens Lane Philadelphia PA 19119

MILLER, IRWIN b New York City, NY, July 3, 28; m 52; c 3. STATISTICS, RESEARCH ADMINISTRATION. Educ: Alfred Univ, BA, 50; Purdue Univ, MS, 52; Va Polytech Inst, PhD, 56. Prof Exp: Mathematician, Appl Res Lab, US Steel Corp, 56-58; prof statist, Ariz State Univ, 58-65; mem prof staff, 65-73, V PRES, ARTHUR D LITTLE, INC, 73-; PRES, OPINION RES CORP, 75- Concurrent Pos: Adj prof math & dir comput ctr, Wesleyan Univ, 69-71. Mem: AAAS; Am Statist Asn; Inst Math Statist; Biomet Soc. Res: Mathematical statistics; continuous stochastic processes; inference. Mailing Add: Arthur D Little Inc 35 Acorn Park Cambridge MA 02140

MILLER, IVAN KEITH, b Rapid City, SDak, July 28, 21; m 45. POLYMER CHEMISTRY. Educ: SDak State Col, BS, 43; Univ Minn, PhD, 50. Prof Exp: Asst, SDak State Col, 42, instr chem, 43; civilian with Off Rubber Res, Univ Minn, 45-50; res chemist, Rayon Dept, 50-58, res assoc, 58-67, RES FEL, TEXTILE FIBERS DEPT, E I DU PONT DE NEMOURS & CO, INC, 67- Mem: AAAS; Am Chem Soc. Res: Reaction mechanisms concerning polymerization reactions; rubber; synthetic resins; viscose rayon. Mailing Add: Textile Fibers Dept E I du Pont de Nemours & Co Inc Wilmington DE 19803

MILLER, JACK CULBERTSON, b Pomona, Calif, Sept 29, 25; m 54; c 2. THEORETICAL PHYSICS. Educ: Pomona Col, BA, 47; Univ Calif, MA, 49; Oxford Univ, DPhil, 55. Prof Exp: Mathematician, Radiation Lab, Univ Calif, 48-49; from instr to assoc prof, 52-66, PROF PHYSICS, POMONA COL, 66- Concurrent Pos: NSF fac fel, 61-62 & 68-69. Mem: Am Phys Soc; Am Geophys Union. Res: Mathematical physics; physical oceanography. Mailing Add: Dept of Physics Pomona Col Claremont CA 91711

MILLER, JACK MARTIN, b Cornwall, Ont, Feb 20, 40. INORGANIC CHEMISTRY. Educ: McGill Univ, BSc, 61, PhD(chem), 64; Cambridge Univ, PhD(chem), 66. Prof Exp: From asst prof to assoc prof, 66-75, PROF CHEM & CHMN DEPT, BROCK UNIV, 75- Concurrent Pos: Nat Res Coun Can overseas fel, Cambridge Univ, 64-66. Mem: Chem Inst Can; Am Chem Soc; The Chem Soc; Am Soc Mass Spectrometry. Res: Nuclear magnetic resonance and mass spectra of donor-acceptor complexes; mass spectra of organometallic and coordination compounds. Mailing Add: Dept of Chem Brock Univ St Catharines ON Can

MILLER, JACK W, b Knoxville, Tenn, Sept 26, 25; m 52; c 1. PHARMACOLOGY. Educ: San Diego State Col, AB, 49; Univ Calif, MS, 52, PhD(pharmacol), 54. Prof Exp: Asst pharmacol, Univ Calif, 52-53, lectr, 53-54; from instr to assoc prof, Univ Wis, 54-62; assoc prof, 62-67, PROF PHARMACOL, UNIV MINN, MINNEAPOLIS, 67- Concurrent Pos: Mem pharmacol-toxicol comt, Nat Inst Gen Med Sci. Mem: Am Soc Pharmacol & Exp Therapeut; Soc Exp Biol & Med; NY Acad Sci. Res: Pharmacology of morphine-type drugs; uterine drugs; adrenergic receptors; catecholamines; drug receptor theory. Mailing Add: Dept of Pharmacol Univ of Minn Minneapolis MN 55455

MILLER, JAMES ALBERT, JR, b Peitaiho, China, June 21, 07; US citizen; m 35; c 2. ANATOMY. Educ: Col Wooster, AB, 28, DSc, 61; Univ Chicago, PhD(zool), 37. Prof Exp: Instr biol, Assiut Col, Egypt, 28-31; instr, Univ Ohio, 35-37; instr anat, Med Sch, Univ Mich, 37-42; asst prof, Univ Tenn, 42-46; from assoc prof to prof, Sch Dent, Emory Univ, 46-54, prof div basic health sci, Univ, 54-60; prof anat & chmn dept, 60-72, asst dean basic med sci, 72-73, EMER PROF ANAT, SCH MED, TULANE UNIV, 73- Concurrent Pos: NSF sr fel, 57-58; Fulbright fels, Finland, 62 & Ger, 72; Alexander von Humboldt sr res award, 73-74. Honors & Awards: Res Prize, Asn Southeast Biol, 59; Res Citation, Sigma Xi, 59. Mem: Am Soc Zool; Soc Cryobiol; Soc Develop Biol; Am Physiol Soc; Am Asn Anat. Res: Hypothermia in the resuscitation of asphyxiated neonates, effects on heart and brain of hypothermia and metabolic depressants; cooling and resuscitating animals from zero degrees centigrade; hyperbaric oxygen and blockage of differentiation; physiology and histochemistry of development in coelenterates. Mailing Add: Dept of Anat Tulane Univ New Orleans LA 70112

MILLER, JAMES ALEXANDER, b Dormont, Pa, May 27, 15; m 42; c 2. ONCOLOGY. Educ: Univ Pittsburgh, BS, 39; Univ Wis, MS, 41, PhD(biochem), 43. Prof Exp: From instr to assoc prof, 44-52, PROF ONCOL, MED CTR, UNIV WIS-MADISON, 52- Concurrent Pos: Finney-Howell Found Med res fel, Med Ctr, Univ Wis-Madison, 43-44. Honors & Awards: Co-recipient, Teplitz-Langer Award, Ann Langer Cancer Res Found, 63 & Lucy Wortham James Award, James Ewing Soc, 65; G H A Clowes Award, Am Asn Cancer Res, 69; Bertner Award, M D Anderson Hosp & Tumor Inst, 71; Wis Nat Div Award, Am Cancer Soc, 73; co-recipient, Papanicolaou Res Award, Papanicolaou Cancer Res Inst, 75 & Lewis S Rosenstiel Award, Brandeis Univ, 76. Mem: Soc Toxicol; Am Chem Soc; Am Soc Biol Chem; Am Asn Cancer Res; NY Acad Sci. Res: Experimental chemical carcinogenesis. Mailing Add: McArdie Lab Univ of Wis Med Ctr Madison WI 53706

MILLER, JAMES E, b McCune, Kans, Sept 19, 16; m 40 & 60; c 4. METEOROLOGY. Educ: Cent Col, Mo, AB, 37; NY Univ, MS, 41. Prof Exp: Weather forecaster, US Weather Bur, Mont, 40; from instr to assoc prof, 40-51, PROF METEOROL, NY UNIV, 51-, CHMN DEPT METEOROL & OCEANOG, 61- Concurrent Pos: Actg chmn dept meteorol & oceanog, NY Univ, 59-61; trustee, Univ Corp Atmos Res, 65-67, mem rep, 67- Honors & Awards: Meisinger Award, Am Meteorol Soc, 48. Mem: Am Meteorol Soc (secy, 61-65); Am Geophys Union; Royal Meteorol Soc. Res: Synoptic and dynamic meteorology; climatology; weather forecasting. Mailing Add: Dept of Meteorol & Oceanog NY Univ New York NY 10453

MILLER, JAMES EDWARD, b Hanover, Pa, Apr 28, 42; c 2. BIOCHEMISTRY. Educ: Shippensburg State Col, BS, 63; Univ NDak, MS, 65, PhD(biochem), 68. Prof Exp: USPHS grant, Sch Med, Temple Univ, 68-70; RES INVESTR BIOCHEM, G D SEARLE & CO, 70- Mem: AAAS; Am Chem Soc. Res: Enzyme chemistry; protein purification; assay development; lipid metabolism; nutrition; hormonal control of metabolism. Mailing Add: Dept of Biol Res Searle Labs Chicago IL 60680

MILLER, JAMES EUGENE, b Loudonville, Ohio, Aug 12, 38; m 63; c 2. BIOCHEMISTRY, MICROBIOLOGY. Educ: Denison Univ, BA, 60; Harvard Univ, MA, 63; Amherst Col, PhD(biochem), 65. Prof Exp: Fel bact, Univ Calif, Los Angeles, 65-66; res assoc, NASA, 66-69; res biochemist, Corp Res Lab, Allied Chem Corp, 69-71; ASST PROF BIOL, DEL VALLEY COL, 71- Res: Bacterial physiology. Mailing Add: Dept of Biol Del Valley Col Doylestown PA 18901

MILLER, JAMES FRANKLIN, agronomy, see 12th edition

MILLER, JAMES FRANKLIN, b Lancaster, Pa, July 18, 12; m 38; c 1. ANALYTICAL CHEMISTRY. Educ: Franklin & Marshall Col, BS, 35; Pa State Univ, MS, 37, PhD(anal chem), 39. Prof Exp: Asst, Pa State Univ, 35-39, instr chem, Altoona Undergrad Ctr, 39-42; asst prof chem, The Citadel, 42; res fel, Res Found, Purdue Univ, 43-44; fel rubber res, Mellon Inst, 44-46, sr fel insecticides, 46-48, coal tar constituents, 48-49, arsenic, 49-51, head anal chem sect, 51-59; lab dir gen chem cent res & eng, Div, Continental Can Co, 59-64, mgr appl res, Corp Res & Develop Dept, 64-68; SECY-TREAS, ALPHA CHI SIGMA FRATERNITY, 68- Concurrent Pos: Civilian with AEC; Off Sci Res & Develop; US Rubber Reserve Corp. Mem: Am Chem Soc; Am Inst Chem Eng. Res: Inorganic non-ferrous analysis; analysis of organic halogen compounds and alcohol; butadiene coverter products; physical properties and behavior of emulsions; air elutriation of particular matter; behavior of fractionated particulate matter on falling utilization of arsenic; organic micro-analysis; polarography; spectrophotometry. Mailing Add: Alpha Chi Sigma Fraternity 5503 E Washington St Indianapolis IN 46219

MILLER, JAMES FREDERICK, b Lancaster, Ohio, June 14, 19; m 56; c 1. FUEL SCIENCE. Educ: Ohio State Univ, BS, 51. Prof Exp: Res chemist, Titanium Div, Nat Lead Co, 51-53; res chemist, 53-58, res proj leader, 58-60, asst chief phys chem div, 60-69, assoc chief phys & inorg chem div, 69-72, sr res chemist, 72-74, SR RES CHEMIST, PROCESS DEVELOP SECT, COLUMBUS LABS, BATTELLE, 74- Mem: Am Inst Mining & Metall Engrs; Am Chem Soc; Metall Soc. Res: Solid state materials synthesis and crystal growth; synthesis and process reactions and studies of phase equilibria in aqueous, molten salt, and other inorganic systems; coal processing and conversion chemistry. Mailing Add: Columbus Labs Battelle 505 King Ave Columbus OH 43201

MILLER, JAMES FREDERICK, b Davenport, Iowa, Feb 18, 43; m 67. MICROPALEONTOLOGY, INVERTEBRATE PALEONTOLOGY. Educ: Augustana Col, Ill, AB, 65; Univ Wis-Madison, MA, 68, PhD(geol), 71. Prof Exp: Asst prof geol, Univ Utah, 70-74; ASST PROF GEOL, SOUTHWEST MO STATE UNIV, 74- Mem: AAAS; Geol Soc Am; Paleont Soc; Pander Soc. Res: Taxonomy, evolution, and biostratigraphy of Cambrian and Lower Ordovician conodonts; stratigraphic position of Cambrian-Ordovician boundary; Paleozoic crinoids. Mailing Add: Dept of Geol & Geog Southwest Mo State Univ Springfield MO 65802

MILLER, JAMES GEGAN, b St Louis Mo, Nov 11, 42; m 66. ULTRASOUND, MEDICAL BIOPHYSICS. Educ: St Louis Univ, AB, 64; Washington Univ, MA, 66, PhD(physics), 69. Prof Exp: Res assoc physics, 69-70, asst prof, 70-72, ASSOC PROF PHYSICS, WASHINGTON UNIV, 72-, ASSOC DIR BIOMED PHYSICS, LAB FOR ULTRASONICS, 74- Mem: Inst Elec & Electronic Engrs; Am Phys Soc. Res: Ultrasonics, biomedical physics, ultrasonic tissue characterization; acoustic magnetic resonance; ultrasonic resonators and transducers; phonon-charge carrier interactions. Mailing Add: Dept of Physics Washington Univ St Louis MO 63130

MILLER, JAMES KINCHELOE, b Elkton, Md, June 16, 32; m 60; c 2. ANIMAL NUTRITION, PHYSIOLOGY. Educ: Berry Col, BS, 53; Univ Ga, MS, 59, PhD(animal nutrit), 62. Prof Exp: Tech asst dairy nutrit, Univ Ga, 57-58, asst, 58-60; asst prof, 61-67, ASSOC PROF DAIRY PHYSIOL, AGR RES LAB, UNIV TENN, 67-, ASSOC PROF, COMPARATIVE ANIMAL RES LAB, UNIV TENN-ENERGY RES & DEVELOP ADMIN, 73- Honors & Awards: Gustav Bohstedt Mineral & Trace Mineral Award, Am Soc of Animal Sci, 74. Mem: Am Dairy Sci Asn; Am Inst Nutrit; Am Soc Animal Sci. Res: Nutrition and physiology of the dairy cow; mineral metabolism; fission product metabolism; effects of internal and external irradiation on cattle. Mailing Add: UT-ERDA Comparative Animal Res Lab 1299 Bethel Valley Rd Oak Ridge TN 37830

MILLER, JAMES L, b Chicago, Ill, May 10, 35; m 58; c 3. ORGANIC CHEMISTRY. Educ: Eastern Ill Univ, BS, 57; Univ Iowa, MS, 62, PhD(chem), 63. Prof Exp: From asst prof to assoc prof, 63-74, PROF CHEM, EAST TENN STATE UNIV, 74- Mem: Am Chem Soc. Res: Mechanisms concerning the bromination of stilbene and tolan; cycloaddition reactions which involve benzyne intermediates. Mailing Add: Dept of Chem East Tenn State Univ Johnson City TN 37601

MILLER, JAMES MILTON, b Austin, Tex, Sept 22, 25. COMPUTER SCIENCE, STATISTICS. Educ: Univ Tex, BS, 52, MA, 54, PhD(chem), 57. Prof Exp: Res scientist chem, Univ Tex, 56-57; engr, Esso Res & Eng Co, 58-60; RES STAFF MEM, THOMAS J WATSON RES CTR, IBM CORP, NEW YORK, 60- Mem: Am Chem Soc; Am Phys Soc; Am Math Soc; Am Statist Asn; NY Acad Sci. Res: Probability and statistics; design of user-oriented, interactive data processing systems; medical research data analysis. Mailing Add: 10 Franklin Ave Apt 4J White Plains NY 10601

MILLER, JAMES MONROE, b Lancaster, Pa, Aug 7, 33; m 55; c 2. ANALYTICAL CHEMISTRY. Educ: Elizabethtown Col, BS, 55; Purdue Univ, MS, 58, PhD(anal chem), 60. Prof Exp: From asst prof to assoc prof, 59-69, PROF ANAL CHEM, DREW UNIV, 69-, CHMN DEPT CHEM, 71- Concurrent Pos: Vis lectr, Univ Ill, Urbana, 64-65; indust consult, 63-64; mem gas chromatography discussion group, Brit Inst Petrol, 65; dir Col Sci Improv Prog, NSF, 67-70; vis prof, Univ Amsterdam, 71. Mem: AAAS; Am Chem Soc. Res: Gas chromatography; applications in teaching; studies of thermal conductivity detector response; determination of non-ionic detergents by column liquid chromatography. Mailing Add: Dept of Chem Drew Univ Madison NJ 07940

MILLER, JAMES NATHANIEL, b Detroit, Mich, Mar 16, 26; m 51; c 2. INFECTIOUS DISEASES. Educ: Univ Calif, Los Angeles, BA, 50, MA, 51, PhD(infectious dis), 56. Prof Exp: Jr res microbiologist, 56-58, asst prof infectious dis, 58-64, from asst prof to assoc prof microbiol & immunol, 66-72, PROF MICROBIOL & IMMUNOL, SCH MED, UNIV CALIF, LOS ANGELES, 72- Mem: Am Soc Microbiol; Am Asn Immunologists; Am Veneral Dis Asn. Res: Venereal diseases; serology and immunology of syphilis; serology of cancer; murine leprosy; tuberculosis; brucellosis. Mailing Add: Dept of Microbiol & Immunol Univ of Calif Sch of Med Los Angeles CA 90024

MILLER, JAMES Q, b Lakewood, Ohio, July 6, 26; m 50; c 4. NEUROLOGY, CYTOGENETICS. Educ: Haverford Col, BA, 49; Columbia Univ, MD, 53. Prof Exp: Nat Inst Neurol Dis & Blindness spec fel neuropath, Harvard Univ, 60-62; asst prof, 62-67, asst dean sch med, 62-70, ASSOC PROF NEUROL, SCH MED, UNIV VA, 67-, NEUROLOGIST, UNIV HOSP, 62- Mem: Am Acad Neurol; Am Epilepsy Soc; AMA. Res: Chromosome disorders and anomalies of the central nervous system. Mailing Add: Dept of Neurol Univ of Va Hosp Charlottesville VA 22901

MILLER, JAMES REGINALD, b Mimico, Ont, Nov 6, 28; m 54; c 5. GENETICS. Educ: Univ Toronto, BA, 51, MA, 53; McGill Univ, PhD, 59. Prof Exp: Asst develop physiol, Jackson Mem Lab, Maine, 54-56; res assoc genetics, Dept Neurol Res, 58-60, from asst prof to assoc prof pediat, 60-67, head div med genetics, 67-73, PROF PAEDIAT, UNIV BC, 67-, PROF MED GENETICS & HEAD DEPT, 73- Mem: Genetics Soc Am; Teratology Soc; Can Col Med Geneticists; Am Soc Human Genetics; Genetics Soc Can. Res: Developmental and population genetics of human beings and other mammals. Mailing Add: Dept of Med Genetics Univ of BC Vancouver BC Can

MILLER, JAMES RICHARD, b St Louis, Mo, June 11, 22; m 45; c 2. PHYSICAL CHEMISTRY, ORGANIC CHEMISTRY. Educ: Mo Sch Mines, BS, 44; Wash Univ, PhD(chem), 51. Prof Exp: Asst chem, Mo Sch Mines, 43; chemist, Ralston-Purina Co, 46-47; asst chem, Washington Univ, 47-50; res chemist, Shell Oil Co, 51-58, sr res chemist, Wood River Res Lab, 58-69; sr res scientist, Shell Res Ltd, Thorton Res Ctr, Chester, Eng, 69-70; MEM STAFF, SHELL DEVELOP CO, WESTHOLLOW RES CTR, HOUSTON, 75- Mem: Am Chem Soc; AAAS; fel Am Inst Chem. Res: Free radical chemistry; electron spin resonance; application of radiochemical techniques to problems in manufacture and use of petroleum. Mailing Add: Westhollow Res Ctr Shell Develop Co Box 1380 Houston TX 77001

MILLER, JAMES ROBERT, b Milford, Nebr, July 2, 22; m 45; c 5. ORGANIC CHEMISTRY. Educ: Iowa State Univ, BS, 43; Syracuse Univ, PhD(chem), 50. Prof Exp: Res asst & jr res chemist, Parke, Davis & Co, 43-47; from asst prof to assoc prof, 50-53, dept head, 52-65, PROF CHEM, HARTWICK COL, 54- Concurrent Pos: Consult lab, Fox Hosp, 58-61. Mem: AAAS; Am Chem Soc; Wilson Ornith Soc; Am Ornith Union. Res: Synthetic organic medicinals and heterocyclic compounds. Mailing Add: Dept of Chem Hartwick Col Oneonta NY 13820

MILLER, JAMES ROBERT, b Holcomb, Mo, Jan 7, 41; m 59; c 3. SOLID STATE PHYSICS. Educ: Mo Sch Mines, BS, 62; Tex Christian Univ, MS, 64, PhD(physics), 66. Prof Exp: Assoc prof, 66-70, PROF PHYSICS, EAST TENN STATE UNIV, 70- Res: Electron spin resonance and nuclear magnetic resonance. Mailing Add: Dept of Physics East Tenn State Univ Johnson City TN 37601

MILLER, JAMES ROLAND, b Millington, Md, May 19, 29; m 54; c 1. SOIL CHEMISTRY, AGRONOMY. Educ: Univ Md, BS, 51, MS, 53, PhD(soil chem), 56. Hon Degrees: State Farmer Degree, FFA, 60. Prof Exp: Asst, Univ Md, 51-56; soil scientist chem, Soils & Plant Relationship Sect, Agr Res Serv, USDA, 56-58; from asst prof to assoc prof soils, 58-63, PROF & HEAD DEPT AGRON, UNIV MD, COLLEGE PARK, 63- Concurrent Pos: Consult, NASA, 65- Mem: Soil Conserv Soc Am; Soil Sci Soc Am; fel Am Soc Agron. Res: Soil test methods for determining available nutrients in soils; fission products; reactions in soils and uptake by plants. Mailing Add: Dept of Agron Univ of Md College Park MD 20742

MILLER, JAMES ROSCOE, b Murray, Utah, Oct 26, 05; m 28; c 3. MEDICINE. Educ: Univ Utah, BA, 25; Northwestern Univ, MD, 30, MS, 31. Hon Degrees: Twelve from US cols & univs. Prof Exp: Instr med, Sch Med, 34-37, from assoc to prof, 37-74, from asst dean to dean med sch, 33-49, pres, Univ, 49-70, chancellor, 70-74, EMER PROF MED, MED SCH & EMER CHANCELLOR, NORTHWESTERN UNIV, EVANSTON, 74- Concurrent Pos: Montgomery Ward fel, Med Sch, Northwestern Univ, 31-33; trustee, Northwestern Mem & Evanston Hosps; dir, Prentice Women's Hosp & Maternity Ctr; consult. Mem: AMA; Am Heart Asn; Asn Am Med Cols (pres, 48-49); fel Am Col Physicians. Res: Cardiology; use of tracer methods in study of arteriosclerosis. Mailing Add: Off of the Chancellor Northwestern Univ Evanston IL 60201

MILLER, JANE ALSOBROOK, b New Orleans, La, Feb 21, 28; c 2. CHEMISTRY. Educ: Agnes Scott Col, AB, 48; Tulane Univ, MS, 50, PhD(hist of chem), 60. Prof Exp: Instr chem, Tulane Univ, 50-52; res asst pharmacol, Washington Univ, 53-54, orthop surg, 63-65, res instr, 65; instr, 65-67, ASST PROF CHEM, UNIV MO-ST LOUIS, 67- Mem: Am Chem Soc; Hist Sci Soc; AAAS. Res: History of chemistry; chemical education. Mailing Add: Dept of Chem Univ of Mo-St Louis St Louis MO 63121

MILLER, JANICE MARGARET, b McPherson, Kans, Nov 11, 38; m 62; c 2. VETERINARY PATHOLOGY. Educ: Kans State Univ, BS, 60, DVM, 62, MS, 63; Univ Wis-Madison, PhD(vet sci), 69. Prof Exp: Res assoc animal nutrit, Mass Inst Technol, 64-65; Leukemia Soc Am spec fel, Univ Wis-Madison, 70-71; spec fel, 71-72, RES VET PATH, NAT ANIMAL DIS LAB, 72- Mem: Am Vet Med Asn; Am Col Vet Path. Res: Viral oncogenesis; leukemia of cattle. Mailing Add: Dept of Path Nat Animal Dis Lab Ames IA 50010

MILLER, JARRELL E, b San Antonio, Tex, Nov 14, 13; m 39; c 4. MEDICINE. Educ: Baylor Univ, MD, 38; Am Bd Radiol, dipl, 42. Prof Exp: Intern, Robert B Green Mem Hosp, San Antonio, Tex, 38-39; resident radiol, Cleveland City Hosp, Ohio, 39-42; assoc prof, 47-56, clin prof, 56-66, PROF RADIOL, UNIV TEX HEALTH SCI CTR, DALLAS, 66- Concurrent Pos: Radiologist, Parkland Hosp, Dallas, 46-49, Children's Med Ctr, 47-65 & St Paul Hosp, Dallas, 67-; consult, Vet Hosps, Lisbon & McKinney, Tex, 46-57; dir dept radiol, Med Ctr, Baylor Univ, 49-66; lectr, Univ Tex Med Br, Galveston, 59-; deleg, AMA, 71- Honors & Awards: Distinguished Serv Award, Tex Med Asn, 71. Mem: Fel Am Col Radiol (pres, 67-); Radiol Soc NAm; Am Roentgen Ray Soc (vpres, 63); Nat Tuberc & Respiratory Dis Asn; Am Cancer Soc. Res: Hypertrophic phyloric stenosis; angiocardiography; anatomy of the heart and great vessels; childhood malignancies. Mailing Add: 6115D Averill Way Dallas TX 75225

MILLER, JERRY K, b Valley City, NDak, Sept 4, 34. ANALYTICAL CHEMISTRY. Educ: Univ Minn, BChem, 57, PhD(anal chem), 66; Univ Mo, MA, 63. Prof Exp: RES CHEMIST, AM CYANAMID CO, 66- Mem: Am Chem Soc. Res: Light scattering; polymer characterization. Mailing Add: Am Cyanamid Co 1937 W Main St Stamford CT 06904

MILLER, JERRY ROLAND, b Tripp, SDak, Feb 5, 22; m 45; c 4. MEDICINE. Educ: Univ SDak, BA, 45; Temple Univ, MD, 47; Univ Iowa, MS, 58; Am Bd Anesthesiol, dipl, 61. Prof Exp: Lab instr physiol, Univ SDak, 43-45; from instr to assoc prof, 58-68, PROF ANESTHESIOL, MED CTR, IND UNIV, INDIANAPOLIS, 68- Concurrent Pos: Attend, Vet Admin Hosp, Indianapolis. Mem: Fel Am Col Anesthesiol; Acad Anesthesiol; Am Soc Anesthesiol; AMA. Res: Pharmacology and physiology pertaining to anesthesiology. Mailing Add: Dept of Anesthesiol Ind Univ Med Ctr Indianapolis IN 46202

MILLER, JESSE WILLIAM, JR, b Mitchell AFB, NY, Aug 19, 41; m 64; c 2. PHYSICAL GEOGRAPHY, GEOMORPHOLOGY. Educ: Pa State Univ, BS, 63; Univ Wis-Madison, MS, 64; Syracuse Univ, PhD(geog), 70. Prof Exp: Asst prof Geog, State Univ NY Col Oswego, 69-72; CONSULT FAC & MILITARY GEOGR, US ARMY COMMAND & GEN STAFF COL, 73- Concurrent Pos: NSF Inst Stipend, Univ Denver, 71; vis lectr geog, Univ Calif, Los Angeles, 72. Mem: AAAS; Asn Am Geog; Am Geog Soc; Am Geol Soc; World Future Soc. Res: Glacial geomorphology, especially drumlins; historical military geography; military terrain analysis at the tactical level. Mailing Add: US Army Command & Gen Staff Col Leavenworth KS 66027

MILLER, JOEL STEVEN, b Detroit, Mich, Oct 14, 44; m 70; c 2. INORGANIC CHEMISTRY. Educ: Wayne State Univ, BS, 67; Univ Calif, Los Angeles, PhD(inorg chem), 71. Prof Exp: Res assoc inorg chem, Stanford Univ, 71-72; assoc scientist, 72-73, SCIENTIST, WEBSTER RES CTR, XEROX CORP, 73- Mem: Am Chem Soc; fel The Chem Soc. Res: Synthetic and physical inorganic chemistry; anisotropic inorganic and organic complexes; structure-function relationships. Mailing Add: 919 S Grosvenor Rd Rochester NY 14618

MILLER, JOHN ALLEN, b Ashland, Ohio, Oct 18, 05; m 51; c 3. ZOOLOGY. Educ: Ohio State Univ, MSc, 27, PhD(zool), 32. Hon Degrees: ScD, Ashland Col, 63. Prof Exp: Teacher high sch, Ohio, 27-28; asst, 28-29, instr, 29-33, from asst prof to prof, 33-63, EMER PROF ZOOL, OHIO STATE UNIV, 63- Concurrent Pos: Trustee, Ashland Col, 48-68, dean acad affairs, 68-72; prof, Univ Miami, 63-68. Mem: AAAS; fel Am Soc Zoologists; Am Inst Biol Sci. Res: Electronic stimulation using balanced square waves; nervous system and its relation to behavior: developmental anatomy. Mailing Add: 5100 N Bayview Dr Ft Lauderdale FL 33308

MILLER, JOHN CLARK, b Pittsburgh, Pa, Mar 31, 28; m 52; c 3. PHYSICAL CHEMISTRY. Educ: Lehigh Univ, BS, 52, MS, 54, PhD(chem), 57. Prof Exp: PROJ SCIENTIST RES & DEVELOP, PLASTICS DIV, UNION CARBIDE CORP, 56- Mem: Am Chem Soc; Soc Rheol (secy-treas); NY Acad Sci. Res: Rheology of dispersions and polymers. Mailing Add: Res & Develop Dept Plastics Div Union Carbide Corp Bound Brook NJ 08805

MILLER, JOHN DAVID, b Todd, NC, Aug 9, 23; m 46; c 3. PLANT BREEDING, PLANT GENETICS. Educ: NC State Col, BS, 48, MS, 50; Univ Minn, PhD(plant breeding), 53. Prof Exp: Res fel, seedstocks prod, Univ Minn, 53; asst prof cereal breeding, Kans State Univ, 53-57; assoc prof, 57-61, ADJ PROF AGRON AGR EXP STA, VA POLYTECH INST & STATE UNIV, 61-; RES LEADER, AGR RES SERV, USDA, 72-, SR AGRONOMIST, 75- Concurrent Pos: Res agronomist, Agr Res Serv, USDA, 57-72. Mem: Am Soc Agron; Am Genetics Asn. Res: Forage breeding; statistical techniques in crops research; improved breeding methods for forages; improved small plot machinery; breeding for tolerance to acid soils. Mailing Add: Dept of Agron Va Polytech Inst & St Univ Blacksburg VA 24061

MILLER, JOHN EDWARD, b McKeesport, Pa, Dec 9, 21; m 44; c 1. MAGNETIC RESONANCE. Educ: Randolph-Macon Col, BS, 48; Univ Va, MA, 50, PhD(physics), 52. Prof Exp: From asst prof to prof physics, Clemson Col, 52-67; VPRES ACAD AFFAIRS, FLA STATE INST TECHNOL, 67- Mem: Am Phys Soc. Res: High speed centrifuges; thin film studies by electron microscope. Mailing Add: Fla Inst of Technol Melbourne FL 32901

MILLER, JOHN FREDERICK, b Los Angeles, Calif, Mar 24, 28; m 54; c 4. METEOROLOGY. Educ: Univ Calif, Los Angeles, AB, 51. Prof Exp: Meteorologist, Coop Studies Sect, Hydrologic Serv Div, US Weather Bur, 53-59 & Dept Navy, 59-62; asst chief coop studies, Off Hydrol, Weather Bur, 62-64, chief spec studies br, 64-71, CHIEF WATER MGT INFO DIV, OFF HYDROL, NAT WEATHER SERV, NAT OCEANIC & ATMOSPHERIC ADMIN, 71- Mem: Am Geophys Union. Res: Investigation of rainfall with respect to cause, frequency, magnitude and estimating the limiting amounts. Mailing Add: 13420 Oriental St Rockville MD 20853

MILLER, JOHN GEORGE, b Philadelphia, Pa, Oct 18, 08; m 40; c 2. PHYSICAL CHEMISTRY. Educ: Univ Pa, AB, 29, MSc, 30, PhD(chem), 32. Prof Exp: Res chemist, Dermat Res Labs, Pa, 29-30; asst, 30-31, from asst instr to assoc prof, 31-52, PROF CHEM, UNIV PA, 52- Concurrent Pos: Proj leader, Thermodyn Res Lab, Univ Pa, 45-51; vis examr, Swarthmore Col, 50-52; consult, Englehard Minerals & Chems Corp, NJ, 44-, Smith, Kline & French Labs, 46-49, Pennwalt Corp, 49-71 & Eastern Regional Res Labs, USDA, 53-57. Mem: Fel AAAS; Am Phys Soc; Clay Minerals Soc; Am Chem Soc; Am Inst Chemists. Res: Molecular structure; dielectric constant measurements; homogeneous catalysis; reaction mechanisms; surface chemistry; gas properties; calorimetry. Mailing Add: Dept of Chem Univ of Pa Philadelphia PA 19174

MILLER, JOHN GRIER, b Boston, Mass, Feb 5, 43. MATHEMATICS. Educ: Univ Chicago, SB, 63, SM, 64; Rice Univ, PhD(math), 67. Prof Exp: Asst prof math, Univ Calif, Los Angeles, 67-69; asst prof, Columbia Univ, 69-72; ASST PROF MATH, ILL STATE UNIV, 72- Mem: Am Math Soc. Res: Topology. Mailing Add: Dept of Math Ill State Univ Normal IL 61761

MILLER, JOHN HENRY, b Washington, DC, Mar 16, 33; m 54. PLANT PHYSIOLOGY. Educ: Yale Univ, BS, 54, MS, 57, PhD(bot), 59. Prof Exp: Nat Cancer Inst fel bot, Yale Univ, 59-60, instr, 60-62; from asst prof to assoc prof, 62-70, PROF BOT, SYRACUSE UNIV, 70- Mem: Am Soc Plant Physiol; Bot Soc Am; Scand Soc Plant Physiol; Japanese Soc Plant Physiol. Res: Photophysiology; developmental physiology. Mailing Add: Dept of Biol Syracuse Univ Syracuse NY 13210

MILLER, JOHN HENRY, III, b Drexel Hill, Pa, Aug 28, 29. ELECTRONIC PHYSICS. Educ: Univ Pa, BA, 51; Princeton Univ, MA, 54, PhD(nuclear physics), 61. Prof Exp: Instr physics, Princeton Univ, 55-58; asst prof, 58-66, ASSOC PROF PHYSICS, UNIV DEL, 66-, ASST CHMN DEPT, 68- Mem: Am Phys Soc; Am Asn Physics Teachers. Res: Beta decay of isotopes involving 0-0 transitions; cathode-ray-excited stimulated emission in zinc sulfide crystals. Mailing Add: Dept of Physics Univ of Del Newark DE 19711

MILLER, JOHN HOWARD, b Columbus, Ohio, Oct 13, 43; m 65; c 1. RADIATION PHYSICS. Educ: Davidson Col, BS, 66; Univ Va, PhD(physics), 71. Prof Exp: Res assoc physics, Univ Fla, 72-74 & Los Alamos Sci Lab, 74-75; STAFF SCIENTIST PHYSICS, PAC NORTHWEST LAB, BATTELLE MEM INST, 75- Mem: Am Phys Soc; Radiation Res Soc. Res: Theory of physical and chemical events resulting from energy deposition by ionizing radiation and the relationship of these events to biological effects of the radiation. Mailing Add: Pac Northwest Lab Battelle Mem Inst Richland WA 99352

MILLER, JOHN IVAN, b Prescott, Kans, Oct 16, 11; m 37; c 3. ANIMAL HUSBANDRY. Educ: Kans State Col, BS, 33; Cornell Univ, MS, 34, PhD(animal husb), 36. Prof Exp: Asst, 33-36, from instr to assoc prof, 37-44, PROF ANIMAL HUSB, CORNELL UNIV, 44- Honors & Awards: NY Farmers Achievement Award; Distinguished Serv Award, Am Soc Animal Sci, 67. Mem: Am Soc Animal Sci (secy-treas, 51-53, vpres, 54, pres, 55). Res: Beef cattle production and nutrition of ruminants, especially protein; energy levels; beef quality; feeding and management. Mailing Add: 135 Bush Lane Ithaca NY 14850

MILLER, JOHN JAMES, b Schreiber, Ont, Oct 13, 18; m 51. MICROBIOLOGY. Educ: Univ Toronto, BA, 41, PhD(bot), 44. Prof Exp: Agr asst, Can Dept Agr, 44-46, asst plant pathologist, 46-47; from asst prof to assoc bot, 47-60, PROF BIOL, McMASTER UNIV, 60- Mem: Can Soc Microbiol; Can Bot Asn. Res: Mycology; yeast sporulation and spore germination. Mailing Add: Dept of Biol McMaster Univ Hamilton ON Can

MILLER, JOHN JOHNSTON, III, b San Francisco, Calif, Apr 9, 34; m 58; c 4. PEDIATRICS, IMMUNOBIOLOGY. Educ: Wesleyan Univ, BA, 55; Univ Rochester, MD, 60; Univ Melbourne, PhD(immunol), 65. Prof Exp: Intern pediat, Univ Calif, San Francisco, 60-61, resident, 61-62; resident, 65, clin teaching asst, 65-67, asst prof pediat, 67-73, DIR RHEUMATIC DIS SERV, CHILDREN'S HOSP, MED CTR, STANFORD UNIV, 67-, SR ATTEND PHYSICIAN, 73- Concurrent Pos: Consult, US Naval Radiol Defense Lab, 67-69. Mem: Fel Am Acad Pediat; Am Rheumatism Asn; Am Asn Immunol; Soc Pediat Res; Am Fedn Clin Res. Res: Pediatric rheumatology; drug-induced systemic lupus; behavior of long-lived lymphocytes in situ; antigen handling in neonatal rats. Mailing Add: Children's Hosp Stanford Univ Med Ctr Stanford CA 94305

MILLER, JOHN JOSEPH, b Denver, Colo, Nov 24, 31; m 61; c 5. ORGANIC CHEMISTRY. Educ: Regis Col, Colo, BS, 53; Univ Ill, PhD(org chem), 57. Prof Exp: Control chemist, McMurtry Mfg Co, Colo, 52; asst, Univ Ill, 53-56; res chemist, Rohm & Haas Co, Pa, 56, sr res chemist, Redstone Arsenal Res Div, Ala, 57-61, Res Labs, 61-73, PROJ LEADER, ROHM & HAAS CO RES LABS, 73- Mem: Am Chem Soc; The Chem Soc. Res: Metallo-organic chemistry; rocket fuels; agricultural chemicals, coatings. Mailing Add: Rohm & Haas Co Spring House PA 19477

MILLER, JOHN MEISTER, organic chemistry, see 12th edition

MILLER, JOHN MELVILLE, III, b Cordova, Ala, Dec 3, 31; m 59; c 2. REHABILITATION MEDICINE. Educ: Vanderbilt Univ, AB, 52; Med Col Ala, MD, 56; Am Bd Phys Med & Rehab, dipl, 68. Prof Exp: Intern, Brooklyn Hosp, NY, 56-57; gen practr, Ala, 59-64; resident phys med & rehab, Columbia Presby Med Ctr, 64-66, instr, Columbia Univ, 66-67, asst prof, 67-70; PROF REHAB MED & CHMN DEPT, SCH MED, UNIV ALA, BIRMINGHAM, 70- Mem: AAAS; Am Acad Phys Med & Rehab; Am Asn Acad Psychiatrists; Int Med Soc Paraplegia. Res: Paraplegia; autonomic function of spinal cord; temperature control mechanisms. Mailing Add: Spain Rehab Ctr 1717 Sixth Ave S Birmingham AL 35233

MILLER, JOHN ROBERT, b Berwyn, Ill, June 22, 44; m 67; c 1. PHYSICAL CHEMISTRY, RADIATION CHEMISTRY. Educ: Ore State Univ, BS, 66, Univ Wis, PhD(phys chem), 71. Prof Exp: Appointee phys & radiation chem, 71-74, ASST CHEMIST, ARGONNE NAT LAB, 74- Mem: Am Chem Soc. Res: Electron transfer over long distances by quantum mechanical tunneling; picosecond pulse radiolysis. Mailing Add: Chem 200 Argonne Nat Lab Argonne IL 60439

MILLER, JOHN WALCOTT, b Royal Oak, Mich, Nov 10, 30; m 72; c 2. ANALYTICAL CHEMISTRY. Educ: Wesleyan Univ, BA, 53; Northwestern Univ, PhD(anal chem), 56. Prof Exp: Methods develop chemist, 56-60, mgr chem methods sect, 60-75, SUPVR CHROMATOGR SECT, PHILLIPS PETROL CO, 75- Concurrent Pos: Nat Acad Sci exchange fel, Prague, 68; chmn, Gordon Res Conf Anal Chem, 75 & Anal Chem Div, Am Chem Soc, 76-77. Res: Liquid and gas chromatography; coulometric analysis; photometric titrations; organic sulfur functional groups analysis; ultraviolet spectroscopy; redox reactions; polarography of coordination compounds; crude oil source identification; water pollution analysis. Mailing Add: Res & Develop Res Bldg 1 Phillips Petrol Co Bartlesville OK 74004

MILLER, JOHN WESLEY, b Dilley, Tex, Feb 23, 37; m 66; c 2. PLANT PATHOLOGY. Educ: Tex A&M Univ, BS, 60, MS, 63; Univ Fla, PhD(plant path), 65. Prof Exp: Asst prof plant path, Mo Delta Ctr, 65-68; PLANT PATHOLOGIST, DIV PLANT INDUST, FLA DEPT AGR, 68- Mem: Am Phytopath Soc. Res: Cotton leaf spot and boll rot diseases; diseases of ornamentals. Mailing Add: Div of Plant Indust Fla Dept Agr PO Box 1269 Gainesville FL 32601

MILLER, JOHN WESLEY, JR, b Philadelphia, Pa, Aug 30, 35; m 58; c 2. MARINE ZOOLOGY, ENTOMOLOGY. Educ: Dickinson Col, BS, 57; Pa State Univ, MS, 60, PhD(entom), 62. Prof Exp: Fel entom & parasitol, Univ Calif, Berkeley, 62-63; assoc prof, 63-74, PROF BIOL, BALDWIN-WALLACE COL, 74- Mem: AAAS; Am Inst Biol Sci; Entom Soc Am. Res: Laboratory culture and use of marine organisms in the undergraduate curriculum; insects as vectors of plant diseases, specifically the relationship of certain plant viruses to their aphid vectors. Mailing Add: Dept of Biol Baldwin-Wallace Col Berea OH 44017

MILLER, JON PHILIP, b Moline, Ill, Mar 30, 44; m 65; c 2. BIOCHEMISTRY, MOLECULAR PHARMACOLOGY. Educ: Augustana Col, Ill, AB, 66; St Louis Univ, PhD(biochem), 70. Prof Exp: Fel biophys chem, 70-71, biochemist, 72, head molecular pharm, 72-73, head drug metab, 73-74, HEAD BIOL DIV, ICN NUCLEIC ACID RES INST, 74- Res: Structure-activity relationships of c-adenosine monophosphate analogs; mechanism of action and metabolism of drugs; mechanism of inhibition of cyclic-adenosine monophosphate phosphodiesterases. Mailing Add: Nucleic Acid Res Inst ICN Pharm Inc 2727 Campus Irvine CA 92664

MILLER, JONATHAN PARIS, biochemistry, see 12th edition

MILLER, JOSEF MAYER, b Philadelphia, Pa, Nov 29, 37; m 60; c 1. PHYSIOLOGY, PSYCHOLOGY. Educ: Univ Calif, Berkeley, BA, 61; Univ Wash, PhD(physiol), 65. Prof Exp: Asst prof psychol, Univ Mich, 67-68; asst prof, 68-75, ASSOC PROF PHYSIOL & OTOLARYNGOL, UNIV WASH, 75- Concurrent Pos: USPHS fel, Univ Mich, 65-67; Deafness Res Found grant, Univ Wash, 69-71, NIH res grant, 69-73. Mem: AAAS; Am Physiol Soc; Acoust Soc Am; Soc Neurosci; Psychonomic Soc. Res: Physiological and behavioral acoustics; animal psychophysics; sensory neurophysiology. Mailing Add: Dept of Otolaryngol Univ of Wash Seattle WA 98105

MILLER, JOSEPH EDWIN, b Carrollton, Mo, Nov 4, 42; m 62; c 2. PLANT BIOCHEMISTRY, PLANT PHYSIOLOGY. Educ: Colo State Univ, BS, 64, MS, 66;

Utah State Univ, PhD(plant biochem), 69. Prof Exp: Asst prof plant physiol, Univ Colo, Denver, 69-74; RES ASSOC, ARGONNE NAT LAB, 74- Mem: Am Soc Plant Physiologists. Res: Plant growth and development, especially physiology of flowering; fluoride effects on oxidative phosphorylation and plant metabolism. Mailing Add: Argonne Nat Lab 9700 S Cass Ave Argonne IL 60439

MILLER, JOSEPH HENRY, b Yonkers, NY, May 27, 24; m 48; c 2. MEDICAL PARASITOLOGY. Educ: Univ Mich, BS, 48, MS, 49; NY Univ, PhD(biol), 53. Prof Exp: Asst biol, NY Univ, 51-53; instr, 53-55, from asst prof to assoc prof, 55-70, PROF MED PARASITOL, SCH MED, LA STATE UNIV, NEW ORLEANS, 70- Concurrent Pos: Fel, China Med Bd, Cent Am, 56; scientist, vis staff, Charity Hosp, New Orleans, 53-; vis prof, Fac Med, Nat Univ Mex, 66 & Col Med, Univ Ariz, 75-76. Mem: Am Soc Trop Med & Hyg; Am Soc Parasitol; fel Royal Soc Trop Med & Hyg; Electron Micros Soc Am (treas, 70-71); Soc Protozool. Res: Medical parasitology, especially electron microscopy of parasites. Mailing Add: Dept of Trop Med & Med Parasitol La State Univ Sch of Med New Orleans LA 70112

MILLER, JOSEPH MORTON, b Boston, Mass, Nov 9, 21; m 48; c 3. CLINICAL MEDICINE, PREVENTIVE MEDICINE. Educ: Harvard Univ, AB, 42, MD, 45, MPH, 60. Prof Exp: Intern med, Mt Sinai Hosp, New York, 45-46; res fel biochem, Trudeau Found, 48; NIH fel, Med Sch, Harvard Univ, 48-50; instr, 50-64, RES ASSOC MED, SCH PUB HEALTH, HARVARD UNIV, 65- Concurrent Pos: Sr assoc, Peter Bent Brigham Hosp, Boston. Mem: Am Fedn Clin Res; fel Am Col Physicians. Res: Rheumatic fever; streptococcal disease; epidemiology of coronary artery disease. Mailing Add: One Boylston Plaza Prudential Ctr Boston MA 02199

MILLER, JOSEPH NELSON, b Hillsboro, Ohio, July 25, 03; m 43; c 1. PARASITOLOGY. Educ: Univ Miami, AB, 25; Ohio State Univ, AM, 26, PhD, 35. Prof Exp: Asst zool, 25-27, from instr to asst prof, 27-48, from assoc prof zool & entom to prof zool, 48-73, EMER PROF ZOOL, OHIO STATE UNIV, 73- Mem: Am Soc Parasitol. Res: Entomology. Mailing Add: Dept of Zool Ohio State Univ Columbus OH 43210

MILLER, JOYCE MARY, b Belton, Tex, Mar 12, 45; m 72. BIOCHEMISTRY. Educ: Southwest Tex State Univ, BS, 67, MA, 68; Tex A&M Univ, PhD(biochem), 72. Prof Exp: Fel lipoprotein chem, Sch Med, Univ Southern Calif, 72-74; res assoc, Sch Med, Wash Univ, 74-75; ASST PROF BIOL, MARYVILLE COL, MO, 75- Concurrent Pos: NIH fel, 73. Mem: AAAS; Sigma Xi. Res: Determination of qualitative and/or quantitative variations in the peptide composition of various serum lipoprotein classes with increasing age of an individual by protein characterization and immunological techniques. Mailing Add: Maryville Col 13550 Conway Rd St Louis MO 63141

MILLER, JULIAN MALCOLM, b Berkeley, Calif, Aug 5, 22; m 60; c 2. CHEMISTRY, PHYSICS. Educ: Univ Calif, BS, 44; Columbia Univ, PhD(chem), 49. Prof Exp: Jr scientist, Off Sci Res & Develop, Calif Inst Technol, 44-45, asst, 45-47; asst, 47-48, from instr to assoc prof, 49-61, PROF CHEM, COLUMBIA UNIV, 61- Concurrent Pos: NSF sr fel, 65-66; chmn dept chem, Columbia Univ, 70-73; Guggenheim fel, 74. Mem: Fel Am Phys Soc; Am Chem Soc. Res: Study of nuclear reactions; radiochemistry. Mailing Add: Dept of Chem Columbia Univ New York NY 10027

MILLER, JULIUS SUMNER, b Billerica, Mass, May 17, 09; m 34. PHYSICS. Educ: Boston Univ, BS, 32, MA, 33; Univ Idaho, MS, 40. Prof Exp: Master math, Cheshire Acad, 36-37; instr physics, Dillard Univ, 37-38, prof, 41-51; PROF PHYSICS, EL CAMINO COL, 53- Concurrent Pos: Chmn dept physics, Mich Col Mining & Technol, 48-49; Carnegie grant, Inst Advan Studies, 50-51; physics ed, Sch Sci & Math, 50-; assoc res physicist, Univ Calif, Los Angeles, 56-; lectr, Uppsala Univ, Sweden, Oslo Univ, Norway & Univ Milan, Italy, 64; Isaac Newton mem lectr, Brit Broadcasting Co, 65; consult, Walt Disney Prods, 60-, Orgn Econ Coop & Develop, Paris, France, 62 & TV progs, Norway & NZ, 63 & Australia, 64-65; lectr physics, TV progs, Australia, 62-74, South Africa, 75. Honors & Awards: TV Soc Australia Award, 66; Hon Tenure Prof, US Air Force Acad. Mem: Am Asn Physics Teachers. Res: Electron configuration in the elements; analytical mechanics; demonstration experiments in physics; physics of toys; history of science. Mailing Add: 16711 Cranbrook Ave Torrance CA 90504

MILLER, KENNETH JAY, b New York, NY, Oct 12, 24; m 48; c 5. PHYSICAL CHEMISTRY. Educ: Eastern Nazarene Col, BS, 49; Johns Hopkins Univ, MA, 50, PhD(phys chem), 52. Prof Exp: Develop engr semiconductors, Westinghouse Elec Corp, Pa, 52-55; asst prof chem, Mt Union Col, 55-58; sr scientist phys chem, Res & Advan Develop Div, Avco Corp, 58-59; mem tech staff semiconductors, Bell Tel Labs, 59-68; PROF CHEM, NORTHEAST LA UNIV, 68- Mem: Am Chem Soc; Electrochem Soc. Res: Thermodynamics and electrochemistry; semiconductors. Mailing Add: Dept of Chem Northeast La Univ Monroe LA 71201

MILLER, KENNETH JOHN, b Chicago, Ill, Mar 24, 39; m 63, 75; c 2. THEORETICAL CHEMISTRY. Educ: Ill Inst Technol, BS, 60; Johns Hopkins Univ, MA, 64; Iowa State Univ, PhD(chem), 66. Prof Exp: Nat Acad Sci-Nat Res Coun resident res assoc, Nat Bur Stand, 66-67; asst prof, 67-75, ASSOC PROF THEORET CHEM, RENSSELAER POLYTECH INST, 75- Mem: Am Phys Soc; Am Inst Physics. Res: Inelastic electron scattering of atoms and molecules with high and low energy electrons; chemical reactivities of mesoionic molecules; quantum mechanics; molecular orbital theory; chemical and physical properties of biological macromolecules; the interaction of drugs with DNA. Mailing Add: Dept of Chem Rensselaer Polytech Inst Troy NY 12181

MILLER, KENNETH LERON, b Magrath, Alta, Sept 6, 24. POLYMER CHEMISTRY. Educ: Brigham Young Univ, BA, 48; Ore State Univ, PhD, 52. Prof Exp: RES & DEVELOP CHEMIST, E I DU PONT DE NEMOURS & CO, INC, 52- Mailing Add: 113 Wood Rd Louisville KY 40222

MILLER, KENNETH MELVIN, b Indianapolis, Ind, Aug 17, 43; m 69; c 1. FRESH WATER ECOLOGY, FOOD SCIENCE. Educ: Ind Univ, AB, 65, AM, 67, PhD(zool), 75. Prof Exp: Instr biol, Purdue Univ, NCent Campus, 70-74; LAB MGR, AM HOME FOODS, 75- Mem: Am Inst Biol Sci; Soc Study Evol; Ecol Soc Am. Res: Species association and diversity in aquatic coleoptera and odonata; heat processing parameters in food canning. Mailing Add: Am Home Foods LaPorte IN 46350

MILLER, KENNETH PHILIP, b Northfield, Minn, Sept 17, 15; m 41; c 4. ANIMAL BREEDING. Educ: Univ Minn, BS, 39, MS, 40; Ohio State Univ, PhD(dairy husb), 56. Prof Exp: Instr, 41-47, asst prof, 47-63, ASSOC PROF ANIMAL BREEDING, SOUTHERN EXP STA, UNIV MINN, WASECA, 63- Mem: AAAS; Am Soc Animal Sci; Am Dairy Sci Asn. Res: Animal nutrition. Mailing Add: 808 Fifth Ave SE Waseca MN 56093

MILLER, KENNETH RAYMOND, b Rahway, NJ, July 14, 48; m 72. CELL BIOLOGY. Educ: Brown Univ, ScB, 70; Univ Colo, PhD(biol), 74. Prof Exp: LECTR BIOL, HARVARD UNIV, 74- Mem: AAAS; Am Soc Cell Biol; Am Soc Photobiol. Res: Structure, biochemistry, and function of biological membranes; most importantly, the photosynthetic membrane. Mailing Add: Biol Labs Harvard Univ 16 Divinity Ave Cambridge MA 02138

MILLER, KENNETH SIELKE, b New York, NY, June 4, 22; m 53; c 2. MATHEMATICS. Educ: Columbia Univ, BS, 43, AM, 47, PhD(math), 50. Prof Exp: Lectr, Columbia Univ, 49; from inst to prof math, NY Univ, 50-64; SR STAFF SCIENTIST, ELECTRONICS RES LABS, COLUMBIA UNIV, 64-; ADJ PROF MATH, FORDHAM UNIV, 64- Concurrent Pos: Mem staff, Inst Advan Study, 50 & 58-59; consult, Army Res Off, Systs Res Labs & Fed Sci Corp. Mem: Am Math Soc; Inst Elec & Electronics Engrs. Res: Differential operators; mathematical machines; linear systems, random noise. Mailing Add: 25 Bonwit Rd Town of Rye Port Chester NY 10573

MILLER, KENNETH WAYNE, b Milwaukee, Wis, Dec 18, 41; m 64; c 2. PHARMACOLOGY, BIOPHARMACEUTICS. Educ: Univ Wis-Madison, BS, 64, MS, 67, PhD(pharmacol), 69. Prof Exp: Res assoc, Vanderbilt Univ, 68-70; ASSOC PROF PHARMACEUT, COL PHARM, UNIV MINN, MINNEAPOLIS, 71- Mem: AAAS; Am Asn Col Pharmacists. Res: Drug metabolism both in vitro and in vivo; investigation of pathways of drug disposition. Mailing Add: Col of Pharm Univ of Minn Minneapolis MN 55455

MILLER, KENT D, b Detroit, Mich, May 9, 25; m 50; c 3. BIOCHEMISTRY. Educ: Oberlin Col, AB, 49; Wayne State Univ, MS, 51, PhD, 54; Albany Med Col, MD, 62. Prof Exp: Res scientist, Div Labs & Res, NY State Dept Health, 54-57, sr res scientist, 57-62, asst dir, 62-69; PROF MED, SCH MED, UNIV MIAMI, 69- Mem: Am Chem Soc; Am Soc Biol Chemists; Am Soc Hemat; Soc Exp Biol & Med. Res: Blood proteins; coagulation; bacterial enzymes; immunology. Mailing Add: Univ Miami Dept of Med Biscayne Annex PO Box 520875 Miami FL 33152

MILLER, KIM IRVING, b Boone, NC, July 26, 36; m 59. PLANT TAXONOMY. Educ: Appalachian State Teachers Col, BS, 58; Purdue Univ, MS, 61, PhD(bot), 64. Prof Exp: Vis asst prof biol, Purdue Univ, 64-65, res assoc & cur herbarium, 65-66; asst prof biol, Eastern Ky Univ, 66-67 & Appalachian State Univ, 67-68; asst prof, 68-71, ASSOC PROF BIOL, JACKSONVILLE UNIV, 72- Mem: Bot Soc Am; Am Soc Plant Taxon. Res: Evolution and systematics of Euphorbiaceae and related families. Mailing Add: Dept of Biol Jacksonville Univ Jacksonville FL 32211

MILLER, KNUDT JOHN, b Forestville, Wis, Sept 29, 40; m 59; c 2. HORTICULTURE, PLANT PHYSIOLOGY. Educ: Univ Wis-Madison, BS, 62, MS, 63; Mich State Univ, PhD(hort), 70. Prof Exp: Asst supvr, Hort Exp Sta, Mich State Univ, 63-69; PLANT PHYSIOLOGIST, AGR RES CTR, LIBBY, McNEILL & LIBBY, 69- Mem: AAAS; Am Soc Hort Sci; Am Soc Plant Physiol; Weed Sci Soc Am. Res: Weed control, growth regulators, environmental factors and cultural practices to improve production and quality of vegetables and fruits for processing. Mailing Add: Agr Res Ctr Libby McNeill & Libby Janesville WI 53545

MILLER, LARRY GENE, b Corsicana, Tex, Dec 30, 46; m 69; c 2. PHARMACEUTICS. Educ: Univ Miss, BS, 69, PhD(pharmaceut), 72. Prof Exp: Sr scientist prod develop, Mead Johnson & Co, 72-73; SR PHARMACIST PHARMACEUT RES, A H ROBINS CO, INC, 73- Mem: Am Pharmaceut Asn; Acad Pharmaceut Sci. Res: Design of dosage forms for pharmaceutical compounds so as to meet the necessary requirements of stability, delivery and bioavailability; new delivery systems for long term therapy. Mailing Add: A H Robins Co Inc 1211 Sherwood Ave Richmond VA 23220

MILLER, LARRY LEE, b Waterloo, Iowa, June 29, 39; m 60; c 2. ORGANIC CHEMISTRY. Educ: Colo State Col, AB, 61; Univ Ill, PhD(chem), 64. Prof Exp: Res chemist, Am Cyanamid Co, 64-66; from asst prof to assoc prof, 66-74, PROF CHEM, COLO STATE UNIV, 74- Mem: Am Chem Soc. Res: Physical organic chemistry. Mailing Add: Dept of Chem Colo State Univ Ft Collins CO 80521

MILLER, LARRY O'DELL, b Los Angeles, Calif, Feb 26, 39; m 72. DEVELOPMENTAL BIOLOGY, NEUROPHYSIOLOGY. Educ: Univ Calif, Santa Barbara, BA, 61, MA, 64, PhD(biol), 67. Prof Exp: Res biologist, Firestone Tire & Rubber Co, 67-69; lectr biol, Univ Calif, Santa Barbara, 69-71; INSTR BIOL, MOORPARK COL, 71- Mem: AAAS. Res: Biochemical control of eukaryotic cellular differentiation. Mailing Add: Dept of Biol Moorpark Col Ventura CA 93021

MILLER, LAURENCE HERBERT, b Newark, NJ, Oct 11, 34; m 59; c 2. DERMATOLOGY. Educ: Muhlenberg Col, BS, 56; Univ Lausanne, MD, 61. Prof Exp: Intern, Newark Beth Israel Hosp, 62-63; house physician, East Orange Gen Hosp, 63; resident, NY Univ Med Ctr, 63-66; DERMAT PROG DIR, NAT INST ARTHRITIS, METAB & DIGESTIVE DIS, 66- Concurrent Pos: Mem, Dermat Found, 68; asst prof, Sch Med, Johns Hopkins Univ, 69-; mem, Sci Adv Bd, Nat Psoriasis Found, 71- Mem: Fel Am Acad Dermat; Soc Invest Dermat. Res: Varicella-Zoster virus; cell controls in psoriasis.

MILLER, LAWRENCE INGRAM, b Jackson Center, Ohio, May 12, 14; m 39; c 2. PHYTOPATHOLOGY. Educ: Oberlin Col, AB, 36; Va Polytech Inst, MS, 38; Univ Minn, PhD(plant path), 53. Prof Exp: Freeport Sulphur Co fel plant path, 38-40, asst plant pathologist, Agr Exp Sta, 40-42, prof plant path, 49-74, ASSOC PLANT PATHOLOGIST, AGR EXP STA, VA POLYTECH INST & STATE UNIV, 49-, PROF PLANT PSYCHOL, 74- Concurrent Pos: Mem comt biol control of soilborne plant pathogens, Nat Acad Sci-Nat Res Coun, 57-65; NATO res grant plant nematol, 73-74. Honors & Awards: J Shelton Horsley Res Award, Va Acad Sci, 60; Golden Peanut Res Award, Nat Peanut Coun, 63. Mem: Int Soc Plant Path; Soc Europ Nematol; Brit Asn Appl Biol; Am Phytopath Soc; Am Soc Nematol. Res: Plant nematology, particularly Heterodera species; diseases of the peanut. Mailing Add: Dept of Plant Path & Physiol Va Polytech Inst & State Univ Blacksburg VA 24061

MILLER, LAWRENCE PETER, b Old Zionsville, Pa, Apr 17, 01; m 35; c 2. PLANT CHEMISTRY, TOXICOLOGY. Educ: Pa State Univ, BS, 23; Univ Wis, MS, 25; Purdue Univ, PhD(agr biochem), 33. Prof Exp: Process operator, Nat Aniline & Chem Co, NY, 23; chemist, WVa Pulp & Paper Co, Md, 23-24; asst, Univ Wis, 24-25; asst biochemist, Boyce Thompson Inst, 25-26; chemist, Exp Sta, Purdue Univ, 26-29; asst biochemist, Boyce Thompson Inst, 29-38, biochemist, 38-62, prog dir sci educ, 62-66; TECH WRITER & ED, VAN NOSTRAND REINHOLD PUBL CO, 67- Concurrent Pos: Consult, Bowey's Inc, Chicago, 42-45, Carbide & Carbon Chem Corp, 45-60, W R Grace & Co, 60-63 & Int Atomic Energy Agency, Vienna, 64-66; mem, Bd Mgrs, Am Inst City New York, 45-50 & Yonkers Health Labs, 58-66; sect ed, Biochem Sect, Chem Abstracts Serv, 67-; mem, Bd Mgrs, NY Bot Garden, 64-68, mem corp, 68-; del, Int Sci Congs, Paris, 54, Oxford Univ, 54, Scheveningen, Netherlands, 55, Geneva, 55, Mondorf-les-Bains, Luxembourg, 55, Hamburg, Ger, 57, Montreal, 59, Tokyo, 67. Mem: AAAS; Am Chem Soc; Am Phytopath Soc; fel Am Inst Chemists; hon mem, NY Acad Sci. Res: Induced information of glycosides in plants; mechanism of fungicidal action; radioisotopes in agricultural research. Mailing Add: 180 West End Ave Apt 27-D New York NY 10023

MILLER, LEE ALAN, organic chemistry, see 12th edition

MILLER, LEE ANTON, invertebrate physiology, animal behavior, see 12th edition

MILLER, LEE NORMAN, b Decatur, Ill, July 9, 30; m 53; c 2. PLANT ECOLOGY, PLANT PHYSIOLOGY. Educ: Southern Methodist Univ, BBA, 53; Yale Univ, MF, 61; Duke Univ, PhD(plant physiol), 66. Prof Exp: Timestudy engr, Bell & Howell Corp, 53-54; methods engr, Borg Warner Corp, 54; mgr, Ben Miller's, 55-59; res assoc radiation ecol, Brookhaven Nat Lab, 61-62; asst prof plant ecol, 66-73, ADMIN ASST TO CHMN DEPT ENTOM, CORNELL UNIV, 74- Mem: Am Soc Plant Physiol; Scand Soc Plant Physiol; Ecol Soc Am; Brit Ecol Soc. Res: Water relations to plants; effects of water stress on plant growth and productivity; physiological basis of niche specialization in plants. Mailing Add: Dept of Entom Cornell Univ Ithaca NY 14850

MILLER, LEON LEE, b Rochester, NY, Dec 7, 12; m 35, 58; c 6. BIOCHEMISTRY. Educ: Cornell Univ, BA & MA, 34, PhD(org chem), 37; Univ Rochester, MD, 45. Prof Exp: Res org chemist, Calco Chem Co, NJ, 37-38; instr path & pharmacol, Sch Med & Dent, Univ Rochester, 45-46; asst prof biochem, Jefferson Med Col, 46-48; assoc prof radiation biol & biochem, 48-59, PROF RADIATION BIOL, BIOPHYS & BIOCHEM, UNIV ROCHESTER, 59- Mem: AAAS; Am Chem Soc; Am Soc Biol Chem; NY Acad Sci. Res: Organic synthesis; blood and liver proteins and their functions; liver injury; enzyme, protein and amino acid metabolism; isolated liver perfusion. Mailing Add: Dept of Radiation Biol & Biophys Univ of Rochester Sch of Med Rochester NY 14642

MILLER, LEONARD DAVID, b Jersey City, NJ, July 8, 30; m 67; c 2. SURGERY. Educ: Yale Univ, AB, 51; Univ Pa, MD, 55. Prof Exp: Asst instr surg, 56-57 & 59-64, assoc in surg, 64-66, from asst prof to assoc prof, 66-72, PROF SURG & VCHMN DEPT, UNIV PA, 72-, J WILLIAM WHITE PROF SURG RES, 70-, DIR HARRISON DEPT SURG RES, 72- Concurrent Pos: NIH sr clin trainee, Univ Pa, 64-65, NIH grants, 65-, John A Hartford Found, Inc grant, 66-; intern, Hosp Univ Pa, 55-56, resident surg, 56-57 & 59-64, dir, Shock & Trauma Clin Res Unit, 67-72; exec officer, Harrison Dept Surg Res, Univ Pa, 67-68, dir NIH training grant, 67-72; vis surgeon, Vet Admin Hosp, 71-; consult, Children's Hosp Philadelphia, 73-; chmn comt surg educ, Soc Univ Surg, 68-; actg chmn dept surg, Am Surg Asn. Honors & Awards: Distinguished Teaching Award, Lindback Found, 70. Mem: AAAS; Soc Univ Surg; Soc Surg Alimentary Tract; Am Soc Surg Trauma; Am Surg Asn. Mailing Add: Dept of Surg 1000 Ravdin Hosp of the Univ of Pa Philadelphia PA 19104

MILLER, LEONARD EDWARD, b New York, NY, Aug 24, 19; m 46; c 4. ORGANIC CHEMISTRY. Educ: Univ Mich, BS, 40, MS, 42, PhD(org chem), 43. Prof Exp: H H Rackham asst, Univ Mich, 43-44; instr chem, Univ Ill, 46-48, instr & admin asst to head dept, 48-51; prof & head dept, Univ NDak, 51-52; assoc prof & dir labs, Univ Ill, 52-56; sr res chemist & sr staff asst, Calif Res Corp, Standard Oil Co Calif, 56-59, res assoc, 59-64; dir explor res, 64-72, HEAD, CHEM PROD RES DEPT, LUBRIZOL CORP, 72- Concurrent Pos: Instr exten, Univ Calif, 58-64. Mem: AAAS; Am Chem Soc. Res: New monomers and polymers; synthesis and applications. Mailing Add: Chem Prod Res Dept Lubrizol Corp Cleveland OH 44117

MILLER, LEONARD ROBERT, b New York, NY, Oct 31, 33; m 57; c 3. PATHOLOGY, CELL BIOLOGY. Educ: Bethany Col, BS, 54; Univ Pittsburgh, MS, 55; Albany Med Col, MD, 59; Am Bd Path, dipl, 65. Prof Exp: Intern, Beverly Hosp, 59-60; resident path, Sch Med, Yale Univ, 61-64, instr, 64-67; asst prof path, Sch Med, Univ Ariz, 67-68; assoc prof, 68-70, VIS ASSOC PROF PHYSIOL & BIOPHYS, MED CTR, UNIV OKLA, 70- Concurrent Pos: Res fel, 60-61; USPHS trainee path, 61-63, spec fel, 63-64; Off Naval Res grant, 68-; consult, Tissue Bank, Naval Med Res Inst, 64-67; sect chief histopath, Armed Forces Radiobiol Res Inst, 64-67. Mem: AAAS; Am Asn Pathologists & Bacteriologists; Int Acad Path; Soc Cryobiol. Res: In vivo, in vitro mammalian and bacterial cell function; cellular control mechanisms; phage induction; mammalian cell transformation; adaptation; immunology; cryobiology; ultrastructure of immune response; inhibition and enhancement of cell repair. Mailing Add: Dept of Physiol & Biophys Univ of Okla Med Ctr Oklahoma City OK

MILLER, LEROY JESSE, b Lebanon, Pa, Aug 12, 33; m 54; c 3. ORGANIC CHEMISTRY. Educ: Elizabethtown Col, BS, 54; Univ Del, MS, 57, PhD(chem), 59. Prof Exp: Sr res engr, Atomics Int Div, NAm Aviation Inc, 58-60; sr res chemist, Sundstrand Corp, 60-62; mem tech staff, 62-71, STAFF ENGR & GROUP HEAD, HUGHES AIRCRAFT CO, 71- Mem: AAAS; Am Chem Soc; Soc Photog Sci & Eng. Res: Polymer chemistry; photochemistry; reaction mechanisms. Mailing Add: 8313 Hillary Dr Canoga Park CA 91304

MILLER, LESTON WAYNE, b Seattle, Wash, Apr 10, 28; m 56; c 2. ELECTRODYNAMICS. Educ: Univ Wash, BS, 51, PhD(physics), 59. Prof Exp: MEM STAFF, LOS ALAMOS SCI LAB, UNIV CALIF, 59- Mem: Am Geophys Union; Am Phys Soc. Res: Radio frequency electromagnetic radiations from nuclear explosions; magnetospheric physics; ionospheric physics; weapon testing. Mailing Add: Los Alamos Sci Lab Univ of Calif Los Alamos NM 87545

MILLER, LEWIS DUDLEY, b Cunningham, Tenn, May 4, 44; m 67; c 1. THEORETICAL NUCLEAR PHYSICS. Educ: Austin Peay State Univ, BA, 66; Univ Fla, PhD(physics), 71. Prof Exp: Res assoc physics, Univ Md, College Park, 71-73 & Mass Inst Technol, 73-75; VIS ASST PROF PHYSICS, UNIV VA, 75- Mem: Am Phys Soc; Am Asn Physics Teachers. Res: Relativistic effects in nuclei; relativistic theories of various problems in nuclear structure. Mailing Add: Dept of Physics Univ of Va Charlottesville VA 22901

MILLER, LEWIS F, b Brooklyn, NY, Apr 30, 33; m 58. POLYMER CHEMISTRY, COLLOID CHEMISTRY. Educ: Brooklyn Col, BS, 55; State Univ NY Col Forestry, Syracuse, MS, 62. Prof Exp: Chemist, 55-73, SR CHEMIST, IBM CORP, 73- Honors & Awards: IBM Corp Invention Awards, 66, 67, 68, 69, 75. Mem: Am Chem Soc; Sigma Xi; fel Am Inst Chemists; Int Soc Hybrid Microelectronics (treas). Res: Material development-inks for magnetic sensing, materials and processes for computer components. Mailing Add: Emans Rd LaGrangeville NY 12540

MILLER, LEWIS SAMUEL, b Portland, Ore, Dec 7, 17; m 42; c 3. ORGANIC POLYMER CHEMISTRY. Educ: Reed Col, BA, 39; Ore State Col, MS, 41; Iowa State Col, PhD(org chem), 50. Prof Exp: Res chemist, Nitrogen Div, Solvay Process Co, 41-45; chg spec prods res, Am-Marietta Co, 50-54, res dir, Adhesive Chem & Resin Div, 54-62; prof specialist pioneering res, 62-70, PROF SPECIALIST NEW BUSINESS RES, WEYERHAEUSER CO, 70- Mem: Am Chem Soc. Res: Organometallic and organosilicon compounds; urea, phenolic, melamine and epoxy resins for adhesives and plastics; coating, adhesive and reprographic polymers; cellulose derivative polymers. Mailing Add: 10260 SE 21st St Bellevue WA 98004

MILLER, LILA, b Cape Girardeau, Mo, Feb 11, 02. BIOCHEMISTRY. Educ: Univ Wis, BA, 26, MS, 27; Univ Mich, PhD(biol chem), 36. Prof Exp: Instr chem, Miss State Col Women, 27-28; res assoc nutrit lab, Battle Creek Sanitarium, Mich, 28-31; instr, 31-36, from instr to assoc prof, 37-70, ASSOC EMER PROF BIOCHEM, MED SCH, UNIV MICH, ANN ARBOR, 70- Concurrent Pos: Rackham fel, Univ Mich, Carlsberg Labs, Copenhagen, 36-37. Mem: AAAS; Am Chem Soc; Am Soc Biol Chem. Res: Nutritional anemia; minerals in plants; microkjeldahl determination applied to amino acids and proteins; enzymatic hydrolysis of native and modified protein; secretin test as diagnostic aid. Mailing Add: 2042 Charlton St Apt 204 Ann Arbor MI 48103

MILLER, LLOYD FREDERICK, preventive medicine, see 12th edition

MILLER, LORRAINE THERESA, b Dane, Wis, Mar 12, 31. NUTRITION, BIOCHEMISTRY. Educ: Univ Wis, BS, 53, MS, 58, PhD(nutrit, biochem), 67. Prof Exp: Dietitian, Med Ctr, Univ Mich, 54-56; instr foods & nutrit, Mich State Univ, 58-63; asst prof, 66-69, ASSOC PROF FOODS & NUTRIT, ORE STATE UNIV, 69- Mem: Am Dietetic Asn. Res: Metabolism of tryptophan in vitamin B6 deficiency; metabolism of taurine; determination of vitamin B6. Mailing Add: Dept of Foods & Nutrit Ore State Univ Corvallis OR 97331

MILLER, LOUIS HOWARD, b Baltimore, Md, Feb 4, 35; m 59; c 1. TROPICAL MEDICINE, PARASITOLOGY. Educ: Haverford Col, BA, 56; Wash Univ, MD, 60; Columbia Univ, MS, 64. Prof Exp: Intern, Mt Sinai Hosp, New York, 60-61; resident med, Montefiore Hosp, Bronx, 61-62 & Mt Sinai Hosp, 62-63; res physician, SEATO Med Res Lab, Bangkok, Thailand, 65-67; asst prof trop med, Col Physicians & Surgeons, Columbia Univ, 67-71, assoc prof, 71; HEAD SECT MALARIA, LAB PARASITIC DIS, NAT INST ALLERGY & INFECTIOUS DIS, 71- Concurrent Pos: NIH fel, Cedar-Sinai Med Ctr, 64-65. Mem: AAAS; fel Am Col Physicians; Am Soc Trop Med & Hyg; Infectious Dis Soc Am; Royal Soc Trop Med & Hyg. Res: Malaria; ultrastructure; immunology; physiology. Mailing Add: Lab of Parasitic Dis Bldg 5 Nat Inst Allergy & Infect Dis Bethesda MD 20014

MILLER, LOWELL D, b Chicago, Ill, Jan 20, 33; m 59; c 2. BIOCHEMISTRY, PHARMACOLOGY. Educ: Univ Mo, BS, 57, MS, 58, PhD(biochem), 60. Prof Exp: Res asst biochem, Univ Mo, 57-60; res biochemist, Neisler Labs, Inc, 60-61, sr res biochemist, 61-64, dir biol res, 64-66, dir toxicol & biochem, Union Carbide Develop Dept, Ill, 66-69; assoc dir biomed res, Warren-Teed Pharmaceut, 69-71; tech dir clin labs & pres, Lab Exp Biol, 71-73; MEM STAFF, MARION LABS, INC, 73- Concurrent Pos: Mem fac, Millikin Univ, 63-68. Mem: Am Chem Soc; NY Acad Sci; Am Asn Clin Chem; Soc Toxicol. Res: Pharmaceutical research; development of screening tests for biological activity; drug metabolism and toxicology. Mailing Add: Marion Labs Inc 10236 Bunker Ridge Rd Kansas City MO 64137

MILLER, LYLE DEVON, b Lebanon, Ind, Dec 8, 38; m 62; c 2. VETERINARY PATHOLOGY. Educ: Kans State Univ, BS, 61, DVM, 63; Univ Wis-Madison, MS, 68, PhD, 71. Prof Exp: PATHOLOGIST, VET SERV, NAT ANIMAL DIS CTR, USDA, 71- Mem: Am Vet Med Asn; Am Asn Vet Lab Diagnosticians; Am Col Vet Pathologists; US Animal Health Asn; Nat Asn Fed Vets. Mailing Add: Nat Animal Dis Ctr Ames IA 50010

MILLER, LYLE HERBERT, b Des Moines, Iowa, Nov 29, 27; m 47; c 3. NEUROPSYCHOLOGY, PSYCHOPHYSIOLOGY. Educ: Drake Univ, BA, 62; Duke Univ, PhD(psychol), 67. Prof Exp: Res asst prof neurosurg, Univ Wash Hosp, 67-68; assoc prof psychiat & head div psychol, La State Univ Med Ctr, 68-73; PROF MED PSYCHOL, MED SCH, TEMPLE UNIV, 73- Mem: AAAS; Am Psychol Asn; Soc Psychophysiol Res. Res: Polypeptide influences on brain-behavior mechanisms. Mailing Add: Dept of Psychiat Temple Univ Med Sch Philadelphia PA 19129

MILLER, LYNN, b McCook, Nebr, Nov 6, 32; m 57; c 2. MICROBIAL GENETICS. Educ: San Francisco State Col, BS, 57; Stanford Univ, PhD(biol), 62. Prof Exp: NIH fel microbiol, Hopkins Marine Sta, Stanford Univ, 62-64; NIH fel genetics, Univ Wash, 64-65; asst prof biol, Am Univ, Beirut, 65-68 & Adelphi Univ, 68-70; assoc prof, 70-74, PROF BIOL, HAMPSHIRE COL, 74- Mem: AAAS; Genetics Soc Am; Am Inst Biol Sci; Am Soc Microbiol. Res: Sterol metabolism and genetics in Saccharomyces; human population genetics. Mailing Add: Sch of Natural Sci Hampshire Col Amherst MA 01002

MILLER, LYSTER KEITH, b Tonapah, Nev, Aug 8, 32; m 60; c 3. ENVIRONMENTAL PHYSIOLOGY, COMPARATIVE PHYSIOLOGY. Educ: Univ Nev, BS, 55, MS, 57; Univ Alaska, PhD(zoophysiol), 66. Prof Exp: Res physiologist, Arctic Health Res Ctr, USPHS, Alaska, 60-62; instr, 62-66, asst prof, 66-69, ASSOC PROF ZOOPHYSIOL, INST ARCTIC BIOL, UNIV ALASKA, 69- Mem: AAAS; Am Soc Zoologists; Am Physiol Soc; Entom Soc Am; Soc Cryobiol. Res: Comparative neurophysiology insect freezing tolerance and physiology of marine mammals; temperature regulation; biometeorology of cold regions. Mailing Add: Inst of Arctic Biol Univ of Alaska Fairbanks AK 99701

MILLER, MALCOLM RAY, b Salt Lake City, Utah, Dec 31, 15; m 42; c 1. ANATOMY. Educ: Univ Wash, BA, 37; Harvard Univ, MA, 39; Univ Calif, Los Angeles, PhD(zool), 42; Univ Calif, MD, 45. Prof Exp: Asst zool, Univ Calif, Los Angeles, 39-42; resident med, Vet Admin Hosp, San Francisco, 48-49; from asst prof to assoc prof anat, Sch Med, Stanford Univ, 49-58; assoc prof, 58-60, PROF ANAT, SCH MED, UNIV CALIF, SAN FRANCISCO, 60- Concurrent Pos: Guggenheim fel, Ger, 55-56 & 66-67; prof & chmn dept anat, Univ Zambia, 68-70. Mem: Am Soc Ichthyol & Herpet; Ecol Soc Am; Am Soc Zool; Am Asn Anat. Res: Comparative endocrinology; neuroanatomy. Mailing Add: Dept of Anat Univ of Calif Sch of Med San Francisco CA 94143

MILLER, MARCIA ANN, microbiology, see 12th edition

MILLER, MARIA G, b Rosshaupten, WGer, Oct 16, 37; m 71. ANIMAL BEHAVIOR. Educ: Univ Munich, WGer, Staatsexamen, 61; City Univ New York, PhD(biol), 74. Prof Exp: Studienrat biol & chem, Wilhelm-Gymnasium, Hamburg, WGer, 62-69; ASST PROF BIOL SCI, BARNARD COL, COLUMBIA UNIV, 74- Concurrent Pos: NIMH fel, 74. Mem: AAAS; Ger Scientists' Union; NY Acad Sci; Animal Behav Soc; Am Soc Zoologists. Res: Role of oral-pharyngeal sensation in the adjustment of intake to metabolic requirements and environmental conditions as it contributes to the regulation of body weight in birds and mammals. Mailing Add: Dept of Biol Sci Barnard Col Columbia Univ New York NY 10027

MILLER, MARTIN WESLEY, b Belden, Nebr, Jan 8, 25; m 48; c 3. MICROBIOLOGY. Educ: Univ Calif, AB, 50, MS, 52, PhD(microbiol), 58. Prof Exp: Res food technologist, 57-59, from asst prof to assoc prof, 59-70, PROF FOOD TECHNOL, UNIV CALIF, DAVIS, 70- Concurrent Pos: Consult, Univ Calif, 57-; Fulbright sr res scientist award, Univ Australia, 64-65. Mem: Am Soc Microbiol; Inst Food Technologists; fel Am Acad Microbiol. Res: Ecology and taxonomy of yeasts;

dehydration and drying of fruits; food science and technology; food fermentations. Mailing Add: Dept of Food Sci Technol Univ of Calif Davis CA 95616

MILLER, MARVIN E, b Convoy, Ohio, Mar 29, 34; m 59; c 2. METEOROLOGY. Educ: Bowling Green Univ, BS, 56; Univ Mich, MS, 64. Prof Exp: Forecaster, US Air Force, 56-59; forecaster & briefer, Indianapolis, 60-61, res meteorologist, Cincinnati, 61-66, state climatologist, Columbus, Ohio, 66-71, meteorologist-in- chg, Columbus, 71-72 & Wilmington, NC, 72-74, METEOROLOGIST-IN-CHG, FORECAST OFF, NAT WEATHER SERV, 74- Honors & Awards: Bronze Medal, Dept of Com, 74; Professionalism Award, Nat Weather Serv, 75. Mem: Am Meteorol Soc. Res: Applied meteorology; climatology. Mailing Add: Nat Weather Serv Forecast Off Kanawha Airport Charleston WV 25311

MILLER, MARVIN FRED, b Chicago, Ill, Sept 16, 24; m 44; c 4. PSYCHIATRY. Educ: State Univ Iowa, MD, 49. Prof Exp: Resident psychiat, USPHS, 50-53, staff psychiatrist, 53-57; from asst prof to assoc prof, 57-69, PROF PSYCHIAT, SCH MED, LA STATE UNIV, NEW ORLEANS, 69- Mem: Am Psychiat Asn. Mailing Add: Dept of Psychiat La State Univ Sch of Med New Orleans LA 70112

MILLER, MARVIN KAY, plant breeding, genetics, see 12th edition

MILLER, MARY H, b Cascade, Mo, Nov 5, 26; m 52; c 1. GEOLOGY. Educ: Univ Mo, AB, 47, MA, 55. Prof Exp: Instr geol, Univ Mo, 51-52; asst commodity geologist, US Geol Surv, 52-55; ed asst, Geol Soc Am, 55-56; asst commodity geologist, 56-63, GEOLOGIST, US GEOL SURV, 63- Mem: Geol Soc Am. Res: Economic geology of antimony and bismuth deposits; plains base-metal mineralization. Mailing Add: US Geol Surv Fed Ctr Lakewood CO 80225

MILLER, MATTHEW WILLIAM, b Columbus, Mont, Mar 14, 15. PHOTOGRAPHIC CHEMISTRY. Educ: Mont State Col, BS, 36; Univ Ill, PhD(org chem), 40. Prof Exp: Asst chem, Univ Ill, 36-39; sr chemist, Resinous Prods & Chem Co, Pa, 40-42; chief, Sci Br, Field Info Agencies Technol, Ger, 46-47, sr chemist, Org Sect, 47-49; group leader, Cent Res Dept, 3M Co, 49-51, asst dir, 51-53, bus mgr, 53-54, tech dir, Abrasives Div, 54-62, dir sci & tech commun, 62-66, CHMN, MINN 3M RES, LTD, 3M CO, 66-, GROUP TECH DIR, PHOTOG PROD, 71- Concurrent Pos: Tech dir, Ferrania SpA, Ferrania, Italy, 67-70; photo prod lab, 70-71; liason off, Off Sci Res & Develop. Mem: AAAS; Am Chem Soc; Soc Chem Indust; The Chem Soc. Res: Resins and polymers; synthetic organic chemistry. Mailing Add: 3M Co 3M Ctr St Paul MN 55101

MILLER, MAURICE MAX, b New Albany, Ind, Feb 18, 29. NUCLEAR PHYSICS. Educ: Ind Univ, AB, 48, MS, 50, PhD(nuclear physics), 52. Prof Exp: Asst, Off Naval Res, Ind, 48-52; sr nuclear engr, Convair Div, Gen Dynamics Corp, 52-53; mem staff physics, Los Alamos Sci Lab, 53, instr nuclear weapons, Armed Forces Spec Weapons Proj, 53-55; mgr, Nuclear Lab Div, Lockheed Aircraft Corp, 55-62, mgr nuclear aerospace div, Lockheed, Ga Co, 62-72, CONSULT ENGR, LOCKHEED MISSILES & SPACE CO, 72- Mem: Am Phys Soc; Am Nuclear Soc. Res: Effects of nuclear radiation on organic and metallic materials; scintillation spectroscopy and measurement of nuclear decay schemes; experimental analysis of beta decay matrix elements; theory of strong interactions. Mailing Add: Lockheed Missiles & Space Co Dept 81-10 Bldg 154 PO Box 504 Sunnyvale CA 94088

MILLER, MAX, b New Haven, Conn, June 22, 10; m 40; c 2. MEDICINE. Educ: Yale Univ, BS, 31, MD, 35; Am Bd Internal Med, dipl, 43. Prof Exp: Intern Trudeau Sanatorium, 34; intern med, New Haven Hosp, 35-36, asst resident, 36-37; from instr to assoc prof, 40-67, PROF MED, SCH MED, CASE WESTERN RESERVE UNIV, 67- Concurrent Pos: Teaching fel, Sch Med, Case Western Reserve Univ, 37-40; asst, Sch Med, Yale Univ, 36-37; from asst physician to assoc physician, Case Western Reserve Univ Hosps, 40-67, physician, 67-; mem gen med study sect, NIH, 57-59, chmn univ group diabetes prog, 59-; consult coop study oral hypoglycemic agents in diabetes mellitus, Vet Admin, 58-65, Vet Admin Hosp, Cleveland, 63-; dean's rep health coun planning bd, Welfare Fedn Cleveland, 62-65; spec consult, Diabetes & Arthritis Control Prog, Bur State Serv, USPHS, 63-69, Nat Inst Arthritis & Metab Dis, 65-, ed adv, Diabetes Lit Index, 66-, mem ad hoc adv comt diabetes, Inst, 67-; spec adv comt diabetic retinopathy, Coun, Nat Inst Neurol Dis & Stroke, 68-; diabetes consult, Off of Dir, Nat Eye Inst, 71-; mem ed bd, Diabetes, 55-70; mem policy adv group, Diabetic Retinopathy Study, Nat Eye Inst, 72- Mem: AAAS; Am Diabetes Asn; Am Fedn Clin Res; Am Soc Clin Invest; Cent Soc Clin Res. Res: Metabolic diseases, especially diabetes and renal disease; intermediary carbohydrates metabolism; epidemiology of diabetes. Mailing Add: Dept of Med Case Western Reserve Univ Cleveland OH 44106

MILLER, MAX JOSEPH, b Saskatoon, Sask, Feb 10, 15; m 47; c 3. MEDICAL PARASITOLOGY, TROPICAL MEDICINE. Educ: Univ Sask, BS, 34; McGill Univ, MS, 36; Tulane Univ, PhD(parasitol), 39; Queen's Univ, MDCM, 43. Prof Exp: Assoc prof parasitol, Inst Parasitol, McGill Univ, 46-52; dir trop med, Liberian Inst Trop Med, 53-56; pres, Am Found Trop Med, New York, 56-62; consult pract trop med, Montreal, Que, 63-67; assoc prof med, Med Fac, McGill Univ, 68-71; PROF TROP MED & CHMN DEPT, TULANE UNIV, 69- Concurrent Pos: Merck sr fel, Sch Trop Med, Calcutta, India, 49-50; consult, Indian Health Serv, Can, 67-72; mem adv panel on parasitic dis, WHO, 68-, mem adv panel on onchocerciasis control, 74- Mem: Royal Soc Trop Med & Hyg; Am Soc Trop Med & Hyg. Res: Medical parasitology with emphasis on clinical and epidemiological studies. Mailing Add: Dept of Trop Med Tulane Univ New Orleans LA 70112

MILLER, MAX K, b Cleburne, Tex, Oct 25, 34; m 58; c 2. MATHEMATICS, GEOPHYSICS. Educ: Univ Tex, Austin, BS & BA, 57, MA, 63, PhD(math), 66. Prof Exp: Res scientist, Defense Res Lab, Univ Tex, Austin, 60-65; RES GEOPHYSICIST, TEX INSTRUMENTS, INC, 65- Concurrent Pos: Teaching asst math, Univ Tex, Austin, 63-65. Mem: Am Math Soc; Acoust Soc Am; Soc Indust & Appl Math; Soc Explor Geophys. Res: Numerical inversion of Laplace transforms; acoustic and seismic wave propagation; numerical analysis. Mailing Add: 7407 Valburn Dr Austin TX 78731

MILLER, MAX WARREN, organic chemistry, see 12th edition

MILLER, MAYNARD MALCOLM, b Seattle, Wash, Jan 23, 21; m 51; c 2. GEOLOGY. Educ: Harvard Univ, SB, 43; Columbia Univ, MA, 48; Cambridge Univ, PhD(geol), 56. Prof Exp: Asst prof navig, Princeton Univ, 46; geologist, Gulf Oil Corp, Cuba, 47; geologist, Off Naval Res, 49-51; vis staff mem, Fed Inst Snow & Avalanche Res, Switz, 52; demonstr phys geog, Cambridge Univ, 54; res assoc geol, Lamont Geol Observ, NY, 55-57; sr scientist, Columbia Univ, 58-59; from asst prof to assoc prof, 59-63, PROF GEOL, MICH STATE UNIV, 64-, DIR GLACIOL & ARCTIC SCI INST, 60- Concurrent Pos: Mem expeds, Alaska & West Can, 40-42, 46-53, 58-72, Argentine Patagonia, Chile & Peru, 49, Mt Rainier & Cascade Mts, 50, 58-60 & 70-72, Greenland, 51-52, Ellesmere Land, Arctic Ocean & N Pole, 51, 58, 67 & 70, Arabia & India, 54, Southeast Asia & Japan, 55 & Norway & Switz, 62; dir & geologist, Juneau Icefield Res Prog, Alaska, 46-; exec dir, Found Glacier Res, Wash, 55-; consult, Boeing Co & US Air Force, 59-60, State of Alaska, 61-63 & US Forest Serv, 63-; geologist, Am Mt Everest Exped, 63; dir, Alaskan Commemorative Proj, Nat Geog Soc, 64-70; field dir, Mt Kennedy-Yukon Exped, 65; dir Lemon Glacier Proj, Int Hydrol Decade Prog, Alaska, 65-72; dir US Navy Oceanog Off Sea Ice Proj, 67-68; chmn World Explor Ctr Found, 68-70. Honors & Awards: Co-recipient, Hubbard Medal, Nat Geog Soc, 63, Franklin L Burr Prize, 67; Elisha Kent Kane Medal, Philadelphia Geog Surv, 64, Karo Award, Soc Am Mil Eng, 65. Mem: Fel Geol Soc Am; Am Meteorol Soc; Am Soc Photogram; Am Geophys Union; Royal Geog Soc. Res: Glacial geology; glaciology; glacio-meteorology; hydrology; geomorphology; volcanology; photogrammetry; inter-disciplinary factors in variation of existing glaciers and sea ice, and related problems in Pleistocene geology and climatology; programming for motivational development and career orientation in the field and exploration sciences. Mailing Add: Dept of Geol Mich State Univ East Lansing MI 48823

MILLER, MEARLE MARION, hydrology, see 12th edition

MILLER, MELVIN P, b Baltimore, Md, May 17, 35; m 64; c 1. PHYSICAL CHEMISTRY. Educ: Loyola Col, Md, BS, 57; Princeton Univ, PhD(molten salts), 62. Prof Exp: From instr to assoc prof, 60-69, PROF CHEM, LOYOLA COL, MD, 69- Concurrent Pos: NSF grant, Conf Surface Colloid & Macromolecular Chem, Lehigh Univ, 65; res partic for col teachers, Boston Univ, 69; sci fac fel, Johns Hopkins Univ, 71-72, courtesy fel, 72-, prof eve col, 75- Mem: Am Chem Soc. Res: Transport properties of molten salts; heterogeneous catalysis; electroanalytical techniques; surface chemistry; infrared studies of carbon monoxide on metal catalysts. Mailing Add: Dept of Chem Loyola Col Baltimore MD 21210

MILLER, MEREDITH, b Murfreesboro, Tenn, Jan 2, 22; m 46; c 3. PHYSICAL CHEMISTRY. Educ: Vanderbilt Univ, BA, 43, MS, 44; Univ Wis, PhD(chem), 50. Prof Exp: Instr chem & physics, Middle Tenn State Col, 46-47; res chemist, Film Dept, E I du Pont de Nemours & Co, NY, 50-54, Va, 54-59, staff scientist, 59-63; prin chemist, 63-67, PROG MGR, HUNTSVILLE DIV, THIOKOL CHEM CORP, 67- Mem: Am Chem Soc. Res: Mechanical and physical properties of elastomers; solid propellant rocket motors. Mailing Add: Hunstville Div Thiokol Chem Corp Huntsville AL 35807

MILLER, MILLAGE CLINTON, III, b Enid, Okla, Aug 28, 32; m 65. BIOSTATISTICS. Educ: Univ Okla, BS, 54, MA, 60, PhD(biostatist), 61. Prof Exp: NIH trainee, Med Ctr, Univ Okla, 59-61; grant, Okla State Univ, 61-62; assoc prof prev med, Univ Okla, 62-67; assoc prof biostatist, Tulane Univ, 67-69; PROF BIOMET & CHMN DEPT, MED UNIV SC, 69- Concurrent Pos: Consult, Fed & State Govts & various acad & res insts, 59-; joint appointment, Biostatist Unit & Med Res Comput Ctr, Univ Okla Med Ctr, 62-66, assoc prof prev med & pub health, Sch Med, 66-67. Mem: Biomet Soc; Am Statist Asn; Inst Math Statist; Asn Comput Mach; Am Pub Health Asn. Res: Experimental design; multivariate analysis; medical application of statistics; biomedical applications of computers. Mailing Add: Dept of Biomet Med Univ of SC Charleston SC 29401

MILLER, MILTON ALBERT, b Pittsburg, Mo, Dec 2, 07; m 41; c 2. ZOOLOGY. Educ: Univ Ill, BA, 29; Univ Calif, PhD(zool), 34. Prof Exp: Asst, Univ Calif, 32-33, assoc, Col Agr, 34, instr, 34-35, instr, Exten Div, 35; from instr to asst prof, Univ Hawaii, 35-41; marine biologist, US Naval Biol Lab, Calif, 41-43 & Oceanog Inst, Woods Hole, Mass, 43-44; from asst prof & asst zoologist to assoc prof & assoc zoologist, Agr Exp Sta, 44-56, prof, 56-74, zoologist, 56-71, chmn dept zool, 59-64, EMER PROF ZOOL, UNIV CALIF, DAVIS, 74- Concurrent Pos: Instr, Golden Gate Jr Col, 35 & San Diego State Col, 42. Mem: Fel AAAS; Soc Syst Zool. Res: Systematics, ecology and distribution of isopod and tanaidaceon Crustacea; biology and prevention of marine fouling; biology and control of pocket gophers. Mailing Add: Dept of Zool Univ of Calif Davis CA 95616

MILLER, MILTON H, b Indianapolis, Ind, Sept 1, 27; m; c 3. PSYCHIATRY. Educ: Ind Univ, BS, 46, MD, 50. Prof Exp: Intern, Indianapolis Gen Hosp, 51; resident, Menninger Sch Psychiat, 53; from instr to prof, 55-72, chmn dept, 62-72, PROF PSYCHIAT & HEAD DEPT, UNIV BC, 72- Concurrent Pos: Examr, Royal Col Psychiat, Can; mem exec bd, World Fedn Ment Health. Mem: Fel Royal Col Psychiat; fel Am Psychiat Asn; Can Psychiat Asn. Mailing Add: Dept of Psychiat Univ of BC Health Sci Centre Hosp Vancouver BC Can

MILLER, MILTON LEONARD, b McKeesport, Pa, Feb 12, 04; m 38; c 1. PSYCHIATRY. Educ: Harvard Univ, AB, 25, Harvard Med Sch, MD, 29. Prof Exp: Asst, Dept Psychiat & Neurol, Cornell Med Sch-New York Hosp, 35-36; staff mem, Chicago Inst Psychoanal, 39-42; pres, Southern Calif Psychoanal Inst, 50-59; PROF PSYCHIAT, SCH MED, UNIV NC, CHAPEL HILL, 59- Concurrent Pos: Rockefeller fel neurol, Nat Hosp, Queens Sq, London, 34-35 & psychiat, Payne-Whitney Psychiat Clin, New York Hosp, 35-36; assoc, Dept Criminol & Jurisp, Univ Ill Med Sch, 39-40; training analyst, Los Angeles Psychoanal Inst, 46-50; pres & training analyst, Inst Psychoanal Med Southern Calif, 50-59; assoc clin prof psychiat, Med Sch, Univ Southern Calif, 53-59; sr attend psychiatrist, Los Angeles County Gen Hosp, 53-59; mem & training analyst, Washington Psychoanal Inst, 60-65; dir & training analyst, Ilniv NC-Duke Univ Psychoanal Training Prog, 60-; mem, Bd Dirs, Am Fund Psychiat, 61-69; consult, Bur Hearings & Appeals, Dept Health, Educ & Welfare, 66- Mem: AAAS; Am Psychosom Soc; Int Psychoanal Asn; fel Am Psychiat Asn; fel Am Psychoanal Asn. Res: Psychosomatic research; psychobiography; psychoanalysis. Mailing Add: Dept of Psychiat Univ of NC Sch of Med Chapel Hill NC 27514

MILLER, MORTON W, b Neptune, NJ, Aug 4, 36; m 59. RADIOBIOLOGY, CYTOGENETICS. Educ: Drew Univ, BA, 58; Univ Chicago, MS, 60, PhD(bot), 62. Prof Exp: Res assoc radiobiol, Brookhaven Nat Lab, 63-65; second officer, Int Atomic Energy Agency, 65-67; MEM STAFF, DEPT RADIATION BIOL & BIOPHYS, SCH MED & DENT, UNIV ROCHESTER, 67-, ASST DIR ATOMIC ENERGY PROJ, 69- Concurrent Pos: NATO fel, Oxford Univ, 62-63. Mem: AAAS; Radiation Res Soc; Am Inst Biol Sci; Environ Mutagen Soc. Res: Radiation genetics and sensitivity. Mailing Add: Dept of Radiation Biol & Biophys Univ Rochester Sch of Med & Dent Rochester NY 14642

MILLER, MURRAY HENRY, b Ont, Can, July 10, 31; m 54; c 2. SOIL FERTILITY. Educ: Ont Agr Col, BSA, 53; Purdue Univ, MS, 55, PhD(agr), 57. Prof Exp: From asst prof to assoc prof, 57-66, PROF SOIL SCI, ONT AGR COL, UNIV GUELPH, 66- Concurrent Pos: Head dept soil sci, Univ Guelph, 66-71. Mem: Am Soc Agron; Agr Inst Can; Can Soc Soil Sci. Res: Soil fertility, especially chemistry of nutrient elements in soils and their absorption by plants; plant nutrients and environmental quality. Mailing Add: Dept of Land Resource Sci Ont Agr Col Univ of Guelph Guelph ON Can

MILLER, MYRON, b Rochester, NY, Mar 31, 33; m 56; c 3. INTERNAL MEDICINE, ENDOCRINOLOGY. Educ: Univ Ill, Urbana, BS, 55; State Univ NY Upstate Med Ctr, MD, 59. Prof Exp: Intern & resident internal med, Univ Wis-

Madison, 59-63; instr med, 65-67, asst prof med & clin investr endocrinol, 67-71, assoc prof med, 71-75, PROF MED, STATE UNIV NY UPSTATE MED CTR, 75- Concurrent Pos: NIH fel endocrinol, Univ Wis-Madison, 62-63; res assoc, Vet Admin Hosp, Syracuse, NY, 65-67, clin investr, 67-71, chief med serv, 72- Mem: Int Soc Neuroendocrinol; Endocrine Soc; Am Col Physicians; Am Fedn Clin Res; Am Physiol Soc. Res: Hypothalamic-pituitary regulation with special interest in the regulation of posterior pituitary function and the role of altered posterior pituitary function in disease states associated with abnormal water regulation. Mailing Add: Dept of Med State Univ of NY Upstate Med Ctr Syracuse NY 13210

MILLER, NATHAN C, b Winnfield, La, Dec 13, 37; m 67. ORGANIC CHEMISTRY. Educ: Emory Univ, BA, 59; Fla State Univ, PhD(org chem), 64. Prof Exp: Res assoc, Univ Vt, 64-65; asst prof, Winthrop Col, 65-66; asst prof, 66-74, ASSOC PROF CHEM, UNIV S ALA, 74- Mem: Am Chem Soc. Res: Synthesis of small, bicyclic ring systems and study of decarboxylation of bridgehead B-keto acids; synthesis of alkaloids; stereoselective reactions. Mailing Add: Dept of Chem Univ of SAla Mobile AL 36608

MILLER, NEAL ELGAR, b Milwaukee, Wis, Aug 3, 09; m 48; c 2. PSYCHOPHYSIOLOGY. Educ: Univ Wash, BS, 31; Stanford Univ, MA, 32; Yale Univ, PhD(psychol), 35. Hon Degrees: DSc, Univ Mich, 65, Univ Pa, 68 & St Lawrence Univ. Prof Exp: Fel, Vienna Psychoanal Inst, Austria, 35-36; asst psychol, Inst Human Rels, Yale Univ, 36-41, res assoc, 41-42 & 46-50, prof, 50-52, James Rowland Angell prof, 52-66; PROF PSYCHOL & HEAD LAB PHYSIOL PSYCHOL, ROCKEFELLER UNIV, 66- Concurrent Pos: Mem fel comt, Found Fund Res Psychiat, 56-61; bd sci counrs, NIMH, 57-61; chmn div anthrop & psychol, Nat Res Coun, 58-60, chmn comt brain sci, 69-71; mem bd sci overseers, Jackson Mem Lab, 50-, chmn, 62-; Langfield lectr & Sigma Xi-Sigma Xi lectr, 68; bd sci counrs, Nat Inst Child Health & Human Develop, 69- Honors & Awards: Warren Medal, Soc Exp Psychol, 54; Cleveland Prize, AAAS, 57; Am Psychol Asn Award, 59; Nat Medal of Sci, 65; Gold Medal, Am Psychol Found, 75. Mem: Nat Acad Sci; Am Acad Arts & Sci; AAAS; Soc Exp Psychol; Am Psychol Asn (pres, 60-61). Res: Learning and behavior theory; conflict, fear and stress; mechanisms of psychosomatic effects; physiological and behavioral studies of motivation; electrical and chemical stimulation of the brain; instrumental learning of visceral responses. Mailing Add: Rockefeller Univ New York NY 10021

MILLER, NEIL AUSTIN, b Grand Rapids, Mich, Apr 9, 32; m 61; c 2. FOREST ECOLOGY. Educ: Mich State Univ, BSF, 58; Memphis State Univ, MS, 64; Southern Ill Univ, PhD(bot), 68; Oak Ridge Radiation Inst, grad, 71. Prof Exp: Teacher high sch & chmn dept biol & physics, Grand Rapids, Mich, 59-62; teacher high sch, Memphis, Tenn, 62-64; instr bot & forestry, Western Ky Univ, 64-65; res asst bot, Southern Ill Univ, 65-66, fel, 66-68; ASSOC PROF BOT, MEMPHIS STATE UNIV, 68- Concurrent Pos: Ill Acad Sci grant, 67-68; consult forestry. Mem: Soc Am Foresters; Am Soc Plant Physiol; Ecol Soc Am; Am Inst Biol Sci. Res: Plant-water balance research dealing with transpiration retardants and soil water conditions related to soil textures and structure characteristics; microclimatic investigations. Mailing Add: 1439 Whitewater Rd Memphis TN 38117

MILLER, NORMAN E, b Tinley Park, Ill, Aug 14, 31; m 52; c 7. INORGANIC CHEMISTRY. Educ: Northern Ill State Teachers Col, BS, 53; Univ Nebr, MS, 55, PhD, 58. Prof Exp: Chemist, Cent Res Dept, E I du Pont de Nemours & Co, 58-63; assoc prof, 63-66, PROF CHEM, UNIV S DAK, 66- Concurrent Pos: Mem fac, Assoc Western Univs, Los Alamos Sci Lab, 72. Mem: Am Chem Soc; Sigma Xi. Res: Lewis acid-base phenomena; inorganic synthesis; boronhydride synthesis and reactivity studies. Mailing Add: Dept of Chem Univ of SDak Vermillion SD 57069

MILLER, NORMAN GUSTAV, b Thermopolis, Wyo, Mar 20, 25; m 54; c 2. MICROBIOLOGY. Educ: Western Reserve Univ, BS, 48; Wash State Univ, MS, 50, PhD(bact), 53; Am Bd Microbiol, Dipl. Prof Exp: From instr to assoc prof, 55-67, PROF MICROBIOL, COL MED, UNIV NEBR, OMAHA, 67- Concurrent Pos: Secy, West-Northcent Interprof Seminar Dis Common to Animals & Man. Mem: AAAS; Am Soc Microbiol; Wildlife Dis Asn; Med Mycol Soc Am; Int Soc Human & Animal Mycol. Res: Infectious diseases of animals transmissible to man; medical mycology. Mailing Add: Dept of Microbiol Univ of Nebr Col of Med Omaha NE 68105

MILLER, NORMAN LEE, organic chemistry, see 12th edition

MILLER, NORTON GEORGE, b Buffalo, NY, Feb 4, 42; m 64. BOTANY. Educ: State Univ NY Buffalo, BA, 63; Mich State Univ, PhD(bot), 69. Prof Exp: Asst cur, Arnold Arboretum, Harvard Univ, 69-70; vis asst prof bot, Univ NC, Chapel Hill, 70-71, asst prof, 71-74; ASSOC PROF BOT, HARVARD UNIV & ASSOC CUR, ARNOLD ARBORETUM & GRAY HERBARIUM, 75- Mem: AAAS; Am Bryol & Lichenol Soc; Am Inst Biol Sci; Am Quaternary Asn; Am Soc Plant Taxon. Res: Systematics of the Bryophyta, especially Hepaticae; Quaternary paleoecology; pollen and plant macrofossil analysis. Mailing Add: Gray Herbarium Harvard Univ 22 Divinity Ave Cambridge MA 02138

MILLER, ORLANDO JACK, b Oklahoma City, Okla, May 11, 27; m 54; c 3. HUMAN GENETICS, OBSTETRICS & GYNECOLOGY. Educ: Yale Univ, BS, 46, MD, 50. Prof Exp: Intern, St Anthony Hosp, Oklahoma City, Okla, 50-51; asst resident obstet & gynec, Grace-New Haven Community Hosp, 54-57, resident & instr, 57-58; res asst human genetics, Univ Col, Univ London, 58-60; from instr to assoc prof obstet & gynec, 60-70, PROF HUMAN GENETICS & DEVELOP, OBSTET & GYNEC, COL PHYSICIANS & SURGEONS, COLUMBIA UNIV, 70- Concurrent Pos: Nat Res Coun fel anat, Yale Univ, 53-54; Pop Coun fel human genetics, Galton Lab, Univ Col, Univ London, 58-60; Josiah Macy, Jr fel obstet & gynec, Columbia Univ, 60-61; NSF sr fel, Oxford Univ, 68-69; career scientist, Health Res Coun, New York, 61-71. Mem: AAAS; Am Soc Human Genetics; Am Soc Cell Biol; Genetics Soc Am; Soc Gynec Invest. Res: Cytogenetics. Mailing Add: Dept of Human Genetics & Develop Col Physicians & Surgeons Columbia Univ New York NY 10032

MILLER, ORSON K, JR, b Cambridge, Mass, Dec 19, 30; m 54; c 3. MYCOLOGY. Educ: Univ Mass, BS, 52; Univ Mich, MF, 57, PhD(bot), 63. Prof Exp: Res forester, Northeastern Forest Exp Sta, USDA, 56-57; asst bot, Univ Mich, 58-59, fel, 60-61; plant pathologist, Intermt Forest & Range Exp Sta, USDA, 61-65 & Forest Dis Lab, 65-70; ASSOC PROF BOT, VA POLYTECH INST & STATE UNIV, 70-, CUR FUNGI, 74- Mem: AAAS; Mycol Soc Am; Bot Soc Am; Am Soc Plant Taxon; Arctic Inst NAm. Res: Taxonomy and physiology of Homobasidiomycetes; arctic and northern fungi. Mailing Add: Dept of Biol Va Polytech Inst & State Univ Blacksburg VA 24061

MILLER, ORVILLE H, b Portland, Ore, July 14, 13; m 39; c 6. PHARMACEUTICAL CHEMISTRY. Educ: Univ Wash, BS, 38, MS, 39, PhD(pharmaceut chem), 45. Prof Exp: Lab asst, Univ Wash, 38-41; drug analyst, Food & Drug Admin, Fed Security Agency, Calif, 41-47; dir res, Groner Labs, 47-48; res chemist, Consol Dairy Prod, 48-50; PROF PHARM, SCH PHARM, UNIV SOUTHERN CALIF, 50- Mem: Am Chem Soc; Am Soc Hosp Pharmacists; Am Pharmaceut Asn. Res: Imino barbiturates; alkaloids of polyploid stramonium; lactose production using ion exchange; protein hydrolyzates; solanaceous alkaloids; ointment bases; prolonged release dosage forms; chemical sterilization of pharmaceuticals. Mailing Add: Univ of Southern Calif Sch of Pharm Los Angeles CA 90007

MILLER, OSCAR L, JR, b Gastonia, NC, Apr 12, 25; m 48; c 2. CELL BIOLOGY. Educ: NC State Univ, BS, 48, MS, 50; Univ Minn, Minneapolis, PhD, 60. Prof Exp: Nat Inst Cancer fel, 60-61; res assoc, Biol Div, Oak Ridge Nat Lab, 61-63, res staff mem, 63-73; PROF BIOL & CHMN DEPT, UNIV VA, 73- Concurrent Pos: Prof, Univ Tenn-Oak Ridge Grad Sch Biomed Sci, 67-73. Mem: AAAS; Am Soc Cell Biol. Res: Correlation of fine structure and genetic activity in chromosomes of prokaryotic and eukaryotic cells; identification of specific genes in action. Mailing Add: Dept of Biol Univ of Va Charlottesville VA 22901

MILLER, OSCAR NEAL, b Canton, Mo, Feb 5, 19; m 47; c 2. NUTRITION, BIOCHEMISTRY. Educ: Univ Mo, BS, 41, MA, 42; Harvard Univ, PhD(biol chem), 50; Am Bd Nutrit, Dipl. Prof Exp: Asst state chemist, Mo Dept Agr, 42-43; res assoc, Children's Fund Mich, 46-47; instr biochem & nutrit, Med Br, Univ Tex, 50-51; from asst prof to prof biochem & med, Sch Med, Tulane Univ, 53-68; DIR DEPT BIOCHEM NUTRIT, HOFFMANN-LA ROCHE INC, 68-, ASSOC DIR BIOL RES, 70- Mem: Fel AAAS; Am Soc Clin Nutrit; Am Inst Nutrit; fel Am Col Clin Pharmacol; fel NY Acad Sci. Res: Enzymology; pharmacology. Mailing Add: Dept of Biochem Nutrit Hoffmann-La Roche Inc Nutley NJ 07110

MILLER, PARK HAYS, JR, b Philadelphia, Pa, Jan 22, 16; m 42; c 3. PHYSICS. Educ: Haverford Col, BS, 36; Calif Inst Technol, PhD(physics), 40. Prof Exp: Asst physics, Calif Inst Technol, 36-39; from instr to prof, Univ Pa, 39-56; chmn exp physics dept, Gen Atomic Div, Gen Dynamics Corp, 56-62, asst dir lab, 60-69; prof — physics & chmn dept, US Int Univ, Calif Western Campus, 69-74; SR PHYSICIST, GEN ATOMIC CO, 74- Concurrent Pos: Chmn dept physics, Univ Pa, 45-46; consult, US Naval Ord Lab, 51-65, US Dept Defense, 52-55 & 60-62, NASA, 58-60 & Mat Adv Bd, Nat Acad Sci, 56 & 60-61; actg ed, Rev Sci Instruments, Am Inst Physics, 53-55. Mem: Fel Am Phys Soc; Soc Explor Geophys; Am Geophys Union; Am Nuclear Soc; Am Asn Physics Teachers. Res: Electrical properties of solids; semiconductor devices; radiation effects; x-ray diffraction; optical instruments; energy conversion; design of fusion power reactors. Mailing Add: Gen Atomic Co PO Box 81608 San Diego CA 92138

MILLER, PATRICK MARTIN, b Springfield, Tenn, Mar 13, 24; m 54; c 2. PLANT PATHOLOGY. Educ: Univ Ill, BS, 50, MS, 52, PhD(plant path), 54. Prof Exp: From asst plant pathologist to assoc plant pathologist, 54-69, PLANT PATHOLOGIST, CONN AGR EXP STA, 69- Mem: Nematol Soc; Am Phytopath Soc. Res: Fruit diseases; nematology; fungicides. Mailing Add: Dept of Plant Path & Bot Conn Agr Exp Sta PO Box 1106 New Haven CT 06504

MILLER, PAUL, b Philadelphia, Pa, Jan 15, 22. SOLID STATE PHYSICS. Educ: George Washington Univ, BS, 43; Univ Pa, MS, 49, PhD(physics), 55. Prof Exp: MEM TECH STAFF, BELL LABS, INC, 54- Mem: AAAS; Am Phys Soc. Res: Surface and transistor physics; microelectronics; semiconductors. Mailing Add: Bell Labs 555 Union Blvd Allentown PA 18103

MILLER, PAUL DEAN, b Cedar Falls, Iowa, Apr 4, 41; m 65; c 1. ANIMAL BREEDING. Educ: Iowa State Univ, BS, 63; Cornell Univ, MS, 66, PhD(animal genetics), 68. Prof Exp: Asst prof animal breeding, Cornell Univ, 67-71; DIR BREEDING PROGS, AM BREEDERS SERV, 71- Concurrent Pos: Chmn, Beef Sire Eval Comt, Am Asn Animal Breeders. Mem: Biomet Soc; Am Soc Animal Sci; Am Dairy Sci Asn; Am Asn Animal Breeders. Res: Population genetics; statistical estimation and linear models; computer simulation. Mailing Add: Am Breeders Serv De Forest WI 53532

MILLER, PAUL GEORGE, b Milwaukee, Wis, Dec 5, 41. APPLIED MATHEMATICS, GENETICS. Educ: Marquette Univ, BSc, 63; Case Inst Technol, PhD(math), 68. Prof Exp: Asst math, Case Inst Technol, 63-68; asst prof, Okla State Univ, 68-73 & Langston Univ, 73-74; TECH REP, BURROUGHS CORP, 74- Mem: Am Math Soc; Genetics Soc Am; Am Genetics Asn. Res: Complex variable theory; boundary behavior of analytic functions; Riemann surfaces; statistical genetics; theory of inbreeding; development of special purpose testing stocks and improved commercial breeds of domestic avian species. Mailing Add: 2228 N 52 St Milwaukee WI 53208

MILLER, PAUL GILBERT, food technology, see 12th edition

MILLER, PAUL R, b Rockport, Ind, Apr 30, 05; m 33; c 3. PLANT PATHOLOGY. Educ: Univ Ind, BS, 29; Purdue Univ, MS, 31; George Washington Univ, PhD(plant path), 38. Prof Exp: Teacher sci, Burritt Col, 25-26 & high sch, Ark, 26-28; asst bot, Agr Exp Sta, Purdue Univ, 29-31; jr pathologist, Bur Plant Indust, USDA, 31-35, from asst pathologist to pathologist, 35-43, Bur Plant Indust, Soils & Agr Eng, 43-46, sr pathologist in-chg, 46-51, leader epidemiol invests, Plant Indust Sta, Beltsville, 51-71, liason, USDA & Fort Valley State Col, 71-72; RETIRED. Mem: Fel Am Phytopath Soc (past pres). Res: Plant disease epidemiology and forecasting. Mailing Add: 3407 Hallaton Ct Silver Spring MD 20906

MILLER, PAUL SCOTT, b Brooklyn, NY, Oct 12, 43; m 71. BIO-ORGANIC CHEMISTRY. Educ: State Univ NY Buffalo, BA, 65; Northwestern Univ, Ill, PhD(chem), 69. Prof Exp: Am Cancer Soc fel nucleic acid chem, 69-73, ASST PROF BIOCHEM & BIOPHYS SCI, BIOPHYS DIV, JOHNS HOPKINS UNIV, 73- Mem: AAAS; Am Chem Soc. Res: Chemical and enzymatic synthesis of nucleic acids and nucleic acid derivatives; interaction of nucleic acids with proteins and nucleic acids; chemical modification of nucleic acids. Mailing Add: Dept of Biochem & Biophys Johns Hopkins Univ Baltimore MD 21205

MILLER, PAUL THEODORE, b Conway, Iowa, Dec 31, 05; m 32; c 1. GEOLOGY. Educ: Simpson Col, BS, 27; Univ Iowa, MS, 30, PhD(geol), 32. Prof Exp: Asst lab, Univ Iowa, 27-29, geol, 29-32; res assoc, 32-34; field geologist, Amerada Petrol Corp, Tex, 35; prof geol & geog, NDak State Teachers Col, 35-36 & Wis State Teachers Col, Superior, 36-47; PROF GEOL & GEOG, ARIZ STATE UNIV, 47- Concurrent Pos: Head dept geol & geog, Ariz State Univ, 50-65. Mem: AAAS; Am Asn Petrol Geol; Nat Asn Geol Teachers. Res: Geography; sedimentation; petrography; water resources. Mailing Add: Dept of Geol Ariz State Univ Tempe AZ 85281

MILLER, PAUL THOMAS, b Atlanta, Ga, May 12, 44; m 66. INORGANIC CHEMISTRY. Educ: Birmingham-Southern Col, BS, 66; Vanderbilt Univ, MA & PhD(chem), 71. Prof Exp: Instr, 71-72, asst prof, 72-75, ASSOC PROF CHEM & PHYS SCI, VOLUNTEER STATE COMMUNITY COL, 75- Concurrent Pos: NSF res grant, 66; res grant, Tulane Univ Scholars & Fels Prog, 66; grad teaching fel, Vanderbilt Univ, 66-69 & 70-71. Mem: AAAS; Am Chem Soc; Sigma Xi. Res: Structure and properties of coordination compounds; chemical education in the two year college; analytical methods for environmental pollution studies; water pollution by heavy metals. Mailing Add: Dept of Chem Volunteer State Community Col Gallatin TN 37066

MILLER, PAUL WILLIAM, b Mt Vernon, Ind, May 2, 01; m 28. PLANT PATHOLOGY. Educ: Univ Ky, BS, 23, MS, 24; Univ Wis, PhD(plant path), 29. Prof Exp: Instr plant path, Univ Wis, 27-29; agent, 29-30, assoc plant pathologist, 30-44, plant pathologist, 44-70, EMER RES PLANT PATHOLOGIST, USDA, ORE STATE UNIV, 70- Res: Fire blight disease of apples; prune russet; vegetable seed diseases; walnut blight; filbert blight; strawberry root rot; strawberry virus diseases; control of rose mildew and rose rust. Mailing Add: 703 NW 30th St Corvallis OR 97330

MILLER, PAULINE MONZ, b Harrisburg, Pa, Apr 2, 31; div. BOTANY, INFORMATION SCIENCE. Educ: Pa State Univ, BS, 52; Univ Pa, PhD(bot), 56; Syracuse Univ, MSLS, 76. Prof Exp: Instr bot, Wheaton Col, 56-57; instr, Conn Col, 57-59, res assoc, 59-61; res assoc, Yale Univ, 61-62; res assoc, Syracuse Univ, 62-71, ADJ PROF BOT, UNIV COL, SYRACUSE UNIV, 71- Mem: Bot Soc Am; Am Libr Asn. Res: Light effects on plant development; organization and dissemination of science information. Mailing Add: 911 Euclid Ave Syracuse NY 13210

MILLER, PETER S, b Middleboro, Mass, June 14, 37; c 2. BIOLOGICAL ANTHROPOLOGY. Educ: Univ Nebr, Lincoln, BA, 62; Univ Ariz, MA, 67, PhD(phys anthrop), 69. Prof Exp: Instr anthrop, State Univ NY Albany, 67-69; asst prof anthrop, Univ Calif, Los Angeles, 69-72; ASST PROF ANTHROP, DREW UNIV, 72- Mem: AAAS; Am Anthrop Asn; Am Asn Phys Anthrop. Res: Biology and culture relationships; human ecology; American Indians—culture change and biological change; human growth. Mailing Add: Dept of Anthrop Drew Univ Madison NJ 07940

MILLER, PHILIP, b Bronx, NY, Aug 26, 38; m 70; c 1. ORGANIC CHEMISTRY, PESTICIDE CHEMISTRY. Educ: Rutgers Univ, BA, 61; Pa State Univ, MS, 63; Univ Mass, PhD(chem), 68. Prof Exp: Res chemist pharmaceut develop, 68-73, RES CHEMIST PESTICIDE METAB, AM CYANAMID CO, 73- Mem: Am Chem Soc. Res: Plant and animal metabolism of compounds potentially useful as pesticides and herbicides; environmental chemistry, including soil metabolism, photolysis, and hydrolysis of these potential new products; radiosynthesis. Mailing Add: Am Cyanamid Co PO Box 400 Princeton NJ 08540

MILLER, PHILIP ARTHUR, b Hastings, Nebr, Feb 1, 23; m 45; c 2. PLANT BREEDING. Educ: Univ Nebr, BSc, 43, MSc, 47; Iowa State Col, PhD(plant breeding), 50. Prof Exp: Res assoc agron, Iowa State Col, 49-50, asst prof, 50-52; assoc prof, 52-59, PROF AGRON, NC STATE UNIV, 59- Concurrent Pos: Dir, NC State Univ Agr Mission, Peru, 59-61. Mem: Fel Am Soc Agron; Sigma Xi. Res: Genetics; corn; cotton; soybeans; quantitative inheritance. Mailing Add: Dept of Crop Sci NC State Univ Raleigh NC 27607

MILLER, PHILIP CLEMENT, b Chicago, Ill, Mar 23, 33; m 58; c 2. ECOLOGY. Educ: Oberlin Col, BA, 54; Iowa State Univ, MS, 59; Univ Colo, PhD(plant ecol), 64. Prof Exp: Res asst ecol, Inst Arctic-Alpine Res, 60-64, res assoc, 64; asst prof ecol & biostatist, Univ Notre Dame, 64-65; asst prof ecol, 65-71, PROF BIOL, SAN DIEGO STATE UNIV, 71- Mem: AAAS; Ecol Soc Am; Am Inst Biol Sci; Am Meteorol Soc. Res: Mathematical models of ecological systems. Mailing Add: Dept of Biol San Diego State Univ San Diego CA 92115

MILLER, PHILIP DIXON, b Albuquerque, NMex, June 7, 32; m 64; c 4. NUCLEAR PHYSICS. Educ: Calif Inst Technol, BS, 54; Rice Inst, MA, 56, PhD(physics), 58. Prof Exp: PHYSICIST, OAK RIDGE NAT LAB, 58-, ACTG DIR, VAN DE GRAAFF LAB, 74- Mem: Fel Am Phys Soc. Res: Low energy physics using Van de Graaff accelerator, including charged particle scattering and reactions and neutron cross sections; measurement of electric dipole moment of neutron; atomic collisions physics. Mailing Add: 5500 Bldg Oak Ridge Nat Lab PO Box X Oak Ridge TN 37830

MILLER, PHILLIP ALLEN, b Ithaca, NY, Sept 5, 08; m 50; c 1. INORGANIC CHEMISTRY. Educ: Cornell Univ, BA, 30, PhD(phys chem), 36; Columbia Univ, MA, 44. Prof Exp: Res asst to Dr W C Geer, NY, 36; phys chemist, US Rubber Co, NJ, 36-37 & Latex Control Lab, Sumatra, 37-40; develop chemist, RI, 40-41; res chemist, Iowa, 41-43; res chemist, Off Naval Res, 46-47, phys sci coordr, 47-66, inorg chemist, Area Off, Calif, 66-68; RETIRED. Mem: Am Chem Soc. Res: Scientific administration and planning of chemistry contract research program. Mailing Add: 736 Leavenworth San Francisco CA 94109

MILLER, RALPH ALBERT, b Bethlehem, Pa, Feb 12, 28; m 47; c 3. FOOD SCIENCE. Educ: Moravian Col, BS, 49; Univ Del, MS, 50. Prof Exp: Res chemist, 50-55, mgr prod res, 55-57, assoc dir, 57-61, dir, 61-66, VPRES PROD DEVELOP, CAMPBELL SOUP CO, CAMDEN, 66- Mem: Inst Food Technologists. Res: Product research and development; research management. Mailing Add: 585 Tarrington Rd Haddonfield NJ 08034

MILLER, RALPH ENGLISH, b Hanover, NH, Sept 23, 33; m 62; c 3. PHARMACOLOGY. Educ: Dartmouth Col, AB, 58; Harvard Univ, MD, 61, MS, 66, DSc, 70. Prof Exp: Intern, Mary Hitchcock Mem Hosp, Hanover, NH, 61-62; NIMH fel, Walter Reed Army Inst Res, 62-64; NASA res & spec fel, Dept Physiol, Harvard Sch Pub Health, 65-69; spec fel physiol, Stanford Univ, 69-70; asst prof, 70-75, ASSOC PROF PHARMACOL, UNIV KY, 75- Concurrent Pos: Chief physiol sect, Dept Neuroendocrinol, Walter Reed Army Inst Res, 64-65. Mem: Am Diabetes Asn; Am Physiol Soc; Soc Neurosci; Endocrine Soc. Res: Neuro-endocrinol; autonomic neuroendocrinol; role of autonomic nervous system in regulation of hormone secretion from the pancreas and kidney. Mailing Add: Dept Pharmacol Col Med MN 504 Med Ctr Univ Ky Lexington KY 40506

MILLER, RALPH J, b Prattville, Mich, Mar 6, 11; m 37; c 3. PHYSICS. Educ: Greenville Col, AB, 36; Univ Mich, MA, 40; Washington Univ, MA, 49; Syracuse Univ, PhD(physics), 58. Prof Exp: Teacher & dean, Spring Arbor Jr Col, 36-45; teacher & dean men, Greenville Col, 45-51; asst prof physics, Wells Col, 51-55; asst, Syracuse Univ, 55-56; PROF PHYSICS, GREENVILLE COL, 56- Concurrent Pos: Teacher, Albion Col, 42-44; Fulbright-Hays lectr, Univ Aleppo, 64-6S. Mem: Am Asn Physics Teachers. Res: History of physics, especially the discovery of the electron; solid state physics, especially the growth of cadmium sulfide crystals. Mailing Add: Dept of Physics Greenville Col Greenville IL 62246

MILLER, RALPH LEROY, b Fountain Hill, Pa, Jan 28, 09; m 39; c 2. ECONOMIC GEOLOGY. Educ: Haverford Col, BS, 29; Columbia Univ, PhD(geol), 37. Prof Exp: Lab asst geol, Columbia Univ, 32-34, lectr, 34-37, instr, 37-46; assoc geologist, US Geol Surv, 42-43, geologist, 44-47, sr geologist, 47-49, chief, Navy Oil Unit, 48-51, Fuels Br, 51-57, staff geologist, 58, tech adv, For Geol Br, Afghanistan, 58, Mex, 59-61, Colombia, 62, Cent Am, 63-70, SR RES GEOLOGIST, OFF INT GEOL, US GEOL SURV, 70- Mem: Fel Geol Soc Am; Am Asn Petrol Geol; Am Inst Mining, Metall & Petrol Engrs. Res: Areal geology and stratigraphy of southeast Utah; areal geology, stratigraphy and structure of Appalachian Mountains from southern New York to Tennessee; manganese deposits of Appalachians; oil geology of southern Appalachians; geology and oil resources of the Arctic slope of Alaska; geology of Central America. Mailing Add: Off of Int Geol US Geol Surv Reston VA 22092

MILLER, RAYMOND ANTHONY, plant biochemistry, see 12th edition

MILLER, RAYMOND EDWIN, b Cincinnati, Ohio, July 6, 37; m 61; c 5. PHYSICS. Educ: Xavier Univ, Ohio, BS, 59; Johns Hopkins Univ, PhD(physics), 65. Prof Exp: Instr physics, Johns Hopkins Univ, 62-65, NASA fel, 65-66; from asst prof to assoc prof, 66-71, PROF PHYSICS & CHMN DEPT, XAVIER UNIV, OHIO, 71- Concurrent Pos: NASA grant, Johns Hopkins Univ, 68-71. Mem: AAAS; Am Asn Physics Teachers; Optical Soc Am; Am Asn Physicists in Med. Res: Atmospheric physics; atomic and molecular physics; biophysics. Mailing Add: Dept of Physics Xavier Univ Cincinnati OH 45207

MILLER, RAYMOND JARVIS, b Claresholm, Alta, Mar 19, 34; m 56; c 3. SOIL CHEMISTRY, PHYSICAL CHEMISTRY. Educ: Univ Alta, BS, 57; Washington State Univ, MS, 60; Purdue Univ, PhD(soil chem), 62. Prof Exp: From asst prof to assoc prof soil & phys chem, NC State Univ, 62-65; from assoc prof to prof, Univ Ill, Urbana, 65-73; DIR, IDAHO AGR EXP STA, 73-; ASSOC DEAN, COL AGR, UNIV IDAHO, 73- Concurrent Pos: Asst dir, Agr Exp Sta, Univ Ill, Urbana, 69, assoc dir, 70, coordr, Col Agr Coun Environ Qual, 70. Mem: Soil Sci Soc Am; Clay Minerals Soc; Am Soc Plant Physiol; Am Soc Agron; AAAS. Res: Clay-water interactions; structure of water in porous media and the effects of water structure on biological activity; agricultural research. Mailing Add: Col of Agr Univ of Idaho Moscow ID 83843

MILLER, RAYMOND SUMNER, b Rochester, NY, Oct 15, 19; m 45; c 2. ANALYTICAL CHEMISTRY. Educ: McMaster Univ, BA, 41. Prof Exp: Asst still operator, Eastman Kodak Co, 41-42; chemist, Manhattan Proj, Los Alamos, NMex, 44-45, asst to area engr, Oak Ridge, Tenn, 45-46; SR RES CHEMIST, EASTMAN KODAK CO, 46- Mem: Am Chem Soc; Am Nuclear Soc. Res: Photo theory; micro analytical technology; radio tracer technology. Mailing Add: 3 Woodview Dr Rochester NY 14624

MILLER, RAYMOND WOODRUFF, b St David, Ariz, Jan 13, 28; m 51; c 5. SOIL FERTILITY, ENVIRONMENTAL SCIENCE. Educ: Univ Ariz, BS, 52, MS, 53; Wash State Univ, PhD(agron, soil chem), 56. Prof Exp: From asst prof to assoc prof, 56-69, PROF SOILS, UTAH STATE UNIV, 69- Concurrent Pos: Consult, Centro Interamericano de Desarollo Integral de Aguas y Tierras, Venezuela, 69-71. Mem: Am Soc Agron; Soil Conserv Soc Am; Soil Sci Soc Am; Int Soil Sci Soc. Res: Soil mineralogy; soil genesis; solid waste management. Mailing Add: Dept of Soil Sci & Biometeorol Utah State Univ Logan UT 84321

MILLER, REGIS BOLDEN, b Meyesdale, Pa, Aug 29, 43; m 68; c 2. PLANT ANATOMY. Educ: WVa Univ, BS, 66; Univ Wis, MS, 68; Univ Md, PhD(bot), 73. Prof Exp: RES BOTANIST, FOREST PROD LAB, US FOREST SERV, USDA, 70- Mem: Int Asn Wood Anatomists; Int Asn Plant Taxon; Bot Soc Am; Asn Trop Biol. Res: Systematic wood anatomy of Flacourtiaceae, Juglandaceae and Boraginaceae; scanning and transmission electron microscopy of structures found on vessel walls of some Juglans. Mailing Add: US Forest Prod Lab Box 5130 Madison WI 53705

MILLER, RICHARD A, organic chemistry, chemical engineering, see 12th edition

MILLER, RICHARD ALAN, organic chemistry, see 12th edition

MILLER, RICHARD ALBERT, b Heber Springs, Ark, Mar 8, 12; m 42; c 2. MATHEMATICS. Educ: Univ Miss, AB, 33, MA, 34. Prof Exp: Asst prof math, Univ Miss, 34-35; asst, Univ Iowa, 35-38 & Univ Ill, 38-41; asst prof, Univ Miss, 46-56; proj nuclear physicist, Ft Worth Div, Gen Dynamics Corp, 56-67, proj opers res analyst, 67-70; MATHEMATICIAN, FREESE & NICHOLS, CONSULT ENGRS, 70- Concurrent Pos: Adj prof, Eve Col, Tex Christian Univ, 57- Mem: Math Asn Am; Soc Indust & Appl Math. Res: Algebraic and differential geometry; games; mathematical programming; hydrological studies; water resource studies. Mailing Add: 4071 W Seventh St Ft Worth TX 76107

MILLER, RICHARD ANTHONY, solar physics, spectroscopy, deceased

MILLER, RICHARD AVERY, b Erie, Pa, June 23, 11. ENDOCRINOLOGY. Educ: Univ Pittsburgh, BS, 32; Univ Iowa, MS, 34, PhD(endocrinol), 37. Prof Exp: Res cytologist, Dept Genetics, Carnegie Inst, 37-46; asst prof anat, Univ Minn, 46-48; from asst prof to assoc prof, 48-56, PROF ANAT, ALBANY MED COL, 56- Concurrent Pos: Vis scientist, NIH, 59-60. Mem: Am Soc Zool; Am Asn Anat. Res: Pituitary adrenal relations; secretory phenomena in cells; neural pathways in brain stem and thalamus. Mailing Add: Dept of Anat Albany Med Col Albany NY 12208

MILLER, RICHARD EDWARD, b Hollis, NY, May 31, 37; m 58. PHYSICAL CHEMISTRY, CHEMICAL PHYSICS. Educ: Stevens Inst Technol, BE, 59; Univ Wash, PhD(phys chem), 66. Prof Exp: Design engr, Boeing Co, 59-62; asst phys chem, Univ Wash, 62-66; res fel, Princeton Univ, 66-67; res fel, Mich State Univ, 67-68, asst prof, 68-69, lab mgr, 69-74; MEM FAC, OHIO STATE UNIV, 74- Concurrent Pos: Consult, Spectra Physics, Inc, 68-71. Res: Molecular and crystal structure using infrared and Raman spectroscopic techniques. Mailing Add: Dept of Chem Ohio State Univ Columbus OH 43210

MILLER, RICHARD FROMAN, organic chemistry, see 12th edition

MILLER, RICHARD GORDON, b Denison, Iowa, July 21, 13; m 39; c 2. VERTEBRATE ZOOLOGY. Educ: Principia Col, AB, 36; Cornell Univ, MS, 41; Stanford Univ, PhD(zool), 51. Prof Exp: Ranger-naturalist, Nat Park Serv, 41-42; dir, State Mus, Nev, 45-47; instr biol, Univ Nev, 47-48; instr, Stanford Univ, 50; assoc prof zool, Long Beach State Col, 51-60; DIR ENVIRON PROJS & INFO CTR, FORESTA INST OCEAN & MOUNTAIN STUDIES, 60- Concurrent Pos: Res grant, Sport Fishing Inst, 54-57; res grant dir, Nat Wildlife Fedn, 51 & 53; proj dir, Fish Res, Antarctic Prog, US Nat Comt, NZ Oceanog Inst, Int Geophys Yr, 58-59; observer, Arg Antarctic Exped, US Antarctic Projs Off, 59-60; mem, Survival Serv Comn & Permanent Comn Educ, Int Union Conserv Nature & Natural Resources. Mem: AAAS; Am Soc Mammal; Ecol Soc Am; Fisheries Soc Am; Am Soc Ichthyologists & Herpetologists. Res: Natural history; food habits; distribution of fishes; marine and continental wildlife conservation and management; ecological and environmental sciences; world protection of native flora and fauna and scenic, park and outdoor recreation resources; environmental education curriculum guidelines. Mailing Add: Foresta Inst Ocean & Mt Studies 6205 Franktown Rd Carson City NV 89701

MILLER, RICHARD GRAHAM, b St Catharines, Ont, Oct 2, 38; m 63; c 2.

BIOPHYSICS, IMMUNOBIOLOGY. Educ: Univ Alta, BS, 60, MS, 61; Calif Inst Technol, PhD(physics & biol), 66. Prof Exp: Res asst physics, Calif Inst Technol, 61-66, fel, 66; fel, 66-67, SR SCIENTIST BIOPHYS, ONT CANCER INST, 67- Concurrent Pos: Vis scientist, Walter & Eliza Hall Inst, Melbourne, Australia, 72-73; asst prof med biophys, Univ Toronto, 67-71, assoc prof, 71-, mem, Inst Med Sci, 70- & Inst Immunol, 73- Mem: Am Asn Immunologists; Can Soc Cell Biol; Can Soc Immunol. Res: Development of procedures for the analysis and sorting of viable mammalian cells; study of initiation of and cellular mechanisms in cell mediated immune responses. Mailing Add: Ont Cancer Inst 500 Sherbourne St Toronto ON Can

MILLER, RICHARD HENRY, b Aurora, Ill, Aug 31, 26; m 52. ASTROPHYSICS. Educ: Iowa State Col, BS, 46; Univ Chicago, PhD(physics), 57. Prof Exp: Engr construct cyclotron, Univ Chicago, 47-51, Calif Res & Develop Co div, Standard Oil Co Calif, 52 & Brazilian Nat Res Coun, 52-54; asst physics, 54-57, res assoc, 57-59, asst prof, 59-62, ASSOC PROF ASTROPHYSICS, UNIV CHICAGO, 63- Concurrent Pos: Assoc dir, Inst Comput Res, Univ Chicago, 62-63, dir, 63-66, actg chmn, Comt Info Sci, 65-66; mem summer study group, Stanford Linear Accelerator Ctr, 64; Nat Res Coun-NASA sr resident res assoc, Goddard Inst Space Studies, 67 & 68; consult astronr, Kitt Peak Nat Observ, 70 & 71; sr resident res assoc, NASA-Ames Res Ctr, 75. Mem: Am Phys Soc; Am Astron Soc; Asn Comput Mach. Res: Structure of extragalactic nebulae; computers on-line for experiments; stellar dynamics; astronomical instrumentation. Mailing Add: Inst for Comput Res 5640 Ellis Ave Chicago IL 60637

MILLER, RICHARD J, b Schenectady, NY, Aug 20, 37. PHYSICAL CHEMISTRY. Educ: Union Col, NY, BS, 59; Lehigh Univ, PhD(phys chem), 64. Prof Exp: Asst chem Lehigh Univ, 59-60 & 63-64; from asst prof to assoc prof, 64-72, PROF CHEM, STATE UNIV NY COL CORTLAND, 72- Mem: Am Chem Soc; AAAS. Res: Thermodynamics. Mailing Add: Dept of Chem State Univ of NY Col Cortland NY 13045

MILLER, RICHARD KEITH, b Clarinda, Iowa, Apr 19, 39; m 63; c 1. MATHEMATICS. Educ: Iowa State Univ, BS, 61; Univ Wis, MS, 62, PhD(math), 64. Prof Exp: Asst prof math, Univ Minn, Minneapolis, 64-66; from asst prof to assoc prof appl math, Brown Univ, 66-71; assoc prof math, 72-74, PROF MATH, IOWA STATE UNIV, 74- Mem: Am Math Soc; Soc Indust & Appl Math. Res: Asymptotic behavior of ordinary differential equations and Volterra integral equations. Mailing Add: Dept of Math Iowa State Univ Ames IA 50010

MILLER, RICHARD LEE, b Glendale, Calif, May 30, 42; m 64; c 2. BIOCHEMISTRY. Educ: San Jose State Col, BS, 64; Ariz State Univ, PhD(biochem), 69. Prof Exp: Res biochemist, 69-72, SR RES BIOCHEMIST, BURROUGHS WELLCOME & CO, 72- Mem: Am Chem Soc. Res: Purine metabolism; nucleotide interconversion; enzymology. Mailing Add: Wellcome Res Labs Burroughs Wellcome & Co Research Triangle Park NC 27709

MILLER, RICHARD LEE, b Boston, Mass, Apr 9, 40. INVERTEBRATE ZOOLOGY, DEVELOPMENTAL BIOLOGY. Educ: Univ Chicago, BS, 62, PhD(zool), 65. Prof Exp: NIH res fel, Calif Inst Technol & Univ Calif, Berkeley, 65-66; asst prof zool, Ore State Univ, 66-68; asst prof, 68-74, ASSOC PROF BIOL, TEMPLE UNIV, 74- Mem: AAAS; Am Soc Zoologists. Res: Chemical mediation of fertilization in the low coelenterates, including sperm attraction; chemical nature of the attractants and their mode of action. Mailing Add: Dept of Biol Temple Univ Philadelphia PA 19122

MILLER, RICHARD LLOYD, b Mishawaka, Ind, Jan 30, 31; m 52; c 7. ENTOMOLOGY, HORTICULTURE. Educ: Purdue Univ, BS, 57; Iowa State Univ, MS, 59, PhD(entom), 62. Prof Exp: Exten entomologist, Univ Ky, 62-67; assoc prof entom, 70-74, PROF ENTOM, OHIO STATE UNIV, 74-, EXTEN ENTOMOLOGIST, 67- Mem: Entom Soc Am. Res: Insect and mite control on fruits, vegetables, ornamentals, turf, yard and garden. Mailing Add: Ohio Coop Exten Serv Ohio State Univ Columbus OH 43210

MILLER, RICHARD PRESSLY, b Troy, Ohio, Feb 18, 22; m 44; c 3. BIOCHEMISTRY. Educ: Miami Univ, BA, 47, MA, 48; Univ Ill, PhD(chem), 51. Prof Exp: RES SCIENTIST, ELI LILLY & CO, 51- Mem: AAAS; Am Chem Soc. Res: Biochemistry of hypertension; separation methods and analytical biochemistry. Mailing Add: Eli Lilly & Co Indianapolis IN 46206

MILLER, RICHARD ROY, b Salt Lake City, Utah, Aug 3, 41. MATHEMATICS, CHEMISTRY. Educ: Univ Utah, BS, 63, PhD(math), 69. Prof Exp: Asst prof, 69-74, ASSOC PROF MATH, WEBER STATE COL, 74- Mem: Am Math Soc; Math Asn Am. Res: Functional analysis. Mailing Add: Dept of Math Weber State Col Ogden UT 84403

MILLER, RICHARD SAMUEL, b Cleveland, Ohio, July 4, 22; m 46; c 2. ANIMAL ECOLOGY. Educ: Univ Colo, BA, 49; Oxford Univ, PhD(animal ecol), 51. Prof Exp: Instr biol, Harvard Univ, 52-55; assoc biologist, Colo State Univ, 55-58; from asst prof to assoc prof, Univ Sask, 59-67; prof forestry, 67-68, OASTLER PROF ECOL, SCH FORESTRY, YALE UNIV, 68- Mem: Am Soc Mammal; Ecol Soc Am; Am Soc Naturalists; Brit Ecol Soc. Res: Population ecology. Mailing Add: Sch of Forestry & Environ Stud Yale Univ New Haven CT 06511

MILLER, RICHARD WILSON, b Miami, Fla, May 14, 34; m 57; c 2. BIOCHEMISTRY, ENZYMOLOGY. Educ: Mass Inst Technol, SB, 56, PhD(biochem), 61. Prof Exp: Res assoc enzymol, Mass Inst Technol, 61; res assoc, Sheffield Sci Sch, 62-63 & Univ Mich, 63-64; biochemist, New Eng Inst Med Res, 64-68; RES SCIENTIST, CHEM & BIOL RES INST, RES BR, CAN AGR, 68- Concurrent Pos: NIH fel, 62-64; USPHS grant, 65-68. Mem: AAAS; Am Soc Biol Chemists; Fedn Am Socs Exp Biol. Res: Respiratory control over pyrimidine biosynthesis; role of free radicals and superoxide anion in biosynthesis, biodegradation and toxicity mechanisms in microorganisms and plants; mechanism of metalloprotein catalysis. Mailing Add: Chem & Biol Res Inst Res Br Can Agr Ottawa ON Can

MILLER, ROBERT, organic chemistry, see 12th edition

MILLER, ROBERT BRUCE, b Darlow, Kans, Oct 28, 10; m 57. PHYSICS. Educ: Manchester Col, AB, 33; Mich State Univ, MS, 50, PhD(physics), 56. Prof Exp: Teacher pub schs, Mich, 33-48; instr physics, Mich State Univ, 50-56; from asst prof to assoc prof, 56-74, EMER ASSOC PROF PHYSICS, WESTERN MICH UNIV, 74- Res: Optical methods in ultrasonics. Mailing Add: 817 Boswell Kalamazoo MI 49007

MILLER, ROBERT BURNHAM, b Dallas, Tex, Dec 14, 42; m 66; c 2. APPLIED STATISTICS. Educ: Univ Iowa, BA, 64, MS, 65, PhD(statist), 68. Prof Exp: Asst prof, 68-73, ASSOC PROF STATIST & BUS, UNIV WIS-MADISON, 73- Mem: AAAS; Am Statist Asn. Res: Statistical problems in risk theory; time series analysis applied to business and economic problems. Mailing Add: 2750 Chamberlain Ave Madison WI 53705

MILLER, ROBERT CARL, b Chicago, Ill, Oct 26, 38; m 69. EXPERIMENTAL HIGH ENERGY PHYSICS, CRYOGENICS. Educ: Ill Inst Technol, BS, 61; Northern Ill Univ, MS, 65, CAS, 72. Prof Exp: RESEARCHER, ARGONNE NAT LAB, 61- Mem: Am Phys Soc; Am Nuclear Soc; Inst Elec & Electronics Engrs; Am Asn Physics Teachers; Instrument Soc Am. Res: Spin dependence in proton-proton scattering; phenomenology of pion-proton, proton-proton scattering including amplitude analysis; polarized proton targets. Mailing Add: Bldg 362 Rm G-216 Argonne Nat Lab Argonne IL 60439

MILLER, ROBERT CHARLES, physical chemistry, see 12th edition

MILLER, ROBERT CHARLES, b State College, Pa, Feb 2, 25; m 52; c 3. OPTICAL PHYSICS, SOLID STATE PHYSICS. Educ: Columbia Univ, AB, 48, MA, 52, PhD(physics), 56. Prof Exp: Asst physics, Columbia Univ, 49-51, lectr, 51-53; mem tech staff, Bell Tel Labs, 54-67; mem, Inst Defense Anal, 67-68; HEAD OPTICAL ELECTRONICS RES DEPT, BELL LABS, INC, 68- Mem: Fel Am Phys Soc; NY Acad Sci. Res: Ferroelectricity; nonlinear optics. Mailing Add: Dept 1155 Rm 1C-325 Bell Labs Murray Hill NJ 07974

MILLER, ROBERT CLAY, b Quincy, Mass, Feb 26, 23; m 49; c 2. ORGANIC CHEMISTRY. Educ: Northeastern Univ, BS, 47; Columbia Univ, AM, 48; Temple Univ, PhD(org chem), 56. Prof Exp: Res chemist, Socony Mobil Oil Co, Inc, 48-56 & exp sta, E I du Pont de Nemours & Co, 56-58; asst prof chem, St Vincent Col, 58-59; from asst prof to assoc prof, DePaul Univ, 59-68; chmn sci div, 70-74, PROF CHEM & CHMN DEPT, ADRIAN COL, 68- Mem: Am Chem Soc; The Chem Soc. Res: Synthesis of organophosphorus compounds; new polymerization systems; reaction mechanisms; correlation of structure of compounds with infrared spectra; polymer chemistry. Mailing Add: Dept of Chem Adrian Col Adrian MI 49221

MILLER, ROBERT CUNNINGHAM, b Blairsville, Pa, July 3, 99; m 37. ZOOLOGY. Educ: Greenville Col, AB, 20; Univ Calif, AM, 21, PhD(zool), 23. Prof Exp: Asst zool, Univ Calif, 20-21, res assoc, 23-24; from asst prof to prof, Univ Wash, 24-38; dir, 38-63, sr scientist, 63-75, EMER DIR, CALIF ACAD SCI, 75- Concurrent Pos: Assoc biologist, San Francisco Bay Marine Piling Surv, 21-24; collabr, Marine Piling Invests Comt, Nat Res Coun, 23-24; vis prof, Lingnan Univ, 29-31; mem staff, Oceanog Labs, Univ Wash, 31-38; secy, Pac Div, AAAS, 44-72, vpres & chmn, Sect Info & Commun, 65, pres, 73-74. Mem: AAAS; Am Soc Zoologists; Am Geophys Union; Am Ornith Union; Am Meteorol Soc. Res: Bird behavior and flight; biology of marine wood-boring organisms; biological and general oceanography; photobiology. Mailing Add: Calif Acad of Sci Golden Gate Park San Francisco CA 94118

MILLER, ROBERT DEMOREST, b Omaha, Nebr, Sept 25, 19; m 41; c 3. SOIL PHYSICS. Educ: Univ Mo, BS, 40; Univ Nebr, MS, 42; Cornell Univ, PhD(soil physics), 48. Prof Exp: Asst soil physicist, Univ Calif, 48-52; assoc prof, 52-59, PROF SOIL PHYSICS, CORNELL UNIV, 59- Concurrent Pos: Fulbright res fel, Norway, 65-66; Royal Norweg Coun Sci & Indust Res fel, 65-66; dean fac, Cornell Univ, 67-71. Mem: Soil Sci Soc Am; fel Am Soc Agron; Am Geophys Union. Res: Soil-water interactions; freezing and heaving of soil; freezing of water in porous media. Mailing Add: Dept of Agron Cornell Univ Ithaca NY 14853

MILLER, ROBERT DENNIS, b Philadelphia, Pa, Sept 23, 41; m 63; c 3. ORGANIC CHEMISTRY. Educ: Lafayette Col, BS, 63; Cornell Univ, PhD(org chem), 68. Prof Exp: RES SCIENTIST, IBM CORP, 68- Mem: AAAS; Am Chem Soc. Res: Organic photochemistry dealing with the production of highly strained, theoretically interesting molecules; high temperature thermal fragmentation reactions; synthetic methods. Mailing Add: IBM Res 5800 Cattle Rd San Jose CA 95193

MILLER, ROBERT DUWAYNE, b Galesburg, Ill, Nov 27, 12; m 37; c 3. PHYSICS. Educ: Knox Col, AB, 34; Univ Ill, MS, 35; Washington Univ, PhD(physics), 37. Prof Exp: Res geophysicist, Subterrex, Tex, 37-38; geophysicist, Shell Oil Co, 38-58, mgr tech info, Shell Develop Co, 58-69, CONSULT, SHELL DEVELOP CO, 69- Concurrent Pos: Physicist, Carnegie Inst Dept Terrestrial Magnetism, 42-43; physicist, Appl Physics Lab, Johns Hopkins Univ, 43-45. Mem: Am Phys Soc; Soc Explor Geophysicists; Am Geophys Union. Res: Diffuse scattering of x-rays; electrical methods of oil exploration. Mailing Add: 6150 Cedar Creek Houston TX 77027

MILLER, ROBERT ERNEST, b Des Moines, Iowa, July 31, 36; m 57; c 3. PLANT PATHOLOGY, SOIL MICROBIOLOGY. Educ: Simpson Col, BA, 58; Cornell Univ, MS, 62, PhD(plant path), 63. Prof Exp: Res assoc, 63-70, RES SCIENTIST, CAMPBELL INST AGR RES, 70- Concurrent Pos: Res grant plant path, Univ Calif, Berkeley, 65-66. Mem: Am Phytopath Soc; Am Soc Microbiol. Res: Ecology and physiology of soil-borne plant pathogens; biological control of soil-borne plant pathogens; genetics of fungi. Mailing Add: Campbell Soup Co Pioneer Plant Res Lab Riverton NJ 08077

MILLER, ROBERT FREDERICK, b Fredonia, NY, July 26, 21; m 46; c 1. ANIMAL NUTRITION. Educ: Cornell Univ, BS, 48, MNS, 49, PhD(animal nutrit), 51. Prof Exp: Asst dir res, Kasco Mills, 51-54; dir res, Park & Pollard, 55-56; DIR TECH SERV, MERCK & CO, 56- Mem: Poultry Sci Asn; Am Soc Animal Sci; Animal Nutrit Res Coun; Am Inst Biol Sci. Res: Animal health; physiology; parasitology. Mailing Add: Animal Health & Feed Prod Merck & Co Rahway NJ 07065

MILLER, ROBERT HAROLD, b Philadelphia, Pa, Feb 10, 19; m 42; c 2. BOTANY. Educ: Univ Calif, Berkeley, AB, 50; Ore State Univ, PhD(bot), 54. Prof Exp: Asst prof biol, Univ Nev, Reno, 53-59 & Univ Wichita, 59-60; assoc prof bot, Univ Ky, 60-63; PLANT SCIENTIST, PLANT INDUST STA, AGR RES SERV, USDA, 63-, PLANT SCIENTIST & REMOTE SENSING SPECIALIST, WASHINGTON, DC, 67-, NAT PROG STAFF SCIENTIST, 72- Concurrent Pos: Sigma Xi res grant, Univ Nev, Reno, 55-56; AID assoc prof, Inst Technol, Univ Indonesia, 60-63. Mem: AAAS; Bot Soc Am; Int Soc Plant Morphol; Soc Econ Bot; Asn Trop Biologists. Res: Plant anatomy and morphology; systematic botany; remote sensing technology; analytical and morphological studies of tropical vegetation. Mailing Add: 10104 Phoebe Lane Adelphi MD 20783

MILLER, ROBERT HAROLD, b Fremont, Wis, Sept 19, 33; m 57; c 3. SOIL SCIENCE, MICROBIOLOGY. Educ: Wis State Univ, River Falls, BS, 58; Univ Minn, MS, 61, PhD(soil microbiol), 64. Prof Exp: From asst prof to assoc prof, 64-70, PROF AGRON, OHIO STATE UNIV, 70- Concurrent Pos: Fulbright lectr, 74-75. Mem: Am Soc Agron; Soil Sci Soc Am; Am Soc Microbiol. Res: Plant rhizosphere microorganisms and their interactions with plants; ecology and physiology of Rhizobium japonicum; chemistry of soil organic matter; recycling of organic wastes in soil. Mailing Add: Dept of Agron Ohio State Univ Columbus OH 43210

MILLER, ROBERT HOOVER, b Salisbury, NY, June 8, 34. GENETICS, BIOMETRY. Educ: NC State Col, BS, 56, PhD(animal genetics), 62; Mich State Univ, MS, 58. Prof Exp: Res geneticist, 60-65, BIOMETRICIAN, AGR RES SERV, USDA, 65- Mem: Biomet Soc; Am Soc Animal Sci; Am Dairy Sci Asn; Am Genetic Asn (secy, 74-75). Mailing Add: Agr Res Serv US Dept of Agr Beltsville MD 20705

MILLER, ROBERT JAMES, b Detroit, Mich, Sept 18, 23; m 43; c 3. ANTHROPOLOGY. Educ: Univ Mich, AB, 48; Univ Wash, PhD(anthrop), 55. Prof Exp: Res anthropologist, Inner Asia Proj, Univ Wash, 55-56; asst prof anthrop, Washington Univ, 56-59; from asst prof to assoc prof anthrop & Indian studies, 59-64, PROF ANTHROP, INDIA & BUDDHIST STUDIES, UNIV WIS-MADISON, 65- Concurrent Pos: Archaeologist-consult, Mo Hist Bldgs Comn, St Louis, Mo, 58-59; NSF & Am Inst Indian Studies fel, Deccan Col Post-Grad & Res Inst Poona, India, 63-64; consult, Am Coun Learned Soc-Conf Buddhist Studies, 65-66; spec asst, Int Progs Off, Univ Wis-Madison, 66-67, chmn comt educ & social applns satellites, 68-70; consult, NDEA Fels Comt, US Dept Health, Educ & Welfare, 67-68; Univ Res Comt fel, Univ Wis-Madison, 68-69; Smithsonian Inst foreign fel, Am Inst Indian Studies, India, 70-72; chmn Fels Comt, Am Inst Indian Studies, 68-69, resident dir, New Delhi, India, 70-72; mem bd, Fulbright Prog Comt, New Delhi, 71-72; consult, Nat Endowment Humanities, 72. Mem: Fel AAAS; Am Folklore Soc; Asn Asian Studies; Soc Gen Systs Res; Am Anthrop Asn. Res: Religious organization; biocultural studies; Buddhist studies, especially Tibet, Mongolia and the Himalayas; comparative minority policies; general systems analysis; anthropological theory; cultural stability and the technological transfer. Mailing Add: Dept of Anthrop Soc Sci Bldg Univ of Wis Madison WI 53706

MILLER, ROBERT JAMES, II, b Dunn, NC, Jan 14, 33; m 59; c 3. PLANT PHYSIOLOGY, ECOLOGY. Educ: NC State Univ, BS, 56; Yale Univ, MF, 62, MS, 65, PhD(biol), 67. Prof Exp: From assoc prof to prof biol, Radford Col, 65-72; PROF BIOL & ACAD DEAN, ST MARY'S COL, 73- Concurrent Pos: Chmn dept biol, Radford Col, 67-68, dean sch natural sci, 68-71, vpres acad affairs, 71-72. Mem: Am Soc Plant Physiol; Ecol Soc Am; Soc Am Foresters. Res: Nitrogen relations of higher plants; ecology of wetlands. Mailing Add: Dept of Biol St Mary's Col Raleigh NC 27609

MILLER, ROBERT JOSEPH, b Ironton, Ohio, June 10, 41; m 73; c 1. EXPERIMENTAL HIGH ENERGY PHYSICS. Educ: Univ Detroit, BS, 63; Purdue Univ, MS, 66, PhD(physics), 69. Prof Exp: Fel exp high energy physics, Rutherford Lab, Sch Res Coun UK, 68-72; res assoc, 72-75, ASST PHYSICIST, ARGONNE NAT LAB, 75- Mem: Am Phys Soc. Res: From factors of K meson; anti-neutrino interaction with protons; search for fluctuations in the pi meson-proton interaction. Mailing Add: Argonne Nat Lab 9700 S Cass Ave Argonne IL 60439

MILLER, ROBERT JOSEPH, b Keokuk, Iowa, Nov 17, 39; m 65; c 2. MARINE ECOLOGY. Educ: William Jewell Col, AB, 61; Col William & Mary, MA, 64; NC State Univ, PhD(zool), 70. Prof Exp: Nat Res Coun Can fel, Marine Ecol Lab, 69-71, RES SCIENTIST, ST JOHN'S BIOL STA, FISHERIES RES BD CAN, 71- Mem: Am Soc Limnol & Oceanog; Am Fisheries Soc. Res: Energy flow in marine populations; yield prediction and conservation in marine fisheries. Mailing Add: Nfld Biol Sta Fisheries Res Bd of Can St John's NF Can

MILLER, ROBERT KENNETH, organic chemistry, see 12th edition

MILLER, ROBERT L, b Chicago, Ill, Jan 26, 26; m 47; c 4. PHYSICAL CHEMISTRY, ACADEMIC ADMINISTRATION. Educ: Univ Chicago, PhB, 48, BS, 50, MS, 51; Ill Inst Technol, PhD(chem), 63. Prof Exp: From instr to assoc prof chem & assoc dean, Univ Ill, Chicago, 51-68; PROF CHEM & DEAN COL ARTS & SCI, UNIV NC, GREENSBORO, 68- Concurrent Pos: Asst dean, Univ Ill, Chicago, 62-65. Mem: AAAS. Res: Applications of quantum mechanics to desription of chemical compounds of biological interest. Mailing Add: Col of Arts & Sci Univ of NC Greensboro NC 27412

MILLER, ROBERT LEE, b Chicago, Ill, Apr 6, 20; m 42; c 4. GEOLOGY. Educ: Univ Ill, AB, 42; Univ Chicago, PhD(geol, paleozool), 50. Prof Exp: Instr phys sci, Univ Ill, 48-49; res assoc, 50-52, from asst prof to assoc prof geol, 52-65, PROF GEOPHYS SCI, UNIV CHICAGO, 65- Concurrent Pos: Vis prof, Univ Frankfurt, 58 & Brown Univ, 63; assoc, Woods Hole Oceanog Inst; ed, J Geol. Mem: Am Geophys Union; Marine Technol Soc; NY Acad Sci. Res: Fluid mechanics; oceanography; sediment transport; marine geophysics and geology. Mailing Add: Dept of Geophys Sci Univ of Chicago Chicago IL 60637

MILLER, ROBERT LLEWELLYN, b Chicago, Ill, Jan 19, 29; m 53; c 1. POLYMER SCIENCE. Educ: Mass Inst Technol, BS, 50; Brown Univ, PhD(chem), 54. Prof Exp: Res chemist, Monsanto Chem Co, 53-59, res specialist, 59-62, group leader, Chemstrand Res Ctr, Inc, Monsanto Co, 62-64, scientist, NC, 64-69, Mo, 69-71; SR RES SCIENTIST, MIDLAND MACROMOLECULAR INST, 72- Concurrent Pos: Sr vis scholar, Univ Manchester, 68-69. Mem: AAAS; Am Chem Soc; Am Phys Soc; Mat Res Soc; Am Crystallog Asn. Res: Solid state physics as applied to polymers, particularly semicrystalline polymers. Mailing Add: Midland Macromolecular Inst 1910 W St Andrews Dr Midland MI 48640

MILLER, ROBERT OGDEN, JR, floriculture, see 12th edition

MILLER, ROBERT RUSH, b Colorado Springs, Colo, Apr 23, 16; m 40; c 5. ICHTHYOLOGY. Educ: Univ Calif, AB, 38; Univ Mich, MA, 43, PhD(zool), 44. Prof Exp: Asst ichthyol surv, Nev, Univ Mich, 38 & Div Fishes, Mus Zool, 39-44; assoc cur fishes, US Nat Mus, Smithsonian Inst, 44-48; from asst prof to assoc prof, 48-60, PROF ZOOL, UNIV MICH, ANN ARBOR, 60- CUR FISHES, MUS ZOOL, 56- Concurrent Pos: Mem, Univ Mich expeds, 38-42, 50 & 59-, Mex, 39, 50 & 55-71, Ichthyol Surv, Guatemala, US Dept State, Smithsonian Inst & Govt Guatemala, 46-47 & Biol Surv, Arnhem Land, Govt Australia, Nat Geol Soc & Smithsonian Inst, 48; assoc cur, Mus Zool, Univ Mich, 48-56; ichthyol ed, Copeia, Am Soc Ichthyol & Herpet, 50-55; collabr, US Nat Park Serv, 60-; Guggenheim fel, 73-74. Honors & Awards: Award Excellence, Am Fisheries Soc, 75. Mem: AAAS; Am Soc Ichthyol & Herpet (vpres, 61, pres, 65); Soc Syst Zool; Soc Study Evolution; Soc Vert Paleont. Res: Taxonomy, distribution, variation, hybridization, ecology, life history and evolution of fishes; paleoichthyology Mailing Add: Univ Mus Bldg Univ of Mich Ann Arbor MI 48104

MILLER, ROBERT STEPHEN, b Perth Amboy, NJ, Jan 6, 34; m 55; c 3. POLYMER CHEMISTRY, POLYMER ENGINEERING. Educ: Rutgers Univ, BSChem, 58. Prof Exp: Polymer chemist, Tex-US Chem Co, 58-62; proj leader plastics & polymers, Atlantic Refining Co, 62-65; sr chemist & engr, 65-69, proj leader acrylics, 69-70, group leader polymers, 70-72, supvr polymer res & develop, 72-75, MGR POLYMER RES & DEVELOP, TENNECO CHEM INC, 75- Mem: Soc Plastics Engrs. Res: Relationship of polymer chemistry and engineering to the modern industrial scene, especially environmental, industrial hygiene, automation, economics of scale, and resource conservation matters. Mailing Add: Tenneco Chem Inc Turner Pl Piscataway NJ 08854

MILLER, ROBERT VANCE, b Enid, Okla, Nov 30, 25; m 55; c 3. APPLIED PHYSICS. Educ: Univ Okla, BS, 46; Univ Calif, Berkeley, MA, 50; Univ Tenn, PhD(physics), 62. Prof Exp: Physicist, Radioisotope Unit, Vet Admin Med Teaching Group Hosp, Memphis, Tenn, 53-56; assoc res engr, Boeing Airplane Co, 56-57; res assoc physics, Oak Ridge Inst Nuclear Studies, 57-59; asst, Univ Tenn, 59-61; tech specialist physics res, Rocketdyne Div, NAm Aviation, Inc, Calif, 62-66; sr systs analyst, Comput & Software, Inc, Calif, 66-70; RES ENGR, SIERRACIN CORP, 71- Mem: AAAS; Am Phys Soc; Inst Elec & Electronics Engrs. Res: Theory and calculation of spectral absorption coefficients; nuclear radiation detectors; mathematical analysis; thin films for applied optics. Mailing Add: 23203 Via Calisero Valencia CA 91355

MILLER, ROBERT VERNE, b Modesto, Calif, Dec 27, 45; m 68. MICROBIAL GENETICS, MOLECULAR GENETICS. Educ: Univ Calif, Davis, BA, 67; Univ Ill, Urbana, MS, 69, PhD(microbiol), 72. Prof Exp: Res assoc molecular genetics, Univ Calif, Berkeley, 72-74; ASST PROF MICROBIOL, UNIV TENN, KNOXVILLE, 74- Concurrent Pos: Dernham fel, Am Cancer Soc, Calif, 72-74. Mem: Sigma Xi; Am Soc Microbiol; Genetics Soc Am. Res: Genetic and biochemical mechanisms of host-viral interactions in Pseudomonas aeruginosa, particularly the host's contribution to vegetive-temperate response of phage and phage conversion of the host. Mailing Add: Dept of Microbiol Univ of Tenn Knoxville TX 37916

MILLER, ROBERT VICTOR, b Batavia, NY, Apr 30, 36; m 56; c 4. SYSTEMATIC ICTHYOLOGY, RESEARCH ADMINISTRATION. Educ: Cornell Univ, BS, 58, PhD(vert zool), 64; Univ Ark, MS, 61. Prof Exp: Asst prof fisheries res, Univ Md, 63-65; zoologist, US Bur Com Fisheries, 65-71, fishery biologist & staff asst, 71-73, SR RES SPECIALIST, NAT MARINE FISHERIES SERV, 73- Concurrent Pos: Chmn, US Del, US/USSR Marine Mammal Prog, Environ Protection Agency, 73-; legis corresp, Am Inst Fishery Res Biologists, 75. Mem: Am Soc Icthyol & Herpet; Soc Syst Zool; Am Soc Zoologists; Am Inst Fishery Res Biologists; Sigma Xi. Res: Comparative morphology and systematics of fishes; research management; marine mammals. Mailing Add: Nat Marine Fisheries Serv Washington DC 20235

MILLER, ROBERT W, b Warrensville, NC, Aug 7, 31; m 53; c 2. AGRONOMY. Educ: Berea Col, BS, 53; Ohio State Univ, MSc, 60, PhD(agron), 63. Prof Exp: Asst county agent, NC State Univ, 55-57; asst prof plant breeding, Cornell Univ, 62-63; from asst prof to assoc prof turfgrass mgt, 63-71, PROF AGRON & TURFGRASS MGT, OHIO STATE UNIV, 71- Mem: Am Soc Agron; Crop Sci Soc Am. Res: Crop physiology; turfgrass management. Mailing Add: Dept of Agron Ohio State Univ 1827 Neil Ave Columbus OH 43210

MILLER, ROBERT WALKER, JR, b Philadelphia, Pa, Nov 3, 41; m 66; c 1. PLANT PATHOLOGY. Educ: Univ Del, BS, 64, PhD(plant path & ecol), 71; Univ Ariz, MS, 70. Prof Exp: Exten specialist, 71-72, ASST PROF PLANT PATH, CLEMSON UNIV, 72- Mem: Am Phytopath Soc; Soc Nematologists. Res: Epidemiology, cultural and chemical controls of the foliar diseases of pecans. Mailing Add: Dept of Plant Path & Physiol Clemson Univ Clemson SC 29631

MILLER, ROBERT WARWICK, b Brooklyn, NY, Sept 29, 21; m 55. PEDIATRICS, EPIDEMIOLOGY. Educ: Univ Pa, AB, 42, MD, 46; Univ Mich, MPH, 58, DrPH, 61. Prof Exp: Mem atomic energy proj, Univ Rochester, 51-53; chief pediat, Atomic Bomb Casualty Comn, Hiroshima, Japan, 53-55, chief child health surv, Hiroshima & Nagasaki, 58-60; prof assoc, Nat Acad Sci, 55-57; CHIEF EPIDEMIOL BR, NAT CANCER INST, 61- Mem: Soc Pediat Res; Am Pediat Soc. Res: Epidemiology of cancer; congenital malformations and radiation effects. Mailing Add: Epidemiol Br A-521 Landow Bldg Nat Cancer Inst Bethesda MD 20014

MILLER, ROBERT WITHERSPOON, b Chester, SC, Oct 29, 18; m 43; c 4. ORGANIC CHEMISTRY. Educ: Erskine Col, AB, 39; Univ NC, PhD(org chem), 48. Prof Exp: Asst instr chem, Clemson Univ, 39-40; chemist, Tenn Eastman Corp, 48-52, sr chemist, Eastman Chem Prod Inc, 52-53, sales rep, 53-56, prod mgr, 56-57, dist sales mgr, 57-58, chief sales develop rep, 58-65, mgr new prod sales, 65-73, RES ASSOC, TENN EASTMAN CO, EASTMAN KODAK CO, 73- Mem: Am Chem Soc; Sigma Xi. Mailing Add: 4531 Stagecoach Rd Kingsport TN 37664

MILLER, ROLAND DREW, b Chicago, Ill, Mar 16, 22; m 43; c 3. MEDICINE. Educ: ePauw Univ, AB, 43; Northwestern Univ, MD, 46; Univ Minn, MS, 51. Prof Exp: From instr to asst prof med, Mayo Clin & Found, Univ Minn, 53-61, from asst dir to assoc dir, 59-73, PROF MED, MAYO MED SCH, UNIV MINN, 61- Concurrent Pos: Consult internal med & pulmonary dis, Mayo Clin; chmn pulmonary bd, Am Bd Internal Med, 70- Mem: AMA; fel Am Col Physicians; Am Col Chest Physicians; Am Fedn Clin Res. Res: Pulmonary diseases as they affect the function of the respiratory system. Mailing Add: 200 SW First St Rochester MN 55901

MILLER, RONALD KENT, b Jersey City, NJ, May 30, 46; m 73. IMMUNOBIOLOGY, ALLERGY. Educ: Rutgers Univ, BS, 68, PhD(zool), 75. Prof Exp: Teaching asst biochem, Rutgers Univ, 69; res scientist biochem, immunol & microbiol, 68-74, SCI PROG ASSOC ALLERGY & ASTHMA, SCHERING PLOUGH CORP, 75- Res: Developments in asthma drug therapy and other related pulmonary diseases; advances in interferon research. Mailing Add: Schering Plough Corp Galloping Hill Rd Kenilworth NJ 07033

MILLER, RONALD LEE, b Magnolia, Ky, Feb 2, 36; m 59; c 3. BIOCHEMISTRY. Educ: Western Ky State Col, BS, 58; Univ Ky, PhD(biochem), 67. Prof Exp: Instr biol sci, Cornell Univ, 67-68; fel, Roche Inst Molecular Biol, 68-70, sr investr, 70-72; ASST PROF BIOCHEM, MED UNIV SC, 72- Mem: AAAS; Am Chem Soc. Res: Connective tissue metabolism; protein synthesis, particularly isolation of protein initiation factors from rabbit reticulocytes. Mailing Add: Dept of Biochem Med Univ of SC Charleston SC 29401

MILLER, ROSCOE EARL, b Shelby Co, Ind, Jan 6, 18; m 52; c 3. MEDICINE. Educ: Ind Univ, BS, 48, MD, 51; Am Bd Radiol, dipl, 55. Prof Exp: Intern, Univ Chicago Hosp, 52, resident radiol, 55, instr, Univ, 55-56; from asst prof to assoc prof, 56-66, PROF RADIOL, MED CTR, IND UNIV, INDIANAPOLIS, 66- Concurrent Pos: USPHS fel, Univ Lund, 64-65; mem adv panel radiologic contrast media, US Pharmacopeia; consult, Sta Hosp, Ft Benjamin Harrison, Cent State Hosp & Vet Admin Hosp; mem ed adv bd, J Radiol; assoc ed, J Gastrointestinal Radiol; mem comt cancer detection, Am Col Radiol; mem coun cancer, Am Gastroenterol Asn, 75-78; mem adv panel radiologic contrast media, US Pharmacopeia, 70-80. Mem: Radiol Soc NAm; Am Col Radiol; Am Gastroenterol Asn; Am Roentgen Ray Soc; Soc Gastrointestinal Radiol. Res: Radiology of the abdomen and gastrointestinal diseases; design and development of radiological equipment; radiological contrast media. Mailing Add: Dept of Radiol Ind Univ Med Ctr Indianapolis IN 46202

MILLER, ROSWELL KENFIELD, b Glen Cove, NY, Aug 4, 32; m 55; c 3. FOREST MANAGEMENT. Educ: State Univ NY Col Forestry, Syracuse Univ, BS, 58, MF, 59; Univ Mich, Ann Arbor, PhD(forest mgt), 72. Prof Exp: Forester, US Forest Serv, Ore, 59-60; forest engr, Crown Zellerbach Corp, Ore, 60-64; logging engr, Navajo Forest Prod Industs, NMex, 64; chief of surv, NMex State Hwy Dept, Gallup, 64-65; asst prof, 65-67 & 69-72, ASSOC PROF FORESTRY, MICH TECHNOL UNIV, 72- Mem: Soc Am Foresters; Am Soc Photogram. Res: Cost control; planning natural resource use; small business management. Mailing Add: Forestry Dept Mich Technol Univ Houghton MI 49931

MILLER, ROY G, JR, b Columbus, Ohio, Dec 23, 33; m 55; c 4. ORGANIC CHEMISTRY. Educ: Ohio Wesleyan Univ, BA, 55; Univ Mich, MS, 60, PhD(chem), 63. Prof Exp: Res chemist, Elastomer Chem Dept, Exp Sta, E I du Pont de Nemours & Co, 63-65; from asst prof to assoc prof, 65-72, PROF CHEM, UNIV NDAK, 72- Mem: Am Chem Soc; Sigma Xi. Res: Organic reaction mechanisms; organometallic chemistry and homogeneous catalysis. Mailing Add: Dept of Chem Univ of NDak Grand Forks ND 58202

MILLER, ROY M, physical science, see 12th edition

MILLER, ROY RICHARD, b Dayton, Ohio, Dec 17, 36; m 58; c 2. PHYSICAL CHEMISTRY. Educ: Mass Inst Technol, BS, 58; Univ Chicago, PhD(chem), 63. Prof Exp: Chemist, Cent Res Labs, Am Cyanamid Co, 58-59; sr res chemist, 62-67, staff chemist, 67-71, DEPT HEAD, ALLEGANY BALLISTICS LAB, HERCULES INC, 71- Mem: AAAS; Am Chem Soc. Res: Influence of fillers on properties of filled plastics; kinetics of polymerization reactions; combustion processes. Mailing Add: Allegany Ballistics Lab Hercules Inc Box 210 Cumberland MD 21502

MILLER, RUDOLPH J, b Gbley, Czech, Sept 25, 34; US citizen; m 57; c 3. ETHOLOGY, ICHTHYOLOGY. Educ: Cornell Univ, BS, 56, PhD(vert zool), 61; Tulane Univ, MS, 58. Prof Exp: NIH fel, Univ Groningen, 61-62; from asst prof to assoc prof, 62-69, PROF ZOOL, OKLA STATE UNIV, 69- Concurrent Pos: Vis investr, Univ Hawaii, 71-72. Mem: AAAS; Am Soc Ichthyol & Herpet; Soc Study Evolution; Soc Syst Zool; Animal Behav Soc. Res: Fish behavior; comparative aspects and motivation analysis, primarily on anabantid, centrarchid and cyprinid fishes; fish morphology; correlatuve studies of brain, sense organs and behavior. Mailing Add: Sch of Biol Sci Okla State Univ Stillwater OK 74075

MILLER, RUPERT GRIEL, b Lancaster, Pa, Jan 31, 33; m 59; c 2. STATISTICS. Educ: Princeton Univ, AB, 54; Stanford Univ, PhD(statist), 58. Prof Exp: Assoc prof, Univ Calif, 58-59, Stanford Univ, 59-63 & Johns Hopkins Univ, 63-64; assoc prof, 64-71, PROF STATIST, STANFORD UNIV, 71- Res: Biostatistics and stochastic processes. Mailing Add: Sch of Med Stanford Univ Stanford CA 94305

MILLER, RUSSEL BRYAN, b Tyler, Tex, May 31, 40; m 69; c 2. ORGANIC CHEMISTRY. Educ: Washington & Lee Univ, BS, 62; Rice Univ, PhD(chem), 67. Prof Exp: Res fel, Columbia Univ, 66-68; asst prof, 68-75, ASSOC PROF CHEM, UNIV CALIF, DAVIS, 75- Mem: AAAS; Am Chem Soc; The Chem Soc. Res: Synthetic natural product chemistry; new synthetic methods; conformational analysis. Mailing Add: Dept of Chem Univ of Calif Davis CA 95616

MILLER, RUSSELL LEE, b Cairo, Ga, Dec 23, 22; m 53; c 4. CROP SCIENCE. Educ: Univ Ga, BSA, 50, MS, 52; La State Univ, PhD, 58. Prof Exp: Instr agron, Univ Ga, 51-52; from instr to assoc prof, 52-68, PROF AGRON, LA STATE UNIV, BATON ROUGE, 68- Concurrent Pos: Mem, Bd Dirs, Lee Col; nat mem chmn, Am Soc Agron, 71-76. Mem: Am Soc Agron. Res: Agronomic education. Mailing Add: Dept of Agron La State Univ Baton Rouge LA 70803

MILLER, SANFORD ARTHUR, b Brookly, NY, May 12, 31; m 58; c 2. BIOCHEMISTRY, NUTRITION. Educ: City Col New York, BS, 52; Rutgers Univ, MS, 56, PhD(physiol, biochem), 57. Prof Exp: Chemist, Appl Res Br, Army Chem Ctr, Md, 52; asst, Bur Biol Res, Rutgers Univ, 54-55, Dept Physiol & Biochem, 55-57; res assoc & supvr animal labs, Dept Food Technol, 57-59, from asst prof to assoc prof, 59-70, PROF NUTRIT BIOCHEM, MASS INST TECHNOL, 70-, DIR, TRAINING PROG ORAL SCI, 70- Concurrent Pos: Mem, Expert Comt GRAS Substances, Fedn Am Socs Exp Biol, Food & Drug Admin, 72- & Comt Maternal & Child Health & Comt Contraceptive Steroids, Nat Inst Child Health Develop, 73- Mem: AAAS; Perinatal Res Soc; Am Soc Pediat Res; Inst Food Technologists; Am Inst Nutrit. Res: Nutrition and development; infant nutrition; synthetic dietary energy sources; oral biology. Mailing Add: Dept of Nutrit & Food Sci Mass Inst of Technol Cambridge MA 02139

MILLER, SANFORD STUART, b Paterson, NJ, June 1, 38; m 67; c 2. MATHEMATICAL ANALYSIS. Educ: Mass Inst Technol, BS, 60; Wash Univ, MA, 66; Univ Ky, PhD(math), 71. Prof Exp: Asst prof math, Ball State Univ, 66-68; from asst prof to assoc prof, 71-76, PROF MATH, STATE UNIV NY COL BROCKPORT, 76- Concurrent Pos: Fel, Int Res Exchange Bd, Poland & Romania, 73-74; Nat Acad Sci exchange scientist, Romania, 74. Mem: Am Math Soc; Math Asn Am. Res: Theory of functions of a complex variable, univalent function theory and differential inequalities in the complex plane. Mailing Add: Dept of Math State Univ of NY Col Brockport NY 14420

MILLER, SHERWOOD ROBERT, b Lamont, Alta, Apr 16, 32; m 55; c 4. POMOLOGY. Educ: Univ Alta, BSc, 54, MSc, 56; Cornell Univ, PhD(pomol), 65. Prof Exp: RES SCIENTIST HORT, CAN DEPT AGR, 56- Mem: Am Soc Hort Sci; Can Soc Hort. Res: Use of synthetic and endogenous growth regulators in apple production; evaluation of dwarfing mechanism in apple rootstocks; spacing trials with emphasis on tree walls and high density plantings. Mailing Add: RR 3 Brighton ON Can

MILLER, SOLOMON, b Philadelphia, Pa, Jan 9, 22; m 56; c 1. CULTURAL ANTHROPOLOGY. Educ: Temple Univ, BS, 49; Columbia Univ, PhD(anthrop), 64. Prof Exp: Asst cur, Mus Am Indian, 60-62; lectr anthrop, Brooklyn Col, 62-64, asst prof, 64-67; from asst prof to assoc prof, New Sch Social Res, 67-74; ASSOC PROF ANTHROP & HEAD DEPT, HOFSTRA UNIV, 74- Mem: Fel AAAS; fel Am Anthrop Asn; fel Brit Anthrop Asn; Am Ethnol Asn. Res: Peasant societies, Peru and Bolivia. Mailing Add: Dept of Anthrop Hofstra Univ 1000 Fulton Ave Hempstead NY 11550

MILLER, STANLEY CUSTER, JR, b Kansas City, Mo, July 30, 26; m 57; c 3. THEORETICAL PHYSICS. Educ: Univ Colo, BS, 48; Univ Calif, PhD(physics), 53. Prof Exp: From asst prof to assoc prof, 53-61, PROF PHYSICS, UNIV COLO, BOULDER, 61- Mem: Am Phys Soc; Am Asn Physics Teachers. Res: Theoretical work in quantum mechanics and in galvanomagnetic effects in semiconductors. Mailing Add: Dept of Physics & Astrophys Univ of Colo Boulder CO 80302

MILLER, STANLEY JOHNSON, b Dayton, Va, June 26, 18; m 41. ORGANIC CHEMISTRY. Educ: Bridgewater Col, AB, 40; Purdue Univ, MS, 42; Univ Southern Calif, PhD(org chem), 47. Prof Exp: Asst chem, Purdue Univ, 40-44; res assoc, Univ Southern Calif, 44-46; res chemist, William T Thompson Co & Thompson Hort Chem Corp, 46-52 & Electrocircuits, Inc, 52-55; consult, 55-60; gen mgr, Los Angeles Div, Gulton Industs, Inc, 60-62; mgr, Ceramic Dept, Scionics Corp, Calif, 62-67; dir res, CAL-R Corp, 67-69; DIR RES & DEVELOP, CIRCUIT FUNCTIONS, INC, 69- Mem: Am Chem Soc. Res: Irradiation of abietic acid with ultraviolet rays; synthesis of antimalarial drugs and vitamins and plant hormones; ultrasonic measurements, barium titanate for electrostrictive and capacitor applications; solid state studies; x-ray diffraction. Mailing Add: 2569 Creston Dr Los Angeles CA 90068

MILLER, STANLEY LLOYD, b Oakland, Calif, Mar 7, 30. CHEMISTRY. Educ: Univ Calif, BS, 51; Univ Chicago, PhD(chem), 54. Prof Exp: Jewett fel chem, Calif Inst Technol, 54-55; instr biochem, Col Physicians & Surgeons, Columbia Univ, 55-58, asst prof, 58-60; from asst prof to assoc prof, 60-68, PROF CHEM, UNIV CALIF, SAN DIEGO, 68- Concurrent Pos: Hon counr, High Coun Sci Res, Spain, 73- Mem: AAAS; Am Chem Soc; Am Soc Biol Chem; Nat Acad Sci. Res: Origin of life; general occurrance of clathrate hydrates; general anesthesia mechanisms. Mailing Add: Dept of Chem Univ of Calif at San Diego La Jolla CA 92093

MILLER, STEPHEN DOUGLAS, b Greeley, Colo, Mar 27, 46; m 69; c 1. WEED SCIENCE. Educ: Colo State Univ, BS, 68; NDak State Univ, MS, 70, PhD(agron), 73. Prof Exp: Asst agron, 73-75, ASST PROF AGRON, NDAK STATE UNIV, 75- Mem: Weed Sci Soc Am; Agron Soc Am; Crop Sci Soc Am. Res: Biology and control of wild oats in field crops; effect of reduced tillage systems on crop yield and crop pests. Mailing Add: Dept of Agron NDak State Univ Fargo ND 58102

MILLER, STEVEN RALPH, b Cleveland, Ohio, Feb 26, 36; m 58; c 2. PHYSICAL CHEMISTRY. Educ: Case Western Reserve Univ, BS, 58; Mass Inst Technol, PhD(phys chem), 62. Prof Exp: Res assoc, Mass Inst Technol, 62; asst prof chem, 62-68, ASSOC PROF CHEM, OAKLAND UNIV, 68- Mem: AAAS; Am Chem Soc. Res: Nuclear magnetic resonance relaxation phenomena in ferroelectric solids; gas phase reaction kinetics; oxidation of sulfur dioxide; aerosol formation. Mailing Add: Dept of Chem Oakland Univ Rochester MI 48063

MILLER, SYDNEY ISRAEL, b Saskatoon, Sask, Can, May 22, 23; nat US; m 50; c 3. PHYSICAL ORGANIC CHEMISTRY. Educ: Univ Man, BSc, 45, MSc, 46; Columbia Univ, PhD(chem), 51. Prof Exp: Instr chem, Univ Man, 46 & Univ Mich, 50-51; from instr to assoc prof, 51-64, PROF CHEM, ILL INST TECHNOL, 64- Concurrent Pos: NSF sr fel, Univ Col, Univ London, 63-64; vis scientist, Argonne Nat Lab, 71-72 & Japan Soc Prom Sci, 73. Mem: Am Chem Soc; Am Asn Univ Profs. Res: Solution kinetics; mechanisms; stereochemistry; acetylene chemistry; heterocyclics; coal. Mailing Add: Dept of Chem Ill Inst of Technol Chicago IL 60616

MILLER, TERRY ALAN, b Girard, Kans, Dec 18, 43; m 66; c 2. CHEMICAL PHYSICS. Educ: Univ Kans, BA, 65; Cambridge Univ, PhD(chem), 68. Prof Exp: SUPVR, BELL LABS, INC, 68- Concurrent Pos: Asst prof, Princeton Univ, 68-71 & Stanford Univ, 72. Mem: Am Phys Soc. Res: Microwave and radio-frequency spectroscopy of transient molecular species, free radicals and excited states; chemical reactions and kinetics of atoms and simple molecules. Mailing Add: Bell Labs Inc 600 Mountain Ave Murray Hill NJ 07974

MILLER, TERRY LEE, b Aberdeen, SDak, Dec 14, 40; m 64. BIOCHEMISTRY. Educ: San Diego State Col, AB, 64, MS, 69; Ore State Univ, PhD(biochem), 69. Prof Exp: NIH fel, Univ Colo, 68-70; ASST PROF BIOCHEM, ORE STATE UNIV, 70- Concurrent Pos: Consult, Salem Mem Hosp, 71- Mem: Am Chem Soc. Res: Protein chemistry; membrane chemistry; mechanisms of action of membrane active compounds; effects of selected environmental toxicants on biological and synthetic membranes. Mailing Add: Dept of Agr Chem Ore State Univ Corvallis OR 97331

MILLER, THEODORE CHARLES, b Troy, NY, July 23, 33; m 59; c 3. PHARMACEUTICAL CHEMISTRY, RESEARCH ADMINISTRATION. Educ: Princeton Univ, AB, 55; Univ Ill, PhD(chem), 59. Prof Exp: Res chemist, 59-69, patent agent trainee, 69-70, PATENT AGENT, STERLING-WINTHROP RES INST, 70- Mem: Am Chem Soc. Res: Free radical rearrangements; synthesis of steroid hormones and heterocyclic compounds. Mailing Add: Sterling-Winthrop Res Inst Rensselaer NY 12144

MILLER, THEODORE LEE, b Crab Orchard, WVa, May 25, 40; m 70; c 1. PHYSICAL CHEMISTRY. Educ: Concord Col, BS, 66; Marshall Univ, MS, 70; Univ Cincinnati, PhD(chem), 74. Prof Exp: Instr chem, Univ Va, 74-75; ASST PROF CHEM, KING'S COL, PA, 75- Mem: Am Chem Soc. Res: Analyzing time-dependent phosphoresence spectra using a computer controlled laser spectrometer; interaction of metal ions with compounds of biological interest. Mailing Add: Dept of Chem King's Col Wilkes-Barre PA 18702

MILLER, THOMAS, b Asheville, NC, Apr 17, 32; m 54; c 1. PLANT PATHOLOGY. Educ: NC State Univ, BS, 62, MS, 64, PhD(plant path), 72. Prof Exp: PLANT PATHOLOGIST, SOUTHEASTERN FOREST EXP STA, US FOREST SERV, 64- Mem: Am Phytopath Soc. Res: Mechanisms of resistance in southern pines to Cronartium fusiforme; control of tree diseases through silviculture. Mailing Add: SE Forest Exp Sta Forestry Sci Lab Athens GA 30602

MILLER, THOMAS ALBERT, b Sharon, Pa, Jan 5, 40; m 65; c 2. ENTOMOLOGY. Educ: Univ Calif, Riverside, BA, 62, PhD(entom), 67. Prof Exp: USPHS fel & res assoc insect physiol, Univ Ill, Urbana, 67-68; NATO fel insect physiol, Glasgow Univ, 68-69; from asst prof & asst entomologist to assoc prof & assoc entomologist, 69-76, PROF ENTOM & ENTOMOLOGIST, UNIV CALIF, RIVERSIDE, 76- Mem: Entom Soc Am; Am Soc Zoologists; Brit Soc Exp Biol. Res: Insect neurophysiology; insect toxicology; mode of action of insecticides; insect visceral muscle physiology. Mailing Add: Dept of Entom Univ of Calif Riverside CA 92502

MILLER, THOMAS BRYAN, JR, physiology, see 12th edition

MILLER, THOMAS EDWARD, b Minneapolis, Minn, Apr 3, 39; m 62; c 2. IMMUNOCHEMISTRY. Educ: Univ Minn, BA, 62, MS, 64, PhD(microbiol), 67. Prof Exp: RES ASSOC BIOMED RES, TRUDEAU RES INST, INC, 67- Mem: Am Soc Microbiol. Res: Biochemical studies on the host-parasite relationship, especially mechanisms responsible for the death of microorganisms within phagocytic cells of the host. Mailing Add: Trudeau Res Inst Inc PO Box 59 Saranac Lake NY 12983

MILLER, THOMAS GILL, nuclear physics, see 12th edition

MILLER, THOMAS GORE, b Greenfield, Ohio, Nov 3, 24; m 53; c 3. ORGANIC CHEMISTRY. Educ: Miami Univ, AB, 48; Univ Ill, MS, 49, PhD(chem), 51. Prof Exp: Res chemist, E I du Pont de Nemours & Co, 51-57; from asst prof to assoc prof, 57-69, PROF CHEM & HEAD DEPT, LAFAYETTE COL, 69- Concurrent Pos: NSF sci fac fel & vis res fel, Princeton Univ, 70-71. Mem: Am Chem Soc. Res: Molecular rearrangements; clathrate compounds; general organic chemistry. Mailing Add: Dept of Chem Lafayette Col Easton PA 18042

MILLER, THOMAS LEE, b Elkhart, Ind, Nov 24, 35; m 62; c 3. BIOCHEMISTRY, MICROBIOLOGY. Educ: Ind State Univ, AB, 61; Univ Wis, MS, 64, PhD(biochem), 66. Prof Exp: Res asst biochem, Univ Wis, 61-66; res assoc microbiol, 66-67, head microbiol sect, 67-70, RES MGR FERMENTATION MICROBIOL, UPJOHN CO, 70- Mem: Am Chem Soc; Am Soc Microbiol. Res: Hydrocarbon fermentations;

steroid bioconversions; microbiological processes; measurement and control of fermentation variables; antibiotic fermentations. Mailing Add: 7599 Orchard Hill Ave Kalamazoo MI 49001

MILLER, THOMAS MARSHALL, b Ft Worth, Tex, Aug 20, 40; m 71. ATOMIC PHYSICS. Educ: Ga Inst Technol, BS, 62, MS, 64, PhD(physics), 68. Prof Exp: Asst prof physics, NY Univ, 68-74; PHYSICIST, STANFORD RES INST, 74- Mem: Am Phys Soc. Res: Transport properties of low energy ions and electrons in gases; ion-molecule reactions; interactions of low energy electrons with thermal atom beams; low-energy atom-atom collisions; atomic polarizabilities; photodissociation. Mailing Add: Molecular Physics 106B Stanford Res Inst Menlo Park CA 94025

MILLER, THOMAS WILLIAM, b Providence, RI, June 12, 29; m 52; c 3. NATURAL PRODUCTS CHEMISTRY. Educ: Univ RI, BS, 50. Prof Exp: Res chemist, 51-61, sr res chemist, 61-69, res fel, 69-70, asst dir, 70-74, SR RES FEL, MERCK SHARP & DOHME RES LABS, 74- Mem: Am Soc Microbiol; Am Chem Soc. Res: Natural products; isolation of antibiotics, vitamins and other fermentation products. Mailing Add: Merck Sharp & Dohme Res Labs Rahway NJ 07065

MILLER, TILFORD DAY, biochemistry, see 12th edition

MILLER, TONY JASPER, b Davidson Co, NC, Jan 4, 34; m 59; c 1. DEVELOPMENTAL BIOLOGY. Educ: Univ NC, AB, 56; Univ Tenn, MS, 58; Rutgers Univ, Newark, PhD, 70. Prof Exp: Instr biol, King Col, 58-59, asst prof, 59-61; sr biologist, Dept Animal Res, St Barnabas Med Ctr, 63-68, assoc dir animal res, 64-68; instr, 65-70, ASST PROF ZOOL, RUTGERS UNIV, NEWARK, 70- Mem: Am Chem Soc; Am Soc Zoologists; Soc Develop Biol. Res: Biochemistry, physiology and morphology of development of vertebrates. Mailing Add: Dept of Zool Rutgers Univ 195 University Ave Newark NJ 07102

MILLER, TRACY BERTRAM, b Syracuse, NY, Oct 19, 27; m 54; c 3. PHARMACOLOGY. Educ: Cornell Univ, AB, 48; Univ Buffalo, MA, 53, PhD, 59. Prof Exp: Instr pharmacol, Univ Buffalo, 53-54; res assoc physiol, Harvard Med Sch, 64-66; assoc prof pharmacol & surg res, State Univ NY Upstate Med Ctr, 66-71; PROF PHYSIOL & PHARMACOL, MED SCH, UNIV MASS, 71- Concurrent Pos: Am Heart Asn res fel, Col Med, State Univ NY Upstate Med Ctr, 59-64, Am Heart Asn estab investr, 62-67. Res: Renal pharmacology; cerebro-spinal fluid physiology. Mailing Add: Dept of Pharmacol Univ of Mass Med Sch Worcester MA 01605

MILLER, VERNA JEAN, b Lebanon, Pa, Aug 17, 48. ETHOLOGY, ANIMAL BEHAVIOR. Educ: Pa State Univ, BS, 70, MS, 72, PhD(zool), 74. Prof Exp: INSTR BIOL, DAVIDSON COL, 74- Mem: Animal Behav Soc; Am Inst Biol Sci; Am Soc Zoologists; AAAS. Res: Development and regulation of vertebrate social systems; relationship between social systems, the ecological setting in which they occur and the evolutionary history of the species. Mailing Add: Dept of Biol Davidson Col Davidson NC 28036

MILLER, VERNON R, inorganic chemistry, see 12th edition

MILLER, VICTOR CHARLES, b Stamford, Conn, Mar 17, 22; c 2. GEOMORPHOLOGY. Educ: Columbia Univ, BA, 43, MA, 48, PhD(geomorphol), 53. Prof Exp: Photogeologist, Sinclair Wyo Oil & Gas Co, 47-49; lectr, Dept Geol, Columbia Univ, 49-51; photogeologist, Geophoto Serv, 51-52; partner, Miller-McCulloch, Alta, 52-54; pres, Miller-McCulloch, Ltd, 54-56, V C Miller & Assocs, Photogeologists, Ltd, Can, 56-57 & Miller & Assocs, Inc, Colo, 57-62; prof geol & chmn dept, Univ Libya, 62-64 & C W Post Col, Long Island Univ, 64-67; PROF GEOL, IND STATE UNIV, TERRE HAUTE, 67- Concurrent Pos: Geol panel mem, Comt Remote Sensing for Earth Resource Surv, Nat Res Coun, 73- Mem: AAAS; Geol Soc Am; Am Soc Photogram; Am Asn Petrol Geologists. Res: Photogeology; photogeomorphology; remote sensing, air photos, radar, infra-red, Skylab and Earth Resources Technol Satellite imagery. Mailing Add: Dept of Geog & Geol Ind State Univ Terre Haute IN 47809

MILLER, VICTOR JAY, b Geff, Ill, Jan 12, 21; m 45; c 3. HORTICULTURE. Educ: Univ Ill, BS, 42, MS, 47, PhD(hort), 49. Prof Exp: Asst bot, Univ Ill, 45-47, asst hort, 47-49; from asst prof to assoc prof & chmn dept, Univ Nebr, 49-58; PROF HORT, ARIZ STATE UNIV, 58- Mem: Am Soc Hort Sci. Res: Fruits and woody ornamentals; turf. Mailing Add: Div of Agr Ariz State Univ Tempe AZ 85281

MILLER, VINCENT PAUL, JR, b Swissvale, Pa, May 11, 32; m 60; c 1. GEOGRAPHY. Educ: Muskingum Col, BSc, 54; Pa State Univ, MSc, 57; Univ Oslo, dipl, 60; Mich State Univ, PhD(geog), 70. Prof Exp: Social sci asst, US Army Res & Eng Command, 57-59; instr goeg, Col Wooster, 59-61; asst instr, Mich State Univ, 61-62; assoc prof, 62-71, PROF GEOG, INDIANA UNIV PA, 71- Concurrent Pos: Teaching assoc, Danforth Found, Mo, 65-; ed, Pa Geogr, 65-; Rural Develop Act res grant, 75. Mem: Asn Am Geogr; Sigma Xi; Arctic Asn NAm; Nat Coun Geog Educ. Res: Philosophy and thought of geography; geography of economic development and urban-rural interaction. Mailing Add: Dept of Geog Indiana Univ of Pa Indiana PA 15701

MILLER, WADE ELLIOTT, II, b Los Angeles, Calif, Oct 20, 32; m 60; c 3. VERTEBRATE PALEONTOLOGY, GEOLOGY. Educ: Brigham Young Univ, BS, 60; Univ Ariz, MS, 63; Univ Calif, Berkeley, PhD(paleont), 68. Prof Exp: Instr geol & phys sci, Santa Ana Col, 61-64; instr geol, Fullerton Jr Col, 68-71; ASSOC PROF ZOOL & GEOL, BRIGHAM YOUNG UNIV, 71- Concurrent Pos: Geologist & paleontologist, Los Angeles County Mus, 69- Mem: Soc Vert Paleont; Paleont Soc; Soc Mammal. Res: Pleistocene mammals of Southern California; fossil vertebrates of Utah, especially mammals. Mailing Add: Dept of Zool Brigham Young Univ Provo UT 84601

MILLER, WALLACE E, b Winona, Minn, Oct 24, 19; m; c 2. ORTHOPEDIC SURGERY. Educ: Wabash Col, AB, 41; Harvard Med Sch, MD, 44; Am Bd Orthop Surg, dipl, 54. Prof Exp: Chmn dept orthop, Jackson Mem Hosp, 57-71; chief orthop serv, Vet Miami Hosp, 71-74; PROF ORTHOP & REHAB, SCH MED, UNIV MIAMI, 74- Concurrent Pos: Prof orthop & chief div, Sch Med, Univ Miami, 57-69; examr, Am Bd Orthop Surg, 55-74. Mem: AMA; Am Acad Orthop Surg. Res: Orthopedics. Mailing Add: Univ of Miami Sch of Med PO Box 520875 Biscayne Annex Miami FL 33152

MILLER, WALTER BERNARD, III, b Pensacola, Fla, Mar 10, 42; m 64. PHYSICAL CHEMISTRY. Educ: Univ Calif, Los Angeles, BS, 63; Harvard Univ, PhD(phys chem), 69. Prof Exp: ASST PROF CHEM, UNIV ARIZ, 68- Res: Chemical kinetics in crossed molecular beams. Mailing Add: Dept of Chem Univ of Ariz Tucson AZ 85721

MILLER, WALTER CHARLES, b Philadelphia, Pa, Dec 9, 18; m 44; c 7. EXPERIMENTAL NUCLEAR PHYSICS. Educ: St Joseph's Col, Pa, BS, 40; Univ Notre Dame, MS, 42, PhD(physics), 48. Prof Exp: Asst physics, Univ Notre Dame, 40-43, instr & asst, Manhattan Proj, 43-44; res assoc, Los Alamos Sci Lab, 44-45; from instr to assoc prof, 45-64, PROF PHYSICS, UNIV NOTRE DAME, 64-, CHMN DEPT, 75- Mem: Am Phys Soc. Res: Electrostatic accelerators; electronics; energy levels; x-rays; nuclear excitation. Mailing Add: Dept of Physics Univ of Notre Dame Notre Dame IN 46556

MILLER, WALTER E, b New York, NY, Jan 28, 14; m 43; c 3. PHYSICAL CHEMISTRY. Educ: City Col New York, BS, 35, ChE, 36; NY Univ, PhD(chem), 41. Prof Exp: Instr, 41-42 & 46-49, from asst prof to assoc prof, 49-66, PROF CHEM, CITY COL NEW YORK, 66- Concurrent Pos: Asst instr, NY Univ, 41-42; mem, US Army Chem Corps Adv Coun, 64-69. Mem: AAAS; Am Chem Soc; NY Acad Sci. Res: Cryogenics; rocket fuels; photo-sensitization; atomic physics; design of radio transmitters and receivers; high polymers; ion exchange; water demineralization; electrolytic treatment of water. Mailing Add: Dept of Chem City Col of New York New York NY 10031

MILLER, WALTER JOHN, astronomy, deceased

MILLER, WALTER PETER, b Dickinson, NDak, Jan 7, 32; m 60; c 4. ORGANIC POLYMER CHEMISTRY. Educ: Univ Minn, BA, 53, PhD, 57. Prof Exp: Fel, Max Planck Inst Coal Res, Ger, 57-58; proj chemist, Union Carbide Chem Co, 58-64, GROUP LEADER, CHEM DIV, UNION CARBIDE CORP, 65- Concurrent Pos: Fulbright travel grant, 57-58. Mem: Am Chem Soc. Res: Emulsion polymerization; organic chemistry. Mailing Add: 1207 Shady Way South Charleston WV 25309

MILLER, WARREN JAMES, b New Kensington, Pa, Oct 12, 31; m 58; c 3. PHYSICAL CHEMISTRY. Educ: Pa State Univ, BS, 57; Fla State Univ, PhD(chem), 62. Prof Exp: Res chemist, 62-64, sr res chem- chemist, 64-69, RES ASSOC, RES LABS, EASTMAN KODAK CO, 69- Mem: Am Chem Soc; fel Am Inst Chemists. Res: Photographic theory; colloid and surface chemistry. Mailing Add: Eastman Kodak Co Res Labs 343 State St Rochester NY 14650

MILLER, WARREN WIDMER, b Bluffton, Ind, Mar 23, 15; m 41; c 3. CHEMISTRY. Educ: Ohio State Univ, BS, 41; Univ Calif, PhD(chem), 44. Prof Exp: Res chemist, Md Res Lab, Washington, DC, 43-44; res assoc chem, Harvard Univ, 44-45; res assoc physics, Mass Inst Technol, 45-47; assoc scientist, Brookhaven Nat Lab, 47-50; assoc prof chem, 50-58, PROF CHEM & HUMANITIES, PA STATE UNIV, 58- Concurrent Pos: Consult, AEC & US Geol Surv; mem, Atoms for Peace Mission, Latin Am, 56-57. Mem: AAAS. Res: Recoil chemistry; activation analysis. Mailing Add: Dept of Chem Pa State Univ University Park PA 16802

MILLER, WATKINS WILFORD, b Hawthorne, Calif, Feb 21, 47; m 69; c 2. SOIL FERTILITY, WATER POLLUTION. Educ: Calif Polytech State Univ, BS, 68; Univ Calif, PhD, 73. Prof Exp: Agr chemist, Soil Testing Serv, Nelson Labs, Stockton, Calif, 69-70; res asst soil water repellency, Univ Calif, Riverside, 70-73; exten specialist natural resource develop, 73-75, ASST PROF SOIL & WATER, UNIV NEV, RENO, 75- Mem: Am Soc Agron; Soil Sci Soc Am; AAAS. Res: Soil fertility analysis and calibration; water quality of irrigation return flows; rural development resource inventories; soil and water testing service for Nevada residents; water management, especially Humbolt River system. Mailing Add: Div of Plant Soil & Water Sci Univ of Nev Col of Agr Reno NV 89507

MILLER, WESLEY LAMAR, organic chemistry, polymer chemistry, see 12th edition

MILLER, WILBUR HOBART, b Boston, Mass, Feb 15, 15; m 41; c 3. ORGANIC CHEMISTRY, BIOCHEMISTRY. Educ: Univ NH, BS, 36, MS, 38; Columbia Univ, PhD(chem), 42. Prof Exp: From asst to instr chem, Univ NH, 36-38; asst, Columbia Univ, 38-39, statutory asst, 39-40; res chemist, Stamford Res Lab, Am Cyanamid Co, 41-49, tech rep, Washington, DC, 49-53, dir custom sales, Lederle Labs, 53-54, dir indust appln, Fine Chem Div, 54-55, dir food indust develop, Farm & Home Div, 55-57; tech dir prod for agr, Cyanamid Int, 57-60; sr scientist, Dunlap & Assocs, Inc, Conn, 60-66; coordr new prod develop, 66-67, mgr commercial res, 67-69, DIR DIVERSIFICATION DEVELOP, CELANESE CORP, NEW YORK, 69- Honors & Awards: Am Design Award, 48. Mem: Fel AAAS; fel Am Inst Chemists; Am Chem Soc; Inst Food Technologists; Tech Asn Pulp & Paper Indust. Res: Agricultural chemicals; animal health and food industry products; chemotherapy; enzyme, organic and general industrial chemistry. Mailing Add: 19 Crestview Ave Stamford CT 06907

MILLER, WILLARD, JR, b Ft Wayne, Ind, Sept 17, 37; m 65; c 2. APPLIED MATHEMATICS, MATHEMATICAL PHYSICS. Educ: Univ Chicago, SB, 58; Univ Calif, Berkeley, PhD(appl math), 63. Prof Exp: NSF fel, Courant Inst, NY Univ, 63-64, vis mem, 64-65; from asst prof to assoc prof, 65-72, PROF MATH, UNIV MINN, MINNEAPOLIS, 72- Concurrent Pos: Vis mem, Ctr Math Res, Univ Montreal, 73-74; assoc ed, J Math Physics, 73-75; managing ed, J Math Anal, 75- Mem: Am Math Soc; Soc Indust & Appl Math. Res: Applications of group theory to special functions. Mailing Add: Sch of Math Univ of Minn Minneapolis MN 55455

MILLER, WILLIAM, b New York, NY, Sept 1, 22; m 57; c 3. EXPERIMENTAL SOLID STATE PHYSICS. Educ: City Col New York, BBA, 43; Univ Pa, PhD(physics), 48. Prof Exp: Physicist, Nat Bur Standards, 48-56; from asst prof to assoc prof, 57-67, PROF PHYSICS, CITY COL NEW YORK, 67- Concurrent Pos: Opers analyst, Opers Res Off, Johns Hopkins Univ, 52-53; sr physicist & consult, Am Mach & Foundry Co, 56- Mem: AAAS; Am Phys Soc; Am Asn Physics Teachers. Res: Noise theory; electromagnetic waves; x-rays; solid state. Mailing Add: Dept of Physics City Col of New York New York NY 10031

MILLER, WILLIAM ALFONSO, b Portland, Ore, Mar 16, 12; m 38; c 6. PHYSICS. Educ: Ore State Col, BS, 32, MS, 36; Purdue Univ, PhD(physics), 41. Prof Exp: Instr physics, Univ Ore, 36-37; asst, Purdue Univ, 37-39; res assoc, Radiation Lab, Mass Inst Technol, 41-42; res & design engr, RCA Labs, 42-61; sr staff scientist, Elec Syst Div, Fairchild-Hiller Corp, 61-65 & Radiometrics Div, Polarad Electronics Corp, 65-67; consult, Sanders Assocs, 67-68; staff scientist, Loral Electronic Systs, 68-69, chief scientist, 69-74; DISPLAY CONSULT, 74- Mem: Am Phys Soc; Am Astron Soc; Am Geophys Union; Soc Photo-Optical Instrument Engrs; sr mem Inst Elec & Electronics Engrs. Res: New semiconducting materials; radio transmitter and antenna design; improvement of frequency stability of radio transmitters; special electron tubes; uranium and thorium fission; radio propagation; solar physics; communication systems design; satellite systems, precision radar and optical systems engineering; special CRT display design. Mailing Add: RR 2 Dogwood Lane Miller Place NY 11764

MILLER, WILLIAM ANTON, b Cedar, Mich, Apr 16, 35; m 60; c 4. MATHEMATICS EDUCATION. Educ: Mich State Univ, BS, 56, MAT, 61; Univ Ill, Urbana, MA, 63; Univ Wis, Madison, PhD(math educ), 68. Prof Exp: Teacher, Sunfield Community Schs, 56-60, Oak Park Schs, 60-61 & Waverly Schs, 61-62; asst prof math educ, Wis State Univ, Whitewater, 65-67, assoc prof math, 67-68; assoc prof, 68-71, PROF MATH, CENT MICH UNIV, 71- Honors & Awards: Awards, Outstanding Educr Am, 75 & Am Biog Inst, 74. Mem: Math Asn Am. Res: Learning theory as it relates to mathematics. Mailing Add: 407 E Grand Ave Mt Pleasant MI 48858

MILLER, WILLIAM BOYNTON, JR, b Atlanta, Ga, Aug 7, 23; m 49; c 2. ACADEMIC ADMINISTRATION, RADIATION PHYSICS. Educ: Emory Univ, BS, 47. Prof Exp: Res asst electronics, 47-52, instr electronics, consult electronics engr & radiation safety officer, 53-56, from instr to asst prof radiation physics & electronics, 57-68, dir clin radioisotope lab, 50-56, ASSOC PROF RADIATION PHYSICS & ELECTRONICS, SCHMED, EMORY UNIV, 68-, ASSOC DIR, DIV ALLIED HEALTH PROFESSIONS, 73- Concurrent Pos: Electronics consult, Grady Mem Hosp, 57-60; consult, Div Radiol Health, USPHS, Md, 64. Mem: Sr mem Inst Elec & Electronics Engrs; NY Acad Sci. Res: Radiation detection instrumentation; radiation exposure incident to x-ray procedures; electrical hazards and safety; electronic radiography. Mailing Add: Emory Univ Sch of Med Atlanta GA 30322

MILLER, WILLIAM BRUNNER, b Bethlehem, Pa, July 27, 23; m 48; c 4. MATHEMATICS. Educ: Lehigh Univ, BS, 47, MA, 55, PhD(math), 62. Prof Exp: Elec engr, Western Elec Co, Inc, 47-49; engr, Laros Textiles Co, 50-53; prof math, Moravian Col, 53-62; ASSOC PROF MATH, WORCESTER POLYTECH INST, 63- Concurrent Pos: Instr, Lehigh Univ, 59-60. Mem: Math Asn Am; Soc Indust & Appl Math. Res: Separation and oscillation theorems of linear differential equations. Mailing Add: Dept of Math Worcester Polytech Inst Worcester MA 01609

MILLER, WILLIAM DONALD, geology, see 12th edition

MILLER, WILLIAM ELDON, b McAllen, Tex, July 13, 30; div; c 4. FOREST ENTOMOLOGY. Educ: La State Univ, BS, 50; Ohio State Univ, MSc, 51, PhD(entom), 55; Mich State Univ, MS, 61. Prof Exp: WGer Govt fel, Univ Göttingen, 56-57; entomologist, 56-64, PRIN INSECT ECOLOGIST, US FOREST SERV, 64- Concurrent Pos: Ed, Forest Sci, 71- Mem: Entom Soc Am; Soc Am Foresters; Lepidop Soc. Res: Population dynamics; insect control techniques; taxonomy of Microlepidoptera; energetics of forest ecosystems. Mailing Add: NCent Forest Exp Sta Fowell Ave St Paul MN 55101

MILLER, WILLIAM FRANKLIN, b Stone Creek, Ohio, Jan 16, 20; m 42, 58; c 5. MEDICINE. Educ: Wittenberg Univ, BA, 42; Western Reserve Univ, MD, 45; Am Bd Internal Med, dipl, 56. Prof Exp: Intern, City Hosp, Cleveland, Ohio, 45-46; resident med, neurol & radiol, Dayton Vet Admin Hosp, 46-48; resident med, Dallas Vet Admin Hosp, Tex, 48-51; clin instr, 51-53, from asst prof to assoc prof, 53-67, PROF MED, UNIV TEX HEALTH SCI CTR DALLAS, 67- Concurrent Pos: Dir cardio-respiratory lab, McKinney Vet Admin Hosp, Tex, 51-53; dir pulmonary div, Parkland Mem Hosp & Woodlawn Hosps, 53-67; dir, Pulmonary Div, Methodist Hosp, 67-; consult, Surgeon Gen, Lackland Air Force & Brooke Army Hosps, San Antonio, US Surgeon Gen, Comt Health Aspects Tobacco, Parkland Mem, Vet Admin, St Paul, Baylor, Methodist & Presby Hosps, Dallas; consult chronic pulmonary dis sect, NIH, 65-67, task force respiratory dis, Nat Heart & Lung Inst, 71 & comt clin training physicians assts, Sch Allied Health Professions, Univ Tex, 72; chmn med adv bd, Am Asn Inhalation Ther; mem med adv bd, Cystic Fibrosis Found; med ed, Respiratory Care, 66- Honors & Awards: Med Movie Award, Am Col Chest Physicians, 67. Mem: AAAS; Am Thoracic Soc; fel Am Col Chest Physicians; Am Fedn Clin Res; Am Heart Asn. Res: Clinical pulmonary asthma; chronic bronchitis and emphysema. Mailing Add: Pulmonary Div Methodist Hosp Dallas Box 5999 Dallas TX 75222

MILLER, WILLIAM FREDERICK, b Vincennes, Ind, Nov 19, 25; m 49; c 1. COMPUTER SCIENCE, ACADEMIC ADMINISTRATION. Educ: Purdue Univ, BS, 49, MS, 51, PhD(physics, math), 56. Hon Degrees: DSc, Purdue Univ, 72. Prof Exp: Assoc physicist, Argonne Nat Lab, 56-59, dir appl math div, 59-64; PROF COMPUT SCI, STANFORD UNIV, 65-, VPRES & PROVOST, 71- Concurrent Pos: Vis prof, Purdue Univ, 62-63; prof lectr, Univ Chicago, 62-64; mem math & comput sci res adv comt, AEC & US deleg study group digital technol, Europ Nuclear Energy technol, Europ Nuclear Energy Agency, 62-; consult, Argonne Nat Lab, 65- & Comput Usage Corp, 69-; mem comput sci & eng bd, Nat Acad Sci, 68-71; mem sci info coun, NSF, 71- Mem: AAAS; Am Phys Soc; Am Math Soc; Asn Comput Mach; Soc Indust & Appl Math. Res: Computer science and applications; computational physics; nuclear scattering; computing machinery; university administration. Mailing Add: Stanford Univ Stanford CA 94305

MILLER, WILLIAM HENRY, b Baltimore, Md, Aug 7, 26; m 57; c 3. BIOLOGY. Educ: Haverford Col, BA, 49; Johns Hopkins Univ, MD, 54. Prof Exp: Intern med, Baltimore City Hosps, 54-55; res assoc, Rockefeller Inst, 55-58, asst prof, 58-64; assoc prof physiol & ophthalmol, 64-69, PROF OPHTHALMOL & VISUAL SCI, SCH MED, YALE UNIV, 69- Res: Anatomy and physiology of invertebrate eyes. Mailing Add: Dept of Ophthal & Vis Sci Yale Univ Sch of Med New Haven CT 06510

MILLER, WILLIAM HUGHES, b Kosciusko, Miss, Mar 16, 41; m 66; c 1. CHEMICAL PHYSICS. Educ: Ga Inst Technol, BS, 63; Harvard Univ, AM, 64, PhD(chem physics), 67. Prof Exp: NATO fel, Univ Freiburg, 67-68; Soc Fels jr fel, Harvard Univ, 68-69; from asst prof to assoc prof, 69-74, PROF CHEM, UNIV CALIF, BERKELEY, 74-, PRIN INVESTR, INORG MAT RES DIV, LAWRENCE BERKELEY LAB, 70- Concurrent Pos: Sloan Found fel, 70; Guggenheim Mem fel, 75-76; fel, Churchill Col, Cambridge Univ, 75-76. Honors & Awards: Int Acad Quantum Molecular Sci Ann Prize, Paris, 74. Res: Quantum theory of atomic and molecular collisions; semiclassical theories and quantum effects in inelastic and reactive scattering of atoms and molecules; collisional transfer of electronic energy. Mailing Add: Dept of Chem Univ of Calif Berkeley CA 94720

MILLER, WILLIAM JACK, b Nathans Creek, NC, Feb 7, 27; m 50; c 4. ANIMAL NUTRITION. Educ: NC State Col, BS, 48, MS, 50; Univ Wis, PhD(animal nutrit), 52. Prof Exp: Wis Alumni Res Found asst, Univ Wis, 50-52; res assoc dairy physiol, Univ Ill, 52-53; from asst prof to prof, 53-73, ALUMNI FOUND DISTINGUISHED PROF DAIRY SCI, UNIV GA, 73- Honors & Awards: Nutrit Res Award, Am Feed Mfrs Asn, 63; Distinguished Serv Award, Univ Ga-Gamma Sigma Delta, 65, Distinguished Serv to Agr Award, 70; Mineral Travel Res Award, Calcium Carbonate Co, 69; Excellence in Res Award, Sigma Xi, 69; Outstanding Scientist of Ga Award, 69; Borden Award, Am Dairy Sci Asn, 71; Gustav Bohstedt Mineral Res Award, Am Soc Animal Sci, 71. Mem: AAAS; Soc Environ Geochem & Health; Am Inst Nutrit; Am Dairy Sci Asn; Am Soc Animal Sci. Res: Zinc, manganese, cadmium, nickel and mercury nutrition and metabolism; mineral nutrition of animals; ruminant nutrition; forage evaluation and utilization. Mailing Add: Dept of Animal & Dairy Sci Univ of Ga Athens GA 30602

MILLER, WILLIAM JOHN, b Malden, Mass, June 13, 36; m 57; c 3. PHYSICAL CHEMISTRY. Educ: Suffolk Univ, BA, 58; Pa State Univ, PhD(chem), 63. Prof Exp: PHYS CHEMIST, AEROCHEM RES LABS, INC, 63- Mem: Am Soc Mass Spectrometry; Optical Soc Am; Combustion Inst. Res: Ions in flames; mass and flame spectroscopy. Mailing Add: AeroChem Res Labs Inc PO Box 12 Princeton NJ 08540

MILLER, WILLIAM KNIGHT, b Salisbury, NC, Nov 14, 18; m 46; c 2. ANALYTICAL CHEMISTRY. Educ: Catawba Col, AB, 40; Univ NC, PhD(anal chem), 50. Prof Exp: Chemist, Carbide & Carbon Chem Co, Oak Ridge Nat Lab, 50-52; ANAL SUPVR, SC JOHNSON & SON, INC, 52- Mem: Am Chem Soc; Soc Appl Spectros. Res: Infrared spectrophotometry; gas chromatography. Mailing Add: 3427 N St Clair St Racine WI 53402

MILLER, WILLIAM LAWRENCE, b Medford, Ore, May 12, 37; m 61; c 2. METALLURGY, GEOLOGY. Educ: Univ Calif, Davis, BS, 63, MD, 67. Prof Exp: Assoc chemist nitroplasticizers develop, Aerojet Gen Corp, Gen Tire & Rubber Co, 63-64; chemist minerals res, Bur Mines, 66-70, prog analyst minerals, Off Secy, 70-71, PHYS SCIENTIST METALL RES, BUR MINES, US DEPT INTERIOR, 71- Concurrent Pos: Alt mem coord comt mat res & develop, Fed Coun Sci & Technol, 69. Mem: Am Chem Soc. Res: Development and application of new and improved process technology to extract, recover, purify, fabricate and recycle metallic and non-metallic minerals with minimum waste and environmental degradation. Mailing Add: US Dept Interior Bur Mines 2401 E St NW Washington DC 20241

MILLER, WILLIAM LLOYD, b Waterman, Ill, Mar 22, 35; m 57; c 4. AGRICULTURAL ECONOMICS. Educ: Univ Ill, Urbana, BS, 57, MS, 60; Mich State Univ, PhD(agr econ), 65. Prof Exp: Asst prof, 65-70, ASSOC PROF AGR ECON, PURDUE UNIV, 70- Mem: Am Agr Econ Asn; Am Econ Asn; Am Water Resources Asn. Res: Resource development. Mailing Add: Dept of Agr Econ Purdue Univ West Lafayette IN 47906

MILLER, WILLIAM LOUIS, b Springville, Utah, June 28, 25; m 52; c 3. BIOCHEMISTRY. Educ: Brigham Young Univ, BS, 48; Univ Wis, MS, 50, PhD(biochem), 52. Prof Exp: Asst biochem, Univ Wis, 48-52; res scientist pharmacol, 52-59, res assoc endocrinol, 59-68, SR RES SCIENTIST, DEPT METAB DIS RES, UPJOHN CO, 68- Mem: Am Soc Pharmacol & Exp Therapeut; Soc Exp Biol & Med. Res: Fate studies on drugs, especially metabolic detoxification of drugs by mammals; skin sterols and azo-dye carcinogenesis; diabetes; hypoglycemic agents; biochemistry of hormone action; lipid metabolism; reproductive physiology. Mailing Add: Upjohn Co 301 Henrietta Kalamazoo MI 49001

MILLER, WILLIAM MARTIN, b Wichita, Kans, Nov 23, 23; m 57. ORGANIC CHEMISTRY. Educ: Univ Ill, BS, 44; Univ Iowa, MS, 50, PhD(org chem), 51. Prof Exp: Asst, Univ Iowa, 46-50; prof chem, Tarkio Col, 51-52; asst, Univ Calif, 52-53; asst prof, Washburn Univ, 53-56; from asst prof to assoc prof, 56-69, PROF CHEM, FRESNO STATE UNIV, 69- Res: Isoflavones; anthrylium salts. Mailing Add: Dept of Chem Fresno State Univ Fresno CA 93710

MILLER, WILLIAM REYNOLDS, JR, b Philadelphia, Pa, Dec 29, 39. POLYMER PHYSICS, PHYSICAL CHEMISTRY. Educ: Princeton Univ, AB, 61; Columbia Univ, MA, 62, PhD(chem), 65. Prof Exp: Fel, Wash Univ, 65-66; proj scientist plastics, Koppers Co, Inc, 68-74, PROJ SCIENTIST PLASTICS, ARCO/POLYMERS, INC, 74- Mem: Am Chem Soc; Soc Rheol; Soc Plastics Eng. Res: Rheology, especially plastics solids and melts; polymer evaluation; dielectric properties of plastics; electron spin resonance spectroscopy of organic free radicals in solution. Mailing Add: Res Dept ARCO/Polymers Inc 440 College Park Dr Monroeville PA 15146

MILLER, WILLIAM RIEDEL, b Cedar Falls, Iowa, Nov 3, 23. ORGANIC CHEMISTRY. Educ: Iowa State Col, BSc, 46; Univ Ill, MS, 48, PhD(chem), 50. Prof Exp: Fel chem, Univ Pittsburgh, 50-51; chemist, Electrochem Dept, E I du Pont de Nemours & Co, 51-57; RES CHEMIST, OILSEEDS CROPS LAB, NORTHERN REGIONAL RES CTR, AGR RES SERV, USDA, 57- Mem: Am Chem Soc; Am Oil Chemists Soc. Res: Organic chemical research; fatty acid derivatives; polymers. Mailing Add: 2707 Randan Ct Peoria IL 61604

MILLER, WILLIAM ROBERT, b Norwalk, Ohio, Mar 11, 43; m 64; c 3. GEOCHEMISTRY. Educ: Ohio State Univ, BCE, 68; Univ Wyo, MS, 72, PhD(geol), 74. Prof Exp: Prod engr, Mobil Oil Corp, 68-69; teaching asst geochem, Univ Wyo, 69-72; GEOLOGIST, BR EXPLOR RES, US GEOL SURV, 74- Mem: Geochem Soc; Geol Soc Am; Sigma Xi. Res: Low temperature geochemistry, particularly reactions involving natural waters and solid phases; clay mineralogy of soils; weathering and controls on the partitioning of trace elements between natural waters and solid phases. Mailing Add: US Geol Surv Mail Stop 932 Fed Ctr Bldg 25 Lakewood CO 80225

MILLER, WILLIAM ROBERT, JR, b Baltimore, Md, June 17, 34; m 58; c 3. ELECTRON PHYSICS, SOLID STATE PHYSICS. Educ: Gettysburg Col, BA, 56; Univ Del, MA, 61, PhD(physics), 65. Prof Exp: Engr, Westinghouse Elec Corp, 58-59, sr engr, 65-67; teaching asst, Univ Del, 59-61; engr, RCA Corp, 67-68; instr, 68, ASST PROF PHYSICS, PA STATE UNIV, 69- Concurrent Pos: Instr physics, York Col Pa, 69-75. Mem: Inst Elec & Electronics Engrs; Am Phys Soc; Am Asn Physics Teachers. Res: Solid state and semiconductor devices; positron physics. Mailing Add: 1029 Preston Rd Lancaster PA 17601

MILLER, WILLIAM SCHUYLER, b East Orange, NJ, Mar 15, 10; m 35; c 2. CHEMISTRY. Educ: Lehigh Univ, BS, 30, MS, 32; Syracuse Univ, PhD(chem), 36. Prof Exp: Asst chem, Lehigh Univ, 30-32 & Syracuse Univ, 32-36; instr, Bethany Col, 36-40; from asst prof to assoc prof, 40-44, chmn div sci, 47-74, head dept chem, 63-75, PROF CHEM, RANDOLPH-MACON COL, 44- Concurrent Pos: Prof, Univ Richmond, 44-45; vis prof, Mary Washington Col, Univ Va, 56-59; vis res assoc, Pa State Univ, 62-63. Mem: AAAS; Am Chem Soc. Res: High pressure chemistry; effects of x-rays on living organisms; x-ray study of coagulation process in proteins; crystal structures of peroxides, peroxide hydrates and polysulfides; certain aspects of chemical history; valence problems; science curricula for undergraduates. Mailing Add: Dept of Chem Randolph-Macon Col Ashland VA 23005

MILLER, WILLIAM SINKABINE, microbiology, see 12th edition

MILLER, WILLIAM TAYLOR, b Winston-Salem, NC, Aug 24, 11; m 51. ORGANIC CHEMISTRY, FLUORINE CHEMISTRY. Educ: Duke Univ, AB, 32, PhD(org chem), 35. Prof Exp: Postdoctoral appt, Stanford Univ, 35-36; from instr to assoc prof, 36-47, PROF CHEM, CORNELL UNIV, 47- Concurrent Pos: Off investr & consult, Nat Defense Res Comt, 41-43; mem fluorocarbon adv comt, Manhattan Prof, 43-45; head fluorocarbon res, SAM Labs, Manhattan Proj, Columbia Univ & Carbide & Carbon Chem Corp, 43-46; consult var govt & indust labs, 47-; consult, E I du Pont de Nemours & Co, Inc, 57- Honors & Awards: Award for Creative Work in Fluorine Chem, Am Chem Soc, 74. Mem: Fel AAAS; Am Chem Soc; The Chem Soc. Res: Chemistry of carbon-fluorine and related highly halogenated compounds; chemistry of haloorgano-metallic compounds. Mailing Add: Baker Lab Dept of Chem Cornell Univ Ithaca NY 14853

MILLER, WILLIAM THEODORE, b Belleville, Ill, Feb 8, 25. ORGANIC CHEMISTRY, BIOLOGICAL CHEMISTRY. Educ: St Louis Univ, AB, 47, BS, 51;

St Mary's Col, Kans, STL, 56; Univ Calif, Berkeley, PhD(chem), 61. Prof Exp: Teacher, Marquette Univ High Sch, 51-52; asst prof, 61-66, ASSOC PROF CHEM, REGIS COL, COLO, 66-, CHMN DEPT, 69-, DIR DIV NATURAL SCI & MATH, 75- Concurrent Pos: Res grants, NIH, 62-, Am Chem Soc Petrol Res Fund, 63- & NSF, 64; res fel, Lab Nuclear Med & Radiation Biol, Univ Calif, Los Angeles, 67-69. Mem: AAAS; Am Chem Soc; Sigma Xi. Res: Comparative study of lipids isolated from in vitro and in vivo grown tubercle bacillus; synthetic organic chemistry; polynuclear aromatic hydrocarbons. Mailing Add: Dept of Chem Regis Col W 50th & Lowell Blvd Denver CO 80221

MILLER, WILLIAM WADD, III, b Starkville, Miss, Oct 4, 32; m 57; c 2. REPRODUCTIVE PHYSIOLOGY. Educ: Miss State Univ, BS, 54, MS, 58; Auburn Univ, PhD(reprod physiol), 62. Prof Exp: Assoc prof biol, Howard Col, 62-67; ASSOC PROF BIOL, NORTHEAST LA UNIV, 67- Concurrent Pos: Sigma Xi grant, 65-66. Mem: AAAS; Soc Study Reproduction; Am Soc Zoologists. Res: Physiology of reproduction; endocrinology. Mailing Add: Dept of Biol Northeast La Univ Monroe LA 71201

MILLER, WILLIAM WALTER, biochemistry, see 12th edition

MILLER, WILLIAM WEAVER, b Winchester, Va, Sept 1, 33; m 58; c 2. PEDIATRICS, CARDIOLOGY. Educ: Va Mil Inst, BA, 54; Univ Pa, MD, 58; Am Bd Pediat, dipl, 66, cert pediat cardiol, 70. Prof Exp: Intern, Univ Pa Hosp, 59; intern, Children's Hosp Philadelphia, 62-63; instr pediat, Med Sch, Univ Pa, 63-65; instr, Harvard Med Sch, 65-66; assoc, Med Sch, Univ Pa, 66-69, asst prof, 69; ASSOC PROF PEDIAT, UNIV TEX HEALTH SCI CTR DALLAS, 69-; DIR PEDIAT CARDIOL DIV, CHILDREN'S MED CTR, 69- Concurrent Pos: Fel pediat cardiol, Children's Hosp, Philadelphia, 64-65, res fel, 66-67; assoc cardiologist & assoc dir cardiovasc labs, 67-69; res fel pediat cardiol, Children's Hosp Med Ctr, Boston, 65-66. Honors & Awards: Am Acad Pediat Award, 69. Mem: Am Acad Pediat; Am Heart Asn. Res: Oxygen transport; myocardial chemistry; congenital heart disease. Mailing Add: Div Pediat Cardiol Children's Med Ctr Dallas TX 75235

MILLER, WILLIE, b Bolivar, Tenn, Aug 24, 42; m 61; c 1. PLANT SCIENCE, SOIL SCIENCE. Educ: Tenn State Univ, BS, 71, MS, 72; Univ Tenn, PhD(plant & soil sci), 76. Prof Exp: RES REP AGR CHEM, ELI LILLY & CO, 76- Mem: Soil Sci Soc Am. Res: Development of new agricultural chemicals; promoting and expanding the use of existing agricultural chemicals. Mailing Add: Suite 209 2950 Metro Dr Minneapolis MN 55420

MILLER, WILLWAM ROBERT, b Arlington, Ala, Sept 1, 24; m 47; c 2. VIROLOGY, MICROBIOLOGY. Educ: Auburn Univ, DVM, 50, MS, 63; Purdue Univ, PhD(virol), 68; Am Bd Vet Pub Health, dipl, 73. Prof Exp: From instr to asst prof, 60-68, ASSOC PROF MICROBIOL, SCH VET MED, AUBURN UNIV, 68- Concurrent Pos: Collabr, USDA, 61-65; Nat Inst Neurol Dis & Blindness spec fel, 65-67; consult virol, Kans State Univ-AID, India, 71. Mem: Am Vet Med Asn; Sigma Xi; Asn Am Vet Cols; Am Asn Food Hyg Vet; Am Soc Microbiol. Res: Pathogenesis of animal virus diseases, specifically respiratory and neurological diseases. Mailing Add: Dept of Path & Parasitol Auburn Univ Sch of Vet Med Auburn AL 36830

MILLER, WILMER GLENN, b Mt Orab, Ohio, Aug 28, 32. POLYMER CHEMISTRY. Educ: Capitol Univ, BS, 54; Univ Wis, PhD(chem), 58. Prof Exp: Res fel chem, Harvard Univ, 58-59 & Univ Minn, 59-60; from asst prof to assoc prof, Univ Iowa, 60-67; assoc prof, 67-70, PROF CHEM, UNIV MINN, MINNEAPOLIS, 70- Concurrent Pos: Guggenheim fel, 64. Mem: Am Chem Soc. Res: Physical chemical studies of synthetic and biopolymers; polymer conformation; thermodynamics and dynamics of polymer liquid crystals; motion of polymers at or near an interface. Mailing Add: Dept of Chem Univ of Minn Minneapolis MN 55455

MILLER, WILMER JAY, b Lawton, Okla, July 15, 25; m 52; c 2. IMMUNOLOGY, GENETICS. Educ: Univ Okla, BA, 48; Univ Wis, PhD(genetics, zool), 54. Prof Exp: Proj assoc immunogenetics, Univ Wis, 53-55; assoc specialist, Sch Vet Med, Univ Calif, 55-62, lectr, 56-57; ASSOC PROF GENETICS, IOWA STATE UNIV, 62- Mem: Fel AAAS; Am Inst Biol Sci; Genetics Soc Am; Am Genetic Asn; NY Acad Sci. Res: Immunogenetics of birds and bovines. Mailing Add: Dept of Genetics Iowa State Univ Ames IA 50010

MILLER (GILBERT), CAROL ANN, b Greenville, Ohio, June 27, 43; m 66; c 1. MICROBIOLOGY, IMMUNOLOGY. Educ: Defiance Col, BA, 65; Ariz State Univ, MS, 69; Ore State Univ, PhD(microbiol), 70. Prof Exp: Clin immunologist, Wilson Mem Hosp, Johnston City, NY, 70-73; RES SCIENTIST MICRO-IMMUNOL, AMES RES LAB, MILES LABS, INC, 73- Mem: Am Soc Microbiol. Res: Development of immunological and chemical test systems which will be used to detect pathognomonic levels of hormones and microbial products in human biological fluids. Mailing Add: Ames Res Lab Miles Labs Inc 1127 Myrtle St Elkhart IN 46514

MILLERO, FRANK JOSEPH, JR, b Greenville, Pa, Mar 16, 39; m 65; c 1. PHYSICAL CHEMISTRY. Educ: Ohio State Univ, BS, 61; Carnegie-Mellon Univ, 64, PhD(phys chem), 65. Prof Exp: Asst chem, Carnegie-Mellon Univ, 61-63, asst thermochem, 63-65, fel, 65; phys chemist, Esso Res & Eng Co, 65-66; res scientist, 66-68, from asst prof to assoc prof, 68-73, PROF CHEM, OCEANOG & PHYS CHEM, ROSENSTIEL SCH OF MARINE & ATMOSPHERIC SCI, UNIV MIAMI, 73- Concurrent Pos: Mem oceanog panel, NSF, 73-75; mem, SCOR Panel on Oceanog Standards, UNESCO, 75. Mem: AAAS; Am Chem Soc; Am Geophys Union. Res: Solution thermodynamics; electrolyte solutions; thermochemistry; chemical oceanography; physical chemistry of aqueous solutions including seawater. Mailing Add: Rosenstiel Sch of Marine & Atmospheric Sci Univ of Miami Miami FL 33149

MILLERS, IMANTS, forest entomology, see 12th edition

MILLET, MARION PAUL, b Ponchatoula, La, July 19, 43; m 71; c 2. HUMAN ANATOMY. Educ: Southeastern La Univ, BS, 67; La State Univ, PhD(anat), 75. Prof Exp: Res asst path, Delta Regional Primate Res Ctr, Tulane Univ, 65-67; res assoc endocrinol, Med Ctr, La State Univ, 69-71; ASST PROF HUMAN ANAT, HEALTH SCI CTR, UNIV OKLA, 75- Mem: Am Asn Anatomists; Electron Soc Am. Res: Functional neuroanatomy; neuropathology. Mailing Add: Dept of Anat Univ Health Sci Ctr 801 NE 13th Oklahoma City OK 73190

MILLETT, FRANCIS SPENCER, b Madison, Wis, Aug 2, 43; m 68. BIOCHEMISTRY. Educ: Univ Wis, BS, 65; Columbia Univ, PhD(chem physics), 70. Prof Exp: NIH fel biochem, Calif Inst Technol, 70-72; ASST PROF BIOCHEM, UNIV ARK, FAYETTEVILLE, 72- Mem: Am Chem Soc (secy-treas, 73-75). Res: Interaction of proteins with biological membranes utilizing nuclear magnetic resonance methods; function of cytochrome C in mitochondria utilizing F-19 nuclear magnetic resonance methods. Mailing Add: Dept of Chem Univ of Ark Fayetteville AR 72701

MILLETT, KENNETH CARY, b Hustiford, Wis, Nov 16, 41. MATHEMATICS. Educ: Mass Inst Technol, BS, 63; Univ Wis-Madison, MS, 64, PhD(math), 67. Prof Exp: Instr math, Mass Inst Technol, 67-69; asst prof, 69-74, ASSOC PROF MATH, UNIV CALIF, SANTA BARBARA, 74- Mem: Am Math Soc. Res: Geometric and algebraic topology. Mailing Add: Dept of Math Univ of Calif Santa Barbara CA 93106

MILLETT, MARION T, physical geography, see 12th edition

MILLETT, MERRILL ALBERT, b Lake Mills, Wis, Nov 17, 15; m 42; c 3. WOOD CHEMISTRY. Educ: Univ Wis, BA, 38, MA, 39, PhD(phys chem), 43. Prof Exp: Asst chemist, 42-43, CHEMIST, US FOREST PROD LAB, 43- Mem: Am Soc Testing & Mat; Am Chem Soc; Forest Prod Res Soc; Tech Asn Pulp & Paper Indust. Res: Modified woods; cellulose and wood chemistry; molecular properties of celluloses; chromatographic analysis of woods and pulps; cellulose and cellulose esters; characterization and chemical utilization of wood residues; kinetics of aging of wood and cellulose; wood and pulping residues as animal feedstuffs. Mailing Add: US Forest Prod Lab Walnut St Madison WI 53705

MILLETT, WALTER ELMER, b Hampton, Ill, July 26, 17; m 44. PHYSICS. Educ: Univ Fla, BS, 40, MS, 42; Harvard Univ, PhD(physics), 49. Prof Exp: Res assoc, Radiation Lab, Mass Inst Technol, 42-45; AEC fel, Calif Inst Technol, 49-50; from instr to asst prof, Univ Fla, 50-52; from asst prof to assoc prof, 52-61, PROF PHYSICS, UNIV TEX, AUSTIN, 61- Mem: Fel Am Phys Soc. Res: Positron decay; electron and ion optics; scintillation spectrometry. Mailing Add: Dept of Physics Univ of Tex Austin TX 78712

MILLETT, WILLIAM HENRY, chemistry, see 12th edition

MILLETTE, CLARKE FRANCIS, b Bridgeport, Conn, Oct 22, 47. REPRODUCTIVE BIOLOGY, IMMUNOLOGY. Educ: Johns Hopkins Univ, BS, 69; Rockefeller Univ, PhD(biochem), 75. Prof Exp: FEL PHYSIOL, HARVARD MED SCH, 75- Mem: Soc Develop Biol. Res: Biochemistry and immunology of gametes and gametogenesis. Mailing Add: Lab Human Reprod & Reprod Biol Harvard Med Sch Boston MA 02115

MILLETTE, GERARD J F, b Montreal, Que, Feb 17, 21; m 45; c 5. SOIL CHEMISTRY. Educ: McGill Univ, BSc, 45, MSc, 48; Pa State Univ, PhD(agron), 55. Prof Exp: Jr pedologist, Can Dept Agr, Que, 45-49; sr pedologist, Ont, 49-53, officer-in-chg soil surv, NB, 53-61; proj mgr land & water surv, UN Spec Fund, Togo, Africa, 61-64; ASSOC PROF SOIL SCI, MacDONALD COL, McGILL UNIV, 64- Concurrent Pos: Mem, Nat Soil Surv Comt Can, 53-; indust consult. Mem: Agr Inst Can; Can Soil Sci Soc; Brit Soc Soil Sci; Int Soc Soil Sci. Res: Soil mapping and classification; interpretive surveys, land use, integrated area planning, land assessment; soil genesis, study of soil-plant-climate ecosystem. Mailing Add: Dept of Renewable Resources MacDonald Col of McGill Univ Ste Anne de Bellevue PQ Can

MILLETTE, ROBERT LOOMIS, b Rockville Centre, NY, May 17, 33; m 57; c 2. BIOCHEMISTRY, VIROLOGY. Educ: Ore State Col, BS, 54; Calif Inst Technol, PhD(biochem), 65. Prof Exp: USPHS res fel, Max Planck Inst Biochem, Ger, 64-67; ASST PROF PATH, MED CTR, UNIV COLO, DENVER, 67- Res: Mechanism and control of RNA biosynthesis; structure and function of ribonucleic acid polymerase; oncogenesis by herpesvirus. Mailing Add: Dept of Path Univ Colo Med Ctr Denver CO 80220

MILLHISER, FREDERICK ROY, b Buffalo, NY, Oct 27, 10; m 37; c 3. INDUSTRIAL CHEMISTRY, POLYMER CHEMISTRY. Educ: Carnegie Inst Technol, BS, 32, MS, 33; Harvard Univ, MS, 36, PhD(chem), 37. Prof Exp: Res chemist, E I du Pont de Nemours & Co, Inc, Del, 37-43, res supvr, 43-47, res mgr, 47-50, tech supt, 50-51, dir, Benger Lab, 51-67, tech mgr, Spunbonded-Nomex Prod, 67-75; RETIRED. Mem: Am Chem Soc. Res: Conductance and viscosity of electrolytes in water; electrochemistry of metals; industrial research on viscose rayon, acetate, acrylic spandex and nylon fibers. Mailing Add: 921 S Spigel Dr Virginia Beach VA 23454

MILLHOUSE, EDWARD W, JR, b West Hartford, Conn, Oct 28, 22. ELECTRON MICROSCOPY, CYTOCHEMISTRY. Educ: Univ Ill, BS, 49, MS, 54, PhD(cytol, anat), 60. Prof Exp: NIH fel anat, Sch Med, Ind Univ, 60-61; from instr to asst prof, 61-71, ASSOC PROF ANAT, CHICAGO MED SCH, 71- Mem: Electron Micros Soc Am; Am Asn Anatomists. Res: Electron microscopy and cyto- and histochemistry of the endocrine organs. Mailing Add: Dept of Anat Chicago Med Sch 2020 W Ogden Ave Chicago IL 60612

MILLHOUSE, OLIVER EUGENE, b Westerville, Ohio, Aug 21, 41. NEUROLOGY, ANATOMY. Educ: Ohio State Univ, BS, 63; Univ Calif, Los Angeles, PhD(anat), 67. Prof Exp: From instr to asst prof, 69-75, ASSOC PROF NEUROL & ANAT, COL MED, UNIV UTAH, 75- Concurrent Pos: Res fel, Dept Anat, Harvard Med Sch, 67-69. Mem: Am Asn Anat; Pan-Am Asn Anat. Res: Structural organization of the mammalian hypothalamus by light and electron microscopy, especially intrinsic connections and relations with other areas of the central nervous system. Mailing Add: Dept of Neurol Univ of Utah Salt Lake City UT 84132

MILLIAN, STEPHEN JERRY, b Okeechobee, Fla, Feb 15, 27; m 56; c 4. VIROLOGY. Educ: Brooklyn Col, BS, 49; Ohio State Univ, MS, 50, PhD(bact), 53; Am Bd Microbiol, dipl pub health & med lab virol, 63; Columbia Univ, MDPE, 75. Prof Exp: Asst, Ohio State Univ, 50-51; bacteriologist-virologist, Res Div, Armour Pharmaceut Co, 53-57; virologist biol res, Charles Pfizer & Co, Inc, 57-59; assoc cancer res scientist, Roswell Park Mem Inst, 60-61; chief virus unit, 61-72, ASST DIR VIROL & SEROL, BUR LABS, NEW YORK CITY DEPT HEALTH, 72- Concurrent Pos: WHO travel-study award, 72; assoc mem, Pub Health Res Inst New York, 61-; lectr pub health, Hunter Col, 66-; lab consult, First Army Med Lab, 66-; res assoc, Mt Sinai Hosp, 66-; consult, Prof Exam Serv, 66- Mem: Harvey Soc; Am Soc Microbiol; fel Am Pub Health Asn; fel NY Acad Sci; fel NY Acad Med. Res: Public health and laboratory virology; immune responses of individuals with neoplasias; murine tumor viruses, mycoplasmas; human and veterinary biologics. Mailing Add: Bur Labs NYC Dept of Health 445 First Ave New York NY 10016

MILLICAN, TROY BEA, b Rosedale, Miss, Aug 15, 37; m 60; c 2. LIMNOLOGY. Educ: Delta State Univ, BS, 63; Univ Miss, MS, 68, PhD(limnol), 71. Prof Exp: Instr zool, Univ Miss, 70-71; ASST PROF BIOL, DELTA STATE UNIV, 71- Concurrent Pos: Limnologist, US Army Corps Engrs, 73-75. Mem: Am Fisheries Soc; Am Soc Limnol & Oceanog; Am Micros Soc; Am Inst Biol Sci. Res: Limnological and ecological studies of flood control impoundments and their major tributaries in northern Mississippi; macro-invertebrates, specifically, of heavy and trace metal content; relationship of benthos to biological productivity of man-made impoundments. Mailing Add: 302 Bolling Ave Cleveland MS 38732

MILLICH, FRANK, b New York, NY, Jan 31, 28; m 60; c 2. POLYMER CHEMISTRY, PHOTOCHEMISTRY. Educ: City Col New York, BS, 49; Polytech Inst Brooklyn, 56, PhD(polymer chem), 59. Prof Exp: Chemist, Norda Essential Oils

& Chem Co, 50-55; Am Cancer Soc res fel, Cambridge Univ, 58-59 & Univ Calif, Berkeley, 59-60; from asst prof to assoc prof, 60-67, res assoc, Syst Space Sci Res Ctr, 66-71, PROF POLYMER CHEM, UNIV MO-KANSAS CITY, 67- Concurrent Pos: Consult, Missiles & Space Div, Lockheed Aircraft Corp, Calif, 59-60, Gulf Oil Corp, Kans, 70- & Midwest Res Inst, Mo, 70- Mem: AAAS; Am Chem Soc; The Chem Soc. Res: New polymer synthesis and characterization; chemical evolution; kinetics of nonenzymatic synthesis of polypeptides and nucleic acids; kinetics of dye-sensitive photochemical reactions; synthesis of new chemotherapeutic drugs; interfacial synthesis; luminescence; polyisocyanides. Mailing Add: Dept of Chem Univ of Mo 5100 Rockhill Rd Kansas City MO 64110

MILLICHAP, JOSEPH GORDON, b Wellington, Eng, Dec 18, 18; US citizen; m 46, 70; c 3. NEUROLOGY, PEDIATRICS. Educ: Univ London, MB, 46, MD, 50; Am Bd Pediat, dipl, 58; Am Bd Psychiat & Neurol, dipl, 60, cert child neurol, 68; FRCP, 71. Prof Exp: House physician med & pediat units, St Bartholomew's Hosp, London, 46-47, demonstr physiol, 47-48, chief asst pediat, 51-53; house physician, Great Ormond St Hosp, 50-51; assoc res prof pharmacol, Univ Utah, 54-55; asst prof neurol, George Washington Univ, 55-56; asst prof pediat & pharmacol, Albert Einstein Col Med, 56-58, assoc prof, 58-59; assoc prof pediat neurol & pharmacol, Mayo Grad Sch Med, Univ Minn, 61-63, pediat neurologist, Mayo Clin, 60-63; PROF NEUROL & PEDIAT, NORTHWESTERN UNIV, CHICAGO, 63- Concurrent Pos: Traveling fel, Brit Med Res Coun, 53-54; fel pediat, Harvard Univ, 53-54, Nat Inst Neurol Dis spec fel neurol, 58-60; pediat neurologist, Children's Mem Hosp, Wesley Hosp & Passavant Hosp; vis physician, DC Gen Hosp, 55-56; vis scientist, Clin Ctr, Nat Inst Neurol Dis, 55-56; vis pediatrician, Bronx Munic Hosp Ctr, NY, 56-58; resident, Mass Gen Hosp, 58-60; mem, Gov Adv Coun Develop Disabilities, 71- Mem: AAAS; Soc Pediat Res; Am Soc Pharmacol & Exp Therapeut; Soc Exp Biol & Med; Am Epilepsy Soc. Res: Pediatric neurology; neuropharmacology of anticonvulsant drugs; biochemistry of developing nervous system; etiology of neurological disorders of children; behavior and learning disabilities; epilepsy; electroencephalography. Mailing Add: 720 N Michigan Ave Chicago IL 60611

MILLIGAN, BARTON, b Cincinnati, Ohio, Oct 22, 29; m 54; c 1. ORGANIC CHEMISTRY. Educ: Haverford Col, AB, 51; Univ NC, MA, 53, PhD(chem), 55. Prof Exp: Lectr org chem, Univ Sydney, 55; Fulbright fel chem, Univ Adelaide, 56; assoc prof, Univ Miss, 56-64; from assoc prof to prof, Fla Atlantic Univ, 64-67; res chemist, 67-75, MGR RES, NITRATION PROD, HOUDRY LABS, AIR PROD & CHEM INC, 75- Concurrent Pos: NSF fac fel, 63-64. Mem: AAAS; Am Chem Soc. Res: Free radical reactions; process research. Mailing Add: Houdry Labs Air Prod & Chem Inc PO Box 427 Marcus Hook PA 19061

MILLIGAN, CARL W, analytical chemistry, see 12th edition

MILLIGAN, DOLPHUS EDWARD, physical chemistry, see 12th edition

MILLIGAN, GEORGE CLINTON, b PEI, Can, Sept 6, 19; m 42; c 3. ECONOMIC GEOLOGY, STRUCTURAL GEOLOGY. Educ: Dalhousie Univ, MSc, 48; Harvard Univ, AM, 50, PhD(struct geol), 60. Prof Exp: Geologist, Man Dept Mines, 50-57; assoc prof, 57-66, PROF GEOL, DALHOUSE UNIV, 66- Concurrent Pos: Nat Res Coun Can fel & guest prof, Swiss Fed Inst Technol, 64-65; Nat Res Coun Can exchange lectr, Univ Fed Pernambuco, Brazil, 74. Mem: Can Inst Mining & Metall; Mineral Asn Can; Geol Asn Can; Geol Soc Am. Mailing Add: Dept of Geol Dalhousie Univ Halifax NS Can

MILLIGAN, JOHN VORLEY, b Edmonton, Alta, Feb 21, 36; m 58; c 3. MEDICAL PHYSIOLOGY. Educ: Univ Alta, BSc, 58, MSc, 60; Univ Minn, PhD(physiol), 64. Prof Exp: Lectr physiol, McGill Univ, 64-66; asst prof, 66-70, ASSOC PROF PHYSIOL, QUEEN'S UNIV, ONT, 70- Concurrent Pos: Assoc ed, Can J Physiol & Pharmacol, 73-; fel, Int Brain Res Orgn/UNESCO, 74. Mem: Soc Neurosci; Int Soc Neuroendocrinol; Can Physiol Soc. Res: Mechanisms of excitation-contraction coupling in muscle; computer analysis of micrographs; ion interactions with plasma membrane; stimulus-secretion coupling in pituitary. Mailing Add: Dept of Physiol Queen's Univ Kingston ON Can

MILLIGAN, LARRY PATRICK, b Innisfail, Alta, Dec 12, 40; m 62; c 2. BIOCHEMISTRY, NUTRITION. Educ: Univ Alta, BSc, 61, MSc, 63; Univ Calif, Davis, PhD(nutrit), 66. Prof Exp: From asst prof to assoc prof, 66-74, PROF ANIMAL SCIENCE, UNIV ALTA, 74-, CHMN DEPT, 73- Mem: Can Biochem Soc; Agr Inst Can; Can Soc Animal Sci. Res: Nitrogen metabolism in animals, particularly ruminants; energy metabolism in animals. Mailing Add: Dept of Animal Sci Univ of Alta Edmonton AB Can

MILLIGAN, MERLE WALLACE, b Des Moines, Iowa, Mar 7, 22; m 48; c 2. MATHEMATICS. Educ: Monmouth Col, BS, 47; Univ Ill, MA, 49; Okla State Univ, EdD(higher educ, math), 60. Prof Exp: Asst math, Univ Ill, 47-52; from asst prof to assoc prof, Adams State Col, 53-62; prof, Albion Col, 62-66; dean arts & sci, 66-74, PROF MATH, METROP STATE COL, 66- Concurrent Pos: NSF grant, 63. Mem: Math Asn Am. Res: Analog computation. Mailing Add: Metrop State Col 250 W 14th Ave Denver CO 80204

MILLIGAN, MORRIS FRANKLIN, chemistry, see 12th edition

MILLIGAN, TERRY WILSON, b Hackensack, NJ, Aug 29, 35; m 57; c 3. ORGANIC CHEMISTRY, PHOTOGRAPHIC CHEMISTRY. Educ: Marietta Col, BS, 56; Univ Ill, PhD(org chem), 59. Prof Exp: Res scientist, 59-64, asst proj mgrq 64-66, mgr color photog res, 66-72, asst to pres, 72-73, MGR INT TECHNOL, POLAROID CORP, CAMBRIDGE, 74- Mem: AAAS; Am Chem Soc. Res: Organic synthesis of dyes, hetero-cyclics, photographic developers; spectroscopic interactions between functional groups; reduction-oxidation reactions; coating technology; diffusion transfer photography; polyester films, polymer coatings, paper and plastic laminations; sensitometry and color analysis. Mailing Add: 51 Prentiss Lane Belmont MA 02178

MILLIGAN, WILBERT HARVEY, III, b Pittsburgh, Pa, Jan 17, 45; m 71; c 1. VIROLOGY. Educ: Washington & Jefferson Col, BA, 66; Univ Pittsburgh, PhD(microbiol), 72. Prof Exp: Lab instr microbiol & physiol, Washington & Jefferson Col, 65-66, lab instr gen biol, 66; teaching asst microbiol, Univ Pittsburgh, 66-67; lab instr, Allegheny Community Col, 67-69; ASST PROF MICROBIOL, SCH DENT MED, SOUTHERN ILL UNIV, EDWARDSVILLE, 72- Concurrent Pos: Abstractor, Am Dent Asn, 75-; consult gastroenterol, Sch Med, Washington Univ, 75- Mem: Sigma Xi; Am Soc Microbiol; AAAS; Int Asn Dent Res; Am Asn Dent Schs. Res: Hepatitis transmission in dentistry; methods for sterilization in dentistry; herpes-simplex latency; immunization against dental caries; immunopathology; viral-mycoplasma interactions; dental caries and peridontal disease; self-instructional learning modules. Mailing Add: Dept of Microbiol Southern Ill Univ Sch Dent Med Edwardsville IL 62025

MILLIGAN, WINFRED OLIVER, b Coulterville, Ill, Nov 5, 08; m 36, 60. PHYSICAL INORGANIC CHEMISTRY. Educ: Ill Col, AB, 30; Rice Univ, MA, 32, PhD(chem), 34. Hon Degrees: ScD, Ill Col, 46; DSc, Tex Christian Univ, 60. Prof Exp: From res chemist to prof chem, Rice Univ, 34-63; vchancellor res, mem bd dirs & pres, Res Found, Tex Christian Univ, 63-65; DISTINGUISHED RES PROF CHEM, BAYLOR UNIV, 65-; DIR RES, ROBERT A WELCH FOUND, 55- Concurrent Pos: Res chemist, Harshaw Chem Co, Ohio, 34; Nat Acad Sci-Nat Res Coun fels, 54-59; consult, Houdry Process Corp, 36-45, Humble Oil & Ref Co, 45-62 & Oak Ridge Nat Lab, 50-; mem comt appln x-ray & electron diffraction, Nat Acad Sci-Nat Res Coun, 38-41, mem panel permanent magnetic mat, 52, mem comt clay mineral, 53-; mem Am Chem Soc adv comt, US Army Chem Corps, 54-62, chmn, 58-62; mem, Tex Adv Comt Atomic Energy, 55-; ed, Res Bull, Robert A Welch Found, 53-, Proc Conf Chem Res, 57- & Clay Minerals, Nat Acad Sci-Nat Res Coun, 55. Mem: Fel Am Inst Chem; fel Am Phys Soc; Am Crystallog Asn; Faraday Soc; Am Chem Soc. Res: X-ray and electron diffraction; electron microscopy; adsorption of gases in solids; magnetic susceptibility; oxides; hydroxides; colloids; general catalysis studies; heavy metal cyano-complexes; inorganic catalysts. Mailing Add: Robert A Welch Found 2010 Bank of Southwest Bldg Houston TX 77002

MILLIGER, LARRY EDWARD, aquatic biology, phycology, see 12th edition

MILLIKAN, ALLAN G, b Charleston, WVa, July 31, 27; wid; c 4. PHOTOGRAPHY, ASTRONOMY. Educ: Oberlin Col, AB, 49; Purdue Univ, MS, 51. Prof Exp: Photog engr, Color Technol, Eastman Kodak Co, 51-57, res assoc, Res Labs, 58-71; vis scientist astron, Kitt Peak Nat Observ, 71-72; SR RES ASSOC SCI PHOTOG, EASTMAN KODAK RES LABS, 72- Concurrent Pos: Mem working group photog mat, Am Astron Soc, 66- Mem: Am Astron Soc. Res: Photographic emulsion as a scientific data recorder; improvement of plates and films used in scientific photography. Mailing Add: Eastman Kodak Co Res Labs B-59 1669 Lake Ave Rochester NY 14650

MILLIKAN, CLARK HAROLD, b Freeport, Ill, Mar 2, 15; m 66; c 3. MEDICINE, NEUROLOGY. Educ: Univ Kans, MD, 39; Am Bd Psychiat & Neurol, dipl, 46. Prof Exp: Intern, St Luke's Hosp, Cleveland, 39-40, from asst resident to resident med, 40-41; resident neurol, Univ Iowa, 41-44, from instr to asst prof, 44-49; from assoc prof to prof, Mayo Found, Univ Minn, 49-65, PROF NEUROL, MAYO GRAD SCH MED, UNIV MINN, 65-; CONSULT, MAYO CLIN, 49-, HEAD NEUROL SECT, 55- Concurrent Pos: Chmn comt cerebrovascular dis, USPHS & mem adv coun, Nat Inst Neurol Dis & Blindness, 61-65; chmn, Joint Coun Subcomt Cerebrovascular Dis, Nat Inst Neurol Dis & Blindness & Nat Heart Inst, mem nat adv comt regional med progs, NIH; past chmn, Coun Cerebrovascular Dis, Am Heart Asn, ed, Stroke; ed, J Cerebral Circulation, 70- Mem: AMA; Asn Res Nerv & Ment Dis (pres, 61); Am Neurol Asn (pres, 74); fel Am Col Physicians; fel Am Acad Neurol. Res: Cerebrovascular disease. Mailing Add: Dept of Neurol Mayo Clin 200 First St SW Rochester MN 55901

MILLIKAN, DANIEL FRANKLIN, JR, b Lyndon, Ill, May 31, 18. BOTANY. Educ: Iowa State Univ, BS, 47; Univ Mo, PhD, 54. Prof Exp: Asst bot, 47-52, instr, 52-54, asst prof hort, 54-58, assoc prof, 58-68, PROF PLANT PATH, UNIV MO-COLUMBIA, 68- Concurrent Pos: USDA assignment, Poland, 66; Polish Minister Agr grant, 70 & 72; Nat Acad Sci-Polish Acad Sci grant, 74. Mem: Fel AAAS; Am Phytopath Soc; Am Soc Hort Sci; Bot Soc Am; foreign mem Polish Acad Sci. Res: Pathology of fruit, vegetable and woody ornamental crops; virology; virus diseases of stone and pome fruit crops. Mailing Add: Dept of Plant Path Rm 108 Waters Hall Univ of Mo Columbia MO 65201

MILLIKAN, LARRY EDWARD, b Sterlin, Ill, May 12, 36; m 62; c 2. DERMATOLOGY, IMMUNOLOGY. Educ: Monmouth Col, AB, 58; Univ Mo, MD, 62. Prof Exp: Physician intern & med officer, Great Lakes Naval Training Ctr, US Navy, 62, med officer aviation med, Naval Air Sta, Pensacola, Fla, 62-64, flight surgeon, Quonset Point, McGuire AFB, 64-67; resident dermat, Univ Hosp, Ann Arbor, Mich, 67-70; asst prof, 70-73, ASSOC PROF DERMAT, MED CTR, UNIV MO-COLUMBIA, 73- Concurrent Pos: Consult physiciant, Student Health Serv, 70-, Vet Admin Hosp, 72- & Ellis Fischell State Cancer Hosp, 72-; mem, Eczema Task Force, Nat Prog Dermat, 73; contrib ed, Int J Dermat, 75- Mem: Sigma Xi; Soc Invest Dermat; AAAS; Am Acad Dermat. Res: Cellular immunity; immunity in neoplasms; melanoma; immune surveillance. Mailing Add: Dept Dermat M752 Med Ctr Univ Mo Columbia MO 65201

MILLIKAN, ROGER CONANT, b Tiffin, Ohio, Jan 27, 31; m 53; c 5. CHEMICAL PHYSICS. Educ: Oberlin Col, BS, 53; Univ Calif, PhD, 57. Prof Exp: Phys chemist, Res Lab, Gen Elec Co, 56-67; PROF CHEM, UNIV CALIF, SANTA BARBARA, 67- Mem: Am Chem Soc; fel Am Phys Soc; fel Optical Soc Am. Res: Infrared spectroscopy; combustion; high temperature reactions; vibrational relaxation; fluorescence; shock tubes; lasers. Mailing Add: Dept of Chem Univ of Calif Santa Barbara CA 93106

MILLIKEN, JOHN ANDREW, b Saskatoon, Sask, May 15, 23; m 46; c 7. INTERNAL MEDICINE. Educ: Queen's Univ, Ont, MD, CM, 46; FRCP(C), 54; Am Bd Internal Med, dipl, 60. Prof Exp: From asst prof to assoc prof, 56-71, PROF MED QUEEN'S UNIV, ONT, 71-; HEAD DEPT MED, HOTEL DIEU HOSP, 56- Mem: Am Col Physicians; NY Acad Med; fel Am Col Chest Physicians; fel Am Col Cardiol; Can Med Asn. Res: Cardiology. Mailing Add: Dept of Med Queen's Univ Kingston ON Can

MILLIKEN, SPENCER RANKIN, b Dallas, Tex, Dec 5, 24; m 45; c 4. PHYSICAL CHEMISTRY. Educ: Ga Inst Technol, BS, 50; Emory Univ, MS, 51; Pa State Univ, PhD, 54. Prof Exp: Asst fuel tech, Pa State Univ, 50-53; res engr lubricants, Aluminum Co Am, 53-57, asst chief, 57-58; res & sales coordr, Foote Mineral Co, 58-60, mgr metall res, 60; dir res, Northern Ill Gas Co, 60-65; mgr appl physics, Roy C Ingersoll Res Ctr, Borg-Warner Corp, 65-66; dir corp eng, Electronic Assistance Corp, 66-74, VPRES RES & DEVELOP, WELCO INDUSTS INC, DIV ELECTRONICS ASSISTANCE CORP, 74- Mem: Am Chem Soc; Am Inst Chem; Am Inst Mining, Metall & Petrol Eng; NY Acad Sci. Res: Chemical constitution of coal; nature and structure of carbons; mineral preparation; lubricants; high purity metal preparation; nonferrous metals fabrication; metal films. Mailing Add: Welco Industs Inc 9027 Shell Rd Cincinnati OH 45236

MILLIMAN, GEORGE ELMER, b New York, NY, Nov 15, 37; m 62; c 2. ANALYTICAL CHEMISTRY. Educ: Univ Rochester, BS, 59; Carnegie Inst Technol, PhD(chem), 64. Prof Exp: Res chemist, Jersey Prod Res Co, 64 & Esso Prod Res Co, Tex, 64-69; SR RES CHEMIST, EXXON RES & ENG CO, 69- Mem: Am Chem Soc. Res: Synthesis; spectroscopy; trace analysis. Mailing Add: Exxon Res & Eng Co PO Box 121 Linden NJ 07036

MILLIMAN, JOHN D, b Rochester, NY, May 5, 38; m 63; c 2. OCEANOGRAPHY, GEOLOGY. Educ: Univ Rochester, BS, 60; Univ Wash, MS, 63; Univ Miami, PhD(oeceanog), 66. Prof Exp: Res asst radiation biol lab, Univ Wash, 61; res asst, Inst Marine Sci, Univ Miami, 63-66, res fel, 66; asst scientist, 66-71, ASSOC SCIENTIST, WOODS HOLE OCEANOG INST, 71- Concurrent Pos: Alexander von Humboldt

Found scholar, Lab Sedimentology, Univ Heidelberg, 69-70. Mem: AAAS; Geol Soc Am; Soc Econ Paleontologists & Mineralogists. Res: Deposition and diagenesis of marine sediments; continental shelf sedimentation; Holocene history and shallow structure; submarine precipitation and lithification of marine carbonates. Mailing Add: Woods Hole Oceanog Inst Woods Hole MA 02543

MILLING, MARCUS EUGENE, b Galveston, Tex, Oct 8, 38; m 59; c 1. SEDIMENTOLOGY, GEOMORPHOLOGY. Educ: Lamar Univ, BS, 61; Univ Iowa, MS, 64, PhD(geol), 68. Prof Exp: SR RES GEOLOGIST, EXXON PROD RES CO, 68- Mem: Geol Soc Am; Am Asn Petrol Geol; Soc Econ Paleontologists & Mineralogists; Int Asn Sedimentol. Res: Environmental facies analysis of Pleistocene and recent sediments. Mailing Add: Exxon Prod Res Co PO Box 2189 Houston TX 77001

MILLINGTON, JAMES E, b New York, NY, Mar 13, 30; m 65. ORGANIC CHEMISTRY. Educ: Lincoln Univ, Pa, AB, 51; Univ Western Ont, MSc, 53, PhD(chem), 56. Prof Exp: Asst, Defence Res Bd Can, Univ Western Ont, 53-55; res chemist, Allis-Chalmers Mfg Co, 56-61, proj leader, 61-63, sect head, 63-67, mgr org chem, 67-69; SECT HEAD, FOSTER GRANT CO INC, 69- Mem: Am Chem Soc. Res: Polymers; structure-properties relationships; polymer characterization. Mailing Add: Res & Develop Foster Grant Co Inc 289 N Main St Leominster MA 01453

MILLINGTON, WILLIAM FRANK, b Ridgewood, NJ, June 16, 22; m 47; c 2. PLANT MORPHOGENESIS. Educ: Rutgers Univ, BSc, 47, MSc, 49; Univ Wis, PhD(bot hort), 52. Prof Exp: Res assoc, Roscoe B Jackson Mem Lab, Maine, 51-53; from instr to asst prof bot, Univ Wis, 53-59; from asst prof to assoc prof, 59-68, PROF BOT, MARQUETTE UNIV, 68- Mem: AAAS; Am Inst Biol Sci; Bot Soc Am; Soc Develop Biol; Phycol Soc Am. Res: Shoot morphogenesis in plants; plant development; regulation of form and pattern. Mailing Add: Dept of Biol Marquette Univ Milwaukee WI 53233

MILLION, RODNEY REIFF, b Idaville, Ind, Apr 3, 29; m 55; c 4. RADIOTHERAPY. Educ: Ind Univ, BS, 51, MD, 54; Am Bd Radiol, dipl, 63. Prof Exp: Intern, Harbor Gen Hosp, Torrance, Calif, 54-55; resident radiol, Ind Univ, 58-60; resident, Univ Tex M D Anderson Hosp & Tumor Inst, 60-62; assoc prof radiol & chief radiother sect, Ind Univ, 62-64; from assoc prof to prof radiol, 64-74, AM CANCER SOC PROF CLIN ONCOL, UNIV FLA, 74- Res: Radiation therapy. Mailing Add: Div of Radiother Univ of Fla Gainesville FL 32610

MILLIS, ALBERT JASON TAYLOR, b Philadelphia, Pa, Oct 4, 41; m 65; c 2. CELL BIOLOGY. Educ: Univ Pa, PhD(biol), 71. Prof Exp: Instr biol, Univ Pa, 67-69; res assoc pediat, Sch Med, Univ Wash, 71-72, res instr, 73, res asst prof, 74; ASST PROF BIOL, STATE UNIV NY ALBANY, 74- Concurrent Pos: Res asst prof, Dept Pediat, Albany Med Col, 76- Mem: Am Soc Cell Biol. Res: Regulation of the mammalian cell cycle and cellular proliferation; effects of growth promoting polypeptides on the mammalian tissue culture cell surface; somatic cell genetics. Mailing Add: 226 Biol Bldg Dept of Biol State Univ of NY Albany NY 12222

MILLIS, JOHN SCHOFF, b Palo Alto, Calif, Nov 22, 03; m 29; c 3. MEDICAL EDUCATION. Educ: Univ Chicago, BS, 24, MS, 27, PhD(physics), 31. Hon Degrees: Fourteen from var US cols & univs. Prof Exp: Master, Howe Sch, Ind, 24-26; from instr to prof physics, Lawrence Col, 27-41, dir, Underwood Astron Observ, 27-37, res assoc inst paper chem, 31-37, dean col, 36-41, dean admin, 38-41; pres, Univ Vt, 41-49; pres, Western Reserve Univ, 49-67; pres, Nat Fund Med Educ, 71-75; chancellor, 67-69, EMER CHANCELLOR, CASE WESTERN RESERVE UNIV, 69-; CHMN, NAT FUND MED EDUC, 75- Concurrent Pos: Consult sci info coun, NSF, 58-61; chmn citizens comn grad med educ, AMA, 62-66; mem nat adv coun med, dent, optom & podiatric educ, USPHS, 66-68; ad hoc consult, Group Continuing Med, Nat Libr Med, 66; mem, Nat Adv Coun Dent Res, 69-73; vpres, Nat Fund Med Educ, 69-71; mem, Comn Foreign Med Grads, 70-; chmn study comn dietetics, Am Dietetic Asn, 70-72; mem, Nat Bd Med Examr, 71-; past trustee, Carnegie Found Advan Teaching; vpres, Nat League Nursing, Inc; chmn, President's Adv Panel Heart Dis, 72; chmn study comn pharm, Am Asn Cols Pharm, 73-76; trustee, Am Nurses' Found, 73- Mem: Hon mem Am Hosp Asn. Res: Atomic and molecular spectroscopy; physical properties of paper; higher education. Mailing Add: Nat Fund for Med Educ 212 Univ Circle Res Ctr Cleveland OH 44106

MILLIS, ROBERT LOWELL, b Martinsville, Ill, Sept 12, 41; m 65. ASTRONOMY. Educ: Eastern Ill Univ, BA, 63; Univ Wis, PhD(astron), 68. Prof Exp: ASTRONR, LOWELL OBSERV, 67- Mem: Am Astron Soc; Int Astron Union. Res: Broad-band photometry of short-period variable stars; photometry of satellites; photometry of asteroids. Mailing Add: Lowell Observ PO Box 1269 Flagstaff AZ 86001

MILLISER, RUSSELL VON, b Leiters Ford, Ind, Feb 27, 06; m 44; c 2. PATHOLOGY. Educ: DePauw Univ, BA, 28; Cornell Univ, PhD, 32; Northwestern Univ, MD, 38; Am Bd Path, dipl, 47. Prof Exp: Instr anat, Med Col, Cornell Univ, 28-31; instr, Med Sch, Northwestern Univ, 31-32; asst prof, Chicago Med Sch, 32-37; intern, White Cross Hosp, Columbus, Ohio, 37-38; from instr to asst prof path, Col Med, Ohio State Univ, 38-48; assoc prof, 48-54, PROF PATH, CHICAGO MED SCH, 55- Concurrent Pos: Assoc, Cook County Hosp, Chicago, 55-; consult, Mt Sinai Hosp. Res: Neuroanatomy; neuropathology; clinical pathology; hematology; immunology; pathologic anatomy. Mailing Add: Dept of Path Chicago Med Sch Chicago IL 60612

MILLMAN, BARRY MACKENZIE, b Toronto, Ont, Oct 17, 34; m 59; c 3. BIOPHYSICS. Educ: Carleton Univ, BSc, 57; King's Col, Univ London, PhD(biophys), 63. Prof Exp: Mem sci staff, Brit Med Res Coun, Biophys Res Unit, King's Col, Univ London, 61-66; from asst prof to prof biol sci, Brock Univ, 67-74, chmn dept, 66-71; PROF PHYSICS, UNIV GUELPH, 74-, CHMN BIOPHYS INTERDEPT COMT, 75- Concurrent Pos: Res grants, Can Med Res Coun, 66-68 & Can Nat Res Coun, 68- Mem: Biophys Soc; Brit Biophys Soc; Brit Physiol Soc; Can Soc Cell Biologists (treas, 68-71); NY Acad Sci. Res: Muscle structure as determined by x-ray diffraction; contraction mechanisms and muscle physiology, especially invertebrate muscles. Mailing Add: Dept of Physics Univ of Guelph Guelph ON Can

MILLMAN, GEORGE HAROLD, b Boston, Mass, June 2, 19; m 43; c 1. IONOSPHERIC PHYSICS. Educ: Univ Mass, BS, 47; Pa State Univ, MS, 49, PhD(physics), 52. Prof Exp: Asst physics, Pa State Univ, 47-50, instr eng res, 50-52; engr, 52-54, res liaison scientist, 54-55, consult physicist, 62-70, SPECIALIST ELECTROMAGNETIC PROPAGATION, GEN ELEC CO, 55-, SR CONSULT PHYSICIST, 70- Concurrent Pos: Mem ionospheric radio propagation comn, Int Union Radio Sci, 58. Mem: Am Phys Soc; fel Inst Elec & Electronics Engrs; Am Geophys Union; NY Acad Sci. Res: Electromagnetic-atmospheric propagation; ionospheric and space physics; radar and radio astronomy; atmospheric effects on radio wave propagation. Mailing Add: 504 Hillsboro Pkwy Syracuse NY 13214

MILLMAN, IRVING, b New York, NY, May 12, 23; m 49; c 2. IMMUNOLOGY, MEDICAL MICROBIOLOGY. Educ: City Col New York, BS, 48; Univ Ky, MS, 51; Northwestern Univ, PhD, 54. Prof Exp: Res bacteriologist, Armour & Co, 49-52; instr bact, Northwestern Univ, 54-55, asst prof, 55-58; asst, Pub Health Res Inst New York, 58-61; res fel, Merck Inst Therapeut Res, 61-67; ASSOC MEM, CLIN RES UNIT, INST CANCER RES, 67-; ASSOC PROF, HAHNEMAN MED COL, 71- Mem: NY Acad Sci; Soc Exp Biol & Med; Am Soc Microbiol; Am Asn Cancer Res. Res: Immunology, hepatitis and cancer research. Mailing Add: Clin Res Unit Inst for Cancer Res Philadelphia PA 19111

MILLMAN, NATHAN, physiology, biochemistry, deceased

MILLMAN, PETER MACKENZIE, astronomy, see 12th edition

MILLMAN, RICHARD STEVEN, b Boston, Mass, Apr 15, 45; c 2. GEOMETRY. Educ: Mass Inst Technol, BS, 66; Cornell Univ, MS, 69, PhD(math), 71. Prof Exp: Asst prof math, Ithaca Col, 70-71; asst prof, 71-74, ASSOC PROF MATH, SOUTHERN ILL UNIV, CARBONDALE, 74- Mem: Math Asn Am; Am Math Soc. Res: Eigenvalues of Laplace operator on Riemannian manifolds; holomorphic connections on fiber bundles. Mailing Add: Dept of Math Southern Ill Univ Carbondale IL 62901

MILLMAN, ROBERT BARNET, b New York, NY, Aug 25, 39. PUBLIC HEALTH. Educ: Cornell Univ, BA, 61; State Univ NY, MD, 65. Prof Exp: Intern, Bellevue Hosp, NY, 65-66; asst physician, New York Hosp, 68-70; asst prof, Rockefeller Univ, 70-72; DIR, ADOLESCENT DEVELOP PROG, MED CTR & ASST PROF PUB HEALTH, MED COL, CORNELL UNIV, 70- Concurrent Pos: Mem methadone maintenance adv comt, Food & Drug Admin, 70-72; consult, NY Dept Corrections, 71-72, Manhattan Borough Presidents' Comt Drug Abuse, 71-74 & NY Methadone Maintenance Treatment Prog, 71-; adj asst prof, Rockefeller Univ, 72-; assoc ed, Millbank Mem Fund Quart, 72-; mem comnr adv comt, NY Addiction Serv Agency, 74-; mem adv comt drug abuse, NY Comnr Health, 75- Mem: Am Pub Health Asn; Am Med Soc Alcoholism. Res: Pathogenesis and patterns of drug and alcohol abuse; characterization of the addictive process, with particular respect to abstinence syndromes. Mailing Add: Med Col Cornell Univ 411 E 69th St New York NY 10021

MILLMAN, SIDNEY, b Dawid-Gorodok, Poland, Mar 15, 08; nat US; m; c 1. NUCLEAR MAGNETIC RESONANCE, MICROWAVE ELECTRONICS. Educ: City Col New York, BS, 31; Columbia Univ, AM, 32, PhD(physics), 35. Hon Degrees: ScD, Lehigh Univ, 74. Prof Exp: Asst physics, Columbia Univ, 33-35, Tyndall fel, 35-36, Barnard fel, 36-37, asst physics, 37-39; instr, City Col New York, 39-41 & Queen's Col, NY, 41-42; mem sci staff, Radiation Lab, Columbia Univ, 42-45; res physicist, Bell Tel Labs, NJ, 45-52, dir phys res, 52-65, exec dir res, physics & acad affairs, 65-74; SECY, AM INST PHYSICS, 74- Mem: AAAS; Am Asn Physics Teachers; fel Am Phys Soc; fel Inst Elec & Electronics Engrs. Res: Nuclear spins and magnetic moments; vacuum tubes such as magnetrons and traveling wave tubes; solid state physics. Mailing Add: Am Inst of Physics 335 E 45th St New York NY 10017

MILLON, RENE, b New York, NY, July 12, 21; m; c 3. ANTHROPOLOGY, ARCHAEOLOGY. Educ: Columbia Univ, AB, 48, AM, 52, PhD(anthrop), 55. Prof Exp: Asst prof anthrop, Univ Calif, Berkeley, 57-61; assoc prof, 61-64, PROF ANTHROP, UNIV ROCHESTER, 64- Concurrent Pos: Henry & Grace Doherty fel, Mex, 55-56; NSF fel, Mex, 56-57; NSF grants, Mex, 59 & Univ Calif, Berkeley, 60-61; dir, Teotihuacan Mapping Proj, 62-; fel, Ctr Advan Study Behav Sci, 62-63; NSF grants, Mex, 62-72. Mem: Fel Am Anthrop Asn; Soc Am Archaeol; Mex Anthrop Soc; Int Cong Americanists; Am Ethnol Soc. Res: Urbanism at Teotihuacan and elsewhere in pre-Hispanic Central Mexico; comparative study of early civilizations; problems of method and interpretation in archaeology; ecology; urbanism and urbanization in nonindustrialized societies. Mailing Add: Dept of Anthrop Univ of Rochester Rochester NY 14627

MILLONIG, ROBERT C, microbiology, see 12th edition

MILLS, ALFRED PRESTON, b Fallon, Nev, Jan 8, 22; m 46; c 2. PHYSICAL CHEMISTRY. Educ: Univ Nev, BS, 43; Tulane Univ, PhD(phys chem), 49. Prof Exp: From instr to asst prof, 49-56, chmn div natural sci & math, 60-61, actg asst dean grad sch, 64-65, ASSOC PROF PHYS CHEM, UNIV MIAMI, 56- Mem: AAAS; Am Chem Soc; Am Phys Soc; Am Inst Chemists; Sigma Xi. Res: Thermodynamics of solutions; prediction of physical properties of liquid solutions; dielectric constant and dipole moment, viscosity; physical properties of silicon compounds and nitro compounds. Mailing Add: Dept of Chem Univ of Miami Coral Gables FL 33124

MILLS, ALLEN PAINE, JR, b Apr 21, 40; US citizen. ATOMIC PHYSICS. Educ: Princeton Univ, BA, 62; Brandeis Univ, MA, 64, PhD(physics), 67. Prof Exp: From instr to assoc prof physics, Brandeis Univ, 67-75; MEM TECH STAFF, BELL LABS, 75- Mem: Am Phys Soc. Res: Atomic physics of positronium. Mailing Add: Bell Labs Murray Hill NJ 07974

MILLS, DALLICE IVAN, b Endeavor, Wis, July 27, 39; m 61; c 1. BOTANY, GENETICS. Educ: Wis State Univ, Stevens Point, BS, 61; Syracuse Univ, MS, 64; Mich State Univ, PhD(bot, plant path), 69. Prof Exp: Teacher high schs, Wis & Ariz, 61-65; spec res asst bot & plant path, Mich State Univ, 65-69; NIH res fel genetics, Univ Wash, 69-74; ASST PROF BIOL SCI, UNIV ILL CHICAGO CIRCLE, 74- Mem: Genetics Soc Am; Bot Soc Am. Res: Fungal and microbial genetics; mechanisms of somatic recombination; genetics and physiology of incompatibility factors in fungi. Mailing Add: Dept of Biol Sci Univ Ill Chicago Cir Chicago IL 60680

MILLS, DANIEL A, biology, chemistry, see 12th edition

MILLS, DON HARPER, b Peking, China, July 29, 27; US citizen; m 49; c 2. FORENSIC MEDICINE. Educ: Univ Cincinnati, BS, 50, MD, 53; Univ Southern Calif, JD, 58. Prof Exp: CONSULT FORENSIC MED, 58- Concurrent Pos: Fel path, Univ Southern Calif, 54-55, instr path, 58-62, from asst clin prof to assoc clin prof path, 62-69, clin prof, 69-; instr humanities, Loma Linda Univ, 60-66, assoc clin prof, 66-; exec ed, Trauma, 64-; mem ed bd, J Forensic Sci, 65-; dep med examr, Off Los Angeles County Coroner, 57-61; mem attend staff, Los Angeles County Hosp, 59-; affil staff, Hosp Good Samaritan, Los Angeles, 67-; res consult, Secy Comn Med Malpract, Dept Health, Educ & Welfare, 72-73; mem adv coun, Assembly Comt Med Malpract, State Calif, 73-75; mem ed bd, J Legal Med, 73- & J AMA, 73-; expert consult, Off Secy, Dept Health, Educ & Welfare, 75-; consult, Armed Forces Inst Path, Dept Army, 74- & Health Resources Admin, Dept Health, Educ & Welfare, 75- Mem: AAAS; fel Am Acad Forensic Sci; AMA; Am Bar Asn; fel Am Col Legal Med (pres, 74-). Res: Forensic medicine, primarily sub-field of legal rights and responsibilities of physicians and hospitals. Mailing Add: 1141 Los Altos Ave Long Beach CA 90815

MILLS, DONALD GRANT, b Wadena, Sask, Apr 19, 43; m 69; c 2. PHARMACOLOGY. Educ: Univ Toronto, BScPhm, 68; Univ Western Ont, PhD(pharmacol), 74. Prof Exp: Med Res Coun fel drug metab, Univ Sask, 73-74; ASST PROF PHARMACOL, UNIV WESTERN ONT, 74- Concurrent Pos: Med

Res Coun fel, 74. Res: Platelet pharmacology; gastrointestinal ulceration and bleeding; effects of diving and decompression on drug metabolism. Mailing Add: Dept of Pharmacol Univ Western Ont Health Sci Ctr London ON Can

MILLS, DOUGLAS LEON, b Berkeley, Calif, Apr 2, 40; m 61. SOLID STATE PHYSICS. Educ: Univ Calif, Berkeley, BS, 61, PhD(physics), 65. Prof Exp: NSF fel physics, Paris, 65-66; from asst prof to assoc prof, 66-74, PROF PHYSICS, UNIV CALIF, IRVINE, 74- Mem: Am Phys Soc. Res: Theoretical investigations of magnetic materials, lattice vibrations, surface effects, light scattering from solids and properties of alloys. Mailing Add: Dept of Physics Univ of Calif Irvine CA 92664

MILLS, EDWARD JAMES, JR, organic chemistry, see 12th edition

MILLS, ELLIOTT, b New York, NY, July 17, 35; m 60; c 1. PHARMACOLOGY, PHYSIOLOGY. Educ: City Col New York, BS, 57; Columbia Univ, PhD(pharmacol), 64. Prof Exp: Nat Inst Neurol Dis & Blindness trainee, Col Physicians & Surgeons, Columbia Univ, 64-65; Nat Heart Inst fel physiol, Middlesex Hosp Med Sch, London, Eng, 65-67; spec fel cardiovasc res inst, Med Ctr, Univ Calif, San Francisco, 67-68; asst prof, 68-74, ASSOC PROF PHYSIOL & PHARMACOL, MED CTR, DUKE UNIV, 74- Concurrent Pos: Estab investr, Am Heart Asn, 74- Mem: Am Phys Soc. Res: Chemoreceptor mechanisms; brainstem and reflex control of cardiovascular function; cardiovascular physiology. Mailing Add: Dept of Physiol & Pharmacol Duke Univ Med Ctr Durham NC 27701

MILLS, ERIC LEONARD, b Toronto, Ont, July 7, 36; m 62; c 2. BIOLOGICAL OCEANOGRAPHY. Educ: Carleton Univ, Can, BSc, 59; Yale Univ, MS, 62, PhD(marine biol), 64. Prof Exp: Asst prof biol, Queen's Univ, Ont, 63-67; assoc prof, 67-71, PROF OCEANOG & BIOL, DALHOUSIE UNIV, 71- Concurrent Pos: Sessional lectr, Carleton Univ, Can, 60; instr, Marine Biol Lab, Woods Hole, Mass, 64-67, mem corp, 65-; instr, Huntsman Marine Lab, NB, 71-73; vis scholar, Corpus Christi Col, Cambridge Univ, 74-75. Mem: AAAS; Am Ornithologists Union; Am Soc Limnol & Oceanog; Can Soc Zoologists; Marine Biol Asn UK. Res: Taxonomy and ecology of amphipod crustaceans; marine ecology; deep-sea biology; history of oceanography. Mailing Add: Dept of Oceanog Dalhousie Univ Halifax NS Can

MILLS, FRANK D, b Cleveland, Ohio, July 29, 37; m 61; c 2. ORGANIC CHEMISTRY. Educ: Western Reserve Univ, BA, 60, MS, 61, PhD, 66. Prof Exp: PRIN CHEMIST CEREAL CARBOHYDRATES, NORTHERN REGIONAL RES CTR, USDA, 66- Honors & Awards: Citation, USDA, 70. Mem: Am Chem Soc. Res: Carbohydrates, structure and synthesis. Mailing Add: Northern Regional Res Ctr USDA 1815 N University St Peoria IL 61604

MILLS, FREDERICK EUGENE, b Streator, Ill, Nov 12, 28; m 50; c 3. PHYSICS. Educ: Univ Ill, BS, 49, MS, 50, PhD(physics), 55. Prof Exp: Res assoc physics, Cornell Univ, 54-56; scientist, Midwestern Univs Res Asn, 56-64, assoc dir sci, 64-65, dir, 65-66; prof physics, Univ Wis Phys Sci Lab, 66-70, dir, 67-70; chmn accelerator dept, Brookhaven Nat Lab, 70-73; SCIENTIST, FERMI NAT ACCELERATOR LAB, 73- Concurrent Pos: Physicist, Saclay Nuclear Res Ctr, France, 61-62; adj prof, Nuclear Eng Dept, Univ Wis, 74- Mem: Fel Am Phys Soc. Res: Accelerators; high energy and plasma physics; energy loss of fast particles in motion; photoproduction of pi mesons; advanced accelerators. Mailing Add: Fermi Nat Lab PO Box 500 Batavia IL 60510

MILLS, GEORGE ALEXANDER, b Saskatoon, Sask, Mar 20, 14; nat US; m 40; c 4. PHYSICAL CHEMISTRY, FUEL SCIENCE. Educ: Univ Sask, BSc, 34, MSc, 36; Columbia Univ, PhD(chem), 40. Prof Exp: Asst chem, Columbia Univ, 36-39; instr, Dartmouth Col, 39-40; res chemist, Houdry Process & Chem Co, 40-47, asst dir res, 47-52, dir, 52-67; dir res, Houdry Lab, Air Prod & Chem Inc, 67-68; asst dir coal res, US Bur Mines, Washington, DC, 68-70, chief div coal, 70-74, asst dir, Off Coal Res, 74-75; DIR FOSSIL ENERGY RES, ENERGY RES & DEVELOP ADMIN, 75- Honors & Awards: Storch Award, Am Chem Soc, 75. Mem: AAAS; Am Chem Soc; Am Inst Chem Engrs; Am Inst Mining, Metall & Petrol Engrs; Catalysis Soc (pres). Res: Active hydrogen; separation of isotopes; exchange reactions with oxygen isotopes; reaction kinetics; petroleum refining; clay minerals; mechanism of catalytic reactions; high polymers; organic nitrogen chemicals; polyurethanes; synthetic fuels from coal; materials research; combustion and power. Mailing Add: Energy Res & Develop Admin 20 Massachusetts Ave NW Washington DC 20545

MILLS, GEORGE SCOTT, b Ft Worth, Tex, Sept 24, 39; m 62. APPLIED PHYSICS. Educ: Univ Tex, BS, 61, PhD(physics), 66. Prof Exp: Physicist, Tracor Inc, 61-64; assoc plasma res, Environ Sci Serv Admin, US Dept Com, 66-68; res physicist, Rocky Flats Div, Dow Chem USA, 68-75; RES PHYSICIST, ROCKY FLATS DIV, ROCKWELL INT, 75- Mem: Am Phys Soc. Res: Experimental investigation of electromagnetic wave absorption and transport phenomena in plasmas; sound propagation and signal processing studies applicable to ultrasonic testing; emission spectroscopy of welding arcs. Mailing Add: 5375 Kewanee Boulder CO 80303

MILLS, GORDON CANDEE, b Fallon, Nev, Feb 13, 24; m 47; c 3. BIOCHEMISTRY. Educ: Univ Nev, BS, 46; Univ Mich, MS, 48, PhD(biochem), 51. Prof Exp: Res assoc biochem, Col Med, Univ Tenn, 50-55; PROF BIOCHEM, UNIV TEX MED BR GALVESTON, 55- Mem: Am Chem Soc; Am Soc Biol Chem; Sigma Xi; Am Asn Univ Prof. Res: Erythrcyte metabolism; genetic disorders of erythrocytes; metabolic control mechanisms. Mailing Add: Div of Biochem Univ of Tex Med Br Galveston TX 77550

MILLS, GORDON FREDERICK, b Cleveland, Ohio, Sept 21, 11; m 43; c 2. PHYSICAL CHEMISTRY, ORGANIC CHEMISTRY. Educ: Oberlin Col, BA, 33; Syracuse Univ, MS, 35; Stanford Univ, PhD(chem), 39. Prof Exp: Res dir, Chem Process Co, 38-49; STAFF, RES & DEVELOP DIV, OAK RIDGE GASEOUS DIFFUSION PLANT, NUCLEAR DIV, UNION CARBIDE CORP, 49- Mem: Am Chem Soc. Res: Colloid chemistry; properties of surfaces; ion exchange resins; absolute adsorption at the water air interface. Mailing Add: Keller Bend Rd Rte 1 Concord TN 37720

MILLS, HARLAN DUNCAN, applied mathematics, see 12th edition

MILLS, HARRY ARVIN, b Paintsville, Ky, Dec 11, 46; m 68; c 2. VEGETABLE CROPS. Educ: Univ Ky, BS, 69; Univ Mass, MS, 72, PhD(plant sci, soil sci), 75. Prof Exp: ASST PROF HORT, UNIV GA, 74- Mem: AAAS; Sigma Xi; Am Soc Hort Sci; Am Soc Agron. Res: Soil fertility and plant nutrition of vegetable crops; nitrogen utilization. Mailing Add: Plant & Soil Sci Dept of Hort Univ of Ga Athens GA 30601

MILLS, HOWARD LEONARD, b Huntington, WVa, May 8, 20. PLANT PHYSIOLOGY, PLANT MORPHOLOGY. Educ: Marshall Col, BS, 44, MS, 49; Univ Iowa, PhD(plant physiol), 51. Prof Exp: Instr bot, Univ Iowa, 49-51; from asst prof to assoc prof bot, 51-61, PROF BIOL SCI, MARSHALL COL, 61- Concurrent Pos: NSF fel, Univ Wyo, 55; NSF-AEC fel, Univ Mich, 59; res assoc, NMex Highlands Univ, 59-60; consult, Div Radiol Health, USPHS, 65-; res consult, Environ Protection Agency, 72- Mem: Bot Soc Am; Am Soc Plant Physiol. Res: Physiology of growth and floral initiation; anthocyanin production and localization; algal nutrition; bactericides; radiobotany; fission product uptake by plant roots; physiognomic analyses of vegetation. Mailing Add: 1234 Ninth St Huntington WV 25701

MILLS, IRA KELLY, b Richmond, Kans, Oct 31, 21; m 43; c 2. PLANT PHYSIOLOGY. Educ: Univ Southern Calif, AB, 52, MS, 53; Ore State Col, PhD(bot), 56. Prof Exp: From asst prof to assoc prof bot, 56-68, asst plant pathologist, 56-62, PROF BOT, MONT STATE UNIV, 68- Concurrent Pos: Fulbright-Hays lectr, Chung Hsing Univ, Taiwan, 66-67; consult, Gen Biol & Genetics Rev Panel, NIH, 68-70. Mem: AAAS; Am Soc Plant Physiol. Res: Effects of virus infection on host plant physiology; metabolism of aquatic plants and systems. Mailing Add: Dept of Biol Mont State Univ Bozeman MT 59715

MILLS, JACK, chemistry, see 12th edition

MILLS, JACK F, b Galesburg, Ill, Feb 3, 28; m 53; c 5. ORGANIC CHEMISTRY. Educ: Knox Col, BA, 50; Univ Iowa, PhD(chem), 53. Prof Exp: Fel, Univ Ill, 53-54; res chemist, 56-62, sr res chemist, 62-74, SR RES SPECIALIST, DOW CHEM USA, 74- Concurrent Pos: Instr, Delta Col. Res: Halogen and polyhalogen compounds and their reactions; study of basic complexes; disinfection and pollution studies; spectrometric studies; synthesis of biological active compounds. Mailing Add: 4524Andre St Midland MI 48640

MILLS, JAMES HERBERT LAWRENCE, b Guelph, Ont, Jan 14, 33; m 59; c 4. VETERINARY PATHOLOGY. Educ: Univ Toronto, BSA, 55, DVM, 61; Univ Conn, MS, 64, PhD(virol), 66; Am Col Vet Path, dipl, 66. Prof Exp: Vet practitioner, Ont, 61-62; instr animal dis, Univ Conn, 62-66, assoc prof vet path, 66-67; assoc prof, 67-71, PROF VET PATH, UNIV SASK, 71- Mem: Am Vet Med Asn; US Animal Health Asn; NY Acad Sci; Can Vet Med Asn; Int Acad Path. Res: Veterinary virology and pathology, particularly bovine mucosal disease; veterinary parasitology, especially lungworms. Mailing Add: Dept Vet Path Col Vet Med Univ of Sask Saskatoon SK Can

MILLS, JAMES SIDNEY, b Winters, Tex, Mar 11, 30; m 53; c 3. PHYSICS. Educ: Univ Calif, Berkeley, AB, 57. Prof Exp: Physicist, Broadview Res Corp, 56-58; physicist, Stanford Res Inst, 58-67, CHMN RADIATION MEASUREMENTS DEPT, STANFORD RES INST, 67- Mem: Am Phys Soc; Sigma Xi. Res: Magnetic resonance of liquids, solids, and gaseous states of matter; mathematical analysis of radiation induced changes in liquids; development of specialized magnetic resonance techniques. Mailing Add: Stanford Res Inst 333 Ravenswood Ave Menlo Park CA 94025

MILLS, JAMES WILSON, b Dayton, Ohio, June 7, 42; m 64; c 1. PHYSICAL CHEMISTRY. Educ: Earlham Col, AB, 63; Brown Univ, PhD(chem), 68. Prof Exp: Res assoc, Joint Inst Lab Astrophys & Univ Colo, 68-69; asst prof chem, Drew Univ, 69-73; ASST PROF CHEM, FT LEWIS COL, 73- Concurrent Pos: Cotrrell Res Corp grant, Drew Univ, 70-71 & Petrol Res Fund grant, 72-73; consult, 70-71 & Pet Four Corners Environ Res Inst, 74- Mem: Am Chem Soc; Am Phys Soc. Res: Molecular spectroscopy; quantum mechanics of small molecules; instrumentation. Mailing Add: Dept of Chem Ft Lewis Col Durango CO 81301

MILLS, JERRY LEE, b Midland, Tex, Mar 6, 43. INORGANIC CHEMISTRY. Educ: Univ Tex, Austin, BS, 65, PhD(inorg chem), 69. Prof Exp: Fel, Ohio State Univ, 69-70; ASST PROF CHEM, TEX TECH UNIV, 70- Mem: Am Chem Soc; The Chem Soc. Res: Preparation, reactivity, and structure of non-transition metal compounds. Mailing Add: Dept of Chem Tex Tech Univ Lubbock TX 79409

MILLS, JOHN BLAKELY, III, b Griffin, Ga, June 15, 39; m 64; c 2. BIOCHEMISTRY. Educ: Ga Inst Technol, BS, 61; Emory Univ, PhD(biochem), 65. Prof Exp: NSF fel, Cambridge Univ, 65-66; Whitehead fel, 66-67, from instr to asst prof, 67-74, ASSOC PROF BIOCHEM, EMORY UNIV, 74- Res: Chemistry of protein hormones. Mailing Add: Dept of Biochem Emory Univ Atlanta GA 30322

MILLS, JOHN NORMAN, b Neenah, Wis, Sept 29, 32; m 58; c 4. BIOCHEMISTRY. Educ: Wis State Col, BA, 54; Okla State Univ, MS, 56; Univ Okla, PhD, 65. Prof Exp: Asst chem, Okla State Univ, 54-58; from instr to asst prof, Okla Baptist Univ, 58-65; sr investr, Okla Med Res Found, 65-67; assoc prof, 67-71, PROF CHEM, OKLA BAPTIST UNIV, 71- Concurrent Pos: NIH res grant, 68- Mem: Am Chem Soc. Res: Human gastric proteolytic enzymes and zymogens; protein structure; enzyme activity. Mailing Add: Dept of Chem Okla Baptist Univ Shawnee OK 74801

MILLS, JOHN T, b Redhill, Eng, July 31, 37; m 65; c 2. PLANT PATHOLOGY. Educ: Univ Sheffield, BSc, 59; Univ London, PhD(plant path), dipl, Imp Col, 62. Prof Exp: Plant pathologist, Tate & Lyle Cent Agr Res Sta, Trinidad, WI, 63-67; RES SCIENTIST, RES STA, CAN DEPT AGR, 67- Concurrent Pos: Mem seed treatment subcomt, Can Comt Grain Dis. Mem: Can Phytopath Soc. Res: Ecology of soil microfloral and microfaunal components as affected by pesticides. Mailing Add: Can Dept Agr Res Sta 25 Dafoe Rd Winnipeg MB Can

MILLS, JOSEPH WILLIAM, b Toronto, Ont, June 9, 17; m 42; c 2. ECONOMIC GEOLOGY. Educ: Univ Toronto, BA, 39; Mass Inst Technol, PhD(geol), 42. Prof Exp: Asst, Mass Inst Technol, 39-42; mine geologist, Consol Mining & Smelting Co, Can, 42-43 & Mic Mac Mines, Que, 43-45; res geologist, O'Brien Gold Mines, Que, 45-50; from asst prof to assoc prof, 50-61, chmn dept, 62-71, PROF GEOL, WASH STATE UNIV, 62- Mem: Soc Econ Geologists; Geol Asn Can. Res: Geology of Canadian gold deposits; geology of eastern Washington; application of stereographic solutions to structural geology problems. Mailing Add: Dept of Geol Wash State Univ Pullman WA 99163

MILLS, KENNETH SELBY, b Loomis, Nebr, Jan 16, 17; m 43; c 1. PHYSIOLOGY. Educ: Univ NMex, BS, 45; Univ Tex, MA, 48, PhD(physiol), 53. Prof Exp: Asst prof biol, ETex State Teachers Col, 53-54; instr biophys, Univ Calif, Los Angeles, 54-57, assoc biol, 57-58; from asst prof to prof zool, Univ Okla, 58-69; PROF PHYSIOL, SOUTHERN CALIF COL OPTOM, 69- Mem: AAAS; Biophys Soc; Am Physiol Soc. Res: Cellular physiology and metabolism; cell and tissue culture; biophysics; bio-electric potentials; comparative neurophysiology. Mailing Add: Visual & Basic Sci Southern Calif Col Optom Fullerton CA 92631

MILLS, KING LOUIS, JR, b Leslie, Ark, Nov 14, 16; m 42, 58; c 1. PETROLEUM CHEMISTRY. Educ: Ark State Teachers Col, BS, 38; Univ Ark, MS, 42. Prof Exp: Res chemist, 43-54, group leader, 54-65, SECT MGR, RES & DEVELOP DEPT, PHILLIPS PETROL CO, 65- Mem: Am Chem Soc; Am Inst Chem Engrs. Res: Petroleum refining processes; hydrocarbon conversions; heterogeneous catalysis; petrochemicals; carbon black technology. Mailing Add: Res & Develop Dept D-5 PRC Phillips Petrol Co Bartlesville OK 74004

MILLS, LEWIS CRAIG, JR, b Chicago, Ill, May 19, 23; m 47; c 4. INTERNAL

MEDICINE. Educ: Baylor Univ, MD, 46; Am Bd Internal Med, dipl, 57. Prof Exp: Intern, John Sealey Hosp, Galveston, Tex, 46-47; resident, Jefferson Davis Hosp, Houston, 49-51 & Methodist Hosp, Houston, 52; from instr to asst prof internal med, Baylor Col Med, 53-57; from asst prof to assoc prof, 57-61, clin prof, 61-64, dir endocrinol & metab dis, Hosp, 69-74, PROF MED, HAHNEMANN MED COL, 64-, ASSOC DEAN AFFIL, 74- Concurrent Pos: Fel cardiol, Jefferson Davis Hosp, Houston, 51; res fel endocrinol, Peter Bent Brigham Hosp, Boston, 52-53; sr attend physician, Hahnemann Hosp, 58-, assoc vpres med affairs. Honors & Awards: Lindback Found Award, 64. Mem: Am Soc Pharmacol & Exp Therapeut; Soc Exp Biol & Med; Am Diabetes Asn; Am Fedn Clin Res; fel Am Col Physicians. Res: Endocrinology; metabolism; diabetes; vasopressor drugs; shock. Mailing Add: Dept of Med Hahnemann Med Col 230 N Broad St Philadelphia PA 19102

MILLS, ROBERT BARNEY, b Lane, Kans, Feb 10, 22; m 45; c 1. ENTOMOLOGY. Educ: Kans State Univ, BS, 49, PhD(entom), 64; Univ Colo, MEd, 53. Prof Exp: High sch teacher, Kans, 49-61; asst prof, 63-70, ASSOC PROF ENTOM, KANS STATE UNIV, 70- Mem: AAAS; Entom Soc Am. Res: Stored product entomology. Mailing Add: Dept of Entom Kans State Univ Manhattan KS 66506

MILLS, ROBERT GAIL, b Effingham, Ill, Jan 20, 24; m 46; c 2. NUCLEAR PHYSICS. Educ: Princeton Univ, BSE, 44; Univ Mich, MA, 47; Univ Calif, PhD(nuclear physics), 52. Prof Exp: Instr elec eng, Princeton Univ, 43-44, res assoc elec eng & physics, 45-46; res assoc, Univ Mich, 46-47; Nat Res Coun res fel physics, Univ Zurich, 52-53, instr, 53-54; MEM SR STAFF, PRINCETON UNIV, 54- Mem: Am Nuclear Soc; Am Phys Soc; Inst Elec & Electronics Engrs. Res: Controlled thermonuclear research; research machines. Mailing Add: 150 Prospect Ave Princeton NJ 08540

MILLS, ROBERT LAURENCE, b Englewood, NJ, Apr 15, 27; m 48; c 5. THEORETICAL PHYSICS. Educ: Columbia Univ, AB, 48, PhD(physics), 55; Cambridge Univ, BA, 50, MA, 54. Prof Exp: Res assoc physics, Brookhaven Nat Lab, 53-55; mem sch math, Inst Advan Study, 55-56; from asst prof to assoc prof, 56-62, PROF PHYSICS, OHIO STATE UNIV, 62- Mem: Am Phys Soc. Res: Quantum field theory; many-body theory; theory of alloys. Mailing Add: Dept of Physics Ohio State Univ Columbus OH 43210

MILLS, ROBERT LEROY, b Canton, Ohio, June 6, 22; m 45; c 4. PHYSICAL CHEMISTRY. Educ: Washington & Jefferson Col, BS, 43; Calif Inst Technol, MS, 48; Stanford Univ, PhD(chem), 50. Prof Exp: Asst chem, Calif Inst Technol, 43-48 & Stanford Univ, 48-49; mem staff, 50-75, ASST GROUP LEADER, LOS ALAMOS SCI LAB, UNIV CALIF, 75- Mem: AAAS. Res: High pressure physics; equation of state; low temperature physics; studies of light molecules to 20 kbar. Mailing Add: Los Alamos Sci Lab Univ Calif PO Box 1663 Los Alamos NM 87545

MILLS, ROGER EDWARD, b Cleveland, Ohio, Nov 19, 30; m 58; c 2. PHYSICS. Educ: Ohio State Univ, BSc & MSc, 52, PhD(physics), 63. Prof Exp: Physicist, Battelle Mem Inst, 60-63, sr physicist, 63-67, assoc div chief, 67-69; assoc prof, 69-75, PROF PHYSICS, UNIV LOUISVILLE, 75-, ASST TO VPRES ACAD AFFAIRS, 74- Mem: Am Phys Soc. Res: Ferrimagnetism; critical phenomena. Mailing Add: Dept of Physics Univ of Louisville Louisville KY 40208

MILLS, RUSSELL CLARENCE, b Milwaukee, Wis, Nov 13, 18; m 40; c 5. MEDICAL EDUCATION. Educ: Univ Wis, BS, 40, MS, 42, PhD(biochem), 44. Prof Exp: Asst biochem, Univ Wis, 40-44; from asst prof to assoc prof, 46-51, assoc dean grad sch, 63-70 & sch med, 63-72, assoc vchancellor, 72-75, PROF BIOCHEM, UNIV KANS, 51-, ASST TO CHANCELLOR, 75- Concurrent Pos: Alan Gregg traveling scholar med educ, Far East, 70-71; consult, Asn Am Med Cols, 75- Mem: AAAS; Am Chem Soc; Am Soc Microbiol; Am Soc Biol Chemists. Res: Intermediary metabolism; oxidative pathways. Mailing Add: 5407 Mission Dr Shawnee Mission KS 66208

MILLS, THOMAS MARSHALL, b Des Moines, Iowa, Nov 2, 38; m 60; c 2. REPRODUCTIVE ENDOCRINOLOGY. Educ: Univ Iowa, BA, 61, MS, 64, PhD(zool), 67. Prof Exp: Trainee steroid biochem, Ohio State Univ, 67-68; res assoc, Endocrine Lab, Univ Miami, 68-71; ASST PROF ENDOCRINOL, MED COL GA, 71- Mem: Endocrine Soc; Soc Study Reproduction. Res: Control of ovulation; ovarian steroid synthesis; metabolism of ovarian tissues. Mailing Add: 3220 Crane Ferry Rd Augusta GA 30907

MILLS, WILFORD RICHARD, b Fillmore, NY, Jan 15, 09; m 38; c 2. PLANT PATHOLOGY. Educ: Cornell Univ, BS, 31, PhD(plant path), 39. Prof Exp: Asst plant path, Cornell Univ, 31-35; geneticist & head dept genetics, Recife Inst Agr Res, Brazil, 35-37; asst plant path, Cornell Univ, 37-41; from asst prof to prof, 41-74, EMER PROF PLANT PATH, PA STATE UNIV, 74- Concurrent Pos: Spec sci aide, Rockefeller Found, Mex, 52; plant indust specialist, Inter-Am Inst Agr Sci, Costa Rica, 58. Mem: AAAS; Am Phytopath Soc; Potato Asn Am. Res: Breeding potatoes for disease resistance; genetics of blight resistance in potatoes. Mailing Add: Dept of Plant Path Pa State Univ University Park PA 16802

MILLS, WILLIAM ANDY, radiation physics, molecular biology, see 12th edition

MILLS, WILLIAM CARLOS, b Cairo, Ga, Aug 26, 13; m 42; c 2. FOOD TECHNOLOGY. Educ: Univ Ga, BSA, 43, MSA, 55. Prof Exp: Teacher, Coffee County Bd Ed & prin pub sch, Ga, 37-41, instr, 43-45; head dept food technol, Univ Ga, 45-51; res food technologist, Kingan, Inc, 51-56; supvr prod develop, Kroger Co, 56-62; ASST PLANT MGR, MURRAY BISCUIT CO, 62- Mem: AAAS; Inst Food Technol. Res: Food products; development of new products and processes in meats, salad dressing, jellies, preserves, peanut butter, canning, cookies, and marshmallow specialities. Mailing Add: Murray Biscuit Co Box 690 Augusta GA 30903

MILLS, WILLIAM CLEARON, JR, b Wake Co, NC, Oct 24, 25; m; c 1. POULTRY SCIENCE, MARKETING. Educ: NC State Univ, BS, 50; Mich State Univ, 60, PhD(poultry mkt), 63. Prof Exp: Sales & serv work, Siler City Mills, NC, 50-52; exten turkey specialist, 52-53, assoc prof poultry sci & proj leader, Poultry Sci Exten, 63-66, PROF POULTRY SCI & CHG POULTRY SCI EXTEN, AGR EXTEN SERV, NC STATE UNIV, 66- Concurrent Pos: Consult mkt mgt. Mem: Poultry Sci Asn. Res: Market research in turkey grades and pricing. Mailing Add: 208 Scott Hall NC State Univ Raleigh NC 27607

MILLS, WILLIAM HAROLD, b New York, NY, Nov 9, 21; m 49; c 3. MATHEMATICS. Educ: Swarthmore Col, AB, 43; Princeton Univ, MA, 47, PhD(math), 49. Prof Exp: Physicist, Aberdeen Proving Ground, 43-44; asst in instr, math, Princeton Univ, 48-49; from instr to assoc prof, Yale Univ, 49-64; MATHEMATICIAN, INST DEFENSE ANAL, 63- Mem: Am Math Soc. Res: Number theory; algebra; theory of games; combinatorial theory. Mailing Add: Inst for Defense Anal Thanet Rd Princeton NJ 08540

MILLS, WILLIAM RAYMOND, JR, b Dallas, Tex, Feb 14, 30; m 52; c 3. NUCLEAR SCIENCE. Educ: Rice Inst, BA, 51; Calif Inst Technol, PhD(physics), 55. Prof Exp: Res asst, Knolls Atomic Power Lab, Gen Elec Co, 55-56; sr res technologist, Socony Mobil Oil Co, Inc, 56-63; RES ASSOC, FIELD RES LAB, MOBIL RES & DEVELOP CORP, 63- Concurrent Pos: Adj prof, Southern Methodist Univ, 69- Mem: Am Phys Soc; Sigma Xi; Am Nuclear Soc. Res: Gamma-ray spectroscopy, especially of common earth elements; neutron physics; pulsed neutron phenomena. Mailing Add: Mobil Res & Develop Corp PO Box 900 Dallas TX 75221

MILLSAPS, KNOX, b Birmingham, Ala, Sept 10, 21; m 56; c 4. APPLIED MATHEMATICS. Educ: Auburn Univ, BS, 40; Calif Inst Technol, PhD, 43. Prof Exp: Assoc prof aeronaut eng, Ohio State Univ, 47-48; mathematician, Off Air Res, US Dept Air Force, 48-49 & 50-51, chief mathematician, Wright Air Develop Ctr, 52-55, chief scientist, Missile Develop Ctr, 56-60, chief scientist, Off Aerospace Res & exec dir, Off Sci Res, 60-63; res prof aerospace eng, Univ Fla, 63-68; head prof mech eng, Colo State Univ, 68-73; PROF ENG SCI & CHMN DEPT, UNIV FLA, 73- Concurrent Pos: Prof, Auburn Univ, 49-52 & Mass Inst Technol, 55-56. Mem: Am Inst Aeronaut & Astronaut; Soc Indust & Appl Math; Am Math Soc; Math Asn Am; Am Phys Soc. Res: Fluid mechanics; heat transfer. Mailing Add: PO Box 13857 Gainesville FL 32604

MILLSON, MAURICE FREDERICK, organic chemistry, see 12th edition

MILLSPAUGH, DICK DARWIN, b Wapello, Iowa, Oct 20, 15; m 43; c 5. ZOOLOGY. Educ: Iowa Wesleyan Col, BA, 37; Southern Methodist Univ, MS, 39. Prof Exp: Instr, 39-41, head dept, 47-59, chmn div natural sci, 50-59, OF BIOL, IOWA WESLEYAN COL, 47- Concurrent Pos: Cur, State Insect Surv Collection, 59-; instr, Iowa State Univ, 59-62. Mem: AAAS; Entom Soc Am; Nat Asn Biol Teachers; Am Ornith Union. Res: Entomology. Mailing Add: Dept of Biol Iowa Wesleyan Col Mt Pleasant IA 52641

MILLSTEIN, LLOYD GILBERT, b Brooklyn, NY, Jan 2, 32; m 53; c 2. MEDICAL PHYSIOLOGY, INFORMATION SCIENCE. Educ: NY Univ, BA, 53; Rutgers Univ, MS, 60, PhD(physiol, biochem), 64. Prof Exp: Microbiologist, Univ Hosp, Bellevue Med Ctr, 53; med writer pharmaceut, Squibb Inst Med Res Div, 55-57, toxicologist, 57-64; sr scientist, Smith Kline & French Labs, 64-67; DIR SCI INFO & COMMUN, McNEIL LABS, 67- Mem: AAAS; Am Heart Asn; NY Acad Sci. Res: Drug regulatory affairs; gerontology; lipid metabolism; arteriosclerosis; pharmacology and drug evaluation. Mailing Add: Sci Info Div McNeil Labs Ft Washington PA 19034

MILLY, GEORGE HARWOOD, b Petersburg, Va, Oct 10, 20; m 43; c 6. EARTH SCIENCE. Educ: Niagara Univ, BS, 40; Univ Mich, PhD(meteorol), 60. Prof Exp: Chem engr, Armour & Co, 41-42, res biochemist, 46-47; chief planning & eval br, Chem Res & Develop Labs, US Dept Army, 47-54, phys scientist, 54-57, sr phys scientist, 57-59, sci dir opers res group, 59-62, dir, 62-63; vpres & dir math sci dept, Travelers Res Ctr, Inc, Conn, 63-67; PRES & CHMN BD, GEOMET, INC, 67- Concurrent Pos: Consult, Environ Protection Agency. Honors & Awards: Am Inst Chemists Medal, 40. Mem: AAAS; Am Chem Soc; Am Meteorol Soc; Opers Res Soc Am; fel Am Inst Chemists. Res: Environmental and mathematical sciences; turbulent diffusion in atmosphere; advanced techniques of minerals exploration involving geometeorological methods applied to uranium, oil, gas, mercury and associated precious and base metals; United States and foreign patents in these areas. Mailing Add: Geomet Inc 15 Firstfield Rd Gaithersburg MD 20760

MILMAN, DORIS H, b New York, NY, Nov 17, 17; m 41; c 1. MEDICINE, PSYCHIATRY. Educ: Bernard Col, BA, 38; NY Univ, MD, 42; Am Bd Pediat, dipl. Prof Exp: From asst prof to assoc prof pediat psychiat, 64-73, PROF PEDIAT & ACTG CHMN DEPT, STATE UNIV NY DOWNSTATE MED CTR, 73-, CHIEF-OF-SERV PEDIAT, STATE UNIV HOSP, 73-; CHIEF-OF-SERV, KINGS COUNTY HOSP CTR, 73- Mem: AMA; Am Psychiat Asn; Am Acad Pediat. Res: Minimal brain impairment; group work with parents of handicapped children; school phobia; adolescent phenomena; drug abuse. Mailing Add: 126 Westminster Rd Brooklyn NY 11218

MILMAN, HARRY ABRAHAM, b Cairo, Egypt, May 16, 43; US citizen; m 68; c 1. BIOCHEMICAL PHARMACOLOGY. Educ: Columbia Univ, BS, 66; St John's Univ, NY, MS, 68; George Washington Univ, PhD(pharmacol), 77. Prof Exp: SCIENTIST TOXICOL, NAT CANCER INST, 70- Res: Metabolism and homeostasis of L-asparagine; biochemical characterization and inhibition of L-asparagine synthetase in pancrease and experimental tumors. Mailing Add: Lab of Toxicol Bldg 37 Rm 5B-22 Nat Cancer Inst Bethesda MD 20014

MILMORE, BENNO KARL, b Schwerin, Ger, Jan 27, 14; nat US; m 37; c 4. EPIDEMIOLOGY. Educ: Univ Calif, AB, 35, MD, 39; Johns Hopkins Univ, MPH, 42; Am Bd Prev Med & Pub Health, dipl. Prof Exp: Intern, US Marine Hosp, USPHS, Calif, 38-39, ward surgeon, Calif & Ga, 39-40, med officer, Typhus Control Unit, Ga, 40-43, chief med sect, Off Malaria Control in War Areas, 43-44, outpatient med serv, US Marine Hosp, Calif, 45-48, diabetes consult, State Dept Pub Health, Calif, 49-53, chronic dis consult, Region IX, 53-54, epidemiologist & head gen field studies sect, Nat Cancer Inst, 54-61; chief bur chronic dis, 61-63, epidemiol ctr, 63-66, ASST CHIEF, CONTRACT COUNTIES HEALTH SERV SECT, STATE DEPT HEALTH, CALIF, 66- Concurrent Pos: Lectr, Univ Calif, 54. Mem: Am Soc Trop Med & Hyg; fel AMA; fel Am Pub Health Asn; Am Diabetes Asn; Am Col Prev Med. Res: Epidemiology of chronic diseases. Mailing Add: 36 Kenyon Ave Kensington CA 94708

MILMORE, JOHN EDWARD, b Brooklyn, NY, Oct 31, 43; m 66; c 2. NEUROENDOCRINOLOGY, PHARMACOLOGY. Educ: Fordham Univ, BS, 65; Long Island Univ, MS, 68; Rutgers Univ, PhD(physiol), 74. Prof Exp: Res asst pharmacol, US Vitamin Inc, 65-66; res asst endocrinol, Hoffmann-La Roche Inc, 66-68; res assoc pharmacol, Squibb Inst Med Res, 68-75; RES ASSOC NUTRIT, AM HEALTH FOUND, 75- Mem: AAAS. Res: Neuroendocrine control mechanisms; nutrition and cancer; central nervous system and cardiovascular pharmacology. Mailing Add: Naylor Dana Inst Am Health Found Valhalla NY 10595

MILNE, ALLEN RITCHIE, b Can, Oct 27, 21; m 45; c 3. PHYSICS, GEOPHYSICS. Educ: Univ Toronto, BSc, 50; McGill Univ, MSc, 53. Prof Exp: Eng asst microwave syst eng, Nat Res Coun Can, 50-52; defence sci res officer chg arctic acoust res prog, Defense Res Estab Pac, Defense Res Bd, Can, 53-73; proj mgr Beaufort sea proj, 73-76, HEAD ARCTIC MARINE GROUP, OCEAN & AQUATIC SCI, DEPT OF ENVIRON, CAN, 76- Mem: Fel Arctic Inst NAm. Res: Oceanography in arctic regions; sea ice related research. Mailing Add: Dept of Environ 512-1230 Government St Victoria BC Can

MILNE, DAVID BAYARD, b Evanston, Ill, Oct 24, 40; m 66; c 2. NUTRITIONAL BIOCHEMISTRY, BIOINORGANIC CHEMISTRY. Educ: Wash State Univ, BS, 62; Ore State Univ, MS, 65, PhD(biochem), 68. Prof Exp: Res assoc biochem, NC State

Univ, 67-69; res chemist, Vet Admin Hosp, Long Beach, Calif, 69-73; RES CHEMIST, LETTERMAN ARMY INST RES, 73- Mem: Am Inst Nutrit; Soc Environ Geochem & Health; Am Chem Soc. Res: Copper metabolism; metabolism and function of new essential trace elements. Mailing Add: Biochem Div Dept of Nutrit Letterman Army Inst of Res San Francisco CA 94129

MILNE, DAVID HALL, b Highland Park, Mich, Dec 15, 39; m 64. ENTOMOLOGY. Educ: Dartmouth Col, BA, 61; Purdue Univ, PhD(entom), 68. Prof Exp: Asst prof gen sci, Ore State Univ, 67-71; MEM FAC, EVERGREEN STATE COL, 71- Mem: AAAS; Am Inst Biol Sci. Res: Mathematics of predation, competition processes; computer simulation of ecosystem dynamics. Mailing Add: Dept of Biol Evergreen State Col Olympia WA 98505

MILNE, EDMUND ALEXANDER, b Eugene, Ore, Mar 3, 27; m 54; c 2. NUCLEAR PHYSICS. Educ: Ore State Col, BA, 49; Calif Inst Technol, MS, 50, PhD(nuclear physics), 53. Prof Exp: Asst, Calif Inst Technol, 52-53, res fel, 53-54; asst prof, 54-58, ASSOC PROF PHYSICS, NAVY POSTGRAD SCH, 58- Mem: Am Phys Soc. Res: Energy levels of light nuclei. Mailing Add: 2 Shady Lane Monterey CA 93940

MILNE, EDWIN LOUIS, chemical physics, solid state physics, see 12th edition

MILNE, ERIC CAMPBELL, b Perth, Scotland, Feb 8, 29; Can citizen; m 55; c 5. RADIOLOGY. Educ: Univ Edinburgh, MB, ChB, 56, DMRD, 60; Royal Col Radiologists, Eng, FRCR, 62. Prof Exp: Intern med surg, Monmouth Med Ctr, NJ, 56-57; resident chest dis, Tulare-Kings Counties Hosp, Calif, 57-58; sr house officer radiol, Royal Infirmary, Univ Edinburgh, 58-60; radiologist, McKellar Gen Hosp, Ft William, Ont, 61-65; asst prof radiol, Univ Western Ont, 65-66 & Peter Bent Brigham Hosp, Harvard Med Sch, 66-68; prof & dir exp radiol, Radiol Res Labs, Univ Toronto, 68-75, consult, Dept Lab Animal Serv, 69-75; PROF RADIOL SCI & CHMN DEPT, UNIV CALIF COL MED, IRVINE, 75- Concurrent Pos: UK Med Res Found fel, Depts Radiol & Med, Royal Infirmary, Univ Edinburgh, Edinburgh, 60-61; Ont Cancer Found Gordon Richards fel, Cardiovasc Res Inst, Univ Calif, San Francisco, 65-66. Mem: Am Col Radiologists; Soc Photo-Optical Instrument Engrs; Can Microcirc Soc; Asn Univ Radiologists. Res: Tumor circulation; radiologic diagnosis of early pulmonary diseases; analysis of x-ray image formation; radiologic magnification techniques; lung water pulmonary microcirculation; x-ray tube construction; x-ray holography; image analysis. Mailing Add: Dept of Radiol Sci Univ of Calif Col of Med Irvine CA 92664

MILNE, FRANK JAMES, b Aberdeen, Scotland, Mar 30, 19; Can citizen; m 44; c 3. VETERINARY MEDICINE. Educ: Royal (Dick) Vet Col, MRCVS, 41; Colo Agr & Mech Col, DVM, 52; Univ Zurich, Dr Med Vet, 62; Am Col Vet Surg, dipl, 66. Prof Exp: Lectr surg, Royal Vet Col, London, 48-49; asst prof, Colo Agr & Mech Col, 50-53; prof obstet, 53-71, PROF SURG, ONT VET COL, UNIV GUELPH, 53- Concurrent Pos: Ed, Am Asn Equine Practitioners, 56- Mem: Am Vet Med ASn; Am Asn Equine Practitioners; Can Vet Med Asn; Brit Vet Asn; Brit Equine Vet Asn. Res: Equine surgery including orthopedics. Mailing Add: Dept of Clin Studies Ont Vet Col Univ Guelph Guelph ON Can

MILNE, GEORGE MCLEAN, JR, b Port Chester, NY, Dec 29, 43; m 65; c 2. MEDICINAL CHEMISTRY, PHARMACOLOGY. Educ: Yale Univ, BSc, 65; Mass Inst Technol, PhD(org chem), 69. Prof Exp: NIH fel chem, Stanford Univ, 69-70; SR RES SCIENTIST, PFIZER CENT RES, PFIZER, INC, 70- Mem: Am Chem Soc. Res: Synthetic medicinal agents; design, synthesis and pharmacological evaluation of central nervous system drugs; biological structure activity relationships; heterocyclic synthesis. Mailing Add: Dept of Pharmacol Pfizer Cent Res Pfizer Inc Groton CT 06340

MILNE, GEORGE WILLIAM ANTHONY, b Stockport, Eng, May, 1937; US citizen. CHEMISTRY. Educ: Univ Manchester, BSc, 57, MS, 58, PhD, 60. Prof Exp: Res fel, Univ Wis, 60-61; vis fel, Lab Chem, Nat Inst Arthritis & Metab Dis, 62-63, vis assoc, 63-64; CHEMIST, LAB CHEM, NAT HEART & LUNG INST, 65- Concurrent Pos: Adj prof chem, Georgetown Univ, 67- Mem: AAAS; Am Chem Soc; The Chem Soc. Res: Chemistry of steroids, terpenes, alkaloids, amino acids, nucleosides, nucleotides, carbohydrates, and application of nuclear magnetic resonance spectroscopy and mass spectrometry to studies in these fields, particularly biological function of various members of these classes. Mailing Add: Lab of Chem Nat Heart & Lung Inst Bethesda MD 20014

MILNE, GORDON GLADSTONE, b Deland, Fla, July 13, 16; m 44; c 3. OPTICS. Educ: Univ Sask, BA, 38, MA, 39; Univ Rochester, PhD(optics, physics), 50. Prof Exp: Res physicist, Inst Optics, Univ Rochester, 42-45, res assoc optics, 45-60, sr res assoc, 60-66; PHYSICIST, TROPEL, INC, 66- Mem: AAAS; Optical Soc Am; Soc Photog Scientists & Engrs. Res: Optical instrumentation; manufacture of precision optics. Mailing Add: Tropel Inc 1000 Fairport Pl Fairport NY 14450

MILNE, HENRY BAYARD, b Walla Walla, Wash, Aug 15, 15; m 37; c 4. ORGANIC CHEMISTRY. Educ: Ore State Col, BS, 37; Northwestern Univ, MS, 39, PhD(org chem), 41. Prof Exp: Instr chem, Univ Idaho, 41-42; fel, Calif Inst Technol, 46; from asst prof to assoc prof, 46-65, PROF CHEM, WASH STATE UNIV, 65- Mem: Am Chem Soc. Res: Peptide synthesis; chemistry of indene derivatives; enzymatic reactions of acylated amino acids; chemical effects of ionizing radiations. Mailing Add: Dept of Chem Wash State Univ Pullman WA 99163

MILNE, IVAN HERBERT, mineralogy, geology, see 12th edition

MILNE, JOHN B, b Vancouver, BC, Mar 24, 34; m 65. INORGANIC CHEMISTRY. Educ: Univ BC, BA, 56, MSc, 60; McMaster Univ, PhD(chem), 65. Prof Exp: Chemist, Govt Can, 56-58; biochemist, Wellcome Found, 60-62; ASSOC PROF CHEM, UNIV OTTAWA, 67- Concurrent Pos: Nat Res Coun fel, 65-67 & operating grant, 67-; Ont Govt grant-in-aid of res, 68- Mem: The Chem Soc; Chem Inst Can. Res: Study of inorganic chemistry in nonaqueous solvents by cryoscopy, conductivity and spectroscopy. Mailing Add: Dept of Chem Univ of Ottawa Ottawa ON Can

MILNE, LORUS JOHNSON, b Toronto, Ont, Sept 12, 12; nat US; m 38. BIOLOGY. Educ: Univ Toronto, BA, 33; Harvard Univ, MA, 34, PhD(biol), 36. Prof Exp: Asst zool, Harvard Univ, 34-36; prof biol, Southwestern Univ, 36-37; adj prof, Randolph-Macon Woman's Col, 37-39, assoc prof, 39-42; war res, Aviation Med, Johnson Found, Univ Pa, 42-47; assoc prof zool, Univ Vt, 47-48; assoc prof, 48-51, PROF ZOOL, UNIV NH, 51- Concurrent Pos: Res grants, Carnegie Corp, Sigma Xi, Am Acad Arts & Sci, Am Philos Soc, Cranbrook Inst Sci & Explorers Club; Fund Advan Educ fac fel, 53-54; exchange lectr, Univs, US-SAfrica Leader Exchange Prog, 59; vis prof environ technol, Fla Int Univ, 74; mem, Marine Biol Lab, Woods Hole. Consult-writer, Biol Sci Curriculum Study, 60-61; Univ NH-Explorers Club deleg, Nairobi Meetings, Int Union Conserv Nature, 63; UNESCO biol consult, NZ, 66. Mem exped, Panama, 51, Cent Am, 53-54, BWI & SAm, 56-57, Equatorial Africa, 59 & 63, NAfrica, Near East, Southeast Africa & Australia, 66. Honors & Awards: Nash Award, 54. Mem: Fel AAAS; Am Soc Zoologists; Animal Behav Soc. Res: Behavioral ecology; natural history. Mailing Add: 1 Garden Lane Durham NH 03824

MILNE, MARGERY (JOAN) (GREENE), b New York, NY, Jan 18, 15; m 38. BIOLOGY. Educ: Hunter Col, BA, 33; Columbia Univ, MA, 34; Radcliffe Col, MA, 37, PhD(zool), 39. Prof Exp: Instr zool, Univ Maine, 36-37; instr biol, Randolph-Macon Woman's Col, 39-40; asst prof biol & bact, Richmond Prof Inst, Col William & Mary, 40-42 & Beaver Col, 42-47; asst prof bot, Univ Vt, 47-48; asst prof zool, Univ NH, 48-50; assoc prof biol, Mass State Teachers Col, Fitchburg, 56; vis prof, Northeastern Univ, 58; consult biologist, Biol Sci Curric Study, Am Inst Biol Sci, Univ Colo, 60; res assoc, Univ NH, 65-74; VIS PROF ENVIRON TECHNOL, FLA INT UNIV, 74- Concurrent Pos: Grantee, Am Acad Arts & Sci, Sigma Xi & Am Philos Soc; mem exped, Panama, 51, Cent Am, 53-54, BWI & SAm, 56-57, Equatorial Africa, 63 & NAfrica, Near East, Southeast Asia & Australia, 66; exchangee, US-SAfrica Leader Exchange Prog, 59; UNESCO biol consult, NZ, 66. Honors & Awards: George Westinghouse Award, 47; Nash Conserv Award, 54; Saxton Mem Award, 54. Mem: AAAS; Nat Audubon Soc; Wilderness Soc. Res: Animal behavior; ecology; science writing. Mailing Add: 1 Garden Lane Durham NH 03824

MILNE, THOMAS ANDERSON, b Winfield, Kans, Dec 29, 27; m 54; c 3. PHYSICAL CHEMISTRY. Educ: Univ Kans, AB, 50, PhD(chem), 55. Prof Exp: Res engr high temperature chem, Atomics Int Div, NAm Aviation, Inc, 54-57; sr chemist, Stanford Res Inst, 57-60; sr physicist, 60-63, prin physicist, 63-68, SR ADV CHEM, MIDWEST RES INST, 68- Mem: AAAS; Am Soc Mass Spectros; Am Combustion Inst; Am Chem Soc; Fedn Am Scientists. Res: Molecular beam formation and sampling at high pressures; combustion processes; thermodynamic behavior of systems at high tempreature; mass spectroscopic study of high temperature reactions; homogeneous nucleation; coal combustion; inhibition chemistry. Mailing Add: Midwest Res Inst 425 Volker Blvd Kansas City KS 64110

MILNE, WALTER LEROY, b Santa Rosa, Calif, Aug 29, 25; m 50; c 2. HEALTH PHYSICS, RADIOLOGICAL HEALTH. Educ: Sacramento State Col, AB, 50. Prof Exp: Radiol biologist, US Naval Radiol Defense Lab, 50-60; phys sci adminr, 60-63, RADIOL SAFETY OFFICER, PAC MISSILE TEST CTR, 63- Concurrent Pos: Lectr, Univ Calif, Santa Barbara; mem, Ventura County Indust-Educ Coun. Mem: Health Physics Soc; Int Radiation Protection Asn; Sigma Xi. Res: Environmental radiation effects; dosimetry; laser radiation effects and safety analyses; fission product uptake and metabolism. Mailing Add: Code 3032 Pac Missile Test Ctr Point Mugu CA 93042

MILNER, ALICE N, b Bay City, Tex, Sept 5, 25. BIOCHEMISTRY. Educ: Tex Woman's Univ, BS, 46, MA, 47; Baylor Univ, PhD(biochem), 59. Prof Exp: Instr chem, Centenary Col La, 47-50; head clin chem, Baylor Hosp, Dallas, Tex, 50-52; phys scientist, Vet Admin Hosp, Dallas, 52-53; Nat Cancer Inst fel biochem & oncol, Univ Tex M D Anderson Hosp & Tumor Inst, 59-62, res assoc biochem, 62-65; dir biochem, Moody Clin Res Lab, Col Med, Baylor Univ, 65-67; ASSOC PROF NUTRIT, TEX WOMAN'S UNIV, 67-, PROG DIR COORD UNDERGRAD PROGS IN DIETETICS-SPEC CLIN DIETETICS, 74- Mem: AAAS; Am Chem Soc; Am Asn Cancer Res; Am Dietetic Asn. Res: Mechanism of action of carcinostatic compounds; bionutritional aspects of mental retardation and obesity. Mailing Add: TWU Res Inst Tex Woman's Univ Denton TX 76204

MILNER, BRENDA (ATKINSON), b Manchester, Eng, July 15, 18; m 44. NEUROPSYCHOLOGY. Educ: Cambridge Univ, BA, 39, MA, 49; McGill Univ, PhD(psychol), 52. Hon Degrees: ScD, McGill Univ, 72. Prof Exp: Exp officer, Ministry of Supply, UK, 41-44; asst prof psychol, Univ Montreal, 45-52; res assoc, Univ, 52-53, from lectr neurol & neurosurg & psychologist, Montreal Neurol Inst to assoc prof psychol, Univ, 53-69, PROF NEUROL & NEUROSURG, McGILL UNIV, 69- Honors & Awards: Distinguished Sci Contrib Award, Am Psychol Asn, 73. Mem: Am Psychol Asn; Am Acad Neurol; Brit Exp Psychol Soc. Res: Brain function; perception and learning in human patients undergoing brain operation for focal cortical epilepsy. Mailing Add: Montreal Neurol Inst McGill Univ Montreal PQ Can

MILNER, CLIFFORD E, b Concord, NH, Aug 10, 28; m 51; c 5. PHYSICAL CHEMISTRY. Educ: Wesleyan Univ, BA, 50, MA, 52; Yale Univ, PhD(phys chem), 55. Prof Exp: Res chemist, 55-71, SR RES CHEMIST, E I DU PONT DE NEMOURS & CO, INC, 71- Mem: Am Chem Soc; Soc Photog Scientists & Engrs. Res: Pressure dependence of the dielectric constant of water; photographic emulsions and processing solutions. Mailing Add: 1763 Winton Rd N Rochester NY 14609

MILNER, DAVID, b Birkenhead, Eng, July 23, 38; US citizen; m 61; c 2. ANALYTICAL CHEMISTRY. Educ: LRIC, 75. Prof Exp: Tech asst chem, Distiller Co Ltd, Eng, 56-61; tech asst, Eli Lilly Ltd, Eng, 61-62, control chemist, 62-64; SECT HEAD PHARMACEUT ANAL, SYNTEX CORP, 64- Mem: Royal Inst Chem; fel The Chem Soc. Res: Development of analytical methods for pharmaceutical dosage forms. Mailing Add: Syntex Res Div 3401 Hillview Ave Palo Alto CA 94304

MILNER, ERIC CHARLES, b London, Eng, May 17, 28; m 54; c 4. MATHEMATICS. Educ: Univ London, BSc, 49, MSc, 50, PhD(math), 63. Prof Exp: Lectr math, Univ Malaya, 52-61 & Univ Reading, 61-67; PROF MATH, UNIV CALGARY, 67-, CHMN DIV PURE MATH, DEPT MATH, STATIST & COMPUT SCI, 74- Mem: Am Math Soc; Math Asn Am; Can Math Cong; London Math Soc. Res: Set theory; combinatorics; graph theory. Mailing Add: DeptMath Statist & Comput Sci Univ of Calgary Calgary AB Can

MILNER, JOHN AUSTIN, b Pine Bluff, Ark, June 11, 47. NUTRITION. Educ: Okla State Univ, BS, 69; Cornell Univ, PhD(nutrit), 74. Prof Exp: Res assoc animal sci, Cornell Univ, 74-75; ASST PROF NUTRIT, UNIV ILL, URBANA-CHAMPAIGN, 75- Mem: Sigma Xi; Nutrit Today Soc; Inst Food Technol; Am Asn Animal Sci. Res: Relationship of dietary factors effecting ammonia detoxification and the synthesis of arginine, carbohydrates, pyrimidines or polyamines in normal and diseased states. Mailing Add: Dept of Food Sci Univ of Ill Urbana IL 61801

MILNER, KELSEY CHARLES, b Kansas City, Kans, Mar 21, 13; m 42; c 4. MEDICAL MICROBIOLOGY. Educ: Univ Chicago, PhB, 34; Tulane Univ, PhD(bact), 50. Prof Exp: Asst bact, Univ Chicago, 39; bacteriologist, Charity Hosp, New Orleans, La, 40-44; from asst to assoc bact, Tulane Univ, 44-50; bacteriologist, Joint Dysentery Unit, USPHS, Korea, 51, sr asst, 51-71, SCIENTIST DIR, ROCKY MOUNTAIN LAB, NAT INST ALLERGY & INFECTIOUS DIS, 51- Concurrent Pos: Vis scientist, Karolinska Inst, Sweden, 61-62; fac affil lectr, Univ Mont, 64- Honors & Awards: Commendation Medal, Dept Health, Educ & Welfare, USPHS, 75. Mem: AAAS; fel Am Acad Microbiol; Am Soc Microbiol; NY Acad Sci. Res: Enteric infectious; endotoxins; structure of antigens; biological assays; immunotherapy of cancer. Mailing Add: Pine Hill Range Rte 1 Box 1410 Hamilton MT 59840

MILNER, MAX, b Edmonton, Alta, Jan 24, 14; nat US; m 42; c 2. FOOD SCIENCE. Educ: Univ Sask, BSc, 38; Univ Minn, MS 41, PhD(biochem), 45. Prof Exp: Res chemist, Pillsbury Mills, Inc, Minn, 41-42; res assoc, Univ Minn, 45-46; prof cereal chem, Kans State Univ, 47-59; sr food technologist, UNICEF, 59-71; dir secretariat, Protein-Calorie Adv Group UN Syst, 71-75; coordr, NSF-Mass Inst Technol Protein Resources Study, 75; NUTRIT COORDR, OFF TECHNOL ASSESSMENT, US CONG, 76- Concurrent Pos: Consult, Food & Agr Orgn, UN, 54-58; adj prof, Columbia Univ, 64-; lectr, Mass Inst Technol, 75-; mem panel world food supply, President's Sci Adv Comt, 68-; chmn, Gordon Res Conf Food & Nutrit, 68. Honors & Awards: Inst Food Technologists Int Award. Mem: Fel AAAS; Am Chem Soc; Am Asn Cereal Chem; Inst Food Technologists. Res: Cereal chemistry; nutrition, protein technology and development. Mailing Add: Office of Technol Assessment US Congress Washington DC 20510

MILNER, PAUL CHAMBERS, b Washington, DC, Aug 23, 31. PHYSICAL CHEMISTRY. Educ: Haverford Col, BS, 52; Princeton Univ, MA, 54, PhD(chem), 56. Prof Exp: MEM TECH STAFF, BELL TEL LABS, INC, 57- Mem: Am Chem Soc; Electrochem Soc. Res: Electrochemistry; kinetics; thermodynamics. Mailing Add: Bell Tel Labs Inc Murray Hill NJ 07974

MILNER, REID THOMPSON, b Carbondale, Ill, Aug 13, 03; m 28, 39; c 1. Educ: Univ Ill, BS, 24, MS, 25; Univ Calif, PhD(phys chem), 28. Prof Exp: From asst to assoc chemist, US Bur Mines, 29-30; assoc chemist, USDA, 30-36, sr chemist, Regional Soybean Indust Prod Lab, Bur Agr Chem & Eng, 36-39, dir, 39-41, head anal & phys chem div, Northern Regional Res Lab, 41-48, dir, 48-54; prof food sci & head dept, 54-71, EMER PROF FOOD SCI, UNIV ILL, URBANA, 71- Mem: AAAS; Am Chem Soc; Am Oil Chem Soc (vpres, 46, pres, 47); Inst Food Technologists (pres, 74). Res: Low temperature specific heats; microanalysis; gas analysis; agricultural and food chemistry. Mailing Add: 614 W Florida Ave Urbana IL 61801

MILNES, HAROLD WILLIS, mathematics, see 12th edition

MILNOR, JOHN WILLARD, b Orange, NJ, Feb 20, 31; m 54 & 68; c 3. MATHEMATICS. Educ: Princeton Univ, AB, 51, PhD(math), 54. Hon Degrees: ScD, Syracuse Univ, 65; DSc, Univ Chicago, 67. Prof Exp: Higgins res asst math, Princeton Univ, 53-54, Higgins lectr, 54-55, from asst prof to prof, 55-62, Henry Putnam Univ prof, 62-67, chmn dept, 63-66; prof, Mass Inst Technol, 68-70; PROF MATH, INST ADVAN STUDY, 70- Concurrent Pos: Alfred P Sloan fel, 55-59; vis prof, Univ Calif, Berkeley, 59-60; mem, Inst Advan Study, 63-; vis prof, Univ Calif, Los Angeles, 67-68. Mem: Int Cong Math. Honors & Awards: Fields Medal, Univ Stockholm, 62; Nat Medal Sci, 66. Mem: Nat Acad Sci; Am Math Soc; Am Acad Arts & Sci; Am Philos Soc. Res: Topology of manifolds. Mailing Add: Inst for Advan Study Princeton NJ 08540

MILNOR, TILLA SAVANUCK KLOTZ, b New York, NY, Sept 29, 34; m 53 & 68; c 2. MATHEMATICS. Educ: NY Univ, BA, 55, MS, 56, PhD(math), 59. Prof Exp: NSF fel, 58-59; instr math, Univ Calif, Los Angeles, 59-60, lectr, 60-61, from asst prof to assoc prof, 61-69; assoc prof, Boston Col, 69-71; PROF MATH & CHMN DEPT, DOUGLASS COL, RUTGERS UNIV, 70- Concurrent Pos: Vis mem, Courant Inst Math Sci, NY Univ, 64-65. Mem: Am Math Soc. Res: Differential geometry of surfaces in 3-space, especially questions involving ordinary or nonstandard conformal structures on surfaces; geometric application of methods form Riemann surface theory. Mailing Add: Dept of Math Douglass Col Rutgers Univ New Brunswick NJ 08903

MILNOR, WILLIAM ROBERT, b Wilmington, Del, May 4, 20; m 44; c 2. MEDICAL PHYSIOLOGY. Educ: Princeton Univ, AB, 41; Johns Hopkins Univ, MD, 44. Prof Exp: From instr to assoc prof med, 51-69, PROF PHYSIOL, SCH MED, JOHNS HOPKINS UNIV, 69- Concurrent Pos: Nat Heart Inst res fel, 49-51; physician, Johns Hopkins Hosp, 52-, physician-in-chg heart sta, 51-60. Mem: Fel Am Col Physicians; Am Physiol Soc; Am Fedn Clin Res. Res: Cardiovascular disease and physiology; cardiac electrophysiology; pulmonary vascular responses; blood volume regulation and distribution; circulatory biophysics. Mailing Add: Dept of Physiol Johns Hopkins Univ Sch of Med Baltimore MD 21205

MILO, GEORGE EDWARD, b Montpelier, Vt, Nov 6, 32; m 56; c 4. VIROLOGY, BIOCHEMISTRY. Educ: Univ Vt, BA, 58, MS, 61; State Univ NY Buffalo, PhD(virol), 68. Prof Exp: Instr biol, Rosary Hill Col, 63-65; NIH fel, Roswell Park Mem Inst, 67-69; sr res virologist, Battelle Mem Inst, 69; ASST PROF VET PATHOBIOL, COL VET MED, OHIO STATE UNIV, 69- Mem: Phytochem Soc NAm; Tissue Cult Asn; NY Acad Sci; Am Soc Microbiol; Am Soc Biol Chemists. Res: Aging and viral oncogenesis at the membrane level; carcinogen, carcinogen and steroid administration to human cell systems in vitro. Mailing Add: Dept of Vet Pathobiol Ohio State Univ Col of Vet Med Columbus OH 43210

MILONE, CHARLES ROBERT, b Uhrichsville, Ohio, Feb 13, 13; m 40; c 2. ORGANIC POLYMER CHEMISTRY. Educ: Mass Inst Technol, BS, 36, PhD(org chem), 39. Prof Exp: Res chemist, Goodyear Tire & Rubber Co, 39-45, sect head, 45-52; supt develop lab, Goodyear Atomic Corp, 52-57, mgr tech div, 57-60, dep gen mgr, 60-67; dir gen prod develop, 67-68, dir res & gen prod develop, 68-70, VPRES, GOODYEAR TIRE & RUBBER CO, 70- Mem: AAAS; Am Chem Soc. Res: Synthetic rubber; polymerization; new plastics and their applications; gaseous diffusion; atomic energy. Mailing Add: Goodyear Tire & Rubber Co 1144 E Market St Akron OH 44316

MILONE, EUGENE FRANK, b New York, NY, June 26, 39; m 59; c 2. ASTRONOMY, PHYSICS. Educ: Columbia Univ, AB, 61; Yale Univ, MS, 63, PhD(astron), 67. Prof Exp: From instr to asst prof physics, Gettysburg Col, 66-71; dir, Hatter Planetarium, 66-71; asst prof, 71-75, ASSOC PROF PHYSICS, UNIV CALGARY, 75-; ASTRONR, US NAVAL RES LAB, 67- Concurrent Pos: Lutheran Church Am & Gettysburg Col res & creativity grant, 67-68, Gettysburg Col fac fel, 68-68 & 71-72; Nat Res Coun grant, 71-76. Mem: AAAS; Am Astron Soc; Optical Soc Am; NY Acad Sci; Can Astron Soc. Res: Photoelectric photometry and spectroscopy of variable stars; radio, infrared and x-ray sources; Cepheids; eclipsing binaries; O'Connell effect in variable asymmetric light curves; rocket-ultraviolet solar spectroscopy. Mailing Add: Dept of Physics Univ of Calgary Calgary AB Can

MILONE, NICHOLAS ARTHUR, b Italy, Nov 20, 03; nat US; m 34; c 1. PUBLIC HEALTH. Educ: Cornell Univ, BS, 29, MS, 51. Prof Exp: Instr bact, Cornell Univ, 29-30; state bacteriologist & sanitarian, State Dept Health, NY, 30-49; res bacteriologist, 49-51, res lectr, 51-59, from assoc prof to prof, 60-74, EMER PROF ENVIRON HEALTH, SCH PUB HEALTH, UNIV MICH, ANN ARBOR, 74- Concurrent Pos: Consult, Nat Sanit Found, 51-58; proj dir microbiol study foods, Comn Environ Hyg, Armed Forces Epidemiol Bd, Washington, DC, 59-, assoc mem comn, 62-; consult, Panama CZ Health Bur, 61, WHO, 62 & NY State Dept, 67- Mem: Fel AAAS; Am Soc Microbiol; Inst Food Technologists; fel Am Pub Health Asn; NY Acad Sci. Res: Environmental health; food sanitation and technology; toxic properties of plastic pipe for potable water supplied; bacteriological studies of automatic clothes washing; bacterial toxins; photography; thermometry; microbicides; blood; milk; water; automatic food vending. Mailing Add: 2329 Yorkshire Rd Ann Arbor MI 48104

MILOSOVICH, GEORGE, JR, pharmaceutical chemistry, see 12th edition

MILSTEAD, WAYNE LAVINE, b Washington, DC, Feb 1, 32; m 57. SYSTEMATIC BOTANY. Educ: Univ Md, BS, 58; Purdue Univ, MS, 61, PhD(plant sci), 64. Prof Exp: From asst prof to assoc prof biol, 64-69, head dept biol sci, 68-73, PROF BIOL, EASTERN MONT COL, 69- Concurrent Pos: Vis fel sci educ, Cornell Univ, 71-72. Res: Taxonomy of the family Compositae, particularly the genus Prenanthes. Mailing Add: Dept of Biol Sci Eastern Mont Col Billings MN 59101

MILSTEAD, WILLIAM WRIGHT, vertebrate zoology, deceased

MILSTED, JOHN, b Bristol, Eng, Sept 11, 17; m 42; c 3. NUCLEAR CHEMISTRY. Educ: Bristol Univ, BSc, 38, PhD(chem), 40. Prof Exp: Chemist, Magnesium Metal Corp, Gt Brit, 40-46; sr sci officer UK staff, Chalk River Nuclear Labs, Can, 46-50; prin sci officer, Atomic Energy Res Estab, Eng, 50-59; ASSOC CHEMIST, ARGONNE NAT LAB, 59- Mem: AAAS; Sigma Xi; The Chem Soc. Res: Nuclear chemistry of heavy elements; alpha spectrometry; assessment of chemical and other environmental impacts of nuclear power and the nuclear fuel cycle. Mailing Add: Environ Statement Proj Argonne Nat Lab Argonne IL 60439

MILSTEIN, NORMAN, organic chemistry, see 12th edition

MILSTIEN, JULIE BLOCK, b Waynesboro, Va, Oct 11, 42; m 66; c 2. BIOCHEMISTRY. Educ: Randolph-Macon Woman's Col, AB, 64; Univ Southern Calif, PhD(biochem), 68. Prof Exp: Res biochemist, Viral Leukemia & Lymphoma Br, Nat Cancer Inst, 72; consult, Litton Biomet, Inc, 73; res chemist, 74-75, CHIEF, MOLECULAR BIOL SECT, BUR BIOLOGICS, FOOD & DRUG ADMIN, 75- Concurrent Pos: Staff fel biophys, Lab Phys Biol, Nat Inst Arthritis & Metab Dis, 68-70, sr staff fel, 70-72; spec fel, Nat Cancer Inst, 72-73. Mem: Biophys Soc; Am Soc Microbiol. Res: Effects of exogenous nucleic acids on eukaryotes; detection of tumor viruses. Mailing Add: Bldg 29 Rm 507 8800 Rockville Pike Bethesda MD 20014

MILSTOC, MAYER, b Iasy, Rumania, Dec 14, 20; US citizen; m 45. MEDICINE, PATHOLOGY. Educ: Univ Bucharest, MD, 52; State Univ NY, MD, 65; Am Bd Path, cert anat & clin path, 69. Prof Exp: Asst in res, Inst Res Antibiotics, Bucharest, 51-55, chief lab, 55-57; chief lab clin path, Colentina Hosp, 57-61; asst path, Montefiore-Morrisania Hosp, New York, 66; asst prof path, 67-73, ASSOC PROF CLIN PATH, MED CTR, NY UNIV, 73-; DIR LABS, GOLDWATER MEM HOSP, 69- Concurrent Pos: Asst prof microbiol, Inst Medico Pharmaceut, Sch Med, Bucharest. Mem: Col Am Path; Am Soc Clin Path. Res: Biology of microorganisms, especially antibiotic problems; enzymes, especially cholinesterase in the normal and the diseased. Mailing Add: 370 E 76th St New York NY 10021

MILSTONE, JACOB HASKELL, b St Louis, Mo, June 30, 12; m 42; c 2. BIOCHEMISTRY, PATHOLOGY. Educ: Johns Hopkins Univ, AB, 33, MD, 37. Hon Degrees: AM, Yale Univ, 67. Prof Exp: Asst bact, NY Univ, 37-40; res asst, 48-49, from asst prof to assoc prof, 49-67, PROF PATH, SCH MED, YALE UNIV, 67- Concurrent Pos: Commonwealth fel blood coagulation, Rockefeller Inst, Princeton, NJ, 40-42; Life Ins Med Res fel, Yale Univ, 46-48; pathologist, Griffin Hosp, 49-52; assoc pathologist, Yale-New Haven Hosp, 49-; mem subcomt blood coagulation, Nat Res Coun, 50-54. Mem: AAAS; Am Asn Path & Bact; Soc Exp Biol & Med; Am Soc Exp Path; NY Acad Sci. Res: Biochemistry of blood coagulation and fibrinolysis. Mailing Add: Dept of Path Yale Univ Sch of Med New Haven CT 06510

MILTON, ALBERT FENNER, b New York, NY, Oct 16, 40; m 65. SOLID STATE PHYSICS. Educ: Williams Col, BA, 62; Harvard Univ, MA, 63, PhD(appl physics), 68. Prof Exp: Staff mem, Inst Defense Anal, 68-71; RES PHYSICIST, NAVAL RES LAB, 71- Mem: Am Phys Soc. Res: Integrated optics; quantum optics; photoconductivity; infrared detection; photoemission. Mailing Add: 2939 28th St NW Washington DC 20008

MILTON, CHARLES, b New York, NY, Apr 25, 96; m 33; c 2. GEOLOGY. Educ: Univ Iowa, BA, 23; Johns Hopkins Univ, PhD(geol), 29. Prof Exp: Geologist, Gulf Oil Co, 25-26, Sinclair Explor Co, 27-28 & US Geol Surv, 31-65; adj prof, 65-66, res prof, 66-74, QUONDAM RES PROF GEOL, GEORGE WASHINGTON UNIV, 74-; GEOLOGIST, US GEOL SURV, 70- Concurrent Pos: Consult, US Dept Interior, 67-69; vis res geologist, Univ Calif, Berkeley, 75- Mem: AAAS; fel Geol Soc Am; fel Mineral Soc Am; Soc Econ Geologists; Can Mineral Soc. Res: Petrology; mineralogy; economic geology. Mailing Add: Beechbank Rd Forest Glen Silver Spring MD 20910

MILTON, DANIEL JEREMY, b Washington, DC, July 28, 34; m 65. GEOLOGY. Educ: Harvard Univ, AB, 54, PhD(geol), 61; Calif Inst Technol, MS, 56. Prof Exp: Geologist geochem & petrol br, 54, fuels br, 56 & regional geophys br, 59-61, GEOLOGIST, BR ASTROGEOL, US GEOL SURV, 61- Mem: Geol Soc Am; Mineral Soc Am; Meteoritical Soc. Res: Astrogeology; geology of the moon and terrestrial impact structures; shock metamorphism. Mailing Add: US Geol Surv 345 Middlefield Rd Menlo Park CA 94025

MILTON, JOHN CHARLES DOUGLAS, b Regina, Sask, June 1, 24; m 53; c 4. NUCLEAR PHYSICS. Educ: Univ Man, BSc, 47; Princeton Univ, MA, 49, PhD(physics), 51. Prof Exp: From asst res officer to sr res officer physics, 51-67, HEAD NUCLEAR PHYSICS, CHALK RIVER NUCLEAR LABS, ATOMIC ENERGY CAN, LTD, 67- Concurrent Pos: Vis physicist, Lawrence Radiation Lab, Univ Calif, 60-61; dir res, Ctr Nuclear Res, Strasbourg, 75; vis physicist, Ctr Study, Bruyeres-le-Chatel, France, 75-76. Mem: Royal Soc Can; Am Phys Soc; Can Asn Physicists. Res: Fission physics; directional correlation of radiations in radioactive decay; production of very high thermal neutron fluxes; intermediate energy physics; high voltage electrostatic accelerators. Mailing Add: Box 459 Deep River ON Can

MILTON, KIRBY MITCHELL, b St Joseph, Mich, May 4, 23; m 45, 65; c 4. CHEMISTRY. Educ: Harvard Univ, SB, 43; Univ Mich, MS, 48, PhD(org chem), 51. Prof Exp: Res chemist, Am Cyanamid Co, Conn, 43-44 & Manhattan Dist, 44-46; res & develop chemist, 50-59, RES ASSOC, EASTMAN KODAK CO, 59- Mem: AAAS; Am Chem Soc; Am Inst Chemists; Soc Photog Scientists & Engrs. Res: Photographic emulsions and supports; gelatin; plasticizers; polymers. Mailing Add: 309 Fishers Rd Fishers NY 14453

MILTON, ROBERT MITCHELL, physical chemistry, see 12th edition

MILTON, ROY CHARLES, b St Paul, Minn, Mar 10, 34; m 55; c 2. STATISTICS. Educ: Univ Minn, BA, 55, MA, 63, PhD(statist), 65. Prof Exp: Statistician, Atomic Bomb Casualty Comn, 64-66; sr scientist, Comput Ctr, Univ Wis-Madison, 66-70; statistician, Atomic Bomb Casualty Comn, 70-71; RES MATH STATISTICIAN, NAT EYE INST, 72- Mem: Asn Comput Mach; Inst Math Statist; Am Statist Asn; Biomet Soc. Res: Application of digital computers to applied and theoretical statistical research; biostatistics. Mailing Add: Off of Biomet & Epidemiol Nat Eye Inst Bldg 31/6A18 Bethesda MD 20014

MILVY, PAUL, b New York, NY, Oct 21, 31; m 56; c 2. BIOPHYSICS, RADIATION BIOLOGY. Educ: Cornell Univ, BA, 53, MA, 61, PhD(biophys), 65. Prof Exp: NIH fel, Rockefeller Univ, 65-66; asst prof physics, City Univ New York, 66-70; ASST PROF PATH, MED CTR, NY UNIV, 70- Mem: Fel NY Acad Sci; Radiation Res Soc; Biophys Soc. Res: Electron spin resonance spectroscopy of biological free radicals; radiation biophysics, chemical protection and sensitization of irradiated biomacromolecules; free radicals in biological systems. Mailing Add: Dept of Path NY Univ Med Ctr New York NY 10016

MILZ, WENDELL COLLINS, b Bedford, Ohio, Feb 21, 18; m 42; c 3. PHYSICAL CHEMISTRY. Educ: Hiram Col, BA, 40. Prof Exp: Asst chief chemist, Cleveland Plant, Aluminum Co Am, 40-48, res engr, Alcoa Res Labs, 48-57, asst chief lubricants div, 57-67; SCI ASSOC, ALCOA TECH CTR, 67- Mem: Am Chem Soc; Sigma Xi; Am Soc Testing & Mat. Res: Lubricants; friction; wear; forging. Mailing Add: Alcoa Tech Ctr Alcoa Center PA 15069

MILZER, ALBERT, microbiology, deceased

MIMEAULT, VICTOR JOSEPH, b Kenogami, Que, July 18, 37; m 60; c 2. PLASTICS CHEMISTRY, COLOR SCIENCE. Educ: Loyola Col, Que, BSc, 60; Iowa State Univ, PhD(chem), 65. Prof Exp: Res chemist, Gen Elec Res & Develop Ctr, 65-70; sr res scientist, 70-72, group leader, Tech Ctr, 72-73, TECH MGR, COLOR DIV, FERRO CORP, 73- Mem: AAAS; Am Chem Soc; Soc Plastics Engrs; Inter-Soc Color Coun; NY Acad Sci. Res: Plastics technology; pigments; applied polymer technology. Mailing Add: 1823 N Monitor Ave Chicago IL 60639

MIMMACK, WILLIAM EDWARD, b Eaton, Colo, Aug 22, 26. OPTICS. Educ: Univ Colo, BA, 56; Univ Rochester, PhD(optics), 73. Prof Exp: Res physicist, White Sands Missile Range, Dept of Army, 51-73; STAFF OPTICAL ENGR, KEUFFEL & ESSER CO, 74- Honors & Awards: Karl Fairbanks Award, Soc Photo-Optical Instrumentation Engrs, 61; Army Res & Develop Achievement Award, 68. Mem: Optical Soc Am; Am Phys Soc; Soc Photo-Optical Instrumentation Engrs. Res: Geometrical optics; lens design; lens design methods; optimization. Mailing Add: Keuffel & Esser Co 20 Whippany Rd Morristown NJ 07960

MIMS, CHARLES WAYNE, b Waukegan, Ill, May 3, 44; m 68; c 1. MYCOLOGY. Educ: McNeese State Col, BS, 66; Univ Tex, Austin, PhD(bot), 69. Prof Exp: ASST PROF BIOL, STEPHEN F AUSTIN STATE UNIV, 69- Mem: Bot Soc Am; Mycol Soc Am. Res: Morphogenesis in fungi, ultrastructure. Mailing Add: Dept of Biol Stephen F Austin State Univ Nacogdoches TX 75961

MIMS, SAM STEWART, physical chemistry, see 12th edition

MIN, DAVID BYONG, b Chong Ju, Korea, Sept 12, 42; US citizen; m 69; c 1. FOOD CHEMISTRY. Educ: Seoul Nat Univ, BS, 65; Univ Minn, MS, 69; Rutgers Univ, PhD(food sci), 73. Prof Exp: Res asst food sci, Seoul Nat Univ, 59-65 & Univ Minn, 66-69; res chemist, Healy Labs, St Paul, Minn, 69; Gen Foods Co res assoc food sci, Rutgers Univ, 73; SR FLAVOR CHEMIST, JOHN STUART RES LABS, QUAKER OATS CO, 73- Mem: Am Oil Chemists Soc; Inst Food Technol; Am Soc Mass Spectrometry. Res: Mechanism of autoxidation and thermaloxidation of fats and oils; chemistry of food emulsifiers; chemistry of food flavors, isolation and identification of flavor compounds. Mailing Add: John Stuart Res Labs Quaker Oats Co 617 W Main St Barrington IL 60010

MIN, HONG SHIK, b Seoul, Korea, July 18, 32; m 64; c 3. CELL PHYSIOLOGY, COMPARATIVE PHYSIOLOGY. Educ: WVa Univ, BA, 57, MS, 58; Univ Ga, PhD(physiol), 63. Prof Exp: From asst prof to assoc prof biol, Ga Inst Technol, 63-72; PROF ZOOL, CLEMSON UNIV, 72- Mem: AAAS; Am Soc Zoologists; Am Inst Biol Sci; Am Soc Cell Biologists; NY Acad Sci. Res: Transport of carbohydrate in protozoans; membrane transport system; kinetics of transport; molecular basis for transfer of memory; effects of cotton dust on byssinosis. Mailing Add: Dept of Zool Clemson Univ Clemson SC 29631

MIN, KONGKI, b Seoul, Korea, May 24, 31; m 58; c 1. NUCLEAR PHYSICS. Educ: Amherst Col, BA, 57; Univ Ill, PhD(physics), 63. Prof Exp: Res assoc physics, Univ Va, 62-63, asst prof, 63-68; ASSOC PROF PHYSICS, RENSSELAER POLYTECH INST, 68- Mem: Am Phys Soc. Res: Photonuclear reactions; nuclear spectroscopy. Mailing Add: Dept of Physics Rensselaer Polytech Inst Troy NY 12181

MIN, KWANG-SHIK, b Seoul, Korea, Sept 25, 27; m 56; c 2. MATHEMATICAL PHYSICS, NUCLEAR SCIENCE. Educ: Seoul Nat Univ, BS, 51; Univ Minn, MS, 59, PhD(physics), 61. Prof Exp: Instr physics, Seoul Nat Univ, 55-57; res assoc reactor physics, Argonne Nat Lab, 61-62; asst prof physics, Seoul Nat Univ, 62-64; assoc prof, 64-71, UNIV FAC RES GRANT, E TEX STATE UNIV, 65-, PROF PHYSICS, 71- Mem: Am Phys Soc; Korean Phys Soc. Res: Neutron transport theory; reactor theory; thermal neutron scattering; application of stochastic processes in physics. Mailing Add: 2802 Rix Commerce TX 75428

MINARD, DAVID, b Fargo, NDak, May 23, 13; m 48; c 4. PHYSIOLOGY. Educ: Univ Chicago, BS, 35, PhD(physiol), 37, MD, 43; Harvard Univ, MPH, 54. Prof Exp: Instr physiol, Med Sch, Univ Louisville, 37-38; instr & assoc, Col Med, Univ Ill, 38-43; actg asst surgeon, US Naval Hosp, Bethesda, Md, 43-44; in chg physiol, Naval Med Res Inst, 46-52, in chg bioenergetics div, 54-63, head bur med, Thermal Stress Control Br, 57-63; PROF OCCUP HEALTH & CHMN DEPT, GRAD SCH PUB HEALTH, UNIV PITTSBURGH, 63- Concurrent Pos: Navy rep physiol study sect, Div Res Grants, NIH, 47-52; mem physiol panel, Res & Develop Bd, US Defense Dept, 51-53; mem Nat Acad Sci-Nat Res Coun comt sanit eng & environ, 63-65, chmn comt environ physiol, 65- & mem comt naval med res, 67-71; mem comn environ hyg, Armed Forces Epidemiol Bd, 65-66, dep dir comn, 66-71, dir, 71-; mem adv comt, Nat Inst Environ Health Sci, 67-70 & lab adv bd air warfare, Naval Res Adv Comt, 71- Mem: Am Physiol Soc; Indust Med Asn; NY Acad Sci; fel Am Acad Occup Med (secy, 70-72); fel Am Col Prev Med. Res: Physiology of histamine in blood and tissues; mechanism of pain sensation; circulatory and environmental physiology; human calorimetry. Mailing Add: Dept of Occup Health Univ Pittsburgh Grad Sch Pub Hlth Pittsburgh PA 15213

MINARD, EDWIN LINCOLN, b Fargo, NDak, Sept 7, 09; m 38; c 3. BACTERIOLOGY. Educ: NDak State Col, BS, 31; St Louis Univ, MS, 33; Yale Univ, PhD, 39. Prof Exp: From instr to asst prof animal dis, Univ Conn, 39-45; from sr instr to asst prof bact, 52-58, ASSOC PROF MICROBIOL, ST LOUIS UNIV, 58- Mem: Am Soc Microbiol. Res: Staphylococcal toxins; avian encephalomyelitis; Newcastle disease; anaphylaxis. Mailing Add: Dept of Microbiol St Louis Univ St Louis MO 63103

MINARD, FREDERICK NELSON, b Iowa, July 2, 22; m 44; c 2. BIOCHEMISTRY. Educ: Iowa State Col, BS, 43, PhD(biochem), 49. Prof Exp: Res chemist, 49-59, res assoc, 59-65, ASSOC RES FEL, ABBOTT LABS, 65- Mem: Am Chem Soc; Am Soc Biol Chemists. Res: Neurochemistry. Mailing Add: Abbott Labs North Chicago IL 60064

MINARD, ROBERT DAVID, b Buffalo, NY, Mar 21, 41; m 66; c 2. ORGANIC CHEMISTRY. Educ: St Olaf Col, BA, 63; Univ Wis, PhD, 68. Prof Exp: Asst prof chem, Col of the Virgin Islands, 68-69; fel, Univ Ill, Chicago Circle, 69-70, asst prof, 70-73; LECTR CHEM, PA STATE UNIV, 73- Mem: AAAS; Am Chem Soc. Res: Polyamino acids; reactions of hydrogen cyanide; chemical evolution; origin of proteins; mass spectrometry; pesticide degradation. Mailing Add: Dept of Chem Pa State Univ University Park PA 16802

MINARIK, CHARLES EDWIN, b Westfield, Mass, Nov 20, 11; m 35; c 2. PLANT PHYSIOLOGY. Educ: Univ Mass, BS, 33, MS, 35; Rutgers Univ, PhD(plant physiol), 39. Prof Exp: Asst plant physiol, Rutgers Univ, 35-39; plant physiologist, Exp Sta, Agr & Mech Col Tex, 39-46; chief, Chem Br, Chem Corps, US Army, 46-52, crops div, 52-67, dir plant sci lab, Ft Detrick, 67-71, chief vegetation control div, 71-72; RETIRED. Concurrent Pos: Ed, Weeds, Weed Sci Soc Am, 53-54. Honors & Awards: Amos A Fries Gold Medal, Am Ord Asn, 67; Meritorious Civilian Serv Award, US Dept Army, 72. Mem: AAAS; Bot Soc Am; Am Soc Plant Physiol; Weed Sci Soc Am; Am Inst Biol Sci. Res: Plant growth regulators; weed control; radioactive tracers in plant physiology; nutrition. Mailing Add: PO Box 682 West Harwich MA 02671

MINASSIAN, DONALD PAUL, b New York, NY, Dec 8, 35; m 64; c 2. MATHEMATICS. Educ: Fresno State Col, BA, 57; Brown Univ, MA, 64; Univ Mich, Ann Arbor, MS, 65, EdD(math), 67. Prof Exp: Assoc prof, 67-73, PROF MATH, BUTLER UNIV, 73- Mem: AAAS; Am Math Soc; Math Asn Am; Soc Indust & Appl Math; Asn Study Grants Econ. Res: Ordered algebraic structures, particularly ordered groups of modern algebra; foreign aid; industrial concentration. Mailing Add: 410 Blue Ridge Rd Indianapolis IN 46208

MINATOYA, HIROAKI, b Japan, Nov 8, 11; US citizen; m 45; c 2. PHARMACOLOGY, PHYSIOLOGY. Educ: Univ Ore, BA, 38; Univ Ill, MS, 42; Nara Med Col, Japan, PhD(pharmacol), 64. Prof Exp: Res asst physiol, Col Med, Univ Ill, 41-42 & 44-48; res pharmacologist, 48-71, SR RES PHARMACOLOGIST, STERLING-WINTHROP RES INST, 71- Mem: Am Soc Pharmacol & Exp Therapeut. Res: Conditioned reflex and insulin hypoglycemia; experimental hypertension in dogs; diuretics; antihypertensive compounds in the renal hypertensive rat; catecholamines, especially absorption, metabolism and elimination; bronchodilators. Mailing Add: 10 Van Buren Ave East Greenbush NY 12061

MINC, HENRYK, b Lodz, Poland, Nov 12, 19; m 43; c 3. ALGEBRA. Educ: Univ Edinburgh, MA, 55, PhD(math), 59. Prof Exp: Lectr math, Univ Dundee, Scotland, 56-58; lectr, Univ BC, 58-59, asst prof, 59-60; assoc prof, Univ Fla, 60-63; PROF MATH, UNIV CALIF, SANTA BARBARA, 63- Concurrent Pos: Vis prof, Israel Inst Technol, 69- Honors & Awards: Ford Award, Math Asn Am, 66. Mem: Am Math Soc; Math Asn Am; Edinburgh Math Soc; Israel Math Union. Res: Linear and multilinear algebra; matrix theory; combinatorial analysis. Mailing Add: Dept of Math Univ of Calif Santa Barbara CA 93106

MINCH, MICHAEL JOSEPH, b Klamath Falls, Ore, Apr 7, 43. PHYSICAL ORGANIC CHEMISTRY. Educ: Ore State Univ, BS, 65; Univ Wash, PhD(chem), 70. Prof Exp: NIH fel chem, Univ Calif, Santa Barbara, 70-72; asst prof chem, Tulane Univ, 72-74; ASST PROF CHEM, UNIV OF THE PAC, 74- Mem: Am Chem Soc. Res: Synthesis and characterization of surfactants; micellar catalysis; models of biochemically significant complexes. Mailing Add: Dept of Chem Univ of the Pac Stockton CA 95211

MINCKLER, JEFF, b Knox, NDak, June 4, 12; m 33; c 6. PATHOLOGY. Educ: Univ Mont, AB, 37; Univ Minn, MA & PhD(neuroanat), 39; St Louis Univ, MD, 44. Prof Exp: Asst anat, Univ Minn, 37-39; from instr to asst prof, Creighton Univ, 39-41; instr, St Louis Univ, 41-43, instr path, 43-45; asst prof, Med Sch, Univ Ore, 45-46, prof gen path & actg head dept, Sch Dent, 49-59; assoc clin prof path, Med Sch, Univ Colo, 60-71; dir labs, Century City Hosp, 73-75; ASSOC RES PROF NEUROSURG, MED CTR, LOMA LINDA UNIV, 71-; DIR LABS, MAD RIVER COMMUNITY HOSP, ARCATA, CALIF, 75- Concurrent Pos: Dir labs, Gen Rose Mem Hosp, 60-71 & Eisenhower Med Ctr, 71-73; lectr radiol, Univ Calif, Los Angeles; adj prof, Univ Denver, 65-71. Mem: AAAS; Am Asn Anat; Am Asn Neuropath; Am Soc Clin Path; Int Acad Path. Res: General and speech pathology; neuropathology; neuroanatomy. Mailing Add: Labs 3800 Janes Rd Mad River Community Hosp Arcata CA 95521

MINCKLER, LEON SHERWOOD, b New Milford, NY, May 7, 06; m 29, 47; c 5. FORESTRY, ECOLOGY. Educ: Syracuse Univ, BS, 28, PhD(plant physiol), 36. Prof Exp: Jr forester, Lake States Forest Exp Sta, US Forest Serv, 35-36, assoc silviculturist, Appalachian Forest Exp Sta, 36-45, silviculturist, Cent States Forest Exp Sta, 45-54, res forester, 54-65, NCent Forest Exp Sta, 65-68; prof forestry & wildlife, Va Polytech Inst, 68-70; ADJ PROF FORESTRY, STATE UNIV NY COL ENVIRON SCI & FORESTRY, 70- Concurrent Pos: Lectr, Univ Mich, 57-58; lectr, Southern Ill Univ, 58-59, adj prof, 64-; consult environ forestry, 71- Honors & Awards: Cert of Merit, US Dept Agr, 62. Mem: Fel AAAS; Soc Am Foresters; Ecol Soc Am. Res: Silviculture of hardwood forests; silvics of individual species; forest ecology and regeneration; yield and quality of forest stands; conservation and forest management. Mailing Add: State Univ of NY Col of Environ Sci & Forestry Syracuse NY 13210

MINCKLER, LEON SHERWOOD, JR, b Lockport, NY, Apr 4, 30; m 51; c 2. ORGANIC CHEMISTRY. Educ: Univ Southern Ill, BA, 51; Northwestern Univ, PhD(org chem), 55. Prof Exp: Asst org chem, Northwestern Univ, 51-53; sr chemist, Esso Res & Eng Co, 55-67, res assoc, 67-74, RES ASSOC, EXXON CHEM CO, 74- Mem: Am Chem Soc; Am Inst Chemists. Res: Polymers; butyl rubber; cationic polymerization; stereochemistry. Mailing Add: 78 Crestwood Dr Watchung NJ 07060

MINCKLER, TATE MULDOWN, b Kalispell, Mont, Apr 1, 34; m 56; c 5. PATHOLOGY. Educ: Reed Col, BA, 55; Univ Ore, MD, 59. Prof Exp: Pathologist, Nat Cancer Inst, Washington, DC, 63-65, head tissue path unit, 64-65; asst pathologist & med systs analyst, Univ Tex M D Anderson Hosp & Tumor Inst, Houston, 65-69, chief sect med info mgt systs, Dept Biomath, 67-69; head dept med automation, Presby Med Ctr, 69-71; assoc prof lab med & dir comput div, Sch Med, Univ Wash, 71-75; ASSOC PATHOLOGIST & ADMINR, MAD RIVER COMMUNITY HOSP, ARCATA, CALIF, 75- Mem: AAAS; Soc Cryobiol; Am Soc Clin Path; fel Col Am Path; Soc Comput Med. Res: Development of medical information systems, especially pathology data automation, application of computers to medical problems; utilization of human tissues for research and patient care; clinical and anatomical pathology; cryobiology; medical records. Mailing Add: Mad River Community Hosp 3800 Janes Rd Arcata CA 95521

MINCKLEY, WENDELL LEE, b Ottawa, Kans, Nov 13, 35; m 56; c 4.

MINCKLEY

ICHTHYOLOGY, AQUATIC ECOLOGY. Educ: Kans State Univ, BS, 57; Univ Kans, MA, 59; Univ Louisville, PhD, 62. Prof Exp: Asst prof zool, Western Mich Univ, 62-63; asst prof, 63-70, ASSOC PROF ZOOL, ARIZ STATE UNIV, 70-, DIR LOWER COLO RIVER BASIN RES LAB, 72- Concurrent Pos: Fac res grant, Western Mich, 62-63; NSF grant, 63-65; Fac Res Comt awards, Ariz State Univ, 63-65; Sport Fishery Inst res grant, 65-66. Mem: Am Soc Ichthyologists & Herpetologists; Am Fisheries Soc; Wildlife Soc; Am Soc Limnol & Oceanog; Am Inst Biol Sci. Res: Systematic and ecological ichthyology; radiation, stream, crustacean and algal ecology; crustacean taxonomy. Mailing Add: Dept of Zool Ariz State Univ Tempe AZ 85281

MINDA, CARL DAVID, b Cincinnati, Ohio, Nov 5, 43; m 73. MATHEMATICAL ANALYSIS. Educ: Univ Cincinnati, BS, 65, MS, 66; Univ Calif, San Diego, PhD(math), 70. Prof Exp: Asst prof math, Univ Minn, 70-71; asst prof, 71-75, ASSOC PROF MATH, UNIV CINCINNATI, 75- Mem: Am Math Soc; Math Asn Am. Res: Complex analysis; Riemann surfaces. Mailing Add: Dept of Math Sci Univ of Cincinnati Cincinnati OH 45221

MINDE, KARL KLAUS, b Leipzig, Ger, Dec 27, 33; Can citizen; c 3. PSYCHIATRY. Educ: Columbia Univ, MA, 60; Munich Univ, MD, 57. Prof Exp: Staff physician, Queen Elizabeth Hosp, 65-71; WHO sr lectr psychiat, Makerere Univ, Uganda, 71-73; staff physician, 73, DIR PSYCHIAT RES, HOSP SICK CHILDREN, 73-; ASSOC PROF PSYCHIAT, UNIV TORONTO, 73- Concurrent Pos: Staff physician, Montreal Childrens Hosp, 65-73; asst prof psychiat, McGill Univ, 66-73; consult psychiat, Baird Residential Treatment Ctr, Burlington, Vt, 65-67. Mem: Can Psychiat Asn; Am Psychiat Asn; Am Acad Child Psychiat; Can Med Asn. Res: Bonding between premature babies and their mothers, and a follow-up study of kindergarten children to determine whether the screening instrument was effective in picking up children at risk for behavior disorders. Mailing Add: Hosp for Sick Children 555 University Ave Toronto ON Can

MINDEL, JOSEPH, b New York, NY, May 3, 12; m 34, 75; c 1. HISTORY OF SCIENCE, SCIENCE EDUCATION. Educ: City Col New York, BS, 32; Columbia Univ, MA, 37; NY Univ, PhD(chem), 43. Prof Exp: Instr high sch, NY, 32-46, head dept sci, 46-60; secy to pres, Bd Ed, NY, 60-61; staff mem, 61-69, Secy steering comt, 69-75, educ dir, Lincoln Lab, Mass Inst Technol, 71-75, CONSULT, LINCOLN LAB, MASS INST TECHNOL, 75-; CONSULT, BETH ISRAEL HOSP, BOSTON, 75- Mem: AAAS; NY Acad Sci. Mailing Add: 18624 Walkers Choice Rd Gaithersburg MD 20760

MINDELL, EUGENE R, b Chicago, Ill, Feb 24, 22; m 45; c 4. ORTHOPEDIC SURGERY. Educ: Univ Chicago, BS, 43, MD, 45. Prof Exp: PROF ORTHOP SURG & HEAD DEPT, SCH MED, STATE UNIV NY BUFFALO, 64- Concurrent Pos: Nat Res Coun fel orthop surg, Univ Chicago Clin, 48-49; Orthop Res & Educ Found res grant, 58-61; NIH res grant, 63-; examr bone path, Am Bd Orthop Surg, 57-. Mem: Am Orthop Asn; Am Soc Surg Trauma; Orthop Res Soc (pres elect, 71); Am Acad Orthop Surg; fel Am Col Surg. Res: Mechanisms by which chondrogenesis occurs; bone pathology; structure and function of cartilage; chemotherapy in bone sarcoma. Mailing Add: Dept of Orthop Surg State Univ of NY Sch of Med Buffalo NY 14214

MINDEN, HENRY THOMAS, b New York, NY, Aug 23, 23; m 53; c 3. CHEMICAL PHYSICS. Educ: Johns Hopkins Univ, BA, 43; Columbia Univ, PhD(chem), 52. Prof Exp: Mem tech staff, Bell Tel Labs, Inc, 48 & RCA Labs, 50-52; sr physicist, Midway Labs, Chicago, 52-56; eng specialist, Res Labs, Sylvania Elec Prod, Inc, 56-60; physicist, Semiconductor Prod Dept, Gen Elec Co, 60-62; TECH STAFF MEM, SPERRY RAND RES CTR, 62- Mem: Am Phys Soc; Sigma Xi. Res: Semiconductor materials and devices. Mailing Add: Sperry Rand Res Ctr Sudbury MA 01776

MINDICH, LEONARD EUGENE, b New York, NY, May 24, 36; m 59; c 3. MICROBIAL PHYSIOLOGY. Educ: Cornell Univ, BS, 57; Rockefeller Univ, PhD, 62. Prof Exp: Asst virol, 62-64, assoc, 64-70, ASSOC MEM MICROBIOL, PUB HEALTH RES INST OF CITY OF NEW YORK, INC, 70- Concurrent Pos: Mem microbial chem study sect, NIH, 72-76. Mem: AAAS; Am Soc Microbiol. Res: Genetics of bacteriocin production, fractionation of desoxyribonucleic acid and the synthesis of bacterial membranes. Mailing Add: Public Health Res Inst 455 First Ave New York NY 10016

MINDLIN, ROWLAND L, b New York, NY, Jan 30, 12; m 40; c 2. PEDIATRICS, COMMUNITY HEALTH. Educ: Harvard Univ, BS, 33, MD, 37, MPH, 62. Prof Exp: Dir maternal & child health, Boston Dept Health & Hosps, 71-74; DIR AMBULATORY CARE, ST MARY'S HOSP, BROOKLYN, 74- Concurrent Pos: Lectr, Harvard Sch Pub Health; chmn comt infant & presch child, Am Acad Pediat; mem comt med in Soc, NY Acad Med. Mem: Am Acad Pediat; NY Acad Med; Am Pub Health Asn. Res: Medical care in maternal and child health. Mailing Add: St Mary's Hosp 1298 St Mark's Ave Brooklyn NY 11213

MINEAR, JOHN W, b Odessa, Tex, Dec 30, 37; m 64; c 2. GEOPHYSICS. Educ: Rice Univ, BA, 60, PhD(geophys), 64. Prof Exp: Res geophysicist, Res Triangle Inst, 65-71, mgr geosci dept, 71-74; PLANETARY SCIENTIST, NASA-JOHNSON SPACE CTR, 74- Concurrent Pos: Adj prof, Univ NC, 65-74; fel, Dept Earth & Planetary Sci, Mass Inst Technol, 68-69. Mem: Seismol Soc Am; Am Geophys Union. Res: Thermal evolution of the terrestrial planets and the moon; physical environment of crustal evolution, magma migration and chemical differentiation of planetary objects; comparative planetology of solid solar system objects. Mailing Add: Johnson Space Ctr TN6 Houston TX 77058

MINEHART, RALPH CONRAD, b Mitchell, SDak, Jan 25, 35; m 59; c 3. EXPERIMENTAL NUCLEAR PHYSICS. Educ: Yale Univ, BS, 56; Harvard Univ, MA, 57, PhD(physics), 62. Prof Exp: Res assoc & lectr physics, Yale Univ, 62-66; asst prof, 66-68, ASSOC PROF, UNIV VA, 68- Concurrent Pos: Vis staff mem, Swiss Inst Nuclear Res, 75-76. Mem: Am Phys Soc. Res: High energy experimental physics; polarized nucleon-nucleon scattering with bubble chamber analysis; high energy photoproduction of charged particles; medium energy nuclear physics; interactions of pi mesons with nucleons and nuclei. Mailing Add: Dept of Physics Univ of Va Charlottesville VA 22901

MINER, BRYANT ALBERT, b Moroni, Utah, Aug 9, 34; m 60; c 6. PHYSICAL CHEMISTRY. Educ: Univ Utah, BA, 61, PhD(phys chem), 65. Prof Exp: From asst prof to assoc prof, 64-73, PROF CHEM, WEBER STATE COL, 73- Mem: Am Chem Soc. Res: Significant structure theory of liquids and electrochemistry; reactions and kinetics. Mailing Add: 4260 Jefferson Ave Ogden UT 84403

MINER, CARL SHELLEY, JR, b Evanston, Ill, May 7, 15; m 41; c 2. ORGANIC CHEMISTRY. Educ: Northwestern Univ, BS, 37; Pa State Univ, PhD, 40. Prof Exp: Res chemist, Miner Labs, 40-47, assoc dir, 47-51, exec dir, 51-56; tech dir, Midwest Div, Arthur D Little, Inc, Ill, 56-64; CHEM CONSULT, 64- Mem: Am Chem Soc; fel Am Inst Chem. Res: Research administration; product development; patents. Mailing Add: 717 Rosewood Ave Winnetka IL 60093

MINER, FLOYD DUANE, b Ft Wayne, Ind, June 25, 15; m 38; c 3. ENTOMOLOGY. Educ: Okla State Univ, BS, 40, MS, 41; Kans State Univ, PhD, 52. Prof Exp: Agent, US Dept Agr, 38; asst entom, Okla State Univ, 39-41 & Cornell Univ 41-42; from instr to assoc prof, 42-69, PROF ENTOM, UNIV ARK, FAYETTEVILLE, 69-, HEAD DEPT, 69- Concurrent Pos: Agency Int Develop adv to Minister Agr, Nicaragua, 54-56. Mem: Entom Soc Am. Res: Biology of insects; control of insects of forage crops. Mailing Add: Dept of Entom Univ of Ark Fayetteville AR 72701

MINER, FREND JOHN, b Loveland, Colo, Dec 8, 28; m 58; c 2. INORGANIC CHEMISTRY, SOIL CHEMISTRY. Educ: Univ Colo, BA, 50; Ore State Univ, MS, 52, PhD(chem), 55. Prof Exp: Asst chem, Ore State Univ, 50-54; from prof chemist to res chemist, 54-67, sr res chemist, 67-68, ASSOC SCIENTIST, ROCKY FLATS PLANT, ROCKWELL INT, 68- Mem: Am Chem Soc; Am Soc Test & Mat; Sigma Xi. Res: Ion exchange separations; complex ions; chemistry of plutonium. Mailing Add: 2445 Dartmouth Ave Boulder CO 80303

MINER, GEORGE KENNETH, b Asheville, NC, Dec 16, 36; m 62; c 3. MAGNETIC RESONANCE. Educ: Thomas More Col, AB, 58; Univ Notre Dame, MS, 60; Univ Cincinnati, PhD(physics), 65. Prof Exp: From instr to assoc prof, 64-74, PROF PHYSICS, THOMAS MORE COL, 74-, CHMN DEPT, 66- Concurrent Pos: NSF sci fac fel, Univ Dayton, 71. Mem: Am Phys Soc; Am Asn Physics Teachers. Res: Radiation damage; electron paramagnetic resonance of rare earths in fluorites. Mailing Add: Dept of Physics Thomas More Col Box 85 Covington KY 41017

MINER, GORDON STANLEY, b Howell, Mich, Apr 25, 40; m 64; c 2. SOIL FERTILITY, AGRONOMY. Educ: Mich State Univ, BS, 62, MS, 64; NC State Univ, PhD(soil sci), 69. Prof Exp: Res agronomist, Rockefeller Found, 69-71; regional dir, Int Soil Fertil Eval & Improv Proj, Costa Rica & Nicaragua, 71-74; ASST PROF SOIL SCI, NC STATE UNIV, 74- Mem: Sigma Xi; Am Soc Agron. Res: Plant bed and field management of tobacco for mechanized production. Mailing Add: Dept of Soil Sci NC State Univ Raleigh NC 27607

MINER, JAMES JOSHUA, b Waldo, Ark, June 11, 28; m 57; c 4. NUTRITION, BIOCHEMISTRY. Educ: Univ Ark, Fayetteville, BS, 53, MS, 54; La State Univ, Baton Rouge, PhD(nutrit), 62. Prof Exp: Poultry husbandman, Poultry Res Br, Agr Res Serv, USDA, 56-58; dir res, Ala Flour Mills, Nebr, 60-62; nutritionist, Loret Mills, Seed Feed Supply Co, 62-65; exten serv poultry man, Univ Ark, 65-66; VPRES FARM PROD, PILGRIM INDUSTS, INC, 66- Mem: Am Soc Animal Sci; Poultry Sci Asn. Res: Biological evaluation of protein supplements for non-ruminants. Mailing Add: Pilgrim Industs Inc 110 S Texas St Pittsburgh TX 75686

MINER, MERTHYR LEILANI, b Honolulu, Hawaii, Apr 13, 12; m 37; c 4. VETERINARY PATHOLOGY. Educ: Utah State Univ, BS, 37; Iowa State Univ, DVM, 41. Prof Exp: Asst vet bact, Mich State Univ, 41-43; from asst prof to assoc prof, 43-54, head dept, 54-73, PROF VET SCI, UTAH STATE UNIV, 54- Concurrent Pos: Fel, Ralston Purina Co, Univ Minn, 53-54. Mem: Am Vet Med Asn; US Animal Health Asn; Am Asn Avian Path. Res: Microbiology; staphylococci; salmonellae; avian skeletal diseases; bovine coccidia. Mailing Add: Dept of Vet Sci Utah State Univ Logan UT 84322

MINER, NORMAN ALLEN, b Ft Wayne, Ind, June 24, 37; m 59; c 5. MOLECULAR BIOLOGY, VIROLOGY. Educ: Carleton Col, BA, 59; Purdue Univ, MS, 65, PhD(molecular biol), 67. Prof Exp: Instr biol, Culver Mil Acad, 59-61; ed, Lea & Febiger Publ Co, 61-62; antiviral proj leader, Human Health Res Ctr, Dow Chem USA, 67-71; MGR BIOL SCI, ARBROOK, INC, JOHNSON & JOHNSON, 71- Concurrent Pos: Vis assoc prof, Purdue Univ, 69-71. Mem: AAAS; Am Soc Microbiol; Soc Indust Microbiol. Res: Animal virology; interferon; microbiology; chemo-sterilants and disinfectants. Mailing Add: Dept of Biol Sci Arbrook Inc 2500 Arbrook Dr Arlington TX 76010

MINER, ROBERT SCOTT, JR, b Chicago, Ill, June 16, 18; m 42; c 3. ORGANIC CHEMISTRY. Educ: Univ Chicago, BS, 40; Polytech Inst Brooklyn, MS, 53; Princeton Univ, MA, 55, PhD(org chem), 56. Prof Exp: Develop chemist, Merck & Co, Inc, NJ, 40-44; chief res chemist, Tung-Sol Lamp Works, Inc, 45-47; asst mfg chemist, Ciba Pharmaceut Co, 47-49, mfg chemist, 49-58, mgr chem mfg div, 58-59, dir, 59-69; asst to chmn dept physics, 70-74, res assoc fac chem, 74-75, MEM PROF RES STAFF, DEPT OF CHEM, PRINCETON UNIV, 75- Concurrent Pos: Fel & eve instr, Union Jr Col, 56-58, guest lectr, 60; guest lectr, Westfield High Sch, 60-; mem, Westfield Bd Educ, 62-65, pres, 64-65; guest lectr, MacMurray Col, 63, trustee, 65-70; chem consult, Haemodialysis Team, Overlook Hosp, 64-; consult chemist to industs, 70- Honors & Awards: Honor Scroll Award, Am Inst Chem, 71. Mem: AAAS; Am Chem Soc; fel Am Inst Chem (treas, 70-75). Res: Synthesis of carbohydrates; vitamins; sulfonamides; steroid hormones; pharmaceuticals; chemical education; catalysis and electrode phenomena involving precious metals. Mailing Add: 1139 Lawrence Ave Westfield NJ 07090

MINERBO, GERALD N, b Alexandria, UAR, Nov 21, 39; US citizen; m 66; c 1. THEORETICAL PHYSICS. Educ: Polytech Inst Brooklyn, 60; Cambridge Univ, PhD(theoret physics), 65. Prof Exp: Res assoc theoret physics, Atomic Energy Res Estab, Harwell, Eng, 64-65; proj physicist, Vitro Lab, Vitro Corp Am, 65-66; asst prof physics, Adelphi Univ, 66-71; STAFF MEM, LOS ALAMOS SCI LAB, 71- Concurrent Pos: Consult, Vitro Corp Am, 66-68; vis prof, Univ Buenos Aires, 68-69. Mem: Am Phys Soc. Res: Theory of elementary particles; formal theory of scattering; atomic physics; radiation transfer; laser theory. Mailing Add: Los Alamos Sci Lab PO Box 1663 Los Alamos NM 87545

MINES, ALLAN HOWARD, b New York, NY, Apr 11, 36; m 59; c 2. PHYSIOLOGY. Educ: Univ Ill, Urbana, BS, 61, MS, 62; Univ Calif, San Francisco, PhD(physiol), 68. Prof Exp: From lectr to asst prof, 68-74, ASSOC PROF PHYSIOL, MED CTR, UNIV CALIF, SAN FRANCISCO, 74- Mem: Am Physiol Soc. Res: Comparative physiology; regulation of respiration, particularly the mechanism responsible for ventilatory acclimatization to altitude. Mailing Add: Dept of Physiol Univ of Calif Med Ctr San Francisco CA 94143

MINES, MATTISON, b Seattle, Wash, Jan 18, 41; m 62; c 3. ANTHROPOLOGY. Educ: Univ Wash, BA, 63, MA, 65; Cornell Univ, PhD(anthrop), 70. Prof Exp: Asst prof anthrop, Hamilton Col, 69-70; asst prof, 70-75, ASSOC PROF ANTHROP, UNIV CALIF, SANTA BARBARA, 75- Mem: Am Anthrop Asn. Res: South Asia; social anthropology; urbanization; Islam. Mailing Add: Dept of Anthrop Univ of Calif Santa Barbara CA 93106

MINESINGER, RICHARD ROCKWELL, b Takoma Park, Md, Nov 25, 39; m 59; c 2. ORGANIC CHEMISTRY. Educ: Columbia Union Col, BA, 61; Univ Md, College Park, PhD(org chem), 66. Prof Exp: Res chemist, E I du Pont de Nemours & Co, Inc,

66-67; sr res assoc chem, US Naval Ord Lab, 67-71; assoc prof, 71-75, PROF CHEM, ANDREWS UNIV, 75- Concurrent Pos: Consult, Lawrence Livermore Labs, 73-; cong sci counr, American Chem Soc appt for Rep E Hutchinson, 74- Mem: AAAS; Am Chem Soc. Res: Physical organic chemistry, specifically photochemistry, spectra-structure relationships, and hydrogen-bonding. Mailing Add: Dept of Chem Andrews Univ Berrien Springs MI 49104

MINETT, ERNEST EVERET, physical chemistry, see 12th edition

MINFORD, JAMES DEAN, b Clairton, Pa, Feb 27, 23; m 44; c 2. METALLURGICAL CHEMISTRY. Educ: Carnegie Inst Technol, BS, 43; Univ Pittsburgh, MLitt, 48, PhD(chem), 51. Prof Exp: Res chemist org chem, Goodyear Tire & Rubber Co, 44; asst chem, Univ Pittsburgh, 46-50, res assoc biochem, Sch Pub Health, 50-52, instr biochem & nutrit, 52-53; sect head, 53-72, GROUP LEADER CHEM METALL DIV, ALCOA LABS, 72- Mem: AAAS; Am Chem Soc; Sigma Xi. Res: Porphyrin synthesis in microorganisms; cancer in experimental animals; nutritional factors in disease; corrosive action of cooling waters on metals; reactions of aluminum and halogenated hydrocarbons; adhesives; high polymers; surface cleaning aluminum. Mailing Add: Chem Metall Div Alcoa Labs Alcoa Ctr PA 15069

MING, SI-CHUN, b Shanghai, China, Nov 10, 22; US citizen; m 57; c 6. PATHOLOGY. Educ: Nat Cent Univ, China, MD, 47. Prof Exp: Resident path, Mass Gen Hosp, Boston, 52-56; from instr to asst prof, Harvard Med Sch, 56-67; assoc prof, Med Sch, Univ Md, 67-71; PROF PATH, MED SCH, TEMPLE UNIV, 71- Concurrent Pos: Assoc pathologist, Beth Israel Hosp, Boston, 56-67; Nat Cancer Inst sr fel, Dept Tumor Biol, Karolinska Inst, Sweden, 64-65. Mem: Int Acad Path; Am Asn Pathologists & Bacteriologists; Am Soc Exp Path; NY Acad Sci; AAAS. Res: Digestive tract disease, gastrointestinal oncology, carcinogenesis and tumor-host relationship. Mailing Add: Dept of Path Temple Univ Med Sch Philadelphia PA 19110

MING, TAO K, chemistry, computer science, see 12th edition

MINGES, PHILIP ADAMS, b Battle Creek, Mich, Jan 10, 13; m 38; c 2. VEGETABLE CROPS. Educ: Mich State Col, BS, 34; Iowa State Univ, MS, 37, PhD(hort), 41. Prof Exp: Asst pomol, Iowa State Univ, 35-37, asst veg crops, 38-40; instr truck crops & jr olericulturist, Col Agr, Univ Calif, 41-42, exten veg crops specialist, 42-55; PROF VEG CROPS & EXTEN PROF, NY STATE COL AGR & LIFE SCI, CORNELL UNIV, 55- Mem: AAAS; Am Soc Hort Sci; Int Soc Hort Sci. Res: Variety testing and adaptation; soils and plant physiology dealing with vegetables; growth regulating substances; cultural practices of vegetables; extension programs and procedures. Mailing Add: NY State Col Agr & Life Sci Cornell Univ Ithaca NY 14850

MINGHI, JULIAN VINCENT, b London, Eng, July 26, 33; m 60; c 1. POLITICAL GEOGRAPHY, GEOGRAPHY OF EUROPE. Educ: Durham Univ, BA, 57; Univ Wash, MA, 59, PhD(geog), 62. Prof Exp: Asst prof geog, Univ Conn, 61-64; from asst prof to assoc prof, Univ BC, 64-73; PROF GEOG & HEAD DEPT, UNIV OF SC, 73- Concurrent Pos: Vis asst prof geog, Univ RI, 64; Can Coun fel impact of boundary changes, Europe, 66; Geog Br fel US-Can boundary, Univ BC, 66-67; consult ed high sch geog proj, Asn Am Geog, 66-69; mem fac, NDEA Inst Polit Geog, Clark Univ, 67; vis lectr geog, Univ Newcastle, 69; Brit Coun vis award, Tour of UK Univs, 69; Can Coun sr leave fel polit geog S Tyrol, 69-70. Mem: Asn Am Geog; Am Geog Soc; Nat Coun Geog Educ; Can Asn Geog. Res: Problems of frontier communities; competition for sea-space; urban political-territorial systems; changing boundaries and border functions in Alpine Europe; cultural conflict in shared space. Mailing Add: Dept of Geog Univ of SC Columbia SC 29208

MINGRONE, LOUIS V, b Pittsburgh, Pa, Jan 27, 40; m 64; c 3. BOTANY. Educ: Slippery Rock State Col, BS, 62; Ohio Univ, MS, 64; Wash State Univ, PhD(bot, taxon), 68. Prof Exp: Assoc prof, 68-73, PROF BIOL & BOT, BLOOMSBURG STATE COL, 73- Mem: Bot Soc Am; Am Soc Plant Taxon; Am Inst Biol Sci. Res: Plant systematics; cytotaxonomy and phytochemistry of the secondary plant constituents; floristic studies. Mailing Add: Dept of Biol Bloomsburg State Col Bloomsburg PA 17815

MINGS, ROBERT CHARLES, b Chicago, Ill, Feb 25, 35; m 61; c 3. ECONOMIC GEOGRAPHY. Educ: Ind Univ, BS, 57, MAT, 58; Ohio State Univ, PhD(geog), 66. Prof Exp: Asst prof geog, Univ Miami, 66-71; ASSOC PROF GEOG, ARIZ STATE UNIV, 73- Concurrent Pos: Ariz State Univ fac grant-in-aid, Tonto Nat Forest, Ariz, 71-72, Eisenhower Consortium grant, 72-73. Mem: AAAS; Asn Am Geog; Conf Latin Americanist Geog; Latin Am Studies Asn; Nat Coun Geog Educ. Res: Spatial analysis of solid waste problems relating to outdoor recreation areas; development of tourism in developing areas; tourist industry in Latin America; improving solid waste collection in recreation areas of the Tonto National Forest, Arizona. Mailing Add: Dept of Geog Ariz State Univ Tempe AZ 85281

MINHAS, KAREEM, b Nairobi, Kenya, Oct 24, 18; US citizen. CARDIOLOGY, PEDIATRIC CARDIOLOGY. Educ: Univ Panjab, WPakistan, MB & BS, 46, MD, 52. Prof Exp: House physician, Mayo Hosp & King Edward Med Col, Lahore, WPakistan, 46-52; asst internal med & cardiol, Hammersmith Hosp & Inst Cardiol, London, Eng, 52-53; chief resident internal med, Faulkner Hosp, Boston, 54-55; chief resident rheumatic heart dis & fever, House Good Samaritan Children's Hosp, 55-56; from asst prof to assoc prof pediat cardiol, Res & Educ Hosp, Univ Ill, 57-64; assoc prof, 64-70, PROF PEDIAT CARDIOL, SCH MED, UNIV LOUISVILLE, 70- Concurrent Pos: Res scholar, Mayo Hosp & King Edward Med Col, Lahore, WPakistan, 46-52; Fulbright scholar cardiovasc dis, Mass Gen Hosp & Harvard Med Sch, 53-54; fel pediat cardiol, Harvard Med Sch, 55-56; res fel, Cook County Children's Hosp, Chicago & Hektoen Inst Med Res, 56-58; grants, Crusade for Children, Ky State Health Dept, Ky & Jefferson County Heart Asn & Children's Bur, DC, 64-74 & Ohio Valley Regional Med Prog, 71-74; resident, Cook County Children's Hosp, Chicago & Hektoen Inst Med Res, 56-58; intern, Presby-St Luke's Hosp, 58-59, asst attend physician, 60-64; attend physician, Hines Mem Vet Admin Hosp, 62-64; dir cardiol, Children's Hosp, Louisville, Ky, 64-; dir cardiac care prog, Comn Handicapped Children, 64-; dir WHAS-TV Crusade Heart Lab, 66- Mem: AMA; Am Heart Asn; fel Am Col Cardiol. Res: Hemodynamics of heart sounds and murmurs; echocardiography. Mailing Add: Univ of Louisville Sch of Med Norton-Children's Hosps Louisville KY 40202

MINIATAS, BIRUTE ONA, b Erzvilkas, Lithuania, Sept 14, 30; US citizen. PHYSICAL CHEMISTRY, INORGANIC CHEMISTRY. Educ: Webster Col, BS, 55; St Louis Univ, MS, 59; Ill Inst Technol, PhD(phys inorg chem), 70. Prof Exp: From instr to asst prof, 60-70, ASSOC PROF CHEM, MUNDELEIN COL, 70- Mem: Am Chem Soc; Sirma Xi. Res: Transition metal chemistry; reaction mechanisms. Mailing Add: Dept of Chem Mundelein Col 6363 N Sheridan Rd Chicago IL 60660

MINIATS, OLGERTS PAULS, b Besagola, Lithuania, Sept 13, 23; Can citizen; m 47; c 3. VETERINARY MEDICINE. Educ: Univ Toronto, DVM, 55; Univ Guelph, MSc, 66. Prof Exp: Clinician, High River Vet Clin, 55-64; asst prof, 66-72, ASSOC PROF CLIN RES, UNIV GUELPH, 72- Mem: Can Vet Med Asn; Asn Gnotobiotics; Asn Advan Baltic Studies; Am Asn Swine Practitioners. Res: Development of techniques for procurement and rearing of gnotobiotic research animals; investigation of the characteristics of gnotobiotic animals; application of gnotobiotic animals for the investigation of infectious diseases. Mailing Add: Dept of Clin Studies Ont Vet Col Univ of Guelph Guelph ON Can

MINICHINO, CAMILLE, b Revere, Mass, June 3, 37. PHYSICS. Educ: Emmanuel Col, BA, 58; Fordham Univ, MS, 65, PhD(physics), 68. Prof Exp: Aide eng, United Aircraft Corp, 58-59; lab asst, 62-63, ASST PROF PHYSICS, EMMANUEL COL, MASS, 68- Mem: Am Asn Physics Teachers; Optical Soc Am. Res: Brillouin and Raman scattering; philosophy of science. Mailing Add: Dept of Physics Emmanuel Col 400 The Fenway Boston MA 02115

MINICK, CHARLES RICHARD, b Sheridan, Wyo, Feb 28, 36; m 57; c 3. PATHOLOGY. Educ: Univ Wyo, BS, 57; Cornell Univ, MD, 60. Prof Exp: Intern path, New York Hosp, 60-61; from asst to asst prof, 61-70, ASSOC PROF PATH, MED COL, CORNELL UNIV, 70- Concurrent Pos: USPHS trainee, Med Ctr, New York Hosp, 61-65, asst researcher, 62-63, chief resident, 63-64, asst attend pathologist, 64-70, assoc attend pathologist, 70-; mem coun arteriosclerosis, Am Heart Asn. Mem: AAAS; Harvey Soc; Am Asn Path & Bact; Am Soc Exp Path; Am Heart Asn. Res: Pathology of arteriosclerosis and hypertension; immunology; culture of cells derived from blood vessels; scanning and transmission electron microscopy. Mailing Add: Dept of Path Cornell Univ Med Col New York NY 10021

MINIERI, P PAUL, b Boston, Mass, Nov 30, 15; m 41; c 2. CHEMISTRY. Educ: Long Island Univ, BS, 37; NY Univ, MS, 41, PhD(chem), 48. Prof Exp: Chemist, NY Stain Co, 37-39; asst chemist, Pharma Chem Corp, NY, 41-42; anal chemist, Gen Chem Co, 42-43; res chemist, Off Sci Res & Develop proj, NY Univ, 43-44, Chas Pfizer & Co, Inc, 44-49, Wallace & Tiernan Prod, Inc, 49-50 & Heyden Chem Corp, 50-53; group leader, Isolation Process Develop, Am Cyanamid Co, 53-55, Organic Synthesis Group, Heyden-Newport Chem Corp, 55-70; MEM STAFF, TENNECO INTERMEDIATES, DIV TENNECO, INC, 70- Mem: Am Chem Soc. Res: Antibiotics; organic synthesis; vitamins; dyestuffs; amides; agricultural pesticides; process development; paint biocides; fire retardents for polyurethane. Mailing Add: 69-21 32nd Ave Woodside NY 11377

MINION, GERALD DOUGLAS, plant breeding, pathology, see 12th edition

MINK, GAYLORD IRA, b Lafayette, Ind, Sept 23, 31; m 54; c 4. PLANT VIROLOGY. Educ: Purdue Univ, BS, 56, MS, 59, PhD(plant path), 62. Prof Exp: Asst prof plant path, Purdue Univ, 62; from asst to assoc plant pathologist, 62-72, PROF, IRRIGATED AGR RES & EXTEN CTR, WASH STATE UNIV, 73-, PLANT PATHOLOGIST, 73- Mem: Am Phytopath Soc. Res: Identification, purification and serology of plant viruses; nature of inactivation of plant virus. Mailing Add: Irrigated Agr Res & Exten Ctr, Wash State Univ Prosser WA 99350

MINK, IRVING BERNARD, b New York, NY, Sept 23, 27; m 50; c 4. HEMATOLOGY. Educ: Univ Buffalo, BA, 48; State Univ NY Buffalo, MS, 69. Prof Exp: Asst cancer res scientist pharmacol, 61-62, cancer res scientist hematol, 62-69, SR CANCER RES SCIENTIST HEMATOL & HEMOSTASIS, ROSWELL PARK MEM INST, 69- Concurrent Pos: Instr pharmacol, Sch Pharm, Univ Buffalo, 56-62; mem bd, Oper Comt & 2nd vpres, Western NY Br, Nat Hemophilia Ctr, 69-; asst res prof, Roswell Park Div, Grad Sch, State Univ NY Buffalo, 72- Mem: AAAS; Int Soc Thrombosis & Hemostasis. Res: Hemostatic mechanism; cancer chemotherapy; bleeding and clotting disorders associated with malignancy; platelet function; development of hemostatic mechanism methodological procedures. Mailing Add: Roswell Park Mem Inst 666 Elm St Buffalo NY 14263

MINK, JOHN F, b Nesquehoning, Pa, June 3, 24; m 51; c 1. HYDROLOGY, GEOLOGY. Educ: Pa State Univ, BS, 49; Univ Chicago, MS, 51. Prof Exp: Res engr, US Steel Co, Ill, 51-52 & Crane Co, 52; asst geologist, Hawaiian Sugar Planters Asn, 52-53; chem supt, Pac Chem & Fertilizer Co, Hawaii, 53-56; geologist, US Geol Surv, 56-60; hydrologist-geologist, Honolulu Bd Water Supply, 60-68; sr analyst, CONSAD Res Corp, 68-69; VPRES, EARTH SCI GROUP, INC, 69- Concurrent Pos: Consult sugar co, Ryukyu Islands, 63- & Govt Guam, 64-; res affil, Water Resources Ctr, Univ Hawaii, 64-; USPHS grant, Johns Hopkins Univ, 65-67; consult, Honolulu Bd Water Supply & Singer-Layne Co; res affil, Univ Guam, 75- Mem: AAAS; Am Geophys Union; Am Water Resources Asn; Geochem Soc; Geol Soc Am. Res: Applications of operations research to water supply; water development and supply, particularly development of ground water in islands and along coasts; groundwater flow; contamination of groundwater; stochastic hydrology of streams on tropical islands. Mailing Add: 611 6th Pl SW Washington DC 20024

MINK, JOHN R, b Peru, Ill, Sept 8, 27; m 52; c 6. DENTISTRY. Educ: Ind Univ, AB, 51, DDS, 56, MSD, 61. Prof Exp: Asst prof pedodont, Ind Univ, 57-62; assoc prof, 62-65, PROF PEDODONT, UNIV KY, 65- Mem: Am Acad Pedodont; Int Asn Dent Res. Res: Pedodontics; replanting teeth; pulp of the tooth root resorption. Mailing Add: Dept of Pedodont Univ of Ky Lexington KY 40506

MINK, LAWRENCE ALBRIGHT, b Birmingham, Ala, Nov 10, 36; m 60; c 1. EXPERIMENTAL PHYSICS. Educ: NC State Univ, BE, 58, MS, 61, PhD(nuclear physics), 66. Prof Exp: Instr physics, NC State Univ, 65-66; asst prof, 66-69, ASSOC PROF PHYSICS, ARK STATE UNIV, 69- Mem: AAAS; Am Phys Soc. Res: Thin liquid films. Mailing Add: Div of Math & Physics Ark State Univ State University AR 72467

MINKEL, CLARENCE WILBERT, b Austin, Minn, Feb 9, 28; m 47; c 5. GEOGRAPHY OF LATIN AMERICA, ECONOMIC GEOGRAPHY. Educ: Colo State Col, BA, 53, MA, 55; Syracuse Univ, PhD(geog), 60. Prof Exp: Instr hist & geog & head coach, College High Sch, Greeley, Colo, 53-55; asst phys geog, Syracuse Univ, 55-56; asst prof, Colo State Col, 60-63, assoc prof, 64-65; assoc prof geog & assoc dir, Latin Am Studies Ctr, 65-67, asst dean, Col Soc Sci, prof geog & dir soc sci res bur, 67-68, assoc dean, Grad Sch, 68-74, ACTG DEAN, GRAD SCH, MICH STATE UNIV, 74-, PROF GEOG, 68- Concurrent Pos: Consult, Encycl Britannica, Inc, 61-; adv planning, USAID, Guatemala, 63-65; pres comt appl geog, Pan-Am Inst Geog & Hist, 65- & US nat mem, Comn Geog, 70-; consult, US Off Educ, 69-71; chmn exec comt, Conf Latin Am Geogrs, 70-71 & secy-treas, Conf, 72-; Fulbright lectr-consult, US Comt Int Exchange Persons, Colombia, 71 & 72. Mem: Am Geog Soc; Asn Am Geog; Nat Coun Geog Educ. Res: Geography of Latin America; economic geography and development; national and regional planning. Mailing Add: Grad Sch Mich State Univ East Lansing MI 48823

MINKER, JACK, b Brooklyn, NY, July 4, 27; m 51; c 2. MATHEMATICS. Educ: Brooklyn Col, BA, 49; Univ Wis, MS, 50; Univ Pa, PhD(math), 59. Prof Exp: Dynamics engr, Bell Aircraft Corp, 51-52; engr, Radio Corp Am, 52-57, mgr info tech, Data Systs Ctr, 57-63; mgr Wash systs sect, Auerbach Corp, 63-67; PROF

MINKER

COMPUT SCI, UNIV MD, 67- Concurrent Pos: Invited lectr, Gordon Res Conf Info Storage & Retrieval, 60; co-chmn Nat Conf Info Storage & Retrieval Asn Comput Mach, 60; mem Nat Acad Sci-US Nat Comt Fedn Info Documentationalists, 69-72; mem adv comt, Annual Rev Info Sci & Technol, 70-; vchmn, Jerusalem Conf Info Technol, 71; co-chmn conf & prog, Nat Info Storage & Retrieval Conf, 71; mem adv comt, Encycl Comput Sci, 71-; mem staff grad sch, NIH. Mem: Asn Comput Mach; Inst Math Statist; Am Math Soc; Am Soc Info Sci; Soc Indust & Appl Math. Res: Computer application; information storage and retrieval; operations research; artificial intelligence. Mailing Add: Dept of Comput Sci Univ of Md College Park MD 20742

MINKIEWICZ, VINCENT JOSEPH, b Shenandoah, Pa, Oct 24, 38; m 65; c 2. SOLID STATE PHYSICS. Educ: Villanova Univ, BS, 60; Univ Calif, Berkeley, PhD(physics), 65. Prof Exp: Assoc scientist solid state physics, Brookhaven Nat Lab, 65-72; assoc prof, Univ Md, College Park, 72-74; RES STAFF MEM, IBM CORP, 75- Mem: Am Phys Soc. Res: Magnetic materials; magnetic bubbles. Mailing Add: IBM Res Lab San Jose CA 95193

MINKLER, WILLIAM S, nuclear physics, reactor physics, see 12th edition

MINKOFF, ELI COOPERMAN, b New York, NY, Sept 5, 43; m 68; c 2. EVOLUTIONARY BIOLOGY, COMPARATIVE ANATOMY. Educ: Columbia Univ, AB, 63; Harvard Univ, AM, 67, PhD(biol), 69. Prof Exp: Res asst primate anat, New Eng Regional Primate Res Ctr, Southboro, Mass, 67-68; asst prof, 68-75, actg chmn dept, 76-77, ASSOC PROF BIOL, BATES COL, 76- Mem: Soc Study Evolution; Am Soc Zoologists; Soc Vert Paleont; Soc Syst Zool; Am Soc Mammalogists. Res: Anatomy and evolutionary biology of primates and other mammals; neuromuscular anatomy, especially of facial nerve and facial muscles, muscle classification, vertebrate paleontology. Mailing Add: Dept of Biol Bates Col Lewiston ME 04240

MINKOFF, JOHN, b June 16, 31; US citizen; m 66; c 2. COMMUNICATION SCIENCE, PHYSICS. Educ: Columbia Univ, BS, 62, MS, 63, PhD(elec eng), 67. Prof Exp: Mem res staff, Columbia Univ Electronics Res Lab, 64-67; mem res staff, 68-69, ASST MGR ANAL ACTIV, RIVERSIDE RES INST, 69- Honors & Awards: NSF res grant, 75. Mem: Am Phys Soc; Inst Elec & Electronics Eng; Am Geophys Union; Int Union Radio Sci. Res: Communications; digital signal and data processing; radio wave propagation and scattering; interpretation of infrared atmospheric and stellar measurements. Mailing Add: Riverside Res Inst 80 West End Ave New York NY 10023

MINKOWITZ, STANLEY, b Brooklyn, NY, July 1, 28; m 57; c 3. PATHOLOGY. Educ: City Col New York, BS, 48; Univ Colo, MS, 50; Univ Geneva, MD, 56. Prof Exp: Chief surg path, Kings County Hosp Med Ctr, 62-71; CLIN ASSOC PROF PATH, STATE UNIV NY DOWNSTATE MED CTR, 71-; DIR LABS, MAIMONIDES MED CTR, 71- Concurrent Pos: Fel pediat med, Jewish Chronic Dis Hosp, 57; asst prof path, State Univ NY Downstate Med Ctr, 63-67, assoc prof, 67-71, consult vet pathologist, 64-; consult pathologist, Brooklyn Women's Hosp, 63- & Unity Hosp, Brooklyn, 65-; consult, Brooklyn State Hosp, 67- Res: Organic chemistry, especially amino acids synthesis; experimental study of the use of gold leaf in skin graft; heterotopic liver transplantation; coxsackie B-virus infection in adult mice; homologous lung transplantation. Mailing Add: Maimonides Med Ctr 4802 Tenth Ave Brooklyn NY 11219

MINKOWSKI, JAN MICHAEL, b Zurich, Switz, Mar 7, 16; nat US; m 51; c 4. SOLID STATE PHYSICS. Educ: Swiss Fed Inst Technol, Dipl Physicist, 49; Johns Hopkins Univ, PhD(physics), 63. Prof Exp: Res physicist, Inst Tehoret Physics, Switz, 49-50; res physicist solid state physics, Erie Resistor Corp, 50-52; head phys res, Carlyle Barton Lab, 52-63, ASSOC PROF ELEC ENG, JOHNS HOPKINS UNIV, 63- Mem: Am Phys Soc. Res: Quantum electronics. Mailing Add: Dept of Elec Eng 308 Barton Hall Johns Hopkins Univ Baltimore MD 21218

MINKOWSKI, RUDOLPH LEO BERNHARD, astronomy, physics, deceased

MINN, FREDRICK LOUIS, physical chemistry, see 12th edition

MINN, JAMES, b Pittsburgh, Pa, Jan 25, 22; m 43; c 3. ORGANIC CHEMISTRY. Educ: Univ Pittsburgh, BS, 48, PhD(org chem), 53. Prof Exp: Res chemist, Hercules Powder Co, 52-71, res chemist res ctr, 71-73, SR RES CHEMIST, RES CTR, HERCULES, INC, 73- Mem: Am Chem Soc; Sigma Xi. Res: Oxidation of hydrocarbons; addition of free radicals to olefins; catalytic studies of isocyanate reactions; addition of carbon monoxide to carbonium ions; process development of agricultural chemicals. Mailing Add: 6 Mallbora Dr Brookside Newark DE 19711

MINNE, RONN N, b Menominee, Mich, Oct 3, 24. INORGANIC CHEMISTRY. Educ: Northwestern Univ, BS, 50, AM, 51; Harvard Univ, PhD(chem), 60. Prof Exp: Teacher high sch, Ill & NMex, 51-54; teacher chem, Culver Mil Acad, 54-56, chmn dept, 60-65; TEACHER CHEM, PHILLIPS ACAD, 65-, CHMN SCI DIV, 72- Concurrent Pos: Off Naval Res grant, 62-65; vis scholar, Cambridge Univ, 71-72. Mem: AAAS; Am Chem Soc; Am Inst Chem; NY Acad Sci. Res: Inorganic polymers; silicon chemistry. Mailing Add: Dept of Chem Phillips Acad Andover MA 01810

MINNEMEYER, HARRY JOSEPH, b Buffalo, NY, July 12, 32; m 66; c 3. BIO-ORGANIC CHEMISTRY. Educ: Univ Buffalo, BA, 58, PhD(chem), 62. Prof Exp: Res assoc chem, State Univ NY Buffalo, 61-66, lectr, Millard Fillmore Col, 63-64; chemist, Starks Assocs, Inc, 66-67, supvr, 67-68, sect mgr, 68-69; sr res chemist, Lorillard Res Ctr, 70, supvr org chem, 70-76, MGR RES, LORILLARD, DIV LOEWS THEATERS, INC, 76- Mem: Am Chem Soc. Res: Organic synthesis; heterocyclic and medicinal chemistry; reaction mechanisms; tobacco science. Mailing Add: Lorillard Res Ctr 420 English St Greensboro NC 27420

MINNERS, HOWARD ALYN, b Rockville Centre, NY, Sept 1, 31; m 58; c 2. MEDICINE. Educ: Princeton Univ, AB, 53; Yale Univ, MD, 57; Harvard Univ, MPH, 60; Am Bd Prev Med, cert aerospace med, 65. Prof Exp: Intern, Wilford Hall US Air Force Hosp, 57-58, flight surgeon, Langley Air Force Base, Va, 61-62; head flight med br, Manned Spacecraft Ctr, NASA, 62-66; spec asst to chief, Off Int Res, NIH, 66-68; from asst chief to chief, Geog Med Br, 68-72, assoc dir collab res, 72-76, ASSOC DIR INT RES, NAT INST ALLERGY & INFECTIOUS DIS, NIH, 76- Concurrent Pos: Am Bd Prev Med fel, 66. Mem: AAAS; AMA; Aerospace Med Asn; Am Pub Health Asn; Am Soc Trop Med & Hyg. Res: Ecology, distribution and determinants of disease prevalence in man; career motivation for international biomedical research; tropical medicine. Mailing Add: Bldg 31 Rm 7A03 Nat Inst Allergy & Infect Dis Bethesda MD 20014

MINNICH, JOHN EDWIN, b Long Beach, CAlif, Oct 23, 42; m 68. ENVIRONMENTAL PHYSIOLOGY, VERTEBRATE ZOOLOGY. Educ: Univ Calif, Riverside, AB, 64; Univ Mich, Ann Arbor, PhD(zool), 68. Prof Exp: ASST PROF ZOOL, UNIV WIS-MILWAUKEE, 68- Concurrent Pos: Wis Alumni Res Found fel physiol, 69-71; environ consult, Environ Analysts, Inc, 74-; textbook consult physiol, MacMillan & Co, Inc, 74- Mem: AAAS; Am Soc Ichthyol & Herpet; Am Soc Zool; Ecol Soc Am; Soc Study Amphibians & Reptiles. Res: Environmental physiology of animals; water, electrolyte, energy and nitrogen metabolism of reptiles; excretion and osmoregulation of vertebrates. Mailing Add: Dept of Zool Univ of Wis Milwaukee WI 53201

MINNICH, VIRGINIA, b Zanesville, Ohio, Jan 24, 10. HEMATOLOGY, NUTRITION. Educ: Ohio State Univ, BSc, 37; Iowa State Col, MS, 38. Hon Degrees: DSc, William Woods Col, 72. Prof Exp: From res asst to res assoc, 39-58, from asst res prof to assoc res prof, 68-74, PROF MED, SCH MED, WASH UNIV, 74-; DIR CLIN MICROS LAB, BARNES HOSP, 75- Concurrent Pos: Fulbright award, Turkey, 64-65. Mem: Am Fedn Clin Res; Am Soc Hemat; Int Soc Hemat. Res: Iron metabolism; platelet agglutinins; abnormal hemoglobins; thalassemia. Mailing Add: Div of Hemat Wash Univ Sch of Med St Louis MO 63110

MINNICK, DANNY RICHARD, b Export, Pa, July 17, 37; m 71. ENTOMOLOGY. Educ: Southern Missionary Col, BA, 61; Univ Fla, MS, 67, PhD(entom), 70. Prof Exp: Asst entom, Univ Fla, 65-70; asst prof biol, Fla Keys Community Col, 70-71; ASST PROF ENTOM, UNIV FLA, 71- Honors & Awards: Outstanding Grad Stud, Entom Soc Am, 70; Serv Award, Am Beekeeping Fedn. Mem: Entom Soc Am; Am Beekeeping Fedn; Sigma Xi. Res: Insect behavior and ecology; pest management of citrus. Mailing Add: Dept of Entom McCarty Hall Univ of Fla Gainesville FL 32611

MINNICK, ROBERT FLETCHER, b Logansport, Ind, Sept 16, 26; m 46. CULTURAL GEOGRAPHY. Educ: Ind State Teachers Col, BS, 50; Univ Nebr, MA, 58. Prof Exp: Instr geog, Cent Mich Univ, 59-63, asst prof, 63-65; ASSOC PROF GEOG, CALIF STATE COL, PA, 65- Mem: Pop Asn Am; Nat Coun Geog Educ; Asn Am Geographers. Res: Historical-geographic study of demographic, industrialization and urbanization processes. Mailing Add: Dept of Geog Calif State Col California PA 15419

MINNIX, RICHARD BRYANT, b Salem, Va, June 20, 33; m 55; c 3. PHYSICS. Educ: Roanoke Col, BS, 54; Univ Va, MS, 57; Univ NC, Chapel Hill, PhD(physics), 65. Prof Exp: From instr to assoc prof, 56-69, PROF PHYSICS, VA MIL INST, 69-, HEAD DEPT, 74- Mem: Am Phys Soc; Am Asn Physics Teachers. Res: Defects in solids, particularly by radiation damage; lecture demonstrations as method of instruction and/or stimulation of public interest in science. Mailing Add: Dept of Physics Va Mil Inst Lexington VA 24450

MINO, GUIDO, b London, Eng, May 14, 20; nat US; m 58. POLYMER CHEMISTRY. Educ: Torino Univ, Italy, PhD(org chem), 46. Prof Exp: Res chemist, Imp Chem Industs, 47-49; res chemist, 51-54, group leader, 54-60, sect mgr, 60-72, dir Bound Brook Lab, 72-74, DIR INT RES & DEVELOP, AM CYANAMID CO, 74- Mem: Am Chem Soc. Res: High polymer chemistry; synthesis of graft and block polymers; management of research and development in the area of agricultural and industrial chemicals. Mailing Add: Admin Hq Am Cyanamid Co Berdan Ave Wayne NJ 07470

MINO, PASCHAL MICHAEL, mathematics, see 12th edition

MINOCHA, HARISH C, b Aug 31, 32; m 55; c 4. BIOCHEMICAL VIROLOGY. Educ: Punjab Univ, India, BVSc, 55; Kans State Univ, MS, 63, PhD(bact), 67. Prof Exp: Res asst microbiol, Indian Vet Res Inst, 55-61; res asst, Kans State Univ, 61-66, res assoc, 66-67; asst prof, NC State Univ, 67-69; ASSOC PROF MICROBIOL, KANS STATE UNIV, 69- Concurrent Pos: Nat Cancer Soc res grant, 69-70; res grant, USDA, 72- Mem: Am Soc Microbiol. Res: Biochemical studies on polyoma and shope fibroma virus infected tissue cultures; immunological studies on swine and equine influenza viruses; antigenic analysis of influenza viruses; bovine respiratory disease. Mailing Add: Dept of Infectious Dis Kans State Univ Manhattan KS 66502

MINOCK, MICHAEL EDWARD, b Los Angeles, Calif, Dec 6, 37. VERTEBRATE ECOLOGY. Educ: Stanford Univ, AB, 60; Calif State Univ, Northridge, MA, 65; Univ Nebr, Omaha, MS, 66; Utah State Univ, PhD(zool), 70. Prof Exp: Teacher biol, Los Angeles City Schs, 62-65; instr, State Univ NY Col, Brockport, 69-70; NIH fel behav biol & ecol, Univ Minn, 70-71; ASST PROF BIOL, UNIV WIS CTR-FOX VALLEY, 71- Mem: AAAS; Am Ornithologists Union; Animal Behav Soc; Ecol Soc Am; Cooper Ornith Soc. Res: Bird and mammal social organizations and their ecological significance. Mailing Add: Univ of Wis Ctr-Fox Valley Midway Rd Menasha WI 54952

MINOR, CHARLES OSCAR, b Churdan, Iowa, June 22, 20; m 43; c 3. FORESTRY. Educ: Iowa State Col, BS, 41; Duke Univ, MF, 42, DF, 58. Prof Exp: Asst prof forestry, La State Univ, 46-54; admin forester, Kirby Lumber Corp, 54-56; assoc prof forestry, Clemson Col, 58; PROF FORESTRY & DEAN SCH FORESTRY, NORTHERN ARIZ UNIV, 58- Mem: Soc Am Foresters. Res: Forest management and mensuration, including aerial photo interpretation. Mailing Add: PO Box 966 Flagstaff AZ 86001

MINOR, GEORGE RIDGWAY, b Cannel City, Ky, Oct 18, 13. SURGERY. Educ: Univ Va, BS, 36, MD, 40; Univ Mich, MS, 48. Prof Exp: Instr thoracic surg, Univ Mich, 45-46, instr gen surg, 47-48; asst prof surg, Col Med, Univ Ill, 48-49; from asst prof to assoc prof, 49-64, PROF SURG & ASST DEAN SCH MED, UNIV VA, 64- Concurrent Pos: Attend surgeon, Chicago Munic Tuberc Sanitarium & Hines Vet Admin Hosp, Ill, 48-49; consult thoracic surgeon, Oak Ridge Assoc Univs, 49-; consult thoracic surgeon, Va State Dept Health, 49-, consult radiologist, 55-60; vis prof surg, Sch Med, Univ Tunis & attend surgeon, Charles Nicolle Hosp, 72-74. Mem: Soc Thoracic Surg; Soc Univ Surg; Am Asn Thoracic Surg; Am Col Surg; Int Soc Surg. Res: Fungous infections; pulmonary metastases; alveolar cell cancer of lung. Mailing Add: Dept of Surg Univ of Va Med Ctr Charlottesville VA 22901

MINOR, HARRY CAMERON, agronomy, see 12th edition

MINOR, JOHN THREECIVELOUS, b Marshall, Mo, Feb 11, 21; m 46; c 4. ORGANIC CHEMISTRY, INFORMATION SCIENCE. Educ: Mo Valley Col, AB, 42; Univ Pittsburgh, MS, 46; Univ Kans, PhD(chem), 50. Prof Exp: Res assoc, Univ Rochester, 44-46; asst prof, Westminster Col, 49-51; res chemist, Commercial Solvents Corp, 51-54; group leader, John Deere Chem Co, 54-61; group leader, 61-70, SR INFO SPECIALIST, CONTINENTAL OIL CO, 70- Mem: Am Chem Soc; fel Am Inst Chem. Res: Aromatic fluorides; nitroaliphatics; ammonia and urea chemistry. Mailing Add: Continental Oil Co PO Box 1267 Library Ponca City OK 74601

MINOR, RONALD R, b Donora, Pa, Sept 13, 36; m 58; c 4. DEVELOPMENTAL ANATOMY, EXPERIMENTAL PATHOLOGY. Educ: Univ Pa, VMD, 66, PhD(path), 71. Prof Exp: Assoc, 71-72, ASST PROF ANAT, MED, PATH & VET MED, UNIV PA, 72- Mem: Am Vet Med Asn; Soc Develop Biol; Soc Cell Biol; AAAS. Res: Regulation of mesodermal differentiation and connective tissue development; regulation of the synthesis, deposition and maintenance of the structural components of basement membrans. Mailing Add: Clin Res Ctr Philadelphia Gen Hosp Philadelphia PA 19104

MINORE, DON, b Chicago, Ill, Oct 31, 31; m 63; c 2. ECOLOGY. Educ: Univ Minn, BS, 53; Univ Calif, Berkeley, PhD(bot), 66. Prof Exp: Res forester, Pac Northwest Forest & Range Exp Sta, US Forest Serv, 55, 57-60; asst bot, Univ Calif, Berkeley, 61-63, teaching fel, 63-65; PLANT ECOLOGIST, FORESTRY SCI LAB, US FOREST SERV, 65- Mem: Ecol Soc Am; Soc Am Foresters. Res: Species-site relationships in the mixed conifer forests of Oregon and Washington; vaccinium ecology and physiology. Mailing Add: Forestry Sci Lab 3200 Jefferson Way Corvallis OR 97331

MINOTTI, PETER LEE, b Burlington, Vt, Aug 20, 35; m 62; c 2. PLANT NUTRITION. Educ: Univ Vt, BS, 57, MS, 62; NC State Univ, PhD(soil sci, plant physiol), 65. Prof Exp: Crop physiologist, Int Minerals & Chem Corp, 65-66; ASSOC PROF VEG CROPS, CORNELL UNIV, 66- Mem: Am Soc Agron; Am Soc Hort Sci. Res: Nutrition of vegetables. Mailing Add: Dept of Veg Crops Cornell Univ Ithaca NY 14850

MINOWADA, JUN, b Kyoto, Japan, Nov 5, 27; m 59; c 3. PATHOLOGY, VIROLOGY. Educ: Mie Med Col, Japan, BS, 48; Kyoto Univ, MD, 52, DMedSci, 59. Prof Exp: Sr res assoc, 61-64, sr res scientist, 66-67, assoc res scientist, 68-72, prin res scientist, 73-74, ASSOC CHIEF SCIENTIST, ROSWELL PARK MEM INST, 75-; ASSOC RES PROF, ROSWELL PARK DIV, GRAD SCH, STATE UNIV NY BUFFALO, 69- Concurrent Pos: Int Atomic Energy Agency res fel, Roswell Park Mem Inst, 60; vis fel, Karolinska Inst, Sweden, 64-66. Mem: Am Asn Cancer Res; Am Soc Microbiol. Res: Viral oncology, immunology. Mailing Add: 376 Countryside Lane Williamsville NY 14221

MINSAVAGE, EDWARD JOSEPH, b Nanticoke, Pa, June 1, 18; m 51. MEDICAL MICROBIOLOGY. Educ: Univ Scranton, BS, 49; Univ Pa, MS, 51, PhD(microbiol), 55. Prof Exp: Ed asst, Biol Abstracts, 50-51; res bacteriologist, Eaton Labs, Inc, 51-52; instr, Med Sch, Univ Pa, 54-55, res assoc, 55-57; from asst prof to assoc prof biol, 57-67, PROF BIOL, KING'S COL, PA, 67- Concurrent Pos: USPHS res grant bact physiol, 63-66. Mem: AAAS; Am Soc Microbiol. Res: Cytology and cytochemistry; bacterial respiration and oxygen transport systems; analysis and biochemistry of nucleic acids and their derivatives; electrophoresis and chromatographic analysis; microphotography; organ deficiency in experimental animals; loss of essential biochemical intermediates by normal cells. Mailing Add: 221 S Hanover St Nanticoke PA 18634

MINSHALL, GERRY WAYNE, b Billings, Mont, Aug 30, 38; m 63; c 2. AQUATIC ECOLOGY. Educ: Mont State Univ, BS, 61; Univ Louisville, PhD(zool), 65. Prof Exp: NATO fel, Freshwater Biol Asn, Eng, 65-66; from asst prof to assoc prof, 66-74, PROF ZOOL, IDAHO STATE UNIV, 74- Concurrent Pos: Fac res grants, Idaho State Univ, 66, 68 & 70; US Dept Health, Educ & Welfare res grant, 68-69; NSF-Int Biol Prog grant, 70-71. Mem: Am Soc Limnol & Oceanog; Brit Ecol Soc; Inst Asn Theoret & Appl Limnol; Ecol Soc Am. Res: Ecology of flowing waters, benthic invertebrates; productivity; detritus; invertebrate drift; pollution. Mailing Add: Dept of Biol Idaho State Univ Pocatello ID 83201

MINSHALL, WILLIAM HAROLD, b Brantford, Ont, Dec 6, 11; m 39; c 2. PLANT PHYSIOLOGY. Educ: Ont Agr Col, BSA, 33; McGill Univ, MSc, 38, PhD(plant physiol), 41. Prof Exp: Asst bot, 33-41, from jr botanist to botanist, 41-51, SR PLANT PHYSIOLOGIST, CAN DEPT AGR, 51- Concurrent Pos: Hon lectr, Univ Western Ont, 52- Mem: AAAS; Am Soc Plant Physiol; Bot Soc Am; fel Weed Sci Soc Am; Can Soc Plant Physiol. Res: Physiology of herbicidal action; metabolic root pressure and uptake of solutes; plant phenology. Mailing Add: 91 Huron St London ON Can

MINSKER, DAVID HARRY, b Huntingdon, Pa, Jan 17, 38; m 65; c 1. PHYSIOLOGY, PHARMACOLOGY. Educ: Juniata Col, BS, 63; Univ Wis, PhD(physiol), 67. Prof Exp: Res pharmacologist thrombosis, 67-69, res pharmacologist hypertension, 69-71, RES FEL THROMBOSIS, MERCK SHARP & DOHME RES LABS, 71- Mem: Am Heart Asn; Int Soc Thrombosis & Hemostasis. Res: Development of anti-thrombotic drugs. Mailing Add: Merck Sharp & Dohme Res Labs West Point PA 19486

MINSKY, MARVIN LEE, b New York, NY, Aug 9, 27; m 52; c 3. MATHEMATICS. Educ: Harvard Univ, BA, 50; Princeton Univ, PhD(math), 54. Prof Exp: Asst math, Princeton Univ, 50-53; res assoc, Tufts Univ, 53-54; jr fel, Soc Fels, Harvard Univ, 54-57; mem staff, Lincoln Lab, 57-58, from asst prof to prof math, 58-64, PROF ELEC ENG, MASS INST TECHNOL, 64- Mem: Nat Acad Sci; fel Am Acad Arts & Sci; fel Inst Elec & Electonics Eng; fel NY Acad Sci. Res: Artificial intelligence; theory of computation; psychology; engineering. Mailing Add: Dept of Elec Eng Mass Inst of Technol Cambridge MA 02139

MINTA, JOE ODURO, b Fomena, Ghana, Mar 16, 42; m 70; c 1. IMMUNOLOGY. Educ: Univ Ghana, BSc, 66; Univ Guelph, MSc, 68; Univ Toronto, PhD(biochem), 71. Prof Exp: Res asst chem, Univ Guelph, 66-68; res asst biochem, Univ Toronto, 68-71; fel immunol, Med Res Coun Can, 71-73; ASST PROF IMMUNOL, UNIV TORONTO, 73-, MEM INST IMMUNOL, 74- Concurrent Pos: Res scholar, Can Heart Found, 73- Mem: AAAS; NY Acad Sci; Can Soc Immunol; Am Asn Exp Path; Am Asn Immunologists. Res: Structure and mechanism of interaction of components of the complement system and their role in immunopathology. Mailing Add: Med Sci Bldg Rm 6308 Univ Toronto Dept Path Toronto ON Can

MINTHORN, MARTIN LLOYD, JR, b Grand Rapids, Mich, Aug 8, 22; m 55, 64; c 4. BIOCHEMISTRY. Educ: Univ Nebr, BS, 44, MS, 49; Univ Ill, PhD(biochem), 53. Prof Exp: Jr chemist, Nat Bur Stand, 44-46; from res assoc to instr biochem, Univ Ill, 52-55; instr, Wash Univ, 53-54; res assoc, Univ Tenn, 55-56, asst prof, 57-64; BIOCHEMIST, ASST BR CHIEF & DEP PROG MGR, DIV BIOMED & ENVIRON RES, AEC, US ENERGY RES & DEVELOP ADMIN, 64-; PROF LECTR, GEORGE WASHINGTON UNIV, 75- Concurrent Pos: USPHS sr res fel, 57-61. Mem: AAAS; Sigma Xi; Am Chem Soc; fel Am Inst Chem. Res: Biochemistry of amino acids; stereochemistry; isotopic tracers; health impacts of energy technologies. Mailing Add: Div Biomed & Environ Res US Energy Res & Develop Admin Washington DC 20545

MINTON, ALLEN PAUL, b Takoma Park, Md, July 5, 43; m 70; c 2. BIOPHYSICAL CHEMISTRY. Educ: Univ Calif, Los Angeles, BS, 64, PhD(phys chem), 68. Prof Exp: Guest scientist, Polymer Dept, Weizmann Inst Sci, 68-70, Chaim Weismann jr fel, 68-70; staff fel, 70-74, SR STAFF FEL, LAB BIOPHYS CHEM, NAT INST ARTHRITIS, METAB & DIGESTIVE DIS, 74- Res: Intermolecular interactions in biochemistry; aggregation of biological macromolecules in solution; structure-function relations in hemoglobin; gelation of sickle-cell hemoglobin; non-ideality in concentrated protein solutions. Mailing Add: Lab of Biophys Chem Nat Inst Arthritis Metab & Digestive Dis Bethesda MD 20014

MINTON, GEORGE HAROLD, nuclear physics, see 12th edition

MINTON, NORMAN A, b Spring Garden, Ala, Oct 12, 24; m 44; c 1. PLANT NEMATOLOGY. Educ: Auburn Univ, BS, 50, MS, 51, PhD(zool), 60. Prof Exp: Asst county agr agent, Coop Exten Serv, Ala, 51-53; horticulturist, Berry Col, 53-55; nematologist, Auburn Univ, 55-64, NEMATOLOGIST, COASTAL PLAIN EXP STA, US DEPT AGR, 64- Mem: Am Phytopath Soc; Soc Nematol; Orgn Trop Am Nematologists. Res: Host-parasite relationship of nematodes to plants; nematode-fungus relationship to plant diseases; nematode population dynamics as influenced by crops, cultural practices, and chemicals; development of nematode resistant varieties. Mailing Add: Coastal Plain Exp Sta US Dept of Agr Tifton GA 31774

MINTON, PAUL DIXON, b Dallas, Tex, Aug 4, 18; m 43; c 2. STATISTICS. Educ: Southern Methodist Univ, BSc, 41, MSc, 48; NC State Col, PhD(exp statist), 57. Prof Exp: Instr math, Univ NC, 48, asst, Inst Statist, 49-52, statist asst, Inst Res Social Sci, 52-53; asst prof math, Southern Methodist Univ, 53-55; assoc prof statist, Va Polytech Inst, 55-56; assoc prof math & dir comput lab, Southern Methodist Univ, 57-61, prof statist, 61-72; DEAN SCH ARTS & SCI, VA COMMONWEALTH UNIV, 72- Mem: Biomet Soc; Inst Math Statist; fel Am Statist Asn; fel Am Soc Qual Control. Res: Distribution theory; experimental statistics. Mailing Add: Va Commonwealth Univ 901 W Franklin Richmond VA 23284

MINTON, ROBERT GEORGE, organic chemistry, see 12th edition

MINTON, SHERMAN ANTHONY, b New Albany, Ind, Feb 24, 19; m 43; c 3. HERPETOLOGY, MICROBIOLOGY. Educ: Ind Univ, AB, 39, MD, 42. Prof Exp: From asst prof to assoc prof microbiol, Sch Med, Ind Univ, Indianapolis, 47-48; vis prof, Postgrad Med Ctr, Karachi, Pakistan, 58-62; assoc prof, 62-67, PROF MICROBIOL, SCH MED, IND UNIV, INDIANAPOLIS, 67- Concurrent Pos: Res assoc, Am Mus Natural Hist, 57-64. Mem: Am Soc Trop Med & Hyg; Am Soc Ichthyol & Herpet; Int Soc Toxinology (pres, 66-68); Soc Study Amphibians & Reptiles. Res: Venomous animals and the injuries they cause; geographic distribution and taxonomy of amphibians and reptiles; serological techniques in taxonomy; arthropods as human parasites and disease vectors. Mailing Add: Dept of Microbiol Ind Univ Med Ctr Indianapolis IN 46202

MINTURN, ROBERT EDWARD, b Deadwood, SDak, May 11, 28; m 51; c 4. PHYSICAL CHEMISTRY. Educ: Ore State Col, BS, 50; Iowa State Col, PhD(chem), 55. Prof Exp: Asst, Ames Lab, Atomic Energy Comn, 50-55; RES CHEMIST, OAK RIDGE NAT LAB, 57- Mem: AAAS; Am Chem Soc; Sigma Xi. Res: Investigation of the treatment of industrial liquid wastes by dynamic membrane filtration. Mailing Add: Oak Ridge Nat Lab PO Box X Oak Ridge TN 37831

MINTY, GEORGE JAMES, JR, b Detroit, Mich, Sept 16, 29; m 59; c 1. MATHEMATICS. Educ: Wayne State Univ, BS, 49, MA, 51; Univ Mich, PhD(math), 59. Prof Exp: Res assoc math, Res Ctr, Univ Mich, 57-58, asst prof 60-64, assoc prof, 64-65; res instr, Duke Univ, 58-59; instr, Univ Wash, Seattle, 59-60; PROF, IND UNIV, BLOOMINGTON, 65- Concurrent Pos: Vis mem, Courant Inst Math Sci, NY Univ, 64-65; Sloan fel, 65-67; vis prof, Univ Calif, Berkeley, 71; US sr scientist awardee, Alexander von Humboldt, 73- 74; vis prof, Inst Appli Math, Univ Hamburg, 73-74. Mem: AAAS; Am Math Soc; Math Asn Am; Soc Indust & Appl Math. Res: Functional analysis; convexity theory; mathematical physics; graph and network theory; mathematical programming. Mailing Add: Dept of Math Ind Univ Bloomington IN 47401

MINTZ, A AARON, b Houston, Tex, July 29, 22; m 47; c 3. MEDICINE. Educ: Rice Inst, BA, 48; Univ Tex, MD, 48. Prof Exp: From instr to asst prof, 52-65, ASSOC PROF PEDIAT, BAYLOR COL MED, 65- Res: Pediatrics. Mailing Add: Dept of Pediat Baylor Col of Med Houston TX 77025

MINTZ, BEATRICE, b New York, NY, Jan 24, 21. BIOLOGY. Educ: Hunter Col, AB, 41; Univ Iowa, MS, 44, PhD(zool), 46. Prof Exp: Res asst zool, Univ Iowa, 42-46, instr, 46; from instr to assoc prof biol sci, Univ Chicago, 46-60; assoc mem, 60-65, SR MEM, INST CANCER RES, 65-; PROF MED GENETICS, UNIV PA, 65- Mem: AAAS; Am Soc Zoologists; Genetics Soc Am; Soc Develop Biol. Res: Gene control of differentiation and disease in mammals; embryo-maternal relationships. Mailing Add: Inst for Cancer Res 7701 Burholme Ave Philadelphia PA 19111

MINTZ, ESTHER URESS, b New York, NY, May 18, 07; m 29; c 1. PHYSICS. Educ: Hunter Col, BA, 28; Columbia Univ, MA, 31; NY Univ, PhD, 36. Prof Exp: Instr physics, Hunter Col, 28-33; res asst biophys, Columbia Univ, 36-39; instr physics, Hunter Col, 39-42; physicist, Signal Corps Labs, US Dept Army, NJ, 42-44; instr math, Fieldston Sch, 54-57; from asst prof to prof, 57-73, EMER PROF PHYSICS & ASTRON, BARUCH COL, CITY UNIV NEW YORK, 73- Mem: Am Phys Soc. Res: Optics; spectroscopy; astronomy for non-science student. Mailing Add: RFD 2 Box 56 Quarry Lane Bedford NY 10506

MINTZ, JEROME RICHARD, b Brooklyn, NY, Mar 29, 30; m 55; c 2. ANTHROPOLOGY. Educ: Brooklyn Col, BA, 52, MA, 55; Ind Univ, PhD(folklore, anthrop), 61. Prof Exp: Lectr folklore, New Sch Social Res, 60; instr Eng, Ohio State Univ, 60-61; from asst prof to assoc prof folklore, 62-68, assoc prof anthrop & folklore, 68-72, PROF ANTHROP & FOLKLORE & CUR MEDITER ETHNOL, IND UNIV, BLOOMINGTON, 72- Concurrent Pos: Am Philos Soc grants, 62 & 68; Ind Univ res grants, 63, 64, 67 & 68; Ford Found fel, 65-66; Guggenheim Found fel, Littauer grant & Nat Endowment for Humanities grant, 69-70; Am Coun Learned Socs grant, 73-74. Mem: AAAS; Am Anthrop Asn; Am Folklore Soc. Res: Anthropological folklore; religious sects and revitalization movements; Mediterranean culture; peasant society; documentary and ethnographic filmmaking. Mailing Add: Dept of Anthrop Ind Univ Bloomington IN 47401

MINTZ, LEIGH WAYNE, b Cleveland, Ohio, June 12, 39; m 62. PALEONTOLOGY, STRATIGRAPHY. Educ: Univ Mich, BS, 61, MS, 62; Univ Calif, Berkeley, PhD(paleont), 66. Prof Exp: From asst prof to assoc prof, 65-75, assoc dean instr, 69-70, actg dean instr, 72-73, PROF EARTH SCI, CALIF STATE UNIV, HAYWARD, 75-, DEAN UNDERGRAD STUDIES, 73- Mem: AAAS; Paleont Soc; Geol Soc Am; Sigma Xi; Paleont Res Inst. Res: Paleozoic crinozoan and Mesozoic echinozoan echinoderms; fossil and recent irregular echinoids; Cenozoic biostratigraphy; historical geology, particularly the significance of plate tectonics in earth history. Mailing Add: Dean of Undergrad Studies Calif State Univ Hayward CA 94542

MINTZ, MICHAEL JEROLD, organic chemistry, see 12th edition

MINTZ, SHELDON, b Toronto, Ont, Jan 3, 40; m 64; c 2. MEDICAL RESEARCH. Educ: Univ Toronto, MD, 63; FRCP(C), 68. Prof Exp: Lectr, 70-71, ASST PROF MED, SCH MED, UNIV TORONTO, 71- Concurrent Pos: Assoc dir, Gage Res Inst, 75- Honors & Awards: Cecile Layman Mayor Award, Am Col Chest Physicians, 71. Mem: Am Thoracic Soc; Can Soc Clin Invest. Res: Alveolar macrophage metabolism;

adaptations to hypoxia; long-term continuous respiratory monitoring; environmental diseases. Mailing Add: Gage Res Inst 223 College St Toronto ON Can

MINTZ, YALE, b New York, NY, Mar 30, 16; m 42; c 3. METEOROLOGY. Educ: Dartmouth Col, AB, 37; Columbia Univ, MA, 42; Univ Calif, Los Angeles, PhD(meteorol), 49. Prof Exp: Instr meteorol, NY Univ, 43-44; from instr to assoc prof, 44-54, PROF METEOROL, UNIV CALIF, LOS ANGELES, 54- Concurrent Pos: Vis prof, Hebrew Univ, Jerusalem, 46-48 & 70- Honors & Awards: Meisinger Award, Am Meteorol Soc, 67. Mem: Fel Am Meteorol Soc. Res: Numerical calculation of global climates. Mailing Add: Dept of Meteorol Univ of Calif Los Angeles CA 90024

MINTZER, DAVID, b New York, NY, May 4, 26; m 49; c 2. GAS DYNAMICS, UNDERWATER ACOUSTICS. Educ: Mass Inst Technol, BS, 45, PhD(physics), 49. Prof Exp: Res assoc, US Navy Opers Eval Group, Mass Inst Technol, 46-48, res assoc, Acoust Lab, 48-49, mem staff, 49; asst prof physics, Brown Univ, 49-55; res assoc, Yale Univ, 55-56, assoc prof & dir lab marine physics, 56-62; assoc dean, 70-73, actg dean, 71-72, PROF MECH ENG & ASTRONAUT SCI, NORTHWESTERN UNIV, 62-, PROF ASTROPHYS, 68-, VPRES RES & DEAN SCI, 73- Concurrent Pos: Mem mine adv comt, Nat Acad Sci-Nat Res Coun, 63-73; mem bd trustees, EDUCOM, 75- Mem: Fel Am Phys Soc; fel Acoust Soc Am; Am Soc Mech Eng; Am Soc Eng Educ. Res: Kinetic theory; plasma physics; wave propagation. Mailing Add: Northwestern Univ Rebecca Crown Ctr 633 Clark St Evanston IL 60201

MINYARD, JAMES PATRICK, b Greenwood, Miss, May 11, 29; m 56; c 5. ORGANIC CHEMISTRY. Educ: Miss State Univ, BS, 51, PhD(org chem), 67. Prof Exp: From asst chemist to chemist, Miss State Chem Labs, 58-64; res chemist, Boll Weevil Res Lab, Agr Res Serv, US Dept Agr, 64-67; PROF CHEM, MISS STATE UNIV, 67-; STATE CHEMIST, MISS STATE CHEM LABS, 67- Concurrent Pos: Instr, Miss State Univ, 59-64. Mem: AAAS; fel Am Chem Soc; Am Oil Chem Soc; Asn Am Feed Control Off; Asn Am Plant Food Control Off (pres, 76). Res: Instrumental analysis; pesticide residues; natural products; organic reaction mechanisms; pesticide photochemistry; insect pheremones. Mailing Add: Box CR Miss State Univ Mississippi State MS 39762

MINZNER, RAYMOND ARTHUR, b Lawrence, Mass, June 9, 15; m 40; c 4. AERONOMY, METEOROLOGY. Educ: Mass State Col, BS, 37, MS, 40. Prof Exp: Lab instr physics, Mass State Col, 38-41; instr, Univ Ariz, 41-42; res assoc electronics, Mass Inst Technol, 42-45; electronic engr, Air Force Cambridge Res Ctr, 46-50, physicist & unit chief, 50-53, sect chief atmospheric standards, 53-56, chief compos sect, 56-58; staff physicist, Geophys Corp Am, 59-64, prin scientist, GCA Tech Div, GCA Corp, Mass, 65-67; aerospace technologist, Electronic Res Ctr, 67-70, AEROSPACE TECHNOLOGIST, GODDARD SPACE FLIGHT CTR, NASA, 70- Concurrent Pos: Mem comt for exten, US Standard Atmosphere, 53-, co-ed, 75. Mem: Am Meteorol Soc; Am Geophys Union; Am Inst Aeronaut & Astronaut; Sigma Xi. Res: Statistical studies and uncertainty of inferred meteorological parameters; stratospheric climatology; cloud motion vs wind relationships; cloud height from aircraft photos and satellite imagery; standard and model atmospheres. Mailing Add: Goddard Space Flight Ctr Greenbelt MD 20771

MIOVIC, MARGARET LANCEFIELD, b Salem, Ore, Apr 15, 43; m 64; c 3. MICROBIOLOGY. Educ: Radcliffe Col, BA, 65; Univ Pa, PhD(microbiol), 70. Prof Exp: Instr microbiol, Cornell Univ, 70-72 & biochem, 73-74; ASST PROF BIOL, SWARTHMORE COL, 75- Mem: Am Soc Microbiol. Res: Metabolic regulation in photosynthetic bacteria. Mailing Add: Martin Biol Bldg Swarthmore Col Swarthmore PA 19081

MIQUEL, JAIME, b Agres, Spain, Jan 7, 29; US citizen; m 55; c 2. GERONTOLOGY. Educ: Univ Granada, MSc, 50, PhD(pharmacol), 52. Prof Exp: NIH res assoc, Nat Inst Neurol Dis & Blindness, 58-61; res scientist, 61-65, CHIEF BR EXP PATH, AMES RES CTR, NASA, 65- Concurrent Pos: Span Inst Pharmacol fel, Span Ministry of Educ, Madrid, 52-54; Cajal Inst fel, Span Ministry of Educ, Valencia, 55-58. Mem: Geront Soc; Am Asn Neuropath; Span Soc Geront; Soc Invert Path. Res: Experimental gerontology; experimental neuropathology; radiobiology. Mailing Add: NASA Ames Res Ctr Moffett Field CA 94035

MIRABELLA, FRANCIS MICHAEL, JR, b Dec 27, 43; m 67. POLYMER CHEMISTRY. Educ: Univ Bridgeport, BA, 66; Univ Conn, MS, 74, PhD(polymer chem), 75. Prof Exp: Res chemist, Technichem Co, 66-69; analyst, Olin Corp, 69-72; teaching asst anal chem, Univ Conn, 72-75; RES SCIENTIST POLYMER EVAL, ARCO/POLYMERS INC, ATLANTIC RICHFIELD CO, 75- Mem: Am Chem Soc. Res: Theory of composition of copolymers and relationship of copolymer composition to molecular weight. Mailing Add: ARCO/Polymers Inc 440 College Park Dr Monroeville PA 15146

MIRABILE, FRANK ANDREW, organic chemistry, see 12th edition

MIRABITO, JOHN A, b Somerville, Mass, May 16, 17; m 39. METEOROLOGY. Educ: Wake Forest Col, BS, 41; Mass Inst Technol, cert, 43. Prof Exp: Meteorologist, Pac Fleet & Naval Air Reserve Training Prog, US Navy, 42-52, personnel & training officer meteorol, Off Chief Naval Oper, 52-54, staff meteorologist & oceanographer, Naval Support Forces, Antarctica, 54-59, exec officer, Fleet Weather Cent, DC, 59-61, staff meteorologist & oceanographer, 6th Fleet, 61-63; marine serv coordr, Environ Sci Serv Admin, 63-70; res prog mgr, Nat Marine Fisheries Serv, 71-72, PROG ANALYST, NAT OCEANIC & ATMOSPHERIC ADMIN, 72- Mem: AAAS; Am Meteorol Soc; Am Geophys Union. Res: Polar meteorology in Antarctica; oceanographic forecasts; marine science, air-sea interaction; fisheries research management. Mailing Add: Nat Oceanic & Atmospheric Admin 6010 Executive Blvd Rockville MD 20853

MIRACLE, CHESTER LEE, b Barbourville, Ky, Apr 30, 34; m 57; c 3. MATHEMATICS. Educ: Berea Col, 54, BA; Auburn Univ, MS, 56; Univ Ky, PhD(math), 59. Prof Exp: From instr to asst prof, 59-63, ASSOC PROF MATH, UNIV MINN, MINNEAPOLIS, 63- Mem: Am Math Soc; Math Asn. Res: Analytical continuation and summability of series. Mailing Add: Dept of Math Univ of Minn Minneapolis MN 55455

MIRAGLIA, GENNARO J, b Italy, Jan 18, 29; US citizen; m 54; c 3. MEDICAL MICROBIOLOGY, PHYSIOLOGY. Educ: St Bonaventure Univ, BS, 51; Univ NH, MS, 57; Univ Tenn, PhD(bact), 60. Prof Exp: Fel bact, Bryn Mawr Col, 61-62; asst prof bact, Seton Hall Col Med & Dent, 62-63; RES FEL, SQUIBB INST MED RES, 63- Mem: AAAS; Am Soc Microbiol; NY Acad Sci. Res: Host-bacteria interrelationships; gram-negative infections; effect of low ambient temperature on infections; chemotherapy of a number of bacterial infections. Mailing Add: Squibb Inst for Med Res Princeton NJ 08540

MIRALDI, FLORO D, b Lorain, Ohio, Mar 21, 31; m 57; c 1. BIOENGINEERING, NUCLEAR MEDICINE. Educ: Col Wooster, AB, 53; Mass Inst Technol, SB, 53, SM, 55, ScD(nuclear eng), 59; Case Western Reserve Univ, MD, 70. Prof Exp: Engr, Atomic Power Div, Westinghouse Elec Corp, 55; asst prof elec eng, Purdue Univ, 58-59; from asst prof to assoc prof nuclear eng, 59-69, assoc prof biomed eng, Schs Eng & Med, 69-72, PROF BIOMED ENG, SCHS ENG & MED, CASE WESTERN RESERVE UNIV, 72- Concurrent Pos: Asst radiologist, Univ Hosps of Cleveland, 71-72, radiologist, 72-; chief nuclear med, Cleveland Metrop Gen Hosp, 73-; mem adv comt biotechnol & human resources, NASA, 63-65; pres, Am Radiation Res Corp, 63-; mem res adv bd, Euclid Clin Found, 66-69. Mem: AAAS; Am Nuclear Soc; Soc Nuclear Med. Res: Radiological health; applications of radiation and radioisotopes; nuclear reactor physics; nuclear medical instrumentation. Mailing Add: Dept of Radiol Univ Hosps Cleveland OH 44106

MIRAND, EDWIN ALBERT, b Buffalo, NY, July 18, 26. BIOLOGY. Educ: Univ Buffalo, BA, 47, MA, 49; Syracuse Univ, PhD, 51. Hon Degrees: DSc, Niagara Univ, 70; DSc, D'Youville Col, 74. Prof Exp: Asst, Univ Buffalo, 47, instr biol, 48; instr, Utica Col, Syracuse, 50; assoc cancer res scientist, 51-60, asst to Inst Dir, 60-67, PRIN CANCER RES SCIENTIST, DIR CANCER RES, ROSWELL PARK MEM INST, 67-, ASSOC DIR, INST, 67- Concurrent Pos: Res prof, Grad Sch, State Univ NY Buffalo, 54-, prof & dir grad studies, 67-, head Roswell Park Grad Div, Grad Sch, 67-; prof, Niagara Univ, 68-, coun mem, 71-; prof, Canisius Col, 68-; mem comt tech guid, Nat Acad Sci-Nat Res Coun, Human Cancer Virus Task Force, NIH, 63-; consult, Cornell Aeronaut Labs, 62-; prof clin cancer comt mem, Nat Cancer Inst, 74- Honors & Awards: Billings Silver Medal, AMA, 60. Mem: Fel AAAS; Radiation Res Soc; Am Soc Zool; Soc Exp Biol & Med; Am Asn Cancer Res. Res: Abnormality of iron metabolism in iron deficiency anemias; erythropoietin; relationships of viruses to cancer; gnotobiology. Mailing Add: Roswell Park Mem Inst 666 Elm St Buffalo NY 14203

MIRANDA, GILBERT A, b Los Angeles, Calif, Oct 21, 43; m 63; c 4. CRYOGENICS, NUCLEAR MAGNETIC RESONANCE. Educ: Calif State Col, Los Angeles, BS, 65; Univ Calif, Los Angeles, MS, 66, PhD(solid state physics), 72. Prof Exp: APPL PHYSICIST, LOS ALAMOS SCI LAB, UNIV CALIF, 72- Res: Critical current measurements of 10-50 kiloampere superconducting cables, braids and monolithic wires. Mailing Add: Los Alamos Sci Labs MS-464 Los Alamos NM 87545

MIRANDA, HENRY A, JR, b New York, NY, Sept 20, 24; m 50; c 2. GEOENVIRONMENTAL SCIENCE. Educ: Iona Col, BS, 52; Fordham Univ, MS, 53, PhD(physics), 56. Prof Exp: Sr scientist, Hudson Labs, Columbia Univ, 56-60; sr consult engr, Braddock, Dunn & McDonald, Inc, Tex, 60-62; sr scientist & group leader radiation physics, Lowell Tech Inst Res Found, 62-65; sr scientist & dept mgr atmospheric sci, GCA Technol Div, 65-67, dir space sci lab, 67-72; VPRES & DIR RES, EPSILON LABS, INC, 72- Mem: Am Geophys Union; Am Phys Soc; Inst Elec & Electronics Eng. Res: Earth and atmospheric radiation; turbulent oceanic diffusion; scintillation counting; optical and infrared tracking systems; nuclear effects; spectroscopy; densitometry; optical instrumentation; laser development; aeronomy; upper-atmospheric winds and diffusion; stratospheric aerosols; aerosol physics; imaging systems; photogrammetry. Mailing Add: Epsilon Labs, Inc 4 Preston Ct Bedford MA 01730

MIRANDA, THOMAS JOSEPH, b Ewa Mill, Hawaii, Nov 18, 27; m 53; c 5. ORGANIC POLYMER CHEMISTRY. Educ: San Jose State Col, AB, 51, MA, 53; Univ Notre Dame, PhD(org chem), 59. Prof Exp: Instr chem, San Jose State Col, 52-53; chemist, Eitel-McCullough Inc, 53; dir res chem, O'Brien Corp, 59-69; sr res mat scientist, 69-71, STAFF SCIENTIST, WHIRLPOOL CORP, 71- Concurrent Pos: Tech ed, J Paint Technol. Honors & Awards: Ernest T Trigg Award, Fedn Socs Coatings Technol, 67. Mem: AAAS; Am Chem Soc; Am Oil Chem Soc. Res: Stereospecific polymerization of olefins; radiation polymerization of olefins; rocket fuels; emulsion polymerization; thermosetting acrylics; epoxy resins; water soluble polymers; thermal analysis; polymer stabilization. Mailing Add: Elisha Gray II Res & Eng Ctr Whirlpool Corp Monte Rd Benton Harbor MI 49022

MIRANKER, WILLARD LEE, b Brooklyn, NY, Mar 8, 32; m 52; c 3. MATHEMATICS. Educ: NY Univ, BA, 52, MS, 53, PhD(math), 57. Prof Exp: Asst math, NY Univ, 53-56; mem staff, Bell Tel Labs, Inc, 56-58; staff mathematician, Res Ctr, Int Bus Mach Corp, 59-63; sr res fel, Calif Inst Technol, 63-64; asst to dir res, 65, STAFF MATHEMATICIAN, RES CTR, IBM CORP, 65-, ASST DIR, MATH SCI DEPT, 72- Concurrent Pos: Adj prof, City Univ New York, 66-67; vis prof dept math, Hebrew Univ, Israel, 68-69; adj prof, Dept Math, NY Univ, 70-; vis prof, Univ Paris, 74-75; vis lectr, Yale Univ, 73-74. Mem: Fel AAAS; Am Math Soc; Soc Indust & Appl Math. Res: Applied mathematics; numerical analysis. Mailing Add: IBM Corp PO Box 218 Yorktown Heights NY 10598

MIRCETICH, SRECKO M, b Skela, Yugoslavia, Sept 2, 26; US citizen; m 56; c 1. PLANT PATHOLOGY. Educ: Univ Sarajevo, BS, 52; Univ Belgrade, MS, 54; Univ Calif, Riverside, PhD(plant path), 66. Prof Exp: Teaching asst plant path, Sch Agr, Univ Sarajevo, 49-52; asst plant pathologist, Exp Sta Subtrop Agr, Bar, Yugoslavia, 52-54, head plant protection dept, 54-57; res asst plant path, Univ Calif, Riverside, 58-66; res plant pathologist, Plant Sci Res Div, 66-73, PLANT PATHOLOGIST, AGR RES SERV, US DEPT AGR, WESTERN REGION, 73- Mem: Am Phytopath Soc. Res: Soilborne pathogens and diseases; soil ecology of Phytophthora and Pythium; soil populations; diseases of stone fruits; Phytophthora root and crown rot of deciduous fruit and nut trees. Mailing Add: Dept of Plant Path Univ of Calif Davis CA 95616

MIRES, RAYMOND WILLIAM, b Mansfield, Tex, Mar 16, 33; m 54; c 3. PHYSICS. Educ: Tex Tech Univ, BS, 55, MS, 60; Univ Okla, PhD(physics), 64. Prof Exp: Mathematician, Holloman Air Develop Ctr, Holloman AFB, NMex, 55-56; engr, Martin Co, 56-57; instr physics, Tex Tech Univ, 57-60; asst, Univ Okla, 60-64; from asst prof to assoc prof, 64-71, PROF PHYSICS, TEX TECH UNIV, 71- Concurrent Pos: Physicist, LTV Electrosysts, Inc, Tex, 66-67; Advan Res Projs Agency res grant, Tex Tech Univ, 67-72. Mem: AAAS; Am Phys Soc; Am Asn Physics Teachers. Res: Optical and magnetic properties of solids; atomic structure. Mailing Add: Dept of Physics Tex Tech Univ Lubbock TX 79409

MIRHEJ, MICHAEL EDWARD, b Dhour-Shweir, Lebanon, Dec 25, 31; Can citizen; m 60; c 2. TEXTILE TECHNOLOGY. Educ: McGill Univ, BSc, 58; Univ Western Ont, MSc, 59; Univ BC, PhD(chem oceanog, chem physics), 62. Prof Exp: Fel physics, Univ BC, 62-63; sr res chemist, Polymer Corp Ltd, Ont, 63-66; res chemist, 66-75, SR RES CHEMIST, CHATTANOOGA NYLON PLANT, E I DU PONT DE NEMOURS & CO, INC, 75- Mem: Am Chem Soc. Res: Nuclear magnetic resonance spin echo technique to study influence of paramagnetic oxygen on proton relaxation; polymer synthesis and characterization; auto regenerating dehydrogenation catalysts; development of new textile products. Mailing Add: Forest Park Dr Signal Mountain TN 37377

MIRKIN, BERNARD LEO, b New York, NY, Mar 31, 28; m 54; c 2. PEDIATRICS, CLINICAL PHARMACOLOGY. Educ: NY Univ, AB, 49; Yale Univ,

PhD(pharmacol), 53; Univ Minn, MD, 64. Prof Exp: Instr physiol & pharmacol, State Univ NY Downstate Med Ctr, 52-54, 56-57, asst prof pharmacol, 57-58, res assoc, 58-60; from asst prof to assoc prof, 66-72, PROF PEDIAT & PHARMACOL, MED SCH, UNIV MINN, MINNEAPOLIS, 72-, DIR DIV CLIN PHARMACOL, 66- Concurrent Pos: Consult, NIH & Nat Res Coun-Nat Acad Sci. Mem: Am Soc Pharmacol & Exp Therapeut; Am Soc Clin Pharmacol & Therapeut; Soc Pediat Res. Res: Pharmacology of neurohumoral mediators; developmental pharmacology. Mailing Add: Dept of Pediat & Pharmacol Univ of Minn Med Sch Minneapolis MN 55455

MIROCHA, CHESTER JOSEPH, b Cudahy, Wis, Feb 7, 30; m 52; c 6. PLANT PATHOLOGY. Educ: Marquette Univ, BS, 55; Univ Calif, PhD(plant path), 60. Prof Exp: Lab technologist, Univ Calif, 57-60; res plant pathologist, Union Carbide Chem Co, 60-63; from asst prof to assoc prof plant path & physiol, 63-71, assoc prof, 71-72, PROF PLANT PATH, UNIV MINN, ST PAUL, 72- Mem: Bot Soc Am; Am Phytopath Soc; Am Soc Plant Physiol; Japanese Soc Plant Physiol. Res: Host-parasite physiology and relationships; mycotoxins; fungus physiology; analytical chemistry. Mailing Add: Dept of Plant Path Univ of Minn St Paul MN 55101

MIRON, JERRY, polymer chemistry, see 12th edition

MIRON, SIMON, b Long Island City, NY, Sept 4, 15; m 46; c 3. ORGANIC CHEMISTRY. Educ: Rice Inst, BA, 36, MA, 38; Univ Pittsburgh, PhD(org chem), 41. Prof Exp: Org res chemist, Lion Oil Co, Ark, 41-44; sr res chemist, Int Minerals & Chem Corp, Ohio, 45-46; res chemist, Pan Am Ref Corp, 46-50, sr chemist, 50-56; res assoc, Am Oil Co, 56-60; res specialist, 60-74, SR RES SPECIALIST, DOW CHEMICAL USA, 74- Mem: Am Chem Soc; fel Am Inst Chem; Am Soc Test & Mat. Res: Organic analytical research; petroleum chemistry; characterization of high-boiling hydrocarbon materials; structure of conjunct polymers; styrene. Mailing Add: Dow Chem Co Bldg B-250 Freeport TX 77541

MIRONE, LEONORA, b New York, NY, May 15, 13. BIOCHEMISTRY. Educ: Col New Rochelle, BA, 34; Fordham Univ, MS, 46, PhD(biochem), 48. Prof Exp: From assoc prof to prof nutrit, Univ Ga, 48-57, head nutrit dept, 51-57, mem grad fac, 53-57; assoc prof org chem, 57-65, PROF CHEM, MANHATTAN COL, 65- Mem: Fel AAAS; Am Chem Soc; Am Pub Health Asn; NY Acad Sci. Res: Metabolism; organic mechanisms.

MIRONESCU, STEFAN GHEORGHE DAN, b Bucharest, Romania, Dec 26, 32; m 65; c 1. CELL BIOLOGY. Educ: Acad Med Sci, Bucharest, Romania, MD, 57, Inst Virol, PhD(chem & viral oncol), 71. Prof Exp: Asst prof anat, Fac Med, Acad Med Sci, Bucharest, Romania, 58, from sci investr to sr sci investr exp oncol, Inst Oncol, 61-67, sr sci investr chem carcinogenesis, Inst Endocrinol, 68-72; vis prof cell biol, Thomas Jefferson Univ, 72-74; SCI INVESTR CELL BIOL, BLOOD RES LAB, AM NAT RED CROSS, 74- Concurrent Pos: Vis sr scientist, NSF, 72. Honors & Awards: Stefan Nicolau Medal Sci Merit, Inst Virol, Acad Med Sci, Bucharest, Romania, 71. Mem: Am Asn Cancer Res; Sigma Xi; Biophys Soc. Res: Control of cell replication in normal, carcinogen-treated and tumor cells; factors controlling iniation of DNA synthesis, chemical carcinogenesis and cell cycle; preservation of cells by freezing. Mailing Add: Am Nat Red Cross Blood Res Lab 9312 Old Georgetown Rd Bethesda MD 20014

MIRSKY, ALFRED EZRA, cell biology, deceased

MIRSKY, ARTHUR, b Philadelphia, Pa, Feb 8, 27; m 61; c 1. ENVIRONMENTAL GEOLOGY, STRATIGRAPHY. Educ: Univ Calif, Los Angeles, BA, 50; Univ Ariz, MS, 55; Ohio State Univ, PhD(stratig geol), 60. Prof Exp: Field geologist, Atomic Energy Comn, 51-53; consult uranium geologist, 55-56; asst dir Inst Polar Studies, Ohio State Univ, 60-67, asst prof geol, 64-67; from asst to assoc prof, 67-72, PROF GEOL, IND UNIV-PURDUE UNIV, INDIANAPOLIS, 73-, CHMN DEPT, 69- Concurrent Pos: NSF res grant, 62-63. Mem: Geol Soc Am; Soc Econ Paleont & Mineral; Nat Asn Geol Teachers; Am Inst Prof Geologists. Res: Environmental and urban geology; geologic factors in urban planning; history of Man's use of geologic resources; medical geology; archaeological geology; stratigraphy-sedimentation and geologic hazards in urbanization. Mailing Add: Dept of Geol Ind Univ-Purdue Univ Indianapolis IN 46202

MIRSKY, I ARTHUR, experimental medicine, deceased

MIRSKY, ISRAEL, biomathematics, cardiology, see 12th edition

MIRSKY, JOSEPH HERBERT, b Mt Vernon, NY, Nov 15, 20; m 49; c 4. PHARMACOLOGY. Educ: City Col New York, BS, 42; Yale Univ, PhD(pharmacol), 54. Prof Exp: Asst pharmacologist, Nepera Chem Co, 46-50; res pharmacologist, Irwin Neisler & Co, 54-59; sr scientist, Smith Kline & French Labs, 59-67; sr scientist, 67-68, SR GROUP LEADER, USV PHARMACEUT CO, 68- Mem: AAAS; Am Soc Pharmacol & Exp Therapeut; NY Acad Sci. Res: Neuropharmacology. Mailing Add: USV Pharmaceut Co Yonkers NY 10701

MIRSKY, RHONA PEARSON, biochemistry, see 12th edition

MIRVISH, SIDNEY SOLOMON, b Cape Town, SAfrica, Mar 12, 29; m 60; c 2. CANCER. Educ: Univ Cape Town, BSc, 48, MSc, 50; Cambridge Univ, PhD(org chem), 55. Prof Exp: Lectr physiol, Univ Witwatersrand, 55-60; res assoc chem carcinogenesis, Weizmann Inst Sci, 61-69; assoc prof, 69-72, PROF CHEM CARCINOGENESIS, EPPLEY INST RES CANCER, UNIV NEBR MED CTR, OMAHA, 72- Concurrent Pos: Res fel med, Hadassah Med Sch, Hebrew Univ, Israel, 60-61; Eleanor Roosevelt Inst Cancer fel, McArdle Labs, Med Sch, Univ Wis-Madison, 65-66, res fel, Univ, 65-67. Mem: Am Chem Soc; Am Asn Cancer Res. Res: Chemical carcinogenesis; metabolism, formation and carcinogenic action of urethane and N-nitroso compounds. Mailing Add: Eppley Inst for Res in Cancer Univ of Nebr Med Ctr Omaha NE 68105

MIRVISS, STANLEY BURTON, b Minneapolis, Minn, Sept 15, 22; m 49; c 2. SYNTHETIC ORGANIC CHEMISTRY. Educ: Univ Wis, BS, 44, PhD(org chem), 50. Prof Exp: Asst org chem, Univ Wis, 46-49; from res chemist to proj leader, Chems Res Div, Esso Res & Eng Co, 50-60, res assoc, 60-63; SUPVR ORG CHEM RES, EASTERN RES CTR, STAUFFER CHEM CO, 63- Mem: AAAS; Am Chem Soc; NY Acad Sci; Am Inst Chem. Res: Organometallic chemistry; polymerization; catalysis; high pressure reactions; organic synthesis; organophorus chemistry. Mailing Add: Eastern Res Ctr Stauffer Chem Co Dobbs Ferry NY 10522

MIRZA, JOHN, b Baghdad, Iraq, Dec 1, 22; m 45; c 2. ORGANIC CHEMISTRY. Educ: DePauw Univ, AB, 44; Univ Ill, PhD(org chem, biochem & physiol), 49. Prof Exp: Res chemist, Rohm & Haas Co, 49-51; RES CHEMIST, MILES LABS, INC, 51-, VPRES, 70- Mem: Am Chem Soc. Res: Syntheses from 2-vinylpyridine; applications of dialdehyde polysaccharides. Mailing Add: Miles Labs, Inc 1127 Myrtle St Elkhart IN 46514

MISCH, DONALD WILLIAM, b Providence, RI, Jan 1, 29; m 59; c 3. CELL BIOLOGY. Educ: Northeastern Univ, BS, 53; Univ Mich, MS, 58, PhD(zool), 63. Prof Exp: Trainee electron micros, Biol Labs, Harvard Univ, 62-63; asst prof, 63-68, ASSOC PROF ZOOL, UNIV NC, CHAPEL HILL, 68- Concurrent Pos: Vis scholar, Dept Zool, Duke Univ, 75. Mem: AAAS; Am Soc Zool. Res: Properties of lysosomes; intestinal transport of sugar and amino acids; permeability properties of dimethylsulfoxide. Mailing Add: Dept of Zool Univ of NC Chapel Hill NC 27514

MISCH, PETER HANS, b Berlin, Ger, Aug 30, 09; nat US; m 47; c 3. GEOLOGY. Educ: Univ Göttingen, PhD(geol), 32. Prof Exp: Asst geol, Univ Göttingen, 32-33; geologist, Himalaya Exped to Nanga Parbat, 34-35; prof geol, Sun Yat-Sen Univ, 36-40 & Peking Univ, 40-46; from asst prof to assoc prof, 47-50, PROF GEOL, UNIV WASH, 50- Concurrent Pos: Guggenheim fel, 54; vis prof, Univ Bonn & Univ Göttingen, 59. Mem: Geol Soc Am; Mineral Soc Am; Geochem Soc; Am Asn Petrol Geol; Am Geophys Union. Res: Structural geology; metamorphic petrology. Mailing Add: Dept of Geol Sci Univ of Wash Seattle WA 98195

MISCH, ROBERT DAVID, physical chemistry, see 12th edition

MISCHKE, RICHARD E, b Bristol, VAa, Aug 19, 40; m 62; c 2. PARTICLE PHYSICS, NUCLEAR PHYSICS. Educ: Univ Tenn, BS, 61; Univ Ill, MS, 62, PhD(physics), 66. Prof Exp: From instr to asst prof physics, Princeton Univ, 66-71; staff mem, 71-75, ASSOC GROUP LEADER, MP DIV, LOS ALAMOS SCI LAB, 75- Mem: Am Phys Soc. Res: Experimental high energy physics; experimental research in particle and nuclear physics at medium energies. Mailing Add: MP Div Los Alamos Sci Lab Los Alamos NM 87544

MISCHLER, TERRENCE WYNN, b Milwaukee, Wis, Oct 12, 40; m 63; c 2. ENDOCRINOLOGY, CLINICAL PHARMACOLOGY. Educ: Mich State Univ, BS, 62, MS, 65, PhD(physiol), 67. Prof Exp: Res scientist, Warner-Lambert Res Inst, 67-71, clin res assoc, 71-72; ASSOC DIR CLIN PHARMACOL, SQUIBB INST MED RES, 72- Mem: Soc Clin Pharmacol & Therapeut; Soc Study Reproduction; NY Acad Sci. Res: Clinical pharmacology studies in the United States; anti-inflammatory, psychotropic and pharmacokinetic research. Mailing Add: Dept of Clin Pharmacol Squibb Inst for Med Res Princeton NJ 08540

MISEK, BERNARD, b New York, NY, Mar 29, 30; m 56; c 4. PHARMACEUTICAL CHEMISTRY, COSMETIC CHEMISTRY. Educ: Columbia Univ, BS, 51; Univ Md, MS, 53; Univ Conn, PhD(pharm), 56. Prof Exp: Pharmaceut chemist, Nepera Chem Co, 53-54; asst prof pharm, Univ Houston, 56-57; head pharm res & develop, Lloyd Brothers Inc, 57-59; dir med prod develop, Nopco Chem Co, 59-60; sr scientist explor res, Richardson-Merrell Inc, 60-64, asst dir, 64-66, dir, 66-72; MGR RES & DEVELOP, BEECHAM PRODS, 72- Mem: Am Chem Soc; Am Pharmaceut Asn; Soc Cosmetic Chem. Res: Pharmaceutical, toiletry and cosmetic product development; dermatological products; dentifrices; proprietary drugs; femine hygiene; hair care; personal products. Mailing Add: Tara Dr Pomona NY 10970

MISELIS, RICHARD ROBERT, b Boston, Mass, Mar 13, 45; m 65; c 1. NEUROBIOLOGY. Educ: Tufts Univ, BS, 67; Univ Pa, VMD, 73 & PhD(biol), 73. Prof Exp: Fel, Dept Neurophysiol, Col France, 73-75; ASST PROF ANIMAL BIOL, SCH VET MED, UNIV PA, 75- Mem: AAAS; Animal Behav Soc. Res: Neurological and physiological basis of feeding and drinking behavior. Mailing Add: Lab Anat Sch of Vet Med Univ of Pa Philadelphia PA 19174

MISENER, AUSTIN DONALD, b Toronto, Ont, Jan 19, 11; m 36; c 3. PHYSICS. Educ: Univ Toronto, BA & MA, 34; Cambridge Univ, PhD(physics), 38. Prof Exp: Res fel, Bristol Univ, 38-39; lectr & demonstr physics, Univ Toronto, 39-46, asst prof, 46-49; prof & head dept, Univ Western Ont, 49-60; dir, Ont Res Found, 60-65; assoc dir, Great Lakes Inst, 65-66, dir, 66-72, MEM STAFF, INST ENVIRON STUDIES, UNIV TORONTO, 72- Mem: Am Asn Physics Teachers; Am Geophys Union; assoc Arctic Inst NAm; Royal Soc Can; Can Asn Physicists; Eng Inst Can; Int Asn Gt Lakes Res; Can Inst Water Pollution Control. Res: Heat flow and geothermal gradients; physical limnology; water management; environmental engineering. Mailing Add: Inst for Environ Studies Univ of Toronto Toronto ON Can

MISENHIMER, HAROLD ROBERT, b Pocatello, Idaho, Feb 13, 31; m 53; c 4. OBSTETRICS & GYNECOLOGY. Educ: George Washington Univ, MD, 56. Prof Exp: Intern, LDS Hosp, Salt Lake City, Utah, 56-57, resident obstet & gynec, 57-60; staff physician, US Air Force Med Corps, Bergstrom AFB, Tex, 60-62, mem teaching staff, Wilford Hall Hosp, Lackland AFB, Tex, 62-64; asst prof, Sch Med, Univ Md, Baltimore, 64-70; from assoc prof to prof obstet & gynec & dir perinatal biol, Rush Med Col, 70-75; PROF OBSTET & GYNEC, SCH MED, TEX TECH UNIV, 75- Concurrent Pos: From asst chief to assoc chief obstet & gynec, Baltimore City Hosps, 64-70; instr, Sch Med, Johns Hopkins Univ, 67-70; fel placental physiol, Carnegie Inst Washington, 67-70. Mem: Am Bd Obstet & Gynec; Am Col Obstetricians & Gynecologists. Res: Placental structure and function in subhuman primates and human beings; studies of high risk pregnancy in general and the specific causes. Mailing Add: R E Thomason Gen Hosp 4815 Alameda Ave El Paso TX 79998

MISH, LAWRENCE BRONISLAW, b Stamford, Conn, Feb 27, 23; m 50; c 4. BOTANY. Educ: Univ Conn, AB, 50; Harvard Univ, AM, 50, PhD(bot), 53. Prof Exp: From instr to asst prof biol, Wheaton Col, 53-59; PROF BOT, BRIDGEWATER STATE COL, 59- Concurrent Pos: Res asst, United Fruit Co, 60-63; ecol consult, 71-; mem, SShore Nature Ctr. Mem: AAAS; Bot Soc Am. Res: Biology of lichens, their algae and fungi; marine and fresh water plankton; taxonomy of flowering plants. Mailing Add: Dept of Biol Bridgewater State Col Bridgewater MA 02324

MISHELEVICH, DAVID JACOB, b Pittsburgh, Pa, Jan 26, 42; m 63. COMPUTER SCIENCES, BIOMEDICAL ENGINEERING. Educ: Univ Pittsburgh, BS, 62; Johns Hopkins Univ, MD, 66, PhD(biomed eng), 70. Prof Exp: Intern med, Baltimore City Hosps, 66-67; staff assoc neurophysiol, Nat Inst Neurol Dis & Stroke, 67-69; exec vpres, Nat Educ Consults, 71-72; asst prof med comput sci, 72-74, ASSOC PROF MED COMPUT SCI & CHMN DEPT, SOUTHWESTERN MED SCH, UNIV TEX HEALTH SCI CTR DALLAS, 74-, ASST PROF INTERNAL MED & DIR MED COMPUT RES CTR, 72-; TECH ADV & ACTG DIR INFO SERV, DALLAS COUNTY HOSP DIST, 75- Concurrent Pos: NIH spec fel, Johns Hopkins Univ, 69-71, fel med & biomed eng, Sch Med, 69-71; attending physician, Dallas County Hosp Dist, 73-; adj assoc prof math sci, Univ Tex, Dallas, 74-; adj assoc prof math, Univ Tex, Arlington, 74-, adj assoc prof biomed eng, 75- Mem: Am Hosp Asn; AAAS; Am Soc Info Sci; Asn Comput Mach; Inst Elec & Electronics Eng. Res: Computer applications in medicine specializing in clinical information and laboratory automation systems; semantic analysis of natural language; computer-based text processing and publication; health evaluation; iconic communication and computer graphics. Mailing Add: Med Comput Resources Ctr Univ of Tex Health Sci Ctr Dallas TX 75235

MISHELOFF, MICHAEL NORMAN, b Brooklyn, NY, Feb 15, 44; m 71. PHYSICS. Educ: Calif Inst Technol, BS, 65; Univ Calif, Berkeley, PhD(physics), 70. Prof Exp: Instr physics, Princeton Univ, 70-75; RES ASSOC PHYSICS, BEHLEN LAB

PHYSICS, UNIV NEBR, LINCOLN, 75- Mem: Am Phys Soc. Res: Particle physics; strong interactions; s-matrix theory; multiperipheral dynamics. Mailing Add: Behlen Lab of Physics Univ of Nebr Lincoln NE 68508

MISHER, ALLEN, b New York, NY, Feb 12, 33; m 56; c 4. PHYSIOLOGY, PHARMACOLOGY. Educ: Philadelphia Col Pharm, BSc, 59; Univ Pa, PhD(physiol), 64. Prof Exp: Sr pharmacologist, Menley & James Labs, Ltd, 64-65; asst sect head pharmacol, 65-66, sect head, 66-67, dir sect pharmacol, 67-71, VPRES RES US, PHARMACEUT PROD, SMITH KLINE & FRENCH LABS, 71- Mem: AAAS; Am Fedn Clin Res; Am Col Neuropsychopharmacol; Am Soc Pharmacol & Exp Therapeut. Mailing Add: Res Dept F-30 Smith Kline & French Labs Philadelphia PA 19101

MISHKIN, ABRAHAM RUDOLPH, biochemistry, see 12th edition

MISHKIN, ELI ABSALOM, b Poland, Apr 25, 17; m 47; c 2. APPLIED PHYSICS. Educ: Israel Inst Technol, Ingenieur, 42, ScD, 52. Prof Exp: Lectr elec eng, Israel Inst Technol, 43-53; asst prof, Mass Inst Technol, 53-55; assoc prof, 55-60, PROF ELEC ENG, POLYTECH INST BROOKLYN, 60- Concurrent Pos: Res fel physics, Harvard Univ, 68; consult, Lawrence Livermore Lab, 73- Mem: Am Phys Soc; NY Acad Sci. Res: Electromagnetic theory; automatic control systems; quantum and nonlinear optics; laser induced fusion. Mailing Add: Dept of Elec Eng & Appl Physics Polytech Inst of Brooklyn Brooklyn NY 11201

MISHKIN, FREDERICK SEYMORE, b Elkhart, Ind, Dec 6, 38; m 64; c 2. NUCLEAR MEDICINE. Educ: Univ Chicago, MD, 62. Prof Exp: Resident radiol, Med Ctr, Ind Univ, Indianapolis, 63-66, assoc prof, 68-71; PROF RADIOL, CHARLES R DREW POSTGRAD MED SCH & ADJ PROF, SCH MED, UNIV CALIF, LOS ANGELES, 71- Concurrent Pos: Fel nuclear med, Johns Hopkins Univ, 66; consult, Vet Admin, Indianapolis, 69-71; dir nuclear med div, Martin Luther King, Jr Hosp, 71- Mem: AAAS; Soc Nuclear Med; AMA; Asn Univ Radiol. Res: Clinical nuclear medicine; application of techniques of radioisotope tracers to the diagnosis and treatment of disease. Mailing Add: Dept of Radiol King Hosp 12021 Wilmington Ave Los Angeles CA 90059

MISHMASH, HAROLD EDWARD, b Richmond, Calif, Nov 8, 42; m 64; c 3. ANALYTICAL CHEMISTRY. Educ: Iowa State Univ, BS, 64; Kans State Univ, PhD(anal chem), 68. Prof Exp: RES SPECIALIST, CENT RES LABS, 3M CO, 68- Mem: Microbeam Anal Soc. Res: Electron microprobe detectors; scanning electron microscopy of material defects; emission spectroscopy sources; surface studies using ion scattering and secondary ion mass spectrometry. Mailing Add: Cent Res Labs 3M Co PO Box 33221 St Paul MN 55133

MISHOE, LUNA I, b Bucksport, SC, Jan 5, 17; m 44; c 4. MATHEMATICS, MATHEMATICAL PHYSICS. Educ: Allen Univ, BSc, 38; Univ Mich, MSc, 42; NY Univ, PhD(math), 53. Prof Exp: Prof math & physics & head dept, Kittrell Col, 39-42 & Del State Col, 46-48; from assoc prof to prof physics, Morgan State Col, 48-60, chmn div natural sci, 57-60; PRES, DEL STATE COL, 60- Concurrent Pos: Consult, Ballistics Res Lab, Aberdeen Proving Ground, 57-60; mem bd dirs, Univ City Sci Ctr, Pa; mem bd trustees, Univ Del. Mem: Am Math Soc; Soc Indust & Appl Math; Math Asn Am. Res: Ordinary differential equations; eigenfunction expansions; boundary value problems; ballistics; physics of upper atmosphere. Mailing Add: Off of the President Delaware State Col Dover DE 19901

MISHUCK, ELI, b Buenos Aires, Arg, July 27, 23; nat US; m 44; c 2. PHYSICAL CHEMISTRY. Educ: Brooklyn Col, BA, 44; Polytech Inst Brooklyn, MS, 46, PhD(chem), 50. Prof Exp: Res assoc chem, Manhattan Proj, SAM Labs, Columbia Univ, 44-46; chief chemist, Bingham Bros Co, NJ, 46-48; res chemist polymers, Polytech Inst Brooklyn, 48-50; head dept solid propellant res, Aerojet-Gen Corp, Gen Tire & Rubber Co, 50-63, sr mgr chem & biol opers, Space-Gen Corp, 63-66, dir chem & biol systs, 66-70, pres, Aerojet Med & Biol Systs, 70-74; GEN MGR, ORGANON DIAGNOSTICS, INC, 74- Concurrent Pos: Mem tech adv bd to pres, Gen Tire & Rubber Co. Res: Biological and chemical detection; medical diagnostic equipment. Mailing Add: Organon Diagnostics Inc 9060 E Flair Dr El Monte CA 91731

MISIEK, MARTIN, b Buffalo, NY, Sept 6, 19; m 45; c 2. MICROBIOLOGY. Educ: Univ Buffalo, BA, 43; Syracuse Univ, MS, 49, PhD(microbiol), 55. Prof Exp: SR MICROBIOLOGIST & PROJ SUPVR ANTIBIOTICS, BRISTOL LABS, SYRACUSE, 43- Mem: Am Soc Microbiol. Res: In vitro evaluation of antibacterial agents. Mailing Add: Bristol Labs Syracuse NY 13201

MISKEL, JOHN ALBERT, b San Francisco, Calif, Aug 21, 19; m 49. CHEMISTRY. Educ: Univ Calif, BS, 43; Washington Univ, PhD(chem), 49. Prof Exp: Jr scientist, Manhattan Proj, Univ Calif, 43-46; asst phys chem, Washington Univ, 46-48, AEC fel phys sci, 48-49; from res assoc to chemist, Brookhaven Nat Lab, 49-55; CHEMIST, LAWRENCE LIVERMORE LAB, UNIV CALIF, 55- Mem: Am Phys Soc. Res: Chemistry of heavy elements; fission fragments; ranges of fission fragments; neutron cross sections; absolute beta counting; nuclear excitation functions; radiochemistry. Mailing Add: Lawrence Livermore Lab Univ Calif PO Box 808 Livermore CA 94550

MISKEL, JOHN J, (SR), b New York, NY, Sept 14, 11; m 33; c 1. ORGANIC CHEMISTRY, PHARMACEUTICAL CHEMISTRY. Educ: Brooklyn Col, BS, 34. Prof Exp: Res chemist, Allied Chem Corp, 34-38; tech dir & asst vpres pharmaceut div, Nopco Chem Co, 38-49; dir pharmaceut prod, Chas Pfizer & Co, 49-60; vpres res & develop, White Labs, Inc Div, Schering Corp, 60-68, DIR LABS, RES, DEVELOP & QUAL CONTROL, R P SCHERER CORP, 68- Mem: Fel AAAS; fel Am Inst Chemists; Am Chem Soc; fel Int Acad Law & Sci; Pharmaceut Mfrs Asn. Mailing Add: R P Scherer Corp 9425 Grinnell Ave Detroit MI 48213

MISKEL, JOHN JOSEPH, JR, b Brooklyn, NY, Aug 2, 33; m 55; c 7. POLYMER CHEMISTRY, ORGANIC CHEMISTRY. Educ: Univ Notre Dame, BS, 55; Univ Pa, PhD(org chem), 60. Prof Exp: Polymer chemist, Atlantic Ref Co, 59-60; supvr polymer develop, Rexall Chem Co, 60-68; DIR RES & DEVELOP SPEC CHEM, NOPCO DIV, DIAMOND SHAMROCK CHEM CO, 60- Mem: Am Chem Soc; Soc Chem Indust; Am Oil Chem Soc. Res: Polyolefin process and product research; specialty polymers and surfactants. Mailing Add: Nopco Chem Div Diamond Shamrock Chem Co Box 2386R Morristown NJ 07960

MISKIMEN, CARMEN RIVERA, b Mayaguez, PR, Mar 15, 33; m 63; c 4. PLANT PATHOLOGY, VIROLOGY. Educ: Univ PR, Mayaguez, BS, 53; Univ Wis-Madison, PhD(plant path), 62. Prof Exp: Instr sci, Commonwealth of PR Dept Educ, 53-55; technician III, Territorial Exp Sta Div, US Dept Agr, PR, 55-58, res plant pathologist, Crops Res Div, 62-63; from asst prof to assoc prof, 63-73, PROF BIOL, UNIV PR, MAYAGUEZ, 73-, COORDR MED TECHNOL, 73- Concurrent Pos: Sugarcane grant, 73- Mem: Am Phytopath Soc. Res: Mosaic and other viruses of economically important crops; morphology and histology of insect sensory structures. Mailing Add: Dept of Biol Univ of PR Mayaguez PR 00708

MISKIMEN, GEORGE WILLIAM, b Appleton, Wis, May 21, 30; m 63; c 4. NEUROPHYSIOLOGY, POPULATION ECOLOGY. Educ: Ohio Univ, BS, 53, MS, 55; Univ Fla, PhD(biol, entom), 66. Prof Exp: Entomologist, VI Agr Prog, US Dept Agr, 58-61, invests leader, Grain & Forage Insects, Agr Res Serv, PR, 62-66; dir entomol pioneering res lab, 66-72, PROF BIOL, UNIV PR, 66- Concurrent Pos: USDA Hatch grant, Univ PR, 64-67; USDA res grant, 72-; NIH res grant, 73- Mem: Entom Soc Am; Asn Trop Biol; Soc Syst Zool; Int Soc Sugarcane Technol; Coleopterists Soc. Res: Ecology of insects; electrophysiology of insect sensory structures; population dynamics; relationships between agriculturally important pests, climate, and biological control organisms; new approaches to insect control; electrophysiology of insects. Mailing Add: Dept of Biol Univ of PR Mayaguez PR 00708

MISKIMEN, MILDRED, b Coshocton, Co, Ohio, Aug 16, 01. ZOOLOGY. Educ: Muskingum Col, BA, 23; Ohio State Univ, MS, 49, PhD, 52. Prof Exp: Orphanage supt & sch adminr, Methodist Mission, Dhulia, India, 23-41; teacher pub sch, Ohio, 43-47; asst prof physiol, Miami Univ, 52-58; vis lectr, Rutgers Univ, 59-60, from asst prof to assoc prof zool, Douglas Col, 60-68; assoc prof, 68-73, ADJ ASSOC PROF ZOOL, FRANZ THEODORE STONE LAB, OHIO STATE UNIV, 73- Mem: AAAS; Wilson Ornith Soc; Cooper Ornith Soc; Am Ornith Union; Acad Zool. Mailing Add: F T Stone Lab Col Biol Sci Ohio State Univ Columbus OH 43210

MISKIN, KOY ELDRIDGE, plant breeding, plant physiology, see 12th edition

MISKOVSKY, NICHOLAS MATTHEW, b Passaic, NJ, Oct 4, 40; m 71; c 1. PHYSICS. Educ: Rutgers Univ, New Brunswick, AB, 62; Pa State Univ, PhD(physics), 70. Prof Exp: ASST PROF PHYSICS, PA STATE UNIV, 70- Res: Theoretical solid state physics; optical properties and band structure of solids; calculation of phonon spectra. Mailing Add: Dept of Physics Pa State Univ Altoona PA 16603

MISKUS, RAYMOND P, b Westville, Ill, June 11, 26; m 47. ORGANIC CHEMISTRY, ANALYTICAL CHEMISTRY. Educ: Univ Ill, BS, 50. Prof Exp: Bacteriologist, US Standard Pharmaceut, 50-51; food bacteriologist, US Army, Calif, 51-52; lab technician, Univ Calif, Berkeley, 52-64; CHEMIST, US DEPT AGR FORESTRY, 64- Mem: Am Chem Soc; Entom Soc Am. Res: Chemistry of insecticides including development of analytical methods and metabolic fate of pesticides in and on insects and plants. Mailing Add: 10 Las Palomas Orinda CA 94563

MISLEVY, PAUL, b Scranton, Pa, May 5, 41; m 64; c 1. AGRONOMY. Educ: Pa State Univ, BS, 66, MS, 69, PhD(agron), 71. Prof Exp: ASST PROF AGRON, UNIV FLA, 71- Mem: Am Soc Agron; Am Forage & Grassland Coun; Weed Sci Soc Am. Res: Forage management; crop physiology; herbicides; extension. Mailing Add: Agr Res Ctr Univ of Fla Ona FL 33865

MISLIVEC, PHILIP BRIAN, b Danville, Ill, Feb 12, 39. MYCOLOGY, PLANT PATHOLOGY. Educ: St Meinrad Col, BS, 61; Ind State Univ, Terre Haute, MA, 65; Purdue Univ, PhD(plant path), 68. Prof Exp: Asst prof bot, Ind State Univ, Terre Haute, 68-69; RES MYCOLOGIST, US FOOD & DRUG ADMIN, 69- Concurrent Pos: Prof, US Dept Agr Grad Sch, DC, 70- Mem: Mycol Soc Am; Soc Indust Microbiol; Am Inst Biol Sci; Am Soc Microbiol. Res: Effects of environment and microbial competition on growth and toxin production by mycetoxin-producing mold species. Mailing Add: Food & Drug Admin 200 C St SW Washington DC 20204

MISLOVE, MICHAEL WILLIAM, b Washington, DC, Feb 8, 44. MATHEMATICS. Educ: Univ of the South, BA, 65; Univ Tenn, Knoxville, PhD(math), 69. Prof Exp: Asst prof math, Univ Fla, 70; asst prof, 70-75, ASSOC PROF MATH, TULANE UNIV LA, 75- Concurrent Pos: NSF res grant, Tulane Univ La, 71-75; res fel, Alexander von Humboldt Found, 76. Mem: AAAS; Am Math Soc; Sigma Xi. Res: Topological semigroups, abstract harmonic analysis on groups and semigroups. Mailing Add: Dept of Math Tulane Univ New Orleans LA 70118

MISLOW, JACQUELINE FORD, cytology, endocrinology, see 12th edition

MISLOW, KURT MARTIN, b Berlin, Ger, June 5, 23; nat US; m 48; c 2. CHEMISTRY. Educ: Tulane Univ, BS, 44; Calif Inst Technol, PhD(org chem), 47. Hon Degrees: Dr H C, Free Univ Brussels, 74; DSc, Tulane Univ, 75. Prof Exp: From instr to prof chem, NY Univ, 47-64; chmn dept, 68-74, H S TAYLOR PROF CHEM, PRINCETON UNIV, 64- Concurrent Pos: Guggenheim fel, 56, 75; Sloan fel, 59; FMC lectr, Princeton, 63; mem adv panel, NSF, 63-66; mem med and org chem panel, NIH, 63-66; univ lectr, Univ London, 65; J A McRae mem lectr, Queen's Univ, 67; mem bd eds, J Org Chem, 65-70; H A Iddles lectr, Univ NH & Solvay lectr, Univ Brussels, 72; E C Lee lectr, Univ Chicago, 73; Churchill fel, Univ Cambridge, 75. Honors & Awards: James Flack Norris Award, Am Chem Soc, 75. Mem: Nat Acad Sci; AAAS; Am Chem Soc; The Chem Soc; fel Am Acad Arts & Sci. Res: Stereochemistry; mechanisms of organic reactions. Mailing Add: Dept of Chem Princeton Univ Princeton NJ 08540

MISNER, CHARLES WILLIAM, b Jackson, Mich, June 13, 32; m 59; c 4. THEORETICAL PHYSICS, COSMOLOGY. Educ: Univ Notre Dame, BS, 52; Princeton Univ, MA, 54, PhD(physics), 57. Prof Exp: Instr, Princeton Univ, 56-59, asst prof, 59-63; assoc prof physics & astron, 63-66, PROF PHYSICS & ASTRON, UNIV MD, COLLEGE PARK, 66- Concurrent Pos: Sloan res fel, 58-62; lectr, Les Houches, Univ Grenoble, 63; NSF sr fel, Univ Cambridge, 66-67 & Niels Bohr Inst, Copenhagen, 67; vis prof, Princeton Univ, 69-70 & Calif Inst Technol, 72; Guggenheim fel, 72; vis fel, All Souls Col, Oxford, 73; mem, NSF Adv Comt Res, 73- Honors & Awards: Notre Dame Sci Centennial Award, 65. Mem: AAAS; Am Phys Soc; Am Math Soc; Royal Astron Soc; Philos Sci Asn. Res: General relativity; relativistic astrophysics; philosophy of physics. Mailing Add: Dept of Physics & Astron Univ of Md College Park MD 20742

MISNER, ROBERT DAVID, b Waynesville, Ill, May 1, 20; m 49; c 2. PHYSICS. Educ: George Washington Univ, BS, 46. Prof Exp: Physicist, 42-44, electronic engr, 45-54, sect head data processing, 54-66, BR HEAD SIGNAL EXPLOITATION, US NAVAL RES LAB, 66- Honors & Awards: Cert Commendation, US Navy, 59, Distinguished Civilian Serv Award, 63. Mem: Sigma Xi. Res: Radio frequency intercept and data handling, especially data storage and processing techniques; magnetic tape storage. Mailing Add: 5815 Blackhawk Dr Washington DC 20021

MISNER, ROBERT E, b Yonkers, NY, May 25, 41; m 68. ORGANIC CHEMISTRY. Educ: Manhattan Col, BS, 63; Fordham Univ, PhD(org chem), 68. Prof Exp: Teaching asst org chem, Fordham Univ, 63-65, res asst, 65-67; SR RES CHEMIST, AM CYANAMID CO, 67- Mem: Am Chem Soc. Res: Organic Synthesis of basic organic chemicals; heterocyclic synthesis; research and development of dyes and organic intermediates. Mailing Add: Am Cyanamid Co Bound Brook NJ 08854

MISRA, DWARIKA NATH, b Sarai-Miran, India, Mar 17, 33; m 54; c 3. PHYSICAL

CHEMISTRY, SURFACE CHEMISTRY. Educ: Univ Lucknow, BS, 51, MSc, 53; Howard Univ, PhD(phys chem), 63. Prof Exp: Res asst phys chem, Regional Res Lab, Hyderabad, India, 54-58; NSF fel, Pa State Univ, 63-66; sr scientist, Itek Corp, Mass, 66-68; lectr, Howard Univ, 70-72; RES ASSOC, AM DENT ASN HEALTH FOUND, 72- Mem: Am Chem Soc; Sigma Xi. Res: Adsorption and catalysis on heterogeneous surfaces; surface chemistry of semiconducting oxides and hydroxyapatite, behavior and kinetics of adsorbed organic molecules; chemical bond between bone material and resins. Mailing Add: Am Dent Asn Health Found Res Unit Nat Bur of Standards Washington DC 20234

MISRA, HARA PRASAD, b Khallikote, Orissa, India, June 1, 40; m 62; c 3. BIOCHEMISTRY, VETERINARY MEDICINE. Educ: Utkal Univ, BVSc & AH, 62; Va Polytech Inst & State Univ, MS, 68, PhD(biochem), 70. Prof Exp: Vet asst surgeon, Orissa Govt, India, 62-64; instr biochem, Orissa Univ Agr & Technol, 64-66; res asst poultry sci, Va Polytech Inst & State Univ, 66-68, res asst biochem, 68-70; res assoc, Duke Univ, 70-73; ASST PROF MICROBIOL, UNIV ALA, BIRMINGHAM, 73- Concurrent Pos: Assoc scientist, Cancer Res Ctr, Univ Ala, Birmingham, 73- Mem: Am Soc Microbiol; Poultry Sci Asn. Res: Superoxide free radical and superoxide dismutase; free radical pathology; aflatoxin and liver cancer; ionizing radiation; aging. Mailing Add: Dept of Microbiol Univ of Ala Med Ctr Birmingham AL 35294

MISRA, KRISHNA PRASAD, biochemistry, see 12th edition

MISRA, RAJENDRA KUMAR, b Nov 1, 31. STATISTICS. Educ: Univ Edinburgh, PhD(biostatist), 63. Prof Exp: Res fel, Univ Sask, 61-63; res assoc, Univ Man, 64-65; asst prof statist, Va Polytech Inst, 65-66; ASSOC PROF ZOOL, UNIV WESTERN ONT, 66- Mem: Biomet Soc. Res: Statistical methods in biometric research and experimental designs associated with inheritance studies in genetics. Mailing Add: Dept of Zool Univ of Western Ont London ON Can

MISRA, SUSHIL, b Budaun, India, Sept 5, 40. NUCLEAR PHYSICS, SOLID STATE PHYSICS. Educ: Agra Univ, BSc, 58; Gorakhpur Univ, MSc, 60; Univ St Louis, PhD(physics), 64. Prof Exp: Jr res scholar physics, Indian Asn Cultivation Sci, Calcutta, 60-61; fel, Univ Toronto, 64-67; asst prof, 67-70, ASSOC PROF PHYSICS, CONCORDIA UNIV, 70- Concurrent Pos: Consult, Electronics & Equip Div, McDonnell Aircraft Corp, Mo, 64. Mem: Am Phys Soc. Res: Electron paramagnetic resonance; nuclear quadrupole resonance; reorientation of oriented nuclei; Mössbauer detection of dynamically oriented nuclei; low-temperature magnetic ordering; spin-diffusion; nuclear rotational spectrum. Mailing Add: Dept of Physics Concordia Univ Montreal PQ Can

MISSERT, RAYMOND FRANCIS, physics, see 12th edition

MISSIMER, JOHN HERTEL, b Audubon, NJ, June 2, 44. PARTICLE PHYSICS, NUCLEAR PHYSICS. Educ: Carnegie-Mellon Univ, BSc, 66; State Univ NY Stony Brook, PhD(physics), 72. Prof Exp: Sr res asst theoret physics, Rudjer Boskovic Inst, Zagreb, Yugoslavia, 72-74; RES PHYSICIST, CARNEGIE-MELLON UNIV, 74- Mem: Am Phys Soc. Res: Nature of weak interactions as manifested in parity violating effects in nuclei. Mailing Add: Dept of Physics Carnegie-Mellon Univ Pittsburgh PA 15213

MISTRETTA, CHARLOTTE MAE, b Washington, DC, Feb 16, 44; m 68. NEUROPHYSIOLOGY, DEVELOPMENTAL PHYSIOLOGY. Educ: Trinity Col, Washington, DC, BA, 66; Fla State Univ, MS, 68, PhD(sensory physiol), 70. Prof Exp: Am Asn Univ Women fel, Nuffield Inst Med Res, Oxford Univ, 70-72; sr res assoc, 72-74, asst res scientist, 74-76, ASSOC RES SCIENTIST, DEPT ORAL BIOL, SCH DENT, UNIV MICH, ANN ARBOR, 76-, ASSOC RES SCIENTIST, CTR HUMAN GROWTH & DEVELOP, 75-, ASSOC PROF RES, SCH NURSING, GRAD STUDIES, 75- Concurrent Pos: Lectr, Ctr Human Growth & Develop, Univ Mich, Ann Arbor, 73-75; mem int comn olfaction & taste, Int Union Physiol Sci, 74- Mem: AAAS; Am Physiol Soc; Europ Chemoreception Res Orgn; Int Soc Develop Psychobiol; Soc Neurosci. Res: Investigation of the development of the sense of taste using anatomical, behavioral and electrophysiological techniques, including study of fetal swallowing activity; study of nerve-epithelium tissue interactions in development. Mailing Add: Dept of Oral Biol Univ of Mich Sch of Dent Ann Arbor MI 48104

MISTRIC, WALTER J, JR, b Opelousas, La, Dec 3, 24; m 46; c 2. ENTOMOLOGY. Educ: La State Univ, BS, 49; Tex A&M Univ, MS, 50, PhD(entom), 54. Prof Exp: Agr inspector, State Dept Agr, La, 46; asst plant path, La State Univ, 48-49; field aide, Bur Entom & Plant Quarantine, US Dept Agr, 49-50, agent, 50; from instr to asst prof entom, Tex A&M Univ, 50-55; from asst prof to assoc prof, 55-65, PROF ENTOM, NC STATE UNIV, 65- Mem: Entom Soc Am. Res: Insect control. Mailing Add: Dept of Entom NC State Univ Box 5215 Raleigh NC 27607

MISTRY, NARIMAN BURJOR, b Bombay, India, Oct 21, 37; m 64; c 2. PHYSICS. Educ: Univ Bombay, BS, 56; Columbia Univ, MA, 60, PhD(neutrino physics), 63. Prof Exp: Res asst high-energy physics, Columbia Univ, 63-64; instr & res assoc, 64-67, SR RES ASSOC HIGH-ENERGY PHYSICS, CORNELL UNIV, 67- Mem: Am Phys Soc. Res: High energy particle physics; cosmic rays; spark chambers; high energy neutrino interactions; weak interaction physics; high energy photon and electron physics. Mailing Add: Dept of Physics Cornell Univ Ithaca NY 14850

MISTRY, SORAB PIROZSHAH, b Bombay, India, Dec 18, 20; nat US; m 53; c 2. BIOCHEMISTRY. Educ: Univ Bombay, BSc, 42; Indian Inst Sci, Bangalore, MSc, 46; Cambridge Univ, PhD(biochem), 51. Prof Exp: Sr res asst, Indian Inst Sci, Bangalore, 45-47; instr, Dunn Nutrit Lab, Cambridge Univ, 48-50, Med Res Coun fel, 50-52; fel, 52-54, res assoc, 54, from asst prof to assoc prof animal nutrit, 54-62, assoc prof, 62-66, PROF BIOCHEM, UNIV ILL, URBANA, 66- Concurrent Pos: Sreenivasaya res award, 45-46; vis prof, Univ Amsterdam, 56 & Univ Zurich, 56-57, 63-64; NIH spec fel, 63-64; vis prof, Bangalore Univ, 70 & Univ Madrid, 71. Mem: Am Soc Biol Chem; Am Inst Nutrit; Brit Biochem Soc. Res: Nutritional and comparative biochemistry; regulation of metabolic processes. Mailing Add: Dept of Animal Sci Univ of Ill Urbana IL 61801

MITACEK, EUGENE JAROSLAV, b Hluk, Moravia, Czech, Apr 22, 35; US citizen; m 64; c 1. BIOCHEMISTRY. Educ: Palacky Univ, Czech, MA, 58; Charles Univ, Prague, PhD(chem & chem educ), 66. Prof Exp: Instr chem, State Lyceum, Veseli na Morave, 58-60, asst prof chem & res assoc chem educ, Inst Educ, Charles Univ, Prague, 60-66, sr res assoc prof chem & chem educ, 66-67; head dept earth sci, UN Int Sch, NY, 67-70; asst prof chem, 70-74, ASSOC PROF CLIN CHEM, DEPT LIFE SCI, NY INST TECHNOL, 74-; ASSOC PROF BIOCHEM, DEPT BASIC HEALTH SCI, COLUMBIA INST CHIROPRACTIC, 75- Concurrent Pos: Lectr, Inst Educ, Univ Warsaw, 61 & 64; consult, Prague Br, UNESCO, Paris, 64-67; res fel, Inst Environ Med, NY Univ Med Sch, 74-75. Mem: AAAS; Am Chem Soc; Nat Sci Teachers Asn; NY Acad Sci. Res: Geochemistry; polarographic analysis; structural chemistry; comparative chemical education; effectiveness of teaching methods in chemistry; motivation and learning in chemistry; biochemistry of cancer; clinical chemistry; environmental chemistry. Mailing Add: Dept of Life Sci NY Inst of Technol Old Westbury NY 11568

MITACEK, PAUL, JR, b Johnson City, NY, Jan 1, 32; m 67. PHYSICAL CHEMISTRY. Educ: Oberlin Col, AB, 54; Pa State Univ, PhD(chem), 62. Prof Exp: Res asst chem, Cryogenic Lab, Pa State Univ, 61-62; instr, Bucknell Univ, 62-63; resident res assoc radiation physics, Argonne Nat Lab, 63-67; asst prof, 67-71, ASSOC PROF CHEM, ST JOHN FISHER COL, 71-, CHMN DEPT, 70- Mem: Am Chem Soc. Res: Cryogenics; radiation dosimetry of mixed radiations of neutrons and gamma rays. Mailing Add: 164 Willowbend Rd Rochester NY 14618

MITALAS, ROMAS, b Kaunas, Lithuania, Feb 28, 33; nat Can. THEORETICAL PHYSICS, ASTROPHYSICS. Educ: Univ Toronto, BA, 57, MA, 58; Cornell Univ, PhD(physics), 64. Prof Exp: Asst prof physics, 64-68, asst prof astron, 68-72, ASSOC PROF ASTRON, UNIV WESTERN ONT, 73- Mem: Am Phys Soc; Am Astron Soc; Can Astron Soc. Res: Stellar structure and evolution. Mailing Add: Dept of Astron Univ of Western Ont London ON Can

MITCH, EUGENE LEONARD, organic chemistry, see 12th edition

MITCH, FRANK ALLAN, b State College, Pa, Apr 21, 20; m 67. ORGANIC CHEMISTRY. Educ: Pa State Col, AB, 41, MS, 48. Prof Exp: Researcher, Photo Prod, E I du Pont de Nemours & Co, Inc, 41-45; control, Floridin Co, 49-51, researcher, 51-58; mem staff res & develop, Ariz Chem Co, 58-61; MGR TECH SERV, GLIDDEN ORGANICS, 61- Mem: AAAS; Am Chem Soc; Soc Soft Drink Technol. Res: Soft drinks; tall oil; terpenes; essential oils. Mailing Add: 4134 Rogero Rd Jacksonville FL 32211

MITCH, WILLIAM EVANS, b Birmingham, Ala, July 22, 41; m 65; c 2. NEPHROLOGY. Educ: Harvard Univ, BA, 63, MD, 67. Prof Exp: Intern med, Peter Bent Brigham Hosp, 67-68, asst resident, 68-69; clin assoc oncol, Nat Cancer Inst, 69-71; fel med, Johns Hopkins Univ, 71-73; chief resident, Peter Bent Brigham Hosp, 73-74; ASST PROF MED & PHARMACOL, JOHNS HOPKINS UNIV, 74- Mem: Am Fedn Clin Res. Res: Nutritional therapy of chronic renal insufficiency. Mailing Add: Sch of Med Johns Hopkins Univ Baltimore MD 21205

MITCHAM, DONALD, b Hazlehurst, Miss, Nov 15, 21. PHYSICS. Educ: Tulane Univ, BS, 48. Prof Exp: Physicist, 48-68, RES PHYSICIST, SOUTHERN REGIONAL RES LAB, US DEPT AGR, 68- Mem: Am Crystallog Asn; Am Asn Textile Chem & Colorists; Soc Appl Spectros; Am Oil Chem Soc; Sigma Xi. Res: X-ray diffraction of cotton cellulose; physical properties of textiles; structure of fatty acids by x-ray diffraction. Mailing Add: US Dept of Agr PO Box 19687 New Orleans LA 70119

MITCHEL, RONALD EDWARD JOHN, b Estevan, Sask, Nov 18, 41; m 68; c 2. BIOCHEMISTRY. Educ: Univ BC, BSc, 63, MSc, 65, PhD(biochem), 68. Prof Exp: Fel, Univ Calif, Los Angeles, 68-70; asst res off, 70-73, ASSOC RES OFFICER BIOCHEM, CHALK RIVER NUCLEAR LABS, ATOMIC ENERGY CAN LTD, 73- Res: Radiation biochemistry, especially damage and repair systems; enzymology, particularly nuclease structure and function. Mailing Add: Biol Br Chalk River Nuclear Labs Atomic Energy of Canada Ltd Chalk River ON Can

MITCHELL, A RICHARD, b Pine Bluff, Ark, Feb 27, 39; m 59; c 2. APPLIED MATHEMATICS. Educ: Southern Methodist Univ, BS, 60; NMex State Univ, MS, 62, PhD(math), 64. Prof Exp: Assoc prof math, Hendrix Col, 64-65; asst prof, 65-67, ASSOC PROF MATH, UNIV TEX, ARLINGTON, 67- Mem: Am Math Soc; Math Asn Am. Res: Differential equations. Mailing Add: Dept of Math Univ of Tex Arlington TX 76010

MITCHELL, ALBERT HAYWOOD, b Magnolia, NJ, Aug 9, 26; m 71; c 1. PHYSICAL CHEMISTRY. Educ: Lincoln Univ, BA, 50; Temple Univ, MA, 53, PhD(phys chem), 66. Prof Exp: Asst chem, Temple Univ, 50-53; anal chemist, 53-59, phys chemist, 59-73, PHYS SCIENTIST RES ADMIN, PITMAN-DUNN LAB, FRANKFORD ARSENAL, 73- Concurrent Pos: Adj asst prof chem, Temple Univ, 69-71; adj fac, Community Col Philadelphia, 70- Mem: AAAS; Sigma Xi. Res: Ignition and combustion of solid and liquid propellants; ignition theory of metals; diffusion in liquids and solids. Mailing Add: Pitman-Dunn Lab Frankford Arsenal Philadelphia PA 19137

MITCHELL, ALEXANDER REBAR, b San Pedro, Calif, June 13, 38; m 63; c 1. BIOCHEMISTRY. Educ: Univ Calif, Berkeley, BA, 61; Ind Univ, PhD(biochem), 69. Prof Exp: Asst biochem, Edgewood Arsenal, US Army, 62-63; res assoc, 69-75, ASST PROF BIOCHEM, ROCKEFELLER UNIV, 75- Mem: Am Chem Soc. Res: Chemical synthesis of peptides; solid phase peptide synthesis; enzyme models; peptide antibiotics and hormones. Mailing Add: Rockefeller Univ 1230 York Ave New York NY 10021

MITCHELL, ANN DENMAN, b Nashville, Tenn, Oct 29, 39. CELL BIOLOGY, BIOCHEMICAL GENETICS. Educ: Univ Tex, Austin, BA, 60, PhD(zool), 71. Prof Exp: Teacher biol & chem, Tex Independent Sch Dists, Ft Worth & Austin, 61-66; instr biol, Huston-Tillotson Col, 68; res assoc path, Sch Med, Stanford Univ, 70-72, NIH fel develop biol, 72-73; CELL BIOLOGIST & PROG MGR BIOCHEM CYTOGENETICS, STANFORD RES INST, 73- Mem: AAAS; Am Soc Cell Biol; Environ Mutagen Soc; Genetics Soc Am; Tissue Cult Asn. Res: Mutagenesis; carcinogenesis; toxicology; DNA repair; cytogenetics; DNA biosynthetic pathways; metabolic activation systems; cell synchronization; drug screening; in vitro mutagenesis and cell transformation. Mailing Add: Biochem Cytogenetics Prog 333 Ravenswood Ave Menlo Park CA 94025

MITCHELL, ARTHUR EDWIN, horticulture, see 12th edition

MITCHELL, BARRY MILLER, b Toronto, Ont, June 25, 34. MATHEMATICS. Educ: Univ Toronto, BA, 55, MA, 56; Brown Univ, PhD(math), 60. Prof Exp: Instr math, Columbia Univ, 60-64; asst prof, Bowdoin Col, 65-68; acad guest, Swiss Fed Inst Technol, 68-69; sr res fel, Dalhousie Univ, 69-71; assoc prof, 71-73, PROF MATH, RUTGERS UNIV, NEW BRUNSWICK, 73- Mem: Am Math Soc; Math Asn Am. Res: Homological algebra. Mailing Add: Dept of Math Rutgers Univ New Brunswick NJ 08903

MITCHELL, BENJAMIN EVANS, b Delhi, La, Feb 15, 20; m 48; c 3. MATHEMATICS. Educ: La State Univ, AB, 41, AM, 48; Univ Wis, PhD(math), 51. Prof Exp: Asst prof math, Ala Polytech Inst, 51-52; from asst prof to assoc prof, 52-64, PROF MATH, LA STATE UNIV, BATON ROUGE, 64- Mem: Am Math Soc; Math Asn Am. Res: Matric theory; algebra. Mailing Add: Dept of Math La State Univ Baton Rouge LA 70803

MITCHELL, BRIAN JAMES, b Minneapolis, Minn, July 25, 36; m 60; c 2. GEOPHYSICS. Educ: Univ Minn, BA, 62, MS, 65; Southern Methodist Univ, PhD(geophys), 70. Prof Exp: Physicist rock properties, US Bur Mines Res Ctr, 62-65;

res fel seismol, Calif Inst Technol, 71-72; geophysicist mining res, Newmont Explor Ltd, 72-73; asst prof, 74-76, ASSOC PROF GEOPHYS, ST LOUIS UNIV, 76- Mem: Am Geophys Union; Seismol Soc Am; Soc Explor Geophysicists; Royal Astron Soc. Res: Crust and upper mantle structure; seismic surface wave propagation. Mailing Add: Dept of Earth & Atmospheric Sci St Louis Univ St Louis MO 63156

MITCHELL, C DEAN, organic chemistry, see 12th edition

MITCHELL, CHARLES VERNON, organic chemistry, see 12th edition

MITCHELL, CLIFFORD L, b Ottumwa, Iowa, Dec 7, 30; m 54; c 4. PHARMACOLOGY. Educ: Univ Iowa, BA, 52, BS, 54, MS, 58, PhD(pharmacol), 59. Prof Exp: Asst prof, Col Med, Univ Iowa, 59-60 & 62-66, from assoc prof to prof pharmacol, 66-73; sr res scientist, 73-74, MGR CENT NERV SYST & CARDIOPULMONARY PHARMACOL, RIKER LABS, 74- Concurrent Pos: Res fel, Stanford Univ, 60-62. Mem: AAAS; Am Soc Pharmacol & Exp Therapeut; Soc Neurosci; Soc Exp Biol & Med. Res: Neuropharmacology; primarily analgesic fields. Mailing Add: Riker Labs 3M Co St Paul MN 55101

MITCHELL, DAVID FARRAR, b Arkansas City, Kans, Dec 15, 18; m 43; c 3. DENTISTRY. Educ: Univ Ill, BS, 40, DDS, 42; Univ Rochester, PhD(path), 48; Am Bd Oral Path, dipl, 69. Prof Exp: Instr dent, Univ Rochester, 48; assoc prof dent & chmn div oral histol & path, Sch Dent, Univ Minn, 48-55, actg chmn div periodontia, 49-51, chmn div oral diag, 49-54; PROF & CHMN DEPT ORAL DIAG, SCH DENT, IND UNIV, INDIANAPOLIS, 55- Concurrent Pos: Pres, Am Bd Oral Path, 69. Mem: AAAS; Am Acad Oral Path (pres, 61); Am Asn Dent Res (pres, 75); Int Asn Dent Res; Am Dent Asn. Res: Aviation chemistry; production and prevention of experimental dental caries and periodontal disease; tissue reactions to dental materials; oral pathology. Mailing Add: Ind Univ Sch of Dent Indianapolis IN 46202

MITCHELL, DAVID LAWRENCE, organic chemistry, see 12th edition

MITCHELL, DEAN LEWIS, b Montour Falls, NY, Apr 9, 29; m 51; c 4. SOLID STATE PHYSICS. Educ: Syracuse Univ, BS, 52, PhD(physics), 59. Prof Exp: Asst prof physics, Utica Col, Syracuse Univ, 58-61; res physicist, Naval Res Lab, 61-70, head solid state physics br, 70-75; MEM STAFF MAT RES, NSF, 75- Concurrent Pos: Nat Acad Sci-Acad Sci USSR exchange prog fel, A F Ioffe Phys Tech Inst, Leningrad, 67-68. Mem: Am Phys Soc. Res: Optical and magneto-optical properties; electronic band structure of solids; semiconductors; infrared sources and detectors. Mailing Add: Div of Mat Sci NSF 1800 G St NW Washington DC 20550

MITCHELL, DONALD FOLK, b Santa Monica, Calif, Aug 16, 25; m 48; c 4. MEDICAL GENETICS. Educ: Univ Calif, Los Angeles, BA, 48, PhD(zool), 52. Prof Exp: Res botanist, Univ Calif, Los Angeles, 54; asst prof genetics, Pa State Univ, 54-59; mem res staff, Northrop Space Labs, 59-64; proj mgr biol systs, Life Sci Dept, NAm Aviation, Inc, 64-66; gen mgr biol systs div, Nus Corp, 66-68; sr scientist, Syst Develop Corp, 68-69; pres, Biome Co, Inc, 69-74; DIR RES, NEURO PSYCHIAT CTR, INC, 74- Concurrent Pos: USPHS fel, 52; vis radiobiologists prog, Am Inst Biol Sci, 62-66; mem, Regional Water Qual Control Bd, State of Calif. Mem: Soc Study Evolution; Genetics Soc Am; Am Astronaut Soc. Res: Population genetics; space biology; water pollution and reclamation; high protein food sources; environmental systems; psychiatric genetics. Mailing Add: Neuro Psychiat Ctr 3029 Deoder Ave Costa Mesa CA 92626

MITCHELL, DONALD GILMAN, b Somerville, Mass, Sept 17, 17; m 42; c 2. FOOD TECHNOLOGY. Educ: Mass Inst Technol, SB, 38. Prof Exp: Chemist, Walter Baker & Co, Inc, 38-41, res chemist, 46-50, asst to res dir, 50-54, qual control mgr, 54-59, mgr chocolate develop, 59-62, MGR TECH SERV, BAKER'S CHOCOLATE & COCOA DIV, GEN FOODS CORP, DOVER, 62- Mem: Am Chem Soc; Inst Food Technologists; Am Asn Candy Technologists; Am Soc Bakery Engrs. Res: Quality control and research and development on chocolate products and their processing. Mailing Add: 706 North Shore Dr Milford DE 19963

MITCHELL, DONALD J, b New Castle, Pa, May 12, 38; m 68. PHYSICAL CHEMISTRY, CRYSTALLOGRAPHY. Educ: Westminster Col, Pa, BS, 60; Vanderbilt Univ, PhD(phys chem), 64. Prof Exp: Res chemist with Dr Jerome Karle, Naval Res Labs, US Govt, DC, 64-67; ASST PROF CHEM, JUNIATA COL, 67- Concurrent Pos: NSF fel, 75. Mem: Am Chem Soc; Am Crystallog Asn. Res: Crystal structure analysis by x-ray diffraction, particularly gas bearing shales, coals and polymers. Mailing Add: Dept of Chem Juniata Col Huntingdon PA 16652

MITCHELL, DONALD JOHN, b Saskatoon, Sask, May 20, 39; m 61. PLANT PHYSIOLOGY. Educ: Univ Sask, BA, 60, MA, 64; Case Western Reserve Univ, PhD(plant physiol), 68. Prof Exp: Lab technician plant physiol, Univ Sask, 60-61; fel, McMaster Univ, 68-69; LECTR II BIOL, SELKIRK COL, 69-, CHMN DEPT ENVIRON SCI, 73- Mem: Can Soc Plant Physiol; Am Soc Plant Physiol; Can Bot Asn. Res: Amino and organic acid metabolism and compartmentation in germinating seedlings; cellulase activity in growing seedlings. Mailing Add: Dept of Environ Sci Selkirk Col Box 1200 Castlegar BC Can

MITCHELL, EARL BRUCE, b Louisville, Miss, Sept 1, 27; m 50; c 3. ENTOMOLOGY. Educ: Miss State Univ, BS, 50, MS, 51, PhD(entom), 71. Prof Exp: Entomologist, La State Univ, 51-52 & Food Mach Corp, 53-55; farmer, 55-62; ENTOMOLOGIST, BOLL WEEVIL RES LAB, AGR RES SERV, USDA, MISSISSIPPI STATE, 63- Mem: Entom Soc Am. Res: Use of pheromones, traps, insecticides, sterile males, and combinations of these for the purpose of suppressing or eradicating the boll weevil. Mailing Add: Rte 2 Box 284 Louisville MS 39339

MITCHELL, EARL DOUGLASS, JR, b New Orleans, La, May 16, 38; m 59; c 3. BIOCHEMISTRY. Educ: Xavier Univ La, BS, 60; Mich State Univ, MS, 63, PhD(biochem), 66. Prof Exp: Res assoc, 67-69, asst prof, 69-74, ASSOC PROF BIOCHEM, OKLA STATE UNIV, 74- Mem: Am Chem Soc. Res: Chemical and physical studies of barley and amylase; plant glycosidases; cell-free biosynthesis of monoterpenoid compounds. Mailing Add: Dept of Biochem Okla State Univ Stillwater OK 74074

MITCHELL, EARL NELSON, b Centerville, Iowa, Aug 30, 26; m 55. PHYSICS. Educ: Univ Iowa, BA, 49, MS, 51; Univ Minn, PhD(physics), 55. Prof Exp: Res physicist, Univac Div, Sperry Rand Corp, 55-58; from asst prof to assoc prof physics, Univ NDak, 58-62; vis assoc prof, 62-65, assoc prof, 65-69, PROF PHYSICS, UNIV NC, CHAPEL HILL, 69-, ASST CHMN DEPT, 68- Mem: Am Phys Soc; Am Asn Physics Teachers. Res: Solid state physics; structure, electric and magnetic properties of thin films. Mailing Add: Dept of Physics Univ of NC Chapel Hill NC 27514

MITCHELL, EDWARD DICK, JR, zoology, paleontology, see 12th edition

MITCHELL, EVERETT ROYAL, b Itasca, Tex, Sept 17, 36; m 57; c 1. ENTOMOLOGY. Educ: Tex Tech Univ, BS, 59; NC State Univ, MS, 61, PhD(entom, bot), 63. Prof Exp: Res entomologist, Cotton Insects Br, USDA, Florence, SC, 63-65; asst prof entom & head dept, Coastal Plain Exp Sta, Univ Ga, Tifton, 65-67; prof biol sci & chmn div sci & math, Tarrant County Community Col, Ft Worth, Tex, 67-69; sr res biologist, Monsanto Co, St Louis, 69-70; res entomologist, 71, actg dir lab, 71-72, RES LEADER ECOL GROUP, INSECT ATTRACTANTS LAB, 72- Concurrent Pos: Assoc prof entom & nematol, Univ Fla, 71- Mem: Entom Soc Am. Res: Biology, ecology, behavior and control of economic pests of field and vegetable crops, particularly chemical messengers produced by insects. Mailing Add: Insect Attractants Lab PO Box 14565 Gainesville FL 32604

MITCHELL, FERDINAND H, b Mobile, Ala, Sept 11, 08; m 35; c 2. PHYSICS. Educ: Univ Ala, BS, 32, AM, 37; Univ Va, PhD(physics), 41. Prof Exp: Instr physics & astron, Univ Ala, 35-42, from asst prof to prof physics, 42-65; actg dean arts & sci, 66-67, PROF PHYSICS, UNIV S ALA, 65-, CHMN DEPT, 66-, DEAN GRAD SCH, 74- Mem: Am Phys Soc; Am Astron Soc; Am Asn Physics Teachers. Res: Electrical discharge in gases; electronics; astronomy. Mailing Add: Dept of Physics Univ of S Ala Mobile AL 36608

MITCHELL, FERN WOOD, chemistry, see 12th edition

MITCHELL, GARY EARL, b Louisville, Ky, July 5, 35; m 57; c 2. PHYSICS. Educ: Univ Louisville, BS, 56; Duke Univ, MA, 58; Fla State Univ, PhD(physics), 62. Prof Exp: Res assoc physics, Columbia Univ, 62-64, asst prof, 64-68; ASSOC PROF PHYSICS, NC STATE UNIV, 68- Mem: AAAS; Am Phys Soc. Res: Nuclear structure physics, including electromagnetic transitions; high resolution proton scattering; fine structure of analog states. Mailing Add: Dept of Physics NC State Univ Raleigh NC 27607

MITCHELL, GARY F, b Chicago, Ill, Apr 16, 40; m 65. PHOTOGRAPHIC CHEMISTRY. Educ: Calif Inst Technol, BS, 62; Mass Inst Technol, PhD(org chem), 67. Prof Exp: Sr res chemist, 67-75, RES ASSOC, EASTMAN KODAK CO, 75- Res: Photographic film design. Mailing Add: Res Labs Eastman Kodak Co 343 State St Rochester NY 14650

MITCHELL, GEORGE ERNEST, JR, b Duoro, NMex, June 7, 30; m 52; c 3. ANIMAL SCIENCE. Educ: Univ Mo, BS, 51, MS, 54; Univ Ill, PhD(animal sci), 56. Prof Exp: Asst animal sci, Univ Ill, 54-56, asst prof, 56-60; assoc prof animal husb, 60-67, PROF ANIMAL SCI, UNIV KY, 67- Concurrent Pos: Sr Fulbright Res Scholar, NZ, 73-74. Mem: AAAS; Am Soc Animal Sci; Am Dairy Sci Asn; Am Inst Nutrit. Res: Vitamin A metabolism; digestive physiology; carbohydrate utilization; rumen function; ruminant-nonruminant comparisons; effect of hormones on nutritive requirements; beef cattle feeding and management. Mailing Add: Dept of Animal Sci Univ of Ky Lexington KY 40506

MITCHELL, GEORGE REDMOND, JR, b Cleveland, Ohio, Mar 3, 17; m 53; c 2. ORGANIC POLYMER CHEMISTRY. Educ: Mass Inst Technol, BS, 39, MS, 40; Case Inst Technol, PhD(org chem), 53. Prof Exp: Shift foreman, Synthetic Rubber Plant, B F Goodrich Chem Co, 41-42, process engr, 42-43, plant chem engr, 43-44, group leader, 44-49; group leader dispersion coating, film res & develop dept, Olin Mathieson Chem Corp, 51-55, asst sect chief, 55-58, chem develop mgr, 58-59; SECT LEADER MAT RES & DEVELOP, GLASTIC CORP, 59- Mem: Am Chem Soc; fel Am Inst Chemists; Am Soc Testing & Mat; Am Inst Chem Engrs; Inst Elec & Electronics Engrs. Res: Thermosetting reinforced polyester laminates and molding compounds; power and electronic electrical properties of reinforced thermosetting plastics; film forming thermoplastics; oxidation of hydrocarbons; synthetic rubber; chemical engineering process development; pilot plant design and operation. Mailing Add: Gates Rd Gates Mills OH 44040

MITCHELL, GEORGE W, JR, b Baltimore, Md, Apr 30, 17; m 42, 57; c 6. OBSTETRICS & GYNECOLOGY. Educ: Johns Hopkins Univ, AB, 38, MD, 42; Am Bd Obstet & Gynec, dipl, 50. Prof Exp: Asst path, Johns Hopkins Univ, 46, instr gynec, 47-49; instr, 50-52, assoc prof, 53-54, prof, 55-56, PROF OBSTET & GYNEC & CHMN DEPT, SCH MED, TUFTS UNIV, 57- Concurrent Pos: Consult, St Margaret's Hosp, 54-, Boston City Hosp, 56- & Chelsea Naval Hosp, 59-; assoc dir, Am Bd Obstet & Gynec; assoc ed, Obstet & Gynec; consult, Surgeon Gen, US Navy. Mem: Am Gynec Soc; Am Asn Obstetricians & Gynecologists; fel Am Col Surgeons; Am Col Obstet & Gynec; AMA. Res: Cytology; female urology. Mailing Add: Dept of Obstet & Gynec Tufts Univ Sch of Med Boston MA 02111

MITCHELL, GEORGE WESTON, b Gainesville, Fla, Dec 13, 42; m 67. BIOCHEMISTRY, ELECTRONICS. Educ: Johns Hopkins Univ, AB, 64; Univ Ill, MS, 65; Harvard Univ, PhD(biol), 69. Prof Exp: Res assoc biol, Harvard Univ, 69-70; res assoc biochem, Univ Ill, 70-74; VPRES ELECTRONICS, SLIM INSTRUMENTS INC, 75- Mem: Am Soc Photochem & Photobiol. Res: Application of advanced spectroscopic techniques to biological problems; bioluminescence and fluorescence spectroscopy. Mailing Add: 910 W Green St Champaign IL 61820

MITCHELL, HAROLD HUGH, b New York, NY, Apr 10, 16. MEDICINE. Educ: Univ Ariz, BS, 36; Univ Southern Calif, MS, 38; Wash Univ, MD, 45. Prof Exp: Lectr bact, Univ Southern Calif, 38-41; asst med dir, Calif Physicians Serv, 48-52; from assoc scientist to sr scientist, Physics Div, Rand Corp, 52-71; SR SCIENTIST, R&D ASSOCS, 71- Concurrent Pos: Chmn, Gov's Adv Comt Radiol Defense, Calif, 62-68. Mem: AAAS; AMA. Res: Biological and environmental consequences of nuclear war radiation effects and radiobiology; mass casualty; civil defense. Mailing Add: R&D Assocs 4640 Admiralty Way Marina del Ray CA 90291

MITCHELL, HENRY ANDREW, b Joplin, Mo, Oct 6, 36; m 61; c 3. MAMMALIAN PHYSIOLOGY. Educ: Southwest Mo State Col, BS, 58; Univ Ariz, MS, 60, PhD(zool), 63. Prof Exp: Asst zool, Univ Ariz, 58-63; asst prof biol, 63-68, asst dean sch grad studies, 69, assoc prof biol & assoc dean col arts & sci, 68-73, lectr med & assoc dean sch med, 72-73, actg dean col arts & sci, 74-75, ASSOC PROVOST HEALTH SCI & PROF BIOL, MED & PHARM, UNIV MO-KANSAS CITY, 73- Concurrent Pos: Field ctr coordr, NSF-AAAS Col Level Short Courses Prog, 71-72; mem, Am Conf Acad Deans; field ctr coordr, NSF Chautauqua-Type Short Courses Prog for Col Teachers, 71-76; proj adminr, Western Mo Area Health Educ Ctr, 75- Mem: Fel AAAS; Am Soc Mammal; Am Inst Biol Sci; Mex Natural Hist Soc. Res: Hematological studies of bats. Mailing Add: Health Sci Univ of Mo 2220 Holmes St Kansas City MO 64108

MITCHELL, HENRY COOPER, b Louisville, Miss, Feb 23, 36; m 57; c 3. ENTOMOLOGY. Educ: Miss State Univ, BS, 58, MS, 67, PhD(entom), 71. Prof Exp: Res entomologist, Boll Weevil Res Lab, Agr Res Serv, USDA, 62-71; asst county agent, 60-62, EXTEN ENTOMOLOGIST, MISS COOP EXTEN SERV, 71- Mem: Entom Soc Am; Nat County Agents Asn; Sigma Xi. Mailing Add: Drawer EM Mississippi State MS 39762

MITCHELL, HENRY REES, b New London, Conn, Oct 13, 08; m 37; c 4. PHYSICS. Educ: Trinity Col, Conn, BS, 31; Johns Hopkins Univ, PhD(physics), 38. Prof Exp:

Instr math, Gettysburg Col, 37-38; asst prof physics, The Citadel, 38-44; physicist, Appl Physics Lab, Johns Hopkins Univ, 44-48; asst prof, Georgetown Univ, 48-51; assoc prof, 51-58, PROF PHYSICS, MICH TECHNOL UNIV, 58- Mem: AAAS; Am Phys Soc; Am Asn Physics Teachers. Res: Thin metallic films servomechanisms; computing mechanism of gun fire control and guided missiles. Mailing Add: 401 Second St Houghton MI 49931

MITCHELL, HERSCHEL KENWORTHY, b Los Nietos, Calif, Nov 27, 13; m 34; c 4. BIOCHEMISTRY. Educ: Pomona Col, AB, 36; Ore State Col, MS, 38; Univ Tex, PhD(org chem, biochem), 41. Prof Exp: Res assoc chem, Ore State Col, 38-39; res biochemist, Univ Tex, 41-43; res assoc biochem, Stanford Univ, 43-46; sr res fel, 46-48, assoc prof biol, 48-53, PROF BIOL, CALIF INST TECHNOL, 53- Concurrent Pos: NSF sr fel zool, Univ Zurich, 59-60. Mem: Am Soc Biol Chemists; Soc Gen Physiol; Am Chem Soc; Genetics Soc Am; Am Soc Naturalists. Res: Growth factors for microorganisms; microchemical methods; biochemical genetics in neurospora; synthesis of compounds of biological significance; biochemistry of Drosophila; developmental enzymology. Mailing Add: Dept of Biol Calif Inst of Technol Pasadena CA 91109

MITCHELL, HOWARD LEE, b Dewey, Okla, Nov 17, 14; m 41; c 2. AGRICULTURAL BIOCHEMISTRY. Educ: Okla Agr & Mech Col, BS, 38; Purdue Univ, PhD(agr biochem), 46. Prof Exp: Asst chemist, Purdue Univ, 40-46; from asst prof to prof chem, 46-61, head dept biochem, 61-73, PROF BIOCHEM, KANS STATE UNIV, 61- Mem: AAAS; Am Chem Soc; Am Soc Plant Physiol. Res: Plant chemistry. Mailing Add: Dept of Biochem Kans State Univ Manhattan KS 66506

MITCHELL, HUGH BERTRON, b Bolton, Miss, Dec 8, 23; m 45; c 4. RADIOBIOLOGY, AEROSPACE MEDICINE. Educ: La State Univ, MD, 47; Univ Calif, Berkeley, MBioradiol, 65. Prof Exp: Physician, Tulsa Clin, Okla, 48-49; pvt pract, Richton, Miss, 49-50; med corps, US Navy, 50-52; physician, Baton Rouge Clin, La, 52-57; med corps, US Air Force, 57-75, med officer, Barksdale Air Force Base Hosp, 57-58, med officer, Sch Aviation Med, Randolph Air Force Base, 58, med officer spec weapons prog & radiobiol, Off Surgeon, Strategic Air Command Hq, 59-62, from dep dir to dir, Armed Forces Radiobiol Res Inst, Nat Naval Med Ctr, Md, 65-71, surgeon, 314th Air Div & comdr, Osan Air Force Base Hosp, Korea, 71-72, staff, Radiobiol Div, US Air Force Sch Aerospace Med, 72-73, comdr, US Air Force Hosp, Plattsburgh Air Force Base, NY, 73-75, res scientist nuclear med, Div Aerospace & Aviation Med, Med Corps, US Air Force, 73-75; AREA MED DIR, SPRINGS MILLS, INC, FT MILL, SC, 75- Mem: AMA; Aerospace Med Asn; Asn Mil Surg US; Am Occup Med Asn. Res: General medicine and surgery; occupational medicine; medical aspects of special weapons and of disaster operations; combined effects of radiation plus other stresses. Mailing Add: Springs Mills Inc Ft Mill SC 29715

MITCHELL, IAN ALASTAIR, b Fernie, BC, May 15, 25; US citizen; m 51; c 1. INTERNAL MEDICINE, MEDICAL ADMINISTRATION. Educ: Univ BC, AB, 47; Univ Toronto, MD, 51. Prof Exp: Jr internship, Vancouver Gen Hosp, BC, 51-52; resident, Med Serv, Sunnybrook Hosp, Toronto, Ont, 52-53, 55-56 & Med Serv, Toronto Gen Hosp, 53-54; res assoc & instr internal med, Thomas Henry Simpson Mem Inst Med Res, Univ Mich, 58-61; head hemat group & actg head bioassay sect, Dept Biol Sci, Defense Res Labs, Gen Motors Corp, 61-63; spec asst to assoc dir field studies, Nat Cancer Inst, 63-66, assoc sci dir planning & anal etiology, 66-67, asst dir, 67-70; SPEC ASST TO ASST SECY HEALTH & SCI AFFAIRS, DEPT HEALTH, EDUC & WELFARE, 70- Concurrent Pos: Teaching fel, McGill Univ & Montreal Gen Hosp, Que, 54-55; fel hemat, Dept Path, Hosp Sick Children, Toronto, 56-57; Gen Motors vis scholar, Inst Indust Health, Univ Mich, 61-62; consult, Fac Periodic Health Unit, Univ Mich, 59-61. Mem: AAAS. Mailing Add: Dept of Health Educ & Welfare Washington DC 20201

MITCHELL, J ANDREW, b Dallas, Tex, May 8, 41; m 71. REPRODUCTIVE PHYSIOLOGY, NEUROENDOCRINOLOGY. Educ: Southern Methodist Univ, BS, 63; Univ Kans, PhD(physiol), 69. Prof Exp: ASST PROF ANAT, SCH MED, WAYNE STATE UNIV, 72- Concurrent Pos: NIH fel, Dept Anat, Baylor Col Med, 69-72. Mem: AAAS; Am Asn Anat; Soc Study Reproduction; Am Asn Hist Med. Mailing Add: Dept of Anat Wayne State Univ Sch of Med Detroit MI 48201

MITCHELL, JACK HARRIS, JR, b Auburn, Ala, Sept 15, 11; m 44; c 4. BIOCHEMISTRY. Educ: Clemson Col, BS, 33; Purdue Univ, PhD(biochem), 41. Prof Exp: Asst chemist, State Chem Lab, SC, 33-36; asst, Columbia Univ, 36-37; asst, Purdue Univ, 37-40, asst chemist, 40-41; res chemist, Am Meat Inst Found, Univ Chicago, 41-42 & Petrol Chem Co, Md, 46-47; head biochem sect, Southern Res Inst, Ala, 47-50; asst chief, Stability Div, Qm Food & Container Inst, Univ Ill, 50-55, chief chem & microbiol div, 55-57; head dept food tech & human nutrit, 57-64, PROF FOOD SCI, CLEMSON UNIV, 64- Concurrent Pos: Mem bd dirs, Res & Develop Assocs for Military Food & Packaging Systs, Inc, 64-67. Mem: Inst Food Technol; fel Am Inst Chem. Res: Research administration; food biochemistry; lipids; food processing and product development. Mailing Add: Dept of Food Sci Clemson Univ Clemson SC 29631

MITCHELL, JAMES DAVID, food science, see 12th edition

MITCHELL, JAMES GEORGE, b Kitchener, Ont, Apr 25, 43; m 67; c 2. COMPUTER SCIENCE. Educ: Univ Waterloo, BSc, 66; Carnegie-Mellon Univ, PhD(comput sci), 70. Prof Exp: Programmer, Berkeley Comput Corp, 70-71; MEM RES STAFF COMPUT SCI, PALO ALTO RES CTR, XEROX CORP, 71- Mem: Asn Comput Mach; Inst Elec & Electronics Engrs; Brit Comput Soc. Res: Programming systems and methodologies; reliable software; programming languages; personal computer systems; computer networks; computer architecture; distributed computing. Mailing Add: Xerox Palo Alto Res Ctr 3333 Coyote Hill Rd Palo Alto CA 94304

MITCHELL, JAMES WINFIELD, b Durham, NC, Nov 16, 43; m 65; c 2. ANALYTICAL CHEMISTRY. Educ: NC A&T State Univ, BS, 65; Iowa State Univ, PhD(anal chem), 70. Prof Exp: SUPVRANAL CHEM, BELL LABS, 70- Mem: Am Chem Soc. Res: Quantitative analysis of submicrogram amounts of inorganic species by radiochemical methods; ultrapurification of reagents for trace analysis. Mailing Add: MH 1A 302A Bell Labs Mountain Ave Murray Hill NJ 07974

MITCHELL, JERE HOLLOWAY, b Longview, Tex, Oct 17, 28; m 60; c 3. CARDIOVASCULAR PHYSIOLOGY. Educ: Va Mil Inst, BS, 50; Univ Tex Southwestern Med Sch, MD, 54. Prof Exp: From intern to resident med, Parkland Mem Hosp, Dallas, 54-56; from sr asst surgeon to surgeon, Lab Cardiovasc Physiol, Nat Heart Inst, 58-62; from asst prof to assoc prof med & physiol, 62-69, PROF MED & PHYSIOL, SOUTHWESTERN MED SCH, UNIV TEX HEALTH SCI CTR, DALLAS, 69- Concurrent Pos: Nat Heart Inst cardiac trainee, Univ Tex Southwestern Med Sch, Dallas, 56-57, res fel, 57-58, USPHS career develop award, 68-73; estab investr, Am Heart Asn, 62-67, chmn coun basic sci, 69-71, fel, Coun Clin Cardiol & Coun Circulation; vis sr scientist, Lab Physiol, Oxford Univ, 70-71; mem ed bd, Am J Cardiol, Am J Physiol & J Appl Physiol; dir, Pauline & Adolph Weinberger Lab Cardiopulmonary Res, 66- Honors & Awards: Young Investr Award, Am Col Cardiol, 61. Mem: Am Physiol Soc; Am Soc Clin Invest; fel Am Col Cardiol; Am Asn Physicians. Res: Cardiovascular and exercise physiology. Mailing Add: Southwestern Med Sch Univ of Tex Health Sci Ctr Dallas TX 75235

MITCHELL, JERRY CHARLES, physical chemistry, see 12th edition

MITCHELL, JOHN, JR, b Catonsville, Md, Oct 5, 13; m 35; c 2. ANALYTICAL CHEMISTRY. Educ: Johns Hopkins Univ, BE, 35; Univ Del, MS, 45. Prof Exp: Control chemist, Consol Edison Co, NY, 35; chem engr, 35-36, chemist, 36-37, anal chemist, 37-45, res supvr, 45-51, sr supvr, 51-66, RES MGR, E I DU PONT DE NEMOURS & CO, INC, 66- Honors & Awards: Fisher Award Anal Chem, 64; Anachem Award, Asn Anal Chemists, 74. Mem: Am Chem Soc. Res: Organic and physical chemistry; chemical engineering. Mailing Add: E I du Pont de Nemours & Co Inc Plastics Dept Exp Sta Wilmington DE 19898

MITCHELL, JOHN CAMPBELL, b Redditch, Eng, Mar 19, 23; Can citizen; m 45, 71; c 4. DERMATOLOGY, MEDICINE. Educ: Univ London, MB & BS, 48, MD, 50; FRCP(C), 61; FRCP, 72. Prof Exp: Asst prof, 57-72, ASSOC PROF DERMAT, UNIV BC, 72- Concurrent Pos: Med Res Coun Can fel, Univ BC, 65-72; consult, Dept Vet Affairs, BC Cancer Inst & Vancouver Gen Hosp, 55- Mem: Can Dermat Asn; Am Acad Dermat. Res: Investigation of allergic contact dermatitis, especially from plants. Mailing Add: 614-750 W Broadway Vancouver BC Can

MITCHELL, JOHN EDWARDS, b San Francisco, Calif, Mar 27, 17; m 42; c 3. PLANT PATHOLOGY. Educ: Univ Minn, BS, 39; Univ Wis, PhD(biochem), 48. Prof Exp: Asst plant path, Univ Minn, 39-40 & 41-42; asst plant path, Univ Wis, 46-48; plant pathologist, Camp Detrick, Md, 48-51, chief biol br, C Div, 51-56; assoc prof plant path, 56-63, PROF PLANT PATH, UNIV WIS-MADISON, 63-, CHMN DEPT, 75- Mem: Am Phytopath Soc; Am Soc Plant Physiol; Sigma Xi. Res: Ecology and control of soil borne plant pathogens; resistance of roots to disease; soil environment and plant disease. Mailing Add: Dept of Plant Path Univ of Wis 1630 Linden Dr Madison WI 53706

MITCHELL, JOHN JACOB, b Schenectady, NY, Mar 4, 17; m 42; c 3. PHYSICAL CHEMISTRY. Educ: Johns Hopkins Univ, PhD(chem), 41. Prof Exp: Asst chem, Johns Hopkins Univ, 40-41; phys chemist, Tex Co, 41-53, staff chemist, 53-57, res assoc, Texaco, Inc, 57-60, SR RES TECHNOLOGIST, TEXACO INC, 60- Concurrent Pos: Supvr lab, Manhattan Dist, Fercleve Corp, Tenn, 44-45. Mem: AAAS; Am Chem Soc; Am Phys Soc; Health Physics Soc. Res: Mass spectroscopy; use of radioactive tracers; radiation chemistry; use of isotope tracers in studying reaction mechanisms; research planning. Mailing Add: Texaco Res Ctr Beacon NY 12508

MITCHELL, JOHN LAURIN AMOS, b Lincoln, Nebr, July 18, 44; m 68; c 2. CELL PHYSIOLOGY, BIOLOGICAL RHYTHMS. Educ: Oberlin Col, BA, 66; Princeton Univ, PhD(biol), 70. Prof Exp: Fel cancer res, McArdle Labs Cancer Res, Univ Wis-Madison, 70-73; ASST PROF BIOL, NORTHERN ILL UNIV, 73- Mem: Am Soc Exp Biologists; Sigma Xi. Res: Regulation of polyamine synthesis in Physarum polycephalum as well as in normal, cancer and cystic fibrosis tissues; post-translational enzyme modifications; biochemistry of cell cycle. Mailing Add: Dept of Biol Sci Northern Ill Univ De Kalb IL 60115

MITCHELL, JOHN MURRAY, JR, b New York, NY, Sept 17, 28; m 56; c 4. METEOROLOGY. Educ: Mass Inst Technol, BS, 51, MS, 52; Pa State Univ, PhD(meteorol), 60. Prof Exp: Weather observer, Blue Hill Meteorol Observ, Harvard Univ, 46-52; mem staff, Div Indust Coop Meteorol Res, Mass Inst Technol, 52; weather officer, Air Weather Serv, US Air Force, 52-55; meteorologist, US Weather Bur, 55-65; PROJ SCIENTIST, NAT OCEANIC & ATMOSPHERIC ADMIN, 65- Concurrent Pos: Ed, Meteorol Monogr, Am Meteorol Soc, 65-; lectr, Univ Calif, Berkeley, 65; vis assoc prof, Univ Wash, 69; chmn, US Nat Comt for Int Union Quaternary Res, 70-; mem glaciol panel, Comt Polar Res & Rev Panel on Arctic Ice Dynamics Joint Exp, Nat Acad Sci & panel on climatic variation, Global Atmospheric Res Prog. Honors & Awards: Silver Medal, US Dept Com, 64. Mem: Am Meteorol Soc; Am Geophys Union; Glaciol Soc; Arctic Inst NAm; Royal Meteorol Soc. Res: Climatic change; climate modification; time series analysis; paleoclimatology. Mailing Add: Nat Oceanic & Atmospheric Admin Rm 608 Gramay Bldg 8060 13th St Silver Springs MD 20910

MITCHELL, JOHN PETER, b Toronto, Ont, June 28, 32; m 56; c 2. PHYSICS. Educ: Univ Toronto, BA, 55, MA, 57, PhD(physics), 60. Prof Exp: Engr, Bell Tel Co, Can, 55-56; teacher math & physics, York Mem Col Inst, Toronto, 60-63; MEM TECH STAFF, BELL LABS, INC, 63- Mem: Am Phys Soc; Inst Elec & Electronics Eng. Res: Surface effects of radiation on semiconductor devices; resistivity of dilute alloys of lead at low temperatures; thin insulating films for cryogenic use. Mailing Add: Bell Labs Inc Room 5E-125 Whippany NJ 07981

MITCHELL, JOHN RICHARD, b Marysville, Ohio, Sept 18, 27; m 49; c 3. MICROBIOLOGY, PUBLIC HEALTH. Educ: Ohio State Univ, DVM, 53; Univ Mich, Ann Arbor, MPH, 60, DrPH, 63. Prof Exp: Pvt pract vet med, 53-56; pub health vet, Dept Health, City of Columbus, Ohio, 55-60; sr vet officer, USPHS, 63-65; PUB HEALTH LAB SCIENTIST, DEPT PUB HEALTH, STATE OF MICH, 65- Mem: Am Vet Med Asn; Am Pub Health Asn. Res: Virology, especially development of viral vaccines. Mailing Add: Bur of Labs Mich Dept of Pub Health Lansing MI 48914

MITCHELL, JOHN TAYLOR, b Buffalo, NY, Aug 16, 31; m 62; c 3. DEVELOPMENTAL BIOLOGY, EXPERIMENTAL EMBRYOLOGY. Educ: Amherst Col, BA, 53; NY Univ, MS, 64, PhD(biol), 66. Prof Exp: Asst cancer res scientist, Roswell Park Mem Inst, 56-60; res asst biol, NY Univ, 60-62; NIH staff fel, Nat Cancer Inst, Md, 66-68; asst prof anat, State Univ NY Upstate Med Ctr, 68-75; ASSOC PROF BIOL, COLGATE UNIV, 75- Concurrent Pos: State Univ NY Res Found grant, 68-70. Mem: AAAS; Tissue Cult Asn; Am Soc Cell Biol; Am Asn Anat; Am Soc Zoologists. Res: Developmental biology, especially cytogenetic, cytological and in vitro approaches; effect of chemical compounds on early developing mammalian embryos; neoplastic conversion of cells in vitro; teratology. Mailing Add: Dept of Biol Colgate Univ Hamilton NY 13346

MITCHELL, JOHN WESLEY, b Christchurch, NZ, Dec 3, 13; m 68. PHYSICS. Educ: Univ NZ, BSc, 33, MSc, 34; Oxford Univ, DrPhil, 38, DSc, 60. Prof Exp: Res physicist, Brit Ministry Supply, 39-45; reader exp physics, Bristol Univ, 45-59; prof, 59-65, WILLIAM BARTON ROGERS PROF PHYSICS, UNIV VA, 65- Honors & Awards: Boys Prize, Brit Inst Physics & Phys Soc, 55; Renwick Mem Medal, Royal Photog Soc, 56. Mem: Hon mem Soc Photog Sci & Eng; fel Am Phys Soc; fel Royal Soc; fel Brit Inst Physics & Phys Soc; fel Royal Photog Soc. Res: Physics of crystals; surface properties of crystalline materials; dislocations and plastic deformation of ionic crystals and metals; theory of photographic sensitivity. Mailing Add: Dept of Physics Univ of Va Charlottesville VA 22903

MITCHELL, JOHN WILLIAM, b Hornell, NY, Sept 16, 05. PLANT PHYSIOLOGY. Educ: Univ Idaho, BS, 28, MS, 29; Univ Chicago, PhD, 32. Hon Degrees: DSc, Univ Chicago, 63. Prof Exp: Asst plant physiol, Univ Chicago, 31-38; physiologist bur plant indust, USDA, 38-42, physiologist guayule res bur plant indust, Soils & Agr Eng, 42-44, in charge res growth regulating substances, 44-58, head growth regulator & antibiotic lab, Agr Res Serv, 58-62, leader plant hormone & regulator pioneering res lab, 62-74; RETIRED. Res: Plant hormones & growth regulating substances, antibiotics and related chemicals. Mailing Add: Plant Hormone & Regulator Pioneering Res Lab USDA Beltsville MD 20705

MITCHELL, JOSEPH CHRISTOPHER, b Albany, Ga, Oct 8, 22; m 45; c 3. PARASITOLOGY, ENTOMOLOGY. Educ: Ft Valley State Col, BS, 43; Atlanta Univ, MS, 49. Hon Degrees: DSc, London Inst Appl Res, 72. Prof Exp: Chmn dept sci high schs, Ga, 43-44, 44-45, 47 & 48-49; scholar, Atlanta Col Mortuary Sci, 45-46, instr physiol, 47-48; asst prof biol, Ft Valley State Col, 49-52; chmn dept phys sci & zool, Ala State Jr Col, 52-54; asst prof biol, Albany State Col, 54-63, assoc prof, Montgomery & chmn sci dept, Mobile Ctr, 63-70; ASSOC PROF BIOL & CHMN DIV NATURAL SCI & MATH, S D BISHOP STATE COL, 70- Concurrent Pos: Carnegie res grant-in-aid & sci fac grant, 50-52; res technician, Cornell Univ, 70; Elem & Sec Educ Act title III field reader, US Off Educ; consult sec sci pub schs, Southwest Ga; proj dir sci improv, NSF, 75- Mem: AAAS; Am Soc Microbiol; Nat Sci Teachers Asn; Nat Asn Biol Teachers; Nat Inst Sci. Res: Mammalian physiology. Mailing Add: S D Bishop State Col 351 N Broad St Mobile AL 36603

MITCHELL, JOSEPHINE MARGARET, b Edmonton, Alta; nat US; m 53. MATHEMATICS. Educ: Univ Alta, BS, 34; Bryn Mawr Col, MA, 41, PhD(math), 42. Prof Exp: Teacher pub sch, Alta, 35-38; instr math, Hollins Col, 42-44; instr, Conn Col, 44-45; assoc prof, Winthrop Col, 45-46; assoc prof, Tex State Col Women, 46-47; asst prof, Okla State Univ, 47-48; asst prof, Univ Ill, 48-54; with adv electronics lab, Gen Elec Co, 55-56; with res lab, Westinghouse Elec Corp, 56-57; assoc prof math, Univ Pittsburgh, 57-58; assoc prof, Pa State Univ, 58-61, prof, 61-69; PROF MATH, STATE UNIV NY BUFFALO, 69- Concurrent Pos: Grantee, Am Philos Soc, 48; NSF grant, 52-53; sr fel, 64-65; mem, Inst Adv Study & Am Asn Univ Women fel, 54-55; Air Force contract, 60-63; mem, US Army Math Res Ctr, Univ Wis, 64-65; Math Asn Am Lectr, 65-72; NSF contract, 68-71. Mem: AAAS; Am Math Soc; Math Asn Am. Res: Multiple Fourier and orthogonal series; bounded symmetric domains in space of several complex variables; solutions of partial differential equations. Mailing Add: Dept of Math State Univ NY 4246 Ridge Lea Rd Amherst NY 14226

MITCHELL, KENNETH FRANK, b Hornchurch, Eng, Apr 18, 40; m 61; c 2. IMMUNOBIOLOGY. Educ: Heriot Watt Univ, BSc, 69; Univ Pa, PhD(immunol), 75. Prof Exp: Technician biochem, Animal Health Trust, Eng, 56-59; technician, May & Baker, Eng, 59-62 & Brit Vitamin Prod Ltd, 62-66; instr, 69-72, res specialist immunol, 72-75, FEL IMMUNOL, UNIV PA, 75- Mem: Fel Royal Soc Arts; Brit Inst Biol. Res: The immunobiology and immunogenetics of the laboratory rat; the immunochemistry of receptors on T and B cells. Mailing Add: Dept of Path Sch Med Univ of Pa Philadelphia PA 19174

MITCHELL, KENNETH JOHN, b Bralorne, BC, Sept 8, 38; m 68; c 2. FOREST MENSURATION. Educ: Univ BC, BSF, 61; Yale Univ, MF, 64, PhD(forest mensuration), 67. Prof Exp: Res officer forest mensuration, Can Forestry Serv, Govt of Can, 61-63, res scientist, 67-70; ASST PROF FOREST MGT, YALE UNIV, 70- Mem: Soc Am Foresters; Can Inst Forestry. Res: Dynamics and simulated yield of Douglas Fir. Mailing Add: Yale Univ Sch of Forestry 205 Prospect New Haven CT 06511

MITCHELL, LAWRENCE CRAIG, organic chemistry, see 12th edition

MITCHELL, LAWRENCE GUSTAVE, b West Chester, Pa, Oct 15, 42; m 65; c 2. ANIMAL PARASITOLOGY, PROTOZOOLOGY. Educ: Pa State Univ, BS, 64; Univ Mont, PhD(zool), 70. Prof Exp: NIH fel, Stella Duncan Mem Res Inst, Univ Mont, 70, asst prof zool, 70-71; ASST PROF ZOOL & BIOL, IOWA STATE UNIV, 71- Concurrent Pos: Instrnl sci equip prog grant, NSF, 74. Mem: Soc Protozoologists; Am Inst Biol Sci. Res: Systematics and ecology of Myxosporida. Mailing Add: Dept of Zool Iowa State Univ Ames IA 50010

MITCHELL, LISLE SERLES, b Mansfield, Ohio, Feb 11, 34; m 58; c 2. URBAN GEOGRAPHY, ECONOMIC GEOGRAPHY. Educ: Ohio State Univ, BS, 61, MA, 63, PhD(geog), 67. Prof Exp: Draftsman, Martin Steel Prods Corp, 52-54, Dom Elec Corp, 56-57 & Porter, Urquart, McCrary & O'Brien, 57-58; asst geog, Ohio State Univ, 62-63, from asst instr to instr, 63-65; from instr to asst prof, 65-70, ASSOC PROF GEOG, UNIV SC, 70- Concurrent Pos: Consult, Battelle Mem Inst, 62-63 & Eastern WVa Res & Develop Ctr, 63; res assoc, Bur Urban & Regional Affairs, Univ SC, 68-69; consult, City of Belton, SC, 72. Mem: Asn Am Geog; Nat Coun Geog Educ (exec secy, 70-71); Regional Sci Asn. Res: Public urban recreation; methods in geographic education; evaluation of central place theory in a recreation context. Mailing Add: Dept of Geog Univ of SC Columbia SC 29208

MITCHELL, LLOYD VERNON, b St Louis, Mo, Aug 17, 20; m 41; c 2. METEOROLOGY, GEOGRAPHY. Educ: Southern Ill Univ, BEd, 42; Univ Chicago, cert, 43. Prof Exp: Teacher pub sch, Ill, 42; salesman, J B Rice Co, Calif, 46; meteorologist, Weather Sta, Hamilton AFB, Calif, 46-49, instr & supvr, Dept Weather, Chanute AFB, Ill, 49-50, meteorologist, Air Force Weather Ctr, Air Weather Serv, DC, 50-51; chief geog & climatol unit, Intel Br, Off Qm Gen, US Dept Army, 51-54, dep chief res & develop support div, Meteorol Dept, Elec Proving Ground, Ariz, 54-59; chief synoptic & dynamic studies br, Sci Serv Hq, Air Weather Serv, Scott AFB, Ill, 59-62, chief spec studies br, Phys Scientist's Off, Aerospace Sci Directorate, 62-65, res meteorologist, Aerospace Sci Div, Air Force Environ Tech Appln Ctr, 65-67, chief rocketsonde br, 67-70, consult, 70-74; RETIRED. Concurrent Pos: Mem meteorol working group & inter-range instrumentation group, Range Meteorol Data Standardization Comt, 66- & comt standardization format pub rocketsonde data, Comt Space Res, 67- Mem: Am Meteorol Soc; Asn Am Geog. Res: Upper atmosphere; synoptic meteorology; climatology; military geography. Mailing Add: 7530 E Marc Pl Tucson AZ 85710

MITCHELL, MADELEINE ENID, b Jamaica, WI, Dec 14, 41. NUTRITION. Educ: McGill Univ, BSc, 63; Cornell Univ, MS, 65, PhD(nutrit), 68. Prof Exp: ASST PROF NUTRIT, WASH STATE UNIV, 69- Mem: Am Dietetic Asn; Nutrit Today Soc. Res: Protein and amino acid nutrition; carnitine; nutritional status of human subjects. Mailing Add: White Hall 308 Home Econ Res Ctr Wash State Univ Pullman WA 99163

MITCHELL, MAURICE MCCLELLAN, JR, b Landsdowne, Pa, Nov 27, 29; m 52. PHYSICAL CHEMISTRY. Educ: Carnegie Inst Technol, BS, 51, MS, 57, PhD(phys chem), 60. Prof Exp: Group leader process chem group, Coal Coke & Coal Chem Div, Appl Res Lab, US Steel Corp, 51-56; asst, Carnegie Inst Technol, 56-58; sr technologist & group leader phys chem group, Coal Chem Div, Appl Res Lab, US Steel Corp, 60-61, actg head process chem sect, 61; supvr phys chem br, Res Div, Melpar, Inc, 61-64; group leader, Res & Develop Dept, Atlantic-Richfield Co, 64-73; DIR RES, HOUDRY DIV, AIR PROD & CHEM , INC, 73- Concurrent Pos: Instr eve sch, Carnegie Inst Technol, 60-61; consult, Blaw-Knox Co, 52-53 & Melpar, Inc, 64-66. Mem: AAAS; Am Chem Soc; Catalysis Soc (ed, Newsletter, 70-). Res: Chemical kinetics; heterogeneous and homogeneous catalysis; catalytic processes; preparation of catalysts; transition metal chemistry; photo, radiation and coal chemistry; chemical vapor deposition. Mailing Add: Houdry Labs Air Prod & Chem Inc Linwood PA 19061

MITCHELL, MERLE, b Dallas, Tex, Apr 1, 21. MATHEMATICS. Educ: Southern Methodist Univ, BA, 42; Univ NMex, MA, 43; Peabody Col, PhD, 58. Prof Exp: Instr math, Univ NMex, 44-46; asst, Univ Wis, 46-48; instr, Arlington State Col, 48-49; instr, 49-58, from asst prof to assoc prof, 58-71, PROF MATH, UNIV NMEX, 71- Mem: Math Asn Am. Mailing Add: Dept of Math Univ of NMex Albuquerque NM 87131

MITCHELL, MICHAEL A, b Austin, Tex, Dec 23, 41; m 65. PHYSICS. Educ: Univ Calif, Riverside, BA, 64; Univ Conn, MS, 66, PhD(physics), 70. Prof Exp: Res physicist, US Naval Ord Lab, 70-74, RES PHYSICIST, NAVAL SURFACE WEAPONS CTR, WHITE OAK LAB, 74- Concurrent Pos: Air Force Off Sci Res-Nat Res Coun fel, 70-72. Mem: AAAS; Am Phys Soc. Res: Transport and magnetic properties of metal alloys; transport properties and structure of metallic materials. Mailing Add: Naval Surface Weapons Ctr White Oak Lab Code WR-32 Silver Spring MD 20910

MITCHELL, MYRON JAMES, b Denver, Colo, Apr 11, 47; m 73; c 1. ECOLOGY. Educ: Lake Forest Col, BA, 69; Univ Calgary, PhD(soil ecol), 74. Prof Exp: Res asst soil ecol, Cornell Univ, 69-70 & Univ Calgary, 70-74; Nat Res Coun Can fel math modelling, Univ BC, 74-75; ASST PROF INVERT ZOOL & ECOL ENERGETICS, STATE UNIV NY COL ENVIRON SCI & FORESTRY, 75- Mem: Ecol Soc Am; AAAS; Acarological Soc Am; Entom Soc Can. Res: Study of floral-faunal interactions and how these interactions affect decomposition, mineral cycling, and energy transformation in terrestrial and aquatic ecosystems. Mailing Add: Dept Forest Zool State Univ of NY Col of Environ Sci & Forestry Syracuse NY 13210

MITCHELL, NORMAN L, b Jamaica, WI, Nov 21, 28; m 68. PLANT PATHOLOGY. Educ: Univ London, BSc, 62; Univ Western Ont, PhD(bot), 67. Prof Exp: Instr biol, WI Col, Jamaica, 62-64; asst prof, 67-69, ASSOC PROF BIOL, LOMA LINDA UNIV, 69- Mem: AAAS; Bot Soc Am. Res: Electron microscopic studies on powdery mildews mainly Sphaerotheca macularis of strawberry. Mailing Add: Dept of Biol Loma Linda Univ Riverside CA 92505

MITCHELL, OLGA MARY MRACEK, b Montreal, Que, Aug 5, 33; m 56; c 2. PHYSICS. Educ: Univ Toronto, BA, 55, MA, 58, PhD(physics), 62. Prof Exp: Res asst textile & metal physics, Ont Res Found, 55-57, res assoc metal physics, 62-63; MEM TECH STAFF, BELL TEL LABS, 63- Mem: Am Phys Soc; Acoust Soc Am. Res: Acoustics and signal processing; interaction of ultrasonic waves in crystals with dislocations, phonons, electrons; nuclear photodisintegration. Mailing Add: Bell Tel Labs Rm 4F 531 Holmdel NJ 07733

MITCHELL, ORMOND GLENN, b Long Beach, Calif, Sept 17, 27. HUMAN ANATOMY, HISTOLOGY. Educ: San Diego State Univ, AB, 49; Univ Southern Calif, MS, 53, PhD(zool), 57. Prof Exp: Instr zool, Univ Southern Calif, 57; asst prof biol, Calif State Polytech Col, 57-62; res assoc, Univ Southern Calif, 62-63; from asst prof to assoc prof anat, Col Dent, NY Univ, 63-69; prof biol, Adelphi Univ, 69-71; assoc prof, 71-75, PROF ANAT, COL DENT, NY UNIV, 75- & CHMN DEPT, 71- Mem: AAAS; Am Soc Mammal; Int Asn Dent Res; NY Acad Sci; Harvey Soc. Res: Growth and development of skin and hair, glands and control of secretion. Mailing Add: Dept of Anat NY Univ Col Dent 421 First Ave New York NY 10010

MITCHELL, PHILLIP VICKERS, solid state physics, see 12th edition

MITCHELL, RALPH, b Dublin, Ireland, Nov 26, 34; m 57; c 3. MICROBIOLOGY. Educ: Trinity Col, Dublin, BA, 56; Cornell Univ, MS, 59, PhD(soil microbiol), 61. Prof Exp: Res assoc soil microbiol, Cornell Univ, 61-62; sr scientist, Weizmann Inst, 62-65; from asst prof to assoc prof appl microbiol, 65-70, GORDON McKAY PROF APPL BIOL, HARVARD UNIV, 70- Mem: Am Soc Microbiol; Brit Soc Gen Microbiol; Water Pollution Control Fedn. Res: Microbiological problems associated with water resources and water pollution; microbial ecology; microbial predator-prey systems; chemoreception in microorganisms; marine fouling. Mailing Add: Div of Eng & Appl Physics Harvard Univ Cambridge MA 02138

MITCHELL, RALPH GERALD, b Waltham, Mass, Sept 9, 25; m 52; c 4. DAIRY HUSBANDRY, GENETICS. Educ: Univ Mass, BS, 50; Univ Wis, MS, 54, PhD(dairy sci), 58. Prof Exp: Asst farm mgr, Univ Mass, 50-51, instr animal husb, 51-53; asst, Univ Wis, 53-57; from asst prof to assoc prof dairy sci, Univ WVa, 57-64; assoc prof, 64-69, PROF ANIMAL GENETICS, RUTGERS UNIV, NEW BRUNSWICK, 69- Mem: AAAS; Am Soc Animal Sci; Am Genetics Asn; Am Dairy Sci Asn. Res: Population genetics and animal breeding; dairy cattle; farm management. Mailing Add: Dept of Animal Sci Rutgers Univ New Brunswick NJ 08903

MITCHELL, REID LOGAN, chemistry, see 12th edition

MITCHELL, RICHARD SCOTT, b Longmont, Colo, Jan 28, 29. MINERALOGY, CRYSTALLOGRAPHY. Educ: Univ Mich, BS, 50, MS, 51, PhD(mineral), 56. Prof Exp: From asst prof to assoc prof geol, 53-63, prof, 63-69, chmn dept, 64-69, PROF ENVIRON SCI, UNIV VA, 69- Mem: AAAS; fel Mineral Soc Am; fel Geol Soc Am; Clay Minerals Soc; Am Crystallog Asn. Res: Morphology and crystal structures of inorganic compounds; structural polytypism; metamict state; mineralogy of coal ash. Mailing Add: Dept of Environ Sci Clark Hall Univ of Va Charlottesville VA 22903

MITCHELL, RICHARD SHEPPARD, b Indianapolis, Ind, Mar 30, 38; m 59; c 2. PLANT TAXONOMY, PLANT MORPHOLOGY. Educ: Fla State Univ, BS, 60, MS, 62; Univ Calif, Berkeley, PhD(bot), 67. Prof Exp: Res botanist, White Mt Res Sta, Univ Calif, Berkeley, 64; res botanist, Tex Res Found, 67; asst prof bot, Va Polytech Inst & State Univ, 67-75; STATE BOTANIST, NY MUS & SCI SERV, 75- Concurrent Pos: Off Water Resources Res, US Dept Interior grant, 68- Mem: Bot Soc Am; Am Inst Biol Sci; Int Asn Aquatic Vascular Plant Biologists. Res: Floristic work; plant distribution; modification of the features of aquatic plants by changes in environment and the implications to their systematics; vascular aquatic plants of the Eastern United States; flora of New York State; smartweeds, Polygonum. Mailing Add: State Mus & Sci Serv Albany NY 12234

MITCHELL, RICHARD SIBLEY, b Barnsdall, Okla, Sept 19, 32; m 59; c 4.

ANALYTICAL CHEMISTRY. Educ: Austin Col, AB, 58; Univ Okla, MS, 63, PhD(chem), 64. Prof Exp: Res chemist, Lion Oil Co Div, Monsanto Chem Co, 57 & 59-61; from asst prof to assoc prof chem, 64-71, PROF CHEM, ARK STATE UNIV, 71- Mem: Am Chem Soc. Res: Gas chromatography and electroanalytical methods. Mailing Add: Drawer TT Div of Phys Sci Ark State Univ State University AR 72467

MITCHELL, RICHARD WARREN, b Lynchburg, Va, Aug 15, 23; m 45 & 62; c 3. PHYSICS. Educ: Lynchburg Col, BS, 44; Agr & Mech Col, Tex, MS, 53, PhD(physics), 60. Prof Exp: Instr math & physics, Lynchburg Col, 45-47; from instr to asst prof physics, Agr & Mech Col, Tex, 47-62, res assoc, Res Found, 54-60; ASSOC PROF PHYSICS, UNIV S FLA, 62- Mem: Am Asn Physics Teachers. Res: Raman and infrared spectroscopy; nuclear magnetic resonance studies relating relaxation times of protons to physical properties of solutions. Mailing Add: Dept of Physics Univ of S Fla Tampa FL 33620

MITCHELL, ROBERT A, b Oakland, Calif, Sept 11, 22; m 51; c 4. PHYSIOLOGY. Educ: Univ Calif, Berkeley, BS, 47; Creighton Univ, MS, 52, MD, 53. Prof Exp: Intern Hosp, Univ Nebr, 53-54, resident internal med, 54-55; resident, Hosp, 55-57, asst res physician, Univ, 59-61, from instr to assoc prof med, 60-74, PROF PHYSIOL & MED, UNIV CALIF, SAN FRANCISCO, 74-, ASSOC STAFF MEM, UNIV CALIF HOSP, 60- Concurrent Pos: Bank of Am Giannini fel pulmonary physiol, Cardiovasc Res Inst, Univ Calif, San Francisco, 57-59; USPHS res career develop award, 63- Mem: Am Physiol Soc. Res: Regulation of respiration; chemoreceptors. Mailing Add: 338 Marietta Dr San Francisco CA 94127

MITCHELL, ROBERT ALEXANDER, b Belfast, Northern Ireland, Apr 3, 35. BIOCHEMISTRY. Educ: Queen's Univ Belfast, BSc, 55, PhD(chem), 60. Prof Exp: Teacher chem, Col Technol, Belfast, 57-60; asst prof, 65-72, ASSOC PROF BIOCHEM, WAYNE STATE UNIV, 72- Concurrent Pos: Fel biochem, Okla State Univ, 60-62, Univ Minn, 63 & Univ Calif, Los Angeles, 63-65. Mem: Am Chem Soc; Am Soc Biol Chem. Res: Mitochondrial energy metabolism; enzyme kinetics; use of substrate analogs to study enzyme catalysis and regulation. Mailing Add: Dept of Biochem Wayne State Univ Sch of Med Detroit MI 48201

MITCHELL, ROBERT BRUCE, b Rochester, Pa, Sept 24, 42; m 65; c 1. PHYSIOLOGY. Educ: Denison Univ, BS, 64; Ohio Univ, MS, 66; Pa State Univ, PhD(physiol), 69. Prof Exp: ASST PROF BIOL, PA STATE UNIV, 69- Mem: AAAS. Res: Reproductive endocrinology; interuterinely transferred blastocysts and ova in the rat; histochemistry; cytophotometric and interferometric analyses of nucleic acids and protein in erythroid cells from hypoxic rats; biology of aging. Mailing Add: 208 Life Sci Bldg I Pa State Univ University Park PA 16802

MITCHELL, ROBERT CURTIS, b Ft Dodge, Iowa, Mar 29, 28; m 49; c 3. ASTROPHYSICS. Educ: NMex State Univ, BS, 49, PhD(physics), 66; Univ Wash, MS, 52. Prof Exp: Physicist ultrasonics, Anderson Labs, West Hartford, Conn, 52-53; physicist atmosphere, Univ Conn, 53; solar observer, Harvard Col Observ, 53-54; teacher math & sci, Colo Rocky Mountain Sch, 54-56 & Gadsden High Sch, Anthony, NMex, 56-62; INSTR PHYSICS, CENT WASH STATE COL, 66- Mem: Am Asn Physics Teachers; Am Asn Variable Star Observers. Res: Photometry and photography of variable stars. Mailing Add: Dept of Geol & Physics Cent Wash State Col Ellensburg WA 98926

MITCHELL, ROBERT DALTON, b Bellingham, Wash, June 2, 23; m 45; c 3. MEDICINE. Educ: La Sierra Col, BS, 44; Col Med Evangelists, MD, 47, MSc, 60; Am Bd Internal Med, dipl & cert gastroenterol. Prof Exp: Asst clin prof, 56-64, asst prof, 64-66, ASSOC PROF MED, SCH MED, LOMA LINDA UNIV, 66- Mem: AMA; fel Am Col Physicians; Am Gastroenterol Asn. Res: Gastroenterology. Mailing Add: Loma Linda Univ Med Ctr Loma Linda CA 92354

MITCHELL, ROBERT DAVIS, b Motherwell, Scotland, Apr 26, 40; m 65; c 2. GEOGRAPHY. Educ: Glasgow Univ, MA, 62; Univ Wis-Madison, PhD(geog), 69. Prof Exp: Instr geog, Calif State Univ, Northridge, 65-66, asst prof, 68-69; asst prof, 69-73, ASSOC PROF GEOG, UNIV MD, COLLEGE PARK, 73- Concurrent Pos: Lectr, Univ Md, College Park, 66; vis lectr, Univ Calif, Irvine, 68; Univ Md Gen Res Bd fel, Univ Md, College Park, 70-71, Asn Am Geog Gibson Fund fel, 71- Mem: Asn Am Geog; Am Geog Soc; Agr Hist Soc; Inst Early Am Hist & Cult. Res: American westward expansion; comparative frontiers; colonization and colonialism. Mailing Add: Dept of Geog Univ of Md College Park MD 20742

MITCHELL, ROBERT W, b Wellington, Tex, Apr 25, 33; m 54; c 3. BIOSPELEOLOGY, ECOLOGY. Educ: Tex Tech Col, BS, 54, MS, 55; Univ Tex, PhD(zool), 65. Prof Exp: From instr to asst prof biol, Lamar State Col, 57-61; from asst prof to assoc prof, 65-74, PROF BIOL SCI, TEX TECH UNIV, 74- Mem: Soc Study Evolution; Nat Speleol Soc; Ecol Soc Am. Res: Distributional, ecological and evolutionary study of cave dwelling animals, particularly invertebrates. Mailing Add: Dept of Biol Sci Tex Tech Univ Lubbock TX 79409

MITCHELL, RODGER (DAVID), b Wheaton, Ill, July 22, 26. POPULATION BIOLOGY. Educ: Univ Mich, PhD(zool), 54. Prof Exp: Instr zool, Univ Vt, 54-57; asst prof biol, Univ Fla, 57-61, assoc prof zool, 61-69; PROF ZOOL, OHIO STATE UNIV, 69- Concurrent Pos: NSF fac fel, Univ Calif, Berkeley, 59-60; Fulbright res fel, Ibaraki Univ, Japan, 65-66. Mem: AAAS; Am Micros Soc; Soc Syst Zool; Soc Study Evolution; Am Soc Naturalists. Res: Morphology and biology of water mites; population biology. Mailing Add: Fac of Zool Ohio State Univ Columbus OH 43210

MITCHELL, ROGER L, b Grinnell, Iowa, Sept 13, 32; m 55; c 4. AGRONOMY, CROP PHYSIOLOGY. Educ: Iowa State Univ, BS, 54, PhD(agron), 61; Cornell Univ, MS, 58. Prof Exp: Asst prof agron, Iowa State Univ, 61-62, from assoc prof to prof agron & farm oper, 62-69; chmn dept agron, Univ Mo-Columbia, 69-72, dean exten div, 72-75; VPRES AGR, KANS STATE UNIV, 75- Concurrent Pos: Am Coun Educ acad admin intern, Univ Calif, Irvine, 66-67. Mem: AAAS; fel Am Soc Agron; Crop Sci Soc Am (pres, 75-76). Res: Soybean physiology; rooting patterns under field conditions; sorghum physiology. Mailing Add: 114 Waters Hall Kans State Univ Manhattan KS 66506

MITCHELL, ROGER SHERMAN, b Wayne, Pa, Sept 26, 07; m 32; c 3. MEDICINE. Educ: Harvard Univ, BA, 30, MD, 34; Am Bd Internal Med, dipl. Prof Exp: Intern med serv, Boston City Hosp, 34-36; asst neurol, Mass Gen Hosp, 36; pvt pract internal med, NY, 37-42; resident tuberc, State Tuberc Sanatorium, NC, 45-46; assoc med dir, Trudeau Sanatorium, 46-53, clin dir, 53-55; dir, Webb-Waring Lung Inst Med, 55-73, assoc prof, 55-67, PROF MED, UNIV COLO MED CTR, DENVER, 67- Concurrent Pos: Asst clin prof, Sch Med, Univ Vt, 50-54; chief of staff, Vet Admin Hosp, Denver, Colo, 71- Mem: Am Thoracic Soc; AMA; Am Clin & Climat Asn; fel Am Col Physicians; fel Am Col Chest Physicians. Res: Chest disease therapy. Mailing Add: 4200 E Ninth Ave Denver CO 80220

MITCHELL, ROGER W, b Ft Worth, Tex, Oct 20, 37; m 59; c 2. MATHEMATICS. Educ: Hendrix Col, AB, 59; Southern Methodist Univ, MS, 61; NMex State Univ, PhD(math), 64. Prof Exp: Asst prof, 64-71, ASSOC PROF MATH, UNIV TEX, ARLINGTON, 71- Mem: Am Math Soc; Math Asn Am. Res: Infinite Abelian groups; algebra. Mailing Add: Dept of Math Univ of Tex Arlington TX 76010

MITCHELL, ROY ERNEST, b Ft Worth, Tex, Aug 27, 36; m 58; c 1. INORGANIC CHEMISTRY, PHYSICAL CHEMISTRY. Educ: Tex A&M Univ, BS, 58; Purdue Univ, PhD(inorg chem), 64. Prof Exp: Fel, Tex A&M Univ, 64-65; asst prof chem, Purdue Univ, 65-66; ASST PROF CHEM, TEX TECH UNIV, 66- Mem: Am Chem Soc. Res: Thermodynamic and kinetic relationships of carbohydrates; ionic solutions; thermochemistry; apparent molar volumes. Mailing Add: Dept of Chem Tex Tech Univ Lubbock TX 79409

MITCHELL, RUSSEL GENE, b Portland, Ore, Feb 14, 30; m 56. FOREST ENTOMOLOGY. Educ: Ore State Col, BS, 56, PhD(entom), 60; State Univ NY, MS, 57. Prof Exp: Res entomologist, 57-71, INSECT ECOLOGIST, PAC NORTHWEST FOREST & RANGE EXP STA, US FOREST SERV, 71- Mem: Entom Soc Am; Soc Am Foresters. Res: Biology, ecology and management of forest insects. Mailing Add: Pac NW Forest & Range Exp Sta Forestry Sci Lab 3200 Jefferson Way Corvallis OR 97331

MITCHELL, SHIELA CRAIG, b Toronto, Ont, Nov 13, 26; m 51; c 1. PEDIATRIC CARDIOLOGY, EPIDEMIOLOGY. Educ: Univ Toronto, MD, 51; Univ Mich, MPH, 60; Am Bd Pediat, dipl, 58. Prof Exp: Examr community health study, Sch Pub Health, Univ Mich, 59-60; res assoc obstet, Mich Med Ctr, 60-61; res assoc epidemiol, Univ Mich, 61-63; med officer, 63-68, asst to dir, 68-74, CHIEF ATHEROGENESIS BR, NAT HEART & LUNG INST, 74- Mem: Am Acad Pediat; Teratology Soc; Royal Soc Med. Mailing Add: C808 Landow Bldg Nat Heart & Lung Inst Bethesda MD 20014

MITCHELL, THEODORE, b Chicago, Ill, Apr 30, 26; m 57; c 2. MATHEMATICS, OPERATIONS RESEARCH. Educ: Ill Inst Technol, BS, 51, MS, 57, PhD(math), 64. Prof Exp: Jr mathematician inst air weapons res, Univ Chicago, 52-54, mathematician, 54-58, sr staff mem labs appl sci, 59-62; staff mathematician, Weapons Systs Eval Group, Inst Defense Anal, 58-59; sr mathematician, Acad Intersci Methodology, 63-64; lectr math, State Univ NY Buffalo, 64-65, assoc prof, 65-67; assoc prof, 67-69, PROF MATH, TEMPLE UNIV, 69- Mem: AAAS; Am Math Soc; Opers Res Soc Am; Math Asn Am. Res: Functional analysis; invariant means; fixed point theorems; game theory; probability theory. Mailing Add: Dept of Math Temple Univ Philadelphia PA 19122

MITCHELL, THOMAS EDGAR, physical chemistry, see 12th edition

MITCHELL, THOMAS F, SR, b Milwaukee, Wis, July 27, 23; m 46; c 6. PROTEIN CHEMISTRY. Educ: Marquette Univ, BS, 49. Prof Exp: Chemist, Bio-Process Co, 50-55; sr res chemist, Lakeside Labs, 55-62; consult protein synthesis, Toni Co, 62-63, res chemist, 63-64; res chemist, 64-67, asst dir res, 67-68, DIR RES, GLUE DIV, DARLING & CO, 68- Mem: Am Chem Soc; Am Inst Chem; Tech Asn Pulp & Paper Indust; Forest Prod Res Soc; Am Soc Testing & Mat. Res: Collagen-proteins and aliphatic amines and their reaction products; vegetable proteins; natural polymer extractions. Mailing Add: 1250 W 46th St Chicago IL 60609

MITCHELL, THOMAS GEORGE, b Philadelphia, Pa, Feb 27, 27; m 46; c 6. PHYSIOLOGY, NUCLEAR MEDICINE. Educ: St Joseph's Col, Pa, BS, 50; Univ Rochester, MS, 56; Georgetown Univ, PhD(physiol), 63; Am Bd Radiol, dipl, 58. Prof Exp: Med Serv Corpsman, US Navy, 50-69, asst supvr radioisotope lab, Naval Hosp, St Albans, NY, 51-55, med nuclear physicist, Bethesda Naval Hosp, Md, 56-60, physiologist & med nuclear physicist, Naval Med Res Unit 2, Taipei, Taiwan, 63-65, med nuclear physicist, Radiation Exposure Eval Lab, Naval Hosp, 65-69; assoc prof radiol sci, Sch Hyg & Pub Health, Johns Hopkins Univ, 69-74; DIR RADIATION CONTROL, ASSOC PROF PHYSIOL & BIOPHYS & ASSOC PROF RADIOL, MED CTR, GEORGETOWN UNIV, 74- Concurrent Pos: Lectr, Naval Med Sch, Bethesda, 56-63 & 65-; adj acad staff, Children's Hosp, DC, 59-63; lectr, Sch Hyg & Pub Health, Johns Hopkins Univ, 65-70; instr, Sch Med & Dent, Georgetown Univ, 65-69; consult, Naval Med Res Inst, 65-70. Mem: Am Col Radiol; Soc Nuclear Med; Health Physics Soc; Am Asn Physicists in Med. Res: Medical nuclear physics; clinical applications of radioisotopes; regional blood flow. Mailing Add: 305 Kober-Cogan Bldg Georgetown Univ Med Ctr Washington DC 20007

MITCHELL, THOMAS OWEN, b New York, NY, Oct 18, 44; m 70. ORGANIC CHEMISTRY, PETROLEUM CHEMISTRY. Educ: Trinity Col, Conn, BS, 66; Northwestern Univ, Ill, PhD(org chem), 70. Prof Exp: Res chemist, 70-73, SR RES CHEMIST, MOBIL RES & DEVELOP CORP, 73- Mem: Am Chem Soc. Res: Catalysis in petrochemistry; catalytic chemistry; organosilicon chemistry; coal chemistry and conversion. Mailing Add: Mobil Res & Develop Corp PO Box 1025 Princeton NJ 08540

MITCHELL, VAL LEONARD, b Salt Lake City, Utah, Sept 9, 38; m 61; c 4. CLIMATOLOGY. Educ: Univ Utah, BS, 64; Univ Wis-Madison, MS, 67, PhD(meteorol), 69. Prof Exp: Res asst meteorol, Univ Wis-Madison, 64-69; from asst prof to assoc prof earth sci, Mont State Univ, 69-74; STATE CLIMATOLOGIST, WIS GEOL & NATURAL HIST SURV, UNIV WIS-EXTEN, 74-, ASST PROF METEOROL, UNIV WIS-MADISON, 74- Mem: Am Meteorol Soc; Asn Am Geogr. Res: Applied climatology; interaction between the atmosphere and the biosphere; climatology of mountainous regions. Mailing Add: Dept of Meteorol Univ of Wis Madison WI 53706

MITCHELL, WALLACE CLARK, b Ames, Iowa, Nov 12, 20; m 58; c 3. ENTOMOLOGY. Educ: Iowa State Col, BS, 47, PhD(zool, entom), 55. Prof Exp: Sr engr aide & entomologist malaria control in war areas, USPHS, Fla, 42-43; asst to state entomologist, Nursery Inspection, State Dept Agr, Iowa, 47-49; instr zool & entom & asst entom, Agr Exp Sta, Univ Hawaii, 49-53; field entomologist midwestern area, Geigy Agr Chem Inc, 53-54; asst prof zool & entom, SDak State Col, 55-56; entomologist fruit insect sect, Agr Res Serv, USDA, 56-62; assoc prof entom & assoc entomologist, 62-69, actg dir, Hawaii Agr Exp Sta, 69-70, PROF ENTOM, CHMN DEPT & ENTOMOLOGIST, UNIV HAWAII, 68-, ACTG DEAN COL TROP AGR, 75- Mem: Entom Soc Am; Am Registry Cert Entomologists. Res: Economic entomology; field evaluation of fruit fly lures, attractants, insecticide sprays, sterilization and eradication procedures for control of fruit and vegetable insects; tropical economic entomology; insect behavior, pest management; insect ecology. Mailing Add: Col of Trop Agr Univ of Hawaii Room 23 Krauss Hall 2500 Dole St Honolulu HI 96822

MITCHELL, WALTER EDMUND, JR, b Franklin, Mass, Nov 16, 25. ASTRONOMY. Educ: Tufts Univ, BS, 49; Univ Va, MS, 51; Univ Mich, PhD, 58. Prof Exp: Asst astron, Univ Mich, 51-52, observer infrared proj, Mt Wilson Observ, 53-55; vis asst prof astron, Brown Univ, 56-57; from instr to assoc prof, 57-69, PROF ASTRON, OHIO STATE UNIV, 69- Mem: Am Astron Soc; Royal Astron Soc. Res:

Solar spectroscopy. Mailing Add: Dept of Astron Ohio State Univ Columbus OH 43210

MITCHELL, WILBUR LEONARD, b Holden, Mass, Nov 19, 15; m 40; c 4. MATHEMATICS. Educ: Colgate Univ, BA, 36; Univ Wis, MA, 38, PhD(math), 40. Prof Exp: Instr math, Univ Wis-Milwaukee, 40-42; asst prof, Mich State Col, 42-44; sr aerodynamicist, Bell Aircraft Corp, 44-48; mathematician, Holloman AFB, NMex, 48-51; eng specialist & mgr eng, NAm Aviation, 51-66; mgr advan prog, Litton Industs, 66-71; asst prof math, Tougaloo Col, 71-72; vpres, Mary Holmes Col, 72-74; PROF MATH, DILLARD UNIV, 74- Mem: Math Asn Am. Mailing Add: Dept of Math Dillard Univ New Orleans LA 70122

MITCHELL, WILLIAM ALEXANDER, b Raymond, Minn, Oct 21, 11; m 38; c 7. FOOD SCIENCE, CARBOHYDRATE CHEMISTRY. Educ: Nebr Wesleyan Univ, BA, 35; Univ Nebr, MS, 38. Prof Exp: Teacher high sch, Nebr, 35-36; asst chem, Univ Nebr, 37-38; org res chemist, Eastman Kodak Co, NY, 38-41; head biochem sect, 41-53, head biocolloids sect, 53-56, sect head chem res, 56-62, res specialist, 62-68, SR RES SPECIALIST, GEN FOODS CORP, 68- Mem: Am Chem Soc. Res: Synthetic organic biochemistry; wheat flour; starch; proteins; fats and oils; emulsifiers; gas reactions; enzymes; gelatin; food processing and products; sugars; pectins; gums; colloid chemistry; coffee; food flavors; cake mixes; eggs; carbonation systems; Kodachrome couplers. Mailing Add: 175 Jacksonville Rd Lincoln Park NJ 07035

MITCHELL, WILLIAM COBBEY, b Rochester, NY, Aug 2, 39; m 62; c 4. SOLID STATE PHYSICS, ENERGY CONVERSION. Educ: Oberlin Col, AB, 61; Wash Univ, PhD(physics), 67. Prof Exp: Sr res physicist, 3M Co, 67-69; Nat Bur Standards-Nat Res Coun fel, Nat Bur Standards, 69-71; consult, Thermoelec Syst Sect, 71-72, sr res physicist, 72-75, RES SPECIALIST, 3M CO, 75- Mem: Am Phys Soc. Res: Non-equilibrium statistical mechanics; solid state transport theory; thermoelectric materials; energy storage. Mailing Add: Energy Systs Proj Bldg 260 3M Ctr St Paul MN 55101

MITCHELL, WILLIAM EDWARD, b Wichita, Kans, Jan 8, 27; m 59; c 2. CULTURAL ANTHROPOLOGY. Educ: Wichita State Univ, BA, 50; Columbia Univ, MA, 54 & 57, PhD(anthrop), 69. Prof Exp: Res asst anthrop, Cornell Med Sch, NY Hosp-Cornell Med Ctr, 55-56; res assoc, Jewish Family Serv & Russell Sage Found, NY, 57-62; proj dir, relationship Ther Proj, Hope Found, 62-65; res assoc social psychiat & anthrop, Col Med, 65-67, asst prof, 67-70, assoc prof social psychiat & anthrop, 70-73, ASSOC PROF ANTHROP, COL LIBERAL ARTS & SCI, UNIV VT, 73- Concurrent Pos: Lectr, Cooper Union Advan Sci & Art, 61; NIMH grant, Hope Found, Vt, 62-65; Wenner-Gren Found travel grant, Columbia Univ & New Guinea, 67; NIMH human behav training grant, Univ Vt Col Med, 67-70, New Guinea Therapeut Systs Res grant, 70-74; dir, Univ Vt New Guinea Proj, 70- Mem: Am Anthrop Asn; Am Ethnol Soc; Royal Anthrop Inst Gt Brit & Ireland; Soc Appl Anthrop; Soc Med Anthrop. Res: Medical anthropology; culturally contrasting therapeutic systems; kinship and the family; deviant behavior; semiotics; cultures studies; American Jews and Chinese; three New Guinea tribes; psychiatric ward. Mailing Add: Williams Sci Hall Univ of Vt Burlington VT 05401

MITCHELL, WILLIAM H, b Acworth, NH, Dec 12, 21; m 46; c 2. AGRONOMY, BOTANY. Educ: Univ NH, BS, 46, MS, 49; Pa State Univ, PhD(agron, bot), 60. Prof Exp: Headmaster & teacher high sch, NH, 45-47; from asst res prof to assoc res prof agron, 49-71, PROF PLANT SCI, UNIV DEL, 71- Concurrent Pos: Dir, Int Coop Improv Asn, 53-66; chmn collabrs regional pasture res lab, Pa State Univ, 60-66; tech comt, Northeastern Regional Res Group, 66-68; mem, Am Grassland Coun. Mem: Soil Sci Soc Am; Am Soc Agron; Crop Sci Soc Am. Res: Management of forage crops and turf; soil fertility. Mailing Add: Dept of Plant Sci Univ of Del Newark DE 19711

MITCHELL, WILLIAM HINCKLEY, b East Orange, NJ, June 5, 03; m 30; c 3. PHYSIOLOGY. Educ: Rutgers Univ, BS, 25; Harvard Univ, MA, 26, PhD(biol), 29; Columbia Univ, BLS, 39. Prof Exp: Asst zool, Harvard Univ & Radcliffe Col, 28-29; agent physiol res bur entom, USDA, Fla, 29-30, assoc physiologist & histologist, Hawaii, 30-33; instr zool, Rutgers Univ, 33-35, library asst, 35-38; library consult, Union Club, NY, 39-43; bibliographer library, USDA, 43-51; supvry librn, Chief of Index & Ref Unit, Pub Br, Adj General's Off, US Dept Army, 51-64; chief tech processing, Fairfax County Pub Library, 64-72; RETIRED. Concurrent Pos: Indexer, H W Wilson Co, 40-43. Res: Physiology of protozoa and Ceratitis capitata; tropisms; mathematical analysis of living processes; entomological bibliography; computerized library book-catalogue preparation. Mailing Add: 4421 Upland Dr Alexandria VA 22310

MITCHELL, WILLIAM MARVIN, b Atlanta, Ga, Mar 3, 35; m 59; c 3. ONCOLOGY, PATHOLOGY. Educ: Vanderbilt Univ, BA, 57, MD, 60; Johns Hopkins Univ, PhD(biol, biochem), 66. Prof Exp: Mem house staff med, Johns Hopkins Hosp, 60-61; from asst prof to assoc prof microbiol, 66-74, ASSOC PROF PATH, VANDERBILT UNIV, 74-, ASST PROF MED, 69-, PLANNING DIR, VANDERBILT CANCER CTR, 71- Concurrent Pos: USPHS fel biol, McCullom-Pratt Inst, Johns Hopkins Univ, 61-66; USPHS grant & Vanderbilt Inst award, 66-67; Nat Inst Arthritis & Metab Dis grant, 66-74; Nat Cancer Inst grant, 71-73; staff physician, Vet Admin Hosp, 66-67; consult, NIH, 71- Mem: AAAS; Am Chem Soc; Am Soc Microbiol; AMA; Am Soc Biol Chem. Res: Structure and function of proteins, including cell collagen, proteolytic enzymes and bacterial exotoxins; oncology, including cell control mechanisms and oncogenic virology; macromolecular chemistry; protein structure-function; molecular pathology. Mailing Add: 251 Vaughn Gap Rd Nashville TN 37205

MITCHELL, WILLIAM WARREN, b Butte, Mont, Mar 10, 23; m 55; c 2. AGRONOMY, BOTANY. Educ: Univ Mont, BA, 57, MA, 58; Iowa State Univ, PhD(bot), 62. Prof Exp: Instr biol sci, Western Mont Col, 58-59; asst prof, Chadron State Col, 62-63; PROF AGRON & HEAD DEPT, INST AGR SCI, UNIV ALASKA, 63- Mem: Am Soc Plant Taxon; Ecol Soc Am; Soc Range Mgt; Am Agron Soc; Am Inst Biol Sci. Res: Applications of introduced and indigenous grass taxa; varietal selection and development of grasses; plant community analysis and management; biosystematics of grasses. Mailing Add: Dept Agron Inst Agr Sci Univ of Alaska Palmer AK 99645

MITCHELL-KERNAN, CLAUDIA I, b Gary, Ind, Aug 29, 41; m 68; c 1. ANTHROPOLOGY. Educ: Ind Univ, Bloomington, BA, 63, MA, 64; Univ Calif, Berkeley, PhD(anthrop), 69. Prof Exp: Lectr anthrop, Harvard Univ, 69-70, asst prof, 70-74; ASST PROF ANTHROP, UNIV CALIF, LOS ANGELES, 74- Concurrent Pos: Res fel urban anthrop, Peabody Mus, Harvard Univ, 69-; mem, Social Probs Rev Comt, NIMH, 71-75; consult, Nat Insts Educ, 71-; Harvard Univ Spencer Found grant, Mass & Vt, 72-73. Mem: Am Anthrop Asn. Res: Social anthropology, especially sociolinguistics and urban anthropology, Afro-American language and culture. Mailing Add: Dept of Anthrop Univ of Calif Los Angeles CA 90024

MITCHEM, JOHN ALAN, b Sterling, Colo, July 28, 40; m 62; c 2. MATHEMATICS. Educ: Univ Nebr, Lincoln, BS, 62; Western Mich Univ, MA, 67, PhD(math), 70. Prof Exp: Asst prof math, 70-74, ASSOC PROF MATH, SAN JOSE STATE UNIV, 74- Mem: Math Asn Am; Am Math Soc. Res: Graph theory, especially partioning problems. Mailing Add: Dept of Math San Jose State Univ San Jose CA 95192

MITCHEN, JOEL RAMON, b Chicago, Ill, Mar 25, 42; m 74; c 1. IMMUNOLOGY, CANCER. Educ: Carthage Col, BA, 64; Purdue Univ, MS, 69, PhD(molecular biol), 71. Prof Exp: Cancer res scientist, Roswell Park Mem Inst, 70-72; assoc cancer res scientist, Denver Gen Hosp, 72-73; ASSOC IMMUNOLOGIST, LIFE SCI RES LAB, 73- Concurrent Pos: Vis scientist, Argonne Nat Lab, 72; vis lectr, Canisius Col, 72. Mem: Am Soc Microbiol. Res: Molecular biology; virology; immunology of cancer. Mailing Add: Life Sci Res Lab 2900 72nd St N St Petersburg FL 33710

MITCHNER, HYMAN, b Vancouver, BC, Nov 23, 30; m 53; c 4. PHARMACEUTICAL CHEMISTRY. Educ: Univ BC, BA, 51, MSc, 53; Univ Wis, Madison, PhD(pharmaceut chem), 56. Prof Exp: Asst prof pharmaceut chem, Sch Pharm, Univ Wis, Madison, 56-59; group leader anal res, Miles Lab, 59-61, sect head, 61-63; dir qual control, Barnes-Hind Labs, 63-66; dir, 66-69, VPRES QUAL CONTROL, SYNTEX LABS INC, 69- Mem: Am Pharmaceut Asn; Am Chem Soc; Am Soc Qual Control. Res: Analytical research on drug products; physical chemical studies in drug systems. Mailing Add: Qual Control Div Syntex Labs Inc 3401 Hillview St Palo Alto CA 94302

MITCHNER, MORTON, b Vancouver, BC, Jan 17, 26; US citizen; m 60; c 2. PHYSICS. Educ: Univ BC, BA, 47, MA, 48; Harvard Univ, PhD(physics), 52. Prof Exp: Sheldon traveling fel, Harvard Univ, 52-53, res fel appl sci, 53-54; staff mem opers res, Arthur D Little, Inc, 54-58; staff scientist, Lockheed Missiles & Space Co, 58-63; vis lectr mech eng, 63-64, assoc prof, 64-69, PROF MECH ENG, STANFORD UNIV, 69- Concurrent Pos: Vis prof eng sci, Columbia Univ, 61-62; consult, United Tech Ctr, United Aircraft Corp, 65-68; mem steering comt, Eng Appln Magnetohydrodynamics, 67-; consult, Nat Comt Nuclear Energy, Italy, 70-71. Mem: Am Phys Soc; Am Inst Aeronaut & Astronaut; Opers Res Soc Am. Res: High temperature gas dynamics; plasma physics; magnetohydrodynamics; kinetic theory; turbulence; combustion aerodynamics; energy conversion; operations research. Mailing Add: Dept of Mech Eng Stanford Univ Stanford CA 94305

MITCHUM, RONALD KEM, b Elk City, Okla, Dec 2, 46; m 65; c 2. MASS SPECTROMETRY. Educ: Southwestern Okla State Univ, BS, 68; Okla State Univ, PhD(chem), 73. Prof Exp: Fel mass spectrometry, Univ Houston, 72-74 & Univ Warwick, 74-75; FEL MASS SPECTROMETRY, UNIV NEBR-LINCOLN, 75- Mem: Am Chem Soc. Res: High pressure negative and positive ion molecule reactions and rate and equilibrium constant determinations in the gas phase of organic compounds. Mailing Add: Dept of Chem Univ of Nebr Lincoln NE 68588

MITESCU, CATALIN DAN, b Bucharest, Roumania, May 2, 38; Can citizen. PHYSICS. Educ: McGill Univ, BEng, 58; Calif Inst Technol, PhD(physics), 66. Prof Exp: From instr to asst prof, 65-69, ASSOC PROF PHYSICS, POMONA COL, 69- Mem: Am Phys Soc; Can Asn Physicists. Res: Low temperature physics-superconductivity, thin films; liquid helium. Mailing Add: Dept of Physics Pomona Col Claremont CA 91713

MITHOEFER, JOHN CALDWELL, b Cincinnati, Ohio, Feb 12, 20; m 42; c 3. MEDICINE. Educ: Brown Univ, BA, 41; Harvard Univ, MD, 44. Prof Exp: From instr to asst clin prof med, Col Physicians & Surgeons, Columbia Univ, 52-68; prof med & chief cardiopulmonary div, Dartmouth Med Sch, 68-72; PROF MED & DIR PULMONARY DIV, MED UNIV SC, 72- Concurrent Pos: Assoc physician & dir cardiopulmonary lab, Mary Imogene Bassett Hosp; mem staff, Fac Med, Paris, 54 & postgrad sch med, Hammersmith Hosp, London, Eng, 62; mem pulmonary training comt & chmn rev comt nat res & demonstration ctrs, Nat Heart & Lung Inst; fel coun clin cardiol, Am Heart Asn. Mem: Am Soc Clin Invest; Am Fedn Clin Res; fel Am Col Cardiol; fel Am Col Chest Physicians; Asn Univ Cardiol. Res: Cardiopulmonary physiology. Mailing Add: Pulmonary Div Med Univ of SC Charleston SC 29401

MITHUN, JACQUELINE STEARNS, b Boston, Mass, July 2, 41. APPLIED ANTHROPOLOGY. Educ: Cornell Univ, BA, 63; Univ Minn, MA, 65; State Univ NY Buffalo, PhD(anthrop), 74. Prof Exp: Instr anthrop, State Univ NY Buffalo, 69-71; instr sociol, State Univ Col Buffalo, 69-70; res anthrop, Ford Found & NSF grants, 72-73; community liaison specialist, Albion Correctional Facil, 73-74; asst prof anthrop, Sangamon State Univ, 74-75; vis asst prof anthrop, Col Old Westbury, State Univ NY, 75-76; NAT ENDOWMENT FOR THE HUMANITIES FEL, JOHNS HOPKINS UNIV, 76- Mem: Fel Am Anthrop Asn; Asn Appl Anthrop; Am Asn Humanistic Psychol; AAAS. Res: Urban minorities, especially Blacks, in psychological aspects and social change processes; applied anthropology in urban area centering on conceptual aspects of social change implementation and processes. Mailing Add: Dept of Anthrop Johns Hopkins Univ Baltimore MD 21218

MITLER, HENRI EMMANUEL, b Paris, France, Oct 26, 30; US citizen; div; c 2. ASTROPHYSICS, NUCLEAR PHYSICS. Educ: City Col New York, BS, 53; Princeton Univ, PhD(nuclear structure), 60. Prof Exp: Jr physicist, Nuclear Develop Assocs, 53; instr physics, Princeton Univ, 57-58; res assoc physics & adj lectr, Brandeis Univ, 59-60; SR STAFF PHYSICIST, SMITHSONIAN ASTROPHYS OBSERV, 61-; RES ASSOC, HARVARD COL OBSERV, 62-, LECTR ASTRON, HARVARD UNIV, 63- Concurrent Pos: Lectr, Brandeis Univ, 67-68 & 71-72. Mem: Am Phys Soc; Am Astron Soc. Res: Quantum theory; nuclear structure and reactions; origin of the elements; radiochemical production in meteoroids; nucleosynthesis by cosmic rays; electron-ion screening in dense plasmas origin of the moon. Mailing Add: Smithsonian Astrophys Observ 60 Garden St Cambridge MA 02138

MITOMA, CHOZO, b San Francisco, Calif, July 21, 22; m 50; c 4. BIOCHEMICAL PHARMACOLOGY. Educ: Univ Calif, BA, 48, PhD(biochem), 51. Prof Exp: Asst, Univ Calif, 48-51; biochemist, NIH, 52-59; sr biochemist, 59-69, DIR DEPT BIOMED RES, STANFORD RES INST, 69- Concurrent Pos: Hite fel, Univ Tex, 51-52. Mem: Am Soc Biol Chem; Am Soc Pharmacol & Exp Therapeut; Soc Exp Biol & Med. Res: Intermediary metabolism of amino acids; mechanism of action of drugs; drug metabolism; neurochemistry. Mailing Add: Dept of Biomed Res Stanford Res Inst Menlo Park CA 94025

MITRA, GRIHAPATI, b Oct 29, 27; m 58; c 1. INORGANIC CHEMISTRY. Educ: Univ Calcutta, BSc, 47, MSc, 49, DSc, 54. Prof Exp: Fel chem, Univ Wash, Seattle, 55-57; res officer, AEC, India, 58; lectr chem, Dum Dum Col, 59-60; fel physics, Pa State Univ, 60-61; asst prof chem, 61-64, chmn dept, 64-67, PROF CHEM, KING'S COL, PA, 67- Concurrent Pos: Vis lectr, Sylvania Elec Prods, Inc, Pa, 62; dir rocket fuel res, Adv Res Projs Agency, 63-67; res dir, Beryllium Corp, 67. Mem: AAAS; Am Chem Soc; Indian Chem Soc. Res: Preparation and properties of compounds containing halogens. Mailing Add: Dept of Chem King's Col Wilkes-Barre PA 18702

MITRA, RATHIN, b Calcutta, India, Jan 1, 37. MICROBIOLOGY. Educ: Univ Calcutta, BVSc & AH, 57; Univ Fla, MS, 63; La State Univ, PhD(microbiol), 66. Prof Exp: Vet officer, Govt of West Bengal, India, 58-62; assoc prof, 66-73, PROF BIOL, WEST LIBERTY STATE COL, 73- Mem: Am Soc Microbiol. Res:

Desoxyribonucleic acid taxonomy in bacteria; mechanism of desoxyribonucleic acid replication of single stranded bacterial virus; study of uptake and role of streptomycin in Escherichia coli protoplast system. Mailing Add: Dept of Biol West Liberty State Col West Liberty WV 26074

MITRA, SANKAR, b Calcutta, India, July 7, 37; m 66; c 2. MOLECULAR BIOLOGY, BIOCHEMISTRY. Educ: Univ Calcutta, BSc, 57, MSc, 59; Univ Wis-Madison, PhD(biochem), 64. Prof Exp: Res asst biochem, Univ Calcutta, 59-60; res asst, Univ Wis-Madison, 60-62, res fel, 62-63; USPHS grant & res assoc, Stanford Univ, 64-65; Indian Govt Coun Sci & Indust Res sci officer, Bose Inst, India, 66-67, sr res fel, 67-70, reader, 71; BIOCHEMIST, BIOL DIV, OAK RIDGE NAT LAB, 71- Concurrent Pos: Mem panel V, Int Cell Res Orgn, 69-; lectr sch biomed sci, Univ Tenn, 72- Mem: Indian Soc Biol Chem; Am Soc Biol Chemists. Res: Molecular biology of viruses and nucleic acids; synthesis of nucleic acids in vivo and in vitro. Mailing Add: Biol Div Oak Ridge Nat Lab Oak Ridge TN 37830

MITROVIC, MILAN, b Yugoslavia, Sept 11, 20; nat US; m 51; c 4. VETERINARY MEDICINE. Educ: Univ Perugia, DVM, 45; Univ Bologna, dipl, 47. Prof Exp: Asst prof clin path, Univ Perugia, 45-46, asst dir animal exp sta, 46-51; head vet lab, State Livestock Sanit Bd, Ark, 52-54; asst mgr vet path, Res Div, Salsbury Labs, 54-56, res assoc & proj leader, 56-57, head microbiol dept, 57-61; asst prof vet sci, Pa State Univ, 61-62; mgr biol lab, Agr Res Dept, 62-67, res group chief, Animal Health Res Dept, 68-73, RES SECT HEAD, ANIMAL HEALTH RES DEPT, HOFFMANN-LA ROCHE INC, 74- Mem: Am Soc Microbiol; Am Asn Avian Path; Am Vet Med Asn; Indust Vet Asn; NY Acad Sci. Res: Bacteriology; virology; pathology; diagnosis; product development; biological production; experimental chemotherapy. Mailing Add: Animal Health Res Dept Hoffmann-La Roche Inc Nutley NJ 07110

MITRUKA, BRIJ MOHAN, b Hanuman Garh Town, Rajasthan, India, May 12, 37; m 57; c 3. MICROBIOLOGY, VETERINARY MEDICINE. Educ: Rajasthan Vet Col, BVSc & AH, 59; Mich State Univ, MS, 62, PhD(microbiol, pub health), 65. Prof Exp: Vet asst surg, Vet Hosp, Hanamangarh County, Rajasthan, 59-60; USPHS fel & res assoc microbiol, Mich State Univ, 65-66; res assoc, Cornell Univ, 66-68; clin pathologist, Sch Med, Yale Univ, 68-69, asst prof lab animal sci & lab med, 69-74; ASSOC PROF, SCH MED, UNIV PA, 74- Concurrent Pos: USPHS grants, Yale Univ, 69-73; prof microbiol & head dept, Punjab Agr Univ, 74- Mem: AAAS; Am Soc Microbiol; Am Asn Clin Chem; Am Vet Med Asn. Res: Bases of microbial pathogenicity; microbial metabolites detection and identification in tissues and body fluids; microbial metabolism in vitro and in vivo; study of pathogenic mechanisms involved in infectious diseases of man and animals; rapid, automated diagnosis in infectious and non infections diseases. Mailing Add: Sch of Med 324 Johnson Pav Univ of Pa Philadelphia PA 19174

MITSCH, RONALD ALLEN, organic chemistry, see 12th edition

MITSCHER, LESTER ALLAN, b Detroit, Mich, Aug 20, 31; m 53; c 3. BIO-ORGANIC CHEMISTRY. Educ: Wayne State Univ, BS, 53, PhD(chem), 59. Prof Exp: Spec instr pharm, Wayne State Univ, 56-58; res scientist bio-org chem, Lederle Labs, Am Cyanamid Co, 58-61, group leader fermentation biochem, 61-67; prof pharmacog & natural prod, Col Pharm, Ohio State Univ, 67-75; UNIV DISTINGUISHED PROF MEDICINAL CHEM & CHMN DEPT, UNIV KANS, 75- Mem: Am Chem Soc; Am Soc Pharmacog; The Chem Soc; Am Soc Microbiol. Res: Chemistry of organic compounds of natural origin, especially alkaloids, terpenes, steroids and antibiotics. Mailing Add: Col of Pharm Univ of Kans Malott Hall Lawrence KS 66044

MITSUI, AKIRA, b Japan, Jan 25, 29; m 64; c 3. BIOLOGICAL OCEANOGRAPHY. Educ: Univ Tokyo, BS, 51, MA, 55, PhD(plant physiol), 58. Prof Exp: PROF BIOL OCEANOG, SCH MARINE & ATMOSPHERIC SCI, UNIV MIAMI, 72- Mem: Am Soc Microbiol; Am Soc Plant Physiologists; Phycol Soc Am. Res: Marine biochemistry; bioenergetics; bioconversion of solar energy. Mailing Add: Div of Biol & Living Resources Univ of Miami Miami FL 33149

MITTAG, LAURENCE, theoretical physics, see 12th edition

MITTAG, THOMAS WALDEMAR, b Pecs, Hungary, Mar 14, 37; m 59; c 2. PHARMACOLOGY, BIOCHEMISTRY. Educ: Univ Cape Town, BS, 59, Hons, 61, PhD(org chem), 64. Prof Exp: Jr lectr chem, Univ Cape Town, 63-65; staff scientist, Worcester Found Exp Biol, 66-67; instr pharmacol, New York Med Col, Flower & Fifth Ave Hosps, 68-69, asst prof, 69-71; ASSOC PROF PHARMACOL, MT SINAI SCH MED, 71- Concurrent Pos: Res fel biochem, Purdue Univ, 65-66; res fel pharmacol, Georgetown Univ, 67-69; NIH grants, 69-77. Mem: AAAS; Am Chem Soc; NY Acad Sci; Am Soc Pharmacol & Exp Therapeut; The Chem Soc. Res: Molecular pharmacology of neurohormone receptors; active site directed reagent synthesis for enzymes and receptors. Mailing Add: Dept of Pharmacol Mt Sinai Sch of Med New York NY 10029

MITTAL, KAMAL KANT, b Pihani, India, July 1, 35; c 1. IMMUNOGENETICS, TRANSPLANTATION IMMUNOLOGY. Educ: Agra Univ, BS, 54, DVM, 58; Univ Ill, Urbana, MS, 62, PhD(immunogenetics), 65. Prof Exp: Res assoc animal sci, Univ Ill, Urbana, 65-66; res fel biol, Calif Inst Technol, 66-67; res geneticist II, Dept Surg, Univ Calif, Los Angeles, 67-69; asst prof microbiol, Baylor Col Med, Houston, 69-70; asst prof immunogenetics, Univ Calif, Los Angeles, 70-72; asst prof surg & physiol, Northwestern Univ, Chicago, 72-75; RES MICROBIOLOGIST, BUR BIOLOGICS, FOOD & DRUG ADMIN, DEPT HEALTH, EDUC & WELFARE, 75- Mem: AAAS; Transplantation Soc; Am Asn Immunol; Am Soc Hemat. Res: Human histocompatibility systems—genetics, immunology, serology and immunoreproduction. Mailing Add: Bur Biologics FDA/DHEW Bldg 29 Rm 402 8800 Rockville Pike Bethesda MD 20014

MITTAL, KASHMIRI LAL, b Kilrodh, India, Oct 15, 45; m 70; c 2. PHYSICAL CHEMISTRY. Educ: Panjab Univ, Chandigarh, BSc, 64; Indian Inst Technol, New Delhi, MSc, 66; Univ Southern Calif, PhD(phys chem), 70. Prof Exp: Res assoc, Pa State Univ, 70-71; fel chem, Univ Pa, 71-72; fel, IBM Corp, San Jose, Calif, 72-74, STAFF ENGR, IBM CORP, POUGHKEEPSIE, 74- Mem: Am Chem Soc; Electrochem Soc; Am Vacuum Soc. Res: Surface, colloid, polymer and electrochemistry; surface properties of materials; adhesion and corrosion. Mailing Add: Dept L04 Bldg 052 IBM Corp Box 390 Poughkeepsie NY 12602

MITTAL, YASHASWINI DEVAL, b Poona, India, Oct 1, 41. STATISTICS. Educ: Poona Univ, BSc, 61; Univ Ill, Urbana, MS, 66; Univ Calif, Los Angeles, PhD(math), 72. Prof Exp: Teaching asst math, Univ Ill, Urbana, 64-66 & Univ Calif, Los Angeles, 66-71; asst prof, Northwestern Univ, Evanston, 72-73; vis mem, Inst Advan Study, Princeton, 73-74; ASST PROF STATIST, STANFORD UNIV, 74- Mem: Am Math Soc; Inst Math Statist. Res: Convergence properties of maxima of stationary Gaussian processes will be extended to isotropic Gaussian random fields in n-dimensions. Mailing Add: Dept of Statist Stanford Univ Stanford CA 94305

MITTELMAN, ARNOLD, b New York, NY, Dec 21, 24; m 56; c 2. SURGERY, MEDICINE. Educ: Columbia Univ, AB, 49, MD, 54; Am Bd Surg, dipl, 66. Prof Exp: Instr surg, Columbia-Presby Med Ctr, 59-61; assoc cancer res surgeon, 61-65, ASSOC CHIEF RES SURGEON & DIR SURG DEVELOP ONCOL, ROSWELL PARK MEM INST, 65-, ASSOC RES PROF BIOCHEM, 69- Concurrent Pos: Asst attend & asst vis surgeon, Presby Hosp, NY, 61. Mem: AAAS; Am Inst Chem; Am Asn Cancer Res. Res: Nucleic acid and steroid biochemistry; acid base physiology. Mailing Add: Surg Develop Oncol Roswell Park Mem Inst Buffalo NY 14263

MITTELMAN, PHILLIP SIDNEY, b New York, NY, Sept 28, 25; m 48; c 2. NUCLEAR PHYSICS. Educ: Rensselaer Polytech Inst, BS, 45, PhD(physics), 53; Harvard Univ, MA, 47. Prof Exp: Jr scientist, Cosmotron Proj, Brookhaven Nat Lab, 47-48; instr physics, Rensselaer Polytech Inst, 48-53; mgr physics & math, Develop Div, United Nuclear Corp, 53-66; pres, White Plains, 66-70, PRES, MATH APPLN GROUP, INC, ELMSFORD, NY, 70- Mem: Am Phys Soc; Am Nuclear Soc. Res: Shielding of nuclear reactors; reactor physics; computer applications. Mailing Add: 9 Hardscrabble Circle Armonk NY 10504

MITTELSTAEDT, STANLEY GEORGE, b Connell, Wash, Oct 15, 09; m 40; c 4. PHARMACEUTICAL CHEMISTRY, PHARMACY. Educ: Northwest Nazarene Col, 34; State Univ Wash, BS & MS, 38; Purdue Univ, PhD(pharm, pharmaceut chem), 48. Prof Exp: Asst, State Univ Wash, 37-38; asst, Purdue Univ, 38-40; actg head dept chem, Boise Jr Col, 40-42; assoc prof, Univ Tex, 48-51; asst dean, Sch Pharm, 51-53, PROF PHARM & PHARMACEUT CHEM, SCH PHARM, UNIV ARK, 51-, DEAN SCH, 53- Concurrent Pos: Co-ed, Vet Drug Encyclop; mem nat adv coun health educ, Dept Health, Educ & Welfare, Washington, DC, 72-76. Mem: AAAS; Am Chem Soc; Am Pharmaceut Asn. Res: Iodo-radio opaques. Mailing Add: Sch of Pharm 4301 W Markham Univ of Ark Med Ctr Little Rock AR 72201

MITTENTHAL, JAY EDWARD, b Boston, Mass, July 28, 41; m 68; c 2. NEUROBIOLOGY. Educ: Amherst Col, BA, 62; Johns Hopkins Univ, PhD(biophys), 70. Prof Exp: Fel neurobiol, Stanford Univ, 70-72; ASST PROF BIOL, PURDUE UNIV, 73- Mem: Sigma Xi; Soc Neurosci. Res: Patterns of synaptic connection among individually identifiable neurons, and developmental processes generating these patterns. Mailing Add: Dept of Biol Sci Purdue Univ West Lafayette IN 47907

MITTERER, RICHARD MAX, b Lancaster, Pa, Sept 8, 38; m; c 2. GEOCHEMISTRY. Educ: Franklin & Marshall Col, BS, 60; Fla State Univ, PhD(geol), 66. Prof Exp: Fel geophys lab, Carnegie Inst, 66-67; asst prof geosci, Southwest Ctr Advan Studies, 67-69; asst prof, 69-73, ASSOC PROF GEOSCI, UNIV TEX, DALLAS, 73- Mem: AAAS; Geol Soc Am; Soc Econ Paleont & Mineral; Geochem Soc. Res: Amino acid diagenesis; carbonate geochemistry. Mailing Add: Div of Geosci Univ of Tex at Dallas Box 688 Richardson TX 75080

MITTERLING, LLOYD ALFRED, b Goshen, Ind, July 22, 23; m 46; c 2. POMOLOGY. Educ: Mich State Univ, BS, 53, MS, 54; Rutgers Univ, PhD(hort), 63. Prof Exp: Asst county agr agent, Mich State Univ, 54-55, field agent, Nat Grape Coop, Mich, 55-56; instr hort, Rutgers Univ, 56-62; asst prof pomol, 62-71, ASSOC PROF POMOL, UNIV CONN, 71- Mem: AAAS; Bot Soc Am; Am Soc Hort Sci. Res: Detailed studies of bird behavior, particularly Cyanocitta cristata; movement and feeding habits. Mailing Add: RT 32 Box 113 Merrow CT 06253

MITTIER, JAMES CARLTON, b Denver, Colo, Nov 7, 35. PHYSIOLOGY, ENDOCRINOLOGY. Educ: Univ Nmex, BS, 56; Univ Ill, Urbana, MS, 61; Mich State Univ, PhD(physiol), 66. Prof Exp: Res fel med, Sch Med, Tulane Univ, 67-69; instr physiol, Rutgers Univ, 65-67; res physiologist, Vet Admin, 69-70; RES FEL ENDOCRINOL, CONEY ISLAND HOSP, MAIMONIDES MED CTR, 70- Res: Neuroendocrinology; physiology of reproduction and growth; hormone assays; radioimmunoassays. Mailing Add: Med Serv Coney Island Hosp Ocean & Shore Pkwy Brooklyn NY 11235

MITTLEMAN, DON, b New York, NY, Apr 7, 19; m 45; c 3. MATHEMATICS, COMPUTER SCIENCE. Educ: Columbia Univ, BS, 39, AM, 40, PhD(math), 51. Prof Exp: Instr math, Hofstra Col, 41-42; physicist, US Dept Navy, 42-43; instr math, Columbia Univ, 46-51; mathematician, Nat Bur Standards, 51-53; mathematician, Diamond Ord Fuze Labs, 53-56, chief math consult, 56-59; chief comput lab, Nat Bur Standards, 59-64; prof comput sci & dir comput ctr, Univ Notre Dame, 64-71; PROF COMPUT SCI & DIR COMPUT CTR, OBERLIN COL, 71- Mem: Fel AAAS; Am Math Soc; Asn Comput Mach; NY Acad Sci. Res: Numerical analysis; systems analysis; computer art. Mailing Add: 80 Parkwood Lane Oberlin OH 44074

MITTLEMAN, MARVIN HAROLD, b New York, NY, Mar 13, 28; m 55; c 3. PHYSICS. Educ: Polytech Inst Brooklyn, BS, 49; Mass Inst Technol, PhD(physics), 53. Prof Exp: Instr physics, Columbia Univ, 52-55; staff scientist, Lawrence Radiation Lab, Livermore, Calif, 55-65; staff scientist space sci lab, Univ Calif, Berkeley, 65-68; assoc prof physics, 68-69, PROF PHYSICS, CITY COL NEW YORK, 69-, EXEC OFFICER PhD PROG, 70- Concurrent Pos: Div Sci & Indust Res, Brit Govt fel, Univ Col, Univ London, 62-63; NASA grant, Univ Calif, Berkeley, 69-70; consult, Lockheed Aircraft Corp, Calif, Convair, Inst Defense Anal, DC & Goddard Space Flight Ctr, NASA. Mem: Fel Am Phys Soc. Res: Atomic scattering and structure; quantum optics. Mailing Add: Dept of Physics City Col of New York New York NY 10031

MITTLER, ARTHUR, b Paterson, NJ, July 15, 43; m 66. PHYSICS. Educ: Drew Univ, BA, 65; Univ Ky, MS, 67, PhD(physics), 70. Prof Exp: ASST PROF PHYSICS, LOWELL TECHNOL INST, 69- Mem: AAAS; Am Phys Soc; Am Nuclear Soc. Res: Low energy nuclear physics; neutron cross section measurements. Mailing Add: Dept of Physics & Appl Physics Lowell Technol Inst Lowell MA 01854

MITTLER, SIDNEY, b Detroit, Mich, Aug 2, 17; m 42, 69; c 3. GENETICS, ZOOLOGY. Educ: Wayne State Univ, BS, 38, MS, 39; Univ Mich, PhD(zool), 44. Prof Exp: Asst, Wayne State Univ, 38-39; asst, Univ Mich, 39-42; instr biol, Bowling Green State Univ, 45-46; instr zool, genetics, nutrit & physiol, Ill Inst Technol, 46-47, asst prof, 47-52, res biologist, IIT Res Found, 52-60; PROF BIOL SCI, NORTHERN ILL UNIV, 60- Honors & Awards: IIT Res Found Award, 55. Mem: Fel AAAS; Genetics Soc Am; Radiation Res Soc; Environ Mutagen Soc. Res: Genetics, radiation and chemical mutation. Mailing Add: Dept of Biol Sci Northern Ill Univ DeKalb IL 60115

MITTLER, THOMAS E, b Vienna, Austria, Sept 18, 28; m 54; c 2. INSECT PHYSIOLOGY. Educ: Univ London, BSc, 49, Imp Col, dipl, 50; Cambridge Univ, PhD(insect physiol), 54. Prof Exp: Lectr zool, Birkbeck Col, London, 53-56; Nat Res Coun Can fel entom, 56-58; asst res entomologist, Univ Calif, Berkeley, 58-59; res scientist, Danish State Bee Inst, 59-61; from asst entomologist to assoc entomologist, 61-73, LECTR ENTOM, UNIV CALIF, BERKELEY, 61-, ENTOMOLOGIST, 73- Concurrent Pos: Ed, Annual Rev Entom, 61- & Entomologia Experimentalis et Applicata, 67- Mem: AAAS; Entom Soc Am; Brit Soc Exp Biol; Royal Entom Soc London. Res: Feeding behavior, metabolism, growth and development of plant sucking insects, especially aphids, in relation to composition of artificial diets and to physiology of host plants. Mailing Add: Div of Entom & Parasitol Wellman Hall Univ of Calif Berkeley CA 94720

MITTON, CARL GREGER, physical chemistry, organic chemistry, see 12th edition

MITTON, JEFFRY BOND, b Glen Ridge, NJ, Mar 16, 47; m 69. POPULATION GENETICS. Educ: Univ Conn, BA, 69; State Univ NY Stony Brook, PhD(econ, evolution), 73. Prof Exp: NIH fel genetics, Univ Calif, Davis, 73-74; ASST PROF BIOL, DEPT ENVIRON POP & ORGANISMIC BIOL, UNIV COLO, BOULDER, 74-, RES ASSOC, INST BEHAV GENETICS, 75- Mem: Genetics Soc Am; Soc Study Evolution; AAAS; Sigma Xi; Soc Syst Biol. Res: Processes of natural selection resulting in population structuring and geographic variation of gene frequencies; protein polymorphisms; multi locus systems; human evolution. Mailing Add: Dept of Environ Pop & Org Biol Univ of Colo Boulder CO 80302

MITTWER, TOD EDWIN, b Minneapolis, Minn, Nov 21, 18; m 42; c 1. BACTERIOLOGY. Educ: Univ Southern Calif, AB, 49, MS, 50, PhD(bact), 52; Univ Nancy, cert, 51. Prof Exp: Asst bact, Univ Southern Calif, 49-52, asst prof, 54; bacteriologist, Gooch Labs, 54-55; tech dir, Terminal Testing Lab, 55; vis asst prof bact, Univ Southern Calif, 55-60; DIR RES, ARDEN-MAYFAIR, INC, 60- Concurrent Pos: Lectr, Univ Southern Calif, 50-53; sr scientist & proj leader, Polaroid Corp, 52-53. Mem: Am Soc Microbiol; Am Chem Soc; Inst Food Technol. Res: Bacterial cytology and physiology; industrial and food bacteriology; photomicrography and optics. Mailing Add: Arden-Mayfair Inc 1900 W Slauson Los Angeles CA 90054

MITUS, WLADYSLAW J, b Zywiec, Poland, May 14, 20; US citizen; m 52; c 2. PATHOLOGY, HEMATOLOGY. Educ: Univ Edinburgh, MB, ChB, 46. Prof Exp: Intern med, Weymouth & Dist Hosp, Eng, 48-49; registr, Kilton Hosp Workshop, 49-52; res pathologist, Children's Hosp, Sheffield, 52-53; res assoc, New Eng Ctr Hosp, Boston, 57-69; CHIEF HEMAT & DIR RES, CARNEY HOSP, 69-; ASSOC PROF MED, TUFTS UNIV, 65- Concurrent Pos: Fel path, City Hosp, Cleveland, Ohio, 54-55; fel hemat, Blood Res Lab, New Eng Ctr Hosp, Boston, 55-57; USPHS grant, 57-58; asst prof, Tufts Univ, 60-65; consult, Med Found, Boston, 61-63. Mem: Am Soc Exp Path; Am Soc Hemat; sr mem Am Fedn Clin Res. Mailing Add: Carney Hosp 2100 Dorchester Ave Boston MA 02124

MITZ, MILTON AARON, b Milwaukee, Wis, May 24, 21; m 46; c 4. BIOCHEMISTRY. Educ: Univ Wis, BS, 44, MS, 45; Univ Pittsburgh, PhD(biochem, org chem), 49. Prof Exp: Asst org chem, Univ Wis, 43-45; group leader chem res, Ciba Pharmaceut Prods, Inc, 45-46; head biochem sect, Res Div, Armour & Co, 49-60; head life sci labs, Melpar Inc, 60-66, dir bioeng, 66-67; chief scientist surface lab syst, Voyager Prog, 67-68, scientist, Viking Mars Prog, 68-70, scientist, Grand Tour Prog, 70-72, CHIEF, ADVAN SCI PLANNING, PLANETARY PROGS OFF, NASA, 68-, PROG SCIENTIST, OUTER PLANETS MISSIONS, 72- Concurrent Pos: Sr investr, World War II Penicillin Prog; proj mgr, Army Rapid Detection Bact & Virus Prog; prin investr, Chem Warfare Detection Proj; consult, Nat Artifical Heart Prog, NIH, Res Appl to Nat Needs, NSF & WHO. Mem: AAAS; Am Chem Soc; Am Soc Biol Chem. Res: Chemistry of natural products; ion exchange; model membranes; study of the chemistry of intermediate metabolism; isolation and synthesis of lipids; amino acids; peptides; isolation of coenzymes; study of mechanism of action of anticholinesterases; hydrazines; virus-cell interaction; drug metabolism. Mailing Add: Nat Aeronaut & Space Admin Code SL 400 Maryland Ave SW Washington DC 20546

MIURA, CAROLE K MASUTANI, b Hilo, Hawaii, June 8, 38; m 62; c 2. MATHEMATICAL STATISTICS. Educ: Cornell Univ, BA, 60; Univ Hawaii, MA, 62; Boston Univ, PhD(math), 73. Prof Exp: Teaching asst, Dept Math, 61-62, instr, 62-65, ASST PROF MATH DISCIPLINE, UNIV HAWAII, HILO, 73- Mem: Am Statist Asn. Res: Theoretical studies of inverse gaussian distribution; statistical studies of remedial education. Mailing Add: Univ of Hawaii PO Box 1357 Hilo HI 96720

MIURA, ROBERT MITSURU, b Selma, Calif, Sept 12, 38; m 61; c 2. APPLIED MATHEMATICS. Educ: Univ Calif, Berkeley, BS, 60, MS, 62; Princeton Univ, MA, 64, PhD(aerospace eng), 66. Prof Exp: Res assoc nonlinear wave propagation, Plasma Physics Lab, Princeton Univ, 65-67; asst prof math, NY Univ, 68-71; assoc prof, Vanderbilt Univ, 71-75; vis assoc prof, 75-76, ASSOC PROF MATH, UNIV BC, 76- Concurrent Pos: Assoc res scientist, Courant Inst Math, NY Univ, 67-69. Mem: Am Math Soc; Soc Indust & Appl Math. Res: Nonlinear partial differential equations; nonlinear wave propagation; fluid mechanics; asymptotic methods; kinetic theory; reaction-diffusion problems. Mailing Add: Dept of Math Univ of BC Vancouver BC Can

MIURA, TAKESHI, b Fukuoka, Japan, Apr 10, 25. MEDICAL ENTOMOLOGY. Educ: Utah State Univ, BSc, 55, MSc, 56; NC State Univ, PhD(entom), 67. Prof Exp: Assoc vector control specialist mosquito res, Calif State Pub Health Dept, 61, 62 & 63-64; asst specialist, Sch Pub Health, 64-65, lab technician, 66-67, ASSOC SPECIALIST, MOSQUITO RES LAB, UNIV CALIF, 67- Mem: AAAS; Entom Soc Am; Am Mosquito Control Asn; Am Inst Biol Sci. Res: Biology and ethology of mosquitoes. Mailing Add: Mosquito Res Lab Univ Calif 5545 E Shields Ave Fresno CA 93727

MIWA, THOMAS KANJI, b Honolulu, Hawaii, Apr 14, 27; m 60; c 3. ORGANIC CHEMISTRY, BIOCHEMISTRY. Educ: Univ Hawaii, BA, 52, MS, 54; Univ Wis, PhD(biochem), 58. Prof Exp: Supvr prep civil censorship detachment, Far East Command, 46-47, chief clerk watch list, 47; asst chem, Univ Hawaii, 52-54; asst biochem, Univ Wis, 54-58; CHEMIST, AGR RES SERV, USDA, 58- Concurrent Pos: Chmn, Int Comt Jojoba Res & Develop, 73-; co-dir, Consejo Int Jojoba, 75- Honors & Awards: Superior Serv Team Award, USDA, 66. Mem: Am Chem Soc; Am Oil Chem Soc. Res: Biological and chemical synthesis of proteins, polypeptides, amino and fatty acids; waxes; plasticization of vinyl polymers; borohydride reactions; gas chromatography; conservation of electron spin; lubricants; allylic plastics; jojoba. Mailing Add: USDA 1815 N University Peoria IL 61604

MIX, LEWELLYN STANLEY, animal nutrition, see 12th edition

MIX, MICHAEL CARY, b Deer Park, Wash, June 27, 41; m 62; c 2. INVERTEBRATE PATHOLOGY, RADIATION BIOLOGY. Educ: Wash State Univ, BS, 63; Univ Wash, PhD(fisheries), 70. Prof Exp: Asst prof biol, 70-74, ASSOC PROF BIOL, ORE STATE UNIV, 74- Concurrent Pos: Consult, Monterey Abalone Farms, 74- Mem: AAAS; Nat Shellfisheries Asn; Soc Invert Path; NY Acad Sci; Sigma Xi. Res: Histopathological effects of irradiation on higher invertebrates; experimental invertebrate pathobiology; diseases of invertebrates; cell renewal systems of mollusks; invertebrate oncology; environmental carcinogens. Mailing Add: Dept of Gen Sci Ore State Univ Corvallis OR 97331

MIXER, ROBERT YOLLAND, organic chemistry, see 12th edition

MIXNER, JOHN PAULDING, b NJ, Nov 1, 15; m 42; c 2. ANIMAL PHYSIOLOGY. Educ: Rutgers Univ, BSc, 36, MSc, 38; Univ Mo, PhD, 43. Prof Exp: Instr zool, La State Univ, 43-47; assoc prof dairy husb & assoc res specialist, Exp Sta, 47-56, chmn dept dairy sci, 61-63, PROF ANIMAL PHYSIOL, RUTGERS UNIV, NEW BRUNSWICK, 56-, CHMN DEPT ANIMAL SCI, 63- Mem: AAAS; Am Physiol Soc; Soc Exp Biol & Med; Am Soc Animal Sci; Endocrine Soc. Res: Artifical insemination in dairy cattle; semen and thyroid physiology; lactation and reproduction. Mailing Add: Dept of Animal Sci Rutgers Univ New Brunswick NJ 08903

MIXON, AUBREY CLIFTON, b Tifton, Ga, Sept 20, 24; m 43. AGRONOMY, PLANT BREEDING. Educ: Univ Ga, BSA, 49; NC State Univ, MSA, 53; Auburn Univ, PhD(plant path), 66. Prof Exp: Asst agronomist, Fla Agr Exten Serv, 53-57; res agronomist, Coop Ala Agr Exp Sta & USDA, 57-73, RES AGRONOMIST, COL AGR, COOP UNIV GA COASTAL PLAIN STA & USDA, 73- Concurrent Pos: Res agronomist, Auburn Agr Exp Sta, 57-73; coordr, USDA Nat Winter Peanut Nursery, Mayaguez, PR, 72-; adj res assoc, Col Agr, Univ Ga, 73-; recorder, Nat Peanut Prod Workshop, 74. Honors & Awards: Twenty Year Serv Award, USDA, 73. Mem: Am Peanut Res & Educ Asn; Am Soc Agron. Res: Breeding, agronomic, physiological, ecological and pathological investigations associated with developing peanut varieties that are resistant to toxin-producing fungi. Mailing Add: Coastal Plain Exp Sta Tifton GA 31794

MIXON, LAWSON WESTON, chemistry, see 12th edition

MIXTER, GEORGE, JR, b Swampscott, Mass, June 25, 15; m 42, 74; c 1. SURGERY. Educ: Harvard Univ, BS, 38, MD, 42. Prof Exp: Instr surg, Univ Rochester, 51-54; from asst prof to assoc prof, Sch Med, NY Univ, 55-61; assoc prof, New York Med Col, 61-63; ASSOC DIR GRAD EDUC, DEPT MED EDUC, AMA, 64-, CHMN COOP COMT ON TEACHING SCI & MATH, 71- Concurrent Pos: Fel surg, Mass Mem Hosp, 46; fel physiol, Western Reserve Univ, 50; consult, US Naval Mat Lab, NY, 55-60 & US Naval Appl Sci Lab, 60-71; mem thermal subpanel, Biomed Div, Defense Atomic Support Agency, DC, 60-71. Mem: Fel Am Col Surg; AMA; Optical Soc Am. Res: Thermal burns; vascular physiology; pancreatitis; medical education. Mailing Add: Coun on Med Educ Am Med Asn 535 N Dearborn St Chicago IL 60610

MIXTER, RUSSELL LOWELL, b Williamston, Mich, Aug 7, 06; m 31; c 4. ANATOMY. Educ: Wheaton Col, Ill, AB, 28; Mich State Col, MS, 30; Univ Ill, PhD(anat), 39. Prof Exp: Instr zool, 28-30, from asst prof to assoc prof, 30-45, PROF ZOOL, WHEATON COL, ILL, 45- Concurrent Pos: Instr, Univ Ill, 35-36. Mem: Am Sci Affil (pres, 51-54, ed jour, 64-68). Res: Macrophages of connective tissue; flexed tail in mice; evolution; spiders of Black Hills. Mailing Add: 1006 N President St Wheaton IL 60187

MIYA, TOM SABURO, b Hanford, Calif, Apr 6, 23; m 48; c 1. PHARMACOLOGY. Educ: Univ Nebr, BSc, 47, MSc, 48; Purdue Univ, PhD(pharmacol), 52. Prof Exp: Asst, Univ Nebr, 47-48; from instr to asst prof pharmacol, Purdue Univ, 48-56; chmn dept, Univ Nebr, 56-57; PROF PHARMACOL, PURDUE UNIV, WEST LAFAYETTE, 58-, HEAD DEPT, 64- Concurrent Pos: Mem rev comt, US Pharmacopoeia, 70-80 & pharmacol-toxicol prog comt, Nat Inst Gen Med Sci; assoc ed, Toxicol & Appl Pharmacol; coun mem, Soc Toxicol, 75-76; chmn chem & biol info handling panel, Res Resources Div, Dept Health, Educ & Welfare, NIH, 75-76. Honors & Awards: Award, Am Pharmaceut Asn, 64. Mem: AAAS; Am Soc Pharmacol & Exp Therapeut; Soc Exp Biol & Med; Am Chem Soc; Am Asn Cols Pharm (pres, 75-76). Res: Hormonal determinants of drug metabolism; factors modifying the normal disposition of drugs. Mailing Add: Dept of Pharmacol & Toxicol Purdue Univ West Lafayette IN 47906

MIYADA, DON SHUSO, b Oceanside, Calif, May 21, 25; m 60; c 4. BIOCHEMISTRY. Educ: Univ Calif, Los Angeles, BS, 49; Mich State Univ, PhD, 53. Prof Exp: Res assoc dept food tech, Univ Calif, 53-55; res assoc dept chem, Ohio State Univ, 55-56; res assoc, McArdle mem lab, Univ Wis, 56-57; asst res biochemist dept med, Univ Calif, Los Angeles, 57-61; biochemist, Long Beach Vet Admin Hosp, 61-67; biochemist, Biochem Procedures, Inc, 67-69; BIOCHEMIST, ORANGE COUNTY MED CTR, ORANGE, 69- Concurrent Pos: Adj asst prof depts path & biochem, Univ Calif, Irvine. Mem: AAAS; Am Chem Soc; Am Asn Clin Chem. Res: Clinical chemistry. Mailing Add: 15152 Temple St Westminster CA 92683

MIYAGAWA, ICHIRO, b Hiratsuka, Japan, Mar 5, 22; m 49; c 3. CHEMICAL PHYSICS. Educ: Nagoya Univ, BS, 45; Univ Tokyo, DrS(chem physics), 54. Prof Exp: Res assoc chem, Nagoya Univ, 48-49; res assoc chem, Univ Tokyo, 50-55, asst prof, Inst Solid State Physics, 60-62; res assoc, Duke Univ, 56-59, vis asst prof, 63-64; from asst prof to assoc prof, 65-71, PROF PHYSICS, UNIV ALA, 71- Mem: Am Phys Soc. Res: Dielectric constant of liquids; electron spin resonance of irradiated molecular crystals. Mailing Add: Dept of Physics Univ of Ala University AL 35486

MIYAI, KATSUMI, b Yokosuka, Japan, Oct 17, 31; m 66; c 3. PATHOLOGY. Educ: Keio Univ, Japan, MD, 56; Univ Toronto, PhD(path), 67; Am Bd Path, dipl, 62. Prof Exp: Intern med, US Naval Hosp, Yokosuka, Japan, 56-57; intern surg, Barnes Hosp, St Louis, Mo, 57-58; resident path, Jewish Hosp St Louis, 58-60; from lectr to asst prof, Fac Med, Univ Toronto, 68-70; ASST PROF PATH, SCH MED, UNIV CALIF, SAN DIEGO, 70- Concurrent Pos: Nat Cancer Inst Can res fel, Banting Inst, Fac Med, Univ Toronto, 63-68; asst res pathologist, Univ Hosp San Diego County, 70- Mem: Am Soc Cell Biol; Int Acad Path; Am Asn Study Liver Dis; Electron Micros Soc Am; Can Asn Path. Res: Study of correlation between ultrastructural and functional alterations in human and experimental diseases, mainly those of the hepatobillary system. Mailing Add: Dept of Path Sch of Med Univ of Calif San Diego La Jolla CA 92093

MIYAKE, MIKIO, b Hiroshima, Japan, Dec 24, 29; US citizen; m 58; c 1. OCEANOGRAPHY, METEOROLOGY. Educ: Drexel Inst, BS, 58; Univ Wash, Seattle, MS, 61, PhD(meteorol), 65. Prof Exp: Res asst, C W Thornthwaite Assoc, 54-58; res asst prof atmospheric sci, Univ Wash, 65-67; asst prof, 67-71, ASSOC PROF ATMOSPHERIC SCI, INST OCEANOG, UNIV BC, 71- Mem: Am Meteorol Soc; Am Geophys Union; Inst Elec & Electronics Eng; Meteorol Soc Japan; Oceanog Soc Japan. Res: Interaction of atmosphere, ocean and earth surface through turbulent transfer process. Mailing Add: Inst of Oceanog Univ of BC Vancouver BC Can

MIYAMOTO, MICHAEL DWIGHT, b Honolulu, Hawaii, Apr 22, 45; m 73. NEUROPHARMACOLOGY. Educ: Northwestern Univ, Evanston, BA, 66, PhD(biol), 71. Prof Exp: Instr pharmacol & USPHS fel, Rutgers Med Sch, Col Med & Dent, NJ, 70-72; ASST PROF PHARMACOL, HEALTH CTR, UNIV CONN, 72- Concurrent Pos: Pharm Mfrs Asn Found grant, 75; USPHS grant neurol dis & stroke, 75-78; Epilepsy Found Am award, 76. Res: Neuromuscular transmitter release; receptor mechanisms. Mailing Add: Dept of Pharmacol Univ of Conn Health Ctr Farmington CT 06032

MIYAMOTO, SEIICHI, b Nagasaki, Japan, Oct 1, 44; US citizen; c 1. SOIL PHYSICS, SOIL CHEMISTRY. Educ: Gifu Univ, BS, 67; Kyushu Univ, MS, 69; Univ Calif, Riverside, PhD(soil sci), 71. Prof Exp: Res assoc soil sci, Univ Ariz, 71-75; RES ASSOC & ASST PROF SOIL SCI, NMEX STATE UNIV, 75- Mem: Soil Sci Soc Am; Am Soc Agron. Res: Mine-spoil reclamation; irrigation and drainage; water quality. Mailing Add: Dept of Agron NMex State Univ Las Cruces NM 88003

MIYASHIRO, AKIHO, b Okayama, Japan, Oct 30, 20. PETROLOGY. Educ: Univ Tokyo, BSc, 43, PhD(petrol), 53. Prof Exp: Asst instr petrol, Univ Tokyo, 43-58, assoc prof, 58-67; vis prof, Lamont-Doherty Geol Observ, Columbia Univ, 67-70; PROF GEOL, STATE UNIV NY ALBANY, 70-, NSF RES GRANT, 71- Concurrent Pos: Vis int scientist, Am Geol Inst, 65; mem working group 9, Inter-union Comn Geodynamics, Int Coun Sci Unions, 71- Honors & Awards: Prize, Geol Soc Japan, 58. Mem: Mineral Soc Am; Geochem Soc. Res: Metamorphic and igneous petrology; petrology of deep ocean floors and oceanic islands; tectonic and geochemical evolution of the earth. Mailing Add: Dept of Geol Sci State Univ of NY 1400 Washington Ave Albany NY 12222

MIZE, CHARLES EDWARD, b Smithville, Tex, Mar 3, 34; m 62; c 1. PEDIATRICS, BIOCHEMISTRY. Educ: Rice Inst, BA, 55; Johns Hopkins Univ, PhD(biochem), 61, MD, 62. Prof Exp: From intern pediat to resident, Johns Hopkins Hosp, 62-64; staff assoc metab, Nat Heart Inst, 64-67; asst prof, 67-73, ASSOC PROF PEDIAT & BIOCHEM, UNIV TEX HEALTH SCI CTR DALLAS, 74- Mem: Soc Pediat Res; Am Soc Neurochem; Am Soc Human Genetics; Am Fedn Clin Res; NY Acad Sci. Res: Biochemistry of membranes in growth and development; metabolic and neurologic disorders of childhood. Mailing Add: Univ of Tex Health Sci Ctr 5323 Harry Hines Blvd Dallas TX 75235

MIZE, JACK PITTS, b Kansas City, Mo, July 27, 23; m 49; c 3. NUCLEAR PHYSICS. Educ: Duke Univ, BS, 47; Univ Rochester, MS, 49; Iowa State Col, PhD(physics), 53. Prof Exp: Res assoc physics, Inst Atomic Res, Iowa State Col, 53; mem staff, Los Alamos Sci Lab, 53-60; TECH STAFF MEM PHYSICS, TEX INSTRUMENTS, INC, DALLAS, 60- Mem: Fel Am Phys Soc. Res: Nuclear spectroscopy; plasma and solid state physics. Mailing Add: 918 Beechwood Richardson TX 75080

MIZEJEWSKI, GERALD JUDE, b Pittsburgh, Pa, Aug 1, 39; m 65; c 5. EMBRYOLOGY, IMMUNOLOGY. Educ: Duquesne Univ, BS, 61; Univ Md, MS, 65, PhD(zool), 68. Prof Exp: From asst zool to res asst immunol, Univ Md, 61-68; from res assoc to lectr, Med Sch, Univ Mich, 68-71; asst prof immunol, Univ SC, 71-74; SR RES SCIENTIST, DIV LABS & RES, NY STATE DEPT HEALTH, 74- Concurrent Pos: Am Cancer Soc grant, Univ Mich, 69-70; Upjohn res grant, 69-70; Abbott radio-pharmaceut res gift, 69-71. Mem: AAAS; Am Soc Zool; Am Inst Biol Sci; Reticuloendothelial Soc; NY Acad Sci. Res: Cancer immunology; choriocarcinoma, thyroid, liver and ovarian carcinoma; lacrimal, salivary and thyroid autoimmunity; fluorescent microscopy; hepatoma; alpha-feto-protein; carcinoembryonic antigen, tumor transplantation; radiolabeled antibodies and antigens; hepatoma and lymphoma cell culture; cytotoxic antibodies; pregnancy interruption. Mailing Add: Div of Labs & Res NY State Dept of Health Albany NY 12201

MIZELL, LOUIS RICHARD, b Gettysburg, Pa, Jan 25, 18; m 43; c 4. TEXTILE CHEMISTRY. Educ: Gettysburg Col, AB, 38; Georgetown Univ, MS, 42. Prof Exp: Res assoc, Textile Found, Inc, 39-42; mem staff, Harris Res Labs, Inc, 46-58, asst dir, 58-67; WOOL MGR NEW MKT OUTLETS, INT SECRETARIAT, 67- Mem: Am Chem Soc; Fiber Soc; Am Asn Textile Chem & Colorists; Am Inst Chem; World Future Soc. Res: Chemical and engineering research on fibrous materials; conception and development of new products and processes and taking them to commercial fruition on a world-wide scale. Mailing Add: 108 Sharon Lane Greenlawn NY 11740

MIZELL, MERLE, b Chicago, Ill, Apr 25, 27; m 58; c 2. DEVELOPMENTAL GENETICS, ONCOLOGY. Educ: Univ Ill, Urbana, BS, 50, MS, 52, PhD(zool), 57. Prof Exp: Asst, Univ Ill, Urbana, 54-57; instr zool, 57-60, from asst prof to assoc prof biol, 60-70, assoc prof anat, Med Sch, 69-70, PROF ANAT, MED SCH & PROF BIOL, TULANE UNIV, 70-, DIR CHAPMAN H HYAMS III LAB TUMOR CELL BIOL, 69- Concurrent Pos: Am Cancer Soc & Cancer Asn Greater New Orleans grant, Tulane Univ, 68-70, 75-76; NSF grant, 69-71; Damon Runyon Mem Fund grant, 70-71, NIH grant, 70-77; consult, Spec Virus Cancer Prog, Nat Cancer Inst & proj site visitor, Cancer Res Ctrs, 69-; corp mem, Marine Biol Lab, Honors & Awards: Sigma Xi Fac Res Award, Tulane Univ, 69. Mem: AAAS; Am Soc Zool; Am Inst Biol Sci; Soc Exp Biol & Med (secy, 69-71); Soc Develop Biol. Res: Cellular differentiation; mechanism of limb regeneration in spontaneous regeneration and induced regeneration; role of viruses as agents of normal and neoplastic differentiation; effects of the regeneration environment on neoplastic growths; amphibian tumor biology; genetic engineering; oncogenic herpesviruses. Mailing Add: Lab of Tumor Cell Biol Tulane Univ New Orleans LA 70118

MIZELL, SHERWIN, b Chicago, Ill, Apr 27, 31; m 57; c 3. GROSS ANATOMY. Educ: Univ Ill, BS, 52, MS, 54, PhD(physiol), 58. Prof Exp: Res fel, Med Col SC, 58-59, instr anat, 59-60, asst prof, 60-64; assoc prof physiol sch med, Creighton Univ, 64-65; assoc prof anat & physiol, 65-75, PROF ANAT, MED SCI PROG, IND UNIV, BLOOMINGTON, 75- Mem: AAAS; Am Asn Anat; Am Soc Zool; Am Physiol Soc; Am Soc Photobiol. Res: Biological rhythms; synchronization and mechanisms responsible for changes in physiology and behavior; physiology and endocrinology of amphibians. Mailing Add: Med Sci Prog Ind Univ Bloomington IN 47401

MIZELLE, JOHN DARY, zoology, see 12th edition

MIZERES, NICHOLAS JAMES, b Pittsburgh, Pa, Nov 13, 24; m 52; c 1. ANATOMY. Educ: Kent State Univ, BS, 48; Mich State Univ, MS, 51; Univ Mich, PhD(anat), 54. Prof Exp: From instr to assoc prof, 54-66, PROF ANAT, SCH MED, WAYNE STATE UNIV, 66- Honors & Awards: Lamp Award, 59. Mem: AAAS; Asn Am Med Cols; Am Asn Anat. Res: Human anatomy; cardiovascular research, especially heart coronary circulation and the autonomic nervous system; descriptive anatomy related to surgery. Mailing Add: Dept of Anat Wayne State Univ Sch Med Detroit MI 48201

MIZUBA, SETH SETSUO, b Hilo, Hawaii, Oct 31, 26; m 63; c 2. MYCOLOGY, BACTERIOLOGY. Educ: Ind State Univ, BS, 50; Mich State Univ, MS, 53, PhD(mycol), 56. Prof Exp: Bacteriologist, State Dept Health, Mich, 51-53; bacteriologist, Kendall Co, 55-57; MYCOLOGIST, G D SEARLE & CO, 57- Concurrent Pos: Instr night sch, Loyola Univ Chicago, 59- Mem: Mycol Soc Am; Am Soc Microbiol; Soc Indust Microbiol; Med Microbiol Soc Am; Int Soc Human & Animal Mycoses. Mailing Add: Searle Labs PO Box 5110 Chicago IL 60680

MIZUKAMI, HIROSHI, b Otaru-Shi, Japan, Oct 11, 32; m 59. BIOPHYSICS, HEMATOLOGY. Educ: Int Christian Univ, Tokyo, BA, 57; Univ Ill, PhD(biophys), 63. Prof Exp: Res fel phys chem, Univ Minn, 62-65, res fel med, Univ Hosps, 64-65; from asst prof to assoc prof biol, 65-74, PROF BIOL, WAYNE STATE UNIV, 74- Concurrent Pos: Vis res prof, Tokyo Med & Dent Univ, 73-74. Mem: AAAS; Biophys Soc; Am Chem Soc. Res: Structure and function of proteins, gas transport and cardiovascular diseases; sickle cells and sickle hemoglobin. Mailing Add: Dept of Biol Wayne State Univ Detroit MI 48202

MIZUKI, MIKISO, mathematical statistics, operations research, see 12th edition

MIZUNO, NOBUKO SHIMOTORI, b Oakland, Calif, Apr 20, 16; wid. BIOCHEMISTRY. Educ: Univ Calif, Berkeley, AB, 37, MA, 39; Univ Minn, PhD(biochem), 56. Prof Exp: Instr med technol, Macalester Col, 46-51; res assoc biochem, Univ Minn, 57-62; RES BIOCHEMIST, VET ADMIN HOSP, 62- Concurrent Pos: Res fel biochem, Univ Minn, 56-57; USPHS grants, 56-62 & 64- Mem: AAAS; Am Chem Soc; Am Asn Cancer Res; Am Soc Biol Chem. Res: Hypervitaminoses A and D; vitamin E and muscular dystrophy; biochemistry of aplastic anemia; metabolism of anticancer drugs; nucleic acids of neoplasm. Mailing Add: Dept of Exp Surg Vet Admin Hosp Minneapolis MN 55417

MIZUNO, SHIGEKI, b Tsinan, China, Aug 17, 36; m 64; c 1. MOLECULAR BIOLOGY. Educ: Univ Tokyo, BAgrSci, 59, MAgrSci, 61, PhD(microbiol), 64. Prof Exp: Asst microbiol biochem, Lab Radiation Genetics & Inst of Appl Microbiol, Univ Tokyo, 64-70; res assoc develop biol, Friday Harbor Lab, Univ Wash, 70-72; res fel molecular cytogenetics, Dept Zool, Univ Leicester, 72-75; RES ASSOC BIOCHEM, DEPT MOLECULAR MED, MAYO CLIN, 75- Honors & Awards: Lalor Found Res Award, 72-73. Mem: Am Soc Microbiol; Soc Develop Biol; Am Soc Cell Biol. Res: Molecular organization, evolution, gene expression and regulation in eukaryotic chromosomes. Mailing Add: Dept of Molecular Med Mayo Clin Rochester MN 55901

MIZUNO, WILLIAM GEORGE, b Ocean Falls, BC; nat US; m 48; c 3. BACTERIOLOGY. Educ: Univ Minn, BA, 48, MS, 50, PhD, 56. Prof Exp: Bacteriologist, 43-47, res bacteriologist, 48-65, sr res scientist, 65-71, MGR CORP TECH SERV, ECON LAB, INC, 71- Concurrent Pos: Hon fel, Univ Minn, 58. Mem: Am Soc Microbiol; Am Pub Health Asn; Sigma Xi. Res: Endamoeba histolytica; biocidal agents; cysticides; bactericides; fungicides; sanitation; bacterial metabolism; radioactive tracers; surface active agents; detergents; enzymes. Mailing Add: 1541 E Sixth St St Paul MN 55106

MIZUSHIMA, MASATAKA, b Tokyo, Japan, Mar 30, 23; US citizen; m 55; c 5. MOLECULAR PHYSICS. Educ: Univ Tokyo, BA, 46, DrSc(physics), 51. Prof Exp: Res assoc physics, Duke Univ, 52-55; from asst prof to assoc prof, 55-60, PROF PHYSICS, UNIV COLO, BOULDER, 60- Concurrent Pos: Mem staff, Nat Bur Standards, 55-69; vis prof univ & inst solid state physics, Univ Tokyo, 62-63, fac sci, Univ Rennes, 64, inst atomic physics, Univ Bucharest, 69-70 & Cath Univ Nijmegen, 72. Mem: Am Phys Soc. Res: Theory of microwave and laser spectroscopy; molecular structure, particularly hyperfine structure and Zeeman effect; theory of radiation processes. Mailing Add: Dept of Physics & Astrophys Univ of Colo Boulder CO 80302

MIZUTANI, SATOSHI, b Yokohama, Japan, Nov 19, 37; m 66. VIROLOGY. Educ: Tokyo Univ Agr & Tech, BS, 62; Univ Kans, PhD(microbiol), 69. Prof Exp: Res scientist antibiotics, Nippon Kayaku Co, Ltd, 62-65; instr tumor virol, 71-72, asst scientist, 72-75, ASSOC SCIENTIST TUMOR VIROL, McARDLE LAB CANCER RES, UNIV WIS, 75- Concurrent Pos: Scholar, Leukemia Soc Am, Inc, 73-78. Mem: Am Soc Microbiol. Res: Molecular mechanism of replication of RNA tumor viruses and their relatives; mechanism of tumor formation by RNA tumor viruses. Mailing Add: McArdle Lab Cancer Res Univ of Wis Madison WI 53706

MIZZONI, RENAT HERBERT, b Morristown, NJ, May 27, 14; m 42; c 2. ORGANIC CHEMISTRY. Educ: Newark Col Eng, BS, 36; Polytech Inst Brooklyn, MS, 44, PhD(chem), 52. Prof Exp: Chemist, Maltbie Chem Co, 36-42; CHEMIST, CIBA-GEIGY CORP, 42- Mem: Am Chem Soc. Res: Chemistry of pyridazines. Mailing Add: Valley Brook Rd Long Valley NJ 07853

MJOLSNESS, RAYMOND C, b Chicago, Ill, Apr 22, 33; m 58; c 3. FLUID DYNAMICS, ATOMIC PHYSICS. Educ: Reed Col, BA, 53; Oxford Univ, BA, 55; Princeton Univ, PhD(math physics), 63. Prof Exp: Asst physics, Los Alamos Sci Lab, 58-61; asst prof math, Reed Col, 61-62; theoret physicist space sci lab, Gen Elec Co, 62-64, consult, 64; staff mem, Los Alamos Sci Lab, 64-67; assoc prof astron, Pa State Univ, 67-69; STAFF MEM, LOS ALAMOS SCI LAB, 69- Concurrent Pos: State secy, Rhodes Scholar Trust, 74- Mem: Am Phys Soc; Math Asn Am. Res: Plasma stability; collisional relaxation of plasmas; scattering of electrons on atoms and molecules; cosmology and galaxy formation; laser energy absorption; fluid dynamics and hydrodynamic turbulence theory. Mailing Add: Los Alamos Sci Lab Los Alamos NM 87544

MLAVSKY, ABRAHAM ISAAC, b London, Eng, July 13, 29; US citizen; m 57; c 3. PHYSICAL CHEMISTRY, MATERIALS SCIENCE. Educ: Univ London, BSc, 50, PhD(phys chem), 54. Prof Exp: Staff scientist, Hirst Res Ctr, Gen Elec Co, Eng, 53-56; staff scientist, Transitron Electronic Corp, Mass, 56-58, mgr thermoelec res, 58-60; mgr energy conversion res, Tyco Labs, Inc, 60-62, vpres & asst dir mat res, 62-65, exec vpres & dir, 65-66, sr vpres technol & dir corp technol ctr, 66-74, actg chief operating officer, 72-73, EXEC VPRES, MOBIL TYCO SOLAR ENERGY CORP, 75- Concurrent Pos: Mem, US-Israel Bi-Nat Adv Coun Res & Develop, 75-; mem bd visitors, Sch Eng, Duke Univ, 76-79. Mem: Int Solar Energy Soc; World Future Soc; AAAS; Electrochem Soc. Res: Solid state materials research, including crystal growth and characterization; solid state devices and device phenomena; electronic and structural high temperature materials; high temperature composites; energy conversion; solar energy; photovoltaics; silicon solar cells; EFG crystal growth. Mailing Add: Mobil Tyco Solar Energy Corp 16 Hickory Dr Waltham MA 02154

MLODOZENIEC, ARTHUR ROMAN, b Buffalo, NY, Mar 29, 37. PHYSICAL CHEMISTRY, PHARMACY. Educ: Fordham Univ, BS, 59; Univ Wis, PhD, 64. Prof Exp: Res assoc prod develop, Upjohn Co, 64-68; sr phys chemist, Solid Surfaces Lab, 68-74, GROUP LEADER, HOFFMANN-La ROCHE, INC, 74- Concurrent Pos: Nat chmn, Indust Pharmaceut Technol, 74-78. Mem: AAAS; Am Pharmaceut Asn; Am Chem Soc; Instrument Soc Am; fel Acad Pharmaceut Sci. Res: Thermal analysis; molecular organic solid physics; scanning electron microscopy and surface analysis; small particle technology; particle flow and cohesion; phase transitions; liquid crystal behavior; microencapsulation of drugs; dosage form design. Mailing Add: Advan Technol Dept Hoffmann-La Roche Inc Nutley NJ 07110

MO, LUKE WEI, b Shantung, China, June 3, 34; m 60; c 2. PHYSICS. Educ: Nat Taiwan Univ, BS, 56; Tsing Hua Univ, Taiwan, MS, 59; Columbia Univ, PhD(physics), 63. Prof Exp: Res assoc physics, Columbia Univ, 63-64; res physicist linear accelerator ctr, Stanford Univ, 65-69; ASST PROF PHYSICS, UNIV CHICAGO, 69- Mem: Am Phys Soc. Res: Electromagnetic and weak interactions in high energy physics; experiments on conserved-vector current theorem; electron scatterings; time-reversal invariance; muon-nucleon and neutrino-electron scatterings at Fermilab; theory on radiative corrections. Mailing Add: Enrico Fermi Inst Univ of Chicago Chicago IL 60637

MOACANIN, JOVAN, physical chemistry, see 12th edition

MOAK, CHARLES DEXTER, b Marshall, Tex, Feb 24, 22; m 43; c 2. PHYSICS. Educ: Univ Tenn, BS, 43, PhD(physics), 54. Prof Exp: Asst, Univ Chicago, 44; PHYSICIST, OAK RIDGE NAT LAB, 44- Mem: Am Phys Soc. Res: Alpha particles accompanying fission; slow neutron cross-sections; neutron capture gamma-ray studies; high-voltage accelerator research on light element charged-particle reactions; interactions of heavy particles with matter. Mailing Add: 332 Louisiana Ave Oak Ridge TN 37830

MOAK, JAMES EMANUEL, b Norfield, Miss, Oct 26, 16; m 43; c 3. FOREST ECONOMICS. Educ: Univ Fla, BSF, S2; Ala Polytech Inst, MSF, 53; State Univ NY Col Forestry, Syracuse, PhD(forestry econ), 65. Prof Exp: Instr forestry, 53-54, from asst prof to assoc prof, 54-66, PROF FORESTRY, MISS STATE UNIV, 66- Concurrent Pos: Sci fac fel, 59. Mem: Soc Am Foresters. Mailing Add: Dept of Forestry Miss State Univ Mississippi State MS 39762

MOAT, ALBERT GROOMBRIDGE, b Nyack, NY, Apr 23, 26; m 49; c 3. MICROBIOLOGY. Educ: Cornell Univ, BS, 49, MS, 50; Univ Minn, PhD(bact), 53; Am Bd Microbiol, dipl. Prof Exp: From asst prof to assoc prof, 52-66, PROF BACT, MICROBIOL & IMMUNOL, HAHNEMANN MED COL, 66- Concurrent Pos: USPHS spec res fel & vis prof, Cornell Univ, 71-72. Mem: AAAS; Am Soc Microbiol; Am Chem Soc; Am Soc Biol Chem; fel Am Acad Microbiol. Res: Nutrition, metabolism and physiology of microorganisms. Mailing Add: Dept of Microbiol Hahnemann Med Col 235 N 15th St Philadelphia PA 19102

MOATES, ROBERT FRANKLIN, b Birmingham, Ala, May 16, 38; m 62; c 2. ORGANIC CHEMISTRY. Educ: Duke Univ, BS, 60; Univ SC, PhD(org chem), 66. Prof Exp: RES CHEMIST, R J REYNOLDS TOBACCO CO, 65- Mem: Am Chem Soc; The Chem Soc. Res: Alkaloid isolation; synthesis of alkaloid systems; synthetic organic chemistry; synthesis of natural products; isolation and identification of natural products. Mailing Add: Res Dept R J Reynolds Tobacco Co Winston-Salem NC 27102

MOATS, WILLIAM ALDEN, b Des Moines, Iowa, Nov 30, 28; m 58; c 2. FOOD BIOCHEMISTRY. Educ: Iowa State Univ, BS, 50; Univ Md, PhD(chem), 57. Prof Exp: Asst chem, Univ Md, 50-55; res chemist, Field Crops & Animal Prod Br, Mkt Qual Res Div, 57-72, RES CHEMIST, AGR MKT RES INST, AGR RES SERV, USDA, 72- Mem: Am Chem Soc; Am Soc Microbiol; NY Acad Sci; Am Dairy Sci Asn. Res: Quality tests for dairy products; staining of bacteria for microscopic examination; determination of pesticide residues in foods; heat resistance of bacteria; improved media for salmonella detection. Mailing Add: Agr Res Ctr Bldg 309 USDA Beltsville MD 20705

MOAWAD, ATEF H, b Dec 2, 35; Can citizen; m 66; c 2. OBSTETRICS & GYNECOLOGY, PHARMACOLOGY. Educ: Cairo Univ, MD, 58; Jefferson Med Col, MS, 63; Am Bd Obstet & Gynec, dipl, 68; FRCS(C), 69. Prof Exp: Fel obstet & gynec, Case Western Reserve Univ, 64-65; vis investr, Univ Lund, 65-66; from lectr pharmacol to assoc prof obstet & gynec & pharmacol, Univ Alta, 66-72; PROF OBSTET & GYNEC & PHARMACOL, DIV BIOL SCI & PRITZKER SCH MED, UNIV CHICAGO, 72- Concurrent Pos: Brush Found scholar, 66-67. Mem: Fel Am Col Obstet & Gynec; Soc Gynec Invest; Pharmacol Soc Can; NY Acad Sci; Can Med Asn. Res: Reproductive physiology and pharmacology, chiefly of the structure and function of uterine and fallopian tube smooth muscle. Mailing Add: Chicago Lying-In Hosp 5841 S Maryland Ave Chicago IL 60637

MOAZED, CYRUS, b Meshed, Iran, Aug 17, 34; US citizen; m 61; c 2. PHYSICS. Educ: Harvard Univ, AB, 57; Purdue Univ, MS, 60; Univ Md, College Park, PhD(physics), 66. Prof Exp: Res assoc nuclear physics, Lab Nuclear Sci, Mass Inst Technol, 66-68; res fel, Bartol Res Found, Franklin Inst, 68-70; ASST PROF NUCLEAR PHYSICS, WAYNE STATE UNIV, 70- Mem: Am Phys Soc. Res: Experimental nuclear physics dealing with the study of nuclear structure. Mailing Add: Dept of Physics Wayne State Univ Detroit MI 48202

MOBERG, GARY PHILIP, b Monmouth, Ill, Feb 14, 41; m 67; c 2. PHYSIOLOGY, NEUROENDOCRINOLOGY. Educ: Monmouth Col, Ill, BA, 63; Univ Ill, Urbana, MS, 65, PhD(physiol), 68. Prof Exp: NIH fel med ctr, Univ Calif, San Francisco, 68-70; ASST PROF ANIMAL SCI & ANIMAL PHYSIOL, UNIV CALIF, DAVIS, 70- Concurrent Pos: NIH res grant, Univ Calif, Davis, 71- Mem: AAAS; Am Soc Animal Sci; Soc Neurosci. Res: Neural control of the anterior pituitary; effects of stress on endocrine system regulation. Mailing Add: Dept of Animal Sci Univ of Calif Davis CA 95606

MOBERLY, LAWRENCE ALLAN, b Topeka, Kans, Aug 9, 47; m 74. LOW TEMPERATURE PHYSICS. Educ: Univ Pa, BS, 69; Dartmouth Col, PhD(physics), 76. Prof Exp: Design engr, Kulicke & Soffa, Inc, Ft Washington, Pa, 71; RES ASSOC & ASSOC INSTR PHYSICS, UNIV UTAH, 75- Res: Experimental measurements of the nuclear magnetic properties of solid helium-3; nuclear cooling experiments in conjunction with measuring properties of magnetic alloy systems and liquid helium-3. Mailing Add: Dept of Physics Univ of Utah Salt Lake City UT 84112

MOBERLY, LAWRENCE ERVIN, b Harris, Kans, Dec 14, 12; m 38; c 3. PHYSICAL CHEMISTRY, INORGANIC CHEMISTRY. Educ: Ottawa Univ, Kans, BS, 34; Univ Kans, MA, 37. Prof Exp: Res chemist, 35-55, supv chemist, 55-65, mgr inorg chem & elec ceramics, 65-69, MGR INORG CHEM, LUBRICANTS & SLIDING CONTACTS, RES LABS, WESTINGHOUSE ELEC CORP, 69- Mem: AAAS; Am Chem Soc. Res: Carbon brush development; chemical aspects of sliding contacts; graphite technology; lubricants; high temperature chemistry. Mailing Add: Westinghouse Res Labs Churchill Borough Pittsburgh PA 15235

MOBERLY, RALPH M, b St Louis, Mo, Apr 17, 29; m 54; c 2. GEOLOGY. Educ: Princeton Univ, AB, 50, PhD(geol), 56. Prof Exp: Geologist, Standard Oil Co, Calif, 56-59; from asst prof to assoc prof geol, 59-70, PROF GEOL, UNIV HAWAII, 70-, CHMN DEPT GEOL & GEOPHYS, 75- Mem: Geol Soc Am; Am Asn Petrol Geol; Am Geophys Union; Soc Econ Paleont & Mineral; Int Asn Sedimentol. Res: Marine geology of the Pacific; sedimentology, tectonics and stratigraphy. Mailing Add: Hawaii Inst of Geophys Univ of Hawaii Honolulu HI 96822

MOBLEY, HAROLD MORTON, b Houston, Tex, Jan 5, 18; m 47; c 4. PLANT PHYSIOLOGY. Educ: Univ Houston, BS, 46, MS, 48; Univ Tex, PhD(bot), 62. Prof Exp: Instr photog, physics & Math, Univ Houston, 46-52; teacher, Sr High Sch, Tex, 52-55; asst prof eng drawing, Univ Tex, 55-60, bot & zool, 61-67; prof biol, McMurry Col, 67-68; assoc prof biol, 69-74, PROF BOT, WESTERN STATE COL, 74- Mem: AAAS; Bot Soc Am. Res: Circadian rhythms. Mailing Add: Div of Natural Sci & Math Western State Col Gunnison CO 81230

MOBLEY, HARRIS W, b Hinesville, Ga, June 19, 29; m 50; c 3. ANTHROPOLOGY. Educ: Mercer Univ, AB, 55; Wake Forest Univ, BD, 59; Hartford Sem Found, MA, 65, PhD(anthrop), 66. Prof Exp: ASSOC PROF SOCIOL & ANTHROP, GA SOUTHERN COL, 66- Mem: Fel Am Anthrop Asn; fel African Studies Asn. Res: African ethnology; social change in folk societies in the southeastern United States. Mailing Add: Dept of Sociol & Anthrop Ga Southern Col Statesboro GA 30458

MOBLEY, JACK ERVIN, b Little Rock, Ark, Nov 12, 25; m 47; c 4. UROLOGY. Educ: Univ Ark, BS, 46; Vanderbilt Univ, MD, 48; Mayo Grad Sch Med, Univ Minn, MS, 56; FRCS(C). Prof Exp: Asst prof urol, Sch Med, Tulane Univ, 66-67; from assoc prof to prof, Sch Med, Univ Ark, Little Rock, 66-72, head dept, 66-72; prof urol & chmn dept, Rush-Presby-St Luke's Med Ctr, 72-74; PROF UROL & ASSOC DEAN CLIN SCI, SCH MED, UNIV S DAK, 74- Mem: Am Col Surgeons; AMA; NY Acad Sci. Res: Renal physiology, lymphatics and transplantation. Mailing Add: 2605 W 22nd Sioux Falls SD 57105

MOBLEY, JEAN BELLINGRATH, b Norfolk, Va, Mar 13, 27; m 49; c 2. MATHEMATICS. Educ: Duke Univ, AB, 48; Univ NC, MA, 54, PhD, 70. Prof Exp: Teacher, Pub Schs, NC, 48-56; asst prof math, Flora MacDonald Col, 56-59, assoc prof & head dept, 59-61; assoc prof, St Andrews Presby Col, 61-63; PROF MATH, PFEIFFER COL, 63- Mem: Math Asn Am. Res: Geometry; mathematics education. Mailing Add: Dept of Math Pfeiffer Col Misenheimer NC 28109

MOBLEY, RALPH CLAUDE, b Buffalo, NY; m. NUCLEAR PHYSICS. Educ: Univ Wis, PhD(physics), 50. Prof Exp: Res assoc physics, Univ Wis, 50-51; res assoc, Duke Univ, 51-53; from asst prof to assoc prof, La State Univ, 53-61; chmn dept, 61-72, PROF PHYSICS, OAKLAND UNIV, 61- Mem: Fel Am Phys Soc; Inst Elec & Electronics Eng. Res: Photoneutron thresholds; charged particle scattering; neutron scattering; ion buncher and accelerator development and neutron scattering by time-of-flight method; mass spectrometry; monopole and macromolecule mass spectrometer development. Mailing Add: Dept of Physics Oakland Univ Rochester MI 48063

MOBLEY, RICHARD MORRIS, physics, see 12th edition

MOBRAATEN, LARRY EDWARD, b Fergus Falls, Minn, Sept 6, 38; m 67; c 2. IMMUNOGENETICS. Educ: Univ Calif, Berkeley, AB, 62; Univ Maine, Orono, PhD(zool), 72. Prof Exp: Curie Found fel immunogenetics, 72-74; ASSOC STAFF SCIENTIST IMMUNOGENETICS, JACKSON LAB, 74- Concurrent Pos: Mem comt genetic stand, Nat Res Coun-Inst Lab Animal Resources & lectr zool, Univ Maine, Orono, 75- Mem: Assoc Sigma Xi; Genetics Soc Am. Res: Genetics and function of histocompatibility genes in mice; analysis of histocompatibility mutations in mice. Mailing Add: Jackson Lab Bar Harbor ME 04609

MOCHEL, JACK MCKINNEY, b Boston, Mass, Jan 27, 39; m 62; c 3. PHYSICS. Educ: Cornell Univ, BA, 61; Univ Rochester, PhD(physics), 65. Prof Exp: Fel physics, Univ Rochester, 65-66; asst prof, 66-72, ASSOC PROF PHYSICS, UNIV ILL, 72- Concurrent Pos: A P Sloan fel, 68-74; assoc, Ctr Advan Study, 73-74. Mem: Am Phys Soc; Sigma Xi. Res: Low temperature physics; properties of helium in two and three dimension; superconductivity and phase transitions. Mailing Add: Dept of Physics Univ of Ill Urbana IL 61801

MOCHEL, VIRGIL DALE, b Woodland, Ind, Oct 29, 30; m 51; c 3. PHYSICAL CHEMISTRY, PHYSICS. Educ: Purdue Univ, BS, 52, MS, 54; Univ Ill, PhD(nuclear magnetic resonance), 60. Prof Exp: Chief chemist, Globe Am Corp, 53-54; res chemist, US Army Chem Ctr, 54-56; sr chemist, Corning Glass Works, 59-64; res chemist, 64-66, sr res chemist, 66-67, RES ASSOC POLYMER NUCLEAR MAGNETIC RESONANCE, FIRESTONE TIRE & RUBBER CO, AKRON, 67- Mem: Am Chem Soc; Am Phys Soc. Res: High resolution nuclear magnetic resonance; kinetics; polymer structure studies; electroluminescence; photoluminescence; semiconduction; photoconduction; glass composition. Mailing Add: 1804 Wall Rd Wadsworth OH 44281

MOCHEL, WALTER EDWIN, b Chicago, Ill, Apr 11, 13; m 37; c 3. PHYSICAL CHEMISTRY, ORGANIC CHEMISTRY. Educ: Univ Chicago, MS, 35, PhD(chem), 37. Prof Exp: Res chemist, 37-49, res supvr, 49-66, HEAD SPECTROS DIV, CENT RES DEPT, EXP STA, E I DU PONT DE NEMOURS & CO, 66- Mem: AAAS; Am Chem Soc; Am Phys Soc. Res: Structure of high polymers; synthetic elastomers; polymerization; photochemistry; photography; radiation chemistry; spectroscopic determination of molecular structure. Mailing Add: Cent Res Dept Exp Sta E I du Pont de Nemours & Co Wilmington DE 19898

MOCHON, MARION JOHNSON, b Saratoga Springs, NY, June 6, 29; div; c 3. ANTHROPOLOGY. Educ: Univ Tex, BA, 50; Univ Wis-Milwaukee, MA, 66, PhD(anthrop), 72. Prof Exp: Elem sch teacher, Charlottesville, Va, 52-53; sci asst anthrop, Milwaukee Pub Mus, 64; instr, Univ Wis Ctr Syst, 66-68; chmn div soc sci, 73-74, ASSOC PROF ANTHROP, UNIV WIS-PARKSIDE, 68-, SPEC ASST TO VCHANCELLOR, 75- Concurrent Pos: Fel, Am Coun Educ, 74-75. Mem: Fel AAAS; fel Am Anthrop Asn; Soc Am Archaeol; Soc Appl Anthrop. Res: Community study; community change, utilizing archival materials with field research; American working class; a taxonomy for the construction of anthropological theory. Mailing Add: Off of Chancellor Univ of Wis-Parkside Kenosha WI 53140

MOCHRIE, RICHARD DOUGLAS, b Lowell, Mass, Feb 17, 28; m 50; c 3. DAIRY SCIENCE. Educ: Univ Conn, BS, 50, MS, 53; NC State Col, PhD(rumen nutrit), 58. Prof Exp: Res technician, Univ Conn, 50-53; asst animal sci, 53-54, res instr, 54-58, from asst prof to assoc prof, 58-72, PROF ANIMAL SCI, NC STATE UNIV, 72- Concurrent Pos: Mem comt abnormal milk control & adv to Coun I, Nat Conf Interstate Milk Shipments, 71- Mem: Am Dairy Sci Asn; Nat Mastitis Coun (vpres, 75-); Int Asn Milk, Food & Environ Sanitarians, Inc. Res: Dairy cattle nutrition, physiology of lactation and forage utilization. Mailing Add: Dept of Animal Sci NC State Univ Raleigh NC 27607

MOCK, DAVID CLINTON, JR, b Redlands, Calif, May 6, 22; m 52. INTERNAL MEDICINE. Educ: Univ Southern Calif, AB, 44; Hahnemann Med Col, MD, 48; Am Bd Internal Med, dipl, 58. Prof Exp: Intern, Hahnemann Hosp, Philadelphia, 48-49; resident, Community Hosp of San Mateo County, Calif, 49-50, resident med, 50-51 & 54; chief med serv, Navajo Base Hosp, Ft Defiance, Ariz, 51-53, med officer in chg, 52-53; resident med, Vet Admin Hosp, Oklahoma City, Okla, 54-55; pvt pract, Calif, 55-56; from clin asst to assoc prof, 56-72, PROF MED, COL MED, UNIV OKLA, 72-, ASSOC DEAN STUDENT AFFAIRS, 69- Concurrent Pos: Res fel exp therapeut, Col Med, Univ Okla, 56-57; Upjohn fel, 57-59; attend physician, Vet Admin Hosp, Oklahoma City, 56- & Univ Okla Hosp, 59-; dir therapeut, Univ Okla, 58-60, asst dean student affairs, 66-69. Mem: Am Fedn Clin Res; fel Am Col Physicians; NY Acad Sci. Res: Clinical drug investigation. Mailing Add: Univ of Okla Col of Med PO Box 26901 Oklahoma City OK 73190

MOCK, GENE VERNON, b Davenport, Iowa, Jan 9, 21; m 42; c 2. ORGANIC CHEMISTRY. Educ: Iowa State Col, BS, 43; Univ Iowa, PhD(chem), 49. Prof Exp: Res chemist, Phillips Petrol Co, Okla, 43-46; res chemist chem dept, E I du Pont de Nemours & Co, 49-51, res chemist explosives dept, 51-52, org sect head, 52-56, res assoc, 56-59; res adminstr, Inst Defense Anal, Washington, DC, 59-60; dir chem off, Adv Res Proj Agency, 60-65; br mgr, Phillips Petrol Co, 65-67; PROF CHEM, MIAMI-DADE COL, 67- Concurrent Pos: Consult, Off Secy Defense. Res: Solid propellant research; organic synthesis; free radical polymerization initiators; nitrogen oxide reactions. Mailing Add: 7525 SW 134th St Miami FL 33156

MOCK, GORDON DUANE, b Bloomington, Ill, Oct 21, 27; m 54; c 2. MATHEMATICS. Educ: Univ Ill, BS, 50, MS, 51; Univ Wis, PhD(educ), 59. Prof Exp: Teacher high sch, Ill, 51-53; teacher lab sch, Univ Wis-Madison, 54-55; teacher lab sch, Univ Southern Ill-Carbondale, 55-56; teacher lab sch, Univ Wis-Madison, 56-58; teacher math, State Univ NY Col Oswego, 58-60; assoc prof, 60-67, PROF MATH, WESTERN ILL UNIV, 67- Concurrent Pos: NSF sci fac fel, 63-64. Mem: Am Math Soc; Math Asn Am. Res: Mathematics education. Mailing Add: Dept of Math Western Ill Univ Macomb IL 61455

MOCK, JAMES JOSEPH, b Geneseo, Ill, Feb 15, 43; m 66; c 3. AGRONOMY, PLANT BREEDING. Educ: Monmouth Col, Ill, BA, 65; Iowa State Univ, PhD(agron), 70. Prof Exp: Asst prof plant breeding, 70-74, ASSOC PROF PLANT BREEDING, IOWA STATE UNIV, 74- Mem: Am Soc Agron; Crop Sci Soc Am. Res: Physiological corn breeding; breeding maize genotypes that will efficiently intercept and convert solar energy into grain. Mailing Add: 111 Agron Bldg Iowa State Univ Ames IA 50011

MOCK, NANCY H, inorganic chemistry, biophysics, see 12th edition

MOCK, ORIN BAILEY, b Elmer, Mo, Oct 22, 38; m 67; c 3. REPRODUCTIVE PHYSIOLOGY, MAMMALOGY. Educ: Northeast Mo State Col, BS, 60, MA, 65; Univ Mo-Columbia, PhD(zool), 70. Prof Exp: Teacher high schs, Mo, 59-60 & 61-62; vol, Peace Corps, Philippines, 62-64; asst prof, 69-74, ASSOC PROF ZOOL, NORTHEAST MO STATE UNIV, 74- Mem: AAAS; Am Soc Zoologists; Am Soc Mammalogists; Am Inst Biol Sci. Res: Reproduction in the least shrew, Cryptotis parva; taxonomy of North American shrews. Mailing Add: 1215 S Florence Kirksville MO 63501

MOCK, RICHARD ARMITAGE, b Blair Co, Pa, Nov 6, 23; m 52; c 4. CHEMISTRY. Educ: Pa State Col, BS, 43; Univ Mich, MS, 45; Yale Univ, PhD(phys chem), 51. Prof Exp: Chemist, Hercules Powder Co, 46-49; res chemist, Dow Chem Co, 51-58, assoc scientist, 58-68; lectr, 66-68, DIR RES, SAGINAW VALLEY COL, 68-, PROF ENVIRON SCI, 72-; CONSULT, DOW CHEM CO, 68- Mem: Am Chem Soc. Res: Physical properties of cellulose and rosin derivatives; physics and chemistry of addition polymers and polymeric electrolytes; human ecology and environmental science. Mailing Add: Sch of Technol Saginaw Valley Col University Center MI 48710

MOCK, WILLIAM L, b Los Angeles, Calif, Aug 5, 38. ORGANIC CHEMISTRY. Educ: Calif Inst Technol, BS, 60; Harvard Univ, PhD(org chem), 65. Prof Exp: ASSOC PROF CHEM, UNIV ILL, CHICAGO CIRCLE, 72- Concurrent Pos: Fel, A P Sloan Found, 72. Mem: Am Chem Soc. Res: Synthetic methods; reaction mechanisms; pericyclic reactions; organosulfur chemistry; bioorganic chemistry. Mailing Add: Dept of Chem Univ of Ill Box 4348 Chicago IL 60680

MOCKETT, PAUL M, b San Francisco, Calif, Apr 9, 36; m 65; c 3. EXPERIMENTAL HIGH ENERGY PHYSICS, PARTICLE PHYSICS. Educ: Reed Col, BA, 59; Mass Inst Technol, PhD(physics), 65. Prof Exp: Res assoc physics, Mass Inst Technol, 65-67; asst physicist, Brookhaven Nat Lab, 67-70, assoc physicist, Div Particles & Fields, 70-72; sr res assoc, 72-75, RES ASSOC PROF PHYSICS, UNIV WASH, 75- Mem: Am Phys Soc. Res: Experimental particle physics. Mailing Add: Dept of Physics FM-15 Univ of Wash Seattle WA 98195

MOCKFORD, EDWARD LEE, b Indianapolis, Ind, June 16, 30. TAXONOMY, ENTOMOLOGY. Educ: Ind Univ, AB, 52; Univ Fla, MS, 54; Univ Ill, PhD, 60. Prof Exp: Asst limnol, Lake & Stream Surv, Ind, 48-52; asst biol, Univ Fla, 52-54; tech asst entom, Nat Hist Surv, Ill, 50-60; from asst prof to assoc prof biol sci, 60-66, PROF BIOL SCI, ILL STATE UNIV, 66- Mem: Soc Syst Zool; Wilson Ornith Soc; Entom Soc Am; Asn Trop Biol. Res: Taxonomic entomology; taxonomy and evolution of insects, especially order Psocoptera. Mailing Add: Dept of Biol Sci Ill State Univ Normal IL 61761

MOCKLE, JERRY AUGUSTE, b Doucet, Que, Sept 16, 26; m 51; c 1. PHARMACOGNOSY, PHARMACOLOGY. Educ: Univ Montreal, BScPharm, 52; Univ Paris, PharmD, 55. Prof Exp: From asst prof to prof pharmacog, Univ Montreal, 55-71, secy fac, 61-65, vdean, 65-69; dir, Opers Br-Drugs, 71-73, DIR, OPERS BR-PROF SERV, QUE HEALTH INS BD, 73- Mem: AAAS; Am Soc Pharmacog; Soc Econ Bot; Am Chem Soc; Pharmacol Soc Can. Res: Chemistry and pharmacology of medicinal plants of various origin, particularly those used in popular medicine in Canada. Mailing Add: Opers Br-Prof Serv Que Health Ins Bd PO Box 6600 Quebec PQ Can

MOCZYGEMBA, GEORGE A, b Panna Maria, Tex, Jan 7, 39; m 66; c 2. POLYMER CHEMISTRY, INORGANIC CHEMISTRY. Educ: Univ Tex, Austin, BS, 62, PhD(chem), 69. Prof Exp: Instr chem, St Edward's Univ, 65-67; SR CHEMIST, PHILLIPS PETROL CO, 68- Mem: Am Chem Soc; Sigma Xi. Res: Polymerization mechanisms, kinetics and catalysis; polymer research for rubber and plastic applications. Mailing Add: 824 SE Crestland Bartlesville OK 74003

MODAK, ARVIND T, b Bombay, India; US citizen; m 65; c 2. NEUROCHEMISTRY, TOXICOLOGY. Educ: Univ Bombay, BSc, 61, BSc, 63, MSc, 65; Univ Tex, Austin, MS, 68, PhD(pharmacol), 70. Prof Exp: Demonstr pharmacol, Univ Bombay, 63-64, lectr, 64-65; teaching asst pharm, Univ Tex, Austin, 65-70; fel pharmacol, Health Sci Ctr, Univ Tex, San Antonio, 70-72; pharmacologist & toxicologist, Pharma Corp, 72-74; res coordr, 73-75, ASST PROF PHARMACOL, HEALTH SCI CTR, UNIV TEX, 75- Concurrent Pos: Adj instr pharmacol, Health Sci Ctr, Univ Tex, 72-74. Mem: Am Soc Neurochem. Res: Study of labile metabolites in the central nervous system—acetylcholine, cyclic nucleotides, indoleamines and catecholamines through the use of microwave irradiation technique; modification of these metabolites by drugs and heavy metals. Mailing Add: Dept of Pharmacol Health Sci Ctr Univ Tex San Antonio TX 78284

MODDERMAN, JOHN PHILIP, b Grand Rapids, Mich, Dec 4, 44. ANALYTICAL CHEMISTRY, FOOD CHEMISTRY. Educ: Calvin Col, AB, 67; Wayne State Univ, PhD(anal chem), 71. Prof Exp: Fel chemiluminescence res, Univ Ga, 71-73; CHEMIST, DIV CHEM & PHYSICS, FOOD & DRUG ADMIN, 73- Mem: Am Chem Soc; AAAS. Res: Analytical chemistry of additives and contaminants in food. Mailing Add: Div Chem & Physics HFF-144 FDA 200 C St SW Washington DC 20204

MODE, CHARLES J, b Bismarck, NDak, Dec 29, 27; m 60; c 1. MATHEMATICS, POPULATION STUDIES. Educ: NDak State Univ, BS, 52; Kans State Univ, MS, 53; Univ Calif, Davis, PhD(genetics), 56. Prof Exp: Res fel statist, NC State Univ, 56-57; prof math, Mont State Univ, 57-66; assoc prof statist, State Univ NY Buffalo, 66-70; PROF MATH, DREXEL UNIV, 70- Concurrent Pos: Mem, Inst Pop Studies, Drexel Univ. Mem: Biomet Soc; Inst Math Statist; Am Math Soc. Res: Probability theory; application of mathematics and statistics to biology and medicine, particularly mathematical genetics and epidemiology; stochastic processes; branching processes, models of population growth; mathematical demography; mathematical ecology. Mailing Add: Dept of Math Drexel Univ 32nd & Chestnut Sts Philadelphia PA 19104

MODE, VINCENT ALAN, b Gilroy, Calif, May 25, 40. INORGANIC CHEMISTRY, COMPUTER SCIENCE. Educ: Whitman Col, AB, 62; Univ Ill, Urbana, PhD(inorg chem), 65. Prof Exp: Res chemist, 65-71, group leader, 71-75, SECT LEADER, LAWRENCE LIVERMORE LAB, UNIV CALIF, 75- Mem: Am Chem Soc; The Chem Soc; Soc Comput Simulation. Res: Fluorescence of rare earth chelates; high pressure liquid chromatography of inorganic species; development of interactive computer programs and large data base analysis. Mailing Add: Lawrence Livermore Lab Box 808-L234 Livermore CA 94550

MODEER, ELMER, biochemistry, see 12th edition

MODEL, FRANK STEVEN, b New York, NY, May 5, 42; m 65; c 2. CHEMISTRY, POLYMER SCIENCE. Educ: Mass Inst Technol, BS, 63; Harvard Univ, MA, 65, PhD(chem), 68. Prof Exp: Res chemist, 67-72, proj leader, 72-75, SR RES CHEMIST, CELANESE RES CO, SUMMIT, 72- Mem: Am Chem Soc; The Chem Soc; Am Soc Artificial Internal Organs; Am Electroplaters Soc. Res: Physical chemistry of formed polymers; dyeing of synthetic and natural fibers; membrane science and technology; reverse osmosis; hemodialysis; polymer characterization; biomedical applications of polymers; organosilicon polymers; inorganic photochromism. Mailing Add: 11 Castle Way Basking Ridge NJ 07920

MODEL, PETER, b Frankfurt, Ger, May 17, 33; US citizen; m 61; c 1. BIOCHEMISTRY, GENETICS. Educ: Stanford Univ, AB, 53; Columbia Univ, PhD(biochem), 65. Prof Exp: Res assoc biochem, Columbia Univ, 65-67; NSF fel, 67-69, asst prof, 69-75, ASSOC PROF BIOCHEM, ROCKEFELLER UNIV, 75- Mem: Am Soc Microbiol. Res: In vitro protein synthesis; bacteriophage genetics and physiology. Mailing Add: Rockefeller Univ New York NY 10021

MODELL, JEROME HERBERT, b St Paul, Minn, Sept 9, 32; m 52; c 3. ANESTHESIOLOGY. Educ: Univ Minn, BA, 54, BS & MD, 57; Am Bd Anesthesiol, dipl, 64. Prof Exp: Jr scientist internal med, Sch Med, Univ Minn, 57-59; from instr to assoc prof anesthesiol, Sch Med, Univ Miami, 63-69; PROF ANESTHESIOL & CHMN DEPT, COL MED, UNIV FLA, 69- Concurrent Pos: NIH res career develop award, 67-69. Mem: AAAS; Am Soc Anesthesiol. Res: Pathophysiology and treatment of near-drowning; physiologic applications of liquid breathing; intensive pulmonary therapy. Mailing Add: Dept of Anesthesiol Univ of Fla Col of Med Gainesville FL 32601

MODELL, WALTER, b Waterbury, Conn, July 18, 07; m 33; c 1. PHARMACOLOGY. Educ: City Col New York, BS, 28; Cornell Univ, MD, 32. Prof Exp: From instr clin pharmacol to prof pharmacol, 32-73, dir clin pharmacol, 56-73, EMER PROF PHARMACOL, MED COL, CORNELL UNIV, 73- Concurrent Pos: Mem bd dirs, US Pharmacopoeia. Mem: Am Soc Pharmacol & Exp Therapeut; Am Soc Clin Pharmacol & Therapeut; fel AMA; fel Am Col Physicians; NY Acad Sci. Res: Clinical pharmacology. Mailing Add: PO Box 119 Larchmont NY 10538

MODESITT, GEORGE (EDWARD), physics, see 12th edition

MODEST, EDWARD JULIAN, chemistry, see 12th edition

MODIC, FRANK JOSEPH, b Cleveland, Ohio, Sept 20, 22; m 56; c 4. PHYSICAL CHEMISTRY, ORGANIC CHEMISTRY. Educ: Case Western Reserve Univ, BS, 43; Iowa State Col, PhD(phys org chem), 51. Prof Exp: Jr chem engr, Kellex Corp, 43-45; prod supvr, Carbide & Carbon Chem Co, 45-46; instr chem, Iowa State Col, 46-48; CHEMIST, SILICONE PROD DEPT, GEN ELEC CO, 51- Mem: Am Chem Soc; The Chem Soc. Res: Physical-organic research in organosilicon chemistry; reaction mechanisms and synthesis of silicones and silicone copolymers. Mailing Add: Silicone Prod Dept Gen Elec Co Waterford NY 12188

MODICA, ANTHONY PETER, physical chemistry, gas dynamics, see 12th edition

MODISETTE, JERRY LEE, b Minden, La, July 28, 34; m 59. SPACE PHYSICS. Educ: La Polytech Inst, BS, 56; Va Polytech Inst, MS, 60; Rice Univ, PhD, 67. Prof Exp: Aeronaut res engr, Langley Res Ctr, Nat Adv Comt Aeronaut, 56-62, aerospace technologist, manned spacecraft ctr, NASA, 62-63, chief radiation & fields br, 63-66, space physics div, 66-69; chmn div sci & math, 69-74, ASSOC DEAN COL SCI & HEALTH PROF, HOUSTON BAPTIST UNIV, 74- Concurrent Pos: Vis lectr, Univ Tex, 65. Mem: AAAS; Am Geophys Union; Am Inst Aeronaut & Astronaut; Am Astron Soc; Am Asn Physics Teachers. Res: Interplanetary medium; solar physics; high temperature chemistry; kinetics of surface reactions; solar energy; air pollution control systems; stellar ultraviolet spectroscopy. Mailing Add: Col of Sci & Health Prof Houston Baptist Col Houston TX 77036

MODLIN, HERBERT CHARLES, b Chicago, Ill, Jan 12, 13; m 33; c 1. PSYCHIATRY. Educ: Univ Nebr, BSc, 35, MD, 38, MA, 40. Prof Exp: Intern, Univ Nebr Hosp, 38-39; resident neuropsychiat, Clarkson Hosp, 39-40; fel psychiat, Pa Hosp, Philadelphia, 40-41 & Adams House, Boston, 41-42; resident neurol, Montreal Neurol Inst, 42-43; chief neuropsychiat serv, Winter Vet Admin Hosp, 46-49; SR PSYCHIATRIST, MENNINGER FOUND, 49-; ASSOC PROF PSYCHIAT, SCH MED, UNIV KANS, 60-, PROF COMMUNITY & FORENSIC PSYCHIAT, 76- Concurrent Pos: Lectr, Sch Law, Univ Kans, 47- Mem: Am Psychiat Asn. Res: Community and social psychiatry; psychiatric education. Mailing Add: Menninger Found Topeka KS 66601

MODRAK, JOHN BRUCE, b New Britain, Conn, June 14, 43; m 68; c 2. PHARMACOLOGY. Educ: Univ Conn, BS, 67, PhD(pharmacol), 75. Prof Exp: ASST PROF PHARMACOL, COL PHARM, UNIV NEBR, 75- Mem: NY Acad Sci. Res: Role of connective tissue, collagen, elastin and mucopolysaccharides in the progression and regression of atherosclerosis. Mailing Add: Col of Pharm Univ of Nebr Lincoln NE 68588

MODRESKI, PETER JOHN, b New Brunswick, NJ, Dec 22, 46; m 68; c 2. GEOCHEMISTRY. Educ: Rutgers Univ, BA, 68; Pa State Univ, MS, 71, PhD(geochem), 72. Prof Exp: Res chemist laser chem, Air Force Weapons Lab,

MODRESKI

Kirtland AFB, 72-75; MEM TECH STAFF GEOCHEM, SANDIA LABS, 75- Mem: Geol Soc Am; Mineral Soc Am. Res: Extraction of energy from molten magma in the earth; chemistry and physics of silicate melts at high pressures; corrosion of metals by molten rock and by volcanic gases. Mailing Add: Sandia Labs Orgn 5831 Albuquerque NM 87115

MOE, CHESNEY RUDOLPH, b Rainy River, Ont, Oct 6, 08; US citizen; m 35, 51; c 2. PHYSICS. Educ: Stanford Univ, AB, 29, AM, 31; Univ Southern Calif, PhD(physics), 41. Prof Exp: Asst, 29-30, instr, 31-35, from asst prof to prof, 35-73, assoc chmn div phys sci, 56-62, chmn dept physics, 62-65, EMER PROF PHYSICS, SAN DIEGO UNIV, 73- Concurrent Pos: Mem bd rev, Marine Phys Lab, Univ Calif, 46-51; consult, US Navy Electronics Lab, 53-56, Tracor Corp, Tex & Calif, 67-69 & Jet Propulsion Lab, 69 & 70. Mem: Fel Acoust Soc Am. Res: Acoustics. Mailing Add: 4669 E Talmadge Dr San Diego CA 92116

MOE, GEORGE, b Portland, Ore, Dec 31, 25; m 52; c 2. PHYSICAL CHEMISTRY, INORGANIC CHEMISTRY. Educ: Reed Col, BA, 47; Univ Rochester, PhD(chem), 50. Prof Exp: Asst chem, Univ Rochester, 47-49; instr, Univ Buffalo, 50-51; from res chemist to head astronaut dept, Aerojet-Gen Corp Div, Gen Tire & Rubber Co, 51-60; vpres res, Astropower, Inc, Douglas Aircraft Corp, 60-64, dir astropower lab, Missiles & Space Syst Div, 64-70, dep dir res & develop, 70-72, DIR RES & DEVELOP, McDONNELL-DOUGLAS ASTRONAUT CO, 72- Mem: Am Chem Soc; Am Phys Soc; Am Inst Aeronaut & Astronaut. Res: Aerospace technology; electrochemical systems; advanced computers and advanced materials; synthesis and evaluation of high-energy propellants; photochemistry and radiation chemistry of free radicals; solid state reactions. Mailing Add: 12231 Afton Lane Santa Ana CA 92705

MOE, GEORGE WYLBUR, b Opportunity, Wash, Apr 24, 42. EXPERIMENTAL ATOMIC PHYSICS. Educ: Gonzaga Univ, BS, 66; Univ Wash, PhD(physics), 74. Prof Exp: RES ASSOC PHYSICS, COLUMBIA RADIATION, COLUMBIA UNIV, 73- Mem: AAAS; Am Phys Soc; Am Inst Physics. Res: Atomic and molecular spectroscopy; alkali-noble gas excimers of possible relevance for laser applications. Mailing Add: Columbia Radiation Lab Columbia Univ 538 W 120th St New York NY 10027

MOE, GORDON KENNETH, b Fairchild, Wis, May 30, 15; m 38; c 6. PHYSIOLOGY. Educ: Univ Minn, BS, 37, MS, 39, PhD(physiol), 40; Harvard Univ, MD, 43. Prof Exp: From asst to instr physiol, Univ Minn, 36-40; Porter fel, Western Reserve Univ, 40-41; instr pharmacol, Harvard Med Sch, 41-44, assoc, 44; from asst prof to assoc prof, Med Sch, Univ Mich, 44-50; prof & chmn dept, 50-60, RES PROF PHYSIOL, COL MED, STATE UNIV NY UPSTATE MED CTR, 60-; DIR RES, MASONIC MED RES LAB, 60- Concurrent Pos: Consult, Walter Reed Army Med Ctr, 46-53; Am Physiol Soc traveling fel, Int Physiol Cong, Oxford, 47; USPHS spec fel, Nat Inst Cardiol, Mex, 48; mem teaching mission, WHO, UN, Israel & Iran, 51; George Fahr lectr, Univ Minn, 60; chmn basic sci coun, Am Heart Asn, 60-62, chmn res comt, 60-61; mem nat heart & lung adv coun, NIH, 70-74. Mem: AAAS (secy sect med, 49-53, chmn, 58); Am Physiol Soc; hon mem Mex Nat Acad Med. Res: Physiology and pharmacology of the cardiovascular system. Mailing Add: Masonic Med Res Lab Utica NY 13503

MOE, JOHN, b Grafton, NDak, Aug 14, 05; m 36; c 2. ORTHOPEDIC SURGERY. Educ: Univ NDak, BS, 27; Northwestern Univ, MB, 29, MD, 30; Am Bd Orthop Surg, dipl, 36. Prof Exp: Chief of staff, Gillette Hosp Crippled Children, St Paul, 58-73; prof & chmn dept, 57-73, EMER PROF ORTHOP SURG, UNIV MINN, MINNEAPOLIS, 73-; EMER STAFF, GILLETTE HOSP CRIPPLED CHILDREN, ST PAUL, 73- Concurrent Pos: Dir, Twin Cities Scoliosis Ctr & John Moe Scoliosis Fel, Fairview Hosp, 73- Mem: Am Clin Orthop Soc (pres, 55); Am Orthop Asn (pres, 72); Am Acad Orthop Surg; Am Col Surg; Int Soc Orthop Surg & Traumatol. Res: Etiology and treatment of scoliosis. Mailing Add: Suite 500 Fairview-St Marys Med Bldg 606 24th Ave S Minneapolis MN 55454

MOE, MICHAEL K, b Milwaukee, Wis, Nov 17, 37; m 61; c 1. ELEMENTARY PARTICLE PHYSICS. Educ: Stanford Univ, BS, 59; Case Western Reserve Univ, MS, 61, PhD(physics), 65. Prof Exp: Res fel cosmic ray physics, Calif Inst Technol, 65-66; asst res physicist, 66-68, asst prof physics, 68-73, asst res physicist, 73-75, ASSOC RES PHYSICIST, UNIV CALIF, IRVINE, 75- Res: Low level scintillation spectrometry; cosmic rays; experimental search for double beta decay. Mailing Add: Sch of Phys Sci Univ of Calif Irvine CA 92664

MOE, MILDRED MINASIAN, b Philadelphia, Pa, Nov 18, 29; m 51; c 2. ATMOSPHERIC PHYSICS. Educ: Univ Calif, Los Angeles, BA, 51, MA, 53, PhD(physics), 57. Prof Exp: Mem tech staff, Ramo-Wooldridge Corp, Calif, 56-60; asst prof physics, Loyola Univ Los Angeles, 67-70, res assoc, 70-72; ASST RES PHYSICIST, PHYSICS DEPT, UNIV CALIF, IRVINE, 73- Mem: Am Phys Soc; Am Asn Physics Teachers. Res: Quantum mechanical scattering theory; supersonic jet flows; satellite orbital and attitude motions; gas-surface interactions; measurement of upper-atmospheric properties; atmospheric modeling. Mailing Add: Dept of Physics Univ of Calif Irvine CA 92664

MOE, OSBORNE KENNETH, b Los Angeles, Calif, Dec 29, 15; m 51; c 2. PHYSICS. Educ: Univ Calif, Los Angeles, BA, 51, MA, 53, PhD(planetary & space sci), 66. Prof Exp: Analyst airborne radar facil, Radio Corp Am, 54-55; mem tech staff, Ramo-Wooldridge Corp, 56-59, consult satellite orbits & model atmospheres, Space Tech Labs, 60-63, consult orbital predictions & model atmospheres, TRW, Inc, 63-67; SR RES SCIENTIST, McDONNELL DOUGLAS ASTRONAUT CO, 67- Concurrent Pos: Asst res geophysicist, Univ Calif, Los Angeles, 66-67; consult, Aerospace Corp, 65-67 & Northrop Corp, 67. Mem: Am Geophys Union; Am Meteorol Soc; Am Inst Aeronaut & Astronaut. Res: Theoretical analysis of the various types of density and composition measurements in the thermosphere; adsorption and energy accommodation at artificial satellite surfaces; thermospheric models; solar-terrestrial relationships. Mailing Add: 2 Jade Cove Corona del Mar CA 92625

MOE, OWEN ARNOLD, b Echo, Minn, Oct 12, 12; m 42; c 3. CHEMISTRY. Educ: St Olaf Col, BA, 35; Univ Minn, PhD(org chem), 42. Prof Exp: Instr chem, St Olaf Col, 37-38; Nat Defense Res Comt fel & res assoc, Northwestern Univ, 42-43; res chemist, Gen Mills, Inc, 43-53, head chem mkt develop dept, 53-58, mgr tech serv, 58-60, dir steroid sales, 60-65, mgr sales & tech serv fine chem, Ill, 65-70, dir mkt & develop, Gen Mills Chem, Inc, 70-73; ASST PROF CHEM, LEBANON VALLEY COL, 73- Mem: AAAS; Am Chem Soc. Res: Organic synthesis; carbohydrate chemistry; amino acids; protein chemistry; rearrangement of phenyl allyl ethers. Mailing Add: Dept of Chem Lebanon Valley Col Annville PA 17003

MOE, PAUL G, b Rochester, Minn, Apr 6, 31; m 55; c 2. BACTERIOLOGY. Educ: Univ Ill, BS, 53, MS, 54; Rutgers Univ, PhD(soil chem), 59. Prof Exp: Agronomist, Found Agr Res, 54-55; chemist, Chem Process Co, 55-56; asst prof, Purdue Univ, 59-67; vis asst prof, Ore State Univ, 67-68; assoc prof bacteriol, 68-72, PROF BACTERIOL & BACTERIOLOGIST, WVA UNIV, 72- Concurrent Pos: Sr lectr, WVa Univ-AID contract, Makerere Univ Col, Uganda, 68-72; vis sr lectr, Univ Nairobi, 72. Mem: Am Soc Agron; Soil Sci Soc Am; Am Chem Soc; Am Soc Microbiol. Res: Waste disposal; environmental microbiology; biodegradation of environmental pollutants. Mailing Add: Div of Plant Sci WVa Univ Morgantown WV 26506

MOE, PAUL WAYNE, animal nutrition, metabolism, see 12th edition

MOE, ROBERT ANTHONY, b Jersey City, NJ, Mar 26, 23; m 49; c 3. PHYSIOLOGY. Educ: Seton Hall Univ, BS, 48; Fordham Univ, MS, 49, PhD(physiol), 51. Prof Exp: Asst biol, Grad Sch, Fordham Univ, 48-51; head physiol testing sect, Toxicol Res Div, Inst Med Res, E R Squibb & Sons, 51-59; head cardiovasc sect, Pharmacol Dept, Hoffmann-La Roche, Inc, 59-65, from asst dir pharmacol dept to asst dir biol res div, 65-70; dir biol res, 70-72, vpres res & develop, 72-73, SR V PRES SCI AFFAIRS, SEARLE LABS, 73- Mem: AAAS; Am Soc Pharmacol & Exp Therapeut; assoc fel Am Soc Clin Pharmacol & Therapeut. Res: Toxicology and physiology of biological preparations; cardiovascular and autonomic systems. Mailing Add: Searle Labs PO Box 5110 Chicago IL 60680

MOEDE, JEROME ALBERT, physical chemistry, see 12th edition

MOEDRITZER, KURT, b Prague, Czech, June 20, 29; nat US; m 57; c 3. ORGANOMETALLIC CHEMISTRY. Educ: Univ Munich, dipl, 53, PhD(inorg chem), 55. Prof Exp: Asst inorg chem, Univ Munich, 53-56, asst & instr, 57-59; Fulbright fel, Univ Southern Calif, 56-57; instr, Munich Br, Univ Md, 58-59; SCI FEL, CORP RES DEPT, MONSANTO CO, 59- Concurrent Pos: Ed, Synthesis & Reactivity in Inorg & Metal-org Chem. Mem: Am Chem Soc; Sigma Xi; Gesellschaft Deutscher Chemiker. Res: Inorganic and organic phosphorus compounds; inorganic polymers; organometallics; metal hydrides; nuclear magnetic resonance; organometallic chemistry. Mailing Add: Corp Res Dept Monsanto Co St Louis MO 63166

MOEHLMAN, ROBERT STEVENS, b Rochester, NY, Feb 23, 10; m 34; c 2. GEOLOGY. Educ: Univ Rochester, AB, 31; Harvard Univ, AM, 32, PhD, 35. Prof Exp: Teaching asst & instr, Harvard Univ, 32-35; field asst, US Geol Surv, 35; mine geologist, Anaconda Copper Mining Co, Mont, 36-38; explor geologist, Ariz, 38-40, Nev & Calif, 40-44; field geologist, SAm Mines Co, NY, 45-46, chief geologist, 47-50; vpres, Austral Oil Co, 51-53, exec vpres, 54-62; PRES, NEWMONT OIL CO, 62- Concurrent Pos: Chmn of bd, Can Export Gas & Oil Ltd, 74- Mem: Soc Econ Geol; fel Geol Soc Am; Am Asn Petrol Geol; Soc Petrol Eng; Nat Petrol Coun. Res: Policy for energy supplies; economics of production of minerals and hydrocarbons. Mailing Add: 242 Maple Valley Rd Houston TX 77027

MOEHRING, DAVID MARION, b Sidney, Ohio, Feb 21, 35; m 59; c 2. FOREST SOILS, FOREST ECOLOGY. Educ: Mont State Univ, BSF, 58; Duke Univ, MF, 59, DF(soils), 65. Prof Exp: Res soil scientist, Southern Forest Exp Sta, US Forest Serv, 58-65; asst prof, 65-70; ASSOC PROF FORESTRY, TEX A&M UNIV, 70- Mem: Soil Sci Soc Am; Ecol Soc Am; Soc Am Foresters. Res: Soil-plant water relations; silviculture; mensuration; climatology; tree physiology; hydrology. Mailing Add: Dept of Forest Sci Tex A&M Univ College Station TX 77840

MOEHRING, JOAN MARQUART, b Orchard Park, NY, Sept 23, 35; m 74. CELL BIOLOGY. Educ: Syracuse Univ, BS, 61; Rutgers Univ, MS, 63, PhD(microbiol), 65. Prof Exp: Fel microbiol, Stanford Univ, 65-68; res assoc cell & tissue cult, Dept Med Microbiol, 68-70, res assoc, Dept Med, 70-71, res assoc, Dept Path, 71-73, RES ASSOC & ASST PROF CELL & TISSUE CULT, DEPT MED MICROBIOL, UNIV VT, 73- Concurrent Pos: Consult & assoc cell & tissue cult, Spec Ctr Res Pulmonary Fibrosis, Vt Lung Ctr, Sch Med, Univ Vt, 74-76. Mem: Am Soc Microbiol; Tissue Cult Asn; Sigma Xi. Res: Application of cell and tissue culture to biomedical research; use of cultured mammalian cells to study the molecular action of bacterial toxins and the genetics of resistance to toxins. Mailing Add: Dept of Med Microbiol Univ of Vt Burlington VT 05401

MOEHRING, THOMAS JOHN, b New York, NY, Aug 15, 36; m 64. MEDICAL MICROBIOLOGY. Educ: Fairleigh Dickinson Univ, BS, 61; Rutgers Univ, MS, 63, PhD(microbiol), 66. Prof Exp: Fel, Stanford Univ, 65-68; asst prof, 68-72, ASSOC PROF MED MICROBIOL, UNIV VT, 72- Mem: Sigma Xi; Am Soc Microbiol; Tissue Cult Asn; AAAS. Res: Cell culture in biomedical research; mechanisms of pathogenesis; in vitro action of microbial toxins; biochemical genetics of cultured cells; protein synthesis. Mailing Add: Dept of Med Microbiol Univ of Vt Col of Med Burlington VT 05401

MOEHS, PETER JOHN, b Yonkers, NY, Apr 4, 04; m 65; c 2. CHEMISTRY. Educ: Norwich Univ, BS, 62; Univ NH, PhD(inorg chem), 67. Prof Exp: Asst prof chem, 69-74, chmn div sci, 71-72, ASSOC PROF CHEM, SAGINAW VALLEY COL, 74- Mem: Am Chem Soc. Res: Chemistry of Group IV organometallics; stabilities of high oxidation states of metal inorganic compounds. Mailing Add: Dept of Chem Saginaw Valley Col 2250 Pierce Rd University Center MI 48710

MOELLER, CARL WILLIAM, JR, b Carroll, Iowa, Mar 2, 24; m 52; c 2. INORGANIC CHEMISTRY. Educ: Harvard Univ, BS, 49; Univ Southern Calif, PhD(chem), 54. Prof Exp: Asst chem, Univ Southern Calif, 49-53; Fulbright fel, Univ Tübingen, 54-55; from instr to asst prof inorg chem, 55-66, ASSOC PROF INORG CHEM, UNIV CONN, 66-, DEP DEPT HEAD BR, 54- Mem: AAAS; Am Chem Soc. Res: Magnetochemical studies of free radicals and solid ternary oxides; photochemistry of coordination compounds; organoboron chemistry; chemical bonding. Mailing Add: Dept of Chem Univ of Conn Storrs CT 06268

MOELLER, FLOYD EDWARD, b Milwaukee, Wis, July 26, 19; m 47; c 3. APICULTURE. Educ: Univ Wis, PhD(entom), 52. Prof Exp: Apiculturist, Div Bee Cult & Biol Control, Bur Entom & Plant Quarantine, USDA, 49-53 & Entom Res Br, Agr Res Serv, 53-73; from instr to assoc prof, 49-75, PROF ENTOM, UNIV WIS-MADISON, 75-; RES LEADER ENTOM & APICULT, AGR RES SERV, USDA, 73- Mem: Entom Soc Am; Sigma Xi. Res: Bee management, diseases and behavior; pollination; stock testing and breeding. Mailing Add: 121 Island Dr Madison WI 53705

MOELLER, GUIDO KONSTANTIN, physics, see 12th edition

MOELLER, HENRY WILLIAM, b Woodbury, NJ, Aug 4, 37; c 3. MARINE BIOLOGY. Educ: Drew Univ, AB, 59; Rutgers Univ, MS, 65, PhD(bot), 69. Prof Exp: Physiologist, Wallace Pharmaceut Corp, NJ, 61-62; from instr to asst prof marine sci, Southampton Col, 65-69; asst prof, 69-72, ASSOC PROF & COORD MARINE SCI, DOWLING COL, 72- Concurrent Pos: Res assoc, NY Ocean Sci Lab, Montauk, 70. Mem: AmPhycol Soc; Brit Phycol Soc; Int Phycol Soc; Am Soc Limnol & Oceanog; NY Acad Sci. Res: Marine botany; experimental farming of Chondrus crispus. Mailing Add: Dept of Biol Dowling Col Oakdale NY 11769

MOELLER, KARL DIETER, b Hamburg, Ger, June 15, 27; m 59; c 2. SPECTROSCOPY, OPTICS. Educ: Univ Hamburg, PhD(physics), 57. Prof Exp: Fel physics, Sorbonne, 58-60; staff mem, US Army Res & Develop Labs, NJ, 60-63; PROF PHYSICS, FAIRLEIGH DICKINSON UNIV, 63- Mem: Optical Soc Am. Res: Far infrared optics and spectroscopy; spectra of torsional vibrations of polyatomic molecules; solid state spectra. Mailing Add: Dept of Physics Fairleigh Dickinson Univ Teaneck NJ 07666

MOELLER, MAX WALTER, poultry nutrition, see 12th edition

MOELLER, THEODORE WILLIAM, b Cincinnati, Ohio, Jan 26, 43; m 66; c 2. FOOD SCIENCE. Educ: Ohio State Univ, BS, 65; Mich State Univ, MS, 67, PhD(food sci), 71. Prof Exp: Group leader food res, 71, MGR PET FOODS RES, JOHN STUART RES LAB, QUAKER OATS CO, 71- Mem: Inst Food Technologists. Res: Cost reductions, product improvements and plant implementation of pet food products. Mailing Add: Quaker Oats Co 617 W Main St Barrington IL 60090

MOELLER, THERALD, b North Bend, Ore, Apr 3, 13; m 35; c 3. INORGANIC CHEMISTRY. Educ: Ore State Col, BS, 34; Univ Wis, PhD, 38. Prof Exp: Instr, Mich State Univ, 38-40; instr inorg chem, Univ Ill, 40-42, assoc, 42-43, from asst prof to assoc prof, 43-53, prof, 53-69, mem, Ctr Advan Study, 65-66; chmn dept, 69-75, PROF CHEM, ARIZ STATE UNIV, 69- Concurrent Pos: Mem, Bd Dir, Inorganic Syntheses, Inc, 71- Mem: Am Chem Soc; The Chem Soc. Res: Lanthanide and actinide chemistry; coordination compounds; synthesis, isomerism, substitution patterns, reactions of phosphonitriles and sulfur-nitrogen compounds; nonaqueous systems. Mailing Add: Dept of Chem Ariz State Univ Tempe AZ 85281

MOELLMANN, GISELA E BIELITZ, b Dessau, Ger, Feb 15, 29; US citizen; m; c 1. CELL BIOLOGY, CYTOCHEMISTRY. Educ: Univ Dayton, BS, 53; Yale Univ, PhD(anat), 67. Prof Exp: Am Cancer Soc fel, 66-68, res assoc anat, 68-69, res assoc dermat & lectr anat, 69-73, ASST PROF DERMAT & CYTOL, SCH MED, YALE UNIV, 73- Mem: AAAS; Am Soc Cell Biol; Soc Invest Dermat; Sigma Xi. Res: Cell biology, especially the fine structure and cytochemistry of mammalian and amphibian pigment cells; intracellular transport of particles; topography of peptide hormone receptor sites. Mailing Add: LCI 500 Dept Dermat Yale Univ Sch of Med New Haven CT 06510

MOEN, AARON NATHAN, b Minn, Nov 15, 36; m 59; c 4. ECOLOGY. Educ: Gustavus Adolphus Col, BS, 58; St Cloud State Col, MS, 63; Univ Minn, PhD(wildlife mgt), 66. Prof Exp: Teacher high schs, Minn, 58-62, 63-64; asst prof zool, Univ Minn, 66-67; ASSOC PROF WILDLIFE ECOL, CORNELL UNIV, 67- Mem: Ecol Soc Am; Wildlife Soc; Am Soc Mammal; Am Inst Biol Sci; Soc Range Mgt. Res: Nutrition, physiology and behavior of free-ranging animals. Mailing Add: Fernow Hall Cornell Univ Ithaca NY 14850

MOEN, ALLEN LEROY, b Badger, Minn, June 15, 33; m 59; c 3. SURFACE PHYSICS. Educ: Pac Lutheran Univ, BA, 55; Wash State Univ, MS, 61, PhD(physics), 68. Prof Exp: Asst physics, 55-57, 59-60 & 61-63; asst prof, 63-68, ASSOC PROF PHYSICS & CHMN DEPT, CENT COL IOWA, 68- Concurrent Pos: Res asst, Wash State Univ, 66-68. Mem: Am Asn Physics Teachers; Am Vacuum Soc. Res: Ion bombardment of metals; sputtering; surface physics; ultra-high vacuum physics. Mailing Add: Dept of Physics Cent Col Pella IA 50219

MOEN, CARL JOHAN, physics, see 12th edition

MOENCH, ROBERT HADLEY, b Boston, Mass, Oct 23, 26; m 55; c 3. GEOLOGY. Educ: Boston Univ, AB, 50, AM, 51, PhD(geol), 54. Prof Exp: GEOLOGIST, US GEOL SURV, 52- Mem: AAAS; Geol Soc Am; Soc Econ Geol. Res: Geology of uranium deposits; stratigraphy, structure and tectonics of metamorphic rocks, mineral resources and environmental geology, especially in New England. Mailing Add: US Geol Surv Bldg 25 Denver Fed Ctr Denver CO 80225

MOENS, PETER B, b Neth, May 15, 31; Can citizen; m 53; c 5. CELL BIOLOGY, CYTOGENETICS. Educ: Univ Toronto, BScF, 59, MA, 61, PhD(biol), 63. Prof Exp: Assoc prof, 64-72, PROF BIOL, YORK UNIV, 72-, CHMN DEPT NATURAL SCI, ATKINSON COL, 74- Concurrent Pos: Assoc ed, Genetic Soc Can, 70. Mem: AAAS; Genetics Soc Am; Genetics Soc Can; Can Soc Cell Biol. Res: Electron microscopy; meiosis; development of germ cells in fungi, plants and animals with emphasis on genetically significant aspects; specifically electron microscopy of the synaptonemal complex. Mailing Add: Dept of Biol York Univ Downsview ON Can

MOENTER, RICHARD LUTHER, b Pemberville, Ohio, Apr 9, 12; m 38; c 7. MATHEMATICS. Educ: Capital Univ, BS, 34; Ohio State Univ, MA, 38. Prof Exp: Asst prof math, Luther Col, Sask, 34-43 & Capital Univ, 43-46; assoc prof, Dana Col, 46-47 & Midland Col, 47-64, ASSOC PROF MATH, BUENA VISTA COL, 64- Concurrent Pos: Instr, Can Air Cadets, 40-43 & pre-flight sch, Capital Univ, 43-44. Mem: Math Asn Am. Res: Reimann surfaces. Mailing Add: Dept of Math Buena Vista Col Storm Lake IA 50588

MOERMAN, MICHAEL, b New York, NY, Dec 6, 34; c 3. ETHNOGRAPHY, ANTHROPOLOGICAL LINGUISTICS. Educ: Columbia Col, AB, 56; Yale Univ, PhD(anthrop), 64. Prof Exp: Asst prof Southern & Southeast Asia studies, Am Univ, 62-64; from asst prof to assoc prof, 64-72, PROF ANTHROP, UNIV CALIF, LOS ANGELES, 72- Concurrent Pos: Consult, AID, 64-70; nat secy, Acad Adv Coun on Thailand, 65-67; Advan Res Proj Agency grant, Univ Calif, Los Angeles, 67-68; NSF grant, Thailand, 68-69. Mem: Asn Asian Studies; Am Anthrop Asn; Thailand Siam Soc; Am Ethnol Soc. Res: Naturalistic observation and microanalysis of social interaction with special emphasis on conversation; Southeast Asia, especially Thailand; ethnographic film and visual anthropology. Mailing Add: Dept of Anthrop Univ of Calif Los Angeles CA 90024

MOERMOND, TIMOTHY CREIGHTON, b Sioux City, Iowa, Apr 4, 47; m 68; c 1. ECOLOGY. Educ: Univ Ill, Urbana-Champaign, BS, 69; Harvard Univ, PhD(biol), 74. Prof Exp: ASST PROF ZOOL, UNIV WIS-MADISON, 73- Mem: Ecol Soc Am; Am Ornithologists Union; Soc Study Evolution; Am Soc Ichthyologists & Herpetologists; Wilson Ornith Soc. Res: Field and theoretical studies of foraging strategies and habitat use patterns. Mailing Add: Dept of Zool Univ of Wis Madison WI 53706

MOERSCH, GEORGE WILLIAM, b Escanaba, Mich, July 23, 15; m 43; c 1. ORGANIC CHEMISTRY. Educ: Lawrence Col, AB, 37; Pa State Col, MS, 40, PhD(org chem), 42. Prof Exp: Instr org chem, Univ Idaho, 40-41; res chemist develop dept, WVa Pulp & Paper Co, SC, 42-44; instr org chem, Pa State Col, 44-46; sr res chemist org med, 46-57, res leader, 57-61, lab dir org chem, 61-69, SECT DIR, PARKE, DAVIS & CO, 69- Concurrent Pos: Admin fel, Mellon Inst, 56-57. Mem: Am Chem Soc; The Chem Soc. Res: Chemistry of aliphatic compounds, pulp by-products, antimalarial drugs, chloramphenicol, steroids and organic medicinals. Mailing Add: Parke Davis & Co Res Div 2800 Plymouth Rd Ann Arbor MI 48106

MOERSCH, HERMAN JOHN, b St Paul, Minn, Mar 14, 95; m 25; c 2. CLINICAL MEDICINE. Educ: Univ Minn, BS, 18, MD, 21, MS, 25. Prof Exp: From instr to assoc prof, 25-47, PROF MED, MAYO GRAD SCH MED, UNIV MINN, 47-; HEAD CLIN SECT, MAYO CLIN, 37- Concurrent Pos: Dir educ & res, Am Col Chest Physicians, 60, sci consult, 60-, award, 65. Honors & Awards: Great Cross Merit, Fed Repub Ger, 58; Chevalier Jackson Award, Am Broncho-Esophagol Asn, 60. Mem: Am Soc Gastrointestinal Endoscopy; Am Thoracic Soc; Am Gastroenterol Asn; AMA; Soc Thoracic Surg. Res: Diseases of the esophagus and chest. Mailing Add: Mayo Clin 200 First SW Rochester MN 55901

MOERTEL, CHARLES GEORGE, b Milwaukee, Wis, Oct 17, 27; c 4. MEDICINE. Educ: Univ Ill, BS, 52, MD, 53; Univ Minn, MS, 58. Prof Exp: From instr to assoc prof, Mayo Grad Sch Med, Univ Minn, 60-71, PROF MED & CHMN DEPT ONCOL, MAYO MED SCH, 71-; CONSULT INTERNAL MED, MAYO CLIN, 58- Mem: Am Col Physicians; Am Asn Cancer Res; Am Soc Clin Oncol; Am Gastroenterol Asn. Res: Gastrointestinal cancer. Mailing Add: Mayo Clin Rochester MN 55901

MOEWS, PAUL CHARLES, JR, inorganic chemistry, see 12th edition

MOFFA, DAVID JOSEPH, b Fairmont, WVa, Dec 6, 42; m 64; c 2. BIOCHEMISTRY, CLINICAL CHEMISTRY. Educ: WVa Univ, AB, 64, MS, 66, PhD(biochem), 68. Prof Exp: Res asst biochem, 64-68, instr, Sch Med, 68-70, ASST PROF BIOCHEM, SCH MED, WVA UNIV, 70-; DIR, BIOPREPS LABS, 69- Concurrent Pos: NIH res fel, WVa Univ, 68-70. Mem: Am Asn Clin Chem; Am Soc Clin Pathologists; Am Med Technologists. Res: Metabolism of vitamin A; lipid metabolism; enzymology; clinical methodology. Mailing Add: BioPreps Labs Box 888 Fairmont WV 26554

MOFFAT, ARLO JAY, physical organic chemistry, see 12th edition

MOFFAT, GRACE H, cytochemistry, see 12th edition

MOFFAT, JAMES, b Turtle Creek, Pa, Feb 24, 21; m 46; c 1. CHEMISTRY. Educ: Allegheny Col, BS, 42; Northwestern Univ, PhD(org chem), 48. Prof Exp: Asst prof chem, Univ Miami Fla, 47-49; res fel, Calif Inst Technol, 49-50; sr res chemist, Nepera Chem Co, 51; res assoc, Northwestern Univ, 51-53; res asst prof chem, Univ Louisville, 53-58; asst prof, 58-62, ASSOC PROF CHEM, UNIV MO-KANSAS CITY, 62- Mem: AAAS; Am Chem Soc. Res: Heterocyclic compounds; organic isocyanides; hydrogen bonding; tautomerism. Mailing Add: 5244 Rockhill Rd Kansas City MO 64110

MOFFAT, JOHN BLAIN, b Owen Sound, Ont, Aug 7, 30; m 56; c 3. PHYSICAL CHEMISTRY, QUANTUM CHEMISTRY. Educ: Univ Toronto, BA, 53, PhD(phys chem), 56. Prof Exp: Res chemist, Res & Develop Lab, Du Pont Can Ltd, 56-61; from asst prof to assoc prof, 61-74, PROF CHEM, UNIV WATERLOO, 74- Mem: Am Chem Soc; The Chem Soc; Am Phys Soc; fel Chem Inst Can. Res: Molecular orbital theory; molecular structure and chemical bonding; surface chemistry and catalysis; adsorption, decomposition, exchange and spectroscopic studies. Mailing Add: 226 Lincoln Rd Waterloo ON Can

MOFFAT, JOHN KEITH, b Edinburgh, Scotland, Apr 3, 43. BIOPHYSICS. Educ: Univ Edinburgh, BSc, 65; Cambridge Univ, PhD(protein crystallog), 71. Prof Exp: Sci staff mem protein crystallog, Med Res Coun Lab Molecular Biol, 68-69; res assoc reaction kinetics, 69-70, ASST PROF BIOCHEM & MOLECULAR BIOL, CORNELL UNIV, 70- Concurrent Pos: Merck, Sharp & Dohme fac develop award, Cornell Univ, 70. Honors & Awards: Medal, Royal Scottish Soc Arts, 64. Mem: Am Crystallog Asn. Res: Protein structure determination by physico-chemical techniques; the relation between structure and function in hemoglobin; protein crystallography. Mailing Add: Sect Biochem & Molecular Biol Cornell Univ Ithaca NY 14853

MOFFAT, ROBERT DOUGLAS, nuclear physics, see 12th edition

MOFFATT, JOHN GILBERT, b Victoria, BC, Sept 19, 30; m 53; c 4. ORGANIC CHEMISTRY. Educ: Univ BC, BA, 52, MSc, 53, PhD(org chem), 56. Prof Exp: Tech officer, Defence Res Bd Can, 53-54; res assoc, BC Res Coun, 56-60; group leader, Calif Corp Biochem Res, 60-61; head org chem, 61-65, assoc dir, 65-67, DIR, SYNTEX INST MOLECULAR BIOL, 67- Mem: Am Chem Soc; The Chem Soc; Am Soc Biol Chemists. Res: Chemistry of nucleosides, nucleotides, nucleoside polyphosphates, sugar phosphates, carbohydrates and nucleic acids. Mailing Add: Inst of Molecular Biol Syntex Res 3401 Hillview Ave Palo Alto CA 94304

MOFFET, ALAN THEODORE, b St Paul, Minn, Mar 24, 36. RADIO ASTRONOMY. Educ: Wesleyan Univ, BA, 57; Calif Inst Technol, PhD(physics, astron), 61. Prof Exp: Fulbright scholar, Univ Bonn, 61-62; res fel radio astron, 62-66, Sloan fel, 63-67, from asst prof to assoc prof, 66-71, PROF RADIO ASTRON, CALIF INST TECHNOL, 71-, DIR, OWENS VALLEY RADIO OBSERV, 75- Concurrent Pos: Mem US Nat Comn 5, Int Sci Radio Union. Mem: Int Astron Union; Am Astron Soc; Am Phys Soc; Inst Elec & Electronics Eng; Int Sci Radio Union. Res: Structure and physical properties of extragalactic radio sources; design of radio telescopes and astronomical instruments. Mailing Add: Owens Valley Radio Observ Calif Inst of Technol Pasadena CA 91125

MOFFET, HUGH L, b Monmouth, Ill, Jan 6, 32; m 54; c 3. PEDIATRICS, INFECTIOUS DISEASES. Educ: Harvard Univ, AB, 53; Yale Univ, MD, 57. Prof Exp: Instr pediat, Bowman Gray Sch Med, 62-63; from asst prof to assoc prof, Med Sch, Northwestern Univ, Chicago, 63-71; assoc prof, 71-74, PROF PEDIAT, MED SCH, UNIV WIS-MADISON, 74- Concurrent Pos: Nat Inst Allergy & Infectious Dis fel, 60-63. Res: Diagnostic microbiology; epidemiology of pediatric infections. Mailing Add: 202 S Park St Madison WI 53715

MOFFETT, BENJAMIN CHARLES, JR, b Spring Lake, NJ, Oct 28, 23; m 49; c 2. DENTAL RESEARCH. Educ: Syracuse Univ, BA, 48; NY Univ, PhD(anat), 52. Prof Exp: From asst prof to assoc prof anat, Med Col, Univ Ala, 52-63; assoc prof, Col Med, Wayne State Univ, 63-64; assoc prof, 64-67, PROF ORTHOD, SCH DENT, UNIV WASH, 67- Concurrent Pos: Res fel anat, Med Sch, Gothenburg Univ, 59-60; res fel anat, Armed Forces Inst Path, 60-61; vis prof, Sch Dent, Cath Univ Nijmegen, 71-72. Honors & Awards: Jerome Schweitzer Res Award, NY Acad Prosthodont, 69. Mem: Am Asn Anat; Int Asn Dent Res. Res: Arthrology; history of anatomy; cranio-facial morphogenesis. Mailing Add: Dept of Orthod Univ of Wash Sch of Dent Seattle WA 98105

MOFFETT, EUGENE WILKIN, b Cherryvale, Kans, Feb 28, 08; m 36; c 1. ORGANIC CHEMISTRY. Educ: Monmouth Col, BS, 29; Northwestern Univ, MS, 30, PhD(chem), 32. Prof Exp: Asst, Northwestern Univ Ill, 29-31; chemist, Marbon Corp, 33-37 & Pittsburgh Plate Glass Co, 37-50, tech dir res & develop, 50-59, mgr resin develop, 59-64, develop dir, PPG Industs, Inc, Springdale, 64-73; RETIRED. Mem: Am Inst Chem; Am Chem Soc. Res: Paint, varnishes and alkyds; polyesters as plasticizers for urea resins; unsaturated polyester resins; polyurethane rigid foams. Mailing Add: 906 Whitley Dr Pittsburgh PA 15237

MOFFETT, JOSEPH ORR, b Peabody, Kans, Jan 9, 26; m 44; c 7. ENTOMOLOGY. Educ: Kans State Univ, BS, 49, MS, 50; Univ Wyo, PhD(entom), 74. Prof Exp: From instr to asst prof apicult, Colo State Univ, 49-59; ed & secy-treas, Am Beekeeping Fedn, 59-64; ENTOMOLOGIST, BEE RES LAB, AGR RES SERV, USDA, 67- Mem: Entom Soc Am; Sigma Xi. Res: Pollination of citrus and hybrid cotton; greenhouse pollination by bees; effect of herbicides on honeybees. Mailing Add: Bee Res Lab Agr Res Serv USDA 2000 E Allen Rd Tucson AZ 85719

MOFFETT, ROBERT BRUCE, b Madison, Ind, June 8, 14; c 2. ORGANIC CHEMISTRY, MEDICINAL CHEMISTRY. Educ: Hanover Col, AB, 37, Univ Ill, AM, 39, PhD(org chem), 41. Hon Degrees: DSc, Hanover Col, 59. Prof Exp: Asst, Univ Ill, 37-39; Abbott, Upjohn & Glidden fel, Northwestern Univ, 41-43; sr res chemist, George A Breon & Co, Mo, 43-44, asst dir labs, 44-47; RES CHEMIST, UPJOHN CO, 47- Mem: Am Chem Soc; Int Soc Heterocyclic Chem. Res: Chemiluminescence; benzopyrylium salts; steroids; analgesics; antispasmodics; drugs for mental diseases. Mailing Add: Upjohn Co Unit 7251-209-7 Kalamazoo MI 49001

MOFFETT, ROBERT PREYER, physical chemistry, see 12th edition

MOFFETT, SAMUEL MCKEE, b Madison, Ind, Jan 23, 16; m 42; c 2. ORGANIC CHEMISTRY. Educ: Hanover Col, AB, 38; Ohio State Univ, PhD(org chem), 42. Prof Exp: Asst chem, Ohio State Univ, 38-42; instr Alma Col, 42-45; asst prof, Ohio Wesleyan Univ, 45-46; asst prof, Park Col, 46-51, from actg head dept to head dept, 46-51; from assoc res chemist to sr res chemist, Midwest Res Inst, 51-54; sr chemist res dept, Grand River Chem Div, Deere & Co, 54-56, res group leader, John Deere Chem Co, 57-61; sr res chemist, Nitrogen Div, Allied Chem Corp, 61-64; prof chem, Va Union Univ, 64-65; assoc prof, 65-67, from actg head dept to head dept, 67-71, PROF CHEM, VA STATE COL, 67- Mem: AAAS; Am Chem Soc; Sigma Xi. Res: Chemiluminescence; natural products; structure determination; waxes; detergents; nitrogen chemistry. Mailing Add: 3001 Riverside Ave Hopewell VA 23860

MOFFITT, EMERSON AMOS, b McAdam, NB, Sept 9, 24; US citizen; m 51; c 3. ANESTHESIOLOGY. Educ: Dalhousie Univ, MD, CM, 51; Univ Minn, MS, 58; Am Bd Anesthesiol, dipl, 60. Prof Exp: Pvt pract, 51-54; resident anesthesiol, Mayo Grad Sch Med, Univ Minn, 54-57, from instr to assoc prof, 59-72; PROF ANESTHESIOL & HEAD DEPT, DALHOUSIE UNIV, 73- Concurrent Pos: Consult, Mayo Clin, 57-72, sect head anesthesiol, 66-72, NIH grant gen med sci, Mayo Found, 67-72. Mem: Am Soc Anesthesiol; Int Anesthesia Res Soc; Can Anaesthetists Soc; Am Heart Asn; Asn Univ Anesthetists. Res: Cardiovascular and metabolic effects of anesthetic state and cardiac surgery with whole body perfusion; metabolism in acute stress and shock. Mailing Add: Dept of Anesthesia Victoria Gen Hosp Halifax NS Can

MOFFITT, HAROLD ROGER, b Ukiah, Calif, Aug 8, 34; m 54; c 3. ACAROLOGY. Educ: Univ Calif, Davis, BS, 57, Univ Calif, Riverside, MS, 63, PhD(entom), 67. Prof Exp: Lab technician entom, Univ Calif, Riverside, 57-63, res asst, 63-67; res entomologist, Agr Res Serv, 67-72, RES LEADER, AGR RES SERV, USDA, 72- Mem: AAAS; Entom Soc Am; Entom Soc Can; Int Orgn Biol Control. Res: Biology and control of insects and mites of agricultural importance; integrated control and applied ecology; biological and genetic control; acarology. Mailing Add: Yakima Agr Res Lab Agr Res Serv USDA 3706 Nob Hill Blvd Yakima WA 98902

MOFFITT, ROBERT ALLAN, b Gillette, Wyo, June 17, 18; m 44; c 4. CHEMISTRY, BIOCHEMISTRY. Educ: Univ Calif, Los Angeles, BA, 40; Univ Southern Calif, MS, 54. Prof Exp: Lab mgr, Fernando Valley Milling & Supply Co, 40-44 & Ralston Purina Co, 46-60; head anal lab, Res Labs, 60-67, MGR ANAL SERV, CARNATION CO, 67- Mem: Am Chem Soc; Am Asn Cereal Chem; Inst Food Technol. Res: Accurate analysis of chemical components of food products; contamination of foods; analytical chemistry; instrumentation. Mailing Add: 1960 Escarpa Dr Los Angeles CA 90041

MOFFITT, ROBERT LEVERE, b Pocatello, Idaho, May 17, 24; m 47; c 3. PHARMACOLOGY. Educ: Univ SDak, Bs, 47, MA, 60; Univ Louisville, PhD(pharmacol), 65. Prof Exp: Res pharmacologist, Miles Labs, Inc, 50-61; sr pharmacologist, Neisler Labs, Inc, 65-66; prof pharmacol & chmn dept, Col Pharm, Ohio Northern Univ, 66-69; DIR SCI INFO & REGULATORY AFFAIRS, MEAD JOHNSON RES CTR, 69- Mem: AAAS; NY Acad Sci. Res: Neuropharmacology; cytopharmacology; bioenergetics. Mailing Add: Sci Info & Regulatory Affairs Mead Johnson Res Ctr Evansville IN 47721

MOFJELD, HAROLD OSWALD, b San Francisco, Calif, Nov 2, 40. PHYSICAL OCEANOGRAPHY. Educ: Univ Wash, BS, 63, MS, 65, PhD(geophys), 70. Prof Exp: Fel oceanog, Univ Wash, 70-71; RES OCEANOGR, ATLANTIC OCEANOG & METEOROL LABS, NAT OCEANIC & ATMOSPHERIC ADMIN, 71- Concurrent Pos: Adj prof oceanog, Univ Miami, 72- Mem: Sigma Xi; Am Geophys Union. Res: Tides and tidal currents both on the continental shelves and in the open oceans; theoretical studies of time-dependent currents. Mailing Add: Atlantic Oceanog & Meteorol Labs 15 Rickenbacker Causeway Miami FL 33149

MOGABGAB, WILLIAM JOSEPH, b Durant, Okla, Nov 2, 21; m 48; c 5. INTERNAL MEDICINE. Educ: Tulane Univ, BS, 42, MD, 44; Am Bd Internal Med, dipl, 51. Prof Exp: Intern, Charity Hosp La, New Orleans, 44-45; from asst to instr med, Sch Med, Tulane Univ, 46-49, instr, Div Infectious Dis, 49-51; Nat Found Infantile Paralysis fel, 51-52; asst prof, Baylor Col Med, 52-53; head virol div, Naval Med Res Unit 4, Great Lakes, Ill, 53-55; assoc prof, 56-62, PROF MED, SCH MED, TULANE UNIV, 62- Concurrent Pos: From resident to chief resident, Tulane Serv, Charity Hosp, La, 46-49, vis physician, 49-51 & sr vis physician, 61-; vis investr & asst physician, Hosp, Rockefeller Inst, 51-52; chief infectious dis, Vet Admin Hosp, Houston, 52-53; consult, Vet Admin Hosp, New Orleans, 56- Mem: Soc Exp Biol & Med; Am Soc Microbiol; Tissue Cult Asn; fel Am Col Physicians; fel Am Acad Microbiol. Res: Virology; tissue culture; infectious disease; bacteriology. Mailing Add: Sect of Infect Dis Tulane Univ Sch of Med New Orleans LA 70112

MOGENSEN, HANS LLOYD, b Price, Utah, Dec 16, 38; m 58; c 2. PLANT ANATOMY, PLANT MORPHOLOGY. Educ: Utah State Univ, BS, 61; Iowa State Univ, MS, 63, PhD(plant anat), 65. Prof Exp: Assoc prof, 65-74, PROF BOT, NORTHERN ARIZ UNIV, 74- Concurrent Pos: Vis prof, Carlsberg Res Lab, Copenhagen, Denmark, 76-77. Mem: Bot Soc Am; Int Soc Plant Morphologists. Res: Ultrastructure of fertilization in flowering plants; plant development. Mailing Add: Dept of Biol Northern Ariz Univ Flagstaff AZ 86001

MOGENSON, GORDON JAMES, b Delisle, Sask, Jan 24, 31; m 54; c 2. PSYCHOPHYSIOLOGY. Educ: Univ Sask, BA, 55, MA, 56; McGill Univ, PhD(physiol psychol), 59. Prof Exp: From asst prof to assoc prof psychol, Univ Sask, 58-65; assoc prof psychol & physiol, 65-68, PROF PSYCHOL & PHYSIOL, UNIV WESTERN ONT, 68- Concurrent Pos: Ed, Can J Psychol, 69-74 & Newsletter, Int Comn Physiol Food & Fluid Intake, 74- Mem: AAAS; Am Psychol Asn; Animal Behav Soc; Can Psychol Asn; Can Physiol Soc. Res: Physiology of behavior; control systems for water and food intake, temperature regulation and self-stimulation; neural mechanisms for reinforcement and learning. Mailing Add: Dept of Physiol Univ of Western Ont London ON Can

MOGFORD, JAMES A, b McCamey, Tex, Dec 6, 30; m 53; c 4. PHYSICS. Educ: Tex Tech Col, BS, 56; Univ Wis, MS, 59. Prof Exp: Physicist, Midwestern Univs Res Asn, 59-61; staff mem, Sandia Corp, NMex, 61-67, supvr, Anal Div, Sandia Labs, Livermore, 67-74, supvr, Inductive Energy Storage Div, 74-75, MEM MGT STAFF, SANDIA LABS, ALBUQUERQUE, 75- Res: Theoretical studies of orbit dynamics in particle accelerators; effects of radiation energy deposition; effects of photon-electron irradiation of materials and systems; design and development of pulsed-power systems. Mailing Add: Sandia Labs Albuquerque NM 87115

MOGGIO, WILLIAM ALDO, b New York, NY, June 4, 17; m 42; c 4. APPLIED CHEMISTRY, ENVIRONMENTAL SCIENCE. Educ: Rutgers Univ, 38, MS, 39. Prof Exp: Asst, Sch Pub Health, Univ NC, 39-41; sanit engr-chemist, Piatt & Davis Eng, 41-42; res engr & proj leader, Nat Coun Stream Improv, 46-54; chief chemist, ETex Pulp & Paper Co, 54-58; prod mgr, Lake States Yeast & Chem Div, St Regis Paper Co, 58-60; RES MGR, ARMSTRONG CORK CO, 60- Concurrent Pos: Res assoc, Eng Exp Sta, La State Univ, 46-54. Honors & Awards: Indust Wastes Award, Water Pollution Control Fedn, 61; Am Inst Chemists Cert. Mem: Tech Asn Pulp & Paper Indust; Am Acad Environ Eng; fel Am Inst Chemists; Am Asn Textile Chemists & Colorists. Res: Pulp and paper technology; nonwoven textile technology; microencapsulation; microbiological fermentation; chemical and microbiological treatment of industrial and domestic wastes; stream sanitation; process water conservation; process and domestic water treatment; carpet technology; printing and dyeing. Mailing Add: Res & Develop Ctr Armstrong Cork Co Lancaster PA 17604

MOGHISSI, KAMRAN S, b Tehran, Iran, Sept 11, 25; US citizen; m 52; c 2. OBSTETRICS & GYNECOLOGY. Educ: Univ Tehran, BS, 47; Univ Geneva, MB & ChB, 51, MD, 52; Am Bd Obstet & Gynec, dipl, 67, dipl reprod endocrinol, 75. Prof Exp: Intern & resident obstet & gynec var hosps, Eng, 52-56; assoc prof, Med Sch & Nemazee Hosp, Univ Shiraz, 56-59; res assoc biochem, Sch Med, Wayne State Univ, 59-61; sr resident obstet & gynec, Detroit Gen Hosp, 61; from asst po assoc prof, 62-71, PROF OBSTET & GYNEC & CHIEF DIV REPROD BIOL, SCH MED, WAYNE STATE UNIV, 71- Concurrent Pos: Attend obstetrician & gynecologist, Detroit Gen Hosp, 62; sr attend, Hutzel Hosp, 62; examr, Am Bd Obstet & Gynec; surgeon, Harper Hosp, 63; consult, Grace Hosp, Detroit; gynecologist, Childrens Hosp Mich. Mem: AMA; fel Am Col Obstet & Gynec; fel Am Col Surg; Asn Planned Parenthood Physicians; Am Soc Androl. Res: Human reproduction and reproductive endocrinology; infertility and conception control. Mailing Add: Dept of Gynec & Obstet Wayne State Univ Sch of Med Detroit MI 48201

MOGREN, EDWIN WALFRED, b Minn, Sept 16, 21; m 44; c 2. FORESTRY. Educ: Univ Minn, BSF, 47, MF, 48; Univ Mich, PhD(forest ecol), 55. Prof Exp: Asst, Lake States Forest Exp Sta, US Forest Serv, 41; asst, Univ Minn, 47-48; from instr to asst prof forestry, 48-55, assoc prof forest mgt, 55-61, PROF FOREST MGT, COLO STATE UNIV, 61- Concurrent Pos: Collabr, Rocky Mt Forest & Range Exp Sta, 52-60. Mem: Fel AAAS; Soc Am Foresters; Ecol Soc Am. Res: Forest management; ecology; siliviculture. Mailing Add: Colo State Univ Col Forest & Natural Resources Ft Collins CO 80521

MOGRO-CAMPERO, ANTONIO, b Liverpool, Eng, Aug 25, 40; Bolivian citizen; m 66; c 2. SPACE PHYSICS, GEOPHYSICS. Educ: Columbia Univ, BS, 63; Univ Chicago, MS, 67, PhD(physics), 71. Prof Exp: Res engr, Lab Cosmic Physics, Univ San Andres, Bolivia, 63-65; res assoc physics, Enrico Fermi Inst, Univ Chicago, 70-74; res physicist & lectr, Univ Calif, San Diego, 74-75; PHYSICIST, GEN ELEC RES & DEVELOP CTR, 75- Mem: Am Geophys Union. Res: Health and environmental physics; gas flow in the Earth; earthquake prediction; Jupiter's and Earth's radiation belts; solar and galactic cosmic rays. Mailing Add: Gen Elec Res & Develop Ctr PO Box 8 Schenectady NY 12301

MOH, CARL CRAIG, b Canton, China, Nov 27, 20; nat US; m 50; c 2. AGRONOMY, RADIATION BOTANY. Educ: Nanking Univ, BS, 43; Wash State Univ, MS, 50, PhD(agron), 53. Prof Exp: Asst cytogenetics, Nanking Univ, 43-47; geneticist, Smithsonian Inst, 53-58; CYTOGENETICIST, TROP AGR RES & TRAINING CTR, INTER-AM INST AGR SCI, 58- Mem: Genetics Soc Am; Am Genetic Asn. Res: Cytogenetics; radiobotany. Mailing Add: Trop Agr Res & Training Ctr Inter-Am Inst of Agr Sci Turrialba Costa Rica

MOHACSI, ERNO, b Zalaegerszeg, Hungary, Jan 26, 29; US citizen. ORGANIC CHEMISTRY, PHYSICAL CHEMISTRY. Educ: Eötvös Lorand Univ, Budapest, dipl org chem, 56; Columbia Univ, MA, 60, PhD(org chem), 62. PXFel, Columbia Univ, 63-64 & Harvard Univ, 64-66; SR RES CHEMIST, HOFFMANN-LA ROCHE, INC, 66- Mem: Am Chem Soc. Res: Synthetic organic chemistry, especially structure elucidation and synthesis of alkaloids and other natural products. Mailing Add: 85 Hillcrest Ave Summit NJ 07901

MOHAMED, ALY HAMED, b Cairo, Egypt, Aug 29, 24; m 43; c 2. GENETICS. Educ: Univ Alexandria, BS, 46; Univ Minn, MS, 53, PhD(genetics), 54. Prof Exp: Asst genetics, Univ Alexandria, 46-48, lectr, 54-60, assoc prof, 60-63; res assoc range sci, Tex A&M Univ, 63-64, res fel genetics & air pollution trainee, 64-66; assoc prof, 66-69, PROF BIOL, UNIV MO-KANSAS CITY, 69- Concurrent Pos: Dept Health, Educ & Welfare res grant, 67-71. Mem: AAAS; Bot Soc Am; Genetics Soc Am; Am Genetic Asn; Int Soc Fluoride Res (vpres). Res: Plant cytogenetics and the inheritance of quantitative characters; studying the effect of some air pollutants on chromosomes. Mailing Add: Dept of Biol Univ Of Mo Kansas City MO 64110

MOHAMED, HAROLD, b Sioux Falls, SDak, Jan 23, 32; m 61; c 3. URBAN GEOGRAPHY, TRANSPORTATION GEOGRAPHY. Educ: Ind State Univ, BS, 57; Northwestern Univ, MS, 58, PhD(geog), 67. Prof Exp: Cartog asst, Off Naval Res Study, 58-59; instr geog, Elmhurst Col, 60-64, asst prof & chmn dept, 64-67; assoc prof geog & dir instnl res, 67-70, PROF GEOG, ASST V PRES ADMIN AFFAIRS & ACTG DIR RES & DEVELOP, NORTHEASTERN ILL UNIV, 70- Concurrent Pos: Kroger Co Study, 58-59; teaching asst, Northwestern Univ, 58-61; Fair Store Corp at Old Orchard study, 60-61; bd chmn, Coop Comput Ctr. Mem: Asn Am Geog; Pop Coun Am; Asn Instnl Res. Res: Research on present-day housing patterns and projection for the future for the city of Elmhurst, Illinois. Mailing Add: Northeastern Ill Univ 5500 N St Louis Ave Chicago IL 60625

MOHAMMED, AUYUAB, b Trinidad, WI, Jan 11, 28; m 55; c 4. APPLIED MATHEMATICS. Educ: Univ Man, BSc, 54, MSc, 56; Univ BC, PhD(appl math), 65. Prof Exp: Sci officer, 54-60, HEAD APPL MATH SECT, DEFENCE RES ESTAB ATLANTIC, DEFENCE RES BD CAN, 61- Mem: Fel Acoust Soc Am; Asn Comput Mach; Inst Elec & Electronics Engrs; Brit Comput Soc. Res: Underwater acoustics; scientific applications of computers; non-linear and linear control theory;

communication theory. Mailing Add: Defence Res Estab Atlantic Defence Res Bd Can Grove St Dartmouth NS Can

MOHAMMED, CLIVE IMRAM, b Trinidad, May 30, 29; US citizen; m; c 4. DENTISTRY, ANATOMY. Educ: Nair Hosp Dent Col, India, LDS, 49; Tufts Univ, cert periodont, 51; Univ Ill, MS, 52, PhD(anat), 55. Prof Exp: Asst histol, Sch Dent, Univ Ill, 54-55, instr, 55-57; from asst prof to prof histopath & periodont, Sch Dent, Univ PR, 57-67, chmn dept, 59-67; chief dent res & educ, US Vet Admin, 69-70; from assoc dean to dean, Sch Dent, Univ Detroit, 70-75. Concurrent Pos: Int Col Dentists fel, 59; Am Col Dentists fel, 63; spec fel award, Univ Tex Dent Br, Houston, 64-65. Honors & Awards: Am Soc Dent Award, Univ Detroit, 73. Mem: Int Asn Dent Res; Int Col Dent; Am Col Dent; Am Dent Asn; Soc Exp Biol & Med. Res: Growth and development of calcified tissues; histopathology and histochemistry of pulpal connective tissue; role of vitamin C in dental tissues; electron microscopy of oral structures. Mailing Add: 1265 Three Mile Dr Grosse Point Park MI 48230

MOHAMMED, KASHEED, b Trinidad, WI, Apr 27, 30; m 64; c 2. NUTRITIONAL BIOCHEMISTRY, HUMAN PHYSIOLOGY. Educ: Univ Ariz, BS, 62, MS, 63, PhD(biochem), 67. Prof Exp: Scientist, Angostura Bitters, Trinidad, 52-56; NIH fel, Univ Ill Med Ctr, 67-69, res fel, 69; NUTRIT BIOCHEMIST, PHARMACEUT DIV, JOHNSON & JOHNSON RES CTR, 69- Concurrent Pos: Sci adv, State Univ NY Agr & Tech Col, Canton, 70- Mem: AAAS; Am Chem Soc. Res: Research and development of enteral and parenteral products for therapeutic usages. Mailing Add: Johnson & Johnson Res Ctr New Brunswick NJ 08902

MOHAMMED, M HAMDI A, b Minia, Egypt, May 10, 40; m 64; c 2. DENTAL MATERIALS, PROSTHODONTICS. Educ: Univ Alexandria, DDS, 63, MSD, 65; Northwestern Univ, MScD, 67; Univ Mich, PhD(dent mat), 71. Prof Exp: Res asst dent mat, Northwestern Univ, 66-67; res assoc, Marquette Univ, 67-68; lectr dent, Washtenaw Community Col, 68-69; res assoc dent mat, Univ Mich, 69-71, scholar, 70-71; from asst prof to assoc prof dent, Sch Dent Med, Univ Conn, 71-74; PROF DENT MAT & CHMN DEPT, COL DENT, UNIV FLA, 74- Concurrent Pos: Consult, Vet Admin Hosp, Gainesville, 74- & Williams Gold Refining Co, Inc, 74- Honors & Awards: Medal Sci Achievement, Repub Egypt, 63. Mem: Am Dent Asn; Am Soc Metals; Int Asn Dent Res; Am Asn Dent Sch. Res: Alloy systems for medical applications; structural design for dental appliances; development of implant materials; adaptation of laser beam holography to dental research. Mailing Add: Col of Dent Dept Dent Mat Univ of Fla Health Ctr Gainesville FL 32610

MOHAN, ARTHUR G, b Trenton, NJ, Mar 26, 35; m 59; c 4. ORGANIC CHEMISTRY. Educ: St Bonaventure Univ, BS, 57, MS, 59; Seton Hall Univ, PhD(org chem), 68. Prof Exp: Chemist, Nopco Chem Co, NJ, 59-65; res chemist, 66-73, SR RES CHEMIST, ORG CHEM DIV, AM CYANAMID CO, 73- Concurrent Pos: Adj assoc prof chem, Seton Hall Univ, 72- Mem: Am Chem Soc. Res: Organic reaction mechanisms; catalysis and organic process research; photochemistry and chemiluminescence. Mailing Add: 34 Windy Willow Way Somerville NJ 08876

MOHAN, RAAM RAMANUJA, b Mysore, India, Mar 13, 22; nat US; m 58; c 2. MICROBIOLOGY. Educ: Univ Bombay, BS, 45; Indian Inst Sci, Bangalore, MS, 48; Rutgers Univ, PhD, 52. Prof Exp: Asst, Indian Inst Sci, Bangalore, 45-49; asst microbiol, Rutgers Univ, 50-52; scientist, Publicker Industs, 52-54; asst, Nat Agr Col, 54- 56; scientist, Warner-Chilcott Labs Div, Warner-Hudnut, Inc, 56-58, sr scientist, Warner-Lambert Res Inst Div, 58-63, sr res assoc, 63-67; RES ASSOC, RES LAB, EXXON RES & ENG CO, 67- Mem: AAAS; Am Soc Microbiol; NY Acad Sci. Res: Chemical microbiology; microbiological transformation; petroleum microbiology; bioengineering; mode of action of anti-biotics in micro-organisms; biochemistry of bacterial cell wall synthesis. Mailing Add: Exxon Res & Eng Co Corp Res PO Box 45 Linden NJ 07036

MOHANDAS, THULUVANCHERI, b Guruvayur, India, Feb 26, 46; Can citizen. HUMAN GENETICS, CYTOGENETICS. Educ: Univ Kerala, India, BS, 66; Ind Agr Res Inst, MS, 69; McGill Univ, PhD(genetics), 72. Prof Exp: Fel human cytogenetics, Dept Pediat, Univ Man, 72-75; ASST RES PROF HUMAN CYTOGENETICS, HARBOR GEN HOSP, SCH MED, UNIV CALIF, LOS ANGELES, 75- Mem: Genetics Soc Am; Genetics Soc Can; Am Soc Human Genetics. Res: Human cytogenetics; human gene mapping using somatic cell hybrids. Mailing Add: Div of Med Genetics (E-4) Harbor Gen Hosp Torrance CA 90509

MOHANTY, SASHI B, b India, Sept 4, 32; m 57; c 2. MICROBIOLOGY, VIROLOGY. Educ: Univ Bihar, BVSc & AH, 56; Univ Md, MS, 61, PhD(microbiol), 63. Prof Exp: Vet asst surgeon, Civil Vet Dept, Govt Orissa, India, 56-60; asst microbiol, 60-63, asst prof vet sci & microbiol, 63-69, assoc prof vet sci, 69-74, PROF VET SCI, UNIV MD, 74- Concurrent Pos: NIH grant, 63-65; mem, Md State Proj Bovine Respiratory viruses, 63-; head working team on bovine & equine picorna viruses, WHO, Food & Agr Orgn, UN, 73- Mem: Am Soc Microbiol; Electron micros Soc Am; Soc Exp Biol & Med; Am Vet Med Asn. Res: Animal virus diseases; experimental infection of cattle with viruses; viral growth; electron microscopy; prevention and control of animal diseases. Mailing Add: 4306 Kenny St Beltsville MD 20705

MOHANTY, SRI GOPAL, b Soro, India, Feb 11, 33; m 63; c 2. MATHEMATICS, STATISTICS. Educ: Utkal Univ, India, BA, 51; Indian Coun Agr Res, dipl agr & animal husb statist, 57; Punjab Univ, MA, 57; Univ Alta, PhD(math statist), 61. Prof Exp: Tech asst, Ministry Food & Agr, Govt of India, 54-56; asst statistician, Indian Coun Agr Res, 58-59; teaching asst math, Univ Alta, 59-61, sessional lectr, 61-62; from asst prof to assoc prof, State Univ NY Buffalo, 62-64; assoc prof, 64-72, PROF MATH, McMASTER UNIV, 72- Concurrent Pos: Asst prof, Indian Inst Technol, New Delhi, 66-68. Mem: Inst Math Statist; Am Statist Asn; Can Math Cong; Math Asn Am; Statist Sci Asn Can. Res: Combinatorial probability; random walk; discrete probability distributions; nonparametric methods in inferences; theory of queues; fluctuation theory; enumeration of trees and certain finite structures. Mailing Add: Dept of Math McMaster Univ Hamilton ON Can

MOHAPATRA, RABINDRA NATH, b Musagadia, India, Sept 1, 44; m 69; c 1. HIGH ENERGY PHYSICS. Educ: Utkal Univ, India, BSc, 64; Delhi Univ, MSc, 66; Univ Rochester, PhD(physics), 69. Prof Exp: Res assoc physics, Inst Theoret Physics, State Univ NY Stony Brook, 69-71 & Univ Md, College Park, 71-74; ASST PROF PHYSICS, CITY COL NEW YORK, 74- Res: Gauge theories of weak, electromagnetic and strong interactions; approximate hadronic symmetries; quark models; neutrino interactions; selection rules in weak and strong interactions; field theories; radiative corrections to weak transitions; mass differences among elementary particles. Mailing Add: City Col of NY Dept of Physics 138th St & Convent Ave New York NY 10031

MOHAT, JOHN THEODORE, b El Paso, Tex, Apr 8, 24; m 45; c 2. MATHEMATICS. Educ: Tex Western Col, BA, S0; Univ Tex, PhD(math), 55. Prof Exp: Instr math, Univ Tex, 51-55 & Duke Univ, 55-57; chief math br, Math Sci Div, Off Ord Res, US Dept Army, 57-59; from asst prof to assoc prof, 59-64, dir dept, 65-69, PROF MATH, N TEX STATE UNIV, 64-, ACTG CHMN DEPT, 75- Concurrent Pos: Vis asst prof, Duke Univ, 58-59. Mem: Am Math Soc; Math Asn Am. Res: General topology; points sets and transformations. Mailing Add: Dept of Math North Tex State Univ Denton TX 76203

MOHBERG, NOEL ROSS, b Britton, SDak, Dec 16, 39; m 63; c 2. BIOSTATISTICS. Educ: NDak State Univ, BS, 61; Va Polytech Inst, MS, 62; Univ NC, PhD(biostatist), 72. Prof Exp: Statistician, NIH, 63-65 & Sandia Corp, 65-68; BIOSTATISTICIAN, UPJOHN CO, 72- Mem: Biomet Soc; Am Statist Asn. Res: Methods of analysis of categorized data; evaluation of cervical cytology screening data. Mailing Add: Upjohn Co 301 Henrietta St Kalamazoo MI 49001

MOHEREK, EMIL ANTHONY, JR, entomology, see 12th edition

MOHILNER, DAVID MORRIS, b Wichita, Kans, Jan 3, 30; m 58. ELECTROCHEMISTRY, ANALYTICAL CHEMISTRY. Educ: Univ Kans, BS, 55, PhD(chem), 61. Prof Exp: Fel chem, La State Univ, 61-62; asst prof, Univ Pittsburgh, 62-64; fel, Univ Tex, Austin, 64-65; from asst prof to assoc prof, 65-71, PROF CHEM, COLO STATE UNIV, 71- Concurrent Pos: Mem electrochem comn, Int Union Pure & Appl Chem, 71- Mem: Am Chem Soc; Electrochem Soc; Int Soc Electrochem (nat secy US, 72-76). Res: Electrochemical thermodynamics and kinetics, especially theory of electrical double layer and its influence on kinetics of electrode reactions; application of digital computers to electrochemistry; mechanisms and electrode reactions; chemical thermodynamics; bioelectrochemistry. Mailing Add: Dept of Chem Colo State Univ Ft Collins CO 80521

MOHILNER, PATRICIA R, analytical chemistry, electrochemistry, see 12th edition

MOHIUDDIN, SYED M, b Hyderabad, India, Nov 14, 34; m 61; c 3. CARDIOLOGY, INTERNAL MEDICINE. Educ: Osmania Univ, MD, 60; Creighton Univ, MS, 67; Laval Univ, DSc(med), 70. Prof Exp: Fel cardiol, Sch Med, Creighton Univ, 65-67; fel res cardiol, Laval Univ, 68-70, adj prof med, 69-70; asst prof, 70-74, ASSOC PROF MED, SCH MED, CREIGHTON UNIV, 74-, DIR CARDIAC GRAPHIC LAB, 73- Mem: Can Cardiovasc Soc; Am Fedn Clin Res; fel Am Col Cardiol; fel Am Col Physicians. Res: Clinical cardiology; coronary flow and cardiac metabolism; cardiomyopathies; graphic methods in cardiology. Mailing Add: 2305 S Tenth St Omaha NE 68108

MOHLENBROCK, ROBERT H, JR, b Murphysboro, Ill, Sept 26, 31; m 57; c 3. SYSTEMATIC BOTANY. Educ: Southern Ill Univ, BS, 53, MA, 54; Wash Univ, PhD(bot), 57. Prof Exp: From asst prof to assoc prof, 57-66, PROF BOT & CHMN DEPT, SOUTHERN ILL UNIV, CARBONDALE, 66- Mem: Am Soc Plant Taxon; Am Fern Soc; Int Asn Plant Taxon. Res: Flora of Midwest and Illinois; tropical legumes. Mailing Add: Dept of Bot Southern Ill Univ Carbondale IL 62901

MOHLENKAMP, MARVIN JOSEPH, JR, b Louisville, Ky, Apr 22, 40; m 63; c 3. FOOD CHEMISTRY. Educ: Univ Notre Dame, BS, 62; Univ Wis, MS, 65, PhD(biochem), 68. Prof Exp: CHEMIST, PROCTER & GAMBLE CO, 68- Mem: Am Chem Soc; Inst Food Technologists. Res: Chemical and organoleptic aspects of food flavors with emphasis on thermally induced flavors. Mailing Add: Procter & Gamble Co Miami Valley Labs Box 39175 Cincinnati OH 45239

MOHLER, IRVIN C, JR, b Lancaster, Pa, Nov 4, 25; m 56; c 2. BACTERIOLOGY. Educ: Franklin & Marshall Col, BS, 49; Pa State Univ, MS, 52. Prof Exp: With US govt, 52-55; asst exec dir biol, Am Inst Biol Sci, 55-59; exec off, McCollum Pratt Inst, Johns Hopkins Univ, 59-61; asst to dir, Am Type Cult Collection, 61-67; asst dir biol sci commun proj, 67-75, ASST RES PROF, DEPT MED & PUB AFFAIRS, GEORGE WASHINGTON UNIV MED CTR, 75- Concurrent Pos: Managing ed, Environ Biol & Med. Mem: Am Soc Microbiol; Am Inst Biol Sci; Am Soc Info Sci. Res: Science administration. Mailing Add: 6 Stratton Ct Potomac MD 20854

MOHLER, JAMES DAWSON, b Liberal, Mo, June 2, 26; m 51; c 3. ZOOLOGY. Educ: Univ Mo, AB, 49, AM, 50; Univ Calif, PhD(zool), 55. Prof Exp: Instr zool, Univ Mo, 50-51; asst, Univ Calif, 51-53; from asst prof to assoc prof, Ore State Univ, 55-66; assoc prof, 66-69, PROF ZOOL, UNIV IOWA, 69- Concurrent Pos: USPHS trainee, Syracuse Univ, 63-64. Mem: Genetics Soc Am. Mailing Add: Dept of Zool Univ of Iowa Iowa City IA 52240

MOHLER, ORREN (CUTHBERT), b Indianapolis, Ind, July 28, 08; m 35; c 2. ASTRONOMY. Educ: Eastern Mich Univ, AB, 29; Univ Mich, AM, 30, PhD, 33. Prof Exp: Observer, McMath-Hulbert Observ, Univ Mich, 33; astronomer & instr astron, Swarthmore Col, 33-40; asst astronomer, 40-46, asst dir, 46-56, from asst prof to assoc prof, 45-56, chmn dept & dir univ observ, 62-70, DIR, McMATH-HULBERT OBSERV, UNIV MICH, ANN ARBOR, 61-, PROF ASTRON, 56- Concurrent Pos: Bd Gov, Cranbrook Inst Sci, 60-; Fulbright res scholar, Inst Astrophys, Univ Liege, 60-61. Honors & Awards: Naval Ord Develop Award, 46. Mem: Am Astron Soc. Res: Astronomical spectroscopy and cinematography; solar astronomy; solar astronomical instruments. Mailing Add: 405 Awixa Rd Ann Arbor MI 48104

MOHLER, STANLEY ROSS, b Amarillo, Tex, Sept 30, 27; m 53; c 1. AEROSPACE MEDICINE. Educ: Univ Tex, BA & MA, 53, MD, 56. Prof Exp: Chem analyst, Longhorn Tin Smelter, Tex, 49-50; intern, USPHS, 56-57, med officer, Div Gen Med Sci, Ctr Aging Res, NIH, 57-61; dir, Civil Aeromed Res Inst, Oklahoma City, Okla, 61-65; CHIEF AEROMED APPLN DIV, FED AVIATION ADMIN, 65- Concurrent Pos: Assoc prof prev med & pub health, Univ Okla, 61-65. Honors & Awards: Meritorious Serv Award, Fed Aviation Admin, 71. Mem: AAAS; Geront Soc; AMA. Res: Gerontology; general medicine; blood clotting; aviation medicine; aircraft accident research. Mailing Add: Aeromed Appln Div Fed Aviation Admin Washington DC 20591

MOHLER, WILLIAM C, b Bridgeton, NJ, Nov 16, 27; m 56; c 3. COMPUTER SCIENCE, MEDICINE. Educ: Yale Univ, BA, 49; Columbia Univ, MD, 53. Prof Exp: Intern & resident, Presby Hosp, New York, 54-55; investr, Nat Cancer Inst, 56-65, asst dir labs & clins, 65-67, ASSOC DIR, DIV COMPUT RES & TECHNOL, NIH, 67- Mem: AAAS. Res: Management science and computing in support of biomedical research. Mailing Add: 5609 Grove Chevy Chase MD 20015

MOHLING, FRANZ, b Jersey City, NJ, July 22, 30; m 62; c 2. THEORETICAL PHYSICS. Educ: Rensselaer Polytech Inst, BS, 51; Univ Wash, PhD(physics), 58. Prof Exp: Fel physics, Columbia Univ, 58-59, res assoc, 59-60; res assoc, Cornell Univ, 60-61; from asst prof to assoc prof, 61-69, PROF PHYSICS, UNIV COLO, BOULDER, 69- Concurrent Pos: US Govt Fulbright scholar, India, 63-64. Mem: Am Phys Soc. Res: Quantum statistics; many body problem; low-temperature physics, especially liquid helium. Mailing Add: Dept of Physics & Astrophys Univ of Colo Boulder CO 80302

MOHLMAN, JOHN WILLIAM, petroleum chemistry, see 12th edition

MOHN, JAMES FREDERIC, b Buffalo, NY, Apr 11, 22; m 45; c 4. IMMUNOLOGY, MEDICAL BACTERIOLOGY. Educ: Univ Buffalo, MD, 44. Prof Exp: From instr to assoc prof, 45-55, PROF BACT & IMMUNOL, SCH MED, STATE UNIV NY BUFFALO, 55- Concurrent Pos: Buswell fel, 59-; asst bacteriologist & serologist, Niagara Sanitorium, NY, 46-48, bacteriologist & dir lab, 48-53; asst bacteriologist, serologist & asst dir blood bank, Buffalo Gen Hosp, 47-58, assoc bacteriologist, serologist & assoc dir blood bank, 58-; consult, Blood Bank, Deaconess Hosp, Buffalo, 57- & Walter Reed Army Inst Res, 58-; mem, Subcomt Transfusion Probs, Nat Acad Sci-Nat Res Coun, 58- Mem: AAAS; Soc Exp Biol & Med; Am Soc Hemat; fel Inst Soc Hemat; Am Asn Immunologists. Res: Investigation of Rh substances and antibodies; blood group specific substances; characterization of blood group isoagglutins; immunologic aspects of hemolytic agents anemia; immunohematologic blood transfusion studies. Mailing Add: Dept of Microbiol State Univ of NY Sch of Med Buffalo NY 14214

MOHN, MELVIN P, b Cleveland, Ohio, June 19, 26; m 52; c 2. ANATOMY, HISTOLOGY. Educ: Marietta Col, AB, 50; Brown Univ, ScM, 52, PhD(biol), 55. Prof Exp: From instr to asst prof anat, State Univ NY Downstate Med Ctr, 55-63; from asst prof to assoc prof, 63-71, PROF ANAT, UNIV KANS MED CTR, KANSAS CITY, 71- Concurrent Pos: Vis prof, Nat Med AV Ctr, 72. Mem: Fel AAAS; Am Asn Anat; Am Soc Zool; Am Inst Biol Sci; Electron Micros Soc Am. Res: Structure, function and embryology of skin and its appendages, particularly hair, nail and ceruminous glands; histochemistry; electron microscopy; comparative anatomy; effects of nutritional deficiencies; endocrine imbalances; irradiations; transplantations; wound healing and circulatory changes. Mailing Add: Dept of Anat Univ of Kans Med Ctr Kansas City KS 66103

MOHNEN, VOLKER A, b Stuttgard, WGer, Mar 11, 37; m 63; c 2. ATMOSPHERIC SCIENCES, PHYSICS. Educ: Univ Karlsruhe, BS, 59; Univ Munich, MS, 63, PhD(physics, meteorol astrophys), 66. Prof Exp: Res assoc, Univ Munich, 62-67; sr res assoc, Atmospheric Sci Res Ctr, 67-75, assoc dir, 72-75, ASSOC PROF ATMOSPHERIC SCI, STATE UNIV NY ALBANY, 67-, DIR, ATMOSPHERIC SCI RES CTR, 75- Mem: Fel AAAS; Am Meteorol Soc; Am Geophys Union; Air Pollution Control Asn; Am Phys Soc. Res: Air pollution; aerosol physics; solar energy. Mailing Add: Atmospheric Sci Res Ctr State Univ NY E S 319 Albany NY 12222

MOHOLY-NAGY, HATTULA, b Berlin, Ger, Oct 11, 33; US citizen; m 65; c 2. ANTHROPOLOGY, ARCHAEOLOGY. Educ: Univ Mich, AB, 55; Univ Chicago, AM, 58. Prof Exp: Field lab head, Joint Casas Grandes Exped, 58-59; vol, Am Mus Natural Hist, 59-60; from mem staff to head field lab, 60-65, ASSOC, UNIV MUS TIKAL PROJ, UNIV PA, 60- Concurrent Pos: Asst & instr, Ethnol Sem, Univ Zurich, 71- Mem: AAAS; Am Anthrop Asn; Soc Am Archaeol. Res: Old world early village-farming communities; Mesoamerican archaeology; museology; primitive technology; prehistoric trade; natural science identifications of archaeological materials. Mailing Add: Sem für Ethnolgie Univ Zürich Rämistrasse 44 CH-8001 Zurich Switzerland

MOHOS, STEVEN CHARLES, b Sopron, Hungary, Jan 20, 18; div; c 2. PATHOLOGY, IMMUNOLOGY. Educ: Pazmany Peter Univ, Budapest, MD, 41; Am Bd Path, dipl, 56. Prof Exp: Res fel exp path, Med Sch, Pazmany Peter Univ, 41-43, asst prof path, 42-43; med staff chief serv, Hosp of Int Red Cross, Ger, 46-50; resident path, Polyclin Hosp, Harrisburg, Pa, 51-52, 53-54, intern, 52-53; instr & res assoc, Med Col, Cornell Univ, 54-56; asst prof path, State Univ NY Downstate Med Ctr, 56-63; ASSOC PROF PATH, NEW YORK MED COL, 63- Concurrent Pos: Asst attend pathologist, New York Hosp, 54-56; Life Ins Res Fund, Off Naval Res & NIH res grants. Honors & Awards: Chinoin Award, 42. Mem: AMA; Col Am Path; Am Soc Exp Path; NY Acad Sci; Int Acad Path. Res: Tissue immunology; filtration membrane; immunological aspects of cancer research; transplantation immunity; experimental nephritis; in vivo effects of complement and application of electron microscopy to these problems. Mailing Add: Dept of Path New York Med Col Valhalla NY 10595

MOHR, HUBERT CHARLES, b Cleveland, Ohio, Jan 30, 15; m 41; c 4. HORTICULTURE. Educ: Ohio State Univ, BS, 38, MS, 39; Agr & Mech Col Tex, PhD(hort), 55. Prof Exp: Instr bot, Agr & Mech Col Tex, 46-47, asst prof hort, 47-54, from assoc prof to prof, 55-62; prof & chmn dept, 62-73, RES PROF HORT, UNIV KY, 73- Mem: Am Soc Hort Sci; Int Soc Hort Sci. Res: Vegetable breeding; plant breeding; development of dwarf varieties of the cultivated cucurbitaceae. Mailing Add: Dept of Hort Univ of Ky Lexington KY 40506

MOHR, JAY PRESTON, b Philadelphia, Pa, Mar 5, 37; m 62; c 2. NEUROLOGY. Educ: Haverford Col, AB, 58; Univ Va, MS & MD, 63. Prof Exp: USPHS fel pharmacol, Univ Va, 60-63; from intern to asst resident internal med, Mary Imogene Bassett Hosp, Cooperstown, NY, 63-65; asst resident neurol, NY Neurol Inst, Columbia-Presby Med Ctr, New York, 65-66; fel neurol, Mass Gen Hosp, Boston, 66-68, Nat Inst Neurol Dis & Stroke fel, 67-69; instr, Univ Md Hosp, Baltimore, 69-71; ASST PROF NEUROL, HARVARD MED SCH, 72- Concurrent Pos: Asst neurologist, Johns Hopkins Hosp, 69-71 & Mass Gen Hosp, 72- Mem: Acad Aphasia; Am Acad Neurol. Res: Behavioral neurology; cerebrovascular disease; aphasiology. Mailing Add: Dept of Neurol Mass Gen Hosp Boston MA 02114

MOHR, JOHN LUTHER, b Reading, Pa, Dec 1, 11; m 39; c 2. MARINE BIOLOGY, PROTOZOOLOGY. Educ: Bucknell Univ, AB, 33; Univ Calif, PhD(zool), 39. Prof Exp: Asst zool, Univ Calif, 34-38, technician, 38-42; res assoc, Pac Islands, Stanford Univ, 42-44; asst prof & res assoc, Allan Hancock Found, 44-47, vis asst prof zool, 47-48, from asst prof to assoc prof, 48-54, assoc prof biol, 54-57, head dept, 59-62, PROF BIOL, UNIV SOUTHERN CALIF, 57- Concurrent Pos: Vis prof, Univ Wash Friday Harbor Labs, 56-57; Guggenheim fel, Plymouth Lab, Marine Biol Asn UK; 57-58; chief marine zool group, Antarctic Ship Eltanin, 62 & 65; mem gen invert comt, Smithsonian Oceanog Sorting Ctr, 63-66; trustee, Biol Stain Comn, 72. Mem: Am Soc Parasitologists; Am Soc Zoologists; Ecol Soc Am; Soc Syst Zoologists; Marine Biol Asn UK. Res: Protozoology and parasitology, especially opalinida, chonotrichs and ciliates of elephants; biological consequences of man on sea; biology of polar seas; philosophy and folkways of biologists; biological stains. Mailing Add: Dept of Biol Sci Univ of Southern Calif Los Angeles CA 90007

MOHR, PAUL HERMAN, inorganic chemistry, see 12th edition

MOHR, SCOTT CHALMERS, b Jamestown, NY, Aug 30, 40; m 64; c 1. BIOCHEMISTRY. Educ: Williams Col, BA, 62; Harvard Univ, MA, 66, PhD(chem), 68. Prof Exp: NIH fel, Cornell Univ, 68-69; asst prof, 69-75, ASSOC PROF CHEM, BOSTON UNIV, 75- Mem: AAAS; Am Chem Soc. Res: Fast kinetics in biochemical systems; allosteric proteins; transfer RNA; nucleic acid-protein interactions; protein synthesis elongation factor I (Tu); compact states of nucleic acids; carcinogen-nucleic acid interactions. Mailing Add: Dept of Chem Boston Univ 675 Commwealth Ave Boston MA 02215

MOHR, WILLARD PHILLIP, b Morden, Man, Nov 14, 28; m 56; c 3. FOOD SCIENCE. Educ: Univ Man, BSA, 49; Univ BC, MSA, 51; Univ Nottingham, PhD(food sci, cell biol), 68. Prof Exp: Food technologist, Can Packers Ltd, Ont, 51-52 & Bulman's Ltd, BC, 52-55; SCIENTIST, RES BR, CAN DEPT AGR, 55- Mem: Agr Inst Can; Can Inst Food Sci & Technol. Res: Color and texture of fruit and vegetable crops and their processed products using ultrastructural, biochemical and food preservation techniques. Mailing Add: RR 3 Brighton ON Can

MOHRBACHER, RICHARD JOSEPH, organic chemistry, medicinal chemistry, see 12th edition

MOHRENWEISER, HARVEY WALTER, b Mora, Minn, Oct 12, 40; m 61; c 2. BIOCHEMISTRY. Educ: Univ Minn, BS, 62, MS, 66; Mich State Univ, PhD(biochem), 70. Prof Exp: NIH fel, McArdle Lab, Univ Wis, 70-73; RES CHEMIST MUTAGENESIS, NAT CTR TOXICOL RES, FOOD & DRUG ADMIN, DEPT HEALTH, EDUC & WELFARE, 73- Concurrent Pos: Asst prof biochem, Univ Ark, 73- Res: Biochemical mechanisms of mutagenesis and carcinogenesis. Mailing Add: Div of Mutagenic Res Nat Ctr Toxicol Res Jefferson AR 72079

MOHRI, HITOSHI, b Taihoku City, Japan, July 28, 30; m 63; c 3. SURGERY. Educ: Tohoku Univ, Japan, BA, 51, MD, 55, DrMedSci(surg), 62. Prof Exp: Asst surg, Sch Med, Tohoku Univ, Japan, 60-63; chief thoracic surg, Katta Gen Hosp, Shiroishi, Japan, 63-64; from vis scientist to asst prof, 64-71, ASSOC PROF SURG, SCH MED, UNIV WASH, 71- Concurrent Pos: NIH fel, Sch Med, Univ Wash, 64-66, Wash State Asn grants, 66-70; estab investr, Am Heart Asn, 70-75. Mem: AMA; Japanese Asn Thoracic Surg; Japanese Surg Soc; Asn Acad Surgeons. Res: Cardiac surgery. Mailing Add: Dept of Surg Univ of Wash Seattle WA 98195

MOHRIG, JERRY R, b Grand Rapids, Mich, Feb 24, 36; m 60; c 3. ORGANIC CHEMISTRY. Educ: Univ Mich, BS, 57; Univ Colo, PhD(chem), 63. Prof Exp: Asst prof, Hope Col, 64-67; from asst prof to assoc prof, 67-75, PROF CHEM, CARLETON COL, 75- Res: Organic reaction mechanisms; reactions of alkanediazonium ions; enzymic catalysis. Mailing Add: Dept of Chem Carleton Col Northfield MN 55057

MOHRMAN, HAROLD W, b Quincy, Ill, Oct 1, 17; m 39; c 4. CHEMISTRY. Educ: Univ Ill, BS, 39. Prof Exp: Res chemist, 39, group leader, 40-46, asst res dir, 46-50, dir res, Plastics Div, 50-59, assoc interests, 59-60 & overseas div, 60-63, dir polymer sect & overseas res, 68-68, SECT DIR, CORP RES DEPT, MONSANTO CO, 68- Mem: AAAS; Am Chem Soc. Res: Condensation resin and vinyl polymers; manufacture of phenolic and melaminealdehyde condensation products; polyelectrolytes and research administration. Mailing Add: Corp Res Dept Monsanto Co 800 N Lindbergh Blvd St Louis MO 63166

MOHRMANN, WILBURN GLENN, physical chemistry, see 12th edition

MOHS, FREDERIC EDWARD, b Burlington, Wis, Mar 1, 10; m 34; c 3. SURGERY. Educ: Univ Wis, BS, 32, MD, 34. Prof Exp: Bowman cancer res fel, 35-38, assoc cancer res & instr surg, 39-42, from asst prof to assoc prof, 42-67, CLIN PROF CHEMOSURG, MED SCH, UNIV WIS-MADISON, 67-, DIR CHEMOSURG CLIN, UNIV HOSPS, 48- Mem: AAAS; AMA; Am Asn Cancer Res. Res: Chemosurgery for the microscopically controlled excision of cancer of the skin, lip, parotid gland and other external structures. Mailing Add: 1300 University Ave Madison WI 53706

MOHUN, WILLIAM ARTHUR, b Toronto, Ont, Nov 11, 13; m 38; c 2. CHEMISTRY. Educ: Univ Toronto, BASc, 35, MASc, 41, ChemE, 49. Prof Exp: Control chemist, Can Industs, Ltd, 35-36; demonstr appl physics & asst photog sensitometry, 37-41; jr res physicist, Optics Sect, Nat Res Coun Can, 41-42, assoc res chemist, Indust Org Sect, 42-44, design engr, Chem Eng Sect, 44-46; from asst to vpres res & develop, Standard Chem Co, Ltd, 47-48; works mgr, Goderich Salt Co, Ltd, 48, from asst to mgr indust, Chem Div, 49-50, mgr res & develop div, 51-52; dir res, Gen Abrasive Co, Inc, NY & vpres opers, Nilok Chem, Inc, 52-60; indust consult, 61-68; TEACHING MASTER MATH & CHEM, SENECA COL APPL ARTS & TECHNOL, 68- Honors & Awards: Blackall Award, Am Soc Mech Eng, 62. Res: Commercial chemical development from research laboratory through all phases to commercial operations; function of products; economic theory. Mailing Add: 1750 Finch Ave E Willowdale ON Can

MOINAT, ARTHUR DAVID, b Pueblo, Colo, Oct 24, 01; m 45; c 3. PLANT PHYSIOLOGY. Educ: Colo State Univ, BS, 24; Ore State Univ, MS, 25; Univ Ill, PhD(bot), 30. Prof Exp: Asst hort, Colo State Univ, 25-27; asst bot, Univ Ill, 27-29, instr, 30-32; from instr to prof biol, Ft Lewis Br, Colo State Univ, 34-50; from asst prof to prof, 50-67, EMER PROF BOT, UNIV NORTHERN COLO, 67- Concurrent Pos: Prof bot, Aims Col, 68-71. Mem: Fel AAAS; Bot Soc Am; Am Soc Plant Physiol; Am Inst Biol Sci. Res: Water relations; plant nutrition; respiration; photosynthesis; rhythmic behavior; control of dwarf mistletoe of pine. Mailing Add: 1816 13th Ave Greeley CO 80631

MOIOLA, RICHARD JAMES, b Reno, Nev, Oct 5, 37; m 70; c 2. SEDIMENTOLOGY. Educ: Univ Calif, Berkeley, AB, 59, PhD(geol), 69. Prof Exp: Res asst geol, Univ Calif, Berkeley, 61-63; res geologist, Field Res Lab, Mobil Oil Corp, Tex, 63-66, sr res geologist, Mobil Res & Develop Corp, 67-74, RES ASSOC, MOBIL RES & DEVELOP CORP, 74- Concurrent Pos: Lectr, Univ Tex, Dallas, 75. Mem: Fel Geol Soc Am; Soc Econ Paleontologists & Mineralogists. Res: Sedimentology of modern and ancient clastic sediments; physical stratigraphy; zeolitic diagenesis. Mailing Add: Mobil Res & Develop Corp Res Dept PO Box 900 Dallas TX 75221

MOIR, DAVID ROSS, b Winnipeg, Man, Sept 25, 18; m 44; c 5. BOTANY. Educ: Univ Man, BS, 40, MS, 42; Univ Minn, PhD, 58. Prof Exp: Asst plant physiol, Univ Man, 42-44; biologist, Govt Man, 44-46; lectr bot, Univ Man, 46-49; instr, Univ Minn, 49-52; from asst prof to prof, NDak State Univ, 52-66; prof & head dept, Brandon Univ, 66-72, dean sci, 67-72; PROF BOT, UNIV MASS, AMHERST, 72- Concurrent Pos: Regional consult, Biol Sci Curric Study, 63-66. Mem: Arctic Inst NAm. Res: Phytogeography and floristics of Hudson Bay area; arctic and sub-arctic floras; systematic botany; plant geography and anatomy. Mailing Add: Dept of Bot Univ of Mass Amherst MA 01002

MOIR, ROBERT YOUNG, b Estevan, Sask, Oct 30, 20; m 46; c 3. ORGANIC CHEMISTRY. Educ: Queen's Univ, Ont, BA, 41, MA, 42; McGill Univ, PhD(org chem), 46. Prof Exp: Jr chemist, Inspection Bd, Can, 41; res chemist, Indust, Org Chem, Res Labs, Dom Rubber Co, Ltd, 43-44, 46-49; from asst prof to assoc prof, 49-64, PROF CHEM, QUEEN'S UNIV, ONT, 64- Mem: Am Chem Soc; Chem Inst Can. Res: Steric effects in cyclohexanes and in diphenyl ethers; general synthesis. Mailing Add: Dept of Chem Queen's Univ Kingston ON Can

MOISE, EDWIN EVARISTE, b New Orleans, La, Dec 22, 18; m 42; c 2.

MATHEMATICS. Educ: Tulane Univ La, BA, 40; Univ Tex, PhD(math), 47. Hon Degrees: MA, Harvard Univ, 60. Prof Exp: Instr math, Univ Tex, 46-47; from instr to prof, Univ Mich, 47-60; prof math & educ, Harvard Univ, 60-71; DISTINGUISHED PROF MATH, QUEENS COL NY, 71- Concurrent Pos: Nat Res Coun fel, Inst Advan Study, 49-50, asst, 50-51, Guggenheim fel, 56-57; vis prof, Res Ctr, Nat Polytech Inst, Mex, 70-71. Mem: Fel Am Acad Arts & Sci; Math Asn Am (pres, 67-68); Am Math Soc (vpres, 73-74). Res: Topology. Mailing Add: Dept of Math Queens Col Flushing NY 11367

MOISE, NORTON L, physics, mathematics, see 12th edition

MOISEYEV, ALEXIS N, b Paris, France, June 27, 32; US citizen. GEOCHEMISTRY, GEOLOGY. Educ: Sorbonne, Lic es Sci, 55, Univ Paris, Dr, 59; Stanford Univ, PhD(geochem), 66. Prof Exp: Mining geologist, Compagnie Royale Asturienne des Mines, 56-58 & 60-61; lectr geol, Univ Calif, Davis, 66-67; asst prof, San Jose State Col, 67-68; asst prof, 68-71, ASSOC PROF GEOL, CALIF STATE COL, HAYWARD, 71- Mem: Am Geophys Union; Geol Soc Am. Res: Geochemistry of hydrothermal processes and low temperature sedimentary deposits; plate tectonics. Mailing Add: Dept of Earth Sci Calif State Col Hayward CA 94542

MOISSIDES-HINES, LYDIA ELIZABETH, b Newton, Mass, Mar 11, 48; m 74. MEDICINAL CHEMISTRY, INDUSTRIAL CHEMISTRY. Educ: Aurora Col, BS, 67; Univ Ill, Urbana-Champaign, MS, 69, PhD(org chem), 71. Prof Exp: Sr scientist med chem, Mead Johnson & Co, 71-75; MEM STAFF PATENT LIAISON DEPT, UPJOHN CO, 75- Mem: AAAS; Am Chem Soc; Metric Asn; Sigma Xi. Mailing Add: Patent Liaison Dept 7264 Upjohn Co Kalamazoo MI 49001

MOJONNIER, LOUISE, nutrition, see 12th edition

MOK, GERALD CHI-HUNG, applied mechanics, applied mathematics, see 12th edition

MOKE, CHARLES BURDETTE, b Pittsburgh, Pa, Mar 13, 10; m 38; c 2. GEOLOGY. Educ: Col Wooster, BA, 31; Harvard Univ, MA, 35, PhD(geol), 48. Prof Exp: From instr to prof geol, 36-72, chmn dept geol & geog, 53-72, EMER PROF GEOL, COL WOOSTER, 72- Mem: Geol Soc Am; Am Asn Petrol Geol; Nat Asn Geol Teachers. Mailing Add: 331 Dorchester Rd Wooster OH 44691

MOKLER, BRIAN VICTOR, b Los Angeles, Calif, May 1, 36; m 68; c 2. INHALATION TOXICOLOGY, AIR POLLUTION. Educ: Pomona Col, BA, 58; Mass Inst Technol, SM, 60; Harvard Univ, SM, 68, ScD(environ health sci), 73; Am Bd Indust Hyg, dipl & cert air pollution, 75. Prof Exp: Chem engr, Arthur D Little Inc, 60-67; teaching asst environ health sci, Sch Pub Health, Harvard Univ, 69-70, teaching fel, 70-71; consult environ sci & air pollution, 73; SR STAFF AEROSOL SCI, LOVELACE BIOMED & ENVIRON RES INST, 74- Mem: Am Chem Soc; Am Indust Hyg Asn; Sigma Xi; Am Acad Indust Hyg. Res: Chemistry and physics of disperse systems; aerosol generation and characterization methodology; condensation aerosols; toxicology of inhaled materials; air pollution control; industrial hygiene. Mailing Add: Lovelace Biomed & Environ Res Inst Box 5890 Albuquerque NM 87115

MOKLER, CORWIN MORRIS, b Forsythe, Ill, Dec 10, 25; m 50; c 2. CARDIOVASCULAR PHYSIOLOGY. Educ: Colo Col, BA, 50; Univ Nev, MS, 52; Univ Ill, PhD(physiol), 58. Prof Exp: Technician virol, Harvard Med Sch, 52-54; biologist, NIH, 54; asst anat, Univ Ill, 54-58; investr cardiac pharmacol & physiol, G D Searle & Co, Ill, 58-61; asst prof pharmacol, Univ Fla, 61-67; ASSOC PROF PHARMACOL, SCH PHARM, UNIV GA, 67- Mem: Am Soc Pharmacol & Exp Therapeut; Sigma Xi. Res: Cardiovascular physiology and pharmacology; cardiac excitability; anti-arrhythmic drugs. Mailing Add: Dept of Pharmacol Univ of Ga Sch of Pharm Athens GA 30602

MOKMA, DELBERT LEWIS, b Holland, Mich, Sept 21, 42. SOIL SCIENCE. Educ: Mich State Univ, BS, 64, MS, 66; Univ Wis-Madison, PhD(soil PhD(soil sci), 71. Prof Exp: From res asst to res assoc soil sci, 71-75, ASST PROF CROP & SOIL SCI, MICH STATE UNIV, 75- Mem: Soil Sci Soc Am; Am Soc Agron; Soil Conserve Soc Am. Res: Soil classification and mapping; use of remote sensing and soil surveys in land use planning; soil mineralogy. Mailing Add: Dept of Crop & Soil Sci Mich State Univ East Lansing MI 48824

MOKOTOFF, MICHAEL, b Brooklyn, NY, Jan 23, 39; m 67; c 3. ORGANIC CHEMISTRY, MEDICINAL CHEMISTRY. Educ: Columbia Univ, BS, 60; Univ Wis, MS, 63, PhD(med chem), 66. Prof Exp: NIH staff fel med chem, Lab Chem, Nat Inst Arthritis & Metab Dis, 66-68; asst prof, 68-72, ASSOC PROF MED CHEM, SCH PHARM, UNIV PITTSBURGH, 72- Concurrent Pos: Health Res Serv Found grant, 69-70; Nat Cancer Inst grant, 70-73; res grant, Am Cancer Soc, 73-75. Mem: Am Chem Soc; Am Pharmaceut Asn; Int Narcotic Res Club. Res: Azabicyclo chemistry; potential inhibitors of asparagine biosynthesis; opiate receptor purification; potential anti-sickly agents. Mailing Add: Sch of Pharm Univ of Pittsburgh Pittsburgh PA 15261

MOKRASCH, LEWIS CARL, b St Paul, Minn, May 9, 30. BIOCHEMISTRY. Educ: Col St Thomas, BS, 52; Univ Wis, PhD(physiol chem), 55; Am Bd Clin Chem, cert. Prof Exp: Res assoc psychiat & neurochem, Sch Med, La State Univ, 56-57; sr res assoc biochem & instr med, Univ Kans Med Ctr, Kansas City, 57-62, dir neurochem lab & assoc med, 59-62; from assoc to asst prof biol chem, Harvard Med Sch, 60-71; ASSOC PROF BIOCHEM, SCH DENT, LA STATE UNIV MED CTR, NEW ORLEANS, 71- Concurrent Pos: Nat Inst Neurol Dis & Blindness spec fel, 60-62; assoc biochemist, McLean Hosp, Mass, 60-71; resident scientist, Neurosci Res Prog, Mass Inst Technol, 70-71. Mem: Am Chem Soc; Am Asn Biol Chem; Am Inst Chem; Am Soc Neurochem; Int Soc Neurochem. Res: Neurochemistry; clinical chemistry; metabolism and chemistry of brain proteins in relation to development; genetic abnormalities and disease; biochemical methods of analysis. Mailing Add: Dept of Biochem La State Univ Med Ctr New Orleans LA 70112

MOLANDER, CHARLES W, microbiology, see 12th edition

MOLAU, GUNTHER ERICH, b Leipzig, Ger, Oct 15, 32; m 58; c 3. POLYMER CHEMISTRY. Educ: Carolo-Wilhelmina Inst Technol, BS, 56, MS, 59, PhD(chem), 61. Prof Exp: Chemist, Dow Chem Co, 61-63, res chemist, 63-65, sr res chemist, 65-72, ASSOC SCIENTIST, DOW CHEM USA, 69- Mem: Am Chem Soc. Res: Organic polymer chemistry; membranes; colloidal and heterogeneous polymers. Mailing Add: Dow Chem Co 2800 Mitchell Dr Walnut Creek CA 94598

MOLD, JAMES DAVIS, b Carlton, Minn, Sept 26, 20; m 46; c 3. BIO-ORGANIC CHEMISTRY. Educ: Univ Minn, BCh, 42; Northwestern Univ, MS, 44, PhD(org chem), 47. Prof Exp: Asst, Northwestern Univ, 42-44, asst chem, 44-46, res assoc, 46-47; biochemist, Parke, Davis & Co, 47-49; org chemist, Allied Sci Div, Biol Lab, US Army Chem Corps, 49-55; chief org chem res, Liggett & Myers Tobacco Co, 55-64, ASST DIR RES, LIGGETT & MYERS INC, 64- Mem: AAAS; Am Chem Soc; Phytochem Soc NAm. Res: Isolation of natural organic substances; synthesis of organic chemicals; degradation and structure proof of natural organic products; chemistry of tobacco and smoke. Mailing Add: Liggett & Myers Inc Res Dept W Main & Fuller Sts Durham NC 27702

MOLDAUER, PETER ARNOLD, b Vienna, Austria, June 18, 23; US citizen; m 49; c 3. NUCLEAR PHYSICS, QUANTUM MECHANICS. Educ: Northeastern Univ, BS, 44; Harvard Univ, MA, 47; Univ Mich, PhD(physics), 56. Prof Exp: Mem res staff, Gordon McKay Lab, Harvard Univ, 47-48 & Instrumentation Lab, Mass Inst Technol, 48-50; instr physics, Univ Conn, 55-57; assoc physicist, 57-67, SR PHYSICIST, ARGONNE NAT LAB, 67- Concurrent Pos: Res assoc, Lab Nuclear Sci, Mass Inst Technol, 64-65. Mem: Am Phys Soc. Res: Relativistic wave equations; theory of measurement; nuclear structure theory; nuclear reaction theory; neutron physics. Mailing Add: Argonne Nat Lab 9700 S Cass Ave Argonne IL 60439

MOLDAVE, KIVIE, b Kiev, Russia, Oct 22, 23; nat US; m 49; c 1. BIOCHEMISTRY. Educ: Univ Calif, AB, 47; Univ Southern Calif, MS, 50, PhD(biochem), 52. Prof Exp: USPHS res fels, Univ Wis, 52-53 & Fac Sci, Univ Paris, 53-54; from asst prof to prof biochem, Sch Med, Tufts Univ, 54-66; prof & chmn dept, Sch Med, Univ Pittsburgh, 66-70; PROF BIOCHEM & CHMN DEPT BIOL CHEM, UNIV CALIF, IRVINE-CALIF COL MED, 70- Concurrent Pos: Mem physiol chem study sect, NIH, 67-71 & res serv merit rev bd basic sci, Vet Admin, 72- Mem: Am Soc Biol Chem; Am Chem Soc. Res: Nucleic acid and protein biosynthesis. Mailing Add: Dept of Biol Chem Univ of Calif-Calif Col of Med Irvine CA 92664

MOLDAWER, MARC, b Philadelphia, Pa, June 4, 22; m 63; c 3. MEDICINE, ENDOCRINOLOGY. Educ: Univ Pa, 39-42; Harvard Univ, MD, 50. Prof Exp: Intern med, Presby Hosp, Columbia Univ, 50-51, resident, 51-52; clin res fel endocrinol, Mass Gen Hosp, 52-55; res fel biochem, Cambridge Univ, 55-56; res fel endocrinol, Harvard Med Sch, 56-57; asst prof, 57-63, ASSOC PROF MED, BAYLOR COL MED, 63-; CHIEF ENDOCRINE SECT, METHODIST HOSP, HOUSTON, 63- Concurrent Pos: Consult, Tex Inst Rehab & Res, Ben Taub Gen, Methodist & Vet Admin Hosps, Houston, 58- Mem: AAAS; Endocrine Soc; Am Fedn Clin Res. Res: Gynecomastia and estrogen metabolism in the male; human growth hormone; immunology and physiology. Mailing Add: Methodist Hosp Houston TX 77025

MOLDENHAUER, RALPH ROY, b Detroit, Mich, Apr 8, 35; m 59; c 2. PHYSIOLOGY, ECOLOGY. Educ: Mich State Univ, BS, 60; Ore State Univ, MS, 65, PhD(zool), 69. Prof Exp: ASSOC PROF BIOL, SAM HOUSTON STATE UNIV, 68- Concurrent Pos: Soc Sigma Xi res grant-in-aid, Sam Houston State Univ, 69-71, Am Mus Natural Hist Frank M Chapman Mem Fund grant, 70-72 & Am Philos Soc res grant, 71-72. Mem: Am Inst Biol Sci; Am Ornith Union; Cooper Ornith Soc; Ecol Soc Am. Res: Physiology and ecology of birds; salt and water balance; thermoregulation; behavioral adjustments in stressful environments. Mailing Add: Dept of Biol Sam Houston State Univ Huntsville TX 77340

MOLDENHAUER, WILLIAM C, b New Underwood, SDak, Oct 27, 23; m 47; c 5. SOIL CONSERVATION. Educ: SDak State Col, BS, 49; Univ Wis, MS, 51, PhD, 56. Prof Exp: Soil surveyor, Exp Sta, SDak State Col, 49-53, asst prof, 53-54; assoc prof, Exp Sta, Iowa State Univ, 56-66, prof, 66-72; prof, Univ Minn, St Paul, 72-75; PROF AGRON, PURDUE UNIV, 75- Concurrent Pos: Soil scientist, Agr Res Serv, USDA, 54- Mem: Am Soc Agron; Soil Sci Soc Am; fel Soil Conserv Soc Am(2nd vpres, 76); Int Soc Soil Sci. Res: Water runoff and erosion of soil by water; tillage; soil management. Mailing Add: Dept of Agron Purdue Univ Lafayette IN 47907

MOLDENKE, HAROLD NORMAN, natural science, see 12th edition

MOLDOVANU, GRAZIELLA, b Bucharest, Romania, May 28, 20; US citizen. HEMATOLOGY, ONCOLOGY. Educ: Univ Bucharest, MD, 46; Educ Coun for Med Grad, cert, 61. Prof Exp: Resident asst prof med & hemat, Inst Physiol, Bucharest, Romania, 46-57; res assoc transplantation, Mary Imogene Bassett Hosp, Cooperstown, NY, 59-60; res assoc lymphoma, Sloan-Kettering Inst Cancer Res, 63-69; ASST PROF MICROBIOL, FAC MED, LAVAL UNIV, 69-, CHIEF VIROL SERV, MED SCH, 73- Concurrent Pos: Rockefeller res fel hemat, Innsbruck Univ, 57-59; res fel clin chemother, Sloan-Kettering Inst Cancer Res, 60-63. Mem: Transplantation Soc; NY Acad Sci; Am Asn Cancer Res; Tissue Cult Asn; Europ Soc Hemat. Res: Virology; transplantation, particularly tumor transplantation; immunology; viral oncology. Mailing Add: Virol Serv Laval Univ Quebec PQ Can

MOLDOVER, MICHAEL ROBERT, b New York, NY, July 19, 40. PHYSICS. Educ: Rensselaer Polytech Inst, BS, 61; Stanford Univ, MS, 62, PhD(physics), 66. Prof Exp: Res assoc physics, Stanford Univ, 66; asst prof, Univ Minn, Minneapolis, 67-72; PHYSICIST, NAT BUR STANDARDS, 72- Res: Thermodynamic properties of liquids and solids, especially near phase transitions; low temperature physics. Mailing Add: Heat Div Nat Bur of Standards Washington DC 20234

MOLE, PAUL ANGELO, b Jamestown, NY, Mar 13, 38; m 57; c3. MUSCULAR PHYSIOLOGY, EXERCISE PHYSIOLOGY. Educ: Univ Ill, Urbana, BS, 60, MS, 62, PhD(physiol), 69. Prof Exp: NIH trainee nutrit, Sch Med, Wash Univ, 69-71; asst prof phys educ, Temple Univ, 71-74; ASST PROF PHYSIOL, LA STATE UNIV MED CTR, 74- Mem: AAAS; NY Acad Sci; fel Am col Sports Med. Res: Biochemical adaptations to exercise and nutrition; biochemical and contractile properties of skeletal and heart muscle. Mailing Add: Dept of Physiol La State Univ Med Ctr New Orleans LA 70112

MOLER, CLEVE B, b Salt Lake City, Utah, Aug 17, 39; m 60; c 2. MATHEMATICS. Educ: Calif Inst Technol, BS, 61; Stanford Univ, PhD(math), 65. Prof Exp: Instr comput sci, Stanford Univ, 65; from asst prof to assoc prof math, Univ Mich, Ann Arbor, 66-72; assoc prof, 72-74, PROF MATH, UNIV N MEX, 74- Concurrent Pos: Off Naval Res assoc, Swiss Fed Inst Technol, 65-66; vis assoc prof, Stanford Univ, 70-71. Mem: Am Math Soc; Asn Comput Mach; Soc Indust & Appl Math. Res: Numerical analysis; computer science; linear algebra; partial differential equations. Mailing Add: Dept of Math Univ of NMex Albuquerque NM 87106

MOLHOLT, BRUCE, microbial genetics, see 12th edition

MOLINARI, PIETRO FILIPPO, b Mestre-Venice, Italy, Sept 9, 23; US citizen; m 56; c 2. ENDOCRINOLOGY, HEMATOLOGY. Educ: Univ Milan, DVM, 52, PhD(clin path), 60. Prof Exp: From asst prof clin vet med to assoc prof clin methodology, Univ Milan, 52-61, dir res lab, 61-64; endocrinologist, Mason Res Inst, 64-67, sr investr endocrinol, 67-71; ASST DIR HEMAT, ST VINCENT HOSP, 71- Concurrent Pos: Ital Res Coun fel physiol, Vet Sch, Cornell Univ, 61; asst prof med, Med Sch, Univ Mass. Mem: AAAS; NY Acad Sci; Endocrine Soc. Res: Study of erythropoietic activity of hormones, particularly steroid hormones in laboratory animals and human cells in vitro. Mailing Add: St Vincent Hosp Dept of Hemat 25 Winthrop St Worcester MA 01610

MOLINARY, SAMUEL VICTOR, b Dante, Va, Jan 15, 39; m 59; c 3. BIOCHEMISTRY, GENETICS. Educ: Univ Va, BA, 61; Med Col Va, PhD(biochem genetics), 68. Prof Exp: Teacher high sch, Va, 61-62; res assoc biochem, Med Units, 68-69, asst prof biochem & sr staff biochemist, Child Develop Ctr, 69-72, ASSOC PROF BIOCHEM & CHILD DEVELOP & CHIEF CLIN BIOCHEM, CHILD DEVELOP CTR, UNIV TENN HEALTH SCI CTR, MEMPHIS, 72- Mem: AAAS; Am Asn Clin Chemists; Genetics Soc Am; Soc Neurosci; Am Asn Ment Deficiency. Res: Biochemical and genetics regulation of metabolism in bacteria and tissue culture cells. Mailing Add: Dept of Biochem Univ of Tenn Med Ctr Memphis TN 38103

MOLIN-CASE, JO ANN, physical chemistry, crystallography, see 12th edition

MOLINE, HAROLD EMIL, b Frederic, Wis, Nov 13, 39; m 65; c 2. PHYTOPATHOLOGY, PLANT VIROLOGY. Educ: Univ Wis-River Falls, BSc, 67, MS, 69; Iowa State Univ, PhD(plant path), 72. Prof Exp: Plant pathologist, Northern Grain Insect Res Lab, Brookings, SDak, 72-73, RES PLANT PATHOLOGIST, HORT CROPS MKT LAB, AGR MKT RES INST, AGR RES SERV, USDA, 74- Mem: Am Phytopath Soc; Bot Soc Am; Sigma Xi; Am Soc Plant Physiologists. Res: Epidemiology of post harvest diseases of fresh market fruits and vegetables; ultrastructural and histochemical modifications of host cells invaded by bacteria, fungi or viruses; physiological disorders. Mailing Add: Agr Mkt Res Inst Hort Crops Lab Beltsville Agr Res Ctr-W Beltsville MD 20705

MOLINE, NORMAN THEODORE, b Gary, Ind, Oct 30, 42; m 66; c 2. CULTURAL GEOGRAPHY, RESOURCE GEOGRAPHY. Educ: Augustana Col, BA, 64; Univ Chicago, MA, 66, PhD(geog), 70. Prof Exp: ASST PROF GEOG, AUGUSTANA COL, 68- Concurrent Pos: Lutheran Church in Am Bd Col Educ res & creativity grant, Southeast Asia, 70; contrib mem, Natural Hazards Res Group, Comn Man & Environ, Int Geog Union, 70- Mem: Am Geog Soc; Asn Am Geog. Res: Environmental perception; flood plain management; settlement history. Mailing Add: Dept of Geog Augustana Col 639 38th St Rock Island IL 61201

MOLINE, ROBERT ALAN, physics, see 12th edition

MOLINE, SHELDON WALTER, b Chicago, Ill, Feb 15, 31; m 52; c 4. BIOCHEMISTRY. Educ: Roosevelt Univ, BS, 52; Univ Chicago, PhD, 58; State Univ NY Buffalo, MBA, 69. Prof Exp: Res technician, Argonne Nat Lab, 52-56; res chemist, Linde Div, 58-63, sr staff biochemist, 64-67, proj scientist chem & plastics, 67-68, tech mgr fermentation, Chem Div, 68-71, dir res, Creative Agr Systs, 71-74, SR GROUP LEADER, CORP RES DEPT, UNION CARBIDE CORP, 74- Mem: AAAS; Am Chem Soc; Inst Food Technol; Sigma Xi; Am Soc Hort Sci. Res: Thermal properties and metabolic processes of biological systems at low temperatures; cryosurgery; cell culture; preservation of foods at low temperatures; shipment of produce and meats; seed and plant physiology. Mailing Add: Union Carbide Corp Tarrytown Tech Ctr Tarrytown NY 10591

MOLINE, WALDEMAR JOHN, b Fredric, Wis, Oct 29, 34; m 57; c 3. AGRONOMY. Educ: Wis State Univ, Riverfalls, BS, 59; Univ Minn, MS, 61; Iowa State Univ, PhD(agron), 65. Prof Exp: Res asst agron, Univ Minn, 59-61; res assoc, Iowa State Univ, 61-65; asst prof, Univ Md, 65-66; assoc prof, 66-71, PROF AGRON, UNIV NEBR, LINCOLN, 71-, VCHMN EXTEN & OUTSTATE PROGS, DEPT AGRON, 74- Concurrent Pos: Chmn mem comt & mem exec comt, Am Forage & Grassland Coun, 60-70; chmn pub rels & info comt, Am Soc Agron, Crop Sci Soc Am & Soil Sci Soc Am, 75. Mem: Am Soc Agron; Soc Range Mgt; Am Forage & Grassland Coun (pres, 75). Res: Forage crops management, production and utilization. Mailing Add: Dept of Agron Keim Hall 236 Univ of Nebr Lincoln NE 68503

MOLINOFF, PERRY BROWN, b Smithtown, NY, June 3, 40; m 63; c 2. NEUROPHARMACOLOGY. Educ: Harvard Univ, BS, 62, MD, 67. Prof Exp: Intern med, Univ Chicago Hosps & Clins, 67-68; res assoc, NIMH, 68-70; vis fel biophys, Univ Col, London, 70-72; ASST PROF PHARMACOL, UNIV COLO MED CTR, 72- Mem: Am Soc Pharmacol & Exp Therapeut; Am Soc Neurochem; Am Heart Asn; AAAS. Res: Regulation of norepinephrine synthesis; isolation and characterization of receptors; storage of neurotransmitters. Mailing Add: Dept of Pharmacol Univ of Colo Med Ctr Denver CO 80220

MOLINSKI, VICTOR JOSEPH, b Ticonderoga, NY, Nov 16, 23. NUCLEAR CHEMISTRY. Educ: Seton Hall Univ, BS, 52. Prof Exp: Chemist, Allen B Dumont Labs, 49-51, F W Berk & Co, 51-57 & Union Carbide Nuclear Co, 57-62, res chemist, Develop Dept, Union Carbide Corp, 62-73, DEVELOP MGR NUCLEONICS, UNION CARBIDE CORP, 73- Mem: Am Chem Soc; Soc Nuclear Med. Res: Analytical chemistry, both wet chemistry and instrumental analysis; radiochemistry, particularly neutron activation analysis; development of radiochemicals and radiodiagnostic agents and instruments for use in nuclear medicine. Mailing Add: Union Carbide Res Ctr Box 324 Tuxedo NY 10987

MOLL, HAROLD WESTBROOK, b Detroit, Mich, Apr 2, 14; m 38; c 3. ORGANIC CHEMISTRY, CHEMICAL ENGINEERING. Educ: Andrews Univ, BS, 37. Prof Exp: Res chemist, 37-48, supt latex pilot plant, 48-58, supvr, Instrument Lab, 59-63, tech expert, E C Britton Res Lab, 63-71, TECH EXPERT, PHYS RES LAB, DOW CHEM, USA, 72- Mem: AAAS; Am Chem Soc; Instrument Soc Am. Res: Organic synthesis; polymers; latexes; chemical engineering; instrumentation science. Mailing Add: 1755 E Isabella Rd RFD 6 Midland MI 48640

MOLL, KENNETH LEON, b Jackson, Mo, Oct 16, 32; m 57; c 2. SPEECH PATHOLOGY. Educ: Southeast Mo State Col, BS, 54; Univ Iowa, MA, 59, PhD(speech path), 60. Prof Exp: Res assoc speech sci, 59-61, res asst prof, 61-64, assoc prof, 64-68, PROF SPEECH SCI & CHMN DEPT SPEECH PATH & AUDIOL, UNIV IOWA, 68- Concurrent Pos: Nat Inst Neurol Dis & Blindness spec fel, Univ Mich, 65-66. Mem: AAAS; Am Speech & Hearing Asn; Acoust Soc Am. Res: Physiological aspects of human speech production through use of x-ray techniques, electromyography and air pressure and air flow recordings. Mailing Add: Dept of Speech Path & Audiol Univ of Iowa Iowa City IA 52242

MOLL, NORMAN GLENN, physical chemistry, see 12th edition

MOLL, ROBERT HARRY, b Lackawanna, NY, July 17, 27; m 50; c 3. GENETICS. Educ: Cornell Univ, BS, 51; Univ Idaho, MS, 53; NC State Col, PhD(plant breeding), 57. Prof Exp: From asst prof to assoc prof, 57-65, PROF GENETICS, NC STATE UNIV, 65- Mem: AAAS; Am Soc Agron; Genetics Soc Am; Am Soc Naturalists. Res: Quantitative genetics; statistics; plant breeding; horticulture. Mailing Add: Dept of Genetics NC State Univ Raleigh NC 27607

MOLL, RUSSELL ADDISON, b Bound Brook, NJ, Aug 12, 46; m 69. LIMNOLOGY. Educ: Univ Vt, BA, 68; Long Island Univ, MS, 71; State Univ NY Stony Brook, PhD(biol), 74. Prof Exp: Jr res assoc, Dept Biol, Brookhaven Nat Lab, 72-73; RES INVESTR LIMNOL, GREAT LAKES RES DIV, UNIV MICH, 74- Mem: Am Soc Limnol & Oceanog; Ecol Soc Am; Int Asn Great Lakes Res. Res: Ecology and community structure of aquatic ecosystems; productivity and carbon in phytoplankton and bacteria; data analysis and biostatistics applied to ecosystems. Mailing Add: Great Lakes Res Div IST Bldg Univ of Mich Ann Arbor MI 48109

MOLL, TORBJORN, b Norway, June 25, 19; m 48; c 2. MICROBIOLOGY. Educ: Vet Col Norway, DVM, 45; Univ Wis, MS, 50, PhD(microbiol), 52. Prof Exp: Asst prof obstet, Vet Col Norway, 47-48; asst & instr, Univ Wis, 48-52; assoc prof, 52-59, PROF VET MICROBIOL, WASH STATE UNIV, 59- Concurrent Pos: Fulbright res fel, Vet Col Norway, 61-62. Res: Enteric infections in the newborn; enteroviruses. Mailing Add: Vet Col Wash State Univ Pullman WA 99163

MOLL, WILLIAM FRANCIS, JR, b Jacksonville Fla, Feb 20, 31; m 60; c 1. GEOCHEMISTRY, MINERALOGY. Educ: Univ Fla, BS, 54; Ind Univ, Bloomington, MA, 58; Wash Univ, PhD(geol), 63. Prof Exp: Instr mineral, Wash Univ, 59-60; technician, Emerson Elec Mfg Co, 61-62; fel mat res, Pa State Univ, 63-65; technologist, Baroid Div, Nat Lead Co, 65-67; mineralogist, 67-74, HEAD, COLLOIDAL MINERALS LAB, GEORGIA KAOLIN CO, ELIZABETH, 74- Mem: Am Chem Soc; Mineral Soc Am; Mineral Soc Gt Brit & Ireland; Clay Minerals Soc; Soc Photog Sci & Eng. Res: Paragenesis of clay minerals and other layer silicates, their chemical properties, modification of their surfaces. Mailing Add: 29 Montrose Ave Summit NJ 07901

MOLLARD, WESLEY N, physics, see 12th edition

MOLLENAUER, JAMES FREDERICK, b Pittsburgh, Pa, Apr 23, 36; m 61; c 2. NUCLEAR PHYSICS, COMPUTER SCIENCE. Educ: Amherst Col, BA, 57; Univ Calif, Berkeley, PhD(nuclear chem), 61. Prof Exp: Res assoc nuclear chem, Brookhaven Nat Lab, 61-63, assoc chemist, 63-64; MEM TECH STAFF, BELL TEL LABS, 64- Mem: Am Phys Soc. Res: Nuclear reactions with multiparticle breakup, or at high angular momentum; applications of computers to nuclear studies and to electronic circuit interconnections. Mailing Add: Bell Labs PO Box 261 Murray Hill NJ 07974

MOLLENAUER, LINN F, b Washington, Pa, Jan 6, 37; m 62; c 1. PHYSICS. Educ: Cornell Univ, BEngPhys, 59; Stanford Univ, PhD(physics), 65. Prof Exp: Asst prof physics, Univ Calif, Berkeley, 65-72; RES STAFF MEM, BELL LABS, 72- Mem: Am Phys Soc; Optical Soc Am. Res: Optical spectroscopy of solids; lasers; light scattering; infrared tunable lasers; color centers. Mailing Add: Bell Labs Holmdel NJ 07733

MOLLER, GOTTFRIED IRVING, b Baltic, SDak, Nov 29, 14; m 41; c 2. PHYSICS. Educ: Augustana Col, AB, 38; Univ SDak, AM, 50. Prof Exp: Teacher high schs, SDak, 38-47; instr physics, Gen Beadle State Teachers Col, 47-49; assoc physicist, Argonne Nat Lab, 50-54; assoc prof, 54-63, PROF PHYSICS, UNIV SDAK, 63- Mem: AAAS; Am Phys Soc; Am Asn Physics Teachers. Res: Thin films; radiological environment. Mailing Add: Dept of Earth Sci & Physics Univ of SDak Vermillion SD 57069

MOLLER, KARLIND THEODORE, b Chisago City, Minn, May 25, 42; m 65; c 1. SPEECH PATHOLOGY. Educ: Univ Minn, Minneapolis, BS, 64, MA, 67, PhD(speech path), 70. Prof Exp: Nat Inst Dent Res spec res fel, 70-72, ASSOC PROF SPEECH PATH, SCH DENT UNIV MINN, MINNEAPOLIS, 70- Mem: Am Speech & Hearing Asn; Am Cleft Palate Asn. Res: Oral physiology; speech production; speech in persons with orofacial anomalies; modification of speech and oral structures. Mailing Add: 686 Arbogast Rd St Paul MN 55112

MOLLER, PALMI, b Iceland, Nov 4, 22; m 45; c 3. DENTISTRY. Educ: Tufts Univ, DMD, 48; Univ Ala, Birmingham, MS, 62; Univ Iceland, Dr Odontol, 71. Prof Exp: From instr to assoc prof, 58-70, PROF CLIN DENT, SCH DENT, UNIV ALA, BIRMINGHAM, 70- Mem: Int Asn Dent Res. Res: Clinical dentistry; oral health survey of Icelandic people; incidence and familial occurrence of cleft lip and palate. Mailing Add: Univ of Ala Sch of Dent 1919 Seventh Ave S Birmingham AL 35233

MOLLER, PETER, b Hamburg, Ger, Nov 19, 41; m 67; c 1. ETHOLOGY. Educ: Free Univ Berlin, dipl biol, 64, PhD(zool), 67. Prof Exp: Asst prof, 70-75, ASSOC PROF PSYCHOL, HUNTER COL, 75-; RES ASSOC, AM MUS NATURAL HIST, 72- Concurrent Pos: NATO res fel, Nat Ctr Sci Res, Paris, 68-69, Dr Carl Duisberg fel, 69-70; City Univ New York fac res grant, Hunter Col, 72-73. Mem: AAAS; Am Soc Zoologists; NY Acad Sci; Ger Zool Soc. Res: Sensory and behavioral physiology, ethology including significance of electric organ discharges in weakly electric fish and orientation mechanisms in arthropods. Mailing Add: Dept of Psychol Hunter Col 695 Park Ave New York NY 10021

MOLLER, RAYMOND WILLIAM, b Brooklyn, NY, Jan 28, 20; m 48; c 8. MATHEMATICS. Educ: Manhattan Col, BS, 41; Cath Univ, PhD(math), 51. Prof Exp: Instr math, Trinity Col, 43-45; from instr to asst prof, 46-57, dept, 57-70, ASSOC PROF MATH, CATH UNIV AM, 57- Mem: Am Math Soc; Math Asn Am. Res: Congruences; primitive roots; cyclotomic polynomials. Mailing Add: Dept of Math Cath Univ of Am Box 105 Washington DC 20017

MOLLES, MANUEL CARL, JR, b Gustine, Calif, July 30, 48. ECOLOGY, ICHTHYOLOGY. Educ: Humboldt State Univ, BS, 70; Univ Ariz, PhD(zool), 76. Prof Exp: ASST PROF ECOL & CUR FISHES, UNIV NMEX, 75- Mem: AAAS; Ecol Soc Am; Am Soc Ichthyologists & Herpetologists. Res: Community structure; island biogeography; diversity-stability relationships. Mailing Add: Dept of Biol Univ of NMex Albuquerque NM 87131

MOLLICA, JOSEPH ANTHONY, b Providence, RI, Oct 24, 40; m 64; c 3. PHARMACEUTICAL CHEMISTRY, ANALYTICAL CHEMISTRY. Educ: Univ RI, BS, 62; Univ Wis, MS, 65, PhD, 66. Prof Exp: Sr chemist, Ciba Pharmaceut Co, 66-68, mgr anal res & develop, 68-70, asst dir, 70-75, DIR ANAL RES & DEVELOP, CIBA-GEIGY CORP, 75- Mem: NY Acad Sci; Am Chem Soc; Am Pharmaceut Asn; Acad Pharmaceut Sci. Res: Developmeht of analytical methods for pharmaceuticals; investigation of rates and mechanisms of organic reactions. Mailing Add: Pharmaceut Div Ciba-Geigy Corp Suffern NY 10901

MOLLO-CHRISTENSEN, ERIK LEONARD, b Bergen, Norway, Jan 10, 23; nat US; m 48; c 3. FLUID DYNAMICS, OCEANOGRAPHY. Educ: Mass Inst Technol, SB, 48, SM, 49, ScD, 54. Prof Exp: Fel, Norweg Defense Res Estab, 48-49, sci officer, 49-51; res assoc, Mass Inst Technol, 54-55, asst prof aeronaut eng, 54-57; sr res fel, Calif Inst Technol, 57-58; from assoc prof to prof aeronaut, 58-64, PROF METEOROL, MASS INST TECHNOL, 64- Honors & Awards: Von Karman Award, Am Inst Aeronaut & Astronaut, 70. Mem: Am Inst Aeronaut & Astronaut; Am Meteorol Soc; Am Phys Soc; Am Acad Arts & Sci; Am Geophys Union. Res: Fluid mechanics. Mailing Add: Dept of Meteorol Mass Inst of Technol Cambridge MA 02139

MOLLOW, BENJAMIN R, b Trenton, NJ, Nov 26, 38; m 67. THEORETICAL

PHYSICS. Educ: Cornell Univ, AB, 60; Harvard Univ, PhD(physics), 66. Prof Exp: NSF fel physics, Brandeis Univ, 66-68, asst prof, 68-69; ASST PROF PHYSICS, UNIV MASS, BOSTON, 69- Res: Quantum optics; light scattering; parametric processes. Mailing Add: Dept of Physics Univ of Mass Boston MA 02125

MOLLOY, ANDREW A, b New York, NY, Mar 19, 30; m 66; c 4. CHEMISTRY. Educ: Marist Col, BA, 51; Cath Univ, PhD(chem), 61. Prof Exp: Teacher parochial schs, 51-56; asst prof chem & chmn dept, Marist Col, 60-66; ASSOC PROF CHEM, ELMIRA COL, 66-, DIR CAREER SERVS, 74- Mem: AAAS; Am Chem Soc. Res: Organotin chemistry; mechanism of paper chromatographic separation of lower fatty acids. Mailing Add: Dept of Chem Elmira Col Elmira NY 14901

MOLLOY, CHARLES THOMAS, b New York, NY, Nov 22, 14; m 36; c 2. PHYSICS. Educ: Cooper Union, BS, 35; NY Univ, MS, 38, PhD(physics), 48. Prof Exp: Res chemist, Muralo Co, NY, 35-38; res engr, Johns Manville Co, NJ, 38-42; physicist, Brooklyn Navy Yard, 42-45 & Bell Tel Labs, Inc, 45-50; head, Physics Res Dept, Labs Div, Vitro Corp Am, 50-54; staff engr sound & vibration, Lockheed Aircraft Corp, 54-59; sr staff engr, TRW Systs Group, Calif, 59-70, SR STAFF ENGR, WASH OPERS, TRW SYSTS GROUP, 70- Concurrent Pos: Instr, Long Island Univ, 42-45 & Brooklyn Col, 39-42; adj prof, Polytech Inst Brooklyn, 50-53, Univ Southern Calif, 57-70 & Univ Calif, Los Angeles, 59- Honors & Awards: Cert Appreciation, Soc Automotive Eng, 61; Spec Merit Award, Inst Environ Sci, 62, Irwin Vigness Award, 71. Mem: Fel Acoust Soc Am; Am Math Soc; Math Asn Am; Audio Eng Soc; sr mem Inst Elec & Electronics Eng. Res: Underwater acoustics; applied mathematics; propagation of sound in lined tubes; radiation theory; loud speaker design; hearing; sound absorbing materials; transients; sound and vibration problems associated with aircraft and missiles. Mailing Add: 2400 Claremont Dr Falls Church VA 22043

MOLLOY, MARILYN, b Caney, Kans, Apr 24, 31. MATHEMATICS. Educ: Our Lady of the Lake Col, BS, 55; Univ Tex, MA, 62, PhD(math) 66. Prof Exp: From instr to asst prof, 59-70, ASSOC PROF MATH, OUR LADY OF THE LAKE COL, 70-, ASSOC ACAD DEAN, 72- Mem: Math Asn Am; Am Math Soc. Res: Mathematical analysis; Stieltjes integral. Mailing Add: Dept of Math Our Lady of the Lake Col San Antonio TX 78207

MOLMUD, PAUL, b Brooklyn, NY, May 29, 23; m 57; c 2. PHYSICS. Educ: Brooklyn Col, AB, 43; Ohio State Univ, PhD(physics), 51. Prof Exp: Jr engr, Nat Union Radio Corp, NJ, 43-45; asst prof physics, Clarkson Col Technol, 51-53; sr res scientist, 54-68, MGR THEORET PHYSICS DEPT, TRW SYSTS INC, 68- Mem: Sr mem Inst Elec & Electronics Eng; Am Phys Soc. Res: Gaseous electronics; ionospheric disturbances; plasma physics; electromagnetic properties of rocket exhausts; electromagnetic fields from nuclear explosions; rarefied gas dynamics. Mailing Add: Theoret Physics Dept TRW Systs Inc Bldg R1/1196 1 Space Park Redondo Beach CA 90278

MOLNAR, CHARLES EDWIN, b Newark, NJ, Mar 14, 35; m 57; c 2. NEUROPHYSIOLOGY, COMPUTER SCIENCE. Educ: Rutgers Univ, BS, 56, MS, 57; Mass Inst Technol, ScD(elec eng), 66. Prof Exp: Staff assoc, Lincoln Lab, Mass Inst Technol, 57-61; res electronics engr, Air Force Cambridge Res Labs, 64-65; assoc prof physiol & biophys, 65-71, assoc prof elec eng, 67-71, assoc dir comput systs lab, 67-72, PROF PHYSIOL & BIOPHYS & ELEC ENG, WASH UNIV, 71-, DIR COMPUT SYSTS LAB, 72- Concurrent Pos: Vis prof, Int Brain Res Orgn Vis Sem, Univ Chile, 67; mem comput & biomath study sect, NIH, 71-74, chmn, 74-75. Res: Experimental studies of auditory system; models for activity of single neurons and neural networks; design of computer systems for biological research; macromodular computer systems. Mailing Add: Comput Systs Lab 724 S Euclid Ave St Louis MO 63110

MOLNAR, GEORGE D, b Szekesfehervar, Hungary, July 30, 22; US citizen; m 47; c 2. ENDOCRINOLOGY, INTERNAL MEDICINE. Educ: Univ Alta, BSc, 49, MD, 51; Univ Minn, PhD(med), 56; Am Bd Internal Med, dipl, 59. Prof Exp: From instr to prof, Mayo Grad Sch Med, 56-73, prof med, Mayo Med Sch, Univ Minn, 73-75; PROF MED & CHMN DEPT, UNIV ALTA, 75- Mem: AMA; Am Diabetes Asn; Endocrine Soc; fel Am Col Physicians. Res: Unstable diabetes; endocrine correlates of the diabetic state; insulin resistance and insulin binding antibodies; correlations of plasma growth hormone levels with diabetes; adrenal and thyroid cancer; hyperfunctioning thyroid nodule; diabetes mellitus, applied physiologic, biochemical and clinical aspects. Mailing Add: Dept of Med 8-121 Clin Sci Bldg Univ Alta Edmonton AB Can

MOLNAR, GEORGE WILLIAM, b Detroit, Mich, Feb 14, 14; m 37; c 2. PHYSIOLOGY. Educ: Oberlin Col, AB, 36; Yale Univ, PhD(zool), 40. Prof Exp: Asst biol, Yale Univ, 36-39; instr zool, Miami Univ, 40-42; asst prof, RI State Col, 42-44; instr physiol, Sch Med & Dent, Univ Rochester, 44-46; physiologist, Army Med Res Lab, Ft Knox, Ky, 46-62; physiologist, Res Lab, Kerrville, Tex, 62-66, coord prof serv, Southern Res Support Ctr, Little Rock, 66-68, RES PHYSIOLOGIST, VET ADMIN HOSP, LITTLE ROCK, 68- Mem: Am Physiol Soc; NY Acad Sci. Res: Thermal physiology of man. Mailing Add: Vet Admin Hosp Little Rock AR 72206

MOLNAR, JANOS, b Budapest, Hungary, Nov 28, 27; US citizen; m 55; c 2. BIOCHEMISTRY. Educ: Eötvös Lorand Univ, Budapest, dipl, 53; Northwestern Univ, PhD(biochem), 60. Prof Exp: Instr chem, Med Univ Budapest, 53-56; res asst biochem, Northwestern Univ, 60-62; res asst, 62-65, asst prof, 65-68, ASSOC PROF BIOCHEM, UNIV ILL COL MED, 68- Mem: AAAS; Am Soc Biol Chemists; Reticuloendothelial Soc; Am Chem Soc. Res: Metabolism of glycoproteins; mechanism of phagocytosis; cell membranes topography and surface antigens as related to cancer cells. Mailing Add: Dept of Biochem Univ of Ill Col of Med Chicago IL 60612

MOLNAR, JULIUS PAUL, physics, see 12th edition

MOLNAR, NICHOLAS M, b Miskolc, Hungary, Sept 23, 03; US citizen; m 30; c 2. ORGANIC CHEMISTRY. Educ: Cooper Union, BSc, 26; NY Univ, MSc, 36. Prof Exp: Pres, 39-74, CHMN, FINE ORGANICS, INC, 74-; DIR, MOLNAR LABS, 27- Honors & Awards: Gano Dunn Medal, Cooper Union, 66. Mem: Am Chem Soc; fel Am Inst Chem; Soc Plastics Eng; Asn Consult Chemists & Chem Eng (pres, 46-48); Tech Asn Pulp & Paper Indust. Mailing Add: Fine Organics Inc 205 Main St Lodi NJ 07644

MOLNAR, PETER HALE, b Pittsburgh, Pa, Aug 25, 43; c 1. SEISMOLOGY, MARINE GEOPHYSICS. Educ: Oberlin Col, AB, 65; Columbia Univ, PhD(geol & seismol), 70. Prof Exp: Res scientist seismol, Lamont-Doherty Geol Observ, Columbia Univ, 70-71; asst res scientist, Scripps Inst Oceanog, Univ Calif, San Diego, 71-73; exchange scientist, Acad Sci, USSR, 73; ASST PROF EARTH SCI, MASS INST TECHNOL, 74- Concurrent Pos: Fel, Alfred P Sloan Found, 75-77. Res: Plate tectonics; earthquake source mechanism; structure of the earth; evolution of the ocean floor; geophysics; structural geology; tectonophysics. Mailing Add: Dept of Earth & Planetary Sci Mass Inst Technol Cambridge MA 02139

MOLNAR, STEPHEN P, b Toledo, Ohio, July 8, 35; m 64; c 2. ORGANIC CHEMISTRY, INORGANIC CHEMISTRY. Educ: Univ Toledo, BS, 57; Purdue Univ, MS, 63; Univ Cincinnati, PhD(chem), 67. Prof Exp: Res chemist, Owens Ill Glass Co, 59-60; teaching asst chem, Purdue Univ, 60-63; res fel, Univ Cincinnati, 63-65, teaching asst, 65-67; asst prof, Miami Univ, 67-75; RES CHEMIST, ARMCO STEEL CORP, 75- Mem: AAAS; Am Chem Soc; The Chem Soc. Res: Chemical kinetics; complexes of Group VIII elements; electronic and chemical spectroscopy; polymer chemistry. Mailing Add: Armoco Steel Corp Res & Technol Middletown OH 45043

MOLNAR, ZELMA VILLANYI, b Komotau, Czech, Jan 29, 31; US citizen; div; c 2. PATHOLOGY. Educ: Med Univ Budapest, MD, 56; Univ Chicago, PhD(path), 73. Prof Exp: Res asst, Dept Surg, Univ Chicago, 57-58 & Dept Anat, 58, from res asst to res assoc, Depts Anat & Physiol, 58-61, resident trainee, Dept Path, 61-65, from instr to asst prof path, 65-73; ASSOC PROF PATH, LOYOLA UNIV CHICAGO, 73- Honors & Awards: Hektoen Award, Chicago Path Soc, 65. Mem: AAAS; Int Acad Path; Am Asn Path & Bact; Am Asn Clin Path; Am Soc Hemat. Res: Lymphohematopoietic tissue and tumors. Mailing Add: Vet Admin Hosp Box 1216 Hines IL 60141

MOLNIA, BRUCE FRANKLIN, b Bronx, NY, Oct 17, 45. MARINE GEOLOGY. Educ: State Univ NY Binghamton, BA, 67; Duke Univ, MA, 69; Univ SC, PhD(geol), 72. Prof Exp: Res fel geol, Duke Univ, 67-69 & Univ SC, 70-72; teaching fel, Cornell Univ, 69-70; asst prof, Amherst Col & Mt Holyoke Col, 72-73; geol oceanogr, US Bur Land Mgt, 73-74, MARINE GEOLOGIST, US GEOL SURV, US DEPT INTERIOR, 74- Concurrent Pos: Sci ed, SC Educ TV Network, 71-73; consult, Environ Defense Fund, 73. Mem: Geol Soc Am; Soc Econ Paleontologists & Mineralogists. Res: Environmental hazards in the marine continental shelf system; marine pleistocene sedimentation; ice rafting; fluvial systems. Mailing Add: US Geol Surv 345 Middlefield Rd Menlo Park CA 94025

MOLO, WILLIAM L, b Dubuque, Iowa, Jan 11, 19; m 45; c 3. METEOROLOGY, OCEANOGRAPHY. Educ: Rockhust Col, BA, 40. Prof Exp: Climatic analyst, Air Force Weather Serv, 46-51, data intel specialist, 51-62; dir info div, Nat Oceanog Data Ctr, 62-68, dir, Serv Div, 68-74, from assoc dir to dir, World Data Ctr A, Oceanog, Nat Oceanog Data Ctr, 70-74; RETIRED. Mem: Am Geophys Union; Am Meteorol Soc; Marine Technol Soc. Res: Climatology; environmental data retrieval. Mailing Add: 5844 Upton St McLean VA 22107

MOLOMUT, NORMAN, b Newark, NJ, Jan 21, 13; m 38; c 3. ONCOLOGY. Educ: Brooklyn Col, AB, 34; Univ Mich, MA, 35; Columbia Univ, PhD(bact), 39. Prof Exp: Asst virus res, Pasteur Inst, Mich, 34-35; from asst bacteriologist to assoc bacteriologist, Dept Med, Col Physicians & Surgeons, Columbia Univ, 39-46; dir, Biol Labs, Inc, NY, 46-54; DIR RES, WALDEMAR MED RES FOUND, INC, 47- Mem: Am Soc Microbiol; Am Asn Immunol; Am Soc Cell Biol; NY Acad Sci. Res: Cell growth; wound healing stimulation with immune sera; cancer biotherapy; effect of immune factors and extracts of living organisms; physiological factors and mechanism of response in resistance to infectious disease agents; cancer immunology. Mailing Add: 246 Wynsum Ave Merrick NY 11566

MOLONEY, JOHN BROMLEY, b Lowell, Mass, Jan 18, 24; m 49; c 2. BIOLOGY. Educ: Tufts Col, BS, 47; George Washington Univ, MS, 53, PhD, 59. Prof Exp: Biologist, 47-60, supvry res biologist, 60-64, head viral leukemia sect, 64-66, assoc chief viral biol br, 66-67, chief, Viral Leukemia & Lymphoma Br, 67-70, ASSOC SCI DIR VIRAL ONCOL & CHMN, SPEC VIRUS CANCER PROG, NAT CANCER INST, 70- Mem: AAAS; Am Asn Cancer Res; Soc Exp Biol & Med; Am Soc Microbiol. Res: Biological and biochemical properties of tumor viruses, especially the leukemia agents, a murine sarcoma virus and the Rous virus. Mailing Add: Viral Oncol Etiol Nat Cancer Inst NIH Bldg 37 Rm 1A13 Bethesda MD 20014

MOLONEY, MICHAEL J, b Albany, NY, Nov 16, 36; m 64; c 2. MICROWAVE PHYSICS. Educ: Ill Inst Technol, BS, 58; Univ Md, PhD(physics), 66. Prof Exp: Asst prof physics, Rose-Hulman Inst Technol, 66-68 & Lafayette Col, 68-70; ASSOC PROF PHYSICS & CHMN DEPT, ROSE-HULMAN INST TECHNOL, 70- Concurrent Pos: NSF res partic, Univ Md, 68; electronics engr, Naval Weapons Support Ctr, Crane, Ind, 72- Mem: Am Asn Physics Teachers. Res: Computer assisted analysis of microwave spectra involving quadropole or centrifugal distortion effects. Mailing Add: Dept of Physics Rose-Hulman Inst of Technol Terre Haute IN 47803

MOLONEY, WILLIAM CURRY, b Boston, Mass, Dec 19, 07; m; c 4. HEMATOLOGY. Educ: Tufts Col, MD, 32. Hon Degrees: DSc, Col of the Holy Cross, 61. Prof Exp: From asst to clin prof med, Med Sch, Tufts Univ, 34-67; clin prof, 67-71, prof, 71-74, EMER PROF MED, HARVARD MED SCH, 74-; PHYSICIAN & CHIEF HEMAT DIV, PETER BENT BRIGHAM HOSP, 67- Concurrent Pos: Consult, Boston Hosps, 38-; consult, Boston City Hosp, 48-, dir clin labs; dir res, Atomic Bomb Casualty Comn, Hiroshima, Japan, 52-54. Mem: AAAS; fel Am Fedn Clin Res; fel AMA; fel Am Col Physicians; Asn Am Physicians. Res: Leukemia. Mailing Add: Peter Bent Brigham Hosp 721 Huntington Ave Boston MA 02115

MOLOTSKY, HYMAN MAX, b Russia, Nov 1, 19; nat US; m 52; c 3. CARBOHYDRATE CHEMISTRY. Educ: Univ Man, BSc, 43; Univ Mo, AM, 49, PhD(chem), 53. Prof Exp: Asst chem, McGill Univ, 45-46 & Univ Mo, 47-52; sr res chemist, Velsicol Corp, 52-55, proj leader, 55-57; res chemist, Richardson Co, 57-58, proj leader, 58-59; res chemist, Corn Prod Co, 59-64, sect leader basic & appl res, Moffett Tech Ctr, 64-66, patent liaison & tech coordr, Moffett Res, 66-74, ASST TO EXEC DIR RES & DEVELOP, INDUST DIV, MOFFETT TECH CTR, CPC INT INC, 74- Mem: Am Chem Soc. Res: Derivatives of sugars and starch for industrial applications; intermediates for urethane foams; synthetic organic intermediates; intermediates for resins and plasticisers. Mailing Add: Moffett Tech Ctr CPC INT INC Box 345 Argo IL 60501

MOLTENI, AGOSTINO, b Como, Italy, Nov 12, 33; US citizen; m; c 2. EXPERIMENTAL PATHOLOGY. Educ: Univ Milan, MD, 57; Ital Bd Internal Med, cert, 63; State Univ NY Buffalo, PhD(path), 70. Prof Exp: Asst prof med, Univ Milano, 58-62; sr investr med res, Farmitalia, Milan, Italy, 63-65; Henry C & Bertha Buswell fel & res asst prof path, State Univ NY Buffalo, 70-72; ASSOC PROF PATH, UNIV KANS MED CTR, KANSAS CITY, 73- Concurrent Pos: NIH res career develop award, 72; consult path, Vet Admin Hosp, Kansas City, 73- Mem: AMA; Am Soc Exp Path; Endocrine Soc; Am Soc Exp Biol & Med; Reticuloendothelial Soc. Res: Cardiovascular diseases, especially systemic and pulmonary hypertension. Mailing Add: Dept of Path Univ of Kans Med Ctr Kansas City KS 66103

MOLTENI-BRIZIO, LOREDANA, see Brizio-Molteni, Loredana, see 12th edition

MOLYNEUX, RUSSELL JOHN, b Luton, Eng, Aug 3, 38; m 63; c 2. ORGANIC CHEMISTRY. Educ: Univ Nottingham, BSc, 60, PhD(org chem), 63. Prof Exp: NSF fel org chem, Univ Ore, 63-65; asst prof forest prod chem, Ore State Univ, 65-67; res assoc, 67-74, RES CHEMIST, WESTERN REGIONAL RES CTR, AGR RES SERV, USDA, 74- Concurrent Pos: Consult, US Brewers Asn, 68-74. Mem: Am Chem Soc; The Chem Soc; Phytochem Soc. Res: Naturally occurring quinones; naturally occurring phenolic compounds. Mailing Add: Western Regional Res Ctr Agr Res Serv 800 Buchanan St Albany CA 94710

MOLZ, FRED JOHN, b Mays Landing, NJ, Aug 13, 43; m 66; c 2. HYDROLOGY, SOIL PHYSICS. Educ: Drexel Univ, BS, 66, MSCE, 68; Stanford Univ, PhD(hydrol), 70. Prof Exp: ASST PROF CIVIL ENG, AUBURN UNIV, 70- Concurrent Pos: Off Water Resources res grant, Auburn Univ, 71-73; hydrologist, US Geol Surv, 71- Mem: Am Geophys Union; Am Soc Agron; Soil Sci Soc Am; Am Soc Civil Eng. Res: Moisture movement in soil, plant and soil-plant systems, including both physical, mathematical, biological and engineering aspects. Mailing Add: Dept of Civil Eng Auburn Univ Auburn AL 36830

MOMBERG, HAROLD LESLIE, b Sedalia, Mo, Mar 24, 29. ZOOLOGY, HISTOLOGY. Educ: Cent Mo State Col, BS, 51; Univ Mo, MA, 55, PhD(zool), 61. Prof Exp: Head dept biol, Hannibal-LaGrange Col, 55-57; instr zool, Univ Mo, 57-58; from assoc prof to prof biol, histol & embryol, William Jewell Col, 60-67, NSF res grant, 64-66; assoc prof biol, Cent Mo State Col, 67-69; assoc prof biol, 70-75, PROF BIOL, CENT METHODIST COL, 75-, HEAD, DEPT BIOL & GEOL, 74- Mem: AAAS; Am Soc Zoologists. Res: Mammalian implantation patterns and cycles; histochemistry and physiology of reproductive tract. Mailing Add: Dept of Biol Cent Methodist Col Fayette MO 65248

MOMENT, GAIRDNER BOSTWICK, b New York, NY, May 4, 05; m 37; c 5. DEVELOPMENTAL BIOLOGY. Educ: Princeton Univ, AB, 28; Yale Univ, PhD(zool), 32. Prof Exp: From instr to prof biol, 32-70, chmn dept, 45-58, EMER PROF BIOL SCI, GOUCHER COL, 70- Concurrent Pos: Lectr, Mt Desert Biol Lab, 38-39; vis assoc prof, Johns Hopkins Univ, 44-45; assoc prog dir inst sect, NSF, 60-61, chmn exec comt, Comn Undergrad Educ in Biol Sci, 63-; prog coordr biol series, Voice of Am, 60-61; secy-gen, Int Cong Zool, DC, 60-63; NIH guest scientist, Geront Res Ctr, City Hosps, Baltimore, Md, 70-; ed-in-chief, Growth, 72-; grants, Am Philos Soc, Am Cancer Soc, NIH & NSF; investr, Woods Hole Biol Lab& Beauford Biol Lab; investr & mem corp, Bermuda Biol Lab; mem, Gov Sci Adv Coun, 74- Mem: Fel AAAS; Am Soc Zoologists (secy, 57-60); Soc Develop Biol; Soc Gen Physiol; Am Inst Biol Sci. Res: Animal development and behavior; biochemical, electrical and anatomical study of animal growth; general biology; annelids; protein electrophoresis; regeneration; tissue culture. Mailing Add: Dept of Biosci Goucher Col Baltimore MD 21204

MOMMAERTS, WILFRIED, b Broechem, Belg, Mar 4, 17; nat US; m 44; c 3. PHYSIOLOGY. Educ: State Univ Leiden, BA, 37, MA, 39; Kolozsvar Univ, Hungary, PhD(biol, phys chem), 43. Prof Exp: Vis assoc prof biochem, Am Univ Beirut, 45-46, adj prof biochem & lectr phys chem, 46-47, assoc prof physiol, 47-48; res assoc biochem, Duke Univ, 48-53; assoc prof, Case Western Reserve Univ, 53-56; PROF MED & PHYSIOL, SCH MED, UNIV CALIF, LOS ANGELES, 56-, CHMN DEPT PHYSIOL, 66- Concurrent Pos: Estab investr, Am Heart Asn, 49-59; dir res lab, Los Angeles Heart Asn, 55- Mem: Biophys Soc; Am Physiol Soc; Am Heart Asn; fel Am Acad Arts & Sci; fel Am Col Cardiol. Res: Molecular physiology and biochemistry of contractile tissues; physical chemistry of tissue proteins and cellular processes. Mailing Add: Dept of Physiol Univ of Calif Sch of Med Los Angeles CA 90024

MOMOT, WALTER THOMAS, b Hamtramck, Mich, Oct 12, 38; m 66; c 1. FISH BIOLOGY. Educ: Wayne State Univ, BS, 60; Univ Mich, MS, 61, PhD(fisheries), 64. Prof Exp: Asst prof zool, Univ Okla, 64; from instr to assoc prof, Ohio State Univ, 64-75; PROF ZOOL, LAKEHEAD UNIV, 75- Mem: AAAS; Am Fisheries Soc; Am Soc Limnol & Oceanog; Ecol Soc Am; Am Inst Fishery Res Biol. Res: Production, population dynamics, trophic ecology of fish and crayfish populations. Mailing Add: Dept of Biol Lakehead Univ Thunder Bay ON Can

MOMPARLER, RICHARD LEWIS, b New York, NY, Jan 6, 35; m 66. PHARMACOLOGY, BIOCHEMISTRY. Educ: Mich State Univ, BS, 57; Univ Vt, PhD(pharmacol), 66. Prof Exp: USPHS fel, Yale Univ, 64-65 & Int Lab Genetics, Italy, 66; asst prof biochem, McGill Univ, 67-74; ASSOC PROF PHARMACOL, SCH MED, UNIV SOUTHERN CALIF, 74- Mem: AAAS. Res: Cancer chemotherapy; enzymes; cell cycle; DNA replication. Mailing Add: Div Hemat-Oncol Children's Hosp Los Angeles CA 90027

MOMSEN, JANET HENSHALL, b Stockport, Eng, Dec 10, 38; m 66; c 2. GEOGRAPHY. Educ: Oxford Univ, BA, 61, dipl educ, 62, MA, 65, BLitt, 67; McGill Univ, MSc, 64; London Univ, PhD(geog), 69. Prof Exp: Lectr geog, Kings Col, Univ London, 64-67; RES ASSOC GEOG, UNIV CALGARY, 72- Concurrent Pos: Vis prof, Fed Univ Rio de Janeiro, 74; vis lectr, Univ Leeds, 75-76. Mem: Inst Brit Geogrs; Am Asn Geogrs; Can Asn Geogrs. Res: Small scale agriculture in the Caribbean; the interaction of agriculture and tourism in the Caribbean and Rio de Janeiro; colonization of the Amazon. Mailing Add: Dept of Geog Univ of Calgary Calgary AB Can

MOMSEN, RICHARD PAUL, JR, b New York, NY, June 14, 23; m 66; c 5. GEOGRAPHY. Educ: Dartmouth Col, AB, 50; Univ Minn, MA, 52, PhD, 60. Prof Exp: Rep, Lindsay Light & Chem Co, Brazil, 48-49; instr geog, Univ Minn, 52-54, asst, Bur Pub Rds Proj, 58-59; field mgr, Brasilia Site Selection Proj, Donald J Belcher & Assocs, Brazil, 54-55; asst prof geog, Univ Colo, 59-60 & Rollins Col, 60-63; chief Guyas Basin Proj, Orgn Am States, 63; assoc prof geog & coordr int studies, Ball State Univ, 63-66; sr photointerpreter geog, Inter-Am Inst Agr Sci, 67-69; ASSOC PROF GEOG, UNIV CALGARY, 69- Concurrent Pos: Vis prof, Univ Coimbra, 62; Ford Found travel grant, Brazil & Caribbean, 65; mem exec comt, Conf Latin Am Geographers, 71-74; consult remote sensing & regional analysis, Superintendency for the Development of the State of Ceara, Brazil, 73. Mem: Am Geog Soc; Asn Am Geog; Brazilian Geog Soc; Can Asn Geog; Conf Latin Am Geographers. Res: Regional geography of Latin America, especially Brazil; underdeveloped areas; interpretation of remote sensing images for physical and human resources analysis and development. Mailing Add: Dept of Geog Univ of Calgary Calgary AB Can

MON, GORDON RUSSELL, applied mechanics, see 12th edition

MONACELLA, VINCENT JOSEPH, b Erie, Pa, May 2, 26; m 61; c 4. HYDRODYNAMICS. Educ: Gannon Col, BS, 48. Prof Exp: Mathematician, David Taylor Model Basin, 53-54, physicist, 54-67, physicist, Naval Ship Res & Develop Ctr, Washington, DC, 67-75, PHYSICIST, DAVID W TAYLOR NAVAL SHIP RES & DEVELOP CTR, BETHESDA, MD, 75- Mem: Soc Naval Archit & Marine Eng. Res: Ship hydrodynamics; water waves; analysis of motions, forces and moments of arbitrary bodies in proximity of a fluid surface. Mailing Add: 2908 Hideaway Rd Fairfax VA 22030

MONACK, LOUISE CHARLOTTE, b Charleroi, Pa, June 16, 19. CHEMISTRY. Educ: WVa Univ, AB, 42, MS, 44; Bryn Mawr Col, PhD(chem), 51. Prof Exp: Chemist, E I du Pont de Nemours & Co, 43-45; instr chem, Sweet Briar Col, 45-47; from instr to assoc prof, 50-67, PROF CHEM, WILSON COL, 67- Mem: Am Chem Soc. Res: Organic chemistry. Mailing Add: Dept of Chem Wilson Col Chambersburg PA 17201

MONACO, ANTHONY L, b Pa, Nov 9, 22; m 46; c 4. PHARMACY. Educ: Temple Univ, BS, 50. Prof Exp: From jr res pharmacist, 51-58, admin asst to mgr prod develop lab, 58-59, asst mgr prod develop lab, 59-63, asst mgr pharm develop sect, 63-69, ASST TO DIR PHARMACEUT RES & DEVELOP, WYETH LABS, AM HOME PROD CORP, 69- Mem: AAAS; Am Pharmaceut Asn; Acad Pharmaceut Sci. Res: Formulae development and stability of injectables, oral liquids and suspensions. Mailing Add: 1426 Harding Blvd Norristown PA 19401

MONACO, ANTHONY PETER, b Philadelphia, Pa, Mar 12, 32; m 60; c 4. SURGERY, IMMUNOLOGY. Educ: Univ Pa, BA, 52; Harvard Med Sch, MD, 56. Prof Exp: From intern to resident surg, Mass Gen Hosp, 56-63; Am Cancer Soc fel, 63-67, from instr to asst prof, 63-69, ASSOC PROF SURG & CHIEF TRANSPLANTATION DIV, HARVARD MED SCH, 69- Concurrent Pos: Lederly fac award, 67-70; ed, Transplantation, 69-; chief transplantation unit, Harvard Surg Serv, New Eng Deaconess Hosp; pres, Interhosp Organ Bank New Eng, 73- Mem: Am Asn Immunol; Transplantation Soc (treas, 70). Res: Transplantation immunology and immunobiology; experimental and clinical transplantation. Mailing Add: Dept of Surg Harvard Med Sch Boston MA 02115

MONACO, LAWRENCE HENRY, b Philadelphia, Pa, Mar 31, 25; m 54; c 5. ZOOLOGY. Educ: LaSalle Col, BA, 49; Univ Notre Dame, MS, 52, PhD(parasitol), 54. Prof Exp: Instr biol sci, Del Mar Col, 54-57 & Villa Madonna Col, 57-58; from instr to assoc prof, 58-61, head dept, 61-66, PROF BIOL SCI, DUTCHESS COMMUNITY COL, 61-, DEAN, 66- Mem: Am Inst Biol Sci. Res: Gill parasites of fish. Mailing Add: 5 High Ridge Rd Poughkeepsie NY 12603

MONAGHAN, PATRICK HENRY, b Memphis, Tenn, July 25, 22; m 43; c 2. GEOCHEMISTRY, PETROLEUM ENGINEERING. Educ: La Polytech Inst, BS, 43; La State Univ, MS, 49, PhD(chem), 50. Prof Exp: Field engr, Sperry Gyroscope Co, 46-47; instr chem, La State Univ, 49-50; asst res engr, Prod Res Div, Humble Oil & Ref Co, 50-51, from res engr to sr res engr, 51-59, from res specialist to sr res specialist, 59-65; res assoc, 65-73, RES ADV, EXXON PROD RES CO, 73- Mem: AAAS; Am Asn Petrol Geologists; Am Chem Soc; Soc Econ Paleontologists & Mineralogists; Am Inst Mining, Metall & Petrol Eng. Res: Oil well drilling and completion techniques; geochemistry; instrumental analysis; organic geochemistry; petroleum geology; environmental management. Mailing Add: Exxon Prod Res Co Box 2189 Houston TX 77001

MONAGLE, DANIEL J, b Eddystone, Pa, Oct 14, 36; m 61; c 5. ORGANIC POLYMER CHEMISTRY. Educ: Mt St Mary's Col Md, BS, 58; Duquesne Univ, MS, 60; Univ Del, PhD(org polymer chem), 68. Prof Exp: Chemist, Hercules Res Ctr, Del, 60-65; lectr anal & gen chem, Univ Del, 65-66; res chemist, 66-70, RES SUPVR COATINGS & SPECIALTY PROD, 70- Mem: Am Chem Soc; Sigma Xi. Res: Synthesis process development, characterization and applications of synthetic water-soluble polymers, particularly polyelectrolytes; coatings; pigments; magnetic materials. Mailing Add: 605 Halstead Rd Sharpley Wilmington DE 19803

MONAGLE, JOHN JOSEPH, JR, b Chester, Pa, Feb 2, 29; m 54; c 5. ORGANIC CHEMISTRY. Educ: Villanova Univ, BS, 50; Polytech Inst Brooklyn, PhD(chem), 54. Prof Exp: Res chemist, Sinclair Res Labs, Inc, 54-56 & Jackson Lab, E I du Pont de Nemours & Co, 56-61; from asst prof to assoc prof chem, NMex State Univ, 61-63, prof & head dept, 63-67; prof, Univ Ala, Tuscaloosa, 67-68; dean col arts & sci, 68-70, assoc dir res ctr arts & sci, 70-71, PROF CHEM, NMEX STATE UNIV, 68-, ASSOC DEAN ARTS & SCI & ASSOC DIR ARTS RES CTR, 75- Concurrent Pos: Consult, Melpar Corp, 64. Mem: Am Chem Soc; NY Acad Sci. Res: Synthetic and physical organic chemistry; polymer chemistry; synthesis and properties of organophosphorus compounds; organic chemistry of polymers. Mailing Add: Col of Art & Sci NMex State Univ Box RC Las Cruces NM 88003

MONAHAN, ALAN RICHARD, b Schenectady, NY, June 17, 39; m 62; c 6. PHYSICAL CHEMISTRY, POLYMER CHEMISTRY. Educ: Rensselaer Polytech Inst, BChE, 61, PhD(phys chem), 64. Prof Exp: Sr chemist gaseous electronics res br, 64-66, scientist chem physics res br, 66-69, sr scientist org solid state physics res br, 69-71, prin scientist, Corp Physics Lab, 71-72, mgr org solid state physics area, 72-75, MGR-DEVELOPER, MAT TECHNOL CTR, XEROX CORP, 75- Mem: Am Chem Soc. Res: Thermal and photochemical conversions in polymers; photochromics; organic photoconductors; molecular spectroscopy; physical properties of dyes and pigments; research, development and engineering of xerographic developer materials (toners and carriers). Mailing Add: 340 Filbert Pl East Rochester NY 14445

MONAHAN, AUDREY SMALL, b Danville, Pa, Nov 15, 37; m. ORGANIC CHEMISTRY. Educ: Univ Rochester, BS, 59; Univ Ill, MS, 60; Columbia Univ, PhD(org chem), 62. Prof Exp: Res assoc org chem, NMex Highlands Univ, 63-64; asst prof chem, 64-70, ASSOC PROF CHEM, UNIV CONN, 70- Concurrent Pos: NSF fel, 63-64, sci fac fel, 71-72. Res: Small ring compounds; kinetics and reaction mechanisms. Mailing Add: Dept of Chem Univ of Conn Storrs CT 06268

MONAHAN, EDWARD CHARLES, b Bayonne, NJ, July 25, 36; m 60; c 2. OCEANOGRAPHY. Educ: Cornell Univ, BEP, 59; Univ Tex, Austin, MA, 61; Mass Inst Technol, PhD(oceanog), 66. Prof Exp: Res asst, Woods Hole Oceanog Inst, 64-65; asst prof physics, Northern Mich Univ, 65-68; asst prof oceanog, Hobart & William Smith Cols, 68-69; from asst prof to assoc prof oceanog, Univ Mich, Ann Arbor, 69-75; DIR EDUC & RES, SEA EDUC ASN, 75- Concurrent Pos: Adj assoc prof, Boston Univ, 75- Mem: AAAS; Int Asn Theoret & Appl Limnol; Am Meteorol Soc; Am Geophys Union; Am Soc Limnol & Oceanog. Res: Air-sea interaction; marine aerosols; physical limnology; teaching of oceanography and meteorology; oceanographic instrumentation; design of drogues and drifters. Mailing Add: 55 Braeside Rd Falmouth MA 02540

MONAHAN, JAMES EMMETT, b Kansas City, Mo, Jan 10, 25; m 48. THEORETICAL PHYSICS. Educ: Rockhurst Col, BS, 48; St Louis Univ, MS, 50, PhD(physics, math), 53. Prof Exp: SR PHYSICIST, ARGONNE NAT LAB, 51- Concurrent Pos: Weizmann fel, 66-67; prof lectr, St Louis Univ, 66-; vis scientist, Univ Ohio, 71- Mem: Am Phys Soc. Res: Theoretical nuclear physics. Mailing Add: Physics Div-203 Argonne Nat Lab 9700 S Cass Ave Argonne IL 60439

MONAHAN, ROBERT LEONARD, b Hoquiam, Wash, July 26, 26; m 50; c 3. GEOGRAPHY. Educ: Univ Wash, BA, 50; Univ Mich, MA, 51; McGill Univ, PhD(geog), 59. Prof Exp: Instr geog, Sir George Williams Col, 52-53; instr, Univ Wis-Exten, 53-54; instr, Univ Wash, 54-55; from instr to assoc prof, 55-68, actg dean, Col Arts & Sci, 72-74, PROF GEOG, WESTERN WASH STATE COL, 68- Concurrent Pos: Res assoc, Kauppakorkeakoulu, Helsinki, Finland, 66-67. Mem: Asn Am Geog; Arctic Inst NAm; Asn Pac Coast Geog (secy-treas, 64-66). Res: Locational aspects of forest industry and forests; resource geography; economic geography, Canada and Norden. Mailing Add: Col of Arts & Sci Western Wash State Col Bellingham WA 98225

MONAHAN, WAYNE GORDON, b Martins Ferry, Ohio, Oct 4, 40; m 62; c 2. BIOPHYSICS, NUCLEAR MEDICINE. Educ: Ohio State Univ, BS, 63, PhD(physics), 68. Prof Exp: Lectr physics, Ohio State Univ, 66-69; asst prof biophys, Grad Sch MedSci, Cornell Univ, 69-73; ASST PROF RADIOL, MT SINAI SCH MED, 73- Concurrent Pos: Assoc biophys, Sloan-Kettering Inst, 69-73; assoc attend physicist, Mt Sinai Hosp, 73- Mem: Soc Nuclear Med. Res: Nuclear medicine instrumentation including positron annihilation and coincidence detection schemes; computer interface electronics and operating system integration; transverse axial tomography. Mailing Add: Dept of Nuclear Med Mt Sinai Hosp New York NY 10029

MONALOY, STEPHEN EMANUEL, b Paterson, NJ, Dec 13, 41; m 70; c 1. GENETICS. Educ: Univ NMex, BS, 65; Kans State Col Pittsburg, MS, 67; Univ Okla, PhD(zool), 71. Prof Exp: ASST PROF BIOL, UNIV TAMPA, 71- Mem: AAAS; Am Genetic Asn. Res: Gene frequencies in the cat population of the Tampa Bay area; morphological variation of melongena species in the Tampa Bay area; heritability of the shell characteristics in melongena. Mailing Add: Div of Sci & Math Univ of Tampa Tampa FL 33606

MONARD, JOYCE ANNE, b Bethlehem, Pa, Nov 5, 46; m 73; c 1. NUCLEAR PHYSICS, NUCLEAR ENGINEERING. Educ: Bryn Mawr Col, AB, 68; Univ Tenn, Knoxville, PhD(physics), 72. Prof Exp: Fel nuclear physics, Lawrence Berkeley Lab, 72-75; ENGR, NUCLEAR ENERGY DIV, GEN ELEC CO, 75- Mem: Am Phys Soc; Am Nuclear Soc. Res: Gamma ray spectroscopy of reactor fuel; nuclear muon capture; Moessbauer effect in the actinides. Mailing Add: Gen Elec Co Nuclear Energy Div 175 Curtner Ave San Jose CA 95125

MONARO, SERGIO, b Padua, Italy, Apr 29, 32; m 66. NUCLEAR PHYSICS. Educ: Univ Milan, Doctorate, 59; Univ Rome, Libera Docenza, 67. Prof Exp: Fel, Nat Inst Nuclear Physics, Naples, 60-61, Inst Nuclear Physics Res, Amsterdam, 61, Nat Inst Nuclear Physics, Naples, 61-62 & Brookhaven Nat Lab, 62-63; physicist, Euratom, Ispra-Varese, Italy, 64-65; PROF PHYSICS, UNIV MONTREAL, 65- Mem: Am Phys Soc. Res: Nuclear spectroscopy; nuclear reactions at low energy. Mailing Add: Dept of Physics Univ of Montreal Box 6128 Montreal PQ Can

MONCHAMP, ROCH ROBERT, b Manchester, NH, Sept 27, 31; m 69; c 2. SOLID STATE CHEMISTRY. Educ: St Anselms Col, AB, 53; Mass Inst Technol, PhD(chem), 59. Prof Exp: Sr res chemist, Res Lab, Merck & Co, 59-63; mgr crystal growth res & develop, Airtron Div, Litton Precision Prod, Inc, NJ, 64-67 & Raytheon Co, 67-73; chief engr, Saphikon Div, Tyco Labs, Inc, 73-75; MGR CRYSTAL GROWTH DEPT, ADOLF MELLER CO, 75- Mem: AAAS; Am Asn Crystal Growth. Res: Semiconductors; lasers and acoustic crystals; crystal growth; hydrothermal synthesis. Mailing Add: Adolf Meller Co PO Box 6001 Providence RI 02904

MONCHICK, LOUIS, b Brooklyn, NY, Dec 27, 27; m 66. CHEMICAL PHYSICS. Educ: Boston Univ, AB, 48, MA, 51, PhD(chem), 54. Prof Exp: Cloud physicist, Air Force Cambridge Res Ctr, 53-54; fel radiation chem, Univ Notre Dame, 54-56; res assoc fused salts, Knolls Atomic Power Lab, Gen Elec Co, 56-57; CHEMIST, APPL PHYSICS LAB, JOHNS HOPKINS UNIV, 57- Concurrent Pos: Assoc prof chem, Johns Hopkins Univ, 68-69, vis prof chem, 75-76. Mem: Am Chem Soc; Am Phys Soc. Res: Diffusion controlled reactions; kinetic theory of gases; intermolecular forces; molecular collisions. Mailing Add: Johns Hopkins Univ Appl Phys Lab Johns Hopkins Rd Laurel MD 20810

MONCRIEF, JOHN A, b Manila, Philippines, July 22, 24; US citizen; m 49; c 3. SURGERY. Educ: Emory Univ, MD, 48; Am Bd Surg, dipl, 54. Prof Exp: Intern, Brooke Gen Hosp, Med Corps, US Army, 48-49, resident gen surg, Barnes Hosp, St Louis, 49-50 & 51-54, battalion surgeon, Korea, 50, chief gen surg, 50-51, chief gen surg, Army Hosp, Ft Sill, Okla, 54-55, chief clin div, Surg Res Unit, 55-57, instr surg, Sch Med, Emory Univ, 57-60, chief surg res br, Med Res & Develop Command, 60-61, comdr & dir surg res unit, Brooke Army Med Ctr, San Antonio, Tex, 61-68; PROF SURG, MED UNIV SC, 68- Concurrent Pos: Instr, Sch Med, Washington Univ, 54-57; mem, Attend Staff, Grady Mem Hosp, Henrietta Egleston Children's Hosp, Piedmont Hosp & Emory Univ Hosp, 57-60; physician pvt pract, 57-60; consult to Surgeon Gen, 60-; asst prof, Univ Tex, 61-68; mem, Surg Study Sect, NIH, 61-68. Mem: Am Asn Surgeons Trauma (pres); Am Col Surgeons; Am Fedn Clin Res; Am Surg Asn; Am Burn Asn (pres, 70-71). Res: Thermal trauma. Mailing Add: Dept of Surg Med Univ of SC Charleston SC 29401

MONCRIEF, JOHN WILLIAM, b Brunswick, Ga, Jan 23, 41; m 63; c 3. CHEMISTRY. Educ: Emory Univ, BS, 63; Harvard Univ, PhD(phys chem), 66. Prof Exp: Asst prof chem, Amherst Col, 66-68; asst prof, 68-71, ASSOC PROF CHEM, EMORY UNIV, 71- Concurrent Pos: Res grants, NIH, 66-70, Eli Lilly, 68-69, NSF, 71- & Sloan fel, Emory Univ, 68-70. Mem: Am Chem Soc; Am Crystallog Asn. Res: X-ray crystallography; structures of organic and inorganic molecules; relation of mechanism to structures. Mailing Add: Dept of Chem Emory Univ Atlanta GA 30322

MONCURE, HENRY, JR, b Stafford, Va, Feb 16, 30; m 55; c 4. ORGANIC POLYMER CHEMISTRY. Educ: Univ Va, BS, 51, PhD(org chem), 58. Prof Exp: NIH fel, Cambridge Univ, 58-59; from chemist to sr res chemist, 59-68, from res supvr to sr supvr, 68-74, RES MGR, PLASTICS DEPT, E I DU PONT DE NEMOURS & CO, INC, 74- Mem: Am Chem Soc. Res: Ring chain tautomerism; aromatic polymides; polymer stability; composite polymers; fluorocarbon polymers, plastics for use in harsh environments. Mailing Add: 1601 Woodsdale Rd Bellvue Manor Wilmington DE 19809

MOND, BERTRAM, b New York, NY, Aug 24, 31; m 57; c 2. MATHEMATICS. Educ: Yeshiva Univ, BA, 51; Bucknell Univ, MA, 59; Univ Cincinnati, PhD(math), 63. Prof Exp: Mech comput analyst, Gen Elec Co, 59-60; res assoc biomet, Col Med & instr math, Univ Cincinnati, 62-63; res mathematician, Aerospace Res Labs, Wright-Patterson AFB, Ohio, 63-69; dean, Sch Phys Sci, 76-78, PROF MATH, LA TROBE UNIV, 69-, CHMN DEPT, 70- Concurrent Pos: Lectr eve col, Univ Cincinnati, 61-64; ed, J Australian Math Soc, 69-74, mem coun, 69-74. Mem: Am Math Soc; Australian Math Soc; Math Prog Soc. Res: Operations research; linear and nonlinear programming; approximation theory. Mailing Add: Dept of Math La Trobe Univ Bundoora 3083 Melbourne Australia

MONDER, CARL, b US, Aug 24, 28; m 59; c 2. BIOCHEMISTRY. Educ: City Col New York, BS, 50; Cornell Univ, MS, 52; Univ Wis, PhD(biochem), 56. Prof Exp: Assoc technologist, Res Lab, Gen Foods Corp, 52-54; USPHS fel, Sch Med, Tufts Univ, 56-57; from instr to asst prof biochem, Albert Einstein Col Med, 58-69; RES ASSOC PROF BIOMED SCI, MT SINAI SCH MED, 69-; HEAD SECT STEROID STUDIES, RES INST SKELETOMUSCULAR DIS, HOSP FOR JOINT DIS, 64- Concurrent Pos: Vis prof chem, Stevens Inst Technol, 63-69; USPHS career develop award, 69- Mem: AAAS; Am Soc Biol Chem; Am Chem Soc; Endocrine Soc; NY Acad Sci. Res: Amino acid and vitamin D metabolism; transamination; steroid chemistry and metabolism; hormone action. Mailing Add: Hosp for Joint Dis 1919 Madison Ave New York NY 10035

MONDY, NELL IRENE, b Pocahontas, Ark, Oct 27, 21. AGRICULTURAL BIOCHEMISTRY. Educ: Ouachita Univ, BS & BA, 43; Univ Tex, MA, 45; Cornell Univ, PhD(biochem), 53. Prof Exp: Asst prof chem, Ouachita Univ, 43; asst biochem, Univ Tex, 43-45; res assoc, Cornell Univ, 45-46; from instr to asst prof chem, Sampson Col, 46-48; from instr to assoc prof food & Nutrit, Cornell Univ, 48-68; prof food chem, Fla State Univ, 68-71; ASSOC PROF HUMAN NUTRIT & FOOD, CORNELL UNIV, 71- Mem: AAAS; Am Chem Soc; fel Am Inst chem; Inst Food Technol; NY Acad Sci. Res: Vitamin B-6 group; choline and betaine aldehyde dehydrogenase in rats; enzymes; phenols; ascorbic acid; alkaloids and lipids in potatoes. Mailing Add: Div Nutrit Sci N-204A Van Rnslr Cornell Univ Ithaca NY 14850

MONER, JOHN GEORGE, b Bayonne, NJ, Sept 4, 28; m 55; c 3. CELL PHYSIOLOGY, BIOCHEMISTRY. Educ: Johns Hopkins Univ, AB, 49; Princeton Univ, MA, 51, PhD(biol), 53. Prof Exp: From instr to assoc prof physiol, 55-73, PROF ZOOL, UNIV MASS, AMHERST, 73- Concurrent Pos: NSF sci fac fel, Biol Inst, Carlsberg Found, Copenhagen, 61-62; pub health spec fel biol, Univ Calif, San Diego. Mem: Soc Protozool; Am Soc Cell Biol. Res: Physiology and biochemistry of cell division, with emphasis on synchronized tetrahymena. Mailing Add: Dept of Zool Univ of Mass Amherst MA 01003

MONES, ARTHUR HARVEY, physical chemistry, inorganic chemistry, see 12th edition

MONES, ROBERT J, b New York, NY, June 18, 29; m 59; c 5. NEUROLOGY. Educ: Ind Univ, BA, 49; NY Univ, MD, 53. Prof Exp: ASSOC PROF NEUROL, MT SINAI SCH MED, 69- Mem: Soc Neurosci; Am Neurol Asn; Am Acad Neurol. Res: Parkinson's disease; dyskinesias; tremor. Mailing Add: 1160 Fifth Ave New York NY 10029

MONETI, GIANCARLO, b Rome, Italy, Nov 2, 31; m 55; c 5. ELEMENTARY PARTICLE PHYSICS. Educ: Univ Rome, Dr, 54, Libero Docente, 63. Prof Exp: Asst prof physics, Univ Rome, 54-62, Nat Res Coun Italy fel, 54-55; physicist nat comt nuclear energy, Frascati Labs, Italy, 57-60; vis assoc physicist, Brookhaven Nat Lab, 61-62; assoc prof physics, Univ Rome, 62-64 & hist of physics, 64-68; PROF PHYSICS, SYRACUSE UNIV, 68- Concurrent Pos: Mem bd dirs, Nat Agency Nuclear Physics, Italy, 66-68; vis scientist, Europ Orgn Nuclear Res, 74-75. Honors & Awards: Montecatini Prize, Ital Phys Soc, 63. Mem: Ital Phys Soc; Europ Phys Soc; fel Am Phys Soc. Res: History of physics; experimental elementary particle physics; meson and baryon resonances; antiproton annihilation. Mailing Add: Dept Physics 323 Physics Bldg Syracuse Univ Syracuse NY 13210

MONETTE, FRANCIS C, b Lowell Mass, Aug 9, 41; m 68. HEMATOLOGY, CELL PHYSIOLOGY. Educ: St Anselm's Col, BA, 62; NY Univ, MS, 65, PhD(biol), 68. Prof Exp: Teaching fel biol, NY Univ, 62-63, res asst, 63-65, asst res scientist, 65-68; NIH trainee hemat, St Elizabeth's Hosp-Tufts Med Sch, 68-71; ASST PROF BIOL, BOSTON UNIV, 71-, ASST PROF HEALTH SCI, 74- Concurrent Pos: Lectr, Iona Col, 66-67. Mem: AAAS; Am Soc Hemat; Soc Exp Hemat; NY Acad Sci; Am Phys Soc. Res: Experimental hematology; kinetics of cellular proliferation; regulation of cell production and differentiation; erythropoietic physiology and biochemistry; aging processes in mammals. Mailing Add: Dept of Biol Sci Boston Univ 2 Cummington St Boston MA 02215

MONEY, KENNETH ERIC, b Toronto, Ont, Jan 4, 35; m 58; c 1. PHYSIOLOGY, BIOLOGY. Educ: Univ Toronto, BA, 58, MA, 59, PhD(physiol), 61. Prof Exp: Res scientist, Defence Res Med Labs, 61-66, sect head vestibular physiol, 66-76, DIR BIOSCI, DEFENCE & CIVIL INST ENVIRON MED, 76-; ASSOC PROF PHYSIOL, UNIV TORONTO, 72- Concurrent Pos: Res assoc, Univ Toronto, 61-62, lectr, 62-68, asst prof, 69-72. Mem: Can Physiol Soc; fel Royal Soc Health; Aerospace Med Asn. Res: Vestibular physiology; motion sickness; histology of the inner ear; eye movements. Mailing Add: Defence & Civil Inst of Environ Med PO Box 2000 Downsview ON Can

MONEY, WILLIAM LANG, b Centerville, RI, Mar 15, 14; m 46; c 2. ENDOCRINOLOGY. Educ: Brown Univ, AB, 41; Harvard Univ, PhD(zool), 47. Prof Exp: Am Cancer Soc fel, Harvard Univ, 47-49; res asst, Med Col, Cornell Univ, 49-50, res assoc med, 50-51, asst prof physiol, 51-52, from asst prof to assoc prof biol, Sloan-Kettering Div, 52-66; PROF ZOOL, UNIV ARK, FAYETTEVILLE, 66- Concurrent Pos: Res fel med, Harvard Univ & Mass Gen Hosp, 47-48; from asst to assoc, Sloan-Kettering Inst, 49-60, assoc mem, 60-; spec fel, NIH, 51 & Nat Cancer Inst, 52. Mem: AAAS; Harvey Soc; Endocrine Soc; Am Thyroid Asn. Res: Endocrine cancer. Mailing Add: Dept of Zool Univ of Ark Fayetteville AR 72701

MONFORE, GERVAISE EDWIN, b Waverly, Kans, Feb 26, 10; m 36; c 1. INSTRUMENTATION. Educ: Col of Emporia, AB, 32; Univ Denver, MS, 50. Prof Exp: Jr physicist, Nat Bur Stand, 36-41; res physicist, B F Goodrich Co, 42-45; res engr, US Bur Reclamation, 45-55; res engr, Portland Cement Asn, 55-60, from sr res physicist to prin res physicist, 67-73; RETIRED. Mem: Am Soc Testing & Mat; Am Concrete Inst. Res: Properties and behavior of portland cement and concrete, including strength, stress, elasticity, creep, shrinkage, electrical and thermal conductivity, heat of hydration and corrosion of embedded metals; fiber reinforcement; ice pressure. Mailing Add: 4083 S Eaton Ave Springfield MO 65801

MONGEAU, J DENIS, b St Hyacinthe, Que, Dec 1, 30; m 56; c 2. VETERINARY MEDICINE, BACTERIOLOGY. Educ: Univ Montreal, BA, 52, DVM, 56; Univ Toronto, MSc, 58. Prof Exp: Lectr poultry path, Ont Vet Col, 58-59; prof bact, Sch Vet Med, Univ Montreal, 59-67; SCI ADV, BUR VET MED, DRUG DIRECTORATE, HEALTH PROTECTION BR, DEPT HEALTH & WELFARE, CAN, 67- Mem: Am Soc Microbiol; US Animal Health Asn; Can Soc Microbiol. Res: Pathology; infectious diseases. Mailing Add: Bur Vet Med Health Protection Br Dept of Health & Welfare Can Ottawa ON Can

MONGEAU, MAURICE, b Montreal, Que, Oct 17, 20; m 49; c 6. PHYSICAL MEDICINE. Educ: Univ Montreal, BA, 42, MD, 48; Am Bd Phys Med & Rehab, dipl, 55. Prof Exp: Asst med dir, Rehab Inst Montreal, 53-62; asst dir, Sch Rehab, 55-70, ASSOC PROF MED, FAC MED, UNIV MONTREAL, 62-; CHIEF SERV PHYS MED & REHAB, REHAB INST MONTREAL, 62- Concurrent Pos: Consult, Dept Vet Affairs, St Anne's Hosp, 54-60 & Pasteur Hosp, 54-; secy, Can Asn Univ Schs Phys Med & Rehab, 59-65. Mem: Can Med Asn; Can Asn Phys Med & Rehab (secy-treas, 56-61, pres, 65-66). Res: Medico-psycho-social aspects in rehabilitation of hemiplegia, cerebral palsy, poliomyelitis, paraplegia and amputee; bioengineering in prosthetics for upper extremity amputee and lower extremity amputee; congenital malformations of the extremities. Mailing Add: Rehab Inst of Montreal 6300 Darlington Ave Montreal PQ Can

MONGELARD, JOSEPH CYRIL, b Mauritius, Apr 18, 33; m 57; c 2. PLANT PHYSIOLOGY, PLANT ECOLOGY. Educ: Univ London, BSc, 60, Imp Col, dipl, 64, MSc, 65, PhD(plant physiol), 68. Prof Exp: Sci asst, Mauritius Dept Agr, 53-56; sci asst, Mauritius Sugar Indust Res Inst, 56-59, asst botanist, 60-64, head dept plant physiol, 66-69; PLANT PHYSIOLOGIST, HAWAIIAN SUGAR PLANTERS' ASN, 69- Concurrent Pos: Lectr, Col Agr Mauritius, 64; mem, London Inst Biologists, 65; sr lectr, Univ Mauritius, 68-69; affil grad fac mem, Univ Hawaii, 75- Mem: Int Soc Sugar Cane Technologists; Crop Sci Soc Am; Am Soc Agron. Res: Plant-water relations; irrigation; photosynthesis; temperature effects on plant growth; screening of herbicides and weed control; environmental biology; psychosociology. Mailing Add: Hawaiian Sugar Planters' Asn Phys & Biochem Dept 99-193 Aiea Heights Dr Aiea HI 96701

MONGINI, PATRICIA KATHERINE ANN, b Cottonwood, Ariz, Dec 13, 50. IMMUNOLOGY. Educ: Northern Ariz Univ, BS, 72; Stanford Univ, PhD(med microbiol), 76. Prof Exp: FEL IMMUNOL, SCH MED, TUFTS UNIV, 76- Mem: Am Soc Microbiol. Res: The role of lymphocyte trapping in the induction of immune responses to antigens; polyclonal activation of T lymphocytes by heterologous anti-immunoglobulin. Mailing Add: Dept of Path Sch Med Tufts Univ Boston MA 02111

MONHEIMER, RICHARD HERBERT, microbial ecology, see 12th edition

MONIE, IAN WHITELAW, b Paisley, Scotland, May 24, 18; nat US; m 42; c 4. ANATOMY. Educ: Glasgow Univ, MB, ChB, 40, MD, 72. Prof Exp: From demonstr to lectr anat, Glasgow Univ, 42-47; from asst prof to assoc prof, Univ Man, 47-52; from asst prof to prof anat, 52-70, vchmn dept, 58-63, chmn dept, 63-70, PROF ANAT & EMBRYOL, UNIV CALIF, SAN FRANCISCO, 70- Concurrent Pos: Cur of unclaimed dead, Northern Calif, 60-66; mem study sect human develop, NIH, 65-69; Guggenheim fel, 67-68. Mem: AAAS; Am Asn Anat; Anat Soc Gt Brit & Ireland; Teratology Soc (pres, 64-65). Res: Mammalian embryology and teratology; human gross anatomy; comparative embryology. Mailing Add: Dept of Anat Univ of of Calif San Francisco CA 94143

MONIZ, WILLIAM BETTENCOURT, b New Bedford, Mass, Feb 12, 32; m 55; c 6. PHYSICAL CHEMISTRY. Educ: Brown Univ, BS, 53; Pa State Univ, PhD(org chem), 60. Prof Exp: Petrol Res Fund fel, Pa State Univ, 60-61; NIH vis scientist, Univ Ill, 61-62; head nuclear magnetic resonance spectros sect, Chem Div, 62-74, HEAD ORG MECHANISMS SECT, CHEM DIV, NAVAL RES LAB, 74- Concurrent Pos: Mem admis comt, Am Chem Soc, 65-66. Mem: Am Chem Soc; NY Acad Sci; Sigma Xi. Res: Magnetic resonance; materials degradation; polymer characterization. Mailing Add: Code 6120 Naval Res Lab Washington DC 20375

MONJAN, ANDREW ARTHUR, b New York, NY, Feb 9, 38; m 69; c 2. NEUROSCIENCES, IMMUNOPATHOLOGY. Educ: Rensselaer Polytech Inst, BS, 60; Univ Rochester, PhD(psychol), 65; Johns Hopkins Univ, MPH, 71. Prof Exp: USPHS res fel, Univ Rochester, 64-66; asst prof psychol & physiol, Univ Western Ont, 66-69; USPHS community health trainee, 69-70, asst prof, 71-75, ASSOC PROF EPIDEMIOL, SCH HYG & PUB HEALTH, JOHNS HOPKINS UNIV, 75- Mem: Am Psychol Asn; Asn Res Vision & Ophthal; Can Physiol Asn; Psychonomic Soc; Soc Neurosci. Res: Effects of viruses upon the nervous system; developmental pathogenesis, mechanisms of immunopathology and long term psychological sequela. Mailing Add: Dept of Epidemiol Sch of Hyg & Pub Health Johns Hopkins Univ Baltimore MD 21205

MONK, CARL DOUGLAS, b Hurdles Mill, NC, July 28, 33; m 57; c 2. PLANT ECOLOGY. Educ: Duke Univ, AB, 55; Rutgers Univ, MS, 58, PhD(bot), 59. Prof Exp: Asst prof bot, Univ Fla, 59-64; from asst prof to assoc prof, 64-71, PROF BOT, UNIV GA, 71- Mem: Bot Soc Am; Ecol Soc Am; Torrey Bot Club; Am Soc Naturalists; Brit Ecol Soc. Res: Vegetation analysis; mineral cycling. Mailing Add: Dept of Bot Univ of Ga Athens GA 30602

MONK, DONALD WAYNE, physical chemistry, see 12th edition

MONK, JAMES DONALD, b Childress, Tex, Sept 27, 30; m 53; c 2. MATHEMATICS. Educ: Univ Chicago, AB, 51; Univ NMex, BS, 56; Univ Calif, Berkeley, MA, 59, PhD(math), 61. Prof Exp: Math analyst, Los Alamos Sci Labs, 51-53; instr math, Univ Calif, Berkeley, 61-62, asst res mathematician, 63-64; asst prof, 62-63 & 64-65, assoc prof, 65-67, PROF MATH, UNIV COLO, BOULDER, 67- Concurrent Pos: NSF res grants, 63-; vis prof, Univ Calif, Berkeley, 67-68. Mem: Am Math Soc; Asn Symbolic Logic. Res: Algebraic logic; general algebra; model theory; foundations of set theory. Mailing Add: Dept of Math Univ of Colo Boulder CO 80302

MONK, MARY ALICE, b Racine, Wis, May 23, 26; m 67. EPIDEMIOLOGY, PUBLIC HEALTH. Educ: Oberlin Col, AB, 48; Univ Mich, MA, 51, PhD(psychol), 54. Prof Exp: Instr prev med, Univ Buffalo, 53-58; asst prof epidemiol, Med Sch, Tulane Univ, 58-60; from asst prof to assoc prof chronic dis, Sch Hyg & Pub Health, Johns Hopkins Univ, 60-69; assoc prof environ med, State Univ NY Downstate Med Sch, 69-70; PROF COMMUNITY & PREV MED, NEW YORK MED COL, 70- Concurrent Pos: Mem epidemiol & dis control study sect, NIH, 72-75. Mem: AAAS; Am Pub Health Asn; Am Psychol Asn; Am Statist Asn. Res: Epidemiology of chronic diseases such as mental illness, gastrointestinal disease and cancer; medical education and medical careers. Mailing Add: Dept of Community & Prev Med New York Med Col New York NY 10029

MONK, RALPH WARNER, b Spanish Fork, Utah, Apr 12, 13; m 40; c 3. PLANT PHYSIOLOGY. Educ: Brigham Young Univ, AB, 37; Colo State Univ, MS, 39; Utah State Univ, PhD, 60. Prof Exp: Jr soil scientist, USDA, 39, soil technologist, Soil Conserv Serv, NMex, 40-42; chemist, Kalunite, Inc, Utah, 42-43; instr pub sch, 44-46; instr & microbiologist, 47-58, assoc prof bot, 60-62, dir acad res & prog develop, 70-74, head dept bot, 60-70, asst dean sch arts, letters & sci, 67-70, PROF BOT, WEBER STATE COL, 62- Mem: AAAS; Am Soc Plant Physiol; Am Soc Hort Sci. Res: Physiology of salt tolerance. Mailing Add: Dept of Bot Weber State Col Ogden UT 84403

MONKHOUSE, FRANK C, b Sask, Can, Dec 1, 13; m 39; c 2. PHYSIOLOGY. Educ: Univ Sask, BA, 43; Univ Toronto, PhD, 52. Prof Exp: Instr physiol, Univ Sask, 43-49; res assoc, Univ Toronto, 49-52, asst prof, 52-54; vis asst prof, Wayne State Univ, 54-55; assoc prof, 55-58, PROF PHYSIOL, UNIV TORONTO, 58- Mem: Am Physiol Soc; Soc Exp Biol & Med; Can Physiol Soc; Int Soc Hemat. Res: Blood coagulation and cardiovascular diseases. Mailing Add: Dept of Physiol Univ of Toronto Toronto ON Can

MONLUX, ANDREW W, b Algona, Iowa, Jan 29, 20; m 50; c 2. VETERINARY PATHOLOGY. Educ: Iowa State Univ, DVM, 42, MS, 47; George Washington Univ, PhD(comp path), 51. Prof Exp: Pvt pract, Iowa, 42; asst, Iowa State Univ, 46-47; vet, USDA, Colo & DC, 51-56; head dept, 56-72, PROF VET PATH, OKLA STATE UNIV, 56-, REGENTS PROF, 72- Mem: Am Vet Med Asn; US Animal Health Asn; Conf Res Workers Animal Dis; Vet Cancer Soc; Vet Urol Soc. Res: Pathology of neoplastic and kidney diseases; lead and photosensitivity syndromes of animals. Mailing Add: Dept of Vet Path Okla State Univ Stillwater OK 74074

MONLUX, WILLIAM S, b Algona, Iowa, Jan 8, 15; m 56. PATHOLOGY, MICROBIOLOGY. Educ: Iowa State Univ, DVM, 37; Cornell Univ, PhD, 48. Prof Exp: Instr path, Cornell Univ, 37-41; fel, Royal Vet Col Sweden, 48; pathologist, Bur Animal Indust, Washington, DC, 49; fel path, Univ Pretoria, 49-50; prof, Tex A&M Univ, 50-53 & Iowa State Univ, 53-62; leader path invest, Nat Animal Dis Lab, Iowa, 62-63, asst dir res, 63-72; RES, FOOD & DRUG ADMIN, 72- Mem: Am Vet Med Asn; Conf Res Workers Animal Dis; Am Col Vet Path; US Animal Health Asn; Am Asn Vet Lab Diagnosticians. Res: Canine leptospirosis; toxoplasmosis; oncology, especially pulmonary neoplasms; porcine and bovine nutrition; pulmonary adenomatosis; erysipelas. Mailing Add: Food & Drug Admin 200 C St HFF134 Washington DC 20204

MONN, DONALD EDGAR, b Chambersburg, Pa, June 21, 38; m 63; c 1. ANALYTICAL CHEMISTRY. Educ: Elizabethtown Col, BS, 59; Univ Del, PhD(anal chem), 64. RES GROUP LEADER, CONTINENTAL OIL CO, 64- Mem: Am Chem Soc. Res: Thermometric and phase titrations; gas chromatography; interfacing analytical instruments to computers and data acquisition equipment; writing computer programs to perform analytical calculations; techniques of data evaluation. Mailing Add: Continental Oil Co PO Drawer 1267 Ponca City OK 74601

MONNETT, VICTOR BROWN, b Ithaca, NY, Dec 6, 15; m 40; c 2. GEOLOGY. Educ: Univ Okla, BS, 37; Univ Mich, PhD(geol), 47. Prof Exp: Asst geologist, Carter Oil Co, 38-40; head dept geol, 50-74, geog, 55-74, PROF GEOL, OKLA STATE UNIV, 47-, ASSOC DEAN COL ARTS & SCI, 66- Mem: Geol Soc Am; Am Asn Petrol Geol. Res: Subsurface stratigraphy; marsh formation of Michigan. Mailing Add: Col of Arts & Sci Okla State Univ Stillwater OK 74074

MONNETT, VICTOR ELVERT, geology, deceased

MONNIER, DWIGHT CHAPIN, b Hartford, Conn, Mar 30, 18; m 38; c 3. ACADEMIC ADMINISTRATION. Educ: Univ Bridgeport, BS, 40; Columbia Univ, MA, 47; Univ Buffalo, EdD, 52. Prof Exp: Dir phys educ, St Paul's Sch, NH, 45-47; dir health & phys educ, Orchard Park Cent Sch, NY, 47-50; exec dir vol agency, Western NY Comt Alcoholism, 50-55; chief health educ adv, Ministry of Health, Govt Pakistan, 55-57; dir div training & res, Mass Dept Pub Health, 57-59; supvr med res prog, Nat Inst Neurol Dis & Blindness, 59-60, prog coordr hearing & speech, 60-62, exec dir grants assocs prog, Div Res Grants, NIH, 62-64, asst chief training grants, Career Develop Rev Br, 64-67; vpres for admin, Am Univ Beirut, 67-73; ASST EXEC DIR, AM COL CARDIOL, 73- Concurrent Pos: Guest lectr, Univ Buffalo, 54. Honors & Awards: Knight, Order of Cedars, Govt of Lebanon, 73. Mem: Lebanese Mgt Asn; NY Acad Sci. Res: Application of economic laws and management principles to institutions of higher education. Mailing Add: Am Col of Cardiol 9650 Rockville Pike Bethesda MD 20014

MONOSON, HERBERT L, b Chicago, Ill, Dec 23, 36; m 65; c 2. MYCOLOGY. Educ: Western Ill Univ, BS, 58, MS, 60; Univ Ill, Urbana, PhD(bot), 67. Prof Exp: Asst prof, 66-71, ASSOC PROF BIOL, BRADLEY UNIV, 71- Concurrent Pos: Res Corp grant, Bradley Univ, 71-72. Mem: Mycol Soc Am; Am Phytopath Soc. Res: Phycology; nematology; hormonal regulation of fungi; rust fungi taxonomy. Mailing Add: Dept of Biol Bradley Univ Peoria IL 61606

MONOSTORI, BENEDICT JOSEPH, b Kövagoörs, Hungary, July 4, 19; US citizen. PHYSICS. Educ: Pazmany Peter Univ, Budapest, BS, 45; Pontifical Univ, St Anselm, Rome, MA, 51; Fordham Univ, PhD(physics), 64. Prof Exp: Instr math & physics, St Stephen's Acad, Hungary, 45-48; prof philos, St Bernard's Col, Hungary, 48-50; instr math, Acad Mary Immaculate, Wichita Falls, Tex, 55-56; instr math & philos, 56-60, asst prof physics, 60-64, ASSOC PROF PHYSICS, UNIV DALLAS, 64-, CHMN DEPT, 71- Mem: Am Phys Soc; Am Asn Physics Teachers. Res: Molecular spectra and structure; Raman spectroscopy; philosophy of science. Mailing Add: Dept of Physics Univ of Dallas Irving TX 75061

MONROE, BARBARA SAMSON GRANGER, b New York, NY, Aug 23, 13; m 51; c 1. ANATOMY. Educ: Mt Holyoke Col, AB, 35, AM, 37; Cornell Univ, PhD(zool), 43. Prof Exp: Instr zool, Wilson Col, 40-41; instr, Univ Ariz, 43-45; from instr to assoc prof, 45-66, PROF ANAT, SCH MED, UNIV SOUTHERN CALIF, 66- Mem: Am Soc Zool; Am Asn Anat; Electron Micros Soc Am. Res: Endocrinology; histology; electron microscopy of the intestine, thyroid, neurohypophysis and hypothalamus. Mailing Add: Dept of Anat Sch of Med Univ of Southern Calif Los Angeles CA 90033

MONROE, BRUCE MALCOLM, b Indianapolis, Ind, July 7, 40; m 68. ORGANIC CHEMISTRY. Educ: Wabash Col, AB, 62; Univ Ill, Urbana, MS, 64, PhD(org chem), 67. Prof Exp: NIH fels, Calif Inst Technol, 67-69; res chemist, Explosives Dept, 69-71, RES CHEMIST, CENT RES DEPT, EXP STA, E I DU PONT DE NEMOURS & CO, INC, 71- Mem: Am Chem Soc; The Chem Soc; Sigma Xi. Res: Organic photochemistry. Mailing Add: Cent Res Dept Exp Sta E I du Pont de Nemours & Co Inc Wilmington DE 19898

MONROE, BURT LEAVELLE, JR, b Louisville, Ky, Aug 25, 30; m 60. ORNITHOLOGY. Educ: Univ Louisville, BS, 53; La State Univ, PhD(zool), 65. Prof Exp: Vis instr zool, La State Univ, 65; from asst prof to assoc prof, 65-74, PROF VERT ZOOL, UNIV LOUISVILLE, 74-, CHMN DEPT BIOL, 71- Mem: Soc Syst Zool; Am Ornith Union; Wilson Ornith Soc; Cooper Ornith Soc. Res: Distribution and systematics of birds, especially Neotropical; entomology, especially Lepidoptera and Coleoptera; zoogeography. Mailing Add: Dept of Biol Univ of Louisville Louisville KY 40208

MONROE, ELIZABETH MCLEISTER, b Pittsburgh, Pa, Dec 11, 40; m 68. ORGANIC CHEMISTRY. Educ: Bucknell Univ, BS, 62; Univ Ill, Urbana, MS, 64, PhD(org chem), 68. Prof Exp: Info chemist, 69-71, SR INFO CHEMIST, E I DU PONT DE NEMOURS & CO, INC, 71- Mem: Am Chem Soc. Res: Chemical information. Mailing Add: Info Systs Dept E I du Pont de Nemours & Co Inc Wilmington DE 19898

MONROE, EUGENE ALAN, b Kansas City, Kans, May 31, 34; m 54; c 3. MINERALOGY, CRYSTALLOGRAPHY. Educ: Univ Wis, BS, 55; Univ Ill, MS, 59, PhD(geol-mineral), 61; Columbia Univ, DDS(dent), 73. Prof Exp: Asst prof, 61-73, ASSOC PROF X-RAY DIFFRACTION CRYSTALLOG, STATE UNIV NY COL CERAMICS, ALFRED UNIV, 73- Mem: Am Crystallog Asn; Sigma Xi; NY Acad Sci; Am Dental Asn; Calcified Tissue Group. Res: Electron microscopy and crystallography of materials; biomedical material development; mineralogical studies;

dental research. Mailing Add: State Univ of NY Col of Ceramics Alfred Univ Alfred NY 14802

MONROE, EZRA, b Soda Springs, Idaho, Jan 21, 10; m 44. PHARMACEUTICAL CHEMISTRY. Educ: Idaho State Univ, BS, 33; Univ Mich, MS, 34, PhD(pharmaceut chem), 36. Prof Exp: Chemist & group leader, Dow Chem Co, 36-54, coordr screening & testing, Exec Res Dept, 54-55, dir appln liaison ctr, 55-66, admin dir res info serv, 66-68, staff asst to vpres & dir corp res & develop, 68-69; assoc dir res, Saginaw Valley State Col, 69-72, asst to dir community rels, 72, dir contracts & govt grants, 72-75; RETIRED. Concurrent Pos: Consult, Saginaw-Midland Water Line, 75- Mem: Am Chem Soc. Res: Organo-arsenic compounds, antispasmodics; insecticides; Friedel-Craft reactions; antimalarials; organic process development; organic synthesis; screening and testing; research publicity; technical writing and information services; environmental research and studies; water quality research. Mailing Add: 1012 S Balfour St Midland MI 48640

MONROE, PEARLE ARVEL, b Bowdoin, Mont, Jan 6, 21; m 46; c 2. ORGANIC CHEMISTRY, BIOCHEMISTRY. Educ: Univ Idaho, BS, 47, MS, 49; Ind Univ, PhD(org chem), 52. Prof Exp: Asst chem, Univ Idaho, 47-48, actg instr, 48-49; res chemist, Goodyear-Sumatra Plantations Co, 52-58; PROF ORG CHEM & BIOCHEM, STATE UNIV NY COL OSWEGO, 58- Concurrent Pos: NIH spec fac fel, Univ Calif, San Francisco, 64. Mem: Am Chem Soc. Res: Synthesis and properties of amino-thiophenes and their derivatives; natural rubber; thiophene derivatives of biological interest. Mailing Add: Dept of Chem State Univ of NY Col at Oswego Oswego NY 13126

MONROE, ROBERT ADAMS, b Weymouth, Mass, Jan 5, 22; m 54; c 2. LABORATORY MEDICINE. Educ: Univ Mass, BS, 44; Univ Mo, MA, 46, PhD(endocrinol), 49. Prof Exp: Asst dairying, Univ Mo, 45-47, instr, 48-49; asst prof & res asst, AEC Prog, Univ Tenn, 50-51, assoc prof & res assoc, 52-55; chief fission prod metab sect, AEC Proj, Univ Calif, Los Angeles, 55-57; asst prof physiol, Cornell Univ, 57-60; group leader, Nat Dairy Prod Corp, 60-65; res biochemist, Nuclear Chicago Corp, Ill, 65-70, group leader, 70-76; GROUP LEADER, SEARLE MED INSTRUMENT GROUP RES, DES PLAINES, 76- Res: Instrumentation and tracer methods in laboratory diagnostics. Mailing Add: 1106 W Clarendon Rd Arlington Heights IL 60004

MONROE, ROBERT JAMES, b Dysart, Iowa, Dec 28, 18; m 42. STATISTICS. Educ: Iowa State Univ, BS, 39; NC State Col, PhD, 49. Prof Exp: Res collabr & supvr, Statist Lab, USDA, Iowa State Univ, 39-41; instr mach comput, NC State Col, 41-42, instr statist, 46-49, asst prof statist & plant sci statistician, 48-52, assoc prof esp statist, 52-53; chief, Oper Anal Off, Air Force Missile Test Ctr, 53-S4; PROF EXP STATIST, SCH PHYS & MATH SCI, NC STATE UNIV, 54- Concurrent Pos: Agent, USDA, 41-42; vis prof, Med Col Va, 61-62. Mem: Biomet Soc; fel Am Statist Asn. Res: Estimation of nutrition requirements; statistical methodology; biometry; operations research. Mailing Add: Dept of Statist NC State Univ Box 5457 Raleigh NC 27607

MONROE, RONALD EUGENE, b Porterville, Calif, Jan 17, 33; m 58; c 1. ENTOMOLOGY. Educ: Fresno State Col, BA, 56; Ore State Col, MS, 58; Kans State Univ, PhD(entom), 64. Prof Exp: Jr vector control officer, Bur Vector Control, State Dept Pub Health, Calif, 55-56; med entomologist insects affecting man & animals, Agr Res Serv, USDA, Ore, 57-58, gen entomologist, Insect Physiol Lab, 58-61, from asst prof to prof entom, Mich State Univ, 64-73; PROF ZOOL, SAN DIEGO STATE UNIV, 73- Mem: Entom Soc Am. Res: Biochemistry and physiology of insects, chiefly lipid, carbohydrate and amino acid metabolism and insect nutrition and reproduction. Mailing Add: Dept of Zool San Diego State Univ San Diego CA 92182

MONROE, RUSSELL RONALD, b Des Moines, Iowa, June 7, 20; m 45; c 3. MEDICINE. Educ: Yale Univ, BS, 42, MD, 44. Prof Exp: Asst med, Yale Univ, 45-46; from asst prof to assoc prof psychiat, Sch Med, Tulane Univ, 50-60; PROF PSYCHIAT, SCH MED, UNIV MD, BALTIMORE, 60- Concurrent Pos: Assoc psychoanalyst, Col Physicians & Surgeons, Columbia Univ, 51-54. Mem: Am Psychosom Soc; Am Psychiat Asn; Acad Psychoanal. Res: Schizophrenia; psychotherapy. Mailing Add: Inst of Psychiat & Human Behav Univ of Md Sch of Med Baltimore MD 21201

MONROE, STUART BENTON, b Manassas, Va, Oct 26, 34; m 60; c 2. ORGANIC CHEMISTRY, POLYMER CHEMISTRY. Educ: Randolph-Macon Col, BS, 56; Univ Fla, PhD(org chem), 62. Prof Exp: Res chemist, Hercules Res Ctr, Del, 61-65; assoc prof, 65-67, PROF ORG CHEM, RANDOLPH-MACON COL, 67-, CHMN DEPT CHEM, 75- Mem: Am Chem Soc. Res: Flame retardant and thermally stable polymers; chemistry of isocyanates. Mailing Add: Dept of Chem Randolph-Macon Col Ashland VA 23005

MONROE, WATSON HINER, b Parkersburg, WVa, Dec 1, 07; m 33; c 1. GEOLOGY. Prof Exp: From jr geologist to geologist, 30-49, chief, Eastern Field Invests Sect, Fuels Br, 49-53, staff geologist, 53-55, chief, P R Coop Invest, Gen Geol Br, 55-66, res geologist, Atlantic Environ Br, 66-72, CONSULT GEOLOGIST, US GEOL SURV, 72- Concurrent Pos: Lectr, Univ PR, 60-65 & 68-69. Mem: Fel Geol Soc Am; Am Asn Petrol Geol. Res: Stratigraphy of Cretaceous and Tertiary rocks of the Gulf Coast Plain; karst geology; stratigraphy, structure and economic geology of Puerto Rico. Mailing Add: 126 E North St Leesburg VA 22075

MONSE, ERNST ULRICH, b Bautzen, Ger, Jan 1, 27. PHYSICAL CHEMISTRY. Educ: Univ Mainz, MS, 53, PhD(phys chem), 57. Prof Exp: Res assoc phys chem, Columbia Univ, 57-59; res assoc, 59-64, assoc prof, 64-74, PROF PHYS CHEM, RUTGERS UNIV, NEWARK, 74- Mem: NY Acad Sci; Sigma Xi. Res: Isotope effects and their correlation with molecular structures and force fields. Mailing Add: Dept of Chem Rutgers Univ 73 Warren St Newark NJ 07102

MONSEN, HARRY, b Trondheim, Norway, Aug 24, 24; nat US; m 50; c 2. ANATOMY. Educ: Univ Minn, MS, 51; Univ Ill, PhD(anat), 54. Prof Exp: Asst anat, Univ Minn, 49-51; asst cancer biol, 51-54, from instr to assoc prof, 54-70, PROF ANAT, UNIV ILL COL MED, 70- Res: Cancer biology; carcinogenesis; pituitary-adrenal-gonadal interrelationships. Mailing Add: Dept of Anat Univ of Ill Col of Med Chicago IL 60680

MONSIMER, HAROLD GENE, b Las Vegas, NMex, Feb 5, 28; m 63; c 2. ORGANIC CHEMISTRY. Educ: Univ Calif, BS, 52; Wayne State Univ, MS, 54, PhD(chem), 57. Prof Exp: Sr res chemist, Nat Drug Co, 56-65; head org chem, MacAndrews & Forbes Co, 65-66; sr res chemist, 66-73, PROJ LEADER, PENNWALT CORP, 73- Mem: Am Chem Soc; The Chem Soc. Res: Medicinal chemistry; organic synthesis. Mailing Add: 2916 Toll Gate Dr Norristown PA 19403

MONSON, ARVID MONROE, b Pequot Lakes, Minn, Feb 9, 38; m 62; c 1. MICROBIOLOGY, GENETICS. Educ: Univ Minn, BS, 59, MS, 64, PhD(genetics), 68. Prof Exp: ASST PROF BIOL, ALLEGHENY COL, 67- Concurrent Pos: Fel, Univ NC-Chapel Hill, 72-73. Mem: AAAS; Am Phytopath Soc; Am Soc Microbiol; Am Inst Biol Sci. Res: Genetic homologies among microorganisms; mutagenesis in microorganisms. Mailing Add: Dept of Biol Allegheny Col Meadville PA 16335

MONSON, FREDERICK CARLTON, b Philadelphia, Pa, Aug 3, 39; m 65; c 2. HISTOLOGY, REPRODUCTIVE BIOLOGY. Educ: Lehigh Univ, BA, 65, MS, 67, PhD(biol), 71. Prof Exp: ASST PROF BIOL, ST JOSEPH'S COL, PA, 71- Mem: Soc Study Reproduction; Soc Develop Biol; Am Soc Zoologists; AAAS. Res: Development of the Sertoli cell population in the mouse testis; contractility, metabolism and general histochemistry of the seminiferous tubule in mammalian testes. Mailing Add: Dept of Biol St Joseph's Col Philadelphia PA 19131

MONSON, GALE (WENDELL), b Munich, NDak, Aug 1, 12; m 41; c 5. ORNITHOLOGY, WILDLIFE MANAGEMENT. Educ: NDak State Univ, BS, 34. Prof Exp: Range exam, US Indian Serv, 34-35; biologist, US Soil Conserv Serv, 35-40; biologist, US Fish & Wildlife Serv, 40-42, refuge mgr, 42-62, staff asst, Div Wildlife Refuges, Wash, DC, 62-69; ed, Atlantic Naturalist, 69-71; WEEKEND SUPVR, ARIZ-SONORA DESERT MUS, 71- Honors & Awards: Am Motors Conserv Award, 59; Distinguished Serv Award, US Interior Dept, 70. Mem: Wildlife Soc; Am Soc Mammal; Cooper Ornith Soc; Wilson Ornith Soc; Am Ornith Union. Res: Distribution and migration of birds in Arizona and New Mexico; desert bighorn sheep; botany, distribution of desert plants; ecology, desert relationships. Mailing Add: 8831 N Riviera Dr Tucson AZ 85704

MONSON, PAUL HERMAN, b Fargo, NDak, Sept 29, 25; m 50; c 3. PLANT TAXONOMY. Educ: Luther Col, BA, 50; Iowa State Univ, MS, 52, PhD(plant taxon), 59. Prof Exp: Instr biol, Luther Col, 52-55; instr bot, Iowa State Univ, 57-58; from asst prof to assoc prof, 58-68, PROF BIOL, UNIV MINN, DULUTH, 68-, CUR, OLGA LEKELA HERBARIUM, 74- Concurrent Pos: Partic, NSF Acad Year Inst, Brown Univ, 65-66. Mem: Bot Soc Am; Am Soc Plant Taxon; Nat Asn Biol Teachers. Res: Flora of the midwest and aquatic plants; Nymphaea. Mailing Add: Dept of Biol Univ of Minn Duluth MN 55812

MONSON, RICHARD STANLEY, b Los Angeles, Calif, May 28, 37; m 66; c 2. CHEMISTRY. Educ: Univ Calif, Los Angeles, BS, 59; Univ Calif, Berkeley, PhD(chem), 64. Prof Exp: From asst prof to assoc prof, 63-72, PROF CHEM, CALIF STATE UNIV, HAYWARD, 72-, CHMN DEPT, 70- Concurrent Pos: Fulbright-Hays fel & lectr, Univ Sarajevo, Yugoslavia, 72-73. Mem: Am Chem Soc. Res: Organic reaction mechanisms; novel methods of organic synthesis; reactions of organophosphorus intermediates. Mailing Add: Dept of Chem Calif State Univ Hayward CA 94542

MONSON, WARREN GLENN, b Clay Center, Nebr, Dec 24, 26; m 58; c 2. AGRONOMY. Educ: Univ Nebr, BSc, 51, MSc, 55, PhD, 58. Prof Exp: Asst agron, Univ Nebr, 57-58; res agronomist, NY, 58-66, RES AGRONOMIST, AGR RES SERV, USDA, GA, 66- Mem: Am Soc Agron; Crop Sci Soc Am; Am Forage & Grassland Coun. Res: Physiology and quality of pasture and forage crops. Mailing Add: Agr Res Serv USDA Tifton GA 31794

MONSON, WILLIAM JOYE, b Menomonie, Wis, Jan 12, 27; m 49; c 1. POULTRY NUTRITION, ANIMAL NUTRITION. Educ: Beloit Col, BS, 49; Univ Wis, MS, 51, PhD(biochem), 53. Prof Exp: Tech adv, Nutrit Res Lab, Chem Div, 53-58, dir tech serv, Feed Suppl Div, 58-62, TECH DIR, NUTRIT RES LAB, CHEM DIV, BORDEN, INC, 62- Mem: AAAS; Poultry Sci Asn; Animal Nutrit Res Coun (pres, 74); Am Inst Nutrit; NY Acad Sci. Res: General nutrition of poultry, large animals and pets with particular emphasis on unknown factors and minerals. Mailing Add: Nutrit Res Lab Chem Div Borden Inc RR 1 Elgin IL 60120

MONSOUR, VICTOR, b Shreveport, La, Aug 28, 22; m 50; c 2. MICROBIOLOGY. Educ: La State Univ, Baton Rouge, BS, 48, MS, 50; Univ Tex, Austin, PhD(microbiol), 54. Prof Exp: Bacteriologist, Shreveport, Charity Hosp, La, 50-51; microbiologist, Confederate Med Ctr, Shreveport, La, 54-57; asst dir bur of lab, Div Health Mo, 57-59; PROF MICROBIOL & HEAD DEPT, MCNEESE STATE UNIV, 59- Concurrent Pos: Consult, Ark-La-Tex area hosps & clins, 54-56; on loan from Div Health Mo to sch Med, Wash Univ, 58-59; consult, area industs, La, 68- Mem: Am Soc Microbiol; Am Chem Soc; Am Pub Health Asn; Nat Environ Health Asn. Res: Microbial metabolism; environmental science; chemical andand/or biological pollution. Mailing Add: Dept of Microbiol McNeese State Univ Lake Charles LA 70601

MONT, GEORGE EDWARD, b New Bedford, Mass, Aug 6, 35; m 57; c 4. POLYMER CHEMISTRY. Educ: Brown Univ, BS, 57; Clark Univ, MA, 59, PhD(chem), 64. Prof Exp: Res chemist, Shawiningan Resins Corp, 61-63; RES SPECIALIST CHEM, MONSANTO CO, 63- Res: Polymer structure-property relationships, poly(vinyl butyral) chemistry. Mailing Add: Monsanto Co 730 Worcester St Indian Orchard MA 01151

MONTAGNA, AMELIO EMIDIO, b Roccacasale Abruzii, Italy, Sept 16, 15; nat US; m 42; c 3. CHEMISTRY. Educ: Bethany Col, BS, 36. Prof Exp: Proj chemist, Chem Div, Union Carbide Corp, 37-49, group leader, 49-55, asst dir, 55-57, assoc dir, 57-64, dir tech progs, 64-65, vpres, Fibers & Fabrics Div, 65-70, vpres & gen mgr process chem div, 70-74; MEM STAFF, GULF RES & DEVELOP CO, 74- Mem: Am Chem Soc. Res: Mathematics; physics; synthetic organic chemicals and polymers. Mailing Add: Gulf Res & Develop Co PO Box 2038 Pittsburgh PA 15230

MONTAGNA, WILLIAM, b Roccacasale, Italy, July 6, 13; nat US; m 39; c 4. CYTOLOGY. Educ: Bethany Col, WVa, AB, 36; Cornell Univ, PhD(histol, embryol), 44. Hon Degrees: DSc, Bethany Col, WVa, 60; DBiolSci, Univ Sardinia, 64. Prof Exp: Instr zool, Cornell Univ, 44-45; from instr to asst prof anat, Long Island Col Med, 45-47; from asst prof to Herbert L Ballou prof biol, Brown Univ, 48-63; DIR, ORE REGIONAL PRIMATE RES CTR & PROF EXP BIOL & HEAD DIV, MED SCH, UNIV ORE, 63- Concurrent Pos: Spec lectr, Univ London, 53; vis prof, Univ Cincinnati, 58; mem sci comt, Int Cong Dermat, 62; consult, Nat Inst Child Health & Human Develop, 65-; counr, Japan Monkey Ctr, Aichi, 65-; mem adv comt, Washington County Child Develop Prog, Ore, 65-; mem comn natural sci, Nat Bd Fels, Bethany Col, 66- Honors & Awards: Soc Cosmetic Chemists Award, 57; Gold Award, Am Acad Dermat, 58; Gold Medal, Decorated Cavaliere Order of Merit, Ital Repub, 63; Cavaliere Ufficiale, Italian Repub, 69, Commendatore, 75; Stephen Rothman Award in Dermat, 72. Mem: AAAS; Am Soc Zool; Histochem Soc; Am Asn Anat; Sigma Xi (vpres, 57, pres, 60). Res: Cytophysiology; histophysiology and comparative anatomy of the skin; skin of primates; prosimians; primate reproduction. Mailing Add: Ore Regional Primate Res Ctr 505 NW 185th Ave Beaverton OR 97005

MONTAGNE, JOHN M, b White Plains, NY, Apr 17, 20; m 42; c 2. PHYSICAL GEOLOGY, ENVIRONMENTAL GEOLOGY. Educ: Dartmouth Col, BA, 42; Univ Wyo, MA, 51, PhD(geol), 55. Prof Exp: From instr to asst prof geol, Colo Sch Mines,

53-57; asst prof, 57-62, PROF GEOL, MONT STATE UNIV, 62- Concurrent Pos: Supply instr, Univ Wyo, 53-54; mem, Int Field Inst Geol, Italy, 64; mem curric panel, Am Geol Inst; coordr environ study, Ski Yellowstone, Inc, 73-74. Mem: Geol Soc Am; Am Inst Prof Geol; Glaciol Soc; Am Quarternary Asn (treas, 70-76); Am Asn Petrol Geol. Res: Cenozoic history of the Rocky Mountain region, particularly structural, stratigraphic and geomorphologic aspects; Pleistocene glacial geology and geomorphology; field geology in undergraduate geological education; snow dynamics; geology applied to land use planning. Mailing Add: Dept of Earth Sci Mont State Univ Bozeman MT 59715

MONTAGU, ASHLEY, b London, Eng, June 28, 05; US citizen; m 41; c 3. PHYSICAL ANTHROPOLOGY, CULTURAL ANTHROPOLOGY. Educ: Columbia Univ, PhD(anthrop), 37. Hon Degrees: DSc, Grinnell Col, 68; DLitt, Ursinus Col, 72. Prof Exp: Res worker, Brit Mus Natural Hist, 26-27; conservator phys anthrop, Wellcome Hist Med Mus London, 29-30; asst prof anat, NY Univ, 31-38; assoc prof, Hahnemann Med Col & Hosp, 38-49; prof anthrop & chmn dept, Rutgers Univ, 49-55; WRITER, 55- Mem: AAAS; Am Asn Anatomists; Soc Study Child Growth & Develop. Res: Bio-social evolution of man. Mailing Add: 321 Cherry Hill Rd Princeton NJ 08540

MONTAGUE, BARBARA ANN, b Hagerstown, Md, Aug 29, 29. INFORMATION SCIENCE. Educ: Randolph-Macon Woman's Col, AB, 51. Prof Exp: Anal chemist, Res Div, Plastics Dept, 51-60, info chemist, 60-61, head plastic dept info syst, 61-64, develop coord, Cent Report Index, Info Syst Div, 64-67, supvr tech opers, 67-74, MGR, INFO SERV PHOTO PRODS INT OPERS DIV, E I DU PONT DE NEMOURS & CO, INC, 74- Mem: Am Chem Soc; Am Soc Info Sci. Res: Quantitative organic analysis; spectra-structure correlations qualitative and quantitative using infrared spectroscopy; designing, installation, operation and testing of coordinate indexing systems for storage and retrieval of scientific and technical information. Mailing Add: RD 2 Box 26A Friends Meeting House Rd Hockessin DE 19707

MONTAGUE, DANIEL GROVER, b Yakima, Wash, July 7, 37; m 57; c 4. PHYSICS. Educ: Ore State Col, BS, 59; Univ Wash, MS, 63; Univ Southern Calif, PhD(physics), 66. Prof Exp: Reactor physicist, Gen Elec Co, Wash, 59-61; sr sci officer, Rutherford High Energy Lab, Sci Res Coun, Chilton, Eng, 66-69; asst prof, 69-74, ASSOC PROF PHYSICS, WILLAMETTE UNIV, 74- Mem: AAAS; Am Phys Soc; Am Asn Physics Teachers. Res: Charged particle scattering and induced reactions for projectiles in the 10 to 50 million electron volts range. Mailing Add: Dept of Physics Willamette Univ Salem OR 97301

MONTAGUE, ELEANOR D, b Genoa, Italy, Feb 11, 26; US citizen; m 53; c 4. RADIOTHERAPY. Educ: Univ Ala, BA, 47; Woman's Med Col Pa, MD, 50. Prof Exp: Resident path, Kings County Hosp, Brooklyn, 52-53; resident radiol, Columbia-Presby Med Ctr, 53-55; radiologist, 6160th US Air Force Hosp, Japan, 55-56; staff physician, Am Tel & Tel Co, NY, 56-57; Am Cancer Soc fel radiother, Univ Tex M D Anderson Hosp & Tumor Inst, 59-61, from asst radiotherapist to radiotherapist, 61-69, assoc prof radiother, 66-69; assoc clin prof radiol, Baylor Col Med, 69-72; PROF RADIATION THER, UNIV TEX M D ANDERSON HOSP & TUMOR INST, 73- Mem: Am Col Radiol; Am Radium Soc; Radiol Soc NAm; Am Soc Therapeut Radiol; AMA. Res: Clinical use of radiation therapy for treatment of neoplasia. Mailing Add: Dept of Radiation Ther M D Anderson Hosp & Tumor Inst Houston TX 77025

MONTAGUE, FREDRICK HOWARD, JR, b Lafayette, Ind, May 31, 45; m 68; c 1. WILDLIFE ECOLOGY. Educ: Purdue Univ, BS, 67, PhD(vert ecol), 75. Prof Exp: From res asst wildlife ecol to teaching asst, 70-75, ASST PROF WILDLIFE ECOL & DIR OFF STUDENT SERV, PURDUE UNIV, 75- Mem: Wildlife Soc; Am Soc Mammalogists. Res: Ecology of wild and domestic canines in Midwest; nesting ecology of birds; farm game management; farm habitat improvement projects. Mailing Add: Dept Forestry & Natural Resources Purdue Univ West Lafayette IN 47907

MONTAGUE, HARRIET FRANCES, b Buffalo, NY, June 9, 05. MATHEMATICS. Educ: Univ Buffalo, BS, 27, MA, 29; Cornell Univ, PhD(math), 35. Prof Exp: Asst, 27-29, from instr to prof, 29-73, dir, NSF Inst Math, 57-70, actg chmn dept math, 61-64, dir undergrad studies, Dept Math, 70-73, EMER PROF MATH, STATE UNIV NY BUFFALO, 73- Mem: Am Math Soc; Math Asn Am. Res: Mathematics education. Mailing Add: 236 Fayette Ave Kenmore NY 14223

MONTAGUE, JOHN H, b Man, July 16, 25; m 55; c 2. NUCLEAR PHYSICS. Educ: Univ Man, BSc, 46; Univ Chicago, SM, 48, PhD(physics), 50. Prof Exp: Fel nuclear physics, Nat Res Coun Can, Chalk River, Ont, 50-52; physicist, Assoc Elec Industs, Eng, 52-54; lectr physics, Queen's Univ, Ont, 54-55; sr Harwell fel nuclear physics, Atomic Energy Res Estab, Eng, 55-58, prin scientist, 58-66; PROF PHYSICS, QUEEN'S UNIV, ONT, 66- Concurrent Pos: Prin res fel, Atomic Energy Res Estab, Harwell, Eng, 71-73. Mem: Am Phys Soc; Can Asn Physicists. Res: Nuclear physics, mainly at low energies with electrostatic generators. Mailing Add: Dept of Physics Queen's Univ Kingston ON Can

MONTAGUE, PATRICIA TUCKER, b Emporia, Kans, Nov 4, 37; m 65; c 2. MATHEMATICS. Educ: Kans State Univ, BS, 57; Univ Wis-Madison, MS, 58, PhD(math), 61. Prof Exp: From instr to asst prof math, Univ Ill, Urbana, 61-67; assoc prof, Univ Tenn, Knoxville, 67-75. Mem: Am Math Soc; Math Asn Am. Res: Algebra; representations of finite groups. Mailing Add: 304 S 50th Ave Omaha NE 68132

MONTAGUE, STEPHEN, b Los Angeles, Calif, July 17, 40; m 65; c 2. MATHEMATICS. Educ: Pomona Col, BS, 62; Univ Ill, Urbana, PhD(math), 67. Prof Exp: Asst prof math, Univ Tenn, Knoxville, 67-74; ASST PROF MATH, UNIV NEBR OMAHA, 75- Mem: Am Math Soc; Math Asn Am. Res: Algebra; transitive extentions of finite permutation groups; operations research; optimization techniques; non-linear programming. Mailing Add: Dept of Math Univ of Nebr at Omaha Omaha NE 68101

MONTALBETTI, RAYMON, b Cranbrook, BC, Can, Feb 7, 24; m 49; c 3. PHYSICS. Educ: Univ Alta, BSc, 46; Univ Sask, PhD(nuclear physics), 52. Prof Exp: Res officer physics, Nat Res Coun Can, 46-49; sci officer upper atmospheric physics, Defense Res Bd, 52-64, officer in charge, Defense Res North Lab, 58-60; from asst prof to assoc prof, 60-68, PROF PHYSICS, UNIV SASK, 68- Mem: Can Asn Physicists. Res: Photonuclear reactions; auroral and upper atmospheric physics. Mailing Add: Dept of Physics Univ of Sask Saskatoon SK Can

MONTALVO, FRANCISCO EMILIO, b Laredo, Tex, Sept 7, 42; m 71. BIOCHEMISTRY, PHYSIOLOGY. Educ: Univ Tex, Austin, BS, 65; La State Univ, New Orleans, MS, 67, PhD(biochem), 70. Prof Exp: Res assoc biochem, Univ Ill, 70-71; ASST PROF AGR BIOCHEM, UNIV HAWAII, 71- Mem: Am Chem Soc. Res: Enzymology; carbohydrates; intermediary metabolism; cell surface studies. Mailing Add: Dept of Agr Biochem Univ of Hawaii Honolulu HI 96822

MONTALVO, JOSE MIGUEL, b Cali, Colombia, June 30, 28; m 51; c 5. MEDICINE, PEDIATRICS. Educ: Univ Tenn, BS, 51, MD, 57. Prof Exp: From intern pediat to chief resident, Frank T Toby Children's Hosp, Univ Tenn, 57-59, Rockefeller Found fel, 58-59; asst prof, Univ Valle, Colombia, 59-60; chief resident, Frank T Toby Children's Hosp, Univ Tenn, 60-61; from instr to assoc prof, 62-74, PROF PEDIAT, MED CTR, UNIV MISS, 74- Concurrent Pos: Consult, Miss State Ment Hosp, Whitfield & USPHS Indian Hosp, Philadelphia, 61- Mem: Endocrine Soc; Am Soc Pediat Nephrology; Am Fedn Clin Res; Int Soc Nephrology; Am Acad Pediat. Res: Pediatric endocrine and metabolic disorders. Mailing Add: Dept of Pediat Univ of Miss Med Ctr Jackson MS 39216

MONTALVO, JOSEPH G, b Cottonport, La, Oct 30, 37; m 61; c 1. ANALYTICAL CHEMISTRY. Educ: Univ Southwestern La, BS, 59, MS, 61; La State Univ, New Orleans, PhD(anal chem), 68. Prof Exp: Anal chemist, Shell Oil Co, 61-64; NSF res fel, La State Univ, New Orleans, 65-68, res fel, 68-69; ANAL CHEMIST, GULF SOUTH RES INST, 69-, MGR, DEPT ANAL CHEM, 71- Concurrent Pos: Res fel anal biochem, Greater New Orleans Cancer Soc, La State Univ, 66. Mem: Am Chem Soc; Am Welding Soc; Am Soc Testing & Mat. Res: Ion-selective electrodes; dipsticks; personnel badges; trace metals in waters; sediments and biota. Mailing Add: Gulf South Res Inst PO Box 26500 New Orleans LA 70186

MONTALVO, RAMIRO A, b Monterrey, Mex, Dec 18, 37; US citizen; m 67; c 1. SOLID STATE PHYSICS. Educ: Ill Inst Technol, BS, 60; Northwestern Univ, PhD(physics), 67. Prof Exp: Asst physics, Northwestern Univ, 60-66; physicist, Aerospace Res Labs, Off Aerospace Res, US Air Force, 66-69; res scientist, Geomet Inc, 69-73; PRIN STAFF MEM, OPERS RES INC, 73- Mem: Am Phys Soc. Res: Magnetic properties of metals at low temperatures; optical properties of semiconductors; underwater acoustics; computer modeling. Mailing Add: Opers Res Inc 1400 Spring St Silver Spring MD 20910

MONTANA, ANDREW FREDERICK, b Oil City, Pa, Jan 15, 30; m 67. ORGANIC CHEMISTRY. Educ: Seattle Pac Col, BS, 51; Univ Wash, PhD(org chem), 57. Prof Exp: Asst prof chem, Seattle Pac Col, 55-61 & Univ Hawaii, 61-63; assoc prof & dept chmn, 63-70, PROF CHEM, CALIF STATE COL, FULLERTON, 70- Mem: AAAS; Am Chem Soc; NY Acad Sci. Res: Pseudoaromatics. Mailing Add: Dept of Chem Calif State Col 800 N State College Blvd Fullerton CA 92631

MONTASIR, MAGDA MOUSTAFA, b Alexandria, Egypt, Oct 18, 31; m 57; c 2. INFECTIOUS DISEASES, PATHOLOGY. Educ: Univ Alexandria, MB, ChB, 55, dipl med, 57, dipl trop med & hyg, 59, Dr Trop Med, 63. Prof Exp: Intern & resident, Univ Alexandria, 55-58, clin asst trop med, 58-63 & Univ Heidelberg, 63-64; NIH fel path, Baylor Col Med, 64-65; asst prof trop med, Univ Alexandria, 65-68; specialist infectious dis, Mininstry Health, Govt Libya, 68-69; asst researcher trop & infectious dis, Sch Pub Health, Univ Calif, Los Angeles, 70; resident path, Hamot Med Ctr, Erie, Pa, 70-71; ASST RESEARCHER, SCH PUB HEALTH, UNIV CALIF, LOS ANGELES, 71-, RESIDENT PATH, SCH MED, 72- Mem: Am Soc Trop Med & Hyg; Royal Soc Trop Med & Hyg; Am Mosquito Control Asn; Am Soc Clin Path; Int Filariasis Asn. Res: Pathology and pathogenesis of tropical and infectious disease, particularly the arthropod-borne parasites and viruses. Mailing Add: Dept of Trop & Infectious Dis Sch of Pub Health Univ Calif Los Angeles CA 90026

MONTEAN, JOHN J, b Welland, Ont, Jan, 6, 16; US citizen; m 43; c 3. SCIENCE EDUCATION. Educ: Eastern Mich Univ, BSc, 43; Columbia Univ, AM, 53; Syracuse Univ, PhD(sci ed), 59. Prof Exp: Chemist, Naugatuck Chem Div, US Rubber Co, 43-48; dept chmn sci, secondary schs, NY & instr math, Hobart Col, 49-50; asst instr chem, Syracuse Univ, 51-53; prof chem & chmn dept sci, Auburn Community Col, 54-59; HEAD SCI EDUC, UNIV ROCHESTER, 59-, PROF, 71- Concurrent Pos: NY State Dept Educ res grants, 61-67; assoc prog dir pre-col sci educ, NSF, 66-67; consult, US Off Educ & NY State Educ Dept, 66-; Stanford Univ Shell Merit fel, 70. Mem: Fel AAAS; Sigma Xi; Nat Asn Res Sci Teaching; Nat Sci Teachers Asn; Asn Educ Teachers Sci (nat pres, 70-71). Res: Pre-college science and mathematics; large-group instruction in chemistry; team teaching in biology; interdisciplinary models for environmental education; curriculum and instructional technology. Mailing Add: 226 Space Sci Ctr Univ of Rochester Rochester NY 14627

MONTELARO, JAMES, b Melville, La, Mar 3, 21; m 55; c 1. HORTICULTURE. Educ: Southwest La Inst, BS, 41; La State Univ, MS, 50; Univ Fla, PhD(hort), 52. Prof Exp: Asst hort, Univ Fla, 52-55; horticulturist, Minute Maid Corp, 55-57; assoc veg specialist, 58-65, VEG CROPS SPECIALIST, UNIV FLA, 65-, PROF, INST FOOD & AGR SCI, 74- Mem: Am Soc Hort Sci. Res: Nutrition and physiology of vegetable crops. Mailing Add: 1605 SW 56th Place Gainesville FL 32601

MONTEMURRO, DONALD GILBERT, b North Bay, Ont, May 27, 30; m 54; c 2. PHYSIOLOGY, NEUROENDOCRINOLOGY. Educ: Univ Western Ont, BA, 51, MSc, 54, PhD(physiol), 57. Prof Exp: Sr res asst, Med Sch, Univ Western Ont, 57-58; Brit Empire Cancer Campaign exchange fel, Chester Beatty Res Inst, Royal Cancer Hosp, London, 58-60; Cancer Inst Can res fel, Sch Med, Yale Univ, 60-61; assoc prof physiol, 61-68, anat, Health Sci Centre, 68-72, PROF ANAT, HEALTH SCI CENTRE, UNIV WESTERN ONT, 72-, CHMN DEPT, 73- Mem: Am Asn Cancer Res; Endocrine Soc; Am Asn Anatomists; Can Asn Anat; Can Physiol Soc. Res: Endocrinology; histology and physiology of the adeno- and neurohypophysis; role of the hypothalamus in water and energy metabolism; electron microscopy; neuroanatomy and neuroendocrine function of the hypothalamus. Mailing Add: Dept of Anat Univ Western Ont Health Sci Ctr London ON Can

MONTENYOHL, VICTOR IRL, b Akron, Ohio, Mar 18, 21; m 46; c 3. CHEMISTRY. Educ: Stanford Univ, AB, 42; Princeton Univ, MA, 47, PhD(chem), 50. Prof Exp: Res assoc, Princeton Univ, 42-46; res chemist, 47-53, res supvr, 53-63, CHIEF SUPVR, BUDGET & PLANNING DIV, E I DU PONT DE NEMOURS & CO, INC, 74- Mem: AAAS; Am Chem Soc; Am Soc Metals. Res: Surface chemistry; corrosion; non-destructive testing; metallurgy; research planning and funding. Mailing Add: 1050 Two Notch Rd Aiken SC 29801

MONTERMOSO, JUAN CARGADO, organic chemistry, see 12th edition

MONTES, LEOPOLDO F, b Buenos Aires, Arg, Nov 22, 29; m 61; c 6. DERMATOLOGY, MYCOLOGY. Educ: Univ Buenos Aires, MD, 54; Univ Mich, MS, 59; FRCPS(C). Prof Exp: Resident dermat, Pirovano Hosp, Buenos Aires, 55; resident, Pa Hosp, Philadelphia, 55-56; resident, Med Ctr, Univ Mich, 56-57, from jr clin instr to sr clin instr, 57-60, res assoc anat, Med Sch, 59, NSF grant & res assoc zool, 60-61, USPHS grant & res assoc anat & dermat, 62; asst prof dermat, Baylor Col Med, 63-66; assoc prof, 66-70, PROF DERMAT, MED CTR, UNIV ALA, BIRMINGHAM, 70-, ASSOC PROF MICROBIOL, 69- Concurrent Pos: USPHS res career develop award, 65-71; ed, J Cutaneous Path. Mem: Int Soc Trop Dermat; Am Acad Dermat; Soc Invest Dermat; fel Am Acad Microbiol; Am Dermat Asn. Res: Cutaneous manifestations of systemic diseases; histochemistry, mainly oxidative enzymes; cutaneous infections caused by yeast; electron microscopy; fine structure of psoriatic skin; acne vulgaris; cytochemistry of the axillary sweat glands. Mailing Add: Dept of Dermat Univ of Ala Med Ctr Birmingham AL 35294

MONTES DE OCA, HECTOR, b Buenos Aires, Arg, Nov 10, 22; m 53; c 2. CELL BIOLOGY. Educ: Univ Buenos Aires, MD, 50. Prof Exp: Teaching asst, Inst Micros Anat & Embryol, Sch Med, Univ Buenos Aires, 45-46; chief blood bank, Ezeiza Hosp, Buenos Aires, Arg, 52-54; head tissue cult sect, Res Labs, E R Squibb & Sons, Martinez, 54-56; chief tissue cult div, Dept Radiol, AEC, Arg, 56-64; vis scientist, Div Biol Stand, NIH, Md, 65-68; assoc prof anat, Sch Med, Univ Sask, 68-69; actg dir, Am Found Biol Res, Md, 69-71; ASSOC PROF SURG, SCH MED, UNIV MIAMI, 71- Concurrent Pos: Nat Res Coun Arg fels, Tissue Cult Lab, Univ Tex Med Br, 59-60 & Tissue Cult Sect, Lab Biol, Nat Cancer Inst, Md, 60-61, vis fel, 61-62; chief teaching labs, Sch Med, Univ Buenos Aires, 55-57, actg prof histol, Vet Sch, 63-64. Mem: Tissue Cult Asn; Soc Cryobiol; Arg Med Asn. Res: Radiobiology; tissue culture. Mailing Add: Dept of Surg Univ of Miami Sch of Med Miami FL 33152

MONTET, GEORGE LOUIS, b Ventress, La, Dec 10, 19; m 46; c 2. CHEMICAL PHYSICS. Educ: La State Univ, BS, 40; Univ Chicago, MS, 49, PhD(chem), 51. Prof Exp: Plant engr, Bird & Son, Inc, 40-42 & 46; shift supt, Huntsville Arsenal, 42; assoc chemist, 51-74, CHEMIST, ARGONNE NAT LAB, 74-, TEAM LEADER, ENVIRONMENTAL STATEMENT PROJ, 72- Concurrent Pos: Fulbright lectr, Univ Ankara, 62-63; adj prof, Northern Ill Univ, 68-71. Mem: AAAS; Am Phys Soc; Sigma Xi. Res: Solid state physics; quantum chemistry; statistical mechanics; environmental assessment. Mailing Add: 424 Bunning Dr Downers Grove IL 60515

MONTGOMERY, ANDREW HARRISON, b Leavenworth, Kans, Dec 3, 18; m 58; c 2. PHYSICAL CHEMISTRY. Educ: Iowa State Univ, BS, 40, MS, 52. Prof Exp: Instr, Boone Jr Col, 46-47; assoc prof, Phillips Univ, 47-49; chmn dept chem, 50-59, chmn dept chem physics & math, 59-67, chmn dept phys sci, 67-74, ASSOC PROF CHEM, CONCORD COL, 50-, CHMN DIV NATURAL SCI, 74- Mem: Am Chem Soc; Nat Educ Asn; Am Asn Higher Educ. Res: Vapor pressure equations; states of matter; coal. Mailing Add: Div of Natural Sci Concord Col Athens WV 24712

MONTGOMERY, ANTHONY JOHN, b Lincoln, Eng, July 13, 37; m 60; c 4. OPTICS. Educ: Univ London, BSc, 58, PhD(optics), 61, DIC, 61. Prof Exp: Sr scientist optics, IIT Res Inst, 61-69; mgr imaging systs, Xerox Corp, NY, 69-73, dept mgr, Xerox Res, Eng, 73-75, MGR ADVAN DEVELOP, XEROX CORP, CT, 75- Mem: Optical Soc Am; Soc Photog Sci & Eng; fel Brit Inst Physics. Res: Optical instrumentation; image evaluation; analysis of imaging systems; psychophysics of images. Mailing Add: Xerox Corp Stamford CT 06904

MONTGOMERY, ARTHUR, mineralogy, see 12th edition

MONTGOMERY, ARTHUR VERNON, b Sinton, Tex, Nov 17, 22; m 56; c 4. PHYSIOLOGY. Educ: Univ Ariz, BS, 46; Univ Colo, PhD(physiol), 53, MD, 57. Prof Exp: Instr physiol, Univ Tex Med Br Galveston, 49-50; res assoc, Med Sch, Univ Colo, 50-52, instr, 52-54, instr surg, 54-57, dir, Halsted Lab Exp Surg, 57-59; chief exp toxicol, US Air Force Sch Aerospace Med, 59-60, chief physiol & biophys, 60-61; mgr life sci depts, McDonnell Aircraft Corp, 61-70, DIR LIFE SCI SUBDIV, McDONNELL DOUGLAS ASTRONAUT CO, 70- Concurrent Pos: Am Heart Asn fel, 57-59. Honors & Awards: Hektoen Gold Medal, AMA, 55. Mem: Am Physiol Soc. Res: Physiology of intrarenal pressure; satiation of thirst; applications of general hypothermia to cardiac surgery; environmental physiology. Mailing Add: McDonnell Douglas Astronaut Co PO Box 516 St Louis MO 63166

MONTGOMERY, CHARLES GRAY, b Philadelphia, Pa, Apr 9, 37; m 66; c 1. THEORETICAL PHYSICS. Educ: Yale Univ, BA, 59; Calif Inst Technol, MS, 61, PhD(physics), 65. Prof Exp: Vis lectr physics, Hollins Col, 65; from asst prof to assoc prof, 65-75, PROF PHYSICS, UNIV TOLEDO, 75- Concurrent Pos: Consult, Owens-Ill, Inc, 65-70 & GTE Labs, Inc, 75-76. Mem: Am Phys Soc; Am Asn physics Teachers. Res: Statistical and solid state physics. Mailing Add: Dept of Physics Univ of Toledo Toledo OH 43606

MONTGOMERY, CHRISTINE ANNE, b Washington, DC. ANTHROPOLOGY, COMPUTER SCIENCE. Educ: Georgetown Univ, BS, 56, MS, 58; Univ Calif, Los Angeles, PhD(anthrop), 66. Prof Exp: Mem tech staff comput ling & info sci, Ramo-Wooldridge Div, Thompson-Ramo-Wooldridge, Inc, 59-63, mem tech staff, Bunker-Ramo Corp, 64-68, mgr natural lang data processing, 68-70; MGR SOFTWARE SYSTS, OPERATING SYSTS, INC, 70- Concurrent Pos: Lectr ling, Cath Univ Am, 58-60, consult, Ctr Appl Ling, 69-71; sr lectr ling, Univ Southern Calif, 69-72; consult, Bunker-Ramo Corp, 70-71; ed Info Retrieval & Lang Processing Communications, Asn Comput Mach, 73- Honors & Awards: Best Paper, Am Soc Info Sci, 72. Mem: Ling Soc Am; Asn Comput Mach; Am Soc Info Sci; African Studies Asn; Asn Computational Ling. Res: Computational linguistics; African linguistics, theoretical and applied; information science. Mailing Add: Operating Systs Inc 21031 Ventura Blvd Woodland Hills CA 91364

MONTGOMERY, DANIEL MICHAEL, b Indianapolis, Ind, Nov 2, 43; m 66; c 1. RADIATION ECOLOGY, RADIOCHEMISTRY. Educ: St Martin's Col, BS, 65; Purdue Univ, Lafayette, PhD(nuclear chem), 69. Prof Exp: Res assoc nuclear chem, Univ Marburg; NSF res assoc, Carnegie-Mellon Univ, 70-71; res chemist, 71-74, HEAD RADIOECOL SECT, NAT ENVIRON RES CTR, ENVIRON PROTECTION AGENCY, 74- Concurrent Pos: Vis scientist, Europ Orgn Nuclear Res, Geneva, Switz, 69-70. Res: Radiochemical analysis of environmental samples; radioecology; neutron activation analysis; radiological surveillance at nuclear facilities. Mailing Add: Nat Environ Res Ctr Environ Protection Agency Cincinnati OH 45269

MONTGOMERY, DAVID CAMPBELL, b Milan, Mo, Mar 5, 36; m 57; c 2. PLASMA PHYSICS, STATISTICAL MECHANICS. Educ: Univ Wis, BS, 56; Princeton Univ, MA, 58, PhD(physics), 59. Prof Exp: Assoc, Proj Matterhorn, Princeton Univ, 59-60; res assoc physics, Univ Wis, 61, instr, 61-62; res asst prof, Univ Md, 62-64; vis researcher, State Univ Utrecht, 64-65; assoc prof, 65-70, PROF PHYSICS, UNIV IOWA, 70- Concurrent Pos: Consult, Oak Ridge Nat Lab, 62-70, Goddard Space Flight Ctr, NASA, 63-64 & Los Alamos Sci Lab, 69-71; vis assoc prof, Univ Calif, Berkeley, 69-70; assoc ed, Physics of Fluids, 71, 72 & 73; assoc ed, Int J Eng Sci, 71-; vis prof, Hunter Col, 73-74, adj prof, 74-7S. Mem: Fel Am Phys Soc; Sigma Xi. Res: Theoretical plasma physics; kinetic theory; strongly magnetized plasmas; turbulence theory; non-equilibrium statistical mechanics and transport theory. Mailing Add: Dept of Physics & Astron Univ of Iowa Iowa City IA 52242

MONTGOMERY, DAVID CAREY, b Elmhurst, Ill, Aug 21, 38; m 68. PHYSICS, INSTITUTIONAL RESEARCH. Educ: Mass Inst Technol, BS, 60; Univ Ill, MS, 61, PhD(physics), 67. Prof Exp: From instr to asst prof physics, Oberlin Col, 66-71, registrar & asst provost, 71-73, dir, Inst Res & Planning, 73-74; COORDR PLANNING & ANAL, STATE UNIV SYST FLA, 75- Mem: Am Phys Soc; Am Asn Higher Educ; Asn Instit Res. Res: Student flow; projections of enrollment, faculty size and mix; analysis of funding formulas. Mailing Add: State Univ Syst of Fla 107 W Gaines St Tallahassee FL 32304

MONTGOMERY, DEANE, b Weaver, Minn, Sept 2, 09; m 33; c 2. TOPOLOGY. Educ: Hamline Univ, BA, 29; Univ Iowa, MS, 30, PhD(math), 33. Prof Exp: Nat Res Coun fel, Harvard Univ, 33-34 & Inst Adv Study, 34-35; from asst prof to prof math, Smith Col, 35-46; assoc prof, Yale Univ, 46-48; Guggenheim fel, 41-42, mem, Nat Defense Res Comt Proj, 45-46, mem, Inst Advan Study, 48-51, PROF MATH, INST ADVAN STUDY, 51- Concurrent Pos: Vis assoc prof, Princeton Univ, 43-45. Mem: Am Philos Soc; Int Math Union (pres, 75-78). Res: Topology; topological groups. Mailing Add: Inst for Advan Study Princeton NJ 08540

MONTGOMERY, DONALD BRUCE, b Hartford, Conn, July 1, 33; m 57; c 2. PHYSICS, ELECTRICAL ENGINEERING. Educ: Williams Col, BA, 57; Mass Inst Technol, BS & MS, 57; Univ Lausanne, DSc, 68. Prof Exp: Mem staff eng, Arthur D Little, Inc, 57-59; mem staff magnet develop, Lincoln Lab, 59-61, GROUP LEADER MAGNET DEVELOP, NAT MAGNET LAB, MASS INST TECHNOL, 61- Res: High field magnet design; cryogenics and superconductivity. Mailing Add: Nat Magnet Lab Mass Inst of Technol Cambridge MA 02139

MONTGOMERY, DONALD JOSEPH, b Cincinnati, Ohio, June 11, 17; m 42; c 4. PHYSICS. Educ: Univ Cincinnati, ChE, 39, PhD(theoret physics), 45. Prof Exp: Instr, Univ Cincinnati, 42-44; res assoc, Princeton Univ, 45-46, asst prof, 46-47; physicist, London Br, Off Naval Res, 47-48, Ballistic Res Labs, 48-50 & Textile Res Inst, 50-53; assoc prof physics, 53-56, prof, 56-60, res prof, 60-65, prof metall, mech & mat sci & chmn dept, 66-71, PROF PHYSICS, MICH STATE UNIV, 66-, RES PROF ENG, 71- Concurrent Pos: Fulbright lectr, Univ Grenoble, 59-60, Guggenheim fel, 60; spec asst to dir, Off Grants & Res Contracts, NASA, Washington, DC, 64-65; vis res physicist, Space Sci Lab, Univ Calif, Berkeley, 65-66; mem, Textile Res Inst. Mem: AAAS; Am Acad Mech; Am Soc Metals; Am Acad Polit & Soc Sci; Am Phys Soc. Res: Materials science, solid state, chemical physics; technology assessment. Mailing Add: Col of Eng Mich State Univ East Lansing MI 48823

MONTGOMERY, DONALD REX, polymer chemistry, organic chemistry, see 12th edition

MONTGOMERY, EDWARD BENJAMIN, b Louisville, Ky, June 15, 15; m 42; c 4. PHYSICS. Educ: Univ Louisville, BA, 42. Prof Exp: Ballistic physicist, Okla Ord Works, 42-43; asst, Metall Lab, Univ Chicago, 43-44; physicist, Hanford Works, Gen Elec Co, 44-50, group head exp physics, 49-51, sect chief pile applns, 51-52, physicist adv tech, 52-54, data processing specialist, 54-56, mgr systs anal, Comput Dept, 56-57, consult prod planning, 57-59, consult physicist comput applns, 59-61, proj engr educ tech proj, 61-63; res consult, Syracuse Univ, 63-65, dean sch libr sci, 65-68; PROF PHYSICS & COORDR, COUN OF PRESIDENTS OF UNIV OF TEX COMPONENTS IN NTEX, UNIV TEX HEALTH SCI CTR DALLAS, 68- Mem: AAAS; Am Phys Soc; Inst Elec & Electronics Eng. Res: Information systems; mass data handling; automation of research process; higher education analyst. Mailing Add: 6720 Greenwich Lane Dallas TX 75230

MONTGOMERY, EDWARD HARRY, b Houston, Tex, July 8, 39; m 63; c 3. PHARMACOLOGY, PHYSIOLOGY. Educ: Univ Houston, BS, 61, MS, 63; Univ Tex, PhD(pharmacol), 67. Prof Exp: Teaching fel, Univ Houston, 61-63, Nat Inst Dent Res training grant, 63-67; USPHS gen res serv fund grant, Univ Ore, 67-69, Nat Inst Dent Res grant, 69-72, from asst prof to assoc prof pharmacol, Dent Sch, 67-72; ASSOC PROF PHARMACOL, UNIV TEX DENT BR HOUSTON, 72- Honors & Awards: Lehn & Fink Award, 61. Mem: AAAS; Sigma Xi; Int Asn Dent Res. Res: Role of vasoactive polypeptides as inflammatory mediators; mechanism of action of anti-inflammatory drugs; release of catecholamines by bradykinin and other vasoactive polypeptides; studies of the mediator systems involved in gingival inflammation. Mailing Add: Dept of Pharmacol Univ of Tex Dent Br Houston TX 77025

MONTGOMERY, ERROL LEE, b Roseburg, Ore, May 1, 39; m 60; c 2. HYDROGEOLOGY. Educ: Ore State Univ, BS, 62; Univ Ariz, MS, 63, PhD(hydrogeol), 71. Prof Exp: Groundwater geologist, Wyo State Engrs Off, 63-65; geohydrologist, Wright Water Engrs, Colo, 65-67; ASST PROF GEOL, NORTHERN ARIZ UNIV, 70- Concurrent Pos: Hydrogeologist, Harshbarger & Assocs, 68- Mem: Am Geophys Union; Am Water Resources Asn; Asn Eng Geol. Res: Hydrogeology; applications of geophysics to hydrogeology; aquifer and aquifer systems analysis; engineering geology. Mailing Add: 2012 N Crescent Flagstaff AZ 86001

MONTGOMERY, FRANKLIN LUCKENBILL, organic chemistry, polymer chemistry, see 12th edition

MONTGOMERY, GEORGE EDWARD, b Manhattan, Kans, Jan 1, 41; m 71. ANTHROPOLOGY. Educ: Stanford Univ, AB, 64; Columbia Univ, PhD(anthrop), 72. Prof Exp: ASST PROF ANTHROP, WASHINGTON UNIV, 71- Mem: AAAS; Am Anthrop Asn; Pop Asn Am; Asn Asian Studies. Res: Infection and disease in relation to population organization and dynamics; human energetics; systems anthropology; methods in human ecological inquiry. Mailing Add: Dept of Anthrop Washington Univ St Louis MO 63130

MONTGOMERY, GEORGE PAUL, JR, b Atlanta, Ga, Jan 5, 43; m 68; c 1. LASERS, SEMICONDUCTORS. Educ: Loyola Col, Md, BS, 64; Univ Ill, Urbana, MS, 66, PhD(physics), 71. Prof Exp: Nat Acad Sci-Nat Res Coun resident res assoc, US Naval Res Lab, 70-72; ASSOC SR RES PHYSICIST, GEN MOTORS RES LABS, 72- Mem: Am Phys Soc. Res: Semiconductor lasers; pollutant trace gas monitoring. Mailing Add: Gen Motors Res Labs GM Tech Ctr Warren MI 48090

MONTGOMERY, HUGH, b Austin, Tex, Apr 17, 04; m 30; c 3. MEDICINE. Educ: Haverford Col, BS, 25; Harvard Univ, MD, 30. Prof Exp: Res, Harvard Med Sch & Marine Biol Lab, Woods Hole, 27-28; intern, Mass Gen Hosp, 31-32; res fel pharmacol, 32-35, instr clin med, 35-41, Heckscher fel, 37-38, Thompson fel, 38-39, assoc med, 41-47, asst prof clin med, 47-52, from assoc prof to prof med, 52-72, EMER PROF MED, SCH MED, UNIV PA, 72- Mem: AAAS; Am Soc Clin Invest; Am Med Asn; Asn Am Physicians; fel Am Col Physicians. Res: Peripheral circulation; metabolism and oxygen tension of tissue; chemical constitution of glomerular and tubular fluids. Mailing Add: 932 Merion Square Rd Gladwyne PA 19035

MONTGOMERY, JAMES DOUGLAS, b Morristown, NJ, July 28, 37. BOTANY, TAXONOMY. Educ: Bucknell Univ, BS, 59; Rutgers Univ, MS, 61, PhD(bot), 64. Prof Exp: From instr to assoc prof biol, Upsala Col, 64-74; RES BIOLOGIST, ICHTHYOL ASSOCS, 74- Mem: Am Soc Plant Taxonomists; Am Fern Soc; Torrey Bot Club. Res: Taxonomy of flowering plants and ferns; biosystematics; floristics; hybridization and distribution of ferns. Mailing Add: Ichthyol Assocs 24 Main St Stamford NY 12167

MONTGOMERY, JOHN ATTERBURY, b Greenville, Miss, Mar 29, 24; m 47; c 4. ORGANIC CHEMISTRY, MEDICINAL CHEMISTRY. Educ: Vanderbilt Univ, BA, 46, MS, 47; Univ NC, PhD(chem), 51. Prof Exp: Fel, Univ NC, 51-52; chemist, 52-56, head org div, 56-62, dir org chem res, 61-74, VPRES, SOUTHERN RES INST, 74- Concurrent Pos: Adj prof, Birmingham Southern Col, 57-62; mem chem adv panel, Cancer Chemother Nat Serv Ctr, Nat Cancer Inst, 60-61, consult, 62-70; consult, Health Res Facilities Br, NIH, 62-63, mem med chem study sect, 64-68 & 71, mem exp therapeut study sect, 75-79. Honors & Awards: Herty Award, Am Chem Soc, 74. Mem: AAAS; Am Chem Soc; Int Soc Heterocyclic Chem; Am Asn Cancer Res; NY Acad Sci. Res: Organic syntheses; biochemistry; chemotherapy. Mailing Add: Southern Res Inst 2000 Ninth Ave S Birmingham AL 35205

MONTGOMERY, JOHN C. MATHEMATICS. Educ: Univ Calif, Los Angeles, BA, 31; Yale Univ, PhD, 40. Prof Exp: From assoc prof to prof, 41-74, head dept, 60-66, EMER PROF MATH, UNIV CONN, 74- Mem: Am Math Soc. Res: Mathematical education. Mailing Add: Dept of Math Univ of Conn Storrs CT 06268

MONTGOMERY, JOHN RICHARD, b Saltsburg, Pa, Sept 23, 13; m 38; c 3. SPEECH PATHOLOGY. Educ: Grove City Col, AB, 33; Univ Mich, MA, 45. Hon Degrees: DSc, London Inst Appl Res, 73. Prof Exp: Instr speech, Univ Pittsburgh, 33-34; pvt pract speech ther, New Kensington, Pa, 34-36; teacher county schs, 36-38; head, Div Speech, Polytech Inst PR, 38-41; pvt pract, 41-43; therapist pub schs, 43-45; dir speech & hearing clin, 45-70, EMER PROF SPEECH, KENT STATE UNIV, 71- Concurrent Pos: Mem speech clin, Western Mich Col, 43 & Univ Mich, 44; dir speech clin, Mercy Hosp, 48-70; consult, Children's Hosp & Rehab Ctr, Akron, Ohio, 53-70. Mem: Am Speech & Hearing Asn. Res: Speech and hearing science. Mailing Add: 1282 Blossom Lane Palm Bay FL 32905

MONTGOMERY, KENNETH O, b Indianapolis, Ind, Sept 6, 30; m 53; c 2. PHARMACY, PHARMACEUTICAL CHEMISTRY. Educ: Butler Univ, BS, 58, MS, 66. Prof Exp: Technician, Allied Labs, Pitman-Moore Div, Dow Chem Co, 47-58, pharmacist, 58-61, chief chemist, 61-63, asst mgr pharmaceut qual control, 63-66; tech dir, 66-72, VPRES TECH AFFAIRS, CENT PHARMACAL CO, 72- Mem: Am Pharmaceut Asn; Pharmaceut Mfrs Asn. Res: Pharmaceutical analysis using ion-exchange resins, also using radioisotopes; quality control of pharmaceuticals; new compounds for iron deficiency. Mailing Add: Cent Pharmacal Co 116-128 E Third St Seymour IN 47274

MONTGOMERY, LAWRENCE KERNAN, b Denver, Colo, May 6, 35; m 58; c 2. ORGANIC CHEMISTRY, PHYSICAL CHEMISTRY. Educ: Colo State Univ, BS, 57; Calif Inst Technol, PhD(chem), 61. Prof Exp: Fel, Harvard Univ, 60-62; from instr to asst prof, 62-67, ASSOC PROF CHEM, IND UNIV, BLOOMINGTON, 67- Concurrent Pos: Consult ed org chem, Holt, Reinhart & Winston, 67- Mem: Am Chem Soc. Res: Reaction mechanisms; deuteron nuclear magnetic resonance; electron diffraction studies on short-lived molecular species. Mailing Add: Dept of Chem Ind Univ Bloomington IN 47401

MONTGOMERY, MABEL D, b Delevan, NY, Aug 18, 19. MATHEMATICS. Educ: Houghton Col, AB, 39; Univ Buffalo, MA, 49, PhD(math), 53. Prof Exp: Teacher pub schs, NY, 39-46; instr math, Univ Buffalo, 46-54, supvr credentials, 54-58; assoc prof, 58-62, PROF MATH, STATE UNIV NY COL BUFFALO, 62- Mem: Math Asn Am; Am Math Soc; Nat Coun Teachers Math. Res: Topology; mathematics education. Mailing Add: Dept of Math State Univ of NY Col Buffalo NY 14222

MONTGOMERY, MAX MALCOLM, b Williamsburg, Iowa, June 14, 04; m 29; MEDICINE. Educ: Univ Iowa, BS, 27; Univ Ill, MD, 31, MS, 40; Am Bd Internal Med, dipl, 41. Prof Exp: Res fel path, Cook County Hosp, 37-39; from asst prof to prof med, 46-71, EMER PROF MED, UNIV ILL MED CTR, 71- Concurrent Pos: Attend physician, Cook County & West Side Vet Admin Hosps, 60- Mem: Am Heart Asn; Am Rheumatism Asn; Am Fedn Clin Res. Res: Rheumatic diseases. Mailing Add: 2052 N Lincoln Pkwy Chicago IL 60614

MONTGOMERY, MICHAEL DAVIS, b San Luis Obispo, Calif, June 4, 36; m 58; c 4. PHYSICS. Educ: Stanford Univ, BS, 58, MS, 59; Univ NMex, PhD(physics), 67. Prof Exp: Staff physicist, Los Alamos Sci Lab, 62-75; MEM STAFF, MAX PLANCK INST EXTRATERRESTRIAL PHYSICS, 75- Concurrent Pos: Assoc ed, J Geophys Res, Am Geophys Union, 72-74. Mem: Am Phys Soc; Am Geophys Union; NY Acad Sci. Res: Space plasma physics; solar wind; stellar wind. Mailing Add:

MONTGOMERY, MONTY J, b Longview, Miss, Dec 26, 39; m 61; c 3. DAIRY SCIENCE. Educ: Miss State Univ, BS, 61; Univ Wis, MS, 63, PhD(dairy sci), 65. Prof Exp: Asst prof, 65-69, ASSOC PROF ANIMAL SCI, UNIV TENN, KNOXVILLE, 69- Mem: Am Dairy Sci Asn. Res: Dairy cattle nutrition especially feed intake regulation and forage evaluation. Mailing Add: Dept of Animal Sci Box 1071 Univ of Tenn Knoxville TN 37916

MONTGOMERY, MORRIS WILLIAM, b Fargo, NDak, Mar 24, 29; m 50; c 4. FOOD SCIENCE, BIOCHEMISTRY. Educ: NDak State Univ, BS, 51, MS, 57; Wash State Univ, PhD(dairy sci, biochem), 61. Prof Exp: Qual control supvr dairy tech, Nat Dairy Prod Corp, Wis, 53-54, asst prod mgr, 54-55, prod mgr, 55-57; res assoc food sci, 61-63, asst prof, 63-69, ASSOC PROF FOOD SCI, ORE STATE UNIV, 69- Concurrent Pos: US Dept Health, Educ & Welfare grant, 65-67; vis prof, State Univ Campinas, Brazil, 73-74. Mem: Inst Food Technologists; Am Chem Soc. Res: Enzymic browning of fruits; polyphenol oxidase; fish muscle enzymes. Mailing Add: Dept of Food Sci & Technol Ore State Univ Corvallis OR 97331

MONTGOMERY, PAUL CHARLES, b Philadelphia, Pa, Jan 29, 44; m 64; c 3. IMMUNOLOGY, MICROBIOLOGY. Educ: Dickinson Col, BSc, 65; Univ Pa, PhD(microbiol), 69. Prof Exp: Smith Kline & French traveling fel, Nat Inst Med Res, London, 69-70; asst prof, 70-73, ASSOC PROF MICROBIOL, SCH DENT MED, UNIV PA, 74- Mem: Am Soc Microbiol; Brit Soc Immunol; Am Asn Immunol. Res: Localized immunity, induction and structural characterization of secretory antibody; immune response, homogeneous antibody, relationship of antigenic structure to antibody combining site structure. Mailing Add: Dept of Microbiol Sch Dent Med Univ of Pa 4001 Spruce St Philadelphia PA 19174

MONTGOMERY, PETER WILLIAMS, b Denver, Colo, May 27, 35; m 60; c 3. PHYSICAL CHEMISTRY. Educ: Univ Colo, BA, 57; Univ Calif, Berkeley, PhD(phys chem), 61. Prof Exp: Sr res chemist, Cent Res Labs, Minn Mining & Mfg Co, 61-67 & Isotope Power Lab, 67-69; asst prof chem, St Cloud State Col, 69-71; consult & itinerant lectr, 71-74; dir, HMO feasibility study, 74, DIR RESOURCE DEVELOP, RAMSEY ACTION PROGS, 75- Mem: AAAS. Res: High pressure physics and chemistry; thermodynamics. Mailing Add: 1477 Goodrich Ave St Paul MN 55105

MONTGOMERY, PHILIP O'BRYAN, JR, b Dallas, Tex, Aug 16, 21; m 53; c 4. PATHOLOGY. Educ: Southern Methodist Univ, BS, 42; Columbia Univ, MD, 45; Am Bd Path, dipl. Prof Exp: Intern, Mary Imogene Bassett Hosp, 46; fel path, Univ Tex Southwest Med Sch Dallas, 50-51; asst path & cancer, Cancer Res Inst, New Eng Deaconess Hosp, 51-52; from asst prof to assoc prof, 52-61, NIH career develop award, 62-68, assoc dean, 68-70, PROF PATH, UNIV TEX HEALTH SCI CTR DALLAS, 61- Concurrent Pos: Consult path var hosps, 52-; mem sci adv comt, Damon Runyan Mem Fund Cancer Res, 66-72, mem bd dir, 74-; pres bd dirs, Damon Runyon-Walter Winchell Cancer Fund, 74-; spec asst to chancellor, Univ Tex Syst, 71-75; mem bd regents, Uniformed Serv Univ of Health Sci, 74-; pres, Biol Humanics Found, Dallas, 74- Honors & Awards: Astronauts' Silver Snoopy Award for Prof Excellence, 70. Mem: Fel Am Soc Clin Path; AMA; fel Col Am Path; fel NY Acad Sci; fel Royal Micros Soc. Res: Pathological aspects of medicolegal cases; ultraviolet irradiation and microscopy; time-lapse photography; cell ultrastructure; carcinogenesis and nucleolar structure and function. Mailing Add: 6343 Kalani Pl Dallas TX 75240

MONTGOMERY, RAYMOND BRAISLIN, b Philadelphia, Pa, May 5, 10; m 44; c 4. PHYSICAL OCEANOGRAPHY. Educ: Harvard Univ, AB, 32; Mass Inst Technol, SM, 34, ScD(oceanog), 38. Prof Exp: Asst meteorol, Mass Inst Technol, 35-36, mem staff, 44-45; jr meteorol statistician, Bur Agr Econ, USDA, 36-37; jr oceanogr, Woods Hole Oceanog Inst, 38-40, phys oceanogr, 40-42 & 45-49; assoc prof meteorol, NY Univ, 43-44; vis prof oceanog, Brown Univ, 49-54; from assoc prof to prof, 54-75, EMER PROF OCEANOG, JOHNS HOPKINS UNIV, 75- Concurrent Pos: Nat Res Coun fel, Univ Berlin, Ger & Univ Helsinki, Finland, 38-39; vis prof, Scripps Inst, Univ Calif, San Diego, 48; Fulbright res scholar, Univ Melbourne, 58; mem corp, Woods Hole Oceanog Inst, 70-; vis prof, Univ Hawaii, 71. Mem: Oceanog Soc Japan. Res: Analysis of water characteristics and oceanic flow patterns; oceanic leveling. Mailing Add: Dept of Earth & Planetary Sci Johns Hopkins Univ Baltimore MD 21218

MONTGOMERY, REX, b Birmingham, Eng, Sept 4, 23; nat US; m 48; c 4. BIOCHEMISTRY. Educ: Univ Birmingham, BSc, 43, PhD(chem), 46, DSc(chem), 63. Prof Exp: Res chemist, Colonial Prod Res Coun, Eng, 43-46 & Dunlop Rubber Co, 46-47; sci off, Ministry of Supply, Brit Govt, 47-48; fel, Ohio State Univ, 48-49; Sugar Res Found fel, USDA, 49-51; res assoc, Univ Minn, 51-55; from asst prof to assoc prof, 55-64, PROF BIOCHEM, UNIV IOWA, 64-, PROG DIR, PHYSICIAN'S ASST PROG, COL MED, 73-, ASSOC DEAN, COL MED, 74- Concurrent Pos: Mem staff, Physiol Chem Study Sect, NIH, 68-72; USPHS sr fel, Australian Nat Univ, 69-70. Mem: Am Chem Soc; Biochem Soc; Am Soc Biol Chem; The Chem Soc. Res: Carbohydrates; protein-carbohydrate complexes; natural products; glycoproteins; carbohydrases; metal complexes of biological import; carbohydrate synthesis; membrane biochemistry; polypeptide antitumor agents. Mailing Add: Dept of Biochem Univ of Iowa Iowa City IA 52240

MONTGOMERY, RICHARD C, b Stark, Kans, Jan 21, 22; m 52; c 2. GEOLOGY. Educ: Univ Idaho, BS, 49; Univ Nebr, MA, 51, PhD(geog, geol), 62. Prof Exp: Intel analyst geog, Eng Strategic Intel Div, Army Map Serv, 51-53; prof earth sci, Willamette Univ, 53-66; PROF GEOG, CALIF STATE UNIV, FRESNO, 66- Mem: Asn Am Geog; Am Geol Inst. Res: Arid land morphology; volcanic landforms; late Pleistocene paleoclimatology. Mailing Add: Dept of Geog Calif State Univ Fresno CA 93726

MONTGOMERY, RICHARD GLEE, b Grayslake, Ill, Feb 9, 38; m 61; c 3. MATHEMATICS. Educ: San Francisco State Col, AB, 60; Brown Univ, MAT, 65; Clark Univ, MA, 68, PhD(math), 69. Prof Exp: Asst prof math, Humboldt State Col, 69-70; ASST PROF MATH, SOUTHERN ORE STATE COL, 70- Mem: Am Math Soc; Math Asn Am; Opers Res Soc Am. Res: Algebraic, topological and categorical structures of rings of continuous functions. Mailing Add: Dept of Math Southern Ore State Col Ashland OR 97520

MONTGOMERY, RONALD EUGENE, b Rural Valley, Pa, Feb 17, 37; m 61; c 4. ORGANIC CHEMISTRY, PESTICIDE CHEMISTRY. Educ: Waynesburg Col, BS, 59; Duke Univ, MA, 62, PhD(org chem), 64. Prof Exp: Res assoc, Duke Univ, 63-64; res chemist, Niagara Chem Div, 64-69, sr res chemist, 69, MGR ORG SYNTHESIS, AGR CHEM DIV, FMC CORP, 69- Mem: Am Chem Soc. Res: Pesticides. Mailing Add: Res Dept Agr Chem Div FMC Corp 100 Niagara St Middleport NY 14105

MONTGOMERY, ROYCE LEE, b Hartsville, Tenn, Nov 8, 33; m 67; c 1. GROSS ANATOMY, NEUROANATOMY. Educ: Univ Va, BA, 55; WVa Univ, MS, 60, PhD(gross anat), 63. Prof Exp: Instr gross anat & neuroanat, Med Sch, WVa Univ, 63-65; from instr to asst prof, 68-72, ASSOC PROF GROSS ANAT & NEUROANAT, SCH MED, UNIV NC, CHAPEL HILL, 72- Mem: Am Asn Anat. Res: Behavior and adrenal status following septal and amygdaloid lesions in the rat; hormonal influence on hippocampal neurons. Mailing Add: Dept of Anat Univ of NC Sch of Med Chapel Hill NC 27514

MONTGOMERY, STEWART ROBERT, b Pottsville, Pa, July 16, 24; m 60. INDUSTRIAL CHEMISTRY. Educ: Pa State Univ, BS, 49; Univ Rochester, PhD(org chem), 55. Prof Exp: Res chemist org synthesis, Cent Res Lab, Allied Chem Corp, 49-51; res chemist petrochem, Res Dept, Lion Oil Co, Monsanto Co, 55-61; sr res chemist, Indust Catalysts Res Dept, 61-73, SR RES CHEMIST, INDUST CHEM DEPT, DAVISON DIV, W R GRACE & CO, 73- Mem: Am Chem Soc; Sigma Xi. Res: Synthesis of amino acids, organosilicon compounds and aromatic hydrocarbons; heterogeneous, vaporphase catalysis; synthetic resins, chemistry of asphalt and bitumens; oxidation of hydrocarbons; preparation and evaluation of industrial catalysts. Mailing Add: W R Grace & Co Indust Chem Dept 5500 Chemical Rd Baltimore MD 21226

MONTGOMERY, SUSAN, b Tampa, Fla, Apr 2, 43. ALGEBRA. Educ: Univ Mich, Ann Arbor, BA, 65; Univ Chicago, MS, 66, PhD(math), 69. Prof Exp: Asst prof math, DePaul Univ, 69-70; asst prof, 70-75, ASSOC PROF MATH, UNIV SOUTHERN CALIF, 75- Concurrent Pos: Vis asst prof, Hebrew Univ Jerusalem, 73. Mem: Am Math Soc; Math Asn Am. Res: Non-commutative rings with involution, and fixed points of automorphisms. Mailing Add: Dept of Math Univ of Southern Calif Los Angeles CA 90007

MONTGOMERY, THEODORE ASHTON, b Los Angeles, Calif, Oct 27, 23. PUBLIC HEALTH, PEDIATRICS. Educ: Univ Southern Calif, MD, 46; Harvard Univ, MPH, 52; Am Bd Pediat & Am Bd Prev Med, dipl. Prof Exp: Consult pediat, Calif Dept Pub Health, 52-53; pvt pract, 52-54; consult child health, 54-58, chief maternal & perinatal health, 58-61 & bur maternal & child health, 61-62, from asst chief to chief div prev med serv, 62-72, dep dir, Calif Dept Pub Health, 69-72, chief div prev med, Alameda County Health Agency, 73-74, MED CONSULT, CALIF DEPT HEALTH, BERKELEY, 74- Concurrent Pos: Mem, Surg Gen Adv Comt Immunization Practices, 64-68, President's Adv Comt Ment Retarded, 65 & Ment Retarded Proj Rev Comt, USPHS, 65-66. Mem: Fel Am Acad Pediat; fel Am Pub Health Asn. Res: Maternal and child health morbidity and mortality causation and procedures for reducing these factors, including designing and implementing approaches to preventive measures. Mailing Add: 85 Wildwood Gardens Piedmont CA 94611

MONTGOMERY, WILLIAM WAYNE, b Proctor, Vt, Aug 20, 23. OTOLARYNGOLOGY. Educ: Middlebury Col, AB, 44; Univ Vt, MD, 47; Am Bd Otolaryngol, dipl, 56. Prof Exp: Intern, Mary Fletcher Hosp, Burlington, Vt, 47-48; physician pvt pract, Vt, 48-50; resident otolaryngol, Mass Eye & Ear Infirmary, 52-55, asst, 55-56; asst otol, 56-58, from instr to assoc prof, 59-70, PROF OTOLARYNGOL, HARVARD MED SCH, 70-; SR SURGEON & ASST TO

CHIEF, DEPT OTOLARYNGOL, MASS EYE & EAR INFIRMARY, 69- Concurrent Pos: Asst surgeon, Mass Eye & Ear Infirmary, 58-60, assoc surgeon, 60-69. Honors & Awards: Harris P Mosher Award, 63. Mem: Fel Am Col Surgeons; AMA; Am Acad Ophthal & Otolaryngol; Am Broncho-Esophagol Asn; Am Laryngol Asn. Res: Dysfunctions of the human larynx; reconstruction of the cervical respiratory areas; carcinoma of the head and neck; radical surgery of the nose and sinuses; surgery of the upper respiratory system. Mailing Add: Mass Eye & Ear Infirmary 243 Charles St Boston MA 02114

MONTI, STEPHEN ARION, b Ross, Calif, Nov 23, 39; m 64. ORGANIC CHEMISTRY. Educ: Univ Calif, Berkeley, BS, 61; Mass Inst Technol, PhD(org chem), 64. Prof Exp: Asst prof chem, Mich State Univ, 64-66; NSF fel, Harvard Univ, 66-67; asst prof, 67-71, ASSOC PROF CHEM, UNIV TEX, AUSTIN, 71- Res: Synthetic organic chemistry; natural products and related areas. Mailing Add: Dept of Chem Univ of Tex Austin TX 78712

MONTIE, THOMAS C, b Cleveland, Ohio, Oct 8, 34; m 56; c 2. BIOCHEMISTRY, MICROBIOLOGY. Educ: Oberlin Col, AB, 56; Univ Md, MS, 58, PhD(microbial physiol), 60. Prof Exp: Asst bot, Univ Md, 56-58, fungus physiol, 59-60; res assoc, Inst Cancer Res, 60-62; asst mem, Albert Einstein Med Ctr, 62-67, assoc mem, 67-69; ASSOC PROF MICROBIOL, UNIV TENN, KNOXVILLE, 69- Mem: AAAS; Am Soc Microbiol; Am Chem Soc; fel Am Inst Chem; Am Inst Biol Sci. Res: Location, synthesis, regulation, mode of action and chemistry of toxic proteins from bacteria; amino acid transport and metabolism. Mailing Add: Dept of Microbiol Univ of Tenn Knoxville TN 37916

MONTIEL, FRANCISCO, b Santiago, Chile, June 30, 34; m 59; c 2. METABOLISM, MICROBIOLOGY. Educ: Univ Chile, MD, 61; Inst Biochem, Curitiba, Brazil, 61. Prof Exp: Asst, 57-64, ASST PROF MICROBIOL, SCH MED, CATH UNIV CHILE, 64-, PROF, SCH NURSING, 60- Concurrent Pos: Henry L & Grace Doherty Charitable Found fel, 64-65; vis asst prof, Stritch Sch Med, Loyola Univ, Ill, 65-66 & 73-74. Mem: Am Soc Microbiol; Chilean Soc Microbiol & Hyg; Med Col Chile. Res: Phage-typing; antibiotics sensitivity; microbial and carbohydrate metabolism; virulence of staphylococcus. Mailing Add: Clin Microbiol Unit Cath Univ of Chile Sch of Med Santiago Chile

MONTIETH, RICHARD VOORHEES, b Indianapolis, Ind, Aug 13, 13; m 41; c 4. CHEMISTRY. Educ: Butler Univ, AB, 35. Prof Exp: Chemist, Allison Div, 40-44, supvr bearing chem lab, 44-54, gen supvr prod metall lab, 54-58, supvr prod metall sect, 58-64, sr exp metallurgist foundry technol, 64-66, Sr Exp Metallurgist, Detroit Diesel Allison Div, Gen Motors Corp, 66-74; RETIRED. Mem: Am Soc Metals. Res: Metallurgical, nonferrous and ferrous and high temperature alloys, selection and properties. Mailing Add: 2350 W 65th St Indianapolis IN 46260

MONTJAR, MONTY JACK, b Latrobe, Pa, Aug 12, 24. PHYSICAL CHEMISTRY. Educ: St Vincent Col, BS, 49; Univ Notre Dame, MS, 50; Carnegie Inst Technol, PhD(phys chem), 55. Prof Exp: Chemist, Callery Chem Co, 54-57; asst prof chem, Pa State Univ, 57-60; from asst prof to assoc prof, St Vincent Col, 60-65; assoc scientist, Brookhaven Nat Lab, 66-68; assoc prof chem, Mt Sinai Sch Med, 68-70; ASSOC PROF CHEM, PA STATE UNIV, HAZLETON, 70- Res: RNA; protein biosynthesis; tissue transplantation; fractionation; peptide synthesis. Mailing Add: Dept of Chem Pa State Univ Hazleton PA 18201

MONTO, ARNOLD SIMON, b Brooklyn, NY, Mar 22, 33; m 58; c 4. EPIDEMIOLOGY, INFECTIOUS DISEASES. Educ: Cornell Univ, BA, 54, MD, 58. Prof Exp: Intern & asst resident med, Vanderbilt Univ Hosp, 58-60; USPHS fel, Med Sch, Stanford Univ, 60-62; mem virus dis sect, Nat Inst Allergy & Infectious Dis, 62-65; from res assoc to asst prof, 65-71, ASSOC PROF EPIDEMIOL, SCH PUB HEALTH, UNIV MICH, ANN ARBOR, 71- Concurrent Pos: NIH career develop award, 68; consult, NIH. Mem: Am Epidemiol Soc; Soc Exp Biol & Med; Am Asn Immunologists; Infectious Dis Soc Am; Soc Epidemiol Res. Res: Epidemiology of respiratory disease in the community; viral diseases and diagnosis; evaluations of vaccines. Mailing Add: Dept of Epidemiol Univ of Mich Sch Pub Health Ann Arbor MI 48104

MONTOURE, JOHN ERNEST, b Shawano, Wis, May 10, 27; m 48; c 4. DAIRY SCIENCE, AGRICULTURAL BIOCHEMISTRY. Educ: Univ Wis, BS, 54, MS, 55; Wash State Univ, PhD(dairy sci), 61. Prof Exp: Asst prof dairy sci, 61-70, head dept food sci, 70-75, ASSOC PROF FOOD SCI, UNIV IDAHO, 70-, FOOD SCIENTIST, 73- Mem: Am Dairy Sci Asn; Inst Food Technol. Res: Pesticide detection and analytical methods; fate of pesticides in dairy products; enzyme isolation; bacteriology. Mailing Add: Dept of Food Sci Univ of Idaho Moscow ID 83843

MONTREUIL, FERNAND, b Quebec City, Que, Apr 20, 16; m 39; c 2. MEDICINE, OTOLARYNGOLOGY. Educ: Laval Univ, MB, 36, MD, 39; FRCS(C). Prof Exp: Instr otolaryngol, Col Physicians & Surgeons, Columbia Univ, 47-52; PROF OTOLARYNGOL, UNIV MONTREAL, 52- Concurrent Pos: Asst otolaryngologist, Columbia-Presby Med Ctr, 47-52; chief otolaryngol, Queen Mary Vet Hosp, Montreal, 52-; chief otolaryngol, Notre Dame Hosp, 52-, dir speech & hearing, 57-; Can mem, Int Comn Study Cancer Larynx, 57. Mem: Fel Am Col Surgeons; fel Am Acad Ophthal & Otolaryngol; fel Am Broncho-Esophagol Asn; fel Am Laryngol Asn; Can Otolaryngol Soc (pres, 61). Res: Inner ear function; problems dealing with hearing and equilibrium. Mailing Add: 1560 Sherbrooke St E Montreal PQ Can

MONTROLL, ELLIOTT WATERS, b Pittsburgh, Pa, May 4, 16; m 43; c 10. MATHEMATICS, PHYSICS. Educ: Univ Pittsburgh, BS, 37, PhD(math), 40. Prof Exp: Asst math, Univ Pittsburgh, 37-39; asst chem, Columbia Univ, 39-40; Sterling res fel, Yale Univ, 40-41; res assoc, Cornell Univ, 41-42; instr physics, Princeton Univ, 42-43; head math res group, Kellex Corp, NY, 43-45; adj prof chem, Polytech Inst Brooklyn, 44-46; from asst prof to assoc prof physics & math, Univ Pittsburgh, 46-48; vis prof, NY Univ, 50; res prof, Inst Fluid Dynamics & Appl Math, Univ Md, 51-61; Lorentz prof, Univ Leiden, 61; dir gen sci, IBM Tech Ctr, 61-63; vpres, Inst Defense Anal, 63-66; EINSTEIN PROF PHYSICS & CHEM & DIR INST FUNDAMENTAL STUDIES, UNIV ROCHESTER, 66- Concurrent Pos: Head physics br, Off Naval Res, Wash, DC, 48-50, dir phys sci div, 52-54; Guggenheim Mem fel, Univ Brussels, 58-59; Fulbright lectr, Univ Grenoble, 59; ed, J Math Physics, 59-70. Honors & Awards: Lancaster Proze, Opers Res Soc Am, 61. Mem: Nat Acad Sci; Am Phys Soc. Res: Statistical mechanics; theory of probability; mathematical physics; mathematical modeling of biological and sociological phenomenon. Mailing Add: Dept of Physics Univ of Rochester Rochester NY 14627

MONTROSE, CHARLES JOSEPH, JR, b Pittsburgh, Pa, Jan 3, 42; m 63; c 3. PHYSICS. Educ: John Carroll Univ, BS, 62, MS, 64; Cath Univ Am, PhD(physics), 67. Prof Exp: Asst prof, 67-70, ASSOC PROF PHYSICS, CATH UNIV AM, 70- Mem: Am Phys Soc; Acoust Soc Am. Res: Study of the structure and dynamics of the liquid state using ultrasonic and optical techniques; laser spectroscopy. Mailing Add: Dept of Physics Cath Univ Am Washington DC 20017

MONTROY, LEO DENNIS, b Wyandotte, Mich, May 9, 47; m 69; c 2. FRESH WATER ECOLOGY. Educ: Univ Windsor, BSc, 69; Univ Notre Dame, PhD(biol), 73. Prof Exp: Fel, Dept Civil Eng, Univ Notre Dame, 73-74, Sch Advan Studies, 74; ASST PROF BIOL, VA COMMONWEALTH UNIV, 74- Mem: Am Soc Limnol & Oceanog; Int Soc Limnol. Res: Extracellular products of phytoplankton; mechanics of filter feeding; mine waste effects. Mailing Add: Dept of Biol Va Commonwealth Univ 816 Park Ave Richmond VA 23184

MONTY, KENNETH JAMES, b Sanford, Maine, Sept 11, 30; m 52; c 2. BIOCHEMISTRY, CELL BIOLOGY. Educ: Bowdoin Col, BA, 51; Univ Rochester, PhD(biochem), 56. Prof Exp: Fel biochem, McCollum-Pratt Inst, Johns Hopkins Univ, 55-57, asst prof biol, 57-63; head dept, 63-75, PROF BIOCHEM, UNIV TENN, KNOXVILLE, 63-, COORDR BIOL SCI, 73- Mem: AAAS; Am Chem Soc; Am Soc Biol Chem; NY Acad Sci. Res: Cellular physiology; sulfur metabolism; microbial genetics and metabolic control; biological adaptation. Mailing Add: Biol Coord 620 Hesler Biol Bldg Univ of Tenn Knoxville TN 37916

MONTZINGO, LLOYD J, JR, b Chicago, Ill, Nov 29, 27; m 48; c 6. MATHEMATICS. Educ: Houghton Col, AB, 49; Univ Buffalo, MA, 51, PhD, 61. Prof Exp: From instr to asst prof math, Roberts Wesleyan Col, 51-56; from instr to asst prof, Univ Buffalo, 56-62; assoc prof, 62-66, PROF MATH, SEATTLE PAC COL, 66-, CHMN DEPT, 62-, DIR, SCH SCI, 73- Concurrent Pos: NSF sci fac fel, Univ Wash, 70-71. Mem: Am Math Soc; Math Asn Am (assoc secy, 58-62). Res: Mathematical statistics. Mailing Add: Sch of Sci Seattle Pac Col Seattle WA 98119

MOOBERRY, JARED BEN, b Pekin, Ill, Mar 6, 42; m 75. SYNTHETIC ORGANIC CHEMISTRY. Educ: Univ Ill, BS, 64; Cornell Univ, PhD(chem), 69. Prof Exp: Fel chem, Swiss Fed Inst Technol, 69-70; res assoc, Cornell Univ, 70-71; RES CHEMIST, EASTMAN KODAK RES LABS, 71- Mem: Am Chem Soc. Res: Synthesis of organic chemicals of photographic utility. Mailing Add: Eastman Kodak Res Labs Rochester NY 14650

MOOD, ALEXANDER MCFARLANE, b Amarillo, Tex, May 31, 13; m 36; c 3. APPLIED MATHEMATICS, PUBLIC POLICY. Educ: Univ Tex, BA, 34; Princeton Univ, PhD(math statist), 40. Prof Exp: Instr appl math, Univ Tex, 40-42; statistician, Bur Labor Statist, 42-44; res assoc, Princeton Univ, 45; prof math statist, Iowa State Col, 45-48; dep chief, Math Div, Rand Corp, 48-55; pres, Gen Anal Corp, 55-60; vpres in charge, Los Angeles Res Ctr, C-E-I-R, Inc, 60-64; asst commissioner ed, US Off Ed, 64-67; dir, Pub Policy Res Orgn, 67-73, PROF ADMIN & POLICY ANALYST, UNIV CALIF, IRVINE, 73- Concurrent Pos: Guest lectr, Army War Col & Air War Col; consult, US Dept Defense; del, NATO Opers Res Conf, Paris, 58, Int Opers Res Conf, Aix-en-Provence, 60 & Oslo, 63. Mem: Am Math Soc; Opers Res Soc Am (pres, 63); Soc Indust & Appl Math; fel Inst Math Statist (pres, 56). Res: Mathematical statistics; operations research; game theory; administration; policy analysis. Mailing Add: Pub Policy Res Orgn Univ of Calif Irvine CA 92664

MOODY, ARNOLD RALPH, b Augusta, Maine, Oct 8, 41; m 65. PLANT PATHOLOGY. Educ: Univ Maine, BS, 63; Univ NH, MS, 65; Univ Calif, Berkeley, PhD(plant path), 71. Prof Exp: Asst res plant pathologist, Univ Calif, Berkeley, 71-72; res plant pathologist, Changins Fed Agr Exp Sta, Nyon, Switz, 72-74; ASSOC PROF PLANT PATH, VA STATE COL, 74- Mem: Am Phytopath Soc. Res: Plant diseases; control of plant disease through natural resistance mechanisms; physiology of disease resistance. Mailing Add: Va State Col Box 476 Petersburg VA 23803

MOODY, DAVID BURRITT, b Rochester, NY, June 20, 40; m 64; c 2. PSYCHOACOUSTICS. Educ: Hamilton Col, BA, 62; Columbia Univ, MA, 64, PhD(psychol), 67. Prof Exp: Dept Health, Educ & Welfare res fel, Kresge Hearing Res Inst, Med Sch, 67-68, res assoc psychoacoustics, 68-71, ASST PROF OTORHINOLARYNGOL, KRESGE HEARING RES INST, MED SCH, UNIV MICH, ANN ARBOR, 71-, ASST PROF PSYCHOL, UNIV, 69- Mem: AAAS; Acoust Soc Am; Asn Res Otorhinolaryngol; Psychonomic Soc. Res: Psychoacoustics; hearing in non-human primates; experimentally produced hearing loss; computers in psychology. Mailing Add: Kresge Hearing Res Inst Univ of Mich Ann Arbor MI 48104

MOODY, DAVID COIT, III, b Florence, SC, Dec 23, 48; m 68. INORGANIC CHEMISTRY. Educ: Univ SC, BS, 67; Ind Univ, PhD(chem), 75. Prof Exp: Res asst chem, Ind Univ, 72-75; FEL CHEM, LOS ALAMOS SCI LAB, UNIV CALIF, 75- Mem: Am Chem Soc; Sigma Xi Res: Synthetic and structural inorganic chemistry as related to the binding of small molecule noxious gases such as sulfur dioxide, nitric oxide, and carbon monoxide, by transition metals. Mailing Add: Los Alamos Sci Lab Los Alamos NM 87545

MOODY, DAVID WRIGHT, b Boston, Mass, Nov 23, 37; m 63; c 1. HYDROLOGY. Educ: Harvard Univ, AB, 60; Johns Hopkins Univ, PhD(geog), 64. Prof Exp: Hydrologist, Water Resources Div, 64-74, PROG ANALYST, OFF OF DIR, US GEOL SURV, 74- Mem: AAAS; Geol Soc Am; Am Geophys Union; World Future Soc. Res: Information systems; applications of computers to water resources planning and management; data network design. Mailing Add: Nat Ctr Stop 105 US Geol Surv Reston VA 22092

MOODY, EDWARD GRANT, b Thatcher, Ariz, Dec 23, 19; m 43; c 7. ANIMAL NUTRITION. Educ: Univ Ariz, BS, 41; Kans State Col, MS, 47; Purdue Univ, PhD(nutrit), 51. Prof Exp: Asst dairy husb, Kans State Col, 46-47; asst, Purdue Univ, 47-49, instr, 49-51; from asst prof to assoc prof, 51-63, PROF ANIMAL SCI, ARIZ STATE UNIV, 63- Mem: Am Soc Animal Sci; Am Dairy Sci Asn; Soc Nutrit Educ. Res: Fat metabolism in ruminants; dairy herd management and production. Mailing Add: Div of Agr Ariz State Univ Tempe AZ 85281

MOODY, ERIC EDWARD MARSHALL, b Neath, Gt Brit, Dec 15, 38; m 67; c 2. MOLECULAR GENETICS, VENEREAL DISEASES. Educ: Univ London, BSc, 64; Univ Edinburgh, PhD(molecular biol), 71. Prof Exp: Res scientist molecular genetics, Southwest Ctr Advan Studies, 65-68; tech officer, Med Res Coun Gt Brit, 68-71; ASST PROF MOLECULAR GENETICS, UNIV TEX HEALTH SCI CTR, 72- Concurrent Pos: Fel, Col Med, Univ Ariz, 71-72. Mem: Soc Gen Microbiol; Am Venereal Dis Asn. Res: Molecular genetics of plasmids; genetics of virulence and antibiotic resistance in Neisseria Gonorrhoeae. Mailing Add: Dept of Microbiol Univ of Tex Health Sci Ctr San Antonio TX 78284

MOODY, FRANK BALDWIN, b Cranston, RI, June 12, 13; m 41; c 1. CHEMISTRY. Educ: RI State Col, BS, 35; Ohio State Univ, MS, 37, PhD(org chem), 40. Prof Exp: Atlas Powder Co fel, Ohio State Univ, 40-41; res chemist, 41-51, SR RES CHEMIST, EXP STA, E I DU PONT DE NEMOURS & CO, INC, 51- Mem: AAAS; Am Chem Soc. Res: Synthetic fibers; characterization of macromolecules; composites; macromolecular synthesis; dyeing processes. Mailing Add: Bldg 262 Exp Sta E I du Pont de Nemours & Co Inc Wilmington DE 19898

MOODY, FRANK GORDON, b Franklin, NH, May 3, 28; m 64; c 3. SURGERY.

Educ: Dartmouth Col, BA, 52; Cornell Univ, MD, 56. Prof Exp: Surg intern, NY Hosp-Cornell Med Ctr, 56-57, surg residency, 57-63; clin instr surg, Med Ctr, Univ Calif, San Francisco, 63-65, asst prof, 65-66; assoc prof, Univ Ala, 66-69, prof, 69-71; PROF SURG & CHMN DEPT, UNIV UTAH, 71- Concurrent Pos: Cardiovasc Res Inst fel, 63-65; Am Heart Asn advan res fel, 63-65; consult, Vet Admin Hosp, Birmingham, Ala, 66-; mem surg training comt, Nat Inst Gen Med Sci, 68-72; mem med sect A, NIH, 72-75 & surg sect B, 75- Mem: Biophys Soc; Am Col Surg; Am Gastroenterol Asn; Am Surg Asn; Soc Surg Alimentary Tract. Res: Gastric acid secretion; pancreatic function; portal hypertension. Mailing Add: Dept of Surg Univ of Utah Med Ctr Salt Lake City UT 84112

MOODY, HARRY JOHN, b Birsay, Sask, May 8, 26; m 52; c 3. MICROWAVE PHYSICS, COMMUNICATIONS ENGINEERING. Educ: Univ Sask, BEng, 48, MS, 50; McGill Univ, PhD(physics), 55. Prof Exp: Asst, Univ Ill, 50-51; res asst, Nat Res Coun Can, 51-55; sr physicist, Can Marconi Co, 55-61; sr mem sci staff, 61-73, RES FEL, RCA LIMITED, MONTREAL, 73- Mem: Can Asn Physicists. Res: Microwave; millimeter wave; infrared; characterizing and optimizing a space system for SCPC communications. Mailing Add: RCA Limited 21001 Trans Canada Hwy Ste Anne de Bellevue PQ Can

MOODY, JOHN ROBERT, b Richmond, Va, Feb 6, 42. ANALYTICAL CHEMISTRY. Educ: Univ Richmond, BS, 64; Univ Md, College Park, MS, 67, PhD(anal chem), 70. Prof Exp: Asst anal chem, Univ Md, 70; RES CHEMIST, US NAT BUR STANDARDS, WASHINGTON, DC, 71- Mem: Am Chem Soc; Soc Appl Spectros. Res: General analytical procedures, including wet, electroanalytical, spectrophotometric, radiochemical and chromatographic procedures as applied to isotope dilution mass spectrometry; high accuracy trace analysis of rare earths; investigation of stoichiometry and separations. Mailing Add: Washington DC

MOODY, JOSEPH E, JR, b Pleasantville, NJ, Sept, 22, 23; m 51; c 3. PHARMACEUTICAL CHEMISTRY. Educ: Univ Conn, BS, 50, MS, 55, PhD(pharmaceut chem), 60. Prof Exp: From asst prof to assoc prof pharm, Fla Agr & Mech Univ, 54-57; chemist, Div Pharmaceut Chem, US Food & Drug Admin, 59-62, res chemist, 62-67; group leader anal res, White Labs, Schering Corp, NJ, 67-70, mgr, 70-72, DIR ANAL RES & DEVELOP, USV PHARMACEUT CORP, 72- Mem: Am Chem Soc; Am Pharmaceut Asn; Asn Off Anal Chem; Acad Pharmaceut Sci. Res: Development of new analytical procedure for official amphetamine preparations; chemical assay for thyroid powder and tablets; analytical profile of chlorthalidone. Mailing Add: 41 Yale Terr West Orange NJ 07052

MOODY, JULIUS REYNARD, b Louisville, Miss, Sept 15, 45; m 67; c 2. ENTOMOLOGY. Educ: Miss State Univ, BS, 67, MS, 69, PhD(entom), 72. Prof Exp: Cotton pest mgt specialist, Miss Coop Exten Serv, 72-75; TECH REP, ICI US INC, 75- Mem: Entom Soc Am. Res: Field screening of candidate pesticides determining whether these materials are of significance to advance to the market place. Mailing Add: 842 Windsor Rd Grenada MS 38901

MOODY, LEONARD ELLSWORTH, chemistry, see 12th edition

MOODY, LEROY STEPHEN, b Elyria, Ohio, Nov 18, 18; m 45; c 3. PHYSICAL CHEMISTRY. Educ: Wesleyan Univ, BA, 41; Univ Wis, PhD(phys chem), 44. Prof Exp: Res chemist, Mass Inst Technol, 44-45; res chemist, 45-51, supvr prod eval, 51-52, specialist mkt res & prod planning, 52-54 & proj eval, 55-57, mgr, New Prod Develop Lab, 57-59 & polycarbonate eng, 59-64, managing dir, N V Polychemie, AKU-GE, Netherlands, 64-67, gen mgr irradiation processing oper, Vallecitos Nuclear Ctr, Calif, 67-69 & reactor fuels & reprocessing dept, 69-71, mgr strategy planning oper, 71-74, RES & DEVELOP MGR MAT SCI & ENG, CORP RES & DEVELOP CTR, GEN ELEC CO, 74- Mem: AAAS; Am Chem Soc; Am Nuclear Soc. Res: General management; administration of scientific programs; high performance materials, especially metals and ceramics for electrically-related applications; plastic materials for structural and engineering purposes. Mailing Add: Corp Res & Develop Ctr Gen Elec Co PO Box 8 Schenectady NY 12309

MOODY, MAX DALE, b Onaga, Kans, Sept 29, 24; m 50. MEDICAL BACTERIOLOGY. Educ: Univ Kans, AB, 48, MA, 49, PhD(bact), 53; Am Bd Med Microbiol, dipl. Prof Exp: Asst med bact, Univ Kans, 46-53; in chg spec projs lab, Nat Commun Dis Ctr, USPHS, 53-62, chief staphylococcus & streptococcus unit, 62-67, dir, Nat Streptococcal Dis Ctr, 67-70, chief reagents eval unit, Ctr Dis Control, 70-71; TECH DIR, WELLCOME REAGENTS DIV, BURROUGHS WELLCOME CO, 71- Concurrent Pos: Mem expert adv panel coccal infections, WHO, 67-; secy subcomt streptococci & pneumococci, Int Cong Microbiol; mem comt on rheumatic fever, Am Heart Asn; assoc mem, Comn on Staphylococcal & Streptococcal Dis, Armed Forces Epidemiol Bd. Honors & Awards: Kimble Methodology Res Award, 67; USPHS Commendation Medal, 68. Mem: Am Soc Microbiol; Sigma Xi; fel Am Acad Med Microbiol. Res: Pathogenesis and immunity of Tularemia; rapid detection and identification of pathogenic microorganisms; fluorescent antibody identification of bacterial pathogens. Mailing Add: Wellcome Reagents Div Burroughs Wellcome Co Research Triangle Park NC 27709

MOODY, PAUL AMOS, b Randolph Center, Vt, Jan 13, 03; m 27; c 2. ZOOLOGY. Educ: Morningside Col, AB, 24; Univ Mich, PhD(zool), 27. Prof Exp: Asst zool, Univ Mich, 25-26; from asst prof to prof, 27-60, dir grad study, 42-49, HOWARD PROF NATURAL HIST & ZOOL, UNIV VT, 60-, EMER PROF, 68- Mem: Am Philos Soc grant-in-aid, 41. Mem: AAAS; Am Soc Mammal; Soc Study Evolution; Am Soc Zool; Am Soc Nat. Res: Evolution; serological investigations of mammalian relationships; human genetics. Mailing Add: Dept of Zool Univ of Vt Burlington VT 05401

MOODY, ROBERT ADAMS, b Swampscott, Mass, Oct 1, 34; m; c 3. NEUROSURGERY. Educ: Univ Chicago, BA, 55 & 56, MD, 60. Prof Exp: Intern, Royal Victoria Hosp, Mon, 60-61; resident surg, Univ Vt, 61-62 & resident neurosurg, 62-66; asst prof neurosurg, Univ Chicago, 66-71; asst prof, Sch Med, Tufts Univ, 71-74; CHMN DIV NEUROSURG, COOK COUNTY HOSP, 74-; PROF NEUROSURG, ABRAHAM LINCOLN SCH MED, UNIV ILL MED CTR, 74- Concurrent Pos: Fel neurosurg, Lahey Clin, 63-64. Mem: Am Asn Neurol Surg; Am Col Surgeons; Asn Acad Surg; Sigma Xi. Res: Head injury; blood brain barrier; brain scanning; microvascular surgery. Mailing Add: Cook County Hosp 1835 W Harrison St Chicago IL 60612

MOODY, ROBERT AUSTIN, analytical chemistry, inorganic chemistry, see 12th edition

MOOERS, CALVIN NORTHRUP, b Minneapolis, Minn, Oct 24, 19; m 45; c 2. MATHEMATICS. Educ: Univ Minn, AB, 41; Mass Inst Technol, MS, 48. Prof Exp: Physicist, Naval Ord Lab, Md, 41-46; PROPRIETOR & DIR RES, ZATOR CO, 47-; PRES & DIR RES, ROCKFORD RES, INC, 61- Mem: Int Fedn Doc; Asn Comput Mach; Am Soc Info Sci; Inst Elec & Electronics Eng. Res: Programming languages; information retrieval and processing, including theory, systems and machines; reactive typewriter. Mailing Add: 13 Bowdoin St Cambridge MA 02138

MOOERS, CHRISTOPHER NORTHRUP KENNARD, b Hagerstown, Md, Nov 11, 35; m 60; c 2. PHYSICAL OCEANOGRAPHY, FLUID DYNAMICS. Educ: US Naval Acad, BS, 57; Univ Conn, MS, 64; Ore State Univ, PhD(phys oceanog), 70. Prof Exp: NATO fel, Univ Liverpool, 69-70; from asst prof to assoc prof phys oceanog, Sch Marine & Atmospheric Sci, Univ Miami, 70-76; ASSOC PROF PHYS OCEANOG, COL MARINE STUDIES, UNIV DEL, 76- Mem: AAAS; Am Geophys Union; Oceanog Soc Japan; Am Meteorol Soc; Sigma Xi. Res: Fluid dynamics of continental margins; oceanic fronts, coastal upwelling, transient circulations due to storms, tidal and other long waves and wave-mean current interactions; analysis of observations; geophysical fluid dynamics. Mailing Add: Col Marine Studies Univ of Del Lewes DE 19958

MOOG, FLORENCE, b Brooklyn, NY, Jan 24, 16. DEVELOPMENTAL BIOLOGY. Educ: NY Univ, AB, 36; Columbia Univ, AM, 38, PhD(zool), 44. Prof Exp: Med records clerk, US Dept Labor, 37-38; instr, Univ Del, 40; res assoc, 42-45, from instr to prof zool, 45-74, REBSTOCK PROF BIOL, WASH UNIV, 74- Concurrent Pos: Merck fel, Cambridge Univ, 54-55; mem study sect human embryol & develop, NIH, 66-70; Walker-Ames prof, Univ Wash, 73. Honors & Awards: AAAS-Westinghouse Award, 48. Mem: Am Soc Zool; Soc Develop Biol; Am Soc Cell Biol; Histochem Soc. Res: Functional development of digestive tract; enzymogenesis in vertebrate development; developmental endocrinology of vertebrates. Mailing Add: Dept of Biol Wash Univ St Louis MO 63130

MOOI, JOHN, physical chemistry, see 12th edition

MOOK, DELO EMERSON, II, b Cleveland, Ohio, Apr 28, 42; m 65. ASTRONOMY. Educ: Case Inst Technol, BS, 64; Univ Mich, PhD(astron), 69. Prof Exp: Instr, Lawrence Univ, 67; res assoc astron, Univ Chicago, 69-70; ASST PROF ASTRON, DARTMOUTH COL, 70- Mem: Am Astron Soc; Royal Astron Soc; Int Astron Union; Astron Soc Pac. Res: Astronomical photometry; polarimetry; celestial x-ray sources. Mailing Add: Dept of Physics & Astron Dartmouth Col Hanover NH 03755

MOOK, DONALD ELTON, dairy industry, see 12th edition

MOOK, HERBERT ARTHUR, JR, b Meadville, Pa, Apr 17, 39; m 65; c 1. APPLIED PHYSICS, SOLID STATE PHYSICS. Educ: Williams Col, BA, 60; Harvard Univ, MA, 61, PhD(physics), 65. Prof Exp: PHYSICIST, OAK RIDGE NAT LAB, 65- Mem: Am Phys Soc. Res: Neutron diffraction studies of solid state physics, especially magnetic materials; polarized and inelastic neutron scattering investigations of the spin density and magnetic excitations in the rare earths and transition metals. Mailing Add: Solid State Div PO Box X Oak Ridge TN 37830

MOOK, LEONARD JAN, b Groningen, Netherlands, Jan 29, 32; Can citizen; m 56; c 2. ANIMAL ECOLOGY. Educ: Univ Groningen, BSc, 54, MSc, 56; McMaster Univ, PhD(biol), 61. Prof Exp: Res officer forestry, Maritime Region, Can Dept Forestry, 60-67; asst prof, 67-69, ASSOC PROF BIOL, UNIV SASK, REGINA, 69- Mem: Entom Soc Can; Brit Ecol Soc. Res: Influence of predation on populations of forest insects. Mailing Add: Dept of Biol Div of Natural Sci Univ of Sask Regina SK Can

MOOK, MAURICE ALLISON, ethnology, deceased

MOOKERJEA, SAILEN, b Jamalpur, India, Sept 1, 30; m 56; c 2. BIOCHEMISTRY, PHYSIOLOGY. Educ: Univ Calcutta, BSc, 49, MSc, 51, PhD(physiol), 56. Prof Exp: Lectr biochem, Univ Nagpur, India, 53-61; lectr physiol, Univ Calcutta, 61-62; res assoc, Banting & Best Dept Med Res, Univ Toronto, 62-65, from asst prof to assoc prof med, 65-75; PROF BIOCHEM, MEM UNIV NFLD, 75- Concurrent Pos: Nat Res Coun Can fel, 56-57; res fel, Banting & Best Dept Med Res, Univ Toronto, 57-58; Med Res Coun scholar, 65-70. Mem: Can Biochem Soc; Can Physiol Soc; Brit Biochem Soc; Am Inst Nutrit. Res: Lipoprotein biosynthesis; role of glycosyltransferases in lipoprotein and membrane metabolism; role of choline in membrane function and lipoprotein biogenesis; glycoprotein biosynthesis. Mailing Add: Dept of Biochem Mem Univ of Nfld St Johns NF Can

MOOKHERJEE, DEBNATH, b Darjeeling, India, Jan 1, 33. URBAN GEOGRAPHY, GEOGRAPHY OF SOUTH ASIA. Educ: Univ Calcutta, BS, 51, MS, 53; Univ Fla, PhD(geog), 61. Prof Exp: Lectr geog, Maharaja Manindra Chandra Col, India, 55-58; asst, Univ Fla, 58-61; from asst prof to assoc prof, 61-71, PROF GEOG, WESTERN WASH STATE COL, 71- Mem: Asn Am Geog; Geog Soc India. Res: Contemporary Asia; urbanization and associated problems; economic growth and development of developing countries. Mailing Add: Dept of Geog Western Wash State Col Bellingham WA 98225

MOOKHERJI, TRIPTY KUMAR, b Calcutta, India, Sept 15, 38; m 65; c 1. SOLID STATE PHYSICS. Educ: Agra Univ, BSc, 57, MSc, 59; Univ Burdwan, DPhil(physics), 66. Prof Exp: Res assoc, Marshall Space Flight Ctr, NASA, 66-68; SR PHYSICIST SOLID STATE, TELEDYNE-BROWN ENG, 68- Concurrent Pos: Mem, Indian Asn Cultivation of Sci. Honors & Awards: New Technol Utilization Award, NASA, 72. Mem: Am Phys Soc; Optical Soc Am. Res: Electronic excitation; color center; optical band structure; surface physics; growing and characterization of metallic and nonmetallic crystals; manufacturing in space; photovoltaic solar energy conversion; amniotic fluid study. Mailing Add: 600 Mountain Gap Dr Huntsville AL 35803

MOOLENAAR, ROBERT JOHN, b DeMotte, Ind, Nov 30, 31; m 55; c 3. PHYSICAL CHEMISTRY, ENVIRONMENTAL SCIENCES. Educ: Hope Col, BA, 53; Univ Ill, PhD, 57. Prof Exp: Instr phys chem, Univ Ill, 56-57; chemist, 57-65, sr res chemist, 65-69, assoc scientist, 69-74, DIR ENVIRON SCI RES, DOW CHEM CO, 74- Mem: Am Chem Soc. Res: Inorganic chemistry; biodegradation; photodegradation; toxicology. Mailing Add: Dow Chem Co 1702 Bldg Midland MI 48640

MOOLGAVKAR, SURESH HIRAJI, b Bombay, India, Jan 3, 43; m 68. GEOMETRY, TOPOLOGY. Educ: Univ Bombay, MB & BS, 65; Johns Hopkins Univ, PhD(math), 73. Prof Exp: Instr math, Johns Hopkins Univ, 72-73; ASST PROF MATH, IND UNIV, BLOOMINGTON, 73- Res: Analysis on manifolds, particularly differential operators and pseudogroup structures. Mailing Add: Dept of Math Ind Univ Bloomington IN 47401

MOOLICK, RICHARD TERRENCE, geology, see 12th edition

MOOLTEN, FREDERICK LONDON, b New York, NY, Oct 11, 32; m 60; c 3. CANCER. Educ: Harvard Col, BA, 53; Harvard Med Sch, MD, 63. Prof Exp: Internship med, Mass Gen Hosp, Boston, 63-64, residency, 64-65 & 67-68, res fel, 65-67, 68-69; res assoc, 69-71, asst prof, 71-74, ASSOC PROF MICROBIOL, SCH MED, BOSTON UNIV, 74- Concurrent Pos: Fel, Am Cancer Soc, 65-67 & NIH, 69-71. Honors & Awards: Cancer Res Scholar Award, Am Cancer Soc, 71. Mem: Am Asn Cancer Res. Res: Tumor immunotherapy studies aimed at eradicating tumor cells selectively with antitumor antibodies conjugated to potent cytotoxins; studies of alterations of the biochemistry and function of cell surfaces of tumor cells. Mailing Add: Sch of Med Boston Univ 80 E Concord St Boston MA 02118

MOOLTEN, SYLVAN E, b New York, NY, Sept 1, 04; m 28; c 2. INTERNAL MEDICINE, PATHOLOGY. Educ: Columbia Univ, AB, 24, MD, 28; Am Bd Path, dipl, 38; Am Bd Internal Med, dipl, 41. Prof Exp: Atten physician, Mt Sinai Hosp, New York, 34-42; dir lab, Middlesex Gen Hosp, 46-70, dir lab res, 56-70; DIR LAB, ROOSEVELT HOSP, 50-; ASSOC PROF, COL PHYSICIANS & SURGEONS, COLUMBIA UNIV, 63- Concurrent Pos: Consult, St Peter's Gen Hosp, 41-, dir lab, 46-56; res grants, USPHS, 50-55 & 61-66, NJ Heart Asn, 56-58 & Wesson Fund for Med Res, 61; consult, Middlesex Gen Hosp, 56-; clin prof, Rutgers Med Sch, Col Med & Dent NJ, 63- Mem: AAAS; Col Am Path; Am Col Physicians; Am Soc Clin Oncol; Asn Hosp Med Educ. Res: Lung pathology; blood platelets and thrombosis; Hodgkin's disease; pulmonary lymphatics; adsorption of virus-like agents to erythrocytes; systemic lupus erythematous. Mailing Add: Dept of Lab & Med Educ Roosevelt Hosp Metuchen NJ 08840

MOOMAW, JAMES CURTIS, b Dickinson, NDak, Jan 6, 28; m 56; c 4. AGRONOMY. Educ: Carleton Col, BA, 49; Univ Idaho, MS, 51; Wash State Univ, AEC, 51, PhD(bot, plant ecol), 57. Prof Exp: Asst agronomist, Univ Hawaii, 56-61, actg asst dir exp sta, 60; Rockefeller Found agronomist, Int Rice Res Inst, Philippines, 61-67 & Ceylon, 67-69; Rockefeller Found agronomist, Int Inst Trop Agr, Nigeria, 70-75, leader, Farming Systs Prog, 72, dir Outreach, 73-75; DIR, ASIAN VEGETABLE RES & DEVELOP CTR, 75- Concurrent Pos: Fulbright adv res grant, Kenya, 58-59; NSF grant, 60; sabbatical leave, Univ Calif, Davis, 69-70. Mem: AAAS; Am Soc Agron; Ecol Soc Am. Res: Range management and tropical forage crops; rice cultural practices and crop ecology; fertilizers, weed control and water management for tropical rice; farming systems; tropical horticulture. Mailing Add: Asian Vegetable Res & Develop Ctr PO Box 42 Shanhua Tainan Taiwan

MOOMAW, WILLIAM RENKEN, b Kansas City, Mo, Feb 18, 38; m 64; c 2. MOLECULAR SPECTROSCOPY, ENVIRONMENTAL SCIENCES. Educ: Williams Col, BA, 59; Mass Inst Technol, PhD(phys chem), 65. Prof Exp: Asst prof chem, Williams Col, 64-65; researcher, Univ Calif, Los Angeles, 65-66; asst prof, 66-71, ASSOC PROF CHEM, WILLIAMS COL, 72- Concurrent Pos: Froman distinguished prof, Russell Sage Col, 71; vis scholar, Univ Calif, Los Angeles, 71-72; AAAS Cong Sci fel, 75-76. Mem: AAAS; Am Phys Soc; Am Chem Soc. Res: Electronic spectra of small organic molecules; nature of the triplet states of these molecules; application of physical science to environmental problems. Mailing Add: Dept of Chem Williams Col Williamstown MA 02167

MOON, BYONG HOON, b Seoul, Korea, Jan 15, 26; US citizen; m 57; c 4. PHARMACOLOGY, MEDICINAL CHEMISTRY. Educ: Seoul Nat Univ, BSc, 51; Univ Nebr, BPharm, 60, MS, 64; Wash State Univ, PhD(pharmacol), 69. Prof Exp: Res assoc pharmacol, Univ Ill Col Med, 69-72; instr pharmacol, 72-73, ASST PROF PHARMACOL & ASST SCIENTIST INTERNAL MED, RUSH MED COL, 73- Mem: NY Acad Sci; Am Chem Soc. Res: The synthesis of 3-aminopiperidone derivatives to provide a basis for study of structure versus potential anticonvulsant activity relationshop; norepinephrine-adenosine triphosphate complex formation, spectroscopical proof of a complex in vitro and in adrenal medulary granules; establishing probable mechanism of hemolysis induced by various oxidant drugs. Mailing Add: Rush-Presby-St Luke's Med Ctr 1725 W Harrison Chicago IL 61612

MOON, FRANCIS C, b Brooklyn, NY, May 1, 39; m 62; c 3. SOLID MECHANICS. Educ: Pratt Inst, BSME, 62; Cornell Univ, MS, 64, PhD(theoret & appl mech), 67. Prof Exp: Asst prof mech & aerospace eng, Univ Del, 66-67; asst prof aerospace & mech sci, Princeton Univ, 67-74; ASSOC PROF, DEPT THEORET & APPL MECH, CORNELL UNIV, 75- Concurrent Pos: Consult, Rand Corp, Calif, 67-69 & Boeing Vertol, 70- Mem: AAAS; Am Soc Mech Engrs; Am Acad Mech. Res: Wave propagation in solids; composite materials; magneto-elasticity; magnetic levitation of trains; superconducting magnets. Mailing Add: Dept of Theoret & Appl Mech Thurston Hall Cornell Univ Ithaca NY 14850

MOON, HARLEY W, b Tracy, Minn, Mar 1, 36; m 56; c 4. VETERINARY PATHOLOGY. Educ: Univ Minn, BS, 58, DVM, 60, PhD(vet path), 65. Prof Exp: Instr vet path, Univ Minn, 60-62, NIH res fel, 62-65; asst scientist, Brookhaven Nat Lab, 65-66; assoc prof vet path, Univ Sask, 66-68; res veterinarian, Nat Animal Dis Lab, Agr Res Serv, USDA, 68-73; PROF VET PATH, IOWA STATE UNIV, 74- Concurrent Pos: Res collabr, Brookhaven Nat Lab, 66- Mem: Am Vet Med Asn; Am Col Vet Path. Res: Infectious diseases of the intestinal tract of animals; role of the lymphocyte in the immune response. Mailing Add: Dept of Vet Path Iowa State Univ Ames IA 50010

MOON, HENRY D, pathology, deceased

MOON, JAY RYONG, b Jeon Buk, Korea, Oct 20, 42; m 74. VETERINARY PUBLIC HEALTH. Educ: Joen Buk Nat Univ, Korea, DVM, 65; Seoul Nat Univ, MPH, 67; Univ Tex, DrPH, 75. Prof Exp: Asst epidemiol res, Korean Hemorrhagic Fever Res Ctr, Korean Army, 66-69; sr health statistician, Korean Med Team, USAID Pub Health, SVietnam, 69-72; fel zoonosis, Univ Tex Sch Pub Health Houston, 72-75; VIS ASST PROF VET PUB HEALTH, COL VET MED, IOWA STATE UNIV, 75- Res: Epidemiology of zoonosis in the United States, especially Iowa. Mailing Add: Dept of Microbiol & Prev Med Iowa State Univ Col of Vet Med Ames IA 50010

MOON, JOHN WESLEY, b Hornell, NY. MATHEMATICS. Educ: Bethany Nazarene Col, BA, 59; Mich State Univ, MS, 60; Univ Alta, PhD(math), 62. Prof Exp: Nat Res Coun Can fel, Univ Col, Univ London, 62-64; from asst prof to assoc prof, 64-69, PROF MATH, UNIV ALTA, 69- Concurrent Pos: Vis prof, Univ Cape Town, 70-71; vis, Math Inst, Oxford Univ, 71. Mem: Am Math Soc; Math Asn Am; Can Math Cong. Res: Graph theory; combinatorial analysis. Mailing Add: Dept of Math Univ of Alta Edmonton AB Can

MOON, KENNETH ANTHONY, physical chemistry, see 12th edition

MOON, MILTON LEWIS, b New Providence, Iowa, Dec 16, 22; m 47; c 4. PHYSICS, ENVIRONMENTAL SCIENCES. Educ: Iowa State Teachers Col, BA, 43; Univ Iowa, MS, 48, PhD(physics), 51. Prof Exp: Physicist, 51-72, POWER PLANT SITE EVAL PROG MGR, APPL PHYSICS LAB, JOHNS HOPKINS UNIV, 72- Concurrent Pos: Mem, Environ Res Guid Comt, 72-75. Res: Communications and navigation systems; environmental effects; cooling systems; air pollution; impact modelling. Mailing Add: 10510 Greenacres Dr Silver Spring MD 20903

MOON, NEIL SENNETT, b Morgantown, WVa, July 18, 13; m 41. SYNTHETIC ORGANIC CHEMISTRY. Educ: Pa State Univ, BS, 35; Univ Ill, PhD(org chem), 39. Prof Exp: Asst chem, Univ Ill, 35-38; chemist, 39-46, chemist & supvr, 47-58, ASST SUPT SYNTHETIC CHEM DIV, EASTMAN KODAK CO, 59- Mem: Am Chem Soc; Soc Photog Sci & Eng. Res: Synthetic organic chemicals. Mailing Add: Eastman Kodak Co Bldg 151 Kodak Park Rochester NY 14650

MOON, PETER CLAYTON, b St Louis, Mo, May 17, 40. DENTAL MATERIALS. Educ: Univ Toledo, BS, 64; Univ Va, MS, 67, PhD(mat sci), 71. Prof Exp: Mat scientist, Texaco Exp, Inc, 66-68; ASST PROF DENT MAT, SCH DENT, MED COL VA, VA COMMONWEALTH UNIV, 71- Mem: Int Asn Dent Res; Am Asn Dent Res; Am Chem Soc. Res: Mechanical properties; fracture; mineralization of bone and teeth; metallurgy; polymers; interfacial bonding. Mailing Add: Dept of Restorative Dent Med Col Va Sch of Dent Richmond VA 23298

MOON, RALPH MARKS, JR, b Bombay, India, Oct 11, 29; US citizen; m 50; c 4. PHYSICS. Educ: Univ Kans, BA, 50, MA, 52; Mass Inst Technol, PhD(physics), 63. Prof Exp: Physicist, Naval Ord Test Sta, Calif, 52-54; PHYSICIST, OAK RIDGE NAT LAB, 63- Mem: Am Phys Soc. Res: Neutron diffraction; magnetism. Mailing Add: Solid State Div Oak Ridge Nat Lab Oak Ridge TN 37830

MOON, RICHARD C, b Beech Grove, Ind, July 25, 26; m 48; c 3. ENDOCRINOLOGY. Educ: Butler Univ, BS, 50; Univ Cincinnati, MS, 52, PhD(zool), 55. Prof Exp: Instr physiol, Col Pharm, Univ Cincinnati, 53-54, instr zool, 54-55; asst prof biol, Duquesne Univ, 55-58; asst prof physiol, Med Units, Univ Tenn, Memphis, 59-63, from assoc prof to prof physiol & biophys, 63-73; head endocrinol res, 73-74, SR SCI ADV, RES INST, ILL INST TECHNOL, 74- Concurrent Pos: Nat Cancer Inst res fel endocrinol, Univ Mo, 58-59; Lederle med fac award, 63-66; Nat Cancer Inst spec res fel, Imp Cancer Res Fund, London, Eng, 65-66; guest investr, Ben May Lab Cancer Res, Univ Chicago, 61. Mem: AAAS; Am Physiol Soc; Endocrine Soc; Soc Exp Biol & Med; Am Asn Cancer Res; Soc Study Reproduction. Res: Hormonal control of mammary gland growth and hormonal influence upon mammary tumor formation; chemical carcinogenesis. Mailing Add: Ill Inst of Technol Res Inst 10 W 35 St Chicago IL 60616

MOON, ROBERT JOHN, b Carbondale, Pa, Aug 6, 42; m 66; c 1. MICROBIOLOGY. Educ: Eastern Col, AB, 64; Bryn Mawr Col, PhD(biol), 69. Prof Exp: From instr to asst prof, 68-75, ASSOC PROF MICROBIOL, MICH STATE UNIV, 75- Concurrent Pos: NIH fel, Harvard Univ, 72-73. Mem: AAAS; Am Soc Microbiol. Res: Medical microbiology and immunology; host responses to infectious agents. Mailing Add: Dept of Microbiol & Pub Health Mich State Univ East Lansing MI 48824

MOON, SUNG, b Seoul, Korea, Oct 6, 28; US citizen; m 61; c 2. ORGANIC CHEMISTRY. Educ: Univ Ill, BS, 56; Mass Inst Technol, PhD(org chem), 59. Prof Exp: Res assoc org chem, Mass Inst Technol, 59-62; from asst prof to assoc prof, 62-71, PROF ORG CHEM, ADELPHI UNIV, 71- Concurrent Pos: Am Chem Soc Petrol Res Fund grant, 65-68. Mem: Am Chem Soc; Brit Chem Soc. Res: Organic synthesis and reaction mechanisms. Mailing Add: Dept of Chem Adelphi Univ Garden City NY 11530

MOON, TAG YOUNG, b Seoul, Korea, May 11, 31; US citizen; m 59; c 4. PHYSICAL CHEMISTRY, STATISTICAL MECHANICS. Educ: Seoul Nat Univ, BS, 59; Yale Univ, PhD(chem), 65. Prof Exp: Lectr & NSF grants, Yale Univ, 65-66; RES CHEMIST, DOW CHEM USA, 66- Mem: Am Chem Soc; Am Phys Soc; Sigma Xi; NY Acad Sci. Res: Statistical thermodynamics of surfaces; interaction of colloidal particles; physics and chemistry of concrete; corrosion of steel. Mailing Add: 4109 Moorland Dr Midland MI 48640

MOON, THOMAS CHARLES, b Detroit, Mich, Apr 1, 40; m 63; c 2. SCIENCE EDUCATION, ECOLOGY. Educ: Kalamazoo Col, BA, 62; Oberlin Col, MAT, 63; Mich State Univ, PhD(sci educ, biol), 69. Prof Exp: Teacher, Detroit Country Day Sch, 63-64 & Farmington High Sch, 64-66; sr lectr sci educ, Mich State Univ, 66-69; PROF SCI EDUC & BIOL, CALIF STATE COL, PA, 69-, DIR ENVIRON STUDIES, 74- Concurrent Pos: Sci consult, Rand McNally & Co, Ill, 68-72. Mem: AAAS. Res: Interdisciplinary environmental science and its effects on science education endeavors. Mailing Add: Dept of Biol Calif State Col California PA 15419

MOON, THOMAS EDWARD, b Pontiac, Mich, Aug 16, 43; m 64. BIOMETRICS. Educ: Northern Ill Univ, BS, 65; Univ Chicago, MS, 67; Univ Calif, Berkeley, PhD(biostatist), 73. Prof Exp: Mathematician, US Army Aberdeen Proving Grounds, 67; statistician, NIH, 67-70; biostatistician, Gorgas Mem Inst, Mid Am Res Unit, 73-74; ASST BIOMETRICIAN, M D ANDERSON HOSP & TUMOR INST, 74- Concurrent Pos: Asst biometrician, Southwest Oncol Group, 74; asst prof biomet, Grad Sch Biomed Sci, Univ Tex, 75. Mem: Am Statist Asn; Biomet Soc. Res: Applications of statistical methods to biomedical research. Mailing Add: M D Anderson Hosp 6723 Bertner Dr Houston TX 77030

MOON, THOMAS WILLIAM, b Portland, Ore, June 10, 44; m 65; c 2. COMPARATIVE PHYSIOLOGY. Educ: Ore State Univ, BSc, 66, MA, 68; Univ BC, PhD(zool), 71. Prof Exp: Fel, Marine Sci Res Lab, Mem Univ, 71-72; asst prof, 72-76, ASSOC PROF BIOL, UNIV OTTAWA, 76- Concurrent Pos: Partic exped, Galapagos Island, 70, Kona Coast, Hawaii, 73 & Amazon, 76; vis prof zool, Univ Toronto, 75. Mem: AAAS; Am Soc Zoologists; Can Soc Zoologists; Can Soc Physiologists. Res: Organism-environment interactions using the tools of biochemistry to probe molecular strategies associated with the adaptation process. Mailing Add: Dept of Biol Univ of Ottawa Ottawa ON Can

MOON, WILCHOR DAVID, b Rose Bud, Ark, June 2, 33; m 53; c 2. MATHEMATICS. Educ: State Col Ark, BS, 58; Univ Tenn, Knoxville, MA, 62; Okla State Univ, PhD(math), 67. Prof Exp: Instr math, Univ Tenn, Knoxville, 60-62; asst prof, Southern State Col, Ark, 62-63; assoc prof, Ark Col, 63-65, prof math & chmn sci div, 66-68; PROF MATH, UNIV CENT ARK, 68-, DEAN UNDERGRAD STUDIES, 72- Concurrent Pos: Consult, Health Physics Div, Union Carbide Nuclear, Oak Ridge Nuclear Co, Tenn, 60-62; vis lectr, Ark Acad Sci, 66-68. Mem: Math Asn Am (1st vpres, 71-72); Am Math Soc. Res: Topology; geometry. Mailing Add: Dept of Math PO Box 1771 Univ of Cent Ark Conway AR 72032

MOONAN, WILLIAM JEANE, b Austin, Minn, Oct 29, 23; m 52; c 2. STATISTICS. Educ: Univ Minn, BS, 48, MS, 50, PhD(statist), 52. Prof Exp: Asst prof statist, Univ Minn, 53-54; dir statist dept, US Naval Personnel Res Activity, 54-69; vpres prog develop, Nat Comput Systs, 69-74; MEM STAFF, NAVAL PERSONNEL RES & TRAINING LAB, 74- Mem: Am Statist Asn; Asn Comput Mach; Inst Math Statist; Opers Res Soc Am. Res: Design of experiments; multivariate analysis. Mailing Add: Naval Personnel Res & Training Lab San Diego CA 92152

MOONEY, CHARLES FRANK, b Baltimore, Md, June 15, 21; m 46, 64. PHYSICS. Educ: Drew Univ, AB, 42; Johns Hopkins Univ, PhD(physics), 52. Prof Exp: Physicist, Bausch & Lomb, Inc, 51-64, res scientist, 64-67, sect head grating res, 62-67; ASSOC PROF PHYSICS, COMMUNITY COL FINGER LAKES, 68- Mem: Am Asn Physics Teachers. Res: Classical and optical physics; polarization; infrared filters; aspheric surfaces; glass making; extended range and surface contour interferometry;

diffraction grating profile measurement; ellipsoid for exciting lasers. Mailing Add: Dept of Physics Community Col of the Finger Lakes Canandaigua NY 14424

MOONEY, DAVID SAMUEL, b Albany, NY, Oct 13, 28; m 60; c 2. ORGANIC CHEMISTRY. Educ: State Univ NY, BA, 50, MA, 51; Rensselaer Polytech Inst, PhD(org chem), 55. Prof Exp: Asst org chem, Rensselaer Polytech Inst, 51-55; from asst prof to assoc prof, 55-65, PROF ORG CHEM, WASHINGTON & JEFFERSON COL, 65- Mem: AAAS; Am Chem Soc. Res: Organo lithium compounds; organic reaction mechanisms. Mailing Add: Dept of Chem Washington & Jefferson Col Washington PA 15301

MOONEY, HAROLD MORTON, b Northfield, Mass, Dec 15, 22; m 46; c 2. GEOPHYSICS. Educ: Harvard Univ, BS, 43; Calif Inst Technol, MS, 48, PhD(geophys), 50. Prof Exp: Engr, Gen Elec Co, 43-44, Manhattan Proj, 44-46 & Geophys Explors, Ltd, 46; from asst prof to assoc prof, 50-65, PROF GEOPHYS, UNIV MINN, MINNEAPOLIS, 65-, GEOL, 74- Concurrent Pos: Vis lectr, Fed Inst Tech & Inst for Geophys, Zurich, 59 & Victoria Univ, Wellington, 68; consult, US Army, US AEC & oil & mining indust. Mem: Soc Explor Geophys; Seismol Soc Am; Acoust Soc Am; Am Geophys Union; Europ Asn Explor Geophys. Res: Geophysical instrumentation; electrical exploration; magnetic properties of rocks; theory of seismic wave propagation. Mailing Add: 4331 Longfellow Ave Minneapolis MN 55407

MOONEY, JOHN BERNARD, b Coalinga, Calif, June 9, 26; m 49; c 2. ANALYTICAL CHEMISTRY, MATERIALS RESEARCH. Educ: Univ Santa Clara, BS, 50; Stanford Univ, MS, 53. Prof Exp: Jr develop chemist, Int Minerals & Chem Corp, 51-54; instr chem, Univ Santa Clara, 54-57; res chemist, Kaiser Aluminum & Chem Corp, 57-59; chief chemist, Thermatest Labs, Inc, 59-60; sr engr chem, Varian Assocs, 60-69; mgr mat technol & corp vpres, Photophys, Inc, 69-75; SR RES CHEMIST, STANFORD RES INST, 76- Concurrent Pos: Vis assoc, Calif Inst Technol, 75-76. Mem: Electrochem Soc. Res: Emission spectroscopy; gas chromatography; materials research; chemical vapor deposition. Mailing Add: Stanford Res Inst Menlo Park CA 94025

MOONEY, LARRY ALBERT, b Idaho Falls, Idaho, Jan 28, 36; m 58; c 7. CLINICAL CHEMISTRY. Educ: Willamette Univ, BS, 58; Univ Wash, PhD(biochem), 64; Am Bd Clin Chem, dipl. Prof Exp: Fel lipid biochem, Va Mason Res Ctr, 64-66; res fel, 66-68; clin biochemist, Mason Clin, 68-69; ASSOC DIR CHEM LAB, SACRED HEART HOSP, 69- Mem: AAAS; assoc Am Soc Clin Path; Am Asn Clin Chem; NY Acad Sci; Am Chem Soc. Res: Biosynthesis of essential fatty acids and biological elongation of fatty acids; development of analytical methods for clinical chemistry; application of biochemical knowledge to problems in medicine. Mailing Add: Chem Lab Sacred Heart Hosp 1200 Adler St Eugene OR 97401

MOONEY, PATRICIA MAY, b Bryn Mawr, Pa, July 12, 45. SOLID STATE PHYSICS. Educ: Wilson Col, AB, 67; Bryn Mawr Col, MA, 69, PhD(physics), 72. Prof Exp: Asst prof physics, Hiram Col, 72-74; ASST PROF PHYSICS, VASSAR COL, 74- Mem: Am Phys Soc; Am Asn Physics Teachers; Sigma Xi. Mailing Add: Dept of Physics Vassar Col Poughkeepsie NY 12601

MOONEY, PAUL DAVID, b Stroud, Okla, Aug 10, 43; m 66; c 4. MEDICINAL CHEMISTRY, BIOCHEMISTRY. Educ: Okla State Univ, BS, 66, MS, 69, PhD(chem), 71. Prof Exp: Res staff, Dept Pharmacol, Sch Med, Yale Univ, 71-74; ASST PROF BIOCHEM, OKLA COL OSTEOP MED & SURG, 74- Concurrent Pos: Adj asst prof chem, Quinnipiac Col, 72-74. Mem: Am Chem Soc. Res: Development of ribonucleoside diphosphate reductase inhibitors that are potential antineoplastic agents; also the chemistry, reactions and biological activity of pyridine, quinoline, isoquinoline and indole derivatives. Mailing Add: Okla Col Osteop Med & Surg 9th & Cincinnati Sts Tulsa OK 74119

MOONEY, RICHARD T, b New York, NY, Jan 12, 25; m 50; c 2. RADIOLOGICAL HEALTH. Educ: Pratt Inst, BS, 44; NY Univ, MS, 52; Am Bd Radiol, dipl. Prof Exp: Jr physicist, Physics Lab, New York, City Dept Hosps, 50-51; SR PHYSICIST, FRANCIS DELAFIELD HOSP, 51-; DIR PHYSICS SERV, NEW YORK CITY HEALTH & HOSPS CORP, 67-; STAFF PHYSICIST, ST CLARE'S HOSP, 70- Concurrent Pos: Staff scientist, AEC contracts, Columbia Univ; staff physicist, Westchester County Med Ctr, 69-; clin assoc, NY Med Col; consult, Vet Admin, hosps & indust; mem, Sci Comt Number 7 & 9, mem task group M-2 & consult subcomt M-2, Nat Coun Radiation Protection. Mem: Assoc fel Am Col Radiol; Radiol Soc NAm; Am Asn Physicists in Med; Am Phys Soc; Health Physics Soc. Res: Radiation protection and dose distribution in patients. Mailing Add: 4 Edgemont Circle Scarsdale NY 10583

MOONEY, RICHARD WARREN, b Lynn, Mass, Aug 2, 23; m 44; c 3. PHYSICAL CHEMISTRY, INORGANIC CHEMISTRY. Educ: Tufts Univ, BS, 44; Cornell Univ, PhD(phys chem), 51. Prof Exp: Engr, Bakelite Corp, NJ, 46-47; sr engr, Lighting Div, Sylvania Elec Prod Inc, Gen Tel & Electronics Corp, Mass, 53-56, engr-in-chg, Chem & Metall Div, 56-58, sect head, 58-59, eng mgr, 59-64, mgr res & develop, 65-66, chief engr, 66-69, vpres & gen mgr, 69-74, PRES, WILBUR B DRIVER DIV, G T E CORP, 74- Concurrent Pos: Ed, J Electrochem Soc, 63-64; Fulbright-Hays res scholar, Tech Univ Norway, 64-65. Mem: Am Chem Soc; Electrochem Soc; fel Am Inst Chemists. Res: Phosphors; phosphates; tungsten and molybdenum; x-ray crystallography; theoretical spectroscopy. Mailing Add: Wilbur B Driver Div GTE Corp 1895 McCarter Hwy Newark NJ 07104

MOONEY, ROBERT ARTHUR, b Rochester, NY, May 29, 33; m 58; c 5. APPLIED CHEMISTRY. Educ: Univ Rochester, BS, 55; Univ Ill, PhD(org chem), 59. Prof Exp: RES CHEMIST PHOTOG SCI, EASTMAN KODAK CO, 58- Mem: Soc Photog Sci & Eng. Res: Silver and non-silver photography; photographic support manufacture. Mailing Add: Roll Coating Div Eastman Kodak Co 343 State St Rochester NY 14650

MOONEY, THOMAS FAULKNER, JR, b Highland Park, Mich, Jan 25, 23; m 48; c 2. ANALYTICAL CHEMISTRY, INDUSTRIAL HYGIENE. Educ: Univ Mich, Ann Arbor, BS, 48; Wayne State Univ, PhD(anal chem), 68. Prof Exp: Gen mgr & chief chemist, Med-Fac Labs, Inc, Mich, 50-61; ASST PROF INDUST HYG, SCH MED, WAYNE STATE UNIV, 61- Mem: AAAS; Am Chem Soc; Am Indust Hyg Asn. Res: Trace metal analysis in biological tissue; air analysis. Mailing Add: Sch of Med Wayne State Univ Detroit MI 48201

MOORCROFT, DONALD ROSS, b Toronto, Ont, June 2, 35; m 64; c 2. PHYSICS. Educ: Univ Toronto, BASc, 57; Univ Sask, MSc, 60, PhD(physics), 62. Prof Exp: NATO sci fel & Rutherford fel, Radiophysics Lab, Stanford Univ, 62-63; from asst prof to assoc prof, 63-74, PROF PHYSICS, UNIV WESTERN ONT, 74- Mem: Can Asn Physicists; Am Asn Physics Teachers; Am Geophys Union. Res: Physics of the upper atmosphere; propagation and scattering of radio waves in ionized media; small-scale ionospheric structure. Mailing Add: Dept of Physics Univ of Western Ont London ON Can

MOORCROFT, WILLIAM HERBERT, b Detroit, Mich, Feb 1, 44; m 71; c 3. PSYCHOBIOLOGY. Educ: Augustana Col, BA, 66; Princeton Univ, PhD(psychol), 70. Prof Exp: USPHS fel ment retardation, Nebr Psychiat Inst & instr psychol, Col Med, Univ Nebr, 70-71; ASST PROF PSYCHOL, LUTHER COL, IOWA, 71- Concurrent Pos: Vis res psychologist, Ment Retardation Ctr, Univ Calif, Los Angeles, 72. Mem: Biofeedback Res Soc; Am Electroencephalog Soc; Am Psychol Asn; Asn Psychophysiol Study Sleep. Res: Developmental psychobiology; ontogeny of anatomical, pharmacological and electrophysiological correlates of behavior. Mailing Add: Dept of Psychol Luther Col Decorah IA 52101

MOORE, AIMEE N, b Conway, SC, Nov 8, 18. NUTRITION. Educ: Univ NC, BS, 39; Columbia Univ, MA, 47; Mich State Univ, PhD(higher educ), 59. Prof Exp: Intern hosp dietetics, Univ Mich, 40; hosp dietitian, Miss State Sanatorium & Stuart Circle Hosp, Richmond, Va, 40-43; teacher inst mgt, Col Human Ecol, Cornell Univ, 47-61; PROF FOOD SYSTS MGT & DIR DEPT NUTRIT & DIETETICS, MED CTR, UNIV MO-COLUMBIA, 61- Concurrent Pos: Rose fel, Am Dietetic Asn, 56 & 58; consult, Vet Admin Dietetic Adv Coun, 61-66; NIH res grant, 66- Mem: Am Dietetic Asn; Am Home Econ Asn; Am Inst Decision Sci; Asn Schs Allied Health Prof; Sigma Xi. Res: Applications of computer technology in hospital departments of nutrition and dietetics. Mailing Add: Dept Nutrit & Dietetics Univ of Mo-Columbia Med Ctr W128 Columbia MO 65201

MOORE, ALEXANDER MAZYCK, b Charleston, SC, Nov 6, 17; m 45; c 4. ORGANIC CHEMISTRY, INFORMATION SCIENCE. Educ: Col Charleston, BS, 38; Johns Hopkins Univ, PhD(org chem), 42. Prof Exp: Lab asst biol, Col Charleston, 36-38; lab asst chem, Johns Hopkins Univ, 38-42; res chemist, Socony-Vacuum Oil Co, Inc, NJ, 42-44 & Surv Antimalarial Drugs, Md, 44-46; res chemist, Parke, Davis & Co, 46-52, lab dir org chem, 52; admin fel, Mellon Inst, 52-56; asst to vpres, 56-66, DIR RES INFO, PARKE, DAVIS & CO, 66- Mem: Am Chem Soc; NY Acad Sci. Res: Chemicals from petroleum; classification and nomenclature of organic compounds; chemotherapy; pharmaceuticals; science information. Mailing Add: Parke Davis & Co 2800 Plymouth Rd Ann Arbor MI 48105

MOORE, ALLEN CHARLTON, b Houston, Tex, Nov 22, 15; m 43; c 5. CHEMISTRY, RESEARCH ADMINISTRATION. Educ: Univ Tex, BA, 37, MA, 43; Univ Ill, PhD(org chem), 48. Prof Exp: Teacher high sch, Tex, 39-41; instr chem, Univ Tex, 41-43; asst prof, Army Specialized Training Prog, The Citadel, 43-44; res chemist, Develop Dept, WVa Pulp & Paper Co, SC, 44-45; asst, Rubber Res Prog, War Prod Bd, Ill, 45-48; res chemist, Parke, Davis & Co, Mich, 48-51, asst mgr res dept, 51-56, coordr res personnel, 56-63; from assoc dir to dir univ off res, 63-67, DIR RES ADMIN, CASE WESTERN RESERVE UNIV, 67- Concurrent Pos: Instr, Eve Sch, Detroit, 48-49. Mem: AAAS; Am Chem Soc. Res: Synthesis of acetylenic alcohols and substituted cyclohexanols; synthesis involving terpene hydrocarbons; chemical study of chloromycetin; research administration. Mailing Add: Off of Res Admin Case Western Reserve Univ Cleveland OH 44106

MOORE, ALLEN MURDOCH, b Ithaca, NY, Mar 15, 40; m 69. ECOLOGY, ZOOLOGY. Educ: Cornell Univ, AB, 61; Univ Tex, Austin, PhD(zool), 68. Prof Exp: Fel, Univ NC, 68-69; ASST PROF BIOL, WESTERN CAROLINA UNIV, 68- Concurrent Pos: Cooperator, Southeastern Forest Exp Sta, US Forest Serv, 73- Mem: AAAS; Ecol Soc Am; Soc Gen Systs Res; Am Inst Biol Sci; Fedn Am Scientists. Res: Systems ecology, especially ecological modelling; energy flow in ecosystems; human impact on ecosystems. Mailing Add: Dept of Biol Western Carolina Univ Cullowhee NC 28723

MOORE, ALTON WALLACE, b Lewiston, Idaho, Sept 16, 16; m 45; c 3. DENTISTRY. Educ: Univ Calif, DDS, 41; Univ Ill, MS, 48; Am Bd Orthod, dipl, 55. Prof Exp: Intern dent, Univ Chicago, 41-42; instr oral path & diag, Univ Louisville, 42-43, asst prof, 43-44; from instr to asst prof, Univ Ill, 45-48; assoc prof, 48-50, head dept orthod, 48-66, dir grad dent educ, 51-58, actg asst dean, Sch Dent, 55-56, head dept dent sci & lit, 66-70, PROF ORTHOD, UNIV WASH, 50-, ASSOC DEAN SCH DENT, 66- Concurrent Pos: Grieve Mem lectr, 56; dir, Am Bd Orthod, 59-60, secy, 60-65, pres, 65-66; Wylie Mem lectr, 67; mem dent study sec, Nat Inst Dent Res, 70-74, chmn, 71-74. Honors & Awards: Award, Am Asn Orthod, 49; Albert H Ketcham Mem Award, Am Bd Orthod, 73. Mem: Am Asn Orthod; Am Dent Asn; fel Am Col Dent; Int Asn Dent Res. Res: Orthodontics; human facial growth and cephalometric appraisal of orthodontic treatment. Mailing Add: Off of the Dean Univ of Wash Sch of Dent Seattle WA 98195

MOORE, ARNOLD ROBERT, b New York, NY, Jan 14, 23; m 46; c 2. SOLID STATE PHYSICS. Educ: Polytech Inst Brooklyn, BS, 42; Cornell Univ, PhD(physics), 49. Prof Exp: Res engr, Victor Div, Radio Corp Am, Pa, 42-45, RES PHYSICIST, RCA LABS, DAVID SARNOFF RES CTR, 49- Concurrent Pos: Vis prof, Brown Univ, 70-71. Mem: Fel Am Phys Soc; NY Acad Sci. Res: Electrical, optical and magnetic properties of semi-conducting and insulating crystals; photoelectricity; acoustoelectric interactions; semi-conducting devices. Mailing Add: 61 Random Rd Princeton NJ 08540

MOORE, ARTHUR DUDLEY, entomology, see 12th edition

MOORE, BARRY NEWTON, b San Antonio, Tex, Jan 27, 41. PLASMA PHYSICS. Educ: Univ Tex, Austin, BS, 62, MA, 69, PhD(physics), 72. Prof Exp: Engr, Missiles & Space Div, LTV Aerospace Corp, 62-67; RES ASSOC PLASMA PHYSICS, FUSION RES CTR, UNIV TEX, AUSTIN, 73- Mem: Sigma Xi; Am Phys Soc; Inst Elec & Electronic Engrs. Res: Feasibility studies of fuel cycles for advanced fusion reactors; applications of radio-frequency heating to plasmas; linear and nonlinear plasma waves. Mailing Add: Dept of Physics Univ of Tex Fusion Res Ctr Austin TX 78712

MOORE, BENJAMIN LABREE, b Louisville, Ky, Feb 10, 15; m 43; c 2. NUCLEAR PHYSICS. Educ: Davidson Col, AB, 34; Vanderbilt Univ, MA, 35; Cornell Univ, PhD(physics), 40. Prof Exp: Asst, Cornell Univ, 35-40; physicist, Bur Ord, Navy Dept, 40-42 & Naval Mine Warfare Sch, 42-44; sr physicist, Tenn Eastman Corp, 44-46; res fel, Harvard Univ, 46-49, asst dir comput lab, 49-50; mem staff, Los Alamos Sci Lab, Univ Calif, 50-51, asst div leader, 51-60, assoc div leader, 60-73; SCI CONSULT, 73- Concurrent Pos: Extra-mural prof, Washington Univ, 57-59; consult, Los Alamos Sci Lab, 73- Mem: Fel AAAS; Am Phys Soc. Res: Nuclear physics; isotope separation by electromagnetic means; design of large scale digital computing machines; atomic weapons. Mailing Add: 3504 Arizona Ave Los Alamos NM 87544

MOORE, BERRIEN, III, b Atlanta, Ga, Nov 12, 41; m 67. MATHEMATICS. Educ: Univ NC, BS, 63; Univ Va, PhD(math), 69. Prof Exp: Asst prof, 69-75, ASSOC PROF MATH, UNIV NH, 75- Mem: Am Math Soc. Res: Hilbert space; operator theory. Mailing Add: Dept of Math Univ of NH Durham NH 03824

MOORE, BETTY CLARK, b Dedham, Mass, Jan 1, 15; m 38; c 1. EMBRYOLOGY. Educ: Radcliff Col, AB, 36; Columbia Univ, MA, 37, PhD(zool), 49. Prof Exp: Asst zool, Manhattanville Col, 37-40; res assoc embryol, Columbia Univ, 49-50; instr biol, Queens Col, 51-52; res assoc cytochem, Columbia Univ, 54-69, lectr, 64-69; RES

ASSOC BIOL, UNIV CALIF, RIVERSIDE, 69- Mem: Am Soc Zool; Am Soc Cell Biol; Am Soc Nat. Res: Biological sciences; population genetics of drosophila. Mailing Add: Dept of Biol Univ of Calif Riverside CA 92502

MOORE, BILL C, b Kansas City, Mo, Sept 12, 07; m 42; c 2. MATHEMATICS, OPERATIONS RESEARCH. Educ: Univ Kans, AB, 29; Princeton Univ, AM, 37. Prof Exp: Assoc prof math, Tex A&M Univ, 48-74; RETIRED. Concurrent Pos: Consult, Humble Oil Co, 54 & 55 & Dow Chem Co, 56; guest worker, Nat Bur Standards, 59. Mem: AAAS; Math Asn Am. Res: Optimization; digital computers. Mailing Add: Dept of Math Tex A&M Univ 1000 Munson College Station TX 77840

MOORE, BLAKE WILLIAM, b Ohio, Sept 18, 26; m 55. BIOCHEMISTRY. Educ: Univ Akron, BS, 48; Northwestern Univ, PhD(biochem), 52. Prof Exp: Asst biochem, Northwestern Univ, 48-52; asst prof, Sch Dent, 52-57, instr cancer res, 57-58, asst prof biochem in cancer res, 58-59, from asst prof to assoc prof, 59-70, PROF BIOCHEM IN PSYCHIAT, SCH MED, WASHINGTON UNIV, 70- Mem: AAAS; Am Chem Soc; Am Soc Biol Chemists. Res: Neurochemistry; proteins; biochemistry of cancer; protein chromatography; enzymes. Mailing Add: Dept of Psychiat Washington Univ Sch of Med St Louis MO 63110

MOORE, BOBBY GRAHAM, b Fulton, Miss, June 22, 40; m 60; c 2. CELL BIOLOGY. Educ: Miss State Univ, BS, 62, MS, 64; Auburn Univ, PhD(biochem), 68. Prof Exp: Res technologist microbiol, Oak Ridge Nat Lab, 64-65; instr, Auburn Univ, 65-68; asst prof, 68-72, ASSOC PROF BIOL, UNIV ALA, TUSCALOOSA, 72- Concurrent Pos: Grant-in-aid, Univ Ala, 69-71. Mem: AAAS. Res: Biochemistry of nucleic acid methylation in diverse organisms and environmental toxicology of naturally occurring chemical substances. Mailing Add: Dept of Biol Univ of Ala University AL 35486

MOORE, BURTON HAROLD, biochemistry, see 12th edition

MOORE, CALVIN C, b New York, NY, Nov 2, 36; m 74. MATHEMATICS. Educ: Harvard Univ, AB, 58, MA, 59, PhD(math), 60. Prof Exp: From asst prof to assoc prof, 61-66, PROF MATH, UNIV CALIF, BERKELEY, 66-, DEAN PHYS SCI, 71- Concurrent Pos: Mem, Inst Adv Study, 64-65; Alfred P Sloan Found fel, 65-67. Mem: Am Math Soc; fel Am Acad Arts & Sci. Res: Group representations. Mailing Add: Dept of Math Univ of Calif Berkeley CA 94720

MOORE, CARL, b Cleveland, Ohio, Aug 28, 19; m 56; c 2. POLYMER CHEMISTRY, COLLOID CHEMISTRY. Educ: Ohio State Univ, BA, 41; Oberlin Col, MA, 43; Case Inst, PhD(phys chem), 51. Prof Exp: Chemist, Standard Oil Co Ohio, 41; asst gen chem, Oberlin Col, 41-43; res assoc colloid sci, Govt Synthetic Rubber Prog, Case Inst, 43-52; chemist, 52-62, sr res chemist, 62-73, RES SPECIALIST, DOW CHEM CO, 73- Mem: Am Chem Soc; Sigma Xi. Res: Physical chemistry of latexes; emulsion polymerization; colloid science; radiation grafting of polymers; organic basic research; complexing resins; plastics and latex foam; rheology of polymer solutions; latex paints; adhesion of coatings; organic coatings. Mailing Add: 1105 Evamar Dr Midland MI 48640

MOORE, CARL ALLPHIN, geology, see 12th edition

MOORE, CARL EDWARD, b Frankfort, Ky, Sept 25, 15; m 40; c 4. ANALYTICAL CHEMISTRY. Educ: Eastern Ky State Col, BS, 39; Univ Louisville, MS, 47; Ohio State Univ, PhD(anal chem), 52. Prof Exp: Chemist, Nat Distillers Prod Corp, 39-41 & E I du Pont de Nemours & Co, 41-45; instr, Univ Louisville, 47-50; PROF CHEM, LOYOLA UNIV CHICAGO, 52- Mem: Am Chem Soc; Soc Appl Spectros. Res: Organic reagents for inorganic analysis; analytical methods for detection and determination of atmospheric contaminants. Mailing Add: Dept of Chem Loyola Univ Chicago IL 60626

MOORE, CARL VERNON, internal medicine, hematology, deceased

MOORE, CARLTON BRYANT, b New York, NY, Sept 1, 32; m 59; c 2. GEOCHEMISTRY, METEORITES. Educ: Alfred Univ, BS, 54; Calif Inst Technol, PhD(chem, geol), 60. Prof Exp: Asst prof geol, Wesleyan Univ, 59-61; asst prof geol, 61-66, assoc prof chem & geol, 66-70, PROF CHEM & GEOL, ARIZ STATE UNIV, 70-, DIR CTR METEORITE STUDIES, 61- Concurrent Pos: Prin investr, Apollo 11-17, mem preliminary exam team, Apollo 12-17; ed, Meteoritics. Mem: Geochem Soc; Geol Soc Am; Am Chem Soc; Am Geophys Union; Mineral Soc Am; Meteoritical Soc (pres, 66). Res: Chemistry and mineralogy of meteorites; geochemistry of organogenic elements; analysis of lunar samples. Mailing Add: Ctr for Meteorite Studies Ariz State Univ Tempe AZ 85281

MOORE, CAROL WOOD, b Columbus, Ohio, Sept 23, 43; m 66; c 1. GENETICS. Educ: Ohio State Univ, BS, 65; Pa State Univ, MS, 69, PhD(genetics), 70. Prof Exp: Teacher pub schs, NY, 65-67; asst human genetics, 70-71, res assoc, 71-74, ASST PROF BIOL, UNIV ROCHESTER, 74- Mem: Genetics Soc Am; AAAS. Res: Recombination and mutation in yeast cells. Mailing Add: Dept of Biol Univ of Rochester Rochester NY 14627

MOORE, CECILIA LOUISE, b Phoenix, Ariz, Apr 17, 28. PHYSICAL ORGANIC CHEMISTRY. Educ: Mt St Mary's Col, Calif, BS, 53; St Louis Univ, PhD(chem), 58. Prof Exp: From instr to assoc prof chem, 57-66, chmn dept phys sci, 60-64, dean fac, 64-67, PROF CHEM, MT ST MARY'S COL CALIF, 66-, PRES, 67- Mem: Am Chem Soc. Res: Kinetics; chymotrypsin-catalyzed hydrolyses of amino acid derivatives; acid-catalyzed dehydrations of sterically-hindered alcohols. Mailing Add: Mt St Mary's Col 12001 Chalon Rd Los Angeles CA 90049

MOORE, CHARLEEN MORIZOT, b Shreveport, La, July 29, 44; div. CYTOGENETICS. Educ: Northeast La Univ, BS, 66; Ind Univ, Bloomington, MA, 67; Univ Tenn, Knoxville, PhD(zool), 71. Prof Exp: Fel med cytogenetics, John F Kennedy Inst, Sch Med, Johns Hopkins Univ, 71-74; ASST PROF PEDIAT & DIR MED CYTOGENETICS LAB, UNIV TEX HEALTH SCI CTR HOUSTON, 74- Mem: Am Soc Human Genetics. Res: Chromosome behavior. Mailing Add: Dept of Pediat Univ of Tex Health Sci Ctr Houston TX 77025

MOORE, CHARLES BACHMAN, JR, b Maryville, Tenn, Oct 28, 20; m 46; c 3. ATMOSPHERIC PHYSICS. Educ: Ga Inst Technol, BChE, 46. Prof Exp: Asst physics, NY Univ, 47-48; head balloon opers, Aero Res Lab, Gen Mills, Inc, 49-53; res investr atmospheric physics, Arthur D Little, Inc, 53-65; assoc prof physics, 65-69, PROF ATMOSPHERIC PHYSICS, NMEX INST MINING & TECHNOL, 69- Honors & Awards: Am Meteorol Soc Award, 60. Mem: AAAS; Am Meteorol Soc; Am Geophys Union; Royal Meteorol Soc. Res: Thunderstorm electricity. Mailing Add: Dept of Physics NMex Inst of Mining & Technol Socorro NM 87801

MOORE, CHARLES GODAT, b Linn, Mo, July 13, 27; m 57; c 3. MATHEMATICS. Educ: Cent Mo State Col, BS, 51, MS, 54; Univ Mich, MA, 60, PhD, 67. Prof Exp: Teacher high schs, Mo, 51-53, 54-59 & Kemper Mil Sch, 53-54; from asst prof to assoc prof, 60-74, PROF MATH, NORTHERN ARIZ UNIV, 74- Res: Mathematics education; continued fractions. Mailing Add: Dept of Math Fac Box 4015 Northern Ariz Univ Flagstaff AZ 86001

MOORE, CHARLES HENKEL, b New Market, Va, Oct 25, 15; m 62; c 1. CHEMISTRY. Educ: Univ Va, BS, 36, MS, 37; Cornell Univ, PhD(geol, chem), 40. Prof Exp: Instr mineral & optics, Cornell Univ, 37-40; chief petrographer, Carborundum Co, NY, 40-43; asst prof mineral, Pa State Col, 43-45; from sect head to dept head, Res Lab, Titanium Div, Nat Lead Co, 45-49, asst to exec res dir, 49-51, tech dir, Ohio, 51; mgr metals, Ceramic & Rectifier Divs, P R Mallory & Co, 51-55, exec dir corp res & develop, 56-60; dir, Copper Prods Develop Asn, 60-63, EXEC VPRES, INT COPPER RES ASN, INC, 63- Concurrent Pos: Vis prof, Rutgers Univ, 46-49. Mem: AAAS; Am Soc Metals; Am Inst Mining, Metall & Petrol Eng. Res: Crystal and high temperature chemistry; single crystal growth; chemistry of metals. Mailing Add: Int Copper Res Asn Inc 825 Third Ave New York NY 10022

MOORE, CHARLES WAYNE, b Seattle, Wash, Apr 14, 43. ECONOMIC GEOGRAPHY. Educ: Univ Wash, BA, 65, MUP, 67, PhD(urban planning), 71. Prof Exp: Asst prof geog, Univ Sask, 70-75; sci adv, Sci Coun Can, 75-76. Concurrent Pos: Consult, Sask Dept Natural Resources, 72-73; res fel, Dept Soc & Econ Res, Univ Glasgow, Scotland, 73-74. Mem: Regional Sci Asn; Can Asn Geog; Am Asn Geographers. Res: Regional development and planning; geography of enterprise; location theory. Mailing Add: Sci Coun of Can 150 Kent Ottawa ON Can

MOORE, CLARENCE L, b Britton, SDak, Dec 6, 31; m 55; c 3. DAIRY SCIENCE. Educ: SDak State Univ, BS, 53, MS, 57, PhD(dairy sci), 59. Prof Exp: Area livestock specialist, Univ Hawaii, 59-61; assoc prof, 61-67, PROF DAIRY SCI, ILL STATE UNIV, 67- Concurrent Pos: Sabbatical, Ruakura Animal Res Sta, Hamilton, NZ, 72. Mem: Am Dairy Sci Asn. Res: Calf nutrition; physiology of milk secretion; dairy cattle management; animal behavior. Mailing Add: Dept of Agr Ill State Univ Normal IL 61761

MOORE, CLAUDE HENRY, b Greensboro, Ala, May 8, 23; m 56; c 5. POULTRY GENETICS. Educ: Auburn Univ, BS, 47; Kans State Col, MS, 48; Purdue Univ, PhD(poultry genetics), 52. Prof Exp: Asst, Kans State Col, 46-48 & Purdue Univ, 48-50; asst nat coordr, Poultry Div, Agr Res Serv, USDA, 50-56; assoc prof, 56-59, PROF POULTRY SCI & HEAD DEPT, AUBURN UNIV, 59- Mem: AAAS; Poultry Sci Asn. Res: Poultry breeding and physiology. Mailing Add: Dept of Poultry Sci Auburn Univ Auburn AL 36830

MOORE, CLYDE H, JR, b Jacksonville, Fla, June 10, 33; m 53; c 3. GEOLOGY. Educ: La State Univ, BS, 55; Univ Tex, MS, 59, PhD(geol), 61. Prof Exp: Res geologist, Shell Develop Co, 61-66; asst prof, 66-69, ASSOC PROF GEOL, LA STATE UNIV, BATON ROUGE, 69- Mem: Am Asn Petrol Geol; Soc Econ Paleont & Mineral. Res: Carbonate petrology; recent carbonate sedimentation and stratigraphy in Caribbean and Gulf Coastal plain. Mailing Add: Dept of Geol La State Univ Baton Rouge LA 70803

MOORE, CONDICT, b Essex Fells, NJ, Apr 29, 16; m 43; c 2. SURGERY. Educ: Princeton Univ, BA, 38; Columbia Univ, MD, 42. Prof Exp: Resident path, St Luke's Hosp, 46-47; resident surg, Methodist Hosp, Brooklyn, 47-49; resident, Mem Hosp, New York, 49-52; from instr to asst prof, 52-62, clin assoc prof, 62-65, assoc prof, 65-69, PROF SURG, SCH MED, UNIV LOUISVILLE, 69- Concurrent Pos: Fel surg, Mem Hosp, New York, 49-51. Mem: Fel Am Col Surgeons; James Ewing Soc; Am Radium Soc. Res: Cancer research; carcinogenesis; tobacco and oral cancer. Mailing Add: Dept of Surg Univ of Louisville Sch of Med Louisville KY 40201

MOORE, CONRAD TAYLOR, b Dormont, Pa, Jan 16, 38; m 67. GEOGRAPHY, HISTORICAL GEOGRAPHY. Educ: Univ Calif, Los Angeles, BA, 64, MA, 67, PhD(geog), 72. Prof Exp: Asst prof, 69-74, ASSOC PROF GEOG, GEORGE PEABODY COL, 74- Mem: AAAS; Asn Am Geogrs; Am Geog Soc. Res: Historical-cultural geography of North America; biogeography. Mailing Add: Dept of Geog George Peabody Col Nashville TN 37203

MOORE, CORNELIUS FRED, b Louisville, Ky, Mar 18, 36; m 61; c 4. NUCLEAR PHYSICS, ATOMIC PHYSICS. Educ: Univ Notre Dame, BS, 59; Univ Louisville, MS, 61; Fla State Univ, PhD(physics), 64. Prof Exp: Asst, Univ Louisville, 60-61; asst, Fla State Univ, 61-63, instr, 63-64, res assoc, 64-65; res scientist, Univ Tex, Austin, 65, asst prof physics, 65-68; vis prof, Univ Heidelberg, 68-69; assoc prof, 69-71, PROF PHYSICS, UNIV TEX, AUSTIN, 71- Res: Isobaric analogue states as compound nuclear resonances; fission studies; low energy electron scattering on ions; atoms and molecules. Mailing Add: Dept of Physics Univ of Tex Austin TX 78712

MOORE, CRAIG DAMON, b Youngstown, Ohio, July 13, 42; m 72. HEALTH PHYSICS, EXPERIMENTAL HIGH ENERGY PHYSICS. Educ: Ohio Univ, BS, 64; Univ Wis, PhD(physics), 70. Prof Exp: Res assoc physics, State Univ NY Stony Brook, 70-73; PHYSICIST, FERMI NAT LAB, 73- Mem: Am Phys Soc. Res: Radiation shielding experiments; anti-proton interactions in the thirty inch hydrogen bubble chamber. Mailing Add: Fermi Nat Lab PO Box 500 Batavia IL 60510

MOORE, CYRIL L, b Trinidad, WIndies, Feb 14, 28; US citizen; m 54; c 7. BIOCHEMISTRY, BIOPHYSICS. Educ: Brooklyn Col, BA, 53, MA, 58; Albert Einstein Col Med, PhD(biochem), 63. Prof Exp: Johnson Res Found fel, Univ Pa, 63-65; univ fel, Albert Einstein Col Med, 65-66; asst prof, 66-69, ASSOC PROF BIOCHEM & NEUROL, ALBERT EINSTEIN COL MED, 70- Mem: Soc Exp Biol & Med; Int Soc Neurochem; Biophys Soc; Am Chem Soc; AAAS; Am Soc Biol Chem; Can Biochem Soc. Res: Ion transport; energy metabolism; development of the central nervous system; mitochondrial biogenesis; neurologic diseases, their origin and causes. Mailing Add: Dept of Neurol & Biochem Albert Einstein Col of Med Bronx NY 10461

MOORE, DAN HOUSTON, b Elk Creek, Va, Apr 7, 09; m 41; c 2. BIOPHYSICS. Educ: Duke Univ, AB, 32, MA, 33; Univ Va, PhD(physics), 36. Hon Degrees: DS, Emory & Henry Col, 74. Prof Exp: Instr physics, Duke Univ, 33-34 & Univ Va, 35-36; asst, Columbia Univ, 36-38; res assoc, Lederle Labs, 38-39; assoc anat, Col Physicians & Surgeons, Columbia Univ, 39-43, from asst prof to assoc prof, 43-53, assoc prof microbiol, 53-58; assoc mem & prof biophys, Rockefeller Inst, 58-66; MEM & HEAD DEPT BIOPHYS CYTOL, INST MED RES, 66- Concurrent Pos: Mem div war res, Off Sci Res & Develop, 40-45; sci liaison officer, Off Naval Res, London, 49-51; Am Cancer Soc res prof, 73. Mem: AAAS; Soc Exp Biol & Med; Am Physiol Soc; Harvey Soc; Electron Micros Soc. Res: Tumor viruses; electron microscopy; characterization of proteins by electrophoresis and ultracentrifugation. Mailing Add: Dept of Biophys Cytol Inst for Med Res Copewood St Camden NJ 08103

MOORE, DAN HOUSTON, II, b New York, NY, Sept 24, 41; m 66; c 1. BIOSTATISTICS. Educ: Univ Calif, Santa Barbara, BA, 63; Univ Calif, Berkeley, PhD(biostatist), 70. Prof Exp: Biostatistician, Univ Pa, 70-73; BIOSTATISTICIAN, LAWRENCE LIVERMORE LAB, UNIV CALIF, 73- Mem: Biomet Soc; Sigma Xi.

Res: Application of statistical methods to chromosome data; Monte Carlo methods; non-linear curve fitting; discriminant analysis. Mailing Add: Biomed Div L-523 Livermore Lab Univ of Calif Livermore CA 94550

MOORE, DANIEL CHARLES, b Cincinnati, Ohio, Sept 9, 18; m 45; c 4. ANESTHESIOLOGY. Educ: Amherst Col, BA, 40; Northwestern Univ, BM & MD, 44. Prof Exp: Clin assoc prof, 56-64, CLIN PROF ANESTHESIOL, SCH MED, UNIV WASH, 65- Concurrent Pos: Mem attend staff, Children's Orthop Hosp, 62-; dir anesthesiol, Mason Clin & chmn dept, Virginia Mason Hosp, 47-72, sr consult, 72-; NIH grant anesthesiol, 67-71. Mem: AMA; Am Soc Anesthesiol (1st vpres, 53-54, 2nd vpres, 54-55, pres-elect, 57-58, pres, 58-59); Acad Anesthesiol; Pan-Am Med Asn. Res: Local anesthetic agents and their distribution and fate in man. Mailing Add: Virginia Mason Res Ctr 1000 Seneca St Seattle WA 98101

MOORE, DAVID GILLIS, b Long Beach, Calif, July 11, 25; m 45; c 5. MARINE GEOLOGY, MARINE GEOPHYSICS. Educ: Univ Southern Calif, AB, 50, MS, 52; State Univ Groningen, PhD, 66. Prof Exp: Jr res geologist, Scripps Inst, Univ Calif, San Diego, 52-55; oceanographer, Electronics Lab, US Navy, 55-69, OCEANOGRAPHER, NAVAL UNDERSEA CTR, 69- Concurrent Pos: Consult, Gen Oceanog, Inc; res assoc, Scripps Inst Oceanog, Univ Calif, San Diego. Honors & Awards: Francis P Shepard Medal, Soc Econ Paleont & Mineral, 67. Mem: Geol Soc Am; Soc Econ Paleont & Mineral; Am Asn Petrol Geol; Soc Explor Geophys; Am Geophys Union. Res: Continental margin tectonics, structure and sedimentation; marine geology; sedimentology. Mailing Add: Naval Undersea Ctr San Diego CA 92132

MOORE, DAVID JAY, b Cecil, Pa, Sept 8, 36; m 58; c 2. ANIMAL ECOLOGY. Educ: Clarion State Col, BS, 59; Ohio Univ, MS, 61; NC State Univ, PhD(animal ecol), 69. Prof Exp: Asst prof biol, SMacomb Community Col, Mich, 62-63; from asst prof to assoc prof, 63-73, dean, Sch Natural Sci, 71-72, PROF BIOL, RADFORD COL, 73-, VPRES ACAD AFFAIRS, 72- Concurrent Pos: Teacher pub schs, Mich, 61-63. Res: Water pollution, especially effects of fluoride pollution on the blue crab. Mailing Add: Off of VPres for Acad Affairs Radford Col Radford VA 24141

MOORE, DAVID SHELDON, b Plattsburg, NY, Jan 28, 40; m 64; c 2. STATISTICS. Educ: Princeton Univ, AB, 62; Cornell Univ, PhD(math), 67. Prof Exp: Asst prof math & statist, 67-71, ASSOC PROF STATIST, PURDUE UNIV, 71- Concurrent Pos: Assoc ed, J Am Statist Asn, 74- Mem: Inst Math Statist; Math Asn Am; Am Statist Asn. Res: Mathematical statistics, especially order statistics, large sample theory and nonparametric statistics. Mailing Add: Dept of Statist Purdue Univ Lafayette IN 47907

MOORE, DONALD BAKER, b Little Rock, Ark, Aug 14, 23; m 64; c 3. EXPERIMENTAL PHYSICS. Educ: Univ Calif, AB, 47. Prof Exp: Jr physicist, Naval Radiol Defense Lab, 50-51; physicist, Calif Res & Develop Co, 51-54, Missile Systs Div, Lockheed Aircraft Corp, 54-55 & Stanford Res Inst, 55-68; STAFF PHYSICIST, EXPLOSIVE TECHNOL, INC, 68- Mem: Soc Motion Picture & TV Eng; Am Phys Soc; Am Statist Asn. Res: Theory, application, and effects of explosives; electrical discharge phenomenology; ion optics; statistical analysis; computer applications to experiment and theory; ultra high speed optical and electronic instrumentation. Mailing Add: Explosive Technol Inc Box KK Fairfield CA 94533

MOORE, DONALD CLARK, b Omaha, Nebr, Oct 14, 20; m 43; c 3. PHYSICS. Educ: Univ Nebr, AB, 42; Univ Calif, PhD(physics), 48. Prof Exp: Electronics engr, Standard Oil Co Calif, 44-45; electrician, Interocean Steamship Co, 45-46; asst, Univ Calif, 46-47, physicist, Radiation Lab, 47-48; asst prof physics, Rensselaer Polytech Inst, 48-51; asst prof, Univ Nebr, 51-56, actg chmn dept, 53-56; sr res proj physicist, Schlumberger Well Surv Corp, 56-67; dean, Housatonic Community Col, 67-68; DEAN FAC, OTERO JR COL, 68- Concurrent Pos: Fulbright prof, Cuyo, 65. Mem: Am Phys Soc; Inst Elec & Electronic Engrs. Res: Electronic instrumentation; nuclear physics. Mailing Add: Otero Jr Col La Junta CO 81050

MOORE, DONALD FLOYD, b Winona, Minn, July 6, 14; m 36; c 4. PSYCHIATRY. Educ: Univ Mich, MD, 38. Prof Exp: Staff psychiatrist, Ypsilanti State Hosp, Mich, 39-41, sr psychiatrist, 45-47; chief psychiat, Sta Hosp, Camp Wallace, Tex, 41-42; chief neuropsychiat serv, 69th Gen Hosp, Ledo, Assam, 44-45; chief neuropsychiat, Vet Admin Hosp, Louisville, Ky, 47-55; MED DIR, LARUE D CARTER MEM HOSP, INDIANAPOLIS, 55-, PROF PSYCHIAT, SCH MED, IND UNIV, INDIANAPOLIS, 55- Concurrent Pos: Chief ment hyg clin, Anti-Aircraft Replacement Training Ctr, Camp Wallace, Tex, 43-44, consult, Spec Training Battalions, 41-42; staff psychiatrist, Louisville Child Guid Clin, 47-55; from assoc to assoc prof, Sch Med, lectr, Kent Sch Social Work & div adult educ, Grad Sch, Louisville, 47-55, southern police inst, 50-62; consult, Norton Psychiat Clin, Louisville, 47-55 & Our Lady Peace Hosp, 51-55; med dir, Chrysler Corp, Detroit, 52, Cent State Hosp, Lakeland, Ky, 55 & Marion County Gen Hosp, Indianapolis, 55- Mem: Fel Am Psychiat Asn; Am Col Psychiat; Am Asn Med Supt Ment Hosps (pres, 66-67). Res: Administrative psychiatry, particularly inter-professional relationships; psychiatric education, especially medical student, psychiatric resident, postgraduate and education of allied disciplines, particularly psychiatric social workers and chaplains. Mailing Add: Sch of Med Ind Univ 1100 W Michigan St Indianapolis IN 46202

MOORE, DONALD FREEMAN, pathology, see 12th edition

MOORE, DONALD GALE, soil science, see 12th edition

MOORE, DONALD R, b Reading, Pa, Dec 29, 33; m 60; c 3. ORGANIC CHEMISTRY, POLYMER CHEMISTRY. Educ: Lafayette Col, BS, 54; Harvard Univ, AM, 55, PhD(org chem), 58, Harvard Bus Sch, AMP, 75. Prof Exp: Chemist, Trubek Labs, 58-62; group leader, Cent Res Lab, J P Stevens & Co, Inc, 62-63, sect leader, 63-68, mgr, 68-69; mgr org fiber chem, Burlington Industs Res Ctr, 70-72, mgr tech, 72-74; DIR, ABRASIVE TECHNOL CTR, CARBORUNDUM CO, 75- Mem: Am Chem Soc; Am Inst Chem; Am Asn Textile Chemists & Colorists. Res: Organic synthesis; reaction mechanisms. Mailing Add: 347 Tracey Lane Grand Island NY 14072

MOORE, DONALD RICHARD, b West Palm Beach, Fla, Feb 16, 21; m 70; c 1. MALACOLOGY. Educ: Univ Miami, BS, 54, PhD, 64; Miss Southern Col, MS, 60. Prof Exp: Field biologist, Oyster Div, State Bd Conserv, Fla, 53 & Shell Develop Co, 54; res scientist marine invert, Inst Marine Sci, Univ Tex, 55; asst marine biologist, Gulf Coast Res Lab, Miss, 55-60; res instr, Inst Marine Sci, 60-64, asst prof, 64-73, ASSOC PROF MARINE INVERT, UNIV MIAMI, 73- Concurrent Pos: Consult, Shell Oil Co, Tex, 55-64 & Freeport Sulphur Co, La, 58-60. Mem: Am Malacol Union (pres, 74-75); Paleont Soc; Soc Syst Zool. Res: Ahermatypic corals; distribution of marine invertebrates, especially mollusks; systematics; ecology and zoogeography; marine geology; systematics, distribution and biology of marine micromollusca. Mailing Add: Sch of Marine Sci Univ of Miami Miami FL 33149

MOORE, DONALD VINCENT, b York, Nebr, Dec 29, 15; m 42; c 2. PARASITOLOGY. Educ: Hastings Col, BA, 37; Univ Nebr, MA, 39; Rice Inst, PhD(parasitol), 42. Prof Exp: Asst, Univ Nebr, 37-39; sr parasitologist, Bur Labs, State Health Dept, Texas, 42-46; asst prof prev med, Col Med, NY Univ, 46-52, assoc prof, 52-55; ASST PROF MICROBIOL, UNIV TEX SOUTHWESTERN MED SCH DALLAS, 55-, ASST PROF PATH, 74- Concurrent Pos: Lectr, Univ Tex, 43-46; asst parasitologist, Bellevue Hosp, 47-55; lectr, Shelton Col, 50; assoc attend microbiologist, Univ Hosp, Post-Grad Med Sch, NY Univ, 50-55; mem Nat Res Coun-NSF biol sci fel panel, 61-63, Nat Res Coun res associateships panel, 72- Mem: Am Micros Soc; Am Soc Parasitol (secy-treas, 66-); Am Soc Trop Med & Hyg. Res: Acanthocephala life histories; schistosomiasis; helminthology; trypanosomiasis medical parasitology; drug resistance in malaria. Mailing Add: Dept of Microbiol Univ of Tex Southwestern Med Sch Dallas TX 75235

MOORE, DOUGLAS HOUSTON, b Los Angeles, Calif, Apr 22, 20; m 47; c 2. APPLIED MATHEMATICS. Educ: Univ Calif, Berkeley, AB, 42; Univ Calif, Los Angeles, MA, 48, PhD(eng), 62. Prof Exp: Instr math, West Coast Univ, 49-53; res engr, NAm Aviation, Inc, 53-54 & Hughes Aircraft Co, 55-56; instr math, West Coast Univ, 56-58; from asst prof to prof, Calif State Polytech Col, 58-68; assoc prof, 68-70, actg dir concentration in environ control, 69-70, PROF MATH, UNIV WIS-GREEN BAY, 70- Mem: Math Asn Am; Inst Elec & Electronics Engrs. Res: Heaviside operational calculus; operational calculus of sequences; scalar dimensional analysis. Mailing Add: 555 Seventh St Santa Monica CA 90402

MOORE, DUANE GREY, b Barron, Wis, Nov 18, 29; m 54; c 4. FOREST SOILS. Educ: Univ Wis, BS, 53, MS, 55, PhD(soils), 60. Prof Exp: Proj assoc bot & chem, Univ Wis, 59-62; asst prof soil sci, Univ Hawaii, 62-65; SOIL SCIENTIST, FORESTRY SCI LAB, PAC NORTHWEST FOREST & RANGE EXP STA, 65- Concurrent Pos: Soil-herbicide chemist, Hawaiian Sugar Planter's Asn Exp Sta, Honolulu, 63. Mem: AAAS; Am Soc Agron; Soil Sci Soc Am; Int Soc Soil Sci. Res: Forest fertilization and water quality; pesticide residues in the forest-soil ecosystem; micronutrient fertility of range forage and soil herbicide interactions as related to water quality. Mailing Add: 1065 SW Stamm Pl Corvallis OR 97330

MOORE, DUANE MILTON, b Rochelle, Ill, Apr 17, 33; m 53; c 3. MINERALOGY, GEOCHEMISTRY. Educ: Univ Ill, BA, 58, MS, 61, PhD(geol), 63. Prof Exp: Asst geologist, Ill State Geol Surv, 62-63; asst prof geol, Marshall Univ, 63-64; asst prof, 64-70, ASSOC PROF GEOL, KNOX COL, ILL, 70- Mem: Mineral Soc Am; Geochem Soc; Geol Soc Am. Res: Mineralogical and geochemical investigations of sedimentary rocks, especially distribution of trace metallic elements; man in geologic perspective. Mailing Add: Dept of Geol Knox Col Galesburg IL 61401

MOORE, DUNCAN THOMAS, b Biddeford, Maine, Dec 7, 46; m 69. OPTICS, OPTICAL ENGINEERING. Educ: Univ Maine, Orono, BA, 69; Univ Rochester, MS, 71, PhD(optics), 74. Prof Exp: Optical engr, Western Elec Co, Inc, Princeton, NJ, 69-71; ASST PROF OPTICS, INST OPTICS, UNIV ROCHESTER, 74- Concurrent Pos: Consult optical eng, 71- Mem: Optical Soc Am; Am Ceramic Soc; Soc Photog & Instrumentation Engrs. Res: Design of optical systems with gradient index materials; design of optical instruments for metrology. Mailing Add: Inst of Optics Univ of Rochester Rochester NY 14627

MOORE, EARL NEIL, b Warner, Ohio, June 8, 04; m 31; c 4. VETERINARY MEDICINE. Educ: Ohio State Univ, BSc, 27, DVM, 30. Prof Exp: Hatchery mgr, Portsmouth Accredited Hatchery, Ohio, 27; asst vet, WVa Dept Agr, 30-34, vet in chg, 34-37; asst prof animal path, WVa Univ, 37-44; assoc prof poultry path, Univ Del, 44-46; prof poultry dis, State Univ NY Vet Col, Cornell Univ, 46-51; prof poultry sci & assoc chmn dept, Ohio Exp Sta, 51-56; with Int Coop Admin-Kans State Univ Team, India, 56-62; consult poultry & livestock, Ford Found, India, 63-65; AID-Kans State Univ adv to dean, Vet Cols, Andhra Pradesh Agr Univ, Hyderabad & Tirunathi, India, 66-67; prof microbiol & poultry sci, Ahmadu Bello Univ, Nigeria, 67-70; mem staff, Int Exec Serv Corps, 70-76; RETIRED. Honors & Awards: Nat Turkey Res Award, 56; Int Vet Cong Prize, Am Vet Med Asn, 73. Mem: Poultry Sci Asn; Am Vet Med Asn; Am Asn Avian Path. Res: Animal and poultry diseases; mastitis; fowl typhoid; coccidiosis; pullorum; artificial insemination of turkeys. Mailing Add: 636 Beall Ave Wooster OH 44691

MOORE, EARL NEIL, b Morgantown, WVa, Dec 19, 32; m 59; c 2. PHYSIOLOGY, CARDIOLOGY. Educ: Cornell Univ, DVM, 56; State Univ NY, PhD, 62; Univ Pa, MA, 71. Prof Exp: Pvt pract, 56; from asst prof to assoc prof physiol animal biol, 62-70, PROF PHYSIOL, SCH VET MED & GRAD SCH ARTS & SCI, UNIV PA, 70-, PROF PHYSIOL IN MED, HOSP, 71- Concurrent Pos: William Stroud estab investr, Am Heart Asn, 66- Mem: Biophys Soc; Am Physiol Soc; Am Heart Asn; Cardiac Muscle Soc; Am Soc Vet Physiologists & Pharmacologists. Res: Cardiac arrhythmias; cardiac electrophysiology; His' bundle electrocardiology; electrocardiology of congenital heart disease; Wolf-Parkinson-White syndrome; cardiac pharmacology. Mailing Add: Comp Cardiovasc Studies Unit Univ of Pa Sch of Vet Med Philadelphia PA 19104

MOORE, EARL PHILLIP, b Jan 16, 27; US citizen; m 53; c 3. INDUSTRIAL CHEMISTRY. Educ: Univ Miami, BS, 50, MS, 52; Ohio State Univ, PhD(org chem), 57. Prof Exp: Res chemist, Plastics Dept, 57-63, sales res chemist, Indust Chem Dept, Tech Serv Labs, 63-67, sr sales res chemist, 67-73, STAFF CHEMIST, INDUST CHEM DEPT, EXP STA, E I DU PONT DE NEMOURS & CO, INC, 73- Mem: Soc Plastics Eng. Res: Industrial inorganic and organic chemical research and development. Mailing Add: Indust Chem Dept Exp Sta E I du Pont de Nemours & Co Inc Wilmington DE 19898

MOORE, EDGAR TILDEN, JR, b Middlesboro, Ky, Jan 18, 37; m 58; c 3. APPLIED PHYSICS. Educ: Univ Ky, BS, 58. Prof Exp: Physicist, Lawrence Radiation Lab, 59-62; physicist, 63-67, mgr shock dynamics dept, 67-70, dep dir appl sci div, 70-72, dir appl sci div, 72-75, VPRES, PHYSICS INT CO, SAN LEANDRO, 75- Mem: Soc Explosive Engrs. Res: Design and development of new methods to enhance the recovery of oil, gas and other natural resources using chemical explosives; use of one and two dimensional, finite-differencing computer programs. Mailing Add: 2845 Russell St Berkeley CA 94705

MOORE, EDWARD FORREST, b Baltimore, Md, Nov 23, 25; m 50; c 3. APPLIED MATHEMATICS. Educ: Va Polytech Inst, BS, 47; Brown Univ, MS, 49, PhD(math), 50. Prof Exp: Asst prof math, Electronic Digital Comput Proj, Univ Ill, 50-51; mem tech staff, Bell Tel Labs, Inc, 51-66; PROF COMPUT SCI & MATH, UNIV WIS-MADISON, 66- Concurrent Pos: Vis lectr, Harvard Univ, 61-62; vis prof, Mass Inst Technol, 61-62 & Stevens Inst Technol, 65-66. Mem: AAAS; Am Math Soc; Nat Speleol Soc; Soc Indust & Appl Math; Math Asn Am. Res: Logical design of switching circuits; automata theory; information retrieval; binary codes; graph theory; diagnostic identification problems. Mailing Add: 4337 Keating Terr Madison WI 53711

MOORE, EDWARD LEE, b Springfield, Mo, Dec 12, 29; m 55; c 3. INORGANIC

CHEMISTRY. Educ: Drury Col, BS, 51; Washington Univ, MS, 56. Prof Exp: Res chemist, Monsanto Co, 56-61, sr res chemist, 61-64, res group leader nitrogen chem, 64-67, mgr mfg technol, Inorg Div, 67-69, mgr mfg, Spec Chem Systs, 69-71; MGR MFG TECHNOL, MONSANTO INDUST CHEM CO, 71- Mem: Am Chem Soc. Res: Chlorinated cyanurates; inorganic phosphates; sulfuric acid and sulfuric acid catalyst. Mailing Add: Inorg Div Monsanto Co 800 N Lindbergh St Louis MO 63166

MOORE, EDWARD WELDON, b Madisonville, Ky, July 6, 30; m 63; c 5. GASTROENTEROLOGY, ELECTROCHEMISTRY. Educ: Vanderbilt Univ, BA, 52, MD, 55. Prof Exp: Intern med, Harvard Med Serv, Boston City Hosp, 55-56; resident, Lemuel Shattuck Hosp, 56-57; clin assoc cancer, Nat Cancer Inst, 57-59; resident med, Harvard Med Serv, Boston City Hosp, 59-60; from asst prof to assoc prof, Med Sch, Tufts Univ, 64-70, instr physiol, 66-70; PROF MED, MED COL VA, VA COMMONWEALTH UNIV, 70- Concurrent Pos: USPHS res fel, Harvard Med Sch, 60-62; Med Found Boston fel, Med Sch, Tufts Univ, 62-65 & NIH res career develop award, 65-70; consult, Gen Med Study Sect, NIH, 70-; informal consult, Electrochem Sect, Nat Bur Stand, 70-; consult, NASA, 71-; consult, Rev Panel Over-the-counter Antacids, Food & Drug Admin, 72- Mem: AAAS; Am Fedn Clin Res; Am Gastroenterol Asn; NY Acad Sci; Am Soc Clin Invest. Res: Gastrointestinal physiology; ion-exchange electrodes in biomedical research. Mailing Add: Health Sci Ctr Med Col of Va Richmond VA 23219

MOORE, EDWIN FORREST, b Dallas, Tex, Nov 16, 12; m 39. ECONOMIC STATISTICS. Educ: Howard Payne Col, AB, 38; Univ Tex, MA, 41; Univ Mo, PhD(statist), 50. Prof Exp: Instr math, Howard Payne Col, 39-41; prof, Hannibal-LaGrange Col, 41-42; civilian instr aerial navig, Bur Aeronaut, US Dept Navy, 42-44; dean, Hannibal-LaGrange Col, 45-48; instr statist, Univ Mo, 48-50; chmn dept statist & dir div res, Sch Bus, Baylor Univ, 50-54; chief statist qual control, Rocket Fuels Div, Phillips Petrol Co, 54-58; chief statist design exp, Astrodyne, Inc, 58-59; chief qual assurance dept, Rocketdyne Solid Rocket Div, NAm Aviation, Inc, 59-62, res specialist, 62-67; mem tech staff, Autonetics Div, NAm Rockwell Corp, 67-72; CHMN DEPT BUS, E TEX BAPTIST COL, 72- Mem: Am Statist Asn; sr mem Am Soc Qual Control; Math Asn Am; Am Inst Aeronaut & Astronaut. Res: Reliability analysis; probability inference; statistical experiment design; statistical quality control; business and economic indices. Mailing Add: Dept of Bus E Tex Baptist Col Marshall TX 75670

MOORE, EDWIN LEWIS, b Springfield, Mass, May 26, 16; m. FOOD SCIENCE. Educ: Mass State Col, BS, 38, MS, 40, PhD(food tech), 42. Prof Exp: Fel, Fla Citrus Comn & collabr, Bur Agr & Indust Chem, USDA, 42-48; Fla Citrus Comn fel, Citrus Exp Sta, 47-67, RES CHEMIST, FLA DEPT CITRUS, UNIV FLA, 67- Honors & Awards: Award, USDA, 52. Mem: Am Chem Soc; Inst Food Technol; fel Am Inst Chem. Res: Nutrition; citrus by-products and processing. Mailing Add: Agr Res & Educ Ctr Univ of Fla Lake Alfred FL 33850

MOORE, EDWIN NEAL, b Dallas, Tex, Aug 14, 34; m 62; c 2. ATOMIC PHYSICS. Educ: Southern Methodist Univ, BS, 57; Yale Univ, MS, 58, PhD(physics), 62. Prof Exp: Lectr physics, Univ Calif, Santa Barbara, 61-62; asst prof, 62-67, NASA res grant, 65-71, ASSOC PROF PHYSICS, UNIV NEV, RENO, 67-, CHMN DEPT, 69- Mem: Am Phys Soc; Am Asn Physics Teachers. Res: Theoretical calculations of accurate atomic wave functions and photoionization cross-sections and other continuum problems. Mailing Add: Dept of Physics Univ of Nev Reno NV 89507

MOORE, EMMETT BURRIS, JR, chemical physics, see 12th edition

MOORE, ERIC G, b Dartford, Eng, Nov 11, 39; m 62; c 3. URBAN GEOGRAPHY, POPULATION GEOGRAPHY. Educ: Univ Cambridge, BA, 62; Univ Queensland, PhD(geog), 66. Prof Exp: Asst prof geog, Northwestern Univ, 66-70, assoc prof, 70-74; MEM FAC GEOG, QUEEN'S UNIV, ONT, 74- Concurrent Pos: Vis lectr, London Sch Econ, 70; consult, H W Lochner & Sons, 70-71; consult, Div Res Epidemiol & Commun Sci, WHO, Geneva, 71-72; consult, Proj Simu-Sch, Chicago Bd Educ, 71- Mem: Asn Am Geogrs; Regional Sci Asn. Res: Micro-level approaches to residential mobility and neighborhood change and the implications of small-area population changes for provision of public services. Mailing Add: Dept of Geog Queen's Univ Kingston ON Can

MOORE, ERIN COLLEEN, b Arlington, Tex, Nov 16, 24. BIOCHEMISTRY, ONCOLOGY. Educ: Univ Tex, BA, 45, MA, 50; Univ Wis, PhD(oncol), 58. Prof Exp: Anal chemist, Univ Tex, 45-50; asst biochem, Dept Genetics, Carnegie Inst, 50-52; jr res chemist, Int Minerals & Chem Corp, 52-54; asst biochem & oncol, Univ Wis, 55-58; sr fel biochem, 58-59, asst, 59-66, ASSOC BIOCHEMIST, UNIV TEX M D ANDERSON HOSP & TUMOR INST HOUSTON, 66-; ASSOC PROF BIOCHEM, UNIV TEX GRAD SCH BIOMED SCI HOUSTON, 66- Concurrent Pos: USPHS fel, Univ Uppsala, 61-63. Mem: AAAS; Am Chem Soc; Am Asn Cancer Res; Am Soc Biol Chem. Res: Biosynthesis of deoxyribonucleotides; intermediary metabolism of nucleotides. Mailing Add: Dept of Biochem Univ of Tex M D Anderson Hosp & Tumor Inst Houston TX 77030

MOORE, ERMER LEON, b Starke, Fla, May 8, 15; m 45; c 2. PLANT PATHOLOGY, GENETICS. Educ: Univ Ga, BSA, 40, MSA, 41; Univ Wis, PhD(plant path, genetics), 47. Prof Exp: Assoc pathologist, Tobacco Exp Sta, NC, 47-50, from agronomist to prin agronomist, 50-63, asst chief, Tobacco & Sugar Crops Res Br, Md, 63-71, chief, 71-72, STAFF SCIENTIST PLANT & ENTOM SCI, NAT PROG STAFF, AGR RES SERV, USDA, 72- Concurrent Pos: From asst prof to prof, NC State Col, 47-63. Mem: AAAS; Am Phytopath Soc; Am Genetic Asn; Am Inst Biol Sci. Res: Genetics of disease resistance in tobacco; varietal development. Mailing Add: Nat Prog Staff Plant & Entom Sci Agr Res Serv USDA Beltsville MD 20705

MOORE, EUGENE ROGER, b Saginaw, Mich, Oct 20, 33; m 58; c 4. POLYMER CHEMISTRY, CHEMICAL ENGINEERING. Educ: Mich Tech Univ, BS(chem) & BS(chem eng), 56; Case Inst, PhD(chem eng), 62. Prof Exp: Chem engr, Dow Chem Co, 56-58; asst, Case Inst, 58-59; proj leader, Nuclear & Basic Res Lab, 61-67, group leader, 67-70, sr res engr phys res, 70-72, res specialist styrene molding polymers res & develop, 72-74, MFG REP PROCESS DEVELOP, DOW CHEM CO, 74- Honors & Awards: Union Carbide Corp Award, Am Chem Soc, 61. Mem: Am Chem Soc. Res: Chemistry and physics of alkyd resins; exploration of the chemistry of reactive copolymers; process design and development. Mailing Add: 5600 Woodview Pass Midland MI 48640

MOORE, EUNICE MARTHA, b Naugatuck, Conn, Sept 23, 09. PHYSICAL CHEMISTRY. Educ: Johns Hopkins Univ, MA, 36; Duke Univ, PhD(chem), 39. Prof Exp: Dir chem, Labs, Mass Gen Hosp, 39-43; res chemist, Kendall Co, 43-47; asst prof chem, Navy Pier, Univ Ill, 47-50; assoc prof, Mt Holyoke Col, 50-56; chief chemist, 56-65, DIR RES & DEVELOP, ELEC UTILITIES CO, 65- Concurrent Pos: Mem adv comt isotope & radiation develop, AEC, 64-; mem, Int Conf Large Elec Systs; trustee, Ill Valley Community Col. Mem: AAAS; Am Chem Soc; Am Soc Testing & Mat; Electrochem Soc; NY Acad Sci. Res: Dielectrics; capacitors. Mailing Add: 1353 Campbell Ave La Salle IL 61301

MOORE, EVON LAMAR, b George Co, Miss, Dec 23, 14; m 39; c 2. HORTICULTURE. Educ: Miss State Col, BS, 40, Mich State Col, PhD, 55. Prof Exp: Agent sweet potato invests, USDA, 40, tung invests, 40-42; asst horticulturist, Veg Invests, Agr Exp Sta, 42-45, assoc horticulturist, 47-59, PROF HORT, MISS STATE UNIV, 59- Mem: Am Soc Hort Sci. Res: Nutrient requirements of vegetables; storage of root crops; physiology and morphology of flowering; plastic film research; tomato breeding. Mailing Add: Dept of Hort Miss State Univ PO Drawer T Mississippi State MS 39762

MOORE, F STANLEY, geography, see 12th edition

MOORE, FELIX E, b Ashland, Ore, Mar 15, 12; m 42; c 2. BIOSTATISTICS. Educ: Univ Wash, BA, 34. Prof Exp: Statistician, Area Statist Off, US Works Proj Admin, Calif & Wash, 35-37; instr sociol, Univ Wash, 40-41; statistician, US Bur Census, Washington, DC, 41 & Info & Educ Div, US War Dept, 41-44; chief surv div, Res Serv, US Vet Admin, 46-47, Natality Anal Sect, Nat Off Vital Statist, USPHS, 47-48 & Biomet Res Sect, Nat Heart Inst, NIH, 48-57; chmn dept, 57-71, PROF BIOSTATIST, UNIV MICH, ANN ARBOR, 57- Concurrent Pos: Mem tech adv comt, US Bur Census, 48-49; consult, Atomic Bomb Casualty Comn, Nat Res Coun, 55-, WHO, 59 & Mich State Health Dept, 71-; consult, NIH, 57-, mem human ecol study sect, 60-64; fel coun epidemiol, Am Heart Asn. Mem: Biomet Soc; Am Epidemiol Soc; fel Am Statist Asn; fel Am Pub Health Asn; Pop Asn Am. Res: Medical and public health biostatistics; demography; epidemiology of cardiovascular diseases. Mailing Add: Dept of Biostatist Univ of Mich Sch of Pub Health Ann Arbor MI 48104

MOORE, FRANCIS BERTRAM, b Des Moines, Iowa, July 31, 05; m. PHYSICAL CHEMISTRY. Educ: Des Moines Univ, BA, 26; Iowa State Col, PhD(phys chem), 40. Prof Exp: Instr & asst coach high sch, Iowa, 26-27, 29-32, prin & coach, 27-29; instr chem, Iowa State Col, 33-40; assoc prof & head dept, Phillips Univ, 40-43; prof, Southeast Mo State Col, 43-52; assoc prof, 52-61, prof, 61-74, head dept, 54-74, EMER PROF CHEM, UNIV MINN, DULUTH, 74- Mem: Am Chem Soc. Res: Electron sharing ability of organic radicals; qualitative separation of copper and cadmium; condensation of mercaptans with chloral in the gaseous phase; chemical composition of western Lake Superior waters; extension of studies of the bathophanthroline determination of iron. Mailing Add: Dept of Chem Univ of Minn Duluth MN 55812

MOORE, FRANCIS DANIELS, b Evanston, Ill, Aug 17, 13; m 35; c 5. SURGERY. Educ: Harvard Univ, AB, 35, MD, 39; Suffolk Univ, DSc, 66. Hon Degrees: MCh, Nat Univ Ireland, 61; LLD, Glasgow Univ, 65; FRCS, 67, FRCS(E), 68, FRCS(C), 70 & FRCS(I), 72. Prof Exp: Nat Res Coun fel med, Harvard Univ, 41-42; instr, 43-46, assoc, 46-47, asst prof & tutor, 47-48, MOSELEY PROF SURG, HARVARD MED SCH, 48- Concurrent Pos: Asst resident surgeon, Mass Gen Hosp, Boston, 42-43, resident surgeon, 43, asst surg, 43-46, asst surgeon, 46; surgeon-in-chief, Peter Bent Brigham Hosp, 48-; consult, Surgeon Gen, Korea, 51; vis prof, Univ Edinburgh, 52 & Univ London, 55; chmn surg study sect, USPHS, 56-59; vis prof, Univ Colo, 58 & Univ Otago, NZ, 67; mem, Exec Comt, Nat Res Coun; chmn, Adv Comt Metab Trauma, Off Surgeon Gen. Mem: Soc Univ Surgeons (pres, 58-); Soc Clin Surg; Am Surg Asn (pres, 71-; AMA; Am Col Surgeons. Res: Clinical surgery as related to gastrointestinal tracts; biochemistry of cellular changes in surgical as revealed by radioactive and stable isotopes; metabolic care in trauma; transplantation of tissues and organs; cancer of the breast; surgical manpower and health care delivery. Mailing Add: Peter Bent Brigham Hosp 721 Hungtington Ave Boston MA 02115

MOORE, FRANK ARCHER, b Tribune, Kans, Mar 30, 20; m 48; c 3. BIOCHEMISTRY. Educ: Ft Hays Kans State Col, BS, 42; Kans State Univ, MS, 52, PhD(chem), 60. Prof Exp: Jr inspector powder & explosives, Weldon Spring Ord Works, 42-43; chemist & analyst, Phillips Petrol Co, 47-49; asst chem, Univ Calif, 51-52, lab technician food technol, 52; asst chem, Kans State Univ, 52-56; instr, NMex State Univ, 56-59; from asst prof to assoc prof, 59-67, PROF CHEM, ADAMS STATE COL, 67- Mem: Am Chem Soc. Res: Interaction of proteins with small molecules; organic chemistry; blood lipids. Mailing Add: Dept of Chem Adams State Col Alamosa CO 81102

MOORE, FRANK DEVITT, III, b Philadelphia, Pa, June 16, 31; m 59; c 2. HORTICULTURE. Educ: Pa State Univ, BS, 58; Univ Del, MS, 60; Univ Md, PhD(hort), 64. Prof Exp: Asst hort, Univ Del, 58-60 & Univ Md, 60-64; asst horticulturist, 64-70, ASSOC PROF HORT, COLO STATE UNIV, 70- Concurrent Pos: Sr res horticulturist, 3M Co, 69; mem, Inst Soc Hort Sci. Mem: AAAS; Am Soc Hort Sci; Am Inst Biol Sci. Res: Effect of environment and nutrition on plant growth and physiology; vegetable crops. Mailing Add: Dept of Hort Plant Sci Bldg Colo State Univ Ft Collins CO 80521

MOORE, FRANK LESLIE, JR, b Brooklyn, NY, Mar 14, 17; m 42; c 3. PHYSICS. Educ: Union Univ, NY, BS, 39; Lafayette Col, MS, 41; Princeton Univ, MA, 47, PhD(spectros), 49. Prof Exp: Asst, Lafayette Col, 39-41; assoc physicist, Div War Res, Univ Calif, Los Angeles, 44-45; asst, Princeton Univ, 41-42 & 45-46, instr physics, 42-44 & 46-48; from instr to asst prof, Dartmouth Col, 48-51; asst prof, Cornell Univ, 51-55; from asst prof to assoc prof, 55-63, PROF PHYSICS, CLARKSON COL TECHNOL, 63- Mem: Am Phys Soc; Am Asn Physics Teachers; Optical Soc Am. Res: Ionic and molecular spectra; plasma physics. Mailing Add: Dept of Physics Clarkson Col of Technol Potsdam NY 13676

MOORE, FRANK LUDWIG, b Fremont, Ohio, Mar 22, 45; m 68; c 1. ENDOCRINOLOGY. Educ: Col Wooster, BA, 67; Univ Colo, MA & PhD(biol), 74. Prof Exp: Teacher biol, Pub Schs, Lakewood, Ohio & Denver, Colo, 67-71; instr, Univ Colo, 71-72; ASST PROF BIOL, ORE STATE UNIV, 73- Mem: Am Soc Zoologists; Sigma Xi; Am Inst Biol Sci. Res: Reproductive endocrinology of amphibia. Mailing Add: Dept of Gen Sci Ore State Univ Corvallis OR 97331

MOORE, FRANK WILLIAM, b Yale, Iowa, Mar 14, 25; m 55. CULTURAL ANTHROPOLOGY. Educ: Cent Col, BA, 48; Univ of the Americas, MS, 50; Univ Mich, MA, 52; Columbia Univ, PhD(anthrop), 56. Prof Exp: Res asst, 54-56, exec dir, 58-74, RES ASSOC ANTHROP, HUMAN RELS AREA FILES, YALE UNIV, 56-, TREAS, 74- Mem: Fel Am Anthrop Asn. Res: Cross-cultural research methodology; documentation; information science. Mailing Add: 2119 Evelyn St Perry IA 50220

MOORE, FRED EDWARD, b Lafayette, Colo, Mar 16, 23. GEOLOGY. Educ: Univ Colo, BA, 48; St Louis Univ, MS, 50; Univ Wyo, PhD(geol), 59. Prof Exp: Oceanogr, Hydrographic Off, US Navy, 50; photogeologist, Geophoto Serv, Colo, 51-54; instr, 55-59, from asst prof to assoc prof, 59-71, PROF GEOL, COLO SCH MINES, 71- Mem: Geol Soc Am; Soc Econ Paleont & Mineral; Am Asn Petrol Geol; Am Geophys Union. Res: Geomorphology, particularly Pleistocene history of the Front Range. Mailing Add: Dept of Geol Colo Sch of Mines Golden CO 80401

MOORE, FREDERICK JESSUP, medicine, information science, deceased

MOORE, G ALEXANDER, JR, b Manila, Philippines, Oct 8, 37; US citizen. ANTHROPOLOGY. Educ: Harvard Univ, AB, 58; Columbia Univ, MA, 63, PhD(anthrop), 66. Prof Exp: Asst prof anthrop & educ, Emory Univ, 65-69; actg chmn dept, 70-71, grad coordr, Dept Anthrop, 71-73, ASSOC PROF ANTHROP, UNIV FLA, 69- Concurrent Pos: Lectr, NDEA Inst Advan Study Teaching Disadvantaged Youth, 67; consult, Regional Curric Study Proj, Fed & State, 67. Mem: Fel Am Anthrop Asn; Am Ethnol Soc. Res: Anthropology and education; Latin American anthropology; tribal republican politics, Central American Indians. Mailing Add: Dept of Anthrop Univ of Fla Gainesville FL 32611

MOORE, GEORGE EDMUND, b Syracuse, NY, Dec 14, 13; m 41; c 3. PHYSICAL CHEMISTRY. Educ: Syracuse Univ, BS, 35; Stanford Univ, PhD(chem), 40. Prof Exp: Asst synthetic resins, Am Cyanamid Co, 35-37, res chemist phys chem, 40-41; chemist thermochem & thermodynamics, US Bur Mines, 41-43; scientist, cryogeny & explosives, Manhattan Proj, Los Alamos Sci Lab, 43-46; res assoc phys chem, Gen Elec Co, 46-56, mgr combustion applns, Corporate Res & Develop, 56-76; RETIRED. Concurrent Pos: Mem subcomt rocket engines, Nat Adv Comt Aeronaut, 51-52. Mem: Am Chem Soc; Combustion Inst; AAAS. Res: Thermochemistry and thermodynamics; combustion and flame phenomena; emissions and air pollution. Mailing Add: 816 Dongan Ave Scotia NY 12302

MOORE, GEORGE EDWARD, b Evanston, Ill, Sept 28, 22; m 49; c 2. PHYSICAL CHEMISTRY. Educ: Northwestern Univ, BS, 43; Univ Tenn, PhD(chem), 61. Prof Exp: Asst chem, Univ Minn, 43-44 & Univ Chicago, 44; res chemist, Clinton Labs, Oak Ridge Nat Lab, 44-48 & Union Carbide Nuclear Co, 48-71; res chemist, Chem Div, 71-74, RES STAFF DIR ADMIN, OAK RIDGE NAT LAB, 74- Concurrent Pos: Chemist, Metall Lab, Manhattan Eng Dist, 44. Mem: AAAS; Am Chem Soc; Sigma Xi; Am Nuclear Soc. Res: Heterogeneous catalysis; radiation chemistry; ion exchange; corrosion; chemistry of actinide elements; molecular beams; gas-surface interactions; water pollution research; radiological protection. Mailing Add: Oak Ridge Nat Lab PO Box X Oak Ridge TN 37830

MOORE, GEORGE EDWARD, b Pittsburgh, Pa, Oct 28, 03; m 31; c 3. PHYSICS. Educ: Calif Inst Technol, BS, 27; Columbia Univ, MA, 32. Prof Exp: Develop engr, Bell Tel Labs, Inc, 27-36, res physics, 37-40, develop, 40-45, mem res staff, 46-66; prof, 66-74, EMER PROF PHYSICS, STATE UNIV NY BINGHAMTON, 74- Mem: Fel Am Phys Soc; Inst Elec & Electronics Eng. Res: Thermionics; gas discharges; chemical reactions; recombination in mercury; catalysis; surface physics. Mailing Add: Dept of Physics State Univ of NY Binghamton NY 13901

MOORE, GEORGE EMERSON, JR, b Lebanon, Mo, Jan 2, 14; m 39; c 3. GEOLOGY. Educ: Univ Mo, AB, 36, MA, 38; Harvard Univ, MA, 41, PhD(geol), 47. Prof Exp: Instr geol, Univ Mo, 38-39; rodman, US Geol Surv, Washington, DC, 39; asst, Harvard Univ, 40-42; geologist, A P Green Fire Brick Co, Mo, 42-46; asst, Harvard Univ, 46-47; from instr to assoc prof, 47-64, PROF GEOL, OHIO STATE UNIV, 64- Mem: Geol Soc Am. Res: Structure and metamorphism of southwestern New Hampshire, bedrock geology of Rhode Island. Mailing Add: Dept of Geol Ohio State Univ Columbus OH 43210

MOORE, GEORGE EUGENE, b Minneapolis, Minn, Feb 22, 20; m 44; c 5. SURGERY, ONCOLOGY. Educ: Univ Minn, BA, 42, MA, 43, BS, 44, BM, 46, MD, 47, PhD, 50; Am Bd Surg, dipl. Prof Exp: Lab asst physiol, Univ Minn, 41-42, asst histol & zool, 42-43, intern surg, Univ Hosps, 46-47, clin instr, Med Sch, 48-50, asst prof & cancer coordr, 51, assoc prof, 51-52; clin prof surg, Sch Med, State Univ NY Buffalo, 52-73, res prof biol & dir, Roswell Park Div, Grad Sch, 62-73; PROF SURG, UNIV COLO, DENVER, 73-; CHIEF DIV SURG ONCOL, DENVER GEN HOSP, 73- Concurrent Pos: Markle Fund scholar, 48-53; dir & chief surg, Roswell Park Mem Inst, 53-67; consult, Nat Cancer Chemother Ctr, NIH, 56; dir pub health res, NY State Dept Health, 67- Honors & Awards: Gross Award, 50; Chilean Iodine Educ Bur Award, Am Pharmaceut Asn, 51. Mem: Soc Exp Biol & Med; Soc Univ Surgeons; Am Surg Asn; NY Acad Sci. Res: Localization of brain tumors; carcinogenesis; experimental biology and oncology; cell culture chemotherapy; cancer surgery. Mailing Add: Dept of Surg Denver Gen Hosp W Eighth Ave & Cherokee St Denver CO 80204

MOORE, GEORGE THOMAS, b Chicago, Ill, Oct 15, 29; m 57; c 1. MARINE GEOLOGY, STRATIGRAPHY. Educ: Univ Notre Dame, BS, 52; Ind Univ, MA, 54, PhD(geol), 56. Prof Exp: Geologist, Chevron Oil Co, 56-67; asst prof, Northern Ill State Col, 67-69; SR RES GEOLOGIST, CHEVRON OIL FIELD RES CO, 69- Concurrent Pos: Indust observer adv bd, Nat Oceanog Data Ctr. Mem: Geol Soc Am; Am Asn Petrol Geologists. Res: Deltaic and deep sea sedimentary accumulations along continental margins associated with major river systems of the world; relationship of petroleum accumulation to this environment of sedimentation. Mailing Add: Chevron Oil Field Res Co PO Box 446 La Habra CA 90631

MOORE, GEORGE WILLIAM, b Palo Alto, Calif, June 7, 28; m 60; c 2. GEOLOGY. Educ: Stanford Univ, BS, 50, MS, 51; Yale Univ, PhD(geol), 60. Prof Exp: Geologist, US Geol Surv, 51-66; geologist-in-chg, La Jolla Marine Geol Lab, 66-75; GEOLOGIST, US GEOL SURV, 75- Concurrent Pos: Res assoc, Scripps Inst Oceanog, 70-75. Mem: AAAS; fel Geol Soc Am; Nat Speleol Soc (pres, 63); Am Asn Petrol Geologists. Res: Stratigraphy and geochemistry of sedimentary rocks; structural geology; cave mineralogy; marine geophysics. Mailing Add: US Geol Surv 345 Middlefield Rd Menlo Park CA 94025

MOORE, GLEN, b Spanish Fork, Utah, Feb 8, 17; m 38; c 7. PLANT PHYSIOLOGY. Educ: Brigham Young Univ, BS, 49; Univ Chicago, PhD(bot), 54. Prof Exp: Asst prof bot, Brigham Young Univ, 54-55; head div biol & phys sci, Church Col Hawaii, 55-58; from asst prof to assoc prof, 58-69, PROF BOT, BRIGHAM YOUNG UNIV, 69- Res: Plant morphogenesis and physiology of reproduction; conservation. Mailing Add: Dept of Bot Brigham Young Univ Provo UT 84601

MOORE, GLENN D, b Barron, Wis, Apr 9, 23; m 47; c 3. ENTOMOLOGY, PLANT BREEDING. Educ: Univ Wis, BS, 50, MS, 51; Univ Minn, PhD, 68. Prof Exp: RES AGRONOMIST & ASST TO MGR RES SERV DEPT, NORTHRUP, KING & CO, 52- Mem: AAAS; Entom Soc Am; Weed Sci Soc Am; Am Seed Trade Asn; Plant Resistance to Insects Asn. Res: Insect resistance in crop plants; alfalfa pollination for seed production; fungicide and fungicide-insecticide seed treatments; herbicides and their application in field crops and vegetables; forage crop management studies. Mailing Add: Northrup King & Co 13410 Research Rd Eden Prairie MN 55343

MOORE, GORDON EARLE, b San Francisco, Calif, Jan 3, 29; m 50. PHYSICAL CHEMISTRY. Educ: Univ Calif, BS, 50; Calif Inst Technol, PhD(chem), 54. Prof Exp: Asst chem, Calif Inst Technol, 50-52; res chemist phys chem, Johns Hopkins Univ, 53-56; mem tech staff, Shockley Semiconductor Labs, Beckman Instruments Corp, 56-57; head eng, Fairchild Semiconductor Corp, 57-58, dir res & develop, 58-68; exec vpres, 68-75, PRES, INTEL CORP, 75- Mem: Electrochem Soc; Am Phys Soc; Inst Elec & Electronics Engrs. Res: Molecular spectroscopy and structure; semiconductors; transistors; microcircuits. Mailing Add: Intel Corp 3065 Bowers Ave Santa Clara CA 95051

MOORE, GORDON GEORGE, b Des Moines, Iowa, Mar 18, 35; m 56; c 3. ORGANIC CHEMISTRY. Educ: Iowa State Univ, BS, 56; Yale Univ, MS, 58, PhD(org chem), 62. Prof Exp: Res assoc, Brookhaven Nat Lab, 60-62; asst prof org chem, Marshall Univ, 62-65; asst prof, 65-71, ASSOC PROF ORG CHEM, PA STATE UNIV, OGONTZ CAMPUS, 71- Concurrent Pos: Res chemist, USDA, Pa, 68- Mem: Am Chem Soc. Res: Organic mechanisms; organic synthesis; compounds of biochemical interest. Mailing Add: Dept of Chem Pa State Univ Ogontz Campus Abington PA 19001

MOORE, HAL G, b Vernal, Utah, Aug 14, 29; m 56; c 4. MATHEMATICS. Educ: Univ Utah, BS, 52, MS, 57; Univ Calif, Santa Barbara, PhD(math), 67. Prof Exp: Jr high sch teacher, Utah, 52-53; instr math, Carbon Jr Col & Carbon High Sch, 53-55; instr, Purdue Univ, 57-61, admin asst to chmn dept, 60-61; exec asst to chmn dept, 61-64, from asst prof to assoc prof, 61-71, PROF MATH, BRIGHAM YOUNG UNIV, 71- Mem: AAAS; Sigma Xi; Am Math Soc; Math Asn Am. Res: Structure of rings and universal algebras. Mailing Add: Dept of Math Brigham Young Univ Provo UT 84602

MOORE, HAROLD ARTHUR, b Brackettville, Tex, Feb 4, 25. NUCLEAR PHYSICS. Educ: St Mary's Univ, Tex, BS, 44; Wash Univ, MS, 48; Univ Fla, PhD, 56. Prof Exp: Asst, Wash Univ, 46-48; physicist, Mo Res Labs, Inc, 48-51; asst, Univ Fla, 51-52, 55-56, instr phys sci, 52-55; from asst prof to assoc prof, 56-62, PROF PHYSICS, BRADLEY UNIV, 62- Concurrent Pos: Consult, Metric Photo Br, US Naval Ord Test Sta, Calif, 57-58. Mem: Am Phys Soc; Am Asn Physics Teachers. Res: Fast resolving time coincidence circuits; deuteron stripping reactions on carbon and oxygen; unified field theory. Mailing Add: Dept of Physics Bradley Univ Peoria IL 61606

MOORE, HAROLD BEVERIDGE, b Alix, Ark, Sept 11, 28; m 50; c 2. MEDICAL BACTERIOLOGY. Educ: San Diego State Col, AB, 51; Univ Calif, Los Angeles, MA, 55, PhD(microbiol), 57; Am Bd Med Microbiol, dipl pub health & med, 65. Prof Exp: Chief microbiologist, Donald N Sharp Mem Community Hosp, San Diego, 57-60; from asst prof to assoc prof, 60-67, PROF MICROBIOL, SAN DIEGO STATE UNIV, 67- Concurrent Pos: Consult, Donald N Sharp Mem Community Hosp, San Diego, 60- & Palomar Hosp, Escondido, 64- Mem: AAAS; Am Soc Microbiol; Am Pub Health Asn. Res: Isolation, identification, taxonomy and clinical significance of poorly defined gram negative rods; clincally significant anaerobic bacteria. Mailing Add: Dept of Microbiol San Diego State Univ San Diego CA 92182

MOORE, HAROLD EMERY, JR, b Winthrop, Mass, July 7, 17. BOTANY. Educ: Univ Mass, BS, 39; Harvard Univ, MA, 40, PhD(biol), 42. Prof Exp: Tech asst, Gray Herbarium, Harvard Univ, 47-48; from asst prof to assoc prof, 48-60, dir, L H Bailey Hortorium, 60-69, PROF BOT, L H BAILEY HORTORIUM, CORNELL UNIV, 60- Concurrent Pos: Guggenheim fel, 46-47 & 55-56; mem bd dirs, Fairchild Trop Garden, 63-71; vis comt, Arnold Arboretum, Harvard Univ, 64-70; bd dirs, Orgn Trop Studies, 70- Honors & Awards: Founder's Medal, Fairchild Trop Garden, 54. Mem: Bot Soc Am; Soc Study Evolution; Am Soc Plant Taxonomists; Am Inst Biol Sci; Int Asn Plant Taxon. Res: Plant taxonomy; Palmae; cultivated Gesneriaceae; Commelinaceae; cultivated plants; tropical botany. Mailing Add: L H Bailey Hortorium 467 Mann Libr Ithaca NY 14853

MOORE, HAROLD W, b Ft Collins, Colo, May 21, 36; m 59; c 2. ORGANIC CHEMISTRY. Educ: Colo State Univ, BS, 59; Univ Ill, PhD(chem), 63. Prof Exp: Asst prof, 65-69, assoc prof, 69-74, chmn dept, 70-74, PROF CHEM, UNIV CALIF, IRVINE, 74- Mem: Am Chem Soc. Res: Organic chemistry, especially synthesis and mechanistic studies. Mailing Add: Dept of Chem Univ of Calif Irvine CA 92650

MOORE, HARRY BALLARD, JR, b Rocky Mount, NC, Sept 28, 28; m 55; c 5. ENTOMOLOGY. Educ: E Carolina Col, AB, 51; Purdue Univ, MS, 55; NC State Univ, PhD(entom), 64. Prof Exp: Chief inspector, NC Struct Pest Control Comn, 58-60; from instr to asst prof, 60-68, ASSOC PROF ENTOM, NC STATE UNIV, 68- Mem: Entom Soc Am; Sigma Xi. Res: Wood-destroying insects. Mailing Add: Dept of Entom NC State Univ Raleigh NC 27607

MOORE, HARVEY CLEAVER, b Port Penn, Del, Mar 13, 18; m 48. ANTHROPOLOGY, CULTURAL ANTHROPOLOGY. Educ: Univ Del, AB, 38; Univ NMex, PhD(anthrop), 50. Prof Exp: From asst prof to assoc prof, 51-58, chmn dept, 65-70, assoc dean grad studies & res, Col Arts & Sci, 70-73, actg dean, Col Arts & Sci, 73-74, PROF ANTHROP, AM UNIV, 58- Concurrent Pos: Res assoc soc sci, Bur Soc Sci Res, DC, 51-54; NIMH grant cult change, Navajo Reservation, 61-63; Am Univ field supvr, Grad Students, Africa, 63. Mem: Fel AAAS; fel Am Anthrop Asn; fel Am Ethnol Soc; fel Royal Anthrop Inst; assoc Current Anthrop. Res: Theory and method of culture change; psychological anthropology; urbanization; applied anthropology; history of anthropological method and theory. Mailing Add: Dept of Anthrop Am Univ Washington DC 20016

MOORE, HENRY J, II, b Albuquerque, NMex, Sept 2, 28; m 59; c 2. GEOLOGY. Educ: Univ Utah, BS, 51; Stanford Univ, MS, 59, PhD(geol), 65. Prof Exp: Prospector-geologist, 54-55; geologist, US Geol Surv, 55-57; asst geol, Stanford Univ, 57-59; GEOLOGIST ASTROGEOL, US GEOL SURV, 60- Concurrent Pos: Prin investr, NASA Exp S-222, 72-76. Honors & Awards: Spec Commendation for Astronaut Training, Geol Soc Am, 73. Mem: AAAS; Sigma Xi; Geol Soc Am; Am Inst Mining, Metall & Petrol Engrs; Am Geophys Union. Res: Application of the principles of geology to lunar problems; investigation of experimental craters produced by projectile impacts with rocks; application of remote sensing data to lunar problems. Mailing Add: 528 Jackson Dr Palo Alto CA 94303

MOORE, HILARY BROOKE, b Arnside, Eng, Aug 23, 07; m. ZOOLOGY. Educ: Univ London, BSc, 29, PhD, 33. Prof Exp: Asst naturalist, Millport Marine Sta, 29-30, Port Erin Biol Sta, 30-34, Plymouth Marine Lab, 34-37 & Bermuda Biol Sta, 37-47; prof marine biol & assoc dir marine lab, 51-74, res assoc, 49-51, EMER PROF MARINE BIOL, UNIV MIAMI, 74-; RES ASSOC, WOODS HOLE OCEANOG INST, 47- Mem: AAAS; Marine Biol Asn UK; Am Soc Limnol & Oceanog. Res: Marine ecology and plankton. Mailing Add: Inst of Marine Sci 1 Rickenbacker Causeway Virginia Keys Miami FL 33149

MOORE, HOWARD EARL, nuclear chemistry, geochemistry, see 12th edition

MOORE, J STROTHER, b Seminole, Okla, Sept 11, 47; m 68; c 1. COMPUTER SCIENCE. Educ: Mass Inst Technol, BS, 70; Univ Edinburgh, PhD(comput logic), 73. Prof Exp: Mem res staff comput sci, Xerox Palo Alto Res Ctr, 73-76; RES MATHEMATICIAN COMPUT SCI, STANFORD RES INST, 76- Res: Automatic computer program verification and advanced debugging aids. Mailing Add: Comput Sci Group Stanford Res Inst Menlo Park CA 94025

MOORE, JAMES A, b Mooresville, NC, Jan 1, 09; m 38; c 3. MEDICINE. Educ: Davidson Col, BS, 30; Harvard Med Sch, MD, 34; Am Bd Otolaryngol, dipl, 40. Prof Exp: House officer gen surg, Jersey City Med Ctr, 35-37 & otolaryngol, Mass Eye & Ear Infirmary, Boston, 37-40; asst bronchoscopy, Chevalier Jackson Bronchial Clin, Philadelphia, 40; mem attend staff otolaryngol, New York Hosp, 40-47, attend surgeon in chg otolaryngol, 47-69; clin assoc prof surg & otolaryngol, 61-69, PROF OTORHINOL & ACTG CHMN DEPT, MED COL, CORNELL UNIV, 69-; ATTEND SURGEON IN CHG OTORHINOL, NEW YORK HOSP, 69- Concurrent Pos: Mem, Bd Dirs & examr, Am Bd Otolaryngol, 58- Mem: Am Otol Soc (secy-treas, 60-65, vpres, 65-66, pres, 66-67); Am Laryngol, Rhinol & Otol Soc. Res: Otolaryngology. Mailing Add: 504 E 87th St New York NY 10028

MOORE, JAMES ALEXANDER, b Johnstown, Pa, Nov 8, 23; m 51; c 3. ORGANIC CHEMISTRY. Educ: Washington & Jefferson Col, BS, 43; Purdue Univ, MS, 44; Pa State Col, PhD(chem), 49. Prof Exp: Res chemist, Parke, Davis & Co, 49-55; from asst prof to assoc prof, 55-63, PROF CHEM, UNIV DEL, 63- Concurrent Pos: Asst, Univ Basel, 52-53; sr ed, J Org Chem, 63-; NIH spec fel, 64-65. Mem: Am Chem Soc; Swiss Chem Soc. Res: Novel heterocyclic systems; new forms of chemical publications. Mailing Add: Dept of Chem Univ of Del Newark DE 19711

MOORE, JAMES ALFRED, b Brooklyn, NY, Aug 30, 39; m 66; c 2. ORGANIC CHEMISTRY, POLYMER CHEMISTRY. Educ: St John's Univ, NY, BS, 61; Polytech Inst Brooklyn, PhD(chem), 67. Prof Exp: NIH fel, Univ Mainz, 67-68; res assoc, Univ Mich, Ann Arbor, 68-69; asst prof, 69-75, ASSOC PROF CHEM, RENSSELAER POLYTECH INST, 75-, COORDR POLYMER SCI & ENG, 73- Concurrent Pos: Assoc ed, Org Preparations & Procedures Int, 73- Mem: Am Chem Soc. Res: Synthesis, characterization and reactions of novel polymers; structure-property relationships; polymer reagents. Mailing Add: Rensselaer Polytech Inst Troy NY 12181

MOORE, JAMES CARLTON, b Harrodsburg, Ky, May 31, 23; m 46; c 3. PHYSIOLOGY. Educ: Univ Ky, BS, 44; Univ Louisville, MD, 46. Prof Exp: Commonwealth Fund fel, Mass Inst Technol, 49-50; from asst prof to assoc prof physiol, 53-70, asst dean student affairs, 73-74, PROF PHYSIOL & BIOPHYS, SCH MED, UNIV LOUISVILLE, 70-, ASSOC DEAN ADMIS, 74- Mem: AAAS; Biophys Soc; Am Physiol Soc; Soc Exp Biol & Med. Res: Blood volume measurement; mechanics of respiration; pulmonary edema; pulmonary circulation; microcirculation. Mailing Add: Health Sci Ctr Louisville KY 40202

MOORE, JAMES ELTON, b Whittier, Calif, Aug 4, 42; m 63; c 1. CHEMICAL PHYSICS. Educ: Calif State Polytech Univ, BS, 67; Univ Southern Calif, PhD(chem physics), 71. Prof Exp: Engr, Xerox Data Syst, 66-68; PROF PHYSICS, UNIV CAMPINAS, BRAZIL, 71- Concurrent Pos: Consult, Telebras, 65- Res: Fiber optics for telecommunications; nonlinear optics in glasses. Mailing Add: Univ of Campinas Campinas SP Brazil

MOORE, JAMES FREDERICK, JR, b Barbourville, Ky, July 19, 38; m 65; c 2. PLANT PATHOLOGY. Educ: Western Ky Univ, BS, 65; Clemson Univ, MS, 68; Univ Ariz, PhD(plant path), 76. Prof Exp: Res asst plant path, Clemson Univ, 65-69; plant pathologist, Dept Agr, Guam, 69-72; res assoc, Univ Ariz, 72-75; RES PLANT PATHOLOGIST, STATE FRUIT EXP STA, SOUTHWEST MO STATE UNIV, 75- Mem: Am Phytopath Soc. Res: Disease of fruit and nut crops; ecology of soil-borne plant pathogens and soil microorganisms; antibiosis; chemical, cultural and biological control; epiphytology. Mailing Add: State Fruit Exp Sta Southwest Mo State Univ Mountain Grove MO 65711

MOORE, JAMES GREGORY, b Palo Alto, Calif, Apr 30, 30; m 52; c 2. GEOLOGY. Educ: Stanford Univ, BS, 51; Univ Wash, MS, 52; Johns Hopkins Univ, PhD(geol), 54. Prof Exp: Instr geol, Johns Hopkins Univ, 52-53; GEOLOGIST, US GEOL SURV, 56- Concurrent Pos: Scientist-in-charge, Hawaiian Volcano Observ, US Geol Surv, 62-64. Mem: Geol Soc Am; Geochem Soc. Res: Petrology and geochemistry of igneous rocks; structural geology; volcanology; submarine geology. Mailing Add: US Geol Surv 345 Middlefield Rd Menlo Park CA 94025

MOORE, JAMES MARVIN, b Van Buren, Ark, Jan 28, 12; m 46. BOTANY, ZOOLOGY. Educ: Univ Ark, BA, 36, BS, 41, MA & MS, 48; Univ Okla, PhD(bot, zool), 58. Prof Exp: Instr high sch, Ark, 36-38; prin, 38-39; instr, 39-42; inspector, Ark Ord Plant, 42-43; machinist, Am Can Co, Calif, 43 & Pac Can Co, 43-44; steel worker, Bethlehem Steel Co, 44; inspector, San Francisco Ord Dist, 44-45; instr high sch, Ark, 45-47; lab instr bot, Univ Ark, 47-48 & Univ Calif, 48-49; instr high sch, Ark, 51-53; instr chem, anat & microbiol, St Edwards Hosp, Sch of Nursing, Ft Smith, Ark, 53-56; asst morphol, plant physiol & genetics, Univ Okla, 56-58; head sci dept, Ft Smith Jr Col, 52-56, 58-59; chmn biol div, 59-69, PROF PLANT ANAT & COMP MORPHOL, UNIV TENN, MARTIN, 59- Mem: AAAS; Am Bot Soc. Res: Developmental plant anatomy and algology. Mailing Add: Dept of Biol Univ of Tenn Martin TN 38237

MOORE, JAMES NORMAN, b Vilonia, Ark, June 10, 31; m 53; c 2. HORTICULTURE. Educ: Univ Ark, BSA, 56, MS, 57; Rutgers Univ, PhD(hort), 61. Prof Exp: Instr hort, Univ Ark, 57; res assoc pomol, Rutgers Univ, 57-61; res horticulturist, USDA, 61-64; assoc prof, 64-69, PROF HORT, UNIV ARK, FAYETTEVILLE, 69- Concurrent Pos: Hon asst prof, Rutgers Univ, 61-64. Honors & Awards: Woodbury Award, Am Soc Hort Sci, 58, Gourley Award, 63 & Ware Res Award, 72. Mem: Am Soc Hort Sci; Am Genetic Asn; Am Pomol Soc; Int Soc Hort Sci. Res: Fruit breeding and genetics; physiology of fruit crops. Mailing Add: Dept of Hort Univ of Ark Fayetteville AR 72701

MOORE, JAY WINSTON, b Madison, Wis, Apr 20, 42; m 65. DEVELOPMENTAL GENETICS, POULTRY GENETICS. Educ: Cedarville Col, BS, 64; Univ Nebr, MS, 67; Univ Mass, PhD(genetics), 70. Prof Exp: ASST PROF BIOL, EASTERN COL, 70- Mem: AAAS; Genetics Soc Am; Am Genetic Asn; Poultry Sci Asn; Am Inst Biol Sci. Res: Plumage color inheritance; genetic factors controlling pigment and keratin development. Mailing Add: Dept of Biol Eastern Col St Davids PA 19087

MOORE, JERRY ARNOLD, b Fairmont, WVa, June 18, 42; m 65; c 1. WILDLIFE ECOLOGY, PESTICIDE MANAGEMENT. Educ: Fairmont State Col, BS, 64; WVa Univ, MS, 67; Am Univ, cert environ syst mgt, 76. Prof Exp: Biol teacher, Marion County Bd Educ, 67-68; biologist wildlife, US Dept Interior, 68-70; biologist pesticide wildlife eval, USDA, 70-71; BIOLOGIST WILDLIFE, US ENVIRON PROTECTION AGENCY, 71- Mem: Wildlife Soc; Sigma Xi; Ecol Soc Am. Res: Effects of pesticides on wildlife resources in the United States and the effects of label terminology in making labels more understandable. Mailing Add: US Environ Protection Agency 401 M St SW Washington DC 20460

MOORE, JERRY LAMAR, b Anderson, SC, Feb 28, 42; m 66; c 2. NUTRITION. Educ: Clemson Univ, BS, 63; Univ Wis, MS, 65, PhD(food sci, biochem), 68. Prof Exp: Clin lab officer, Wilford Hall US Air Force Med Ctr, 68-69, biomed lab officer, Nutrit Br, US Air Force Sch Aerospace Med, 69-72; assoc dir nutrit, 72-74, ASSOC DIR CORP RES & DEVELOP, PILLSBURY CO, 74- Mem: Am Dietetic Asn; Inst Food Technol; Soc Nutrit Educ (secy, 74-77). Res: Applied human nutrition leading to new or nutritionally improved food products, new health related services and more effective informational and educational materials. Mailing Add: Pillsbury Co Res & Develop Labs 311 Second St SE Minneapolis MN 55414

MOORE, JEWEL ELIZABETH, b Hot Springs, Ark, June 5, 18. BOTANY. Educ: Henderson State Col, AB, 40; Univ Ark, MS, 48; Univ Tenn, PhD(bot), 58. Prof Exp: Teacher high sch, Ark, 40-42; instr sci, Beebe Jr Col, 42-47; PROF BIOL, STATE COL ARK, 47- Mem: AAAS; Bot Soc Am; Am Fern Soc; Am Bryol & Lichenological Soc; Nat Audobon Soc. Res: Arkansas plants; bryophytes; ecology. Mailing Add: Dept of Biol State Col of Ark Conway AR 72032

MOORE, JOANNE IWEITA, b Greenville, Ohio, July 23, 28. PHARMACOLOGY. Educ: Univ Cincinnati, AB, 50; Univ Mich, PhD(pharmacol), 59. Prof Exp: Asst, Christ Hosp Inst Med Res, Cincinnati, 50-55; asst pharmacol, Univ Mich, 55-57; fel cardiovasc pharmacol, Basic Sci Div, Emory Univ, 59-61; asst prof, 61-66, assoc prof, 66-71, actg chmn dept, 69-70, interim chmn dept, 71-73, PROF PHARMACOL, COL MED, UNIV OKLA, 71-, CHMN DEPT, 73- Mem: Am Soc Pharmacol & Exp Therapeut; Soc Exp Biol & Med; Am Soc Clin Res; Soc Neurosci; NY Acad Sci. Res: Cardiovascular pharmacology; mechanism of action of drugs upon the contractile and/or conduction system of the heart. Mailing Add: Dept of Pharmacol Univ of Okla Col of Med Oklahoma City OK 73190

MOORE, JOHN ALEXANDER, b Charles Town, WVa, June 27, 15; m 38; c 1. EVOLUTIONARY BIOLOGY. Educ: Columbia Univ, AB, 36, MS, 39, PhD(zool), 40. Prof Exp: Asst zool, Columbia Univ, 36-39; tutor biol, Brooklyn Col, 39-41; instr, Queens Col, NY, 41-43; from asst prof to prof zool, Barnard Col & Columbia Univ, 43-69, chmn dept, Col, 48-52, 53-54 & 60-66, chmn dept, Univ, 49-52; PROF BIOL, UNIV CALIF, RIVERSIDE, 69- Concurrent Pos: Res assoc, Am Mus Natural Hist, 42-, Fulbright res scholar, Australia, 52-53; managing ed, J Morphol, 55-60; Guggenheim fel, 59-60; partic, Biol Sci Curric Study, 59- Mem: Nat Acad Sci; Am Acad Arts & Sci; Am Soc Zoologists (pres, 74); Am Soc Naturalists (pres, 72); Int Soc Develop Biologists. Res: Evolution of amphibians and Drosophila. Mailing Add: Dept of Biol Univ of Calif Riverside CA 92502

MOORE, JOHN CARMAN GAILEY, b Belleville, Ont, Aug 15, 16; m 52; c 3. ECONOMIC GEOLOGY. Educ: Univ Toronto, BA, 38; Cornell Univ, MS, 40; Harvard Univ, PhD, 55. Prof Exp: Field asst, Ont Dept Mines, 37-39; geologist, Int Petrol Co, Ecuador, 40-41, Labrador Mining & Explor Co, 46 & Oliver Iron Mining Co, Venezuela, 47; tech officer, Geol Surv Can, 48-51; geologist, Photog Surv Corp, Ltd, 51-52, Am Metal Co, Ltd, 52-54 & Dom Gulf Co, 54-56; div supvr, McIntyre Porcupine Mines, Ltd, 56-61; vis lectr geol, Univ NB, 61-62; from asst prof to assoc prof, 62-68, head dept, 66-73, PROF GEOL, MT ALLISON UNIV, 68- Mem: Geol Soc Am; Geol Asn Can; Can Inst Mining & Metall; Asn Explor Geochemists; Soc Econ Geologists. Res: Dispersion haloes around ore deposits. Mailing Add: 1 Bennett St PO Box 471 Sackville NB Can

MOORE, JOHN COLEMAN, b Staten Island, NY, May 27, 23. MATHEMATICS. Educ: Mass Inst Technol, BS, 48; Brown Univ, PhD(math), 52. Prof Exp: Fine instr, 53, NSF fel, 53-55, from asst prof to assoc prof, 55-60, PROF MATH, PRINCETON UNIV, 60- Mem: Am Math Soc; Math Asn Am; London Math Soc; Math Soc France; Mex Math Soc. Res: Algebraic topology; homological algebra. Mailing Add: Fine Hall Dept of Math Box 37 Princeton Univ Princeton NJ 08540

MOORE, JOHN CRISWELL, b Matador, Tex, Sept 24, 99; m 23; c 2. CHEMISTRY. Educ: Agr & Mech Col, Tex, BS, 22. Prof Exp: Res chem engr, Tex Co, 23-27; asst supt, Tex Pac Coat & Oil Co, 27-29 & Richardson Ref Co, 29-31; asst supt patent res, Sinclair Ref Co, 31-34, supt paint dept, 34-47; tech dir, Nat Paint Varnish & Lacquer Asn, 47-56, tech dir, Coatings Res Group, Inc, 56-60; PRES, MOORE RES LABS, INC, 60- Concurrent Pos: Pres, Fedn Socs Paint Tech, 46-47. Mem: Am Chem Soc; Am Soc Test & Mat. Res: Coatings; petrochemicals and related fields. Mailing Add: Moore Res Labs Inc 6306 Winston Blvd Bethesda MD 20034

MOORE, JOHN DAVID, b Granite Falls, NC, Aug 12, 29; m 60; c 3. ANIMAL PHYSIOLOGY, PARASITOLOGY. Educ: Wake Forest Col, BS, 51; Univ NC, MSPH, 52, MA, 56; Emory Univ, PhD(biol), 65. Prof Exp: Asst prof biol, Charleston Col, 57-59; from asst prof to assoc prof, 62-69, PROF BIOL, TENN STATE UNIV, 69- Concurrent Pos: Examr physiol, Governor's Bd Basic Sci Exam, Tenn, 63- Res: Physiology of animal parasites, especially their in vitro cultivation. Mailing Add: Dept of Biol East Tenn State Univ Johnson City TN 37601

MOORE, JOHN DOUGLAS, b Austin, Tex, Oct 5, 39; m 63. MATHEMATICS. Educ: Idaho State Univ, BS, 61, MS, 63; Syracuse Univ, PhD(math), 69. Prof Exp: Asst prof math, State Univ NY Col Oswego, 67-69; ASST PROF MATH, ARIZ STATE UNIV, 69- Mem: Am Math Soc; Math Asn Am. Res: Theory of Abelian groups. Mailing Add: Dept of Math Ariz State Univ Tempe AZ 85281

MOORE, JOHN DOUGLAS, b Chicago, Ill, July 11, 43; m 65; c 1. MATHEMATICS. Educ: Univ Calif, Berkeley, BA, 65, PhD(math), 69. Prof Exp: Asst prof, 69-75, ASSOC PROF MATH, UNIV CALIF, SANTA BARBARA, 75- Mem: Am Math Soc; Math Asn Am. Res: Isometric immersions of Riemannian manifolds. Mailing Add: Dept of Math Univ of Calif Santa Barbara CA 93106

MOORE, JOHN DUAIN, b Lancaster, Pa, Dec 11, 13; m 40; c 2. PLANT PATHOLOGY. Educ: Pa State Univ, BS, 39; Univ Wis, PhD(plant path), 45. Prof Exp: From asst prof to assoc prof, 45-54, PROF PLANT PATH, UNIV WIS-MADISON, 54-, DIR UNIV EXP FARMS, 74- Concurrent Pos: Plant pathologist, USDA, 56-65; prof plant sci & head dept, Fac Agr, Univ Ife, Nigeria, 68-70, dean grad studies, 69, dean agr, 69-70. Mem: AAAS; Am Phytopath Soc; Bot Soc Am; Am Inst Biol Sci. Res: Fungus, bacterial and virus diseases of sour cherries and apples; effects of fungicides on quality of cherries and apples. Mailing Add: Dept of Plant Path Univ of Wis Madison WI 53706

MOORE, JOHN EDWARD, b Kirkersville, Ohio, Mar 7, 35; m 57; c 1. RUMINANT NUTRITION. Educ: Ohio State Univ, BS, 57, MS, 58, PhD(animal sci), 61. Prof Exp: PROF ANIMAL SCI, UNIV FLA, 61- Mem: Am Dairy Sci Asn; Am Soc Animal Sci; Am Forage & Grassland Coun. Res: Forage evaluation and utilization. Mailing Add: Nutrit Lab Univ of Fla Gainesville FL 32611

MOORE, JOHN EZRA, b Columbus, Ohio, Jan 25, 31; m 56. HYDROLOGY, GROUNDWATER GEOLOGY. Educ: Ohio Wesleyan Univ, BA, 53; Univ Ill, MS, 58, PhD(geol), 60. Prof Exp: Asst geologist, Am Zinc Co, Mo, 53-54; teaching asst geol, Univ Ill, 57-60; asst geologist, Nev Test Site, Nev & Colo, 60-63, proj chief, Groundwater Br, Water Resources Div, 63-67, supvry hydrologist, Colo Dist, 67-71, STAFF HYDROLOGIST, ROCKY MT REGION, WATER RESOURCES DIV, US GEOL SURV, 71- Mem: Soc Econ Paleont & Mineral; fel Geol Soc Am; Am Geophys Union. Res: Clay mineralogy; sedimentation; groundwater hydrology;

application of electrical analog and digital models to the study of groundwater and surfacewater systems; evaluations of water resources for planning and management. Mailing Add: US Geol Surv Bldg 25 Denver Fed Ctr Denver CO 80225

MOORE, JOHN GARBER, mathematics, see 12th edition

MOORE, JOHN GEORGE, JR, b Berkeley, Calif, Sept 17, 17; m 46; c 4. OBSTETRICS & GYNECOLOGY. Educ: Univ Calif, AB, 39, MD, 42. Prof Exp: From intern to resident obstet & gynec, Univ Calif, 42-49; from instr to asst prof, Univ Iowa, 50-51; from asst prof to prof, Univ Calif, Los Angeles, 51-65; prof & chmn dept, Col Physicians & Surgeons, Columbia Univ, 65-69; PROF OBSTET & GYNEC & CHMN DEPT, SCH MED, UNIV CALIF, LOS ANGELES, 69- Concurrent Pos: Dir, Am Bd Obstet & Gynec, 67-73, pres, 74-78. Mem: Soc Gynec Invest (pres, 66); Asn Profs Gynec & Obstet (pres, 73); Soc Gynec Oncol; Am Asn Obstet & Gynec; Am Gynec Soc. Res: Carcinoma of the uterine cervix; tissue culture growth of genital tissues. Mailing Add: Dept of Obstet & Gynec Univ of Calif Sch of Med Los Angeles CA 90024

MOORE, JOHN HAYS, JR, b Pittsburgh, Pa, Nov 6, 41; m 63; c 2. PHYSICAL CHEMISTRY, MOLECULAR PHYSICS. Educ: Carnegie Inst Technol, BS, 63; Johns Hopkins Univ, MA, 65, PhD(chem), 67. Prof Exp: Res assoc chem, Johns Hopkins Univ, 67-69; asst prof, 69-74, ASSOC PROF CHEM, UNIV MD, COLLEGE PARK, 74- Concurrent Pos: Vis fel, Joint Inst Lab Astrophys, Colo, 75-76. Mem: Am Phys Soc. Res: Molecular spectroscopy; ion-molecule collisions; collision induced vibrational and electronic excitation of molecules; atomic physics. Mailing Add: Dept of Chem Univ of Md College Park MD 20742

MOORE, JOHN MARSHALL, JR, b Winnipeg, Man, Aug 14, 35; m 58; c 3. GEOLOGY, PETROLOGY. Educ: Univ Man, BSc, 56; Mass Inst Technol, PhD(petrol), 60. Prof Exp: Queen Elizabeth II fel, 60-62, asst prof, 62-67, ASSOC PROF GEOL, CARLETON UNIV, 67- Concurrent Pos: Guest prof, Inst Mineral, Univ Geneva, 68-69. Mem: Geol Asn Can; Mineral Asn Can; Mineral Soc Am. Res: Precambrian geology; metamorphic zoning; exchange equilibria in metamorphic rocks. Mailing Add: Dept of Geol Carleton Univ Ottawa ON Can

MOORE, JOHN (NEWTON), b Columbus, Ohio, Apr 2, 20; m 41; c 2. PLANT PATHOLOGY. Educ: Denison Univ, AB, 41; Mich State Univ, MS, 43, EdD, 52. Prof Exp: Instr math, 43-44, instr biol sci, 46-53, from asst prof to assoc prof natural sci, 53-70, PROF NATURAL SCI, MICH STATE UNIV, 70- Res: Study of philosophy of science, especially teaching of organic evolution, natural selection and related topics. Mailing Add: Dept of Natural Sci Mich State Univ East Lansing MI 48823

MOORE, JOHN OLIVER, biochemistry, physiology, see 12th edition

MOORE, JOHN ROBERT, b Bridgeville, Pa, Mar 3, 34; m 55; c 2. PLANT ECOLOGY. Educ: Clarion State Col, BS, 57; Univ Pittsburgh, 62, PhD(biol), 65. Prof Exp: Pub sch teacher, Pa, 57-62; asst biol, Univ Pittsburgh, 62-65; assoc prof, 65-67, PROF BIOL, CLARION STATE COL, 67- Mem: Ecol Soc Am. Res: Primary production of vascular aquatic plants. Mailing Add: Dept of Biol Clarion State Col Clarion PA 16214

MOORE, JOHN THOMAS, b Wallacetown, Ont, Sept 21, 15; nat US. MATHEMATICS. Educ: Univ Western Ont, BA, 36; Univ Wis, MA, 38; Univ Chicago, PhD(math), 52. Prof Exp: Instr math, Milwaukee Voc Jr Col, 39-43, Univ Wis, 43-45 & Univ Western Ont, 47-49; asst prof, Ga Inst Technol, 53; assoc prof, Univ Fla, 53-65; PROF MATH, UNIV WESTERN ONT, 65- Concurrent Pos: Vis prof, Univ Western Ont, 63-64. Mem: Am Math Soc; Math Asn Am; Can Math Cong; NY Acad Sci. Res: Modern and abstract algebra. Mailing Add: Dept of Math Univ of Western Ont London ON Can

MOORE, JOHN V, b Grants, NMex, Nov 8, 41. ANIMAL BREEDING. Educ: Colo State Univ, BS, 62; NMex State Univ, MS, 65; Iowa State Univ, PhD(animal sci), 69. Prof Exp: Res geneticist, Dairy Cattle Res Br, Agr Res Serv, USDA, 69; CATTLE GENETICIST IN-CHG BEEF BREEDING PROG, FARMERS HYBRID CO, INC, 69- Mem: Am Soc Animal Sci; Am Soc Dairy Sci. Res: Analysis of dairy cattle breeding and management data; beef cattle breeding program. Mailing Add: Farmers Hybrid Co Inc Hampton IA 50441

MOORE, JOHN WARD, b Lancaster, Pa, July 17, 39; m 61. INORGANIC CHEMISTRY, PHYSICAL CHEMISTRY. Educ: Franklin & Marshall Col, AB, 61; Northwestern Univ, PhD(phys chem), 65. Prof Exp: NSF fel phys chem, Univ Copenhagen, 64-65; asst prof inorg chem, Ind Univ, Bloomington, 65-71; ASSOC PROF CHEM, EASTERN MICH UNIV, 71- Concurrent Pos: Fel, Ctr Study Contemp Issues, Eastern Mich Univ, 74-75. Mem: AAAS; Am Chem Soc. Res: Coordination chemistry; molecular orbital theory; chemical education; environmental chemistry. Mailing Add: Dept of Chem Eastern Mich Univ Ypsilanti MI 48197

MOORE, JOHN WILLIAMSON, b Houston, Tex, Jan 16, 21; m 43; c 2. PHYSICAL CHEMISTRY. Educ: Univ Tex, BS, 41, AM, 42. Prof Exp: Tutor phys chem, Univ Tex, 41-42; jr chem engr, Tenn Valley Authority, 42-43, asst chem engr, 43-44; chem engr, Nat Cotton Coun, 44-45; chem engr, Southern Alkali Corp, 45-48, supvr res labs, 48-55, dir res, 55-70, dir develop, 67-70, MGR CHEM DIV TECH CTR, PPG INDUSTS, INC, 70- Mem: Am Chem Soc. Res: Kinetics, industrial process development. Mailing Add: Chem Div Tech Ctr PPG Industs Inc Box 31 Barberton OH 44203

MOORE, JOHN WILSON, b Winston-Salem, NC, Nov 1, 20; m 44; c 3. NEUROPHYSIOLOGY. Educ: Davidson Col, BS, 41; Univ Va, MA, 42, PhD(physics), 45. Prof Exp: Researcher, RCA Labs, 45-46; asst prof physics, Med Col Va, 46-50; biophysicist, Naval Med Res Inst, 50-54; biophysicist, Nat Inst Neurol Dis & Blindness, 54-56, assoc chief lab biophys, 56-61; assoc prof physiol, 61-65, PROF PHYSIOL, DUKE UNIV, 65- Concurrent Pos: Nat Neurol Res Found fel, 61-66; trustee, Marine Biol Lab, Woods Hole, 71-79. Mem: AAAS; Am Physiol Soc; Biophys Soc; Inst Elec & Electronic Engrs. Res: Physiological computation and instrumentation. Mailing Add: Dept of Physiol & Pharmacol Duke Univ Med Ctr Durham NC 27710

MOORE, JOHNES KITTELLE, b Washington, DC, Apr 20, 31; m 59; c 3. MARINE ECOLOGY. Educ: Bowdoin Col, AB, 53; Univ RI, PhD(oceanog), 66. Prof Exp: Asst acct exec, Batton, Barton, Durstine & Osborn, Inc, 55-60; res asst oceanog, Narragansett Marine Lab, Univ RI, 61-66; from asst prof to assoc prof, 66-74, PROF BIOL, SALEM STATE COL, 74- Concurrent Pos: Consult, NSF, 67-69. Mem: Am Soc Limnol & Oceanog; Sigma Xi. Res: Estuarine ecology. Mailing Add: Dept of Biol Salem State Col Salem MA 01970

MOORE, JON THOMAS, b Kingsport, Tenn, Nov 25, 43. POLYMER PHYSICS. Educ: Ga Inst Technol, BEE, 65; Univ Tenn, MS, 66; Univ Va, PhD(physics), 70. Prof Exp: RES PHYSICIST, TENN EASTMAN CO, EASTMAN KODAK CO, 70- Mailing Add: Tenn Eastman Co PO Box 511 Kingsport TN 37662

MOORE, JOSEPH B, b Potchefstroom, SAfrica, Dec 17, 10; m 37; c 1. ECONOMIC ENTOMOLOGY. Educ: Cornell Univ, BS, 33, MS, 35, PhD(entom), 37. Prof Exp: Asst res entom, NY Exp Sta, Geneva, 35-39; entomologist, State Col Wash, 39; entomologist, 39-71, vpres & dir res & develop, 54-71, SR STAFF VPRES, McLAUGHLIN GORMLEY KING CO, 71-, DIR RES & DEVELOP, 74- Concurrent Pos: Dir, Hardwicke Chem Co. Mem: Am Chem Soc; Entom Soc Am. Res: New insecticides and their use in the field. Mailing Add: McLaughlin Gormley King Co 8810 10th Ave N Minneapolis MN 55427

MOORE, JOSEPH CURTIS, b Washington, DC, June 5, 14; m 40; c 3. MAMMALOGY. Educ: Univ Ky, BS, 39; Univ Fla, MS, 42, PhD(biol), 53. Prof Exp: Asst bot, Univ Ky, 37-39; asst biol, Univ Fla, 39-41; park biologist, Everglades Nat Park, 49-55; res fel, Dept Mammals, Am Mus Natural Hist, 55-61; cur mammals, Field Mus Natural Hist, 62-71, res assoc, 71-73; ECOL & SYSTEMATICS, FLA SOUTHERN COL, 72- Concurrent Pos: Hon lectr, Dept Anat, Univ Chicago, 65-71. Mem: AAAS; Am Soc Mammal; Zool Soc London. Res: Systematics and zoogeography of mammals, especially Cetacea and Sciuridae; ecology of vertebrates, especially Sciuridae and Sirenia in Florida; local mid-peninsular Florida ecology, especially vertebrate, terrestrial and volant, with emphasis on vegetation emphasizing relationships to the seasons, temperature, and rainfall. Mailing Add: Fla Southern Col Lakeland FL 33802

MOORE, JOSEPH GRAESSLE, b Dubuque, Iowa, Feb 19, 04; m 30; c 3. ANTHROPOLOGY. Educ: Hamline Univ, BA, 29; Seabury-West Theol Sem, BD, 31; Northwestern Univ, MA, 47, PhD(anthrop), 53. Prof Exp: Dir, Western Surv Staff, 28-30; curate, St Mark's Church, Evanston, Ill, 30-31; rector, St Paul's Church, Evansville, Ind, 31-42; dir training, prof practical theol & lectr social sci, Seabury-Western Theol Sem, 46-55; exec dir gen div res & field studies, Nat Coun, Protestant Episcopal Church, 48-62, exec officer, Strategic Adv Comt to Presiding Bishop, 62-65; planning officer, Prov of Caribbean, Episcopal Church USA, 65-67; prof anthrop & chmn dept anthrop & sociol, 68-73, dir div behav sci, 70-73, EMER PROF ANTHROP & SOCIOL, 73- Concurrent Pos: Viking Found fel, BWI, 50. Mem: Fel AAAS; fel Am Anthrop Asn; fel Am Sociol Asn; fel Soc Appl Anthrop. Res: Institutional and acculturation studies of Central America, South America, Alaska, Philippines, South East Asia and of American Indians; population; housing and institutional studies; the Negro in the New World and in Africa. Mailing Add: PO Box 1396 Frederiksted St Croix VI 00840

MOORE, JOSEPHINE CARROLL, b Ann Arbor, Mich, Sept 20, 25. NEUROANATOMY, ELECTROMYOGRAPHY. Educ: Univ Mich, BA, 47, MS, 59, PhD(anat), 64; Eastern Mich Univ, BS, 54. Prof Exp: From instr to asst prof occup ther, Eastern Mich Univ, 55-60; instr anat, Med Sch, Univ Mich, 64-66; asst prof, 66-69, ASSOC PROF ANAT, MED SCH, UNIV S DAK, 69- Mem: AAAS; Am Asn Anat; Int Soc Electromyographic Kinesiology; World Fedn Occup Ther; Am Occup Ther Asn. Res: Rehabilitation of handicapped individuals, especially dealing with neuroanatomical and neurophysiological concepts; electromyography of normal individuals performing prescribed actions under normal conditions. Mailing Add: Dept of Anat Univ of SDak Med Sch Vermillion SD 57069

MOORE, KEITH LEON, b Brantford, Ont, Oct 5, 25; m 49; c 5. ANATOMY. Educ: Univ Western Ont, 49, MSc, 51, PhD(micros anat), 54. Prof Exp: Lectr anat, Univ Western Ont, 54-56; from asst prof to assoc prof, 56-65, PROF ANAT & HEAD DEPT, UNIV MAN, 65- Concurrent Pos: Consult, Children's Hosp, Winnipeg, Man, 59- Mem: Am Asn Anat; Anat Soc Gt Brit & Ireland; Can Asn Anat (secy, 62-65, pres, 66-68); Can Fedn Biol Socs; Can Cytol Coun. Res: Embryology and teratology. Mailing Add: Dept of Anat Univ of Man Winnipeg MB Can

MOORE, KENNETH BOYD, b Pratt, Kans, Jan 31, 17; c 2. PSYCHIATRY. Educ: Univ Kans, AB, 38, MA, 40, PhD(psychol), 43, MD, 47; Am Bd Psychiat & Neurol, dipl psychiat, 54. Prof Exp: Clin dir, Ypsilanti State Hosp, Mich, 58-60, asst med supt, 60-61; assoc chief staff, 61-72, CHIEF MENT HYG CLIN, VET ADMIN HOSP, 72- Concurrent Pos: Mem, Ment Health Res Inst, Univ Mich, 59-60; asst prof clin psychiat, Sch Med, Univ Ky, 61- Mem: Fel Am Psychiat Asn; Am Psychol Asn. Res: Treatment of schizophrenia; out-patient psychiatric treatment methods. Mailing Add: Vet Admin Hosp Ment Hyg Clin Lexington KY 40507

MOORE, KENNETH EDWIN, b Edmonton, Alta, Aug 8, 33; m 53; c 3. PSYCHOPHARMACOLOGY. Educ: Univ Alta, BS, 55, MS, 57; Univ Mich, PhD(pharmacol), 60. Prof Exp: Instr pharmacol, Dartmouth Med Sch, 60-61, asst prof, 62-65; assoc prof, 66-69, PROF PHARMACOL, MICH STATE UNIV, 70- Concurrent Pos: Mem rev comt pharmacol & endocrinol, NIH, 68-70, mem pharmacol study sect, 75-79; vis scholar, Cambridge Univ, Eng, 74. Mem: Am Soc Pharmacol & Exp Therapeut; Pharmacol Soc Can; Soc Exp Biol & Med; Soc Neurosci; Am Col Neuropsychopharmacol. Res: Biochemical pharmacology, neuropharmacology and toxicology related to the role of endocrine and nervous systems in drug and environmental induced stress; central nervous system transmitters; catecholamines. Mailing Add: Dept of Pharmacol Mich State Univ East Lansing MI 48823

MOORE, KENNETH HOWARD, b Leek, Eng, June 20, 07; nat US; m 41; c 2. PHYSICS. Educ: Rensselaer Polytech Inst, BS, 30, MS, 33, PhD(physics), 45. Prof Exp: Instr physics & elec eng, 30-36, from instr to prof physics, 36-73, EMER PROF PHYSICS, RENSSELAER POLYTECH INST, 73- Mem: Am Phys Soc; Am Soc Eng Educ; Am Asn Physics Teachers. Res: X-ray and electron diffraction studies, particularly of treated surfaces and of epitaxy propagated through overlayers; development of demonstration, teaching and grading techniques. Mailing Add: RD 1 Box 37 Johnsonville NY 12094

MOORE, KENNETH VIRGIL, b Emporia, Kans, May 16, 33; m 52; c 2. NUCLEAR SCIENCE. Educ: Kans State Teachers Col, BA, 54; Stanford Univ, MS, 64. Prof Exp: Jr geophysicist, Gulf Oil Corp, 56; nuclear analyst, Phillips Petrol Co, Atomic Energy Div, 57-63, group leader nuclear res, 64-69; group leader thermal hydraulics, Idaho Nuclear Corp, 69-71, Aerojet Nuclear Corp, 71-74; STAFF CONSULT REACTOR SAFETY, ENERGY INC, 74- Mem: Am Soc Mech Engrs. Res: Major director and author of thermal hydraulic (transient, two-phase water) computer codes used by the United States government and the nuclear reactor industry for safety analysis and experimental predictions. Mailing Add: Energy Inc PO Box 736 Idaho Falls ID 83401

MOORE, LARRY WALLACE, b Menan, Idaho, Aug 24, 37; m 61; c 4. PLANT PATHOLOGY. Educ: Univ Idaho, BS, 62, MS, 64; Univ Calif, Berkeley, PhD(plant path), 70. Prof Exp: ASST PROF PLANT PATH, ORE STATE UNIV, 69- Mem: Am Phytopath Soc; Am Inst Biol Sci. Res: Biology of phytopathogenic bacteria and bacterial diseases of nursery and ornamental plants. Mailing Add: Dept of Bot & Plant Path Ore State Univ Corvallis OR 97331

MOORE, LAURENCE DALE, b Danville, Ill, July 12, 37; m 58; c 4. PLANT PATHOLOGY. Educ: Univ Ill, BS, 59; Pa State Univ, MS, 61, PhD(plant path), 65. Prof Exp: Asst, Pa State Univ, 59-65; asst prof, 65-70, ASSOC PROF PLANT PATH, VA POLYTECH INST & STATE UNIV, 70- Mem: Am Phytopath Soc; Am Soc Plant Path; Int Soc Plant Path. Res: Physiology of disease; role of enzymes and plant nutrition in disease development; air pollution effects on the physiology and biochemistry of plants. Mailing Add: Dept of Plant Path & Physiol Va Polytech Inst & State Univ Blacksburg VA 24061

MOORE, LAWRENCE EDWARD, b Conway, SC, May 28, 38; m 64; c 2. INORGANIC CHEMISTRY. Educ: Davidson Col, BS, 60; Univ Tenn, PhD(chem), 64. Prof Exp: Asst prof chem, Wofford Col, 66-70; asst prof, 70-73, ASSOC PROF CHEM, UNIV SC, SPARTANBURG, 73- Mem: Am Chem Soc. Res: Coordination compounds of nickel in which the perchlorate ion is a ligand; aromatic substituent effects in coordination compounds of nickel with substituted pyridines. Mailing Add: Dept of Chem Univ of SC Spartanburg SC 29303

MOORE, LEE E, b Chelsea, Okla, Mar 6, 38; m 64; c 2. NEUROPHYSIOLOGY, BIOPHYSICS. Educ: Univ Okla, BS, 60; Duke Univ, PhD(physiol), 66. Prof Exp: Asst prof physiol, Ohio Univ, 66-68; asst prof, 68-72, ASSOC PROF PHYSIOL, MED SCH, CASE WESTERN RESERVE UNIV, 72- Concurrent Pos: Nat Inst Neurol Dis & Stroke res grant, 67-75. Mem: Neurosci Soc; Biophys Soc; Soc Gen Physiol; Am Physiol Soc. Res: Cellular neurophysiology; biophysics of membrane transport; effect of low temperature on the ionic conductance of myelinated nerve; ion permeability of skeletal muscle. Mailing Add: Dept of Physiol Case Western Reserve Univ Cleveland OH 44106

MOORE, LEON, b Waldron, Ark, Oct 2, 31; m 51; c 4. ENTOMOLOGY. Educ: Univ Ark, BSA, 57, MS, 59; Kans State Univ, PhD, 72. Prof Exp: Surv entomologist, 59-62, EXTEN ENTOMOLOGIST, AGR EXP STA, UNIV ARIZ, 62-, ASSOC PROF ENTOM & ASSOC ENTOMOLOGIST, 74- Mem: Entom Soc Am. Res: Extension entomology with emphasis on educational programs on cotton and vegetable insects, cereal and forage crop insects and pesticides. Mailing Add: Dept of Entom Univ of Ariz Tucson AZ 85721

MOORE, LEONARD ORO, b Payson, Utah, Sept 20, 31; m 54; c 2. RESEARCH ADMINISTRATION, ORGANIC CHEMISTRY. Educ: Brigham Young Univ, BS, 53; Iowa State Univ, PhD(chem), 57. Prof Exp: Europ Res Assocs fel, Swiss Fed Inst Technol, 57-58; res chemist, Union Carbide Chem Co, 58-64 & Olefins Div, Union Carbide Corp, 64-71; mgr res, 71-76, MGR RES & DEVELOP, ANSUL CO, 76- Concurrent Pos: Chmn adv comt, Kanawha Valley Grad Ctr. Mem: Am Chem Soc; fel Am Inst Chemists; Sigma Xi. Res: Halogen chemicals; fluorocarbons; pesticides; free radicals; kinetics; environmental chemistry; organometallics. Mailing Add: Ansul Res Ctr PO Drawer 1165 Weslaco TX 78596

MOORE, LEONARD PATRICK, organic chemistry, see 12th edition

MOORE, LESLIE DAVID, b Mt Hope, WVa, Nov 9, 31; m 56; c 4. ORGANIC CHEMISTRY. Educ: WVa Univ, BS, 53; Univ Akron, MS, 56; Purdue Univ, PhD(org chem), 59. Prof Exp: Proj engr, Mat Div, Eng Res Labs, US Dept Army, Va, 55; res chemist, Gulf Res & Develop Co, 58-59, group leader, 59-62; group leader, Nalco Chem Co, 62-63, sect head, 64-67, admin tech coordr, 67-68, dir res, 68-70, corp dir technol, 70-72, vpres petrol div, 72-73; VPRES & DIR, CHEMETRON CORP, 73- Mem: Am Chem Soc. Res: Fluorocarbon, polymer, petroleum, paper and surface chemistry; ion exchange; industrial microbiocides; pigments; plastic additives. Mailing Add: Chemetron Corp 111 E Wacker Dr Chicago IL 60601

MOORE, LOUIS DOYLE, JR, b Royston, Ga, Sept 16, 26; m 48; c 2. POLYMER CHEMISTRY. Educ: Duke Univ, BS, 47; Mass Inst Technol, PhD(phys chem), 51. Prof Exp: Res chemist, 51-57, sr res chemist, 57-67, res assoc, 67-71, SR RES ASSOC, TENN EASTMAN CO, 71- Mem: Am Chem Soc. Res: Molecular characterization of polymers; polymer rheology; polymer morphology. Mailing Add: 4411 Leedy Rd Kingsport TN 37664

MOORE, MARIAN ALEASE, b Vincennes, Ind, May 18, 07. MATHEMATICAL ANALYSIS. Educ: Greenville Col, AB, 29; Univ Ill, AM, 35; Purdue Univ, PhD(math), 53. Prof Exp: Teacher high sch, Ill, 29-46, chmn dept math, 53-55; from asst prof to assoc prof, Southern Ill Univ, 55-62; teacher high sch, Ill, 62-67; assoc prof, Houston Baptist Col, 67-73; RETIRED. Concurrent Pos: Instr, Purdue Univ, 46-53; NSF vis prof, Hiram Col, 58-60. Mem: Am Math Soc. Res: Theory of integration as evolved from set theory. Mailing Add: 1017 10th St Lawrenceville IL 62439

MOORE, MARION E, b Boise City, Okla, May 22, 34; m 53; c 1. MATHEMATICS. Educ: W Tex State Univ, BS, 57; Tex Tech Col, MS, 60; Univ NMex, PhD(math), 68. Prof Exp: Instr math, W Tex State Univ, 58-61 & Univ NMex, 61-66; asst prof, 66-70, ASSOC PROF MATH, UNIV TEX, ARLINGTON, 70- Mem: Math Asn Am; Am Math Soc. Res: Ring theory. Mailing Add: Dept of Math Univ of Tex Arlington TX 76010

MOORE, MARVIN G, b Harrison, Ark, Dec 8, 08; m 35; c 1. MATHEMATICS. Educ: Univ Ark, AB, 31; Univ Ill, MA, 32, PhD(math), 38. Prof Exp: Instr math, Ala Polytech Inst, 36-37 & Ind Univ, 37-40; prof, Tri-State Col, 40-43; from asst prof to prof & chmn dept, 43-74, EMER PROF MATH, BRADLEY UNIV, 74- Mem: Am Math Soc; Math Asn Am; Soc Indust & Appl Math. Res: Generalizations of Fourier series in complex plane; relativity; expansions in series of exponential functions; residual stresses. Mailing Add: 708 Henryetta Springdale AR 72764

MOORE, MARY ELIZABETH, b New London, Conn, Dec 7, 30; m 74. RHEUMATOLOGY. Educ: Douglass Col, AB, 52; Rutgers Univ, MS, 58, PhD(psychol), 60; Temple Univ, MD, 67. Prof Exp: Res assoc sociol, Rutgers Univ, 52-60; assoc psychol & psychiat, Univ Pa, 60-63; asst prof, 72-75, ASSOC PROF MED, SCH MED, TEMPLE UNIV, 76- Concurrent Pos: Attend staff, Temple Univ Hosp & Episcopal Hosp, 70-, St Mary's Hosp, 76-; Fel, Philadelphia Found, 72-; mem, Pain Conf, Temple Univ Hosp, 73- Mem: Fel Am Col Physicians; Am Rheumatologic Asn. Res: Methods of pain relief. Mailing Add: Temple Univ Hosp 3401 N Broad St Philadelphia PA 19140

MOORE, MARYALICE CONLEY, b Stoneham, Mass, June 1, 18; m 42; c 2. ORGANIC CHEMISTRY. Educ: Simmons Col, BS, 39; Mass Inst Technol, PhD(org chem), 43. Prof Exp: Chemist, E I du Pont de Nemours & Co, 43-47; res assoc civil eng, Mass Inst Technol, 47-48; from lectr to assoc prof, 56-66, PROF ORG CHEM, STONEHILL COL, 66- Concurrent Pos: NSF col teachers res participation grant, 64-65. Mem: Am Chem Soc. Res: Organic synthesis of compounds of biochemical interest. Mailing Add: Dept of Chem Stonehill Col North Easton MA 02356

MOORE, MATTHEW THIBAUD, b Philadelphia, Pa, Sept 12, 01; m 27. NEUROPSYCHIATRY. Educ: Temple Univ, MD, 27; Am Bd Psychiat & Neurol, dipl, 39. Prof Exp: Intern, Univ Hosp, Temple Univ, 27-28, clin asst neurol, Med Sch, 28-29, asst, Neurol Dispensary & in-chg convulsive state clin, Hosp, 29-35; from instr to prof & neuropathologist, 36-70, EMER PROF NEUROPATH, GRAD SCH MED, UNIV PA, 70- Concurrent Pos: Clin asst, Jewish Hosp, 28-29, assoc, 33-54; encephalogr, Norristown State Hosp, 29-33; instr & demonstr, Med Sch, Temple Univ, 33-35; attend chief psychiatrist, Philadelphia Psychiat Ctr, 37-67, emer chief psychiatrist, 67-; neuropsychiatrist-in-chief, Doctors Hosp, Philadelphia, 40-70; med examr, US Army Induction Bd, Philadelphia, 42-44; vis neuropsychiatrist, Del State Hosp, Farnhurst, 52-70; prof, Hahnemann Med Col, 53-58; sr attend neurologist, Albert Einstein Med Ctr, 54-67, emer sr attend neurologist, 67-; consult neuropsychiatrist, Home Jewish Aged, 58-70; archivist, Int Soc & Congs Neuropath, 70- Mem: AAAS; fel Royal Soc Med; fel Am Psychiat Asn; Soc Biol Psychiat (pres, 67-); fel Am Acad Neurol. Res: Cerebral vascular disease; neurological and psychiatric problems of the aged; schizophrenia; degenerative disorders. Mailing Add: 1813 Delancey Pl Philadelphia PA 19103

MOORE, MAURICE LEE, b Laurel Hill, Fla, Sept 11, 09; m 33; c 3. MEDICINAL CHEMISTRY. Educ: Univ Fla, BS, 30, MS, 31; Northwestern Univ, PhD(org chem), 34. Prof Exp: Asst chem, Univ Fla, 27-31; asst instr gen chem, Dental Sch, Northwestern Univ, 31-34; Smith fel org chem, Yale Univ, 34-36; res chemist, Sharp & Dohme, Inc, Pa, 36-43; dir org res, Frederick Stearns & Co, Mich, 43-45, asst dir res, 45-47; dir, Smith, Kline & French Labs, Pa, 47-51; vpres, Vick Chem Co, 51-59; dir new prod develop, 59-65, exec vpres, Winthrop Labs Div, 59-70, CORP SCI OFFICER, STERLING DRUG INC, 70- Mem: AAAS; Am Chem Soc; Soc Chem Indust; Am Pharmaceut Asn; fel Am Inst Chem. Res: Research and development of new therapeutic agents. Mailing Add: Sterling Drug Inc 90 Park Ave New York NY 10016

MOORE, MELVYN WILLIAM, organic chemistry, chemical engineering, see 12th edition

MOORE, MICHAEL CABOT, b Lincoln, Ill, Sept 23, 40; m 67; c 2. INORGANIC CHEMISTRY. Educ: Knox Col, BA, 62; Univ Colo, Boulder, MS, 65, PhD(chem), 67. Prof Exp: US Army, 65-, res chemist, Harry Diamond Labs, 69-71, instr & assoc prof gen chem, US Mil Acad, 71-74. Res: Reaction kinetics. Mailing Add: HHC 197th Ordnance Battalion APO New York NY 09144

MOORE, MICHAEL STANLEY, b Grass Creek, Wyo, Oct 24, 30; m 54; c 4. NUCLEAR PHYSICS. Educ: Rice Univ, BA, 52, MA, 53, PhD(physics), 56. Prof Exp: Physicist, Atomic Energy Div, Phillips Petrol Co, 56-66 & Idaho Nuclear Corp, 66-68; STAFF MEM, LOS ALAMOS SCI LAB, 68- Mem: Fel Am Phys Soc; fel Am Nuclear Soc. Res: Neutron cross sections by time of flight; fission physics. Mailing Add: 104 Tesuque Los Alamos NM 87544

MOORE, MORRIS, b Natick, Mass, Feb 14, 08; m 40; c 2. MYCOLOGY, DERMATOLOGY. Educ: Boston Univ, BS, 28; Harvard Univ, MA, 29; Washington Univ, PhD(mycol), 33. Prof Exp: Res assoc mycol, Washington Univ, 33-34; resident, Barnard Free Skin & Cancer Hosp, 34-35; Guggenheim fel, SAm, 35-36; MYCOLOGIST, BARNARD FREE SKIN & CANCER HOSP, 36-; CHIEF BACTERIOLOGIST, HOMER G PHILLIPS HOSP, 59- Concurrent Pos: Nat Tuberc Asn grant, 41; clin asst prof mycol & dermat, Washington Univ, 36-; mem staff, Barnes Hosp, 36-; collabr, Mycopathologia et Mycologia Applicata, 38. Honors & Awards: Bronze Medal, AMA, 39, Mem: AAAS; AMA; Mycol Soc Am; Soc Invest Dermat; Am Soc Trop Med & Hyg. Res: Mycotic infections; clinical and laboratory diagnosis; antibiotics; histopathology of fungus diseases; dermatological investigations. Mailing Add: Microbiol Lab Homer G Phillips Hosp St Louis MO 63113

MOORE, MORTIMER NORMAN, b Los Angeles, Calif, Mar 16, 27; m 58; c 2. MATHEMATICAL PHYSICS. Educ: Univ Calif, Los Angeles, BA, 50; Univ London, PhD(math physics), 55. Prof Exp: Sr res scientist, Atomics Int, NAm Aviation, 55-57; lectr theoret physics, Birkbeck Col, London, 57-61; assoc prof physics, 61-66, PROF PHYSICS, CALIF STATE UNIV, NORTHRIDGE, 66-, PROF ASTRON, 74- Concurrent Pos: Consult, Inst Nuclear Res, Stuttgart, Ger, 58, Atomics Int, 59-64 & Fr AEC fast reactor prog, 64-65; vis prof, Univ Ariz, 69. Mem: Am Phys Soc; Am Nuclear Soc; Brit Inst Physics. Res: Stochastic processes; transport processes; partial coherence theory; low energy neutron physics; non-equilibrium processes; laser physics. Mailing Add: Dept Physics Calif State Univ 18111 Nordoff St Northridge CA 91324

MOORE, NADINE HANSON, b Idaho Falls, Idaho, Feb 10, 41; m 63. MATHEMATICS. Educ: Idaho State Univ, BS, 63; Syracuse Univ, MA, 65, PhD(math), 69. Prof Exp: Instr math, State Univ NY Col Oswego, 68-69; ASST PROF MATH, ARIZ STATE UNIV, 69- Mem: Am Math Soc. Res: Homological algebra; structure of projective modules. Mailing Add: 2322 E Geneva Dr Tempe AZ 85253

MOORE, NELSON JAY, b Greenville, Ohio, Aug 29, 41; m 64; c 2. ETHOLOGY, ORNITHOLOGY. Educ: Manchester Col, BA, 63; Ohio State Univ, MS, 68; Univ Ariz, PhD(zool), 72. Prof Exp: ASSOC PROF BIOL, OHIO NORTHERN UNIV, 72- Mem: Am Ornithologists Union; Cooper Ornith Soc. Res: Description and analysis of the behaviors of the Yellow-eyed Junco; investigation of avian fossils of southeastern Arizona; wintering bird ecology of northwestern Ohio. Mailing Add: Dept of Biol Ohio Northern Univ Ada OH 45810

MOORE, PATRICIA ANN, b Dallas, Tex, June 18, 37. BIOCHEMISTRY, PHYSIOLOGY. Educ: Rice Univ, BA, 60; Univ NC, Chapel Hill, MS, 63, PhD(biochem), 67. Prof Exp: Res asst biochem, Univ NC, Chapel Hill, 61-64; asst ed biochem, 66-69, ASSOC ED BIOCHEM, CHEM ABSTR SERV, 69- Mem: Sigma Xi; Am Soc Info Sci; Am Chem Soc. Res: Indexing biochemical literature with emphasis on nucleic acid and protein structure and metabolism; microbiological genetic systems; research in mammalian nucleic acid and protein metabolism and regulatory systems. Mailing Add: Chem Abstr Serv 2540 Olentangy River Rd Columbus OH 43202

MOORE, PAUL BRIAN, b Stamford, Conn, Nov 24, 40. MINERALOGY, CRYSTALLOGRAPHY. Educ: Mich Technol Univ, BS, 62; Univ Chicago, SM, 63, PhD(geophys), 65. Prof Exp: NSF fel, Swed Natural Hist Mus, Stockholm, 65-66; from instr to assoc prof, 66-71, PROF MINERAL & CRYSTALLOG, UNIV CHICAGO, 72- Honors & Awards: Dreyfus Found Award, 71. Mem: Am Crystallog Asn; Mineral Soc Am; Mineral Asn Can; Am Geophys Union. Res: Descriptive, paragenetic and taxonomic mineralogy; crystal structure analysis of silicates, phosphates and arsenates; crystallochemical classifications; dense-packed structures; polyhedra theory. Mailing Add: Dept of Geophys Sci Univ of Chicago Chicago IL 60637

MOORE, PAUL HARRIS, plant physiology, see 12th edition

MOORE, PERRY ALLDREDGE, b Kent, NY, Jan 10, 03; m 27; c 2. ORGANIC CHEMISTRY, PHYSICAL CHEMISTRY. Educ: Des Moines Univ, BS, 25; Iowa State Col, MS, 26, PhD(chem), 36. Prof Exp: Dean & prof, Webster City Jr Col, 26-

29; grad instr chem, Iowa State Col, 29-36, instr, 36-39; from assoc prof to prof & head dept, 39-73, EMER PROF CHEM, HAMLINE UNIV, 73- Mem: Am Chem Soc. Res: Electron-sharing of organic radicals. Mailing Add: Dept of Chem Hamline Univ St Paul MN 55101

MOORE, PETER BARTLETT, b Boston, Mass, Oct 15, 39; m 66; c 2. BIOCHEMISTRY. Educ: Yale Univ, BS, 61; Harvard Univ, PhD(biophys), 66. Prof Exp: NSF fel, Univ Geneva, 66-67; US Air Force Off Sci Res fel, Med Res Coun Lab, Cambridge, Eng, 67-69; asst prof, 69-73, ASSOC PROF MOLECULAR BIOPHYS & BIOCHEM, YALE UNIV, 73-, NIH & NSF RES GRANTS, 69- Mem: AAAS; Am Soc Biol Chem; Biophys Soc. Res: Structure and function of ribosomes; application of neutron scattering to study of quaternary structure; macromolecular structure. Mailing Add: Dept Molecular Biophys & Biochem Yale Univ New Haven CT 06520

MOORE, PETER FRANCIS, b New York, NY, July 24, 36; m 60; c 5. PHARMACOLOGY. Educ: Fordham Univ, BS, 56; Purdue Univ, MS, 59, PhD(pharmacol), 61. Prof Exp: Mem staff, 61-73, SR RES INVESTR, PHARMACOL DEPT, PFIZER, INC, 73- Mem: AAAS; Am Heart Asn. Res: Pulmonary asthma, allergy, immediate hypersensitivity, bronchodilators, mucolytics; cyclic mononucleotides; hypolipemics; diuretics. Mailing Add: Pharmacol Dept Pfizer Inc Groton CT 06340

MOORE, PETER WARREN, chemistry, physics, see 12th edition

MOORE, RALPH BISHOP, b Carbonear, Nfld, Mar 23, 41; m 65; c 2. INDUSTRIAL CHEMISTRY, ANALYTICAL CHEMISTRY. Educ: Mem Univ Nfld, BSc, 62; Univ Alta, PhD(phys org chem), 67. Prof Exp: Res chemist, 67-69, process chemist, 69-74, SR PROCESS CHEMIST, DEPT ORG CHEM, E I DU PONT DE NEMOURS & CO INC, 74- Res: Organolead chemistry; photochemistry; reaction kinetics, sodium borohydride chemistry; trace metal analysis. Mailing Add: 8 Candlewick Ct New Castle DE 19720

MOORE, RALPH GOWER DAVIES, b Charlottetown, PEI, Mar 30, 11; nat US; m; c 4. ORGANIC CHEMISTRY. Educ: Univ BC, BA, 32, MA, 34; McGill Univ, PhD(org chem), 36. Prof Exp: Lab asst chem, Univ BC, 33-34; fel, Rainier Pulp & Paper Co, McGill Univ, 36-37; res chemist, Arthur D Little, Inc, Mass, 38-45, J T Baker Chem Co, 45-51 & Gen Aniline & Film Corp, 51-62; sr tech assoc, GAF Corp, NY, 62-76; RETIRED. Mem: AAAS; Am Chem Soc. Res: Structure of lignin; synthesis of rubber chemicals; pharmaceuticals and intermediates; synthesis of insecticides, herbicides and rodenticides; chemicals for photoreproductive processes, especially diazotypy. Mailing Add: PO Box 188 Chenango Forks NY 13746

MOORE, RAMON EDGAR, b Sacramento, Calif, Dec 27, 29. MATHEMATICS, COMPUTER SCIENCE. Educ: Univ Calif, AB, 50; Stanford Univ, PhD(math), 63. Prof Exp: PROF COMPUT SCI, UNIV WIS-MADISON, 68- Concurrent Pos: Alexander von Humboldt Found sr scientist award, 75. Mem: Soc Neurosci. Res: Interval analysis; neural modeling; celestial mechanics; computing methods. Mailing Add: Dept of Comput Sci Univ of Wis Madison WI 53706

MOORE, RAYMOND A, b Britton, SDak, Nov 16, 27; m 51; c 3. AGRONOMY, PLANT PHYSIOLOGY. Educ: SDak State Univ, BS, 51, MS, 56; Purdue Univ, PhD(agron), 63. Prof Exp: High sch voc agr instr, SDak, 51-56; instr agron, 56-58, asst agronomist field crops, 58-62, assoc prof, 63-67, head dept plant sci, 69-73, PROF FIELD CROPS, SDAK STATE UNIV, 67-, DIR SDAK EXP STA, 73- Mem: Am Soc Agron; Soc Range Mgt. Res: Administration; university teaching; pasture crops. Mailing Add: 207 17th Ave Brookings SD 57006

MOORE, RAYMOND CECIL, geology, paleontology, deceased

MOORE, RAYMOND F, JR, b Fishersville, Va, Dec 17, 27; m 51; c 3. ENTOMOLOGY. Educ: Bridgewater Col, BS, 51; Univ Richmond, MA, 56; Rutgers Univ, PhD(entom), 59. Prof Exp: Entomologist, USDA, 59-65; asst prof biol, Univ SC, 65-66; ENTOMOLOGIST, USDA, 66- Mem: AAAS; Entom Soc Am; Am Chem Soc. Res: Insect physiology and nutrition. Mailing Add: PO Box 271 Florence SC 29501

MOORE, RAYMOND JOHN, b Hamilton, Ont, Oct 26, 18. CYTOTAXONOMY. Educ: McMaster Univ, BA, 41; Univ Va, MA, 43, PhD(bot), 46. Prof Exp: RES SCIENTIST, CAN DEPT AGR, 43- Mem: Genetics Soc Can; Can Bot Asn; Int Asn Plant Taxon. Res: Cytotaxonomy of Buddleia, Medicago, Cardueae-Compositae. Mailing Add: Biosyst Res Inst Can Dept of Agr Cent Exp Farm Ottawa ON Can

MOORE, RAYMOND KENWORTHY, b Meriden, Conn, Jan 9, 42; m 61; c 4. GEOCHEMISTRY, MINERALOGY. Educ: Univ Fla, BS, 63, MS, 65; Pa State Univ, PhD(geochem), 69. Prof Exp: Res assoc, Mat Res Labs, Pa State Univ, 69-70; ASSOC PROF GEOL, RADFORD COL, 70- Mem: Mineral Asn Can. Res: Petrology; spectroscopic studies of minerals; geomorphological processes; spectroscopic properties of naturally occurring solids; landform development. Mailing Add: 341 Riggs St Dublin VA 24084

MOORE, REGINALD GEORGE, b Brantford, Ont, Mar 16, 32; m 57; c 3. INVERTEBRATE PALEONTOLOGY. Educ: Univ Western Ont, BSc, 54; Univ Mich, MS, 55, PhD(geol), 60. Prof Exp: Instr geol, Oberlin Col, 57-58; assoc prof, 60-71, PROF GEOL, ACADIA UNIV, 71- Res: Non-marine invertebrate paleontology and paleoecology; malacology. Mailing Add: Dept of Geol Acadia Univ Wolfville NS Can

MOORE, RICHARD, b Los Angeles, Calif, Jan 19, 27; m 57, 69; c 2. MEDICAL PHYSICS, BIOMEDICAL ENGINEERING. Educ: Univ Mo, BS, 49; Univ Rochester, PhD(biophys), 56; George Washington Univ, DSc(bioeng), 70; Am Bd Health Physics, cert; Am Bd Radiol, cert diag radiol physics, 75. Prof Exp: Asst biophys, Univ Rochester, 52-53, res assoc, 53-56; biophysicist, Metab Dis Br, Nat Inst Arthritis & Metab Dis, 57-60; res biophysicist, Blood Prog Res Lab, Am Nat Red Cross, 60-69; ASSOC PROF RADIIOL, SCH MED, UNIV MINN, MINNEAPOLIS, 69- Concurrent Pos: Sr asst engr, Div Sanit Eng Serv, USPHS, 55-57, lectr, R A Taft Sanit Eng Ctr, 56-57; vis prof, Howard Univ, 58-; consult, Comt Data Logging Systs, NIH, 59-60 & Dept Biophys, Div Nuclear Med, Walter Reed Army Inst Res, 61-65; from vis asst prof to vis prof, Sch Med, George Washington Univ, 64-69; assoc ed, Pattern Recognition, 68- & Comput in Biol & Med, 69-; contrib ed, Med Electronics & Data, 70-; consult ed, Measurements & Data, 71-; mem coun cardiovasc radiol, Am Heart Asn. Mem: Fel AAAS; fel Soc Advan Med Systs; Am Col Radiol; Am Asn Physicists in Med; Radiol Soc NAm. Res: Radioisotopic tracing; compartmental analysis; radiological physics; ultracentrifugation; radiation biology; diagnostic radiologic sciences; membrane permeability; mathematical modeling; computer simulation. Mailing Add: Dept of Radiol Univ of Minn Sch of Med Minneapolis MN 55455

MOORE, RICHARD ALLAN, b Mansfield, Ohio, Jan 11, 24; m 49; c 3. MATHEMATICS. Educ: Washington Univ, AB, 48, AM, 50, PhD, 53. Prof Exp: Instr math, Univ Nebr, 53-54 & Yale Univ, 54-56; from asst prof to assoc prof, 56-67, assoc head dept, 65-71, chmn dept, 71-75, PROF MATH, CARNEGIE-MELLON UNIV, 67-, ASSOC HEAD DEPT, 75- Mem: Am Math Soc; Math Asn Am. Res: Ordinary differential equations, especially second order equations. Mailing Add: Dept of Math Carnegie-Mellon Univ Pittsburgh PA 15213

MOORE, RICHARD ANTHONY, b Pittsburgh, Pa, Oct 29, 28; m 55; c 8. ORGANIC CHEMISTRY. Educ: Duquesne Univ, BS, 51; Purdue Univ, MS, 56; Univ Pittsburgh, PhD(org chem), 62. Prof Exp: Proj officer, Wright Air Develop Ctr, Ohio, 51-53; teaching asst chem, Purdue Univ, 53-56; res chemist, Develop Dept, Koppers Co, Inc, Pa, 56-57; teaching asst chem, Univ Pittsburgh, 57-58; res chemist, Wyandotte Chem Corp, 61-72, SR RES CHEMIST, BASF WYANDOTTE CORP, 72- Mem: Am Chem Soc; fel Am Inst Chem; Sigma Xi. Res: Heterocyclic and organic fluorine chemistry; polyethers research. Mailing Add: Res Dept BASF Wyandotte Corp Wyandotte MI 48192

MOORE, RICHARD BYRON, b Kansas City, Mo, July 17, 33; m 53; c 4. MICROBIOLOGY, MARINE ECOLOGY. Educ: Tex Christian Univ, BA, 58; Col William & Mary, MA, 61; Univ Tex, Austin, PhD(microbiol), 66. Prof Exp: Res asst marine biol, Tex Fish & Game Comn, 57 & Va Inst Marine Sci, 58-60; res asst microbiol, Univ Tex, Austin, 60-62; asst res scientist, Life Sci Div, Syracuse Univ Res Corp, 66-67, assoc res scientist & mgr microbiol res, 68-70; DIR LAKE ONT ENVIRON LAB, STATE UNIV NY COL OSWEGO, 70- Mem: AAAS; Am Soc Microbiol; Phycol Soc Am; NY Acad Sci. Res: Environmental problems; oceanography, physiology and biochemistry of microorganisms; interaction of microorganisms and their environment; pesticides; residues; environmental impact to toxic chemicals in various ecosystems. Mailing Add: Lake Ont Environ Lab State Univ of NY Col Oswego NY 13126

MOORE, RICHARD DANA, b Battle Creek, Mich, Feb 11, 26. ANATOMY. Educ: Olivet Col, BS, 48; Mich State Univ, MS, 52, PhD(anat), 56. Prof Exp: Asst biol, Olivet Col, 48-49; asst anat, Mich State Univ, 51-55; from asst prof to prof & actg head dept biol, Hardin-Simmons Univ, 55-66; assoc prof, Albright Col, 66-67; assoc prof, 67-70, chmn dept, 68-70, PROF BIOL, McMURRY COL, 70- Concurrent Pos: Am Physiol Asn fels, N Tex State Col, 59- Mem: NY Acad Sci. Res: Histology of urinary system of domestic animals; human histology. Mailing Add: Dept of Biol McMurry Col Abilene TX 79605

MOORE, RICHARD DAVIS, b Salina, Kans, Mar 17, 32. BIOPHYSICS. Educ: Purdue Univ, PhD(oxytocin action), 63; Ind Univ, MD, 57. Prof Exp: NIH fel biophys, Sch Med, Univ Md, 63-64, res asst prof, 64; assoc prof, 64-67, PROF BIOPHYS, STATE UNIV NY COL PLATTSBURGH, 67- Mem: AAAS; Biophys Soc. Res: Cellular control mechanisms, especially ionic regulation; role of ions in hormone action and gene activation; electrical interactions between hormones and receptor sites; active transport. Mailing Add: Div of Sci & Math State Univ of NY Col Plattsburgh NY 12901

MOORE, RICHARD DONALD, b Spokane, Wash, Mar 7, 24; m 46; c 4. PATHOLOGY. Educ: Western Reserve Univ, MD, 47; Am Bd Path, dipl, 55. Prof Exp: Resident path, Univ Hosps, Western Reserve Univ, 49-52, from instr to sr instr, Sch Med, 52-55, asst prof, 55-56; asst prof, Sch Med, Univ Rochester, 56-57; assoc prof, Sch Med, Western Reserve Univ, 57-67, prof, 67-69; PROF PATH & CHMN DEPT, MED SCH, UNIV ORE, 69- Mem: Am Soc Exp Path; Reticuloendothelial Soc; Am Asn Path & Bact. Res: Structure and function of connective tissue and the reticuloendothelial system. Mailing Add: Dept of Path Univ of Ore Med Sch Portland OR 97201

MOORE, RICHARD E, b San Francisco, Calif, July 30, 33; m 60; c 4. ORGANIC CHEMISTRY. Educ: Univ San Francisco, BS, 57, MS, 59; Univ Calif, Berkeley, PhD(org chem), 62. Prof Exp: Res assoc, 63-66, asst prof, 66-70, ASSOC PROF ORG CHEM, UNIV HAWAII, HONOLULU, 70- Mem: Am Chem Soc. Res: Structure determination, chemistry and synthesis of natural products from marine animals and plants. Mailing Add: Dept of Chem Univ of Hawaii Honolulu HI 96822

MOORE, RICHARD GERALD, JR, physics, see 12th edition

MOORE, RICHARD LEE, b Philadelphia, Pa, June 20, 18; m 42; c 3. PLASMA PHYSICS. Educ: Univ Calif, Los Angeles, AB, 41, MA, 42; Ohio State Univ, PhD(physics), 53. Prof Exp: Meteorologist & res physicist, US Army & US Air Force, 41-51; res asst physics, Los Alamos Sci Lab, 51-53, mem staff, 54, staff asst, Weapon Systs Anal Dept, Northrop Aircraft, Inc, Calif, 54-56; sr assoc, Planning Res Corp, Calif, 56-57; consult physicist, R L Moore Consults, 57-62; head appl & theoret physics, Douglas Aircraft Co, Inc, Calif, 62-66, mgr phys sci & math res, 66-70; phys scientist, US Army Weapons Command, 71-73, PHYS SCIENTIST, US ARMY ARMAMENT COMMAND, 73- Mem: Am Phys Soc; assoc fel Am Inst Aeronaut & Astronaut. Res: Operations analysis; plasma physics; variational methods in continuum theory; data reduction principles. Mailing Add: US Army Armament Command Rock Island Arsenal Rock Island IL 61201

MOORE, RICHARD NEWTON, b Bernice, La, Mar 4, 26; m 60; c 3. ORGANIC CHEMISTRY. Educ: Miss Col, BS, 49. Prof Exp: Chemist, Southern Regional Res Lab, USDA, 49-57; res chemist, 57-59, sr res chemist, 59-67, res specialist, 67-68, res group leader, 68-70, SR RES SPECIALIST, RES DEPT, HYDROCARBONS & POLYMERS DIV, MONSANTO CO, 70- Mem: Am Chem Soc; Sigma Xi. Res: Chemistry of hydrocarbons, free radical reactions, and especially pyrolytic processes. Mailing Add: Monsanto Co 800 N Lindbergh Blvd St Louis MO 63166

MOORE, RICHARD OWEN, b Zanesville, Ohio, Apr 30, 20; m 49. BIOCHEMISTRY. Educ: DePauw Univ, AB, 42; Cornell Univ, PhD(biochem), 51. Prof Exp: Res assoc biochem, Sch Med, Ind Univ, 46-48; PROF BIOCHEM, OHIO STATE UNIV, 51-, ASSOC DEAN BIOL SCI, 71- Concurrent Pos: Vis prof, Harvard Med Sch, 59-60. Mem: AAAS; Am Chem Soc. Res: Lipogenesis; mammary gland metabolism; hormone action; isoenzymes. Mailing Add: Col of Biol Sci Ohio State Univ Columbus OH 43210

MOORE, RICHARD WAYNE, b Lamar, Mo, Aug 15, 26; m 49; c 3. VETERINARY MICROBIOLOGY. Educ: Tex A&M Univ, DVM, 55, MS, 56. Prof Exp: Res vet, DeKalb Agr Asn Inc, 56-58; from asst prof to assoc prof, 58-67, PROF VET MICROBIOL, AGR EXP STA, TEX A&M UNIV, 67- Concurrent Pos: McLaughlin fel, Med Br, Univ Tex, 59-60. Mem: Am Vet Med Asn; Am Soc Microbiol; Am Asn Avian Path; NY Acad Sci. Res: Viral and mycoplasma diseases of poultry and swine; enteroviruses and new species of mycoplasma of swine; mycoplasma of poultry and of rheumatoid arthritis; equine diseases, especially equine infectious anemia and Venezuelan Equine Encephalomyelitis. Mailing Add: Dept of Vet Microbiol Tex A&M Univ College Station TX 77843

MOORE, ROBERT ALONZO, b Indianapolis, Ind, June 15, 31; m 58; c 3. MATHEMATICS. Educ: Hanover Col, BA, 53; Ind Univ, Bloomington, PhD(math), 60. Prof Exp: Asst prof math, Pa State Univ, 61-64; ASSOC PROF MATH, SOUTHERN ILL UNIV, CARBONDALE, 64- Mem: Am Math Soc; Math Asn Am. Res: Class field theory. Mailing Add: Dept of Math Southern Ill Univ Carbondale IL 62901

MOORE, ROBERT B, b Windsor, Nfld, June 28, 35; m 58. NUCLEAR PHYSICS. Educ: McGill Univ, BEngPhys, 57, MSc, 59, PhD(physics), 61. Prof Exp: Lectr, 61-63, asst prof, 63-66, ASSOC PROF PHYSICS, McGILL UNIV, 66- Mem: Am Phys Soc; Can Asn Physicists; Eng Inst Can. Res: Nuclear reactions; nuclear spectroscopy; fission and short-lived isomeric states; heavy ion accelerators; proton and heavy ion irradiations. Mailing Add: Dept of Physics McGill Univ Montreal PQ Can

MOORE, ROBERT CONLEY, b Rock Hill, SC, Nov 24, 41; m 73; c 1. PHARMACY ADMINISTRATION, PUBLIC HEALTH ADMINISTRATION. Educ: Univ SC, AB, 66; Univ Ga, BS, 70; Univ Tenn, Memphis, PharmD, 71. Prof Exp: Asst prof pharm, Col Pharm, Univ Fla, 71-72; ASST PROF PHARM ADMIN, INST COMMUNITY & AREA DEVELOP, SCH PHARM, UNIV GA, 72- Concurrent Pos: Consult pharm, Ga Dept Human Resources & Ga Medicaid Prog, 75- Mem: AAAS; Sigma Xi. Res: Applied research for development and implementation of demonstration projects directed toward community health, especially rural health delivery. Mailing Add: Inst of Community & Area Develop Univ of Ga Sch of Pharm Athens GA 30602

MOORE, ROBERT D, experimental physics, see 12th edition

MOORE, ROBERT EARL, b South Bend, Ind, Dec 26, 23; m 47, 56; c 3. PHYSICAL INORGANIC CHEMISTRY. Educ: Purdue Univ, BS, 45; Univ Chicago, SM, 48, PhD(chem), 50. Prof Exp: Chemist, Navy Inorg Chem Res Proj, Chicago, 47-50; CHEMIST, OAK RIDGE NAT LAB, 50- Mem: AAAS; Am Chem Soc. Res: Radiation damage to high temperature materials; high temperature phase equilibria; ultrapure materials and solvent extraction; molten salts; boron chemistry; radiological effects of releases of radionuclides from nuclear facilities and power plants. Mailing Add: 107 E Irving Lane Oak Ridge TN 37830

MOORE, ROBERT EMMETT, b Wichita Falls, Tex, Nov 17, 31; m 53; c 3. VERTEBRATE ECOLOGY. Educ: N Tex State Univ, BA, 52; Ore State Univ, MS, 59; Univ Tex, PhD(mammalian speciation), 62. Prof Exp: From asst prof to assoc prof, 62-72, PROF ZOOL, MONT STATE UNIV, 72- Mem: Am Soc Naturalists; Ecol Soc Am; Am Soc Mammal; Sigma Xi. Res: Isolating mechanisms in the speciation of small mammals. Mailing Add: Dept of Biol Mont State Univ Bozeman MT 59715

MOORE, ROBERT H, b Baltimore, Md, Dec 19, 30; m 58; c 4. MATHEMATICS. Educ: Univ Md, BS, 53, MA, 55; Univ Mich, PhD(math), 59. Prof Exp: NSF fel, Munich Tech Univ, 59-60; res mathematician, Math Res Ctr, US Army, Univ Wis, 60-63; asst prof, 63-68, ASSOC PROF MATH, UNIV WIS-MILWAUKEE, 68- Mem: Am Math Soc; Math Asn Am. Res: Functional analytic study of certain approximation methods in numerical analysis, especially collectively compact operators and Newton's method, generalized inverses of linear operators; numerical solution of hyperbolic partial differential equations. Mailing Add: Dept of Math Univ of Wis Milwaukee WI 53201

MOORE, ROBERT J, physics, see 12th edition

MOORE, ROBERT LEE, mathematics, deceased

MOORE, ROBERT LEE, b Gainesville, Tex, Dec 28, 20; m 53. PHYSICAL CHEMISTRY. Educ: N Tex State Univ, BA, 42; Univ Tex, MA, 44, PhD(phys chem), 47. Prof Exp: Instr math, N Tex State Univ, 41-42; instr chem, Univ Tex, 42-45; chemist, Hanford Atomic Prod Oper, Gen Elec Co, 47-56, mgr fission prod chem, 56-65; mgr, Pac Northwest Lab, Battelle Mem Inst, 65-70; MGR APPL CHEM & ANAL, WESTINGHOUSE HANFORD CORP, 70- Mem: Am Chem Soc; Am Nuclear Soc; fel Am Inst Chemists. Res: Nuclear fuel processing; separations chemistry; waste treatment; analytical chemistry. Mailing Add: Westinghouse Hanford Corp PO Box 1970 Richland WA 99352

MOORE, ROBERT PARKER, b Blackburn, Okla, Jan 30, 12; m 36; c 2. AGRONOMY. Educ: Okla State Univ, BS, 34; Iowa State Univ, MS, 35; Ohio State Univ, PhD(field crops), 40. Prof Exp: Instr agron, Okla State Univ, 35-37; Ohio Crop Improve Asn asst, Ohio State Univ, 37-39; asst prof agron, Univ Tenn, 39-43; assoc prof agron, 43-46, dir seed improv, 46-53, PROF FIELD CROPS & IN CHG CROP STAND RES, NC STATE UNIV, 53- Mem: Am Soc Agron. Res: Seed evaluation and improvement; seedling establishment. Mailing Add: Dept of Crop Sci NC State Univ Raleigh NC 27607

MOORE, ROBERT STEPHENS, b Dubuque, Iowa, Sept 12, 33; m 56; c 3. POLYMER SCIENCE. Educ: Univ Wis, BS, 55, PhD(phys chem), 62. Prof Exp: Mem tech staff, Bell Tel Labs, Inc, 62-69; res assoc, 69-71, HEAD POLYMER PHYS CHEM LAB, EASTMAN KODAK CO, 71- Mem: Am Chem Soc; Am Phys Soc; Soc Rheol; Am Inst Chemists. Res: Viscoelastic properties of polymers; dilute solution properties of polymers; rheological and rheo-optical properties of polymers; light scattering from polymeric systems. Mailing Add: Polymer Phys Chem Lab Eastman Kodak Co Rochester NY 14650

MOORE, ROBERT VERNON, b Columbus, Ga, Nov 17, 20; m 44; c 3. RADIOCHEMISTRY. Educ: Col Charleston, BS, 42; Columbia Univ, MA, 49; Univ NC, PhD(chem), 53. Prof Exp: From instr to assoc prof chem, The Citadel, 46-57; res assoc, Med Col SC, 57-59, asst prof, 59-63; chief chemist, Ecol Field & Training Sta, USPHS, 63-70; RES CHEMIST, ENVIRON RES LAB, ENVIRON PROTECTION AGENCY, 70- Concurrent Pos: Mem coun arteriosclerosis, Am Heart Asn. Mem: Am Chem Soc; Am Nuclear Soc. Res: Elemental concentrations in water and sediments. Mailing Add: EPA Nuclear Anal Facil 900 Atlantic Dr NW Atlanta GA 30318

MOORE, ROBERT YATES, b Harvey, Ill, Dec 5, 31; m 69; c 4. NEUROLOGY, PEDIATRIC NEUROLOGY. Educ: Lawrence Univ, BA, 53; Univ Chicago, MD, 57, PhD(psychol), 62; Am Bd Psychiat & Neurol, dipl & cert neurol, 66. Hon Degrees: MD, Univ Lund, 74. Prof Exp: Intern, Univ Mich Hosp, 58-59; instr anat & res neurol, Sch Med, Univ Chicago, 59-64, from asst prof to prof pediat, neurol & anat, 64-74; PROF NEUROSCI, UNIV CALIF, SAN DIEGO, 74- Concurrent Pos: Markle scholar, 65-70; consult, NIH, 71-75 & Food & Drug Admin, 72-; assoc, Neurosci Res Prog, 74- Mem: Am Acad Neurol; Am Neurol Asn; Child Neurol Soc; Am Soc Clin Invest; Soc Pediat Res. Res: Organization and function of monoamine neuron systems in the mammalian brain; central neural regulation of diurnal rhythms. Mailing Add: Dept of Neurosci Univ of Calif at San Diego La Jolla CA 92037

MOORE, ROGER HUGHES, mathematical statistics, see 12th edition

MOORE, ROSCOE MICHAEL, JR, b Richmond, Va, Dec 2, 44; m 69; c 2. EPIDEMIOLOGY. Educ: Tuskegee Inst, BS, 68, DVM, 69; Univ Mich, MPH, 70. Prof Exp: Researcher gnotobiotics, NIH, 70-71; epidemiologist, Ctr Dis Control, 71-73 & Food & Drug Admin, 73-74; EPIDEMIOLOGIST, NAT INST OCCUP SAFETY & HEALTH, 74- Concurrent Pos: Consult, Dade County, Fla Health Dept & Ala State Health Dept, 72-73; adv, Washington Tech Inst, 74- Mem: Am Pub Health Asn; Soc Occup & Environ Health; assoc fel Am Col Vet Toxicologist; Soc Epidemiol Res; Vet Cancer Soc. Res: Epidemiology of chemical carcinogenesis in man and animals. Mailing Add: Nat Inst Occup Safety & Health 5600 Fishers Ln Rockville MD 20852

MOORE, ROYALL TYLER, b Lexington, Va, Oct 11, 30. MYCOLOGY. Educ: Mich State Univ, BS, 51; Univ Iowa, MS, 53; Harvard Univ, PhD(biol), 59. Prof Exp: Vis scientist mycol, Marine Lab, Univ Miami, 59; NIH fel, Cornell Univ, 59-61; fel, Univ Calif, Berkeley, 61-63, asst res biophysicist, 63-64; from asst prof to assoc prof bot, NC State Univ, 64-74; MEM FAC BIOL & ENVIRON SCI, COLERAINE COL, NEW ULSTER UNIV, 74- Concurrent Pos: Neth Orgn Advan Pure Res fel, State Univ Groningen, 71-72. Mem: AAAS; Mycol Soc Am; Bot Soc Am; Am Soc Cell Biol. Res: Ultracytology of fungi; cell biology. Mailing Add: Dept of Biol & Environ Stud Coleraine Col New Ulster Univ Londonderry Northern Ireland

MOORE, RUFUS ADOLPHUS, b Brackettville, Tex, Feb 8, 23. PHYSICS. Educ: St Mary's Univ, Tex, BS, 43; Univ Tex, MA, 49, PhD(physics), 58. Prof Exp: Asst math & physics, Univ Tex, 48-50, res scientist physics, Defense Res Lab, 51-52, asst math & physics, Univ, 53; mathematician, Sch Aviation Med, 54-56; asst math & physics, Univ Tex, 56-58; mathematician, Personnel Lab, Wright Air Develop Ctr, 58-59; asst prof physics, St Mary's Univ, Tex, 59-62; assoc prof, 62-66, PROF PHYSICS, STATE UNIV NY COL OSWEGO, 66- Mem: AAAS; Am Math Soc; Inst Elec & Electronics Engrs; Am Phys Soc; Am Asn Physics Teachers. Res: Molecular theory; electronic communications. Mailing Add: Dept of Physics State Univ of NY Oswego NY 13126

MOORE, RUSSELL THOMAS, b Las Animas, Colo, Dec 30, 43; m 64; c 2. PLANT ECOLOGY. Educ: Univ Idaho, BS, 66; Utah State Univ, PhD(plant ecol), 72. Prof Exp: Fel plant ecophysiol, San Diego State Univ, 71-73; proj mgr plant ecol, 73-75, SECT MGR MINING & RECLAMATION, ECOL CONSULTS INC, 75- Mem: Ecol Soc Am; Soc Range Mgt; Am Soc Plant Physiologists. Res: Selection of plant species and development of planting techniques suitable for revegetation of western surface mined lands. Mailing Add: Ecol Consults Inc Box 2105 Ft Collins CO 80522

MOORE, SOPHIE (OLEKSA), physics, see 12th edition

MOORE, STANFORD, b Chicago, Ill, Sept 4, 13. BIOCHEMISTRY. Educ: Vanderbilt Univ, AB, 35; Univ Wis, PhD(org chem), 38. Hon Degrees: MD, Free Univ Brussels, 54; Dr, Univ Paris, 64; DSc, Univ Wis, 74. Prof Exp: Asst, 39-42, assoc, 42 & 45-49, assoc mem, 49-52, PROF BIOCHEM & MEM, ROCKEFELLER UNIV, 52- Concurrent Pos: Chmn, Panel on Proteins, Comt on Growth, Nat Res Coun, 47-49; vis Franqui prof, Free Univ Brussels, 50-51; vis investr, Cambridge Univ, 51; trustee, Vanderbilt Univ, 74- Honors & Awards: Nobel Prize in Chem, 72; Linderstrom-Lang Medal, 72; Richards Medal, Am Chem Soc, 72. Mem: Nat Acad Sci; AAAS; Am Chem Soc; Am Soc Biol Chem (treas, 57-59, pres, 66); Am Acad Arts & Sci. Res: Chromatography; chemistry of carbohydrates, proteins and amino acids. Mailing Add: Rockefeller Univ York Ave & 66th St New York NY 10021

MOORE, STEVENSON, III, b Chicopee, Mass, May 13, 28; m 48; c 4. ENTOMOLOGY. Educ: Am Int Col, AB, 49; Univ Mass, MS, 51; Cornell Univ, PhD(entom), 53. Prof Exp: Asst entom, Cornell Univ, 51-53; from asst prof to assoc prof entom, 53-72, PROF AGR ENTOM & ENTOMOLOGIST, UNIV ILL, URBANA, 72- Mem: AAAS; Entom Soc Am. Res: Livestock, field crop and household insect control. Mailing Add: Dept of Entom Univ of Ill Urbana IL 61801

MOORE, THEODORE CARLTON, JR, b Kinston, NC, Feb 16, 38; m 60; c 2. OCEANOGRAPHY, MARINE GEOLOGY. Educ: Univ NC, BS, 60; Univ Calif, San Diego, PhD(oceanog), 68. Prof Exp: Res assoc oceanog, Ore State Univ, 68-69, asst prof, 69-75; ASSOC PROF OCEANOG, UNIV RI, 75- Mem: AAAS; Am Geophys Union. Res: Stratigraphy and sedimentation in the deep-sea; micropaleontological studies of Radiolaria and calcareous nanoplankton. Mailing Add: Grad Sch of Oceanog Univ of RI Kingston RI 02881

MOORE, THERAL ORVIS, b Emerson, Ark, Oct 16, 27; m 62; c 2. MATHEMATICS. Educ: Univ Ark, BA, 49, MA, 51; Univ Mo, PhD(math), 55. Prof Exp: Asst math, Univ Mo, 51-55; asst prof, 55-66, ASSOC PROF MATH, UNIV FLA, 66- Mem: Am Math Soc; Math Asn Am. Res: Topology; lattice theory; abstract algebra. Mailing Add: Dept of Math Univ of Fla Gainesville FL 32601

MOORE, THERON LANGFORD, b Mayo, Va, Nov 24, 34; m 59; c 3. ORGANIC CHEMISTRY. Educ: Yale Univ, BS, 56; Univ Calif, Los Angeles, PhD(org chem), 61. Prof Exp: Chemist, Procter & Gamble Co, 61-66; assoc prof, 66-68, PROF CHEM, NORFOLK STATE COL, 68- Mem: AAAS; Am Chem Soc. Mailing Add: Dept of Chem Norfolk State Col Norfolk VA 23504

MOORE, THOMAS CARLETON, b Muncie, Ind, Apr 7, 21; m 49; c 4. MEDICINE. Educ: Dartmouth Col, AB, 42; Harvard Med Sch, MD, 45. Prof Exp: Jr asst resident surg, Children's Hosp, Boston, 48-49; asst resident, Ind Univ Hosp, 49-50, chief resident, 50-51, instr, Sch Med, 51-52, assoc, 52-56, from asst prof to assoc prof, 56-62; prof, Univ Ky, 62-63, prof clin surg, 63-68; PROF SURG, SCH MED, UNIV CALIF, LOS ANGELES & CHIEF RENAL TRANSPLANTATION & PEDIAT SURG, UNIV CALIF-HARBOR MED CTR, 68- Mem: Soc Univ Surgeons (treas, 56-59); Soc Vascular Surg; Am Surg Asn; Transplantation Soc; Am Col Surgeons. Res: Thoracic pediatric and cardiovascular surgery. Mailing Add: Univ of Calif-Harbor Med Ctr 1000 W Carson St Torrance CA 90509

MOORE, THOMAS CARROL, b Sanger, Tex, Sept 22, 36; m 56; c 3. PLANT PHYSIOLOGY. Educ: N Tex State Univ, BA, 56; Univ Colo, MA, 58, PhD(bot), 61. Prof Exp: Instr biol, Univ Colo, 58-59; asst prof bot, Ariz State Col, 61-63; from asst prof to assoc prof, 63-71, PROF BOT, ORE STATE UNIV, 71-, CHMN DEPT BOT & PLANT PATH, 73- Mem: Am Soc Plant Physiologists; Bot Soc Am; Int Plant Growth Substances Asn; Sigma Xi. Res: Hormonal regulation of growth and flowering in angiosperms, including metabolic control mechanisms; modes of action of plant growth-regulating chemicals; physiological ecology of seed plants. Mailing Add: Dept of Bot & Plant Path Ore State Univ Corvallis OR 97331

MOORE, THOMAS D, b Philadelphia, Pa, Dec 17, 28. BACTERIOLOGY. Educ: Lincoln Univ, Pa, AB, 50; Pa State Univ, MS, 56, PhD(bact), 59. Prof Exp: Res assoc dairy bact, Iowa State Univ, 58-62; BACTERIOLOGIST, NIH, 62- Mem: Am Soc Microbiol; Am Pub Health Asn; Am Asn Lab Animal Sci. Res: Food poisoning staphylococci; diseases of laboratory animals. Mailing Add: 1301 Delaware SW Washington DC 20024

MOORE, THOMAS D, b Cleveland, Okla, Sept 7, 30; m 50; c 5. MEDICINE. Educ: Okla State Univ, BS, 51; Univ Okla, MD, 54; Am Bd Pediat, dipl, 60. Prof Exp: Intern, Letterman Gen Hosp, 54-55; pediat resident, New York Hosp-Cornell Med Ctr, 55-57; chief pediat serv, Ramey AFB, PR, 57-60 & Hamilton AFB, Calif, 60-61; dir clin res, Ross Labs, Ohio, 62-65, med dir, 65-67, vpres sci affairs, 67-69; assoc prof, 69-75, PROF PEDIAT, UNIV TEX HEALTH SCI CTR DALLAS, 75- Mem: AAAS; Am Acad Pediat; AMA. Mailing Add: Suite 802 Stemmons Tower North 2710 Stemmons Freeway Dallas TX 75207

MOORE, THOMAS EDWIN, b Amarillo, Tex, Jan 15, 18; m 42; c 4. INORGANIC CHEMISTRY. Educ: Univ Tex, BA, 40, MS, 42, PhD(chem), 46. Prof Exp: Res assoc, Radio Res Lab, Harvard Univ, 43-45; US Signal Corps fel, Northwestern Univ, 46-47; from asst prof to assoc prof, 47-57, PROF CHEM, OKLA STATE UNIV, 57- Mem: Am Chem Soc. Res: Solvent extraction of inorganic salts; nonaqueous solutions; thermodynamics of electrolyte mixtures. Mailing Add: Dept of Chem Okla State Univ Stillwater OK 74074

MOORE, THOMAS EDWIN, b Champaign, Ill, Mar 10, 30; m 51; c 2. ZOOLOGY, ENTOMOLOGY. Educ: Univ Ill, BS, 51, MS, 52, PhD, 56. Prof Exp: Asst, Ill Natural Hist Surv, 52-56; from instr to assoc prof, 56-67, PROF ZOOL, UNIV MICH, ANN ARBOR, 67-, CUR INSECTS, MUS ZOOL, 59-, CHMN COMN TROP STUDIES, 74- Concurrent Pos: Mem, Orgn Trop Studies, vis prof, 70 & 72; mem, Nat Acad Sci-Nat Res Coun panel on NSF grad fels. Mem: AAAS; Animal Behav Soc; Soc Study Evolution; Soc Syst Zool; Asn Trop Biol. Res: Acoustical behavior of insects; evolution and systematics of cicadas. Mailing Add: Mus of Zool Univ of Mich Ann Arbor MI 48104

MOORE, THOMAS FRANCIS, b Camden, NJ, Oct 1, 22; m 48; c 2. GEOCHEMISTRY. Educ: Southern Methodist Univ, BS, 44; Johns Hopkins Univ, MS & PhD(chem), 50. Prof Exp: From group leader petrol recovery res to sr res engr, Atlantic Refining Co, 42-67; PRIN RES ENGR, ATLANTIC RICHFIELD CO, DALLAS, 67- Concurrent Pos: Res chemist, Tenn Eastman Div, Oak Ridge. Mem: AAAS; Soc Petrol Engrs. Res: Uranium purification and extraction; surface chemistry of silica; petroleum production and exploration research; economic feasibility analysis; thermal oil recovery methods; in-situ solution mining of uranium. Mailing Add: 100 West Shore Pl Richardson TX 75080

MOORE, THOMAS WARNER, b New Haven, Conn, Mar 22, 28; m 51; c 3. SOLID STATE PHYSICS. Educ: Calif Inst Technol, BS, 50; Univ Calif, Berkeley, PhD(physics), 61. Prof Exp: Staff mem, Gen Elec Res & Develop Ctr, 61-67; assoc prof physics, 67-75, PROF PHYSICS, MT HOLYOKE COL, 75- Mem: Am Phys Soc. Res: Metal physics; transport phenomena in metals at cryogenic temperatures. Mailing Add: Dept of Physics Mt Holyoke Col South Hadley MA 01075

MOORE, VAUGHN CLAYTON, b Osawatomie, Kans, May 16, 34; m 56; c 3. RADIOLOGICAL PHYSICS, BIOPHYSICS. Educ: Univ Kans, BS, 56, MS, 57; Univ Minn, PhD(biophys), 68; Am Bd Radiol, dipl, 64. Prof Exp: Health physicist, Univ Chicago, 57-58, asst dir health physics, 58-59; asst prof radiol, Health Sci Ctr, Univ Minn, 68-70, asst prof therapeut radiol, 70-75; DIR RADIATION PHYSICS, NEUROPSYCHIAT INST, FARGO, 75- Concurrent Pos: Attend, Vet Admin Hosp, Minneapolis, 68-75; consult, St Joseph's Hosp, St Paul, 69-75; adj prof physics & bionucleonics, NDak State Univ, 75- Mem: Am Asn Physicists in Med. Res: Applications of electron linear accelerators and computers to radiation therapy. Mailing Add: Neuropsychiat Inst Radiation Physics Fifth & Mills Ave N Fargo ND 58102

MOORE, WALTER EDWARD CLADEK, b Rahway, NJ, Oct 12, 27; m 49; c 3. MEDICAL MICROBIOLOGY, BACTERIOLOGY. Educ: Univ NH, BS, 51; Univ Wis, MS, 52, PhD(dairy husb, bact), 55. Prof Exp: Asst, Alumni Res Found, Univ Wis, 51-52; assoc prof, 54-61, PROF BACT, VA POLYTECH INST & STATE UNIV, 61-, SCIENTIST-IN-CHG ANAEROBE LAB, 71- Concurrent Pos: Mem judicial comn, Int Comn Bact Nomenclature, 75- Honors & Awards: Kimble Methodology Res Award, Conf Pub Health Lab Dirs, 73. Mem: AAAS; Am Soc Microbiol; Am Acad Microbiol. Res: Intestinal and anaerobic microbiology. Mailing Add: Anaerobe Lab Va Polytech Inst & State Univ Blacksburg VA 24061

MOORE, WALTER GUY, b Detroit, Mich, June 21, 13; m 39; c 6. AQUATIC BIOLOGY, ECOLOGY. Educ: Wayne State Univ, BA, 34; Univ Minn, MA, 38, PhD(zool), 40. Prof Exp: Instr, Wayne State Univ, 34-35; asst, Univ Minn, 35-40; from instr to assoc prof, 40-51, actg chmn depts biol & med technol, 42-44, PROF BIOL, LOYOLA UNIV, LA, 51- Mem: AAAS; Ecol Soc Am; Am Micros Soc; Am Soc Limnol & Oceanog; Int Soc Limnol. Res: Limnology; ecology of Anostraca; temporary ponds. Mailing Add: Dept of Biol Sci Loyola Univ New Orleans LA 70118

MOORE, WALTER JOHN, b New York, NY, Mar 25, 18; m 43; c 3. PHYSICAL CHEMISTRY. Educ: NY Univ, BS, 37; Princeton Univ, PhD(phys chem), 40. Prof Exp: Nat Res Coun fel, Calif Inst Technol, 40-41; from instr to assoc prof chem, Cath Univ Am, 41-51; Guggenheim & Fulbright fels, Bristol Univ, 51-52; prof chem, Ind Univ, Bloomington, 52-63, res prof, 63-74; PROF PHYS CHEM, UNIV SYDNEY, 74- Concurrent Pos: NSF sr fel, Paris, 58-59; vis prof, Harvard Univ, 60 & Univ Brazil, 62 & 63; chmn comn phys chem, Nat Acad Sci-Nat Res Coun, 64-66; Austalian Am Educ Found prof, Univ Queensland, 66, vis prof, 68- Honors & Awards: James F Norris Award, Am Chem Soc, 65. Mem: Am Chem Soc; Biophys Soc; Am Soc Biol Chemists; Am Soc Neurochem; Int Soc Neurochem. Res: Solid state chemistry; neurochemistry; biophysical chemistry of brain function. Mailing Add: Dept of Phys Chem Univ of Sydney Sydney NSW 2006 Australia

MOORE, WALTER LEROY, b Omaha, Nebr, Oct 24, 25; m 46; c 2. GEOLOGY. Educ: Utah State Univ, BS, 50; Univ Wis, MS, 54, PhD(geol), 59. Prof Exp: Mem staff geol, Gulf Oil Corp, 54-60; assoc prof, 60-66, PROF GEOL, UNIV NDAK, 66- Concurrent Pos: In-serv inst dir, NSF, 63-66. Mem: Am Asn Petrol Geol; Soc Econ Paleont & Mineral. Res: Stratigraphy; paleontology; sedimentary petrology. Mailing Add: Dept of Geol Univ of NDak Grand Forks ND 58201

MOORE, WARD WILFRED, b Cowden, Ill, Feb 12, 24; m 49; c 4. PHYSIOLOGY. Educ: Univ Ill, AB, 48, MS, 51, PhD(physiol), 52. Prof Exp: Asst animal physiol, Univ Ill, 50-52, res assoc, 52-54; asst prof physiol, Okla State Univ, 54-55; from asst prof to assoc prof physiol, 55-66, actg chmn dept anat, 71-73, PROF PHYSIOL, IND UNIV, INDIANAPOLIS, 66-, ASSOC DEAN BASIC MED SCI, 71- Concurrent Pos: Vis prof, Jinnah Postgrad Med Ctr, Karachi, 63-64; mem staff, Rockefeller Found & vis prof & chmn dept physiol, Fac Sci, Mahidol Univ, Thailand, 68-71. Mem: AAAS; Am Physiol Soc; Am Asn Anat; Endocrine Soc; Am Soc Nephrology. Res: Neuroendocrinology; stress; regulation of antidiuretic hormone secretion. Mailing Add: Dept of Physiol Ind Univ Indianapolis IN 46202

MOORE, WARREN KEITH, b Wellington, Kans, Feb 11, 23; m 44; c 5. MATHEMATICS. Educ: Southwestern Col, Kans, AB, 47; Univ Kans, MA, 48, PhD(math), 51. Prof Exp: Cottrell asst, Univ Kans, 48-49, instr math, 51-52; from asst prof to assoc prof, 52-63, PROF MATH, ALBION COL, 63-, CHMN DEPT, 61- Mem: Am Math Soc; Math Asn Am. Res: Mathematical analysis. Mailing Add: Dept of Math Albion Col Albion MI 49224

MOORE, WAYNE ELDEN, b McLeansboro, Ill, Sept 2, 19; m 44; c 4. PETROLEUM GEOLOGY. Educ: Univ Ill, BS, 46; Cornell Univ, MS, 48, PhD(geol), 50. Prof Exp: Assoc prof geol, Va Polytech Inst, 50-56; paleontologist, Chevron Oil Co, Stand Oil Co, Calif, 56-70; PROF GEOL & CHMN DEPT, CENT MICH UNIV, 71- Mem: Am Asn Petrol Geologists; Geol Soc Am; Paleont Soc; Am Asn Stratig Palynologists; AAAS. Res: Origin, maturation and migration of Paleozoic petroleum and subsurface brines, especially in Michigan; micropaleontology of the source beds. Mailing Add: Dept of Geol Cent Mich Univ Mt Pleasant MI 48859

MOORE, WELLINGTON, JR, b Lexington, Ky, Sept 24, 28; m 56; c 5. VETERINARY TOXICOLOGY, INHALATION TOXICOLOGY. Educ: Univ Ky, BS, 50; Auburn Univ, DVM, 55; Cornell Univ, PhD(radiation biol), 61. Prof Exp: Pvt pract, 55-56; epidemic intel officer, Commun Dis Ctr, USPHS, 56-58, res officer, Div Radiol Health, 58-66, chief radiation cytol lab, Nat Ctr Radiol Health, 66-68; dep dir, Fed Dept Vet Res, Vom, Nigeria, 68-70; chief biol effects br, 71-73, CHIEF EXP TOXICOL BR, ENVIRON TOXICOL RES LAB, NAT ENVIRON RES CTR, 74- Mem: Am Vet Med Asn. Res: Radiobiology; cytogenetics; toxicology; toxicology of environmental pollutants. Mailing Add: Exp Toxicol Br Nat Environ Res Ctr 1055 Laidlaw Ave Cincinnati OH 45237

MOORE, WESLEY SANFORD, b San Bernardino, Calif, Aug 1, 35; m 60; c 2. SURGERY. Educ: Univ Southern Calif, BS, 55; Univ Calif, San Francisco, MD, 59. Prof Exp: Intern surg, Univ Calif Hosps, San Francisco, 59-60; asst resident surg, Vet Admin Hosp, San Francisco, 60-63, chief resident, 63-64; clin instr, 66-68, asst prof, 68-73, ASSOC PROF SURG, SCH MED, UNIV CALIF, SAN FRANCISCO, 73-; CHIEF VASCULAR SURG SECT, VET ADMIN HOSP, 66- Concurrent Pos: NIH fel cerebrovasc insufficiency, 66-67; mem comt prosthetics & orthotics, Nat Acad Sci. Mem: Soc Vascular Surg; AMA; fel Am Col Surg; Soc Univ Surgeons. Res: Vascular surgery; circulation research; stroke prevention. Mailing Add: Dept of Surg Vet Admin Hosp San Francisco CA 94121

MOORE, WILLARD S, b Jackson, Miss, Sept 27, 41. MARINE GEOCHEMISTRY. Educ: Millsaps Col, BS, 62; Columbia Univ, MA, 65; State Univ NY Stony Brook, PhD(earth & space sci), 69. Prof Exp: OCEANOGR, OCEAN FLOOR ANAL DIV, US NAVAL OCEANOG OFF, 69- Concurrent Pos: Vis asst prof, State Univ NY Stony Brook, 70; vis res fel, Tata Inst Fundamental Res, India, 71; vis adj prof, Univ SC, 70 & 71. Mem: AAAS; Am Geophys Union. Res: Radioisotopes in the ocean; age determinations of fossil corals; coral growth rates; ocean mixing. Mailing Add: Ocean Floor Anal Div US Naval Oceanog Off Code 038 Chesapeake Beach MD 20732

MOORE, WILLIAM EARL, b Tuscaloosa, Ala, Nov 10, 41; m 63; c 2. PROTEIN CHEMISTRY. Educ: Southern Univ, BS, 63; Purdue Univ, PhD(protein chem), 67. Prof Exp: Assoc prof, 67-71, PROF CHEM, SOUTHERN UNIV, 71-; DIR, INST SERV EDUC, 75- Concurrent Pos: Charles Pfizer Pfizer res fel med, La State Univ, 68; consult, Oxford Univ Press, 71; NIH res fel biomed, Naperville Tech Ctr, Standard Oil Co, 72; guest lectr, Purdue Univ, 73; prog assoc, Inst Serv Educ, 74-75; consult interdisciplinary sci, Clark Col, 75. Mem: AAAS; Sigma Xi; Am Chem Soc; Inst Food Technologists; Nat Sci Teachers Asn. Res: Physicochemical studies of protein stability by examining denaturation in various media. Mailing Add: Inst for Serv to Educ 2001 S St NW Washington DC 20009

MOORE, WILLIAM MARSHALL, b Lincoln, Nebr, Dec 25, 30; m 58, 68; c 4. PHYSICAL CHEMISTRY. Educ: Colo Col, BA, 52; Iowa State Univ, PhD(phys chem), 59. Prof Exp: Res chemist, Monsanto Chem Co, 52-53; NIH fel, Cambridge Univ, 59-60; from asst prof to assoc prof, 60-69, PROF CHEM, UTAH STATE UNIV, 69- Mem: AAAS; Am Chem Soc; The Chem Soc; Am Soc Photobiol. Res: Mechanisms for photochemical reactions; atmospheric chemistry and photochemistry; emission spectroscopy. Mailing Add: Dept of Chem Utah State Univ Logan UT 84321

MOORE, WILLIAM ROBERT, b Minneapolis, Minn, July 18, 28; m 56; c 3. ORGANIC CHEMISTRY. Educ: Univ Calif, Los Angeles, BS; Univ Minn, PhD, 54. Prof Exp: Res assoc org chem, Mass Inst Technol, 54-55, from instr to assoc prof, 55-72; PROF CHEM & CHMN DEPT, WVA UNIV, 72- Mem: AAAS; Am Chem Soc. Res: Mechanisms of organic reactions; unsaturated cyclic hydrocarbons; highly-strained compounds; carbenes. Mailing Add: Dept of Chem WVa Univ Morgantown WV 26506

MOORE, WILLIAM SAMUEL, b Manhatten, Kans, June 8, 42; m 63; c 2. POPULATION BIOLOGY. Educ: Mich State Univ, BS, 66; Univ Conn, PhD(biol), 71. Prof Exp: Asst prof, 71-75, ASSOC PROF BIOL, WAYNE STATE UNIV, 75- Mem: Soc Study Evolution; Ecol Soc Am; Genetics Soc Am. Res: Evolution and adaptive strategies of various genetic systems. Mailing Add: Dept of Biol Wayne State Univ Detroit MI 48202

MOORE, WILLIS EUGENE, b Plain City, Ohio, Dec 28, 24; m 45; c 4. PHARMACY. Educ: Ohio State Univ, BSc, 49, PhD(pharmaceut chem), 53. Prof Exp: Asst prof pharmaceut chem, George Washington Univ, 53-56; res assoc, Sterling Winthrop Res Inst, 56-59, assoc mem, 59-60, mem & sect head, 60-63, asst dir pharm div, 64-68; assoc prof, 68-73, asst dean col pharm, 71-72, actg dean col pharm, 73-74, assoc dean col pharm & allied health prof, 74-76, PROF PHARMACEUT, COL PHARM, WAYNE STATE UNIV, 73- Mem: Am Pharmaceut Asn. Res: Biopharmaceutics; research management; product development. Mailing Add: 7148 Osage Ave Allen Park MI 48101

MOOREFIELD, HERBERT HUGHES, b Baltimore, Md, July 25, 18; m 53; c 1. TOXICOLOGY. Educ: Univ Md, BS, 51; Univ Ill, MS, 52, PhD, 53. Prof Exp: Res assoc, Univ Ill, 53-54; entomologist, Boyce Thompson Inst, 54-60; dir agr res, Res Sta, Union Carbide Chem Co, 60-64; asst dir res & develop, NC, 64-67, mgr agr prod mkt develop, Calif, 67-72, TECHNOL MGR AGR PROD, RES & DEVELOP DEPT, TECH CTR, UNION CARBIDE CORP, 72- Res: Pesticides; insect physiology-toxicology. Mailing Add: Res & Develop Dept Tech Ctr Union Carbide Corp PO Box 8361 South Charleston WV 25303

MOOREHEAD, THOMAS J, b Jersey City, NJ, Nov 1, 47. CHEMISTRY. Educ: St Peter's Col, NJ, BS, 69; Univ Notre Dame, PhD(chem), 75. Prof Exp: RES ASST BIO-ORG CHEM, PA STATE UNIV, UNIVERSITY PARK, 74- Mem: Am Chem Soc. Res: Structure and mechanism of catalysis of the enzyme 4-hydroxyphenylpyruvic acid oxygenase. Mailing Add: Dept of Chem 152 Davey Lab Pa State Univ University Park PA 16802

MOOREHEAD, WELLS RUFUS, b Hickory Flat, Miss, July 20, 31; m 62; c 1.

CLINICAL CHEMISTRY, BIOCHEMISTRY. Educ: Miss State Univ, BS, 53, MS, 60; Univ Tenn, PhD(biochem), 65. Prof Exp: ASSOC PROF CLIN PATH & ASSOC DIR CLIN CHEM, MED CTR, IND UNIV, INDIANAPOLIS, 70- Mem: AAAS; Am Asn Clin Chemists; NY Acad Sci; Sigma Xi. Res: Serum enzyme activity levels in different disease conditions. Mailing Add: Dept of Path Ind Univ Med Ctr Indianapolis IN 46202

MOOREHEAD, WILLIAM DOUGLAS, b Black Mountain, NC, May 19, 29; m 57; c 3. BIOLOGY, SCIENCE EDUCATION. Educ: NC Col Durham, BS, 53, MS, 54; Univ Okla, PhD, 65. Prof Exp: From instr to asst prof, 54-62, actg chmn div sci & math, 59-62, PROF BIOL & CHMN DIV SCI & MATH, 65- Mem: Nat Sci Teachers Asn. Res: Elementary school science; zoology. Mailing Add: Dept of Biol Ft Valley State Col Ft Valley GA 31030

MOORE-LANDECKER, ELIZABETH JANE, mycology, see 12th edition

MOORES, ELDRIDGE MORTON, b Phoenix, Ariz, Oct 13, 38; m 65; c 2. PETROLOGY, STRUCTURAL GEOLOGY. Educ: Calif Inst Technol, BS, 59; Princeton Univ, MA, 61, PhD(geol), 63. Prof Exp: Teaching asst geol, Princeton Univ, 62-63, NSF vis res fel, 63-65, res assoc, 65-66; lectr, 66-67, from asst prof to assoc prof, 67-75, chmn dept, 71-72, PROF GEOL, UNIV CALIF, DAVIS, 75-, CHMN DEPT, 73- Mem: Geol Soc Am; Mineral Soc Am; Am Geophys Union. Res: Ophiolites and plate tectonics; plate tectonics of deformed belts; history of plate interactions. Mailing Add: Dept of Geol Univ of Calif Davis CA 95616

MOORES, EUGENE ALBERT, b Waterville, Maine, Oct 27, 31; m 59; c 2. EARTH SCIENCES, RESEARCH ADMINISTRATION. Educ: Univ Alaska, BS, 61, MS, 62; Iowa State Univ, PhD(geol), 66. Prof Exp: Instr geol, Iowa State Univ, 62-66; oper res analyst, Sunray DX Oil Co, 66-67, supvr regional studies, 67-70; supvr oceanog res, 70-72, sr sci assoc gen res, 72, mgr explor res, 72-74, chief scientist, 74-75, MGR GEOL, SUN OIL CO, 75- Concurrent Pos: Lectr geol, Univ Tulsa, 67-69; mem site selection comt, Int Geodynamics Proj, 74- Mem: Am Asn Petrol Geologists; Soc Econ Paleontologists & Mineralogists. Res: Sedimentary tectonics as applied to the search for new methods to locate pooled hydrocarbons, including use of all exploration tools, particularly seismic, organic geochemistry, geology and remote sensing. Mailing Add: Sun Oil Co 12850 Hillcrest Rd Dallas TX 75230

MOORES, MEAD STEPHEN, b Pittsburgh, Pa, Jan 4, 27; m 48; c 4. ORGANIC CHEMISTRY. Educ: Carnegie Inst Technol, BS, 52, MS, 57, PhD(org chem), 59. Prof Exp: Res chemist resins & adhesives, Koppers Co, Inc, Pa, 52-55, res chemist organometallics, 58-62; res chemist dyes, NJ, 62-69, sr supvr, 69-71, chief supvr, 71-73, tech mgr, PR, 73-75, CHIEF SUPVR, E I DU PONT DE NEMOURS & CO, INC, 75- Mem: Am Chem Soc; Am Asn Textile Chemists & Colorists. Res: Organic synthesis; dye chemistry. Mailing Add: Azo Lab Chambers Works E I du Pont de Nemours & Co Inc Deepwater NJ 08023

MOORES, RUSSELL R, b St Louis, Mo, Feb 25, 35; m 57; c 7. MEDICINE. Educ: Ark State Univ, BS, 55; Univ Ark, MD, 58; Am Bd Internal Med, dipl, 65. Prof Exp: From intern to resident med, Strong Mem Hosp, Rochester, NY, 58-60; resident, Barnes Hosp, St Louis, Mo, 60-61; NIH fel hemat, 61-63; staff hematologist, US Naval Hosp, Oakland, Calif, 63-65; from asst prof to assoc prof med, 65-71, assoc dean curriculum, 72-74, ASSOC DEAN SPEC PROGS, MED COL GA, 74-, PROF HUMANITIES & MED, MED COL GA, 71- Mem: Am Soc Hemat; AMA; fel Am Col Physicians; Am Fedn Clin Res; fel Int Soc Hemat. Res: Erythropoiesis. Mailing Add: Off of Humanities Med Col of Ga Augusta GA 30904

MOORHEAD, EDWARD DARRELL, analytical chemistry, electrochemistry, see 12th edition

MOORHEAD, PAUL SIDNEY, b El Dorado, Ark, Apr 18, 24; m 49; c 3. CYTOGENETICS. Educ: Univ NC, AB, 48, MA, 50; Univ Tex, PhD(zool), 54. Prof Exp: Res assoc cytol, Med Br, Univ Tex, 54-56, Runyan Mem fel, 55; res assoc, Sch Med, Univ Pittsburgh, 56-58; assoc mem, Wistar Inst Anat & Biol, 59-69; ASSOC PROF HUMAN GENETICS, SCH MED, UNIV PA, 69- Mem: Genetics Soc Am; Am Soc Human Genetics; Environ Mutagen Soc; Am Asn Cancer Res; Tissue Cult Asn. Res: Mammalian chromosomes. Mailing Add: Dept of Human Genetics Univ of Pa Sch of Med Philadelphia PA 19104

MOORHEAD, PHILIP DARWIN, b Pratt, Kans, Nov 21, 33; m 56; c 3. VETERINARY PATHOLOGY. Educ: Kans State Univ, BS & DVM, 57; Purdue Univ, MS, 64, PhD(vet path), 66. Prof Exp: Vet, gen pract, 57-62; instr vet path, Col Vet Med, Purdue Univ, 63-66; ASSOC PROF VET PATH, OHIO AGR RES & DEVELOP CTR, 66- Concurrent Pos: Secy, NC-65 Comt Poultry Respiratory Dis, Coop Res Serv, USDA, 69- Mem: Am Vet Med Asn; Am Asn Avian Path; Conf Res Workers Animal Dis. Res: Sporadic toxocologic environmental infections and non-specific problems in chickens and turkeys. Mailing Add: Dept of Vet Sci Ohio Agr Res & Develop Ctr Wooster OH 44691

MOORHEAD, WILLIAM DEAN, b Youngstown, Ohio, Nov 2, 36; m 65; c 1. THEORETICAL PHYSICS. Educ: Ohio Wesleyan Univ, BA, 58; Ohio State Univ, PhD(physics), 68. Prof Exp: Asst prof, 68-72, ASSOC PROF PHYSICS & ASTRON, YOUNGSTOWN STATE UNIV, 68- Mem: Am Phys Soc. Res: Theories of angular momentum, magnetism, and the many body problem; statistical physics. Mailing Add: Dept of Physics & Astron Youngstown State Univ Youngstown OH 44503

MOORHOUSE, JOHN A, b Winnipeg, Man, Oct 4, 26. MEDICINE, ENDOCRINOLOGY. Educ: Univ Man, MD, 50, MSc, 55; FRCP(C). Prof Exp: Fel path, Winnipeg Gen Hosp, Man, 50-51, asst resident med, 51-53, chief resident, 53-54; res fel metab, Res & Educ Hosp & Presby Hosp, Univ Ill, 54-56; res fel & clin asst endocrinol & metab, Univ Mich Hosp, 56-58; res assoc, Clin Invest Univ, Winnipeg Gen Hosp, 58-61; from asst prof to assoc prof, 60-72, PROF PHYSIOL, UNIV MAN, 72-, ASSOC PROF MED, 70-; DIR ENDOCRINE & METAB LAB, HEALTH SCI CTR, 61-, ASST PHYSICIAN, 63- Mem: Fel Am Col Physicians; Endocrine Soc; Am Diabetes Asn; Can Soc Clin Invest. Res: Carbohydrate and fat metabolism, particularly as related to diabetes mellitus. Mailing Add: Health Sci Ctr G449 700 William Ave Winnipeg MB Can

MOORING, FRANCIS PAUL, b Stokes, NC, Feb 6, 21; m 48; c 3. NUCLEAR PHYSICS. Educ: Duke Univ, BA, 44; Univ Wis, PhD(physics), 51. Prof Exp: Instr physics, Duke Univ, 44-46; assoc physicist, 51-75, PHYSICIST, ARGONNE NAT LAB, 75- Concurrent Pos: Fulbright res fel, Univ Helsinki, Finland, 62-63; adj prof, St Louis Univ, 66- Mem: Am Phys Soc; AAAS. Res: Particle accelerators; nuclear physics-nuclear cross sections from 0.1-3.0 MeV. Mailing Add: 295 Abbotsford Ct Glen Ellyn IL 60137

MOORING, JOHN STUART, b Long Beach, Calif, July 14, 26; m 55; c 4. BOTANY. Educ: Univ Calif, Santa Barbara, AB, 50; Univ Calif, Los Angeles, PhD, 56. Prof Exp: Instr biol, Occidental Col, 55; from instr to asst prof bot, Wash State Univ, 55-61; vis asst prof, Univ Calif, Riverside, 61-62, lectr, 62-63; from asst prof to assoc prof, 63-69, PROF BOT, UNIV SANTA CLARA, 69- Concurrent Pos: NSF res grants, 60-63, 65-66 & 66-69. Mem: Bot Soc Am; Soc Study Evolution. Res: Biosystematics. Mailing Add: Dept of Biol Univ of Santa Clara Santa Clara CA 95053

MOORING, PAUL K, b Houston, Tex, Sept 16, 23; m 62; c 2. PEDIATRIC CARDIOLOGY, PEDIATRICS. Educ: NY Univ, BS, 55; Columbia Univ, MD, 57. Prof Exp: Intern med, Bellevue Hosp, NY, 57-58; res pediat, Presby Med Ctr-Babies Hosp, 58-60; NIH vis fel pediat cardiol, Presby Med Ctr, NY, 60-62; from asst prof to assoc prof pediat, 62-71, dir pediat cardiol & cardiovasc labs, 64-71, PROF PEDIAT, SCH MED, UNIV NEBR AT OMAHA, 71- Mem: Am Acad Pediat; AMA; fel Am Col Cardiol; Am Heart Asn. Res: Radiotelemetry electrocardiology at maximum stress; educational television as a technique for postgraduate physician education; heart disease in the retarded. Mailing Add: Dept of Pediat Univ of Nebr at Omaha Omaha NE 68105

MOORJANI, KISHIN, b Karachi, Pakistan, Apr 9, 35; m 62. PHYSICS. Educ: Univ Delhi, BSc, 55, MSc, 57; Cath Univ Am, PhD(statist mech), 64. Prof Exp: Prin physicist, Melpar, Inc, Westinghouse Air Brake Co, 61-63; res assoc statist mech & solid state physics, Cath Univ Am, 64-65, asst prof, 65-66; vis scientist, Nat Ctr Sci Res, France, 66-67; sr physicist, 67-74, PRIN PHYSICIST, APPL PHYSICS LAB, JOHNS HOPKINS UNIV, 74- Concurrent Pos: Consult, Melpar, Inc, Westinghouse Air Brake Co, 63-67 & Goddard Space Flight Ctr, NASA, 64-65; vis prof, Univ Grenoble, 74-75; vis scientist, Nat Ctr Sci Res, Grenoble, France, 74-75. Mem: Am Phys Soc. Res: Disordered solids; semiconductor physics; statistical mechanics; optical properties of solids. Mailing Add: Appl Physics Lab Johns Hopkins Univ Silver Spring MD 20910

MOOR-JANKOWSKI, JAN K, b Czestochowa, Poland, Feb 5, 24; US citizen. IMMUNOGENETICS, PRIMATOLOGY. Educ: Univ Bern, MD, 54. Prof Exp: Prin investr, Alpine Isolates Genetic Res Prog, Swiss Nat Fund Sci Res, 53-57; res asst, Inst Human Genetics, Univ Geneva, 55-57; res assoc, Inst Study Human Variation, Columbia Univ, 57-58; NIH spec res fel, Blood Group Res Unit, Lister Inst Prev Med, London & Dept Path, Cambridge Univ, 59-62; vis scientist, Human Genetics Br, NIH, 62-64; chief div exp immunogenetics oncol, Yerkes Primate Ctr, Emory Univ, 64-65; res assoc prof forensic med, 65-69, RES PROF FORENSIC MED & DIR LAB EXP MED & SURG PRIMATES, MED SCH, NY UNIV, 69- Concurrent Pos: Ed-in-chief, Primates in Med & J Med Primatol. Mem: AAAS; Am Soc Human Genetics; Am Soc Immunol; Tissue Cult Asn; Soc Cryobiol. Res: Immunological specificity of tissue cells in culture; primate immunogenetics; experimental medicine in nonhuman primates. Mailing Add: Lab Exp Med & Surg Primates NY Univ Med Ctr New York NY 10016

MOORMAN, ROBERT BRUCE, b Chadron, Nebr, Oct 21, 16; m 43; c 3. FISHERIES. Educ: Iowa State Col, BS, 39, MS, 42, PhD(fisheries mgt), 53. Prof Exp: Asst, Iowa State Univ, 39-42, exten wildlife conservationist, 48-53; asst prof zool, Kans State Univ, 53-56; from asst prof to assoc prof, 56-69, PROF ZOOL & ENTOM, IOWA STATE UNIV, 69-, EXTEN WILDLIFE CONSERVATIONIST, 56- Concurrent Pos: With Iowa Conserv Comn, 46-48. Mem: Am Fisheries Soc; Wildlife Soc. Res: Fisheries management in farm ponds; bobwhite quail. Mailing Add: Dept of Animal Ecol Iowa State Univ Ames IA 50011

MOORREES, COENRAAD FRANS AUGUST, b Hague, Neth, Oct 23, 16; nat US; m 39; c 2. ORTHODONTICS. Educ: State Univ Utrecht, dipl, 39; Univ Pa, DDS, 41. Hon Degrees: AM, Harvard Univ, 59; Dr Med, Univ Utrecht, 71. Prof Exp: Intern, Eastman Dent Dispensary, 41; sr fel orthod, 47-48, actg chief dept orthod, 48-56, assoc prof, 59-64, PROF ORTHOD, FORSYTH DENT CTR, HARVARD UNIV, 64-, CHIEF DEPT, 56- Concurrent Pos: Mem sci exped, Aleutian Islands, 48; res fel ondontol, Peabody Mus, 51-66, res assoc odontol, 66-68, hon assoc, 68- Mem: Am Asn Orthod; Am Asn Phys Anthrop; Int Asn Dent Res; Neth Soc Study Orthod; Neth Dent Soc. Res: Child growth and development, especially face and dentition; dental anthropology; evolutionary and racial aspects of the dentition. Mailing Add: Forsyth Dent Ctr 140 The Fenway Boston MA 02115

MOORTI, VARAHUR R GURU, b Poona, India, Oct 6, 32; m 63; c 1. PHYSICAL CHEMISTRY, APPLIED MATHEMATICS. Educ: NWadia Col, India, BS, 52; Univ Poona, MS, 59, BA, 61; NY Univ, PhD(chem physics), 68. Prof Exp: Sci asst phys chem, Nat Chem Lab, India, 53-62; asst res scientist, NY Univ, 62-64, res assoc, 67-68; asst prof, 68-71, ASSOC PROF PHYS CHEM, GRAMBLING COL, 71- Mem: Am Chem Soc; Am Math Soc. Res: Quantum mechanics; thin film metallography; electron microscopy; statistics. Mailing Add: Dept of Chem Grambling Col Grambling LA 71245

MOOS, ANTHONY MANUEL, b Madrid, Spain, June 14, 13; m 39; c 2. PHYSICAL CHEMISTRY. Educ: Univ Geneva, Chem Eng, 33; Univ Paris, ScD(phys chem), 37. Prof Exp: Chief chemist, Lederle Labs, Am Cyanamid Co, NY, 37-44 & R H Macy & Co, 44-46; dir chem res, Eversharp, Inc, 46-48; vpres, Patterson, Moos & Co, Inc, 48-53, pres, 54; gen mgr, Patterson, Moos Div, Universal Winding Co, 54-59; VPRES, LEESONA CORP, WARWICK, 59- Mem: AAAS; Am Chem Soc; NY Acad Sci; Inst Elec & Electronics Engrs; Am Phys Soc. Res: Industrial and organic physical chemistry; nucleonics; ordnance engineering; electronic instrumentation. Mailing Add: 14 Cindy Lane Portsmouth RI 02871

MOOS, CARL, b New York, NY, Mar 3, 30; m 51; c 3. BIOCHEMISTRY, CELL PHYSIOLOGY. Educ: Mass Inst Technol, SB, 50; Columbia Univ, PhD(biophys), 57. Prof Exp: Res assoc chem, Northwestern Univ, 55-57; res assoc physiol, Col Med, Univ Ill, 57-58, instr, 58-59; assoc biophys, Sch Med, State Univ NY Buffalo, 59-61, asst prof, 61-66; assoc prof biol sci, 66-69, ASSOC PROF BIOCHEM, STATE UNIV NY STONY BROOK, 69- Concurrent Pos: NIH spec fel, King's Col, Univ London, 70-71. Mem: Fel AAAS; Biophys Soc; Soc Gen Physiologists; Am Soc Biol Chemists. Res: Muscle contraction; molecular mechanisms of contraction and relaxation; role of muscle proteins. Mailing Add: Dept of Biochem State Univ of NY Stony Brook NY 11794

MOOS, FELIX, b Konstanz, Baden, Ger, Sept 28, 29; US citizen; m 55; c 1. ANTHROPOLOGY. Educ: Univ Cincinnati, BA, 55; Univ Wash, MA, 57, PhD(anthrop), 63. Prof Exp: Lectr anthrop, Inst Oriental Cult, Univ Tokyo, 59-60; lectr anthrop & Asian studies, Univ Md in Far East, 60-61; PROF ANTHROP, UNIV KANS, 61-, DIR CTR EAST ASIAN STUDIES, 68- Concurrent Pos: Lectr anthrop, Korea Univ, 64-65; consult, AID, Korea, 65-67; mem, Comt Korean studies, Soc Sci Res Coun, 67-69 & East Asia Comt, Am Coun Learned Socs, 68-70; Claude V Ricketts Chair comp cult, US Naval War Col, 72-73; tech specialist, World Bank, Kasetsart Univ, Bangkok Develop Proj, 75-76. Mem: Fel Am Anthrop Asn; fel Soc Appl Anthrop; Royal Asiatic Soc Gt Brit & Ireland. Res: Applied anthropology; rapid acculturation and culture change in Asia and Micronesia; modernization through culture contact with the West. Mailing Add: Dept of Anthrop Univ of Kans Lawrence KS 66044

MOOS, GILBERT ELLSWORTH, b Hasbrouck Heights, NJ, May 1, 15; m 44; c 2. ORGANIC CHEMISTRY, BIOCHEMISTRY. Educ: St Lawrence Univ, BS, 36; Mass Inst Technol, SM, 37, PhD(org chem), 39. Prof Exp: Instr chem, Rollins Col, 39-40; res assoc vitamins, Mass Inst Technol, 40; res chemist, Am OakLeather Co, Ohio, 41-42; res chemist, Celanese Corp Am, NJ & Md, 42-49, head spinning res, 49-52; from asst prof to assoc prof chem, St Lawrence Univ, 52-63; PROF CHEM, STATE UNIV NY COL FREDONIA, 63- Concurrent Pos: Chmn dept chem, State Univ NY Col Fredonia, 63-68, sabbatical leave, 69-70; consult, Howard Smith Paper Mills, Ltd. Mem: Fel AAAS; Am Chem Soc; fel Am Inst Chemists; assoc NY Acad Sci. Res: Syntheses of proteins and lipoproteins; cancer chemotherapy. Mailing Add: 34 Middlesex Rd Fredonia NY 14063

MOOS, HENRY WARREN, b New York, NY, Mar 26, 36; m 57; c 2. PHYSICS. Educ: Brown Univ, BA, 57; Univ Mich, MA, 59, PhD(physics), 61. Prof Exp: Res assoc physics, Stanford Univ, 61-63, actg asst prof, 63-64; from asst prof to assoc prof, 64-71, PROF PHYSICS, JOHNS HOPKINS UNIV, 71- Concurrent Pos: Sloan Found fel, 65-67; vis fel, Joint Inst Lab Astrophys & Lab Atmospheric & Space Physics, Univ Colo, 72-73. Mem: Am Phys Soc; Am Astron Soc. Res: Astrophysics; atomic spectroscopy; ultraviolet spectroscopy from space vehicles of objects of astrophysical interest and of very high temperature plasmas. Mailing Add: Dept of Physics Johns Hopkins Univ Baltimore MD 21218

MOOS, WALTER SAM, b Konstanz, Ger, May 1, 24; nat US; m 49; c 2. RADIOLOGY. Educ: Univ Fribourg, PhD(physics, physiol), 48. Prof Exp: Res assoc, Radiation Lab, Mass Inst Technol, 49-50; assoc prof radiol & radiation physics, Univ Ill Col Med, 52-61, res prof radiol & radiation physicist, 61-68; head dosemetry sect, Int Atomic Energy Agency, 68-73. Concurrent Pos: Nat Cancer Inst fel, 50-52. Mem: AAAS; Health Physics Soc; Europ Phys Soc; Am Asn Physicists in Med; Radiation Res Soc. Res: Radiation effects on biological objects; x-ray dosimetry; dosecomputer. Mailing Add: Waldrainstrasse 16 Köniz CH3098 Switzerland

MOOSE, MANUEL FOSTER, chemistry, see 12th edition

MOOSMAN, DARVAN ALBERT, b Toledo, Ohio. ANATOMY, SURGERY. Educ: Bowling Green Univ, AB, 34; Univ Mich, MD, 37, MS, 47; Am Bd Surg, dipl, 52. Prof Exp: Intern & asst resident surg, Univ Mich Hosp, Ann Arbor, 37-40; resident, St Joseph Mercy Hosp, Pontiac, 40-42; clin instr surg, Hosp, 50-52, PROF CLIN ANAT, MED CTR, UNIV MICH, ANN ARBOR, 52- Concurrent Pos: Vis prof clin anat, Med Sch, Univ Hawaii, 70 & Med Ctr, Univ Ala, Birmingham, 73. Mem: Am Soc Abdominal Surg. Res: Biliary duct system; blood supply liver; thyroid gland. Mailing Add: Dept of Anat Univ of Mich Med Ctr Ann Arbor MI 48104

MOOSSY, JOHN, b Shreveport, La, Aug 24, 25; m 51; c 2. NEUROPATHOLOGY, NEUROLOGY. Educ: Tulane Univ, MD, 50. Prof Exp: USPHS fel neuropath, Columbia Univ, 53-54; lectr, Sch Med, Tulane Univ, 54-60; from asst prof to prof neurol & path, Sch Med, La State Univ, 57-65; PROF PATH, SCH MED, UNIV PITTSBURGH, 65- Mem: Am Asn Neuropath; Am Neurol Asn. Res: Neuropathology and neurology, especially those areas related to vascular disease in the nervous system. Mailing Add: Dept of Path Univ of Pittsburgh Sch of Med Pittsburgh PA 15261

MOOZ, ELIZABETH DODD, b Middletown, Conn, Nov 22, 39; m 64; c 2. BIOCHEMISTRY. Educ: Hollins Col, BA, 61; Tufts Univ, PhD(biochem), 67. Prof Exp: Instr res surg, Univ Pa Grad Hosp, 67-68; fel biochem, Univ Del, 68-71, asst prof health sci, 71-73; RES ASSOC CHEM, BOWDOIN COL, 73- Concurrent Pos: Mem, Adv Comt Maine, New Eng Bd Higher Educ, 74- Mem: AAAS; Am Chem Soc. Res: Biosynthesis of gluthathione; amino acid transport, enzyme purification; substrate specificity. Mailing Add: Dept of Chem Bowdoin Col Brunswick ME 04011

MOPPETT, CHARLES EDWARD, b London, Eng, Sept 23, 41; m 65; c 2. ORGANIC CHEMISTRY. Educ: Univ London, BSc & ARCS, 63, PhD(org chem) & dipl, Imp Col, 66. Prof Exp: Res grant org chem, Res Inst Med & Chem, Cambridge, Mass, 66-67 & Harvard Univ, 67-69; res chemist, Dow Chem USA, 69-73; RES CHEMIST, PFIZER, USA, 73- Mem: The Chem Soc; Am Chem Soc. Res: Organic chemistry, especially structure determination of natural products and their synthesis; general synthetic chemistry; new synthetic methods; discovery isolation; physico-chemical characterization; chemistry and biology of microbial metabolites. Mailing Add: Cent Res Pfizer Inc Groton CT 06340

MOPSIK, FREDERICK ISRAEL, b New York, NY, May 20, 38; m 65; c 1. PHYSICAL CHEMISTRY. Educ: Queen's Col, NY, BS, 59; Brown Univ, PhD(chem), 64. Prof Exp: PHYS CHEMIST, POLYMERS DIV, NAT BUR STANDARDS, 63- Mem: Am Phys Soc. Res: Dielectrics research; equations of state; polymer research. Mailing Add: Polymers Div Nat Bur of Standards Washington DC 20234

MORA, EMILIO CHAVEZ, b Valedon, NMex, Aug 14, 28; m 52; c 4. BACTERIOLOGY. Educ: Univ NMex, BS, 51; NMex State Univ, MS, 54; Kans State Univ, PhD(bact), 59. Prof Exp: Assoc prof, 58-70, PROF POULTRY SCI, AUBURN UNIV, 70- Mem: Am Soc Microbiol. Res: Chemistry of virus-host cell relationships; virus vaccines; electron microscopy. Mailing Add: Dept of Poultry Sci Auburn Univ Auburn AL 36830

MORA, PETER TIBOR, biochemistry, see 12th edition

MORABITO, JOSEPH MICHAEL, b Asbury Park, NJ, Feb 26, 41; m 68; c 4. ELECTRONIC SPECTROSCOPY. Educ: Univ Notre Dame, BSMet, 63; Univ Pa, PhD(mat sci), 67. Prof Exp: Fel low energy electron diffraction, Univ Calif, Berkeley, 68-69; vis scientist thin film res & develop, Phillips Res, Neth, 69-70; mem tech staff, 70-74, SUPVR THIN FILM RES & DEVELOP, BELL LABS, AM TEL & TEL CO, 75- Concurrent Pos: Consult, NSF & Energy Res & Develop Admin, 74- Mem: Am Vacuum Soc. Res: Thin film research and development using auger electron spectroscopy and secondary ion mass spectrometry; bonding techniques; interdiffusion studies; solar cells. Mailing Add: Bell Labs 555 Union Blvd Allentown PA 18103

MORACK, JOHN LUDWIG, b Schenectady, NY. PHYSICS. Educ: Union Col, NY, BS, 61; Ore State Univ, PhD(physics), 67. Prof Exp: Asst prof, 67-70, ASSOC PROF PHYSICS, UNIV ALASKA, 70- Mem: Am Phys Soc; Am Asn Physics Teachers. Res: Atomic and molecular physics. Mailing Add: Dept of Physics Univ of Alaska Fairbanks AK 99701

MORACZEWSKI, ALBERT STEPHAN, psychopharmacology, see 12th edition

MORAGNE, EDWARD LEVERNE, physics, mathematics, see 12th edition

MORAHAN, PAGE SMITH, b Newport News, Va, Jan 7, 40; m 63. VIROLOGY, IMMUNOLOGY. Educ: Agnes Scott Col, BA, 61; Hunter Col, MA, 64; Marquette Univ, PhD(microbiol), 69. Prof Exp: Res technician, Rockefeller Univ, 61-65; NIH trainee, 69-70, A D Williams Jr acad fel, 70, asst prof, 71-74, ASSOC PROF MICROBIOL, MED COL VA, 74- Concurrent Pos: NIH res career development award, 74. Mem: AAAS; Soc Exp Biol & Med; Am Soc Microbiol; Am Chem Soc; Sigma Xi. Res: Host resistance to viruses and tumors; interferon inducers and antitumor drugs; interferon and immunity in age-related resistance to viruses; immunomodulators. Mailing Add: Dept of Microbiol Med Col of Va Richmond VA 23298

MORAIS, REJEAN, b Montreal, Que, Oct 26, 38; m 63; c 2. BIOCHEMISTRY. Educ: Univ Montreal, BSc, 60, MSc, 62, PhD(biochem), 65. Prof Exp: Res fel med, Harvard Med Sch, 65-67; asst prof, 67-73, ASSOC PROF BIOCHEM, UNIV MONTREAL, 73-; RES ASST CANCER BIOCHEM, INST CANCER MONTREAL, NOTRE-DAME HOSP, 67- Mem: AAAS. Res: Interrelation between cellular nuclease activities, mitochondria and the viral and chemical transformation of normal cells to tumor cells. Mailing Add: Inst of Cancer of Montreal Notre-Dame Hosp Montreal PQ Can

MORALES, DANIEL RICHARD, b San Francisco, Calif, Jan 1, 29; m 57. CLINICAL BIOCHEMISTRY. Educ: Univ San Francisco, BS, 55; Calif Inst Technol, MS, 56; Univ Calif, PhD(biochem), 62. Prof Exp: Fel pharmacol, Yale Univ, 62-65, res assoc, 65-66; asst prof, Sch Med, Univ Kans, 66-68; res clin chemist, Calif Dept Pub Health, 68-72, CHIEF CLIN CHEM LAB, CALIF DEPT HEALTH, 72- Mailing Add: Clin Chem Lab Calif Dept of Health Berkeley CA 94704

MORALES, GUSTAVO ADOLFO, b Mexico City, Mex, Mar 7, 35; m 61; c 2. GEOLOGY, MICROPALEONTOLOGY. Educ: Baylor Univ, BS, 60; Univ Mo, MA, 62; La State Univ, PhD(geol), 65. Prof Exp: Res geologist, Esso Prod Res Co, 65-67; assoc prof paleont & micropaleont, 67-74, ASSOC PROF GEOL, BAYLOR UNIV, 74- Concurrent Pos: Consult, Inst Geol, Nat Univ Mex, 65-69 & Mex Petrol Inst, 69-70. Mem: Nat Asn Geol Teachers. Res: Ecology, taxonomy and distribution of Recent and Tertiary ostracodes and foraminifers; taxonomy and distribution of charophytes. Mailing Add: Dept of Geol Baylor Univ Waco TX 76703

MORALES, MANUEL FRANK, b San Pedro, Honduras, July 23, 19; US citizen. BIOPHYSICS. Educ: Univ Calif, AB, 39, PhD(physiol), 42; Harvard Univ, AM, 41. Prof Exp: Teaching fel physiol, Univ Calif, 41-42; instr physics, Western Reserve Univ, 42-43; instr math biophys & asst prof physiol, Univ Chicago, 46-48; head phys biochem div, Naval Med Res Inst, 48-57; prof biochem & chmn dept, Dartmouth Med Sch, 57-60; prof biochem, 60-69, PROF BIOPHYS, DEPT BIOCHEM & BIOPHYS & CARDIOVASC RES INST, SCH MED, UNIV CALIF, SAN FRANCISCO, 69- Concurrent Pos: Mem panel physiol, Comt Undersea Warfare, Nat Res Coun, 49; mem US cultural mission, Honduras, 51 & physiol study sect, USPHS, 52-; mem, Nat Adv Res Resources Coun, 67-, mem sr fel selection comt; mem molecular biol panel, NSF; career investr, Am Heart Asn, 60-; ed, Ann Rev Biophys & Bioeng. Honors & Awards: Flemming Award, US Fed Serv. Mem: Nat Acad Sci; Am Soc Biol Chem; Soc Gen Physiol; Biophys Soc (pres, 68); Am Physiol Soc. Res: Biochemical thermodynamics and kinetics; physical chemistry of muscle contraction. Mailing Add: Cardiovasc Res Inst Univ of Calif San Francisco CA 94143

MORALES, RAUL, b San Pedro Sula, Honduras, Sept 27, 35; US citizen; m 59; c 4. ANALYTICAL CHEMISTRY. Educ: Univ Southwestern La, BS, 61; La State Univ, PhD(anal chem), 66. Prof Exp: Asst prof chem, Nicholls State Col, 65-66; res chemist, Indust & Biochem Dept, E I du Pont de Nemours & Co, 66-75; MEM STAFF, LOS ALAMOS SCI LAB, UNIV CALIF, 75- Mem: Am Chem Soc. Res: Precipitations from homogeneous solution; titrations in non-aqueous solvents; metabolism and analytical chemistry of agricultural chemicals; sampling and trace analytical methods for carcinogenic substances in the occupational environment. Mailing Add: Los Alamos Sci Lab Univ of Calif Los Alamos NM 87545

MORAN, CHARLES HENRY, agronomy, see 12th edition

MORAN, CHARLES WILLIAM, b Chicago, Ill, June 15, 07; m 37; c 2. MATHEMATICS. Educ: Lewis Inst, BS, 33; Loyola Univ, Ill, MA, 36; Univ Ill, PhD(math), 40. Prof Exp: Teacher, Lane Tech Sch, 31-45; instr math, Ill Inst Technol, 45-47; dean, 62-67, vpres admin affairs, 67-70, PROF MATH, NORTHEASTERN ILL UNIV, 54- Concurrent Pos: Instr, Loyola Univ, Chicago, 37 & Wright Jr Col, 46-54. Mem: Fel AAAS; Am Math Soc; Math Asn Am. Res: Analysis; hypercomplex variables; asymptotic theory of linear differential equations singular in several parameters. Mailing Add: Dept of Math Northeastern Ill Univ Chicago IL 60625

MORAN, DANIEL AUSTIN, b Chicago, Ill, Feb 17, 36; m 68; c 2. TOPOLOGY. Educ: St Mary's Col, Tex, BS, 57; Univ Ill, MS, 58, PhD(math), 62. Prof Exp: Res instr math, Univ Chicago, 62-64; asst prof, 64-67, ASSOC PROF MATH, MICH STATE UNIV, 67- Concurrent Pos: Assoc chmn dept math, Mich State Univ, 69-70; vis scholar, Cambridge Univ, 70-71. Mem: Am Math Soc; Math Asn Am. Res: Theory of topological manifolds. Mailing Add: Dept of Math Mich State Univ East Lansing MI 48823

MORAN, DAVID TAYLOR, b New York, NY, June 30, 40; m 63; c 2. NEUROBIOLOGY, CELL BIOLOGY. Educ: Princeton Univ, AB, 62; Brown Univ, PhD(biol), 69. Prof Exp: NIH fel, Harvard Univ, 69-70; NIH fel, 70-71, ASST PROF ANAT, MED SCH, UNIV COLO, DENVER, 71- Mem: Am Soc Cell Biol; AAAS; Am Soc Zoologists; Sigma Xi. Res: Neurobiology of sensory transduction in mechanoreceptors. Mailing Add: Dept of Anat Univ of Colo Med Sch Denver CO 80220

MORAN, EDWARD FRANCIS, JR, b Lowell, Mass, July 1, 32; m 58; c 4. INORGANIC CHEMISTRY. Educ: Villanova Univ, BS, 54; Univ Pa, PhD(inorg chem), 61. Prof Exp: Asst, Yale Univ, 60-62; SR RES CHEMIST INORG RES, EXP STA, E I DU PONT DE NEMOURS & CO, 62- Mem: Am Chem Soc; Am Inst Mining, Metall & Petrol Engrs. Res: Synthesis and structure determination of inorganic polymers; low-temperature spectroscopy; hydrometallurgy of copper concentrates and ores. Mailing Add: Exp Sta E I du Pont de Nemours & Co Wilmington DE 19898

MORAN, EDWIN THORNE, JR, animal science, nutrition, see 12th edition

MORAN, EMILIO FEDERICO, b Habana, Cuba, July 21, 46; US citizen; m 72. Educ: Spring Hill Col, BA, 68; Univ Fla, MA, 69, PhD(anthrop), 75. Prof Exp: Field supvr, Trop SAm Prog, Univ Fla, 74; ASST PROF ANTHROP, IND UNIV, BLOOMINGTON, 75- Mem: Am Anthrop Asn; Am Ethnol Soc; AAAS; Soc Appl Anthrop; Latin Am Studies Asn. Res: Development of a comprehensive model for the development of tropical rain forest ecosystems, particularly that of the Amazon Basin; identification of cultural and ecological bottlenecks to development, institutional malfunctioning and use of tropical plants. Mailing Add: Dept of Anthrop Ind Univ Bloomington IN 47401

MORAN, JAMES CHRISTOPHER, organic chemistry, see 12th edition

MORAN, JAMES HERBERT, b Great Falls, Mont, Feb 12, 25; m 48; c 3. THEORETICAL PHYSICS, APPLIED MATHEMATICS. Educ: Calif Inst Technol, BS, 45; Case Inst Technol, MS, 49, PhD(elec eng), 52. Prof Exp: Elec engr, Anaconda Copper Mining Co, 46-48; instr electronics Case Inst Technol, 48-52, asst prof, 52-53; res physicist, Schlumberger Well Surv Corp, 53-63, head physics res, Conn, 63-67, dir res, Schlumberger Res Ctr, 67-69, vpres eng, Schlumberger Well Serv, Tex, 69-72, VPRES RES, SCHLUMBERGER DOLL RES CTR, 72- Mem: Am Math Soc; Am Phys Soc; Asn Comput Mach; Inst Elec & Electronics Engrs; Am Inst Mining, Metall & Petrol Engrs. Res: Low frequency and microwave antennas; application of physics and mathematical analysis to oil well logging; information theory and computers. Mailing Add: Schlumberger Doll Res Ctr PO Box 307 Ridgefield CT 06877

MORAN, JAMES PAUL, organic chemistry, see 12th edition

MORAN, JOHN F, biochemistry, pharmacology, see 12th edition

MORAN, JOHN J, b Scranton, Pa, Jan 11, 27; m 52; c 3. MEDICINE. Educ: Univ Scranton, BS, 48; Jefferson Med Col, MD, 52. Prof Exp: Assoc path, Univ Pa & assoc pathologist, Univ Hosp, 57-61; asst prof path, 61-64, ASSOC PROF PATH, JEFFERSON MED COL, 64-, ASST DIR CLIN LABS, HOSP, 61- Concurrent Pos: Consult, Philadelphia Vet Admin Hosp, 57-59. Mem: AMA; Col Am Path; Am Soc Clin Path; Int Acad Path. Res: Pathology. Mailing Add: Dept of Path Jefferson Med Col Philadelphia PA 19107

MORAN, JOSEPH FRANCIS, JR, b Springfield, Mass, July 4, 30; m 55; c 5. PHYSIOLOGY. Educ: Stonehill Col, BS, 52; Univ Notre Dame, MS, 54, PhD(zool), 56. Prof Exp: Instr anat & physiol, Exten, Ind Univ, 54-56; asst prof biol, Duquesne Univ, 56-57; from instr to assoc prof, Russell Sage Col, 57-67; PROF BIOL & CHMN DEPT, SACRED HEART UNIV, 67- Concurrent Pos: Nat Cancer Inst spec fel, 64-65; mem, Marine Biol Lab, Woods Hole. Mem: Am Inst Biol Sci; Am Soc Parasitol. Res: Intermediary metabolism of islet tissue in fish; intermediary metabolism and phagocytosis in leukocytes. Mailing Add: Dept of Biol Sacred Heart Univ 5229 Park Ave Bridgeport CT 06604

MORAN, JOSEPH MICHAEL, b Boston, Mass, Feb 14, 44; m 71. CLIMATOLOGY. Educ: Boston Col, BS, 65, MS, 67; Univ Wis-Madison, PhD(meteorol), 72. Prof Exp: From instr to assoc prof earth sci, Univ Wis-Green Bay, 69-75; VIS ASSOC PROF CLIMAT, UNIV ILL, URBANA, 75- Mem: Am Meteorol Soc; AAAS; Sigma Xi; Nat Asn Geol Teachers; Nat Sci Teachers Asn. Res: Pleistocene climatology; nature of climatic change; environmental science education. Mailing Add: Dept of Geog Univ of Ill Urbana IL 61801

MORAN, JULIETTE MAY, b New York, NY, June 12, 17. ORGANIC CHEMISTRY. Educ: Columbia Univ, BS, 39; NY Univ, MS, 48. Prof Exp: Asst chem, Columbia Univ, 41; jr engr, Signal Corps Lab, US Army, 42-43; jr chemist, Process Develop Dept, Gen Aniline & Film Corp, 43-44, tech asst, 44-48, tech asst to dir cent res lab, 49-52 & com develop, 52-55, supvr tech serv, Com Develop Dept, 55-59, sr develop specialist, Develop Dept, 59-60, mgr planning, 61, asst to pres, 62-67, vpres, GAF Corp, 67-71, SR VPRES, GAF CORP, 71- Mem: Fel AAAS; Am Chem Soc; Com Develop Asn; Am Asn Textile Chem & Colorists; fel Am Inst Chem. Res: Structure of hemocyanine; synthesis and application of dyestuffs, detergents and acetylene derivatives; technical information systems; commercial development; administration. Mailing Add: GAF Corp 140 W 51st St New York NY 10020

MORAN, MARIUS ROBERT, biology, see 12th edition

MORAN, MICHAEL J, b Butte, Mont, Aug 10, 41; m 65; c 3. PHARMACEUTICAL CHEMISTRY. Educ: Mont State Univ, BS, 63; Univ Cincinnati, PhD(biochem), 72. Prof Exp: Res asst bee venom anal, Mont State Univ, 63-65; anal chemist, Mat Lab, Wright-Patterson AFB, Ohio, 67-68; lab chemist, Radiol Health Lab, US Air Force, Ohio, 58-70; ASSOC PROF CHEM, US AIR FORCE ACAD, 72- Concurrent Pos: Co-chmn drug educ, US Air Force Acad, 75- Res: Control of cholesterol levels in blood using lecithin trace metal; analysis using atomic absorption. Mailing Add: Dept of Chem US Air Force Academy CO 80846

MORAN, NEIL CLYMER, b Phoenix, Ariz, Oct 12, 24; m 48; c 3. PHARMACOLOGY. Educ: Stanford Univ, AB, 49, MD, 50. Prof Exp: Irving fel physiol, Stanford Univ, 50-51; med officer, USPHS Hop, Savannah, Ga, 51-52; Nat Heart Inst pharmacologist, Emory Univ, 52-54; head sect pharmacodynamics, Nat Heart Inst, Bethesda, Md, 54-56; assoc prof pharmacol, 56-62, PROF PHARMACOL & CHMN DEPT, EMORY UNIV, 62- Concurrent Pos: USPHS sr res fel, 57-60, res career develop award, 60-62; asst ed, J Pharmacol & Exp Therapeut, 58-60, ed, 61-65; vis scientist, Karolinska Inst, Sweden, 60-61; mem res career award comt, Nat Inst Gen Med Sci, 64-68; mem res comt, Am Heart Asn, 68-73, chmn, 72-73; mem res comt A, Nat Heart & Lung Inst, 75- Honors & Awards: Citation for Distinguished Serv to Res, Am Heart Asn, 74. Mem: AAAS; Am Soc Pharmacol & Exp Therapeut; Soc Exp Biol & Med; Am Heart Asn; Am Asn Univ Prof. Res: Cardiovascular and autonomic pharmacology and physiology; allergy. Mailing Add: Dept of Pharmacol Emory Univ Atlanta GA 30322

MORAN, PAUL RICHARD, b Buffalo, NY, June 1, 36; m 58; c 4. SOLID STATE PHYSICS. Educ: Univ Notre Dame, BS, 58; Cornell Univ, PhD(physics), 63. Prof Exp: Staff consult scientist, Kaman Nuclear Co, Colo, 63; NSF fel physics, Univ Ill, Urbana, 63-65; from asst prof to assoc prof, 65-73, PROF PHYSICS, UNIV WIS-MADISON, 73- Mem: AAAS; Am Phys Soc. Res: Magnetic resonance and optical studies of defect states in insulating solids. Mailing Add: Dept of Physics Sterling Hall Univ of Wis Madison WI 53706

MORAN, REID VENABLE, b Los Angeles, Calif, June 30, 16. PLANT TAXONOMY. Educ: Stanford Univ, AB, 39; Cornell Univ, MS, 42; Univ Calif, PhD(bot), 51. Prof Exp: Botanist, Santa Barbara Bot Garden, 47-48; instr bot, Bailey Hortorium, Cornell Univ, 51-53; lectr Far East prog, Univ Calif, 53-56; CUR BOT, SAN DIEGO MUS NATURAL HIST, 57- Concurrent Pos: Ed, San Diego Mus Natural Hist, 57-62, actg dir, 65-66. Mem: Am Fern Soc; Am Soc Plant Taxon; Bot Soc Am; Int Asn Plant Taxon. Res; Taxonomy of Crassulaceae and Cactaceae; vascular flora of Baja California. Mailing Add: Natural Hist Mus San Diego CA 92112

MORAN, STEPHEN ROYSE, geology, see 12th edition

MORAN, THOMAS FRANCIS, b Manchester, NH, Dec 11, 36; m 60; c 4. PHYSICAL CHEMISTRY. Educ: St Anselm's Col, BA, 58; Univ Notre Dame, PhD(chem), 62. Prof Exp: Asst instr chem, Univ Notre Dame, 58-59; AEC fel, Brookhaven Nat Lab, 62-64, assoc scientist, 64-66; asst prof, 66-68, ASSOC PROF CHEM, GA INST TECHNOL, 68- Concurrent Pos: Danforth fel, 71- Mem: AAAS; Am Chem Soc; Am Phys Soc. Res: Collisions of electrons and ions with molecules; mass spectrometry; energy transfer processes; kinetics of chemical reactions. Mailing Add: Sch of Chem Ga Inst of Technol Atlanta GA 30332

MORAN, THOMAS IRVING, b Amsterdam, NY, Nov 8, 30; m 53; c 4. PHYSICS. Educ: Union Univ, NY, BS, 53; Yale Univ, PhD, 57. Prof Exp: Res physicist, Gen Elec Co, 58-59; res assoc physics, Syracuse Univ, 59, asst prof, 60; res assoc, Brookhaven Nat Lab, 61-63; Fulbright sr scholar, Univ Heidelberg, 63-64; asst prof, 64-74, ASSOC PROF PHYSICS, UNIV CONN, 74- Mem: Am Phys Soc. Res: Thermal diffusion; atomic beams; atomic physics; optical biophysics. Mailing Add: Dept of Physics Univ of Conn Storrs CT 06268

MORAN, THOMAS JAMES, b Rennerdale, Pa, Oct 14, 12; m 41; c 4. PATHOLOGY, MEDICINE. Educ: Univ Pittsburgh, BS & MD, 36; Am Bd Path, dipl, 45. Prof Exp: Dir labs, City Hosp, Pittsburgh, Pa, 39-42; pathologist, Welborn Hosp, Ind, 46; mem staff, Hosp, Danville, Va, 46-50 & St Margaret Hosp, 50-54; assoc prof path, Univ Pittsburgh, 54-58, prof, 58-62; DIR LABS, MEM HOSP, DANVILLE, 62- Mem: Am Soc Clin Path; Am Soc Exp Path; Am Asn Path & Bact; Col Am Path; Path Soc Gt Brit & Ireland. Res: Aspiration pneumonia; cortisone effects. Mailing Add: Creekside Danville VA 24541

MORAN, THOMAS PATRICK, b Detroit, Mich, Nov 6, 41; m 71; c 1. COMMUNICATION SCIENCE. Educ: Univ Detroit, BArch, 65; Carnegie-Mellon Univ, PhD(comput sci), 74. Prof Exp: SCIENTIST PSYCHOL & COMPUT SCI, PALO ALTO RES CTR, XEROX CORP, 74- Mem: Asn Comput Mach; Am Psychol Asn. Res: Information processing analysis of human cognitive behavior; human mental representation of spatial information; artificial intelligence. Mailing Add: Xerox Palo Alto Res Ctr 3333 Coyote Hill Rd Palo Alto CA 94304

MORAN, WALTER HARRISON, JR, b Grand Forks, NDak, Nov 16, 30; m 52; c 2. SURGERY, PHYSIOLOGY. Educ: Univ NDak, BA, 52, BS, 53; Harvard Univ, MD, 55; Am Bd Surg, dipl, 63. Prof Exp: Intern, Dept Surg, Univ Minn Hosps, 55-56; from instr to assoc prof surg, 60-70, dir surg res labs, 60-73, PROF PHYSIOL, SURG & BIOPHYS, SCH MED, WVA UNIV, 70-, COORDR DIV & ACTG DIR EMERGENCY ROOM, 73- Concurrent Pos: Med fel, Dept Surg, Univ Minn Hosps, 56-58, med fel specialist, 58-59, Nat Heart Inst fel, 59-60; USPHS fel, Sch Med, WVa Univ, 60-63; Nat Inst Arthritis & Metab Dis res career develop award, 62-67; co-dir metab unit, Univ Hosp, WVa Univ, 61-70. Mem: Endocrine Soc; AMA; Soc Univ Surgeons; Am Physiol Soc. Res: Biophysics; surgical endocrinology, especially vasopressin physiology; burn surgery. Mailing Add: Dept of Surg WVa Univ Med Ctr Morgantown WV 26506

MORAN, WILLIAM JOSEPH, b Amityville, NY, May 30, 12; m 40; c 2. ORGANIC CHEMISTRY. Educ: Washington & Lee Univ, BS, 34; Polytech Inst Brooklyn, MS, 40, PhD(chem), 49. Prof Exp: Chemist, Mabrand Prod, 35-37 & Dixon Crucible Co, 37-39; develop chemist, Merck & Co, Inc, 39-45; sr org chemist, Vick Chem Co, 45-47, Gen Chem Co, 48 & Nat Dairy Res Lab, Inc, 48-51; develop chemist, Ciba Pharmaceut Prod Corp, 51-53, mfg chemist, 53-68, mgr chem mfg, 68-70, DIR CHEM MFG, PHARMACEUT DIV, CIBA-GEIGY CORP, 70- Mem: Am Chem Soc; NY Acad Sci. Res: Friedel-Craft reaction on aromatic substituted aliphatic amides; nuclear methylation of substituted phenols; resolution of synthetic menthol; penicillin. Mailing Add: 211 Longwood Ave Chatham NJ 07928

MORAN, WILLIAM RODES, b Los Angeles, Calif, July 29, 19. GEOLOGY. Educ: Stanford Univ, AB, 42. Prof Exp: Geologist, Union Oil Co, Calif, 43-46, Paraguay, 46-50, sr geologist, Nev, 50-51, Cia Petrol de Costa Rica, 51-52, mem spec explor staff, 52-59, actg mgr, Union Oil Develop Corp, Australia, 59-60, for opers, Los Angeles, 60-63, VPRES & MGR, MINERALS EXPLOR CO, UNION OIL CO, CALIF, 63- Concurrent Pos: Consult, Stanford Archive Recorded Sound; assoc ed, Bull, Am Asn Petrol Geologists, 56- Mem: Fel AAAS; fel Am Geog Soc; Am Inst Mining, Metall & Petrol Engrs; fel Geol Soc Am; fel Geol Soc London. Res: Historical sound recording; mining administration; mining and petroleum geology. Mailing Add: 1335 Olive Lane La Canada CA 91011

MORANA, SIMON JOSEPH, b St Louis, Mo, Apr 22, 17; m 45; c 4. ORGANIC CHEMISTRY. Educ: Western Reserve Univ, BS, 41; Duke Univ, MA, 42. Prof Exp: Asst, Duke Univ, 42; res chemist, Beacon Res Lab, Tex Co, 42-47; res engr, Brush Beryllium Co, Ohio, 47-49; res engr, Clifton Prod, Inc, 49-51; supvry engr, Crane Co, Ill, 51-52; chief res engr, Beryllium Corp, Pa, 52-56, dir res, 56-69, VPRES, AG-MET INC, 69- Mem: Am Chem Soc; Electrochem Soc; Am Soc Metals; fel Am Inst Chemists. Res: Extractive metallurgy of non-ferrous metals and alloys, synthesis of complex metallo-organic compounds: fused salt bath electrolyses; beryllium and titanium metal extraction; electrowinning and electrorefining; plastics. Mailing Add: Ag-Met Inc Box 523 Hazleton PA 18201

MORAND, PETER, b Montreal, Can, Feb 11, 35; m 57; c 2. ORGANIC CHEMISTRY. Educ: Bishop's Univ, Can, BSc, 56; McGill Univ, PhD(org chem), 59. Prof Exp: NATO fel, Imp Col, Univ London, 59-61; sr chemist, Ayerst Labs, Can, 61-63; asst prof, 63-67, asst vice rector acad, 69-71, ASSOC PROF ORG CHEM, UNIV OTTAWA, 67- Concurrent Pos: Nat Res Coun Can grants, 63-; Ont Res Found grant, 64-65. Mem: Am Chem Soc; The Chem Soc; fel Chem Inst Can. Res: Chemistry of natural products; especially steroids; conformational transmission effects and reaction mechanisms; biosynthesis of estrogens; stereochemistry. Mailing Add: Dept of Chem Univ of Ottawa Ottawa ON Can

MORATH, RICHARD JOSEPH, b St Paul, Minn, July 13, 25; m 52; c 2. ORGANIC CHEMISTRY. Educ: Univ Minn, BChem, 49; Wash State Univ, MS, 51, PhD(chem), 54. Prof Exp: Res assoc, Univ NC, 53-54 & Univ Iowa, 54-55; asst prof org chem, Univ Dayton, 55-57; from asst prof to assoc prof, 57-70, PROF CHEM, COL ST THOMAS, 70- Mem: Am Chem Soc; The Chem Soc. Res: Organic syntheses; reaction mechanisms. Mailing Add: Dept of Chem Col of St Thomas St Paul MN 55105

MORATO, TOMAS, b Tampico, Mex, Dec 21, 29; m 57; c 3. ENDOCRINOLOGY, REPRODUCTIVE PHYSIOLOGY. Educ: Nat Polytech Inst, Mex, MD, 54. Prof Exp: Asst prof physiol, Nat Polytech Inst, Mex, 54-59; ASSOC RESEARCHER ENDOCRINOL, NAT INST NUTRIT, 63-; ASSOC PROF ENDOCRINOL, MED SCH, NAT UNIV MEX, 69- Concurrent Pos: Fel endocrinol, diabetes & nutrit, Nat Inst Nutrit, Mexico City, 59-60; NIH fel, Worcester Found, 61-63; asst prof, Med Sch, Nat Univ Mex & Nat Polytech Inst, 65- Mem: Endocrine Soc; Mex Acad Sci Res; Mex Soc Nutrit & Endocrinol; Mex Soc Biochem. Res: Steroid metabolism in gonadal tissue; metabolic fate of steroid hormones; steroid metabolism in target tissues; steroid hormones quantification. Mailing Add: Dept of Biol Reproduction Nat Inst of Nutrit Mexico DF Mexico

MORAVCSIK, MICHAEL JULIUS, b Budapest, Hungary, June 25, 28; US citizen; m 56; c 2. HIGH ENERGY PHYSICS, SCIENCE POLICY. Educ: Harvard Col, AB, 51; Cornell Univ, PhD(theoret physics), 56. Prof Exp: Res assoc theoret physics, Brookhaven Nat Lab, 56-58; physicist & head elem particle & nuclear theory group,

Lawrence Radiation Lab, Univ Calif, 58-67; PROF PHYSICS & RES ASSOC, INST THEORET SCI, UNIV ORE, 67-, DIR INST, 69- Concurrent Pos: Vis prof, Purdue Univ, 57 & Int Atomic Energy Agency, Atomic Energy Ctr, Pakistan AEC, Lahore, 62-63; vis lectr, Harvard Univ, 66-67; SEED grant, Nigeria, 75; NATO sr fel sci, Sci Policy Res Unit, Univ Sussex, Falmer, Brighton, Sussex, Eng, 75. Mem: Fel Am Phys Soc. Res: Theoretical elementary particle physics, especially as it is related to experiments; photoproduction processes; nuclear forces; spin structure of particle reactions; assignments of intrinsic quantum numbers; problems of science policy, organizattion and management in developing countries. Mailing Add: Inst of Theoret Sci Univ of Ore Eugene OR 97403

MORAVEK, PAUL HAYNAL, theoretical physics, see 12th edition

MORAVEK, RICHARD THOMAS, analytical chemistry, see 12th edition

MORAWA, ARNOLD PETER, b Detroit, Mich, Feb 14, 40; m 65; c 3. PEDODONTICS, ORAL BIOLOGY. Educ: Univ Mich, DDS, 64, MS, 66 & 68, PhD(anat), 73. Prof Exp: ASST PROF DENT & HOSP DENT, UNIV MICH, ANN ARBOR, 73- Concurrent Pos: Consult, Hawthorne Ctr, State Mich, 66-72, Plymouth State Home, 68-72 & Ctr Study Ment Retardation, Univ Mich, 75-; pedodontist, pvt pract, Ann Arbor, 66- Mem: Int Asn Dent Res. Res: Biochemical and ultrastructural evaluation of dilute formocresol upon dental pulp; concurrent studies of clinical application of this medicament; biochemical and ultrastructural effects of hypoxia upon protein secreting cells. Mailing Add: 3920 Waldenwood Ann Arbor MI 48105

MORAWETZ, CATHLEEN SYNGE, b Toronto, Ont, May 5, 23; nat US; m 45; c 4. APPLIED MATHEMATICS. Educ: Univ Toronto, BS, 44; Mass Inst Technol, MS, 46; NY Univ, PhD(math), 51. Prof Exp: Res assoc, 51-57, from asst prof to assoc prof, 57-66, PROF MATH, COURANT INST MATH SCI, NY UNIV, 66- Concurrent Pos: Guggenheim fel, 66-67; term trustee, Princeton Univ. Mem: Am Math Soc; Soc Indust & Appl Math. Res: Applications of partial differential equations. Mailing Add: Courant Inst Math Sci NY Univ 251 Mercer St New York NY 10012

MORAWETZ, HERBERT, b Prague, Czech, Oct 16, 15; nat US; m 45; c 4. PHYSICAL CHEMISTRY. Educ: Univ Toronto, BASc, 43, MASc, 44; Polytech Inst Brooklyn, PhD(chem), 50. Prof Exp: Res chemist, Bakelite Co, 45-49; fel, NIH, 50-51; from asst prof to assoc prof, 51-58, PROF POLYMER CHEM, POLYTECH INST BROOKLYN, 58- Concurrent Pos: NIH fel, 67-68. Mem: AAAS; Am Chem Soc. Res: Molecular association of polymers; reaction kinetics in polymer solutions; polyelectrolyte chelates; solid state polymerization; enzyme models. Mailing Add: Dept of Chem Polytech Inst of Brooklyn Brooklyn NY 11201

MORAWITZ, HANS, b Neustadt, Austria, Feb 6, 35; m 63; c 1. THEORETICAL PHYSICS. Educ: Stanford Univ, BS, 56, PhD(physics), 63. Prof Exp: Staff physicist, IBM Res Lab, 63-65; Australian Dept Supply fel & sr lectr physics, Monash Univ, Australia, 65-66; Austrian Ministry Educ res assoc, Univ Vienna, 66-67; STAFF PHYSICIST, IBM RES LAB, 67- Mem: Am Phys Soc. Res: Theoretical atomic, molecular and solid state physics; surface physics; cooperative radiative processes; phase transition in organic solids and magnetic systems; macroscopic quantum states. Mailing Add: Dept K32 IBM Res Lab Monterey & Cottle Rds San Jose CA 95193

MORBEY, GRAHAM KENNETH, b Birmingham, Eng, Apr 5, 35; m 60; c 2. TEXTILE CHEMISTRY, CHEMICAL ENGINEERING. Educ: Univ Birmingham, BSc, 56; Univ Toronto, MASc, 57; Princeton Univ, MA, 58, PhD(chem eng), 61. Prof Exp: Res scientist, Dunlop Res Ctr, Can, 60-61, sect leader, Dunlop Tyre Co, Eng, 61-62, asst to tech dir, 62-63; group leader, Celanese Fibers Mkt Co, 63-64, develop mgr, 64-67, develop dir, 67-69; tech dir, Hoechst Fibers Inc, 69-72; DIR APPL RES, PERSONAL PRODS CO, 72- Mem: Textile Res Inst; Fiber Soc; AAAS. Res: Polymer chemistry; statistics; fiber physics and textiles. Mailing Add: Personal Prods Co Milltown NJ 08850

MORCH, ERNST TRIER, b Slagelse, Denmark, May 14, 08; nat US; m 40; c 4. ANESTHESIOLOGY. Educ: Copenhagen Univ, MD, 35, PhD(human genetics), 42; FRCS, 54; Am Bd Anesthesiol, dipl, 56. Prof Exp: Sci asst, Inst Human Genetics, Copenhagen Univ, 38-40, privatdocent anesthesia, 43-49; asst prof, Univ Kans, 50-52; prof surg & dir anesthesia, Univ Chicago, 53-58; PROF SURG, UNIV ILL MED CTR, 59- Concurrent Pos: Assoc, Gentofte Hosp, Denmark, 42-46; Brit Coun scholar, Oxford Univ, 46-47; dir anesthesia, Cook County Hosp, Chicago, 59-61 & Rush-Presby Hosp, St Luke's Med Ctr, 67- Honors & Awards: Decorated, Kings of Denmark, Norway & Sweden. Mem: Int Anesthesia Res Soc; corresp mem Asn Anaesthetists Gt Brit & Northern Ireland; corresp mem Fr Soc Anesthesia & Analgesia; fel Int Col Anesthetists; Am Col Anesthesiol. Res: Anesthesia; artificial respiration; poliomyelitis; respiratory physiology. Mailing Add: Apt 5009 300 N State St Chicago IL 60610

MORCK, ROLAND ANTON, b Crookston, Minn, July 11, 13; m 39; c 4. BIOCHEMISTRY. Educ: St Olaf Col, BA, 35; Pa State Univ, MS, 37, PhD(biochem), 39. Prof Exp: Nutrit specialist, R B Davis Co, 39-41, chemist, 41-43, chief chemist, 44-55; res chemist, 55-58, asst dir res, 58-72, DI RES, NAT BISCUIT CO, 72- Mem: AAAS; Am Chem Soc; Am Oil Chem Soc; Am Asn Cereal Chem; Inst Food Technol. Res: Food research. Mailing Add: 32 Abbington Terr Glen Rock NJ 07452

MORCOCK, ROBERT EDWARD, b Washington, DC, Aug 26, 38; m 67; c 2. PHYSIOLOGICAL ECOLOGY. Educ: NC Wesleyan Col, BA, 66; Wake Forest Univ, MA, 70, PhD(physiol ecol), 74. Prof Exp: RES ASSOC PARASITOL, UNIV MASS, AMHERST, 74- Concurrent Pos: NIH trainee, Univ Mass, Amherst, 74-76. Mem: AAAS; Am Inst Biol Sci; Am Soc Parasitologist. Res: Physiology and biochemistry of symbiotic helminths; carbon dioxide fixation by developing cestodes; effects of malnutrition on host-parasite relations. Mailing Add: Dept of Zool Univ of Mass Morrill Sci Ctr Amherst MA 01002

MORDESON, JOHN N, b Council Bluffs, Iowa, Apr 22, 34; m 60; c 5. MATHEMATICS. Educ: Iowa State Univ, BS, 59, MS, 61, PhD(math), 63. Prof Exp: PROF MATH, CREIGHTON UNIV, 63- Mem: Am Math Soc. Res: Field theory and ring theory. Mailing Add: Dept of Math Creighton Univ Omaha NE 68178

MORDUCHOWITZ, ABRAHAM, b New York, NY, Aug 17, 33; m 56; c 3. ORGANIC POLYMER CHEMISTRY. Educ: Yeshiva Col, BA, 54; Univ Chicago, MS, 58, PhD(org photochem), 62. Prof Exp: Chemist, 62-63, sr chemist, 63-68, res chemist, 68-75, SR RES CHEMIST, BEACON RES LAB, TEXACO INC, 75- Mem: Am Chem Soc. Res: Synthesis and study of solution properties of macromolecules. Mailing Add: 3 Crown Rd Monsey NY 10952

MORDUE, DALE LEWIS, b Colchester, Ill, Apr 26, 33; m 60; c 3. PHYSICS, MATHEMATICS. Educ: Western Ill Univ, BS, 54; Univ Ill, Urbana, MS, 59; Tex A&M Univ, PhD(physics), 65. Prof Exp: Teacher high sch, Ill, 56-58; instr phys sci & chem, Evansville Col, 59-61; asst physics, Tex A&M Univ, 61-63; from asst prof to assoc prof, 65-71, PROF PHYSICS, MANKATO STATE COL, 71- Mem: Am Asn Physics Teachers. Res: Biophysics using electrophysiological methods to determine spectral sensitivity of insects. Mailing Add: Dept of Physics Mankato State Col Mankato MN 56001

MORDY, WENDELL ALLEN, meteorology, physics, see 12th edition

MORE, KENNETH RIDDELL, b Vancouver, BC, Jan 9, 10; nat US; m 42; c 2. PHYSICS. Educ: Univ BC, BA, 29, MA, 31; Univ Calif, PhD(physics), 34. Prof Exp: Asst physics, Univ BC, 29-31; asst, Univ Calif, 31-34, fel, 34-35; Royal Soc Can fel, Mass Inst Technol, 35-36; Sterling fel, Yale Univ, 36-38; instr physics, Ohio State Univ, 38-44, asst prof, 44-45; res physicist, Phillips Petrol Co, Okla, 46-47; prof physics, Univ BC, 47-50; opers analyst, US Dept Air Force, 50-61; sr physicist, Stanford Res Inst, 61-63; opers res scientist, Ctr Naval Anal, 63-64, dir naval objectives anal group, 64-67, naval implications technol & tech group, Naval Res Lab, 67-68; sci & eng adv, US Bur Mines, 68-70, CHIEF OFF OPERS RES, US BUR MINES, 70- Concurrent Pos: Mem staff, Radiation Lab, Mass Inst Technol, 42-45. Mem: AAAS; fel Am Phys Soc; Opers Res Soc Am. Res: Atomic and molecular spectroscopy; microwave magnetron design; nuclear physics; operations research. Mailing Add: US Bur Mines 2401 E Street NW Washington DC 20241

MORE, RICHARD MICHAEL, b Kenosha, Wis, July 27, 42; m 60; c 2. THEORETICAL PHYSICS. Educ: Univ Calif, Riverside, BA, 63; Univ Calif, San Diego, MS, 64, PhD(physics), 68. Prof Exp: Res assoc, 68-69, ASST PROF PHYSICS, UNIV PITTSBURGH, 69- Mem: Am Phys Soc. Res: Theoretical solid state physics; impurities in metals. Mailing Add: Dept of Physics Univ of Pittsburgh Pittsburgh PA 15213

MORE, ROBERT HALL, b Kitchener, Ont, Dec 16, 12; m 43; c 3. PATHOLOGY. Educ: Univ Toronto, MD, 39; McGill Univ, MS, 42; FRCPS(C), 61. Prof Exp: Jr intern, Toronto Gen Hosp, 39-40, sr intern surg, 42-43; lectr path, McGill Univ, 43-46, Fraser asst prof comp path, 47-49, Fraser prof, 50-51; prof path & head dept, Queen's Univ, Ont, 51-67; STRATHCONA PROF PATH & CHMN DEPT, PATH INST, McGILL UNIV, 67- Concurrent Pos: Fel, Univ Toronto, 42-43; prosector, Royal Victoria Hosp, Montreal, 43-46; pathologist, Women's Gen Hosp, 45-46; asst pathologist, NY Hosp, 46-47; res assoc, Med Col, Cornell Univ, 46-47; consult pathologist & pathologist, Kingston Gen Hosp, 51-; consult pathologist, Hotel Dieu Hosp, 51- Mem: Am Asn Path & Bact; Can Soc Exp Path; Can Asn Path; Path Soc Gt Brit & Ireland. Res: Relation of hypersensitivity to cardiovascular renal diseases; arteriosclerosis. Mailing Add: Dept of Path McGill Univ Path Inst Montreal PQ Can

MOREAU, JEAN RAYMOND, b Village des Aulnaies, Que, June 27, 24; m 5-; c 4. FOOD SCIENCE. Educ: Laval Univ, BA, 44, BSc, 48; Mass Inst Technol, PhD(food sci & eng), 57. Prof Exp: Res asst biol, Dept Fisheries, Que, 48-49; supt fisheries plant, 49-51; res scientist, Can Packers Ltd, 57-64; prof food sci, Univ Toronto, 64-65; prof FOOD SCI, LAVAL UNIV, 65- Concurrent Pos: Sci consult, 67- Honors & Awards: United Inventors & Scientists Am Achievement Award, 73. Mem: Can Inst Food Sci & Technol. Res: Food science and engineering; hygiene, quality and nutritive value of meat, meat products, vitamins and beverages. Mailing Add: Dept of Chem Eng Laval Univ Ste-Foy PQ Can

MOREE, RAY, b Ellensburg, Wash, Oct 14, 13; m 41. GENETICS, ZOOLOGY. Educ: Wash State Univ, BS, 37; Univ Mich, MS, 39, PhD(zool), 45. Prof Exp: Asst zool, Wash State Univ, 35-37; asst, Univ Mich, 37-42, asst genetics, 42; from instr to assoc prof zool, 42-65, assoc prof genetics, 65-73, PROF ZOOL, WASH STATE UNIV, 73- Mem: AAAS; Am Genetic Asn; Soc Study Evolution; Genetics Soc Am; Am Soc Human Genetics; Am Soc Nat; Am Soc Zool. Res: General and population genetics; fitness components. Mailing Add: Dept of Zool Wash State Univ Pullman WA 99163

MOREHART, ALLEN L, b Williamsport, Pa, Apr 1, 33; m 57; c 5. PHYTOPATHOLOGY, MEDICAL MYCOLOGY. Educ: Lycoming Col, AB, 59; Univ Del, MS, 61, PhD(biol sci), 64. Prof Exp: Res asst plant path, Univ Del, 59-61, res assoc, 61-64; fel med mycol & res assoc microbiol, Univ Okla, 64-65; asst prof, WVa Univ, 65-68, assoc prof biol, Lycoming Col, 68-71, chmn dept, 69-71; ACTG CHMN DEPT PLANT SCI, UNIV DEL, 72- Concurrent Pos: Danforth Assoc, 69. Mem: AAAS; Mycol Soc Am; Am Soc Microbiol; Int Soc Human & Animal Mycol; Am Phytopath Soc. Res: Mechanism of fungicidal action; fungal physiology. Mailing Add: Dept of Plant Sci Univ of Del Newark DE 19711

MOREHEAD, FREDERICK FERGUSON, JR, b Roanoke, Va, July 30, 29; m 54; c 5. PHYSICAL CHEMISTRY, EXPERIMENTAL SOLID STATE PHYSICS. Educ: Swarthmore Col, BA, 50; Univ Wis, MS, 51, PhD(phys chem), 53. Prof Exp: Asst prof chem, Union Col, NY, 53-54; mem staff, Lamp Develop Lab, Gen Elec Co, Ohio, 54-59; MEM STAFF, T J WATSON RES CTR, IBM CORP, 59- Mem: Am Phys Soc; Sigma Xi. Res: Luminescence and photoconductivity in II-IV compounds; electroluminescence; ion implantation. Mailing Add: RFD Baldwin Rd Yorktown Heights NY 10598

MOREHEAD, ROBERT P, b Lasker, NC, Sept 4, 10; m 46; c 3. PATHOLOGY. Educ: Wake Forest Col, BS, 31, MA, 32, BSMed, 34; Jefferson Med Col, MD, 36. Prof Exp: Instr path, 36-37, asst prof, Sch Med Sci, 38-41, assoc prof, Bowman Gray Sch Med, 41-46, PROF PATH & CHMN DEPT, BOWMAN GRAY SCH MED, WAKE FOREST UNIV, 46- Concurrent Pos: Pathologist, NC Baptist Hosp, 41-; consult, Nat Cancer Inst. Mem: Am Soc Clin Path; Am Asn Path & Bact; Am Asn Cancer Res; Col Am Path; Am Col Physicians. Res: Histogenesis of tumors; pathogenesis of appendicitis. Mailing Add: Dept of Path Bowman Gray Sch of Med Winston-Salem NC 27103

MOREHOUSE, ALPHA L, b Lafayette, Ind, Sept 27, 23; m 47; c 4. BIOCHEMISTRY. Educ: Purdue Univ, BS, 48; Pa State Univ, MS, 50, PhD(biochem), 52. Prof Exp: SR BIOCHEMIST, GRAIN PROCESSING CORP, 52- Mem: Am Chem Soc; Poultry Sci Asn. Res: Poultry nutrition; recovery of vitamins and antibiotics from fermentation liquors; chemical modification of corn starch; enzymatic hydrolysis of starch; biological value of proteins; extraction and purification of plant proteins. Mailing Add: Tech Dept Grain Processing Corp Muscatine IA 52761

MOREHOUSE, CLARENCE KOPPERL, b Boston, Mass, Apr 8, 17; m 42; c 2. ELECTROCHEMISTRY. Educ: Tufts Col, BS, 39; McGill Univ, MS, 40; Mass Inst Technol, PhD(chem), 47. Prof Exp: Res chemist, Naval Ord Lab, Washington, DC, 42-43 & Nat Bur Standards, 43-45; leader battery res & develop, Olin Industs, Inc, Ill, 47-49, asst dir res, Elec Div, Conn, 49-52, mgr, Res & Develop, 52-53; chemist, Res Labs, Radio Corp Am, 53-58, mgr battery & capacitor develop & eng, Semiconductor & Mat Div, 58-59, component develop, 59-60; vpres eng, Globe Battery Co div, 60-67, vpres res & eng, 67-70, VPRES & GEN MGR, INT DIV, GLOBE-UNION INC, 70- Honors & Awards: Achievement Award, Labs, Radio Corp Am, 57. Mem: Am Chem Soc; Electrochem Soc. Res: Inorganic and physical chemistry, specifically batteries and capacitors; corrosion and chemistry of less familiar elements. Mailing Add: 5757 N Green Bay Ave Milwaukee WI 53201

MOREHOUSE, DONALD S, JR, b Owosso, Mich, Jan 12, 28; m 52; c 5. INDUSTRIAL CHEMISTRY. Educ: Mich State Univ, BA, 58. Prof Exp: Group leader, Polymer Sci Lab, 58-68, sect head tech serv & develop, 68-69, group leader org chem prod res, 69-70, res mgr, Designed Polymers Res Lab, 70-72, tech dir, 72-73, TECH DIR, ORG CHEM RES LAB, DOW CHEM CO, 73- Res: Process and new product research, particularly herbicides, insecticides, fine organics, oxide and phenolic derivatives. Mailing Add: 608 Columbia Rd Midland MI 48640

MOREHOUSE, LAURENCE ENGLEMOHR, b Danbury, Conn, July 13, 13; m 39; c 2. PHYSIOLOGY. Educ: Springfield Col, BS, 36, MEd, 37; Univ Iowa, PhD(phys ed), 41. Prof Exp: Asst, Univ Iowa, 37-40; head dept health & phys ed, Univ Wichita, 41-42; asst prof ed, Univ Kans, 42; fel, Harvard Univ, 45-46; assoc prof phys educ, Univ Southern Calif, 46-54; prof phys educ, 54-74, PROF KINESIOL, UNIV CALIF, LOS ANGELES, 74- Concurrent Pos: Mem exped, Nat Acad Sci, NH, 46; researcher, Off Naval Res, 47-49; chief performance physiol sect, Care of Flyer Dept, Randolph Air Force Sch Aviation Med, 49; mem port study comt, Nat Res Coun, 58-; specialist, US State Dept, 60-61, consult to minister sci res, UAR, 63; consult, Henry Dryfuss, 60-63, Douglas Aircraft Co, 63- & Mayor's Space Adv Comt, 65-; Nat Res Coun vis scientist, Manned Spacecraft Ctr, NASA, 68-69. Mem: Am Physiol Soc; Ergonomics Res Soc; Human Factors Soc; Aerospace Med Asn; Am Asn Health, Phys Educ & Recreation. Res: Physiology of exercise; industrial physiology; sports medicine; aerospace medicine; kinesiology; fatigue. Mailing Add: Dept of Phys Educ Univ of Calif Los Angeles CA 90024

MOREHOUSE, LAWRENCE G, b Manchester, Kans, July 21, 25; m; c 2. VETERINARY PATHOLOGY, VETERINARY MICROBIOLOGY. Educ: Kans State Univ, BS & DVM, 52; Purdue Univ, MS, 56, PhD(animal path), 60. Prof Exp: Supvr brucellosis labs, USDA, Purdue Univ, 53-60, staff vet, Agr Res Serv, 60-61, discipline leader path & toxicol, Animal Health Div, Nat Animal Dis Lab, Iowa, 61-64; chmn dept, 64-71, PROF VET PATH, UNIV MO-COLUMBIA, 64-, DIR VET MED DIAG LAB, 71- Concurrent Pos: Consult, Agr Res Serv, USDA, 64-; mem, Comt Salmonellosis, NCent US Poultry Dis Conf, 62-, studies enteric dis in young swine, NCent State Tech Comt, 64- & Nat Conf Vet Lab Diagnosticians; chmn Accreditation Bd, Am Asn Vet Lab Diagnosticians. Honors & Awards: Cert Merit, USDA, 59, 63 & 64. Mem: AAAS; Am Vet Med Asn; Am Asn Avian Path; Conf Res Workers Animal Dis; fel Royal Soc Health. Res: Virus-host cell relationships in tissue culture systems; respiratory diseases of poultry; brucellosis and tuberculosis in domestic animals; enteric diseases of young swine; rabies in swine; streptococcic lymphadenitis of swine. Mailing Add: Vet Med Diagn Lab Univ of Mo-Columbia Sch of Vet Med Columbia MO 65201

MOREHOUSE, MARGARET GULICK, b Champaign, Ill, Aug 22, 04; m 33. BIOCHEMISTRY. Educ: Univ Calif, AB, 27; Univ Southern Calif, PhD(biochem), 39. Prof Exp: Asst biochem, 29-33 & 36-40, instr, 40-46, asst prof, 46-54, assoc prof, 54-72, EMER ASSOC PROF BIOCHEM, SCH MED, UNIV SOUTHERN CALIF, 73- Mem: Am Soc Biol Chem; Soc Exp Biol & Med; Am Oil Chem Soc. Res: Beta oxidation; deuterium as an indicator in study of oxidation of fat; tracer in glyceride absorption and phospholipid formation in normal and irradiated animals; digestibility of fats; effects of dietary fat levels; heart lipids under normal and stress conditions; rat heart lipid changes under diet and aging. Mailing Add: Dept of Biol Sci Univ of Southern Calif Los Angeles CA 90007

MOREHOUSE, NEAL FRANCIS, b Emmett, Idaho, Sept 7, 08; m 36; c 1. POULTRY SCIENCE. Educ: Kans State Col, BS, 33; Iowa State Col, MS, 35, PhD(parasitol), 42. Prof Exp: Instr zool, Univ Omaha, 36-37; scientist protozool, Salsbury Labs, 37-73, mgr, Poultry Res Farm, 44-60, mgr appl res, 60-70; RES CONSULT, 71- Mem: AAAS; Am Soc Parasitol; Soc Protozool; Poultry Sci Asn. Res: Helminthology; protozoology; life cycle of Capillaria caudinflata; therapeutics of poultry parasites; growth stimulants, especially arsenic acids. Mailing Add: 2763 Stagecoach Dr Fayetteville AR 72701

MOREHOUSE, ROGER LYMAN, molecular physics, see 12th edition

MOREHOUSE, SHEILA MCENNESS, b Auburn, NY. INORGANIC CHEMISTRY. Educ: Salve Regina Col, BA, 59; Cornell Univ, MS, 62; Imp Col, Univ London, dipl, 63; Columbia Univ, PhD(inorg chem), 70. Prof Exp: Res asst inorg chem, Mass Inst Technol, 63-65; ASST PROF CHEM, MANHATTANVILLE COL, 70- Mem: AAAS; Am Chem Soc; NY Acad Sci. Res: Coordination chemistry; organometallic compounds. Mailing Add: Dept of Chem Manhattanville Col Purchase NY 10577

MOREJOHN, G VICTOR, b Havana, Cuba, Mar 10, 23; US citizen; m 49; c 9. ZOOLOGY, GENETICS. Educ: Univ Calif, Davis, AB, 53, MA, 54, PhD(zool), 60. Prof Exp: Asst cur ornith & mammal, Los Angeles County Mus, 46-49; asst genetics, Univ Calif, Davis, 52-53, asst zool, 58-60, teaching assoc, 54-58; from asst prof to prof zool, 60-73, PROF BIOL, SAN JOSE STATE UNIV, 73- Concurrent Pos: NSF lectr, 60; Soc Sigma Xi-Sci Res Soc Am Found grant, 61; Am Philos Soc grant, 61. Mem: Cooper Ornith Soc; Am Ornith Union; Am Soc Mammal; Soc Study Evolution. Res: Morphology and evolution of marine birds and mammals; comparative anatomy, genetics and evolution, especially gallinaceous birds and evolution of domestic animals. Mailing Add: Dept of Biol San Jose State Univ San Jose CA 95114

MOREJON, CLARA BAEZ, b Matanzas, Cuba, Nov 30, 40; US citizen; m 65; c 2. CHEMISTRY. Educ: Univ Miami, BS, 64, MS, 67. Prof Exp: Scientist II spec chem, 66-70, supvr spec chem & chem res & develop, 70-72, SR RES SCIENTIST IMMUNOCHEM RES & DEVELOP, DADE DIV, AM HOSP SUPPLY CORP, 72-, RADIATION SAFETY OFFICER, 70- Mem: Am Chem Soc. Res: Radioimmunoassay and competitive protein binding procedures for compounds of clinical significance mostly in thyroid function testing. Mailing Add: Dade Div Am Hosp Supply Corp 1851 Delaware Pkwy Miami FL 33152

MORELAND, ALVIN FRANKLIN, b Morven, Ga, Sept 5, 31; m 55; c 3. VETERINARY MEDICINE, COMPARATIVE MEDICINE. Educ: Ga Teachers Col, BS, 51; Univ Ga, MS, 52, DVM, 60; Am Col Lab Animal Med, Dipl. Prof Exp: NIH fel lab animal med, Bowman Gray Sch Med, 60-62; asst prof exp path, Univ Va, 62-63; from asst prof to assoc prof, 63-72, PROF COMP MED, COL MED, UNIV FLA, 72-, HEAD DIV, 63- Concurrent Pos: Vchmn, Coun Accreditation, Am Asn Accreditation of Lab Animal Care, 68-72, chmn, 72-74; consult field lab animal med. Mem: AAAS; Am Vet Med Asn; Am Asn Lab Animal Sci; NY Acad Sci. Res: Atherosclerosis; subhuman primate medicine. Mailing Add: Div of Comp Med Univ of Fla Col of Med Gainesville FL 32601

MORELAND, CHARLES GLEN, b St Petersburg, Fla, Nov 24, 36; m 60; c 3. PHYSICAL CHEMISTRY. Educ: Univ Fla, BS, 60, MS, 62, PhD(phys chem), 64. Prof Exp: Asst prof, 64-69, ASSOC PROF CHEM, NC STATE UNIV, 69- Honors & Awards: Sigma Xi Res Award. Mem: Am Chem Soc. Res: Nuclear magnetic resonance; thermodynamics and kinetics of redistribution reactions of organo-arsenic (V) and antimony (V) derivatives. Mailing Add: Dept of Chem NC State Univ PO Box 5247 Raleigh NC 27607

MORELAND, DONALD EDWIN, b Enfield, Conn, Oct 12, 19; m 54; c 3. PLANT PHYSIOLOGY, WEED SCIENCE. Educ: NC State Univ, BS, 49, MS, 50, PhD(plant physiol), 53. Prof Exp: Asst, 50, res asst prof field crops, 53-61, assoc prof crop sci, 61-65, PROF CROP SCI, BOT & FORESTRY, NC STATE UNIV, 65-; PLANT PHYSIOLOGIST, SOUTHERN REGION, AGR RES SERV, USDA, 53-, RES LEADER, 73- Concurrent Pos: Asst, State Univ NY Col Forestry, Syracuse, 52-53; mem toxicol study sect, NIH, 63-67. Mem: Fel AAAS; Am Soc Plant Physiol; Bot Soc Am; Weed Sci Soc Am. Res: Mechanism of action of herbicides and growth regulators; photosynthesis; respiration; enzymology; translocation. Mailing Add: Dept of Crop Sci NC State Univ Raleigh NC 27607

MORELAND, FERRIN BATES, b Portland, Ore, Aug 12, 09; m 37; c 2. FORENSIC TOXICOLOGY, CLINICAL CHEMISTRY. Educ: Ore State Univ, BS, 30; Rice Univ, MA, 32; Vanderbilt Univ, PhD(biochem), 36; Am Bd Clin Chem, Dipl, 51; Am Bd Toxicol Chem, Dipl, 72. Prof Exp: Asst biochem, Vanderbilt Univ, 32-36; instr, Tulane Univ, 36; chemist-bacteriologist, Tenn Dept Pub Health, 36; instr biochem, Univ Iowa, 36-42; chief chemist, Kans Dept Pub Health Labs, 45-47; assoc prof biochem, Col Med, Baylor Univ, 47-65; biochemist & dir clin lab, Tex Inst Rehab & Res, 59-65; chief, Biomed Support Sect, Crew Systs Labs, Brown & Root-Northrop, Manned Spacecraft Ctr, Tex, 65-67, mgr, Bio-Med Support Labs, 67-68; dir clin chem & qual control, Metab Res Found, 68-70; CHIEF TOXICOLOGIST, OFF HARRIS COUNTY MED EXAMR, 70- Concurrent Pos: Biochemist, Methodist Hosp, Houston, 47-52 & US Vet Admin Hosp, 49-65. Mem: Am Inst Chemists; Am Chem Soc; Am Asn Clin Chemists (vpres, 57-58, pres, 58-59); Am Acad Forensic Sci; Int Asn Forensic Toxicol. Res: Methods for toxicology. Mailing Add: 3752 Jardin Houston TX 77005

MORELAND, PARKER ELBERT, JR, b Ft Worth, Tex, Nov 5, 31; m 55; c 3. APPLIED PHYSICS, ENGINEERING MANAGEMENT. Educ: Baylor Univ, BS, 54; Harvard Univ, AM, 55, PhD(physics), 62. Prof Exp: Asst physicist, Argonne Nat Lab, 62-70; sr physicist, Packard Instrument Co, Ill, 70-73; VPRES-TECH DIR, SPERRY DIV, AUTOMATION INDUSTS, INC, 73- Mem: Am Phys Soc; Am Soc Nondestructive Testing. Res: Mass spectroscopy; atomic masses; ion-molecule reactions; nuclear instrumentation; x-ray fluorescence spectrometry; ultrasonic transducers and instrumentation; materials evaluation. Mailing Add: 13 Strawberry Hill Rd Danbury CT 06810

MORELAND, WALTER THOMAS, JR, b New London, NH, Apr 30, 26; m 48; c 4. MEDICINAL CHEMISTRY. Educ: Univ NH, BS, 48, MS, 50; Mass Inst Technol, PhD(chem), 52. Prof Exp: Res chemist, Chas Pfizer & Co, Inc, Groton, 52-59, proj leader, 59-61, group supvr, 61-63, sect mgr, 63-65, asst dir med chem res, 65-68, dir, 68-72, EXEC DIR MED CHEM RES, PFIZER, INC, 72- Mem: Am Chem Soc; AAAS; NY Acad Sci. Res: Cardiovascular drugs; adrenergic agents; medicinal products; antibiotics; central nervous system; cardiopulmonary and metabolic diseases. Mailing Add: Pfizer Inc Med Res Labs Groton CT 06340

MORELL, PIERRE, b Dominican Repub, Dec 10, 41; US citizen; m 65; c 2. BIOCHEMISTRY. Educ: Columbia Univ, AB, 63; Albert Einstein Col Med, PhD(biochem), 68. Prof Exp: ASST PROF BIOCHEM IN NEUROL, ALBERT EINSTEIN COL MED, 69- Concurrent Pos: Nat Inst Ment Health fel, Ment Health Res Inst, Univ Mich, Ann Arbor, 68-69. Mem: Am Soc Neurochem; Int Soc Neurochem; Am Chem Soc. Res: Sphingolipid metabolism; myelin proteins; cultured neuroblastoma cells. Mailing Add: Dept of Neurol Albert Einstein Col of Med Bronx NY 10461

MORELL, SAMUEL ALLAN, b Worcester, Mass, Feb 5, 09; m 36; c 1. BIOCHEMISTRY. Educ: Univ Wis, BA, 30, MS, 31, PhD(biochem), 34. Prof Exp: Asst org chem, Calif Inst Technol, 38-39; assoc biochemist, Bur Plant Indust, USDA, 39-40, assoc chemist, Div Agr Residues, Northern Regional Res Lab, Bur Agr & Indust Chem, 41-42, chemist, 42-45; head fine chem dept, Pabst Brewing Co, Wis, 48-58; PROF BIOCHEM IN PATH, MED COL WIS, 58-; DIR BIOCHEM DEPT, MILWAUKEE BLOOD CTR, 58- Concurrent Pos: Fel biochem, Univ Wis, 34-36; Lilly fel, Mt Sinai Hosp, New York, 36-38. Mem: Am Chem Soc; Am Asn Clin Chem; Am Soc Biol Chem. Res: Enzymes and coenzymes; nucleotides; fine chemicals derived from yeast; erythrocyte metabolism; hemoglobin; thiol functions in erythrocytes. Mailing Add: Dept of Path Med Col of Wis Milwaukee WI 53233

MORELLO, EDWIN FRANCIS, b Marseilles, Ill, Jan 12, 28. ORGANIC CHEMISTRY. Educ: Univ Ill, BS, 48; Univ Minn, PhD(org chem), 52. Prof Exp: Res chemist, Standard Oil Co, Ind, 52-61; res chemist, 61-72, RES ASSOC, AMOCO CHEM CORP, 72- Mem: Am Chem Soc. Res: High-temperature polymers; organic chemicals; solid propellants. Mailing Add: 19 Olympus Dr Apt 1A Naperville IL 60540

MORELLO, JOSEPHINE A (MRS ROBERT E BUTZ), b Boston, Mass, May 2, 36; m 71. MEDICAL MICROBIOLOGY. Educ: Simmons Col, BS, 57; Boston Univ, AM, 60, PhD(microbiol), 62; Am Bd Microbiol, Cert med microbiol. Prof Exp: Instr microbiol, Boston Univ, 62-64; res assoc, Rockefeller Univ, 64-66; resident med microbiol, Col Physicians & Surgeons, Columbia Univ, 66-68, asst prof microbiol, 68-69; asst prof, 70-73, ASSOC PROF PATH & MED, UNIV CHICAGO, 73-, DIR CLIN MICROBIOL, 70- Concurrent Pos: Dir microbiol, Harlem Hosp Ctr, 68-69. Mem: AAAS; Am Soc Microbiol; fel Am Acad Microbiol; Am Soc Clin Path; Sigma Xi. Res: Phagocytosis and cellular immunology; salmonella infection; improved methods of clinical microbiology; gonococcal pathogenesis. Mailing Add: Box 290 Univ of Chicago Chicago IL 60637

MORELOCK, JACK, b Houston, Tex, Nov 27, 28; m 58. OCEANOGRAPHY. Educ: Univ Houston, BS, 50 & 53; Tex A&M Univ, PhD(oceanog), 67. Prof Exp: Teaching asst geol, Univ Houston, 50-53; geologist, Magnolia Petrol Co, Tex, 53; teaching asst geol, Univ Kans, 53-55; explor geologist, Continental Oil Co, Wyo, 55-59 & Oasis Oil Co, Libya, 59-62; teaching asst oceanog, Tex A&M Univ, 62-66; prof & head dept, Fla Inst Technol, 66-69 & Inst Oceanog, Univ Oriente, Venezuela, 69-72; ASSOC PROF MARINE SCI, UNIV PR, MAYAGUEZ, 72- Concurrent Pos: Consult to secy natural resources, Dept Natural Resources, Book Study Asn, PR, 74- Mem: Soc Econ Paleont & Mineral. Res: Geological oceanography in areas of marine sedimentation and geophysics; beach and coastal processes; estuarine geology. Mailing Add: Dept of Marine Sci Univ of PR Mayaguez PR 00708

MORELOCK, JAMES CRUTCHFIELD, b Martin, Tenn, Feb 7, 20; m 45; c 3. MATHEMATICS. Educ: Memphis State Col, BS, 41; Univ Mo, MA, 48; Univ Fla, PhD(math), 52. Prof Exp: Asst instr math, Univ Mo, 46-48; instr astron & math, Univ Fla, 49-52; asst prof math, Ala Polytech Inst, 52-56; prof, King Col, 56-60; mathematician, US Naval Comput Lab, 60-61; mem staff, Huntsville Comput Ctr, Gen Elec Co, 61-63; MEM STAFF COMPUT LAB, GEORGE C MARSHALL

MORELOCK

SPACE FLIGHT CTR, NASA, 63- Mem: Am Math Soc; Math Asn Am; Asn Comput Mach. Res: Algebraic geometry; electronics; satellite applications. Mailing Add: 2917 Garth Rd SE Huntsville AL 35801

MORENG, ROBERT EDWARD, b New York, NY, Jan 29, 22; m 50; c 7. POULTRY PHYSIOLOGY. Educ: Univ Md, BS, 44, MS, 48, PhD(poultry physiol), 50. Prof Exp: Asst poultry husb, Univ Md, 47-50; asst prof, NDak Agr Col, 50-55; prof avian sci, Colo State Univ, 55-72, DIR RES, COL AGR SCI, COLO STATE UNIV, 72- Mem: AAAS; Soc Exp Biol & Med; Radiation Res Soc; Poultry Sci Asn; NY Acad Sci. Res: Effects of low temperature on the chicken and the chicken embryo; x-ray irradiation and avian embryo; environment and growth; physiological genetics of reproductive adaptation to high altitude in turkeys. Mailing Add: Col of Agr Sci Colo State Univ Ft Collins CO 80521

MORENO, CARLOS JULIO, b Sevilla, Colombia, July 30, 46; US citizen; m 72. MATHEMATICS. Educ: NY Univ, BA, 68, PhD(math), 71. Prof Exp: ASST PROF MATH, UNIV ILL, URBANA-CHAMPAIGN, 71- Concurrent Pos: Assoc mem, Ctr Advan Study, Univ Ill, 75-76; vis mem, Inst Advan Study, Princeton, 75-76. Mem: Am Math Soc; Math Asn Am. Res: Applications of group representations and automorphic forms to problems in number theory and algebraic geometry. Mailing Add: 273 Altgeld Hall Univ Ill Dept Math Urbana IL 61801

MORENO, EDGARD CAMACHO, physical chemistry, see 12th edition

MORENO, ESTEBAN, b San Juan, PR, Aug 3, 26; US citizen. MEDICINE, PATHOLOGY. Educ: Columbia Univ, BA, 48; Temple Univ, MD, 51. Prof Exp: Attend pathologist, Univ Hosp, PR, 57-63; assoc prof, 61-73, PROF PATH, SCH MED, UNIV PR, SAN JUAN, 73- Concurrent Pos: Attend path, Oncol Hosp, PR Med Ctr, 66-73. Mem: Fel Am Soc Clin Path. Res: Dermatopathology; training of physicians in pathology. Mailing Add: Dept of Path GPO Box 5067 Univ of Pr Sch of Med San Juan PR 00936

MORENO, HERNAN, b Medellin, Colombia, Sept 9, 39; m 64; c 3. PEDIATRICS, HEMATOLOGY. Educ: Berchmans Col, Colombia, BS, 56; Univ Valle, Colombia, MD, 63; Am Bd Pediat, dipl, 67; Educ Coun for Med Grad, cert. Prof Exp: Intern, Univ Hosp, Cali, Colombia, 61-62, jr pediat resident, Dept Pediat, 62-63, sr pediat resident, 63-64; pediat resident, Children's Hosp, Birmingham, Ala, 64-65; instr pediat, Col Med, Univ Cincinnati, 65-67; from instr to asst prof, 67-71, ASSOC PROF PEDIAT, SCH MED, UNIV ALA, BIRMINGHAM, 71-, DIR PEDIAT HEMAT & ONCOL DIV, 69- Concurrent Pos: Fel pediat hemat, Children's Hosp Res Found, Cincinnati, Ohio, 65-67; consult, Children's Hosp, Birmingham, 67-; mem hosps & clins staff, Sch Med, Univ Ala, 68- Mem: Am Acad Pediat; Am Soc Hemat; NY Acad Sci. Res: Cancer chemotherapy. Mailing Add: Dept of Pediat Univ of Ala Sch of Med Birmingham AL 35233

MORENO-BLACK, GERALDINE S, b Brooklyn, NY, Aug 25, 46. PHYSICAL ANTHROPOLOGY. Educ: State Univ NY Buffalo, BA, 67; Univ Ariz, MA, 70; Univ Fla, PhD(anthrop), 74. Prof Exp: ASST PROF ANTHROP, UNIV ORE, 74- Mem: Am Asn Phys Anthropologists; Soc Med Anthrop; Am Anthrop Asn; AAAS. Res: Affect of disturbed habitat conditions on diet, behavior and social structure of non-human primates with emphasis on African Colobinae; natural diets of non-human primates; human ecology. Mailing Add: Dept of Anthrop Univ of Ore Eugene OR 97403

MORENZONI, RICHARD ANTHONY, b Sonoma, Calif, Sept 30, 46; m 74. MICROBIOLOGY. Educ: Univ Calif, Davis, BS, 68, PhD(microbiol), 73. Prof Exp: RES & STAFF MICROBIOLOGIST, E & J GALLO WINERY, 73- Mem: Sigma Xi. Res: Industrial research related to microbiological practices or problems encountered in the production of wines and spirits. Mailing Add: Res Microbiol Dept E & J Gallo Winery PO Box 1130 Modesto CA 95353

MOREST, DONALD KENT, b Kansas City, Mo, Oct 4, 34; m 63; c 2. ANATOMY. Educ: Univ Chicago, BA, 55; Yale Univ, MD, 60. Prof Exp: Sr asst surgeon, NIH, 60-63; asst prof anat, Univ Chicago, 63-65; assoc, 65-67, asst prof, 67-70, ASSOC PROF ANAT, HARVARD MED SCH, 70- Concurrent Pos: Res assoc otolaryngol, Mass Eye & Ear Infirmary, 65- Honors & Awards: Herrick Award Comp Neurol, 66. Mem: AAAS; Am Asn Anatomists; Soc Neurosci; Asn Res Otolaryngol. Res: Neuroembryology; neuroanatomy; connections of central neural pathways and neurocytology; autonomic and sensory systems; auditory system; neurobiology. Mailing Add: Dept of Anat Harvard Med Sch Boston MA 02115

MORET, CASSARD LAWRENCE, physical chemistry, see 12th edition

MORETON, ROBERT DULANEY, b Brookhaven, Miss, Sept 24, 13; m 45. RADIOLOGY. Educ: Millsaps Col, BS, 34; Univ Miss, cert, 36; Univ Tenn, MD, 38; Am Bd Radiol, dipl, 43. Prof Exp: Intern, Lloyd Noland Mem Hosp, Fairfield, Ala, 38-39; instr, Sch Med, Univ Miss, 40; lectr, Univ Tex Med Br, Galveston, 45-50; from instr to assoc prof clin radiol, Univ Tex Southwestern Med Sch, Dallas, 51-65; PROF RADIOL, UNIV TEX M D ANDERSON HOSP & TUMOR INST, HOUSTON, 65-, VPRES PROF & PUB AFFAIRS, 69- Concurrent Pos: Fel radiol, Mayo Found, 40-42; staff radiologist, Scott & White Hosp & Clin & Santa Fe Hosp, Temple, Tex, 42-50; chmn & partner, Bond Radiol Group, Tex, 50-65; consult var hosps, companies & railroads, Tex, 52-65; chmn & dir dept radiol, Harris Hosp, Ft Worth & Ft Worth Childrens Hosp, 61-65; vpres, Univ Cancer Found, Tex, 65-; founding mem bd, Charter Blood Bank & Bd Radiation & Res Found Southwest; mem bd & exec comt, Radiation Ctr, Ft Worth. Mem: Am Col Radiol; AMA; Am Roentgen Ray Soc; Am Geriat Soc; Indust Med Asn. Mailing Add: Univ of Tex M D Anderson Hosp & Tumor Inst Houston TX 77025

MORETTI, RICHARD LEO, b Ft Collins, Colo, Feb 15, 29; m 54; c 2. DEVELOPMENTAL BIOLOGY. Educ: Univ Calif, Riverside, AB, 57, Berkeley, MA, 60, PhD(zool), 64. Prof Exp: Lectr embryol, Univ Calif, Riverside, 64-65, asst prof embryol & asst res zoologist, Air Pollution Res Ctr, 65-73; RES BIOLOGIST, BRUCE LYON MEM RES LAB, CHILDREN'S HOSP, OAKLAND, 73- Res: Mechanisms regulating coronary blood flow. Mailing Add: Bruce Lyon Mem Res Lab Children's Hosp Oakland CA 94609

MORETZ, WILLIAM HENRY, b Hickory, NC, Oct 23, 14; m 47; c 6. SURGERY. Educ: Lenoir-Rhyne Col, BS, 35; Harvard Univ, MD, 39. Hon Degrees: DSc, Lenoir-Rhyne Col, 61. Prof Exp: Instr surg, Sch Med, Univ Rochester, 44-47; from asst prof to assoc prof, Sch Med, Univ Utah, 47-55; prof & chmn dept, 55-72, PRES, MED COL GA, 72- Concurrent Pos: Attend surgeon, North Bench Vet Admin Hosp & assoc surgeon, Salt Lake Gen Hosp, Salt Lake City, Utah, 47-55; consult surgeon, Tooele Army Hosp, 51-55; chief surg, Eugene Talmadge Mem Hosp, Ga, 55-72; consult, US Vet Admin Hosp, 55- Mem: Soc Univ Surgeons; Am Surg Asn; Int Cardiovasc Soc; Int Soc Surgeons; Soc Surg Alimentary Tract. Res: Thromboembolism and arterial diseases. Mailing Add: Dept of Surg Med Col of Ga Augusta GA 30902

MOREWITZ, HARRY ALAN, b Newport News, Va, June 2, 23; m 48; c 2. NUCLEAR PHYSICS. Educ: Col William & Mary, BS, 43; Columbia Univ, AM, 49; NY Univ, PhD(physics), 53. Prof Exp: Res assoc cosmic rays, NY Univ, 49-53; supvr reactor physics, Westinghouse Elec Corp, 53-59; PROJ MGR, ATOMICS INT DIV, ROCKWELL INT CORP, 59- Concurrent Pos: Proj engr, Metavac, Inc, 52-53; lectr, Dept Radiol, Ctr Health Sci, Univ Calif, Los Angeles, 68-72; mem, Adv Comt Reactor Physics, US AEC, 69-73. Mem: Am Phys Soc; Am Nuclear Soc; Sigma Xi; NY Acad Sci; Health Physics Soc. Res: Reactor physics; reactor safety; aerosol physics; transient liquid metal heat transfer; radiation dosimetry. Mailing Add: 5300 Bothwell Rd Tarzana CA 91356

MOREY, DARRELL DORR, b Manhattan, Kans, Dec 6, 14; m 40; c 6. AGRONOMY. Educ: Kans State Col, BS, 37; Tex Tech Col, MS, 38; Iowa State Col, PhD(agron), 47. Prof Exp: Asst, Tex Tech Col, 37-38, instr crops, 38-39; jr supvr grain inspection, Grain Br, USDA, Minn, 39-44; assoc, Exp Sta, Iowa State Col, 44-47; assoc prof agron, Univ Ga, 48-49; assoc agronomist, Fla Agr Exp Sta, 49-53; assoc plant breeder, 53-65, PLANT BREEDER RYE TRITICALE, WHEAT BREEDING INVESTS, GA COASTAL PLAIN EXP STA, CROP RES DIV, AGR RES SERV, USDA & PROF AGRON, UNIV GA, 65- Concurrent Pos: Mem, Nat Oat Conf Comt, 51-52 & Nat Wheat Improv Coun, 73-; coordr, Uniform Southern Soft Wheat Exps, 73- Mem: Am Plant Life Soc; Am Forestry Asn; Am Soc Agron. Res: Breeding disease resistance in small grains; genetics and cytology of cereals; rye breeding. Mailing Add: Ga Coastal Plain Exp Sta PO Box 748 Tifton GA 31794

MOREY, GLENN BERNHARDT, b Duluth, Minn, Oct 17, 35; m 67; c 2. GEOLOGY. Educ: Univ Minn, Duluth, BA, 53, Minneapolis, MS, 62, PhD(geol), 65. Prof Exp: Instr geol, Univ Minn, Minneapolis, 64-6S, asst prof, Minn Geol Surv, 65-69, ASSOC PROF GEOL, UNIV MINN, ST PAUL, 69-, PRIN GEOLOGIST PRECAMBRIAN, MINN GEOL SURV, 73- Mem: Soc Econ Paleont & Mineral; fel Geol Soc Am; Soc Econ Geol. Res: Field and laboratory studies of Precambrian rocks, principally in Minnesota and the Lake Superior Region. Mailing Add: Minn Geol Surv Univ of Minn St Paul MN 55108

MOREY, PHILIP RICHARD, b Cleveland, Ohio, July 3, 40; m 67. PLANT ANATOMY, INDUSTRIAL HYGIENE. Educ: Univ Dayton, BS, 62; Yale Univ, MS, 64, PhD(biol), 67. Prof Exp: NIH fel, Univ Calif, Santa Barbara, 67; lectr biol, Harvard Univ, 68-70, forest botanist, 67-70; ASSOC PROF BIOL, TEX TECH UNIV, 70- Mem: Bot Soc Am; Am Soc Plant Physiol; Weed Sci Soc Am; Am Soc Agron; Sigma Xi. Res: Anatomy of agronomic plants; herbicide effects on plant growth; identification and analysis of vegetable dusts relative to industrial hygiene. Mailing Add: Dept of Biol Tex Tech Univ Lubbock TX 79409

MOREY, PHILIP STOCKTON, JR, mathematics, see 12th edition

MOREY, RICHARD CARL, b Syracuse, NY, Nov 3, 38; m 64. OPERATIONS RESEARCH. Educ: Syracuse Univ, BS, 60; Univ Calif, Berkeley, MA, 62, PhD(opers res), 65. Prof Exp: Staff mem prog, Eastman Kodak, 61-62, Opers Res Ctr, Univ Calif, 62-65 & Weapons Systs Eval, Inst Defense Anal, Va, 65-67; opers analyst, Stanford Res Inst, 67-68; vpres, Decisions Studies Group Inc, Calif, 68-71; PRES, CONTROL ANAL CORP, PALO ALTO, 71- Concurrent Pos: Lectr, George Washington Univ, 65-68 & Stanford Univ, 68-69. Mem: Sigma Xi. Res: Reliability theory; queueing theory; nonparametric results concerning first passage time problems; inventory control; logistics. Mailing Add: Control Anal Corp 800 Welch Rd Palo Alto CA 94304

MOREY, ROBERT V, b Perry, Iowa, Feb 6, 38; m 62. CULTURAL ANTHROPOLOGY, ETHNOLOGY. Educ: Univ Iowa, BA, 62; Univ Pittsburgh, PhD(anthrop), 70. Prof Exp: Asst instr anthrop, Univ Pittsburgh, Titusville, 67; asst prof, Univ Utah, 67-71; ASSOC PROF ANTHROP, WESTERN ILL UNIV, 71- Concurrent Pos: Co-prin investr, Nat Geog Soc res grant, Colombia, 73-74. Mem: AAAS; Am Anthrop Asn; Soc Am Archaeol; Sigma Xi; Am Soc Ethnohist. Res: Social organization, ecology and culture change in lowland, especially tropical savanna regions of South America; South American lowland-highland developments and their interrelationship; influences. Mailing Add: Dept of Sociol & Anthrop Western Ill Univ Macomb IL 61455

MORF, MAX, organic chemistry, see 12th edition

MORFIT, HENRY MASON, b Baltimore, Md, Jan 19, 13; m 40; c 3. SURGERY. Educ: Johns Hopkins Univ, AB, 35, MD, 39. Prof Exp: Rockefeller fel, Mem Hosp Cancer & Allied Dis, 39-48; from asst prof to assoc prof, 48-67, PROF SURG, UNIV COLO MED CTR, DENVER, 67-, DIR, BONFILS TUMOR CLIN, 48- Concurrent Pos: Consult & attend surgeon, Hosps, Denver; proj dir, Colo Regional Cancer Ctr. Mem: Fel Am Col Surgeons; Soc Head & Neck Surgeons (pres); James Ewing Soc. Res: Cancer surgery of the head and neck. Mailing Add: 3303 E Kentucky Ave Denver CO 80209

MORFITT, JOHN WINSLOW, nuclear physics, see 12th edition

MORGAN, ANTONY RICHARD, b Mombasa, Kenya, Jan 5, 40. MOLECULAR BIOLOGY. Educ: Cambridge Univ, BA, 61; Univ Alta, PhD(chem), 64. Prof Exp: Proj assoc DNA chem & enzymol, Enzyme Inst, Univ Wis-Madison, 65-67, asst prof, 67-69; ASST PROF DNA CHEM & ENZYMOL, UNIV ALTA, 69- Mem: AAAS; Can Fedn Biol Socs. Res: Chemical mechanism of DNA replication in vitro; three-stranded complexes between DNA and RNA and their biological implications for transcription. Mailing Add: Dept of Biochem Univ of Alta Edmonton AB Can

MORGAN, ASHLEY GRANTHAM, JR, b Marianna, Fla, Apr 2, 30; m 56; c 3. SCIENCE EDUCATION. Educ: Univ Ga, BS, 55, EdD, 64; Univ Va, MEd, 58. Prof Exp: Teacher pub schs, Ga, 55-59; sci consult, Ga Dept Educ, 59-62; prof sci educ & phys sci & dept chmn, Southern Miss Univ, 64-69; PROF SCI EDUC, GA STATE UNIV, 69-, DIR, CTR IMPROV ELEM SCH SCI, 73- Concurrent Pos: Sci consult, Miss Authority Instructional TV, 67-68; inst disadvantaged youth, Miss, Nat Defense Educ Act, 68-69; instnl dir, Ga Sci Teacher Proj, NSF, 69-72. Mem: Fel AAAS; Nat Sci Teachers Asn; fel Am Inst Chemists. Res: Cognitive style and inquiry processes; language development through science. Mailing Add: Dept of Curric & Instr Ga State Univ Atlanta GA 30303

MORGAN, BEVERLY CARVER, b New York, NY, May 29, 27; m 54; c 3. PEDIATRIC CARDIOLOGY. Educ: Duke Univ, MD, 55; Am Bd Pediat, dipl, 60, cert cardiol, 61. Prof Exp: Intern & asst resident pediat, Stanford Univ, 55-56; trainee pediat cardiol, Babies Hosp, Columbia Univ, 57-60; dir cardiol, Heart Sta, Robert Green Hosp, San Antonio, Tex, 60-62; from instr to assoc prof, 62-73, PROF PEDIAT & CHMN DEPT, SCH MED, UNIV WASH, 73- Concurrent Pos: Clin fel, Babies Hosp, Columbia Univ, 56-57, res fel, Cardiovasc Lab, Presby Hosp, 59-60; NIH res career prog award, 66-71; consult to Surgeon Gen, Brooke Army Med Ctr, Tex, 60-62; clin lectr, Sch Med, Univ Tex, 60-62; mem pulmonary training comt, Nat Heart & Lung Inst, 72, mem pulmonary acad award panel, 72-; mem comt NIH training & fel progs, Nat Res Coun, 72-74. Mem: Am Fedn Clin Res; Am Heart Asn; Am Acad Pediat; Soc Pediat Res; Am Col Cardiol. Res: Clinical research; effects of respiration on circulation. Mailing Add: 1031 Evergreen Point Rd Bellevue WA 98004

MORGAN, BRUCE HARRY, b Sharon, Pa, Sept 30, 31; m 58; c 2. PHYSICS. Educ: Harvard Univ, AB, 53; Calif Inst Technol, MS, 54; George Washington Univ, JD, 68. Prof Exp: Assoc engr systs anal, Westinghouse Elec Corp, 56-57; ASSOC PROF GEN PHYSICS, US NAVAL ACAD, 57- Mem: Am Asn Physics Teachers. Res: General physics. Mailing Add: Dept of Physics Michelson Hall US Naval Acad Annapolis MD 21402

MORGAN, BRUCE HENRY, b Chicago, Ill, Jan 21, 20; m 75; c 2. FOOD MICROBIOLOGY. Educ: Purdue Univ, BS, 41; Rutgers Univ, PhD(microbiol), 53. Prof Exp: Res bacteriologist, Continental Can Co, Inc, Ill, 41-42 & Baxter Labs, 43; bacteriologist, US Biol Warfare Res Labs, 47-49; res bacteriologist, Nat Canners Asn Res Labs, DC, 52-53; dep for radiation preservation, Off Sci Dir, Qm Food & Container Inst, Ill, 54-57; mgr packaging eng, Continental Can Co, Inc, 58-64, dir corp new prod res, 64-67; dir res & eng, Lamb Weston, Inc, 67-68, vpres res & eng, 68-73; SR VPRES CORP DEVELOP, AMFAC FOODS, INC, 74- Concurrent Pos: Asst ed, Food Res, 59-61; mem, Adv Bd, Food Res Inst, 59-63; instr, Ill Inst Technol, 63-64; mem, Bd Dirs, Qm Res & Develop Assocs, 64- & Bd Govs, Food Update, Food & Drug Law Inst, 69-; mem, Food Industs Adv Comt, Nutrit Found, 66-67 & Indust Action Comt, Am Health Found, 71- Mem: Am Soc Microbiol; Inst Food Technologists (treas, 70-74); fel Royal Soc Health. Res: Radiation biology of microbiological population; radiation chemistry of foods; vegetable freezing and dehydration. Mailing Add: PO Box 23564 Portland OR 97223

MORGAN, CARL ROBERT, b Dubuque, Iowa, Nov 25, 29; m 50; c 2. ANATOMY. Educ: Wartburg Col, BA, 50; Univ Nebr, MA, 52; Univ Minn, PhD(anat), 63. Prof Exp: Jr scientist bact, Univ Minn, 52-56; instr biol, Villa Madonna Col, 56-57; asst prof, Concordia Col, 57-58; instr anat, Univ Minn, 63-64; asst prof, 64-70, PROF ANAT, MED CTR, IND UNIV, INDIANAPOLIS, 70- Concurrent Pos: Res fel exp diabetes & anat, Univ Minn, 58-63; Am Diabetes Asn res fel, 63-64. Mem: AAAS; Am Diabetes Asn; Am Asn Anatomists; Soc Exp Biol & Med. Res: Experimental diabetes. Mailing Add: Dept of Anat Ind Univ Med Ctr Indianapolis IN 46207

MORGAN, CHARLES O, b Fairfield, Iowa, Nov 28, 31; m 52; c 2. HYDROLOGY. Educ: Univ Iowa, BA, 54, MS, 56. Prof Exp: Geologist, US Geol Surv, Mich, 56, La, 56-63, Kans, 63-67, hydrologist, 67-70, HYDROLOGIST, US GEOL SURV, CALIF, 70- Concurrent Pos: US AID consult, Pakistan, 69 & 70. Honors & Awards: Superior Performance Award, US Geol Surv, 65. Mem: Geol Soc Am; Int Asn Math Geol. Res: Salt-water encroachment and chemical facies changes in coastal aquifers; digital computer techniques as applied in the field of hydrology. Mailing Add: Laguna Fed Bldg 24000 Avila Rd Laguna Niguel CA 92677

MORGAN, CHARLES ROBERT, b Kingston, Pa, July 18, 34; m 68. PHYSICAL ORGANIC CHEMISTRY. Educ: Pa State Univ, BS, 56; Mass Inst Technol, PhD(org chem), 63. Prof Exp: Res assoc, Mass Inst Technol, 63-65; res chemist, Org Res Dept, 65-73, sr res chemist, Photopolymer Systs, 73-74, MGR PROD DEVELOP, PHOTOPOLYMER SYSTS, W R GRACE & CO, 74- Mem: Am Chem Soc; Soc Chem Coaters. Res: Kinetics and mechanisms of organic reactions; cyanide and sulfur chemistry; nucleophilicity and solvent isotope effects; organometallic reactions; polymer stabilization; photochemistry and photopolymers. Mailing Add: Photopolymer Systs W R Grace & Co Columbia MD 21044

MORGAN, CLARENCE RICHARD, organic chemistry, see 12th edition

MORGAN, COUNCILMAN, b Boston, Mass, Sept 6, 20; m 45; c 4. MICROBIOLOGY, VIROLOGY. Educ: Harvard Univ, BS, 43; Columbia Univ, MD, 46. Prof Exp: From intern to resident med, Bellevue Hosp, New York, 46-53; from asst prof med to assoc prof microbiol, 54-63, assoc dean students & curric, 70-72, PROF MICROBIOL, COL PHYSICIANS & SURGEONS, COLUMBIA UNIV, 63-, ASSOC DEAN CURRIC AFFAIRS, 72- Mem: Am Soc Microbiol; Am Asn Clin Invest; Am Asn Immunol; Am Soc Cell Biol; hon mem Fr Soc Electron Microscope. Res: Structure and development of viruses; curriculum design. Mailing Add: Columbia Univ Col of Phys & Surg 630 W 168th St New York NY 10032

MORGAN, DAVID WILLIAM, b Nashville, Tenn, June 11, 44; m 68; c 1. BIOLOGICAL OCEANOGRAPHY, AQUATIC BIOLOGY. Educ: Univ Ga, BS, 66; Univ RI, PhD(biol oceanog), 73. Prof Exp: ASST PROF BIOL, UNIV NOTRE DAME, 72- Mem: AAAS; Am Fisheries Soc; Animal Behav Soc; Am Soc Limnol & Oceanog; Marine Technol Soc. Res: Behavior of aquatic vertebrates, especially as affected by human environmental perturbation; wetland resource allocation and management; fisheries dynamics; systems for processing natural resources data. Mailing Add: Dept of Biol Univ of Notre Dame Notre Dame IN 46556

MORGAN, DAVID ZACKQUILL, b Fairmont, WVa, Mar 27, 25; m 48; c 2. INTERNAL MEDICINE. Educ: WVa Univ, AB, 48, BS, 50; Med Col Va, MD, 52; Am Bd Internal Med, dipl, 64. Prof Exp: From instr to assoc prof, 63-73, asst dean, 65-72, PROF MED, SCH MED, WVA UNIV, 73-, ASSOC DEAN, 72- Mem: Fel Am Col Physicians. Res: Cardiology. Mailing Add: Dept of Med WVa Univ Sch of Med Morgantown WV 26506

MORGAN, DEE R, dairy industry, food technology, see 12th edition

MORGAN, DELBERT THOMAS, JR, b Cuyahoga Falls, Ohio, Mar 3, 18; m 47; c 1. BOTANY. Educ: Kent State Univ, BS, 40; Columbia Univ, MS, 42, PhD(cytogenetics), 48. Prof Exp: From asst prof to assoc prof, 47-59, PROF BOT, UNIV MD, COLLEGE PARK, 59- Mem: Am Inst Biol Sci; Genetics Soc Am; Am Genetic Asn. Res: Cytogenetics of maize and vegetable crops; haploidy and polyembryony in plants; plant cytogenetics and breeding. Mailing Add: Dept of Bot Univ of Md College Park MD 20742

MORGAN, DONALD O'QUINN, b Star, NC, Mar 24, 34; m 59; c 2. VETERINARY IMMUNOLOGY. Educ: NC State Univ, BS, 55, MS, 63; Univ Ga, DVM, 59; Univ Ill, PhD(physiol), 67. Prof Exp: Res asst microbiol, NC State Univ, 59-62, instr animal sci, 62-63; diagnostician, Dept Agr, NC, 63-64; NIH fel, Univ Ill, 64-67, asst prof vet physiol, 67; vet med officer, Plum Island Animal Dis Lab, Agr Res Serv, USDA, 67-69; from asst prof to assoc prof vet sci, Univ Ky, 69-74; ASSOC PROF VET MED, COL VET MED, UNIV ILL, URBANA, 74- Mem: Am Soc Vet Physiol & Pharmacol; Am Vet Med Asn; Am Soc Microbiol; Conf Res Workers Clin Dis. Res: Equine serum proteins; equine viral diseases and their immunology; immunological competence of the foal and the equine fetus; equine immunoglobulins. Mailing Add: 2313 Barberry Dr Champaign IL 61820

MORGAN, DONALD PRYSE, b Indianapolis, Ind; m 52; c 3. PREVENTIVE MEDICINE. Educ: Ind Univ, MD, 47; Northwestern Univ, PhD(physiol), 53. Prof Exp: Asst prof physiol, Sch Med, Ind Univ, 53-54; asst prof, Med Sch, Northwestern Univ, 54-60; pvt pract, Ariz, 60-67; epidemiologist, Community Pesticide Proj, Univ Ariz, 67-73; ASST PROF PREV MED, MED SCH, UNIV IOWA, 73- Mem: Soc Toxicol; Am Physiol Soc. Res: Respiration; body fluids; pesticide toxicology. Mailing Add: Community Pesticide Proj Univ of Iowa Med Sch Oakdale IA 52319

MORGAN, DONALD R, b Boston, Mass, May 26, 33; m 61; c 2. SOLID STATE PHYSICS. Educ: Boston Col, BS, 55, MS, 57; Univ Notre Dame, PhD(physics), 61. Prof Exp: Res engr, Melpar Inc, Mass, 56-57; instr physics, Univ Notre Dame, 60-61; from asst prof to assoc prof, 61-70, PROF PHYSICS, ST MARY'S COL, MINN, 70-, CHMN DIV NATURAL SCI & MATH, 62- Mem: Am Phys Soc; Am Asn Physics Teachers. Res: Physical electronics; thermionic and field emission with special emphasis on the effects of absorbed molecules on the work function. Mailing Add: Dept of Physics St Mary's Col Winona MN 55987

MORGAN, EVAN, b Spokane, Wash, Feb 26, 30; m 59; c 1. ANALYTICAL CHEMISTRY. Educ: Gonzaga Univ, BS, 52; Univ Wash, MS, 54, PhD(anal chem), 56. Prof Exp: Staff chemist, Int Bus Mach Corp, 56-60; group supvr, Olin Mathieson Chem Corp, 60-64; assoc prof chem, High Point Col, 64-65; sr chemist, Metall Res Dept, Reynolds Metals Co, Va, 65-72; group supvr, 72, SR RES CHEMIST, BABCOCK & WILCOX, 72- Mem: Am Chem Soc; Am Soc Testing & Mat. Res: Electrochemistry; gas chromatography; evolved gas analysis; mass spectrometry. Mailing Add: 3114 Bute Lane Richmond VA 23221

MORGAN, FREDERIC L, plant pathology, botany, see 12th edition

MORGAN, GEORGE L, b Cleveland, Ohio, Aug 8, 37. INORGANIC CHEMISTRY. Educ: Case Inst, BS, 58; Univ Ill, MS, 60, PhD(inorg chem), 63. Prof Exp: From asst prof to assoc prof, 62-72, PROF INORG CHEM, UNIV WYO, 72- Mem: Am Chem Soc; The Chem Soc. Res: Electron-deficient bonding; structures of organometallic compounds; infrared and nuclear magnetic resonance spectroscopy. Mailing Add: Dept of Chem Univ of Wyo Laramie WY 82070

MORGAN, GEORGE WALLACE, b Shreveport, La, Aug 14, 41; m 62; c 4. POULTRY PHYSIOLOGY. Educ: Miss State Univ, BS, 64, MS, 66, PhD(animal physiol), 70. Prof Exp: Staff fel immunol, Div Biologics Standards, 70-73; physiologist, Bur Biologics, Food & Drug Admin, 73-74; ASST PROF IMMUNOL & PHYSIOLOGIST, DEPT POULTRY SCI, NC STATE UNIV, 74- Mem: Poultry Sci Asn; Sigma Xi. Res: Inter-relationships of physiological stress and heavy metal toxicity of immune responsiveness. Mailing Add: NC State Univ PO Box 5307 Raleigh NC 27607

MORGAN, HAROLD EUGENE, b New Madison, Ohio, Apr 18, 08; m 39; c 3. PHYSICS. Educ: Miami Univ, AB, 29; Pa State Univ, MS, 31, PhD(physics), 37. Prof Exp: Asst physics, Pa State Univ, 29-37; from instr to asst prof, Fenn Col, 37-43; physicist, US War Dept, Wright Field, Ohio, 43-46; from assoc prof to prof, 46-74, chmn dept, 53-70, EMER PROF PHYSICS, CLEVELAND STATE UNIV, 74- Mem: Am Asn Physics Teachers. Res: Electronics; magnetron tubes and circuits; high pressure viscosity measurements; effect of temperature and pressure on viscosity. Mailing Add: Dept Phys Cleveland State Univ Euclid Ave at 24th St Cleveland OH 44115

MORGAN, HARRY CLARK, b Kalamazoo, Mich, Dec 17, 16; m; c 3. PHYSICS. Educ: Mich State Univ, BS, 38, MS, 39. Prof Exp: Physicist, Moraine Prod Div, Gen Motors Corp, 39-41; partner, Optron Labs, Ohio, 41-42; proj engr, Curtis Eng Co, 42-47; nuclear physicist, Mound Labs, Monsanto Chem Co, 47-49; owner, Morgan Instruments Co, 49-52; proj engr, Electronics Div, Century Metalcraft Corp, 52-54; lead engr instrumentation data handling, Rocketdyne Div, NAm Aviation Inc, 54-57; sr engr, Res Div, Cohu Electronics, Inc, 57-58; mem tech staff, Ramo Wooldridge Div, TRW, Inc, 58-60; mem tech staff, Autonetics Div, NAm Rockwell Corp, 60-70, space div, 70-71, sr staff assoc, Sci Ctr, 71-74; MEM STAFF, EON INSTRUMENTATION INC, 74- Mem: Am Phys Soc; Inst Elec & Electronics Engrs; Am Inst Aeronaut & Astronaut. Res: Computers and digital data handling systems; optics and nuclear physics; analog electronic systems. Mailing Add: Eon Instrumentation Inc 15547 Cabrito Rd Van Nuys CA 91406

MORGAN, HENRY MERRIAM, polymer physics, see 12th edition

MORGAN, HERBERT ROY, b St Paul, Minn, July 19, 14; m 48; c 3. MICROBIOLOGY, MEDICINE. Educ: Univ Calif, BA, 36, MA, 38; Harvard Med Sch, MD, 42; Am Bd Prev Med & Pub Health, dipl; Am Bd Med Microbiol, dipl. Prof Exp: Asst bact, Univ Calif, 36-38, intern, Hosp, 42-43; assoc prof epidemiol, Sch Pub Health & asst prof med, Med Sch, Univ Mich, 48-50; PROF MICROBIOL & ASSOC PROF MED, SCH MED & DENT, UNIV ROCHESTER, 50-, DIR INDEPENDENT STUDIES PROG, 71- Concurrent Pos: Sr fel med sci, Harvard Med Sch, 46-48; Fulbright res scholar, Univ Oslo, 60-; Commonwealth prof fel, Inst Cancer Res, France, 67-68; dir, City Health Bur Labs, Rochester, 50-59, Bact Labs, Strong Mem Hosp, 50-70 & Microbiol Labs, Monroe County Health Dept, 59-70; consult, Highland & Genesee Hosps, Rochester, 50- & WHO, 56; ed, Infection & Immunity; mem adv panel regulatory biol, NSF, 58, mem test comt bact, Nat Bd Med Examr, 59-63, mem virol & rickettsiology study sect, NIH, 59-63 & Microbiol Training Comt, 64-68, mem spec virus-cancer-leukemia effort adv comt, Nat Cancer Inst, 64-72 & adv comt res etiology of cancer, Am Cancer Soc, 65-71. Mem: Am Soc Microbiol; Soc Exp Biol & Med; Am Soc Clin Invest; Am Soc Cell Biol; Infectious Dis Soc Am. Res: Antigens of Salmonella organisms; latent infections; tumor viruses. Mailing Add: Univ of Rochester Sch of Med & Dent Rochester NY 14620

MORGAN, HORACE C, b Piedmont, Ala, July 3, 28; m 56; c 2. CLINICAL PATHOLOGY. Educ: Auburn Univ, DVM, 55, MS, 58. Prof Exp: Instr physiol, Auburn Univ, 55-58, asst prof, 58-59; asst prof clin path, Univ Ga, 60-61, from asst prof to assoc prof physiol, 61-65, assoc prof clin path, 65-70; prof & dir continuing educ & learning resources, 70-73, ASST DEAN, SCH VET MED, AUBURN UNIV, 73- Concurrent Pos: Consult, WHO, 73- Mem: Am Vet Med Asn; Am Soc Vet Physiol & Pharmacol; Am Soc Vet Clin Path; Am Physiol Soc; Biol Photog Asn. Res: Liver function; heartworm disease; renal function; medical photography; hematology of domestic animals; fetal electrocardiography. Mailing Add: Dept of Admin Sch of Vet Med Auburn Univ Auburn AL 36830

MORGAN, HOWARD E, b Bloomington, Ill, Oct 8, 27; m 47; c 3. PHYSIOLOGY, BIOCHEMISTRY. Educ: Johns Hopkins Univ, MD, 49. Prof Exp: Intern obstet & gynec, Vanderbilt Univ, 49-50, asst resident, 50-54, instr, 53-54, res assoc physiol, 54-55, from instr to prof, 57-67; PROF PHYSIOL & CHMN DEPT, PA STATE UNIV, HERSHEY MED CTR, 67- Concurrent Pos: Fel, Howard Hughes Med Inst, 54-55, investr, 57-; vis scientist, Cambridge Univ, 60-61. Mem: Am Soc Biol Chemists; Am Physiol Soc; Cardiac Muscle Soc; Biophys Soc; Brit Biochem Soc. Res: Mechanism of hormone action; regulation of glucose and glycogen metabolism; membrane transport; regulation of protein turnover. Mailing Add: Dept of Physiol Pa State Univ Hershey Med Ctr Hershey PA 17033

MORGAN, HOWARD LEE, computer science, operations research, see 12th edition

MORGAN, IRA LON, b Ft Worth, Tex, Aug 3, 26; m 48; c 3. NUCLEAR PHYSICS. Educ: Tex Christian Univ, BA, 49, MA, 51; Univ Tex, PhD(physics), 54. Prof Exp: Instr physics, Tex Christian Univ, 48-51; res scientist, Univ Tex, 51-56; vpres nuclear physics, Tex Nuclear Corp, 56-60, exec vpres & dir res, 60-66, pres, 66-68; prof physics & dir ctr nuclear studies, Univ Tex, Austin, 68-73; PRES, COLUMBIA SCI INDUSTS CORP, 68- Concurrent Pos: AEC fel, 54-56; vpres, Nuclear-Chicago Corp, 65-68; consult, Los Alamos Sci Lab; mem, Bd Dirs, Capital Area Radiation & Res Found; chmn, Adv Comt Isotopes & Radiation Develop, AEC, 70-72; mem, Comn Nuclear Sci, Nat Res Coun & Tech Electronic Prod Radiation Safety Standards Comn, Food & Drug Admin, 74-77. Mem: Fel Am Phys Soc; Am Nuclear Soc; Am Inst Mgt. Res: Inelastic neutron scattering; activation analysis and radiation interaction in biological systems; nucleon-nucleon interactions. Mailing Add: Columbia Sci Industs Corp PO Box 9908 Austin TX 78766

MORGAN, JAMES EBENEZER, b Mayland, Eng, May 22, 35; m 60; c 4. PHYSICAL CHEMISTRY. Educ: Univ London, BSc, 57; McGill Univ, PhD(gas kinetics), 62. Prof Exp: Res chemist, Indust Cellulose Res Ltd, Ont, 57-58; res fel gas kinetics, McGill Univ, 62-63, res assoc, 63-65; PRES, MORGAN-SCHAFFER CORP, 65- Concurrent Pos: Asst prof, McGill Univ, 63-64. Mem: Chem Inst Can; Am Soc Mass Spectrometry. Res: Degradation of insulating materials in extra high voltage equipment under conditions of abnormal stress; gas kinetics; mass spectrometry. Mailing Add: Morgan-Schaffer Corp 5110 Courtrai Ave Montreal PQ Can

MORGAN, JAMES FREDERICK, b Gretna, La, Oct 21, 15; m 43; c 6. INDUSTRIAL HYGIENE. Educ: George Washington Univ, BS, 39. Prof Exp: Asst chemist, US Food & Drug Admin, 39-42, biochemist, 46-47; head chem hygienist, Indust Hyg Found, Inc, 47-55; dir indust hyg, Pa RR Co, 55-61; asst to dir, Haskell Labs, 61-73, CONSULT INDUST HYG, E I DU PONT DE NEMOURS & CO, INC, 73- Concurrent Pos: Lectr, Grad Sch Pub Health, Univ Pittsburgh, 50-55; consult, Threshold Limit Values Comt, Am Cong Govt Indust Hyg, 72- Mem: Am Chem Soc; Am Indust Hyg Asn. Res: Physical and chemical environmental health factors; industrial toxicology; design of company operated industrial hygiene services; direction of company-wide industrial hygiene program; composition, analysis and health effects of diesel exhaust; scientific information storage and retrieval; industrial chemical carinogenic agents. Mailing Add: 612 Merion Ave Havertown PA 19083

MORGAN, JAMES FREDERICK, b Minneapolis, Minn, June 20, 41; m 68; c 2. NUCLEAR PHYSICS. Educ: St Mary's Col, Minn, BA, 63; Univ Minn, MA, 66, PhD(physics), 68. Prof Exp: Res fel, Calif Inst Technol, 68-70; res assoc, 70-74, ASST PROF NUCLEAR PHYSICS, OHIO STATE UNIV, 74- Mem: Am Phys Soc. Res: Nuclear astrophysics; understanding of the origin and abundance of the elements, nuclear spectroscopy via analog resonances. Mailing Add: Van de Graaff Lab 1302 Kinnear Rd Columbus OH 43212

MORGAN, JAMES JOHN, b New York, NY, June 23, 32; m 57; c 6. ENVIRONMENTAL SCIENCE, WATER CHEMISTRY. Educ: Manhattan Col, BCE, 54; Univ Mich, MSE, 56; Harvard Univ, AM, 62, PhD(water chem), 64. Prof Exp: Instr, Univ Ill, 56-60; Danforth Teacher, Harvard Univ, 60-61; assoc prof chem & eng, Univ Fla, 63-65; ASSOC PROF ENG, CALIF INST TECHNOL, 65- Concurrent Pos: Ed, Environ Sci & Technol, Am Chem Soc, 66-74; dean students, Calif Inst Technol, 72-75. Honors & Awards: Water Purification Award, Am Water Works Asn, 63. Mem: AAAS; Am Chem Soc; Am Water Works Asn; Am Soc Limnol & Oceanog. Res: Chemistry of natural water systems; coagulation processes in aqueous solutions; mineral equilibria. Mailing Add: Div of Eng & Appl Sci Calif Inst of Technol Pasadena CA 91109

MORGAN, JAMES PLUMMER, b Beaumont, Tex, Dec 2, 19; m 45; c 3. GEOLOGY. Educ: Univ Calif, BA, 42; La State Univ, PhD(geol), 51. Prof Exp: From instr to assoc prof, 46-62, PROF GEOL, LA STATE UNIV, BATON ROUGE, 62-, CHMN DEPT, 70- Concurrent Pos: Contract grant, Corps Engrs, 48-50 & Off Naval Res, 50-53; managing dir, Coastal Studies Inst, 54-66; US Geol Surv grant, 68-71; mem, Sci Planning Coun, Gulf Univ Res Consortium, 70- Mem: AAAS; Geol Soc Am; Am Asn Petrol Geologists; Soc Econ Paleont & Mineral. Res: Deltaic geology; coastal morphology and processes; marine geology. Mailing Add: Dept of Geol La State Univ Baton Rouge LA 70803

MORGAN, JASPER EUGENE, b Waynesville, NC, June 5, 11; m 33; c 1. PHYSICS. Educ: Wake Forest Col, BS, 31; Duke Univ, AM, 32, PhD(physics), 35. Prof Exp: Assoc physics, Duke Univ, 32-35, instr radiol & consult physicist, Univ Hosp, 35-41; RADIATION PHYSICIST, MED CTR, UNIV CALIF, LOS ANGELES, 58-, CLIN PROF RADIOL, 62- Res: Radiation physics. Mailing Add: Dept of Radiol Univ of Calif Hosp Los Angeles CA 90024

MORGAN, JEAN MCNEIL, b Dover, Del, Aug 29, 26; m 61. INTERNAL MEDICINE. Educ: Univ Pa, BA, 46; Women's Med Col, MD, 54. Prof Exp: NIH fel metab, Med Ctr, Univ Ala, 56-57 & renal med, Harvard Med Sch & Peter Bent Brigham Hosp, 58-59; from instr to asst prof med, Med Col Ala, 60-63; asst prof, Med Col SC, 63-64; from asst prof to prof med, Sch Med, Univ Ala, Birmingham, 64-75; RETIRED. Concurrent Pos: Mem staff, Vet Admin Hosp, Birmingham. Mem: Am Col Physicians; Am Fedn Clin Res; Am Soc Nephrology. Res: Lead poisoning; immunology of the kidney; transplantation. Mailing Add: Univ of Ala Sch of Med Univ Sta Birmingham AL 35294

MORGAN, JOE PETER, b Olds, Iowa, July 30, 31; m 52; c 4. VETERINARY RADIOLOGY. Educ: Iowa State Teachers Col, BA, 52; Colo State Univ, DVM, 60, MS, 62; Royal Vet Col, Sweden, Vet med dr, 67. Prof Exp: Instr radiol, Colo State Univ, 60-63, asst prof, 63-64 & 67-68; PROF RADIOL, SCH VET MED, UNIV CALIF, DAVIS, 68-, VET MED DIR & CHMN DEPT RADIOL SCI, 74- Res: Comparative orthopedics. Mailing Add: Dept of Radiol Univ of Calif Sch of Vet Med Davis CA 95616

MORGAN, JOHN CLIFFORD, II, b Darby, Pa, Oct 29, 38; m 61; c 4. MATHEMATICAL STATISTICS, PURE MATHEMATICS. Educ: San Diego State Col, AB, 64; Univ Calif, Berkeley, MA, 68, PhD(statist), 72. Prof Exp: ASST PROF MATH, SYRACUSE UNIV, 72- Mem: Inst Math Statist; Am Statist Asn. Res: Methods of classifying point sets, including Baire category, measure theory and dimension theory; nonparametric statistics. Mailing Add: Dept of Math Syracuse Univ Syracuse NY 13210

MORGAN, JOHN WALTER, b Walsall, Eng, Jan 27, 32; m 59; c 2. GEOCHEMISTRY, RADIOCHEMISTRY. Educ: Univ Birmingham, BSc, 55; Australian Nat Univ, PhD(geochem), 66. Prof Exp: Sci asst electronics, Atomic Energy Res Estab, UK, 48-51, asst exp off, 51-52, anal chem, 55-59; exp off, Australian Atomic Energy Comn Res Estab, 59-61, sr res scientist, 66-68; res asst geochem, Australian Nat Univ, 61-66, res fel, 66; res assoc, Univ Ky, 68-69, chem, 69-70; SR RES ASSOC GEOCHEM, ENRICO FERMI INST, UNIV CHICAGO, 70- Mem: AAAS; Inst Asn Geochem & Cosmochem; Am Geophys Union; Meteoritical Soc. Res: Application of analytical chemistry, particularly those aspects involving isotope and radiochemistry to the problems of earth and space science. Mailing Add: Enrico Fermi Inst Univ of Chicago Chicago IL 60637

MORGAN, JOSEPH, b Kiev, Russia, Mar 4, 09; nat US; m 42; c 2. PHYSICS, X-RAY CRYSTALLOGRAPHY. Educ: Temple Univ, AB, 31, MA, 33; Mass Inst Technol, PhD(physics), 37. Prof Exp: Asst physics, Temple Univ, 29-31; supvr, Wave Length Proj, Mass Inst Technol, 37-38; instr physics, Tex Agr & Mech Col, 38-41, asst prof, 41; from asst prof to assoc prof, 41-45, PROF PHYSICS, TEX CHRISTIAN UNIV, 45-, DIR RES COORD, 69- Concurrent Pos: Dir eng, Tex Christian Univ, 52-69, chmn, Div Natural Sci, 53-55 & Dept Physics, 58-59, vpres, Univ Res Found, 66-73. Mem: AAAS; Am Phys Soc; Am Asn Physics Teachers; Am Crystallog Asn; Am Geophys Union. Res: Optical spectroscopy and physical optics; x-ray diffraction; neutron diffraction; nuclear physics. Mailing Add: 3833 S Hills Circle Ft Worth TX 76109

MORGAN, JOSEPH FRANCIS, biochemistry, deceased

MORGAN, JULIET, b Clacton-on-Sea, Eng, Jan 30, 37; m 59; c 2. CELL BIOLOGY, ZOOLOGY. Educ: Australian Nat Univ, BSc, 66; MacQuarie Univ, MS, 68; Univ Ky, PhD(cell biol), 70. Prof Exp: Asst zool, Australian Nat Univ, 64-66; exp officer, Australian AEC, 66-68; res asst cell biol, T H Morgan Sch Biol Sci, Univ Ky, 68-70; RES ASSOC CELL BIOL, UNIV CHICAGO, 70- Res: Basic research in muscle disease; cell structure and function; cell recognition, aggregation and adhesion; cell ultrastructure. Mailing Add: Dept of Med Univ of Chicago Chicago IL 60637

MORGAN, KARL ZIEGLER, b Enochsville, NC, Sept 27, 07; m; c 4. HEALTH PHYSICS. Educ: Univ NC, AB, 29, MS, 30; Duke Univ, PhD(physics), 34. Hon Degrees: DS, Lenoir Rhyne Col, 71. Prof Exp: Asst, Duke Univ, 31-34; prof physics & head dept, Lenoir-Rhyne Col, 34-43; mem res staff, Metall Lab, Univ Chicago, 43; from mem staff to dir health physics div, Oak Ridge Nat Lab, 43-72; PROF NUCLEAR ENERGY, SCH OF NUCLEAR ENERGY, GA INST TECHNOL, 72- Concurrent Pos: Mem, Nat Comt Radiation Protection & Int Comn Radiol Protection; ed, J Health Physics, Health Physics Soc; mem subcomt, Adv Comt Reactor Safeguards-Nuclear Reg Comn, 75-, Transportation of Radioactive Mat Comt to Joint Comt Cong, 74- & Adv Comt, Bur Radiol Health, Dept Health, Educ & Welfare, Pub Health Serv, Food & Drug Admin, 75- Honors & Awards: First Gold Medal Work Radiation Protection, Royal Acad Sci Sweden, 62; First Distinguished Award, Health Physics Soc, 68; Distinguished Achievement Award, Health Physics Soc, 73. Mem: Health Physics Soc (pres, 55-57); fel Am Nuclear Soc; Am Asn Physicists in Med; Am Asn Physics Teachers; hon mem Ger Asn Experts & Radiation Protection. Res: Cosmic ray experiments; health physics; radiation protection; neutron dosimetry cataract studies; low level exposure of man to ionizing radiation; non-ionizing radiation effects on man. Mailing Add: 1984 Castleway Dr Atlanta GA 30345

MORGAN, KATHRYN A, b Riverside, Calif, Nov 18, 22. MATHEMATICS. Educ: Stanford Univ, AB, 42, AM, 43, PhD(math), 46. Prof Exp: From instr to asst prof, 46-57, ASSOC PROF MATH, SYRACUSE UNIV, 57- Res: Number theory. Mailing Add: Dept of Math Syracuse Univ Syracuse NY 13210

MORGAN, LEE ROY, JR, b New Orleans, La, Nov 5, 36; m 57; c 4. PHARMACOLOGY, CHEMOTHERAPY. Educ: Tulane Univ, BS, 58, MS, 59, PhD, 60; Imp Col, Univ London, DIC, 61; La State Univ, MD, 71. Prof Exp: Instr, 61-63, asst prof, 63-72, ASSOC PROF PHARMACOL, SCH MED, LA STATE UNIV, MED CTR, 72- Concurrent Pos: Res assoc, Imp Col, Univ London, 60-61; asst prof, Loyola Univ, 61-63; vis prof pharmacol, Sch Med, Univ Costa Rica, 63; consult clin biochemist, Charity Hosp, New Orleans, 67- Honors & Awards: Sigma Xi Award, 58. Mem: AAAS; Am Chem Soc; The Chem Soc; Soc Exp Med & Biol; Am Asn Cancer Res. Res: Drug profiles; enzyme kinetics; immunology and chemotherapy of cancer. Mailing Add: Dept of Pharmacol La State Univ Med Ctr New Orleans LA 70112

MORGAN, LEON OWEN, b Oklahoma City, Okla, Oct 25, 19; m 42; c 3. CHEMISTRY. Educ: Oklahoma City Univ, BA, 41; Univ Tex, MS, 43; Univ Calif, PhD(chem), 48. Prof Exp: Anal chemist, Okla Gas & Elec Co, 41; instr chem, Univ Tex, 41-44; res assoc, Metall Lab, Univ Chicago, 44-45; chemist, Radiation Labs, Univ Calif, 45-47; from asst prof to assoc prof, 47-61, PROF CHEM, UNIV TEX, AUSTIN, 62- Mem: Am Chem Soc; Am Phys Soc. Res: Nuclear and electron paramagnetic resonance and relaxation; fast reaction rate processes; radiation effects; inorganic biochemistry. Mailing Add: Dept of Chem Univ of Tex Austin TX 78712

MORGAN, LYMAN WALLACE, physical chemistry, see 12th edition

MORGAN, MARCUS S, b Pittsburgh, Pa, Feb 16, 14; m 40; c 1. ORGANIC CHEMISTRY. Educ: Univ Pittsburgh, BSc, 35, PhD(org chem), 43. Prof Exp: Fel, Dept Res Pure Chem, 43-52, sr fel med chem, 52-57, sr fel & head coal chem res proj, 57-67, ASSOC PROF CHEM, MELLON INST SCI, CARNEGIE-MELLON UNIV, 67- Mem: AAAS; Am Chem Soc. Honors & Awards: Phillips Medal, 35. Res: Teaching general, organic and biochemistry; biomedical materials; chemotherapy. Mailing Add: Mellon Inst of Sci Carnegie-Mellon Univ 4400 Fifth Ave Pittsburgh PA 15213

MORGAN, MAX EUGENE, b Bellingham, Wash, Dec 28, 13; m 37; c 2. DAIRY BACTERIOLOGY. Educ: Wash State Univ, BS, 39; Univ Conn, MS, 41; Iowa State Univ, PhD(dairy bact), 48. Prof Exp: Assoc prof dairy mfg, Univ Conn, 48-58, prof dairy mfg, 58-70, bact, Grad Sch, 54-70; PROF FOOD SCI, ORE STATE UNIV, 70- Honors & Awards: Borden Award, 67. Mem: Am Dairy Sci Asn; Am Soc Microbiol; Am Acad Microbiol. Res: Metabolism of microorganisms of importance in dairy products; chemistry of the flavor of dairy products. Mailing Add: Dept of Food Sci & Technol Ore State Univ Corvallis OR 97331

MORGAN, MEREDITH WALTER, b Kingman, Ariz, Mar 22, 12; m 37; c 1. OPTOMETRY, VISUAL PHYSIOLOGY. Educ: Univ Calif, AB, 34, MA, 39, PhD(physiol), 41. Hon Degrees: DOS, Ill Col Optom, 68. Prof Exp: Clin asst optom, 36-42, from instr to assoc prof, 42-51, dean sch optom, 60-73, PROF PHYSIOL OPTICS & OPTOM, UNIV CALIF, BERKELEY, 51- Concurrent Pos: Pvt pract, 34-60; mem ed coun, Am J Optom, 55-; mem, Coun Optom Educ, 60-73; mem rev comt construct schs optom, USPHS, 64-65, nat adv coun med, dent, optom, podiatric & vet educ, 65-67; mem adv coun, Nat Eye Inst, 69-71. Honors & Awards: Nelson Achievement Award, 59; Prentice Medal, Am Acad Optom, 67. Mem: AAAS; Geront Soc; Am Acad Optom (pres, 53-54). Res: Accommodation and convergence; binocular vision. Mailing Add: Sch of Optom Univ of Calif Berkeley CA 94720

MORGAN, MICHAEL DEAN, b Marion, Ind, Oct 27, 41; m 66; c 1. PLANT ECOLOGY. Educ: Butler Univ, BA, 63; Univ Ill, Urbana-Champaign, MS, 65, PhD(bot), 68. Prof Exp: Asst prof, 68-72, ASSOC PROF ECOSYSTS ANAL, UNIV

WIS-GREEN BAY, 72- Concurrent Pos: Wis Alumni Res Fund grant, Univ Wis-Green Bay, 72. Mem: Ecol Soc Am; Am Inst Biol Sci; AAAS. Res: Relationships between climatic change and plant distribution and production; ecological relationships during late Pleistocene; plant phenology; interactions between plants and air pollutants. Mailing Add: Col of Environ Sci Univ of Wis Green Bay WI 54302

MORGAN, MILLETT GRANGER, b Hanover, NH, Mar 17, 41; m 63; c 2. SCIENCE POLICY, ENVIRONMENTAL MANAGEMENT. Educ: Harvard Univ, BA, 63; Cornell Univ, MS, 65; Univ Calif, San Diego, PhD(appl physics), 69. Prof Exp: Lectr appl physics & info sci, Univ Calif, San Diego, 70-71, actg asst prof info sci, 71-72, dir comput jobs through training proj, 69-72; prog dir, Comput Impact on Soc, NSF, 72-74; assoc res physicist, Biomed & Environ Assessment Group, Brookhaven Nat Lab, 74-75; ASST PROF ELEC ENG & PUB AFFAIRS & COORDR EDPA GRAD PROG, CARNEGIE-MELLON UNIV, 75- Concurrent Pos: Consult, NSF, 74- & Brookhaven Nat Lab, 75- Mem: AAAS; Am Geophys Union; Inst Elec & Electronics Engrs; Air Pollution Control Asn. Res: Problems in the science-society interface area, especially energy use and computers and society. Mailing Add: Elec Eng & Pub Affairs Carnegie-Mellon Univ Pittsburgh PA 15213

MORGAN, MONROE TALTON, SR, b Mars Hill, NC, June 29, 33; m 60; c 2. ENVIRONMENTAL HEALTH, PUBLIC HEALTH. Educ: ETenn State Univ, BA & CPHS, 60; Univ NC, Chapel Hill, MSPH, 62; Tulane Univ, La, DrPh(environ & pub health), 69. Prof Exp: Sanitarian, Fairfax County Health Dept, Va, 60-61; training officer pub health, Va State Health Dept, 62-63; PROF ENVIRON HEALTH & CHMN DEPT, E TENN STATE UNIV, 63- Concurrent Pos: Consult ed, Nat J Environ Health, 69-; consult environ health educ, Nat Environ Health Asn, 70-; mem, Pub Health Rev Comt, 71-75. Mem: Am Pub Health Asn; Nat Environ Health Asn (2nd vpres, 71-). Res: Survival of aerobic sporeformers and mycobacterium tuberculosis vas hominis; human ecology as it relates to environmental stresses. Mailing Add: Dept of Environ Health ETenn State Univ Johnson City TN 37601

MORGAN, NEAL O, b Los Angeles, Calif, May 13, 28; m 52; c 3. MEDICAL ENTOMOLOGY, ECOLOGY. Educ: SDak State Col, BS, 56, MS, 57; Va Polytech Inst, PhD(entom), 62. Prof Exp: Entomologist, Entom Res Div, Agr Res Serv, USDA, Tex, 61-65, invest leader fly control, Md, 65-72, RES ENTOMOLOGIST, CHEM & BIOPHYS CONTROL LAB, AGR ENVIRON QUAL INST, ENTOM RES DIV, AGR RES SERV, USDA, 72- Mem: Entom Soc Am; Ecol Soc Am. Res: Insects affecting man and animals, their ecology, physiology and control methods; electromagnetic forces as insect attractants and repellants. Mailing Add: Bldg 177A Beltsville Agr Res Ctr-E Beltsville MD 20705

MORGAN, OMAR DRENNAN, JR, b Shelbyville, Mo, Mar 5, 13; m 42; c 2. PLANT PATHOLOGY. Educ: Ill State Univ, BEd, 40; Univ Ill, PhD(plant path), 50. Prof Exp: Teacher high sch, 41-42; asst plant path, Univ Ill, 48-49; from asst prof to assoc prof, 49-74, PROF PLANT PATH, UNIV MD, 74- Mem: Am Phytopath Soc. Res: Soil fumigation for control of soil borne diseases; developing tobacco resistance to major diseases. Mailing Add: 3108 Gumwood Dr Hyattsville MD 20783

MORGAN, PAGE WESLEY, b Phoenix, Ariz, Apr 3, 33; m 55; c 3. PLANT PHYSIOLOGY, BIOCHEMISTRY. Educ: Tex A&M UNiv, BS, 55, MS, 58, PhD(plant physiol), 61. Prof Exp: Asst plant physiol & range mgt, 56-58, Anderson-Clayton fel plant physiol, 58-60, from asst prof to assoc prof, 60-69, PROF PLANT PHYSIOL, TEX A&M UNIV, 69- Concurrent Pos: Cotton Producers Inst grants, 62-73; NSF grants, 64-74; mem, Plant Growth Regulator Working Group, Am Soc Plant Physiol. Mem: Am Soc Plant Physiol; Japanese Soc Plant Physiol. Res: Plant regulatory mechanisms; phytohormones; enzymes; destruction of auxin; synthesis of ethylene; interaction of phytohormones; auxin transport; floral initiation; apical dominance in grain sorghum. Mailing Add: Dept of Plant Sci Tex A&M Univ College Station TX 77843

MORGAN, PAUL HARPER, b Paris, Tenn, Aug 15, 40. BIOCHEMISTRY. Educ: Bethel Col, Tenn, BS, 63; Vanderbilt Univ, PhD(biochem), 70. Prof Exp: Instr chem, David Lipscomb High Sch, Nashville, Tenn, 63-65; NIH sr fel biochem, Univ Wash, 70-72; ASST PROF BIOCHEM, UNIV S ALA, 72- Mem: Am Chem Soc; Sigma Xi. Res: Mechanism of activation of the first component of complement; development of quantitative models for the interpretation of nutritional responses and animal growth curves. Mailing Add: Dept of Biochem Univ of SAla Col of Med Mobile AL 36688

MORGAN, PAUL NOLAN, b Konawa, Okla, Jan 15, 27; m 52; c 3. MICROBIOLOGY, IMMUNOLOGY. Educ: Univ Ark, BS, 52, MS, 56; Univ Okla, PhD(microbiol), 63. Prof Exp: RES MICROBIOLOGIST, MED RES SECT, VET ADMIN HOSP, 63- Concurrent Pos: From asst to assoc prof microbiol, Med Ctr, Univ Ark, Little Rock, 63-73. Mem: AAAS; Am Soc Microbiol; Am Soc Trop Med & Hyg; Int Soc Toxinology. Res: Viral replication in cell cultures; basic immunochemistry; spider venoms. Mailing Add: Vet Admin Hosp Med Res Sect 300 E Roosevelt Rd Little Rock AR 72206

MORGAN, PAUL VINCENT, b Pittsburgh, Pa, June 10, 26; m 51; c 3. MICROBIOLOGY. Educ: Univ Pittsburgh, BS, 50, MS, 58 & 61. Prof Exp: Asst biol sci, Univ Pittsburgh, 50-52, res assoc dent res, 52-54; jr fel water resources res, Mellon Inst Indust Res, 54-61; res fel environ health, Univ Pittsburgh, 61-67; mgr sci serv, 67-68, vpres, 68-72, V PRES & GEN MGR, CYRUS WILLIAM RICE DIV, NUS CORP, 72- Mem: Am Chem Soc; Am Inst Biol Sci; Am Water Resources Asn; Water Pollution Control Asn; Int Asn Theoret & Appl Limnol. Res: The ecological impact of large utilities, industries and municipalities on the environment, which involves siting studies, pre- and post-construction comparisons, modeling, thermal effects and water quality monitoring studies. Mailing Add: 6909 Meade St Pittsburgh PA 15208

MORGAN, PAUL WINTHROP, b West Chesterfield, NH, Aug 30, 11; m 39; c 2. CHEMISTRY. Educ: Univ Maine, BS, 37; Ohio State Univ, PhD(org chem), 40. Prof Exp: Asst chem, Ohio State Univ, 37-40, Du Pont fel, 40-41; res chemist, E I du Pont de Nemours & Co, 41-46, res assoc, 46-50, Pioneering Res Labs, Textile Fibers Dept, Exp Sta, 50-57, res fel, 57-73, SR RES FEL, PIONEERING RES LABS, TEXTILE FIBERS DEPT, EXP STA, E I DU PONT DE NEMOURS & CO, 73- Concurrent Pos: Mem, Polymer Nomenclature Comt, Am Chem Soc, 70-; chmn, Gordon Conf Polymers, 74. Honors & Awards: Am Chem Soc Award, 60, Polymer Chem Award, 76. Mem: AAAS; Am Chem Soc. Res: Cellulose derivatives; moisture permeability of polymers; low temperature and interfacial polycondensations; condensation polymers; thermally stable polymers; extended chain polymers and their liquid crystalline solutions; high strength-high modulus fibers; tire cord and reinforcement fibers. Mailing Add: Exp Sta Bldg 302 E I du Pont de Nemours & Co Wilmington DE 19898

MORGAN, PETER ERNEST DAVID, ceramics, chemistry, see 12th edition

MORGAN, RAYMOND VICTOR, JR, b Brownwood, Tex, May 10, 42; m 67. MATHEMATICS. Educ: Howard Payne Col, BA, 64; Vanderbilt Univ, MA, 65; Univ Mo, PhD(math), 69. Prof Exp: Instr math, Univ Mo, 69; ASST PROF MATH, SOUTHERN METHODIST UNIV, 69- Mem: Am Math Soc; Math Asn Am. Res: Non-associative algebra; generalizations of alternative algebras and their structures relative to Peirce decompositions and Wedderburn decompositions. Mailing Add: Dept of Math Southern Methodist Univ Dallas TX 75222

MORGAN, RELBUE MARVIN, b Oxford, Miss, May 7, 39; m 58. PHYSICS. Educ: Christian Bros Col, BS, 62; Iowa State Univ, PhD(physics), 67. Prof Exp: Asst prof, 67-69, ASSOC PROF PHYSICS, CHRISTIAN BROS COL, 69- Mem: Am Phys Soc; Am Asn Physics Teachers. Res: Electronic properties of metals and alloys; electrical and optical properties of thin metal films. Mailing Add: Dept of Physics Christian Bros Col Memphis TN 38104

MORGAN, RICHARD ALAN, organic chemistry, polymer chemistry, see 12th edition

MORGAN, RICHARD C, b Jamaica, NY, June 9, 38; m 61; c 3. APPLIED MATHEMATICS. Educ: Stevens Inst Technol, BE, 59; NY Univ, MS, 62, PhD(math), 65. Prof Exp: From instr to assoc prof, 60-69, chmn dept, 65-70, PROF MATH, ST JOHN'S UNIV, NY, 69- Mem: Am Math Soc; Math Asn Am; Inst Elec & Electronics Eng. Res: Electromagnetic diffraction theory; partial differential equations; fluid dynamics. Mailing Add: Dept of Math St John's Univ Grand Central & Utopia Pkwys Jamaica NY 11432

MORGAN, ROBERT LEE, b Clawson, Mich, Nov 26, 29; m 52; c 2. ORGANIC CHEMISTRY. Educ: Antioch Col, BS, 52; Univ Chicago, PhD(chem), 60. Prof Exp: Asst chemist, Univ of Chicago, 52-53; org res chemist, Cent Res Dept, E I du Pont de Nemours & Co, 60-62, res chemist, Elastomers Dept, 62-67, prod develop chemist, 67-71, market develop rep, 71-72, supvr wire & cable indust develop & serv, 72-74, DIV HEAD, PROD DEVELOP, ELASTOMERS DEPT, E I DU PONT DE NEMOURS & CO, 74- Concurrent Pos: Inst, Salem Col, 52. Res: Steroids; organo-inorganic chemistry; homogeneous catalysis; elastomers. Mailing Add: Elastomer Dept E I du Pont de Nemours & Co Wilmington DE 19898

MORGAN, RUSSELL HEDLEY, b London, Ont, Oct 9, 11; nat US; m 38; c 2. RADIOLOGY. Educ: Univ Western Ont, BA, 34, MD, 37. Prof Exp: From instr to assoc prof roentgenol, Univ Chicago, 42-46; med officer-in-chg radiol sect, Tuberc Control Div, USPHS, 44-46; prof radiol, Sch Med & radiologist-in-chief, Univ Hosp, 46-71, dean sch med, 71-75, vpres health divs, 73-75, PROF MED, JOHNS HOPKINS UNIV, 75- Concurrent Pos: Sr consult radiation, Off Surgeon Gen, USPHS. Mem: Am Roentgen Ray Soc; Am Thoracic Soc; Radiol Soc NAm; Am Col Radiol. Res: Electronic image tubes; antiocardiographic methods for detection of gastric cancer; mass photofluorographic examinations of the chest for tuberculosis; design of x-ray automatic timers. Mailing Add: Off of Dean Sch Med Johns Hopkins Univ Baltimore MD 21205

MORGAN, SAMUEL POPE, b San Diego, Calif, July 14, 23; m 48; c 4. MATHEMATICAL PHYSICS, COMPUTER SCIENCE. Educ: Calif Inst Technol, BS, 43, MS, 44, PhD(physics), 47. Prof Exp: Asst physics, Univ Calif, 43-44 & Calif Inst Technol, 44-47; mem tech staff, Bell Tel Labs, 47-59, head, Math Physics Dept, 59-67, dir comput technol, 69-70, DIR COMPUT SCI RES, BELL LABS, INC, NJ, 67- Mem: AAAS; fel Inst Elec & Electronics Engrs; Am Phys Soc; Soc Indust & Appl Math; Asn Comput Mach. Res: Electromagnetic theory; mechanics of continua; wave propagation; special mathematical functions; numerical methods; research administration. Mailing Add: Comput Sci Res Ctr Bell Labs Inc Murray Hill NJ 07974

MORGAN, STANLEY L, b Sandyville, Ohio, Jan 28, 18; m 41; c 3. PHARMACEUTICS, CHEMICAL ENGINEERING. Educ: Case Inst Technol, BSChE, 39. Prof Exp: Chem engr, 40-42, mgr blood plasma labs, 42-43, gen mgr, 43-63, vpres, 63-64, EXEC VPRES, BEN VENUE LABS, INC, 64- Mem: AAAS; Am Chem Soc; Am Inst Chem Eng; Asn Off Racing Chemists; fel Am Inst Chem. Res: Pharmaceuticals under aseptic conditions; ethylene oxide sterilization; freeze drying; racing chemistry; preservation of viable organisms; cryobiology. Mailing Add: 270 Northfield Rd Bedford OH 44146

MORGAN, THOMAS ANTHONY, b New York, NY, Aug 4, 37. PHYSICS. Educ: Mass Inst Technol, BS, 58; Syracuse Univ, PhD(theoret physics), 64. Prof Exp: Physicist, Gen Elec Co, NY, 63-64; asst prof, 64-72, ASSOC PROF PHYSICS, UNIV NEBR, LINCOLN, 72- Concurrent Pos: Res assoc math dept, Kings Col, Univ London, 65 & Kellogg Radiation Lab, Calif Inst Technol, 69-71. Mem: Am Astron Soc; Am Math Soc; Am Phys Soc. Res: General relativity; relativistic astrophysics. Mailing Add: Dept of Physics Univ of Nebr Lincoln NE 68508

MORGAN, THOMAS DAVID, b Cushing, Okla, Aug 4, 16; m 40; c 3. PHYSICS. Educ: Southwestern Col, Kans, AB, 39. Prof Exp: Asst physics, Univ Ky, 39-40; physicist, Res Div, Phillips Petrol Co, 40-51, opers supvr, Atomic Energy Div, 51-53, spectral anal supvr, 53-63, sec leader anal res, 63-67, SECT MGR SPECTROS, PHILLIPS PETROL CO, 67- Mem: Am Chem Soc; Soc Appl Spectros; Am Soc Mass Spectrometry. Res: Mass spectrometry; analytical chemistry; physical science instrumentation. Mailing Add: 224 RB 1 Phillips Petrol Co Bartlesville OK 74004

MORGAN, THOMAS EDWARD, b Bay Shore, NY, Dec 27, 43. ASTROPHYSICS. Educ: Hofstra Univ, BA, 65; Ind Univ, Bloomington, MA, 67, PhD(astrophys), 71. Prof Exp: Instr, 70-71, ASST PROF ASTRON, STATE UNIV NY COL OSWEGO, 71- Mem: Am Astron Soc; Sigma Xi. Res: Line blanketing in stellar atmospheres, primarily Epsilon Virginis and Arcturus; ancient solar cycle; astronomy education. Mailing Add: Dept of Earth Sci State Univ of NY Col Oswego NY 13126

MORGAN, THOMAS EDWARD, JR, b Jacksonville, Fla, Nov 9, 29; m 54; c 3. INTERNAL MEDICINE, BIOCHEMISTRY. Educ: Duke Univ, BS, 50, MD, 54. Prof Exp: Intern & resident med, Hosps, Stanford & Columbia Univs, 54-57; asst physician, Presby Hosp, New York, 57-60; res assoc biochem, Sch Med, Univ Wash, 62-64, asst prof internal med, 64-68, assoc prof med, 68-73, assoc dean acad affairs, 68-74, prof med, 73-74; DIR DIV BIOMED RES & DEP DIR DEPT ACAD AFFAIRS, ASN AM MED COLS, 75- Concurrent Pos: Nat Found res fel, Columbia Univ, 57-60; NIH spec res fel, 62-64; USPHS res career award, 65-; asst physician, Francis Delafield Hosp, New York, 59-60; mem med scientist comt, Nat Inst Gen Med, 69-72, mem nat heart & lung adv coun, Nat Heart & Lung Inst, 72-75. Mem: Am Fedn Clin Res; Am Soc Clin Invest; Am Col Physicians; Am Thoracic Soc. Res: Transport mechanisms; pulmonary surface active lipids; lipid biosynthesis. Mailing Add: Asn Am Med Cols 1 Dupont Circle NW Washington DC 20036

MORGAN, THOMAS HARLOW, b Jacksonville, Fla, May 31, 45. ASTROPHYSICS, PLANETARY SCIENCES. Educ: Univ Fla, BS, 66, PhD(physics), 72. Prof Exp: Fel, Nat Res Coun, NASA Johnson Space Ctr, 73-74; ASST PROF PHYSICS ASTRON, HOUSTON BAPTIST UNIV, 75- Mem: Am Astron Soc; Am Phys Soc; Astron Soc of Pac; Soc Photo Optical Instrumentation Engrs. Res: Ultraviolet observations of stars and planets; ground based observations in the infrared one to three microns of emission-line objects and planetary atmospheres; high spatial resolution planetary imagery. Mailing Add: Houston Baptist Univ 7502 Fondren Houston TX 77036

MORGAN, THOMAS JOSEPH, b Brooklyn, NY, Oct 20, 43; m 68; c 2. ATOMIC PHYSICS. Educ: Carroll Col, BA, 65; Mont State Univ, BSc, 66; Univ Calif, Berkeley, MSc, 68, PhD(eng physics), 71. Prof Exp: Univ fel pure & appl physics, Queen's Univ Belfast, 71-73; ASST PROF PHYSICS, WESLEYAN UNIV, 73- Mem: Am Phys Soc. Res: Heavy particle collision phenomena; atomic beams; collisional properties of excited states. Mailing Add: Dept of Physics Wesleyan Univ Middletown CT 06457

MORGAN, WALTER CLIFFORD, b Ledyard, Conn, Dec 22, 21; m 48; c 3. ANIMAL GENETICS. Educ: Univ Conn, BSc, 46, PhD(genetics), 53; George Washington Univ, MSc, 49. Prof Exp: Animal geneticist, Nat Cancer Inst, 46-49; res assoc mammalian genetics, Columbia Univ, 51-53; asst prof poultry genetics, Univ Tenn, 53-54; assoc prof, 54-58, PROF GENETICS & PHYSIOL, SDAK STATE UNIV, 58- Concurrent Pos: Vis scientist, Radiobiol Dept, Nuclear Energy Ctr, Mol, Belg, 68-69. Mem: Am Genetic Asn; World Poultry Sci Asn; Nat Asn Biol Teachers; Radiation Res Soc; NY Acad Sci. Res: Developmental genetics of poultry and mice; physiology of reproduction; new mutations; lethals; irradiation effects on chick embryos and on wheat. Mailing Add: Dept of Animal Sci SDak State Univ Brookings SD 57006

MORGAN, WILLIAM JASON, b Savannah, Ga, Oct 10, 35; m 59; c 2. GEOPHYSICS. Educ: Ga Inst Technol, BS, 57; Princeton Univ, PhD(physics), 64. Prof Exp: Res assoc, 64-66, asst prof, 66-71, ASSOC PROF GEOPHYS, PRINCETON UNIV, 71- Mem: Am Geophys Union. Res: Mantle convection; heat flow; plate tectonics; marine geophysics. Mailing Add: Dept of Geol & Geophys Sci Princeton Univ Princeton NJ 08540

MORGAN, WILLIAM KEITH C, b July 1, 29; US citizen; m 53; c 3. INTERNAL MEDICINE. Educ: Univ Sheffield, MB, ChB, 53, MD, 61; MRCP(D), 58, FRCP(E), 71, FACP, 72. Prof Exp: From instr to assoc prof, Univ Md, 59-67; assoc prof, 67-70, PROF MED, MED CTR, WVA UNIV, 70-, HEAD DIV PULMONARY DIS, 74- Concurrent Pos: Chief, Appalachian Lab Occup Respiratory Dis, Nat Inst Occup Safety & Health, 67-71, dir, 71-74. Mem: Am Thoracic Soc; Brit Med Asn; Am Col Chest Physicians. Res: Chest diseases and occupational medicine. Mailing Add: Dept of Med WVa Univ Med Ctr Morgantown WV 26505

MORGAN, WILLIAM L, JR, b Honolulu, Hawaii, Nov 18, 27; m 54; c 2. INTERNAL MEDICINE, CARDIOVASCULAR DISEASES. Educ: Yale Univ, BA, 48; Harvard Med Sch, MD, 52; Am Bd Internal Med, dipl, 62. Prof Exp: Instr med, Harvard Univ, 56-58, tutor med sci, 57-58; assoc, Div Cardiovasc Dis, Henry Ford Hosp, 58-62; assoc prof, 62-66, PROF MED, SCH MED & DENT, UNIV ROCHESTER, 66-, ASSOC CHMN DEPT MED & DIR EDUC PROGS, 69- Concurrent Pos: Clin assoc, Nat Heart Inst; mem, Am Bd Internal Med, 73- Mem: Fel Am Col Physicians. Res: Cardiovascular hemodynamics; cardiac arrhythmias; medical education; physical diagnosis. Mailing Add: Dept of Med Strong Mem Hosp Rochester NY 14642

MORGAN, WILLIAM WILSON, b Bethesda, Tenn, Jan 3, 06; m 28, 66; c 2. ASTRONOMY. Educ: Univ Chicago, BS, 27, PhD, 31. Hon Degrees: Dr, Nat Univ Cordoba. Prof Exp: Asst astron, Yerkes Observ, Univ Chicago, 26-32, from instr to prof, 32-66, chmn dept, 60-66, dir, Yerkes & McDonald Observs, 60-63, Bernard E & Ellen C Sunny distinguished serv prof, 66-74, EMER PROF, UNIV CHICAGO, 74- Honors & Awards: Bruce Gold Medal Award, Astron Soc Pac, 58. Mem: Nat Acad Sci; Am Acad Arts & Sci; Royal Danish Acad; Royal Soc Sci Liege; cor mem Arg Nat Acad Sci. Res: Stellar spectroscopy; galaxies. Mailing Add: Yerkes Observ Williams Bay WI 53191

MORGAN, WINFIELD SCOTT, b Takoma Park, Md, Jan 9, 21; m 48; c 5. PATHOLOGY. Educ: Albright Col, BS, 42; Temple Univ, MD, 45. Prof Exp: Resident path, Mass Gen Hosp, 48-51, asst, 52-53, from asst pathologist to assoc pathologist, 53-62; dir path, Cleveland Metrop Gen Hosp, 62-67; dir labs, Aultman Hosp, Canton, Ohio, 67-74; PROF PATH & DIR SURG PATH, MED CTR, WVA UNIV, 74- Concurrent Pos: Fel med, Guthrie Clin, Sayre, Pa, 46; Am Cancer Soc-Brit Empire Cancer Campaign exchange fel, Oxford Univ, 51-52; Nat Cancer Inst spec res fel cellular physiol, Wenner-Gren Inst, Univ Stockholm, 58-60; consult, Health Res Facil Br, NIH, 65-67; instr, Med Sch, Tufts Univ, 52-60; Harvard Med Sch, 60-61, assoc, 61-62; prof, Sch Med, Case Western Reserve Univ, 62-67. Mem: Am Asn Path & Bact; Am Soc Exp Path; NY Acad Med; Int Acad Path. Res: Biochemical pathology. Mailing Add: Dept of Path WVa Univ Med Ctr Morgantown WV 26506

MORGAN, WYMAN, b Russellville, Ala, Jan 7, 41; m 59. INORGANIC CHEMISTRY. Educ: Florence State Col, BS, 62; Univ Fla, PhD(chem), 67. Prof Exp: Res chemist, Inorg Div, Monsanto Co, 67-71, sr res chemist, Monsanto Indust Chem Co, 71-75, SR RES SPECIALIST, MONSANTO INDUST CHEM CO, 75- Mem: Am Chem Soc. Res: Chemistry of metallic elements; exploratory process development. Mailing Add: Monsanto Indust Chem Co 800 N Lindbergh Blvd St Louis MO 63166

MORGANA, NORLEY, organic chemistry, see 12th edition

MORGANE, PETER J, b Atlanta, Ga, May 14, 27. NEUROPHYSIOLOGY, PSYCHOPHARMACOLOGY. Educ: Tulane Univ, BS, 48; Northwestern Univ, MS, 57, PhD(physiol), 59. Prof Exp: Resident res, Northwestern Univ, 56-59; instr physiol, Col Med, Univ Tenn, 59, asst prof, 59-61; sr physiologist, Life Sci Br, Goodyear Aircraft Corp, Ohio, 61-62; sr vis scientist, Brain Res Unit, Mex, 62-63; chmn div neurol sci, Commun Res Inst, 63-68; SR SCIENTIST, WORCESTER FOUND EXP BIOL, 68- Concurrent Pos: Affil prof, Worcester Polytech Inst, Clark Univ & Boston Univ, 69-; mem panel neurobiol, NSF, 71-74; mem study sect, Nat Inst Neurol Dis & Stroke, 74-; dir post-doctoral training prog neurobiol, NIMH. Mem: Am Physiol Soc; Am Asn Anatomists; Asn Psychophysiol Study Sleep; Am Psychol Asn; NY Acad Sci. Res: Comparative morphology of the brains of mammals; neural regulation of food and water intake; hypothalamic-limbic interactions in behavior; physiological, pharmacological and neurobiochemical studies on the sleep states; quantitative electroencephalographic analysis following pharmacological manipulation of the biogenic amines; neurophysiological role of serotonin; effects of protein malnutrition on developing brain. Mailing Add: Worcester Found for Exp Biol 222 Maple Ave Shrewsbury MA 01545

MORGANROTH, JOEL, b Detroit, Mich, Oct 29, 45; m 72. CARDIOLOGY, INTERNAL MEDICINE. Educ: Univ Mich, BS, 68, MD, 70; Am Bd Internal Med, dipl, 73, Am Bd Cardiovasc Dis, dipl, 75. Prof Exp: From intern to resident, Beth Israel Hosp, Harvard Med Sch, 70-72; clin assoc cardiol, Nat Heart & Lung Inst, 72-74; clin instr med, Georgetown Univ, 74; clin fel, 74-75, ASST PROF MED, MED SCH, UNIV PA, 75-, DIR, ECG-EXERCISE LAB & ASSOC DIR, NON-INVASIVE LAB, UNIV HOSP, 75- Honors & Awards: Physician Recognition Award, AMA, 73-76. Mem: Am Fedn Clin Res; Am Col Physicians; assoc fel, Am Col Cardiol; Am Heart Asn. Res: Pathophysiology of clinical cardiology utilizing primarily non-invasive techniques; echocardiography and exercise testing with particular attention to computerized models. Mailing Add: Hosp Univ Pa 937 Gates Pavil 3400 Spruce St Philadelphia PA 19104

MORGANS, LELAND FOSTER, b Stuttgart, Ark, June 30, 39; m 60; c 2. HISTOLOGY, ZOOLOGY. Educ: Tex Lutheran Col, BS, 61; Univ Ark, Fayetteville, MS, 65; Okla State Univ, PhD(zool), 68. Prof Exp: Asst prof biol, Phillips Univ, 68-69; ASST PROF BIOL, UNIV ARK, LITTLE ROCK, 69- Mem: AAAS; Sigma Xi; Am Asn Anatomists. Res: Comparative vertebrate physiology. Mailing Add: Dept of Biol Univ of Ark Little Rock AR 72204

MORGANSTERN, KENNARD H, b St Louis, Mo, Nov 24, 24; m 54; c 4. PHYSICS. Educ: Washington Univ, AB, 47, MS, 48, PhD, 51. Prof Exp: Asst nuclear physics, Washington Univ, 48-50; pres, Nuclear Consult, Inc, 50-53; exec vpres, Nuclear Corp Am, 54-58; PRES, RADIATION DYNAMICS, INC, 58- Res: Scattering of positrons from protons. Mailing Add: Radiation Dynamics Inc 1800 Shames Dr Westbury NY 11590

MORGAREIDGE, KENNETH GOODRICH, biochemistry, see 12th edition

MORGENSTERN, ALAN LAWRENCE, b Brooklyn, NY, Dec 21, 33; m 60; c 2. PSYCHIATRY. Educ: Cornell Univ, BA, 54; Duke Univ, MD, 59. Prof Exp: Intern internal med, Med Sch, Duke Univ, 59-60; resident psychiat, Med Sch, Univ Colo, 60-63; from instr to asst prof, 65-69, ASSOC PROF PSYCHIAT, MED SCH, UNIV ORE, 69- Concurrent Pos: WHO travel-study fel, Inst Psychiat, London, 72-73; consult, Training Br, NIMH & prog develop, Good Samaritan Hosp, Portland, 74- Mem: Fel Am Psychiat Asn; AMA. Res: Tests of clinical competence for psychiatrists; mental health services for disadvantaged adolescents; psychotherapy of borderline states; marital psychotherapy. Mailing Add: Suite 525 1220 SW Morrison St Portland OR 97205

MORGENSTERN, ERWIN KRISTIAN, b Heiligenkreutz, Ger, Dec 20, 28; Can citizen; m 61; c 2. FOREST GENETICS. Educ: Univ NB, BScF, 59; Univ Toronto, MScF, 61; Univ Hamburgh, Dr rer nat(forest genetics), 66. Prof Exp: Forest res officer forest genetics, Can Dept Forestry, 61-66; RES SCIENTIST FOREST GENETICS, CAN DEPT ENVIRON, 67- Concurrent Pos: Secy, Study Group Forest Tree Improv, NAm Forestry Comn, Food & Arg Orgn, UN, 73- Mem: Can Tree Improv Asn; Biomet Soc; Genetics Soc Can; Can Inst Forestry. Res: Population genetics of boreal tree species as a basis for the planning of breeding programs. Mailing Add: Petawawa Forest Exp Sta Chalk River ON Can

MORGENSTERN, LEON, b Pittsburgh, Pa, July 14, 19. SURGERY. Educ: Brooklyn Col, BA, 40; NY Univ, MD, 43. Prof Exp: Pvt pract, 52-59; asst prof surg, Albert Einstein Col Med, 59-60; asst clin prof, 64-70, assoc clin prof, 70-74, CLIN PROF SURG, SCH MED, UNIV CALIF, LOS ANGELES, 74-; DIR SURG, CEDARS OF LEBANON HOSP, 60-; DIR SURG, CEDARSSINAI MED CTR, 73- Mem: Am Col Surgeons; Soc Surg Alimentary Tract; Am Gastroenterol Asn; Int Soc Surgeons. Res: Gastric physiology and cancer; carcinoma gastric stump; vagotomy and gastric carcinoma; partial splenectomy; splenic hemostasis; postoperative jaundice; inflammatory bowel disease; malignant diverticulitis; wound healing; carcinoma of breast. Mailing Add: Dept of Surg Cedars-Sinai Med Ctr Los Angeles CA 90048

MORGENSTERN, PAUL, b Vienna, Austria, Jan 28, 31; US citizen; m 55; c 3. METEOROLOGY, AIR POLLUTION. Educ: St Louis Univ, BS, 53; Mass Inst Technol, SM, 55. Prof Exp: Mem staff radar meteorol, Lincoln Lab, Mass Inst Technol, 55; eng scientist, Missile Electronics & Controls Div, Radio Corp Am, 57-63; staff Scientist, Tech Div, GCA Corp, 63-68; PRIN METEOROLOGIST & DIR ENVIRON SYSTS DEPT, WALDEN RES DIV, ABCOR INC, 68- Mem: Air Pollution Control Asn; Am Meteorol Soc; Opers Res Soc Am. Res: Radar meteorology; meteorological forecasting; communication system studies; radar target simulation; ionospheric winds; meteorological data collection systems; atmospheric diffusion; air pollution; meteorological instrumentation; diffusion modelling; air pollution emission inventories; statistical and numerical analysis; data processing. Mailing Add: Walden Res Div Abcor Inc 201 Vassar St Cambridge MA 02139

MORGENTHALER, GEORGE WILLIAM, b Chicago, Ill, Dec 16, 26; m 49; c 4. MATHEMATICS. Educ: Concordia Col, BS, 46; Univ Chicago, MS, 48, PhD, 53; Univ Denver, MS, 63; Mass Inst Technol, MS, 70. Prof Exp: Mathematician, Inst Air Weapons Res, Chicago, 51-55, group leader, 55-58; assoc prof math & head dept, Chicago Undergrad Div, Univ Ill, 58-60; mgr electronics & math res dept, Martin Co, 60-65, dir res & develop, Martin Marietta Corp, 66-69, corp dir res & develop, 70-74, VPRES TECH OPERS, MARTIN MARIETTA AEROSPACE, DENVER DIV, 74- Concurrent Pos: Instr, Ill Inst Technol, 51-58; adj prof, Univ Colo, 60-65, vis prof, 70-; bk rev ed, J Astronaut Sci. Mem: Fel AAAS; Am Math Soc; Opers Res Soc Am; Inst Math Statist; fel Am Astronaut Soc (pres, 64-66). Res: Statistics; operations research; applied mathematics; astronautics. Mailing Add: Martin Marietta Aerospace PO Box 179 Mail Stop 1002 Denver CO 80201

MORGENTHALER, LAWRENCE P, analytical chemistry, see 12th edition

MORHARDT, J EMIL, b Bishop, Calif, Aug 19, 42; m 65. COMPARATIVE PHYSIOLOGY, ECOLOGY. Educ: Pomona Col, AB, 64; Rice Univ, PhD(vert physiol), 68. Prof Exp: Asst prof biol, Wash Univ, 67-75; DIR BIOL SERV & CHIEF SCIENTIST, HENNINGSON, DURHAM & RICHARDSON, ECOSCI DIV, 75- Concurrent Pos: Environ consult, Mo Bot Garden, 73-75 & Harland Bartholomew & Assocs, 74-75. Mem: AAAS; Am Soc Zoologists; Ecol Soc Am; Am Soc Naturalists; Am Soc Mammalogists. Res: Physiology and behavioral temperature regulation in the vertebrates; relationships between animals and environmental energy, food and water; environmental analyses. Mailing Add: HDR Ecosci 804 Anacapa Santa Barbara CA 93101

MORHARDT, SYLVIA STAEHLE, b Rochester, NY, June 21, 43; m 65. ECOLOGY, ENVIRONMENTAL MANAGEMENT. Educ: Pomona Col, BA, 65; Rice Univ, MA, 68; Wash Univ, PhD(biol), 71. Prof Exp: Asst prof biol, Wash Univ, 71-72; ecologist, Harland Bartholomew & Assocs, 72-75; SR ECOLOGIST, HENNINGSON, DURHAM & RICHARDSON, ECOSCI, 75- Mem: Am Inst Biol Sci; Ecol Soc Am. Res: Quantitative analysis of energy exchange between organisms and the environment; physiological and behavioral adaptations of animals to the environment; environmental planning and management of natural resources. Mailing Add: HDR Ecosci 804 Anacapa Santa Barbara CA 93101

MORI, KANAKA FRED, b Tokyo, Japan, Feb 4, 25; Can citizen; m 64; c 1. BIOCHEMISTRY, ENDOCRINOLOGY. Educ: Univ Tokyo, DVM, 48; Univ Montreal, PhD(exp med), 63. Prof Exp: Lectr biochem, Univ Tottori, 50-55, asst prof, 55-57; asst prof, Col Osteop Med & Surg, 59-61; fel, Cardiovasc Res Inst, Sch Med, Univ Calif, San Francisco, 61; res scientist, Animal Dis Res Inst, Can Dept Agr, 62-64; res scientist, 64-69, SECT HEAD, FOOD & DRUG RES LABS, CAN DEPT NAT HEALTH & WELFARE, 69- Mem: Am Chem Soc; Endocrine Soc. Res:

Structure-activity relationship of protein hormones. Mailing Add: Health Protection Br Can Dept of Health & Welfare Ottawa ON Can

MORI, KEN, b Tobata, Japan, July 14, 25; US citizen; m 62. PATHOLOGY. Educ: Univ Tokyo, 51. Prof Exp: Clin instr path, Col Med, State Univ NY Downstate Med Ctr, 63-66; assoc pathologist, 66-68, PATHOLOGIST, BETH ISRAEL HOSP, 68-; ASST PROF PATH, MT SINAI SCH MED, 67- Concurrent Pos: Asst pathologist, Maimonides Hosp, Brooklyn, NY, 63-66. Mem: Col Am Path. Res: Human pathology; electromicroscopy. Mailing Add: 33 Greenwich Ave New York NY 10014

MORI, PETER TAKETOSHI, b Montebello, Calif, Feb 16, 25; m 57; c 3. ORGANIC CHEMISTRY. Educ: Park Col, BS, 45; Okla State Univ, MS, 48; Purdue Univ, PhD, 54. Prof Exp: Res chemist, Dearborn Chem Co, 52-55, Am Potash & Chem Corp, 55-57, Turco Prod, Inc, 57 & Stuart Co, 58-62; RES CHEMIST, ICI UNITED STATES, INC, 62- Mem: AAAS; Am Chem Soc; Sigma Xi. Res: Medicinals, insecticides; detergents and corrosion inhibitors. Mailing Add: 1309 Quincy Dr Green Acres Wilmington DE 19803

MORI, RAYMOND I, b Hawaii, Oct 7, 26; m 51; c 4. ORGANIC CHEMISTRY. Educ: Univ Hawaii, BA, 49; Northwestern Univ, PhD(chem), 55. Prof Exp: Org chemist, Hawaiian Pineapple Co, Ltd, 54-61; asst dir qual assurance, Dole Co, 61-64, dir qual assurance, 64-69; CORP DIR QUAL ASSURANCE, CASTLE & COOKE FOODS, INC, 69- Mem: Am Chem Soc; Inst Food Technologists; Asn Offs Agr. Res: Quality control; enzymes. Mailing Add: Castle & Cooke Foods 50 California St San Francisco CA 94111

MORIARTY, C MICHAEL, b Schenectady, NY, Apr 12, 41; m 66; c 1. PHYSIOLOGY, BIOPHYSICS. Educ: Carnegie Inst Technol, BS, 62; Cornell Univ, MS, 65; Univ Rochester, PhD(biophys), 68. Prof Exp: Instr physiol & biophys, Sch Med, Univ Iowa, 68-70; asst prof, 70-73, ASSOC PROF PHYSIOL & BIOPHYS, UNIV NEBR MED CTR, OMAHA, 73- Concurrent Pos: USPHS fel, Sch Med, Univ Iowa, 68-70; NSF res grant, 71-74. Mem: Am Physiol Soc; Biophys Soc. Res: Cell physiology, membrane transport; calcium; hormone regulation. Mailing Add: Dept of Physiol & Biophys Univ of Nebr Med Ctr Omaha NE 68105

MORIARTY, DANIEL DELMAR, JR, b New Orleans, La, June 16, 46; m 69; c 1. ANIMAL BEHAVIOR. Educ: La State Univ, New Orleans, BA, 68; Tulane Univ, MS, 72, PhD(psychol), 73. Prof Exp: ASST PROF EXP PSYCHOL, UNIV SAN DIEGO, 73- Mem: Animal Behav Soc; Am Psychol Asn. Res: Genetic analysis of the behavioral response of frustration in rats and mice; effects of food deprivation and partial reinforcement on estrus cycling in female rats; effects of irrelevant drive on aversively motivated instrumental responses; Hippocampal chemical stimulation and delay of reinforcement. Mailing Add: Dept of Behav Sci Univ of San Diego San Diego CA 92110

MORIARTY, JOHN ALAN, b Chicago, Ill, Jan 17, 44; m 69. THEORETICAL SOLID STATE PHYSICS. Educ: Univ Calif, Berkeley, AB, 65; Stanford Univ, PhD(appl physics), 71. Prof Exp: Res assoc, Los Alamos Sci Lab, NMex, 71-73; res assoc, Univ Cambridge, 73-74; RES ASSOC PHYSICS, COL WILLIAM & MARY, 74- Concurrent Pos: Contractor, NASA Langley Res Ctr, 74- Mem: Am Phys Soc. Res: Electronic structure and properties of metals and semiconductors; pseudopotential, tight-binding, and resonance methods. Mailing Add: Dept of Physics Col of William & Mary Williamsburg VA 23185

MORIARTY, JOHN HENRY, b Methuen, Mass, July 3, 20; m 49; c 5. FOOD SCIENCE. Educ: Tufts Col, BS, 42. Prof Exp: Asst, Dept Food Technol, Mass Inst Technol, 46-51; food technologist & mem staff, Flavor Lab, Arthur D Little, Inc, 51-59, proj dir & bus mgr, Food & Biol Div, 59-60, PROJ DIR LIFE SCI DIV, ARTHUR D LITTLE, INC, 60- Concurrent Pos: Dir Res & Develop Assocs, 57-60. Mem: AAAS; fel Soc Advan Food Serv Res (vpres, 65-66, pres, 66-67); Inst Food Technologists; Am Inst Chemists. Res: Flavor evaluation; new products concepts and development; group administration. Mailing Add: 172 Lincoln St Melrose MA 02176

MORIARTY, ROBERT M, b New York, NY, Oct 9, 33; m 57; c 3. ORGANIC CHEMISTRY. Educ: Fordham Univ, BS, 55; Princeton Univ, PhD(org chem), 59. Prof Exp: Res worker org chem, Merck, Sharpe & Dohme, 55-56; NSF, NATO, Fulbright & NIH fels, Harvard Univ & Univ Munich, 59-61; from assoc prof to prof org chem, Cath Univ Am, 61-68; PROF ORG CHEM, UNIV ILL, CHICAGO CIRCLE, 68- Mem: Am Chem Soc; The Chem Soc. Res: Biochemistry; physical methods. Mailing Add: Dept of Chem Univ of Ill at Chicago Circle Chicago IL 60680

MORIBER, LOUIS G, b New York, NY, July 12, 17; m 41; c 2. CELL BIOLOGY. Educ: Brooklyn Col, AB, 38; Columbia Univ, MA, 52, PhD(zool, cytol), 56. Prof Exp: Jr aquatic physiologist, US Fish & Wildlife Serv, 41-43; assoc prof, 50-67, PROF BIOL, BROOKLYN COL, 67- Concurrent Pos: Exec officer, PhD prog biol, City Univ New York, 71- Mem: Am Soc Cell Biol. Res: Cytochemistry; cellular endocrinology. Mailing Add: Dept of Biol Brooklyn Col Brooklyn NY 11210

MORICONI, EMIL JOHN, organic chemistry, see 12th edition

MORIE, GERALD PRESCOTT, b St Louis, Mo, Mar 20, 39; m 61; c 2. ANALYTICAL CHEMISTRY. Educ: Cent Mo State Col, BS, 61; Ohio State Univ, MS, 63, PhD(anal chem), 66. Prof Exp: Asst instr chem, Ohio State Univ, 65; res chemist, 66-67, sr res chemist, 68-74, RES ASSOC, TENN EASTMAN CO, EASTMAN KODAK CO, 74- Mem: Am Chem Soc; Sigma Xi. Res: Solvent extraction of metal chelates; chemical separations especially liquid chromatography; selective ion electrodes; gas chromatography; analysis of tobacco smoke. Mailing Add: Tenn Eastman Co Cent Res Bldg Kingsport TN 37660

MORIGI, EUGENE MARIO EDMUND, b Istanbul, Turkey, Jan 1, 22; US citizen; m 51; c 2. BACTERIOLOGY, IMMUNOLOGY. Educ: Univ Bologna, MD, 47; McGill Univ, PhD(bact, immunol), 59. Prof Exp: Ship's surgeon, 49-51; intern, St Boniface Hosp, Man, 52-53; staff physician, Sanatorium Bd Man Cent Tuberc Clin, Winnipeg, 53-56; asst dir clin invest, Bristol Labs, NY, 59-63; assoc dir med res, William S Merrell Co, Ohio, 63-66; DIR CLIN RES, INT ORTHO RES FOUND, 66- Res: Medicine; metabolism of tubercle bacillus; clinical pharmacology of penicillins and tetracyclines; physiology of sleep. Mailing Add: Int Ortho Res Found US Hwy 202 Raritan NJ 08869

MORIMOTO, EDWARD MAMORU, radiochemistry, physical chemistry, see 12th edition

MORIMOTO, HIDEO, b Auburn, Calif, Jan 1, 39; m 64; c 2. MICROBIOLOGY, MOLECULAR BIOLOGY. Educ: Univ Calif, Los Angeles, AB, 60, PhD(cell physiol), 67. Prof Exp: Res zoologist, Univ Calif, Los Angeles, 66-67; USPHS fel, Univ Wis-Madison, 67-69, trainee, 69-70, res specialist molecular biol, 70-71; ASST PROF BIOL SCI, CARNEGIE-MELLON UNIV, 71- Res: Biochemistry and molecular biology of mitochondria development in yeast. Mailing Add: Dept of Biol Carnegie Mellon Univ Pittsburgh PA 15213

MORIN, DORNIS CLINTON, b Detroit, Mich, Oct 9, 23; m 48; c 3. PLASMA PHYSICS, ELECTRICAL ENGINEERING. Educ: Wayne State Univ, BS, 48, MSE, 51; Univ Wis, PhD(physics), 62. Prof Exp: Instr elec eng, Wayne State Univ, 48-51; res assoc, Univ Mich, 50-52; test engr, Vickers, Inc, 52-54; res assoc, Univ Mich, 54-55; mgr res dept, Dynex, Inc, 55-58; engr, Midwestern Univs Res Asn, 59-62, physicist, 62-63; res assoc, Univ Wis, 63-64, asst prof physics, Univ Wis-Milwaukee, 64-68, assoc prof, Univ Wis-Whitewater, 68-71; PROJ ASSOC, PLASMA PHYSICS GROUP, UNIV WIS-MADISON, 71- Concurrent Pos: Consult, Dynex, Inc, 58-60. Mem: AAAS; Am Phys Soc; Inst Elec & Electronics Eng. Res: Mechanics; fluid mechanics; plasma physics; electromagnetic theory; electronics; automatic control systems; applied mathematics. Mailing Add: 622 Jacobson Ave Madison WI 53714

MORIN, FRANCIS JOSEPH, b Laconia, NH, Oct 10, 17; m 46; c 7. SOLID STATE PHYSICS. Educ: Univ NH, BS, 39, MS, 40. Prof Exp: Mem tech staff, Bell Tel Labs, Inc, 41-62; assoc dir, Sci Ctr, NAm Aviation, Inc, 62-67, dir, Space Sci, Space & Info Systs Div, 65-67, dir, Sci Ctr, NAm Rockwell Corp, 67-70, mem tech staff, Sci Ctr, 70-74, DISTINGUISHED FEL, SCI CTR, ROCKWELL INT CORP, 74- Concurrent Pos: Vis prof physics, Univ Mo, 75. Mem: Am Chem Soc; Am Phys Soc. Res: Semiconductors; low temperature; electronic transport and energy-band structure in solids; transition metal oxides; diffusion and dislocations in solids; transition metal superconductors; surface physics; mechanism of catalysis of d-band surface states. Mailing Add: Sci Ctr Rockwell Int Corp Thousand Oaks CA 91360

MORIN, JAMES GUNNAR, b Minneapolis, Minn, Sept 13, 42. INVERTEBRATE ZOOLOGY, INVERTEBRATE PHYSIOLOGY. Educ: Univ Calif, Santa Barbara, BA, 65; Harvard Univ, MA, 67, PhD(biol), 69. Prof Exp: ASST PROF ZOOL, UNIV CALIF, LOS ANGELES, 69- Concurrent Pos: Nat Inst Neurol Dis & Stroke fel, Univ Calif, Los Angeles, 69-72. Mem: AAAS; Am Soc Zoologists; Soc Exp Biol & Med. Res: Bioluminescence in marine invertebrate, especially cnidaria; physiology of primitive nervous systems; community structure of subtidal sand bottoms. Mailing Add: Dept of Zool Univ of Calif Los Angeles CA 90024

MORIN, LEO GREGORY, b Berlin, NH, May 9, 41; m 68; c 2. CLINICAL BIOCHEMISTRY. Educ: Spring Hill Col, BS, 65; Boston Col, PhD(molecular biol), 68. Prof Exp: Asst prof biol, Rollins Col, 68-70; res biochemist, Sunland Hosp, Orlando, Fla, 70-71 & Kiess Instruments, Inc, 71-73; CLIN BIOCHEMIST & DIR CLIN CHEM, VET ADMIN HOSP, ATLANTA, 73- Concurrent Pos: Consult, E I du Pont de Nemours & Co, Inc, 76- Mem: AAAS; Am Chem Soc; Am Asn Clin Chem. Res: Metabolic regulation and disorders; analytical methodology; creatine kinase enzymology. Mailing Add: Lab Serv Vet Adm Hosp 1670 Clairemont Rd Decatur GA 30033

MORIN, RICHARD DUDLEY, b Quincy, Ill, Oct 5, 18; m 42; c 4. MEDICINAL CHEMISTRY. Educ: Univ Mich, BS, 40, MS, 42, PhD(org chem), 43. Prof Exp: Res chemist, Battelle Mem Inst, 43-52, asst div chief, 52-62, res assoc, 62-63; assoc prof, 63-70, PROF MED CHEM, SCH MED, UNIV ALA, BIRMINGHAM, 70- Mem: AAAS; Am Chem Soc. Res: Alkaloid synthesis; synthesis and biochemical and psychopharmacological action of psychotomimetic agents related to mescaline and psilocybin; structure-activity relationships among excitant and depressant compounds. Mailing Add: Neurosci Prog Sch of Med Univ of Ala Birmingham AL 35294

MORIN, ROBERT BENNETT, b Canton, Ohio, Dec 29, 31; m 56; c 4. ORGANIC CHEMISTRY. Educ: Swarthmore Col, BA, 53; Rice Univ, PhD(chem), 59. Prof Exp: Asst chemist, Rocket Fuels Div, Phillips Petrol Co, 54-55; sr res chemist, Chem Res Div, Eli Lilly & Co, Ind, 59-66, res assoc, 66-70; assoc prof pharmaceut chem, Sch Pharm, Univ Wis-Madison, 70-75; MEM FAC CHEM, UNIV ALTA, 75- Mem: AAAS; Am Chem Soc; The Chem Soc; Swiss Chem Soc; Am Soc Microbiol. Res: Structure and synthesis of certain diterpenes, antibiotics and lipid hormones. Mailing Add: Dept of Chem Univ of Alta Edmonton AB Can

MORIN, THOMAS LEE, b Rahway, NJ, Aug 27, 43; m 65; c 1. OPERATIONS RESEARCH. Educ: Rutgers Univ, BS, 65; Univ NMex, MS, 67; Case Western Reserve Univ, MS, 70, PhD(opers res), 71. Prof Exp: Spec lectr opers res, Case Western Reserve Univ, 69-70; opers res analyst, Univ Assocs, Inc, 70-71; ASST PROF OPERS RES, DEPT INDUST ENG & MGT SCI, TECHNOL INST, NORTHWESTERN UNIV, EVANSTON, 71- Concurrent Pos: NSF grant, Urban Systs Eng Ctr, Northwestern Univ, Evanston, 71-72. Mem: Opers Res Soc Am; Inst Mgt Sci; Am Soc Civil Engrs. Res: Dynamic programming; optimization of stochastic processes; application of operations research to urban and water resources systems. Mailing Add: Dept of Indust Eng & Mgt Sci Northwestern Univ Evanston IL 60202

MORIN, WALTER ARTHUR, b Salem, Mass, Oct 31, 33; m 57; c 7. NEUROPHYSIOLOGY. Educ: Merrimack Col, AB, 58; Boston Col, MS, 60; Clark Univ, PhD(physiol), 66. Prof Exp: Assoc prof, 61-74, PROF ZOOL, BRIDGEWATER STATE COL, 74- Concurrent Pos: NIH res grants, 66-67 & 69-71; Nat Res Coun Can Fel, Univ Toronto, 67-68; Sigma Xi res grant, 69-70. Mem: Am Soc Zool. Res: Excitatory and inhibitory synaptic vesicles in crustaceans. Mailing Add: Dept of Zool Bridgewater State Col Bridgewater MA 02324

MORIN, YVES, b Quebec, Que, Nov 28, 29; m 59; c 4. CARDIOLOGY. Educ: Laval Univ, BA, 48, MD, 53; FRCP(C), 58. Prof Exp: Chmn dept & asst dean res, 71-75, PROF MED, LAVAL UNIV, 71-, DEAN FAC MED, 75- Concurrent Pos: Chmn, Que Med Res Coun, 69-; vpres, Can Med Res Coun, 71-; pres, Comn Univ Res, 73- Mem: Fel Am Col Cardiol; fel Am Heart Asn; Can Cardiovasc Soc (secy-treas, 67); Can Soc Clin Invest. Res: Toxic cardiomyopathies; effect of alcohol on the heart; coronary circulation in man. Mailing Add: Fac of Med Laval Univ Quebec PQ Can

MORINIGO, FERNANDO BERNARDINO, b Parana, Arg, June 1, 36; m 63. PHYSICS. Educ: Univ Southern Calif, BS, 57; Calif Inst Technol, PhD(physics), 63. Prof Exp: Asst prof physics, Calif State Col, Los Angeles, 63-64; vis prof, Univ Freiburg, 64-65; from asst prof to assoc prof, 65-70, PROF PHYSICS, CALIF STATE UNIV, LOS ANGELES, 71- Concurrent Pos: Res fel, Calif Inst Technol, 63-64, sr res fel, 65; res physicist, Lab Nuclear Spectrometry, Nat Ctr Sci Res, Strasbourg, France, 68-69. Mem: Am Phys Soc. Res: Nuclear beta decay; nuclear spectroscopy; theoretical nuclear physics; nuclear reaction theory; quantum theory of gravitation. Mailing Add: Dept of Physics Calif State Univ 5151 State College Dr Los Angeles CA 90032

MORISAWA, MARIE, b Toledo, Ohio, Nov 2, 19. GEOLOGY, GEOMORPHOLOGY. Educ: Hunter Col, AB, 41; Union Theol Sem, MA, 45; Univ Wyo, MA, 52; Columbia Univ, PhD(geol), 60. Prof Exp: Instr geol, Bryn Mawr Col, 55-59; asst prof, Mont State Univ, 59-61; assoc prof, Antioch Col, 63-69; assoc prof, 69-75, PROF GEOL, STATE UNIV NY BINGHAMTON, 75- Mem: Geol Soc Am; Am Quaternary Asn; Am Geophys Union. Res: Environmental geomorphology, geology and planning; geological aesthetics. Mailing Add: Dept of Geol State Univ of NY Binghamton NY 13901

MORISHIMA, AKIRA, b Tokyo, Japan, Apr 18, 30; US citizen; m 61; c 2. PEDIATRICS, CYTOGENETICS. Educ: Keio Univ, Japan, MD, 54, PhD(med), 61. Prof Exp: Fulbright travel grant, 55-57; fel pediat endocrinol, Col Physicians & Surgeons, Columbia Univ, 58-61; instr pediat, 61-63, assoc, 63-65, asst prof pediat endocrinol, 65-66; asst prof, Med Ctr, Univ Calif, San Francisco, 66-68; ASSOC PROF PEDIAT ENDOCRINOL, COL PHYSICIANS & SURGEONS, COLUMBIA UNIV, 68- Concurrent Pos: From asst attend to assoc attend pediatrician, Babies Hosp, New York, 63-; NIH res career develop award, 66-68. Mem: NY Acad Sci; Soc Study Reproduction; Endocrine Soc; Soc Pediat Res; Am Soc Pediat. Res: Human cytogenetics; endocrinology. Mailing Add: Babies Hosp 622 W 168th St New York NY 10032

MORISHIMA, HISAYO ODA, b Ito-shi, Japan, July 27, 29; m 61; c 2. MEDICINE. Educ: Toho Univ, MD, 51; Tokyo Univ, PhD(med), 60. Prof Exp: Resident anesthesiol, Sch Med, Tokyo Univ, 54-56; dir anesthesiol, Izu Teishin Hosp, Shizuoka-ken, Japan, 56-59; resident, DC Gen Hosp, 59-60 & Washington Hosp Ctr, DC, 60-61; res assoc, Col Physicians & Surgeons, Columbia Univ, 61-66; clin instr, Med Ctr, Univ Calif, San Francisco, 66-68; asst prof anesthesiol, 68-74, ASSOC PROF ANESTHESIOL, COL PHYSICIANS & SURGEONS, COLUMBIA UNIV, 74- Mem: Am Soc Anesthesiol; Soc Obstet Anesthesia & Perinatology. Res: Fetal and neonatal physiology and pharmacology; obstetric anesthesia. Mailing Add: Dept of Anesthesiol Columbia Univ New York NY 10032

MORISON, IAN GEORGE, b Perth, Australia, Jan 31, 28; m 72; c 4. FORESTRY. Educ: Univ Western Australia, BScFor, 51; Australian Forestry Sch, DipFor, 52; Univ Wash, PhD(forestry), 70. Prof Exp: Sr scientist soils, 69-72, ASSOC PROF FORESTRY, COL FOREST RESOURCES, UNIV WASH, DIR INST FOREST PROD & DIR DIV CONTINUING EDUC FOREST RESOURCES, 72- Concurrent Pos: Proj mgr regional forest nutrit res prog & land use res appl nat needs, NSF, 69-72. Mem: Am Soc Agron; AAAS; Soc Am Foresters; Am Soc Plant Physiologists; Int Soc Soil Sci. Res: Growth response of Douglas fir to application of nitrogenous fertilizers and the relationship of response to stand, soil and foliar characteristics. Mailing Add: Col of Forest Resources Univ of Wash Seattle WA 98195

MORISON, ROBERT SWAIN, b Milwaukee, Wis, Nov 25, 06; m 36; c 2. NEUROPHYSIOLOGY. Educ: Harvard Univ, AB, 30, MD, 35. Hon Degrees: DSc, Loyola Univ Ill, 70 & Univ Rochester, 73. Prof Exp: Res physician, Collis P Huntington Mem Hosp, Boston, 34-35; Austin teaching fel, Harvard Med Sch, 35-36, instr physiol, 36-38, assoc anat, 38-41, asst prof, 41-44; from asst dir to assoc dir med sci, Rockefeller Found, 44-51, med & pub health, 51-55, dir biol & med res, 55-59, med & natural sci, 59-64; prof biol & dir div biol sci, 64-70, Richard J Schwartz prof sci & soc, 70-75, EMER PROF SCI & SOC, CORNELL UNIV, 75- Concurrent Pos: Class 1949 vis prof, Mass Inst Technol, 75- Mem: AAAS; Am Physiol Soc; Am Asn Anat; Am Acad Arts & Sci. Res: Electrophysiology; neuromuscular junction; thalamocortical relations; science policy; biomedical ethics. Mailing Add: Box 277 Peterborough NH 03458

MORITA, EIICHI, b Waipahu, Hawaii, Oct 16, 16; m 25. RUBBER CHEMISTRY. Educ: WVa State Col, BS, 61; WVa Univ, MS, 64; Osaka Univ, PhD(chem), 68. Prof Exp: Anal res chemist, Dunlop Rubber Co Ltd, Japan, 40-55 & Honolulu Gas Co, Hawaii, 55-56; sr res chemist, Monsanto Co, 56-74, RES SPECIALIST, MONSANTO INDUST CHEM CO, 74- Mem: Am Chem Soc; Soc Rubber Indust Japan. Res: Bonding rubber to other materials; synthesis and application of chemicals in elastomers; vulcanization of various elastomers; mechanism of vulcanization. Mailing Add: Monsanto Indust Chem Co 260 Springside Dr Akron OH 44313

MORITA, HIROKAZU, b Steveston, BC, July 17, 26; m 65. ORGANIC CHEMISTRY. Educ: Univ Man, BSc, 48, MSc, 49; Univ Notre Dame, PhD(org chem), 51. Prof Exp: US Off Naval Res fel, Northwestern Univ, 51-52; agr res off, Chem Div, Sci Serv, Can Dept Agr, 52-57, res off chem, Soil Res Inst, 67, RES SCIENTIST CHEM, SOIL RES INST, CAN DEPT AGR, 67- Concurrent Pos: Goodyear res fel, Princeton Univ, 56-57; Commonwealth Sci & Indust Res Orgn vis scientist, Univ Melbourne, 66-67. Mem: Fel AAAS; Phytochem Soc NAm; Am Chem Soc; fel Chem Inst Can; Can Soc Soil Sci. Res: Natural products in soil organic matter; organic mass spectrometry; gas chromatography; synthesis and properties of polyphenols; differential thermal analysis of organic polymers. Mailing Add: Soil Res Inst Can Dept Agr Ottawa ON Can

MORITA, RICHARD YUKIO, b Pasadena, Calif, Mar 27, 23; m 53; c 3. MICROBIOLOGY, OCEANOGRAPHY. Educ: Univ Nebr, BS, 47; Univ Southern Calif, MS, 49; Univ Calif, PhD(microbiol, oceanog), 54. Prof Exp: Asst, Univ Southern Calif, 47-49; asst, Univ Calif, 49-54, res microbiologist, Scripps Inst Oceanog, 54-55; asst prof bact, Univ Houston, 55-58; assoc prof, Univ Nebr, 58-62; PROF MICROBIOL & OCEANOG, ORE STATE UNIV, 62- Concurrent Pos: Microbiologist, Mid-Pac Exped, 50 & 53 & Trans-Pac Exped, 53; vis investr, Danish Galathea Deep-Sea Exped, 52 & Dodo Exped, 64; prog dir biochem, NSF, 68-69, mem, Panel Molecular Biol, 69-70; consult, Nat Inst Gen Med Sci, 68-71; proj reviewer, Environ Protection Agency, 72; mem, Panel Biol Oceanog, NSF, 73; Queen Elizabeth II sr fel, Govt Australia, 73-74. Mem: AAAS; Am Soc Microbiol; Am Soc Limnol & Oceanog; Can Soc Microbiol; fel Am Acad Microbiol. Res: Effect of hydrostatic pressure on the physiology of microorganisms; study of Beggiatoa, marine psychophilic bacteria; eutrophication. Mailing Add: Dept of Microbiol Ore State Univ Corvallis OR 97331

MORITA, TOSHIKO N, b Los Angeles, Calif, May 29, 26; m 53; c 3. FOOD MICROBIOLOGY. Educ: Univ Calif, Los Angeles, BS, 48, MA, 50, PhD(zool), 52. Prof Exp: Res assoc chemother, Barlow Sanitarium, 51-52; serologist, Vet Admin, 52-53; res assoc biochem, Scripps Metab Clin, 53-55; asst prof biol, Univ Houston, 57-58; asst prof immunol, Ore State Univ, 65-67; res assoc biol, Georgetown Univ, 68-69; ASST PROF FOOD & NUTRIT, ORE STATE UNIV, 70- Res: Staph toxins. Mailing Add: Dept of Food & Nutrit Ore State Univ Corvallis OR 97330

MORITA, YOSHIKAZU, b San Francisco, Calif, Feb 17, 21; m 46. NEPHROLOGY, INTERNAL MEDICINE. Educ: Univ Calif, AB, 42; Wayne State Univ, MD, 46, MS, 49; Am Bd Internal Med, dipl, 55, cert nephrol, 72. Prof Exp: Intern, Detroit Gen Hosp, Mich, 46-47, resident internal med, 49-51; from instr to asst prof, 53-65, clin assoc prof, 65-66, assoc prof, 66-70, adj assoc prof med, 70-75, CLIN PROF MED, SCH MED, WAYNE STATE UNIV, 75-; CHIEF MED, WILLIAM BEAUMONT HOSP, 70- Concurrent Pos: Consult, Dearborn Vet Hosp, Mich, 55-; staff mem, Hutzel Hosp, Detroit, 58-, chief med, 66-70; staff mem, Grace Hosp, Detroit, 61-; St Joseph Mercy Hosp, Pontiac, 64- Mem: Am Fedn Clin Res; Am Soc Nephrology. Res: Fluid and electrolyte problems and renal disease. Mailing Add: Dept of Med William Beaumont Hosp Royal Oak MI 48072

MORITSUGU, TOSHIO, b Honolulu, Hawaii, Apr 2, 25; m 59; c 2. SUGAR CHEMISTRY. Educ: Univ Louisville, BA, 49; Ohio State Univ, MS, 51, PhD(org chem), 54. Prof Exp: Res fel, Ohio State Univ, 54-55; assoc technologist sugar cane res, 55-68, SUGAR TECHNOLOGIST, EXP STA, HAWAIIAN SUGAR PLANTERS' ASN, 68- Concurrent Pos: Mem, US Nat Comt Sugar Anal & Int Comn Uniform Methods Sugar Anal. Mem: AAAS; fel Am Inst Chemists; Am Chem Soc; Int Soc Sugar Cane Technol. Res: Carbohydrate chemistry; crystallization, clarification and ion exchange in sugar cane processing; steric effects in organic chemistry; cane factory analysis, control and calculations. Mailing Add: Hawaiian Sugar Planters' Asn Exp Sta 99-193 Aiea Hts Honolulu HI 96701

MORITZ, ALAN RICHARDS, b Hastings, Nebr, Dec 25, 99; m 27; c 3. PATHOLOGY. Educ: Univ Nebr, BSc, 20, MA, 21, MD, 23; Am Bd Path, dipl, 51. Hon Degrees: DSc, Univ Nebr, 50; MA, Harvard Univ, 45. Prof Exp: Intern, Lakeside Hosp, Cleveland, 23-24; from instr to assoc prof path, Sch Med, Western Reserve Univ, 29-37, pathologist in chg univ hosps, 30-37; prof legal med, Harvard Med Sch, 37-49; dir inst path, Med Sch, 49-65, PROF PATH, CASE WESTERN RESERVE UNIV, 49-, PROVOST, 65- Concurrent Pos: Hanna res fel path, Western Reserve Univ, 24-26; pathologist, Mass State Dept Pub Safety, 49; consult, Sci Adv Bd, Armed Forces Inst Path, 55-; chmn, Intersoc Comt Res Potential Path, 54-57; mem, Comt Path, Nat Res Coun, 54- & Path Training Comt, USPHS, 58- Mem: Am Soc Cancer Res; Soc Exp Biol & Med; Am Soc Clin Path; AMA; Col Am Path. Res: Pathogenesis of lesions of cardiovascular system; mechanically induced injuries; hypo and hyperthermic injuries. Mailing Add: 2085 Adelbert Rd Cleveland OH 44106

MORITZ, BARRY KYLER, b Elizabeth, NJ, Apr 16, 41; m 65; c 2. PHYSICS, COMPUTER SCIENCE. Educ: Calif Inst Technol, BS, 63; Univ Md, PhD(physics), 69. Prof Exp: Asst physics, Univ Md, 63-69; NSF fel, E O Hulburt Ctr Space Res, US Naval Res Lab, Washington, DC, 69-71; sr assoc, 71-75, PRIN, PRC INFO SCI CO, 75- Concurrent Pos: Lectr, The Heights, Washington, DC, 69-70. Mem: AAAS; Am Phys Soc. Res: General relativity, astrophysics and cosmology; solar corona and streamers; digital image analysis and pattern recognition; distributed logic systems; associative and parallel techniques; computer information and science. Mailing Add: 201 W Pine St Rome NY 13440

MORITZ, CARL ALBERT, b Bellevue, Ky, Jan 27, 14; m 43; c 5. GEOLOGY. Educ: Univ Kans, AB, 40; Harvard Univ, MA, 48, PhD(geol), 50. Prof Exp: Jr geologist, St Louis Smelting & Refining, 41; topog engr, US Coast & Geod Surv, 42-43; jr geologist, Union Mines Develop Corp, 44; sr geologist, Phillips Petrol Co, 44-46; instr, Univ Tenn, 46; asst prof, Dartmouth Col, 48-53; CONSULT GEOLOGIST, ALEX W McCOY ASSOCS, INC, 53- Mem: Fel Geol Soc Am; Paleont Soc; Am Asn Petrol Geologists; Am Inst Mining, Metall & Petrol Engrs. Res: Stratigraphy; sedimentation and petroleum geology. Mailing Add: 5223 S Birmingham Pl Tulsa OK 74105

MORITZ, ROGER HOMER, b Cleveland, Ohio, Mar 11, 37; m 59; c 4. MATHEMATICS, ENGINEERING STATISTICS. Educ: Valparaiso Univ, BS, 59; Univ Pittsburgh, MS, 61, PhD(math), 64. Prof Exp: Asst math, Univ Pittsburgh, 59-61; sr engr, Goodyear Atomic Corp, 62-64; res mathematician, Cornell Aeronaut Lab, Inc, Cornell Univ, 64-70; ASSOC PROF MATH, ALFRED UNIV, 70- Concurrent Pos: Instr math, State Univ NY Buffalo, 65-69. Mem: Am Math Soc; Math Asn Am. Res: Summability and infinite series; number theory; applied mathematics; Egyptian fractions. Mailing Add: Dept of Math Alfred Univ Alfred NY 14802

MORIYAMA, IWAO MILTON, b San Francisco, Calif, Jan 26, 09; m 46; c 2. PUBLIC HEALTH. Educ: Univ Calif, BS, 31; Yale Univ, MPH, 34, PhD(pub health statist), 37. Prof Exp: Sanit engr, George Williams Hooper Found Med Res, Calif, 31-32; asst, Pierce Lab Hyg, Conn, 33-39; tech secy comt hyg housing, Am Pub Health Asn, 39-40; jr biometrician, US Bur Census, Washington, DC, 40-46; chief mortality anal sect, Nat Off Vital Statist, 47-61, dir off health statist anal, Nat Ctr Health Statist, 61-71, chief statist dept, Atomic Bomb Casualty Comn, 71-73, ASSOC DIR INT STATIST, NAT CTR HEALTH STATIST, USPHS, 74- Concurrent Pos: Mem, Int Union for Sci Study Pop; secy, US Nat Comt, Vital & Health Statist. Mem: Fel AAAS; fel Am Statist Asn; fel Am Pub Health Asn. Res: Vital and health statistics; demography. Mailing Add: Nat Ctr for Health Statist 5600 Fishers Lane Rockville MD 20852

MORIYASU, KEIHACHIRO, b Tacoma, Wash, Jan 26, 40. EXPERIMENTAL HIGH ENERGY PHYSICS. Educ: Mass Inst Technol, BS, 62; Univ Calif, Berkeley, PhD(physics), 67. Prof Exp: NSF fel, 67-68; res assoc physics, Stanford Linear Accelerator Ctr, 68-71; sr res assoc, 71-73, ASST PROF PHYSICS, UNIV WASH, 71- Concurrent Pos: NSF foreign travel grant, 74. Mem: Am Inst Physics. Res: Experimental investigation of elementary particle interactions at high energy. Mailing Add: Dept of Physics Univ of Wash Seattle WA 98195

MORK, BYRON O, b Hills, Minn, May 11, 08; m 33. MEDICINE. Educ: Univ Minn, BS, 29, MB, 30, MD, 31, MPH, 50; Am Bd Prev Med, dipl, 53. Prof Exp: Pvt pract, Worthington Clin, Minn, 32-49; dist health officer, City Health Dept, Los Angeles, 50-55; assoc clin prof prev med & pub health, Univ Calif, Los Angeles, 55-65; med consult, Bur Chronic Dis, State Dept Pub Health, Calif, 65-68; med consult, State Dept Voc Rehab, Calif, 68-70; dist med consult, State Dept Health Care Servs, Calif, 70-74. Concurrent Pos: Dist health officer, Minn State Dept Health, 47-50; lectr, Univ Calif, Los Angeles, 51-65; med consult, State Voc Rehab Serv, Calif, 56-59; regional med coordr pub health, State Dept Pub Health, 59-64. Mem: Fel Am Pub Health Asn. Res: Health administration; preventive medicine; rehabilitation. Mailing Add: 6021 Vista De La Mesa La Jolla CA 92037

MORK, DAVID PETER SOGN, b Thief River Falls, Minn, Sept 25, 42; m 66. BIOLOGY. Educ: Moorhead State Col, BS, 64; Purdue Univ, MS, 66, PhD(bionucleonics), 69. Prof Exp: Asst prof, 68-75, ASSOC PROF BIOL, ST CLOUD STATE COL, 75-, LECTR, SCH NURSING, 73- Mem: AAAS; Am Soc Health Physics; Am Soc Zoologists. Mailing Add: Dept of Biol Sci St Cloud State Univ St Cloud MN 56301

MORKEN, DONALD A, b Crookston, Minn, Feb 2, 22; m 47; c 2. BIOPHYSICS. Educ: Cornell Univ, BEE, 49; Univ Rochester, PhD(biophys), 54. Prof Exp: From instr to asst prof, 54-61, ASSOC PROF RADIATION BIOL, SCH MED & DENT, UNIV ROCHESTER, 61- Mem: Radiation Res Soc; Inst Elec & Electronics Engrs; Biophys Soc; Health Physics Soc. Res: Effects of ionizing radiation on physiological functions of organisms. Mailing Add: Univ Rochester Sch of Med & Dent PO Box 287 Sta 3 Rochester NY 14642

MORKOVIN, DIMITRY, b Prague, Austria, May 11, 11; nat US. RADIOLOGY. Educ: Univ Southern Calif, BS, 34, MS, 35, BS, 37; Univ Ill, PhD(eng), 44, MD, 47; Am Bd Radiol, dipl. Prof Exp: Instr theoret & appl mech, Univ Ill, 42-44; instr radiol, Med Ctr, Univ Colo, 56-57, clin asst prof, 57-58, asst prof, 58-59; clin assoc prof, 59-70, CLIN PROF RADIOL, UNIV TEX HEALTH SCI CTR, DALLAS, 70-; CHIEF RADIOL SERV, VET ADMIN HOSP, 60- Concurrent Pos: Radiation therapist, Parkland Mem Hosp, 59-60. Mem: AMA; Brit Inst Radiol. Res: Radiation biology and therapy; cancer; diagnostic radiology. Mailing Add: Dept of Radiol Univ of Tex Health Sci Ctr Dallas TX 75235

MORLANG, BARBARA LOUISE, b Danbury, Conn, May 24, 44; m 64. NUTRITION. Educ: Brigham Young Univ, BS, 60; Columbia Univ, MS, 64; Univ

Mass, PhD(food & nutrit), 69. Prof Exp: Therapeut dietitian, Yale-New Haven Community Hosp, 61-62 & St Luke's Hosp, New York, 62-64; nutrit consult, Bur Nutrit, City New York, 64-65; dir & nutritionist, Springfield Dairy Coun, Mass, 65-66; STATE NUTRIT CONSULT, WVA DEPT HEALTH, 70- Mem: Am Dietetic Asn; Am Home Econ Asn; Am Pub Health Asn. Res: Nutritional status of local populations. Mailing Add: Va Dept of Health 703 Townside Rd Roanoke VA 24014

MORLANG, CHARLES, JR, b Apr 21, 35; US citizen; m 64. PLANT MORPHOLOGY. Educ: City Col New York, BS, 56; Columbia Univ, PhD(bot), 65. Prof Exp: Lab instr gen bot, Columbia Univ, 56-59, Grad Morphol Lab, 59-60, lectr bot, 60-63, coordr, Biol Methods Sect, Univ High Sch Sci Honors Prog, 61-64; lectr biol, City Col New York, 64-65; asst prof, Mass State Col Westfield, 65-67; ASSOC PROF BIOL, HOLLINS COL, 67- Concurrent Pos: Consult ecol curric, Am Inst Biol Sci, 73-75; consult, Blue Ridge Parkway Nat Park, Roanoke City Schs & Va State Dept Health. Mem: AAAS; Bot Soc Am; Torrey Bot Club; Am Inst Biol Sci; Sigma Xi. Res: Plant morphogenesis; neoplasms in ferns; ecological education; morphogenesis in ferns; electron microscopy; ecological problems. Mailing Add: Dept of Biol Hollins Col Hollins VA 24020

MORLEY, COLIN GODFREY DENNIS, b Sittingbourne, Eng, Nov 12, 41; m 63; c 2. BIOCHEMISTRY. Educ: Univ Nottingham, BSc, 63; Australian Nat Univ, PhD(biochem), 69. Prof Exp: Res scientist, Imp Chem Industs, 63-64; asst lectr chem, Woolwich Col Further Educ, London, 64-65; NIH fel, Nat Heart & Lung Inst, 68-70; ASST PROF, DEPT MED, UNIV CHICAGO, 70- Concurrent Pos: Guest lectr, Purdue Univ, Calumet campus, 74-75. Mem: AAAS; assoc The Chem Soc. Res: Cell biology, particularly control of cell growth in mammalian systems, hormonal aspects of such control; relationship between nor...-¹ and cancer cells for growth control. Mailing Add: Dept of Med Univ of Chicago Chicago IL 60637

MORLEY, DOUROSSOFF EDMUND, b Edmonton, Alta, Nov 9, 11; nat US; m 59. SPEECH PATHOLOGY. Educ: Eastern Mich Univ, AB, 33; Univ Mich, MA, 38, PhD(speech path, hearing), 49. Prof Exp: Teacher pub sch, Mich, 34-39; instr Eng & speech, Univ Pa, 39-40; instr speech, Pa State Teachers Col, Calif, 40-44; supvr speech correction, Commonwealth of Pa, 44-45; from instr to prof speech path, 49-69, PROF PHYS MED & REHAB & SPEECH, MED SCH, UNIV MICH, ANN ARBOR, 69- Concurrent Pos: Consult, United Cerebral Palsy Asn Mich, 55- & US Vet Admin Hosp, 55-; Fulbright lectr, Univ Oslo, 56-57; teacher, Am Acad, Am Community Schs, Athens, Greece, 64-65. Mem: Fel Am Speech & Hearing Asn. Res: Neurological disorders; audiology. Mailing Add: Univ of Mich Speech Clin 1111 E Catherine St Ann Arbor MI 48104

MORLEY, GAYLE L, b Moroni, Utah, Feb 29, 36; m 64; c 3. THEORETICAL SOLID STATE PHYSICS. Educ: Brigham Young Univ, BS, 58; Univ Calif, Los Angeles, MS, 60; Iowa State Univ, PhD(physics), 67. Prof Exp: Microwave engr, Hughes Aircraft Co, 58-60; fel, Tex A&M Univ, 67-68; ASSOC PROF PHYSICS, MANKATO STATE UNIV, 68- Mem: Am Phys Soc. Res: Theory of lattice vibrations in solids. Mailing Add: Dept of Physics Mankato State Univ Mankato MN 56001

MORLEY, GEORGE W, b Toledo, Ohio, June 6, 23; m 46; c 3. MEDICINE. Educ: Univ Mich, BS, 44, MD, 49, MS, 55; Am Bd Obstet & Gynec, dipl, 53, cert gynec oncol, 74. Prof Exp: Intern, 49-50, from asst resident to resident, 50-54, from instr to assoc prof, 56-70, PROF OBSTET & GYNEC, MED CTR, UNIV MICH, ANN ARBOR, 70- Concurrent Pos: Consult, US Vet Admin Hosp, Ann Arbor, Mich, 56- & Wayne County Gen Hosp, Eloise, 60- Mem: Fel Am Col Surg; fel Am Col Obstet & Gynec; Soc Gynec Oncol; Soc Pelvic Surg; Int Soc Study Vulvar Dis. Res: Malignancy of the female genital tract. Mailing Add: Dept of Obstet & Gynec Univ of Mich Med Ctr Ann Arbor MI 48104

MORLEY, HAROLD VICTOR, b Buenos Aires, Arg, July 21, 27; m 53; c 1. PESTICIDE CHEMISTRY, ENVIRONMENTAL CHEMISTRY. Educ: Univ London, BSc, 48, PhD(org chem), 55. Prof Exp: Asst lectr chem, Royal Free Hosp, Sch Med, Univ London, 51-57; fel, Nat Res Coun Can, 57-59 & McMaster Univ, 59-60; res scientist, Anal Chem Res Serv, Can Dept Agr, 60-71, Chem Biol Res Inst, 71-72, RES COORDR, ENVIRON & RESOURCES, RES BR, CAN DEPT AGR, OTTAWA, 72- Concurrent Pos: Mem, Reference Group Land Use Activities, Int Joint Comn, 73-77. Mem: The Chem Soc; Chem Inst Can. Res: Determination and function of ergothioneine in biological fluids; goitrogenic compounds, especially imidazole-2-thiols; porphyrins, especially structure of chlorophylls; pesticide residue chemistry. Mailing Add: Res Br Agr Can Ottawa ON Can

MORLEY, LAWRENCE WHITAKER, b Toronto, Ont, Feb 19, 20; m 50; c 4. GEOPHYSICS. Educ: Univ Toronto, BA, 46, MA, 49, PhD(geophys), 52. Hon Degrees: DSc, York Univ, 74. Prof Exp: Geophysicist, Fairchild Aerial Surv, Inc, Calif, 46-48; chief geophysicist, Dom Gulf Co, 48-49. chief geophys div, Geol Surv Can, 52-69, DIR CAN CTR REMOTE SENSING, DEPT ENERGY, MINES & RESOURCES, GEOL SURV CAN, 69- Concurrent Pos: Lectr, Carleton Univ, 57 & 59. Honors & Awards: McCurdy Medal, Can Aeronaut & Space Admin, 74. Mem: Soc Explor Geophys; Am Geophys Union; Can Inst Mining & Metall. Res: Paleomagnetic research; geophysical instrumentation and interpretation; aerogeophysics. Mailing Add: Can Ctr for Rem Sens DEMR Geol Surv of Can Ottawa ON Can

MORLEY, MICHAEL DARWIN, b Youngstown, Ohio, Sept 29, 30; m 54. MATHEMATICAL LOGIC. Educ: Case Inst Technol, BS, 51; Univ Chicago, MS, 53, PhD, 62. Prof Exp: Sr mathematician, Labs Appl Math, Univ Chicago, 55-61; instr math, Univ Calif, Berkeley, 62-63; asst prof, Univ Wis-Madison, 63-66; assoc prof, 67-70, PROF MATH, CORNELL UNIV, 70- Mem: Am Math Soc; Asn Symbolic Logic. Res: Foundations of mathematics. Mailing Add: 325 Highland Rd Ithaca NY 14850

MORLEY, NINA HOPE, b Kamloops, BC, Dec 14, 09. NUTRITION. Educ: Univ Toronto, BA, 41, MA, 44; Ore State Col, PhD(nutrit), 57. Prof Exp: Asst nutrit, Univ Toronto, 42-44; from asst prof to assoc prof foods & nutrit, Univ BC, 44-53; asst nutrit, Ore State Col, 54-57; res assoc, Dept Ophthal, Banting Res Inst, Univ Toronto, 57-61, RES ASSOC BIOCHEM, CHARLES H BEST INST, UNIV TORONTO, 61- Mem: Can Biochem Soc. Res: Metabolism of riboflavin and niacin in women; insulin; enzyme stereospecificity. Mailing Add: Charles H Best Inst 112 College St Toronto ON Can

MORLEY, THOMAS, b Berkeley, Calif, Oct 26, 17. SYSTEMATIC BOTANY. Educ: Univ Calif, AB, 40, MA, 41, PhD(bot), 49. Prof Exp: Asst, Univ Calif, 41-42 & 46-48; from instr to assoc prof, 49-62, PROF BOT, UNIV MINN, ST PAUL, 62- Concurrent Pos: NSF grant, 58-59. Mem: Bot Soc Am; Am Soc Plant Taxon; Soc Study Evolution; Int Soc Plant Morphol. Res: Taxonomy, phylogeny, comparative morphology and anatomy of Memecyleae; Minnesota flora. Mailing Add: Dept of Bot Univ of Minn St Paul MN 55108

MORLEY, THOMAS PATERSON, b Manchester, Eng, June 13, 20; Can citizen; m 43; c 3. NEUROSURGERY. Educ: Oxford Univ, BA, 41, BM, BCh, 43; FRCS, 49; FRCPS(C), 53. Prof Exp: Consult, Sunnybrook Hosp, Dept Vet Affairs, Govt of Can, 54-60; PROF SURG, UNIV TORONTO, 64-, CHMN DIV NEUROSURG, UNIV & TORONTO GEN HOSP, 74- Concurrent Pos: Ont Cancer Treatment & Res Found res fel, 55-65; consult, Toronto E Gen Hosp, 62-69, Queen Elizabeth Hosp, Toronto, 63-, Princess Margaret Hosp, 63-, Ont Cancer Inst, 63- & Wellesley Hosp, Toronto, 64-69. Mem: Am Asn Neurol Surg; Soc Neurol Surg; Neurosurg Soc Am; Can Neurosurg Soc (secy, 60-64, pres, 71-72); Soc Brit Neurol Surg. Res: Diagnostic use of radio-isotopes in neurosurgery; echoencephalography; recovery of tumor cells from blood in glioma cases; radiotherapy in gliomas. Mailing Add: Toronto Gen Hosp Toronto ON Can

MORLOCK, CARL G, b Crediton, Ont, Sept 11, 06; nat US; m 37; c 2. MEDICINE. Educ: Univ Western Ont, BA, 29, MD, 32; Univ Minn, MS, 37. Prof Exp: Intern, Victoria Hosp, London, Ont, 32-33, resident physician, 33-34; from instr to prof clin med, Mayo Grad Sch Med, 39-72, prof med, 72-75, EMER PROF MED, MAYO MED SCH, UNIV MINN, 75-, CONSULT, MAYO CLIN, 39- Mem: AMA; Am Gastroenterol Asn; fel Am Col Physicians. Res: Peptic ulcer and its complications; gastric carcinoma; anorexia nervosa; regional enteritis; arteriolar pathology of hypertension; blood pressure in renal tumors; liver disease; hemochromatosis; suprarenal insufficiency. Mailing Add: Mayo Clin 200 First St SW Rochester MN 55901

MORNEWECK, SAMUEL, b Meadville, Pa, Sept 3, 39; m 61. ORGANIC BIOCHEMISTRY. Educ: Allegheny Col, BS, 60; Case Inst Technol, PhD(org chem), 65. Prof Exp: Chemist, Agr Prod Labs, Esso Res & Eng Co, 65-67, res chemist, 67-70; asst prof, 70-75, ASSOC PROF CHEM, ST PETER'S COL, NJ, 75- Res: Determination of steric substituent constants. Mailing Add: Dept of Chem St Peter's Col Jersey City NJ 07306

MOROI, DAVID S, b Tokyo, Japan, Oct 15, 26; m 59; c 3. THEORETICAL PHYSICS. Educ: St Paul's Univ, Tokyo, BSc, 53; Johns Hopkins Univ, PhD(physics), 59. Prof Exp: Res assoc physics, Iowa State Univ, 59-61, instr, 60-61; res assoc, Univ Notre Dame, 61-63; asst prof, 63-68, ASSOC PROF PHYSICS, KENT STATE UNIV, 68- Concurrent Pos: US Air Force Off Sci Res grant, 64-69. Mem: Am Phys Soc. Res: Quantum electrodynamics; field theory; general relativity; electromagnetic and other physical properties of liquid crystals; elementary particles in intense laser beams. Mailing Add: Dept of Physics Kent State Univ Kent OH 44242

MOROKUMA, KEIJI, b Kagoshima, Japan, July 12, 34; m 59; c 4. QUANTUM CHEMISTRY, PHYSICAL CHEMISTRY. Educ: Kyoto Univ, BS, 57, MS, 59, PhD(chem), 63. Prof Exp: Asst chem, Kyoto Univ, 62-66; res fel, Harvard Univ, 66-67; from asst prof to assoc prof, 67-71, PROF CHEM, UNIV ROCHESTER, 71- Concurrent Pos: Res assoc, Columbia Univ, 64-65, vis res asst prof, 65-66; Quincy Boese fel, 64-65; Fulbright travel grant, 64-67. Mem: Am Phys Soc; Am Chem Soc. Res: Electronic structures of molecules; reactivities of organic molecule; electronic structures of biological systems; theoretical studies on electron spin resonance spectra; collision theories of elementary reactions; elastic and inelastic collisions. Mailing Add: Dept of Chem Univ of Rochester Rochester NY 14627

MORONI, ENEO C, b Fiume, Italy, Feb 6, 23; US citizen; m 60; c 1. INDUSTRIAL ORGANIC CHEMISTRY. Educ: Univ Milan, PhD(phys org chem), 49. Prof Exp: Res chemist, Ledoga SpA, Italy, 49-55; prod chemist, Industria Saccarifera Parmense, 55-56; prod chemist, Pitt-Consol Chem Co Div, Consol Coal Co, 56-57, res chemist, 57-64; sr res chemist, Jones & Laughlin Steel Corp, 64-67; res chemist, Pittsburgh Energy Res Ctr, US Bur Mines, 67-75; PROG MGR, ENERGY RES & DEVELOP ADMIN, 75- Mem: Am Chem Soc. Res: Activated carbons; phenolic and cresylic acids resins; catalytic alkylation, dealkylation and isomerization of phenols and thiophenols; corrosion and coating surface phenomena; organometallics; coal desulfurization and liquefaction. Mailing Add: Apt 124S 1600 S Eads St S Eads St Arlington VA 22202

MOROS, STEPHEN ANDREW, b New York, NY, July 29, 28; m 57; c 2. ANALYTICAL CHEMISTRY. Educ: Polytech Inst Brooklyn, BS, 48, PhD(electroanal chem), 61; Cornell Univ, MS, 50. Prof Exp: Proj group leader, Foster D Snell, Inc, NY, 50-53; from proj chemist to res chemist, Am Cyanamid Co, NJ, 58-71; sr chemist, 71-73, GROUP LEADER, HOFFMANN-LA ROCHE INC, 73- Mem: AAAS; Am Chem Soc; Electrochem Soc; Am Pharmaceut Asn; Am Soc Testing & Mat. Res: Electroanalytical chemistry, including coulometry and potentiometry; organic electrosynthesis; instrumentation; thermal methods of analysis; spectroscopic methods; radiotracers; instrumental methods of analysis; thermoanalytical chemistry; spectrochemical analysis. Mailing Add: Hoffmann-La Roche Inc Nutley NJ 07110

MOROSIN, BRUNO, b Klamath Falls, Ore, Feb 10, 34; m 58; c 4. PHYSICAL CHEMISTRY. Educ: Univ Ore, BA, 56; Univ Wash, PhD, 59. Prof Exp: Asst, Univ Wash, 56-59; mem tech staff, Hughes Aircraft Co, 60-61; staff mem, Sandia Labs, 61-67, supvr chem physics div, 67-73, SUPVR, SOLID STATE MAT, SANDIA LABS, 73- Mem: Am Chem Soc; Am Phys Soc; Am Crystallog Asn. Res: Diffraction studies; structural properties related to magnetic and dielectric behavior. Mailing Add: 12317 Eastridge Dr NE Albuquerque NM 87112

MOROSOFF, NICHOLAS, b New York, NY, Jan 27, 37; m 67; c 1. PHYSICAL CHEMISTRY, POLYMER CHEMISTRY. Educ: Queens Col, NY, BS, 58; Polytech Inst Brooklyn, PhD(phys chem), 65. Prof Exp: NIH fel, 65-67; PHYS CHEMIST, CAMILLE DREYFUS LAB, RES TRIANGLE INST, 67- Mem: AAAS; Am Chem Soc; Am Phys Soc; Am Crystallog Asn. Res: X-ray diffraction from polymers; polymer crystal structure and morphology; solid state polymerization. Mailing Add: Camille Dreyfus Lab Res Triangle Inst PO Box 12194 Research Triangle Park NC 27709

MOROSON, HAROLD, b New York, NY, Aug 9, 27; m 52; c 2. RADIATION IMMUNOLOGY, RADIOBIOLOGY. Educ: Columbia Univ, BS, 50; Univ Del, MS, 55; Polytech Inst Brooklyn, PhD(chem), 58. Prof Exp: Res chemist, Reichhold Chem Inc, 52-56; asst prof chem, Newark Col Eng, 59-61; res assoc biophys, 61-63, assoc radiobiol & sect head, Sloan-Kettering Inst Cancer Res, 63-72, asst prof biophys, Sloan-Kettering Div, Cornell Univ, 63-72; ASSOC PROF RADIOL, NY MED COL, 72-, DIR DIV RADIOBIOL, 72- Concurrent Pos: NIH fels radiobiol, Chester Beatty Res Inst, London, Eng, 58-59; Damon Runyon grant, 69-72; Nat Cancer Inst res grants, 72-78. Mem: Radiation Res Soc; Biophys Soc; Am Soc Exp Path; Am Asn Cancer Res. Res: Effects of ultraviolet and ionizing radiation on bacterial and mammalian cells; changes in macromolecular properties of DNA; radiation sensitization of tumor cells; radiation immunology; tumor immunology; experimental radiotherapy. Mailing Add: Div of Radiobiol NY Med Col Flower & 5th Ave Hosp New York NY 10029

MOROWITZ, HAROLD JOSEPH, b Poughkeepsie, NY, Dec 4, 27; m 49; c 5. BIOPHYSICS, BIOLOGY. Educ: Yale Univ, BS, 47, MS, 50, PhD(biophys), 51. Prof

MOROWITZ

Exp: Biophysicist, Nat Bur Standards, 51-53 & Nat Heart Inst, 53-55; from asst prof to assoc prof, 55-67, PROF BIOPHYS & BIOCHEM, YALE UNIV, 67- Concurrent Pos: Mem planetary biol subcomt, NASA, 66-72; mem adv bd, J Theoret Biol, 68-; assoc ed, J Biomed Computing, 69-; mem eval panel phys chem, Nat Bur Standards, 69-74; columnist, Hosp Pract, 74- Mem: AAAS; fel NY Acad Sci; Biophys Soc; Am Inst Biol Scientists. Res: Energy transduction in biological systems; structure and function of biological membranes; thermodynamic foundations of biology. Mailing Add: 56 Ox Bow Lane Woodbridge CT 06525

MOROZOWICH, WALTER, b Irwin, Pa, Oct 27, 33; m 69; c 1. MEDICINAL CHEMISTRY. Educ: Duquesne Univ, BS, 55; Ohio State Univ, MS, 56, PhD(pharmaceut chem), 59. Prof Exp: SR RES SCIENTIST, UPJOHN CO, 59- Mem: Am Pharmaceut Asn; Am Chem Soc. Res: Analogs and in vivo reversible derivatives of steroids, sulfonylureas, antibiotics and prostaglandins. Mailing Add: Upjohn Co 301 Henrietta St Kalamazoo MI 49001

MORR, CHARLES VERNON, b Ashland, Ohio, Oct 7, 27; m 51; c 2. FOOD CHEMISTRY. Educ: Ohio State Univ, BS, 52, MS, 55, PhD(dairy technol), 59. Prof Exp: Res assoc food, Carnation Res Lab, Calif, 59-61; asst prof milk protein res, Ohio State Univ, 61-64; from asst prof to prof, Univ Minn, St Paul, 64-73; DIR PROTEIN RES, RALSTON PURINA CO, 73- Honors & Awards: Dairy Res Award, Am Dairy Sci Asn, 73. Mem: Am Chem Soc; Am Dairy Sci Asn; Inst Food Technologists; Am Asn Cereal Chemists. Res: Chemistry and functional properties of plant proteins. Mailing Add: Ralston Purina Co St Louis MO 63188

MORRE, D JAMES, b Drake, Mo, Oct 20, 35; m 56; c 3. BIOCHEMISTRY. Educ: Univ Mo, BS, 57; Purdue Univ, MS, 59; Calif Inst Technol, PhD(biochem), 63. Prof Exp: From asst prof to assoc prof, 63-71, PROF BIOCHEM, PURDUE UNIV, LAFAYETTE, 71- Mem: Am Soc Cell Biol. Res: Cell growth; Golgi apparatus structure-function; membrane formation; surface coats; secretion. Mailing Add: Dept of Biol Sci Purdue Univ Lafayette IN 47906

MORREAL, CHARLES EDWARD, organic chemistry, see 12th edition

MORREL, BERNARD BALDWIN, b Lynchburg, Va, Nov 28, 40; m 64; c 1. MATHEMATICS. Educ: Univ Va, BA, 62, MA, 66, PhD(math), 68. Prof Exp: Teaching asst math, Johns Hopkins Univ, 62-64; from teaching asst appl math to jr instr math, Univ Va, 65-68; asst prof math, Univ Ga, 68-75; VIS ASST PROF, IND UNIV, Bloomington, 75- Mem: Am Math Soc; Math Asn Am; Soc Indust & Appl Math. Res: Functional analysis; theory of operators in Hilbert Space. Mailing Add: Dept of Math Swain Hall-East Ind Univ Bloomington IN 47101

MORRELL, FRANK, b New York, NY, June 4, 26; c 4. NEUROLOGY, NEUROPHYSIOLOGY. Educ: Columbia Univ, AB, 48, MD, 51; McGill Univ, MSc, 55; Am Bd Psychiat & Neurol, dipl, 58. Prof Exp: Med intern, Montefiore Hosp, New York, 51-52, chief resident neurol, 53-54; from instr to assoc prof neurol, Med Sch, Univ Minn, 55-61; prof & chmn dept, Sch Med, Stanford Univ, 61-69; prof neurol & psychiat, New York Med Col, 69-72; PROF NEUROL SCI, MED COL, RUSH UNIV, 72- Concurrent Pos: Rosenthal fel, Nat Hosp, London, Eng, 52-53; fel neurophysiol, Montreal Neurol Inst, 54-55; consult, Epilepsy Found Am, 59-; consult, NIH & NSF, 61-; assoc neurosci res prog, Mass Inst Technol, 62-; mem brain sci comt, Nat Acad Sci, 66-70; Wall Mem lectr, Children's Hosp, DC, 67. Mem: Fel Royal Soc Health; Am Electroencephalog Soc; Am Epilepsy Soc; Am Acad Neurol; Soc Neurosci. Res: Pathophysiology of epilepsy and neural mechanisms of learning. Mailing Add: Dept of Neurol Sci Rush Univ Med Col Chicago IL 60612

MORRELL, JOSEPH SALVADOR, b Maringouin, La, Dec 31, 38. MATHEMATICS. Educ: Univ Southwestern La, BS, 60; La State Univ, Baton Rouge, MA, 63; Fla State Univ, PhD(math), 70. Prof Exp: Teaching asst math, Univ Southwestern La, 59-60 & La State Univ, 62-63; instr, Fla State Univ, 63-68, part-time instr & asst, 68-70; ASSOC PROF MATH, UNIV SOUTHERN MISS, 70- Concurrent Pos: NATO fel advan studies, 72. Mem: Sigma Xi; London Math Soc; Am Math Soc; Math Asn Am. Res: Functional analysis; operator theory in Banach spaces; statistical pattern recognition. Mailing Add: Dept of Math Univ of Southern Miss Box 5265 Hattiesburg MS 39401

MORRELL, WILLIAM EGBERT, b Logan, Utah, July 30, 09; m 33; c 3. CHEMISTRY, SCIENCE EDUCATION. Educ: Utah State Univ, BS, 33; Univ Calif, PhD(phys chem), 38. Hon Degrees: LHD, Suffolk Univ, 73. Prof Exp: Instr phys sci, Chicago City Col, 38-42; from instr to prof phys sci, Univ Ill, 42-59; PROG DIR SUMMER STUDY, NSF, 59- Concurrent Pos: Asst prof dir summer study, NSF, 58-59. Honors & Awards: Meritorious Serv Award, NSF, 75. Mem: AAAS; Am Chem Soc. Mailing Add: Nat Sci Found Washington DC 20550

MORREY, CHARLES BRADFIELD, JR, b Columbus, Ohio, July 23, 07; m 37; c 3. MATHEMATICS. Educ: Ohio State Univ, AB, 27, MA, 28; Harvard Univ, Phd(math), 31. Prof Exp: Asst, Ohio State Univ, 27-28; instr math, Harvard Univ, 29-31; Nat Res Coun fel, Princeton Univ, 31-32 & Rice Inst, 32-33; from instr to assoc prof, 33-45, chmn dept, 49-54, PROF MATH, UNIV CALIF, BERKELEY, 45- Concurrent Pos: Tutor, Harvard Univ, 30-31; vis prof, Univ Calif & Inst Advan Study, 37-38; mathematician, Aberdeen Proving Ground, Md, 42-45; mem math div, Nat Res Coun, 53- & Inst Advan Study, 54-55. Mem: Nat Acad Sci; Am Acad Arts & Sci; Am Math Soc (pres elect, 66, pres, 67 & 68); Math Asn Am. Res: Area of surfaces; calculus of variations; elliptic partial differential equations. Mailing Add: Dept of Math Univ of Calif Berkeley CA 94720

MORREY, JOHN ROLPH, b Joseph, Utah, May 30, 30; m 52; c 5. PHYSICAL CHEMISTRY, INORGANIC CHEMISTRY. Educ: Brigham Young Univ, BA, 53; Univ Utah, PhD(phys chem), 58. Prof Exp: Sr scientist, Gen Elec Co, 58-63, tech specialist, 63-64; RES ASSOC CHEM, PAC NORTHWEST LABS, BATTELLE MEM INST, 65- Concurrent Pos: Adj assoc prof, Wash State Univ, 67- Mem: AAAS; Am Chem Soc; NY Acad Sci. Res: Boron hydride fused salt and actinide element chemistry; chemical kinetics; molecular spectroscopy; computer applications to chemistry; laser chemistry. Mailing Add: Dept of Chem 325 Bldg 300 Area Northwest Labs Battelle Mem Inst Richland WA 99352

MORRILL, CALLIS GARY, b Tridell, Utah, Nov 6, 38; m 64; c 5. PHYSIOLOGY. Educ: Brigham Young Univ, BA, 64; Univ Calif, San Francisco, PhD(physiol), 70. Prof Exp: NIH res fel physiol, Univ Colo Med Ctr, 71-73; RES ASSOC PHYSIOL, NAT ASTHMA CTR, DENVER, 73- Concurrent Pos: Am Lung Asn fel, Nat Asthma Ctr, 74-75; NIH res grant, 75. Mem: Am Thoracic Soc. Res: Ventilatory control of asthmatic children; effects of low levels of carbon monoxide on exercise tolerance. Mailing Add: Dept of Clin Physiol Nat Asthma Ctr 1999 Julian St Denver CO 80204

MORRILL, GENE A, b Bend, Ore, Aug 5, 31; m 60; c 2. BIOCHEMISTRY, DEVELOPMENTAL BIOLOGY. Educ: Univ Portland, BS, 54; Univ Utah, PhD(biochem), 59. Prof Exp: Res asst prof, 63-64, asst prof, 65-71, ASSOC PROF PHYSIOL, ALBERT EINSTEIN COL MED, 71- Concurrent Pos: Fel biochem, Albert Einstein Col Med, 58-60, sr fel physiol, 62; fel, Inst Training Res Behav & Neurol Sci, 60-61; City of New York Health Res Coun career scientist award, 69- Mem: Biophys Soc; Am Physiol Soc; Am Soc Biol Chemists; Soc Develop Biol. Res: Cell physiology; developmental biology; biophysics. Mailing Add: Albert Einstein Col of Med Yeshiva Univ New York NY 10461

MORRILL, JAMES LAWRENCE, JR, b Graves Co, Ky, Nov 23, 30; m 52; c 5. DAIRY SCIENCE. Educ: Murray State Col, BS, 58; Univ Ky, MS, 59; Iowa State Univ, PhD(dairy cattle nutrit), 63. Prof Exp: From instr to asst prof, 62-69, ASSOC PROF DAIRY SCI, KANS STATE UNIV, 69-, ASSOC PROF POULTRY SCI & DAIRY CATTLE RES NUTRITIONIST, AGR EXP STA, 74- Mem: Am Dairy Sci Asn; Am Soc Animal Sci. Res: Dairy cattle nutrition, especially nutrition of young. Mailing Add: Dept of Dairy Sci Kans State Univ Manhattan KS 66502

MORRILL, JOHN BARSTOW, JR, b Chicago, Ill, Nov 20, 29; m 53; c 2. DEVELOPMENTAL BIOLOGY. Educ: Grinnell Col, BA, 51; Iowa State Col, MS, 53; Fla State Univ, PhD(zool), 58. Prof Exp: From instr to asst prof biol, Wesleyan Univ, 58-65; assoc prof, Col William & Mary, 65-67; assoc prof, 67-69, PROF BIOL, NEW COL, FLA, 69-, COORDR ENVIRON STUDIES PROG, 74- Concurrent Pos: NSF fel, 58; NIH special fel, 64; mem corp, Marine Biol Lab. Mem: Am Soc Zoologists; Soc Develop Biol. Res: Development of mollusk eggs; experimental analyses of molluscan development; effects of chemicals on mollusk eggs. Mailing Add: Div Natural Sci New Col Sarasota, FL 33578

MORRILL, JOHN ELLIOTT, b Oak Park, Ill, Nov 4, 35; m 58; c 3. MATHEMATICS. Educ: DePauw Univ, BA, 57; Univ Mich, MA, 60, PhD(math), 64. Prof Exp: From asst prof to assoc prof math, DePauw Univ, 64-70; vis assoc prof, Univ Mich, 70-71; ASSOC PROF MATH, DePAUW UNIV, 71- Concurrent Pos: Acad guest, Res Inst Math, ETH, Zurich, 72-73. Mem: Assoc Soc Actuaries; Am Math Soc; Math Asn Am. Res: Actuarial mathematics; mathematical economics. Mailing Add: Dept of Math DePauw Univ Greencastle IN 46135

MORRILL, LAWRENCE GEORGE, b Tridell, Utah, July 21, 29; m 49; c 4. SOIL CHEMISTRY, FERTILITY. Educ: Utah State Univ, BS, 55, MS, 56; Cornell Univ, PhD(soil chem), 59. Prof Exp: From assoc chem to sr chemist, Thiokol Chem Corp, Utah, 59-66; res specialist, Cornell Univ, 60-61; ASSOC PROF SOIL CHEM & FERTIL, OKLA STATE UNIV, 66- Mem: Am Soc Agron; Soil Sci Soc Am. Res: Soil chemistry and fertility research and teaching. Mailing Add: Dept of Agron Okla State Univ Stillwater OK 74074

MORRILL, RICHARD LELAND, b Santa Monica, Calif, Feb 15, 34; m 58; c 3. GEOGRAPHY, REGIONAL ECONOMICS. Educ: Dartmouth Col, BA, 55; Univ Wash, MA, 57, PhD(geog), 59. Prof Exp: Asst prof geog, Northwestern Univ, 59-60; NSF fel, Univ Lund, Sweden, 60-61; from asst prof to assoc prof, 61-68, PROF GEOG, UNIV WASH, 69-, CHMN DEPT, 73- & ASSOC DIR ENVIRON STUDIES, 74- Concurrent Pos: Vis assoc prof geog, Univ Chicago, 66-67; NIH fel, 66-68, NIH fel, Univ Wash, 66-68; Sir John McTaggart fel & vis prof, Univ Glasgow, 70. Honors & Awards: Meritorious Contrib Award, Asn Am Geogr, 71. Mem: Asn Am Geogr; Regional Sci Asn; Int Union Sci Study Pop. Res: Location theory; social and economic inequality; regional development and planning; migration and diffusion. Mailing Add: Dept of Geog Univ of Wash Seattle WA 98105

MORRILL, TERENCE CLARK, b Albany, NY, Mar 1, 40; m 65; c 2. ORGANIC CHEMISTRY. Educ: Syracuse Univ, BS, 61; San Jose State Col, MS, 64; Univ Colo, PhD(org chem), 66. Prof Exp: NSF fel org chem, Yale Univ, 66-67; from asst prof to assoc prof, 67-75, PROF ORG CHEM, ROCHESTER INST TECHNOL, 75- Mem: Am Chem Soc; Sigma Xi. Res: Stereochemistry of addition and solvolysis reactions; mechanisms of additions, solvolyses and oxidation reactions: alkylations; hydrazone chemistry; use of lanthanide shift reagents. Mailing Add: Dept of Chem Rochester Inst of Technol Rochester NY 14623

MORRILL, WARREN THOMAS, b Portland, Maine, Nov 5, 29; m 53; c 3. ETHNOLOGY, ANTHROPOLOGY. Educ: Univ Mich, AB, 50; Univ Chicago, MA, 52, PhD(anthrop), 61. Prof Exp: Instr anthrop, Univ Chicago, 61; from asst prof to assoc prof, Bucknell Univ, 61-70; PROF ANTHROP & HEAD DEPT, PA STATE UNIV, UNIVERSITY PARK, 70- Mem: Fel AAAS; fel Am Anthrop Asn; fel Royal Anthrop Inst Gt Brit & Ireland. Res: Demography and peasant societies. Mailing Add: Dept of Anthrop Pa State Univ University Park PA 16802

MORRIN, PETER ARTHUR FRANCIS, b Dublin, Ireland, Oct 8, 31; m 60; c 2. MEDICINE. Educ: Nat Univ Ireland, MB, BCh & BAO, 54, BSc, 55; FRCP(C), 61. Prof Exp: Instr med, Wash Univ, 60-61; lectr, 62-63, asst prof, 63-69, ASSOC PROF MED, QUEEN'S UNIV, ONT, 69- Mem: Am Fedn Clin Res; Can Med Asn; Can Soc Clin Invest. Res: Internal medicine; renal disease. Mailing Add: Etherington Hall Queen's Univ Kingston ON Can

MORRIS, ALBERT GREGORY, b Philadelphia, Pa, Sept 28, 23; m 47; c 1. INORGANIC CHEMISTRY. Educ: Haverford Col, AB, 48; Univ Del, MS, 50, PhD(chem), 53. Prof Exp: Asst mech drawing & descriptive geometry, Univ Del, 48 & 50; res engr labs div, Radio Corp Am, 52-55; sr res chemist, Foote Mineral Co, Pa, 55-59, group leader semiconductor res, 59-60; dir res, United Mineral & Chem Corp, 60-62; asst prof chem, Villanova Univ, 64-67; ASSOC PROF CHEM, PA STATE UNIV, DELAWARE COUNTY CAMPUS, 67- Concurrent Pos: Consult, Technol Data, Pa, 62- Mem: AAAS; Am Chem Soc; Am Phys Soc; Am Soc Testing & Mat. Res: Inorganic fluoride chemistry; infrared photoconductivity of germanium; preparation of silicon, germanium, intermetallic semiconductor materials, chemical literature. Mailing Add: Dept of Chem Pa State Univ Delaware County Media PA 19063

MORRIS, ALLAN J, b Linn Grove, Iowa, June 26, 26; m 56; c 2. BIOCHEMISTRY. Educ: Iowa State Univ, BA, 55; Univ Utah, MA, 57, PhD(biochem), 59. Prof Exp: From asst prof to assoc prof, 63-72, PROF BIOCHEM, MICH STATE UNIV, 72- Concurrent Pos: NIH fels, City of Hope Hosp, Duarte, Calif, 59-60, Med Sch, Univ Ky, 60-62 & Nat Inst Med Res, London, 62-63. Mem: AAAS; Am Soc Biol Chem; Brit Biochem Soc. Res: Genetics and molecular biology of hemoglobin biosynthesis; metabolism of nucleotides in red blood cells. Mailing Add: Dept of Biochem Mich State Univ East Lansing MI 48824

MORRIS, ALVIN LEONARD, b Detroit, Mich, July 2, 27; wid; c 3. DENTISTRY. Educ: Univ Mich, DDS, 51; Univ Rochester, PhD(path), 57. Prof Exp: Asst prof oral med & actg head dept oral diag, Sch Dent, Univ Pa, 57-60, asst prof, Grad Sch Med, 58-61, assoc prof & head dept, 60-61; dean, Sch Dent, Univ Ky, 61-68, asst vpres, Med Ctr, 68-69, spec asst to pres admin, 69-70, prof oral diag & oral med, 61-75, vpres admin, 70-75; EXEC DIR, ASN ACAD HEALTH CTRS, 75- Concurrent Pos: Consult, Vet Admin, Lexington, Ky, 62; consult, USPHS, Ky, 63, chmn dent study sect, Nat Inst Dent Res, 65-67; mem army med serv adv comt prev dent, Off Surgeon Gen, 67-74; mem nat adv coun educ health professions, NIH, 68-72; mem dent adv comt, Dept Defense, 70-73; pres, Am Fund Dent Educ, 70-74. Mem: Nat Inst Med; AAAS; Int Asn Dent Res. Res: Experimental oral cancer with emphasis on the histochemistry and biochemistry of carcinogenesis; normal and abnormal keratinization of oral mucosa. Mailing Add: Asn Acad Health Ctrs Rm 302 1625 Massachusetts Ave NW Washington DC 20036

MORRIS, ARTHUR PEEBLES, entomology, see 12th edition

MORRIS, BERT MILLER, physical chemistry, see 12th edition

MORRIS, BYRON FREDERICK, b Montevideo, Minn, July 29, 43; m 65; c 2. MARINE ECOLOGY, OCEANOGRAPHY. Educ: Calif State Univ, Long Beach, BS, 66, MA, 68; Dalhousie Univ, PhD(oceanog), 75. Prof Exp: Res assoc oil pollution, 71-75, SUPVR SHIP OPER, BERMUDA BIOL STA, 73-, PROJ LEADER BERMUDA INSHORE WATERS INVEST & ASST DIR ADMIN, 76- Concurrent Pos: Mem, Bermuda Govt Law of Sea Comt, 74- Mem: AAAS; Am Inst Biol Sci. Res: Environmental ecology of marine organisms, zooplankton ecology, and the effects of man on natural marine ecosystems. Mailing Add: Bermuda Biol Sta Res Inc St George's West Bermuda

MORRIS, CHARLES EDWARD, b Detroit, Mich, Feb 17, 41; m 68. SOLID STATE PHYSICS. Educ: Iowa State Univ, BS, 63, PhD(physics), 68. Prof Exp: MEM STAFF, LOS ALAMOS SCI LAB, UNIV CALIF, 68- Mem: Am Phys Soc. Res: Mechanical properties of solids and liquids; optical properties of solids. Mailing Add: Los Alamos Sci Lab PO Box 1663 Los Alamos NM 87544

MORRIS, CHARLES ELLIOT, b Denver, Colo, Mar 30, 29; m 51; c 2. NEUROLOGY. Educ: Univ Denver, BA, 50, MA, 51; Univ Colo, MD, 55. Prof Exp: Teaching fel neurol, Harvard Med Sch, 56-59; asst prof, 61-65, ASSOC PROF NEUROL, SCH MED, UNIV NC, CHAPEL HILL, 65- Concurrent Pos: Consult coun drugs, AMA, 66; dir, Nat Insts Neurol Dis & Stroke Res Ctr, Agana, Guam, 70-71. Mem: Fel Am Acad Neurol; Asn Res Nerv & Ment Dis; AMA; Am Epilepsy Soc. Res: Neuroimmunology, especially investigations into the pathogenesis and diagnosis of autoimmune diseases of the central and peripheral nervous system; diseases of muscle; Parkinsonism and neurogenic amines; amyotropic lateral sclerosis; movement disorders. Mailing Add: Dept of Nuerol Univ of NC Sch of Med Chapel Hill NC 27514

MORRIS, CLETUS EUGENE, b Alcorn County, Miss, Jan 30, 35; m 62; c 1. ORGANIC CHEMISTRY. Educ: Auburn Univ, BS, 59, PhD(org chem), 66. Prof Exp: RES CHEMIST, SOUTHERN REGIONAL RES CTR, AGR RES SERV, USDA, 65- Mem: Am Chem Soc; Am Asn Textile Chemists & Colorists. Res: Chemical modification of cotton; organophosphorus chemistry; flame retardants for textiles. Mailing Add: Southern Regional Res Ctr USDA PO Box 19687 New Orleans LA 70179

MORRIS, DALE DUANE, biophysics, see 12th edition

MORRIS, DANIEL LUZON, b Newtown, Conn, July 29, 07; m 29, 61. CHEMISTRY. Educ: Yale Univ, AB, 29, PhD(org chem), 34. Prof Exp: Field dir speed surv, State Hwy Dept, Conn, 33-34; fel physiol, Sch Med, Yale Univ, 34-35; teacher, Putney Sch, 35-44; res chemist, Mead Johnson & 44-48; lab supvr, Food, Chem & Res Labs, Inc, 48-51; head dept sci, 51-69, teacher math & sci Lakeside Sch, 51- Concurrent Pos: Fulbright exchange teacher, Eng, 58-59. Mem: AAAS; Am Chem Soc. Res: Descriptive geometry of four dimensions; lipids from animal sources; effects of colloids on crystallization; isolation, determination and physiological effects of carbohydrates; isolation of tryptophane; Christian theology. Mailing Add: Lakeside Sch 14050 First Ave NE Seattle WA 98125

MORRIS, DANIEL W, b New York, NY, Mar 21, 39. MOLECULAR BIOLOGY, BIOCHEMISTRY. Educ: Univ Calif, Berkeley, BA, 60; Univ Calif, San Diego, PhD(cellular biol), 66. Prof Exp: NIH fel, Inst Microbiol, Copenhagen Univ, 66-68; ASST PROF MICROBIOL, SCH MED, ST LOUIS UNIV, 68- Concurrent Pos: NIH res grant, 69-72. Mem: Am Soc Microbiol. Res: Regulation of RNA synthesis in synthesis in bacteria; biosynthesis of bacterial ribosomes; analysis of the RNA control lucus in Escherichia coli. Mailing Add: Dept of Microbiol St Louis Univ Sch of Med St Louis MO 63104

MORRIS, DAVID ALBERT, b Marietta, Ohio, July 30, 36; m 58; c 2. GEOCHEMISTRY, EXPLORATION GEOLOGY. Educ: Marietta Col, BS, 58; Univ Kans, MS, 61, PhD(geol), 67. Prof Exp: From res geologist to sr res geologist, 67-69, mgr appl projs sect, 70-75, MGR GEOCHEM BR, EXPLOR PROD RES DIV, PHILLIPS PETROL CO, 75- Mem: AAAS; Am Asn Petrol Geologists; Soc Econ Paleontologists & Mineralogists; Geol Soc Am. Res: Stratigraphy, sedimentation and depositional environments of Pennsylvania rocks in North America and Europe; paleogeomorphology of coal swamps and formation of coal splits; origin and geology of petroliferous source rocks and migration of oil; diagenesis and compaction of rocks. Mailing Add: Res & Develop Div Phillips Petrol Co 142 RB 1 Bartlesville OK 74004

MORRIS, DAVID EUGENE, organic chemistry, see 12th edition

MORRIS, DAVID JULIAN, b Ramsgate, Eng, May 17, 39; m 65; c 2. ORGANIC CHEMISTRY, ENDOCRINOLOGY. Educ: Oxford Univ, BA, 60, MA & DPhil(org chem), 63. Prof Exp: Fel with Prof F W Barnes, Brown Univ, 63-66; sect leader med chem, Beecham Res Labs, 66-68; asst prof biochem pharmacol, 68-75, ASSOC PROF BIOCHEM PHARMACOL, BROWN UNIV, 75-; CHIEF BIOCHEMIST, DEPTS MED & LAB MED, MIRIAM HOSP, PROVIDENCE, 68- Mem: Endocrine Soc; Am Asn Clin Chem; Am Chem Soc. Res: Mechanism of action of glucocorticoids in liver and aldosterone in kidney; primary receptors of steroid hormones in target tissues; characterization and physiological role of aldosterone metabolites in the kidney; steroid metabolism. Mailing Add: Dept of Med Miriam Hosp Providence RI 02990

MORRIS, DAVID MARKLAND, JR, anatomy, physiology, deceased

MORRIS, DAVID ROBERT, b Whittier, Calif, June 25, 39; m 61; c 2. BIOCHEMISTRY. Educ: Univ Calif, Los Angeles, BA, 61; Univ Ill, PhD(chem), 64. Prof Exp: NIH fel, 64-66; asst prof biochem, 66-70, ASSOC PROF BIOCHEM, UNIV WASH, 70- Concurrent Pos: John Simon Guggenheim fel, 71-72. Mem: AAAS; Am Chem Soc; Am Soc Biol Chem; Am Soc Microbiol. Res: Regulation of cell growth and division; biological function of polyamines. Mailing Add: Dept of Biochem Univ of Wash Seattle WA 98195

MORRIS, DEREK, b Hove, Eng, May 6, 30; m 53; c 3. METROLOGY, QUANTUM PHYSICS. Educ: London Univ, BSc, 50, PhD(physics), 53. Prof Exp: Fel, Div Pure Physics, Nat Res Coun, Can, 53-55; sci officer, Guided Weapons Dept, Royal Aircraft Estab, Eng, 55-57; from asst res officer to assoc res officer div appl physics, 57-75, SR RES OFFICER, NAT RES COUN, CAN, 75- Res: Atomic frequency standards. Mailing Add: Time & Frequency Sect Div of Physics Nat Res Coun Montreal Rd Ottawa ON Can

MORRIS, DONALD EUGENE, b Tulsa, Okla, July 9, 40; m 63; c 4. ORGANOMETALLIC CHEMISTRY. Educ: Univ Tulsa, BS, 63; Northwestern Univ, PhD(inorg chem), 67. Prof Exp: NSF res grant, Stanford Univ, 67-68; res chemist, 68-73, RES SPECIALIST, CORP RES DEPT, MONSANTO CO, 73- Mem: Am Chem Soc. Res: Metal complexes and their application in homogeneous catalysis. Mailing Add: Corp Res Dept Monsanto Co 800 N Lindbergh Blvd St Louis MO 63166

MORRIS, EDWARD C, b West Branch, Iowa, Apr 4, 16; m 44; c 5. CHEMISTRY. Educ: William Penn Col, BS, 40; Friends Univ, AB, 49; Univ Iowa, MS, 57. Prof Exp: Prof physicist x-ray sect, Nat Bur Stand, Washington, DC, 42-43; assoc prof physics & chem, William Penn Col, 56-63; PROF CHEM & PHYS SCI, ASBURY COL, 63- Mem: Am Chem Soc; Am Asn Physics Teachers. Res: Solubility of metal chlorides in nonaqueous acetic acid, especially metals that complex spectrophotometric studies of species. Mailing Add: Asbury Col Wilmore KY 40390

MORRIS, EDWARD CRAIG, b Murray, Ky, Oct 7, 39. ANTHROPOLOGY. Educ: Vanderbilt Univ, BA, 61; Univ Chicago, MA, 64, PhD(anthrop), 67. Prof Exp: Asst prof anthrop, Northern Ill Univ, 67-68; asst prof anthrop, Brandeis Univ, 68-75; ASST CUR ANTHROP, AM MUS NATURAL HIST, 75- Mem: Am Anthrop Asn; Soc Am Archaeol; Inst Andean Studies. Res: Andean archaeology and ethnohistory; Inca urbanism; methods of urban archaeology; development of civilization in the central Andes. Mailing Add: Dept of Anthrop Am Mus Natural Hist New York NY 10024

MORRIS, ELLIOT COBIA, b Ely, Nev, June 24, 26; m 50; c 3. ASTROGEOLOGY, MINERALOGY. Educ: Univ Utah, BS, 50, MS, 53; Stanford Univ, PhD(geol), 62. Prof Exp: Seismic comput, Seismic Explor, Inc, Wyo, 53-54; asst explor geologist, Phillips Petrol Co, Utah, 54-56; geologist & coordr Surveyor TV invests, Astrogeol Br, 61-69, SUPVRY GEOLOGIST, VIKING MARS LANDERS IMAGING TEAM, US GEOL SURV, 69- Concurrent Pos: Staff scientist planetary progs, NASA hq, 70-71. Mem: AAAS; Geol Soc Am. Res: Structural and stratigraphic geology of Western Uinta and Wasatch Mountains, Utah and southern Alaskan areas; sedimentary mineralogy of central California; astrogeologic studies. Mailing Add: 515 N Bertrand Flagstaff AZ 86001

MORRIS, ESTELL E, b Pellville, Ky, Jan 20, 20; m 42; c 3. ORAL SURGERY. Educ: Ind Univ, AB, 49, DDS, 53, MS, 60. Prof Exp: From intern to resident, Med Sch, 54-55, instr oral surg, Sch Dent, 55-61, ASST PROF ORAL SURG, SCH DENT, IND UNIV, INDIANAPOLIS, 61- Mem: Am Dent Asn; Am Soc Oral Surg. Res: Ilosone, the propionyl ester of erythromycin; management of oral and dental infections; unusual luxation of mandibular condyle. Mailing Add: Dept of Oral Surg Ind Univ Sch of Dent Indianapolis IN 46202

MORRIS, EUGENE RAY, b Albion, Nebr, Aug 26, 30; m 52; c 2. NUTRITION, PHYSIOLOGICAL CHEMISTRY. Educ: Univ Mo, BS, 52, MS, 56, PhD(agr chem), 62. Prof Exp: Assoc chemist, Midwest Res Inst, 62-64; instr agr chem, Univ Mo, 64-68; RES CHEMIST, HUMAN NUTRIT RES DIV, AGR RES SERV, USDA, 68- Mem: Am Chem Soc; Am Sci Affil; Am Inst Nutrit. Res: Animal and human nutrition; physiological chemistry of magnesium; unidentified growth factors; iron availability. Mailing Add: Agr Res Ctr Nutrit Inst Agr Res Serv Beltsville MD 20705

MORRIS, EVERETT FRANKLIN, b Bellmont, Ill, May 23, 24; m 46; c 2. BOTANY. Educ: Eastern Ill Univ, BS, 50; Univ Wyo, MS, 52; Univ Iowa, PhD(bot), 55. Prof Exp: Asst prof biol, Millikin Univ, 55-57; assoc prof, Martin Br, Univ Tenn, 57-58; asst mycologist, Ill Natural Hist Surv, 58; from asst prof to assoc prof, 58-66, PROF BIOL, WESTERN ILL UNIV, 66-, CO-ADMINR, A L KIBBE LIFE SCI STA, 65-, CHMN DEPT BIOL SCI, 69- Mem: Mycol Soc Am. Res: Taxonomy of myxomycetes and Fungi Imperfecti. Mailing Add: Dept of Biol Sci Western Ill Univ Macomb IL 61455

MORRIS, FRED JOHN, b Chicago, Ill, Dec 6, 19; m 42; c 2. PHYSICS. Educ: Tex Col Arts & Indust, BS, 42; Univ Tex, MS, 44, PhD(physics), 51. Prof Exp: Instr physics, Univ Tex, 46-51; PRES & DIR RES, ELECTRO-MECH CO, 51- Concurrent Pos: Sci adv, Joint Spectrum Eval Group, Joint Chiefs of Staff, DC, 57-58; consult, Dept of Defense, 59- Mem: AAAS; Soc Am Mil Eng. Res: Electronics and magnetics, particularly development of measurement techniques. Mailing Add: Electro-Mech Co PO Box 1546 Austin TX 78767

MORRIS, GENE FRANKLIN, b Cedar Rapids, Iowa, Nov 22, 34; m 59; c 4. ORGANIC CHEMISTRY. Educ: Iowa State Univ, 55; Kans State Univ, PhD(org chem), 61. Prof Exp: USPHS fel, Iowa State Univ, 61-63, instr chem, 63-66; from asst prof to assoc prof, Wis State Univ, Eau Claire, 66-69; vis prof, Univ Calgary, 69; ASSOC PROF CHEM, WESTERN CAROLINA UNIV, 69- Mem: Am Chem Soc. Res: Physical organic chemistry; mechanisms of reactions, epoxidations, hydrogenations. Mailing Add: Dept of Chem Western Carolina Univ Cullowhee NC 28723

MORRIS, GEORGE COOPER, JR, b Evanston, Ill, Feb 15, 24; m 46; c 5. SURGERY. Educ: Univ Pa, MD, 48. Prof Exp: Instr surg, Sch Med, Univ Pa, 49-50; from instr to assoc prof, 50-68, PROF SURG, BAYLOR COL MED, 68- Concurrent Pos: Markle scholar. Mem: Soc Vascular Surg; Soc Thoracic Surgeons; Am Col Chest Physicians; Am Asn Surg Trauma; Int Soc Surgeons. Res: Cardiovascular research. Mailing Add: Tex Med Ctr Baylor Col of Med Houston TX 77030

MORRIS, GEORGE V, b Providence, RI, Nov 18, 30; m 59; c 2. PHYSICAL CHEMISTRY, ANALYTICAL CHEMISTRY. Educ: Providence Col, BS, 52; Univ RI, MS, 57, PhD(phys chem), 62. Prof Exp: Asst chem, Univ RI, 55-56; res chemist, Nat Res Corp, Mass, 56-57 & Eltex Res Corp, RI, 57-59; asst chem, Univ RI, 59-61; res chemist, US Naval Underwater Ord Sta, 62-65; from instr to assoc prof, 63-72, PROF PHYS CHEM, SALVE REGINA COL, 72- Concurrent Pos: Sr engr, Raytheon Co, RI, 66- Mem: Am Chem Soc. Res: Research and development of continuous total organic carbon analyzer; development of stable, long deployment life, dissolved oxygen probe. Mailing Add: Dept of Physics & Chem Salve-Regina Col Newport RI 02840

MORRIS, GEORGE WILLIAM, b Granite, Okla, Apr 23, 21; m 59; c 2. MATHEMATICS, MATHEMATICAL PHYSICS. Educ: Southest Inst Technol, BA, 42; Univ Okla, MA, 48; Univ Calif, Los Angeles, PhD(math), 57. Prof Exp: Instr math, Univ Tulsa, 47-48; engr, Northrop Aircraft, Inc, 51-53; sr engr, NAm Aviation, Inc, 53-58; proj engr, Aerolab Develop Co, 58-60; mem tech staff, Land-Air, Inc, Point Mugu, 60-62 & Douglas Aircraft Co, 62-68; PROF MATH, NAVAL POSTGRAD SCH, 68- Mem: Soc Indust & Appl Math. Res: Numerical analysis; celestial mechanics; exterior ballistics; elasticity; aerodynamics; electromagnetic wave propagation; thermodynamics; operations research. Mailing Add: Dept of Math Naval Postgrad Sch Monterey CA 93940

MORRIS, GERALD BROOKS, b Decatur, Tex, July 2, 33; m 59. GEOPHYSICS. Educ: Tex A&M Univ, 56, MS, 62; Univ Calif, San Diego, PhD(earth sci), 69. Prof Exp: Res engr, Res Lab, Carter Oil Co, 57-58 & Jersey Prod Res Co, 58-63; asst res geophysicist, Marine Phys Lab, Univ Calif, San Diego, 69-70; ASST PROF GEOPHYS, UNIV HAWAII, 70- Mem: Soc Explor Geophysicists; Am Geophys Union. Res: Underwater acoustics and sound propagation; explosion seismology, particularly marine seismic refraction studies; elastic properties of earth materials. Mailing Add: Dept of Geol & Geophys Univ of Hawaii 2525 Correa Rd Honolulu HI 96822

MORRIS, GERALD PATRICK, b Edmonton, Alta, June 13, 39; m 64. CELL BIOLOGY, PARASITOLOGY. Educ: Univ BC, BSc, 64, MSc, 66; Queens Univ, Belfast, PhD(zool), 68. Prof Exp: Lectr zool, Univ BC, 65-66; NIH fel electron micros, Univ Kans, 68-69; asst prof biol, 69-74, ASSOC PROF BIOL, QUEENS UNIV, ONT, 74- Mem: Am Soc Cell Biol; Am Soc Parasitol; Can Soc Zoologists; Brit Soc Parasitol. Res: Ultrastructure and biochemistry of parasitic flatworms; insect neurosecretion and experimental gastric ulcers. Mailing Add: Dept of Biol Queens Univ Kingston ON Can

MORRIS, GLEN JEFFS, b Salt Lake City, Utah, Jan 14, 24; m 46; c 2. PLASMA PHYSICS, ELECTROOPTICS. Educ: Univ Utah, BS, 49, PhD(physics), 56. Prof Exp: Asst math & physics, Univ Utah, 50-56; mem staff, Los Alamos Sci Lab, NMex, 56-59; asst prof physics, San Diego State Col, 59-60; staff mem, Boeing Sci Res Labs, Wash, 60-69; SR SCIENTIFIC SPECIALIST, E G & G, INC, 69- Mem: Am Phys Soc; Optical Soc Am. Res: Non-linear optics, including interaction of light with plasmas. Mailing Add: 7110 Natalie NE Albuquerque NM 87110

MORRIS, GLENN KARL, b Toronto, Ont, Oct 26, 38; m 64; c 6. ENTOMOLOGY, ANIMAL BEHAVIOR. Educ: Ont Agr Col, Univ Guelph, BSA, 62; Cornell Univ, MS, 65, PhD(entom), 67. Prof Exp: Lectr, 67-68, asst prof, 68-74, ASSOC PROF ZOOL, UNIV TORONTO, 74- Mem: Animal Behav Soc; Entom Soc Can; Royal Can Inst. Res: Sound communication of long horned grasshoppers. Mailing Add: Dept of Zool Erindale Col Univ of Toronto Toronto ON Can

MORRIS, HAL TRYON, b Salt Lake City, Utah, Oct 24, 20; m 42; c 3. GEOLOGY. Educ: Univ Utah, BS, 42, MS, 47. Prof Exp: Geologist, 46-69, RES GEOLOGIST, US GEOL SURV, 69- Mem: Fel AAAS; fel Geol Soc Am; Soc Econ Geologists. Res: Mineral deposits; structural geology; stratigraphy; geochemical pros- pecting; detection and discovery of concealed ore deposits. Mailing Add: 345 Middlefield Rd Menlo Park CA 94025

MORRIS, HALCYON ELLEN MCNEIL, b Delphos, Kans, May 24, 27; m 59. MATHEMATICS. Educ: Kans State Univ, BS, 51; Univ Tulsa, MS, 58. Prof Exp: Mathematician, Carter Oil Co, 51-58, Jersey Prod Res Co, 58-59, Pan Am Petrol Co, 60-63 & Naval Electronics Lab, 64-68; SUPVRY MATHEMATICIAN, NAVAL UNDERSEA CTR, 68- Mem: Seismol Soc Am; Acoust Soc Am. Res: Detection of submarines by surveillance systems; development of theory for new sonar systems; opportunities in science for women; underwater acoustics, especially sound propagation and botton reflection loss. Mailing Add: Code 409 Naval Undersea Ctr San Diego CA 92132

MORRIS, HAROLD DONALD, b Crawfordville, Ga, Nov 24, 12; m 39; c 3. AGRONOMY. Educ: NC State Col, BS, 38, MS, 41; Iowa State Univ, PhD(soil fertil), 47. Prof Exp: Asst agronomist, Exp Sta, NC State Col, 38-41; assoc prof, 48-52, PROF AGRON, UNIV GA, 52- Mem: Am Soc Agron. Res: Soil fertility; minor elements; soil acidity; potassium; soluble manganese as a factor affecting growth of various legumes in culture solutions and in acid soils. Mailing Add: Dept of Agron Univ of Ga Athens GA 30601

MORRIS, HAROLD H, b Lincoln, Nebr, Mar 24, 43; m 61; c 3. NEUROLOGY. Educ: Baylor Col Med, MD, 68. Prof Exp: Intern med, Methodist Hosp, Houston, 68-69; med officer, USPHS, 69-71; resident, 71-74, ASST PROF NEUROL, UNIV TEX MED BR, 74- Mem: Am Acad Neurol; Am Epilepsy Soc; AMA. Res: Neuromuscular transmission defects; muscular dystrophy. Mailing Add: Dept of Neurol Univ of Tex Med Br Galveston TX 77550

MORRIS, HAROLD HOLLINGSWORTH, JR, b Shanghai, China, Sept 23, 17; US citizen; m 44; c 2. MEDICINE, PSYCHIATRY. Educ: Haverford Col, BS, 39; Tulane Univ, MD, 43. Prof Exp: ASSOC PROF PSYCHIAT, SCH MED, UNIV PA, 56- Concurrent Pos: Clin dir psychiat, Inst Pa Hosp, 48-56; consult, Episcopal Diocese Pa, 54-, Vet Admin Hosp, Coatesville, 56-, Am Friends Serv Comt, 62 & Peace Corps, 64; dir, Mercy-Douglass Hosp, Philadelphia, 56-63 & Misericordia Hosp, 66- Mem: Am Psychiat Asn. Res: Effect of phenothiazines on schizophrenia; long-term follow-up studies on 3,000 patients, including 225 children; effect of psychological mileu on organic illness. Mailing Add: Misericordia Hosp 54th & Cedar Ave Philadelphia PA 19143

MORRIS, HAROLD PAUL, b Salem, Ind, May 8, 00; m 28; c 3. NUTRITION, BIOCHEMISTRY. Educ: Univ Minn, BS, 25, PhD(biochem), 30; Kans State Col, MS, 26. Prof Exp: Asst animal genetics, Kans State Col, 25-26; asst animal nutrit, Exp Sta, Univ Ill, 26-28; asst biochem & dairy chem, Exp Sta, Univ Minn, 28-30; res assoc, Bur Fisheries, US Dept Commerce, 31-32; jr bacteriologist, Bur Home Econ, USDA, 33-34, assoc biochemist, Food & Drug Admin, 34-38; biochemist, Nat Cancer Inst, 38-41, sr nutrit chemist, 41-48, prin biochemist nutrit, 48-51, head nutrit & carcinogenesis sect, Lab Biochem, 51-68; PROF BIOCHEM, COL MED, HOWARD UNIV, 68- Concurrent Pos: Mem incentive awards bd, NIH, 57-61, chmn, 60-61; US organizer, US-Japanese Exchange Mission, Conf Biol & Biochem Eval Malignancy in Exp Hepatomas, Kyoto, Japan, 65. Honors & Awards: Superior Serv Award, Nat Cancer Inst, 56. Mem: Am Chem Soc; Am Soc Biol Chemists; Soc Exp Biol & Med; Am Asn Cancer Res; Am Inst Nutrit. Res: Mouse nutrition; radio and chemical isotopic tracers; hereditary factors influencing food utilization in the rat; development of a spectrum of transplantable rat hepatocellular carcinomas of different growth rate and possessing many variable biological and biochemical characteristics. Mailing Add: Dept of Biochem Howard Univ Col of Med Washington DC 20059

MORRIS, HARRIS LEE, b Syracuse, NY, June 15, 38; m 65; c 1. PHYSICAL CHEMISTRY, CHEMICAL ENGINEERING. Educ: Stevens Inst Technol, BEng, 60, MS, 62; Univ Mich, PhD(phys chem), 65. Prof Exp: RES CHEMIST, CHEM DIV, UNIROYAL, INC, NAUGATUCK, 66- Concurrent Pos: Instr, Southern Conn State Col, 67-69; adj assoc prof, Univ New Haven, 69- Mem: Am Chem Soc; Soc Rheol. Res: Physical chemistry and physics of plastics and rubbers; structural and molecular characterization of polymeric materials; relationship of physical, mechanical and rheological properties of polymers to their molecular structure and morphology. Mailing Add: 14 Old Quarry Rd Woodbridge CT 06525

MORRIS, HARRY DUNLAP, b El Dorado, Okla, Dec 15, 06; m 31. ORTHOPEDIC SURGERY. Educ: Ore State Col, BS, 26; St Louis Univ, MD, 33; Am Bd Orthop Surg, dipl, 42. Prof Exp: From asst resident surgeon to resident surgeon, St Louis City Hosp, Mo, 34-36; resident surgeon, Mass Gen Hosp, Boston, 36 & Children's Hosp, 37; clin asst, Washington Univ, 37-39; instr orthop surg, Sch Med, Tulane Univ, 39-42; chief amputation ctr, McClosky Gen Hosp, US Army, 43-46 & chief orthop serv, Percy Jones Gen Hosp, 46-47; assoc orthop surg, 48-57, head dept, 57-67, chief med staff, 64, SR CONSULT, DEPT ORTHOP SURG, OCHSNER CLIN, 67- Concurrent Pos: Res surgeon, Shriners' Hosp, St Louis, 37-39; orthop surgeon, Div Crippled Children, La State Dept Health, 39-53; sr vis orthop surgeon, Charity Hosp, New Orleans, 39-67, consult, 67-; sr vis orthop surgeon, Touro Infirmary, 39-; from asst clin prof to assoc clin prof orthop surg, Sch Med, Tulane Univ, 46-60, clin prof, 60-; area consult, Vet Admin, 47-53; vis orthop surgeon, Ochsner Found Hosp, New Orleans, 48-57, chief dept orthop surg, 57-67; mem, Am Bd Orthop Surg, 59-65. Mem: Clin Orthop Soc; Am Orthop Asn; fel Am Col Surgeons; Am Acad Orthop Surg (vpres, 65); Int Soc Orthop Surg & Traumatol. Res: Surgical treatment and prosthetic fitting of congenital amputees; reconstruction of foot deformities due to paralytic involvement. Mailing Add: Ochsner Clin 1514 Jefferson Hwy New Orleans LA 70121

MORRIS, HERBERT ALLEN, b Okla, Sept 15, 19; m 45; c 3. MATHEMATICS. Educ: Southeastern State Col, BA, 46; Univ Tex, MA, 51. Prof Exp: Asst prof math, Lamar State Col, 51-55 & Colo Sch Mines, 55-57; mathematician, Atomic Energy Div, Phillips Petrol Co, 57-59, sr mathematician, Comput Dept, 59-65; DIR COMPUT CTR & CHMN DEPT COMPUT SCI, BRADLEY UNIV, 65-, ASSOC PROF COMPUT SCI, 74- Concurrent Pos: Adj prof, Okla State Univ, 60-65. Mem: Soc Indust & Appl Math; Asn Comput Mach. Res: Numerical analysis; point set topology; operations research. Mailing Add: Comput Ctr Bradley Univ Peoria IL 61606

MORRIS, HORTON HAROLD, b Post, Tex, May 26, 22; m 45; c 3. CLAY MINERALOGY, PULP AND PAPER TECHNOLOGY. Educ: Tex Tech Col, BS, 49; Univ Maine, MS, 52. Prof Exp: From instr to assoc prof chem, Univ Maine, 52-57; res dir, South Clays, Inc, 57-63; vpres res & develop, Freeport Kaolin Co, 63-74; PRES, SSI CONSULTS, 74- Mem: AAAS; Am Chem Soc; Tech Asn Pulp & Paper Indust; Am Soc Test & Mat; NY Acad Sci. Res: Molecular rearrangements; Glycidic esters; 2, 3-dihydroxy esters and halohydrins; kaolin clay studies; mineral benefaction; delaminated clays; calcined clays. Mailing Add: 4684 Twin Oak Dr Macon GA 31204

MORRIS, HOWARD ARTHUR, b Draper, Utah, Feb 9, 19; m 41; c 4. FOOD SCIENCE. Educ: Univ Minn, MS, 49, PhD(dairy technol), 52. Prof Exp: Asst & instr, 46-51, asst prof dairy technol, 52-55, assoc prof dairy indust, 55-60, PROF FOOD SCI, UNIV MINN, ST PAUL, 60- Mem: Am Soc Microbiol; Brit Soc Appl Bact; Am Dairy Sci Asn. Res: Chemistry, bacteriology and enzymology applied to food processing. Mailing Add: Dept of Food sci & Nutrit Univ of Minn St Paul MN 55101

MORRIS, HUGHLETT LEWIS, b Big Rock, Tenn, Mar 18, 31; m 50; c 3. SPEECH PATHOLOGY. Educ: Univ Iowa, BA, 52, MA, 57, PhD(speech path), 60. Prof Exp: Clinician speech & hearing, Pub Schs, Iowa, 54-56; res assoc otolaryngol & maxillofacial surg, 58-61, coordr cleft palate clin, 59-64, res asst prof otolaryngol & maxillofacial surg, 61-64, assoc prof otolaryngol, maxillofacial surg, speech path & audiol, 65-67, PROF OTOLARYNGOL & MAXILLOFACIAL SURG, SPEECH PATH & AUDIOL & ORTHOD, UNIV IOWA, 68-, DIR DIV SPEECH & HEARING, DEPT OTOLARYNGOL & MAXILLOFACIAL SURG & DIR CLEFT PALATE RES PROG, 65- Concurrent Pos: Ed, Cleft Palate J, Am Cleft Palate Asn, 64-70; prin investr, Proj Res Grant, Nat Inst Dent Res Prog, 65-; mem, Am Bds Examrs Speech Path & Audiol, 73-76; vpres, Am Cleft Palate Educ Found, 74-75. Mem: AAAS; fel Am Speech & Hearing Asn; Am Cleft Palate Asn (pres, 73-74). Res: Cleft lip and palate; disorders of the voice. Mailing Add: Dept Otolaryng-Maxillofac Surg Univ of Iowa Hosps Iowa City IA 62242

MORRIS, HUMBERT, b Chicago, Ill, Jan 30, 17; m 36; c 2. PHYSICAL CHEMISTRY. Educ: Ohio Univ, BS, 39; Northwestern Univ, PhD(inorg chem), 43. Prof Exp: Chemist, Stand Oil Co, Ind, 43-46; res engr, Douglas Aircraft Co, 46-49; chemist, Rand Corp, 49-53; mem, Off Ord Res, 53-55, mgr western off, 55-60; tech dir, Army Weapons Command, 60-63, staff mem, Inst Defense Anal, 63-68, chem engr, Defense Special Proj Group, 68-72; SR SCIENTIST, RAFF ASSOCS, 72- Mem: Am Chem Soc; Am Ord Asn. Res: Catalysis; radio chemistry; thermodynamics; ballistics. Mailing Add: 6705 Pawtucket Rd Bethesda MD 20034

MORRIS, JAMES ALBERT, b Crawfordsville, Ind, June 4, 42; m 67; c 2. ENZYMOLOGY, MICROBIAL BIOCHEMISTRY. Educ: Wabash Col, BA, 64; Purdue Univ, PhD(microbiol), 69. Prof Exp: Res assoc gustation, Monell Chem Senses Ctr, Univ Pa, 70-73; SR MICROBIOLOGIST, INT FLAVORS & FRAGRANCES, INC, 73- Mem: Am Chem Soc. Res: Production or modification of flavors by microbial and/or enzymatic processes. Mailing Add: 1515 Hwy 36 Union Beach NJ 07735

MORRIS, JAMES F, b New York, NY, Mar 22, 22. MEDICINE. Educ: Ohio Wesleyan Univ, AB, 43; Univ Rochester, MD, 48. Prof Exp: Fel med bact, Sch Med, Univ Rochester, 50-51; instr med, Col Med, Univ Utah, 53-54; assoc prof, 57-71, PROF MED, UNIV ORE, 71- Concurrent Pos: Chief pulmonary & infectious dis, Vet Admin Hosp, Portland, 57- Mem: Am Thoracic Soc. Res: Clinical pulmonary physiology, primarily obstructive airway diseases. Mailing Add: Dept of Med Vet Admin Hosp Portland OR 97207

MORRIS, JAMES GRANT, b Brisbane, Australia, Aug 30, 30; m 59; c 3. ANIMAL NUTRITION. Educ: Univ Queensland, BAgrSc, 53, Hons, 55, BSc, 58, MAgrSci, 59; Utah State Univ, PhD(nutrit & biochem), 61. Prof Exp: Dir husb res, Animal Res Inst, Brisbane, Australia, 65-69; assoc prof ruminant nutrit & assoc nutritionist, 69-75, PROF ANIMAL SCI & PHYSIOL CHEM & NUTRITIONIST, EXP STA, UNIV CALIF, DAVIS, 75- Mem: Brit Nutrit Soc; Australian Inst Agr Sci; Am Inst Nutrit; Am Soc Animal Sci. Res: Physiology and biochemistry. Mailing Add: Dept of Animal Sci Univ of Calif Davis CA 95616

MORRIS, JAMES JOSEPH, JR, b Jersey City, NJ, Aug 16, 33; m 54; c 3. INTERNAL MEDICINE, CARDIOLOGY. Educ: Hofstra Univ, BA, 55; State Univ NY, MD, 59. Prof Exp: Intern, 59-60, instr, 60-61, resident, 61-62, instr, 62-63, chief resident, 63-64, assoc, 64-66, asst prof, 66-70, ASSOC PROF MED, DUKE UNIV, 70- Concurrent Pos: USPHS fels, 60-61, 62-63 & spec fel, 64-67. Mem: Am Heart Asn; Am Col Physicians. Res: Electrocardiology; arrhythmia; cardiac catherization; hemodynamics. Mailing Add: Dept of Med Cardiovasc Lab Duke Univ Med Ctr Durham NC 27706

MORRIS, JESSE CLENDENIN, petroleum chemistry, see 12th edition

MORRIS, JOHN CARRELL, b Philadelphia, Pa, May 30, 14; m 40, 67 & 73; c 3. SANITARY CHEMISTRY. Educ: Rutgers Univ, BS, 34; Princeton Univ, AM, 35, PhD(phys chem), 38. Hon Degrees: AM, Harvard Univ, 49. Prof Exp: Instr chem, Harvard Univ, 38-41; asst prof, Bucknell Univ, 41-42 & Worcester Polytech Inst, 42-

43; from instr to assoc prof sanit chem, 44-58, Allston Burr Sr Tutor, 59-64, GORDON McKAY PROF, SANIT CHEM, HARVARD UNIV, 58- Concurrent Pos: WHO expert consult, Univ Alexandria, 58-59; vis lectr, Int Courses Sanit Eng, Delft, Neth, 64-; mem safe drinking water comt, Nat Acad Sci, 75-76. Honors & Awards: Buswell-Porges Award, Inst Advan Sanit Res Int, 73. Mem: Am Water Works Asn; Int Water Supply Asn; Am Chem Soc; Water Pollution Control Fedn. Res: Kinetics and equilibria of reactions in water and sewage treatment; dynamics of water disinfection; reactions of aqueous chlorine with organic and nitrogenous compounds. Mailing Add: 127 Pierce Hall Harvard Univ Cambridge MA 02138

MORRIS, JOHN EDWARD, b Pasadena, Calif, July 9, 36; m 58; c 2. DEVELOPMENTAL BIOLOGY. Educ: Stanford Univ, BA, 58; Univ Hawaii, MS, 60; Univ Calif, Los Angeles, PhD(zool), 66. Prof Exp: Teaching asst, Univ Hawaii, 60 & Univ Calif, Los Angeles, 60-64; investr, Wenner-Gren Inst, Sweden, 65-67 & Univ Chicago, 67-68; asst prof zool, 68-74, ASSOC PROF ZOOL, ORE STATE UNIV, 74- Concurrent Pos: NIH fel, 65-67, trainee, 68; NSF grant, 70-; vis asst prof pediat, Univ Chicago, 74-75. Mem: AAAS; Am Inst Biol Sci; Am Soc Zoologists; Am Soc Cell Biol; Soc Develop Biol. Res: Tissue and cell interactions during embryonic differentiation and growth; control mechanisms in differentiation; reversibility of development during dehydration in cryptobiotic embryos. Mailing Add: Dept of Zool Ore State Univ, Corvallis OR 97331

MORRIS, JOHN EMORY, b Takoma Park, Md, June 15, 37. ONCOLOGY, BIOCHEMISTRY. Educ: Cornell Univ, BA, 59; Univ Wis, MS, 62, PhD(oncol), 66. Prof Exp: Res asst clin oncol, Univ Wis-Madison, 61-65; Am Peace Corps vis asst prof biochem, Fac Med, Pahlavi Univ, Iran, 66-67; asst prof, 67-70, ASSOC PROF CHEM, STATE UNIV NY COL BROCKPORT, 70- Mem: Am Chem Soc. Res: Carcinogenic activity of nitrofurans; synthesis of substituted furans; protein structure; mechanism of enzyme regulation. Mailing Add: Dept of Chem State Univ of NY Col Brockport NY 14420

MORRIS, JOHN LEONARD, b Des Moines, Iowa, Dec 12, 29; m 52; c 2. PLANT BREEDING, PLANT PATHOLOGY. Educ: Iowa State Univ, BS, 61; Utah State Univ, PhD(plant breeding & path), 67. Prof Exp: Fieldman, 61-63, res asst seed develop, 63-68, DIR PEA & BEAN RES, ROGERS BROS CO, 68- Mem: Am Soc Agron; Sci Res Soc Am. Res: Breeding, development and research of snap beans and garden peas; seed quality research. Mailing Add: Rogers Bros Co PO Box 104 Twin Falls ID 83301

MORRIS, JOHN LLEWELYN, b Newtown, Wales, Sept 19, 43; m 65; c 3. NUMERICAL ANALYSIS. Educ: Univ Leicester, BS, 65; Univ St Andrews, PhD(math), 68. Prof Exp: Nat Cash Register Co res fel numerical anal, Univ Dundee, 67-69, lectr comput sci, 67-75; ASSOC PROF COMPUT SCI, UNIV WATERLOO, 75- Mem: Brit Inst Math & Applns. Res: Numerical solution of partial differential equations. Mailing Add: Comput Sci Dept Univ of Waterloo Waterloo ON Can

MORRIS, JOHN MCLEAN, b Kuling, China, Sept 1, 14; m 51; c 5. MEDICINE. Educ: Princeton Univ, AB, 36; Harvard Univ, MD, 40; Yale Univ, MA, 62; Am Bd Surg, dipl, 50; Am Bd Obstet & Gynec, dipl, 58. Prof Exp: Asst surg, Mass Gen Hosp, Boston, 47-52; assoc prof, 52-61, prof, 61-69, JOHN SLADE ELY PROF GYNEC, SCH MED, YALE UNIV, 69- Concurrent Pos: Am Cancer Soc fel, Radiumhemmet, Stockholm, Sweden, 51-52; consult gynecologist, Hosps, Conn, 52-; chief obstet & gynec, Yale-New Haven Hosp, 65-66; vis prof gynec & obstet, Stanford Univ, 66-67; vis prof, Univ Tex M D Anderson Hosp & Tumor Inst, 70; consult, Walter Reed Hosp, Gorgas Hosp & Tripler Gen Hosp. Mem: Fel Am Col Surg; fel Am Col Obstet & Gynec; Am Gynec Soc; Am Fertil Soc; Soc Pelvic Surg. Res: Gynecology; surgery; endocrinology; radiation biology; intersexuality; agents affecting ovum development. Mailing Add: 333 Cedar St New Haven CT 06511

MORRIS, JOHN WESLEY, b Billings, Okla, Nov 14, 07; m 32; c 2. GEOGRAPHY. Educ: Univ Okla, BS, 30; Okla State Univ, MS, 34; George Peabody Col, PhD(geog), 41. Prof Exp: Instr geog, Seminole Jr Col, 31-38; instr, George Peabody Col, 38-39; from assoc prof to prof, Southeastern State Col, 39-48; prof, 48-73, EMER PROF GEOG, UNIV OKLA, 73- Honors & Awards: Distinguished Serv Award, Nat Coun Geog Educ 66. Mem: Nat Coun Geog Educ (secy, 55-58, pres, 59-60); Asn Am Geog. Res: Historical geography; geographic education. Mailing Add: Dept of Geog Univ of Okla 455 Lindsey Ave 804 Norman OK 73069

MORRIS, JOSEPH ANTHONY, b Prince Georges Co, Md, Sept 6, 18; m 42; c 4. BACTERIOLOGY. Educ: Cath Univ Am, BS, 40, MS, 42, PhD(bact), 47. Prof Exp: Bacteriologist, Josiah Macy Jr Found, NY, 43-44, US Dept Interior & USDA, 44-47, Walter Reed Army Inst Res, DC, 47-56 & US Army Med Command, Japan, 56-59; virologist, NIH, 59-72; DIR SLOW, LATENT & TEMPERATE VIRUS BR, BUR BIOLOGICS, FOOD & DRUG ADMIN, 72- Concurrent Pos: Instr, Am Univ, 43-46. Mem: Am Soc Microbiol; Am Soc Trop Med & Hyg; Soc Exp Biol & Med; Am Asn Immunol; NY Acad Sci. Res: Virus and rickettsial diseases. Mailing Add: Lab of Virol Bur of Biologics Food & Drug Admin Bethesda MD 20014

MORRIS, JOSEPH BURTON, b Del, Jan 25, 25. ANALYTICAL CHEMISTRY. Educ: Howard Univ, BS, 49, MS, 51; Pa State Univ, PhD(anal chem), 56. Prof Exp: Instr chem, Howard Univ, 51-53; asst, Pa State Univ, 53-56; res & develop chemist, E I du Pont de Nemours & Co, 56-57; from asst prof to assoc prof, 57-70, PROF ANAL CHEM, HOWARD UNIV, 70- Mem: AAAS; Am Chem Soc. Res: Polarography; chronopotentiometry; voltammetry at solid electrodes; coulometry; instrumental methods of analysis. Mailing Add: Dept of Chem Howard Univ Washington DC 20001

MORRIS, JOSEPH RICHARD, b Richmond, Va, Aug 3, 35; m 59; c 1. TOPOLOGY. Educ: Va Polytech Inst, BS, 57, MS, 60; Univ Ala, MA, 65, PhD(math), 69. Prof Exp: Asst prof math, Samford Univ, 59-64; ASST PROF MATH, VA COMMONWEALTH UNIV, 69- Mem: Am Math Soc; Math Asn Am. Res: The existence of invariant means of Banach spaces and the common fixed point property for a family of functions. Mailing Add: Dept of Math Sci Va Commonwealth Univ Richmond VA 23284

MORRIS, JUSTIN ROY, b Nashville, Ark, Feb 20, 37; m 56; c 2. HORTICULTURE, PLANT PHYSIOLOGY. Educ: Univ Ark, BSA, 57, MS, 61; Rutgers Univ, PhD(hort), 64. Prof Exp: Asst hort, Univ Ark, 57-61; instr pomol, Rutgers Univ, 61-64; exten horticulturist & assoc prof food sci, 64-75, PROF HORT & FOOD SCI, UNIV ARK, FAYETTEVILLE, 75- Concurrent Pos: Consult, Int Exten, Fed Exten Serv, USDA, 71-72 & US AID, 74. Mem: Am Soc Hort Sci; Inst Food Technol. Res: Preharvest production and handling of mechanically harvested fruits. Mailing Add: Hort Food Sci Rte 6 Fayetteville AR 72701

MORRIS, KELSO BRONSON, b Beaumont, Tex; m 61; c 4. PHYSICAL INORGANIC CHEMISTRY. Educ: Wiley Col, BSc, 30; Cornell Univ, MSc, 37, PhD(inorg chem), 40. Prof Exp: Instr chem & math, Wiley Col, 30-37, from assoc prof to prof chem & head dept, 37-46; assoc prof, 46-60, head dept, 65-69, PROF CHEM, HOWARD UNIV, 61- Concurrent Pos: From assoc prof to prof & head chem sect, Air Force Inst Technol, 59-61. Mem: Fel AAAS; Am Chem Soc; Nat Asn Res Sci Teaching; fel Am Inst Chem. Res: Electrochemistry; chemistry of hydroxylamine; complexing tendencies of metal ions; chemistry of fused salts. Mailing Add: Dept of Chem Howard Univ Washington DC 20001

MORRIS, LEO RAYMOND, b South Whitley, Ind, June 19, 22; m 45; c 3. INDUSTRIAL ORGANIC CHEMISTRY. Educ: Manchester Col, BA, 47; Univ Wis, PhD(org chem), 52. Prof Exp: Asst chemist, Univ Wis, 47-50; from org chemist to sr res chemist, 51-71, RES SPECIALIST, DOW CHEM USA, 71- Mem: Am Chem Soc. Res: Dehydrohalogenation, synthesis of new monomers and bioactive compounds; free-radical additions. Mailing Add: Dow Chem USA Midland MI 48640

MORRIS, LEONARD LESLIE, b Terre Haute, Ind, Aug 5, 14; m 40. VEGETABLE CROPS, PLANT PHYSIOLOGY. Educ: Purdue Univ, BS, 37; Cornell Univ, MS, 39, PhD(veg crops), 41. Prof Exp: Asst veg crops, Cornell Univ, 37-41; instr truck crops, 41-45, from jr olericulturist to assoc olericulturist, 41-60, from asst prof to assoc prof veg crops, 45-60, PROF VEG CROPS, UNIV CALIF, DAVIS & OLERICULTURIST, EXP STA, 60- Honors & Awards: Hauck Award, Produce Packaging Asn, 57; Vaughan Award, Am Soc Hort Sci, 58. Mem: Fel AAAS; fel Am Soc Hort Sci; Am Soc Plant Physiol; Am Inst Biol Sci. Res: Physiology of vegetable crops as related to transit, storage and marketing. Mailing Add: Dept of Veg Crops Mann Lab Univ of Calif Davis CA 95616

MORRIS, LUCIEN ELLIS, b Mattoon, Ill, Nov 30, 14; m 42; c 5. ANESTHESIOLOGY. Educ: Oberlin Col, AB, 36; Case Western Reserve Univ, MD, 43; Am Bd Anesthesiol, dipl, 49. Prof Exp: Intern, Grasslands Hosp, Valhalla, NY, 43; resident anesthesia, Wis Gen Hosp, Madison, 46-48; instr anesthesiol, Univ Wis, 48-49; from asst prof to assoc prof, Univ Iowa, 49-54; prof, Univ Wash, 54-60, clin prof, 61-68; prof anaesthesia, Fac Med, Univ Toronto, 68-70; PROF ANESTHESIA & CHMN DEPT, MED COL OHIO, 70- Concurrent Pos: Mem traveling med fac, WHO & Unitarian Serv Comt, Israel & Iran, 51; mem subcomt anesthesia, Nat Res Coun, 56-61; dir anesthesia res labs, Providence Hosp, Seattle, 60-68, dir med educ & res, 65-68; chief anaesthetist, St Michael's Hosp Unit, Toronto, 68-70. Mem: Am Soc Anesthesiol; Am Soc Pharmacol & Exp Therapeut; Soc Exp Biol & Med; AMA; Asn Anaesthetists Gt Brit & Northern Ireland. Res: Cardiac conduction; placental transmission of drugs; anesthetic apparatus; fundamental neurophysiologic mechanism in anesthesia; carbon dioxide homeostasis; cardiac output; acid-base status with cardiopulmonary bypass and hypothermia; liver function with various anesthetic agents; medical education. Mailing Add: 3425 Bentley Blvd Toledo OH 43606

MORRIS, MANFORD D, b Kamiah, Idaho, Apr 18, 26; m 51; c 3. BIOCHEMISTRY. Educ: Univ San Francisco, BS, 49, MS, 51; Univ Calif, PhD(biochem), 58. Prof Exp: Asst res biochemist, Sch Med, Univ Calif, 58-61; asst prof, 61-65, assoc prof, 65-72, PROF BIOCHEM, SCH MED, UNIV ARK, LITTLE ROCK, 72- Concurrent Pos: Asst res biochemist, Clin Invest Ctr, US Naval Hosp, Oakland, 58-61. Res: Cholesterol metabolism; sterol methodology; primate lipid and lipoprotein metabolism; heritable diplipoproteinemia. Mailing Add: Dept of Biochem Univ of Ark Med Ctr Little Rock AR 72201

MORRIS, MARION CLYDE, b Akron, Ohio, Oct 14, 32; m 54; c 3. POLYMER CHEMISTRY. Educ: Univ Akron, BS, 54, MS, 60, PhD(polymer chem), 63. Prof Exp: Sr res chemist, 62-71, SECT HEAD PHYS CHEM, RES DIV, GOODYEAR TIRE & RUBBER CO, 71- Mem: Am Chem Soc. Res: Rubberlike elasticity; effect on physical properties due to blending, crystallization and molecular weight distribution; characterization of polymers by gel permeation chromatography and thermal analysis; rheology and processing properties. Mailing Add: Dept 455B Res Div Goodyear Tire & Rubber Co Akron OH 44316

MORRIS, MARK ROOT, b Aberdeen, Wash, Sept 2, 47. RADIO ASTRONOMY. Educ: Univ Calif, Riverside, BA, 69; Univ Chicago, PhD(physics), 75. Prof Exp: RES FEL RADIO ASTRON, OWENS VALLEY RADIO OBSERV, CALIF INST TECHNOL, 74- Mem: Am Astron Soc. Res: Spectral lines of molecular species in interstellar medium and surrounding young and old stars; spectral line radio astronomy. Mailing Add: Owens Valley Radio Observ Calif Inst of Technol Pasadena CA 91125

MORRIS, MARY ROSALIND, b Ruthin, Wales, May 8, 20; US citizen. PLANT CYTOGENETICS. Educ: Ont Agr Col, BSA, 42; Cornell Univ, PhD(plant breeding), 47. Prof Exp: Asst agron, 47-51, from asst prof to assoc prof, 51-58, PROF CYTOGENETICS, UNIV NEBR, LINCOLN, 58- Concurrent Pos: Univ Nebr Johnson fel, Calif Inst Technol, 49-50; Guggenheim fel Sweden & Eng, 56-57. Mem: AAAS; Genetics Soc Am; Crop Sci Soc Am; Genetics Soc Can; Am Inst Biol Sci. Res: Wheat cytogenetics; assignment of genes for important wheat characters to specific chromosomes by use of aneuploids and chromosome substitutions. Mailing Add: Dept of Agron Univ of Nebr Lincoln NE 68583

MORRIS, MAURICE F, b Blanca, Colo, Apr 29, 19; m 42; c 3. APPLIED PHYSICS. Educ: Adams State Col, AB, 53; Univ Colo, MA, 55, EdD(phys sci), 63. Prof Exp: From instr to assoc prof, 55-66, PROF PHYSICS, 66-75, CHMN DIV SCI & TECHNOL STUDIES, ADAMS STATE COL, 75- Mem: Fel AAAS; Am Asn Physics Teachers; Nat Sci Teachers Asn. Res: Improvement of demonstration apparatus for college physics use. Mailing Add: Chmn Div Sci & Technol Studies Adams State Col Alamosa, CO 81102

MORRIS, MELVIN L, b Cincinnati, Ohio, Mar 27, 29. INORGANIC CHEMISTRY. Educ: Ohio State Univ, BSc, 51, MSc, 55, PhD(chem), 58. Prof Exp: Fel, Northwestern Univ, 58-59 & Ohio State Univ, 59-60; asst prof chem, Tex Tech Col, 60-61; from asst prof to assoc prof, NDak State Univ, 63-68; educ sci adminr, NSF, 68-69; assoc prof chem, 69-74, PROF CHEM, N DAK STATE UNIV, 74- Concurrent Pos: Res chemist, Wright Patterson AFB, Ohio, 61. Mem: Am Chem Soc. Res: Mass spectrometry and synthesis of inorganic compounds; inorganic synthesis of beta- diketone complexes. Mailing Add: Dept of Chem NDak State Univ, Fargo ND 58102

MORRIS, MELVIN LEWIS, b New York, NY, Nov 28, 14; m 43; c 3. DENTISTRY. Educ: City Col New York, BS, 34; Columbia Univ, MA, 37, DDS, 41; Am Bd periodont, dipl, 51. Prof Exp: From instr to clin prof, 48-70, ADJ PROF DENT, SCH DENT & ORAL SURG, COLUMBIA UNIV, 70- Concurrent Pos: NIH res grant, 66-68; consult, Vet Admin Hosp, Castle Point, 53-56 & Franklin Delano Hosp, 53-59. Mem: Am Dent Asn; Int Asn Dent Res; Am Acad Periodont. Res: Experimental wound healing of periodontal tissues. Mailing Add: Columbia Univ Sch of Dent & Oral Surg 630 W 168th St New York NY 10032

MORRIS, MELVIN SOLOMON, b Denver, Colo; m 33; c 2. RANGE SCIENCE. Educ: Colo State Univ, BS, 30, MS, 32. Prof Exp: From instr to asst prof bot, Colo State Univ, 30-36; assoc prof range mgt, 36; from asst prof to prof, lectr, 36-72, EMER PROF FORESTRY, UNIV MONT, 72- Concurrent Pos: Spec State Univ NY

Col Forestry, 49, 50; collabr, US Forest Serv, 50-58 & Fish & Wildlife Serv, 50-62; dir bison site, US Grassland Biome, Int Biol Prog, 70-72; consult, rangeland ecol & range mgt, 72; comn mem, Environ Protection Comn, The Navajo Nation, Window Rock, Ariz, 74- Honors & Awards: Outstanding Achievement & Sci Award, Soc Range Mgt, 74. Mem: Fel AAAS; Soc Range Mgt(treas, 49, pres-elec, 65, pres, 66); Soil Conserv Soc Am; Am Soc Agron; Ecol Soc Am. Res: Range ecology; land classification; natural resources and ecology of big sagebrush in Montana; primary productivity; grazing land management; grassland ecology. Mailing Add: 211 Mary Ave Missoula MT 59801

MORRIS, MEREDITH JAMES, biostatistics, see 12th edition

MORRIS, MICHAEL D, b New York, NY, Mar 27, 39; m 61; c 4. ANALYTICAL CHEMISTRY. Educ: Reed Col, BA, 60; Harvard Univ, MA, 62, PhD(chem), 64. Prof Exp: Asst prof chem, Pa State Univ, 64-69; ASSOC PROF CHEM, UNIV MICH, ANN ARBOR, 69- Concurrent Pos: Res grants, USPHS, 66-68 & NSF, 66-72. Mem: AAAS; Am Chem Soc. Res: Applications of Raman spectroscopy and laser spectroscopy to analytical chemistry. Mailing Add: Dept of Chem Univ of Mich Ann Arbor MI 48104

MORRIS, MICHAEL I, systematic botany, see 12th edition

MORRIS, MURRELL PINCKNEY, organic chemistry, see 12th edition

MORRIS, N RONALD, b New York, NY, July 22, 33; m 57; c 1. PHARMACOLOGY. Educ: Yale Univ, BS, 55, MD, 59. Prof Exp: Asst prof pharmacol, Sch Med, Yale Univ, 63-67; asst prof, 67-68, ASSOC PROF PHARMACOL, RUTGERS MED SCH, COL MED & DENT NJ, 68- Mem: AAAS. Res: Desoxyribonucleic acid synthesis and methylation; effects of halogenated pyrimidines. Mailing Add: Dept Pharmacol Rutgers Med Sch Col of Med & Dent of NJ New Brunswick NJ 08903

MORRIS, OSWALD NATHANIEL, entomology, microbiology, see 12th edition

MORRIS, PETER ALAN, b Oakland, Calif, Oct 6, 45; m 69; c 1. OPERATIONS RESEARCH, SYSTEMS ANALYSIS. Educ: Univ Calif, Berkeley, BS, 68; Stanford Univ, MS, 70, PhD(eng, econ syst), 71. Prof Exp: Opers res analyst, Off Systs Anal, Dept Defense, 71-74, opers res mgr, Modeling & Anal Off, Manpower & Reserve Affairs, 74; RES SCIENTIST ANAL RES GROUP, PALO ALTO RES CTR, XEROX CORP, 74- Concurrent Pos: Assoc ed, Mgt Sci, 71-; consult asst prof eng-econ syst, Stanford Univ, 75- Mem: Inst Mgt Sci; Opers Res Soc Am; Inst Elec & Electronics Engrs. Res: Decision analysis; systems modeling; probabilistic models; use of experts. Mailing Add: Anal Res Group Xerox Res Ctr 3333 Coyote Hill Rd Palo Alto CA 94304

MORRIS, PETER CRAIG, b Kansas City, Mo, Sept 5, 37; m 60; c 3. MATHEMATICS. Educ: Univ Southern Ill, BA, 59; Univ Iowa, MS, 61; Okla State Univ, PhD(math), 67. Prof Exp: Asst prof math, State Col Iowa, 63-65; Belg-Am Educ Found fel, 67-68; asst prof math, Fla State Univ, 68-72; ASSOC PROF MATH & HEAD DEPT, SHEPHERD COL, 72- Mem: Am Math Soc; Math Asn Am; London Math Soc; Math Soc Belg; Math Soc France. Res: Homological algebra. Mailing Add: Dept of Math Shepherd Col Shepherdstown WV 25443

MORRIS, QUENTIN L, b Ridgeway, Tex, Oct 6, 18; m 42. ORGANIC CHEMISTRY. Educ: ETex State Teachers Col, BS, 38, MS, 49; Univ Okla, PhD(chem), 53. Prof Exp: Asst prof chem, ETex State Teachers Col, 51-53; assoc prof, Northwestern State Col, La, 53-54; asst prof, ETex State Teachers Col, 54-56; assoc prof, Northwestern State Col, 57-59; explor res chemist, Phillips Petrol Co, 56-57; res chemist, Rocketdyne Div, NAm Aviation, Inc, 59-67; SCI SPECIALIST, VOUGHT CORP, 67- Mem: Am Chem Soc. Res: Corrosion control. Mailing Add: Rte 4 Sulphur Springs TX 75482

MORRIS, RALPH DENNIS, b Humboldt, Sask, Feb 13, 40; m 63; c 2. ECOLOGY. Educ: Univ Sask, BSc, 61, PhD(ecol), 69; Univ Colo, Boulder, BEd, 63. Prof Exp: Nat Res Coun Can fel biol, McGill Univ, 69-70; asst prof, 70-73, ASSOC PROF BIOL, BROCK UNIV, 73- Mem: AAAS; Can Soc Zool; Can Soc Wildlife & Fishery Biol; Am Soc Mammalogists. Res: Population dynamics of colonial seabirds and small mammals. Mailing Add: Dept of Biosci Brock Univ St Catherines ON Can

MORRIS, RALPH WILLIAM, b Cleveland Heights, Ohio, July 30, 28; m 55; c 5. PHARMACOLOGY. Educ: Ohio Univ, BA, 50, MS, 53; Univ Iowa, PhD(pharmacol), 55. Prof Exp: Asst pharmacol, Univ Iowa, 52-53; instr, Col Med, 55-56, from asst prof to assoc prof, Col Pharm, 56-69, PROF PHARMACOG & PHARMACOL, COL PHARM, UNIV ILL, 69- Mem: AAAS; Drug Info Asn; Am Soc Pharmacol & Exp Therapeut; Am Pharmaceut Asn; Int Soc Chronobiol. Res: Chronopharmacology; drug abuse education; drug interactions; drug information retrieval; drug screening. Mailing Add: Dept of Pharmacog & Pharmacol Univ of Ill Col of Pharm Chicago IL 60612

MORRIS, RICHARD HERBERT, b Oakland, Calif, Nov 22, 28; m 56; c 2. ELECTROMAGNETICS. Educ: Univ Calif, Berkeley, AB, 50, PhD(nuclear physics), 57. Prof Exp: Asst physics, Univ Calif, Berkeley, 50-53, asst nuclear physics, Lawrence Radiation Lab, 53- 55; instr physics, Sacramento State Col, 56-57; from asst prof to assoc prof, 57-66, PROF PHYSICS, SAN DIEGO STATE COL, 66- Concurrent Pos: Consult, Naval Electronics Lab, 62-63. Mem: Am Asn Physics Teachers. Res: Theoretical physics; modern optics; teaching of physics. Mailing Add: Dept of Physics San Diego State Col San Diego CA 92182

MORRIS, RICHARD KNOWLES, b Somerville, Mass, Nov 16, 15; m 46. ANTHROPOLOGY. Educ: Trinity Col, Conn, AB, 40; Yale Univ, MA, 49, PhD(philos educ), 51. Prof Exp: Master, Loomis Inst, Conn, 42-44; teacher, Deep River High Sch, Conn, 45-46 & Chester High Sch, 46-47; mem exec staff, Conn Educ Asn, 47-49; instr educ, 51-52, from asst prof to prof educ & anthrop, 52-75, chmn dept educ, 71-74, EMER PROF EDUC, TRINITY COL, CONN, 75-, EDUC CONSULT, 74- Mem: AAAS; fel Philos Educ Soc; Humane Soc US. Res: History and philosophy of education; naval history; inventions in culture; scientific, philosophical and theological foundations for a humane ethic. Mailing Add: 120 Cherry Hill Dr Newington CT 06111

MORRIS, RITA, M L, b Boston, Mass. GEOGRAPHY. Educ: Teachers Col, City of Boston, BSE, 42, EdM, 43; Harvard Univ, EdD(geog), 62. Prof Exp: Teacher hist, Boston Pub Schs, 43-46; instr hist & psychol, Boston Teachers Col, 46-48; instr hist, Farmington Teacher Col, 48-51; instr hist & geog, Worcester State Col, 51-59, from asst prof to assoc prof geog, 59-65, PROF GEOG, WORCESTER STATE COL, 65- Concurrent Pos: NSF fel copper mining & steel mfg, Mich Col Mining & Technol, 63, fel water resources, NMex Univ, 65 & fel polit geog, Clark Univ, 67; US deleg, Int Geog Cong, New Delhi, 68 & Montreal, 72; NSF fel comput & syst anal, Rutgers Univ, 69; distinguished scholar, Rutgers Univ, 70. Mem: Asn Am Geogrs; Nat Coun Geog Educ. Res: Water resources, especially at the North Pole and Canadian high artic; geography as an academic discipline at Harvard from 1636 to 1920. Mailing Add: Dept of Geog & Geol Worcester State Col Worcester MA 01602

MORRIS, ROBERT CARTER, b Richmond, Va, Oct 3, 43; m 63; c 1. EXPERIMENTAL SOLID STATE PHYSICS. Educ: Hampden-Sydney Col, BS, 66; Univ Va, PhD(physics), 70. Prof Exp: Res assoc physics, Univ Va, 70-71, asst prof, 71-73; ASST PROF PHYSICS, FLA STATE UNIV, 73- Mem: Am Phys Soc. Res: Experimental studies of the normal and superconducting state properties of layer-structure, transition-metal dichalcogenide compounds and tungsten bronze compounds, both pure and doped with impurity atoms. Mailing Add: Dept of Physics Fla State Univ Tallahassee FL 32302

MORRIS, ROBERT CLARENCE, b Electra, Tex, Feb 19, 28; m 54; c 4. GEOLOGY. Educ: Tex Technol Col, BS, 52; Univ Wis, MS, 62, PhD(geol), 65. Prof Exp: Subsurface geologist, Int Petrol Co, Ltd, 52-56, explor geologist, 56-61; asst prof, 64-67, ASSOC PROF GEOL, NORTHERN ILL UNIV, 67- Concurrent Pos: NSF fel, Northern Ill Univ, 67-69. Mem: Am Asn Petrol Geologists; Soc Econ Paleontologists & Mineralogists; Geol Soc Am; Int Asn Sedimentol. Res: Stratigraphy and sedimentology of Carboniferous, Ouachita Mountains, Arkansas; petrology of cabonate banks; modern evaporite deposition; classification of disturbed bedding; sedimentary and tectonic history of northwest Peru. Mailing Add: Dept of Geol Northern Ill Univ De Kalb IL 60115

MORRIS, ROBERT FRANKLIN, ecology, see 12th edition

MORRIS, ROBERT GEMMILL, b Des Moines, Iowa, July 20, 29; m 55; c 3. SOLID STATE PHYSICS. Educ: Iowa State Univ, BS, 51, PhD(physics), 57; Calif Inst Technol, MS, 54. Prof Exp: Am-Swiss Found Sci exchange fel, Swiss Fed Inst Technol, 57-58; from asst prof to prof physics & head dept, SDak Sch Mines & Technol, 58-68; physicist, Physics Prog, Off Naval Res, Va, 68-72, actg dir, Electronics Prog, 72-73, dir, Electronics Prog, 73-74; DEP DIR, OFF TECHNOL POLICY & SPACE AFFAIRS, BUR OCEANS & INT ENVIRON & SCI AFFAIRS, US DEPT STATE, 74- Concurrent Pos: Vis prof, Swiss Fed Inst Technol, 63-64. Mem: Fel Am Phys Soc; Swiss Phys Soc. Res: Electrical, thermal and magnetic properties of semiconductor elements and compounds; technology transfer. Mailing Add: Off of Technol Policy & Space Affairs US Dept of State Washington DC 20520

MORRIS, ROBERT HAMILTON, b Lackawanna, NY, Aug 3, 21; m 48; c 3. GEOLOGY. Educ: Upsala Col, BA, 48; Rutgers Univ, MS, 53. Prof Exp: Instr geol, Upsala Col, 48-50; geologist, Navy Oil Unit, US Geol Surv, DC, 50-54, photogeol sect, Colo, 54-58; Int Coop Admin prof geomorphol & photogeol, Univ Rio Grande do Sul, Brazil, 58-60; geologist & photogeologist, Ky Br, 60-64, photgeologist, Spec Projs Br, 65-72, GEOLOGIST, NUCLEAR REACTOR SITE INVESTS, ENG GEOL BR, US GEOL SURV, 72- Mem: Asn Eng Geol; Am Asn Petrol Geologists; Am Soc Photogram; Geol Soc Am. Res: Sedimentation and heavy minerals; photogeology. Mailing Add: US Geol Surv Box 25046 Mail Stop 903 KAE Denver CO 80225

MORRIS, ROBERT JAMES, b New Auburn, Wis, Mar 10, 15; m 40; c 2. CHEMISTRY. Educ: Univ Idaho, BS, 36, MS, 38; Ohio State Univ, PhD(chem), 47. Prof Exp: Res worker, State Planning Bd, 36; asst chem, Univ Idaho, 36-38, instr, 38-41; asst, Ohio State Univ, 41-43, spec asst, 43-44, res assoc, 45-47; from asst prof to prof org chem & biochem, 47-74, PROF BIOCHEM, FLEISCHMANN COL AGR, UNIV NEV, RENO, 61-, BIOCHEMIST, 74- Concurrent Pos: Consult, AEC, 54-58; consult & res dir, Cosmetic Div, Sea & Ski Corp, 61-; assoc, Desert Res Inst, 61-66, dir anal lab, Water Resource Ctr, 66-67, head indust div, 67- Mem: Am Chem Soc. Res: Ultraviolet, visible and infrared absorption spectroscopy; synthetic organic chemistry and biochemistry; filter applications; chromatography and analytical methods; chemistry of alfalfa saponins; synthesis and isolation of natural products; analytical controls, air and water pollution studies. Mailing Add: Biochem Div Fleischmann Col Agr Univ of Nev Reno NV 89507

MORRIS, ROBERT LYLE, b Oelwein, Iowa, Dec 6, 14; m; c 1. CHEMISTRY. Educ: Univ Iowa, BS, 40, MS, 50, PhD, 59. Prof Exp: Res chemist, Nat Aluminate Corp, 40-48; from instr to asst prof hyg & prev med, 52-69, chief chemist, 50-65, ASSOC PROF PREV MED & ENVIRON HEALTH, UNIV IOWA, 69-, PRIN CHEMIST, STATE BACT LAB, 65-, ASSOC DIR, STATE HYG LAB, 67- Mem: Am Chem Soc; Am Water Works Asn; Am Pub Health Asn. Res: Instrumental analysis; environmental chemistry; corrosion control; industrial hygiene. Mailing Add: State Hyg Lab Univ of Iowa Iowa City IA 52240

MORRIS, ROBERT NICHOLAS, b Chicago, Ill, Jan 11, 37; m 60; c 7. PHARMACOLOGY. Educ: St Mary's Col, Minn, BA, 59; Loyola Univ Chicago, MS, 63, PhD(pharmacol), 66. Prof Exp: Asst prof pharmacol, Sch Pharm, Univ Wis, 66-71; assoc prof & chmn dept, Col Pharm, Univ Nebr, 71-72; head sect pharmacol, 72-74, HEAD PHARMACOL DEPT, ARNAR-STONE LABS, 74- Concurrent Pos: Vis assoc prof, Col Pharm, Univ Nebr, 72- Mem: AAAS. Res: Cardiovascular pharmacology; drug metabolism reproductive pharmacology. Mailing Add: Pharmacol Dept Arnar-Stone Labs 601 E Kensington Rd Mt Prospect IL 60056

MORRIS, ROBERT WHARTON, b Liberal, Mo, Aug 27, 20; m 45; c 2. ICHTHYOLOGY. Educ: Wichita State Univ, AB, 42; Ore State Col, MS, 48; Stanford Univ, PhD(biol), 54. Prof Exp: Biologist agr exp sta, Ore State Col, 48-49 & US Fish & Wildlife Serv, 51-55; from instr to assoc prof, 55-68, PROF BIOL, UNIV ORE, 68- Concurrent Pos: Guggenheim fel, 62-63. Res: Biology of fishes and lower vertebrates. Mailing Add: Dept of Biol Univ of Ore Eugene OR 97403

MORRIS, ROBERT WILLIAM, b Staten Island, NY, Sept 28, 41; m 68; c 2. GEOLOGY, INVERTEBRATE PALEONTOLOGY. Educ: Duke Univ, AB, 63; Columbia Univ, MA, 65, PhD(geol), 69. Prof Exp: Teaching asst geol, Columbia Univ, 63-66; asst instr, Rutgers Univ, 66-67; asst prof, 68-74, ASSOC PROF GEOL, WITTENBERG UNIV, 75- Mem: Sigma Xi; Paleont Soc; Geol Soc Am; Int Palaeont Union; Nat Asn Geol Teachers. Res: Paleoecology; micropaleontology. Mailing Add: Dept of Geol Wittenberg Univ Springfield OH 45501

MORRIS, ROGER EARL, organic chemistry, see 12th edition

MORRIS, ROSEMARY SHULL, b Los Angeles, Calif, Aug 11, 29. SCIENCE ADMINISTRATION, NUTRITION. Educ: Univ Calif, Berkeley, 50; Univ Southern Calif, BS, 53, MS, 56, PhD(biochem, nutrit), 59. Prof Exp: Res assoc biochem, Univ Southern Calif, 59; jr res biochemist, Univ Calif, Los Angeles, 59-61; res biochemist, Eastern Utilization Res & Develop Div, Agr Res Serv, USDA, Washington, DC, 61-66; res chemist, Div Nutrit, US Food & Drug Admin, 66-67 & Human Nutrit Res Div, Agr Res Serv, USDA, Md, 67-72; health scientist adminr, Nat Heart & Lung Inst, NIH, 72-75, ASST CHIEF RES, REFERRAL BR DIV RES GRANTS, NIH, 75- Mem: Am Inst Nutrit. Res: Lipid and cholesterol metabolism; vitamin E; essential fatty acids. Mailing Add: Div of Res Grants Nat Inst Health Bethesda MD 20014

MORRIS, ROY OWEN, b Kingston-on-Thames, Eng, May 24, 34; m 63; c 2. BIOCHEMISTRY. Educ: Univ London, BSc, 55, PhD(chem), 59. Prof Exp: Asst lectr chem & biochem, St Thomas Hosp Med Sch, London, Eng, 59-61; res assoc biochem, Sci Res Inst, 61-64, asst prof agr chem, 64-70, ASSOC PROF AGR CHEM, ORE STATE UNIV, 70-, ASSOC PROF CHEM, 74- Mem: AAAS; Am Soc Plant Physiol. Res: Biochemistry of plant development; protein and nucleic acid biosynthesis. Mailing Add: Dept of Agr Chem Ore State Univ Corvallis OR 97331

MORRIS, RUPERT CLARKE, b Hattiesburg, Miss, July 15, 11; m 47. ORGANIC CHEMISTRY. Educ: Miss State Col, BS, 32; Univ Ill, PhD(chem), 36. Prof Exp: Res asst org chem, Univ Ill, 36-38; res supvr, 38-73, DEPT MGR CHEM RES & APPL, SHELL DEVELOP CO, 73- Mem: AAAS; Am Chem Soc. Res: Structure of gossypol; purification and separation of unsaturated hydrocarbons; chemistry of 3-sulfolene; synthetic lubricants; chemistry of organophosphorus compounds; process research; applications. Mailing Add: 1002 Hollow Tree Ave La Porte TX 77571

MORRIS, SIDNEY MACHEN, JR, b Tyler, Tex, Mar 31, 46. BIOLOGICAL CHEMISTRY. Educ: Univ Tex, Austin, BS, 68; Univ Calif, Berkeley, PhD(biochem), 75. Prof Exp: NIH FEL PHYSIOL CHEM, UNIV WIS MED CTR, 75- Res: Molecular processes involved in eucaryotic gene regulation and development; amphibian metamorphosis; hormone action. Mailing Add: Dept of Physiol Chem Univ of Wis Med Ctr Madison WI 53706

MORRIS, STANLEY P, b Montreal, Que, Nov 23, 37; m 63; c 4. THEORETICAL SOLID STATE PHYSICS. Educ: McGill Univ, BSc, 58, PhD(physics), 64. Prof Exp: Lectr physics, Loyola Col, Can, 63-64; asst prof, Sir George Williams Univ, 64-69; ASSOC PROF PHYSICS, CONCORDIA UNIV, 69-, CHMN DEPT, 74- Mem: Am Phys Soc; Am Asn Physics Teachers; Can Asn Physicists. Res: Electron interactions in the presence of a uniform magnetic field; Bloch electrons in a magnetic field. Mailing Add: Dept of Physics Concordia Univ SGW Campus 1455 de Maisonneuve Montreal QB Can

MORRIS, STEPHEN CHARLES, astronomy, see 12th edition

MORRIS, THOMAS WENDELL, b Emory, Ga, Jan 31, 30; m 54; c 2. PHYSICS. Educ: Duke Univ, BS, 51; Yale Univ, MS, 53, PhD(physics), 55. Prof Exp: PHYSICIST, BROOKHAVEN NAT LAB, 55- Concurrent Pos: Vis physicist, Saclay Nuclear Res Ctr, France, 59-60. Mem: Am Phys Soc. Res: Particle physics; computer applications; nuclear instrumentation. Mailing Add: 20 Pennsylvania St Upton NY 11973

MORRIS, WILFORD ERNST, b Hamilton, Ohio, Mar 7, 14; m 41; c 4. PHYSICS. Educ: Miami Univ, AB, 36; Univ Mich, PhD(physics), 42. Prof Exp: Res physicist, Goodyear Tire & Rubber Co, Ohio, 41-50; chief, Air-Ground Explosions Div, Naval Ord Lab, Md, 50-58; physicist, Advan Systs Studies, Radio Corp Am, 58-62; CHIEF SCIENTIST, ELEC COMPONENTS DIV, BENDIX CORP, 62- Honors & Awards: Civilian Meritorious Award, US Navy, 56. Mem: Am Phys Soc. Res: Physical effects of explosions; aerodynamics and fluid dynamics; supersonics; Geiger-Muller tubes; radioactivity; nuclear physics; physical properties of rubber; electromagnetic wave propagation; ignition systems; electronics. Mailing Add: Bendix Corp Eng & Electron Comp Div Sidney NY 13838

MORRIS, WILLIAM COLLINS, b Port Gibson, Miss, Sept 23, 09; m 35; c 2. CHEMISTRY. Educ: Washington & Lee Univ, BS, 31; Western Reserve Univ, MA, 33, PhD(chem), 35. Prof Exp: Res chemist, Ferro Enamel Corp, Ohio, 35-37 & Chandler Chem Co, 37-38; res dir, Promat Div, Poor & Co, Ill, 38-40; res chemist, Harshaw Chem Co, 40-44; res assoc, Fla Eng Exp Sta, 44-45; res chemist, 45-58, RES DIR, GLASS ENAMEL DEPT, HARSHAW CHEM CO, 58- Mem: Am Chem Soc; Am Ceramic Soc. Res: Glass enamel and related ceramic fields; inorganic fluorides; detergents; corrosion control. Mailing Add: 17113 Sunset Dr Chagrin Falls OH 44022

MORRIS, WILLIAM JOSEPH, b Baltimore, Md, Oct 14, 23; m 45; c 2. GEOLOGY. Educ: Syracuse Univ, BA, 48; Princeton Univ, MA & PhD(geol), 51. Prof Exp: Prof geol, Agr & Mech Col, Tex, 51-55; PROF GEOL, OCCIDENTAL COL, 55-, CHMN DEPT, 71- Concurrent Pos: Res assoc, Mus Natural Hist, Los Angeles. Honors & Awards: Arnold Guyot Mem Award, Nat Geog Soc, 68. Mem: Fel Geol Soc Am; Soc Vert Paleont; Am Asn Petrol Geologists; Soc Study Evolution. Res: Vertebrate paleontology; sedimentary petrology; invertebrate paleontology. Mailing Add: Dept of Geol Occidental Col 1600 Campus Rd Los Angeles CA 90041

MORRIS, WILLIAM LEWIS, b Hamilton, Ohio, Aug 19, 31; m 56; c 7. MATHEMATICS. Educ: Univ Cincinnati, AB, 58, AM, 60; Univ Tenn, PhD(math), 67. Prof Exp: Engr, Honeywell, Inc, 60-61; sr engr, Gen Dynamics, Inc, 61-62; consult comput, Aerospace Corp, 63-65; from instr to asst prof, Univ Tenn, 62-66; res mathematician, Oak Ridge Nat Lab, 66-68; asst prof math, 68-71, ASSOC PROF MATH, UNIV HOUSTON, 71- Mem: Am Math Soc. Res: Numerical analysis; numerical solutions of functional equations; number theory. Mailing Add: Rte 6 Box 66D Center TX 75935

MORRISETT, JOEL DAVID, b Winston-Salem, NC, May 2, 42; m 67; c 2. BIOPHYSICS, BIOCHEMISTRY. Educ: Davidson Col, BS, 64; Univ NC, Chapel Hill, PhD(org chem), 68. Prof Exp: NIH fel biophys, Stanford Univ, 70-71; ASST PROF EXP MED, BAYLOR COL MED, 71- Concurrent Pos: Estab investr, Am Heart Asn, 74-79. Mem: Am Heart Asn; Biophys Soc; AAAS; Fedn Am Soc Exp Biol; NY Acad Sci. Res: Correlation and determination of structure and function of proteins, particularly enzymes and lipoproteins. Mailing Add: Div Atherosci & Lipoprotein Res Baylor Col of Med Houston TX 77025

MORRISETT, PETER EDWARD, analytical chemistry, see 12th edition

MORRISH, ALLAN HENRY, b Winnipeg, Man, Apr 18, 21; m 52; c 2. PHYSICS. Educ: Univ Manitoba, BSc, 43; Univ Toronto, MA, 46; Univ Chicago, PhD(physics), 49. Prof Exp: Asst physics, Univ Chicago, 48-49; lectr, Univ BC, 49-50, res assoc 51-52; physicist, McGill Univ, 52-53; from res assoc to prof elec eng, Univ n, Minneapolis, 53-64; PROF PHYSICS, UNIV MAN, 64-, HEAD DEPT, 66- Concurrent Pos: Nat Res Coun Can fel, 50-51; Guggenheim fel, 57-58; pres, Can Asn Physicists, 74-75. Mem: Fel Am Phys Soc; Royal Soc Can; fel Brit Inst Physics; Can Asn Physicists. Res: Particle accelerators; elementary and small magnetic particles; nuclear emulsions; ferromagnetism; magnetic resonance; low temperatures; Mössbauer effect in magnetic materials. Mailing Add: Dept of Physics Univ of Man Winnipeg MB Can

MORRISON, ADRIAN RUSSEL, b Philadelphia, Pa, Nov 5, 35; m 58; c 5. NEUROANATOMY, NEUROPHYSIOLOGY. Educ: Cornell Univ, DVM, 60, MS, 62; Univ Pa, PhD(anat), 64. Prof Exp: NIH spec fel neurophysiol, Univ Pisa, 64-65; asst prof, 66-70, ASSOC PROF ANAT, SCH VET MED, UNIV PA, 70- Mem: Am Vet Med Asn; Am Asn Vet Anat; Am Asn Anat; Asn Psychophysiol Study Sleep. Res: Neuroanatomical and neurophysiological bases of mammalian behavior. Mailing Add: Dept of Animal Biol Sch Vet Med Univ of Pa Philadelphia PA 19104

MORRISON, ALEXANDER BAILLIE, b Edmonton, Alta, Dec 22, 30; m 51; c 6. TOXICOLOGY, NUTRITION. Educ: Univ Alta, BSc, 51, MSc, 52; Cornell Univ, PhD(nutrit), 56; Univ Mich, MS, 66; Am Bd Nutrit, dipl. Prof Exp: From chemist to sr chemist, Mead Johnson & Co, Ind, 56-58, group leader, 58-59; chemist, Vitamins & Nutrit Sect, Food & Drug Directorate, 59-63, chief nutrit div, 63-66, chief pharmacol div, 66-68, dir res labs, 68-69, from dep dir-gen to dir-gen, 69-71, asst dep minister, Food & Drug Br, 71, ASST DEP MINISTER, HEALTH PROTECTION BR, DEPT NAT HEALTH & WELFARE, 71- Concurrent Pos: Vis prof, Univ Toronto, 72. Honors & Awards: Borden Award, Nutrit Soc Can, 63. Mem: Pharmacol Soc Can; Am Inst Nutrit; Nutrit Soc Can (treas, 61-65, pres, 71-72). Res: Protein nutrition; evaluation of protein quality; toxic substances in foods; B-vitamins; effects of pharmaceutical dosage form on physiological availability of vitamins and drugs. Mailing Add: Health Protection Br Dept of Nat Hlth & Welf Tunney's Pasture Ottawa ON Can

MORRISON, ASHTON BYROM, b Belfast, Ireland, Oct 13, 22; m 50; c 1. PATHOLOGY. Educ: Queens Univ Belfast, MD, 46, PhD, 50; Duke Univ, MD, 46. Prof Exp: Asst lectr biochem, Queens Univ Belfast, 47-50 & anat, 50-51; mem sci staff exp med, Univ Cambridge, 52-55; assoc path, Duke Univ, 55-58; asst prof, Sch Med, Univ Pa, 58-61; assoc prof, Sch Med, Univ Rochester, 61-65; PROF PATH & CHMN DEPT, RUTGERS MED SCH, 65- Concurrent Pos: Markle scholar, 56-61. Mem: Am Soc Exp Path; Am Fedn Clin Res; Am Physiol Soc; Soc Exp Biol & Med; Brit Biochem Soc. Res: Experimental chronic renal insufficiency; experimental nephropathies. Mailing Add: Dept of Path Rutgers Med Sch Piscataway NJ 08854

MORRISON, BARBARA ANN, b Providence, RI; m 55. APPLIED MATHEMATICS. Educ: Brown Univ, AB, 44, ScM, 53, PhD(appl math), 56. Prof Exp: Mem tech staff, Bell Labs, NJ, 56-60; from asst prof to assoc prof & chmn dept, 64-73, PROF MATH, COL ST ELIZABETH, NJ, 73- Concurrent Pos: Mem, Edwin Aldrin Adv Panel, 71-73; vis fel, Dept Statist, Princeton Univ, 74-75. Mem: Am Math Soc; Math Asn Am; Sigma Xi. Mailing Add: 28 Ashwood Rd New Providence NJ 07974

MORRISON, CHARLES FREEMAN, JR, b Yakima, Wash, Sept 24, 29; m 52; c 4. ANALYTICAL CHEMISTRY, MEDICAL TECHNOLOGY. Educ: Univ Puget Sound, BS, 53; Mass Inst Technol, PhD(anal chem), 57. Prof Exp: From instr to asst prof anal chem, Wash State Univ, 57-62; res scientist, Granville-Phillips Co, Colo, 62-70; sr scientist, Universal Instruments Corp, 70-71; eng-opers mgr, 71-75, CHIEF RES, VALLEYLAB, INC, 75- Concurrent Pos: NSF fel, 61-63. Mem: Am Chem Soc; Am Vacuum Soc; Am Asn Advan Med Instrumentation. Res: Instrumental methods of analysis; analytical instruments; vacuum instrumentation; low pressure measurement methods; calibration techniques; biomedical instrumentation; electrosurgical mechanisms and methods. Mailing Add: 4790 Sioux Dr Boulder CO 80303

MORRISON, CLYDE ARTHUR, b Lachine, Mich, July 17, 26; m 52; c 4. PHYSICS, ELECTRICAL ENGINEERING. Educ: Mich State Col, BS, 50, MS, 51; Univ Mich, PhD(nuclear sci), 68. Prof Exp: Res assoc comput, Willow Run Res Ctr, Univ Mich, 51-53; physicist, Nat Bur Standards, 53-54; PHYSICIST, HARRY DIAMOND LABS, 54- Res: Interaction of radiation and solids with particular application of angular momentum interactions. Mailing Add: Harry Diamond Labs Washington DC 20438

MORRISON, COHN L, b Cloverdale, Ind, July 20, 09; m 41; c 1. PHYSICS. Educ: DePauw Univ, AB, 31; Ind Univ, MA, 36; Univ Cincinnati, PhD(physics), 51. Prof Exp: Dir physics, Dept Res & Develop, US Air Force Aerial Navig, 40-45, asst chief mech br, Aeronaut Res Lab, 51-56; chief reactor eng, US Air Force Nuclear Eng Test Facil, 56-60; head physics dept, Alexandria Div, Am Mach & Foundry Co, 60-63, ASST GEN MGR & TECH DIR, 63-71, AMF, INC, 71- Concurrent Pos: Assoc prof physics, Northern Va Community Col, 72- Mem: AAAS; Inst Elec & Electronics Engr; Am Inst Aeronaut & Astronaut. Res: Reactor physics and system dynamics. Mailing Add: 1225 Tudor Place Alexandria VA 22307

MORRISON, DAVID CAMPBELL, b Stoneham, Mass, Sept 1, 41; m 66. IMMUNOLOGY. Educ: Univ Mass, BS, 63; Yale Univ, MS, 66, PhD(molecular biol), 69. Prof Exp: Fel, Lab Biochem Pharm, Nat Inst Allergy & Infectious Dis, NIH, 69-71; fel, Dept Exp Path, 71-74, asst, 74-75, ASSOC, DEPT IMMUNOPATH, SCRIPPS CLIN & RES FOUND, 75- Concurrent Pos: Res career develop award, Nat Inst Allergy & Infectious Dis, NIH, 75. Mem: Am Asn Immunol; Am Soc Exp Path; Am Soc Microbiol. Res: Interaction of bacterial lipopolysaccharides with cellular and humoral mediation systems; mechanisms of activation of peritoneal mast cells by various stimuli. Mailing Add: Dept of Immunopath Scripps Clin & Res Found La Jolla CA 92037

MORRISON, DAVID DOUGLAS, b Danville, Ill, June 26, 40; m 66. ASTRONOMY, PLANETARY SCIENCE. Educ: Univ Ill, Urbana, BA, 62; Harvard Univ, AM, 64, PhD(astron), 69. Prof Exp: Res assoc astron, Cornell Univ, 68-69; asst astronr, 69-74, ASSOC ASTRONR, INST ASTRON, UNIV HAWAII, 74- Concurrent Pos: Co-investr, Mariner Venus/Mercury Mission, infrared radiometer, 71-75; vis assoc, Calif Inst Technol, 72; vis assoc prof, Lunar & Planetary Lab, Univ Ariz, 75-76; mem, NASA Mariner Jupiter Orbiter Sci Working Group, 75-76; vis scientist, Kitt Peak Nat Observ, 76. Mem: Int Astron Union; Am Astron Soc (secy-treas, 71-77); AAAS. Res: Planetary surfaces and atmospheres; exploration of the planets by spacecraft; infrared astronomy and infrared observational techniques; radio astronomy. Mailing Add: Inst for Astron Univ of Hawaii Honolulu HI 96822

MORRISON, DAVID LEE, b Butler, Pa, Jan 25, 33; m 54; c 2. NUCLEAR CHEMISTRY, PHYSICAL CHEMISTRY. Educ: Grove City Col, BS, 54; Carnegie Inst Technol, MS, 60, PhD(chem), 61. Prof Exp: Chemist, Callery Chem Co, 54; sr chemist, Inst, 61-65, from assoc chief to chief chem physics div, 65-70, mgr environ systs & processes sect, 70-74, mgr energy & environ prog off, 74-75; DIR PROG DEVELOP & MGT, BATTELLE MEM INST, 75- Mem: AAAS; Am Chem Soc; Am Nuclear Soc. Res: Energy and environmental research; technology and environmental impact assessment; research and development planning; nuclear reactor safety analysis; radiochemistry; research management. Mailing Add: Battelle Mem Inst 505 King Ave Columbus OH 43201

MORRISON, DONALD ALLEN, b Mt Forest, Ont, July 20, 36; US citizen; m 69; c 1. PETROLOGY, GEOLOGY. Educ: State Univ NY Buffalo, BS, 62; Univ Alaska, MS, 64; Univ Idaho, PhD(geol), 68. Prof Exp: Aerospace scientist, Manned Spacecraft Ctr, NASA, 68-74; ASST PROF BIOL, UNIV ILL CHICAGO CIRCLE, 74- Mem: AAAS; Geol Soc Am. Res: Lunar geology; petrology of lunar rocks; meteorite and micrometeorite studies as related to lunar surface processes. Mailing Add: Dept of Biol Sci Univ of Ill at Chicago Circle Chicago IL 60680

MORRISON, DONALD FRANKLIN, b Stoneham, Mass, Feb 10, 31. MATHEMATICAL STATISTICS. Educ: Boston Univ, BS, 53, AM, 54; Univ NC,

MS, 57; Va Polytech Inst, PhD, 60. Prof Exp: Asst math, Boston Univ, 53-54 & Univ NC, 54-56; res math statistician, Biomet Br, NIMH, 56-63; PROF STATIST, WHARTON SCH, UNIV PA, 63- Concurrent Pos: Instr, Found Advan Educ in Sci, 60-63; mem staff, Lincoln Lab, Mass Inst Technol, 56, consult, 56-57; div comput sci, NIH, 63-65; mem tech staff, Bell Tel Labs, NJ, 67; ed, Am Statistician, 72-75; assoc ed, Biometrics, 72-75. Mem: Fel Am Statist Asn; Fel Inst Math Statist; Biomet Soc; Psychomet Soc: Royal Statist Soc. Res: Statistical theory and methodology; multivariate analysis. Mailing Add: Dept of Statist The Wharton Sch Univ of Pa Philadelphia PA 19174

MORRISON, DONALD ROSS, b Tacoma, Wash, May 3, 22; m 43; c 3. MATHEMATICS, COMPUTER SCIENCE. Educ: Northern Ill State Teachers Col, BE, 42; Univ Wis, PhM, 46, PhD(math), 50. Prof Exp: Instr math, Univ Wis, 50; asst prof, Tulane Univ, 50-55; from mem staff to mgr, Sandia Corp, 55-71; PROF MATH & COMPUT SCI, UNIV N MEX, 71- Concurrent Pos: Consult, Los Alamos Sci Lab, Univ Calif, 71- Mem: Am Math Soc; Math Asn Am; Asn Comput Mach. Res: Pattern recognition; information retrieval; graph theory; abstract algebra. Mailing Add: 712 Laguayra Dr NE Albuquerque NM 87108

MORRISON, EDWARD JOSEPH, b Bridgeport, Conn, Apr 18, 27; m 56; c 6. TERATOLOGY, ANATOMY. Educ: Univ Bridgeport, BA, 50, MS, 59; Boston Univ, AM, 63; Univ Fla, PhD(anat), 68. Prof Exp: Teacher high sch, Conn, 58-59, sci coordr, 59-60; asst prof biol, Dutchess Community Col, 64-66; ASST PROF ANAT, MEHARRY MED COL, 68- Mem: AAAS. Res: Teratological effects of antisera on chick embryos. Mailing Add: Dept of Anat Meharry Med Col Nashville TN 37208

MORRISON, EDWARD RANZ, organic chemistry, see 12th edition

MORRISON, ESTON ODELL, b Sabinal, Tex, Sept 18, 32; m 58; c 3. ENTOMOLOGY, PARASITOLOGY. Educ: Tex Col Arts & Indust, BS, 57; Tex A&M Univ, MS, 60, PhD(entom), 63. Prof Exp: Instr biol, Tex Col Arts & Indust, 60-61; asst prof, Lamar State Col, 63-66; assoc prof, 66-68, PROF BIOL, TARLETON STATE COL, 68- Concurrent Pos: Lamar Res Ctr grant, 65-66. Mem: Am Soc Parasitol. Res: Lung flukes of salientia; helminthology. Mailing Add: Box 219 Tarleton Sta Stephenville TX 76401

MORRISON, FRANK ORVILLE, b Wetaskiwin, Alta, May 23, 10; m 39; c 2. ENTOMOLOGY, TOXICOLOGY. Educ: Univ Alta, BA, 34, BScAgr, 36, MSc, 38; McGill Univ, PhD(entom), 39. Prof Exp: Teacher pub sch, Alta, 30-32; asst entom, Univ Alta, 36-37 & Macdonald Col, McGill Univ, 37-39; fieldman, Rocky Mountain spotted fever surv, Dept Health, Alta, 39; assoc prof econ entom, 49-62, PROF ECON ENTOM, MACDONALD COL, MCGILL UNIV, 62- Concurrent Pos: Chmn dept entom, Macdonald Col, McGill Univ, 73-75. Mem: Fel AAAS; Am Entom Soc; Entom Soc Can (pres, 66-67); fel Agr Inst Can; Can Zool Soc. Res: Control of agricultural and household pests; biology of economically important pest species and their toxicology. Mailing Add: Dept of Entom Macdonald Col Quebec PQ Can

MORRISON, GARRETT LOUIS, b Port Jefferson, NY, Jan 18, 41; m 64; c 3. GEOCEHMISTRY. Educ: Univ Maine, BA, 64; Univ SC, MS, 68; Univ Okla, PhD(geol), 73. Prof Exp: RES GEOLOGIST, KANS DEPT TRANSP, 72- Mem: Asn Eng Geologists (treas, 75-); Geol Soc Am; Clay Mineral Soc; AAAS. Res: Electrochemical modification of concrete-polymer concretes; clay mineralogy; material analysis. Mailing Add: Res & Mat Lab 2300 Van Buren Topeka KS 66611

MORRISON, GEORGE HAROLD, b New York, NY, Aug 24, 21; m 52; c 3. ANALYTICAL CHEMISTRY. Educ: Brooklyn Col, BA, 42; Princeton Univ, MA & PhD, 48. Prof Exp: Instr chem, Rutgers Univ, 48-50; head inorg & anal chem, Gen Tel & Electronics Labs, 51-61; PROF CHEM, CORNELL UNIV, 61- Concurrent Pos: Res chemist, US AEC, 49-51; mem chem adv panel, NSF, 62-65; chmn comt anal chem, Nat Acad Sci, Nat Res Coun, 66-; NSF sr fel, Univ Calif, San Diego, 67-68; Guggenheim fel, Univ Paris, Orsay, 74-75. Honors & Awards: Anal Chem Award, Am Chem Soc, 71. Mem: Fel AAAS; Am Chem Soc; Soc Appl Spectros. Res: Ion microprobe; microscopy; mass spectroscopy; radiochemistry; atomic spectroscopy; trace and microanalysis. Mailing Add: Dept of Chem Cornell Univ Ithaca NY 14853

MORRISON, GEORGE ROBERT, b Camden, NJ, Dec 2, 25; m 53; c 3. BIOCHEMISTRY, INTERNAL MEDICINE. Educ: Holy Cross Col, BNSc, 46; Univ Rochester, AB, 48, MD, 54; Am Bd Internal Med, dipl, 62. Prof Exp: Asst biochem, Univ Rochester, 48-50; intern med, Ward Med Serv, Barnes Hosp, 54-55, asst resident med, 59-61; from instr to asst prof prev med, 61-67, ASSOC PROF MED & PREV MED, WASH UNIV, 67- Concurrent Pos: Res fel biochem, Dept Prev Med, Wash Univ, 57-59; Markel scholar med sci, 62-67. Res: Biochemical study of liver and its diseases, including microchemical analyses of different parts of liver lobule. Mailing Add: Dept of Prev Med Wash Univ Sch of Med St Louis MO 63110

MORRISON, GEORGE SAMUEL, biochemistry, see 12th edition

MORRISON, GLENN C, b New Haven, Conn, Mar 24, 33; m 55; c 4. ORGANIC CHEMISTRY. Educ: Brown Univ, ScB, 54; Univ Rochester, PhD(org chem), 58. Prof Exp: Res chemist org chem, Am Cyanamid Co, 57-70; sr scientist, 60-70, SR RES ASSOC ORG CHEM, WARNER-LAMBERT RES INST, 70- Mem: Am Chem Soc. Res: Synthetic organic medicinals. Mailing Add: Warner-Lambert Res Inst 170 Tabor Rd Morris Plains NJ 07950

MORRISON, HARRY, b New York, NY, Apr 25, 37; m 58; c 3. ORGANIC CHEMISTRY. Educ: Brandeis Univ, BS, 57; Harvard Univ, PhD(org chem), 61. Prof Exp: NSF-NATO fel, Swiss Fed Inst Technol, 61-62; res fel org chem, Univ Wis, 62-63; asst prof chem, 63-69, ASSOC PROF CHEM, PURDUE UNIV, 70- Concurrent Pos: Mem bd fels, Brandeis Univ, 65-; vis scientist, Weizman Inst, Rehovot, Israel, 72; consult, Sun Chem Corp, 72- Mem: Am Soc Photobiol; Am Chem Soc; The Chem Soc. Res: Organic photochemistry; organic reaction mechanisms. Mailing Add: Dept of Chem Purdue Univ West Lafayette IN 47907

MORRISON, HUGH MACGREGOR, b Liverpool, Eng, Aug 28, 36; m 63; c 3. PHYSICS. Educ: Univ Edinburgh, BSc, 59, PhD(physics), 65. Prof Exp: Asst lectr physics, Univ Edinburgh, 62-65; asst prof, 65-70, ASSOC PROF PHYSICS, UNIV WATERLOO, 70- Mem: Brit Inst Physics. Res: Dislocation enhanced diffusion at relatively low temperatures; diffusion in metals. Mailing Add: Dept of Physics Univ of Waterloo Waterloo ON Can

MORRISON, HUNTLY FRANK, b Montreal, Que, May 16, 38; m 70; c 2. GEOPHYSICS. Educ: McGill Univ, BSc, 59, MSc, 61; Univ Calif, Berkeley, PhD(eng geosci), 67. Prof Exp: Asst prof, 67-70, ASSOC PROF GEOPHYS ENG, UNIV CALIF, BERKELEY, 70- Mem: AAAS; Am Geophys Union; Soc Explor Geophys; Europ Asn Explor Geophys; Archaeol Inst Am. Res: Applied geophysics; electromagnetic and electrical prospecting methods. Mailing Add: Dept of Mat Sci & Eng Univ of Calif Berkeley CA 94720

MORRISON, JACK WILLIAM, b Alta, Can, Aug 8, 22; m 43; c 2. CYTOGENETICS. Educ: Univ Alta, BSc, 50, MSc, 51; Univ London, PhD(bot, cytol), 53. Prof Exp: Cytogeneticist, Cereal Crops Div, Cent Exp Farm, 53-60, supt exp farm, 60-65, RES COORDR, RES BR, CAN DEPT AGR, 65- Mem: Agr Inst Can. Res: Cytogenetics in cereals; research management. Mailing Add: Res Br Can Dept of Agr Ottawa ON Can

MORRISON, JAMES ALEXANDER, b Medicine Hat, Alta, June 11, 18: m 43; c 5. PHYSICAL CHEMISTRY. Educ: Univ Alta, BSc, 40, MSc, 41; McGill Univ, PhD(phys chem), 43. Prof Exp: Chemist, Dept Nat Defense, 43-45; Nat Res Coun fel, Pa State Col, 45-46; from asst res chemist to prin res chemist, Nat Res Coun Can, 46-66, dir div pure chem, 66-69; PROF CHEM & DIR INST MAT RES MCMASTER UNIV, 69- Mem: Am Phys Soc; Am Chem Soc; The Chem Soc; Chem Inst Can. Res: Phase equilibria; solutions of high polymers; surface chemistry; adsorption; low temperature calorimetry; imperfections in solids; thermodynamics; solid state. Mailing Add: Inst for Mat Res McMaster Univ Hamilton ON Can

MORRISON, JAMES DANIEL, b Bryn Mawr, Pa, Mar 28, 36; m 58; c 3. ORGANIC CHEMISTRY. Educ: Franklin & Marshall Col, BS, 58; Northwestern Univ, PhD(org chem), 63. Prof Exp: From teaching asst to teaching assoc gen chem, Northwestern Univ, 58-62; NSF fel, Stanford Univ, 62-63; asst prof org chem, Wake Forest Col, 63-65; from asst prof to assoc prof, 65-72, PROF ORG CHEM, UNIV NH, 72- Concurrent Pos: NSF sci fac fel, Univ NC, 71-72. Res: Asymmetric organic reactions; chiral organometallics; homogeneous hydrogenation; epoxides; Grignard reagents; novel peptides. Mailing Add: Dept of Chem Parsons Hall Univ of NH Durham NH 03824

MORRISON, JAMES GEORGE, organic chemistry, see 12th edition

MORRISON, JAMES LESLIE, b Pittsburgh, Pa, June 27, 34; m 58; c 2. THEORETICAL PHYSICS. Educ: Carnegie Inst Technol, BS, 56; Cornell Univ, PhD(physics), 64. Prof Exp: Asst prof, 64-70, SR RES ASSOC PHYSICS, UNIV UTAH, 70- Mem: Am Phys Soc. Res: High energy cosmic ray particle physics. Mailing Add: Dept of Physics Univ of Utah Salt Lake City UT 84112

MORRISON, JOHN AGNEW, b Ridgefield Park, NJ, Mar 13, 32; m 54; c 3. CHEMISTRY. Educ: Fairleigh Dickinson Univ, BS, 59; Rutgers Univ, PhD(org chem), 70. Prof Exp: From chemist to res chemist, 57-73, GROUP LEADER, LEDERLE LABS, AM CYANAMID CO, 73- Concurrent Pos: Lectr, Fairleigh Dickinson Univ, 60- Mem: AAAS; Am Chem Soc; Can Asn Res Toxicol. Res: Pharmacokinetics; drug metabolism. Mailing Add: Lederle Lags Am Cyanamid Co Pearl River NY 10965

MORRISON, JOHN ALBERT, b Wichita, Kans, Dec 1, 24; m 50; c 3. WILDLIFE MANAGEMENT, ECOLOGY. Educ: Mont State Univ, BS, 55, MS, 57; Wash State Univ, PhD(zool), 65. Prof Exp: Wildlife biologist, Idaho Fish & Game Dept, 57-61; res asst zool, Wash State Univ, 61-65; res biologist, lab perinatal physiol & chief sect primate ecol, US Dept Health, Educ & Welfare, PR, 65-67; leader, Okla Coop Wildlife Res Unit, Okla State Univ , 67-75; TERRESTRIAL ECOLOGIST, WESTERN ENERGY & LAND USE TEAM, US FISH & WILDLIFE SERV, 75- Mem: Wildlife Soc; Am Soc Mammalogists; Ecol Soc Am; Wildlife Dis Asn. Res: Behavior, reproductive and nutritional physiology and general ecology of birds and mammals; environmental protection and reclamation of disturbed energy development locations. Mailing Add: US Fish & Wildlife Serv Rm 208 Federal Bldg Ft Collins CO 80521

MORRISON, JOHN ALLAN, b Beckenham, Eng, June 10. 27; nat US; m 55. APPLIED MATHEMATICS. Educ: Univ London, BSc, 52, ScM, 54, PhD(appl math), 56. Prof Exp: Asst appl math, Brown Univ, 52-56; MEM TECH STAFF, MATH, PHYSICS & NETWORKS DEPT, BELL LABS, INC, 56- Mem: Am Math Soc; Soc Indust & Appl Math. Res: Mathematical physics; nonlinear oscillations; methods of averaging; stochastic differential equations; propagation in random media. Mailing Add: Math of Physics & Networks Dept Bell Telephone Labs Inc Mountain Ave Murray Hill NJ 07974

MORRISON, JOHN COULTER, b Hickman, Ky, Sept 11, 43; m 67; c 3. OBSTETRICS & GYNECOLOGY, BIOCHEMISTRY. Educ: Memphis State Univ, BS, 65, Univ Tenn, MD, 68. Prof Exp: Res asst biochem, Univ Tenn, Memphis, 65-66, res assoc, 67-68; intern, City of Memphis Hosps, 68-69, resident, 69-72; from instr to asst prof obstet & gynec, 71-75, ASSOC PROF OBSTET & GYNEC, UNIV TENN, MEMPHIS, 75- Concurrent Pos: Chief resident, City of Memphis Hosps, 71-72, asst prof, 72- Mem: Am Chem Soc; Am Col Obstet & Gynec; AMA; Am Fertil Soc. Res: Clinical research in obstetrics and gynecology; basic research in carcinogenesis and protein biosynthesis. Mailing Add: Dept of Obstet & Gynec Univ of Tenn Memphis TN 38163

MORRISON, JOHN JOSEPH, b Holyoke, Mass, Oct 23, 20; m 47; c 3. MATHEMATICS. Educ: Univ Notre Dame, AB, 44, PhD(philos, math), 51, MS, 58. Prof Exp: From instr to asst prof philos, Univ Notre Dame, 47-55, res asst, 56-58; consult, Opers Model Eval Group, US Air Force, 58; specialist mathematician, Fairchild-Hiller Corp, 59-66; aerospace engr, Electronics Res Ctr, 66-70, RES MATHEMATICIAN, DEPT TRANSP, TRANSP SYSTS CTR, NASA, 70- Concurrent Pos: Res asst, Off Naval Res, 56-58. Mem: Am Inst Aeronaut & Astronaut. Res: Philosophy; celestial mechanics; cybernetics. Mailing Add: 100 Memorial Dr Cambridge MA 02142

MORRISON, JOHN OWEN, chemistry, see 12th edition

MORRISON, JOSEPH LOUIS, b Brooklyn, NY, Aug 3, 36; m 59; c 3. BIOCHEMISTRY, DRUG METABOLISM. Educ: Brooklyn Col, BS, 57; Purdue Univ, MS, 59, PhD(biochem), 62. Prof Exp: Res chemist, Am Mach & Foundry Co, 62-64; mgr biochem & anal chem dept, Salsbury Labs, 64-70; dir assay lab, Sigma Chem Co, Mo, 70-72; CHIEF BIOCHEM SECT, NORWICH PHARMACAL CO, 72- Mem: AAAS; Am Chem Soc. Res: Pharmacokinetics; drug metabolism; drug residues; method development. Mailing Add: Norwich Pharmacal Co Norwich NY 13815

MORRISON, KENNETH JESS, b Rudy, Ark, Feb 14, 21; m 46; c 2. AGRONOMY. Educ: Kans State Univ, BS, 48; Purdue Univ, Lafayette, MS, 50, PhD(agr), 67. Prof Exp: Assoc agronomist, 66-68, AGRONOMIST, WASH STATE UNIV, 68-, EXTEN AGRONOMIST, 52- Mem: Soc Agron; Soc Range Mgt; Crop Sci Soc Am. Res: Effect of environment on cultivars of wheat and barley; interrelation of emergence in dry soils and cold hardiness in common winter wheats. Mailing Add: Dept of Agron & Soils Wash State Univ Pullman WA 99163

MORRISON, KENNETH N, b Guelph, Ont, Nov 14, 17; US citizen; m 48; c 4. DENTISTRY. Educ: Univ Toronto, DDS, 43; Univ Wash, MSD, 52. Prof Exp: Instr oper dent, 48-52, asst prof, 52-57, assoc prof fixed partial dentures, 57-65, prof, 65-70, chmn dept, 57-70, PROF RESTORATIVE DENT & CHMN DEPT, SCH DENT, UNIV WASH, 70- Mem: Am Acad Crown & Bridge Prosthodont (pres, 73-74). Res: Restorative dentistry; recording mandibular movement. Mailing Add: Univ of Wash Sch of Dent Seattle WA 98195

MORRISON, LESTER MARVIN, b London, Eng, Sept 18, 07; US citizen; m 38. INTERNAL MEDICINE. Educ: Temple Univ, MD, 33. Prof Exp: F Prof Exp: Fel gastroenterol, Guy's Hosp, London, Eng, 34-35; instr med & asst gastroenterologist, Med Sch, Temple Univ, 35-44; asst prof med, Med Sch & dir, Arteriosclerosis Res Unit, Loma Linda Univ, 45-65; PRES & DIR, INST ARTERIOSCLEROSIS RES, 65- Concurrent Pos: Sr attend physician, Los Angeles County Gen Hosp, 45-; pres, Crenshaw Hosp, 53-61, Whittier Hosp, 54-62, Doctors' Hosp, Santa Ana, 54-67, Sun Valley Hosp, 59-62, Pasadena Community Hosp, 63-65 & Covina Convalescent Hosp, 64-65; consult med, Univ Calif Med Ctr, Los Angeles, 65- Mem: Fel Am Col Physicians; fel Am Col Gastroenterol; fel Am Col Cardiol; fel Royal Soc Med. Res: Arteriosclerosis; clinical medicine. Mailing Add: 9331 Venice Blvd Culver City CA 90230

MORRISON, MARCUS EUGENE, b Corinth, Miss, Sept 6, 10; m 39; c 1. BACTERIOLOGY, OPHTHALMOLOGY. Educ: Univ Miss, BS, 36; Univ Tenn, MD, 38. Prof Exp: Intern, St Vincent's Hosp, Jacksonville, 38-39; asst prof path & bact, Sch Med, Univ Miss, 46, from assoc prof to prof bact, 46-57; pvt pract, Miss, 57-67; PROF BIOL, UNIV MISS, 68- Mem: Am Soc Microbiol; AMA. Res: Ocular infections. Mailing Add: Dept of Biol Univ of Miss University MS 38677

MORRISON, MARTIN, b Detroit, Mich, Dec 9, 22; m 47; c 4. BIOCHEMISTRY. Educ: Univ Mich, BS, 47; Wayne State Univ, PhD(biochem), 52. Prof Exp: From instr to asst prof biochem, Sch Med & Dent, Univ Rochester, 52-60; head sect respiratory enzym, Dept Biochem, City of Hope Med Ctr, Duarte, Calif, 61-67; PROF BIOCHEM, UNIV TENN, MEMPHIS, 67-; CHMN BIOCHEM DEPT, ST JUDE CHILDREN'S RES HOSP, 67- Concurrent Pos: USPHS career develop award, 56-61; res fel, Molteno Inst, Cambridge Univ, 60-61. Mem: Am Soc Biol Chemists; Am Chem Soc; Am Inst Biol Scientists; Am Soc Microbiol; Am Phys Soc. Res: Membrane structure and function; mammalian peroxidases, thyroid peroxidase and lactoperoxidase. Mailing Add: St Jude Children's Res Hosp 332 N Lauderdale Memphis TN 38101

MORRISON, NANCY DUNLAP, b Schenectady, NY, Dec 14, 46; m 66. ASTRONOMY. Educ: Radcliffe Col, BA, 67; Univ Hawaii, MS, 71, PhD(astron), 75. Prof Exp: RES ASSOC ASTRON, JOINT INST LAB ASTROPHYS, UNIV COLO, BOULDER, 75- Mem: Am Astron Soc; AAAS. Res: Photometry and spectroscopy of O-type stars, utilizing these techniques to determine the masses of stars in binary systems. Mailing Add: Joint Inst for Lab Astrophys Univ of Colo Boulder CO 80309

MORRISON, NATHAN, b RI, Dec 4, 12; m; c 1. MATHEMATICAL STATISTICS. Educ: Brooklyn Col, AB, 32. Prof Exp: Prin actuary, State Dept Labor, NY, 44-61; exec assoc, Assoc Hosp Serv, 61-73; assoc dir res dept, 73-74, CONSULT, INT LADIES GARMENT WORKERS UNION, 74- Concurrent Pos: Consult, US War Dept, 43-46; exec secy, State Adv Coun Employ & Unemploy Ins, NY, 43-70; vis prof, Cornell Univ, 56-; lectr, Teachers Col, Columbia Univ, 59-61; adj prof, Grad Div, Brooklyn Col, 65-67. Mem: Fel Am Statist Asn; Economet Soc; Am Math Soc; Soc Indust & Appl Math; Inst Math Statist. Res: Labor market analysis and unemployment insurance; mathematical physics; social insurance; hospital and medical care. Mailing Add: 196 Elm St New Rochelle NY 10805

MORRISON, PETER REED, b Washington, DC, Nov 11, 19; m 45; c 6. COMPARATIVE PHYSIOLOGY, ENVIRONMENTAL PHYSIOLOGY. Educ: Swarthmore Col, BA, 40; Harvard Univ, PhD(biol), 47. Prof Exp: Asst physiol, Harvard Univ, 42; asst phys chem, Harvard Med Sch, 42-46; from asst prof to prof zool & physiol, Univ Wis, 47-64; PROF ZOOPHYSIOL, INST ARCTIC BIOL, UNIV ALASKA, 63-, ADV SCI DIR, INST, 66- Concurrent Pos: Guggenheim & Fulbright fel, Australia, 54-55; NSF sr fel, SAm, 59-60. Mem: Am Soc Zool; Am Soc Biol Chem; Am Physiol Soc; Am Soc Mammal; Int Soc Biometeorol. Res: Respiratory function of blood; energy metabolism and temperature regulation in mammals; physical chemistry of plasma proteins and of blood coagulation; comparative cold and high altitude physiology; hibernation. Mailing Add: Inst of Arctic Biol Univ of Alaska Fairbanks AK 99701

MORRISON, PHILIP, b Somerville, NJ, Nov 7, 15; m 38; PHYSICS. Educ: Carnegie Inst Technol, BS, 36; Univ Calif, PhD(theoret physics), 40. Prof Exp: Instr physics, San Francisco State Col, Calif, 41 & Univ Ill, 41-42; physicist, metall lab, Univ Chicago, 43-44; physicist & group leader, Los Alamos Sci Lab, Univ Calif, 44-46; from assoc prof to prof physics, Cornell Univ, 46-65; PROF PHYSICS, MASS INST TECHNOL, 75- Concurrent Pos: Book ed, Sci Am, 65. Honors & Awards: Pregel Prize, 55; Babson Prize, 57; Oerstad Medal, 65. Mem: Nat Acad Sci; Am Phys Soc; Am Astron Soc; Fedn Am Scientists (chmn, 72-). Res: Applications of physics in astronomy. Mailing Add: Dept of Physics Mass Inst Technol Cambridge MA 02139

MORRISON, RALPH M, b Annapolis, Md, June 23, 32; m 60. PLANT PHYSIOLOGY, MICROBIOLOGY. Educ: Col William & Mary, BS, 55; Ind Univ, PhD(bot), 60. Prof Exp: From instr to asst prof biol, 60-66, ASSOC PROF BIOL, UNIV NC, GREENSBORO, 66- Mem: AAAS; Mycol Soc Am; Bot Soc Am; Am Soc Plant Physiologists. Res: Mathematical method for studying the rate of seed germination; botany; mycology. Mailing Add: Dept of Biol Univ NC Greensboro NC 27412

MORRISON, REGINALD GRAHAM, JR, physical chemistry, see 12th edition

MORRISON, RICHARD CHARLES, b Lowell, Mass, Jan 24, 38; m 60; c 4. ENERGY CONVERSION. Educ: Princeton Univ, AB, 59; Yale Univ, MS, 61, PhD(physics), 65. Prof Exp: From instr to asst prof physics, New Haven Col, 63-67; asst prof, Iowa State Univ, 67-74; PROF PHYSICS, UNIV NEW HAVEN, 74- Concurrent Pos: Assoc physicist, Ames Lab, AEC, 67-74; sr partner, Enercon Assocs, 75- Mem: Am Nuclear Soc; Am Phys Soc; Am Asn Physics Teachers. Res: Nuclear instrumentation applied to environmental research; energy production, conversion, and consumption; alternative energy systems. Mailing Add: Dept of Physics Univ of New Haven West Haven CT 06516

MORRISON, RICHARD DONALD, b Kansas City, Mo, Aug 5, 27; m 55; c 1. DENTISTRY. Educ: Washington Univ, AB, 49, DDS, 54. Prof Exp: Instr oper dent, Sch Dent, Washington Univ, 54-56, instr pharmacol, 56-58, USPHS fel, 59-60, asst prof oper dent & pharmacol, 60-62, assoc prof oper dent, 62-66; Fulbright prof dent, Univ Baghdad, 66-67; asst dean instr, 71-73, PROF PREV MED & COMMUNITY DENT, SCH DENT, WASHINGTON UNIV, 70- Mailing Add: Washington Univ Sch of Dent St Louis MO 63110

MORRISON, RICHARD H, physical inorganic chemistry, see 12th edition

MORRISON, ROBERT DEAN, b Wetumka, Okla, Sept 27, 15; m 39; c 1. BIOSTATISTICS. Educ: Okla State Univ, BS, 38, MS, 42; NC State Col, PhD(exp statist), 57. Prof Exp: From asst prof to assoc prof math, 46-61, PROF STATIST, OKLA STATE UNIV, 61-. STATISTICIAN, AGR EXP STA, 57- Mem: Biomet Soc; Am Statist Asn. Res: Technometrics; experimental statistics; statistics for agricultural research. Mailing Add: Dept of Statist Okla State Univ Stillwater OK 74074

MORRISON, ROBERT THOMAS, b Vermilion, Alta, July 5, 33; m 56; c 2. NUCLEAR MEDICINE. Educ: Univ Alta, BSc, 54, MD, 58, MSc, 61; Univ Iowa, PhD(radiation biol), 65. Prof Exp: Asst prof radiobiol & nuclear med, Col Med, Univ Iowa, 65-67; ASSOC PROF MED, FAC MED, UNIV BC, 67-; HEAD SECT NUCLEAR MED, VANCOUVER GEN HOSP, 67- Concurrent Pos: Fel gastroenterol, Univ Alta, 59-61; consult, Vet Admin Hosp, Iowa City, 66-67. Mem: AAAS; Soc Nuclear Med. Res: Radiation biology, particularly organ visualization and data readout systems. Mailing Add: Dept of Med Univ of BC Vancouver BC Can

MORRISON, ROBERT W, JR, b Columbia, SC, Dec 5, 38; m 61; c 2. ORGANIC CHEMISTRY. Educ: Davidson Col, BS, 60, MA, 62, PhD(org chem), 64. Prof Exp: Res chemist, Chemstrand Res Ctr, Inc, Monsanto Co, 64-69; sr res chemist, 69-73, GROUP LEADER, BURROUGHS WELLCOME & CO, USA, 73- Concurrent Pos: From vis asst prof to adj assoc prof chem, NC State Univ, 68-71. Mem: Am Chem Soc. Res: Synthesis of nitrogen heterocyclic compounds; enzyme inhibitors. Mailing Add: Burroughs Wellcome & Co USA Inc 3030 Cornwallis Rd Research Triangle Park NC 27709

MORRISON, ROBERT WILLIAM, b Lena, SC, Aug 28, 06; m 32, 60; c 1. PHARMACOLOGY, PHARMACY. Educ: Univ SC, PhC, 28, BS, 29; Univ Tenn, MS, 31. Prof Exp: Instr pharm & pharmacol, 31-33, adj prof, 35-40, assoc prof, 40-47, actg dean, 52-57, DEAN SCH PHARM, UNIV SC, 57-, PROF PHARM & PHARMACOL, 47- Concurrent Pos: Teaching fel pharmacol, Univ Tenn, 28-32. Mem: Am Col Apothecaries; Am Pharmaceut Asn; Am Soc Hosp Pharmacists. Res: Hospital pharmacy. Mailing Add: Univ of SC Sch of Pharm Columbia SC 29208

MORRISON, ROGER ALBERT, dairy husbandry, see 12th edition

MORRISON, ROGER BARRON, b Madison, Wis, Mar 26,. 14; c 3. GEOLOGY. Educ: Cornell Univ, BA, 33, MS, 34; Univ Nev, PhD(geol), 64. Prof Exp: GEOLOGIST, US GEOL SURV, 39- Concurrent Pos: Mem, Friends Pleistocene. Mem: Geol Soc Am; Int Asn Quaternary Res; Soil Sci Soc Am; Am Soc Photogram; Am Geog Soc. Res: Cenozoic geology, especially Quaternary stratigraphy and geomorphology; soil stratigraphy; glacial geology; geochronology; climatic history; hydrogeology and remote sensing; photo interpretation of geological terrain features. Mailing Add: US Geol Surv Fed Ctr Denver CO 80225

MORRISON, ROLLIN JOHN, b Akron, Ohio, Oct 8, 37; m 64; c 2. HIGH ENERGY PHYSICS. Educ: Ohio Wesleyan Univ, BA, 59; Univ Ill, MA, 61, PhD(physics), 64. Prof Exp: Volkswagen fel, Ger Electron Syncrotron, 64-66; res physicist, 67, asst prof, 67-71, ASSOC PROF PHYSICS, UNIV CALIF, SANTA BARBARA, 71- Mem: Am Phys Soc. Res: Nuclear physics using the Mössbauer effect; high energy experimental physics. Mailing Add: Dept of Physics Univ of Calif Santa Barbara CA 93106

MORRISON, SPENCER HORTON, b Madison, Wis, Apr 17, 19; m 46; c 6. VETERINARY MEDICINE, ANIMAL NUTRITION. Educ: Cornell Univ, BS, 39, MS, 46, PhD(animal nutrit & physiol, biochem), 49; Univ Ga, DVM, 54. Prof Exp: Asst animal husb, Cornell Univ, 46-48; assoc, Univ Calif, 48-49; assoc prof dairying, Univ Ga, 49-54; tech sales dir, Feed & Soy Div, Pillsbury Mills, Inc, 54-57, res dir, 57-58; DIR, AGRICON, 58-; ED & MGR, MORRISON PUB CO, 58-, PARNTNER, 40- Mem: Fel AAAS; Am Soc Animal Sci; Poultry Sci Asn; Am Dairy Sci Asn; Am Vet Med Asn. Res: Animal physiology; biochemistry; feeding value of concentrates and roughages; antibiotics, vitamins; physiology of reproduction; protein synthesis in ruminants. Mailing Add: RR 6 Orangeville ON Can

MORRISON, STANLEY ROY, b Saskatoon, Sask, Sept 24, 26; m 49; c 3. SURFACE CHEMISTRY, SURFACE PHYSICS. Educ: Univ BC, BA, 48, MA, 49; Univ Pa, PhD(physics), 52. Prof Exp: Res assoc solid state physics, Univ Ill, 52-54; sr scientist, Sylvania Elec Prod Inc, 54-55; from staff scientist to asst dir res, Res Ctr, Minneapolis-Honeywell Regulator Co, 54-64; SR PHYSICIST, STANFORD RES INST, 64- Concurrent Pos: Guest prof, Inst Physics & Chem, Göttingen Univ, 71-72. Honors & Awards: Cert of Recognition, NASA, 73 & 74. Mem: Am Phys Soc. Res: Adsorption and catalysis; semiconductor electrochemistry; electrical and chemical properties of surfaces; influence of adsorbed gases and other surface imperfections on the electrical properties of solid state materials and devices. Mailing Add: Stanford Res Inst Ravenswood Ave Menlo Park CA 94025

MORRISON, SUMNER MARTIN, b Boston, Mass, Jan 23, 19; m 46; c 3. MICROBIOLOGY. Educ: Univ Mass, BS, 41; Univ Purdue, MS, 42; Ohio State Univ, PhD(bact), 50. Prof Exp: Bacteriologist, Upjohn Co, 43-46; from asst to prof microbiol, 50-70, PROF MICROBIOL & CIVIL ENG & DIR ENVIRON HEALTH SERV, COLO STATE UNIV, 70- Concurrent Pos: Consult, US Pub Health Serv, 65-, NIH, 69-, off water progs, Environ Protection Agency, 68- & industry. Mem: AAAS; Am Soc Microbiol; Am Pub Health Asn; Water Pollution Control Fedn. Res: Enteric bacteria, epidemiology staphylococcus aureus; ice sanitation; surface and ground water; feed lot wastes; low temperature waste treatment; microbiology and geologic formations; industrial waste. Mailing Add: Dept of Microbiol Colo State Univ Ft Collins CO 80523

MORRISON, WILEY HERBERT, III, b Statesville, NC, Aug 1943; m 67. ORGANIC CHEMISTRY. Educ: Wake Forest Univ, BS, 66; Univ Ga, PhD(chem), 70. Prof Exp: Res assoc & fel org chem, Fla State Univ, 70-71; RES CHEMIST, RICHARD B RUSSELL AGR RES CTR, AGR RES SERV, USDA, 71- Mem: Am Chem Soc; Am Oil Chemists Soc. Res: Organic reaction mechanisms and stereo-chemistry; oxidative stability of edible oils. Mailing Add: Richard B Russell Agr Res Ctr PO Box 5677 Athens GA 30604

MORRISON, WILLIAM ALFRED, b Chicago, Ill, Mar 27, 48; m 71. INORGANIC CHEMISTRY. Educ: Ill Wesleyan Univ, BA, 70; Univ Kans, PhD(inorg chem), 74. Prof Exp: Asst prof chem, Monmouth Col, 74-75; VIS ASST PROF CHEM, UNIV EVANSVILLE, 75- Mem: Am Chem Soc. Res: Transition metal organometallics; carbenes; metal-metal bonds; nonaqueous solvents. Mailing Add: Dept of Chem Univ of Evansville Evansville IN 47702

MORRISON, WILLIAM D, b Provost, Alta, Oct 16, 27; m 49; c 4. ANIMAL NUTRITION, ANIMAL PHYSIOLOGY. Educ: Ont Agr Col, Toronto, BSA, 49; Univ Ill, MSc, 54, PhD(animal nutrit), 55. Prof Exp: Territory mgr, Master Feeds Div, Maple Leaf Mills Ltd, 49-52, nutritionist, 55-57, dir nutrit & res, 57-71; CHMN DEPT ANIMAL & POULTRY SCI, UNIV GUELPH, 71- Mem: AAAS; Am Soc Animal Sci Asn; Poultry Sci Asn; Nutrit Soc Can; World Poultry Sci Asn. Res:

Amino acid utilization by the chick, with special emphasis on the D isomer; protein and energy requirements of chickens and turkeys; control and prevention of certain diseases in poultry. Mailing Add: RR 4 Fergus ON Can

MORRISON, WILLIAM HARVEY, JR, b Memphis, Tenn, Sept 9, 48; m 75. INORGANIC CHEMISTRY. Educ: Vanderbilt Univ, BE, 70; Univ Ill, PhD(chem), 74. Prof Exp: RES CHEMIST, E I DU PONT DE NEMOURS & CO, INC, 74- Mem: Am Chem Soc; The Chem Soc. Res: Development and synthesis of new refractory or ceramic materials; high temperature synthesis. Mailing Add: E I du Pont de Nemours & Co Inc Exp Sta Bldg 335 Wilmington DE 19898

MORRISON, WILLIAM JOSEPH, b Plainfield, NJ, Feb 17, 42. GENETICS. Educ: Clemson Univ, BS, 65; Pa State Univ, PhD(genetics), 69. Prof Exp: NIH fel genetics, Hershey Med Ctr, Pa State Univ, 69-70; NIH fel, Cornell Univ, 70-73; ASST PROF BIOL, SHIPPENSBURG STATE COL, PA, 73- Mem: AAAS; Genetics Soc Am; Am Inst Biol Sci. Res: Biochemical and development genetics; Drosophila genetics. Mailing Add: Dept of Biol Shippensburg State Col Shippensburg PA 17257

MORRISON-CLEATOR, IAIN GOESTA, b Edinburgh, Scotland, Oct 18, 39; m 61; c 3. SURGERY. Educ: Edinburgh Univ, MB, ChB, 62; FRCS(E), 66; FRCS(London), 67; FRCS(C), 72. Prof Exp: Residency surg, Edinburgh Teaching Hosps, 62-71; fel, 71-72, ASST PROF SURG, UNIV BC, 72- Concurrent Pos: Active staff mem, St Paul's Hosp, Vancouver, 73-, dir, Woodward Gastrointestinal Clin, 74- Mem: Can Asn Gastroenterol; fel Am Col Surgeons. Res: Role of gastro-intestinal hormones in normal and pathological conditions. Mailing Add: Univ of BC 1081 Burrard St Vancouver BC Can

MORRISS, FRANCIS V, organic chemistry, see 12th edition

MORRISSETTE, HUGUES, b Arthabaska, Can, Feb 1, 37; m 62; c 3. GEOGRAPHY. Educ: Univ Montreal, BA, 58; Laval Univ, BA, 60, MA, 62; Univ Strasbourg, PhD(geog), 65. Prof Exp: Researcher geog, Govt PQ, 62-66; asst prof geog, Laval Univ, 66-67; chmn dept geog, Univ Ottawa, 67-74; CHIEF, NORTHERN RES DIV, DEPT INDIAN AFFAIRS & NORTHERN DEVELOP, GOVT CAN, 74- Mem: Asn Am Geogrs; Can Asn Geog; Fr Am Geog Asn. Res: Regional planning; agricultural geography. Mailing Add: 31 Lessard Hull PQ Can

MORRISSETTE, MAURICE CORLETTE, b Clyde, Kans, Aug 27, 21; m 45; c 3. PHYSIOLOGY, PHARMACOLOGY. Educ: Kans State Univ, BS & DVM, 54; Okla State Univ, MS, 56, PhD(reprod physiol), 64. Prof Exp: Instr physiol & pharmacol, Sch Vet Med, Okla State Univ, 54-56, asst prof, 56-57; asst prof physiol, Kans State Univ, 57-59; from asst prof to assoc prof physiol & pharmacol, Okla State Univ, 59-69; PROF & HEAD DEPT, SCH VET MED, LA STATE UNIV, BATON ROUGE, 69- Mem: Am Soc Vet Physiologists & Pharmacologists (pres, 72-73); Am Vet Med Asn; Am Fertil Soc; Soc Study Reproduction; Am Soc Animal Sci. Res: Reproductive physiology; veterinary medicine; female reproduction in swine and cattle; veterinary physiology, pharmacology and toxicology. Mailing Add: Dept of Physiol & Pharmacol Sch of Vet Med La State Univ Baton Rouge LA 70803

MORRISSEY, ARTHUR CHARLES, b Westerly, RI, Mar 7, 41; m 67; c 2. SCIENCE POLICY. Educ: Washington & Jefferson Col, AB, 63; Univ SC, PhD(phys chem), 67. Prof Exp: Teaching chem & res spectros, Univ SC, 63-67; analyst, Cent Intel Agency, 67-70, br chief anal technol, 70-74; technol staff position, Intel Community Staff, 74-75; POLIT-MIL SCI OFFICER FOREIGN TECH DEVELOP, DEPT OF STATE, 75- Mem: AAAS; Am Chem Soc; Fedn Am Scientists; Sigma Xi. Res: Role of technology in foreign relations, a tool and/or determinate. Mailing Add: Dept of State Rm 6524A Washington DC 20520

MORRISSEY, BRUCE WILLIAM, b Danbury, Conn, Apr 18, 42; m 67; c 1. PHYSICAL CHEMISTRY, BIOCHEMISTRY. Educ: Rensselaer Polytech Inst, BS, 64, PhD(theoret chem), 70; Yale Univ, MS, 66. Prof Exp: Nat Res Coun res assoc , 70-72, RES CHEMIST, POLYMERS DIV, NAT BUR STAND, 72- Mem: Am Chem Soc; Am Soc Artificial Internal Organs. Res: Polymer surface chemistry; relation between the conformation and conformational changes of adsorbed protein and their reactivity and function at surfaces. Mailing Add: Polymers Div Nat Bur Stand Washington DC 20234

MORRISSEY, J EDWARD, b Grinnell, Iowa, Aug 7, 32; m 57; c 2. ZOOLOGY, PHYSIOLOGY. Educ: St Ambrose Col, BA, 56; Northwestern Univ, Ill, MS, 58; Univ Mo-Columbia, PhD(zool), 68. Hon Degrees: MHL, Ottawa Univ, 72. Prof Exp: Instr biol, Stevens Col, 60-65; asst prof, MacMurray Col, 68; asst prof, 68-72, ASSOC PROF BIOL, OTTAWA UNIV, 72- Mem: AAAS. Res: Immunological development; studies of immune responses in chickens. Mailing Add: Dept of Biol Ottawa Univ Ottawa KS 66067

MORRISSEY, JOHN F, b Brookline, Mass, June 16, 24; m 50; c 2. MEDICINE, GASTROENTEROLOGY. Educ: Dartmouth Col, AB, 46; Harvard Univ, MD, 49. Prof Exp: Asst prof med, Univ Wis, 56-60; asst prof, Univ Wash, 60-62; from asst prof to assoc prof, 62-71, PROF MED, UNIV WIS-MADISON, 71- Mem: Am Gastroenterol Asn; Am Col Physicians; Am Fedn Clin Res; Am Soc Gastrointestinal Endoscopy. Res: Evaluation of new instruments for digestive tract endoscopy; effects of drugs on gastrointestinal mucosa. Mailing Add: Univ of Wis Hosps 1300 University Ave Madison WI 53706

MORRISSEY, JOYCE ANN, biochemistry, see 12th edition

MORRISSEY, RONALD JAMES, physical chemistry, see 12th edition

MORRITZ, FRED LEONARD, b Chicago, Ill, Aug 13, 22; m 49; c 4. ORGANIC CHEMISTRY. Educ: Univ Chicago, BS, 42; Ill Inst Technol, PhD, 53. Prof Exp: Res chemist, Sinclair Res Labs, Inc, 52-55; sr chemist, Armour Res Found, 55-62; chief thin films res, Autonetics Div, NAm Aviation, Inc, Calif, 62-69; TECH DIR, INDUST & SCI CONF MGT, INC, 69- Mem: AAAS; Am Chem Soc. Res: Catalytic reactions; thin films; semiconductors. Mailing Add: Indust & Sci Conf Mgt Inc 222 W Adams Chicago IL 60606

MORRONE, TERRY, b New York, NY, May 30, 36; m 62; c 2. PLASMA PHYSICS. Educ: Columbia Univ, BS, 57, MS, 58; Polytech Inst Brooklyn, PhD(electrophys), 64. Prof Exp: Asst prof, 64-69, ASSOC PROF PHYSICS, ADELPHI UNIV, 69- Mem: Am Phys Soc. Mailing Add: Dept of Physics Adelphi Univ Garden City NY 11530

MORROW, ANDREW GLENN, b Indianapolis, Ind, Nov 3, 22; m 45; c 2. SURGERY. Educ: Wabash Col, AB, 43; Johns Hopkins Univ, MD, 46. Prof Exp: Asst, 50-51, from instr to asst prof, 51-60, ASSOC PROF SURG, JOHNS HOPKINS UNIV, 60-; CHIEF CLIN SURG, NAT HEART & LUNG INST, 53- Mem: Soc Vascular Surg (pres, 71-72); Soc Univ Surg; Am Asn Thoracic Surg; fel Am Col Surg; Am Fedn Clin Res. Res: Cardiovascular surgery; diagnostic methods and allied physiology. Mailing Add: Surg Br Nat Heart & Lung Inst Bethesda MD 20014

MORROW, BARRY ALBERT, b Regina, Sask, Aug, 39. PHYSICAL CHEMISTRY. Educ: Univ BC, BSc, 61, MSc, 62; Cambridge Univ, PhD(chem), 65. Prof Exp: Nat Res Coun fel, 65-66; lectr chem, Univ West Indies, 66-67; ASSOC PROF CHEM, UNIV OTTAWA, 67- Mem: The Chem Soc. Res: Spectroscopic studies of adsorption and catalysis; spectroscopic studies of small molecules. Mailing Add: Dept of Chem Univ of Ottawa Ottawa ON Can

MORROW, CHARLES TABOR, b Gloucester, Mass, May 3, 17; m 49; c 2. ACOUSTICS. Educ: Harvard Univ, AB, 37, SM, 38, ScD, 46. Prof Exp: Engr, Harvard Univ, 37-40, res assoc physics, 40-41, spec res assoc underwater sound, 41, instr radio & radar officers training, 42-44; lectr indust electronics, Northeastern Univ, 44-45; sr proj engr, Sperry Gyroscopy Co, 46-51; res physicist, Hughes Aircraft Co, Inc, 51-55; mem sr staff, Space Technol Labs, Inc, 55-60 & Aerospace Corp, Calif, 60-67; from mem staff to staff scientist, Western Div, LTV Res Ctr, 67-69; STAFF SCIENTIST, ADVAN TECHNOL CTR, INC, 69- Honors & Awards: Vigness Award, Inst Environ Sci, 71. Mem: AAAS; Acoust Soc Am; Am Inst Aeronaut & Astronaut; Inst Elec & Electronics Eng; Am Soc Eng Educ. Res: Vibratory gyroscopes; silencing of diving masks; study of speech in gas oxygen and diving masks and in helium atmospheres; development of microphone for divers; shock and vibration analysis; prediction of structural response to aerodynamic turbulence and rocket noise; vibration instrumentation. Mailing Add: Advan Technol Ctr Inc PO Box 6144 Dallas TX 75222

MORROW, DARRELL ROY, b Pen Yan, NY, Feb 28, 37; m 60; c 2. SOLID STATE PHYSICS, POLYMER PHYSICS. Educ: Lehigh Univ, BA, 59; Clarkson Col Technol, MS, 61; Pa State Univ, PhD, 65. Prof Exp: Asst, Clarkson Col Technol, 59-60; asst, Pa State Univ, 61-63; asst res specialist, 64-65, asst prof, 65-67, assoc prof mech, 67-71, PROF MECH, RUTGERS UNIV, NEW BRUNSWICK, 71-, CHMN GRAD PROG PACKAGING SCI & ENG, 71- Mem: AAAS; Am Crystallog Asn; Am Chem Soc; Soc Packaging & Handling Engrs; Packaging Inst US; NY Acad Sci. Res: Structure and properties of polymers; development and design of instrumentation; packaging materials testing; package design and testing; design and analysis of packaging machinery. Mailing Add: Packaging Sci & Eng Rutgers Univ Col of Eng New Brunswick NJ 08903

MORROW, DAVID AUSTIN, b Tyrone, Pa, Jan 14, 35; m 65; c 2. THERIOENOLOGY. Educ: Pa State Univ, BS, 56; Cornell Univ, DVM, 60, PhD(obstet), 67. Prof Exp: Pvt pract vet med, 60-61; intern, Cornell Univ, 61-62, asst, 62-64, NIH fel, 64-67, res assoc, 67-68; ASSOC PROF VET MED, MICH STATE UNIV, 68- Mem: Soc Study Reprod; Am Col Theriogenologists. Res: Bovine reproductive physiology; bovine theriogenology. Mailing Add: Large Animal Surg & Med Mich State Univ Col of Vet Med East Lansing MI 48824

MORROW, DAVID CLARENCE, b La Riviere, Man, Jan 12, 00; nat US; m 29; c 2. MATHEMATICS. Educ: Univ Man, BA, 24; Univ Toronto, MA, 25; Univ Chicago, PhD(math), 28. Prof Exp: Instr math, Northwestern Univ, 28-29; instr, Col City of Detroit, 29-30, from asst prof to prof, 30-62, EMER PROF MATH, WAYNE STATE UNIV, 62- Concurrent Pos: Chmn dept math, Mercy Col, Mich, 62-68. Mem: Am Math Soc; Math Asn Am. Res: Representation of positive integers by quaternary quadratic forms. Mailing Add: 1704 Cherrywood Dr Holiday FL 33589

MORROW, DEAN HUSTON, b Indianapolis, Ind, June 11, 31; m 53; c 3. ANESTHESIOLOGY, CARDIOVASCULAR PHYSIOLOGY. Educ: Butler Univ, BS, 53; Ind Univ, MD, 56; Am Bd Anesthesiol, dipl, 63. Prof Exp: Staff anesthesiologist, Clin Ctr, NIH, Md, 59-61, res anesthesiologist, 62-64; assoc prof anesthesiol, 64-66, PROF ANESTHESIOL & DIR DEPT RES, COL MED, UNIV KY, 66-, PROF PHARMACOL, 71- Concurrent Pos: Nat Inst Gen Med Sci res career award, 67-71; res affil, Nat Heart Inst, 61; mem comt anesthesia, Nat Res Coun-Nat Acad Sci, 63-65. Mem: Am Soc Anesthesiol; Int Anesthesia Res Soc; Asn Univ Anesthetists; fel Am Col Anesthesiol; Am Soc Pharmacol & Exp Therapeut. Res: Pharmacology. Mailing Add: Dept of Anesthesiol Univ of Ky Col of Med Lexington KY 40506

MORROW, DUANE FRANCIS, b Detroit, Mich, June 23, 33; m 58; c 3. PHARMACEUTICAL CHEMISTRY. Educ: Wayne State Univ, BS, 54; Univ Ill, PhD(chem), 57. Prof Exp: From assoc res chemist to sr res chemist, Parke, Davis, & Co, 57-69; from from group leader to sr investr, 69-74, PRIN INVESTR, RES CTR, MEAD JOHNSON & CO, 74- Mem: Am Chem Soc. Res: Pharmaceutical chemistry; steroids; anti-fertility. Mailing Add: Mead Johnson Res Ctr Evansville IN 47721

MORROW, HOMER NICHOLAS, JR, b Houston, Tex, Jan 6, 13; m 42; c 1. ORGANIC CHEMISTRY. Educ: Agr & Mech Col Tex, BS, 34, MS, 38. Prof Exp: Res chemist, Baroid Sales Div, Nat Lead Co, 38-41; asst prof org chem, Agr & Mech Col Tex, 46-48, fel & res chemist, Eng Exp Sta, 48-50; chief chemist refinery, Delhi-Taylor Oil Corp, 50-56, mgr develop lab, 56-63 & Hess Oil & Chem Corp, 63-74, MGR DEVELOP LAB, AMERADA HESS CORP, 74- Mem: Am Chem Soc. Res: Product and process development in organic and petroleum chemistry. Mailing Add: Amerada Hess Corp PO Box 4936 Corpus Christi TX 78408

MORROW, JACK I, b New York, NY, Jan 30, 33; m 58; c 1. PHYSICAL INORGANIC CHEMISTRY. Educ: NY Univ, BA, 54, PhD(su- rface chem), 59. Prof Exp: From asst prof to assoc prof chem, 57-73, PROF CHEM, CITY COL, NEW YORK, 73- Mem: Fel Am Inst Chemists; Am Chem Soc. Res: Inorganic reaction mechanisms and instrumentation for the study of fast reactions. Mailing Add: Dept of Chem City Col of New York Convent Ave & 138th St New York NY 10031

MORROW, JAMES ALLEN, JR, b Little Rock, Ark, Sept 14, 41; m 68; c 1. MATHEMATICS. Educ: Calif Inst Technol, BSc, 63; Stanford Univ, PhD(math), 67. Prof Exp: Teaching asst math, Stanford Univ, 63-67; instr, Univ Calif, 67-68, lectr, 68-69; asst prof, 69-73, ASSOC PROF MATH, UNIV WASH, 73- Mem: Am Math Soc; Math Asn Am. Res: Complex manifolds; singularities. Mailing Add: Dept of Math Univ of Wash Seattle WA 98105

MORROW, JAMES EDWIN, JR, b Brooklyn, NY, Nov 7, 18; m 50; c 1. SYSTEMATIC ICHTHYOLOGY. Educ: Middlebury Col, AB, 40, MS, 42; Yale Univ, MS, 44, PhD(zool), 49. Prof Exp: From res asst to res assoc, Bingham Oceanog Lab, Yale Univ, 49-54; from assoc prof to prof fisheries, 60-66, head dept biol sci, 66-69, PROF ZOOL, UNIV ALASKA, 66- Concurrent Pos: Vis res assoc, Div Fishes, US Nat Mus, 69-70; ichthyol ed, Am Soc Ichthyologists & Herpetologists, 72- Mem: Fel, AAAS; Am Soc Ichthyologists & Herpetologists; Am Fisheries Soc; Soc Syst Zool; fel Am Inst Fishery Res Biol. Res: Systematic ichthyology. Mailing Add: Div Life Sci Univ of Alaska Fairbanks AK 99701

MORROW, JOHN CHARLES, III, b Hendersonville, NC, Sept 20, 24; m 50; c 3. PHYSICAL CHEMISTRY. Educ: Univ NC, BS, 44; Mass Inst Technol, PhD(phys chem), 49. Prof Exp: From asst prof to assoc prof, 49-59, PROF CHEM, UNIV NC, CHAPEL HILL, 59-, PROVOST, 68- Concurrent Pos: NSF fac fel, Univ Heidelberg, 62-63; dean, Col Arts & Sci & Gen Col, 66-68; NATO sr fel, Univ Hamburg, 71.

Mem: Am Chem Soc; Am Phys Soc; Am Crystallog Asn. Res: X-ray crystallography. Mailing Add: 263 Venable Hall Univ of NC Chapel Hill NC 27514

MORROW, KENNETH JOHN, JR, b Wallace, Idaho, Nov 2, 38; m 60. GENETICS. Educ: Whitman Col, AB, 60; Univ Wash, MS, 62, PhD(genetics), 64. Prof Exp: Fulbright fel, 64; NIH fel, Inst Genetics, Univ Pavia, 64-66; res assoc biol, Inst Cancer Res, Philadelphia, 66-68; asst prof physiol & cell biol, Univ Kans, 68-73; ASSOC PROF BIOCHEM, SCH MED, TEX TECH UNIV, 73- Mem: Am Soc Cell Biol; Am Tissue Cult Asn. Res: Somatic cell genetics; genetics of mammalian somatic cell cultivated in vitro; isolation of variants and their characterization; cell culture hybridization. Mailing Add: Dept of Biochem Sch of Med Tex Tech Univ Lubbock TX 79409

MORROW, LARRY ALAN, b Boise, Idaho, Oct 3, 38; m 57; c 4. RANGE SCIENCE, WEED SCIENCE. Educ: Utah State Univ, BS, 65; Univ Nebr, MS, 71, PhD(weed sci), 74. Prof Exp: Biologist, 66-74, RES AGRONOMIST, AGR RES SERV, USDA, 74- Mem: Soc Range Mgt; Weed Sci Soc Am. Res: Improvement of forage production and grazing lands through management, weed control and fertilization. Mailing Add: 405 13th St NW Mandan ND 58554

MORROW, LEONARD OWEN, b Richmond, Va, May 28, 35; m 56; c 3. BOTANY. Educ: Univ Richmond, BS, 57, MS, 60; Cornell Univ, PhD(bot), 65. Prof Exp: Chemist, E R Carpenter Co, Va, 58-59; asst prof biol, Richmond Prof Inst, 64-67; from asst prof to assoc prof, Randolph-Macon Col, 67-71; sci specialist, Math & Sci Ctr, Mitre Corp, 71-74; res supvr, Charles County Community Col, 74; MEM TECH STAFF, MITRE CORP, 75- Mem: Bot Soc Am; AAAS; Nature Conservancy. Res: Plant morphology; vegetation analysis; impact assessment. Mailing Add: Mitre Corp Westgate Research Park McLean VA 22101

MORROW, LEROY WAGGONER, nuclear physics, see 12th edition

MORROW, NORMAN LOUIS, b Brooklyn, NY, Feb 1, 42; m 64; c 2. PHYSICAL CHEMISTRY, ANALYTICAL CHEMISTRY. Educ: Stevens Inst Technol, BS, 63; Univ Conn, PhD(chem), 67. Prof Exp: Res chemist, Exxon Res & Eng Co, NJ, 67-70; group head anal chem, Exxon Chem Co, USA, 70-73, chief chemist, 73-75; STAFF CHEMIST, EXXON CHEM CO, 75- Res: Process and analytical chemistry of resins, adhesives and hot melts. Mailing Add: Exxon Chem Co Chem Specialities Technol Div PO Box 241 Baton Rouge LA 70821

MORROW, NORMAN ROBERT, chemical engineering, surface chemistry, see 12th edition

MORROW, PAUL EDWARD, b Fairmont, WVa, Dec 27, 22; m 47; c 2. TOXICOLOGY, PHARMACOLOGY. Educ: Univ Ga, BS, 42, MS, 47; Univ Rochester, PhD(pharmacol), 51. Prof Exp: Indust hygienist, Holston Ord Works, Tenn Eastman Corp, 42-43; asst, Univ Ga, 56-57; res assoc, 47-52, from instr to prof radiation biol & pharmacol, 52-69, PROF RADIATION BIOL & BIOPHYS, SCH MED & DENT, UNIV ROCHESTER, 69- Concurrent Pos: NIH spec res fel, Univ Göttingen, 59-60; mem, Int Comn Radiol Protection, 65- & Nat Coun Radiation Protection & Measurements, 69-; mem subcomt airborne particles & sulfur oxide, Nat Res Coun-Nat Acad Sci. Mem: Health Physics Soc; Radiation Res Soc; Am Soc Pharmacol & Exp Therapeut; Am Indust Hyg Asn; Am Thoracic Soc. Res: Radiation toxicology; deposition and retention of inhaled dusts; dust clearance mechanisms in the lung; radioactive dust hazards; respiratory physiology in uncooperative subjects; aerosols; lung models. Mailing Add: Sch of Med & Dent Univ of Rochester Rochester NY 14642

MORROW, RICHARD ALEXANDER, b Powassan, Ont, Apr 19, 37; m 64; c 2. PARTICLE PHYSICS. Educ: Queens Univ, Ont, BSc, 58; Univ BC, MSc, 59; Princeton Univ, PhD(physics), 63. Prof Exp: Instr physics, Princeton Univ, 63-64; asst prof, Dartmouth Col, 64-70; ASSOC PROF PHYSICS, UNIV MAINE, ORONO, 70- Mem: Am Phys Soc. Res: Theoretical high energy physics. Mailing Add: Dept of Physics Univ of Maine Orono ME 04473

MORROW, RICHARD JOSEPH, b Portland, Ore, Aug 28, 28; m 58; c 5. NUCLEAR CHEMISTRY, PHYSICAL CHEMISTRY. Educ: Reed Col, BA, 52; Univ Idaho, MS, 58. Prof Exp: Res chemist, Reed Inst, 52 & Gen Elec Co, Hanford, Wash, 53-55, opers chemist, 56-58; nuclear chemist, Lawrence Livermore Lab, Univ Calif, 58-71; STAFF SCIENTIST ATMOSPHERIC RES, TECH APPLNS CTR, US AIR FORCE, 71- Mem: Am Chem Soc. Res: Nuclear decay schemes, actinide and rare earth separation chemistry; Raman and infrared studies of metal oxide systems; fast neutron reactions; radioactivity in the environment. Mailing Add: Air Force Tech Appln Ctr Patrick AFB FL 32925

MORROW, ROY WAYNE, b Hopkinsville, Ky, Sept 28, 42; m 66; c 1. ANALYTICAL CHEMISTRY. Educ: Murray State Univ, BS, 64; Univ Tenn, MS, 67, PhD(chem), 70. Prof Exp: DEVELOP CHEMIST, OAK RIDGE Y-12 PLANT, NUCLEAR DIV, UNION CARBIDE CORP, 70- Mem: Am Chem Soc; Soc Appl Spectros; Sigma Xi. Res: Flame emission and atomic absorption spectroscopy; gas chromatography. Mailing Add: Oak Ridge Y-12 Plant Union Carbide Corp Oak Ridge TN 37830

MORROW, SCOTT, b Oklahoma City, Okla, Sept 11, 20; m 45; c 3. MICROSCOPY, INORGANIC CHEMISTRY. Educ: Case Western Reserve Univ, BS, 47, MS, 49, PhD(inorg chem), 51. Prof Exp: Res chemist, Mound Lab, Monsanto Chem Co, Ohio, 51-53, Mass, 53-54 & Socony Mobil Co, NJ, 54-56; proj chemist, Thiokol Chem Co, 56-66; RES CHEMIST PROPELLANTS, PICATINNY ARSENAL, DOVER, 66- Mem: Am Chem Soc; Am Inst Physics. Res: Solid state propellants and explosives; catalysis and combustion research; mixed crystals and solid solutions; microscopy of thin films; thermal analysis by microscopy; ignition and combustion of gun propellants; electric ignition of guns; microscopical characterization of nitrocellulose. Mailing Add: 36 East Shore Rd Denville NJ 07834

MORROW, TERRY ORAN, b Latrobe, Pa, May 24, 47; m 75; c 1. MICROBIAL GENETICS. Educ: Grove City Col, BS, 69; Bowling Green State Univ, MA, 71, PhD(microbial genetics), 73. Prof Exp: Asst prof biol, Univ Wis-River Falls, 73-75; ASST PROF BIOL, CLARION STATE STATE UNIV, 75- Mem: Am Soc Microbiol; Sigma Xi; Am Inst Biol Sci. Res: Genetics of Staphylococcus aureus. Mailing Add: Dept of Biol Clarion State Col Clarion PA 16214

MORS, WALTER B, b Sao Paulo, Brazil, Nov 23, 20; m 44; c 3. NATURAL PRODUCTS CHEMISTRY. Educ: Univ Sao Paulo, MS, 42; Univ Brazil, Dr(chem), 60. Prof Exp: Res chemist, N Inst Agr, 43-46 & Inst Agr Chem, 47-62; PROF PHYTOCHEM, NATURAL PROD RES CTR, FED UNIV RIO DE JANEIRO, 63- Concurrent Pos: Ed annals, Brazilian Chem Soc, 59-62; privat-docent, Nat Sch Chem, Fed Univ Rio de Janeiro, 60; res chemist, Div Agr Technol, Ministry Agr, 63-66, dir, 66-72. Mem: Brazilian Acad Sci (gen secy, 65-69); Brazilian Soc Advan Sci; Brazilian Chem Soc (secy, 52-54); Soc Econ Bot. Res: Chemistry of natural products. Mailing Add: Estrada de Jacarepagua 6784 Rio de Janeiro Brazil

MORSE, ANTHONY PERRY, b Ithaca, NY, Aug 21, 11; m 34, 56; c 6. MATHEMATICS. Educ: Cornell Univ, AB, 33; Brown Univ, PhD(math), 37. Prof Exp: Instr math, Brown Univ, 36-37 & Princeton Univ, 37-38; from instr to prof, 39-73, EMER PROF MATH, UNIV CALIF, BERKELEY, 73- Concurrent Pos: Fel, Inst Advan Study, 37-39; mathematician, Theory Sect, Aberdeen Proving Ground, 44-45. Mem: Am Math Soc; NY Acad Sci. Res: Analysis; real function and measure theory. Mailing Add: Dept of Math Univ of Calif Berkeley CA 94720

MORSE, BURT JULES, b New York, NY, June 17, 26. APPLIED MATHEMATICS. Educ: City Col New York, BS, 49; Columbia Univ, AM, 51; NY Univ, PhD(math), 63. Prof Exp: Mathematician, Vitro Corp Am, 52-54; mathematician, Int Bus Mach Corp, 54-58; res assoc electromagnetic theory, NY Univ, 58-63; asst prof math, St John's Univ, NY, 63 & Univ NMex, 63-66; res mathematician, Gen Elec Co, 66-67 & Philco-Ford Corp, Pa, 67-68; mathematician, Nat Hurricane Res Lab, 68-73, MATHEMATICIAN, SATELLITE EXP LAB, NAT OCEANIC & ATMOSPHERIC ADMIN, 73- Mem: Am Math Soc; Soc Indust & Appl Math. Res: Geophysical fluid dynamics; numerical analysis; meteorology; electromagnetic theory. Mailing Add: Nat Oceanic & Atmospheric Admin NESS SEL FOB-4 S-321-B Suitland MD 20233

MORSE, DAN FRANKLIN, b Wheeling, WVa, Mar 10, 35; m 60; c 3. ANTHROPOLOGY, ARCHAEOLOGY. Educ: Univ Mich, BA, 56, MA, 59, PhD(anthrop), 67. Prof Exp: State hwy archeologist, Ga Hist Comn, 59-60; asst prof anthrop, Univ Tenn, 62-64; instr anthrop & cur, State Hwy Arch, Idaho State Univ, 66-67; asst prof, 67-71, ASSOC PROF ANTHROP, UNIV ARK, 71-; ARCHEOLOGIST, ARK ARCHEOL SURV, 67- Mem: AAAS; Soc Am Archeol; Prehist Soc Gt Brit; Am Quaternary Asn; fel Am Anthrop Asn. Res: Archeology of eastern United States; early man; beginnings and development of Mississippi culture; lithic technology; settlement pattern. Mailing Add: Ark Archeol Surv Drawer 820 State University AR 72467

MORSE, DANIEL EDWARD, molecular biology, genetics, see 12th edition

MORSE, DENNIS ERVIN, b Loup City, Nebr, Mar 21, 47; m 67; c 3. HUMAN ANATOMY. Educ: Hastings Col, BA, 69; Univ NDak, MS, 71, PhD(anat), 73. Prof Exp: Asst prof anat, George Washington Univ, 73-76; ASST PROF ANAT, MED COL OHIO, 76- Mem: Am Asn Anatomists; Am Soc Cell Biol; Sigma Xi; Electron Micros Soc Am. Res: Scanning and transmission electron microscopy of cardiovascular embryology; extracellular connective tissue ultrastructure; meningeal ultrastructure. Mailing Add: Dept of Anat Med Col of Ohio Toledo OH 43616

MORSE, DOUGLASS HATHAWAY, b Lewiston, Maine, July 20, 38; m 64. ECOLOGY. Educ: Bates Col, BS, 60; Univ Mich, MS, 62; La State Univ, PhD(zool), 65. Prof Exp: Res assoc, 65-66, from asst prof to assoc prof, 66-75, PROF ZOOL, UNIV MD, COLLEGE PARK, 75- Mem: Ecol Soc Am; Am Soc Naturalists; Soc Study Evolution. Res: Foraging ecology; interspecific relationships; ecology of social groups. Mailing Add: Dept of Zool Univ of Md College Park MD 20742

MORSE, EDWARD EVERETT, b Gardner, Mass, June 7, 32; m 54; c 4. MEDICINE, HEMATOLOGY. Educ: Harvard Univ, AB, 54, MD, 58; Am Bd Path, dipl hemat, 69, cert blood banking, 73. Prof Exp: Asst surgeon, Nat Cancer Inst, 60-62; USPHS spec fel hemat, Johns Hopkins Hosp, 62-63, from instr to asst prof med, Johns Hopkins Univ, 63-68; assoc prof, 68-74, PROF LAB MED, SCH MED, UNIV CONN, FARMINGTON, 74-, DIR HEMAT DIV, 68- Concurrent Pos: Med dir, Conn Red Cross Blood Prog, 68-74, consult, 74-; consult, Newington Vet Admin Hosp, Hartford & Bristol Hosps. Mem: Fel Am Col Physicians; fel Am Soc Clin Path; fel Asn Clin Scientists. Res: Laboratory medicine; cell preservation; transfusion therapy; platelet fibrinogen and physiology; granulocyte function. Mailing Add: Dept of Lab Med Univ of Conn Health Ctr Farmington CT 06032

MORSE, ELLEN HASTINGS, b Durham, NH, Oct 10, 08. NUTRITION. Educ: Wellesley Col, BA, 30; Smith Col, MA, 38; Univ Mass, MS, 49; Univ Conn, PhD(nutrit), 60. Prof Exp: Asst cur, Dept Art, Smith Col, 31-46; res fel home econ & nutrit, Univ Mass, 46-49; nutritionist, Colo Exp Sta, 49-51 & NY State Health Dept, 51-52; res asst home econ & nutrit, Univ Conn, 52-60; from assoc prof to prof, 60-74, EMER PROF NUTRIT, UNIV VT, 74- Mem: AAAS; Am Home Econ Asn; Am Dietetic Asn; Am Inst Nutrit. Mailing Add: 38-A University Heights Burlington VT 05401

MORSE, ERSKINE VANCE, b Peoria, Ill, June 25, 21; m 45; c 4. VETERINARY MICROBIOLOGY, VETERINARY PUBLIC HEALTH. Educ: Cornell Univ, DVM, 44, MS, 48, PhD(vet bact), 49; Am Col Vet Prev Med, dipl & cert vet pub health, 75. Prof Exp: Asst pathogenic bact, Cornell Univ, 47-48, Am Vet Med Asn fel, 48-49; from asst prof to assoc prof vet sci, Univ Wis, 49-55; from assoc prof to prof microbiol & pub health, Mich State Univ, 55-58; prof & assoc dir vet med res inst, Iowa State Univ, 58-60; prof vet sci, head dept & dean sch agr exp sta, 60-70, H W HANDLEY PROF VET MED & ENVIRON HEALTH & ASSOC DIR ENVIRON HEALTH INST, SCH VET MED, PURDUE UNIV, 70- Concurrent Pos: Consult, Nat Asn Stand Med Vocab, 62-, Vet Admin, 66-, Surgeon Gen, US Air Force, 68-70, USPHS & AID; alt chmn US deleg, Conf Vet Med Educ, Food & Agr Orgn-WHO, Copenhagen, 65; judge, Int Sci Fair, 65, 68, 69, 74 & 75; mem, Nat Bd Vet Med Examr, 65-74; nat counr, Purdue Res Found, 65-; mem bd trustees, Am Asn Accreditation of Lab Animal Care, 68-75, vchmn bd, 72-73, chmn bd, 74-75; mem nat coun health prof educ assistance, Dept Health, Educ & Welfare, 69-72; chmn comt animal health, Nat Res Coun-Nat Acad Sci, 69-72; evaluator-consult, NCent Asn Cols & Schs-Comn Insts Higher Educ, 72-; Purdue Univ liaison rep for vet serv to Asst Surgeon Gen, US Army, 73-; reviewer, Jour, Am Vet Med Asn, 74- & Am J Vet Res, 75- Mem: Soc Exp Biol & Med; Conf Res Workers Animal Dis; Am Vet Med Asn (secy, 66-69); US Animal Health Asn; Am Asn Lab Animal Sci. Res: College administration; environmental health; pathogenic bacteriology; infectious diseases of animals; leptospirosis, vibriosis, brucellosis and corynebacterial infections; laboratory animal medicine; epidemiology, microbiology and treatment of salmonellosis and zoonotic diseases; monitoring environmental quality and water pollution. Mailing Add: Sch of Vet Med Purdue Univ Lafayette IN 47907

MORSE, ERWIN EMERSON, b Willimantic, Conn, Oct 6, 15; m 44; c 3. PHYSICAL CHEMISTRY. Educ: Bowdoin Col, BS, 36; Univ Calif, PhD(chem), 39. Prof Exp: Res chemist, Spreckels Sugar Co, 40-48; res mgr, 48-65, gen mgr res & develop div, 65-71, DIR TECH SERV, BROWN CO, 71- Mem: AAAS; Tech Asn Pulp & Paper Indust; Am Chem Soc; Inst Food Technologists. Res: Cellulose chemistry and derivatives, including cellulose ion-exchangers; pulp and paper technology. Mailing Add: 26 Jewell St Gorham NH 03581

MORSE, FRED A, b Colorado Springs, Colo, Jan 11, 37; m 57; c 3. SPACE PHYSICS. Educ: Univ Idaho, BS, 58; Univ Mich, MS, 60, PhD(phys chem), 62. Prof Exp: Instr phys chem, Univ Del, 61-62; fel, Univ Mich, 62; mem tech staff, 64-68, staff scientist, 68-72, ASSOC DEPT HEAD ATMOSPHERIC PHYSICS, SPACE PHYSICS LAB, AEROSPACE CORP, 72- Mem: AAAS; Am Geophys Union. Res: Molecular and ion beam collision; atomic spectroscopy; atmospheric and auroral physics;

chemiluminescent reaction rate; aeronomy; ionosphere. Mailing Add: Space Physics Lab Aerospace Corp PO Box 95085 Los Angeles CA 90045

MORSE, GARTH EDWIN, b Dell Rapids, SDak, Dec 16, 21; m 48; c 2. PHYSICS. Educ: Pasadena Col, AB, 50; Univ Southern Calif, MS, 58; Univ Calif, Riverside, PhD(physics), 66. Prof Exp: Physicist, Naval Ord Test Sta, 51-52; from asst prof to prof physics, Pasadena Col, 52-73; PROF PHYSICS, POINT LOMA COL, 73- Concurrent Pos: Consult, DeMornay-Bonardi, Inc, Calif, 66-67. Mem: Am Asn Physics Teachers. Res: Lattice dynamics, especially microwave frequencies. Mailing Add: Dept of Physics Point Loma Col San Diego CA 92106

MORSE, GUY EMERY, b Harford, NY, Jan 11, 17; m 47; c 3. VETERINARY BACTERIOLOGY. Educ: Cornell Univ, DVM, 42. Prof Exp: Dir lab bovine mastitis, State Univ NY Vet Col, Cornell Univ, 54-61; ASST PROF VET MED & DIR MASTITIS RES, SCH VET MED, UNIV PA, 61- Concurrent Pos: Mem, Nat Mastitis Res Worker's Conf. Mem: AAAS; Am Dairy Sci Asn; US Animal Health Asn; Am Vet Med Asn; Am Soc Microbiol. Res: Bovine mastitis, pathogenesis and control. Mailing Add: Dept of Clin Studies Univ of Pa Sch of Vet Med Kennett Square PA 19348

MORSE, HERBERT CARPENTER, III, b Washington, DC, May 7, 43; m 69; c 2. IMMUNOLOGY. Educ: Oberlin Col, BA, 65; Harvard Univ, MD, 70. Prof Exp: From intern to resident, Peter Bent Brigham Hosp, 70-72; res assoc, 72-75, SR INVESTR IMMUNOL, NAT INST ALLERGY & INFECTIOUS DIS, 75- Concurrent Pos: Attend physician, Arthritis & Rheumatism Br, Nat Inst Arthritis, Metab & Digestive Dis, 76- Mem: Am Asn Immunologists. Res: Interactions of viruses and the immune response; regulation of antibody formation. Mailing Add: Lab of Microbial Immunity Nat Inst Allergy & Infectious Dis Bethesda MD 20014

MORSE, HOWARD CURTIS, cell biology, reproductive physiology, see 12th edition

MORSE, JANE H, b Grosse Pointe, Mich, Aug 27, 29; m 56; c 2. IMMUNOLOGY. Educ: Smith Col, BA, 51; Columbia Univ, MD, 55. Prof Exp: Res assoc immunol, Rockefeller Univ, 60-62; instr & asst prof med, 62-75, ASSOC PROF CLIN MED, COL PHYSICIANS & SURGEONS, COLUMBIA UNIV, 75- Concurrent Pos: Mem, Allergy & Immunol Study Sect, USPHS, 73-77. Mem: Am Asn Immunologists; Am Rheumatism Asn. Res: Immune complexes and complement in autoimmune diseases; the role of inhibitors in the immune response. Mailing Add: 630 W 168th St Dept of Med New York NY 10032

MORSE, JOHN THOMAS, b Oakland, Calif, Apr 30, 35; m 60; c 2. ENVIRONMENTAL PHYSIOLOGY. Educ: Ore State Univ, BS, 56; Univ Calif, Davis, PhD(physiol), 68. Prof Exp: Pvt indust res grant, 68-69, asst prof, 68-72, ASSOC PROF PHYSIOL, CALIF STATE UNIV, SACRAMENTO, 72- Mem: Am Physiol Soc; Am Inst Biol Sci. Res: Physiological adaptations favoring physical work performance in varying environments. Mailing Add: Dept of Biol Sci Calif State Univ Sacramento CA 95819

MORSE, JOHN WILBUR, b Ft Dodge, Iowa, Nov 11, 46. GEOCHEMISTRY. Educ: Univ Minn, BS, 69; Yale Univ, MPhil, 71, PhD(geol), 73. Prof Exp: ASST PROF OCEANOG, FLA STATE UNIV, 73- Mem: Am Geophys Union; Int Asn Geochemists & Cosmochemists. Res: Application of chemical kinetics and surface chemistry to diagenetic reactions in marine sediment. Mailing Add: Dept of Oceanog Fla State Univ Tallahassee FL 32306

MORSE, JOSEPH GRANT, b Colorado Springs, Colo, Oct 16, 39; m 63; c 2. INORGANIC CHEMISTRY. Educ: SDak State Col, BS, 61; Univ Mich, MS, 63, PhD(inorg chem), 67. Prof Exp: Asst prof, 68-74, ASSOC PROF CHEM, UTAH STATE UNIV, 74- Concurrent Pos: Lectr, Univ Mich, 65-66. Mem: AAAS; Am Chem Soc. Res: Synthesis and properties of new ligands, especially of phosphorus group; chemistry of pi-acid chelates; photochemistry of fluorophosphines. Mailing Add: Dept of Chem & Biochem Utah State Univ Logan UT 84332

MORSE, KAREN W, b Monroe, Mich, May 8, 40; m 63; c 1. INORGANIC CHEMISTRY. Educ: Denison Univ, BSc, 62; Univ Mich, MSc, 64, PhD(chem), 67. Prof Exp: Res scientist, Ballistic Res Inst, 67-68; ASST PROF CHEM, UTAH STATE UNIV, 68- Mem: Am Chem Soc. Res: Synthesis and behavior of fluorophosphine derivatives; synthesis and behavior of transition metal complexes of borano carbonate derivatives. Mailing Add: Dept of Chem & Biochem Utah State Univ Logan UT 84322

MORSE, LEWIS DAVID, b Brooklyn, NY, Oct 29, 24; m 46; c 1. BIOCHEMISTRY. Educ: NY Univ, BA, 48; Brooklyn Col, MA, 52. Prof Exp: Res chemist, Col Physicians & Surgeons, Columbia Univ, 50-51 & Stein Hall Corp, 51-55; chief chemist & prod supvr, Myer 1890 Beverages, Inc, 55-59; proj leader, Am Sugar Refining Co, Inc, 59-63 & Nat Cash Register Co, 63-64; mgr prod res, Ionac Chem Co, NJ, 64-67; prod develop fel, 67, SECT HEAD FINE CHEM, PROD DEVELOP, MERCK & CO, INC, RAHWAY, 67- Mem: AAAS; Am Chem Soc; Am Inst Chemists; Inst Food Technologists; Am Asn Cereal Chemists. Res: Polymer chemistry and synthesis; microencapsulation; ion exchange; carbonated beverage technology; flavor compounding; enzyme, sugar, starch, adhesive and microbiological chemistry; statistical design and analysis of experiments; vitamin technology in foods and medicinals; nutrition; antiseptics. Mailing Add: 133 Snowden Lane Princeton NJ 08540

MORSE, LURA MYRA, b Aberdeen, Wash, Dec 1, 13. NUTRITION, BIOCHEMISTRY. Educ: Univ Calif, Berkeley, BA, 36, MA, 39, PhD(biochem), 46. Prof Exp: Assoc biochem, Univ Calif, Berkeley, 45-46; instr foods & nutrit, Univ Calif, Davis, 46-48; from instr to asst prof nutrit, Med Ctr, Univ Calif, San Francisco, 48-58; assoc prof, 58-64, PROF NUTRIT, UNIV MINN, ST PAUL, 64-, PROF FOOD SCI, 74- Concurrent Pos: USPHS spec res fel folic acid, Univ Calif, Berkeley, 67-68. Mem: AAAS; Am Inst Nutrit; Am Dietetic Asn; Teratology Soc. Res: Folic acid metabolism; congenital defects in folic acid deficiency; vitamin C metabolism in pregnant rats. Mailing Add: Dept of Nutrit Univ of Minn St Paul MN 55101

MORSE, MARSTON, b Waterville, Maine, Mar 24, 92; m 22, 40; c 7. MATHEMATICS. Educ: Colby Col, AB, 14; Harvard Univ, AM, 15, PhD(math), 17. Hon Degrees: Many from US and foreign univs, 35- Prof Exp: Benjamin Pierce instr math, Harvard Univ, 19-20; from instr to asst prof, Cornell Univ, 20-25; assoc prof, Brown Univ, 25-26; from asst prof to prof, Harvard Univ, 26-35; prof, 35-75, EMER PROF MATH, INST ADVAN STUDY, 75- Concurrent Pos: Lectr, Int Math Cong, 32, Sorbonne & Inst Henri Poincare, 38, Gibbs lectr, 52; vis prof, City Univ New York, 65-66. Consult, Off Chief Ord, US Coast & Geod Surv, Nat Defense Res Comt & Off Sci Res & Develop, 44. Chmn div math, Nat Res Coun, 50-52; mem bd, NSF, 50-54; vpres, Int Math Union, 58-62, chmn US nat comn math, 59-63. Honors & Awards: Bocher Prize, 33; Presidential Cert of Merit, 47; Nat Medal of Sci, 65; Croix de Guerre; Chevalier, Legion d'Honneur. Mem: Nat Acad Sci; AAAS (vpres, 39); fel Am Acad Arts & Sci; Am Math Soc (vpres, 33-35, pres, 40-42); Am Philos Soc. Res: Analysis and differential geometry in the large; variational theory; differential topology. Mailing Add: Inst for Advan Study Princeton NJ 08540

MORSE, MARY PATRICIA, b Hyannis, Mass, Aug 29, 38. MALACOLOGY. Educ: Bates Col, BS, 60; Univ NH, MS, 62, PhD(zool), 66. Prof Exp: Instr biol, Suffolk Univ, 62-63; from instr to asst prof, 64-70, ASSOC PROF BIOL, MARINE SCI INST, NORTHEASTERN UNIV, 70- Concurrent Pos: Trustee, Bates Col & Charles River Acad; res assoc malacol, Harvard Univ, 74-; Brasilian grant to study interstitial molluscs, 75. Mem: Am Soc Zoologists; Malacol Soc London; Sigma Xi. Res: Systematics and biology of interstitial molluscs; fine structural studies of molluscan epithelia. Mailing Add: Marine Sci Inst Northeastern Univ Nahant MA 01908

MORSE, MELVIN LAURANCE, b Hopkinton, Mass, Feb 23, 21; m 49; c 2. GENETICS. Educ: Univ NH, BS, 44; Univ Ky, MS, 47; Univ Wis, PhD(genetics), 55. Prof Exp: Jr biologist radiation res, Biol Div, Oak Ridge Nat Lab, 47-51; res assoc genetics, Univ Wis, 55-56; res microbiologist, Webb-Waring Lung Inst, 56-58, asst dir inst, 59-71, actg dir, 71-72, from asst prof to assoc prof, Med Ctr, 56-66, vchmn dept, 71-73, PROF BIOPHYS & GENETICS, MED CTR, UNIV COLO, 66-, HEAD DIV GENETICS, WEBB-WARING LUNG INST, 56-, JAMES J WARING CHAIR BIOL, 60- Concurrent Pos: USPHS sr res fel, 61, career develop award, 62-; consult, Army Med Res & Nutrit Lab, Fitzsimons Gen Hosp, 60-70; foreign res, Inst Molecular Biol, Univ Geneva, 62-63. Mem: Am Soc Microbiol; Genetics Soc Am; Biophys Soc; Am Genetic Asn; Soc Human Genetics. Res: Biochemical genetics; biophysics; microbiology. Mailing Add: Webb-Waring Lung Inst Univ of Colo Med Ctr Denver CO 80220

MORSE, PHILIP DEXTER, II, b Bakersfield, Calif, Oct 17, 44; m 66; c 2. MOLECULAR BIOLOGY. Educ: Univ Calif, Davis, BA, 67, PhD(zool), 72. Prof Exp: Res assoc pharmacol, Univ Berne, 71-73; res assoc biophys, Pa State Univ, 73-75; ASST PROF BIOL, WAYNE STATE UNIV, 75- Mem: AAAS; Biophys Soc. Res: Structure and function of biological membranes using electron spin resonance and freeze-fracture electron microscopy; interrelationships between intracellular water order and membrane structure. Mailing Add: Dept of Biol Wayne State Univ Detroit MI 48202

MORSE, PHILIP MCCORD, b Shreveport, La, Aug 6, 03; m; c 2. THEORETICAL PHYSICS, OPERATIONS RESEARCH. Educ: Case Inst Technol, BS, 26; Princeton Univ, AM, 27, PhD(physics), 29. Hon Degrees: ScD, Case Inst Technol, 40. Prof Exp: Instr physics, Princeton Univ, 29-30; Int Res fel, Univ Munich & Cambridge Univ, 30-31; from asst prof to prof physics, 31-69, dir underwater sound lab, 39-42, dir comput ctr, 56-67, chmn fac, 58-60, dir opers res ctr, 58-69, EMER LECTR PHYSICS, MASS INST TECHNOL, 69- Concurrent Pos: Supvr sound control lab, Harvard Univ, 39-45; dir opers res group, Off Sci Res & Develop, US Navy, 42-46 & Brookhaven Nat Lab, 46-48; dep dir weapons systs eval group, Off Secy Defense, Nat Mil Estab, 49-50; ed, Annals of Physics, 57-; dir, Control Data Corp, 64-; mem bd trustees, New Eng Regional Comput Net, 68- Trustee, Rand Corp, 48-62, Inst Defense Anal, 55-60, Adage Inc, 58-64, Anal Serv Inc, 62-72 & Coun Libr Resources, 64- Chmn comt revision math tables, Nat Res Coun, 54-64, mem comt uses comput, 60-65, comt natural resources, 62-65; chmn adv panel opers res, Sci Adv, NATO, 59-65; chmn adv panel opers res, Orgn Econ Coop & Develop, 62-; mem telecommun sci panel, Dept Com, 65- Honors & Awards: Presidential Medal for Merit, 46; Silver Medal, Brit Oper Res Soc, 65; Lanchester Prize, Opers Res Soc Am, 69, Gold Medal, 74. Mem: Fel Nat Acad Sci; AAAS; fel Am Acad Arts & Sci; fel Acoust Soc Am (vpres, 46, pres, 50); fel Am Phys Soc (vpres, 70-71, pres, 71-72). Res: Acoustics; search theory; library operation; computer networking. Mailing Add: 126 Wildwood St Winchester MA 01890

MORSE, RICHARD KENNETH, b Seattle, Wash, Oct 31, 41; m 66. DEVELOPMENTAL PHYSIOLOGY. Educ: Johns Hopkins Univ, AB, 63; Case Western Reserve Univ, PhD(biol), 68. Prof Exp: Res asst biol, Case Western Reserve Univ, 63-67; genetics & cell biol, Univ Conn, 68-70; asst prof biol, Point Park Col, 70-74; ASST PROF BIOL, UNIV NC, GREENSBORO, 74- Mem: Soc Develop Biol; Am Soc Zoologists; Am Inst Biol Sci; Sigma Xi (secy, 75-76); Soc Behavioral Kinesiology. Educ: Myoblast orientation; affinity chromatography. Mailing Add: Dept of Biol Univ of NC Greensboro NC 27412

MORSE, RICHARD STETSON, b North Abington, Mass, Aug 19, 11; m 35; c 2. PHYSICS. Educ: Mass Inst Technol, SB, 33. Hon Degrees: DSc, Polytech Inst Brooklyn, 59; PhD(eng), Clark Univ, 60. Prof Exp: Mem staff, Mass Inst Technol, 35; physicist, Eastman Kodak Co & Distillation Prod, Inc, NY, 35-40; pres, Nat Res Corp, 40-59; dir res & Asst Secy Army, 59-61; sr lectr, Sloan Sch Mgt, Mass Inst Technol, 62-72; PRES, MIT DEVELOP FOUND, INC, 72- Concurrent Pos: Dir, Dresser Industs, Inc, Res Anal Corp & Japan Fund Inc; trustee, Midwest Res Inst, Boston Mus Sci, Marine Biol Lab, Woods Hole Oceanog Inst & Aerospace Corp, 74-; chmn bd vis, Air Force Systs Command; chmn & dir, Sci Energy Systs Corp; former mem, Defense Sci Bd & tech adv, Panel Biol & Chem Warfare, US Dept Defense; vchmn tech adv bd, US Dept Com; chmn, Army Sci Adv Panel; mem gen adv comt, Energy Res & Develop Admin, 75- Honors & Awards: Dept of Army Distinguished Civilian Serv Medal. Mem: Am Chem Soc. Res: High vacuum technology; vacuum metallurgy; dehydration; evaporation of metals; vacuum pumps and gauges; photo processes. Mailing Add: MIT Develop Found Inc 50 Memorial Dr Cambridge MA 02139

MORSE, ROBERT IRA, physical chemistry, see 12th edition

MORSE, ROBERT MALCOLM, b Haverhill, Mass, Dec 21, 38; m 61; c 2. HIGH ENERGY PHYSICS. Educ: San Jose State Col, BA, 63; Univ Wis-Madison, MA, 65, PhD(physics), 69. Prof Exp: Res assoc physics, Univ Wis, 69-70; RES ASSOC PHYSICS, UNIV COLO, BOULDER, 70-, ASST PROF PHYSICS & ASTROPHYSICS, 74- Mem: Am Phys Soc. Res: Weak interactions of strange particles; meson spectroscopy of strongly interacting particles; meson production and scattering processes at high energies. Mailing Add: Dept of Physics Univ of Colo Boulder CO 80302

MORSE, ROBERT WARREN, b Boston, Mass, May 25, 21; m 43; c 3. PHYSICS. Educ: Bowdoin Col, BS, 43; Brown Univ, ScM, 47, PhD(physics), 49. Hon Degrees: ScD, Bowdoin Col, 66. Prof Exp: From asst prof to prof physics & head dept, Brown Univ, 49-62, dean col, 62-64; Asst Secy Navy for Res & Develop, 64-66; pres, Case Western Reserve Univ, 66-71; dir res, 71-73, ASSOC DIR & DEAN GRAD STUDIES, WOODS HOLE OCEANOG INST, 73- Concurrent Pos: Howard Found fel, Cambridge Univ, 54-55; mem comt undersea warfare, Nat Acad Sci, 57-64, chmn, 62-64, chmn bd human resources, 70-74, chmn ocean affairs bd, 71-75; chmn interagency comt oceanog, Fed Coun Sci & Technol, 64-66; mem, Naval Res Adv Comt, 71-74. Mem: Fel Am Acad Arts & Sci; fel Am Phys Soc; fel Acoust Soc Am (pres, 65-66). Res: Ultrasonics; elastic waves; metals; cryogenics; underwater sound. Mailing Add: Woods Hole Oceanog Inst Woods Hole MA 02543

MORSE, ROGER ALFRED, b Saugerties, NY, July 5, 27; m 51; c 3.

ENTOMOLOGY, APICULTURE. Educ: Cornell Univ, BS, 50, MS, 53, PhD(entom), 55. Prof Exp: Entomologist, State Plant Bd, Fla, 55-57; asst prof hort, Univ Mass, 57; from asst prof to assoc prof, 57-72, PROF APICULT, CORNELL UNIV, 72- Concurrent Pos: Vis prof, Col Agr, Univ Philippines, 68. Mem: Fel AAAS; Entom Soc Am; Bee Res Asn. Res: Evolution of the Apoidea; toxicity of insecticides to honey bees, honey wine, honey production and handling; social structure of honey bee colony. Mailing Add: Dept of Entom Cornell Univ Ithaca NY 14850

MORSE, RONALD LOYD, b Kearney, Nebr, May 15, 40; m 63; c 2. INDUSTRIAL ORGANIC CHEMISTRY. Educ: Univ Nebr, BS, 63; Univ Wis, PhD(org chem), 68. Prof Exp: SR RES CHEMIST, MONSANTO INDUST CHEM CO, 67- Mem: Am Chem Soc. Mailing Add: Monsanto Indust Chem Co 800 N Lindbergh Blvd St Louis MO 63166

MORSE, ROY EARL, b Boston, Mass, Nov 3, 16; m 46; c 3. FOOD SCIENCE. Educ: Univ Mass, BS, 40, MS, 41, PhD(food technol), 48. Prof Exp: Control food technologist, Hills Bros Co, NY, 41; instr food technol, Ore State Col, 41-42; packaging researcher, Owens-Ill Co, Calif, 42-43; prod mgr, Featherweight Foods, Maine, 43-44; instr food technol, Univ Mass, 46-48; assoc prof, Univ Ga, 48-49; group leader, Food Lab, Monsanto Chem Co, 49-51; dir res, Kingan & Co, Ind, 51-53 & Wm J Stange Co, 53-55; chmn dept food sci, Rutgers Univ, 55-59; vpres tech res, Thomas J Lipton, Inc, 59-66; vpres res, Pepsi Co, Inc, NY, 66-69; PROF FOOD SCI, RUTGERS UNIV, NEW BRUNSWICK, 69- Concurrent Pos: Mem nutrit adv comt, Govt Tunisia; ed, Biol Abstracts; vis prof, Univ New South Wales, 75-76 & Univ Iceland, 76. Mem: Fel AAAS; Am Chem Soc; Inst Food Technologists; Am Mgt Soc. Res: Effect of organic acids on yeasts and molds; factors in production and storage of dried bakers' yeast; antagonisms in yeast growth factors; animal fat stability and rendering; meat product processing; food preservation and dehydration; international feeding problems; bean protein digestibility; mycotoxins; fish protein compression; food foam stability. Mailing Add: Dept of Food Sci Rutgers Univ New Brunswick NJ 08903

MORSE, RUSSELL W, physiology, biophysics, deceased

MORSE, STEARNS ANTHONY, b Hanover, NH, Jan 3, 31; m 60; c 3. PETROLOGY. Educ: Dartmouth Col, AB, 52; McGill Univ, MSc, 58, PhD(geol), 62. Prof Exp: Mem, Blue Dolphin Labrador Exped, 49, 51, 52 & 54; petrologist, Brit Nfld Explor Ltd, Can, 59-61; geologist, Cold Regions Res & Eng Lab, US Army, 62; from asst prof to assoc prof geol, Franklin & Marshall Col, 62-71; assoc prof, 71-74, PROF GEOL, UNIV MASS, AMHERST, 74- Concurrent Pos: Carnegie Corp fel, Carnegie Inst Geophys Lab, Washington, DC, 67-68. Honors & Awards: Peacock Mem Prize, 62. Mem: AAAS; Geol Soc Am; Mineral Soc Am; Am Geophys Union; Mineral Asn Can. Res: Geochemistry; igneous and metamorphic petrology; layered intrusions and magma evolution; feldspars; anorthosites. Mailing Add: Dept of Geol Univ of Mass Amherst MA 01002

MORSE, STEPHEN ALLEN, b Los Angeles, Calif, Apr 11, 42; m 74. MICROBIOLOGY. Educ: San Jose State Col, BA, 64; Univ NC, MSPH, 66, PhD(microbiol), 69. Prof Exp: NSF fel, Univ Ga, 69-70; asst prof biol, Southeastern Mass Univ, 70-71; res assoc microbiol, Sch Pub Health, Harvard Univ, 71-72, asst prof, 72-74; asst prof microbiol, 74-75, ASSOC PROF MICROBIOL, HEALTH SCI CTR, UNIV ORE, 75- Concurrent Pos: Vis lectr, Sch Pub Health, Univ Calif, Los Angeles, 72; sr instr clin bact, Harvard Med Sch, 72-74. Honors & Awards: Mary Poston Award, NC Br, Am Soc Microbiol, 65. Mem: AAAS; Am Soc Microbiol; Soc Gen Microbiol; Am Veneral Dis Asn; Soc Exp Biol Med. Res: Physiology and metabolism of infectious agents. Mailing Add: Dept of Microbiol & Immunol Univ of Ore Health Sci Ctr Portland OR 97201

MORSE, STEPHEN IVOR, b New York, NY, Nov 3, 30; m 56; c 2. MICROBIOLOGY, MEDICINE. Educ: Yale Univ, BA, 51; Wash Univ, MD, 55; Rockefeller Univ, PhD, 60. Prof Exp: From intern to asst resident med, Columbia-Presby Hosp, 55-57; from asst prof to assoc prof, Rockefeller Univ, 60-69; PROF MICROBIOL & IMMUNOL & CHMN DEPT, STATE UNIV NY DOWNSTATE MED CTR, 69- Concurrent Pos: Guggenheim fel, Nat Inst Med Res, Mill Hill, London, Eng, 68-69. Mem: Am Soc Clin Invests; Am Asn Immunol; Harvey Soc; Am Soc Microbiol. Res: Lymphocytosis induced by B pertussis; cellular immunology. Mailing Add: Dept of Microbiol & Immunol State Univ NY Downstate Med Ctr Brooklyn NY 11203

MORSE, WILLIAM HERBERT, b Yorktown, Va, May 30, 28; m 58; c 4. PHARMACOLOGY. Educ: Univ Va, BA, 50, MA, 52; Harvard Univ, PhD(psychol), 55. Prof Exp: Res fels psychol, 55-58, from instr to asst prof, Med Sch, 58-67, ASSOC PROF PSYCHOL, MED SCH, HARVARD UNIV, 67- Mem: Am Soc Pharmacol & Exp Therapeut. Res: Behavioral pharmacology and physiology. Mailing Add: Dept of Pharmacol Harvard Med Sch Boston MA 02115

MORSS, LESTER ROBERT, b Boston, Mass, Apr 6, 40; m 66; c 4. PHYSICAL INORGANIC CHEMISTRY. Educ: Harvard Univ, BA, 61; Univ Calif, Berkeley, PhD(chem), 69. Prof Exp: Res assoc chem, Purdue Univ, 69-70; ASST PROF CHEM, RUTGERS UNIV, NEW BRUNSWICK, 71- Concurrent Pos: Vis asst prof & NSF fel, Purdue Univ, 70-71. Mem: AAAS; Am Chem Soc. Res: Chemistry of lanthanides and actinides; thermochemistry of electrolytes; inorganic complex compounds. Mailing Add: Sch of Chem Rutgers Univ New Brunswick NJ 08903

MORTADA, MOHAMED, b Alexandria, Egypt, Mar 14, 25; m 58; c 1. INDUSTRIAL CHEMISTRY. Educ: Univ Cairo, BSc, 46; Univ Calif, MS, 49, PhD(petrol eng), 52. Prof Exp: Instr, Univ Cairo, 46-47; petrol engr, Mobil Oil Corp, 52-53, sr res technologist, Field Res Lab, 54-59, from res assoc to sr opers res assoc, 60-63; consult petrol reservoir eng, 63-67; OWNER, MORTADA INT, 69- Concurrent Pos: Distinguished lectr, Soc Petrol Engrs, 70-71. Honors & Awards: Noble Prize, Founder Eng Soc, 56; Raymond Award, Am Inst Mining, Metall & Petrol Engrs, 57. Mem: Sigma Xi; Soc Petrol Engrs. Res: Flow of fluids through porous media; petroleum chemistry and technology; computer science. Mailing Add: Mortada Int Suite 406 7616 LBJ Freeway Dallas TX 75240

MORTARA, LORNE B, b Chicago, Ill, Sept 29, 32; m 57; c 4. HIGH ENERGY PHYSICS. Educ: Purdue Univ, BS, 53, PhD(physics), 63. Prof Exp: Asst physics, Purdue Univ, 56-63; res assoc, Univ Ariz, 63-65; sr scientist physics instrumentation, Albuquerque Lab, Edgerton, Germeshausen & Grier, Inc, 65-68; SR ENG PHYSICIST, AURA, INC, 68- Mem: AAAS; Am Phys Soc. Res: Instrumentation design; application of computer systems. Mailing Add: Kitt Peak Nat Observ Box 26732 Tucson AZ 85726

MORTEL, RODRIGUE, b Saint-Marc, Haiti, Dec 3, 33; m 71. OBSTETRICS & GYNECOLOGY. Educ: Univ Haiti, MD, 60; Am Bd Obstet & Gynec, dipl, 70. Prof Exp: Asst clin instr obstet & gynec, Hahnemann Med Col & Hosp, 67-68, instr, 68-70, sr instr, 70-71, asst prof, 71-72; asst prof obstet & gynec, 72-74, ASSOC PROF OBSTET & GYNEC, HERSHEY MED CTR, PA STATE UNIV, 74-, CHIEF DIV GYNEC ONCOL, 74- Concurrent Pos: USPHS grant gynec oncol, Hahnemann Med Col & Hosp, 68-69; USPHS grant, Mem Hosp Cancer & Allied Dis, New York, 69-70. Mem: Fel Am Col Obstet & Gynec; fel Am Col Surgeons; James Ewing Soc; NY Acad Sci. Res: Clinical research in gynecologic oncology. Mailing Add: Dept of Obstet & Gynec Hershey Med Ctr Pa State Univ Hershey PA 17033

MORTENSEN, EARL MILLER, b Salt Lake City, Utah, June 25, 33; m 62; c 4. PHYSICAL CHEMISTRY. Educ: Univ Utah, BA, 5S, PhD(chem), 59. Prof Exp: NSF fel chem, Univ Calif, Berkeley, 59-60, lectr, 60-61, chemist, Radiation Lab, 60-62; asst prof chem, Univ Mass, Amherst, 62-69; ASSOC PROF CHEM, CLEVELAND STATE UNIV, 69- Concurrent Pos: NSF grant, 64-66; NSF grant, Univ Mass, Amherst, 68-69. Mem: Am Phys Soc; Am Chem Soc. Res: Theoretical reaction kinetics; polarizabilities and anisotropies of molecules; computer applications to chemical education. Mailing Add: Dept of Chem Cleveland State Univ Cleveland OH 44115

MORTENSEN, EDITH (ELIZABETH), b Odense, Denmark, July 16, 03; US citizen. ZOOLOGY. Educ: Carleton Col, BA, 25; Univ Minn, MA, 27; George Washington Univ, PhD(zool), 45. Prof Exp: Asst zool, Mt Holyoke Col, 25-26 & Univ Minn, 26-27; from instr to asst prof, St Paul-Luther Col, 27-33; asst prof, Univ Maine, 33-36; from asst prof to prof, 36-72, EMER PROF ZOOL, GEORGE WASHINGTON UNIV, 72- Mem: Am Soc Zoologists. Res: Invertebrate zoology; protozoology. Mailing Add: 4444 Faraday Pl NW Washington DC 20016

MORTENSEN, HARLEY EUGENE, b Albuquerque, NMex, Mar 1, 31; m 52; c 5. ORGANIC BIOCHEMISTRY. Educ: Regis Col, BS, 54; Kans State Univ, PhD(org chem), 61. Prof Exp: Res chemist, Benger Lab, E I du Pont de Nemours & Co, 61-67; from asst prof to assoc prof, 67-75, PROF CHEM, SOUTHWEST MO STATE UNIV, 75- Concurrent Pos: NSF res partic, Acad Year Exten, Southwest Mo State Univ, 70-72. Mem: Am Chem Soc. Res: Organic synthesis; enzymology. Mailing Add: Dept of Chem Southwest Mo State Univ Springfield MO 65802

MORTENSEN, JOHN ALAN, b San Antonio, Tex, May 11, 29; m 57; c 1. HORTICULTURE, GENETICS. Educ: Tex A&M Univ, BS, 50, MS, 51; Cornell Univ, PhD(plant breeding), 58. Prof Exp: Plant breeder, Birds Eye Div, Gen Foods Corp, 57-60; asst geneticist, Agr Exp Sta, 60-68, ASSOC PROF & ASSOC GENETICIST, WATERMELON & GRAPE INVESTS LAB, INST FOOD & AGR SCI, AGR RES CTR, UNIV FLA, 68- Mem: Am Genetic Asn; Am Soc Hort Sci; Am Pomol Soc. Res: Disease and insect resistance in grapes; development of improved varieties of scions and rootstocks in grapes through breeding and testing; nutritional and inheritance studies in grapes. Mailing Add: Agr Res Ctr Univ of Fla PO Box 388 Leesburg FL 32748

MORTENSEN, OTTO AXEL, b Milwaukee, Wis, June 4, 02; m 31; c 3. ANATOMY. Educ: Univ Wis, BS, 27, MS, 28, MD, 29. Prof Exp: From instr to prof anat, 30-72, chmn dept, 50-67, from asst dean to assoc dean sch med, 48-67, EMER PROF ANAT, SCH MED, UNIV WIS-MADISON, 72-; VIS PROF ANAT, MED CTR, STANFORD UNIV, 72- Concurrent Pos: Mem, Gov Adv Comt Educ, 55-; consult, Int Coop Admin, Peru, 57-58; mem res career awards comt, NIH, 62-65, mem anat training grants comt, 65-69. Mem: Am Asn Anatomists. Res: Cerebrospinal fluid; tissue fluid space of peripheral nerves; phagocytosis by reticuloendothelial cells; electromyography. Mailing Add: Dept of Anat Stanford Univ Med Ctr Stanford CA 94305

MORTENSEN, RAYMOND ARCHIE, b Marquam, Ore, Sept 17, 96; m 23; c 1. CHEMISTRY. Educ: Pac Union Col, AB, 19; Univ Southern Calif, MS, 25; Stanford Univ, PhD, 32. Prof Exp: Instr physics & chem, Pac Union Col, 19-24, prof chem, 25-37; prof biochem, 38-74, DISTINGUISHED SERV PROF BIOCHEM, SCH MED, LOMA LINDA UNIV, 74- Mem: Am Chem Soc. Res: Photochemical reactions; behavior of lead in the animal body; metabolism of cholesterol, glutathione, pyruvate and glycine. Mailing Add: Dept of Biochem Loma Linda Univ Sch of Med Loma Linda CA 92354

MORTENSEN, WALTER PETER, soils, see 12th edition

MORTENSON, LEONARD EARL, b Melrose, Mass, June 24, 28; m 52; c 4. BIOCHEMISTRY. Educ: RI State Col, BS, 50; Univ Wis, MS, 52, PhD(bact, biochem), 54. Prof Exp: Asst bact, Univ Wis, 50-52, asst, Enzyme Inst, 52-53, NSF fel, 53-54; res biochemist, E I du Pont de Nemours & Co, 54-61; assoc prof, 62-66, PROF BIOL, PURDUE UNIV, WEST LAFAYETTE, 66- Honors & Awards: Hoblitzelle Nat Award, 65. Mem: AAAS; Am Soc Microbiol; Am Soc Biol Chemists; Am Chem Soc; NY Acad Sci. Res: Biological nitrogen fixation; electron transport; energy and carbohydrate metabolism; biosynthetic reactions; ferredoxin biochemistry; hydrogenase. Mailing Add: Dept of Biol Sci Purdue Univ West Lafayette IN 47907

MORTER, RAYMOND LIONE, b Arlington, Wis, Sept 7, 20; m 46; c 2. MEDICAL MICROBIOLOGY, IMMUNOPATHOLOGY. Educ: Iowa State Univ, BS, 54, DVM, 57; Mich State Univ, MS, 58, PhD(microbiol & path), 60. Prof Exp: Asst vet anat, Iowa State Univ, 55-57; NSF fel vet microbiol, Mich State Univ, 57-59; asst prof, Vet Med Res Inst, Iowa State Univ, 59-60; assoc prof vet microbiol, path & pub health, 60-64, PROF VET MICROBIOL, SCH VET SCI & MED, PURDUE UNIV, WEST LAFAYETTE, 64-, HEAD DEPT VET SCI & ASSOC DEAN RES, 66- Concurrent Pos: AID staff, Philippines, 66; consult, Food & Agr Orgn, UN; bd dirs, Lab Supply Co, 71- Mem: AAAS; Am Asn Lab Animal Sci; Am Soc Microbiol; Am Vet Med Asn; Am Soc Exp Path. Res: Mechanism and course of infectious diseases and health management of feedlot cattle. Mailing Add: Rm 109 Lynn Hall Purdue Univ Sch of Vet Med West Lafayette IN 47907

MORTHLAND, FRANCIS WILLIAM, biochemistry, see 12th edition

MORTIMER, CHARLES EDGAR, b Allentown, Pa, Nov 21, 21; m 60; c 1. ORGANIC CHEMISTRY, HISTORY OF SCIENCE. Educ: Muhlenberg Col, BS, 42; Purdue Univ, MS, 48, PhD(chem), 50. Prof Exp: Line shift supvr, Hercules Powder Co, Va & Kans, 42-44; res assoc, Carbide & Carbon Chem Co, NY, 44-46; from asst prof to assoc prof, 50-59, PROF CHEM, MUHLENBERG COL, 59- Mem: Am Chem Soc. Res: Chemical education. Mailing Add: Dept of Chem Muhlenberg Col Allentown PA 18104

MORTIMER, CLIFFORD HILEY, b Whitchurch, Eng, Feb 27, 11; m 36; c 2. LIMNOLOGY, PHYSICAL OCEANOGRAPHY. Educ: Univ Manchester, BS, 32, DSc, 46; Univ Berlin, PhD(zool), 35. Prof Exp: Sci officer, Freshwater Biol Asn, Eng, 35-41, Royal Naval Sci Serv, 41-46 & Freshwater Biol Asn, 46-56; dir, Scottish Marine Biol Asn, 56-66; DISTINGUISHED PROF ZOOL & DIR CTR GREAT LAKES STUDIES, UNIV WIS-MILWAUKEE, 66- Concurrent Pos: Vis prof, Univ Wis-Madison, 62-63; lectr, Glasgow Univ, 63-66. Honors & Awards: Naumann Medal, Int Asn Limnol, 65. Mem: Am Soc Limnol & Oceanog (pres, 70-71); Marine Biol Asn UK; Brit Freshwater Biol Asn (vpres, 66-); Int Asn Limnol; Int Asn Great Lakes Res (pres, 73-74). Res: Physics, biology and chemistry of lakes and oceans, in particular water motion in large basins and coastal marine waters, including internal waves in the Laurentian Great Lakes. Mailing Add: Ctr for Great Lakes Studies Univ of Wis Milwaukee WI 53201

MORTIMER, DONALD CHARLES, b Didsbury, Alta, Mar 4, 24; m 47; c 7. PLANT PHYSIOLOGY. Educ: Univ Alta, BSc, 45; Univ Wis, MS, 47, PhD(biochem), 50. Prof Exp: Asst res officer, 50-59, ASSOC RES OFFICER APPL BIOL, RES LABS, NAT RES COUN CAN, 59- Res: Ecology and physiology of freshwater plants; photosynthesis; translocation. Mailing Add: Div of Biol Sci Nat Res Coun Ottawa ON Can

MORTIMER, EDWARD ALBERT, JR, b Chicago, Ill, Mar 22, 22; m 44; c 3. MEDICINE. Educ: Dartmouth Col, AB, 43; Dartmouth Med Sch, dipl, 44; Northwestern Univ, MD, 47. Prof Exp: Resident pediat, Boston Children's Hosp, 50-52; from sr instr to prof, Case Western Reserve Univ, 52-66; prof & chmn dept, Sch Med, Univ NMex, 66-75; PROF COMMUNITY HEALTH & PEDIAT & CHMN DEPT COMMUNITY HEALTH, SCH MED, CASE WESTERN RESERVE UNIV, 75- Concurrent Pos: Markle scholar, Case Western Reserve Univ, 61-66; asst dir dept pediat, Cleveland Metrop Gen Hosp, 52-66; chief pediat, Bernalillo County Med Ctr, 66-75; mem comn streptococcal & staphylococcal dis, Armed Forces Epidemiol Bd, 69-72; mem epidemiol & dis control study sect, NIH, 69-73; chmn, Joint Coun Nat Pediat Soc, 72-74; vis prof epidemiol, Sch Pub Health, Harvard Univ, 73. Mem: Soc Pediat Res; Am Pediat Soc; Am Soc Microbiol. Res: Pediatrics; epidemiology; rheumatic fever; streptococcal and staphylococcal diseases and infections. Mailing Add: Dept of Community Health Case Western Reserve Univ Cleveland OH 44106

MORTIMER, FORREST SPENCER, chemistry, see 12th edition

MORTIMER, GEORGE ALLAN, polymer chemistry, physical organic chemistry, see 12th edition

MORTIMER, ROBERT GEORGE, b Provo, Utah, Aug 25, 33; m 60; c 5. PHYSICAL CHEMISTRY, THEORETICAL CHEMISTRY. Educ: Utah State Univ, BS, 58, MS, 59; Calif Inst Technol, PhD(chem), 63. Prof Exp: Res chemist, Univ Calif, San Diego, 62-64; asst prof chem, Ind Univ, Bloomington, 64-70; asst prof, 70-72, ASSOC PROF CHEM, SOUTHWESTERN AT MEMPHIS, 72- Res: Statistical mechanics; irreversible thermodynamics; experimental study of transport processes in liquids. Mailing Add: Dept of Chem Southwestern at Memphis Memphis TN 38112

MORTIMER, ROBERT KEITH, b Didsbury, Alta, Nov 1, 27; nat US; m 49; c 4. GENETICS, BIOPHYSICS. Educ: Univ Alta, BSc, 49; Univ Calif, Berkeley, PhD(biophys), 53. Prof Exp: From instr to assoc prof, 53-66, PROF MED PHYSICS, UNIV CALIF, BERKELEY, 66-, CHMN MED, 72- Mem: AAAS; Genetics Soc Am; Radiation Res Soc. Res: Genetics and radiation biology of microorganisms. Mailing Add: Dept of Med Physics 103 Donner Lab Univ of Calif Berkeley CA 94720

MORTIMORE, DONALD MERTON, b Portland, Ore, July 4, 17; m 40; c 3. ANALYTICAL CHEMISTRY. Educ: Seattle Pac Col, BS, 40. Prof Exp: Anal chemist, Eastern Regional Res Lab, USDA, 41-43; spectrographer, Aluminum Co Am, Washington, 43; head physics sect, US Bur Mines, 46-56; mgr tech control, Ore Metall Co, 56-61; proj coordr chg anal lab, Albany Metall Res Ctr, US Bur Mines, 61-74; RETIRED. Mem: AAAS; Am Chem Soc; Optical Soc Am; Sigma Xi; Am Soc Metals. Res: Development of new analytical methods using both emission and x-ray spectroscopic techniques. Mailing Add: 715 Clarmount St NW Salem OR 97304

MORTIMORE, GLENN EDWARD, b Portland, Ore, Apr 13, 25; m 59; c 2. PHYSIOLOGY, BIOCHEMISTRY. Educ: Ore State Col, BS, 49; Univ Ore, MD, 52. Prof Exp: NSF fel, 57-58; sr investr, Nat Inst Arthritis & Metab Dis, 58-67; assoc prof, 67-71, PROF PHYSIOL, MILTON S HERSHEY MED CTR, PA STATE UNIV, 71- Mem: AAAS; Endocrine Soc; Am Fedn Clin Res; Am Physiol Soc; Am Soc Biol Chemists. Res: Mechanism of hormone action; effect of insulin of liver metabolism; regulation of metabolism and protein turnover; lysosomes. Mailing Add: Dept of Physiol Hershey Med Ctr Pa State Univ Hershey PA 17033

MORTLAND, MAX MERLE, b Streator, Ill, Mar 30, 23; m 47; c 4. SOIL CHEMISTRY. Educ: Univ Ill, BS, 46, MS, 47 & 50, PhD(agron), 51. Prof Exp: Asst prof soils, Univ Wyo, 51-53; from asst prof to assoc prof, 53-69, PROF SOIL SCI, MICH STATE UNIV, 69-, CROP SCI, 74- Concurrent Pos: Fulbright sr res scholar, Cath Univ Louvain, 61-62. Mem: Am Soc Agron; Soil Sci Soc Am. Res: Physical chemical reactions of soils; reactions of ammonia in soils; rate controlling processes in potassium release from minerals; clay-organic complexes. Mailing Add: Dept of Crop & Soil Sci Mich State Univ East Lansing MI 48823

MORTLOCK, ROBERT PAUL, b Bronxville, NY, May 12, 31; m 54; c 3. MICROBIAL PHYSIOLOGY. Educ: Rensselaer Polytech Inst, BS, 53; Univ Ill, PhD(bact), 58. Prof Exp: Bacteriologist, US Army Chem Corps Res & Develop Labs, 59-61; res assoc biochem, Mich State Univ, 61-63; from asst prof to assoc prof, 68-73, PROF MICROBIOL, UNIV MASS, AMHERST, 73-, HEAD DEPT, 72- Concurrent Pos: USPHS fel, 61-63. Mem: AAAS; Am Soc Microbiol; Am Acad Microbiol. Res: Physiological bacteriology; microbial physiology and metabolism; cellular regulatory mechanisms; carbohydrate metabolism and enzyme regulation in microorganisms; the utilization of uncommon and unnatural carbohydrates by microorganisms. Mailing Add: Dept of Microbiol Univ of Mass Amherst MA 01002

MORTON, BRUCE ELDINE, b Loma Linda, Calif, May 9, 38; m 60; c 2. BIOCHEMISTRY, REPRODUCTIVE BIOLOGY. Educ: La Sierra Col, BS, 60; Univ Wis, MS, 63, PhD(biochem), 65. Prof Exp: Fel, Inst Enzyme Res, Univ Wis, 65-66; NIH fel, Mass Inst Technol, 66-67; Harvard Univ res fel med, Beth Israel Hosp, 67-69; asst prof, 69-74, ASSOC PROF BIOCHEM, UNIV HAWAII, MANOA, 74- Concurrent Pos: Consult, New Eng Mem Hosp, Stoneham, Mass, 67-69 & St Francis Hosp, Honolulu, 70-71. Mem: Soc Study Reproduction; Am Soc Biol Chemists; Am Soc Andrology. Res: Biochemical mechanisms of sperm maturation and egg penetration; energy metabolism and control of cell movement; structure, function and control of ribosomes. Mailing Add: Dept of Biochem Univ of Hawaii Manoa Campus Honolulu HI 96822

MORTON, CONRAD VERNON, systematic botany, deceased

MORTON, DONALD CHARLES, b Kapuskasing, Ont, June 12, 33; m 70. ASTROPHYSICS. Educ: Univ Toronto, BA, 56; Princeton Univ, PhD, 59. Prof Exp: Astronr, US Naval Res Lab, 59-61; res assoc, 61-63, res staff mem, 63-65, res astronr, 65-68, SR RES ASTRONR & LECTR ASTROPHYS SCI, PRINCETON UNIV, 68- Mem: Int Astron Union; Am Astron Soc; Royal Astron Soc; Royal Astron Soc Can. Res: Stellar spectroscopy; space instrumentation and observations; galaxies and quasars. Mailing Add: Dept of Astrophys Sci Princeton Univ Observ Princeton NJ 08540

MORTON, DONALD JOHN, b Brooklyn, NY, Jan 11, 31; m 53; c 3. INFORMATION SCIENCE. Educ: Univ Del, BS, 52; La State Univ, MS, 54; Univ Calif, Berkeley, PhD(plant path), 57; Simmons Col, MLS, 69, DA(libr sci), 76. Prof Exp: Asst prof plant nematol, NMex State Univ, 57-58; asst plant path, NDak State Univ, 59-61; sr res plant pathologist, USDA, Ga, 61-65; assoc prof plant path, Univ Del, 65-68; dir sci libr, Northeastern Univ, 69-70; asst prof hist of med, 70-74, ASSOC PROF LIBR SCI, MED SCH, UNIV MASS, 74-, LIBR DIR, 70- Concurrent Pos: Prof libr sci, Worcester State Col, 74- Mem: Mycol Soc Am; Med Libr Asn; Am Soc Info Sci; Am Libr Asn; Spec Libr Asn. Res: Air pollution effects on plants; serological studies of plant pathogens; organization and retrieval of scientific information. Mailing Add: Libr Univ of Mass Med Sch Worcester MA 01605

MORTON, DONALD LEE, b Richwood, WVa, Sept 12, 34; m 57; c 4. SURGERY, ONCOLOGY. Educ: Univ Calif, BA, 55, MD, 58; Am Bd Surg, dipl, 67; Am Bd Thoracic Surg, dipl, 69. Prof Exp: Intern med, Med Ctr, Univ Calif, 58-59, resident surg, 59-60; clin assoc, Nat Cancer Inst, 60-62; resident surg, Med Ctr, Univ Calif, 62-66; sr surgeon, Nat Cancer Inst, 66-69, head tumor immunol sect, 69-71; assoc prof surg, Sch Med, Johns Hopkins Univ, 70-71; PROF SURG & CHIEF DIV ONCOL, SCH MED, UNIV CALIF, LOS ANGELES, 71- Concurrent Pos: Fel, Cancer Res Inst, Med Ctr, Univ Calif, 62-66; immunol adv mem, Spec Virus Cancer Prog, Nat Cancer Inst, 69-71, mem bd sci counr, 74-; mem comt for objective 6, Nat Cancer Plan, 71-; chief surg, Sepulveda Vet Admin Hosp, Calif, 71-74, chief oncol sect, Surg Serv, 74-; mem sci adv coun, Cancer Res Inst, Inc, 74- Honors & Awards: US Dept Health, Educ & Welfare Superior Serv Award, 70; Cancer Res Inst Inc Award, 75. Mem: Am Asn Cancer Res; Am Surg Asn; Am Soc Clin Oncol; Soc of Surg Oncol; Soc Univ Surgeons. Res: Immunologic and virologic aspects of neoplastic disease, including immunotherapy of melanoma, skeletal and soft tissue sarcoma and mammary carcinoma; surgical oncology; thoracic surgery. Mailing Add: Ctr of Health Sci Univ of Calif Los Angeles CA 90024

MORTON, HARRISON LEON, b St Paul, Minn, Oct 19, 38; m 62; c 4. FOREST PATHOLOGY. Educ: Univ Minn, BS, 61, MS, 64, PhD(plant path), 67. Prof Exp: Asst prof forest path, 66-72, ASSOC PROF FOREST PATH & CHMN FISHERIES, FORESTRY & WILDLIFE PROG, SCH NATURAL RESOURCES, UNIV MICH, ANN ARBOR, 72- Mem: Am Phytopath Soc; Soc Am Foresters. Res: Biodegradation of wood preservatives; air pollution damage to trees. Mailing Add: Dept of Forest Path Univ of Mich Sch of Natural Resources Ann Arbor MI 48104

MORTON, HARRY E, b Wayne, Mich, Aug 4, 06; m 34; c 2. BACTERIOLOGY. Educ: Univ Mich, BS, 30, MS, 31, ScD(bact), 36. Prof Exp: Instr, 31-35, assoc, 35-37, from asst prof to prof bact, 37-75, EMER PROF BACT, SCH MED, UNIV PA, 75- Concurrent Pos: Chief microbiol div, Pepper Lab, Univ Hosp, 67-75. Honors & Awards: Kimble Methodology Res Award, 68. Mem: Electron Micros Soc Am; Am Asn Path & Bact; Am Pub Health Asn; Biol Photog Asn; Am Acad Microbiol. Res: Bacterial variation; electron microscopy; chemical disinfectants; antibiotics; filtration; cultivation of the gonococcus; bacteriophage; variation of diphtheria bacillus; anaerobic microorganisms; spirochetes; pleuropneumonia-like organisms; visual instruction. Mailing Add: 4114 School Lane Drexel Hill PA 19026

MORTON, HELEN JANET, b Liverpool, Eng, Oct 27, 25; Can citizen; div; c 1. BIOCHEMISTRY, CYTOLOGY. Educ: Univ Liverpool, BSc, 47; Univ Toronto, MA, 59; Univ Ottawa, PhD(biochem), 65. Prof Exp: Res asst tissue cult, Connaught Med Res Labs, Ont, 47-52; chemist, Lab Hyg, Dept Nat Health & Welfare, Can, 54-62; ASSOC RES OFFICER, DIV BIOL, NAT RES COUN CAN, 65- Concurrent Pos: Lectr, Univ Ottawa, 67- Mem: Tissue Cult Asn; Soc Exp Biol & Med; Can Soc Cell Biol; NY Acad Sci. Res: Synthetic media for tissue culture studies; comparative nutritional and enzymatic studies on normal and malignant cells in vitro; methods of cultivation of, and radiation effects on, bone marrow cells; haematology. Mailing Add: Div of Biol Nat Res Coun of Can Ottawa ON Can

MORTON, HOWARD LEROY, b Moscow, Idaho, Dec 13, 24; m 50; c 2. PLANT PHYSIOLOGY. Educ: Univ Idaho, BS, 50, MS, 52; Agr & Mech Col, Tex, PhD(plant physiol), 61. Prof Exp: Asst agronomist, Univ Idaho, 52-57; res agronomist, Tex, 57-66, plant physiologist, 66-68, PLANT PHYSIOLOGIST, CROPS RES DIV, AGR RES SERV, USDA, ARIZ, 68- Mem: Weed Sci Soc Am; Am Soc Plant Physiol; Soc Range Mgt. Res: Absorption, translocation and metabolism of herbicides; weed control on range lands; range revegetation; poisonous weeds; pesticide residues; plant growth and development; controlled environment systems; effects of herbicides on honey bees. Mailing Add: Agr Res Serv USDA 2000 E Allen Rd Tucson AZ 85719

MORTON, JOHN DUDLEY, b Southampton, Eng, July 25, 14; US citizen. ENVIRONMENTAL HEALTH. Educ: Cambridge Univ, BA, 36, MA, 40. Prof Exp: Sect leader appl chem, Exp Sta, Eng, 36-47; asst dir aerobiol, Microbiol Res Estab, Eng, 47-62; lab mgr meteorol, Melpar Inc, 63-70; PRIN SCIENTIST ENVIRON HEALTH, ENVIRO CONTROL INC, 70- Concurrent Pos: Consult, Nat Acad Eng, 69-70. Honors & Awards: Officer, Order of the Brit Empire, 56. Res: Studies of human health in relation to environmental and occupational exposure to toxic substances. Mailing Add: Enviro Control Inc 11300 Rockville Pike Rockville MD 20852

MORTON, JOHN HENDERSON, b New Haven, Conn, Jan 15, 23; m 49; c 4. SURGERY. Educ: Amherst Col, BA, 45; Yale Univ, MD, 46. Prof Exp: Intern surg, gynec & obstet, Strong Mem Hosp & Rochester Munic Hosp, 46-47, asst resident surg, 47 & 49-52, resident, 53; from instr to assoc prof surg, 53-69, from instr to asst prof surg anat, 55-67, PROF HEALTH SERV, SCH MED & DENT, UNIV ROCHESTER, 69- Concurrent Pos: From asst surgeon to assoc surgeon, Med Ctr, Univ Rochester, 54-62, sr assoc surgeon, 62- Mem: AMA; fel Am Col Surg; Am Asn Surg of Trauma; Am Burn Asn. Res: Liver and gastrointestinal tract. Mailing Add: 260 Crittenden Blvd Rochester NY 14642

MORTON, JOHN KENNETH, b Tamworth, Eng, Jan 3, 28; m 51; c 2. BOTANY. Educ: Univ Durham, BSc, 44, PhD(bot), 53. Prof Exp: Lectr bot, Univ Ghana, 51-60, sr lectr & cur, Ghana Herbarium, 60-61; lectr, Birkbeck Col, London, 61-63; prof & chmn dept, Fourah Bay Col, Sierra Leone, 63-67; PROF BIOL, UNIV WATERLOO, 68-, CHMN DEPT, 74- Mem: Can Bot Soc (pres, 74-75); Bot Soc Brit Isles; Bot Soc Am; Am Soc Plant Taxonomists. Res: Experimental taxonomy and biogeography of North American and tropical African vascular plants; palynology; evolution. Mailing Add: Dept of Biol Univ of Waterloo Waterloo ON Can

MORTON, JOHN ROBERT, III, b Palestine, Tex, June 5, 29; m 53; c 2. NUCLEAR SCIENCE. Educ: Univ Ala, BS, 50; Univ Calif, Berkeley, PhD(chem), 61. Prof Exp: Tech grad, Hanford Atomic Prod Oper, Gen Elec Co, 50-51, supvr health physics, 51-56; PHYSICIST, LAWRENCE LIVERMORE LAB, UNIV CALIF, 61- Mem: AAAS; Am Phys Soc; Am Nuclear Soc. Res: Nuclear weapons test diagnostic techniques; pinhole imagery; reactor and critical assembly physics; seeking evidence for neutrino decay from fission explosions. Mailing Add: L Div Lawrence Livermore Lab Univ of Calif PO Box 808 Livermore CA 94550

MORTON, JOHN WEST, JR, b Dallas, Tex, Mar 3, 25; m 50; c 2. ORGANIC CHEMISTRY. Educ: Southern Methodist Univ, BS, 46; Iowa State Univ, PhD(org chem), 52. Prof Exp: Res chemist, Procter & Gamble Co, Ohio, 52-54; from assoc prof to prof chem, La Polytech Inst, 54-62; assoc prof chem, 62-74, PROF PHYS SCI, WESTERN NMEX UNIV, 74- Mem: Am Chem Soc. Res: Organolithium compounds. Mailing Add: Dept of Phys Sci Western NMex Univ Silver City NM 88061

MORTON, JOSEPH JAMES PANDOZZI, b Hartford, Conn, May 9, 41; m 68; c 2. PHARMACOLOGY, BIOLOGY. Educ: Univ Hartford, BS, 63; Univ Conn, MS, 66, PhD(pharmacol), 68. Prof Exp: Actg dir pharmacol, Natural Prod Res, Amazon Natural Drug Co, 67-68, dir, 68-69; toxicologist, Med Eval Dept, Gillette Res Inst, 69-70, med rev officer, 70-73, CHIEF OFF MED REV, GILLETTE MED EVAL LABS, 74- Mem: Soc Toxicol; Soc Cosmetic Chemists; Am Soc Pharmacog; Am Chem Soc. Res: Medical safety/toxicity evaluation of drugs; drugs, cosmetics and household chemical products; screening of natural products for potential therapeutic activity. Mailing Add: Gillette Med Eval Labs 1413 Research Blvd Rockville MD 20850

MORTON, MARTIN LEWIS, b Tony, Wis, May 1, 34; m 53; c 4. ZOOLOGY, PHYSIOLOGY. Educ: San Jose State Col, BA, 59, MA, 61; Wash State Univ, PhD(zoophysiol), 66. Prof Exp: Res asst zool, Wash State Univ, 63-65; res assoc, Univ Wash, 65-66, asst prof, 66-67; asst prof, 67-70, ASSOC PROF BIOL, OCCIDENTAL COL, 70- Mem: Am Ornith Union; Am Soc Mammal; Am Soc Zool; Ecol Soc Am; Am Inst Biol Sci. Res: Bioenergetics, orientation, phenology, endocrinology, annual cycles and biological clocks of migratory birds. Mailing Add: Dept of Biol Occidental Col 1600 Campus Rd Los Angeles CA 90041

MORTON, MAURICE, b Latvia, June 3, 13; nat US; m 33; c 3. POLYMER CHEMISTRY. Educ: McGill Univ, BSc, 34, PhD(chem), 45. Prof Exp: Chief chemist, Johns-Manville Co, Can, 36-41; chemist, Congoleum of Can, Ltd, 41-44; from asst prof to prof chem, Sir George Williams Col, 45-48; prof, 53-68, asst dir inst rubber res, 48-53, dir, 53-65, REGENTS PROF POLYMER CHEM, UNIV AKRON, 68-, DIR INST POLYMER SCI, 65-, HEAD DEPT POLYMER SCI, 67- Concurrent Pos: Lectr, McGill Univ, 46-48; chmn, Gordon Conf Elastomers, 53; chmn comt macromolecular chem, Nat Acad Sci-Nat Res Coun, 64-67. Mem: AAAS; Am Chem Soc. Res: Polymerization kinetics; emulsion polymerization; synthetic rubber; anionic polymerization. Mailing Add: Inst of Polymer Sci Univ of Akron Akron OH 44325

MORTON, NEWTON ENNIS, b Camden, NJ, Dec 21, 29; m 49; c 5. POPULATION GENETICS. Educ: Univ Hawaii, BA, 51; Univ Wis, MS, 52, PhD, 55. Prof Exp: Geneticist, Atomic Bomb Casualty Comn, Japan, 52-53; fel, Nat Cancer Inst, 55-56; asst prof med genetics, Univ Wis, 56-60, assoc prof, 60-61; dir genetics res proj, 58-59, prof genetics, 61-69, chmn dept, 62-65, DIR POP GENETICS LAB, UNIV HAWAII, HONOLULU, 69-, PROF, SCH PUB HEALTH, 75- Concurrent Pos: Consult, NIH, 59 & genetics training comt, 61-65; mem expert adv comt human genetics, WHO, 61-; dir med genetics proj, Immigrants Hosp, Sao Paulo, Brazil, 62-63. Honors & Awards: Lederle Award, 58; Allan Award, 63. Mem: AAAS; Genetics Soc Am; Am Soc Human Genetics; Am Soc Naturalists; Brazilian Acad Sci. Res: Human and population genetics. Mailing Add: Pop Genetics Lab Univ of Hawaii Honolulu HI 96822

MORTON, PERRY WILKES, JR, b Strong, Ark, Jan 19, 23; m 58; c 3. PHYSICS. Educ: Rice Univ, BS, 47; Miss State Univ, MS, 51; Duke Univ, PhD(physics), 57. Prof Exp: Asst math & instr physics, Miss State Univ, 52-53; asst, Duke Univ, 53-57; from assoc prof to prof, Miss State Univ, 57-63; prof physics & math & chmn div natural sci, Ky Southern Col, 63-69; CHMN DEPT PHYSICS, SAMFORD UNIV, 69- Mem: Am Phys Soc. Res: Nuclear and classical physics; beta-ray spectroscopy. Mailing Add: Dept of Physics Samford Univ Birmingham AL 35209

MORTON, RICHARD ALAN, b Chicago, Ill, Dec 14, 38; m 62; c 4. BIOPHYSICS. Educ: Univ Chicago, SB, 61, SM, 62, PhD(biophys), 65. Prof Exp: Res assoc biophys, Johns Hopkins Univ, 65-67; NSF res fel, Univ Calif, Santa Barbara, 68, res assoc, 69; asst prof, 69-73, ASSOC PROF BIOL, McMASTER UNIV, 73- Mem: Sigma Xi. Res: Structure and function of cytochromes; electron-transfer reactions; bioenergetics. Mailing Add: Dept of Biol McMaster Univ Hamilton ON Can

MORTON, RICHARD FREEMAN, b Boston, Mass, Aug 8, 14; m 41; c 3. PHYSICS. Educ: Mass Inst Technol, SB, 36; Harvard Univ, AM, 48. Prof Exp: Teacher, Monson Acad, 36-39 & Moses Brown Sch, 39-41; dean, Monson Acad, 46-49; from instr to assoc prof, 49-59, assoc dean fac, 62-71, PROF PHYSICS, WORCESTER POLYTECH INST, 59- Concurrent Pos: Phys sci adminr, NSF, 58-59. Mem: Am Asn Physics Teachers. Res: Optics; thermodynamics. Mailing Add: Dept of Physics Worcester Polytech Inst Worcester MA 01609

MORTON, ROBERT ALEX, b Cincinnati, Ohio, Oct 17, 42; m 68; c 2. SEDIMENTOLOGY. Educ: Univ Chattanooga, BA, 65; WVa Univ, MS, 66, PhD(geol), 72. Prof Exp: Petrol geologist, Chevron Oil Co, 66-69; ASSOC RES SCIENTIST GEOL, BUR ECON GEOL, UNIV TEX, AUSTIN, 72- Mem: Am Asn Petrol Geologists; Geol Soc Am; Soc Econ Paleontologists & Mineralogists. Res: Coastal processes; ancient and modern clastic depositional systems; marine geology; environmental geology. Mailing Add: Bur of Econ Geol Univ of Tex Box X Austin TX 78712

MORTON, ROBERT C, electrical engineering, see 12th edition

MORTON, ROGER DAVID, b Nottingham, Eng, Oct 20, 35; m 61; c 2. MINERALOGY. Educ: Univ Nottingham, BSc, 56, PhD(geol), 59. Prof Exp: Sci asst, Univ Oslo, 59-61; lectr geol, Univ Nottingham, 61-66; assoc prof, 66-73, PROF GEOL, UNIV ALTA, 73- Concurrent Pos: G V Hobson Bequest Fund, Brit Inst Mining & Metall, 62; consult, Can Int Develop Agency, 75- Mem: Fel Geol Asn Can; Can Inst Mining & Metall; Soc Econ Geologists; Mineral Soc Am. Res: Investigation of uranium deposits in northwest Canada and mineral resources of Indonesia. Mailing Add: Rm 366 Dept of Geol Univ of Alta Edmonton AB Can

MORTON, STEPHEN DANA, b Madison, Wis, Sept 7, 32. WATER CHEMISTRY, PHYSICAL CHEMISTRY. Educ: Univ Wis, BS, 54, PhD(chem), 62. Prof Exp: Asst prof chem, Otterbein Col, 62-66; fel water chem, Univ Wis, 66-67; res chemist, 67-73, HEAD ENVIRON QUAL DEPT, WARF INST, 73- Mem: Am Chem Soc; Am Soc Limnol & Oceanog; Am Water Works Asn; Water Pollution Control Fedn; fel Am Inst Chemists. Res: Water pollution; lake and stream studies; waste treatment. Mailing Add: WARF Inst PO Box 2599 Madison WI 53701

MORTON, SUE BRAKEBILL, textile chemistry, see 12th edition

MORTON, THOMAS HARLOW, b Plymouth, Mass, Apr 29, 06; m 30; c 1. BIOLOGY. Educ: Univ Vt, BS, 29; Niagara Univ, MS, 36, PhD(histol), 38. Prof Exp: Instr math & chem, 29-34, from instr to asst prof biol, 34-42, chmn dept, 42-74, prof physics, Air Force Detachment, 43-44 & math, Army Spec Training Prog, 44-45, PROF BIOL, NIAGARA UNIV, 42- Mem: AAAS; Biol Photog Asn; Am Asn Anat; NY Acad Sci. Res: Histology; histological staining; color photomicrography; bile capillary wall structure. Mailing Add: Dept of Biol PO Box 215 Niagara Univ Niagara University NY 14109

MORTON, THOMAS HELLMAN, b Los Angeles, Calif, Feb 10, 47; m 75. ORGANIC CHEMISTRY. Educ: Harvard Univ, AB, 68; Calif Inst Technol, PhD(chem), 73. Prof Exp: ASST PROF CHEM, BROWN UNIV, 72- Mem: Am Chem Soc. Res: Molecular rearrangements and the interaction of radiation with matter. Mailing Add: Dept of Chem Box H Brown Univ Providence RI 02912

MORTON, WILLIAM EDWARDS, b Boston, Mass, June 30, 29; m 56; c 3. EPIDEMIOLOGY. Educ: Univ Puget Sound, BS, 52; Univ Wash, MD, 55; Univ Mich, MPH, 60, DrPH, 62. Prof Exp: Intern med, Doctors Hosp, Seattle, Wash, 55-56; USPHS heart dis control officer, Colo Dept Pub Health, 56-58; sr resident med, San Mateo County Hosp, Calif, 58-59; trainee epidemiol, Sch Pub Health, Univ Mich, 59-62; res epidemiologist, Colo Heart Asn & asst clin prof prev med, Med Sch, Univ Colo, 62-67; assoc prof, 67-70, PROF PUB HEALTH & PREV MED, MED SCH, UNIV ORE, 70-, HEAD DIV ENVIRON MED, 72- Concurrent Pos: Med res consult, Selective Serv, Colo, 64-67 & Ore, 70-; consult, Environ Health Sci Ctr, Ore State Univ, 72- Mem: Am Heart Asn; Am Pub Health Asn; Am Cancer Soc; Am Occup Med Asn; Asn Teachers Prev Med. Res: Cancer epidemiology; hypertension; other cardiovascular diseases; electrocardiography; streptococcosis; screening method evaluation; environmental and occupational health hazards. Mailing Add: Dept of Pub Health & Prev Med Univ of Ore Med Sch Portland OR 97201

MORTVEDT, JOHN JACOB, b Dell Rapids, SDak, Jan 25, 32; m 55; c 3. SOIL CHEMISTRY. Educ: SDak State Col, BS, 53, MS, 59; Univ Wis, PhD(soil), 62. Prof Exp: SOIL CHEMIST, SOILS & FERTIL RES BR, TENN VALLEY AUTHORITY, 62- Mem: Am Soc Agron; Soil Sci Soc Am. Res: Agronomic effectiveness of micronutrients when applied with macronutrient carriers, including plant uptake and soil-fertilizer reactions. Mailing Add: Soils & Fertilizer Res Br Tenn Valley Authority Muscle Shoals AL 35660

MOSAK, RICHARD DAVID, b Washington, DC, Oct 8, 45; m 67; c 3. MATHEMATICAL ANALYSIS. Educ: Columbia Univ, AB, 66, PhD(math), 70. Prof Exp: Instr math, Yale Univ, 70-72 & Univ Chicago, 72-73; ASST PROF MATH, UNIV ROCHESTER, 73- Mem: Am Math Soc. Res: Harmonic analysis and representations of topological groups. Mailing Add: Dept of Math Univ of Rochester Rochester NY 14627

MOSBACH, ERWIN HEINZ, b Ger, Feb 18, 20; nat US; m 44. BIOCHEMISTRY. Educ: Columbia Univ, BA, 43, MA, 48, PhD(chem), 50. Prof Exp: Tutor chem, Brooklyn Col, 42-46; asst, Columbia Univ, 46-50; biochemist, Biol Div, Oak Ridge Nat Lab, 50-51; res assoc biochem, Col Physicians & Surgeons, Columbia Univ, 51-54, from asst prof to assoc prof, 54-61; assoc mem & chief biochem sect, Dept Lab Diag, 61-71, MEM & CHIEF DEPT LIPID RES, PUB HEALTH RES INST NEW YORK, 72-; ASST DIR BUR LABS, DEPT HEALTH, NEW YORK, 61- Concurrent Pos: Lectr, Hunter Col, 51-54; consult, NY Infirmary, 57-; consult, Metab Study Sect, NIH, 67-71, mem lipid metab adv comt, 74-; adj assoc prof med, Med Sch, NY Univ, 61-; assoc ed, J Lipid Res, 68-72, ed, 76; consult, Manhattan Vet Admin Hosp, 71-; fel coun arteriosclerosis, Am Heart Asn. Mem: Am Soc Biol Chemists; Soc Exp Biol & Med; Am Inst Nutrit; Am Asn Study Liver Dis; Am Gastroenterol Asn. Res: Metabolism of sterols and bile acids. Mailing Add: Pub Health Res Inst 455 First Ave New York NY 10016

MOSBO, JOHN ALVIN, b Davenport, Iowa, June 11, 47; m 68; c 1. INORGANIC CHEMISTRY. Educ: Univ Northern Colo, BA, 69; Iowa State Univ, PhD(inorg chem), 73. Prof Exp: ASST PROF CHEM, BALL STATE UNIV, 73- Mem: Am Chem Soc. Res: Reaction mechanisms of phosphorus and boron compounds; heteroatom configurations and ring conformations of cyclic organophosphorus and organoboron compounds. Mailing Add: Dept of Chem Ball State Univ Muncie IN 47306

MOSBURG, EARL R, JR, b Frederick, Md, Jan 23, 28; m 58; c 1. ATOMIC PHYSICS. Educ: Yale Univ, BS, 52, PhD(physics), 56. Prof Exp: PHYSICIST, NAT BUR STANDARDS, 56- Concurrent Pos: Mem subcomt neutron standards & measurements, Comt Nuclear Sci, Nat Res Coun, 59-61. Mem: Am Phys Soc. Res: Plasma physics; gas discharges; gaseous electronics. Mailing Add: Nat Bur of Standards Boulder CO 80302

MOSBY, HENRY SACKETT, b Lynchburg, Va, Oct 28, 13; m 41; c 3. WILDLIFE MANAGEMENT, FORESTRY. Educ: Hampden-Sydney Col, BS, 35; Univ Mich, BS & MF, 37, PhD(wildlife), 41. Prof Exp: Asst, Va Coop Wildlife Res Unit, Va Polytech Inst, 38; field biologist, State Comn Game & Inland Fisheries, Va, 39-42, actg supt game, 42-43, dist game technician, 46-47; dir, Va Coop Wildlife Sta, 47-55, PROF WILDLIFE MGT, VA POLYTECH INST & STATE UNIV, 55-, HEAD DEPT FISHERIES & WILDLIFE SCI, 74- Concurrent Pos: Ed, Wildlife Soc Newslett, 49-55; regional dir, Nat Wild Turkey Fedn, 72- Honors & Awards: Wine Award, Va Polytech Inst & State Univ, 63. Mem: Hon mem Wildlife Soc (vpres, 59, pres, 65). Res: Forest and wildlife management; wild turkey; wildlife techniques; population dynamics; population dynamics of hunting on deer, squirrel, quail, and wild turkey; influence of habitat disturbance of forest wildlife. Mailing Add: Dept of Fisheries & Wildlife Sci Va Polytech Inst & State Univ Blacksburg VA 24061

MOSBY, WILLIAM LINDSAY, b Rockford, Ill, Nov 30, 21; m 49. ORGANIC CHEMISTRY. Educ: Harvard Univ, BSc, 43; Ohio State Univ, PhD(org chem), 49. Prof Exp: Res chemist, Gen Aniline & Film Corp, 49-52; res chemist, Res Dept, 52-54, group leader, 55-58, RES ASSOC, AM CYANAMID CO, 59- Concurrent Pos: Am Cyanamid fel, Univ Munich, 64-65. Mem: Am Chem Soc. Res: Synthetic and theoretical organic chemistry; intermediates for dyes and pharmaceuticals; vat dyes; polyester dyes; antioxidants; aromatic and polycyclic compounds and heterocyclic system with bridgehead nitrogen atoms. Mailing Add: Am Cyanamid Co Bound Brook NJ 08805

MOSCARELLO, MARIO ANTONIO, b Timmins, Ont, Sept 20, 29; m 55; c 2. MEDICINE, BIOCHEMISTRY. Educ: Univ Toronto, BA, 51, MD, 55, PhD(biochem), 62. Prof Exp: Intern, St Michael's Hosp, 55-56; investr, Res Inst, Hosp for Sick Children, 64-65, asst scientist, 65-69; asst prof biochem & path, 65-69, ASSOC PROF BIOCHEM, UNIV TORONTO, 69-; ASSOC SCIENTIST BIOCHEM, RES INST, HOSP FOR SICK CHILDREN, 69- Concurrent Pos: Ont Cancer Inst fel virol, 62-64. Mem: Can Biochem Soc. Res: Glycoprotein biosynthesis in nephrosis; protein constituents of myelin membrane; conformational changes of proteins on solid surfaces. Mailing Add: Res Inst Hosp for Sick Children 555 University Ave Toronto ON Can

MOSCATELLI, EZIO ANTHONY, b New York, NY, Nov 17, 26; c 1.

BIOCHEMISTRY, NEUROCHEMISTRY. Educ: Columbia Univ, AB, 48; Univ Ill, MS, 49, PhD(biochem), 58. Prof Exp: Assoc chemist, Merck & Co, Inc, 49-55; chemist, Nat Heart Inst, 58-59; sr chemist, Merck, Sharp & Dohme Res Labs, 59-62; asst prof biochem, Univ Tex Southwest Med Sch, Dallas, 62-70; assoc prof psychiat & biochem, Mo Inst Psychiat, 70-74; ASSOC PROF BIOCHEM, SCH MED & INVESTR, DALTON RES CTR, UNIV MO-COLUMBIA, 74- Mem: Am Chem Soc; Am Oil Chem Soc; Am Soc Biol Chemists; Am Soc Neurochem; Int Soc Neurochem. Res: Biochemistry of sphingolipids; biochemistry of alcohol abuse. Mailing Add: Dalton Res Ctr Research Park Columbia MO 65201

MOSCHERA, JOHN ANTHONY, b New York, NY, June 10, 43. BIOCHEMISTRY. Educ: Polytech Inst Brooklyn, BS, 64; New York Med Col, PhD(biochem), 75. Prof Exp: Instr, 72-75, ASST PROF BIOCHEM, NEW YORK MED COL, 75- Concurrent Pos: Fel, Roche Inst Molecular Biol, 76- Mem: Am Chem Soc; AAAS; Soc Complex Carbohydrates. Res: Structures and functions of the epithelial mucus glycoproteins derived from mammalian submaxillary and sublingual glands. Mailing Add: Dept of Biochem Basic Sci Bldg New York Med Col Valhalla NY 10595

MOSCHOPEDIS, SPEROS E, b Piraeus, Greece, June 1, 26; Can citizen; m 56; c 3. ORGANIC CHEMISTRY. Educ: Nat Univ Athens, BSc, 54, PhD(chem), 69. Prof Exp: Teacher chem, Archimides Inst Technol, Greece, 54-56; chemist, Sherritt Gordon Mines, Ltd, Alta, 56-57; RES OFFICER ORG CHEM, RES COUN ALTA, 57- Mem: Chem Inst Can; Greek Chem Asn. Res: Humic acids, lignites, coals and asphaltic type bituminous materials; water-soluble derivatives of humic acids; synthesis of polypeptides. Mailing Add: Res Coun of Alta 114th St & 87th Ave Edmonton AB Can

MOSCHOVAKIS, JOAN RAND, b Glendale, Calif, Dec 24, 37; m 63. MATHEMATICS. Educ: Univ Calif, Berkeley, AB, 59; Univ Wis, MS, 61, PhD(math), 65. Prof Exp: Instr math, Oberlin Col, 63-64; asst prof, 65-67, 69-74, ASSOC PROF MATH, OCCIDENTAL COL, 74- Mem: Asn Symbolic Logic; Am Math Soc; Math Asn Am. Res: Foundations of mathematics; formal and symbolic logic; intuitionism; point set-topology. Mailing Add: Dept of Math Occidental Col Los Angeles CA 90041

MOSCHOVAKIS, YIANNIS N, b Athens, Greece, Jan 18, 38; m 63. MATHEMATICS. Educ: Mass Inst Technol, SB & SM, 60; Univ Wis, PhD(math), 63. Prof Exp: Actg instr math, Univ Wis, 62-63; Benjamin Peirce instr, Harvard Univ, 63-64; from asst prof to assoc prof, 64-74, PROF MATH, UNIV CALIF, LOS ANGELES, 74- Mem: Am Math Soc; Asn Symbolic Logic. Res: Foundations of mathematics; recursive functions; hierarchy theory. Mailing Add: Dept of Math Univ of Calif Los Angeles CA 90024

MOSCONA, ARON ARTHUR, b Haifa, Israel, July 4, 22; m 49; c 1. DEVELOPMENTAL BIOLOGY. Educ: Hebrew Univ Jerusalem, MSc, 47, PhD(zool), 50. Prof Exp: Res fel embryol, Strangeways Res Lab, Cambridge Univ, 50-52; assoc prof physiol, Sch Med, Hebrew Univ Jerusalem, 53-55; vis investr develop biol, Rockefeller Inst, 55-57; from assoc prof to prof zool, 58-69, prof biol, 69-74, LOUIS BLOCK PROF BIOL SCI & MEM COMT DEVELOP BIOL & GENETICS, UNIV CHICAGO, 74-, DIR TRAINING PROG IN DEVELOP BIOL, 69- Concurrent Pos: Vis assoc prof, Stanford Univ, 59; Claude Bernard vis prof, Univ Montreal, 60; Lillie fel, Marine Biol Lab, Woods Hole, 60 & vis prof, Univ Palermo, 66; ed, Exp Cell Res, Current Topics Develop Biol & Univ Chicago Pubs Biol. Mem: Am Soc Zool; Am Asn Anat; Int Inst Embryol; fel NY Acad Sci; Int Soc Cell Biol. Res: Mechanisms of embryonic differentiation; tissue culture; cell physiology; neoplasia. Mailing Add: Dept of Biol Univ of Chicago Chicago IL 60637

MOSCONY, JOHN JOSEPH, b Philadelphia, Pa, Aug 26, 29. CHEMISTRY. Educ: St Joseph's Col, Pa, BS, 51, MS, 58; Univ Pa, PhD(chem), 65. Prof Exp: Chemist, Elec Storage Battery Co, 51-54 & Waterman Prod Co, 54-57; ENG LEADER, RCA CORP, 57- Concurrent Pos: Lectr, RCA Corp Eng Serv, 71- Mem: Am Chem Soc; Am Vacuum Soc. Res: High pressure synthesis of silicon fluorides; thermoelectric and thermionic energy conversion; materials and processes related to vacuum and color television picture tubes. Mailing Add: 936 Martha Ave Lancaster PA 17601

MOSCOVICI, CARLO, b Cairo, Egypt, July 27, 25; US citizen; m 55; c 2. VIROLOGY. Educ: Univ Rome, PhD(microbiol), 52. Prof Exp: Asst prof pediat, Med Sch, Univ Colo, 57-67; ASSOC PROF IMMUNOL & MICROBIOL, MED SCH, UNIV FLA, 67-; CHIEF VIROL RES LAB, VET ADMIN HOSP, 67- Honors & Awards: USPHS-Carrier Award, 61. Res: Tumor virology; avian tumor viruses; RNA tumor viruses; cell differentiation. Mailing Add: Virus Res Lab Vet Admin Hosp Gainesville FL 32601

MOSCOVICI, MAURICIO, b Sao Paulo, Brazil, Apr 26, 25; m 48; c 3. ANATOMY, SURGERY. Educ: Univ Brazil, DDS, 47, PhD(anat), 65; Univ Fluminense, Brazil, PhD(anat), 54. Prof Exp: From asst prof to prof anat, Fac Odontol, State of Rio de Janeiro, 49-60; prof, Fac Odontol, 60-64, prof, Fac Pharm, 64-74, hon prof, 65, PROF TITULAR ANAT, FAC MED, FED UNIV FLUMINENSE, RIO DE JANEIRO, 74-, HEAD DEPT, 67- Concurrent Pos: Fel, Neurol Inst, Univ Brazil, 59-60; fel anat, Tech Coop Prog, US Dept State-Agency Int Develop, 60-61; dent surgeon Welfare Inst, Labor Ministry, Brazil, 51-66, head clin oral-facial surg, Ipanema Hosp, 66-; ed, J Brazilian Dent Asn, 60-61; partic teacher's training prog oral biol, Chicago Col Dent Surg, Loyola Univ, 61; vis prof & lectr, Med Sch, Northwestern Univ, 67; vis prof, Med Col Ohio, 72. Mem: AAAS; Am Asn Anat; Int Asn Dent Res; Brazilian Soc Anat; Port-Brazilian Soc Anat (pres, 74-76). Res: Oral, head and neck surgery; human anatomy; oral anatomy; mandibular nerve; facial nerve; ear anatomy; gastric-esophagus junction; human embryology; temporo mandibular joint. Mailing Add: Av Atlantica 720 Apt 1002 Leme ZC-07 Rio de Janeiro Brazil

MOSCOVITZ, HOWARD, b New York, NY, Apr 6, 23; m 58; c 3. CARDIOVASCULAR PHYSIOLOGY. Educ: City Col New York, BS, 43; NY Univ, MD, 46. Prof Exp: Asst attend physician, 58-64, ASSOC PHYSICIAN, MT SINAI HOSP, NEW YORK, 64-, SR MEM CARDIAC CATHETERIZATION TEAM, 58- Concurrent Pos: Exec med officer, Jewish Hosp for Aged, 60-; assoc clin prof, Mt Sinai Sch Med, 66-75, clin prof med, 75-; attend cardiologist, Kingsbridge Vet Admin Hosp, 70-; fel coun clin cardiol, Am Heart Asn. Mem: Sr mem Am Fedn Clin Res; fel Am Col Physicians; fel Am Col Cardiol; fel Am Col Chest Physicians. Res: Intracardiac phonocardiography; cardiac catheterization. Mailing Add: 7 E 87th St New York NY 10028

MOSCOWITZ, ALBERT, b Manchester, NH, Aug 20, 29. PHYSICAL CHEMISTRY. Educ: City Col New York, BS, 50; Harvard Univ, MA, 54, PhD, 57. Prof Exp: Nat Res Coun-Am Chem Soc fel petrol res, Harvard Univ, 57-58 & Wash Univ, 58-59; from asst prof to assoc prof, 59-65, PROF PHYS CHEM, UNIV MINN, MINNEAPOLIS, 65- Concurrent Pos: Fulbright lectr & vis prof, Copenhagen Univ, 61-62, vis prof, 67-68; Alfred P Sloan Found fel, 62-66; mem nat screening comt, Fulbright Awards to Scandinavia, 65, chmn, 66; Seydel-Woolley vis prof, Ga Inst Technol, 66; adv ed, Chem Physics Lett, 67-; vchmn, Gordon Conf Theoret Chem, 68, chmn, 70; assoc ed, J Chem Physics, 70-73. Mem: AAAS; Am Chem Soc; Am Phys Soc; The Chem Soc; fel NY Acad Sci. Res: Electronic structure of molecules; optical activity; Faraday effect; stereochemistry. Mailing Add: Dept of Chem Univ of Minn Minneapolis MN 55455

MOSE, DOUGLAS GEORGE, b Chicago, Ill, July 18, 42; m 69. GEOCHEMISTRY, GEOCHRONOLOGY. Educ: Univ Ill, Urbana, BS, 65; Univ Kans, MS, 68, PhD(geol), 71. Prof Exp: Asst prof geol, Brooklyn Col, 71-75; ASSOC PROF GEOL, GEORGE MASON UNIV, 75- Mem: Geol Soc Am; Geol Soc Can; Nat Asn Geol Teachers; Sigma Xi. Res: Evolution of igneous and metamorphic rocks in North American Precambrian and Paleozoic terranes. Mailing Add: Dept of Chem George Mason Univ Fairfax VA 22030

MOSELEY, DONALD STILLMAN, b Albany, NY, June 3, 19; m 46; c 7. UNDERWATER ACOUSTICS. Educ: Rensselaer Polytech Inst, BS, 41, MS, 42, PhD(physics), 54. Prof Exp: Jr physicist, US Naval Ord Lab, 42-47; asst instr, Rensselaer Polytech Inst, 47-52; res engr, Brush Develop Co, 52-55; res physicist, Southwest Res Inst, Tex, 55-58; ACOUST PHYSICIST, VITRO LABS DIV, AUTOMATION INDUSTS, INC, 58- Mem: Acoust Soc Am. Res: Underwater sound and its production, propagation and reception; sonic imaging; piezoelectricity; dynamics of elastic bodies; pulsative flow. Mailing Add: Vitro Labs Div 14000 Georgia Ave Silver Spring MD 20910

MOSELEY, HARRISON MILLER, b Dundee, Tex, Dec 14, 21. PHYSICS. Educ: Tex Christian Univ, AB, 43; Univ NC, PhD(physics), 50. Prof Exp: From asst prof to assoc prof, 50-65, PROF PHYSICS, TEX CHRISTIAN UNIV, 65- Mem: Am Phys Soc; Am Asn Physics Teachers. Res: Thermal diffusion; fundamental particle theory. Mailing Add: Dept of Physics Tex Christian Univ Ft Worth TX 76129

MOSELEY, HARRY EDWARD, b New Iberia, La, Oct 18, 29; m 55; c 2. CHEMISTRY. Educ: La State Univ, BS, 51, MS, 52, PhD(chem), 69. Prof Exp: Res chemist, Monsanto Chem Co, 54-61; from instr to assoc prof, 61-75, PROF CHEM, LA TECH UNIV, 75- Mem: AAAS; Am Chem Soc. Res: Separation and determination of the platinum metals. Mailing Add: Dept of Chem La Tech Univ Ruston LA 71270

MOSELEY, JOHN MARSHALL, chemistry, see 12th edition

MOSELEY, JOHN TRAVIS, b New Orleans, La, Feb 26, 42; m 61; c 2. ATOMIC PHYSICS, MOLECULAR PHYSICS. Educ: Ga Inst Technol, BS, 64, MS, 66, PhD(physics), 69. Prof Exp: Asst res physicist, Eng Exp Sta, Ga Inst Technol, 64-65; asst prof physics, Univ W Fla, 68-69; physicist, 69-75, SR PHYSICIST, STANFORD RES INST, 75- Concurrent Pos: Vis scientist, Univ Paris, 75-76. Mem: Am Phys Soc. Res: Transport properties; reaction rates of gaseous ions; ion-ion mutual neutralization; metastable ions and molecules; photodissociation and photodetachment of ions. Mailing Add: Molecular Physics Ctr Stanford Res Inst Menlo Park CA 94025

MOSELEY, MAYNARD FOWLE, JR, b Boston, Mass, July 15, 18; m 49; c 2. PLANT ANATOMY. Educ: Univ Mass, BS, 40; Univ Ill, MS, 42, PhD(bot), 47. Prof Exp: Instr bot, Cornell Univ, 47-49; from instr to assoc prof, 49-63, PROF BOT, UNIV CALIF, SANTA BARBARA, 63- Mem: Bot Soc Am; Int Soc Plant Morphol; Int Asn Wood Anat; Am Inst Biol Sci. Mailing Add: Dept of Biol Sci Univ of Calif Santa Barbara CA 93106

MOSELEY, PATTERSON B, b Holland, Mo, May 27, 18; m 42; c 3. CHROMATOGRAPHY. Educ: Ouachita Col, BS, 43; La State Univ, MS, 49, PhD, 51. Prof Exp: Res chemist, Hercules Powder Co, Del, 51-57, res supvr, 57-64; assoc prof, 64-69, PROF CHEM, LA TECH UNIV, 69-, DIR RES, COL ARTS & SCI & DIR DIV ALLIED HEALTH, 68-, ASSOC DEAN COL ARTS & SCI, 70- Mem: AAAS; Am Chem Soc. Res: Adsorption chromatography; ion exchange. Mailing Add: Col of Arts & Sci La Tech Univ Ruston LA 71271

MOSELEY, ROBERT DAVID, JR, b Minden, La, Feb 29, 24; m 47; c 3. RADIOLOGY. Educ: La State Univ, MD, 47; Am Bd Radiol, dipl, 55. Prof Exp: Asst resident radiol, Univ Chicago, 49-50; spec res proj, Los Alamos Sci Lab, 50-51, staff mem, Univ Calif, 51-52; from instr to prof radiol & chmn dept, Sch Med, Univ Chicago, 54-71, dir radiation protection serv, 60-71; PROF RADIOL, ASST CHMN DEPT & CHIEF DIAG RADIOL DIV, UNIV NMEX, 71- Concurrent Pos: Assoc chief of staff, Los Alamos Med Ctr, 51-52; Am Cancer Soc clin fel, 55-56; tech adv, US deleg, Int Conf Peaceful Uses of Atomic Energy, Geneva, 58; prof radiol, Argonne Cancer Res Hosp, Chicago, 58-71; vis prof & res scholar, Roentgen Diag Dept, Univ Lund, 62-63; chief of staff, Bernalillo County Med Ctr, NMex, 71-72; mem bd dirs, Nat Coun Radiation Protection; mem radiation study sect, Div Res Grants, NIH, 71-75. Mem: Asn Univ Radiol (pres, 61-62); Radiol Soc NAm (1st vpres, 75-76); Am Col Radiol (pres, 72-73); Int Soc Radiol; hon fel Royal Col Radiologists. Res: Radiation protection; bone neoplasm; diagnosis of diseases of pancreas and upper gastrointestinal tract; diagnostic radiologic instrumentation. Mailing Add: Dept of Radiol Univ of NMex Sch of Med Albuquerque NM 87131

MOSELEY, VINCE, b Orangeburg, SC, Oct 29, 12; m 38; c 8. INTERNAL MEDICINE. Educ: Duke Univ, AB, 33, MD, 36; Am Bd Internal Med, dipl, 45. Prof Exp: Asst med, Duke Univ, 38-39, asst physiol & bact, 39-40; assoc med, Univ Pa, 40-41; from assoc to assoc prof, 47-59, PROF MED, MED UNIV SC, 59-, DIR DIV CONTINUING EDUC & COORDR SC REGIONAL MED PROGS, 69- Concurrent Pos: Fel dermat & syphil, Duke Univ, 39-40; fel gastroenterol, Univ Pa, 40-41; co-chmn dept med & dir outpatient clin, Med Univ SC, 47-66, dean clin med, 60-66, med dir, Hosp; chief med serv, Vet Admin Hosp, Charleston, SC, 66-69; physician-in-chief, Roper Hosp; consult, US Depts Army & Navy. Mem: Soc Exp Biol & Med; Am Geriat Soc; Am Clin & Climat Asn; fel Am Col Physicians; Am Fedn Clin Res. Res: Absorption from the gastrointestinal tract; gastrointestinal motility, tonus and secretory activity; pancreatic enzyme studies. Mailing Add: Div of Continuing Educ Med Univ of SC Charleston SC 29401

MOSELEY, WILLIAM DAVID, JR, b Cleveland, Ohio, Nov 27, 36. PHYSICAL CHEMISTRY. Educ: Williams Col, BA, 58; Wash State Univ, PhD(chem), 63. Prof Exp: Jr scientist quantum chem, Univ Uppsala, 63-66; asst prof chem, Howard Univ, 66-69; ASST PROF CHEM, WASH STATE UNIV, 69- Res: Quantum theory; valence theory; collisions. Mailing Add: Dept of Chem Wash State Univ Pullman WA 99163

MOSELY, ROBERT BRUCE, physical chemistry, organic chemistry, see 12th edition

MOSEMAN, ALBERT HENRY, b Oakland, Nebr, Jan 27, 14; m 42; c 3. AGRICULTURE, RESEARCH ADMINISTRATION. Educ: Univ Nebr, BSc, 38, MS, 40; Univ Minn, PhD(plant breeding, genetics), 44. Hon Degrees: DSc, Univ Nebr, 65. Prof Exp: Agt, Bur Plant Indust, Soils & Agr Eng, USDA, 36-42, assoc agronomist, 42-44, asst to chief bur, 44-51, chief bur, 51-53, dir crops res, Agr Res Serv, 53-56; from assoc dir to dir agr sci, Rockefeller Found, 56-65; asst adminr tech coop & res, AID, Washington, DC, 65-67; assoc, Agr Develop Ccun, 67-74; CONSULT AGR DEVELOP, ROCKEFELLER FOUND, 74- Concurrent Pos: Consult to Secy Agr, 63; dir, Malaysian Agr Res & Develop Inst, Ministry Agr & Coop, Kuala Lumpur, 69-71, consult, 71-72. Mem: Fel AAAS (vpres agr sect, 63); fel Am Soc Agron; Am Genetic Asn. Res: International agricultural development; agricultural research organization administration. Mailing Add: 34 Shadblow Hill Ridgefield CT 06877

MOSEMAN, JOHN GUSTAV, b Oakland, Nebr, Dec 7, 21; m 48; c 3. PLANT PATHOLOGY, AGRONOMY. Educ: Univ Nebr, BS, 43; Wash State Univ, MS, 48; Iowa State Univ, PhD(agron, plant path), 50. Prof Exp: Res plant pathologist, NC, 50-54, res plant pathologist, Cereal Crops Res Br, 54-69, leader barley invest, 69-72, CHMN, PLANT GENETICS & GERMPLASM INST, NORTHEASTERN REGION, AGR RES SERV, USDA, 72- Mem: Am Phytopath Soc; Am Soc Agron. Res: Genetics; host-pathogen relationship obligate pathogens; diseases of barley. Mailing Add: Beltsville Agr Res Ctr Northeastern Region USDA Beltsville MD 20705

MOSEMAN, ROBERT FREDRICK, b Indianapolis, Ind, Dec 18, 41; m 65; c 3. PESTICIDE CHEMISTRY. Educ: Marian Col, BS, 63; Univ NC, MS, 65; Univ Mo, PhD(agr chem), 71. Prof Exp: Chemist, NC State Bd Health, 65-66; chemist, USPHS, 66-68; RES CHEMIST, US ENVIRON PROTECTION AGENCY, 71- Mem: Sigma Xi; Am Chem Soc. Res: Development of analytical methodology for the determination of trace levels of pesticides and other toxic substances and their transformation products in various types of biological and environmental samples. Mailing Add: Health Effects Res Lab US Environ Protection Agency Res Triangle Park NC 27711

MOSEN, ARTHUR WALTER, b Bemidji, Minn, July 11, 22; m 46; c 2. ANALYTICAL CHEMISTRY. Educ: Ore State Col, BS, 49, MS, 51. Prof Exp: Instr chem, San Diego State Col, 50-51; res asst anal chem, Los Alamos Sci Lab, 51-53, staff mem, 53-55; chemist, Rohr Corp, 55-56; staff mem, John Jay Hopkins Lab Pure & Appl Sci, Gen Atomic Div, Gen Dynamics Corp, 56-59, group leader anal chem, 59-66, asst chmn chem dept, 66-70; mgr anal chem br & sr staff mem, Mat Sci Dept, Gulf Gen Atomic Co, 70-73; MGR ANAL CHEM DEPT, GEN ATOMIC CO, 73- Concurrent Pos: Lectr, San Diego State Col, 55-57 & 59-60. Mem: AAAS; Am Chem Soc; Am Soc Testing & Mat. Res: Analytical chemistry methods development applied particularly to nuclear reactor materials; gases in metals; rare earth elements. Mailing Add: Anal Chem Dept Gen Atomic Co PO Box 81608 San Diego CA 92138

MOSER, BRUNO CARL, b Elmhurst, Ill, Mar 31, 40; m 62; c 3. HORTICULTURE. Educ: Mich State Univ, BS, 62, MS, 64; Rutgers Univ, PhD(hort), 69. Prof Exp: From asst prof to assoc prof hort, Rutgers Univ, New Brunswick, 69-75; PROF HORT & HEAD DEPT, PURDUE UNIV, WEST LAFAYETTE, 75- Honors & Awards: Kenneth Post Award, Am Soc Hort Sci, 69. Mem: Am Soc Hort Sci; Int Plant Propagation Soc. Res: Physiology of root regeneration; bud dormancy; tuberization. Mailing Add: Dept of Hort Purdue Univ West Lafayette IN 47907

MOSER, CHARLES EDWIN, b Farmville, Va, Feb 15, 13; m 41; c 4. CHEMISTRY. Educ: Univ Idaho, BS, 33, MS, 35; Northwestern Univ, PhD(phys chem), 39. Prof Exp: Asst, Univ Idaho, 33-35 & Northwestern Univ, 35-38; chemist, Coleman Elec Co, Ill, 38; res chemist, Texas Co, 39-53, assoc dir res, 53-54; asst mgr res & develop div, 54-60, dir res planning fuels, 60-67, asst to vpres res & technol dept, 67-71, ASST TO VPRES ENVIRON PROTECTION DEPT, TEXACO INC, 71- Concurrent Pos: Chmn res adv comt, Coord Res Coun Air Pollution, 74- Honors & Awards: Cert of Appreciation, Am Petrol Inst, 62. Mem: Am Chem Soc; Air Pollution Control Asn; Sigma Xi; Am Inst Chemists; Am Inst Chem Engrs. Res: Thermodynamics of solutions; dielectric constants; electrochemistry; petroleum chemistry; environmental chemistry; automotive emissions. Mailing Add: Environ Protection Dept Texaco Inc Beacon NY 12508

MOSER, CHARLES R, b Woodland, Calif, Oct 8, 39; m 60; c 1. DEVELOPMENTAL BIOLOGY. Educ: Humboldt State Col, AB, 61; State Univ NY Buffalo, PhD(biol), 67. Prof Exp: Asst prof, 66-71, ASSOC PROF BIOL SCI, CALIF STATE UNIV, SACRAMENTO, 71- Res: DNA, RNA, protein synthesis and their interrelationships in developing frog embryos. Mailing Add: Dept of Biol Sci Calif State Univ Sacramento CA 95819

MOSER, DONALD EUGENE, b Steubenville, Ohio, Jan 22, 25; m 46; c 3. MATHEMATICS. Educ: Amherst Col, AB, 47; Brown Univ, AM, 49; Univ Pittsburgh, PhD(math), 56. Prof Exp: From instr to asst prof math, Univ Mass, 49-60; assoc prof, 60-70, PROF MATH, UNIV VT, 70- Mem: Math Asn Am. Res: Algebra; analysis. Mailing Add: Dept of Math Univ of Vt Burlington VT 05401

MOSER, FRANK, b Winnipeg, Man, Sept 5, 27; US citizen; m 49; c 3. SOLID STATE PHYSICS. Educ: Univ Man, BSc hons, 49; Univ Minn, Minneapolis, MSc, 52. Prof Exp: RES PHYSICIST, EASTMAN KODAK CO RES LABS, 52- Concurrent Pos: Vis scientists, Oxford Univ, 65-66 & Israel Inst Technol, 73-74. Mem: Am Phys Soc. Res: Electronic properties of semiconductors; photoeffects and imaging phenomena in semiconductors and other solids and thin films. Mailing Add: Eastman Kodak Co Res Labs 1669 Lake Ave Rochester NY 14650

MOSER, FRANK HANS, b Chicago, Ill, Aug 4, 07; m 30, 69; c 2. POLLUTION CHEMISTRY. Educ: Hope Col, AB, 28; Univ Mich, MS, 29, PhD(chem), 31. Prof Exp: Asst, Univ Mich, 28-31, chief analyst, Eng Res Dept, 31-32; chief analyst, Nat Aniline Co, NY, 32-38; supt, Intermediate Dept, Standard Ultramarine Co, 38-53; supt, Intermediate & Phthalocyanine Depts, Standard Ultramarine & Color Co, 53-59, res dir, 59-64; res dir, Holland Suco Color Co, 65-68; tech dir pigments div, Chemetron Corp, 68-72; CONSULT POLLUTION CONTROL & SAFETY SYSTS, 72- Mem: Am Chem Soc; fel Am Inst Chemists; NY Acad Sci. Res: Phthalocyanine compounds; pigments; intermediates for pigments and dyes; pigments application; process, product and raw material safety. Mailing Add: 112 E 28th St Holland MI 49423

MOSER, H GEOFFREY, b Philadelphia, Pa, Dec 5, 38; m 61; c 2. MARINE BIOLOGY, ICHTHYOLOGY. Educ: Dartmouth Col, AB, 60; Univ Southern Calif, PhD(biol), 66. Prof Exp: FISHERIES BIOLOGIST, NAT MARINE FISHERIES SERV, 64- Prof Exp: Instr, Mesa Co, 68-69. Honors & Awards: Johannes Schmidt Stipendium in Oceanog, Carlsberg Found, 70; Wildlife Soc Res Award, 72. Mem: AAAS; Am Soc Ichthyologists & Herpetologists. Res: Marine zoology; reproduction and development of marine fishes; functional anatomy of vertebrates. Mailing Add: Nat Marine Fisheries Serv PO Box 271 La Jolla CA 92037

MOSER, HERBERT CHARLES, b Camp Verde, Ariz, Mar 5, 29; m 51; c 2. PHYSICAL CHEMISTRY. Educ: San Jose State Col, BS, 52; Iowa State Univ, PhD(chem), 57. Prof Exp: From asst prof to assoc prof, 57-69, PROF CHEM, KANS STATE UNIV, 69- Mem: Am Chem Soc. Res: Atomic and free radical reactions; radiation chemistry; use of radioisotopes as tracers. Mailing Add: Dept of Chem Kans State Univ Manhattan KS 66502

MOSER, JAMES HOWARD, b Santa Rosa, Calif, Apr 29, 28; m 50; c 4. CHEMISTRY, CHEMICAL ENGINEERING. Educ: Univ Calif, BS, 50; Ore State Col, MS, 52, PhD(chem), 54. Prof Exp: Asst food technol, Ore State Col, 50-51, asst chem, 51-52; res chemist & technologist, Shell Oil Co, Calif, 54-60, sr res chemist, Houston Res Lab, 60-68, STAFF RES ENGR, HOUSTON RES LAB, SHELL DEVELOP CO, 68- Mem: Am Inst Chem Eng. Res: Chemical and petroleum process design and development; applied mathematics and statistics; mathematical simulation of chemical and petroleum processes; control systems, environmental engineering; operations research; analytical and physical chemistry; reaction kinetics. Mailing Add: 12511 Ravensway Cypress TX 77429

MOSER, JOHN BENEDIKT, b Salzburg, Austria, Oct 17, 23; US citizen; m 54; c 1. MATERIALS SCIENCE, DENTAL MATERIALS. Educ: Sir George Williams Univ, BS, 50; Northwestern Univ, MS, 58, PhD(mat sci), 61. Prof Exp: Res assoc mat sci, Northwestern Univ, 61-62; assoc ceramist, Argonne Nat Lab, 62-70; fel, 70-71, ASST PROF BIOL MAT, NORTHWESTERN UNIV, CHICAGO, 71- Mem: Am Dent Asn; Int Asn Dent Res. Res: Equilibrium and kinetic behavior of inorganic compounds; mechanical properties of dental materials; properties and structure of reinforced resins and dental cements. Mailing Add: Dept of Biol Mat Northwestern Univ Chicago IL 60611

MOSER, JOHN C, b Columbus, Ohio, Mar 29, 29. ACAROLOGY. Educ: Ohio State Univ, BS, 51, MS, 54; Cornell Univ, PhD(insect ecol), 58. Prof Exp: ENTOMOLOGIST, ALEXANDRIA FORESTRY CTR, US FOREST SERV, 58- Mem: Entom Soc Am; Royal Entom Soc London. Res: Biology of leaf cutting ants; Celtis gallmaker parasites; bark beetle mites. Mailing Add: Southern Forest Exp Sta US Forest Serv Pineville LA 71360

MOSER, JOHN WILLIAM, JR, b Hagerstown, Md, Oct 8, 36; m 64; c 3. FOREST BIOMETRY. Educ: WVa Univ, BS, 58; Pa State Univ, MS, 61; Purdue Univ, PhD(forest biomet), 67. Prof Exp: Forester, US Forest Serv, 61-63 & Coop Exten Serv, WVa Univ, 63-64; asst, 64-66, from instr to asst prof, 66-71, ASSOC PROF FORESTRY & CONSERV, PURDUE UNIV, WEST LAFAYETTE, 71- Mem: Biomet Soc; Soc Am Foresters. Res: Modeling the dynamics of forest stands; computer applications to forest management. Mailing Add: Dept of Forestry & Conserv Purdue Univ West Lafayette IN 47906

MOSER, JOSEPH M, b Spring Hill, Minn, Apr 13, 30. MATHEMATICAL STATISTICS. Educ: St John's Univ, Minn, BA, 54; St Louis Univ, MA, 55, PhD(math statist), 59. Prof Exp: Statistician, Aberdeen Proving Ground, Md, 55-56; from asst prof to assoc prof, 59-69, PROF MATH STATIST, SAN DIEGO STATE UNIV, 69- Concurrent Pos: Consult, Navy Electronics Lab, Calif, 64-71. Mem: Inst Math Statist; Math Asn Am; Am Statist Asn. Res: Distribution-free statistics. Mailing Add: Dept of Math San Diego State Univ San Diego CA 92115

MOSER, JÜRGEN (KURT), b Königsberg, Ger, July 4, 28; nat US; m 55; c 3. MATHEMATICS. Educ: Univ Göttingen, Dr rer nat, 52. Prof Exp: Asst math, Univ Göttingen, 53; res assoc, NY Univ, 53-54; asst, Univ Göttingen, 54-55; res assoc, NY Univ, 55-56, asst prof, 56-57; from asst prof to assoc prof, Mass Inst Technol, 57-60; dir, Courant Inst Math Sci, 67-70, PROF MATH, NY UNIV, 60- Concurrent Pos: Sloan fel, 62 & 63; Am Acad Arts & Sci fel, 64; Guggenheim fel, 70. Honors & Awards: G B Birkhoff Prize, 68; J Craig Watson Medal, 69. Mem: Nat Acad Sci; Am Math Soc; corresp mem Int Astron Union. Res: Ordinary and partial differential equations; spectral theory; celestial mechanics. Mailing Add: Courant Inst of Math Sci NY Univ New York NY 10012

MOSER, KENNETH BRUCE, b Malverne, NY, Mar 27, 33; m 58; c 5. CARBOHYDRATE CHEMISTRY. Educ: Tusculum Col, BS, 54; Duke Univ, PhD(chem), 59. Prof Exp: Res chemist, 58-60, sr res chemist, 60-70, lab head, 70-73, GROUP LEADER, A E STALEY MFG CO, 73- Mem: Am Chem Soc. Res: Synthesis of polycyclics aromatic systems containing quarternary nitrogen at the bridgehead position; preparation of acrylic polymer emulsions; carbohydrates; nitrogen heterocyclics; starch modification. Mailing Add: Res Div AE Staley Co Bldg 63 Decatur IL 62525

MOSER, KENNETH MILES, b Baltimore, Md, Apr 12, 29; m 51; c 4. MEDICINE. Educ: Haverford Col, AB, 50; Johns Hopkins Univ, MD, 54. Prof Exp: From instr to assoc prof med, Georgetown Univ, 58-68, chief pulmonary div, Univ Hosp, 61-68; assoc prof, 68-73, PROF MED, UNIV CALIF, SAN DIEGO, 73-, DIR PULMONARY DIV, 68-, DIR PULMONARY SPEC CTR RES, 70- Concurrent Pos: Dir head, chest & contagious dis div, Nat Naval Med Ctr, Md, 59-61; chief pulmonary sect, Georgetown Clin Res Inst, Bur Aviation Med, Fed Aviation Agency, 61-66; consult, US Naval Hosp, Md, 61 & NIH, 65- Mem: Am Thoracic Soc; Am Heart Asn; Am Fedn Clin Res; fel Am Col Physicians; fel Am Col Chest Physicians. Res: Pulmonary and cardiac physiology; blood coagulation; pathogenesis and therapy of thromboembolism; clinical pulmonary disease. Mailing Add: Pulmonary Div Univ Hosp 225 W Dickinson St San Diego CA 92103

MOSER, LOUISE ELIZABETH, b Racine, Wis, July 24, 43. MATHEMATICS. Educ: Univ Wis-Madison, BS, 65, MS, 66, PhD(math), 70. Prof Exp: Asst prof, 70-74, ASSOC PROF MATH, CALIF STATE UNIV, HAYWARD, 74- Mem: Am Math Soc; Math Asn Am. Res: Topology of 3-manifolds; knot theory; group theory; ring theory. Mailing Add: Dept Math Calif State Univ 25800 Hillary St Hayward CA 94542

MOSER, LOWELL E, b Akron, Ohio, Mar 19, 40; m 64; c 2. AGRONOMY. Educ: Ohio State Univ, BS, 62, PhD(agron), 67; Kans State Univ, MS, 64. Prof Exp: Asst prof agron, Ohio State Univ, 67-70; assoc prof, 70-75, PROF AGRON, UNIV NEBR-LINCOLN, 75- Honors & Awards: Teaching Award of Merit, Gamma Sigma Delta, 73; Distinguished Teaching Award, Univ Nebr, 74. Mem: Am Soc Agron; Crop Sci Soc Am; Am Soc Range Mgt. Res: Forage physiology; physiological investigation into cool and warm season grasses in Nebraska. Mailing Add: Dept of Agron Univ of Nebr Lincoln NE 68503

MOSER, MARVIN, b Newark, NJ, Jan 24, 24; m 54; c 3. INTERNAL MEDICINE, CARDIOLOGY. Educ: Cornell Univ, AB, 43; State Univ NY Downstate Med Ctr, MD, 47. Prof Exp: From intern to resident med, Univ Div, Kings County Hosp, Brooklyn, 47-49; resident, Montefiore Hosp, NY, 49-50; in-chg vascular dis & asst chief cardiol, Walter Reed Army Med Ctr, DC, 51-53; assoc, Albert Einstein Col Med, 65-71, asst prof clin med, 71-74; CLIN PROF MED, NEW YORK MED COL, 74-; PHYSICIAN & CHIEF CARDIOL, WHITE PLAINS HOSP, 71- Concurrent Pos: Nat Heart Asn fel cardiol, Mat Sinai Hosp, 50-51; NIH res grants cardiol & hypertensive vascular dis, 55-57 & 60-61; consult, St Agnes Hosp, White Plains, NY, 55-62; assoc attend med, White Plains Hosp, 56-71; chief hypertension clin, Montefiore Hosp, 56-73; fel coun clin cardiol, Am Heart Asn, 60-, mem med adv bd coun high blood pressure res, 74-; attend physician in-chg-hypertension, Westchester County Med Ctr, Valhalla, 74-; sr med consult, Nat High Blood Pressure Educ Prog, Nat Heart & Lung Inst, 75- Honors & Awards: Am Col Angiol Honors Award, 59.

Mem: AMA; Am Fedn Clin Res; fel Am Col Cardiol; fel Am Col Physicians. Res: Hypertension; hypertensive vascular disease. Mailing Add: 33 Davis Ave White Plains NY 10605

MOSER, PAUL E, b Auburn, Wyo, Jan 18, 42; m 63; c 4. PLANT BREEDING, PLANT PATHOLOGY. Educ: Utah State Univ, MS, 68, PhD(plant path & breeding), 71. Prof Exp: Fel plant path, McGill Univ, 71-72; PATHOLOGIST & PLANT BREEDER, GALLATIN VALLEY SEED CO, 72- Mem: Am Phytopath Soc; Sigma Xi. Res: Breeding of new pea and bean varieties; indexing and screening for disease resistance of peas and beans. Mailing Add: Gallatin Valley Seed Co Box 167 Twin Falls ID 83301

MOSER, ROBERT E, b Defiance, Ohio, Feb 18, 39; m 61; c 2. BIO-ORGANIC CHEMISTRY. Educ: Bowling Green State Univ, BS, 61; Yale Univ, MS, 63, PhD(org chem), 65. Prof Exp: Res specialist, Cent Res Dept, Monsanto Co, Mo, 65-69, group leader, 69-71; sr res chemist, 71-73, RES GROUP LEADER, T R EVANS RES CTR, DIAMOND SHAMROCK CORP, 73- Mem: AAAS; Am Chem Soc. Res: Oxidation-reduction polymers and charge transfer interactions; chemical evolution; protein food and nutrition; design and synthesis of herbicides and animal health products; biological molecular interactions; structure-activity correlation. Mailing Add: T R Evans Res Ctr Diamond Shamrock Corp PO Box 348 Painesville OH 44077

MOSER, ROY EDGAR, b Steubenville, Ohio, Sept 30, 22; m 42; c 2. FOOD TECHNOLOGY. Educ: Univ Mass, Amherst, BS, 47, MS, 49. Prof Exp: Food technologist, USDA, DC, 48-50; exten food technologist, Va Polytech Inst, 50-57; assoc prof food technol, Ore State Univ, 58-67, Univ Hawaii, Manoa Campus, 67-68 & Ore State Univ, 68-69; chmn dept food sci & technol, 70-73, PROF FOOD TECHNOL, UNIV HAWAII, MANOA CAMPUS, 69- Mem: Inst Food Technol. Res: Food processing. Mailing Add: Dept of Food Sci & Technol Univ of Hawaii Manoa Campus Honolulu HI 96822

MOSER, RUSSEL JOHN, organic chemistry, see 12th edition

MOSER, WILLIAM O J, b Winnipeg, Man, Sept 5, 27; m 53; c 3. MATHEMATICS. Educ: Univ Man, BSc, 49; Univ Minn, MA, 51; Univ Toronto, PhD(math), 57. Prof Exp: Lectr math, Univ Sask, 55-57, asst prof, 57-59; assoc prof, Univ Man, 59-64; assoc prof, 64-66, PROF MATH, McGILL UNIV, 66- Concurrent Pos: Ed, Can Math Bull, 61-69. Mem: Am Math Soc; Math Asn Am; Can Math Cong (pres, 75-77); Soc Indust & Appl Math. Res: Finite groups; combinatorial mathematics. Mailing Add: Dept of Math McGill Univ Montreal PQ Can

MOSER, WILLIAM RAY, b Old Hickory, Tenn, Aug 3, 35; m 60; c 3. ORGANOMETALLIC CHEMISTRY. Educ: Mid Tenn State Univ, BS, 59; Mass Inst Technol, PhD(org chem), 64. Prof Exp: Res chemist, Monsanto Res SA, Switz, 66-67, staff scientist, 67-69; sr res chemist, Corp Res Labs, Esso Res & Eng Co, NJ, 69-75; RES ASSOC, BADGER CO, 75- Mem: Am Chem Soc; NY Acad Sci; NAm Catalysis Soc. Res: Homogeneous and heterogeneous catalysis in areas of chemical and petrochemical processes; organometallic synthesis and reaction mechanisms. Mailing Add: Badger Co One Broadway Cambridge MA 02142

MOSES, ALFRED JAMES, b Loerrach, Ger, Feb 5, 21; nat US; m 50; c 1. FORENSIC SCIENCE. Educ: NY Univ, BS, 48; Iowa State Univ, MS, 50. Prof Exp: Asst physics, NY Univ, 47-48; asst chem, Iowa State Univ, 48-49, jr radiochemist, Ames Lab, 49-51; radiochemist, Isotope Unit, Air Force Cambridge Res Ctr, 51-52; chief radioisotope sect, Watertown Arsenal Lab, 52-55; sr engr, Aircraft Nuclear Propulsion Dept, Gen Elec Co, 55-57; sr scientist, Bettis Atomic Power Div, Westinghouse Elec Corp, 57-59; mgr hot lab opers, Isotopes Specialties Co Div, Nuclear Corp Am, 59-60, opers mgr, 60-61; opers mgr, Hazleton-Nuclear Sci Corp, 61; chemist, Atomics Int Div, NAm Aviation Inc, 61-65; eng specialist flight & electronics systs, Aires Mfg Co Div, Garrett Corp, 65-67; mem tech staff, Electronic Properties Info Ctr, Hughes Aircraft Co, 67-70; criminalist, City of West Covina, Calif, 72-73; CRIMINALIST, WEST COVINA SATELLITE LAB, INVEST SERV BR, CALIF DEPT JUSTICE, 73- Concurrent Pos: Abstractor, Chem Abstr, 50; radiation protection officer, City of West Covina, 75- Mem: Am Chem Soc. Res: Forensic chemistry; radioisotope technology. Mailing Add: 609 S Sunset Ave West Covina CA 91790

MOSES, CAMPBELL, JR, b Pittsburgh, Pa, Feb 12, 17; m 40; c 4. MEDICINE. Educ: Univ Pittsburgh, BS, 39, MD, 41. Prof Exp: Instr physiol & pharmacol, Sch Med, Univ Pittsburgh, 41-46, from asst prof to assoc prof med, 46-48, dir, Addison H Gibson Lab, 48-68, dir postgrad educ, 60-68; med dir, Am Heart Asn, 68-73. Concurrent Pos: Mem coun arteriosclerosis, Am Heart Asn. Mem: Am Physiol Soc; Soc Exp Biol & Med; fel AMA; fel Am Col Physicians; Am Diabetes Asn. Res: Thrombosis and embolism; anticoagulants; liver and kidney function tests; nutrition, liver and kidney function in arteriosclerosis. Mailing Add: Medicus Commun Inc 909 Third Ave New York NY 10022

MOSES, EDWARD JOEL, b Newark, NJ, Oct 9, 38; m 65; c 2. UNDERWATER ACOUSTICS, OPERATIONS RESEARCH. Educ: Rensselaer Polytech Inst, BS, 60; Johns Hopkins Univ, PhD(physics), 67. Prof Exp: Res assoc physics, Vanderbilt Univ, 67-68, instr, 68-71; res scientist, Raff Assocs, Inc, Md, 71-75; dept head, Ocean Systs Dept, Gen Res Co, 75; SR PROJ STAFF, OPERS RES, INC, 75- Mem: Am Phys Soc. Res: Sound propagation and noise background in the oceans; statistical character of oceanic noise; statistics of signal detection; tactical analysis and operations research for naval operations and systems. Mailing Add: Opers Res Inc 1400 Spring St Silver Spring MD 20910

MOSES, FRANCIS GUY, b Baltimore, Md, Nov 15, 37. PHYSICAL ORGANIC CHEMISTRY. Educ: Univ Del, BA, 59; Calif Inst Technol, PhD(chem), 67. Prof Exp: Fel chem, Iowa State Univ, 64-65; SUPV CHEMIST, E I DU PONT DE NEMOURS & CO, INC, 66- Mem: Am Chem Soc. Res: High pressure chemistry; reaction mechanisms; organic photochemistry. Mailing Add: Cent Res Dept Exp Sta E I du Pont de Nemours & Co Inc Wilmington DE 19898

MOSES, GERALD ROBERT, b Chicago, Ill, June 7, 38; m 61; c 2. SPEECH PATHOLOGY. Educ: Loyola Univ Chicago, BS, 61; Western MIch Univ, MA, 65; Ohio State Univ, PhD(speech sci), 69. Prof Exp: Instr speech path, Miami Univ, 65-67; ASSOC PROF SPEECH PATH, EASTERN MICH UNIV, 69- Concurrent Pos: Clin dir, Preble County Speech & Hearing Clin, Ohio, 65-67; ed, J Commun Path, 75- Mem: Am Speech & Hearing Asn; Am Asn Higher Educ. Res: Stuttering; experimental phonetics; supervision of public school speech therapy. Mailing Add: Speech Clin Eastern Mich Univ Ypsilanti MI 48197

MOSES, HAROLD EUGENE, b Troy, Ohio, Dec 3, 12; m 39; c 4. VETERINARY MEDICINE. Educ: Ohio State Univ, DVM, 36, MSc, 39. Prof Exp: Jr vet, Bur Animal Indust, USDA, 36; instr bact, Ohio State Univ, 36-39; spec res asst, Harvard Univ, 43-45; PROF VET SCI, PURDUE UNIV, 46- Res: Experimental chemotherapy of tuberculosis; naturally-occurring tuberculosis in animals; general veterinary microbiology. Mailing Add: Sch of Vet Sci & Med Purdue Univ West Lafayette IN 47907

MOSES, HARRY, meteorology, see 12th edition

MOSES, HARRY ELECKS, b Canton, Ohio, Aug 30, 22; m 58; c 1. THEORETICAL PHYSICS. Educ: Univ Mich, BS, 44, MS, 47; Columbia Univ, PhD(physics), 50. Prof Exp: Res scientist aerodyn, Nat Adv Comt Aeronaut, 44-46 & Univ Mich, 46-47; asst physics, Columbia Univ, 47-49; asst wave propagation, NY Univ, 49-50, res assoc upper atmosphere res, 50-60; assoc prof physics, Polytech Inst Brooklyn, 60-61; staff mem, Geophys Corp Am, Mass, 61-62 & Lincoln Lab, Mass Inst Technol, 62-69; STAFF MEM, AIR FORCE CAMBRIDGE RES LABS, 69- Mem: Am Phys Soc. Res: Aerodynamics; quantum theory of scattering; upper atmosphere physics. Mailing Add: Aeronomy Lab Air Force Cambridge Res Labs Bedford MA 01730

MOSES, HENRY A, b Gastonia, NC, Sept 8, 39. BIOCHEMISTRY, PHYSIOLOGY. Educ: Livingstone Col, BS, 59; Purdue Univ, MS, 62, PhD(biochem), 64. Prof Exp: Asst prof, 64-70, ASSOC PROF BIOCHEM, MEHARRY MED COL, 70- Concurrent Pos: Tenn Heart Asn res grant, 65; vis lectr, Wheaton Col, Mass, 65; consult biochemist, Vet Admin Hosp, Tuskegee, Ala, 65; asst prof, Tenn State Univ, 65-66; consult, George W Hubbard Hosp, Nashville. Mem: AAAS; Am Chem Soc. Res: Zinc metabolism; biochemical effects of oral contraceptives; anti-cariogenic effects of vanadium compounds; trace minerals in the post-myocardial infracted patient; toxic metals in the environment. Mailing Add: Dept of Biochem Meharry Med Col Nashville TN 37208

MOSES, HERBERT A, b Hartford, Conn, July 22, 29; m 54; c 3. ATOMIC PHYSICS. Educ: Mich State Univ, BS, 51, MS, 53; Univ Conn, PhD(physics), 63. Prof Exp: From asst to instr physics, Mich State Univ, 51-56; res asst, Univ Conn, 56-59; from asst prof to assoc prof, 59-66, PROF PHYSICS, TRENTON STATE COL, 66- Mem: Am Asn Physics Teachers. Res: Experimental work in nuclear magnetic resonance in liquid crystals total cross sections for multiple electron stripping in atomic collisions at one hundred kiloelectron volts and five billion electron volts c-muon interactions; theoretical calculations of atomic wave functions for cesium and electron distribution in a deuterium plasma; nuclear magnetic resonance signals in liquid crystals. Mailing Add: Dept of Physics Trenton State Col Trenton NJ 08625

MOSES, JOEL, b Petach Tikvah, Israel, Nov 25, 41; US citizen; m 70. COMPUTER SCIENCE. Educ: Columbia Univ, BA, 62, MA, 63; Mass Inst Technol, PhD(math), 67. Prof Exp: Asst prof comput sci, 67-71, ASSOC PROF COMPUT SCI, MASS INST TECHNOL, 71-, ASSOC DIR, LAB COMPUT SCI, 74- Concurrent Pos: Res assoc, Electro Tech Lab, Tokyo, Japan, 75. Mem: Sigma Xi; Asn Comput Mach; Math Asn Am. Res: Symbolic formula manipulation; symbolic integration algorithms; artificial intelligence. Mailing Add: Lab for Comput Sci Mass Inst of Technol Cambridge MA 02139

MOSES, JOHN HERRICK, b New York, NY, Feb 1, 09; m 34; c 3. GEOLOGY. Educ: Harvard Univ, BS, 31, AM, 32, PhD(mining geol), 36. Prof Exp: Asst instr econ geol, Harvard Univ, 33-35; geologist, Cerro de Pasco Copper Corp, 35-41, asst chief geologist, 41-48, chief geologist, 48-51; chief geologist, Reynolds Metals Co, 51-74; RETIRED. Concurrent Pos: Field asst, US Geol Surv, 32 & 34. Mem: AAAS; Soc Mining Eng; fel Geol Soc Am; Soc Econ Geol. Res: Mining geology; ore microscopy; exploration and development of mineral resources; identification of opaque minerals by their reflecting power as measured photoelectrically. Mailing Add: 100 Penshurst Rd Richmond VA 23221

MOSES, LINCOLN ELLSWORTH, b Kansas City, Mo, Dec 21, 21; m 42, 68; c 5. STATISTICS. Educ: Stanford Univ, AB, 41, PhD(statist), 50. Prof Exp: Asst prof educ, Teachers Col, Columbia Univ, 50-52; from asst prof to assoc prof, 52-59, exec head dept, 64-68, assoc dean humanities & sci, 65-68, dean grad div, 69-75, PROF STATIST, UNIV & SCH MED, STANFORD UNIV, 59- Concurrent Pos: Guggenheim fel, 60-61; fel, Ctr Advan Study in Behav Sci, 75-76. Mem: Nat Inst Med; Biomet Soc; Am Statist Asn; Inst Math Statist. Res: Experimental design; biological and psychological applications of statistical methods; data analysis. Mailing Add: Dept of Statist Stanford Univ Stanford CA 94305

MOSES, MONTROSE JAMES, b New York, NY, June 26, 19; m 49; c 2. CELL BIOLOGY, ELECTRON MICROSCOPY. Educ: Bates Col, BS, 41; Columbia Univ, AM, 42, PhD(zool), 49. Prof Exp: Assoc cytochemist, Brookhaven Nat Lab, 48-52, cytochemist, 52-55; assoc prof anat, 59-66, PROF ANAT, SCH MED, DUKE UNIV, 66- Concurrent Pos: Vis investr, Rockefeller Inst, 54-55, asst, 55-, assoc, 55-56, asst prof cytol, 56-59; mem, Nat Res Coun, 62-64 & 70-74; mem molecular biol study sect, NIH, 66-69; adv ed, Int Rev Cytol, 71-76. Mem: Am Soc Zoologists; Genetics Soc Am; Am Soc Naturalists; Am Asn Anatomists; Am Soc Cell Biol (secy, 61-67, pres, 68-69). Res: Cytology; fine structure and cytochemistry of nucleus and chromosomes; synaptonemal complex in meiosis; microtubules in motility and cell differentiation; light and electron microscopic techniques for investigating cell structure and function. Mailing Add: Dept of Anat Duke Univ Sch of Med Durham NC 27710

MOSES, RAY NAPOLEON, JR, b Clinton, NC, Jan 23, 36; m 62; c 1. ASTRONOMY. Educ: Ga Inst Technol, BS, 64; Ohio State Univ, PhD(astron), 74. Prof Exp: Astronaut engr, Boeing Co, 64-67 & Lockheed Missiles & Space Co, Inc, 67-68; systs analyst, US Air Force, Wright-Patterson AFB, Ohio, 69-70; ASST PROF PHYSICS, FURMAN UNIV, 74- Mem: Am Astron Soc. Res: Solar system studies, especially those related to interplanetary exploration. Mailing Add: Dept of Physics Furman Univ Greenville SC 29613

MOSES, RONALD ELLIOT, b Chelsea, Mass, Dec 29, 30; m 52; c 2. ORGANIC CHEMISTRY. Educ: Harvard Univ, AB, 52; Northeastern Univ, MS, 59. Prof Exp: From jr res chemist to sr res chemist, Atlantic Gelatin Div, Gen Foods Corp, Mass, 54-60; sr prof chemist, Gillette Safety Razor Co, 60-66; proj chemist, 66-71, sr mgr res, 71-72, dir prod develop, 72-75, DIR TECH & ADMIN SERV, TOILETRIES DIV, GILLETTE CO, 75- Mem: Fel Am Inst Chemists; Am Chem Soc; Soc Cosmetic Chemists. Res: Gelatin, processing and properties; organic synthesis, particularly of heterocyclic compounds, especially pyridines, quinolines and pyrimidines. Mailing Add: 1039 Shirley St Winthrop MA 02152

MOSES, WILLIAM, b New York, NY, Dec 27, 10; m 46; c 3. COMPARATIVE BIOCHEMISTRY. Educ: City Col New York, BS, 41; Univ Calif, PhD(comp biochem), 52. Prof Exp: Lab asst, Wallerstein Labs, 35-41, res chemist, 41-42, group leader, 46-49; sect head, 55-72, DIR MICROBIOL RES, S B PENICK, CPC INT, INC, NEW YORK, 72- Res: Biochemistry; microbiology; fermentation; antibiotics; enzymes; tissue culture research. Mailing Add: 375 Harland Ave Haworth NJ 07641

MOSESMAN, MAX ABE, b Dallas, Tex, June 9, 15; m 39; c 2. PHYSICAL CHEMISTRY. Educ: Agr & Mech Col Tex, BS, 36, MS, 38; Univ Calif, PhD(phys chem), 41. Prof Exp: Asst, Agr & Mech Col Tex, 36-38; res chemist, Humble Oil &

Refining Co, 45-52, sr res chemist, 52-56, res specialist, 56-57, res sect head, 57-65; dept head, Esso Res & Eng Co, 65-66; MGR RES SERV, EXXON RES & ENG CO, 66- Mem: Am Chem Soc; Am Inst Chem Engrs. Res: Chemical thermodynamics; catalysis; thermodynamic properties of the crystalline forms of silica; x-ray diffraction; adsorption; analytical chemistry. Mailing Add: Exxon Res & Eng Co Box 4255 Baytown TX 77520

MOSESSON, MICHAEL W, b New York, NY, Dec 31, 34; m 67; c 2. BIOCHEMISTRY, INTERNAL MEDICINE. Educ: Brooklyn Col, BS, 55; State Univ NY, 59. Prof Exp: Intern, II & IV Med Serv, Boston City Hosp, Mass, 59-60; asst resident, Ward Med Serv, Barnes Hosp, St Louis, Mo, 63-64, instr, Dept Med, 65-67; from asst prof to assoc prof med, 67-75, PROF MED, COL MED, STATE UNIV NY DOWNSTATE MED CTR, 75- Concurrent Pos: NIH res career develop award, 67; mem exec comt thrombosis, Am Heart Asn, 71- Mem: Am Fedn Clin Res; Int Soc Hemat; Am Soc Clin Invest; Am Soc Biol Chem; Int Soc Thrombosis & Haemostasis. Res: Structure of fibrinogen and related proteins; metabolism of coagulation proteins; hemostasis. Mailing Add: Dept of Med State Univ NY Downstate Med Ctr Brooklyn NY 11203

MOSESSON, ZEHMAN I, b Brownsville, Pa, Oct 11, 11. MATHEMATICS. Educ: Harvard Univ, AB, 31, MA, 34, PhD(math), 37. Prof Exp: Instr math, Harvard Univ, 32-36; actuarial student, Prudential Ins Co Am, 37-47, asst mathematician, 47-49, sr actuarial asst, 49-50, chief actuarial asst, 50-54, asst actuarial dir, 54-55, assoc actuarial dir, 55-59, asst actuary, 59-72; ADJ ASSOC PROF, COL INS, 73- Mem: Math Asn Am; fel Soc Actuaries; Am Acad Actuaries. Res: Actuarial science. Mailing Add: 1403-B Troy Towers Bloomfield NJ 07003

MOSEVICH, JACK WALTER, b Melrose Park, Ill, Oct 9, 44; m 65; c 2. APPLIED MATHEMATICS. Educ: Univ Ill, BSc, 65; Northern Ill Univ, MSc, 67; Univ BC, PhD(math), 72. Prof Exp: Asst prof math, Univ Alta, 72-73; ASST PROF MATH, MT ALLISON UNIV, 73- Mem: Math Asn Am. Res: Computing unknown functions or parameters occurring in a differential equation whose solution is known. Mailing Add: Dept of Math Mt Allison Univ Sackville NB Can

MOSEY, LOIS MARGOT, b Evanston, Wyo, May 1, 22. OBSTETRICS & GYNECOLOGY. Educ: Univ Wyo, BS, 43; Boston Univ, MA, 51, MD, 53. Prof Exp: Intern, Bellevue Hosp, New York, 53-54, asst resident med, 54-55; asst resident obstet & gynec, Grace-New Haven Hosp, Conn, 55-57; resident, New Brit Gen Hosp, 57-58; asst prof, 58-66, ASSOC PROF OBSTET & GYNEC, SCH MED, UNIV MISS, 66- Mem: AMA. Res: Pathologic physiology of reproduction. Mailing Add: Dept of Obstet & Gynec Univ of Miss Sch of Med Jackson MS 39216

MOSHER, CAROL WALKER, b Loveland, Colo, June 23, 21; m 44; c 3. BIO-ORGANIC CHEMISTRY. Educ: Colo State Col, BS, 42; Pa State Col, MS, 43, PhD(chem), 47. Prof Exp: RES CHEMIST, STANFORD RES INST, 47- Mem: Am Chem Soc. Res: Organic syntheses; drug-DNA interactions. Mailing Add: 713 Mayfield Stanford CA 94305

MOSHER, HAROLD ELWOOD, b Sterling, Mass, Aug 6, 20; m 46; c 3. HORTICULTURE. Educ: Mass State Col, BS, 42; Univ Mass, BLA, 47, MLA, 57. Prof Exp: Landscape architect & supt grounds, Lake Placid Club, NY, 47-50; from instr to asst prof hort, Univ Mo, 50-58; assoc exten prof, 58-66, assoc prof landscape archit, 66-72, PROF LANDSCAPE ARCHIT, UNIV MASS, AMHERST, 72- Res: Ornamental plants and their uses in landscape architecture; ecological determinants in land use and landscaping. Mailing Add: Clark Hall Univ of Mass Amherst MA 01002

MOSHER, HARRY STONE, b Salem, Ore, Aug 31, 15; m 44; c 3. ORGANIC CHEMISTRY. Educ: Willamette Univ, AB, 37; Ore State Col, MS, 39; Pa State Col, PhD(org chem), 42. Prof Exp: Asst prof chem, Willamette Univ, 39-40; from instr to asst prof, Pa State Col, 43-46; from asst prof to assoc prof, 47-56, PROF CHEM, STANFORD UNIV, 56- Concurrent Pos: US sr res fel, Univ London, 59-60; Am Chem Soc fel, Univ Zurich, 67-68; vis prof, Univ Amsterdam, 75. Mem: AAAS; Am Chem Soc. Res: Stereochemistry; synthetic drugs; mechanisms of organic reactions; chemistry of pyridine compounds; peroxides; animal toxins. Mailing Add: Dept of Chem Stanford Univ Stanford CA 94305

MOSHER, JAMES ARTHUR, b Green, NY, Oct 25, 42; m 68; c 1. PHYSIOLOGICAL ECOLOGY, ORNITHOLOGY. Educ: Utica Col, BS, 65; State Univ NY Col Environ Sci & Forestry, MS, 73; Brigham Young Univ, PhD(zool), 75. Prof Exp: RES ASSOC PHYSIOL ECOL, NAVAL ARCTIC RES LAB, UNIV ALASKA, 75- Mem: Ecol Soc Am; AAAS; Am Ornithologists Union; Cooper Ornith Soc. Res: Study of metabolic and thermoregulatory adaptations of arctic vertebrates; physiological ecology of birds of prey and avian ecology. Mailing Add: Naval Arctic Res Lab Barrow AK 99723

MOSHER, JOHN IVAN, b Waterloo, NY, Sept 26, 33; m 60; c 3. HUMAN ECOLOGY. Educ: Hobart Col, BA, 56; Western State Col Colo, MA, 61; Utah State Univ, PhD, 72. Prof Exp: Chem analyst, NY Agr Sta, Cornell Univ, 58-59; instr biol, Lyndon Inst, Vt, 59-60; asst & preparator, Univ Rochester, 60-61; ASSOC PROF BIOL, STATE UNIV NY COL BROCKPORT, 61- Mem: Ecol Soc Am; Am Soc Zoologists; Zool Soc London. Res: Human ecology, connected with understanding the practical applications of basic ecological principles to living a life style harmonious with the environment; experimentations with shelter, biodynamic food production, and human reactions to employing environmentally sound living practices. Mailing Add: Dept of Biol State Univ of NY Brockport NY 14420

MOSHER, LOREN CAMERON, b Phoenix, Ariz, June 20, 38; m 63; c 4. PETROLEUM GEOLOGY, PALEONTOLOGY. Educ: Calif Inst Technol, BS, 60; Univ Wis, MS, 64, PhD(micropaleont), 67. Prof Exp: Asst prof geol, Fla State Univ, 67-71; assoc prof geosci, Univ Ariz, 71-75; RES GEOLOGIST, PHILLIPS PETROL CO, 75- Concurrent Pos: NSF grant, Fla State Univ, 68-71. Mem: AAAS; Soc Econ Paleontologists & Mineralogists; Geol Soc Am. Res: Petroleum and alternate energy; stratigraphic and zoologic studies of conodonts. Mailing Add: Phillips Petrol Co Phillips Res Ctr Bartlesville OK 74004

MOSHER, MELVYN WAYNE, b Palo Alto, Calif, June 10, 40; m 63; c 3. PHYSICAL ORGANIC CHEMISTRY. Educ: Univ Washington, BA, 62; Univ Idaho, MS, 64, PhD(org chem), 68. Prof Exp: Fel, Univ Alta, 67-69; asst prof chem, Marshall Univ, 69-74; ASST PROF CHEM & ASST DIR, REGIONAL CRIME LAB, MO SOUTHERN STATE COL, 74- Mem: Am Chem Soc. Res: Free radical reactions and mechanisms; polar effects in free radical reactions; free radical introduction of functional groups into alkanes. Mailing Add: Dept of Chem Mo Southern State Col Joplin MO 64801

MOSHER, MILTON MONROE, b Nonquitt, Mass, Apr 24, 14; m 39; c 2. FOREST MANAGEMENT. Educ: Syracuse Univ, BS, 35, MF, 38. Prof Exp: From instr to assoc prof forestry & range mgt & assoc forester, 37-61, PROF FORESTRY & RANGE MGT & FOR SCIENTIST, WASH STATE UNIV, 61- Mem: Soc Am Foresters. Res: Forest policy; farm woodland management practices; forest irrigation and fertilization. Mailing Add: Dept of Forestry & Range Mgt Wash State Univ Pullman WA 99163

MOSHER, ROBERT ALDEN, organic chemistry, see 12th edition

MOSHER, ROBERT E, b Mt Vernon, NY, July 15, 37; m 71. MATHEMATICS. Educ: Kenyon Col, AB, 58; Mass Inst Technol, PhD(math), 62. Prof Exp: Res assoc math, Brandeis Univ, 62-63; asst prof, Northwestern Univ, 63-66; assoc prof, 66-71, PROF MATH, CALIF STATE COL, LONG BEACH, 71- Concurrent Pos: NSF res grant, 68-70. Mem: Am Math Soc. Res: Algebraic topology, including cohomology operations, homotopy groups, fibre spaces, and characteristic classes. Mailing Add: Dept of Math Calif State Col Long Beach CA 90801

MOSHER, ROBERT EUGENE, b Detroit, Mich, Sept 27, 20; m 43; c 6. ANALYTICAL CHEMISTRY. Educ: Wayne State Univ, BS, 42, MS, 49, PhD(anal chem), 50. Prof Exp: Chemist rubber develop, US Rubber Co, 42-45 & 46-47; instr, Wayne State Univ, 47-50; DIR DEPT PHYSIOL & RES, PROVIDENCE HOSP, 50-; ASSOC PROF GEOL, WAYNE STATE UNIV, 70- Mem: AAAS; Am Chem Soc; Am Fedn Clin Res; Am Asn Clin Chemists; Am Asn Lab Animal Sci. Res: Application of analytical chemistry to research in clinical and geochemical research. Mailing Add: Dept of Res Providence Hosp Southfield MI 48075

MOSHER, WILLIAM ALLISON, organic chemistry, deceased

MOSHIRI, GERALD ALEXANDER, b Teheran, Iran, June 1, 29; US citizen. LIMNOLOGY, PHYSIOLOGICAL ECOLOGY. Educ: Oberlin Col, BA, 52, MA, 54; Univ Pittsburgh, PhD(biol), 68. Prof Exp: Instr biol sci, Cent Fla Jr Col, 64-66; asst prof, 69-73, ASSOC PROF BIOL, UNIV W FLA, 73- Concurrent Pos: Fed Water Pollution Control Admin sr res fel, Inst Ecol, Univ Calif, Davis, 68-69; consult, NASA, 71-72, Environ Protection Agency, 71-73 & Escambia & Santa Rosa Counties Regional Planning Coun, 71-73; chief consult, Theta Anal, Inc, 73- Mem: AAAS; Brit Freshwater Biol Asn; Ecol Soc Am; Am Soc Limnol & Oceanog; Int Asn Theoret & Appl Limnol. Res: Aquatic ecology; energetics of aquatic ecosystems, including problems involving cycling of nutrients and eutrophication of inland waters and estuaries. Mailing Add: Fac of Biol Univ of W Fla Pensacola FL 32504

MOSHMAN, JACK, b Richmond Hill, NY, Aug 12, 24; m 47; c 4. STATISTICAL ANALYSIS. Educ: NY Univ, BA, 46; Columbia Univ, MA, 47; Univ Tenn, PhD(math), 53. Prof Exp: Tutor math, Queens Col, NY, 47; instr, Univ Tenn, 47-50; sr statistician, Oak Ridge Nat Lab, 50-54; mem tech staff, Bell Tel Labs, Inc, 54-57; mem tech staff statist, Serv Div, C-E-I-R, Inc, 57-60, vpres & dir tech sci, 60-65, vpres & gen mgr appl res & mgt sci div, 65-66; managing dir, EBS Mgt Consults, Inc, Washington, DC, 66-67; sr vpres, Leasco Systs & Res Corp, Md, 67-69; PRES, MOSHMAN ASSOCS, INC, 69- Concurrent Pos: Statistician, AEC, 48-50; mem div math, Nat Res Coun-Nat Acad Sci, 53-56; lectr & mem adv comt math & statist, USDA Grad Sch, 59-75; prof lectr, George Washington Univ, 60-61; vis prof, Eagleton Inst Polit, Rutgers Univ; exec secy, Comt to Evaluate Nat Ctr Health Statist, 71-73; mem, Comt Nat Info Syst in Math Sci; mem adv comt statist policy, Off Mgt & Budget, 74-; chmn, Inst Safety Anal, 75- Mem: Opers Res Soc Am; Inst Mgt Sci; Asn Comput Mach (secy, 56-60, vpres, 60-62); fel Am Statist Asn; Inst Math Statist. Res: High speed electronic digital computers; Monte Carlo methods; operations research; mathematical models of political behavior; information systems; statistical problems of compliance with safety and other regulatory requirements. Mailing Add: Moshman Assocs Inc Suite 304 6400 Goldsboro Rd Washington DC 20034

MOSHY, RAYMOND JOSEPH, b Brooklyn, NY, Aug 12, 25; m 48; c 5. FOOD SCIENCE. Educ: St John's Univ, NY, BS, 48; Fordham Univ, MS, 49, PhD(chem), 53. Prof Exp: Chemist res & develop, Am Lecithin Co, 50-52; synthetic org chem, Heyden Chem Corp, 52-55; proj leader, Res Ctr, Gen Foods Corp, 55-59; group mgr chem lab, Am Mach & Foundry Co, 59-60, sect mgr, 60-65, mgr chem develop lab, Res & Develop Div, 65-66, staff vpres & dir res div, 66-70; vpres res & develop, 70-75, VPRES & GROUP EXEC, HUNT-WESSON FOODS, INC, 75- Mem: Am Chem Soc; Sigma Xi; Am Inst Chemists; Inst Food Technologists; Indust Res Inst. Res: Agricultural chemicals; proteins; starches; food processing; nutrition; agricultural research. Mailing Add: Hunt-Wesson Foods Inc 1645 W Valencia Dr Fullerton CA 92634

MOSIER, ARVIN RAY, b Olney Springs, Colo, June 11, 45; m 65; c 2. AGRICULTURAL CHEMISTRY. Educ: Colo State Univ, BS, 67, MS, 68, PhD(soil sci), 74. Prof Exp: RES CHEMIST, AGR RES SERV, USDA, 67- Mem: AAAS; Am Soc Agron; Soil Sci Soc Am; Int Soc Soil Sci. Res: Distribution of nitrogen and organic compounds emanating from agricultural sources and the effect of these chemicals on soil, water and plant systems; nitrogen metabolism and nutrition of algae and higher plants. Mailing Add: USDA Agr Res Serv PO Box E Ft Collins CO 80522

MOSIER, BENJAMIN, b Corsicana, Tex, July 15, 26; m 54; c 3. CHEMISTRY. Educ: Tex A&M Univ, BS, 49, MS, 52; Univ Ill, PhD(chem), 57. Prof Exp: Instr chem, Kilgore Col, 49-50; asst, Tex A&M Univ, 50-51; res chemist, Gen Dynamics Corp, 51-52; res scientist, Humble Oil & Refining Co, 57-60; owner & res dir, 60-69, PRES, INST RES, INC, 69- Concurrent Pos: Lectr, Univ Houston, Rice Univ, Baylor Col Med, Univ Tex M D Anderson Hosp & Tumor Inst & Tex Res Inst Ment Sci; mem, Am Coun Independent Labs; res asst prof path, Baylor Col Med, 74- Mem: Am Chem Soc; Nat Asn Corrosion Eng; Am Inst Mining, Metall & Petrol Engrs; fel Am Inst Chemists; The Chem Soc. Res: Electrochemistry; surface and colloidal phenomena; microencapsulation; instrumental methods including polarography, infrared, ultraviolet and visible spectrometry, gas chromatography, nuclear radiation methods, x-ray diffraction and fluorescence; microprobe analysis; electron microscopy; mass spectrometry; nuclear magnetic spectroscopy; differential thermal, thermogravimetry, emission spectrographic and neutron activation analysis. Mailing Add: Inst for Res Inc 8330 Westglen Dr Houston TX 77042

MOSIER, H DAVID, JR, b Topeka, Kans, May 22, 25. PEDIATRICS, ENDOCRINOLOGY. Educ: Notre Dame Univ, BS, 48; Johns Hopkins Univ, MD, 52; Am Bd Pediat, dipl, 57. Prof Exp: Intern pediat, Johns Hopkins Hosp, 52-53; asst path, Univ Southern Calif, 54-55; fel pediat endocrinol, Johns Hopkins Hosp, 55-57; from asst prof to assoc prof pediat, Sch Med, Univ Calif, Los Angeles, 57-63; assoc prof, Univ Ill Col Med, 63-67; PROF PEDIAT & HEAD DIV ENDOCRINOL & METAB, UNIV CALIF, IRVINE-CALIF COL MED, 67- Concurrent Pos: Asst resident, Los Angeles Children's Hosp, Calif, 53-54, resident pediat path, 54-55; consult, Pac State Hosp, Pomona, Calif, 57- & Tichenor Orthop Clin, Long Beach, 57-63 & 74-; dir res, Ill State Pediat Inst, 63-67; mem staff, Childrens Hosp Med Ctr, Long Beach. Mem: Endocrine Soc; Soc Pediat Res; Soc Exp Biol & Med; Lawson Wilkins Pediat Endocrine Soc; Sigma Xi. Res: Somatic growth and development. Mailing Add: Childrens Hosp Med Ctr 2801 Atlantic Ave Long Beach CA 90801

MOSIER, JACOB EUGENE, b Hoxie, Kans, Feb 5, 24; m 45; c 4. VETERINARY

MEDICINE. Educ: Kans State Univ, DVM, 45, MS, 48. Prof Exp: Instr anat, surg & med, Kans State Univ, 45-47, instr surg & med, 47-48, asst prof large animal med, 48-49; asst prof large animal surg, Univ Ill, 49-50; assoc prof surg & med, 50-54, prof small animal med, 54-61, PROF SURG & MED & HEAD DEPT, KANS STATE UNIV, 61- Concurrent Pos: Mem nat adv comt, Food & Drug Admin, 71-, consult, Bur Vet Med, 74- Honors & Awards: Am Animal Hosp Asn Award, 73. Mem: Am Vet Med Asn; NY Acad Sci. Res: Canine pediatrics; internal medicine of the dog and cat; cause and effect of perinatal disease in the cat and dog. Mailing Add: Dept of Surg & Med Kans State Univ Manhattan KS 66502

MOSIG, GISELA, b Sehmorkau, Ger, Nov 29, 30. GENETICS. Educ: Univ Cologne, Dr rer nat(bot), 59. Prof Exp: Res assoc phage genetics, Vanderbilt Univ, 59-62; NIH fel, Carnegie Inst Genetics Res Unit, 62-63, res assoc, 63-65; from asst prof to assoc prof, 65-71, PROF MOLECULAR BIOL, VANDERBILT UNIV, 71- Mem: AAAS; Genetics Soc Am; Am Soc Microbiol; NY Acad Sci. Res: Mechanism of genetic recombination and replication of DNA in bacteriophage. Mailing Add: Dept of Molecular Biol Vanderbilt Univ Nashville TN 37203

MOSIMANN, JAMES EMILE, b Charleston, SC, Oct 26, 30; m 53; c 8. BIOSTATISTICS. Educ: Univ Mich, BA, 52, MS, 53, PhD(zool), 56; Johns Hopkins Univ, MS, 61. Prof Exp: Res assoc, Willow Run Labs, Univ Mich, 55; asst prof biol, Univ Montreal, 55-61; NIH res fel, 61-62; res assoc, Univ Ariz, 62-63; math statistician, 63-75, CHIEF LAB STATIST & MATH METHODOLOGY, DIV COMPUT RES & TECHNOL, NIH, 75- Concurrent Pos: NIH res fel, 59-60. Mem: Am Statist Asn; Biomet Soc; Am Soc Ichthyologists & Herpetologists; Royal Statist Soc. Res: Discrete probability models in biology; statistical distribution theory; biometry; ecological and population statistics. Mailing Add: Div of Comput Res & Technol NIH Bethesda MD 20014

MOSKALYK, RICHARD EDWARD, b Hafford, Sask, Apr 17, 36; m 57; c 3. MEDICINAL CHEMISTRY. Educ: Univ Sask, BS, 56, MSc, 59; Univ Alta, PhD(pharmaceut chem), 65. Prof Exp: Control chemist, Merck, Sharp & Dohme Ltd, Can, 56-57; res chemist, Food & Drug Directorate, Dept Nat Health & Welfare, 58-61; from asst prof to assoc prof, 63-75, PROF PHARMACEUT CHEM, FACULTY PHARM & PHARMACEUT SCI, UNIV ALTA, 75- Concurrent Pos: Invited prof, Univ Geneva, 71-72. Mem: Can Pharmaceut Asn; Am Chem Soc; Chem Inst Can. Res: Development of methodology for the determination of the bioavailability of drugs and their metabolites in biological fluids. Mailing Add: Fac of Pharm & Pharmaceut Sci Univ of Alta Edmonton AB Can

MOSKOVITS, MARTIN, b Apr 13, 43; Can citizen. SPECTROCHEMISTRY, SURFACE CHEMISTRY. Educ: Univ Toronto, BSc, 65, PhD(chem), 70. Prof Exp: Res scientist chem, Alcan Int, 70-71; ASST PROF CHEM, UNIV TORONTO, 72- Mem: Can Inst Chemists; Optical Soc Am. Res: Chemistry and spectroscopy of metal surfaces and metal particles, both bulk and isolated in rare gas solids, raman spectroscopy, difference raman, differential raman scattering, circular and linear. Mailing Add: Dept of Chem Univ of Toronto Toronto ON Can

MOSKOVITZ, DAVID, b Ungvar, Austria-Hungary, Apr 1, 03; US citizen; m 27; c 2. MATHEMATICAL ANALYSIS. Educ: Carnegie Inst Technol, BS, 25, MS, 27; Brown Univ, PhD(math), 32. Prof Exp: Instr math, Carnegie Inst Technol, 25-29 & Brown Univ, 29-32; from asst prof to assoc prof, 32-50, actg & assoc head dept, 52-65, PROF MATH, CARNEGIE-MELLON UNIV, 50- Mem: Am Math Soc; Math Asn Am. Res: Difference equations; convexity in linear spaces; numerical solution of Laplace's and Poisson's equations; analysis. Mailing Add: Dept of Math Carnegie-Mellon Univ Pittsburgh PA 15213

MOSKOWITZ, HARVEY D, b Philadelphia, Pa, Oct 24, 41; m 69; c 2. BIOMEDICAL ENGINEERING, DENTAL MATERIALS. Educ: Drexel Univ, BS, 64, PhD(biomed eng), 72; NY Univ, MS, 68. Prof Exp: Asst prof dent mat & chmn dept, Col Dent, NY Univ, 71-73; ASSOC PROF DENT MAT, COL DENT NJ, 74- Mailing Add: Col Dent NJ Div Dent Mat 100 Bergan St Newark NJ 07103

MOSKOWITZ, JULES WARREN, b Newark, NJ, June 11, 34; m 57; c 1. PHYSICAL CHEMISTRY. Educ: Princeton Univ, AB, 56; Mass Inst Technol, PhD(chem), 61. Prof Exp: From asst prof to assoc prof, 63-72, PROF CHEM, NY UNIV, 72- Concurrent Pos: Consult, Bell Tel Labs, NJ, 65- Mem: Am Phys Soc. Res: Quantum mechanics of solid state and molecular systems; application of digital computers to problems of chemical interest. Mailing Add: Dept of Chem NY Univ New York NY 10003

MOSKOWITZ, MARK LEWIS, b Brooklyn, NY, Dec 5, 25; m 49; c 2. PHOTOGRAPHIC CHEMISTRY. Educ: City Col New York, BS, 50; Syracuse Univ, PhD(org chem), 54. Prof Exp: Org chemist, 54-59, supvr prod develop lab, 59-65, mgr reproduction prod res & develop, 65-69, tech dir off systs div, 69-74, MGR, ADV REPRODUCTION MAT RES & DEVELOP, GAF CORP, 74- Mem: AAAS; Am Chem Soc; Soc Photog Sci & Eng; Am Soc Test & Mat. Res: Diazotype coatings chemistry and sensitometry; light sensitive coatings; electrophotography. Mailing Add: 37 Colfax Dr Pequannock NJ 07440

MOSKOWITZ, MARTIN A, b New York, NY, June 25, 35; m 58; c 2. MATHEMATICS. Educ: Brooklyn Col, BA, 57; Univ Calif, Berkeley, MA, 59, PhD(math), 64. Prof Exp: Instr math, Univ Chicago, 64-66; asst prof, Columbia Univ, 66-69; assoc prof, 69-76, PROF MATH, GRAD CTR, CITY UNIV NEW YORK, 76- Concurrent Pos: Prin investr, NSF contract, 75. Mem: Am Math Soc. Res: Topological groups; harmonic analysis and representation theory. Mailing Add: Dept of Math City Univ of NY Grad Ctr New York NY 10036

MOSKOWITZ, MERWIN, b New York, NY, May 26, 21; m 44; c 3. MICROBIOLOGY, IMMUNOCHEMISTRY. Educ: Univ Mich, BS, 44; Univ Calif, PhD(bact), 49. Prof Exp: Asst chem, Univ Calif, 46-49; fel, USPHS, 49-50; instr microbiol, Yale Univ, 50; from asst prof to assoc prof bact, 51-61, prof biol, 61-73, PROF BIOL SCI, PURDUE UNIV, 73- Mem: AAAS; Am Soc Microbiol; Am Soc Cell Biol; Am Asn Immunol. Res: Medical microbiology; pathogenesis of rheumatic fever; autosensitization; red cell structure; enzymes of Clostridia; cell biology. Mailing Add: Dept of biol Sci Purdue Univ West Lafayette IN 47907

MOSKOWITZ, MICHAEL ARTHUR, b New York, NY, May 26, 42; m 65; c 1. NEUROSCIENCE. Educ: Johns Hopkins Univ, AB, 64; Tufts Univ, MD, 68. Prof Exp: Chief resident, Dept Neurol, 72-73, instr, 74-76, ASST PROF NEUROL, SCH MED, HARVARD UNIV, 76- Concurrent Pos: Res assoc, Mass Inst Technol, 73-, lectr neurosci, Dept Nutrit, 75-; res fel, Found Fund Res Psychiat, 73-75, Alfred P Sloan Found, 76. Mem: Neurosci Soc; Am Acad Neurol. Res: Biogenic amines and neurological diseases; the pineal gland and its pathological implications. Mailing Add: Dept of Nutrit Mass Inst of Technol Cambridge MA 02139

MOSKOWITZ, NORMAN, b Trenton, NJ, Jan 25, 22. ANATOMY. Educ: Rutgers Univ, BS, 43; Univ Pa, MS, 47, PhD(zool), 51. Prof Exp: Asst instr zool, Univ Pa, 48-50; lectr biol, Rutgers Univ, 58-59; from asst prof to assoc prof, 62-74, PROF ANAT, JEFFERSON MED COL, 74- Concurrent Pos: Fel neuroanat, Col Physicians & Surgeons, Columbia Univ, 59-62. Mem: Am Asn Anat; Harvey Soc. Res: Neuroanatomy; central auditory system. Mailing Add: 1020 Locust St Philadelphia PA 19107

MOSKOWSKI, ERICA F, b Clij, Rumania, Apr 14, 29; m 54; c 2. OBSTETRICS & GYNECOLOGY, ENDOCRINOLOGY. Educ: Univ Buenos Aires, MD, 54. Prof Exp: Rotating intern, Lincoln Hosp, New York, 54-55; jr resident surg & gynec, Tucson Med Ctr, 55-56; asst resident obstet & gynec, Quincy City Hosp, Boston, 56-57; resident, New Eng Hosp, 57-58; asst resident gynec & gynec endocrinol, Johns Hopkins Hosp, 59-60; instr, 60-68, ASST PROF OBSTET & GYNEC, UNIV MD HOSP, 68- Concurrent Pos: Fel gynec & gynec endocrinol, Johns Hopkins Hosp, 59-60; consult, Church Home Hosp, 65-; lectureships, Prince George Hosp, 69- Res: Gynecological endocrinology and infertility. Mailing Add: Dept of Gynec Univ of Md Hosp Baltimore MD 21201

MOSLEMI, ALI A, b Kermanshah, Iran, Aug 26, 35; m 62; c 1. FOREST PRODUCTS. Educ: Univ Tehran, BS, 57; Mich State Univ, MS, 60, PhD(forest prod), 64. Prof Exp: Asst instr wood technol, Mich State Univ, 61-62, instr, 62-64; from asst prof to prof, Southern Ill Univ, Carbondale, 65-75, chmn dept forestry, 73-75; ASSOC DEAN & ASSOC DIR, COL FORESTRY, WILDLIFE & RANGE SCI, UNIV OF IDAHO, 75- Mem: Forest Prod Res Soc (chmn, wood particle & fiber process div); Soc Wood Sci & Technol. Res: Wood mechanics; particle composites. Mailing Add: Col Forestry Wildlife & Range Sci Univ of Idaho Moscow ID 83843

MOSLEY, JAMES W, b Temple, Tex, Aug 8, 29; m 53; c 2. EPIDEMIOLOGY. Educ: Univ Tex, BA, 50; Cornell Univ, MD, 54. Prof Exp: Intern med, New York Hosp, 54-55; chief hepatitis unit, Commun Dis Ctr, USPHS, Ga, 55-58; resident med, Peter Bent Brigham Hosp, 58-59; fel virol, Harvard Med Sch, 59-61; resident, New Eng Ctr Hosp, 61-62; chief hepatitis unit, Commun Dis Ctr, 62-66, chief viral dis sect, 66-70; assoc prof med, 70-75, PROF MED, UNIV SOUTHERN CALIF, 75-; DIR, HEPATIC EPIDEMIOL LAB, JOHN WESLEY COUNTY HOSP, 73- Res: Infectious diseases; epidemiology of viral hepatitis. Mailing Add: John Wesley County Hosp 2826 S Hope St Los Angeles CA 90007

MOSLEY, JOHN ROSS, b Wichita, Kans, Oct 18, 22; m 50; c 2. PHYSICAL CHEMISTRY. Educ: Stanford Univ, AB, 44, PhD(chem), 49. Prof Exp: Mem staff, 48-68, assoc group leader, 68-73, ALT GROUP LEADER, LOS ALAMOS SCI LAB, 73- Mem: Am Chem Soc; Am Phys Soc; fel Am Inst Chemists. Res: Materials science; chemical kinetics; high vacuum. Mailing Add: PO Box 1663 Los Alamos NM 87544

MOSLEY, KIRK THORNTON, epidemiology, see 12th edition

MOSLEY, RONALD BRUCE, b Hyden, Ky, Mar 10, 43; m 66; c 2. SOLID STATE PHYSICS. Educ: Berea Col, AB, 65; Auburn Univ, MS, 68, PhD(physics), 73. Prof Exp: ASST PROF PHYSICS, VA COMMONWEALTH UNIV, 73- Honors & Awards: Outstanding Res Award, Sigma Xi, 74. Mem: Sigma Xi. Res: Electrical transport properties of thin films of semiconducting materials. Mailing Add: Dept of Physics Va Commonwealth Univ Richmond VA 23284

MOSLEY, WILBUR CLANTON, JR, b Birmingham, Ala, Oct 30, 38; m 60; c 2. SOLID STATE PHYSICS, MATERIALS SCIENCE. Educ: Auburn Univ, BEP, 60, MS, 62; Univ Ala, Tuscaloosa, PhD(physics), 65. Prof Exp: RES PHYSICIST, NUCLEAR MAT DIV, SAVANNAH RIVER LAB, E I DU PONT DE NEMOURS & CO, INC, 65- Mem: Am Phys Soc. Res: Chemical and radiation stability of compounds of actinides. Mailing Add: Nuclear Mat Div Savanah Rivr Lab E I du Pont de Nemours & Co Inc Aiken SC 29801

MOSLEY, WILEY HENRY, b Tsingkiangpu, China, Oct 17, 33; US citizen; m 56; c 3. EPIDEMIOLOGY. Educ: Southwestern at Memphis, BA, 55; Univ Okla, MD, 59; Johns Hopkins Univ, MPH, 65. Prof Exp: Intern med, Johns Hopkins Hosp, 59-60, asst resident, 60-61, resident, 63-64; epidemiologist epidemic intel serv, Commun Dis Ctr, USPHS, 61-63; head epidemiol div, Pakistan-SEATO Cholera Res Lab, 65-71; PROF POP & CHMN DEPT POP DYNAMICS, SCH HYG & PUB HEALTH, JOHNS HOPKINS UNIV, 71- Concurrent Pos: Consult, WHO, 69-71, mem adv comt int quarantine, 70-75. Mem: AAAS; Pop Asn Am; Am Pub Health Asn; Am Epidemiol Soc; Int Epidemiol Asn. Res: Population dynamics; infectious diseases. Mailing Add: Dept of Pop Dynamics Sch of Hyg & Pub Health Johns Hopkins Univ Baltimore MD 21205

MOSQUIN, THEODORE, b Brokenhead, Man, July 8, 32; m 61; c 1. BOTANY, TAXONOMY. Educ: Univ Man, BSc, 56; Univ Calif, Los Angeles, PhD(bot), 61. Prof Exp: Asst prof bot, Univ Alta, 61-62; res officer taxon, Plant Res Inst, Can Dept Agr, 62-67, res scientist, 67-74; MEM STAFF, CAN NATURE FEDN, 74- Concurrent Pos: Grants, Univ Alta Fac Res Fund, 61-62 & Am Philos Soc, 62-63; sessional lectr, Carleton Univ, 65-66; vis lectr, Univ Calif, Berkeley, 68-69; exec dir, Can Nature Fedn, 72-73. Mem: Can Bot Asn; Genetics Soc Can; Int Asn Plant Taxon. Res: Evolution and biological classification of Linum; taxonomy of Epilobium angustifolium; general systematics of Canadian flora. Mailing Add: Can Nature Fedn 46 Elgin St Ottawa ON Can

MOSS, ALFRED JEFFERSON, JR, b Little Rock, Ark, Nov 22, 40; m 65. BIOPHYSICS, PHYSIOLOGY. Educ: Univ Ark, Little Rock, BS, 62, MS, 64, PhD(biophys, physiol), 70. Prof Exp: Res chemist, Dow Chem Co, 65-68; ASST PROF RADIOL & PHYSIOL, MED CTR, UNIV ARK, LITTLE ROCK, 70- Mem: Biophys Soc; Radiation Res Soc. Res: Radiation biophysics; molecular biology. Mailing Add: Div Nuclear Med & Radiation Biol Univ of Ark Med Ctr Little Rock AR 72201

MOSS, BERNARD, b New York, NY, July 26, 37; m 60; c 3. BIOCHEMISTRY, VIROLOGY. Educ: NY Univ, BA, 57, MD, 61; Mass Inst Technol, PhD(biochem), 67. Prof Exp: Intern med, Children's Hosp Med Ctr, Boston, Mass, 61-62; USPHS basic sci training fel, 62-66; investr, lab biol viruses, 66-71, HEAD MACROMOLECULAR BIOL SECT, LAB BIOL VIRUSES, NIH, 71- Mem: Am Soc Microbiol; Am Soc Biol Chemists; AAAS; Fedn Am Scientists. Res: Animal viruses; proteins, enzymes; assembly of virus particles; anti-viral substances. Mailing Add: Lab of Biol of Viruses Nat Inst of Health Bethesda MD 20014

MOSS, BUELON REXFORD, b Columbia, Ky, Oct 24, 37; m 59; c 3. ANIMAL NUTRITION, DAIRY SCIENCE. Educ: Berea Col, BS, 60; Univ Tenn, Knoxville, PhD(animal sci), 68. Prof Exp: Instr high sch, Ky, 61-63; res technician, Univ Tenn-AEC Agr Res Lab, Oak Ridge, 67-68; res assoc animal sci, Univ Tenn, Knoxville, 68-69; asst prof, 69-74, ASSOC PROF ANIMAL SCI & DAIRY NUTRIT, MONT STATE UNIV, 74- Mem: Am Dairy Sci Asn; Am Inst Nutrit; Poultry Sci Asn. Res: Forage evaluation; HiProly-High Lysine barley evaluation; barleys as feeds for dairy calves, lactating cows and poultry; amino acid supplements to poultry rations. Mailing Add: Dept of Animal & Range Sci Mont State Univ Bozeman MT 59715

MOSS, CALVIN E, b Richmond, Va, Nov 27, 39; m 61. EXPERIMENTAL NUCLEAR PHYSICS. Educ: Univ Va, BS, 61; Calif Inst Technol, MS, 63, PhD(physics), 68. Prof Exp: Res assoc nuclear physics, Duke Univ, 67-69 & Univ Colo, 69-71; res fel, Australian Nat Univ, 71-72; res assoc, Univ Colo, 72-73; STAFF MEM, LOS ALAMOS SCI LAB, 73- Mem: Am Phys Soc. Res: Nuclear spectroscopy of light nuclei and accelerator development. Mailing Add: P-9 MS 480 Los Alamos Sci Lab Los Alamos NM 87545

MOSS, CHARLES NORMAN, b Los Angeles, Calif, June 13, 14; m 63; c 3. MEDICINE, SURGERY. Educ: Stanford Univ, AB, 40; Harvard Med Sch, MD, 44; Univ Vienna, 47; Univ Calif, Berkeley, MPH, 55; Univ Calif, Los Angeles, DrPH, 70; Am Bd Prev Med, cert aerospace med, 58 & cert occup med, 68. Prof Exp: Intern surg, Peter Bent Brigham Hosp, Boston, 44-45, asst, 47; med dir, NAm Rockwell Corp, 69-70; physician, Los Angeles County Occup Health Serv, 70-73; CHIEF MED ADV UNIT, BD RETIREMENT, LOS ANGELES COUNTY, 73- Concurrent Pos: Aviation med examr, Fed Aviation Admin, 70-73. Honors & Awards: AMA Physician's Recognition Award, 69 & 72. Mem: Fel Am Col Prev Med; fel Am Pub Health Asn; fel Royal Soc Health; fel Am Acad Occup Med; Asn Mil Surgeons US. Res: Venous pump. Mailing Add: 7714 Cowan Ave Los Angeles CA 90045

MOSS, CLAUDE WAYNE, b Rural Hall, NC, Mar 20, 35; m 58; c 4. MICROBIOLOGY, BIOCHEMISTRY. Educ: NC State Univ, BS, 57, MS, 62, PhD(microbiol), 65. Prof Exp: Res assoc microbiol, NC State Univ, 63-65; RES MICROBIOLOGIST, CTR DIS CONTROL, USPHS, 65- Honors & Awards: Superior Performance Award, USPHS, 66. Mem: Sci Res Soc Am; Am Soc Microbiol. Res: Physiology and metabolism of pathogenic bacteria; application of chromatographic techniques to diagnostic bacteriology and clinical chemistry; cryobiology of microorganisms and animal cells; biochemistry of drug metabolism. Mailing Add: Anal Bacteriol Unit Ctr for Dis Control Atlanta GA 30333

MOSS, DALE NELSON, b Thornton, Idaho, Mar 27, 30; m 53; c 8. AGRONOMY, PLANT PHYSIOLOGY. Educ: Ricks Col, BS, 55; Cornell Univ, MS, 55, PhD(crop physiol), 59. Prof Exp: Asst prof chem, Ricks Col, 55-56; asst agr scientist, Conn Agr Exp Sta, 59-61, assoc agr scientist, 61-63, agr scientist, 63-67; PROF CROP PHYSIOL, UNIV MINN, ST PAUL, 67- Mem: Fel Am Soc Agron; Crop Sci Soc Am (pres-elect, 75-76, pres, 76-78); Am Soc Plant Physiol. Res: Effects on photosynthesis, respiration and transpiration of higher plants of light intensity, temperature, carbon dioxide concentration, nutrition, removal of storage organs, air turbulence, planting patterns, and age of leaves. Mailing Add: Dept of Agron & Plant Genetics Univ of Minn St Paul MN 55108

MOSS, DONOVAN DEAN, b Bunker Hill, Ind, Feb 28, 26; m 48; c 2. FISH BIOLOGY. Educ: Auburn Univ, BS, 49, MS, 50; Univ Ga, PhD(zool), 62. Prof Exp: Fisheries biologist, Ala State Dept Conserv, 51-56, asst chief fisheries, 56-57; res asst zool, Univ Ga, 57-61; from assoc prof to prof, Univ Ky, 62-65; assoc prof biol, Tenn Technol Univ, 65-67; assoc prof, 67-72, PROF FISHERIES, AUBURN UNIV, 72-, ASST DIR INT CTR AQUACULT, 70- Mem: Am Fisheries Soc; Am Soc Limnol & Oceanog. Res: Fish management and biology of fishes including ecological requirements of fish species. Mailing Add: Fisheries Bldg Auburn Univ Auburn AL 36830

MOSS, ERNEST KENT, b Woodlake, Calif, Aug 19, 38; m 64. PHYSICAL ORGANIC CHEMISTRY, POLYMER CHEMISTRY. Educ: Univ Ore, BS, 59; Univ Ariz, PhD(chem), 63. Prof Exp: Develop chemist, Fibers Div, Am Cyanamid Co, 63-64, sr develop chemist, 64-67; sr res chemist, 67-70, group leader, 70-74, RES ASSOC, JIM WALTERS RES CORP, 74- Mem: Int Solar Energy Soc. Res: Polyurethane and modified polyisocyanurate foams for insulation applications; computer programming and applications; thermally stable low flammability plastics; epoxy resin applications to building products. Mailing Add: Jim Walter Res Corp 10301 Ninth St N St Petersburg FL 33702

MOSS, FRANK ANTHONY JAMES, b Sept 5, 24; US citizen; m 51; c 3. INORGANIC CHEMISTRY, PHYSICAL CHEMISTRY. Educ: Univ London, BSc, 45; Univ Ill, PhD(inorg & phys chem), 52. Prof Exp: Res chemist, Delanium Ltd, Eng, 45-46 & Thorium Ltd, Imp Chem Industs, Eng, 46-49; process engr, Esso Standard Oil Co, La, 52-56; asst dir res, Mallinckrodt Chem Co, Mo, 56-63; corp mgr inorg res, Glidden Co, Ohio, 63-65, dir res, Chem Group, Md, 65-69; DIR, EASTERN RES CTR, STAUFFER CHEM CO, 69- Mem: Indust Res Inst; Am Chem Soc; Electrochem Soc; Am Mgt Asn. Res: Rare earths; solvent hydrofining; lube oils; ore processing; columbium-tantalum; uranium; semiconductor silicon; fungicides; fine organics; metal powders; ceramic coatings; inorganic pigments. Mailing Add: Eastern Res Ctr Stauffer Chem Co Dobbs Ferry NY 10522

MOSS, GERALD, b New York, NY, Feb 1, 31; m 63; c 1. BIOMEDICAL ENGINEERING, SURGERY. Educ: NY Univ, BA, 51, MS, 56; Union Univ, NY, PhD(biochem), 61; Albany Med Col, MD, 61. Prof Exp: Res phys chemist, Vitro Labs, Inc, 56; resident surg, Albany Med Ctr Hosp, 61-68; RES PROF BIOMED ENG, RENSSELAER POLYTECH INST, 68- Concurrent Pos: Clin asst prof surg, Albany Med Col, 68- Res: Biochemistry; postoperative protein metabolism and wound healing; cerebral glucose metabolism; circulatory physiology; gastrointestinal physiology. Mailing Add: Biomed Eng Lab Rensselaer Polytech Inst Troy NY 12181

MOSS, GERALD ALLEN, b Milwaukee, Wis, Jan 24, 40; m 62; c 2. NUCLEAR PHYSICS. Educ: Univ Wis, BS, 61; Univ Ore, MS, 63, PhD(physics), 66. Prof Exp: Fel nuclear physics, Univ Man, 67-69; asst prof, 69-72, ASSOC PROF NUCLEAR PHYSICS, UNIV ALTA, 72- Concurrent Pos: Vis scientist, Saclay Nuclear Res Ctr, France, 75-76. Mem: Am Asn Physics Teachers. Res: Nuclear reactions; spectroscopy; intermediate energy experimental nuclear physics. Mailing Add: Nuclear Res Ctr Univ of Alta Edmonton AB Can

MOSS, GERALD S, b Cleveland, Ohio, Mar 4, 35; m; c 3. SURGERY, EXPERIMENTAL SURGERY. Educ: Ohio State Univ, BA, 56, MD, 60. Prof Exp: Teaching fel anat, Sch Med, Harvard Univ, 62; tutor surg, Manchester Royal Infirmary, Eng, 64; head exp surg, US Naval Res Inst, 66-68; from asst prof to assoc prof, 68-72, PROF SURG, UNIV ILL, CHICAGO, 73- Concurrent Pos: Asst chief surg, Vet Admin West Side Hosp, Chicago, 68-70; attend surg, Cook County Hosp, 70-72, chmn dept, 72-; dir surg res, Hektoen Inst Med Res, Chicago, 72- Mem: Soc Univ Surgeons; fel Am Col Surgeons; Asn Acad Surg (pres, 76-); Nat Soc Med Res; Int Cardiovasc Soc. Res: Shock and resuscitation; blood preservation; blood component therapy and blood substitutes; cardiopulmonary physiology. Mailing Add: Dept Surg Cook County Hosp 1835 W Harrison St Chicago IL 60612

MOSS, HERBERT IRWIN, b Brooklyn, NY, Mar 8, 32; m 60; c 3. INORGANIC CHEMISTRY, CERAMICS. Educ: Univ Louisville, BS, 53; Indiana Univ, PhD(chem), 60. Prof Exp: MEM TECH STAFF, DAVID SARNOFF RES CTR, RCA LABS, 59- Mem: Am Chem Soc; Electrochem Soc; Am Ceramic Soc. Res: Pressure sintering of magnetic, electronic and optically active materials; synthesis and properties of electronically active materials; thin films; materials for magnetic recording heads. Mailing Add: David Sarnoff Res Ctr RCA Labs Princeton NJ 08540

MOSS, JACK N, b Philadelphia, Pa, Jan 25, 26; m 47; c 1. PHARMACOLOGY, CHEMOTHERAPY. Educ: Philadelphia Col Pharm, BSc, 46. Prof Exp: Lab head bact, Nat Drug Co, 46-56 & cell physiol, 56-60; lab head pharmacol, William H Rorer Pharmaceut Co, 60-67; LAB HEAD PHARMACOL, ROHM AND HAAS CO, 67- Concurrent Pos: Vis prof, Gwynedd Mercy Col, 65-67. Mem: AAAS. Res: Metabolite antagonists; anti-inflammatory enzymes; gastrointestinal pharmacology; bacterial and viral diseases. Mailing Add: 777 W Germantown Pike Plymouth Meeting PA 19462

MOSS, JAMES MERCER, b Bradley, Ga, Dec 15, 17; m 41; c 4. INTERNAL MEDICINE. Educ: Univ Va, MD, 41; Am Bd Internal Med, dipl, 52. Prof Exp: Intern med, Univ Va Hosp, 41-42, from asst resident physician to resident physician, 47-49; instr endocrinol, Duke Univ, 46-47; instr, 49-52, attend physician & dir diabetic clin, Univ Hosp, 49-75, from clin asst prof to clin assoc prof, 52-62, CLIN PROF MED, GEORGETOWN UNIV, 62- Concurrent Pos: Vis physician, DC Gen Hosp, 50-66, dir diabetic clin, 50-55; chief med, Circle Terrace Hosp, Alexandria, Va, 64-68, pres med staff, 65-68; consult, USPHS, 65-67; ed consult, Am Family Pract, 66-; mem consult staff, Arlington Hosp, Va, Fairfax Hosp & Nat Orthop Hosp; mem pub adv comt endocrine & metab drugs, Food & Drug Admin, 74-; mem health manpower comn, Commonwealth of Va, 74-76. Mem: Am Heart Asn; Am Diabetes Asn; fel Am Col Physicians; fel Am Col Cardiol; Am Med Writers' Asn. Res: Treatment of patients with diabetes mellitus; evaluation of drugs and methods used in treating diabetes. Mailing Add: 1707 Osage St Alexandria VA 22302

MOSS, JOHN HALL, b Philadelphia, Pa, Oct 25, 18; m 42; c 6. GEOLOGY. Educ: Princeton Univ, AB, 41; Mass Inst Technol, SM, 43; Harvard Univ, AM, 47, PhD(geol), 49. Prof Exp: Geologist, Mil Geol Unit, US Geol Surv, 43-45; from asst prof to assoc prof, 48-59, chmn dept, 59-71, PROF GEOL, FRANKLIN & MARSHALL COL, 59-, DIR ENVIRON STUDIES, 71- Concurrent Pos: Chmn high sch earth & space sci adv comt, Pa Dept Pub Instr, 58-63, mem state sci adv comt, 63-; dir, Pa Environ Coun, 71- Mem: AAAS; Geol Soc Am; Soc Econ Paleont & Mineral; Nat Asn Geol Teachers (pres, 64). Res: Pleistocene glaciation and river terrace formation in Rocky Mountains, New York and Pennsylvania; arid region geomorphology; geologic studies of early man sites; environmental geology. Mailing Add: Dept of Geol Franklin & Marshall Col Lancaster PA 17604

MOSS, JOHN SEABORN, physics, see 12th edition

MOSS, L HOWARD, III, b Jamaica, NY, Apr 12, 33; m 59; c 2. MICROBIOLOGY, VIROLOGY. Educ: Univ Tenn, BS, 60, MS, 61, PhD(microbiol), 67. Prof Exp: Asst prof microbiol, Sch Med, Wayne State Univ, 69; ASST PROF MICROBIOL, MED COL OHIO, 69- Concurrent Pos: USPHS res fel virol, St Jude Children's Res Hosp, 67-69. Mem: Tissue Cult Asn; Am Soc Microbiol. Res: Herpesviruses; host-cell interaction; structure and function of virus envelope and capsid; sequential development of herpesviruses by electron microscopy. Mailing Add: Dept of Microbiol Med Col of Ohio PO Box 6190 Toledo OH 43614

MOSS, LEO (DAVID), b Berlin-Spandau, Ger, June 25, 11; nat US; m 35; c 3. PATHOLOGY. Educ: Univ Berne, MD, 34. Prof Exp: Fel path, Mt Sinai Hosp, New York, 37-42; pathologist & dir labs, Bradford Hosp, Pa, 42-58; PATHOLOGIST & DIR LABS, OLEAN GEN & ST FRANCIS HOSPS, 42-56 & 58-; DIR CATTARAUGUS COUNTY LABS, 60-, DEP COMNR HEALTH, COUNTY DEPT HEALTH, 72- Concurrent Pos: Adj prof, St Bonaventure Univ, 59- Mem: Fel Am Soc Clin Path; Am Asn Path & Bact; fel Am Col Physicians; fel Col Am Path; Int Acad Path. Res: Pathological anatomy; clinical pathology; arteriosclerosis; hypertension. Mailing Add: 477 Vermont St Olean NY 14760

MOSS, LEONARD WALLACE, b Detroit, Mich, Sept 7, 23; m 45; c 1. ANTHROPOLOGY. Educ: Wayne State Univ, BSc, 47, MA, 50; Univ Mich, PhD(sociol), 55. Prof Exp: From instr to prof sociol & anthrop, Wayne State Univ, 52-69, chmn dept, 62-68, dir soc prog, 72-73, secy liberal arts col fac & fac coun, 70-74, PROF ANTHROP, WAYNE STATE UNIV, 69- Concurrent Pos: Consult, Fac Agr Econ, Sch Soc Serv & Univ Naples & Fulbright res scholar, Univ Rome & Univ Naples, 55-56; guest lectr, Int Univ Soc Sci, Rome, 61-62; Fulbright sr lectr anthrop, Univ Rome & Mus Pop Art & Tradition, 61-62; Am Coun Learned Socs travel grants, Italy & Israel, 59 & 63; Wenner-Gren Found Anthrop Res grant to estab, Italian Ctr Cult Anthrop, 62, travel grant, 63 & travel grant, World Cong Anthrop, USSR, 64; mem, Mich Regional Adv Bd, Anti-Defamation League, 61-63, 64-69 & 70-; Fulbright-Hays sr .es scholar, Univ Rome & Mus Pop Art & Tradition, 68-69; mem nat adv bd, Am-Yugoslav Proj Regional & Urban Planning Studies, Ljubljana, 68-73; Mead lectr, Trinity Col, Conn, 71, Barbieri Lectureship, 74. Honors & Awards: Knight, Order of Merit, Repub of Italy, 65. Mem: Fel AAAS; fel Am Anthrop Asn; fel Royal Anthrop Inst Gr Brit & Ireland; Italian Asn Soc Sci; Cent States Anthrop Soc (pres, 69-70). Res: Impact of technology and cultural change on South Italian peasant villages; adjustment of peasant in-migrants in cities; pre-Roman archaeology. Mailing Add: Dept Anthrop 137 Manoogian Hall Wayne State Univ Detroit MI 48202

MOSS, LLOYD KENT, b Los Angeles, Calif, Aug 8, 24; m 50; c 4. BIOCHEMISTRY. Educ: Univ Calif, Los Angeles, BS, 50; Stanford Univ, PhD(chem), 57. Prof Exp: Chemist, Aerojet-Gen Corp, Gen Tire & Rubber Co, 50-52; assoc chemist, Stanford Res Inst, 56-58, biochemist, 58-66; from asst prof to assoc prof, 66-70, PROF CHEM, FOOTHILL COL, 70- Mem: AAAS; Am Chem Soc; NY Acad Sci. Res: Protein characterization; preparative chromatography; continuous flow electrophoresis; periodate oxidations; photosynthetic energy transfer mechanisms; allergens, hemagglutinins and toxins. Mailing Add: Dept of Chem Foothill Col 12345 El Monte Rd Los Altos CA 94022

MOSS, MARVIN KENT, b Burlington, NC, Feb 3, 31; m 56; c 1. PHYSICS, THEORETICAL MECHANICS. Educ: Elon Col, BA, 54; NC State Univ, MS, 57, PhD(physics), 61. Prof Exp: Instr physics, 57-58, asst eng physics, 60-61, asst prof physics, 61-66, ASSOC PROF PHYSICS, NC STATE UNIV, 66- Concurrent Pos: NSF fac fel, Imp Col, Univ London & Univ Edinburgh, 66-67. Mem: Am Phys Soc; Am Asn Physics Teachers. Res: Relativity; theoretical physics, especially general theory of relativity and theoretical mechanics; development of lasers and laser systems; operations research. Mailing Add: Dept of Physics NC State Univ Raleigh NC 27607

MOSS, MELVIN LANE, b Deerfield, Ohio, July 3, 15; m 52; c 1. BIOCHEMISTRY, SCIENCE ADMINISTRATION. Educ: Mt Union Col, BS, 38; Purdue Univ, MS, 40, PhD(anal chem), 42. Prof Exp: Asst instr chem, Purdue Univ, 38-41; res chemist, Hercules Powder Co, 42-48; asst div head, Alcoa Res Labs, Aluminum Co Am, 48-62; NIH fel, Inst Neurobiol, Univ Gothenburg, 63-64; sr scientist staff mem, Oak Ridge Nat Lab, 69-73; assoc mem, Inst Muscle Dis, Inc, 64-69, actg dir, 73-74, DIR RES &

DEVELOP, MUSCULAR DYSTROPHY ASN, INC, 73- Mem: AAAS; Am Chem Soc; Histochem Soc; Am Micrchem Soc. Res: Analytical methods and instrumentation; chemical and metallurgical process control; analysis of isolated nerve, muscle and amniotic cells; prenatal detection of genetic disorders; biochemistry of muscle disease. Mailing Add: Muscular Dystrophy Asn Inc 810 7th Ave New York NY 10019

MOSS, MELVIN LIONEL, b New York, NY, Jan 3, 23; m 70; c 2. ANATOMY. Educ: NY Univ, AB, 42; Columbia Univ, DDS, 46, PhD, 54. Prof Exp: From asst anat to assoc prof, 52-67, dean, Sch Dent & Oral Surg, 68-73, PROF ANAT, COLUMBIA UNIV, 68-, PROF ORAL BIOL, COL PHYSICIANS & SURGEONS, 67- Concurrent Pos: Lederle med fac award, 54-56; mem int comt standardization human biol. Mem: AAAS; Am Soc Zoologists; Am Asn Anatomists; Am Asn Phys Anthrop; Int Asn Dent Res. Res: Skeletal morphology and physiology; physical anthropology. Mailing Add: Dept of Anat Columbia Univ 630 W 168th St New York NY 10032

MOSS, PHILIP H, organic chemistry, see 12th edition

MOSS, ROBERT ALLEN, b Brooklyn, NY, May 27, 40; m 67. ORGANIC CHEMISTRY. Educ: Brooklyn Col, BS, 60; Univ Chicago, MS, 62, PhD(chem), 63. Prof Exp: Nat Acad Sci-Nat Res Coun res fel chem, Columbia Univ, 63-64; from asst prof to assoc prof, 64-73, PROF CHEM, RUTGERS UNIV, 73- Concurrent Pos: Vis scientist, Mass Inst Technol, 71-72; Alfred P Sloan Found fel, 71-73. Mem: Am Chem Soc; The Chem Soc. Res: Organic and bioorganic chemistry in the micellar phase; deamination; chemistry of alkyl diazotate salts and alkyldiazonium ions; carbenes and carbenoids. Mailing Add: Sch of Chem Rutgers Univ New Brunswick NJ 08903

MOSS, ROBERT HENRY, b New York, NY, July 26, 22; m 52; c 4. PHYSICAL INORGANIC CHEMISTRY, SOLID STATE CHEMISTRY. Educ: Univ NH, BS, 43; Univ Ark, MS, 48; Univ Conn, PhD(phys chem), 55. Prof Exp: Res asst indust eng, Eng Exp Sta, Univ NH, 43-46; res asst chem, Univ Conn, 53-54, asst instr, 54-55; prin chemist, Mine Safety Appliances Co, 55-56; engr, Westinghouse Elec Corp, 56-58, sr engr, 58-64; sr scientist, 64-69, SECT HEAD RES & DEVELOP PURIFICATION & SUPT PROD PURIFICATION, CRYSTALS & ELECTRONIC PARTS DEPT, HARSHAW CHEM CO, 69- Mem: Am Chem Soc; Electrochem Soc. Res: Materials preparation and purification of soild state materials; semiconductor, thermoelectric, electro-optical, scintillation and optical materials. Mailing Add: Crystal & Electronic Parts Dept Harshaw Chem Co 6801 Cochran Rd Solon OH 44139

MOSS, ROBERT L, b Brooklyn, NY, Aug 24, 40. NEUROPHYSIOLOGY, NEUROENDOCRINOLOGY. Educ: Villanova Univ, BS, 62; Claremont Grad Sch & Univ Ctr, MA, 67; PhD(neurophysiol, neuropsychol), 69. Prof Exp: Res asst, Inst Behav Res, Silver Spring, Md, 63-64; res asst grade I operant behav, Patton State Hosp, Calif, 64-65; res assoc anat, Med Sch, Univ Bristol, 69-71; ASST PROF PHYSIOL, SOUTHWESTERN MED SCH, UNIV TEX HEALTH SCI CTR DALLAS, 71- Concurrent Pos: NIMH fel, Dept Anat, Med Sch, Univ Bristol, 69-71; Instnl grant, Dept Physiol, Southwestern Med Sch, Univ Tex Health Sci Ctr, Dallas, 71-72, NIH grant, 72-78, NSF grant, 74-76 & Ayerst Labs Inc grant, 74- Honors & Awards: Young Scientist Award, Am Psychol Asn, 69. Mem: AAAS; Am Physiol Soc; Endocrine Soc; Int Soc Neuroendocrinol; Soc Neurosci. Res: Neural and biochemical mechanisms involved in hypothalamic control over pituitary function(s) and reproductive behavior. Mailing Add: Dept Physiol Southwestern Med Sch Univ of Tex Health Sci Ctr Dallas TX 75235

MOSS, RODNEY DALE, b Oakdale, Nebr, Apr 9, 27; m 50; c 2. ORGANIC CHEMISTRY, ANALYTICAL CHEMISTRY. Educ: Univ Nebr, BS, 48, MS, 49; Indiana Univ, PhD(chem), 51. Prof Exp: Anal chemist, Eastman Kodak Co, 48; res chemist, 51-60, proj leader, 60-63, group leader, 63-65, head chem res dept, 65-68, DIR CHEM, AGR PROD CTR, DOW CHEM, USA, 68- Mem: Am Chem Soc. Res: Organic sulfur and phosphorus chemistry; residue analysis and environmental studies. Mailing Add: Agr Prod Ctr 9008 Bldg Dow Chem USA Midland MI 48640

MOSS, RONNIE LEE, b Ft Sill, Okla, Oct 16, 38; m 66; c 1. STATISTICS, OPERATIONS RESEARCH. Educ: Okla State Univ, BS, 60, MS, 61; Univ Okla, PhD(statist), 66. Prof Exp: Instr math sci, Glencoe Pub Schs, Okla, 60-61; instr math, Enid Pub Schs, Okla, 61-63; asst statist, Univ Okla, 63-66; asst prof math, Northwest Mo State Univ, 66-69, PROF STATIST & COMPUT SCI & CHMN DEPT, NORTHWEST MO STATE UNIV, 69- Mem: Asn Comput Mach; Asn Educ Data Systs (pres, 71-). Mailing Add: Dept of Statist & Comput Sci Northwest Mo State Univ Maryville MO 64468

MOSS, SAMUEL, b Czech, July 2, 14; nat US; m 44; c 2. PHYSIOLOGY. Educ: NY Univ, PhD(physiol), 50. Prof Exp: Physiologist & histochemist, Dairy Husb Res Br, Agr Res Serv, USDA, 50-56; EXEC SECY, HUMAN EMBRYOL & DEVELOP STUDY SECT, DIV RES GRANTS, NIH, 56- Mem: Teratology Soc. Res: Physiology of reproduction; embryology; neonatology; congenital anomalies; science administration. Mailing Add: Div of Res Grants Nat Inst Health Bethesda MD 20014

MOSS, SAMUEL J, applied physics, mathematics, see 12th edition

MOSS, SANFORD ALEXANDER, III, b Ridley Park, Pa, Apr 8, 39; m 65; c 1. ENVIRONMENTAL PHYSIOLOGY, MARINE BIOLOGY. Educ: Yale Univ, BS, 61; Cornell Univ, PhD(zool), 65. Prof Exp: Instr biol, Yale Univ, 65-67; asst prof, 67-71, ASSOC PROF BIOL, SOUTHEASTERN MASS UNIV, 71- Mem: AAAS; Am Fisheries Soc. Res: Schooling behavior and respiratory physiology of fish; feeding mechanisms of sharks and rays. Mailing Add: Dept of Biol Southeastern Mass Univ North Dartmouth MA 02747

MOSS, THOMAS HENRY, b Cleveland, Ohio, June 27, 39; m. BIOPHYSICS. Educ: Harvard Univ, AB, 61; Cornell Univ, PhD(physics), 65. Prof Exp: NSF fel, Univ Calif, San Diego, 65-66 & Nobel Inst, Royal Caroline Inst, Sweden, 66-67; RES PHYSICIST, IBM WATSON RES CTR, 67-74, 75- Concurrent Pos: Analyst, Fed Support Acad Res, US Bur Budget, DC, 63-66; adj asst prof, Columbia Univ, 67-; consult, Sci & Technol Comt, US House of Reps, 75- Mem: Cong fel Am Phys Soc; Biophys Soc; NY Acad Sci. Res: Mössbauer spectroscopy of heme and non heme iron proteins and model compounds; electron spin resonance, nuclear magnetic resonance and magnetic susceptibility studies of metalloproteins; copper binding in blue copper proteins. Mailing Add: IBM Watson Res Ctr Yorktown Heights NY 10598

MOSS, VALENTIN G, b Oct 16, 13; US citizen; m 41; c 1. FOOD CHEMISTRY. Educ: Munich Univ, BS, 33; Univ Debrecen, PhD(org chem), 42. Prof Exp: RES BIOCHEMIST, OSCAR MAYER & CO, 56- Mem: Am Chem Soc; Inst Food Technologists. Res: Meat Proteins; enzymes; biochemicals. Mailing Add: Oscar Mayer & Co Packers Ave Madison WI 53701

MOSS, W WAYNE, b Toronto, Ont, Mar 14, 37; m 67; c 2. SYSTEMATICS, ENTOMOLOGY. Educ: Carleton Univ, Can, BSc, 59, MSc, 61; Univ Kans, PhD(entom), 66. Prof Exp: Fel entom, 66-68, asst curator, 68-70, ASSOC CURATOR & CHMN ACAD NATURAL SCI PHILADELPHIA, 70- Concurrent Pos: Lectr biol, Univ Pa, 66-68, adj assoc prof, 70-; NSF res grant, 67-; Miller res fel, Miller Inst for Basic Res in Sci, Div Entom, Univ Calif, Berkeley, 68-70, lectr entom & parasitol, 70; Am Philos Soc res grant, 75. Honors & Awards: Hungerford Mem Award, Univ Kans, 67. Mem: AAAS; Soc Study Evolution; Soc Syst Zool; Entom Soc Am. Res: Numerical taxonomy, theory and methods with application to Acari and other taxonomic groups in collaboration; computers in biology, data processing, analysis and storage; host-parasite relations of Acari on birds. Mailing Add: Acad of Natural Sci 19th & Pkwy Philadelphia PA 19103

MOSS, WOODROW GLEN, b Red Fork, Okla, June 11, 14; m 40; c 2. PHYSIOLOGY, SCIENCE ADMINISTRATION. Educ: Univ Wichita, BA, 37; Univ Chicago, MS, 42; Univ Ill, PhD(physiol), 48. Prof Exp: Instr zool, Univ Wichita, 37-39; asst, Univ Mo, 39-40; asst physiol, Univ Ill, 42-47, from instr to asst prof, 47-53; assoc prof pharmacol, Temple Univ, 48-50; from assoc prof to prof physiol, Sch Med, Miami Univ, 53-61; exec secy cardiovasc study sect, NIH, 61-63, chief training grants & awards br, Nat Heart & Lung Inst, 63-64, chief res grants br, 64-69, chief hypertension & kidney dis br & dep assoc dir extramural res & training, 69-72, DEP DIR DIV EXTRAMURAL AFFAIRS, NAT HEART & LUNG INST, 72- Res: Cardiac dynamics; hypertension; vitamin A metabolism; local anesthesia; blocking of pressor amines; atherosclerosis. Mailing Add: Div of Extramural Affairs Nat Heart & Lung Inst Bethesda MD 20014

MOSSBERG, HOWARD E, b Chicago, Ill, Sept 1, 32; m 55; c 2. PHARMACOLOGY, PHARMACY. Educ: Univ Fla, BS, 54, PhD(pharmacol), 58. Prof Exp: Fel pharmacol, Univ Fla, 54-58; from assoc prof to prof pharmacol, Sch Pharm, Southwestern State Col, Okla, 58-65, assoc dean, 65-66; DEAN SCH PHARM, UNIV KANS, 66- Mailing Add: Off of the Dean Univ of Kans Sch of Pharm Lawrence KS 66045

MOSSER, JOHN SNAVELY, b Canton, Ohio, Apr 7, 28. PHYSICAL CHEMISTRY, ENVIRONMENTAL MANAGEMENT. Educ: Case Western Reserve Univ, BS, 50, MS, 52, PhD(phys chem), 63. Prof Exp: Res engr, Indust Rayon Corp, 52-54; sr res chemist, Gen Tire & Rubber Co, 62-70; WASTEWATER QUAL COORDR, CITY OF AKRON, 72- Mem: Am Chem Soc; Sigma Xi. Res: Cryogenic measurement and calculation of thermodynamic properties; preparation and characterization of polymers and latices. Mailing Add: 565 Stevenson Ave Akron OH 44312

MOSSMAN, ARCHIE STANTON, b Madison, Wis, Feb 5, 26; m; c 3. ZOOLOGY. Educ: Univ Wis, BA, 49, PhD(zool, wildlife mgt), 55; Univ Calif, MA, 51. Prof Exp: Biologist, Dept Fish & Game, Alaska, 55-57; instr, Exten Div, Univ Wis, 57-58; asst prof, Univ Wyo, 58-59; Fulbright res scholar, Nat Mus Southern Rhodesia, 59-61; from asst prof to assoc prof, 61-71, PROF, SCH NATURAL RESOURCES, HUMBOLDT STATE UNIV, 71- Concurrent Pos: Sr lectr, Univ Col Rhodesia & Nyasaland, 63-65; Food & Agr Orgn consult, Malawi, 69; prin investr evaluation game ranching in southern Africa, Int Union Conserv Nature & Natural Resources & World Wildlife Found Joint Proj, 74-75. Mem: Ecol Soc Am; Wildlife Soc; Am Soc Mammal; Cooper Ornith Soc; Am Inst Biol Sci. Res: Animal behavioral response to environment; food production from wildlife; predation. Mailing Add: Sch of Natural Resources Humboldt State Univ Arcata CA 95521

MOSSMAN, DAVID JOHN, b Mar 9, 38; Can citizen; m 63; c 1. MINERALOGY, GEOLOGY. Educ: Dalhousie Univ, BSc, 59, MSc, 63; Univ Otago, NZ, PhD(geol), 70. Prof Exp: Field geologist, Anglo Am Corp, SAfrica, 59-62; party chief groundwater res, NS Govt, Can, 64; lectr appl geol, Univ Otago, NZ, 70; geologist, Dept Natural Resources, NB, Can, 71; ASST PROF ECON GEOL, UNIV SASK, 71- Mem: Geol Soc NZ; Mineral Asn Can; Can Inst Mining & Metall. Res: Economic geology; petrology of basic and ultrabasic rocks; petrology of coal. Mailing Add: Dept of Geol Sci Univ of Sask Saskatoon SK Can

MOSSMAN, REUEL WALLACE, b Fresno, Calif, Aug 18, 14; m 40. GEOPHYSICS. Educ: Columbia Univ, AB, 35, MA, 39. Prof Exp: County Supvr, Okla State Mineral Surv, 35-36; survr-comput, Geophys Res Corp, 37; supvr, 39-50, chief geophysicist, 50-55, ASST V PRES, SEISMOGRAPH SERV CORP, 55- Mem: Am Asn Petrol Geologists; Soc Explor Geophys; Am Geophys Union; Europ Asn Explor Geophys. Mailing Add: Seismograph Serv Corp Box 1590 Tulsa OK 74102

MOSSOP, GRANT DILWORTH, b Calgary, Alta, Apr 15, 48; m 69; c 1. SEDIMENTOLOGY, SEDIMENTARY PETROLOGY. Educ: Univ Calgary, BSc, 70, MSc, 71; Univ London, PhD(sedimentary geol), & DIC, 73. Prof Exp: Fel geol, Univ Calgary, 74; RES OFFICER OIL SAND GEOL, ALTA RES COUN, 75- Mem: Can Soc Petrol Geologists; Geol Asn Can; Soc Econ Paleontologists & Mineralogists. Res: Detailed sedimentology and petrology of the Athabasca Oil Sands, Alberta. Mailing Add: Alta Res Coun 11315-87 Ave Edmonton AB Can

MOSSOTTI, VICTOR GIOVONI, b St Louis, Mo, May 20, 38; m 61. PHYSICAL CHEMISTRY, ANALYTICAL CHEMISTRY. Educ: Southeast Mo State Col, BS, 60; Iowa State Univ, PhD(phys chem), 64. Prof Exp: Asst, Iowa State Univ, 60-64; Alexander von Humboldt Found res fel, Inst Spectrog Chem, Univ Münster, 64-65, sr res chemist, Mat Res Lab, Univ Ill, Urbana, 65-70; ASSOC PROF CHEM, UNIV MINN, MINNEAPOLIS, 70- Concurrent Pos: Alexander von Humboldt Found grant & cert German, Goethe Inst, 64. Mem: AAAS; Am Chem Soc; Soc Appl Spectros; Am Soc Test & Mat. Res: Physical analytical chemistry, including atomic spectroscopy, laser applications in chemical instrumentation, information theory and theory of stochastic processes. Mailing Add: Dept of Chem Univ of Minn Minneapolis MN 55455

MOSS-SALENTIJN, LETTY, b Amsterdam, Netherlands, Apr 14, 43; m 70. ANATOMY, DENTAL RESEARCH. Educ: State Univ Utrecht, DDS, 67. Prof Exp: Chief instr oral hist growth & develop, Holland Lab Hist & Micros Anat, State Univ Utrecht, 67-68; asst prof anat, 68-74, ASSOC PROF ORAL BIOL, COLUMBIA UNIV, 74- Mem: AAAS; Int Asn Dent Res; Am Soc Zool; Int Soc Stereology; fel Royal Micros Soc. Res: Growth and development of skeletal tissues; growth of cartilages; orofacial embryology; comparative odontology. Mailing Add: Dept of Anat Columbia Univ 630 W 168th St New York NY 10032

MOST, DAVID S, b Boston, Mass, Feb 7, 29; m 52; c 4. PAPER CHEMISTRY. Educ: Boston Univ, AB, 52; Lawrence Col, MS, 54, PhD(chem), 57. Prof Exp: Group leader, Res Dept, Albemarle Paper Co Div, Ethyl Corp, 57-60; dept mgr appl res, Itek Corp, 60-62; vpres & gen mgr, New Eng Labs, Inc, Rahn Corp, 62-65; consult paper chem, 65-70; pres, M/K Systs, Inc, 70-75; CONSULT, 75- Mem: AAAS; Am Chem Soc; Tech Asn Pulp & Paper Indust; Soc Photog Sci & Eng. Res: Reproduction technology; development of special papers for use in recording and communications applications including copy papers and facsimile. Mailing Add: 199 W Share Dr Marblehead MA 01945

MOST, ELMER EDWIN, JR, organic chemistry, see 12th edition

MOST, HARRY, b New York, NY, Sept 18, 07; m 38; c 2. TROPICAL MEDICINE. Educ: NY Univ, BS, 27, MD, 31, DMedSc, 39; London Sch Trop Med, DTM & H, 36. Prof Exp: Asst prof med, 41-46, assoc prof prev med, 46-49, prof trop med, 49-54, BIGGS PROF PREV MED & CHMN DEPT, NY UNIV, 54- Concurrent Pos: Res fel, Int Health Div, Rockefeller Found, PR & Haiti, 41; lectr, Columbia Univ, 42-44 & Army Med Ctr, 46 & 47; chief consult trop med, Vet Admin, 46-; consult, Surgeon Gen, US Army, 46-; comnr health, State Health Dept, NY, 46-; comnr health, USPHS, 46-, Commun Dis Ctr, Ga, 50-; Clin Ctr, 55-; mem, Nat Res Coun, 46; dir trop dis diag clin, City Health Dept, New York, 49-; mem parasitic dis comn, US Armed Forces Epidemiol Bd, 52-, dir, 60-; vis lectr, Harvard Univ, 56-; vis physician, Bellevue Hosp; consult, Meadowbrook, Phelps Mem & NShore Hosps. Mem: Am Soc Parasitol; Am Soc Clin Invest; Harvey Soc; AMA; NY Acad Sci. Res: Malaria; amebiasis; schistosomiasis; trichinosis; hookworm. Mailing Add: Dept of Prev Med NY Univ Sch of Med New York NY 10016

MOST, JOSEPH MORRIS, b New York, NY, Apr 24, 43; m 65; c 2. PHYSICAL CHEMISTRY. Educ: Rutgers Univ, AB, 64, PhD(inorg chem), 74. Prof Exp: Chemist, NL Industs, Inc, 64-66; instr chem, Rutgers Col, Rutgers Univ, 71-74; ASST PROF CHEM, UPSALA COL, 74- Mem: Am Chem Soc. Res: Chemical education, especially course development. Mailing Add: Dept of Chem Upsala Col East Orange NJ 07019

MOSTARDI, RICHARD ALBERT, b Bryn Mawr, Pa, July 1, 38; m 62; c 4. PHYSIOLOGY. Educ: Kent State Univ, BS, 60, MEd, 64; Ohio State Univ, PhD(physiol), 68. Prof Exp: Res asst physiol, Aviation Med Lab, Ohio State Univ, 66-68; ASSOC PROF PHYSIOL, UNIV AKRON, 68- Concurrent Pos: NIH fel, Milan, Italy, 72-73. Mem: NY Acad Sci; Sigma Xi. Res: Acoustic diagnosis of arthritis; effects of drag reducing polymers in the vertebrate system; exercise in humans. Mailing Add: Dept of Biol Univ of Akron Akron OH 44325

MOSTELLER, C FREDERICK, b Clarksburg, WVa, Dec 24, 16; m 41; c 2. MATHEMATICAL STATISTICS. Educ: Carnegie Inst Technol, BS, 38, MS, 39; Princeton Univ, AM, 42, PhD(math), 46. Hon Degrees: DSc, Univ Chicago, 73 & Carnegie-Mellon Univ, 74. Prof Exp: Instr math, Princeton Univ, 42-44; res mathematician, Statist Res Group, 44-45; MEM FAC, DEPT SOCIAL RELS, HARVARD UNIV, 46-, PROF MATH STATIST, 51-, CHMN DEPT STATIST, 57-69, 75-, MEM FAC, J F K SCH GOVT, 70- Concurrent Pos: Fund Advan Educ fel, Univ Chicago, 54-55; mem staff probability & statist, NBC's Continental Classroom TV Course, 60-61; fel, Ctr Advan Study Behav Sci, 62-63; chmn bd dirs, Soc Sci Res Coun, 66-68; Guggenheim fel, 69-70; vchmn, President's Comn Fed Statist, 71; Miller res prof, Univ Calif, 74-75. Mem: Nat Acad Sci; Inst of Med of Nat Acad Sci; AAAS; Am Philos Soc; Am Acad Arts & Sci. Res: Theoretical statistics and its applications to social science, medicine, public policy and industry. Mailing Add: Rm 603 Sci Ctr Harvard Univ One Oxford St Cambridge MA 02138

MOSTELLER, RAYMOND DEE, b Austin, Tex, Dec 30, 41. BIOCHEMISTRY. Educ: Univ Tex, Austin, BA, 64, PhD(biochem), 68. Prof Exp: ASST PROF BIOCHEM, SCH MED, UNIV SOUTHERN CALIF, 70- Concurrent Pos: Fel molecular biol, Stanford Univ, 68-70; NSF res grants, 71-73 & 73-75; NIH res grant, 73-76. Mem: Am Soc Microbiol; AAAS. Res: Mechanism of protein biosynthesis; control of gene expression; regulation of protein degradation. Mailing Add: Dept of Biochem Univ of Southern Calif Sch Med Los Angeles CA 90033

MOSTELLER, ROBERT COBB, b Lynchburg, Va, Oct 14, 38; m 60; c 2. BIOMETRICS. Educ: Randolph-Macon Col, BA, 61; Emory Univ, MS, 67, PhD(statist), 76. Prof Exp: Res assoc biostatist, Biomet Unit, Dept Plant Breeding, Cornell Univ, 65; instr biometry, Univ Kans Med Ctr, 68-70; assoc, Dept Biometry & Statist, 70-73, CHIEF STATIST UNIT BIOMETRY, MAMMOGRAPHY SECT, DEPT RADIOL, EMORY UNIV, 73- Res: Statistical evaluation of breast cancer data aimed at the identification of women with a high risk of either current or future breast cancer. Mailing Add: Mammography Sect Emory Univ Atlanta GA 30322

MOSTERT, PAUL STALLINGS, b Morrilton, Ark, Nov 27, 27; m 47; c 4. MATHEMATICS. Educ: Southwestern at Memphis, BS, 50; Univ Chicago, MS, 51; Purdue Univ, PhD(math), 53. Prof Exp: Asst math, Purdue Univ, 51-53; res instr, Tulane Univ La, 53-54, from asst prof to prof, 54-70, chmn dept, 68-70; chmn dept, 70-73, PROF MATH, UNIV KANS, 70- Concurrent Pos: Vis prof, Univ Tübingen, 62-63 & 66; NSF sr fel & mem, Inst Advan Study, 67-68; mem selection of postdoctoral fels panel, Nat Res Coun, 69-71; managing ed & co-founder, Semigroup Forum, exec ed, 74-; chmn comt acad freedom, tenure and employment security, Agr Mkt Serv. Mem: Am Math Soc. Res: Topological semigroups; transformation groups; category theory. Mailing Add: Dept of Math Univ of Kans Lawrence KS 66044

MOSTOFI, FATHOLLAH KESHVAR, b Teheran, Iran, Aug 10, 11; nat US; m 40; c 1. PATHOLOGY. Educ: Univ Nebr, BA, 34, BS, 38; Harvard Med Sch, MD, 39. Prof Exp: Intern, St Luke's Hosp, Pa, 39-40; house officer path, Peter Bent Brigham Hosp, Boston, 40-41; resident, Boston Lying-in-Hosp, 41-42, Free Hosp Women, 42 & Children's Hosp, 42-43; asst pathologist, Mass Gen Hosp, 43-44; consult pathologist, Mass Eye & Ear Infirmary, 44; spec cancer fel, Nat Cancer Inst, 47-48; pathologist, Vet Admin Cent Lab, 48-62, ASSOC CHMN CTR ADVAN PATH & CHMN DEPT GENITOURINARY PATH, ARMED FORCES INST PATH, 62- Concurrent Pos: Consult, Nat Cancer Inst, 48-50; sci dir, Am Registry Path, 57-59; assoc prof, Johns Hopkins Univ, 60-; clin prof, Med Ctr, Georgetown Univ, 61- Mem: Fel Am Soc Clin Path; Am Cancer Soc; fel AMA; Asn Mil Surgeons US; fel Am Col Path. Res: Cancer. Mailing Add: Armed Forces Inst of Path Washington DC 20306

MOSTOLLER, MARK ELLSWORTH, b Somerset, Pa, Sept 13, 41; m 63; c 2. SOLID STATE PHYSICS. Educ: Harvard Col, AB, 62; Harvard Univ, SM, 63, PhD(appl physics), 69. Prof Exp: MEM RES STAFF, OAK RIDGE NAT LAB, 69- Mem: Am Phys Soc. Res: Lattice dynamics; disordered systems; defects in solids; electronic structure and properties. Mailing Add: Solid State Div Oak Ridge Nat Lab PO Box X Oak Ridge TN 37830

MOSTOW, GEORGE DANIEL, b Boston, Mass, July 4, 23; m 47; c 4. MATHEMATICS. Educ: Harvard Univ, BA, 43, MA, 46, PhD(math), 48. Prof Exp: Instr math, Princeton Univ, 47-48; asst prof, Syracuse Univ, 49-52; from asst prof to prof, Johns Hopkins Univ, 52-61; prof, 61-63, chmn dept, 71-74, JAMES E ENGLISH PROF MATH, YALE UNIV, 63- Concurrent Pos: Mem staff, Inst Advan Study, 47-49, 56-57, 75; vis prof, Inst Pure & Appl Math, Brazil, 53-54; Guggenheim fel & Fulbright res scholar, State Univ Utrecht, 57-58; ed, Am J Math, 63-67, assoc ed, 67-; exchange prof, Univ Paris, 66; vis prof, Hebrew Univ, Israel, 67; Tata Inst Fundamental Res, India, 69; Inst Advan Study Sci, 71 & 75; chmn, US Nat Comt Math, 73-74; chmn, Off Math Sci, Nat Acad Sci-Nat Res Coun, 75-78. Mem: Nat Acad Sci. Res: Lie groups; discrete subgroups of algebraic groups. Mailing Add: Dept of Math Yale Univ New Haven CT 06520

MOSZKOWSKI, STEVEN ALEXANDER, b Berlin, Ger, Mar 13, 27; nat US; m 52; c 3. THEORETICAL PHYSICS. Educ: Univ Chicago, BS, 46, MS, 50, PhD(physics), 52. Prof Exp: Jr physicist, Argonne Nat Lab, 50-51; res asst, Columbia Univ, 52-53; from asst prof to assoc prof physics, 53-63, PROF PHYSICS, UNIV CALIF, LOS ANGELES, 63- Concurrent Pos: Consult, Rand Corp, 53-71 & Oak Ridge Nat Lab, 62-68; Guggenheim fel, 61-62. Mem: Fel Am Phys Soc. Res: Nuclear shell structure; many-body problem. Mailing Add: Dept of Physics Univ of Calif Los Angeles CA 90024

MOTA, ANA CELIA, b Maria Grande, Arg, June 10, 35; US citizen; div; c 2. LOW TEMPERATURE PHYSICS. Educ: Univ Cuyo, Arg, MS, 60, PhD(physics), 67. Prof Exp: From asst to assoc physics res, Univ Calif, San Diego, 68-75, lectr physics, 72-74; SR RES SCIENTIST, UNIV COLOGNE, 75- Concurrent Pos: Consult, Superconducting Helium Electronics, San Diego, Calif, 75- Mem: Am Phys Soc. Res: Experimental ultra low temperature physics. Mailing Add: II Phys Inst Univ of Cologne 5 Cologne 41 Germany

MOTAWI, KAMAL EL-DIN HUSSEIN, b Singerg, Egypt, July 22, 34; US citizen; m 64; c 5. FOOD SCIENCE. Educ: Ain Shams Univ, Cairo, BSc, 56; Mich State Univ, MS, 62, PhD(food sci), 66. Prof Exp: Food technologist, Ministry Agr, Egypt, 58-60; res assoc human nutrit, Mich State Univ, 65-66; tech adv div food technol, Ministry Indust, Egypt, 66-67; asst prof food sci & human nutrit, Ain Shams Univ, Cairo, 67-68; asst mgr prod develop, 68-69, MGR PROD DEVELOP, GERBER PROD CO, 69- Mem: Inst Food Technologists. Res: Development of food systems and processes to improve nutritional contribution and acceptability of natural foods; transform plant, animal and synthetic materials into wholesome, stable, safe and nutritious foods for human consumption. Mailing Add: Gerber Prod Co 445 State St Fremont MI 49412

MOTE, MICHAEL ISNARDI, b San Francisco, Calif, Feb 5, 35; m 65; c 2. NEUROPHYSIOLOGY. Educ: Univ Calif, Berkeley, AB, 58; San Francisco State Col, MA, 63; Univ Calif, Los Angeles, PhD(zool), 68. Prof Exp: NIH fel biol, Yale Univ, 68-70; asst prof, 70-74, ASSOC PROF BIOL, TEMPLE UNIV, 74-, NIH FEL, EYE INST & NSF FEL PSYCHOBIOL, 71- Mem: Fel AAAS; Am Soc Zool; Soc Gen Physiol; Asn Res Vision & Opthal. Res: Integrative neurophysiology of invertebrate nervous systems with emphasis on vision in arthropods. Mailing Add: Dept of Biol Temple Univ Philadelphia PA 19122

MOTE, VICTOR LEE, b Corpus Christi, Tex, Nov 2, 41; m 64; c 2. GEOGRAPHY, ENVIRONMENTAL SCIENCES. Educ: Univ Denver, BA, 64; Univ Wash, MA, 69, PhD(geog), 71. Prof Exp: ASST PROF GEOG, UNIV HOUSTON, 71- Concurrent Pos: Vis prof, Univ St Thomas, Houston, Tex, 75. Mem: Am Asn Advan Slavic Studies; Asn Am Geog; Nat Coun Geog Educ. Res: Environmental quality, especially in the Soviet Union; historical and geographical evolution of Siberia; Baykal-Amur Mainline Railroad, East Siberia; city size in USSR; air and noise pollution in USSR. Mailing Add: Dept of Geog Univ of Houston Houston TX 77004

MOTEKAITIS, RAMUNAS JUOZAS, organic chemistry, inorganic chemistry, see 12th edition

MOTHERSILL, JOHN SYDNEY, b Ottawa, Ont, Mar 24, 31. GEOLOGY. Educ: Carleton Univ, Can, BSc, 53; Queen's Univ, Ont, BSc, 56, PhD(geol), 67. Prof Exp: Geologist, Esso Standard Turkey, Inc, Stand Oil Co NJ, 56-58; geologist, Mobil Explor Nigeria, Inc, Mobil Int Oil Co, Inc, 58-61, sect head admin stratig, Colombia Petrol Co, 61-62, sr geologist, Mobil Explor Nigeria, Inc, 62-64; asst prof, 66-70, ASSOC PROF STRATIG & SEDIMENTATION, LAKEHEAD UNIV, 70- Concurrent Pos: Reader, Univ Nigeria, 72-73; mem staff, Limnol Surv Lake Superior, Can Ctr Inland Waters. Mem: Am Asn Petrol Geologists. Res: Stratigraphy; sedimentation; nearshore clastic sedimentation. Mailing Add: Dept of Geol Lakehead Univ Thunder Bay ON Can

MOTICKA, EDWARD JAMES, b Oak Park, Ill, May 21, 44; m 69; c 1. IMMUNOLOGY. Educ: Kalamazoo Col, BA, 66; Univ Ill, PhD(anat), 70. Prof Exp: Vis scientist immunol, Czech Acad Sci, Inst Microbiol, 71-72; ASST PROF CELL BIOL, UNIV TEX HEALTH SCI CTR, DALLAS, 72- Mem: Am Asn Immunologists; Am Soc Zoologists. Res: Development and control of polyclonal immunoglobulin synthesis and the controlling functions of thymus-derived lymphocytes in immune responses. Mailing Add: Dept of Cell Biol Univ Tex Health Sci Ctr Dallas TX 75235

MOTILL, RONALD ALLEN, b Feb 5, 41; US citizen. THEORETICAL PHYSICS, COSMOLOGY. Educ: Syracuse Univ, BS, 61, MS, 63, PhD(physics), 75. Prof Exp: Technician physics, Syracuse Univ, 58, Bausch & Lomb Optical Co, 60; ASST PROF PHYSICS, FURMAN UNIV, 74- Mem: Am Phys Soc; Am Asn Physics Teachers; AAAS. Res: Construction of a quantum field theory of the gravitational field based on commutators within characteristic, or null, hypersurfaces; general relativity; quantum field theory. Mailing Add: Dept of Physics Furman Univ Greenville SC 29613

MOTLEY, HURLEY LEE, b Silex, Mo, July 23, 04; m 41; c 3. PHYSIOLOGY. Educ: Univ Mo, AB, 30, BS & AM, 32, PhD(physiol), 34; Harvard Univ, MD, 36. Prof Exp: Asst zool, Univ Mo, 28-30, asst physiol & pharmacol, 30-34, from asst prof to assoc prof, Sch Med, 36-47; assoc prof med & dir cardio-respiratory lab, Jefferson Med Col, 47-52; PROF MED & DIR CARDIO-RESPIRATORY LAB, SCH MED, UNIV SOUTHERN CALIF, 52- Concurrent Pos: Consult med to Army Surgeon Gen; staff mem, Hollywood Presby Hosp; fel, Columbia Univ, 45-46; staff mem & dir res, Good Samaritan Hosp, Los Angeles, 62-; mem sci comt, Los Angeles Air Pollution Control Dist & nat med & res adv coun, City of Hope Med Ctr. Mem: Fel Am Col Physicians; Am Indust Hyg Asn; Am Physiol Soc; Am Thoracic Soc; fel AMA. Res: Physiology of fresh water mussel heart; intermittent positive pressure breathing; pulmonary function studies in man; air pollution; pneumoconiosis; emphysema; clinical pulmonary physiology; internal medicine, especially the chest. Mailing Add: 2003 N Serrano Ave Los Angeles CA 90027

MOTLEY, ROBERT W, b Pottsville, Pa, Aug 23, 31. PLASMA PHYSICS. Educ: Pa State Univ, BS, 53; Princeton Univ, PhD(physics), 58. Prof Exp: RES PHYSICIST, PLASMA PHYSICS LAB, PRINCETON UNIV, 58- Res: High energy particle detection; plasma recombination and diffusion; wave propagation in plasma; RF heating of plasma. Mailing Add: Plasma Physics Lab Princeton Univ Princeton NJ 08540

MOTOYAMA, ETSURO K, b Japan, Apr 11, 32; US citizen. ANESTHESIOLOGY, RESPIRATORY PHYSIOLOGY. Educ: Chiba Univ, Japan, BS, 53; Chiba Univ Med Sch, Japan, MD, 57. Prof Exp: Res assoc pediat, 64-66, asst prof anesthesiol & pediat, 66-70, ASSOC PROF ANESTHESIOL & PEDIAT, MED SCH, YALE UNIV, 70- Concurrent Pos: Fel respiratory physiol, Harvard Med Sch, 62-64; attend anesthesiologist, Yale-New Haven Hosp, 66-; prin investr lung res ctr, Yale Univ, 71- Mem: Am Physiol Soc; Am Acad Pediat; Soc Pediat Res; Am Soc Anesthesiol. Res: Perinatal and pediatric respiratory physiology; pediatric anesthesiology. Mailing Add: Dept of Anesthesiol Yale Univ Med Sch New Haven CT 06510

MOTSAVAGE, VINCENT ANDREW, b Scranton, Pa, May 10, 34; m 57; c 5.

MOTSAVAGE

PHARMACY, PHYSICAL CHEMISTRY. Educ: Philadelphia Col Pharm, BS, 55; Temple Univ, MS, 57, PhD(pharm), 62. Prof Exp: Res assoc pharmaceut develop, Merck Sharp & Dohme Res Labs, Merck & Co, Inc, 57-59, pharm res, 61-63; asst instr pharm, Temple Univ, 59-60, res asst, 60-61; sr phys chemist, Avon Prod, Inc, 64-67, sect head chem res, 67-69; HEAD DEPT COSMETIC & TOILETRIES DEVELOP, MENLEY & JAMES LABS, INC, SMITH KLINE & FRENCH LABS, 69- Mem: Am Chem Soc; Am Pharmaceut Asn; Soc Cosmetic Chemists. Res: Physical and colloid chemistry as applied to the research and development of cosmetics and pharmaceuticals. Mailing Add: Gwynedd Manor Rd North Wales PA 19454

MOTSINGER, RALPH E, b Carthage, NC, May 10, 31; m 52; c 2. PLANT PATHOLOGY. Educ: NC State Univ, BS, 56; Univ Md, MS, 60; Auburn Univ, PhD(plant path), 64. Prof Exp: Asst county agent, NC Agr Exten Serv, 56-58; asst county agent, Md Agr Exten Serv, 58-59, exten tobacco specialist, 59-61; exten plant pathologist, La Coop Exten Serv, 64-70; EXTEN PLANT PATHOLOGIST, GA COOP EXTEN SERV, UNIV GA, 70- Concurrent Pos: Ford Found vis prof, Col Agr, Selangor Malaysia, 69-70. Mem: Am Phytopath Soc; Soc Nematol. Res: Nematode population dynamics; grower acceptance of plant disease control information. Mailing Add: Ga Coop Exten Serv Univ of Ga Athens GA 30601

MOTT, DAVID GORDON, b Dalhousie, NB, Feb 9, 32; m 53; c 4. ECOLOGY, FOREST ENTOMOLOGY. Educ: Univ NB, BSc, 54; Yale Univ, MF, 57. Prof Exp: Res officer forest entom, Forest Entom & Path Lab, Can Dept Forestry, 54-62; ecologist, Forest Insect & Dis Lab, Northeastern Forest Exp Sta, Conn, 62-72, RES FORESTER, FORESTRY SCI LAB, US FOREST SERV, 72- Res: Animal population ecology; biometrics; mathematical modelling; forest insect control and population regulation; resource management and optimization. Mailing Add: US Forest Serv Forestry Sci Lab Concord-Mast Rds PO Box 640 Durham NH 03824

MOTT, DAVID LOWE, b Springfield, Vt, Apr 28, 25; m 45; c 3. PHYSICS. Educ: Tex Western Col, BS, 58; NMex State Univ, MS, 60, PhD(physics), 63. Prof Exp: PHYSICIST, PHYS SCI LAB, N MEX STATE UNIV, 63-, ASST PROF PHYSICS, N MEX STATE UNIV, 65- Mem: Am Asn Physics Teachers. Res: X-ray spectroscopy; classical mechanics. Mailing Add: Dept of Physics NMex State Univ Las Cruces NM 88001

MOTT, FREDERICK DODGE, b Wooster, Ohio, Aug 3, 04; m 30; c 3. PREVENTIVE MEDICINE. Educ: Princeton Univ, AB, 27; McGill Univ, MD & CM, 32. Hon Degrees: LLD, Univ Sask, 55. Prof Exp: From med officer to chief med officer, US Farm Security Admin, 37-46; chmn, Health Serv Planning Comn, Sask, 46-51; dep minister, Dept Pub Health, Sask, 51; med adminr, Miners Mem Hosp Asn, Washington, DC, 52-57; exec dir, Community Health Asn, 57-64; consult med care, NY Acad Med, 64-66; prof med care, Sch Hyg, Univ Toronto, 66-72; CONSULT, 72- Concurrent Pos: Chief health serv br, Off of Labor, War Food Admin, 43-45; actg dept minister, Dept Pub Health, Sask, 49-51; mem expert adv panel on orgn of med care, WHO, 51- Mem: Fel Am Pub Health Asn. Res: Rural health and medical care; organization and financing of health services; planning of hospital and health facilities; medical administration. Mailing Add: 19 E Jefferson Circle Pittsford NY 14534

MOTT, GEORGE ROBSON, b Syracuse, NY, Oct 8, 21; m 44; c 3. PHYSICS. Educ: Ohio Wesleyan Univ, AB, 43; Univ Rochester, PhD(physics), 51. Prof Exp: Res assoc electronics, Strong Mem Hosp, 51-52; PHYSICIST XEROGRAPHIC RES, XEROX CORP, 52- Mem: Am Phys Soc; Inst Elec & Electronics Engrs. Res: High energy nuclear physics. Mailing Add: 113 Shirewood Dr Rochester NY 14625

MOTT, GERALD O, b Hastings, Nebr, Mar 31, 12; m 37; c 4. AGRONOMY. Educ: Univ Nebr, BS, 34; Cornell Univ, PhD, 40. Prof Exp: Agent & jr agronomist, Soil Conserv Serv, USDA, 34-36; asst, Cornell Univ, 36-39; prof agron, Purdue Univ, 39-63; dir res, IRI Res Inst, 64-66; prof agron, Purdue Univ, 66-68; PROF AGRON, UNIV FLA, 69- Concurrent Pos: Vis prof, NC State Univ, 52 & Cornell Univ, 56; consult, IRI Res Inst, 56-64; sr Fulbright-Hays grant, Univ Queensland & Cunningham Lab, Commonwealth Sci & Indust Res Orgn, Australia, 70. Honors & Awards: Crop Sci Award, Am Soc Agron, 68. Mem: Am Soc Agron; Crop Sci Soc Am (pres, 55-56); Am Soc Animal Sci; Sigma Xi. Res: Tropical forage and pasture research and development programs for international agriculture. Mailing Add: Dept of Agron 2183 McCarty Hall Univ of Fla Gainesville FL 32601

MOTT, JACK EDWARD, b Hammond, Ind, May 4, 37; m 59; c 2. ENGINEERING PHYSICS. Educ: Univ Chicago, MS, 60; Northwestern Univ, PhD(physics), 67. Prof Exp: Assoc scientist, Bettis Atomic Power Div, Westinghouse Elec Co, 60-63; res asst physics, Northwestern Univ, Evanston, 63-67, res assoc, 67-68; from asst prof to assoc prof, Indiana Univ, 68-75; SR PHYSICIST, ENERGY SYSTS & TECHNOL DIV, GEN ELEC CO, 75- Mem: Am Phys Soc; Sigma Xi; Am Nuclear Soc. Res: Fundamental particle research and teaching; nuclear physics research and teaching; nuclear reactor and energy systems research. Mailing Add: Energy Systs & Technol Div GE Co 310 DeGuinge Dr Sunnyvale CA 94086

MOTT, JOE LEONARD, b Linden, Tex, Apr 1, 37; m 60; c 2. ALGEBRA. Educ: E Tex Baptist Col, BS, 58; La State Univ, MS, 60, PhD(math), 63. Prof Exp: Asst prof math, Univ Kans, 63-65; assoc prof, 65-75, PROF MATH, FLA STATE UNIV, 75- Concurrent Pos: Vis prof, Mich State Univ, 73-74. Mem: Am Math Soc; Math Asn Am. Res: Ideal theory of commutative rings; partially ordered Abelian groups. Mailing Add: Dept of Math Fla State Univ Tallahassee FL 32306

MOTT, THOMAS, b Oswego, NY, Feb 14, 26; m 52; c 7. MATHEMATICS. Educ: Union Col, AB, 50; Univ Pa, AM, 52; Pa State Univ, PhD, 67. Prof Exp: Jr mathematician, Cornell Aeronaut Lab, Buffalo, NY, 52-53; instr math, Pa State Univ, Erie, 53; instr, Clarkson Inst Technol, 53-55; instr, Pa State Univ, Hazleton, 55-57, University Park, 57-62; assoc prof, State Univ NY Col Fredonia, 62-67; PROF MATH, STATE UNIV NY COL BUFFALO, 67- Concurrent Pos: Math Asn Am lectr, 70- Mem: Math Asn Am. Res: Analysis; integration theory; limit theorems. Mailing Add: Dept of Math State Univ NY Col Buffalo NY 14222

MOTTA, JEROME J, b Los Angeles, Calif, July 6, 33. MYCOLOGY. Educ: San Francisco State Col, AB, 58, MA, 64; Univ Calif, Berkeley, PhD(bot), 68. Prof Exp: Res plant pathologist, Univ Calif, Berkeley, 68-69; ASSOC PROF BOT, UNIV MD, COLLEGE PARK, 69- Mem: AAAS; Mycol Soc Am; Bot Soc Am. Res: Cytology and ultrastructure of fungi. Mailing Add: Dept of Bot Univ of MD College Park MD 20742

MOTTELER, ZANE CLINTON, b Wenatchee, Wash, July 4, 35; m 60; c 4. MATHEMATICS, COMPUTER SCIENCE. Educ: Stanford Univ, BS, 57, MA, 62, PhD(math), 64. Prof Exp: NSF fel, Univ Minn, 57-58 & Univ NMex, 58-59; res asst math, Los Alamos Sci Lab, Univ Calif, 58-60, staff mem, 60-65; from asst prof to assoc prof math, Gonzaga Univ, 65-72, chmn dept, 66-71; PROF & HEAD DEPT MATH, MICH TECHNOL UNIV, 72- Concurrent Pos: NSF vis scientist prog lectr, 60-61, 65-66; consult, Northwest Col & Univ Asn Sci, 70; nat lectr, Soc Indust & Appl Math, 70-72. Mem: Soc Indust & Appl Math; Math Asn Am. Res: Existence theory for non-linear elliptic partial differential equations of second order; quantitative and qualitative behavior of polynomials near roots; computer systems. Mailing Add: Dept of Math Mich Technol Univ Houghton MI 49931

MOTTER, ROBERT FRANKLIN, b Early, Iowa, June 29, 33; m 55; c 3. PHOTOGRAPHY, CHEMISTRY. Educ: Morningside Col, BS, 55; Univ Minn, PhD(org chem), 58. Prof Exp: Res chemist, Dow Chem Co, 55; asst, Univ Minn, 55-58; RES CHEMIST, EASTMAN KODAK CO, 58- Mem: Am Chem Soc; Soc Photog Sci & Eng; Soc Motion Picture & TV Engrs. Res: Organic synthesis in heterocyclics and sulfur chemistry. Mailing Add: 142 El Mar Dr Rochester NY 14616

MOTTERN, HENRY ORVILLE, organic chemistry, see 12th edition

MOTTET, N KARLE, b Renton, Wash, Jan 8, 24; m 52; c 3. PATHOLOGY, TERATOLOGY. Educ: Wash State Univ, BS, 47; Yale Univ, MD, 52; Am Bd Path, dipl, 57. Prof Exp: Instr physiol, Yale Univ, 51-52 & path, 55-59; from asst prof to assoc prof, 59-66, PROF PATH, SCH MED, UNIV WASH, 66-, DIR HOSP PATH, 59- Concurrent Pos: Nat Found Infantile Paralysis fel, Cambridge Univ, 52-53; USPHS trainee path, 54-55; pathologist & dir lab, Griffin Hosp, Derby, Conn, 55-59; vis scientist, Strangeways Res Lab, Eng, 69-70; consult ed, McGraw-Hill Encycl Sci & Technol & Yearbk Sci. Mem: Fel Col Am Path; Am Asn Path & Bact; fel Am Soc Clin Path; Am Soc Exp Path; NY Acad Sci. Res: Embryonic induction by connective tissue; metals and birth defects. Mailing Add: BB228 Univ Hosp Seattle WA 98195

MOTTINGER, JOHN P, b Detroit, Mich, Nov 28, 38. CYTOGENETICS. Educ: Ohio Wesleyan Univ, BA, 61, Indiana Univ, PhD(cytogenetics), 68. Prof Exp: Instr, 67-68, ASST PROF BOT, UNIV R I, 68- Mem: AAAS; Genetics Soc Am. Res: Cytogenetics of maize, particularly induced mutations and mutable systems. Mailing Add: Dept of Bot Univ of R I Kingston RI 02881

MOTTLEY, CAROLYN, b Palestine, Tex, Oct 29, 47. PHYSICAL CHEMISTRY. Educ: Wayland Col, BS, 69; Univ NC, Chapel Hill, PhD(chem), 73. Prof Exp: Res assoc chem, Univ Ala, Tuscaloosa, 74-75; ASST PROF CHEM, LUTHER COL, 75- Mem: Am Chem Soc; Am Phys Soc. Res: Study of molecular motion and structure through the use of magnetic resonance techniques. Mailing Add: Dept of Chem Luther Col Decorah IA 52101

MOTTO, HARRY LEE, soil chemistry, agronomy. see 12th edition

MOTTO, JEROME (ARTHUR), b Kansas City, Mo, Oct 16, 21. PSYCHIATRY. Educ: Univ Calif, AB, 48, MD, 51. Prof Exp: Intern, San Francisco Gen Hosp, 51-52; resident psychiat, Henry Phipps Psychiat Clin, Johns Hopkins Hosp, 52-55; sr resident, Langley Porter Neuropsychiat Inst, 55-56; from instr to asst prof, 56-64, lectr, 64-67, assoc clin prof, 67-69, assoc prof, 69-73, PROF PSYCHIAT, SCH MED, UNIV CALIF, SAN FRANCISCO, 73- Concurrent Pos: Attend psychiatrist, Langley Porter Neuropsychiat Inst, 56-; chief psychiat serv, San Francisco Gen Hosp, 71. Res: Clinical psychiatry. Mailing Add: Dept of Psychiat Univ of Calif Med Ctr San Francisco CA 94143

MOTTOLA, HORACIO ANTONIO, b Buenos Aires, Arg, Mar 22, 30; m 58; c 2. ANALYTICAL CHEMISTRY. Educ: Indust Tech, Indust Nat Sch, Arg, 49; Univ Buenos Aires, MS, 57, PhD(chem), 62. Prof Exp: Teaching asst chem, Univ Buenos Aires, 56-57, instr, 57-58, 60-63; res assoc, Univ Ariz, 63-64; asst prof, Elbert Covell Col, Univ of the Pac, 64-67; from asst prof to assoc prof, 67-75, PROF CHEM, OKLA STATE UNIV, 75- Concurrent Pos: Lectr, Univ Ariz, 66-67. Mem: Am Chem Soc; Sigma Xi. Res: Separation and determination of traces of metals; mechanisms of liquid-liquid distribution; chemistry of metal chelates; reaction rate methods; fast analyzers. Mailing Add: Dept of Chem Okla State Univ Stillwater OK 74074

MOTTS, WARD SUNDT, b Cleveland, Ohio, Oct 31, 24; m 51; c 2. GEOLOGY. Educ: Columbia Univ, BA, 49; Univ Minn, MS, 51; Univ Ill, PhD(geol), 57. Prof Exp: Geologist, US Bur Reclamation, 51-53; geologist, US Geol Surv, 53-60 & Okla Geol Surv, 60-61; ASSOC PROF GEOL, UNIV MASS, AMHERST, 61- Concurrent Pos: Asst, Univ Ill, 55-57. Mem: Fel Geol Soc Am; Am Geophys Union. Res: Hydrogeology; environmental geology; geomorphology; engineering geology. Mailing Add: Dept of Geol Univ of Mass Amherst MA 01003

MOTTUS, EDWARD HUGO, b Eckville, Alta, June 12, 22; m 45; c 3. ORGANIC CHEMISTRY. Educ: Univ Alta, BSc, 49; Univ Ill, PhD(chem), 52. Prof Exp: Teacher pub schs, Can, 41-42; asst, Univ Ill, 49-51; fel, Nat Res Coun Can, 52-53; res chemist, 53-67, scientist, 67-71, SR SCI FEL, MONSANTO CO, 71- Mem: AAAS; Am Chem Soc; The Chem Soc. Res: Electrolytic reduction of bicyclic aminoketones; polarographic reduction of diketones; lycopodine; protopine; polyethylene, ionic ringopening polymerizations; polyethers; polyamides; catalysis; polymer syntheses. Mailing Add: 850 Claymont Dr Ballwin MO 63011

MOTULSKY, ARNO GUNTHER, b Fischhausen, Ger, July 5, 23; nat US; m 45; c 3. INTERNAL MEDICINE, MEDICAL GENETICS. Educ: Univ Ill, BS, 45, MD, 47. Prof Exp: Res assoc internal med, Sch Med, George Washington Univ, 52-53; from instr to assoc prof, 53-61, PROF MED & GENETICS, SCH MED, UNIV WASH, 61- Concurrent Pos: Clin investr, Army Med Serv Grad Sch, Walter Reed Army Med Ctr, DC, 52-53; attend physician, King County & Vet Admin Hosps, Seattle, 54-; consult, Madigan Army Hosp, Tacoma, 55-74; Commonwealth Fund fel, Univ London, 57-58; Markle scholar, 57-62; mem subcomt transfusion probs, Nat Res Coun, 58-63; attend physician, Univ Wash Hosp, 59-; mem human ecol study sect, NIH, 61-65 & hemat study, 69-72; mem, US Panel Methods Eval Environ Mutagenesis & Carcinogenesis, Nat Inst Allergy & Infectious Dis, 72- Mem: Nat Acad Sci; Nat Inst Med; Asn Am Physicians; fel Am Col Physicians; Am Fedn Clin Res. Res: Role of genetic factors in disease etiology; genetics of coronary heart disease; hereditary hemolytic anemias; abnormal hemoglobins; genetics of drug reaction and response; human population genetics. Mailing Add: Div of Med Genetics Univ of Wash Sch of Med Seattle WA 98195

MOTZ, HENRY THOMAS, b St Louis, Mo, June 10, 23; m 47; c 3. NUCLEAR PHYSICS. Educ: Yale Univ, BS, 44, MS, 48, PhD, 49. Prof Exp: Physicist, Brookhaven Nat Lab, 49-56; physicist, 56-61, group leader res reactor group, 61-65, assoc physics div leader, 65-71, PHYSICS DIV LEADER, LOS ALAMOS SCI LAB, 71- Concurrent Pos: Res fel, Univ Zurich, 53-54; guest lectr, Netherlands-Norweg Reactor Sch, 63; secy & chmn nuclear cross sect adv comt & controlled thermonuclear res standing comt, USAEC; mem, Europ-Am Nuclear Data Comt; adv to US mem, Int Nuclear Data Comt, Int Atomic Energy Agency & US mem, Nuclear Energy Agency-Nuclear Data Comt, 73-; Univ Calif contractor to Energy Res & Develop Admin. Mem: Am Phys Soc; NY Acad Sci. Res: Cyclotron bombardment; si reactions; slow neutron capture gamma rays. Mailing Add: Los Alamos Sci Lab PO Box 1663 Los Alamos NM 87545

MOTZ, JOSEPH WILLIAM, b Binghamton, NY, Nov 11, 18; m 45; c 2. PHSYICS.

Educ: Univ Wis, BS, 41; Cornell Univ, MS, 42; Ind Univ, PhD(physics), 49. Prof Exp: Physicist, Armour Res Found, 43-46; PHYSICIST, X-RAY DIV, NAT BUR STAND, 49- Mem: Fel Am Phys Soc. Res: Radiation physics; photon and electron scattering processes. Mailing Add: 11306 Cushman Rd Rockville MD 20852

MOTZ, KAYE LA MARR, b Bluffton, Ind, Aug 10, 32; m 59. INDUSTRIAL ORGANIC CHEMISTRY. Educ: Univ Colo, BA, 54; Univ Ill, PhD(chem), 58. Prof Exp: Instr chem, Mich State Univ, 58-59; res assoc, Univ Mich, 59-60; RES ASSOC, CONTINENTAL OIL CO, 60- Mem: Am Chem Soc; The Chem Soc. Res: Reactions of aluminum alkyls; antioxidants; tertiary oil recovery. Mailing Add: Res & Develop Dept Continental Oil Co Drawer 1267 Ponca City OK 74602

MOTZ, LLOYD, b Susquehanna, Pa, June 5, 10; m 34; c 2. ASTROPHYSICS, NUCLEAR PHYSICS. Educ: City Col NY, BS, 30; Columbia Univ, PhD(physics), 36. Prof Exp: Instr physics, City Col New York, 31-40; dir res & optical design, Dome Precision Corp, NY, 42-46; dir, Park Instrument Co, NJ, 46-49; from asst prof to assoc prof, 50-62, PROF ASTRON, COLUMBIA UNIV, 62- Concurrent Pos: Lectr, Columbia Univ, 35-49; adj prof, Polytech Inst Brooklyn, 50-; consult, AMF, Inc, Grumman Aerospace Corp, Polarad Electronics Corp & Razdow Labs; mem bd dirs, Geosci Instrument Corp & Thexon Corp. Honors & Awards: Award, Gravity Res Found, 60; Boris Pregel Award in Astron & Physics, NY Acad Sci, 72. Mem: AAAS; fel Am Phys Soc; Am Astron Soc; Royal Astron Soc; NY Acad Sci (pres, 70-71). Res: Internal constitution of stars; unified field theory; design of optical instruments; geometrical optics; structure of elementary particles. Mailing Add: Dept of Astron Columbia Univ New York NY 10027

MOTZ, ROBIN OWEN, b New York, NY, Mar 9, 39; m 59. PLASMA PHYSICS, INTERNAL MEDICINE. Educ: Columbia Univ, AB, 59, AM, 60, PhD(physics), 65, MD, 75. Prof Exp: Asst astron, Columbia Univ, 58-63; lectr physics, City Col New York, 63-65; asst prof, Stevens Inst Technol, 65-71; lectr, Columbia Univ, 71-75; MED HOUSE STAFF MEM, DEPT MED, PRESBY HOSP, NEW YORK, 75- Concurrent Pos: Astron ed, Am Oxford Encyclop, 62-63; assoc ed, Am J Physics, 69-72. Mem: Fel NY Acad Sci; AAAS; Am Phys Soc; AMA. Res: Magnetohydrodynamic drag; Alfven waves; plasma radiation; optical diagnostics. Mailing Add: Dept of Med Presby Hosp 622 W 168th St New York NY 10032

MOTZKIN, SHIRLEY M, b New York, NY, Jan 12, 27; m 52; c 3. ANATOMY, DEVELOPMENTAL BIOLOGY. Educ: Brooklyn Col, BS, 47; Columbia Univ, AM, 49; NY Univ, PhD(anat), 58. Prof Exp: Instr biol, Brooklyn Col, 47-52; instr histol, NY Univ, 51-59, asst prof, 59-66; assoc prof biol, 66-73, DIR PRE-MED CURRICULUM, POLYTECH INST NEW YORK, 66- Concurrent Pos: Adj instr & prof, Brooklyn Col, 52-; guest lectr, Guggenheim Dent Clin, 60-; partic, interdisciplinary prog, NIH basic res prog, 62-66. Mem: Sigma Xi; NY Acad Sci; Int Asn Dent Res. Res: Development teratologic studies of bone formation, palate development and tissue interactions using biochemical embryonics, histological, cytological, histochemical, cytochemical and radioisotopic techniques. Mailing Add: Polytech Inst of New York 333 Jay St Brooklyn NY 11201

MOTZOK, ILARY, b Stoney Mountain, Man, Oct 11, 11; m 48. NUTRITION. Educ: Ont Agr Col, BSA, 36; Univ Toronto, MSA, 39, PhD(biochem), 46. Prof Exp: Instr chem & asst nutrit, 39-44, from asst prof to assoc prof, 45-62, PROF NUTRIT, COL BIOL SCI, UNIV GUELPH, 62- Mem: Am Soc Animal Sci; Animal Nutrit Res Coun; Am Inst Nutrit; Can Biochem Soc; Can Soc Animal Prod. Res: Enzymology; minerals in nutrition. Mailing Add: Dept of Nutrit Col of Biol Sci Univ Guelph Guelph ON Can

MOU, THOMAS WILLIAM, b Philadelphia, Pa, May 17, 20; m 45; c 2. MEDICINE, PREVENTIVE MEDICINE. Educ: Philadelphia Col Pharm, BS, 41; Univ Rochester, MD, 50. Prof Exp: Intern & resident internal med, Univ Rochester, 50-53, instr med & microbiol, Sch Med, 54-56; asst prof, 56-61, assoc prof prev med, 61-68, PROF COMMUNITY HEALTH, STATE UNIV NY UPSTATE MED CTR, 68-, PROVOST HEALTH SCI, STATE UNIV NY CENT ADMIN, 70- Concurrent Pos: Mem bd trustees, Philadelphia Col Pharm & Sci; res fel infectious dis, Thorndike Lab, Harvard Med Sch, 53-54. Mem: Am Fedn Clin Res; Infectious Dis Soc. Res: Infectious disease, especially preventive aspects; urinary tract infections; respiratory infections; toxoplasmic uveitis; rheumatic fever prophylaxis. Mailing Add: 99 Washington Ave Albany NY 12210

MOUALIN, RICHARD JOSEPH, chemistry, see 12th edition

MOUK, ROBERT WATTS, b Trenton, NJ, June 23, 40; m 65. POLYMER CHEMISTRY, SYNTHETIC ORGANIC CHEMISTRY. Educ: Wittenberg Univ, BS, 63; Bowling Green State Univ, MA, 67; Mich State Univ, PhD(chem), 70. Prof Exp: SR RES CHEMIST, ASHLAND CHEM CO, ASHLAND OIL INC, 69- Mem: Am Chem Soc. Res: Emulsion polymerization; adhesives. Mailing Add: Res & Develop Lab Ashland Chem Co PO Box 2219 Columbus OH 43216

MOULD, RICHARD A, b Reading, Pa, Mar 4, 27. THEORETICAL PHYSICS. Educ: Lehigh Univ, BS, 51; Yale Univ, MS, 55, PhD(physics), 57. Prof Exp: Asst prof, 57-64, ASSOC PROF PHYSICS, STATE UNIV NY STONY BROOK, 64-, MASTER, LEARNED HAND COL, 70- Concurrent Pos: Master, Washington Irving Col, 68, dir . 68. Mem: Am Asn Physics Teachers. Res: General relativ ity; quantum theory of measurements. Mailing Add: Dept of Physics State Univ of NY Stony Brook NY 11790

MOULD, RICHARD EVERETT, b Toledo, Ohio, Mar 29, 26; m 49; c 3. SOLID STATE PHYSICS, GLASS TECHNOLOGY. Educ: Mass Inst Technol, SB, 48; Univ Ill, MS, 52, PhD(physics), 54. Prof Exp: Res physicist, Preston Labs, Inc, 48-51, chief physicist, 54-57, vpres res, Am Glass Res, Inc, 57-60, pres, 61-72, CHMN BD, AM GLASS RES, INC, 73- Mem: Am Ceramic Soc; Am Phys Soc; Am Soc Test & Mat; Brit Soc Glass Technol. Res: chemical, physical and mechanical properties of glass; strength of brittle materials; cryogenics and superconductivity. Mailing Add: Am Glass Res Inc Box 149 Butler PA 16001

MOULDER, JAMES WILLIAM, b Burgin, Ky, Mar 28, 21; m 42; c 4. MICROBIOLOGY. Educ: Univ Chicago, SB, 41, PhD(biochem), 44. Prof Exp: Res assoc malaria, Off Sci Res & Develop Proj, 44-45, Logan fel, 46, instr biochem, Dept Microbiol & Biochem, 46-47, from asst prof to assoc prof, 47-57, chmn dept, 60-69, PROF MICROBIOL, UNIV CHICAGO, 57- Concurrent Pos: Fulbright scholar & Guggenheim fel, Oxford Univ, 52-53; Ciba lectr microbial biochem, 63; ed, J Infectious Dis, 57-68. Honors & Awards: Lilly Award, 54. Mem: AAAS; Am Soc Biol Chem; Am Soc Microbiol; Am Acad Microbiol. Res: Biochemistry of intracellular parasitism. Mailing Add: Dept of Microbiol Univ of Chicago Chicago IL 60637

MOULDER, JERRY WRIGHT, b Bowling Green, Ky, Sept 2, 42; m 67; c 1. PHYSICS. Educ: Western Ky Univ, BS, 64; Univ Tenn, PhD(physics), 70. Prof Exp: Asst prof physics, WVa Inst Technol, 70-75; ASST PROF PHYSICS, TRI-STATE COL, 75- Concurrent Pos: NSF Acad Yr exten grant, WVa Inst Technol, 71-73. Mem: Am Asn Physics Teachers. Res: The interaction of low energy K mesons with nuclei. Mailing Add: Dept of Physics Tri-State Col Angola IN 46703

MOULDER, PETER VINCENT, JR, b Jackson, Mich, Jan 26, 21; m 46; c 4. CARDIOVASCULAR SURGERY. Educ: Univ Notre Dame, BS, 42; Univ Chicago, MD, 45; Am Bd Surg, dipl, 54; Am Bd Thoracic Surg, dipl, 56. Hon Degrees: MS, Univ Pa, 72. Prof Exp: Intern surg, Univ Chicago Clins, 45-46, resident gen surg, 48-51; resident, Univ Ill, 52; from instr to prof surg, Univ Chicago Clins, 52-68; prof thoracic & cardiovasc surg & dir dept, Pa Hosp, Sch Med, Univ Pa, 68-72; PROF THORACIC & CARDIOVASC SURG, UNIV FLA, 72- Concurrent Pos: Resident thoracic surg, Univ Chicago Clins, 52-53, chief resident surgeon, 53-54, secy dept surg, 59-64; consult & lectr, Great Lakes Naval Hosp, Chicago, 59-68; consult, Philadelphia Naval Hosp, 68-; med investr, Vet Admin Hosp, 73-74 & 76-82. Mem: Am Physiol Soc; Am Asn Thoracic Surg; Am Surg Asn; Am Col Surg; Soc Clin Surg. Res: Biochemical and physiological studies on the heart; pulmonary hypertension; myocardial hypertrophy; time series analysis of cardiovascular phenomena; computer science. Mailing Add: Div Thoracic & Cardiovasc Surg Univ of Fla & Vet Admin Hosp Gainesville FL 32610

MOULDS, GORDON MARS, chemistry, see 12th edition

MOULE, DAVID, b Hamilton, Ont, Nov 17, 33; m 62; c 2. PHYSICAL CHEMISTRY, MOLECULAR SPECTROSCOPY. Educ: McMaster Univ, BSc, 58, PhD(chem), 62. Prof Exp: Asst res officer chem, Atomic Energy Comn Can, 64-66; asst prof, 66-70, ASSOC PROF CHEM, BROCK UNIV, 70- Concurrent Pos: NATO fel, 62-64. Res: spectroscopy of polyatomic molecules in excited electronic states; isotope equilibria. Mailing Add: Dept of Chem Brock Univ St Catherines ON Can

MOULIS, EDWARD JEAN, JR, b Natchitoches, La, July 26, 40; m 67; c 2. MATHEMATICS. Educ: Harvard Univ, BA, 62; Univ Del, MS, 67, PhD(math), 71. Prof Exp: Instr math, US Navy Nuclear Power Sch, Md, 62-66; asst prof, Frostburg State Col, 71-75; ASST PROF MATH, US NAVAL ACAD, 75- Mem: Math Asn Am; Am Math Soc. Res: Complex analysis; univalent function theory; conformal mapping. Mailing Add: Dept of Math US Naval Acad Annapolis MD 21402

MOULT, ROY HEPWORTH, b Long Island, NY, Nov 18, 13; m 42; c 4. ORGANIC POLYMER CHEMISTRY. Educ: Univ Mass, BS, 38. Prof Exp: Anal res chemist, Colgate-Palmolive-Peet Co, 40-42; res chemist, Stamford Res Lab, Am Cyanamid Co, 43-46; res chemist, Everett Res Lab, Monsanto Chem Co, 46-49; anal chemist, Lever Bros Co, 49-50; sr chemist, Chem Div, 50-52, lab supvr resins & adhesives develop, 52-55, sr scientist, Res Dept, 55-65, sr proj leader resins & adhesives, 65-69, GROUP MGR, RESINS & ADHESIVES RES, RES DEPT, KOPPERS CO INC, 69- Mem: Am Chem Soc. Res: Synthetic resin chemistry, resorcinol, phenolic, epoxides; adhesion; adhesives for wood, rubber, metals and textiles; tire cord bonding; reinforcement of rubber. Mailing Add: Koppers Co Inc 440 College Park Dr Monroeville Res Ctr Monroeville PA 15146

MOULTHROP, PETER HILL, b Berkeley, Calif, Apr 7, 26; div; c 4. ENERGY CONVERSION. Educ: Univ Calif, AB, 50, PhD(physics), 55. Prof Exp: Physicist, Lawrence Berkeley Lab, 50-55 & Lawrence Livermore, Lab, 55-67, A-Div leader, 67-73, asst assoc dir, 73-74, PHYSICIST, LAWRENCE LIVERMORE LAB, UNIV CALIF, 74- Mem: AAAS; Am Phys Soc; Sigma Xi. Res: Lasers and neutron physics; astrophysics; hydrodynamics; energy and resurce planning. Mailing Add: 4128 Graham Pleasanton CA 94566

MOULTON, BENJAMIN, b Medford, Mass, Aug 25, 17; m 44; c 3. GEOGRAPHY, GEOLOGY. Educ: Clark Univ, AB, 39; Butler Univ, MS, 41; Indiana Univ, MA, 45, PhD(geog, geol), 50. Prof Exp: Instr geog, Butler Univ, 43-44, from asst prof to assoc prof, 48-55; asst prof, Fla State Univ, 46-47; instr, Case Western Reserve Univ, 47-48; Ballenger chair earth sci, Flint Community Jr Col, 55-61; PROF GEOG & CHMN DEPT GEOG & GEOL, IND STATE UNIV, 61- Mem: Asn Am Geog. Res: Climatology; geomorphology. Mailing Add: Dept of Geog & Geol Indiana State Univ Terre Haute IN 47809

MOULTON, BRUCE CARL, b Oneida, NY, Nov 6, 40; m 67; c 2. ENDOCRINOLOGY, BIOCHEMISTRY. Educ: Hamilton Col, AB, 62; Cornell Univ, MS, 65, PhD(endocrinol), 68. Prof Exp: Fel, Reprod Biol Training Prog, Col Med, Univ Nebr, Omaha, 68-70, res asst prof obstet-gynec & biochem, 70; asst prof obstet-gynec & biol chem, 70-75, ASSOC PROF OBSTET-GYNEC, COL MED, UNIV CINCINNATI, 75- Mem: AAAS; Endocrine Soc; Soc Study Reprod; Am Physiol Soc. Res: Biochemical mechanisms of hormone action in reproductive physiology. Mailing Add: Dept of Obstet-Gynec Univ of Cincinnati Col of Med Cincinnati OH 45267

MOULTON, CHARLES WESLEY, physical chemistry, see 12th edition

MOULTON, DAVID GILLMAN, b Bombay, India, Nov 29, 28; m 55; c 3. PHYSIOLOGY. Educ: Glasgow Univ, BSc, 54; Univ Birmingham, PhD(anat), 58. Prof Exp: Thomas Welton Stanford fel psychol, Stanford Univ, 58-59; res assoc physiol, Fla State Univ, 60-63, vis assoc prof, 63-65; from assoc prof to prof, Clark Univ, 65-69; ASSOC PROF PHYSIOL, SCH MED & MEM MONELL CHEM SENSES CTR, UNIV PA, 69-; RES PHYSIOLOGIST, VET ADMIN HOSP, 69- Mem: AAAS; Brit Soc Exp Biol; Am Physiol Soc; Brit Inst Biol; Soc Neurosci. Res: Olfaction; sensory physiology; neurophysiology. Mailing Add: Monell Chem Senses Ctr Univ of Pa Philadelphia PA 19104

MOULTON, GEORGE HERBERT, b Minn, Nov 25, 12; m 34; c 2. DENTISTRY. Educ: Gustavus Adolphus Col, BA, 38; Univ Minn, DDS, 35. Prof Exp: Chief crown, bridge & operative dent, Walter Reed Army Med Ctr, Washington, DC, 48-53; pvt pract, Tex, 54-55; PROF CROWN & BRIDGE DENT & CHMN DEPT, SCH DENT, EMORY UNIV, 55-, DIR CLINS, 58-, DEAN, 61- Concurrent Pos: Consult, US Army, Ft Benning, Ga, 58- Mem: Am Dent Asn; fel Am Col Dent. Res: Occlusion of the natural dentition. Mailing Add: Dept of Crown & Bridge Dent Emory Univ Sch of Dent Atlanta GA 30322

MOULTON, GRACE CHARBONNET, b New Orleans, La, Nov 1, 23; m 47; c 2. BIOPHYSICS, EXPERIMENTAL SOLID STATE PHYSICS. Educ: Tulane Univ, BA, 44; Univ Ill, MS, 48; Univ Ala, PhD(physics), 62. Prof Exp: Asst biophys, Univ Ill, 50-52; physicist, Argonne Cancer Res Hosp, Ill, 52; asst physics, Univ Ala, 59-61, asst prof, 61-65; asst prof, 65-74, ASSOC PROF PHYSICS, FLA STATE UNIV, 74- Mem: Am Phys Soc. Res: Radiation effects in materials of biological importance with emphasis on the mechanisms involved, as studied by electron spin resonance and electron nuclear double resonance. Mailing Add: Dept of Physics Fla State Univ Tallahassee FL 32306

MOULTON, JACK E, b Seattle, Wash, Mar 4, 22; m 49; c 1. VETERINARY PATHOLOGY. Educ: Wash State Univ, BS, 47, DVM, 49; Univ Minn, PhD(path), 53. Prof Exp: Instr vet path, Univ Minn, 49-52; from asst prof to assoc prof, 52-64,

PROF VET PATH, COL VET MED, UNIV CALIF, DAVIS, 64- Mem: Am Vet Med Asn. Res: Immunopathology of trypanosomiasis. Mailing Add: Dept of Vet Path Univ of Calif Davis CA 95616

MOULTON, JAMES EDWARD, b Evanston, Ill, Apr 6, 13; m 42; c 5. BOTANY. Educ: Northwestern Univ, BS, 36; Univ Chicago, PhD(bot), 41. Prof Exp: Instr bot, Univ Mo, 42-43, physics, 43-44; from asst prof to assoc prof, 44-70, PROF HORT, MICH STATE UNIV, 70- Mem: Am Soc Hort Sci; Am Pomol Soc. Res: Fruit breeding. Mailing Add: Dept of Hort Mich State Univ East Lansing MI 48823

MOULTON, JAMES FRANK, JR, b Wash, DC, Nov 9, 21; m 44; c 2. RESEARCH ADMINISTRATION. Educ: Georgetown Univ, BS, 43. Prof Exp: Res physicist, Underwater & Air Explosion Effects, US Dept Navy, 43-46, sr res assoc, Shock Wave Phenomena in Air, Naval Ord Lab, 46-58, chief, Air-Ground Explosions Div, 58-65, chief, Naval Effects Br, Defense Atomic Support Agency, 65-67, CHIEF, AEROSPACE SYSTS DIV, DEFENSE NUCLEAR AGENCY, 67- Concurrent Pos: Sci consult, Energy Res & Develop Admin, Dept Navy, Armed Serv Explosives Safety Bd & Nat Mat Adv Bd; mem working group, S2-54 Atmospheric Blast Effects, Am Nat Standards Comt, 71- Honors & Awards: Newmann Award, 40; Meritorious Civilian Serv Award, US Navy, 51, 59; Sustained Superior Performance, Defense Nuclear Agency, 74. Mem: AAAS; Am Phys Soc. Res: Detection and measurement of blast and shock phenomena in high explosive and nuclear explosion environments; impulsive irradiation and response of aerospace systems materials and structures; research and development resource management. Mailing Add: 4105 Glenrose St Kensington MD 20795

MOULTON, JAMES MALCOLM, b West Haven, Conn, July 25, 21; m 49; c 3. VERTEBRATE MORPHOLOGY. Educ: Univ Mass, BS, 47; Harvard Univ, MA, 50, PhD(zool), 52. Prof Exp: Asst biol, Williams Col, 47-48; instr, Brown Univ, 51; instr anat, Sch Med, Johns Hopkins Univ, 51-52; from instr to assoc prof, 52-65, actg chmn dept, 59-60 & 66-67, chmn, 70-73, PROF BIOL, BOWDOIN COL, 65-, PREMED ADV, 69- Concurrent Pos: Res assoc, Woods Hole Oceanog Inst, 55-; NSF grant, 57-63; Fulbright res scholar & Guggenheim fel, Univ Queensland, 60-61; mem conf animal orientation, Univ Munich, 62; Wenner-Gren Conf Animal Communication, 65; partic, Southeast Pac Biol Oceanog Prog, cruise 18A of Anton Bruun, 66; vis scientist, Inst Animal Genetics, Edinburgh Univ, 67; vis prof, Mus Comp Zool, Harvard Univ, 74. Mem: Fel AAAS; Soc Ichthyl & Herpet; Am Soc Zool; Am Fisheries Soc; fel Am Inst Fishery Res Biologists. Res: Animal morphology and development; acoustical biology of animals. Mailing Add: Dept of Biol Bowdoin Col Brunswick ME 04011

MOULTON, JOHN MAXIM, b Camp Hill, Pa, Dec 28, 05; m 32; c 3. ZOOLOGY. Educ: Bates Col, AB, 28; Univ Ill, MA, 30. Prof Exp: From instr to assoc prof biol, 30-37, instr geog, 37-46, PROF BIOL & GEOG, HASTINGS COL, 46-, CHMN DEPT GEOG & GEOL, 59- Mem: Asn Am Geogrs; Nat Coun Geog Educ. Res: Land utilization; anatomy of nemathelminthes; irrigation in the high plains of Kansas and Nebraska. Mailing Add: Dept of Geog & Geol Hastings Col Hastings NE 68901

MOULTON, WILBUR NORTON, b Winner, SDak, June 16, 26; m 49, 72; c 3. ORGANIC CHEMISTRY. Educ: Sioux Falls Col, BS, 49; Univ Minn, MS, 52, PhD(org chem), 54. Prof Exp: Cereal chemist, Pillsbury Mills, Inc, 49-50; instr chem, Col St Thomas, 53-54; asst prof, Morningside Col, 54-56; from asst prof to prof, Southern Ill Univ, Carbondale, 56-72, asst dean col liberal arts & sci, 63-66, assoc dean int serv div, 66-67, dean students, 67-71; ASST TO PRES, SANGAMON STATE UNIV, 72- Concurrent Pos: Smith-Mundt vis lectr, Univ Baghdad, 61-62; Ellis L Phillips Found intern acad admin, Brown Univ, 65-66. Mem: AAAS. Res: Organic synthesis; reaction mechanisms. Mailing Add: Off of the Pres Sangamon State Univ Springfield IL 62708

MOULTON, WILLIAM G, b Waverly, Ill, Jan 4, 25; m 47; c 2. SOLID STATE PHYSICS, LOW TEMPERATURE PHYSICS. Educ: Western Ill State Col, BS, 46; Univ Ill, MS, 48, PhD(physics), 52. Prof Exp: Asst physics, Univ Ill, 46-51; instr, Chicago Div, Univ Ill, 51-53, asst prof, 53-56; from asst prof to prof, Univ Ala, 56-65; PROF PHYSICS, FLA STATE UNIV, 65- Concurrent Pos: Consult, Army Missile Command, 61-64; consult & dir, Recon Inc, 65- Mem: Am Phys Soc. Res: Superconductivity; magnetic resonance; magnetic ordered states; solid state and low temperature physics. Mailing Add: Dept of Physics Fla State Univ Tallahassee FL 32306

MOULTRIE, FRED, genetics, see 12th edition

MOUNIB, M SAID, b Cairo, Egypt, July 5, 29; m 66; c 2. REPRODUCTIVE PHYSIOLOGY, ENDOCRINOLOGY. Educ: Univ Alexandria, BSc, 50, MSc, 54; Aberdeen Univ, PhD(physiol), 56. Prof Exp: Teaching asst endocrinol, Univ Alexandria, 50-54, from asst prof to assoc prof reproductive physiol, 56-64; SR SCIENTIST, ENDOCRINOL SECT, FISHERIES RES BD CAN, 64- Concurrent Pos: Res fel reproductive physiol, Worcester Found Exp Biol, 62-63, Population Coun res fel, 63-64. Mem: Fel AAAS; Am Physiol Soc; Endocrine Soc; Soc Study Reproduction; NY Acad Sci. Res: Estrous cycle in sheep; establishing biochemical and physiological differences between sheep of different levels of potassium and sodium in their erythrocytes; metabolic changes in sperm after irradiation and capacitation; metabolism of eggs, sperm, developing embryo, endometrium, fallopian tube, ovaries and testes; freezing of sperm. Mailing Add: Environ Can Halifax Lab Res & Develop Directorate Box 429 Halifax NS Can

MOUNT, BENJAMIN HARRISON, JR, mathematics, see 12th edition

MOUNT, BERTHA LAURITZEN, b Valparaiso, Ind, Mar 26, 40; m 66; c 2. MATHEMATICS. Educ: Carleton Col, BA, 62; Northwestern Univ, MA, 64, PhD(math), 70. Prof Exp: Mathematician, Rock Island Arsenal, 62; SPECIALIST, PROJ SPEC ELEM EDUC FOR DISADVANTAGED, INC, 75- Mem: Sigma Xi. Mailing Add: 2705 Noyes St Evanston IL 60201

MOUNT, DAVID WILLIAM ALEXANDER, b Bromley, Eng, Jan 15, 38; m 60; c 2. GENETICS, MOLECULAR BIOLOGY. Educ: Univ Alta, BSc, 60; Univ Toronto, MA, 63, PhD(med biophys), 66. Prof Exp: ASST PROF MICROBIOL, COL MED, UNIV ARIZ, 69- Concurrent Pos: Fel genetics, Univ Alta, 67-68; USPHS fel molecular biol, Univ Calif, Berkeley, 68-69; lectr, Univ Ottawa, 66, Carleton Univ, 66-67 & Univ Alta, 67-68. Mem: AAAS; Genetics Soc Am; Am Soc Microbiol. Res: Genetics of bacteria and bacterial viruses; radiation biology; biophysics. Mailing Add: Dept of Microbiol Univ of Ariz Col of Med Tucson AZ 85274

MOUNT, DONALD I, b Miamisburg, Ohio, Sept 20, 31; m 53; c 2. FISH BIOLOGY. Educ: Ohio State Univ, BS, 53, MS, 57, PhD(zool, fish physiol), 60. Prof Exp: Fisheries res biologist, R A Taft Sanit Eng Ctr, Fed Water Pollution Control Admin, 60-67, DIR NAT WATER QUAL LAB, US ENVIRON PROTECTION AGENCY, 67- Honors & Awards: Superior Serv Award, US Dept Health, Educ & Welfare, 65; Gold Medal, US Environ Protection Agency, 73. Mem: AAAS; Am Fisheries Soc; Water Pollution Control Fedn; Inst Fishery Res Biologists. Res: Fish toxicology and physiology; effects of water pollution on fishes, especially the chronic effects of pollutants. Mailing Add: Nat Water Qual Lab 6201 Congdon US Environ Protection Agency Duluth MN 55804

MOUNT, GARY A, b Bristow, Okla, Oct 8, 36; m 63, 73; c 2. MEDICAL ENTOMOLOGY. Educ: Okla State Univ, BS, 58, MS, 60, PhD(entom), 63. Prof Exp: RES ENTOMOLOGIST, INSECTS AFFECTING MAN LAB, USDA, 63- Mem: Entom Soc Am; Am Mosquito Control Asn. Res: Control methods for mosquitoes, biting flies, ticks and chigger mites. Mailing Add: USDA PO Box 14565 Gainesville FL 32604

MOUNT, JOSEPH F, mathematics, biology, see 12th edition

MOUNT, KENNETH R, b Champaign, Ill, Apr 29, 33; m 66; c 1. MATHEMATICS. Educ: Univ Ill, BA, 54, MA, 55; Univ Calif, Berkeley, PhD(math), 60. Prof Exp: From instr to asst prof, 60-66, ASSOC PROF MATH, NORTHWESTERN UNIV, ILL, 66- Concurrent Pos: NSF res grant, France, 64-65; NATO grant, 74. Mem: Am Math Soc. Res: Algebraic geometry and commutative algebra; economics. Mailing Add: Dept of Math Northwestern Univ Evanston IL 60201

MOUNT, LESTER ADRAN, b Lebanon, Ohio, Mar 23, 10; m 34; c 3. NEUROSURGERY. Educ: Univ Cincinnati, BS, 32, BM, 34, MD, 35. Prof Exp: Intern, Henry Ford Hosp, 34-35, resident, 35-36, resident neurosurgeon, 36-38; asst resident neurologist, Neurol Inst, Presby Hosp, 38-39, from asst resident neurosurgeon to resident neurosurgeon, 39-41; asst neurol, 41-43, instr, 45-47, assoc, 47-52, from asst prof to assoc prof clin neurol surg, 52-70, PROF CLIN NEUROL SURG, COL PHYSICIANS & SURGEONS, COLUMBIA UNIV, 70- Concurrent Pos: Asst neurosurgeon, Vanderbilt Clin, Neurol Inst, Presby Hosp, 41-45, from asst attend neurosurgeon to assoc attend neurosurgeon, 45-60, attend neurosurgeon, 60-; chief neurosurg & mem med bd, Proj Hope, 65; founding mem, trustee, vpres & mem exec comt, Found Int Educ Neurosurg. Mem: Neurosurg Soc Am (secy, 53-56, vpres, 56-57, pres, 61-62); Am Asn Neurol Surgeons (pres elect, 75-76, pres, 76-77); Soc Neurol Surg (vpres, 69-70); hon mem Soc Peruvian Psychiat Neurol & Neurosurg; fel NY Acad Sci. Res: Intracranial aneurysms; premature closure of sutures of cranial vault; craniophyaryngioma; pituitary tumors; primary suprasellar teratoma. Mailing Add: Neurol Inst 710 W 168th St New York NY 10032

MOUNT, LLOYD GORDON, b Central Square, NY, Mar 29, 16; m 41; c 3. INDUSTRIAL ORGANIC CHEMISTRY. Educ: Cornell Univ, AB, 37; Yale Univ, PhD(org chem), 40. Prof Exp: Res chemist, Calco Chem Div, Am Cyanamid Co, NJ, 40-41, group leader process develop, 45-51; chief, Pyrotech Res & Develop Sect, Picatinny Arsenal, 51-52; tech liason mkt res, Chemstrand Corp, 52-55; commercial develop, Food Mach & Chem Corp, 55-58; head chem res & develop dept, Vitro Labs, 58-60; pres, Carnegies Fine Chem of Kearny, NJ, 60-62; dir bus res, Thiokol Chem Corp, 62-65; vpres planning & develop, Clarkson Col Technol, 65-67; dir res & develop, 67-68, tech dir, 68-71, dir mfg, 71-72, VPRES PROD, NEASE CHEM CO, INC, 72- Concurrent Pos: Lectr, Rutgers Univ, 46-47. Mem: Am Chem Soc; Chem Mkt Res Asn; Am Inst Chem. Res: Sulfa compounds; vat dyes; military explosives; synthetic fibers; planning, development and marketing research. Mailing Add: Nease Chem Co Inc Box 221 State College PA 16801

MOUNT, MARK SAMUEL, b Crawfordsville, Ind, Nov 18, 40; m 63; c 1. PLANT PATHOLOGY, PLANT PHYSIOLOGY. Educ: Ill Wesleyan Univ, BS, 63; Mich State Univ, MS, 65, PhD(bot, plant path), 68. Prof Exp: Res assoc, Cornell Univ, 68-69; ASST PROF PLANT PATH, UNIV MASS, AMHERST, 69- Concurrent Pos: NSF res grant, Univ Mass, Amherst, 70-72, NIH res grant, 73-76. Mem: Am Phytopath Soc; Am Soc Plant Physiol. Res: Physiology of plant disease development; nucleic acid metabolism in diseased plants. Mailing Add: Dept of Plant Path Univ of Mass Amherst MA 01002

MOUNT, RAMON ALBERT, b Lohrville, Iowa, May 4, 39; m 62; c 2. ORGANIC CHEMISTRY. Educ: Ariz State Univ, BS, 61; Mich State Univ, PhD(org chem), 67; St Louis Univ, MBA, 72. Prof Exp: Chemist, Dow Chem Co, 61-64; res specialist, 67-75, SR RES SPECIALIST, MONSANTO CO, 75- Mem: Am Chem Soc. Res: Catalytic oxidation of hydrocarbons; carbonium ion reaction mechanisms; industrial process research in organic chemistry. Mailing Add: 4810 Broad Oak Dr St Louis MO 63128

MOUNT, ROBERT HUGHES, b Lewisburg, Tenn, Dec 25, 31; m 61; c 2. VERTEBRATE ZOOLOGY. Educ: Auburn Univ, BS, 54, MS, 56; Univ Fla, PhD(biol), 61. Prof Exp: From asst prof to assoc prof biol, Ala Col, 61-66; assoc prof, 66-72, PROF ZOOL, AUBURN UNIV, 72- Mem: Am Soc Ichthyol & Herpet; Soc Study Amphibians & Reptiles. Res: Herpetology; natural history of reptiles and amphibians of southern United States. Mailing Add: Dept of Zool & Entom Auburn Univ Auburn AL 36830

MOUNT, WAYNE DELANO, b West Allis, Wis, Dec 15, 27; m 49; c 3. METEOROLOGY, ENVIRONMENTAL MANAGEMENT. Educ: Mass Inst Technol, BS, 52, MS, 53, PhD(meteorol), 58. Prof Exp: Res meteorologist, Geophys Res Directorate, Air Force Cambridge Res Ctr, 53-57, proj scientist, 57-58, chief technique develop sect, 58-60, technique appl & develop br, 60-62; head atmospheric physics dept, Sperry Rand Res Ctr, 62-70, dir, Atmospheric Sci Lab, 70-75; PRES, GEO-ATMOSPHERICS CORP, 75- Concurrent Pos: Chmn objective anal session, Int Symposium Numerical Weather Prediction, Tokyo, 60; mem comt atmospheric dynamics & comt atmospheric structures & circulations for planning & writing pub, Atmospheric Sci, 61-71, Nat Acad Sci-Nat Res Coun, 61; partic meterol panel, White House Conf Int Coop, 65. Mem: AAAS; Am Meteorol Soc; Am Geophys Union; Royal Meteorol Soc; Am Inst Aeronaut & Astronaut. Res: Air-sea interactions; oceanography; environmental modeling; techniques for directly and indirectly probing the atmosphere; atmospheric sciences; air pollution. Mailing Add: Geo-Atmospherics Corp Box 177 Lincoln MA 01773

MOUNTAIN, CLIFTON FLETCHER, b Toledo, Ohio, Apr 15, 24; m 45; c 3. SURGERY. Educ: Harvard Col, AB, 47; Boston Univ, MD, 54; Am Bd Surg, dipl, 62. Prof Exp: Dir dept statist res, Univ Boston, 47-50; consult & res analyst, Mass Dept Pub Health, 51-53; resident surgeon, Univ Chicago Clins, 54-58; instr surg, Univ Chicago, 58-59; asst prof, 60-63, ASSOC PROF SURG, UNIV TEX GRAD SCH BIOMED SCI & M D ANDERSON HOSP & TUMOR INST, 63- Concurrent Pos: Fel surg physiol, Univ Chicago, 55-58; sr fel thoracic surg, Univ Tex M D Anderson Hosp & Tumor Inst, 59-60; prin investr & mem solid tumor study group, Cancer Chemother Nat Serv Ctr, NIH, 61-, prin investr, Cancer Res Progs, Nat Cancer Inst, 63-; sect ed chest dis, Yearbk Cancer, 61-; sr investr & chmn prog biomath & comput sci, Univ Tex, 62-64; consult med sci adv comt, Systs Develop Corp, Calif, 63-; consult mem, Am Joint Comt Cancer Staging & End Result Reporting; mem, Task Force Lung & Esophageal Cancer & Nat Working Party Lung Cancer; chmn, Task Force Surg. Mem: AAAS; Am Col Chest Physicians; Am Asn Cancer Res; Am Thoracic Soc; AMA. Res: Thoracic malignant diseases; surgical techniques and adjunctive therapeutic programs in cancer chemotherapy and supervoltage irradiation;

quantitative biology through biomathematics and computer sciences. Mailing Add: Univ of Tex M D Anderson Hosp & Tumor Inst Houston TX 77025

MOUNTAIN, ISABEL MORGAN, b New Bedford, Mass, Aug 20, 11; m 49; c 1. IMMUNOLOGY, VIROLOGY. Educ: Stanford Univ, AB, 32; Cornell Univ, MA, 36; Univ Pa, PhD(bact), 38; Columbia Univ, MS, 61. Prof Exp: From fel to assoc, Rockefeller Inst, 38-44; from assoc to asst prof, Poliomyelitis Res Ctr, Johns Hopkins Univ, 44-49; res assoc, Westchester County Dept Lab & Res, 50-53; from lectr to res assoc, Columbia Univ, 53-61; ASSOC, SLOAN-KETTERING RES INST, 62- Concurrent Pos: With Pub Health Inst New York, 61-64; res assoc, Med Col, Cornell Univ, 64-69. Mem: Harvey Soc; fel NY Acad Sci. Res: Physiological basis of immune response to poliovirus; effectiveness of immunization of monkeys with formalin-inactivated poliovirus. Mailing Add: 17 Brookfield Pl Pleasantville NY 10570

MOUNTAIN, RAYMOND DALE, b Great Falls, Mont, Mar 28, 37; m 61; c 3. THEORETICAL PHYSICS. Educ: Mont State Col, BS, 59; Case Western Reserve Univ, MS, 61, PhD(physics), 63. Prof Exp: Physicist, 63-68, CHIEF STATIST PHYSICS SECT, NAT BUR STANDARDS, 68- Concurrent Pos: Nat Acad Sci-Nat Res Coun fel, 63-65; John Simon Guggenheim Mem Found fel, 74. Mem: Am Phys Soc. Res: Statistical mechanics; physics of liquids. Mailing Add: Nat Bur Standards Statist Physics Washington DC 20234

MOUNTAIN, WILLIAM BUCKINGHAM, b Kamsack, Sask, Dec 3, 22; m 48; c 2. NEMATOLOGY. Educ: Univ Western Ont, BSc, 50; Univ Toronto, PhD(plant path), 53. Prof Exp: Nematologist, Harrow Res Sta, 50-59, head nematol sect, 59-64; dir, Vineland Res Sta, 64-69; dir, Entom Res Inst, 69-73, ASST DIR-GEN, RES BR, AGR CAN, 73- Mem: Soc Nematol (vpres, 63, pres, 64); Agr Inst Can; Soc Europ Nematol. Res: Phytonematology; management and administration of research. Mailing Add: Res Br Agr Can Rm 2121 Neatby Bldg Cent Exp Farm Ottawa ON Can

MOUNTCASTLE, VERNON BENJAMIN, b Shelbyville, Ky, July 15, 18; m 45; c 3. PHYSIOLOGY. Educ: Roanoke Col, BS, 38; Johns Hopkins Univ, MD, 42. Hon Degrees: DSc, Roanoke Col, 68. Prof Exp: House officer surg, Johns Hopkins Hosp, 42-43; resident fel physiol, 46-48, from asst prof to assoc prof, 48-59, PROF PHYSIOL, SCH MED, JOHNS HOPKINS UNIV, 59-, DIR DEPT, 64- Concurrent Pos: Mem physiol study sect, NIH, 57-58, chmn, 58-63, chmn physiol training comt, 58-63; Lilly lectr & spec univ lectr, Univ London, 59; vis lectr, Col France, 59; chief ed, J Neurophysiol, 62-64; mem vis comt psychol & mem neurosci res prog, Mass Inst Technol, 66-; mem bd biol & med, NSF, 70-73; Wilder Penfield Mem lectr, Am Univ Beirut, 71; mem coun, Nat Eye Inst, 71-74; chmn sect physiol, Nat Acad Sci, 71-74; mem comn neurophysiol, Int Union Phys Sci; Sherrington lectr, Univ Liverpool, 74. Honors & Awards: Lashley Prize, Am Philos Soc, 74. Mem: Nat Acad Sci; AAAS; Am Acad Arts & Sci; Am Physiol Soc; Soc Neurosci (pres, 70-71). Res: Central nervous mechanisms in emotion; neurophysiology of the great afferent systems; sensation. Mailing Add: Dept of Physiol Johns Hopkins Univ Sch of Med Baltimore MD 21205

MOUNTCASTLE, WILLIAM R, JR, b Smyrna, Ga, Oct 31, 21; m 50; c 2. PHYSICAL CHEMISTRY, ANALYTICAL CHEMISTRY. Educ: Ga Inst Technol, BS, 43; Univ Ala, MS, 56, PhD(chem), 58. Prof Exp: Rubber chemist, Goodyear Tire & Rubber Co, Ala, 43-53; chem engr, Southeastern Exp Sta, US Bur Mines, 54-56; instr chem, Univ Ala, 57-58; assoc prof anal & phys chem, Birmingham-Southern Col, 58-65, prof, 65-66; ASST PROF CHEM, AUBURN UNIV, 66- Concurrent Pos: Consult res ctr, Med Col, Univ Ala, 58-60; participation contract, Oak Ridge Assoc Univs, 61-; res assoc, Union Carbide Nuclear Corp, 63; dir, NSF-Undergrad Res Chem, 64-66. Mem: Am Chem Soc. Res: Solvent extraction using phenyl phosphate diester and applications of electrochemical and spectrographic methods to this study; coulometry; spectroscopy. Mailing Add: Dept of Chem Auburn Univ Auburn AL 36830

MOUNTJOY, ERIC W, b Calgary, Alta, Nov 28, 31; m 58. GEOLOGY, STRATIGRAPHY. Educ: Univ BC, BASc, 55; Univ Toronto, PhD(stratig, struct geol), 60. Prof Exp: Tech officer field geol, Geol Surv Can, 57-60, geologist, 60-63; from asst prof to assoc prof, 63-74, PROF SEDIMENTATION STRATIG, McGILL UNIV, 74- Mem: Soc Econ Paleont & Mineral; Am Asn Petrol Geol; fel Geol Soc Am; Can Soc Petrol Geol; Int Asn Sedimentologists. Res: Sedimentation; structural geology of Alberta Rocky Mountains; Devonian Reef complexes; carbonate sedimentology; recent carbonates. Mailing Add: Dept of Geol Sci McGill Univ PO Box 6070 Montreal PQ Can

MOUNTJOY, JOSEPH BODE, b Lincoln, Ill. ANTHROPOLOGY, ARCHEOLOGY. Educ: Univ Ill, BA, 63; Southern Ill Univ, PhD(anthrop), 70. Prof Exp: Lectr, 69-70, asst prof, 70-73, ASSOC PROF ANTHROP, UNIV NC, GREENSBORO, 73- Mem: Am Anthrop Asn; Soc Am Archaeol. Res: Archeological research in Mexico focusing on cultural ecology, rise of civilization, longrange contacts and archeo-ethnography. Mailing Add: Dept of Anthrop Univ of NC Greensboro NC 27412

MOUNTNEY, GEORGE JOSEPH, b Plainfield Township, Pa, Aug 1, 21; m 42; c 3. POULTRY SCIENCE, FOOD TECHNOLOGY. Educ: Univ Md, BS, 48; Pa State Univ, MS, 49; Agr & Mech Col, Tex, PhD(poultry sci, food tech), 57. Prof Exp: Instr poultry sci, Univ Tenn, 49-50; asst prof, Agr & Mech Col, Tex, 50-58; asst prof, Ohio State Univ, 58-67, prof, 67-69; RES MGT SPECIALIST, COOP STATE RES SERV, USDA, 69- Mem: AAAS; Poultry Sci Asn; Inst Food Technol; Am Dairy Sci Asn. Res: Physical, bacterial, biochemical and organoleptic breakdown of dairy and poultry products and pesticide residues in these products. Mailing Add: Coop State Res Serv USDA Washington DC 20250

MOUNTS, RICHARD DUANE, b San Diego, Calif, Nov 15, 41; m 75; c 2. ANALYTICAL CHEMISTRY. Educ: Wheaton Col, Ill, BS, 64; Ariz State Univ, MS, 68; Univ Ariz, PhD(anal chem), 74. Prof Exp: Res chemist, J T Baker Chem Co, 67-69 & Grefco, Inc, 69-70; asst prof chem, Wake Forest Univ, 74-75; ASST PROF CHEM, FLA INST TECHNOL, 75- Mem: Am Chem Soc. Res: Steric effects in analytical reagents and electronic modules for instruction in analytical chemistry instrumentation. Mailing Add: Dept of Chem Fla Inst of Technol Melbourne FL 32901

MOUNTS, TIMOTHY LEE, b Peoria, Ill, Sept 14, 37; m 58; c 3. AGRICULTURAL CHEMISTRY, RADIOCHEMISTRY. Educ: Bradley Univ, BS, 59, MS, 68. Prof Exp: Res chemist, 57-75, LEADER, EDIBLE OILS PROD & PROCESSES, OILSEEDS CROPS LAB, NORTHERN REGIONAL RES CTR, AGR RES SERV, USDA, 75- Honors & Awards: Bond Award, Am Oil Chemists Soc, 69 & 71. Mem: Am Oil Chemists Soc; Am Chem Soc; Inst Food Technologists. Res: Development of improved edible oil products and processes so as to maintain a safe and nutritious food supply; hydrogenation and refining of edible oils, organoleptic evaluation of oils for flavor and stability, oils from damaged soybeans. Mailing Add: 1815 N University Peoria IL 61604

MOURAD, A GEORGE, b Bludan, Syria, Nov 6, 31; US citizen; m 58; c 1. GEODESY, PETROLEUM ENGINEERING. Educ: Ohio State Univ, BSc, 57, MSc, 59. Prof Exp: Res asst gravity & geod, Inst Geod, Ohio State Univ Res Found, 56-59, res assoc, 59-62; sr engr, NAm Aviation, 62-64; res geodesist, 64-66, SR GEODESIST, BATTELLE MEM INST, 66-, PROG DIR MARINE GEOD, 68-, PROJ MGR GEOD & OCEAN PHYSICS, 72- Concurrent Pos: Vchmn navig subcomt antisubmarine warfare, Nat Security Indust Asn, 70-; chmn spec study group on marine geod, Int Asn Geod, 70-; chmn comt marine geod, Int Asn Geod, 75-79. Mem: Marine Technol Soc; Am Geophys Union; Am Inst Navig. Res: Research and management ain in programs on satellite applications to earth and ocean dynamics disciplines; satellite altimetry for determining mean sea level; radar techniques for sea state measurements; satellite interferometry techniques for navigation; traffic control; data transfer; search and rescue applications. Mailing Add: Battelle Mem Inst 505 King Ave Columbus OH 43201

MOURANT, WALTER ARTHUR, b Cleveland, Ohio, Apr 5, 13; m 42; c 2. GEOLOGY. Educ: Rutgers Univ, BS, 50. Prof Exp: Indust photographer, Otis Elevator Co, 36-43; geologist, 50-67, hydrologist, 67-71, HYDROLOGY-GROUND WATER SPECIALIST, WATER RESOURCES DIV, US GEOL SURV, 71- Mem: Fel Geol Soc Am; Am Inst Prof Geol. Res: Occurrence and movement of ground water; ground water geology: hydrology. Mailing Add: 1716 Vassar NE Albuquerque NM 87106

MOURATOFF, GEORGE J, internal medicine, clinical pharmacology, see 12th edition

MOURER, KERMIT L, b Spearville, Kans, May 1, 13; m 35; c 2. PHARMACEUTICAL CHEMISTRY. Educ: Univ Nebr, BS, 34. Prof Exp: Res pharmacist vet med, 43-60, dir pharmaceut div, 60-65, VPRES MFG DIV, NORDEN LABS, INC, SMITH KLINE & FRENCH LAB, INC, 65- Mem: Am Chem Soc; Am Pharmaceut Asn. Res: Pharmaceutical production. Mailing Add: 2200 S 46th St Lincoln NE 68506

MOURNING, MICHAEL CHARLES, b Jerseyville, Ill, Oct 6, 40; m 67; c 2. CHEMISTRY. Educ: Univ Ill, Urbana, BS, 63; Univ NC, Chapel Hill, PhD(chem), 68. Prof Exp: SR CHEMIST, GAF CORP, 68- Mem: Am Chem Soc; Soc Photog Sci & Eng. Res: Organic synthesis. Mailing Add: GAF Corp 44-4 Binghamton NY 13902

MOURSHED, FAROUK ALI, b Cairo, Egypt, Sept 1, 33; m 66; c 1. DENTAL RADIOLOGY. Educ: Cairo Univ, BDS, 58; Univ Pa, MSc, 63. Prof Exp: Instr dent radiol, Cairo Univ, 58-60; lectr, 60-68, from asst prof to assoc prof dent radiol, 68-74, PROF ORAL DIAG & RADIOL & CHMN DEPT, COL DENT, HOWARD UNIV, 74- Mem: Am Acad Dent Radiol; Int Asn Dent Res. Mailing Add: Dept of Oral Diag & Radiol Howard Univ 600 W St NW Washington DC 20059

MOURSUND, ANDREW FLEMING, mathematics, see 12th edition

MOURSUND, ANNE LOREEN, physical chemistry, atomic physics, see 12th edition

MOURSUND, DAVID G, b Eugene, Ore, Nov 3, 36; m 61; c 4. COMPUTER SCIENCE, MATHEMATICS EDUCATION. Educ: Univ Ore, BA, 58; Univ Wis, MS, 60, PhD(math), 63. Prof Exp: From asst prof to assoc prof math, Mich State Univ, 63-67; res assoc, Comput Ctr, 67-70, head dept comput sci, 69-75, ASSOC PROF COMPUT SCI, UNIV ORE, 67- Mem: Am Math Asn; Asn Comput Mach. Res: Computers in education, with major emphasis upon uses of computers in pre-college education; computer literacy. Mailing Add: Dept of Comput Sci Univ of Ore Eugene OR 94703

MOURY, DANIEL NORMAN, b Greensboro, NC, Mar 18, 35; m 60; c 2. BIOCHEMISTRY. Educ: Wake Forest Col, BS, 60; Purdue Univ, PhD(biochem), 63. Prof Exp: Cardiovasc training prog fel biochem, Bowman Gray Sch Med, 63-64, instr, 64-66; assoc prof chem & chmn dept, Tusculum Col, 66-70; dean fac natural sci & math, Stockton State Col, 70-75; ASSOC DIR, OFF STATE COLS, NJ DEPT HIGHER EDUC, 75- Mem: AAAS; Am Asn Higher Educ. Res: Effects of thyroxin on mitochondrial electron transport system; electron transport particle from rat liver. Mailing Add: Dept of Higher Educ 225 W State Trenton NJ 08625

MOUSER, GILBERT WARREN, b St Louis, Mo, Jan 10, 11; m 34; c 4. BIOLOGY. Educ: Greenville Col, BS, 33; Cornell Univ, PhD, 50. Prof Exp: Teacher high sch, 36-44; asst, Cornell Univ, 44-45, zool, 45, rural educ, 46; instr biol, Iowa State Teachers Col, 47-51; asst prof, 51-52, actg head dept land & water conserv, 52-53, ASSOC PROF FISHERIES & WILDLIFE, 53- Mem: Soil Conserv Soc Am; Am Nature Study Soc; Nat Asn Biol Teachers. Res: Leadership training for outdoor education; comparative performance of high school and university freshmen on a test of biological misconception; field techniques for teaching conservation. Mailing Add: Dept of Fisheries & Wildlife Mich State Univ East Lansing MI 48823

MOUSHEGIAN, GEORGE, b Detroit, Mich, Jan 19, 23; m 52; c 3. PHYSIOLOGICAL PSYCHOLOGY. Educ: Wayne State Univ, BS, 47, MA, 51; Univ Tex, PhD, 57. Prof Exp: Res scientist, Defense Res Lab, Univ Tex, 56-59; res fel hearing, neurophysiol & psychol, Walter Reed Army Inst Res, 59-64; prof physiol psychol, Lab Sensory Commun, Syracuse Univ, 64-68; DIR RES, CALLIER CTR COMMUN DIS, 68- Concurrent Pos: Adj prof, Dept Physiol, Univ Tex Southwest Med Br, 69- Mem: AAAS; Am Psychol Asn; Acoust Soc Am; Am Physiol Soc; Soc Neurosci. Res: Electrophysiological study of responses from the brain stem to acoustic stimulation, using micro and macro electrodes; study of human responses to sounds; neurophysiology; psychophysics. Mailing Add: Callier Ctr for Commun Dis 1966 Inwood Rd Dallas TX 75235

MOUSSA, MOUFIED ABDEL-AZIZ, entomology, see 12th edition

MOUSSA, MOUNIR, TAWFIK, b Cairo, UAR, June 22, 33; m 65. SEDIMENTARY PETROLOGY, STRATIGRAPHY. Educ: Cairo Univ, BSc, 58; Univ Utah, PhD(geol), 65. Prof Exp: Chemist, Egyptian Soc Fertilizers & Chem Indust, 58; asst res worker geol, Desert Inst, Nat Res Ctr, Egypt, 58-60; from asst prof to assoc prof, 65-75, PROF GEOL, UNIV PR, MAYAGUEZ, 75- Concurrent Pos: Scientist, PR Nuclear Ctr, USAEC, 72-75. Mem: Geol Soc Am; Am Asn Petrol Geol; Soc Econ Paleont & Mineral; Sigma Xi. Res: Green River formation in Uinta Basin; upper cretaceous stratigraphy in the Wasatch Plateau, Utah; mid tertiary stratigraphy of southwestern Puerto Rico; rock weathering; carbonate petrology and diagenesis, biostratigraphy and paleoenvironments of middle tertiary of southwestern and northern Puerto Rico; bioclastic turbidite sedimentation; grain-size distribution in sediments. Mailing Add: Dept of Geol Univ of PR Mayaguez PR 00708

MOUSTAFA, LAILA AHMED, b Cairo, Egypt, June 5, 37; m 69; c 1. DEVELOPMENTAL BIOLOGY, GENETICS. Educ: Cairo Univ, BScAgEng, 59; Calif State Polytech Col, San Luis Obispo, MA, 63; Wash State Univ, PhD(animal sci, genetics), 69. Prof Exp: Agr engr, Agr Dept, Egyptian Govt, 59-60; agr engr, Cairo Munic Govt, 60-61; res asst animal physiol, Wash State Univ, 67-68, electron micros, Electron Micros Lab, 69; NIH training grant & res assoc tissue & embryol, Lab

Reproduction Physiol, Sch Vet Med, Univ Pa, 69-71; staff fel, 71-72, SR STAFF FEL, ENVIRON TOXICOL BR, NAT INST ENVIRON HEALTH SCI, 72- Mem: Egyptian Agr Eng Asn; Am Soc Animal Sci; Genetics Soc Am; Soc Exp Biol & Med; Soc Study Reproduction. Res: Mammalian embryo development; control of differentiation and retrodifferentiation; use of cell transplantation and micrurgy to study the pattern of organ development and gene action; in vitro differentiation of mammalian blastocysts during implantation phase and effect of terata on ontogenesis. Mailing Add: Environ Toxicol Br PO Box 12233 Nat Inst Environ Health Sci Research Triangle Park NC 27709

MOUW, DAVID RICHARD, b Carlisle, Pa, Aug 22, 42; c 2. PHYSIOLOGY. Educ: Hope Col, BA, 64; Univ Mich, Ann Arbor, PhD(physiol), 69. Prof Exp: Instr biol, Hampton Inst, 67-68; NIH fel exp physiol, Howard Florey Labs Exp Physiol, Melbourne, Australia, 70-71; asst prof physiol, 71-75, ASSOC PROF PHYSIOL, UNIV MICH, ANN AROBR, 75- Mem: Am Physiol Soc; AAAS. Res: Lead handling by the kidney; lead toxicity and the kidney; renal electrolyte metabolism. Mailing Add: 6812 Med Sci II Univ of Mich Ann Arbor MI 48104

MOUZON, JAMES CARLISLE, physics, deceased

MOVAT, HENRY ZOLTAN, b Temesvar, Romania, Aug 11, 23; nat US; m 56; c 2. PATHOLOGY. Educ: Innsbruck Univ, MD, 48; Queen's Univ, Ont, MSc, 54, PhD, 56; Royal Col Physicians & Surgeons Can, cert path, 59; FRCP(C), 67. Prof Exp: From asst prof to assoc prof, 57-65, PROF PATH, UNIV TORONTO, 65-, HEAD DIV EXP PATH, 68-, MEM INST IMMUNOL, 71- Concurrent Pos: Res assoc, Med Res Coun, 60- Mem: Am Soc Exp Path; Am Asn Path & Bact; Am Asn Immunol; Soc Exp Biol & Med; Int Acad Path. Res: Acute inflammatory reaction; chemical mediators of acute inflammation and hypersensitivity. Mailing Add: Div of Exp Path Med Sci Bldg Univ of Toronto Toronto ON Can

MOVIUS, WILLIAM GUST, b Portland, Ore, Jan 15, 43. INORGANIC CHEMISTRY. Educ: Univ Ore, BA, 65; Pa State Univ, PhD(chem), 68. Prof Exp: Fel, Univ Calif, San Diego, 68-69; ASST PROF CHEM, KENT STATE UNIV, 70- Res: Oxidation-reduction reactions, especially those involving uncommon oxidation states; coordination compounds in nonaqueous electrolyte solutions, especially those incompatible with water. Mailing Add: Dept of Chem Kent State Univ Kent OH 44242

MOWAT, DAVID NAIRN, animal nutrition, crop breeding, see 12th edition

MOWAT, JOHN GORDON, b Honolulu, Hawaii, May 24, 29; m 57; c 2. PHYSICS. Educ: Stanford Univ, BS, 51, MS, 53; Univ Va, PhD(physics), 61. Prof Exp: Mathematician, Naval Ord Testing Sta, Calif, 52-54; temp instr physics, Ala Polytech Inst, 57-58; asst prof, Univ Ky, 61-64; ASSOC PROF PHYSICS, AUBURN UNIV, 64- Mem: Am Phys Soc. Res: Theoretical physics; teaching of quantum mechanics; connections between the sciences and the humanities; non-Western cultures. Mailing Add: Dept of Physics Auburn Univ Auburn AL 36830

MOWAT, JOHN HALLEY, b Houlton, Maine, May 20, 10; m 43. ORGANIC CHEMISTRY. Educ: Univ Maine, BSc, 32, MSc, 33; McGill Univ, PhD(org chem), 42. Prof Exp: Chemist, Lederle Labs, Am Cyanamid Co, 41-68, group leader, 44-50, 54-58; ASSOC RES SCIENTIST, RES CTR, ROCKLAND STATE HOSP, 68- Mem: Am Chem Soc. Res: Vitamins; antibiotics; isolation; structure; synthesis; metabolism of drugs; mass spectrometry. Mailing Add: Res Ctr Rockland State Hosp Orangeburg NY 10962

MOWATT, THOMAS C, b Orange, NJ, Apr 24, 36; m 59; c 1. GEOCHEMISTRY, GEOLOGY. Educ: Rutgers Univ, BA, 59; Univ Mont, PhD(geol), 65. Prof Exp: Res scientist, Pan Am Petrol Corp, Okla, 65-67; asst prof geol, Winona State Col, 67-68 & Univ SDak, 68-70; supvr minerals anal & res, Alaska Geol Surv, 70-74; geologist, Bur Land Mgt, 74-75, GEOLOGIST, BUR MINES, US DEPT INTERIOR, 75- Concurrent Pos: Lectr, Univ Tulsa, 67; res assoc, SDak State Geol Surv, 68-70; adj assoc prof, Inst Marine Sci, Univ Alaska, 73- Mem: Geochem Soc; Mineral Soc Am; Clay Minerals Soc; Mineral Asn Can; Soc Econ Paleontologists & Mineralogists. Res: Petrology; sedimentology; environmental studies; economic geology; geochemistry, mineralogy and petrology in the contexts of economic geology, environmental science and marine science. Mailing Add: US Dept Interior Bur of Mines PO Box 2259 Anchorage AK 99510

MOWBRAY, THOMAS BRUCE, b Duluth, Minn, Mar 1, 40; m 66. BOTANY, PLANT ECOLOGY. Educ: Univ Minn, Duluth, BA, 62; Duke Univ, MA, 64, PhD(bot), 67. Prof Exp: Instr biol, Duke Univ, 67-68; asst prof, 68-74, ASSOC PROF BIOL & CHAIRPERSON POP DYNAMICS, UNIV WIS-GREEN BAY, 74- Mem: Ecol Soc Am; Am Inst Biol Sci. Res: Plant community analysis; vegetation gradient analysis. Mailing Add: Dept of Biol Univ of Wis Green Bay WI 54305

MOWER, HOWARD FREDERICK, b Chicago, Ill, Aug 25, 29; m; c 2. ORGANIC CHEMISTRY. Educ: Calif Inst Technol, BS, 51, PhD(org chem), 56. Prof Exp: Res chemist, Cent Res Dept, E I du Pont de Nemours & Co, Del, 56; assoc prof, 65-69, PROF BIOCHEM, UNIV HAWAII, 69- Res: Ferredoxins; hydrogenase enzymes; biological nitrogen fixation. Mailing Add: Dept of Biochem Univ of Hawaii Honolulu HI 96822

MOWER, LYMAN, b Berkeley, Calif, June 15, 27; m 48; c 3. PHYSICS. Educ: Univ Calif, BS, 49; Mass Inst Technol, PhD(physics), 53. Prof Exp: Eng specialist, Sylvania Elec Prod Inc, 53-57; from asst prof to assoc prof, 57-64, PROF PHYSICS, UNIV NH, 64- Concurrent Pos: Vis fel, Joint Inst Lab Astrophys, 64-65. Mem: AAAS; Am Phys Soc. Res: Atomic and plasma physics; quantum electronics. Mailing Add: Dept of Physics Univ of NH Durham NH 03824

MOWER, ROBERT G, b Gasport, NY, Sept 27, 28. FLORICULTURE, ORNAMENTAL HORTICULTURE. Educ: Cornell Univ, BS, 56, MS, 59, PhD(turf dis), 61. Prof Exp: Asst prof, 61-67, ASSOC PROF WOODY ORNAMENTALS, CORNELL UNIV, 67- Mem: AAAS; Am Soc Hort Sci; Int Soc Hort Sci. Res: Taxonomy, evaluation of woody plants for landscape use. Mailing Add: Dept of Floriculture & Ornamental Hort Plant Sci Bldg Cornell Univ Ithaca NY 14853

MOWERY, DWIGHT FAY, JR, b Moorehead, Minn, May 1, 15; m 43. CARBOHYDRATE CHEMISTRY, CHEMICAL KINETICS. Educ: Harvard Univ, AB, 37; Mass Inst Technol, PhD(org chem), 40. Prof Exp: Res chemist, E I du Pont de Nemours & Co, Del, 40-42; res chemist, Hercules Powder Co, 42-43; head chem dept, Elms Col, 43-46; head dept, Franklin Tech Inst, 46-49; asst prof, Trinity Col, Conn, 49-53; chmn dept chem, Ripon Col, 53-57; prof chem & dir grad prog, New Bedford Inst Tech, 57-64; chmn dept, 64-70, COMMONWEALTH PROF CHEM, SOUTHEASTERN MASS UNIV, 65- Concurrent Pos: Researcher, J B Williams Co, Conn, 52-53, WTM Mfg Co, Wis, 54-56, Aerovox Corp, Mass, 62 & Acushnet Process Corp, Mass, 65-67. Mem: Am Chem Soc. Res: Seed disinfectants and bactericides; carbohydrate chemistry; chromatographic adsorption; gas chromatography; organic microanalysis; chemical kinetics and computer programming. Mailing Add: Dept of Chem Southeastern Mass Univ North Dartmouth MA 02747

MOWERY, RICHARD ALLEN, JR, b Newboston, Ohio, June 2, 38. ANALYTICAL CHEMISTRY. Educ: Univ Calif, Los Angeles, BS, 63; Univ Southern Calif, MAOM, 69; Ariz State Univ, PhD(chem), 74. Prof Exp: RES CHEMIST, APPL AUTOMATION, INC, PHILLIPS PETROL CO, 74- Mem: Am Chem Soc; The Chem Soc. Mailing Add: Appl Automation Inc Phillips Res Ctr Rm 125 RB 2 Bartlesville OK 74004

MOWITZ, ARNOLD MARTIN, b New York, NY, Jan 14, 23; m 46; c 2. ANALYTICAL CHEMISTRY. Educ: Univ Buffalo, MA, 53. Prof Exp: Supvr control analysts, Nat Aniline Div, Allied Chem & Dye Corp, 46-48, chief analyst res & develop, 48-50, anal res chemist, 50-53, chief anal res, 53-55; group leader, Anal Dept, Interchem Corp, 55-65, prog mgr, 66-67, mgr res serv dept, 67-70, MGR OPERS DEPT, CENT RES LABS, INMONT CORP, 70- Mem: Am Chem Soc; Am Microchem Soc; Soc Appl Spectros; Am Soc Test & Mat; NY Acad Sci. Res: Instrumental analysis; infrared and ultraviolet absorption analysis; spectrographic and microchemical analysis; x-ray diffraction; gas chromatography; light and electron microscopy; physical testing; nuclear magnetic resonance spectroscopy; research management; environmental chemistry. Mailing Add: Res Serv Dept Cent Res Labs Inmont Corp 1255 Broad St Clifton NJ 07015

MOWLES, THOMAS FRANCIS, b Boston, Mass, Feb 26, 34; m 56; c 4. ENDOCRINOLOGY, BIOCHEMISTRY. Educ: Boston Univ, BA, 55; NY Univ, MS, 64; Rutgers Univ, PhD(zool), 68. Prof Exp: Lab supvr, Ciba Pharmaceut Co, 56-68; sr res biochemist, 68-74, RES GROUP CHIEF, HOFFMANN-LA ROCHE INC, 75- Mem: AAAS; Am Chem Soc. Res: Mechanism of hormone action; control of fertility; biochemical pharmacology; steroid biosynthesis; releasing factors. Mailing Add: Biochem Pharmacol Div Hoffmann-La Roche Inc Nutley NJ 07110

MOWRY, DAVID THOMAS, b Pyengyang, Korea, Mar 11, 17; US citizen; m 38; c 3. INDUSTRIAL CHEMISTRY. Educ: Col Wooster, BS, 38; Ohio State Univ, MSc, 40, PhD(org chem), 41. Prof Exp: Chemist, Ohio State Univ, 38-41; res chemist, Cent Res Labs, Monsanto Co, 41-44, res group leader, 45-52, mgr chem develop, Phosphate Div, 52-53, fine chem, Org Div, Develop Dept, 54-57, asst dir, 57-58, dir res & eng div, 58-61, mgr plastics div, 61-64, mgr planning E Asia, Int Div, 64-68, dir, Monsanto Japan Ltd & Ryoko Chemstrand Ltd, 68-74; PRIN ENGR, INT OPERS DIV, NUS CORP, 74- Mem: AAAS; Am Chem Soc; Commercial Develop Asn; Am Nuclear Soc. Res: Structure of natural products; synthesis and reactions of nitriles; exploratory organic synthesis; high polymers; agricultural chemicals; commercial development; patent and know-how licensing; energy economics. Mailing Add: NUS Corp 4 Research Place Rockville MD 20850

MOWRY, JAMES B, b Peoria, Ill, Oct 3, 20; m 46; c 2. HORTICULTURE. Educ: Univ Ill, BS, 48; Purdue Univ, MS, 49; Rutgers Univ, PhD(genetics, fruit breeding), 51. Prof Exp: Asst fruit breeding, Univ Ill, 47-48; asst bot & plant path, Purdue Univ, 48-49; asst fruit breeding, Rutgers Univ, 49-51; from asst prof to assoc prof, 51-62, PROF HORT, UNIV ILL, URBANA & SOUTHERN ILL UNIV, CARBONDALE, 62-; SUPT, ILL HORT EXP STA, 51- Honors & Awards: Shepard Award, 60. Mem: Am Pomol Soc (secy-treas, 65-); AAAS; Genetics Soc Am; Am Soc Hort Sci; Am Phytopath Soc. Res: Cross inoculation of prunus with coccomyces; embryology and cytology of pear; peach and apple genetics and breeding for disease resistance; climatic adaptation, phenology, bud hardiness and disease susceptibility of peaches and apples; root stock-scion interactions affecting apples. Mailing Add: Dept of Plant Indust Southern Ill Univ Carbondale IL 62901

MOWRY, ROBERT WILBUR, b Griffin, Ga, Jan 10, 23; m 49; c 3. PATHOLOGY. Educ: Birmingham Southern Col, BS, 44; Johns Hopkins Univ, MD, 46. Prof Exp: Intern, Med Col Ala, 46-47, asst resident path, 47-48; sr asst surgeon, NIH, 48-52; asst prof path, Sch Med, Washington Univ, 52-53; from asst prof to assoc prof path, Med Ctr, 53-57, dir grad progs path, 64-72, sr scientist, Inst Dent Res, 67-72, PROF PATH, MED CTR, UNIV ALA, BIRMINGHAM, 58- Concurrent Pos: Fel, Mallory Inst Path, Boston Univ, 49-50; dir, Anat Path Lab, Univ Ala Hosp, 60-64 & 75-; assoc ed, J Histochem & Cytochem, 61- & Stain Technol, 65-; mem, Path A Study Sect, USPHS, 64-68, trustee, Biol Stain Comn, 66-, vpres, 74-; vis scientist, Dept Path, Cambridge Univ, 72-73. Mem: Am Soc Exp Path; Histochem Soc; Am Asn Path & Bact; Biol Stain Comn; Int Acad Path. Res: Histochemistry and its applications to pathology; histopathologic technic; methods for detection and characterization of complex carbohydrates, microbial agents, amyloids and insulin in cells and tissues. Mailing Add: Dept of Path Univ of Ala Med Ctr Birmingham AL 35233

MOWSHOWITZ, ABBE, b Liberty, NY, Nov 13, 39; m 64. COMPUTER SCIENCE. Educ: Univ Chicago, SB, 61; Univ Mich, Ann Arbor, MA, 65, MS, 66, PhD(comput sci), 67. Prof Exp: Res assoc methodology, Human Sci Res, Inc, 62-63; res asst appl math, Ment Health Res Inst, Univ Mich, 63-67, asst res mathematician, 67-68; asst prof comput sci & indust eng, Univ Toronto, 68-69; asst prof, 69-74, ASSOC PROF COMPUT SCI, UNIV BC, 74- Concurrent Pos: Res assoc, Inst Social Res & lectr, Dept Commun & Comput Sci, Univ Mich, 67-68; vis res assoc, Dept Comput Sci, Cornell Univ, 75-76. Mem: Asn Comput Mach; Am Math Soc; Math Asn Am; Can Math Cong. Res: Combinatorial mathematics and graph theory; social impact of science and technology; complexity of algebraic systems. Mailing Add: Dept of Comput Sci Univ of BC Vancouver BC Can

MOXHAM, ROBERT LYNN, b Burlington, Ont, July 2, 33; m 61; c 3. GEOCHEMISTRY. Educ: McMaster Univ, BA, 55, MSc, 58; Univ Chicago, PhD(geochem), 63. Prof Exp: Field geologist, Int Nickel Co, Can, 55-57, 58-59; Nat Res Coun Can fel, Univ Man, 63-64; geochemist, NY State Mus & Sci Serv, 64-69; geochemist, Off Tech Coop, UN, 69-75; GEOCHEMIST, INT ATOMIC ENERGY AGENCY, 75- Mem: Geol Soc Am; Soc Econ Geol; Mineral Asn Can; Am Inst Prof Geol; Asn Explor Geochemists. Res: Geochemical exploration; geochemistry of ore deposits; instrumental analysis of rocks; trace element abundances.

MOXHAM, ROBERT MORGAN, b Columbus, Ohio, Sept 15, 19; m 49; c 3. GEOLOGY. Educ: Ohio State Univ, BS, 42. Prof Exp: Geologist, Alaska, 42-51, chief radiation sect, Geophys Br, 53-55, staff geologist, 56-61, chief br theoret geophys, 62-65, RES GEOPHYSICIST, GEOPHYS BR, US GEOL SURV, 66- Concurrent Pos: UN lectr & consult. Mem: Geol Soc Am; Soc Explor Geophys; Am Asn Petrol Geol; Am Geophys Union. Res: Radioactive and industrial minerals in Alaska; nuclear geology and aeroradiometry; infrared techniques in geophysical exploration; neutron activation analysis for mineral exploration. Mailing Add: 6404 Fairland St Alexandria VA 22312

MOXON, ALVIN LLOYD, b Flandreau, SDak, July 25, 09; m 38; c 1. BIOCHEMISTRY, NUTRITION. Educ: SDak State Col, BS, 34, MS, 37; Univ Wis, PhD(biochem), 41. Prof Exp: Anal chemist, Exp Sta, SDak State Col, 34-37, actg chemist, 37-40, chemist, 40-45, prof chem, 45-51; PROF ANIMAL SCI & ASSOC

CHMN DEPT, OHIO AGR RES & DEVELOP CTR & OHIO STATE UNIV, 51- Concurrent Pos: Collabr, Northern Region Res Lab, Bur Agr & Indust Chem, USDA, Ill, 44-; head dept chem, SDak State Col, 47-51; prof, Univ Sao Paulo, 64-66; consult, USPHS, 37-38 & Punjab Agr Col, India, 60-61. Mem: AAAS; Am Soc Animal Sci; Am Chem Soc; Am Soc Biol Chem; Am Inst Nutrit. Res: Distribution of selenium in nature; mechanism of selenium toxicity; nutritional significance of trace elements; forms of carbohydrates in feeds; ruminant nutrition. Mailing Add: Ohio Agr Res & Develop Ctr Wooster OH 44691

MOYA, FRANK, b New York, NY, Jan 20, 29; c 6. ANESTHESIOLOGY. Educ: NY Univ, BA, 49; State Univ NY Downstate Med Ctr, MD, 53; Am Bd Anesthesiol, dipl, 59. Prof Exp: Attend-in-chg anesthesiol, Sloane Hosp Women, NY, 59-62; assoc dean hosp affairs, 67-68, actg dean, Sch Med, 68-69, PROF ANESTHESIOL & CHMN DEPT, SCH MED, UNIV MIAMI, 62-, ASSOC DEAN SCH, 69- Concurrent Pos: Assoc, Col Physicians & Surgeons, Columbia Univ, 58-60, asst prof, 60-62; attend physician, Jackson Mem Hosp, Miami, Fla, 62-, chmn med bd, 69; vis prof, Stanford Univ, 64, Columbia Univ & Albert Einstein Col Med, 65, Sch Med, Case Western Reserve Univ, 66, Univ Va & Univ Wash, 67, Univ Man & Queen's Univ, Ont, 70, Univ Ky, Univ Mo & Univ Tenn, 71; assoc exam, Am Bd Anesthesiol, 64-71; spec consult, Nat Inst Gen Med Sci, 67; mem anesthesiol training comt, NIH, 67-71; William Washington Graves lectr, St Louis Univ, 69; A William Friend Mem lectr, Univ Ont, 70; consult mgr acad ctrs. Mem: Am Soc Pharmacol & Exp Therapeut; fel Am Col Anesthesiol; NY Acad Sci; NY Acad Med; fel Royal Soc Med. Res: Maternal and neonatal physiology and pharmacology, obstetric anesthesia and newborn resuscitation. Mailing Add: Dept of Anesthesiol Univ of Miami Sch of Med Miami FL 33136

MOYE, ALFRED LEON, b New Windsor, Md, May 30, 38. CHEMISTRY. Educ: WVa Wesleyan Univ, BS, 60; Univ Pittsburgh, PhD(chem), 68. Prof Exp: Instr chem, WVa Wesleyan Col, 62-63; ASST PROF CHEM, UNIV PITTSBURGH, 68-, DEAN STUDENT AFFAIRS, 72- Mem: AAAS; Am Chem Soc. Res: Reactions of carbon monoxide borane with Lewis bases, especially organometallic species. Mailing Add: Dept of Chem Univ of Pittsburgh Pittsburgh PA 15213

MOYE, ANTHONY JOSEPH, b McAdoo, Pa, Oct 15, 33; m 57; c 3. ACADEMIC ADMINISTRATION, PHYSICAL ORGANIC CHEMISTRY. Educ: Upsala Col, BS, 55; Iowa State Univ, MS, 57, PhD(org chem), 62. Prof Exp: Prof chem & dean acad planning & grad studies, Calif State Col, Los Angeles, 62-71; prof chem & vpres acad affairs, Quinnipiac Col, 71-72; STATE UNIV DEAN EDUC PROGS & RESOURCES, CALIF STATE UNIV & COLS, 72- Mem: AAAS; Am Chem Soc; The Chem Soc. Res: Free radicals in solution; chemiluminescence. Mailing Add: Calif State Univ 5670 Wilshire Blvd Los Angeles CA 90032

MOYE, HUGH ANSON, b Mobile, Ala, Oct 18, 38. ANALYTICAL CHEMISTRY. Educ: Spring Hill Col, BS, 61; Univ Fla, PhD(chem), 65. Prof Exp: Asst prof, 65-75, ASSOC PROF CHEM & ASSOC CHEMIST PESTICIDE RES, PESTICIDE RES LAB, SOUTHERN EXP STA, UNIV FLA, 75- Mem: Am Chem Soc; Asn Offs Anal Chem. Res: Analaytical methods for pesticides; reaction gas chromatography of pesticides; gas chromatography detectors. Mailing Add: Southern Exp Sta Pesticide Res Lab Univ Fla Gainesville FL 32601

MOYED, HARRIS S, b Philadelphia, Pa, May 15, 25; m 54; c 2. BACTERIOLOGY, BIOCHEMISTRY. Educ: Nat Found res fel biochem, Mass Gen Hosp, 54-55 & bact, Harvard Med Sch, 55-57; from instr to asst prof bact, Harvard Med Sch, 57-63; Hastings prof microbiol, Sch Med, Univ Southern Calif, 63-69; PROF MED MICROBIOL, COL MED, UNIV CALIF, IRVINE, 69-, ASSOC DEAN ACAD AFFAIRS, 74- Concurrent Pos: Lederle award, 58-60. Mem: Am Soc Microbiol; Am Soc Biol Chem. Res: Biochemistry of bacteria; regulation of biosynthetic reactions; action of plant auxin. Mailing Add: Dept of Med Microbiol Univ of Calif Irvine CA 92717

MOYER, ARDEN WESLEY, b Frederick, Ill, Jan 16, 09; m 43; c 1. PHYSIOLOGICAL CHEMISTRY. Educ: Univ Ill, AB, 37, MS, 38; Cornell Univ, PhD(biochem), 42. Prof Exp: Instr biochem, Med Col, Cornell Univ, 42-46; res biochemist, Lederle Labs Div, Am Cyanamid Co, 46-58, head virus biol res dept, 58-70, sr res virologist, 70-74; RETIRED. Concurrent Pos: Res assoc, Med Col, Cornell Univ, 43-46. Mem: AAAS; Harvey Soc; Am Asn Immunol; NY Acad Sci. Res: Amino acid metabolism; transmethylation; specificity of choline in transmethylation; structure of biotin; immunology and allergies; cancer, viral and rickettsial research; lipid metabolism. Mailing Add: 138 Fremont Ave Park Ridge NJ 07656

MOYER, BURTON JONES, physics, see 12th edition

MOYER, CALVIN LYLE, b Philadelphia, Pa, Nov 2, 41; m 63; c 2. ORGANIC CHEMISTRY. Educ: Ursinus Col, BS, 63; Harvard Univ, MA, 65, PhD(chem), 68. Prof Exp: RES CHEMIST, BENGER LAB, E I DU PONT DE NEMOURS & CO, INC, 68- Mem: AAAS; Am Chem Soc; Am Inst Chemists; Am Asn Textile Chemists & Colorists; The Chem Soc. Res: Synthetics; polymer chemistry; textile fibers; dyeing. Mailing Add: 617 Shore Rd Waynesboro VA 22980

MOYER, CARL EDWARD, b Dayton, Ohio, Dec 24, 26; m 50; c 4. PHYSIOLOGICAL CHEMISTRY. Educ: Univ Dayton, BS, 53; Ohio State Univ, MS, 57, PhD, 59. Prof Exp: CLIN BIOCHEMIST & HEAD CLIN LAB, RES LABS, PARKE, DAVIS & CO, 63- Concurrent Pos: Supvr clin labs, Riverside Methodist Hosp, Columbus, Ohio, 59-63. Mem: Am Chem Soc; Am Asn Clin Chem; Asn Clin Sci. Res: Clinical biochemistry. Mailing Add: Head Clin Lab Parke Davis & Co 2800 Plymouth Rd Ann Arbor MI 48106

MOYER, DEAN LA ROCHE, b Pa, May 17, 25; m 53; c 2. PATHOLOGY. Educ: Lehigh Univ, BA, 48; Univ Rochester, MD, 52. Prof Exp: From instr to prof path, Med Ctr, Univ Calif, Los Angeles, 56-69; PROF PATH, OBSTET & GYNEC, MED SCH, UNIV SOUTHERN CALIF, 69-, HEAD SECT EXP PATH, 69- Concurrent Pos: Fel oncol, Mass Gen Hosp, 55-56; dir labs, Harbor Gen Hosp, Torrance, 61-69. Res: Early reproduction. Mailing Add: Sect of Exp Path Univ of Southern Calif Med Sch Los Angeles CA 90033

MOYER, ELIZABETH KING, anatomy, deceased

MOYER, FRANK H, b Topeka, Kans, Apr 5, 27; div; c 2. CYTOLOGY, ENVIRONMENTAL SCIENCE. Educ: Univ Md, BS, 56; Johns Hopkins Univ, PhD(biol), 61. Prof Exp: Lectr biol, Goucher Col, 61; instr anat, Sch Med, Johns Hopkins Univ, 61-62; asst prof zool, Univ Ill, 62-65; assoc prof, Washington Univ, 65-67; chmn dept, 67-70, PROF BIOL, UNIV MO-ST LOUIS, 67- Concurrent Pos: Vis lectr, Univ Miami, 65; mem, St Louis Planned Parenthood Prog & Res Comt, 67-; mem, Bd Dirs, St Louis Coalition for Environ, 70-, chmn, Standing Comt Pop, 70-72. Mem: AAAS; Am Soc Cell Biol; Am Soc Zool; Electron Micros Soc Am; Soc Develop Biol. Res: Factors regulating cytodifferentiation in mammalian pigment cells; echinoderm and ascidian eggs and embryos; amphibian embryos studies by means of electron microscopy; cell fractionation techniques; biochemical analysis. Mailing Add: Dept of Biol Univ of Mo 8001 Natural Bridge Rd St Louis MO 63121

MOYER, JAMES CHARLES, b Guelph, Ont, Feb 24, 14; nat US; m 42; c 3. FOOD TECHNOLOGY. Educ: Ont Agr Col, BSA, 36; Univ Toronto, MSA, 38; Cornell Univ, PhD, 42. Prof Exp: Asst hort, Ont Agr Col, 36-38; fel, Nutrit Found, 42-44; from asst prof to assoc prof chem, 44-54, PROF FOOD SCI & TECHNOL, AGR EXP STA, CORNELL UNIV, 54- Concurrent Pos: Fulbright scholar, Australia, 54-55. Mem: Am Chem Soc; Inst Food Technologists. Res: Vitamin content of vegetables; experimental canning, freezing and dehydration of fruits and vegetables; design of processing equipment. Mailing Add: Dept of Food Sci Cornell Univ Agr Exp Sta Geneva NY 14456

MOYER, JAMES EARL, b Altoona, Pa, July 9, 28; m 52; c 2. MICROBIOLOGY. Educ: Southwest Tex State Col, BS, 50, MA, 51; Univ Tex, PhD(bact), 60. Prof Exp: Clin lab instr, US Air Force Sch Aviation Med, 51-55, res microbiologist, Sch Aerospace Med, 58-66; chief biol sci, 66-67, CHIEF RES RESOURCES PROG, ROBERT S KERR WATER RES CTR, 67- Mem: Am Soc Microbiol. Res: Water Pollution. Mailing Add: Robert S Kerr Water Res Ctr PO Box 1198 Ada OK 74820

MOYER, JAMES WARD, b Chicago, Ill, June 29, 44; m 69. INORGANIC CHEMISTRY. Educ: Univ Rochester, BS, 66; Univ Wis-Madison, PhD(inorg chem), 71. Prof Exp: RES CHEMIST, LAMP BUS DIV, GEN ELEC CO, 70- Mem: Am Chem Soc; AAAS; Sigma Xi. Res: High purity inorganic chemicals; reduced oxidation state of transition metals; metal-halide chemistry; chemical transport phenomenon; molten salt chemistry. Mailing Add: LR & TSO Bldg 336 Lamp Bus Div Gen Elec Co Nela Park East Cleveland OH 44112

MOYER, JOHN CLARENCE, b Chicago, Ill, Jan 9, 46; m 75. MATHEMATICS. Educ: Christian Bros Col, BS, 67; Northwestern Univ, MS, 69, PhD(math educ), 74. Prof Exp: Teacher math, St Patrick High Sch, Chicago, 67-69 & St Joseph High Sch, Chicago, 69-72; ASST PROF MATH, MARQUETTE UNIV, 74- Mem: Math Asn Am; Am Math Soc; Nat Coun Teachers Math. Res: Relationship between cognitive, mathematical, and instructional structures as exhibited in children's geometric development. Mailing Add: Dept of Math & Statist Marquette Univ Milwaukee WI 53233

MOYER, JOHN HENRY, b Hershey, Pa, Apr 1, 17; m; c 7. MEDICINE. Educ: Lebanon Valley Col, BS, 39; Univ Pa, MD, 43; Am Bd Internal Med, dipl. Hon Degrees: DSc, Lebanon Valley Col, 68. Prof Exp: Intern, Pa Hosp, 43; resident, Belmont Hosp, Worcester, Mass, 44-45; asst instr tuberc & contagious dis, Univ Vt, 44-45; chief resident med, Brooke Gen Hosp, 47; fel pharmacol & med, Sch Med, Univ Pa, 48-50; from asst prof to prof pharmacol, Col Med, Baylor Univ, 50-57; prof med, Hahnemann Med Col & Hosp, 57-74, chmn dept med, 57-71, vpres acad affairs, 71-73; VPRES, DIR PROF & EDUC AFFAIRS, CONEMAUGH VALLEY MEM HOSP, 74- Concurrent Pos: From attend physician to sr attend, Jefferson Davis Hosp, Houston, 50-57; consult, Vet Admin Hosp, Houston & Houston Tuberc Hosp, 50-57; vis prof, Sch Med, La State Univ, 52; consult, Vet Admin Hosp, Philadelphia, 58-68, Philadelphia Naval Hosp, 58-, Bd Vet Appeals, 63- & comn drugs, AMA, 68-; deleg at large, AMA, 70-75; adv & consult, Hypertension Info & Educ Adv Comt, US Dept Health, 72-75; pres bd trustees, US Pharmacopeia, 72-75; adv, Gov Task Force Hypertension, State of Pa, 74-; ed consult, Am J Cardiol; ed cardiovasc sect, Cyclopedia Med, Surg & Specialties; Milliken lect, Pa Hosp, 58. Honors & Awards: Hunter Award, Am Therapeut Soc, 59; Clyde M Fish Mem Lect, 60; Mayo Found Honor Lect, 60; Susan & Theodore Cummings Humanitarian Award, 62, 65 & 66; Presidential Citation, 64. Mem: Fel Am Col Cardiol; Am Soc Clin Pharmacol & Therapeut (pres, 65); Am Acad Tuberc Physicians (pres, 61); fel Am Col Clin Pharmacol & Chemother (pres, 64-66); fel NY Acad Sci. Res: Hypertension and pharmacodynamics of the cardiovascular system; renal function. Mailing Add: Conemaugh Valley Mem Hosp 1086 Franklin St Johnstown PA 15905

MOYER, JOHN RAYMOND, b Buffalo, NY, June 9, 31; m 52; c 4. INORGANIC CHEMISTRY. Educ: Eastern Mich Univ, AB, 52; Univ Mich, PhD(phys inorg chem), 58. Prof Exp: Res chemist, Electro-inorg Res Lab, 59-63, sr res chemist, 63-68, assoc scientist, 68-70, environ res lab, 70-74, ASSOC SCIENTIST, CENT RES-INORG LAB, DOW CHEM, USA, 74- Mem: Am Chem Soc. Res: Aqueous chemistry of halogens; peroxide chemistry. Mailing Add: 2704 Swede Rd Midland MI 48640

MOYER, JOSEPH DONALD, b Dunbar, Pa, Jan 15, 20; m 46; c 3. ORGANIC CHEMISTRY. Educ: Pa State Col, BS, 48, MS, 49; Univ Md, PhD(org chem), 58. Prof Exp: Chemist, Nat Bur Standards, 49-60; CHEMIST, W R GRACE & CO, 60- Mem: Am Chem Soc. Res: Carbohydrates; isotopic tracers in organic chemistry; polymer research; synthesis of monomers for photopolymers. Mailing Add: W R Grace & Co Wash Res Ctr Columbia MD 21044

MOYER, MARY PAT SUTTER, b Arlington, Mass, Apr 27, 51; m 75; c 1. VIROLOGY, ONCOLOGY. Educ: Fla Atlantic Univ, BS, 72, MS, 74. Prof Exp: Dir tissue culture res virol, Equine Res Inst, 70-73; CANCER RES SCIENTIST VIRAL ONCOL, THORMAN CANCER RES LAB, TRINITY UNIV, 74- Mem: Am Soc Microbiol; AAAS; Tissue Culture Asn. Res: Biological activity of simian virus 40 DNA fragments; characterization of BK virus; tumor immunology. Mailing Add: Trinity Univ Box 295 San Antonio TX 78284

MOYER, MELVIN ISAAC, b Newton, Kans, June 30, 21; m 61. ORGANIC CHEMISTRY. Educ: Bethel Col, AB, 42; Univ Okla, MS, 44; Univ Kans, PhD(chem), 52. Prof Exp: Asst, Univ Okla, 42-44; res chemist, Cities Serv Oil Co, 46-48; asst instr, Univ Kans, 48-50; develop chemist, 52-57, chief chemist mfg, 57-62, sr chemist, NJ, 62-73, SR CHEMIST, AM CYANAMID CO, W VA, 73- Mem: AAAS; Am Chem Soc. Res: Manufacturing. Mailing Add: Am Cyanamid Co Willow Island WV 26190

MOYER, PATRICIA HELEN, b Greensboro, NC, Sept 30, 27; m 50; c 3. ORGANIC CHEMISTRY. Educ: Northwestern Univ, BA, 49; Univ Wis, PhD(chem), 54. Prof Exp: Res chemist, Phillips Petrol Co, 53; sr chemist, Clevite Corp, 55-56; sr res chemist, Res Ctr, B F Goodrich Co, 56-63; head biochem lab, Midwest Med Res Found, 65-66; sr res chemist, Frontier Chem Co, 66-67; chem div, Vulcan Mat Co, 67-68, group leader chem res, 68-73; INSTR VIS STAFF, PHOENIX COL, 74- Mem: AAAS; fel Am Inst Chem; Am Chem Soc. Res: Rates and mechanisms of organic reactions; polymerization; organometallics. Mailing Add: 8102 N 6th St Phoenix AZ 85020

MOYER, RALPH OWEN, JR, b New Bedford, Mass, May 19, 36. INORGANIC CHEMISTRY. Educ: Southeastern Mass Univ, BS, 57; Univ Toledo, MS, 63; Univ Conn, PhD(inorg chem), 69. Prof Exp: Develop engr, Union Carbide Corp, 57-64; asst prof, 69-75, ASSOC PROF CHEM, TRINITY COL, CONN, 75- Mem: Am Chem Soc; Sigma Xi. Res: Preparation, structure and properties of ternary hydrides. Mailing Add: Clement Chem Lab Trinity Col Hartford CT 06106

MOYER, REX CARLTON, b Elkhart, Ind, Dec 8, 35; m 58; c 3. CANCER, MICROBIOLOGY. Educ: Purdue Univ, BS, 57; Univ Nebr, MS, 61; Univ Tex, PhD(microbiol), 65. Prof Exp: Asst bacteriologist, Miles-Ames Res Labs, 57-58; lab instr gen microbiol, Univ Nebr, 58-61; trainee molecular biol, Univ Tex, 61-65; Nat Acad Sci-Nat Res Coun res fel microbial genetics & bacteriophagy, Ft Detrick, 65-66; res microbiologist, Ft Detrick, 66-69; co-dir, Bettye Thorman Cancer Res Lab, 70-75, ASST PROF BIOL, TRINITY UNIV, 69-, DIR, BETTYE THORMAN CANCER RES LAB, 75- Mem: Am Soc Microbiol; Tissue Cult Asn; NY Acad Sci. Res: Cancer virology; microbial genetics; nucleic acids. Mailing Add: Dept of Biol Trinity Univ San Antonio TX 78284

MOYER, ROBERT DALE, b Allentown, Pa, Sept 5, 38; m 63; c 2. MATHEMATICAL ANALYSIS. Educ: Pa State Univ, BS, 60; Univ Calif, Berkeley, MS, 62, PhD(appl math), 64. Prof Exp: Mem tech staff, Bellcomm, Inc, 64-65; asst prof math, Pa State Univ, 65-67; ASSOC PROF MATH, UNIV KANS, 67- Concurrent Pos: Assoc prof, Purdue Univ, 73-74. Mem: Am Math Soc; Inst Advan Study. Res: Partial differential equations; functional analysis; global analysis; complex analysis. Mailing Add: Dept of Math Univ of Kans Lawrence KS 66044

MOYER, ROBERT (FINDLEY), b New York, NY, May 12, 37; m 61; c 3. RADIATION PHYSICS. Educ: Pa State Univ, BS, 59, MS, 61; Univ Calif, Los Angeles, PhD(med physics), 65. Prof Exp: Instr radiol physics, 65-70, ASST PROF RADIOL, STATE UNIV NY UPSTATE MED CTR, 70- Mem: Am Asn Physicists in Med. Res: Radiological physics and biology. Mailing Add: Dept of Radiol State Univ NY Upstate Med Ctr Syracuse NY 13210

MOYER, RONALD CLARENCE, organic chemistry, see 12th edition

MOYER, RUDOLPH HENRY, b Sask, June 1, 35; m 57; c 2. BIOCHEMISTRY. Educ: Univ BC, BSA, 58, MSc, 62; Univ Calif, Los Angeles, PhD(biochem), 66. Prof Exp: Food technologist, Fisheries Res Bd, Can, 58-60; sr biochemist, Aerojet-Gen Corp Div, 65-69; SR SCIENTIST, GEOMET, INC, 69- Mem: AAAS; Am Chem Soc. Res: Development of instrument systems for bio-medical, geochemical and environmental applications; detection and quantitation of aerosols and other atmospheric pollutants. Mailing Add: Geomet Inc 2814A Metropolitan Pl Pomona CA 91767

MOYER, SAMUEL EDWARD, b Hershey, Pa, Oct 5, 34; m 59; c 2. POPULATION GENETICS. Educ: Pa State Univ, BS, 56; Univ NH, MS, 59; Univ Minn, PhD(genetics), 64. Prof Exp: Asst geneticist, NC State Univ, 64-65; asst prof genetics, Northeastern Univ, 66-71; ASSOC PROF BIOL, BURLINGTON COUNTY COL, 71- Mem: Genetics Soc Am; Am Genetic Asn. Res: Genetic traits of economic importance in poultry; effects of linkage on survival; genetic loads of populations. Mailing Add: Dept of Biol Burlington County Col Pemberton NJ 08068

MOYER, VANCE EDWARDS, b Orwigsburg, Pa, Nov 22, 14; m 53; c 2. METEOROLOGY. Educ: Pa State Univ, BS, 50, MS, 51, PhD(meteorol), 54. Prof Exp: Res asst meteorol, Pa State Univ, 51, res assoc, 52-54; asst prof, Univ Tex, 54-58; assoc prof, 58-61, chmn instruct meteorol, 60-66, actg head dept, 66-67, head dept, 67-75, PROF METEOROL, TEX A&M UNIV, 61- Concurrent Pos: NSF lectr, 58-64; mem earth sci curriculum proj, 65. Mem: Am Meteorol Soc; Am Geophys Union. Res: Cloud and precipitation physics; physical and radar meteorology. Mailing Add: Dept of Meteorol Tex A&M Univ College Station TX 77843

MOYER, WALTER ALLEN, JR, b Philadelphia, Pa, Nov 16, 22; m 46; c 2. ORGANIC CHEMISTRY. Educ: Philadelphia Col Pharm, BSc, 43; Middlebury Col, MSc, 48; Univ Del, PhD(org chem), 51. Prof Exp: From instr to assoc prof, 51-67, PROF CHEM, MIDDLEBURY COL, 67-, ASSOC DEAN, COL INST RES & SPEC ADMIS, 71- Mem: Am Chem Soc. Res: Carbohydrates; natural products; organic synthesis. Mailing Add: 9 Adirondack View Middlebury VT 05735

MOYER, WENDELL WILLIAM, JR, organic chemistry, see 12th edition

MOYER, WILLIAM C, JR, b Dallas, Tex, Apr 5, 37. UNDERWATER ACOUSTICS. Educ: Southern Methodist Univ, BS, 59; Univ Tex, PhD(mech eng), 66. Prof Exp: Mem tech staff, Hughes Aircraft Co, 59; sr scientist, 66-68, asst dir res dept, 68-69, dir anal dept, 69-71, asst vpres appl technol div, 71-74, VPRES ANAL & APPL RES DIV, TRACOR INC, 74- Mem: Acoust Soc Am. Res: Sonar system performance analysis; radiation associated with underwater arrays; transducer and baffle interactions; specialized underwater sensor systems. Mailing Add: Tracor Inc 6500 Tracor Lane Austin TX 78721

MOYERMAN, ROBERT MAX, b Atlantic City, NJ, Sept 14, 25; m 51; c 3. ORGANIC CHEMISTRY, ANALYTICAL CHEMISTRY. Educ: Rutgers Univ, BS, 49; Univ Ala, MS, 51. Prof Exp: Chemist high temperature nuclear reactors, Nuclear Develop Assocs, Inc, 51-52; assoc chemist carbohydrate & anal chem, Johns Hopkins Univ, 52-53, chem kinetics, Appl Physics Lab, 53-55; res investr, Am Smelting & Ref Co, 55-58; res chemist anal chem & org separations, Ansul Co, Wis, 58-63, sr res chemist process res, Org Res & Sect Head Anal Dept, 63-65; sr chemist, Scholler Bros, Inc, Pa, 65-68; group leader, Chem Div, Sun Chem Corp, RI, 68-70; owner & new prod mgr, Warwick Labs, 70-74; MGR RES & DEVELOP, HYDROLABS, INC, PATERSON, 74- Concurrent Pos: Tech serv dir, Org Chem Corp, 70-72. Mem: Am Chem Soc; Am Microchem Soc; Am Asn Textile Chemists & Colorists; Sigma Xi. Res: Organic syntheses; organic arsenic. Mailing Add: 118 Edmond Dr Warwick RI 02886

MOYERS, JACK, b Sidney, Iowa, Dec 7, 21; m 45; c 2. ANESTHESIOLOGY. Educ: Univ Iowa, BS, 43, MD, 45; Am Bd Anesthesiol, dipl, 53. Prof Exp: Intern med, Mt Carmel Mercy Hosp, Detroit, 45-46; resident anesthesiol, Col Med, Univ Iowa, 48-50; instr, WHO Anesthesiol Training Ctr, Univ Copenhagen, 50-51; instr, 51-52, assoc, 52-53, from asst prof to assoc prof, 53-66, actg head dept, 67-68, PROF ANESTHESIA, COL MED, UNIV IOWA, 66-, HEAD DEPT, 68-; ATTEND ANESTHESIOLOGIST, VET ADMIN HOSP, 52- Mem: Am Soc Anesthesiol; Asn Univ Anesthetists; fel Am Col Anesthesiol; Am Heart Asn; NY Acad Sci. Res: Clinical and laboratory investigation in field of anesthesiology. Mailing Add: Dept of Anesthesia Univ of Iowa Col of Med Iowa City IA 52240

MOYERS, JARVIS LEE, b Houston, Tex, Sept 7, 43; m 66; c 1. CHEMISTRY. Educ: Marshall Univ, BS, 65; Univ Hawaii, PhD(chem), 70. Prof Exp: LAB DIR, DEPT CHEM, UNIV ARIZ, 71- Mem: AAAS; Am Chem Soc; Am Geophys Union; Am Soc Testing & Mat. Res: Analytic environmental chemistry; atmospheric chemistry. Mailing Add: Atmospheric Anal Lab Dept Chem Univ of Ariz Tucson AZ 85721

MOYERS, ROBERT EDISON, b Sidney, Iowa, Nov 19, 19; m 56; c 2. ORTHODONTICS, HUMAN DEVELOPMENT. Educ: Univ Iowa, BS & DDS, 42, MS, 47, PhD(physiol), 49. Prof Exp: Assoc orthod, Col Dent, Univ Iowa, 45-47, instr, 47-49; prof & head dept, Fac Dent, Univ Toronto, 49-53; head dept orthod, 53-65, PROF DENT, SCH DENT, UNIV MICH, ANN ARBOR, 53-, DIR CTR HUMAN GROWTH & DEVELOP, 65- Concurrent Pos: Fulbright scholar, Nat Univ Athens, 51, vis prof, 64; consult, WHO, 57- Honors & Awards: Milo Hellman Res Award, 50; mem, Order Brit Empire; Order of Phoenix. Mem: AAAS; Am Asn Orthod; Am Dent Asn; Int Asn Dent Res; Int Soc Craniofacial Biol. Res: Electromyography; facial growth. Mailing Add: Ctr for Human Growth & Develop Univ of Mich 1111 E Catherine St Ann Arbor MI 48109

MOYLE, CLARENCE LLEWELLYN, b Hibbing, Minn, Apr 25, 09; m 43, 58; c 1. ORGANIC CHEMISTRY. Educ: Univ Minn, BS, 30, MS, 32, PhD(org chem), 35. Prof Exp: Asst chem, Univ Minn, 32-33 & 34-35; anal chemist, US Steel Corp, Minn, 34; sr chemist, State Hwy Dept, 35; group leader org res, Dow Chem Co, 35-43, process develop mgr, Dow Corning Corp, 43-45; chief chem div, Deering Milliken Res Trust, 45-46; bio-org res, Dow Chem USA, 46-55, res scientist, 55-73; RETIRED. Mem: Am Chem Soc; NY Acad Sci. Res: Synthesis of biologically active structures. Mailing Add: 40 Hibiscus Drive Punta Gorda FL 33950

MOYLE, JOHN BRIGGS, biology, see 12th edition

MOYLE, PETER BRIGGS, b Minneapolis, Minn, May 29, 42; m 66. ICHTHYOLOGY, AQUATIC ECOLOGY. Educ: Univ Minn, BA, 64, PhD(zool), 69; Cornell Univ, MS, 66. Prof Exp: Asst prof biol, Fresno State Col, 69-72; ASST PROF BIOL, DIV WILDLIFE & FISHERIES BIOL, UNIV CALIF, DAVIS, 72- Mem: AAAS; Ecol Soc Am; Am Fisheries Soc; Am Soc Ichthyol & Herpet. Res: Ecology of freshwater fishes and amphibians; distribution and ecology of freshwater fishes of California. Mailing Add: Div of Wildlife & Fisheries Biol Univ Calif Davis CA 95616

MOYLE, RICHARD W, b American Fork, Utah, Mar 22, 30; m 57; c 2. PALEONTOLOGY, GEOLOGY. Educ: Brigham Young Univ, BS, 53, MS, 58; Univ Iowa, PhD(gen geol), 63. Prof Exp: Instr geol, Western State Col Colo, 61-63, asst prof, 63-65; from asst prof to assoc prof, 65-71, actg head dept geol & geog, 68-69, chmn dept, 69-75, PROF GEOL, WEBER STATE COL, 71- Mem: Geol Soc Am; Nat Asn Geol Teachers; Soc Econ Paleont & Mineral; Paleont Asn. Res: Mississippian and Pennsylvanian sponges; paleoecology of Upper Mississippian and Lower Pennsylvanian sediments in west central Utah; ammonoids of Wolfcampian from the Glass Mountains of west Texas and continguous areas; microcrystals and photography of same upper; Mississippian Blastoids of Utah. Mailing Add: Dept of Geol & Geog Weber State Col Ogden UT 84403

MOYLE, SUSAN MARY, b St Paul, Minn, May 4, 44. BRYOLOGY. Educ: Carleton Col, BA, 66; Univ Tenn, Knoxville, PhD(bot), 73. Prof Exp: Instr biol, Wellesley Col, 71-72; asst prof, Va Commonwealth Univ, 72-74; ASST PROF BIOL, CENTRE COL KY, 74- Mem: AAAS; Am Bryol & Lichenological Soc; Am Inst Biol Sci; Bot Soc Am; Int Asn Plant Taxon. Res: Floristics and ecology of bryophytes of Kentucky and Virginia; culturing of bryophytes. Mailing Add: Div of Sci & Math Centre Col of Ky Danville KY 40422

MOYLS, BENJAMIN NELSON, b Vancouver, BC, May 1, 19; m 42; c 2. ALGEBRA. Educ: Univ BC, BA, 40, MA, 41; Harvard Univ. AM, 42, PhD(math), 47. Prof Exp: From instr to assoc prof, 47-59, PROF MATH, UNIV BC, 59-, ASST DEAN GRAD STUDIES, 67- Mem: Am Math Soc; Math Asn Am; Soc Indust & Appl Math; Edinburgh Math Soc; Can Math Cong. Res: Linear algebra. Mailing Add: Dept of Math Univ of BC Vancouver BC Can

MOYNIHAN, CORNELIUS TIMOTHY, b Inglewood, Calif, Feb 2, 39; m 63; c 2. PHYSICAL CHEMISTRY, GLASS TECHNOLOGY. Educ: Univ Santa Clara, BS, 60; Princeton Univ, MA, 62, PhD(chem), 65. Prof Exp: From asst prof to assoc prof chem, Calif State Col Los Angeles, 64-69; assoc prof chem, 68-69, assoc prof chem eng, 69-75, PROF CHEM ENG, CATH UNIV AM, 75- Concurrent Pos: Res assoc, Purdue Univ, 68-69. Mem: Am Chem Soc; Am Ceramic Soc. Res: Physical chemistry of molten salts, electrolyte solutions and glasses; molecular engineering of glasses. Mailing Add: Dept of Chem Eng & Mat Sci Cath Univ of Am Washington DC 20064

MOYNIHAN, MARTIN HUMPHREY, b Chicago, Ill, Feb 5, 28. ANIMAL BEHAVIOR. Educ: Princeton Univ, AB, 50, DPhil(zool), Oxford Univ, 53. Prof Exp: Vis fel, Cornell Univ, 53-55; res fel, Harvard Univ, 55-57; DIR SMITHSONIAN TROP RES INST, 57- Mem: Soc Study Evolution; Am Ornith Union; Am Soc Naturalists; Asn Trop Biol. Res: Behavior, ecology and evaluation. Mailing Add: Smithsonian Trop Res Inst PO Box 2072 Balboa CZ

MOYNIHAN, ROBERT EDWARD, b Batavia, NY, Aug 3, 28; m 50; c 8. PHYSICAL CHEMISTRY. Educ: Canisius Col, BS, 50; Purdue Univ, PhD(chem), 54. Prof Exp: RES ASSOC, WASHINGTON LAB, PLASTICS DEPT, E I DU PONT DE NEMOURS & CO, INC, 54- Mem: Am Chem Soc; Am Inst Chemists. Res: Chemical physics; polymer chemistry; automotive safety glazing. Mailing Add: RD 2 Lowell OH 45744

MOZELL, MAXWELL MARK, b Brooklyn, NY, May 20, 29; m 55; c 4. SENSORY PHYSIOLOGY, PSYCHOPHYSIOLOGY. Educ: Brown Univ, AB, 51, MSc, 53, PhD(phsiol psychol), 56. Prof Exp: Fel physiol, Fla State Univ, 59-61; from asst prof to assoc prof, 61-70, PROF PHYSIOL, STATE UNIV NY UPSTATE MED CTR, 70-, ASSOC DEAN, 71- Concurrent Pos: Mem study panel, NSF, 74- Mem: Am Physiol Soc; Am Psychol Asn; Soc Neurosci; Sigma Xi. Res: Sensory psychophysiology; oflaction; electrophysiology; determine the physical, chemical, physiological mechanisms basic to olfactory discriminations. Mailing Add: Dept of Physiol State Univ of NY Upstate Med Ctr Syracuse NY 13210

MOZEN, MILTON (MICHAEL), biochemistry, see 12th edition

MOZER, BERNARD, b Denver, Colo, Dec 2, 25; m 57; c 3. PHYSICS. Educ: Univ Denver, BS, 50; Univ Colo, MS, 52; Carnegie Inst Technol, PhD(physics), 60. Prof Exp: Res assoc physics, Univ Denver, 52-53; jr physicist, Brookhaven Nat Lab, 53-55; res asst physics, Carnegie Inst Technol, 55-59; res assoc, Brookhaven Nat Lab, 59-61, from asst physicist to assoc physicist, 61-67; PHYSICIST, NAT BUR STAND, 67- Mem: Am Phys Soc. Res: Theoretical plasma physics; theoretical and experimental aspects of Mössbauer effect; inelastic neutron scattering experiments on liquids, metals and alloys; vibrational and electronic effects of impurities in solids. Mailing Add: Reactor Bld Nat Bur of Stand Washington DC 20234

MOZER, FORREST S, b Lincoln, Nebr, Feb 13, 29; m 58; c 3. SPACE PHYSICS. Educ: Univ Nebr, BS, 51; Calif Inst Technol, MS, 53, PhD(physics), 56. Prof Exp: Fel, Calif Inst Technol, 56-57; res scientist, Lockheed Res Lab, 57-61 & Aerospace Corp, 62-63; res dir space physics, Univ Paris, 63-66; from asst prof to assoc prof, 66-70, PROF PHYSICS, UNIV CALIF, BERKELEY, 70- Mem: Am Geophys Union; Am Phys Soc. Mailing Add: Dept of Physics Univ of Calif Berkeley CA 94720

MOZERSKY, SAMUEL M, b Sask, Can, Sept 19, 24; US citizen; m 55; c 2. BIOCHEMISTRY, ENZYMOLOGY. Educ: Univ Calif, Los Angeles, BA, 47; Univ

Southern Calif, PhD(biochem), 57. Prof Exp: Sr lab asst turnover of plasma proteins, Col Med, Univ Ill, 52-55; grad res physiol chemist, Med Ctr, Univ Calif, Los Angeles, 56-57, asst res physiol chemist, 57-61; RES BIOCHEMIST, EASTERN REGIONAL LAB, USDA, WYNDMOOR, 63- Concurrent Pos: USPHS spec fel, Dept Biol, Brookhaven Nat Lab, 61-63. Mem: Biophys Soc; Am Chem Soc; NY Acad Sci. Res: Biosynthesis of plasma proteins; purification and characterization of hyaluronidases; protein modification; mechanisms of enzyme action; structure and enzymatic properties of contractile proteins. Mailing Add: Eastern Regional Lab USDA 600 E Mermaid Lane Wyndmoor PA 19118

MOZINGO, HUGH NELSON, b Monongahela, Pa, Apr 23, 25; m 49. BOTANY. Educ: Univ Pittsburgh, BS, 46, MS, 47; Columbia Univ, PhD(bot), 50. Prof Exp: Asst biol, Univ Pittsburgh, 46-47; asst bot, Columbia Univ, 47-50; instr, Univ Tenn, 50-51; assoc prof biol & bot & chmn sci div, Fla Southern Col, 51-55; asst prof, Mich State Univ, 55-59; assoc prof, 59-68, PROF BIOL, UNIV NEV, RENO, 68-, CHMN DEPT, 69- Concurrent Pos: Chmn Rocky Mt Sci coun, 72-73. Mem: AAAS; Bot Soc Am; Am Soc Plant Physiol; Am Bryol & Lichenological Soc; Electron Micros Soc Am. Res: Plant morphogenesis; electron microscopy; cytochemistry; anaerobic effects on plants, Nevada bryophytes. Mailing Add: Dept of Biol Univ of Nev Reno NV 89507

MOZINGO, JAMES ROBERT, JR, medicinal chemistry, organic chemistry, see 12th edition

MOZLEY, JAMES MARSHALL, JR, b Marion, Ill, Nov 1, 22; m 44; c 1. BIOMEDICAL ENGINEERING. Educ: Washington Univ, BS, 43, MS, 47, PhD(chem & elec eng), 50. Prof Exp: Instr chem eng, Washington Univ, 47-49; assoc chemist sec oil recovery, Atlantic Refining Co, Tex, 49-51; sr instr chem eng, Polytech Inst Brooklyn, 51-52; res engr, automatic control, E I du Pont de Nemours & Co, Del, 51-57; assoc prof radiol, Johns Hopkins Univ, 57-65, dir div radiation chem, Sch Pub Health & Hyg, 59-65; PROF RADIOL, UNIV HOSP, STATE UNIV NY UPSTATE MED CTR, 65-, DIR DIV RADIOL PHYSICS & ENG, 75-; PROF CHEM ENG, SYRACUSE UNIV, 67- Concurrent Pos: Pres, Radiation Assocs Md, Inc, 59-; ed, Trans, Instrument Soc Am, 62- Mem: AAAS; Soc Nuclear Med; Am Chem Soc; Am Soc Mech Eng; sr mem Instrument Soc Am. Res: Automatic control of chemical processes; radiological instrumentation; data processing; computation; nuclear medicine. Mailing Add: Dept of Radiol State Univ NY Upstate Med Ctr Syracuse NY 13210

MOZLEY, ROBERT FRED, b Boston, Mass, Apr 18, 17; c 2. NUCLEAR PHYSICS. Educ: Harvard Univ, AB, 38; Univ Calif, MS & PhD(physics), 50. Prof Exp: Elec engr radar, Sperry Gyroscope Co, 41-45; asst radiation lab, Univ Calif, 45-50; from instr to asst prof physics, Princeton Univ, 50-53; assoc prof, 53-62, PROF PHYSICS, STANFORD UNIV, 62- Mem: Fel Am Phys Soc. Res: Elementary particle physics. Mailing Add: Stanford Linear Accelerator Ctr Stanford Univ Stanford CA 94305

MOZLEY, SAMUEL CLIFFORD, b Atlanta, Ga, Aug 13, 43; m 64; c 2. AQUATIC ECOLOGY. Educ: Emory Univ, BS, 64, MS, 66, PhD(animal ecol), 68. Prof Exp: NATO fel, Max Planck Inst Limnol, Ger, 68-69, NSF fel, 69-70; Nat Res Coun Can fel, Univ Toronto, 70; ASST RES SCIENTIST, GREAT LAKES RES DIV, UNIV MICH, ANN ARBOR, 70- Mem: Am Soc Limnol & Oceanog; Int Asn Theoret & Appl Limnol; Int Asn Great Lakes Res. Res: Taxonomy and morphology of Chironomidae; community structure of benthic animals; benthos and pollution in fresh waters; methodology of benthic sampling. Mailing Add: Great Lakes Res Div Univ of Mich Ann Arbor MI 48105

MOZOLA, ANDREW JOHN, b Pittston, Pa, Dec 6, 15; m 64; c 2. GEOLOGY. Educ: Wayne State Univ, BA, 36; Syracuse Univ, MSc, 38, PhD, 54. Prof Exp: Spec instr geol, 46-47, instr gen geol, 47-51, from asst prof to assoc prof, 51-70, PROF GEN GEOL, WAYNE STATE UNIV, 70- Mem: Fel AAAS; fel Geol Soc Am; Am Asn Petrol Geologists; Nat Asn Geol Teachers. Res: Engineering, groundwater and glacial geology; geomorphology; economic geology of the nonmetallic resources; environmental geology of southeastern Michigan. Mailing Add: Dept of Geol Wayne State Univ Detroit MI 48202

MOZUMDER, ASOKENDU, b Baherok, India, June 2, 31; m 61. RADIATION CHEMISTRY, RADIATION PHYSICS. Educ: Univ Calcutta, BSc, 50, MSc, 53; Indian Inst Technol, Kharagpur, PhD(physics). 61. Prof Exp: From assoc lectr to lectr physics, Indian Inst Technol, Kharagpur, 54-62; assoc, 62-65, from assoc res scientist to res scientist, 65-69, ASSOC FAC FEL, RADIATION LAB, UNIV NOTRE DAME, 69- Mem: Am Phys Soc; Radiation Res Soc. Res: Theoretical radiation chemistry; application of the methods of theoretical physics to problems involving interaction of radiation with matter. Mailing Add: Radiation Lab Univ of Notre Dame Notre Dame IN 46556

MOZZI, ROBERT LEWIS, b Meriden, Conn, Dec 8, 31; m 56; c 1. EXPERIMENTAL SOLID STATE PHYSICS. Educ: Villanova Univ, BS, 53; Univ Pittsburgh, MS, 56; Mass Inst Technol, PhD(physics), 68. Prof Exp: Physicist, Pratt & Whitney Aircraft Div, United Aircraft Corp, 55-57; PRIN SCIENTIST, RES DIV, RAYTHEON CO, 57- Mem: Am Crystallog Asn; Am Phys Soc. Res: X-ray diffraction studies of the structure of glass and imperfections in crystals; ion implantation in semiconductors. Mailing Add: Res Div Raytheon Co Waltham MA 02154

MPELKAS, CHRISTOS C, plant physiology, horticultural sciences, see 12th edition

MRAZ, FRANK RUDOLPH, b New York, NY, Feb 25, 25. NUTRITION, BIOCHEMISTRY. Educ: Rutgers Univ, BS, 51; Univ Pa, MS, 53, PhD(agr & biol chem, poultry husb), 54. Prof Exp: From asst scientist to assoc scientist, 55-65, assoc prof nutrit biochem, 55-65, SCIENTIST & PROF METAB & NUTRIT, UNIV TENN-ATOMIC ENERGY COMN COMP ANIMAL RES LAB, OAK RIDGE, 65- Mem: Soc Exp Biol & Med; Poultry Sci Asn; Am Inst Nutrit; Radiation Res Soc. Res: Comparative metabolism and influence of diet on metabolic pathways of fission products and industrial nuclides in mammalian species; mechanisms of discrimination between the elements at physiological interfaces. Mailing Add: 1299 Bethel Valley Rd Oak Ridge TN 37830

MRAZ, RICHARD GEORGE, organic chemistry, see 12th edition

MRAZEK, RUDOLPH G, b Chicago, Ill, May 23, 22; m 44; c 3. SURGERY. Educ: Univ Ill, BA, 41, MD, 44, MS, 45. Prof Exp: Resident surg, MacNeal Hosp, 45-46 & Hines Vet Admin Hosp, 49-52; from instr to clin assoc prof, 52-73, PROF SURG, UNIV ILL COL MED, 73-; DIR MED EDUC, MACNEAL HOSP, 71- Mem: AMA; Am Asn Cancer Res; fel Am Col Surg; Soc Surg Alimentary Tract; James Ewing Soc. Res: Cancer chemotherapy. Mailing Add: 3237 Oak Park Ave Berwyn IL 60402

MROCHEK, JOHN EDWARD, chemistry, analytical biochemistry, see 12th edition

MROCZKOWSKI, STANLEY, b Poland, Mar 29, 25; US citizen; m 54; c 1. SOLID STATE CHEMISTRY. Educ: Adam Mickiewicz Univ, Poznan, MPH, 52; Univ Warsaw, Cand Sci & PhD(inorg chem), 56. Prof Exp: Group leader rare earth compounds, Inst Electron Technol, Polish Acad Sci, Warsaw, 56-58; head anal sect, Weitzman Inst, Albar Kvar-Saba, 59-61; res assoc chem, Columbia Univ, 61-63; SR RES ASSOC CHEM RARE EARTH COMPOUNDS, YALE UNIV, 63- Concurrent Pos: Tech consult, David Sarnoff Res Ctr, RCA Corp, 67- & Autoclave Engrs, Inc, Erie, Pa, 74- Mem: Am Chem Soc; Inst Elec & Electronics Engrs; Am Crystal Growth Asn. Res: Crystallization processes under high pressure and high temperature; crystal growth from high temperature solution; vapor transport reaction. Mailing Add: Yale Univ Dept of Eng & Appl Sci 427 Becton Ctr New Haven CT 06520

MROWCA, ADALBERT, b Chicago, Ill, Mar 1, 13. PHYSICS. Educ: Univ Notre Dame, PhD, 44. Prof Exp: ASSOC PROF PHYSICS, UNIV NOTRE DAME, 44- Mem: Am Phys Soc. Res: Polymer physics; nuclear magnetic resonance. Mailing Add: Dept of Physics Univ of Notre Dame Notre Dame IN 46556

MROWCA, JOSEPH J, b Taylor, Pa, Jan 25, 39; m 60; c 3. ORGANOMETALLIC CHEMISTRY. Educ: Univ Scranton, BS, 60; Columbia Univ, MA, 62, PhD(chem), 65. Prof Exp: RES CHEMIST, CENT RES DEPT, E I DU PONT DE NEMOURS & CO, INC, 65- Mem: Am Chem Soc. Res: Transition metal catalysis. Mailing Add: 1513 Forsythia Ave Wilmington DE 19810

MROWKA, STANISLAW GRZEGORZ, mathematics, see 12th edition

MROZIK, HELMUT, b Habelschwerdt, Ger, Oct 23, 31; m 57; c 2. ORGANIC CHEMISTRY. Educ: Univ Basel, PhD(chem), 58. Prof Exp: Asst chem, Columbia Univ, 59-60; sr chemist, 60-68, SR RES FEL, MERCK & CO, 68- Mem: Am Chem Soc. Res: Chemotherapy of parasitic diseases; aromatic and heterocyclic chemistry. Mailing Add: 159 Idlebrook Lane Matawan NJ 07747

MROZINSKI, PETER MATTHEW, b Chicago, Ill, Apr 22, 47; m 70; c 1. APPLIED PHYSICS, LOW TEMPERATURE PHYSICS. Educ: St Mary's Col, Minn, BA, 69; Ohio State Univ, MS, 72, PhD(physics), 75- Prof Exp: RES PHYSICIST, E I DU PONT DE NEMOURS & CO, INC, 74- Mem: Am Phys Soc. Res: Applications of x-ray fluorescence. Mailing Add: Exp Sta B-357 E I du Pont de Nemours & Co Inc Wilmington DE 19898

MRTEK, MARSHA BEDFORD, b Crawfordsville, Ind, Dec 18, 42; m 66. PHARMACY. Educ: Univ Ill, BS, 66, MS, 69, PhD(pharm), 75. Prof Exp: Instr, 70-75, RES ASSOC PHARM, COL PHARM, UNIV ILL MED CTR, 75- Concurrent Pos: Consult, Walter Reed Army Inst Res, 66-69. Mem: Am Pharmaceut Asn; Am Inst Hist Pharm; Acad Pharmaceut Sci. Res: Physicochemical properties of drug-drug complexes; molecular complexes as models for drug-receptor interactions; statistical evaluation of tablet manufacturing parameters. Mailing Add: Univ of Ill Col of Pharm 833 SWood St Chicago IL 60612

MRTEK, ROBERT GEORGE, b Oak Park, Ill, Sept 2, 40; m 66. PHARMACY, HISTORY OF PHARMACY. Educ: Univ Ill, BS, 62, PhD(pharm), 67. Prof Exp: Resident res assoc chem, Argonne Nat Lab, 64-66; from asst prof to assoc prof pharm, 67-73, coordr educ res develop, 70-73, PROF PHARM, COL PHARM, UNIV ILL, CHICAGO, 73-, ASST DEAN EDUC DEVELOP, 73- Concurrent Pos: Consult, Walter Reed Army Inst Res, 66-69; Univ Ill res bd grants, 68-69; consult & examr, Civil Serv Comn, City of Chicago, 69; US Vitamin & Pharmaceut Co res grant, 70; spec proj prog, Health Professions, Bur Health Manpower Educ grant, 72-74; vis prof, Health Sci Ctr, Univ Wis, Madison, 75-76. Honors & Awards: C P van Schaak Chem Award, Lehn & Fink Gold Medal Award & Elich Prize, 62; Citation Hist Res, Am Inst Hist Pharm, 72. Mem: AAAS; Am Pharmaceut Asn; Am Inst Hist Pharm. Res: History of pharmacy in Illinois; drug partitioning in complex coacervate systems; self study resources for pharmaceutical education; continuing education for pharmacists, systems models; admissions criteria in higher education. Mailing Add: Dept of Pharm Prac Col of Pharm Univ of Ill Med Ctr PO Box 6998 Chicago IL 60680

MRUK, NORBERT JOHN, organic chemistry, polymer chemistry, see 12th edition

MRUK, WALTER FREDERICK, physics, see 12th edition

MUAN, ARNULF, b Lökken Verk, Norway, Apr 19, 23; nat US; m 60; c 2. GEOCHEMISTRY. Educ: Tech Univ Norway, dipl, 48; Pa State Univ, PhD(geochem), 55. Prof Exp: Instr anal chem, Tech Univ Norway, 48-49; asst geochem, 50-51, res assoc, 52-55, from asst prof to prof metall, 55-66, PROF MINERAL SCI, PA STATE UNIV, 66-, HEAD DEPT GEOSCI, 71- Concurrent Pos: Head dept geochem & mineral, Pa State Univ, 66-71. Honors & Awards: Purdy Award, Am Ceramic Soc, 59. Mem: Am Chem Soc; Am Ceramic Soc; Am Inst Mining, Metall & Petrol Engrs; Mineral Soc Am; Am Soc Metals. Res: Hetergeneous equilibria at high temperatures; thermodynamics at high temperatures, oxide systems; application of these principles to petrology, ore deposits, slag and refractory problems. Mailing Add: Dept of Geosci Pa State Univ University Park PA 16802

MUCCI, JOSEPH FRANCIS, b Southington, Conn, Apr m 53; c 3. PHYSICAL CHEMISTRY. Educ: Cent Conn State Col, BS, 50; Wesleyan Univ, MA, 53; Yale Univ, PhD, 57. Prof Exp: Asst, Yale Univ, 54-57; from instr to assoc prof, 57-66, PROF CHEM, VASSAR COL, 66-, CHMN DEPT, 75- Mem: AAAS; Am Chem Soc; NY Acad Sci; fel Am Inst Chem. Res: Complexions in various media by spectrophotometric, conductometric, polarographic, ion exchange, extraction and radiochemical methods; quantum chemical studies employing self consistent field-linear combination of atomic orbitals-molecular orbitals; statsitical mechanics. Mailing Add: Dept of Chem Vassar Col Poughkeepsie NY 12601

MUCCINI, GIUSEPPE ANTONIO, physical chemistry, see 12th edition

MUCENIEKS, PAUL RAYMOND, b Riga, Latvia, Feb 3, 21; US citizen; m 56; c 1. PHYSICAL CHEMISTRY, ELECTROCHEMISTRY. Educ: Johns Hopkins Univ, MA, 61, PhD(phys chem), 64. Prof Exp: Prin physicist, Litton Systs, Litton Indust, Inc, 64; SR RES CHEMIST, FMC CORP, 64- Mem: Am Chem Soc; Electrochem Soc; Am Inst Chemists. Res: Studies of reaction kinetics and mechanisms; x-ray diffraction in heavy metal salt solutions; synthesis of stable free radicals; electrochemical synthesis and electrodialysis; chemical instrumentation; elimination of industrial pollutants. Mailing Add: 338 Glenn Ave Lawrenceville NJ 08648

MUCHMORE, HAROLD GORDON, b Ponca City, Okla, Mar 8, 20; m 54; c 4. INTERNAL MEDICINE. Educ: Rice Univ, BA, 43; Univ Okla, MD, 46, MS, 56; Am Bd Internal Med, dipl, 62. Prof Exp: Intern, Jersey City Med Ctr, 46-47; univ fel, Univ Okla, 47-48, instr pharmacol, Sch Med, 48-49, instr med, 49-52, resident, Univ Hosps, 54-56, asst prof, 57-62; assoc prof, Med Sch, Univ Minn, 62-66; assoc prof med, microbiol & immunol, 66-70, chief infectious dis sect, 66-68, Carl Puckett assoc prof pulmonary dis, 68-70, CARL PUCKETT PROF PULMONARY DIS & PROF MED, MED SCH, UNIV OKLA, 70-, PROF MICROBIOL & IMMUNOL, 71-

Concurrent Pos: Clin investr, Vet Admin, 57-60, chief tuberc & infectious dis sect, Vet Admin Hosp, Oklahoma City, 60-62 & 66-; chief infectious dis sect, Ancker Hosp, St Paul, Minn, 62-66. Mem: Fel Am Col Physicians; Am Fedn Clin Res; Am Thoracic Soc; Med Mycol Soc Americas; Int Soc Human & Animal Mycol. Res: Clinical research; infectious diseases; fungus disease and tuberculosis. Mailing Add: Univ Okla Health Sci Ctr 800 NE 13th St Oklahoma City OK 73104

MUCHMORE, WILLIAM BREULEUX, b Cincinnati, Ohio, July 7, 20; m 43; c 2. ARACHNOLOGY. Educ: Oberlin Col, AB, 42; Washington Univ, PhD(zool), 50. Prof Exp: From instr to assoc prof, 50-70, PROF BIOL, UNIV ROCHESTER, 70- Concurrent Pos: Res assoc, Fla State Collection Arthropods, 74. Mem: Micros Soc Am; Am Soc Zool; Am Arachnolog Soc; Brit Arachnolog Soc. Res: Systematics and biogeography of pseudoscorpions. Mailing Add: Dept of Biol Univ of Rochester Rochester NY 14627

MUCHOW, GORDON MARK, b Evanston, Ill, June 15, 21; m 44; c 3. PHYSICAL CHEMISTRY. Educ: Northwestern Univ, Ill, BS, 42, MS, 51; St Louis Univ, PhD(chem), 54. Prof Exp: Res chemist, Pure Oil Co, 46-48; chemist, Graymills Corp, 48; res chemist, Monsanto Co, 53-62; from res scientist to sr res scientist, Owens-Ill, Inc. 62-73; STAFF TECHNOLOGIST, BRUSH WELLMAN, INC, 73- Mem: Am Chem Soc; Am Inst Chem; Am Crystallog Asn. Res: Silicates; x-ray crystallography; crystal chemistry. Mailing Add: 5923 Winding Way Sylvania OH 43560

MUCHOWSKI, JOSEPH MARTIN, b Odessa, Sask, Jan 30, 37; m 65; c 2. ORGANIC CHEMISTRY. Educ: Univ Sask, BSc, 58, MSc, 59; Univ Ottawa, PhD(org chem), 59. Prof Exp: Nat Res Coun Can overseas fel with Prof A Eschenmoser, Swiss Fed Inst Technol, 62-63; sr res chemist, Bristol Labs of Can, Que, 63-71; sr chemist, Syntex, SA, Mexico City, 71-72, asst dir res, 72-73, DIR CHEM RES, SYNTEX, SA, MEXICO CITY, 73-, ASST DIR RES, SYNTEX RES CENTRE, PALO ALTO, 75- Concurrent Pos: Prof extraordinary, Iberoamerican Univ, Mex, 74- Mem: Am Chem Soc; Chem Inst Can. Res: Mechanistic and synthetic organic chemistry; medicinal chemistry. Mailing Add: Div of Investigation Syntex SA Apartado Postal 10-820 Mexico DF Mexico

MUCK, DARREL LEE, b Larned, Kans, Jan 26, 38; m 60; c 2. PHYSICAL ORGANIC CHEMISTRY. Educ: Wichita State Univ, BS, 59, MS, 62; Univ Fla, PhD(phys org chem), 65. Prof Exp: Res chemist, Procter & Gamble Co, 65-71 & Pfizer, Inc, 71-72; tech mgr detergents res, 72-74, PROD MGR DETERGENT CHEM, PHILADELPHIA QUARTZ CO, VALLEY FORGE, PA, 74- Mem: Am Oil Chemists Soc. Res: Chelating and/or sequestering tendencies of organic hydroxy acids and inorganic polymers to heavy metal ions. Mailing Add: 19 Oak Hill Circle Malvern PA 19355

MUCK, GEORGE A, b Fillmore, Ill, Sept 28, 37; m 59; c 2. DAIRY SCIENCE, BIOCHEMISTRY. Educ: Univ Ill, BS, 59, MS, 61, PhD(dairy tech), 62. Prof Exp: Res asst food tech, Univ Ill, 59-62; head prod develop sect, Res Dept, Dean Foods Co, 62-67, dir res, 67-70, V PRES, RES & DEVELOP, RES DEPT, DEAN FOODS CO, 70- Mem: Am Dairy Sci Asn; Inst Food Technol; Am Chem Soc; Am Oil Chemist Soc. Res: Dairy technology; isolation and identification of flavors; effect of high heat treatment on model milk systems; development of new dairy and food products. Mailing Add: Res Dept Dean Foods Co 1126 Kilburn Ave Rockford IL 61101

MUCKENFUSS, CHARLES, b Cleveland, Ohio, May 2, 27; m 54; c 2. THEORETICAL CHEMISTRY. Educ: Univ Wis, PhD(chem), 57. Prof Exp: Nat Res Coun fel, Nat Bur Standards, DC, 57-58; res assoc, Gen Elec Res & Develop Ctr, NY, 58-67; ASSOC PROF CHEM ENG, RENSSELAER POLYTECH INST, 67- Res: Kinetic theory; statistical mechanics; irreversible thermodynamics; transport phenomena. Mailing Add: Div of Chem Eng Rensselaer Polytech Inst Troy NY 12181

MUCKENHOUPT, BENJAMIN, b Newton, Mass, Dec 22, 33; m 64; c 2. MATHEMATICAL ANALYSIS. Educ: Harvard Univ, AB, 54; Univ Chicago, MS, 55, PhD(math), 58; from instr to asst prof math, DePaul Univ, 58-60; from asst prof to assoc prof, 60-70, PROF MATH, RUTGERS UNIV, 70- Concurrent Pos: Vis assoc prof, Mt Holyoke Col, 63-65; visitor, Inst Advan Study, 68-69 & 75-76; vis prof, Stat Univ NY, Albany, 70-71. Mem: Am Math Soc; Math Asn Am. Res: Singular transformations; Fourier series. Mailing Add: Dept of Math Rutgers Univ New Brunswick NJ 08903

MUCKENTHALER, FLORIAN AUGUST, b McFarland, Kans, July 31, 33; m 69; c 2. ZOOLOGY. Educ: Spring Hill Col, BS, 59; Catholic Univ, PhD(zool), 64. Prof Exp: USPHS fel cell biol, Johns Hopkins Univ, 64-65; asst prof biol, State Univ NY Albany, 65-71; asst prof, 71-75, ASSOC PROF BIOL, BRIDGEWATER STATE COL, 75- Mem: AAAS; Am Soc Cell Biol; Am Genetics Asn; Am Inst Biol Sci; Am Soc Zool. Res: Developmental genetics; cell biology; nucleic acid synthesis in spermatogenesis and oogenesis; mechanisms in meiosis; accumulation of materials in oogenesis and utilization in early development. Mailing Add: Dept of Biol Sci Bridgewater State Col Bridgewater MA 02324

MUCKERMAN, JAMES TERRY, physical chemistry, theoretical chemistry, see 12th edition

MUDD, J GERARD, b St Louis, Mo, Apr 7, 21; m 46; c 5. CARDIOLOGY. Educ: Col Holy Cross, BS, 43; St Louis Univ, MD, 45. Prof Exp: Intern med, 45-46, resident, 48-50, instr, 51-55, sr instr, 55-56, asst prof, 56-64, ASSOC PROF MED, ST LOUIS UNIV, 64- Concurrent Pos: Fel, Johns Hopkins Univ, 50-51; St Louis Heart Asn fel cardiol, 51-52; Nat Heart Inst fel, 52-54. Mem: Am Heart Asn; Am Col Cardiol; Am Fedn Clin Res. Res: Cardiac catheterization, including right heart, left, retrograde and coronary arteriography. Mailing Add: Dept of Internal Med St Louis Univ Sch of Med St Louis MO 63104

MUDD, JOHN BRIAN, b Darlington, Eng, Aug 31, 29. BIOCHEMISTRY. Educ: Cambridge Univ, BA, 52; Univ Alta, MSc, 55; Univ Wis, PhD(biochem), 58. Prof Exp: Jane Coffin Childs Mem Fund Med Res fel, Univ Calif, Davis, 59-60; from asst prof to assoc prof, 61-69, PROF BIOCHEM, UNIV CALIF, RIVERSIDE, 69- Mem: Am Chem Soc; Am Soc Plant Physiol; Brit Biochem Soc; Am Soc Biol Chem. Res: Lipid metabolism in plants; mechanism of enzyme action; biochemical effects of toxic oxidants. Mailing Add: Dept of Biochem Univ of Calif Riverside CA 92502

MUDD, STUART, pathology, microbiology, deceased

MUDD, STUART HARVEY, b Bryn Mawr, Pa, Apr 29, 27; m 55; c 3. BIOCHEMISTRY. Educ: Harvard Univ, BS, 49, MD, 53. Prof Exp: Intern med, Mass Gen Hosp, 53-54; MED DIR, LAB GEN & COMP BIOCHEM, NAT INST MENT HEALTH, 56- Concurrent Pos: NSF res fel, Biochem Res Lab, Mass Gen Hosp, Boston, 54-56. Mem: Am Soc Biol Chem. Res: Oxidative phosphorylation; transmethylation; plant metabolism; mechanism of enzyme action. Mailing Add: Lab of Gen & Comp Biochem Nat Inst of Ment Health Bethesda MD 20014

MUDGE, GILBERT HORTON, b Brooklyn, NY, Apr 19, 15; m 41; c 4. PHYSIOLOGY, MEDICINE. Educ: Amherst Col, BA, 36; Columbia Univ, MD, 41, Med Sci Dr, 45. Prof Exp: Instr med, Columbia Univ, 48-49, assoc, 49-51, from asst prof to assoc prof, 51-55; prof pharmacol & exp therapeut & dir dept, Johns Hopkins Univ, 55-62, dean, 62-65, actg chmn dept med, 65-66, chmn dept, 66-67; PROF MED, DARTMOUTH MED SCH, 65- Concurrent Pos: Mem pharmacol study sect, USPHS, 57-60, Life Ins Med Res Fund, 60-64 & Nat Res Coun, 60-62; assoc dean pharmacol & exp therapeut, Johns Hopkins Univ, 60-62, prof exp therapeut, 62-66; mem regulatory biol panel, NSF, 61-65. Mem: Soc Exp Biol & Med; Am Physiol Soc; Am Soc Clin Invest; Am Soc Pharmacol & Exp Therapeut; Asn Am Physicians. Res: Renal function; electrolyte physiology. Mailing Add: Lyme NH 03768

MUDGE, JOSEPH WILLIAM, b Gridley, Kans, Apr 12, 21; m 47; c 2. ANIMAL BREEDING, GENETICS. Educ: Kans State Univ, BS, 42, MS, 57, PhD(animal breeding), 66. Prof Exp: Instr dairy husb, Kans State Univ, 46-47; dairy farmer, 48-55; res asst dairy husb, Kans State Univ, 56-60; from instr to assoc prof, 60-72, PROF DAIRY HUSB, UNIV MINN, ST PAUL, 72-, EXTEN DAIRYMAN, 60- Mem: Am Dairy Sci Asn; Am Genetic Asn. Res: Dairy cattle breeding; extension activities in dairy management. Mailing Add: Dept of Animal Sci Univ of Minn St Paul MN 55101

MUDGETT, MEREDITH, b Springfield, Mass, Jan 9, 45. IMMUNOCHEMISTRY. Educ: Mt Holyoke Col, BA, 67; Univ Ill, Chicago, PhD(biochem), 73. Prof Exp: FEL IMMUNOL, ROCKEFELLER UNIV, 73- Concurrent Pos: USPHS trainee, NIH, 67-75; James N Jarvie Mem fel, New York Heart Asn, 75- Mem: Harvey Soc; Sigma Xi. Res: Regulation of rabbit allotype and idiotype expression; protein and peptide sequencing by mass spectrometry. Mailing Add: Rockefeller Univ York Ave & 66th St New York NY 10021

MUDHOLKAR, GOVIND S, b Aurangabad, India, Jan 5, 34. MATHEMATICAL STATISTICS. Educ: Univ Poona, BSc, 56, MSc, 57 & 58; Univ NC, PhD(statist), 63. Prof Exp: Lectr math & statist,SP Col, Poona, 57-60; from asst prof to assoc prof, 63-75, PROF STATIST & BIOSTATIST, MED SCH, UNIV ROCHESTER, 75- Concurrent Pos: NSF grant, 66-67. Mem: Inst Math Statist; Am Statist Asn; Am Math Soc; Math Asn Am. Res: Power functions of multivariate procedures in inequalities and their sharpness properties; estimation in sample surveys; topics in statistical inference. Mailing Add: Dept of Math Univ of Rochester Rochester NY 14627

MUDIE, JOHN DAVID, b Durban, SAfrica, Jan 9, 38; m 63; c 2. GEOPHYSICS. Educ: Univ Cape Town, BSc, 57, Hons, 58, MSc, 59; Cambridge Univ, PhD(geophys), 63. Prof Exp: Res geophysicist, 63-65, asst res geophysicist, 65-69, asst prof geophys, 69-75, ASSOC PROF GEOPHYS, SCRIPPS INST OCEANOG, UNIV CALIF, SAN DIEGO, 75- Mem: Am Geophys Union. Res: Magnetism; magnetometers; geophysics in a marine environment; geomagnetism; archeomagnetism; electronic instrumentation; observations near the sea floor. Mailing Add: Scripps Inst of Oceanog Univ of Calif San Diego La Jolla CA 92038

MUDRAK, ANTON, b Stickney, Ill, Jan 21, 16. ORGANIC CHEMISTRY. Educ: Univ Ill, BS, 38, MS, 47, PhD(chem), 49. Prof Exp: Res chemist org, Colgate-Palmolive Co, 49-52; group leader, 52-69, DIR ORG RES, HARSHAW CHEM CO, 69- Mem: Am Chem Soc. Res: Dyes and pigments; electroplating addition agents. Mailing Add: Cent Res & Develop Dept Harshaw Chem Co 1945 E 97th St Cleveland OH 44106

MUDRICK, STEPHEN EDWARD, b Washington, DC, Dec 1, 44; m 70; c 1. DYNAMIC METEOROLOGY. Educ: Univ Md, BS, 66; Mass Inst Technol, PhD(meteorol), 73. Prof Exp: PHYSICIST, METEOROL DIV, AIR FORCE GEOPHYS LAB, 72- Mem: Am Meteorol Soc. Res: An atmospheric simulation model to study the effects of pollutants upon the stratosphere; also higher order filtering effects on frontal scale numerical integrations. Mailing Add: Air Force Geophys Lab Code LYD Hanscom AFB Bedford MA 01731

MUECKE, HERBERT OSCAR, b Kenedy, Tex, Jan 14, 40; m 61. MATHEMATICS. Educ: Univ Tex, Austin, BS & MA, 62, PhD(math), 68. Prof Exp: Asst prof, 68-74, ASSOC PROF MATH, SAM HOUSTON STATE UNIV, 74- Res: Complex analysis; symbolic logic; history and philosophy of science. Mailing Add: Dept of Math Sam Houston State Univ Huntsville TX 77340

MUEGGLER, WALTER FRANK, b Enterprise, Ore, May 22, 26; m 58; c 6. PLANT ECOLOGY. Educ: Univ Idaho, BS, 49, Univ Wis, MS, 53; Duke Univ, PhD, 61. Prof Exp: Plant ecologist, 49-74, PROJ LEADER, INTERMOUNTAIN FOREST & RANGE EXP STA, US FOREST SERV, 74- Mem: Ecol Soc Am; Soc Range Mgt. Res: Range and wildlife habitat research; plant synecology and autecology. Mailing Add: Forestry Sci Lab US Forest Serv 860 N 12 E Logan UT 84321

MUEHLBAECHER-HEISE, CLARA A, biochemistry, pharmacology, see 12th edition

MUEHLBAUER, FREDERICK JOSEPH, b Buffalo, NY, Feb 22, 40; m 62; c 4. PLANT BREEDING. Educ: Univ Ga, BS, 63; Pa State Univ, MS, 65, PhD(genetics), 69. Prof Exp: Instr agron, Pa State Univ, 68-69; RES GENETICIST PLANT BREEDING, AGR RES SERV, USDA, 69- Mem: Crop Sci Soc Am; Pisum Genetics Asn; Nat Pea Improv Asn; Lentil Res Orgn; Coun Agr Sci & Technol. Res: Genetics and breeding of dry-edible legumes. Mailing Add: Dept of Agron & Soils Wash State Univ Pullman WA 99163

MUEHLBERGER, WILLIAM RUDOLF, b New York, NY, Sept 26, 23; m 49; c 2. GEOLOGY. Educ: Calif Inst Technol, BS & MS, 49, PhD(geol), 54. Prof Exp: From asst prof to assoc prof geol, 54-62, chmn dept geol sci, 66-70, PROF GEOL, UNIV TEX, AUSTIN, 62- Concurrent Pos: Geologic field asst, US Geol Surv, 48-49, geologist, 49 & 71-; geologist, State Bur Mines & Mineral Resources, NMex, 53-61, dir crustal studies lab, 61-66; prin investr, Apollo Field Geol Invests, Apollo 16 & 17, 71-74. Honors & Awards: Medal for Except Sci Achievement, NASA, 73. Mem: AAAS; Geol Soc Am; Am Asn Petrol Geol; Am Geophys Union; Nat Asn Geol Teachers. Res: Structural, areal and lunar geology; analysis of global tectonics. Mailing Add: Dept of Geol Sci Univ of Tex Austin TX 78712

MUEHLHAUSE, CARL OLIVER, physics, see 12th edition

MUELLER, ALBERT JOSEPH, organic chemistry, physical organic chemistry, see 12th edition

MUELLER, ARTHUR JACOB, b Evansville, Ind, July 10, 12; m 41; c 2. BIOCHEMISTRY. Educ: Ala Polytech Inst, BS, 35. Prof Exp: From res chemist to sr res chemist, 35-62, group leader, 62-69, SR INVESTR, BIOCHEM DEPT, MEAD JOHNSON & CO, 69- Mem: AAAS; Am Chem Soc; NY Acad Sci. Res: Amino acids; bio-assays; arteriosclerosis; antithrombosis; anti-diabetes. Mailing Add: Biochem Dept Mead Johnson & Co 2404 Pennsylvania St Evansville IN 47221

MUELLER, AUGUST P, b Fargo, NDak, July 30, 33; m 58; c 2. IMMUNOLOGY, SEROLOGY. Educ: Moorhead State Col, BSc, 55; Univ Wis, MSc, 57, PhD(zool, chem), 60. Prof Exp: Res assoc zool, Univ Wis, 60 & 62; fel, Univ Edinburgh, 61-62; asst prof, 62-67, ASSOC PROF BIOL, STATE UNIV NY BINGHAMPTON, 67- Concurrent Pos: NIH res grant, 63-66. Mem: AAAS; Am Soc Zool; Reticuloendothelial Soc; Am Inst Biol Sci. Res: Immune unresponsiveness in juvenile and adult animals; development of the immune system; bursa of Fabricus in chickens and the thymus in mammals; genetics; biometry. Mailing Add: Dept of Biol State Univ of NY Binghamton NY 13901

MUELLER, C BARBER, b Carlinville, Ill, Jan 22, 17; m 40; c 4. SURGERY. Educ: Univ Ill, AB, 38; Washington Univ, MD, 42; FRCS(C). Prof Exp: Rockefeller asst, 46-49; from instr to asst prof surg, Washington Univ, 51-56; prof & chmn dept, State Univ NY Upstate Med Ctr, 56-67; chmn dept, 67-72, PROF SURG, McMASTER UNIV, 67- Concurrent Pos: Markle scholar med sci, 49- Mem: Am Soc Univ Surg; Am Surg Asn; Am Col Surg; Am Fedn Clin Res. Res: Renal physiology and disease; survivorship in breast carcinoma. Mailing Add: Dept of Surg McMaster Univ Hamilton ON Can

MUELLER, CHARLES FREDERICK, b Sharon, Pa, Oct 3, 39; m 64; c 2. ECOLOGY. Educ: Ind Univ Pa, BSEd, 62; Ohio Univ, MS, 65; Mont State Univ, PhD(zool), 67. Prof Exp: ASSOC PROF BIOL, SLIPPERY ROCK STATE COL, 67-, CHMN DEPT, 75- Concurrent Pos: NSF res grant, Slippery Rock State Col, 69-71. Mem: Soc Study Amphibians & Reptiles; Am Soc Ichthyol & Herpet. Res: Temperature and energy characteristics; bioenergetics of poikilotherms. Mailing Add: Dept of Biol Slippery Rock State Col Slippery Rock PA 16057

MUELLER, CHARLES RICHARD, b St Louis, Mo, June 22, 25; m 47; c 2. THEORETICAL CHEMISTRY, PHYSICAL CHEMISTRY. Educ: Washington Univ, AB, 48; Univ Utah, PhD(chem), 51. Prof Exp: From instr to assoc prof, 51-64, PROF CHEM, PURDUE UNIV. 64- Mem: Am Chem Soc; Sigma Xi. Res: Intermolecular forces and molecular beam methods; scattering theory. Mailing Add: Dept of Chem Purdue Univ Lafayette IN 47907

MUELLER, CLYDE DEWEY, b Sawyer, Kans, Apr 23, 18; m 40; c 8. GENETICS. Educ: Kans State Col, BS, 39; Cornell Univ, MS, 40, PhD(animal breeding), 43. Prof Exp: Asst, Cornell Univ, 39-43; geneticist, Westhill Farms, NY, 43-48, poultry dept, Kans State Col, 48-55, Arbor Acres Farm, Inc, Conn, 55-64 & Kimber Farms, Inc, Calif, 64-68; INDEPENDENT GENETIC & COMPUT CONSULT, 68- Mem: Poultry Sci Asn; Am Genetic Asn. Res: Poultry breeding and genetics; genetic variability in the fowl; computer programming; systems analysis. Mailing Add: Box 457 Fayetteville AR 72701

MUELLER, DAVID CLARK, organic chemistry, see 12th edition

MUELLER, DELBERT DEAN, b Claremore, Okla, Oct 22, 33; m 59; c 3. PHYSICAL BIOCHEMISTRY. Educ: Univ Okla, BS, 62, PhD(phys chem), 66. Prof Exp: Res assoc phys biochem, Northwestern Univ, 66-68; asst prof, 68-75, ASSOC PROF BIOCHEM, KANS STATE UNIV, 75- Concurrent Pos: Vis staff mem, Los Alamos Sci Lab, 74- Mem: AAAS; Am Chem Soc. Res: Physical studies on biopolymers with emphasis on the applications of carbon-13 nuclear magnetic resonance spectroscopy and hydrogen exchange techniques to conformational change and binding problems. Mailing Add: Dept of Biochem Kans State Univ Manhattan KS 66506

MUELLER, DONALD SCOTT, b Cleveland, Ohio, May 8, 47; m 69; c 1. PLASTICS CHEMISTRY, POLYMER CHEMISTRY. Educ: Hiram Col, BA, 69; Univ Ill, PhD(org chem), 73. Prof Exp: Teaching asst org chem, Univ Ill, 69-73; CHEMIST, ROHM AND HAAS CO, 73- Mem: Am Chem Soc. Res: Synthesis and development of new polymeric systems, especially in reference to the production of new plastics. Mailing Add: Rohm and Haas Co PO Box 219 Bldg 78 Bristol PA 19007

MUELLER, ERWIN W, b Berlin, Ger, June 13, 11; nat US; m 39; c 1. ELECTRON PHYSICS. Educ: Berlin Tech Univ, dipl, 35, Dr Ing, 36, Dr habil, 50. Hon Degrees: Dr rer nat, Free Univ Berlin, 68; Dr, Claude Bernard Univ, Lyon, France, 75. Prof Exp: Res physicist, Siemens Res Lab, Ger, 35-37 & Stabilovolt Co, 37-45; prof phys chem, Tech Inst, Altenburg, 45-47; div chief, Kaiser-Wilhelm Inst, Berlin, 47-52; prof, 52-55, res prof, 55-68, EVAN-PUGH RES PROF PHYSICS, PA STATE UNIV, UNIVERSITY PARK, 68- Concurrent Pos: Privat-docent, Tech Univ W Berlin, 50; extraordinary prof, Free Univ W Berlin, 51-52; sci mem-at-lg, Max-Planck Soc Advan Sci, 57- Honors & Awards: Gauss Medal, 52; Award, Instrument Soc Am, 60; Potts Medal, Franklin Inst, 64; Centenary Lect Medal, The Chem Soc, 69; John Scott Medal, Philadelphia City Trusts, 70; M V Welch Award, Am Vacuum Soc, 70; Davisson-Germer Prize, Am Phys Soc, 72. Mem: Nat Acad Sci; Nat Acad Eng; fel Am Phys Soc; Am Vacuum Soc; hon fel Royal Micros Soc. Res: Field emission research; field electron and ion microscopes; field desorption and ion emission; adsorption; surfaces; metal physics; radiation damage; atom probe field-ion microscopy. Mailing Add: 659 Glenn Rd State College PA 16801

MUELLER, FRED MICHAEL, b Chicago, Ill, Oct 8, 38; div. SOLID STATE PHYSICS. Educ: Univ Chicago, SB, 61, SM, 62, PhD(physics), 66. Prof Exp: Assoc physicist, 66-73, SR SCIENTIST, ARGONNE NAT LAB, 73-; PROF PHYSICS, NORTHERN ILL UNIV, 69- Concurrent Pos: Consult, Northwestern Univ, 69-, Univ Chicago, 69-, Stanford Univ, 70- & Am Photo Copying Equip Co, 71-; prof dr, Physics Lab, Nijmegen, Netherlands, 75- Mem: Am Phys Soc. Res: Electronic structure of transition metals; phonon spectra; superconductivity. Mailing Add: Argonne Nat Lab Argonne IL 60439

MUELLER, GEORGE, b Budapest, Hungary, Jan 19, 16; m 63; c 1. ORGANIC GEOCHEMISTRY, COSMOCHEMISTRY. Educ: Oxford Univ, BSc, 44; Univ London, PhD(mineral), 51. Prof Exp: Asst org chem, Oxford Univ, 43-44; consult & industrialist chem & dir, Durplate Co, London, 44-48; res asst geochem, Univ Col, Univ London, 48-53; prof geol & chmn dept, Concepcion Univ, 54-68; prof geochem, Inst Molecular Evolution, Univ Miami, 68-73; ADJ PROF CHEM & GEOL, FLA ATLANTIC UNIV, 74- Concurrent Pos: Vis prof & Nuffield Found grant, Birkbeck Col, Univ London, 62-65; vis prof, Inst Molecular Evolution, Univ Miami, 66-68, NASA grant, univ, 66-; vis prof, Concepcion Univ, 68-70. Mem: AAAS; Am Chem Soc; Geol Soc Am; Meteoritical Soc. Res: Organic chemical studies on carbonaceous meteorites; hydrothermally associated bitumens; comparative studies of sprays from terrestrial, meteoritic and lunar sources; crystal growth; mineral deposits; evolutional philosophy; inclusions of oils and bitumens in crystals of hydrothermal minerals.

MUELLER, GEORGE L, b Alton, Ill, Dec 1, 17; m; c 1. VETERINARY MEDICINE. Educ: Tex A&M Univ, DVM, 42. Prof Exp: Pvt pract vet med, 46-52; assoc dir vet field serv, Am Cyanamid Co, 52-60; ASST DIR VET REGULATORY & PROF AFFAIRS, E R SQUIBB & SONS, INC, 60- Mem: NY Acad Sci; Am Vet Med Asn; World Asn Advan Vet Parasitol; Am Asn Equine Practr; Am Vet Radiol Soc. Res: Chemotherapy of animal and poultry diseases. Mailing Add: E R Squibb & Sons Inc PO Box 4000 Princeton NJ 08540

MUELLER, GEORGE PETER, b Atchison, Kans, Aug 7, 18; m 46; c 1. ORGANIC CHEMISTRY. Educ: Univ Nebr, BSc, 40, MSc, 41; Univ Ill, PhD(org chem), 43. Prof Exp: Chemist, State Hwy Testing Lab, Nebr, 38-40 & Eastman Kodak Co, 42; asst, Univ Ill, 44; res assoc, Harvard Univ, 44-46; chemist. Wyeth Inst Appl Biochem, Pa, 46-47; assoc prof chem, Univ Tenn, 47-52; res supvr, G D Searle & Co, 52-62, coordr, Searle Chem Inc, Ill & Mex, 59-62; dir res, Marine Colloids, Inc, Maine, 62-68; CONSULT, 68- Concurrent Pos: Fel, Univ Ill, 44-45; consult, Oak Ridge Nat Lab & Oak Ridge Inst Nuclear Studies, 49-52. Mem: AAAS; Am Chem Soc; NY Acad Sci. Res: Synthetic and steroidal estrogens; barbiturates; insecticides; alicyclic synthesis; androgens; corticoids; isolation, pharmacology; structure of botanical isolates; sources, chemistry steroidal sapogenins; seaweed sources; structure, modification, utilization of seaweed colloids; polysaccharides. Mailing Add: 5 Cedar St Camden ME 04843

MUELLER, GERALD CONRAD, b Centuria, Wis, May 22, 20; m 44; c 3. BIOCHEMISTRY. Educ: Univ Wis, BS, 43, MD, 46, PhD(biochem, physiol), 50. Prof Exp: Intern, Med Col, Va Hosp, 47; from instr to assoc prof oncol, 50-58, asst prof acad affairs, Univ, 63-67, PROF ONCOL, McARDLE LAB CANCER RES, UNIV WIS-MADISON, 58- Concurrent Pos: Schering scholar, Max Planck Inst Virus Res, 58; mem drug eval panel, Cancer Chemother Nat Serv Ctr, 60-61, chmn, 61-62, mem biochem comt, 60-62; USPHS res career award, Univ Wis-Madison, 62-; mem bd sci coun, Nat Cancer Inst, 65-69, mem organizational task force, 68, mem chemother adv comt, 69-; vis prof, Univ Sao Paulo, 71. Mem: Am Soc Biol Chem; Am Asn Cancer Res. Res: Mechanism of action of estrogenic hormones; molecular processes regulating animal cell replication; intermediate metabolism of growth regulation. Mailing Add: McArdle Lab for Cancer Res Univ of Wis Madison WI 53706

MUELLER, GERHARD W, physical chemistry, physics, see 12th edition

MUELLER, HELMUT, b Schneeberg, Ger, Jan 2, 26; US citizen; m 53; c 2. BIOCHEMISTRY. Educ: Univ Würzburg, MD, 52; Univ Birmingham, PhD(biochem), 61. Prof Exp: Univ fel, Univ Colo, 53-54; resident physician, Northwestern Univ, 54-56; Am Heart Asn res fel, Inst Muscle Res, Mass, 57-58, adv res fel, 58-60; asst res prof biochem, Univ Pittsburgh, 61-64, assoc res prof, 64-65; sr fel, Mellon Inst, 65-68; HEALTH SCI ADMINR & MED OFFICER DRUG SURVEILLANCE, CTR POP RES, NAT INST CHILD HEALTH & HUMAN DEVELOP, 68- Concurrent Pos: Am Heart Asn estab investr, Univ Birmingham, 60-61 & Univ Pittsburgh, 61-65. Mem: AAAS; Am Soc Biol Chem; Soc Gen Physiol; Biophys Soc. Res: Structure and function of contractile proteins; cardiovascular disease. Mailing Add: Ctr for Pop Res Nat Inst of Child Health & Human Develop Bethesda MD 20014

MUELLER, HELMUT CHARLES, b Milwaukee, Wis, Mar 20, 31; m 59; c 1. ZOOLOGY, ANIMAL BEHAVIOR. Educ: Univ Wis-Madison, BS, 53, MS, 58, PhD(zool), 62. Prof Exp: Res assoc zool, Univ Wis-Madison, 62-65, lectr, 66; asst prof, 66-70, ASSOC PROF ZOOL, UNIV NC, CHAPEL HILL, 70- Mem: Fel AAAS; fel Am Ornithologists Union; Wilson Ornith Soc; Ecol Soc Am; Animal Behav Soc. Res: Behavioral aspects of the predator-prey interaction; hawk behavior; bird migration; behavioral ecology. Mailing Add: Dept of Zool Univ of NC Chapel Hill NC 27514

MUELLER, HERBERT J, b Vienna, Austria, Dec 1, 23; US citizen; m 50; c 1. PHYSICS, MATERIAL SCIENCE. Educ: Univ Vienna, PhD(physics), 49. Prof Exp: Asst prof physics, II Inst Physics, Univ Vienna, 49-58; asst dir basic res lab, US Army Eng Res & Develop Lab, Va, 58-67; RES ADMINR, NAVAL AIR SYSTS COMMAND, 67- Mem: Sigma Xi (pres, Sci Res Soc Am, 65-66); Austrian Phys Soc; Austrian Phys-Chem Soc (secy, 56-58). Res: Solid state and metal physics; x-ray diffraction; exo-electron emission; luminescence. Mailing Add: Naval Air Systs Command AIR-310 Washington DC 20360

MUELLER, IRENE MARIAN, b St Libory, Nebr, July 12, 04. PLANT ECOLOGY. Educ: Nebr Cent Col, AB, 27; Univ Nebr, AM, 37, PhD(plant ecol), 40. Prof Exp: Teacher high schs, Nebr, 27-35; asst bot, Univ Nebr, 36-39; instr biol sci, Wis State Teachers Col, Platteville, 40-43; from assoc prof to prof, 43-75, EMER PROF BIOL, NORTHWEST MO STATE UNIV, 75- Mem: AAAS; Ecol Soc Am; Am Asn Univ Profs. Res: Rhizomes of prairie plants; drought resistance in prarie plants. Mailing Add: 728 West Third St Maryville MO 64468

MUELLER, IVAN I, b Budapest, Hungary, Jan 9, 30; US citizen; m 50; c 2. GEODESY, GEOPHYSICS. Educ: Budapest Tech Univ, dipl eng, 52; Ohio State Univ, PhD(geod sci), 60. Prof Exp: Asst prof geod, Budapest Tech Univ, 52-56; design engr, C H Sells Consult Engr, NY, 57-58; from instr to assoc prof, 59-66, PROF GEOD, OHIO STATE UNIV, 66- Concurrent Pos: Mem geod-cartog working group, Manned Space Sci Coord Comt, NASA, 65-66, prin investr, Nat Geod Satellite Prog, 65-74, mem ad hoc adv group satellite geod, 66-67, geod & cartog subcomt & space sci & appln steering comt, 67-68; panel solid earth geophys & earthquake eng, comt adv to Environ Sci Serv Admin, Nat Acad Sci-Nat Acad Eng, 67-69; assoc ed J geophys res, 67-74; consult, UN Develop Prog, Ctr Survey Training & Map Prod, Hyderabad, India, 71 & 72; Ed in chief, Bull Geodesique, Paris, 75- Mem: Am Soc Photogram; Am Geophys Union (pres/geod, 74). Res: Geodetic astronomy; gravimetric and satellite geodesy. Mailing Add: Dept of Geod Sci Ohio State Univ Columbus OH 43210

MUELLER, JOHN FREDERICK, b Goshen, Ind, June 15, 22; m 45; c 4. INTERNAL MEDICINE. Educ: Capital Univ, BA, 44; Univ Cincinnati, MD, 46. Prof Exp: Intern, King's County Hosp, Brooklyn, NY, 46-47; fel hemat & nutrit, Univ Cincinnati, 48-49; res fel hemat, Western Reserve Univ, 50-51; from instr to assoc prof med, Sch Med, Univ Cincinnati, 51-62, assoc dir lab hemat & nutrit, 55, co-dir, 57-62; prof med, Univ Colo, 62-64; prof, State Univ NY Downstate Med Ctr, 64-73; PROF MED, UNIV COLO MED CTR, DENVER, 73-; DIR INTERNAL MED, ST LUKES HOSP, 73- Concurrent Pos: Resident, Cincinnati Gen Hosp, 47-50, chief clinician, Out Patient Dept, 57-62; chief med, Denver Vet Admin Hosp, 62-64; physician-in-chief, Brooklyn-Cumberland Med Ctr, 64-73; consult, Surgeon Gen, NIH. Mem: Am Soc Clin Invest; Am Fedn Clin Res; Am Soc Clin Nutrit (secy-treas, 63-66); Am Heart Asn. Res: General field of lipid chemistry of blood and tissues as related to various disease states; nutrition and hematology. Mailing Add: St Lukes Hosp Dept of Med 601 E 19th Ave Denver CO 80203

MUELLER, JOSEPH ROBERT, b Appleton, Wis, Oct 23, 42. BIOCHEMISTRY, ENDOCRINOLOGY. Educ: St Procopius Col, BS, 64; Med Univ SC, PhD(biochem), 75. Prof Exp: Res assoc endocrinol, Univ Iowa, 67-70; assoc fac biochem & endocrinol, Med Univ SC, 74-75; NIH res fel, 75-76, RES INSTR, WASHINGTON UNIV, 76- Honors & Awards: Cert Merit, Am Chem Soc, 64. Mem: Am Chem Soc; AAAS. Res: The elucidation of the topography of steroid binding sites of steroid converting enzymes and steroid receptor proteins via affinity labeling steroid derivatives. Mailing Add: Dept Obstet-Gynec Sch Med Washington Univ St Louis MO 63110

MUELLER, JUSTUS FREDERICK, b Baltimore, Md, Nov 20, 02. ZOOLOGY. Educ: Johns Hopkins Univ, AB, 23; Univ Ill, MA, 26, PhD(zool), 28. Prof Exp: Sci asst, Bur Fisheries, 23-24; asst zool, Univ Ill, 24-28; from instr to assoc prof zool, State Univ NY Col Forestry, Syracuse Univ, 28-42, assoc prof parasitol, Col Med, Univ, 42-50; assoc prof parasitol, 50-56, actg chmn dept microbiol, 54-57, prof microbiol, 56-72, EMER PROF, STATE UNIV NY UPSTATE MED CTR, 72- Concurrent Pos: Field naturalist, Roosevelt Wild Life Forest Exp Sta, 28-35; mem trop med & parasitol study sect, NIH, 62-66; leader Ore Biol Colloquium, 65; fel trop med, Cent Am Prog, La State Univ, 65; consult Merck Inst Therapeut Res, 68-69; lectr, Col Med, Syracuse Univ, 30-42, Med Sch, Marquette Univ, 56-60, Univ Pittsburgh, 65 & Yale Univ, 67; ed, J Parasitol, 62- Mem: Am Soc Parasitol (pres elect, 72, pres, 73); Am Micros Soc; Am Soc Trop Med & Hyg. Res: Invertebrates; fish and human parasites; pseudophyllidean tapeworms; sparganosis; visual education; models; in vitro culture of cestodes; growth-promoting substances in cestodes; parasite-induced obesity. Mailing Add: Dept of Microbiol State Univ of NY Upstate Med Ctr Syracuse NY 13210

MUELLER, KARL HUGO, JR, b Ft Worth, Tex, May 27, 43; m 65; c 2. LOW TEMPERATURE PHYSICS. Educ: Rice Univ, BA, 65; Duke Univ, PhD(physics), 72. Prof Exp: Res assoc physics, Duke Univ, 71-73; staff mem, Kernforschungsanlage Julich GMBH, 73-75; STAFF MEM PHYSICS, LOS ALAMOS SCI LAB, 75- Mem: Am Phys Soc. Res: Intrinsic dissipation in flowing helium films in the temperature range below one degree kelvin; precision thermodynamic measurements near the lambda point in pure helium 4. Mailing Add: MS 764 PO Box 1663 Los Alamos NM 875444

MUELLER, MANFRED ERNST, b Orange, Calif, Mar 4, 11; m 36; c 3. PHYSICAL CHEMISTRY. Educ: Univ Calif, AB, 33, MA, 35, PhD(phys chem), 37. Prof Exp: Instr chem, San Francisco Jr Col, 35-42; phys chemist, Radiation Lab, Univ Calif, 42-45; instr chem, City Col San Francisco, 45-74; PARTNER, SEMANS-MUELLER ASSOCS, 74- Concurrent Pos: Lectr, Univ San Francisco, 56-71. Mem: Am Chem Soc. Res: Compressibility of gases; vapor pressures of solids; inorganic chemistry synthesis; basic chemistry; statistical analysis of demographic data for California Community College. Mailing Add: 1045 Cole St San Francisco CA 94117

MUELLER, MARVIN MARTIN, b Broken Arrow, Okla, Sept 29, 28; m 54, 67. PHYSICS. Educ: Univ Okla, BS, 51, MS, 54, PhD(physics), 59. Prof Exp: STAFF MEM, LOS ALAMOS SCI LAB, UNIV CALIF, 59- Mem: AAAS; Am Asn Physics Teachers; Am Phys Soc. Res: Laser-generated plasmas; x-ray plasma diagnostics; radiation physics. Mailing Add: 409 Estante Way Los Alamos NM 87544

MUELLER, MARY CASIMIRA, b Cincinnati, Ohio, July 16, 18. PHYSICAL CHEMISTRY, THEORETICAL CHEMISTRY. Educ: Villa Madonna Col, BA, 41; Cath Univ Am, MS, 42; Univ Cincinnati, PhD(chem), 62. Prof Exp: From asst prof to assoc prof, 42-52, PROF CHEM, THOMAS MORE COL, 52-, CHMN DEPT, 73- Mem: Am Chem Soc. Res: Physics and mathematics. Mailing Add: Dept of Chem Thomas More Col Box 85 Ft Mitchell KY 41017

MUELLER, MAX BEST, b Freeport, Ill, May 2, 15; m 39; c 4. ORGANIC CHEMISTRY. Educ: Univ Ill, AB, 36, MS, 37, PhD(chem), 39. Prof Exp: Chemist org chem, Am Cyanamid Co, 39-41; chemist org chem, 41-52, supvr, 52-69, MGR RES & DEVELOP, SPECIALTY CHEM DIV, ALLIED CHEM CORP, 69- Mem: Am Chem Soc; Am Soc Qual Control; Am Statist Asn. Res: Organic synthesis. Mailing Add: Specialty Chem Div Allied Chem Corp Box 1087 R Morristown NJ 07960

MUELLER, NANCY SCHNEIDER, b Wooster, Ohio, Mar 8, 33; m 59; c 1. DEVELOPMENTAL BIOLOGY, REPRODUCTIVE BIOLOGY. Educ: Col Wooster, AB, 55; Univ Wis, MS, 47, PhD(zool), 62. Prof Exp: Hon fel, Univ Wis, Madison, 65-66, instr develop biol, 66; vis asst prof zool, NC State Univ, 68 & zool & poultry sci, 68-71; vis prof, 71-72, ASSOC PROF BIOL, NC CENT UNIV, 72- Mem: Am Soc Cell Biol; Am Soc Zool. Res: Genetic and hormonal influences on sexual dichromatism in plumage of birds; control of molting in insect embryos; sex determination and differentiation in bird and mammal embryos. Mailing Add: Dept of Biol NC Cent Univ Durham NC 27707

MUELLER, PAUL ALLEN, b Anniston, Ala, Sept 9, 45. GEOCHEMISTRY. Educ: Washington Univ, AB, 67; Rice Univ, MA & PhD(geol), 71. Prof Exp: Advan res projs agency res assoc geochem, Mat Res Ctr, Univ NC, Chapel Hill, 71-73; ASST PROF GEOL, UNIV FLA, 73- Honors & Awards: Nininger Prize, 69. Mem: Geol Soc Am; Am Geophys Union; Geochem Soc. Res: Petrology, geochemistry, isotopic geochemistry and geochronology of igneous and metamorphic rocks. Mailing Add: Dept of Geol Univ of Fla Gainesville FL 32611

MUELLER, PETER STERLING, b New York, NY, Dec 28, 30; m 58; c 4. METABOLISM, PSYCHIATRY. Educ: Princeton Univ, AB, 52; Univ Rochester, MD, 56. Prof Exp: Intern Bellevue Hosp, 56-57; clin assoc metab serv, Gen Med Br, Nat Cancer Inst, 57-59; clin assoc, Off Chief, Unit Psychosom, Lab Clin Sci, NIMH, 59-63; asst resident psychiat, Phipps Psychiat Clin, Johns Hopkins Univ, 63-66; asst prof, Sch Med, Yale Univ, 66-72; ASSOC PROF PSYCHIAT, RUTGERS MED SCH, COL MED & DENT NJ, 72- Concurrent Pos: Consult to resident supvr psychother, Conn Valley Hosp, 66-72. Mem: Am Psychosom Soc. Res: Free fatty acid and neurohormone metabolism in cancer and mental illness; psychoendocrinology; family therapy of schizophrenia; psychosomatic medicine. Mailing Add: Dept of Psychiat Rutgers Med Sch PO Box 101 Piscataway NJ 08854

MUELLER, RAYMOND KARL, b East St Louis, Ill, Oct 18, 41; m 67. APPLIED MATHEMATICS. Educ: Washington Univ, BS, 63, MS, 65, DSc(appl math), 67. Prof Exp: Mem tech staff, Bell Tel Labs, 67-70; ASST PROF MATH, COLO SCH MINES, 70- Mem: Opers Res Soc Am; Inst Math Statist. Res: Applied probability; probability and stochastic processes; operations research. Mailing Add: Dept of Math Colo Sch of Mines Golden CO 80401

MUELLER, RICHARD AUGUST, organic chemistry, see 12th edition

MUELLER, ROBERT ARTHUR, b Fond du Lac, Wis, July 24, 38; m 62; c 3. ANESTHESIOLOGY, PHARMACOLOGY. Educ: Univ Wis-Madison, BS, 60, MS, 63; Univ Minn, Minneapolis, MD, 65, PhD(pharmacol), 66. Prof Exp: Am Cancer Soc fel, Univ Minn, Minneapolis, 65-66, intern surg, Hosps, 66-67, resident anesthesiol, 67; res assoc pharmacol & toxicol, Lab Clin Sci, NIMH, 67-69; resident anesthesia, Med Sch, Northwestern Univ, Ill, 69-70; assoc prof anesthesiol & asst prof pharmacol, 70-75, PROF ANESTHESIOL & ASSOC PROF PHARMACOL, MED SCH, UNIV NC, CHAPEL HILL, 75- Mem: Am Soc Pharmacol & Exp Therapeut; Fedn Am Socs Exp Biol. Res: Adrenergic pharmacology. Mailing Add: Dept of Pharmacol Univ of NC Med Sch Chapel Hill NC 27514

MUELLER, ROBERT FRANCIS, petrology, geochemistry, see 12th edition

MUELLER, ROLF KARL, b Zurich, Switz, Aug 30, 14; nat US; m 42; c 3. Educ: Munich Tech Univ, Dipl phys, 39, Dr rer nat, 42, habil, 50. Prof Exp: Asst electronics, Univ Jena, 39-45; asst prof theoret physics, Stuttgart Tech, 47-48; dozent, Munich Tech Univ, 48-52; consult, Air Force Cambridge Res Ctr, Mass, 52-55; sr tech specialist & head solid state res sect, Gen Mills, Inc, 56-63; mgr, Gen Sci & Technol Lab, Bendix Corp, 63, Lab dir, Bendix Ctr, 63-74; PROF ELEC ENG, UNIV MINN, 74- Mem: AAAS; fel Am Phys Soc; Am Vacuum Soc; Optical Soc Am; Inst sr mem Inst Elec & Electronics Engrs. Res: Acoustics; acousto-optics; electronics; electromagnetic theory; laser physics. Mailing Add: 9707 Manning Ave Stillwater MN 55082

MUELLER, SABINA GERTRUDE, b Binghamton, NY, Apr 29, 40; m 70; c 2. BOTANY. Educ: Swarthmore Col, BA, 61; Univ NC, Chapel Hill, PhD(bot), 68. Prof Exp: From asst prof to assoc prof bot, Shippensburg State Col, 66-70; staff mem plant records, Cox Arboretum, Dayton, 72-73, botanist, Educ Serv, 73-75; BOT, FULLMER'S LANDSCAPE SERV, DAYTON, 75- Mem: Am Asn Bot Gardens & Arboreta; Am Hort Soc. Res: Plant nomenclature. Mailing Add: 5512 Woodbridge Lane Dayton OH 45429

MUELLER, THEODORE ARNOLD, b St Louis, Mo, Jan 29, 38; m 63; c 1. BIOPHYSICS. Educ: Cent Methodist Col, AB, 59; NMex Highlands Univ, MS, 62, PhD(biophys chem), 65. Prof Exp: PROF SCI, ADAMS STATE COL, 65- Mem: Asn Physics Teachers; Biophys Soc; Air Pollution Control Asn. Res: Cellular and photochemical ultraviolet effects; photobiology; environmental pollution; pollution monitoring. Mailing Add: Div of Sci & Math Adams State Col Alamosa CO 81102

MUELLER, THEODORE ROLF, b Ft Wayne, Ind, Dec 14, 28; m 51; c 3. ANALYTICAL CHEMISTRY, INSTRUMENTATION. Educ: Valparaiso Univ, BA, 50; Univ Kans, PhD(anal chem), 63. Prof Exp: Instr math & sci, Lutheran High Sch, Houston, Tex, 50-51; instr chem, Concordia Col Inst, Bronxville, NY, 51-53; CHEMIST, OAK RIDGE NAT LAB, 61- Mem: AAAS; Am Chem Soc; Sigma Xi. Res: Application of electroanalytical techniques to environmental problems; research and development in analytical procedures; instruments for automated analyses. Mailing Add: Oak Ridge Nat Lab PO Box X Oak Ridge TN 37830

MUELLER, WALTER A, b Koenigshofen, Ger, Jan 16, 10; Can citizen; m 38; c 3. PHYSICAL CHEMISTRY, METALLURGY. Educ: Univ Heidelberg, PhD(phys chem), 34. Prof Exp: Fel metall, Univ Göttingen, 34-36; head lab, Siemens-Schuckertwerke, Ger, 36-47; consult, Brit Mil Forces, 47-50; corrosion specialist, Mannesman Roehren Werke, AG, Ger, 50-51; sr scientist, Pulp & Paper Res Inst Can, 51-72, prin scientist, 72-75; CONSULT CORROSION & ENERGY CONSERV, 75- Honors & Awards: Weldon Medal, Can Pulp & Paper Asn, 60. Mem: Electrochem Soc; Nat Asn Corrosion Eng; Tech Asn Pulp & Paper Indust; Can Pulp & Paper Asn. Res: Corrosion; polarization curves; anodic protection; corrosion studies in: Kraft liquor recovery, bleaching solutions and paper machines. Mailing Add: 138 Mimosa Ave Dorval PQ Can

MUELLER, WALTER CARL, b Newark, NJ, Nov 29, 34; m 56; c 3. PLANT PATHOLOGY. Educ: Rutgers Univ, BS, 56; Cornell Univ, PhD(plant path), 61. Prof Exp: From asst prof to assoc prof, 61-74, PROF PATH-ENTOM, UNIV RI, 74- Mem: AAAS; Am Phytopath Soc; Electron Micros Soc Am. Res: Viruses and virus diseases of plants; electron microscopy. Mailing Add: Dept of Plant Path-Entom Univ of RI Kingston RI 02881

MUELLER, WALTER E, b Olten, Switz, Apr 9, 38; m 67; c 1. SOLID STATE PHYSICS. Educ: Swiss Fed Inst Technol, BS, 60, MS, 62, PhD(physics), 67. Prof Exp: Res assoc basic res, IBM Res Lab, Zurich, Switz, 62-67; R A Welch Found fel, Univ Tex, Austin, 67-69; sr res physicist, Photo Prod Dept, E I Du Pont de Nemours & Co, Inc, 69-75; SR RES PHYSICIST, CIBA-GEIGY AG, BASLE, SWITZ, 75- Mem: Am Phys Soc; Swiss Phys Soc; Soc Photog Sci & Eng. Res: Optical properties of metals and metal ammonia solutions; photographic silver halide systems; photographic development mechanisms; research planning; marketing. Mailing Add: CIBA-GEIGY AG, R-1001 B 1 79 CH-4002 Basle Switzerland

MUELLER, WAYNE PAUL, b Evansville, Ind, July 27, 33; m 57; c 2. Educ: Univ Evansville, BA, 56; Ind Univ, MA, 61, PhD(zool), 62. Prof Exp: From asst prof to assoc prof, 62-72, PROF BIOL, UNIV EVANSVILLE, 72-, DIR ENVIRON STUDIES, 75- Mem: Nat Asn Biol Teachers. Res: Protozoan ecology. Mailing Add: 2020 Vogel Rd Evansville IN 47711

MUELLER, WERNER JULIUS, b Zurich, Switzerland, Oct 7, 26; nat US; m 53. POULTRY SCIENCE, PHYSIOLOGY. Educ: Swiss Fed Inst Technol, Agr Eng, 50, DSc(animal nutrit), 53. Prof Exp: Fel poultry nutrit, Pa State Univ, 53-54, from asst prof to assoc prof, 54-63, PROF POULTRY SCI, PA STATE UNIV, 64- Concurrent Pos: Poultry prod officer, Food & Agr Orgn, UN, 62-63. Mem: Poultry Sci Asn; Am Soc Zool; Am Inst Nutrit. Res: Endocrine control of calcium metabolism; avian physiology; poultry nutrition. Mailing Add: Dept of Poultry Sci Pa State Univ University Park PA 16802

MUELLER, WILLIAM ALEX, physical organic chemistry, see 12th edition

MUELLER, WILLIAM H, b Chicago, Ill, Mar 25, 26; m 49; c 2. PHARMACEUTICAL CHEMISTRY, PHARMACOLOGY. Educ: Univ Ill, BS, 50, MS, 55. Prof Exp: Pharm asst, Armour Pharmaceut Co, 50-53; develop chemist, Toni Res Labs, 53-58, res dir pharmaceut prod, Gillette Labs, 58-62, assoc res dir cosmetics, Toni Res Labs, 62-66, lab dir, Toni Div, 66-68, V PRES RES & DEVELOP, PERSONAL CARE DIV, GILLETTE CO, 68- Mem: Am Pharmaceut Asn; Soc Cosmetic Chemists (pres, 66). Res: Cosmetics. Mailing Add: Personal Care Div Gillette Co 458 Merchandise Mart Chicago IL 60654

MUELLER, WILLIAM SAMUEL, b Union, Mo, May 24, 01; m 27; c 2. MICROBIOLOGY. Educ: Univ Ill, BS, 27; Rutgers Univ, MS, 28; Mass State Col, PhD(food tech), 39. Prof Exp: Asst dairy prod, Rutgers Univ, 27-28; bacteriologist, State Bd Health, Wis, 29; asst res prof dairy indust, Univ Mass, Amherst, 32-50, assoc prof, 50-71, EMER PROF ENVIRON SCI, UNIV MASS, AMHERST, 71- Mem: Am Dairy Sci Asn; Inst Food Technol. Res: Paper mill process water; aerosol germicides; preservation of live lobsters. Mailing Add: 128 E Pleasant St Amherst MA 01002

MUELLER-DOMBOIS, DIETER, b Bethel, Ger, July 26, 25; nat Can; m 51; c 5. PLANT ECOLOGY. Educ: Univ Göttingen, BScA, 51; Hohenheim Agr Univ, dipl, 51; Univ BC, BScF, 55, PhD, 60. Prof Exp: Asst biol & bot, Univ BC, 55-57; forest ecologist, Res Div, Can Dept Forestry, 58-63; forest ecologist, Dept Bot, Univ Hawaii, 63-66; prin field investr, Smithsonian-Ceylon Ecol Proj, Univ Ceylon, 67-69; PRIN FIELD INVESTR, UNIV HAWAII, 69-, PROF BOT & ECOL, 71- Concurrent Pos: Co-dir & sci coordr, Island Ecosysts Integrated Res Prog, Int Biol Prog, 70-71, dir,

72- Mem: Fel AAAS; Asn Trop Biol; Ecol Soc Am; Int Soc Trop Ecol. Res: Agriculture; botany; soil science; climatology; forest site classification; vegetation and environmental studies; synecology and autecology; tree physiology; soil water-plant growth relations; ecology of vegetation on recent volcanic matter; tropical and ecosystems ecology; animal-vegetation interactions. Mailing Add: Dept of Bot Univ of Hawaii Honolulu HI 96822

MUENCH, DONALD LEO, b Rochester, NY, Jan 31, 34; m 60; c 4. MATHEMATICS. Educ: St John Fisher Col, BS, 55; St John's Univ, NY, MS, 60; Idaho State Univ, DA(math), 74. Prof Exp: Asst prof math, US Naval Acad, 60-66; ASSOC PROF MATH, ST JOHN FISHER COL, 66- Mem: Math Asn Am. Res: Linear Algebra; matrix theory. Mailing Add: Dept of Math St John Fisher Col Rochester NY 14618

MUENCH, JOHN, JR, b Annapolis, Md, July 26, 30; m 54; c 2. FORESTRY, FOREST ECONOMICS. Educ: Pa State Univ, BS, 53, MF, 58; Duke Univ, DFor, 64. Prof Exp: From instr to asst prof forestry, Pa State Univ, 57-65; forest economist, 65-73, DIR ECON, NAT FOREST PRODS ASN, 73- Mem: Soc Am Foresters; Am Econ Asn; Nat Asn Bus Econ. Res: Public forestry programs for private landowners; impact of public versus private ownership of forest land on a rural economy; management of even-aged forest stands. Mailing Add: Nat Forest Prods Asn 1619 Massachusetts Ave NW Washington DC 20036

MUENCH, KARL HUGO, b St Louis, Mo, May 3, 34; m 57; c 3. BIOCHEMISTRY, GENETICS. Educ: Princeton Univ, AB, 56; Washington Univ, MD, 60. Prof Exp: Intern med, Barnes Hosp, St Louis, 60-61; USPHS fel biochem, Stanford Univ, 61-65; from instr to assoc prof med & biochem, 65-73, PROF MED, SCH MED, UNIV MIAMI, 73-, CHIEF GENETIC MED, 68- Concurrent Pos: Am Cancer Soc fac res assoc, Univ Miami, 65-70; Markle scholar acad med, 69-74; Leukemia Soc Am scholar, 71-76; USPHS res career develop award, 71-76. Mem: Am Soc Biol Chem; Am Chem Soc; Biochem Soc; Am Soc Human Genetics. Res: Amino acyl-transfer RNA synthetases; transfer RNA; protein synthesis; interaction of drugs with nucleic acids. Mailing Add: Dept of Med Univ of Miami Sch of Med Miami FL 33152

MUENCH, NILS LILIENBERG, b Houston, Tex, Feb 27, 28; m 50; c 1. PHYSICS. Educ: Rice Univ, BA, 49, MA, 50, PhD(physics), 55; S Tex Col, LLB, 59. Prof Exp: Sr res engr,, Humble Oil & Ref Co, 55-59; chief scientist, Army Rocket & Guided Missile Agency, 59-62; mem Inst Defense Anal, 62-63; head physics dept, 63-69, TECH DIR RES LABS, GEN MOTORS CORP, 69- Mem: Am Phys Soc; Am Inst Aeronaut & Astronaut; Am Bar Asn. Res: Solid state physics; management of research. Mailing Add: Res Labs Gen Motors Corp Warren MI 48090

MUENCH, ROBIN DAVIE, b North Conway, NH, Sept 16, 42; m 66; c 2. PHYSICAL OCEANOGRAPHY. Educ: Bowdoin Col. AB. 64; Dartmouth Col, MA, 66; Univ Washington, PhD(oceanog), 70. Prof Exp: From res asst oceanog to oceanogr, Univ Washington, 68-70; asst prof, 70-75, ASSOC PROF OCEANOG. INST MARINE SCI, UNIV ALASKA, 75- Mem: AAAS; Am Geophys Union; Am Meteorol Soc. Res: Physical oceanography of estuarine and inshore waters; physical oceanography in the Arctic; air-sea interaction. Mailing Add: Inst of Marine Sci Univ of Alaska Fairbanks AK 99701

MUENOW, DAVID W, b Chicago, Ill, May 28, 39. PHYSICAL CHEMISTRY, GEOCHEMISTRY. Educ: Carleton Col, BA, 61; Purdue Univ, PhD(chem), 67. Prof Exp: Welch fel mass spectrometry, Rice Univ, 67-70; asst prof, 70-74, ASSOC PROF CHEM, UNIV HAWAII, 75- Mem: Am Chem Soc. Res: High temperature mass spectrometry; thermodynamics; meteoritics; silicate chemistry; geochemistry. Mailing Add: 2545 The Mall Dept of Chem Univ of Hawaii Honolulu HI 96822

MUENSTERBERGER, WERNER, US citizen; m 64. PSYCHIATRY, PSYCHOANALYSIS. Educ: Univ Basel, PhD(anthrop), 38. Prof Exp: Res assoc, Royal Inst Indies, Amsterdam, 39-46; consult, Proj Contemp Cults, Columbia Univ, 47-50; lectr psychiat, 50-56, clin asst prof, 56-61, ASSOC PROF PSYCHIAT, STATE UNIV NY DOWNSTATE MED CTR, 61- Concurrent Pos: Lectr, Sch Appl Psychoanal, NY Psychoanal Inst, 53- & NJ Neuropsychiat Inst, 58-; Guggenheim fel, 69-70. Mem: Fel Am Anthrop Asn; Soc Appl Anthrop; Am Psychol Asn; Royal Anthrop Inst Gt Brit & Ireland. Res: Sociopsychoanalytic research in personality, social structure and culture; motivation and human interaction; psychoanalysis. Mailing Add: Dept of Psychiat State Univ NY Downstate Med Ctr Brooklyn NY 11203

MUENTENER, DONALD ARTHUR, b Pigeon, Mich, Apr 24, 26; m 51; c 2. BACTERIOLOGY. Educ: Alma Col, BSc, 50; Mich State Univ, MS, 53. Prof Exp: Instr biol, Alma Col, 50; asst, Mich State Univ, 53-55; microbiologist, 56-60, asst dir, 61-75, ACTG DIR LAB DIV, MICH DEPT AGR, 75- Mem: Regist Nat Registry Microbiol; Am Soc Microbiol; Brit Soc Appl Bact; Asn Food & Drug Off. Res: Laboratory administration; agricultural, industrial, food, dairy and sanitation microbiology. Mailing Add: 3312 Inverary Dr Lansing MI 48910

MUENTER, ANNABEL ADAMS, b New York, NY, Dec 3, 44; m 68. PHOTOGRAPHIC CHEMISTRY, PHYSICAL CHEMISTRY. Educ: Univ Mich, BSChem, 66; Harvard Univ, PhD(chem physics), 72. Prof Exp: SR RES CHEMIST, EASTMAN KODAK RES LABS, 70- Mem: Soc Photog Scientists & Engrs. Res: Dyes and spectral sensitization of silver halide using various spectroscopic techniques, including sub nanosecond fluorescence lifetime measurements. Mailing Add: Eastman Kodak Res Labs Kodak Park Rochester NY 14650

MUENTER, JOHN STUART, b Cleveland, Ohio, May 10, 38; m 68. PHYSICAL CHEMISTRY, SPECTROSCOPY. Educ: Kenyon Col, BA, 60; Stanford Univ, PhD(chem). 65. Prof Exp: Fel Stanford Univ, 65-66; NIH fel, Harvard Univ, 66-68; asst prof, 69-75, ASSOC PROF PHYS CHEM, UNIV ROCHESTER, 75- Mem: Am Inst Physics. Res: Spectroscopic studies of the electronic structure of small molecules utilizing microwave and molecular beam electric resonance spectroscopy. Mailing Add: Dept of Chem Univ of Rochester Rochester NY 14627

MUESSIG, SIEGFRIED JOSEPH, b Freiburg, Ger, Jan 19, 22; nat US; m 49; c 2. ECONOMIC GEOLOGY. Educ: Ohio State Univ, BSc, 47, PhD(geol), 51. Prof Exp: Instr field geol, Ohio State Univ, 50; geologist, Mineral Deposits Br, US Geol Surv, 51-59; chief geologist, US Borax & Chem Corp, 59-66; minerals explor mgr, Tidewater Oil Co, Calif, 66-67; minerals explor mgr, Getty Oil Co, 67-73, PRES, GETTY OIL DEVELOPMENT CO, LTD, 73- Concurrent Pos: Consult, Borax Consol, Ltd, 55; regional dir, Nat Defense Exec Reserve, 71-; dir region 9, Emergency Minerals Admin, Dept Interior, 71- Mem: AAAS; Am Inst Mining, Metall & Petrol Eng; Geol Soc Am; Soc Econ Geologists (vpres, 73-); Am Mining Cong. Res: Stratigraphy and structure; geology of saline deposits; geology of northeastern Washington; economic geology; geology of northwest Argentina. Mailing Add: Getty Oil Development Co Ltd 3810 Wilshire Blvd Los Angeles CA 90010

MUETHER, HERBERT ROBERT, b Winfield, NY, Sept 27, 21; m 51; c 6. NUCLEAR PHYSICS. Educ: Queen's Col, NY, BS, 42; Princeton Univ, AM, 47, PhD(physics), 51. Prof Exp: Instr physics, Princeton Univ, 49-50; lectr, Queens Col, NY, 50-52, instr, 52-55, asst prof, 55-59; assoc prof, 59-61, assoc chmn dept, 68-75, PROF PHYSICS, STATE UNIV NY STONY BROOK, 61-, DIR UNDERGRAD PROG PHYSICS, 75- Concurrent Pos: Res collabr, Brookhaven Nat Lab, 51-; consult, Frankford Arsenal, Pa, 58- Mem: AAAS; Am Phys Soc; Am Asn Physics Teachers. Res: Neutron physics. Mailing Add: Dept of Physics State Univ of NY Stony Brook NY 11790

MUETTERTIES, EARL LEONARD, b Elgin, Ill, June 23, 27; m 56; c 6. INORGANIC CHEMISTRY. Educ: Northwestern Univ, BS, 49; Harvard Univ, AM, 51, PhD(inorg chem), 52. Prof Exp: Res chemist, Cent Res Dept, E I Du Pont de Nemours & Co, Inc, 52-57, res supvr, 57-65, assoc res dir, 65-73; PROF CHEM, CORNELL UNIV, 73- Concurrent Pos: Adj prof dept chem & assoc mem Monell Chem Senses Ctr, Univ Pa, 69- Honors & Awards: Award, Am Chem Soc, 65. Mem: Nat Acad Sci; Am Chem Soc; Am Acad Arts & Sci; The Chem Soc; Am Phys Soc. Res: organometallics; catalysis; stereochemistry; pheromones; animal behavior as affected by olfactory signals; chemoreception. Mailing Add: 201 Christopher Lane Ithaca NY 14850

MUFFLER, LEROY JOHN PATRICK, b Alhambra, Calif, Sept 19, 37; m 66; c 1. GEOLOGY, GEOCHEMISTRY. Educ: Pomona Col, BA, 58; Princeton Univ, MA, 61, PhD(geol), 62. Prof Exp: Geologist, Southwest States Br, 62, Alaska Geol Br, 62-64, GEOLOGIST, FIELD GEOCHEM & PETROL BR, US GEOL SURV, 64- Mem: Geol Soc Am; Mineral Soc Am; Am Geophys Union. Res: Geothermal resources; hydrothermal alteration; hot springs; Cenozoic volcanic rocks. Mailing Add: Geol Div US Geol Surv 345 Middlefield Rd Menlo Park CA 94025

MUFFLEY, HARRY CHILTON, b Urbana, Ill, Dec 2, 21; m 50. BIOLOGY, ORGANIC CHEMISTRY. Educ: Milliken Univ, BS, 49. Prof Exp: Res chemist, Fine Chem Div, Glidden Co, 50-52; res chemist, Rock Island Arsenal, 52-63, res phys scientist, US Army Weapons Command, 63-70; RES PHYS SCIENTIST, GEN THOMAS J RODMAN LAB, 70-. Honors & Awards: Wilbur Deutsch Mem Award, Am Soc Lubrication Eng, 67. Mem: AAAS; Am Chem Soc (pres, 61); Am Soc Lubrication Eng; Am Ord Asn. Res: Biomechanics; study of the unique or unusual characteristics of animal leading to new concepts in weapons or weapons systems; hydraulic fluids; corrosion preventives. Mailing Add: 1711 29th St Rock Island IL 61201

MUFSON, DANIEL, b Bronx, NY, Dec 24, 42; m 64; c 2. PHARMACEUTICAL CHEMISTRY. Educ: Columbia Univ, BS, 63, MS, 65; Univ Mich, PhD(pharmaceut chem), 68. Prof Exp: Res pharmacist, Parke Davis & Co, 68-71; sr scientist pharmaceut chem, Smith Kline & French Labs, 71-73, group leader, 73-74; MGR BIOPHARMACEUT & DRUG METAB, USV PHARMACEUT CORP, DIV REVLON, 74- Mem: Am Pharmaceut Asn; Acad Pharmaceut Sci. Res: Solubilization and dissolution studies on pharmaceuticals and endogenous lipids; biopharmaceutics and drug metabolism; pharmaceutical development. Mailing Add: USV Pharmaceut Corp 1 Scarsdale Rd Tuckahoe NY 10707

MUFTI, IZHAR-UL HAQ, b Batala, India, June 15, 31; Can citizen; m 61; c 2. APPLIED MATHEMATICS. Educ: D J Col, Univ Karachi, Pakistan, BSc, 51, MSc, 53; Univ BC, PhD(appl math), 60. Prof Exp: Lectr math, DJ Col, Univ Karachi, 51-56; asst, Univ BC, 56-60; asst res officer, Nat Res Coun Can, Ont, 60-66, ASSOC RES OFFICER MATH, ANAL LAB, DIV MECH ENG, NAT RES COUN CAN, ONT, 66- Mem: Soc Indust & Appl Math. Res: System theory; control engineering; stability; numerical mathematics. Mailing Add: Anal Lab Div of Mech Eng Nat Res Coun of Can Ottawa ON Can

MUGA, MARVIN LUIS, b Dallas, Tex, Mar 1, 32. NUCLEAR CHEMISTRY. Educ: Southern Methodist Univ, BS, 53, MS, 54; Univ Tex, PhD, 57. Prof Exp: Res nuclear chemist, Lawrence Radiation Lab, Univ Calif, 57-60; asst prof, 60-68, ASSOC PROF NUCLEAR CHEM, UNIV FLA, 68- Concurrent Pos: Fulbright lectr, 60. Mem: Am Chem Soc; Am Phys Soc. Res: Thin film scintillator detectors for dE dx measurements of energetic heavy ions; fission decay phenomena. Mailing Add: Dept of Chem & Physics Univ of Fla Gainesville FL 32601

MUGGENBURG, BRUCE AL, b St Paul, Minn, May 2. 37; m 60; c 3. VETERINARY PHYSIOLOGY. Educ: Univ Minn, BS, 59, DVM, 61; Univ Wis, Madison, MS, 64, PhD(vet sci), 66. Prof Exp: From instr to asst prof vet sci, Univ Wis, Madison, 64-69; VET PHYSIOLOGIST, INHALATION TOXICOL RES INST, LOVELACE FOUND, 69- Res: Am Vet Med Asn; Am Phsiol Soc; Health Physics Soc. Res: Therapy of radiation induced disease; cardiopulmonary physiology. Mailing Add: Inhalation Toxicol Res Inst Lovelace Found PO Box 5890 Albuquerque NM 87115

MUGGLI, JOANNE, b Richardton, NDak, Aug 21, 08. MATHEMATICS. Educ: Univ Minn, BS, 29; Univ NDak, MA, 33; Univ Wash, PhD(math), 45. Prof Exp: Teacher schs, 29-44; prof math & chmn dept, Col St Benedict, 44-70; prof, St John's Univ, Minn, 70-75; MEM FAC, DEPT NATURAL SCI, COL ST BENEDICT, 75- Concurrent Pos: Dir, NSF Inst, Col St Benedict, 65-66. Mailing Add: Dept of Natural Sci Col of St Benedict St Joseph MN 56374

MUGGLI, ROBERT ZENO, b Richardton, NDak, Dec 6, 29; m 54; c 8. BIO-ORGANIC CHEMISTRY, CHEMICAL MICROSCOPY. Educ: St John's Univ, Minn, BA, 51; NDak State Univ, MS, 56; Kans State Univ, PhD(org chem). 60. Prof Exp: Paint chemist, Western Paint & Varnish, Minn, 53-54; anal chemist, Standard Oil Co (Ind), NDak, 54-55; res chemist, Sinclair Res Inc, Ill, 60-65; SR RES CHEMIST, WALTER C McCRONE ASSOCS, INC, CHICAGO, 65- Mem: Am Chem Soc. Res: Paint, petroleum, biological and organic chemistry; electron microscopy; x-ray diffraction; non-routine microanalytical chemistry; optical microscopy. Mailing Add: 17938 Homewood Ave Homewood IL 60430

MUGHABGHAB, SAID F, b Beirut, Lebanon, July 4, 34; m 63. NUCLEAR PHYSICS. Educ: Am Univ Beirut, BSc, 56, MSc, 59; Univ Pa, PhD(nuclear physics), 63. Prof Exp: Res assoc physics, 63-65, asst physicist, 65-67, ASSOC PHYSICIST, BROOKHAVEN NAT LAB, 67- Concurrent Pos: Consult, Nuclear Data Proj, Oak Ridge Nat Lab, 66-; adj assoc prof, Dowling Col, 70- Mem: Am Phys Soc. Res: Compilation of nuclear data in the neutron field; photonuclear reactions; neutron total cross section measurements, study of nuclear structure with n,lambda reactions. Mailing Add: Dept of Appl Sci Brookhaven Nat Lab Upton NY 11973

MUGLER, DALE H, b Denver, Colo, Nov 8, 48; m 72. MATHEMATICAL ANALYSIS. Educ: Univ Colo, BA, 70; Northwestern Univ, MA, 71, PhD(math), 74. Prof Exp: Asst prof math, Syracuse Univ, 74-75; ASST PROF MATH, UNIV SANTA CLARA, 75- Mem: Am Math Soc; Math Asn Am; Soc Indust & Appl Math. Res: Complex function theory and related parts of applicable mathematics, especially differential and integral equations. Mailing Add: Dept of Math Univ of Santa Clara Santa Clara CA 95053

MUGNAINI, ENRICO, b Siena, Italy, Dec 10, 37; m 61; c 2. NEUROANATOMY. Educ: Univ Pisa, MD, 62. Prof Exp: Trainee neuroanat, Univ Oslo, 63; asst prof anat, Med Sch, Univ Bergen, 64-66; assoc prof, Med Sch, Univ Oslo, 67-69; PROF BIOBEHAV SCI & HEAD LAB NEUROMORPHOL, UNIV CONN, 69- Concurrent Pos: Vis prof, Harvard Med Sch, 70. Mem: AAAS; Royal Soc Med; Scand Electron Microscope Soc; Am Asn Cell Biol. Res: Neurohistology. Mailing Add: Lab of Neuromorphol Univ of Conn U-154 Storrs CT 06268

MUGWIRA, LUKE MAKORE, b Selukwe, Rhodesia, Mar 21, 40; m 65; c 2. SOIL CHEMISTRY. Educ: Lewis & Clark Col, BS, 65; Mich State Univ, MS, 67, PhD(soil chem), 70. Prof Exp: Teaching asst soil chem, Mich State Univ, 65-67, res asst, 67-70, fel, 70-71; ASSOC PROF SOIL CHEM, ALA A&M UNIV, 71- Concurrent Pos: Consult, Biochem Labs, 74 & Hayes Int Corp, City Investing Co, 75- Mem: Am Soc Agron; Soil Sci Soc Am. Res: Triticale adaptation to acid soils, its mineral nutrition and physiological characteristics related to fertilizer efficiency utilization; organic waste evaluation for crop production and role in environmental pollution. Mailing Add: Ala A&M Univ PO Box 137 Normal AL 35762

MUHLER, JOSEPH CHARLES, b Ft Wayne, Ind, Dec 22, 23; m 49; c 2. BIOCHEMISTRY. Educ: Ind Univ, BS, 45, DDS, 48, PhD(chem), 51. Prof Exp: From asst prof to prof chem, Ind Univ, Indianapolis, 51-61, res prof basic sci, Sch Dent, 61-72; RES PROF DENT SCI & DIR PREV DENT RES INST, SCH DENT, IND UNIV, FT WAYNE, 72- Concurrent Pos: Consult, Procter & Gamble Co, 49-; chmn biochem sect, Am Asn Dent Schs, 58; chmn dept prev dent, Sch Dent, Ind Univ, Indianapolis, 58-72, dir prev dent res inst, 68-72; consult, US Air Force Sch Aviation Med, 59-61, Off Surgeon Gen, US Army, 61-, Ft Knox, 62-, Mead Johnson Co, 61-, Gen Foods Corp, 64-69, Bur Med & Surg, US Navy, 64-, White Labs, 66-, & Dentsply Corp, 70- Honors & Awards: Award, Int Asn Dent Res, 68. Mem: Fel AAAS; fel Am Col Dent; fel Am Inst Chem; Am Chem Soc; Am Dent Asn. Res: Essentiality of trace elements; lipid metabolism. Mailing Add: Ind Univ at Ft Wayne 2101 Coliseum Blvd E Ft Wayne IN 46805

MUHLHAUSER, RICHARD OREN, pharmacy, pharmaceutical chemistry, see 12th edition

MUHRER, MERLE E, b Kahoka, Mo, Aug 5, 13; m 39; c 4. BIOCHEMISTRY. Educ: Northeast Mo State Univ, BS, 35; Mo Univ, AM, 40, PhD(agr chem), 44. Prof Exp: Teacher high sch, Mo, 35-39; asst instr agr chem, Mo Univ, 39-40; actg prof chem, Northeast Mo State Univ, 42-43; from asst prof to prof agr chem, 44-74, chmn dept, 55-68, PROF BIOCHEM, UNIV MO-COLUMBIA, 74- Concurrent Pos: Collabr USDA Lab, Peoria, Ill, 50-; mem thrombosis coun, Scientific Coun, Am Heart Asn, 74. Mem: Fel AAAS; Am Chem Soc; Am Soc Animal Sci. Res: Hemostasis in farm animals; giotrogenic effects in farm animals; ruminant biochemistry; nutrition; science education; hemostasis in animals found to be models of the same defect found in human beings; nonprotein nitrogen substitutes in animal nutrition to spare protein for human nutrition. Mailing Add: 105 Schweitzer Hall Dept of Biochem Univ of Mo Columbia MO 65201

MUHS, MERRILL ARTHUR, b San Francisco, Calif, May 9, 26; m 52; c 3. ORGANIC CHEMISTRY, ANALYTICAL CHEMISTRY. Educ: Univ Calif, BS, 49; Univ Wash, PhD(chem), 54. Prof Exp: Res chemist, 54-73, SUPVR SR STAFF, SHELL DEVELOP CO, 73- Mem: Am Chem Soc. Res: Gas chromatography; applied spectroscopy; characterization of odors; analysis of air and water pollutants; liquid chromatography; polymer analysis. Mailing Add: Shell Develop Co PO Box 1380 Houston TX 77001

MUI, PAUL TING-KAI, b Tai-Sen, China, July 19, 46. BIOCHEMISTRY. Educ: Southern Missionary Col, BA, 68; Mid Tenn State Univ, MS, 70; Univ Tenn, PhD(biochem), 75. Prof Exp: Teaching asst chem, Mid Tenn State Univ, 69-70; res asst med chem, Ctr Health Sci, Univ Tenn, 70-72; technologist hemat, 72-73, trainee biochem, 73-75, TRAINEE HEMAT, ST JUDE CHILDREN'S RES HOSP, 75- Mem: Am Chem Soc; Am Asn Univ Profs. Res: Biochemistry of blood coagulation and fibrinolysis as well as their interrelationship with respect to the process of hemostasis. Mailing Add: St Jude Children's Hosp 332 N Lauderdale Memphis TN 38101

MUIR, ARTHUR H, JR, b San Antonio, Tex, Aug 26, 31; div; c 2. SOLID STATE SCIENCE. Educ: Williams Col, BA, 53; Calif Inst Technol, MS, 55, PhD(physics), 60. Prof Exp: Sr physicist, Atomics Int Div, NAm Rockwell Corp, 60-62, specialist Mössbauer effect, 62-63, mem tech staff, Sci Ctr, 63-69, mem technol advan staff, 69-70, mgr int technol prog, 70-74, DIR PHYSICS & CHEM DEPT, ROCKWELL INT SCI CTR, 74- Mem: Am Phys Soc; Sigma Xi; Int Solar Energy Soc. Res: Applications of Mössbauer effect to solid state physics; nuclear spectroscopy; reactor physics; low energy nuclear physics; solar energy; research administration. Mailing Add: Rockwell Int Sci Ctr PO Box 1085 Thousand Oaks CA 91360

MUIR, BARRY SINCLAIR, b Belleville, Ont, July 21, 32; m 61; c 2. ZOOLOGY. Educ: Univ Toronto, BA, 56, MA, 58, PhD(zool), 61. Prof Exp: From asst prof to assoc prof zool, Univ Hawaii, 61-67; sr res scientist, Hydronautics, Inc, Md, 67-68; scientist, 68-71, ASST DIR RES, FISHERIES RES BD CAN, MARINE ECOL LAB, BEDFORD INST, 71- Concurrent Pos: Prin investr, NSF grant, 63-68. Mem: AAAS; Ecol Soc Am; Am Soc Zool; Am Fisheries Soc; Am Soc Ichthyologists & Herpteologists; Can Sos Zoologists. Res: Dynamics of fish populations; environmental influence on energy metabolism of fish, especially processes of growth and reproduction. Mailing Add: Marine Ecol Lab Bedford Inst Dartmouth NS Can

MUIR, DONALD EARL, b Seattle, Wash, Apr 15, 33; m 56; c 2. MATHEMATICS. Educ: Univ Idaho, BS, 56; Stanford Univ, MS, 74. Prof Exp: Instr math, Univ Idaho, 56-57; design specialist, Martin Co, 57-61; sr engr, Raytheon Co, 61-62; res engr, Boeing Co, 62; opers analyst, Tech Opers, Inc, 62-64; MEM PROF STAFF, CTR NAVAL ANAL, 64- Mem: Am Math Soc; Am Astron Soc; Math Soc Am; Am Geophys Union; Opers Res Soc Am. Res: Military operations analysis; computer sciences. Mailing Add: Ctr for Naval Anal 1401 Wilson Blvd Arlington VA 22209

MUIR, DONALD RIDLEY, b Toronto, Ont, Jan 3, 29; m 54; c 2. PHYSICAL CHEMISTRY, RESEARCH ADMINISTRATION. Educ: Univ Toronto, BA, 51, MA, 52, PhD(phys chem), 54. Prof Exp: Mem staff, Res & Develop, Johnson & Johnson Ltd, 54-62; dir res & develop div, Columbia Cellulose Co, Ltd, BC, 62-69; dir res & develop, Oxford Paper Co, 69-73; PRES, SULPHUR DEVELOP INST CAN, 73- Mem: Am Chem Soc; Tech Asn Pulp & Paper Indust; Can Pulp & Paper Asn; fel Chem Inst Can. Res: Research and development of new uses for sulphur. Mailing Add: Sulphur Develop Inst Can Box 9505 202 6 Ave SW Calgary AB Can

MUIR, DOUGLAS WILLIAM, b Kalamazoo, Mich, May 18, 40; m 67; c 2. NUCLEAR SCIENCE. Educ: Southern Ill Univ, BA, 62; NMex State Univ, MS, 65, PhD(physics), 68. Prof Exp: STAFF MEM, LOS ALAMOS SCI LAB, 68- Mem: Am Nuclear Soc; Am Phys Soc. Res: Measurement, compilation, processing and testing of nuclear data for calculations of neutron/proton transport and radiation effects. Mailing Add: 4200 Arkansas Los Alamos NM 87544

MUIR, FOREST VERN, b West Frankfort, Ill, Nov 26, 39; m 61. POULTRY SCIENCE. Educ: Southern Ill Univ, BS, 61, MS, 63; Ohio State Univ, PhD(poultry genetics), 67. Prof Exp: EXTEN POULTRY SPECIALIST, UNIV MAINE, ORONO, 68- Mem: Poultry Sci Asn. Mailing Add: Hitchner Hall Univ of Maine Orono ME 04473

MUIR, J LAWRENCE, b Enid, Okla, Apr 4, 03; m 30; c 4. GEOLOGY. Educ: Univ Okla, BS, 30, MS, 33. Prof Exp: Asst geol, North La field party, Gulf Ref Co, La, 30-31; asst, Amerada Petrol Corp, Okla, 33-44, dist exploitation geologist, 44-48; chief geologist, Champlin Ref Co, 48-51, vpres explor & develop, 51-54; consult geologist, 54-62; petrol geologist, Securities & Exchange Comn, 62-68, chief oil & gas geologist & chief sect oil & gas, Washington, DC, 68-72, CHIEF OIL & GAS GEOLOGIST & CHIEF, OFFICE OF OIL & GAS, WASHINGTON, DC, SECURITIES & EXCHANGE COMN, 73- Concurrent Pos: Vis assoc prof, Univ Okla, 55-56. Mem: Fel Geol Soc Am; Am Asn Petrol Geol; Am Inst Mining, Metall & Petrol Eng; Soc Econ Paleont & Mineral; Soc Explor Geophys. Res: Petroleum geology, especially subsurface geology, sedimentation and stratigraphy; oil and gas industry securities regulation. Mailing Add: 533 W Great Falls St Falls Church VA 22046

MUIR, JAMES ALEXANDER, solid state physics, see 12th edition

MUIR, LARRY ALLEN, b Warren, Ohio, May 31, 42; m 65; c 1. ANIMAL NUTRITION & PHYSIOLOGY. Educ: Ohio State Univ, BS, 64, MS, 66, PhD(ruminant nutrit), 70. Prof Exp: Res assoc ruminant nutrit, Ohio State Univ, 64-70; sr res physiologist, 70-75, RES FEL BASIC ANIMAL SCI RES, MERCK & CO, INC, 75- Mem: Am Dairy Sci Asn; Am Soc Animal Sci. Res: Ruminant nutrition and physiology. Mailing Add: Basic Animal Sci Res Merck & Co Inc Rahway NJ 07065

MUIR, MARIEL MEENTS, physical inorganic chemistry, see 12th edition

MUIR, MELVIN K, b Johannesburg, SAfrica, Jan 25, 32; US citizen; m 57; c 2. SOIL CHEMISTRY, PLANT NUTRITION. Educ: Brigham Young Univ, BSc, 61; Pa State Univ, MSc, 63, PhD(agron), 66. Prof Exp: Mine officer, Johannesburg City Deep Mines, SAfrica, 52-57; asst soils, Pa State Univ, 61-66; asst prof, Mont State Univ, Univ, 57-59; ENVIRON ASST SCIENTIST, KENNECOTT COPPER CORP, 69- Mem: Am Soc Agron; Soil Sci Soc Am. Res: Environmental research; amounts of hydrogen fluorine extractable ammonium in Pennsylvania soils; evaluation of slags as soil liming materials; minor element availability for plant growth; correlations of plant growth factors; influence of pollutants on air, water, soil and vegetation; methods development on analytical techniques for environment and biological samples. Mailing Add: Kennecott Res & Develop Ctr 1515 Mineral Sq Salt Lake City UT 84112

MUIR, ROBERT DONALD, b Sharon, Pa, June 27, 14; m 40; c 2. MICROBIOLOGY. Educ: Allegheny Col, AB, 36; Yale Univ, PhD(bact), 42. Prof Exp: Instr med bact, Sch Med, St Louis Univ, 42-43; microbiologist, Bristol Labs, 43-53; microbiologist, G D Searle & Co, 53-69, ASST DIR BIOL RES SEARLE LABS, 69- Mem: AAAS; Am Soc Microbiol; Mycol Soc Am; Soc Indust Microbiol; Am Phytopath Soc. Res: Fermentation; mycology; product development and control; pharmaceutical bacteriology; control microbiology; antibiotics. Mailing Add: G D Searle & Co Box 5110 Chicago IL 60680

MUIR, ROBERT MATHEW, b Laramie, Wyo, Oct 15, 17; m 47; c 3. PLANT PHYSIOLOGY. Educ: Univ Wyo, BA, 38; Univ Mich, MA, 41, PhD(bot), 46. Prof Exp: Asst prof bot, Pomona Col, 46-48; from asst prof to assoc prof, 48-61, PROF BOT, UNIV IOWA, 61- Mem: Bot Soc Am; Am Soc Plant Physiol; Soc Exp Biol & Med. Res: Relation of chemical constitution to growth regulator action; role of hormones in fruit development; vernalization; abscission; gibberellin and auxin physiology. Mailing Add: Dept of Bot Univ of Iowa Iowa City IA 52242

MUIR, THOMAS GUSTAVE, JR, b San Antonio, Tex, Aug 3, 38; m 66; c 2. ACOUSTICS. Educ: Univ Tex, Austin, BS, 61, MA, 65, PhD(mech eng), 71. Prof Exp: RES SCIENTIST ACOUST, APPL RES LABS, UNIV TEX, AUSTIN, 61- Mem: Acoust Soc Am; Brit Acoust Soc. Res: Underwater, sonar and nonlinear acoustics; sound propagation and scattering. Mailing Add: Appl Res Labs Univ of Tex PO Box 8029 Austin TX 78712

MUIR, WILLIAM A, b Pittsburgh, Pa, Dec 8, 37; m 62; c 1. HUMAN GENETICS, ANATOMY. Educ: George Washington Univ, BS, 60, MS, 62; Univ Rochester, PhD(anat), 66. Prof Exp: Instr anat & human genetics, Sch Med, Univ Rochester, 65-66; fel human genetics, 66-74, ASST PROF MED, CASE WESTERN RESERVE UNIV, 74- Mem: Am Soc Human Genetics; NY Acad Sci. Res: Genetic and biochemical investigation of the thalassemias; hematology; biochemistry. Mailing Add: Dept of Med Case Western Reserve Univ Cleveland OH 44106

MUIR, WILLIAM HOWARD, b Sharon, Pa, May 26, 28; m 52; c 4. PLANT PATHOLOGY, PHYSIOLOGY. Educ: Allegheny Col, BS, 49; Johns Hopkins Univ, MA, 51; Univ Wis, PhD(plant path, physiol), 55. Prof Exp: Res assoc plant path, Univ Wis, 55-57; asst prof biol, 57-62, assoc prof biol, 62-66, chmn dept biol, 64-69, PROF BOT, CARLETON COL, 66- Concurrent Pos: NSF res grants, 58-65, eval panelist, 62-69. Mem: AAAS; Bot Soc Am; Am Bryol & Lichenological Soc; Am Phytopath Soc; Am Soc Plant Physiol. Res: Morphogenesis; development of plant tissue cultures from single cells; morphogenesis in tissue cultures; auxin relations; phylogeny and development of cryptogams; mycology; boreal ecology. Mailing Add: Dept of Biol Carleton Col Northfield MN 55057

MUIR, WILSON BURNETT, b Montreal, Que, Can, July 20, 32; m 55; c 3. SOLID STATE PHYSICS. Educ: McGill Univ, BSc, 53; Univ Western Ont, MSc, 55; Ottawa Univ, PhD(solid state physics), 62. Prof Exp: Sci officer, Radioactivity Div, Dept Mines & Tech Surv, Can, 55-57; physicist, Franklin Inst Labs, 61-64 & Noranda Res Centre, 64-66; physicist, 66-68, asst prof, 68-69, ASSOC PROF PHYSICS, EATON LAB, McGILL UNIV, 69- Mem: Am Phys Soc; Can Asn Physicists. Res: Electron transport; magnetization and Mössbauer Effect in metals, alloys, semiconductors and minerals. Mailing Add: Dept of Physics Eaton Lab McGill Univ PO Box 6070 Montreal PQ Can

MUIRHEAD, ERNEST ERIC, b Pernambuco, Brazil, Sept 13, 16; US citizen; m 42; c 5. PATHOLOGY. Educ: Baylor Univ, BA & MD, 39; Am Bd Path, dipl. Prof Exp: Asst path, Col Med, Baylor Univ, 37-39, from intern to resident, Hosp, 39-41, instr clin path, Col Med, 41-43, asst prof path, Col Dent, 46-48; from assoc prof to prof, Univ Tex Southwestern Med Sch, 48-59, chmn dept, 50-56, chief div hemat, Dept Internal Med, 51-59; prof clin path, Col Med, Wayne State Univ, 59-65; DIR LABS & BLOOD BANK, BAPTIST MEM HOSP, MEMPHIS, TENN, 65- Concurrent Pos: Asst dir labs, Baylor Univ Hosp, 41-43 & William Buchanan Blood, Plasma & Serum Ctr, Dallas, 46-48; from instr to asst prof, Univ Tex Southwestern Med Sch, 47-48; consult, Vet Admin Hosps, Dallas, Tex, 48 & US Army Brooke Gen Hosp, Ft Sam

Houston, 48-68; dir labs, Parkland Hosp, 48-56; dir labs & blood bank, Hutzel Hosp, Detroit, Mich, 59-65; consult, Vet Admin Hosp, Dearborn, Mich, 65; prof path & med, Col Med, Univ Tenn, Memphis, 65-74; mem coun high blood pressure res, Am Heart Asn. Mem: Am Soc Exp Path; Am Soc Clin Path; fel Am Col Physicians; fel Col Am Path; fel Int Soc Hemat. Res: Hematology; renal disorders. Mailing Add: Baptist Hosp 849 Madison Memphis TN 38146

MUIRHEAD, JAMES SEGNER, physical chemistry, see 12th edition

MUIRHEAD, ROBB JOHN, b Adelaide, S Australia, July 7, 46; m 70; c 1. MATHEMATICAL STATISTICS. Educ: Univ Adelaide, BSc, 68, PhD(statist), 70. Prof Exp: Asst prof, 70-75, ASSOC PROF STATIST, YALE UNIV, 75- Mem: Am Statist Asn; Inst Math Statist; Royal Statist Soc. Res: Multivariate analysis and distribution theory; asymptotic methods; linear models. Mailing Add: Dept of Statist Yale Univ Box 2179 Yale Station New Haven CT 06520

MUKA, ARTHUR ALLEN, b Adams, Mass, Oct 23, 24; m 52; c 5. ENTOMOLOGY. Educ: Univ Mass, BS, 50; Cornell Univ, MS, 52, PhD(econ entom), 54. Prof Exp: Asst entom, Cornell Univ, 50-54; assoc entomologist, Va Agr Exp Sta, 54-56; from asst prof to assoc prof, 56-65, PROF ENTOM, CORNELL UNIV, 65- Concurrent Pos: Rockefeller Found entomologist, Int Rice Res Inst, 65-66. Mem: Entom Soc Am; Entom Soc Can. Res: Field and forage crop insect pests; vegetable insect pests; pest management; extension entomology; international agricultural development. Mailing Add: Dept of Entom Cornell Univ Ithaca NY 14850

MUKAI, CROMWELL DAISAKU, b Bostonia, Calif, Apr 13, 17; m 44; c 4. CHEMISTRY. Educ: Univ Calif, BS, 43; NY Univ, MS, 49, PhD(org chem), 55. Prof Exp: Res chemist, Gelatin Prods Corp Mich, 44-46; res chem chemist, Boyle-Midway Div, Am Home Prods Corp, 46-67, res assoc, 67-75; MGR, ANAL LAB, POLYCHROME CORP, YONKERS, 75- Mem: AAAS; Am Chem Soc. Res: Synthesis of amino acids; emulsion technology as applied to waxes; petroleum additives; plastics; mechanisms of Grignard reactions; instrumental analysis; aerosol technology; cleaners; detergents; emulsion polymerization. Mailing Add: Polychrome Corp 137 Alexander St Yonkers NY 10702

MUKERJEE, BARID, b Suri, India, Oct 27, 28; m 59. GENETICS. Educ: Univ Calcutta, BSc, 51; Brigham Young Univ, MS, 56; Univ Utah, PhD(genetics), 58. Prof Exp: Demonstr biol, Vidyasagar Col, India, 51-55; asst prof, Westminster Col, 58-59; asst res prof, Univ Utah, 60-61; res assoc genetics, Columbia Univ, 61-63; from asst prof to assoc prof, 63-70, PROF GENETICS, McGILL UNIV, 70- Res: Relationship of embryonic differentiation and malignancy; mechanism of chromosome differentiation. Mailing Add: Dept of Biol McGill Univ PO Box 6070 Montreal PQ Can

MUKERJEE, DEBDAS, human genetics, oncology, see 12th edition

MUKERJEE, PASUPATI, b Calcutta, India, Feb 13, 32; m 64. PHYSICAL CHEMISTRY, COLLOID CHEMISTRY. Educ: Univ Calcutta, BSc, 49, MSc, 51; Univ Southern Calif, PhD(colloid chem), 57. Prof Exp: Lab asst chem, Univ Southern Calif, 52-54, res fel, 54-56, lectr & res fel, 56-57; res assoc, Brookhaven Nat Lab, 57-59; reader phys chem, Indian Asn Cultivation Sci, 59-64; guest scientist chem, Van't Hoff Lab, Univ Utrecht, 64; sr scientist, Univ Southern Calif, 64-66; vis assoc prof, 66-67, PROF CHEM, SCH PHARM, UNIV WIS-MADISON, 67- Concurrent Pos: Vis asst prof, Univ Southern Calif, 57; hon lectr, Univ Calcutta, 61-64. Mem: Fel AAAS; Am Chem Soc; The Chem Soc; Am Pharmaceut Asn; Sigma Xi. Res: Equilibrium and transport properties of micellar systems; dyes; inorganic ions in aqueous solution; hydrophobic bonding; multiple equilibria; electrical double layers. Mailing Add: Sch of Pharm Univ of Wis Madison WI 53706

MUKERJI, AMBUJ, b Calcutta, India, Mar 1, 19; m 54; c 2. NUCLEAR PHYSICS. Educ: Univ Calcutta, BSc, 40, MSc, 42; Swiss Fed Inst Technol, DrScNat(nuclear physics), 52. Prof Exp: Res fel nuclear spectros, Tata Inst Fundamental Res, India, 52-55, fel, 59-61; res assoc nuclear physics, Univ Wis, 55-56; physicist, Bartol Res Found, Franklin Inst, 56-57; assoc prof physics, Tex A&M Univ, 57-59; vis assoc prof, Univ Ala, 61-64, prof, 64-68; PROF PHYSICS, LEHMAN COL, 68-, MEM GRAD FAC, CITY UNIV NEW YORK, 69-, CHMN DEPT PHYSICS & ASTRON, 75- Concurrent Pos: Consult, Phys Sci Lab, US Army Missile Command, Redstone Arsenal, 64- Mem: AAAS; Am Phys Soc; Swiss Phys Soc. Res: Nuclear spectroscopy by the study of the decay of radioactive nuclei; determination of the fluorescent yields in the K and L shells; nuclear and solid state properties with help of Mössbauer effect experiments. Mailing Add: Dept of Physics Herbert H Lehman Col Bronx NY 10468

MUKHERJEA, ARUNAVA, b Calcutta, India, Aug 14, 41; m 71. MATHEMATICS. Educ: Univ Calcutta, MSc, 61; Wayne State Univ, PhD(math), 67. Prof Exp: Asst prof math, Eastern Mich Univ, 67-69; asst prof, 69-70, ASSOC PROF MATH, UNIV S FLA, 70- Mem: Am Math Soc. Res: Functional analysis; measure theory; analysis on semigroups. Mailing Add: Dept of Math Univ of SFla Tampa FL 33620

MUKHERJEE, ANIL B, b Suri, India, Jan 20, 42. CELL BIOLOGY, GENETICS. Educ: Univ Calcutta, BSc, 62; Univ Utah, MS, 64, PhD, 66. Prof Exp: Res assoc human genetics & cytol, Col Physicians & Surgeons, Columbia Univ, 66; asst prof biol & sometic cell genetics, Queen's Univ, Ont, 66-67; asst res prof, 67-68, ASST PROF PEDIAT & HUMAN GENETICS, STATE UNIV NY, BUFFALO, 68- Concurrent Pos: Nat Res Coun Can res grant, 66-67; NIH res grant; Lalor Found fel. Mem: AAAS; AMA; Genetics Soc Am; Am Soc Cell Biol. Res: Human genetics. Mailing Add: Sch of Med State Univ of NY Buffalo NY 14222

MUKHERJEE, ASIT B, b Suri, India, Apr 6, 40. CYTOGENETICS, DEVELOPMENTAL GENETICS. Educ: Univ Utah, BS, 65, MS, 66, PhD(zool), 68. Prof Exp: Teaching asst biol, Univ Utah, 65-67; res assoc human genetics, State Univ NY Upstate Med Ctr, 68-69; fel genetics, Med Ctr, Columbia Univ, 69-70; instr, Albert Einstein Col Med, 70-72; ASST PROF, DEPT BIOL, FORDHAM UNIV, 72- Mem: AAAS; Tissue Cult Asn; Am Soc Cell Biol. Res: Cytogenetics; somatic cell genetics; human genetics; somatic cell hybridization. Mailing Add: Dept of Biol Fordham Univ Bronx NY 10458

MUKHERJEE, JOAN WEINMANN, organic chemistry, see 12th edition

MUKHERJEE, LAL MOHAN, analytical chemistry, physical chemistry, see 12th edition

MUKHERJEE, TAPAN KUMAR, b Gorakhpur, India, Jan '5, 29; m 57; c 2. ORGANIC CHEMISTRY, SOLID STATE SCIENCE. Educ: Patna Univ, BS, 48, MS, 50, DSc(chem), 74; Wayne State Univ, PhD(chem), 56. Prof Exp: Lectr chem, Patna Univ, 50-52; asst, Wayne State Univ, 52-55, res fel, 55-56; lectr, Univ Bihar, 57-58; res chemist, Mass Inst Technol, 58-60, Gen Aniline & Film Corp, 60-61, Retina Found, 61-62 & Air Force Cambridge Res Lab, 62-74; prog mgr, Solar Energy Conversion, 74-75, PROG MGR, ADVAN RESOURCE SYSTS, NSF, 76- Mem: Am Chem Soc; Int Solar Energy Soc. Res: Syntheses of organic compounds; mechanism of reactions; stable free radicals; organic semiconductors; photoconductors; high temperature laser window materials, photovoltaic materials and devices. Mailing Add: Res Appl Directorate NSF 1800 G St NW Washington DC 20550

MUKHERJI, KALYAN KUMAR, b Calcutta, India, May 30, 39; m 65. GEOLOGY, GEOPHYSICS. Educ: Univ Calcutta, BSc, 59, MSc, 61; Univ Leeds, Dipl, 63; Univ Western Ont, PhD(geol), 68. Prof Exp: Demonstr geol, Univ Western Ont, 63-67; asst prof, 68-73, PROF GEOL, CONCORDIA UNIV, LOYOLA, MONTREAL, 73- Concurrent Pos: Fel, Carleton Univ, Ont, 68; Nat Res Coun Can grant, 69-; Geol Surv Can res grant, 69-70. Mem: Can Soc Petrol Geologists; Int Asn Sedimentologists; Indian Soc Earth Sci; Soc Econ Paleont & Mineral; Geol Asn Can. Res: Ordovician stratigraphy and sedimentation in southwest Ontario; carbonate sedimentation and petrology of Sicker Group, Vancouver Islands; thermo luminescence study of Middle Ordovician limestones and recent sediments; exploration geophysics; carbonate trace element geochemistry. Mailing Add: Dept of Geol Concordia Univ Loyola Montreal PQ Can

MULAIK, STANLEY B, entomology, arachnology, see 12th edition

MULARIE, WILLIAM MACK, b Duluth, Minn, Dec 4, 38; m 61; c 2. SOLID STATE PHYSICS. Educ: Univ Minn, Duluth, BA, 61; Univ Minn, Minneapolis, MSEE, 66, PhD(elec eng), 71. Prof Exp: Physicist, Minn Mining & Mfg Co, 61-63; res scientist solid state physics, Res Inst Advan Studies, 71-74; SR RES ENGR, 3M CO, 74- Mem: Inst Elec & Electronics Engrs. Res: Physics of semiconductor surfaces; auger electron spectroscopy; photoemission; infrared detectors; optics. Mailing Add: 3M Co 3M Ctr St Paul MN 55101

MULAS, PABLO MARCELO, b Atlixco, Mex, Apr 26, 39. PHYSICAL CHEMISTRY, NUCLEAR ENGINEERING. Educ: Univ Ottawa, BS, 60; Princeton Univ, PhD(chem eng), 65. Prof Exp: Resident res assoc, Jet Propulsion Lab, Nat Acad Sci, 67-68; asst vis prof, Grad Sect, Sch Eng, Univ Calif, Los Angeles, 68; prof nuclear sci, Nat Polytech Inst, Mex, 67-71, head dept nuclear eng, 69-71; dir, Reactor Lab, Nat Inst Nuclear Energy, Mex, 71-73; PROF PHYS CHEM, CTR RES, NAT POLYTECH INST, MEX, 73- Concurrent Pos: Mem comt, Nat Prog Basic Sci, Nat Coun Sci & Technol, Mex & consult, Nat Energy Comn, Mex, 75- Mem: Am Phys Soc; Am Nuclear Soc; Am Chem Soc; Acad Sci Res Mex; Mex Phys Soc. Res: Microscopic studies of interfacial phenomena, mass and energy transport; direct conversion of solar electromagnetic radiation to chemical fuels. Mailing Add: Mexicali 36 Mexico 11 DF Mexico

MULAY, AMBADAS SHRIHARI, plant physiology, see 12th edition

MULAY, LAXMAN NILAKANTHA, b Rahuri, India, Mar 5, 23; m 45. PHYSICAL CHEMISTRY, INORGANIC CHEMISTRY. Educ: Univ Bombay, MS, 46, PhD(phys chem), 50. Prof Exp: Daxina Merit fel, Karnatak Col, Bombay, 43-45, demonstr & lectr chem, 46-48, lectr, Inst Sci, 48-53 & 57-58; res assoc, Northwestern Univ, 53-55; res fel, Harvard Univ, 55-57; asst prof, Univ Cincinnati, 58-63; assoc prof, 63-67, PROF CHEM & CHMN SOLID STATE SCI PROG, PA STATE UNIV, 67- Mem: Am Chem Soc; Am Phys Soc; assoc Royal Inst Chem. Res: Magneto-chemistry applied to inorganic polymeric systems; nuclear and electron magnetic resonance studies on metallocenes and biological systems; adsorption; coordination compounds; superparamagnetic systems. Mailing Add: Mat Res Lab Pa State Univ University Park PA 16802

MULCAHY, DAVID LOUIS, b Manchester, NH, Oct 16, 37; m 63; c 2. ECOLOGY, EVOLUTION. Educ: Dartmouth Col, BA, 59; Vanderbilt Univ, PhD(bot), 63. Prof Exp: Asst prof bot, Univ Ga, 63-66; vis scientist, Brookhaven Nat Lab, 66-68; asst prof, 68-71, ASSOC PROF BOT, UNIV MASS, AMHERST, 71- Mem: Ecol Soc Am; Genetics Soc Am; Am Genetic Asn; Soc Study Evolution; Bot Soc Am. Res: Gametophytic competition; population structure of trees; evolution of heterostylv; pollination systems. Mailing Add: Dept of Bot Univ of Mass Amherst MA 01002

MULCAHY, GABRIEL MICHAEL, b Jersey City, NJ, Feb 16, 29; m 58; c 7. PATHOLOGY. Educ: St Peter's Col, NJ, AB, 50; Georgetown Univ, MD, 54. Prof Exp: USPHS med officer, Navajo Indian Reservation, 55-57; resident path, USPHS Hosp, Seattle, Wash, 57-59, USPHS Hosp, Staten Island, NY, 59-61; chief path, USPHS Hosp, Detroit, Mich, 61-62; from instr to assoc prof path, Sch of Med, Creighton Univ, 62-69, actg chmn dept, 67; DIR PATH, JERSEY CITY MED CTR, 69-; PROF GEN & ORAL PATH, NJ DENT SCH & ASSOC CLIN PROF PATH, NJ MED SCH, COL MED & DENT NJ, 71- Mem: Am Soc Clin Pathologists; Int Acad Path; Am Soc Human Genetics; AAAS; AMA. Res: Pathology of familial tumors; cytogenetics. Mailing Add: Dept of Path Jersey City Med Ctr 50 Baldwin Ave Jersey City NJ 07304

MULCAHY, JOHN JOSEPH, b New York, NY, Jan 7, 41; m 70; c 2. UROLOGY. Educ: Georgetown Univ, MD, 66; Univ Minn, MS, 74; Univ Mich, PhD(physiol), 72. Prof Exp: ASST PROF SURG, UNIV KY, 74- Concurrent Pos: Consult, Lexington Vet Admin Hosp, 74- & Cardinal Hill Hosp, 75-; mem provisional staff, St Joseph's Hosp & courtesy staff, Good Samaritan Hosp, 75-; mem perm staff, Univ Ky Med Ctr. Mem: AMA. Res: Renal perfusion from a physiologic standpoint; post-obstructive diuresis; compensatory renal hypertrophy. Mailing Add: Div of Urol Univ of Ky Med Ctr Lexington KY 40506

MULCARE, DONALD J, b New York, NY, July 27, 38; m 68; c 2. ZOOLOGY, DEVELOPMENTAL BIOLOGY. Educ: St Procopius Col, BS, 62; Univ Notre Dame, PhD(biol), 68. Prof Exp: Teaching asst biol, Univ Notre Dame, 62-66; asst, St Mary's Col, Ind, 66-67; Nat Cancer Inst fel zool, Univ Mich, Ann Arbor, 68-69, mem staff amphibian facil, 69; ASST PROF BIOL, SOUTHEASTERN MASS UNIV, 69- Mem: AAAS; Am Soc Zoologists; Soc Develop Biol; NY Acad Sci. Res: Oncology in amphibians; transmission of the Lucke renal adenocarcinoma in Rana pipiens; Rana palustris and their hybrids. Mailing Add: Dept of Biol Southeastern Mass Univ North Dartmouth MA 02747

MULCHI, CHARLES LEE, b Warren County, NC, Dec 2, 41; m 62; c 2. AGRONOMY, PLANT PHYSIOLOGY. Educ: NC State Univ, BS(crop sci) & BS(soil sci), 64, MS, 67, PhD(plant physiol), 70. Prof Exp: Instr soil-plant rels & biochem, NC State Univ, 66-70; asst prof, 70-75, ASSOC PROF CROP PHYSIOL, UNIV MD, COLLEGE PARK, 75- Mem: Am Soc Agron; Crop Sci Soc Am. Res: Soil-plant relations; plant physiology with special interest in photosynthesis and photo-respiration; air pollution effects on plants; aerosol salt effects on crops or cooling tower salt drift effects on vegetation. Mailing Add: Dept of Agron Univ of Md College Park MD 20740

MULCRONE, THOMAS FRANCIS, b Chicago, Ill, Aug 5, 12. MATHEMATICS. Educ: Spring Hill Col, BS, 39; Catholic Univ, MS, 42; St Mays Col, Kans, STL, 47. Prof Exp: Instr math, Spring Hill Col, 40-41 & 42-43; spec lectr, St Louis Univ, 43-47; asst prof, Spring Hill Col, 48-54; asst prof Loyola Univ, La, 54-60; assoc prof,

MULCRONE

Spring Hill Col, 61-75; ASSOC PROF MATH, LOYOLA UNIV, LA, 75- Mem: Math Asn Am; Nat Coun Teachers Math. Res: Modern geometry and algebra; semigroups; history of mathematics; matric with same zero-pattern. Mailing Add: Dept Math Loyola Univ 6363 St Charles Ave New Orleans LA 70118

MULDAWER, LEONARD, b Philadelphia, Pa, Aug 6, 20; m 50; c 3. SOLID STATE PHYSICS, MEDICAL PHYSICS. Educ: Temple Univ, AB, 42, AM, 44; Mass Inst Technol, PhD(physics), 48. Prof Exp: Instr eng, Sci & Mgt War Training, Temple Univ, 42-44, from asst prof to assoc prof, 48-61, PROF PHYSICS, TEMPLE UNIV, 61- Concurrent Pos: Consult, Labs, Frankford Arsenal; staff, Diag Radiol Res Lab, Temple Univ Med Sch, 72- Mem: Am Phys Soc; Am Soc Metals; Am Crystallog Asn; Am Asn Physics Teachers. Res: X-ray diffraction studies of order, particle size and strain; electron diffraction studies of oxides; physiological acoustics; optical and transport properties of alloys; phase transformations; science education; radiological physics. Mailing Add: Dept of Physics Temple Univ Philadelphia PA 19122

MULDER, CAREL, b Arnhem, Neth, Mar 19, 28; m 61; c 4. VIROLOGY, MOLECULAR BIOLOGY. Educ: Univ Leiden, BS, 51, Drs, 55; Oxford Univ, DPhil(microbiol), 63. Prof Exp: Instr molecular biol, Sch Med, Leiden Univ, 60-65; res assoc, Dept Chem, Harvard Univ, 65-67; NIH spec fel tumor virol, Sch Med, St Louis Univ, 67-68 & Salk Inst Biol Studies, 68-70; sr staff investr, Cold Spring Harbor Lab, 70-75; ASSOC PROF MICROBIOL & PHARMACOL, SCH MED, UNIV MASS, 75- Concurrent Pos: Vis prof, Sch Med, Univ Leiden, 75. Mem: Am Soc Microbiol; Soc Gen Microbiol; Neth Soc Biochem. Res: Molecular biology of DNA tumor viruses; restriction endonucleases; cloning of genes. Mailing Add: Dept of Pharmacol Sch of Med Univ of Mass Worcester MA 01605

MULDER, DONALD WILLIAM, b Rehoboth, NMex, June 30, 17; m 43. CLINICAL NEUROLOGY. Educ: Calvin Col, AB, 40; Marquette Univ, MD, 43; Univ Mich, MS, 46. Prof Exp: Asst prof neurol, Univ Colo, 49-50; from instr to assoc prof, Mayo Grad Sch Med, Univ Minn, 50-64; chmn sect neurol, 66-71, pres staff, 70-71, SR CONSULT, MAYO CLIN, 50-; PROF NEUROL, MAYO MED SCH, UNIV MINN, 64- Mem: Am Neurol Asn; Am Psychiat Asn; fel Am Acad Neurol. Res: Amyotropic lateral sclerosis; epilepsy; neuromuscular disease. Mailing Add: Mayo Clin 200 First St SW Rochester MN 55901

MULDER, HARVEY DALE, physical chemistry, see 12th edition

MULDERS, GERARD FRANCIS WILLIAM, b Nymegen, Holland, May 30, 08; nat US; m 37. ASTROPHYSICS. Educ: Univ Utrecht, BS, 29, MA, 33, PhD(astrophys), 34. Prof Exp: Dutch-Am Found fel astrophys res, Mt Wilson Observ, Carnegie Inst Technol, 35-38; actg instr math, Stanford Univ, 38-40; asst prof physics, Univ Redlands, 41-42; from asst prof to assoc prof, 42-47; sci res adminr, San Francisco Br, Off Naval Res, 47-50, chief scientist, 50-56, sci dir, London Br, 56-57, chief scientist, Pasadena Br, 57-59; from assoc prog dir to prog dir astron, NSF, Washington, DC, 59-63, head astron sect, 63-69; ED, FEDN AM SOCS EXP BIOL, 69- Mem: AAAS; Int Astron Union; Am Astron Soc. Res: Photographic photometry of solar spectrum; curve of growth for Fraunhofer lines; energy distribution in continuous spectra of sun and corona; administration of research programs in astronomy supported by the government Mailing Add: Fedn of Am Socs for Exp Biol 9650 Rockville Pike Bethesda MD 20014

MULDOON, THOMAS GEORGE, b Brooklyn, NY, May 13, 38. BIOCHEMISTRY. Educ: Queens Col, NY, BS, 60; Univ Louisville, PhD(biochem), 67. Prof Exp: Fel biochem, Med Ctr, Univ Kans, 67-69; asst prof, 69-74, ASSOC PROF ENDOCRINOL, MED COL GA, 74- Mem: Am Soc Biol Chemists; Endocrine Soc; NY Acad Sci; Soc Study Reprod; Soc Exp Biol & Med. Res: Tissue-specific interactions of steroid hormones and proteins; regulation of hormone receptor activity. Mailing Add: Dept of Endocrinol Med Col of Ga Augusta GA 30902

MULDREW, JAMES ARCHIBALD, b Winnipeg, Man, Oct 9, 25; m 58; c 4. FOREST ENTOMOLOGY. Educ: Univ Manitoba, BSc, 49, MSc, 52. Prof Exp: Agr res officer forest entom, Can Dept Agr, 49-61; res officer forestry, Can Dept Forestry, Winnipeg, 61-71; RES SCIENTIST, CAN DEPT ENVIRON, 71- Mem: AAAS; Entom Soc Can; Can Inst Forestry. Res: Biological control of forest insects; resistance of insects to parasites; population dynamics. Mailing Add: Northern Forest Res Centre 5320 122nd St Edmonton AB Can

MULDROW, CHARLES NORMENT, JR, b Washington, DC, 1930; m 58; c 3. POLYMER CHEMISTRY. Educ: Col Charleston, BS, 50; Univ NC, MA, 54; Univ Va, PhD(phys chem), 58. Prof Exp: Instr chem, Univ NC, 51-52; res chemist, Shell Develop Co, 58-59; from res chemist to sr res chemist, Am Enka Corp, 59-65, head polyester develop sect, 65-70, HEAD POLYMER DEVELOP, AM ENKA CORP, 70- Mem: Am Chem Soc. Res: Synthesis of dielectric materials; magnetochemistry; organic and inorganic solution thermodynamics; process development of plastics and synthetic rubber; reaction kinetics; synthetic fibers. Mailing Add: Am Enka Corp Enka NC 28728

MULE, JAMES GASPARE, b Norco, La, Jan 27, 25; m 49; c 2. OBSTETRICS & GYNECOLOGY. Educ: La State Univ, MD, 47. Prof Exp: From asst to res assoc, 52-55, from instr to prof, 55-74, EMER PROF OBSTET & GYNEC, LA STATE UNIV SCH MED, NEW ORLEANS, 74- Mailing Add: Dept of Obstet & Gynec La State Univ Sch of Med New Orleans LA 70112

MULE, SALVATORE JOSEPH, b Trenton, NJ, Apr 7, 32; m 56; c 4. PHARMACOLOGY, BIOCHEMISTRY. Educ: Col Wooster, BA, 54; Rutgers Univ, MS, 55; Univ Mich, PhD(pharmacol), 61. Prof Exp: Fel biochem & pharmacol, Univ Wis, 61-63; res pharmacologist, Addiction Res Ctr, NIMH, Ky, 63-68; DIR, DRUG ABUSE CONTROL COMN TESTING & RES LAB, 68- Mem: AAAS; Am Chem Soc; Am Soc Pharmacol & Exp Therapeut; Am Acad Forensic Sci; fel Am Inst Chem. Res: Drug metabolism as related to biochemical mechanisms associated with the action of narcotic analgesics. Mailing Add: Drug Abuse Control Comn Testing & Res Lab 80 Hanson Pl Brooklyn NY 11217

MULFORD, DWIGHT JAMES, b Greenville, Ill, Feb 9, 11; m 37. BIOCHEMISTRY. Educ: Greenville Col, BS, 33; St Louis Univ, PhD(biochem), 42. Prof Exp: Res assoc phys chem, Harvard Univ, 42-47, assoc, 47-49; from asst prof to assoc prof biochem, Univ Kans, Lawrence, 49-56; actg chmn dept, 57-58, asst dean, 67-71, PROF BIOCHEM, UNIV KANS MED CTR, KANSAS CITY, 56-, DEAN, 71- Concurrent Pos: Tutor, Harvard Univ, 42-46; chief blood processing lab, Mass Dept Pub Health, 45-47, asst dir, 47-49; asst prof, Boston Univ, 46-47. Mem: Am Soc Biol Chem; Soc Exp Biol & Med. Res: Choline metabolism; plasma fractionation; stability of proteins; blood bank. Mailing Add: Dept of Biochem Univ of Kans Med Ctr Kansas City KS 66103

MULFORD, ROBERT NEAL RAMSAY, b US, Oct 2, 22; m 51; c 1. PHYSICAL CHEMISTRY, PHYSICAL METALLURGY. Educ: Hofstra Col, BA, 47; Brown Univ, PhD(chem), 50. Prof Exp: Asst, Brown Univ, 46-49; staff chemist, 50-69, alt group leader, 69-74, GROUP LEADER, LOS ALAMOS SCI LAB, UNIV CALIF, 74- Res: Hydride chemistry; gas-metal equilibria; high temperature chemistry; plutonium chemistry. Mailing Add: Los Alamos Sci Lab Los Alamos NM 87544

MULHARE, MIRTA T, biological anthropology, see 12th edition

MULHAUSEN, HEDY ANN, b Cleveland, Ohio, Dec 5, 40. BIOCHEMISTRY, INFORMATION SCIENCE. Educ: Ursuline Col, Ohio, AB, 62; Ohio State Univ, MS, 65, PhD(biochem), 67. Prof Exp: Res assoc biochem, Ohio State Univ, 68; assoc, Univ Ga, 68-69; BIOCHEM ED ANALYST, CHEM ABSTR SERV, OHIO STATE UNIV, 69- Mem: Am Chem Soc. Res: Mechanisms of control in mammalian carbohydrate metabolism. Mailing Add: Chem Abstracts Serv Ohio State Univ Columbus OH 43210

MULHAUSEN, ROBERT OSCAR, b Chicago, Ill, June 7, 30; m 54; c 4. INTERNAL MEDICINE, MEDICAL ADMINISTRATION. Educ: Univ Ill, BS, 51, MD, 55; Univ Minn, MS, 64. Prof Exp: Fel internal med, Univ Minn, Minneapolis, 56-59; from instr to assoc prof internal med, 59-73, from asst dean to assoc dean med sch, 67-73, PROF INTERNAL MED, UNIV MINN, MINNEAPOLIS, 73-; CHIEF MED, ST PAUL RAMSEY HOSP, 73- Concurrent Pos: Fulbright res award grant, Rigshospitalet, Copenhagen, 65-66. Mem: AAAS; fel Am Col Physicians; Am Fedn Clin Res; NY Acad Sci; Am Soc Nephrol. Res: Fluid, electrolyte and acid-base physiology; renal physiology. Mailing Add: St Paul Ramsey Hosp St Paul MN 55101

MULHERN, JOHN E, JR, b Chicago, Ill, Mar 26, 26; m 50; c 3. PHYSICS. Educ: Okla State Univ, BS, 48; Boston Univ, MS, 49, PhD, 54. Prof Exp: Semiconductor physicist, Gen Elec Co, 51-54; from asst prof to assoc prof, 54-66, PROF PHYSICS, UNIV NH, 66- Concurrent Pos: Vis asst prof, Brandeis Univ, 57-58; consult, NASA Cambridge Res Ctr, 65-69; NIH spec res fel, 69-70; sr vis fel, Cavendish Lab, Eng, 69-70. Mem: Am Phys Soc. Res: Solid state physics; use of an electron microprobe to analyse the elemental content of biological tissue. Mailing Add: Dept of Physics Univ of NH Durham NH 03824

MULHERN, THOMAS PATRICK, b Brooklyn, NY, May 2, 32; m 63; c 5. MATHEMATICS. Educ: Fordham Univ, ScB, 53; Brown Univ, AM, 55. Prof Exp: Asst prof math, Fordham Univ, 58-60; sr mathematician, Shell Oil Co, NY, 60-65; assoc mgr opers anal, Gen Foods Corp, 65-68, mgr corp mgt sci, 68-73, MGR DISTRIB SYSTS & LOGISTICS ANAL, GEN FOODS CORP, 73- Concurrent Pos: Assoc ed, J Systs Mgt, 71-75. Mem: Inst Mgt Sci. Res: Management science; corporate planning; logistics. Mailing Add: 133 Country Club Rd Stamford CT 06903

MULHOLLAND, JOHN DERRAL, b Muncie, Ind, Sept 28, 34; m 57; c 2. ASTRONOMY, ASTRONAUTICS. Educ: Purdue Univ, BSAE, 57; Univ Cincinnati, MS, 61, PhD(celestial mech), 65; Yale Univ, MS, 64. Prof Exp: Asst proj engr, Kett Tech Ctr, US Indust Inc, 57-58; staff engr, Ketco, Inc, 58-59; res asst & lectr aerodynamics, Univ Cincinnati, 59-60, instr astronaut, 60-64, res assoc astron, 64-65, asst prof, 65-66; mem tech staff, Jet Propulsion Lab, Calif Inst Technol, 66-71; RES SCIENTIST, McDONALD OBSERV & UNIV TEX, AUSTIN, 71- Concurrent Pos: Consult, CTL Div, Studebaker-Packard Corp, 61-62; consult, Aerospace Res Lab, Wright-Patterson AFB, 65; consult mem working group on ephemerides for space res, Int Astron Union, 67-70; mem lunar laser ranging panel, Comt Space Res, Int Coun Sci Unions, chmn panel 1D, 75-; res assoc, Res Group Space Geodesy, France, 73-74; consult, Encycl Britannica, 74- Honors & Awards: NASA Group Achievement Award, 73; Bronze Medal, Nat Ctr Space Studies, France, 74. Mem: Am Astron Soc; Int Astron Union. Res: Celestial mechanics; theory of differential correction processes; orbital and rotational dynamics of the moon; astrometry and dynamics of natural satellites; geophysical application of laser ranging. Mailing Add: Dept of Astron Univ of Tex Austin TX 78712

MULHOLLAND, JOHN HUGH, surgery, deceased

MULIERI, BERTHANN SCUBON, b Menticle, Pa, May 4, 37; m 60. PHYSIOLOGY. Educ: Pa State Univ, BS, 58; Univ Vt, PhD(physiol, biophys), 68. Prof Exp: Asst prof biol sci, Hunter Col, 67-70; USPHS fel, pharmacol inst, Univ Lund, 71-72; from instr to asst prof biol, Trinity Col, Vt. 72-74; RES ASSOC PHYSIOL & BIOPHYS, UNIV VT, 74- Concurrent Pos: Health Educ & Welfare res fel physiol & biophys, Univ Vt, 74-75. Mem: Biophys Soc; NY Acad Sci; Biophys Soc; Sigma Xi. Res: Mechanics of muscle contraction; desensitization at neuromuscular junction. Mailing Add: Dept of Physiol & Biophys Univ Vt Col of Med Burlington VT 05401

MULINOS, MICHAEL GEORGE, b Cairo, Egypt, Nov 24, 97; nat US; m 27; c 2. CLINICAL PHARMACOLOGY. Educ: Columbia Univ, AB, 21, AM, 22, MD, 24, PhD(physiol), 29. Prof Exp: Intern, St Vincent's Hosp, Pa, 24-25; instr pediat, Univ Minn, 25-26; vis asst, Univ Chicago, 26; from instr to assoc prof pharmacol, Col Physicians & Surgeons, Columbia Univ, 27-44; assoc prof, NY Med Col, 44-45; dir med res, Interchem Corp, NJ, 45-47; med dir, Com Solvents Corp, 53-63; consult & med dir, McCann-Erickson, Inc, 66-69; MED DIR, ERWIN WASEY, INC, 69- Concurrent Pos: Asst, Inst Child Guid, Minneapolis, Minn, 26; asst med dir, Life Exten Inst New York. Mem: AAAS; Am Soc Pharmacol & Exp Therapeut; Asn Med Dirs (past pres); Soc Exp Biol & Med; Harvey Soc. Res: Pharmacology of gastrointestinal tract; toxicology of pharmaceuticals and irritation; physiology of autonomic nervous systems; clinical testing of drugs. Mailing Add: 869 Standish Ave Westfield NJ 07090

MULKERN, GREGORY BENEDICT, b Tulsa, Okla, Mar 27, 31; m 54; c 6. ENTOMOLOGY. Educ: Univ Ill, BS, 53; Kans State Univ, MS, 54, PhD(entom), 57. Prof Exp: From asst prof to assoc prof, 57-69, PROF ENTOM, N DAK STATE UNIV, 69- Mem: Entom Soc Am. Res: Insect morphology and physiology; insect ecology; host plant selection; biology of orthoptera; acrididae. Mailing Add: Dept of Entom NDak State Univ Fargo ND 58102

MULKEY, GEORGE JACKSON, population genetics, physiology, see 12th edition

MULKEY, JAMES ROBERT, JR, b Alamo, Tex, Oct 30, 35; m 63; c 3. AGRONOMY. Educ: Tex A&I Univ, BS, 57; Okla State Univ, MS, 59; Tex A&M Univ, PhD(soil physics), 64. Prof Exp: Asst soil scientist, Tex Agr Exp Sta, Tex A&M Univ, 64-67, asst prof, 67-73, ASSOC PROF CHG SOILS, TEX AGR EXP STA, TEX A&M UNIV, 73- Mem: Am Soc Agron; Soil Sci Soc Am; Crop Sci Soc Am. Res: Soil physical properties, fertility and plant nutrition; water conservation and utilization. Mailing Add: Texas Agr Exp Sta PO Drawer 1051 Uvalde TX 78801

MULL, LEON EDMUND, b Assumption, Ill, Mar 10, 13; m 42; c 2. BACTERIOLOGY. Educ: Univ Ill, BS, 39; Univ Mo, MS, 40; Iowa State Univ, PhD(microbiol), 50. Prof Exp: PROF MICROBIOL, UNIV FLA, 40- Mem: Am Dairy Sci Asn. Res: Metabolism lactic cultures. Mailing Add: Dept of Dairy Sci Inst of Food & Agr Sci Univ of Fla Gainesville FL 32601

MULL, ROBERT PAUL, chemistry, see 12th edition

MULLA, MIR S, b Kandahar, Afghanistan, Feb 15, 27; nat US; m 54; c 4. ENTOMOLOGY. Educ: Cornell Univ, BS, 52; Univ Calif, Berkeley, PhD(entom), 55. Prof Exp: Jr res entomologist, 56-58, asst res entomologist, 58-63, assoc prof, 63-69, PROF ENTOM, UNIV CALIF, RIVERSIDE, 69- Concurrent Pos: Consult, WHO, 64. Mem: Entom Soc Am; Am Mosquito Asn. Res: Medical entomology; biology, ecology and control of arthropods affecting the health and quality of living of man. Mailing Add: Dept of Entom Univ of Calif Riverside CA 92502

MULLAHY, JOHN HENRY, b Baltimore, Md, June 26, 14. BIOLOGY, PHYCOLOGY. Educ: St Louis Univ, AB, 37, STL, 47; Fordham Univ, MS, 42; Vanderbilt Univ, PhD(biol), 51. Prof Exp: From instr to asst prof biol, Spring Hill Col, 39-48; assoc prof, 52-55, PROF BIOL, LOYOLA UNIV, LA, 55-, CHMN DEPT, 54- Concurrent Pos: Fulbright fel, Univ Manchester, 51-52; guest investr, Sta Zool, Napoli, Italy; vis prof, Univ Washington, 53. Mem: Fel AAAS; Am Phycol Soc; Bot Soc Am; Am Micros Soc; NY Acad Sci. Res: Cytology of Rhodophyceae; microflora of atmosphere. Mailing Add: 6363 St Charles Ave New Orleans LA 70118

MULLAN, CHARLES EDWARD, mathematics, see 12th edition

MULLAN, DERMOTT JOSEPH, b Omagh, Northern Ireland, Jan 10, 44; m 70; c 4. ASTROPHYSICS. Educ: Queen's Univ, Belfast, BS, 64, BS, 65; Univ Md, PhD(astron), 69. Prof Exp: Astronr, Armagh Observ, Northern Ireland, 69-72; presidential intern fel astron, 72-73, ASST PROF ASTROPHYS, BARTOL RES FOUND, 73- Mem: Am Astron Soc. Res: Physics of stellar flares; formation of sunspots and starspots; structure of magnetic fields in late-type stars; magnetic convection. Mailing Add: Bartol Res Found Swarthmore PA 19081

MULLAN, JOHN F, b County Derry, NIreland, May 17, 25; US citizen; m 59; c 4. NEUROSURGERY. Educ: St Columbus Col, Ireland, BAO, 42; Queen's Univ, Belfast, MB & BCh, 47; FRCS, 51; Am Bd Neurol Surg, dipl, 57. Prof Exp: Resident surg, Royal Victoria Hosp, Belfast, Ireland, 50-51; resident neurol surg, Montreal Neurol Inst, 53-55; from asst prof to assoc prof, 55-63, PROF NEUROL SURG, UNIV CHICAGO, 63- Concurrent Pos: Fel, Middlesex Hosp, Univ London, 49-50 & Guys Hosp, 51. Mem: Fel Am Col Surg; Am Asn Neurol Surg; Am Acad Neurol Surg. Res: Head injury, intracranial aneurysm and pain; epilepsy. Mailing Add: Div of Neurol Surg Univ of Chicago Hosp Chicago IL 60637

MULLANEY, HENRY WENDELL, b Boston, Mass, July 6, 43; m 70; c 1. ELECTROMAGNETICS, IONOSPHERIC PHYSICS. Educ: Providence Col, BS, 65; Boston Univ, MA, 67, PhD(physics & astron), 72. Prof Exp: Res asst physics & astron, Boston Univ, 68-71; mem tech staff physics, Sanders Assocs, Inc, 71-75; SCI OFFICER, OFF NAVAL RES, 75- Mem: Am Geophys Union; Am Inst Physics. Res: Electromagnetics, primarily antennas, antenna arrays and adaptive beamforming; radio and electromagnetic propagation including guided and ducted modes, propagation through plasma including irregularity effects, waveguides and optical fibers; communications and radar. Mailing Add: 6316 25th St N Arlington VA 22207

MULLANEY, OWEN CHRISTOPHER, b Boston, Mass, Oct 4, 11; m 42; c 1. MEDICINE. Educ: Boston Col, AB, 33; Boston Univ, MD, 37; Am Bd Obstet & Gynec, dipl, 46. Prof Exp: From instr to asst prof obstet & gynec, Sch Med, Boston Univ, 46-64, assoc clin prof, 64; instr, 54, ASST PROF OBSTET & GYNEC, SCH MED, TUFTS UNIV, 65- Concurrent Pos: Assoc vis obstetrician & gynecologist, Mass Mem Hosp, 54-60, vis obstetrician & gynecologist, Univ Hosp, 60-; sr obstetrician, St Margaret's Hosp, Dorchester, Mass, 55. Mem: Am Col Surg. Res: Obstetrics and gynecology. Mailing Add: 408 Granite Ave Milton MA 02186

MULLANEY, PAUL F, b New York, NY, Jan 26, 38; m 63; c 2. PHYSICS, BIOPHYSICS. Educ: Iona Col, BS, 59; Univ Del, MS, 63, PhD(physics), 65. Prof Exp: Asst prof physics, St Bonaventure Univ, 65-66; staff mem, Bio-Med Res Group, Los Alamos Sci Lab, 66-72, sect leader biophys, 72-73, GROUP LEADER, BIOPHYS & INSTRUMENTATION GROUP H-10, LOS ALAMOS SCI LAB, 73- Concurrent Pos: Consult, Photoconductor Devices Div, Sylvania Elec Prod, Inc, Pa, 65-66; Nat Cancer Inst, Comt Cytol Automation, 73-75. Mem: AAAS; Biophys Soc; Am Asn Physics Teachers. Res: Cellular biophysics; light scattering and fluorescence of cells; high speed cell analysis by flow methods; biomedical engineering. Mailing Add: Group H-10 Los Alamos Sci Lab Univ of Calif Los Alamos NM 87544

MULLEE-JONES, MARY TERESA, parasitology, zoology, see 12th edition

MULLEN, ANTHONY J, b Jermyn, Pa, Sept 2, 27. ELECTROMAGNETISM. Educ: Villanova Univ, BS, 50; Cath Univ Am, MS, 54; Bryn Mawr Col, PhD, 68. Prof Exp: Teacher, Archbishop Carroll High Sch, 54-59; prof elec eng, Villanova Univ, 59-71, chmn dept, 59-67; sr engr flying qual, Vertol Div, 72, ENG SUPVR ELECTROMAGNETIC PULSE ANAL ENG, BOEING AEROSPACE CO, 72- Concurrent Pos: Consult, Reentry Systs Div, Gen Elec Corp, 66-67; res assoc, Geophys Inst, Univ Alaska, 70-71. Mem: Am Phys Soc. Res: Electromagnetic radiation in plasmas; geomagnetic micropulsations. Mailing Add: 13235 SE 160th Pl Renton WA 98055

MULLEN, DAVID ANTHONY, b Pomona, Calif, May 19, 36; m 58; c 2. ZOOLOGY. Educ: Univ Calif, Berkeley, AB, 58, PhD(zool), 65; Univ Calif, Davis, MA, 61. Prof Exp: Lab asst plant biochem, Univ Calif, Riverside, 56, lab technician, Cancer Res Genetics Lab, Univ Calif, Berkeley, 56-58, res asst environ physiol, Univ Calif, Davis, 58-60, lab technician endocrine physiol, Univ Calif, Berkeley, 60-61; investr environ physiol, Arctic Res Lab, Barrow, Alaska, 61-62, resident investr, 62-64; asst prof anat, Univ Calif, San Francisco & consult environ physiol, Hooper Found Med Res, 64-67; mem fac, 67-70, ASSOC PROF BIOL, UNIV SAN FRANCISCO, 70- Concurrent Pos: Consult comp oncol, WHO, 64-; dir, Hist-Tech Labs, Univ Calif, Berkeley, 66-70. Mem: AAAS; Am Soc Zoologists; Am Soc Mammalogists; Arctic Inst NAm. Res: Environmental physiology, especially effects of stressors on the endocrine system and medical repercussions of such effects. Mailing Add: Dept of Biol Univ of San Francisco San Francisco CA 94117

MULLEN, GARY RICHARD, b Ogdensburg, NY, Nov 16, 45; m 69; c 2. MEDICAL ENTOMOLOGY. Educ: Northeastern Univ, BA, 68; Cornell Univ, MS, 70, PhD(entom), 74. Prof Exp: Med entomologist & adminr, Allegheny County Health Dept, Vector Control Prog, Pittsburgh, 74-75; ASST PROF, DEPT ZOOL & ENTOM, AUBURN UNIV, 75- Concurrent Pos: Consult, Environ Sci Div, QLM Labs, Inc, 73-, Ecol Sci Div, NUS Corp, 75- Mem: Entom Soc Am; Am Mosquito Control Asn; Acarological Soc Am; Am Arachnological Soc. Res: Ecology of insects and mites of medical and veterinary importance; general acarology eith emphasis on the Parasitengona, their evolution and host-parasite relationship with arthropods; acarine parasites of mosquitoes. Mailing Add: Dept of Zool Entom Auburn Univ Auburn AL 36830

MULLEN, GEORGE HENRY, b Hackensack, NJ, Nov 10, 34; m 58; c 5. THEORETICAL PHYSICS. Educ: Rutgers Univ, BA, 56; Syracuse Univ, MS, 58, PhD(physics), 61. Prof Exp: Nat Acad Sci-Nat Res Coun res assoc, US Naval Ord Lab, Md, 61-63; asst prof physics, Univ NH, 63-69, PROF PHYSICS & CHMN DEPT, MANSFIELD STATE COL, 69- Mem: Am Phys Soc. Res: Theoretical investigations in planetary atmospheres. Mailing Add: Dept of Physics Grant Sci Bldg Mansfield State Col Mansfield PA 16933

MULLEN, JAMES A, b Malden, Mass, May 28, 28; m 61; c 4. APPLIED PHYSICS. Educ: Providence Col, BS, 50; Harvard Univ, MA, 51, PhD, 55. Prof Exp: Mem res staff, 55-65, prin scientist, 65-69, CONSULT SCIENTIST, RES DIV, RAYTHEON CO, 69- Mem: Am Phys Soc; Soc Indust & Appl Math: Am Math Soc; Inst Elec & Electronics Engrs. Res: Statistical communication theory; noise in non-linear circuits and oscillators; adaptive detection theory; pattern recognition. Mailing Add: 337 S Main St Cohasset MA 02025

MULLEN, JAMES G, b St Louis, Mo, Sept 17, 33; m 58; c 3. SOLID STATE PHYSICS. Educ: Univ Mo-Rolla, BS, 55; Univ Ill, MS, 57, PhD(physics), 60. Prof Exp: From asst physicist to assoc physicist, Argonne Nat Lab, 60-64; from asst prof to assoc prof, 64-75, PROF PHYSICS, PURDUE UNIV, LAFAYETTE, 75- Concurrent Pos: Consult, Argonne Nat Lab, 64-65; mem, Ad Hoc Panel Mössbauer Data. Mem: Am Phys Soc; Am Asn Physics Teachers. Res: Studies of solid state diffusion in ionic and metallic crystals; Mössbauer studies of properties of solids. Mailing Add: Dept of Physics Purdue Univ Lafayette IN 47907

MULLEN, JOSEPH DAVID, b Green Isle, Minn, Jan 6, 34; m 56; c 4. BIOCHEMISTRY, FOOD SCIENCE. Educ: Col St Thomas, BS, 56; Univ Minn, PhD(biochem), 62. Prof Exp: Group head, 61-72, DEPT HEAD, GEN MILLS, INC, 72- Mem: AAAS; Am Chem Soc; Am Asn Cereal Chem; Inst Food Technologists. Res: Protein chemistry; relation of protein structures to function in foods; new food product development including extruded protein and gum systems, edible protein films, gel technology, low calorie foods and nutrition. Mailing Add: Gen Mills Inc Bell Res Ctr 9000 Plymouth Ave N Minneapolis MN 55427

MULLEN, JOSEPH MATTHEW, b Washington, DC, June 1, 44; m 65. CHEMICAL PHYSICS, ASTROPHYSICS. Educ: Old Dom Univ, BS, 66; Univ Fla, PhD(physics), 72. Prof Exp: Asst prof physics, Valdosta State Col, 71-72; fel, Univ Fla, 72-74; PRIN STAFF OPER ANAL, OPER RES INC, 74- Mem: AAAS; Sigma Xi; Am Phys Soc. Res: Determination of intermolecular potentials; development of instrumentation for ion-molecule reaction studies; theory of mass transfer in close binary star systems. Mailing Add: Oper Res Inc 1400 Spring St Silver Spring MD 20910

MULLEN, KENNETH, b London, Eng, Feb 28, 39; US citizen; m 61; c 3. STATISTICS, MATHEMATICS. Educ: Western Reserve Univ, BA, 61; Va Polytech Inst & State Univ, PhD(statist), 66. Prof Exp: Asst prof math, Radford Col, 64-65; asst prof biometry, Med Col Va, 65-67; sr statistician, Ciba, Ltd, 20 Switz, 67-69; asst prof, 69-71, ASSOC PROF STATIST, UNIV GUELPH, 71- Concurrent Pos: Consult, Am Tobacco Co, 65-67 & Albemarle Paper Co, 66-67. Mem: Am Statist Asn. Res: Estimation problems associated with censored and truncated data; development of non-parametric procedures. Mailing Add: Dept of Math Univ of Guelph Guelph ON Can

MULLEN, PATRICIA ANN, b Flushing, NY, July 10, 35. COSMETIC CHEMISTRY. Educ: Seton Hill Col, BA, 57; Mt Holyoke Col, MA, 61. Prof Exp: Res chemist, Charles Bruning Co, 57-59; res asst spectros, Mt Holyoke Col, 59-61; res chemist, 63-74, GROUP LEADER, AM CYANAMID CO, 74- Mem: Am Chem Soc; Soc Cosmetic Chemists. Res: Ultraviolet and vacuum spectroscopy; photochemistry; spectropolarimetry; cosmetic chemistry. Mailing Add: Am Cyanamid Co Clifton NJ 07015

MULLEN, RICHARD JOSEPH, b Leominster, Mass, Aug 15, 41; m 64; c 3. DEVELOPMENTAL GENETICS, NEUROSCIENCES. Educ: Fitchburg State Col, BS, 63; Univ NH, MS, 69, PhD(genetics), 71. Prof Exp: USPHS fel develop genetics, Harvard Med Sch, 71-74, RES ASSOC NEUROSCI, CHILDREN'S HOSP MED CTR, 73-; INSTR NEUROPATHOL, HARVARD MED SCH, 73- Res: Mammalian developmental genetics; culture and manipulation of preimplantation embryos; use of experimental chimeric mice in studies of brain development. Mailing Add: Dept of Neurosci Children's Hosp Med Ctr Boston MA 02115

MULLEN, ROBERT KEECH, b Newark, NJ, Feb 26, 32; m 56; c 2. PHYSIOLOGICAL ECOLOGY, RADIATION ECOLOGY. Educ: Calif State Univ, Northridge, AB, 63; Univ Southern Calif, MS, 65, PhD(biol), 70. Prof Exp: Scientific specialist radioecology, EG&G, Inc, 70-73; RES BIOLOGIST, MISSION RES CORP, 73- Concurrent Pos: Lectr, Univ Calif, Santa Barbara, 73-; vchmn, Environ Qual Adv Bd, Santa Barbara, 73-; consult, Off Emergency Serv, State of Calif, 74- & nuclear safeguards, US Nuclear Regulatory Comn, 75- Mem: Ecol Soc Am; Int Asn Ecol; AAAS; Am Soc Ichthyologists & Herpetologists; Asn Trop Biol. Res: Physiological ecology of vertebrates; energy metabolism of free-living organisms; comparative radioecology of natural populations; effects and uptake of fallout radiation in vertebrates; internal dosimetry; environmental radiation surveillance utilizing dosimetric characteristics of naturally occurring materials. Mailing Add: Mission Res Corp Santa Barbara CA 93102

MULLEN, ROBERT TERRENCE, b Chicago, Ill, Sept 25, 35; m 55; c 3. PHYSICAL CHEMISTRY. Educ: Univ Ill, BS, 57; Univ Calif, PhD(phys chem), 61. Prof Exp: Res assoc chem, Brookhaven Nat Lab, 61-63; sr chemist, Res Lab, Merck & Co, Inc, 63-65, sect leader, 65-67, res fel, 67-68, supvr automation & control dept, 68-70, mgr automation & control lab automation, 70-75, MGR LAB & GEN AUTOMATION, MERCK & CO, INC, 75- Mem: AAAS; Am Chem Soc; Soc Appl Spectros; NY Acad Sci. Res: Hot atom chemistry; photochemistry; Mössbauer spectroscopy; radioisotope tracer applications; laboratory automation. Mailing Add: Automation & Control Dept Merck & Co Inc Rahway NJ 07065

MULLEN, RUSSELL EDWARD, b Atlantic, Iowa, Sept 4, 49; m 71; c 1. CROP PHYSIOLOGY. Educ: Northwest Mo State Univ, BS, 71, MSEd, 72; Purdue Univ, PhD(crop physiol & prod), 75. Prof Exp: Grad teaching asst crops, Northwest Mo State Univ, 71-72; grad teaching asst crop prod, 72-74, ASST PROF CROP PROD, PURDUE UNIV, 75- Mem: Sigma Xi; Am Soc Agron; Crop Sci Soc Am. Res: Forage physiology and production. Mailing Add: Dept of Agron Purdue Univ West Lafayette IN 47907

MULLENAX, CHARLES HOWARD, b Sterling, Colo, Feb 5, 32; m 54; c 4. ANIMAL SCIENCE, ECOLOGY. Educ: Colo State Univ, BS, 53, DVM, 56; Cornell Univ, MS, 61. Hon Degrees: DVM, Cent Univ, Ecuador, 66. Prof Exp: Owner & vet, Mountain Parks Vet Hosp, 56-59; asst vet physiol, NY State Vet Col, Cornell, 59-61; res vet, Nat Animal Dis Lab, USDA, 61-64; lectr vet med & dir clins, Fulbright Binational Educ Comn, US Dept State, Cent Univ, Ecuador, 64-66; assoc pathologist, Agr Sci Prog, Rockefeller Found, Bogota, Colombia, 66-69; animal prod training specialist, Int Ctr Trop Agr, Cali, Colombia, 69-71; tech dir, Colombia Livestock Proj, Int Bank Reconstruction & Develop, 71-73; PRES & GEN MGR, CONCORDIA INT LTDA, 73- Concurrent Pos: Consult, Bahamas Livestock Co, 56-57. Mem: NY Acad Sci; Am

Soc Trop Med & Hyg; Int Asn Ecol; Asn Trop Biol. Res: Ruminant intermediary metabolism; physiopathology; veterinary epidemiology; tropical livestock diseases; livestock production; range ecology. Mailing Add: Concordia Int Ltda Apartado Aereo 21-18 Villavicencio Meta Colombia

MULLENDORE, JAMES MYERS, b Ft Wayne, Ind, Aug 15, 19; m 42; c 4. SPEECH PATHOLOGY, AUDIOLOGY. Educ: Northwestern Univ, BS, 41, MA, 42, PhD(speech path), 48. Prof Exp: Instr speech correction, Northwestern Univ, 44-45; from asst prof to assoc prof speech, Univ Va, 45-61; prof audiol & speech path, Vanderbilt Univ, 61-63 & WVa Univ, 63-67; PROF SPEECH & HEARING SCI & DIR SCH SPEECH & HEARING SCI, BRADLEY UNIV, 67- Concurrent Pos: Ed consult, State Farm Mutual Ins Co, 56-61; prof, George Peabody Col, 61-63; dir, Bill Wilkerson Hearing & Speech Ctr, 61-63. Mem: Fel Am Speech & Hearing Asn. Res: Stuttering; supervision and administration; voice disorders. Mailing Add: 909 W Burnside Dr Peoria IL 61614

MULLER, BURTON HARLOW, b New York, NY, May 11, 24; m 52; c 2. PHYSICS. Educ: Wesleyan Univ, BA, 44; Yale Univ, MS, 45; Univ Ill, PhD(physics), 54. Prof Exp: Asst physics, SAM Labs, Columbia Univ, 45; from asst prof to assoc prof, 53-62, PROF PHYSICS, UNIV WYO,62- Concurrent Pos: NSF fac fel, Univ BC, 59-60 & Univ Nottingham, 65-66; vis prof, Univ Kent, 71-72. Mem: Am Phys Soc; Am Asn Physics Teachers. Res: Nuclear magnetic resonance; molecular motion. Mailing Add: Dept of Physics Univ of Wyo Laramie WY 82070

MULLER, CORNELIUS HERMAN, b Collinsville, Ill, July 22, 09; m 39; c 1. PLANT ECOLOGY. Educ: Univ Tex, BA, 32, MA, 33; Univ Ill, PhD(plant ecol), 38. Prof Exp: Asst bot, Univ Tex, 29-33; asst Univ Ill, 34-38; ecologist, Ill State Natural Hist Surv, 38; asst botanist, Div Plant Explor & Introd, Bur Plant Indust, USDA, 38-42, rubber plant invests, 42, assoc botanist, Spec Guayule Res Proj, 42-45; from asst prof to assoc prof, 45-56, PROF BOT, UNIV CALIF, SANTA BARBARA, 56- Concurrent Pos: Instr grad sch, USDA, 41-42, collabr, 45-46; res assoc, Inst Tech & Plant Indust, Southern Methodist Univ, 45 & Santa Barbara Bot Garden, 48-; fac res lectr, Univ Calif, Santa Barbara, 57, actg dean grad div, 61-62; adj prof bot, Univ Tex, Austin, 74- Mem: Eminent ecologist, Ecol Soc Am; Linnean Soc London. Res: Vegetation of the southwestern United States, Mexico and Central America; basic nature of the biotic community; biochemical inhibition among higher plants; plant competition and community interactions; taxonomy and evolution of American Quercus. Mailing Add: Dept of Biol Sci Univ of Calif Santa Barbara CA 93106

MULLER, DAVID EUGENE, b Austin, Tex, Nov 2, 24; m 44; c 2. APPLIED MATHEMATICS. Educ: Calif Inst Technol, BS, 47, PhD(physics), 51. Prof Exp: Res fel physics, Calif Inst Technol, 51-52; fel electronic digital comput, 52-53, res asst prof appl math, 53-56, res assoc prof, 56-60, res prof, 60-64, PROF MATH, UNIV ILL, URBANA, 64- Concurrent Pos: Consult, IBM Corp, 59-; Fulbright res scholar, Univ Tokyo, 61-62. Mem: AAAS; Am Phys Soc; Am Math Soc. Res: Switching and automata theory; error correcting codes. Mailing Add: Dept of Math 273 Atgeld Hall Univ Ill Urbana IL 61801

MÜLLER, DIETRICH, b Leipzig, Ger, Sept 14, 36; m 68; c 2. PHYSICS, ASTROPHYSICS. Educ: Univ Bonn, Dipl physics, 61, PhD(physics), 64. Prof Exp: Res assoc physics, Univ Bonn, 64-68; res assoc physics, Univ Chicago, 68-70, ASST PROF, ENRICO FERMI INST & DEPT PHYSICS, 70- Mem: Am Phys Soc; Am Geophys Union; Ger Phys Soc. Res: Experimental physics; mass spectroscopy; aeronomy; experimental astrophysics; cosmic ray research. Mailing Add: Enrico Fermi Inst Univ of Chicago 933 E 56th St Chicago IL 60637

MULLER, DONALD EDWARD, b Waterbury, Conn, July 21, 40; m 64; c 2. ANALYTICAL CHEMISTRY. Educ: Yale Univ, BE, 62, MAT, 63; Ind Univ, Bloomington, PhD(phys chem), 69. Prof Exp: Teacher high sch, Mass, 63-64; engr, Refinery & Chem Div, Bechtel Corp, Calif, 64-65; asst phys & gen chem, Ind Univ, Bloomington, 65-69; res assoc Raman spectros, Univ Mass, Amherst, 69-70; asst prof chem, RI Col, 70-72; QUAL CONTROL MGR CHEM PROD, PIERCE CHEM CO, 72- Mem: Am Inst Physics. Res: Organic and clinical reagents analysis; clathrate hydrate crystal growth; laser Raman spectroscopy; far infrared spectroscopy. Mailing Add: Pierce Chemical Co 3747 Meridian Rd Rockford IL 61105

MULLER, ELSIE, b Wakefield, Nebr, Dec 7, 12. MATHEMATICS. Educ: Wayne State Col, BA, 35; Univ Mich, MA, 44; Iowa State Univ, PhD, 63. Prof Exp: Instr high sch, SDak, 42-43; instr math, Waukon High Sch & Jr Col, Iowa, 43-45; instr high sch, Mich, 45-47, Univ Ill, 47-49, Britt High Sch & Jr Col, Iowa, 49-50 & La Salle High Sch & Jr Col, Ill, 50-55; asst prof, Morningside Col, 55-58; Iowa State Univ, 58-63; PROF MATH, MORNINGSIDE COL, 63-, HEAD DEPT, 72- Mem: AAAS; Math Asn Am; Soc Indust & Appl Math; Am Math Soc; Fedn Am Scientists. Res: Algebra. Mailing Add: Dept of Math Morningside Col Sioux City IA 51106

MULLER, ERIC RENE, b Morija, Lesotho, Nov 5, 38; Can citizen; m 65; c 2. MATHEMATICS, PHYSICS. Educ: Univ Natal, BSc, 60, MSc, 62; Univ Sheffield, PhD(theoret physics), 67. Prof Exp: Lectr math, Rhodes Univ, SAfrica, 61-64; lectr col appl arts & technol, Univ Sheffield, 64-67; asst prof math & physics, 67-71, ASSOC PROF MATH & PHYSICS, BROCK UNIV, 71- Concurrent Pos: Consult, Steltner Develop, 70- & Transp Develop Agency, Can Ministry Transport, 73- Mem: Math Soc Can; Am Asn Physics Teachers; Math Asn Am; Soc Indust & Appl Math. Res: Mathematical models and analyses of transportation systems; theoretical solid state physics. Mailing Add: Dept of Math Brock Univ St Catharines ON Can

MULLER, ERNEST HATHAWAY, b Tabriz, Iran, Mar 4, 23; US citizen; m 51; c 3. GEOLOGY. Educ: Col Wooster, AB, 47; Univ Ill, MS, 49, PhD(geol), 52. Prof Exp: Geologist mil geol br, US Geol Surv, Alaska, 48, proj head, Bristol Bay Area, 49-54; asst prof geol, Cornell Univ, 54-59; assoc prof, 59-65, PROF GEOL, SYRACUSE UNIV, 65- Concurrent Pos: Geologist, NY State Sci Serv, 56-; mem SChile exped, Am Geog Soc, 59; mem exped, Katmai Nat Monument, Alaska, 63-64; res assoc, Mus Natural Hist, Reykjavik, Iceland, 68-69 & Churchill Falls Power Proj, Labrador, 70; Erskine fel, Univ Canterbury, Christchurch, NZ, 73-74. Mem: AAAS; Geol Soc Am; Nat Asn Geol Teachers; Am Geog Soc; Glaciol Soc. Res: Geomorphology; glacial, engineering and environmental geology; permafrost; denudation; drumlin origins; glacial geology of New York, Southwestern Alaska, Iceland, and South Island of New Zealand. Mailing Add: Dept of Geol 204 Heroy Geol Lab Syracuse Univ Syracuse NY 13210

MULLER, FREDRIK ARTHUR, physical chemistry, crystallography, see 12th edition

MULLER, GEORGE HEINZ, b Ger, June 6, 19; US citizen; m 49. VETERINARY DERMATOLOGY. Educ: Tex A&M Univ, DVM, 43; Am Col Vet Internal Med, dipl & cert dermat, 74. Prof Exp: Dir, Pittsburg Vet Hosp, 46-56; DIR, MULLER VET HOSP, 56-; CLIN PROF DERMAT, SCH MED, STANFORD UNIV, 58- Concurrent Pos: Ed dermat, Current Vet Ther, 66-75; pres, Dermat Specialty Group, Am Col Vet Internal Med, 74-76. Honors & Awards: McCoy Mem Award, Wash State Univ, 69; Merit Award Dermat, Am Animal Hosp Asn, 70. Mem: Am Acad Vet Dermat (pres); affil Am Acad Dermat; Am Animal Hosp Asn. Res: Small animal veterinary dermatology; comparative dermatology; canine and human demodicosis. Mailing Add: Dept of Dermat Stanford Univ Med Sch Stanford CA 94305

MULLER, HARRY DIERKS, genetics, see 12th edition

MULLER, JAN ENGELBERT, b Leiden, Neth, Aug 7, 17; nat US; m 42; c 4. REGIONAL GEOLOGY. Educ: Univ Groningen, BSc, 38, PhD(geol), 43; Univ Leiden, MSc, 40. Prof Exp: Geologist, Neth Bur Mines, Heerlen, 41-45; geologist, Royal Dutch Shell Oil Co, 45-48; GEOLOGIST, GEOL SURV CAN, 48- Mem: Fel Geol Soc Am; fel Geol Asn Can. Res: sedimentary petrology; stratigraphy; structure research in the Netherlands; Appalachian and Cordilleran geology in Canada; geology, stratigraphy, structure and mineral deposits of Vancouver Island, British Columbia. Mailing Add: Geol Surv of Can 100 W Pender St Vancouver BC Can

MULLER, JOHN HUBERT, solid state physics, see 12th edition

MULLER, JON DAVID, b Salina, Kans, Oct 23, 41; m 63. ANTHROPOLOGY, ARCHAEOLOGY. Educ: Univ Kans, AB, 63; Harvard Univ, PhD(anthrop), 67. Prof Exp: Asst prof, 66-71, ASSOC PROF ANTHROP, SOUTHERN ILL UNIV, CARBONDALE, 71- Mem: AAAS; Soc Am Archaeol; Am Anthrop Asn. Res: The formal description of art styles and the cultural ecology of the prehistoric Southeast and also Africa. Mailing Add: Dept of Anthrop Southern Ill Univ Carbondale IL 62901

MULLER, KARL FREDERICK, b Glen Ridge, NJ, Sept 5, 35; m 61; c 2. APPLIED MATHEMATICS. Educ: Lafayette Col, BSME, 57; Syracuse Univ, MSEE, 65, PhD(appl math), 70. Prof Exp: Field engr, Leeds & Northrup Co, Philadelphia, 57-63; res engr, Syracuse Univ Res Corp, 64-66, group leader appl math, 66-69; MEM TECH STAFF, MITRE CORP, BEDFORD, MASS, 69- Concurrent Pos: Mem weather '85 study group, US Air Force Systs Command, 71-72; consult, PHI Comput Serv, Inc, Arlington, Mass, 72. Res: Detailed mathematical analysis, modeling and computer simulation of physical and probabalistic phenomena associated with seismic, and acoustic propagation, weather predictability, optimal control systems, high-sensitivity electro-optical sensors and detection, especially identification algorithms. Mailing Add: 17 Charles Rd Winchester MA 01890

MULLER, LAWRENCE DEAN, b Peoria, Ill, Nov 26, 41; m 65; c 1. DAIRY HUSBANDRY. Educ: Univ Ill, Urbana, BS, 64, MS, 66; Purdue Univ, Lafayette, PhD(animal sci), 69. Prof Exp: Asst prof animal sci, Purdue Univ, 69-71; ASSOC PROF DAIRY SCI, SDAK STATE UNIV, 71- Mem: Am Dairy Sci Asn; Am Soc Animal Sci. Res: Animal production with emphasis on interrelationships between nutrition, physiology and management on animal productivity. Mailing Add: Dept of Dairy Sci SDak State Univ Brookings SD 57006

MULLER, MARCEL WETTSTEIN, b Vienna, Austria, Nov 1, 22; nat US; m 47; c 3. PHYSICS. Educ: Columbia Univ, BS, 49, AM, 52; Stanford Univ, PhD(physics), 57. Prof Exp: Sr scientist, Varian Assocs, Calif, 52-66; PROF ELEC ENG, WASH UNIV, 66- Concurrent Pos: Lectr, Univ Zurich, 62-63; vis prof, Univ Colo, 68. Mem: Am Phys Soc; Inst Elec & Electronics Eng. Res: Microwave electronics; quantum electronics; solid state physics. Mailing Add: Dept of Elec Eng Wash Univ Lindell & Skinker Blvds St Louis MO 63130

MULLER, MERVIN EDGAR, b Hollywood, Calif, June 1, 28; m 63; c 3. COMPUTER SCIENCE, STATISTICS. Educ: Univ Calif, Los Angeles, PhD(math), 54. Prof Exp: Instr math, Cornell Univ, 54-56; res assoc, Princeton Univ, 56-59; mem staff control planning, Data Processing Div, Int Bus Mach Corp, 59-60, mgr proj weld, 60-64; prof comput sci, Univ Wis-Madison, 64-70, prof comput sci & statist, 70-71, dir comput ctr, 64-70; DIR DEPT COMPUT ACTIV, WORLD BANK, 71- Concurrent Pos: Mem bd dir, Am Fedn Info Processing, 71-73, chmn finance comt, 71-75; chmn comt statist comput, Int Statist Inst, 75. Mem: Am Math Soc; Soc Indust & Appl Math; Math Asn Am; fel Am Statist Asn; Asn Comput Mach. Res: Monte Carlo procedures and simulation; statistical design of experiments; use of computers in statistics, data processing and statistical control procedures; computer information systems and languages; management information systems; data base management systems. Mailing Add: 5303 Mohican Rd Washington DC 20016

MÜLLER, MIKLOS, b Budapest, Hungary, Nov 24, 30; m 73; c 2. BIOLOGICAL CHEMISTRY, PARASITOLOGY. Educ: Med Univ Budapest, MD, 55. Prof Exp: Instr biol & histol, Med Univ Budapest, 50-55, asst prof, 55-64; res assoc biochem cytol, Rockefeller Inst, 64-65; Rask-Orsted fel & guest investr cell biol, Dept Physiol, Carlsberg Lab, Copenhagen, 65-66; asst prof biochem cytol, 66-68, ASSOC PROF BIOCHEM CYTOL, ROCKEFELLER UNIV, 68- Mem: Soc Protozool; Am Soc Parasitol; Am Soc Cell Biol; Am Microbiol Soc. Res: Physiology and biochemistry of parasitic and free living protozoa; peroxisomes; hydrogenosomes; action of antiprotozoal drugs; lysosomes in intracellular digestion and in host-parasite relationships. Mailing Add: Rockefeller Univ 1230 York Ave New York NY 10021

MULLER, NORBERT, b Hamburg, Ger, Jan 25, 29; nat US; m 58; c 7. PHYSICAL CHEMISTRY. Educ: Univ Calif, BS, 49; Harvard Univ, MA, 51, PhD(chem physics), 53. Prof Exp: Instr chem, 53-54, from asst prof to assoc prof, 56-68, PROF CHEM, PURDUE UNIV, LAFAYETTE, 68- Mem: Am Chem Soc; The Chem Soc. Res: Molecular structure and spectra; nuclear magnetic resonance; surfactant chemistry. Mailing Add: Dept of Chem Purdue Univ Lafayette IN 47907

MULLER, OLAF, b Tallinn, Estonia, Jan 14, 38; US citizen. INORGANIC CHEMISTRY, SOLID STATE CHEMISTRY. Educ: Western Reserve Univ, BA, 60, MS, 61; Pa State Univ, PhD(solid state sci), 68. Prof Exp: Res assoc solid state sci, Pa State Univ, 68-72; inorg chemist, Corp Res & Develop Ctr, Gen Elec Co, 72-75; SCIENTIST SOLID STATE SCI, WEBSTER RES CTR, XEROX CORP, 75- Mem: Am Chem Soc; Am Ceramic Soc. Res: Synthesis and crystal chemistry of inorganic materials; property-composition relationships; magnetic materials. Mailing Add: Webster Res Ctr Xerox Corp 800 Phillips Rd Webster NY 14580

MULLER, OTTO HEINRICH, b Esslingen, Ger, Sept 20, 08; nat US; m 41; c 6. PHYSIOLOGY. Educ: Stanford Univ, BA, 33; Charles Univ, Prague, RNDr(phys chem), 35. Prof Exp: Res asst physiol, Stanford Univ, 35-37, res assoc chem, 37-38; res assoc surg & assoc physiol, Med Col, Cornell Univ, 38-40, res assoc anat, 40-45; from instr to asst prof physiol & pharmacol, Col Med, Univ Nebr, 45-47; from asst prof to assoc prof physiol, Syracuse Univ, 47-50; assoc prof, 50-56, PROF PHYSIOL, STATE UNIV NY UPSTATE MED CTR, 56- Concurrent Pos: Upjohn fel, Stanford Univ, 37-38; instr physiol, Med Col, Cornell Univ, 42-45. Mem: Am Chem Soc; Am Physiol Soc; fel NY Acad Sci; Harvey Soc. Res: Polarographic studies of passivity; overvoltage; oxid-red systems; buffer action; proteins; enzymes; respiratory gases; organic chemistry. Mailing Add: Dept of Physiol State Univ of NY Upstate Med Ctr Syracuse NY 13210

MULLER, RICHARD A, b New York, NY, Jan 6, 44; m 66. EXPERIMENTAL

PHYSICS. Educ: Columbia Univ, AB, 64; Univ Calif, Berkeley, PhD(physics), 69. Prof Exp: Asst res physicist, Space Sci Lab, 69-75, ASSOC RES PHYSICIST, SPACE SCI LAB & LAWRENCE BERKELEY LAB, UNIV CALIF, 75- Concurrent Pos: Lectr physics, Univ Calif, Berkeley, 72-74; consult, Stanford Res Inst, 73- Mem: Am Phys Soc; Am Astron Soc; AAAS; Sigma Xi. Res: Astrophysics; cosmology; elementary particles. Mailing Add: Lawrence Berkeley Lab 50/232 Univ of Calif Berkeley CA 94720

MULLER, ROBERT ALBERT, b Passaic, NJ, Dec 5, 28; m 50; c 2. CLIMATOLOGY, PHYSICAL GEOGRAPHY. Educ: Rutgers Univ, BA, 58; Syracuse Univ, MA, 59, PhD(geog), 62. Prof Exp: Phys geogr, Pac Southwest Forest & Range Exp Sta, Calif, 62-64; lectr climat, Univ Calif, Berkeley, 64; from asst prof to assoc prof geog, Rutgers Univ, 64-68; assoc prof, 69-72, PROF GEOG, LA STATE UNIV, 72- Mem: Asn Am Geog; Am Geog Soc; Am Meteorol Soc. Res: Water balance and synoptic climatology evaluations of evapotranspiration loss, water yield, and river basin regimen including flooding. Mailing Add: Dept of Geog La State Univ Baton Rouge LA 70803

MULLER, ROBERT NEIL, b Santa Barbara, Calif, Aug 29, 46; m 70. PLANT ECOLOGY. Educ: Univ Calif, Riverside, BA, 69; Yale Univ, MFS, 72, PhD(plant ecol), 75. Prof Exp: Appointee ecol, 74-76, ASST ECOLOGIST, ARGONNE NAT LAB, 76- Mem: Ecol Soc Am. Res: Adaptations of species to their environment; factors affecting distributions of species within a plant community; structure and function of terrestrial ecosystems. Mailing Add: Radiol & Environ Res Div Argonne Nat Lab Argonne IL 60439

MULLER, ROLF HUGO, b Aarau, Switz, Aug 6, 29; nat US; m 62; c 2. ELECTROCHEMISTRY, CHEMICAL ENGINEERING. Educ: Swiss Fed Inst Technol, Dipl sc nat, 53, PhD(phys chem), 57. Prof Exp: Res chemist, Plastics Div, E I du Pont de Nemours & Co, 57-60; res assoc electrochem eng, Univ Calif, Berkeley, 61-62; asst head inorg mat res div, 70-75, PRIN INVESTR PHYSICS & CHEM OF PHASE BOUNDARIES, LAWRENCE BERKELEY LAB, UNIV CALIF, 62-, LECTR CHEM ENG, 66-, ASST HEAD MAT & MOLECULAR RES DIV, 75- Mem: AAAS; Am Chem Soc; Optical Soc Am; Int Soc Electrochem; Electrochem Soc. Res: Scale-dependent processes in electrochemistry; optical observation of surfaces, thin films and boundary layers; electrolytic metal dissolution and crystallization. Mailing Add: Mat & Molecular Res Div Lawrence Berkeley Lab Bldg 62 Berkeley CA 94720

MULLER, THOMAS C, b Jersey City, NJ, Aug 13, 32. ORGANIC CHEMISTRY. Educ: Holy Cross Col, BS, 53, MS, 54; Univ NH, PhD(chem), 60. Prof Exp: Res assoc org chem, Univ Chicago, 59-61; res scientist, Am Radiator & Standard Sanit Corp, 61-63; proj leader biodegradable detergents, Cent Res Lab, Witco Chem Co, Inc, 63-68. Mem: Am Chem Soc. Res: Reaction mechanisms; electrophilic aromatic substitution; kinetics of electron transfer reactions; organometallic polymers; surfactants. Mailing Add: 312 Park Ave East Orange NJ 07017

MULLER, WALTER HENRY, b New York, NY, Jan 30, 21; m 46; c 4. PLANT PHYSIOLOGY. Educ: Queens Col, NY, BS, 42; Cornell Univ, PhD(plant physiol), 50. Prof Exp: Bacteriologist, Appl Res Labs, Inc, NJ, 42; instr bot, Cornell Univ, 49-50; from instr to assoc prof, 50-65, PROF BOT, UNIV CALIF, SANTA BARBARA, 65- Mem: AAAS; Am Soc Plant Physiol; Bot Soc Am. Res: Natural inhibitors in vascular plants; influence of natural inhibitors on plant growth, development and metabolism. Mailing Add: Dept of Bio-Sci Univ of Calif Santa Barbara CA 93106

MULLER, WILLIAM HENRY, JR, b Dillon, SC, Aug 19, 19; m 46; c 3. SURGERY. Educ: The Citadel, BS, 40; Duke Univ, MD, 43; Am Bd Surg, dipl; Am Bd Thoracic Surg, dipl. Prof Exp: Intern, Johns Hopkins Hosp, 44, asst resident & asst surg, 44-46, instr surg & resident gen surg, 48-49, resident cardiovasc surg, 49; from asst prof to assoc prof surg, Sch Med, Univ Calif, Los Angeles, 49-54; STEPHEN H WATTS PROF SURG & CHMN DEPT, SCH MED, UNIV VA, 54-, SURGEON-IN-CHIEF, HOSP, 54- Concurrent Pos: Attend specialist, Wadsworth Vet Admin Hosp, Los Angeles, chief sect cardiovasc surg, Los Angeles County Gen Hosp, Torrence, consult, St John's Hosp, Santa Monica & Santa Monica Hosp, 49-54; mem exam bd, Am Bd Surg; chmn surg study sect, NIH; mem, President's Panel on Heart Dis, 72; mem Nat Joint Practice Comn of Med & Nursing, 72; mem bd trustees, Duke Univ & Duke Univ Med Ctr. Mem: Soc Vascular Surg (pres, 66-67); Soc Clin Surg; Am Asn Thoracic Surg; AMA; fel Am Col Surg. Res: Surgery of cardiovascular deformities; pulmonary hypertension; enzymatic debridement of wounds. Mailing Add: Dept of Surg Univ of Va Hosp Charlottesville VA 22901

MÜLLER-EBERHARD, HANS JOACHIM, b Magdeburg, Ger, May 5, 27; m 53; c 2. IMMUNOLOGY, BIOCHEMISTRY. Educ: Univ Göttingen, MD, 53. Prof Exp: Asst physician, Dept Med, Univ Göttingen, 53-54; asst & asst physician, Rockefeller Inst, 54-57; fel, Swedish Med Res Coun, Dept Clin Chem, Univ Uppsala, 57-59; from asst prof to assoc prof biochem & immunol, Rockefeller Inst, 59-63; mem dept exp path, 63-74, CECIL H & IDA M GREEN INVESTR MED RES, SCRIPPS CLIN & RES FOUND, 72-, CHMN DEPT MOLECULAR IMMUNOL, 74- Concurrent Pos: Assoc physician, Rockefeller Inst, 59-62; mem allergy & immunol A study sect, NIH, 65-69; adj prof, Univ Calif, San Diego, 68-; Harvey lect, 70. Honors & Awards: Parke Davis Meritorious Award, 66; Squibb Award, Infectious Dis Soc Am, 70; T Duckett Jones Mem Award, Helen Hay Whitney Found, 71; Modern Med Distinguished Achievement Award, 74; Karl Landsteiner Mem Award, Am Asn Blood Banks, 74; Annual Int Award, Gairdner Found, Can, 74. Mem: Nat Acad Sci; Am Soc Clin Invest; Am Asn Immunol; Am Soc Exp Path; Asn Am Physicians. Res: Antibody-complement interaction; reaction mechanisms of complement; membrane pathology. Mailing Add: Dept of Molecular Immunol Scripps Clin & Res Found La Jolla CA 92037

MULLER-EBERHARD, URSULA, b Göttingen, Ger, June 14, 28; US citizen; m 53; c 2. HEMATOLOGY, BIOCHEMISTRY. Educ: Univ Göttingen & Univ Freiburg, MD, 53. Prof Exp: Intern, Wyckoff Heights Hosp, Brooklyn, 54-55, asst res pediat, 55-56; asst res pediat, Univ Hosp, Bellevule Med Ctr, 56, sr asst res pediat, Bellevue Med Ctr, 59-60; asst, Med Ctr, Cornell Univ, 60-62, instr pediat & head div pediat hemat, 62-63; from assoc to assoc mem dept biochem, 63-75, MEM, DEPT BIOCHEM, SCRIPPS CLIN & RES FOUND, 75- Concurrent Pos: Fel pediat hemat, Med Ctr, Cornell Univ, 60-62; Health Res Coun NY career scientist award, 62; USPHS res career develop award, 63-71. Mem: Harvey Soc; Am Soc Hemat; Am Soc Exp Path; Am Soc Clin Invest; Am Asn Study Liver Dis. Res: Biochemical hematology; developmental biochemistry. Mailing Add: Dept of Biochem Scripps Clin & Res Found La Jolla CA 92037

MÜLLER-SCHWARZE, DIETLAND, b Grosshartmannsdorf, Ger, Oct 4, 34; m 65. ANIMAL BEHAVIOR. Educ: Univ Freiburg, PhD(zool), 63. Prof Exp: Asst prof zool, Univ Freiburg, 63-65; asst prof biol, San Francisco State Col, 65-68; assoc prof animal behav, Utah State Univ, 68-73; ASSOC PROF, COL ENVIRON SCI & FORESTRY, STATE UNIV NY, 73- Mem: Animal Behav Soc; Ger Zool Soc; Ger Ornith Soc; Am Soc Mammalogists. Res: Vertebrate pheromones; behavioral adaptations in birds and mammals; predator-prey relations. Mailing Add: State Univ of NY Col of Environ Sci & Forestry Syracuse NY 13210

MULLHAUPT, JOSEPH TIMOTHY, b St Marys, Pa, Feb 25, 32; m 57; c 5. PHYSICAL CHEMISTRY. Educ: Univ Rochester, BS, 54; Brown Univ, PhD(phys chem), 58. Prof Exp: From res chemist to sr res chemist, 58-67, res supvr phys chem, 67-69, SR RES SCIENTIST PHYS CHEM, LINDE DIV, UNION CARBIDE CORP, 69- Mem: AAAS; Am Phys Soc; Sigma Xi; Am Chem Soc. Res: Solid state chemistry; adsorption and surface chemistry; thermodynamics of phase equilibria. Mailing Add: Linde Div Union Carbide Corp Tarrytown Tech Ctr Tarrytown NY 10591

MULLIER, MICHEL E, polymer chemistry, see 12th edition

MULLIGAN, BENJAMIN EDWARD, b Greensboro, NC, May 17, 36; m 63; c 2. SENSORY PSYCHOLOGY. Educ: Univ Ga, BA, 58; Univ Miss, MA, 61, PhD(sensory psychol), 64. Prof Exp: Asst prof, 64-69, ASSOC PROF SENSORY PSYCHOL, UNIV GA, 69- Concurrent Pos: Nat Inst Neurol Dis & Stroke grant, 65-70; partic, Int Cong Physiol Sci, 65. Mem: Acoust Soc Am; Optical Soc Am; Am Soc Cybernet; Soc Neurosci. Res: Sensory processes; psychophysics; mathematical models; noise pollution; communication. Mailing Add: Dept of Psychol Univ of Ga Athens GA 30601

MULLIGAN, BERNARD, b Montgomery, Ala, Aug 31, 34; m 64. THEORETICAL PHYSICS, NUCLEAR PHYSICS. Educ: Univ Ala, BS, 56; Mass Inst Technol, PhD(theoret physics), 62. Prof Exp: Vis asst prof, 61-63, asst prof, 63-66, ASSOC PROF PHYSICS, OHIO STATE UNIV, 66- Res: Theory of nuclear phenomena; mathematical physics. Mailing Add: Dept of Physics Ohio State Univ Columbus OH 43210

MULLIGAN, HUGH FRANCIS, botany, ecology, see 12th edition

MULLIGAN, JAMES ANTHONY, b Denver, Colo, Aug 31, 24. ANIMAL BEHAVIOR. Educ: St Louis Univ, AB, 47, STL, 57; Univ Calif, Berkeley, PhD(zool), 63. Prof Exp: From instr to asst prof, 63-68, ASSOC PROF BIOL, ST LOUIS UNIV, 68- Concurrent Pos: Frank M Chapman Mem Fund res grant, 64; NSF res grant, 65-71; fel, Woodrow Wilson Int Ctr Scholars, 71-72. Mem: Animal Behav Soc; Ecol Soc Am; Am Soc Zoologists; Am Ornith Union. Res: Social behavior and communication in animals by means of sound; field study and physical analysis of avian vocalizations; ontogeny and genetic analysis of bird vocalizations; vertebrate ecology; conservation of natural areas; ethics of the environmental crisis. Mailing Add: St Louis Univ St Louis MO 63103

MULLIGAN, JAMES EDWARD, JR, b Boston, Mass, Apr 26, 27; m 47; c 2. MATHEMATICS. Educ: Univ Mich, AB, 49, AM, 50. Prof Exp: Math statistician, US Census Bur, 50-51; mathematician hydrographic off, US Dept Navy, 51-52, mathematician, Naval Weapons Lab, 53-58, supvry opers res analyst, 59-60; MATHEMATICIAN, GEN ELEC CO, 61- Mem: Fel AAAS; Opers Res Soc Am; Math Asn Am. Res: Computer simulation; numerical analysis; operations research; systems design. Mailing Add: 27 Red Coat Lane Trumbull CT 06611

MULLIGAN, JOSEPH FRANCIS, b New York, NY, Dec 12, 20; m 68. ATOMIC PHYSICS, MOLECULAR PHYSICS. Educ: Boston Col, AB, 45, MA, 46; Cath Univ, PhD(physics), 51. Prof Exp: Instr physics, St Peter's Col, 46-47; instr, Fordham Univ, 55-57, from asst prof to assoc prof, 57-68, chmn dept, 57-64, dean grad sch arts & sci & dean fac, 64-67; PROF PHYSICS & DIR GRAD STUDIES & RES, UNIV MD, BALTIMORE COUNTY, 68- Concurrent Pos: Mem adv comt grad fels, Nat Defense Educ Act, 59-63, mem adv comt grad educ, NY State, 63-68; NSF fac fel, Univ Calif, San Diego, 61-62. Mem: AAAS; Am Phys Soc; Am Asn Physics Teachers; Sigma Xi. Res: Fundamental constants of physics. Mailing Add: Dept of Physics Univ of Md Baltimore County Baltimore MD 21228

MULLIGAN, LEO VIRGIL, b Emmett, Kans, Mar 28, 10; m 46; c 3. SURGERY. Educ: St Mary's Col, AB, 31; St Louis Univ, MD, 37; Am Bd Surg, dipl. Prof Exp: Med dir, St Louis City Hosp, 41-42; from instr to asst prof surg, 42-55, assoc prof clin surg, 55-72, PROF CLIN SURG, ST LOUIS UNIV, 72- Concurrent Pos: Pvt pract, 46-; asst surgeon, St Louis Univ Group Hosps, 46-48, assoc surgeon, 48-; pres, Inst Med Educ & Res, 54-71; chief staff & dir training prog, St Louis City Hosp, 57-73. Mem: Am Col Surg. Res: Clinical surgery. Mailing Add: 665 S Skinker Blvd St Louis MO 63105

MULLIGAN, RICHARD MICHAEL, b Sherburne, NY, July 1, 12; m 38; c 1. PATHOLOGY. Educ: Cornell Univ, AB, 33; Univ Rochester, MD, 37. Prof Exp: Intern path & pediat, Colo Gen Hosp, Denver, 37-38; fel path, Dartmouth Med Sch, 38-39; from instr to prof path, Sch Med, Univ Colo, Denver, 39-76. Mem: Soc Exp Biol & Med; Am Soc Exp Path; Am Asn Cancer Res; Tissue Cult Asn. Res: General and tumor pathology. Mailing Add: Dept of Path Univ of Colo Sch of Med Denver CO 80220

MULLIKEN, ROBERT SANDERSON, b Newburyport, Mass, June 7, 96; m 29; c 2. PHYSICS, CHEMISTRY. Educ: Mass Inst Technol, BS, 17; Univ Chicago, PhD(phys chem), 21. Hon Degrees: ScD, Columbia Univ, 39, Marquette Univ, 67, Cambridge Univ, 67, Gustavus Adolphus Col, 75; PhD, Univ Stockholm, 60. Prof Exp: Jr chem eng bur mines, US Dept Interior, Washington, DC, 17-18; asst rubber res, NJ Zinc Co, Pa, 19; Nat Res Coun fel, Univ Chicago & Harvard Univ, 21-25; asst prof physics, Wash Sq Col, NY Univ, 26-28; from assoc prof to prof, 28-57, dir ed work & info, Plutonium Proj, 42-45, Ernest DeWitt Burton distinguished serv prof, 57-61, DISTINGUISHED SERV PROF PHYSICS & CHEM, UNIV CHICAGO, 61- Concurrent Pos: Guggenheim fel, 30 & 32; Fulbright scholar, Oxford Univ, 52-53; sci attache, London, 55; Baker lectr, Cornell Univ, 60; lectr, Atomic Energy Estab, Trombay, India & Indian Inst Tech, Kanpur, 62; Silliman lectr, Yale Univ, 65; Jan Van Geuns vis prof, Univ Amsterdam, 65; distinguished res prof, Fla State Univ, Winters, 65-71. Honors & Awards: Nobel Prize in Chem, 66; Lewis Gold Medal, Am Chem Soc, 60, Richards Gold Medal, 60, Debye Award, 63, Kirkwood Medal, 64, City Col New York Alumni Asn Gold Medal, 65, Willard Gibbs Gold Medal, 65. Mem: Nat Acad Sci; AAAS; fel Am Phys Soc; fel Am Acad Arts & Sci; hon fel The Chem Soc. Res: Separation of isotopes; molecular spectra; diatomic molecules; theory of molecular spectra and electronic structure of molecules; electron donor-acceptor interactions and charge-transfer spectra. Mailing Add: Dept of Chem Univ of Chicago Chicago IL 60637

MULLIKIN, THOMAS WILSON, b Tenn, Jan 9, 28; m 52; c 3. APPLIED MATHEMATICS. Educ: Univ Tenn, AB, 50; Harvard Univ, MA, 54, PhD(math), 58. Prof Exp: Asst math, Oak Ridge Nat Lab, 47-48; mathematician, Rand Corp, 57-64; PROF MATH, PURDUE UNIV, 64- Mem: AAAS; Am Math Soc; Soc Indust & Appl Math. Res: Differential and integral equations; functional analysis. Mailing Add: Dept of Math Purdue Univ West Lafayette IN 47906

MULLIN, CHARLES JAMES, theoretical physics, see 12th edition

MULLIN, CHARLES R, b Staten Island, NY, Jan 28, 39; m 60; c 5. PHYSICAL CHEMISTRY. Educ: St Vincent Col, BS, 59; Univ Notre Dame, PhD(phys chem), 64. Prof Exp: Phys chemist, Dow Chem Co, 64-69; asst prof chem, 69-72, ASSOC PROF CHEM & PHYSICS, LAKE SUPERIOR STATE COL, 72-, PHYSICS COORDR, 69- Mem: AAAS; Am Inst Physics. Res: Organic photochemistry; environmental sciences. Mailing Add: Dept of Chem Lake Superior State Col Marie MI 49783

MULLIN, MICHAEL MAHLON, b Galveston, Tex, Nov 17, 37; m 64; c 3. BIOLOGICAL OCEANOGRAPHY, ECOLOGY. Educ: Shimer Col, AB, 57; Harvard Univ, AB, 59, MA, 60, PhD(biol), 64. Prof Exp: NSF fel, 64; instr oceanog & res biologist, 64-65, asst prof & asst res biologist, 65-71, ASSOC PROF OCEANOG, SCRIPPS INST OCEANOG, UNIV CALIF, SAN DIEGO, 71-, ASSOC RES BIOLOGIST, INST MARINE RESOURCES, 71- Mem: Am Soc Limnol & Oceanog. Res: Ecology of marine plankton, especially energetics and population dynamics of zooplankton. Mailing Add: Inst of Marine Resources Univ Calif San Diego PO Box 1529 La Jolla CA 92093

MULLIN, ROBERT SPENCER, b Tazewell, Va, May 19, 12; m 38; c 2. PLANT PATHOLOGY. Educ: Hampden-Sydney Col, BS, 34; Va Polytech Inst, MS, 37; Univ Minn, PhD, 50. Prof Exp: Agt directing grain rust control, USDA, Va, 36-41, state leader, 41-44; assoc pathologist, Truck Exp Sta, Univ, Va, 45-46, plant pathologist & head dept, 48-58; assoc prof plant path, physiol & bot, Va Polytech Inst, 46-48; PROF & PLANT PATHOLOGIST, COOP EXTEN SERV, UNIV FLA, 58- Mem: Am Phytopath Soc. Res: Control of plant diseases, especially vegetable, ornamental and fruit crops. Mailing Add: Plant Path Lab Univ of Fla Gainesville FL 32611

MULLIN, RONALD CLEVELAND, b Guelph, Ont, Aug 15, 36. MATHEMATICS. Educ: Univ Western Ont, BA, 59; Univ Waterloo, MA, 60, PhD(math), 64. Prof Exp: Lectr math, Univ Waterloo, 60-64, from asst prof to assoc prof, 64-68; prof, Fla Atlantic Univ, 68-69; assoc dean grad studies, 71-75, PROF MATH, UNIV WATERLOO, 69-, CHMN DEPT COMBINATORICS & OPTIMIZATION, 75- Mem: Math Asn Am; Can Math Cong; Am Math Soc. Res: Combinatorial mathematics, especially enumeration and design theory. Mailing Add: Dept Combinatorics Optimization Univ of Waterloo Waterloo ON Can

MULLIN, WILLIAM JESSE, b Brentwood, Mo, Dec 8, 34; m 61; c 3. THEORETICAL SOLID STATE PHYSICS. Educ: St Louis Univ, BS, 56; Washington Univ, PhD(theoret solid state physics), 65. Prof Exp: Res physicist, Aerospace Res Labs, Wright-Patterson AFB, Ohio, 64-65; res assoc physics, Univ Minn, Minneapolis, 65-67; asst prof theoret solid state physics, 67-71, ASSOC PROF THEORET SOLID STATE PHYSICS, UNIV MASS, AMHERST, 71- Concurrent Pos: Sci Res Coun fel, Univ Sussex, Eng, 73-74. Mem: Am Phys Soc. Res: Many-body theory; analysis of properties of quantum solids and liquids at low temperatures. Mailing Add: Dept of Physics Univ of Mass Amherst MA 01002

MULLINAX, PERRY FRANKLIN, b Quebec, Que, June 7, 31; US citizen; m 57; c 2. RHEUMATOLOGY, IMMUNOLOGY. Educ: Duke Univ, BA, 51; Med Col Va, MD, 55; Am Bd Internal Med, dipl; Subspecialty Bd Rheumatol, dipl. Prof Exp: Clin & res fel med, Mass Geh Hosp, 59-61; Helen Hay Whitney res fel, Sch Med, Washington Univ, 61-62; Helen Hay Whitney res fel, Mass Inst Technol, 62-63; asst prof, 63-67, asst dir, Clin Res Ctr, 70-75, ASSOC PROF MED, MED COL VA, 67- Concurrent Pos: Fel med, Harvard Med Sch, 59-61; res fel, Arthritis Found, 59-61. Mem: AAAS; Am Fedn Clin Res; Am Rheumatism Asn; fel Am Col Physicians. Res: Clinical immunology; immunochemistry; rheumatic diseases. Mailing Add: Dept of Med Med Col of Va Richmond VA 23298

MULLINEAUX, DONAL RAY, b Weed, Calif, Feb 16, 25; m 51; c 3. GEOLOGY. Educ: Univ Wash, Seattle, BS, 47 & 49, MS, 50, PhD(geol), 61. Prof Exp: Field asst, 50, GEOLOGIST, ENG GEOL BR, US GEOL SURV, 50- Mem: Geol Soc Am; Am Quaternary Asn. Res: Geology of Puget Sound Basin; engineering geology; volcanic ash and mudflow deposits. Mailing Add: US Geol Surv Denver Fed Ctr Denver CO 80225

MULLINEAUX, RICHARD DENISON, b Portland, Ore, Feb 23, 23; m 47; c 2. ORGANIC CHEMISTRY. Educ: Univ Wash, Seattle, BS, 48; Univ Wis, PhD(org chem), 51. Prof Exp: Chemist, Shell Develop Co, 51-59, res supvr, 60-63, spec technologist, Wilmington Refinery, Shell Oil Co, 63-64, mgr aromatics dept, Wood River Refinery, 65-66, asst mgr head off tech dept, 66-67, dir, Gen Sci Div, Shell Develop Co, Calif, 67-69, gen mgr, Mfg, Transport & Mkt, Shell Oil Co, 69-74, gen mgr, Mfg, Transport & Mkt/Chem Res & Develop, 74, GEN MGR RES & DEVELOP PROD, SHELL OIL CO, 75- Mem: Am Chem Soc; Catalysis Soc; Am Soc Advan Sci. Res: Hydrocarbon chemistry; organo metallics; engine lubricants; petroleum refining; organic chemical products and processes. Mailing Add: Shell Oil Co One Shell Plaza Houston TX 77002

MULLINGS, MARCUS EVANS, mathematics, see 12th edition

MULLINIX, KATHLEEN PATRICIA, b Boston, Mass, Mar 19, 44; m 66. BIOCHEMISTRY, ENDOCRINOLOGY. Educ: Trinity Col, DC, AB, 65; Columbia Univ, PhD(chem biol), 69. Prof Exp: NIH fel, Harvard Univ, 69-71, res assoc biol, 71-72; staff fel, Nat Inst Arthritis, Metab & Digestive Dis, 72-73, SR STAFF FEL, NAT CANCER INST, NIH, 73- Mem: AAAS; Am Chem Soc; Am Soc Microbiol. Res: Enzymes catalyzing macromolecular biosynthesis; interrelationships among subcellular organelles; mechanisms of hormonal regulation of gene expression. Mailing Add: Nat Cancer Inst NIH Bethesda MD 20014

MULLINS, AUTTIS MARR, b Dyersburg, Tenn, Jan 28, 25; m 46; c 3. MEAT SCIENCES, FOOD TECHNOLOGY. Educ: Univ Ky, BS, 53, MS, 54; Univ Mo, PhD(animal sci), 57. Prof Exp: Lab asst animal sci, Univ Ky, 52-54; instr, Univ Mo, 54-57; from asst prof to prof, La State Univ, 57-70; prof animal industs & head dept, 70-72, DEAN COL AGR, UNIV IDAHO, 72- Mem: Am Meat Sci Asn; Am Soc Animal Sci; Inst Food Technologists. Res: Basic characterization of muscle's fine structure and biochemical composition as they affect muscle as a food, especially tenderness as textural differences and water binding properties. Mailing Add: Col of Agr Univ of Idaho Moscow ID 83843

MULLINS, DONALD EUGENE, b La Junta, Colo, Nov 2, 44; m 68; c 1. INSECT PHYSIOLOGY. Educ: Univ Colo, BA, 66; Colo State Univ, MS, 68; Va Polytech Inst & State Univ, PhD(entom), 72. Prof Exp: Lectr zool, Univ Western Ont, 71-73; INSTR ENTOM, VA POLYTECH INST & STATE UNIV, 73- Mem: Sigma Xi; AAAS; Entom Soc Am. Res: Physiology and biochemistry of nitrogen metabolism in insects as it relates to osmoregulation and excretion, particularly the role of stored urates. Mailing Add: Dept of Entom Va Polytech Inst & State Univ Blacksburg VA 24061

MULLINS, EDGAR RAYMOND, JR, b Champaign, Ill, Aug 16, 23; m 72; c 6. MATHEMATICS. Educ: Grinnell Col, AB, 47; Univ Ill, AM, 49, PhD(math), 52. Prof Exp: Asst math, Univ Ill, 47-52; instr, Swarthmore Col, 52, asst prof, 53-61; assoc prof, Grinnell Col, 61-68, prof math & dir comput serv, 68-70; DIR COMPUT-EDUC ACTIV, SWARTHMORE COL, 70- Res: Computer education; mathematical education. Mailing Add: 11 S Princeton Ave Swarthmore PA 19081

MULLINS, J FRED, b Lampasas, Tex, June 4, 20; m 49; c 2. DERMATOLOGY. Educ: Univ Tex, BA, 42, MD, 46; Am Bd Dermat, dipl. Prof Exp: Instr, 50-52, PROF DERMAT, UNIV TEX MED BR GALVESTON, 53-, CHMN DEPT, 68- Concurrent Pos: Consult, USPHS Hosp, Galveston, 64-; mem exam bd, Am Bd Dermat, 69- Mem: AMA; Am Acad Dermat; Am Dermat Asn. Res: Improvment in techniques of dermatologic teaching; controlled clinical research; basic work in microbiology of the skin. Mailing Add: Dept of Dermat Univ of Tex Med Br Galveston TX 77550

MULLINS, JEANETTE SOMERVILLE, b Salem, Ohio, Aug 1, 32; c 2. MICROBIAL ECOLOGY, PLANT TAXONOMY. Educ: Wayne State Univ, BA, 55, MS, 62; NDak State Univ, PhD(bot), 75. Prof Exp: Sr bacteriologist, Henry Ford Hosp, Detroit, 55-66; bacteriologist, Providence Hosp, Southfield, Mich, 68; plant physiologist & res asst, Metab & Radiation Res Lab, Agr Res Serv, USDA, Fargo, NDak, 69-74; ASST PROF BIOL SCI, CALIFORNIA STATE COL, PA, 75- Mem: AAAS; Am Inst Biol Sci; Am Soc Microbiol; Sigma Xi. Res: Enumeration and identification of leaf surface bacteria; effects of surface applied herbicides on phyllpplane bacteria. Mailing Add: Dept of Biol Sci Calif State Col California PA 15419

MULLINS, JOHN A, b Philadelphia, Pa, Feb 16, 31. CHEMISTRY. Educ: Univ Pa, BS, 58, PhD(chem), 64. Prof Exp: Fel photochem, Brandeis Univ, 64-66; lectr chem, Bucknell Univ, 66-67; asst prof, 67-74, ASSOC PROF NATURAL SCI, MICH STATE UNIV, 74- Res: Biophysical chemistry; history and philosophy of science. Mailing Add: Dept of Natural Sci Kedzie Lab Mich State Univ East Lansing MI 48823

MULLINS, JOHN DOLAN, b Avon, Mass, May 2, 24; m 45. PHARMACY. Educ: Mass Col Pharm, BS, 50, MS, 52; Univ Fla, PhD, 55. Prof Exp: Res assoc, Merck Sharp & Dohme Res Labs, 55-61; asst dir prod develop, Mead Johnson Res Ctr, 61-64; dir prod develop, 64-68, dir develop, 68-71, dep gen mgr sci & technol, 71-74, DIR DERMAT RES & DEVELOP, ALCON LABS, INC, 74- Mem: Am Acad Dermat; Am Chem Soc; Am Pharmaceut Asn; NY Acad Sci. Res: Pharmaceutical dosage forms; colloids; emulsion technology. Mailing Add: 4801 Westlake Dr Ft Worth TX 76132

MULLINS, JOHN THOMAS, b Richmond, Va, Nov 18, 32; m 55; c 3. BOTANY. Educ: Univ Richmond, BS, 55, MS, 57; Univ NC, PhD(bot), 60. Prof Exp: Asst prof bot & biol sci, Univ Fla, 59-64; NIH spec res fel, Harvard Univ, 64-65; assoc prof bot, 65-73, PROF BOT, UNIV FLA, 73-, ASSOC CHMN DEPT, 75- Mem: AAAS; Bot Soc Am; Mycol Soc Am; Sigma Xi. Res: Regulatory mechanisms in hormonal control of sexual morphogenesis in fungi. Mailing Add: Dept of Bot Univ of Fla Gainesville FL 32611

MULLINS, LAWRENCE J, JR, b New York, NY, Nov 7, 21; m 46; c 4. PHYSICAL CHEMISTRY. Educ: Queen's Col, NY, BS, 43; Univ NMex, PhD(chem), 57. Prof Exp: STAFF MEM, LOS ALAMOS SCI LAB, 46- Mem: Fel Am Inst Chem; Am Chem Soc; Am Nuclear Soc. Res: Plutonium chemistry and metallurgy; electrochemistry and electrorefining of plutonium metals; thermodynamic properties of nuclear materials; fused salt chemistry; high temperature chemistry of plutonium, uranium, americum and rare earths; plutonium 238 heat sources. Mailing Add: 30 Manhattan Loop Los Alamos NM 87544

MULLINS, LORIN JOHN, b San Francisco, Calif, Sept 23, 17; m 46; c 2. BIOPHYSICS. Educ: Univ Calif, BS, 37, PhD(biophys), 40. Prof Exp: Asst, Univ Calif, 38-40; asst physiol, Sch Med & Dent, Univ Rochester, 40-41, instr, 41-43; res assoc, Med Sch, Wayne State Univ, 46; Am-Scand Found fel, Inst Theoret Physics, Copenhagen, 47-48; Nat Res Coun Merck fel, Zool Sta, Naples & Johnson Res Found, Sch Med, Univ Pa, 48-49; Nat Res Coun Merck fel biophys, Johns Hopkins Univ, 49-50; assoc prof biol sci, Purdue Univ, 50-58; PROF BIOPHYS & HEAD DEPT, SCH MED, UNIV MD, BALTIMORE CITY, 59- Concurrent Pos: Mem Corp Bermuda Biol Sta, 51-; mem, Marine Biol Lab, Woods Hole, Mass, 56-; USPHS fel, Zoophysiol Lab, Univ Copenhagen, 56-57; mem bd sci counr, Nat Inst Neurol Dis & Stroke, 69-73; chmn, J Neurosci Res, 75-; ed, Ann Rev Biophys & Bioeng, 72- Mem: Am Physiol Soc; Biophys Soc; Am Chem Soc; Soc Gen Physiol; Sigma Xi. Res: Permeability of cells to ions; applications of radioisotopes to biological problems; active transport of ions; modes of anesthetic action. Mailing Add: Dept of Biophys Univ of Md Baltimore MD 21201

MULLINS, ROBERT EMMET, b New York, NY, Sept 24, 37; m 64; c 1. MATHEMATICAL ANALYSIS. Educ: Iona Col, BS, 58; Univ Notre Dame, MS, 60; Northwestern Univ, PhD(function algebras), 65. Prof Exp: Instr, 64-65, asst prof, 65-70, ASSOC PROF MATH, MARQUETTE UNIV, 70- Mem: Am Math Soc. Res: Algebras of functions. Mailing Add: Dept of Math Marquette Univ Milwaukee WI 53233

MULLINS, WILLIAM WILSON, b Boonville, Ind, Mar 5, 27; m 48; c 4. PHYSICS. Educ: Univ Chicago, MS, 51, PhD(physics), 55. Prof Exp: Res physicist, Res Labs, Westinghouse Elec Corp, 55-59, adv physicist, 59-60; assoc prof metall eng, Carnegie Inst Technol, 60-63; prof & head dept, 63-66, dean, 66-70; PROF APPL SCI, CARNEGIE-MELLON UNIV, 70- Mem: Am Phys Soc; Am Inst Mining, Metall & Petrol Engrs. Res: Metallic surfaces and interfaces; physical metallurgy; statistical mechanics of alloys; morphology of solid state transformations; defect structures in crystalline lattices; particle flow and soil mechanics. Mailing Add: Carnegie-Mellon Univ Pittsburgh PA 15213

MULLISON, WENDELL ROXBY, b Philadelphia, Pa, Sept 24, 13; m. PLANT PHYSIOLOGY. Educ: Univ NMex, BA, 34; Univ Chicago, PhD(plant physiol), 38. Prof Exp: Instr biol, Purdue Univ, 40-44; plant physiologist olericult, Curacaosche Petrol Indust Maatschappij, Neth WIndies, 44-46; plant physiologist, 45-50, asst tech dir in chg agr chem, Dow Chem Int, Ltd, 50-59, prod mgr, 59-62, dir info serv, Bioprod Dept, 62-65, mgr govt contract res & develop, 66-72, REGISTR SPECIALIST, DOW CHEM CO, 72- Mem: Am Soc Plant Physiol; Bot Soc Am; Am Soc Hort Sci; Am Soc Agron; Soil Sci Soc Am. Res: Plant nutrition and hormones; herbicides. Mailing Add: Dow Chem Co PO Box 1706 Midland MI 48640

MULLOOLY, JOHN P, b Manhattan, NY, July 8, 37; m 69. BIOSTATISTICS, BIOMATHEMATICS. Educ: St Francis Col, BS, 59; Mich State Univ, MS, 61; Cath Univ Am, PhD(math statist), 66. Prof Exp: Math statistician, NIH, Md, 66-68; prof statist, Ore State Univ, 68-73; BIOSTATISTICIAN, KAISER FOUND HOSP, HEALTH SERV RES CTR, PORTLAND, ORE, 73- Mem: Biomet Soc; Am Statist

Asn; Soc Math Biol; Am Pub Health Asn. Res: Stochastic kinetics; mathematical models in biology; statistical inference; health services research. Mailing Add: Kaiser Found Hosp 4610 SE Belmont Portland OR 97215

MULLOY, WILLIAM, b Salt Lake City, Utah, May 3, 17. ANTHROPOLOGY. Educ: Univ Utah, AB, 39; Univ Chicago, MA, 48, PhD, 53. Prof Exp: Field archaeologist, La State Archaeol Surv, 38-39; state supvr, Mont, 40-42; asst prof anthrop, 48-54, assoc prof anthrop & sociol, 54-58, PROF ANTHROP, UNIV WYO, 58-, CUR ANTHROP MUS, 71- Concurrent Pos: Mem, Norweg Archaeol Exped, EPolynesia, 55-56; mem, Univ Chile Mission, Isla de Pascua, 59-60; Fulbright res grant, Easter Island, 65-66. Honors & Awards: Humphrey Dist Fac Medal, Univ Wo, 64. Mem: Am Anthrop Asn; Soc Am Archaeol; Polynesian Soc. Res: Archaeology of North America and Polynesia; prehistory and conservation of Easter Island. Mailing Add: Dept of Anthrop Univ of Wyo Laramie WY 82070

MULNIX, JOHN ARTHUR, b Colorado Springs, Colo, Jan 17, 39; m 61; c 4. VETERINARY MEDICINE. Educ: Colo State Univ, DVM, 63, MS, 74. Prof Exp: Vet, Lubbock Animal Hosp, 63-64; asst prof med, Dept Small Animal Med & Surg, Cornell Univ, 66-68; from instr to asst prof, 64-74, ASSOC PROF MED, DEPT CLIN SCI, COLO STATE UNIV, 74- Concurrent Pos: Guest lectr, Small Animal Clin, Sch Vet Med, Univ Utrecht, 74-75. Mem: Acad Vet Cardiol; Am Vet Med Asn; Am Animal Hosp Asn; Am Col Vet Int Med; Am Asn Vet Clinicians. Res: Pathophysiology of polydipsia and polyuria in spontaneous hyperadrenocorticism in dogs; role of the hypothalamus in spontaneous hyperadrenocorticism in dogs; classification and identification of spontaneous polyuric disorders in dogs. Mailing Add: Dept of Clin Sci Col Vet Med Colo State Univ Ft Collins CO 80523

MULRENNAN, CECILIA AGNES, b Everett, Mass, Aug 4, 25. BIOLOGY. Educ: Regis Col, Mass, AB, 46; Fordham Univ, MA, 57, PhD(genetics), 59. Prof Exp: Instr biol, 59-63, PROF BIOL & CHMN DEPT, REGIS COL, MASS, 63- Concurrent Pos: NIH res grant, 60-61; Grass Found res grant, 68-71. Res: Genetics of Drosophila; philosophy of science. Mailing Add: Dept of Biol Regis Col Wellesley St Weston MA 02193

MULROW, PATRICK J, b New York, NY, Dec 16, 26; m 53; c 4. MEDICINE. Educ: Colgate Univ, AB, 47; Cornell Univ, MD, 51; Am Bd Internal Med, dipl. Prof Exp: Instr physiol, Med Col, Cornell Univ, 54-55; from instr to assoc prof med, Sch Med, Yale Univ, 57-69, prof internal med, 69-76; PROF MED & CHMN DEPT, MED COL OHIO, 76- Concurrent Pos: USPHS res fel, 54-56, res grant, 57-66; Arthritis Res Found res fel, 56-57; clin investr, Vet Admin Hosp, West Haven, Conn, 58-61; attend, Yale-New Haven Hosp, 60-; mem study sect, NIH. Mem: Am Soc Clin Invest; Am Physiol Soc; Endocrine Soc; Am Fedn Clin Res; Int Soc Cardiol. Res: Hypertension and endocrinology. Mailing Add: Dept of Med Med Col of Ohio Toledo OH 43614

MULROY, MICHAEL JOSEPH, b Wyandotte, Mich, July 26, 31; m 65; c 1. ANATOMY. Educ: Our Lady of the Forest Sem, AB, 57; DePaul Univ, MS, 60; Univ Calif, San Francisco, PhD(anat), 68. Prof Exp: NIH fel auditory physiol, Harvard Med Sch-Mass Inst Technol, 68-70; instr anat, Harvard Med Sch, 72-74; ASST PROF ANAT, MED CTR, UNIV MASS, 74-; RES ASSOC OTOLARYNGOL, MASS EYE & EAR INFIRMARY, EATON PEABODY LAB, 71- Concurrent Pos: Teaching fel gross anat, Harvard Med Sch, 71- Mem: Am Asn Anatomists. Res: Structure and function of comparative vertebrate hearing. Mailing Add: Dept of Anat Univ of Mass Med Ctr Worcester MA 01605

MULSON, JOSEPH F, b Milwaukee, Wis, Feb 6, 29; m 49; c 2. ELECTRON PHYSICS. Educ: Rollins Col, BS, 56; Pa State Univ, MS, 61, PhD(physics), 63. Prof Exp: From asst prof to assoc prof physics, 63-73, chmn sci div, 70-72, PROF PHYSICS, ROLLINS COL, 73- Concurrent Pos: Cotrell grant, 64-65; mem, Nat Sci Stud Res Partic, 65-66. Res: Holography and laser applications. Mailing Add: Dept of Physics A G Bush Sci Ctr Rollins Col Winter Park FL 32789

MULTER, H GRAY, b Syracuse, NY, July 7, 26; m 50; c 2. GEOLOGY. Educ: Syracuse Univ, AB, 49, MS, 51; Ohio State Univ, PhD(geol), 55. Prof Exp: Petrol geologist, Tex Co, Calif, 51-53; prof geol, Col Wooster, 55-69; dir, West Indies Lab, 69-75, CHMN DEPT EARTH SCI, FAIRLEIGH DICKINSON UNIV, MADISON, NJ, 75- Mem: Geol Soc Am; Nat Asn Geol Teachers; Am Asn Petrol Geol. Res: Sedimentation; marine geology; enrivonmental geology. Mailing Add: Dept of Earth Sci Fairleigh Dickinson Univ Madison NJ 07940

MULTHAUF, DELMAR CHARLES, b Hartford, Wis, May 2, 24; m 53; c 2. GEOGRAPHY. Educ: Wis State Univ-Oshkosh, BS, 50; Univ Wis, MS, 52; Columbia Univ, DEd, 58. Prof Exp: Instr geog, Wis State Univ-La Crosse, 54, Whitewater, 54-55, Oshkosh, 55; asst prof, WConn State Col, 58-62; asst prof, Mankato State Col, 62-64; asst prof, Wis State Univ-Oshkosh, 64-66; assoc prof, 66-70, PROF GEOG, UNIV WIS-STEVENS POINT, 70- Mem: Asn Am Geogrs; Nat Coun Geog Educ. Res: Cultural geography of the state of Wisconsin; regional and urban geography in New Jersey and Connecticut. Mailing Add: Dept of Geog Univ of Wis Stevens Point WI 54481

MULVANEY, JAMES EDWARD, b Brooklyn, NY, Aug 4, 29; m 52; c 4. ORGANIC CHEMISTRY. Educ: Polytech Inst Brooklyn, BS, 51, PhD(chem), 59. . Prof Exp: Res chemist gen chem div, Allied Chem & Dye Corp, 51-53; asst, Polytech Inst Brooklyn, 55-59; res assoc chem, Univ Ill, 59-61; from asst prof to assoc prof, 61-71, PROF CHEM, UNIV ARIZ, 71- Mem: AAAS; Am Chem Soc; The Chem Soc. Res: Organic synthesis; synthesis and mechanism of high polymer formation. Mailing Add: Dept of Chem Univ of Ariz Tucson AZ 85721

MULVANEY, JOHN FRANCIS, b Newark, NJ, Oct 3, 16; m 43; c 2. ORGANIC CHEMISTRY. Educ: NY Univ, BA, 38, MSc, 41, PhD(org chem), 42. Prof Exp: Instr chem, Newark State Col, 40-47; res chemist, Evans Res Develop Corp, 42-48; res chemist, Gen Aniline & Film Corp, 48-58, prod mgr, 58-61, plant mgr, 61-62, dir develop, 62-66; from exec vpres to pres, Grestco Dyes & Chem, Inc, NY, 66-71; DIR MFG, AM HOECHST CORP, 71- Mem: AAAS; Am Chem Soc; Com Develop Asn. Res: Sulfonamides; cyclic ureas; esters; properties and derivatives of mercapto acids; physical properties of nitro compounds; method synthesizing cyclic ureas; vat dyes; intermediates; surface active agents. Mailing Add: Am Hoechst Corp 129 Quidnick St Coventry RI 02816

MULVANEY, THOMAS RICHARD, b Bellevue, Mich, July 4, 33; m 54; c 4. FOOD SCIENCE. Educ: Mich State Univ, BS, 56, MS, 59, PhD(food sci), 62. Prof Exp: Asst food eng, Mich State Univ, 56-59, asst food sci, 59-61, asst instr, 61-62, asst, 62; res engr, Alcoa Res Labs, Aluminum Co Am, Pa, 62-67, sr res scientist, 67-68; assoc prof food sci & technol, Univ Mass, Amherst, 68-71; sr scientist, 71-72, CHIEF FOOD PROCESSING SECT, FOOD & DRUG ADMIN, 72- Mem: AAAS; Am Chem Soc; Inst Food Technologists; fel Am Inst Chemists. Res: Thermal processing of foods and beverages; effect of sequestering agents on metals; chemical and physical changes in foods induced by packaging materials; nature of metallic flavors; consumer protection. Mailing Add: 8307 Forrester Blvd Springfield VA 22152

MULVEY, DENNIS MICHAEL, b Lockport, NY, Nov 17, 38; m 63; c 2. ORGANIC CHEMISTRY. Educ: Univ Pa, AB, 60; State Univ NY Buffalo, PhD(org chem), 65. Prof Exp: Res assoc, Columbia Univ, 64-65; sr res chemist, 65-73, RES FEL, MERCK, SHARP & DOHME RES LABS, RAHWAY, 73- Mem: Am Chem Soc. Res: Heterocyclic compounds; reaction mechanisms; synthesis of nonclassical aromatic systems; photochemistry; chemistry of natural products. Mailing Add: RD 2 Milford NJ 08848

MULVEY, PHILIP FRANCIS, JR, b Worcester, Mass, Dec 22, 31; m 55; c 4. RADIOBIOLOGY, PHYSIOLOGY. Educ: Clark Univ, AB, 53; Bowling Green State Univ, MA, 55; Univ Buffalo, PhD(biol), 59. Prof Exp: Asst biol, Bowling Green State Univ, 54-55; res biochemist radioisotope serv, Vet Admin Hosp, 58-65; res physiologist, Electronics Res Ctr, NASA, Mass, 66; res physiologist, US Army Natick Labs, 66-68; lectr, 59-68, PROF BIOL, SUFFOLK INST, 68- Concurrent Pos: Asst, Sch Med, Boston Univ, 59-66. Mem: AAAS; Am Soc Zoologists; Am Physiol Soc. Res: Use of radioisotopes and antithyroid agents to study thyroid gland physiology; use of activation analysis to determine trace element concentrations in biological systems. Mailing Add: Dept of Biol Suffolk Inst Boston MA 02114

MULVEY, RICHARD KNEALE, pharmacology, see 12th edition

MULVIHILL, JOHN JOSEPH, b Washington, DC, Aug 20, 43; m 66; c 2. EPIDEMIOLOGY, PEDIATRICS. Educ: Col of the Holy Cross, BS, 65; Dartmouth Med Sch, BMS, 67; Univ Wash, MD, 69. Prof Exp: Staff assoc epidemiol, 70-74, HEAD CLIN GENETICS SECT, CLIN EPIDEMIOL BR, NAT CANCER INST, 74- Concurrent Pos: Mem, Comt Biol Effects Ionizing Radiation, Nat Acad Sci, 70-73; fel pediat, Sch Med, Johns Hopkins Univ, 72-74. Mem: Teratology Soc; Am Acad Pediat; Am Soc Human Genetics. Res: Epidemiology of cancer and congenital defects in man, especially genetic and familial factors; animal models of congenital and genetic disease; medical genetics. Mailing Add: Clin Epidemiol Br Landow Bldg Nat Cancer Inst Bethesda MD 20014

MULVIHILL, MARY LOU JOLIE, b Chicago, Ill, Sept 28, 28. HUMAN PHYSIOLOGY. Educ: St Xavier Col, Ill, BA, 60; Purdue Univ, PhD(physiol), 67. Prof Exp: From asst prof to assoc prof biol, St Xavier Col, Ill, 67-72, chmn div natural sci, 69-72, vpres student affairs, 70-72; ASSOC PROF BIOL, HARPER COL, 72- Concurrent Pos: Consult physiol, W C Brown, Harper & Row & Scott, Foresman, 73- Mem: Am Inst Biol Sci. Res: Effect of a vitamin A deficiency on the ultrastructure of the mosquito eye. Mailing Add: Dept of Biol William Rainey Harper Col Palatine IL 60067

MUMA, MARTIN HAMMOND, b Topeka, Kans, July 24, 16; m 40; c 6. ARACHNOLOGY, DESERT ECOLOGY. Educ: Univ Md, BS, 39, MS, 40, PhD(entom), 43. Prof Exp: Lab asst, USDA, Md, 37-38; asst, Univ Md, 41-43, instr entom, 43-44, asst entomologist, 44-45; assoc exten entomologist, Univ Nebr, 45-48, assoc prof entom & assoc entomologist, 48-51; from assoc prof entom & assoc entomologist to prof & entomologist, 51-71, EMER PROF ENTOM & EMER ENTOMOLOGIST, CITRUS EXP STA, UNIV FLA, 71-; RES ASSOC, DIV PLANT INDUST, FLA DEPT AGR & CONSUMER SERV & WESTERN NMEX UNIV, 71- Mem: Ecol Soc Am; Soc Syst Zool; Animal Behav Soc; Am Arachnol Soc. Res: Insect control by natural or biological factors; taxonomy, systematics, biology, behavior and ecology of arid-land arachnids. Mailing Add: PO Box 2020 Silver City NM 88061

MUMBACH, NORBERT R, b Buffalo, NY, June 6, 20; m 48; c 5. APPLIED CHEMISTRY. Educ: Canisius Col, BS, 42. Prof Exp: Instr phys chem, Canisius Col, 42-43; instr chem oper, Lake Ont Ord Works, 43; anal chemist, Manhattan Proj, Linde Air Prods Co, 43-46, res chemist, 46-52, mass spectrometrist, Linde Div, Union Carbide Corp, 52-56, spectroscopist, 59-61, RES CHEMIST, LINDE DIV, UNION CARBIDE CORP, 61- Mem: Am Chem Soc; Sigma Xi. Res: Synthetic gemstones; hydrothermal methods; new silicate phases created at high pressures and temperatures. Mailing Add: Genessee Rd East Concord NY 14055

MUMFORD, DAVID LOUIS, b Salt Lake City, Utah, May 2, 32; m 55; c 4. PLANT PATHOLOGY. Educ: Brigham Young Univ, BS, 56, MS, 58; Univ Minn, PhD(plant path), 62. Prof Exp: Plant pathologist, USDA, Mich State Univ, 63-67; PLANT PATHOLOGIST, USDA, CROPS RES LAB, UTAH STATE UNIV, 67- Mem: Am Phytopath Soc. Res: Virus diseases of sugar beets; disease resistance in sugar beets. Mailing Add: Crops Res Lab Utah State Univ Logan UT 84321

MUMFORD, FRANKLIN EDWARD, organic chemistry, see 12th edition

MUMFORD, GEORGE, b Sydney, Australia, Apr 4, 27; US citizen; c 3. PROSTHODONTICS. Educ: Univ Sydney, BDS, 53, MDS, 61, DDSc, 61; Ind Univ, DDS, 64. Prof Exp: Sr lectr oper dent, Univ Sydney, 58-59; from asst prof to assoc prof crown & bridge, Sch Dent, Ind Univ & dir ceramics sect, 65-68; assoc prof restorative dent & chmn dept, 68-72, PROF GEN DENT, HEALTH CTR, UNIV CONN, FARMINGTON, 72- Mem: Int Asn Dent Res; fel Am Col Dent; fel Australian Col Dent Surg. Res: Material sciences and restorative dentistry. Mailing Add: Dept Restorative Dent Sch Dent Univ of Conn Health Ctr Farmington CT 06032

MUMFORD, GEORGE SALTONSTALL, III, b Milton, Mass, Nov 13, 28; m 49; c 4. ASTRONOMY. Educ: Harvard Univ, AB, 50; Ind Univ, MA, 52; Univ Va, PhD(astron), 55. Prof Exp: Instr math, Randolph-Macon Woman's Col, 52-53; instr, 55-56, from asst prof to assoc prof, 56-68, PROF MATH, TUFTS UNIV, 68-, DEAN COL LIBERAL ARTS, 69- Concurrent Pos: Vis astronr, Kitt Peak Nat Observ, 62- Mem: AAAS; Am Astron Soc; Am Asn Variable Star Observers; Am Phys Soc; Int Astron Union. Res: Cataclysmic variable stars; close binaries. Mailing Add: Ballou Hall Tufts Univ Medford MA 02155

MUMFORD, RUSSELL EUGENE, b Casey, Ill, May 26, 22; m 47; c 3. ANIMAL ECOLOGY. Educ: Purdue Univ, BS, 48, MS, 52, PhD, 61. Prof Exp: Res biologist, State Dept Conserv, Ind, 48-50; teacher natural hist, Fla Audubon Soc, 50-51; res biologist, State Dept Conserv, Ind, 52-55; asst mus zool, Univ Mich, 55-57; PROF VERT NATURAL HIST, PURDUE UNIV, LAFAYETTE, 58- Mem: Am Soc Mammal; Wilson Ornith Soc; Am Ornith Union. Res: Life history and distribution of ornithology and mammalogy; bat banding. Mailing Add: Dept Forestry Natural Resources Purdue Univ Lafayette IN 47907

MUMM, ROBERT FRANKLIN, b Urbana, Ill, Oct 16, 35. BIOMETRICS, GENETICS. Educ: Univ Ill, BS, 57, MS, 58; Univ Nebr, PhD(quant genetics), 67. Prof Exp: Assoc dir res, Crow's Hybrid Corn Co, Ill, 61-65; instr biomet, 65-67, from asst prof to assoc prof agron, 67-75, PROF AGRON & CONSULT STATIST LAB, UNIV NEBR, LINCOLN, 75- Mem: Am Soc Agron; Crop Sci Soc Am; Am Statist

Asn. Res: Genetic variance components; computer simulation studies of their distribution and effect of choice of mating deisgn on their estimation. Mailing Add: Rm 103 Miller Hall Univ of Nebr Col of Agr Lincoln NE 68503

MUMM, WALTER JOHN, b Sidney, Ill, Nov 20, 95; m 21; c 1. GENETICS. Educ: Univ Ill, BS, 19, MS, 28, PhD(agron, plant breeding), 40. Prof Exp: Asst agron, Univ Ill, 27-28, instr, 28-31, assoc, 31-38; dir res & plant breeding, 38-67, RES CONSULT, CROW'S HYBRID CORN CO, 67- Mem: Am Soc Agron; AAAS. Res: Genetics of dent corn; breeding multiple-ear and dwarf corn; new approaches to maize inbreeding. Mailing Add: 510 E Jones St Milford IL 60953

MUMMA, MARTIN DALE, b Gideon, Mo, Jan 21, 36; m 56; c 3. GEOLOGY. Educ: Univ Mo, AB, 58, MA, 60; La State Univ, PhD(geol), 65. Prof Exp: Field geologist, Magnolia Petrol Co, 58; geologist, Esso Prod Res Co, Stand Oil Co, NJ, 65-66; sr geologist, Humble Oil & Refining Co, 66-68; assoc prof geol, Eastern Ky Univ, 68-69; asst prof, 69-71, ASSOC PROF GEOL, EASTERN WASH STATE COL, 71- Mem: Soc Econ Paleont & Mineral; Am Asn Petrol Geologists; Paleont Soc. Res: Sedimentary petrology and environments; biostratigraphy; micropaleontology. Mailing Add: Dept of Geol Eastern Wash State Col Cheney WA 99004

MUMMA, MICHAEL JON, b Lancaster, Pa, Dec 3, 41; m 66; c 2. EXPERIMENTAL ATOMIC PHYSICS, MOLECULAR PHYSICS. Educ: Franklin & Marshall Col, AB, 63; Univ Pittsburgh, PhD(physics), 70. Prof Exp: Space scientist, 70-76, HEAD RADIO & INFRARED ASTRON BR, GODDARD SPACE FLIGHT CTR, NASA, 76- Mem: Am Geophys Union; Am Phys Soc; Am Astron Soc; AAAS. Res: Experimental atomic and molecular physics; cometary physics; planetary atmospheres; infrared astronomy; molecular spectroscopy; coherent detection techniques; Doppler limited laser spectroscopy; molecular astrophysics. Mailing Add: Radio Astron Br Goddard Space Flight Ctr Code 693 Greenbelt MD 20771

MUMMA, RALPH O, b Mechanicsburg, Pa, June 20, 34; m 58; c 2. BIOCHEMISTRY, ENTOMOLOGY. Educ: Juniata Col, BS, 56; Pa State Univ, PhD(chem), 60. Prof Exp: Fel biochem, 60-61, asst prof, 61-66, from asst prof to assoc prof chem pesticides, 66-72, PROF CHEM PESTICIDES, 72- Mem: Am Chem Soc; Am Oil Chem Soc; Entom Soc Am. Res: Lipid, sulfur, pesticide and insect metabolism. Mailing Add: Pesticide Res Lab Dept Entom Pa State Univ University Park PA 16802

MUMMA, RICHARD HOWARD, chemistry, see 12th edition

MUMPTON, FREDERICK ALBERT, b Rome, NY, Dec 14, 32; m 54; c 5. MINERALOGY. Educ: St Lawrence Univ, BS, 54; Pa State Univ, MS, 56, PhD(geochem), 58. Prof Exp: Res chemist, Linde Div, Union Carbide Corp, 58-60, res geochemist, Nuclear Div, 60-65; mineral group leader, Mining & Metals Div, NY, 65-69; assoc prof, 69-74, PROF, DEPT EARTH SCI, STATE UNIV NY COL BROCKPORT, 74- Mem: Geochem Soc; fel Mineral Soc Am; Clay Minerals Soc; Sigma Xi. Res: Silicate chemistry; synthetic and clay mineralogy; mineralogy and utilization of zeolite minerals; mineralogy of asbestos and serpentinites; mineral resources. Mailing Add: Dept of the Earth Sci State Univ of NY Col Brockport NY 14420

MUN, ALTON M, b Honolulu, Hawaii, Apr 1, 23; m 55; c 5. DEVELOPMENTAL BIOLOGY. Educ: Univ Southern Calif, BA, 49; Univ Ill, MS, 51; Ind Univ, PhD(zool), 56. Prof Exp: Res assoc avian embryol, Wash State Univ, 56-59; asst invest exp embryol, Carnegie Inst, 59-61; assoc prof zool, 61-70, PROF ZOOL, UNIV MAINE, 70- Mem: Am Soc Zool; Sigma Xi; Soc Develop Biol. Res: Experimental embryology; zoology; effects of antisera on chick embryos; enhancement of growth of chick host spleens by homologous adult organ fragments; homograft reaction in the chick embryo; parthenogenetic development in unfertilized turkey eggs; teratological effects of trypan blue in the chick embryo. Mailing Add: Dept of Zool Murray Hall Univ of Maine Orono ME 04473

MUNAN, LOUIS, b New York, NY, Feb 10, 21; m 47; c 4. EPIDEMIOLOGY. Educ: City Col New York, BA, 48; George Washington Univ, AB, 48, MSc, 50. Prof Exp: Res analyst, Nat Acad Sci-Nat Res Coun, 49-51; res analyst, Prev Med Div, Off Surgeon Gen, Dept Army, 51-55, chief res & develop sect, Med Info & Intel Div, 55-56; Fulbright exchange prof, Schs Med, Lima & Guayaquil, 57-58; statistician, Pan-Am Sanit Bur, WHO, 58-61, res scientist, Off Res Coord, Pan-Am Health Orgn, 61-68; chmn dept, 68-75, ASSOC PROF EPIDEMIOL, FAC MED, UNIV SHERBROOKE, 68- Concurrent Pos: Assoc, Sch Med, George Washington Univ, 50-58; vis prof, Malaria Eradication Training Ctr, Kingston, Jamaica, 62 & 63. Mem: Fel Am Pub Health Asn; Biomet Soc; fel NY Acad Sci; Can Pediat Soc; Int Edpidemiol Asn. Mailing Add: 39 Belvedere Lennoxville PQ Can

MUNAVALLI, SOMASHEKHAR, b Sumpgaon, India, Oct 14, 31; m 61; c 3. ORGANIC CHEMISTRY, BIOCHEMISTRY. Educ: Karnatak Univ, India, MS, 55; Univ Kans, AM, 59; Univ Strasbourg, PhD(org chem), 62, DSc(org chem), 64. Prof Exp: From researcher to sr researcher org chem, Nat Ctr Sci Res, France, 60-65; sr lectr, Karnatak Univ, India, 65-67; staff scientist, Worchester Found Exp Biol, Mass, 67; prof chem, Lane Col, 67-69; PROF CHEM, LIVINGSTONE COL, 69-, CHMN DEPT, 71- Concurrent Pos: Visitor, Inst Org Chem, Gif-sur-Yvette, France, 64; Piedmont Univ Ctr res grant, 70; NSF exten res grant, 71; dir, Col Sci Improv Proj, NSF, prin investr, Res Iniation Grant; coordr, Consortium Health Serv, Dept Health, Educ & Welfare. Mem: Am Chem Soc; The Chem Soc; fel Am Inst Chem; NY Acad Sci; Chem Soc France. Res: Bio-organic chemistry; chemistry of natural products, particularly the terpenoids and alkaloids and their biosynthesis; enzymes; medical chemistry. Mailing Add: Dept of Chem Livingstone Col Salisbury NC 28144

MUNCH, GUIDO, b San Cristobal, Mex, June 9, 21; m 47; c 4. ASTROPHYSICS, ASTRONOMY. Educ: Nat Univ Mex, BS, 38, MS, 44; Univ Chicago, PhD(astron, astrophys), 47. Prof Exp: Instr astrophys, Yerkes Observ, Univ Chicago, 47-48, asst prof, 48-51; from asst prof to assoc prof, 51-59, PROF ASTROPHYS, CALIF INST TECHNOL, 59- Mem: Nat Acad Sci; fel Am Astron Soc; fel Am Acad Arts & Sci; fel Royal Astron Soc; fel Int Astron Union. Res: Stellar atmospheres; interstellar matter; planetary atmospheres. Mailing Add: Hale Observ Calif Inst Technol Pasadena CA 91109

MUNCH, JOHN HOWARD, b St Louis, Mo, Feb 9, 38; m 65. ORGANIC CHEMISTRY. Educ: Swarthmore Col, BA, 60; Univ Wis, Madison, PhD(org chem), 66. Prof Exp: Asst prof chem, Dickinson Col, 65-69; RES CHEMIST, TRETOLITE DIV, PETROLITE CORP, ST LOUIS, 69- Mem: AAAS; Am Chem Soc; The Chem Soc. Res: Organic reaction mechanisms and synthesis; electrophilic substitution reactions; Mannich reactions; surface-active compounds. Mailing Add: 9 Douglass Lane Kirkwood MO 63122

MUNCH, RALPH HOWARD, b Lafayette, Ind, May 5, 11; m 35; c 5. PHYSICAL CHEMISTRY. Educ: Univ NC, BS, 31, MS, 32; Northwestern Univ, PhD(phys chem), 35. Prof Exp: Res asst, Rockefeller Found, Chicago, 35-37; res chemist ord div, Res Dept, Monsanto Chem Co, 37-41, group leader, 41-54, sect leader, 54-56, asst dir res, 56-60, res assoc, 60-64, sr res specialist, Monsanto Co, 64-67, advan scientist, 67-70, sr sci fel, 70-74, DISTINGUISHED SCI FEL, MONSANTO INDUST CHEM CO, 74- Concurrent Pos: Vchmn, Gordon Res Conf Instrumentation, 53, chmn, 54. Mem: AAAS; Am Chem Soc; fel Instrument Soc Am (vpres, 46). Res: Spectroscopy; process control instrumentation; gas chromatography; dielectrics. Mailing Add: Res Dept Monsanto Indust Chem Co 800 N Lindberg Blvd St Louis MO 63166

MUNCH, THEODORE, b Columbus, Ohio, Nov 9, 19; m 60. BACTERIOLOGY, SCIENCE EDUCATION. Educ: Ohio State Univ, BS, 41 & 46; Colo State Univ, MEd, 48; Stanford Univ, EdD(sci educ), 52. Prof Exp: Instr, Balboa High Sch, CZ, 49-50; instr sci & educ, San Francisco State Col, 52; instr life sci, Fullerton Jr Col, 53; asst prof sci educ, Univ Tex, 54-58; assoc prof, 59-63, PROF SCI EDUC, ARIZ STATE UNIV, 63- Mem: Fel AAAS; Nat Sci Teachers Asn. Res: Science books for children. Mailing Add: Dept of Physics Ariz State Univ Tempe AZ 85282

MUNCHAUSEN, LINDA LOU, b New Orleans, La, Aug 30, 46. ORGANIC CHEMISTRY. Educ: Southeastern La Univ, BS, 68; Univ Ark, PhD(chem), 73. Prof Exp: Fel biol, Oak Ridge Nat Lab, 73-75; fel chem, La State Univ, 75; INSTR ORG CHEM, DEPT CHEM, SOUTHEASTERN LA UNIV, 75- Mem: Am Chem Soc; The Chem Soc; Am Soc Photobiol. Mailing Add: Dept of Chem Southeastern La Univ Hammond LA 70402

MUNCK, ALLAN ULF, b Buenos Aires, Arg, July 4, 25; US citizen; m 57; c 3. ENDOCRINOLOGY. Educ: Mass Inst Technol, BS, 48, MS, 49, PhD(biophys), 56. Prof Exp: Nat Cancer Inst fel steroid biochem, Worcester Found Exp Biol, 57-58; from asst prof to assoc prof, 59-67, PROF PHYSIOL, DARTMOUTH MED SCH, 67- Concurrent Pos: Res career develop award, Dartmouth Med Sch, 63-72; assoc ed, J Steroid Biochem, 67- Mem: Biophys Soc; Endocrine Soc; Physiol Soc. Res: Physiological and molecular mechanisms of action of glucocorticoids. Mailing Add: Dept of Physiol Darmouth Med Sch Hanover NH 03755

MUNCY, ROBERT JESS, b Narrows, Va, Apr 13, 29; m 54; c 1. ZOOLOGY. Educ: Va Polytech Inst & State Univ, BS, 50, MS, 54; Iowa State Col, PhD(fishery biol), 57. Prof Exp: Asst zool, Iowa State Col, 54-57; fishery biologist, Chesapeake Biol Lab, Md, 57-59; asst prof wildlife mgt, La State Univ, 59-65; asst prof, Colo State Univ, 65-66; PROF ZOOL & ENTOM & UNIT LEADER, IOWA COOP FISHERY UNIT, BUR SPORT FISHERIES & WILDLIFE, IOWA STATE UNIV, 66- Concurrent Pos: Vis lectr, Univ Md, 58-59 & Mt Lake Sta, Va, 64; fishery biologist, Food & Agr Orgn, UN, Zambia, Africa, 72-73. Mem: Am Fisheries Soc; Am Soc Limnol & Oceanog. Res: Fishery biology and limnology. Mailing Add: Iowa Coop Fishery Unit Sci Hall Iowa State Univ Ames IA 50010

MUNCZEC, HERMAN J, b Buenos Aires, Arg, June 9, 27; m 54; c 2. PHYSICS. Educ: Univ Buenos Aires, Lic, 54, PhD(physics), 58. Prof Exp: Res investr, AEC, Arg, 54-60; from asst prof to assoc prof physics, Univ Buenos Aires, 61-66; vis assoc prof, Northwestern Univ, 66-69; assoc prof, 69-71, PROF PHYSICS, UNIV KANS, 71- Concurrent Pos: Univ Buenos Aires res fel elem particle physics, Univ Rome, 58-60. Mem: Am Phys Soc. Res: Theory of elementary particles. Mailing Add: Dept of Physics & Astron Univ of Kans Lawrence KS 66044

MUNDAY, JOHN CLINGMAN, JR, b Plainfield, NJ, June 10, 40; m 65; c 3. MARINE SCIENCES, REMOTE SENSING. Educ: Cornell Univ, AB, 62; Univ Ill, PhD(biophys), 68. Prof Exp: Res asst photosynthesis, Univ Ill, 65-68, res assoc, 68; physicist, Air Force Missile Develop Ctr, Holloman Air Force Base, NMex, 68-69; assoc marine scientist, Va Inst Marine Sci, 69-71; asst prof geog, Univ Toronto, 71-75; ASSOC MARINE SCIENTIST, VA INST MARINE SCI, 75- Concurrent Pos: Nat Res Coun res associateship, 68-69; asst prof, Univ Va & Col William & Mary, 69-71, assoc prof marine sci, 75- Mem: AAAS. Res: Spectroscopy, photosynthesis and membrane physiology of algae; missile reentry spectral photography; estuarine oil pollution; remote sensing of coastal water quality and circulation. Mailing Add: Va Inst of Marine Sci Gloucester Point VA 23062

MUNDELL, PERCY MELDRUM, b Vancouver, BC, Dec 14, 21; m 50; c 3. ORGANIC CHEMISTRY. Educ: Univ BC, BA, 43, MA, 45; Ohio State Univ, PhD(chem), 53. Prof Exp: Asst chem, Univ BC, 42-45, instr, 45-46; asst, Ohio State Univ, 47-51; from asst prof to assoc prof, 52-67, PROF CHEM, MIAMI UNIV, 67- Mem: Am Chem Soc. Res: Structural studies of natural products. Mailing Add: Dept of Chem Miami Univ Oxford OH 45056

MUNDELL, ROBERT DAVID, b Greensburg, Pa, Aug 30, 36; m 62; c 2. ANATOMY, CELL BIOLOGY. Educ: Waynesburg Col, BS, 57; Univ Pittsburgh, PhD(anat, cell biol), 65. Prof Exp: Instr anat & histol, Sch Med, Tufts Univ, 64-66; asst prof histol, 66-68, head dept, 68-70, assoc prof histol & anat, 68-72, PROF HISTOL & ANAT, SCH DENT MED, UNIV PITTSBURGH, 72-, HEAD DEPT ANAT, 70- Concurrent Pos: NSF res grant, Tufts Univ & Univ Pittsburgh, 66-68. Mem: Am Asn Anatomists; AAAS; Sigma Xi; NY Acad Sci; Am Asn Dent Schs. Res: Craniofacial development. Mailing Add: 630 Salk Hall Sch of Dent Med Univ of Pittsburgh Pittsburgh PA 15261

MUNDEN, BILL J, b Seymour, Ind, Jan 27, 35; m 58; c 2. PHARMACY. Educ: Purdue Univ, BS, 57, MS, 59, PhD(pharm), 62. Prof Exp: Asst pharm, Purdue Univ, 57-59, instr, 59-62, asst prof, 62-64; sr pharmacist, 64-65, HEAD PHARMACEUT PROD DEVELOP, LIFE SCI DIV, DOW CHEM CO, 65- Mem: Fel AAAS; Am Pharmaceut Asn; Am Chem Soc. Res: Application of physical-chemical principles to development of solid dosage forms, controlled drug release and dosage form effect on drug absorption. Mailing Add: Life Sci Div Dow Chem Co PO Box 68511 Indianapolis IN 46268

MUNDIE, J RYLAND, b Shawnee, Okla, Oct 25, 30; m 55; c 3. MEDICINE. Educ: Ouachita Baptist Univ, BS, 51; Univ Ark, MD, 56. Prof Exp: Intern, Med Ctr, Univ Ark, 56-57; CHIEF NEUROPHYSIOL BR, AEROSPACE MED RES LAB, WRIGHT-PATTERSON AFB, 59- Mem: AAAS; Acoust Soc Am. Res: Neurophysiology; bionics; hearing physiology; automatic speech recognition and speech synthesis; biomedical instrumentation; computer applications to biological research. Mailing Add: Aerospace Med Res Lab Wright-Patterson AFB OH 45433

MUNDIE, LLOYD GEORGE, b Udney, Ont, Dec 15, 16; nat US; m 42; c 2. OPTICS. Educ: Univ Sask, BSc, 35, MSc, 37; Purdue Univ, PhD(physics), 43. Prof Exp: Instr physics, Purdue Univ, 39-47; infrared physicist, Naval Ord Lab, 47-51; physicist, Nat Bur Standards, 51-54; physicist, Univ Mich, 54-57; head infrared & optics dept, Bendix Systs Div, 57-61; dir basic res lab, Lockheed-Calif Co, 61-65; PHYSICIST, RAND CORP, 65- Mem: Fel AAAS; fel Optical Soc Am. Res: Spectroscopy; infrared. Mailing Add: Rand Corp 1700 Main St Santa Monica CA 90406

MUNDKUR, BALAJI, b Mangalore, India, Dec 27, 24; nat US; m 46. CYTOLOGY. Educ: Univ Bombay, India, BSc, 45; Washington Univ, PhD(bot, genetics), 51. Prof

Exp: Assoc mycol, Indian Agr Res Inst, 47; res assoc, Univ Southern Ill, 50-53; sr sci officer, Indian Cancer Res Ctr, 53-55; assoc bacteriologist, Univ PR, 55-58; spec res fel anat, USPHS, Chicago, 58-60; ASSOC PROF BIOL SCI, UNIV CONN, 60- Res: Electron microscopy; application of cytological freeze-drying techniques in cytochemistry and electron microscopy. Mailing Add: Biol Sci Group Univ of Conn Storrs CT 06268

MUNDT, JOHN ORVIN, b Wisconsin Rapids, Wis, Oct 6, 12; m 40; c 4. FOOD MICROBIOLOGY. Educ: Univ Wis, BS, 38; Univ Maine, MS, 40; Mich State Univ, PhD(bact), 44. Prof Exp: With qual control, Birdseye Corp, 39-40; lab asst, Mich State Univ, 40-42; asst prof, Hobart Col, 42-47; assoc prof microbiol, 47-56, PROF MICROBIOL, UNIV TENN, KNOXVILLE, 56-, ASSOC, AGR EXP STA, 53- Mem: Am Soc Microbiol; Inst Food Technol; Am Acad Microbiol; fel Am Pub Health Asn; Sigma Xi. Res: Microbiology of foods and dairy products; food plant sanitation; streptococci in frozen and non-sterile foods; water pollution. Mailing Add: Dept of Microbiol Univ of Tenn Knoxville TN 37916

MUNDT, PHILIP A, b Sioux Falls, SDak, Oct 2, 27; m 51; c 3. PETROLEUM GEOLOGY. Educ: SDak Sch Mines & Technol, BS, 51; Washington Univ, MA, 53; Stanford Univ, PhD(geol), 55. Prof Exp: Geologist, Mobil Producing Co, 55-58, subsurface supvr, Mobil Mediterranean Inc, 58-63, staff geologist, Mobil Oil Corp, 63-65, explor supvr, Libya, 65-67, geol supvr, 67-69; explor mgr, US Nat Resources, Inc, 69-72; MGR GEOL & GEOCHEM RES, MOBIL RES DEVELOP CORP, 72- Res: Petroleum exploration. Mailing Add: Mobil Res & Develop Corp PO Box 900 Dallas TX 75221

MUNDY, BELVEY WASHINGTON, b Roanoke, Va, Nov 6, 17. PHYSICAL CHEMISTRY. Educ: Va Mil Inst, BS, 40; Ind Univ, MA, 41, PhD(chem), 48. Prof Exp: Asst chem, Ind Univ, 41-42 & 46-47; from asst prof to assoc prof chem, Va Mil Inst, 48-55; asst prof, 55-56, ASSOC PROF PHYS SCI, UNIV FLA, 56- Concurrent Pos: Ford Found fel, Harvard Unvi, 54-55. Mem: Am Chem Soc. Res: Alkylation; x-ray diffraction; polarography. Mailing Add: 218 SW Tenth St Gainesville FL 32601

MUNDY, BRADFORD PHILIP, b Warrensburg, NY, Nov 9, 38; m 63; c 3. CHEMISTRY. Educ: State Univ NY Albany, BS, 61; Univ Vt, PhD(chem), 65. Prof Exp: Res assoc chem, Univ Calif, Berkeley, 65-66; NIH fel, 66-67; asst prof, 67-71, ASSOC PROF CHEM, MONT STATE UNIV, 71- Mem: Am Chem Soc. Res: Synthesis of natural products; heterocyclic chemistry; biosynthesis. Mailing Add: Dept of Chem Mont State Univ Bozeman MT 59715

MUNDY, RODERICK ASHBY, physical chemistry, organic chemistry, see 12th edition

MUNDY, ROY LEE, b Charlottesville, Va, Mar 4, 22; m 41; c 3. PHARMACOLOGY. Educ: Howard Col, BS, 48; Univ Ala, MS, 50; Univ Va, PhD(pharmacol), 57. Prof Exp: Asst prof pharmacol, Howard Col, 50-51; chief pharmacol dept, Walter Reed Army Inst Res, 55-66; assoc prof, 66-71, PROF PHARMACOL, MED CTR, UNIV ALA, BIRMINGHAM, 71- Mem: Soc Toxicol; Soc Exp Biol & Med; Am Soc Pharmacol & Exp Therapeut. Res: Pharmacology of sulfhydryl radio-protectant chemicals; removal of radiation agents from the animal body; autonomic pharmacology. Mailing Add: Dept of Pharmacol Univ of Ala Med Ctr Birmingham AL 35294

MUNETA, PAUL, b Harlowton, Mont, Apr 21, 31; m 56; c 2. FOOD SCIENCE. Educ: Mont State Col, BS, 53; Cornell Univ, PhD(veg crops), 59. Prof Exp: Asst agr chem, 59-68, asst prof, 61-68, ASSOC PROF FOOD SCI & ASSOC FOOD SCIENTIST, UNIV IDAHO, 68- Mem: Am Chem Soc; Inst Food Technol; Phytochem Soc NAm. Res: Changes in nitrate and nitrite in cured meats. Mailing Add: Food Res Bldg Univ of Idaho Moscow ID 83843

MUNGALL, ALLAN GEORGE, b Vancouver, BC, Mar 12, 28; m 50; c 3. EXPERIMENTAL ATOMIC PHYSICS. Educ: Univ BC, BASc, 49, MASc, 50; McGill Univ, PhD(physics), 54. Prof Exp: Geophysicist, Calif Stand Co, Alta, 50; jr res officer physics, Nat Res Coun Can, 50-52; asst, McGill Univ, 54; from asst res officer to assoc res officer, 54-67, SR RES OFFICER, NAT RES COUN CAN, 67- Mem: Can Asn Physicists; Inst Elec & Electronics Eng. Res: Atomic frequency and time standards. Mailing Add: Physics Div M-36 Nat Res Coun Montreal Rd Ottawa ON Can

MUNGALL, WILLIAM STEWART, b Buffalo, NY, July 24, 45; m 67; c 2. BIO-ORGANIC CHEMISTRY. Educ: State Univ NY Buffalo, BA, 67; Northwestern Univ, PhD(org chem), 70. Prof Exp: Asst prof chem, 71-74, ASSOC PROF CHEM, HOPE COL, 74- Concurrent Pos: Consult, Donnelley Mirrors Inc, 72-75. Mem: Am Chem Soc; Sigma Xi. Res: Synthesis of biologically active organic compounds; phosphoramidate chemistry. Mailing Add: Dept of Chem Hope Col Holland MI 49423

MUNGER, BRYCE LEON, b Everett, Wash, May 20, 33; m 57; c 4. HUMAN ANATOMY, CELL BIOLOGY. Educ: Washington Univ, MD, 58. Prof Exp: Intern path, Johns Hopkins Hosp, 58-59; asst prof anat, Washington Univ, 61-65; assoc prof, Univ Chicago, 65-66; PROF ANAT & CHMN DEPT, HERSHEY MED CTR, PA STATE UNIV, 66- Mem: AAAS; Am Asn Anat; Am Soc Cell Biol; Am Diabetes Asn; NY Acad Sci. Res: Cytology of secretion, especially pancreatic islets and comparative ultrastructure of sensory nerve endings, particularly mechanoreceptors. Mailing Add: Dept of Anat Hershey Med Ctr Pa State Univ Hershey PA 17033

MUNGER, GEORGE DONALD, b Wilmington, Del, May 21, 23; m 50; c 2. PLANT PATHOLOGY. Educ: Univ Del, BS, 48, MS, 49; Ohio State Univ, PhD(bot, plant path), 55. Prof Exp: Tech serv rep, E I du Pont de Nemours & Co, 49-50; sr res biologist, Battelle Mem Inst, 50-54; plant pathologist, B F Goodrich Co, 55-56; plant pathologist, Diamond Shamrock Corp, 56-58; plant pathologist fungicide develop coord, Am Cyanamid Co, 58-66; supvr field res, T R Evans Res Ctr, 66-69, MGR COM DEVELOP, AGR CHEM, DIAMOND SHAMROCK CORP, 69- Mem: Am Phytopath Soc. Res: Agricultural pesticides; field research and development. Mailing Add: Diamond Shamrock Corp Agr Chem Div 1100 Superior Ave Cleveland OH 44114

MUNGER, HENRY MARTIN, b Ames, Iowa, May 10, 16; m 50; c 2. CROP BREEDING. Educ: Cornell Univ, BS, 36, PhD(plant breeding), 41; Ohio State Univ, MS, 37. Prof Exp: From instr to asst prof hort, Univ Wis, 41-42; from asst prof to assoc prof, 42-48, head dept veg crops, 51-66, PROF PLANT BREEDING & VEG CROPS, CORNELL UNIV, 48- Concurrent Pos: Consult, Food & Agr Orgn, UN, Ministry Agr, UAR, 62-63; vis prof, Univ Philippines, 69-70. Mem: Am Soc Hort Sci (pres, 66-67); AAAS; Am Soc Agron; Res: Vegetable breeding; breeding vegetables for disease resistance and use of F1 hybrids. Mailing Add: Dept of Plant Breeding Cornell Univ Ithaca NY 14850

MUNGER, ROBERT SHOOP, b Perry, Okla, Mar 17, 21; m 47; c 2. ANALYTICAL CHEMISTRY. Educ: Okla State Univ, BS, 48, MS, 49. Prof Exp: Jr res chemist, Continental Oil Co, 49-50, assoc res chemist, 51-55, from res chemist to sr res chemist, 55-57; res group leader, Petrol Chem, Inc, 58, head serv sect, 59-60; head anal div, Petrochem Res Lab, Cities Serv Res & Develop Co, 60-63, mgr anal sect, Lake Charles Res Ctr, Columbian Carbon Co, La, 64-67, mgr anal sect, Technol & Planning Div, NJ, 67, MGR ANAL SECT, RES DEPT, PETROCHEM GROUP, CITIES SERV CO, 67- Mem: Am Chem Soc. Res: Infrared and ultraviolet spectrometry; structure of polymers. Mailing Add: Res Dept Petrochem Group Cities Serv Co Drawer 4 Cranbury NJ 08512

MUNI, INDU A, b Amreli, India, Oct 24, 42; m 69. BIOCHEMICAL PHARMACOLOGY. Educ: Univ Nagpur, BS, 64; NDak State Univ, MS, 66; Univ Miss, PhD(biochem, pharmacol), 68. Prof Exp: NIMH res fel, Univ Miss, 68-69; clin biochemist dept path, St Joseph's Hosp, Milwaukee, 69-74; SR RES SCIENTIST, MILES LABS, INC, 74- Mem: Am Asn Clin Chem; Am Chem Soc; Soc Appl Spectros. Res: Drug metabolism; clinical chemistry. Mailing Add: Toxicol Miles Labs Inc Elkhart IN 46514

MUNIER, JOHN HAMMOND, b San Diego, Calif, June 19, 08; m 37; c 2. PHYSICS. Educ: Univ Calif, Los Angeles, AB, 34; Johns Hopkins Univ, PhD(physics), 39. Prof Exp: Res physicist, Corning Glass Works, 39-44, plant mgr, 44-47, sr res physicist, 47-55, mgr gen prod develop lab, 55-61, staff res mgr, Res & Develop, 61-63, mgr tech commun, Tech Staffs Div, 63-73; RETIRED. Mem: AAAS. Res: Physical properties of glass; composite materials. Mailing Add: Tech Staffs Div Corning Glass Works Corning NY 14830

MUNIES, ROBERT, b New York, NY, Oct 19, 35. INDUSTRIAL PHARMACY, RESEARCH ADMINISTRATION. Educ: Columbia Univ, BS, 56; Univ Wis, PhD(pharm), 65. Prof Exp: Sr pharmaceut scientist, Vick Div Res & Develop, Richardson-Merrell, Inc, 65-71, group supvr sci info, 71; MGR REGULATORY AFAIRS, BOEHRINGER INGELHEIM LTD, 71- Mem: Am Pharmaceut Asn; Am Chem Soc; NY Acad Sci. Res: Factors influencing percutaneous absorption through human skin; etiology and treatment of acne; food and drug law; clinical research. Mailing Add: 33 W Tarrytown Rd Elmsford NY 10533

MUNIGLE, JO ANNE, b Los Angeles, Calif, Nov 14, 34. ANATOMY. Educ: Conn Col, BA, 57; Cornell Univ, PhD(anat), 67. Prof Exp: Fel anat, McGill Univ, 66-68; MGR REPRODUCTION & MUTAGENESIS, CIBA-GEIGY CORP, 68- Mem: AAAS; Genetics Soc Am. Res: Developmental biology and genetics. Mailing Add: Dept of Toxicol & Path Ciba-Geigy Corp Ardsley NY 10502

MUNIZ, RAUL A, b Lima, Peru, July 5, 37; m 68; c 2. POULTRY PATHOLOGY. Educ: San Marcos Univ, Lima, BS, 59, DVM, 60; Univ Calif, Davis, MS, 66. Prof Exp: Teaching asst vet path, Sch Vet Med, San Marcos Univ, Lima, 59-60, lectr, 60-61; asst prof animal husb sch, Agrarian Univ Peru, 62-66, assoc prof, 67-68; dir tech serv & res poultry, Merck Sharp & Dohme, Peru, SA, 69-71; res fel develop parasitol, 71-73, REGIONAL DIR, LATIN AM-CARIBBEAN AREA, INT ANIMAL SCI RES, MERCK SHARP & DOHME RES LABS, 73- Mem: Am Vet Med Asn; Latin Am Asn Animal Prod. Res: Coccidiosis; mechanisms of pathogenesis of Eimeria species in chickens; drug sensitivity and evaluation of field strains; drug resistance and immunity; evaluation parameters of new compounds. Mailing Add: Merck Sharp & Dohme Res Labs Rahway NJ 07065

MUNJAL, DEVIDAYAL, Indian citizen. IMMUNOCHEMISTRY. Educ: Poona Univ, India, BS, 63; Panjab Univ, India, BS, 65, MS, 67; State Univ NY Buffalo, PhD(biochem), 71. Prof Exp: Lab asst bot, Delhi Univ, 63-64; asst res officer biochem, Indian Coun Med Res, New Delhi, 67-68; teaching asst, State Univ NY Buffalo, 68-71, res assoc immunochem, Dept Biochem & Ctr Immunol, 71-72; assoc biochem, Dept Med, Sch Med, Harvard Univ, 72-74, prin res assoc immunochem, 74-75; res assoc, Mallory Gastroenterol Lab, Boston City Hosp, 72-75; res assoc, Sch Med, Boston Univ, 74-75; ASST PROF PATH, UNIV KY MED CTR, 75- Concurrent Pos: Fel, Ctr Immunol & Erie County Lab, State Univ NY Buffalo, 71-72. Mem: Am Asn Clin Chemists; Am Fedn Clin Res; Am Chem Soc; Int Soc Toxinology; AAAS. Res: Detection, isolation, purification and immunochemical characterization of tumor-associated macromolecules in blood and tumor tissues of patients with gastrointestinal cancers, especially colonic cancer; routine testing of blood samples from cancer patients for carcinoembryonic antigen using radioimmunoassays. Mailing Add: 2504 Larkin Rd Apt F-188 Lexington KY 40503

MUNK, MINER NELSON, b Napa, Calif, Nov 17, 34; m 55; c 2. APPLIED PHYSICS. Educ: Univ Calif, Berkeley, AB, 57, MA, 59, PhD(physics), 67. Prof Exp: Physicist, Aerojet Gen Corp, Calif, 59-62; sr physicist, Varian Aerograph, 67-75; PHYSICIST, MILTON ROY CO, 75- Mem: Sigma Xi. Res: Physical methods of analysis; mass spectroscopy; liquid chromatography. Mailing Add: Milton Roy Co 5000 Park St N St Petersburg FL 33733

MUNK, PETR, b Praha, Czech, Oct 31, 32; m 61; c 2. POLYMER CHEMISTRY. Educ: Col Chem Technol, Prague, Czech, MS, 56; Inst Macromolecular Chem, Czech Acad Sci, PhD(phys chem macromolecules), 60, DSc(phys chem macrumolecules), 67. Prof Exp: Head dept molecular hydrodyn, Inst Macromolecular Chem, Czech Acad Sci, 56-67; res scientist, Res Triangle Inst, NC, 68; vis assoc prof chem, 69-71, asst prof, 71-72, ASSOC PROF CHEM, UNIV TEX, AUSTIN, 72- Mem: Am Chem Soc. Res: Streaming birefringence; viscometry; sedimentation analysis; light scattering; interactions among macromolecules; interactions between macromolecules and solvent. Mailing Add: Dept of Chem Univ of Tex Austin TX 78712

MUNK, VLADIMIR, b Pardubice, Czech, Feb 27, 25; m 50; c 2. MICROBIOLOGY, BIOCHEMISTRY. Educ: Prague Tech Univ, MS, 50, PhD(biochem), 55. Prof Exp: Chief anal dept, Cent Res Inst Food Indust, 53-57, chief microbiol dept, 58-63; sr microbiologist, Inst Microbiol, Czech Acad Sci, 63-69; PROF BIOL, STATE UNIV NY COL PLATTSBURGH, 69- Concurrent Pos: Univ fel & grant-in-aid, 70-71, Chase Chem Co fel, 71-72. Honors & Awards: State Prize, Govt Czech Socialistic Repub, 68. Mem: AAAS; Am Soc Microbiol. Res: Industrial use of microorganisms; microbial production of vitamins and enzymes; fermentation of hydrocarbons; continuous cultivation of microorganisms. Mailing Add: Fac of Sci State Univ of NY Col Plattsburgh NY 12901

MUNK, WALTER HEINRICH, b Vienna, Austria, Oct 19, 17; nat US; m 53; c 2. PHYSICAL OCEANOGRAPHY. Educ: Calif Inst Technol, BS, 39; Univ Calif, MS, 40; PhD(oceanog), 47. Hon Degrees: Doctor Philosophiae Honoris Causa, Univ Bergen, Norway, 75. Prof Exp: From asst prof to assoc prof geophys, 47-54, PROF GEOPHYS, UNIV CALIF, SAN DIEGO, 54-, DIR LA JOLLA UNIT & ASSOC DIR, INST GEOPHYS, 59- Concurrent Pos: Guggenheim Found fel, Univ Oslo, 48, Cambridge Univ, 55 & 62; mem, Oceanog Adv Comt. Honors & Awards: Arthur L Day medal, Am Geol Soc; Sverdrup Gold Medal; Gold Medal, Royal Astron Soc; co-recipient, Award Ocean Sci & Eng; Josiah Willard Gibbs Lectr. Mem: Nat Acad Sci; Am Acad Arts & Sci; Am Philos Soc; Am Geophys Union; Leopoldina Ger Acad Res Natural Sci. Res: Ocean waves; tides; wind stress and ocean currents; rotation of the earth; ocean acoustics. Mailing Add: Inst of Geophys & Planetary Sci Univ of Calif San Diego La Jolla CA 92093

MUNKACSI, ISTVAN, b Budapest, Hungary, Apr 15, 27; m 54; c 1. ANATOMY. Educ: Med Univ Budapest, Dr med, 53; Univ Khartoum, Dr philos, 64. Prof Exp: Lectr anat, Med Univ Budapest, 53-57, sr lectr, 57-60; sr lectr, Univ Khartoum, 60-64; sr lectr, Med Univ Budapest, 64-66; vis asst prof, 66-68, assoc prof, 68-74, PROF ANAT, UNIV SASK, 74- Mem: Anat Soc Gt Brit & Ireland; Can Asn Anat; Int Soc Nephrology; Pan Am Asn Anat (secy gen, 72-75). Res: Comparative morphology; innervation of the kidney and distribution of the type of nephrons investigated in desert and laboratory mammals; blood vessels and arterio-venous connections of the kidney; dynamics of synapse formation in the sympathetic nervous tissue in tissue culture. Mailing Add: Dept of Anat Univ of Sask Saskatoon SK Can

MUNKRES, JAMES RAYMOND, b Omaha, Nebr, Aug 18, 30; m 64. MATHEMATICS. Educ: Nebr Wesleyan Univ, AB, 51; Univ Mich, AM, 52, PhD(math), 56. Prof Exp: Instr math, Univ Mich, 55-57; instr, Princeton Univ, 57-58, Fine instr, 58-60; from asst prof to assoc prof, 60-66, PROF MATH, MASS INST TECHNOL, 66- Concurrent Pos: Sloan res fel, 65-67. Mem: Am Math Soc; Math Asn Am. Res: Differential and combinatorial topology. Mailing Add: 2-242 Mass Inst of Technol Cambridge MA 02139

MUNKRES, KENNETH DEAN, genetics, biochemistry, see 12th edition

MUNN, GEORGE EDWARD, b Lawrence Co, Pa, Nov 29, 24. ORGANIC CHEMISTRY. Educ: Westminster Col, Pa, BS, 45; Univ Ill, PhD(org chem), 48. Prof Exp: Asst, Univ Ill, 45-47; Du Pont fel, Mass Inst Technol, 48-49; CHEMIST, E I DU PONT DE NEMOURS & CO, INC, 49- Mem: Am Chem Soc. Res: Synthetic organic chemistry; polymer chemistry. Mailing Add: Plastics Dept E I du Pont de Nemours & Co Inc Wilmington DE 19898

MUNN, JOHN IRVIN, b Pittsburgh, Pa, Oct 28, 22; m 48; c 3. PHARMACOLOGY, BIOCHEMISTRY. Educ: Ind Univ Pa, BS, 48; George Washington Univ, MS, 52; Georgetown Univ, PhD(biochem), 57. Prof Exp: Res biochemist, US Naval Med Res Inst, 48-52; res hematologist, Walter Reed Army Inst Res, 52-58; biochemist, Food & Drug Admin, 58-61; scientist adminr-pharmacologist, NIH, Md, 61-71; SR SCIENTIST, WHO, 71- Honors & Awards: Superior Performance Award, Walter Reed Army Inst Res, 57. Mem: AAAS; Am Soc Pharmacol & Exp Therapeut. Res: Metabolism of gallium; biochemical methodology; hemoglobin; fatty acids and surface areas of erythrocytes; toxicity of emulsifiers in foods. Mailing Add: Food Additives Safety Unit Div of Environ Health WHO 1211 Geneva 27 Switzerland

MUNN, NANCY D, b New York, NY, Apr 13, 31. ANTHROPOLOGY. Educ: Univ Okla, BA, 51; Ind Univ, MA, 55; Australian Nat Univ, PhD, 61. Prof Exp: Asst prof anthrop, Bennington Col, 61-63; asst prof, 66-71, ASSOC PROF ANTHROP, UNIV MASS, AMHERST, 71- Concurrent Pos: Soc Sci Res Coun res grant, 60-61; vis fel, Inst Advan Study, 72-73. Mem: Fel Am Anthrop Asn; fel Anthrop Inst Gt Brit & Ireland. Res: Australian aborigines; problems in comparative art; symbol theory; New Guinea; social theory. Mailing Add: Dept of Anthrop Univ of Mass Amherst MA 01002

MUNN, ROBERT EDWARD, b Winnipeg, Man, July 26, 19; m 44; c 4. METEOROLOGY. Educ: McMaster Univ, BA, 41; Univ Toronto, MA, 45; Univ Mich, PhD(meteorol), 62. Prof Exp: METEOROLOGIST, METEOROL SERV CAN, 41- Concurrent Pos: Vis prof, Univ Stockholm, 70; mem comn environ monitoring, Int Coun Sci Unions/Spec Comt Problems Environ. Mem: Fel Am Meteorol Soc; Fel Royal Meteorol Soc. Res: Micrometeorology; regional air pollution; environmental monitoring. Mailing Add: Atmospheric Envir Serv 4905 Dufferin St Downsview ON Can

MUNN, ROBERT JAMES, b Southampton, Eng, Jan 31, 37; m 62; c 2. CHEMICAL PHYSICS. Educ: Bristol Univ, BSc, 57, PhD(chem), 61. Prof Exp: Res fel chem, Bristol Univ, 61-63; Harkness fel of Commonwealth Fund, Univ Md, 63-64; lectr chem, Bristol Univ, 64-65; lectr math, Queen's Univ, Belfast, 65-66; from asst prof to assoc prof chem physics, 66-72, PROF CHEM, UNIV MD, COLLEGE PARK, 72-, DIR, 70-, ASST DEAN, 75- Concurrent Pos: Consult, Lockheed Electronics, 70-71 & NSF, 70- Mem: AAAS; Am Phys Soc. Res: Scattering; educational technology; computer aided instruction. Mailing Add: Dept of Chem Univ of Md College Park MD 20742

MUNNECKE, DONALD EDWIN, b St Paul, Minn, May 30, 20; m 42; c 4. PHYTOPATHOLOGY. Educ: Univ Minn, BA, 42, MS, 49, PhD(plant path), 50. Prof Exp: Instr & jr plant pathologist, Univ Calif, Los Angeles, 51-53, from asst prof & asst plant pathologist to assoc prof & assoc plant pathologist, 53-61; assoc prof & assoc plant pathologist, 61-65, PROF & PLANT PATHOLOGIST, UNIV CALIF, RIVERSIDE, 65- Concurrent Pos: Guggenheim fel & Fulbright res scholar, Univ Göttingen, 65-66. Mem: Am Phytopath Soc; Bot Soc Am. Res: Ornamental plant diseases; chemical soil treatments for disease control; fungicide action in soils; ecological relations in control of Armillaria mellea. Mailing Add: Dept of Plant Path Univ of Calif Riverside CA 92502

MUNNELL, EQUINN W, b Sayville, NY, June 28, 13; m 37; c 3. MEDICINE. Educ: Amherst Col, BA, 35; Cornell Univ, MD, 39; Am Bd Obstet & Gynec, dipl, 48. Prof Exp: Asst instr obstet & gynec, Col Med, NY Univ, 45-47; from asst prof to assoc prof, 50-59, PROF OBSTET & GYNEC, COL PHYSICIANS & SURGEONS, COLUMBIA UNIV, 69- Concurrent Pos: Asst gynecologist, Mem Hosp, New York, 45-47; asst obstetrician & gynecologist, Presby Hosp, New York, 47-57, assoc obstetrician & gynecologist, 57-69, attend obstetrician & gynecologist, 69-; assoc attend gynecologist, Francis Delafield Hosp, 53-62, attend gynecologist, 62- Mem: Fel Am Col Surg; fel Am Col Obstet & Gynec; Am Asn Obstet & Gynec; Am Cancer Soc. Res: Obstetrics and gynecology; gynecologic cancer. Mailing Add: 842 Park Ave New York NY 10021

MUNNS, DONALD NEVILLE, b Sydney, Australia, Sept 6, 31; m 60; c 3. PLANT NUTRITION, SOIL SCIENCE. Educ: Univ Sydney, BScAgr, 54; Univ Calif, Berkeley, PhD(soil sci), 61. Prof Exp: Asst chemist, NSW Dept Agr, Australia, 54-57; res asst, Univ Calif, Berkeley, 57-60; from res officer to sr res officer, Commonwealth Sci & Indust Res Orgn, Canberra, 60-66; asst chemist, 66-68, ASSOC PROF SOILS & PLANT NUTRIT, UNIV CALIF, DAVIS, 68- Concurrent Pos: Vis soil scientist, Univ Hawaii, 73-74. Mem: AAAS; Soil Sci Soc Am; Am Soc Agron. Res: Plant and soil analysis; soil phosphate, acidity, and legume growth; nodulation and nitrogen fixation; salinity; varietal variations in plant response to substrate factors. Mailing Add: Dept of Soils & Plant Nutrit Univ of Calif Davis CA 95616

MUNNS, THEODORE WILLARD, b Peoria, Ill, June 11, 41. BIOCHEMISTRY. Educ: Bradley Univ, BS, 63; St Louis Univ, PhD(biochem), 70. Prof Exp: Res assoc biochem, 70-74, INSTR BIOCHEM, ST LOUIS UNIV, 74- Concurrent Pos: NSF fel, St Louis Univ, 70-71, NIH fel, 72- Mem: Am Chem Soc; Sigma Xi; NY Acad Sci. Res: Nucleic acid and protein metabolism in neoplastic systems. Mailing Add: 1402 S Grand Blvd St Louis MO 63104

MUNOZ, JAMES LOOMIS, b East Orange, NJ, Oct 31, 39. GEOCHEMISTRY. Educ: Princeton Univ, AB, 61; Johns Hopkins Univ, PhD(geol), 66. Prof Exp: Fel, Carnegie Inst Geophys Lab, 66-68; asst prof geol, 68-74, ASSOC PROF GEOL, UNIV COLO, BOULDER, 74- Mem: AAAS; Am Geophys Union; Mineral Soc Am; Geochem Soc. Res: Application of thermodynamics to petrology; geochemistry of fluorine in igneous, metamorphic, and ore-forming processes. Mailing Add: Dept of Geol Sci Univ of Colo Boulder CO 80302

MUNOZ, JOHN JOAQUIN, b Guatemala, Dec 23, 18; nat US; m 47; c 4. IMMUNOLOGY, MICROBIOLOGY. Educ: La State Univ, BS, 42; Univ Ky, MS, 45; Univ Wis, PhD(med bact), 47; Am Bd Microbiol, dipl. Prof Exp: Jr & sr technician bact, Univ Ky, 42-44; asst microbiol, Univ Wis, 44-47; asst prof med bact, Sch Med, Univ Ill, 47-51; res assoc, Merck Sharp & Dohme Res Labs, 51-57; prof microbiol & pub health, chmn dept & dir Stella Duncan Mem Labs, Mont State Univ, 57-61; RES MICROBIOLOGIST, ROCKY MOUNTAIN LAB, NAT INST ALLERGY & INFECTIOUS DIS, 61- Concurrent Pos: Spec assignment, Pasteur Inst, Paris, 66-67. Mem: AAAS; fel Am Acad Microbiol; Am Soc Microbiol; Am Asn Immunol; Soc Exp Biol & Med. Res: Pneumococcal immunity; serology of Pseudomonas genus; precipitin reaction in agar; complexity of antigens; standardization of bacterin; effects of Bordetella pertussis on mice; mouse anaphylaxis and antibody; mechanism of anaphylaxis; serological techniques. Mailing Add: Rocky Mountain Lab Nat Inst Allergy & Infect Dis Hamilton MT 59840

MUNRO, DAVID (AIRD), b Victoria, BC, May 25, 23; m 43; c 4. BIOLOGY. Educ: Univ BC, BA, 47; Univ Toronto, PhD(zool, geog), 56. Prof Exp: Wildlife officer, Can Wildlife Serv, 48-53, chief ornithologist, 53-62, staff specialist, 62-63, chief, 64-66, dir, 66-68; dir community affairs br, Dept Indian Affairs & Northern Develop, 68-69, asst dept minister, 69-70; dir prog coord, Fisheries Res Bd Can, 70, dep chmn policy & planning, 70-71, asst dir gen res & develop, Fisheries Serv, 71, DIR GEN, LIAISON & COORD DIRECTORATE, DEPT ENVIRON, 71- Concurrent Pos: Lectr, Univ Ottawa, Ont, 57- Mem: Wildlife Soc. Res: Migratory bird research and administration; land use in relation to wildlife management and recreation. Mailing Add: Liaison & Coord Directorate Dept of the Environment Ottawa ON Can

MUNRO, DONALD W, JR, b Narberth, Pa, Dec 27, 37; m 61; c 2. PHYSIOLOGY, GENETICS. Educ: Wheaton Col, BS, 59; Pa State Univ, MS, 63, PhD(zool), 66. Prof Exp: Assoc prof zool, 66-71, PROF BIOL & HEAD DEPT, HOUGHTON COL, 71- Mem: Am Soc Zool; Am Physiol Soc; Am Inst Biol Sci; NY Acad Sci; Genetics Soc Am. Res: Effects of cold exposure and hibernation on the respiration rates and oxidative phosphorylation of various tissues of carp, trout, frogs, rats, hamsters and chipmunks. Mailing Add: Dept of Biol Houghton Col Houghton NY 14744

MUNRO, DOUGLAS CARLYLE, b Renown, Sask, Jan 19, 35; m 62; c 2. SOIL FERTILITY, PLANT NUTRITION. Educ: Univ Sask, BSA, 57, MSc, 59. Prof Exp: RES OFFICER SOIL FERTIL, RES BR, CAN DEPT AGR, 59- Mem: Soil Sci Soc Am; Am Soc Agron; Agr Inst Can; Can Soc Soil Sci; Prof Inst Pub Serv Can. Res: Fertility of podzol soils with emphasis on nitrogen and phosphorous availability; nitrogen, phosphorus and potassium nutrition of cole crops and potatoes. Mailing Add: Charlotte Res Sta Can Dept of Agr Box 1210 Charlottetown PE Can

MUNRO, HAMISH N, b Edinburgh, Scotland, July 3, 15; m 46; c 4. BIOCHEMISTRY, NUTRITION. Educ: Glasgow Univ, BSc, 36, MB, 39, DSc(biochem), 56. Prof Exp: Clin tutor med, Victoria Infirmary, Glasgow, Scotland, 40-45, asst dir path, 42-45; lectr physiol, Glasgow Univ, 45-47, sr lectr biochem, 47-56, reader nutrit biochem, 56-64, prof biochem, 64-66; PROF PHYSIOL CHEM, MASS INST TECHNOL, 66- Concurrent Pos: Rockefeller traveling fel, 48; Fleck lectr, Glasgow Univ, 60; mem protein requirements comt, WHO-Food & Agr Orgn, 63; chmn food & nutrit bd, US Nat Res Coun, 75- Honors & Awards: Osborne & Mendel Award, Am Inst Nutrit, 68. Mem: Nat Acad Sci; Am Inst Nutrit; fel Royal Soc Edinburgh; fel, Brit Inst Biol; Brit Biochem Soc. Res: Mammalian protein metabolism, notably nutritional aspects; protein synthesis control mechanisms and actions of hormones on protein and RNA metabolism; tissue analysis of mammals. Mailing Add: Dept of Nutrit & Food Sci Mass Inst of Technol Cambridge MA 02139

MUNRO, HOWARD EVERETT, organic chemistry, see 12th edition

MUNRO, JAMES, b Newcastle-on-Tyne, Eng, Aug 29, 11; m 42; c 1. PLANT PATHOLOGY. Educ: Univ Durham, BSc, 41, MSc, 50. Prof Exp: Lectr, King's Col, Univ Durham, 41-46; plant virus pathologist, Nat Inst Agr Bot, Eng, 46-51; pathologist, 51-68, CHIEF CROP CERTIFICATION, PLANT PROTECTION DIV, CAN DEPT AGR, 68- Mem: Am Phytopath Soc; Potato Asn Am (pres, 69-70); fel NY Acad Sci; Can Phytopath Soc. Res: Potato viruses in relation to forms of resistance and immunity in Solanum tuberosum and other Solanum species. Mailing Add: Plant Protection Div Can Dept of Agr Sir John Carling Bldg Cent Exp Farm Ottawa ON Can

MUNRO, WILLIAM DELMAR, b Cedaredge, Colo, Nov 22, 16; m 51; c 3. NUMERCIAL ANALYSIS. Educ: Univ Colo, BA, 38; Univ Minn, MA, 40, PhD(math), 47. Prof Exp: Asst math, 38-41, instr, 41-43, asst prof math & mech, 45-49, from assoc prof to prof math, 49-69, PROF COMPUT, INFO & CONTROL SCI & ASSOC HEAD DEPT, UNIV MINN, MINNEAPOLIS, 69- Concurrent Pos: Proj engr, Honeywell Inc, 43-; consult, Radio Corp Am, 45-, Maico Corp, 58-, Viron Div, GCA Corp, 65- & North Star Res Corp, 70-; vis res mathematician, Univ Calif, Los Angeles, 57-58; vis prof, Johns Hopkins Univ, 59-60. Mem: Am Math Soc; Math Asn Am; Soc Indust & Appl Math; Asn Comput Mach. Res: Theory of approximation; computers; numerical methods and analysis research; navigation computer for aircraft; orthogonal trigonometric sums with auxiliary conditions; high order precision by exact arithmetic. Mailing Add: Dept Comput Info & Control Sci 120 Lind Hall Univ of Minn Minneapolis MN 55455

MUNROE, EUGENE GORDON, b Detroit, Mich, Sept 8, 19; nat Can; m 44; c 4. ENTOMOLOGY, ECOLOGY. Educ: McGill Univ, BSc, 40, MSc, 41; Cornell Univ, PhD(entom), 48. Prof Exp: Lectr & res asst, Inst Parasitol, Macdonald Col, McGill Univ, 46-50; agr res officer, Can Dept Agr, 50-65; sci adv, Sci Secretariat, Off Privy Counr, 65-67, head studies, 67-68; RES SCIENTIST, BIOSYST RES INST, CAN DEPT AGR, 68- Concurrent Pos: Vis lectr, Univ Calif, Berkeley, 60-61; mem, Steering Comt, Biol Coun Can, 65-66; res assoc, Entom Res Inst, Can Dept Agr, 65-68. Mem: AAAS; hon mem Lepidop Soc (pres, 59-60); Entom Soc Can (pres, 63-64); Soc Syst Zool; Royal Entom Soc London. Res: Science policy; research planning and management; management and conservation of renewable resources; ecology; biogeography; taxonomy. Mailing Add: Entom Res Inst Can Dept of Agr Ottawa ON Can

MUNROE, JOSCELYN SPENCER, b Jamaica, WI, June 10, 18; nat US; m 45; c 1. CANCER. Educ: NY Univ, BA; Univ Toronto, PhD(endocrinol), 54, MD, 55. Prof

Exp: Asst Sloan-Kettering Inst Cancer Res, 56-64; asst prof, Inst Phys Med & Rehab, NY Univ, 65-70; assoc dir clin pharmacol, 67-73, ASSOC DIR CLIN RES, CLIN ONCOL SECT, SCHERING CORP, 73-; RES ASSOC PROF REHAB MED, MED CTR, NY UNIV, 67- Concurrent Pos: Consult, Nat Inst Neurol Dis & Blindness, 63-65; exp pathologist, Lenox Hill Hosp, 64-67, head sect primate oncogenesis & immunol, Dept Path, 64-65. Mem: AAAS; Am Soc Exp Path; Am Asn Cancer Res; AMA; Reticuloendothelial Soc. Res: Clinical oncological studies in new drug development; viral oncogenesis; experimental primate oncogenesis; immunology as related to cancer, endocrinology and metabolism; congenital brain defects. Mailing Add: Schering Corp 60 Orange St Bloomfield NJ 07003

MUNROE, MARIAN HALL, b Albany, NY, June 15, 38. BOTANY. Educ: Stanford Univ, BA, 60; Yale Univ, MS, 61, PhD(biol), 66. Prof Exp: NSF fel & asst bot, Univ Col, London Univ, 65-66; instr bot, 66-69, asst prof biol, 69-73, RESEARCHER BIOL, UNIV CHICAGO, 73- Mem: Bot Soc Am; Int Soc Plant Morphol; Am Soc Plant Physiol. Res: Plant growth and development; ferm organ culture; electron microscopy; genetics. Mailing Add: Dept of Biol Univ of Chicago Chicago IL 60637

MUNROE, MARSHALL EVANS, b Gainesville, Ga, 18; m 47; c 1. MATHEMATICS. Educ: Univ Tex, BA, 40; Brown Univ, ScM, 41, PhD(math), 45. Prof Exp: Instr math, Brown Univ, 43-45; from instr to assoc prof, Univ Ill, 45-58; PROF MATH & CHMN DEPT, UNIV NH, 59- Concurrent Pos: Vis prof, Cairo Univ, 65-66. Mem: Am Math Soc; Math Asn Am. Res: Abstract integration theory; measure theory; modernization of calculus. Mailing Add: Dept of Math Univ of NH Durham NH 03824

MUNSEE, JACK HOWARD, b Niagra Falls, NY, Sept 27, 34; m 62; c 4. PHYSICS. Educ: Col Wooster, BA, 56; Case Western Reserve Univ, MS, 62, PhD(physics), 68. Prof Exp: Jr engr, Aeroprod Opers, 56-57; asst physics, Case Western Reserve Univ, 58-62, part-time instr, 64-68; instr physics & math, Col Wooster, 62-64; asst prof, 68-71, ASSOC PROF PHYSICS, CALIF STATE UNIV, LONG BEACH, 71- Mem: Am Phys Soc; Am Asn Physics Teachers. Res: Low energy nuclear physics; neutrinos; physics education. Mailing Add: Dept of Physics Calif State Univ Long Beach Long Beach CA 90801

MUNSELL, MONROE WALLWORK, b New London, Conn, Jan 8, 25; m 54; c 2. ORGANIC CHEMISTRY, PETROLEUM TECHNOLOGY. Educ: Carnegie-Mellon Univ, BS, 47, MS, 50, PhD, 55. Prof Exp: Res chemist, Esso Res & Eng Co, 55-64, sr chemist & proj leader, 64-68, res assoc, Esso Kagaku kk, Japan, 68-72, RES ASSOC, ESSO RES & ENG CO, 72- Mem: Am Chem Soc; Sigma Xi; Soc Automotive Eng. Res: Lubrication oil and fuel additives; technical service and product application in field of detergents; viscosity improvers and wax crystal modifiers. Mailing Add: 180 Sutton Dr Berkeley Heights NJ 07922

MUNSICK, ROBERT ALLIOT, b Glen Ridge, NJ, Oct 21, 28; m 53, 65; c 3. OBSTETRICS & GYNECOLOGY. Educ: Cornell Univ, AB, 50; Columbia Univ, MD, 54, PhD(pharmacol), 62; Am Bd Obstet & Gynec, dipl, 66. Prof Exp: Intern med, Roosevelt Hosp, New York, 54-55; resident obstet & gynec, Sloane Hosp Women, 59-62; asst prof, Sch Med, Univ Colo, 62-65; prof, Sch Med, Univ NMex, 65-74, chmn dept, 64-73; PROF OBSTET & GYNEC, SCH MED, IND UNIV, INDIANAPOLIS, 74- Concurrent Pos: Josiah Macy, Jr Found fel, 57-62; NIH res grant, 62- Mem: Soc Gynec Invest; fel Am Col Obstet & Gynec; Am Gynec Soc; Am Asn Obstetricians & Gynecologists. Res: Comparative aspects of neurohypophyseal endocrinology; physiology of the chorioamniotic membranes. Mailing Add: Dept of Obstet & Gynec Ind Univ Sch of Med Indianapolis IN 46202

MUNSON, ARVID W, b Paterson, NJ, Aug 22, 33; m 54; c 2. AGRICULTURAL STATISTICS. Educ: Iowa State Univ, BS, 55, MS, 57; Okla State Univ, PhD(animal breeding), 66. Prof Exp: Anal statistician, Biomet Serv Staff, Agr Res Serv, USDA, 65-66, biometrician, 66-67, asst dir, 67-68, actg dir data processing, Data Syst Appln Div, 68; DIR RES SERV DIV, RALSTON PURINA CO, 68- Mem: AAAS; Am Soc Animal Sic; Am Statist Asn; Am Mgt Asn; Inst Food Technologists. Res: Statistical methodology and computer techniques in animal science research and improvement. Mailing Add: Res Serv Div Ralston Purina Co 835 S Eighth St Louis MO 63188

MUNSON, BENJAMIN RAY, b Tonawanda, NY, Sept 19, 37; m 63; c 2. BIOCHEMISTRY, ONCOLOGY. Educ: Houghton Col, BA, 60; State Univ NY Buffalo, PhD(biochem, pharmacol), 68. Prof Exp: Teacher, NY High Schs, 60-62; SR CANCER RES SCIENTIST, SPRINGVILLE LABS, ROSWELL PARK MEM INST, 70- Concurrent Pos: NIH res fel biochem, Springville Labs, Roswell Park Mem Inst, 68-70, fel physiol, Inst, 70-, Nat Cancer Inst res grant, 71- Mem: AAAS; Am Chem Soc; Int Soc Biochem Pharmacol; NY Acad Sci. Res: Viral oncology; biochemical oncology; biochemical pharmacology. Mailing Add: Dept Exp Biol Roswell Park Mem Inst 666 Elm St Buffalo NY 14263

MUNSON, BURNABY, b Wharton, Tex, Mar 20, 33. PHYSICAL CHEMISTRY, ANALYTICAL CHEMISTRY. Educ: Univ Tex, BA, 54, MA, 56, PhD(phys chem), 59. Prof Exp: Res chemist, Humble Oil & Refining Co, 59-62, sr res chemist, 62-64, sr res chemist, Esso Res & Eng Co, 64-66, res specialist, Tex, 66-67; assoc prof chem, 67-72, PROF CHEM, UNIV DEL, 72- Mem: AAAS; Am Chem Soc; Am Inst Chem; Am Soc Mass Spectros; Sigma Xi. Res: Kinetics; mass spectrometry; reactions of gaseous ions and excited species. Mailing Add: Dept of Chem Univ of Del Newark DE 19711

MUNSON, DONALD ALBERT, b New York, NY, May 13, 41; m 71. ZOOLOGY, PARASITOLOGY. Educ: Colgate Univ, AB, 63; Adelphi Univ, MS, 66; Univ NH, PhD(zool), 70. Prof Exp: Teaching asst zool, Univ NH, 66-68, teaching fel, 68-70; NIH fel parasitol, Med Sch, Tulane Univ, 70-72; ASST PROF BIOL, HOOD COL, 72- Mem: Am Soc Parasitol; Am Soc Trop Med & Hyg; Wildlife Dis Asn. Res: Histological, histochemical and histopathological studies of medically important parasites; host-parasite relations of fish parasites. Mailing Add: Dept of Biol Hood Col Frederick MD 21701

MUNSON, EDWIN STERLING, b Akron, Ohio, Dec 29, 33. ANESTHESIOLOGY. Educ: Univ Tenn, MD, 57. Prof Exp: Intern, State Univ Iowa, 57-58; from clin instr to asst clin prof anesthesia, Univ Calif, San Francisco, 63-65; asst prof, Univ Va, Charlottesville, 65-67; asst prof, Univ Calif, Davis, 67-69, assoc prof anesthesia & guest investr, Nat Ctr Primate Biol, 69-71; PROF ANESTHESIOL & DIR RES TRAINING, COL MED, UNIV FLA, 71- Concurrent Pos: NIH trainee anesthesia, Univ Calif, San Francisco, 64; assoc ed, Anesthesia & Analgesia, 72- Mem: Am Soc Anesthesiologists; Int Anesthesia Res Soc; fel Am Col Anesthesiologists; Asn Univ Anesthetists. Res: Pharmacology of anesthetic drugs. Mailing Add: Dept of Anesthesiol Univ of Fla Col of Med Box J-254 Gainesville FL 32610

MUNSON, H RANDALL, JR, b Washington, DC, Sept 2, 34. ORGANIC CHEMISTRY. Educ: Univ Md, College Park, BS, 58; Georgetown Univ, PhD(org chem), 69. Prof Exp: Res asst biochem pharmacol, Med Sch, Georgetown Univ, 62-64; Walter Reed Army Inst Res fel, Univ Va, 68-69; SR RES CHEMIST, A H ROBINS CO, 69- Mem: Am Chem Soc. Res: Medicinal chemistry; synthesis of antimalarials, antivirals and nucleosides; quantitative structure-activity relationships. Mailing Add: A H Robins Co Chem Res Dept 1211 Sherwood Ave Richmond VA 23220

MUNSON, JAMES WILLIAM, b Perrysburg, Ohio, Aug 13, 43; m 66; c 2. ANALYTICAL CHEMISTRY. Educ: Ohio State Univ, BS, 67; Univ Wis, MS, 69, PhD(pharmaceut), 71. Prof Exp: Asst prof pharm, Univ Conn, 71-73; ASST PROF PHARM, UNIV KY, 73- Mem: Am Pharmaceut Asn; Acad Pharmaceut Sci. Res: Analysis of drugs in biological fluids. Mailing Add: Col of Pharm Univ of Ky Lexington KY 40506

MUNSON, JOHN BACON, b Clifton Springs, NY, Nov 15, 32; m 59; c 3. NEUROSCIENCES. Educ: Union Col, NY, AB, 57; Univ Rochester, PhD(neurobiol), 65. Prof Exp: Fel neurophysiol, Inst Physiol, Univ Pisa, 65-66; res assoc physiol, 66-68, from instr to asst prof physiol & psychol, 69-73, asst prof neurosci, 71-73, ASSOC PROF NEUROSCI, PHYSIOL & PSYCHOL, COL MED, UNIV FLA, 73- Concurrent Pos: Vis scientist, Duke Med Ctr, 73 & Neurol Sci Inst, Portland, Ore, 75. Mem: AAAS; Am Physiol Soc; Soc Neurosci; Int Brain Res Orgn; Asn Res Vision & Ophthal. Res: Disfunction and recovery of function in damaged spinal motoneurons; spinal cord damage; effects and recovery; central correlates of eye movements. Mailing Add: Dept of Neurosci Univ of Fla Col of Med Gainesville FL 32610

MUNSON, JOHN HERBERT, computer science, nuclear physics, deceased

MUNSON, PAUL LEWIS, b Washta, Iowa, Aug 21, 10; m 31; c 1; m 48; c 2. PHARMACOLOGY. Educ: Antioch Col, BA, 33; Univ Wis, MA, 37; Univ Chicago, PhD(biochem), 42. Hon Degrees: MA, Harvard Univ, 55. Prof Exp: Asst biochem, Univ Chicago, 39-42; res biochemist, Wm S Merrell Co, Ohio, 42-43; res biochemist & head endocrinol res, Armour & Co, Ill, 43-48; from res asst to res assoc pharmacol, Yale Univ, 48-50; from asst prof to prof, Sch Dent Med, Harvard Univ, 50-65; prof pharmacol, 65-76, SARAH GRAHAM KENAN PROF PHARMACOL & ENDOCRINOL, SCH MED, UNIV NC, CHAPEL HILL, 70-, CHMN DEPT PHARMACOL, 65- Concurrent Pos: Mem, Corticotropin Assay Study Panel, US Pharmacopoeia, 51-55; tutor, Harvard Univ, 55-58, lectr, 65-66; Claude Bernard vis prof, Univ Montreal, 64; mem, Gen Med B Study Sect, USPHS, 66-70, chmn, 69-70; sr adv comn, Laurentian Hormone Conf, 66-; mem pharmacol test comt, Nat Bd Med Exam, 67-71; ed, Vitamins & Hormones, 68-; mem pharmacol-toxicol prog comt, Nat Inst Gen Med Sci, 72-76. Mem: Am Soc Pharmacol & Exp Therapeut (secy-treas, 71-72); Am Chem Soc; Endocrine Soc; Biomet Soc; Asn Med Sch Pharmacol (secy, 72-73, pres, 74-76). Res: Isolation, bioassay and mechanism of action of hormones, especially hypothalamic, pituitary, androgenic, parathyroid and thyrocalcitonin; steroid metabolism; calcium metabolism; mechanism of stimulation of adrenocorticotropic hormone secretion. Mailing Add: Dept of Pharmacol Univ of NC Sch of Med Chapel Hill NC 27514

MUNSON, ROBERT DEAN, b Stockport, Iowa, Mar 14, 27; m 50; c 3. SOIL FERTILITY, PLANT NUTRITION. Educ: Univ Minn, BS, 51; Iowa State Univ, MS, 54, PhD(soil fertil, agr econ), 57. Prof Exp: Instr high sch, Minn, 51-52; agr economist, Tenn Valley Authority, 57-58; agronomist, 58-67, MIDWEST DIR, POTASH INST N AM, INC, 67- Concurrent Pos: Mem panel fertilizer use res needs, NSF, 75. Mem: Fel AAAS; fel Am Soc Agron (assoc ed, J Agron Educ, 73); Soil Sci Soc Am; Int Soil Sci Soc; Crop Sci Soc Am. Res: Potassium availability, movement in soils and physiological function in plants and animals; cation balance and interrelationship of plant nutrients; diagnostic techniques in crop production; soil and plant analysis; availability of soil and fertilizer nitrogen as affected by residues; economics of fertilizer use and research methodology. Mailing Add: Potash Inst of NAm Inc 2147 Doswell Ave St Paul MN 55108

MUNSON, RONALD ALFRED, b Lancaster, Pa, Aug 12, 33; m 67; c 2. PHYSICAL CHEMISTRY. Educ: Franklin & Marshall Col, BS, 55; Northwestern Univ, PhD(phys chem), 59. Prof Exp: NSF fel, Max-Planck-Inst Phys Chem, Göttingen, Ger, 58-59; phys chemist, Gen Elec Co Res & Develop Ctr, 60-67; proj coordr, 68-70, RES CHEMIST, US BUR MINES, 67- Mem: Am Inst Mining, Metall & Petrol Eng; Am Chem Soc. Res: Chemical kinetics; electrochemistry; ultra high pressure synthesis; zeolites. Mailing Add: Div of Metall US Bur of Mines 2410 E St NW Washington DC 20241

MUNSON, SAM CLARK, b Kosciusko, Miss, May 27, 08; m 41; c 3. PHYSIOLOGY, TOXICOLOGY. Educ: Miss State Col, BS, 30, MS, 31; Univ Md, PhD(zool, entom), 52. Prof Exp: Asst zool, Duke Univ, 31-32; asst, US Govt, 34-37; from jr zoologist to assoc entomologist, USDA, Washington, DC, 37-46; asst prof biol, 46-52, from assoc prof to prof, 52-73, EMER PROF BIOL, GEORGE WASHINGTON UNIV, 73- Res: Insect physiology. Mailing Add: Dept of Biol Sci George Washington Univ Washington DC 20006

MUNTER, PAUL ANTHONY, analytical chemistry, see 12th edition

MUNTZ, ALFRED PHILIP, geography, see 12th edition

MUNTZ, JOHN ADOLPH, b New York, NY, Feb 23, 11; m 38; c 2. BIOCHEMISTRY. Educ: NY Univ, AB, 36; Univ Chicago, MS, 41, PhD(med biochem), 45. Prof Exp: Asst, Zoller Dent Biochem Lab, Univ Chicago, 37-43; res assoc, Manhattan Proj, 43-45; from asst prof to assoc prof, 46-55, prof biochem, Western Reserve Univ, 46-55; chmn dept, 55-74, prof, 55-75, EMER PROF BIOCHEM, ALBANY MED COL, 75- Concurrent Pos: Consult & lectr, Albany Med Ctr. Mem: Am Chem Soc; Am Soc Biol Chem. Res: Development of micro methods for biological compounds; etiology of dental caries; toxicity of metals; intermediary metabolism; uranium poisoning; carbohydrate metabolism. Mailing Add: 29 Darroch Rd Delmar NY 12054

MUNTZ, RONALD LEE, b Bonaparte, Iowa, Sept 19, 45; m 66; c 2. ORGANOMETALLIC CHEMISTRY, ORGANIC CHEMISTRY. Educ: Iowa State Univ, BS, 68; Univ Ill, PhD(org chem), 72. Prof Exp: RES CHEMIST, STAUFFER CHEM CO, 72- Mem: Am Chem Soc. Res: Process development for the synthesis of new organo-metallic catalysts and the development of new homogeneous catalysts. Mailing Add: Stauffer Chem Co Eastern Res Ctr Dobbs Ferry NY 10522

MUNYAN, ARTHUR CLAUDE, b Lexington, Ky, May 31, 08; m 32; c 1. GEOLOGY, ENGINEERING GEOLOGY. Educ: Univ Ky, BS, 30; Univ Cincinnati, AM, 31, PhD(geol), 51. Prof Exp: Jr geologist, US Geol Surv, 34-36; asst state geologist, Ky, 36-37; asst state geologist, Ga, 37-41; from asst prof to assoc prof geol, Emory Univ, 41-51; staff geologist, Sohio Petrol Co, Can & US, 51-56, div geologist, 56-60, explor geologist, 60; chmn dept geol, 61-70, prof, 61-73, EMER PROF GEOL & GEOPHYS, OLD DOM UNIV, 73-; CONSULT GEOLOGIST, 60-; PRES, GEOSERV CO, INC, 65- Mem: Fel Geol Soc Am; Soc Econ Paleont & Mineral; Am Asn Petrol Geol; Int Geol Cong; Am Inst Prof Geologists (vpres, 68). Res: Structure;

petroleum; sedimentology. Mailing Add: 3204 Blue Ridge Ct Virginia Beach VA 23452

MUNYER, EDWARD ARNOLD, b Chicago, Ill, May 8, 36; m 60; c 3. ZOOLOGY, SCIENCE EDUCATION. Educ: Ill State Univ, BSEd, 58, MS, 62. Prof Exp: Teacher pub schs, Ill, 58-59; asst biol, Ill State Univ, 59-60; teacher pub schs, Ill, 60-63; instr, Ill State Univ, 63-64; cur zool, Ill State Mus, 64-67; assoc prof sci, Vincennes Univ, 67-70; COORD EDUC, FLA STATE MUS, UNIV FLA, 70- Mem: Am Inst Biol Sci; Am Asn Mus; Wilson Ornith Soc. Res: Museum education; vertebrate natural history; raptor ecology; comparative anatomy; perceptual learning; science education. Mailing Add: Fla State Mus Univ of Fla Gainesville FL 32611

MUNZ, FREDERICK WOLF, b Pomona, Calif, Sept 3, 29; m 54; c 3. VISUAL PHYSIOLOGY. Educ: Pomona Col, BA, 50; Univ Calif, Los Angeles, MA, 52, PhD(zool), 58. Prof Exp: Asst zool, Univ Calif, Los Angeles, 50-52, from actg instr to instr, 57-58; USPHS fel, Inst Ophthal, London, 58-59; from asst prof to assoc prof biol, 59-71, PROF BIOL, UNIV ORE, 71- Concurrent Pos: Guggenheim fel, 67-68. Mem: AAAS; Am Soc Zool. Mailing Add: Dept of Biol Univ of Ore Eugene OR 97403

MURACA, RAFFAELE FRANCESCO, b Easton, Pa, July 27, 21. ANALYTICAL CHEMISTRY, PHYSICS. Educ: Lehigh Univ, BS, 44, MS, 47, PhD(chem physics), 50. Prof Exp: Asst chem cent res labs, Gen Aniline & Film, Inc, 43-44; assoc prof & head dept, Concord Col, 47-48; asst, Lehigh Univ, 48-50, instr, 50-51, asst prof & div head, 51-55; group leader chem & physics, Jet Propulsion Lab, Calif Inst Technol, 55-56, sect chief chem, 56-59; asst div head propulsion, Stanford Res Inst, 59-61, dir space sci, 61-62, asst gen mgr phys sci, 62-64, dir anal & instrumentation, 64-69; PRES, WESTERN APPL RES & DEVELOP, INC, 69-; PROF CHEM & PHYSICS, COL NOTRE DAME, BELMONT, CALIF, 75- Concurrent Pos: Ed, Chemist-Analyst, 53-55; chmn, Joint Army-Navy-Air Force Panel Anal Chem, 58-63. Honors & Awards: C E Heussner Award, 55; Cert of Recognition, NASA, 75. Mem: Am Chem Soc. Res: Analytical chemistry; mass spectroscopy of natural products; high vacuum techniques; electrochemistry; electronics; instrument design; instrumental analysis; space sciences; applications. Mailing Add: Dept of Chem Col of Notre Dame Belmont CA 94002

MURACO, WILLIAM ANTHONY, b Cleveland, Ohio, Dec 23, 40; m 62; c 2. URBAN GEOGRAPHY. Educ: Ohio State Univ, BS, 64, MA, 66, PhD(geog), 71. Prof Exp: Teaching assoc geog, Ohio State Univ, 64-68; instr, Wright State Univ, 68-71; asst prof, 71-75, ASSOC PROF GEOG, UNIV TOLEDO, 75- Concurrent Pos: Fel, NSF, 72; res assoc, Community Res Assoc Inc, 73-76; co-prin investr, NASA Grant, 74-75 & Ohio Real Estate Comn Grant, 74-76. Mem: Asn Am Geogrs; Am Real Estate & Urban Econ Asn. Res: Technological assessment and impact; regional and urban planning decision theory; design of community health delivery systems. Mailing Add: Univ Toledo Dept Geog 2801 W Bancroft Toledo OH 43606

MURAD, EDMOND, b Bagdad, Iraq, Nov 29, 34; US citizen. PHYSICAL CHEMISTRY. Educ: NY Univ, BA, 55; Univ Rochester, PhD(phys chem), 59. Prof Exp: Res assoc phys chem, Nat Bur Standards, DC, 59-60; res assoc, Univ Wis, 60-61; res assoc chem phys, Univ Chicago, 61-63; res assoc phys chem, Cornell Univ, 63-64; res scientist, Aeronutronic Div, Ford Motor Co, 64-66; RES CHEMIST, AIR FORCE CAMBRIDGE RES LABS, 66- Mem: AAAS; Am Chem Soc; Am Phys Soc. Res: Mass spectrometry; high temperature thermodynamics; ion-neutral collision phenomena. Mailing Add: Air Force Cambridge Res Labs LKB Bedford MA 01730

MURAD, EMIL MOISE, b Detroit, Mich, May 10, 26. OPERATIONS RESEARCH. Educ: Univ Southern Calif, AB, 49, MS, 51. Prof Exp: Res chemist, Standard Coil Prod, Inc, 51-53; res chemist, Hydroaire Div, Crane Co, 53-55; chief chemist, Marvelco Electronics Div, Nat Aircraft Corp, 55-57; prin scientist, Stromberg-Carlson Div, Gen Dynamics Corp, 57-59; res specialist, Autonetics Div, NAm Aviation, Inc, 59; dir res, Orbitec Corp, 60-62; sr tech specialist space div, NAm Rockwell Corp, 62-68; PRES, QUANTADYNE ASSOCS, INC, 66- Mem: AAAS; Am Phys Soc; Electrochem Soc; Inst Elec & Electronics Eng; Opers Res Soc Am. Res: Development of computer technology for the application of data processing techniques to industrial electronic systems design. Mailing Add: Exec Off Quantadyne Assocs Inc 1929 Livonia Ave Los Angeles CA 90034

MURAD, FERID, b Whiting, Ind, Sept 14, 36; m 58; c 5. CLINICAL PHARMACOLOGY. Educ: DePauw Univ, BA, 58; Western Reserve Univ, MD & PhD(pharmacol), 65. Prof Exp: From intern to resident med, Mass Gen Hosp, 65-67; sr asst surg, Nat Heart & Lung Inst, 67-69, sr staff fel res, 69-70; assoc prof, 70-75, PROF PHARMACOL & INTERNAL MED, SCH MED, UNIV VA, 75-, DIR CLIN RES CTR, 71-, DIR DIV CLIN PHARMACOL, 73- Concurrent Pos: Nat Inst Arthritis & Metab Dis grant, Sch Med, Univ Va, 71-; USPHS res career develop award, 72-; Nat Heart & Lung Inst res grant, 75- Mem: Am Soc Biol Chem; Am Fedn Clin Res; Endocrine Soc; Am Soc Pharmacol & Exp Therapeut; Am Soc Clin Invest. Res: Cyclic adenosine monophosphate and cyclic guanosine monophosphate metabolism; endocrinology. Mailing Add: Div of Clin Pharmacol Univ of Va Sch of Med Charlottesville VA 22903

MURAD, JOHN LOUIS, b Tyler, Tex, Dec 15, 32; m 58; c 4. ZOOLOGY. Educ: Austin Col, BA, 56; NTex State Univ, MA, 58; Tex A&M Univ, PhD(zool), 65. Prof Exp: Asst microbiol, NTex State Univ, 56-58; teaching res med br, Univ Tex, 58-59; instr biol, Stephen F Austin State Col, 59-61; instr, Tex A&M Univ, 61-65; asst prof zool, 65-71, PROF ZOOL & DIR RES, COL LIFE SCI, LA TECH UNIV, 71- Concurrent Pos: Res partic, NSF-NAtlantic Treaty Orgn, Ger, 72, Eng, 74. Mem: AAAS; Soc Nematol; Am Soc Microbiol; Am Soc Testing & Mat; Am Inst Biol Sci. Res: Parasitology of wild and domestic animals, especially helminthic parasites; nematodes of soil, water and sewage. Mailing Add: Col of Life Sci La Tech Univ Box 5797 Ruston LA 71270

MURAD, TARIQ MOHAMMED, b Karbalaa, Iraq, July 28, 36; US citizen; m 69; c 2. PATHOLOGY. Educ: Univ Baghdad, MB ChB, 59; Ohio State Univ, MSc, 65, PhD(path), 67; Am Bd Path, cert anat path & clin path, 68. Prof Exp: From instr to assoc prof path, Ohio State Univ, 65-72; PROF PATH, MED CTR, UNIV ALA, 72-, DIR, DIV SURG PATH & CYTOL, 75- Concurrent Pos: Dir div clin cytol, Ohio State Univ, 71-72. Mem: Int Acad Path; Am Soc Cytol; Am Asn Pathologists & Bacteriologists; fel, Int Acad Cytol; Soc Surg Oncol. Res: Classification of breast cancers according to cell of origin and application of electron microscopy and identification of tumor. Mailing Add: Med Ctr Univ Ala 619 S 19th St Birmingham AL 35233

MURAD, TURHON ALLEN, b Hammond, Ind, July 27, 44; m 68. BIOLOGICAL ANTHROPOLOGY. Educ: Ind Univ, Bloomington, AB, 68, MA, 71, PhD(bio-anthrop), 75. Prof Exp: ASST PROF PHYS ANTHROP, CALIF STATE UNIV, CHICO, 72- Mem: Am Asn Phys Anthropologists; Sigma Xi. Res: North Alaskan Eskimo intrapopulation variation for palmar dermatoglyphics. Mailing Add: Dept of Anthrop Calif State Univ Chico CA 95929

MURAI, KOTARO, b San Francisco, Calif, Jan 10, 25; m 54; c 1. ORGANIC CHEMISTRY. Educ: Univ Nebr, BSc, 44, MSc, 45; Univ Minn, PhD(org chem), 49. Prof Exp: Asst, Univ Nebr, 44-45; asst, Univ Minn, 45-48; mem res staff, Pfizer, Inc, 49-73, SR RES INVESTR, PFIZER CENT RES, PFIZER MED RES LABS, 73- Mem: AAAS; Am Chem Soc; NY Acad Sci; Am Inst Chem. Res: Physico-organic approaches to kinetics, antibiotics, alkaloids and steroids. Mailing Add: Pfizer Med Res Labs Groton CT 06340

MURAI, MARY MIYEKO, b San Francisco, Calif, Jan 16, 13. NUTRITION. Educ: Univ Calif, BA, 34, MS, 50, MPH, 60, DrPH, 64. Prof Exp: Staff dietitian, St Luke's Int Med Ctr, Tokyo, Japan, 35-39; chief dietitian, Kaukini Hosp, Honolulu, Hawaii, 41-47; lab technician, Univ Calif, 50; technician res proj, US Dept Navy, 51; asst prof home econ, Univ Hawaii, 53-60; fel pub health nutrit, Children's Br, 59-64, lectr, Sch Pub Health, 64-66, ASST CLIN PROF PUB HEALTH NUTRIT, SCH PUB HEALTH, UNIV CALIF, BERKELEY, 66- Concurrent Pos: Nat Res Coun fel, Pac Sci Bd, 51; mem, Food & Agr Orgn-WHO, 55; consult, Off Surg Gen, US Army. Mem: AAAS; Am Pub Health Asn; Am Dietetic Asn; NY Acad Sci. Res: Nutrition survey and food habits of Micronesia; food values of South Pacific foods; training public health nutritionists; development and improvement of the criteria for selection of public health nutrition students. Mailing Add: 423 Earl Warren Hall Sch of Pub Health Univ of Calif Berkeley CA 94720

MURAKAMI, TAKIO, b Kanazawa, Japan, Mar 17, 21; m 49; c 2. METEOROLOGY. Educ: Meteorol Col, dipl, 43 & 49; Univ Tokyo, ScD(meteorol), 60. Prof Exp: Chief gen circulation, Meteorol Res Inst, Tokyo, 53-67; res meteorologist, Meteorol Satellite Lab, Washington, DC, 67-69; PROF METEOROL, UNIV HAWAII, 69- Concurrent Pos: Fel, Sch Advan Study, Mass Inst Technol, 60-62; res meteorologist, Inst Trop Meteorol, Poona, India, 66-67. Honors & Awards: Meteorol Soc Japan Award, 54. Mem: Am Meteorol Soc; Meteorol Soc Japan. Res: Synoptic and theoretical tropical meteorology. Mailing Add: Dept of Meteorol Univ of Hawaii Honolulu HI 96822

MURAKISHI, HARRY HARUO, b San Francisco, Calif, Oct 21, 17; m 48; c 3. PLANT PATHOLOGY. Educ: Univ Calif, BS, 40; Univ NC, MS, 47; Univ Minn, PhD(plant path), 48. Prof Exp: Asst plant path, Univ NC, 44-45; asst, Univ Minn, 46-48; from asst plant pathologist to plant pathologist, Univ Hawaii, 48-56, head dept plant path, 52-56; assoc prof bot & plant path, 56-63, PROF BOT & PLANT PATH, MICH STATE UNIV, 63- Concurrent Pos: Agent, USDA, 44; Guggenheim Mem Found fel & assoc plant path, Univ Calif, 55-56. Mem: Am Phytopath Soc; Tissue Culture Asn. Res: Plant virology; virus diseases of vegetables and orchids; plant tissue culture. Mailing Add: Dept of Bot & Plant Path Mich State Univ East Lansing MI 48824

MURAMOTO, HIROSHI, b Hilo, Hawaii, June 6, 22; m 56; c 2. PLANT BREEDING, GENETICS. Educ: NMex Col Agr & Mech Arts, BS, 55; Univ Ariz, PhD(agron), 58. Prof Exp: Asst plant breeder, 58-64, assoc prof, 64-75, ASSOC PLANT BREEDER, UNIV ARIZ, 64-, PROF PLANT BREEDING, 75- Res: Cotton breeding and genetics. Mailing Add: Dept of Agron Univ of Ariz Tucson AZ 85721

MURANO, GENESIO, b Cairano, Italy, Oct 23, 41; PHYSIOLOGY, BIOCHEMISTRY. Educ: Univ Mass, BA, 64; Wayne State Univ, MS, 66, PhD(physiol), 68. Prof Exp: Res assoc, 68-70, ASST PROF PHYSIOL, WAYNE STATE UNIV, 71- Concurrent Pos: NIH fel, Karolinska Inst, Sweden, 70-71. Mem: NY Acad Sci; Int Soc Thrombosis & Haemostasis. Res: Biochemical interactions of clotting factors; fibrinogen structure. Mailing Add: Dept of Physiol Wayne State Univ Detroit MI 48201

MURANY, ERNEST ELMER, b Avella, Pa, Mar 28, 23; m 57; c 2. EXPLORATION GEOLOGY. Educ: Kent State Univ, BS, 50; Univ Utah, PhD(geol), 63. Prof Exp: Photogrammeter, Corps Engrs, US Army, 51-52; stratigr, US Geol Surv, 52-53; sect chief, Mene Grande Oil Co, Gulf Oil Corp, 53-60; explor mgr & dist geologist, Sinclair Venezuelan Oil Co, Sinclair Oil Co, 63-68; vpres, Collman Indust, 69-70; consult geologist, Dallas, Tex, 70-71 & Venezuelan Petrol Corp, 71-73; SR STAFF GEOLOGIST, BELCO PETROL CORP, 73- Mem: AAAS; Am Asn Petrol Geologists; Asn Venezuelan Mineralogists & Petrologists; Am Geophys Union. Res: Structural, sedimentological and stratigraphic implication of plate tectonics to mountain building as related to generation, migration and accumulation of oil in Venezuela, Colombia, Ecuador and Peru. Mailing Add: 5406 Foresthaven Dr Houston TX 77066

MURASHIGE, TOSHIO, b Kapoho, Hawaii, May 26, 30; m 53; c 5. PLANT PHYSIOLOGY. Educ: Univ Hawaii, BS, 52; Ohio State Univ, MS, 54; Univ Wis, PhD(plant physiol), 58. Prof Exp: Res assoc, Univ Wis, 58-59; asst prof, Univ Hawaii, 59-64; from asst plant physiologist to assoc plant physiologist, 64-72, assoc prof plant sci, 67-72, PROF HORT SCI & PLANT PHYSIOLOGIST, UNIV CALIF, RIVERSIDE, 72- Concurrent Pos: Elvenia Slosson fel ornamental hort, 72-77. Mem: AAAS; Am Soc Plant Physiol; Bot Soc Am; Am Soc Hort Sci; Tissue Culture Asn. Res: Plant tissue culture; experimental morphogenesis; growth regulators. Mailing Add: Dept of Plant Sci Univ of Calif Riverside CA 92502

MURASKIN, MURRAY, b Brooklyn, Ny, Aug 7, 35; m; c 3. PHYSICS. Educ: Mass Inst Technol, BS, 57; Univ Ill, MS, 59, PhD(physics), 61. Prof Exp: Res assoc physics, Univ Minn, 61-63; asst prof, Univ Nebr, Lincoln, 63-69; assoc prof, 69-74, PROF PHYSICS, UNIV NDAK, 74- Mem: Am Phys Soc. Res: Elementary particle physics. Mailing Add: Dept of Physics Univ of NDak Grand Forks ND 58201

MURASUGI, KUNIO, b Tokyo, Japan, Mar 25, 29; m 55; c 3. MATHEMATICS. Educ: Tokyo Univ Educ, BSc, 52, DSc(math), 61; Univ Toronto, MA, 61. Prof Exp: Lectr math, Hosei Univ, 55-59, asst prof, 59-61; res asst, Univ Toronto, 61-62; res asst, Princeton Univ, 62-64; asst prof, 64-66, assoc prof, 66-69, PROF MATH, UNIV TORONTO, 69- Concurrent Pos: Res grants, Nat Res Coun Can, 61-62, NSF, 62-64 & Can Coun grant; vis scientist, Princeton Univ, 71-72; ed, Can J Math, 69-71. Mem: Am Math Soc; Can Math Cong; Math Soc Japan. Res: Knot theory in combinatorial topology; infinite group theory. Mailing Add: 611 Cummer Ave Willowdale ON Can

MURAY, JULIUS J, b Budapest, Hungary, Mar 22, 31; US citizen; m 56; c 2. PHYSICS. Educ: Eötvos Lorand, Budapest, dipl, 53; Univ Calif, Berkeley, MA, 60, PhD(physics), 61. Prof Exp: Asst prof physics, Eötvos Lorand, Budapest, 53-56; asst, Univ Calif, Berkeley, 57-61; physicist, Stanford Univ, 61-65; physicist, Hewlett-Packard Co, Palo Alto, 65-68; exec vpres, Cintra Inc, Physics Int Co, 68-71; exec Sloan fel, Grad Sch Bus, Stanford Univ, 71-72; MGR PHYS ELECTRONICS LAB, CENT ENG LAB, FMC CORP, 72- Concurrent Pos: Consult, Appl Radiation Co, Calif, 60-65; pres, Instruments Int, 71- Mem: Am Phys Soc; sr mem Inst Elec & Electronics Eng. Res: Electron and radiation physics; atomic and molecular physics; electro-optics. Mailing Add: FMC Corp 1185 Coleman Ave Santa Clara CA 95052

MURAYAMA, MAKIO, b San Francisco, Calif, Aug 10, 12; m 45. BIOCHEMISTRY. Educ: Univ Calif, BA, 39, MA, 40, PhD(immunochem), 53. Prof Exp: Asst biochem, Univ Calif, 39-42; res chemist, Bellevue Hosp, NY, 43-45; res biochemist, Univ Hosp,

Univ Mich, 45-48; res biochemist res div, Harper Hosp, 50-54; res fel chem, Calif Inst Technol, 54-56; res assoc biochem, Univ Pa, 55-58; Nat Cancer Inst spec res fel, Cavendish Lab, Cambridge Univ, 58; BIOCHEMIST, NIH, 58- Honors & Awards: Martin Luther King, Jr, Med Achievement Award, Southern Christian Leadership Conf, 72. Mem: AAAS; Am Soc Biol Chem. Res: Protein chemistry; chemistry and structure of hemoglobin, especially electron microscopic studies of human sickle cell hemoglobin cable; molecular mechanism of human red cell sickling with hemoglobin S. Mailing Add: Nat Inst of Health Bethesda MD 20014

MURAYAMA, TAKAYUKI, b Tokyo, Japan, Mar 29, 32; US citizen. POLYMER SCIENCE. Educ: Tokyo Univ Agr & Technol, BS, 54; Lowell Technol Inst, MS, 62; Kyushu Univ, PhD(polymer sci), 68. Prof Exp: Res asst eng, Tokyo Univ Agr & Technol, 54-59; mech engr, Chemstrand Res Ctr, 62-66, from res physicist to sr res physicist, 66-72; RES SPECIALIST, TRIANGLE PARK DEVELOP CTR, MONSANTO CO, 72- Mem: Soc Rheol; Am Chem Soc; Fiber Soc. Res: Dynamic mechanical analyses on material; studies on differential thermoproperties relating to molecular structure on polymers; dynamic anisotropic viscoelasticity on material. Mailing Add: PO Box 12274 Research Triangle Park NC 27709

MURBACH, EARL WESLEY, b Almira, Wash, Oct 10, 22; m 48; c 2. INORGANIC CHEMISTRY. Educ: Gonzaga Univ, BS, 43; Wash State Univ, MS, 49, PhD(chem), 52. Prof Exp: Anal chemist, Kaiser Aluminum Co, 46-47; chem engr div indust res, Wash State Univ, 49-50; chemist, Calif Res & Develop Corp Div, Standard Oil Co, Calif, 52-53; chemist, Nat Carbon Co, 53-54; sr chemist, Phillips Petrol Co, Idaho, 54-56; sr res engr, Atomics Int Div, NAm Aviation, Inc, 56-57, supvr chem develop, 57-70; mgr nuclear process develop, Allied Gulf Nuclear Serv, 70-74, MGR NUCLEAR PROCESS DEVELOP, ALLIED GEN NUCLEAR SERV, 74- Mem: Am Chem Soc. Res: High temperature methods for reprocessing reactor fuel; chemical reaction kinetics at high temperatures. Mailing Add: 877 Sycamore Dr Aiken SC 29801

MURCH, ROBERT MATTHEWS, b Lackawanna, NY, June 27, 24; m 51; c 4. INORGANIC CHEMISTRY, ORGANOMETALLIC CHEMISTRY. Educ: Univ Mich, BS, 48; Pa State Univ, MS, 50; Univ Wis, PhD(inorg chem), 66. Prof Exp: Chem engr, Dow Corning Corp, 50-54, res chemist, 54-58, proj leader, 58-62; sr chemist, 65-67, RES SUPVR, WASH RES CTR, W R GRACE & CO, 67- Mem: Am Chem Soc; Am Ord Asn; Sigma Xi. Res: Organosilicon; organophosphorus; organofluorine; inorganic polymers; fire retardant urethanes and polyester resins; flammability and smoke testing research. Mailing Add: Wash Res Ctr W R Grace & Co Columbia MD 21044

MURCHISON, JOHN TAYNTON, b Ft Niagara, NY, Feb 7, 06; m 32; c 4. ORGANIC CHEMISTRY. Educ: Univ Nebr, AB, 27; Univ Tex, AM, 30, PhD(org chem), 33. Prof Exp: Asst chem, Univ Nebr, 27-28; tutor, Univ Tex, 28-33; prof, 33-75, head dept, 33-67, asst dean sci, 67-71, EMER PROF CHEM, UNIV TEX, ARLINGTON, 75- Mem: Fel AAAS; Am Chem Soc. Res: Derivatives of halogenated ethers. Mailing Add: Univ of Tex Arlington TX 76010

MURCHISON, PAMELA W, b Pittsburgh, Pa, June 14, 43; m 70; c 2. PHYSICAL CHEMISTRY. Educ: Carnegie Inst Technol, BS, 65; Univ Minn, PhD(phys chem), 69. Prof Exp: Vis lectr chem, Hamline Univ, 69-70; ASST PROF PHYS CHEM, CENT MICH UNIV, 70-71, 72-74 & 75- Mem: Am Chem Soc. Res: Effective undergraduate chemical education. Mailing Add: 606 W Meadowbrook Midland MI 48640

MURCHISON, THOMAS EDGAR, b Kingsville, Tex, Aug 7, 32. VETERINARY PATHOLOGY. Tex A&M Univ, DVM, 55; Ohio State Univ, MSc, 57, PhD, 59. Prof Exp: Instr vet path, Ohio State Univ, 55-59; head path dept, Fla Dept Agr, 59-62; head exp path, Orange Mem Hosp, 62-65; head, Dawson Res Inst, 60-65, PRES, DAWSON RES CORP, 65- Concurrent Pos: Pres, Temson Co, 65- Mem: Am Vet Med Asn; Am Col Vet Path; Int Acad Path. Res: Pharmaceutical toxicology; veterinary pathology; comparative pathology; research administration. Mailing Add: Dawson Res Corp 114 W Grant Ave Orlando FL 32806

MURCRAY, DAVID GUY, b Leadville, Colo, Jan 19, 24; m 45; c 2. ATMOSPHERIC PHYSICS. Educ: Univ Denver, BS, 48, PhD, 63; Okla State Univ, MS, 50. Prof Exp: Fel, Univ Kans, 50-51; res mathematician, Phillips Petrol Co, 51-52; res physicist res inst, 52-63, asst prof, 63, assoc prof, 66-69, PROF PHYSICS, UNIV DENVER, 69-, SR RES PHYSICIST, 63- Mem: AAAS; Optical Soc Am; Am Geophys Union; Royal Meteorol Soc. Res: Upper atmospheric physics; infrared transmission in the upper atmosphere; radiation balance of the atmosphere; infrared spectroscopy; operations analysis. Mailing Add: Dept of Physics Univ of Denver Denver CO 80210

MURDAUGH, HERSCHEL VICTOR, JR, b Columbia, SC, Mar 4, 28; 48; c 3. INTERNAL MEDICINE. Educ: Duke Univ, MD, 50. Prof Exp: Intern med, Grady Mem Hosp, Atlanta, Ga, 50-51; from asst resident to sr asst resident, Duke Univ Hosp, 53-56, instr, 56-57, assoc, 57-58; from asst prof to assoc prof med, Med Col Ala, 58-65, dir renal & electrolyte div, 58-61; ASSOC PROF MED & DIR RENAL DIV, SCH MED, UNIV PITTSBURGH, 65- Concurrent Pos: Res fel, Duke Univ & USPHS Hosp, 54-55; chief res, Vet Admin Hosp, Durham, NC, 56-57, clin investr, 57-58; trustee, Mt Desert Island Biol Lab, 62- Mem: Am Physiol Soc; Am Fedn Clin Res; Am Soc Clin Invest; fel Am Col Physicians; Soc Exp Biol & Med. Res: Renal physiology and disease; physiology of aquatic mammals. Mailing Add: Univ of Pittsburgh Sch of Med Pittsburgh PA 15213

MURDAY, JAMES STANLEY, b Trenton, NJ, Sept 16, 42; m 67; c 2. SURFACE PHYSICS. Educ: Case Inst Technol, BS, 64; Cornell Univ, PhD(physics), 70. Prof Exp: Res physicist, 69-75, SECT HEAD SURFACE ANAL, NAVAL RES LAB, 75- Concurrent Pos: Consult, Off Naval Res, 73- Mem: AAAS; Am Chem Soc; Am Phys Soc; Am Vacuum Soc; Int Soc Hybrid Microelectronics. Res: Interaction of energy forms with surfaces, surface chemical analysis, surface reactions. Mailing Add: Code 6170 Naval Res Lab Washington DC 20375

MURDESHWAR, MANGESH GANESH, b Bombay, India, Mar 25, 33; m 61; c 3. TOPOLOGY. Educ: Univ Bombay, BA, 54, MA, 56; Univ Alta, PhD(math), 64. Prof Exp: Lectr math, Wilson Col, Univ Bombay, 56-57, Khalsa Col, 58-59 & Parle Col, 59-61; ASST PROF MATH, UNIV ALTA, 64- Concurrent Pos: Nat Res Coun Can overseas fel, Univ Bombay, 66-68. Mem: Can Math Cong; Math Asn Am; Am Math Soc. Res: Point-set topology. Mailing Add: Dept of Math Univ of Alta Edmonton AB Can

MURDICK, PHILIP W, b Akron, Ohio, Nov 13, 28; m 52; c 4. VETERINARY MEDICINE. Educ: Ohio State Univ, DVM, 52, MS, 58, PhD(physiol), 64. Prof Exp: Instr vet med, 56-64, from asst prof to assoc prof, 64-69, PROF VET MED, OHIO STATE UNIV, 69-, CHMN DEPT VET CLIN SCI, 71- Mem: Am Vet Med Asn. Res: Veterinary obstetrics and diseases of the genitalia; development of methods for the detection of drugs illegally administered to race horses. Mailing Add: Col of Vet Med Ohio State Univ 1935 Coffey Rd Columbus OH 43210

MURDOCH, ARTHUR, b DuBois, Nebr, Aug 25, 34; m 57; c 2. ORGANIC CHEMISTRY. Educ: Westmar Col, BA, 56; Yale Univ, MS, 58, PhD(org chem), 64. Prof Exp: Asst prof chem, Morningside Col, 62-68; ASSOC PROF CHEM & CHMN DEPT, MT UNION COL, 68- Mem: AAAS; Am Chem Soc. Res: Organic reduction-oxidation polymers. Mailing Add: Dept of Chem Mt Union Col Alliance OH 44601

MURDOCH, BRUCE THOMAS, b Prague, Okla, Mar 15, 40; m 69; c 1. EXPERIMENTAL NUCLEAR PHYSICS. Educ: Carleton Col, BA, 62; Rice Univ, MA, 66; Utah State Univ, PhD(physics), 75. Prof Exp: Develop engr, Goodyear Aerospace Corp, 67-70; FEL NUCLEAR PHYSICS, UNIV MAN, 74- Mem: Am Phys Soc. Res: Medium energy experimental nuclear physics; cryogenic targets; nuclear orientation at ultralow temperatures; directional gamma-gamma correlations; magnetic materials; fundamental physical symmetries. Mailing Add: Dept of Physics Univ of Man Winnipeg MB Can

MURDOCH, CHARLES LORAINE, b Atkins, Ark, Aug 23, 32; m 66; c 2. HORTICULTURE. Educ: Univ Ark, BS, 59, MS, 60; Univ Ill, PhD(agron), 66. Prof Exp: Res asst agron, Southwest Br Exp Sta, Univ Ark, 60-62 & Univ Ill, 62-66; res assoc, Univ Ark, 66-70; asst prof, 70-74, ASSOC PROF HORT, UNIV HAWAII, HONOLULU, 74- Mem: Am Soc Agron. Res: Turfgrass management; ecological and physiological aspects of turfgrass growth and development. Mailing Add: Dept of Hort Univ of Hawaii Honolulu HI 96822

MURDOCH, DAVID CARRUTHERS, b Tunbridge Wells, Eng, Mar 31, 12; m 50; c 3. MATHEMATICS. Educ: Univ BC, BA, 31, MA, 33; Univ Toronto, PhD(math), 37. Prof Exp: Sterling fel math, Yale Univ, 37-38, instr, 38-40; instr, Univ Sask, 40-42, asst prof, 42-44; assoc prof, 44-53, PROF MATH, UNIV BC, 53- Concurrent Pos: Vis prof, Ford Found, Mass Inst Technol Proj, Birla Inst Math & Sci, Pilani, India, 66-68, Mem: Am Math Soc; Math Asn Am; Can Math Cong. Res: Non-commutative ideal theory and theory of rings; abstract algebra. Mailing Add: Dept of Math Univ of BC Vancouver BC Can

MURDOCH, JOSEPH RICHARD, b Portland, Ore, Sept 15, 46; m 71. ORGANIC CHEMISTRY. Educ: Univ Calif, Santa Barbara, BA, 68; Univ Calif, Berkeley, PhD(chem), 73. Prof Exp: Damon Runyon Cancer Found fel, Dept Chem, Stanford Univ, 73-75; res assoc chem, Univ Calif, Berkeley, 75-76; ASST PROF CHEM, UNIV CALIF, LOS ANGELES, 76- Mem: Am Chem Soc. Res: Solid state chemistry, dynamics of chemical reactions in solution and condensed phases, applications to biological systems. Mailing Add: Dept of Chem Univ of Calif Los Angeles CA 90024

MURDOCH, RAYMOND LESTER, surgery, see 12th edition

MURDOCH, WALLACE PIERCE, b Farnum, Idaho, Oct 23, 24; m 47; c 4. ENTOMOLOGY. Educ: Utah State Univ, BS, 49, MS, 51; Univ Utah, PhD(entom), 62. Prof Exp: Commanding officer, Malaria Surv Detachment, Med Serv Corps, US Army, Korea, 52-53, instr med entom, Med Field Serv Sch, Tex, 53-54, chief entom div, Environ Hyg Agency, Md, 54-56, chief dept entom, Med Gen Lab, Japan, 57-60, chief div environ hyg, Off Chief Surgeon, CZ, 62-66, chief prev med res br, Hq, Res & Develop Command, Washington, DC, 66-69; EXEC SECY, ENTOM SOC AM, 69- Concurrent Pos: US Army Med Res Command biol grant, Panama, 62-66; entom consult, Off Interoceanic Canal Studies, Balboa, CZ, 63-66; WHO consult to Pan Am Sanit Bur, Paraguay, 64; mem, Int Cong Entom, 74-76. Mem: AAAS (treas, 74-); Sigma Xi. Res: Medical zoology; taxonomy of palearctic Tabanidae; bionomics of Culex tritaeniorhynchus in Japan and Korea; biology of immature Tabanidae; ecology of cutaneous leishmaniasis. Mailing Add: RD 2 Gettysburg PA 17325

MURDOCH, WARREN FRANK, organic chemistry, see 12th edition

MURDOCH, WILLIAM W, b Glassford, Scotland, Jan 28, 39; m 63; c 1. POPULATION BIOLOGY, ECOLOGY. Educ: Univ Glasgow, BSc, 60; Oxford Univ, DPhil(ecol), 63. Prof Exp: Res assoc & instr ecol, Univ Mich, 63-65; from asst prof to assoc prof, 65-75, PROF BIOL SCI, UNIV CALIF, SANTA BARBARA, 75- Concurrent Pos: Vis lectr, Univ BC, 65. Mem: Ecol Soc Am; Brit Ecol Soc; Japanese Soc Pop Biol. Res: Population and community dynamics of organisms. Mailing Add: Dept of Biol Sci Univ of Calif Santa Barbara CA 93106

MURDOCK, ARCHIE LEE, b Arcola, Mo, Nov 5, 33; m 53; c 3. BIOCHEMISTRY. Educ: Southwest Mo State Col, BS, 57; Univ Mo, MS, 60, PhD(biochem), 63. Prof Exp: Res assoc biochem, Brookhaven Nat Lab, 63-65; ASST PROF BIOCHEM, MED CTR, UNIV KANS, 65- Mem: Am Chem Soc. Res: Protein chemistry; molecular mechanisms of thermophily in bacteria; pancreatic acinar cell function; methodology. Mailing Add: Dept of Biochem Univ of Kans Med Ctr Kansas City KS 66103

MURDOCK, FENOI R, b Blackfoot, Idaho, Feb 2, 17; m 42; c 3. ANIMAL NUTRITION. Educ: Univ Idaho, BS, 38; Pa State Univ, PhD(agr), 43. Prof Exp: Res chemist, Borden Co, 43-49; asst prof dairy sci, Western Wash Exp Sta, 49-52, assoc prof dairy sci & assoc dairy scientist, 52-60, PROF DAIRY SCI & DAIRY SCIENTIST, WESTERN WASH RES & EXTEN CTR, WASH STATE UNIV, 60- Mem: Am Dairy Sci Asn. Res: Dairy cattle nutrition, utilization of forages and by-products as feeds, animal waste management to conserve nutrients and prevent air and water pollution. Mailing Add: Western Wash Res & Exten Ctr Puyallup WA 98371

MURDOCK, GEORGE PETER, b Meriden, Conn, May 11, 97; m 25; c 1. ANTHROPOLOGY. Educ: Yale Univ, AB, 19, PhD(sociol), 25. Prof Exp: Instr sociol, Univ Md, 25-27; asst prof, Yale Univ, 28-34, assoc prof ethnol, 34-39, prof anthrop, 39-60, dir cross-cult surv, Inst Human Rels, 37-46; Mellon Prof anthrop, 60-71, ANDREW W MELLON EMER PROF ANTHROP, 71-; RES ASSOC ETHNOL, AFRICA, NAM & OCEANIA, CARNEGIE MUS NATURAL HIST, 75- Concurrent Pos: Ed, Ethnology, 62-73; Am Anthrop Asn rep, Nat Res Coun, 42-45, chmn div behav sci, 64-66. Mem: Nat Acad Sci; fel Am Anthrop Asn (pres, 55); Soc Appl Anthrop (pres, 47); Am Sociol Asn; Am Ethnol Soc (pres, 52). Res: Ethnology of Africa, North America and Micronesia; social theory and organization. Mailing Add: Carnegie Mus of Natural Hist 4400 Forbes Pittsburgh PA 15213

MURDOCK, GORDON ALFRED, b Minneapolis, Minn, Jan 4, 23; m 50; c 5. PHYSICAL CHEMISTRY. Educ: Willamette Univ, BS, 50; Univ Ore, MA, 51, PhD, 54. Prof Exp: Asst, Univ Ore, 49-53; res chemist, 53-57, prof leader, 57-60, supvr printing grades develop, 60-68, supvr reprographic res, 68-69, TECH MGR, REPROGRAPHIC PROD, CROWN ZELLERBACH CORP, 69- Mem: Am Chem Soc; Tech Asn Pulp & Paper Indust; Soc Photog Sci & Eng. Res: Electrochemistry; corrosion studies; polarography; paper products; graphic arts; electrophotography and reproduction papers. Mailing Add: 1906 NW Couch St Camas WA 98607

MURDOCK, GORDON ROBERT, b Redlands, Calif, Jan 4, 43; m 68; c 1. INVERTEBRATE ZOOLOGY. Educ: Reed Col, AB, 65, PhD(zool), 72. Prof Exp: Asst prof zool, Ariz State Univ, 70-75; FEL, DUKE UNIV, 75- Concurrent Pos: Fel, Univ Manchester, 74-75. Mem: AAAS; Am Soc Zoologists. Res: Prey capture and utilization by sessile invertebrates, particularly the influence of water motion and growth form on success of food capture. Mailing Add: Dept of Zool Duke Univ Durham NC 27706

MURDOCK, HAROLD RUSSELL, JR, b Orange, NJ, Oct 15, 19; m 50; c 3. PHARMACOLOGY. Educ: Davidson Col, BS, 43; Univ Rochester, MS, 49; Univ Buffalo, PhD(physiol, pharmacol), 51. Prof Exp: Tech asst, Inst Paper Chem, Lawrence Col, 42-43; sr pharmacologist, Johnson & Johnson Res Found, 52-54; dir biol sci, Lloyd Bros, Inc, 54-59; prin scientist, Vet Admin Hosp, Huntington, WVa, 59-66, res pharmacologist, Livermore, 66-70; PHARMACOLOGIST, FOOD & DRUG ADMIN, WASHINGTON, DC, 70- Mem: AAAS; Am Chem Soc. Res: Biochemical pharmacology; bioavailability of drugs; biopharmaceutics; drug metabolism. Mailing Add: 5011 Macon Rd Rockville MD 20852

MURDOCK, JAMES DAVID, organic chemistry, see 12th edition

MURDOCK, JOHN THOMAS, b Lynn Grove, Ky, Nov 21, 27; m 49; c 3. SOIL SCIENCE. Educ: Univ Ky, BS, 52, MS, 53; Univ Wis, PhD(soils), 56. Prof Exp: Assoc prof, 55-72, asst dir int agr progs, 68-70, PROF SOILS, UNIV WIS-MADISON, 72-, ASSOC DIR INT AGR PROGS, 74- Concurrent Pos: Soils specialist, Univ Wis Contract, Univ Rio Grande do Sul, Brazil, 64-68; proj coordr, Midwestern Univs Consortium Int Activities Higher Agr Educ Proj, Bogor, Indonesia, 70-72; sr res adv spec prog agr res, Brasilia, Brazil, 72-73. Mem: Soil Sci Soc Am. Res: Soil fertility and management. Mailing Add: Int Agr Progs Off Univ of Wis Madison WI 53706

MURDOCK, JOHN WALLACE, physics, information science, see 12th edition

MURDOCK, JOSEPH RICHARD, b Provo, Utah, Apr 25, 21; m 47; c 4. PLANT ECOLOGY. Educ: Brigham Young Univ, BS, 49, MS, 51; Wash State Univ,PhD(bot), 56. Prof Exp: From asst prof to assoc prof, 52-69, PROF BOT, BRIGHAM YOUNG UNIV, 69- Concurrent Pos: Res assoc radiation ecol proj, Nev Test Site, AEC, 59-60. Res: Microenvironmental studies of plant habitats; temperature and moisture. Mailing Add: Dept of Bot Brigham Young Univ Provo UT 84601

MURDOCK, KEITH CHADWICK, b Garfield, Utah, Feb 5, 28; m 53; c 2. SYNTHETIC ORGANIC CHEMISTRY, MEDICINAL CHEMISTRY. Educ: Univ Utah, BA, 48, MA, 50; Univ Ill, PhD(chem), 53. Prof Exp: Asst pharmacol & chem, Univ Utah, 48; asst chem, Univ Ill, 50-53; chemist, Res Div, Am Cyanamid Co, NJ, 53-54 & Army Chem Ctr, Edgewood, Md, 55-56; res chemist, 56-63, SR RES CHEMIST, LEDERLE DIV, AM CYANAMID CO, 63- Mem: Am Chem Soc. Res: Mannich reaction; copolymerization; pharmaceutical and medicinal chemistry; heterocyclics; cancer and antiviral chemotherapy; nucleic acids. Mailing Add: Lederle Div Am Cyanamid Co Pearl River NY 10965

MURDOCK, LARRY LEE, b Linton, Ind, Aug 18, 42; m 63. PHYSIOLOGY, BIOCHEMISTRY. Educ: DePauw Univ, BA, 64; Kans State Univ, MS, 66, PhD(entom), 69. Prof Exp: Res assoc entom, Kans State Univ, 68-69; res assoc invert physiol, Univ Wash, 69; sci res asst invert neurobiochem, Univ Konstanz, 69-74; MEM FAC, DEPT NEUROBIOL, KENNEDY LAB, UNIV WIS MED SCH, 74- Mem: AAAS; Am Chem Soc; Entom Soc Am. Res: Biochemistry of neurotransmitter substances; comparative physiology and biochemistry; invertebrate pharmacology. Mailing Add: Dept of Neurobiol Kennedy Lab Univ Wis Med Sch Madison WI 53706

MURDY, WILLIAM HENRY, b New Bedford, Mass, Dec 25, 28; m 52; c 2. BOTANY. Educ: Univ Mass, BS, 56; Washington Univ, PhD(bot), 59. Prof Exp: From instr to assoc prof, 59-71, chmn Biol Dept, 71-74, PROF BIOL, EMORY UNIV, 71- Concurrent Pos: Scholar, Harvard Univ, 67-68. Mem: Bot Soc Am; Soc Study Evolution. Res: Systematics of plant species of granite outcrop communities in the Southeastern Piedmont; systematics of urban plant species. Mailing Add: Dept of Biol Emory Univ Atlanta GA 30322

MUREIKA, ROMAN A, b Lithuania, Aug 9, 44; Can citizen; m 66; c 2. MATHEMATICS. Educ: Cath Univ Am, BA, 64, MA, 67, PhD(math), 69. Prof Exp: ASST PROF MATH, UNIV ALTA, 68- Mem: Can Math Cong; Inst Math Statist; Am Math Soc. Res: Probability theory; information theory; stochastic processes; entropy. Mailing Add: Dept of Math Univ of Alta Edmonton AB Can

MUREN, JAMES FRANCIS, medicinal chemistry, see 12th edition

MURIE, JAN O, b Okanogan, Wash, July 24, 39; m 61; c 2. ANIMAL BEHAVIOR, ECOLOGY. Educ: Colo State Univ, BS, 59; Univ Mont, MA, 63; Pa State Univ, PhD(zool), 67. Prof Exp: Fel & lectr, 67-69, asst prof, 69-75, ASSOC PROF ZOOL, UNIV ALTA, 75- Mem: Ecol Soc Am; Animal Behav Soc; Am Soc Zoologists; Am Soc Mammal; Can Soc Environ Biologists. Res: Ecology and behavior of small mammals, especially interspecific relationships; rodents. Mailing Add: Dept of Zool Univ of Alta Edmonton AB Can

MURIE, MARTIN L, b Twisp, Wash, July 10, 25; m 52; c 3. VERTEBRATE ZOOLOGY. Educ: Reed Col, BA, 50; Univ Calif, Berkeley, BA, 53, PhD(zool), 60. Prof Exp: Instr biol, Univ Calif, Santa Barbara, 60-61; asst prof, 61-67, ASSOC PROF BIOL, ANTIOCH COL, 67- Mem: Am Soc Mammal; Am Soc Zoologists; Ecol Soc Am. Res: Vertebrate physiology; ecology. Mailing Add: Rte 1 Yellow Springs OH 45387

MURIE, RICHARD A, b Mt Pleasant, Ohio, Oct 3, 23; m 50; c 2. ANALYTICAL CHEMISTRY. Educ: Ohio Univ, BS, 50; Iowa State Univ, MS, 52, PhD(anal chem), 55. Prof Exp: Assoc prof chem, Drake Univ, 51-52; res chemist, Monsanto Chem Co, 55-60; sr res chemist, Res Ctr, Diamond Alkali Chem Co, 60, group leader, 60-63; prin scientist, Allison Div, 63-68, supvr appl math res, Tech Ctr, 68-70, SR RES CHEMIST, GEN MOTORS RES LABS, 70- Concurrent Pos: Instr, Eve Div, WVa State Col, 55-60; vis prof, Univ Guadalajara, 72-75; instr, Eve Div, Lawrence Inst Technol, Southfield, Mich, 72- Mem: Am Chem Soc; Instrument Soc Am; Soc Appl Spectros. Res: Instrumental methods of analysis, especially gas chromatography, differential thermal analysis and ultraviolet and infrared spectroscopy; development of high temperature battery systems and materials to serve as insulators in such systems. Mailing Add: 4661 Barcroft Way Sterling Heights MI 48077

MURINO, CLIFFORD JOHN, b Yonkers, NY, Feb 10, 29; m 54; c 3. METEOROLOGY. Educ: St Louis Univ, BS, 50, MS, 54, PhD(geophys), 57. Prof Exp: Assoc prof, 54-67, PROF GEOPHYS, INST TECHNOL, ST LOUIS UNIV, 67-, VPRES FINANCE & RES, 71- Concurrent Pos: Vpres res, St Louis Univ,69-71; sci consult, Div Environ Sci, NSF, Washington, DC, 67-69. Mem: Am Meteorol Soc. Res: Satellite meteorology; radiation physics; atmospheric energetics; severe storms. Mailing Add: St Louis Univ Off of the Vpres Financial Affs St Louis MO 63103

MURINO, VINCENT S, b New York, NY, July 30, 24; m 50; c 1. METEOROLOGY, RESEARCH ADMINISTRATION. Educ: St Louis Univ, BS, 51, MS, 52; Am Univ, MA, 67. Prof Exp: Meteorologist, 52-64, meteorol syst analyst, 64-73, EXEC OFFICER, SYSTS DEVELOP OFF, NAT WEATHER SERV, NAT OCEANIC & ATMOSPHERIC ADMIN, 73- Concurrent Pos: Environ Sci Serv Admin fel, 66-67. Mem: Am Meteorol Soc; Am Geophys Union. Res: Administration of overall applied research and development program. Mailing Add: Nat Weather Serv 8060 13th St Silver Spring MD 20910

MURISON, GERALD LEONARD, b SAfrica, May 16, 39; US citizen; m 69; c 1. CELL BIOLOGY, DEVELOPMENTAL BIOLOGY. Educ: Univ Witwatersrand, BSc, 61, MSc, 63; Johns Hopkins Univ, PhD(biol), 69. Prof Exp: Jr lectr biochem, Univ Witwatersrand, 61-64; Pa Plan scholar & instr histol, Med Sch, Univ Pa, 69-70; asst prof biol, Univ Miami, 70-73; ASSOC PROF BIOL, FLA INT UNIV, 73- Mem: AAAS; Soc Develop Biol; Am Soc Cell Biologists. Res: Biochemistry of development with emphasis on protein synthesis in tissues before and after birth; regulation of liver metabolism; protein biosynthesis. Mailing Add: Dept of Biol Sci Fla Int Univ Miami FL 33199

MURMANN, ROBERT KENT, b Chicago, Ill, Oct 7, 27; m 55; c 2. INORGANIC CHEMISTRY. Educ: Monmouth Col, BS, 49; Northwestern Univ, MS, 51, PhD(chem), 53. Prof Exp: Res assoc, Univ Chicago, 53-54; from instr to asst prof chem, Univ Conn, 54-57; assoc prof, 58-60, PROF CHEM, UNIV MO-COLUMBIA, 60- Mem: AAAS; Am Chem Soc. Res: Coordination compounds; rhenium chemistry. Mailing Add: Dept of Chem Univ of Mo Columbia MO 65202

MURNAGHAN, FRANCIS DOMINIC, mathematics, see 12th edition

MURNANE, THOMAS GEORGE, b Dallas, Tex, May 5, 26; m 53; c 5. VETERINARY MEDICINE, RESOURCE MANAGEMENT. Educ: Agr & Mech Col Tex, DVM, 47. Prof Exp: Pvt pract, Tex, 47-48; vet, Foot & Mouth Campaign, Joint US-Mex Comn, Mex, 48-49; US Army, 49-, vet lab officer, Area Med Labs, 49-56, vet adv, US Mil Mission, Repub Panama, 56-59, dir defense subsistence testing lab, Ill, 59-63, dep dir div vet med, Walter Reed Army Inst Res, 63-66, chief vet dept, Ninth Med Field Lab, 66-67, chief vet res div, US Army Med Res & Develop Command, 67-72, sr vet corps staff officer, Off Surgeon Gen, 72-74, CHIEF VET CORPS CAREER ACTIVITIES OFF, ARMY MED DEPT PERSONNEL SUPPORT AGENCY, OFF SURGEON GEN, 74- Concurrent Pos: Consult vet pub health, Surgeon Gen, Dept Army, Washington, DC, 74- Mem: Am Vet Med Asn; US Animal Health Asn; Asn Mil Surgeons US; Am Asn Vet Lab Diagnosticians; Conf Res Workers Animal Dis. Res: Food hygiene; rabies; leptospirosis; encephalomyocarditis virus; foot and mouth disease; military participation in emergency animal disease programs; career management of health professionals. Mailing Add: Army Med Dept Personnel Support Agency Washington DC 20314

MURNANE, THOMAS WILLIAM, b Cambridge, Mass, July 18, 36; m 65; c 2. ORAL SURGERY, ANATOMY. Educ: Tufts Univ, BS, 58, DMD, 62, PhD(anat), 68; Am Bd Oral Surg, dipl. Prof Exp: Fel anesthesia, Tufts Univ-Boston City Hosp, 63-64; fel, Queen Victoria Hosp, Eng, 65; NIH fel, Med Col Va, 65-66; sr instr anat, Sch Med, 67-68, asst prof oral surg, Sch Dent Med, 67-71, actg dean, 71-72, ASSOC PROF ORAL SURG, SCH DENT MED, TUFTS UNIV, 71-, LECTR ANAT, SCH MED, 71-, ASSOC DEAN SCH DENT MED & CHMN STEERING COMT, UNIV, 72- Concurrent Pos: Vis assoc oral surgeon, Boston City Hosp & Tufts New Eng Med Ctr, 68; consult, Coun Dent Educ, Am Dent Asn; mem coun oral surg, Pan-Am Med Asn. Mem: Am Dent Asn; Am Soc Oral Surg. Res: Joints; synovial membrane; connective tissue; electron and light microscopy; experimental pathology. Mailing Add: Tufts Univ Sch of Dent Med 136 Harrison Ave Boston MA 02111

MURNIK, MARY RENGO, b Manistee, Mich, Aug 30, 42; m 70. GENETICS. Educ: Mich State Univ, BS, 64, PhD(zool, genetics), 69. Prof Exp: Asst prof biol & genetics, Fitchburg State Col, 68-70; asst prof, 70-73, ASSOC PROF BIOL & GENETICS, WESTERN ILL UNIV, 73- Mem: AAAS; Genetics Soc Am; Environ Mutagen Soc. Res: Environmental mutagenesis, primarily pesticides; Drosophila female behavior in reproductive isolation. Mailing Add: Dept of Biol Sci Western Ill Univ Macomb IL 61455

MUROGA, SABURO, b Numazu, Japan, Mar 15, 25; m 56; c 4. COMPUTER SCIENCE. Educ: Univ Tokyo, BE, 47, PhD(info theory), 58. Prof Exp: Mem res staff, Nat Railway Pub Corp, Japan, 47-49; mem eng staff, Govt Radio Regulatory Comn, Japan, 50-51; mem res staff, Elec Commun Lab, Nippon Tel & Tel Pub Corp, Japan, 51-60 & IBM Res Ctr, Yorktown Heights, NY, 60-64; PROF COMPUT SCI, UNIV ILL, URBANA, 64- Mem: Inst Elec & Electronic Engrs; Asn Comput Mach; Info Processing Soc Japan; Inst Electronics & Commun Engrs Japan. Res: Logical design; switching theory; integrated circuits; mathematical programming; integrated programming. Mailing Add: Dept of Comput Sci Univ of Ill Urbana IL 61801

MUROV, STEVEN LEE, b Los Angeles, Calif, Oct 16, 40; m 66. PHOTOCHEMISTRY. Educ: Harvey Mudd Col, BS, 62; Univ Chicago, PhD(photochem), 67. Prof Exp: NIH fel photochem, Calif Inst Technol, 67-68; asst prof org chem, State Univ NY Stony Brook, 68-73; ASSOC PROF PHYS SCI, SANGAMON STATE UNIV, 73- Mem: Am Chem Soc; The Chem Soc. Res: Organic photochemistry; organic pollutants in water; methods in chemistry education at the elementary and secondary school levels. Mailing Add: Dept of Phys Sci Sangamon State Univ Springfield IL 62708

MURPHEY, BARRY THANE, organic chemistry, see 12th edition

MURPHEY, BYRON FREEZE, b Great Falls, Mont, Aug 12, 18; m 41; c 2. PHYSICS. Educ: Univ Mont, BA, 39; Univ Minn, MA, 41, PhD(physics), 48. Prof Exp: Physicist underwater ord, Naval Ord Lab, 41-45; physicist, Minn Mining & Mfg Co, 48-49, physics sect leader, 53-58; div supvr weapons effects, Sandia Corp, NMex, 49-53, div supvr underground explosions, 58-61, div supvr appl phys sci, 61-62, dept mgr nuclear burst physics, 62-67, dir underground exp, 67-71; DIR APPL RES, SANDIA LABS, 71- Mem: Fel Am Phys Soc; Am Asn Physics Teachers. Res: Magnetism; solid state physics; effects of nuclear weapons. Mailing Add: 1822 Vancouver Way Livermore CA 94550

MURPHEY, FRANK J, b Wilmington, Del, June 8, 20; m 42; c 1. BIOLOGY. Educ: Univ Del, BS, 53, MS, 54, PhD, 61. Prof Exp: Asst entom, 52-55, res analyst, 56-57, res assoc, 57-74, ASSOC RES PROF ENTOM, UNIV DEL, 74- Mem: Am Mosquito Control Asn. Res: Insect physiology, especially blood sucking arthropods; zoology; entomology; environmental toxicology. Mailing Add: Dept of Entom Univ of Del Newark DE 19711

MURPHEY, MILLEDGE, b Augusta, Ga, Dec 15, 12; m 37; c 3. ENTOMOLOGY. Educ: Univ Fla, BSA, 35; Okla State Univ, PhD(entom), 53. Prof Exp: From dist supvr to state supvr insect control proj, USDA, 35-47; entomologist & asst dir, State Dept Entom, Ga, 37-42; from asst prof to assoc prof, 47-65, PROF ENTOM, UNIV FLA, 65- Mem: Entom Soc Am. Res: Bee behavior; use of antibiotics for bee diseases;

insects of importance to fruit and vegetable crops; use of insects in biological control. Mailing Add: 3093 McCarty Hall Univ of Fla Gainesville FL 32611

MURPHEY, RHOADS, geography of China & India, historical geography, see 12th edition

MURPHEY, ROBERT STAFFORD, b Littleton, NC, Oct 29, 21; m 46; c 2. MEDICINAL CHEMISTRY. Educ: Univ Richmond, BS, 42; Univ Va, MS, 47, PhD(org chem), 49. Prof Exp: Res chemist medicinal chem, 48-54, dir chem res, 54-56, asst dir res, 56-58, dir, 58-60, dir int res, 60-66, asst vpres, 67-73, DIR SCI DEVELOP, A H ROBINS & CO, INC, 66-, VPRES, 73- Mem: AAAS; Am Chem Soc. Res: Organic chemistry. Mailing Add: 1407 Cummings Dr Richmond VA 23220

MURPHEY, RODNEY KEITH, b Minneapolis, Minn, May 6, 42; m 64; c 2. NEUROBIOLOGY. Educ: Univ Minn, Minneapolis, BA, 65, MS, 67; Univ Ore, PhD(biol), 70. Prof Exp: NIH fel, Univ Calif, Berkeley, 70-71; asst prof zool, Univ Iowa, 71-74; vis asst prof biol, Univ Ore, 74-75; RES ASSOC, CTR FOR NEUROBIOL, STATE UNIV NY ALBANY, 75- Mem: AAAS; Brit Soc Exp Biol; Soc Neurosci; Am Soc Zoologists. Res: Neural mechanisms of animal behavior; mechanisms of orientation in animals; neurophysiology of invertebrates; developmental neurobiology. Mailing Add: Ctr for Neurobiol State Univ NY Albany NY 12203

MURPHEY, WAYNE K, b Glenolden, Pa, Sept 5, 27; m 52; c 5. WOOD TECHNOLOGY. Educ: Pa State Univ, BS, 52, MF, 53; Univ Mich, PhD, 61. Prof Exp: Engr res, Koppers Co, Inc, 53-55; instr forest prod, Ohio Agr Exp Sta, 55-60; asst prof wood utilization, 60-67, head dept wood sci & technol & in chg forestry res lab, 67-68, actg asst dean resident instr, Col Agr, 68-70, PROF WOOD TECHNOL & ASST DIR SCH FOREST RESOURCES, PA STATE UNIV, UNIVERSITY PARK, 70- Res: Physical and mechanical properties of wood; adhesives; wood preservation and seasoning; wooden component design; extractives; effects of environment on wood and fiber properties. Mailing Add: 309 Forest Resources Lab Pa State Univ University Park PA 16802

MURPHEY, WILBUR ALFORD, b Augusta, Ga, Apr 27, 23; m 45; c 3. RESEARCH ADMINISTRATION, POLYMER CHEMISTRY. Educ: Univ Ga, BS, 44; Purdue Univ, PhD(org chem), 50. Prof Exp: Chemist, Eastman Kodak Co, 44-46; asst, Purdue Univ, 46-48; res chemist, 50-55, RES SUPVR, E I DU PONT DE NEMOURS & CO, INC, 55- Mem: Am Chem Soc. Res: Synthetic, organic and polymer chemistry. Mailing Add: 2533 Colton Dr Richmond VA 23235

MURPHEY, WILLIAM HOWARD, b Detroit, Mich, Aug 18, 35; m 60; c 3. BIOCHEMISTRY, MICROBIOLOGY. Educ: Univ Calif, Los Angeles, BA, 57; Purdue Univ, MS, 60; Univ Tex, PhD(microbiol), 64. Prof Exp: Res assoc, 66-70, RESIDENT ASST PROF PEDIAT, CHILDREN'S HOSP, SCH MED, STATE UNIV NY, BUFFALO, 70- Concurrent Pos: Fel biochem, Brandeis Univ, 64-66. Mem: Am Soc Microbiol. Res: Biochemistry and genetics of mannitol fermentation; biochemistry of microbial malate dehydrogenases; biochemical genetics and inborn errors of metabolism. Mailing Add: Dept of Pediat State Univ of NY Sch of Med Buffalo NY 14214

MURPHREE, HENRY BERNARD SCOTT, b Decatur, Ala, Aug 11, 27; m 53; c 3. CLINICAL PHARMACOLOGY. Educ: Yale Univ, BA, 50; Emory Univ, MD, 59. Prof Exp: Instr pharmacol, Emory Univ, 59-61, intern med, Grady Mem Hosp, 59-61; asst chief pharmacol sect, Bur Res Neurol & Psychiat, NJ Neuropsychiat Inst, 61-68, mem staff, Inst, 62-68; assoc prof psychiat, 68-71, mem prof staff, Rutgers Ctr Alcohol Studies, 68-72, assoc, Grad Fac Psychol, 69-72, PROF PSYCHIAT & PHARMACOL, RUTGERS MED SCH, 71-, MEM GRAD FAC PSYCHOL & DIR LIAISON PSYCHIAT, 72- Concurrent Pos: Consult, Princeton Hosp, NJ, 64-75; lectr, Hahnemann Med Col, 65-; chief psychiat, Raritan Valley Hosp, 72- Mem: AAAS; Am Soc Pharmacol & Exp Therapeut; NY Acad Sci; Soc Biol Psychiat; Am Col Neuropsychopharmacol. Res: Human psychopharmacology, psychophysiology, psychometrics, electronics, computer techniques, as all these come together in the understanding of the biological correlates and determinants of behavior. Mailing Add: Dept of Psychiat Rutgers Med Sch Piscataway NJ 08854

MURPHREE, R L, physiology, see 12th edition

MURPHY, ALAN PEARCE, organic chemistry, physical chemistry, see 12th edition

MURPHY, ALEXANDER JAMES, b New York, NY, May 19, 39; m 60; c 1. BIOCHEMISTRY. Educ: Brooklyn Col, BS, 62; Yale Univ, PhD(biochem), 67. Prof Exp: Am Heart Asn spec fel, Univ Calif, San Francisco, 67-70; asst mem dept contractile proteins, Inst Muscle Dis, 70-72; asst prof physiol, 72-74, ASST PROF BIOCHEM, SCH DENT, UNIV OF THE PAC, 74- Concurrent Pos: NIH career develop award, 72-77. Mem: Biophys Soc; NY Acad Sci. Res: Protein structure; active sites of contractile and membrane proteins; synthesis of nucleotide analogs. Mailing Add: Dept of Biochem Sch of Dent Univ of the Pac 2155 Webster San Francisco CA 94115

MURPHY, ALLAN HUNT, atmospheric sciences, mathematics, see 12th edition

MURPHY, ALLEN EMERSON, b Barnesville, Ohio, Aug 9, 21; m 43; c 3. GEOLOGY. Educ: Mt Union Col, AB, 43; WVa Univ, MS, 48; Syracuse Univ, PhD, 55. Prof Exp: Topog engr, US Coast & Geod Surv, 43; geol engr, Guy B Panero, 48; from asst prof to assoc prof geol, 48-61, PROF GEOL, POTOMAC STATE COL, WVA UNIV, 61-, HEAD DEPT GEOL & GEOG, 48- Mem: Geol Soc Am. Res: Physical, historical geomorphology and general geology. Mailing Add: Dept of Geol Potomac State Col Keyser WV 26726

MURPHY, BERNARD T, b Hull, Eng, May 30, 32; m 59; c 2. PHYSICS. Educ: Univ Leeds, BSc, 53, PhD(physics), 59. Prof Exp: Physicist-engr, Mullard Res Labs, Eng, 56-59; supvry engr, Westinghouse Elec Co, Pa, 59-62; dir develop, Siliconix, Inc, Calif, 62-63; DEPT HEAD, BELL TEL LABS, INC, 63- Mem: Am Phys Soc; Inst Elec & Electronics Engrs. Res: Medical physics; electron beam studies; integrated circuit structures. Mailing Add: Bell Tel Labs Inc Mountain Ave Murray Hill NJ 07974

MURPHY, BEVERLEY (ELAINE) PEARSON, b Toronto, Ont, Mar 15, 29; m 58; c 2. BIOCHEMISTRY, ENDOCRINOLOGY. Educ: Univ Toronto, BA, 52, MD, 56; McGill Univ, MSc, 60, PhD(invest med), 64. Prof Exp: Res assoc & consult endocrinol, 64-75, DIR, ENDOCRINOL LAB, QUEEN MARY VET HOSP, 75- Concurrent Pos: Med Res Coun Can fel, McGill Univ, 64; Med Res Coun Can assoc, 68-; assoc prof med, McGill Univ, 70-75, prof, 75-; asst physician, Royal Victoria Hosp, 70-; consult, Lab Med, Reddy Mem Hosp, 71-; asst physician, Montreal Gen Hosp, 71-, asst obstetrician & gynecologist, 72-75, assoc obstetrician & gynecologist, 75- Mem: Can Soc Clin Invest; Am Soc Clin Invest; Endocrine Soc. Res: Protein-binding of substances in plasma, particularly hormones and application to their determination by competitive protein-binding analysis; hormonal changes in the developing fetus. Mailing Add: Montreal Gen Hosp Res Inst 1650 Cedar Ave Montreal ON Can

MURPHY, BRIAN DONAL, b Dublin, Ireland, May 31, 39; US citizen; m 67; c 2. PHYSICS. Educ: Nat Univ Ireland, BSc, 61, MSc, 63; Univ Va, PhD(physics), 73. Prof Exp: Res officer soil physics, Agr Inst, Dublin, 63-65; res assoc radiation physics, Med Col Va, 65-66; res asst physics, Univ Va, 68-72; res assoc, Univ Wis-Madison, 72-74; TECH COMPUT SPECIALIST, NUCLEAR DIV, UNION CARBIDE CORP, 74- Mem: Am Phys Soc. Res: Applications of computers in physics; computer modeling; environmental physics; nuclear physics. Mailing Add: Union Carbide Corp Nuclear Div PO Box X Oak Ridge TN 37830

MURPHY, BRIAN LOGAN, b Hartford, Conn, Apr 24, 39; m 61; c 6. AIR POLLUTION, SCIENCE POLICY. Educ: Brown Univ, ScB, 61; Yale Univ, MS, 63, PhD(physics), 66. Prof Exp: Physicist, Mt Auburn Res Assocs, Inc, 65-75; CHIEF SCIENTIST & DEP MGR AIR QUAL STUDIES DIV, ENVIRON RES & TECHNOL, INC, CONCORD, 75- Res: Air quality diffusion modeling; fluid mechanics; meteorology; energy and environmental policy analysis. Mailing Add: 101 Avalon Rd Waban MA 02168

MURPHY, BRUCE DANIEL, b Denver, Colo, May 16, 41; m 67. REPRODUCTIVE PHYSIOLOGY. Educ: Colo State Univ, BSc, 65, MSc, 69; Univ Sask, PhD(physiol), 73. Prof Exp: Asst prof zool, Univ Idaho, 72-73; ASST PROF BIOL, UNIV SASK, 73- Concurrent Pos: Res consult, Ctr Nat Sci Invest, Cuba, 73- Mem: Soc Study Reprod; Can Soc Endocrinol & Metab. Res: Reproductive physiology of ovulation, implantation and sexual maturation in mammals. Mailing Add: Dept of Biol Univ of Sask Saskatoon SK Can

MURPHY, CHARLES FRANKLIN, b Des Moines, Iowa, Dec 13, 33; m 61; c 1. PLANT BREEDING. Educ: Iowa State Univ, BS, 56, PhD(crop breeding), 61; Purdue Univ, MS, 57. Prof Exp: Asst prof, 60-67, ASSOC PROF CROP SCI, NC STATE UNIV, 67- Mem: AAAS; Am Soc Agron. Res: Effects of diverse polygenic systems on yield; yield components and other quantitative characters in oats; small grain breeding. Mailing Add: Dept of Crop Sci NC State Univ Raleigh NC 27607

MURPHY, CHARLES FRANKLIN, b Ithaca, NY, June 9, 40; m 63. MEDICINAL CHEMISTRY. Educ: Rochester Inst Technol, BS, 63; Iowa State Univ, PhD(org chem), 66. Prof Exp: NSF fel org chem, Inst Chem, Strasbourg, France, 66-67; sr chemist, 67-71, HEAD ORG CHEM DEPT, LILLY RES LABS, ELI LILLY & CO, 71- Mem: Am Chem Soc. Res: Natural products chemistry, especially in the study of antibiotics. Mailing Add: M705 Org Chem Dept Chem Res Div Lilly Res Labs Eli Lilly & Co Indianapolis IN 46206

MURPHY, CHARLES THORNTON, b Boston, Mass, May 20, 38; m 69; c 4. ELEMENTARY PARTICLE PHYSICS. Educ: Princeton Univ, AB, 59; Univ Wis, MA, 61, PhD(physics), 63. Prof Exp: Res assoc physics, Univ Wis, 63-64; asst prof, Univ Mich, Ann Arbor, 64-68; from asst prof to assoc prof, Carnegie-Mellon Univ, 68-73; PHYSICIST, FERMI NAT ACCELERATOR LAB, 73- Mem: Am Phys Soc; Am Asn Physics Teachers. Res: Experimental high energy physics with bubble chambers; weak interactions; bubble chamber and particle beam technology. Mailing Add: Fermilab Box 500 Batavia IL 60510

MURPHY, CLARENCE JOHN, b Manchester, NH, Apr 20, 34; m 60; c 3. ORGANIC CHEMISTRY, INORGANIC CHEMISTRY. Educ: Univ NH, BS, 55, MS, 57; Univ Buffalo, PhD(organometallic chem), 62. Prof Exp: Res assoc chem, Mass Inst Technol, 60-61; from asst prof to assoc prof & chmn dept, Ithaca Col, 61-69; PROF CHEM & CHMN DEPT, E STROUDSBURG STATE COL, 69- Concurrent Pos: NSF res vis prof, Cornell Univ, 67-69. Mem: AAAS; Am Chem Soc; Soc Appl Spectros; Am Inst Chemists; Coblenz Soc. Res: Absorption spectroscopy of group IIB complexes. Mailing Add: Dept of Chem E Stroudsburg State Col East Stroudsburg PA 18301

MURPHY, CLIFFORD ELYMAN, b Blocher, Ind, Apr 2, 12; m 40; c 2. BIOLOGY. Educ: Hanover Col, AB, 36; Univ Ill, MS, 48; Univ Okla, PhD(limnol), 62. Prof Exp: Pub sch teacher, Ind, 36-44, 46-47; instr zool, Univ Ill, 47-48; from asst prof to assoc prof biol, 48-68, PROF BIOL, TEX CHRISTIAN UNIV, 68- Mem: AAAS; Am Soc Limnol & Oceanog; Am Micros Soc; World Maricult Soc; Am Water Resources Asn. Res: Ecology of impoundments; water pollution, especially industrial effluents and their effects on aquatic ecology. Mailing Add: 5836 Waltham Ave Ft Worth TX 76133

MURPHY, COLLIN GRISSEAU, b Dayton, Ohio, Oct 25, 40; m 62; c 2. DEVELOPMENTAL BIOLOGY. Educ: Ohio State Univ, BS, 62; Univ Calif, Berkeley, MA, 6S, PhD(zool), 66. Prof Exp: Res asst zool, 66-69, asst res zoologist, 69-70, NIH trainee genetics, 70-71, NIH spec fel, 71-73, ASST RES GENETICIST, UNIV CALIF, BERKELEY, 73- Concurrent Pos: Lectr biol, San Francisco State Univ, 74- Mem: AAAS; Genetics Soc Am. Res: Developmental genetics; development of imaginal discs in Drosophila; disc ultrastructure and histochemistry; pattern formation. Mailing Add: Dept of Genetics Univ of Calif Berkeley CA 94720

MURPHY, CORNELIUS BERNARD, b Worcester, Mass, Dec 10, 18; m 43; c 4. CHEMISTRY. Educ: Col of the Holy Cross, BS, 41, MS, 42; Clark Univ, PhD(chem), 52. Prof Exp: From instr to asst prof chem, Col of the Holy Cross, 45-52; res chemist, Stamford Labs, Am Cyanamid Co, 52-55; develop chemist, Gen Eng Lab, Gen Elec Co, 55-57, mgr anal chem, 57-58, mgr anal & phys chem, 58-63, proj engr, 63-65; mgr mat anal, 65-70, prog mgr, 70-72, mgr chem eng, 72-73, mgr toner processing, 73-75, MAT COORDR, XEROX CORP, 76- Concurrent Pos: Mem bd dirs, Delta Labs, NY. Mem: Am Chem Soc; Electrochem Soc; NY Acad Sci; Int Confedn Thermal Anal (pres, 68-71); NAm Thermal Anal Soc. Res: Chelation; phase equilibria; differential thermal analysis. Mailing Add: Webster Res Ctr Xerox Corp 800 Phillips Rd Webster NY 14580

MURPHY, DANIEL BARKER, b Richmond Hill, NY, Apr 7, 28; m 51; c 4. ORGANIC CHEMISTRY. Educ: Fordham Univ, BS, 47, MS, 49; Pa State Univ, PhD(fuel technol), 58. Prof Exp: Asst chem, Fordham Univ, 47-49; instr, Univ Scranton, 49-51; res chemist, Picatinny Arsenal, US Dept Army, 51-54; asst fuel technol, Pa State Univ, 54-57; from instr to assoc prof, 57-70, PROF CHEM, LEHMAN COL, 70- Mem: Am Chem Soc; fel The Chem Soc; Am Carbon Soc. Res: Organic synthesis; nitrogen heterocyclics; propellants and explosives; carbon and graphite. Mailing Add: Dept of Chem Herbert H Lehman Col Bronx NY 10468

MURPHY, DANIEL LAWSON, b Tarrytown, NY, Oct 3, 29; m 54; c 2. GEOLOGY. Educ: Lehigh Univ, BA, 51; Univ Mo, MA, 55; Univ Mich, PhD(geol), 60. Prof Exp: Asst, Univ Mo, 53-55; geologist, Stand Oil Co, Tex, 55-56; from asst prof to assoc prof geol, Univ Wichita, 58-64; sr geologist, Am Metal Climax, 64-69; asst explor mgr, Jefferson Lake Sulphur Co, 69-71; vpres, Geometrics Inc, Houston, Tex, 71-74; VPRES, MINERAL RESOURCES, ANSCHUTZ CORP, DENVER, COLO, 74- Concurrent Pos: Consult geologist, 58-64. Mem: AAAS; Geol Soc Am; Soc Econ

Geol; Geochem Soc; Am Inst Mining, Metall & Petrol Eng. Res: Economic geology; metals and nonmetals, including petroleum; structural geology of Rocky Mountains and Canadian shield regions; mineral economics. Mailing Add: Anschutz Corp Mineral Resources 1110 Denver Club Bldg Denver CO 80202

MURPHY, DON ROBISON, b Long Beach, Calif, Sept 12, 29; m 58; c 4. GEOGRAPHY, GEOLOGY. Educ: Brigham Young Univ, BS, 53, MS, 54; Univ Nebr, Lincoln, PhD(geog), 69. Prof Exp: Geologist, Standard Oil Co, Tex, 54-57; geologist, Texaco, Inc, 57-61; instr geog, Univ Nev, Las Vegas, 63-65; assoc prof geog & geol, 65-74, PROF GEOG, WEBER STATE COL, 74- Concurrent Pos: NSF grant, Univ Natal, 71-72. Mem: Asn Am Geogr; Am Geog Soc; Nat Coun Geog Educ. Res: Regional climatology of western United States. Mailing Add: Dept of Geol & Geog Weber State Col Ogden UT 84403

MURPHY, DONALD G, b New York, NY, July 14, 34; m 56; c 2. MEDICAL RESEARCH, CELL BIOLOGY. Educ: Ore State Univ, BS, 56, PhD(nematol), 61. Prof Exp: NSF grant & asst prof nematol, Ore State Univ, 61-62; NIH fel & spec fel, Univ Hamburg, 62-65; res nematologist, Agr Res Serv, USDA, 65-67; grants assoc, NIH, 67; biologist, Nat Inst Child Health & Human Develop, 68-74; HEALTH SCIENTIST ADMINR, NAT INST AGING, 75- Concurrent Pos: Fel, Dept Path, Johns Hopkins Univ, 71-73. Mem: AAAS; Geront Soc; Tissue Cult Asn. Res: Cellular aging; nematode phylogeny and bionomics; evolution; research administration. Mailing Add: Nat Inst on Aging NIH Bethesda MD 20014

MURPHY, DONALD HENRY, b Plainfield, NJ, Sept 10, 34; m 67. BIOENGINEERING. Educ: Manhattan Col, BEE, 58; Polytech Inst Brooklyn, MS, 68, PhD(bioeng), 71. Prof Exp: Develop engr, Ford Instrument Co, 58-59; res engr, Repub Aviation Corp, 59-65; res engr, Grumman Aircraft Eng Corp, 65-66; RES SCIENTIST PHYSIOL SYSTS, LONG ISLAND COL HOSP, BROOKLYN, 66- Concurrent Pos: Lectr bioeng, Polytech Inst Brooklyn, 72- Res: Cardiovascular dynamics. Mailing Add: 14 Foster Pl Sea Cliff NY 11579

MURPHY, DOUGLAS RICHARD, b Sunapee, NH, Dec 28, 21; m 43; c 3. PLANT PATHOLOGY. Educ: Univ NH, BS, 49, MS, 51; Iowa State Univ, PhD(plant path), 54. Prof Exp: Res asst plant path, Univ NH, 49-51; tech sales mgr, 54-58, PROD MGR, AGR CHEM DIV, STAUFFER CHEM CO, 59- Mem: Weed Sci Soc Am; Am Entom Soc. Res: Control of oak wilt. Mailing Add: Stauffer Chem Co Westport CT 06880

MURPHY, EDUARDO S, b Utica, NY, June 8, 23; m 53; c 5. PATHOLOGY, ONCOLOGY. Educ: Univ Mich, BA, 45, MD, 46. Prof Exp: Intern med, St Joseph Hosp, Ann Arbor, Mich, 47-48; resident path, St Luke & Children's Hosp, Denver, 49-52; Barth fel, Nat Cardiol Inst, Mex, 52-53; chief path, Atomic Bomb Casualty Comn, Hiroshima, 55-57; fel radioisotopes, Roswell Park Mem Inst, 58; PROF PATH, NAT UNIV MEX, 59- Concurrent Pos: Chief path, Nat Cancer Inst, Mex, 58-66 & Hosp Santelena, Mex, 61-; pathologist, Mex Comn Nuclear Energy, 63-; secy-treas, Med Bd Path, 63-; pathologist, Hosp de Jesus, 64-70; mem & secy, Mex Bd Nuclear Med, 73- Mem: Fel Am Soc Clin Path; fel Col Am Path; Mex Soc Nuclear Med (pres, 73-74); Mex Asn Path (secy, 65-66). Res: Radioactive effects on tissue. Mailing Add: Hosp Santelena Queretaro 58 Col Roma Mexico DF Mexico

MURPHY, EDWARD G, b Sheffield, Eng, Dec 6, 21; Can citizen; m 47; c 1. MEDICINE, PEDIATRICS. Educ: Univ London, MB & BS, 45; Royal Col Physicians & Surgeons, dipl, 50; Royal Col Physicians & Surgeons Can, cert, 53; FRCPS(C), 72. Prof Exp: Intern med & surg, Guy's Univ, County Hosp, Pembury, Eng, 45; intern surg, St Mary's Hosp, Roehampton, 46; intern med, Edgware Gen Hosp, 49; intern, Evelina Children's Hosp, 49-50; intern med & pediat, 52-53, ELECTROENCEPHALOGRAPHER, HOSP SICK CHILDREN, TORONTO, ONT, 55-, CONSULT, 56-; ASSOC PROF MED & PEDIAT, UNIV TORONTO, 74- Concurrent Pos: Fel neurol serv, Hosp Sick Children, Toronto, Ont, 53-55; neurol consult, Ont Crippled Children's Ctr, 63-; assoc, Univ Toronto, 67-71, asst prof, 71-74. Mem: Can Med Asn; Can Pediat Soc; fel Can Soc Electroencephalog; assoc Can Neurol Soc; Brit Med Asn. Res: Neuromuscular disorders. Mailing Add: Hosp for Sick Children Toronto ON Can

MURPHY, EDWARD JOSEPH, b Moosomin, Sask, Apr 6, 98; m 34; c 1. BIOPHYSICS. Educ: Univ Sask, BSc, 18. Prof Exp: Asst, Univ Sask, 18-19 & Harvard Univ, 22-23; mem tech staff, Bell Tel Labs, Inc, 23-58; guest investr, Rockefeller Inst, 58-64; RES SCIENTIST, STANLEY-THOMPSON LAB SURFACE STUDIES, COLUMBIA UNIV, 64- Mem: AAAS; Am Phys Soc; Biophys Soc; NY Acad Sci. Res: Studies of dielectric properties of water absorbed on internal surfaces in substances of biological interest; conduction in fibrous proteins and other dielectrics of biological interest. Mailing Add: 217 E 66th St New York NY 10021

MURPHY, EDWIN DANIEL, b Brooklyn, NY, July 30, 17; m 42; c 7. EXPERIMENTAL PATHOLOGY. Educ: St John's Univ, NY, BS, 39; Yale Univ, MD, 43. Prof Exp: Intern surg, New Haven Hosp & Yale Univ, 43; instr path, Col Med, Univ Tenn, 46-48; pathologist, Nat Cancer Inst, 48-53; res assoc, 53-57, sci dir, 56-57, asst dir res, 57-58, staff scientist, 57-70, SR STAFF SCIENTIST, JACKSON LAB, 70- Concurrent Pos: Childs fel clin & exp oncol, Sch Med, Yale Univ, 44-46; Guest prof & Fulbright sr res award, Univ Frankfurt, 63-64; Japan Soc Prom Sci fel, 73; vis scientist, Aichi Cancer Ctr Res Inst, Nagoya, Japan, 73-74; Nat Cancer Inst spec res fel, 74. Mem: Am Asn Cancer Res. Res: Experimental tumorigenesis; pathologic anatomy of inbred mice. Mailing Add: Jackson Lab Bar Harbor ME 04609

MURPHY, ELIAS SMITH, JR, b Ogden, Utah, Apr 19, 26; m 51; c 5. NUCLEAR PHYSICS, ASTRONOMY. Educ: Brigham Young Univ, BS, 47; Colo State Univ, MS, 52, PhD(physics), 61. Prof Exp: Instr physics, Mont State Col, 53-55; instr, Idaho State Col, 55-56; instr, Colo State Univ, 56-60; asst prof, Wash State Univ, 61-65; assoc prof, 65-70, chmn dept, 70-74, PROF PHYSICS, CENT WASH STATE COL, 70- Mem: AAAS; Am Asn Physics Teachers. Res: Nuclear decay scheme studies; beta and gamma ray spectroscopy. Mailing Add: Dept of Physics Cent Wash State Col Ellensburg WA 98926

MURPHY, ELIZABETH WILCOX, b Ionia, Mich, Feb 11, 27; m 55; c 2. NUTRITION, FOOD CHEMISTRY. Educ: Cent Mich Col Educ, BS, 48; Univ Chicago, MS, 54. Prof Exp: Lab technician, Dept Home Econ, Univ Chicago, 48-51 & Nutrit Lab, Quaker Oats Co, Ill, 51-54; nutrit specialist, Human Nutrit Res Div, 54-56, res chemist, 56-65, res chemist, Consumer & Food Econ Inst, Agr Res Serv, 65-76, NUTRITIONIST, MEAT & POULTRY INSPECTION, ANIMAL & PLANT HEALTH INSPECTION SERV, USDA, 76- Mem: Am Inst Nutrit; Am Dietetic Asn; Am Chem Soc; Soc Environ Geochem & Health. Res: Nutritional aspects of policies related to labeling of meat and poultry products. Mailing Add: Prod Labels Packaging & Standards APHIS USDA 300 12th St SW Washington DC 20250

MURPHY, EUGENE VICTOR THOMAS, physical chemistry, see 12th edition

MURPHY, FREDERICK A, b New York, NY, June 14, 34; m 60; c 4. VIROLOGY, IMMUNOLOGY. Educ: Cornell Univ, BS, 57, DVM, 59; Univ Calif, Davis, PhD(comp path), 64. Prof Exp: CHIEF VIRAL PATH BR, CTR DIS CONTROL, US DEPT HEALTH, EDUC & WELFARE, 64- Concurrent Pos: Mem, Int Comt Taxonomy of Viruses; chmn, Am Comt Arthropod-borne Viruses; hon fel, John Curtin Sch Med Res, Australian Nat Univ, 70-71. Mem: Am Vet Med Asn; Am Asn Immunol; Soc Exp Biol & Med; Am Soc Microbiol; Electron Micros Soc Am. Res: Pathogenesis of viral diseases and encephalitis; electron microscopy; viral ultrastructure. Mailing Add: Viral Path Br Ctr for Dis Control Atlanta GA 30333

MURPHY, FREDERICK VERNON, b Washington, DC, Mar 26, 38; m 65; c 2. EXPERIMENTAL HIGH ENERGY PHYSICS. Educ: Georgetown Univ, BS, 59; Princeton Univ, MA, 61, PhD(physics), 67. Prof Exp: Instr physics, Princeton Univ, 66-67; ASST RES PHYSICIST, UNIV CALIF, SANTA BARBARA, 67- Res: Elementary particle physics; K meson scattering and decays; photoproduction and total photon cross sections; counters; spark and streamer chamber techniques; secondary particle beam design. Mailing Add: Dept of Physics Univ of Calif Santa Barbara CA 93106

MURPHY, GARTH IVOR, b Portland, Ore, Nov 7, 22; m 42; c 6. FISH BIOLOGY. Educ: Univ Calif, AB, 42, MA, 47, PhD, 65. Prof Exp: Asst zool, Univ Calif, 42-43; dist fisheries biologist, State Dept Fish & Game, Calif, 46-51; fishery res biologist, Pac Oceanic Fishery Invests, US Fish & Wildlife Serv, Hawaii, 51-58; coordr, Calif Coop Oceanic Fishery Invest, Scripps Inst Oceanog, Univ Calif, San Diego, 58-65; PROF OCEANOG, UNIV HAWAII, 65- Concurrent Pos: Food & Agr Orgn consult, UN, 71-; mem, Animal Species Adv Comn, State Hawaii, 71- Mem: Am Fisheries Soc; Ecol Soc Am. Res: Ecology of marine organisms; fisheries. Mailing Add: Dept of Oceanog Univ of Hawaii Honolulu HI 96822

MURPHY, GEORGE EARL, b Portland, Ore, Oct 17, 22; m 53; c 2. PSYCHIATRY. Educ: Ore State Univ, BS, 49; Washington Univ, MD, 52. Prof Exp: From intern to asst resident med, Highland-Alameda County Hosp, Oakland, Calif, 52-54; fel psychosom med, Sch Med, Washington Univ, 54-55; clin & res fel psychiat, Mass Gen Hosp, 55-56; asst resident, Renard Hosp, Barnes Hosp Group, St Louis, Mo, 56-57; from instr to assoc prof, 57-69, PROF PSYCHIAT, SCH MED, WASHINGTON UNIV, 69- Mem: Fel Am Psychiat Asn; Asn Res Nerv & Ment Dis; Psychiat Res Soc; Am Psychopath Asn; Sigma Xi. Res: Clinical and epidemiologic studies in suicide, alcoholism, drug addiction, affective disorder and life stress; problems of psychotherapy. Mailing Add: Dept of Psychiat Washington Univ Sch of Med St Louis MO 63110

MURPHY, GEORGE EDWARD, b Kansas City, Mo, Aug 22, 18; m 43. EXPERIMENTAL PATHOLOGY. Educ: Univ Kans, AB, 39; Univ Pa, MD, 43. Prof Exp: Intern, Univ Kans Hosp, 43; asst resident & res pathologist, Hosp & asst path, Sch Med, Johns Hopkins Univ, 44-45; asst physician, Rockefeller Inst Hosp, 46-53; assoc prof path, Med Col, Cornell Univ, 53-68; assoc attend pathologist, 61-67, ATTEND PATHOLOGIST, NY HOSP, 68-, PROF PATH, MED COL, CORNELL UNIV, 68- Concurrent Pos: Mem coun on arteriosclerosis & coun cardiovasc dis of the young, Am Heart Asn; Life Ins Med Res Fund fel rheumatic fever, 46-49; Helen Hay Whitney Found fel rheumatic fever, 49-53; Lederle med fac award, 54-57. Honors & Awards: William Osler Medal, Am Asn Hist Med, 43. Mem: Am Soc Exp Path; Soc Exp Biol & Med; Am Asn Path & Bact; NY Acad Med; NY Acad Sci. Res: Experimental and histopathologic studies on nature of rheumatic fever, especially rheumatic heart disease, glomerulonephritis, and arteriosclerosis; medical education. Mailing Add: Dept of Path Cornell Univ Med Col New York NY 10021

MURPHY, GEORGE GRAHAM, b Clarksville, Tenn, Aug 31, 43; m 64; c 1. HERPETOLOGY. Educ: Austin Peay State Univ, BS, 65; Miss State Univ, MS, 67, PhD(zool), 70. Prof Exp: ASSOC PROF BIOL, MID TENN STATE UNIV, 69- Mem: Sigma Xi; Soc Study Amphibians & Reptiles; Herpetologist's League; Am Soc Ichthyologists & Herpetologists. Res: Behavior of turtles; reproductive biology of turtles. Mailing Add: Dept of Biol Mid Tenn State Univ Murfreesboro TN 37130

MURPHY, GEORGE WASHINGTON, b Hot Springs, Ark, Jan 2, 19; m 45, 67; c 4. PHYSICAL CHEMISTRY. Educ: Univ Ark, AB, 40; Univ NC, PhD(phys chem), 46. Prof Exp: Asst chem, Univ NC, 40-42; res chemist, US Naval Res Lab, 42-45; from instr to asst prof chem, Univ Wis, 46-51; assoc chemist, Argonne Nat Lab, 51-53; prof chem & chmn dept, State Univ NY Col Teachers, Albany, 53-56; assoc prof, 56-59, chmn dept, 60-68, PROF CHEM, UNIV OKLA, 59- Mem: Am Chem Soc; Electrochem Soc. Res: Theory of solutions; thermodynamics; irreversible processes; electrochemistry. Mailing Add: Dept of Chem Univ of Okla Norman OK 73069

MURPHY, GRATTAN PATRICK, b Parsons, Kans, Sept 15, 35; m 61; c 3. MATHEMATICS. Educ: Rockhurst Col, BS, 57; St Louis Univ, MS, 62, PhD(math), 66. Prof Exp: Tech analyst comput prog & data reduction, McDonnell Aircraft Co, 59-61; instr math, St Louis Univ, 62-65; asst prof, 65-69, ASSOC PROF MATH, UNIV MAINE, ORONO, 69- Concurrent Pos: Guest prof, Univ Freiburg, 71-72. Mem: Am Math Soc; Math Asn Am; Ger Math Soc. Res: Geometry of generalized metric spaces. Mailing Add: Dept of Math Univ of Maine Orono ME 04473

MURPHY, HENRY BRIAN MEGGET, b Edinburgh, Scotland, Sept 17, 15; m 46; c 5. PSYCHIATRY, SOCIOLOGY. Educ: Univ Edinburgh, MB, 38, MD, 58; London Sch Hyg, dipl pub health, 52; New Sch Social Res, PhD(sociol), 59. Prof Exp: Med supvr refugee care, UN Relief & Rehab Agency, 46-47; med consult, Int Relief Orgn, 47-49; dir student health serv, Univ Malaya, 52-57; Milbank Mem Fund fel, Milbank Mem Fund & New Sch Social Res, 57-59; from asst prof to assoc prof, 59-68, PROF PSYCHIAT, McGILL UNIV, 68- Concurrent Pos: Consult, WHO, Asia, 55, Europe, 56, Caribbean, 65, Brazil, 74 & Geneva, 75; res consult, Laurentides Hosp, 66-71; NIMH fel, Social Sci Res Inst, Univ Hawaii, 70; res dir, Albert Prevost Inst, 71-; chmn, Inter-univ Group for Res in Med Anthrop & Transcult Psychiat, 74- Res: Transcultural psychiatry; social psychiatry; evaluation of health services; medical anthropology; culture and personality studies. Mailing Add: Transcult Psychiat Studies McGill Univ Montreal PQ Can

MURPHY, HENRY D, b Hartshorne, Okla, Mar 21, 29. HISTOLOGY, ANATOMY. Educ: Univ Calif, Berkeley, AB, 58, MA, 60; Univ Calif, San Francisco, PhD(anat), 65. Prof Exp: Asst prof, 65-74, ASSOC PROF ANAT & PHYSIOL, SAN JOSE STATE UNIV, 74- Res: Endocrine research on role of follicle stimulating hormone on the testes of rats; comparative histological study of marine mammals, seals, sea-lions, porpoises and various whales. Mailing Add: Dept of Anat San Jose State Univ San Jose CA 95114

MURPHY, HUBERT WILLIAM, b Circle, Mont, Nov 3, 15; m 50; c 3. PHARMACEUTICAL CHEMISTRY. Educ: Univ Mont, BS, 37, MS, 39; Purdue Univ, PhD(pharmaceut chem), 42. Prof Exp: Mgr, Mission Drug Co, Mont, 37-38; asst pharm, Univ Mont, 38-39; asst pharmacist, Purdue Univ, 39-41; pharmaceut chemist, 43-57, res assoc chem, 57-65, head pharmaceut res dept, 65-68, RES ASSOC ANAL CHEM, ELI LILLY & CO, 68- Mem: Am Chem Soc; Am Pharmaceut Asn. Res: Synthesis of oxytoxics and plant growth hormones; synthesis of local anesthetics;

analysis and stability of vitamins; improved medicinal capsules; erythromycin esters and derivatives. Mailing Add: Dept M-772 Eli Lilly & Co 307 E McCarty St Indianapolis IN 46206

MURPHY, JAMES A, b Philadelphia, Pa, July 28, 35; m 63; c 3. PHYSICAL CHEMISTRY, SURFACE CHEMISTRY. Educ: St Joseph's Col, BS, 57; Iowa State Univ, PhD(phys chem), 63. Prof Exp: Sr chemist, 63-66, res chemist, 66-69, sr res chemist, 69-72, PROJ LEADER, CORNING GLASS WORKS, 72- Mem: Am Chem Soc; Am Inst Chemists. Res: Materials research, especially thin films; surface chemistry and the interaction of solid, liquids and gases with solids. Mailing Add: 106 Fairview Ave Painted Post NY 14870

MURPHY, JAMES CLAIR, b Salt Lake City, Utah, July 29, 31. PATHOLOGY. Educ: Utah State Univ, BS, 57; Wash State Univ, DVM, 61; Colo State Univ, PhD(path), 66. Prof Exp: Fel path, Col Vet Med, Colo State Univ, 62-66; res fel, Harvard Med Sch, 66-67; pathologist, Hazleton Labs, Va, 67-68, supvr teratol sect, 67-68; instr, Sch Med, Tufts Univ, 68-74, asst prof surg, 74-75; VET PATHOLOGIST MED DEPT & DIR RES ANIMAL LAB, DIV LAB ANIMAL MED, MASS INST TECHNOL, 75- Concurrent Pos: Mem spec sci staff, New Eng Med Ctr Hosps, 68-75, vet & dir res animal lab, 70-75. Mem: Am Vet Med Asn; Am Col Vet Path; Int Acad Path. Res: Pathogenesis of infectious diseases. Mailing Add: Div Lab Animal Med Mass Inst Technol Cambridge MA 02139

MURPHY, JAMES FRANCIS, b Lethbridge, Alta, Aug 19, 22; nat US; m 49; c 4. PHYSICAL CHEMISTRY. Educ: Univ Alta, BS, 45, MS, 46; Univ Calif, PhD(chem), 50. Prof Exp: Asst prof chem, Univ Idaho, 49-51; res chemist, Dept Metall Res, Kaiser Aluminum & Chem Corp, 51-55; mgr phys chem, Gen Eng Lab, Gen Elec Co, 55-58; chief chem sect, Metall Labs, Olin Mathieson Chem Corp, 58-66; res assoc, Kaiser chem Res Labs, 66-68, lab mgr, 68-71; MGR CHEM RES, KAISER ALUMINUM & CHEM CORP, 71- Mem: Am Chem Soc; Electrochem Soc; Int Soc Gen Semantics. Res: Surface chemistry and physics; electrochemistry; inorganic chemistry; corrosion and finishing reactions; welding and fusion reactions; adhesion. Mailing Add: Kaiser Aluminum & Chem Corp PO Box 870 Pleasanton CA 94566

MURPHY, JAMES GILBERT, b Brooklyn, NY, July 25, 19; m 47; c 8. ORGANIC CHEMISTRY. Educ: St Francis Col, NY, BS, 47; Polytech Inst Brooklyn, MS, 50; Georgetown Univ, PhD(chem), 59. Prof Exp: Asst chemist, Nat Oil Prod Co, NJ, 40-42; chemist, Evans Res & Develop Corp, NY, 45-51; res chemist, NIH, 52-71; CONSULT CHEMIST, 72- Mem: Am Chem Soc. Res: Organic sulfur compounds; medicinal chemistry; biological substrates. Mailing Add: 5111 Edgemoor Lane Bethesda MD 20014

MURPHY, JAMES JOSEPH, b New York, NY, Apr 29, 38; m 61; c 3. EXPERIMENTAL SOLID STATE PHYSICS. Educ: St Joseph's Col, Pa, BS, 59; Fordham Univ, MS, 61, PhD(physics), 71. Prof Exp: From instr to asst prof, 61-71, chmn dept, 66-75, ASSOC PROF PHYSICS, IONA COL, 71- Concurrent Pos: Mem adj fac, Bergen Community Col, 69- Mem: AAAS; Am Asn Physics Teachers. Res: Magnetism in transition metals; spin echo studies of ferromagnetic alloys. Mailing Add: Dept of Physics Iona Col New Rochelle NY 10801

MURPHY, JAMES L, b Pasadena, Calif, May 21, 27; m 55; c 4. FOREST ECONOMICS, FORESTRY. Educ: Utah State Univ, BS, 58, MS, 59; Univ Mich, PhD(forest mgt-econ), 65. Prof Exp: Fire control aid, US Forest Serv, 45-48, forester, 58-61, res forester, Calif, 61-68, proj leader forest fire sci proj, Wash, 68-71, asst to dep chief res, Washington, DC, 71-73, prog leader forest fire prev proj, 73-74; MEM STAFF, PAC SOUTHWEST EXP STA, US FOREST SERV, 74- Concurrent Pos: Assoc prof, Col Forest Resources, Univ Wash, 68-; consult, US Peace Corps, 71 & Repub Chile, 71-; chmn, Nat Insurgency Wildfire Prev Anal Task Force, 73-74. Mem: AAAS; Soc Am Foresters. Res: Forest fire research and economics; wildfire prevention; evaluation systems; early warning systems; behavioral studies. Mailing Add: Pac Southwest Exp Sta Box 245 Berkeley CA 94701

MURPHY, JAMES LEE, b Grand Ledge, Mich, Aug 29, 40; m 65; c 4. MATHEMATICS. Educ: Univ Detroit, BA, 64; Mich State Univ, MS, 66, PhD(math), 70. Prof Exp: Admin asst math, Mich State Univ, 66-69, instr, 69-70; ASST PROF MATH, CALIF STATE COL, SAN BERNARDINO, 70- Mem: Math Asn Am. Res: Piecewise linear topology in Euclidean four-space. Mailing Add: Dept of Math Calif State Col San Bernardino CA 92407

MURPHY, JAMES SLATER, b New York, NY, June 2, 21; m 48, 64; c 6. MICROBIOLOGY. Educ: Johns Hopkins Univ, MD, 45. Prof Exp: Intern med, Johns Hopkins Univ, 45-46; USPHS fel, 48-50, Am Cancer Soc fel, 50-51; from asst to asst prof, 51-60, ASSOC PROF VIROL & MED, ROCKEFELLER UNIV, 60- Mem: AAAS; Soc Exp Biol & Med; Am Soc Microbiol; Harvey Soc; Am Asn Immunologists. Res: Virology; influenza; virus development cycle; genetics; bacteriophage; aging and nutrition of Crustacea. Mailing Add: 177 E 64th St New York NY 10021

MURPHY, JAMES WALLACE, b St John, NB, Mar 30, 14. PHYSICAL CHEMISTRY. Educ: St Joseph's Univ, BA, 34; Univ Toronto, MA, 43, PhD(chem), 58. Prof Exp: Lectr chem, Loyola Col, Can, 48-49; lectr, St Paul's Col, Man, 49-50; from asst prof to assoc prof, 51-74, PROF CHEM, ST MARY'S UNIV, NS, 74- Mem: AAAS; Chem Inst Can. Res: Properties of ionic melts and fog particles. Mailing Add: Dept of Chem St Mary's Univ Halifax NS Can

MURPHY, JOHN CORNELIUS, b Wilmington, Del, Feb 28, 36; m 58; c 7. PHYSICS. Educ: Cath Univ Am, BA, 57, PhD(physics), 71; Univ Notre Dame, MS, 59. Prof Exp: PHYSICIST, APPL PHYSICS LAB, JOHNS HOPKINS UNIV, 59- Mem: Am Phys Soc. Res: Microwave-optical double resonance experiments on excitation migration in solids; photo acoustic spectroscopy; electron spin resonance of electro generated radical ions in solution, including double resonance in chemiluminescence. Mailing Add: Johns Hopkins Appl Physics Lab Johns Hopkins Rd Laurel MD 20810

MURPHY, JOHN F, b Hazleton, Pa, Jan 29, 25; m 51; c 8. MICROBIOLOGY, BIOCHEMISTRY. Educ: Georgetown Univ,.BS, 49; Pa State Univ, MS, 51, PhD, 53. Prof Exp: Res bacteriologist, 53-55, head poultry res, 55-57, asst to dir res, 57-58, dir res, 58-65, vpres, 65-72, DIR, SWIFT & CO, 72- Mem: Am Soc Microbiol; Inst Food Technol. Res: Applied microbiology; research administration. Mailing Add: Swift & Co 115 W Jackson Blvd Chicago IL 60608

MURPHY, JOHN FRANCIS, b Cranston, RI, Aug 27, 22; m 45; c 2. GEOLOGY. Educ: Dartmouth Col, AB, 47, AM, 49. Prof Exp: Geologist, 51-65, chief org fuels br, Colo, 65-66, geologist, Heavy Metals Br, 67-68, dep assoc chief geologist, 68-72, DEP CHIEF OFF ENERGY RESOURCES, US GEOL SURV, 72- Mem: AAAS; Geol Soc Am; Am Asn Petrol Geologists; Soc Econ Geologists. Res: Structural geology; stratigraphy; petrology. Mailing Add: 8612 Tuckerman Lane Potomac MD 20854

MURPHY, JOHN JOSEPH, physics, see 12th edition

MURPHY, JOHN JOSEPH, b Tucson, Ariz, July 28, 40; m 64; c 4. BIOCHEMISTRY, PLANT PHYSIOLOGY. Educ: Univ Ariz, BS, 62, MS, 63; Purdue Univ, Lafayette, PhD(biochem), 66. Prof Exp: NSF fel, King's Col, Univ London, 66-67; res biochemist, Agr Res Ctr, Stauffer Chem Co, Calif, 67-74; SR RES BIOCHEMIST, CHEMAGRO AGR DIV, MOBAY CHEM CORP, 74- Mem: Am Chem Soc; Am Inst Biol Sci; Am Soc Plant Physiologists; Am Soc Agron; Weed Sci Soc Am. Res: Metabolism and mechanisms of action of pesticides in plants, animals and soil. Mailing Add: Chemagro Agr Div Mobay Chem Corp PO Box 4913 Kansas City MO 64120

MURPHY, JOHN JOSEPH, b New York, NY, July 10, 34; m 65; c 2. ORGANIC POLYMER CHEMISTRY, PHOTOCHEMISTRY. Educ: Manhattan Col, BS, 56; Niagara Univ, MS, 58. Prof Exp: Res chemist, E I du Pont de Nemours & Co, Inc, 62-69; SR RES ASSOC CHEM, INT PAPER CO, 69- Mem: Am Chem Soc. Res: Investigation of polymeric systems, curable by high energy radiation sources, mainly ultraviolet and electron beam with application of these systems to organic coatings, inks. Mailing Add: Int Paper Co Corp Res Ctr Tuxedo Park NY 10987

MURPHY, JOHN JOSEPH, b Scranton, Pa, Oct 2, 20; m 44; c 6. UROLOGY, SURGERY. Educ: Univ Scranton, BS, 42; Univ Pa, MD, 45. Prof Exp: Asst instr surg, Harrison Dept Surg Res, Sch Med, Univ Pa, 48-52; sr instr surg, Dept Urol, Sch Med, Univ Mich, 52-53; assoc urol, Hosp, 53-56, instr, 56-58, from asst prof to assoc prof, 56-64, PROF UROL, SCH MED, UNIV PA, 64- Concurrent Pos: Fel, Harrison Dept Surg Res, Sch Med, Univ Pa, 48-51, Am Cancer Soc fel, 51-52; Harrison fel urol surg & Am Cancer Soc fel, Dept Urol, Sch Med, Univ Mich, 52-53; Ravidin traveling fel, 52-53; consult urologist, Vet Admin Hosp, 53- & Children's Seashore House, Atlantic City, NJ; consult, Univ Hosp, Pa. Mem: Am Soc Exp Path; Am Surg Asn; Am Urol Asn; Am Col Surg; Am Asn Genito-Urinary Surg. Res: Lymphatic system of the kidney; hydrodynamics of the urinary tract; hypertension as related to the kidney; renal healing; pyelonephritis; cineradiography in urology. Mailing Add: Div of Urol Univ of Pa Sch of Med Philadelphia PA 19104

MURPHY, JOHN RIFFE, b Hooker, Okla, Apr 12, 42; m 62; c 3. EXPERIMENTAL STATISTICS. Educ: Panhandle State Col, BS, 64; Okla State Univ, MS, 67, PhD(statist), 73. Prof Exp: Math statistician, Control Systs Div, Environ Protection Agency, Research Triangle Park, NC, 71-72; SR STATISTICIAN, ELI LILLY & CO, 74- Concurrent Pos: Lectr statist, Butler Univ, 75. Mem: Am Statist Asn; Biomet Soc. Res: Development and study of statistical procedures for obtaining an objective grouping in a set of observed means, especially multiple decision procedures. Mailing Add: Eli Lilly & Co Dept MC730 307 EMcCarty St Indianapolis IN 46206

MURPHY, JOHN THOMAS, b Yonkers, NY, Mar 14, 38; c 4. MEDICAL PHYSIOLOGY, NEUROLOGY. Educ: Columbia Univ, MD, 63; McGill Univ, PhD(neurol & neurosurg), 68. Prof Exp: Intern med, Columbia Univ, 63-64, resident surg, 64-65; fel electroencephalography & clin neurophysiol, Montreal Neurol Inst, McGill Univ, 65-66, res fel neurol & neurosurg, 65-68, lectr physiol, 68; asst prof, State Univ NY, 68-70; assoc prof, 70-73, PROF PHYSIOL, UNIV TORONTO, 73-, CHMN DEPT, 75- Concurrent Pos: Invited res lectr, Int Conf Nat Ctr Sci Res, Aix-Marseille, 74. Mem: Am Physiol Soc; Can Physiol Soc; Soc Neurosci; Int Brain Res Orgn; Can Soc Electroencephalographers, Electromyographers & Clin Neurophysiologists. Res: Brain mechanisms in control of voluntary movement. Mailing Add: Dept of Physiol Univ of Toronto Toronto ON Can

MURPHY, JOSEPH, b Montreal, Que, Nov 6, 32; m 58; c 2. SOLID STATE PHYSICS. Educ: McGill Univ, BSc, 56, MSc, 58, PhD(physics), 63. Prof Exp: Sr physicist, US Naval Ord Lab, Calif, 60-63; sr physicist, Solid State Sci Dept, 63-70, SR PHYSICIST, PHYSICS DEPT, WESTINGHOUSE RES & DEVELOP CTR, 70- Mem: Am Phys Soc. Res: Theoretical nuclear physics; microwave and optical properties of solids; theory of interaction of localized defects with each other and with lattice vibrations. Mailing Add: Westinghouse R&D Ctr Physics Dept 401 4621 Beulah Rd Pittsburgh PA 15235

MURPHY, JOSEPH ROBISON, b Salt Lake City, Utah, June 14, 25; m 46; c 5. ZOOLOGY. Educ: Brigham Young Univ, AB, 50, MA, 51; Univ Nebr, PhD(zool), 57. Prof Exp: From instr to asst prof zool, Univ Nebr, 51-60; from asst prof to assoc prof, 60-68, chmn dept, 68-74, PROF ZOOL, BRIGHAM YOUNG UNIV, 68- Mem: Ecol Soc Am; Cooper Ornith Soc; Am Ornithologists Soc. Res: Ecology of predatory birds, especially American eagles. Mailing Add: Dept of Zool 575 WIDB Brigham Young Univ Provo UT 84601

MURPHY, JUNEANN WADSWORTH, b Chickasha, Okla, Mar 13, 37; m 67; c 2. MEDICAL MICROBIOLOGY. Educ: Univ Okla, BS, 59, MSS, 61, MS, 65, PhD(microbiol), 69. Prof Exp: Res asst, Sch Med, Tulane Univ, 59; med technologist, Cent State Hosp, 61, instr, 63-64; res asst med mycol, 62-63, vis asst prof, 69-70, ASST PROF MICROBIOL, UNIV OKLA, 70- Mem: Am Soc Microbiol; Med Mycol Soc Am; Am Soc Med Technol. Res: Host-parasite relationships in systematic mycotic diseases, with a primary interest in host defense mechanisms in Cryptococcosis. Mailing Add: Dept of Bot-Microbiol Univ of Okla 770 Van Vleet Oval Norman OK 73069

MURPHY, LARRY S, b Greenfield, Mo, Dec 15, 37; m 59; c 2. AGRONOMY, PLANT PHYSIOLOGY. Educ: Univ Mo, BS, 59, MS, 60, PhD(agron), 65. Prof Exp: Instr soils, Univ Mo, 60-65; from asst prof to assoc prof, 65-73, PROF AGRON, KANS STATE UNIV, 73- Concurrent Pos: Res assoc, Mich State Univ, 71-72. Honors & Awards: Geigy Award, Am Soc Agron, 73. Mem: Hon mem Nat Fertilizer Solutions Asn; Am Soc Agron; Soil Sci Soc Am. Res: Nitrate accumulation in forage crops and water supplies; wheat, corn, grain sorghum, forage production and quality; micronutrient nutrition of plants; evaluation of P fertilizers; water pollution; animal waste disposal. Mailing Add: Dept of Agron Kans State Univ Manhattan KS 66502

MURPHY, LESLIE CARLTON, b Mercer, NDak, May 28, 13; m 38; c 3. MICROBIOLOGY. Educ: Univ Idaho, BS, 35; State Col Wash, DVM, 39; Am Bd Vet Pub Health, dipl. Prof Exp: Asst vet, Western Wash Exp Sta, 39-40; res vet, Carnation Res Labs, Wis, 40; vis investr, Rockefeller Inst, NY, 49-50, chief dept bact, Vet Div, Walter Reed Army Inst Res, Washington, DC, 53-58, chief br IV, Virus & Rockettsia Div, 58-61; head cancer virol sect, Virol Res Resources Br, Nat Cancer Inst, 61-64; dir res develop, 64-68, ASSOC DEAN RES & DEVELOP, SCH VET MED, UNIV MO-COLUMBIA, 68- Mem: Am Vet Med Asn; Conf Res Workers Animal Dis; Conf Pub Health Vets; Am Col Vet Microbiol; Am Asn Lab Animal Sci. Res: Veterinary medicine. Mailing Add: 100B Connaway Hall Univ of Mo Sch of Vet Med Columbia MO 65201

MURPHY, MARJORY BETH, b Page, Nebr, July 21, 25. CELL PHYSIOLOGY, BIOCHEMISTRY. Educ: Nebr Wesleyan Univ, BA, 47; Univ Colo, MA, 53; Univ Ill, PhD(cell physiol), 61. Prof Exp: PROF CHEM, PHILLIPS UNIV, 53- Mem: AAAS.

Res: Enzymes involved in membrane transport; active sites of enzymes. Mailing Add: Dept of Chem Phillips Univ Enid OK 73701

MURPHY, MARTIN JOSEPH, JR, b Colorado Springs, Colo, Dec 29, 42; m 65; c 5. IMMUNOHEMATOLOGY. Educ: Regis Col, Colo, BS, 64; NY Univ, MS, 67, PhD(physiol, hemat), 69. Prof Exp: Reader & lect asst, Grad Sch Arts & Sci, NY Univ, 65-68; instr biol, Nassau Community Col, 68-69; asst mem, St Jude Children's Res Hosp, Memphis, 73-75; ASSOC, SLOAN-KETTERING INST CANCER RES, 75- Concurrent Pos: Damon Runyon res fel, Inst Cellular Path, Hopital Bicetre, Kremlin-Bicetre, France, 69-70; NIH res fel, Paterson Labs, Christie Hosp & Holt Radium Inst, Manchester, Eng, 70-71; Leukemia Soc Am spec fel immunol, John Curtin Sch Med Res, Australian Nat Univ, 71-73. Mem: NY Acad Sci; Electron Micros Soc Am; Soc Exp Biol & Med; Am Asn Cancer Res; Sigma Xi. Res: Physiology of blood cell production in health and disease. Mailing Add: Sloan-Kettering Inst Cancer Res Labs Exp Hemat 410 E 68th St New York NY 10021

MURPHY, MARY LOIS, b Nebr, Oct 16, 16. MEDICINE. Educ: Univ Nebr, BA, 39, MD, 44; Am Bd Pediat, dipl, 51. Prof Exp: Asst bacteriologist, Minn State Dept of Health, 42-43; intern, Woman's Med Col, Pa, 44-45; resident pediat, St Christopher's Hosp Children, Philadelphia, 45-46; resident path, Children's Hosp, Washington, DC, 46-47; asst instr clin path, Med Sch, Georgetown Univ, 47-48; asst instr pediat, Med Sch, Univ Pa, 49-52; res assoc med, Sloan-Kettering div, Cornell Univ, 52-53, asst prof, 54-57; assoc attend pediatrician, 57-66, ATTEND PEDIATRICIAN & CHMN DEPT PEDIAT, MEM HOSP, 66-; PROF PEDIAT, MED COL, CORNELL UNIV, 70- Concurrent Pos: Res fel, Children's Hosp, Philadelphia, 49-51; res fel, Sloan-Kettering Inst Cancer Res, 51-54; from resident to asst chief resident, Children's Hosp, Philadelphia, 47-49, asst vis physician, 49; chief resident, Camden Munic Hosp Contagious Dis, 49-51; asst, Sloan-Kettering Inst Cancer Res, assoc mem, 60-70, mem, 70-; assoc prof pediat, Med Col Cornell Univ, 67-70. Mem: Harvey Soc; AMA; Am Asn Cancer Res; Am Fedn Clin Res; Am Acad Pediat. Res: Pediatrics; leukemia; cancer; teratogenesis. Mailing Add: Dept of Pediat Mem Hosp 1275 York Ave New York NY 10021

MURPHY, MARY NADINE, b Waucoma, Iowa, June 22, 33. MICROBIOLOGY. Educ: Clarke Col, AB, 54; Purdue Univ, NSF fel, 61-64, MS, 64, PhD(biol sci), 65. Prof Exp: Instr, 59-61 & 65-66, asst prof, 66-70, ASSOC PROF BIOL, MUNDELEIN COL, 70-, CHMN DEPT, 66- Concurrent Pos: Consult-evaluator, NCent Asn Cols & Schs, 73- Mem: AAAS; Am Soc Microbiol. Res: Cellular differentiation; physiology of aquatic fungi. Mailing Add: Dept of Biol Mundelein Col Chicago IL 60660

MURPHY, MARY TERESA JOSEPH, b Hartford, Conn, Sept 10, 28. GEOCHEMISTRY. Educ: St Joseph Col, Conn, BS, 50; Wesleyan Univ, MA, 56; Fordham Univ, PhD(chem), 65. Hon Degrees: DSc, Univ Hartford, 75. Prof Exp: Chemist, Naugatuck Chem Div, US Rubber Co, 50-51 & Travelers Ins Co, 51-54; res chemist, Monsanto Chem Co, 56-58; from asst prof to assoc prof, 65-73, PROF CHEM, ST JOSEPH COL, CONN, 73- Concurrent Pos: NASA fel, Univ Glasgow, 66-67. Mem: AAAS; Am Chem Soc; Geochem Soc. Res: Geochemistry of organic matter in rocks and meteorites; oil shales. Mailing Add: Dept of Chem St Joseph Col West Hartford CT 06117

MURPHY, MICHAEL A, b Spokane, Wash, Mar 11, 25; m 47; c 2. GEOLOGY, STRATIGRAPHY. Educ: Univ Calif, Los Angeles, PhD, 54. Prof Exp: Subsurface geologist, Shell Oil Co, 53-54; from asst prof to assoc prof geol, 54-67, PROF GEOL, UNIV CALIF, RIVERSIDE, 67- Concurrent Pos: NSF grant, 58, 61, 67 & 69; prof, Univ Cenap, Brazil, 59-60. Mem: Paleont Soc; fel Geol Soc Am. Mailing Add: Dept of Geol Sci Univ of Calif Riverside CA 92502

MURPHY, MICHAEL JOSEPH, b Butte, Mont, Feb 12, 23. PHYSICAL GEOLOGY. Educ: Univ Notre Dame, AB, 45, BS, 51; Univ Calif, Berkeley, MS, 53. Prof Exp: From instr to asst prof, 53-65, asst chmn dept, 66-70, ASSOC PROF GEOL, UNIV NOTRE DAME, 65-, CHMN DEPT, 70- Concurrent Pos: NSF fel, Columbia Univ, 60-61. Mem: AAAS; Geol Soc Am; Mineral Soc Am; Nat Asn Geol Teachers; Sigma Xi. Res: Isomorphic mineral systems. Mailing Add: Dept of Earth Sci Univ of Notre Dame Notre Dame IN 46556

MURPHY, PATRICK AIDAN, b Liverpool, Eng, June 4, 37; m 64; c 4. MICROBIOLOGY, MEDICINE. Educ: Univ Liverpool, BSc, 57, MB & ChB, 60; Oxford Univ, DPhil, 66. Prof Exp: UK Med Res Coun fel, Oxford Univ, 64-67; fel microbiol, 67-68, instr, 68-69, ASST PROF MICROBIOL & MED, SCH MED, JOHNS HOPKINS UNIV, 69- Res: Pathogenesis of fever. Mailing Add: Dept of Microbiol Sch of Med Johns Hopkins Univ Baltimore MD 21205

MURPHY, PATRICK JOSEPH, b Chicago, Ill, June 11, 40; m 66; c 3. BIOCHEMISTRY. Educ: Loyola Univ Chicago, BS, 62; San Diego State Col, MS, 64; Univ Calif, Los Angeles, PhD(biochem), 67. Prof Exp: Sr scientist, 67-73, RES SCIENTIST, ELI LILLY & CO, 73- Mem: AAAS; Am Chem Soc; Am Soc Pharmacol & Exp Therapeut. Res: Drug metabolism; biochemical pharmacology; studies of enzymes involved in metabolism of endogenous and exogenous compounds. Mailing Add: Eli Lilly & Co Dept M304 Indianapolis IN 46206

MURPHY, PAUL HENRY, b Boston, Mass, July 7, 42; m 65; c 3. NUCLEAR MEDICINE, MEDICAL PHYSICS. Educ: Univ Kans, MS, 68, PhD(radiation biophys), 70. Prof Exp: ASST PROF NUCLEAR MED & PHYSICS, BAYLOR COL MED, 71- Concurrent Pos: Advan sr fel med physics, Univ Tex M D Anderson Hosp & Tumor Inst Houston, 70-71. Mem: Soc Nuclear Med; Health Physics Soc; Am Asn Physicists in Med. Res: Nuclear medicine imaging; radionuclid pulmonary function tests; computer applications in nuclear medicine. Mailing Add: Dept of Radiol Baylor Col of Med Houston TX 77025

MURPHY, PETER GEORGE, b New York, NY, Feb 23, 42; m 67. ECOLOGY, BOTANY. Educ: Syracuse Univ, BS, 63, MS, 68; Univ NC, Chapel Hill, PhD(plant ecol), 70. Prof Exp: Res assoc trop ecol, PR Nuclear Ctr, Rio Piedras, 63-66; asst prof, 70-75, ASSOC PROF BOT & ECOL, MICH STATE UNIV, 75- Concurrent Pos: AID ecol adv, Indonesia, 73-74; mem directorate, US Man & Biosphere Prog, Trop Forest Prog, 75- Mem: AAAS; Ecol Soc Am; Am Inst Biol Sci; Asn Trop Biol; Int Soc Trop Ecol. Res: Structure and function of tropical ecosystems; primary productivity; sand dune ecosystems; radiation ecology. Mailing Add: Dept of Bot & Plant Path Mich State Univ East Lansing MI 48824

MURPHY, PRESTON V, b East Chicago, Ind, Mar 3, 30; m 58; c 4. SOLID STATE SCIENCE. Educ: Univ Notre Dame, BS, 51; Washington Univ, PhD(radio-chem), 56. Prof Exp: Scientist, US Army Sci Liaison & Adv Group, 56-59; sci dir, Nat Dosimetry Lab, Brazil, 59-62; vpres, Panoramic Res, Inc, 62-65, pres, 65-66; mgr dielectrics prod, Thermo Electron Corp, 66-73, gen mgr, Thermo Electron SA, Spain, 73-75; PRES, LECTRET SA, SWITZ, 76- Concurrent Pos: Assoc prof, Cath Univ, Brazil, 61-62. Mem: AAAS; Am Chem Soc; Am Phys Soc. Res: Persistent polarization and charge conduction in dielectrics; transient radiation effects in insulators; radiation dosimetry; electrostatic aerosol filters; electroacoustic transducers. Mailing Add: 103 Av de Villiers Paris 75017 France

MURPHY, QUILLIAN R, JR, b Birmingham, Ala, Nov 9, 17; m 49. PHYSIOLOGY. Educ: Birmingham-Southern Col, BS, 38; Univ NC, MA, 40; Univ Wis, PhD(physiol), 46, MD, 48. Prof Exp: From instr to assoc prof, 47-59, PROF PHYSIOL, MED SCH, UNIV WIS-MADISON, 59- Mem: Am Physiol Soc. Res: Autonomic nervous system; cardiac arrhythmias. Mailing Add: Dept of Physiol Univ of Wis Med Sch Madison WI 53705

MURPHY, RAY BRADFORD, b USA, June 7, 22; m 54; c 5. MATHEMATICAL STATISTICS, APPLIED STATISTICS. Educ: Princeton Univ, AB, 43, MA, 48, PhD(math), 51. Prof Exp: From instr to asst prof math, Carnegie Inst Technol, 49-52; mem tech staff, 52-58, dept head qual theory, 58-67, DEPT HEAD APPL STATIST, BELL TEL LABS, HOLMDEL, 67- Mem: Am Math Soc; Economet Soc; Am Statist Asn; Inst Math Statist. Res: Quality control. Mailing Add: 39 Borden Pl Silver NJ 07739

MURPHY, RAYMOND EDWARD, b Apple River, Ill, July 24, 98; m 26; c 1. URBAN GEOGRAPHY. Educ: Mo Sch Mines & Metall, BS, 23; Univ Wis-Madison, MS, 26, PhD(geog), 30. Prof Exp: Instr geol, Univ Ky, 26-28; head dept geog, Concord State Teachers Col, 30-31; from asst prof to prof, Pa State Col, 31-45; actg chmn dept, Univ Hawaii, 45-46; prof econ geog, 46-62, ed, Econ Geog, 49-62 & 65-69, dir grad sch geog, 62-65, EMER PROF ECON GEOG, CLARK UNIV, 70- Concurrent Pos: Off Naval Res grant, Cent Bus Dist Res Proj, 52-54; Royal Scottish Geog Soc fel urban & econ geog, 64. Mem: Asn Am Geogr; fel Am Geog Soc. Res: Geography of mineral production; geography of American Micronesia. Mailing Add: 1299 Briarwood Ave Deltona FL 32763

MURPHY, RICHARD ALAN, b Twin Falls, Idaho, July 4, 38; m 61; c 2. PHYSIOLOGY. Educ: Harvard Univ, AB, 60; Columbia Univ, PhD(physiol), 64. Prof Exp: NIH fel physiol, Max Planck Inst Med Res, Heidelberg, 64-66; res assoc, Univ Mich, 66-68; asst prof, 68-71, ASSOC PROF PHYSIOL, SCH MED, UNIV VA, 72- Concurrent Pos: NIH career develop award, 71. Mem: AAAS; Am Physiol Soc; Biophys Soc; Soc Gen Physiologists. Res: Biochemistry of the contractile proteins of vascular smooth muscle; contractile properties of arterial smooth muscle. Mailing Add: Dept of Physiol Univ of Va Sch of Med Charlottesville VA 22901

MURPHY, RICHARD ALLAN, b Evergreen Park, Ill, Feb 23, 41; m 65; c 3. MEDICAL MICROBIOLOGY. Educ: Loyola Univ, BS, 63; Univ Ill, MS, 66, PhD(microbiol), 71. Prof Exp: Teaching asst, 64-70, ASST PROF ORAL DIAGNOSIS, UNIV ILL MED CTR, 70- Mem: Am Soc Microbiol. Res: Role of bacterial enzymes and toxins in pathogenesis. Mailing Add: Dept of Microbiol Univ of Ill Med Ctr Chicago IL 60612

MURPHY, RICHARD ERNEST, b Hibbing, Minn, Sept 21, 20; m 49; c 3. PHYSICAL GEOGRAPHY. Educ: St Lawrence Univ, BA, 43; George Washington Univ, MA, 52; Clark Univ, PhD(geog), 57. Prof Exp: Cartographic aide, US Army Map Serv, 46-48; read reference sect, Map Div, Libr Cong, 49-54; from asst prof to assoc prof geog, George Washington Univ, 55-59; from assoc prof to prof, Univ Wyo, 59-63; vis prof, Univ Hawaii, 63-64; NSF sci fac fel, Inst Geog, Univ Paris, 64-65; PROF GEOG & CHMN DEPT, UNIV NMEX, 65- Concurrent Pos: Consult theory group, George Washington Univ-US Army Logistics Res Proj, 55-56; res scientist, George Washington Univ-US Army Qm Intel Res Proj, 58-59; consult, Univ NMex Technol Appln Ctr, NASA, 67-; Fulbright lectr, Inst Geog, Tohoku Univ, Japan, 68-69. Mem: Asn Am Geogr; Int Geog Union. Res: Classification and distribution of world ethnic groups; the raison d'etre of nation-states; classification and distribution of world landform regions; conservation, especially wilderness areas. Mailing Add: Dept of Geog Univ of NMex Albuquerque NM 87106

MURPHY, ROBERT CARL, b Wheeler, Pa, Dec 18, 19; m 45; c 3. ANATOMY. Educ: Geneva Col, BS, 49; Univ Wis, MS, 52, PhD, 55. Prof Exp: Asst zool, Univ Wis, 49-52, asst anat, 52-54, instr, 54-55; asst prof, Univ Iowa, 55-57; from asst prof to assoc prof, 57-72, PROF ANAT, TERRE HAUTE CTR MED EDUC, IND UNIV SCH MED, 72- Res: Cells and tissues of the lymphoid system, immunology. Mailing Add: Terre Haute Ctr for Med Educ Ind Univ Sch of Med Terre Haute IN 47809

MURPHY, ROBERT CARL, b Seymour, Ind, Dec 15, 44; m 65; c 2. ORGANIC CHEMISTRY, PHARMACOLOGY. Educ: Mt Union Col, BS, 66; Mass Inst Technol, PhD(org chem), 70. Prof Exp: NIH trainee & Harvard Univ fel, Mass Inst Technol & Harvard Univ, 70-71; ASST PROF PHARMACOL, UNIV COLO MED CTR, DENVER, 71- Concurrent Pos: Assoc ed, Org Mass Spectrometry, 74- Mem: Am Chem Soc; Am Soc Mass Spectrometry. Res: Application of stable isotopes and mass spectrometry to biomedical research; drug metabolism; structure determination of phospholipids and pharmacologically active molecules by mass spectrometry. Mailing Add: Dept of Pharmacol Univ of Colo Med Ctr Denver CO 80220

MURPHY, ROBERT CUSHMAN, zoology, ornithology, deceased

MURPHY, ROBERT EARL, astronomy, see 12th edition

MURPHY, ROBERT EMMETT, b Chicago, Ill, Jan 28, 27; m 71; c 4. MEAT SCIENCES. Educ: Univ Ill, BS, 51. Prof Exp: Meat technologist, 53-61, proj leader, 61-75, ASSOC SCIENTIST, RES & DEVELOP CTR, SWIFT & CO, 76- Concurrent Pos: Meat technol consult, Consortium Develop Technol, Brazil, 74. Mem: Inst Food Technologists. Res: International application of antemortem and postmortem enzymatic meat tenderization in Australia, New Zealand, Korea, Brazil, Costa Rica and Canada; fresh meat products and processing. Mailing Add: Res & Develop Ctr Swift & Co 1919 Swift Dr Oak Brook IL 60521

MURPHY, ROBERT FRANCIS, b Rockaway Beach, NY, Mar 3, 24; m 50; c 2. ANTHROPOLOGY. Educ: Columbia Col, BA, 49; Columbia Univ, PhD(anthrop), 54. Prof Exp: Res assoc anthrop, Univ Ill, Urbana, 53-55; asst prof, Univ Calif, Berkeley, 55-61; assoc prof, 61-63; chmn dept anthrop, 69-72, PROF ANTHROP, COLUMBIA UNIV, 63- Concurrent Pos: Soc Sci Res Coun fac res fel, Univ Calif, Berkeley, 57-60; Ford Found foreign area fel, Niger, WAfrica, 59-60; Guggenheim Found fel, Columbia Univ, 68-69. Mem: Fel Am Anthrop Asn; fel Am Ethnol Soc. Res: Theories of social structure; kinship and the family; ethnology of South and North America and northern Africa. Mailing Add: Dept of Anthrop Columbia Univ New York NY 10027

MURPHY, ROBERT T, b Washington, DC, Nov 17, 31; m 54; c 1. ANALYTICAL CHEMISTRY. Educ: Univ Md, BS, 58. Prof Exp: Res chemist, USDA, 58-63; chemist, Univ Calif, Riverside, 63-65; res chemist, 65-68, dir anal chem, NY, 68-75, DIR REGISTRATION & TOXICOL, AG DIV, CIBA-GEIGY CORP, 75- Mem: Am Chem Soc; Entom Soc Am; Asn Off Anal Chemists; Weed Sci Soc Am. Res: Agricultural chemicals; metabolic pathway and dissipation of pesticides; toxicology and safety aspects of pesticides; state and federal registration of pesticides. Mailing

Add: Registration & Toxicol Dept Ag Chem Div Ciba-Geigy Corp Greensboro NC 27409

MURPHY, ROYSE PEAK, b Norton, Kans, May 2, 14; m 41; c 3. PLANT BREEDING. Educ: Kans State Univ, BS, 36; Univ Minn, MS, 38, PhD(plant breeding, genetics), 41. Prof Exp: Asst, Div Agron & Plant Genetics, Univ Minn, 36-37, from instr to asst prof, 37-42; assoc prof, Mont State Univ, 42-44; assoc prof plant breeding, 46-48, head dept, 53-64, dean univ fac, 64-67, PROF PLANT BREEDING, CORNELL UNIV, 48- Mem: AAAS; Am Soc Agron; Genetics Soc Am; Am Inst Biol Sci. Res: Plant genetics and breeding with perennial forage legumes and grasses. Mailing Add: Dept of Plant Breeding & Biometry Cornell Univ Ithaca NY 14850

MURPHY, SAMUEL G, b Long Beach, Calif, Dec 29, 37; m 64; c 3. CANCER, ONCOLOGY. Educ: Duke Univ, AB, 61; NC State Univ, MS, 64; Albany Med Col, PhD(microbiol), 68; Ohio State Univ, MD, 73. Prof Exp: Tech coordr, Pharmacia Fine Chem, Inc, 63-64; biochemist, Div Labs & Res, NY State Dept Health, Albany, 65-67; sr natural prod chemist, Wallace Labs, 67-68; asst prof, Dept Path, 68-70, CLIN INSTR, DEPT MED & DEPT PATH, COL MED, OHIO STATE UNIV, 70- Res: Cancer immunology, immunopathology. Mailing Add: 1826 Elmwood Columbus OH 43212

MURPHY, SHELDON DOUGLAS, b Forestburg, SDak, July 16, 33; m 54; c 1. TOXICOLOGY, PHARMACOLOGY. Educ: SDak State Col, BS, 55; Univ Chicago, PhD(pharmacol), 58. Prof Exp: Technician pharmacol, Univ Chicago, 55-58; asst scientist, Occup Health Field Hqs, USPHS, 58-59, chief pharmacol & toxicol sect, Div Air Pollution, 59-63; asst prof toxicol, 63-67, ASSOC PROF TOXICOL & DIR TOXICOL TRAINING PROG, SCH PUB HEALTH, HARVARD UNIV, 67- Concurrent Pos: Mem toxicol study sect, NIH, 69-73, chmn, 72-73; mem marine food resources comt, Nat Acad Sci-Nat Res Coun; mem pesticide bd, Dept Pub Health, Commonwealth of Mass, 70-; mem WHO expert comt pesticide residues, Food & Agr Orgn-WHO, Rome, 72, chmn, Geneva, Switz, 73, temporary adv, Rome, 74; mem expert adv panel food additives, WHO, 72-; mem bd sci counr, Nat Inst Environ Health Sci, 73-; mem panel oxidants med biol effects environ pollutants, Nat Acad Sci-Nat Res Coun, 73-75; mem toxicol adv comt, US Food & Drug Admin, 75-76; mem hazardous mat adv comt, Environ Protection Agency, 75- Honors & Awards: Achievement Award, Soc Toxicol, 70. Mem: AAAS; Soc Toxicol (pres, 74-75); Am Soc Pharmacol & Exp Therapeut; Soc Exp Biol & Med; NY Acad Sci. Res: Metabolism and interaction of drugs and chemicals by mammalian organisms; environmental toxicology; comparative pesticide toxicology. Mailing Add: Dept of Physiol Sch Pub Health Harvard Univ 665 Huntington Ave Boston MA 02115

MURPHY, SHELDON R, chemical engineering, see 12th edition

MURPHY, STANLEY REED, b Guthrie, Okla, Nov 3, 24; m 57; c 1. PHYSICS. Educ: Fresno State Col, BA, 48; Univ Wash, PhD(physics), 59. Prof Exp: Res engr, Boeing Airplane Co, 50-52; assoc physicist, Appl Physics Lab, 52-54, physicist, 54-57, sr physicist, 57-64, asst dir, 64-68, PROF OCEANOG, COL ARTS & SCI, PROF MECH & OCEAN ENG, COL ENG & DIR DIV, MARINE RESOURCES, UNIV WASH, 68- Concurrent Pos: Mem, Wash Comn Oceanog, 69-, vchmn, 70-71; adj prof, Inst Marine Study, Univ Wash, 73- Mem: Am Phys Soc. Res: Acoustics; oceanography; instrumentation. Mailing Add: Div of Marine Resources Univ of Wash Seattle WA 98195

MURPHY, TED DANIEL, b Stanley, NC, Apr 21, 36; m 57; c 1. BIOLOGY, ZOOLOGY. Educ: Duke Univ, AB, 58, MA, 60, PhD(zool), 63. Prof Exp: Asst prof biol, State Univ NY Binghamton, 63-70; assoc prof, Siena Col, 70-74; PROF BIOL, CALIF STATE COL, BAKERSFIELD, 74- Mem: AAAS; Am Soc Ichthyologists & Herpetologists; Ecol Soc Am; Am Inst Biol Sci. Res: Ecology of amphibians and reptiles. Mailing Add: Dept of Biol Calif State Col Bakersfield CA 93309

MURPHY, TERENCE MARTIN, b Seattle, Wash, July 1, 42; m 69. PLANT BIOCHEMISTRY, PHYSIOLOGY. Educ: Calif Inst Technol, BS, 64; Univ Calif, San Diego, PhD(cell biol), 68. Prof Exp: USPHS fel, Univ Wash, 69-70; ASST PROF BOT, UNIV CALIF, DAVIS, 70- Res: Photochemistry and photobiology of RNA and DNA; plant cell repair mechanisms; immunochemical characterizations of plant enzymes. Mailing Add: Dept of Bot Univ of Calif Davis CA 95616

MURPHY, TERENCE W, b Liverpool, Eng, Feb 19, 31; US citizen; m; c 5. BIOENGINEERING, PUBLIC HEALTH. Educ: Univ Liverpool, BSc, 51, MB & ChB, 56; Harvard Univ, MPH, 73; Am Bd Anesthesiol, dipl, 63. Prof Exp: Assoc anesthesia, Sloan-Kettering Inst Cancer Res, 60-63; USPHS spec fel bioeng, Columbia Univ, 63-64; math consult, Rand Corp, 64-66; assoc prof anesthesiol, Col Med, NY Univ, 66-74, dir biomed comput facil, Med Ctr, 66-70; vpres med affairs, Misericordia Hosp Med Ctr, New York, 74-75; MED DIR, BLUE CROSS BLUE SHIELD GREATER NEW YORK, 75- Concurrent Pos: Consult anesthesiol, Roosevelt Hosp, New York. Mem: Am Soc Anesthesiol; Inst Elec & Electronics Eng; Asn Comput Mach; Int Anesthesia Res Soc; assoc Am Col Legal Med. Res: Computers in medicine; biomathematics; models of the control of ventilation; ambulatory care; anesthesiology; technology in health care; legal medicine. Mailing Add: Blue Cross Blue Shield Great NY 622 3rd Ave New York NY 10017

MURPHY, THOMAS JAMES, b Brooklyn, NY, Feb 17, 42; m 68; c 2. STATISTICAL MECHANICS. Educ: Fordham Univ, BS, 63; Rockefeller Univ, PhD(physics), 68. Prof Exp: Res staff physicist, Yale Univ, 68-69; asst prof, 69-75, ASSOC PROF CHEM, UNIV MD, COLLEGE PARK, 75- Res: Equilibrium and non-equilibrium statistical mechanics of Coulomb systems; Brownian motion of interacting particles; density dependence of transport coefficients of gases. Mailing Add: Dept of Chem Univ of Md College Park MD 20742

MURPHY, THOMAS JOSEPH, b Pittsburgh, Pa, Oct 4, 41; m. WATER CHEMISTRY. Educ: Univ Notre Dame, BS, 63; Iowa State Univ, PhD(photochem), 67. Prof Exp: NIH fel org chem, Ohio State Univ, 67-68; ASST PROF ORG CHEM, DePAUL UNIV, 68- Mem: Am Chem Soc; Int Asn Great Lakes Res; Int Soc Limnol; Am Soc Limnol & Oceanog. Res: Sources, sinks and cycling of materials in bodies of water; atmospheric inputs. Mailing Add: Dept of Chem DePaul Univ Chicago IL 60614

MURPHY, WALTER THOMAS, b Medford, Mass, Oct 5, 28; m 55; c 5. ORGANIC CHEMISTRY, POLYMER CHEMISTRY. Educ: Boston Col, BS, 50, MS, 52. Prof Exp: Lab supvr, Main Plant, 52-53, jr res chemist, Res Ctr, 53-57, res chemist, 57-64, SR RES CHEMIST, RES CTR, B F GOODRICH CO, 64- Mem: Am Chem Soc. Res: Polyurethane polymers; catalysis; structure versus mechanical properties; spandex fiber; adhesives; coatings, poromeric films; formulation and processing of reactive liquid polymers of epoxy, butadiene and acrylonitrile. Mailing Add: 1091 Taft Ave Cuyahoga Falls OH 44223

MURPHY, WILLIAM FREDERICK, b Dunkirk, NY, June 10, 39; m 61; c 3. MOLECULAR SPECTROSCOPY. Educ: Case Inst Technol, BS, 61; Univ Wis, PhD(phys chem), 66. Prof Exp: Fel, 66-68, asst res officer, 68-72, ASSOC RES OFFICER, NAT RES COUN CAN, 72- Mem: Am Phys Soc; Optical Soc Am; Soc Appl Spectros; Coblentz Soc. Res: Raman spectroscopy; Raman gas phase intensities and band contours; instrumentation for Raman spectroscopy. Mailing Add: Div Chem Nat Res Coun of Can 100 Sussex Dr Ottawa ON Can

MURPHY, WILLIAM HENRY, JR, b New York, NY, June 26, 25; m 48; c 4. MICROBIOLOGY. Educ: Pa State Univ, BS, 50, MS, 51; Univ Minn, PhD, 54. Prof Exp: Instr microbiol, Univ Minn, 54-56; from asst prof to assoc prof, 56-67, PROF MICROBIOL, UNIV MICH, ANN ARBOR, 67- Concurrent Pos: Rockefeller exchange prof, Colombia, SAm, 58. Mem: AAAS; Am Soc Microbiol; Am Asn Immunologists. Res: Virology; immunology; leukemia, infectious diseases; immunology and immunopathology of leukemia. Mailing Add: Dept of Microbiol 6706 MS II Univ of Mich Ann Arbor MI 48104

MURPHY, WILLIAM PARRY, JR, b Boston, Mass, Nov 11, 23; c 3. MEDICINE. Educ: Univ Ill, MD, 47. Prof Exp: Instr med, Harvard Med Sch, 49-51, res assoc, 53-55; dir res, Dade Reagents, 55-57; PRES, CORDIS CORP, 57- Concurrent Pos: Res fel med, Peter Bent Brigham Hosp, Boston, 49-51, asst med, 53-55; chief engr, Fenwal Labs, Mass, 49 & 54; with lab biol control, NIH, 51-53; res assoc, Miami Heart Inst, 56-68; res assoc prof biophys & chmn div, Med Sch, Univ Miami, 58-70; pres, Cordis Dow Corp, 70-75. Honors & Awards: Award, Am Roentgen Ray Soc, 48. Mem: AAAS; AMA; Inst Elec & Electronics Eng. Res: Artificial internal organs; transfusion; biology instrumentation; heart disease; biophysics. Mailing Add: 3915 Biscayne Blvd Miami FL 33137

MURPHY, ZATIS LUAIN, b Pascagula, Miss, Dec 31, 34; m 57; c 3. ORGANIC CHEMISTRY. Educ: Wichita State Univ, BS, 64, MS, 66; Univ Wash, PhD(org chem), 69. Prof Exp: Res chemist, Ill, 69-73, SR RES CHEMIST, SHELL DEVELOP CO, 73- Res: Synthesis of heterocycles and aromatic amines; oxidation studies of lubricants. Mailing Add: Shell Develop Co Westhollow Res Ctr Houston TX 77001

MURR, BROWN L, JR, b Atlanta, Ga, Feb 23, 31; m 53, 73; c 2. PHYSICAL ORGANIC CHEMISTRY. Educ: Emory Univ, AB, 52, MS, 53; Ind Univ, PhD(chem), 61. Prof Exp: NSF fel org chem, Mass Inst Technol, 61-62, res assoc, 62; from asst prof to assoc prof, 62-71, PROF ORG CHEM, JOHNS HOPKINS UNIV, 71-, CHMN DEPT CHEM, 73- Mem: Am Chem Soc; The Chem Soc. Res: Reaction mechanisms; kinetics; stereochemistry; kinetic isotope effects. Mailing Add: Dept of Chem Johns Hopkins Univ Baltimore MD 21218

MURR, SANDRA M, developmental genetics, see 12th edition

MURRA, JOHN VICTOR, b Odessa, Ukraine, Aug 24, 16; US citizen. ETHNOLOGY, ETHNOHISTORY. Educ: Univ Chicago, AB, 36, MA, 42, PhD(anthrop), 56. Prof Exp: Instr anthrop, Univ Chicago, 43-46; ed, Encycl Britannica, 46-47; from asst prof to assoc prof, Univ PR, 47-50; from lectr to prof, Vassar Col, 50-61; vis prof, Yale Univ, 61-63; PROF ANTHROP, CORNELL UNIV, 68- Concurrent Pos: Area specialist, UN Secretariat, 51; vis prof, San Marcos Univ, Lima, 58 & 66; NSF grant Inca prov & peasant life, Huanuco, Peru, 63-66; Nat Acad Sci fel, Smithsonian Inst, 66-67; Lewis Henry Morgan lectr, Univ Rochester, 69; vis prof anthrop, Yale Univ, 70-71; organizer, Nispa Ninchis, 70-; mem, Inst Advan Study, Princeton, 74-75; assoc dir, Sch Advan Study Social Sci, Paris, 75-76. Mem: Am Anthrop Asn; Am Ethnol Soc (pres, 72-73); Am Soc Ethnohist (pres, 70-71); Inter-Am Indian Inst; Soc Am Archaeol; Int African Inst; Nat Inst Anthrop & Hist Ecuador. Res: Economic, social and political organization of early states, particularly the Andean ones. Mailing Add: Dept of Anthrop Cornell Univ Ithaca NY 14853

MURRAY, BEATRICE E, b Young, Sask, Jan 27, 19. CYTOGENETICS. Educ: Univ Sask, BSA, 45, MSc, 47; Cornell Univ, PhD(cytogenetics, bot), 55. Prof Exp: Asst genetics, Univ Sask, 45-47; res officer genetics & forage crops, Res Sta, 47-54, RES SCIENTIST CYTOGENETICS, RES BR, CAN DEPT AGR, 55- Mem: Genetics Soc Can; Can Bot Asn; Agr Inst Can. Res: Plant mutations; morphology. Mailing Add: Ottawa Res Sta Can Dept of Agr Ottawa ON Can

MURRAY, BERTRAM GEORGE, JR, b Elizabeth, NJ, Sept 24, 33. ZOOLOGY. Educ: Rutgers Univ, AB, 61; Univ Mich, MS, 63, PhD(zool), 67. Prof Exp: Lectr biol, Cornell Univ, 67-68; asst prof natural sci, Mich State Univ, 68-71; asst prof, 71-74, ASSOC PROF BIOL, RUTGERS UNIV, NEW BRUNSWICK, 74- Mem: Am Ornithologists Union; Ecol Soc Am. Res: Ecology, behavior and evolution of birds; migration; orientation; territoriality; paleontology. Mailing Add: Dept of Sci Rutgers Univ New Brunswick NJ 08903

MURRAY, BRUCE B, radiochemistry, molecular spectroscopy, see 12th edition

MURRAY, BRUCE C, b New York, NY, Nov 30, 31; m 54, 71; c 3. ASTRONOMY, GEOLOGY. Educ: Mass Inst Technol, SB, 53, SM, 54, PhD(geol), 55. Prof Exp: Explor & exploitation geologist, Calif Co, La, 55-58; geophysicist, Geophys Res Directorate, L G Hanscom Field, Mass, 58-60; res fel, 60-63, assoc prof, 63-68, PROF PLANETARY SCI, CALIF INST TECHNOL, 68-, DIR JET PROPULSION LAB, 76- Concurrent Pos: Guest observer, Mt Wilson & Palomar Observ, 60-65, staff assoc, 65-69; consult, Rand Corp, 61-75; co-investr, TV Exp, Mariner 4, 65, Mariner 6 & 7, 69, Mariner 9, 71, leader imaging team, Mariner Venus Mercury 73 Mission; Guggenheim fel, 75-76. Honors & Awards: Except Sci Achievement Award, NASA, 69 & Distinguished Pub Serv Medal, 74. Mem: AAAS; Am Astron Soc; Am Geophys Union. Res: Planetary exploration; geology and geophysics of the surfaces of the moon and planets; techniques of space photography. Mailing Add: Div of Geol & Planetary Sci Calif Inst of Technol Pasadena CA 91109

MURRAY, CALVIN CLYDE, b Oakboro, NC, Aug 5, 07; m 34; c 1. AGRONOMY. Educ: NC State Col, BS, 32; Univ Ga, MS, 38; Cornell Univ, PhD(plant breeding), 45. Prof Exp: High sch teacher, NC, 32-35; asst agronomist, Soil Conserv Serv, USDA, NC, 35; from asst prof to prof agron, Univ Ga, 36-46; prof & agronomist, La State Univ, 46-48; dir exp sta, 48-50, dean col agr, 50-68, regents prof int educ, dir inter-instnl progs, int affairs & exec dir southern consortium int educ, 68-74, EMER DEAN COL AGR & REGENTS PROF INT EDUC, UNIV GA & EMER DIR INT AFFAIRS, UNIV GA SYST, 74- Concurrent Pos: Consult, AID, World Bank, 74- & Inter Am Develop Bank, 75-; mem, Int Exec Serv Corps, 75- Res: International education; administration; institutional development overseas. Mailing Add: 236 West View Dr Athens GA 30601

MURRAY, CHARLES RICHARD, b Trinidad, Colo, Dec 1, 09; m 36; c 3. HYDROLOGY. Educ: Univ Colo, BS, 32, MS, 34. Prof Exp: Asst geol, Univ Colo, 33-34; mining geologist, Ward Big 5 Mining Co, 34-35; petrol geologist, Carter Oil Co, 37-38; surveyor & engr, State Hwy Dept, 38-39; geologist, US Geol Surv, NMex, 39-48; actg instr geol, Stanford Univ, 48-50; asst dist geologist, Iowa, 50-53; chief ground water hydrol, Foreign Opers Admin, Egypt, 53-56; tech adv geol, Int Coop Admin, Manila, 57-61, AID, Jordan, 62 & Turkey, 63-65; HYDROLOGIST, US GEOL SURV, 65- Mem: Geol Soc Am; Am Geophys Union; Soc Explor Geophys;

Am Water Resources Asn; Am Water Works Asn. Res: Ore deposits; mineralogy and petrography; stratigraphy and structure; geology and hydrology of ground water; water resources development; water use; quinquennial study of water use in the United States by category of use, source, and quality; summarizing water resources research projects of US Geological Survey annually. Mailing Add: US Geol Surv Nat Ctr Reston VA 22092

MURRAY, CHRISTOPHER BROCK, b Meadville, Pa, Jan 16, 37; m 61; c 2. MATHEMATICS. Educ: Rice Univ, BA, 58; Univ Tex, PhD(math), 64. Prof Exp: Spec instr math, Univ Tex, 61-63; engr scientist, Tracor, Inc, 64-66; ASST PROF MATH, UNIV HOUSTON, 66- Mem: Am Math Soc; Math Asn Am. Res: Mathematical analysis. Mailing Add: Dept of Math Univ of Houston Houston TX 77004

MURRAY, DONALD SHIPLEY, b Philadelphia, Pa, June 30, 16; m 38; c 3. ACADEMIC ADMINISTRATION, APPLIED STATISTICS. Educ: Univ Pa, BS, 37, AM, 40, PhD(statist), 44. Prof Exp: Instr statist & acct, 37-44, from asst prof to assoc prof statist, 45-57, chmn dept, 64, PROF STATIST, UNIV PA, 57- Concurrent Pos: Treas, Am Inst Indian Studies, 64-73, pres asst for Indian opers, 73-; dir, Nat Conf Admin Res, 66-72; mem, Nat Adv Coun Arthritis & Metab Dis, 69-72; mem, Grants Admin Adv Comt, Dept Health, Educ & Welfare, 69-72, chmn, 70-72. Mem: Am Statist Asn; Am Soc Eng Educ. Res: Development of models for financial management of private institutions of higher education.

MURRAY, EDWARD CONLEY, b Mullen, Nebr, Sept 25, 31; m 53; c 2. INORGANIC CHEMISTRY. Educ: Nebr State Teachers Col, Kearney, AB, 52; Univ Colo, Boulder, MS, 63, PhD(inorg chem), 69. Prof Exp: Jr chemist, Ames Lab, AEC, Iowa, 52-55; proj engr, Aeronaut Res Lab, Wright-Patterson AFB, Ohio, 59-60; sr res chemist, Am Potash & Chem Corp, 69-70; SR RES CHEMIST, KERR-McGEE CORP, 70- Mem: Am Chem Soc. Res: Preparation of high purity inorganic chemicals; surface chemistry of titanium dioxide pigments. Mailing Add: 2204 Reveille Dr Oklahoma City OK 73111

MURRAY, EDWARD DONALD, b Minnedosa, Man, June 23, 36; m 58; c 2. MICROBIAL BIOCHEMISTRY. Educ: Univ Man, BSc, 61, MSc, 62; Univ Western Ont, PhD(biochem), 70. Prof Exp: Res scientist, John Labatt Ltd, 63-72; SR RES SPECIALIST, GEN FOODS, LTD, 72- Concurrent Pos: Ed proc, Soc Indust Microbiol, 70-75; jour bus mgr, Can Inst Food Sci & Technol, 75- Mem: Can Soc Microbiol; Am Soc Microbiol; Am Chem Soc; Soc Indust Microbiol; Can Inst Food Sci & Technol. Res: Protein chemistry; enzymology; fermentations; food microbiology and microscopy. Mailing Add: Res Dept Gen Foods Ltd Cobourg ON Can

MURRAY, FINNIE ARDREY, JR, b Burgaw, NC, May 30, 43; m 64; c 2. REPRODUCTIVE PHYSIOLOGY. Educ: NC State Univ, BS, 66, MS, 68; Univ Fla, PhD(reproductive physiol), 71. Prof Exp: Res assoc, Dept Molecular, Cellular & Develop Biol, Univ Colo, 70-71; instr, Dept Zool, Univ Tenn, 71-72, asst prof, 72-74; ASST PROF, DEPT ANIMAL SCI, OHIO AGR RES & DEVELOP CTR, WOOSTER, 74- Concurrent Pos: Asst prof, Dept Animal Sci, Ohio State Univ, 74-; embryologist, KBJ Ranch Bovine Embryo Transplantation Lab, Xenia, Ohio, 75- Mem: Am Soc Animal Sci; AAAS; Soc Study Reproduction; Int Embryo Transfer Soc. Res: Function of the uterus in embryonic development; endocrine control of uterine and ovarian function; biochemistry of uterine secretions. Mailing Add: Dept of Animal Sci Ohio Agr Res & Develop Ctr Wooster OH 44691

MURRAY, FRANCIS E, b Grande Prairie, Alta, Nov 17, 18; m 39; c 4. PHYSICAL CHEMISTRY. Educ: Univ Alta, BSc, 50; McGill Univ, PhD(phys chem), 53. Prof Exp: Asst prof phys chem, Univ Man, 53-55; asst prof, Can Serv Col, Royal Roads, 55-56; res chemist, BC Res Coun, 56-61; res coord, Consol Paper Corp, 61-62; head div chem, BC Res Coun, 62-68; assoc prof chem eng, 68-70, PROF CHEM ENG & HEAD DEPT, FAC APPL SCI, UNIV BC, 70- Concurrent Pos: Consult, Air Pollution Br, USPHS. Mem: Can Pulp & Paper Asn; Tech Asn Pulp & Paper Indust; Air Pollution Control Asn. Res: Molecular association forces; critical temperature phenomena; Kraft pulping chemical recovery and pulp purification. Mailing Add: Dept of Chem Eng Univ of BC Fac of Appl Sci Vancouver BC Can

MURRAY, FRANCIS JOSEPH, b New York, NY, Feb 3, 11; m 35; c 6. MATHEMATICS. Educ: Columbia Univ, AB, 32, MA, 33, PhD(math), 35. Prof Exp: From instr to prof math, Columbia Univ, 36-60; dir spec res numerical anal, 60-70, PROF MATH, DUKE UNIV, 60- Concurrent Pos: Consult ed, Math Tables & Other Aids to Comput, Div Math, Nat Res Coun, 53-57. Mem: Am Math Soc; Asn Comput Mach. Res: Partial differential equations; linear spaces; rings of operators; Hilbert space; mathematical machines; aids to computation. Mailing Add: Dept of Math Duke Univ Durham NC 27706

MURRAY, FRANCIS JOSEPH, b Jersey City, NJ, Oct 16, 20; m 47; c 5. BACTERIOLOGY. Educ: St Peters Col, BS, 42; Purdue Univ, PhD(bact), 48. Prof Exp: Asst bact, Lederle Labs, Am Cyanamid Co, NY, 44; asst serol, Purdue Univ, 44-46; asst chief bacteriologist, Wm S Merrell Co, 48-51, head dept microbiol, 51-56, exec asst to dir res, 56-59, dir sci rels, 60-67, dir sci & com develop & cent sci serv, 67-69, VPRES, RICHARDSON-MERRELL INC, 69- Mem: AAAS; Am Soc Microbiol; Am Asn Immunol; Am Soc Clin Pharmacol & Therapeut. Res: Antibiotics; immunology; virology; chemotherapy. Mailing Add: Richardson-Merrell Inc 10 Westport Rd Wilton CT 06897

MURRAY, FRANCIS WILLIAM, b San Antonio, Tex, July 29, 21; m 54; c 1. METEOROLOGY. Educ: Univ Tex, BA, 41; Univ Calif, Los Angeles, MA, 48; Mass Inst Technol, PhD(meteorol), 60. Prof Exp: Sr res scientist, Douglas Aircraft Co, Inc, 63-66; PHYS SCIENTIST, RAND CORP, 66- Concurrent Pos: Weather officer, Air Weather Serv, DC, Marshall Islands & Morocco. Mem: Am Meteorol Soc; Am Geophys Union; Royal Meteorol Soc. Res: Atmospheric science; cloud dynamics; dynamic meteorology; numerical weather prediction. Mailing Add: Rand Corp Phys Sci Dept 1700 Main St Santa Monica CA 90406

MURRAY, FREDERICK NELSON, b Tulsa, Okla, Apr 21, 35; m 67. STRUCTURAL GEOLOGY, STRATIGRAPHY. Educ: Univ Tulsa, BS, 57; Univ Wash, BS, 62; Univ Colo, MS, 62, PhD(geol), 66. Prof Exp: Jr geologist, Pan Am Petrol Corp, 57-58; asst geologist, Ill State Geol Surv, 65-67; asst prof geol, Allegheny Col, 68-71; geologist, US Geol Surv, 71-75; MEM STAFF, NAPCO INC, 75- Mem: Am Asn Petrol Geologists; Am Geophys Union; Geol Soc Am. Res: Computer application in geology; geologic mapping; coal geology. Mailing Add: 1211 S College Tulsa OK 74104

MURRAY, GEORGE CLOYD, b Minneapolis, Minn, May 20, 34; m 57; c 2. NEUROPHYSIOLOGY, PHYSICS. Educ: George Washington Univ, BS, 59; Univ Colo, MS, 62; Johns Hopkins Univ, PhD(biophys), 68. Prof Exp: Assoc staff engr, Appl Physics Lab, Johns Hopkins Univ, 54-59; instr & teaching assoc physics, Univ Colo, 59-62; asst prof, George Washington Univ, 62; instr & res assoc biophys, Johns Hopkins Univ, 62-68; staff fel neurophysiol, Lab Neurophysiol, Nat Inst Neurol Dis & Blindness, 68-72, head, Communicative Disorders Sect & Biomed Eng Sect, C & FR, Nat Inst Neurol Dis & Stroke, 72-74, SPEC ASST TO DIR & DEP DIR, DIV BLOOD DIS & RESOURCES, NAT HEART & LUNG INST, NIH, 74- Res: Biophysical study of the mechanisms of excitable biological membrane systems. Mailing Add: 1706 Mark Lane Rockville MD 20852

MURRAY, GEORGE GRAHAM, JR, b Detroit, Mich, Feb 23, 24; m 51; c 3. MATHEMATICS. Educ: Harvard Univ, AB, 44, MA, 48, PhD(math), 51. Prof Exp: Mathematician, Appl Physics Lab, Johns Hopkins Univ, 51-55; systs engr, Radio Corp Am, 55-57, mgr digital systs group, Missile & Surface Radar Div, 57-63; dir adv tech, Librascope Div, Gen Precision, Inc, 63-65, chief engr, Kearfott-San Marcos Div, 65-69; PARTNER, DATA-WARE DEVELOP, 69- Mem: Am Math Soc; Soc Indust & Appl Math; Asn Comput Mach. Res: Analysis; digital computer systems; Walsh and Fourier transforms; special processor design and development. Mailing Add: 8650 Kilbourn Dr La Jolla CA 92037

MURRAY, GLEN A, b Sidney, Mont, Mar 1, 39; m 61; c 2. CROP PHYSIOLOGY, AGRONOMY. Educ: Mont State Univ, BS, 62, MS, 64; Univ Ariz, PhD(agron), 67. Prof Exp: ASSOC PROF PLANT SCI & ASSOC CROP PHYSIOLOGIST, UNIV IDAHO, 67- Mem: Am Soc Agron; Crop Sci Soc Am. Res: Cold hardiness of grass seedlings; photoperiodic studies on legumes; associating changes from vegetative to reproductive states; regulation of nitrogen distribution in wheat plants; protein and yield studies on Austrian winter peas, growth regulators; row spacing, date of seeding and photoperiod studies. Mailing Add: Dept of Plant & Soil Sci Univ of Idaho Moscow ID 83843

MURRAY, GROVER ELMER, b Maiden, NC, Oct 26, 16; m 41; c 2. GEOLOGY. Educ: Univ NC, BS, 37; La State Univ, MSc, 39, PhD(geol), 42. Prof Exp: Res geologist, State Geol Surv, La, 38-41; geologist, Magnolia Petrol Co, 41-48; prof stratig geol, La State Univ, 48-55, chmn dept, 50-53, Boyd prof geol, 55-66, vpres & dean, 63-65, vpres acad affairs, La State Univ Syst, 65-66; PRES, TEX TECH UNIV, 66-, PRES SCH MED, 69- Concurrent Pos: With Ark Fuel Oil Corp, 51-60; dir, Orgn Trop Studies, Inc, 64-65, dir adv coun, 65-; dir, Gulf Univs Res Corp, 64-, pres, 65-66; consult, 49-; mem, Am Comn Stratig Nomenclature, 51-54 & 57-63; ed, J Paleont, 52-54; mem, Int Comn Stratig Nomenclature, 54-; ed, Bull Am Asn Petrol Geologists, 59-63; mem, US Nat Comt Geol, 65-69; deleg, House Soc Reps, Am Geol Inst, 58-63 & 65-68; mem bd gov, ICASALS, Inc, 67-; mem, Nat Sci Bd, 68-; bd dirs, Tex Partners of Americas, 71-; mem, Nat Adv Comt Oceans & Atmosphere, 75- Mem: Geol Soc Am; Soc Econ Paleontologists & Mineralogists (pres, 62-63); Paleont Soc; hon mem Am Asn Petrol Geologists (pres, 64-65); Am Geophys Union. Res: Structural and field geology; geomorphology; geophysics; micropaleontology; stratigraphy of the Gulf coast and southern Appalachians; petroleum geology of coastal plain. Mailing Add: Off of the Pres Tex Tech Univ Lubbock TX 79409

MURRAY, HAROLD DIXON, b Neodesha, Kans, May 25, 31; m 554; c 1. MALACOLOGY. Educ: Ottawa Univ, BA, 52; Kans State Col Pittsburg, 53; Univ Kans, PhD(zool), 60. Prof Exp: Asst biol, zool & parasitol, Univ Kans, 55-60, instr limnol & invert zool, 60-61; from asst prof to assoc prof, 61-74, PROF BIOL, TRINITY UNIV, TEX, 74-, CHMN DEPT, 75- Mem: AAAS; Am Malacol Union (vpres, pres, 74). Res: Geographical distribution of Unionidae and Unionicolidae; biology and ecology of Thiaridae. Mailing Add: Dept of Biol Trinity Univ San Antonio TX 78284

MURRAY, HAYDN HERBERT, b Kewanee, Ill, Aug 31, 24; m 44; c 3. CLAY MINERAOLOGY, ECONOMIC GEOLOGY. Educ: Univ Ill, BS, 48, MS, 50, PhD(geol), 51. Prof Exp: From asst prof to assoc prof geol, Ind Univ, 51-57; dir appl res, Ga Kaolin Co, 57-59, dir res & mfg, 59-63, vpres, 63-64, exec vpres, 64-73; CHMN DEPT GEOL, IND UNIV, 73- Concurrent Pos: Clay mineralogist, State Geol Surv, Ind, 51-57; mem exec comt, Working Comt Genesis & Age of Kaolins, UNESCO, 73- Honors & Awards: Hal Williams Hardinge Award, Am Inst Mining, Metall & Petrol Engr, 76. Mem: Am Ceramic Soc (vpres, 74-75); Geol Soc Am; Mineral Soc Am; Am Inst Mining, Metall & Petrol Engr; Clay Minerals Soc (pres, 65-66). Res: Geology, economic uses and chemistry of clay minerals; beneficiation of metallic, non-metallic, and coals using high intensity magnetic separation. Mailing Add: Dept of Geol Ind Univ Bloomington IN 47401

MURRAY, HERBERT CHARLES, b Indianapolis, Ind, Aug 4, 12; m 46; c 4. MICROBIOLOGY. Educ: Purdue Univ, BS, 38, MS, 40, PhD(bact), 42. Exp:Asst bact, Purdue Univ, 38-42; bacteriologist, 42-54, MICROBIOLOGIST, UPJOHN CO, 54- Mem: AAAS; Am Soc Microbiol. Res: Fermentation of steroids; biochemistry of fungi. Mailing Add: Upjohn Co Unit 7252 301 Henrietta St Kalamazoo MI 49001

MURRAY, IRWIN MACKAY, b NS, Can, Nov 20, 19; m 44; c 4. ANATOMY. Educ: Dalhousie Univ, MD, 44. Prof Exp: From asst prof to assoc prof anat, Dalhousie Univ, 47-50; from asst prof to assoc prof, 51-68, PROF ANAT, COL MED, STATE UNIV NY DOWNSTATE MED CTR, 68- Mem: Gerontol Soc; Am Asn Anat; NY Acad Sci. Res: Physiologic aspects of aging; physiologic functions of reticuloendothelial system; human anatomy. Mailing Add: Dept of Anat State Univ NY Downstate Med Ctr Brooklyn NY 11203

MURRAY, JAMES GORDON, b Flint, Mich, July 8, 27; m 47; c 4. ORGANIC POLYMER CHEMISTRY. Educ: Univ Mich, BS, 50; Dartmouth Univ, MA, 52; Duke Univ, PhD(chem), 55. Prof Exp: Chemist, Monsanto Chem Co, 55-59 & Gen Elec Co, 59-67; group leader polyolefins, 67-74, RES ASSOC, MOBIL CHEM CO, 74- Mem: Am Chem Soc. Res: Polymers; organometallics; catalysis; petroleum chemistry. Mailing Add: Mobil Chem Co PO Box 240 Edison NJ 08817

MURRAY, JAMES W, b Berwyn, Alta, Sept 11, 33; m 60. GEOLOGY. Educ: Univ Alta, BSc, 56; Princeton Univ, MA, 63, PhD(geol), 64. Prof Exp: Geologist, Texaco Explor Co, Alta, 57-61; fel, Inst Oceanog, 64-65, asst prof geol, 65-69, assoc prof geol & oceanog, 69-74, actg head dept geol, 71-72, PROF GEOL SCI, UNIV BC, 74- Concurrent Pos: Mem subcomt stratig, paleont & fossil fuels res, Nat Adv Comt Can, 65-68; co-chmn, Marine Geol Prog Int Geol Cong, Montreal, 72. Mem: Fel Geol Soc Am; Am Asn Petrol Geologists; Soc Econ Paleontologists & Mineralogists. Res: Origin and distribution of recent marine sediments in fjords and delta; origin of continental shelves and slopes. Mailing Add: Dept of Geol Univ of BC Vancouver BC Can

MURRAY, JAY CLARENCE, b Lapoint, Utah, June 27, 29; m 49; c 2. GENETICS, AGRONOMY. Educ: Utah State Univ, BS, 51; Colo Agr & Mech Col, MS, 55; Cornell Univ, PhD, 59. Prof Exp: Asst agron, Colo Agr & Mech Col, 53-55; asst genetics, Cornell Univ, 55; from asst prof to assoc prof, 59-67, PROF AGRON, OKLA STATE UNIV, 67-, ASSOC DIR AGR EXP STA, 68- Res: Physiological genetics of neurospora and photoperiodism in Gossypium; geneome complementation in the tetraploid species of Gossypium; genetics of fiber differentiation in cotton; quantitative genetic studies of fiber properties and earliness of cotton; genetics, cytogenetics, and breeding of forage grasses and legumes. Mailing Add: Okla Agr Exp Sta Okla State Univ Stillwater OK 74074

MURRAY, JOAN BAIRD, b Rochester, NY, Nov 20, 26; m 52; c 2. VERTEBRATE ZOOLOGY. Educ: Alfred Univ, BA, 48; Syracuse Univ, MA, 50, PhD, 53. Prof Exp: Asst prof biol, Western Col, 60-68; TEACHER BIOL, DEKALB COL, 68- Concurrent Pos: Grammar sch teacher & demonstr, Ibadan, 65-66. Mem: AAAS; Am Inst Biol Sci; Am Soc Zoologists. Res: Biogeography. Mailing Add: Dept of Biol DeKalb Col Clarkston GA 30021

MURRAY, JOHN FREDERIC, b Mineola, NY, June 8, 27; m 49; c 3. INTERNAL MEDICINE. Educ: Stanford Univ, AB, 49, MD, 53. Prof Exp: From intern to asst resident med, San Francisco Hosp, 52-54; from resident to sr resident, Kings County Hosp, New York, 54-56; res fel, Am Col Physicians, Post-Grad Sch Med, Univ London, 56-57; from instr to assoc prof med & physiol, Univ Calif, Los Angeles, 57-59; assoc prof, 66-69, PROF MED, UNIV CALIF, SAN FRANCISCO, 69-, STAFF MEM CARDIOVASC RES INST, SCH MED, 66-; CHIEF CHEST SERV, SAN FRANCISCO HOSP, 66- Concurrent Pos: Attend specialist, Vet Admin, 59-; chmn pulmonary training comt, Nat Heart & Lung Inst, 70-72; ed, Am Rev Respiratory Dis; chmn pulmonary acad award comt, Nat Heart & Lung Inst, 74-, mem pulmonary dis adv comt, 75- Mem: Asn Am Physicians; Am Soc Clin Invest; Am Fedn Clin Res; Am Physiol Soc. Res: Cardiopulmonary physiological techniques in clinical medicine. Mailing Add: Chest Serv San Francisco Hosp San Francisco CA 94110

MURRAY, JOHN JOSEPH, b New York, NY, Oct 16, 37; m 60; c 3. ORGANIC CHEMISTRY. Educ: Manhattan Col, BS, 59; Fordham Univ, PhD(org chem), 64. Prof Exp: Res chemist, Gen Chem Div, Allied Chem Corp, NJ, 64-70; asst prof, 70-75, ASSOC PROF CHEM, MIDDLESEX COUNTY COL, 75- Mem: Am Chem Soc. Res: Synthesis of new fluorin fluorinated compounds; chemistry of carbenes and their diazo precursors. Mailing Add: Dept of Chem Middlesex County Col Edison NJ 08817

MURRAY, JOHN RANDOLPH, b PEI, Aug 18, 16; m 42; c 2. PHARMACOLOGY. Educ: Univ Alta, BSc, 40, MSc, 50; Ohio State Univ, PhD(pharmacol), 55. Prof Exp: Asst dispenser, Dunford Drug Co, Ltd, Alta, 40-41; asst dept mgr, Parke, Davis & Co, Ont, 41-42; from lectr pharm to prof, Sch Pharm, Univ Alta, 46-59; dir, Sch Pharm, 59-70, DEAN FAC PHARM, UNIV MAN, 70- Mem: AAAS; Pharmacol Soc Can; Can Pharmaceut Asn; Can Conf Pharmaceut Fac; Can Found Adv Pharm. Res: Hypertension and the antihypertensive drugs; tissue respiration. Mailing Add: Fac of Pharm Univ of Man Winnipeg MB Can

MURRAY, JOHN WOLCOTT, b Flushing, NY, Jan 9, 09; m 38; c 2. CHEMISTRY. Educ: Colgate Univ, AB, 30; Johns Hopkins Univ, PhD(chem), 33. Prof Exp: Asst chem, Johns Hopkins Univ, 32-33, instr, 33-34; asst, Rockefeller Inst, 34-39; chief chemist, Thomasville Stone & Lime Co, 39-42; from asst prof to prof, 42-71, EMER PROF CHEM, VA POLYTECH INST & STATE UNIV, 71- Mem: Am Chem Soc; Nat Speleol Soc. Res: Raman spectra; molecular models; dissociation of phenols; movement of water in cell models; factors controlling nature of cave deposits. Mailing Add: 701 York Dr NE Blacksburg VA 24060

MURRAY, JOSEPH, b Pocatello, Idaho, Dec 14, 16; m 41; c 2. ORGANIC CHEMISTRY. Educ: Univ Puget Sound, BS, 48; Ore State Univ, MS, 50. Prof Exp: Clin chemist, J W Creed Labs, Idaho, 50-51; res chemist, Trojan Powder Co, 51-52; instr chem, Buffalo Univ, 52-56; sr engr, Energy Div, Olin Mathieson Chem Corp, 56-59, dir comput applns, Nuclear Fuels Div, 59-61, corp mgr, 61-62; from asst prof to assoc prof chem, 62-70, PROF CHEM, MONT COL MINERAL SCI & TECHNOL, 70-, ACTG HEAD DEPT, 71- Mem: Am Chem Soc; fel Am Inst Chemists. Res: Synthetic and theoretical organic chemistry; organometallic chemistry; chemistry of parabanic acid; high temperature thermodynamics; high energy fuels; radioactive tracers; applications of computers to science and engineering. Mailing Add: Dept of Chem Mont Col of Mineral Sci & Technol Butte MT 59701

MURRAY, JOSEPH BUFORD, b Birmingham, Ala, July 29, 33; m 55; c 2. GEOLOGY. Educ: Univ Chattanooga, BS, 55; Univ Tenn, Knoxville, MS, 60; Case Western Reserve Univ, PhD(geol), 71. Prof Exp: Asst prof geol & geog, Grove City Col, 60-66; CHIEF GEOLOGIST, DEPT NATURAL RESOURCES, DIV MINES, MINING & GEOL, GA GEOL SURV, 69- Mem: Geol Soc Am; Soc Econ Paleontologists & Mineralogists. Res: Precambrian and Paleozoic stratigraphy of the southern and central Appalachian Mountains. Mailing Add: Dept of Natural Resources Div Mines Mining & Geol Ga Geol Surv Atlanta GA 30334

MURRAY, JOSEPH JAMES, JR, b Lexington, Va, Mar 13, 30; m 57; c 3. ZOOLOGY. Educ: Davidson Col, BS, 51; Oxford Univ, BA, 54, MA, 57, DPhil(zool), 62. Prof Exp: Instr biol, Washington & Lee Univ, 56-58; from asst prof to assoc prof, 62-73, PROF BIOL, UNIV VA, 73-, CO-DIR, MOUNTAIN LAKE BIOL STA, 64- Mem: AAAS; Soc Study Evolution; Am Soc Naturalists; Am Soc Ichthyologists & Herpetologists; Genetics Soc Am. Res: Genetics of populations of gastropods. Mailing Add: Dept of Biol Univ of Va Charlottesville VA 22903

MURRAY, KENNETH JAMES, organic chemistry, see 12th edition

MURRAY, KENNETH MALCOLM, JR, b Philadelphia, Pa, July 17, 25; div; c 2. NUCLEAR PHYSICS. Educ: Univ Miami, BSc, 49; Univ Md, MS, 63; Georgetown Univ, PhD(physics), 69. Prof Exp: Instr physics, Univ Miami, 49-51, oceanogr, Marine Lab, 51-52; oceanogr, 52-56, physicist, 56-70, HEAD X-RAY APPLN SECT, RADIATION TECHNOL DIV, NAVAL RES LAB, 70- Concurrent Pos: Edison mem grad training prog, Naval Res Lab, 64-66. Mem: Am Phys Soc. Res: Measured short period current variations in Florida Current between Miami and Gun Cay; photonuclear reactions in light elements; applications of nuclear physics. Mailing Add: Code 6633 Naval Res Lab Washington DC 20375

MURRAY, LEO THOMAS, b New York, NY, May 15, 37; m 60; c 5. INORGANIC CHEMISTRY, ORGANIC CHEMISTRY. Educ: Manhattan Col, BS, 58; Purdue Univ, PhD(inorg chem), 63. Prof Exp: Res chemist, 62-63, sr res chemist, 63-66, sect head oral prod, 66-70, sect head laundry res, 70-72, tech coordr mkt, 72-73, dir contract purchasing, 73-74, MGR HOUSEHOLD SPECIALTIES RES, COLGATE-PALMOLIVE CO, 74- Mem: Am Chem Soc. Res: Chemistry organoboranes, specifically amine-boranes; chemistry of oxidizing agents, specifically N-kalo and peroxide types; development of household and toiletries consumer products. Mailing Add: Colgate-Palmolive Co 909 River Rd Piscataway NJ 08854

MURRAY, LINWOOD ASA, JR, b Syracuse, NY, July 11, 08; m 34; c 1. PHYSICAL CHEMISTRY. Educ: Augustana Col, AB, 29; Univ Ill, PhD(chem), 34. Prof Exp: Res & develop chemist, Gen Labs, US Rubber Co, 34-44, sr chemist, Providence Plant, 44-46, asst mgr develop dept, 46-52, group leader new prod develop dept, Gen Labs, 52-54; dir, Mellon Group, US Naval Powder Factory, Md, 54-56; chief chem res lab, US Army Eng Res & Develop Lab, 56-60, chief chem br, US Army Sci Liaison & Adv Group, 60-69, chief scientist, 69-72; RETIRED. Mem: AAAS; Am Chem Soc; Am Phys Soc; Inst Elec & Electronics Engrs. Res: Dipole moment; latex technology and applications; rubber technology; resin research and development; radiography. Mailing Add: 6 Cherry Lane Asheville NC 28804

MURRAY, MALCOLM ARTHUR, b Stratford, Ont, Dec 22, 25; US citizen; m 52; c 2. GEOGRAPHY. Educ: Univ Western Ont, BA, 48; Syracuse Univ, MA, 50, PhD(geog), 55. Prof Exp: From asst prof to prof geog, Miami Univ, 52-68, res assoc, Scripps Found Pop Studies, 64-68; prof urban life & chmn dept geog, 68-72, PROF GEOG, GA STATE UNIV, 68- Concurrent Pos: Fulbright travel grant, Univ Southampton, 58-59; Soc Sci Res Coun travel grant, Sweden, 60; Rockefeller Found vis assoc prof, Univ Ibadan, 65-66; Asn Am Geogr & NSF vis geog scientist, Univ Vt, Middlebury Col & Darmouth Col, 67; Australian-Am Educ Found vis sr lectr geog, Univ New Eng, Australia, 75. Mem: Am Geog Soc; Asn Am Geogr; Can Asn Geogr; Int Geog Union; Pop Asn Am. Res: Medical and urban geography; computer graphics as applied to demographic data. Mailing Add: Dept of Geog Ga State Univ 33 Gilmer St SE Atlanta GA 30303

MURRAY, MARC MICHAEL, b Rochester, NY, Jan 4, 46; m 71. PETROLOGY, GEOCHEMISTRY. Educ: Oberlin Col, BA, 67; Rice Univ, MA & PhD(geol), 70. Prof Exp: Welch Found grant geol, Rice Univ, 70-71; asst prof, 71-75, ASSOC PROF GEOL, LA STATE UNIV, BATON ROUGE, 75- Concurrent Pos: Am Chem Soc Petrol Res Fund grant geochem, NMex Inst Mining & Technol, 70-71. Mem: AAAS; Geol Soc Am; Geochem Soc; Am Geophys Union. Res: Granite petrology; igneous and metamorphic petrology; volcanic geochemistry; plate tectonics and metamorphism; crustal evolution. Mailing Add: Dept of Geol La State Univ Baton Rouge LA 70803

MURRAY, MARGARET RANSONE (MRS BURTON LE DOUX), b Mathews Co, Va, Nov 16, 01; m 41. NEUROSCIENCES. Educ: Goucher Col, AB, 22; Washington Univ, MS, 24; Univ Chicago, PhD(zool), 26. Prof Exp: Nat Res Coun fel, Univ Chicago, 26-28; assoc prof biol & physiol, Fla State Col Women, 28-29; from instr to prof, 29-70, EMER PROF SURG, COL PHYSICIANS & SURGEONS, COLUMBIA UNIV, 70-; RES BIOLOGIST & SR SCIENTIST, NIH, 73- Concurrent Pos: Mem fel rev panel, NIH, 60-63; NIH res career award, 62-72; Commonwealth Fund traveling fel, Europe & Asia, 63-64. Honors & Awards: Sci Medal, Univ Brussels, 64. Mem: Am Soc Cell Biologists; hon mem Tissue Cult Asn (secy, 46-50, pres, 54-56); Am Asn Anat; Soc Neurosci; hon mem Am Asn Neuropath. Res: Neurobiology; organotypic culture of the nervous system; functional differentiation of nerve and mucle in vitro; degenerative diseases of the nervous system; active transport, cerebrovascular. Mailing Add: Lab Neuropath & Neuroanat Sci Inst Neurol Commun Dis & Stroke Bethesda MD 20014

MURRAY, MARION, b Evanston, Ill, Feb 27, 37. NEUROBIOLOGY. Educ: McGill Univ, BSc, 59; Harvard Univ, MA, 61; Univ Wis-Madison, PhD(physiol), 64. Prof Exp: Res asst neurobiol, Rockefeller Univ, 67-69; ASST PROF ANAT, PRITZKER SCH MED, UNIV CHICAGO, 69- Concurrent Pos: NIH fel anat, McGill Univ, 64-67. Mem: Am Asn Anat; Am Soc Cell Biol; Soc Neurosci. Res: Structure and function of nerve cells with special emphasis on synthesis and transport of protein. Mailing Add: Dept of Anat Pritzker Sch of Med Univ of Chicago Chicago IL 60637

MURRAY, MARVIN, b Milwaukee, Wis, June 10, 27; m 59; c 2. CLINICAL PATHOLOGY. Educ: Marquette Univ, BS, 48; Mich State Univ, MS, 50, PhD(physiol), 56; Wayne State Univ, MD, 55. Prof Exp: Res path resident path, Univ Wis, 56-59; from asst prof to assoc prof, 59-71, PROF PATH, SCH MED, UNIV LOUISVILLE, 71-, ADJ ASSOC PROF CHEM, 65-, DIR CLIN PATH, 59- Concurrent Pos: Nat Cancer Inst trainee, Univ Wis, 56-59, dir clin labs, Louisville Gen Hosp, 59-; Lederle Med Fac award, 64. Res: Blood coagulation; protein chemistry; clinical analytical chemistry. Mailing Add: 323 E Chestnut St Louisville KY 40202

MURRAY, MARY AILEEN, b Washington, DC, Dec 11, 14. BOTANY. Educ: Univ Ariz, BA, 36, MS, 38; Univ Chicago, PhD(bot), 45. Prof Exp: Pub sch teacher, Ariz, 37-43; asst bot, Univ Chicago, 43-45, res assoc & instr, 45-48; asst prof, 48-55, ASSOC PROF BOT, DePAUL UNIV, 55- Mem: AAAS; Bot Soc Am; Am Inst Biol Sci. Res: Plant morphology and anatomy. Mailing Add: Dept of Biol DePaul Univ Chicago IL 60614

MURRAY, MARY PATRICIA, b Milwaukee, Wis, July 27, 23; m 63. KINESIOLOGY, MEDICAL RESEARCH. Educ: Ripon Col, BA, 45; Marquette Univ, MS, 56, PhD(anat), 61. Prof Exp: Resident phys ther, Mayo Clin, 45; staff phys therapist, Curative Workshop of Milwaukee, 45-48; staff phys therapist, 48-51, supvr, 51-53, asst chief, 53-54, clin training supvr, 55-58, res & educ supvr, 59-64, CHIEF KINESIOLOGY RES LAB, VET ADMIN CTR, 64- Concurrent Pos: From instr to asst prof phys ther, Med Col Wis, 54-67, assoc prof, 67-, assoc prof anat, 75-; grants, NIH, 62-63 & 67-, Vet Admin, 62- & Dept Health, Educ & Welfare, 63-66; mem appl physiol & bioeng study sect, NIH, 71-76. Honors & Awards: Sci Exhibit Award, Am Phys Ther Asn, 64, Marian Williams Res Award, 67. Mem: Am Phys Ther Asn; Am Cong Rehab Med; Soc Behav Kinesiology. Res: Normal and abnormal human motion; locomotion, postural steadiness and upright stability, muscle strength and joint mobility in patients with neuro-musculo-skeletal disabilities. Mailing Add: Kinesiology Res Lab Vet Admin Ctr Wood WI 53193

MURRAY, MURRAY J, b Palmerston North, NZ, Oct 30, 22; m 46; c 4. MEDICINE. Educ: Univ NZ, MB & ChB, 46, MD, 53; Univ Otago, NZ, DSc, 70; Royal Australasian Col Physicians, dipl, 67; Royal Col Physicians Edinburgh, dipl, 69. Prof Exp: Registr, St Stephens Hosp, London, Eng, 50-51; asst physician, Wellington Hosp, NZ, 52-55; from instr to assoc prof, 56-69, PROF MED, UNIV MINN, MINNEAPOLIS, 69-, DIR MED SERV, UNIV HOSP, 70- Concurrent Pos: Hartford Found res grant, 68-69. Mem: Fel Am Col Cardiol; Am Gastroenterol Asn; Royal Col Physicians; fel Royal Soc Med. Res: Coronary artery disease; portal hypertension; iron absorption. Mailing Add: Dept of Med Univ of Minn Minneapolis MN 55455

MURRAY, RAYMOND CARL, b Fitchburg, Mass, July 2, 29; m 55; c 2. SEDIMENTARY PETROLOGY. Educ: Tufts Col, BS, 51; Univ Wis, MS, 52, PhD, 55. Prof Exp: Asst, Tufts Col, 50-51 & Univ Wis, 52-55; res geologist, Shell Develop Co Div, Shell Oil Co, 55-62, sect head prod geol res, 62-66; assoc prof geol, Univ NMex, 66-67; PROF GEOL & CHMN DEPT, RUTGERS UNIV, NEW BRUNSWICK, 67- Concurrent Pos: Sr field geologist, State Geol Surv, Wis, 52, 54 & Buchans Mining Co, Ltd, 53; instr, Edgewood Col, 53 & 55; vis lectr, Univ Calif, 62; vis scientist, Rijswijk, Neth, 63-64; vis prof, Univ Mont, 75. Honors & Awards: Best Paper Award, Soc Econ Paleontologists & Mineralogists, 60. Mem: Fel Geol Soc Am; Soc Econ Paleontologists & Mineralogists; Am Asn Petrol Geologists. Res: Petrology of recent sediments and carbonate rocks; forensic geology. Mailing Add: Dept of Geol Rutgers Univ New Brunswick NJ 08903

MURRAY, RAYMOND GORBOLD, b Tokyo, Japan, May 12, 16; m 38, 56; c 6. HISTOLOGY. Educ: Monmouth Col, SB, 37; Univ Chicago, PhD(histol), 42. Prof Exp: Asst histol, Univ Chicago, 39-43, res assoc, 43-46; instr, Med Sch, Tufts Col, 46-48; asst prof path, Dent Sch, Northwestern Univ, 48-49; assoc prof, 49-65, PROF ANAT, MED SCH, IND UNIV, BLOOMINGTON, 65-, CHMN DEPT ANAT & PHYSIOL, 73- Mem: AAAS; Electron Micros Soc Am; Am Soc Cell Biol; Am Asn Anat; Soc Neurosci. Res: Tissue culture of thymus; parenteral nutrition;

histopathology of x-rays; histopathology and distribution of radioisotopes; morphology and function of lymphocytes; fine structure of taste buds. Mailing Add: 1910 E First St Bloomington IN 47401

MURRAY, RAYMOND HAROLD, b Cambridge, Mass, Aug 17, 25; m 48; c 8. CARDIOLOGY. Educ: Univ Notre Dame, BS, 45; Harvard Univ, MD, 48; Am Bd Internal Med, dipl, 55; Am Bd Cardiovasc Dis, dipl, 60. Prof Exp: Intern med, Peter Bent Brigham Hosp, Boston, Mass, 48-49; resident, Roosevelt Hosp, New York, 49-50; res assoc cardiol, Nat Heart Inst, 50-53; instr med, Univ Mich, 53-54; assoc prof, 62-67, PROF MED, SCH MED, IND UNIV INDIANAPOLIS, 67-, CHMN DEPT COMMUNITY HEALTH SCI, 72- Concurrent Pos: Fel coun clin cardiol, Am Heart Asn, 64; dir, Regenstrief Inst Health Care. Mem: Aerospace Med Asn; Am Fedn Clin Res; Am Col Physicians. Res: Health care delivery. Mailing Add: Marion County Gen Hosp Ind Univ Sch of Med Indianapolis IN 46202

MURRAY, RICHARD BENNETT, b Marietta, Ga, Dec 5, 28; m 56; c 2. SOLID STATE PHYSICS. Educ: Emory Univ, AB, 47; Ohio State Univ, MS, 50; Univ Tenn, PhD(physics), 55. Prof Exp: Asst physics, Oak Ridge Gaseous Diffusion Plant, 47-48; physicist, Oak Ridge Nat Lab, 55-66; assoc prof, 66-68, actg chmn dept, 75-76, PROF PHYSICS, UNIV DEL, 69- Concurrent Pos: Vis assoc prof, Univ Del, 62-63; lectr, Univ Tenn, 63-66. Mem: AAAS; Am Asn Physics Teachers; Sigma Xi; fel Am Phys Soc. Res: Luminescence and scintillation phenomena in solids; color centers; channeling; ion penetration and radiation damage. Mailing Add: Dept of Physics Univ of Del Newark DE 19711

MURRAY, ROBERT EDWARD, b New Haven, Conn, Oct 25, 43; m 65; c 2. BIOCHEMISTRY, VIROLOGY. Educ: Bellarmine Col, Ky, BA, 65; Yale Univ, MS, 67, PhD(biol), 69. Prof Exp: ASST PROF MICROBIOL, MEHARRY MED COL, 70- Concurrent Pos: Fel, Inst Molecular Virol, St Louis Univ, 68-69; Brown-Hazen Found res grant, 71. Mem: Am Soc Microbiol. Res: Biochemisty of viral growth; application of modern virology to routine medical practice. Mailing Add: Dept of Microbiol Meharry Med Col Nashville TN 37208

MURRAY, ROBERT FULTON, JR, b Newburgh, NY, Oct 19, 31; m 56; c 4. MEDICAL GENETICS. Educ: Union Col, NY, BS, 53; Univ Rochester, MD, 58; Am Bd Internal Med, dipl, 66; Univ Wash, MS, 68. Prof Exp: Resident med, Colo Gen Hosp, 59-62; sr surgeon, USPHS, NIH, 62-65; fel med genetics, Univ Wash, 65-67; from asst prof to assoc prof pediat & med, 67-74, PROF PEDIAT & MED, COL MED, HOWARD UNIV, 74- Concurrent Pos: Mem nat adv coun, Nat Inst Gen Med Sci, 71-75; chmn ad hoc comt sickel cell trait, Armed Forces, 72; mem comt inborn errors metab, Nat Res Coun, Nat Acad Sci, 72-75. Mem: Nat Inst Med; Am Soc Human Genetics; fel AAAS; fel Am Col Physicians; fel Inst Soc Ethics & Life Sci. Res: Studies of factors influencing genetic counseling; genetic and developmental variations in isoenzymes; inherited susceptibility to disease. Mailing Add: Div of Med Genetics Box 75 Howard Univ Col of Med Washington DC 20059

MURRAY, ROBERT GEORGE EVERITT, b Ruislip, Eng, May 19, 19; m 44; c 3. BACTERIOLOGY. Educ: Cambridge Univ, BA, 41, MA, 42; McGill Univ, MD, CM, 43. Prof Exp: Lectr, 45-47, from asst prof to assoc prof, 47-49, head dept, 49-74, PROF BACT & IMMUNOL, UNIV WESTERN ONT, 49- Concurrent Pos: Med, Can J Microbiol, 54-60; hon consult, St Joseph's Hosp, London, Ont, 60-; mem, Bergey's Manual Trust, 64-; mem, Int Comt Bact Nomenclature, 66-; gov bd, Biol Coun Can, 66-; ed, Bact Rev, 69- Honors & Awards: Coronation Medal, 53; Harrison Prize, Royal Soc Can, 57 & Award, 60-61; Prize, Can Soc Microbiol, 63; Centennial Medal, Govt Can, 67. Mem: Electron Micros Soc Am; Am Soc Microbiol (vpres, 71-72, pres, 72-73); Am Soc Cell Biologists; fel Royal Soc Can; Can Soc Microbiol (pres, 51-52). Res: Bacterial cytology and physiology; ultrastructure of bacteria and relation of structure to function, with emphasis on the cell wall and macromolecular arrangement. Mailing Add: Fac of Med Univ of Western Ont London ON Can

MURRAY, ROBERT KINCAID, b Glasgow Scotland, Dec 18, 32; m 59; c 4. BIOCHEMISTRY. Educ: Glasgow Univ, MB, ChB, 56; Univ Mich, MS, 58; Univ Toronto, PhD(biochem), 61. Prof Exp: From asst prof to assoc prof, 61-73, PROF BIOCHEM, UNIV TORONTO, 73- Res: Biochemistry of glycosphingolipids; biochemistry of cancer. Mailing Add: Dept of Biochem Univ of Toronto Toronto ON Can

MURRAY, ROBERT MARIE, b New London, Conn, Mar 5, 27. RUBBER CHEMISTRY. Educ: Mass Inst Technol, BS, 48. Prof Exp: Rubber chemist, B F Goodrich Co, 49-53; rubber chemist, 53-63, HEAD RUBBER DIV, ELASTOMERS LAB, E I DU PONT DE NEMOURS & CO, INC, 63- Mem: Am Chem Soc. Res: Development of new synthetic elastomers, especially compounding and applications. Mailing Add: Elastomers Lab Chestnut Run E I du Pont de Nemours & Co Inc Wilmington DE 19898

MURRAY, ROBERT WALLACE, b Brockton, Mass, June 20, 28; m 51; c 7. PHYSICAL ORGANIC CHEMISTRY. Educ: Brown Univ, AB, 51; Wesleyan Univ, MA, 56; Yale Univ, PhD(chem), 60. Prof Exp: Asst chem, Wesleyan Univ, 54-56 & Yale Univ, 56-57; res chemist polymer chem, Olin-Mathieson Chem Corp, 56-57; mem tech staff chem, Bell Tel Labs, Inc, 59-63, res supvr, 63-68; PROF CHEM, UNIV MO-ST LOUIS, 68-, CHMN DEPT, 75- Concurrent Pos: Mem, Nat Adv Comt Air Pollution Res Grants, 70-73; consult, Panel Vapor Phase Org Air Pollutants from Hydrocarbons, Nat Acad Sci. Honors & Awards: Am Chem Soc Award, 74. Mem: AAAS; Am Chem Soc; Air Pollution Control Asn; The Chem Soc; Am Inst Chemists. Res: Oxidation of organic compounds; singlet oxygen and ozone chemistry; air pollution chemistry; chemistry of aging; carbene chemistry; reaction mechanisms. Mailing Add: 1810 Walnutway Dr Creve Coeur MO 63141

MURRAY, ROYCE WILTON, b Birmingham, Ala, Jan 9, 37; m 57; c 5. ANALYTICAL CHEMISTRY. Educ: Birmingham-Southern Col, BS, 57; Northwestern Univ, PhD(anal chem), 60. Prof Exp: From instr to assoc prof, 60-69, actg chmn dept, 70-71, PROF CHEM, UNIV NC, CHAPEL HILL, 70- Concurrent Pos: Alfred P Sloan res fel, 69-72; prog dir chem anal, NSF, 71-72. Mem: Am Chem Soc. Res: Electroanalytical chemistry, including surface chemistry, instrumentation and nonaqueous media; stopped-flow kinetics. Mailing Add: Dept of Chem Univ of NC Chapel Hill NC 27514

MURRAY, STEPHEN PATRICK, b New York, NY, Oct 4, 38; m 62; c 3. PHYSICAL OCEANOGRAPHY. Educ: Rutgers Univ, AB, 60; La State Univ, MS, 63; Univ Chicago, PhD(geophys), 66. Prof Exp: NSF fel, 66-67; asst prof, 67-71, ASSOC PROF MARINE SCI, COASTAL STUDIES INST, LA STATE UNIV, BATON ROUGE, 71-, DIR COASTAL STUDIES INST, 75- Mem: Am Geophys Union; Am Meteorol Soc; Sigma Xi; Estuarine Res Fedn. Res: Coastal oceanography, including generation and trajectories of coastal currents; land-sea interaction and the turbulent diffusion of solid particles under shoaling waves; estuarine dynamics and air. Mailing Add: Coastal Studies Inst La State Univ Baton Rouge LA 70803

MURRAY, STEPHEN S, b New York, NY, Aug 28, 44; m 65; c 2. X-RAY ASTRONOMY. Educ: Columbia Univ, BS, 65; Calif Inst Technol, PhD(physics), 71. Prof Exp: Staff scientist x-ray astron, Am Sci & Eng, Inc, 71-73; ASTROPHYSICIST, CTR ASTROPHYS, SMITHSONIAN ASTROPHYS OBSERV, 73- Concurrent Pos: Assoc, Harvard Col Observ, 74- Mem: Am Astron Soc. Res: Observational x-ray astronomy, particularly extragalactic objects; development and use of high sensitivity, high resolution x-ray imaging detectors for extragalactic observations. Mailing Add: Ctr for Astrophys 60 Garden St Cambridge MA 02138

MURRAY, THOMAS EDMUND, b Albany, NY, July 26, 34. SOLID STATE PHYSICS. Educ: Fordham Univ, AB, 59, MS, 63; Syracuse Univ, PhD(physics), 66. Prof Exp: Instr physics, Loyola Col Md, 68; ASST PROF PHYSICS, LE MOYNE COL, NY, 69-, DIR FED RELS OFF, 71- Concurrent Pos: NSF acad yr inst partic, Univ Md, 68-69. Mem: Am Phys Soc. Res: Magnetic materials, including measurement of antiferromagnetic susceptibilities and resonance at low temperatures. Mailing Add: Dept of Physics Le Moyne Col Syracuse NY 13214

MURRAY, THOMAS J, b Los Angeles, Calif, Oct 21, 29; m 54; c 4. ORGANIC CHEMISTRY. Educ: Univ Notre Dame, BS, 52, MS, 54. Prof Exp: From chemist to sr res chemist, 57-66, RES ASSOC, RES LABS, EASTMAN KODAK CO, 66- Mem: Soc Photog Scientists & Engrs; Am Chem Soc. Res: Color photographic systems. Mailing Add: Eastman Kodak Co Res Labs 343 State St Rochester NY 14650

MURRAY, THOMAS PINKNEY, b Charleston, SC, Oct 8, 42; m 65; c 2. ORGANIC CHEMISTRY. Educ: Western Carolina Univ, BS, 64; Appalachian State Univ, MA, 66; Va Polytech Inst & State Univ, PhD(chem), 69. Prof Exp: Res assoc chem, Univ Alta, 69-71, Vanderbilt Univ, 71-72; ASST PROF CHEM, UNIV N ALA, 72- Concurrent Pos: Consult, Tenn Valley Authority, 76- Mem: Am Chem Soc; Sigma Xi. Res: Biosynthesis of phenolic plant metabolites, isolation and characterization of new metabolites; organic synthesis; beneficiation of phosphate rock by foam floatation. Mailing Add: Dept of Chem Univ of N Ala Florence AL 35630

MURRAY, WALLACE JASPER, b Quantico, Va, July 13, 40; m 64; c 3. MEDICINAL CHEMISTRY. Educ: San Diego State Univ, BS, 64; Univ Calif, San Francisco, PhD(pharmaceut chem), 74. Prof Exp: Instr chem, Mass Col Pharm, 74-75; ASST PROF MED CHEM, UNIV NEBR-LINCOLN, 75- Mem: Sigma Xi; Am Chem Soc; sci assoc Am Pharmaceut Asn. Res: Topological indices in structure-activity relationships; molecular orbital approach to drug-receptor interactions; pharmaceutical analysis. Mailing Add: Dept of Med Chem & Pharmacog Univ of Nebr Col of Pharm Lincoln NE 68588

MURRAY, WILLIAM DONALD, b Shadyside, Ohio, Aug 9, 13; m 35; c 2. MEDICAL ENTOMOLOGY. Educ: Ohio State Univ, BA, 34, MS, 35; Univ Minn, PhD(entom), 40. Prof Exp: Instr bot & zool, Eveleth Jr Col, 41-42; instr biol, Minn State Teachers Col, Bemidji, 42-43; entomologist, State Dept Pub Health, Calif, 46; instr biol, Eastern Ill State Teachers Col, 46-47; MGR & ENTOMOLOGIST, DELTA VECTOR CONTROL DIST, 47- Mem: Entom Soc Am; Am Mosquito Control Asn (treas, 65-). Res: Taxonomic revisions in hymenoptera; filariasis; mosquito biologies. Mailing Add: Delta Vector Control Dist 1737 W Houston Visalia CA 93277

MURRAY, WILLIAM JAMES, anesthesiology, pharmacology, see 12th edition

MURRAY, WILLIAM MOZLEY, JR, b Roswell, NMex, Aug 4, 12; m. ANALYTICAL CHEMISTRY. Educ: Emory Univ, AB, 32, MA, 33; Princeton Univ, PhD(anal chem), 36. Prof Exp: Lab instr chem, Harvard Univ, 36-37; anal chemist, Gen Elec Co, Mass, 37-45; anal chemist, 45-47, asst dir, 47-48, dir, 48-63, pres, 64-74, CONSULT, SOUTHERN RES INST, 74- Mem: AAAS; Am Chem Soc. Res: Spectroscopy; x-ray diffraction; spectrophotometry; high-vacuum analyses of gases; high-purity metals; research administration. Mailing Add: 2250 Highland Ave Birmingham AL 35205

MURRAY, WILLIAM R, b Ottawa, Ont, Dec 4, 24; US citizen; m; c 2. ORTHOPEDIC SURGERY. Educ: St Patrick's Col, Ottawa, BSc, 47; McGill Univ, MD & CM, 52. Prof Exp: From instr to assoc prof, 58-73, PROF ORTHOP SURG & CHMN DEPT, SCH MED, UNIV CALIF, SAN FRANCISCO, 73-, CHIEF ORTHOP SURG CLIN, 58- Concurrent Pos: Adv, Bur Hearings & Appeals, Soc Security Admin, Dept Health, Educ & Welfare, 64-72; mem, Arthritis Found. Mem: AMA; Am Acad Orthop Surg; Am Rheumatism Asn; dipl mem Pan Am Med Asn. Res: Total hip joint replacement arthroplasty; rheumatoid arthritis. Mailing Add: Dept of Orthop Surg Univ of Calif Med Ctr San Francisco CA 94143

MURRAY, WILLIAM SINGLER, b Chicago, Ill, Aug 16, 17; m 50; c 4. CHEMISTRY. Educ: Univ Notre Dame, BS, 39, MS, 40, PhD(org chem), 42. Prof Exp: Res chemist, Jackson Lab, Del, 42-51, chief supvr plant technol sect, Chambers Works, NJ, 51-60, div head Jackson Lab, Del, 60-64, div head process dept, Chambers Works, NJ, 64-70, TECH SUPT PETCHEM DIV, CHAMBERS WORKS, E I DU PONT DE NEMOURS & CO, INC, 70- Mem: AAAS; Am Chem Soc. Res: Fluorinated hydrocarbons; organometallics. Mailing Add: Petchem Div Chambers Works E I du Pont de Nemours & Co Inc Deepwater NJ 08023

MURRAY, WILLIAM SPARROW, b Wilkes Barre, Pa, July 15, 26; m 52; c 2. SCIENCE ADMINISTRATION. Educ: Juniata Col, BS, 50; Univ Md, MS, 52, PhD(entom), 63. Prof Exp: Agt entomologist, USDA, 51; dist entomologist, US Army Corps Engrs, 52-55; entomologist, Norfolk Dist, US Dept Navy, 55-58, dist entomologist, River Commands, Washington, DC, 58-62, entomologist, Dept Navy, 63-64; consult, Nat Pesticide Prob, House of Rep, US Cong, 64-65; asst exec secy, Fed Comt Pest Control, 65-69; exec secy, Working Group Pesticides, President's Cabinet Comt Environ, 69-71; staff dir hazardous mat adv comt, 71-72, PHYS SCI ADMINR, OFF OF PESTICIDE PROGS, ENVIRON PROTECTION AGENCY, 73- Concurrent Pos: Consult, Nat Plant & Animal Dis & Quarantine Probs, House Appropriations Comt, US Cong, 67. Honors & Awards: Outstanding Performance Award, US Dept Navy, 63-64; Qual Increase Award, Off of Pesticide Progs, Environ Protection Agency, 74. Mem: AAAS; Am Inst Biol Sci; Entom Soc Am. Res: National pesticide problem; incidence and effects of pesticides and other pollutants on human health and the environment. Mailing Add: 1281 Bartonshire Way Potomac Woods Rockville MD 20854

MURRELL, JAMES THOMAS, JR, b Dickson, Tenn, Mar 17, 42; m 60; c 2. SYSTEMATIC BOTANY. Educ: Austin Peay State Col, BS, 64; Vanderbilt Univ, PhD(syst bot), 69. Prof Exp: NIH trainee, Univ Miami, 68-69; ASST PROF BIOL, GEORGE PEABODY COL, 69- Res: Chemotaxonomy; allelopathy; pollination biology. Mailing Add: Dept of Biol George Peabody Col Nashville TN 37203

MURRELL, KENNETH DARWIN, b Burley, Idaho, Jan 19, 40; m 65; c 2. PARASITOLOGY, IMMUNOLOGY. Educ: Chico State Col, AB, 62; Univ NC, Chapel Hill, MSPH, 63, PhD(parasitol), 69. Prof Exp: NIH trainee microbiol, Univ Chicago, 69-71; RES ZOOLOGIST, NAVAL MED RES INST, 71- Mem: Am Soc Trop Med & Hyg; Am Soc Parasitol; Am Inst Biol Sci; Wildlife Dis Asn. Res:

Fundamental mechanisms of immunity to animal parasites; immunochemistry of parasite antigens. Mailing Add: Naval Med Res Inst Bethesda MD 20014

MURRELL, LAWRENCE LEE, inorganic chemistry, see 12th edition

MURRELL, LEONARD RICHARD, b Stamford Centre, Ont, June 17, 33. ANATOMY. Educ: McMaster Univ, BSc, 57, MSc, 58; Univ Minn, Minneapolis, PhD(anat), 64. Prof Exp: Asst biol, McMaster Univ, 57-58; from instr to asst prof anat, Univ Minn, Minneapolis, 64-67; from asst prof to assoc prof, 67-74, PROF ANAT, UNIV TENN CTR HEALTH SCI, 74- Concurrent Pos: Am Diabetes Asn res fel, Univ Minn, Minneapolis, 64-66. Mem: Am Asn Anat; Am Diabetes Asn; Brit Soc Cell Biol; Tissue Cult Asn; Can Asn Anat. Res: Human anatomy; experimental diabetes; functional cytodifferentiation; organ culture; carcinogenesis model systems. Mailing Add: Dept of Anat Univ of Tenn Ctr for Health Sci Memphis TN 38163

MURRILL, EVELYN, organic chemistry, biochemistry, see 12th edition

MURRILL, RUPERT IVAN, b Mexico City, Mex, Apr 19, 15; nat US; m 63; c 4. ANTHROPOLOGY. Educ: McGill Univ, BS, 40; Columbia Univ, PhD, 54. Prof Exp: From instr to asst prof anthrop & sociol, Univ Kans, 50-58, dir dept western civilization, 53-56; from asst prof to assoc prof, 58-72, PROF ANTHROP, UNIV MINN, MINNEAPOLIS, 72- Mem: Fel AAAS; fel Am Anthrop Asn; Am Asn Phys Anthrop. Res: Racial blood pressure; primate paleontology. Mailing Add: 1315 June Ave S Minneapolis MN 55416

MURRISH, DAVID EARL, b Glasgow, Mont, Jan 28, 37; m 65. COMPARATIVE PHYSIOLOGY. Educ: Calif State Col, Los Angeles, BA, 53, MA, 65; Univ Mont, PhD(comp physiol), 68. Prof Exp: Res assoc comp physiol, Duke Univ, 68-70; ASST PROF BIOL, CASE WESTERN RESERVE UNIV, 70- Mem: Am Physiol Soc; Am Soc Zoologists. Res: Temperature regulation in birds; transcapillary fluid exchange in birds. Mailing Add: Dept of Biol Case Western Reserve Univ Cleveland OH 44106

MURRMANN, RICHARD P, b South Bend, Ind, Aug 3, 40; m 61; c 2. PHYSICAL CHEMISTRY, SOIL SCIENCE. Educ: Purdue Univ, BS, 62; Cornell Univ, MS, 63, PhD(soil sci chem), 66. Prof Exp: Res assoc soil chem, Cornell Univ, 66; res chemist, US Army Cold Regions Res & Eng Lab, NH, 66-74; ASST AREA DIR, AGR RES SERV, USDA, 74- Mem: Am Soc Agron; Soil Sci Soc Am; Sigma Xi. Res: Inorganic phosphate in soil; electrical conductivity and diffusivity of ions; adsorption of heavy metal ions by minerals; chemistry of trace components in atmosphere, soil and water; land treatment wastewater. Mailing Add: Ala/N Miss Area 209 S Lafayette St PO Box 1486 Starkville MS 39759

MURSKY, GREGORY, b Ukraine, Feb 13, 29; Can citizen; m 52; c 2. GEOLOGY. Educ: Univ BC, BSc, 56; Stanford Univ, MS, 60, PhD(geol), 63. Prof Exp: Geologist, Eldorado Mining & Refining, Ont, 56-57, chief geologist, 57-59; Nat Res Coun Can fel, 63-64; from asst prof to assoc prof geol, 64-68, chmn dept geol sci, 66-68, PROF GEOL, UNIV WIS-MILWAUKEE, 68- Concurrent Pos: Geol Surv Can res grant, 65-66; NSF instructional improv grants, 65-68, in-serv inst grant, 67; Univ Wis res grants, 65-71. Mem: AAAS; Soc Econ Geol; Mineral Soc Am; Mineral Asn Can; Can Inst Mining & Metall. Res: Mineralogy; economic geology; petrology; geochemistry. Mailing Add: Dept of Geol Sci Univ of Wis Milwaukee WI 53201

MURTAGH, FREDERICK, JR, b Philadelphia, Pa, May 16, 17; m 43; c 3. NEUROSURGERY. Educ: Temple Univ, MD, 43, MSc, 51. Prof Exp: From asst prof to prof neurosurg, Med Sch, Temple Univ, 57-75, chmn div neurol & sensory sci, 66-75, chmn dept neurosurg, 71-75; PROF NEUROSURG, UNIV PA, 75-, DIR DEPT, HOSP UNIV PA, 75- Concurrent Pos: Mem staff, Children's Hosp, Philadelphia, Arlington Hosp, Consol Bryn Mawr Hosp, St Cloris & Philadelphia Gen & Pa Hosps; consult & mem staff, Phoenixville Hosp. Mem: Am Asn Neurol Surg; AMA; Am Col Surgeons; Cong Neurol Surg. Res: Neurological research in problems of infants and children. Mailing Add: 3401 Spruce St Philadelphia PA 19140

MURTAUGH, WALTER A, b Providence, RI, Apr 7, 03. NUCLEAR PHYSICS, AERONAUTICS. Educ: Providence Col, BA, 24; Cath Univ Am, MS, 43. Hon Degrees: MA, Providence Col, 53, DSc, 64. Prof Exp: Chmn physics, Aquinas Col High Sch, 33-43; chmn dept, 43-68, PROF PHYSICS, PROVIDENCE COL, 68- Concurrent Pos: Partic, Oper Plumbbob, Nev Test Site, 57; mem staff, Oak Ridge Inst Nuclear Studies, 59; mem adv comt radiation & merit award, Indust Code Comn Safety & Health, 64; mem, RI State Atomic Energy Comn. Mem: Fel AAAS; Am Phys Soc; Am Nuclear Soc; Am Meteorol Soc; Am Optical Soc. Res: Scattering processes; nuclear materials testing; optical methods of material testing; stability phenomena. Mailing Add: Dept of Physics Providence Col Providence RI 02918

MURTHY, GOPALA KRISHNA, b Bangalore, India, Mar 13, 25; nat US; m 57. DAIRYING. Educ: Univ Mysore, BSc, 44; Univ Ill, MS, 53, PhD(dairy technol, biochem), 56. Prof Exp: Chemist, Indian Dairy Res Inst, Bangalore, 44-50; asst dairy technol, Univ Ill, 50-56; res assoc dairy chem, Iowa State Col, 56-57; RES CHEMIST MILK & FOODS RES, ROBERT A TAFT SANIT ENG CTR, 57-; RES CHEMIST, BUR FOODS, US FOOD & DRUG ADMIN, 69- Mem: AAAS; Am Chem Soc; Am Dairy Sci Asn; Am Inst Chemists; Inst Food Technologists; NY Acad Sci. Res: Milk and food biochemical research. Mailing Add: Div of Microbiol Bur of Foods 1090 Tusculum Ave Cincinnati OH 45226

MURTHY, GUMMULURU SATYANARAYANA, b Balijipeta, India, Dec 30, 41; Can citizen; m 71; c 1. PALEOMAGNETISM. Educ: Andhra Univ, India, BS, 59, MS, 61; Mem Univ Nfld, MS, 66; Univ Alta, PhD(geophys), 69. Prof Exp: Lectr geophys, Andhra Univ, India, 62-63; fel, 69-71, res assoc, 71-72, ASST PROF PHYSICS, MEM UNIV NFLD, 72- Mem: Fel Geol Asn Can; Can Geophys Union; Am Geophys Union; Soc Terrestrial Magnetism & Elec Japan. Res: Study of the magnetic properties of rocks with applications to continental drift and plate tectonics; study of the processes by which rocks acquire permanent magnetism. Mailing Add: Dept of Physics Mem Univ St Johns NF Can

MURTHY, KRISHNA A S, b Bangalore, India, Feb 8, 32; m 65; c 1. HISTOCHEMISTRY, EXPERIMENTAL PATHOLOGY. Educ: Univ Mysore, BSc, 50; Univ Bombay, MSc, 55, PhD(biochem), 61. Prof Exp: Res asst histochem & exp path, Indian Cancer Res Ctr, Bombay, 52-57, asst res officer endocrinol, 57-61; res assoc morphol, Chicago Med Sch, 61-63; res assoc histochem, Children's Cancer Found, Boston, 63-65; res officer, Indian Coun Med Res, New Delhi, 65-67; RES ASSOC HISTOCHEM, EXP PATH & BIOCHEM, CHILDREN'S CANCER RES FOUND, BOSTON, 67- Concurrent Pos: Ill Rheumatism & Arthritis Found fel, Chicago Med Sch, 61-62, Chicago Heart Asn fel, 62. Honors & Awards: Dr Khanolkar Prize, Indian Asn Path & Bact, 63. Mem: AAAS; Endocrine Soc. Res: Functional endocrine tumors and their induction in animals; endocrine interrelationships; carcinogens; chemical analysis of tumors. Mailing Add: Children's Cancer Res Found Boston MA 02115

MURTHY, MAHADI RAGHAVANDRARAO V, b Bangalore, India, Nov 3, 29. BIOCHEMISTRY, NEUROCHEMISTRY. Educ: Univ Mysore, BSc, 49; Indian Inst Sci, Bangalore, PhD(biochem), 55. Prof Exp: Res assoc biochem, Med Br, Univ Tex, 61-63; from asst prof to assoc prof, 64-69, PROF BIOCHEM, FAC MED, LAVAL UNIV, 69-, DIR MOLECULAR NEUROBIOL LAB, 69- Concurrent Pos: Fel biochem, Texas A&M Univ, 55-58; fel, Sch Med, Yale Univ, 58-59; Welch Found fel biochem & entom, Med Br, Univ Tex, 59-61. Honors & Awards: M Sreenivasaya Award, Coun Indian Inst Sci, 53. Mem: AAAS; Can Biochem Soc; Brit Biochem Soc; Fedn Europ Biochem Socs; Chem Inst Can. Res: Regulation of protein and nucleic acid synthesis in tissues during growth; metabolism of biological macromolecules; biochemical mechanisms of memory. Mailing Add: Dept of Biochem Laval Univ Fac of Med Quebec PQ Can

MURTHY, VARANASI RAMA, b Visakhapatnam, India, July 2, 33; m 59. GEOCHEMISTRY. Educ: Andhra Univ, India, BSc, 51; Yale Univ, MS, 55, PhD, 57. Prof Exp: Res fel geol, Calif Inst Technol, 57-59; res asst geochem, Univ Calif, San Diego, 59-62, asst prof, 62-65; assoc prof, 65-69, PROF GEOCHEM, UNIV MINN, MINNEAPOLIS, 69-, CHMN DEPT, 71- Mem: Am Geophys Union; Geochem Soc; Geol Soc Am. Res: Petrology; cosmochemistry and lunar investigation; early crystal and mantle evolution in the earth. Mailing Add: Dept of Geol & Geophys Univ of Minn Minneapolis MN 55455

MURTHY, VEERARAGHAVAN KRISHNA, b Pudukottah, India, Feb 27, 34; c 3. BIOCHEMISTRY, PHYSIOLOGY. Educ: Univ Madras, BS, 53; Univ Bombay, MS, 60, PhD(biochem), 64. Prof Exp: Res asst biochem, Vallabhbhai Patel Chest Inst, Univ Delhi, 55-57; sci officer, Indian Cancer Res Ctr, AEC, Govt India, 57-64; ASST PROF MED & BIOCHEM, UNIV NEBR MED CTR, OMAHA, 74- Concurrent Pos: Res fels, Univ Fla, 64-68 & Univ Toronto, 68-74. Mem: Royal Inst Chemists; Can Biochem Soc; Am Diabetes Asn. Res: Diabetes and lipid metabolism; hormones and lipids; cardiac muscle metabolism; muscle contraction; drug metabolism in cancer. Mailing Add: Dept of Internal Med Univ of Nebr Med Ctr Omaha NE 68105

MURTHY, VISHNUBHAKTA SHRINIVAS, b Kanker, India, Jan 1, 42; m 68; c 2. PHARMACOLOGY. Educ: Univ Indore, India, BS & MB, 65, MD, 68; Univ Manitoba, PhD(pharmacol), 72. Prof Exp: Demonstr & lectr pharmacol, Mahatma Gandhi Mem Med Col, Indore, India, 66-69; sr scientist, Warner-Lambert Res Inst, 72-74; SR RES INVESTR, SQUIBB INST MED RES, 74- Concurrent Pos: Adj asst prof physiol, Rutgers Med Sch, Col Med & Dent NJ, 74- Mem: NY Acad Sci. Res: Physiology of cardiovascular homeostasis and its modification in cardiovascular diseases; pharmacological modulation of experimental myocardial ischemia, infarction and oxygen transport to tissue. Mailing Add: Squibb Inst for Med Res Box 4000 Princeton NJ 08540

MURTY, DANGETY SATYANARAYANA, b Visakhapatnam, India, Dec 28, 27; m 52; c 2. PHYSICS. Educ: Govt Arts Col, India, BSc, 48; Presidency Col, MA, 50; Andhra Univ, MSc, 51, DSc(ionosphere physics), 56. Prof Exp: Lectr appl physics, Andhra Univ, India, 52-57; Colombo Plan res scholar, 57-58; lectr appl physics, Andhra Univ, 58-60; assoc prof physics & actg head dept, Tex Southern Univ, 60-63; assoc prof, 63-69, chmn dept, 63-72, PROF PHYSICS, ST MARY'S UNIV, NS, 69- Mem: Can Asn Physicists; Inst Elec & Electronics Engrs; Am Asn Physics Teachers; fel Brit Inst Elec Engrs; fel Brit Inst Electronics & Radio Engrs. Res: Mössbauer effect; low energy nuclear physics. Mailing Add: 1123 Belmont on the Arm Halifax NS Can

MURTY, DASIKA RADHA KRISHNA, b Guntur, India, Dec 13, 31; m 49; c 4. ORGANIC CHEMISTRY. Educ: Andhra Univ, India, BSc, 51, MSc, 52; Fla State Univ, PhD(org chem), 60. Prof Exp: Lab instr chem, Gudivada Col, India, 52-53; sci teacher, AVH Sch, Hyderabad, 53-54; lab instr, Andhra Christian Col, 54-55; asst org chem, Fla State Univ, 55-60; fel, Wayne State Univ, 60; sr chemist, Tracerlab Div, Lab for Electronics, Inc, 61-63; sr res scientist, 63-68, supvr radiomed synthesis sect, 68-70, HEAD RADIOPHARMACEUT RES SECT, SQUIBB INST MED RES, 70-, HEAD IN-VITRO DIAG SECT, E R SQUIBB & SONS, 72- Mem: AAAS; Am Inst Clin Chemists; Am Chem Soc; Soc Nuclear Med. Res: Heterocyclics synthesis, reaction mechanisms; synthesis of radiochemicals and radiopharmaceuticals; clinical radioassay research and development. Mailing Add: 755 Hoover Dr North Brunswick NJ 08902

MURTY, HARI SRIRAM, b Gudivada, India, Oct 15, 38; m 71. BIOCHEMISTRY. Educ: Osmania Univ, India, BS, 58; Univ Delhi, MS, 60, MS, 63, PhD(biochem), 66. Prof Exp: Res assoc, Biochem Res Div, Sinai Hosp, Baltimore, Inc, 67-71; RES INVESTR, RADIOPHARMACEUT RES & DEVELOP, SQUIBB INST MED RES, 71- Concurrent Pos: Fel, Foreign Res & Tech Prog Div, USDA Res Servs, 66-67. Mem: AAAS; Am Chem Soc; Soc Nuclear Med. Res: Immunology; drug metabolism; nuclear medicine. Mailing Add: Radiopharmaceut Res & Develop Squibb Inst for Med Res New Brunswick NJ 08903

MURTY, KATTA GOPALAKRISHNA, b Pandillapalli, India, Sept 9, 36; m 64; c 1. OPERATIONS RESEARCH. Educ: Madras Univ, BSc, 55; Univ Calif, Berkeley, MS, 66, PhD(opers res), 68. Prof Exp: Consult asst prof statist & opers res, Indian Statist Inst, Calcutta, 58-65; ASSOC PROF OPERS RES, UNIV MICH, ANN ARBOR, 68-, GRAD PROG ADV, DEPT INDUST & OPERS ENG, 75- Concurrent Pos: Fulbright travel grant, 61-62. Mem: Assoc mem Opers Res Soc Am. Res: Mathematical programming; branch and bound algorithms; complementarity problem; networkflows; convex polyhedra. Mailing Add: Dept of Indust & Opers Eng Univ of Mich Ann Arbor MI 48104

MURTY, RAMA CHANDRA, b Vizianagaram, India, July 1, 28; Can citizen; m 62; c 1. PHYSICS. Educ: Andhra Univ, India, BSc, 47; Univ Bombay, MSc, 50, dipl librarianship, 51; Univ Western Ont, MSc, 58, PhD(physics), 62. Prof Exp: Demonstr physics, Wilson Col, Bombay, 47-52; librn, Express Newspapers Ltd, India, 52; sr master physics, Harrison Col, Barbados, West Indies, 52-57; demonstr, 57-58, sr demonstr, 58-61, res assoc geophys, 59-61, lectr physics, 61-63, asst prof, 63-68, ASSOC PROF PHYSICS, UNIV WESTERN ONT, 68- Concurrent Pos: Scanner, Tata Inst Fundamental Res, India, 50; Nat Res Coun Can grant, 65-; contract, Meteorol Br, Can Dept Transportation, 65-72; chmn comn VII, Can Div, Int Sci Radio Union. Mem: NY Acad Sci; Am Asn Physics Teachers; Am Meteorol Soc; Can Asn Physicists; Brit Inst Physics. Res: Physics of lightning; atmospheric electricity; sferics and meteorology. Mailing Add: Dept of Physics Univ of Western Ont London ON Can

MURTY, TADEPALLI SATYANARAYANA, b Rambhotlapalem, India, Aug 5, 37; m 67. PHYSICAL OCEANOGRAPHY. Educ: Andhra Univ, India, BSc, 55, MSc, 59; Univ Chicago, MS, 62, PhD(geophys), 67. Prof Exp: Lectr physics, Osmania Col, India, 59-60; res asst geophys, Univ Chicago, 60-67; res scientist I, 67-69, RES SCIENTIST II, CAN DEPT ENVIRON, 69- Concurrent Pos: Can mem int tsunami comt, Int Union Geod & Geophys, 71- Honors & Awards: Distinguished Res Medal, Univ Chicago, 67. Mem: Am Geophys Union; Am Meteorol Soc; Am Soc Limnol & Oceanog; Seismol Soc Am; Am Math Soc. Res: Theoretical research in physical oceanography using numerical integration techniques. Mailing Add: Marine Sci Br Dept of Environ 615 Booth St Ottawa ON Can

MURVOSH, CHAD M, b Toronto, Ohio, Aug 10, 31; m 65; c 3. AQUATIC ECOLOGY. Educ: Kent State Univ, BS, 53; Ohio State Univ, MS, 58, PhD(zool, entom), 60. Prof Exp: Instr entom, Ohio State Univ, 60-61; instr zool, Ohio Wesleyan Univ, 61; med entomologist, Entom Res Div, USDA, 62-64; asst prof, 64-69, ASSOC PROF ZOOL, UNIV NEV, LAS VEGAS, 69- Concurrent Pos: Grant, Desert Res Inst, Univ Nev, 65-66. Mem: Entom Soc Am; Ecol Soc Am; Soc Study Evolution; Soc Syst Zool; Am Mosquito Control Asn. Res: Aquatic insect ecology and medical entomology. Mailing Add: Dept of Biol Sci Univ of Nev Las Vegas NV 89154

MUSA, RAIQ S, b Nazareth, Palestine, July 19, 17; nat US; m 46; c 2. PHYSICS. Educ: Tex A&M Univ, BS, 53, MS, 55, PhD(physics), 57. Prof Exp: Instr math & physics, Am Univ Beirut, 48-51; asst, Tex A&M Univ, 52-57; res physicist, 57-67, MGR ACOUST & NOISE CONTROL RES, WESTINGHOUSE RES LAB, 67- Concurrent Pos: Mem, Sect Comt, Am Nat Standards Inst. Mem: Am Phys Soc; Acoust Soc Am; Sigma Xi; Inst Elec & Electronics Engrs; Am Soc Testing & Mat. Res: Physical and applied acoustics; instrument and apparatus; calibrations and standards in vibration and acoustics; the measurement and control of noise in industrial products. Mailing Add: Westinghouse Res Labs Beulah Rd Churchill Pittsburgh PA 15235

MUSACCHIA, XAVIER JOSEPH, b Brooklyn, NY, Feb 11, 23; m 50; c 4. PHYSIOLOGY, ZOOLOGY. Educ: St Francis Col, NY, BS, 44; Fordham Univ, MS, 47, PhD(biol), 49. Prof Exp: Instr biol, Marymount Col, NY, 47-49; instr comp physiol, St Louis Univ, 49-51, from asst prof to prof, 51-65; PROF PHYSIOL, UNIV MO-COLUMBIA, 65-, ASSOC DEAN, GRAD SCH & ASSOC DIR RES, 72-, DIR, DALTON RES CTR, 74- Concurrent Pos: Co-dir, Arctic Res Projs, St Louis Univ, 49-52, actg dir, Biol Labs, 52-53; vis scientist, Am Physiol Soc, 63-65; sr investr, Dalton Res Ctr, 65-74. Mem: Fel AAAS; Am Physiol Soc; Am Soc Zoologists; Am Micros Soc; Am Soc Mammal. Res: Environmental physiology; biochemistry of hibernation in reptiles and mammals; radiation biology and comparative physiology of intestinal absorption; physiology of depressed metabolism, hypothermia and hibernation; radio-protection in mammals. Mailing Add: Dalton Res Ctr Univ of Mo Columbia MO 65201

MUSCARI, JOSEPH A, b Chicago, Ill, May 13, 35; m 60; c 6. PHYSICS. Educ: Beloit Col, BS, 57; Johns Hopkins Univ, MS, 63; Wash State Univ, PhD(physics), 66. Prof Exp: Teacher high sch, 59-63; chief, Optical Physics Sect, 66-70, RES SCIENTIST, MARTIN MARIETTA CORP, 70- Concurrent Pos: Prin investr, Skylab Prog Exp, Martin Marietta Corp. Mem: Optical Soc Am; Am Vacuum Soc. Res: Nuclear physics; gamma ray spectroscopy; optical physics; vacuum ultraviolet spectroscopy.

MUSCATINE, LEONARD, b Trenton, NJ, Sept 7, 32; m 57; c 4. INVERTEBRATE ZOOLOGY, COMPARATIVE PHYSIOLOGY. Educ: Lafayette Col, BA, 54; Univ Calif, Berkeley, MA, 56, PhD(zool), 61. Prof Exp: Fel biochem, Howard Hughes Med Inst, 61-62; fel plant biochem, Scripps Inst Oceanog, 62-63, res biologist, 63-64; assoc prof, 64-74, PROF ZOOL, UNIV CALIF, LOS ANGELES, 74- Concurrent Pos: NIH fel, 61-63; NSF res grant, 63-64 & 65-75; Guggenheim fel, Oxford Univ, 70-71. Mem: Am Soc Zoologists; Brit Soc Exp Biol; Marine Biol Asn UK. Res: Coelenterate physiology; symbiosis of invertebrates and unicellular algae. Mailing Add: Dept of Biol Univ of Calif Los Angeles CA 90024

MUSCHEK, LAWRENCE DAVID, b Philadelphia, Pa, Apr 28, 43; m 64; c 3. BIOCHEMICAL PHARMACOLOGY. Educ: Philadelphia Col Pharm & Sci, BSc, 65; Mich State Univ, PhD(biochem), 70. Prof Exp: Mich Heart Asn fel cardiovasc pharmacol, Mich State Univ, 70-72; res scientist, 72-73, sr scientist, 73-76, GROUP LEADER CARDIOVASC BIOCHEM, McNEIL LABS, INC, 76- Mem: Am Chem Soc; AAAS. Res: Discovery and development of new agents effective in the treatment and/or prevention of thrombosis; mechanisms responsible for myocardial ischemia and infarction. Mailing Add: Dept of Biochem McNeil Labs Inc Ft Washington PA 19034

MUSCHEL, LOUIS HENRY, b New York, NY, July 4, 16; m 46; c 1. IMMUNOLOGY. Educ: NY Univ, BS, 36; Columbia Univ, AM, 38; Yale Univ, MS, 51, PhD(microbiol), 53. Prof Exp: Asst supvr, Serum Diag Dept, Div Labs & Res, NY State Dept Health, 39-41 & 46; chief, Dept Spec Serol & exec officer, Fourth Area Lab, Brooke Med Ctr, US Army Med Serv Corps, Tex, 46-47, chief lab serv, 20th Sta Hosp, Clark Field, PI, 47-48, Depts Serol & Chem, Second Area Lab, Ft Meade, Md, 48-50, 406th Med Gen Lab, Far East Command, 53-56, Exp Immunol Sect, Dept Appl Immunol, Walter Reed Army Inst Res, Washington, DC, 56-58, Dept Serol, 58-62; from assoc prof to prof microbiol, Med Sch, Univ Minn, Minneapolis, 62-70; MEM STAFF, RES DEPT, AM CANCER SOC, 70- Concurrent Pos: Abstractor, Biol Abstr, 47-48 & Chem Abstr, Am Chem Soc, 55-62; mem, Bact & Mycol Study Sect, Div Res Grants, NIH, 59- & Grants Rev Comt, Minn Chap, Arthritis & Rheumatism Found, 65-70; consult, Walter Reed Army Inst Res, Washington, DC, 63- & Vet Admin Hosp, Minneapolis, 65- Mem: Am Soc Microbiol; Soc Exp Biol & Med; Am Asn Immunol; Am Asn Cancer Res; NY Acad Sci. Res: Immunochemistry; natural resistance; bactericidal reactions; serology of syphilis; immunohematology. Mailing Add: Res Dept Am Cancer Soc 219 E 42nd St New York NY 10017

MUSCHIO, HENRY M, b New York, NY, Apr 25, 31; m 57; c 4. HUMAN GENETICS. Educ: Syracuse Univ, AB, 52; Fordham Univ, MS, 57, PhD(biol), 63. Prof Exp: Instr biol sci, Fairleigh Dickinson Univ, 58-62; asst prof, Montclair State Col, 62-66; assoc prof, 66-68, PROF BIOL SCI, DUTCHESS COMMUNITY COL, 68-, HEAD DEPT, 66- Concurrent Pos: Dir & lectr, NSF Inserv Inst Modern Biol, Montclair State Col, 65-66. Mem: AAAS; NY Acad Sci. Res: Human cytogenetics and cytological research related to the effects of chemical agents and their effects on the human karyotype and various human cell lines in vitro, with consideration of ethical and moral issues and values. Mailing Add: Dept of Biol Sci Dutchess Community Col Poughkeepsie NY 12601

MUSCHLITZ, EARLE EUGENE, JR, b Palmerton, Pa, Apr 23, 21; m 53; c 2. PHYSICAL CHEMISTRY, CHEMICAL PHYSICS. Educ: Pa State Univ, BS, 41, MS, 42, PhD(phys chem), 47. Prof Exp: Asst, Pa State Univ, 43-46; instr phys chem, Cornell Univ, 47-51; asst res prof, Col Eng, Univ Fla, 51-53, assoc prof, Univ Fla, 53-58, PROF CHEM, UNIV FLA, 58-, CHMN DEPT, 73- Concurrent Pos: NSF sr fel, 63-64; vis fel, Joint Inst Lab Astrophys, Boulder, Colo, 68. Mem: AAAS; Am Chem Soc; fel Am Phys Soc. Res: Ion, electron and excited atom scattering in gases; negative ions; molecular beams; mass spectrometry; molecular structure; upper atmosphere phenomena. Mailing Add: Dept of Chem Univ of Fla Gainesville FL 32611

MUSE, JOEL, JR, b Williamston, NC, July 11, 41; m 65; c 2. INDUSTRIAL ORGANIC CHEMISTRY. Educ: Univ NC, Chapel Hill, AB, 63; Univ Md, College Park, PhD(chem), 68. Prof Exp: SR RES CHEMIST, GOODYEAR TIRE & RUBBER CO, AKRON, 68- Mem: Am Chem Soc. Res: Process development involving rubber chemicals. Mailing Add: 4604 Walena Dr Medina OH 44256

MUSEN, PETER, b Nikolaiev, Russia, Jan 29, 12; m 46. ASTRONOMY. Educ: Univ Belgrade, PhD(math), 37. Prof Exp: Observator, Astron Observ, Univ Belgrade, Yugoslavia, 39-42; scientist, Astron Reicheninstitut, Berlin, 42-45; res asst, Observ, Univ Cincinnati, 50-53, instr, 53-56, from asst prof to assoc prof astron, 56-59; physicist, Theoret Div, 59, astronr, Mission Anal & Geodynamics Div, 71-73, STAFF SCIENTIST, EARTH SURV & APPLN DIV, GODDARD SPACE FLIGHT CTR, 73- Concurrent Pos: Part-time prof, Dept Physics & Astron, Univ Md, College Park, 68-72. Mem: Am Astron Soc; Am Astronaut Soc; Am Geophys Union. Res: Celestial mechanics; theoretical astronomy; analytical mechanics; astronomical computations; tides, oceanic and solid earth. Mailing Add: Goddard Space Flight Ctr Greenbelt MD 20771

MUSES, CHARLES ARTHUR, b NJ, Apr 28, 19. MATHEMATICS, CYBERNETICS. Educ: City Col, New York, BSc, 38; Columbia Univ, AM, 47, PhD(philos), 51. Prof Exp: Chemist, Gar-Baker Labs, Inc, 41-43; consult, 45-54; ed-in-chief, Falcon's Wing Press, Colo, 54-59; res dir, Barth Found, 60-62; res dir, Res Ctr Math & Morphol, Switz, 63-69, RES DIR, RES CTR MATH & MORPHOL, CALIF, 69- Concurrent Pos: Ed, J Study Consciousness, 68-73; chmn, Artificial Intelligence Div, 3rd Int Cong Cybernetics & Gen Systs, Bucharest, 75. Mem: NY Acad Sci; Royal Astron Soc Can; Am Math Soc; Math Asn Am. Res: Logic; cybernetics and artificial intelligence; quantum theory; chronotopology; algebraic structures; morphogenesis; alterations of consciousness; noetics; hypernumbers. Mailing Add: Ed Offs Res Ctr for Math & Morphol Santa Barbara CA 93108

MUSGRAVE, ALBERT WAYNE, b Eads, Colo, Jan 22, 23; m 43; c 4. GEOPHYSICS, ENGINEERING. Educ: Colo Sch Mines, ScD(geophys eng), 52. Prof Exp: Geol engr, Colo Sch Mines, 47; from trainee to interpreter, Seismic Surv, Mangolia Petrol Co, Socony Mobil Oil Co, Inc, 47-49, seismologist seismic interpretation, 50, seismic party chief, Seismic Surv, 52-53, res geophysicist geophys explor, 54-60, supt spec probs, 60-65, SR GEOPHYS SCIENTIST, GEOPHYS SERV CTR, MOBIL OIL CORP, 65- Honors & Awards: Van Diest Gold Medal, Colo Sch Mines, 61. Mem: Soc Explor Geophys; Sigma Xi; Am Asn Petrol Geologists. Res: Geophysical engineering; physics; geology; mathematics; electronics; seismology; gravity; magnetism; well logging. Mailing Add: Geophys Serv Ctr Mobil Oil Corp PO Box 900 Dallas TX 75221

MUSGRAVE, ANTHONY JOHN, b London, Eng, Mar 30, 13; m 45; c 6. BIOLOGY. Educ: Univ London, BSc, 34, MSc, 47, Imp Col, dipl, 36; McMaster Univ, PhD, 55. Prof Exp: Researcher & instr entom, Imp Col, London, 36-39, asst, Sch Hyg & Trop Med, London, 43-45; asst, Univ Cambridge, 45; mem sr res staff biol, Brit Leather Mfrs Res Asn, 45-48; from asst prof to prof entom & zool, Ont Agr Col, 48-65, PROF ZOOL, UNIV GUELPH, 65- Mem: Can Soc Microbiol; Can Soc Cell Biol; Entom Soc Can; Can Soc Zoologists. Res: Symbiotes, pathology, histology, cytology and physiology of invertebrates. Mailing Add: Dept of Zool Univ of Guelph Guelph ON Can

MUSGRAVE, BURDON C, physical chemistry, see 12th edition

MUSGRAVE, CAROL ANN, b Riverside, Calif, Sept 30, 48. ENTOMOLOGY. Educ: Boise State Col, BS, 71; Ore State Univ, PhD(entom), 74. Prof Exp: Res aide entom, Ore State Univ, 74; ASST PROF ENTOM & NEMATOL, UNIV FLA, 74- Mem: Entom Soc Am; Am Inst Biol Sci; Sigma Xi; AAAS. Res: Biology and ecology of parasitic insects and agriculturally important arthropod pests; biology, ecology and systematics of leafhoppers (Homoptera Cicadellidae). Mailing Add: Dept of Entom & Nematol 3103 McCarty Hall Univ Fla Gainesville FL 32611

MUSGRAVE, F STORY, b Boston, Mass, Aug 19, 35; m; c 5. PHYSIOLOGY, SURGERY. Educ: Syracuse Univ, BS, 58; Univ Calif, Los Angeles, MBA, 59; Marietta Col, BA, 60; Columbia Univ, MD, 64; Univ Ky, MS, 66. Prof Exp: Intern surg, Med Ctr, Univ Ky, 64-65; SCIENTIST-ASTRONAUT, JOHNSON SPACECRAFT CTR, NASA, 67- Concurrent Pos: US Air Force fel aerospace physiol & med & Nat Heart Inst fel, Univ Ky, 65-67; instr physiol & biophys, Med Ctr, Univ Ky, 69-; fel surg, Denver Gen Hosp, 69- Honors & Awards: Except Serv Medal, NASA, 74. Mem: AAAS; Aerospace Med Asn; Am Inst Aeronaut & Astronaut; AMA; Civil Aviation Med Asn. Res: Design and development of Space Shuttle extravehicular activity equipment and procedures; design and development of Spacelab. Mailing Add: NASA Code DF Houston TX 77058

MUSGRAVE, ORLO LYNN, b Findlay, Ohio, Aug 25, 19; m 41; c 2. SOILS. Educ: Ohio State Univ, BSc, 41, MSc, 53, PhD(agron), 57. Prof Exp: Voc agr instr, High Sch, Ohio, 41-42; county agr exten agt, 46-55, exten agronomist, 55-65, asst dir, Coop Exten Serv, 65-70, PROF AGRON & ASSOC DIR, OHIO COOP EXTEN SERV, COL AGR & HOME ECON, OHIO STATE UNIV, 70- Mem: Am Soc Agron; Soil Sci Soc Am. Res: Soil fertility. Mailing Add: Coop Exten Serv Ohio State Univ 2120 Fyffe Rd Columbus OH 43210

MUSGRAVE, ROBERT BURNS, b Hutsonville, Ill, Apr 15, 13; m 36; c 4. AGRONOMY. Educ: Univ Ill, BS, 36, MS, 38, PhD, 40. Prof Exp: Asst agron, Univ Ill, 36-40; asst prof field crops, 40-50, asst agronomist, 40-50, vis prof, Grad Educ Prog, 64-66, PROF FIELD CROPS, CORNELL UNIV, 59- Concurrent Pos: Vis prof, Univ Philippines, 57-58. Res: Crop ecology; dynamics and energetics of the soil-plant-atmosphere continuum with emphasis on the responses of genotypes of maize with specific morphological and physiological characters to stresses of energy, water, carbon dioxide and oxygen. Mailing Add: Dept of Agron Cornell Univ Ithaca NY 14850

MUSGRAVE, STANLEY DEAN, b Hutsonville, Ill, Jan 26, 19; m 44; c 2. ANIMAL BREEDING, ANIMAL NUTRITION. Educ: Univ Ill, BS, 47, MS, 48; Cornell Univ, PhD(animal breeding), 51. Prof Exp: Asst animal husb, Cornell Univ, 47-50; asst prof dairy prod, Univ Ill, 50-51; from asst prof to prof dairying & head dept, Okla State Univ, 51-68; PROF ANIMAL & VET SCI, UNIV MAINE, ORONO, 68- Concurrent Pos: Prog chmn, Am Dairy Sci Asn, 66; chmn, Dept Animal & Vet Sci, Univ Maine, Orono, 68-73; consult, Mossoro Advan Sch Agr, Brazil, 74- & Univ Mosul, Iraq, 73- Mem: AAAS; Am Soc Animal Sci; Am Genetics Asn; Am Dairy Sci Asn. Res: Milk component analysis; dairy cattle nutrient, health requirements; scanning electron microscopic studies of feed and age effect on gastrointestinal epithelium; dairy management systems development and analysis. Mailing Add: Dept of Animal & Vet Sci Univ of Maine Orono ME 04473

MUSGRAVE, TED RUSSELL, b Sheridan, Wyo, Dec 19, 38; m 57; c 2. INORGANIC CHEMISTRY. Educ: Harvard Univ, AB, 60; Univ Colo, Boulder, PhD(inorg chem), 64. Prof Exp: Instr, 64-65, ASST PROF CHEM, COLO STATE UNIV, 65- Mem: Am Chem Soc. Res: Coordination polymers of nitrogen heterocycles; complex ion equilibria; inorganic syntheses. Mailing Add: Dept of Chem Colo State Univ Ft Collins CO 80521

MUSHAHWAR, ISA KHAMIS, biochemistry, see 12th edition

MUSHAK, PAUL, b Dunmore, Pa, Dec 9, 35. BIOCHEMISTRY, CHEMISTRY. Educ:

Univ Scranton, BS, 61; Univ Fla, PhD(chem), 70. Prof Exp: Res asst clin biochem & toxicol, Clin Res Labs, Sch Med, Univ Fla, 67-69; NIH res assoc metalloenzym, Dept Molecular Biophys & Biochem, Yale Univ, 69-71; ASST PROF METAL BIOCHEM & PATH, UNIV NC, CHAPEL HILL, 71- Concurrent Pos: Sr mem, Nat Inst Environ Health Sci proj prog heavy metal path, Univ NC, Chapel Hill, 71-, Environ Protection Agency & Inter-univ Consortium Environ Studies grants, 71-; consult, Nat Inst Environ Health Studies, 71-; mem & consult, Inter-univ Consortium Environ Studies, 71- Mem: AAAS; Am Chem Soc. Res: Metalloenzymology; trace metal analysis; metabolism and biochemical effects of metal chelating agents; heavy metal toxicology; organometallic chemistry of the nickel triad metals. Mailing Add: Dept of Path Univ of NC Sch of Med Chapel Hill NC 27514

MUSHER, JEREMY ISRAEL, chemical physics, deceased

MUSHETT, CHARLES WILBUR, b Elizabeth, NJ, Apr 1, 14; m 39. PATHOLOGY. Educ: NY Univ, AB, 39, MS, 41, PhD(vert morphol), 44. Prof Exp: Technician, Merck Inst Therapeut Res, 33-35, lab asst, 35-37, sr worker, Bact & Path Dept, 37-40, assoc hemat & path, 40-43, from assoc head to head dept path, 43-56, from asst dir to dir sci rels, Merck Sharp & Dohme Res Labs, 57-66, dir int sci rels, 66-70, DIR SCI INDUST LIAISON, MERCK SHARP & DOHME RES LABS DIV, MERCK & CO, INC, 71- Concurrent Pos: Merck foreign fel, Denmark & Ger, 52-53. Mem: AAAS; Endocrine Soc; Am Soc Exp Path; fel NY Acad Sci; fel Int Soc Hemat. Res: Experimental animal pathology and hematology in relation to nutrition, infection and toxicology of drugs; blood coagulation and anticoagulants. Mailing Add: Merck Sharp & Dohme Res Labs Rahway NJ 07065

MUSHINSKI, JOSEPH FREDERIC, b New Brighton, Pa, Mar 18, 38; m 71. BIOCHEMISTRY, CANCER. Educ: Yale Univ, BA, 59; Harvard Med Sch, MD, 63. Prof Exp: Intern med, Med Ctr, Duke Univ, 63-64; res assoc biochem, 65-70, SR INVESTR, LAB CELL BIOL, NAT CANCER INST, 70- Concurrent Pos: USPHS fel, Res Training Prog, Med Ctr, Duke Univ, 64-65; William O Moseley traveling fel from Harvard Univ, Max Planck Inst Exp Med, 69-70. Mem: Am Asn Cancer Res; AAAS. Res: Molecular biology of cancer; immunology; transfer RNA. Mailing Add: Lab of Cell Biol Nat Cancer Inst Bldg 8 Rm 202 NIH Bethesda MD 20014

MUSHINSKY, HENRY RICHARD, b Passaic, NJ, Oct 3, 43; m 67; c 2. ECOLOGY, BEHAVIORAL BIOLOGY. Educ: Tusculum Col, BS, 67; Tenn State Univ, MS, 69; Clemson Univ, PhD(zool), 73. Instr zool, 73-74, ASST PROF ZOOL, LA STATE UNIV, BATON ROUGE, 74- Concurrent Pos: Sigma Xi res grant, 73-74. Mem: Sigma Xi; Soc Ichthyologists & Herpetologists; Am Soc Zoologists; Animal Behav Soc. Res: Ontogeny of food and habitat preference; resource partitioning; niche development in amphibians and reptiles. Mailing Add: Dept of Zool & Physiol La State Univ Baton Rouge LA 70803

MUSIC, JACK FARRIS, b Childress, Tex, Oct 5, 21; m 42; c 2. PHYSICAL CHEMISTRY, PHYSICS. Educ: Univ Tex, BA, 46, PhD(phys chem), 51. Prof Exp: Res scientist, Gen Elec Co, 51-52, supvr graphite & mat develop, 52-54, process tech, 54-56, mgr, 56-60, proj analyst, 60-61, consult analyst, 61-65, mgr, Div Anal & Planning, 65-68, mgr aerospace anal & planning, Aerospace Group, Valley Forge Space Technol Ctr, 68-69, mgr, Group Planning Oper, Info Systs Group, 69-71; PRES, STRATEGIC MGT, INC, PA, 71- Mem: Am Chem Soc; Am Phys Soc; Inst Mgt Sci. Res: Properties of matter; chemicals, materials, nuclear energy, aerospace, computers, economics; management of research and development; coupling of science and technology to business; strategic management of the enterprise. Mailing Add: 590 Blair Rd Berwyn PA 19312

MUSICK, GERALD JOE, b Ponca City, Okla, May 24, 40; m 62; c 2. ENTOMOLOGY. Educ: Okla State Univ, BS, 62; Iowa State Univ, MS, 64; Univ Mo-Columbia, PhD(entom), 69. Prof Exp: Asst entom, Iowa State Univ, 62-64; instr, Univ Mo-Columbia, 64-69; asst prof, 69-71, ASSOC PROF ENTOM, OHIO AGR RES & DEVELOP CTR, 71- Mem: Entom Soc Am. Res: Biology and control of soil insects in corn; insects associated with no tillage corn. Mailing Add: Dept of Entom Ohio Agr Res & Develop Ctr Wooster OH 44691

MUSICK, JOHN A, b 1940; m. ICHTHYOLOGY, ECOLOGY. Educ: Rutgers Univ, AB, 62; Harvard Univ, MA, 64, PhD, 67. Prof Exp: Fisheries biologist, US Fish & Wildlife Serv, 62; teaching fel comp anat & gen biol, Harvard Univ, 62-63, anthrop, 63-64, ichthyol, 65 & 67; ASST PROF MARINE SCI, COL WILLIAM & MARY & UNIV VA, 67- Concurrent Pos: Assoc marine scientist, Va Inst Marine Sci, 67-; mem, Bd Govs, Am Soc Ichthyol & Herpet, 74-75; sci collabr, Capes Hatteras & Lookout Nat Seashores, US Park Serv. Mem: AAAS; Am Soc Ichthyol & Herpet; Ecol Soc Am; Animal Behav Soc; Am Fisheries Soc. Res: Community ecology of demersal fishes of the continental slope and rise; systematics of fishes and reptiles. Mailing Add: Va Inst of Marine Sci Gloucester Point VA 23062

MUSKAT, JOSEPH BARUCH, b Marietta, Ohio, Sept 20, 35; m 59. NUMBER THEORY, COMPUTER SCIENCE. Educ: Yale Univ, AB, 55; Mass Inst Technol, SM, 56, PhD(math), 61. Prof Exp: From asst prof to assoc prof math, Univ Pittsburgh, 61-69; vis assoc prof, 69-70, ASSOC PROF MATH, BAR-ILAN UNIV, ISRAEL, 70- Concurrent Pos: NSF fel, 63-69; res assoc comput, Univ Pittsburgh, 61-69; chmn dept math, Bar-Ilan Univ, 71-74. Mem: Am Math Soc; Math Asn Am; Asn Comput Mach. Res: Reciprocity laws; cyclotomy; use of computers in number theory. Mailing Add: Dept of Math Bar-Ilan Univ Ramat-Gan Israel

MUSKER, WARREN KENNETH, b Chicago, Ill, Apr 17, 34; m 57; c 2. INORGANIC CHEMISTRY. Educ: Bradley Univ, BS, 55; Univ Ill, PhD(org chem), 59. Prof Exp: Asst boron hydrides, Univ Mich, 61-62; from asst prof to assoc prof, 62-75, PROF INORG CHEM, UNIV CALIF, DAVIS, 75- Concurrent Pos: Von Humboldt fel, 70-71. Mem: Am Chem Soc. Res: Intramolecular donor-acceptor interactions in group IV compounds; influence of ligand structure on the stereochemistry and reactivity of metal complexes; copper II oxidations; medium ring complexes of transition metals. Mailing Add: Dept of Chem Univ of Calif Davis CA 95616

MUSS, DANIEL R, b Birmingham, Ala, Apr 5, 28; m 65; c 2. PHYSICS. Educ: Mass Inst Technol, BS, 48; Calif Inst Technol, MS, 53; Univ Pittsburgh, PhD(physics), 61. Prof Exp: Sr physicist, 48-64, mgr silicon device develop, 64-69, MGR SOLID STATE DEVICE RES, WESTINGHOUSE RES LABS, 69- Mem: Am Phys Soc; sr mem Inst Elec & Electronics Eng. Res: Defect structure of metals; semiconductors; semiconductor devices. Mailing Add: Westinghouse Res Labs Churchill Borough Pittsburgh PA 15235

MUSSELL, HARRY W, b Paterson, NJ, Nov 10, 41; m 64. PLANT PATHOLOGY, PLANT BIOCHEMISTRY. Educ: Drew Univ, AB, 65; Duke Univ, MF, 65; Purdue Univ, PhD(bot), 68. Prof Exp: PLANT PATHOLOGIST, BOYCE THOMPSON INST PLANT RES, INC, 68- Concurrent Pos: Lectr, State Univ NY Col Purchase, 73- Honors & Awards: Ciba Sci Res Award, 62. Mem: AAAS; Am Phytopath Soc; Am Soc Plant Physiol; Bot Soc Am; Am Inst Biol Sci. Res: Physiology of parasitism, particularly enzymology of pathogenesis in plants. Mailing Add: Boyce Thompson Inst for Plant Res 1086 N Broadway Yonkers NY 10701

MUSSELLS, FRANCIS LLOYD, b Montreal, Can, Oct 24, 17; m 50; c 4. HOSPITAL ADMINISTRATION. Educ: McGill Univ, BA, 40, MD CM, 44; Columbia Univ, MS, 49. Prof Exp: Intern, Montreal Gen Hosp, 44-45, admitting off, 46-47; admin asst, Strong Mem Hosp, Rochester, 47-48, asst dir, 48-53; med dir, Philadelphia Gen Hosp, 53-54, exec dir, 54-58; dir, Peter Bent Brigham Hosp, 58-67; asst clin prof, Dept Soc & Prev Med, Univ Sask, 67-71; DIR, DIV HEALTH SERVS, WOODS, GORDON & CO MGT CONSULTS, 71- Concurrent Pos: Dep exec dir, Comt Med Sci, Res & Develop Bd, US Dept Defense, 51-52, exec dir, 52-53; lectr, Harvard Univ, 58-67; exec dir, South Sask Hosp Ctr, Regina, 67-70. Mem: Am Pub Health Asn; Am Hosp Asn. Res: Hospital administration; preventive medicine. Mailing Add: PO Box 253 Toronto-Dom Ctr Toronto ON Can

MUSSELMAN, MERLE MCNEIL, b Topeka, Kans, Sept 19, 15; m 40; c 4. SURGERY. Educ: Univ Nebr, BS, 37, MD, 39; Univ Mich, MS, 49; Am Bd Surg, dipl, 49. Prof Exp: Dir surg, Wayne County Gen Hosp, Mich, 50-54; PROF SURG, COL MED, UNIV NEBR, OMAHA, 54- Mem: AAAS; Am Asn Surg of Trauma; fel Am Col Surg; Am Fedn Clin Res; AMA. Res: Surgical infections and antibiotics; pancreatitis; fat embolism. Mailing Add: Col of Med Univ of Nebr Omaha NE 68105

MUSSELMAN, NELSON PAGE, b Luray, Va, Mar 20, 17. TOXICOLOGY, ANALYTICAL CHEMISTRY. Educ: Western Md Col, AB, 38. Prof Exp: Teacher, Baltimore Dept Educ & social investr, Dept Pub Welfare, 38-42; mem sales & mgt staff, Chrysler Air Temp, Frigidaire div, Gen Motors Corp , 46-51; chemist, Armco Steel Corp, 51-54; RES CHEMIST, US ARMY EDGEWOOD ARSENAL, 54- Mem: AAAS; Am Chem Soc. Res: Vapor and carbon monoxide toxicity; chemical analysis and methods; evaluation of chemical protective devices; air pollution. Mailing Add: 1127 St Paul St Baltimore MD 21202

MUSSER, A WENDELL, b Herrick, Ill, Dec 15, 30; m 53; c 2. PATHOLOGY. Educ: Purdue Univ, BS, 52; Ind Univ, MD, 56; Am Bd Path, dipl, 61. Prof Exp: Res instr, Med Ctr, Ind Univ, 57-61, lectr clin path, 61; from asst prof to assoc prof path, Med Ctr, Duke Univ, 63-74; PROF PATH, MED CTR, UNIV KY, 74-, ASSOC DEAN VET ADMIN AFFAIRS, 74- Concurrent Pos: Nat Cancer Inst fel, Med Ctr, Ind Univ, 58-61; chief lab serv, Vet Admin Hosp, NC, 63-70; dir allied health educ prog, Duke Univ, 67-70; chmn, Nat Coun Med Technol Educ, 68-; asst chief med dir planning & eval, Cent Off, Vet Admin, Washington, DC, 70-74; chief staff, Vet Admin Hosp, Lexington, Ky, 74-; vpres, Bd Dirs, Nat Registry Clin Chem, 75- Mem: AAAS; AMA; Col Am Path; Am Soc Clin Path; Int Acad Path. Res: Clinical pathology and chemistry; medical education. Mailing Add: Vet Admin Hosp Lexington KY 40507

MUSSER, DAVID MUSSELMAN, b Bowmansville, Pa, Apr 30, 09; m 38; c 2. ORGANIC CHEMISTRY. Educ: Pa State Col, BS, 31; Ga Inst Technol, MS, 33; Univ Wis, PhD(org chem), 37. Prof Exp: Indust fel, Mellon Inst, 37-42; res chemist, Pac Mills, NJ, 42-46; sr chemist, Deering Milliken Res Trust, Conn, 46-47; head textile res & develop, Onyx Oil & Chem Co, 47-52; dir res, Refined Prods Corp, 52-62; DIR RES, RAYTEX CHEM CORP, 62- Mem: Am Chem Soc; Am Asn Textile Chemists & Colorists. Res: Cellulose; textile finishing agents. Mailing Add: 821 Lawrence Dr Emmaus PA 18049

MUSSER, GLENN LUTHER, physics, see 12th edition

MUSSER, HARRY ROBERT, polymer chemistry, organic chemistry, see 12th edition

MUSSER, MARC JAMES, (JR), b Terre Haute, Ind, July 3, 10; m 33, 46; c 4. INTERNAL MEDICINE. Educ: Univ Wis, AB, 32, MD, 34. Prof Exp: From instr to asst prof neuropsychiat, Sch Med, Univ Wis, 38-46, from asst prof to assoc prof internal med, 46-53, prof med, 53-58; dir prof serv, Vet Admin Hosp, Houston, 57-59; dir res serv, Vet Admin Cent Off, Washington, DC, 59-62, from asst chief med dir to chief med dir, 62-74; DIR MED RELS, SMITH KLINE & FRENCH LABS, 74- Concurrent Pos: Consult, Vet Admin, Washington, DC, 47-57; exec dir, NC regional med prog, 66-70; prof, Sch Med, Duke Univ, 66-; adj prof, Bowman Gray Sch Med, 66- & Sch Pub Health, Univ NC, 66- Mem: Fel Am Col Physicians; AMA. Res: Psychiatry. Mailing Add: 4538 N 39th St Arlington VA 22207

MUSSER, MICHAEL TUTTLE, b Williamsport, Pa, Jan 31, 42; m 66; c 1. ORGANIC CHEMISTRY. Educ: Purdue Univ, BS, 63, PhD(org chem), 68. Prof Exp: Res chemist, Intermediates Div, Plastics Dept, Exp Sta, E I du Pont De Nemours & Co, 67-72, Nylon Intermediates Div, Polymer Intermediates Dept, 72-74, SR RES CHEMIST, SABINE RIVER WORKS, E I DU PONT DE NEMOURS & CO, TEX, 74- Mem: Am Chem Soc. Res: Homogeneous catalysis as a route to organic intermediates. Mailing Add: 1529 Lindenwood Dr Orange TX 77630

MUSSER, SAMUEL JOHN, b South Haven, Mich, Nov 22, 16; m 51; c 3. BIOCHEMISTRY. Educ: Mich State Univ, BS, 49, MS, 51. Prof Exp: Res assoc virol, Upjohn Co, 52-58; assoc dir biol res, Anchor Serum Co div, Philips Roxane, Inc, 58-64, dir res, 64-69, VPRES, PHILIPS ROXANE, INC, 69- Mem: Soc Cryobiol; NY Acad Sci. Res: Human or veterinary virus vaccines; measles vaccine; blood fractionation; physical properties of viruses. Mailing Add: Philips Roxane Inc 2621 N Belt Highway St Joseph MO 64502

MUSSINAN, CYNTHIA JUNE, b Elizabeth, NJ, Dec 23, 46. ANALYTICAL CHEMISTRY. Educ: Georgian Court Col, BA, 68; Rutgers Univ, MS, 75. Prof Exp: SR CHEMIST, INT FLAVORS & FRAGRANCES, INC, 68- Mem: Am Chem Soc. Res: Isolation, identification and synthesis of the volatile and nonvolatile flavor constituents of foods. Mailing Add: Int Flavors & Fragrances 1515 Hwy 36 Union Beach NJ 07735

MUSSMAN, HARRY CHARLES, veterinary pathology, see 12th edition

MUSSO, ROCCO CARMEN, physical chemistry, organic chemistry, see 12th edition

MUSSON, ALFRED LYMAN, b Honolulu, Hawaii, Aug 31, 11; m 35; c 3. ANIMAL SCIENCE. Educ: Univ Conn, BS, 33; Iowa State Univ, MS, 34, PhD(animal breeding), 51. Prof Exp: Res assoc & asst prof swine breeding, Iowa State Univ, 46-52; prof, 52-73, EMER PROF ANIMAL SCI, SDAK STATE UNIV, 73- Concurrent Pos: Head dept animal sci, SDak State Univ, 52-60, from asst dir to assoc dir, Exp Sta, 59-73. Mem: Am Soc Animal Sci. Res: Swine breeding. Mailing Add: Exp Sta SDak State Univ Brookings SD 57006

MUSTACCHI, PIERO OSCAR, b Cairo, Egypt, May 29, 20; nat US; m 48; c 2. MEDICINE. Educ: Italian Lyceum, Cairo, BS, 37; Faud 1st Univ, Cairo, MB, ChB, 44. Prof Exp: Asst resident path, Med Sch, Univ Calif, 49-51; clin instr med, 53-58, clin asst prof med & prev med, 58-66, consult, Hemat Clin, 54-58, asst dir continuing educ med & health sci, 64-69, assoc dir, 69-74, vchmn dept prev med, 65-66, assoc dir spec serv extended prog med educ, 74-75, CLIN ASSOC PROF MED & PREV MED, MED SCH, UNIV CALIF, SAN FRANCISCO, 66-, MED CONSULT, WORK CLIN, 75- Concurrent Pos: Res fel, Am Cancer Soc, 49-52; fel, Sloan-Kettering Inst, 51-53; vis instr, Fac Med, Cairo Univ, 50; resident, Mem Hosp Cancer & Allied Dis, New York, 51-53; physician-in-chg, Hemat & Lymphoma Clin, St Mary's Hosp, San Francisco, 54-56, consult, 68; mem consult tumor bd, Children's Hosp, San Francisco, 55-56, consult, 58-, head off epidemiol & biomet, 60-; consult, staff, Franklin Hosp, 70-; physician, Ital Consulate, San Francisco, hon vconsul Italy, 71-; chmn, Comt Continuing Educ, Children's Hosp, 72-; med consult, Work Clin, Univ San Francisco Med Ctr, 74-75. Honors & Awards: Knight Officer, Order of Merit, Italy, 71. Mem: AAAS; Am Soc Clin Invest; AMA; fel Am Col Physicians; Am Soc Environ & Occup Health. Res: Epidemiology of cancer; general ecology and education. Mailing Add: 3838 California St San Francisco CA 94118

MUSTAFA, SYED JAMAL, b Lucknow, India, July 10, 46; m 73; c 1. MEDICAL RESEARCH. Educ: Lucknow Univ, BS, 62, MS, 65, PhD(biochem), 70. Prof Exp: Fel, Indust Toxicol Res Ctr, Lucknow Res Ctr, Lucknow, India, 70-71, Dept Physiol, Univ Va, 71-74; ASST PROF PHARMACOL, COL MED, UNIV S ALA, 74- Concurrent Pos: Fel, Coun Sci & Indust Res & Indian Coun Med Res, New Delhi, India, 70-71, NIH, 71-74. Mem: Am Physiol Soc; Am Neurochem Soc; Sigma Xi; NY Acad Sci; Am Heart Asn; Int Study Group Res Cardiac Metab. Res: The field of cardiology; the study of the relationship between vasoactive agents and blood flow. Mailing Add: Dept of Pharmacol Col of Med Univ of SAla Mobile AL 36688

MUSTARD, JAMES FRASER, b Toronto, Ont, Oct 16, 27; m 52; c 6. PATHOLOGY. Educ: Univ Toronto, MD, 53; Cambridge Univ, PhD, 56; FRCP, 65. Prof Exp: Can Heart Found sr res assoc med, Univ Toronto, 58-63, from asst prof to assoc prof path, 61-66, assoc med, 63-66, asst prof med, 65; chmn dept path, 66-72, PROF PATH, MED CTR, McMASTER UNIV, 66-, DEAN FAC MED, 72-, DEAN FAC HEALTH SCI, 74- Concurrent Pos: Mem, Coun Arteriosclerosis, Am Heart Asn, 65-, mem, Coun Thrombosis; mem, Int Comt Haemostasis & Thrombosis; chmn, Health Res Comt, Ont Coun Health, 66-, chmn, Task Force Health Planning, Ont Coun Health, 73-74; chmn, Med Adv Comt, Can Heart Found, 72-; mem, Expert Adv Panel Cardiovasc Dis, WHO, 72. Honors & Awards: Gairdner Found Int Award, 67. Mem: Am Soc Clin Invest; Can Soc Clin Invest (pres, 65-66); Can Physiol Soc; Am Soc Hemat (secy, 64-67, pres, 70); Am Soc Exp Path. Res: Blood and vascular disease. Mailing Add: McMaster Univ Med Ctr 1200 Main St W Hamilton ON Can

MUSTARD, MARGARET JEAN, b Bayfield, Ont, Feb 18, 20; nat US. HORTICULTURE. Educ: Univ Miami, Bs, 42, MS, 50; Ohio State Univ, PhD(hort), 58. Prof Exp: Lab technician, Sub-trop Exp Sta, Univ Fla, 42-45; instr hort, 45-50, from asst prof to assoc prof, 50-68, PROF TROP BOT, UNIV MIAMI, 68- Mem: AAAS; Am Soc Hort Sci. Res: Anatomical and morphological aspects of botany; fruit setting in horticultural plants; chemical analyses of plant tissue; handling and marketing of fruits and vegetables; growth regulators in relation to horticulture. Mailing Add: Dept of Biol Univ of Miami PO Box 9118 Coral Gables FL 33124

MUSULIN, BORIS, theoretical chemistry, see 12th edition

MUT, STUART CREIGHTON, b Dallas, Tex, July 27, 24; m 47; c 5. GEOPHYSICS. Educ: Rice Inst, BS, 47, MS, 48. Prof Exp: Asst physicist, Atlantic Refining Co, 48-49, admin asst, Res Admin, 49-51, sr physicist, 51-56, supvry physicist, 56-59, dir res & develop, Explor Sect, 59-61, mgr, Eng Div, Producing Dept, 61-63, Eastern Dist, 63-66, VPRES EASTERN REGION, N AM PRODUCING DIV, ATLANTIC RICHFIELD CO, 66- Mem: Soc Explor Geophysicists; Soc Petrol Engrs; Inst Elec & Electronics Engrs. Res: Petroleum exploration geophysics; petroleum engineering; geology. Mailing Add: Box 2819 Dallas TX 75221

MUTCH, GEORGE WILLIAM, b Ann Arbor, Mich, June 22, 43; m 66. CHEMICAL KINETICS, PHYSICAL CHEMISTRY. Educ: Andrews Univ, BA, 66; Univ Calif, Davis, PhD(chem), 73. Prof Exp: Teaching asst chem, Univ Calif, Davis, 67-70, res asst, 70-73; ASST PROF CHEM, ANDREWS UNIV, 73- Mem: Am Chem Soc; Am Phys Soc. Res: Chemical dynamics of high energy unimolecular decomposition processes. Mailing Add: Dept of Chem Andrews Univ Berrien Springs MI 49104

MUTCH, THOMAS ANDREW, b Rochester, NY, Aug 26, 31; m 56; c 3. GEOLOGY. Educ: Princeton Univ, AB, 52, PhD(geol), 60; Rutgers Univ, MSc, 57. Prof Exp: From instr to assoc prof, 60-70, PROF GEOL, BROWN UNIV, 70- Concurrent Pos: Chmn dept geol sci, Brown Univ, 68-71; consult, NASA & leader, Landing Imaging Sci Team, Viking Mission to Mars, 70- Mem: Geol Soc Am. Res: Planetary geology; photogeology; stratigraphy. Mailing Add: Dept of Geol Brown Univ Providence RI 02912

MUTCH, WILLIAM WARREN, physics, see 12th edition

MUTCHLER, GORDON SINCLAIR, b Iowa City, Iowa, Mar 18, 38; m 63; c 1. NUCLEAR PHYSICS. Educ: Mass Inst Technol, BS, 60, PhD(physics), 66. Prof Exp: Res assoc nuclear physics, Los Alamos Sci Lab, Univ Calif, 66-68; from res assoc to sr res assoc, 68-73, ASST PROF PHYSICS, T W BONNER NUCLEAR LABS, RICE UNIV, 73- Mem: Am Phys Soc. Res: Investigation of nuclear structure using intermediate energy particles. Mailing Add: T W Bonner Nuclear Labs Rice Univ Houston TX 77001

MUTCHMOR, JOHN A, b Ft William, Ont, Aug 21, 29; m 55; c 1. INSECT PHYSIOLOGY. Educ: Univ Alta, BSc, 50; Univ Minn, MS, 55, PhD(entom), 61. Prof Exp: Tech officer entom, Sci Serv Lab, Can Dept Agr, 50-51 & Chatham Entom Lab, 56-61; from asst prof to assoc prof zool & insect physiol, 62-70, PROF ZOOL & INSECT PHYSIOL, IOWA STATE UNIV, 70- Mem: Entom Soc Am. Res: Low temperature adaption of poikilotherms and the influence of temperature and thermal adaptation on their dispersion and other activities; physiology of diapause in insects. Mailing Add: Dept of Zool & Entom Iowa State Univ Ames IA 50010

MUTH, CHESTER WILLIAM, b Antioch, Ohio, May 23, 22; m 49; c 5. ORGANIC CHEMISTRY. Educ: Ohio Univ, BS, 43; Ohio State Univ, PhD(chem), 49. Prof Exp: Synthetic org chemist, Eastman Kodak Co, 43-44, jr chemist, Tenn Eastman Corp, 44-45; asst, Ohio State Univ, 45-49; from asst prof to assoc prof chem, 49-63, PROF CHEM, WVA UNIV, 63- Mem: Am Chem Soc. Res: Synthetic organic chemistry with emphasis on tertiary amine-N-oxides and antimalarials. Mailing Add: Dept of Chem WVa Univ Morgantown WV 26506

MUTHA, SHANTILAL CHHOTMAL, b Ujjain, India, July 12, 34; m 54; c 4. ANALYTICAL CHEMISTRY. Educ: Birla Col, Pilani, India, BPharm, 56; Banaras Hindu Univ, India, MPharm, 57; Univ Calif, San Francisco, PhD(pharmaceut chem), 68. Prof Exp: Lectr pharmaceut chem, D A V Col, Kanpur, 57-58 & Banaras Hindu Univ, India, 58-64; teaching asst, Sch Pharm, Univ Calif, San Francisco, 64-68, Nat Inst Ment Heallth res chemist, 68; from anal chemist to sr anal chemist, 68-73, SUPVR CHEM ANAL RES, CUTTER LABS, 73- Mem: Am Chem Soc. Res: Pharmaceutical and medicinal chemistry; protein chemistry; intravenous solutions development; applied clinical chemistry. Mailing Add: Cutter Labs Inc 4th & Parker St Berkeley CA 94710

MUTHUKRISHNAN, RAJAGOPAL, physics, see 12th edition

MUTIS-DUPLAT, EMILIO, b Cucuta, Colombia, Sept 6, 32; m 58; c 3. GEOLOGY, PETROLOGY. Educ: Nat Univ Colombia, geologist, 60; Tex A&M Univ, MS, 69; Univ Tex, Austin, PhD(geol), 72. Prof Exp: From asst prof to assoc prof geol, Nat Univ Colombia, 61-73, dean fac, 64-65, chmn dept, 65-67; explor geologist, Tex Land & Trading Co, Austin, 73-74; ASSOC PROF EARTH SCI, UNIV TEX PERMIAN BASIN, 74- Concurrent Pos: Fel, Univ Tex, Austin, 72-73. Mem: Geol Soc Am; Am Geophys Union; Geochem Soc; Mineral Soc Am; Geol Soc London. Res: Origin of migmatites; origin of augen gneiss; amphibolite facies in regional metamorphism; marbles in regional metamorphic rocks. Mailing Add: Fac of Earth Sci Univ of Tex Permian Basin Odessa TX 79762

MUTO, PETER, b Chicago, Ill, Apr 23, 24; m 45; c 5. SCIENCE EDUCATION. Educ: Wis State Univ, Stevens Point, BS, 48; Phillips Univ, MEd, 51. Prof Exp: Teacher high sch, Okla, 49-51 & Iowa, 51-54; asst prof chem, 54-61, ASSOC PROF PHYS SCI, UNIV WIS-RIVER FALLS, 62- Mem: Nat Sci Teachers Asn. Res: Improvement of instruction in science for non-scientists at the college level. Mailing Add: Dept of Chem Univ of Wis River Falls WI 54022

MUTSCH, EDWARD L, b Madelia, Minn, Mar 15, 39; m 61. ORGANIC CHEMISTRY. Educ: St Olaf Col, BS, 61; Univ Minn, PhD(org chem), 65. Prof Exp: Sr res investr, G D Searle & Co, Ill, 65-67; sr res chemist, Biochem Res Cent Res Lab, Minn Mining & Mfg Co, 67-74; MEM STAFF CHEM RES & DEVELOP, RIKER LABS, 74- Mem: Am Chem Soc. Res: Synthesis of nucleoside antimetabolites. Mailing Add: Riker Labs Inc Chem Res & Develop 3M Ctr Bldg 218-1 St Paul MN 55101

MUTSCHLECNER, JOSEPH PAUL, b Ft Wayne, Ind, July 6, 30; m 55; c 3. ASTROPHYSICS, HYDRODYNAMICS. Educ: Ind Univ, AB, 52, MA, 54; Univ Mich, PhD(astron), 63. Prof Exp: Scientist, US Naval Ord Test Sta, 55-57 & 61-63; staff mem scientist, Los Alamos Sci Lab, Univ Calif, 63-67; ASSOC PROF ASTRON, IND UNIV, BLOOMINGTON, 67- Mem: Am Astron Soc; Int Astron Union. Res: Solar and stellar atmosphere and abundances. Mailing Add: Dept of Astron Ind Univ Bloomington IN 47401

MUTTER, WALTER EDWARD, b New York, NY, Nov 13, 21; m 63; c 2. PHYSICS. Educ: Polytech Inst Brooklyn, BS, 42; Mass Inst Technol, PhD(physics), 49. Prof Exp: Engr, Radio Corp Am, 42-46; res assoc physics, Mass Inst Technol, 46-49; engr, 49-59, SR ENGR, IBM CORP, 59- Honors & Awards: Sr Medal, Am Inst Chemists, 42. Mem: Am Phys Soc; Am Chem Soc; Am Inst Chemists; Inst Elec & Electronics Engrs. Res: Electronic processes in crystals and semiconductors; semiconductor devices; vacuum tubes. Mailing Add: 8 Bobrick Rd Poughkeepsie NY 12601

MUTTON, DONALD BARRETT, b New Toronto, Ont, Oct 29, 27; m 53; c 2. CHEMISTRY. Educ: Univ Toronto, BASc, 49, MASc, 51, PhD(cellulose chem), 53. Prof Exp: Asst res chemist, Int Cellulose Res, Ltd, 52-55, asst in-chg, Pioneering Res Div, 55-58, asst mgr, Basic Res Div, 58-60, mgr, 60-62, dir basic res & spec serv, 62-70, dir sci, 70-71, DIR RES, CIP RES LTD, 71-, VPRES, 72- Mem: Can Pulp & Paper Asn; Chem Inst Can (treas, 62-64); Tech Asn Pulp & Paper Indust; Asn Sci, Tech & Eng Community Can. Res: Pulp and paper technology; wood and cellulose chemistry. Mailing Add: CIP Res Ltd Hawkesbury ON Can

MUUL, ILLAR, b Tallinn, Estonia, Feb 18, 38; US citizen; m 61; c 2. ECOLOGY, ANIMAL BEHAVIOR. Educ: Univ Mass, BS, 60; Univ Mich, MS, 62, PhD(zool), 65. Prof Exp: Researcher ecol virus transmission, Walter Reed Army Inst Res, Washington, DC, 65-68, chief, Dept Ecol, Army Med Res Unit, Inst Med Res, Malaysia, 68-75, CHIEF, ENVIRON RES REQUIREMENTS BR, US ARMY MED BIOENG LAB, FT DETRICK, MD, 75- Concurrent Pos: Lectr, Eastern Mich Univ, 65. Mem: Am Soc Mammal; Ecol Soc Am; AAAS; Malaysian Soc Parasitol & Trop Med. Res: Environmental physiology; ethology; systematics of flying squirrels of the world; ecological factors involved in disease transmission; tropical ecology; population dynamics of mammals; zoogeography; environmental quality. Mailing Add: USA Med Bioeng Res & Dev Lab Ft Detrick Frederick MD 21701

MUUS, JYTTE MARIE, b Copenhagen, Denmark, Sept 21, 04. BIOCHEMISTRY. Educ: Univ Copenhagen, Mag Sci, 30. Prof Exp: Instr biochem, Univ Copenhagen, 30-36; Rockefeller fel, Harvard Med Sch, 36-37, asst, 37-39, fel, 39-40; from asst prof to prof, 40-70, EMER PROF PHYSIOL, MT HOLYOKE COL, 70- Concurrent Pos: Instr, Sch Pub Health, Harvard Univ, 42-46; Am Asn Univ Women fel, Carlsberg Lab, Copenhagen, 51-52; vis investr, Rockefeller Inst, 53; Fulbright lectr, Univ Col, Rhodesia & Nyasaland, 58-59 & 65-66. Mem: AAAS; Am Soc Biol Chem. Res: Tissue metabolism; chemical changes in burns; enzyme chemistry. Mailing Add: Dept of Biol Sci Mt Holyoke Col South Hadley MA 01075

MUZIK, THOMAS J, b Lorain, Ohio, Dec 21, 19; m 45; c 3. PLANT PHYSIOLOGY. Educ: Univ Mich, AB, MS, 42, PhD(bot), 50. Prof Exp: Res botanist, Firestone Plantations, Liberia, 42-47; plant physiologist, Fed Exp Sta, PR, 49-56; assoc prof agron, 56-62, PROF AGRON & AGRONOMIST, WASH STATE UNIV, 62- Concurrent Pos: Vis prof, Univ Madrid, 70-71. Mem: AAAS; Bot Soc Am; Am Soc Plant Physiol; Weed Sci Soc Am; Am Soc Agron. Res: Growth and development; environmental relationships; growth-regulators and herbicides. Mailing Add: Dept of Agron Wash State Univ Pullman WA 99163

MUZINICH, IVAN J, b San Francisco, Calif, Oct 26, 36; m 66. THEORETICAL HIGH ENERGY PHYSICS. Educ: Univ Calif, Berkeley, AB, 59, PhD(physics), 62. Prof Exp: Res asst physics, Lawrence Radiation Lab, Univ Calif, 60-62; res assoc, Univ Wash, 62-64, asst prof, 64-66; PHYSICIST, BROOKHAVEN NAT LAB, 68- Concurrent Pos: Consult, Los Alamos Sci Lab, 64-; vis scientist, Mass Inst Technol, 66-67; vis assoc prof, Rockefeller Univ, 67-68, affil, 68-; vis mem, Inst Advan Study, 74-75. Res: Theory of high energy elementary particle reactions and weak interactions; quantum field theory. Mailing Add: Brookhaven Nat Lab Upton NY 11973

MUZYCZKO, THADDEUS MARION, b Chicago, Ill, Jan 14, 36; m 65; c 2. POLYMER CHEMISTRY. Educ: Mich State Univ, BA, 59; Roosevelt Univ, MS, 68. Prof Exp: Proj engr, Chicago Pump Div, FMC Corp, 60-61; chemist & supvr chem, 61-69, MGR & SUPVR GRAPHIC ARTS, RES & DEVELOP DIV, RICHARDSON CO, MELROSE PARK, ILL, 70- Concurrent Pos: Lectr polymer chem, Roosevelt Univ, 69-; instr polymer technol, Col DuPage, 71- Mem: Am Chem Soc; Soc Plastics Engrs; Soc Photog Scientists & Engrs. Res: Polymer morphology; photopolymers and photopolymerizations; polymer characterizations. Mailing Add: 530 W 36th St Downers Grove IL 60515

MYALL, ROBERT WILLIAM T, b London, Eng, Dec 2, 38; m 67; c 2. ORAL

MEDICINE, DENTISTRY. Educ: Univ London, BDS, 65; FRCS(E), 67. Prof Exp: Asst prof oral med, Univ Ky, 67-70; ASST PROF ORAL MED, UNIV BC, 70- Mem: Brit Dent Asn; Can Dent Asn; Am Acad Oral Path; Am Acad Oral Med; Int Asn Dent Res. Res: Prevention of infective endocarditis in susceptible patients receiving dental care. Mailing Add: Dept of Oral Med Univ of BC Vancouver BC Can

MYATT, DEWITT O'KELLY, information science, see 12th edition

MYATT, WILFRED GERVAIS, geography, deceased

MYCEK, MARY J, b Shelton, Conn, Dec 19, 26. BIOCHEMISTRY, PHARMACOLOGY. Educ: Brown Univ, BA, 48; Yale Univ, PhD(biochem), 55. Prof Exp: Instr biochem, Yale Univ, 54-55; sr res biochemist, NY State Psychiat Inst, 57-61; res assoc biochem, Col Physicians & Surgeons, Columbia Univ, 59-61; res assoc, 61-63, asst prof, 63-69, ASSOC PROF PHARMACOL, COL MED & DENT, NJ, 69- Concurrent Pos: USPHS fel, Rockefeller Inst, 55-57; USPHS grants, 61-66, 67-71 & 72- Mem: AAAS; Am Chem Soc; Am Soc Pharmacol & Exp Therapeut; NY Acad Sci. Res: Effect of drugs on nucleic acid metabolism; drug metabolism; transamidation reactions; breakdown or alteration of proteins. Mailing Add: Dept of Pharmacol Col of Med & Dent of NJ Newark NJ 07103

MYCIELSKI, JAN, b Wisniowa, Poland, Feb 7, 32; m 59. MATHEMATICS. Educ: Wroclaw Univ, MA, 55, PhD(math), 57. Prof Exp: Full researcher, Nat Ctr Sci Res, Paris, 57-58; adjunkt, Inst Math, Polish Acad Sci, 58-63, docent, 63-68, prof, 68-69; PROF MATH, UNIV COLO, BOULDER, 69- Concurrent Pos: Vis prof, Univ Calif, Berkeley, 61-62 & 70, Case Western Reserve Univ, 67 & Univ Colo, Boulder, 67. Honors & Awards: Polish Math Soc Award, 56, Stefan Banach Prize, 66. Mem: Am Math Soc; Math Asn Am; Polish Math Soc. Res: Logic and foundations; artificial intelligence; theory of games; varia. Mailing Add: Dept of Math Univ of Colo Boulder CO 80302

MYER, DONAL GENE, b Toledo, Ohio, May 4, 30; m 51; c 3. PARASITOLOGY. Educ: Ohio State Univ, BSc, 51, MSc, 53, PhD(zool), 58. Prof Exp: Asst instr zool, Ohio State Univ, 57-58; from asst prof to assoc prof, 58-70, PROF ZOOL, SOUTHERN ILL UNIV, 70-, CHMN DEPT BIOL SCI, 74- Concurrent Pos: Asst dean, Grad Sch, Southern Ill Univ, 64-70; Am Coun Educ acad admin intern, Fla State Univ, 67-68. Mem: Am Soc Parasitol; Am Micros Soc; Sigma Xi. Res: Invertebrate zoology. Mailing Add: Dept of Biol Sci Southern Ill Univ Edwardsville IL 62025

MYER, GEORGE HENRY, b Bronx, NY, Dec 25, 37; m 63; c 2. GEOLOGY, MINERALOGY. Educ: Univ Calif, Santa Barbara, BA, 59; Yale Univ, PhD(geol), 65. Prof Exp: Asst prof geol, Univ Maine, 65-70; ASST PROF GEOL, TEMPLE UNIV, 70- Mem: AAAS; Mineral Soc Am; Geol Soc Am. Res: Mineralogy and metamorphic petrogenesis. Mailing Add: Dept of Geol Temple Univ Philadelphia PA 19122

MYER, GLENN EVANS, b Kingston, NY, Sept 16, 41; m 61. OCEANOGRAPHY, METEOROLOGY. Educ: State Univ NY Plattsburgh, BS, 65; State Univ NY Albany, MS, 69, PhD(atmospheric physics), 71. Prof Exp: Teacher physics, Plattsburgh High Sch, NY, 65-66; ASST PROF METEOROL, STATE UNIV NY PLATTSBURGH & DIR NORTH COUNTRY PLANETARIUM, 71-, DIR & CHMN LAKES & RIVERS RES LAB, 72- Mem: Am Phys Soc; Am Asn Physics Teachers; Am Soc Limnol & Oceanog; Int Asn Great Lakes Res; Royal Astron Soc Can. Res: Computer modeling and field measurements related to turbulent transport processes; applications of fluid dynamics to problems of physical limnology, atmospheric transport and modeling of planetary atmospheres. Mailing Add: Hudson Hall State Univ of NY Plattsburgh NY 12901

MYER, JON HAROLD, b Heilbronn, Ger, Sept 29, 22; nat US; m 48; c 4. EXPERIMENTAL PHYSICS. Educ: Hebrew Tech Col, BEE, 41. Prof Exp: Instrument maker, Anglo Iranian Oil Co, Iran, 42-44; instrument designer, Hebrew Tech Col, 44-46; eng consult, 46-47; instrumentologist, Dept Chem, Univ Southern Calif, 47-53; sub-lab head, Semiconductor Div, Hughes Aircraft Co, 53-60, mgr laser metall, 60-66, sr staff engr, Theoret Studies Dept, Res Labs, Malibu Beach, 66-70, Chem Physics Dept, 70-74, SR STAFF ENGR, OPTO ELECTRONICS DEPT, HUGHES AIRCRAFT CO, 74- Concurrent Pos: Lectr, Calif State Lutheran Col, 73- Mem: Sigma Xi; Am Phys Soc; sr mem Inst Elec & Electronics Engrs; Optical Soc Am; Magnetics Soc. Res: Physical instrumentation and apparatus design; semiconductor devices; laser applications; bubble domains; forensic science and technology. Mailing Add: 22931 Gershwin Dr Woodland Hills CA 91364

MYER, YASH PAUL, b Jullundur City, India, May 5, 32; m 59; c 2. PHYSICAL BIOCHEMISTRY. Educ: Punjab Univ, BSc, 53, MSc, 55; Univ Ore, PhD(chem), 61. Prof Exp: Lectr chem, SD Col, Punjab, India, 53-55; jr sci officer, Coun Sci & Indust Res, Punjab Univ, 55-57; res assoc biochem, Sch Med, Yale Univ, 61-66; from asst prof to assoc prof, 66-74, PROF CHEM, STATE UNIV NY ALBANY, 74- Mem: Am Chem Soc; Am Soc Biol Chem; Biophys Soc. Res: Macromolecular conformation and structure. Mailing Add: Dept of Chem State Univ of NY Albany NY 12222

MYERHOLTZ, RALPH W, JR, b Bucyrus, Ohio, July 29, 26; m 51; c 2. POLYMER CHEMISTRY. Educ: Purdue Univ, BS, 50; Northwestern Univ, PhD(org chem), 54. Prof Exp: Asst proj chemist, Standard Oil Co, Ind, 54-55, proj chemist, 55-58, group leader high polymers, 58-60; group leader, 60-66, res assoc, 66-69, DIR POLYMER PROPERTIES DIV, AMOCO CHEM CORP, 69- Mem: Am Chem Soc; Soc Plastics Engrs. Res: Structure-property relationships of high polymers; rheology, dynamic mechanical properties, stability and crystallization of high polymers; anionic polymerization processes; catalysis and hydrocarbon isomerization. Mailing Add: Amoco Chem Corp PO Box 400 Naperville IL 60540

MYERLY, RICHARD CREBS, b Westminster, Md, Aug 6, 28; m 49; c 1. ORGANIC CHEMISTRY. Educ: Franklin & Marshall Col, BS, 48; Pa State Col, MS, 50, PhD(chem), 52. Prof Exp: Proj leader, Process Develop Labs, Carbide & Carbon Chem Co Div, Union Carbide & Carbon Corp, 52-58, group leader, Union Carbide Chem Co Div, Union Carbide Corp, 58-63, mgr res & develop detergents & specialty chem, Chem Div, WVa, 63-67, tech mgr soaps & detergents, 67-69, TECH MGR AGR & BIOCHEM INTERMEDIATES, CHEM DIV, UNION CARBIDE CORP, 69- Mem: Am Chem Soc. Res: Production of organic chemicals by biological processes. Mailing Add: 1578 Nottingham Rd Charleston WV 25314

MYERS, ALBERT LEROY, b Church Hill, Tenn, Dec 18, 15; m 41; c 3. PHYSICAL CHEMISTRY. Educ: Carson-Newman Col, BS, 37; Univ Ga, MS, 39; Purdue Univ, PhD(chem), 50. Prof Exp: Asst area supvr, Plum Brook Ord Works, Trojan Powder Co, 41-43; prod supvr & staff chemist, Clinton Eng Works, Tenn Eastman Corp, 43-46; assoc prof chem, Ouachita Col, 46-49; assoc prof, FURMAN UNIV, 50-54, actg head dept, 51-53; from assoc prof to prof, Carson-Newman Col, 54-63; prof chem & chmn div sci & math, Houston Baptist Col, 63-67; PROF CHEM & CHMN DEPT, CARSON-NEWMAN COL, 67- Concurrent Pos: Res chemist, Oak Ridge Nat Lab, 59-62; chemist, Univ Tex Med Br, Galveston, 65. Mem: AAAS; Am Chem Soc. Res: Dipole moments; solubility products; vapor pressure of solutions; infrared spectra. Mailing Add: 901 S Russell Ave Jefferson City TN 37760

MYERS, ARTHUR JOHN, b South Haven, Mich, Aug 27, 18. GEOMORPHOLOGY. Educ: Kalamazoo Col, BA, 41; Mich Col Mining, BS & MS, 49; Univ Mich, PhD(geol), 57. Prof Exp: Asst prof, 51-61, ASSOC PROF GEOL, UNIV OKLA, 61- Mem: AAAS; Geol Soc Am; Am Asn Petrol Geol. Res: Permian and Pleistocene fluviatile deposits of northwestern Oklahoma; geomorphology as it reflects rock types, structure, climate and time; geologic mapping using aerial photographs and field observations; photogrammetry. Mailing Add: Sch of Geol & Geophys Univ of Okla Norman OK 73069

MYERS, BENJAMIN FRANKLIN, JR, b Steelton, Pa, Sept 3, 26; m 56; c 2. PHYSICAL CHEMISTRY. Educ: Pa State Univ, BS, 50; Northwestern Univ, PhD(chem), 55. Prof Exp: Res chemist, Union Oil Co, Calif, 55-59; res assoc & instr phys chem, Princeton Univ, 59-62; staff scientist, Gen Dynamics/Convair, 62-69 & Sci Applns, Inc, 69-74; STAFF CHEMIST, GEN ATOMIC CO, 74- Concurrent Pos: Consult, Sci Applns, Inc, 74- Mem: AAAS; Am Chem Soc; Am Phys Soc; fel Am Inst Chemists; Combustion Inst. Res: Shock phenomena; chemical kinetics; energy exchange processes; atmospheric chemistry; nuclear reactor fission product transport. Mailing Add: 1031 Alexandria Dr San Diego CA 92107

MYERS, BETTY JUNE, b Ashland, Ohio, Apr 18, 28. PARASITOLOGY. Educ: Ashland Col, BA & BSc, 49; Univ Nebr, MA, 51; McGill Univ, PhD(parasitol), 59. Prof Exp: Lab asst biol, Ashland Col, 47-49, instr, 51-52; asst biol, Univ Nebr, 49-51; instr parasitol, McGill Univ, 52-53, res assoc, 53-59, asst prof parasitol, 59-64; parasitologist, 64-69, ASST FOUND SCIENTIST, SOUTHWEST FOUND RES & EDUC, 69- Concurrent Pos: Fisheries Res Bd Can res grant, 54-58; consult, Food & Agr Orgn, 57-; Mem, Can Comt Freshwater Fisheries Res & Sci Adv Bd, Ashland Col, 59-; Nat Res Coun Can fel, 59-60; consult, NIH, 61; parasitologist, Arctic Unit, Fisheries Res Bd Can, 62-63; coun mem at large, Am Soc Parasitol, 74- Mem: AAAS; Am Soc Mammal; Am Soc Parasitol; Am Soc Trop Med & Hyg; Am Micros Soc (pres, 74). Res: Parasites of marine mammals, fish and primates; anisakiasis; schistosomiasis haematobium; primate ecological relationships of host and parasites; host-phylogenetic relationships; zoogeography; systematic experimental studies; helminthology. Mailing Add: SW Found for Res & Educ PO Box 28147 San Antonio TX 78228

MYERS, BRYANT LEE, biochemistry, chemistry, see 12th edition

MYERS, CARROL BRUCE, b Asheville, NC, Sept 6, 43; m 65; c 2. ALGEBRA. Educ: Berea Col, BA, 65; Univ Ky, MA, 67, PhD(math), 70. Prof Exp: Asst prof, 70-72, ASSOC PROF MATH, AUSTIN PEAY STATE UNIV, 72- Mem: Math Asn Am. Res: Ring theory; module theory. Mailing Add: Dept of Math Austin Peay State Univ Clarksville TN 37040

MYERS, CHARLES CHRISTOPHER, b Richwood, WVa, June 12, 34; m 60; c 3. FORESTRY, BIOMETRY. Educ: WVa Univ, BS, 60; Syracuse Univ, MS, 62; Purdue Univ, PhD(forestry), 66. Prof Exp: From instr to asst prof forestry, Purdue Univ, 62-67; technician, US AID, 67-69; asst prof, Univ Vt, 69-73; ASSOC PROF FORESTRY, SOUTHERN ILL UNIV, CARBONDALE, 73- Concurrent Pos: NSF grant, 71. Mem: Soc Am Foresters. Res: Applications of statistics and computers to forest inventory. Mailing Add: Dept of Forestry Southern Ill Univ Carbondale IL 62901

MYERS, CHARLES WILLIAM, b St Louis, Mo, Mar 4, 36. HERPETOLOGY. Educ: Univ Fla, BS, 60; Southern Ill Univ, MA, 62; Univ Kans, PhD(zool), 70. Prof Exp: Vis scientist herpet, Gorgas Mem Lab, Panama, 64-67; asst cur, 68-73, ASSOC CUR HERPET, AM MUS NATURAL HIST, 73- Mem: Am Soc Ichthyologists & Herpetologists; Soc Study Amphibians & Reptiles. Res: Systematics of neotropical amphibians and reptiles. Mailing Add: Dept of Herpet Am Mus Natural Hist New York NY 10024

MYERS, CLAUDE GRENVILLE, b Haverhill, Mass, June 23, 17; m 43; c 3. PHYSICAL CHEMISTRY, ENGINEERING. Educ: Yale Univ, BS, 39; Columbia Univ, PhD(mining, metall), 48. Prof Exp: Res chemist, 39-48, sr res chemist, 48-59, res assoc, 60-66, asst supvr, 66-70, ASST MGR CATALYST RES & DEVELOP, MOBIL RES & DEVELOP CORP, 70- Mem: Am Chem Soc. Res: Catalysis, especially the hydrogenative conversion of petroleum and catalytic cracking; petroleum catalyst research and development; structure and properties of aluminosilicates. Mailing Add: Mobil Res & Develop Corp Paulsboro NJ 08066

MYERS, CLIFFORD ALBERT, JR, b New London, Conn, Oct 19, 20; m 42; c 1. FOREST MANAGEMENT, FOREST MENSURATION. Educ: Colo State Univ, BSF, 42, MF, 47; Yale Univ, PhD(forestry), 59. Prof Exp: Asst prof forestry, Univ WVa, 47-55; res forester, Rocky Mountain Forest & Range Exp Sta, 55-75, RES FORESTER, SOUTHERN FOREST EXP STA, US FOREST SERV, 75- Honors & Awards: Superior Serv Award, USDA, 72; Mgt Improv Award, US President, 75. Mem: Soc Am Foresters; Biometrics Soc; Soc Am Archaeol. Res: Systems analysis in forestry; simulation in forest management. Mailing Add: 1435 Shelton Dr Nacogdoches TX 75961

MYERS, CLIFFORD EARL, b Jefferson City, Tenn, June 1, 29; m 53; c 2. INORGANIC CHEMISTRY, PHYSICAL CHEMISTRY. Educ: Carson-Newman Col, BS & BA, 51; Purdue Univ, MS, 53, PhD(inorg chem), 56. Prof Exp: Asst, Purdue Univ, 51-54; grad res chemist, Inst Eng Res, Univ Calif, 54-55; res assoc chem eng, Univ Ill, 55-56; from asst prof to assoc prof chem, Lynchburg Col, 56-58; asst prof, State Univ NY Col Ceramics, Alfred Univ, 58-63; ASSOC PROF CHEM, STATE UNIV NY BINGHAMTON, 63- Mem: Fel AAAS; Am Chem Soc; fel Am Inst Chemists. Res: High temperature vaporization processes; thermodynamic stabilities of refractory substances. Mailing Add: Dept of Chem State Univ of NY Binghamton NY 13901

MYERS, CLOVIS D, b Bloomington, Ind, Mar 17, 14; m 38; c 3. TEXTILE CHEMISTRY. Educ: Simpson Col, BA, 34; Univ Iowa, MS, 35, PhD(org chem), 37. Prof Exp: Instr chem, Park Col, 37-39, Col Emporia, 39-40 & Boise Jr Col, 40-41; res chemist, Tech Div, Rayon Dept, E I du Pont de Nemours & Co, 41-44, Nylon Div, 44-47, group leader, Process Develop Sect, 47, supvr, 48-52, tech supt, 52-64, planning supt, 65-67, planning-control supt, Process Develop Sect, 67-71; exec asst to pres, 71-72, dir bus affairs, 72-75, ASSOC DIR ESTATE PLANNING, SIMPSON COL, 75- Res: Chemistry of nylon; high polymers for synthetic fibers. Mailing Add: Simpson Col Indianola IA 50125

MYERS, DALE KAMERER, b Prospect, Pa, Apr 30, 38; m 74. ORGANIC CHEMISTRY. Educ: Berea Col, BA, 60; Auburn Univ, PhD(chem), 66. Prof Exp: Asst prof chem, Berea Col, 65-68; res assoc, Duke Univ, 68-71; ASSOC PROF CHEM, UNION COL, KY, 71-, DIR EXP EDUC, 75- Concurrent Pos: Chmn, Sci Div, Union Col, 73-75. Mem: Am Chem Soc; Sigma Xi. Res: Organophosphorus chemistry, particularly esters containing the tropane ring system; reactions of N-bromoamides; organophosphorus-phospholene ring. Mailing Add: Box 448 Union Col Barbourville KY 40906

MYERS, DAVID, b Philadelphia, Pa, Sept 18, 06; m 30; c 2. MEDICINE. Educ: Univ Pa, 27; Temple Univ, MD, 30; Am Bd Otolaryngol, dipl, 35. Prof Exp: Intern, Temple Univ, 30-32, preceptor, Temple Univ & trainee, Temple Univ Hosp, 32-40, mem staff dept otorhinol, Temple Univ Med Sch, 32-55, prof & chmn dept, 55-62; prof otorhinolaryngol, 62-71, chmn dept, 64-71, PROF OTOLARYNGOL, SCH MED, UNIV PA, 71- Concurrent Pos: Dir inst otol, Presby Hosp, Philadelphia, 62- Mem: Fel Am Otol Soc; Am Laryngol, Rhinol & Otol Soc; fel Am Col Surg; fel Am Acad Ophthal & Otolaryngol. Res: Otolaryngology. Mailing Add: Dept of Otorhinolaryngol Univ of Pa Sch of Med Philadelphia PA 19104

MYERS, DAVID DANIEL, b Morris, Minn, Dec 19, 32; m 59; c 1. LABORATORY ANIMAL SCIENCE, ANIMAL PATHOLOGY. Educ: Univ Minn, BS, 55, DVM, 57; Univ Ill, MS, 62, PhD(vet path & microbiol), 65. Prof Exp: Poultry vet, Fla Livestock Bd, 57-58; animal pathologist, Ill Dept Agr, 58-59; instr vet path, Univ Ill, 59-61; assoc staff scientist, 65-69, staff scientist, 69-75, SR STAFF SCIENTIST, JACKSON LAB, 75- Mem: AAAS; Am Vet Med Asn; Am Asn Lab Animal Sci; Am Soc Lab Animal Practrs. Res: Histopathology; pathogenesis of infectious diseases. Mailing Add: Jackson Lab Bar Harbor ME 04609

MYERS, DAVID KENNETH, biochemistry, see 12th edition

MYERS, DIRCK V, b New York, NY, Aug 24, 35; m 59; c 3. BIOCHEMISTRY. Educ: Dartmouth Col, AB, 57; Univ Wash, PhD(biochem), 62. Prof Exp: Res fel biol, Harvard Univ, 62-64; sr res scientist, Squibb Inst Med Res, NJ, 64-68; SR SCIENTIST, COCA-COLA CO, 68- Mem: Am Chem Soc. Res: Protein chemistry; peptide synthesis; enzymology. Mailing Add: 730 Starlight Lane NE Atlanta GA 30342

MYERS, DONALD ALBIN, b Denver, Colo, May 17, 36; m 58; c 2. PHYSIOLOGY, BIOMEDICAL ENGINEERING. Educ: Colo Sch Mines, PE, 58; Univ Colo, PhD(physiol), 73. Prof Exp: Pipeline engr, US Army Corps Engrs, 58-60; design engr, Martin Marietta Corp, 60-65; develop engr, AiResearch Mfg Co, Garrett Corp, 65-67; staff engr, Martin Marietta Corp, 67-69; PHYSIOLOGIST, US GOVT, WASHINGTON, DC, 73- Mem: AAAS; Aerospace Med Asn; Sigma Xi. Res: Effects of environmental stressors on the normal functioning of the human organism. Mailing Add: 8537 Pepperdine Dr Vienna VA 22180

MYERS, DONALD EARL, b Chanute, Kans, Dec 29, 31; m 54; c 2. MATHEMATICS. Educ: Kans State Univ, BS, 53, MS, 55; Univ Ill, PhD, 60. Prof Exp: Asst math, Kans State Univ, 53-55 & Univ Ill, 55-58; assoc prof, Millikin Univ, 58-60; from asst prof to assoc prof, 60-68, PROF MATH, UNIV ARIZ, 68- Concurrent Pos: Co-dir, Sec Sci Training Prog, NSF, Univ Ariz, 61, dir, 63, mem, Adv Panel Judge Proposals, 63-66 & writing team, Minn Math & Sci Teaching Proj, Univ Minn, 63-66; vis lectr, Teachers Col, 66; consult, India Prog, NSF-AID, 67. Mem: AAAS; Math Asn Am; Am Math Soc; Inst Math Statist. Res: Analysis; theory of distributions. Mailing Add: 3322 E Waverly Tucson AZ 85716

MYERS, DONALD ROYAL, b Cleveland, Ohio, Dec 18, 13; m 40; c 1. ORGANIC CHEMISTRY. Educ: Ohio State Univ, AB, 35, PhD(chem), 40. Prof Exp: Asst anal chem, Ohio State Univ, 35-40; res engr, Battelle Mem Inst, Ohio, 40-42; fel, Carnegie Inst Technol, 42-44; res chemist, 44-54, sect head, Chem Dept, 54-60, MGR CHEM RES PREP DEPT, UPJOHN CO, 60- Mem: AAAS; Am Chem Soc; NY Acad Sci. Res: Steroids; medicinal chemicals. Mailing Add: Upjohn Co Unit 7265 Bldg 91 Henrietta St Kalamazoo MI 49001

MYERS, DREWFUS YOUNG, JR, b Corcicana, Tex, Aug 20, 46; m 73. ORGANIC POLYMER CHEMISTRY, COLLOID CHEMISTRY. Educ: Tex Lutheran Col, BS, 68; Univ Utah, PhD(org chem), 74. Prof Exp: RES CHEMIST, EASTMAN KODAK CO, 74- Mem: AAAS; Am Chem Soc. Res: Various aspects of emulsion polymerization, latex polymer application and the colloidal stability of latex polymers. Mailing Add: Eastman Kodak Co Res Labs Bldg 82C Kodak Park Rochester NY 14650

MYERS, EARL EUGENE, b Ruffsdale, Pa, Nov 5, 24; m 54; c 4. PETROLEUM CHEMISTRY. Educ: Thiel Col, BS, 47; Western Reserve Univ, MS, 49, PhD(org chem), 51. Prof Exp: Res chemist, Esso Res & Eng Co, 51-52; res chemist, 53-62, SR RES CHEMIST, GULF RES & DEVELOP CO, 62- Mem: AAAS; Am Chem Soc. Res: Preparation of chemicals for use as petroleum additives; relationships of structure to activity; oil-soluble polymers. Mailing Add: 1526 Sherman St Cheswick PA 15024

MYERS, EUGENE NICHOLAS, b Philadelphia, Pa, Nov 27, 33; m 56; c 2. OTOLARYNGOLOGY. Educ: Univ Pa, BS, 54; Temple Univ, MD, 60. Prof Exp: PROF OTOLARYNGOL & CHMN DEPT, SCH MED, UNIV PITTSBURGH, 72- Concurrent Pos: Consult, US Naval Hosp, Philadelphia, 71- & Children's Hosp & Vet Admin Hosp, Pittsburgh, 72- Mem: Am Acad Facial, Plastic & Reconstruct Surg; Am Acad Ophthal & Otolaryngol; Am Soc Head & Neck Surg; Am Otol Soc; Asn Res Otolaryngol. Res: Histopathology of the temporal bones of children. Mailing Add: Dept of Otolaryngol Eye & Ear Hosp 230 Lothrop St Pittsburgh PA 15213

MYERS, FRANKLIN GUY, b Baltimore, Md, Aug 9, 18; m 37; c 3. MATHEMATICS. Educ: Univ Va, BA, 38, MA, 40, PhD(math), 42. Prof Exp: Instr math, Univ Va, 39-42; res engr, Martin Co, 42-50, supvr automatic comput, 50-53, chief servomech, 53-57, mgr advan reactor systs, 57-62; dir res, Allison Div, 62-68, chief mfg technol, Detroit Diesel Allison Div, 68-74, MGR MFG COMPUT SYSTS, DETROIT DIESEL ALLISON DIV, GEN MOTORS CORP, 74- Res: Gas turbine engines. Mailing Add: Detroit Diesel Div Gen Motors Corp Indianapolis IN 46206

MYERS, GARDINER HUBBARD, b Washington, DC, Jan 16, 39; m 63; c 2. PHYSICAL CHEMISTRY. Educ: Princeton Univ, AB, 59; Univ Calif, Berkeley, PhD(chem), 65. Prof Exp: Asst prof, 65-72, ASSOC PROF CHEM, UNIV FLA, 72- Mem: Am Chem Soc. Res: Gas kinetics; singlet oxygen; chemical education. Mailing Add: Dept of Chem Univ of Fla Gainesville FL 32611

MYERS, GEORGE E, b Detroit, Mich, Aug 9, 26; m 53; c 6. PHYSICAL CHEMISTRY. Educ: Univ Southern Calif, BS, 48, MS, 49; Harvard Univ, PhD, 52. Prof Exp: Res chemist, Oak Ridge Nat Lab, 52-56; proj scientist, Plastics Div, Union Carbide Corp, 56-63; sr tech specialist & sect chief, Lockheed Propulsion Co, Calif, 63-75; RES CHEMIST, FOREST PROD LAB, 75- Mem: Am Chem Soc; Sigma Xi. Res: Physical chemistry of proteins, clays, ion-exchange resins and polymers; solid propellants; wood adhesives. Mailing Add: 120 Grand Canyon Dr Madison WI 53705

MYERS, GEORGE HENRY, b New York, NY, Feb 21, 30; m 56; c 3. ELECTRICAL ENGINEERING, BIOMEDICAL ENGINEERING. Educ: Mass Inst Technol, SB & SM, 52; Columbia Univ EngScD, 59. Prof Exp: Mem tech staff, Bell Tel Labs, Inc, 52-59, supvr guid & control, 59-65; from assoc prof to prof elec eng, NY Univ, 65-69; mgr biomed eng lab, Riverside Res Inst, New York, 69-74; TECH DIR PACEMAKER CTR, NEWARK BETH ISRAEL MED CTR, 74- Mem: Sr mem Inst Elec & Electronics Eng; Am Soc Artificial Internal Organs; Biomed Eng Soc. Res: Ultrasonics; pacemakers and control of respiration; digital and analog computers. Mailing Add: 190 Wyoming Ave Maplewood NJ 07040

MYERS, GEORGE RODNEY, organic chemistry, see 12th edition

MYERS, GEORGE SCOTT, JR, b Monte Vista, Colo, Mar 21, 34; m 61; c 7. ANIMAL NUTRITION. Educ: Colo State Univ, BS, 56; Univ Conn, MS, 58; Cornell Univ, PhD(animal nutrit), 66. Prof Exp: Asst animal nutrit, Univ Conn, 56-58 & Cornell Univ, 60-63; sr nutritionist, Ciba Res Farm, NJ, 63-69; sr res nutritionist, Squibb Agr Res Ctr, 69-70; PRES, MYERS ANIMAL SCI CO, 70- Mem: Am Soc Animal Sci. Res: Carotene; vitamins A and E requirements of cattle, sheep and swine; rumen metabolism; ration formulation; animal husbandry management. Mailing Add: Myers Animal Sci Co PO Box 543 Clovis CA 93612

MYERS, GERALD, molecular biology, genetics, see 12th edition

MYERS, GERALD ANDY, b Boelus, Nebr, Sept 23, 28; m 53; c 4. PLANT ANATOMY, SCIENCE EDUCATION. Educ: Kearney State Col, AB, 51; Colo State Col, AM, 57; SDak State Col, PhD(plant sci), 63. Prof Exp: Teacher elem sch, Nebr, 51-52; teacher, jr high sch, Idaho, 52-55, high sch, 55-56; asst bot & biol, Colo State Col, 56-57; elem teacher adminr, elem sch, Ill, 57-58; instr bot, 58-64, asst prof, 64-68, assoc prof, 68-72, PROF BOT & BIOL & HEAD DEPT, SDAK STATE UNIV, 72- Concurrent Pos: NSF instr, Ind Univ, 59, SDak State Col, 60, Univ Wash, 61, Fla State Univ, 67 & Pa State Univ, 68 & Ohio State Univ, 72; consult, Biol Sci Curric Study, SDak, 63-, chmn, Testing Comt, 66-67; consult sci process approach, AAAS, 68-; consult, Intermediate Sci Curric Study, 69- Mem: Nat Asn Biol Teachers; Bot Soc Am. Res: Confidence testing on computer managed instructional modules; thermoperiodic effects on developmental anatomy; internal consistency, reliability of computer managed instructional tests. Mailing Add: Dept of Bot & Biol SDak State Univ Brookings SD 57006

MYERS, GORDON EDWARD, b Calgary, Alta, Oct 1, 18; m; c 2. MICROBIOLOGY. Educ: Univ Alta, BSc, 46, MSc, 48; McGill Univ, PhD(bact), 51. Prof Exp: Lectr bact, 46-49, from asst prof to prof, 51-63, microbiol, Fac Microbiol, 63-66, PROF MICROBIOL, FAC PHARM, UNIV ALTA, 67-, ASSOC DEAN, FAC PHARM & PHARMACEUT SCI, 71- Concurrent Pos: Asst prov bacteriologist, Alta Dept Pub Health, 46-57; head dept microbiol, Univ Alta, 63-66; consult microbiol, Pharm & Petrol Indust, Prov & Fed Depts, 51-66; Nat Res Coun Can sr res fel, Univ Hawaii, 66-67. Mem: Can Pharmaceut Asn. Res: Antimicrobial agents in medical practice and industry; industrial microbiology, especially pharmaceutical microbiology. Mailing Add: Fac of Pharm & Pharmaceut Sci Univ of Alberta Edmonton AB Can

MYERS, GORDON SHARP, b St John, NB, Jan 23, 21; m 47; c 5. MEDICINAL CHEMISTRY. Educ: Univ NB, BA, 41; Univ Toronto, MA, 42, PhD(org chem), 45. Prof Exp: Res chemist, Dom Rubber Co, 45-47; res assoc, Dept Biochem, Queen's Univ, Ont, 47-49; res chemist, Ayerst, McKenna & Harrison, Ltd. 49-67, mgr chem develop, 67-69, DIR CHEM DEVELOP, AYERST LABS, 69- Mem: Am Chem Soc; fel Chem Inst Can. Res: Medicinal and pharmaceutical chemistry; synthesis of anticonvulsants, antispasmodics, analgesics and steroids. Mailing Add: Ayerst Labs PO Box 6115 Montreal PQ Can

MYERS, HAROLD EDWIN, b Netawaka, Kans, Feb 12, 07; m 32; c 1. AGRONOMY. Educ: Kans State Univ, BS, 28; Univ Ill, MS, 29; Univ Mo, PhD(soils), 37. Prof Exp: Asst, Univ Ill, 28-29; instr agron, Kans State Univ, 29-31, from asst prof to prof, 31-56; dean, Col Agr, 56-73, DIR OVERSEAS PROGS, UNIV ARIZ, 73- Concurrent Pos: Agr adv, US Dept State, MidE, 43-45; head dept agron, Kans State Univ, 46-52, asst dean, Sch Agr & assoc dir, Agr Exp Sta, 52-56; consult, Mo Basin Surv Comn, 52; coordr, Bicentennial Comn, Univ Ariz, 7S- Honors & Awards: Cert Distinguished Serv Agr, Kans State Univ, 70; Citation Merit Agr, Univ Mo Alumni Asn, 70. Mem: Fel Am Soc Agron (pres, 53); Soil Sci Soc Am (pres, 50); Crop Sci Soc Am; Soil Conserv Soc Am; fel AAAS. Res: Soil fertility, nitrogen and moisture; factors influencing soil aggregation. Mailing Add: Off of Overseas Progs Univ of Ariz Tucson AZ 85721

MYERS, HARVEY NATHANIEL, b Tampa, Fla, Aug 26, 46; m 70; c 1. BIO-ORGANIC CHEMISTRY. Educ: Morehouse Col, BS, 69; Univ Ill, Urbana, MS, 71, PhD(chem), 74. Prof Exp: Fel comput based teaching, Univ Ill, Urbana, 74; ASST PROF CHEM, CHICAGO STATE UNIV, 74-, COORDR COMPUT ASSISTED INSTR, 75- Concurrent Pos: Consult, Shannon Furniture Refinishing, 75- Mem: AAAS; Am Chem Soc. Res: Curriculum development; bio-organic chemistry for computer assisted instruction. Mailing Add: Dept of Phys Sci Chicago State Univ Chicago IL 60628

MYERS, HERBERT, organic chemistry, see 12th edition

MYERS, HOWARD, b New York, NY, Jan 27, 28; m 48; c 2. CHEMICAL PHYSICS. Educ: Univ Chicago, PhB, 48, BS, 51, MS, 58. Prof Exp: Mem tech staff, Hughes Res Labs, 54-56; res specialist, Douglas Aircraft Co, Inc, 56-61; mem tech staff, Aerospace Corp, 61-63, mgr reentry physics, 63-66; mem tech staff, TRW Systs Group, 66-68; pres & tech dir, CPRL, Inc, 68-69; mgr plasma physics, 69-70, mgr planetary atmospheric physics, 70-73, STAFF SCIENTIST, PLANETARY PROGS, McDONNELL DOUGLAS ASTRONAUT CO, 73- Mem: AAAS; Am Phys Soc; Am Geophys Union; fel Am Inst Chemists. Res: Thermodynamics and reaction kinetics of solids, gases and plasmas. Mailing Add: 1232 Wissman Ave Manchester MO 63011

MYERS, HOWARD M, b Brooklyn, NY, Dec 12, 23; m 72; c 3. PHARMACOLOGY. Educ: Western Reserve Univ, DDS, 49; Univ Calif, MS, 53; Univ Rochester, PhD(pharmacol), 58; San Francisco State Col, MA, 64. Hon Degrees: MA, Univ Pa, 74. Prof Exp: Asst dent med, Sch Dent, Univ Calif, San Francisco, 4951, from instr to asst prof, 51-59, assoc prof dent med & biochem, 59-65, prof oral biol & lectr biochem, 65-71, vchmn dept biochem, Sch Med, 67-71, prof biochem & biophys, 71-72; prof biochem & chmn dept, Sch Dent, Univ Pac, 71-74; PROF PHARMACOL, SCH DENT MED & DIR CTR ORAL HEALTH RES, UNIV PA, 74- Concurrent Pos: Nat Inst Dent Res spec res fel, Dept Med Physics, Karolinska Inst, Sweden, 64-65; trainee, Advan Seminar Res Educ, Am Col Dentists, 63, mentor, 64; consult, Stanford Res Inst, 63-74; mem, Dent Training Comt, Nat Inst Dent Res, 65-69; dent res consult, Vet Admin Hosp, San Francisco, 66-74; mem, Dent Study Comt, Div Res Grants, NIH, 69-73; ed, Monograms Oral Sci, 70-; consult, Cooper Labs, 74- Mem: AAAS; Am Asn Dent Res (pres, 74); Int Asn Dent Res. Res: Mineral metabolism; composition of saliva; surface properties of tooth and bone mineral. Mailing Add: Ctr for Oral Health Res Univ of Pa Philadelphia PA 19174

MYERS, HUGH IRVIN, anatomy, pathology, see 12th edition

MYERS, IRA LEE, b Madison Co, Ala, Feb 9, 24; m 43; c 4. PREVENTIVE MEDICINE, PUBLIC HEALTH. Educ: Howard Col, BS, 45; Univ Ala, MD, 49; Harvard Univ, MPH, 53; Am Bd Prev Med, dipl pub health, 67. Prof Exp: Chief epidemic intel serv officer & asst to chief epidemiol br, Commun Dis Ctr, USPHS, 49-55; admin officer & asst state health officer, 55-62, STATE HEALTH OFFICER, DEPT PUB HEALTH, ALA, 63-; ASST CLIN PROF PREV MED, MED COL ALA, 57- Concurrent Pos: Secy, Ala Bd Med Examrs, 62-73; chmn, Ala Water Improv Comn, 63- Mem: Asn State & Prov Health Authorities of NAm; Asn State & Territorial Health Offs; Am Thoracic Soc. Res: Epidemiology of acute and chronic disease; problems of the aged, including medical care and nursing. Mailing Add: State Off Bldg Rm 381 Ala Dept Of Pub Health Montgomery AL 36104

MYERS, IRA THOMAS, b Iona, SDak, June 10, 25; m 47; c 8. PHYSICS. Educ: Wash State Univ, BS, 48, MS, 52, PhD(physics), 58. Prof Exp: Actg chief engr, Radio Sta KWSC, 48-49; sr physicist, Gen Elec Co, 49-62, mgr radiation effects sect, 62-70, MGR POWER COMPONENTS SECT, LEWIS RES CTR, NASA, 70- Res: Radiation effects on semiconductor devices; high power electronic components; solar cell arrays for space power. Mailing Add: NASA Lewis Res Ctr 21000 Brookpart Mail Stop 54-4 Cleveland OH 44135

MYERS, JACK DUANE, b New Brighton, Pa, May 24, 13; m 46; c 5. CLINICAL MEDICINE. Educ: Stanford Univ, AB, 33, MD, 37. Prof Exp: From intern to asst resident, Stanford Univ Hosps, 36-38; from asst resident to resident, Peter Bent Brigham Hosp, Brigham 39-52; assoc, Emory Univ, 46-47; assoc prof med, Duke Univ, 47-55; prof & chmn dept, 55-70, UNIV PROF MED, SCH MED, UNIV PITTSBURGH, 70- Concurrent Pos: Chmn, Am Bd Internal Med, 67-70; mem, Nat Adv Coun Arthritis & Metab Dis, 70-74; chmn, Nat Bd Med Examrs, 71-75. Mem: Soc Exp Biol & Med; Am Soc Clin Invest (secy-treas, 55-57); fel Am Col Physicians (pres, 76); Am Physiol Soc; Asn Am Physicians. Res: Clinical investigation of circulatory system of man, particularly the hepatic blood flow. Mailing Add: 1291 Scaife Hall Univ Pittsburgh 3550 Terrace St Pittsburgh PA 15261

MYERS, JACK EDGAR, b Boyds Mills, Pa, July 10, 13; m 37; c 4. PHOTOBIOLOGY. Educ: Juniata Col, BS, 34; Mont State Col, MS, 36; Univ Minn, PhD(bot), 39. Hon Degrees: DSc, Juniata Col, 66. Prof Exp: Nat Res Coun fel, Smithsonian Inst, 39-41; asst prof physiol, 41-46, from assoc prof to prof zool, 46-55, PROF BOT & ZOOL, UNIV TEX, AUSTIN, 55- Concurrent Pos: Guggenheim fel, 60; sci ed, Highlights for Children, 61. Honors & Awards: Kettering Award, Am Soc Plant Physiologists, 74. Mem: Nat Acad Sci; Bot Soc Am; Am Soc Plant Physiologists; Soc Gen Physiol; Am Soc Photobiol (pres, 75). Res: Photosynthesis; plant pigments; biological effects of radiation; physiology of algae. Mailing Add: Dept of Zool Univ of Tex Austin TX 78712

MYERS, JACOB MARTIN, b Mercersburg, Pa, Aug 16, 19; m 45; c 2. PSYCHIATRY. Educ: Princeton Univ, AB, 40; Johns Hopkins Univ, MD, 43; Am Bd Psychiat & Neurol, dipl, 49. Prof Exp: Exec med officer, Pa Hosp, 51-62, med dir, 62-70, from asst prof to assoc prof, 54-70, PROF PSYCHIAT, SCH MED, UNIV PA, 70-; DIR PSYCHIAT, PA HOSP & PSYCHIATRIST-IN-CHIEF, INST, 70- Concurrent Pos: Consult, Vet Admin Hosp, Coatesville, 55-65; US Naval Hosp, Philadelphia, 57-75; mem, Accreditation Coun, Psychiat Facil, Joint Comn Accreditation Hosps, 76- Mem: AAAS; Am Psychopath Asn; AMA; fel Am Psychiat Asn; Am Col Psychiat (pres, 71-72). Res: Clinical evaluation of treatment of hospitalized psychiatric patients. Mailing Add: Inst of the Pa Hosp 111 N 49th St Philadelphia PA 19139

MYERS, JAMES EDWARD, b San Francisco, Calif, Aug 9, 31; m 53; c 2. ANTHROPOLOGY. Educ: Calif State Univ, Sacramento, AB, 56; Cornell Univ, MEd, 57; Univ Calif, Berkeley, PhD(anthrop & educ), 61. Prof Exp: Teacher sci, Oakland Sch Dist, 58-60; assoc prof, 60-71, PROF ANTHROP, CALIF STATE UNIV, CHICO, 71- Concurrent Pos: Consult, Yuba City Unified Sch Dist, 63-65 & Nat Defense Educ Act Inst Cult Disadvantaged Children, Calif State Univ, Chico, 65-67; mem, Coun Anthrop & Educ, 65-; NSF fel, Hoopa Indian Reservation, 67-68, Nat Study Am Indian fel, 68-69. Mem: Fel Am Anthrop Asn; fel Soc Appl Anthrop. Res: Contemporary American Indian education. Mailing Add: Dept of Anthrop Calif State Univ Chico CA 95926

MYERS, JAMES HURLEY, b Memphis, Tenn, Sept 28, 40; m 63; c 2. PHYSIOLOGY. Educ: Memphis State Univ, BS, 63; Univ Tenn, Memphis, PhD(physiol), 69. Prof Exp: Instr physiol, Memphis State Univ, 68-69; ASST PROF PHYSIOL, SOUTHERN ILL UNIV, CARBONDALE, 71-, SCH MED, 71-, CURRIC COORDR, SCH & REP TO AM ASN MED COLS GROUP STUDENT AFFAIRS, 71- Concurrent Pos: USPHS fel biol, Brookhaven Nat Lab, 69-71. Mem: Assoc Am Physiol Soc; Geront Soc. Res: Physiology of circulation; radiation injury in primates; radioisotope techniques; physiology of aging. Mailing Add: Dept of Physiol Southern Ill Univ Sch Med Carbondale IL 62901

MYERS, JEFFERY, b Philadelphia, Pa, Feb 8, 32; div; c 1. PATHOLOGY, INFORMATION SCIENCE. Educ: Univ Pa, AB, 52; Temple Univ, MD, 57, MSc, 62; McGill Univ, PhD(path), 65. Prof Exp: Pathologist, Allentown Hosp, 65-66; pathologist, 66-69, CHIEF SURG PATH, PHILADELPHIA GEN HOSP, 69-; ASST PROF PATH, SCH MED, UNIV PA, 66- Concurrent Pos: Asst prof, Sch Med, Temple Univ, 65-66; consult, Wyeth Labs, 66- Mem: AAAS; fel Col Am Path; NY Acad Sci. Res: Use of computers to analyze pathology data; sources of errors in medical information systems; validity of computerized information. Mailing Add: 1900 J F Kennedy Blvd Apt 1221 Philadelphia PA 19103

MYERS, JOHN ALBERT, b Sandusky, Ohio, Mar 13, 43; c 1. ORGANIC CHEMISTRY. Educ: Carson-Newman Col, BS & BA, 65; Univ Fla, PhD(chem), 70. Prof Exp: Res grant, Mich State Univ, 70-71; ASSOC PROF CHEM, NC CENT UNIV, 71- Mem: Am Chem Soc. Res: Non-benzenoid aromatics; organic reaction mechanisms; heterocyclics. Mailing Add: Dept of Chem NC Cent Univ Durham NC 27707

MYERS, JOHN MARTIN, b Portland, Ore, June 8, 35; m 59; c 3. APPLIED PHYSICS, SYSTEMS THEORY. Educ: Calif Inst Technol, BS, 56; Harvard Univ, MS, 57, PhD(appl physics), 62. Prof Exp: Jr engr, Raytheon Co, 56-57, engr, 57-60, res scientist, 60-62, sr res scientist, 62-65, prin res scientist, 65-67; opers res analyst, Off Asst Secy Defense, 67-68; asst adminr, Model City Admin, Boston, 68-70; CONSULT PROCESS CONTROL & REPORTING SYSTS, 70- Mem: Inst Elec & Electronics Engrs. Res: Boundary value problems; transformational generating principles for biological structure; systems research; concurrency, choice, graphics and simulation. Mailing Add: 18 Joy St Boston MA 02114

MYERS, JOSEPH B, b Indianapolis, Ind, Dec 30, 27; m 49; c 5. PHYSIOLOGY. Educ: WVa State Col, BS, 49; Atlanta Univ, MS, 64; Howard Univ, PhD(zool), 71. Prof Exp: Res assoc exp path, Sch Med, Ind Univ, 51-58; instr biol, Indianapolis Pub Sch Syst, 58-63; asst prof, St Joseph Col, 64-68; ASST PROF BIOL, ATLANTA UNIV, 71- Mem: AAAS. Res: Renal hypertension; physiological variation during hypertensive state; the physiology of insect body fluids. Mailing Add: Dept of Physiol Atlanta Univ Atlanta GA 30314

MYERS, KARL JOHNSON, SR, b Nestorville, WVa, July 17, 99; m 22; c 2. RADIOLOGY. Educ: WVa Univ, BSc, 21; Univ Md, MD, 23; Am Bd Radiol, dipl, 39. Hon Degrees: DSc, Alderson-Broaddus Col, 69. Prof Exp: Part-time instr biol, human biol & radiol technol, 41-58, PROF RADIOL TECHNOL & HEAD DEPT, ALDERSON-BROADDUS COL, 58-; RADIOLOGIST, MYERS CLIN, BROADDUS HOSP, 33-, DIR, 70- Concurrent Pos: Trustee, Alderson-Broaddus Col, 62-, secy, Bd Dirs, 65- Mem: Am Roentgen Ray Soc; Am Thoracic Soc; Soc Nuclear Med; fel Am Col Radiol; Radiol Soc NAm. Res: Clinical radiology; radiologic technology. Mailing Add: 340 S Main St Philippi WV 26416

MYERS, LAWRENCE STANLEY, JR, b Memphis, Tenn, Apr 29, 19; m 42; c 3. RADIATION BIOPHYSICS, ENVIRONMENTAL SCIENCES. Educ: Univ Chicago, 41, PhD(phys chem), 49. Prof Exp: Asst chem, Metall Lab, Manhattan Eng Dist, Chicago, 42-44; assoc chemist, Clinton Lab, Tenn, 44-46; asst, Inst Nuclear Studies, Univ Chicago, 46-48, chemist, Univ, 48-49; assoc chemist, Argonne Nat Lab, 49-52; asst prof biophys, nuclear med & radiol, Sch Med, 53-70, RES RADIOBIOLOGIST, LAB NUCLEAR MED & RADIATION BIOL, UNIV CALIF, LOS ANGELES, 53-, LECTR RADIOL, SCH MED, 70- Concurrent Pos: Biophysicist, Biol Br, Div Biol & Med, AEC, 72-74; assoc ed, Radiation Res, 74- Honors & Awards: AEC Spec Achievement Cert, 74. Mem: Am Soc Photobiol; AAAS; Radiation Res Soc; Biophys Soc. Res: Effects of ionizing radiation and environmental contaminants on nucleic acids, nucleoproteins and simple biological systems. Mailing Add: Lab of Nuclear Med Univ of Calif Los Angeles CA 90024

MYERS, LYLE LESLIE, b Salem, Ore, June 11, 38; m 60; c 3. BIOCHEMISTRY, IMMUNOCHEMISTRY. Educ: Ore State Univ, BS, 60; Mont State Univ, MS, 62; Purdue Univ, PhD(biochem), 66. Prof Exp: Asst prof, 66-71, ASSOC PROF VET BIOCHEM, VET RES LAB, MONT STATE UNIV, 71- Mem: Conf Res Workers Animal Dis. Res: Immunological and biochemical aspects of neonatal enteritis of the bovine; antigenic structure of the vibrio fetus bacteria. Mailing Add: Vet Res Lab Mont State Univ Bozeman MT 58715

MYERS, MARCUS NORVILLE, b Boise, Idaho, May 30, 28; m 50; c 3. ANALYTICAL CHEMISTRY. Educ: Brigham Young Univ, BS, 50, MS, 52; Univ Utah, PhD(phys chem), 65. Prof Exp: Engr, Hanford Works, Gen Elec Co, 51-57, chemist, Idaho, 57-61, Vallecitos Atomic Lab, 61-62; res asst, 62-65, res assoc, 65-67, ASST RES PROF PHYS CHEM, UNIV UTAH, 67- Mem: AAAS; Am Chem Soc. Res: High pressure gas chromatography; theory of all forms of chromatography; field flow fractionization; activation analysis. Mailing Add: Dept of Chem Univ of Utah Salt Lake City UT 84112

MYERS, MARK B, b Winchester, Ind, Oct 14, 38; m 59; c 4. MATERIALS SCIENCE, CERAMICS. Educ: Earlham Col, AB, 60; Pa State Univ, PhD(solid state technol), 64. Prof Exp: Mem sci staff, Xerox Res Labs, NY, 64-68, mgr mat sci br, 68-71, mgr mat res lab, 71-75, MGR, XEROX RES CENTRE OF CAN LTD, 75- Concurrent Pos: Assoc prof, Univ Rochester, 70-75. Mem: Am Phys Soc; Am Ceramic Soc. Res: Thermodynamics and kinetics of glass formation; phase transitions; glass transition phenomena; two phase glass ceramics; chalcogenide materials. Mailing Add: Xerox Res Centre of Can Ltd 2480 Dunwin Dr Mississauga ON Can

MYERS, MARSHALL JAY, nutritional biochemistry, food science, see 12th edition

MYERS, MAX H, b Lynchburg, Va, July 2, 36; m 59; c 2. BIOMETRICS, STATISTICS. Educ: Bridgewater Col, BA, 58; Va Polytech Inst, MS, 60; Univ Minn, PhD(biomet), 71. Prof Exp: USPHS officer, End Results Sect, Biomet Br, 60-62, math statistician, 62-73, HEAD END RESULT SECT, BIOMET BR, FIELD STUDIES & STATIST, DIV CANCER CAUSE & PREV, NAT CANCER INST, 73- Mem: Am Statist Asn; Biomet Soc. Res: Epidemiology of cancer patient survival including detailed study of factors related to prognosis; statistical methodology for evaluating multifactor relationships to survival. Mailing Add: Nat Cancer Inst Landow Bldg C518 7910 Woodmont Ave Bethesda MD 20014

MYERS, MELVIL BERTRAND, JR, b New Orleans, La, Sept 12, 28; m 54; c 3. MEDICINE. Educ: Tulane Univ, MD, 51; Am Bd Surg, dipl, 58. Prof Exp: Clin instr, 56-71, assoc prof, 71-75, PROF SURG, SCH MED, LA STATE UNIV MED CTR, NEW ORLEANS, 75-; PRIN INVESTR SURG RES, VET ADMIN HOSP, NEW ORLEANS, 63-, SR SURGEON, TOURO INFIRMARY, 65- Concurrent Pos: Grants, Am Cancer Soc, La Heart Asn, Southeast Surg Cong, NY Acad Sci, Ethicon, Inc, Warren-Teed, Inc & John A Hartford Found; sr surgeon, Charity Hosp, New Orleans, 61-; attend physician, US Vet Admin Hosp, 66-71, staff physician, 71- Mem: Fel Am Col Surgeons; Am Heart Asn; Plastic Surg Res Coun; Am Soc Plastic & Reconstruct Surgeons. Res: Wound healing and revascularization; cause of tissue necrosis; tissue changes following devascularization and revascularization; mechanism of ventricular fibrillation following ischemia; various clinical surgical problems. Mailing Add: Vet Admin Hosp 1601 Perdido St New Orleans LA 70140

MYERS, MERLE WENTWORTH, b Bellevue, Ill, Nov 18, 13; m 40; c 2. GEOGRAPHY. Educ: Univ Ill, BS, 36; Clark Univ, MA, 37, PhD(geog), 48. Prof Exp: Instr geog, Mich State Univ, 37-38; asst prof, George Peabody Col, 39; prof, Troy State Univ, 39-40; assoc prof, 40-48, PROF GEOG, MISS STATE UNIV, 48- Mem: Nat Coun Geog Educ; Am Geog Soc; Asn Am Geogrs. Res: Historic cultural geography in southern United States, especially Mississippi. Mailing Add: Dept of Geol & Geog Drawer GG Miss State Univ Mississippi State MS 39762

MYERS, MICHAEL KENNETH, applied mechanics, applied mathematics, see 12th edition

MYERS, ORLO EDMUND, chemistry, see 12th edition

MYERS, OVAL, JR, b Roachdale, Ind, July 28, 33; m 59; c 2. PLANT GENETICS. Educ: Wabash Col, BA, 58; Dartmouth Col, MA, 60; Cornell Univ, PhD(genetics), 63. Prof Exp: From instr to asst prof bot & bact, Univ Ark, Fayetteville, 63-68; assoc prof bot & plant indust, 68-75, PROF PLANT & SOIL SCI, SOUTHERN ILL UNIV, CARBONDALE, 75- Concurrent Pos: Educ specialist, Southern Ill Univ & Food & Agr Orgn, Brazil, 72-74. Mem: Fel AAAS; Bot Soc Am; Genetics Soc Am; Ecol Soc Am; Am Soc Agron. Res: Genetics and cytogenetics of Zea mays, sorghum bicolor, Hypericum and Impatiens; developmental plant morphology. Mailing Add: Dept of Plant & Soil Sci Southern Ill Univ Carbondale IL 62901

MYERS, PAUL WALTER, b Schenectady, NY, Jan 15, 23; m 44; c 4. NEUROSURGERY. Educ: Albany Med Col, MD, 46; Am Bd Neurol Surg, dipl, 60. Prof Exp: Intern med, Ellis Hosp, Schenectady, NY, 46-47; US Air Force, 51-, resident surg, Ellis Hosp, 52-53, resident neurol surg, Albany Med Ctr, 53-56, chief neurol surg, 58-71, COMMANDER, WILFORD HALL AIR FORCE MED CTR, 71, PROF NEUROSURG, UNIV TEX MED SCH, SAN ANTONIO, 69-

Concurrent Pos: Hon flight surgeon designation, Govt Chile, 68. Mem: AMA; Cong Neurol Surgeons; Am Asn Neurol Surg; Am Col Surgeons. Res: Cervical injuries; neuroanatomy. Mailing Add: Wilford Hall AF Med Ctr Lackland AFB TX 78236

MYERS, PETER BRIGGS, b Washington, DC, Apr 24, 26; m 48; c 2. PHYSICS. Educ: Worcester Polytech Inst, BSEE, 46; Oxford Univ, DPhil(physics), 50. Prof Exp: Mem tech staff, Switching Res Dept, Bell Tel Labs, Inc, 50-59; exec staff scientist, Motorola, Inc, 59-62; dir res lab, Bunker-Ramo Corp, 62-66, vpres res & develop, 66-68; mgr advan technol, Magnavox Res Labs, Calif, 68-74, DIR, MAGNAVOX CO, MD, 74- Mem: AAAS; Am Soc Qual Control; Electrochem Soc; Philos Sci Asn; Inst Elec & Electronics Engrs. Res: Solid state device physics; solid state circuits; magnetic logic and memory; solid state integrated circuits. Mailing Add: Magnavox Co 8750 Georgia Ave Silver Spring MD 20910

MYERS, PHILIP, b Baltimore, Md, June 10, 47; m 69; c 1. MAMMALOGY. Educ: Swarthmore Col, BA, 69; Univ Calif, Berkeley, PhD(zool), 75. Prof Exp: ASST PROF ZOOL, UNIV MICH, ANN ARBOR, 75- Mem: Assoc Sigma Xi; Am Soc Mammalogists. Res: Ecology and evolution of mammals; population biology; biosystematics of mammals. Mailing Add: Mus of Zool Univ of Mich Ann Arbor MI 48104

MYERS, PHILIP CHERDAK, b Elizabeth, NJ, Nov 18, 44; m 72. RADIO ASTRONOMY. Educ: Columbia Univ, AB, 66; Mass Inst Technol, PhD(physics), 72. Prof Exp: Staff scientist radio physics & astron, Res Lab Electronics, 72-75, ASST PROF PHYSICS, MASS INST TECHNOL, 75- Mem: Am Astron Soc; Int Union Radio Sci; Int Astron Union; AAAS. Res: Radio astronomy of interstellar medium, particularly spectral lines in dark clouds; microwave radiometry of human tissue for purposes of medical diagnosis. Mailing Add: Dept of Physics Mass Inst of Technol Cambridge MA 02139

MYERS, PHILLIP WARD, b Evanston, Ill, Nov 11, 39; m 63; c 3. OTOLARYNGOLOGY, AUDIOLOGY. Educ: Western Ill Univ, BS, 61; Univ Ill, MD, 65; Am Bd Otolaryngol, dipl, 71. Prof Exp: Intern med, St Paul-Ramsey Hosp, Minn, 66; resident otolaryngol, Univ Louisville, 66-67, resident gen surg, 67-68; resident otolaryngol & maxillofacial surg, Med Sch, Northwestern Univ, 68-70; DIR AUDIOL & SPEECH CTR & BIOACOUST RES LAB, WALTER REED ARMY MED CTR, 71- Concurrent Pos: Fel otolaryngol & maxillofacial surg, Med Sch, Northwestern Univ, 70-71; staff physician & asst chief otolaryngol serv, Walter Reed Hosp, 71-73; Army liaison consult, Acoust Soc Am, 71-73; staff otolaryngologist & consult, Va Otolaryngol Clin, Ft Meyer, 71-73. Honors & Awards: Physician Recognition Award, AMA, 71. Mem: Cand mem Am Col Surg. Res: Otology-perilymphatic fistulas in post stapedectomy patients and the relationship of increased central venous pressure to the elevation of perilymphatic fluid pressure; hearing conservation. Mailing Add: Audiol & Speech Ctr Walter Reed Army Med Ctr Washington DC 20012

MYERS, RALPH, b Ft Smith, Ark, Nov 25, 38; m 63; c 1. GEOLOGY. Educ: Okla State Univ, BS, 55; Harvard Univ, AM, 62; Univ Tex, Austin, PhD(biostratig), 65. Prof Exp: NSF fel, Univ Tex, Austin, 65-67; asst prof geol, East Tex State Univ, 67-68; consult, Compania Dominicana de Desarrollo y Fomento, 68; res geologist, 69-71, SR RES GEOLOGIST, MOBIL RES & DEVELOP CORP, 71- Mem: AAAS; Geol Soc Am; Am Asn Petrol Geologists; Paleont Soc. Res: Habitat of oil; mesozoic stratigraphy of Mexico and the Carribean; evolution of rudists; plate tectonics. Mailing Add: Field Res Lab Mobil Res & Develop Corp Dallas TX 75221

MYERS, RALPH ARTHUR, physical chemistry, chemical engineering, see 12th edition

MYERS, RALPH DUANE, b Chambersburg, Pa, Sept 29, 12; m 40; c 3. PHYSICS. Educ: Cornell Univ, AB, 34, AM, 35, PhD(physics), 37. Prof Exp: Asst physics, Cornell Univ, 34-37; instr, Purdue Univ, 37-38; from instr to asst prof, Univ Md, 38-44; assoc physicist, Nat Bur Standards, 44-45; assoc prof physics, 45-47, PROF PHYSICS, UNIV MD, COLLEGE PARK, 47- Concurrent Pos: Consult, Naval Res Lab, DC, 48. Mem: Am Phys Soc. Res: Photoelectricity; solid state theory; angular momentum in nuclear reactions. Mailing Add: Dept of Physics Univ of Md College Park MD 20740

MYERS, RALPH THOMAS, b Maidsville, WVa, Mar 28, 21; m 54; c 2. PHYSICAL INORGANIC CHEMISTRY. Educ: WVa Univ, AB, 41, PhD(phys org chem), 49. Prof Exp: Asst, Manhattan Proj, Columbia Univ, 44-46; assoc prof chem & head dept, Waynesburg Col, 48-51; asst prof phys chem & consult, Res Found, Colo Sch Mines, 51-56; asst prof chem, 56-61, ASSOC PROF CHEM, KENT STATE UNIV, 61- Mem: AAAS; Am Chem Soc. Res: Nonaqueous solvents; dielectric constant; electrical conductance. Mailing Add: Dept of Chem Kent State Univ Kent OH 44242

MYERS, RAYMOND HAROLD, b Charleston, WVa, Oct 13, 37; m 59; c 2. MATHEMATICAL STATISTICS. Educ: Va Polytech Inst, BSc, 59, MSc, 61, PhD(statist), 64. Prof Exp: From asst prof to assoc prof, 63-71, PROF STATIST, VA POLYTECH INST & STATE UNIV, 71- Mem: Am Statist Asn. Res: Experimental design and analysis; response surface techniques. Mailing Add: Dept of Statist Va Polytech Inst & State Univ Blacksburg VA 24060

MYERS, RAYMOND J, b Bellevue, Ohio, Dec 5, 03; m 32; c 1. ZOOLOGY. Educ: Heidelberg Col, AB, 27; Univ Pa, PhD(zool), 33. Prof Exp: Asst, Univ Ind, 27-30; instr, Univ Pa, 30-34; instr, 34-42, from asst prof to assoc prof, 42-56, PROF ZOOL, COLGATE UNIV, 56-, CHMN, DEPT BIOL, 67- Concurrent Pos: Instr, LI Col Med, 44-46. Mem: AAAS; assoc Am Soc Zoologists. Res: Reproduction and taxonomy of Hirudinea; human hair growth. Mailing Add: 13 Payne St Hamilton NY 13346

MYERS, RAYMOND REEVER, b New Oxford, Pa, Jan 23, 20; m 43; c 3. CHEMISTRY. Educ: Lehigh Univ, AB, 41, PhD(chem), 52; Univ Tenn, MS, 42. Prof Exp: Res chemist, Cent Res Labs, Monsanto Co, 42-46 & Jefferson Chem Co, 46-50; asst chem, Lehigh Univ, 50-52, res assoc, 52-53, from res asst prof to res prof, 53-65; PROF & CHMN DEPT CHEM, KENT STATE UNIV, 65- Concurrent Pos: Instr, Univ Dayton, 45-46; res dir, Paint Res Inst, 64-; former consult, Nat Bur Standards, R T Vanderbilt Co & Air Reduction Co; ed, Soc Rheol, 63; Matiello lectr, Fedn Socs Coatings Technol, 75. Honors & Awards: Borden Award, Am Chem Soc, 71; Morrison Award, NY Acad Sci, 58. Mem: Am Chem Soc; fel Am Inst Chemists; Soc Rheol fel NY Acad Sci; Brit Soc Rheol. Res: Rheology of coatings; adhesion; application of spectra in catalysis; structure of matter; research administration. Mailing Add: Dept of Chem Kent State Univ Kent OH 44242

MYERS, RICHARD F, b Hammond, Ind, Feb 1, 31; m 51; c 4. VERTEBRATE ZOOLOGY, ACADEMIC ADMINISTRATION. Educ: Earlham Col, AB, 52; Cornell Univ, MS, 54; Univ Mo, PhD, 64. Prof Exp: From asst prof to assoc prof zool, Cent Mo State Col, 59-67; assoc prof zool & ecol, Univ Mo-Kansas City, 67-72; DIR, NAT WEATHER SERV TECH TRAINING CTR, 72- Concurrent Pos: Inst Int Educ-US AID consult, Bangladesh, 69-70. Mem: AAAS; Am Soc Mammal; Ecol Soc Am. Res: Mammalogy; wildlife biology; movement and migration patterns; population ecology, especially bats. Mailing Add: Nat Weather Serv 617 Hardesty Kansas City MO 64124

MYERS, RICHARD HAROLD, physical chemistry, see 12th edition

MYERS, RICHARD LEE, b Doylestown, Pa, Oct 26, 44; m 66. ANALYTICAL CHEMISTRY, ENVIRONMENTAL SYSTEMS & TECHNOLOGY. Educ: Calif Inst Technol, BS, 66; Univ Wis-Madison, PhD(anal chem), 71. Prof Exp: Mem tech staff, NAm Rockwell Sci Ctr, 71-73, prog mgr, Air Monitoring Ctr, 73-75, CENT REGION MGR, AIR MONITORING CTR, ROCKWELL INT CORP, 75- Mem: Am Chem Soc; Electrochem Soc. Res: Application of mini-computers to chemical and physical measurements; ambient air pollution measurement techniques and instrumentation. Mailing Add: Rockwell Int Corp 11640 Administration Dr Creve Coeur MO 63141

MYERS, RICHARD SHOWSE, b Jackson, Miss, Oct 26, 42; m 65; c 2. PHYSICAL CHEMISTRY. Educ: Miss Col, BS, 64; La State Univ, MS, 66; Emory Univ, PhD(chem), 68. Prof Exp: Asst prof, 68-74, ASSOC PROF CHEM, DELTA STATE COL, 74- Mem: Am Chem Soc. Res: Surface thermodynamics and surface tension of nonelectrolyte solutions; determination of activity coefficients of nonelectrolyte solutions by Rayleigh light scattering techniques. Mailing Add: Dept of Chem Delta State Col Cleveland MS 38732

MYERS, RICHARD THOMAS, b Macon, Ga, Dec 25, 18; m 59; c 4. SURGERY. Educ: Univ NC, AB, 39; Univ Pa, MD, 43. Prof Exp: From asst prof to assoc prof, 50-68, PROF SURG & CHMN DEPT, BOWMAN GRAY SCH MED, 68- Mem: AMA; Am Col Surg. Res: Upper gastrointestinal tract; pancreas and biliary tract. Mailing Add: Dept of Surg Bowman Gray Sch of Med Winston-Salem NC 27103

MYERS, ROBERT ANTHONY, b Brooklyn, NY, Feb 22, 37. SOLID STATE PHYSICS. Educ: Harvard Univ, AB, 58, AM, 59, PhD(appl physics), 64. Prof Exp: Physicist, IBM Watson Res Ctr, NY, 63-68, mem corp tech comt staff, IBM Corp, 68-72, MGR TERMINAL TECHNOL, IBM WATSON RES CTR, 73- Concurrent Pos: Secy, Sci Adv Comt, IBM Corp, 71-72. Mem: Am Phys Soc; Inst Elec & Electronics Engrs. Res: Application of solid state technology to computer input/output devices. Mailing Add: IBM Watson Res Ctr PO Box 218 Yorktown Heights NY 10598

MYERS, ROBERT DURANT, b Philadelphia, Pa, Oct 25, 31; m 53; c 2. PSYCHOBIOLOGY, NEUROBIOLOGY. Educ: Ursinus Col, BS, 53; Purdue Univ, MS, 54, PhD, 56. Prof Exp: Asst psychol, Purdue Univ, 54-55; from asst prof to assoc prof & dir res coun, Colgate Univ, 56-64; fel, Neurol Sci Group, Sch Med, Johns Hopkins Univ, 60-61; prof psychol, 65-72, PROF PSYCHOL & BIOL SCI, PURDUE UNIV, 72-, COORDR NEUROBIOL TRAINING PROG, 70-, DIR, PSYCHOBIOL PROG, 73- Concurrent Pos: Res psychologist, Rome Air Develop Ctr, Griffiss AFB, 57-58; vis scientist, Nat Inst Med Res Eng, 63-65 & 69-70; Sigma Xi res award, Purdue Univ, 71; vis prof, La Trobe Univ, Australia, 75. Mem: AAAS; Am Psychol Asn; Am Physiol Soc. Res: Neural mechanisms controlling feeding, drinking and emotional behavior; transmitter synthesis, turnover and release in brain stem; physiology and pharmacology of hypothalamus. Mailing Add: Dept of Psychol Sci Purdue Univ Lafayette IN 47907

MYERS, ROBERT FREDERICK, b Trenton, NJ, Feb 23, 16; m 40; c 3. METEOROLOGY, CHEMICAL ENGINEERING. Educ: Va Polytech Inst, BS, 36. Prof Exp: Observer, US Weather Bur, 39-41, jr instrument engr, 41-42; sr inspector eng, Bur Ord, US Navy, 42-43; meteorologist, US Weather Bur, 45-57, liaison officer, Ga, 47-48, meteorologist, Tenn, 48-54, meteorologist in charge res, 54-55; chief data handling br, US Air Force Cambridge Res Lab, 55-60, SR RES ENGR, AIR FORCE GEOPHYS LAB, 60- Mem: Am Meteorol Soc. Res: Meteorological instrumentation including data acquisition, communication and display; micrometeorological research; ozone research; meteorological satellite ground stations with integral interactive computer. Mailing Add: 73 Lexington Ave Needham Heights MA 02194

MYERS, ROBERT LEE, organic chemistry, see 12th edition

MYERS, ROLLIE JOHN, JR, b Nebr, July 15, 24; m 50; c 2. PHYSICAL CHEMISTRY. Educ: Calif Inst Technol, BS, 47, MS, 48; Univ Calif, PhD(chem), 51. Prof Exp: From instr to assoc prof, 51-62, PROF CHEM, UNIV CALIF, BERKELEY, 62-, ASST DEAN COL CHEM, 73-, PRIN INVESTR, INORG MAT RES DIV, 73- Concurrent Pos: Guggenheim fel, 57-58; int fac award, Am Chem Soc-Petrol Res Fund, 65-66. Mem: Am Chem Soc; Am Phys Soc. Res: Spectroscopy; magnetic resonance; microwave and molecular structure. Mailing Add: Dept of Chem Univ of Calif Berkeley CA 94720

MYERS, RONALD ELWOOD, b Chicago Heights, Ill, Sept 24, 29; m 57; c 4. NEUROLOGY. Educ: Univ Chicago, AB, 50, PhD(neuroanat), 55, MD, 56. Prof Exp: Intern, Univ Chicago Clins, 56-57; res officer, Walter Reed Army Inst Res, 57-60; dir, Lab Neurol Sci, Spring Grove State Hosp, Baltimore, Md, 63-64; CHIEF LAB PERINATAL PHYSIOL, NAT INST NEUROL DIS & STROKE, 64- Concurrent Pos: Spec fel physiol & neurol med, Sch Med, Johns Hopkins Univ, 60-63. Mem: Am Asn Anat; Am Physiol Soc; Am Acad Neurol; Soc Gynec Invest; Pavlovian Soc NAm. Res: Physiological pyschology; fiber connections of the brain; experimental neuropathology; perinatalogy. Mailing Add: Lab of Perinatal Physiol NIH 9000 Rockville Pike Bethesda MD 20014

MYERS, RONALD FENNER, b East Haven, Conn, July 22, 30; m 59; c 3. NEMATOLOGY. Educ: Univ Conn, BS, 57, MS, 59; Univ Md, PhD(plant path, nematol), 64. Prof Exp: Nematologist, USDA, Md, 59-61; asst bot, Univ Md, 61-63; asst prof, Univ Conn, 64-65; from asst prof to assoc prof nematode physiol, 65-75, PROF PLANT PHYSIOL, RUTGERS UNIV, NEW BRUNSWICK, 75- Concurrent Pos: Res prof, Mem Univ Nfld, 73-74. Mem: Soc Nematol; Soc Europ Nematol. Res: Physiology and biochemistry of nematodes; culture and nutrition of nematodes; nematode detection and control recommendations. Mailing Add: Dept of Plant Biol Cook Col Rutgers Univ New Brunswick NJ 08903

MYERS, ROY MAURICE, b Scottdale, Pa, Sept 24, 11; m 39; c 4. BOTANY. Educ: Ohio State Univ, BSc, 34, MA, 37, PhD(plant physiol), 39. Prof Exp: Teacher high sch, Ohio, 34-35; asst bot, Ohio State Univ, 35-38 & Northwestern Univ, 38-40; instr, Boise Jr Col, 40-42; instr biol, Denison Univ, 42-45; from asst prof to assoc prof, 45-52, PROF BIOL, WESTERN ILL UNIV, 52-, CUR HERBARIUM, 69- Concurrent Pos: Chmn, Dept Biol Sci, Western Ill Univ, 53-69. Mem: AAAS; Am Inst Biol Sci; Soc Econ Bot. Res: Economic botany; plant taxonomy, flora of west-central Illinois. Mailing Add: 1416 Westview Dr Macomb IL 61455

MYERS, SAMUEL MAXWELL, JR, b Florence, SC, Jan 20, 43. SOLID STATE PHYSICS. Educ: Duke Univ, BS, 65, PhD(physics), 70. Prof Exp: Sandia Corp fel, 70-72, MEM STAFF, SANDIA LABS, 72- Mem: AAAS; Am Phys Soc. Res:

Magnetism in solids; nuclear magnetic resonance; solid hydrogen; physical metallurgy; corrosion; ion-solid interactions; surface physics. Mailing Add: Orgn 5111 Sandia Labs Albuquerque NM 87115

MYERS, SARAH KERR, b Rio de Janeiro, Brazil, June 18, 40; US citizen; m 61; c 2. CULTURAL GEOGRAPHY, GEOGRAPHY OF LATIN AMERICA. Educ: Oberlin Col, AB, 61; Stanford Univ, MA, 62; Univ Chicago, MA, 67, PhD(geog), 71. Prof Exp: ED, GEOG REV, AM GEOG SOC, 73- Mem: Asn Am Geogr; Conf Latin Am Geog; Am Anthrop Asn. Res: Problems of culture change experienced by rural-urban migrants; geography of languages; west coast South America. Mailing Add: Am Geog Soc Broadway at 156th St New York NY 10032

MYERS, THOMAS DEWITT, b Wilmington, Del, Apr 8, 38; m 61; c 3. INVERTEBRATE ZOOLOGY, OCEANOGRAPHY. Educ: Bridgewater Col, BA, 59; Univ NC, MA, 65; Duke Univ, PhD(zool), 68. Prof Exp: Fishery biologist, Biol Lab, US Bur Commercial Fisheries, 61-62; ASST PROF BIOL SCI, UNIV DEL, 67-, ADJ ASST PROF MARINE SCI, 73- Mem: AAAS; Am Inst Biol Sci; Am Soc Limnol & Oceanog; Marine Biol Asn UK. Res: Biology and distribution of pelagic molluscs; migration of marine animals; physiological tolerances of zooplankton; marine pollution and ocean waste disposal. Mailing Add: Col of Marine Studies Univ of Del Newark DE 19711

MYERS, VERNON WORK, b New Castle, Pa, Feb 16, 19; m 47; c 5. PHYSICS. Educ: Geneva Col, BS, 40; Syracuse Univ, MA, 42; Yale Univ, PhD(physics), 47. Prof Exp: Instr physics, Yale Univ, 43-44; physicist, Naval Res Lab, DC, 44 & Argonne Nat Lab, 47-48; from asst prof to assoc prof physics, Pa State Univ, 48-63; RES PHYSICIST, REACTOR RADIATION DIV, NAT BUR STANDARDS, 63- Concurrent Pos: Guest scientist, Brookhaven Nat Lab, 52-53, 60-61 & 63-66; Fulbright prof, Univ Philippines, 61-62. Mem: AAAS; Am Phys Soc. Res: Molecular quantum mechanics; experimental neutron physics. Mailing Add: Reactor Radiation Div Nat Bur of Standards Washington DC 20234

MYERS, WALTER LOY, b Joliet, Ill, Mar 13, 33; m 59. IMMUNOLOGY. Educ: Univ Ill, BS, 55, DVM, 57, MS, 59; Univ Wis, PhD(vet sci), 61. Prof Exp: From asst prof to prof vet path & hyg, Univ Ill, Urbana, 61-73; PROF IMMUNOL, SCH MED, SOUTHERN ILL UNIV, SPRINGFIELD, 73- Mem: AAAS; Am Vet Med Asn; Conf Res Workers Animal Dis; Am Asn Immunologists; Am Soc Vet Allergists & Immunologists. Res: The induction of immunological tolerance to transplantation antigens; the role of cellular immunity in neoplasia; clinical immunology. Mailing Add: Southern Ill Univ Sch of Med PO Box 3926 Springfield IL 62708

MYERS, WARREN POWERS LAIRD, b Philadelphia, Pa, May 2, 21; m 44; c 4. INTERNAL MEDICINE, ONCOLOGY. Educ: Yale Univ, BS, 43; Columbia Univ, MD, 45; Univ Minn, MS, 52; Am Bd Internal Med, dipl. Prof Exp: Intern, Philadelphia Gen Hosp, 45-46; intern med, Maimonides Hosp, NY, 48-49; from asst prof to assoc prof, 54-68, PROF MED, MED COL CORNELL UNIV, 68- Concurrent Pos: Fel, Mem Hosp, NY, 48; Eleanor Roosevelt Found fel, Cambridge, Eng, 62-63; clin asst, Mem Hosp, NY, 52-54, from asst attend physician to attend physician, 54-59, assoc chmn dept med, 64-67, chmn dept med, 68-; asst, Sloan-Kettering Inst Cancer Res, 52-56, assoc, 56-60, assoc mem, 60-69, mem, 69-, head metab & renal studies sect, Div Clin Invest, 57-66, head calcium metab lab, 67-; asst attend physician, NY Hosp, 59-68, attend physician, 68-; vis physician, Bellevue Hosp, 60-68; consult, Grasslands Hosp, Valhalla, 66-68; mem clin cancer training comt, Nat Cancer Inst, 70-73, chmn, 71-73, mem & chmn clin cancer educ comt, 75-78. Mem: AAAS; Endocrine Soc; Harvey Soc; AMA; Am Diabetes Asn. Res: Medical oncology; calcium metabolism; clinical endocrinology. Mailing Add: Mem Hosp 1275 York Ave New York NY 10021

MYERS, WAYNE LAWRENCE, b Adrian, Mich, Sept 17, 42; m 62; c 3. FORESTRY, BIOMETRY. Educ: Univ Mich, Ann Arbor, BS, 64, MF, 65, PhD(forestry), 67. Prof Exp: Res scientist, Forest Res Lab, Can Dept Fisheries & Forestry, Ont, 66-69; asst prof, 69-74, ASSOC PROF FORESTRY, MICH STATE UNIV, 74- Mem: Soc Am Foresters; Am Soc Photogram; Am Statist Asn. Res: Forest biometry; quantitative ecology; remote sensing. Mailing Add: Dept of Forestry Mich State Univ East Lansing MI 48823

MYERS, WILLIAM GRAYDON, b Toledo, Ohio, Aug 7, 08; m 40. MEDICAL BIOPHYSICS, CHEMISTRY. Educ: Ohio State Univ, BA, 33, MSc, 37, PhD(phys chem), 39, MD, 41. Prof Exp: Asst chem, 33-37, intern med, 41-42, res assoc chem, Res Found, 42-43, res assoc bact, 43-45, assoc prof med biophys, 49-53, STONE RES PROF MED BIOPHYS, COL MED OHIO STATE UNIV, 53- Concurrent Pos: Stone fel med biophys, Col Med, Ohio State Univ, 45-49; consult, Oak Ridge Nat Lab, 49- & US Naval Med Sch, 52-64; lectr, Int Cong Radiol, Munich, 59; consult, Oak Ridge Inst Nuclear Studies, 64-; consult, Inst Nuclear Med, Heidelberg, 66-; vis prof nuclear med, Univ Calif, Berkeley, 70- Honors & Awards: Lucy Wortham James Award, James Ewing Soc, 66; Aebersold Award, Soc Nuclear Med, 73. Mem: Am Chem Soc; Am Physiol Soc; Am Asn Cancer Res; Soc Nuclear Med (historian); Am Asn Physicists in Med. Res: Development of cobalt-60, gold-198, chromium-51 as substitutes for radium and radon in therapy, and iodine-121, iodine-123, iodine-125, iodine-126, strontium-85m, strontium-87m, potassium-38 and carbon-11 for diagnosis and fundamental studies; synthesis of radioactively labeled compounds for diagnosis. Mailing Add: 2724 Wexford Rd Columbus OH 43221

MYERS, WILLIAM HOWARD, b Dodge Co, Nebr, Nov 17, 08; m 32; c 3. MATHEMATICS. Educ: Stanford Univ, AB, 34, PhD(math), 39; Univ Calif, MA, 35. Prof Exp: Asst math, Stanford Univ, 35-36, instr, 36-39; instr, Univ Utah, 39-40; from asst prof to prof, 40-74, EMER PROF MATH, SAN JOSE STATE UNIV, 74- Mem: Math Asn Am. Res: Linear groups; algebra; analysis. Mailing Add: 2352 Sunny Vista Dr San Jose CA 95128

MYERS, WILLIAM HOWARD, b Oak Ridge, Tenn, Jan 26, 46; m 67; c 2. INORGANIC CHEMISTRY. Educ: Houston Baptist Col, BA, 67; Univ Fla, PhD(chem), 72. Prof Exp: Instr chem, Univ Fla, 68-69; fel, Ohio State Univ, 72-73; ASST PROF CHEM, UNIV RICHMOND, 73- Mem: Am Chem Soc; Sigma Xi. Res: Halogenation reactions of amine-boranes and amine-alanes and investigations of steric effects on reactivity in such systems. Mailing Add: Dept of Chem Univ of Richmond Richmond VA 23173

MYERS, WILLIAM HUNTER, physical oceanography, see 12th edition

MYERSON, ALBERT LEON, b New York, NY, Nov 14, 19; m 53; c 3. PHYSICAL CHEMISTRY. Educ: Pa State Univ, BS, 41; Univ Wis, PhD(phys chem), 48. Prof Exp: Asst org chem, Off Sci Res & Develop & Nat Defense Res Comt, Columbia Univ, 41-42, org chem & phys chem, Manhattan Proj, SAM labs, 42-45; asst chem, Univ Wis, 46-48; mem staff, Franklin Inst, 48-56; mgr phys chem, Missile & Space Vehicle Dept, Gen Elec Co, 56-60; prin res phys chemist, Cornell Aeronaut Lab, Inc, 60-69; RES ASSOC, EXXON RES & ENG RES LABS, 69- Mem: Am Chem Soc; Am Phys Soc; Combustion Inst; fel Am Inst Chemists. Res: Mechanisms and spectroscopy of gaseous explosions; gas-phase chemical kinetics; shock-tube reaction rate studies; non-steady state catalytic recombination mechanisms; physical chemistry of air pollution; isotopic separation. Mailing Add: Corp Res Labs EXXON Res & Engr Co Linden NJ 07036

MYERSON, PAUL GRAVES, b Boston, Mass, Sept 15, 14; m 38; c 4. PSYCHIATRY. Educ: Harvard Univ, AB, 35, MD, 39; Am Bd Psychiat & Neurol, dipl, 46. Prof Exp: Resident neurol, Mt Sinai Hosp, New York, 41-42; resident psychiat, NY State Psychiat Inst, 42-43; instr neurol, 46-47, ASST NEUROL, SCH MED, TUFTS UNIV, 47-, PROF PSYCHIAT, 62-, CHMN DEPT, 63- Concurrent Pos: Res assoc, Boston Psychopath Hosp, 46-47; from instr to assoc prof psychiat, Sch Med, Tufts Univ, 46-62; instr, Harvard Med Sch, 48-55; physician in charge dept psychiat, Boston Dispensary, 48-, consult phys med, 51-; assoc, Beth Israel Hosp, 49-; training analyst, Boston Psychoanal Soc & Inst, 59-, pres, 72-; psychiatrist in chief, New Eng Med Ctr Hosps, 63-; area dir, Tufts Ment Health Ctr, 68. Mem: AAAS; Am Psychiat Asn; Am Psychoanal Asn. Res: Doctor patient relationship; psychotherapeutic process. Mailing Add: Dept of Psychiat Tufts Univ Sch of Med Boston MA 02111

MYERSON, RALPH M, b New Britain, Conn, July 21, 18; m 43; c 2. INTERNAL MEDICINE. Educ: Tufts Univ, BS, 38, MD, 42. Prof Exp: Intern, Boston City Hosp, 42-43, resident med, 46-48; instr, Sch Med, Tufts Univ, 47-48; ward physician, Vet Admin Hosp. Wilmington, Del, 48-52; asst chief med serv, Vet Admin Hosp, Philadelphia, 53-67, chief med serv, 67-72; clin prof, 57-66, prof med, 66-75, CLIN PROF MED, MED COL PA, 75; ASSOC DIR, CLIN SERV DEPT, SMITH, KLINE & FRENCH LABS, PA, 75- Concurrent Pos: Chief staff, Vet Admin Hosp, Philadelphia, 72-75. Mem: Fel Am Col Physicians; Am Fedn Clin Res; Int Soc Internal Med; fel Am Col Gastroenterol; Am Gastroenterol Asn. Res: Alcoholism; liver disease; application of new diagnostic techniques and evaluation in clinical medicine; gastroenterology; hepatology. Mailing Add: Smith Kline & French Labs 1500 Spring Garden St Philadelphia PA 19101

MYHILL, JOHN, b Birmingham, Eng, Aug 11, 23; m 69; c 4. MATHEMATICS. Educ: Harvard Univ, PhD(philos), 49. Prof Exp: Instr philos, Vassar Col, 48-49, Temple Univ, 49-51 & Yale Univ, 51-53; asst prof, Univ Chicago, 53-54; from asst prof to assoc prof, Univ Calif, 54-60; prof philos & math, Stanford Univ, 60-63 & Univ Ill, 63-66; PROF MATH, STATE UNIV NY BUFFALO, 66- Concurrent Pos: Guggenheim fel, 53-54; mem, Inst Advan Study, 57-59 & 63; res assoc, Moore Sch Elec Eng & assoc prof, Univ Pa, 59; vis prof, Univ Mich, 69-70 & Univ Leeds, 71-72; consult, air weapons res, Univ Chicago, 56-57, NSF, 57-59 & 68-75, Stanford Res Inst, 60, Hughes Aircraft Corp, 66-68 & IBM Corp, 68-70. Mem: Am Math Soc. Res: Constructive foundations of mathematics; computer music; automata theory; philosophy of science; mathematical linguistics; cognitive psychology. Mailing Add: Dept of Math State Univ of NY Buffalo NY 14226

MYHR, BRIAN CECIL, molecular biophysics, see 12th edition

MYHRE, BYRON ARNOLD, b Fargo, NDak, Oct 22, 28; m 53; c 2. PATHOLOGY, IMMUNOHEMATOLOGY. Educ: Univ Ill, BS, 50; Northwestern Univ, MS, 52, MD, 53; Univ Wis, PhD, 62. Prof Exp: Resident path, Univ Wis, 56-60; asst prof med microbiol & immunol, Sch Med, Marquette Univ, 62-64, asst prof path, 64-66; assoc clin prof, Univ Southern Calif, 66-69, assoc prof path, 69-72; PROF PATH, UNIV CALIF, LOS ANGELES, 72-; DIR, BLOOD BANK, HARBOR GEN HOSP, 72- Concurrent Pos: Nat Inst Arthritis & Metab Dis fel, Univ Wis, 60-62; res grants, Ortho Found, 63-64, NIH, 65-66 & 72-73, Am Nat Red Cross, 67-68, Nat Heart Inst, 73-75; assoc med dir, Milwaukee Blood Ctr, Wis, 62-65; sci dir, Los Angeles-Orange Counties Res Cross Blood Ctr, 66-72. Mem: Am Soc Clin Path; Am Soc Exp Path; Am Asn Blood Banks. Res: Blood banking; immunopathology; histochemistry; cryobiology. Mailing Add: 1000 W Carson Torrance CA 90509

MYHRE, DAVID V, b Lloydminster, Sask, Jan 4, 32; m 57; c 2. FOOD CHEMISTRY. Educ: Concordia Col, Moorhead, Minn, BA, 54; NDak State Univ, MS, 55; Univ Minn, PhD(biochem), 62. Prof Exp: RES BIOCHEMIST, MIAMI VALLEY LABS, PROCTER & GAMBLE CO, 62- Mem: Am Chem Soc. Res: Flavor and food chemistry. Mailing Add: Miami Valley Labs Procter & Gamble Co Cincinnati OH 45239

MYHRE, DONALD L, soil science, see 12th edition

MYHRE, JANET M, b Tacoma, Wash, Sept 24, 32; m 54; c 1. MATHEMATICAL STATISTICS. Educ: Pac Lutheran Univ, BA, 54; Univ Wash, MA, 56; Univ Stockholm, Fil Lic, 68. Prof Exp: Res engr, Boeing Co, 56-58; lectr math, Harvey Mudd Col, 61-62; PROF MATH, CLAREMONT MENS COL, 62- Concurrent Pos: Consult, US Navy, 68-; assoc ed, Technometrics, 70-75; guest prof math, Univ Stockholm, 71-72; pres, Math Anal Res Corp, 74-; dir, Inst Decision Sci, 75- Mem: Am Statist Asn; Inst Math Statist. Res: Reliability theory. Mailing Add: Claremont Mens Col Dept of Math Claremont CA 91711

MYHRE, PHILIP C, b Tacoma, Wash, Mar 13, 33; m 54; c 1. ORGANIC CHEMISTRY. Educ: Pac Lutheran Univ, BA, 54; Univ Wash, PhD(chem), 58. Prof Exp: NSF fel, Nobel Inst Chem, Stockholm, 58-60; from asst prof to assoc prof, 60-69, PROF CHEM, HARVEY MUDD COL, 69- Concurrent Pos: Vis assoc, Calif Inst Technol, 67-68; guest prof, Swiss Fed Inst Technol, 71-72; chmn dept chem, Harvey Mudd Col, 74-75. Mem: AAAS; Am Chem Soc. Res: Mechanisms of organic reactions; nuclear magnetic resonance spectroscopy. Mailing Add: Dept of Chem Harvey Mudd Col Claremont CA 91711

MYHRE, RICHARD JOHN, b Tacoma, Wash, Dec 11, 20; m 49; c 4. FISHERIES MANAGEMENT. Educ: Univ Wash, BS, 50, MS, 60. Prof Exp: Jr scientist, 49-54, asst scientist, 54-55, assoc scientist, 55-57, biologist, 57-64, sr biologist, 64-70, ASST DIR, RESOURCE MGT, INT PAC HALIBUT COMN, 70- Concurrent Pos: Consult, Int NPac Fisheries Comn, 63-71. Mem: AAAS; Am Inst Fishery Res Biologists; Am Fisheries Soc; Biomet Soc. Res: Population dynamics; tagging studies; biometrics; mortality estimation; population models. Mailing Add: PO Box 9 University Station Seattle WA 98105

MYINT, THAN, b Moulmein, Burma, June 5, 18; US citizen; m 66; c 3. BIOCHEMISTRY. Educ: Univ Rangoon, BS, 41; Utah State Univ, MS, 48, PhD(biochem), 50. Prof Exp: Res assoc nutrit, Wash State Univ, 50-51; liaison personnel, US Tech Corp Admin, Burma, 52-53; pub analyst, Dept Pub Health, Govt of Union of Burma, 53-57; res asst chem, Utah State Univ, 57-58; res assoc radiation biol, Mass Inst Technol, 58-59; res fel radiol, Mass Gen Hosp, Boston, 59; protein chemist, Armour & Co, Ohio, 59-60; biochemist, Case Western Reserve Univ, 60-68; HEAD CLIN CHEM DIV, ST VINCENT CHARITY HOSP, 68-; DIR CLIN CHEM, LINCOLN HOSP, ALBERT EINSTEIN COL MED, 68- Concurrent Pos: Lectr, Mass Ed Ctr, Burma, 52-53; Sch Health Asst, Burma, 57. Honors & Awards: Cleveland Obstet & Gynec Soc Award, 68. Mem: AAAS; Am Chem Soc; Am Asn Clin Chem. Res: Clinical biochemistry; microbiology chemistry; vitamins;

methodology. Mailing Add: Lincoln Hosp Albert Einstein Col of Med New York NY 10454

MYINT-U, TYN, b Mandalay, Burma, Jan 1, 32; m 60; c 3. MATHEMATICS. Educ: Univ Mich, BSE, 55, MSE, 57; NY Univ, PhD(aeronaut, astronaut), 64. Prof Exp: Res asst aeronaut & instr astronaut, NY Univ, 61-64; asst prof, 65-70, ASSOC PROF MATH, MANHATTAN COL, 70- Concurrent Pos: Adj assoc prof, Lehman Col, 71-72. Mem: Am Math Soc; Soc Indust & Appl Math; NY Acad Sci; fel Brit Interplanetary Soc. Res: Space science; applied mathematics; satellite theory and partial differential equations. Mailing Add: Dept of Math Manhattan Col Bronx NY 10471

MYKLEBY, RAY W, b Crookston, Minn, June 20, 18; m 42; c 3. FOOD TECHNOLOGY. Educ: Univ Minn, BS, 41; Pa State Univ, MS, 42. Prof Exp: Milk sanitarian, Pittsburgh Dairy Coun, 42-44; mem staff, Med Dept, US Army, 44-46; lab technician, 46-48, lab supvr, 48-58, lab mgr, 58-65, DIR RES & PROD DEVELOP, LAND O'LAKES, INC, 65-, VPRES RES & TECHNOL, 72- Concurrent Pos: Mem res comn, Am Butter Inst. Mem: Am Dairy Sci Asn; Inst Food Technologists; Dry Milk Inst. Res: Technology of production of dehydrated dairy products for use in food processing; technology of production of sodium caseinate and concentrated milk proteins. Mailing Add: Cent Res & Prod Develop Labs Land O'Lakes Inc PO Box 116 Minneapolis MN 55440

MYKOLAJEWYCZ, ROMAN, physical chemistry, see 12th edition

MYLES, WILLIAM JOHN, b Millerton, NY, Oct 18, 13; m 41; c 2. ORGANIC CHEMISTRY. Educ: City Col New York, BS, 42. Prof Exp: Chemist, Celanese Corp, 43-56, group leader textile finishing, 56-60, head dyeing & finishing div, NC, 60-65, res assoc, 66-70; supvr finishing, Am Cyanamid Co, 70-75; SUPVR APPLNS LAB, APEX CHEM CO, 75- Mem: Am Chem Soc; Am Asn Textile Chemists & Colorists. Res: Dyeing and finishing; flammability; photochemistry; surface chemistry. Mailing Add: Apex Chem Co 200 S First St Elizabeth Port NJ 07206

MYODA, TOSHIO TIMOTHY, b Mukden, Manchuria, Mar 17, 29; m 63; c 2. MICROBIOLOGY, BIOCHEMISTRY. Educ: Hokkaido Univ, BS, 49, MS, 52; Iowa State Univ, PhD(bact), 59. Prof Exp: Asst microbiol, Hokkaido Univ, 52-54, instr, 54-59; asst bact, Iowa State Univ, 56-59; res fel, Nat Res Coun Can, 59-60; res fel microbiol, Western Reserve Univ, 60-64; chief microbiol, Inst Microbial Chem, Japan, 64-66; res assoc & instr, La Rabida-Univ Chicago Inst, 66-67; assoc chief microbiol & immunochemist & affil med staff, 67-73, CHIEF MICROBIOL DEPT, ALFRED I DU PONT INST, 73- Concurrent Pos: Vis prof, Valparaiso Univ, 67. Mem: AAAS; Am Soc MIcrobiol; The Biochem Soc; Am Chem Soc; Am Inst Biol Sci. Res: Molecular biology of group A streptococcal M antigens; bacterial cell wall structures. Mailing Add: Dept of Microbiol Aflred I du Pont Inst Wilmington DE 19899

MYRBERG, ARTHUR AUGUST, JR, b Chicago Heights, Ill, June 28, 33; c 2. ANIMAL BEHAVIOR, MARINE BIOLOGY. Educ: Ripon Col, BS, 54; Univ Ill, MS, 58; Univ Calif, Los Angeles, PhD(zool), 61. Prof Exp: Asst, Ill Natural Hist Surv, 57; asst zool, - Univ Ill, 57-58 & Univ Calif, Los Angeles, 58-61; fel, Max Planck Inst Behav Physiol, Seewieson, Ger, 61-64; from asst prof to assoc prof, 64-71, PROF MARINE SCI, UNIV MIAMI, 71- Mem: Am Soc Zoologists; Am Soc Ichthyol & Herpet; Animal Behav Soc; Am Inst Biol Sci; fel Am Inst Fishery Res Biologists. Res: Ichthyology; comparative behavior of fishes, particularly those of tropical waters with emphasis on the families Cichlidae and Pomacentridae; underwater acoustics and its biological significance; sensory physiology and behavior of sharks. Mailing Add: Div of Biol & Living Resources Rosenstiel Sch Marine & Atmos Sci Univ of Miami Miami FL 33149

MYRES, MILES TIMOTHY, b London, Eng, May 16, 31. ECOLOGY, ORNITHOLOGY. Educ: Univ Cambridge, BA, 53, MA, 58; Univ BC, MA, 57, PhD(zool), 60. Prof Exp: Res officer, Edward Grey Inst Field Ornith, Oxford Univ, 59-61; asst prof biol, Lakehead Col, 62-63; asst prof zool, 63-69, ASSOC PROF ZOOL, UNIV CALGARY, 69- Concurrent Pos: Can Wildlife Serv res grant, 67-68; nat dir, Can Nature Fedn, 72-74; mem coun, Pac Seabird Group, 73-74. Mem: Am Ornithologists Union; Can Soc Environ Biol; Brit Inst Biol; Brit Ornithologists Union. Res: Bird migration; ecology of birds; North Pacific seabirds; man-land-fauna interactions; cultural aspects of environmental conservation at different times in history; relationship between amateur naturalists and professional biologists. Mailing Add: Dept of Biol Univ of Calgary Calgary AB Can

MYRIANTHOPOULOS, NTINOS, b Cyprus, July 24, 21; m 55; c 3. HUMAN GENETICS. Educ: George Washington Univ, BS, 52; Univ Minn, MS, 54, PhD(genetics), 57. Prof Exp: Res assoc neurol, Univ Ill, 55-57; res geneticist, 57-63, proj consult, 55-57, HEAD SECT EPIDEMIOL & GENETICS, DEVELOP NEUROL BR, NAT INST NEUROL & COMMUN DIS & STROKE, 63- Concurrent Pos: Assoc prof neurol, George Washington Univ, 58-; instr grad prog, NIH, 58-; dir, Genetic Counseling Ctr, 58- Mem: Am Soc Human Genetics; NY Acad Sci; Soc Study Social Biol; Teratology Soc. Res: Genetics and epidemiology of neurological disease; congenital malformations, prevalence, incidence and mutation rates; metabolic etiology; clinical and behavioral genetics. Mailing Add: Develop Neurol Br Inst of Neurol & Commun Dis & Stroke Bethesda MD 20014

MYRICK, ALVIN GRANT, b NC, Jan 22, 35; m 56; c 2. MATHEMATICS. Educ: NC State Univ, BS, 62; Univ NC, Chapel Hill, MA, 63; Duke Univ, EdD(math educ), 69. Prof Exp: Instr math & physics, High Point Col, 65-67, assoc prof math & head dept, 69-72; DIR MATH & SCI, BUNCOMBE COUNTY SCHS, 73- Mem: Math Asn Am; Nat Coun Teachers Math; Nat Sci Teachers Asn. Res: Using space mathematics to motivate learning. Mailing Add: 27 Fieldcrest Circle Asheville NC 28806

MYRON, DUANE R, b Birmingham, Ala, Jan 3, 43; m 62; c 2. BIOCHEMISTRY, NUTRITION. Educ: Univ NDak, BS, 65, PhD(biochem), 70. Prof Exp: Fel biochem, St Jude Children's Res Hosp, 70-72 & Mem Univ Nfld, 72-73; fel biochem & nutrit, 73-74, RES ASSOC BIOCHEM & NUTRIT, HUMAN NUTRIT LAB, USDA AGR RES SERV, UNIV NDAK, 74- Mem: Sigma Xi; AAAS; Am Chem Soc. Res: Biochemical and metabolic role of trace mineral nutrients in laboratory animals and in humans. Mailing Add: USDA Human Nutrit Lab Univ of NDak Dept of Biochem Grand Forks ND 58201

MYRON, HAROLD WILLIAM, b New York, NY, Apr 28, 47. THEORETICAL SOLID STATE PHYSICS. Educ: City Univ New York, BA, 67; Iowa State Univ, PhD(physics), 72. Prof Exp: Res asst physics, Ames Lab, US AEC, 69-72; res assoc, Dept Physics, Magnetic Theory Group, Northwestern Univ, 72-74, vis asst prof, 74-75; SCIENTIST PHYSICS, INST METAL PHYSICS, UNIV NIJMEGEN, NETH, 75- Mem: Am Phys Soc. Res: Ab initio calculations of electron and phonon properties in metals including dielectric functions of transition metal compounds. Mailing Add: Inst for Metal Physics Univ of Nijmegen Nijmegen Netherlands

MYRON, JAMES JOHN, physical chemistry, see 12th edition

MYRVIK, QUENTIN NEWELL, b Minneota, Minn, Nov 9, 21; m 44; c 1. MICROBIOLOGY, IMMUNOLOGY. Educ: Univ Wash, BS, 48, MS, 50, PhD(microbiol), 52. Prof Exp: From asst prof to assoc prof microbiol, Univ Va, 52-63; PROF MICROBIOL & CHMN DEPT, BOWMAN GRAY SCH MED, 63- Mem: Am Soc Microbiol; Am Asn Immunol; Reticuloendothelial Soc; Am Acad Microbiol. Res: Mechanisms of natural immunity to infectious agents and emphasis on role of lysozyme; cellular immunity with emphasis on the structure and function of alveolar macrophages. Mailing Add: Dept of Microbiol Bowman Gray Sch of Med Winston-Salem NC 27103

MYSAK, LAWRENCE ALEXANDER, b Saskatoon, Sask, Jan 22, 40. APPLIED MATHEMATICS, PHYSICAL OCEANOGRAPHY. Educ: Univ Alta, BSc, 61; Univ Adelaide, MSc, 63; Harvard Univ, AM, 64, PhD(appl math), 66. Prof Exp: US Navy fel, Harvard Univ, 66-67; asst prof, 67-70, ASSOC PROF MATH, UNIV BC, 70- Concurrent Pos: Vis res assoc, Ore State Univ, 68; sr visitor, Cambridge Univ, 71-72; travel fel, Nat Res Coun Can, 71-; Siam vis lectr, Soc Indust & Appl Math, 75-76. Mem: Soc Indust & Appl Math; Can Meteorol Soc. Res: Analytical and probabilistic methods of applied mathematics; waves and currents in rotating stratified fluids. Mailing Add: Dept of Math Univ of BC Vancouver BC Can

MYSELS, ESTELLA KATZENELLENBOGEN, b Berlin, Ger, Jan 12, 21; nat US; m 53. CHEMISTRY. Educ: Univ Calif, BS, 42, PhD(chem), 46. Prof Exp: Chemist, Richfield Oil Co, 42-44; instr chem, Univ Calif, 46-47 & Univ Southern Calif, 47-50; asst, Sloan-Kettering Inst Cancer Res, 50-53; res assoc chem, Univ Southern Calif, 54-66; lectr chem, Salem Col, 67, prof, 67-70; lectr chem, Univ Calif, San Diego, 75. Concurrent Pos: Chmn dept chem, Salem Col, 68-70. Mem: Am Chem Soc. Res: Infrared and ultraviolet spectroscopy; electrophoresis; surfactant solutions; organic and physical chemistry. Mailing Add: 8327 LaJolla Scenic Dr La Jolla CA 92037

MYSELS, KAROL JOSEPH, b Krakow, Poland, Apr 14, 14; nat US; m 53. COLLOID CHEMISTRY, SURFACE CHEMISTRY. Educ: Univ Lyon, Lic es sc, 37; Harvard Univ, PhD(inorg chem), 41. Prof Exp: Asst chem Stanford Univ, 41-42, res assoc, 43-45; instr, NY Univ, 45-47; from asst prof to prof, Univ Southern Calif, 47-66; assoc dir res, R J Reynolds Industs, 66-70; SR RES ADV, GENERAL ATOMIC CO, 70- Concurrent Pos: Mem staff, Shell Develop Co, 40-42; NSF fac fel, 57-58, sr fel, Strasbourg Ctr Macromolecule Res, France, 62-63; Guggenheim fel, 65-66; Rennebohm lectr, Univ Wis, 64, Pharm Alumni lectr, 67; John Watson Mem Lectr, Va Polytech Inst, 68; assoc mem, Comn Colloid & Surface Chem, Int Union Pure & Appl Chem, 65-69, titular mem, 68-69 & 73, chmn, 73-, mem, Div Comt Phys Chem Div, 75-; Phi Lambda Upsillon Lectr, Univ Okla, 74; Am Chem Soc tour lectr, 71 & 74. Honors & Awards: Kendall Award, Am Chem Soc, 64. Mem: AAAS; Am Chem Soc (counr, 61-62 & 73-); The Chem Soc. Res: Surfactant solutions; surface tension; soap films; evaporation control; reverse osmosis; conductivity; electrophoresis; diffusion; rheology; intermolecular forces; gas-cooled nuclear reactor design and applications. Mailing Add: Gen Atomic Co PO Box 81608 San Diego CA 92138

MYSER, WILLARD C, b Cuyahoga Falls, Ohio, Apr 22, 23; m 43; c 4. ZOOLOGY. Educ: Kent State Univ, BS, 44; Ohio State Univ, MS, 47, PhD, 52. Prof Exp: Asst zool, 45-47, asst instr, 47-48, from instr to assoc prof, 48-61, asst chmn dept, 61-68, PROF ZOOL, OHIO STATE UNIV, 61- Concurrent Pos: Res assoc, Argonne Nat Lab, 56. Mem: Radiation Res Soc; Entom Soc Am. Res: Radiation biology; cytology. Mailing Add: Dept of Zool Ohio State Univ Columbus OH 43210

MYTON, BECKY ANN, b Johnstown, Pa, Jan 14, 42. ETHOLOGY, ECOLOGY. Educ: Allegheny Col, BS, 63; Univ Md, College Park, MS, 65, PhD(zool), 71. Prof Exp: Teaching asst zool, Univ Md, College Park, 63-70, instr, 70-71; ASST PROF BIOL, NAT UNIV HONDURAS, 71- Concurrent Pos: Lectr, USDA, 65-66. Mem: AAAS; Am Soc Mammal; Animal Behav Soc. Res: Utilization of space in small mammal communities; behavior and ecology of Central American mammals. Mailing Add: Dept of Biol Nat Univ of Honduras Tegucigalpa DC Honduras

MYTTON, JAMES W, b Kansas City, Mo, Feb 18, 27; m 69; c 1. GEOLOGY. Educ: Dartmouth Col, AB, 49; Univ Wyoming, MA, 51. Prof Exp: Geologist, US Geol Surv, 51-58; photogeologist, Knox, Bergman & Shearer, 61-63; GEOLOGIST, US GEOL SURV, 63- Mem: AAAS; Geol Soc Am; Am Asn Petrol Geologists; Geochem Soc. Res: Investigation of fissionable materials in sedimentary rocks; regional stratigraphic studies; photogeologic interpretation; evaluation of mineral resources on a regional scale; investigation of phosphate and related commodities; waste disposal studies; studies related to underground nuclear testing; geologic studies related to geothermal energy. Mailing Add: US Geol Surv Fed Ctr Denver CO 80225

N

NAAE, DOUGLAS GENE, b Graettinger, Iowa, Dec 24, 46; m 71; c 2. FLUORINE CHEMISTRY, SOLID STATE CHEMISTRY. Educ: Univ Iowa, BS, 69, MS, 71, PhD(org chem), 72. Prof Exp: Res assoc solid state org chem, Univ Minn, Minneapolis, 73-74; ASST PROF ORG CHEM, UNIV KY, 74- Mem: Am Chem Soc. Res: Solid-state reactions of organo-fluorine compounds, synthesis of polyfluorinated olefins and aromatics, halogenation reactions, x-ray crystallography and structure determinations. Mailing Add: Dept of Chem Univ of Ky Lexington KY 40506

NAAKE, HANS JOACHIM, b Leipzig, Ger, Jan 2, 25; nat; m 56; c 3. ACOUSTICS, SOLID STATE PHYSICS. Educ: Univ Göttingen, Dipl, 51, PhD(physics), 53. Prof Exp: Sci co-worker physics, Physics Inst, Univ Göttingen, 53-57; physicist, 57-69, MGR APPL PHYSICS, MAJOR APPLIANCE LABS, GEN ELEC CO, 69- Concurrent Pos: Adj prof, Univ Louisville, 69- Mem: Acoust Soc Am; Ger Phys Soc. Res: General acoustics; vibrations; sound propagation in liquids and solids; solid state and semiconductor physics; thermoelectricity, especially thermoelectric refrigeration. Mailing Add: Major Appliance Labs Gen Elec Co Louisville KY 40225

NABB, DALE PRESTON, b East St Louis, Ill, Jan 1, 37; m 60; c 4. BIOCHEMISTRY. Educ: Univ Mo, BS, 59, MS, 60, PhD(agr chem), 63. Prof Exp: Res chemist, Commun Dis Ctr, USPHS, 63-66; clin chemist, C W Long Mem Hosp, Atlanta, Ga, 66-71; EXEC DIR, PATHOLOGISTS' SERV, PROF ASSOCS INC, 71- Concurrent Pos: Dipl, Am Asn Clin Chemists, 75 & Nat Registry Clin Chemists. Mem: Am Asn Clin Chemists; fel Am Bd Clin Chemists. Res: Biochemical studies of the transport of dichloro-diphenyl-trichloroethane in mammals; dermal absorption of pesticides; effects of pesticides on amino acid metabolism; kinetic studies of acetylcholinesterase; relationship of nutrition and disease resistance. Mailing Add: PO Box 50122 Fed Annex Atlanta GA 30302

NABER, EDWARD CARL, b Mayville, Wis, Sept 12, 26; m 53; c 2. POULTRY NUTRITION. Educ: Univ Wis, BS, 50, MS, 52, PhD(biochem), 54. Prof Exp: Asst poultry husb, Univ Wis, 53-54; asst poultry nutritionist, Clemson Col, 54-56; from asst prof to assoc prof, 56-63, PROF POULTRY SCI, OHIO STATE UNIV, 63-, CHMN DEPT, 69- Concurrent Pos: Vis prof, Univ Wis, 64-65. Mem: AAAS; Am Chem Soc;

Poutry Sci Asn; Am Inst Nutrit; Am Inst Biol Sci. Res: Nutrition and metabolism in the avian species; vitamin metabolism and protein formation in the avian embryo; energy utilization in the chick; amino acid and lipid metabolism in the laying hen. Mailing Add: Dept of Poultry Sci Ohio State Univ Columbus OH 43210

NABER, JAMES ALLEN, b Buffalo, NY, Sept 6, 36; m 67; c 1. RADIATION PHYSICS, SOLID STATE PHYSICS. Educ: Canisius Col, BS, 58, Purdue Univ, MS, 61, PhD(radiation effects in germanium), 65. Prof Exp: Sr staff physicist & mgr radiation effects physics br, Gulf Radiation Technol, 65-72; mgr physics dept, 72-75, VPRES, IRT CORP, 75- Mem: Am Phys Soc. Res: Radiation effects in materials and systems; experimental solid state physics phenomena. Mailing Add: IRT Corp PO Box 80817 San Diego CA 92138

NABI, HOSNI ABDEL, b Cairo, Egypt, Nov 9, 38; m 63; c 2. PLANT GENETICS, PLANT BREEDING. Educ: Cairo Univ, BS, 59; La State Univ, MS, 63, PhD(plant breeding), 65. Prof Exp: Res assoc plant genetics, Dept Cotton Breeding, Ministry Agr, Egypt, 65-67; res dir plant breeding, Dunn Seed Co, 68-71; prof biol, Odessa Col, 70-75; MEM FAC SCI, MIDLAND COL, 75- Mem: Am Genetic Asn; Am Soc Agron. Res: Industrial research on development and improvement of cotton; the inheritance of quantitative and economic characters in field crops, with major emphasis on disease resistance, insect tolerance and fiber qualities in upland cotton. Mailing Add: Dept of Sci Midland Col Midland TX 79701

NABI, ISIDORE, b Brno, Czech, July 22, 10; m 30; c 6. POPULATION BIOLOGY. Educ: Cochabamba Univ, AB, 30; Nat Univ Mex, MD, 36. Hon Degrees: PhD, Cochabamba Univ, 50; LLB, Nat Univ Mex, 39. Prof Exp: Petrol geologist, Ministerio de Fomento, Venezuela, 40-42; instr biol, Hunter Col, 45-47; resident path, Kings County Hosp, Brooklyn, 47-49; ed & publisher, Boletin de Medicina Forensica, Caracas, 49-51; lectr & res assoc path, Univ Venezuela, 51-56; Guggenheim fel biol, Yeshiva Univ, 56-57; res assoc pharmacol, NY Univ, 62-65; res assoc anat, 65-67, evolutionary biol, 67-71, RES ASSOC BIOL, UNIV CHICAGO, 71- Concurrent Pos: Consult, Standard Oil Co, 45-47 & Kings County Coroner, 47-49; NIH res grant, 65. Mem: Soc Study Evolution; Am Col Legal Med; Int Acad Path. Res: Cytopathology; forensic cytology; paleocytopathology. Mailing Add: Dept of Biol Univ of Chicago Chicago IL 60637

NABIGHIAN, MISAC N, b Bucharest, Rumania, Dec 5, 31; US citizen; m 66; c 2. GEOPHYSICS. Educ: Mining Inst Bucharest, BS, 54; Columbia Univ, PhD(geophys), 67. Prof Exp: Geophysicist, Geol Comt, Bucharest, 55-57; res scientist, Geophys Inst, Rumanian Acad Sci, 57-62; res asst, Lamont Geol Observ, NY, 63-67; GEOPHYSICIST, NEWMONT EXPLOR LTD, 67- Mem: Soc Explor Geophys; Europ Asn Explor Geophys. Res: Theoretical research for the development of new exploration and interpretive techniques in geophysical processing. Mailing Add: Newmont Explor Ltd PO Box 1310 Danbury CT 06810

NABORS, CHARLES J, JR, b Cleveland, Ohio, Jan 11, 34; m 61; c 3. CYTOLOGY, ANATOMY. Educ: Wabash Col, AB, 55; Univ Utah, PhD(anat), 65. Prof Exp: Instr, 65-66, ASST PROF ANAT, COL MED, UNIV UTAH, 66- Concurrent Pos: Markle scholar acad med, 69-74. Mem: AAAS; Endocrine Soc; Reticuloendothelial Soc; Radiation Res Soc; NY Acad Sci. Res: Steroid hormone effects on fibroblasts and reticuloendothelial cells; steroid biochemistry; endocrinology; circadian rhythms; radiobiology. Mailing Add: Dept of Anat Univ of Utah Med Ctr Salt Lake City UT 84112

NABORS, JAMES BALUS, JR, chemistry, see 12th edition

NABORS, MURRAY WAYNE, b Carlisle, Pa, Oct 4, 43; m 66 PLANT PHYSIOLOGY, PLANT BREEDING. Educ: Yale Univ, BS, 65; Mich State Univ, PhD(bot), 70. Prof Exp: NSF grant & vis asst prof biol, Univ Ore, 70; asst prof, Univ Santa Clara, 70-72; ASST PROF BIOL, COLO STATE UNIV, 72- Mem: AAAS; Am Soc Plant Physiol. Res: Phytochrome; water relations; tissue culture. Mailing Add: Dept of Bot Colo State Univ Ft Collins CO 80521

NABRIT, SAMUEL MILTON, b Macon, Ga, Feb 21, 05; m. MORPHOLOGY, PHYSIOLOGY. Educ: Morehouse Col, BS, 25; Brown Univ, MS, 28, PhD(biol), 32. Prof Exp: Instr zool, Morehouse Col, 25-27, prof, 28-31; prof, Atlanta Univ, 32-55; pres, Tex Southern Univ, 55-66; comnr, US AEC, 66-67; EXEC DIR, SOUTHERN FELS FUND, 67- Concurrent Pos: Exchange prof, Atlanta Univ, 30, dean grad sch; Gen Educ Bd fel, Columbia Univ, 43; res fel, Univ Brussels, 50; coordr, Carnegie Exp Grant-in-Aid Res Prog; mem sci bd, NSF, 56-60; mem corp, Marine Biol Lab, Woods Hole. Mem: Inst of Med of Nat Acad Sci; Am Soc Zool; Soc Develop Biol; Nat Asn Res Sci Teaching; Nat Inst Sci (pres, 45). Res: Neuroembryology; role of fin rays in regeneration of tail-fins of fishes. Mailing Add: 686 Beckwith St SW Atlanta GA 30314

NACARRATO, WILLIAM FRANK, b New Kensington, Pa, Jan 23, 47; m 72; c 2. BIOCHEMISTRY, LIPID CHEMISTRY. Educ: Gannon Col, BS, 68; Univ Pittsburgh, PhD(biochem), 72. Prof Exp: Res assoc biochem, Mich State Univ, 72-74 & Ctr Theoret Biol, State Univ NY Buffalo, 74-76; SR RES ASSOC BIOCHEM, DEPT PHARMACOL & PHYSIOL, SCH DENT MED, UNIV PITTSBURGH, 76- Mem: AAAS; Am Chem Soc; Biophys Soc. Res: lipids associated with biological membranes. Mailing Add: 931 Dorchester Ave Pittsburgh PA 15226

NACE, DONALD MILLER, b Hanover, Pa, Nov 28, 24; m 45; c 2. PHYSICAL CHEMISTRY. Educ: Lehigh Univ, BS, 47, MS, 49; Pa State Univ, PhD(chem), 56. Prof Exp: Asst, Nat Printing Ink Res Inst, 47-49; res chemist, 49-58, sr res chemist, 58-74, RES ASSOC, MOBIL RES & DEVELOP CORP, 74- Mem: Am Chem Soc. Res: Catalysis in petroleum processing; kinetics of vapor phase reactions over heterogeneous catalysis. Mailing Add: Mobil Res & Develop Corp Res Dept Paulsboro NJ 08066

NACE, GEORGE WILLIAM, b Connellsville, Pa, Apr 1, 20; m 46; c 4. DEVELOPMENTAL BIOLOGY. Educ: Reed Col, BA, 43; Univ Calif, Los Angeles, MA, 48, PhD(embryol), 50. Prof Exp: Lab asst zool, Reed Col, 41-42; asst, Univ Calif, Los Angeles, 46-49; NIH fel, Brussels, Belg, 50-51; asst prof, Duke Univ, 51-57; assoc prof, 57-61, PROF ZOOL, UNIV MICH, ANN ARBOR, 61-, DIR AMPHIBIAN FACILITY, 68-, FEL, CTR HUMAN GROWTH & DEVELOP, 69- Concurrent Pos: Staff asst, Div Biol & Agr, Nat Res Coun, 52-53; vis prof, Kyoto Univ, 64, adv, 64-; consult, NIH, 65; vis prof, Hiroshima Univ, 66, 68, 70; vis lecturers' trust fund vis prof, Univ Witwatersrand, 71; chmn subcomt amphibian standards, Comt Standards, Inst Lab Animal Resources, Nat Acad Sci-Nat Res Coun. Mem: Soc Develop Biol; Fedn Am Scientists; Am Soc Zoologists; Am Asn Anatomists; fel NY Acad Sci. Res: Protein ontogeny in amphibia; development and maintenance of defined strains of amphibia; transfer of macromolecules from material organism to egg on conceptus; serology of neoplasia in the frog; role of specific macromolecules in amphibian fertilization. Mailing Add: Dept of Zool Univ of Mich Ann Arbor MI 48104

NACE, HAROLD RUSS, b Collingswood, NJ, July 5, 21; m 44. ORGANIC CHEMISTRY. Educ: Lehigh Univ, BS, 43; Mass Inst Technol, PhD(org chem), 48; Brown Univ, MS, 57. Prof Exp: Res chemist, Merck & Co, Inc, 44-45; from instr to assoc prof, 48-59, PROF CHEM, BROWN UNIV, 59- Concurrent Pos: Res chemist, Jackson Lab, E I du Pont de Nemours & Co, 56-57, consult, 57-60; consult, William S Merrell Co div, Richardson-Merrell Inc, 60-68 & Wyeth Labs, 68- Mem: AAAS; Am Chem Soc; fel Am Inst Chemists; fel NY Acad Sci. Res: Stereochemistry and partial synthesis of steroids; dipole moment studies of organic compounds; prostaglandin syntheses. Mailing Add: Dept of Chem Brown Univ Providence RI 02912

NACE, PAUL FOLEY, b Brooklyn, NY, May 6, 17; m 40; c 4. MARINE BIOLOGY, ENDOCRINOLOGY. Educ: Columbia Univ, AB, 38, MA, 46; NY Univ, PhD(histol), 51. Prof Exp: Asst zool, Univ NC, 41-42; asst, Columbia Univ, 46; asst prof biol, St John's Univ, 46-49; assoc anat, NY Med Col, 49-50, from asst prof to assoc prof, 50-56; from assoc prof zool & med res to prof, McMaster Univ, 56-65; dir, Clapp Lab, Battelle Mem Inst, 65-70; prof biol, Ind Univ, Indianapolis, 70-71; PROF BIOL, STATEN ISLAND COMMUNITY COL, 71- Concurrent Pos: Independent investr, Marine Biol Lab, Woods Hole, 60-61; staff mem, Dept Nuclear Med, St Joseph's Hosp, 60- Mem: Am Physiol Soc; Am Soc Zoologists. Res: Liver histochemistry; pathology and biochemistry of diabetes in fish, birds, mammals; inhalation radiocardiography; autoradiography; radiobiology. Mailing Add: Dept of Biol Sci Staten Island Community Col Staten Island NY 10301

NACE, RAYMOND LEE, b Los Angeles, Calif, Oct 13, 07; m 35; c 2. HYDROLOGY. Educ: Univ Wyo, BS, 35, MA, 36; Columbia Univ, PhD(geol), 60. Prof Exp: Asst geol, Geol Surv, Wyo, 35-36; instr, Univ Wyo, 37-38 & Yale Univ, 39-40; jr geologist, US Geol Surv, WVa, 41-42, asst geologist, 42-46, assoc geologist, Nebr, 46, dist geologist, Groundwater Br, Idaho, 46-56, regional coordr, Pac Northwest, 51-56, asst chief, Water Resources Div, 56-57, assoc chief, 57-63, RES HYDROLOGIST, WATER RESOURCES DIV, US GEOL SURV, 64- Concurrent Pos: Chmn, US Nat Comt, Int Hydrol Decade, Nat Acad Sci, 64-66, US rep, UNESCO Coord Coun, 65-69; adj prof civil eng, NC State Univ, 74- Mem: AAAS; Asn Eng Geol; fel Geol Soc Am; Am Water Works Asn; Am Geophys Union. Res: Groundwater hydrology of Idaho; tertiary stratigraphy of Rocky Mountain area and of Snake River Plain; radioelements in natural water; general hydrology of the United States; underground disposal of industrial wastes; history of hydrology; world water balance. Mailing Add: 4715-F Edwards Mill Rd Raleigh NC 27612

NACHBAR, MARTIN STEPHEN, b New York, NY, July 17, 37; m 62; c 1. MEDICAL MICROBIOLOGY. Educ: Union Col, BS, 58; NY Univ, MD, 62. Prof Exp: From intern to resident, Bellevue Hosp, New York, 62-66; USPHS med scientist training fel, 66-69, from instr to asst prof med, 69-72, ASST PROF MED & MICROBIOL, MED CTR, NY UNIV, 72- Concurrent Pos: NY Heart Asn sr investr, 69. Honors & Awards: Career Scientist Award, Irma T Hirschl Trust, 74. Res: Glycoproteins of mammalian cell surfaces—relationship of blood group antigens to disease; use of lectins to elucidate surface morphology, composition and function. Mailing Add: Dept of Med & Microbiol Med Ctr NY Univ 550 First Ave New York NY 10016

NACHBAR, WILLIAM, b Brooklyn, NY, Apr 25, 23; m 52; c 1. APPLIED MATHEMATICS. Educ: Cornell Univ, BME, 44; NY Univ, MS, 48; Brown Univ, PhD(appl math), 51. Prof Exp: Res assoc appl math, Brown Univ, 49-51; staff mem, Math Servs Unit, Boeing Airplane Co, 51-55; sect head mech, Appl Math Dept, Missiles & Space Div, Lockheed Aircraft Corp, 55-60, staff scientist, 60-63; assoc prof, Dept Aeronaut & Astronaut, Stanford, 63-65; PROF APPL MECH & ENG SCI, UNIV CALIF, SAN DIEGO, 65- Concurrent Pos: Res assoc, Stanford Univ, 61-63. Mem: Assoc fel, Am Inst Aeronaut & Astronaut; Am Math Soc; Am Soc Mech Eng; Combustion Inst. Res: Solid mechanics and structural analysis; stability theory; combustion and flame thoery. Mailing Add: Dept of Appl Mech & Eng Sci Univ of Calif PO Box 109 La Jolla CA 92037

NACHBIN, LEOPOLDO, b Recife, Brazil, Jan 7, 22; m 56; c 4. MATHEMATICS. Educ: Univ Brazil, MS, 43, PhD(math), 47; Hon Degrees: Dr, Univ Pernambuco, 66. Prof Exp: Prof math, Univ Brazil, 50-61; vis prof, Univ Paris, 61-63; prof, 63-67, GEORGE EASTMAN PROF MATH, UNIV ROCHESTER, 67- Concurrent Pos: Fels, Guggenheim Found, 49-50, 57-58 & Rockefeller Found, 56-57; chmn, Nat Comt Math Brazil, 54-; head div math res, Nat Res Coun Brazil, 55-56, mem gen bd, 60-61. Honors & Awards: Moinho Santista Found Prize, 62. Mem: Brazilian Acad Sci; Lisbon Acad Sci. Res: Approximation theory; differential operators; function spaces. Mailing Add: Dept of Math Univ of Rochester Rochester NY 14627

NACHLAS, MARVIN MORTON, b Baltimore, Md, Jan 26, 20; m 45; c 5. SURGERY. Educ: Johns Hopkins Univ, AB, 40, MD, 43; Am Bd Surg, dipl, 55. Prof Exp: Asst prof surg, Sch Med, Tufts Univ & asst dir, Tufts Surg Serv, Boston City Hosp, 52-55; asst prof, 55-61, ASSOC PROF SURG, SCH MED, JOHNS HOPKINS UNIV, 61- Concurrent Pos: Res fel surg, Beth Israel Hosp, 47-49; asst, Harvard Med Sch, 48-49; attend surgeon, Sinai Hosp, Baltimore, 55-; surgeon, Johns Hopkins Hosp. Res: Histochemistry of colorimetry of esterases, lipase and dehydrogenases; diagnosis and treatment of upper gastrointestinal hemorrhage; experimental and clinical studies on resuscitation after cardiac arrest. Mailing Add: 6503 Park Heights Ave Baltimore MD 21215

NACHMAN, RALPH LOUIS, b Bayonne, NJ, June 29, 31; m 58; c 2. MEDICINE. Educ: Vanderbilt Univ, AB, 53, MD, 56. Prof Exp: From instr to asst prof, 63-68, ASSOC PROF MED, MED COL, CORNELL UNIV, 68-, ASSOC ATTEND PHYSICIAN & DIR DIV HEMAT, NY HOSP, 70- Concurrent Pos: Res fel med, New York Hosp-Cornell Med Ctr, 62-63; dir labs clin path, NY Hosp, 63-70. Mem: AAAS; Am Fedn Clin Res; Am Soc Hemat; Am Physiol Soc. Res: Immunological aspects of hematologic diseases, particularly platelet abnormalities. Mailing Add: Dept of Med Cornell Univ Col of Med New York NY 10021

NACHMANSOHN, DAVID, b Jekaterinoslaw, Russia, Mar 17, 99; nat US; m 29; c 1. NEUROLOGY. Educ: Univ Berlin, MD, 26. Hon Degrees: MD, Free Univ Berlin, 64. Prof Exp: Res fel, Kaiser Wilhelm Inst Biol, Ger, 26-30; asst med, Univ Frankfurt, 30-31; independent investr, Paris, 33-39; instr, Sch Med, Yale Univ, 39-42; res assoc neurol, 42-47, from asst prof to assoc prof, 47-55, prof biochem, 55-68, EMER PROF BIOCHEM & SPEC LECTR, COL PHYSICIANS & SURGEONS, COLUMBIA UNIV, 68- Concurrent Pos: Hon fel, Weizmann Inst Sci, Israel, 72. Honors & Awards: Pasteur Medal, Paris, 52; Neuberg Medal, 53. Mem: Nat Acad Sci; AAAS; Am Acad Arts & Sci; Soc Exp Biol & Med; Am Physiol Soc. Res: Chemical and molecular basis of nerve activity; role of acetylcholine cycle in control of ion movements across excitable membranes; transduction of chemical into electrical energy, also nerve excitability and bioelectricity; molecular forces in enzymes and proteins associated with function of acetylcholine; electric fish; organophosphorous compounds. Mailing Add: Dept of Neurol Col of Phys & Surgeons Columbia Univ New York NY 10032

NACHOD, FREDERICK CONSTANTINE, b Leipzig, Ger, Oct 4, 13; nat US; m 40; c 2. PHYSICAL CHEMISTRY. Educ: Univ Freiburg, BS, 32; Univ Leipzig, AM, 38; State Univ Utrecht, ScD(phys chem), 38. Prof Exp: Res assoc, State Univ Utrecht, 38-39; fel, Columbia Univ, 39; res chemist, Baker & Co, Inc, NJ, 39; instr chem, City Col New York, 39-40; res chemist, Permutit Co, NJ, 40-44; consult, Smith Kline & French Labs, 44; sr chemist, Atlantic Ref Co, Pa, 44-46; head dept phys chem, 46-64, chem liaison staff dir, 64-74, DIR SPEC PROJS, STERLING-WINTHROP RES INST, 74- Concurrent Pos: Adj prof, Rensselaer Polytech Inst, 52- Mem: AAAS; Am Chem Soc. Res: Absorption spectra; ion exchange; medicinal chemistry; structure of organic compounds; research administration. Mailing Add: Sterling Res Inst Rensselaer NY 12144

NACHREINER, RAYMOND F, b Richland Center, Wis, Apr 29, 42; m 68; c 3. VETERINARY PHYSIOLOGY, ENDOCRINOLOGY. Educ: Iowa State Univ, DVM, 66; Univ Wis, PhD(endocrinol), 72. Prof Exp: Res asst endocrinol, Univ Wis, 68-72; ASST PROF PHYSIOL, AUBURN UNIV, 72- Honors & Awards: Burr Beach Award, Univ Wis, 72. Mem: Soc Study Fertil; Soc Study Reprod; Am Vet Med Asn. Res: Population control in animals; porcine stress syndrome and porcine agalactia syndrome. Mailing Add: 422 Green St Auburn AL 36830

NACHTIGALL, GUENTER WILLI, b Hamburg, Ger, Jan 1, 29; nat US; m 50; c 1. ORGANIC CHEMISTRY. Educ: Columbia Univ, BS, 62; Univ Colo, PhD(org chem), 68. Prof Exp: Chemist, Indust Chem Div, 61, chemist cent res div, 62-63, res chemist, 64-74, SR RES CHEMIST, CHEM RES DIV, AM CYANAMID CO, 74- Mem: Am Chem Soc; AAAS. Res: Carbonium ion reactions, particularly those of bridged polycyclic compounds; asymmetric organic syntheses. Mailing Add: Chem Res Div Am Cyanamid Co 1937 W Main St Stamford CT 06904

NACHTRIEB, NORMAN HARRY, b Chicago, Ill, Mar 4, 16; m 41, 53; c 1. PHYSICAL CHEMISTRY. Educ: Univ Chicago, BS, 36, PhD(chem), 41. Prof Exp: Anal chemist, State Geol Surv, Ill, 37-38; head, Anal Sect, Columbia Chem Div, Pittsburgh Plate Glass Co, Ohio, 41-43; res chemist, Manhattan Dist, Metall Lab, Chicago, 43-44; alternate group leader anal group, Los Alamos Sci Lab, NMex, 44-46; from asst prof to assoc prof, 46-53, chmn dept, 62-71, PROF CHEM, UNIV CHICAGO, 53-, MASTER, PHYS SCI COL DIV & ASSOC DEAN, DIV PHYS SCI, 73-, PROF CHEM, INST STUDY METALS, 46- Concurrent Pos: Adv ed, Encyclopaedia Britannica, 55-; NSF fel, 59-60, mem adv panel, Sci Educ Div, 65-68. Mem: Am Chem Soc; fel Am Phys Soc. Res: Spectrochemical analysis of solutions; fused salts; metals; extraction of metal halides by organic solvents; electrode potentials; diffusion in crystalline solids and liquids; high pressure chemistry; nuclear magnetic resonance; solid state chemistry; magnetic, electrical and optical properties of metal-molten salt solutions. Mailing Add: Dept of Chem Univ of Chicago 5735 Ellis Ave Chicago IL 60637

NACHTWEY, DAVID STUART, b Seattle, Wash, Aug 9, 29; m; c 3. PHOTOBIOLOGY, RADIOBIOLOGY. Educ: Univ Wash, BA, 51; Univ Tex, MA, 56; Stanford Univ, PhD(biol sci), 61. Prof Exp: Nat Cancer Inst fel, Biol Inst, Carlsberg Found, Denmark, 61-62; res biologist, Cellular Radiobiol Br, US Naval Radiol Defense Lab, 62-68; assoc prof radiation biol, Ore State Univ, 68-74; ed, Climatic Impact Assessment Prog Monograph 5, 74-75; MGR, LIFE SCI STRATOSPHERIC RES PROG, NASA, 75- Mem: AAAS; Soc Protozool; Radiation Res Soc; Am Soc Cell Biol; Am Soc Photobiol. Res: Cell division processes in protozoa and unicellular algae; effects of ultraviolet and ionizing radiation on cells; mutagenesis; recovery from radiation damage; ecosystem responses to solar ultraviolet radiation. Mailing Add: Biol Br DD7 Johnson Space Ctr NASA Houston TX 77058

NACKOWSKI, MATTHEW PETER, b Manchester, Conn, Oct 19, 15; m 42; c 6. GEOLOGY. Educ: Univ Calif, BA, 41; Univ Mo, MS, 49, PhD(geol), 52. Prof Exp: Geologist, Consol Coppermines Corp, 42-44 & Callahan Zinc-Lead Co, 44-46; instr geol, Mo Sch Mines, 47-50 & 51-52, from asst prof to assoc prof, 52-55; from assoc prof to prof mining & geol, 55-71, prof econ geol & geol eng, 71-75, asst chmn dept, 72-75, PROF GEOL & GEOPHYSICS, UNIV UTAH, 75- Concurrent Pos: Geol expert, UNESCO, 67-69. Mem: Geol Soc Am; Soc Econ Geol; Geochem Soc; Am Inst Mining, Metall & Petrol Eng. Res: Geochemical exploration; exploration and economic geology; mineral resources and economics; mine valuation. Mailing Add: Dept of Geol & Geophys Univ Utah Salt Lake City UT 84112

NACOZY, PAUL E, b Los Angeles, Calif, Apr 15, 42. ASTRONOMY, CELESTIAL MECHANICS. Educ: San Diego State Col, BA, 64; Yale Univ, MS, 66, PhD(astron), 68. Prof Exp: Asst prof, 68-74, ASSOC PROF AEROSPACE ENG, UNIV TEX, AUSTIN, 74- Mem: Am Astron Soc; Int Astron Soc; Am Inst Aeronaut & Astronaut. Res: Celestial mechanics and aerospace mechanics, especially series-solutions of the motions of space vehicles, asteroids, and comets; earth-moon-space vehicle dynamical system. Mailing Add: Dept of Aerospace Eng Univ of Tex Austin TX 78712

NADALIN, ROBERT JOHN, analytical chemistry, see 12th edition

NADAS, ALEXANDER SANDOR, b Budapest, Hungary, Nov 12, 13; US citizen; m 41; c 3. PEDIATRIC CARDIOLOGY. Educ: Med Univ Budapest, MD, 37; Wayne State Univ, MD, 45; Am Bd Pediat, dipl & cert pediat cardiol. Prof Exp: Intern, Fairview Park Hosp, Cleveland, Ohio, 39-40 & Wilmington Gen Hosp, Del, 40-41; resident pediat, Mass Mem Hosps, Boston, 41-42; vol asst med serv, Children's Hosp, Boston, 42-43, asst resident, 43; chief resident, Children's Hosp, Mich, 43-45; pvt pract, Greenfield, Mass, 45-49; instr pediat, 50-52, clin assoc, 52-55, from asst clin prof to clin prof, 55-69, PROF PEDIAT, HARVARD MED SCH, 60-; CHIEF CARDIOL DEPT, CHILDREN'S HOSP MED CTR, 69- Concurrent Pos: Res fel pediat, Harvard Med Sch, 49; Guggenheim fel, 70; instr, Wayne State Univ, 43-45; asst physician & assoc chief cardiol div, Children's Hosp Med Ctr, 49-50, from assoc physician to physician, Sharon Sanatorium, 50-51; assoc physician & assoc cardiologist, Sharon Cardiovasc Unit, 51-52, cardiologist, 52-66, sr assoc med & Good Samaritan Div, 62-, chief cardiol div, 66-69; Fulbright prof, State Univ Groningen, 56-57; consult, NShore Children's Hosp, Salem, Mass, Boston Lying-in-Hosp, Mass Gen Hosp, Boston & Newton Wellesley Hosp. Mem: Am Heart Asn; Am Acad Pediat; Soc Pediat Res; Am Pediat Soc. Res: Applied cardiovascular physiology and physiology of congenital heart disease; natural history of congenital heart disease with special emphasis on clinical-physiologic correlations. Mailing Add: Children's Hosp Med Ctr 300 Longwood Ave Boston MA 02115

NADAS, ARTHUR JOSEPH, b Budapest, Hungary, Apr 10, 34; US citizen; m 55; c 4. MATHEMATICAL STATISTICS. Educ: Alfred Univ, BA, 59; Univ Ore, MA, 61; Columbia Univ, PhD(math statist), 67. Prof Exp: Asst math, Univ Ore, 59-61; statistician math statist, Int Bus Mach Corp, 61-68; asst prof math, Polytech Inst Brooklyn, 68-70; SR MATHEMATICIAN, IBM COR SYSTS PRODS, IBM CORP, HOPEWELL JCT, 71- Mem: Inst Math Statist; Am Statist Asn. Res: Probability; reliability theory; operations research. Mailing Add: Bull Rd RD Rock Tavern NY 12575

NADDOR, ELIEZER, b Jerusalem, Israel, Sept 23, 20; US citizen; m 54; c 3. OPERATIONS RESEARCH. Educ: Israel Inst Technol, BS, 51, CE, 52; Columbia Univ, MS, 53; Case Inst Technol, PhD(opers res), 57. Prof Exp: Asst opers res, Case Inst Technol, 53-55, instr bus admin, 55-56; from asst prof to assoc prof opers res, 56-64, PROF OPERS RES, JOHNS HOPKINS UNIV, 64- Concurrent Pos: Consult, Cleveland Graphite Bronze Co, 53-55, Am Airlines, Inc, 55-56 & Chesapeake & Potomac Tel Co Md, 57-66; res labs, Gen Motors Corp, 59-64; consult, Res Labs, Gen Motors Corp, 59-64 & Bell Tel Labs, 63-64; Fulbright lectr, Finland, 64-65; consult, Gen Elec Co & US Naval Acad, 66- Mem: Opers Res Soc Am; Inst Mgt Sci (secy-treas, 64-66); Am Inst Indust Eng; Am Statist Asn; Math Asn Am. Res: Inventory systems. Mailing Add: Dept of Math Sci Johns Hopkins Univ Baltimore MD 21218

NADDY, BADIE IHRAHIM, b Haifa, Palestine, Dec 31, 33; US citizen; m 63; c 3. PHYSICAL CHEMISTRY, SOIL CHEMISTRY. Educ: Am Univ Beirut, BS, 57; Kans State Univ, PhD(soil & phys chem), 63. Prof Exp: Chmn div sci, Henderson State Col, 63-65; dir labs, Jordan Govt, 65-67; CHMN DIV MATH & SCI, COLUMBIA STATE COMMUNITY COL, 68- Mem: Am Chem Soc. Res: Effect of cation exchange on the dielectric constants of minerals; x-ray diffraction of minerals saturated with different cations. Mailing Add: Div of Math & Sci Columbia State Community Col Columbia TN 38401

NADEAU, GERARD, b Quebec, Que, Oct 28, 25; m 53; c 5. THEORETICAL PHYSICS. Educ: Laval Univ, BS, 48, MS, 49. Prof Exp: From asst prof to assoc prof, 49-62, PROF PHYSICS, LAVAL UNIV, 63- Mem: French-Can Asn Advan Sci; Can Asn Physicists. Res: Theory of elasticity. Mailing Add: Dept of Physics Fac Sci Laval Univ Quebec PQ Can

NADEAU, HERBERT GERARD, b Cranston, RI, Aug 1, 28; m 51; c 4. ANALYTICAL CHEMISTRY, PHYSICAL CHEMISTRY. Educ: Providence Col, BS, 51. Prof Exp: Anal res chemist, Geigy Chem Corp, 51-55; chief, Sect Anal Chem, Olin Mathieson Chem Corp, 55-65; head anal res serv, 65, mgr cellular plastics, 66-68, mgr anal & phys res, 68-74, MGR PHYS & FLAMMABILITY RES, UPJOHN CO, 74- Concurrent Pos: Mem, Prods Res Comt, 75-80. Honors & Awards: Distinguished Serv Award & Cert Appreciation, Soc Plastics Indust, 75. Mem: Am Chem Soc; Soc Plastics Indust. Res: Organic analysis; infrared; vapor phase chromatography; high vacuum techniques; boron compounds; commercial products; urethane chemistry; polymers; flammability testing and research. Mailing Add: 140 Patten Rd North Haven CT 06473

NADEAU, REGINALD ANTOINE, b St Leonard, NB, Dec 18, 32; m 57; c 2. CARDIOVASCULAR PHYSIOLOGY, CARDIOLOGY. Educ: Loyola Col, Can, BA, 52; Univ Montreal, MD, 57; FRCP(C), 62. Prof Exp: From asst prof to assoc prof, 64-70, prof physiol, 72-75, PROF MED, FAC MED, UNIV MONTREAL, 75- Concurrent Pos: Assoc, Med Res Coun Can, 65; dir res, Cardiol Hosp Sacre Coeur, Montreal. Mem: Can Physiol Soc; Can Cardiovasc Soc; Am Col Cardiol. Res: Cardiovascular pharmacology; clinical cardiology. Mailing Add: Dept of Physiol Univ of Montreal Fac of Med Montreal PQ Can

NADEL, ELI MAURICE, b New York, NY, Oct 9, 18; m 43; c 3. PHYSIOLOGY, PATHOLOGY. Educ: City Col New York, BS, 37, MS, 39; Long Island Col Med, MD, 45; Am Bd Path, dipl, 51. Prof Exp: Intern, Mt Sinai Hosp, New York, 45-46; pathologist, NIH, 46-56, exec secy path sect, 56-59, spec asst assoc dir, 58-60, asst dir, Nat Cancer Inst, 60-61, chief diag res br, 61-65; chief res path, Lab Med & Hemat, dir clin invest, res assoc & sr investr progs, Vet Admin, 65-68; prof path, Sch Med, St Louis Univ, 68-72, prof community med, 70-72, assoc dean, 68-72, actg chmn depts path & physiol, 68-69 & 71-72; prof path & assoc dean, Med Univ SC, 72-75; chief staff, Charleston Vet Admin Hosp, 72-75, actg dir, 74-75; ASST VPRES, MED RES CTR, PROF PATH & CANCER COORDR, ST LOUIS UNIV MED CTR, 75 Concurrent Pos: Fel path, Sch Med, Washington Univ, 48-50; fel biochem, Jefferson Med Col, 50-51; mem, Path Study Sect, Am Bd Path, 60-62; clin prof res path, Med Ctr, Georgetown Univ, 66-; consult, Vet Admin Cent Off, 68-70 & St Louis Vet Admin Hosp, 70-72. Mem: Soc Exp Path; Am Asn Path & Bact; Am Asn Cancer Res; Am Soc Cytol; Am Soc Clin Path. Res: Steroid metabolism in cancer; experimental pathology; diagnostic research; leukemia; malaria; carcinogenesis. Mailing Add: St Louis Univ Med Ctr 1402 S Grand Blvd St Louis MO 63104

NADEL, ETHAN RICHARD, b Washington, DC, Sept 3, 41. PHYSIOLOGY. Educ: Williams Col, BA, 63; Univ Calif, Santa Barbara, MA, 66, PhD(biol), 69. Prof Exp: Asst fel, 70-73, ASSOC FEL, JOHN B PIERCE FOUND LAB, 74-; ASST PROF, DEPT EPIDEMIOL & PUB HEALTH, SCH MED, YALE UNIV, 70 Concurrent Pos: NIH fel, Sch Med, Yale Univ, 69-70, USPHS grant, 70-; partic, US-Japan Prog Human Adaptability, 72-73. Mem: AAAS; Am Physiol Soc; fel Am Col Sports Med. Res: Physiological regulations against hyperthermia. Mailing Add: John B Pierce Found Lab 290 Congress Ave New Haven CT 06519

NADEL, JAY A, b Philadelphia, Pa, Jan 21, 29; m 60; c 3. PULMONARY PHYSIOLOGY, CARDIOVASCULAR PHYSIOLOGY. Educ: Temple Univ, AB, 49; Jefferson Med Col, MD, 53. Prof Exp: Trainee heart & lung res, 58-62, clin instr med, 61-62, asst clin prof, 62-64, from asst prof to assoc prof, 64-70, PROF MED & RADIOL, CARDIOVASC RES INST, MED CTR, UNIV CALIF, SAN FRANCISCO, 70-, STAFF MEM, INST, 64- Concurrent Pos: Mem, Cardiovasc B Study Sect, Nat Heart & Lung Inst, 71-75; adv, Comn State-Wide Air Pollution Res, Univ Calif; mem, Comn Surv Prof Manpower Pulmonary Dis. Mem: Am Physiol Soc; Asn Am Physicians; Thoracic Soc (pres, 73-); Am Soc Clin Invest. Res: Control of airways, mucous secretion, respiration and pulmonary circulation; pulmonary radiology. Mailing Add: Cardiovasc Res Inst Univ of Calif San Francisco CA 94143

NADEL, MARVIN KEITH, b Brookline, Mass, Oct 2, 26; m; c 4. CHEMISTRY. Educ: Univ Mass, AB, 49, MS, 51; Kans State Univ, PhD(virol), 55. Prof Exp: Bacteriologist, Ft Detrick, Md, 51-52; USPHS fel, Kans State Univ, 55-56; vpres & dir biol res, Res Labs, Inc & Anchor Serum Co, 56-58; supt diag agents, Lederle Labs, Am Cyanamid Co, 58-61; prin res scientist & sect head, Bur Labs, New York City Dept Health, 61-64; mgr biol res & develop, Chem & Biol Opers, Space-Gen Corp, Calif, 64-68; exec div mgr, Electro-Optical Systs, Inc, Xerox Corp, Calif, 68, dir res & mgr chem opers, Med Diagnostics, 68-69; pres & chief exec off, Bio-Diagnostics, Inc, Calif, 69-71; vpres & gen mgr chem reagents group, Abbott Labs, Inc, 71-72; DIR HEALTH CARE COMMERCIAL DEVELOP, NEW ENTERPRISE DIV, MONSANTO CO, ST LOUIS, 72- Concurrent Pos: Mem, Pub Health Res Inst New York, 61-64. Mem: AAAS; Am Soc Microbiol; Am Pub Health Asn; Brit Soc Gen Microbiol; Am Asn Clin Chemists. Res: Host-parasite relationships. Mailing Add: 762 Haw Thicket Lane Des Peres MO 63166

NADELHAFT, IRVING, b New York, NY, Nov 4, 28; m 56; c 2. NEUROBIOLOGY. Educ: City Col New York, BS, 49; Syracuse Univ, PhD, 56. Prof Exp: Res assoc, Carnegie Inst Technol, 58-61, asst prof physics, 61-67; ASST PROF NEUROL SURG, MED SCH, UNIV PITTSBURGH, 74-; RES PHYSICIST, VET ADMIN HOSP, PITTSBURGH, 69 Concurrent Pos: Ford Found fel, Europ Orgn Nuclear Res, 56-57; NIH spec fel, Dept Anat & Cell Biol, Sch Med, Univ Pittsburgh & Dept Biol,

Mass Inst Technol, 67-69; guest res physicist, Brookhaven Nat Lab, 62-63; adj asst prof pharmacol, Med Sch, Univ Pittsburgh, 69- Mem: AAAS; Biophys Soc; Soc Neurosci; Am Phys Soc. Res: High energy physics; weak interactions; particle physics; axoplasmic flow; tracer techniques; computer simulation. Mailing Add: Dept of Neurol Surg Univ of Pittsburgh Med Sch Pittsburgh PA 15261

NADELL, JUDITH, b Newark, NJ, Apr 27, 26; m 46; c 2. MEDICINE, HEMATOLOGY. Educ: NY Univ, BA, 46; State Univ NY Downstate Med Ctr, MD, 50. Prof Exp: Intern med, Maimonides Hosp, Brooklyn, 50-51; resident, Columbia Div, Goldwater Hosp, New York, 51-53; teaching asst, Sch Med, Univ Calif, San Francisco, 53-55; res assoc, Sch Med, Stanford Univ, 55-57, from res asst to res assoc hemat, 60-68; ASSOC MED DIR, SYNTEX RES INST CLIN MED, 70- Concurrent Pos: USPHS res fel, Sch Med, Univ Calif, San Francisco, 53-55; Am Heart Asn fel physiol, Stanford Univ, 55-57; teaching asst, Col Physicians & Surgeons, Columbia Univ, 51-53; mem, Leucocyte Cult Conf, 65- Mem: Am Soc Hemat; Am Fedn Clin Res. Res: Erythropoiesis; thrombopoiesis; transplantation immunology; clinical investigation in the anemias of bone marrow failure; hemolytic anemia; sickle cell disease. Mailing Add: Syntex Res Inst for Clin Med 3401 Hillview Ave Palo Alto CA 94304

NADER, ALLAN E, b Chicago, Ill, Dec 24, 37; m 65; c 1. ORGANIC CHEMISTRY. Educ: Ill Inst Technol, BS, 60; Western Mich Univ, MA, 63; Purdue Univ, PhD(org chem), 67. Prof Exp: Res chemist, 66-71, SR RES CHEMIST, POLYMER INTERMEDIATES DEPT, E I DU PONT DE NEMOURS & CO, INC, 71- Mem: Am Chem Soc; Sigma Xi. Res: Reaction mechanism studies; hydrocarbon nitration and oxidation; metal catalysis; polymer degradation; chemical process research. Mailing Add: Polymer Intermed Dept Exp Sta E I du Pont de Nemours & Co Inc Wilmington DE 19898

NADER, LAURA, b Winsted, Conn, Sept 30, 30; m 62; c 3. ANTHROPOLOGY. Educ: Wells Col, BA, 52; Radcliffe Col, PhD(anthrop), 62. Prof Exp: Info analyst, Indonesian Consulate, 53; res asst, Harvard Univ, 55-60; vchmn dept, 68-71, PROF ANTHROP, UNIV CALIF, BERKELEY, 60- Concurrent Pos: Consult anthrop, Arthur D Little, Inc, 57-59; fel, Ctr Advan Study Behav Sci, 63-64; mem, Rincon Zapotec Film Exped, 64; NSF grant, 66-68; trustee, Ctr Study Responsive Law, 68-; chmn cult anthrop comt, NIMH, 68-71, mem comt, 68-; mem, Soc Sci Res Coun, 69-; mem soc sci adv comt, NSF, 71-; mem, Carnegie Coun Children, 72-; Nat Acad Sci, Nat Res Coun rep, Behav Div, 69-71 & 73- Mem: Am Anthrop Asn; Soc Women Geographers. Mailing Add: Dept of Anthrop Univ of Calif Berkeley CA 94720

NADKARNI, MORESHWAR VITHAL, b Bombay, India, July 1, 18; nat US; m 50; c 5. BIOCHEMICAL PHARMACOLOGY. Educ: Univ Bombay, BS, 37; Univ Iowa, MS, 47, PhD(pharmaceut chem), 49. Prof Exp: Asst res prof pharmacol, George Washington Univ, 54-58; pharmacologist, 58-59, head pharmacol sect, Drug Eval Br, 59-65, DIR PHARMACOL PROG, EXTRAMURAL ACTIVITIES, NAT CANCER INST, 65- Concurrent Pos: Fel, Nat Cancer Inst, 49-51; Am Cancer Soc fel, George Washington Univ, 52-54. Mem: Am Chem Soc; Am Pharmaceut Asn; Am Asn Cancer Res; NY Acad Sci. Res. Chemotherapy of cancer; pharmacology and toxicology; drug metabolism; isotope tracer techniques. Mailing Add: Pharmacol Prog Nat Cancer Inst Bethesda MD 20014

NADLER, CHARLES FENGER, b Chicago, Ill, Nov 8, 29; m 53; c 3. INTERNAL MEDICINE, ZOOLOGY. Educ: Dartmouth Col, AB, 51; Northwestern Univ, MD, 55. Prof Exp: From intern to asst resident med, Barnes Hosp, St Louis, Mo, 55-57 & 59-60; instr, 61-63, assoc, 63-67, asst prof, 67-71, ASSOC PROF MED, MED SCH, NORTHWESTERN UNIV, 71; RES ASSOC, DIV MAMMALS, FIELD MUS NATURAL HIST, 65- Concurrent Pos: Fel hemat, Med Ctr, Univ Colo, 6061; attend physician, Passavant Mem Hosp, Chicago, Ill, 61-72; attend physician, Northwestern Mem Hosp, Chicago, 73-; assoc mammal, Mus Natural Hist, Univ Kans, 73- Mem: Am Fedn Clin Res; Soc Exp Biol & Med; Am Soc Mammal; Soc Syst Zool; Am Col Physicians. Res: Hematology; application of cytogenetics and comparative biochemistry of proteins to the evolution of Asian and North American mammals. Mailing Add: Dept of Med Northwestern Univ Med Sch Chicago IL 60611

NADLER, HENRY LOUIS, b New York, NY, Apr 15, 36; m 57; c 4. PEDIATRICS, HUMAN GENETICS. Educ: Colgate Univ, AB, 57; Northwestern Univ, MD, 61; Univ Wis, MS, 65. Prof Exp: From intern to resident pediat, Med Ctr, NY Univ, 61-63, chief resident & inst pediat, 63-64; instr, Sch Med, Univ Wis, 64-65; assoc, Med Sch, 65-66, from asst prof to assoc prof, 67-70, PROF PEDIAT, MED SCH & GRAD SCH & CHMN DEPT PEDIAT, MED SCH, NORTHWESTERN UNIV, 70-; HEAD DIV GENETICS, CHILDREN'S MEM HOSP, 69-, CHIEF OF STAFF PEDIAT, 70- Concurrent Pos: Res fel pediat, Children's Mem Hosp, Chicago, 64-65. Honors & Awards: Irene Heinz Given & John La Porte Given Res Prof Pediat, Children's Mem Hosp, Chicago; E Mead Johnson Award, Am Acad Pediat, 73. Mem: Am Soc Clin Invest; Am Soc Human Genetics; Am Pediat Soc; Soc Pediat Res; Soc Exp Biol & Med. Res: Human biochemical genetics; chromosomal disorders and inborn errors of metabolism; prenatal detection of genetic diseases. Mailing Add: Children's Mem Hosp 2300 Children's Plaza Chicago IL 60614

NADLER, KENNETH DAVID, b Bronx, NY, Sept 18, 42; m 67. PLANT PHYSIOLOGY, BIOCHEMISTRY. Educ: Rensselaer Polytech Inst, BSc, 63; Rockefeller Univ, PhD(life sci), 68. Prof Exp: NIH res fel biol, Revelle Col, Univ Calif, San Diego, 68-70; ASST PROF BOT, MICH STATE UNIV, 70- Concurrent Pos: Lectr, Univ Calif, San Diego, 68. Mem: AAAS; Am Soc Plant Physiologists. Res: Biochemical mechanisms with which subcellular activities are integrated into cellular metabolism; control mechanisms on cellular development and differentiation; biochemistry of chlorophylls and hemes; control of differentiation of plastids, mitochondria and cells in relation to porphyrin biosynthesis. Mailing Add: Dept of Bot Mich State Univ East Lansing MI 48823

NADLER, MELVIN PHILIP, b Malden, Mass, May 20, 40; m 71. PHYSICAL CHEMISTRY. Educ: Northeastern Univ, AB, 63; Cornell Univ, PhD(phys chem), 69. Prof Exp: NSF fel under R Weiner, Northeastern Univ, 61-63; res assoc, State Univ NY Binghamton, 68-70; supvr air & water pollution control, Remington Rand Div, Sperry Rand Corp, 71-72; RES CHEMIST, NAVAL WEAPONS CTR, 72- Mem: Am Chem Soc. Res: Kinetics, spectroscopy and photochemistry of boron systems and flames; infrared emission; combustion of solid propellants. Mailing Add: Naval Weapons Ctr China Lake CA 93555

NADLER, NORMAN JACOB, b Montreal, Que, Dec 24, 27; m 53. ENDOCRINOLOGY, ANATOMY. Educ: McGill Univ, BSc, 47, MD, CM, 51, PhD(thyroid), 55. Prof Exp: Lectr, 57-59, asst prof, 59-64, ASSOC PROF ANAT, McGILL UNIV, 65- Concurrent Pos: Consult med, Jewish Gen Hosp, Montreal, 59- Mem: Endocrine Soc; Am Thyroid Asn; Am Diabetes Asn; Can Asn Anat. Res: Biophysical approach to morphological physiological aspects of thyroid gland. Mailing Add: Dept of Anat McGill Univ Montreal PQ Can

NADLER, RONALD DAVID, b Newark, NJ, Jan 19, 36; m 59; c 3. ANIMAL BEHAVIOR, PRIMATOLOGY. Educ: Univ Calif, Los Angeles, BA, 60, MA, 63, PhD(physiol psychol), 65. Prof Exp: USPHS res fels, Oxford Univ, 65-66 & Univ Wash, 66-67; asst prof psychiat, State Univ NY Downstate Med Ctr, 67-71; DEVELOP BIOLOGIST PRIMATOL, YERKES PRIMATE RES CTR, EMORY UNIV, 71- Concurrent Pos: State Univ NY Res Found grant, State Univ NY Downstate Med Ctr, 68-70, NSF grant, 69-71, USPHS grants, 69-70 & 71-72; NSF grants, Yerkes Primate Res Ctr, Emory Univ, 71-73 & 71-74; NSF res grant, 75-78. Mem: Int Soc Psychoneuroendocrinol; Int Primatol Soc; AAAS; Int Acad Sex Res. Res: Comparative, developmental research on socio-sexual behavior of the great apes, chimpanzee, gorilla and orang-utan; emphasis is placed on physiological correlates of behavior, especially hormonal. Mailing Add: Yerkes Regional Primate Res Ctr Emory Univ Atlanta GA 30322

NADOL, BRONISLAW JOSEPH, JR, b Cambridge, Mass, Oct 2, 43; m 70; c 2. OTOLARYNGOLOGY. Educ: Harvard Col, BA, 66; Johns Hopkins Univ, MD, 70; Am Bd Otolaryngol, dipl, 75. Prof Exp: Intern surg, Beth Israel Hosp, 70-71, residency surg, 71-72, residency otolaryngol, 72-75, ASST OTOLARYNGOL, MASS EYE & EAR INFIRMARY & INSTR, SCH MED, HARVARD UNIV, 75- Mem: Am Acad Ophthal & Otolaryngol. Res: Electron microscopy of the human inner ear; clinical electrocochleography. Mailing Add: Mass Eye & Ear Infirmary 243 Charles St Boston MA 02114

NAEGELE, EDWARD WISTER, JR, b Philadelphia, Pa, Sept 30, 23; m 50; c 1. ORGANIC CHEMISTRY. Educ: Temple Univ, AB, 48, MA, 50, PhD(chem), 55. Prof Exp: Technologist, E I du Pont de Nemours & Co, 55-57; supvr basic res, Acheson Dispersed Pigments Co, 57-58; assoc prof chem, 58-62, PROF CHEM & CHMN DEPT, GROVE CITY COL, 62- Mem: AAAS; Am Chem Soc. Res: Heterocyclic compounds. Mailing Add: Dept of Chem Grove City Col Grove City PA 16127

NAEGELE, JOHN ADAM, b Hackensack, NJ, June 13, 28; m 50; c 4. ENVIRONMENTAL BIOLOGY, ENTOMOLOGY. Educ: Cornell Univ, BS, 49, PhD, 53. Prof Exp: Asst entom, Cornell Univ, 49-53, from asst prof to assoc prof, 53-63; prof environ sci & head suburban exp sta, 63-74, ASSOC DEAN RES & ASSOC DIR MASS AGR EXP STA, COL FOOD & NAT RESOURCES, UNIV MASS, 74- Concurrent Pos: Consult, Jackson & Perkins Co & A N Pierson Co, 62; consult, CLM Systs, Inc & US Dept Educ; vis lectr, Sch Pub Health, Harvard Univ, 71- Mem: Entom Soc Am. Res: Environmental biology, especially behavior, entomology, resistance, acarology and pest control; air pollution effects on plants; total spectrum agriculture research. Mailing Add: Stockbridle Hall Col of Food & Nat Res Univ of Mass Amherst MA 01002

NAEGELE, ROBERT FRANK, virology, see 12th edition

NAEGELI, DAVID W, physical chemistry, see 12th edition

NAEGER, LEONARD L, b St Louis, Mo, July 19, 41; m 66; c 2. PHARMACOLOGY. Educ: St Louis Col Pharm, BS, 63, MS, 65; Univ Fla, PhD(pharmacol), 71. Prof Exp: ASST PROF PHARMACOL, ST LOUIS COL PHARM, 70-; ASST PROF, SCH DENT MED, WASHINGTON UNIV, 72- Concurrent Pos: Lectr, Med Ctr, St Louis Univ, 71- Res: Ethanol metabolism. Mailing Add: Dept of Pharmacol St Louis Col of Pharm St Louis MO 64110

NAESER, CHARLES RUDOLPH, b Mineral Point, Wis, Nov 13, 10; m 36; c 2. INORGANIC CHEMISTRY. Educ: Univ Wis, BS, 31; Univ Ill, MS, 33, PhD(inorg chem), 35. Prof Exp: Asst gen chem, Univ Ill, 32-35; from instr to asst prof inorg chem, 35-42, assoc prof, 45-47, PROF CHEM, GEORGE WASHINGTON UNIV, 47-, CHMN DEPT, 48-50, 51-53 & 56- Concurrent Pos: Chief, Chem Group, US Geol Surv, 53-56, consult, 56-75; consult, Off Saline Water, 62-72. Honors & Awards: Am Inst Chem Honor Award, 62. Mem: AAAS; Am Inst Chem; Am Chem Soc; Geochem Soc. Res: Inorganic chemistry of rare earths, beryllium, uranium, rhenium and selenium; electro reduction of less common metals; fluoroplatinates; geochemistry; radioactive waste disposal; desalinization. Mailing Add: Dept of Chem George Washington Univ Washington DC 20052

NAESER, CHARLES WILBUR, b Washington, DC, July 2, 40; m 63; c 2. GEOLOGY. Educ: Dartmouth Col, AB, 62, MA, 64; Southern Methodist Univ, PhD(geol), 67. Prof Exp: GEOLOGIST, US GEOL SURV, 67- Mem: Fel Geol Soc Am; Am Geophys Union. Res: Geochronology, specifically the use of fission track dating of minerals. Mailing Add: US Geol Surv Box 25046 Denver Fed Ctr Stp 424 Denver CO 80225

NAEYE, RICHARD L, b Rochester, NY, Nov 27, 29; m 55; c 3. PATHOLOGY. Educ: Colgate Univ, AB, 51; Columbia Univ, MD, 55. Prof Exp: Instr, Columbia Univ, 57-58; trainee, Med Col, Univ Vt, 58-60, from asst prof to prof, 60-67; PROF PATH & CHMN DEPT, HERSHEY MED CTR, PA STATE UNIV, 67- Concurrent Pos: Marckle scholar, 60-65. Mem: Col Am Path; Am Soc Exp Path; Am Soc Clin Path; Am Asn Pathologists & Bacteriologists; Int Acad Path. Res: Prenatal and postnatal growth and development; pulmonary vascular disease; systemic vascular disease. Mailing Add: Dept of Path Pa State Univ Hershey Med Ctr Hershey PA 17033

NÄF, ULRICH, b Wülflingen, Switz, June 27, 21; nat US. PLANT PHYSIOLOGY. Educ: Swiss Fed Inst Technol, dipl, 48; Yale Univ, PhD(bot), 53. Prof Exp: Damon Runyon fel, Rockefeller Univ, 53-55, res assoc plant physiol, 55-60, asst prof, 60-62; assoc prof, 62-69, PROF PLANT PHYSIOL & DIR LAB PLANT MORPHOGENESIS, MANHATTAN COL, 69- Mem: AAAS; Bot Soc Am; Am Soc Plant Physiol; Scand Soc Plant Physiol. Res: Developmental physiology of ferns; physiology of Kostoff's tumor. Mailing Add: Dept of Biol Manhattan Col Bronx NY 10471

NAFE, JOHN ELLIOTT, b Seattle, Wash, July 22, 14; m 41; c 2. PHYSICS, GEOPHYSICS. Educ: Univ Mich, BS, 38; Wash Univ, MS, 40; Columbia Univ, PhD(physics), 48. Prof Exp: Asst physics, Wash Univ, 38-39; asst, Columbia Univ, 40-41, instr, 46-49; asst prof, Univ Minn, 49-51; dir res, Hudson Lab, 51-53, res assoc, Lamont-Doherty Geol Observ, 53-55, adj assoc prof geophys, 55-58, chmn dept geol, 62-65, PROF GEOPHYS, LAMONT-DOHERTY GEOL OBSERV, COLUMBIA UNIV, 58- Concurrent Pos: Vis fel, Cambridge Univ, 71-72. Mem: Fel AAAS; fel Am Geophys Union; fel Am Phys Soc; fel Geol Soc Am; Seismol Soc Am. Res: Atomic beams; hyperfine structure of deuterium and hydrogen; seismology; marine geophysics; underwater sound. Mailing Add: Lamont-Doherty Geol Observ Columbia Univ Palisades NY 10964

NAFF, JOHN DAVIS, b Atlanta, Ga, Nov 12, 18; m 42; c 5. GEOLOGY. Educ: Univ Ala, AB, 39, MA, 40; Univ Kans, PhD, 60. Prof Exp: Assoc prof, 50-65, PROF GEOL, OKLA STATE UNIV, 65- Concurrent Pos: Staff geologist, Juneau Icefield Res Prog, NSF lect series; ed, Geol Sect, Okla Acad Sci, 68- Mem: Soc Econ Paleont & Mineral; Am Asn Petrol Geol; Nat Asn Geol Teachers (secy). Res: Invertebrate paleontology; stratigraphy; mountaineering; photography; electronics; miniaturization and packaging of equipment. Mailing Add: Dept of Geol Okla State Univ Stillwater OK 74074

NAFF, MARION BENTON, b Lexington, Ky, Mar 23, 18; m 46. ORGANIC CHEMISTRY. Educ: Univ Ky, BS, 41, MS, 46; Ore State Col, PhD(org chem), 50. Prof Exp: Asst prof chem, Western Carolina State Col, 50-51 & Bowling Green State Univ, 51-55; assoc prof, Loyola Univ, La, 55-58 & Dickinson Col, 58-66; CHEMIST, DRUG RES & DEVELOP, NAT CANCER INST, 66- Concurrent Pos: Vis assoc prof chem, Brown Univ, 64-65. Mem: Fel AAAS; Am Chem Soc. Res: Synthesis of heterocyclics; acetylenic chemistry; chemical kinetics; medicinal chemistry; organic synthesis and analysis; instrumentation. Mailing Add: Drug Res & Develop Nat Cancer Inst NIH Bethesda MD 20014

NAFIE, LAURENCE ALLEN, b Detroit, Mich, Aug 9, 45; m 68; c 1. PHYSICAL CHEMISTRY. Educ: Univ Minn, Minneapolis, BChem, 67; Univ Ore, MS, 69, PhD(chem), 73. Prof Exp: Sci & eng asst nuclear physics, Nuclear Effects Lab, Edgewood Arsenal, Md, 69-71; res assoc infrared circular dichroism, Univ Southern Calif, 73-75; ASST PROF PHYS CHEM, SYRACUSE UNIV, 75- Mem: Am Chem Soc; Optical Soc Am. Res: Vibrational optical activity, including experimental and theoretical research in vibrational circular dichroism and Raman circular intensity differential scattering; resonance Raman spectroscopy and the theory of Raman line shapes. Mailing Add: Dept of Chem Syracuse Univ Syracuse NY 13210

NAFISSI-VARCHEI, MOHAMMAD MEHDI, b Arak, Iran, Sept 23, 36; m 69; c 1. ORGANIC CHEMISTRY. Educ: Tehran Univ, Iran, Licentiate, 60; Miami Univ, Ohio, MSc, 66; Mass Inst Technol, PhD(org chem), 69. Prof Exp: Instr chem, Tehran Univ, 60-65; res assoc spectros, Mass Inst Technol, 69-70; sr scientist med chem, 71-74, PRIN SCIENTIST, SCHERING CORP, 74- Mem: Am Chem Soc; Chem Soc Eng. Res: Medicinal chemistry; anti-infective agents. Mailing Add: Schering Corp 60 Orange St Bloomfield NJ 07003

NAFOOSI, A AZIZ, b Mosul, Iraq, June 2, 22; m 50; c 5. MATHEMATICS. Educ: Univ Baghdad, BA, 44; Univ Mich, MA, 50; Univ Colo, PhD(math), 60. Prof Exp: High sch teacher, Iraq, 44-48; instr, Higher Teachers Col, 50-56; instr, Univ Colo, 56-59; asst prof, Kans State Col, 59-62; assoc prof, Tulsa, 62-63; design specialist, Hayes Int Corp, 63-65, eng specialist, 65-67; PROF MATH, CHICAGO STATE UNIV, 67- Mem: Am Math Soc; Opers Res Soc Am; Iraq Math Soc. Res: Pure mathematics, particularly the theory of numbers and nonlinear differential equations; applied mathematics; guidance; optimization technique; application of calculus of variation for obtaining optimum trajectory equations; operation research, especially mathematical models for lethal area and related areas of research. Mailing Add: Dept of Math Chicago State Univ Chicago IL 60628

NAFPAKTITIS, BASIL G, b Athens, Greece, Dec 23, 29; m 64; c 2. BIOLOGY. Educ: Am Univ Beirut, BSc, 62, MS, 63; Harvard Univ, PhD(biol), 69. Prof Exp: Asst prof, 67-71, ASSOC PROF BIOL, UNIV SOUTHERN CALIF, 71- Mem: Am Soc Ichthyologists & Herpetologists; Am Soc Zoologists; Soc Syst Zool. Res: Ichthyology, particularly distribution, systematics and ecology of deep-sea fishes; bioluminescence. Mailing Add: Dept of Biol Sci Univ of Southern Calif Los Angeles CA 90007

NAFZIGER, RALPH HAMILTON, b Minneapolis, Minn, Aug 9, 37. CHEMICAL METALLURGY. Educ: Univ Wis, BS, 60; Pa State Univ, PhD(geochem), 66. Prof Exp: Geol asst, US Geol Surv, 63; res assoc geochem, Pa State Univ, 66-67; RES CHEMIST, US BUR MINES, 67- Concurrent Pos: Mem electroslag & plasma arc melting, Nat Mat Adv Bd, 74-75. Mem: Geol Soc Am; Mineral Soc Am; Am Inst Mining & Metall Engrs; Am Soc Metals. Res: Thermochemistry of metal-nonmetal systems; electroslag melting of metals; electric furnace smelting of lower-grade titaniferous, chromite and ferrous ores. Mailing Add: US Bur of Mines PO Box 70 Albany OR 97321

NAG, MONI, b India. ANTHROPOLOGY, POPULATION STUDIES. Educ: Univ Calcutta, BSc, 44, MSc, 46; Yale Univ, MA, 59, PhD(anthrop), 61. Prof Exp: Statistician, Anthrop Surv, Govt of India, 48-62; chief urban sociologist, Calcutta Metrop Planning Orgn, 62-64; superintending anthropologist, Anthrop Surv, Govt of India, 64-66; from asst prof to assoc prof anthrop, 66-71, sr lectr, 71-75, ASSOC PROF PUB HEALTH, COLUMBIA UNIV, 75-; CHIEF SOCIAL DEMOG SECT, INT INST STUDY HUMAN REPROD, 73- Mem: Fel Am Anthrop Asn; fel Soc Appl Anthrop; Int Union Sci Study Pop; Pop Asn Am; Indian Asn Study Pop. Res: Cultural factors related to human fertility and family planning; breast feeding in South Asia; consequences of its declining trend in terms of health, economics and fertility. Mailing Add: Int Inst Study of Human Reprod Columbia Univ 78 Haven Ave New York NY 10032

NAGAI, JIRO, b Nagano, Japan, Sept 26, 27; Can citizen; m 59; c 2. ANIMAL BREEDING. Educ: Univ Tokyo, BA, 52, DAgr(animal breeding), 61. Prof Exp: Instr animal breeding, Univ Tokyo, 55-65; RES SCIENTIST ANIMAL GENETICS, AGR CAN, 65- Concurrent Pos: Vis prof, NC State Univ, 74-75. Mem: Genetics Soc Can; Genetics Soc Am; Can Asn Lab Animal Sci; Japan Exp Animal Res Asn; Am Soc Animal Sci. Res: Animal breeding through experiments using mice; selection for nursing ability, mature weight and index combining the two, and long-term performance of crosses from the selected lines of mice. Mailing Add: Animal Genetics Bldg Animal Res Inst Agr Can Ottawa ON Can

NAGAI, TOSHIO, b Osaka, Japan, Jan 23, 22; m 48; c 3. PHYSIOLOGY, ZOOLOGY. Educ: Univ Tokyo, MEng, 45, MSc, 50, DSc(physiol), 60. Prof Exp: Asst prof zool & physiol, Natural Sci Div, Int Christian Univ, Tokyo, 58-63, prof, 63-67; RES SCIENTIST, RES INST, CAN DEPT AGR, 67- Concurrent Pos: Fulbright travel grant, 60-62; res assoc, Dept Physiol, Univ Ill, 60-62, vis scientist, 64-67; hon lectr, Univ Western Ont, 71- Mem: AAAS; Am Inst Biol Sci; NY Acad Sci; Am Soc Zool; Zool Soc Japan. Res: Electrophysiological study of smooth muscle; neuromuscular activity of intestinal muscle. Mailing Add: Univ Sub PO Can Dept of Agr London ON Can

NAGAMATSU, HENRY T, b Garden Grove, Calif, Feb 13, 16; m 42; c 2. FLUID PHYSICS, PLASMA PHYSICS. Educ: Calif Inst Technol, BS, 38 & 39, MS, 40, PhD(aeronaut), 49. Prof Exp: Asst, Calif Inst Technol, 38-41, theoret aerodynamicist, Douglas Aircraft Co, 41-42; theoret aerodynamicist, Curtiss-Wright Aircraft Corp, 42-43, head aeronaut res, Res Lab, 43-46; asst sect head, Jet Propulsion Lab, Calif Inst Technol, 46-49, sr res fel & dir hypersonic res, Aeronaut Dept, 49-55; RES ASSOC, RES & DEVELOP CTR, GEN ELEC CO, 55- Concurrent Pos: Consult, US Naval Ord Test Sta, 49-56, Rand Corp, 50-59, Atlas Proj, Gen Dynamics/Convair, 50-53, Midwest Res Inst, 52-55, Off Sci Res, US Dept Air Force, 57- & Aeronaut Lab, Wright-Patterson AFB, 60-; adj prof, Rensselaer Polytech Inst, 56-64; mem, Eng Noise Subcomt, Nat Acad Sci, 68-71. Mem: AAAS; fel Am Phys Soc; fel Am Inst Aeronaut & Astronaut; NY Acad Sci. Res: High temperature gas dynamics associated with intercontinental ballistic missiles, satellites and space vehicles; applied mathematics; magnetohydrodynamics; high temperature physics; physical chemistry; fluid mechanics; rockets and missiles; jet noise and acoustics; arc physics. Mailing Add: 1046 Cornelius Ave Schenectady NY 12309

NAGASAWA, HERBERT TSUKASA, b Hilo, Hawaii, May 31, 27; m 51; c 2. MEDICINAL CHEMISTRY. Educ: Western Reserve Univ, BS, 50; Univ Minn, PhD(org chem), 55. Prof Exp: Fel biochem, Univ Minn, 55-57; sr chemist, Radioisotope Serv, 57-61, sr scientist, Lab Cancer Res, 61, PRIN SCIENTIST, MED RES LABS, VET ADMIN HOSP, 61- Concurrent Pos: From asst prof to assoc prof, Col Pharm, Univ Minn, 59-72, prof, 73-; assoc ed, J Med Chem, 72-, actg ed, 73. Mem: AAAS; Am Chem Soc; Am Asn Cancer Res; NY Acad Sci; Am Soc Pharmacol & Exp Therapeut. Res: Synthesis on rational biochemical basis of cytotoxic agents and of amino acid antimetabolites, and evaluation of their biochemical pharmacology. Mailing Add: Vet Admin Hosp 54th St & 48th Ave S Minneapolis MN 55417

NAGATA, JUN-ITI, b Osaka, Japan, Mar 4, 25; m 57. TOPOLOGY. Educ: Univ Tokyo, BS, 47; Osaka Univ, DSc(topol), 56. Prof Exp: Lectr math, Osaka City Univ, 49-55, from asst prof to prof, 55-65; PROF MATH, UNIV PITTSBURGH, 63- Concurrent Pos: Vis assoc prof, Univ Wash, 59-60; mem, Inst Adv Study, 63-64; NSF grants, 63-64, 66-68 & 71-72; vis prof, Univ van Amsterdam, 75- Mem: Am Math Soc; Math Soc Japan. Res: Mathematics, especially topology. Mailing Add: Dept of Math Univ of Pittsburgh Pittsburgh PA 15213

NAGATANI, KUNIO, b Manchuria, Jan 23, 36; Japanese citizen; m 59; c 2. NUCLEAR PHYSICS. Educ: Tohoku Univ, BS, 58, MS, 60; Yale Univ, PhD(physics), 65. Prof Exp: Res assoc physics, Mass Inst Technol, 65-66; res fel physics, Calif Inst Technol, 67-68; from asst physicist to physicist, Brookhaven Nat Lab, 68-72; ASSOC PROF PHYSICS, TEX A&M UNIV, 72- Mem: Am Phys Soc; Sigma Xi. Res: Experimental nuclear physics, especially heavy ion reaction studies. Mailing Add: Cyclotron Inst Tex A&M Univ College Station TX 77843

NAGEL, CHARLES WILLIAM, b St Helena, Calif, Dec 8, 26; m 51; c 5. FOOD SCIENCE. Educ: Univ Calif, BA, 50, PhD(microbiol), 60. Prof Exp: Bacteriologist, US Dept War, Dugway Proving Grounds, Utah, 51-52; lab technician food technol, Univ Calif, 52-54; sr lab technician & coop agt food technol & preserv of refrig poultry, USDA & Univ Calif, 54-60; from asst prof to prof fruit & veg processing, Wash State Univ, 60-71; res dir, United Vintners Inc, Calif, 71-73; PROF FOOD SCI & TECHNOL, WASH STATE UNIV, 73- Concurrent Pos: Chmn food sci exec comt, Wash State Univ, 64-68. Mem: Am Soc Microbiol; Inst Food Technologists; Am Soc Enol; Sigma Xi (secy-treas, 75-77). Res: Sanitation; preservation of refrigerated poultry; psychrophilic bacteria; pectic enzymes of bacteria; food fermentations; enology. Mailing Add: Dept of Food Sci & Technol Wash State Univ Pullman WA 99163

NAGEL, CLATUS MARTIN, b Sheldon, NDak, Nov 10, 06. PLANT PATHOLOGY. Educ: NDak State Col, BS, 29; Iowa State Col, MS, 32, PhD(plant path), 38. Prof Exp: Asst plant path, Iowa State Col, 34-41; agt, Div Cereal Crops & Dis, Bur Plant Indust, USDA, 41-42, pathologist, Soils & Agr Eng, 43; pathologist, Naco Fertilizer Co, Fla, 44; asst prof plant path, 44-47, head dept, 47-69, PROF PLANT PATH, SDAK STATE UNIV, 47- Mem: AAAS; Am Phytopath Soc. Res: Microbiological aspects of grain spoilage in storage; biology, host range and control of sugar beet leaf spot; root and stalk rot resistance in corn; seed treatments of cereal crops; rust resistance in Populus; control of streak mosaic virus of wheat. Mailing Add: Dept of Plant Sci SDak State Univ Brookings SD 57006

NAGEL, DONALD LEWIS, b Blue Island, Ill, May 24, 41; m 63; c 1. CANCER, ORGANIC CHEMISTRY. Educ: Knox Col, BA, 67; Univ Nebr-Lincoln, PhD(org chem), 71. Prof Exp: Chemist, Libby, McNeill & Libby, 63-67; asst org chem, Univ Nebr-Lincoln, 67-71; instr, 71-74, ASST PROF CANCER, EPPLEY INST RES CANCER, UNIV NEBR MED CTR, OMAHA, 74- Mem: Am Chem Soc; Sigma Xi. Res: Chemical carcinogenesis with special emphasis on problems relating to organic chemical applications, nuclear magnetic resonance and mass spectrometry. Mailing Add: Eppley Inst Res Cancer Univ Nebr Med Ctr Omaha NE 68105

NAGEL, EDGAR HERBERT, b San Diego, Calif, Mar 24, 38; m 69; c 2. ANALYTICAL CHEMISTRY. Educ: Valparaiso Univ, BS, 60; Northwestern Univ, PhD(chem), 65. Prof Exp: From instr to asst prof, 63-69, ASSOC PROF CHEM, VALPARAISO UNIV, 69- Concurrent Pos: Consult, Argonne Nat Lab, 65- Mem: Am Chem Soc. Res: Gas chromatography; electrochemistry; computer applications in teaching. Mailing Add: Dept of Chem Valparaiso Univ Valparaiso IN 46383

NAGEL, EUGENE L, b Quincy, Ill, Aug 12, 24; c 3. ANESTHESIOLOGY. Educ: Cornell Univ, BEE, 49; Wash Univ, MD, 59. Prof Exp: Intern, St Luke's Hosp, St Louis, Mo, 59-60; resident anesthesiol, Presby Hosp, New York, 60-62; from asst prof to assoc prof, Sch Med, Univ Miami, 65-73; PROF ANESTHESIOL, UNIV CALIF, LOS ANGELES, 74- Concurrent Pos: Attend physician, Jackson Mem Hosp, Miami, Fla, 63-74, clin dir anesthesiol, 66-74; consult, Am Heart Asn, 69, Nat Registry of Emergency Med Technicians, 71-, Am Col Surgeons, Robert Wood Johnson Found & Bur Med Serv, Dept Health, Educ & Welfare, 72-; chmn sect clin care, Am Soc Anesthesiol, 73-74; chmn comn emergency med serv, AMA, 74- Mem: AMA; AM Physiol Soc; Am Soc Anesthesiol; Am Col Cardiol; Soc Critical Care Med. Res: Emergency care; telemetry; sudden cardiac death. Mailing Add: Harbor Gen Hosp Dept of Anesthes 1000 W Carson St Torrence CA 90274

NAGEL, FRITZ JOHN, b Ger, Oct 20, 19; nat US; m 53; c 3. ORGANIC CHEMISTRY. Educ: Univ Notre Dame, BSChE, 41, MS, 42. Prof Exp: Engr mat, Gen Elec Co, 50-52; tech dir, Capac Plastics, Inc, 52-53; dir res, Congoleum-Nairn, Inc, 53-56; vpres, Polymer Processes, Inc, 56-68; res & develop mgr, Signal Oil & Gas Co, 68-70; VPRES, CHAPMAN CHEM CO, 70- Concurrent Pos: Res engr, Res Labs, Westinghouse Elec Corp, 43-68. Mem: Am Chem Soc. Res: High polymers; laminates; insulating varnishes; adhesives; protective coatings; plastics; electrical insulation; fungicides. Mailing Add: 1264 E Massey Rd Memphis TN 38138

NAGEL, GLENN M, b Blue Island, Ill, Apr 16, 44; m 66. BIOCHEMISTRY. Educ: Knox Col, BA, 66; Univ Ill, PhD(biol chem), 71. Prof Exp: Scholar molecular biol & biochem, Univ Calif, Berkeley, 70-72; ASST PROF CHEM & MOLECULAR BIOL, CALIF STATE UNIV, FULLERTON, 72- Concurrent Pos: NIH res fel, 70-72. Mem: Am Chem Soc. Res: Structure and function of oligomeric enzymes; metabolic control; specific interactions between proteins and nucleic acids. Mailing Add: Dept of Chem Calif State Univ Fullerton CA 92634

NAGEL, MAX RICHARD, physics, see 12th edition

NAGEL, ROGER MILES, b Cleveland, Ohio, Nov 25, 25; m 47; c 2. POLYMER CHEMISTRY. Educ: Western Reserve Univ, BS, 49; Univ Buffalo, PhD(chem), 52. Prof Exp: Res chemist, Lubrizol Corp, 52-55; group leader res, Petro-Tex Chem Corp, FMC Corp, 56-61, supvr develop, 61-63; res assoc, Air Reduction Co, Inc, 63-69; mgr low pressure polyethylene group, Koppers Co, Inc, 69-75, MGR NEW PROD

OPPORTUNITIES, ARCO/POLYMERS, INC, 75- Mem: Am Chem Soc; Com Develop Asn. Res: Phosphorus chemistry; oil additives; poly-alpha-olefins; alpha-olefin copolymers; Ziegler catalysts; vinyl chloride homo- and copolymers. Mailing Add: Arco/Polymers Inc 440 College Park Dr Monroeville PA 15146

NAGEL, RONALD LAFUENTE, b Santiago, Chile, Jan 18, 36; US citizen; m 60; c 3. HEMATOLOGY. Educ: Univ Chile, Baccalaureate, 52, MD, 60. Prof Exp: Asst resident, Hosp Salvador, Sch Med, Univ Chile, 60-63; int fel, NIH, 63-64; res fel, 64-67, assoc med, 67-69, asst prof, 69-73, ASSOC PROF MED, ALBERT EINSTEIN COL MED, 73- Concurrent Pos: Mem, Exec Comt, Hemolytic Anemia Study Group, 74-; consult, Watson Lab, IBM Corp, 75-76. Honors & Awards: Award in Black, Found Res & Educ Sickle Cell Dis, 73. Mem: Am Soc Clin Invest; Am Soc Biol Chemists; Am Soc Hemat; Int Soc Hemat. Res: The molecular, cellular and clinical aspects of sickle cell anemia and other hemoglobinopathies; the structural and functional relationships in hemoglobin. Mailing Add: Dept of Med 1300 Morris Park Ave Albert Einstein Col of Med Bronx NY 10461

NAGEL, SIDNEY ROBERT, b New York, NY, Sept 28, 48. SOLID STATE PHYSICS. Educ: Columbia Univ, BA, 69; Princeton Univ, MA, 71, PhD(physics), 74. Prof Exp: RES ASSOC SOLID STATE PHYSICS, DIV ENG, BROWN UNIV, 74- Mem: Am Phys Soc. Res: Electronic structure of metals and alloys; the stability of metallic glasses in terms of its electronic structure; dielectric response of solids at finite wave vector. Mailing Add: Div of Eng Brown Univ Providence RI 02912

NAGEL, TERRY MARVIN, b Rochester, Minn, Mar 25, 43. INORGANIC CHEMISTRY, PHYSICAL CHEMISTRY. Educ: Macalester Col, BA, 65; Univ Minn, Minneapolis, PhD(chem), 70. Prof Exp: ASST PROF CHEM, MONMOUTH COL, 70- Mem: AAAS; Am Chem Soc. Res: Synthesis and characterization of organo-germanes and their transition metal derivatives; mechanisms of electron-transfer reactions. Mailing Add: Dept of Chem Monmouth Col Monmouth IL 61462

NAGELL, RAYMOND H, b Rochester, NY, Apr 10, 27; m 49; c 1. GEOLOGY. Educ: Univ Rochester, BA, 51, MS, 52; Stanford Univ, PhD(geol), 58. Prof Exp: Geologist, Cerro Corp, 52-55 & Cia Minera Cuprum, 56-57; chief geologist, Industria e Comercio de Minerios SA, 57-61; geologist, Shenon & Full, 61-62 & US Geol Surv, 63-70; GEOLOGIST, BETHLEHEM STEEL CORP, 70- Concurrent Pos: Tech consult, Industria e Comercio de Minerios, SA, 72- Mem: Soc Econ Geol; Geol Soc Am; Am Inst Mining, Metall & Petrol Engrs; Brazilian Geol Soc. Res: Economic geology; ore deposits. Mailing Add: Geol Dept Bethlehem Steel Corp Bethlehem PA 18016

NAGER, GEORGE THEODORE, b Zurich, Switz, Dec 1, 17; m 50; c 2. OTOLARYNGOLOGY. Educ: Zurich Univ, MD, 47; Swiss Bd Otolaryngol, dipl, 54; Am Bd Otolaryngol, dipl, 59. Prof Exp: PROF LARYNGOL & OTOL & CHIEF DIV, SCH MED, JOHNS HOPKINS UNIV, 70- Concurrent Pos: Consult Baltimore City Hosps, Greater Baltimore Med Ctr, Good Samaritan Hosp & USPHS Hosp, Baltimore. Mem: Am Acad Ophthal & Otolaryngol; Am Col Surg; Am Laryngol, Rhinol & Otol Soc; Am Otol Soc. Mailing Add: Dept of Laryngol & Otol Johns Hopkins Univ Sch of Med Baltimore MD 21205

NAGER, MAXWELL, organic chemistry, petroleum chemistry, see 12th edition

NAGER, URS FELIX, b Zurich, Switz, May 15, 22; nat US; m 51; c 3. ORGANIC CHEMISTRY, BIOCHEMISTRY. Educ: Swiss Fed Inst Technol, Chem Eng, 45, PhD(org chem), 49. Prof Exp: Res biochemist, Commercial Solvents Corp, Ind, 59-52; sr res chemist, Burke Res Co, Mich, 52-58; res assoc, Squibb Inst Med Res, 58-60, res supvr, 60-69, ASST DIR, SQUIBB CORP, 69- Mem: Am Chem Soc; NY Acad Sci. Res: Antibiotics; enzymes; steroids; designed experimentation. Mailing Add: Bunkerhill Rd RD 1 Princeton NJ 08540

NAGERA, HUMBERTO, b Havana, Cuba, May 23, 27; Brit citizen; m 52; c 3. PSYCHIATRY, PSYCHOANALYSIS. Educ: Maristas Col, BS, 45; Havana Univ, MD, 52. Prof Exp: Mem staff child development, Hampstead Clin, London, 58-68; prof psychiat, 68-73, PROF CHILD PSYCHIAT & CHIEF YOUTH SERV, UNIV MICH, ANN ARBOR, 73-, DIR CHILD PSYCHOANAL STUDY PROG, 68- Mem: Brit Psychoanal Soc; Am Psychoanal Asn; Int Psychoanal Asn; Asn Prof Child Psychiatrists. Res: Child development and psychopathology. Mailing Add: Children's Psychiat Hosp Ann Arbor MI 48104

NAGIN, DANIEL STEVEN, b Philadelphia, Pa, Nov 29, 48; m 72. STATISTICAL ANALYSIS. Educ: Carnegie-Mellon Univ, BS & MS, 71, PhD(urban & pub affairs), 75. Prof Exp: Systs analyst health, 71-72, RES ASSOC URBAN & PUB AFFAIRS, CARNEGIE-MELLON UNIV, 74-, ASST PROF TRANSP, 75- Concurrent Pos: Prin staff mem panel on deterrence & incapacitation, Nat Acad Sci, 75- Res: Determining the deterrent effects of criminal sanctions; factors which influence an individual's choice among alternative transportation modes and alternative sources for receiving primary care. Mailing Add: Sch of Urban & Pub Affairs Carnegie-Mellon Univ Pittsburgh PA 14213

NAGLE, DARRAGH (EDMUND), b New York, NY, Feb 25, 19; m 49; c 3. PHYSICS. Educ: Calif Inst Technol, BS, 40; Columbia Univ, AM, 42; Mass Inst Technol, PhD(physics), 47. Prof Exp: Lectr, Columbia Univ, 41-42; res assoc, Metall Lab, Univ Chicago, 43; group leader, Argonne Nat Lab, 43-44; asst group leader, Los Alamos Sci Lab, 44-45; res assoc, Mass Inst Technol, 45-47, instr, 47-48; Fulbright fel, Cambridge Univ, 48-49; asst prof physics, Univ Chicago, 49-55; mem staff, 55-62, group leader, 62-65, assoc div leader, 65-68, ALT DIV LEADER MEDIUM ENERGY PHYSICS, LOS ALAMOS SCI LAB, 68- Concurrent Pos: Guggenheim fel, 53; mem, Spec Comt Latin Am Coord Policy Bd, 70-73 & Mass Inst Technol Bates Linac Prog Comt, 73- Mem: Fel Am Phys Soc. Res: Physics of particles; nuclear physics; accelerator design; plasma physics. Mailing Add: Los Alamos Sci Lab Los Alamos NM 87545

NAGLE, FRANCIS J, b Lynn, Mass, July 1, 24; m 60; c 11. EXERCISE PHYSIOLOGY, CARDIOVASCULAR PHYSIOLOGY. Educ: Univ Nebr, BS, 51, MA, 53; Boston Univ, EdD(health, phys educ), 59; Univ Okla, PhD(physiol), 66. Prof Exp: Asst prof phys educ, Univ Fla, 56-62; sect chief biodynamics br, Civil Aeromed Res Inst, Fed Aviation Agency, 62-64; from asst prof to assoc prof, 66-75, PROF PHYSIOL & PHYS EDUC, BIODYNAMICS LAB, UNIV WIS-MADISON, 75-, DIR LAB, 73- Mem: AAAS; Am Asn Health Phys Educ & Recreation; Am Asn Univ Profs; Am Physiol Soc; NY Acad Sci. Res: Cardiovascular physiology and metabolism in stress. Mailing Add: Dept of Phys Educ Univ of Wis Sch of Educ Madison WI 53706

NAGLE, FREDERICK, JR, b Queens, NY, Jan 30, 37; m 57; c 2. PETROLOGY, MINERALOGY. Educ: Lafayette Col, BA, 58; Princeton Univ, MA, 61, PhD(geol), 66. Prof Exp: Asst prof geol, Juniata Col, 64-68; asst prof marine geol, 68-72, ASSOC PROF GEOL, UNIV MIAMI, 72- Concurrent Pos: NSF col teacher res fel, 67-68. Mem: AAAS; Mineral Soc Am; Geol Soc Am; Am Geophys Union. Res: Igneous petrology; Caribbean Island arc geology. Mailing Add: Rosensteil Sch Marine & Atmos Sci Univ of Miami Miami FL 33149

NAGLE, JAMES JOHN, b Wilkes-Barre, Pa, Nov 10, 37; m 60. GENETICS, EVOLUTION. Educ: Bloomsburg State Col, BS, 62; NC State Univ, MS, 65, PhD(genetics), 67. Prof Exp: Instr genetics, NC State Univ, 66-67; asst prof genetics, 67-74, ASSOC PROF ZOOL & BOT, DREW UNIV, 74- Concurrent Pos: Adj prof, Hunter Col, City Univ New York, 71- Mem: Soc Study Evolution. Res: Experimental evolution, especially interspecific hybridization, chromosomal polymorphism and population fitness concerning Drosophila species; biological education, especially biology for non-majors; social implications of biology. Mailing Add: Dept of Zool Drew Univ Madison NJ 07940

NAGLE, JOHN F, b Easton, Pa, Sept 29, 39; m 67. PHYSICS, STATISTICAL MECHANICS. Educ: Yale Univ, BA, 60, MS, 62, PhD(physics), 65. Prof Exp: NATO fel physics & statist mech, King's Col, London, 65-66; res assoc statist mech, Cornell Univ, 66-67; asst prof physics, 67-72, ASSOC PROF PHYSICS & BIOL SCI, CARNEGIE MELLON UNIV, 72- Concurrent Pos: A P Sloan Found fel, 69-71. Res: Phase transitions; solid state physics; biophysics. Mailing Add: Dept of Physics Carnegie-Mellon Univ Pittsburgh PA 15213

NAGLE, WILLIAM ARTHUR, b W Reading, Pa, May 16, 43; m 70. RADIATION BIOPHYSICS, MOLECULAR BIOLOGY. Educ: Albright Col, BS, 65; Okla Univ, Norman, MS, 66; Univ Tex Southwestern Med Sch, PhD(radiation biol), 72. Prof Exp: Fel, Harvard Med Sch, 72-74; ASST PROF RADIOL-NUCLEAR MED, UNIV ARK MED SCI & MOLECULAR BIOLOGIST, MED RES SERV, VET ADMIN HOSP, LITTLE ROCK, 74- Honors & Awards: Student Travel Award, Radiation Res Soc, 71. Mem: Biophys Soc; Radiation Res Soc. Mailing Add: Div Nuclear Med Col of Med Univ Ark for Med Sci Little Rock AR 72201

NAGLER, ARNOLD LEON, b New York, NY, Aug 18, 32; m 61; c 2. PATHOLOGY, PHYSIOLOGY. Educ: City Col New York, BS, 53; NY Univ, MD, 58, PhD(path), 60. Prof Exp: Res assoc, Col Med, NY Univ, 58-60; res assoc, Mt Sinai Hosp, 60-61; ASST PROF SURG & PATH, ALBERT EINSTEIN COL MED, 61- Concurrent Pos: NIH fel, NY Univ, 60-61; prin investr, NIH grant, Albert Einstein Col Med, 70- Mem: NY Acad Sci; Fedn Am Socs Exp Biol. Res: Pathophysiology of shock; determination of the mechanisms involved in the pathogenesis of shock and the mortality therefrom; methods of circumventing lethality; choline and nutritional deficiencies and the pathology resulting therefrom. Mailing Add: Dept of Surg Albert Einstein Col of Med Bronx NY 10461

NAGLER, BENEDICT, b Czernowitz, Austria, Mar 14, 00; nat US; m 27; c 2. NEUROLOGY, PSYCHIATRY. Educ: Univ Hamburg, MD, 25. Prof Exp: Resident neurol, psychiat & internal med, St George Gen Hosp, Hamburg, Ger, 25-27 & Munic Hosp, Berlin, 27-31; pvt practr, Berlin, 31-33, Tunis, NAfrica, 34 & Newark, NJ, 35-43; chief neurol & psychiat serv, Vet Admin Hosp, Richmond, Va, 46-53, chief neurol div, Psychiat & Neurol Serv, Vet Admin, DC, 53-57; supt, Lynchburg Training Sch & Hosp, 57-73; RETIRED. Concurrent Pos: Asst prof psychiat & neurol, Med Col Va, 46-67; lectr, Richmond Prof Inst, 46-53; assoc prof, Georgetown Univ, 53-57, prof lectr, 57-67; consult, Neurol Study Sect & Nat Adv Neurol Dis & Blindness Coun, NIH, 53-57, neurol field invests study sect, Div Res Grants, 58-61, Neurol Prog Proj Comt, 61-65 & Perinatal Res Comt, 65-73; lectr, Univ Va, 57-67; consult, Children's Rehab Ctr, Md, 59; mem coun, Int Asn Sci Study Ment Deficiency, 64-67. Mem: Am Epilepsy Soc; Am EEG Soc; fel Am Psychiat Asn; Asn Res Nerv & Ment Dis; fel Am Acad Neurol. Res: Multiple sclerosis; cerebral atherosclerosis; brain damage due to prenatal, perinatal and neonatal causes; electroencephalography; physiological impact of physical disability. Mailing Add: 2404 Langhorne Rd Lynchburg VA 24501

NAGLER, FREDERICK PAUL, medicine, deceased

NAGLER, KENNETH MALCOLM, b Springfield, Mass, June 26, 20; m 47; c 3. METEOROLOGY. Educ: Univ Mass, BS, 42; Univ Chicago, MS, 47. Prof Exp: Asst instr meteorol, Univ Chicago, 46-48; METEOROLOGIST, NAT WEATHER SERV, 48-, CHIEF, SPACE OPERS SUPPORT DIV, 65- Honors & Awards: Meritorious Serv Award, US Dept Com, 59, Gold Medal Award, 73. Mem: Am Meteorol Soc; Am Geophys Union. Res: Transport of atmospheric contaminants; worldwide weather support for space operations. Mailing Add: 3912 Longfellow St Hyattsville MD 20781

NAGLER, ROBERT CARLTON, b Iowa City, Iowa, July 4, 23; m 47; c 5. ORGANIC CHEMISTRY. Educ: William Penn Col, BS, 47; Univ Mo, MA, 49; Univ Iowa, PhD(chem), 53. Prof Exp: Asst chem, Univ Mo, 47-49; asst, Univ Iowa, 49-51, instr, 52-53; asst prof, Purdue Univ, 53-56; from asst prof to assoc prof, 56-69, PROF CHEM, WESTERN MICH UNIV, 69-, ASST CHMN DEPT, 68- Concurrent Pos: USAID sci adv, Nigeria, 62-64. Mem: Am Chem Soc. Res: Nitrogen-magnesium-halide reagents; Grignard reactions on coumarins; synthesis of anti-tumor agents; fluoride analysis. Mailing Add: Dept of Chem Western Mich Univ Kalamazoo MI 49008

NAGLIERI, ANTHONY N, b New York, NY, Apr 15, 30; m 55; c 1. INDUSTRIAL ORGANIC CHEMISTRY. Educ: Fordham Univ, BS, 51, MS, 53; Columbia Univ, PhD(chem), 59. Prof Exp: Proj leader, 59-64, res assoc, 64-68, sect head res & develop, 68-74, ASST DIR RES, HALCON INT, INC, LITTLE FERRY, NJ, 74- Mem: Am Chem Soc. Res: Liquid and vapor phase oxidations; catalysis; free radical chemistry.

NAGODE, LARRY ALLEN, b New Deal, Mont, Nov 15, 38; m 63; c 2. VETERINARY PATHOLOGY. Educ: Colo State Univ, DVM, 63; Ohio State Univ, MSc, 65, PhD(vet path), 68. Prof Exp: Morris Animal Found res fel vet path, Ohio State Univ, 63-65, Nat Cancer Inst res fel, 65-68; Nat Cancer Inst res fel biochem, Med Sch, Univ Pa, 69-70; ASST PROF VET PATH, COL VET MED, OHIO STATE UNIV, 70- Res: Metabolism and mechanism of action of vitamin D in normal and diseased states. Mailing Add: Dept of Vet Path Ohio State Univ Col of Vet Med 1925 Coffey Rd Columbus OH 43210

NAGPAL, TARLOK SINGH, b Ft Saundeman, India, Dec 1, 32; m; c 1. NUCLEAR PHYSICS. Educ: Univ Panjab, India, BA, 52, BT, 53; Aligarh Muslim Univ, India, MSc, 56; Univ BC, PhD(nuclear physics), 64. Prof Exp: Lectr physics, Khalsa Col, Amritsar, 56-59; demonstr, Univ BC, 60-64; asst prof, 64-69, ASSOC PROF PHYSICS, BISHOP'S UNIV, 69- Res: Radioactive decay schemes. Mailing Add: Dept of Physics Bishop's Univ Lennoxville PQ Can

NAGY, ANDREW F, b Budapest, Hungary, May 2, 32; m 65. AERONOMY, ATMOSPHERIC PHYSICS. Educ: Univ New South Wales, BE, 57; Univ Nebr, MSc, 59; Univ Mich, MSE, 60, PhD(elec eng), 63. Prof Exp: Design engr, Elec Control & Eng Co, 56-57; instr elec eng, Univ Nebr, 57-59; asst res engr, 60-63, from asst prof to assoc prof elec eng, 63-71, PROF ATMOSPHERIC SCI & PROF ELEC ENG, UNIV MICH, ANN ARBOR, 71- Mem: Inst Elec & Electronics Eng; Am Geophys

Union. Res: Theoretical and experimental studies of the chemistry and physics of the terrestrial and planetary atmospheres. Mailing Add: Dept of Atmos & Oceanic Sci Univ of Mich Ann Arbor MI 48109

NAGY, BARTHOLOMEW STEPHEN, b Budapest, Hungary, May 11, 27; nat US; m 52; c 2. ORGANIC GEOCHEMISTRY. Educ: Pazmany Peter Univ, Hungary, BA, 48; Columbia Univ, MA, 50; Pa State Univ, PhD(mineral), 53. Prof Exp: Asst, Pa State Univ, 49-53; res engr, Pan Am Oil Co, 53-55; res assoc & supvr geophys res, Cities Serv Res & Develop Co, Okla, 55-57; asst prof chem, Fordham Univ, 57-60, assoc prof, 60-65; assoc res geochemist, Univ Calif, San Diego, 65-68; PROF GEOSCI, UNIV ARIZ, 68- Concurrent Pos: Vis assoc prof, Univ Calif, San Diego, 63-65; mem adv comt, Lunar Sci Inst, 72-; managing ed, Precambrian Res, Elsevier Pub Co, Amsterdam, Neth, 72- Mem: Geochem Soc; Am Chem Soc; NY Acad Sci; fel Am Inst Chemists; Int Soc Study Origin Life. Res: Lunar sample analysis; hydrocarbons; meteorites; clay mineralogy; x-ray crystallography. Mailing Add: 533 Space Sci Dept of Geosci Univ of Ariz Tucson AZ 85721

NAGY, BELA FERENC, b Nagybanhegyes, Hungary, May 15, 26; US citizen; m 58; c 2. BIOCHEMISTRY. Educ: Eötvös Lorand Univ, Budapest, dipl biol & chem, 53; Brandeis Univ, PhD(biochem), 64. Prof Exp: Asst prof biochem, Eötvös Lorand Univ, Budapest, 53-56; Nat Acad Sci res fel, Rockefeller Inst, 57, res assoc, 57-59; res assoc, NY Univ, 59-60; Muscular Dystrophy Asn Am spec fel, Inst Muscle Dis, 60; res assoc, Retina Found, 64-70, STAFF SCIENTIST, BOSTON BIOMED RES INST, 70- Concurrent Pos: NIH res grant, 67-69 & career develop award, 67-72; res assoc neuropath, Harvard Univ, 68-69; assoc, 69- Mem: AAAS; Am Chem Soc; Brit Biochem Soc; Biophys Soc; Am Soc Biol Chem. Res: Chemistry and physiology of muscle contraction; chemistry and physical chemistry of proteins. Mailing Add: Dept of Muscle Res Boston Biomed Res Inst 20 Staniford St Boston MA 02114

NAGY, JULIUS G, b Balatonboglar, Hungary, Aug 7, 25; US citizen; m 49; c 2. NUTRITION, BACTERIOLOGY. Educ: Wayne State Univ, BS, 60; Colo State Univ, MS, 63, PhD(wildlife nutrit), 66. Prof Exp: Animal scientist, 63-65, from instr to asst prof, 65-70, ASSOC PROF WILDLIFE BIOL, COLO STATE UNIV, 70- Mem: Wildlife Soc. Res: Wildlife nutrition and physiology, especially the rumen microbiological digestion of wild ruminants. Mailing Add: Dept of Wildlife Biol Colo State Univ Ft Collins CO 80521

NAGY, KENNETH ALEX, b Santa Monica, Calif, July 1, 43; m 67; c 2. ENVIRONMENTAL PHYSIOLOGY. Educ: Univ Calif, Riverside, AB, 67, PhD(biol), 71. Prof Exp: Actg asst prof zool, 71-72, ADJ ASST PROF BIOL, DEPT BIOL & ASST RES ZOOLOGIST, ENVIRON BIOL, NUCLEAR MED & RADIATION BIOL LAB, UNIV CALIF, LOS ANGELES, 72- Mem: Sigma Xi; AAAS; Am Soc Ichthyologists & Herpetologists; Am Inst Biol Sci; Ecol Soc Am. Res: Physiology and behavior of desert vertebrates as these relate to the animal's survival in nature, including water, electrolyte and energy balance in field animals measured with isotopically-labeled water. Mailing Add: Nuclear Med & Radiation Biol Lab Univ of Calif Los Angeles CA 90024

NAGY, THERESA ANN, b Wheeling, WVa, July 16, 46. ASTROPHYSICS. Educ: West Liberty State Col, BS, 68; Tex A&M Univ, MS, 70; Univ Pa, PhD(astron), 74. Prof Exp: Analyst & programmer, Anal Inc, 72-73; SCI ANALYST, COMPUT SCI CORP, NASA GODDARD SPACE FLIGHT CTR, 73- Mem: Am Astron Soc; Sigma Xi. Res: Computation of synthetic light curves for contact binary star systems; computer generation and manipulation for stellar and nonstellar astronomical data; image processing; cosmic ray muon intensity research. Mailing Add: Code 671 NASA Goddard Space Flight Ctr Greenbelt MD 20771

NAGYLAKI, THOMAS ANDREW, b Budapest, Hungary, Jan 29, 44; Can citizen; m 69. POPULATION GENETICS. Educ: McGill Univ, BS, 64; Calif Inst Technol, PhD(physics), 69. Prof Exp: Res assoc physics, Univ Colo, 69-71; vis asst prof, Ore State Univ, 71-72; proj assoc med genetics, Univ Wis-Madison, 72-74, asst scientist med genetics & math res ctr, 74-75; ASST PROF BIOPHYS & THEORET BIOL, UNIV CHICAGO, 75- Res: Theoretical population genetics; linkage and selection, geographical structure of populations; human behavior genetics. Mailing Add: Dept of Biophys & Theoret Biol Univ of Chicago Chicago IL 60637

NAGYVARY, JOSEPH, b Szeged, Hungary, Apr 18, 34; m 63; c 3. ORGANIC CHEMISTRY. Educ: Univ Zurich, PhD(org chem), 62. Prof Exp: Res asst peptide synthesis, CIBA, Ltd, Switz, 62; fel nucleotides, Univ Cambridge, 62-64; res asst prof nucleotides synthesis, Univ Conn, 64-65; asst prof nucleic acids, Sch Med, Creighton Univ, 65-68; assoc prof biochem, 68-74, PROF BIOCHEM, COL AGR, TEX A&M UNIV, 74- Concurrent Pos: Swiss Regional Scholar fel, 62-63; NIH grant, 64-66; consult, CIBA, Ltd, 64. Mem: Am Chem Soc. Res: Nucleic acid chemistry; synthesis of nucleotide di-and triesters. Mailing Add: Dept of Biochem & Biophys Tex A&M Univ Col of Agr College Station TX 77840

NAHABEDIAN, KEVORK VARTAN, b Boston, Mass, Oct 31, 28; m 57; c 1. CHEMISTRY. Educ: Mass Inst Technol, SB, 52; Univ Vt, MS, 54; Univ NH, PhD(chem), 59. Prof Exp: Asst chem, Univ Vt, 52-54, instr, 62-63; instr, Lafayette Col, 54-55; asst, Univ NH, 55-57, fel, 57-59; res assoc, Brown Univ, 59-61; res chemist, Qm Res & Eng Ctr, 61-62; asst prof chem, Union Col, NY, 63-68; PROF CHEM & CHMN DEPT, STATE UNIV NY COL GENESEO, 68- Mem: Am Chem Soc. Res: Physical organic chemistry; electrophilic and nucleophilic aromatic substitution; acid and base catalysis; donor functions for basic media. Mailing Add: Dept of Chem State Univ of NY Geneseo NY 14454

NAHAS, ALY, b Cairo, Egypt, Dec 29, 29; US citizen; m 58; c 5. BIOCHEMICAL PHARMACOLOGY. Educ: Cairo Univ, BSc, 51, MSc, 56 & 58; Tufts Univ, PhD(pharmacol), 67. Prof Exp: Instr pharmacol, Sch Med, Yale Univ, 69; ASST PROF BIOCHEM & MED, SCH MED, UNIV ROCHESTER, 69- Concurrent Pos: Nat Cancer Inst fel cancer chemother, Sch Med, Yale Univ, 67-69. Mem: AAAS; Am Asn Cancer Res; Am Soc Pharmacol & Exp Therapeut; Am Chem Soc; Am Acad Polit & Soc Sci. Res: Cancer chemotherapy. Mailing Add: Div of Oncol Univ of Rochester Med Ctr Rochester NY 14642

NAHAS, GABRIEL GEORGES, b Alexandria, Egypt, Mar 4, 20; nat US; m 54; c 3. PHARMACOLOGY. Educ: Univ Toulouse, BA, 37, MD, 44; Univ Rochester, MS, 49; Univ Minn, PhD(physiol), 53. Prof Exp: Chief lab exp surg, Marie Lannelongue Hosp, Paris, 53-55; asst prof physiol, Univ Minn, 55-57; chief respiratory sect, Walter Reed Army Inst Res, 57-59; assoc prof anesthesiol, 59-62, RES PROF ANESTHESIOL, COL PHYSICIANS & SURGEONS, COLUMBIA UNIV, 62- Concurrent Pos: Mem, Med Adv Bd, Coun Circulation & Basic Sci, Am Heart Asn; mem, Comt Trauma, Nat Res Coun, 64; consult, Oceanog Inst, Monaco; adj prof anesthesiol res, Univ Paris, 68. Honors & Awards: Mem, Order Brit Empire; Off, Order Orange Nassau; Presidential Medal Freedom with Gold Palm. Mem: Am Physiol Soc; Am Soc Pharmacol & Exp Therapeut; Am Soc Artificial Internal Organs; Harvey Soc. Res: Acid-base equilibrium, catecholamines metabolism; mechanism of action of drugs; marihuana. Mailing Add: Dept of Anesthesiol Columbia Univ Col Phys & Surg New York NY 10032

NAHHAS, FUAD MICHAEL, b Sidon, Lebanon, Jan 29, 27; m 53; c 3. PARASITOLOGY, MEDICAL MICROBIOLOGY. Educ: Univ of the Pac, AB, 58, MA, 60; Purdue Univ, PhD(biol sci), 63. Prof Exp: Res fel parasitol, Fla State Univ, 63-64; from asst prof to assoc prof, 64-71, PROF BIOL, UNIV OF THE PAC, 71- Mem: Am Inst Biol Sci; Am Soc Parasitol; Am Soc Microbiol; NY Acad Sci. Res: Parasites of vertebrates; taxonomy and life history studies; geographic distribution of parasites; the antibiogram as an aid in the identification of bacteria. Mailing Add: Dept of Biol Sci Univ of the Pac Stockton CA 95211

NAHIKIAN, HOWARD MOVESS, b Asheville, NC, June 9, 10; m 36; c 2. MATHEMATICS. Educ: Univ NC, AB, 33, AM, 34, PhD(math), 39. Prof Exp: Asst instr math, Univ NC, 34-35; from instr to asst prof, 35-42, assoc prof, 46-53, PROF MATH, NC STATE UNIV, 53- Mem: Am Math Soc; Math Asn Am. Res: Modern algebra; matrices. Mailing Add: 3116 Leonard St Raleigh NC 27607

NAHIN, PAUL GILBERT, b St Paul, Minn, July 7, 16; m 39; c 5. CHEMISTRY. Educ: Univ Minn, AB, 39; Univ Calif, PhD(radiation chem), 42. Prof Exp: Group leader synthetic rubber res, US Rubber Co, 42-44; res engr, Union Oil Co, Calif, 44-52, sr res chemist, Res Ctr, 52-55, res assoc, 55-69; exec vpres res & develop div, Direct Image Corp, 69-75; MANAGING DIR FOREIGN TECHNOL VENTURES, B&W INT DIV, B&W INC, 75- Concurrent Pos: Mem comt clay minerals, Nat Res Coun, 52-64. Mem: AAAS; Graphic Arts Technol Found; Clay Minerals Soc (vpres, 68, pres, 69); Am Chem Soc. Res: Rubber synthesis; desulfurization, hydroforming and cracking catalysis; electron microscopy; tracer chemistry with radioactive isotopes; chemistry of clay and petroleum reservoirs; radiation chemistry; petrochemicals; chemistry of lithography, aerosols, polymers and adhesives. Mailing Add: 525 Union Pl Brea CA 92621

NAHMIAS, ANDRE JOSEPH, b Alexandria, Egypt, Nov 20, 30; US citizen; m 56; c 3. VIROLOGY, PEDIATRICS. Educ: Univ Tex, Austin, BA, 50, MA, 52; Univ Mich, MPH, 53; George Washington Univ, MD, 57. Prof Exp: Intern med, USPHS Hosp, Staten Island, NY, 57-58; resident pediat, Boston City Hosp, Mass, 60-62, clin assoc, 62-64; asst prof pediat & prev med, 64-67, assoc prof, 67-70, PROF PEDIAT, SCH MED, EMORY UNIV, 70-, CHIEF SECT INFECTIOUS DIS & IMMUNOL, 64- Concurrent Pos: NIH spec res fel virol, Mass Mem Hosp, 62-64; var study grants, USPHS, Am Cancer Soc, Surgeon Gen, US Army & Nat Found, 64-; NIH res career develop award, 66-71; res assoc microbiol, Sch Med, Boston Univ, 62-64; lectr var univs, US, Europe & Australia, 64-; consult, Am Red Cross, 68-72; mem study sect, Grants Rev Bd, Nat Commun Dis Ctr, 69-70. Honors & Awards: Mead Johnson Award Pediat Res, 74. Mem: AAAS; fel Am Pub Health Asn; fel Am Acad Pediat; Soc Pediat Res; Infectious Dis Soc Am. Res: Bacterial infections, particularly staphylococcus, pertussis and leptospirosis; viral infections, particularly herpes viruses and relation to cancer; serological techniques, particularly immunofluorescence; immune mechanisms, particularly cellular immunity; clinical and epidemiological aspects of infectious diseases; fetal and neonatal diseases and neurological diseases. Mailing Add: Emory Univ Sch of Med Atlanta GA 30303

NAHORY, ROBERT EDWARD, b McKeesport, Pa, Mar 1, 38; m 60; c 3. PHYSICS. Educ: Carnegie-Mellon Univ, BS, 60; Purdue Univ, MS, 62, PhD(physics), 67. Prof Exp: MEM TECH STAFF PHYSICS, BELL TEL LABS, 67- Mem: Am Phys Soc. Res: Semiconductor physics, especially optical properties; characteristics of new semiconductor materials; behavior of semiconductors under high excitation; lasers; oscillatory photoconductivity. Mailing Add: Bel Tel Labs Room 4d-409 Holmdel NJ 07733

NAHRWOLD, DAVID LANGE, b St Louis, Mo, Dec 21, 35; m 58; c 4. SURGERY, GASTROENTEROLOGY. Educ: Ind Univ, AB, 57, MD, 60. Prof Exp: Intern surg, Med Ctr, Ind Univ, Indianapolis, 60-61, resident, 61-65, asst prof, Sch Med, 68-70; assoc prof surg, 70-73, PROF SURG, VCHMN DEPT & CHIEF DIV GEN SURG, COL MED, PA STATE UNIV, 73- Concurrent Pos: Scholar, Univ Calif, Los Angeles, 65-66. Mem: Am Gastroenterol Soc; Cent Surg Asn; fel Am Col Surgeons; Soc Surg Alimentary Tract; Soc Univ Surgeons. Res: Gastrointestinal physiology. Mailing Add: Dept of Surg Pa State Univ Hershey Med Ctr Hershey PA 17033

NAHRWOLD, MICHAEL LANGE, b St Louis, Mo, Nov 23, 43; m 71; c 2. ANESTHESIOLOGY, NEUROSCIENCES. Educ: Ind Univ, AB, 65, MD, 69; Am Bd Anesthesiology, dipl, 74. Prof Exp: Intern anesthesia, Univ Colo Med Ctr, 69-70, resident, 70-72, res fel & instr, 72-73; clin assoc, NIH, 73-75; ASST PROF & RES DIR ANESTHESIA & RES ASSOC PHYSIOL, M S HERSHEY MED CTR, 75- Concurrent Pos: Res Initiation grant, Am Soc Anesthesiologists, 76. Mem: Fel Am Col Anesthesiologists; Am Soc Anesthesiologists; Am Physiol Soc; Am Soc Neurochem. Res: The effects of volatile and nonvolatile anesthetics on metabolic parameters of C-6 astrocytoma and C-1300 neuroblastoma cells in culture. Mailing Add: Dept of Anesthesia Milton S Hershey Med Ctr Hershey PA 17033

NAIB, ZUHER M, b Aleppo, Syria, Dec 10, 27; US citizen; m 58; c 3. PATHOLOGY. Educ: Univ Geneva, BS, 49, MD, 52; Am Bd Path, dipl, 60. Prof Exp: From instr to asst prof cytopath, Univ Md, 58-63; assoc prof, 63-67, PROF PATH & PROF GYNEC & OBSTET, EMORY UNIV, 67- Concurrent Pos: Fel path, Univ Va, 54-58; mem staff, Cytopath Div, Grady Mem Hosp, Atlanta, Ga. Mem: Am Cytol Soc. Res: Exfoliative cytopathology. Mailing Add: Dept of Path Emory Univ Atlanta GA 30322

NAIBERT, ZANE ELVIN, b Cedar Rapids, Iowa, Nov 19, 31; m 58. SCIENCE EDUCATION, CHEMISTRY. Educ: Coe Col, BA, 54; Univ Iowa, MS, 57, PhD(sci educ, chem), 64. Prof Exp: Asst anal chem, La State Univ, 54-55; asst, Univ Iowa, 55-57, asst & teacher, Lab Schs, 59-61; teacher, Algona Community Schs, 58-59; asst prof chem, State Univ NY Col Cortland, 61-67; assoc prof & chmn div sci & math, 67-71, PROF CHEM, MONTGOMERY COL, 71- Mem: AAAS; Am Chem Soc. Mailing Add: Dept of Chem Montgomery Col Takoma Park MD 20012

NAIDE, MEYER, b Russia, Mar 13, 07; nat US; m 33; c 2. MEDICINE. Educ: Univ Pa, AB, 29, MD, 32. Prof Exp: In chg vascular clin, Woman's Med Col Pa, 52-58, assoc prof med, 58-67; ASST PROF MED, UNIV PA, 67- DIR PERIPHERAL VASCULAR DIV, GRAD HOSP, 64- Concurrent Pos: In chg vascular clin, Einstein Med Ctr, 36-72; assoc, Univ Pa, 46-; pvt practr. Mem: AAAS; AMA; Am Col Physicians; Am Col Angiol; Int Cardiovasc Soc. Res: Peripheral vascular disease. Mailing Add: 2034 Spruce St Philadelphia PA 19103

NAIDORF, IRVING JOSEPH, b Brooklyn, NY, Oct 19, 15; m 42; c 2. DENTISTRY. Educ: NY Univ, BA, 37; Columbia Univ, DDS, 41; Am Bd Endodont, dipl, 65. Prof Exp: Clin asst dent, 52-54, asst, 54-55, instr, 55-58, asst clin prof, Sch Dent & Oral Surg, 58-67, assoc clin prof, 67-71, CLIN PROF DENT, SCH DENT & ORAL SURG, COLUMBIA UNIV, 71-, ASST DEAN POST GRAD EDUC, 74-, DIR POST-GRAD ENDODONT, 70- Concurrent Pos: Consult, Vet Admin Hosp, Montrose, 59-; dir, Am Bd Endodont, 73-; consult, Vet Admin Hosp, Bronx, 75- &

Columbia Univ, 58-; head microbiol sect, Columbia Univ, 58-71, assoc attend dent surgeon, Med Ctr; vis prof endodont, Univ Pa. Mem: AAAS; Am Soc Microbiol; fel Am Asn Endodont; Am Col Dent; Am Dent Asn. Res: Bacteriology and immunology of root canal and periapical infections. Mailing Add: 30 Central Park South New York NY 10019

NAIDU, ANGI SATYANARAYAN, b Sundargarh, India, Oct 21, 36; m 69; c 2. SEDIMENTOLOGY, MARINE GEOLOGY. Educ: Andhra Univ, India, BSc, 59, MSc, 60, PhD(geol), 68. Prof Exp: Demonstr geol, Andhra Univ, India, 60-61, Univ Grants Comn jr res fel, 66-69; asst prof marine sci, 69-71, MARINE GEOCHEMIST, INST MARINE SCI, UNIV ALASKA, FAIRBANKS, 71- Mem: AAAS; Soc Econ Paleont & Mineral; Geochem Soc; Clay Minerals Soc; Int Asn Study Clays. Res: Marine geochemistry; lithological and chemical facies changes in recent sediments of arctic and tropical deltas and in subarctic fjordal environment, their present and paleoenvironmental implications; Cenozoic sedimentary history of Arctic Ocean. Mailing Add: Inst of Marine Sci Univ of Alaska Fairbanks AK 99701

NAIDUS, HAROLD, b New York, NY, Apr 11, 21; m 43; c 2. CHEMISTRY. Educ: Univ Ill, AB, 41, MS, 42; Polytech Inst Brooklyn, PhD(org chem), 44. Prof Exp: Sr res chemist, Publicker Industs, Pa, 44-48; res dir, Am Polymer Corp Div, Borden Co, 48-55, tech dir & vpres, Polyvinyl Chem, Inc, 55-62; assoc prof chem, 62-69, ASST DEAN & DIR LIB ARTS PROGS, NORTHEASTERN UNIV, 69- Mem: Am Chem Soc. Res: Kinetics of polymerization; physical properties and preparation of polymers; monomer synthesis; organic synthesis. Mailing Add: Northeastern Univ Boston MA 02115

NAIL, BILLY RAY, b Roby, Tex, Jan 19, 33; m 52; c 3. ALGEBRA. Educ: Hardin-Simmons Univ, BA, 56; Univ Ill, Urbana, MA, 62; PhD(math), 67. Prof Exp: Teacher math, High Sch, Tex, 57-61; instr, Wayland Baptist Col, 62-64; from assoc prof to prof, Morehead State Univ, 67-72, chmn div math sci, 67-72; PROF MATH & DEAN, CLAYTON JR COL, 72- Mem: Math Asn Am. Res: Lie algebras. Mailing Add: Clayton Jr Col Morrow GA 30260

NAIMAN, BARNET, b Baltimore, Md, Mar 21, 00; m 34; c 1. ANALYTICAL CHEMISTRY. Educ: Univ NC, BS, 21, MS, 22; Columbia Univ, PhD(org chem), 34. Prof Exp: Asst chemist, State Dept Agr, NC, 22-24; from instr to prof, 25-71, EMER PROF CHEM, CITY COL NEW YORK, 71- Concurrent Pos: Res consult, Lewkowitz Dye Co, 30-31; sci text recorder for blind students, NJ Comn Blind; mem & past pres, Teaneck Student Loan Asn. Mem: AAAS; Am Chem Soc; NY Acad Sci. Res: Synthesis and structure of benzothiazoles; analytical methods; organic reagents in quantitative analysis; okra seed oil; male hormone; quantitative analysis. Mailing Add: 1123 Cambridge Rd Teaneck NJ 07666

NAIMAN, CHARLES S, solid state physics, chemical physics, see 12th edition

NAIMAN, ROBERT JOSEPH, b Pasadena, Calif, July 31, 47; m 72. AQUATIC ECOLOGY. Educ: Calif State Polytech Col, BS, 69; Univ Calif, Los Angeles, MA, 71- Ariz State Univ, PhD(zool), 74. Prof Exp: Nat Res Coun Can fel estuarine ecol, Pac Biol Sta, Fisheries & Marine Serv, Can, 74-76; RES ASSOC STREAM ECOL, DEPT FISHERIES & WILDLIFE, ORE STATE UNIV, 76- Mem: Am Fisheries Soc; Am Soc Ichthyologists & Herpetologists; Am Soc Limnol & Oceanog; Int Soc Limnol. Res: Aquatic ecology with emphasis on community energetics of fishes and stream ecosystems, including feeding habits, fish production and growth, primary production, and detritus sampling. Mailing Add: Dept Fisheries & Wildlife Ore State Univ Corvallis OR 97331

NAIMARK, GEORGE MODELL, b New York, NY, Feb 5, 25; m 46; c 3. BIOCHEMISTRY. Educ: Bucknell Univ, BS, 47, MS, 48; Univ Del, PhD(biochem), 51. Prof Exp: Res fel ultrasonics & biochem, Biochem Res Found, 48-51; res biochemist, Clevite-Brush Develop Co, 51; asst dir biochem labs, Strong, Cobb & Co, Inc, 51-53 & control, 53-54, dir, 54; asst to dir med res dept, White Labs, Inc, 54-58, dir sci serv, 58-60; dir, Burdick & Becker, Inc, 60-61; vpres & dir res & develop, Dean L Burdick Assocs, Inc, New York, 61-66; PRES, NAIMARK & BARBA, INC, NEW YORK, 66- Mem: AAAS; Am Chem Soc; Am Inst Chem; NY Acad Sci. Res: Product development; promotional utilization of scientific and medical information. Mailing Add: 87 Canoe Brook Pkwy Summit NJ 07901

NAIMPALLY, SOMASHEKHAR AMRITH, b Bombay, India, Aug 31, 31; m 55; c 3. TOPOLOGY. Educ: Univ Bombay, BSc, 52, MSc, 54 & 58; Mich State Univ, PhD(math), 64. Prof Exp: Lectr math, Ruparel Col, India, 52-58; prof, Kirti Col, India, 59-61; teaching asst, Mich State Univ, 61-64; asst prof, Iowa State Univ, 64-65; assoc prof, Univ Alta, 65-69; prof, Indian Inst Technol, 69-71; vis prof, 71-74, PROF MATH, LAKEHEAD UNIV, 74- Concurrent Pos: Fel, Inst Sci India, 52-53. Mem: Am Math Soc; Math Asn Am; Indian Math Soc; Can Math Cong. Res: General topology; proximity and uniform spaces; function spaces; semi-metric, developable spaces; compactifications; convexity; proximity. Mailing Add: Dept of Math Lakehead Univ Thunder Bay ON Can

NAIR, GANGADHARAN V M, b Madras, India, Jan 26, 30; m 71. PLANT PATHOLOGY, MYCOLOGY. Educ: Univ Madras, BSc, 51; Aligarh Muslim Univ, MSc, 53; Univ Wis-Madison, PhD(plant path, mycol), 64. Prof Exp: Mycologist, Indian Agr Res Inst, New Delhi, 55-59; res scientist, Univ Wis-Madison, 64-68; asst prof environ sci, Univ Wis-Green Bay, 68-69; UN expert, Food & Agr Orgn, UN Develop Prog, Italy, 69-71; ASSOC PROF ENVIRON CONTROL, UNIV WIS-GREEN BAY, 71-, DIR INT PROGS, 74- Concurrent Pos: NSF fel plant path, Univ Wis-Madison, 64-66, NSF sr fel, 66-68; external exam doctoral thesis, Aligarh Muslim Univ, 69-71; UN expert, Food & Agr Orgn, UN Italy, 69- Mem: Am Phytopath Soc; Am Inst Biol Sci; Indian Phytopath Soc; Forestry Asn Nigeria. Res: International control programs of plant-forest tree diseases; weedicide-Sylvicide applications in the establishment of exotic tree species in developing countries; host parasite interactions of vascular wilt pathogens; electron microscopy. Mailing Add: Col of Environ Sci Univ of Wis Green Bay WI 54301

NAIR, K AIYAPPAN, b Trivandrum, India, Jan 7, 36; m 66; c 1. MATHEMATICAL STATISTICS. Educ: Univ Kerala, BSc, 56, MSc, 58; State Univ NY Buffalo, PhD(statist), 70. Prof Exp: Statist asst, Damodar Valley Corp, India, 59; res off statist, Univ Kerala, 59-63, lectr, 63-66; ASSOC PROF MATH, EDINBORO STATE COL, 70- Mem: Am Statist Asn. Mailing Add: Dept of Math Edinboro State Col Edinboro PA 16412

NAIR, KUTTENAIR GOPINATHAN, b Trivandrum, India, May 20, 37; m 67; c 2. NUCLEAR PHYSICS. Educ: Kerala Univ, India, BSc, 57; Mass Inst Technol, MS, 67; Univ Wash, PhD(nuclear physics), 73. Prof Exp: Jr res officer nuclear physics, Tata Inst Fundamental Res, Bombay, 58-64; RES SCIENTIST NUCLEAR PHYSICS, CYCLOTRON INST, TEX A&M UNIV, 73- Mem: Am Phys Soc; assoc mem Sigma Xi. Res: Gross properties of heavy ion induced nuclear reactions; multinucleon transfer reactions between heavy nuclei and the clustering effect in nuclear matter; theoretical aspects of reaction mechanism and nuclear structure. Mailing Add: Cyclotron Inst Tex A&M Univ College Station TX 77843

NAIR, PADMANABHAN PADMANABHAN, b Singapore, Nov, 9, 31; m 59; c 3. BIOCHEMISTRY. Educ: Univ Travancore, India, BSc, 51; Univ Bombay, MSc, 55, PhD(biochem), 57. Prof Exp: Res off chem path, All India Inst Med Sci, 58-60; res assoc biol, McCollum-Pratt Inst, Johns Hopkins Univ, 60-62; res assoc biochem, 62-64, DIR BIOCHEM RES DIV, SINAI HOSP, BALTIMORE, 64- Concurrent Pos: Fel, Indian Coun Med Res, 57-58; Fulbright res grant, 60-63; instr, Sch Med, Johns Hopkins Univ, 64-69, lectr, 72- Mem: AAAS; Am Inst Nutrit; Am Oil Chem Soc; NY Acad Sci. Res: Biochemistry of fat soluble vitamins; fatty acids and sterols. Mailing Add: Dept of Med Sinai Hosp Biochem Res Div Baltimore MD 21215

NAIR, RAMACHANDRAN MUKUNDALAYAM SIVARAMA, b North Parur, India, Nov 15, 38; m 73. NATURAL PRODUCT CHEMISTRY. Educ: Kerala Univ, India, BSc, 57, MSc, 59; Poona Univ, PhD(chem), 64. Prof Exp: Sr res fel org synthesis & natural prod, Nat Chem Lab, Poona, India, 64-66; res fel fungal metabolites, NY Bot Garden, 66-69; sr researcher org synthesis, Univ Paris, France, 69-70; sci pool officer org synthesis, Indian Inst Technol, Bombay, 70-71; res fel, 71-73, RES ASSOC FUNGAL PRODS, NY BOT GARDEN, NY, 73- Concurrent Pos: Jr res fel, Coun Sci & Indust Res, India, 61-64, sr res fel, 64-66. Mem: Am Chem Soc; Am Inst Chemists. Res: Study of the structure, chemistry, biogenesis and biological activity of secondary metabolites of fungi and of certain higher plants; synthesis of natural products. Mailing Add: NY Bot Garden Bronx Park Bronx NY 10458

NAIR, SREEDHAR, b Trivandrum, India, July 28, 28; nat US; m 54; c 3. PULMONARY PHYSIOLOGY, PHARMACOLOGY. Educ: Univ Travancore, India, ISc, 45; Univ Madras, MB, BS, 51; Am Bd Internal Med & Am Bd Pulmonary Dis, dipl. Prof Exp: Res physician med & pulmonary dis, Metrop Hosp & NY Med Col, 54-57, res assoc med, 57-58; asst prof physiol & pharmacol, 58-64, ASST CLIN PROF MED, NY MED COL, 64- Concurrent Pos: Sr attend physician, Norwalk Hosp, Conn, 68-, dir dept chest dis, 70- Mem: Fel Am Col Physicians; Am Thoracic Soc; fel Am Col Chest Physicians; AMA. Res: Pulmonary diseases; physiology of respiration. Mailing Add: Dept of Chest Dis Norwalk Hosp Norwalk CT 06856

NAIR, SREEKANTAN S, b Trivandrum, India, May 2, 41; m 71. STATISTICS, OPERATIONS RESEARCH. Educ: Univ Kerala, BSc, 61, MSc, 63; Purdue Univ, West Lafayette, MS, 69, PhD(math statist), 70. Prof Exp: Teaching asst statist, Purdue Univ, 66-70; ASSOC PROF MATH, ST AUGUSTINE'S COL, NC, 70- Mem: Inst Math Statist; assoc mem Opers Res Soc Am. Res: Applied probability; theory of queues; mathematical statistics. Mailing Add: Dept of Math St Augustine's Col Raleigh NC 27611

NAIR, VASU, b Suva, Fiji Islands, Jan 7, 39. ORGANIC CHEMISTRY, BIO-ORGANIC CHEMISTRY. Educ: Univ Otago, NZ, BS, 63; Univ Adelaide, PhD(org chem), 66. Prof Exp: USPHS res fel org chem, Univ Sydney, 66-67; res assoc chem, Univ Ill, Urbana, 67-68; res fel, Harvard Univ, 68-69; asst prof, 69-73, ASSOC PROF CHEM, UNIV IOWA, 73- Mem: Am Chem Soc; Royal Australian Chem Inst. Res: Synthetic, structural and mechanistic organic chemistry; chemistry of natural products; nuclear magnetic resonance spectroscopy. Mailing Add: Dept of Chem Univ of Iowa Iowa City IA 52242

NAIR, VELAYUDHAN, b India, Dec 29, 28; US citizen; m 57; c 3. PHARMACOLOGY. Educ: Benares Univ, BPharm, MS, 48; Univ London, PhD(med), 56. Prof Exp: Res assoc pharmacol, Col Med, Univ Ill, Chicago, 56-58; asst prof, Univ Chicago, 58-63; assoc prof, 63-66, PROF PHARMACOL & THERAPEUT, CHICAGO MED SCH, 66-, VCHMN DEPT, 71- Concurrent Pos: Dir lab neuropharmacol & biochem, Psychiat Inst, Michael Reese Hosp, 63-66, dir therapeut res, 66-70. Mem: Am Soc Pharmacol & Exp Therapeut; Radiation Res Soc; Soc Toxicol; Int Soc Chronobiol; Int Brain Res Orgn. Res: Blood-brain barrier; effects of environmental toxicants in pregnancy on biochemical and functional development in the offspring; radiation effects on the nervous system; radiation pharmacology; circadian rhythms in drug action. Mailing Add: Dept of Pharmacol & Exp Therapeut Chicago Med Sch Chicago IL 60612

NAIRN, ALAN EBEN MACKENZIE, b Newcastle on Tyne, Eng, Sept 9, 27. GEOLOGY. Educ: Univ Durham, BSc, 51; Glasgow Univ, PhD(geol), 54. Prof Exp: Asst geophys, Cambridge Univ, 54-55; asst, King's Col, Univ Durham, 56-58, Turner & Newall fel, 58-62, lectr, 62-65; vis prof, 63-64, from assoc prof to prof geol & geophys, Case Western Reserve Univ, 65-73; PROF GEOL, UNIV SC, 73- Concurrent Pos: Brit Geol Soc Lyell Fund grant, 63; guest prof, Univ Bonn, 65-66. Mem: Am Geophys Union; Geol Soc London; Royal Astron Soc; Am Geol Soc. Res: Paleomagnetism of Mediterranean and Mexican rocks for interpretation of continental drift, megatectonic deformation; paleoclimates and origin of glaciations; development and changes of climatic belts. Mailing Add: Dept of Geol Univ SC Columbia SC 29208

NAIRN, JOHN GRAHAM, b Toronto, Ont, Aug 23, 28; m 54; c 4. PHARMACY, PHYSICAL PHARMACY. Educ: Univ Toronto, BScPharm, 52; State Univ NY Buffalo, PhD(chem), 59. Prof Exp: Retail pharmacist, 52-54; from asst prof to assoc prof, 58-72, PROF PHARM, UNIV TORONTO, 73- Concurrent Pos: Grants, Can Found Adv Pharm, 61-65, Nat Res Coun Can, 63-66, Nu Chapter, Rho Phi Fraternity, 62 & 65, Univ Toronto, 64-65, Med Res Coun Can, 67. Mem: Am Chem Soc; Asn Faculties Pharm Can; Can Pharmaceut Asn. Res: Ion exchange resins in pharmacy; surface active agents in pharmaceutical systems; kinetics of drug decomposition and stabilization. Mailing Add: Fac of Pharm Univ of Toronto 19 Russell St Toronto ON Can

NAISH, JOHN MICHAEL, b London, Eng, Sept 29, 14; m 44; c 1. PHYSICS. Educ: Univ London, BSc, 35, MSc, 36, PhD(info theory), 63. Prof Exp: Sr investr spectros, Brit Non-Ferrous Metals Res Asn, 47-50; prin sci officer, Royal Aircraft Estab, Eng, 50-65; SR PRIN SCIENTIST, MCDONNELL DOUGLAS CORP, 65- Mem: Assoc fel, Am Inst Aeronaut & Astronaut. Res: Visual flight theory; combination of information in superimposed visual fields; electronic aircraft displays; visual flight simulators. Mailing Add: McDonnell Douglas Corp 3855 Lakewood Blvd Long Beach CA 90846

NAISTAT, SAMUEL SOLOMON, b Worcester, Mass, Mar 6, 17; m 42; c 3. PHYSICAL CHEMISTRY. Educ: Worcester Polytech Inst, BS, 37, MS, 39; Univ Wis, PhD(chem), 44. Prof Exp: Asst physics, Worcester Polytech Inst, 37-39; asst chem, Univ Wis, 39-41, instr physics, 43-44; res engr, Westinghouse Elec Corp, NJ, 44-45; res chemist, Congoleum-Nairn, Inc, 45-50; res chemist, Buffalo Electro-Chem Co, Inc, 50-55; group leader & asst to mgr, Becco Chem Div, FMC Corp, 55-57, sect mgr, Inorg Res & Develop Dept, 57-61, proj evaluator & tech consult, 61-63; supvr propulsion tech, Rocketdyne Solid Rocket Div, NAm Aviation, Inc, Tex, 63, res specialist, 64-65; from asst prof to assoc prof chem, 65-74, PROF CHEM, STEPHEN F AUSTIN STATE UNIV, 74- Mem: Am Chem Soc. Res: Physical chemistry of macromolecules; manufacture and applications of peroxygen chemicals; chemistry and energetics of propellants; energy conversion processes. Mailing Add: Dept of Chem Stephen F Austin State Univ Nacogdoches TX 75961

NAITOH, PAUL YOSHIMASA, b Japan, Feb 1, 31; US citizen; m 59; c 2. PSYCHOPHYSIOLOGY. Educ: Yamaguchi Univ, BA, 53; Univ Minn, MA, 56, PhD(psychol), 64. Prof Exp: Asst psychol, Univ Minn, 53-56, psychiat, 57-58 & vet radiol, 58-61; head psychophysiol lab, Neuropsychiat Inst, Univ Calif, Los Angeles, 65-67, staff psychologist, Inst & asst prof med psychol, Univ, 66-67; head, Behav Res Br, Psychophysiol Div, US Navy Med Neuropsychiat Res Unit, 67-74, HEAD, BEHAV RES BR, PSYCHOPHYSIOL DIV, NAVAL HEALTH RES CTR, 74- Concurrent Pos: Trainee, Nat Inst Ment Health interdisciplinary res training prog, 64-66; res assoc, Nat Ctr Sci Res, France, 74-75. Mem: Soc Psychophysiol Res; Am Electroencephalographic Soc; Int Soc Chronobiol; Biomet Soc; Asn Psychophysiol Study Sleep. Res: Psychophysiological analyses of electroencephalography; psychophysiological correlates of alcoholism; psychophysiology of sleep and sleep loss. Mailing Add: Behav Res Br Psychophysiol Div Naval Health Res Ctr San Diego CA 92152

NAITOVE, ARTHUR, b New York, NY, Mar 25, 26; m 46; c 5. SURGERY, PHYSIOLOGY. Educ: Dartmouth Col, BA, 45; NY Univ, MD, 48; Am Bd Surg, dipl, 58. Prof Exp: Asst chief surg, White River Junction Vet Admin Hosp, Vt, 58; from instr to asst prof, 59-71, ASSOC PROF SURG, DARTMOUTH MED SCH, 71-, ASST PROF PHYSIOL, 64-, DIR SURG LABS, 71- Concurrent Pos: USPHS fel physiol, Dartmouth Med Sch, 59-61; attend, White River Junction Vet Admin Hosp, 58-, consult surg, 70-; prin investr USPHS res grant, Dartmouth Med Sch, 61-; consult, Hitchcock Clin, Hanover, NH, 70- Mem: Fel Am Col Surg; Am Gastroenterol Asn. Res: Abdominal surgery; gastrointestinal tract and physiology of gastrointestinal tract, particularly relating to hemodynamic events, their control and their influence on secretory processes. Mailing Add: Dept of Surg Dartmouth Med Sch Hanover NH 03755

NAJAR, RUDOLPH MICHAEL, b San Fernando, Calif, June 11, 31; m 70; c 1. MATHEMATICS, PHYSICS. Educ: St Mary's Col Calif, BS, 54; Univ Calif, Berkeley, MA, 61; Univ Notre Dame, MS, 62, PhD(math), 69. Prof Exp: From instr to asst prof math, St Mary's Col Calif, 67-70, chmn dept, 69-70; ASST PROF MATH, UNIV WIS-WHITEWATER, 70- Mem: Am Math Soc; Math Asn Am. Res: Homological algebra and category theory; cohomology of finite groups. Mailing Add: Dept of Math Univ of Wis Whitewater WI 53190

NAJARIAN, HAIG HAGOP, b Nashua, NH, Jan 5, 25; m 57; c 3. PARASITOLOGY. Educ: Univ Mass, BS, 48; Boston Univ, MA, 49; Univ Mich, PhD(zool), 53. Prof Exp: Asst biol, Boston Univ, 49; asst zool, Univ Mich, 49-51; asst prof biol, Northeastern Univ, 53-55; assoc res parasitologist, Parke, Davis & Co, 55-57; scientist, Bilharziasis Control Proj, WHO, Iraq, 58-59; USPHS trainee trop med & parasitol, Med Br, Univ Tex, 59-60, asst prof microbiol, 60-66; assoc prof, 66-68, chmn div sci & math, 67-71, chmn dept biol, 71-75, PROF BIOL, UNIV MAINE, PORTLAND-GORHAM, 68- Mem: AAAS; Am Soc Parasitol; Am Micros Soc; Am Soc Trop Med & Hyg. Res: Life histories of digenetic trematodes; morphology of aspidogastrid trematodes and parasitic copepods; experimental chemotherapy of schistosomiasis, malaria, amebiasis, paragonimiasis and intestinal helminths; biharziasis control; ecology of bulinid snails; filariasis; haemobartonellae; haemoflagellates; experimental amebiasis. Mailing Add: Dept of Biol Univ of Maine at Portland-Gorham Portland ME 04103

NAJARIAN, JOHN SARKIS, b Oakland, Calif, Dec 22, 27; m; c 4. SURGERY. Educ: Univ Calif, AB, 48, MD, 52; Am Bd Surg, dipl. Prof Exp: From intern to resident surg, Sch Med, Univ Calif, 52-60; prof surg & vchmn dept, dir surg res labs & chief transplantation serv, Sch Med, Univ Calif, San Francisco, 63-67; PROF SURG & CHMN DEPT, COL MED SCI, UNIV MINN, MINNEAPOLIS, 67- Concurrent Pos: NIH spec res fel immunopath, Sch Med, Univ Pittsburgh, 60-61; NIH assoc & sr fel tissue transplantation immunol, Scripps Clin & Res Found, La Jolla, Calif, 61-63; Markle Award, 64-69; NIH spec consult clin res training comt, Nat Inst Gen Med Sci, 65-69. Mem: AAAS; fel Am Col Surgeons; Soc Exp Biol & Med; Am Soc Exp Path; Am Fedn Clin Res. Mailing Add: Univ of Minn Health Sci Ctr Mayo Mem Bldg Box 195 Minneapolis MN 55455

NAJJAR, VICTOR ASSAD, b Zalka, Lebanon, Apr 15, 14; nat US; m 48; c 3. PEDIATRICS. Educ: Am Univ, Beirut, MD, 35; Am Bd Nutrit, dipl. Prof Exp: From instr to assoc prof pediat, Harriet Lane Home, Johns Hopkins Univ, 39-57; prof microbiol & head dept, Sch Med, Vanderbilt Univ, 57-68; prof molecular biol, 68, AM CANCER SOC PROF MOLECULAR BIOL, PROF PEDIAT & CHMN DIV PROTEIN CHEM, SCH MED, TUFTS UNIV, 68- Concurrent Pos: Nat Res Coun fels, Sch Med, Washington Univ, 46-48 & Sch Biochem, Cambridge Univ, 48-49; Irving McQuarrie lectr, 57; prof, Boston Floating Hosp Infants & Children, 68; ed-in-chief, Molecular & Cellular Biochem, 72- Honors & Awards: Meade Johnson Award, 51. Mem: AAAS; NY Acad Sci; Am Soc Biol Chemists; Am Soc Microbiol; Am Soc Clin Invest. Res: Vitamin metabolism and human requirement; mammalian and bacterial enzymology; mechanism of enzyme action; immunochemistry. Mailing Add: Div of Protein Chem Tufts Univ Sch of Med Boston MA 02111

NAKACHE, FERNAND ROBERT, physics, see 12th edition

NAKADA, DAISUKE, b Osaka, Japan, July 23, 25; m 58; c 2. BIOCHEMISTRY, MICROBIOLOGY. Educ: Kyoto Univ, BS, 48; Osaka Univ, PhD(bact), 55. Prof Exp: Lectr bact, Osaka Univ, 55-56, asst prof, 56-58; vis asst prof microbial genetics, Columbia Univ, 58-61; vis asst prof biol, Mass Inst Technol, 61-63; res proj leader molecular biol, E I du Pont de Nemours & Co, Inc, 63-67; assoc prof, 67-71, PROF BIOCHEM, SCH MED, UNIV PITTSBURGH, 71- Concurrent Pos: Fulbright exchange fel, Columbia Univ, 58-61; NIH res grant, 68-75. Mem: Am Soc Biol Chem; Am Soc Microbiol; Am Chem Soc. Res: Control of macromolecule synthesis in bacteria and bacteriophage; ribosomal function. Mailing Add: Dept of Biochem Univ of Pittsburgh Sch of Med Pittsburgh PA 15213

NAKADA, HENRY ISAO, b Los Angeles, Calif, Oct 12, 22; m 45; c 3. BIOCHEMISTRY. Educ: Temple Univ, BA, 48, PhD, 53. Prof Exp: Res assoc, Inst Cancer Res, Philadelphia, 50-54; mem, Scripps Clin & Res Found, 54-62; ASSOC PROF BIOCHEM, UNIV CALIF, SANTA BARBARA, 62- Mem: Am Soc Biol Chemists; Soc Exp Biol & Med; Am Chem Soc. Res: Mucopolysaccharide, carbohydrate and amino acid metabolism. Mailing Add: Dept of Biol Sci Univ of Calif Santa Barbara CA 93018

NAKADA, MINORU PAUL, b Los Angeles, Calif, Jan 15, 21; m 53; c 3. PHYSICS. Educ: Univ Calif, AB, 47, PhD(cosmic rays, physics), 52. Prof Exp: Physicist, Lawrence Radiation Lab, Univ Calif, 52-61 & Jet Propulsion Lab, Calif Inst Technol, 61-62; PHYSICIST, GODDARD SPACE FLIGHT CTR, NASA, GREENBELT, 62- Mem: Am Phys Soc; Am Geophys Union. Res: Space plasma physics.

NAKADA, YOSHINAO, b Los Angeles, Calif, Mar 14, 18; m 44; c 2. PHYSICS. Educ: Calif Inst Technol, BS, 40, MS, 41. Prof Exp: Chief chemist, Nobell Res Found, 46-48; res physicist, West Precipitation Corp, 48-55, Magnavox Co, 55-56, Hughes Aircraft Co, 56-59 & A C Spark Plug Div, Gen Motors Corp, 59-61; RES PHYSICIST, HUGHES AIRCRAFT CO, 61- Mem: Am Phys Soc. Res: Real time data processing; electrical discharge; infrared. Mailing Add: 4227 Don Mariano Dr Los Angeles CA 90008

NAKADOMARI, HISAMITSU, b Japan, Sept 15, 35; US citizen; m 67; c 2. ELECTROCHEMISTRY. Educ: Kans State Col of Pittsburgh, BA, 61; Ft Hays Kans State Col, MS, 65; Colo State Univ, PhD(chem), 74. Prof Exp: Asst prof chem, St Gregory's Col, 65-68; SR RES ASSOC BIOELECTROCHEM, BROOKHAVEN NAT LAB, 74- Concurrent Pos: Fel, Colo State Univ, 73-74. Mem: Sigma Xi; Am Chem Soc; AAAS. Res: Ion transport phenomena in lipid bilayer membranes; studies of the electrode-solution interfaces; the electrical double layer. Mailing Add: Dept of Appl Sci Brookhaven Nat Lab Upton NY 11973

NAKAGAWA, T WILLIAM, b San Jose, Calif, Jan 9, 25; m 56; c 2. ORGANIC CHEMISTRY, POLYMER CHEMISTRY. Educ: Univ Calif, Berkeley, BS, 51; Univ Calif, Davis, MS, 56, PhD(chem), 59. Prof Exp: Res chemist, Spreckels Sugar Co, 51-52; sr technician chem, Univ Calif, Davis, 53-55, asst, 56-58; res chemist, Monsanto Co, 59-61; sr chemist, Chem Systs Div, United Technologies, 61-75, SR STAFF SCIENTIST, UNITED AIRCRAFT CORP, 71-, FORMULATION SPECIALIST, ELECTRONIC MEMORIES & MAGNETICS COMPUT PROD DIV, 75- Mem: Am Chem Soc. Res: Development of magnetic coatings for tapes and disc. Mailing Add: 5550 Muir Dr San Jose CA 95124

NAKAI, SHURYO, b Kanazawa, Japan, Dec 13, 26; m 52; c 3. FOOD CHEMISTRY. Educ: Univ Tokyo, BSc, 50, PhD(dairy tech), 62. Prof Exp: Res chemist, Okayama Dairy, Meiji Milk Prod Co Ltd, 50-51, sect head dairy chem, Res Lab, 52-62; res assoc, Univ Ill, 62-66; asst prof dairying, 66-70, assoc prof food chem, 70-75, PROF FOOD CHEM, UNIV BC, 75- Mem: Am Dairy Sci Asn; Can Inst Food Sci & Technol; Inst Food Technol. Res: Chemistry of food proteins; chemical studies on food products. Mailing Add: Dept Food Sci 233 MacMillan Bldg Univ of BC Vancouver BC Can

NAKAJIMA, NOBUYUKI, b Tokyo, Japan, Nov 3, 23; c 3. POLYMER CHEMISTRY, POLYMER PHYSICS. Educ: Univ Tokyo, BS, 45; Polytech Inst Brooklyn, MS, 55; Case Inst Technol, PhD(phys chem), 58. Prof Exp: Asst polymer & phys chem, Naval Air Force Res Ctr, Japan, 44-45; prod engr, Chem Div, Osaka Gas Co, 45-51; from res asst to res assoc, Case Inst Technol, 55-60; res chemist, W R Grace & Co, 60-63, sect leader, Polymer Chem Div, 63-66, asst to vpres chem, Res Div, 65-66; tech supvr plastics div, Allied Chem Corp, NJ, 66-67, mgr polymer physics & anal plastics div, 67-71; MEM STAFF, TECH CTR, B F GOODRICH CHEM CO, 71- Concurrent Pos: Assoc ed, Rubber Chem & Technol. Mem: AAAS; Am Chem Soc; Soc Rheol; Am Phys Soc; Sigma Xi. Res: Elastomer rheology and processing; polymer solution thermodynamics; rheology of polymer melts and solution; molecular weight distribution of polymers; polymer morphology, processing and structural analysis. Mailing Add: B F Goodrich Chem Co Tech Ctr PO Box 122 Avon Lake OH 44012

NAKAJIMA, SHIGEHIRO, b Kobe, Japan, July 13, 31; m 57; c 2. NEUROPHYSIOLOGY. Educ: Univ Tokyo, MD, 55, PhD(physiol), 61. Prof Exp: Instr physiol, Sch Med, Univ Tokyo, 60-65; assoc prof, Sch Med, Juntendo Univ, Japan, 65-69; assoc prof, 69-73, PROF BIOL & NEUROPHYSIOL, PURDUE UNIV, 73- Concurrent Pos: United Cerebral Palsy Res & Educ Found fel neurophysiol, Col Physicians & Surgeons, Columbia Univ, 62-64; asst zoologist, Brain Res Inst, Univ Calif, Los Angeles, 64-65; Wellcome Trust fel physiol, Univ Cambridge, 67-69. Mem: Am Physiol Soc; Biophys Soc; Soc Neurosci; Soc Gen Physiol. Res: Electrophysiology of peripheral nerves and muscle. Mailing Add: Dept of Biol Sci Purdue Univ Lafayette IN 47907

NAKAJIMA, YASUKO, b Osaka, Japan, Jan 8, 32; m 57; c 2. ANATOMY, NEUROBIOLOGY. Educ: Univ Tokyo, MD, 55, PhD(anat), 62. Prof Exp: Instr anat, Univ Tokyo, 62-67; vis res fel zool, Univ Cambridge, 67-69; ASSOC PROF NEUROBIOL, PURDUE UNIV, WEST LAFAYETTE, 69- Concurrent Pos: Vis res fel anat, Col Physicians & Surgeons, Columbia Univ, 62-64; asst researcher, Med Sch, Univ Calif, Los Angeles, 64-65. Mem: Am Asn Anat; Am Soc Cell Biol; Soc Neurosci. Res: Neurobiology at the cellular level; electron microscopy, electrophysiology, and tissue culture. Mailing Add: Dept of Biol Sci Purdue Univ West Lafayette IN 47907

NAKAMOTO, KAZUO, b Kobe, Japan, Mar 1, 22; m 50; c 3. PHYSICAL CHEMISTRY. Educ: Osaka Univ, BS, 45, DSc, 53. Prof Exp: Res asst chem, Osaka Univ, 45-46, res assoc, 46-51, lectr, 51-57, assoc prof, 57; res fel, Clark Univ, 57-58, asst prof, 58-61; from asst prof to prof, Ill Inst Technol, 61-69; WEHR PROF CHEM, MARQUETTE UNIV, 69- Concurrent Pos: Res fel, Iowa State Univ, 53-55. Mem: Am Chem Soc; The Chem Soc; Chem Soc Japan; Spectros Soc Japan. Res: Electronic and vibrational spectra. Mailing Add: Dept of Chem Marquette Univ Milwaukee WI 53233

NAKAMOTO, TOKUMASA, b Kohala, Hawaii, July 8, 28; m 50; c 3. BIOCHEMISTRY. Educ: Univ Chicago, BA, 56, PhD(biochem), 59. Prof Exp: Res assoc biochem, Univ Chicago, 59-62; res assoc, Rockefeller Inst, 62-64, asst prof, 64-65; asst prof, 65-67, ASSOC PROF BIOCHEM, SCH MED, UNIV CHICAGO, 67- Concurrent Pos: Mem, Argonne Cancer Res Hosp. Res: Biosynthesis of RNA and proteins. Mailing Add: Dept of Biochem Univ of Chicago Sch of Med Chicago IL 60637

NAKAMURA, EUGENE LEROY, b San Diego, Calif, June 8, 26; m 61. BIOLOGICAL OCEANOGRAPHY, FISH BIOLOGY. Educ: Univ Ill, BS, 50, MS, 51. Prof Exp: Res asst zool, Univ Hawaii, 51-56; fishery biologist, Biol Lab, US Bur Com Fisheries, Hawaii, 56-70; dir, Eastern Gulf Sport Fisheries Marine Lab, 70-72, OFFICER-IN-CHARGE, GULF COASTAL FISHERIES CTR, PANAMA CITY LAB, NAT MARINE FISHERIES SERV, 72- Mem: AAAS; Am Fisheries Soc; Am Soc Ichthyol & Herpet; Am Soc Limnol & Oceanog; Am Inst Fishery Res Biol. Res: Biology, ecology and sport fishery of marine fishes. Mailing Add: Nat Marine Fish Serv Panama City Lab PO Box 4218 Panama City FL 32401

NAKAMURA, KAZUO, b Fukuoka, Japan, June 30, 29; m 60; c 2. MICROBIAL GENETICS, PHYCOLOGY. Educ: Kyushu Univ, BS, 54, MS, 56, DSc, 67; Univ Mo, PhD(genetics), 64. Prof Exp: Res assoc bot, Univ Mo, 64-65; scholar, Univ Calif, Los Angeles, 65-67; res fel, Kyushu Univ, 67-69; prof, Fukuoka Univ, Japan, 69-70; vis asst prof, 69-70, asst prof, 70-74, ASSOC PROF BOT, UNIV LETHBRIDGE, 74- Concurrent Pos: Vis prof, Univ Mo-Columbia, 75-76; Nat Res Coun Can travel fel, 75. Mem: Genetics Soc Am; Int Phycol Soc; Genetics Soc Can; Can Soc Microbiol; Am Soc Microbiol. Res: Genetics and physiology of Chlamymonas and Neurospora. Mailing Add: Dept of Biol Sci Univ of Lethbridge Lethbridge AB Can

NAKAMURA, MITSURU J, b Los Angeles, Calif, Dec 17, 26; m 51; c 3.

MICROBIOLOGY. Educ: Univ Calif, Los Angeles, AB, 49; Univ Southern Calif, MS, 50, PhD(med sci), 56; Am Bd Med Microbiol, dipl, 62. Prof Exp: Asst, Sch Med, Univ Calif, 50-52; from asst prof to assoc prof, Northeastern Univ, 52-56; assoc prof, 57-63, PROF MICROBIOL & CHMN DEPT, UNIV MONT, 63- Concurrent Pos: Res assoc, Sch Med, Boston Univ, 55-56; responsible investr, Comn Enteric Infections, Armed Forces Epidemiol Bd, Off Surgeon Gen, US Dept Army, 57-62; fel, Sch Med, La State Univ, 59 & 60; fel, Univ Costa Rica, 63; Am Acad Microbiol fel, 67. Mem: Fel AAAS; fel Am Pub Health Asn; Am Soc Microbiol; Am Soc Trop Med & Hyg; Soc Exp Biol & Med. Res: Physiology of Shigella; water pollution; effects of ultraviolet irradiation on bacteria; cultivation of Protozoa; Clostridium perfringens food poisoning. Mailing Add: Dept of Microbiol Univ of Mont Missoula MT 59801

NAKAMURA, ROBERT MASAO, b Los Angeles, Calif, Sept 18, 35; m 66; c 3. REPRODUCTIVE BIOLOGY, BIOPHYSICS. Educ: Occidental Col, BA, 59; Univ Southern Calif, MS, 64, PhD(biochem), 68. Prof Exp: Res asst biochem, Univ Southern Calif, 59-61, res assoc biophys, Allan Hancock Found, 66-67; asst res biochemist, Dept Obstet & Gynec, Harbor Gen Hosp, Torrance, Calif, 67-68, res biochemist, 68-69; ASST PROF REPROD BIOL, SCH MED, UNIV SOUTHERN CALIF, 69- Concurrent Pos: Consult, Ford Found, 69-; mem task force human reproduction, WHO; managing ed, Contraception, 69- Mem: AAAS; The Chem Soc; Am Chem Soc; NY Acad Sci; Endocrine Soc. Res: Hormonal profiles in normal menstruating women; effects of contraceptive steroids on the serum hormone levels; biophysical properties of cervical mucus; properties of protein hormones. Mailing Add: Dept of Obstet & Gynec Univ of Southern Calif Sch Med Los Angeles CA 90033

NAKAMURA, ROBERT MOTOHARU, b Montebello, Calif, June 10, 27; m 57; c 2. PATHOLOGY, IMMUNOLOGY. Educ: Whittier Col, AB, 49; Temple Univ, MD, 54. Prof Exp: Chief clin path, Long Beach Vet Admin Hosp, 59-60; pathologist, Atomic Bomb Casualty Comn, Japan, 60-61; instr path, Sch Med, Univ Calif, Los Angeles, 61-62, asst prof, 62-65; pathologist, St Joseph Hosp, Orange, Calif, 68-69; from assoc prof to prof path & dir clin labs, Orange County Med Ctr, Univ Calif, Irvine, 69-74; HEAD DEPT PATH, HOSP SCRIPPS CLIN, 74-, ASSOC MEM, DEPT MOLECULAR IMMUNOL, SCRIPPS CLIN & RES FOUND, 74-, ADJ PROF PATH, UNIV CALIF, SAN DIEGO, 75- Concurrent Pos: USPHS spec fel exp path, Scripps Clin & Res Found, Univ Calif, 65-68; pathologist, Los Angeles County Harbor Gen Hosp, Calif, 61-65; consult dept path, Orange County Gen Hosp, 62-65; adj prof path, Univ Calif, Irvine, 74- Mem: AAAS; AMA; fel Am Col Path; fel Am Soc Clin Path; Am Soc Exp Path. Res: Clinical pathology; general area of cell proliferation; immunopathology, specifically autoimmune diseases and immunological tolerance. Mailing Add: Dept of Path Hosp of Scripps Clin La Jolla CA 92037

NAKANE, PAUL K, b Yokahama, Japan, Oct 20, 35; m 59; c 3. HISTOCHEMISTRY, ELECTRON MICROSCOPY. Educ: Huntingdon Col, BA, 58; Brown Univ, MS, 61, PhD(cytol), 63. Prof Exp: Res assoc histochem, Sch Med, Stanford Univ, 63-65; instr path, Univ Mich, Ann Arbor, 65-67, asst prof cell biol in path, 67-68; asst prof, 68-69, assoc prof, 69-72, PROF PATH, SCH MED, MED CTR, UNIV COLO, DENVER, 73- Concurrent Pos: Cancer Inst trainee, 59-63. Mem: AAAS. Res: Ultrastructural localization of antigens by enzyme-labeled antibody. Mailing Add: Dept of Path Sch of Med Univ of Colo Med Ctr Denver CO 80220

NAKANISHI, KOJI, b Hong Kong, May 11, 25; m 47; c 2. ORGANIC CHEMISTRY. Educ: Nagoya Univ, BSc, 47, PhD(chem), 54. Prof Exp: Garioa fel, Harvard Univ, 50-52; asst prof chem, Nagoya Univ, 55-58; prof, Tokyo Kyoiku Univ, 58-63 & Tohoku Univ, Japan, 63-69; PROF CHEM, COLUMBIA UNIV, 69- Concurrent Pos: Consult, Syntex/Zoecon Corp, 65- & Lederle Labs, 69- Honors & Awards: Chem Soc Japan Award, 54; Cult Award, Asahi Press, Japan, 68. Mem: Am Chem Soc; Chem Soc France; The Chem Soc; Chem Soc Japan; fel Am Acad Arts & Sci. Res: Isolation and structural studies of physiologically active natural products; applications of spectroscopy to structure determination. Mailing Add: Dept of Chem Columbia Univ New York NY 10027

NAKANISHI, SUSUMU, organic chemistry, see 12th edition

NAKANO, JAMES HIROTO, b Hiroshima, Japan, Jan 17, 22; nat US; m 47; c 3. MICROBIOLOGY, EXPERIMENTAL PATHOLOGY. Educ: Stanford Univ, AB, 47, MA, 51, PhD(bact), 53. Prof Exp: Asst dir labs clin path, Kaiser Found Hosp, 54-59; virologist, Commun Dis Ctr, 59-65, chief vesicular dis virus lab, 65-67, res microbiologist, Enterovirus Infections Unit, 67-71, chief vesicular dis lab & poliovirus strain characterization lab, 71-72, chief viral vaccine invest sect, 72-74, CHIEF VIRAL EXANTHEMS BR, CTR DIS CONTROL, USPHS, 74-; DIR, COLLAB REF CTR SMALLPOX & OTHER POXVIRUS INFECTIONS, WHO, 74- Concurrent Pos: Dir, Regional Ref Ctr Smallpox, WHO, 71-72, dir, Int Ref Ctr Smallpox, 72-74, consult, 71- Mem: AAAS; fel Am Acad Microbiol; Am Soc Microbiol; NY Acad Sci. Res: Enteroviruses; poxviruses; viral immunology. Mailing Add: Rm 230 Bldg 7 Viral Exanthems Br Virol Div USPHS Dis Control Atlanta GA 30333

NAKANO, JIRO, pharmacology, internal medicine, see 12th edition

NAKAO, AKIRA, biochemistry, see 12th edition

NAKASHIMA, TADAYOSHI, b Yokkaichi, Japan, Dec 1, 22; m 47; c 1. BIOCHEMISTRY. Educ: Nagoya Pharmaceut Col, BP, 43; Taihoku Imp Univ, BS, 46; Kyushu Univ, PhD(biochem), 61. Prof Exp: Fel biochem, Univ Hawaii, 62-64; res scientist biochem, Inst Molecular Evolution, 65-73, RES ASST PROF BIOCHEM, INST MOLECULAR & CELLULAR EVOLUTION, UNIV MIAMI, 73- Concurrent Pos: Vis res scientist, Inst Animal Physiol, Univ Bonn, Ger, 66-69. Mem: Int Soc Study Origin Life; Japanese Soc Food & Nutrit; Sigma Xi. Res: Nucleic acid-protein interaction, genetic code, protein synthesis on the model ribosomes, including prebiological chemistry. Mailing Add: Inst of Molecular & Cellular Evol Univ of Miami Coral Gables FL 33134

NAKASONE, HENRY YOSHIKI, b Kauai, Hawaii, July 6, 20; m 48; c 2. HORTICULTURE. Educ: Univ Hawaii, BA, 43, MS, 52, PhD(genetics), 60. Prof Exp: Asst hort, Agr Exp Sta, 48-52, instr plant propagation, Col Agr & jr horticulturist, Agr Exp Sta, 52-58, asst prof plant propagation & trop pomol & asst horticulturist, 58-60, assoc prof & assoc horticulturist, 60-69, PROF HORT, COL AGR & HORTICULTURIST, AGR EXP STA, UNIV HAWAII, 69-, CHMN DEPT, 75- Concurrent Pos: Consult, Heinz Alimentos, Mex, 66-72, Comn Fruit Cult, Mex Govt & DaCosta Bros, Jamaica, 72- Honors & Awards: Best Paper Award, Am Soc Hort Sci, 73. Mem: Fel AAAS; Am Soc Hort Sci. Res: Plant breeding and culture of tropical crops; genetics of tropical crops. Mailing Add: Dept of Hort Univ of Hawaii 3190 Maile Way Honolulu HI 96822

NAKATA, HERBERT MINORU, b Pasadena, Calif, Mar 10, 30; m 60; c 1. BACTERIOLOGY. Educ: Univ Ill, BS, 52, MS, 56, PhD, 59. Prof Exp: From instr to assoc prof, 59-71, PROF BACT, WASH STATE UNIV, 71-, CHMN DEPT BACT & PUB HEALTH, 68- Mem: AAAS; Am Soc Microbiol. Res: Bacterial physiology, particularly the biochemical processes associated with sporulation of aerobic bacilli. Mailing Add: Dept of Bacteriol & Pub Health Wash State Univ Pullman WA 99163

NAKATA, SHIGERU, b Honolulu, Hawaii, Nov 15, 19; m 47; c 2. PLANT PHYSIOLOGY. Educ: Univ Hawaii, BS, 47, MS, 49, PhD(bot), 65. Prof Exp: Asst plant physiologist, 59-67, ASSOC PLANT PHYSIOLOGIST, UNIV HAWAII, 67- Concurrent Pos: Mem staff, Purdue Univ, 65-66. Mem: AAAS; Am Soc Plant Physiol; Bot Soc Am; Am Soc Hort Sci. Res: Growth and development of plants, especially floral initiation in subtropical plants; mineral nutrition and water stress. Mailing Add: Dept of Plant Physiol Univ of Hawaii 3190 Maileway Honolulu HI 96822

NAKATANI, ROY E, b Seattle, Wash, June 8, 18; m 55; c 4. FISH BIOLOGY. Educ: Univ Wash, BS, 47, PhD(fisheries), 60. Prof Exp: Fishery biologist, Fisheries Res Inst, 47-48; res assoc blood chem salmonoids, Univ Wash, 52-58, anal past Alaskan fisheries data, 58-59; biol scientist, Hanford Labs, Gen Elec Co, 59-62, mgr aquatic biol, 62-66; mgr ecol, Pac Northwest Labs, Battelle Mem Inst, 66-70; assoc prof fisheries, 70-73, PROF FISHERIES, UNIV WASH, 73-, ASSOC DIR FISHERIES RES INST & PROG DIR, DIV MARINE RESOURCES, 70- Concurrent Pos: Consult, AEC, Westinghouse Elec Corp, Argonne Nat Lab, ARCO Chem Co & Bio-Test; mem panel radioactivity in marine environ & biol effects of ionizing radiation subcomt environ effects, Nat Acad Sci-Nat Res Coun. Mem: AAAS; Am Inst Fishery Res Biol; Am Soc Limnol & Oceanog; Int Acad Fishery Sci; Am Fisheries Soc. Res: Radiation biology of aquatic organisms; water pollution; nuclear power plant siting; oil pollution. Mailing Add: Fisheries Res Inst Univ of Wash Seattle WA 98195

NAKATSUGAWA, TSUTOMU, b Kochi-Ken, Japan, Apr 17, 33; m 65; c 2. INSECT TOXICOLOGY, INSECTICIDE TOXICOLOGY. Educ: Univ Tokyo, BAgr, 57; Iowa State Univ, MS, 61, PhD(insect toxicol), 64. Prof Exp: Asst entomologist, Nat Inst Agr Sci, Tokyo, 57-60; res assoc insect toxicol, Iowa State Univ, 64-68; asst prof, 68-72, ASSOC PROF INSECT TOXICOL, STATE UNIV NY COL ENVIRON SCI & FORESTRY, 72- Mem: AAAS; Am Chem Soc; Entom Soc Am. Res: Enzymology of detoxication and mode of action of insecticides and pharmacological agents. Mailing Add: Dept of Entom State Univ of NY Col Environ Sci & Forestry Syracuse NY 13210

NAKAUE, HARRY S, agricultural biochemistry, nutrition, see 12th edition

NAKAYAMA, FRANCIS SHIGERU, b Honolulu, Hawaii, July 1, 30. SOIL CHEMISTRY. Educ: Univ Hawaii, BS, 52; Iowa State Univ, MS, 55, PhD(soil fertil), 58. Prof Exp: Asst, Iowa State Univ, 53-58; CHEMIST, US WATER CONSERV LAB, USDA, 58- Mem: AAAS; Am Soc Agron; Am Chem Soc. Res: Solubility of calcium constituents of soil and calcium complex formation; interrelation between water quality and soil water movement. Mailing Add: US Water Conserv Lab 4331 E Broadway Phoenix AZ 85040

NAKAYAMA, ROY MINORU, b Dona Ana, NMex, Sept 11, 23. PLANT BREEDING, HORTICULTURE. Educ: NMex Col, BS, 48; Iowa State Univ, MS, 50, PhD, 60. Prof Exp: Asst plant path, Exp Sta, NMex Col, 51-53; asst plant pathologist, Plant Path Bur, State Dept Agr, Calif, 53-56; asst prof agr sci, 56-60, asst prof hort, 60-69, ASSOC PROF HORT, NMEX STATE UNIV, 69- Mem: Am Hort Soc. Res: Breeding of chile peppers and pecans. Mailing Add: Dept of Hort NMex State Univ Las Cruces NM 88003

NAKAYAMA, TAKAO, b Sacramento, Calif, Sept 19, 13; m 41; c 2. PHYTOPATHOLOGY, HORTICULTURE. Educ: Univ Calif, BS, 37. Prof Exp: Plant pathologist, Ohara Inst Agr Res, 40-46; agriculturist & dir, Chofu Hydroponic Farm, Japan, 46-61; horticulturist, US Army Procurement Agency Japan, 62-74; HORTICULTURIST, PAC AIR FORCE PROCUREMENT CTR JAPAN, 74- Concurrent Pos: Agr consult, Land Auth, Govt PR, 59, Repub Korea, 64-65 & Repub China, Thailand & SVietnam, 66-68. Mem: Am Phytopath Soc; Phytopath Soc Japan. Res: Wheat scab and vegetable diseases; vegetable production; hydroponic vegetable production; technical and economical implications surrounding production of fresh fruits and vegetables in Japan and far eastern countries. Mailing Add: Pac Air Force Procure Ctr Japan FPO Seattle WA 98760

NAKAYAMA, TOMMY, b Ballico, Calif, Mar 15, 28; m 57; c 4. FOOD SCIENCE. Educ: Univ Calif, Berkeley, BS, 51, MS, 52, PhD(agr chem), 57. Prof Exp: Jr specialist, Univ Calif, Berkeley, 54-57, asst specialist, 57-58; asst prof food sci & technol, Univ Calif, Davis, 59-63; res supvr, Miller Brewing Co, 63-66; assoc prof food sci, Univ Ga, 66-70; PROF FOOD SCI, UNIV HAWAII, 70- Mem: Inst Food Technol; Am Chem Soc; Am Soc Brewing Chem; Am Soc Microbiol. Res: Carotenoids and polyphenolic compounds in foods; agricultural waste utilization. Mailing Add: Dept of Food Sci & Technol Univ Hawaii 1920 Edmondson Rd Honolulu HI 96822

NAKON, ROBERT STEVEN, b Brooklyn, NY, May 1, 44. INORGANIC CHEMISTRY. Educ: DePaul Univ, BS, 65; Tex A&M Univ, PhD(inorg chem), 71. Prof Exp: Fel inorg chem, Iowa State Univ, 71-73 & Memphis State Univ, 73-74; ASST PROF INORG CHEM, WVA UNIV, 74- Mem: Am Chem Soc. Res: Metal complexes as catalysts for biomimetic reactions; metal ion interactions with biologically important substances; correlation of reactivity with the structures of metal chelates as intermediates. Mailing Add: Dept of Chem WVa Univ Morgantown WV 26506

NALBANDIAN, JOHN, b Providence, RI, Nov 26, 32; m 70. ORAL PATHOLOGY, ELECTRON MICROSCOPY. Educ: Brown Univ, AB, 54; Harvard Univ, DMD, 58. Prof Exp: Res assoc periodont, Sch Dent Med, Harvard Univ, 62-63, assoc, 63-64, asst prof oper dent, 64-69; prof gen dent, Sch Dent Med, Hosp, Univ Conn, Hartford, 69-74, head dept, 69-71; PROF PERIODONT, HEALTH CTR, UNIV CONN, FARMINGTON, 74- Concurrent Pos: Fel periodont, Sch Dent Med, Harvard Univ, 58-61; USPHS spec fel, Nat Inst Dent Res, 61-62; actg head dept periodont, Health Ctr, Univ Conn, 74-75. Mem: AAAS; Int Asn Dent Res. Res: Dental aspects of aging; dental embryology; ultrastructure of oral tissues and of bone; bone resorption; dental caries; dental plaque. Mailing Add: Dept of Periodont Univ of Conn Health Ctr Farmington CT 06032

NALBANDOV, ANDREW VLADIMIR, b Simferopol, Russia, July 4, 12; nat US; m 36. GENETICS. Educ: Univ Munich, dipl, 33; Okla Agr & Mech Col, MS, 36; Univ Wis, PhD(physiol genetics), 40. Hon Degrees: Dr, Tech Univ, Munich, 74. Prof Exp: Asst genetics, Univ Wis, 37-40; assoc prof, 40-50, PROF ANIMAL PHYSIOL, UNIV ILL, URBANA, 50- Honors & Awards: Borden Award, 59; Endocrinol & Reprod Award, 64; Morrison Award, 69; Leadership Award, Endocrine Soc, 74. Mem: AAAS; Endocrine Soc; Am Soc Zool; Genetics Soc Am; Soc Study Reprod (pres, 73). Res: Physiology of reproduction; sterility; hypothalamo-hypophysial interrelation; pituitary ano gonadal hormones, releasing factors. Mailing Add: Animal Sci 102 Animal Genet Lab Univ of Ill Urbana IL 61801

NALBANDOV, OLGA OLIVER, biochemistry, see 12th edition

NALDRETT, ANTHONY JAMES, b London, Eng, June 23, 33; m 60; c 3. GEOLOGY. Educ: Cambridge Univ, BA, 56, MA, 62; Queen's Univ, MSc, 61, PhD(geol), 64. Prof Exp: Geologist, Falconbridge Nickel Mines Ltd, 57-59; fel geochem, Geophys Lab, Carnegie Inst Wash, 64-67; from asst prof to assoc prof, 67-72, PROF GEOL, UNIV TORONTO, 72- Concurrent Pos: Mem, Comn Exp Petrol, Int Union Geol Sci; ed, J Petrol, 73- Honors & Awards: Barlow Medal, Can Inst Mining & Metall, 74. Mem: Geol Soc Am; Can Inst Mining & Metall; Geol Asn Can; Soc Econ Geol; Mineral Asn Can. Res: Geology and geochemistry of nickel deposits; geology of the Sudbury area; petrology of mafic and ultramafic rocks; experimental geochemistry of sulfide, sulfide-oxide and sulfide-silicate systems. Mailing Add: Dept of Geol Univ of Toronto Toronto ON Can

NALDRETT, STANLEY NORMAN, physical chemistry, see 12th edition

NALEWAJA, JOHN DENNIS, b Browerville, Minn, Oct 7, 30; m 59; c 3. WEED SCIENCE. Educ: Univ Minn, BS, 53, MS, 59, PhD(agron), 62. Prof Exp: PROF WEED SCI, NDAK STATE UNIV, 62- Mem: Weed Sci Soc Am. Res: Basic and applied aspects of weed science. Mailing Add: Dept of Agron NDak State Univ Fargo ND 58102

NALIN, DAVID ROBERT, b New York, NY, Apr 22, 41. MEDICAL RESEARCH. Educ: Cornell Univ, AB, 61; Albany Med Col, MD, 65. Prof Exp: Intern & resident med, Montefiore Hosp, New York, 66-67; res assoc, Cholera Res Hosp, Off Int Res, NIH, 67-70; sr resident, Boston City Hosp, 70-71; instr, Sch Med, Dept Med & res assoc pathobiol, Sch Hyg & Pub Health, 71-72, ASST PROF MED & PATHOBIOL, JOHNS HOPKINS UNIV, 72- Concurrent Pos: Guest scientist, Cholera Res Lab, Dacca, Bangladesh & sr consult, Int Rescue Comt, 72-75. Mem: Royal Soc Trop Med & Hyg. Res: Infectious diseases, particularly enteric, including treatment and pathophysiology; tropical diseases and adaptation of medical technology to needs of developing countries. Mailing Add: Johns Hopkins Ctr for Med Res Suite 115 550 N Broadway Baltimore MD 21205

NALL, DANIEL MAX, organic chemistry, see 12th edition

NALL, JULIAN CLARK, b Memphis, Tenn, Apr 28, 21; m 56; c 1. NUCLEAR PHYSICS. Educ: Southwestern at Memphis, BS, 43; Mass Inst Technol, MS, 48; Vanderbilt Univ, PhD(physics), 58. Prof Exp: Asst prof physics, Southwestern at Memphis, 48-54; PHYSICIST, US GOVT, 58- Concurrent Pos: Assoc prof lectr, George Washington Univ, 60. Mem: Am Phys Soc; Am Asn Physics Teachers. Res: Very low energy electron spectroscopy; beta ray spectrometers; electronics. Mailing Add: 4631 N 27th St Arlington VA 22207

NALL, RAYMOND WILLETT, b Flaherty, Ky, Nov 21, 39; m 58; c 3. LIMNOLOGY, BOTANY. Educ: Western Ky Univ, BS, 61; Univ Louisville, PhD(bot), 65. Prof Exp: Staff biologist, 65, RESOURCE PROJS MGR, LAND BETWEEN THE LAKES, TENN VALLEY AUTHORITY, 65- Concurrent Pos: Asst prof, Murray State Univ, 70-72, adj prof, 70- Mem: Am Soc Limnol & Oceanog. Res: Water pollution ecology; life histories of deer and wild turkey; improved techniques for management of natural resources. Mailing Add: Land Between the Lakes Tenn Valley Authority Golden Pond KY 42231

NALLEY, SAMUEL JOSEPH, b Benton, Ark, May 5, 43; m 62; c 2. ATOMIC PHYSICS, MOLECULAR PHYSICS. Educ: State Col Ark, BS, 65; Univ Tenn, Knoxville, MS, 67, PhD(physics), 71. Prof Exp: Asst prof, 71-74, ASSOC PROF PHYSICS, CHATTANOOGA STATE TECH COMMUNITY COL, 74- Mem: Am Phys Soc; Am Asn Physics Teachers. Res: Atomic and molecular collision. Mailing Add: Dept of Physics Chattanooga State Tech Com Col Chattanooga TN 37406

NALOS, ERVIN JOSEPH, b Prague, Czech, Sept 10, 24; nat US; m 47; c 3. PHYSICS. Educ: Univ BC, BASc, 46, MASc, 47; Stanford Univ, PhD(elec eng), 51. Prof Exp: Res assoc microwave physics, 50-54; group leader microwave tube develop & res, Microwave Lab, Gen Elec Co, Calif, 54-59, sci rep to Europe, Res Lab, 59-62; staff engr, Off Vpres Res & Develop, 62-69, mgr appl technol, Military Airplane Systs Div, 69-71, supvr civil & com systs, 71-74, SUPVR PLANS & ANAL, RES & ENG DIV, AEROSPACE GROUP, BOEING CO, 74- Honors & Awards: Baker Award, Inst Radio Eng, 59. Mem: Sr mem Inst Elec & Electronics Eng; Sigma Xi. Res: Electron physics and electronics; high power microwave devices. Mailing Add: M/S 88-22 Res & Eng Div Aerosp Grp Boeing Co PO Box 3999 Seattle WA 98124

NALWALK, ANDREW JEROME, marine geology, oceanography, deceased

NAMBA, RYOJI, b Honolulu, Hawaii, Jan 31, 22; m 48; c 4. ENTOMOLOGY. Educ: Mich State Col, BS, 48, MS, 50; Univ Minn, PhD(entom), 53. Prof Exp: From asst entomologist to assoc entomologist, 53-68, ENTOMOLOGIST, AGR EXP STA, UNIV HAWAII, 68- Mem: Entom Soc Am. Res: Leafhoppers; insect transmission of plant pathogens. Mailing Add: Dept of Entom Univ of Hawaii Agr Exp Sta Honolulu HI 96822

NAMBA, TATSUJI, b Changchun, China, Jan 29, 27. NEUROLOGY, PHARMACOLOGY. Educ: Okayama Univ, MD, 50, PhD(med), 56. Prof Exp: Asst med, Med Sch, Okayama Univ, 56-57, lectr, 57-62; res assoc, 62-64, from asst attend physician to assoc attend physician, 64-71, dir neuromuscular dis labs, 66-71, ATTEND PHYSICIAN, MAIMONIDES MED CTR, 71-, HEAD ELECTROMYOGRAPHY CLIN, 66-, DIR NEUROMUSCULAR DIS DIV, 71-; ASSOC PROF MED, STATE UNIV NY DOWNSTATE MED CTR, 71- Concurrent Pos: Res fel, Maimonides Med Ctr, 59-62; Fulbright fel, 59-62; consult, Fukuyama Defense Force Hosp, Japan, 57-59; from instr to asst prof, State Univ NY Downstate Med Ctr, 59-71; from asst vis physician to assoc vis physician, Kings County Hosp Ctr, 65-73, vis physician, 73-; from asst attend physician to assoc attend physician, Coney Island Hosp, 66-74, attend physician, 74-; attend physician, State Univ Hosp, 66-; mem, Med Adv Bd, Myasthenia Gravis Found. Mem: Am Col Physicians; Am Acad Neurol; Am Soc Pharmacol & Exp Therapeut; Am Soc Clin Pharmacol & Chemother; AMA. Res: Basic and clinical research of skeletal muscle and neuromuscular diseases; clinical pharmacology of neuromuscular agents. Mailing Add: Maimonides Med Ctr 4802 Tenth Ave Brooklyn NY 11219

NAMBIAR, GOVINDAN KUPPADAKKATH, b Kunhimangalam, India, Jan 20, 32; m 60; c 3. GENETICS. Educ: Univ Madras, BVSc, 54; Univ Tenn, Knoxville, MS, 59; Tex A&M Univ, PhD(genetics), 68. Prof Exp: ASSOC Prof Exp: Assoc prof biol, J C Smith Univ, 68-69; assoc prof, 69-73, PROF BIOL, SAVANNAH STATE COL, 73- Mem: Am Inst Biol Sci; Poultry Sci Asn; Nat Inst Sci; Genetics Soc Am; NY Acad Sci. Res: Transfer of genetic units through DNA transplants; population genetics of Drosophila melanogaster; gene transfer in Drosophila melanogaster. Mailing Add: 2328 Margaret St Savannah GA 31404

NAMBOODIRI, MADASSERY NEELAKANTAN, b Kothamangalam, India, Oct 18, 35; m 63; c 1. NUCLEAR CHEMISTRY. Educ: Kerala Univ, India, BSc, 57; State Univ NY Stony Brook, PhD(chem), 72. Prof Exp: Sci officer radiochem, Bhabha Atomic Res Ctr, Bombay, India, 58-67; RES SCIENTIST NUCLEAR CHEM, CYCLOTRON INST, TEX A&M UNIV, 72- Concurrent Pos: Res affil chem, Argonne Nat Lab, 60-61. Mem: Am Phys Soc; AAAS. Res: Nuclear reactions, especially heavy ion induced reactions, with emphasis on fission, heavy ion fusion, deep inelastic reactions and related phenomena. Mailing Add: Cyclotron Inst Tex A&M Univ College Station TX 77843

NAMBU, YOICHIRO, b Tokyo, Japan, Jan 18, 21; m 45; c 2. THEORETICAL PHYSICS. Educ: Univ Tokyo, BS, 42, ScD, 52. Prof Exp: Asst, Univ Tokyo, 45-49; from asst prof to prof, Osaka City Univ, 49-56; res assoc, 54-56, from assoc prof to prof, 56-71, DISTINGUISHED PROF PHYSICS, ENRICO FERMI INST NUCLEAR STUDIES, UNIV CHICAGO, 71- Concurrent Pos: Mem, Inst Advan Study, 52-54. Honors & Awards: Dannie Heineman Prize Math Physics, Am Phys Soc, 70. Mem: Nat Acad Sci; Am Phys Soc; Phys Soc Japan; Am Acad Arts & Sci. Res: Field theory; theory of elementary particles; theory of superconductivity. Mailing Add: Dept Physics E Fermi Inst for Nuclear Studies Univ of Chicago Chicago IL 60637

NAMEROFF, MARK A, b Philadelphia, Pa, May 16, 39. DEVELOPMENTAL BIOLOGY. Educ: Univ Pa, BA, 60, MD, 65, PhD(anat embryol), 66. Prof Exp: Instr anat, Sch Med, Univ Pa, 66-67; staff mem, Armed Forces Inst Path, Washington, DC, 67-70; asst prof, 70-75, ASSOC PROF BIOL STRUCT, UNIV WASH, 75- Res: Cell differentiation; embryonic chondrocytes and muscle cells. Mailing Add: Dept of Biol Struct Univ of Wash Seattle WA 98195

NAMIAS, JEROME, b Bridgeport, Conn, Mar 19, 10; m 38; c 1. METEOROLOGY. Educ: Mass Inst Technol, MS, 41. Hon Degrees: ScD, Univ RI, 72. Prof Exp: Asst aerology, Blue Hill Meteorol Observ, Harvard Univ, 33-36; res assoc, Mass Inst Technol, 35-40; chief extended forecast div, US Weather Bur, Nat Oceanic & Atmospheric Agency, 41-71; RES METEOROLOGIST, SCRIPPS INST OCEANOG, UNIV CALIF, SAN DIEGO, 72- Concurrent Pos: Meteorologist, Trans World Airlines, Inc, 34; lectr, Univ Stockholm, 50; assoc, Woods Hole Oceanog Inst, 54-, mem sci vis comt, 64-; distinguished lectr, Pa State Univ, 62; part-time res meteorologist, Scripps Inst Oceanog, 68-71. Honors & Awards: Meisinger Award, 38; Citation, Secy Navy, 43; Meritorious Serv Award, US Dept Com, 50, Gold Medal Award, 65; Rockefeller Pub Serv Award, 55; Extraordinary Sci Accomplishment Award, 55. Mem: Fel Am Meteorol Soc; fel Am Geophys Union; fel NY Acad Sci; Royal Meteorol Soc. Res: Long range weather forecasting and general circulation of atmosphere; aerology; large scale air-sea interaction. Mailing Add: Scripps Inst of Oceanog Univ of Calif at San Diego La Jolla CA 92037

NAMKOONG, GENE, b New York, NY, Jan 25, 34; m 56; c 3. FOREST GENETICS, EVOLUTION. Educ: State Univ NY, BS, 56, MS, 58; NC State Univ, PhD(genetics), 63. Prof Exp: Res forester, 58-60, plant geneticist, 63-71, PIONEERING RES SCIENTIST POP GENETICS, FOREST SERV, USDA, 71- Concurrent Pos: From asst prof to assoc prof genetics & forestry, NC State Univ, 63-71, prof, 71-; vis prof, Univ Chicago, 68; consult, BC Forest Serv, 72- & Repub Korea Forest Serv, 74-; adj prof, Shaw Univ, 75- Honors & Awards: Sci Achievement Award, Int Union Forest Res Orgns, 71. Mem: Biomet Soc; Genetics Soc Am; Soc Am Foresters; Am Soc Nat; Soc Study Evolution. Res: Mathematical, population genetics, particularly with respect to forest tree species. Mailing Add: 811 Beaver Dam Rd Raleigh NC 27607

NAMM, DONALD H, b Hamden, Conn, Feb 10, 40; m 63; c 2. PHARMACOLOGY. Educ: Rensselaer Polytech Inst, BS, 61; Albany Med Col, PhD(pharmacol), 65. Prof Exp: Instr pharmacol, Emory Univ, 67-68; asst prof, Sch Med, Univ Okla, 68-69; sr res pharmacologist, 69-75, GROUP LEADER, RES LABS, BURROUGHS-WELLCOME CO, 75- Concurrent Pos: Fel pharmacol, Emory Univ, 65-67. Mem: Am Soc Pharmacol & Exp Therapeut. Res: Heart metabolism; blood vessel metabolism; drug-enzyme interactions. Mailing Add: Dept of Pharmacol Wellcome Res Labs Research Triangle Park NC 27709

NAMMINGA, HAROLD EUGENE, b Scotland, SDak, July 26, 45; m 73. ECOLOGY, LIMNOLOGY. Educ: Univ SDak at Springfield, BS, 67; Univ SDak at Vermillion, MA, 69; Okla State Univ, PhD(zool), 75. Prof Exp: Instr biol, Kearney State Col, 69-71; ENVIRON SCIENTIST-ECOL, TECHNOL RES & DEVELOP, INC DIV, BENHAM-BLAIR & AFFIL, 75- Mem: AAAS; Am Soc Limnol & Oceanog; Am Inst Biol Sci. Res: Heavy metals in aquatic ecosystems and the effects of metals on aquatic community structure; secondary interests include effects of metals in land application of municipal wastewaters. Mailing Add: 9515 N Greystone Oklahoma City OK 73120

NAMY, JEROME NICHOLAS, b Cleveland, Ohio, Aug 11, 38; m 63; c 3. PETROLOGY, STRATIGRAPHY. Educ: Western Reserve Univ, BA, 60; Univ Tex, Austin, PhD(geol), 69. Prof Exp: Explor geologist, Pan Am Petrol Corp, Standard Oil Co, Ind, 67-70; asst prof, 70-73, ASSOC PROF GEOL, BAYLOR UNIV, 73- Mem: Nat Asn Geol Teachers; Geol Soc Am; Am Asn Petrol Geologists; Sigma Xi; Soc Econ Paleontologists & Mineralogists. Res: The stratigraphy and petrology of carbonate and sedimentary rocks. Mailing Add: Dept of Geol Baylor Univ Box 6367 Waco TX 76706

NANCE, CHARLES ROGER, b Berkeley, Calif, Mar 18, 38; m 65; c 2. CULTURAL ANTHROPOLOGY, ARCHAEOLOGY. Educ: Univ Calif, Los Angeles, BA, 62; Wash State Univ, MA, 66; Univ Tex, PhD(anthrop), 71. Prof Exp: Asst prof, 67-73, ASSOC PROF ANTHROP, UNIV ALA, BIRMINGHAM, 73- Concurrent Pos: Comnr, Ala Hist Comn, 70-72. Mem: Am Anthrop Asn; Soc Am Archaeol. Res: North American prehistory; lithic technology; prehistoric human ecology. Mailing Add: Dept Anthrop Univ of Ala Birmingham AL 35233

NANCE, FRANCIS CARTER, b Manila, Philippines, Jan 1, 32; US citizen; m 59; c 4. SURGERY, PHYSIOLOGY. Educ: Univ Tenn, MD & MS, 59. Prof Exp: Instr surg, 65-67, from asst prof to assoc prof surg & physiol, 67-73, PROF SURG & PHYSIOL, LA STATE UNIV SCH MED, NEW ORLEANS, 73- Concurrent Pos: Am Cancer Soc fel, Univ Pa, 63-64. Mem: AAAS; Am Gastroenterol Soc; Am Col Surgeons; Am Asn Surg Trauma; Am Surg Asn. Res: Gastrointestinal physiology; effects of microbial flora on various gastrointestinal functions and diseases; burns. Mailing Add: Dept of Surg La State Univ Sch of Med New Orleans LA 70112

NANCE, JAMES FRANCIS, b San Bernardino, Calif, May 5, 14; m 47; c 4. PLANT PHYSIOLOGY. Educ: Univ Calif, AB, 37, MA, 39, PhD(plant physiol), 44. Prof Exp: Mkt specialist, USDA, Calif, 42-43; instr bot, Univ Wis, 46-47, res assoc, 47-48; from asst prof to prof, 48-75, EMER PROF BOT, UNIV ILL, URBANA, 75- Mem: AAAS; Bot Soc Am; Am Soc Plant Physiol. Res: Biochemical effects of the auxins. Mailing Add: Rt 1 Neoga IL 62447

NANCE, JOHN ARTHUR, b Ralls, Tex, Jan 26, 19; m 44; c 3. NUTRITION. Educ: Tex A&M Univ, BS, 41; Iowa State Univ, MS, 47; Okla State Univ, PhD(animal sci), 52. Prof Exp: Instr animal sci, Navarro Jr Col, 47-48; asst prof, Tex A&I Univ, 48-49; PROF AGR, SAM HOUSTON STATE UNIV, 52- Mem: Am Soc Animal Sci. Res:

Ruminant nutrition; roughage substitutes. Mailing Add: Dept of Agr Sam Houston State Univ Huntsville TX 77340

NANCE, JON ROLAND, b Springfield, Mo, Nov 13, 39; m 60; c 3. PHYSICS. Educ: Univ Mo-Rolla, BS, 60; Univ Ill, Urbana, MS, 61, PhD(physics), 66. Prof Exp: ASST PROF PHYSICS, UNIV GA, 66- Concurrent Pos: Consult, PEC Corp, 66-67. Mem: Am Phys Soc; Am Asn Physics Teachers. Res: Relativistic quantum theory; field theory; space physics. Mailing Add: Dept of Physics Univ of Ga Athens GA 30601

NANCE, OLEN ALVIN, computer science, see 12th edition

NANCE, RICHARD E, b Raleigh, NC, July 22, 40; m 62; c 1. COMPUTER SCIENCE, OPERATIONS RESEARCH. Educ: NC State Univ, BS, 62, MS, 66; Purdue Univ, PhD(opers res), 68. Prof Exp: From asst prof to assoc prof comput sci & opers res, Southern Methodist Univ, 68-73; HEAD, DEPT COMPUT SCI, VA POLYTECH INST & STATE UNIV, 73- Concurrent Pos: Chmn, Col Simulation & Gaming, Inst Mgt Sci, 75- Mem: Asn Comput Mach; Opers Res Soc Am; Inst Mgt Sci; Am Soc Info Sci; Am Inst Indust Engrs. Res: Digital simulation theory; mathematical models of information networks; theory of information retrieval; computer systems modeling and performance evaluation. Mailing Add: Dept of Comput Sci Va Polytech Inst & State Univ Blacksburg VA 24061

NANCE, WALTER ELMORE, b Manila, Philippines, Mar 25, 33; US citizen; m 57; c 2. HUMAN GENETICS, INTERNAL MEDICINE. Educ: Univ of South, SB, 54; Harvard Univ, MD, 58; Univ Wis, PhD(med genetics), 68. Prof Exp: From intern to resident med, Sch Med, Vanderbilt Univ, 58-61, asst prof, Sch Med, Vanderbilt Univ, 64-69; prof genetics & med, Sch Med, Ind Univ, Indianapolis, 69-75; PROF HUMAN GENETICS & CHMN DEPT, MED COL VA, 75- Concurrent Pos: Mem, Genetics Training Comt, Nat Inst Gen Med Sci, 71-74; prin investr, Ind Univ Human Genetics Ctr, 74-75; consult, Nat Inst Neurol Dis & Stroke, Annual Surv Hearing Impaired Children & Youth & Genetics Sect, WHO; mem, Epidemiol & Dis Control Study Sect, NIH, 75-79. Mem: AAAS; Math Asn Am; Am Soc Human Genetics (secy, 71-74); Int Soc Twin Studies; fel Am Col Physicians. Res: Medical genetics; hereditary deafness; human twin studies; population genetics; analysis of human genetic polymorphisms; genetically determined disorders of metabolism. Mailing Add: Dept of Genetics Med Col of Va Box 33 Richmond VA 23298

NANCOLLAS, GEORGE H, b Wales, Brit, Sept 24, 28; m 54; c 2. PHYSICAL CHEMISTRY, INORGANIC CHEMISTRY. Educ: Univ Wales, BSc, 48, PhD(phys chem), 51; Glasgow Univ, DSc, 63. Prof Exp: Res assoc, Univ Manchester, 51-53; lectr, Glasgow Univ, 53-65; provost fac natural sci & math, 70-75, PROF CHEM, STATE UNIV NY BUFFALO, 65- Concurrent Pos: Fel, Univ Wales, 51-52; vis scientist, Brookhaven Nat Lab, 63-64. Mem: AAAS; Faraday Soc; Am Chem Soc; The Chem Soc; fel Royal Inst Chem. Res: Formation of metal complexes and ion-pairs; inorganic ion exchangers; kinetics of crystal growth and dissolution; the electrical double layer. Mailing Add: Dept of Chem State Univ of NY Buffalo NY 14214

NANDA, DEVENDER (DAVE) KUMAR, b Mandibahudin, India, Mar 15, 38; m 66; c 1. PLANT BREEDING, GENETICS. Educ: Cent Col Agr, New Delhi, India, BSc, 58; Indian Agr Res Inst, New Delhi, MSc, 60; Univ Wis-Madison, PhD(agron), 64. Prof Exp: Res asst sorghum breeding, Rockefeller Found, 60, in-charge field collection corn & millets, 60-61; res agronomist, DeKalb AgRes, Inc, 64-65; dir res corn breeding & genetics, Edward J Funk & Sons, Inc, 65-68; DIR RES CORN BREEDING & GENETICS, TROJAN SEED CO, 68- Mem: Am Soc Agron; Crop Sci Soc Am; Soil Sci Soc Am. Res: Botany; cytology; plant pathology; fungicides; herbicides; teaching; extension; seed production; marketing of seeds; quality control in seeds; crop management; fertilizers; statistics; economics; entomology. Mailing Add: Trojan Seed Co Windfall IN 46076

NANDA, JAGDISH L, b Punjab, India, Feb 1, 33; m 63; c 2. MATHEMATICS. Educ: Univ Delhi, BA, 53, MA, 55; Ind Univ, PhD(geom), 61. Prof Exp: Asst math, Govt India, New Delhi, 55-57; res assoc, Wright-Patterson AFB, Ohio, 60-62; asst prof, Univ Dayton, 61-63; assoc prof, Villanova Univ, 63; assoc prof, Univ Delhi, 63-64; PROF MATH, EASTERN ILL UNIV, 64- Mem: Math Asn Am; Am Math Soc. Res: Geometry; relativity. Mailing Add: Dept of Math Eastern Ill Univ Charleston IL 61920

NANDA, RAVINDRA, b Layallpur, India, Feb 19, 43; m 66; c 2. ORTHODONTICS, TERATOLOGY. Educ: Univ Lucknow, BDS, 64, MDS, 66; Roman Cath Univ, Nijmegen, PhD(med), 69. Prof Exp: Res assoc orthod, Roman Cath Univ, Nijmegen, 67-70; asst prof, Col Dent Surg, Loyola Univ, Chicago, 70-73; ASSOC PROF ORTHOD, UNIV CONN HEALTH CTR, FARMINGTON, 73- Mem: Teratol Soc; Int Asn Dent Res; Europ Orthod Soc; Indian Dent Asn. Res: Clinical orthodontics; growth and development of face; tooth development; radiotracer studies. Mailing Add: 3 Wintergreen Lane West Simsbury CT 06092

NANDAN, RAJIVA, b India, Aug 23, 40; m 66; c 2. MICROBIOLOGY, BIOCHEMISTRY. Educ: Univ Lucknow, BS, 58, MS, 61; Ore State Univ, PhD(microbiol), 67. Prof Exp: From res asst to res assoc biochem, Ore State Univ, 66-68; bacteriologist, Goshen Gen Hosp, Ind, 68-69; res biochemist, South Bend Med Found, 68-69; sr res pharmacologist, 69-73, MGR BIOCHEM PHARMACOL, BAXTER LABS, INC, 73- Mem: Am Asn Clin Chem; Am Soc Microbiol. Res: Development and cure of experimental atherosclerosis in animals; study of microaggregates in stored blood and red cell preservation; carbohydrate metabolism in bacteria and animals; various methods for in vitro pyrogen testing in intravenous fluids, blood and urine; hyperalimentation. Mailing Add: Pharmacol Dept Baxter Labs Inc Morton Grove IL 60053

NANDEDKAR, ARVINDKUMAR NARHARI, b Nagpur, India, Apr 8, 37; m 64; c 2. BIOCHEMISTRY, CLINICAL CHEMISTRY. Educ: Univ Nagpur, BSc, 59, MSc, 61; Univ Delhi, PhD(med biochem), 66. Prof Exp: Asst res officer, Indian Coun Med Res, V Patel Chest Inst, Univ Delhi, 66; NIH res assoc biochem, Georgetown Univ, 66-68; from instr to asst prof, 68-74, ASSOC PROF BIOCHEM, HOWARD UNIV, 74-, ASST DIR ACAD REINFORCEMENT PROG, 73- Concurrent Pos: AEC fel, Howard Univ, 68-71; vis assoc prof, Cornell Univ Med Ctr, 75-; consult clin biochem, New York Hosp, 75- Mem: Am Asn Clin Chemists; AAAS; Am Chem Soc; fel Am Inst Chem. Res: Lipid biochemistry, including clinical application to tuberculosis and anaphylaxis; metabolism of fatty acids in mammary gland; protein chemistry; binding and carrier of metals in body fluids. Mailing Add: Dept of Biochem Howard Univ Col of Med Washington DC 20001

NANDI, JEAN, zoology, comparative endocrinology, see 12th edition

NANDI, SATYABRATA, b North Lakhimpur, India, Dec 1, 31; nat US; m 57. ZOOLOGY, ENDOCRINOLOGY. Educ: Univ Calcutta, BSc, 49, MSc, 51; Univ Calif, PhD(zool), 58. Prof Exp: Demonstr zool, Bethune Col, India, 49-53; lectr, City Col Calcutta, 53; asst biophys, Saha Inst Nuclear Physics, Calcutta, 53-54; asst zool, 54-56, res zoologist, 56-57, jr res zoologist & lectr, 58-59, asst res endocrinologist, 59-61, actg asst prof zool, 61-62, from asst prof to prof, 62-70, Miller prof, 70-71, res endocrinologist, Cancer Res Lab, 68-74, chmn dept zool, 71-73, PROF ZOOL, UNIV CALIF, BERKELEY, 71-, DIR CANCER RES LAB, 74- Concurrent Pos: Guggenheim fel, Netherlands Cancer Inst, Amsterdam, 67-68; vis scientist, Virus Res Inst, Kyoto Univ, 65. Mem: Am Soc Zoologists; Endocrine Soc; Am Asn Cancer Res. Res: Tumor biology, including endocrinology, virology, genetics and hormone receptor. Mailing Add: 230 Warren Hall Univ of Calif Cancer Res Lab Berkeley CA 94720

NANDY, KALIDAS, b Calcutta, India, Oct 1, 30; m 61; c 2. GERIATRICS, NEUROANATOMY. Educ: Univ Calcutta, MD, 53; Univ Lucknow, MSurg, 60; Emory Univ, PhD(anat), 63. Prof Exp: Lectr anat, Univ Calcutta, 54-57, asst prof, 57-60, reader, 60-61; from asst prof to prof anat, Emory Univ, 63-75; DEP DIR & DIR RES, GERIATRIC RES EDUC & CLIN CTR, BEDFORD VET ADMIN HOSP, 75-; PROF ANAT, BOSTON UNIV, 75- Concurrent Pos: Indian Coun Med Res-Rockefeller Found fel, 59-60; Tull fel, 62-63. Mem: Am Asn Anat; Geront Soc; Soc Neurosci. Res: Cytochemistry of nervous system; biology of aging; blood-brain barrier. Mailing Add: Geriatric Res Educ & Clin Ctr Vet Admin Hosp 200 Springs Rd Bedford MA 01730

NANES, ROGER, b Brooklyn, NY, Jan 25, 44; m 68; c 2. MOLECULAR SPECTROSCOPY. Educ: Harpur Col, BA, 65; Johns Hopkins Univ, PhD(phys chem), 70. Prof Exp: Nat Res Coun Can fel spectros, Div Pure Physics, Nat Res Coun Can, Ont, 70-71 & Univ Western Ont, 71-72; asst prof, 72-75, ASSOC PROF PHYSICS, CALIF STATE UNIV, FULLERTON, 75- Mem: Am Chem Soc; Am Phys Soc; AAAS; Soc Appl Spectros. Res: Spectroscopy and molecular structure; vapor phase electronic spectroscopy of polyatomic molecules; electric and magnetic field effects on spectra; air pollution. Mailing Add: Dept of Physics Calif State Univ Fullerton CA 92634

NANKERVIS, GEORGE ARTHUR, b Meriden, Conn, Apr 1, 30; m 54; c 2. PEDIATRICS, MICROBIOLOGY. Educ: Princeton Univ, AB, 52; Univ Rochester, PhD(microbiol), 59, MD, 62. Prof Exp: Asst prof, 67-71, ASSOC PROF PEDIAT, CASE WESTERN RESERVE UNIV, 71- Concurrent Pos: Teaching fel pediat, Harvard Univ, 64-65; res fel infectious dis, Case Western Reserve Univ, 65-67. Res: Diagnostic virology and vaccine evaluation; study of pediatric populations with respect to immune status to infectious diseases; congenital viral infections. Mailing Add: Cleveland Metrop Gen Hosp 3395 Scranton Rd Cleveland OH 44109

NANKIVELL, JOHN (ELBERT), b Mt Vernon, NY, Jan 31, 21; m 45; c 3. PHYSICS. Educ: Stevens Inst Technol, ME, 43, MS, 51, PhD, 62. Prof Exp: Mech engr, Gen Elec Co, 43-46; asst prof physics, Stevens Inst Technol, 46-58, res assoc, 58-63; PROF MECH TECHNOL, STATEN ISLAND COMMUNITY COL, 63- Mem: Am Phys Soc; Am Asn Physics Teachers; Am Soc Eng Educ. Res: Plasma physics. Mailing Add: Dept of Mech Technol Staten Island Community Col Staten Island NY 10301

NANNELLI, PIERO, b Montelupo, Italy, Sept 29, 35; m 61; c 2. POLYMER CHEMISTRY. Educ: Univ Florence, DSc(org chem), 61. Prof Exp: Res assoc inorg polymers, Univ Ill, 61-63; asst prof coord chem, Univ Florence, 63-65; sr res chemist, 65-70, proj leader, 70-72, GROUP LEADER, PENNWALT CORP, 72- Mem: Am Chem Soc. Res: Inorganic polymers; coordination compounds; high performance structural materials and coatings. Mailing Add: Pennwalt Corp Technol Ctr 900 First Ave King of Prussia PA 19406

NANNEY, DAVID LEDBETTER, b Abingdon, Va, Oct 10, 25; m 51; c 2. GENETICS. Educ: Okla Baptist Univ, AB, 46; Ind Univ, PhD(zool), 51. Prof Exp: From asst prof to assoc prof zool, Univ Mich, 51-58; fel, Calif Inst Technol, 58-59; PROF ZOOL, UNIV ILL, URBANA, 59- Mem: AAAS; Genetics Soc Am; Soc Protozool; Am Soc Zool; Am Soc Nat. Res: Formal genetics, cytogenetics, developmental genetics and evolutionary genetics of ciliated protozoa. Mailing Add: Dept of Genetics & Develop Univ of Ill Urbana IL 61801

NANNEY, THOMAS RAY, b Concord, NC, Apr 21, 31; m 54; c 2. PHYSICAL CHEMISTRY, COMPUTER SCIENCE. Educ: Univ NC, BS, 53; Univ SC, PhD(phys chem), 62. Prof Exp: Chemist, E I du Pont de Nemours & Co, Inc, 53-54; asst prof chem, 60-66, ASSOC PROF CHEM, FURMAN UNIV, 66-, DIR COMPUT CTR, 67- Concurrent Pos: Vis asst prof & USPHS & Dept Defense Advan Res Projs Agency fel, 64-65; consult, SC Found Independent Cols, 68-; consult, Duke Power Co, NC, 70; consult, Nat Lab Higher Educ, Durham, NC, 70- Mem: AAAS; Am Chem Soc; Asn Comput Mach; Soc Appl Spectros. Res: Information retrieval; programming languages; infrared and Raman spectroscopy; physical chemistry of solutions. Mailing Add: Comput Ctr Furman Univ Greenville SC 29613

NANNI, LUIS FERNANDO, statistical analysis, see 12th edition

NANZ, ROBERT AUGUSTUS ROLLINS, b Baltimore, Md, Apr 3, 15; m 39. FOOD SCIENCE. Educ: Rutgers Univ, BS, 37; Columbia Univ, MS, 39. Prof Exp: Food chemist, Quaker Maid Co, Inc, Great Atlantic & Pac Tea Co, NY, 37-38; biochemist, Watchung Labs, NJ, 38-39; nutrit specialist, Walker-Gordon Lab Co, Inc, 39-43; asst to coordr res, Spec Prod Div, Borden Co, NY, 46-47; dir food tech sect, Foster D Snell, Inc, 47-50; res chemist, Fla Citrus Canners Coop, 50-51; tech rep, Crown Can Co Div, Crown Cork & Seal Co, Fla, 51-53; pres, Fla Chemists & Engrs, Inc, 53-60; pres, Sci Assocs, Inc, 60-62; aerospace technologist, Food & Nutrit Group, Biomed Specialties Br, Biomed Res Off, Manned Spacecraft Ctr, NASA, Tex, 62-67; asst dir, Nat Ctr Fish Protein Concentrate, US Dept Com, DC, 67-68; prog dir tech develop, Aquatic Sci Inc, Fla, 69-70; CONSULT FOODS & NUTRIT, 70- Concurrent Pos: Instr, Col Boca Raton, 71-74. Mem: Inst Food Technologists. Res: Food processing, formulations; marketing; nutritional factors in food. Mailing Add: Apt 701 300 NE 20th St Boca Raton FL 33431

NANZ, ROBERT HAMILTON, JR, b Shelbyville, Ky, Sept 14, 23; m; c 2. GEOLOGY. Educ: Miami Univ, AB, 44; Univ Chicago, PhD(geol), 52. Prof Exp: Res geologist, Shell Develop Co, 47-58, mgr geol dept, 58-59, dir explor res, 59-64, explor mgr, 64-67, vpres explor & prod res div, 67-70, vpres explor, Shell Oil Co, 70-75, VPRES WESTERN E&P REGION, SHELL OIL CO, 75- Mem: Geol Soc Am; Am Asn Petrol Geol. Res: Petroleum geology. Mailing Add: Shell Oil Co PO Box 576 Houston TX 77001

NANZETTA, PHILIP NEWCOMB, b Wilmington, NC, June 4, 40; m 62; c 2. MATHEMATICS, ACADEMIC ADMINISTRATION. Educ: NC State Univ, BS, 62; Univ Ill, MS, 63, PhD(math), 66. Prof Exp: Res assoc math, Case Western Reserve Univ, 66-67; asst prof, Univ Fla, 67-70; assoc prof, St Mary's Col Md, 70-74; DEAN, FAC NATURAL SCI & MATH, STOCKTON STATE COL, 74- Mem: AAAS; Am Math Soc. Res: Topology; structure spaces; lattice theory; genetics. Mailing Add: Stockton State Col Pomona NJ 08240

NAPIER, DONALD RAY, organic chemistry, see 12th edition

NAPIER, ROGER PAUL, b Rochester, NY, Apr 1, 38; m 65. ORGANIC CHEMISTRY. Educ: St John Fisher Col, BS, 59; Univ Rochester, PhD(chem), 63. Prof Exp: Fel organophosphorus chem, Rutgers Univ, 63-65; PROJ LEADER, MOBIL CHEM CO, 65- Mem: Sr mem Am Chem Soc. Res: Natural products; nitrogen heterocycles; organophosphorus chemistry; pesticide chemistry. Mailing Add: Mobil Chem Co Route 27 Edison NJ 08817

NAPKE, EDWARD, b Zahle, Lebanon, Jan 21, 24; Can citizen; m 61; c 3. MEDICINE, PHYSIOLOGY. Educ: Univ NB, Fredericton, BSc, 45; Univ Toronto, MD, 51, dipl pub health, 68. Prof Exp: Res officer aviation physiol, Defence Med Res Labs, 57-58; med officer, Dept Health & Welfare, Food & Drug Directorate, 63-65, MED OFFICER, DRUG ADVERSE REACTION & POISON CONTROL PROGS, DEPT HEALTH & WELFARE, HEALTH PROTECTION BR, GOVT CAN, 65- Concurrent Pos: Res fels, Karolinska Inst, Sweden, 5861 & C H Best Inst, 61-63; consult, Drug Monitoring Prog, WHO, 68- Mem: AAAS; Am Asn Poison Control Ctrs; NY Acad Sci; Can Soc Forensic Sci; Int Soc Biometeorol. Res: G stress and unconsciousness; drug reaction and interreaction detection by program methods; poison control and prevention; human toxicology; hypnosis; fatty tissue studies; biometeorological studies; blood clotting. Mailing Add: Dept of Health & Welfare Health Protection Br Ottawa ON Can

NAPLES, FELIX JOHN, b Quadrelle, Italy, July 7, 12; nat US; m 41; c 3. ORGANIC CHEMISTRY. Educ: Youngstown Col, AB, 33; Univ Vt, MS, 34; Ind Univ, PhD(org chem), 36. Prof Exp: Asst, Ind Univ, 34-36; instr chem & physics, Youngstown Col, 36-37, assoc prof, 40-43; head dept chem, Springfield Jr Col, 37-40; res chemist, 43-45, SR RES CHEMIST, GOODYEAR TIRE & RUBBER CO, 45- Mem: Am Chem Soc. Res: Polymerization; solubility of amino acids; chemical equilibrium; organic synthesis; synthetic rubber; chemical derivatives of diene rubbers. Mailing Add: Goodyear Tire & Rubber Co Dept 455B-2 Goodyear Blvd Akron OH 44316

NAPLES, JOHN OTTO, b Long Beach, Calif, Apr 2, 47; m 69; c 1. ORGANIC CHEMISTRY, POLYMER CHEMISTRY. Educ: Stanford Univ, BS, 69; Univ Calif, Los Angeles, PhD(org chem), 74. Prof Exp: SR SCIENTIST, ROHM AND HAAS CO, 74- Mem: Am Chem Soc. Res: Synthesis of ion exchange resins of unusual selectivity, of superior mechanical, thermal and chemical stability; solid phase organic synthesis; solid phase organic photosensitization. Mailing Add: Rohm and Haas Co Norristown & McKean Rds Spring House PA 19477

NAPOLITANO, JOSEPH J, b New York, NY, Feb 15, 35; m 58; c 2. BIOLOGY, PROTOZOOLOGY. Educ: Iona Col, BS, 56; St John's Univ, NY, MS, 59; NY Univ, PhD(biol), 63. Prof Exp: From instr to asst prof biol, Iona Col, 58-63; asst prof, 63-66, ASSOC PROF BIOL, ADELPHI UNIV, 66- Concurrent Pos: NIH res grant, 64-66. Mem: Soc Protozool; Am Micros Soc. Res: Morphogenesis and taxonomy of amoeboflagellates. Mailing Add: Dept of Biol Adelphi Univ Garden City NY 11530

NAPOLITANO, LEONARD MICHAEL, b Oakland, Calif, Jan 8, 30; m 55; c 3. ANATOMY. Educ: Univ Santa Clara, BS, 51; St Louis Univ, MS, 54, PhD(anat), 56. Prof Exp: Instr anat, Med Col, Cornell Univ, 56-58; from instr to asst prof, Sch Med, Univ Pittsburgh, 58-64; assoc prof, 64-70, PROF ANAT, SCH MED, UNIV N MEX, 70-, DEAN SCH MED, 72- Mem: Am Asn Anat; Am Asn Cell Biol; Electron Micros Soc Am. Res: Autonomic nervous system; fine structure of adipose tissue, heart and myelin. Mailing Add: Deans Off Univ of NMex Sch of Med Albuquerque NM 87131

NAPORA, THEODORE ALEXANDER, b Ridgewood, NJ, Sept 14, 27. BIOLOGICAL OCEANOGRAPHY. Educ: Columbia Univ, BS, 51; Univ RI, MS, 53; Yale Univ, PhD(biol), 64. ASSOC PROF OCEANOG, UNIV RI, 64-, ASST DEAN STUDENTS, GRAD SCH OCEANOG, 71- Mem: AAAS; Am Soc Limnol & Oceanog. Res: Plankton ecology; composition and distribution of oceanic zooplankton; physiology of deep-sea organisms. Mailing Add: Grad Sch of Oceanog Univ of RI Kingston RI 02881

NAPP, DUANE THEODOR, analaytical chemistry, see 12th edition

NAPPI, ANTHONY JOSEPH, b New Britain, Conn, Oct 21, 37; m; c 3. INSECT PHYSIOLOGY, PATHOLOGY. Educ: Cent Conn State Col, BS, 59, MS, 64; Univ Conn, PhD(entom, zool), 68. Prof Exp: Instr biol, Cent Conn State Col, 64-65; res asst entom, Univ Conn, 65-67; univ res grant, 68-69, NSF res grant, 71-72, asst prof, 68-70, ASSOC PROF BIOL, STATE UNIV NY COL OSWEGO, 71- Concurrent Pos: NIH res grant, 74; Am Cancer Soc scholarship, 75. Mem: AAAS; Am Inst Biol Sci; Am Soc Zool; Soc Invertebrate Path; Am Soc Parasitologists. Res: Cellular immune mechanisms of insects against metazoan parasites; insect pathology; parasitology. Mailing Add: Dept of Biol State Univ NY Col Oswego NY 13126

NAPTON, LEWIS KYLE, b Bozeman, Mont, Nov 15, 33; m 60; c 2. ARCHAEOLOGY, PHYSICAL ANTHROPOLOGY. Educ: Mont State Univ, BS, 59; Univ Mont, MA, 65; Univ Calif, Berkeley, PhD(anthrop), 70. Prof Exp: Wenner-Gren fel, Univ Calif, Berkeley, 68-69, asst prof anthrop, 70-71; assoc prof, 72-74, PROF ANTHROP, CALIF STATE COL, STANISLAUS, 74- Concurrent Pos: NSF fel, Univ Calif, Berkeley, 70-; NSF res assoc, Cent Australian Exped, 73-74. Res: North American archaeology; paleoanthropology; environmental archaeology, prehistoric man in arid environments. Mailing Add: Dept of Anthrop Calif State Col Stanislaus Turlock CA 95380

NAQVI, ALI MEHDI, physics, see 12th edition

NAQVI, SAIYID ISHRAT HUSAIN, b Saharanpur, India, June 29, 31; m 58; c 3. NUCLEAR PHYSICS, ASTRONOMY. Educ: Univ Lucknow, BSc, 51, MSc, 53; Univ Man, MSc, 56, PhD(nuclear physics), 61. Prof Exp: Lectr physics, Univ Man, 56-61; asst prof, St Paul's Col, Man, 61-65 & Univ Man, 65-66; asst prof, 66-68, ASSOC PROF PHYSICS, UNIV REGINA, 68- Concurrent Pos: Vis prof, Copenhagen Univ Observ, 72-73. Mem: Am Phys Soc; Royal Astron Soc Can; Can Asn Univ Teachers. Res: Beta and gamma ray spectroscopy; low and medium energy nuclear physics; photoelectric photometry. Mailing Add: Dept of Physics & Astron Univ of Regina Regina SK Can

NAQVI, SAIYID MAHMOODUL HASAN, b Fatehpur, India, Jan 1, 31; m 62; c 2. ORGANIC CHEMISTRY, BIOCHEMISTRY. Educ: Sind Univ, Pakistan, BSc, 51; Ill Inst Technol, MS, 56, PhD(org chem), 60; Fairleigh Dickinson Univ, MBA, 66. Prof Exp: Chief accounts off, Int Islamic Econ Orgn, Pakistan, 51-52; res chemist, Wyandotte Chem Corp, Mich, 60-63; res chemist, Reaction Motors Div, Thiokol Chem Corp, NJ, 63-65, supvr organometallic synthesis, 65-67; sr res scientist, Amoco Chem Corp, Inc, 67-69; supvr biochem dept, Cook County Hosp, Chicago, 69-70; DIR, AM MED LABS, 70- Mem: Am Chem Soc; Am Asn Clin Chemists. Res: General organic synthesis; synthetic lubricants; synthesis of thermally stable ultraviolet absorbers; thermally stable coatings; surfactants; heterocycles; synthesis of monomers; radioimmunoassay; gas chromatography; drug assay; lipids; endocrinology. Mailing Add: Am Med Labs 4817 W 83rd St Oak Lawn IL 60459

NARAGON, ERNEST ASHLEY, b Hanoverton, Ohio, Jan 7, 13; m 40; c 2. PHYSICAL CHEMISTRY, ORGANIC CHEMISTRY. Educ: Mt Union Col, BS, 36; Western Reserve Univ, PhD(chem), 40. Prof Exp: Lab asst, Mt Union Col, 34-36; res chemist, Texas Co, 40-43, proj leader, 43-52, admin asst, 52-54, asst supvr chem res, 54-55, tech assoc, Texaco Develop Corp, 55-62 & Texaco Inc, 62-64, asst mgr tech sect, Patent & Trademark Div, Legal Dept, 64-67, asst mgr, Tech Div, 67-68, mgr, 68-73, ASST TO THE PRES & MGR, TEXACO DEVELOP CORP, 73- Mem: AAAS; Am Chem Soc; Sigma Xi. Res: Petroleum research of fuels; gases; chemicals; lubricants; design of fractionating column; patent-research liaison. Mailing Add: New York NY

NARAHARA, HIROMICHI TSUDA, b Tokyo, Japan, Oct 24, 23; US citizen; m 54; c 4. BIOCHEMISTRY, METABOLISM. Educ: Columbia Univ, BA, 43, MD, 47; Am Bd Internal Med, dipl, 55. Prof Exp: USPHS res fel med, Univ Wash, 53-56, res instr, 56-58; USPHS spec res fel biol chem, Wash Univ, 58-60, from asst prof to assoc prof, 60-71; MEM STAFF, DIV LABS & RES, NY STATE DEPT HEALTH, 71- Mem: Am Diabetes Asn; Endocrine Soc; Am Soc Biol Chem. Res: Intermediary metabolism of carbohydrates; muscle physiology; effect of hormones and muscle contraction on carbohydrate metabolism; cell membranes. Mailing Add: Div of Labs & Res NY State Dept of Health Albany NY 12201

NARAHASHI, TOSHIO, b Fukuoka, Japan, Jan 30, 27; m 56; c 2. NEUROPHYSIOLOGY, NEUROPHARMACOLOGY. Educ: Univ Tokyo, BS, 48, PhD(insect neurotoxicol), 60. Prof Exp: Res assoc physiol, Univ Chicago, 61-62, asst prof, 62; asst prof, 62-63, from asst prof to prof, 65-69, vchmn dept, 73-75, PROF PHYSIOL & PHARMACOL, MED CTR, DUKE UNIV, 69 Concurrent Pos: Res assoc, Fac Agr, Univ Tokyo, 51-65. Honors & Awards: Japanese Soc Appl Entom & Zool Prize, 55. Mem: Soc Neurosci; Am Soc Pharmacol & Exp Therapeut; Am Physiol Soc; NY Acad Sci; Int Soc Toxinology. Res: Electrophysiology and pharmacology of nerve and muscle membrane and synaptic junctions in general; basic insect neurophysiology; neurotoxicology of insects. Mailing Add: Dept of Physiol & Pharmacol Duke Univ Med Ctr Durham NC 27710

NARANG, SARAN A, b Agra, India, Sept 10, 30; m; c 1. ORGANIC CHEMISTRY, MOLECULAR BIOLOGY. Educ: Panjab Univ, India, BSc, 51, MSc, 53; Univ Calcutta, PhD(org chem), 60. Prof Exp: Sr res fel chem, Indian Asn for Cultivation of sci, Calcutta, 59-62; res assoc, Johns Hopkins Univ, 62-63; proj assoc molecular biol, Inst Enzyme Res, Univ Wis-Madison, 63-66; asst res officer, Div Pure Chem, 66-67, assoc res officer, Div Biochem & Molecular Biol, 67-73, SR RES OFFICER, DIV BIOL SCI, NAT RES COUN CAN, 73- Concurrent Pos: Adj prof, Carleton Univ, 74- Honors & Awards: Coochbihar Professorship Mem Award, Indian Asn Cultivation Sci, Calcutta, 74. Res: Chemico-enzymatic synthesis of DNA and RNA and their biological roles; DNA-protein recognition and studies viroids. Mailing Add: Div of Biol Sci Nat Res Coun of Can Ottawa ON Can

NARASIMHAN, KALATUR S V L, b Tirupati, India, Aug 7, 42; m 75. SOLID STATE CHEMISTRY. Educ: Univ Poona, India, PhD(chem), 67. Prof Exp: Res assoc, Mat Sci Lab, Dept Chem Eng, Univ Tex, Austin, 67-70; RES ASST PROF CHEM, UNIV PITTSBURGH, 70- Mem: Inst Elec & Electronics Engrs; Sigma Xi. Res: Physical properties of solids; magnetic, electrical and thermal properties of rare earth intermetallics, intermetallic alloys, and their application to permanent magnet development. Mailing Add: Dept of Chem Univ of Pittsburgh Pittsburgh PA 15260

NARASIMHAN, MYSORE N L, b Mysore City, India, July 7, 28; US citizen; m 49; c 3. APPLIED MATHEMATICS, ENGINEERING SCIENCE. Educ: Univ Mysore, MSc, 51; Indian Inst Technol, Kharagpur, PhD(math), 58. Prof Exp: Lectr math, Lingaraj Col, India, 51-55; asst lectr, Indian Inst Technol, Kharagpur, 55-58; lectr, Indian Inst Technol, Bombay, 58-61, asst prof, 61-62, assoc prof, 64-65; res prof, Math Res Ctr, Univ Wis, 62-64; assoc prof, Univ Calgary, 65-66; PROF MATH, ORE STATE UNIV, 66- Concurrent Pos: Vis prof, Princeton Univ, 72-73. Honors & Awards: Iyengar Mem Prize, Mysore, 51. Mem: Am Math Soc; Soc Indust & Appl Math; US Soc Eng Sci; Indian Soc Theoret & Appl Mech. Res: Non-Newtonian fluid flows; flow through elastic tubes; porous channel and magnetohydrodynamic flows; stability of fluid flows; microcontinuum theory; liquid crystal theory; thermodynamics; nonlinear continuum-mechanics. Mailing Add: Dept of Math Ore State Univ Corvallis OR 97331

NARATH, ALBERT, b Berlin, Ger, Mar 5, 33; US citizen; m 58; c 3. SOLID STATE PHYSICS. Educ: Univ Cincinnati, BS, 55; Univ Calif, Berkeley, PhD(phys chem, molecular spectros), 59. Prof Exp: Dept mgr solid state res, 59-68, dir solid state sci res, 68-71, managing dir phys sci, 71-73, VPRES RES, SANDIA LABS, 73- Mem: Fel Am Phys Soc. Res: Nuclear magnetic resonance in nonmetallic magnetic crystals and in transition metals and intermetallic compounds; properties of ferromagnets and antiferromagnets. Mailing Add: Orgn 50 Sandia Labs Albuquerque NM 87115

NARAYAN, KRISHAMURTHI ANANTH, b Secunderabad, India, Oct 1, 30; m 61; c 2. NUTRITIONAL BIOCHEMISTRY, FOOD SCIENCE. Educ: Madras Univ, BS, 49; Osmania Univ, India, MS, 51; Univ Ill, Urbana, PhD(food technol), 57. Prof Exp: Res assoc phys chem, Wash State Univ, 57-60, res assoc agr chem, 61-62; sci officer, Nutrit Res Lab, 60-61; from asst prof to assoc prof food chem, Univ Ill, Urbana, 62-71; RES NUTRITIONIST, FOOD LABS, US ARMY NATICK LABS, 71- Concurrent Pos: Nat Cancer Inst career develop award, 66- Mem: AAAS; Am Oil Chem Soc; Am Inst Nutrit; NY Acad Sci; Brit Biochem Soc. Res: Lipids in cancer; lipoprotein metabolism; disc electrophoresis and lipoproteins; wheat and serum proteins; oxidized lipid-protein complexes; lipoproteins in cancer; liver plasma membranes; essential fatty acid deficiency; absorption, transport and utilization of lipids. Mailing Add: Nutrit Div Food Lab US Army Natick Labs Natick MA 01760

NARAYAN, TV LAKSHMI, b Udamalpet, India, June 5, 37; m 68; c 2. ORGANIC CHEMISTRY, POLYMER CHEMISTRY. Educ: Univ Madras, BSc, 58; Annamalai Univ, Madras, MSc, 61; Univ 21 Pa, PhD(chem), 65. fel polymer chem, Univ Ariz, 65-66; res chemist, Am Cyanamid Co, Conn, 66-69; SR RES CHEMIST, BASF WYANDOTTE CORP, 69- Mem: Am Chem Soc; The Chem Soc. Res: Organophosphorous and sulfur chemistry; thermostable polymers; engineering thermoplastics; flame retardant polymers; polyurethane chemistry. Mailing Add: Cent Res BASF Wyandotte Corp Wyandotte MI 48192

NARAYANA, TADEPALLI VENKATA, b Madras, India, Apr 23, 30; m 60. MATHEMATICS. Educ: Univ Madras, MA, 50; Univ Bombay, MA, 51; Univ NC, PhD, 53. Prof Exp: Statistician, Indian Coun Agr Res, India, 54; asst prof math, McGill Univ, 55-58; assoc prof, 58-66, PROF MATH, UNIV ALTA, 66- Mem: Can Math Cong. Res: Probability and statistics; number theory. Mailing Add: Dept of Math Univ of Alta Edmonton AB Can

NARAYANAMURTI, VENKATESH, b Bangalore, India, Sept 9, 39; m 61; c 2.

EXPERIMENTAL SOLID STATE PHYSICS. Educ: Univ Delhi, BS, 58, MS, 60; Cornell Univ, PhD(physics), 65. Prof Exp: Res assoc physics, Cornell Univ, 64-65, instr, 67-68; asst prof, Indian Inst Technol, Bombay, 65-66; mem tech staff physics, 68-76, HEAD SEMICONDUCTOR ELECTRONICS RES DEPT, BELL LABS, 76- Mem: Am Phys Soc; Assoc Inst Physics London; Fel Acad Sci India; AAAS. Res: Phonons in solids and liquid helium; second sound and sound propagation in matter; superconductivity; metal-insulator transitions under pressure. Mailing Add: Bell Labs Murray Hill NJ 07974

NARAYANASWAMY, PADMANABHA, b Madras, India, July 25, 36; m 63; c 2. THEORETICAL HIGH ENERGY PHYSICS. Educ: Univ Delhi, BS, 56, MS, 58, PhD(physics), 63. Prof Exp: Res fel theoret physics, Univ Delhi, 63-64; fel, Tata Inst Fundamental Res, India, 64-65; vis scientist, Int Ctr Theoret Physics, Trieste, 65-66 & Ctr Europ Nuclear Res, Switz, 66-67. fel theoret physics, Tata Inst Fundamental Res, India, 67-68; asst prof physics, Am Univ Beirut, 68-69; asst prof, 69-72, ASSOC PROF PHYSICS, SOUTHERN ILL UNIV, EDWARDSVILLE, 72- Concurrent Pos: Vis scientist, Ctr Particle Theory, Univ Tex, Austin, 74-75. Mem: Am Phys Soc; Ital Phys Soc. Res: Field theory; theory of elementary particles. Mailing Add: Dept of Physics Southern Ill Univ Edwardsville IL 62025

NARBAITZ, ROBERTO, b Arg, Oct 1, 26; m 52; c 3. EMBRYOLOGY. Educ: Univ Buenos Aires, MD, 51, DrMed, 61. Prof Exp: From instr to adj prof, Univ Buenos Aires, 57-66; asst prof, Univ Md, Baltimore City, 67-68; asst prof, Univ Pittsburgh, 68-69; assoc prof, 69-71, PROF HISTOL & EMBRYOL, UNIV OTTAWA, 71- Mem: Am Soc Zool; Am Asn Anat; Can Fedn Biol Socs. Res: Experimental embryology; sex differentiation; histochemistry; electromicroscopy. Mailing Add: Dept of Histol Univ of Ottawa Ottawa ON Can

NARCISI, ROCCO S, b Bristol, Pa, Apr 4, 31; m 57; c 1. AERONOMY. Educ: Pa State Univ, BS, 53; Harvard Univ, MS, 55, PhD(physics), 59. Prof Exp: Supvry physicist, 60-74, PROG MGR & PROJ SCIENTIST STRATOSPHERIC ENVIRON PROJ, AIR FORCE CAMBRIDGE RES LABS, 74- Concurrent Pos: Lectr, Int Sch Atmospheric Physics, Erice, Sicily, 70. Honors & Awards: Marcus D O'Day Award, Air Force Cambridge Res Labs, 66 & Guenter Loeser Mem Award, 70; Except Civilian Serv Award, US Air Force, 72. Mem: Sigma Xi; Am Geophys Union; AAAS; NY Acad Sci. Res: Ionospheric and upper atmospheric structure and dynamics; mass spectrometry; rocket and satellite instrumentation to measure composition and density of the neutral and ionized constituents of the atmosphere; experimental and theoretical research on the compsition of the atmosphere and the associated physical and chemical processes; study of the earth's stratosphere. Mailing Add: Code LKD Air Force Cambridge Res Labs Bedford MA 01731

NARDONE, ROLAND MARIO, b Brooklyn, NY, Mar 29, 28; m 51; c 4. PHYSIOLOGY. Educ: Fordham Univ, BS, 47, MS, 49, PhD(biol), 51. Prof Exp: Instr, St Francis Col, 48-51 & St Louis Univ, 51-52; from asst prof to assoc prof, 52-63, PROF BIOL, CATH UNIV AM, 63- Mem: Tissue Cult Asn (secy, 72-76). Res: Cell division; physiology of cells in culture; radiation effects. Mailing Add: Dept of Biol Cath Univ of Am Washington DC 20017

NARDUCCI, LORENZO M, b Torino, Italy, May 25, 42; m 65; c 3. QUANTUM OPTICS. Educ: Univ Milan, PhD(physics), 64. Prof Exp: Asst prof quantum electronics, Univ Milan, 65-66; asst prof physics, 66-71, ASSOC PROF PHYSICS, WORCESTER POLYTECH INST, 71- Concurrent Pos: Consult, Am Optical Corp, 67-68 &Phys Sci Directorate, Redstone Arsenal, 72- Mem: Am Phys Soc; Optical Soc Am. Res: Laser amplifiers; interaction of radiation and matter; quantum statistics; light scattering and phase transitions. Mailing Add: Dept of Physics Worcester Polytech Inst Worcester MA 01609

NARIBOLI, GUNDO A, b Dharwar, India, Sept 2, 25; m 47; c 4. APPLIED MATHEMATICS. Educ: Univ Bombay, BSc, 47, MSc, 52; Karnatak Univ, India, MSc, 54; Indian Inst Technol, Kharagpur, PhD(appl math), 59. Prof Exp: Lectr math, Col Eng & Technol, Hubli, India, 52-55 & Indian Inst Technol, Kharagpur, 56-59; reader, Univ Bombay, 59-62; assoc prof, Iowa State Univ, 62-64; reader, Univ Bombay, 64-66; assoc prof math & eng mech, 66-69, PROF ENG MECH, IOWA STATE UNIV, 69- Concurrent Pos: Reviewer, Appl Math Rev. Mem: Soc Indust & Appl Math; Math Asn Am; Tensor Soc. Res: Linear and nonlinear waves; group-invariant solutions; Baeklund transformations; perturbation methods; method of perturbation for waves in bounded media; traffic flow theory. Mailing Add: 209 N Wilmoth Ave Ames IA 50010

NARICI, LAWRENCE ROBERT, b Brooklyn, NY, Nov 15, 41. MATHEMATICAL ANALYSIS. Educ: Polytech Inst Brooklyn, BS, 62, MS, 63, PhD(math), 66. Prof Exp: From instr to asst prof math, Polytech Inst Brooklyn, 65-67; assoc prof, 67-72, PROF MATH, ST JOHN'S UNIV, NY, 72- Mem: Am Math Soc; Math Asn Am; Mex Math Soc; Math Soc France; Israel Math Union. Res: Non-Archimedean Banach spaces and algebras; topological algebras; functional analysis; valuation theory. Mailing Add: Dept of Math St John's Univ Jamaica NY 11432

NARIN, FRANCIS, b Philadelphia, Pa, May 10, 34; m 58; c 2. RESEARCH ADMINISTRATION, COMPUTER SCIENCES. Educ: Franklin & Marshall Col, BS, 55; NC State Col, MS, 57. Prof Exp: Physicist, IIT Res Inst, 57-59; staff mem, Los Alamos Sci Lab, NMex, 59-63; sr staff assoc, IIT Res Inst, 63-66, group leader space sci, 65-68, sr scientist, 66-68; PRES, COMPUT HORIZONS, INC, 68- Concurrent Pos: Consult, WHO, 74. Mem: Fel AAAS; Asn Comput Mach; World Future Soc; Am Soc Info Sci; NY Acad Sci. Res: Computer applications to commercial, industrial and medical areas; evaluation of scientific research; information science and analysis of scientific literature. Mailing Add: Comput Horizons Inc 1050 Kings Hwy Cherry Hill NJ 08034

NARINS, DORICE MARIE, b Chicago, Ill, Oct 9, 32; m 70. NUTRITIONAL BIOCHEMISTRY. Educ: Northwestern Univ, BS, 54; Mass Inst Technol, PhD(nutrit biochem), 66. Prof Exp: NIH fel overnutrit, Rockefeller Univ, 66-68; asst prof nutrit biochem, Mich State Univ, 68-74; ASSOC PROF NUTRIT BIOCHEM, RUSH UNIV, 74- Mem: Inst Food Technologists; Soc Nutrit Educ; Am Soc Cell Biol; Sigma Xi; AAAS. Res: Undernutrition and overnutrition, particularly effects on brain and liver; iron deficiency anemia in infants with low socioeconomic background; world nutrition problems. Mailing Add: Dept of Clin Nutrit Rush Univ Chicago IL 60612

NARO, PAUL ANTHONY, b Scranton, Pa, Aug 17, 34; m 57; c 2. ORGANIC CHEMISTRY, PHYSICAL CHEMISTRY. Educ: Temple Univ, AB, 56; Pa State Univ, PhD(chem), 60. Prof Exp: Res chemist, Socony Mobil Oil Co, Inc, 59-61, sr res chemist, Mobil Oil Corp, 61-72, asst supvr, 64-66, ADMIN MGR, MOBIL RES & DEVELOP CORP, 72- Mem: Am Chem Soc. Res: Hydrocarbon synthesis; organic sulfur compounds; polymer chemistry; heterogeneous catalysis; computer applications. Mailing Add: Mobil Res & Develop Corp Cent Res Div Box 1025 Princeton NJ 08540

NAROLL, RAOUL, b Toronto, Ont, Sept 10, 20; US citizen; m 41; c 1. CULTURAL ANTHROPOLOGY, SOCIAL ANTHROPOLOGY. Educ: Univ Calif, Los Angeles, AB, 50, MA, 52, PhD(hist), 53. Prof Exp: Fel, Ctr Advan Studies Behav Sci, 54-55; res assoc anthrop, Human Rels Area Files, Washington, DC, 55-57; from asst prof to assoc prof, Calif State Univ, Northridge, 57-62; from assoc prof to prof, Northwestern Univ, 62-67; PROF ANTHROP, STATE UNIV NY BUFFALO, 67- Concurrent Pos: Dir, Inst Cross-Cult Studies, 60-67; pres, Human Rels Area Files, 73- Mem: AAAS; Am Anthrop Asn; Int Studies Asn. Res: Cross-cultural methodology; comparative evolutionary studies; cross-historical studies; peasantry and urbanization. Mailing Add: Dept of Anthrop State Univ of NY at Buffalo Amherst NY 14226

NAROTSKY, SAUL, b Brooklyn, NY, May 19, 22; m 50; c 4. VETERINARY MEDICINE, POULTRY PATHOLOGY. Educ: Univ Conn, BS, 44; Kans State Univ, DVM, 47. Prof Exp: Instr bact & poultry path, Mich State Univ, 47-49; dir regional vet lab, Cornell Univ, 49-71; DIR, AVIAN VET SERV, 71- Mem: AAAS; Am Asn Avian Path; Am Vet Med Asn; Poultry Sci Asn. Res: Prevention and control of poultry disease; poultry immunology and nutrition; clinical evaluation of vaccines, drugs and nutrition in poultry disease. Mailing Add: Avian Vet Serv 239 Capen Blvd Buffalo NY 14226

NARROD, MARIAN FREEDMAN, b Orlando, Fla; m 58; c 2. PHARMACOLOGY. Educ: George Washington Univ, BS & MS, PhD(pharmacol), 53. Prof Exp: Vis lectr, Med Sch, Univ Pa, 68-69; asst prof pharmacol, Med Col, Pa, 69-71, vis asst prof, 71-72; pharmacologist, Inst Cancer Res, 71-74; ASST PROF PHARMACOL, MED SCH, UNIV PA, 72-; PHARMACOLOGIST, DEPT MED, EINSTEIN MED CTR, 74- Concurrent Pos: Res assoc pharmacol, Oxford Univ. Mem: Am Soc Pharmacol & Exp Therapeut; Soc Exp Biol & Med. Res: Pharmacokinetics; metabolism, toxicity and mechanism of action of oncologic drugs. Mailing Add: Dept of Med Einstein Med Ctr ND Philadelphia PA 19141

NARROD, STUART ALLAN, b Chicago, Ill, Apr 9, 25; m 58; c 2. BIOCHEMISTRY. Educ: Univ Ill, BS, 48, PhD(dairy sci, bact), 55. Prof Exp: Asst, Univ Ill, 51-55; res assoc, Am Dent Asn, NIH, 55-58; chemist, NIH, 61-64; ASSOC PROF BIOCHEM, MED COL PA, 64-, ACTG CHMN DEPT, 69- Concurrent Pos: Res fel dept biochem, Brandeis Univ, 58-60 & Oxford Univ, 60-61. Mem: Am Soc Microbiol; Am Chem Soc. Res: Biochemistry and bacterial physiology in relation to metabolic processes and enzyme chemistry. Mailing Add: Dept of Biochem Med Col of Pa Philadelphia PA 19129

NARSKE, RICHARD MARTIN, b Berwyn, Ill, July 4, 42; m 65; c 2. ORGANIC CHEMISTRY. Educ: Augustana Col, BA, 64; Univ Iowa, MS, 66, PhD(chem), 68. Prof Exp: Asst prof, 68-70, ASSOC PROF CHEM, UNIV TAMPA, 70- Concurrent Pos: Consult, Erny Supply Co, 69-; chief chem consult, Intersci Inc, 70- Mem: Am Chem Soc (treas-secy, 69-71); Am Inst Chemists; Sigma Xi. Res: Gas chromatographic analysis of pesticide residues; forensic chemistry; organic metallic complexes. Mailing Add: Dept of Chem Univ Tampa 401 W Kennedy Blvd Tampa FL 33606

NARTEN, PERRY FOOTE, b Cleveland, Ohio, July 24, 21; m 46; c 2. ENVIRONMENTAL MANAGEMENT, HORTICULTURE. Educ: Col Wooster, BA, 43; Wash Univ, MS, 49. Prof Exp: Geologist, Pa State Geol Surv, 47 & US Geol Surv, 48-63; opers analyst, Res Anal Corp, 63-72; staff environmentalist, Gen Res Corp, 72-74; environ consult, 74-75; GEOLOGIST, US GEOL SURV, 75- Concurrent Pos: Dir, Coover Arboretum. Mem: Geol Soc Am; Am Hort Soc; World Future Soc; Int Soc Technol Assessment. Res: Environmental and engineering geology; ecological systems in planning; resource limitations in land use; suburban and inner city environmental-life quality relationships; environmental impact analysis; aesthetics measurement; ornamental horticulture. Mailing Add: 3708 N Randolph St Arlington VA 22207

NARVAEZ, RICHARD, b New York, NY, May 4, 30; m 63; c 5. POLYMER CHEMISTRY. Educ: City Col New York, BS, 51; NY Univ, PhD(phys chem), 63. Prof Exp: Chemist, Savannah River Lab, 51-55, 57-58, res chemist, 63-65, res chemist, Carothers Lab, 65-69, SR RES CHEMIST, CAROTHERS LAB, EXP STA, E I DU PONT DE NEMOURS & CO, INC, 69- Mem: Am Chem Soc. Res: Improved polyamidation and polyesterification processes and their control; polyamide and polyester product development. Mailing Add: Carothers Lab E I du Pont de Nemours & Co Wilmington DE 19898

NARVAEZ MORALES, IGNACIO, phytopathology, agronomy, see 12th edition

NASATIR, MAIMON, b Chicago, Ill, Apr 16, 29; m 52; c 3. CELL BIOLOGY. Educ: Univ Chicago, PhB, 50; Univ Pa, PhD(bot), 58. Prof Exp: Instr biol, Univ Pa, 58-59; lectr, Haverford Col, 59; USPHS fel, Univ Brussels, 59-60 & Univ Ill, 60-61; asst prof bot, Brown Univ, 61-66, asst to dean, Pembroke Col, 63-65; chmn dept biol, 66-70, PROF BIOL, UNIV TOLEDO, 66- Concurrent Pos: Lalor fel, 62 & 63; adj prof dept physiol, Med Col Ohio, 68-71. Mem: AAAS; Bot Soc Am; Am Soc Cell Biol. Res: Biochemical cytology; plant physiology; cellular biology. Mailing Add: Dept of Biol Univ of Toledo Toledo OH 43606

NASELOW, ARTHUR BENSON, organic chemistry, polymer chemistry, see 12th edition

NASH, CARROLL BLUE, b. Louisville, Ky, Jan 29, 14; m 41. BIOLOGY, PARAPSYCHOLOGY. Educ: George Washington Univ, BS, 34; Univ Md, MS, 36, PhD, 39. Prof Exp: Instr zool, Univ Ariz, 39-41; assoc prof biol, Pa Mil Col, 41-44; asst prof, Am Univ, 44-45; prof, Washington Col, Md, 45-48; PROF BIOL, ST JOSEPH'S COL, PA, 48-, DIR PARAPSYCHOL LAB, 56- Honors & Awards: William McDougall Award, Parapsychol Lab, Duke Univ, 60. Mem: AAAS; Parapsychol Asn. Res: Extrasensory perception; precognition; psychokinesis. Mailing Add: Dept of Biol St Joseph's Col Philadelphia PA 19131

NASH, CHARLES PRESLEY, b Sacramento, Calif, Mar 15, 32; m 55; c 3. PHYSICAL CHEMISTRY. Educ: Univ Calif, BS, 52; Univ Calif, Los Angeles, PhD(chem), 58. Prof Exp: Actg instr chem, Univ Calif, Los Angeles, 56; instr, 57-59, assoc prof, 65-70, PROF CHEM, UNIV CALIF, DAVIS, 70- Concurrent Pos: Consult, Lawrence Livermore Lab, Univ Calif, 57-68; vis sr lectr, Imp Col, Univ London, 68. Mem: Am Phys Soc; Am Chem Soc. Res: Exploding wires; solution chemistry; vibrational spectroscopy; amino acids. Mailing Add: Dept of Chem Univ of Calif Davis CA 95616

NASH, CHARLES WILLIAM, b Man, July 6, 15; m 45; c 2. PHARMACOLOGY. Educ: Univ Man, BSc, 42; Univ Minn, MSc, 47, PhD(pharmacol), 53. Prof Exp: Lectr pharm, Univ Man, 43-46, from asst prof to assoc prof, 46-54; assoc prof, 54-56, PROF PHARMACOL, UNIV ALTA, 56- Concurrent Pos: Vis scientist, Nat Heart Inst, 62-63; res fel, Dept Pharmacol, Univ Glasgow, 72-73. Mem: Am Soc Pharmacol; Can Physiol Soc; Pharmacol Soc Can (secy, 64-66, pres, 68-69). Res: Cardiovascular pharmacology. Mailing Add: Dept of Pharmacol Univ of Alta Fac of Med Edmonton AB Can

NASH, CLAUDE HAMILTON, III, b Palestine, Tex, Jan 28, 43; m 62; c 2. MICROBIAL PHYSIOLOGY. Educ: Lamar State Col, BS, 65; Colo State Univ, MS, 67, PhD(microbiol), 69. Prof Exp: SR MICROBIOLOGIST, ANTIBIOTIC DEVELOP DIV, ELI LILLY & CO, 69- Mem: Am Soc Microbiol. Res: Metabolism of psychrophilic microorganisms; protein biosynthesis and ribosomal function; antibiotic biosynthesis, control and fermentation technology. Mailing Add: Antibiotic Develop Div Eli Lilly & Co Indianapolis IN 46206

NASH, CLINTON BROOKS, b Gunnison, Miss, Jan 3, 18; m 46; c 1. PHARMACOLOGY. Educ: Univ Tenn, BS, 50, MS, 52, PhD(pharmacol), 55. Prof Exp: Sr pharmacologist, Res Labs, Mead Johnson & Co, 54-57, group leader pharmacol, 57-58; from asst prof to assoc prof, 58-65, PROF PHARMACOL, UNIV TENN CTR HEALTH SCI, MEMPHIS, 65- Mem: Am Soc Pharmacol & Exp Therapeut; Am Heart Asn; Soc Exp Biol & Med; Soc Toxicol. Res: Cardiovascular effects of anesthetic agents; intraocular pressures; peripheral vasodilators; catecholamine content of various tissues; coronary blood flow; antiarrhythmic agents; cardiovascular actions of vasopressin, reserpine and digitalis. Mailing Add: Dept of Pharmacol Univ of Tenn Ctr for Health Sci Memphis TN 38163

NASH, COLIN EDWARD, b Liverpool, Eng, Jan 23, 37; m 61; c 3. RESOURCE MANAGEMENT. Educ: Univ Leeds, BS, 59, PhD(textile technol), 62. Prof Exp: Res fel textile technol, Int Wood Secretariat, 62-63; res officer aquacult, White Fish Authority, Eng, 63-67, asst prin officer, 67-71; dir, Multunna Lab, Wash, 71-72; DIR, OCEANIC INST, HAWAII, 72- Concurrent Pos: Deleg, S Pac Comn, UN, 71 & 72; consult, US AID, Rockefeller Found, Nat Sea Grant Prog & Int Aquacult Consultancy, 72-; mem comn, Dept Commerce Aquacult Surv, Nat Oceanic & Atmospheric Admin, 72, Marine Resources Subcomt, State of Hawaii, 73; adv, Int Ctr Living Aquatic Resources Mgt, 74- Mem: Challenger Soc; Am Fisheries Soc; World Maricult Soc. Res: Fish culture science and technology, embracing the interdisciplinary fields of reproductive physiology, nutrition, environmental and aquatic biology, engineering and economics, relating the whole to aquatic resources management for increased animal protein production. Mailing Add: Oceanic Inst Waimanalo HI 96795

NASH, DAVID, b London, Eng, Sept 10, 37. BIOCHEMICAL GENETICS, CYTOGENETICS. Educ: Univ London, BSc, 60; Univ Cambridge, PhD(genetics), 63. Prof Exp: Wis Alumni Res Found fel zool, Univ Wis, 63-64, res assoc, 64-65; asst prof genetics, 65-70, ASSOC PROF GENETICS, UNIV ALTA, 70- Mem: Genetics Soc Am; Brit Genetical Soc; Can Soc Cell Biol. Res: Metabolism of nucleic acids in Drosophila, particularly autoradiography of polytene chromosomes; developmental studies on mutants of Drosophila. Mailing Add: Dept of Genetics Univ of Alba Edmonton AB Can

NASH, DAVID HENRY GEORGE, b Ash Vale, Eng, June 19, 43. APPLIED MATHEMATICS. Educ: Univ Calif, Riverside, BA, 65; Univ Calif, Berkeley, MA, 67, PhD(math), 70. Prof Exp: Actg asst prof math, Univ Hawaii, 69-70; Woodrow Wilson intern, Va State Col, 70-71; lectr, Univ Calif, Berkeley, 71; ASSOC SR RES MATHEMATICIAN, RES LABS, GEN MOTORS CORP, 72- Mem: Am Math Soc; Math Asn Am. Res: Functional analysis. Mailing Add: Res Labs Gen Motors Corp Tech Ctr Warren MI 48093

NASH, DONALD JOSEPH, b New York, NY, Dec 20, 30; m 54; c 3. GENETICS, ZOOLOGY. Educ: Univ Mich, BS, 51; Univ Kans, MA, 57; Iowa State Univ, PhD(genetics), 60. Prof Exp: Asst prof genetics, Pa State Univ, 60-62; asst prof zool, Rutgers Univ, 62-65; assoc prof radiation biol & zool, 65-66, assoc prof zool, 66-71, PROF ZOOL, COLO STATE UNIV, 71- Mem: AAAS; Genetics Soc Am; Am Genetic Asn; Am Soc Mammal; Soc Study Evolution. Res: Physiological and quantitative genetics; radiation biology. Mailing Add: Dept of Zool Colo State Univ Ft Collins CO 80521

NASH, DONALD ROBERT, b Pittsfield, Mass, Nov 15, 38; m 63; c 1. IMMUNOBIOLOGY. Educ: Am Int Col, BA, 61; Boston Col, MS, 63; Univ NC, Chapel Hill, PhD(bact, immunol), 67. Prof Exp: Asst prof immunol, Univ Hawaii, 69-70; HEAD IMMUNOBIOL RES, E TEX CHEST HOSP, 72-; CONSULT, M D ANDERSON HOSP & TUMOR INST, 75- Concurrent Pos: Res fel immunol, Univ NC, Chapel Hill, 67-68; Belg Am Educ Fund res fel, Cath Univ Louvain, 68-69; sr res fel immunol, Ref & Training Ctr, WHO, Switz, 70-72. Mem: AAAS; Am Asn Immunologists. Res: Humoral and cellular immunity. Mailing Add: East Tex Chest Hosp Box 2003 Tyler TX 75701

NASH, DOUGLAS B, b Elgin, Ill, Dec 2, 32; m 64. GEOLOGY, SPACE PHYSICS. Educ: Univ Calif, Berkeley, AB, 60, MA, 62. Prof Exp: From assoc scientist to sr scientist, 62-68, res group supvr, 68-70, prin investr, Lunar Sample Anal, 69-74, CONSULT, JET PROPULSION LAB, CALIF INST TECHNOL, 74- Mem: AAAS; Am Geophys Union; Geol Soc Am. Res: Lunar luminescence; lunar surface optical properties; proton irradiation effects on rocks; x-ray diffraction analysis of rock glass; instrument development for lunar and planetary geological analysis; surface properties of Galilean satellites. Mailing Add: Space Sci Div Jet Propulsion Lab Pasadena CA 91103

NASH, EDMUND GARRETT, b Manitowoc, Wis, Nov 19, 36; m 61; c 2. ORGANIC CHEMISTRY. Educ: Lawrence Col, BS, 59; Univ Colo, PhD(chem), 65. Prof Exp: Sr res chemist, Gen Mills, Inc, 65-66; res assoc, Johns Hopkins Univ, 66-67; asst prof, 67-70, ASSOC PROF CHEM, FERRIS STATE COL, 70- Mem: AAAS; Am Chem Soc. Res: Chemistry of organic nitrogen compounds; nuclear magnetic resonance of systems with restricted rotation; organic polymer chemistry. Mailing Add: Dept of Phys Sci Ferris State Col Big Rapids MI 49307

NASH, EDWARD THOMAS, b New York, NY, July 31, 43; m 70. EXPERIMENTAL HIGH ENERGY PHYSICS. Educ: Princeton Univ, AB, 65; Columbia Univ, MA, 67, PhD(physics), 70. Prof Exp: Res assoc physics, Nevis Cyclotron Lab, Columbia Univ, 70 & Lab Nuclear Sci, Mass Inst Technol, 70-71; STAFF PHYSICIST & HEAD INTERNAL TARGET LAB, FERMI NAT ACCELERATOR LAB, 71- Mem: Am Phys Soc. Res: Fundamental forces and symmetries; searches for and studies of the properties of new particles; study of the interaction of photons with matter at very high energy. Mailing Add: Fermi Nat Accelerator Lab PO Box 500 Batavia IL 60510

NASH, FRANKLIN D, b Brooklyn, NY, May 2, 32; c 2. PHYSIOLOGY, NEPHROLOGY. Educ: Ind Univ, AB, 55, MD, 58. Prof Exp: Asst physiol, Ind Univ, 52-55, res asst surg, Med Ctr, 57, instr basic sci, Sch Nursing, 57-58, intern surg, Med Ctr, 59-60; asst resident, Kings County Hosp Ctr-State Univ NY, 60-61; from instr to assoc prof, 62-75, PROF PHYSIOL, SCH MED, IND UNIV, INDIANAPOLIS, 75-, ASSOC PROF MED, 70- Concurrent Pos: Res fel surg, Med Ctr, Ind Univ, 58-59; USPHS res fel renal physiol, Med Col, Cornell Univ, 61-62; dir renal hypertension lab, Sch Med, Ind Univ, Indianapolis, 69-73, assoc dir specialized ctr res in hypertension, 71-73. Mem: Asn Am Med Cols; Am Physiol Soc; Am Soc Nephrology; NY Acad Sci; Int Soc Nephrology. Res: Pathophysiology of renal and cardiovascular diseases; evaluation of the interrelationships between humoral and neural factors in the etiology and maintenance of hypertension; renal diseases, dialysis and transplantation; hypertension. Mailing Add: Ind Univ Med Ctr Indianapolis IN 46202

NASH, FRANKLIN RICHARD, physics, see 12th edition

NASH, HAROLD ANTHONY, b Corvallis, Ore, Sept 28, 18; m 46; c 2. BIOCHEMISTRY. Educ: Ore State Col, BS, 40; Purdue Univ, PhD(biochem), 47. Prof Exp: Asst agr chem, Purdue Univ, 40-44, asst chemist, 42-44 & 46-47; res chemist, Pitman-Moore Co, 47-55, dir chem res, 55-60, dir pharmaceut res, 60-61, asst to tech dir, 61-63; dir res biosci, NStar Res & Develop Inst, 63-64, dir biosci div, 64-70; staff assoc, 70-71, ASSOC DIR, POP COUN, 72- Mem: Am Chem Soc. Res: Chemistry of natural products; medicinal chemicals; mechanism of drug action; biochemistry of chronic diseases; fertility control; contraceptive development. Mailing Add: Pop Coun Rockefeller Univ York Ave & 66th St New York NY 10021

NASH, HAROLD EARL, b Lindsay, Calif, July 14, 14; m 40. PHYSICS. Educ: Univ Calif, Berkeley, BA, 38. Prof Exp: Radio engr, US Signal Corps, McClellan Field, Calif, 42-44; res assoc, Underwater Sound Lab, Harvard Univ, 44-45; sect leader, US Naval Underwater Sound Lab, 45-50, div head, 50-60, assoc tech dir systs develop, 60-63, tech dir, 63-70, TECH DIR, US NAVAL UNDERWATER SYSTS CTR, 70- Mem: Fel Acoust Soc Am; fel Inst Elec & Electronics Engrs. Res: Sonar systems; laboratory administration. Mailing Add: Naval Underwater Systs Ctr New London Lab New London CT 06320

NASH, HARRY CHARLES, b Cleveland, Ohio, Mar 24, 27; m 51; c 12. SOLID STATE PHYSICS, OPTICS. Educ: John Carroll Univ, BS, 50, MS, 51; Case Inst Technol, PhD(physics), 58. Prof Exp: From instr to assoc prof, 51-64, PROF PHYSICS, JOHN CARROLL UNIV, 64-, CHMN DEPT, 71- Mem: Optical Soc Am; Am Asn Physics Teachers. Res: Optical properties of absorbing thin films; elastic constants of single crystals; emission spectroscopy. Mailing Add: Dept of Physics John Carroll Univ Cleveland OH 44118

NASH, J FRANK, b Canton, Ohio, May 19, 24; m 49; c 5. PHARMACEUTICAL CHEMISTRY, PHYSICAL PHARMACY. Educ: Ohio State Univ, BS, 48, PhD(pharmaceut chem), 52. Prof Exp: Res scientist, 52-68, GROUP LEADER BIOAVAILABILITY, ELI LILLY & CO, 68- Mem: Acad Pharmaceut Sci; Am Chem Soc; Am Pharmaceut Asn. Res: Assay of drugs in biological tissues; drug formation; pharmacokinetics; bioavailability; dissolution of drugs. Mailing Add: Eli Lilly & Co 307 E McCarty St Indianapolis IN 46206

NASH, JAMES LEWIS, JR, b Drakesboro, Ky, Sept 24, 26; m 51; c 2. POLYMER CHEMISTRY. Educ: Western Ky State Col, BS, 48; Univ Fla, MS, 50, PhD(chem), 53. Prof Exp: Asst chem, Univ Fla, 48-49, 51-53; sr chemist, 53-56, group supvr, 56-60, sr res chemist, 60-69, TECH SERV SPECIALIST, TEXTILE FIBERS DEPT, E I DU PONT DE NEMOURS & CO, INC, 69- Mem: Am Chem Soc. Res: Textile chemistry. Mailing Add: Textile Fibers Dept E I du Pont de Nemours & Co Wilmington DE 19898

NASH, JOE BERT, b Normangee, Tex, Aug 31, 21; m 43; c 4. PHARMACOLOGY. Educ: Agr & Mech Col, Tex, BS, 43; Univ Tex, BS, 47, MA, 51, PhD(pharmacol), 54. Prof Exp: Instr, Col Pharm, Univ Tex, 47-49; instr, 52-53, asst prof, 53-57, ASSOC PROF PHARMACOL, UNIV TEX MED BR, GALVESTON, 57- Concurrent Pos: Res collabr, Brookhaven Nat Lab, 59-; consult, Oak Ridge Inst Nuclear Studies. Mem: AAAS; NY Acad Sci. Res: Metabolic formation of histamines; factors influencing the voluntary intake of alcohol; metabolic fate of anticonvulsant agents; development of autonomic and ataractic agents; toxicology of thallium compounds; toxicological techniques. Mailing Add: Dept of Pharmacol & Toxicol Univ of Tex Med Br Galveston TX 77550

NASH, JOHN BARRY, radiation chemistry, polymer chemistry, see 12th edition

NASH, JOHN PURCELL, applied mathematics, see 12th edition

NASH, JOHN THOMAS, b Glen Cove, NY, July 30, 41; m 66; c 1. GEOLOGY, GEOCHEMISTRY. Educ: Amherst Col, BA, 63; Columbia Univ, MA, 65, PhD(geol), 67. Prof Exp: GEOLOGIST, US GEOL SURV, 67- Mem: Geol Soc Am; Am Inst Mining, Metall & Petrol Engrs; Mineral Soc Am; Mineral Asn Can. Res: Geochemistry of mineral deposits; fluid inclusions; geology of uranium deposits; clay mineralogy; stable isotopes. Mailing Add: US Geol Surv Fed Ctr Box 25046 Denver CO 80225

NASH, JUNE C, b Salem, Mass, May 30, 27; m 51; c 2. ANTHROPOLOGY. Educ: Columbia Univ, BA, 48; Univ Chicago, MA, 53, PhD(anthrop), 60. Prof Exp: Asst prof anthrop, Chicago Teachers Col-North, 60-64; asst prof, Yale Univ, 64-68; assoc prof, NY Univ, 68-72; PROF ANTHROP, CITY COL NEW YORK, 72- Concurrent Pos: NIMH res fels, 61-62 & 64-67; Soc Sci Res Coun res fel, 69; Fulbright Hays res fel, 70; Guggenheim fel, 71; mem bd, Soc Sci Res Coun Latin Am, 71- Mem: Am Anthrop Asn; Am Ethnol Soc; Sigma Xi. Res: Industrialization; cultural impact; United States corporations abroad. Mailing Add: Dept of Anthrop City Col of New York New York NY 10037

NASH, LEONARD KOLLENDER, b New York, NY, Oct 27, 18; m 45; c 2. PHYSICAL CHEMISTRY. Educ: Harvard Univ, BS, 39, MA, 41, PhD(anal chem), 44. Prof Exp: Asst chem, Harvard Univ, 43-44; res assoc, Columbia Univ, 44-45; instr, Univ Ill, 45-46; from instr to assoc prof, 46-59, chmn dept, 71-74, PROF CHEM, HARVARD UNIV, 59- Mem: Am Acad Arts & Sci. Res: Chemical education; history and philosophy of science. Mailing Add: Dept of Chem Harvard Univ Cambridge MA 02138

NASH, MANNING, b Philadelphia, Pa, May 4, 24. ANTHROPOLOGY. Educ: Temple Univ, BS, 49; Univ Chicago, AM, 52, PhD(anthrop), 55. Prof Exp: Instr anthrop, Univ Calif, Los Angeles, 55-56; asst prof, Univ Wash, 56-57; assoc prof, 57-63, PROF ANTHROP, UNIV CHICAGO, 64- Concurrent Pos: NSF fels, Upper Burma, 60-61 & Malaya, 64, 66 & 68. Mem: Am Anthrop Asn; Royal Anthrop Inst Gt Brit & Ireland; Am Ethnol Soc. Res: Modernization and social change in Southeast Asia; ethnology of Mesoamerica; education and cultural systems. Mailing Add: Dept of Anthrop Univ of Chicago Chicago IL 60637

NASH, MURRAY L, b Brooklyn, NY, Dec 28, 17; m 40; c 3. PHYSICAL CHEMISTRY, CHEMICAL ENGINEERING. Educ: Brooklyn Col, BS, 39. Prof Exp: Phys sci aide, Nat Bur Standards, DC, 41-42; jr chemist, US Bur Mines, Md, 42-44; res assoc, SAM Labs, Columbia Univ, 44; chemist & maj supvr, Carbide & Carbon Chem Corp, Tenn, 44-46; sci analyst, AEC, 46-49, tech asst, 49-55, chief classification analyst, 55-56, chief classification br, 56-64, asst dir opers, Div Classification, 64-67, DEP DIR DIV CLASSIFICATION, US ENERGY RES & DEVELOP ADMIN, 67- Mem: Am Chem Soc; Am Nuclear Soc; NY Acad Sci. Res: Separation of isotopes; mass and emission spectroscopy; spectrographic methods of analysis and production of uranium. Mailing Add: Div of Classification US Energy Res & Develop Admin Washington DC 20545

NASH, NAT H, b New York, NY, Dec 9, 14; m 39; c 2. FOOD SCIENCE. Educ: City Col New York, BS, 37. Prof Exp: Jr pharmacist, Nauheim Pharm, Inc, 31-36; jr chemist, Walcourse Chemists, Inc, 36-38; chemist, Weiner Chemists, Inc, 38-40; chemist, M B Picker Corp, 40-42; chief chemist, Joe Lowe Corp, 42-47; dir res, Lanco Prod Corp, 47-62; sales & prod mgr, 62-74, MKT MGR, SPECIALTY CHEM DIV, PVO INT INC, 74- Mem: Am Asn Cereal Chemists; Inst Food Technologists; Am Oil Chemists Soc; Am Dairy Sci Asn; Am Chem Soc. Res: Fats, oils, cereals and cereal products; gums, hydrocolloids; emulsifiers and surfactants; dairy chemistry and technology of ice cream, milk and milk products; imitation dairy products; structured foods and convenience foods. Mailing Add: 571 Columbia St New Milford NJ 07646

NASH, PETER HOWARD, b Sidcup, Eng, Feb 20, 17; Can citizen; m 51; c 4. RESEARCH ADMINISTRATION. Educ: Cambridge Univ, BA, 38, MB, BCh, 41, MA, 45, MD, 50; Univ London, DPH, 47, DIH, 53. Prof Exp: Intern surg, Middlesex Hosp, London, 41-42; resident internal med, Metrop Hosp, London, 49-50; asst dir, Slough Indust Health Serv, 50-53; regional med dir occup health, Bell Tel Co Can, 54-57; med dir, Abbott Labs, Ltd, 57-64, DIR SCI AFFAIRS, ABBOTT LABS, LTD, 64- Concurrent Pos: Rockefeller fel prev med, 46-48; res fel indust toxicol, Harvard Med Sch, 48; lectr indust health, London Sch Hyg & Trop Med, 50-53; asst physician, Royal Victoria Hosp, Montreal, Can, 57- Mem: Fel Indust Med Asn; NY Acad Sci; Can Med Asn; Brit Med Asn; Pharmacol Soc Can. Res: Pharmaceutical, medical and research development; quality control. Mailing Add: Abbott Labs Ltd PO Box 6150 Montreal PQ Can

NASH, PETER HUGH, SR, b Frankfort on Main, Ger, Sept 18, 21; US citizen; m 55; c 2. URBAN GEOGRAPHY, ACADEMIC ADMINISTRATION. Educ: Univ Calif, Los Angeles, BA, 42, MA, 46; Univ Grenoble, cert d' Etudes, 45; Harvard Univ, MCP, 49, MPA, 56, PhD(archit sci), 58. Prof Exp: Teaching asst geog, Univ Calif, Los Angeles, 42; instr, US Army Prog, Univ Grenoble, 45-46; instr geog, Univ Wis-Madison, 46-47; prin planning asst, Boston City Planning Bd, Mass, 49-50; sr planner, City Worcester Planning Dept, 50-51; asst chief urban redevelop div, Boston Housing Authority, 51-52; dir planning dept, City Medford, 52-56; vis critic, Grad Sch Design, Harvard Univ, 56-57; assoc prof city & regional planning, Univ NC, Chapel Hill & res assoc, Inst Res Soc Sci, 57-59; prof geog & regional planning & head dept, Univ Cincinnati, 59-63; dean grad sch, Univ RI, 63-68, prof geog & regional planning & dir curric community planning & area develop, 63-70; actg dir sch archit, 72-73, PROF ARCHIT, GEOG & PLANNING & DEAN FAC ENVIRON STUDIES, UNIV WATERLOO, 70- Concurrent Pos: Instr, Ctr Adult Educ, Cambridge, Mass, 49-57; lectr, Northeastern Univ, 54-56; lectr & acting asst prof geog, Univ Boston, 56-57; res dir, Citizens Develop Comt, City Cincinnati, 59-60; consult, Action, Inc, 59-61; vis prof, Univ Southern Calif, 59 & 62; adv, Better Housing League, Cincinnati & Hamilton County, 59-63; Soc Sci Res Coun & NSF grants, Sweden, 60; consult metrop probs, Brookings Inst, 60-61; corresp, J Ekistics, 60-73; Western Hemisphere rep, Comn Appl Geog, Int Geog Union, 64-; vis critic, Grad Sch Design, Harvard Univ, 67-68; urban res consult, Battelle Mem Inst, 67-73; Am Coun Learned Socs grant, India, 68 & NZ, 74; vis prof inst human sci, Boston Col, 69-70. Honors & Awards: Medal Appl Geog, Univ Liege, 67. Mem: Fel Am Geog Soc; Asn Am Geog; Am Inst Planners; Am Soc Planning Officers; Can Asn Geogr. Res: Area development and redevelopment; city and regional planning; environmental education and research; future studies; urban growth patterns; problems of administration of environmental studies programs. Mailing Add: 588 Sugarbush Dr Waterloo ON Can

NASH, PHILLEO, b Wisconsin Rapids, Wis, Oct 25, 09; m 35; c 2. APPLIED ANTHROPOLOGY. Educ: Univ Wis, AB, 32; Univ Chicago, PhD(anthrop), 37. Prof Exp: Lectr anthrop, Univ Toronto, 37-41; spec lectr, Univ Wis, 41-42; spec asst to dir, Domestic Br, Off War Info, 42-45; spec asst, Exec Off President, White House, 46-52, admin asst to President of United States, 52-53; lieutenant gov, State Wis, 59-61; spec asst to Asst Secy, US Dept Interior, 61, US Comnr Indian Affairs, 61-66; CONSULT ANTHROPOLOGIST, 66-; PROF ANTHROP, AM UNIV, 71- Concurrent Pos: Asst keeper, Royal Ont Mus Archaeol, Univ Toronto, 37-41; spec consult, US Secy War, 43-44; pres, Georgetown Day Sch, 42-52; hon cur, Milwaukee Pub Mus, Wis, 58-; mem task force Indian affairs, US Secy Interior, 61; partic Am specialist prog, US Dept State, India, 66; res assoc, Smithsonian Inst, 68-71; mem, Coun Anthrop & Educ, 72- Mem: AAAS; fel Am Anthrop Asn (treas, 67-69); fel Soc Appl Anthrop (pres, 71-72); Inst Intercult Studies. Res: Acculturation; applied anthropology; community development; American Indians; developing countries; race relations; economic development; anthropology of education. Mailing Add: Dept of Anthrop Am Univ Washington DC 20016

NASH, RALPH GLEN, b Del Norte, Colo, July 26, 30; m 57; c 3. SOIL SCIENCE, CHEMISTRY. Educ: Colo State Univ, BS, 58, MS, 61, PhD(soil sci), 63. Prof Exp: SOIL SCIENTIST, AGR ENVIRON QUAL INST, AGR RES SERV, USDA, 65- Mem: Am Soc Agron; Soil Sci Soc Am; Weed Sci Soc Am; Am Chem Soc. Res: Weed science; toxic and residual interactions which result from a combination of two or more pesticides added to soils; plant absorption of pesticides; pesticide degradation; persistence and movement in soil and plants; soil and plant analytical pesticide methods. Mailing Add: Agr Environ Qual Inst USDA BARC-West Bldg 050HH1 Beltsville MD 20705

NASH, REGINALD GEORGE, b LaValle, Wis, Nov 20, 22; m 52; c 3. PARASITOLOGY. Educ: William Penn Col, BA, 48; Univ Iowa, MS, 52; Mich State Univ, PhD(zool), 64. Prof Exp: Asst prof biol, Northern Ill Univ, 52-54; instr natural sci, Mich State Univ, 54-58; PROF BIOL, UNIV WIS-WHITEWATER, 58- Mem: Am Soc Parasitol. Res: Immunologic studies involving infections with roundworms Trichinella spiralis, Ascaris lumbricoides and Toxocara canis. Mailing Add: Dept of Biol Univ of Wis Whitewater WI 53190

NASH, RICHARD FULLER, JR, entomology. see 12th edition

NASH, ROBERT ARNOLD, b Brooklyn, NY, July 6, 30; m 52; c 3. PHARMACEUTICAL CHEMISTRY, PHARMACY. Educ: Brooklyn Col Pharm, BS, 52; Rutgers Univ, MS, 54; Univ Conn, PhD, 58. Prof Exp: Asst, Rutgers Univ, 53-54 & Univ Conn, 54-57; res assoc, Merck, Sharp & Dohme Res Labs, 57-60; proj leader, Lederle Labs Div, 60-67, HEAD PHARMACEUT PROD DEVELOP, LEDERLE LABS & CYANAMID INT DIV, AM CYANAMID CO, 67- Concurrent Pos: Mfg pharmacist, Hartford Hosp, Conn, 55-57. Honors & Awards: Richardson Award, 57. Mem: AAAS; Am Chem Soc; Am Pharmaceut Asn. Res: Application of the sciences of physical and analytical chemistry, biochemistry and related disciplines to the solution of industrial pharmaceutical problems; physical pharmacy. Mailing Add: Dept 947 Lederle Labs Pearl River NY 10965

NASH, ROBERT JOSEPH, b Coventry, Eng, Sept 12, 39; m 63; c 2. SURFACE CHEMISTRY. Educ: Univ Wales, BSc, 62; Bristol Univ, PhD(phys chem), 66. Prof Exp: Res assoc surface chem, Amherst Col, 65-66, asst prof chem, 66-67; res assoc surface chem, Case Western Reserve Univ, 67-69; scientist, 70-73, SR SCIENTIST, XEROX CORP, 73- Mem: Am Chem Soc. Res: Palladium-hydrogen system; hysteresis in absorption processes; surface chemistry of metals; uses of gas chromatography in surface chemistry; surface chemistry of pigments and polymers; surface potentials. Mailing Add: Xerox Corp 800 Phillips Rd Bldg 139 Webster NY 14580

NASH, ROBLEY WILSON, b Whitman, Mass, June 24, 08; m 34; c 3. ENTOMOLOGY. Educ: Univ Mass, BSc, 29; Cornell Univ, MSc, 34. Prof Exp: Asst entomologist, State of Maine, 29-45; entomologist, US Forest Serv, 45-47; sr entomologist, 47-56, STATE ENTOMOLOGIST, MAINE FOREST SERV, 56- Concurrent Pos: Consult, Gould Paper Co, NY, 53-64. Mem: Entom Soc Am. Res: Forest entomology; biology; biological and chemical control. Mailing Add: Maine Forest Serv Augusta ME 04330

NASH, STANLEY WILLIAM, b Yakima County, Wash, Oct 8, 15; Can citizen. MATHEMATICAL STATISTICS. Educ: Col Puget Sound, BA, 39; Univ Calif, Berkeley, MA, 46, PhD(math statist), 50. Prof Exp: From asst prof to assoc prof, 50-70, PROF MATH, UNIV BC, 70- Concurrent Pos: Vis assoc prof, Dept Statist, Iowa State Univ, 60-61. Mem: Biomet Soc; Am Statist Asn; Inst Math Statist; Am Math Soc; Math Asn Am. Res: Classification problems; dissection of mixed distributions; contingency tables in several dimensions; least squares estimation in the singular case; growth curves. Mailing Add: Dept of Math Univ of BC Vancouver BC Can

NASH, VICTOR E, b Frankfort, Ky, Sept 27, 28; m 56; c 4. SOIL CHEMISTRY, MINERALOGY. Educ: Univ Ky, BS, 51, MS, 52; Univ Mo, PhD(soils), 55. Prof Exp: Res geochemist, Cities Serv Res & Develop Co, 56-59; from asst prof to assoc prof soils, 59-68, assoc agronomist, 59-68, PROF SOILS & AGRONOMIST, MISS STATE UNIV, 68- Mem: Am Soc Agron; Clay Minerals Soc; Soil Sci Soc Am; Int Soil Sci Soc. Res: Cation exchange of soil colloids and interaction of soil colloids; soil micromorphology and mineralogy. Mailing Add: Dept of Agron Miss State Univ Mississippi State MS 39762

NASH, WILLIAM DONALD, b Shreveport, La, Jan 17, 47; m 69; c 1. SYNTHETIC ORGANIC CHEMISTRY. Educ: McNeese State Univ, La, BS, 70; Tex A&M Univ, PhD(org chem), 74. Prof Exp: RES CHEMIST, EL PASO PROD CO, 74- Mem: Am Chem Soc. Mailing Add: El Paso Prod Co PO Box 3986 Odessa TX 79760

NASH, WILLIAM PURCELL, b Boston, Mass, Mar 20, 44; m 66; c 2. PETROLOGY. Educ: Univ Calif, Berkeley, BA, 65, PhD(geol), 71. Prof Exp: Asst prof, 70-73, ASSOC PROF GEOL, UNIV UTAH, 73- Mem: Am Geophys Union; Mineral Soc Am; Geochem Soc. Res: Theoretical, chemical and thermodynamic methods applied to determining the origin, evolution and crystallization of igneous rocks. Mailing Add: Dept of Geol & Geophys Univ of Utah Salt Lake City UT 84112

NASHED, MOHAMMED ZUHAIR ZAKI, b Aleppo, Syria, May 14, 36; m 59; c 3. MATHEMATICS. Educ: Mass Inst Technol, SB, 57, SM, 58; Univ Mich, MS & PhD(math), 63. Prof Exp: From asst prof to assoc prof, 63-69, PROF MATH, GA INST TECHNOL, 69- Concurrent Pos: Assoc prof, Am Univ Beirut, 67-69; vis prof math res ctr, Univ Wis-Madison, 67, 70-72. Honors & Awards: Lester Ford Award, Math Asn Am, 67. Mem: AAAS; Am Math Soc; Math Asn Am; Soc Indust & Appl Math; Inst Math Statist. Res: Nonlinear functional analysis; iterative methods for operator equations; numerical analysis; optimization; mathematical programming and control theory; generalized inverses; ill-posed problems; random operators. Mailing Add: Sch of Math Ga Inst of Technol Atlanta GA 30332

NASHED, WILSON, b Damanhour, Egypt, Feb 16, 19; nat US; m 54; c 1. PHARMACY. Educ: Univ Cairo, BS, 39; Purdue Univ, MS, 51, PhD(pharm), 54. Prof Exp: Hosp pharmacist, Egyptian Govt, 40-41; tech dir, Delta Labs, Egypt, 41-50; sr res chemist, 54-55, res group leader, 56-62, asst dir prod coord, 62-65, mgr regulatory affairs, 65-69, dir sci info, 69-74, assoc dir res, 74-75, DIR TECH SERV & RES FACIL, JOHNSON & JOHNSON, 75- Mem: Am Pharmaceut Asn. Res: Pharmaceutical research and product development. Mailing Add: Johnson & Johnson New Brunswick NJ 08903

NASHOLD, BLAINE S, b Lennox, SDak, Nov 12, 23; m 48; c 4. NEUROSURGERY, NEUROPHYSIOLOGY. Educ: Ind Univ, AB, 43; Ohio State Univ, MSc, 44; Univ Louisville, MD, 49; McGill Univ, MSc, 54. Prof Exp: Instr neuroanat, McGill Univ, 53; asst neurosurg, Bowman Gray Sch Med, 56-57; from asst prof to assoc prof, 57-75, PROF NEUROSURG, SCH MED, DUKE UNIV, 75- Concurrent Pos: Chief neurosurg sect, Vet Admin Hosp, Durham, NC, 57-59; chmn, Coop Studies Intervertebral Disc Dis & Parkinsonism, Vet Admin, 60-; mem, Cong French Speaking Neurosurgeons, 64. Mem: Am Asn Neurol Surg; Am Acad Neurol; Asn Res Nerv & Ment Dis; Am Acad Cerebral Palsy. Res: Stereotactic neurosurgical problems in relation to extrapyramidal diseases and problems of central pain; neurochemistry of brain function. Mailing Add: Dept of Neurosurg Duke Univ Med Ctr Durham NC 27716

NASIM, MOHAMMED ANWAR, b Pasrur, Pakistan, Dec 7, 35; Can citizen. GENETICS. Educ: Punjab Univ, BS, 55, MS, 57; Univ Edinburgh, PhD(genetics), 66. Prof Exp: Res officer genetics, Atomic Energy Can Ltd, 66-73; RES OFFICER GENETICS, NAT RES COUN CAN, 73- Mem: Environ Mutagen Soc; Genetics Soc Can. Res: Mechanisms of mutagenesis; DNA repair in yeast; genetic hazards. Mailing Add: Div of Biol Sci Nat Res Coun Ottawa ON Can

NASIR-UD-DIN, NASIR, b Amritsar, India, Aug 15, 37; m 74; c 1. BIOCHEMISTRY, CARBOHYDRATE CHEMISTRY. Educ: Panjab Univ, BS, 55, MS, 57; Univ Edinburgh, PhD(chem), 63. Prof Exp: Demonstr anal chem, Panjab Univ, 58-60; res fel biol chem, Sch Med, Harvard Univ, 63-65; lectr org chem, Panjab Univ, 66-67; chief chemist, mgr qual control & tech adv, Johnson & Johnson, Pakistan, 67-69; ASSOC BIOL CHEM, SCH MED, HARVARD UNIV, 70- Mem: The Chem Soc; Am Chem Soc; Pakistan Asn Advan Sci. Res: Chemistry of carbohydrates; chemistry of bacterial cell walls; glycoproteins from wheat smut; glycoproteins from cervical mucus. Mailing Add: Lab for Carbohydrate Res Mass Gen Hosp Boston MA 02114

NASJLETI, CARLOS EDUARDO, b San Juan, Arg, Apr 24, 21; US citizen; m 55; c 3. CYTOGENETICS, TERATOLOGY. Educ: San Juan Nat Col, BA & BS, 41; Nat Univ Cordoba, DDS, 45. Prof Exp: Pvt practr dent, San Juan, 45-55; res asst, Dept Nuclear Med, Med Ctr, 55-63, res biologist, 63-69, RES ASSOC DENT, SCH DENT, UNIV MICH, ANN ARBOR, 63; DIR DENT RES & EDUC TRAINEE PROG, 69-, COORDR PROG, 71- Concurrent Pos: Partic, Nat Inst Dent Res-Am Col Dent Inst Dent Res, Nat Inst Dent Res, Bethesda, Md & Rockefeller Inst, New York, 65. Mem: AAAS; Am Dent Asn; Int Asn Dent Res; Soc Nuclear Med; NY Acad Sci. Res: Human cytogenetics, especially investigation of the effects of ionizing radiation and certain chemotherapeutic agents on human chromosomes, both in vivo and invitro. Mailing Add: Vet Admin Hosp Ann Arbor MI 48105

NASKALI, RICHARD JOHN, b Jefferson, Ohio, Dec 11, 35. BOTANY. Educ: Ohio State Univ, BSc, 57, MSc, 61, PhD(bot), 69. Prof Exp: Instr bot, Ohio State Univ, 60-

67; ASST PROF BOT, UNIV IDAHO, 67- Mem: AAAS; Bot Soc Am; Am Soc Plant Physiol; Int Soc Plant Morphol. Res: Developmental plant anatomy, particularly of flowering and plant chimeras; aquatic macrophytes of Pacific Northwest. Mailing Add: Dept of Biol Sci Univ of Idaho Moscow ID 83843

NASON, ALVIN, b Coatesville, Pa, June 10, 19; m 44; c 5. ENZYMOLOGY. Educ: Cornell Univ, BS, 40; Columbia Univ, AM, 48, PhD(biochem), 49. Prof Exp: Asst genetics, Carnegie Inst, 40; chemist, Chromium Corp Am, 41-42; agent plant physiol & biochem, Field Lab Tung Invest, USDA, 42-43; asst plant physiol chem, Columbia Univ, 46-48; from asst prof to assoc prof, 49-58, PROF BIOL, McCOLLUM PRATT INST, JOHNS HOPKINS UNIV, 58-, ASSOC DIR INST, 59- Concurrent Pos: Travel award, Am Soc Biol Chem, Vienna, 58; ed-in-chief, Anal Biochem, 59-; mem ad hoc panel, Grad Training Fels, NSF, 65, consult, Adv Panel, 71-74; consult, Nat Cancer Inst, 65-69; chmn vis comt biol, Brookhaven Nat Lab, 66-67. Mem: Am Soc Biol Chem; Am Chem Soc; Am Soc Plant Physiol; Am Soc Microbiol. Res: Enzymology and mechanism of terminal electron transport and inorganic nitrogen metabolism; vitamin metabolism; metallo-flavoproteins. Mailing Add: Dept of Biol Johns Hopkins Univ Baltimore MD 21218

NASON, HOWARD KING, b Kansas City, Mo, July 12, 13; m 34. CHEMISTRY. Educ: Univ Kans, AB, 34. Prof Exp: Chief chemist, Anderson-Stolz Corp, 35-36; res chemist, Org Div, Monsanto Chem Co, 36-39, asst dir res plastics div, 39-44, dir develop cent res dept, 44-46, assoc dir, 46-48, dir, 48-50, asst to vpres, 50-51, dir res org div, 51-56, vpres & gen mgr res & eng div, 56-60, PRES, MONSANTO RES CORP, 60- Concurrent Pos: Mem adv comt isotopes & radiation develop, AEC, 64-68, labor-mgt adv comt, 65; mem, President's Comn Patent Syst, 65-68; mem patent adv comt, US Patent Off, 68; trustee-at-large, Univs Res Asn, Inc, 71; vpres & mem exec comt, Atomic Indust Forum, Inc, 71-73; chmn, Aerospace Safety Adv Panel, 72-trustee, Charles F Kettering Found, 73-; mem, Nat Mat Adv Bd, Nat Acad Eng, 73- Mem: AAAS; Am Chem Soc; Am Soc Testing & Mat; Soc Rheol; Am Inst Chem Engrs. Res: Plastics; plasticizers; industrial microbiology; physical testing; water treatment; industrial application of chemicals; protective coatings. Mailing Add: Monsanto Res Corp 800 N Lindbergh Blvd St Louis MO 63166

NASON, JAMES DUANE, b Los Angeles, Calif, July 19, 42; m 65; c 1. CULTURAL ANTHROPOLOGY. Educ: Univ Calif, Riverside, BA, 64; Univ Wash, MA, 67, PhC, 68, PhD(anthrop), 70. Prof Exp: ASST PROF ANTHROP, UNIV WASH, 70-, CUR ETHNOL, THOMAS BURKE MEM WASH STATE MUS, 70- Concurrent Pos: Grad Sch res grant, Univ Wash, 71-72; ed newslett, Asn Social Anthrop Oceania, 72- Mem: Asn Social Anthrop Oceania (secy, 72-); Am Anthrop Asn; Am Ethnol Soc; Soc Appl Anthrop. Res: Ethnology and social change in Oceania; applied anthropology in rural and underdeveloped regions; material culture studies. Mailing Add: Burke Mem Wash State Mus Univ of Wash Seattle WA 98195

NASON, ROBERT DOHRMANN, b San Francisco, Calif, Dec 9, 39; SEISMOLOGY. Educ: Calif Inst Technol, BS, 61; Univ Calif, San Diego, PhD, 71. Prof Exp: Seismologist, Earthquake Mechanism Lab, Nat Oceanic & Atmospheric Admin, Calif, 66-73; EARTHQUAKE GEOPHYSICIST, US GEOL SURV, 73- Mem: Am Geophys Union; Geol Soc Am; Seismol Soc Am; Earthquake Eng Res Inst. Res: Heat-flow and marine tectonics; earthquakes and earthquake tectonics; movement on the San Andreas fault; fault creep. Mailing Add: US Geol Surv 345 Middlefield Rd Menlo Park CA 94025

NASRALLAH, MIKHAIL ELIA, b Kafarmishky, Lebanon, Feb 1, 39; m 63; c 1. BIOLOGY, GENETICS. Educ: Am Univ Beirut, BSc, 60; Univ Vt, MS, 62; Cornell Univ, PhD, 65. Prof Exp: Res assoc physiol genetics, Cornell Univ, 65-66; asst prof plant breeding & biol, Cornell Univ & State Univ NY Col Cortland, 66-67; from asst prof to assoc prof biol, 67-74, PROF BIOL, STATE UNIV NY COL CORTLAND, 74- Res: Physiological genetics of self-incompatible plants; characterization of self-incompatibility antigens; cytochemical, enzymatic and immunogenetic studies with pollen and stigmatic proteins. Mailing Add: Dept of Biol Sci State Univ NY Col Cortland NY 13045

NASS, HANS GEORGE, b Danzig, Ger, Mar 6, 41; Can citizen; m 64; c 1. AGRONOMY, PLANT SCIENCE. Educ: Univ Man, BSA, 64; Colo State Univ, MSc, 66; Purdue Univ, Lafayette, PhD(plant physiol), 69. Prof Exp: NSF grant, Univ Mo-Columbia, 69-70; RES SCIENTIST PHYSIOL & BREEDING, CAN DEPT AGR, 70- Mem: Am Soc Agron; Agr Inst Can; Genetics Soc Can; Soc Cryobiol; Can Soc Agron. Res: Plant breeding; genetics; plant physiology; plant metabolism; winter handiness; biochemistry. Mailing Add: Can Dept of Agr Res Sta PO Box 1210 Charlottetown PE Can

NASS, HAROLD WILLIAM, b Croswell, Mich, May 8, 31; m 52; c 2. NUCLEAR CHEMISTRY, ANALYTICAL CHEMISTRY. Educ: Eastern Mich Univ, AB, 57; State Univ NY, MA, 73. Prof Exp: Res asst nuclear chem, Univ Mich, 58-61, asst res chemist, 61-63; nuclear chemist, Develop Dept, 63-65, supvr nuclear anal serv, 65-71, group leader anal serv, 71-72, MGR QUAL ASSURANCE, UNION CARBIDE CORP, 72- Mem: Am Chem Soc; Am Soc Qual Control. Res: Use of interdisciplinary methods to assure quality of clinical diagnostic products, especially clinical radioisotopes and radioimmunoassays; reagents for blood component testing and neutron activation analysis for trace elements. Mailing Add: Union Carbide Corp PO Box 324 Tuxedo NY 10987

NASSAU, KURT, b Stockerau, Austria, Aug 25, 27; US citizen; m 49. SOLID STATE CHEMISTRY, PHYSICS. Educ: Bristol Univ, BSc, 48; Univ Pittsburgh, PhD(phys chem), 59. Prof Exp: Res chemist, Glyco Prod Co, Inc, Pa, 49-54; chemist, Dept Metab, Walter Reed Army Med Ctr, Washington, DC, 54-56; MEM TECH STAFF, BELL LABS, MURRAY HILL, 59- Mem: Am Chem Soc; Am Crystallog Asn; Am Asn Crystal Growth. Res: Growth of crystals and their physical and chemical properties; solid state and crystal chemistry and physics; crystallography; laser, magnetic, piezoelectric and ferroelectric materials. Mailing Add: Round Top Rd Bernardsville NJ 07924

NASSE, GEORGE NICHOLAS, b Woonsocket, RI, Oct 26, 24; m 61; c 1. GEOGRAPHY. Educ: Clark Univ, AB, 53; Univ Mich, MA, 54, PhD, 60. Prof Exp: Instr geog, Wis State Col, Oshkosh, 57-58; vis lectr, geog & Asian civilization, New Paltz Col Educ, State Univ NY, 59-60; asst prof geog, Western Mich Univ, 60-62; asst prof, Northern Mich Univ, 62-65; assoc prof, 65-71, PROF GEOG, CALIF STATE UNIV, FRESNO, 71- Mem: Asn Am Geogr; Am Geog Soc. Res: Strategic importance of and evolution of governments of the Mediterranean lands and Balkan Peninsula. Mailing Add: Dept of Geog Calif State Univ Fresno CA 93710

NASSER, DELILL, b Terre Haute, Ind, July 17, 29. MICROBIOLOGY, MICROBIAL GENETICS. Educ: Ind State Col, BS, 50; Purdue Univ, MS, 55, PhD(microbiol), 63. Prof Exp: Res asst, Purdue Univ, 55-56; res microbiologist, Eli Lilly & Co, 56-58; instr, Purdue Univ, 58-60, NIH res trainee microbiol, 63-64; NIH res trainee, Univ Wash, 64-67; asst prof bact, Univ Fla, 67-70, assoc prof microbiol, 70-72; ASSOC RES BIOCHEMIST, UNIV CALIF, SAN FRANCISCO, 72- Mem: Am Soc Microbiol. Res: Glucose metabolism of Fusarium; bioconversion of steroids; biosynthesis of flagellin; transformation of the property of flagellation in Bacillus subtilis; genetic and enzymatic studies of the synthesis of aromatic amino acids in Bacillus subtilis; eucaryotic chromosome structure and regulation. Mailing Add: Dept of Biochem Univ of Calif San Francisco CA 94143

NASSET, EDMUND SIGURD, b Willmar, Minn, Feb 24, 00; m 28; c 2. PHYSIOLOGY. Educ: St Olaf Col, AB, 25; Pa State Col, MS, 27; Univ Rochester, PhD(physiol), 31. Hon Degrees: DMSc, Soo do Med Col, Korea, 63; LHD, Dickinson State Col, 66. Prof Exp: Asst biochem, Pa State Col, 26-27; asst, Univ Calif, 27-28; asst physiol, 28-31, from instr to prof, 31-65, EMER PROF PHYSIOL, SCH MED, UNIV ROCHESTER, 65-; SR RES PHYSIOLOGIST, CHILDREN'S HOSP, OAKLAND, 74- Concurrent Pos: Fulbright lectr, Med Schs, India, 54-55; vis prof, Univ Lucknow, 54-55; mem, Fulbright Screening Comt Biol & Agr, 57; Food & Agr Orgn nutrit adv, Govt India, 61-63; res physiologist & lectr, Univ Calif, Berkeley, 67-; res physiologist, Children's Hosp, Oakland, 70-74. Mem: Am Physiol Soc; Soc Exp Biol & Med; fel Am Inst Nutrit. Res: Digestive secretions; biological value of amino acid mixtures and of proteins; physiological effects of high frequency current; enterocrinin; homeostasis of amino acids in gut lumen and blood plasma. Mailing Add: Bruce Lyon Mem Res Lab Children's Hosp Oakland CA 94609

NASTUK, WILLIAM LEO, b Passaic, NJ, June 17, 17; m 50; c 3. PHYSIOLOGY. Educ: Rutgers Univ, BSc, 39, PhD(physiol), 45. Prof Exp: Asst chem, Rutgers Univ, 39-40, asst physiol, 40-43, instr elec & radio eng, 43-44, asst, Off Sci Res & Develop Proj, 44-45; from instr to assoc prof physiol, 45-60, PROF PHYSIOL, COL PHYSICIANS & SURGEONS, COLUMBIA UNIV, 60-, DIR BIOENG INST, 74- Concurrent Pos: Mem, Physiol Study Sect, USPHS, 60-64 & Physiol Training Comt, 64-66; mem, Sci Adv Bd, Myasthenia Gravis Found, 63- Mem: AAAS; Soc Neurosci; Am Physiol Soc; Harvey Soc; NY Acad Sci. Res: Membrane potentials, neuromuscular transmission; instruments for electrophysiology; gravity and hemorrhagic shock; myasthenia gravis; neuromuscular pharmacology; muscle contraction. Mailing Add: Dept of Physiol Columbia Univ Col Phys & Surg New York NY 10032

NATALINI, JOHN JOSEPH, b Norristown, Pa, Apr 27, 44. BIOLOGICAL RHYTHMS, VERTEBRATE BIOLOGY. Educ: Villanova Univ, BS, 66; Northwestern Univ, MS & PhD(biol), 71. Prof Exp: ASST PROF BIOL, QUINCY COL, 71- Mem: Sigma Xi; Int Soc Chroniobiol; Int Soc Biometerol; Animal Behavior Soc; AAAS. Res: Phase response curves to light and the means of entrainment of biological rhythms to various zeitgebers. Mailing Add: Dept of Biol Quincy Col Quincy IL 62301

NATARAJAN, KOTTAYAM VISWANATHAN, b Cochin, India, Apr 21, 33; m; c 3. OCEANOGRAPHY, MICROBIOLOGY. Educ: Univ Travancore, India, BS, 52; Benaras Hindu Univ, MS, 55; Univ Alaska, PhD(marine sci), 65. Prof Exp: Demonstr bot, NSS Col, Kerala, 52-53 & Vivekananda Col, Madras, 55-56; lectr, Mar Ivanios Col, Kerala, 56-57; res asst, Indian Agr Res Inst, 57-60; scientist, Kaiser Found Res Inst, Calif, 60-61; res asst bot, Univ Calif, Berkeley, 61-62; sr res asst marine sci, Univ Alaska, 62-65, asst prof, 65-70; ASSOC PROF SCI, GREATER HARTFORD COMMUNITY COL, 70- Concurrent Pos: Mem sci fac fel panel, NSF, 74. Mem: AAAS; Am Soc Limnol & Oceanog; Int Phycol Soc. Res: Nitrogen fixation by blue-green algae; general physiology of algae; vitamins of the sea. Mailing Add: Dept of Sci Greater Hartford Community Col Hartford CT 06105

NATELSON, SAMUEL, b New York, NY, Feb 28, 09; m 37; c 4. CLINICAL CHEMISTRY. Educ: NY Univ, MS, 30, PhD(chem), 31. Prof Exp: Instr chem, NY Univ, 28-31; res chemist in-chg, NY Testing Lab, 31-32; res biochemist, Jewish Hosp Brooklyn, 33-49; chmn dept biochem, Rockford Mem Hosp, 49-57, St Vincent's Hosp, New York, 57-58 & Roosevelt Hosp, New York, 58-65; CHMN DEPT BIOCHEM, MICHAEL REESE HOSP, 65- Concurrent Pos: Lectr grad sch, Brooklyn Col, 47-49 & 57-65, New York Polyclin Med Sch & Hosp, 62-65 & Ill Inst Technol, 71- Honors & Awards: Van Slyke Award Clin Chem, 61; Ames Award, Am Asn Clin Chemists, 65. Mem: AAAS; Am Microchem Soc; Harvey Soc; Soc Appl Spectros; Am Chem Soc. Res: Citric acid metabolism in humans; infant feeding; radiopaques; surface tension; vapor pressure; sterols; vitamin D; resins; alkaloids; synthetic organic chemistry; organic analysis; microanalysis; instrumentation; nitrogen metabolism. Mailing Add: Michael Reese Hosp & Med Ctr 29th St & Ellis Ave Chicago IL 60616

NATH, AMAR, b Agra, India, Nov 28, 29; m 57; c 1. PHYSICAL CHEMISTRY, SOLID STATE CHEMISTRY. Educ: Agra Univ, MSc, 50, DSc, 70; Moscow State Univ, PhD(chem), 61. Prof Exp: Sci officer, Bhabha Atomic Res Ctr, 51-66; assoc geophysicist, Inst Geophys & Planetary Physics, Univ Calif, Los Angeles, 66-67; res chemist, Lawrence Radiation Lab, Univ Calif, Berkeley, 67-69; PROF CHEM, DREXEL UNIV, 69- Mem: Fel Am Inst Chemists; Am Chem Soc; NY Acad Sci. Res: Hot-atom chemistry of solids; radiolysis of solids using electron paramagnetic resonance, thermoluminescence and semiconductivity measurements; Mössbauer studies of after-effects of Auger events; Mössbauer spectroscopy of vitamin B12 and hemoglobin. Mailing Add: Dept of Chem Drexel Univ Philadelphia PA 19104

NATH, JOGINDER, b Joginder nagar, India, May 12, 32; m 69; c 1. CYTOLOGY. Educ: Panjab Univ, India, BS, 53, MS, 55; Univ Wis, PhD(agron), 60. Prof Exp: Res assoc cryobiol, Am Found Biol Res, Wis, 60-63; asst prof cytol & res grant, Southern Ill Univ, 64-66; asst prof genetics, 66-72, PROF GENETICS & REPROD PHYSIOL, WVA UNIV, 72- Concurrent Pos: NSF res grant, WVa Univ, 67- Mem: Electron Micros Soc Am; Soc Cryobiol; Am Soc Agron; Indian Soc Genetics & Plant Breeding. Res: Cytology of genus Selaginella; cytogenetic studies of some species of Paniceae and Phleum; cryobiological studies on blood cells, plasma, heart, intestines, mesentery and embryo; electron microscopy of frog oocytes; cryobiological studies on semen, pollen grains and bacteria; cryobiology. Mailing Add: Dept of Genetics WVa Univ Morgantown WV 26506

NATH, K RAJINDER, b Ferozepur, India, May 26, 37; m 73; c 1. DAIRY MICROBIOLOGY, FOOD SCIENCE. Educ: Delhi Univ, BSc Hons, 59; Agra Univ, MSc, 61; Cornell Univ, PhD(food sci), 69. Prof Exp: Res assoc dairy technol, Ohio State Univ, 69-70; res assoc food sci, Cornell Univ, 70-75; RES MICROBIOLOGIST, KRAFTCO CORP, 75- Concurrent Pos: Proj leader & secy, Qual Assurance Consumer Foods Proj, NE-83, 72-75. Mem: Am Dairy Sci Asn; Inst Food Technologists. Res: Development of microbial culture for food use; microbial interaction in foods; stimulation and inhibition of lactic acid bacteria; protein hydrolysis and protein modification. Mailing Add: Kraftco Corp Res & Dev 801 Waukegan Rd Glenview IL 60025

NATH, RAVINDER KATYAL, b Jullundar, India, Apr 9, 42; m 71; c 1. NUCLEAR PHYSICS, RADIOLOGICAL PHYSICS. Educ: Univ Delhi, BS, 63, MS, 65; Yale Univ, PhD(physics), 71. Prof Exp: Res staff physicist, 71-73, res assoc, 73-76, ASST PROF PHYSICS, YALE UNIV, 76- Honors & Awards: Med Physics Award, Am Asn Physicists Med, 75. Mem: Am Phys Soc; Am Asn Physicists Med. Res: Radiological physics related to radiation therapy, nuclear physics, especially photonuclear and neutron physics. Mailing Add: Dept of Therapeut Radiol Yale Univ 333 Cedar St New Haven CT 06510

NATHAN, ALAN HART, b New York, NY, Apr 25, 13; m 45; c 2. CHEMISTRY. Educ: Yale Univ, BS, 35; Columbia Univ, MA, 37, PhD(chem), 40. Prof Exp: Chemist, 41-61, staff asst dept chem, 61-68, ADMIN ASST EXP BIOL & CHEM, UPJOHN CO, 68- Mem: Am Chem Soc; AAAS. Mailing Add: 1109 Washburn Ave Kalamazoo MI 49001

NATHAN, ALAN MARC, b Rumford, Maine, Sept 17, 46; m 70. EXPERIMENTAL NUCLEAR PHYSICS. Educ: Univ Md, BS, 68; Princeton Univ, MA, 72, PhD(physics), 75. Prof Exp: RES ASSOC PHYSICS, BROOKHAVEN NAT LAB, 75- Mem: Am Phys Soc. Res: Nuclear spectroscopy and fundamental interactions of nuclei and nucleons. Mailing Add: Bldg 901 A Brookhaven Nat Lab Upton NY 11973

NATHAN, EDWARD C, III, organic chemistry, see 12th edition

NATHAN, HELMUTH M, b Hamburg, Ger, Oct 26, 01; US citizen; m 27; c 1. SURGERY, HISTORY OF MEDICINE. Educ: Univ Frieburg & Univ Hamburg, Med Dr, 25. Prof Exp: EMER PROF SURG, UNIV HAMBURG, 70-; EMER PROF SURG & ANAT, 70-, FIRST PROF HIST MED & CHMN DEPT, ALBERT EINSTEIN COL MED, 71- Concurrent Pos: Asst in surg, Surg Dept, Univ Freiburg, 27; from asst path to asst internal med & surg, Hamburg St Georg Hosp; assoc surg, Jewish Hosp, Hamburg, 43-55; attend surgeon, Sydenham Hosp, New York & Bronx Munic Hosp, 54-72; consult surgeon, Albert Einstein Col Hosp, Montefiore Med Ctr & Bronx Munic Hosp, 65-; prof lectr, Mt Sinai Sch Med, 74. Honors & Awards: Deneke Medal, Hamburg St Georg Hosp, 32; Solomon Heine Medal, Jewish Hosp Hamburg, 36; Officer's Cross, Order Merit 1st Class, Fed Repub Ger, 73. Mem: Int Col Surg; Am Asn Abdominal Surg; Am Col Gastroenterol; Am Col Cardiol; Ger Surg Asn. Res: Septical infections; intestinal granulomatosis; ulcerative colitis; viridans encephalitis; art and medicine; internal hernias; problems of death; tumor of salivary glands; total parotidectomy; intestinal volvulus; biographies of leading physicians and scientists. Mailing Add: Dept of the Hist of Med Albert Einstein Col of Med Bronx NY 10461

NATHAN, HENRY C, b New York, NY, Aug 16, 24; m 57; c 1. BIOLOGY. Educ: NY Univ, BA, 50; Columbia Univ, PhD(histochem, path), 66. Prof Exp: Microbiologist, Personality & Celebrity Printing Co, 51-53; res biologist, Wellcome Res Labs, 53-65; cytopharmacologist, Union Carbide Res Inst, NY, 65-68; instr biol sci, anat & physiol, Hunter Col, 68-74; ADJ ASST PROF BIOL, PACE UNIV, 74- Concurrent Pos: Instr, NY Inst Dietetics, 56-57; asst lab instr, Darwin Animal Labs, Brooklyn, NY, 60-65. Mem: AAAS; Am Soc Zoologists; Am Soc Microbiol; Fedn Am Scientists; NY Acad Sci. Res: Nutrition and disease; antibacterials; experimental cancer chemotherapy; metabolism and disease; chemical suppression of immune response; nature of immune mechanisms. Mailing Add: Dept of Pharmacol Pace Univ Haskins Labs New York NY 10038

NATHAN, LAWRENCE CHARLES, b Corning, Calif, Nov 26, 44; m 66; c 2. INORGANIC CHEMISTRY. Educ: Linfield Col, BA, 66; Univ Utah, PhD(inorg chem), 71. Prof Exp: ASST PROF CHEM, UNIV SANTA CLARA, 70- Concurrent Pos: Petrol Res Fund res grant, 71. Mem: Am Chem Soc. Res: Preparation and characterization of new transition metal coordination complexes. Mailing Add: Dept of Chem Univ of Santa Clara Santa Clara CA 95053

NATHAN, MARC A, b Great Falls, Mont, Sept 14, 37; m 57; c 2. NEUROPHYSIOLOGY, PHYSIOLOGICAL PSYCHOLOGY. Educ: Wash State Univ, BS, 60; Univ Wash, MS, 62, PhD(psychol), 67. Prof Exp: Res assoc physiol psychol, Sch Med, Univ Wash, 67-68, res psychologist radiobiol, Sch Aerospace Med, 68-71; instr, 72-73, ASST PROF NEUROL, MED COL, CORNELL UNIV, 73-, DIR, FIELD NEUROBIOL & BEHAV, 73- Mem: AAAS; Am Physiol Soc; Inst Elec & Electronics Engrs; Soc Neurosci; Am Heart Asn. Res: Neural control of the cardiovascular system; neurogenic hypertension; emotional behavior. Mailing Add: Dept of Neurol Cornell Univ Med Col New York NY 10021

NATHAN, MARSHALL I, b Lakewood, NJ, Jan 22, 33; m 55, 71; c 2. PHYSICS. Educ: Mass Inst Technol, BS, 54; Harvard Univ, PhD(physics), 58. Prof Exp: Staff mem, 58-71, mgr coop phenomena group, 71-74, consult to dir res, 74-75, MGR OPTICAL SOLID STATE TECHNOL GROUP, IBM CORP, 75- Mem: Fel Am Phys Soc; fel Inst Elec & Electronics Engrs. Res: Solid state physics; semiconductor devices; optics. Mailing Add: IBM Corp Res Ctr Box 218 Yorktown Heights NY 10598

NATHAN, PAUL, b Chicago, Ill, June 18, 24; m 53; c 4. PHYSIOLOGY. Educ: Univ Chicago, PhB, 46, PhD(physiol), 53. Prof Exp: Biochemist, Galesburg State Res Hosp, Ill, 53-55; from instr to asst prof physiol, 55-68, ASSOC PROF PHYSIOL, COL MED, UNIV CINCINNATI, 68-, ASST PROF EXP SURG, 66-, DIR DEPT CELL BIOL & IMMUNOL, SHRINERS BURNS INST, 66- Concurrent Pos: Advan res fel, Am Heart Asn, 59-61; res assoc, May Inst Med Res, Cincinnati Jewish Hosp, 55-65; estab investr, Am Heart Asn, 61-66; res collabr, Brookhaven Nat Lab, 64-71. Mem: AAAS; Am Physiol Soc; Int Soc Burn Injuries. Res: Transplantation; physiology of digestion; immunology; burn injury. Mailing Add: Shriners Burns Inst Univ of Cincinnati Col of Med Cincinnati OH 45219

NATHAN, RICHARD ARNOLD, b New York, NY, Sept 25, 44; m 66; c 2. ORGANIC CHEMISTRY. Educ: Mass Inst Technol, BS, 65; Polytech Inst Brooklyn, PhD(org chem), 69. Prof Exp: Chemist, Rohm and Haas Co, 65 & Polaroid Corp, 69; proj leader, Org Chem Div, 70-74, ASSOC MGR ORG & STRUCT CHEM SECT, COLUMBUS LABS, BATTELLE MEM INST, 74- Mem: Am Chem Soc; The Chem Soc; Int Solar Energy Soc; NY Acad Sci. Res: Photochemistry and bioorganic chemistry. Mailing Add: Org & Struct Chem Sect Battelle Mem Inst Columbus OH 43201

NATHAN, ROBERT, applied mathematics, see 12th edition

NATHANIEL, EDWARD J H, b Guntur, India, Apr 21, 28; m 53; c 3. ANATOMY. Educ: Univ Madras, MB, BS, 52; Univ Calif, Los Angeles, MS, 58, PhD(anat), 62. Prof Exp: Demonstr anat, Christian Med Col, Vellore, 52-54; civil asst surgeon & actg chief med officer, Leprosy Treatment & Study Ctr, Turukoilur, 55-56; res anatomist, Sch Med, Univ Calif, Los Angeles, 57-62; chief electron micros lab path, Cedars of Labanon Hosp, 62-64; assoc prof anat & path, Med Col Ga, 64-66; asst prof anat, McGill Univ, 66-68; ASSOC PROF ANAT, SCH MED, UNIV MAN, 68- Concurrent Pos: Med Res Coun Can fel, Sch Med, Univ Man, 60-72, Nat Cancer Inst Can fel, 69-72; USPHS fel, Cedars of Lebanon Hosp, Los Angeles, 63-66; USPHS fel, Med Col Ga, 64-66. Mem: Am Asn Anatomists; Can Asn Anat; Electron Micros Soc Am; Am Soc Cell Biol; Am Soc Exp Path. Res: Electron microscopic research in experimental neurology with emphasis on demyelination and remyelination, postnatal development of nervous system, platelet morphology in artificially induced thrombi in experimental situations and chemotherapy of experimental mouse tumors. Mailing Add: Univ of Man Sch of Med Winnipeg MB Can

NATHANS, DANIEL, b Wilmington, Del, Oct 30, 28; m 56; c 3. MICROBIOLOGY, MOLECULAR BIOLOGY. Educ: Univ Del, BS, 50; Washington Univ, MD, 54. Prof Exp: Resident, Columbia-Presby Med Ctr, 57-59; from asst prof to assoc prof, 62-67, PROF MICROBIOL, JOHNS HOPKINS UNIV, 67-, DIR, 72- Concurrent Pos: USPHS grant, Rockefeller Univ, 59-62. Res: Tumor viruses. Mailing Add: Dept of Microbiol Johns Hopkins Univ Baltimore MD 21205

NATHANS, MARCEL WILLEM, b The Hague, Neth, Apr 16, 22; nat US; m 47; c 2. PHYSICAL CHEMISTRY. Educ: State Univ Leiden, Cand, 41; Univ Calif, PhD(phys chem), 49. Prof Exp: Res chemist, Univ Calif, 49-50 & Great Lakes Carbon Corp, 50-54; asst chemist, Argonne Nat Lab, 54-58; chemist, Lawrence Radiation Lab, Univ Calif, 58-64; supvr anal res, Tracer-Lab, 64-67; mgr res dept, 67-74, TECH DIR, LFE ENVIRON ANAL LABS DIV, LFE CORP, 74- Concurrent Pos: Teacher, Wright Jr Col, 52-57 & Diablo Valley Col, 59-61. Mem: Fel Am Inst Chemists; Am Geophys Union. Res: Mechanical metallurgy; absorption spectroscopy; carbon and graphite; heterogeneous reaction kinetics; phase diagrams; shock-induced reactions; peaceful applications of nuclear explosives; activation analysis; fallout studies; environmental studies; thermal analysis; trace contaminants in environment. Mailing Add: LFE Environ Anal Labs 2030 Wright Ave Richmond CA 94804

NATHANS, ROBERT, b Wilmington, Del, May 27, 27; m 49; c 5. SOLID STATE PHYSICS, URBAN STUDIES. Educ: Univ Del, BS, 49; Univ Minn, MS, 50; Univ Pa, PhD(physics), 54. Prof Exp: Assoc prof physics, Pa State Univ, 54-58; vis prof, Osaka Univ, Japan, 58-59; physicist, Lincoln Lab, Mass Inst Technol, 59-60; physicist, Brookhaven Nat Lab, 60-68; PROF PHYSICS & MAT SCI, STATE UNIV NY STONY BROOK, 68-, CHMN PROG URBAN & POLICY SCI, 71- Concurrent Pos: Fulbright grant, 58. Mem: Fel Am Phys Soc. Res: Neutron diffraction; application of mathematical modeling to urban and environmental problems. Mailing Add: Dept of Physics State Univ of NY Stony Brook NY 11790

NATHANSON, BENJAMIN, b New York, NY, Jan 9, 29. AIR POLLUTION, CHEMICAL INSTRUMENTATION. Educ: City Col New York, BS, 47; Columbia Univ, MA, 49; NY Univ, PhD(chem), 65. Prof Exp: Asst chem, NY Univ, 50-54; lectr, Hunter Col, 54-55; instr, Finch Col, 66-67; lectr, Pace Col, 67-68; SR CHEMIST, DEPT AIR RESOURCES, CITY OF NEW YORK, 68- Concurrent Pos: Adj asst prof, Pratt Inst, 67-68. Mem: Am Chem Soc. Res: Instrumental analysis of air pollutants utilizing atomic absorption; x-ray fluorescence; high-pressure liquid chromatography; gas chromatography-mass spectroscopy; analysis of airborne particulates for trace metals and organic components; analyzing gasoline for lead; trace gas analysis; freon analysis in air. Mailing Add: Dept of Air Resources 51 Astor Pl New York NY 10003

NATHANSON, MELVYN BERNARD, b Philadelphia, Pa, Oct 10, 44. MATHEMATICS. Educ: Univ Pa, BA, 65; Univ Rochester, MA, 68, PhD(math), 72. Prof Exp: ASSOC PROF MATH, SOUTHERN ILL UNIV, CARBONDALE, 71- Concurrent Pos: Asst to Andre Weil, Inst Advan Study, 74-75; assoc prof math, Brooklyn Col, 75-76; guest, Rockefeller Univ, 75-76. Mem: Am Math Soc; Math Asn Am. Res: Number theory; algebra; combinatorial theory. Mailing Add: Dept of Math Southern Ill Univ Carbondale IL 62901

NATHANSON, NEAL, b Boston, Mass, Sept 1, 27; m 54; c 3. VIROLOGY, EPIDEMIOLOGY. Educ: Harvard Univ, AB, 49, MD, 53. Prof Exp: Chief poliomyelitis surveillance unit, USPHS Commun Dis Ctr, 55-57; res assoc anat & asst prof, Sch Med, 57-63, assoc prof epidemiol, 63-70, PROF EPIDEMIOL, JOHNS HOPKINS UNIV, 70- Concurrent Pos: Ed-in-chief, Am J Epidemiol, 64- Mem: Am Epidemiol Soc; Am Asn Immunol; Am Soc Microbiol; Am Soc Trop Med & Hyg; Am Pub Health Asn. Res: Neurotropic viruses; neuropathology. Mailing Add: Dept of Epidemiol Johns Hopkins Univ Sch of Hyg Baltimore MD 21205

NATHANSON, NEIL MARC, b Philadelphia, Pa, Dec 11, 48. NEUROBIOLOGY. Educ: Univ Pa, BA, 70; Brandeis Univ, PhD(biochem), 75. Prof Exp: MUSCULAR DYSTROPHY ASN, INC FEL, LAB BIOCHEM GENETICS, NAT HEART & LUNG INST, 75- Mailing Add: Lab of Biochem Genetics Nat Heart & Lung Inst NIH Bethesda MD 20014

NATHANSON, WESTON IRWIN, b Detroit, Mich, May 2, 38; m 58; c 2. MATHEMATICS. Educ: Univ Calif, Los Angeles, BA, 61, MA, 63, PhD(math), 70. Prof Exp: Aeronaut engr, Douglas Aircraft Co, 61-66; ASSOC PROF MATH, CALIF STATE UNIV, NORTHRIDGE, 66- Mem: AAAS; Am Math Soc; Math Asn Am; Soc Indust & Appl Math. Res: Control theory; calculus of variations and the mathematical theory of control processes. Mailing Add: Dept of Math Calif State Univ Northridge CA 91324

NATHENSON, STANLEY G, b Denver, Colo, Aug 1, 33; m 59; c 2. IMMUNOBIOLOGY, IMMUNOCHEMISTRY. Educ: Reed Col, BA, 55; Washington Univ, MD, 59. Prof Exp: From asst prof to assoc prof microbiol & immunol, 66-73, PROF CELL BIOL, MICROBIOL & IMMUNOL, ALBERT EINSTEIN COL MED, 73- Concurrent Pos: Nat Found fel pharmacol, Washington Univ, 60-62; Helen Hay Whitney Found fel, Queen Victoria Hosp, Sussex, Eng, 64-67; mem, Nat Cancer Inst Immunobiol Contract Study Sect. Mem: Am Asn Immunologists; Transplantation Soc. Res: Immunogenetics of transplantation antigens; immunochemistry and biochemistry of mammalian cell membranes; cellular regulation and biosynthetic mechanisms of membrane macromolecules. Mailing Add: Dept of Microbiol & Immunol Albert Einstein Col of Med Bronx NY 10461

NATHER, ROY EDWARD, b Helena, Mont, Sept 23, 26; m 62; c 3. ASTRONOMY, PHYSICS. Educ: Whitman Col, BA, 49; Univ Cape Town, PhD(astron), 72. Prof Exp: Physicist, Hanford Works, Gen Elec Co, 47-51, Calif Res & Develop Corp, 51-53, Tracerlab, Inc, 53-56 & Gen Atomics Div, Gen Dynamics Corp, 56-60; programmer comput div, Royal McBee, Inc, 60-61 & Packard Bell Comput Co, 61-62; tech dir, Sharp Lab, Beckman Inst, 62-67; spec res assoc astron, 67-73, ASSOC PROF ASTRON, UNIV TEX, AUSTIN, 73- Res: Application of high speed electronic techniques to study of short timescale astronomical phenomena. Mailing Add: Dept of Astron Univ of Tex Austin TX 78712

NATION, JAMES LAMAR, b Webster Co, Miss, Mar 3, 36; m 59; c 3. INSECT PHYSIOLOGY. Educ: Miss State Univ, BS, 57; Cornell Univ, PhD(entom), 60. Prof Exp: Assoc prof biol sci, 60-70, assoc prof entom & nematol, 70-72, PROF ENTOM & NEMATOL, UNIV FLA, 72- Concurrent Pos: NSF grants, 61-65; mem staff, Univ Guelph, 69-70; USDA coop grant, 70-72. Mem: AAAS; Entom Soc Am. Res: Nitrogen metabolism and purine excretion in insects; insect nutrition; sex attractants in insects. Mailing Add: Dept of Entom & Nematol Univ of Fla Gainesville FL 32611

NATION, JOHN, b Bridgwater, Eng, Aug 8, 35; m 61; c 2. PHYSICS, ELECTRICAL ENGINEERING. Educ: Univ London, BSc & ARCS, 57, DIC & PhD(plasma physics), 60. Prof Exp: Consult plasma physics, Nat Comt for Nuclear Energy, Rome, Italy, 60-62; staff scientist, Cen Elec Generating Bd, Eng, 62-65; asst prof, 65-71, ASSOC PROF PLASMA PHYSICS & ELEC ENG, CORNELL UNIV, 71-, ASST DIR, LAB PLASMA STUDIES, 75- Concurrent Pos: Lectr, Chelsea Col Sci & Technol, Eng, 64-65; sr vis fel, Sci Res Coun, London, Eng, 73-76. Mem: Am Phys Soc. Res: Plasma physics; relativistic electron beams; collective ion accelerators; high power microwave generation. Mailing Add: 212 Phillips Hall Cornell Univ Ithaca NY 14853

NATIONS, CLAUDE, b Marlow, Okla, July 9, 29; m 53; c 4. CELL BIOLOGY. Educ: Univ Okla, BS, 53; Okla State Univ, MS, 58, PhD(bot), 67. Prof Exp: Teacher, Okla City Bd Educ, Okla, 55-57; teacher, Wichita Bd Educ, Kans, 58-62; consult sci educ, Mo State Dept Educ, 63-64; instr biol, bot & plant physiol, Okla State Univ, 64-66; asst prof bot, Univ Tex, Arlington, 66-68; ASST PROF BIOL, SOUTHERN METHODIST UNIV, 68- Concurrent Pos: Fel, Lab Cancer Res, Univ Wis, 72-73. Mem: Am Soc Cell Biol. Res: Gene regulation by nuclear proteins. Mailing Add: Dept of Biol Southern Methodist Univ Dallas TX 75275

NATIONS, JACK DALE, b Prairie Grove, Ark, Oct 18, 34; m 57; c 4. PALEOECOLOGY. Educ: Ariz State Univ, BS, 56; Univ Ariz, MS, 62; Univ Calif, Berkeley, PhD(paleont), 69. Prof Exp: Geologist, Stand Oil Co Tex, 61-64; ASST PROF GEOL, NORTHERN ARIZ UNIV, 69- Concurrent Pos: Grant paleoecol & biostratig of Cenozoic Basins, 75-76. Mem: Am Asn Petrol Geol; Paleont Soc; Soc Econ Paleont & Mineral; Int Paleont Union. Res: Paleoecology and biostratigraphy of freshwater mollusks; systematics, phylogeny and paleobiogeography of fossil crabs. Mailing Add: Dept of Geol Northern Ariz Univ Flagstaff AZ 86001

NATOLI, SALVATORE JOHN, b Reading, Pa, Sept 12, 29. GEOGRAPHY. Educ: Kutztown State Col, BS, 51; Clark Univ, AM, 57, PhD(geog), 67. Prof Exp: Teacher soc sci, Chesapeake City High Sch, Md, 53-55; assoc prof geog, Mansfield State Col, 57-63; vis lectr, Clark Univ, 63-65; vis lectr, Univ Conn, 66; prog mgr, Div Educ Personnel Develop, US Off Educ, 66-68, dep chief trainers of teacher trainers br, 68-69; EDUC AFFAIRS DIR, ASN AM GEOGR, 69- Concurrent Pos: Consult, Soc Sci Inst Br, Div Educ Personnel Training, US Off Educ, 66-67, Environ Educ Prog, 72, mem basic studies task force, Nat Ctr Improv Educ Systs, 72-; consult, Sch Col Coop Prog, NSF, 71. Mem: Fel AAAS; Asn Am Geogr; fel Am Geog Soc; Nat Coun Geog Educ; Nat Coun Social Studies. Res: Urban geography; geography teaching and learning; environmental issues and problems. Mailing Add: Asn of Am Geogr 1710 16th St NW Washington DC 20009

NATOWITZ, JOSEPH BERNARD, b Saranac Lake, NY, Dec 24, 36; m 61; c 2. NUCLEAR CHEMISTRY. Educ: Univ Fla, BS, 58; Univ Pittsburgh, PhD(chem), 65. Prof Exp: Asst chem, Univ Pittsburgh, 61-65; res assoc, State Univ NY Stony Brook, 65-67; asst prof, 67-72, ASSOC PROF CHEM, TEX A&M UNIV, 72- Concurrent Pos: Res collabr, Brookhaven Nat Lab, 65-67. Mem: Am Chem Soc; Chem Inst Can; Am Phys Soc. Res: Nuclear reaction studies; fission, light fragment emission, angular momentum effects. Mailing Add: Dept of Chem Tex A&M Univ College Station TX 77840

NATOWSKY, SHELDON, b Brooklyn, NY, May 6, 44; m 68; c 2. INDUSTRIAL ORGANIC CHEMISTRY. Educ: Brooklyn Col, BS, 66; Cornell Univ, PhD(phys org chem), 73. Prof Exp: Res specialist surfactants, GAF Corp, 73-75; DIR RES & DEVELOP LABS, CARSON CHEM INC, DIV QUAD CHEM CORP, 75- Mem: Am Chem Soc; Am Oil Chemists Soc; Sigma Xi. Res: Synthesis and product-process development of surfactants and specialty organic chemicals; surfactant formulation. Mailing Add: 3130 W 187th St Torrance CA 90504

NATT, MICHAEL PHILIP, b New York, NY, Sept 30, 25; m 55; c 4. INFORMATION SCIENCE, COMMUNICATION SCIENCE. Educ: Univ Wis, BA, 47, MA, 48, PhD(zool, biochem), 51. Prof Exp: Res asst, Univ Wis, 48-51; res assoc, Hektoen Inst Med Res, Cook County Hosp, 51-52 & Trop Res Found Chicago & Mex, 52; chief sect parasitol, Eaton Labs Inc Div, Norwich Pharmacal Co, 52-59, sr info scientist, 59-60; sr info scientist, Ciba Pharmaceut Prod, Inc, NJ, 60-64; MGR SCI INFO SECT, WYETH LABS, AM HOME PROD CORP, 64- Mem: AAAS; Am Soc Parasitol; Am Soc Trop Med & Hyg; Am Asn Vet Parasitol; Am Soc Info Sci. Res: Biochemistry, physiology and chemotherapy of parasites; manual and machine documentation; storage and retrieval of scientific information; administration; data processing; data cost analysis; information management. Mailing Add: Sci Info Sect Wyeth Labs Radnor PA 19087

NATTIE, EUGENE EDWARD, b Alexandria, Va, June 15, 44; m 70. PULMONARY PHYSIOLOGY. Educ: Dartmouth Col, BA, 66, BMS, 68; Harvard Univ, MD, 71. Prof Exp: Intern med, Peter Bent Brigham Hosp, 71-72; fel, 72-75, ASST PROF PHYSIOL, SCH MED, DARTMOUTH COL, 75- Concurrent Pos: Young Pulmonary investr, NIH, 75. Res: The relationship of brain intra and extra-cellular acid-base and electrolyte balance and the control of breathing with special emphasis on metabolic acid base disturbances and potassium. Mailing Add: Dept of Physiol Dartmouth Med Sch Hanover NH 03755

NATZKE, ROGER PAUL, b Greenleaf, Wis, June 15, 39; m 64; c 2. DAIRY SCIENCE, PHYSIOLOGY. Educ: Univ Wis, BS, 62, MS, 63, PhD(dairy sci), 66. Prof Exp: Asst prof, 66-71, ASSOC PROF DAIRY SCI, CORNELL UNIV, 71- Mem: Am Dairy Sci Asn; Nat Mastitis Coun; Int Asn Milk, Food & Environ Sanit. Res: Factors affecting screening tests; effect of sanitary practices on mastitis; free stall management factors. Mailing Add: Dept of Animal Sci Cornell Univ Ithaca NY 16850

NAU, CARL AUGUST, b Yorktown, Tex, June 25, 03; m 25; c 2. PREVENTIVE MEDICINE. Educ: Univ Tex, BA, 23, MA, 24; Rush Med Col, MD, 34. Prof Exp: Asst zool, Univ Tex, 22, from asst chem to asst prof, 23-27, asst prof physiol, Sch Med, 27-29; from asst prof to assoc prof, Sch Med, Univ Okla, 29-36; dir div indust hyg, State Dept Health, Tex, 36-41; prof physiol, prev med & pub health, Sch Med, Univ Tex, 41-43, prof prev med & pub health, Med Br, 43-61; prof prev med & pub health & dir inst environ health, 61-73, EMER PROF PREV MED & PUB HEALTH, COL MED, UNIV OKLA, 73-; PROF CLIN TOXICOL & CONSULT TO DEAN, SCH MED, TEX TECH UNIV, 74- Concurrent Pos: Mem, Occup Med & Indust Hyg Adv Comt to Surgeon Gen. Mem: Fel Am Pub Health Asn; fel Indust Med Asn; Am Indust Hyg Asn; Am Conf Govt Indust Hygienists. Res: Occupational medicine; toxicology; public health. Mailing Add: Box A Coleman Hall Tex Tech Univ Lubbock TX 79406

NAU, RICHARD WILLIAM, b Lakefield, Minn, May 28, 41; m 63; c 2. MATHEMATICS. Educ: SDak Sch Mines & Technol, BS, 63, MS, 65; Univ Va, PhD(appl math), 70. Prof Exp: Res assoc, Boeing Co, 65-66; asst prof math, Clarkson Col, 66-67; actg asst prof comput sci, Univ Va, 70; ASST PROF MATH & COMPUT SCI, CARLETON COL, 70- Mem: Math Asn Am; Asn Comput Mach; Soc Indust & Appl Math. Res: Asymptotic methods; optimization methods; computer graphics. Mailing Add: Dept of Math Carleton Col Northfield MN 55057

NAUENBERG, MICHAEL, b Berlin, Ger, Dec 19, 34; US citizen; m 69; c 2. ELEMENTARY PARTICLE PHYSICS. Educ: Mass Inst Technol, BS, 55; Cornell Univ, PhD(physics), 59. Prof Exp: Asst prof physics, Columbia Univ, 61-64; vis res prof, Stanford Univ, 64-66; PROF PHYSICS, UNIV CALIF, SANTA CRUZ, 66- Concurrent Pos: Consult, Brookhaven Nat Lab, 62-64 & NSF; Guggenheim fel, 63-64; A P Sloan fel, 64-66; NSF grant, 66- Mem: Am Phys Soc. Res: Elementary particles and their interactions; astrophysics. Mailing Add: Div of Natural Sci Univ of Calif Santa Cruz CA 95060

NAUGHTEN, JOHN CHARLES, b Chicago, Ill, Jan 29, 42; m 66. ZOOLOGY, EMBRYOLOGY. Educ: Univ Chicago, AB, 64; Univ Iowa, MS, 68, PhD(zool), 71. Prof Exp: Res assoc neuroembryol, Univ Iowa, 71-72; ASST PROF BIOL, UNIV WIS-EAU CLAIRE, 72- Mem: AAAS; Am Inst Biol Sci; Am Soc Zool; Soc Develop Biol. Res: Vertebrate embryology, tissue interactions and differentiation, neuroembryology, metamorphosis. Mailing Add: Dept of Biol Univ of Wis Eau Claire WI 54701

NAUGHTON, JOHN JOSEPH, b New York, NY, June 5, 12; m 40; c 4. CHEMISTRY. Educ: City Col New York, BS, 36; NY Univ, MS, 40, PhD(phys chem), 42. Prof Exp: Chem engr, Rockwood Chocolate Co, NY, 36; teacher chem, Washington Sq Col, NY Univ, 37-40; res chemist, Tagliabue Instrument Co, NY, 40-41; instr chem, Union Col, 41-43; res chemist, Res Labs, Gen Elec Co, NY, 42-47; from asst prof to assoc prof chem, 47-56, chmn dept, 55-59, PROF CHEM, UNIV HAWAII, 56- Concurrent Pos: Vis res assoc, Princeton Univ, 50-51, 53-54; vis chemist, Brookhaven Nat Lab, 60-61 & Oak Ridge Nat Lab, 67. Mem: AAAS; Am Chem Soc; Geochem Soc. Res: Geochemistry; chemical volcanology; geochronology; natural stable and radioactive isotopes; igneous rock and gas equilibria. Mailing Add: Dept of Chem Univ of Hawaii Honolulu HI 96822

NAUGHTON, JOHN PATRICK, b Nanticoke, Pa, May 20, 33; c 6. INTERNAL MEDICINE, CARDIOLOGY. Educ: St Louis Univ, BS, 54; Univ Okla, MD, 58. Prof Exp: Intern, George Washington Univ Hosp, 58-59; resident med, Univ Okla Hosp, 59-61, clin asst, Med Ctr, 61-63, instr, 63-66, asst prof med & physiol, 66-68; assoc prof med & coordr rehab ctr, Univ Ill Col Med, 68-70; prof med & dir rehab med, Hosp & Med Ctr, George Washington Univ, 70-75; DEAN, SCH MED, STATE UNIV NY BUFFALO, 75- Concurrent Pos: Psychosom cardviovasc trainee, Med Ctr, Univ Okla, 61-62, chief resident internal med, 62-63, cardiovasc trainee, 63-64; attend physician, Vet Admin Hosp, Oklahoma City, 63-68; consult, Civil Aeromed Inst, Fed Aviation Agency, 64- & Cent State Hosp, Norman, Okla, 65-68; Dean, Acad Affairs, George Washington Univ, 70-75. Mem: Am Fedn Clin Res: Am Psychosom Soc; Soc Exp Biol & Med; fel Am Col Cardiol; fel Am Col Physicians. Res: Exercise physiology; coronary artery disease. Mailing Add: Sch of Med State Univ of NY Buffalo NY 14214

NAUGHTON, MICHAEL A, b UK, June 23, 26; US citizen; c 2. BIOPHYSICS, OBSTETRICS & GYNECOLOGY. Educ: Univ St Andrews, BSc, 52; Cambridge Univ, PhD, 59. Prof Exp: Res asst, Cambridge Univ, 54-56; res assoc, Mass Inst Technol, 59-62; assoc prof biophys, Johns Hopkins Univ, 62-67; sr prin res scientist, Div Animal Genetics, Commonwealth Sci, Indust & Res Orgn, 67-70; PROF BIOPHYS & OBSTET & GYNEC, SCH MED, UNIV COLO, DENVER, 70- Concurrent Pos: NIH career develop award, 64. Mem: Brit Biochem Soc; Australian Biochem Soc; Am Soc Biol Chemists. Res: Molecular biology and immunology of proteins. Mailing Add: Div of Perinatal Med Box 2302 Univ of Colo Med Ctr Denver CO 80220

NAUGLE, DONALD, b Wetumpka, Okla, Apr 23, 36; m 58; c 1. LOW TEMPERATURE PHYSICS. Educ: Rice Univ, BA, 58; Tex A&M Univ, PhD(physics), 65. Prof Exp: Res fel physics, Tex A&M Univ, 65-66; res assoc, Univ Md, College Park, 67-69; asst prof, 69-75, ASSOC PROF PHYSICS, TEX A&M UNIV, 75- Concurrent Pos: NATO fel, Univ Göttingen, 66-67. Mem: Am Phys Soc. Res: Ultrasonics; fluid transport properties; superconductivity. Mailing Add: Dept of Physics Tex A&M Univ College Station TX 77843

NAUGLE, JOHN EARL, b Belle Fourche, SDak, Feb 9, 23; m 45; c 3. PHYSICS. Educ: Univ Minn, BS, 49, MS, 50, PhD(physics), 53. Prof Exp: Sr staff scientist, Convair Sci Res Lab, 56-59; head nuclear emulsion sect, Goddard Space Flight Ctr, 59-60, head energetic particles prog, satellite & sounding rocket progs, Off Space Flight Progs, 60-61, chief physics, geophys & astron progs, Off Space Sci, 61-62, dir physics & astron progs, Off Space Sci & Appln, 62-66, from dep assoc adminr to assoc adminr space sci & appln, 66-72, ASSOC ADMINR SPACE SCI, NASA, 72- Mem: Am Phys Soc; Am Inst Aeronaut & Astronaut; Am Geophys Union. Res: Cosmic rays; high energy physics; trapped radiation. Mailing Add: 7211 Rollingwood Dr Chevy Chase MD 20015

NAUGLE, NORMAN WAKEFIELD, b Saginaw, Tex, Jan 9, 31; m 55; c 2. NUMERICAL ANALYSIS, APPLIED MATHEMATICS. Educ: Tex A&M Univ, AB, 53, MS, 58, PhD(physics), 65. Prof Exp: Asst math, Tex A&M Univ, 55-57 from instr to asst prof, 57-64; mathematician, Manned Spacecraft Ctr, NASA, 64-68; ASSOC PROF MATH, TEX A&M UNIV, 68- Concurrent Pos: Teacher, Allen Mil Acad, 57-59 & Alvin Jr Col, 65-67; consult, Appl Res Corp, Tex, 68-72. Mem: Math Asn Am; Soc Indust & Appl Math. Res: Numerical analysis; digital picture data processing; photoclinometry; molecular structure calculations for vibrational and rotational analysis. Mailing Add: Dept of Math Tex A&M Univ College Station TX 77843

NAULT, LOWELL RAYMOND, b San Francisco, Calif, Apr 4, 40; m 63; c 2. ENTOMOLOGY. Educ: Univ Calif, Davis, BS, 62; Cornell Univ, MS, 64, PhD(entom), 66. Prof Exp: From asst prof entom to assoc prof entom & plant path, 66-75, PROF ENTOM & PLANT PATH, OHIO AGR RES & DEVELOP CTR, 75- Mem: Entom Soc Am; Am Phytopath Soc; AAAS. Res: Interrelationships among arthropods, plant pathogens and plants, especially on vectors of corn viruses and spiroplasmas; chemical ecology of aphids; phagastimulants, pheromones. Mailing Add: Dept of Entom Ohio Agr Res & Develop Ctr Wooster OH 44691

NAUMAN, CHARLES HARTLEY, b Philadelphia, Pa, June 6, 37; m 68. BIOLOGY, GENETICS. Educ: Davis & Elkins Col, BS, 62; Univ Ark, Fayetteville, MS, 65; Northwestern Univ, Evanston, PhD(biol), 72. Prof Exp: Assoc radiobiol, Brookhaven Nat Lab, 65-68, tech collabr, Brookhaven Nat Lab-Columbia Univ, 71-72, asst biologist, Brookhaven Nat Lab, 73-74, ASSOC BIOLOGIST, BROOKHAVEN NAT LAB, 75- Mem: AAAS; Radiation Res Soc. Res: Mutagenesis by environmental chemical mutagens and ionizing radiation, and application of radiobiological techniques to problems in developmental cytology and genetics. Mailing Add: Biol Dept Brookhaven Nat Lab Upton NY 11973

NAUMAN, LOUIS WILLIAM, b Atlantic City, NJ, Jan 26, 27; m 57; c 2. CLINICAL CHEMISTRY, FORENSIC SCIENCE. Educ: St Joseph's Col, Pa, BS, 51; Agr & Tech Col NC, MS, 66; Univ Alaska, PhD(zool, chem), 71. Prof Exp: Res chemist, Eastern Regional Res Lab, USDA, Pa, 55-62; teacher high sch, Va, 63-65; res assoc protein chem, Inst Arctic Biol, Univ Alaska, 66-70; asst dir clin & crime labs, 70-73, DIR CLIN & CRIME LABS, ALASKA MED LABS, INC, 73- Mem: Am Soc Crime Lab Dirs; Am Chem Soc; Am Acad Forensic Sci. Res: Animal proteins; use of proteins as taxonomic markers; critical-noncritical substitutions of amino acids; development of analytical courses for applied areas in biology. Mailing Add: Alaska Med Labs Inc 207 E Northern Lights Blvd Anchorage AK 99503

NAUMAN, ROBERT KARL, b Allentown, Pa, Feb 26, 41; m 68; c 2. MICROBIOLOGY. Educ: Pa State Univ, BS, 63; Univ Mass, MS, 65, PhD(microbiol), 68. Prof Exp: Res asst microbiol, Univ Mass, 63-68; res assoc, Med Ctr, WVa Univ, 68-69, instr, 69-70; ASST PROF MICROBIOL, SCH DENT, UNIV MD, BALTIMORE, 70- Mem: Am Soc Microbiol; Electron Micros Soc Am. Res: General microbiology; microbial cytology, motility, ultrastructure and function; ultrastructure of mammalian tumors and tumor viruses. Mailing Add: Dept of Microbiol Univ of Md Sch of Dent Baltimore MD 21201

NAUMAN, ROBERT VINCENT, b East Stroudsburg, Pa, Dec 6, 23; m 55; c 4. PHYSICAL CHEMISTRY. Educ: Duke Univ, BS, 44; Univ Calif, PhD(chem), 47. Prof Exp: Res assoc chem, Cornell Univ, 47-52; asst prof, Univ Ark, 52-53; from asst prof to assoc prof, 53-63, PROF CHEM, LA STATE UNIV, BATON ROUGE, 63- Concurrent Pos: Fulbright-Hays lect award, Valparaiso Tech Univ Chile, 66-67; vis prof, Fac Sci, Univ Chile, 71-72. Mem: AAAS; Am Chem Soc; Am Phys Soc. Res: Molecular spectra; internal energy conversion; light scattering; sodium silicates; structure, size and shape of molecules; detergents. Mailing Add: Dept of Chem La State Univ Baton Rouge LA 70803

NAUMANN, ALFRED WAYNE, b Farmington, Iowa, May 8, 28; m 52; c 4. PHYSICAL CHEMISTRY. Educ: Grinnell Col, BA, 51; Iowa State Univ, PhD, 56. Prof Exp: Asst, Iowa State Univ, 51-56; assoc chemist, 56-71, SR RES SCIENTIST, UNION CARBIDE CORP, 71- Mem: AAAS; Am Chem Soc; Am Ceramic Soc. Res: Hydrometallurgical separation processes; surface and colloid chemistry; preparation and properties of high performance ceramics; heterogeneous catalysis. Mailing Add: 14 Shuart Rd Monsey NY 10952

NAUMANN, DOROTHY ETHEL, b Syracuse, NY, Mar 18, 15. MEDICINE. Educ: Syracuse Univ, MD, 40. Prof Exp: Indust physician, Gen Elec Co, 42-43; univ physician, Syracuse Univ, 44-50; physician, US Vet Admin, 51-52; assoc prof health & prev med, Syracuse Univ, 52-63; assoc prof prev med & assoc dir student health in chg women's div, 63-72, ASST PROF COMMUNITY HEALTH SCI, DUKE UNIV, 72-, DIR STUDENT HEALTH, 72- Concurrent Pos: Exam physician, YWCA, Schenectady, NY, 40-43 & Syracuse, 47-63. Mem: Fel AMA; Am Soc Microbiol; Am Acad Family Physicians. Res: Clinical research in student health service. Mailing Add: Box 6635 Col Sta Durham NC 27708

NAUMANN, HANS NORBERT, b Berlin, Ger, Feb 4, 01; US citizen; m 40; c 3. PATHOLOGY, PATHOLOGICAL CHEMISTRY. Educ: Univ Berlin, MD, 26; Am Bd Path, dipl & cert path, 42, cert path anat, 47. Prof Exp: Asst med, Long Island Col Med, 38-39; clin pathologist, Jewish Hosp, Louisville, Ky, 39-40; dir path labs, Taunton State Hosp, Mass, 45-48; chief lab serv, Vet Admin Ctr, Jackson, 48-57; chief biochem serv, Memphis Vet Admin Hosp, 57-69; ASST PROF MED, MED COL, UNIV TENN, MEMPHIS, 59- Concurrent Pos: Res grant, Emergency Comt Ger Sci, 30-33; res fel, Seamen's Hosp, London, 34-36 & Long Island Col Med, 36-37. Honors & Awards: Asn Clin Sci Dipl Hon, 60. Mem: AAAS; Am Chem Soc; Soc Exp Biol & Med; AMA; Am Soc Clin Path. Res: Optical rotation of blood filtrates; sensitive bilirubin test; urobilinogen in spinal fluid; post-mortem chemistry and diagnosis of diabetic and uremic coma; plasma hemoglobin determination; Bence Jones protein differentiation from uroglobulins. Mailing Add: Dept of Hemat Univ of Tenn ICI Bldg Memphis TN 38103

NAUMANN, HUGH DONALD, b Newport, Ark, Oct 26, 23; m 45; c 2. FOOD MICROBIOLOGY. Educ: Univ Mo, BS, 49, MS, 50, PhD(meat technol), 56. Prof Exp: From asst instr to instr animal husb, Univ Mo, 50-53; asst prof, Cornell Univ, 53-55; from asst prof to prof animal husb, 55-68, PROF FOOD SCI & NUTRIT, UNIV MO-COLUMBIA, 68-, CHMN DEPT, 75- Concurrent Pos: Fulbright sr res scholar, Commonwealth Sci & Indust Orgn, 65-66, Fulbright sr lectr, Gida Fermantasyon Teknolojisi Kurusu Ege Univ, Turkey, 74-75. Mem: Inst Food Technologists; Food Distrib Res Soc; Am Soc Animal Prod. Res: Meat technology; evaluation of quality attributes of meat, formulation of meat food products and processing meats, particularly the effects of sanitation, temperature environment, gas environment, and packaging upon the stability of fresh, frozen and cured meats. Mailing Add: 174 Agr Bldg Univ of Mo Columbia MO 65201

NAUMANN, ROBERT ALEXANDER, b Dresden, Ger, June 7, 29; nat US; m 61; c 2. PHYSICAL CHEMISTRY. Educ: Univ Calif, BS, 49; Princeton Univ, MA, 51, PhD(phys chem), 53. Prof Exp: From instr to assoc prof, 52-73, PROF CHEM & PHYSICS, PRINCETON UNIV, 73- Concurrent Pos: Procter & Gamble fac fel, 59-60. Mem: Fel AAAS; Am Chem Soc; fel Am Phys Soc; NY Acad Sci. Res: Radioactivity; inorganic chemistry; nuclear physics. Mailing Add: Dept of Physics Jadwin Hall Princeton Univ PO Box 708 Princeton NJ 08540

NAUNTON, RALPH FREDERICK, b London, Eng, Sept 26, 21; m; c 2. OTOLARYNGOLOGY. Educ: Univ London, MB, BS, 45. Prof Exp: Resident surg & dir audiol, Univ Col Hosp, Univ London, 47-50; Med Res Coun res fel, Cent Inst for Deaf, Mo, 51; sci officer, Med Res Coun Eng, 51-52; res asst, Cent Inst for Deaf, Mo, 52-53; sci officer, Med Res Coun Eng, 53-54; instr surg, 54-57, from asst prof to assoc prof otolaryngol, 57-64, PROF OTOLARYNGOL, SCH MED, UNIV CHICAGO, 64-, CHMN DEPT, 68- Concurrent Pos: Asst surgeon, St John's Hosp, London, 47-50; Univ London fel, Holland, 49; clin asst, Royal Nat Throat, Nose & Ear Hosp, 51-54. Mem: Am Speech & Hearing Asn; fel Royal Soc Med. Res: Otology. Mailing Add: 950 E 59th St Chicago IL 60637

NAUS, JOSEPH IRWIN, mathematical statistics, see 12th edition

NAUTA, WALLE J H, b Medan, Indonesia, June 8, 16; US citizen; m 42; c 3. ANATOMY. Educ: State Univ Utrecht, MD, 42, PhD(anat), 45. Prof Exp: Lectr anat, State Univ Utrecht, 42-46; assoc prof, State Univ Leiden, 46-47 & Univ Zurich, 47-51; neurophysiologist, Walter Reed Army Inst Res, 51-64; PROF NEUROANAT, MASS INST TECHNOL, 64- Concurrent Pos: Mem bd, Found Fund Res Psychiat, 59-62; mem res career develop awards comt, NIMH, 66-; mem Biol Stain Comn, 57- Honors & Awards: Karl Spencer Lashley Award, Am Philos Soc, 64. Mem: Nat Acad Sci; Am Philos Soc; Am Asn Anat; Am Acad Arts & Sci; Swiss Asn Anat. Res: Neuroanatomy; neurophysiology. Mailing Add: Dept of Psychol Mass Inst of Technol Cambridge MA 02139

NAUTIYAL, JAGDISH CHANDRA, b Garhwal, India, Oct 8, 32; m 60; c 2. FOREST ECONOMICS. Educ: DAV Col, Dehra Dun, BSc, 51; Indian Forest Col, AIFC, 54; Univ BC, MF, 65, PhD(forestry, econ), 67. Prof Exp: Asst conservator forests, Forest Dept, India, 54-60; instr forest mensuration & mgt, Northern Forest Rangers Col, Forest Res Inst, India, 60-63; div forest off, Soil Conserv Div, Forest Dept, India, 63-64, working plan off, EAlmora Forest Div, 67-68; asst prof, 68-69, ASSOC PROF FOREST ECON, UNIV TORONTO, 69- Mem: Can Inst Forestry; Am Econ Asn; Can Econ Asn; Can Oper Res Soc; Int Soc Trop Foresters. Res: Application of economic theory and linear programming in the management of forest resources including timber and recreation. Mailing Add: Fac of Forestry Univ of Toronto Toronto ON Can

NAVALKAR, RAM G, b Bombay, India, May 7, 24; m 66; c 2. MICROBIOLOGY, IMMUNOLOGY. Educ: St Xavier's Col, India, BSc, 46; Univ Bombay, PhD(microbiol), 56. Prof Exp: Res officer leprosy, Acworth Leprosy Hosp, Bombay, 58-60; proj assoc tuberc, Sch Med, Univ Wis, 60-63; res assoc, Gothenburg Univ, 64-65; proj assoc tuberc & leprosy, Sch Med, Univ Wis, 66-67; asst prof, 67-72, ASSOC PROF MED MICROBIOL, MEHARRY MED COL, 72- Concurrent Pos: Fel microbiol, Sch Med, Stanford Univ, 56-58; NIH grant, 67-; consult, Stanford Res Inst, 57-58. Mem: AAAS; Am Soc Microbiol; Int Leprosy Asn. Res: Antigenic studies on mycobacteria; study of Mycobacterium leprae by comparative analysis, using various immunochemical techniques; specific antigens of Mycobacterium leprae in relation to their biological activity; immune mechanisms in Mycobacterial infections. Mailing Add: Dept of Microbiol Meharry Med Col Nashville TN 37208

NAVAR, LUIS GABRIEL, b El Paso, Tex, Mar 24, 41; m 65; c 3. PHYSIOLOGY, BIOPHYSICS. Educ: Agr & Mech Col, Tex, BS, 62; Univ Miss, PhD(biophys, physiol), 66. Prof Exp: Instr physiol & biophys, Sch Med, Univ Miss, 66-67, from asst prof to assoc prof, 67-74; ASSOC PROF PHYSIOL & BIOPHYS, SCH MED, UNIV ALA, BIRMINGHAM, 75- Concurrent Pos: Nat Inst Arthritis & Metab Dis fel physiol & biophys, Med Ctr, Univ Miss, 66-69 & spec fel med, Duke Univ, 72; Nat Heart & Lung Inst res career develop award, 74; mem, Coun Kidney & Cardiovasc Dis, Am Heart Asn. Mem: Am Soc Nephrology; Am Heart Asn; Am Physiol Soc; NY Acad Sci; Sigma Xi. Res: Control of renal hemodynamics; regulation of sodium excretion; pathophysiology of high blood pressure; regulation of extracellular fluid volume. Mailing Add: Dept of Physiol & Biophys Univ of Ala Med Sch Birmingham AL 35294

NAVARRA, JOHN GABRIEL, b Bayonne, NJ, July 3, 27; m 47; c 2. EARTH SCIENCES, SCIENCE EDUCATION. Educ: Columbia Univ, AB, 49, MA, 50, EdD(sci educ), 54. Prof Exp: Assoc prof chem, physics & sci educ, ECarolina Univ, 54-58; prof sci & chmn dept, 58-68, PROF GEOSCI, JERSEY CITY STATE COL, 68-; DIR, LEARNING RESOURCES LABS, 67- Concurrent Pos: Consult, NC State Bd Educ Sci Curriculum Study, 58, State of Calif NDEA Workshops Strengthening Elem Sch Sci, 60-61 & State Adv Comt on Sci, 60-62; partic, White House Conf Children & Youth Golden Anniversary, 60; teacher educ study, Nat Asn State Dir Teacher Educ & Cert, 60-61; Am Inst Physics coordr vis scientists prog physics for high schs, NJ, 61-64; sci ed, Arabian-Am Oil Co, Saudi Arabia, 63. Mem: Am Geol Soc; AAAS. Res: Applications of chromatography; development and refinement of broad areas of science in elementary, junior high school and college curriculum. Mailing Add: Learning Resources Labs PO Box 647 Farmingdale NJ 07727

NAVARRO, JOSEPH ANTHONY, b New Britain, Conn, July 6, 27; m 51; c 3. MATHEMATICAL STATISTICS. Educ: Cent Conn State Col, BS, 50; Purdue Univ, MS, 52, PhD(math statist), 55. Prof Exp: Asst math, Purdue Univ, 51, math statist, 51-55; consult statistician, Gen Elec Co, 55-59; mem res staff, IBM Corp, 59-64; mem res staff, Inst Defense Anal, 64-70, asst dir systs eval div, 70-72; EXEC VPRES, SYST PLANNING CORP, 72- Mem: Sigma Xi; Opers Res Soc Am; Am Statist Asn. Res: Use of operations research, probability and mathematical statistics in the area of management decisions; systems analysis. Mailing Add: 7825 Fulbright Ct Bethesda MD 20034

NAVE, CARL R, b Newport, Ark, July 21, 39; m 62; c 2. PHYSICS. Educ: Ga Inst Technol, BEE, 61, MS, 64, PhD(physics), 66. Prof Exp: Fel physics, Univ Col NWales, 66-67; asst prof, 68-71, ASSOC PROF PHYSICS, GA STATE UNIV, 71- Mem: Am Phys Soc; AAAS; Audio Eng Soc; Am Asn Physics Teachers. Res: Determination of molecular structure and study of intramolecular interactions by microwave spectroscopy. Mailing Add: Dept of Physics Ga State Univ University Plaza Atlanta GA 30303

NAVE, FLOYD ROGER, b Moline, Ill, Oct 7, 25; m 49; c 4. GEOMORPHOLOGY, PALEONTOLOGY. Educ: Augustana Col, AB, 49; Univ Iowa, MS, 52; Ohio State Univ, PhD, 68. Prof Exp: Instr geol, Augustana Col, 51-52; geologist, Gen Petrol Corp, Calif, 52-53; asst prof geol, 53-61, assoc prof, 62-69, chmn dept, 64-74, PROF GEOL, WITTENBERG UNIV, 69- Concurrent Pos: NSF sci fac fel, 58-59. Mem: Geol Soc Am; Am Asn Petrol Geol; Nat Asn Geol Teachers; Sigma Xi. Res: Environmental geology. Mailing Add: Dept of Geol Wittenberg Univ Springfield OH 45501

NAVE, PAUL MICHAEL, b Lancaster, Pa, June 3, 43; m 65; c 2. ORGANIC CHEMISTRY. Educ: Memphis State Univ, BS, 65; Iowa State Univ, PhD(org chem), 69. Prof Exp: Asst prof, 69-74, ASSOC PROF CHEM, ARK STATE UNIV, 74- Mem: Am Chem Soc. Res: Metal ion oxidation of organic compounds; ligand transfer oxidation of free radicals. Mailing Add: Dept of Chem Ark State Univ State University AR 72467

NAVES, RENEE G, b Toulouse, France, Nov 18, 27. ORGANIC CHEMISTRY. Educ: Univ Grenoble, BS, 48, PhB, 49; Univ Geneva, MS, 52, PhD(biol), 55. Prof Exp: Instr, Univ Geneva, 52-55; res fel org chem, Mass Gen Hosp & Harvard Med Sch, 55-57; res asst, Univ Fla, 57-58; prof & chmn dept, Newton Col Sacred Heart, 58-72; researcher, Am Univ, 72-74; Am Chem Soc fel chem & pub affairs, serv in House of Rep, 74- Concurrent Pos: Res asst, Harvard Med Sch, 58-59; abstractor, Chem Abstr, 65-69; consult, Sci Curriculum in East Boston Schs, 71; vis scientist, Duke Univ, 74-75. Mem: AAAS; Am Chem Soc; NY Acad Sci. Res: Carbohydrate chemistry and synthesis, specifically examinations for determinations of carbohydrate content of microorganisms cell walls; carbohydrate research through use of nuclear magnetic resonance. Mailing Add: Comt on Sci & Technol Rayburn Off Bldg Rm 2319 Washington DC 20515

NAVIA, JUAN MARCELO, b Havana, Cuba, Jan 16, 27; m 50; c 4. NUTRITION. Educ: Mass Inst Technol, BS & MS, 50, PhD, 65. Prof Exp: Tech dir, Cuba Indust & Com Co, 50-52; assoc prof nutrit & food sci, Univ Villanueva, Cuba, 55-61; res assoc, Mass Inst Technol, 61-65, assoc prof nutrit biochem, 66-69; ASSOC PROF BIOCHEM, UNIV ALA, BIRMINGHAM, PROF DENT & COMP MED & SR SCIENTIST, INST DENT RES, 69-, DIR RES TRAINING, SCH DENT, 73- Concurrent Pos: Dir, FIM Nutrit Lab, 52-55; asst dir, Cuban Inst Tech Invest, 55-61. Mem: Am Chem Soc; Am Soc Microbiol; Am Inst Nutrit; Inst Food Technologists; Am Inst Chemists. Res: Nutritional biochemistry; mineral metabolism; oral biology. Mailing Add: Univ of Ala Sch of Dent University Station Birmingham AL 35294

NAVIDI, MARJORIE HANSON, inorganic chemistry, see 12th edition

NAVON, DAVID H, b New York, NY, Oct 28, 24; m 47; c 2. ELECTRONIC PHYSICS, MICROELECTRONICS. Educ: City Col New York, BEE, 47; NY Univ, MS, 50; Purdue Univ, PhD(physics), 53. Prof Exp: Instr physics, Mohawk Col, 47 & Queen's Univ, NY, 47-50; res assoc, Purdue Univ, 53-54; asst dir res, Transition Electronic Corp, 54-60, dir semiconductor res, 60-65; vis assoc prof elec eng, Mass Inst Technol, 65-68; PROF ELEC ENG, UNIV MASS, AMHERST, 68- Mem: Inst Elec & Electronics Eng; Am Vacuum Soc. Res: Solid state physics; semiconductor electronics. Mailing Add: Dept of Elec Eng Univ of Mass Amherst MA 01002

NAVRATIL, JAMES DALE, b Denver, Colo, Jan 20, 41; m 67; c 4. INDUSTRIAL CHEMISTRY. Educ: Univ Colo, Boulder, BA, 70, MSc, 72, PhD(anal chem), 75. Prof Exp: Anal lab technician anal chem, Dow Chem USA, 61-66, chem res & develop master technician, 66-68, sr chemist, 70-73, res chemist, 73-75; SR RES CHEMIST RES & DEVELOP, ROCKWELL INT, 75- Concurrent Pos: Res assoc, Univ Colo, Boulder, 75-76. Mem: AAAS; Am Chem Soc; Calorimetry Conf. Res: Chemical separations methods; ion exchange chromatography; liquid chromatography; solvent extraction research; chemical synthesis and characterization. Mailing Add: Rockwell Int Rocky Flats Plant PO Box 464 Golden CO 80401

NAVROTSKY, ALEXANDRA, b New York, NY, June 20, 43. CHEMISTRY. Educ: Univ Chicago, BS, 63, MS, 64, PhD(chem), 67. Prof Exp: Res assoc theoret metall, Clausthal Tech Univ, 67-68; res assoc geochem, Pa State Univ, 68-69; asst prof, 69-74, ASSOC PROF CHEM, ARIZ STATE UNIV, 74- Concurrent Pos: Alfred P Sloan Found fel, 73. Mem: Am Geophys Union; Mineral Soc Am; Am Ceramic Soc. Res: Thermodynamics; phase equilibria and high temperature calorimetry; oxides and oxide solid solutions; order-disorder; geochemistry; geothermal fluids. Mailing Add: Dept of Chem Ariz State Univ Tempe AZ 85281

NAWAR, TEWFIK, b Cairo, Egypt, Sept 6, 39; Can citizen. NEPHROLOGY. Educ: Einshams Univ, Cairo, MB, BCh, 63; McGill Univ, MSc, 72; Col Physicians & Surgeons Can, FRCP(C), 72. Prof Exp: Med Res Coun fel, Montreal Clin Res Inst, 70-72; ASST PROF MED, MED SCH, UNIV SHERBROOKE, 72- Concurrent Pos: Med Res Coun fel, Renal Div, Med Sch, Wash Univ, 71-72. Mem: Can Soc Nephrol; Am Soc Nephrol; Int Soc Nephrol. Res: Renal handling of adenosine 3',5' cyclic monophosphate in health and disease. Mailing Add: Univ Hosp Ctr Univ of Sherbrooke Sherbrooke PQ Can

NAWAR, WASSEF W, b Cairo, Egypt, May 17, 26; m 53; c 1. FOOD CHEMISTRY. Educ: Univ Cairo, BSc, 47, MS, 50; Univ Ill, PhD(food sci), 59. Prof Exp: Asst dairy tech, Univ Ill, 50-52 & 57-59; from asst prof to assoc prof, 59-70, PROF FOOD SCI, UNIV MASS, AMHERST, 70- Concurrent Pos: Res grants, Sigma Xi, 61, USPHS, 65-69 & AEC, 65-; mem comt food irradiation, Nat Acad Sci-Nat Res Coun, 74- Mem: Am Chem Soc; Am Oil Chem Soc; AAAS; Inst Food Technol. Res: Flavor chemistry; thermal decomposition of fats; effects of ionizing radiation on fats. Mailing Add: Dept of Food Sci Univ of Mass Amherst MA 01002

NAWN, GEORGE HENRY, organic chemistry, photographic science, see 12th edition

NAYAK, DEBI PROSAD, b West Bengal, India, Apr 1, 37; m 65; c 2. VIROLOGY, ONCOLOGY. Educ: Univ Calcutta, BVSc, 57; Univ Nebr, Lincoln, MS, 63, PhD(virol), 64. Prof Exp: Actg asst prof virol, 64-66, asst res virologist, 66-68, asst prof, 68-71, ASSOC PROF VIROL, UNIV CALIF, LOS ANGELES, 71- Concurrent Pos: Cancer Res Coord Comt & Calif Inst Cancer Res fels, Univ Calif, Los Angeles, 69, Calif Div Am Cancer Soc sr Dernham fel, 69-74, Am Cancer Soc res grant, 72-74, Nat Cancer Inst res grant, 74-, Nat Inst Allergy & Infectious Dis res grant, 75- Mem: AAAS; Am Soc Microbiol. Res: Influenza virus genome and its translation, transcription, and replication; role of viruses in oncogenesis; activation of endogenous viral genes; role of viruses in the induction of neoplasmas in humans. Mailing Add: Dept of Med Microbiol & Immunol Univ of Calif Ctr Health Sci Los Angeles CA 90024

NAYAK, RAMESH KADBET, b Udipi, India, Sept 6, 34; US citizen; m 58; c 2. REPRODUCTIVE PHYSIOLOGY. Educ: Univ Madras, BS, 54; Univ Bombay, MS, 56; Univ RI, MS, 64; Ore State Univ, PhD(physiol), 70. Prof Exp: Lectr zool, Inst Sci, Bombay, 56-61; res asst biol, Childrens Cancer Res Found, 64-65; fel, Univ Nebr, 70-72; res assoc anat, George Washington Univ, 72-75; ASSOC PROF ZOOL, KUWAIT UNIV, 75- Concurrent Pos: Scientist biol, Smithsonian Sci Info Exchange, 72-75. Mem: Electron Micros Soc Am; Am Soc Cell Biol; Am Soc Animal Sci; Am Inst Biol Sci; AAAS. Res: Electron microscopic studies of mammalian oviduct; effect of contraceptive steroids on the cardiovascular system. Mailing Add: 200 Beecher Ave Indian Head MD 20640

NAYAK, RAMNATH V, b Bombay, India, Dec 3, 31; Can citizen; m 57; c 2. MEDICINE, ENDOCRINOLOGY. Educ: Univ Bombay, MB, BS, 56; McGill Univ, PhD(exp med), 66. Prof Exp: Sir Edward Beatty mem scholar, McGill Univ, 67-68; ASST PROF MED, SCH MED, CREIGHTON UNIV, 68- Concurrent Pos: Vet Admin res grant, Omaha Vet Admin Hosp, 70-73. Mem: Endocrine Soc; Am Fedn Clin Res; NY Acad Sci. Res: Study of pituitary cytology in various species. Mailing Add: Dept of Med Creighton Univ Sch of Med Omaha NE 68131

NAYAR, JAI KRISHEN, b Kisumu, East Africa, Jan 3, 33; US citizen; m 64; c 3. INSECT PHYSIOLOGY, MEDICAL ENTOMOLOGY. Educ: Univ Delhi, BSc, 54, MSc, 56; Univ Ill, Urbana, PhD(entom), 62. Prof Exp: Sr res asst entom, Indian Agr Res Inst, Delhi, 56-58; MED ENTOMOLOGIST, DIV HEALTH, FLA MED ENTOM LAB, VERO BEACH, FLA, 63-; ADJ ASSOC PROF ENTOM & NEMATOL, UNIV FLA, GAINESVILLE, 75- Concurrent Pos: Nat Res Coun Can award, Univ Man, 62-63. Mem: Entom Soc Am; Am Soc Trop Med & Hyg. Res: Biology and physiology of mosquitoes of medical and veterinary importance. Mailing Add: Fla Med Entom Lab PO Box 520 Vero Beach FL 32960

NAYFEH, SHIHADEH NASRI, b Merj 'Youn, Lebanon. BIOCHEMISTRY, ENDOCRINOLOGY. Educ: Am Univ Beirut, BS & teaching dipl, 59, MS, 61; Univ NC, Chapel Hill, PhD(biochem), 64. Prof Exp: Investr biochem, Lebanese Agr Res Inst, 65-67 & Univ Pa, 67-68; asst prof biochem & endocrinol & dir endocrinol lab, NC Mem Hosp, 68-72, ASSOC PROF BIOCHEM, NUTRIT & PEDIAT, SCH MED, UNIV NC, CHAPEL HILL, 72- Concurrent Pos: Fel, Harvard Univ, 64-65. Mem: AAAS; Soc Study Reproduction; Endocrine Soc; NY Acad Sci; Am Soc Biol Chemists. Res: Mechanism of action of steroid hormones in target tissue; metabolism of steroid hormones in normal and pseudohermaphrodite male rats; effects of androgens on polyribosomal integrity of rat ventral prostate. Mailing Add: Dept of Pediat Univ of NC Sch of Med Chapel Hill NC 27514

NAYLOR, ALFRED F, b South River, NJ, Oct 17, 27; m 50; c 3. GENETICS. Educ: Univ Chicago, AB, 50, PhD(zool), 57. Prof Exp: Asst, Univ Chicago, 51-54 & 55-57; asst prof zool, Univ Okla, 57-60; asst prof genetics, McGill Univ, 60-64; GENETICIST, NAT INST NEUROL DIS & STROKE, 64- Mem: Am Soc Human Genetics. Res: Population and human genetics; biometry; population ecology. Mailing Add: Perinatal Res Br Nat Inst Neurol Commun Dis & Stroke Bethesda MD 20014

NAYLOR, AUBREY WILLARD, b Union City, Tenn, Feb 5, 15; m 40; c 2. PLANT PHYSIOLOGY. Educ: Univ Chicago, BS, 37, MS, 38, PhD(bot), 40. Prof Exp: Rockefeller asst, Univ Chicago, 38; mem staff, Bur Plant Indust, USDA, 38-40; instr bot, Univ Chicago, 40-44, naval radio, 42-44; instr bot, Northwestern Univ, 44-45; Nat Res Coun fel biol sci, Boyce Thompson Inst, 45-46; asst prof plant physiol, Univ Wash, 46-47 & Yale Univ, 47-52; from assoc prof to prof, 52-72, JAMES B DUKE PROF PLANT PHYSIOL, DUKE UNIV, 72- Concurrent Pos: Res partic & consult, Oak Ridge Inst Nuclear Studies, 54-64; consult, Oak Ridge Nat Lab, 57-58, NSF, 60-65, Res Triangle Inst, 68-, Biol Div, Tenn Valley Authority, 69-75 & Educ Testing Serv, 72-; NSF sr fel & vis prof, Univ Bristol, 58-59, prog dir metab biol, NSF, 61-62; mem, Comn Undergrad Educ in Biol Sci, chmn panel interdisciplinary activ; chmn comt examr, Grad Rec Exam Biol, 66-72; mem bd, Southeastern Plant Environ Labs, 68- Mem: Fel AAAS; Bot Soc Am; Am Soc Plant Physiol (secy, 53-55, vpres, 56, pres, 61); Soc Exp Biol & Med; Scand Soc Plant Physiol. Res: Photophysiology; growth regulation; enzymes; amino acid metabolism. Mailing Add: Dept of Bot Duke Univ Durham NC 27706

NAYLOR, BENJAMIN FRANKLIN, b Gilroy, Calif, Nov 15, 17; m 46; c 3. CHEMISTRY. Educ: San Jose State Col, AB, 40; Stanford Univ, MA & PhD(phys chem), 43. Prof Exp: Phys chemist, US Bur Mines, Calif, 43-45; chemist, Stand Oil Co Calif, 45; from instr to assoc prof chem, 45-61, head dept, 50-61, PROF CHEM, SAN JOSE STATE COL, 61-, COORDR GEN CHEM, 61- Mem: Am Chem Soc. Res: Chemical thermodynamics; high-temperature heat contents of titanium carbide and titanium nitride; specific heats of metals at high temperatures by AC-DC method. Mailing Add: Dept of Chem San Jose State Col San Jose CA 95114

NAYLOR, CARTER GRAHAM, b Denver, Colo, May 22, 42; m 64; c 2. PETROLEUM CHEMISTRY. Educ: Calif Inst Technol, BS, 64; Univ Colo, PhD(org chem), 69. Prof Exp: Res chemist, 69-70, SR RES CHEMIST, JEFFERSON CHEM CO, 70- Mem: Am Chem Soc; Am Oil Chemists Soc. Res: Petrochemicals, especially surfactants. Mailing Add: Jefferson Chem Co Box 4128 Austin TX 78765

NAYLOR, CHARLES MARTIN, mathematics, see 12th edition

NAYLOR, DENNY VE, b Twin Falls, Idaho, Oct 26, 37; m 59; c 2. SOIL CHEMISTRY. Educ: Univ Idaho, BS, 59, MS, 61; Univ Calif, Berkeley, PhD(soil sci), 66. Prof Exp: ASSOC PROF SOILS, UNIV IDAHO, 66- Mem: AAAS; Am Soc Agron; Soil Sci Soc Am. Res: Nutrients in soil-water systems; water quality and agricultural practices; soil-pesticide relationships. Mailing Add: Dept of Plant & Soil Sci Univ of Idaho Moscow ID 83843

NAYLOR, DEREK, b Eng, Nov 9, 29; m 60; c 2. APPLIED MATHEMATICS. Educ: Univ London, BSc, 51, PhD(aerodyn), 53. Prof Exp: Res assoc, Brown Univ, 53-54; aerodynamicist, A V Roe & Co, 54-56; asst prof appl math, Univ Toronto, 56-62; sr lectr, Royal Col Sci, Glasgow, 62-63; assoc prof, 63-65, PROF APPL MATH, UNIV WESTERN ONT, 65- Mem: London Math Soc; fel Brit Inst Math & Appln. Res: Aerodynamics; integral transforms. Mailing Add: Dept of Appl Math Univ of Western Ont London ON Can

NAYLOR, ERNST E, b Shelbina, Mo, Jan 3, 99. BOTANY. Educ: Univ Mo, BS, 22, AM, 24, PhD(morphol), 31. Prof Exp: Asst prof bot, Univ Mo, 31-42; from tech asst to assoc cur ed, NY Bot Garden, 42-48; prof bot, 48-75, EMER PROF SCI, UPPER IOWA UNIV, 75- Mem: AAAS; Bot Soc Am; Torrey Bot Club. Res: Morphology of plants. Mailing Add: Dept of Bot Upper Iowa Univ Fayette IA 52142

NAYLOR, FLOYD EDMOND, b Becker, NMex, Feb 9, 22; m 52; c 4. POLYMER CHEMISTRY. Educ: Wash State Col, BS, 51; Univ Md, PhD(org chem), 56. Prof Exp: RES CHEMIST & GROUP LEADER, PHILLIPS PETROL CO, 55- Mem: Am Chem Soc. Res: Chemistry of high polymers; rubber; plastics. Mailing Add: Res & Develop Dept Phillips Petrol Co Bartlesville OK 74003

NAYLOR, GERALD WAYNE, b Keener, Ala, Feb 15, 22; m 44; c 3. AGRONOMY, PLANT PHYSIOLOGY. Educ: Auburn Univ, BS, 47, MS, 49; NC State Univ, PhD(agron), 53; Southeastern Baptist Theol Sem, BD, 56. Prof Exp: Minister & hosp chaplain, NC Baptist Hosp, 53-60, res fel, 60-61; assoc prof, 61-71, PROF BIOL, CARSON-NEWMAN COL, 71- Concurrent Pos: Partic, NSF res prog, Univ Tex, 64-66 & acad year exten, 64-66, 67-69; proj dir, NSF Student Sci Training Prog, 72- Mem: AAAS; Am Soc Plant Physiol. Res: Population density effects of various species of pollen grains and diffusable compounds which regulate or influence pollen germination and tube growth. Mailing Add: Dept of Biol Carson-Newman Col Jefferson City TN 37760

NAYLOR, HARRY BROOKS, b Minn, Mar 30, 14; m 40; c 3. BACTERIOLOGY. Educ: Univ Minn, BS, 38; Cornell Univ, PhD(bact), 43. Prof Exp: Dairy chemist, Sheffield Farms Co, 46-47; prof dairy indust, 47-50, PROF BACT, CORNELL UNIV, 50- Concurrent Pos: Fulbright-Hays lectureship, Univ Alexandria, 66-67; Orgn Am States lectureship, Univ Campinas, Brazil, 72, 73, sabbatical res leave, Univ Campinas, 75. Mem: Am Soc Microbiol; Am Acad Microbiol; Inst Food Technologists. Res: Bacterial physiology; virology. Mailing Add: 413 Stocking Hall Cornell Univ Ithaca NY 14850

NAYLOR, HARRY LEE, applied anthropology, see 12th edition

NAYLOR, JAMES MAURICE, b Hawarden, Sask, Feb 22, 20; nat US; m 43; c 2. BOTANY. Educ: Univ Wis, PhD(bot), 53. Prof Exp: From asst prof to assoc prof field husb, 53-64, prof biol, 64-74, RAWSON PROF BIOL, UNIV SASK, 74-, HEAD DEPT, 68- Res: Dormancy in seeds and buds; action of endogenous plant growth regulators; fine structure in relation to function in plant cells. Mailing Add: Dept of Biol Univ of Sask Saskatoon SK Can

NAYLOR, JASPER ROSS, b Buffalo, NY, May 7, 08; m 28; c 2. MATHEMATICS. Educ: Cent State Col, BS, 32; Univ Okla, AM, 37. Prof Exp: High sch teacher, Okla, 28-41; dean men, 41-48, bus mgr, 48-50, dean students, 50-57, dir admis, 57-74, PROF MATH, EASTERN NAZARENE COL, 57- Concurrent Pos: Dean, Kiowa County Jr Col, Okla, 34-41. Res: Synthetic geometry. Mailing Add: Dept of Math Eastern Nazarene Col Quincy MA 02170

NAYLOR, MARCUS A, JR, b Oberlin, Ohio, Apr 27, 20; m 43; c 2. ORGANIC CHEMISTRY. Educ: Col Wooster, BA, 42; Johns Hopkins Univ, MA, 43, PhD(org chem), 45. Prof Exp: Lab instr chem, Johns Hopkins Univ, 42-44; res chemist, Plastics Dept, 44-50, supvr res div, Polychem Dept, 50-55, com investr, Planning Div, 55-56, sect mgr res div, 56-58, dir gen prod, res & develop, 58-59, asst dir res div, Indust & Biochem Dept, 59-67, asst dir indust chem sales div, 67-72, DIR LABS, INDUST CHEM DEPT, E I DU PONT DE NEMOURS & CO, INC, 73- Mem: Am Chem Soc. Res: Research administration; organic synthesis; reaction mechanisms; chemistry of high polymers. Mailing Add: Indust Chem Dept E I du Pont de Nemours & Co Inc Wilmington DE 19898

NAYLOR, RICHARD STEVENS, b Lakeland, Fla, July 15, 39. GEOLOGY, GEOCHEMISTRY. Educ: Mass Inst Technol, BS, 61; Calif Inst Technol, PhD(geol), 67. Prof Exp: Asst prof geol, Mass Inst Technol, 67-74; ASSOC PROF EARTH SCI & CHMN DEPT, NORTHEASTERN UNIV, 74- Mem: AAAS; Geol Soc Am; Am Geophys Union; Geochem Soc. Res: Geology and geochronology of northern Appalachian Mountain system; geology and geochronology of mantled gneiss domes. Mailing Add: Dept of Earth Sci 103GR Northeastern Univ Boston MA 02115

NAYLOR, ROBERT ERNEST, JR, b Nashville, Ark, July 14, 32; m 63; c 1. CHEMICAL PHYSICS. Educ: Univ SC, BS, 51; Harvard Univ MA, 54, PhD(chem physics), 56. Prof Exp: Res chemist, Film Dept, 56-62, res supvr, 62-64, res mgr, 64-66, tech supt, 66-67, lab dir, 67-71, tech mgr, 71-74, PROD & TECH DIR, FILM DEPT, E I DU PONT DE NEMOURS & CO, INC, 74- Res: Polymer science. Mailing Add: Film Dept E I du Pont de Nemours & Co Inc Wilmington DE 19898

NAYMIK, DANIEL ALLAN, b Lorain, Ohio, Mar 8, 22; m 44. MATHEMATICS, QUANTUM PHYSICS. Educ: Univ Mich, BS, 47, MS, 48, PhD(physics), 58. Prof Exp: Mem tech staff semiconductor adv develop, Bell Tel Labs, NJ, 58-64; mem sr staff physics of thin films, Gen Dynamics/Electronics, 64-66, prin engr, Comput Sci Dept, 66-69; SR RES SCIENTIST, COMPUT RES DEPT, AMOCO PROD CO, 69- Mem: Am Phys Soc; Asn Comput Mach. Res: Computer and management sciences; thin film physics; high energy and mathematical physics; quantum mechanics; electron scattering; semiconductor device development; computer graphics systems. Mailing Add: Res Ctr Amoco Prod Co Tulsa OK 74102

NAYUDU, Y RAMMOHANROY, b Masulipatam-Andhr, India, Jan 13, 22; US citizen; m 43; c 3. MARINE GEOLOGY, PETROLOGY. Educ: Univ Bombay, BS, 45, MSc, 47; Univ Wash, PhD(geol), 59. Prof Exp: Lectr geol, Univ Rangoon, 51-53; geologist, Burma Geol Dept, Ministry Mines, 54-55; res instr geol oceanog, Univ Wash, 59-61; res asst geologist, NSF fel, Scripps Inst, Univ Calif, 61-63; res asst prof marine geol, Univ Wash, 63-65, res assoc prof, 65-68; dep dir, Inst Marine Sci, Alaska, 68-69; prof marine geol, Univ Alaska, 69-71; SCI ADV TO GOV, ALASKA, 70-; DIR DIV MARINE & COASTAL ZONE MGT, ALASKA DEPT ENVIRON CONSERV, 71- Concurrent Pos: NSF grants, 59-63, 65 & 66-67. Mem: Am Asn Petrol Geologists; Soc Econ Paleontologists & Mineralogists; Geol Soc Am; fel Brit Geol Asn; Int Asn Sedimentol. Res: Deep sea sediments and submarine volcanics. Mailing Add: Div of Marine & Coastal Zone Mgt Dept of Environ Conserv Box 323 Juneau AK 99802

NAYYAR, RAJINDER, b Khanna, India, June 14, 36; m 69; c 1. NEUROLOGY, MICROSCOPIC ANATOMY. Educ: Panjab Univ, India, BSc, 57, MSc, 59; Univ Delhi, PhD(zool), 64. Prof Exp: Asst lectr zool, Univ Delhi, 63-64, chmn dept zool, H R Col, 64-65; ASST PROF NEUROL & ANAT, MED SCH, NORTHWESTERN UNIV, CHICAGO, 67-; RES BIOLOGIST & ELECTRON MICROSCOPIST, NEUROL SERV, VET ADMIN HOSP, 67- Concurrent Pos: Joseph P Kennedy fel anat, Univ Western Ont, 65-66, Med Res Coun fel, 66-67. Mem: Electron Micros Soc Am; Am Asn Anat; Am Soc Cell Biol; Histochem Soc. Res: Fish chromosome studies; histochemistry of fish oocytes; sex-chromatin studies; histochemistry of the diabetic retina; electron microscopy and histochemistry; the effect of antibiotics on the brain; cytogenesis of lysosomes; electromagnetic autoradiography; chemically induced myelopathy; preputial gland secretion; effect of castration. Mailing Add: Lakeside Vet Admin Hosp 333 E Huron St Chicago IL 60611

NAZARETH, JOHN LAWRENCE, b Nairobi, Kenya. NUMERICAL ANALYSIS, COMPUTER SCIENCE. Educ: Univ Cambridge, BA, 67, dipl comput sci, 68; Univ Calif, Berkeley, MS & PhD(comput sci), 73. Prof Exp: Fel, 73-74, ASST COMPUT SCIENTIST, APPL MATH DIV, ARGONNE NAT LAB, 74- Mem: Math Prog Soc; Soc Inst Appl Math; Asn Comput Mach. Res: Mathematical programming and numerical analysis, particularly the invention and analysis of optimization algorithms and the development of mathematical software for solving optimization problems. Mailing Add: Appl Math Div Argonne Nat Lab 9700 S Cass Ave Argonne IL 60439

NAZARIAN, GIRAIR MIHRAN, b Trenton, NJ, Sept 5, 26; m 52; c 4. PHYSICAL CHEMISTRY. Educ: Rutgers Univ, BS, 46, MS, 48; Calif Inst Technol, PhD(chem, physics), 57. Prof Exp: Asst chem, Rutgers Univ, 46-47, Cornell Univ, 47-48 & Calif Inst Technol, 48-49; sr engr, Physics Lab, Sylvania Elec Prod, Inc, 51-55; res fel chem & lectr statist mech, Calif Inst Technol, 57-58; mem tech staff, Res Lab, Thompson Ramo Wooldridge, Inc, 58-61; assoc prof, 61-65, PROF CHEM, CALIF STATE UNIV, NORTHRIDGE, 65- Concurrent Pos: Instr, Univ Southern Calif, 58-61 & exten, Univ Calif, Los Angeles, 59-60; consult, Space Technol Labs, 61-62. Res: Statistical mechanics of liquid-vapor interface, molecular friction constant and gas adsorption; luminescence phenomena; electron diffraction calculations; ultracentrifuge; ionic propulsion; radiolytic gas in nuclear reactors; protein diffusion through liquid-liquid interfaces. Mailing Add: Dept of Chem Calif State Univ Northridge CA 91324

NAZAROFF, GEORGE VASILY, b San Francisco, Calif, Apr 12, 38; m 63; c 1. THEORETICAL CHEMISTRY. Educ: Univ Calif, Berkeley, BS, 59; Univ Wis-Madison, PhD(chem), 65. Prof Exp: NSF fel, 65-66; asst prof chem, Col Natural Sci, Mich State Univ, 66-72; ASSOC PROF CHEM & CHMN DEPT, IND UNIV, SOUTH BEND, 72- Concurrent Pos: Res Corp starter grant, 66- Mem: Am Chem Soc. Res: Perturbation theory; generalized Hartree-Fock formalisms; natural spin orbitals; resonant scattering; electron-diatomic collision theory. Mailing Add: Dept of Chem Ind Univ South Bend IN 46615

NAZERIAN, KEYVAN, b Tehran, Iran, Dec 21, 34; m 59; c 3. VIROLOGY, ELECTRON MICROSCOPY. Educ: Univ Tehran, DVM, 58; Mich State Univ, MS, 60, PhD(virol), 65. Prof Exp: Asst virol, Mich State Univ, 59-60; vis scientist, Pub Health Inst, Padua, 60-62; asst virol, Mich State Univ, 63-65; head electron micros lab, South Jersey Med Res Found, 65-66; MICROBIOLOGIST, REGIONAL POULTRY RES LAB, USDA, 66- Mem: AAAS; Am Soc Microbiol; Electron Micros Soc Am; NY Acad Sci. Res: Biochemical, biophysical and morphological studies of animal viruses, particularly oncogenic viruses and their interaction with susceptible hosts. Mailing Add: Regional Poultry Res Lab USDA 3606 E Mt Hope Rd East Lansing MI 48823

NAZY, JOHN ROBERT, b Alamosa, Colo, July 12, 33; m 56; c 5. INDUSTRIAL ORGANIC CHEMISTRY. Educ: Regis Col, Colo, BS, 54; Northwestern Univ, PhD(org chem), 59. Prof Exp: Chemist, Union Carbide Chem Co, 58-60; proj leader, 60-66, tech supvr, 66-69, sect leader, Tech Serv Dept, 69-73, TECH DIR WATER SOLUBLE POLYMERS, GEN MILLS CHEM, INC, 73- Mem: Tech Asn Pulp & Paper Indust; Am Chem Soc. Res: Organoboron and fatty acid chemistry; polymer synthesis; guar and natural gums. Mailing Add: Gen Mills Chem Inc 2010 E Hennepin Ave Minneapolis MN 55413

NEAGLE, LYLE H, b Mutual, Okla, Nov 6, 31; m 64; c 2. ANIMAL NUTRITION, BIOCHEMISTRY. Educ: Okla State Univ, BS, 53; Iowa State Univ, PhD(animal nutrit), 60. Prof Exp: Asst dir animal nutrit & res, Supersweet Feeds Div, Int Milling Co, Minn, 60-67; mgr res, 67-72, DIR RES, ALLIED MILLS, INC, 72- Mem: Am Soc Animal Sci. Mailing Add: Allied Mills Inc PO Box 459 Libertyville IL 60048

NEAL, ARTHUR HOMER, organic chemistry, see 12th edition

NEAL, ARTHUR LESLIE, b Belmont, Wis, May 3, 11; m 36; c 4. BIOCHEMISTRY. Educ: Monmouth Col, BS, 34; Univ Ill, MS, 35; Univ Wis, PhD(biochem), 43. Prof Exp: Res chemist, Continental Can Co, Ill, 35-37; instr chem, Kans State Col, 37-40; asst biochem, Univ Wis, 40-43; from asst prof to assoc prof, Exp Sta, Mich State Col, 43-46; ASSOC PROF BIOCHEM, CORNELL UNIV, 47- Concurrent Pos: Guggenheim fel, 55; vis prof, Inst Microbiol, Rutgers Univ, 62. Mem: Am Chem Soc; NY Acad Sci. Res: Metabolic inhibitors; biochemical relationship between plant pathogenic organisms and their hosts. Mailing Add: Sect of Biochem & Molecular Biol Cornell Univ Ithaca NY 14850

NEAL, BOBBY J, zoology, see 12th edition

NEAL, HOMER ALFRED, b Franklin, Ky, June 13, 42; m 62; c 2. EXPERIMENTAL HIGH ENERGY PHYSICS. Educ: Ind Univ, BS, 61; Univ Mich, MS, 63, PhD(physics), 66. Prof Exp: NSF fel, Europ Orgn Nuclear Res, 66-67; from asst prof to assoc prof, 67-72, PROF PHYSICS, IND UNIV, BLOOMINGTON, 72- Concurrent Pos: Alfred P Sloan Found fel, Ind Univ, Bloomington, 68-; chmn zero gradient synchrotron accelerator users orgn, mem zero gradient synchrotron prog comt, 70-72; mem bd trustees, Argonne Univs Asn, 71-74. Mem: Fel Am Phys Soc. Res: Application of spark chamber and scintillation counter techniques in the study of elementary particle interactions. Mailing Add: Dept of Physics Ind Univ Bloomington IN 47401

NEAL, JACK LAURANCE, b London, Eng, Jan 18, 27; Can citizen; m 53; c 2. BIOPHYSICS. Educ: McGill Univ, BSc, 49, PhD(phys chem), 69; Yale Univ, MS, 54. Prof Exp: Res chemist, Int Cellulose Res Ltd, Can, 54-62, sr res chemist, 62-69; ASST PROF BIOPHYS, McGILL UNIV, 69-, DIR DEPT, MONTREAL CHILDREN'S HOSP, 72- Mem: Chem Inst Can. Res: Cystic fibrosis. Mailing Add: Biophys Dept Montreal Children's Hosp Montreal PQ Can

NEAL, JAMES THOMAS, b Detroit, Mich, Feb 9, 36; m 60; c 4. GEOLOGY. Educ: Mich State Univ, BS, 57, MS, 59. Prof Exp: Geologist, Can Cliffs Ltd, 57 & Albanel Minerals Ltd, 59; US Air Force, 60-, proj officer geol res, Air Force Cambridge Res Labs, 60-63, proj scientist, 63-66, chief geotech br, 66-68, from instr to assoc prof geog, US Air Force Acad, 68-73, CHIEF GROUND SHOCK & CRATERING, CIVIL ENG RES DIV, US AIR FORCE WEAPONS LAB, US AIR FORCE, 73- Honors & Awards: Outstanding Res & Develop Award, US Air Force, 66. Mem: Sigma Xi; fel Geol Soc Am. Res: Engineering and military geology; geology of playas; remote sensing. Mailing Add: 712 Raton Ave SE Albuquerque NM 87123

NEAL, JOHN ALEXANDER, b Aliquippa, Pa, Aug 7, 40; m 64; c 2. INORGANIC CHEMISTRY. Educ: Eastern Wash State Col, BA, 66; Univ Wash, PhD(inorg chem), 70. Prof Exp: Res assoc chem, Wash State Univ, 71-72; RES CHEMIST, GA-PAC CORP, 72- Mem: Am Chem Soc. Res: Factors which influence formation and stability of transition metal complexes in polydentate systems; sterochemistry of polydentate complexes; oxidative degradation of metal complexes. Mailing Add: Ga-Pac Corp Bellingham WA 98225

NEAL, JOHN LLOYD, JR, b Concordia, Kans, Oct 18, 37; m 62; c 2. MICROBIOLOGY, SOIL SCIENCE. Educ: Ore State Univ, BSc, 60, MSc, 63, PhD(soil microbiol), 67. Prof Exp: Res asst microbiol, Ore State Univ, 60-67; RES SCIENTIST, CAN AGR RES STA, 67- Mem: Can Soc Microbiol; Can Soc Soil Sci. Res: Interrelationship between plant roots and soil microorganisms; soil biochemistry as related to soil fertility; microbial transformations in soil affecting soil fertility. Mailing Add: Can Agr Res Sta Lethbridge AB Can

NEAL, JOHN WILLIAM, JR, b St Louis, Mo, Nov 17, 37. ENTOMOLOGY. Educ: Univ Mo-Columbia, BS, 61, MS, 64; Univ Md, College Park, PhD(entom), 70. Prof Exp: Mus aid & entomologist, Dept Mammals (Iran), US Nat Mus, 63-65; fac res asst alfalfa weevil, Univ Md, College Park, 66-68, instr biol control, 68-70; RES ENTOMOLOGIST, APPL PLANT GENETICS LAB, PLANT GENETICS & GERMPLASM INST, AGR RES SERV, USDA, 70- Mem: Entom Soc Am. Res: Biological and chemical control pests of forage crops; screening pesticides and synthetic insect hormones for biological activity. Mailing Add: Plant Genetics & Germplasm Inst Agr Res Serv USDA Bldg 467-C Beltsville MD 20705

NEAL, LOUISE ADELAIDE, b Caldwell, Kans, Nov 5, 08. BIOLOGY, SCIENCE EDUCATION. Educ: Friends Univ, AB, 36; Univ Colo, MA, 41; Univ Northern Colo, EdD(sci & elem educ), 57. Prof Exp: Teacher & prin pub schs, Kans, 25-46; prof, 46-75, EMER PROF ELEM SCI, UNIV NORTHERN COLO, 75- Concurrent Pos: Asst, Columbia Univ, 50; dir, NSF Elem Sci Inst, 63; mem feasibility conf, NSF-AAAS. Mem: Fel AAAS; Nat Asn Biol Teachers; Nat Sci Teachers Asn; Nat Asn Res Sci Teaching. Res: Development of critical thinking and scientific inquiry abilities of children; laboratory experience; use of quantitative procedures for the study of science; preparation of teachers of science education for children; inquiry learning adapted to pupils' learning styles and sequential levels of achievement. Mailing Add: Dept of Elem Sci Univ of Northern Colo Greeley CO 80631

NEAL, MARCUS PINSON, JR, b Columbia, Mo, Apr 22, 27; m 61; c 3. RADIOLOGY. Educ: Univ Mo, AB, 49, BS, 51; Univ Tenn, MD, 53; Am Bd Radiol, cert radiol, 58 & radiol in nuclear med, 59. Prof Exp: Intern, Med Col Va Hosp, 53-54; res assoc path, Sch Med, Univ Mo, 54; resident radiol, Univ Wis Hosps, 54-57, from instr to asst prof, Sch Med, 57-63; assoc prof radiol, 63-66, chmn, Div Diag Radiol, 65-71, asst dean health sci div & dir, Regional Med Progs, 68-71, dir housestaff educ & dir continuing educ med & grad educ med, 69-71, interim dean sch med, 71, asst vpres health sci, 71-73, PROF RADIOL, MED COL VA, VA COMMONWEALTH UNIV, 66-, PROVOST HEALTH SCI, 73- Concurrent Pos: Radiologist, Cent Wis Colony, Madison, 59-63 & Vet Admin Hosp, 61-63; consult, Wis Diag Ctr, 61-63, US Air Force Hosp, Truax Field, 63 & Vet Admin Hosp, Richmond, Va, 63-; pres, Va Coun Health & Med Care, 70-74. Mem: Am Col Radiol; AMA; Radiol Soc NAm; Brit Inst Radiol. Res: Diagnostic radiology in medical research. Mailing Add: Med Col of Va Va Commonwealth Univ Richmond VA 23219

NEAL, OLIVER MEADER, JR, b North Berwick, Maine, Oct 23, 15; m 45; c 7. HORTICULTURE. Educ: Univ Maine, BS, 38; Mich State Univ, PhD(hort), 56. Prof Exp: Asst, Mich State Univ, 38-41; from instr to assoc prof hort, 42-63, from asst horticulturist to assoc horticulturist, 51-63, PROF HORT & HORTICULTURIST, WVA UNIV, 63- Mem: Am Soc Hort Sci. Res: Propagation and breeding of lilies, azaleas and hollies; transpiration studies. Mailing Add: Div of Plant Sci WVa Univ Morgantown WV 26506

NEAL, RICHARD ALLAN, b Waverly, Iowa, July 27, 39; m 62; c 3. FISHERIES. Educ: Iowa State Univ, BS, 61, MS, 62; Univ Wash, PhD(invert fishery biol), 67. Prof Exp: SUPVRY FISHERY BIOLOGIST, NAT MARINE FISHERIES SERV, 66- Mem: AAAS; Am Fisheries Soc; Wildlife Soc; Ecol Soc Am; Biomet Soc. Res: Freshwater fishery biology; ecological studies of paralytic shellfish poisoning; fishery population dynamics; penaeid shrimp culture. Mailing Add: Galveston Biol Lab Nat Marine Fisheries Serv Galveston TX 77550

NEAL, RICHARD B, b Lawrenceburg, Tenn, Sept 5, 17; m 44; c 2. PHYSICS. Educ: US Naval Acad, BS, 39; Stanford Univ, PhD(physics), 53. Prof Exp: Field serv engr, Sperry Gyroscope Co, 41-42, fire control serv supt, 42-46, res engr, 46-47; res assoc, 50-62, ASSOC DIR TECH DIV, LINEAR ACCELERATOR CTR, STANFORD UNIV, 62- Mem: Am Phys Soc. Res: Microwave and accelerator physics; high energy linear electron accelerators. Mailing Add: 1351 N Lemon Ave Menlo Park CA 94025

NEAL, ROBERT A, b Casper, Wyo, Apr 21, 28; m 58; c 3. TOXICOLOGY, BIOCHEMISTRY. Educ: Univ Denver, BS, 49; Vanderbilt Univ, PhD(biochem), 63. Prof Exp: From asst prof to assoc prof, 64-75, PROF BIOCHEM, SCH MED, VANDERBILT UNIV, 75-, DIR, CTR ENVIRON TOXICOL, 73- Concurrent Pos: NIH res fel toxicol, Univ Chicago, 63-64; mem, Food Protection Comt & Toxciol Subcomt, Nat Acad Sci; mem, Toxicol Study Sect & Prog Comt Multiple Factors Causation Environ Induced Dis, NIH. Mem: AAAS; Am Soc Pharmacol & Exp Therapeut; Soc Toxicol; Am Inst Nutrit; Am Asn Biol Chemists. Res: Natural product chemistry; isolation and identification of natural products; detoxification mechanisms. Mailing Add: Dept of Biochem Vanderbilt Univ Sch of Med Nashville TN 37203

NEAL, SCOTTY RAY, b Redlands, Calif, July 12, 37; m 58; c 2. APPLIED MATHEMATICS, TELECOMMUNICATIONS. Educ: Univ Calif, Riverside, BA, 61, MA, 63, PhD(math), 65. Prof Exp: Res mathematician, US Naval Weapons Ctr, China Lake, Calif, 64-67; mem tech staff, 67-73, SUPVR, TRAFFIC RES GROUP, BELL TEL LABS, 73- Mem: Am Math Soc; Oper Res Soc. Res: Optimal design strategy for stochastic networks. Mailing Add: Bell Tel Labs Crawford Corner Rd Holmdel NJ 07733

NEAL, THOMAS EDWARD, b Royal Oak, Mich, May 2, 42; m 67; c 2. ANALYTICAL CHEMISTRY, TEXTILE CHEMISTRY. Educ: Univ Mich, BS, 64; Univ NC, PhD(anal chem), 70. Prof Exp: Res asst inorg chem, Univ Mich, 64-65; res chemist, 70-73, SR RES CHEMIST, TEXTILE FIBERS DEPT, E I DU PONT DE NEMOURS & CO, INC, 73- Res: Electroanalytical chemistry in non-aqueous systems; end use research of synthetic polymer products in textile applications. Mailing Add: Textile Fibers Dept E I du Pont de Nemours & Co Inc Wilmington DE 19898

NEAL, VICTOR THOMAS, b Dell Rapids, SDak, Nov 1, 24; m 48; c 2. PHYSICAL OCEANOGRAPHY. Educ: Univ Notre Dame, BS, 48; Univ NDak, MEd, 54; Ore State Univ, PhD(phys oceanog), 65. Prof Exp: Geophysicist, Carter Oil Co, 48-49; teacher, Various Sec Schs & Jr Cols, 50-62; instr phys oceanog, Ore State Univ, 64; asst prof, US Naval Postgrad Sch, 64-66; asst prof, 66-74, ASSOC PROF OCEANOG, ORE STATE UNIV, 74-, ASST CHMN INSTRNL PROGS, 70- Mem: Am Geophys Union; Am Soc Oceanog. Res: Estuarine, coastal and arctic oceanography. Mailing Add: Dept of Oceanog Ore State Univ Corvallis OR 97331

NEAL, WILLIAM JOSEPH, b Princeton, Ind, Nov 19, 39; m 59; c 3. SEDIMENTARY PETROLOGY. Educ: Univ Notre Dame, BS, 61; Univ Miss, MA, 64, PhD(geol), 68. Prof Exp: Fel geol, McMaster Univ, 67-68; asst prof, 71-73, ASSOC PROF GEOL, GRAND VALLEY STATE COL, 73-, CHMN DEPT, 75- Concurrent Pos: Adj prof, Skidaway Inst Oceanog, Ga, 69-71. Mem: Am Asn Petrol Geol; Soc Econ Paleont & Mineral; Int Asn Sedimentologists; Nat Asn Geol Teachers. Res: Heavy minerals; recent and Pleistocene deep-sea sediments; carbonate petrology; Pennsylvanian cyclothems; ancient turbidites. Mailing Add: Dept of Geol Grand Valley State Col Allendale MI 49401

NEALE, CLAUDE LINWOOD, b Saluda, Va, Dec 4, 02; m 29; c 2. PSYCHIATRY, NEUROLOGY. Educ: Univ Richmond, BS, 24; Med Col Va, MD, 28; Am Bd Psychiat & Neurol, dipl, Prof Exp: Intern, Walter Reed Gen Hosp, 28-29; med officer, US Army, 29-43; from asst prof to assoc prof psychiat, 43-69, PROF PSYCHIAT, MED COL VA, 69-, DIR INPATIENT PSYCHIAT SERV, HOSP, 59- Mem: Fel Am Psychiat Asn; AMA; Asn Mil Surgeons US. Mailing Add: 3900 Seminary Ave Richmond VA 23227

NEALE, ELAINE ANNE, b Philadelphia, Pa, May 20, 44; m 67; c 2. ELECTRON MICROSCOPY, NEUROCYTOLOGY. Educ: Rosemont Col, AB, 65; Georgetown Univ, PhD(biol), 69. Prof Exp: Asst res neuromorphologist, Ment Health Res Inst, Univ Mich, Ann Arbor, 70-75; NEUROCYTOLOGIST, BEHAV BIOL BR, NEUROBIOL SECT, NAT INST CHILD HEALTH & HUMAN DEVELOP, NIH, 75- Concurrent Pos: NIH staff fel, Nat Cancer Inst, 69-70 & Nat Inst Child Health & Human Develop, 73-; NIH fel, Univ Mich, Ann Arbor, 70-73. Mem: Am Soc Cell Biol; Am Asn Women Sci; Sigma Xi. Res: Structure-function relationships in the nervous system; techniques for the ultrastructural localization of specific macromolecules; ultrastructural anatomy. Mailing Add: Nat Insts of Health Bldg 36 Rm 2A-21 Bethesda MD 20014

NEALE, ERNEST RICHARD WARD, b Montreal, Que, July 3, 23; m 50; c 2. GEOLOGY. Educ: McGill Univ, BSc, 49; Yale Univ, MS, 50, PhD(geol), 52. Prof Exp: Asst, Yale Univ, 51-52; asst prof, Rochester Univ, 52-54; geologist, Geol Surv Can, 54-60, 65-67, head Pre-Cambrian shield sect, 67-68; PROF GEOL & HEAD DEPT, MEM UNIV NEWF, 68- Concurrent Pos: Field geologist, Que Dept Mines, 47-53; actg head, Appalachian Sect, Geol Surv Can 59-60, head, 60-62; Brit Commonwealth Geol Liaison Off, Eng, 63-65; ed, Can J of Earth Sci, 74-; vis prof, Univ BC, 74-75. Honors & Awards: Bancroft Award, Royal Soc Can, 75. Mem: Am Geol Soc; Royal Soc Can; Geol Asn Can (pres, 72); Can Inst Mining & Metall; Can Geosci Coun, (vpres, 72-73, 74-75, pres, 75-76). Res: Appalachian geology and mineral resources; Canadian science policy. Mailing Add: Dept of Geol Mem Univ of Newfoundland St John's NF Can

NEALE, JOSEPH HICKMAN, biochemistry, see 12th edition

NEALE, ROBERT S, b Abington, Pa, Mar 19, 36; m 57; c 3. ORGANIC CHEMISTRY. Educ: Amherst Col, AB, 57; Univ Ill, PhD(org chem), 61. Prof Exp: Org chemist, Union Carbide Res Inst, 60-67, RES SCIENTIST, UNION CARBIDE CHEM & PLASTICS, 67- Mem: Am Chem Soc. Res: Chemistry of nitrogen free radicals, especially those derived from N-Halo compounds; chemistry of hydroperoxide oxidations; synthesis of organosilicon compounds; organosilicon chemistry. Mailing Add: Union Carbide Chem & Plastics Tarrytown Tech Ctr Tarrytown NY 10591

NEALEN, JOSEPH PETER, b Lyons, Ore, July 13, 08. PHYSICS. Educ: Gonzaga Univ, MSc, 34. Prof Exp: Chmn dept physics, Seattle Univ, 35-37; chmn dept, 44-64, prof, 64-73, EMER PROF PHYSICS, GONZAGA UNIV, 73- Concurrent Pos: Physicist, Isotope Comt, Sacred Heart Hosp, 52-66; mem, Wash State Tech Adv Bd Radiation Control, 66- Mem: Soc Nuclear Med; Am Asn Physics Teachers. Res: Nuclear instrumentation; radio communication. Mailing Add: Dept of Physics Gonzaga Univ Spokane WA 99202

NEALEY, RICHARD H, b Lawrence, Mass, May 30, 36; m 60; c 4. ORGANIC CHEMISTRY. Educ: Merrimack Col, BSc, 57; Univ Conn, MSc, 59; Brown Univ, PhD(chem), 63. Prof Exp: Res chemist, Ethyl Corp, Mich, 62-63; sr res chemist, Monsanto Res Corp, 63-68; mgr org chem res, Tech Opers Inc, 68-69; TECH AREA MGR, RES LAB DIV, XEROX CORP, 69- Mem: Am Chem Soc. Res: Organometallic chemistry; heterocyclic synthesis; photo- sensitizing dyes; physical properties and drug action; synthesis of novel imaging materials. Mailing Add: Xerox Corp 800 Phillips Rd W 114 Webster NY 14580

NEALON, THOMAS F, JR, b Jessup, Pa, Feb 24, 20; m 46; c 4. SURGERY, THORACIC SURGERY. Educ: Scranton Univ, BS, 41; Jefferson Med Col, MD, 44. Prof Exp: Am Cancer Soc fel surg, Jefferson Med Col, 51-53; from instr to prof, 55-68; PROF SURG, NY UNIV, 68-; DIR SURG, ST VINCENT'S HOSP & MED CTR, NEW YORK, 68- Concurrent Pos: Consult, Greenwich Hosp, Conn, Holy Name Hosp, Teaneck, NJ, St Agnes Hosp, White Plains, NY & St Vincent's Med Ctr of Richmond, Staten Island. Mem: Am Col Chest Physicians; Am Surg Asn; Am Asn Thoracic Surg; Am Col Surg; Am Soc Artificial Internal Organs. Res: Cancer; cardiorespiratory physiology during operations; gastrointestinal surgery. Mailing Add: Dept of Surg St Vincent's Hosp & Med Ctr New York NY 10011

NEALSON, KENNETH HENRY, b Iowa City, Iowa, Oct 8, 43. MARINE MICROBIOLOGY. Educ: Univ Chicago, BS, 65, MS, 66, PhD(microbiol), 69. Prof Exp: NIH fel, Harvard Univ, 69-71; asst prof biol, Univ Mass, Boston, 71-73; ASST PROF MARINE BIOL, SCRIPPS INST OCEANOG, 73- Mem: AAAS; Am Soc Microbiol. Res: Physiology, biochemistry and genetics of luminous bacteria; study of the symbiotic relationship between luminous bacteria and marine luminous fishes. Mailing Add: Scripps Inst of Oceanog A-002 La Jolla CA 92037

NEALY, CARSON LOUIS, b Natchitoches, La, Dec 24, 38; m; c 1. ANALYTICAL CHEMISTRY. Educ: Northwestern State Col, La, BS, 60; Fla State Univ, MS, 63, PhD(nuclear & inorg chem), 65. Prof Exp: Nuclear chemist, Shell Develop Co, Houston, 65-70, anal chemist, Shell Oil Co, 70-72; MGR ANAL CHEM ATOMICS INT DIV, ROCKWELL INT CORP, 72- Mem: Am Chem Soc; Sigma Xi. Res: Nuclear reaction spectroscopy in nuclear structure studies; neutron activation analysis; gas chromatography; instrument development; nuclear fuel analysis; nuclear methods in analytical chemistry. Mailing Add: Atomics Int Div Rockwell Int Corp 8900 Desoto Ave Canoga Park CA 91304

NEALY, DAVID LEWIS, b Monticello, NY, June 29, 36; m 62; c 2. ORGANIC CHEMISTRY. Educ: Duke Univ, BS, 58; Cornell Univ, PhD(org chem), 63. Prof Exp: NSF fel org chem, Mass Inst Technol, 63-64; res chemist, 64-65, sr res chemist, 65-69, res assoc, 70, head phys & anal chem div, 71-73, SUPT, FIBER DEVELOP DIV, TENN EASTMAN CO, 74- Mem: Am Asn Textile Technologists; Am Chem Soc. Res: Organic polymer chemistry; polymer morphology; chemical kinetics. Mailing Add: Res Labs Tenn Eastman Co Kingsport TN 37662

NEARN, WILLIAM THOMAS, b Middletown, NY, Dec 2, 20; m 46; c 4. FOREST PRODUCTS. Educ: State Univ NY, BS, 43; Yale Univ, MF, 47, DF, 54. Prof Exp: From instr to assoc prof, Pa State Univ, 47-60; prof specialist, 60-73, MGR WOOD SCI & MORPHOL, WEYERHAEUSER CO, 73-, MGR INSTRUMENTATION & TEST, 74- Mem: Soc Wood Sci & Technol. Res: Wood-moisture relations; wood adhesives; composite wood products; scanning and transmission electron microscopy. Mailing Add: Res Div Weyerhaeuser Co Tacoma WA 98401

NEARY, EDWARD R, b Jersey City, NY, Oct 13, 11; m 38; c 2. MEDICINE. Educ: Seton Hall Univ, AB, 32; Jefferson Med Col, MD, 37. Prof Exp: Med dir, White Labs, Inc, 39-71; DIR FOOD & DRUG ADMIN LIAISON, SCHERING CORP, 71- Mem: AMA; Am Heart Asn; Am Fedn Clin Res; fel Am Col Clin Pharmacol & Chemother; NY Acad Sci. Res: Therapeutics. Mailing Add: Food & Drug Admin Liaison Schering Corp Bloomfield NJ 07003

NEAS, ROBERT EDWIN, b Sheldon, Mo, May 7, 35; m 57; c 3. ANALYTICAL CHEMISTRY. Educ: Cent Methodist Col, AB, 57; Southern Ill Univ, Carbondale, MS, 65; Univ Mo-Columbia, PhD(chem), 70. Prof Exp: Analyst chem, Mallinckrodt Chem Works, 57-58, chemist, 58-59, supvr, 59-61, asst to dir qual control, 61-62, res asst chem, 62-63; instr, Univ Mo-Mo Exp Sta Lab, 65-66; asst prof, 69-74, ASSOC PROF CHEM, WESTERN ILL UNIV, 74- Concurrent Pos: Univ res coun grant, Western Ill Univ, 70-71. Mem: Am Chem Soc. Res: Analytical chemistry of anions; spectrophotometry; ion selective electrodes; trace analysis. Mailing Add: Dept of Chem Western Ill Univ Macomb IL 61455

NEATHERY, MILTON WHITE, b Chapel Hill, Tenn, Apr 15, 28; m 50; c 1. ANIMAL NUTRITION. Educ: Univ Tenn, BS, 50, MS, 55; Univ Ga, PhD(animal nutrit), 73. Prof Exp: Farm mgr dairy, Minglewood Farm, Tenn, 50-51; fieldman dairy, Nashville Pure Milk Co, 53-54; asst dairy, Univ Tenn, 55-56; asst animal husbandman animal nutrit, Ga Mountain Exp Sta, 56-59; ASST PROF DAIRY SCI, UNIV GA, 59- Mem: Sigma Xi; Am Dairy Sci Asn; Am Soc Animal Sci. Res: Mineral metabolism in animals; primarily trace mineral metabolism in ruminants using radioisotopes. Mailing Add: Animal & Dairy Sci Dept Univ Ga Livestock-Poultry Bldg Athens GA 30602

NEATHERY, THORNTON LEE, b Atlanta, Ga, Mar 12, 31; m 56; c 3. GEOLOGY. Educ: Univ Ala, BS, 56, MS, 64. Prof Exp: Geologist, Reynolds Metals Co, 56-62; asst vpres, Textile Rubber & Chem Co, 63; geologist, 64-73, CHIEF GEOLOGIST, GEOL SURV ALA, 73- Mem: Sigma Xi; fel Geol Soc Am; Soc Econ Geologist; Mineral Soc Am; Soc Mining Engrs. Res: Regional geologic mapping in southern Piedmont and folded Appalachians with emphasis on sedimentation, metamorphism, and structural evolution as applied to distribution of ore deposits. Mailing Add: Mineral Resources Div Geol Surv of Ala University AL 35486

NEAVEL, RICHARD CHARLES, b Philadelphia, Pa, Oct 21, 31; m 58; c 3. FUEL SCIENCE. Educ: Temple Univ, BA, 54; Pa State Univ, MS, 57, PhD(geol), 66. Prof Exp: Coal petrologist, Ind Geol Surv, 57-61; res assoc, Dept Geol, Pa State Univ, 61-66; staff geologist, Humble Oil & Refining Co, 67; geologist, Synthetic Fuel Lab, 68-69, group leader, 70, res assoc, Gasification Lab, 71-75, SR RES ASSOC, GASIFICATION LAB, EXXON RES & ENG CO, 75- Concurrent Pos: Consult, Inst Gas Technol, 59-63; chmn fuels sci, Gordon Res Conf, 75. Mem: AAAS; Sigma Xi; Am Soc Testing & Mat. Res: Characterization of coals and relationships between coal properties and utilization, emphasizing synthetic fuels processes. Mailing Add: Exxon Res & Eng Co PO Box 4255 Baytown TX 77520

NEAVES, WILLIAM BARLOW, b Spur, Tex, Dec 25, 43; m 65; c 2. ANATOMY, CELL BIOLOGY. Educ: Harvard Univ, AB, 66, PhD(anat), 69. Prof Exp: Lectr vet anat, Univ Nairobi, 70-71; lectr anat, Med Sch, Harvard Univ, 72; asst prof, 72-74, ASSOC PROF CELL BIOL & DIR ANAT, UNIV TEX HEALTH SCI CTR

DALLAS, 74- Concurrent Pos: Rockefeller Found fel anat, Harvard Univ & Univ Nairobi, 70-71; res assoc, Los Angeles County Mus, 70-73; consult, Ford Found, 73-74; assoc ed, Anat Record, 75- Mem: AAAS; Am Asn Anat; Soc Study Reproduction. Res: Reproductive biology; androgenic tumors; testicular endocrinology; contraception. Mailing Add: Dept of Cell Biol Univ of Tex Health Sci Ctr Dallas TX 75235

NEBB, JACK, mathematical analysis, see 12th edition

NEBEKER, ALAN V, b Salt Lake City, Utah, Apr 8, 38; m 60; c 3. ENTOMOLOGY, AQUATIC ECOLOGY. Educ: Univ Utah, BS, 61, MS, 63, PhD(zool), 55. Prof Exp: Res aquatic biologist entom, Nat Water Qual Lab, 66-71, RES AQUATIC BIOLOGIST, ENVIRON PROTECTION AGENCY, ENTOM & FISHERIES, WESTERN FISH TOXICOL LAB, 71- Mem: Entom Soc Am; Am Entom Soc; Am Fisheries Soc. Res: Water pollution toxicology; systematics of aquatic insects; water quality criteria for protection of aquatic life; bioassay analysis. Mailing Add: Western Fish Toxicol Lab Environ Protection Agency Corvallis OR 97330

NEBEKER, THOMAS EVAN, b Richfield, Utah, May 10, 45; m 64; c 3. FOREST ENTOMOLOGY. Educ: Col Southern Utah, BS, 67; Utah State Univ, MS, 70; Ore State Univ, PhD(entom), 74. Prof Exp: Teaching asst zool & entom, Utah State Univ, 67-70; NSF trainee pest pop ecol, Ore State Univ, 70-73; fel pop ecol, Utah State Univ, 73-74; ASST PROF FOREST ENTOM, MISS STATE UNIV, 74- Mem: Entom Soc Am; Ecol Soc Am; Sigma Xi; Can Entom Soc. Res: Population biology of forest insects with emphasis on the dynamics of southern pine beetle populations; parasite and predator efficiency studies utlizing behavior patterns. Mailing Add: Dept of Entom Miss State Univ Drawer EM Mississippi State MS 39762

NEBEL, BERNARD JAMES, biology, plant physiology, see 12th edition

NEBEL, CARL WALTER, organic chemistry, see 12th edition

NEBEL, RICHARD WILSON, b Madison, Wis, July 5, 15; m 43; c 2. CHEMISTRY. Educ: Princeton Univ, AB, 36; Harvard Univ, MA, 37, PhD(org chem), 41. Prof Exp: Res chemist, E I du Pont de Nemours & Co, Inc, 41-46, res supvr, 46-50, res mgr, 50-51, lab dir, 51-69, prod develop mgr, Christina Lab, Del, 69-73; RETIRED. Mem: Am Chem Soc. Res: Chemistry of high polymers; cellulose derivatives; technology of synthetic textiles. Mailing Add: 2 E Third St New Castle DE 19720

NEBENZAHL, LINDA LEVINE, b Duluth, Minn, Oct 4, 49; m 71. PHYSICAL ORGANIC CHEMISTRY, SURFACE CHEMISTRY. Educ: Univ Minn, Minneapolis, BA, 71; Univ Calif, Berkeley, PhD(org chem), 75. Prof Exp: STAFF CHEMIST, IBM CORP, 75- Mem: Am Chem Soc. Res: Fluid flow through porous media. Mailing Add: E21/030 IBM Corp San Jose CA 95193

NEBERGALL, WILLIAM HARRISON, b Cuba, Ill, Dec 21, 14; m 40; c 3. INORGANIC CHEMISTRY. Educ: Western Ill State Teachers Col, BEd, 36; Univ Ill, MS, 39; Univ Minn, PhD(inorg chem), 49. Prof Exp: High sch teacher, Ill, 36-38; instr chem, Tenn State Teachers Col, Johnson City, 39-41, Univ Ky, 41-42, Wis State Teachers Col, Superior, 42-44 & Univ Minn, 44-49; asst prof, 49-55, ASSOC PROF CHEM, IND UNIV, BLOOMINGTON, 55- Concurrent Pos: Guest prof, Brunswick Tech Univ, 59-60. Mem: Am Chem Soc. Res: Organometallic, organosilicon and fluoride chemistry; inorganic phosphates and fluorides. Mailing Add: Dept of Chem Ind Univ Bloomington IN 47403

NEBERT, DANIEL WALTER, b Portland, Ore, Sept 26, 38; m 60; c 2. PEDIATRICS, PHARMACOLOGY. Educ: Univ Ore, BA, 61, MS & MD, 64. Prof Exp: From intern to resident pediat, Ctr Health Sci, Univ Calif, Los Angeles, 64-66; res assoc biochem, Lab Chem, Etiology Br, Nat Cancer Inst, 66-68; res investr pharmacol, Sect Develop Enzym, 68-70, head sect develop pharmacol, Lab Biomed Sci, 70-74, chief neonatal & pediat med br, 74-75, CHIEF DEVELOP PHARMACOL BR, NAT INST CHILD HEALTH & HUMAN DEVELOP, NIH, 75- Mem: AAAS; Am Soc Pharmacol & Exp Therapeut; Am Soc Biol Chemists; Sigma Xi; Am Soc Clin Invest. Res: Application of mammalian cell culture and inbred animal strains as experimental model systems to clinical pharmacology, toxicology, cancer research and pharmacogenetic disorders. Mailing Add: Rm 13 N-256 Bldg 10 Nat Inst Child Health & Human Dev Bethesda MD 20014

NEBESKY, EDWARD ANTHONY, food technology, see 12th edition

NEBGEN, JOHN WILLIAM, b Independence, Mo, May 20, 3S; m 57; c 2. PHYSICAL CHEMISTRY, INORGANIC CHEMISTRY. Educ: Washington Univ, AB, 56; Univ Pa, PhD(chem), 60. Prof Exp: Assoc chemist, 60-65, sr chemist, 65-68, PRIN CHEMIST, MIDWEST RES INST, 68- Mem: Am Chem Soc. Res: Molecular structure; infrared, nuclear magnetic resonance and electron spin resonance spectroscopy; inorganic synthesis; portland cement manufacture; wastewater treatment; water pollution abatement; environmental systems analysis. Mailing Add: Midwest Res Inst 425 Volker Blvd Kansas City MO 64110

NEBLETT, RICHARD FLEMON, organic chemistry, see 12th edition

NEBZYDOSKI, JOHN WALTER, organic chemistry, see 12th edition

NECHAMKIN, HOWARD, b Brooklyn, NY, Aug 18, 18; m 56; c 1. INORGANIC CHEMISTRY. Educ: Brooklyn Col, BA, 39; Polytech Inst Brooklyn, MS, 49; NY Univ, EdD, 61. Prof Exp: Chemist, R H Macy Co, NY, 39-41 & US Naval Supply Depot, 41-42; head res chemist, Hazeltine Electronics Corp, 42-45; instr chem, Brooklyn Col, 45-46; from instr to assoc prof chem, Pratt Inst, 46-61; PROF CHEM, TRENTON STATE COL, 61-, CHMN DEPT, 68- Honors & Awards: Professional Chemist Award, Am Inst Chemists, 74. Mem: Fel Am Inst Chem; Am Chem Soc. Res: Analysis of synthetic fibers; metal coatings research; plastics development; detection of vanillin flavor; chemistry of rhenium; volumetric analysis of sulfate; chemistry of recyclable materials. Mailing Add: Dept of Chem Trenton State Col Trenton NJ 08625

NECHAY, BOHDAN ROMAN, b Prague, Czech, Nov 26, 25; nat US; m; c 2. PHARMACOLOGY, THERAPEUTICS. Educ: Univ Minn, DVM, 53. Prof Exp: Pvt practr, Minn, 53-56; asst prof pharmacol, Col Med, Univ Fla, 61-66; asst prof pharmacol & urol, Med Ctr, Duke Univ, 66-68; ASSOC PROF PHARMACOL & TOXICOL, UNIV TEX MED BR GALVESTON, 68- Concurrent Pos: Fel pharmacol, Col Med, Univ Fla, 56-60; Am Heart Asn fel, 58-60; NIH fel pharmacol, Univ Uppsala, 60-61. Mem: Am Soc Nephrology; Am Soc Pharmacol & Exp Therapeut. Res: Electrolyte physiology and pharmacology; environmental toxicology; kidney. Mailing Add: Dept of Pharmacol & Toxicol Univ of Tex Med Br Galveston TX 77550

NECHELES, THOMAS, b Hamburg, Ger, Oct 1, 33; US citizen; m 55; c 5. PEDIATRICS, PHYSIOLOGY. Educ: Univ Chicago, BA, 53, BS, 57, MS, 58, MD & PhD(biol, physiol), 61. Prof Exp: Instr med, 65-67, asst prof pediat, 67-70, ASSOC PROF PEDIAT, SCH MED, TUFTS UNIV, 70- Concurrent Pos: USPHS hemat fel, Blood Res Lab, New Eng Med Ctr Hosps, 63-66; asst hematologist, Boston Floating Hosp, 65-67; consult & sr instr, Blood Res Lab, New Eng Med Ctr Hosps, 66-, hematologist, 67-70, chief pediat hemat, 70-; estab investr, Am Heart Asn, 6671; consult, Springfield Hosp, Mass, 72-, St Elizabeth Hosp, Brighton, 73- & Kennedy Mem Hosp, 74- Mem: Am Soc Clin Oncol; AAAS; Am Fedn Clin Res; Am Soc Hemat; Am Soc Exp Path. Res: Control of cellular proliferation; hemoglobin synthesis; red cell metabolism. Mailing Add: Dept of Pediat Hemat New Eng Med Ctr Hosps Boston MA 02111

NECKER, WALTER LUDWIG, b Hamburg, Ger, Dec 27, 13; US citizen; m 39, 68. HISTORY OF BIOLOGY, HISTORY OF MEDICINE. Educ: Univ Chicago, BS, 40. Prof Exp: Herpetologist & librn, Chicago Acad Sci, 29-40; cataloger & buyer natural hist mat, Barnes & Noble, Inc, 41; antiqn & bibliogr, Natural Hist Bks, 42-55, chief cataloger, Chicago Hist Soc, 55-57; ref librn, Gary Pub Libr, Ind, 57-59, asst librn, 59-60; chief libr br, Qm Food & Container Inst Armed Forces, 60-63; librn-cur, Wood Libr-Mus Anesthesiol, 63-67; BIO-MED LIBRN, UNIV CHICAGO LIBR, 67- Concurrent Pos: Chg, Smithsonian Inst Exhib, World's Fair, 34; Carnegie grant, Mex, 39-40; historian, Am Soc Ichthyologists & Herpetologists, 38-50; Fulbright sr fel, UK, 52-53; vis prof, Inst Anesthesiol, Univ Mainz, 69; archivist, Med Libr Asn, 75- Mem: Am Asn Hist Med; Hist Sci Soc; Med Libr Asn; Int Soc Hist Pharm. Res: Bibliography and history of biology and medicine; pre-Linnean natural history. Mailing Add: 1415 Hoffman Ave Park Ridge IL 60068

NECKERS, DOUGLAS CARLYLE, organic chemistry, see 12th edition

NEDDENRIEP, RICHARD JOE, b Leipsic, Ohio, June 3, 30; m 57; c 2. PHYSICAL CHEMISTRY. Educ: Miami Univ, BA, 53; Univ Wis, PhD(phys chem), 58. Prof Exp: Res chemist, Linde Div, Union Carbide Corp, NY, 57-65, group leader, 65-69; group leader, 69-72, mgr prod res, 72-73, asst dir res, 73-74, DIR RES & ASST VPRES, BETZ LABS, INC, 74- Mem: Am Chem Soc. Res: Reaction kinetics, particularly free radical reactions and the radiolysis of organic materials; adsorption; catalysis, particularly with molecular sciences; water and air purification; corrosion and scale inhibition. Mailing Add: Res Dept Betz Labs Inc 4636 Somerton Rd Trevose PA 19047

NEDDERMAN, HOWARD CHARLES, b Mishawaka, Ind, Apr 11, 19; m 40; c 4. PHYSICS. Educ: Purdue Univ, BS, 42, MS, 44; Columbia Univ, PhD(physics), 56. Prof Exp: Instr chem, Purdue Univ, 42-46; engr microwave tubes, Radio Corp Am, Pa, 46-48; res assoc, Radiation Lab, Columbia Univ, 48-53; sr physicist thermionics, Gen Elec Co, 53-56; res specialist, Stromberg-Carlson Div, Gen Dynamics Corp, NY, 56-57, mgr res, 57-61, from assoc dir to dir res, Electronics Div, 61-66; pres & gen mgr, Megadyne Industs, Inc, 66-70; indust consult, 70-75; VPRES, INTERSTATE CIRCUITS, INC, 75- Mem: AAAS; Am Phys Soc. Mailing Add: 86 Bastian Rd Rochester NY 14623

NEDDERMEYER, PETER ARTHUR, b Elze, Hannover, Ger, Mar 5, 41; US citizen; m 65; c 3. ANALYTICAL CHEMISTRY. Educ: Union Col, NY, BS, 63; Purdue Univ, PhD(anal chem), 68. Prof Exp: SR RES CHEMIST, EASTMAN KODAK CO, 68- Mem: Am Chem Soc. Res: Liquid chromatography, molecular weight characterization, surface chemistry, photographic science. Mailing Add: Indust Lab Eastman Kodak Co 343 State St Rochester NY 14650

NEDDERMEYER, SETH HENRY, b Richmond, Mich, Sept 16, 07; m 44; c 2. PHYSICS. Educ: Stanford Univ, AB, 29; Calif Inst Technol, PhD(physics), 35. Prof Exp: Res fel physics, Calif Inst Technol, 35-41; physicist, Nat Defense Res Comt, Nat Bur Standards; 41-43; mem staff, Los Alamos Sci Lab, 43-46; from assoc prof to prof, 46-73, EMER PROF PHYSICS, UNIV WASH, 73- Res: Particle physics; electromagnetic interactions; paraphysics; parapsychology. Mailing Add: Dept of Physics BJ 10 Univ of Wash Seattle WA 98105

NEDELSKY, LEO, b Russia, Oct 28, 03; nat US; m 41; c 3. THEORETICAL PHYSICS. Educ: Univ Wash, Seattle, BS, 28; Univ Calif, MS, 31, PhD(theoret physics), 32. Prof Exp: Instr physics, Univ Calif, 32-35 & Hunter Col, 35-40; fel Gen Educ Bd, 40-41, dir res basic nursing educ, 54-55, PROF PHYS SCI & EXAM, UNIV CHICAGO, 41- Concurrent Pos: Lectr, Baker & Ottawa Univs, 48, Univ Pa, 52-53, Univ Wash, Seattle, 56, Univ Mo, 58, Northern Mich Col, 58-59, Univs Colombia, El Salvador, Guatemala, Israel, Mex, Puerto Rico, Venezuela, Cuba & Brazil, 66; consult, Univ Wash, Seattle, 54-57, NY State Univ, 57, Am Bd Radiol, 58, CBS-TV, 58, Univ Ill, 59-60, Michael Reese Hosp, 60; res assoc dent educ, Univ Ill, 65-; consult, WHO, 66- Mem: AAAS; Am Phys Soc; Am Asn Physics Teachers. Res: Physical sciences in general education. Mailing Add: Univ of Chicago 5811 S Ellis Ave Chicago IL 60637

NEDICH, RONALD LEE, b Chicago, Ill. PHYSICAL PHARMACY, INDUSTRIAL PHARMACY. Educ: St Louis Col Pharm, BS, 65; Purdue Univ, MS, 68, PhD(phys pharm), 70. Prof Exp: Sr res pharmacist, 70-73, MGR PHARM DEVELOP, BAXTER LABS, INC, 73- Mem: Acad Pharmaceut Sci; Am Pharmaceut Asn; Am Chem Soc. Res: Dosage form design; physical chemical principles associated with pharmaceutical dosage forms; pharmacokinetics. Mailing Add: Baxter Labs Inc 6301 Lincoln Ave Morton Grove IL 60053

NEDOLUHA, ALFRED K, b Vienna, Austria, Sept 13, 28; m 57; c 1. THEORETICAL SOLID STATE PHYSICS. Educ: Univ Vienna, PhD(physics), 51. Prof Exp: Staff mem, Felten & Guilleaume, A G, Austria, 51-56, head high voltage lab, 56-57; physicist, White Sands Missile Range, NMex, 57-59; Naval Ord Lab, Corona, 59-62, res physicist, 62-70, RES PHYSICIST, NAVAL ELECTRONICS LAB CTR, 70- Mem: Am Phys Soc. Res: Solid state theory; transport and optical phenomena. Mailing Add: 5434 Bothe Ave San Diego CA 92122

NEDROW, WARREN WESLEY, b Falls City, Nebr, Oct 13, 06; m 27. BOTANY. Educ: Nebr State Teachers Col, AB, 28; Univ Denver, MA, 33; Univ Nebr, PhD(bot), 36. Prof Exp: Sch supt Colo, 27-32; asst bot, Univ Denver, 32-33; prof biol & head dept, Ark State Col, 36-52, head dept sci, 53-65, dean grad sch, 65-70; RETIRED. Mem: AAAS; spec affil Am Med Asn; Ecol Soc Am. Res: Ecology of roots. Mailing Add: 5122 S Main Jonesboro AR 72401

NEDWICK, JOHN JOSEPH, b Ranshaw, Pa, Jan 11, 22; m 61; c 3. INDUSTRIAL ORGANIC CHEMISTRY, CHEMICAL ENGINEERING. Educ: Univ Louisville, AB, 47; Univ Pa, MS, 53. Prof Exp: Chemist, 48-56, group leader high pressure chem, 56-63, GROUP LEADER POLYMER CHEM, ROHM & HAAS CO, 63-, PROJ LEADER CHEM PROCESS ENG DEPT, 73- Mem: Am Chem Soc; Am Inst Chem Engrs. Res: High pressure research and process development; acetylene reactions; continuous bench scale pilot plants; polymer chemistry; plastics; coatings; process development of agricultural chemicals and health products. Mailing Add: 133 Forrest Ave Southampton PA 18966

NEDWICKI, EDWARD G, b Detroit, Mich, Mar 12, 16; m 42; c 3. PULMONARY

DISEASES, INTERNAL MEDICINE. Educ: Univ Detroit, BS, 39; Univ Mich, MD, 43. Prof Exp: Instr clin med, Sch Med, 50-53, from asst prof to assoc prof med, Univ, 53-74, ASSOC PROF MED, SCH MED, WAYNE STATE UNIV, 74-; CHIEF PULMONARY DIS SECT, DETROIT GEN HOSP, 74- Concurrent Pos: Chief tuberc serv, Vet Admin Hosp, Allen Park, 50-59, chief pulmonary dis sect, 59-74, actg assoc chief staff res & educ, 71-74. Mem: Fel Am Col Chest Physicians; Am Thoracic Soc; AMA; fel Am Col Physicians; affil Royal Soc Med. Res: Pulmonary tuberculosis; clinical application of drug therapy; diagnosis and treatment of pulmonary diseases. Mailing Add: Dept of Med Wayne State Univ 1326 St Antoine Detroit MI 48226

NEDZEL, V ALEXANDER, b Constantinople, Turkey, Sept 20, 21; nat US; m 45; c 3. PHYSICS. Educ: Univ Chicago, SB, 41, SM, 50, PhD(nuclear physics), 53. Prof Exp: Physicist, Metall Lab, Manhattan Dist, 42-44; assoc scientist, Los Alamos Sci Lab, 44-47; asst physics, Univ Chicago, 47-51; syst anal & eval, 53-55, group leader, 55-65, assoc head data systs div, 65-69, HEAD AEROSPACE DIV, LINCOLN LAB, MASS INST TECHNOL, 69- Concurrent Pos: Asst to dir, Los Alamos Group, Bikini Tests, 46. Mem: AAAS; fel Am Phys Soc; Opers Res Soc Am; Am Inst Aeronaut & Astronaut. Res: Radar and computer systems; defense sciences. Mailing Add: Lincoln Lab Mass Inst of Technol Lexington MA 02173

NEE, M COLEMAN, b Taylor, Pa, Nov 14, 17. MATHEMATICS. Educ: Marywood Col, AB, 39, MA, 43; Univ Notre Dame, MS, 59. Prof Exp: Teacher math & Latin, Marywood Sem, 43-55; asst prof math, 59-70, PRES MARYWOOD COL, 70- Res: Group theory of algebra. Mailing Add: Marywood Col Scranton PA 18509

NEECE, GEORGE A, b Pine Bluff, Ark, Sept 18, 39; m 62. PHYSICAL CHEMISTRY. Educ: Rice Univ, AB, 61; Duke Univ, PhD(phys chem), 64. Prof Exp: Chemist, US Army Res Off-Durham, 64-67; res assoc chem, Cornell Univ, 67-68; asst prof, Univ Ga, 68-71; CHEMIST, OFF NAVAL RES, 71-, DIR CHEM PROG, 72- Res: Statistical mechanics. Mailing Add: Off of Naval Res 800 N Quincy Arlington VA 22217

NEEDELS, THEODORE STANTON, physics, see 12th edition

NEEDHAM, CHARLES D, b Chicago, Ill, Sept 17, 37. POLYMER CHEMISTRY. Educ: Carnegie Inst Technol, BS, 59; Univ Minn, PhD(chem), 65. Prof Exp: Staff fel phys biol, NIH, 65-67; STAFF CHEMIST, COMPONENTS DIV, IBM CORP, 67- Mem: Am Chem Soc; Am Phys Soc. Res: Radiation chemistry and analysis of polymers. Mailing Add: B/300 Syst Prod Div IBM Corp Hopewell Jct NY 12533

NEEDHAM, GERALD MORTON, b Caldwell, Idado, Aug 4, 17; m 42; c 2. BACTERIOLOGY. Educ: Col Idaho, BS, 40; Univ Minn, PhD(bact), 47; Am Bd Med Microbiol, dipl. Prof Exp: Bacteriologist, State Dept Health, Minn, 41-42; instr bact, Med Sch, Univ Minn, 41-46; med bacteriologist, Mayo Clin, 46-68; ASSOC DIR DIV EDUC, MAYO SCH HEALTH REL SCI, MAYO FOUND, 68-: ASSOC DEAN STUDENT AFFAIRS, MAYO MED SCH, UNIV MINN, 71- Concurrent Pos: Consult bacteriologist, Econ Labs, St Paul, 41-46. Mem: AAAS; Am Soc Microbiol; fel Am Acad Microbiol; fel Am Pub Health Asn. Res: Medical bacteriology; antibiotics; tuberculosis; action of a few antibacterial substances on resting cells. Mailing Add: Mayo Clin Rochester MN 55901

NEEDHAM, JOHN W, information science, biochemistry, see 12th edition

NEEDHAM, THOMAS E, JR, b Newton, NJ, Apr 12, 42. PHARMACEUTICS. Educ: Univ RI, BS, 65, MS, 67, PhD(pharmaceut sci), 70. Prof Exp: Pharmacist, Galen Drug, Inc, 65-68; instr pharm, Univ RI, 67-69; pharmacist, Pinault Drug, Inc, 69-70; asst prof, 70-74, ASSOC PROF PHARM, SCH PHARM, UNIV GA, 74- Mem: Am Pharmaceut Asn; Acad Pharmaceut Sci; Am Asn Cols of Pharm. Res: Effects of selected variables on the dissolution and absorption of drugs; particulate and incompatibilities in glass and plastic parenteral containers. Mailing Add: Sch of Pharm Univ of Ga Athens GA 30602

NEEDLEMAN, SAUL BEN, b Chicago, Ill, Sept 25, 27; m 54; c 4. BIOCHEMISTRY, NEUROCHEMISTRY. Educ: Ill Inst Technol, BS, 50, MS, 55; Northwestern Univ, PhD(chem), 57. Prof Exp: Res assoc biochem, Col Med, Univ Ill, 52-53; asst chem, Ill Inst Technol, 53-55; asst chem, Northwestern Univ, 55-57, from instr to asst prof neurobiochem, 57-72; prof biochem & head dept, Roosevelt Univ, 72-74; CLIN DIAG SPECIALIST, ABBOTT LABS, 74- Concurrent Pos: Group leader, Helene Curtis Industs, Inc, 57-58; sr res chemist, Nalco Chem Co, 58-62; sr res biochemist, Abbott Labs, 62-66; asst chief radioisotope serv, Vet Res Hosp, Chicago, 66-71. Mem: Am Chem Soc; Sigma Xi; Am Soc Biol Chemists; Am Soc Cell Biol; Biophys Soc. Res: Protein sequence determination; mechanism of enzyme action; lysosomal enzyme diseases; Wilson's disease; organic synthesis of biochemically active substances; Collagen disease metabolism. Mailing Add: 867 Marion Highland Park IL 60035

NEEDLER, ALFRED WALKER HOLLINSHEAD, b Huntsville, Ont, Aug 14, 06; m 30, 53; c 5. MARINE BIOLOGY. Educ: Univ Toronto, BA, 26, MA, 27, PhD, 30. Hon Degrees: DSc, Univ NB, 54 & Brit Col, 69. Prof Exp: Asst bot, Univ Toronto, 24-25, asst embryol, 25-26; vol investr, Fisheries Res Bd, Can Dept Fisheries, 26-29, sci asst marine biol, 29-30, in-chg, Atlantic Oyster Invests & PEI Biol Sta, 30-41, in-chg oyster cult, 30-42, from asst biologist to chief biologist, 31-41, dir, Atlantic Biol Sta, 41-54 & Pac Biol Sta, 54-63, asst dep minister fisheries, 48-50, dep minister, 63-71, exec dir, Huntsman Marine Lab, St Andrews, NB, 71-76. Concurrent Pos: Asst, Univ Toronto, 26-27; sci adv, NAm Coun Fisheries Invests, 28-29 & Int Passamaquoddy Fisheries Comn, 31-34; mem standing adv comt fisheries, UN Food & Agr Orgn, 46-49; chmn adv comt marine resources res, 63-65; chmn res comt, Int Comn Northwest Atlantic Fisheries, 51-54, head Can deleg, 63-, chmn, 73-75; chmn res comt, Int N Pac Fisheries Comn, 54-57 & Skeena Salmon Mgt Comt, 54-63; chmn, Comt on Fisheries, 66-67; mem, Sci Coun Can, 68-71; sr fisheries adv, Can Deleg UN Law of Sea Conf, 74-76. Honors & Awards: Officer, Order Brit Empire; Gold Medal, Prof Inst Pub Serv Can, 73. Mem: Fel Royal Soc Can; Royal Can Inst. Res: Oceanography; life history; migrations and ecology of haddock; ecology of the oyster of Canadian Atlantic waters; fisheries biology, management, development and international relations.

NEEDLER, GEORGE TREGLOHAN, theoretical physics, see 12th edition

NEEDLES, HOWARD LEE, b Bloomington, Calif, June 26, 37; m 60; c 4. POLYMER CHEMISTRY, TEXTILE CHEMISTRY. Educ: Univ Calif, Riverside, AB, 59; Univ Mo, PhD(org chem), 63. Prof Exp: Res chemist, Western Regional Res Lab, USDA, 63-69; asst prof textile sci & asst textile chemist, Agr Exp Sta, 69-74, ASSOC PROF TEXTILE SCI, UNIV CALIF, DAVIS, 74- Mem: Am Chem Soc; Am Asn Textile Chem & Colorists. Res: Chemical modification and crosslinking of fibers and related models; graft polymerization on fibers and films; photodegradation of fibers, finishes and dyes; physical and chemical properties of modified fibers. Mailing Add: Div of Textiles & Clothing Univ of Calif Davis CA 95616

NEEL, JAMES VAN GUNDIA, b Middletown, Ohio, Mar 22, 15; m 43; c 3. GENETICS. Educ: Col Wooster, AB, 35; Univ Rochester, PhD(genetics), 39, MD, 44. Prof Exp: Asst, Univ Rochester, 35-39; instr zool, Dartmouth Col, 39-41; from intern to asst resident med, Strong Mem Hosp & Rochester Munic Hosp, NY, 44-46; dir field studies, Atomic Bomb Casualty Comn, Nat Res Coun, 47-48; from asst prof to assoc prof internal med, 49-56, assoc geneticist, Inst Human Biol, 48-56, prof human genetics & internal med, Sch Med, 56-66, LEE R DICE UNIV PROF HUMAN GENETICS & INTERNAL MED, SCH MED, UNIV MICH, ANN ARBOR, 66-, CHMN DEPT HUMAN GENETICS, 56- Concurrent Pos: Mem comt res probs sex, Nat Res Coun, 49-56, panel genetics, Comt Growth, 51-56, comt atomic casualties, 51-54, adv comt, Atomic Bomb Casualty Comn, 57-, comt epidemiol & vet follow-up studies, 65-; mem comt genetic effects of atomic radiation, Nat Acad Sci, 55-, comt int biol prog, 65-, selection comt sr res fels, NIH, 56-60 & gen res training grant comt, 57-58, chmn genetic training grant comt, 58-63, mem comt int ctrs for med res & training, 61-65, comt int res, 65-69, expert adv panel radiation, WHO, 57-61, expert adv panel human genetics, 61, mem coun, Nat Acad Sci, 70-; US deleg, US-Japan Coop Med Sci Prog, 71- Honors & Awards: Lasker Award, Am Pub Health Asn, 60; Allen Award, Am Soc Human Genetics, 65; Nat Medal of Sci, 74. Mem: Nat Acad Sci; Am Acad Arts & Sci; Asn Am Physicians; Am Philos Soc; Am Soc Naturalists. Res: Genetics of man. Mailing Add: Dept of Human Genetics Univ of Mich Ann Arbor MI 48104

NEEL, JAMES WILLIAM, b Turlock, Calif, July 20, 25; m 56. SOIL SCIENCE, BOTANY. Educ: Univ Calif, Berkeley, BS, 49, Univ Calif, Los Angeles, PhD(bot sci), 64. Prof Exp: From lab technician to prin lab technician, Atomic Energy Proj, Univ Calif, Los Angeles, 49-59, asst plant physiol, Dept Irrig & Soil Sci, 59-63; from asst prof to assoc prof, 63-69, chmn dept, 69-71, PROF BIOL, SAN DIEGO STATE UNIV, 69- Concurrent Pos: Consult, Rand Corp, Calif, 63-71. Mem: AAAS; Am Soc Agron; Soil Sci Soc Am; NY Acad Sci. Res: Fate of soluble and insoluble forms of radionuclides in soils and availability to plants; effects of heavy metals on plant systems. Mailing Add: Dept of Biol San Diego State Univ San Diego CA 92182

NEEL, JOE KENDALL, SR, b Tacoma, Va, June 12, 15; m 42; c 1. LIMNOLOGY. Educ: Univ Ky, BS, 37, MS, 38; Univ Mich, PhD(limnol), 47. Prof Exp: Instr zool, Univ Ky, 38-39; teaching fel, Univ Mich, 41-42; med entomologist, USPHS, 42-46; Rackham fel, Univ Mich, 46-47; asst prof zool, Univ Ky, 47-50; basin biologist, Div Water Pollution Control, USPHS, 50-53, biologist, 53-58, chief water quality sect, 58-63; dir potamological inst, Louisville, Ky, 63-66; PROF BIOL, UNIV N DAK, 66-, CHMN DEPT, 74- Mem: AAAS; Am Inst Biol Sci; Am Soc Limnol & Oceanog; Am Micros Soc; Water Pollution Control Fedn. Res: Psammon; stream limnology; waste treatment; eutrophication. Mailing Add: 2221 Chestnut St Grand Forks ND 58201

NEEL, PERCY LANDRETH, b Bryn Mawr, Pa, Feb 15, 44; m 70. ORNAMENTAL HORTICULTURE. Educ: Calif State Univ, Fresno, BS, 66; Univ Calif, Davis, MS, 68, PhD(bot), 71. Prof Exp: ASST PROF ORNAMENTAL HORT RES, INST FOOD & AGR SCI, AGR RES CTR, UNIV FLA, FT LAUDERDALE, 71- Mem: Am Soc Hort Sci; Int Plant Propagators Soc. Res: Weed control in ornamentals; care and culture of ornamentals; tissue culture; propagation; plant growth regulators; growing media; new plant introductions. Mailing Add: Inst Food & Agr Sci 3205 SW 70th Ave Ft Lauderdale FL 33314

NEEL, WILLIAM WALLACE, b Thomasville, Ga, Feb 4, 18; m 52; c 2. ENTOMOLOGY. Educ: Emory Univ, BA, 40; Univ Fla, MS, 49; Agr & Mech Col, Tex, PhD(entom), 54. Prof Exp: Field supvr, Disease Control, USPHS, 45-47; asst entomologist, Inter-Am Inst of Agr Sci, 48-49, 52-53; asst prof entom & asst entomologist, Agr Exp Sta, Miss State Univ, 54-62; field res, Chemagro Corp, 62-63; asst prof entom & asst entomologist, Agr Exp Sta, Univ WVa, 63-66; ASSOC PROF ENTOM & ASSOC ENTOMOLOGIST, AGR EXP STA, MISS STATE UNIV, 66- Mem: Entom Soc Am; Sigma Xi. Res: Forest and pecan entomology with special interest in seed and cone insects; toxicant research on pecan and seed and cone insects; pheromone isolation of pecan weevil. Mailing Add: Dept of Entom PO Drawer EM State College MS 39762

NEELAKANTAN, LAKSHMANAN, b Ernakulam, India, Feb 1, 27; m 66; c 1. ORGANIC CHEMISTRY, BIOCHEMISTRY. Educ: Univ Madras, BSc, 46; Indian Inst Sci, Bangalore, MSc, 49; Univ Bombay, PhD(org chem), 54. Prof Exp: Fel anti-malarials, Sch Pharm, Univ Kans, 53-55; fel antitumor agents, Sch Pharm, Univ NC, 55-56; res assoc, Sch Pharm, Med Col Va, 56-62; res assoc reaction mechanisms, Karolinska Inst Sweden, 63-68; mem res staff, Inst Biomed Res, AMA Educ & Res Found, 68-70; RES ASSOC, COL PHARM, UNIV KY, 71- Concurrent Pos: NIH grant, 61-62. Mem: Am Chem Soc; Indian Chem Soc. Res: Antimalarials; antitumor agents; reaction mechanisms; asymetric synthesis; molecular biophysics; gas chromatography; electron capture detection; chemical ionization. Mailing Add: Col of Pharm Univ of Ky Lexington KY 40506

NEELEY, CHARLES MACK, b Pine Bluff, Ark, Mar 7, 42; m 68. PHYSICAL CHEMISTRY, POLYMER CHEMISTRY. Educ: Univ Ark, Fayetteville, BS, 65, PhD(chem), 69. Prof Exp: Lab technician, Houston Chem Corp, 62-63; asst phys chem, Univ Ark, Fayetteville, 65-69; fel, Univ Fla, 69-70; res chemist polymer phys chem, Plastics Lab, 70-75, SR CHEMIST, POLYMER DEVELOP POLYETHYLENE & POLYPROPYLENE DIVS, POLYMER DEVELOP DIV, TEX EASTMAN CO, 75- Mem: AAAS; Am Chem Soc. Res: Pyrolysis gas chromatography; rheology; gas phase photochemistry; gas phase kinetics. Mailing Add: 1600 Pineridge St Longview TX 75601

NEELEY, JOHN CHARLES, b Portland, Ore, Apr 9, 38; m 64; c 1. GENETICS. Educ: Portland State Col, BS, 60; Ore State Univ, MS, 63, PhD(zool), 70. Prof Exp: Asst prof biol, St Benedict's Col, Kans, 65-67; asst prof biol, Univ Portland, 67-75, head dept, 71-73; ASST PROF BIOL, PURDUE UNIV, 75- Mem: AAAS; Genetics Soc Am. Res: Factors affecting viability of metazoan aneuploids. Mailing Add: Dept of Biol Purdue Univ at Calumet Hammond IN 46323

NEELEY, VICTOR ISAAC, b Bell, Calif, Nov 5, 30; m 54; c 3. SOLID STATE PHYSICS. Educ: Univ Minn, BA, 57; Univ Idaho, MS, 60; Univ Ore, PhD(physics), 64. Prof Exp: Physicist, Gen Elec Co, Wash, 57-60; mgr solid state physics, Pac Northwest Labs, Battelle Mem Inst, 64-69; VPRES TECHNOL DEVELOP, HOLOSONICS, INC, 69- Mem: Am Nuclear Soc; Am Phys Soc. Res: Phonon interactions; electron spin resonance; lasers. Mailing Add: Holosonics Inc 2950 George Washington Way Richland WA 99352

NEELIN, JAMES MICHAEL, b London, Ont, Dec 4, 30; m 53; c 3. BIOCHEMISTRY. Educ: Univ Toronto, BA, 53, PhD(biochem), 58. Prof Exp: Jr scientist, Atlantic Tech Sta, Fisheries Res Bd, Can, 53-54, asst scientist, 54-55; Coun fel chem, Div Appl Biol, Nat Res Coun Can, 58-59, asst res officer, 59-62; res assoc chicken histones, Stanford Univ, 62-63; assoc res officer, Div Biosci, Nat Res Coun Can, 64-70, sr res officer, 71; PROF BIOL & CHMN DEPT, CARLETON UNIV, 71- Mem: Can Biochem Soc; Can Fedn Biol Sci. Res: Chemistry and biological implication of histones, especially of avian erythrocytes; mechanisms and exploitation of high resolution electrophoresis; cell separation and differentiation. Mailing Add: Dept of Biol Carleton Univ Ottawa ON Can

NEELY, BROCK WESLEY, b London, Ont, Apr 28, 26; nat US; m 53; c 3. BIOCHEMISTRY, PHYSICAL CHEMISTRY. Educ: Univ Toronto, BS, 48; Mich State Univ, PhD(biochem), 52. Prof Exp: Res Found fel chem, Ohio State Univ, 52-53; Rockefeller fel, Univ Birmingham, 53-54; res assoc, G D Searle & Co, Ill, 54-55; RES ASSOC BIOCHEM, DOW CHEM CO, 57-, RES ASSOC ENVIRON SCI, 73- Mem: Am Chem Soc; Sigma Xi; Am Soc Biol Chem. Res: Structure- activity relationships; environment research; math modeling. Mailing Add: 1702 Bldg Dow Chem Co Midland MI 48640

NEELY, CHARLES LEA, JR, b Memphis, Tenn, Aug 3, 27; m 57; c 2. MEDICINE, HEMATOLOGY. Educ: Princeton Univ, AB, 50; Washington Univ, MD, 54. Prof Exp: Intern, Bellevue Hosp, New York, 54-55; resident, Barnes Hosp, St Louis, 55-57, fel med & NIH trainee chemother, 57-58; from instr to assoc prof, 58-71, PROF MED, COL MED, UNIV TENN, MEMPHIS, 71- Concurrent Pos: Chief sect med oncol, City of Memphis Hosps, 74- Mem: Am Soc Hemat; fel Am Col Physicians; Am Fedn Clin Res; Am Soc Clin Oncol; AMA. Res: Hemolytic anemias and oncology. Mailing Add: Dept of Med Univ of Tenn Memphis TN 38163

NEELY, FLORENCE ELIZABETH, b Bushnell, Ill, Sept 13, 20. BOTANY. Educ: Univ Iowa, AB, 42, MS, 44; Univ Ill, PhD(bot), 51. Prof Exp: Asst prof bot, Univ Kansas City, 52-60; from asst prof to assoc prof, 60-73, PROF BOT, AUGUSTANA COL, ILL, 73- Mem: AAAS; Bot Soc Am; Am Inst Biol Sci. Mailing Add: Dept of Biol Augustana Col Rock Island IL 61201

NEELY, JAMES ALAN, b St Louis, Mo, Nov 7, 36; m 63. ANTHROPOLOGY, ARCHAEOLOGY. Educ: Univ of the Americas, BA, 58; Univ Ariz, MA, 68, PhD(anthrop), 73. Prof Exp: Asst archaeologist, Mus Northern Ariz, 62-63; res assoc archaeol, Rice Univ, 63-64 & Smithsonian Inst, 64; ASST PROF ANTHROP, CTR MID EASTERN STUDIES, UNIV TEX, AUSTIN, 68- Concurrent Pos: Res assoc archaeol, Smithsonian Inst, 66; archaeol field res in Iran, Rice Univ, 68-69. Mem: Fel AAAS; Am Anthrop Asn; Mex Anthrop Soc; Brit Inst Persian Studies. Res: Archaeological research on prehistoric plant and animal domestication and development of village farming communities and irrigation agriculture in Mexico and the Middle East. Mailing Add: Dept of Anthrop Univ Tex Ctr Mid Eastern Studies Austin TX 78712

NEELY, JAMES ROBERT, b Tenn, June 11, 35; m 54; c 5. CARDIOVASCULAR PHYSIOLOGY. Educ: Tenn Tech Univ, BS, 61, MA, 62; Vanderbilt Univ, PhD(physiol), 66. Prof Exp: Tenn Heart Asn fel, Vanderbilt Univ, 66-67; from instr to assoc prof, 67-76, PROF PHYSIOL, M S HERSHEY MED CTR, COL MED, PA STATE UNIV, 76- Concurrent Pos: Mem, Int Study Group Res in Cardiac Metab. Mem: Am Physiol Soc; Cardiac Muscle Soc. Res: Cardiac metabolism. Mailing Add: M S Hershey Med Ctr Pa State Univ Hershey PA 17033

NEELY, JAMES W, b Pittsburgh, Pa, May 23, 45; m 67. PHYSICAL INORGANIC CHEMISTRY. Educ: Carnegie Mellon Univ, BS, 67; Univ Calif, Berkeley, PhD(chem), 71. Prof Exp: Fel phys biochem, Univ Rochester, 71-77; SR SCIENTIST, ROHM AND HAAS CO, 74- Mailing Add: 15 Forest Lane Levittown PA 19055

NEELY, JAMES WINSTON, b Cotton Plant, Ark, Feb 4, 06; m 35; c 1. PLANT BREEDING. Educ: Univ Ark, BS, 28; Cornell Univ, PhD(genetics), 35. Prof Exp: Tech asst plant breeding, Univ Ark, 28-30; asst agron, Cornell Univ, 30-34; asst agronomist, Soil Erosion Serv, USDA, 34-35, assoc geneticist, Div Cotton & Other Fiber Crops & Diseases, Bur Plant Indust, 35-40, geneticist, 40-45; sr geneticist, Bur Plant Indust, Soils & Agr Eng, 45-46; plant breeder, Stoneville Pedigreed Seed Co, 46-51; vpres & dir, Plant Breeding & Agr Res, 51-71, CONSULT, COKER'S PEDIGREED SEED CO, 71- Mem: AAAS; Am Soc Agron. Res: Inheritance and linkage studies in cotton; improvement of field crops through breeding; field plot technique; breeding methods. Mailing Add: Coker's Pedigreed Seed Co Hartsville SC 29550

NEELY, PETER MUNRO, b Los Angeles, Calif, Dec 31, 27; m 65; c 2. STATISTICAL BIOLOGY. Educ: Univ Calif, Los Angeles, BA, 52, PhD(bot), 60. Prof Exp: Asst bot, Univ Calif, Los Angeles, 52-63; asst prof bot & res assoc, Biol Sci Comput Ctr, Chicago, 63-68; assoc prof statist biol, Dept Bot, 68-71, ASSOC PROF STATIST BIOL, DEPT SYSTS & ECOL, UNIV KANS, 71-, ASSOC DIR COMPUT CTR, 69- Concurrent Pos: USPHS fel, Sch Pub Health, Univ Calif, Los Angeles, 61-62; mem comt fortran language standards, Am Nat Standards. Res: Application of computer processing and statistical techniques to biological problems; grouping algorithms and development of general theory of classifications; development of interactive pedagogic programs; computational methodology. Mailing Add: Dept of Systs & Ecol Univ of Kans Lawrence KS 66044

NEELY, ROBERT DAN, b Senath, Mo, Oct 6, 28; m 53; c 2. PLANT PATHOLOGY. Educ: Univ Mo, BS, 50, PhD(bot), 57. Prof Exp: PLANT PATHOLOGIST, ILL NATURAL HIST SURV, 57- Concurrent Pos: Adj prof plant path, Univ Ill, Urbana-Champaign, 72- Honors & Awards: Author's Citation, Int Soc Arboriculture, 74. Mem: Mycol Soc Am; Am Phytopath Soc. Res: Diseases of shade and forest trees and woody ornamentals. Mailing Add: 382 Natural Resources Bldg Ill Natural Hist Surv Urbana IL 61801

NEELY, STANLEY CARRELL, b Abilene, Tex, Sept 11, 37; m 59; c 3. PHYSICAL CHEMISTRY. Educ: Southern Methodist Univ, BS, 60; Yale Univ, PhD(phys chem), 65. Prof Exp: Asst prof, 65-69, ASSOC PROF CHEM, UNIV OKLA, 69- Mem: Am Chem Soc. Res: Molecular and solid state spectroscopy; molecular interactions. Mailing Add: Dept of Chem Univ of Okla Norman OK 73069

NEELY, THOMAS ALEXANDER, physical chemistry, organic chemistry, see 12th edition

NEELY, WILLIAM CHARLES, b Cave City, Ark, Nov 22, 31; m 57; c 2. PHYSICAL CHEMISTRY. Educ: Miss State Col, BS, 53; La State Univ, MS, 60, PhD(chem), 62. Prof Exp: Chemist, Nylon Div, Chemstrand Corp, 53-54, res chemist, Chemstrand Res Ctr, Inc, 62-66; asst prof, 66-70, ASSOC PROF CHEM, AUBURN UNIV, 70- Mem: Am Chem Soc. Res: Molecular spectroscopy and photochemistry of uranyl compounds; amides; azo compounds; anthraquinones; electronic excitation energy transfer processes. Mailing Add: Dept of Chem Auburn Univ Auburn AL 36830

NEEMAN, MOSHE, b Latvia, Apr 1, 19; nat US; m 60; c 4. ORGANIC CHEMISTRY, BIOCHEMISTRY. Educ: Univ London, BSc, 43; Hebrew Univ, MSc, 45, PhD, 47. Prof Exp: Asst, D Sieff Res Inst, Israel, 36-44; with indust, 45-46; indust safety & hyg expert, Govt Israel, 48-52; dep dir, Israel Res Coun & dir, Lab Appl Org Chem, 52-56; sr lectr, Israel Inst Technol, 56; vis res assoc chem, Univ Wis, 56-59; assoc res prof chem, 64-72, co-chmn chem prog, 69-73, ASSOC CANCER RES SCIENTIST, ROSWELL PARK MEM INST, 59-, RES PROF CHEM, 72- Concurrent Pos: Chmn standards comt, Israel Standards Inst, 48-50; dir, Inst Indust Hyg, 50-52; res prof chem, Niagara Univ, 69-; res prof biol, Canisuis Col, 69- Honors & Awards: Szold Award, Israel, 57. Mem: Am Chem Soc; fel Royal Soc Health; NY Acad Sci; fel The Chem Soc; fel Am Inst Chemists. Res: Heterocyclic syntheses; diazoalkanes; medicinal chemistry; mechanism of carcinogenesis and tumor promotion; chemistry and reaction mechanisms of steroids; steroidal alkaloids; steroid hormone metabolism. Mailing Add: Roswell Park Mem Inst 666 Elm St Buffalo NY 14263

NEEPER, DONALD ANDREW, b New York, NY, Aug 9, 37. PHYSICS. Educ: Pomona Col, BA, 58; Univ Wis, MS, 60, PhD(physics), 64. Prof Exp: Res assoc, James Franck Inst, Univ Chicago, 66-68; STAFF MEM, LOS ALAMOS SCI LAB, 68- Mem: Am Phys Soc; AAAS. Res: Thermal transport at cryogenic and stellar temperatures. Mailing Add: 2708 B Walnut St Los Alamos NM 87544

NEER, KEITH LOWELL, b Springfield, Ohio, Feb 18, 49; m 71. MEAT SCIENCE. Educ: Ohio State Univ, BSc, 72, MSc, 73; Univ Nebr, PhD(animal sci), 75. Prof Exp: Technician meats, Ohio State Univ, 72-73; instr animal sci, Univ Nebr, Lincoln, 75-76; ASST PROF ANIMAL SCI, VA POLYTECH INST & STATE UNIV, 76- Mem: Am Meat Sci Asn; Am Soc Animal Sci; Inst Food Sci; Sigma Xi. Res: Investigating various feeding regimes and their effect on beef palatability; evaluating mechanical tenderization, pressing, power cleaving and cooking methods on beef palatability. Mailing Add: Dept of Animal Sci Va Polytech Inst & State Univ Blacksburg VA 24061

NEES, MONICA (ROSE), organic chemistry, information science, see 12th edition

NEESBY, TORBEN EMIL, b Copenhagen, Denmark, Apr 21, 09; nat US; m 39; c 3. CLINICAL CHEMISTRY. Educ: Tech Univ Denmark, MSc, 32. Prof Exp: Consult chemist, 32-40; tech dir, Norsk Sulfo, Norway, 40-43; private bus, Denmark, 43-48; tech dir, Am Sulfo, Inc, NY, 48-51; head org chem, Carroll Dunham Smith Pharmacal Co, 51-56; head lab, E F Drew & Co, 56-58; asst dir tech servs, Bristol Myers, Inc, 58; sr scientist, Schieffelin & Co, 58-61; sr biochemist surg res, Harbor Gen Hosp, Univ Calif, 61-68; CLIN BIOCHEMIST, VALLEY MED CTR, 68- Mem: AAAS; Am Chem Soc. Res: Structure-activity relationship; isotopes; investigation of mediator compounds; research on exogenous and endogenous polypeptide factors with effect on growth of fibroblasts and connective tissue, in vitro. Mailing Add: Valley Med Ctr 445 S Cedar St Fresno CA 93702

NEESON, JOHN FRANCIS, b Buffalo, NY, Dec 9, 36; m 63; c 1. NUCLEAR PHYSICS. Educ: Canisius Col, BS, 58; Univ Mich, MS, 60; State Univ NY Buffalo, PhD(physics), 65. Prof Exp: From asst prof to assoc prof physics, St Bonaventure Univ, 65-68; assoc prof, State Univ NY Col Brockport, 68-69; assoc prof, 69-74, PROF PHYSICS, ST BONAVENTURE UNIV, 74-, ASSOC DEAN COL ARTS & SCI, 69- Concurrent Pos: Consult, Clarke Bros Co Div, Dresser Industs, 65-67 & Dresser-Clark Div, 67-; sr scientist, Western NY Nuclear Res Ctr, 69- Mem: Am Phys Soc; Am Nuclear Soc; Am Asn Physics Teachers; Am Asn Higher Educ. Res: Low energy nuclear physics, especially nuclear structure of spherical nuclei; high pressure shock wave absorption. Mailing Add: Admin Bldg St Bonaventure Univ St Bonaventure NY 14778

NEET, KENNETH EDWARD, b St Petersburg, Fla, Sept 24, 36; m 60; c 4. BIOCHEMISTRY. Educ: Univ Fla, BSCh, 48, MS, 60, PhD(biochem), 65. Prof Exp: Fel biochem, Univ Calif, Berkeley, 65-67; asst prof, 67-72, ASSOC PROF BIOCHEM, SCH MED, CASE WESTERN RESERVE UNIV, 72- Mem: AAAS; Am Chem Soc; Am Soc Biol Chemists; NY Acad Sci. Res: Protein chemistry; enzyme regulation and mechanisms; subunit interactions. Mailing Add: Dept of Biochem Sch of Med Case Western Reserve Univ Cleveland OH 44106

NEFF, ALVEN WILLIAM, b Lafayette, Ind, Sept 13, 23; m 48; c 4. BIOCHEMISTRY. Educ: Kans State Col, BS, 47, MS, 48; Purdue Univ, PhD, 52. BIOCHEMIST, UPJOHN CO, 52- Mem: AAAS; Am Chem Soc. Res: Drug and pesticide metabolism; drug residue in animals; pesticide residues in plants. Mailing Add: Upjohn Co 301 Henrietta St Kalamazoo MI 49001

NEFF, BEVERLY JEAN, b Los Angeles, Calif, Nov 20, 25. VIROLOGY, EPIDEMIOLOGY. Educ: Stanford Univ, AB, 47; Univ Mich, Ann Arbor, PhD(epidemiol sci), 63. Prof Exp: Res asst virol, Dept Pharmacol, Sch Med, Stanford Univ, 47-49; pub health microbiologist, Virus Lab, Calif State Dept Health, 49-52, sr pub health microbiologist, 52-58; res assoc epidemiol-chronic dis, Dept Epidemiol, Sch Pub Health, Univ Mich, 58-70, asst prof nutrit, Dept Community Health Serv, 68-70; SR RES VIROLOGIST, MERCK INST THERAPEUT RES, MERCK & CO, 70- Mem: AAAS; Am Soc Microbiol; Tissue Cult Asn; Soc Exp Biol & Med. Res: Virology and serology of cell-associated Herpesviruses, varicella, cytomegalovirus, Mareks virus; live virus vaccines; epidemiology of heart disease and other chronic diseases; immunochemistry of lipoproteins and collagen. Mailing Add: Virus & Cell Biol MSDRL Merck Inst Therapeut Res West Point PA 19486

NEFF, CARROLL FORSYTH, b Pigeon, Mich, Jan 10, 08; m 30, 57; c 2. NUTRITION. Educ: Washington Univ, AB, 29; Univ Calif, MA, 30. Prof Exp: With res lab, Ralston Purina Co, Mo, 26-28 & 30-31; head biol lab, Anheuser-Busch, Inc, 31-38; mgr plant & labs, Sterling Drug, Inc, Ga, 38-50, plant mgr, SAfrica, 50-51, consult chemist, Ga, 51-56; with res dept, New Prod Liaison, Foremost Dairies, Inc, 57-62; TECH CONSULT FOOD, DRUGS & COSMETICS, 62- Concurrent Pos: Scholar, Food Law Inst, Emory Univ, 54; western rep, Food & Drug Res Labs, Inc, NY, 68-72; lectr-consult, Nat Agr Col Chapingo, Mex, 74- Mem: AAAS; Am Chem Soc; Am Pharmaceut Asn; Am Mkt Asn; Inst Food Technologists. Res: Chemotherapy of poultry coccidiosis; yeast vitamins; diet and immunity; milk derivatives; special dietary use foods; food and drug regulations. Mailing Add: Av Tonanzintla 40 Colonia LaPaz Puebla Mexico

NEFF, DON JOHNSON, wildlife management, see 12th edition

NEFF, JERRY MICHAEL, comparative physiology, electron microscopy, see 12th edition

NEFF, JOHN DAVID, b Cedar Rapids, Iowa, July 30, 26; m 52. MATHEMATICS. Educ: Marquette Univ, BNS, 46; Coe Col, BA, 49; Kans State Univ, MS, 51; Univ Fla, PhD(math), 56. Prof Exp: Mem tech staff, Bell Tel Labs, NY, 52-53; instr math, Univ Fla, 53-55; from instr to asst prof, Case Inst Technol, 56-61; from asst prof to assoc prof, 61-72, actg dir math, 70-72, PROF MATH, GA INST TECHNOL, 72-, DIR MATH, 72- Mem: Am Math Soc; Math Asn Am; Inst Math Statist; Soc Indust & Appl Math; Nat Coun Teachers Math. Res: Differential equations; probability. Mailing Add: Sch of Math Ga Inst of Technol Atlanta GA 30332

NEFF, JOHN S, b Milwaukee, Wis, Nov 24, 34; m 60; c 2. ASTRONOMY, ASTROPHYSICS. Educ: Univ Wis, BS, 57, MS, 58, PhD(astron), 61. Prof Exp: Res assoc, Yerkes Observ, Chicago, 61-64; asst prof, 64-68, ASSOC PROF ASTRON, UNIV IOWA, 68- Concurrent Pos: NSF res grant, 65-67. Mem: Am Astron Soc; fel

Royal Astron Soc; Int Astron Union. Res: Stellar photometry and spectrophotometry; planetary spectrophotometry; design of astronomical instruments; stellar classification and investigation of galactic structure. Mailing Add: Dept of Physics & Astron Univ of Iowa 704 Physics Bldg Iowa City IA 52240

NEFF, LAURENCE D, b Santa Ana, Calif, Jan 11, 38; m 60; c 2. PHYSICAL CHEMISTRY. Educ: John Brown Univ, BA, 59; Univ Ark, MS, 62, PhD(phys chem), 64. Prof Exp: Sr chemist, NAm Aviation, Inc, 64; res chemist, Beckman Instruments, Inc, 64-66; res specialist mat sci, NAm Rockwell Corp, 66-68; asst prof, 68-74, ASSOC PROF CHEM, E TEX STATE UNIV, 74- Mem: AAAS; Am Chem Soc. Res: Heterogeneous catalysis; application of infrared spectroscopy to gas-solid phase interactions. Mailing Add: Dept of Chem ETex State Univ Commerce TX 75428

NEFF, LOREN LEE, b Seattle, Wash, Sept 14, 18; m 40; c 1. CHEMISTRY. Educ: Univ Wash, Seattle, BS, 39, PhD(chem), 43. Prof Exp: Asst chem, Univ Wash, Seattle, 39-43; SUPVR RES DEPT, UNION OIL CO CALIF, 43- Mem: Am Chem Soc. Res: Lubricating oil additives; lubricating oils; activity of colloidal electrolytes; determination of the thermodynamic activity of 1-dodecane sulfonic acid in aqueous solutions at forty degrees Centigrade by electromotive force measurements; petroleum industry corrosion; chemicals. Mailing Add: Res Dept Union Oil Co of Calif PO Box 76 Brea CA 92621

NEFF, MARY MUSKOFF, b Jacksonville, Fla, Jan 20, 30; m 52. MATHEMATICS. Educ: Purdue Univ, BS, 51, MS, 52; Univ Fla, PhD(math), 56. Prof Exp: Instr math, Univ Fla, 55-56; from instr to asst prof, John Carroll Univ, 56-61; asst prof, 61-68, ASSOC PROF MATH, EMORY UNIV, 68- Mem: Am Math Soc. Res: Identities in lattices; groups given by generators and relations. Mailing Add: Dept of Math Emory Univ Atlanta GA 30322

NEFF, RICHARD D, b Elmo, Mo, Oct 7, 32; m 54; c 4. HEALTH PHYSICS. Educ: Northwest Mo State Col, BS, 57; Univ Kans, MS, 59; Univ Calif, Los Angeles, PhD(biophys), 64; Am Bd Health Physics, cert, 73. Prof Exp: Res biophysicist, Lab Nuclear Med & Radiation Biol, Univ Calif, Los Angeles, 59; health physicist, Atomics Int, 59-60; radiation physicist, Tison-Pease Inc, 60-61; asst prof radiol health, Mich State Univ, 64-66; from asst prof to assoc prof, 66-73, RADIOL SAFETY OFF, TEX A&M UNIV, 66-, PROF NUCLEAR ENG, 73- Mem: AAAS; Health Physics Soc; Am Nuclear Soc. Res: Mammalian radiation biology; environmental radioactivity. Mailing Add: Dept of Nuclear Eng Tex A&M Univ College Station TX 77843

NEFF, ROBERT JACK, b Kansas City, Mo, Jan 22, 21; m 48; c 3. BIOLOGY, PHYSIOLOGY. Educ: Univ Mo, AB, 42, MA, 48, PhD(zool), 51. Prof Exp: Instr anat, Sch Med, Johns Hopkins Univ, 51-52; from asst prof to assoc prof biol, 52-64, ASSOC PROF MOLECULAR BIOL, VANDERBILT UNIV, 64- Concurrent Pos: Vis assoc prof, Univ Calif, 57-58; NIH spec fel, Biol Inst, Carlsberg Found, Copenhagen, 65-66. Mem: AAAS; Soc Protozool; Am Soc Cell Biol; NY Acad Sci. Res: Cellular physiology; nuclear-cytoplasmic control mechanisms; cellular osmoregulation; macromolecular organization of protoplasm; cytodifferentiation; encystment; cell growth-division cycle. Mailing Add: Dept of Molecular Biol Vanderbilt Univ Nashville TX 37235

NEFF, RUTH HENSLEY, zoology, see 12th edition

NEFF, STUART EDMUND, b Louisville, Ky, Oct 3, 26; m 48; c 3. BIOLOGY. Educ: Univ Louisville, BS, 54; Cornell Univ, PhD(limnol), 60. Prof Exp: Asst limnol, Cornell Univ, 54-59, instr, 60; from asst prof to assoc prof, Va Polytech Inst, 60-68; assoc res prof, 68-72, RES PROF BIOL, WATER RESOURCES LAB, UNIV LOUISVILLE, 72- Mem: Soc Syst Zool; Entom Soc Am; Royal Entom Soc London. Res: Hydrobiology; immature stages of aquatic insects; taxonomy and biology of acalyptrate Diptera; biology of Chironomidae. Mailing Add: Water Resources Lab Univ of Louisville Louisville KY 40208

NEFF, THOMAS RODNEY, b Salt Lake City, Utah, Sept 22, 37; m 70. GEOLOGY. Educ: Univ Utah, BS, 60, MS, 63; Stanford Univ, PhD(geol), 69. Prof Exp: Field asst, Utah Geol & Mineral Surv, 59-60 & Humble Oil & Ref Co, 63; ASST PROF GEOL, WEBER STATE COL, 68- Concurrent Pos: Vis res prof, Mineral-Geol Mus, Oslo, Norway, 73-74. Mem: Geol Soc Am; Sigma Xi. Res: Structural geology; petrography; petrology of Norwegian granitic rocks. Mailing Add: Dept of Geol & Geog Weber State Col Ogden UT 84403

NEFF, VERNON DUANE, b Rochester, NY, Sept 16, 32; m 55; c 4. PHYSICAL CHEMISTRY. Educ: Syracuse Univ, BS, 53, PhD(phys chem), 60. Prof Exp: Spectroscopist, Gen Tire & Rubber Co, 59-61; asst prof, 61-70, ASSOC PROF PHYS CHEM, KENT STATE UNIV, 70- Concurrent Pos: Consult, Gen Tire & Rubber Co, 61- Mem: Am Chem Soc; Am Phys Soc. Res: Infrared spectroscopy; quantum chemistry. Mailing Add: Dept of Chem Kent State Univ Kent OH 44240

NEFF, WILLIAM DAVID, b Portland, Ore, Mar 25, 45; m 67; c 2. ATMOSPHERIC PHYSICS. Educ: Lewis & Clark Col, BA, 67; Univ Wash, MS, 68. Prof Exp: Commissioned officer, 68-73, PHYS SCIENTIST, ATMOSPHERIC ACOUST, NAT OCEANIC & ATMOSPHERIC ADMIN, 73- Mem: Am Meteorol Soc. Res: Applications of acoustic remote sensing techniques to boundary layer meteorology, particularly of boundary layer structure over Antarctic ice plateau. Mailing Add: Wave Propagation Lab R45X8 Boulder CO 80302

NEFF, WILLIAM DUWAYNE, b Lomax, Ill, Oct 27, 12; m 37; c 2. NEUROSCIENCES. Educ: Univ Ill, AB, 36; Univ Rochester, PhD(psychol), 40. Prof Exp: Res assoc psychol, Swarthmore Col, 40-42; from asst prof to prof, Univ Chicago, 46-59, psychol & physiol, 59-61; dir, Lab Physiol Psychol, Bolt, Beranek, Newman, Inc, Mass, 61-63; prof psychol, 63-64, RES PROF, IND UNIV, BLOOMINGTON, 64-, DIR CTR NEURAL SCI, 65- Concurrent Pos: Consult, Nat Acad Sci, Nat Res Coun, NSF, NIH, & NASA; mem, Otolaryngol Res Group, Int Brain Res Orgn & sci liaison off, London Br, Off Naval Res, 53-54. Honors & Awards: Hearing Res Award, Beltone Inst. Mem: Nat Acad Sci; fel AAAS; fel Am Physiol Soc; fel Acoust Soc Am; Soc Exp Psychol (secy-treas, 52-59). Res: Brain functions; neural mechanisms of sensory discrimination; physiological acoustics. Mailing Add: Ctr for Neural Sci Ind Univ Bloomington IN 47401

NEFF, WILLIAM H, b May 13, 31; US citizen; m 50; c 6. ZOOLOGY, PHYSIOLOGY. Educ: Pa State Univ, BS, 56, MS, 59, PhD(zool), 66. Prof Exp: Instr anat & physiol, 59-65, asst prof zool, 66-75, ASSOC PROF BIOL, PA STATE UNIV, 75-, COORDR BIOL SCI, COMMONWEALTH CAMPUSES, 66- Res: Adaptation to environmental stress; gross metabolic and electrolyte response during cold acclimation; histological changes in reproductive tract of season breeding mammals. Mailing Add: Dept of Biol Pa State Univ University Park PA 16802

NEFF, WILLIAM MEDINA, b San Francisco, Calif, Oct 27, 29; m 52; c 4. EMBRYOLOGY. Educ: Stanford Univ, AB, 51, PhD(biol sci, statist), 58. Prof Exp: From instr to assoc prof biol, Knox Col, 56-68; assoc prof, Chico State Col, 68-70; ASSOC PROF CITY COL SAN FRANCISCO, 70- Mem: Sigma Xi. Res: Development of the skin; cells of the dermis. Mailing Add: Dept of Biol City Col of San Francisco San Francisco CA 94112

NEFZGER, MERL DEAN, statistics, see 12th edition

NEGAS, TAKI, high temperature chemistry, solid state chemistry, see 12th edition

NEGELE, JOHN WILLIAM, b Cleveland, Ohio, Apr 18, 44; m 67. THEORETICAL PHYSICS. Educ: Purdue Univ, Lafayette, BS, 65; Cornell Univ, PhD(theoret physics), 69. Prof Exp: NATO fel & vis physicist, Niels Bohr Inst, Copenhagen, Denmark, 69-70; vis asst prof, 70-71, asst prof, 71-72, ASSOC PROF PHYSICS, MASS INST TECHNOL, 72- Concurrent Pos: Alfred P Sloan Foun res fel, 72; consult, Los Alamos Sci Lab, 73; mem, Brookhaven Nat Lab, Tandem Prog Adv Comt, 73- Mem: Am Phys Soc. Res: Theoretical nuclear physics; many-body theory; microscopic theory of nuclear structure. Mailing Add: Rm 6-302 Dept of Physics Mass Inst of Technol Cambridge MA 02139

NEGGERS, JOSEPH, b Amsterdam, Netherlands, Jan 10, 40; US citizen; m 65. MATHEMATICS. Educ: Fla State Univ, BS, 59, MS, 60, PhD(math), 63. Prof Exp: Asst prof math, Fla State Univ, 63-64; sci asst math, Univ Amsterdam, 64-65; lectr pure math, King's Col, London, 65-66; vis assoc prof, Univ P R, 66-67; asst prof, 67-68, ASSOC PROF MATH, UNIV ALA, 68- Mem: Am Math Soc. Res: Algebra; derivations and automorphisms on local-rings, associated structures; partially ordered sets. Mailing Add: Dept of Math Univ of Ala University AL 35486

NEGHME, AMADOR, b Iquique, Chile, Feb 15, 12; m 60; c 5. PARASITOLOGY, MEDICINE. Prof Exp: Prof parasitol, Med Sch, Univ Chile, 47-58, dean fac med, 63-68; sci dir, Regional Libr Med, Pan-Am Health Orgn, 69-76. Concurrent Pos: Chilean rep, Int Cong Malaria & Trop Med, 48, 58 & 63; mem expert comt parasitic infection, WHO, 56-61; mem coun, Chilean Nat Health Serv, 59-68; pres, Latin Am Fedn Parasitologists, 63-67 & Pan-Am Fedn Med Schs, 64-68; mem, Acad Med, Inst Chile. Mem: AAAS; hon mem Am Soc Trop Med & Hyg; Am Soc Parasitologists. Res: Medical education; information sciences; biomedical librarianship. Mailing Add: Rua Botucatu 862 Vila Clementino 04023 Sao Paulo Brazil

NEGIN, MICHAEL, b Tampa, Fla, Dec 19, 42; m 63; c 3. ELECTRICAL ENGINEERING, BIOMEDICAL ENGINEERING. Educ: Univ Fla, BEE, 64, MSE, 65, PhD(elec eng), 68; Temple Univ, MS, 75. Prof Exp: Asst prof, 68-71, ASSOC PROF ELEC & BIOMED ENG, DREXEL UNIV, 71- Concurrent Pos: Nat Inst Gen Med Sci spec fel, 72-74. Mem: AAAS; Sigma Xi; Inst Elec & Electronics Engrs. Res: Early detection of breast cancer by computerized thermography and radiography; sleep electroencephalographic analysis by computer; digital simulation of biosystems; electromyographic and electroencephalographic studies and the relationship between these bioelectric events; dental radiography by digital computer. Mailing Add: Biomed Eng & Sci Prog Drexel Univ Philadelphia PA 19104

NEGISHI, EI-ICHI, b Shinkyo-City, Japan, July 14, 35; m 60; c 2. ORGANIC CHEMISTRY. Educ: Univ Tokyo, BS, 58; Univ Pa, PhD(chem), 63. Prof Exp: Res chemist, Res Inst, Teijin Ltd, Japan, 58-60, 63-66; res assoc organometallic chem, Purdue Univ, 66-72; ASST PROF ORGANIC & ORGANOMETALLIC CHEM, SYRACUSE UNIV, 72- Mem: Am Chem Soc. Res: Synthesis of organic compounds via organometallic compounds; exploratory organometallic chemistry and its application to organic synthesis. Mailing Add: Dept of Chem Syracuse Univ Syracuse NY 13210

NEGLIA, MARIA TERESA, spectroscopy, see 12th edition

NEGREPONTIS, STYLIANOS, b Thessaloniki, Greece, Feb 22, 39. MATHEMATICS. Educ: Univ Rochester, BSc, 61, MA, 63, PhD(math), 65. Prof Exp: Vis asst prof math, Ind Univ, 65-66; asst prof, 66-70, ASSOC PROF MATH, McGILL UNIV, 70- Concurrent Pos: Nat Res Coun Can res grant, 71-74; prof, Nat Univ Athens, 72-73. Mem: Can Math Cong; Greek Math Soc; Am Math Soc; Asn Symbolic Logic. Res: Model theory and ultraproducts; set theoretical, topological and algebraic aspects in the theory of rings of continuous functions and the Stone-Cech compactification. Mailing Add: Dept of Math McGill Univ Montreal PQ Can

NEGUS, NORMAN CURTISS, b Portland, Ore, Sept 20, 26; m 48; c 4. ZOOLOGY. Educ: Miami Univ, BA, 48, MA, 50; Ohio State Univ, PhD(zool), 56. Prof Exp: Asst zool, Miami Univ, 48-50; res fel, Ohio State Univ, 51-55; from instr to prof, Tulane Univ, 55-70; PROF ZOOL, UNIV UTAH, 70- Concurrent Pos: Vis investr, SEATO Med Res Lab, Thailand, 65-66. Mem: Am Soc Mammalogists; Ecol Soc Am; Am Soc Zoologists; Am Asn Anatomists; Wildlife Soc. Res: Mammalian population dynamics; reproductive physiology; molting in mammals; orientation and movements of mammals. Mailing Add: Dept of Biol Univ of Utah Salt Lake City UT 84112

NEHARI, ZEEV, b Berlin, Ger, Feb 2, 15; nat US; m 38. MATHEMATICS. Educ: Hebrew Univ, Israel, MA, 39, PhD(math), 41. Prof Exp: Instr math, Hebrew Univ, Israel, 46-47; res proj, Harvard Univ, 47-48; from assoc prof to prof math, Wash Univ, 48-54; PROF MATH, CARNEGIE-MELLON UNIV, 54- Mem: Am Math Soc; Math Asn Am. Res: Theory of functions; conformal mapping; differential equations. Mailing Add: Dept of Math Carnegie-Mellon Univ Pittsburgh PA 15213

NEHER, CLARENCE M, b Twin Falls, Idaho, May 14, 16; m 39; c 4. ORGANIC CHEMISTRY. Educ: Manchester Col, AB, 37; Purdue Univ, MS, 39, PhD(chem), 41. Prof Exp: Asst, Purdue Univ, 37-40; res chemist, 41-45, res supvr, 45-51, asst dir res & develop labs, 51-54, proj mgr, 54-57, dir com develop, 57-63, spec assignment, 63-64, vpres & gen mgr, Plastics Div, 64-69, SR VPRES, ETHYL CORP, 69- MEM BD DIRS, 70- Mem: Mfg Chem Asn; Am Chem Soc; Chem Mkt Res Asn; Com Chem Develop Asn; Soc Plastics Indust. Res: Chlorination; polymers; chlorination of aliphatics and aromatics; cracking of chlorocarbons; chlorination methane and ethane. Mailing Add: Ethyl Corp Ethyl Tower 451 Florida Baton Rouge LA 70801

NEHER, DAVID DANIEL, b McCune, Kans, July 12, 23; m 50; c 3. SOILS. Educ: Kans State Univ, BS, 46, MS, 48; Utah State Univ, PhD(soil sci), 59. Prof Exp: Instr soils, Kans State Univ, 48-49; from asst prof to assoc prof, 49-65, PROF SOILS, TEX A&I UNIV, 65- Mem: Soil Sci Soc Am; Am Soc Agron; Crop Sci Soc Am; Soil Conserv Soc Am. Mailing Add: Col of Agr PO Box 156 Tex A&I Univ Kingsville TX 78363

NEHER, DEAN ROYCE, b Enterprise, Kans, Feb 10, 29; m 53; c 4. PHYSICS, COMPUTER SCIENCE. Educ: McPherson Col, BS, 54; Univ Kans, MS, 59, PhD(physics), 64. Prof Exp: From asst prof to assoc prof, 61-68, PROF PHYSICS, BRIDGEWATER COL, 68-, DIR COMPUT CTR, 70- Mem: Am Asn Physics Teachers. Res: Nuclear physics; use of computers in teaching. Mailing Add: 210 E College St Bridgewater VA 22812

NEHER, GALEN HARL, experimental pathology, toxicology, see 12th edition

NEHER, GEORGE MARTIN, b Chicago, Ill, June 4, 21; m 45. VETERINARY SCIENCE. Educ: Purdue Univ, BS, 47, MS, 50, PhD(endocrinol), 53, DVM, 67. Prof Exp: From asst prof to assoc prof vet sci, 54-66, PROF VET PHYSIOL & PHARMACOL, SCH VET SCI & MED, PURDUE UNIV, WEST LAFAYETTE, 66- Mem: Assoc Am Vet Radiol Soc; assoc Am Rheumatism Asn; Conf Res Workers Animal Dis; Am Vet Med Asn; Soc Study Reproduction. Res: Histochemistry and pathogenesis of the arthritides of animals; experimental hypopituitarism in swine; physiologic effects of noise. Mailing Add: Dept of Physiol & Pharmacol Purdue Univ Sch Vet Sci & Med West Lafayette IN 47906

NEHER, LELAND K, b Porterville, Calif, Dec 2, 20; m 43; c 7. NUCLEAR PHYSICS. Educ: Pomona Col, BA, 43; Univ Calif, PhD(physics), 53. Prof Exp: MEM STAFF, LOS ALAMOS SCI LAB, UNIV CALIF, 53- Mem: AAAS; Am Phys Soc. Res: Nuclear science. Mailing Add: 205 Rio Bravo W R Los Alamos NM 87544

NEHER, MAYNARD BRUCE, b Greenville, Ohio, Apr 2, 23; m 44; c 3. CHEMISTRY. Educ: Manchester Col, AB, 44; Purdue Univ, PhD(chem), 47. Prof Exp: Asst chem, Purdue Univ, 44-47; asst prof, Univ Ohio, 47-51; res chemist, 51-56, ASST DIV CHIEF, BATTELLE MEM INST, 56- Mem: Am Chem Soc; AAAS. Res: Nitroparaffin derivatives as insecticidal compounds; aldol condensation of fluoroform with acetone; preparation of selected diaryl nitroalkanes and their derivatives; leather treatments; application of computer techniques in organic analysis; structural organic chemistry; mass spectroscopy; gas-liquid chromatography. Mailing Add: 3911 Bickley Pl Columbus OH 43220

NEHER, ROBERT TROSTLE, b Mt Morris, Ill, Nov 1, 30; m 54; c 3. PLANT TAXONOMY, ENVIRONMENTAL BIOLOGY. Educ: Manchester Col, BS, 53; Univ Ind, MAT, 55, PhD(bot), 66; Bethany Sem, MRE, 57. Prof Exp: From asst prof to assoc prof, 57-67, PROF BIOL, LA VERNE COL, 67-, CHMN DEPT, 71-, CHMN NATURAL SCI DIV, 72- Concurrent Pos: NSF fac fel, 62-63. Mem: AAAS; Am Soc Nat; Am Soc Plant Taxon; Asn Am Med Cols. Res: Systematic studies in Tagetes, stressing chemotaxonomy and cytogenetics; study and development of environemental control models; development of multi-stage aquaculture systems. Mailing Add: Dept of Biol La Verne Col 1950 Third St La Verne CA 91750

NEHLS, JAMES WARWICK, b Memphis, Tenn, June 30, 26; m 53; c 3. INORGANIC CHEMISTRY. Educ: Univ Tenn, BS, 48, MS, 49, PhD(chem), 52. Prof Exp: Asst chem, Univ Tenn, 49-52; chemist, E I du Pont de Nemours & Co Tenn, 52, Ind, 52-53, & SC, 53-60; chemist, Oak Ridge Opers, 60-65, ASST TO DIR, RES & DEVELOP DIV, USAEC, 65- Concurrent Pos: Chemist, Res & Tech Support Div, US Energy Res & Develop Admin, 73- Mem: AAAS; Am Chem Soc. Res: Transplutonium elements; radisotope production; exchange reactions; radiochemistry. Mailing Add: 121 Balboa Circle Oak Ridge TN 37830

NEI, MASATOSHI, b Miyazaki, Japan, Jan 2, 31; m 63; c 2. POPULATION GENETICS, EVOLUTION. Educ: Miyazaki Univ, BS, 53; Kyoto Univ, MS, 55, PhD(genetics), 59. Prof Exp: Instr, Kyoto Univ, 58-62; geneticist, Nat Inst Radiol Sci, Japan, 62-64, chief geneticist, 65, head lab, 65-69; from assoc prof to prof, Brown Univ, 69-72; PROF POP GENETICS, CTR DEMOG & POP GENETICS, UNIV TEX, HOUSTON, 72- Mem: AAAS; Am Soc Naturalists; Am Soc Human Genetics; Soc Study Evolution; Genetics Soc Am. Res: Population dynamics of mutant genes; genetic structure of populations; molecular evolution. Mailing Add: Ctr for Demog & Pop Genetics Univ of Tex Houston TX 77030

NEIBURGER, MORRIS, b Hazleton, Pa, Dec 5, 10; m 32; c 1. METEOROLOGY. Educ: Univ Chicago, BS, 36, PhD(meteorol), 45. Prof Exp: Observer meteorol, US Weather Bur, Ill, 30-38, from jr meteorologist to asst meteorologist, Washington, DC, 39-40; instr meteorol, Mass Inst Technol, 40-41; from instr to assoc prof, 41-54, chmn dept, 56-62, PROF METEOROL, UNIV CALIF, LOS ANGELES, 54- Concurrent Pos: Instr grad sch, USDA, 39-40; sr meteorologist, Air Pollution Found, 54-56; members' rep, Univ Corp for Atmospheric Res, Univ Calif, Los Angeles, 67- Honors & Awards: Meisinger Award, Am Meteorol Soc, 46. Mem: Fel AAAS; Am Meteorol Soc (pres, 62-64); for mem Royal Meteorol Soc. Res: Physics of clouds; air pollution; synoptic meteorology; upper wind dynamics and temperature forecasting; atmospheric radiaton; evaporation. Mailing Add: Dept of Meteorol Univ of Calif Los Angeles CA 90024

NEIDELL, NORMAN SAMSON, b New York, NY, Mar 11, 39; m 63; c 4. GEOPHYSICS. Educ: NY Univ, BA, 59; Univ London, DIC, 61; Cambridge Univ, PhD(geod, geophys), 64. Prof Exp: Res geophysicist, Gulf Res & Develop Co, 64-68; geophys researcher, Seismic Comput Corp, 68-72; CHMN & PRES, N S NEIDELL & ASSOCS, INC, 73- Concurrent Pos: Consult, partner prin, Geoquest Int, Ltd, 73-; assoc ed marine opers, Geophysics, 76-; lectr, Univ Houston. Honors & Awards: Best Presentation Award, Soc Explor Geophysicists, 74. Mem: Soc Explor Geophys; Am Inst Navig; Soc Photo-optical Instrument Eng. Res: Signal analysis and computer processing in the earth sciences; digital computer applications in geology and geophysics. Mailing Add: 13015 Kimberley Houston TX 77024

NEIDERHISER, DEWEY HAROLD, b Masontown, Pa, Jan 22, 35. BIOCHEMISTRY. Educ: Duquesne Univ, BS, 56; Univ Pittsburgh, MS, 59; Univ Wis, PhD(biochem), 63. Prof Exp: ASST PROF BIOCHEM & RES CHEMIST, VET ADMIN HOSP, SCH MED, CASE WESTERN RESERVE UNIV, 62- Res: Biochemistry of human cholesterol gallstone formation. Mailing Add: Res Dept Vet Admin Hosp Cleveland OH 44106

NEIDHARDT, FREDERICK CARL, b Philadelphia, Pa, May 12, 31; m 56; c 2. MICROBIOLOGY, BIOCHEMISTRY. Educ: Kenyon Col, BA, 52; Harvard Univ, PhD(bact), 56. Prof Exp: Am Cancer Soc res fel, Inst Pasteur, Paris, 56-57; Harold C Ernst res fel, Harvard Med Sch, 57-58, instr bact & immunol, 58-59, assoc, 59-61; from assoc prof biol sci to prof & assoc head dept, Purdue Univ, 61-70; PROF MICROBIOL & CHMN DEPT, MED SCH, UNIV MICH, ANN ARBOR, 70- Concurrent Pos: Mem microbial chem study sect, NIH, 65-69; NSF sr fels, Univ Inst Microbiol, Copenhagen, 68-69; mem comn scholars, Ill Bd Higher Educ, 73- Honors & Awards: Award Bact & Immunol, Eli Lilly & Co, 66. Mem: AAAS; Am Soc Microbiol; Am Soc Biol Chemists; Soc Gen Physiol; NY Acad Sci. Res: Regulation of gene expression in bacteria; regulation of bacterial metabolism; regulation of macromolecule synthesis in bacteria. Mailing Add: Dept Microbiol 6643 Med Sci II Univ of Mich Med Sch Ann Arbor MI 48104

NEIDHARDT, WALTER JIM, b Paterson, NJ, June 19, 34; m 62; c 2. LOW TEMPERATURE PHYSICS, QUANTUM PHYSICS. Educ: Stevens Inst Technol, ME, 56, MS, 58, PhD(physics), 62. Prof Exp: Instr, Newark Col Eng, 62-63, asst prof, 63-67, ASSOC PROF PHYSICS, NJ INST TECHNOL, 67- Concurrent Pos: Consult ed, J Am Sci Affil, 68- Mem: Am Phys Soc; Am Inst Physics Teachers; fel Am Sci Affil; Sigma Xi; NY Acad Sci. Res: Application of quantum physics to low temperature phenomena; examination of the nature of science, its proper domains and limits; emphasis on those integrative concepts that are common to science, philosophy and religion. Mailing Add: 146 Park Ave RD 2 Randolph NJ 07801

NEIDIG, HOWARD ANTHONY, b Lemoyne, Pa, Jan 25, 23; m 46; c 2. PHYSICAL CHEMISTRY, ORGANIC CHEMISTRY. Educ: Lebanon Valley Col, BS, 43; Univ Del, MS, 46, PhD(chem), 48. Prof Exp: Instr chem, Univ Del, 46-48; from asst prof to assoc prof, 48-59, HEAD DEPT, LEBANON VALLEY COL, 51-, PROF CHEM, 59- Mem: AAAS; Am Chem Soc. Res: Reaction mechanism; molecular rearrangements; mechanism of oxidation and reduction reactions in organic chemistry. Mailing Add: Dept of Chem Lebanon Valley Col Annville PA 17003

NEIDLE, AMOS, biochemistry, neurochemistry, see 12th edition

NEIDLE, ENID ANNE, b New York, NY, Apr 6, 24; m 49; c 2. PHYSIOLOGY, PHARMACOLOGY. Educ: Vassar Col, AB, 44; Columbia Univ, PhD(physiol), 49. Prof Exp: Assoc pharmacol, Jefferson Med Col, 49-50; instr biol, Brooklyn Col, 50-54; from instr to assoc prof physiol & pharmacol, 55-72, PROF PHYSIOL & PHARMACOL, COL DENT & GRAD FAC ARTS & SCI, NY UNIV, 72- Concurrent Pos: USPHS grant, 60. Mem: AAAS; Am Asn Dent Schs; Sigma Xi; Am Physiol Soc; Harvey Soc. Res: Vasomotor innervation of orofacial structures; circulation in the dental pulp; contribution of the mandibular nerve to growth, development and vasomotion in orofacial structures. Mailing Add: NY Univ Col of Dent 342 E 26th St New York NY 10010

NEIDLEMAN, SAUL L, b New York, NY, Oct 3, 29; m 56; c 4. BIOCHEMISTRY. Educ: Mass Inst Technol, MS, 52; Univ Ariz, PhD(biochem), 59. Prof Exp: Biophysicist, Peter Bent Brigham Hosp, Boston, 53-54; res assoc agr biochem, Univ Ariz, 58-59; sr res microbiologist, 59-67, res assoc, 67-69, SECT HEAD MICROBIAL BIOCHEM, SQUIBB INST MED RES DIV, OLIN CORP, 69- Mem: AAAS; Am Chem Soc. Res: Antibiotic biosynthesis; antimetabolites; amino acid and fungal metabolism; enzyme reactions. Mailing Add: Squibb Inst for Med Res PO Box 400 Princeton NJ 08540

NEIE, VAN ELROY, b Clifton, Tex, Nov 1, 38; m 63; c 2. PHYSICS EDUCATION. Educ: McMurry Col, BA, 61; NTex State Univ, MS, 66; Fla State Univ, PhD(sci educ), 70. Prof Exp: Instr physics, NTex State Univ, 64-67; teacher, Fla High Sch, Tallahassee, 68-69; asst prof, 70-76, ASSOC PROF PHYSICS & EDUC, PURDUE UNIV, WEST LAFAYETTE, 76- Concurrent Pos: Mem sci adv comt, Ind Dept Pub Instr, 71- Mem: Nat Asn Res Sci Teaching; Nat Sci Teachers Asn; Asn Educ Teachers Sci; Am Asn Physics Teachers. Res: Information theory and its application to lexical communication in science. Mailing Add: Dept of Physics Purdue Univ West Lafayette IN 47907

NEIGHBOURS, JOHN ROBERT, b Cleveland, Ohio, Nov 22, 24; m 46; c 4. SOLID STATE PHYSICS. Educ: Case Inst Technol, PhD(physics), 53. Prof Exp: Asst physics, Case Inst Technol, 49-53; asst prof, Rensselaer Polytech Inst, 53-55; res engr, Ford Motor Co, 55-59; assoc prof, 59-64, PROF PHYSICS, US NAVAL POSTGRAD SCH, 64- Concurrent Pos: Chmn dept physics, Colo State Univ, 68-70. Mem: Fel Am Phys Soc. Res: Wave propagation in anisotropic media; electronic and magnetic properties of solids at low temperatures. Mailing Add: Dept of Physics US Naval Postgrad Sch Monterey CA 93940

NEIHOF, REX A, b Ponca City, Okla, Oct 31, 21; m 49; c 3. MARINE CHEMISTRY, MICROBIAL BIOCHEMISTRY. Educ: Tex Technol Col, BS, 43; Univ Minn, PhD(biochem), 50. Prof Exp: Chem engr, Gates Rubber Co, Colo, 43-45; asst, Univ Minn, 46-48, instr physiol chem, 48-49; phys chemist, NIH, 49-55, 57-58; Nat Heart Inst spec fel, Physiol Inst, Univ Uppsala, 55-57; RES CHEMIST, US NAVAL RES LAB, 58- Concurrent Pos: Vis scientist, Univ Calif, San Diego, 70-71. Mem: Am Soc Microbiol; Am Chem Soc; Sigma Xi. Res: Physical chemistry and electrochemistry of membrane processes; permselective membranes; biophysical studies on biological membranes; microbial cell walls; chemistry of seawater; marine fouling; biodeterioration. Mailing Add: Marine Biol & Biochem Br US Naval Res Lab Washington DC 20375

NEIKAM, WILLIAM C, physical chemistry, see 12th edition

NEIL, DONALD E, b Alliance, Nebr, Jan 15, 28; m 51; c 3. PHYSICAL CHEMISTRY. Educ: Univ SDak, AB, 50, MA, 54; Rensselaer Polytech Inst, PhD(phys chem), 59. Prof Exp: Jr res assoc, Brookhaven Nat Lab, 55-58; res fel, Reactor Div, Brookhaven Nat Lab & Rensselaer Polytech Inst, 58-60; res scientist, Union Carbide Olefins Co, 60-63; assoc chemist, Argonne Nat Lab, 63-64; tech sales rep, Instrument Prod Div, 64-68, prod mgr, 68-69, mkt mgr, Sci & Process Div, 73-74, SALES MGR, INSTRUMENT & EQUIP DIV, E I DU PONT DE NEMOURS & CO, 69-, MGR, SCI & PROCESS DIV, 74- Mem: Am Chem Soc. Res: High temperature thermodynamics; fused salts; radiation chemistry; analytical and process instrumentation. Mailing Add: Instrument & Equip Div E I du Pont de Nemours & Co Wilmington DE 19898

NEIL, GARY LAWRENCE, b Regina, Sask, June 13, 40; m 62; c 2. ORGANIC CHEMISTRY, BIOCHEMICAL PHARMACOLOGY. Educ: Queen's Univ, Ont, BSc, 62; Calif Inst Technol, PhD(chem), 66. Prof Exp: Res assoc cancer res, 66-74, RES HEAD CANCER RES, UPJOHN CO, KALAMAZOO, 74- Mem: Am Asn Cancer Res; Am Soc Pharmacol & Exp Therapeut; Am Chem Soc. Res: Enzyme kinetics; mode of action of antitumor agents; drug metabolism and pharmacokinetics. Mailing Add: 10619 Sudan St Portage MI 49081

NEIL, HUGH GROSS, b Gordonville, Tenn, Aug 28, 20; m 44; c 2. PHYSICS. Educ: Univ Tenn, BA, 47. Prof Exp: Jr physicist, Tenn Eastman Corp, 43-44, sr physicist, 44-46; asst physics, Univ Tenn, 47-48, res assoc, 48-49; PRES, SPEC INSTRUMENTS LAB, INC, 49-; VPRES STARLAB, INC, 66- Concurrent Pos: Dir, Fiber Lab, Inc, Gastonia, NC, 68-; dir, Spinlab, Ate, Switzerland, 70-; health physicist, Univ Tenn Hosp, 48-58. Mem: AAAS; Instrument Soc Am; Am Soc Test & Mat; Fiber Soc. Res: Textile testing instruments and controls; design of radioisotope detectors and allied equipment; electronic, medical and color instrumentation. Mailing Add: Special Instruments Lab Inc 312 W Vine Ave Knoxville TN 37902

NEIL, THOMAS C, b Tacoma, Wash, Dec 21, 34; m 63; c 4. ORGANIC CHEMISTRY, ANALYTICAL CHEMISTRY. Educ: Earlham Col, AB, 56; Pa State Univ, MS, 60, PhD(org chem), 64. Prof Exp: Asst prof chem, Baldwin-Wallace Col, 64-66; ASSOC PROF CHEM, KEENE STATE COL, 66- Mem: Am Chem Soc. Res: Photochemistry; reactions of carbenes, waxes and macromolecules. Mailing Add: Dept of Chem Keene State Col Keene NH 03431

NEILAND, BONITA, b Eugene, Ore, June 5, 28; m 55. PLANT ECOLOGY, RESOURCE MANAGEMENT. Educ: Univ Ore, BS, 49; Ore State Univ, BA, 51; Univ Col Wales, dipl rural sci, 52; Univ Wis, PhD(bot), 54. Prof Exp: Instr biol, Univ Ore, 54-55; asst prof, Gen Exten Div, Ore State Univ, 55-60; from asst prof to assoc prof, 61-68, head dept land resources & agr sci, 71-73, PROF BOT, UNIV ALASKA, 68-, ASST DEAN, SCH AGR & LAND RESOURCES MGT, 75- Concurrent Pos: NSF grant, 55-57 & 62-70; mem contract group, Proj Chariot, AEC, Alaska, 61-62; McIntire-Stennis Fund grant, 70-; Bur Land Mgt grant, 71-75; US Forest Serv contracts, 74- Mem: Fel AAAS; Ecol Soc Am; Arctic Inst NAm; Brit Ecol Soc; Soil Conserv Soc Am. Res: Comparisons of burned and unburned forest lands; analysis of forest and bog communities; vegetation, topography and ground ice correlations in the Fairbanks area; revegetation of denuded lands. Mailing Add: Sch of Agr & Land Resources Mgt Univ Ala Fairbanks AK 99701

NEILAND, KENNETH ALFRED, b Portland, Ore, Feb 18, 29; m 55; c 1. INVERTEBRATE ZOOLOGY. Educ: Reed Col, BA, 50; Univ Ore, MA, 53. Prof Exp: Asst parasitol, Reed Col, 50-51; asst physiol, Univ Ore, 51-53; asst zool, physiol & parasitol, Univ Calif, Los Angeles, 53-54; res fel physiol, Univ Ore, 55-56; instr biol sci, Ore Col Educ, 57-59; RES BIOLOGIST PARASITOL & COMP PHYSIOL, ALASKA DEPT FISH & GAME, 59- Mem: Am Soc Parasitologists. Res: Comparative parasitology; biology of helminth parasites; comparative physiology of molting in Crustacea. Mailing Add: Alaskan Dept Fish & Game 1300 College Rd Fairbanks AK 99701

NEILANDS, JOHN BRIAN, b Glen Valley, BC, Sept 11, 21; nat US. Educ: Univ Toronto, BS, 44; Dalhousie Univ, MSc, 46; Univ Wis, PhD(biochem), 49. Prof Exp: Nat Res Coun chemist, SAM Med Inst, Stockholm, 49-51; instr biochem, Univ Wis, 51-52; from asst prof to assoc prof, 52-61, PROF BIOCHEM, UNIV CALIF, BERKELEY, 61- Concurrent Pos: Guggenheim Found fel, 58-59. Mem: Am Chem Soc; Am Soc Biol Chem; Biochem Soc; Bertrand Russell Soc. Res: Bioinorganic chemistry; chemistry and biochemistry of iron compounds; microbial iron transport; membranes; cell surface receptors. Mailing Add: 185 Hill Rd Berkeley CA 94708

NEILD, RALPH E, b Georgetown, Ill, Apr 14, 24; m 49; c 4. HORTICULTURE, AGRICULTURAL CLIMATOLOGY. Educ: Univ Ill, Urbana, BS, 49; Iowa State Univ, MS, 51; Kans State Univ, PhD, 70. Prof Exp: Agr & opers researcher, Libby McNeill & Libby, 51-64; assoc prof, 64-74, PROF HORT, UNIV NEBR-LINCOLN, 74- Concurrent Pos: Consult, Libby McNeil & Libby, Imp Govt, Iran & Hashemite Kingdom, Jordan. Mem: Am Soc Hort Sci. Res: Crop ecology; operations research; crop geography. Mailing Add: Dept of Hort Univ of Nebr Lincoln NE 68502

NEILER, JOHN HENRY, b Mt Oliver, Pa, Dec 21, 22; m 47; c 3. NUCLEAR PHYSICS. Educ: Univ Pittsburgh, BS, 47, MS, 50, PhD(physics), 53. Prof Exp: Instr physics, Univ Pittsburgh, 47-51; physicist, Oak Ridge Nat Lab, 53-62, vpres-tech dir, Oak Ridge Tech Enterprises Corp, 62-67, VPRES-TECH DIR, ORTEC INC, 67- Concurrent Pos: Lectr, Univ Tenn, 57-; vis scientist, Am Inst Physics-Am Asn Physics Teachers Prog, 57-63. Mem: Am Asn Physics Teachers. Res: Neutron and gamma ray spectrometry; nanosecond pulsing and timing techniques; neutron cross section measurements; nuclear spectrometry with semiconductor diode detectors; fission fragment energy correlations. Mailing Add: Ortec Inc PO Box C Oak Ridge TN 37830

NEILL, ALEXANDER BOLD, b Jersey City, NJ, Sept 27, 19; m 47; c 4. ORGANIC CHEMISTRY. Educ: Lehigh Univ, BS, 41, MS, 47, PhD(chem), 49. Prof Exp: Chemist, Hercules Powder Co, 41, shift supvr smokeless powder, 42-44, shift supvr rocket powder, 44-45; develop chemist, Carwin Co, 49-50; sr res chemist, 51-58, admin asst to dir res, 58-62, admin asst to vpres res, 62-65, chief scheduling & control, 65-67, chief document, 67, mgr info serv, 67-74, DIR REGULATORY AFFAIRS, NORWICH PHARMACAL CO, 74- Mem: Am Chem Soc; Sigma Xi. Mailing Add: 17 Ridgeland Rd Norwich NY 13815

NEILL, CATHERINE ANNIE, b London, Eng, Sept 3, 21. PEDIATRICS, CARDIOLOGY. Educ: Royal Free Hosp Sch Med, MB, BS, 44; Univ London, MD, 47. Prof Exp: Intern internal med & pediat, Royal Free Hosp, London, 44, fel, Sch Med, 45-47; resident pediat, Queen Elizabeth Hosp Children, 47-50; fel pediat cardiol, Hosp Sick Children, Ont, 50-51 & Harriet Lane Home Cardiac Clin, Johns Hopkins Univ, 51-53; Am Heart Asn res fel embryol, Carnegie Inst, 53-54; res clinician, Queen Elizabeth Hosp Children, 54-56; asst prof, 56-64, ASSOC PROF PEDIAT, SCH MED, JOHNS HOPKINS UNIV, 64-, PEDIATRICIAN, CARDIAC CLIN, JOHNS HOPKINS HOSP, 65- Mem: Am Heart Asn; Am Pediat Soc. Res: Pediatric cardiology; familial incidence of congenital heart disease. Mailing Add: Dept of Pediat Johns Hopkins Univ Baltimore MD 21218

NEILL, JAMES CLEMENS, statistics, see 12th edition

NEILL, JIMMY DYKE, b Merkel, Tex, Mar 6, 39; m 60; c 2. PHYSIOLOGY, ENDOCRINOLOGY. Educ: Tex Tech Col, BS, 61; Univ Mo, MS, 63, PhD(reproductive physiol), 65. Prof Exp: Nat Inst Child Health & Human Develop res fel physiol, Sch Med, Univ Pittsburgh, 65-67, instr, 67-69; asst prof, 69-71, ASSOC PROF PHYSIOL, DIV BASIC HEALTH SCI, SCH MED, EMORY UNIV, 71- Concurrent Pos: Nat Inst Child Health & Human Develop career develop award, 70-75. Mem: Am Fertil Soc; Soc Study Reproduction; Am Physiol Soc; Endocrine Soc. Res: Pituitary-ovarian relationships; hormone levels in various reproductive states. Mailing Add: Dept of Physiol Sch of Med Emory Univ Div Basic Health Sci Atlanta GA 30029

NEILL, ROBERT H, radiological health, see 12th edition

NEILL, WARREN JOSEPH, b Rochester, NY, June 6, 22; m 44; c 4. PHYSICAL CHEMISTRY, RADIOCHEMISTRY. Educ: Eastern NMex Univ, BS & MS, 53; Univ NMex, PhD(chem), 57. Prof Exp: Res chemist, Oak Ridge Nat Lab, 56-58; assoc prof, 58-60, PROF CHEM, EASTERN NMEX UNIV, 60-, HEAD DEPT, 66- Mem: AAAS; Am Chem Soc. Res: Ion exchange chromatography; chemical kinetics; soil chemistry. Mailing Add: Dept of Chem Eastern NMex Univ Portales NM 88130

NEILL, WILLIAM ALEXANDER, b Nashville, Tenn. MEDICINE, CARDIOLOGY. Educ: Amherst Col, BA, 51; Cornell Univ, MD, 55. Prof Exp: NIH fel, Peter Bent Brigham Hosp, Boston, 59-61; instr med, Mass Mem Hosp, 61-63; ASSOC PROF MED, SCH MED, UNIV ORE, 63-; CHIEF CARDIOL, PORTLAND VET ADMIN HOSP, 70- Concurrent Pos: Mem staff, USPHS Commun Dis Ctr, 56-68; fel, Physiol Inst, Dusseldorf, Ger, 69-70; fel coun clin cardiol, Am Heart Asn. Mem: Am Fedn Clin Res; Am Physiol Soc. Res: Coronary circulation; muscle metabolism; tissue oxygen supply. Mailing Add: Vet Admin Hosp Portland OR 97207

NEILL, WILLIAM HAROLD, b Wynne, Ark, Oct 21, 43; m 64; c 1. FISH BIOLOGY. Educ: Univ Ark, BS, 65, MS, 67; Univ Wis, PhD(zool), 71. Prof Exp: Res fishery biologist, Nat Marine Fisheries Serv, Nat Oceanog & Atmospheric Admin, US Dept Commerce, 71-74; ASSOC PROF FISHERIES, TEX A&M UNIV, 74- Concurrent Pos: Affil prof zool, Univ Hawaii, 73- Honors & Awards: Spec Achievement Award, Nat Marine Fisheries Serv, 74. Mem: Am Fisheries Soc; Am Inst Fishery Res Biologists; Am Inst Biol Sci; AAAS; Sigma Xi. Res: Behavioral and physiolgoical ecology of fishes, with emphasis on behavioral regulation of environment and intra-habitat distribution. Mailing Add: Dept Wildlife & Fisheries Sci Tex A&M Univ College Station TX 77843

NEILSEN, IVAN ROBERT, b Rulison, Colo, Aug 12, 15; m 37; c 2. PHYSICS. Educ: Pac Union Col, AB, 36; Stanford Univ, MS, 48, PhD(physics), 52. Prof Exp: Instr physics, Glendale Union Acad, 36-38, San Diego Union Acad, 38-40 & Modesto Union Acad, 40-43; from instr to asst prof, Pac Union Col, 43-48; res assoc, Microwave Lab, Stanford Univ, 48-51; assoc prof, Pac Union Col, 51-52, prof & chmn phys sci div, 52-64, dir data processing lab, 58-64; prof physiol & biophys, 64-69, PROF BIOMATH & CHMN DEPT, SCH MED, LOMA LINDA UNIV, 69-, COORDR, SCI COMPUT FACIL, 65- Concurrent Pos: Res consult, Hansen Labs, Stanford Univ, 52-; consult, Calif State Dept Educ, 64. Mem: AAAS; Am Phys Soc; Am Asn Physics Teachers; Asn Comput Mach; Inst Elec & Electronics Engrs; Radiation Res Soc. Res: Applied electromagnetic field theory; high power pulsed klystrons; linear electron accelerators; chemical and biological reactions induced by ionizing radiation; computer models of living systems. Mailing Add: Dept of Biomath Loma Linda Univ Sch of Med Loma Linda CA 92354

NEILSON, CLARENCE ALBERT, physical chemistry, see 12th edition

NEILSON, GEORGE CROYDEN, b Vancouver, BC, Apr 4, 28; m; c 3. NUCLEAR PHYSICS. Educ: Univ BC, BA, 50, MA, 52, PhD(physics), 55. Prof Exp: Physicist, Radiation Sect, Defence Res Bd, 55-58, head radiation sect, 58-59; from asst prof to assoc prof, 59-66, PROF PHYSICS, UNIV ALTA, 66- Mem: Am Phys Soc; Can Asn Physicists. Res: Measurement of the energy angular distribution, gamma ray correlation and polarization of neutrons and protons produced by deuteron bombardment of light nuclei; angular correlation of cascade gamma rays. Mailing Add: Nuclear Res Ctr Dept of Physics Univ of Alta Edmonton AB Can

NEILSON, GEORGE FRANCIS, JR, b Portland, Ore, Jan 19, 30; m 55; c 3. PHYSICAL CHEMISTRY. Educ: Ore State Univ, BS, 51, MS, 53; Ohio State Univ, PhD, 62. Prof Exp: Res chemist, Cent Res Dept, E I du Pont de Nemours & Co, 58-62; res scientist, 62-67, SR SCIENTIST, OWENS-ILL, INC, 67- Mem: Am Chem Soc; Am Phys Soc; Am Crystallog Asn. Res: Small-angle x-ray scattering; kinetics and mechanisms of nucleation and crystallization; phase transformation processes in glass systems; microstructure of amorphous and polycrystalline materials. Mailing Add: Tech Ctr Owens-Ill Inc PO Box 1035 Toledo OH 43666

NEILSON, JAMES A, JR, plant ecology, plant morphology, see 12th edition

NEILSON, JAMES MAXWELL, b Shellbrook, Sask, Apr 5, 12; m 45; c 3. GEOLOGY. Educ: Queen's Univ, Ont, BSc, 36; McGill Univ, MSc, 47; Univ Minn, PhD(geol), 50. Prof Exp: Mining engr, Perron Gold Mines, Ltd, 37-39; mine mgr, Senore Gold Mines, Ltd, 39-40; chief geologist, Mistassini Explor, Ltd, 46-47; geologist, Que Dept Mines, 48-50; from asst prof to assoc prof geol & geol eng, Mich Technol Univ, 50-57, prof geol eng, 57-66, asst dean fac, 60-64; PROF GEOL SCI, QUEEN'S UNIV, ONT, 66- Concurrent Pos: Consult geologist, Dames & Moore. Mem: Fel Geol Soc Am; Am Inst Mining, Metall & Petrol Engrs; fel Geol Asn Can; Can Inst Mining & Metall; Soc Econ Geologists. Res: Precambrian regional, glacial and engineering geology; iron ore deposits. Mailing Add: Dept of Geol Sci Queen's Univ Kingston ON Can

NEILSON, JOHN TAYLOR MCLAREN, b Denny, Scotland, Mar 27, 38; m 62; c 2. IMMUNOLOGY, PARASITOLOGY. Educ: Glasgow Univ, BSc, 61, PhD(immunol), 65. Prof Exp: Res asst biochem, Sch Vet Med, Glasgow Univ, 61-65; res scientist parasite immunity, Div Animal Health, Commonwealth Sci & Indust Res Orgn, McMaster Lab, Australia, 65-68; asst prof, 68-73, ASSOC PROF PARASITE IMMUNITY, DEPT VET SCI, UNIV FLA, 73- Mem: Am Soc Parasitol. Res: Parasite immunity and biochemistry. Mailing Add: Col Vet Med Inst Food & Agr Sci Univ of Fla Gainesville FL 32611

NEILSON, JOHN WARRINGTON, b Saskatoon, Sask, Feb 13, 18; m 47; c 3. DENTISTRY. Educ: Univ Sask, BA, 39; Univ Alta, DDS, 41; Univ Mich, MSc, 46; Am Bd Periodont, dipl. Prof Exp: From asst prof to assoc prof periodont & oral path, Univ Alta, 45-47; assoc prof periodont, Univ Wash, 52-57; PROF ORAL BIOL & DEAN FAC DENT, UNIV MAN, 57- Concurrent Pos: Consult, USPHS Hosp, Seattle, Wash, 54-57, Royal Can Dent Corps, 60, Winnipeg Gen & Children's Hosps, 60; examr oral med, Nat Dent Exam Bd Can, 58-60; mem assoc comt dent res, Nat Res Coun Can, 59-60; mem dent adv comt, Nat Health & Welfare, 65-; mem coun higher learning, Prov Man, 65- Mem: Fel Am Col Dent; Am Acad Oral Path; Am Acad Periodont; Can Dent Asn; Can Acad Periodont (pres, 61-62). Res: Oral pathology and medicine; periodontology; effect of irritation on supporting tissues of the dentition. Mailing Add: Univ of Man Fac of Dent 780 Bannatyne Ave Winnipeg MB Can

NEILSON, M M, forest entomology, see 12th edition

NEIMAN, BENJAMIN H, b Toronto, Ont, Oct 12, 10; US citizen; m 31; c 3. PATHOLOGY. Educ: Univ Chicago, BS, 26; Rush Med Col, MD, 30; Univ Ill, MS, 34. Prof Exp: From assoc prof to prof, 46-68, CLIN PROF PATH, CHICAGO MED SCH, 68- Concurrent Pos: Dir labs, MacNeal Mem Hosp, 37-; attend pathologist, Cook County Hosp, 45- Mem: Col Am Path; Am Col Physicians; Am Soc Clin Path; AMA. Mailing Add: 6 N Michigan Ave Chicago IL 60602

NEIMAN, ROBERT R, physical chemistry, see 12th edition

NEIMARK, HAROLD CARL, b Detroit, Mich, July 25, 32; m 69. MICROBIOLOGY, IMMUNOLOGY. Educ: Univ Calif, Los Angeles, BA, 54, PhD(microbiol), 60. Prof Exp: Res assoc, Inst Microbiol, 59-60; from instr to asst prof, 60-71, ASSOC PROF MICROBIOL & IMMUNOL, STATE UNIV NY DOWNSTATE MED CTR, 72- Concurrent Pos: NIH grants, 67-76; collabr, USDA. Mem: AAAS; Am Soc Microbiol; Brit Soc Gen Microbiol. Res: Genetics and physiology of microorganisms; L forms and mycoplasmas; infectious diseases. Mailing Add: Dept of Microbiol & Immunol State Univ NY Downstate Med Ctr Brooklyn NY 11203

NEIMS, ALLEN HOWARD, b Chicago, Ill, Oct 24, 38; m 61; c 3. BIOCHEMISTRY, PEDIATRICS. Educ: Univ Chicago, BA & BS, 57; Johns Hopkins Univ, MD, 61, PhD(physiol chem), 66. Prof Exp: NIH fel, Lab Neurochem, Nat Inst Neurol Dis & Stroke, 68-70; asst prof pediat, Johns Hopkins Univ, 70, asst prof physiol chem, 70-72; asst prof pharmacol, 72-74, ASSOC PROF PEDIAT, McGILL UNIV, 72-, ASSOC PROF PHARMACOL, 74- Concurrent Pos: Physician, Johns Hopkins Hosp & J F Kennedy Inst, Baltimore, 70-72. Honors & Awards: Henry Strong Denison Award, 61. Mem: Soc Pediat Res; Am Fedn Clin Res; Perinatal Res Soc; Am Soc Pharmacol & Exp Therapeut; Can Soc Clin Invest. Res: Developmental pharmacology and therapeutics; developmental biology; clinical pharmacology. Mailing Add: Dept of Pharmacol McGill Univ Montreal PQ Can

NEISH, ARTHUR CHARLES, biochemistry, see 12th edition

NEISSEL, JOHN P, physics, see 12th edition

NEISWANDER, ROBERT SOUTH, aeronautical engineering, see 12th edition

NEISWENDER, DAVID DANIEL, b Palmdale, Pa, Oct 6, 30; m 55; c 2. PETROLEUM CHEMISTRY. Educ: Lebanon Valley Col, BS, 53; Pa State Univ, MS, 55, PhD(chem), 57. Prof Exp: Res chemist, Cent Res Div, Mobil Oil Corp, 57-60, sr res chemist, 60-62, asst supvr, 62-64, RES ASSOC, PAULSBORO LAB, MOBIL RES & DEVELOP CORP, 64- Mem: Am Chem Soc. Res: Development and testing of automotive engine, transmission and gear oils; chemistry of electrical discharges; synthesis of petrochemicals; preparation and reactions of organoboron compounds; hydrocarbon oxidation; design and testing of synthetic lubricants; lubricant contributions to fuel economy. Mailing Add: Paulsboro Lab Mobil Res & Develop Corp Paulsboro NJ 08066

NEITHAMER, RICHARD WALTER, b Wesleyville, Pa, Aug 3, 29; m 58; c 1. INORGANIC CHEMISTRY. Educ: Allegheny Col, BS, 51; Univ Ind, PhD(inorg chem), 57. Prof Exp: Asst prof chem, Lebanon Valley Col, 55-59, East Tex State Univ, 59-61, Rose Polytech Inst, 61-64; assoc prof & coord chem, 64-67, PROF CHEM, ECKERD COL, 67-, CHMN, COLLEGIUM NATURAL SCI, 72- Concurrent Pos: Vis scientist, Ind Acad Sci, 63-64 & Fla Acad Sci, 65-66; consult, US Naval Weapons Lab, 65-66, contract res, 66-69. Mem: AAAS; Am Chem Soc; fel Am Inst Chem. Res: Coordination and metal chelate compounds; polarography. Mailing Add: Collegium of Natural Sci Eckerd Col St Petersburg FL 33733

NEKLUTIN, VADIM CONSTANTIN, b Samara, Russia, Feb 4, 18; nat US; m 41; c 3. ORGANIC CHEMISTRY, POLYMER CHEMISTRY. Educ: Univ Ill, BS, 39; Univ Pa, MS & PhD(org chem), 44. Prof Exp: Instr inorg chem, Univ Pa, 41-44; sr res chemist, Synthetic Rubber Div, US Rubber Co, 44-52, mgr process develop, Chem Div, 52-55, prod supt, 55-59, tech coord, 59-64, opers mgr, Naugatuck Chem Int Div, 64-69, DEVELOP MGR, CHEM DIV, UNIROYAL, INC, 69-, MGR OVERSEAS OPERS, 72- Mem: Am Chem Soc; fel Am Inst Chemists; AAAS. Res: Synthetic rubber; plastics; organic synthesis; rubber and agricultural chemicals; research and development in fields of organic and polymer chemistry leading to design of processing plants. Mailing Add: Chem Div UniRoyal Inc Spencer St Naugatuck CT 06770

NEL, LOUIS DANIEL, b Barkly West, SAfrica, June 5, 34; m 56; c 4. TOPOLOGY. Educ: Univ Stellenbosch, BSc, 54, MSc, 58; Cambridge Univ, PhD(math), 62. Prof Exp: Lectr math, Univ Stellenbosch, 56-62; sr lectr, Univ Cape Town, 62-65; prof, Port Elizabeth Univ, 66-68; ASSOC PROF MATH, CARLETON UNIV, 68- Mem: SAfrican Math Soc (secy, 63-67); Am Math Soc; Res: Categorical topology; Cartesian closed topological categories and their applications to topology, functional analysis, topological algebra. Mailing Add: Dept of Math Carleton Univ Ottawa ON Can

NELAN, DONALD ROYCE, b Wheatland, NY, Oct 12, 22; m 47; c 4. ORGANIC CHEMISTRY. Educ: Univ Rochester, BS, 49. Prof Exp: Lab technician ceramics, Victor Insulators, Inc, 40-41; res chemist, Distillation Prod Indust Div, Eastman Kodak Co, 50-62, sr res chemist, 62-70, SR RES CHEMIST, HEALTH & NUTRIT DIV, TENN EASTMAN CO, 70- Mem: AAAS; Am Chem Soc. Res: Chemistry of vitamin E, including synthesis, resolution of optical isomers, oxidation and related reactions; synthesis of vitamins A and K and related products; synthesis of products for human and animal nutrition. Mailing Add: Hidden Valley Rt 9 Kingsport TN 37663

NELB, ROBERT GILMAN, b Lawrence, Mass, Dec 4, 23; m 45; c 5. ORGANIC CHEMISTRY. Educ: Dartmouth Col, BA, 45, MA, 46; Univ Rochester, PhD(chem), 49. Prof Exp: Instr chem, Dartmouth Col, 45-46; asst, Univ Rochester, 46-48; res chemist, Chem Div, US Rubber Co, 49-52, group leader new prod, 52-53, mgr vibrin polyester develop, 53-57, mgr vibrin polyester & vibrathane polyurethane res & develop, 57-60, mgr kralastic res & develop, 60-65; group mgr plastics res & develop, 65-66, dir res & develop Consumer, Indust & Plastics Div, 66-74, DIR RES & DEVELOP, INDUST PRODS CO, UNIROYAL, INC, 74- Mem: Am Chem Soc. Res: High temperature and reinforced polyesters; polyurethane elastomers; emulsion polymerization; thermoplastics; elastomers and plastics—fabrication of rubber products. Mailing Add: Birchwood Terr Middlebury CT 06762

NELIGAN, ROBERT EMMETT, chemistry, see 12th edition

NELKIN, MARK, b New York, NY, May 12, 31; m 52; c 2. THEORETICAL PHYSICS. Educ: Mass Inst Technol, SB, 51; Cornell Univ, PhD, 55. Prof Exp: Res assoc, Knolls Atomic Power Lab, Gen Elec Co, 55-57; mem res staff, Gen Atomic Div, Gen Dynamics Corp, 57-62; assoc prof eng physics, 62-67, PROF APPL PHYSICS, CORNELL UNIV, 67- Concurrent Pos: Vis res assoc, State Univ Utrecht, 60-61; mem adv comt reactor physics, AEC, 64-67; Guggenheim fel, Orsay, Paris, 68-69. Mem: Am Phys Soc. Res: Statistical physics; physics of fluids; neutron physics. Mailing Add: Sch of Appl & Eng Physics Cornell Univ Ithaca NY 14853

NELLES, MAURICE, b Madison, SDak, Oct 19, 06; m 29; c 3. PHYSICAL CHEMISTRY. Educ: Univ SDak, AB, 27, AM, 28; Harvard Univ, PhD(phys chem), 34. Hon Degrees: DSc, Univ SDak, 55. Prof Exp: Prof physics, Columbus Col, SDak, 28-29; instr chem, Univ SDak, 29-30; res chemist, Nat Aniline Chem Co, 32-34 & Union Oil Co, 34-36; camp dir, Civilian Conserv Corps, Ft Lewis, Wash, 36-37; res engr, Riverside Cement Co, 37-39; engr, Permanente Corp, 39-40; staff asst, Lockheed Aircraft Corp, 40-47; prof aeronaut eng, Univ Southern Calif, 46-49, res mgr, Allan Hancock Found, 49-50; prof & dir eng exp sta, Pa State Univ, 50-51; vpres petromech div & dir res, Borg-Warner Corp, 51-54; dir res & diversification & mgr, Graphic Arts Div, Technicolor Corp, 54-57; vpres eng, Crane Co, 57-59; dir Lamb-Weston Div, 57-71, vpres, Am Electronics Corp, 59-61; mgr mfg res & develop, Westinghouse Defense Ctr, 64-66; prof bus admin, Univ Va, 66-70; CONSULT, NAT RES COUN, 71- Concurrent Pos: Dep dir, War Prod Bd, 44-45; pres, Corwith Co, Crane Co, 57-59; dir, MER & D Corp, 61-, Hydro-Aire Co & Crane, Ltd, Can; mem expeds; chief engr, Off Prod Res & Develop; mem tech adv comt, Calif Air Resources Bd. Mem: AAAS; Am Soc Mech Eng; Soc Automotive Eng; Am Chem Soc; Am Soc Metals. Res: Electronics; radio transmitter design; structures; petroleum; catalytic reactions; design of mechanisms; preliminary design for a mass production factor for frozen stuffed quail. Mailing Add: 5522 Rutgers Rd La Jolla CA 92037

NELLIGAN, WILLIAM BRYON, b Northampton, Mass, Jan 19, 20; m 42. PHYSICS. Educ: Rensselaer Polytech Inst, BS, 50, MS, 51. Prof Exp: Designer elec power, Gen Elec Co, Mass, 42-44; asst physics, Rensselaer Polytech Inst, 50-51; physicist, Res Lab, Schlumberger Well Surv Corp, 51-53, sr res physicist, 53-65, res proj physicist, 65-67, RES PROJ PHYSICIST, SCHLUMBERGER TECH CORP, 67- Mem: AAAS; Am Phys Soc; Am Nuclear Soc; Soc Petrol Engrs. Res: Applied nuclear physics; electronic instrumentation; mathematics. Mailing Add: Candlewood Vista Danbury CT 06810

NELLIS, LOIS FONDA, b Dayton, Ohio, Nov 30, 26. MICROBIOLOGY. Educ: Hobart & William Smith Cols, BA, 46; Smith Col, MA, 48; Purdue Univ, PhD(bact), 62. Prof Exp: From instr to assoc prof, 48-68, chmn dept, 69-74, PROF BIOL, HOBART & WILLIAM SMITH COLS, 68- Concurrent Pos: Res mem, Bergey's Manual Comn, 53-55 & 61; Geneva City bacteriologist, 50-52; United Health Found of western NY grant, 68-70; NIH co-proj dir, Dept Pharmacol, Med Sch, State Univ NY Buffalo, 70-76; vis prof med, Univ Rochester, 75-76. Mem: AAAS; Am Soc Microbiology; Am Inst Biol Sci. Res: Myxobacteria; R factors and tetracycline resistance in Escherichia coli. Mailing Add: Dept of Biol Hobart & William Smith Cols Geneva NY 14456

NELLOR, JOHN ERNEST, b Omaha, Nebr, Oct 31, 22; m 46; c 3. PHYSIOLOGY, ENDOCRINOLOGY. Educ: Univ Calif, BS, 50, PhD(comp physiol), 55. Prof Exp: From instr to assoc prof, 55-64, mem staff, Endocrine Res Unit, 64-69, asst vpres, 69-71, PROF PHYSIOL, COL HUMAN MED & NATURAL SCI, MICH STATE UNIV, 64-, ASSOC VPRES RES & DEVELOP & DIR CTR ENVIRON QUAL, 71- Concurrent Pos: Mem staff, NSF, 66-67, prog dir metals biol, 67-68; mem, US Nat Comn for UNESCO Man & Biosphere Prog, 75- Mem: Am Physiol Soc; Am Asn Anatomists; Soc Exp Biol & Med; Soc Study Reproduction; Sigma Xi. Res: Comparative reproductive physiology; hormones and tissue leucocytogenesis; adrenal-pituitary hormones and aging. Mailing Add: Off for Res & Develop Mich State Univ East Lansing MI 48824

NELMS, GEORGE E, b Ark, Feb 6, 27; m 50; c 3. ANIMAL BREEDING. Educ: Ark State Col, BS, 51; Ore State Col, MS, 54, PhD(genetics), 56. Prof Exp: Instr animal husb, Ore State Col, 55-56; asst prof animal sci, Univ Ariz, 56-59; ASST PROF ANIMAL BREEDING, UNIV WYO, 59- Mem: Am Soc Animal Sci; Am Genetic Asn. Res: Reproductive and environmental physiology; genetics of beef cattle. Mailing Add: Div of Animal Sci Univ of Wyo Box 3354 Univ Sta Laramie WY 82070

NELP, WIL B, b Pittsburgh, Pa, July 30, 29; m 52, 69; c 4. INTERNAL MEDICINE, NUCLEAR MEDICINE. Educ: Franklin Col, BA, 51; Johns Hopkins Univ, MD, 55. Hon Degrees: DSc, Franklin Col, 67. Prof Exp: NIH fel med & radiol, Johns Hopkins Univ, 60-62; from asst prof to assoc prof, 62-71, PROF MED & RADIOL, UNIV WASH, 71-, CHIEF DIV NUCLEAR MED & HEAD CLIN NUCLEAR MED, UNIV HOSP, 62- Concurrent Pos: Instr, Johns Hopkins Univ, 61-62; Nat Inst Arthritis & Metab Dis training grant, 63-69; consult, Providence Hosp, Seattle, Wash, 64-68, Nat Heart Inst, 67-68 & Nat Heart & Lung Inst, 68-; consult adv radiopharmaceut, Food & Drug Admin, 70; consult, Children's Orthop Hosp, Seattle Vet Admin Hosp, Harborview Med Ctr & USPHS Hosp, Seattle. Mem: Am Fedn Clin Res; Soc Nuclear Med (vpres, 69-70, pres, 73-74); fel Am Col Physicians. Res: Physiologic and clinical investigations in nuclear medicine. Mailing Add: BB20 Univ Hosp RC-70 Univ of Wash Seattle WA 98195

NELSEN, OLIN E, b South Lancaster, Mass, Nov 30, 98. ANATOMY, EMBRYOLOGY. Educ: Clark Univ, AB, 22; Brown Univ, AM, 23; Univ Pa, PhD, 29. Prof Exp: Instr zool, Univ Toledo, 24-26; instr, 26-29, Nat Res Coun fel, 29-30, from instr to prof, 30-70, EMER PROF ZOOL, UNIV PA, 70- Mem: Am Soc Zool; Soc Develop Biol. Res: Favors concept that development from fertilization through early body form in chick and frog embryos is the result of a succession of synchronized blasxemata. Mailing Add: 406 Cresswell St Ridley Park PA 19078

NELSEN, ROGER BAIN, b Chicago, Ill, Dec 20, 42. MATHEMATICS. Educ: DePauw Univ, BA, 64; Duke Univ, PhD(math), 69. Prof Exp: ASST PROF MATH, LEWIS & CLARK COL, 69- Mem: Math Asn Am. Res: Stochastic processes. Mailing Add: Dept of Math Lewis & Clark Col Box LC110 Portland OR 97219

NELSEN, STEPHEN FLANDERS, b Chicago, Ill, Apr 17, 40; m 62; c 1. ORGANIC CHEMISTRY. Educ: Univ Mich, BS, 62; Harvard Univ, PhD(chem), 65. Prof Exp: From asst prof to assoc prof, 65-75, PROF ORG CHEM, UNIV WIS-MADISON, 75- Mem: Am Chem Soc. Res: Physical organic chemistry; physical and chemical properties of free radicals; electrochemistry; conformational analysis. Mailing Add: 1101 W University Ave Madison WI 53706

NELSEN, THOMAS ROBERT, b Hicksville, NY, Oct 25, 45; m 70; c 1. ORGANIC CHEMISTRY. Educ: Syracuse Univ, AB, 67; State Univ NY Buffalo, PhD(chem), 73. Prof Exp: Res assoc chem, Syrauscue Syracuse Univ, 72-74; asst prof chem, State Univ NY Col, Geneseo, 74-75; RES ASSOC CHEM, NY STATE AGR EXP STA, 75- Mem: AAAS; Am Chem Soc. Res: Pesticide residue analysis; synthetic and mechanistic organic chemistry, particularly heterocyclic compounds. Mailing Add: Dept of Food Sci & Technol NY State Agr Exp Sta Geneva NY 14456

NELSEN, THOMAS SLOAN, b Tacoma, Wash, Aug 4, 26; m 45; c 2. SURGERY. Educ: Univ Wash, BS, 47, MD, 51; Am Bd Surg, dipl, 59. Prof Exp: From instr to asst prof, Univ Chicago, 57-60; from asst prof to assoc prof, 60-71, PROF SURG, SCH MED, STANFORD UNIV, 71- Mem: Am Col Surgeons; Inst Elec & Electronics Engrs. Res: Surgery of neoplasms; gastrointestinal physiology and surgery. Mailing Add: Dept of Surg Stanford Univ Sch of Med Stanford CA 94305

NELSESTUEN, GARY LEE, b Galesville, Wis, Sept 10, 44; m 67; c 2. BIOCHEMISTRY. Educ: Univ Wis-Madison, BS, 66; Univ Minn, St Paul, PhD(biochem), 70. Prof Exp: NIH fel biochem, Univ Wis, 70-72; ASST PROF BIOCHEM, UNIV MINN, ST PAUL, 72- Concurrent Pos: Estab investr, Am Heart Asn, 75-80. Mem: Am Chem Soc. Res: Function of the carbohydrate portion of glycoproteins; role of vitamin K and the vitamin K-dependent amino acid, gamma-carboxyglutamic acid. Mailing Add: Dept of Biochem Univ of Minn Col of Biol Sci St Paul MN 55108

NELSON, A CARL, JR, b West Chester, Pa, Jan 2, 26; m 50; c 4. MATHEMATICAL STATISTICS, MATHEMATICS. Educ: Mass Inst Technol, SB, 46; Univ Del, MS, 48. Prof Exp: Instr math, Univ Del, 48-50, 51-52 & 53-56; scientist, Bettis Atomic Power Div, Westinghouse Elec Corp, Pa, 56-60, fel scientist, 60; statistician, Res Triangle Inst, 60-63, sr statistician, 63-73; statist consult, 73-75; SR PROF SCIENTIST, PED CO-ENVIRON SPECIALISTS, INC, 75- Concurrent Pos: Mem, Hwy Res Bd, Nat Acad Sci-Nat Res Coun. Mem: Am Statist Asn; Biomet Soc; Inst Math Statist; Am Soc Qual Control. Res: Applied research in application of statistics to physical sciences, particularly the fields of environmental analysis, occupational and highway safety, quality assurance, systems analysis, and the statistical design of experiments for developing mathematical models. Mailing Add: 3219 Ridge Rd Durham NC 27705

NELSON, A GENE, b Galesburg, Ill, Sept 9, 42; m 64; c 2. AGRICULTURAL ECONOMICS. Educ: Western Ill Univ, BS, 64; Purdue Univ, MS, 67, PhD(agr econ), 69. Prof Exp: Asst prof, 69-75, ASSOC PROF AGR ECON, ORE STATE UNIV, 75- Mem: Am Agr Econ Asn; Am Soc Farm Mgrs & Rural Appraisers. Res: Systems analysis of beef and forage production; decision making under risk and uncertainty. Mailing Add: Dept Agr & Resource Econ Ore State Univ Corvallis OR 97331

NELSON, AARON LOUIS, b Deer Lodge, Mont, June 23, 20; m 51; c 3. ORGANIC CHEMISTRY. Educ: Harvard Univ, BS, 42, MS, 46, PhD(org chem), 49. Prof Exp: Asst prof org chem, Case Univ, 49-55; res chemist, E I du Pont de Nemours & Co, Inc, 55-61; mem res staff, Wright Lab, 61-67, ASSOC PROF CHEM, UNIV COL, RUTGERS UNIV. NEW BRUNSWICK, 67- Concurrent Pos: Rutgers res coun vis fel, Princeton Univ, 70-71. Mem: AAAS; Am Chem Soc; The Chem Soc. Res: Heterocyclic and aromatic chemistry and synthesis. Mailing Add: 1055 Sunny Slope Dr Mountainside NJ 07092

NELSON, ALAN R, b Logan, Utah, June 11, 33; m 59; c 3. MEDICAL QUALITY ASSESSMENT. Educ: Northwestern Univ, BS, 55, MD, 58. Prof Exp: Pres, Utah Prof Rev Orgn, 71-75; ASSOC, MEM MED CTR, 64- Concurrent Pos: Mem, Nat Prof Stand Rev Coun, 73- Honors & Awards: Recognition Award, Am Soc Int Med, 73. Mem: Inst Med of Nat Acad Sci. Res: Medical utilization review and quality assessment. Mailing Add: 2000 S Ninth East Salt Lake City UT 84105

NELSON, ALBERT WENDELL, b Boston, Mass, June 2, 35; m 59; c 3. CARDIOVASCULAR DISEASES. Educ: Cornell Univ, DVM, 59; Colo State Univ, MS, 62, PhD(path), 65. Prof Exp: Vet, private practice, 59-60; asst prof, 65-69, ASSOC PROF SURG, COLO STATE UNIV, 69- Concurrent Pos: NIH res grants, Colo State Univ, 65-68, 71-73; Colo Heart Asn res grant, 71-73; Nat Heart & Lung Inst Contract, 73-77. Mem: Am Vet Med Asn. Res: Cardiovascular pathology and physiology, primarily in relation to the microcirculation; reconstructive surgery relative to animal and human problems. Mailing Add: Surg Lab Colo State Univ Ft Collins CO 80521

NELSON, ALLEN CHARLES, b Plum City, Wis, July 13, 32; m 54; c 4. MYCOLOGY, MICROBIOLOGY. Educ: Wis State Univ, River Falls, BS, 54; Univ SDak, MA, 61; Univ Wis, PhD(bot), 64. Prof Exp: From asst prof to assoc prof, 64-68, chmn dept biol, 67-73, PROF BOT, UNIV WIS-LA CROSSE, 68- Mem: Mycol Soc Am; Bot Soc Am; Sigma Xi; Am Soc Microbiol. Res: Ascomycetes; morphological and cytological studies. Mailing Add: Dept of Biol Univ of Wis-LaCrosse La Crosse WI 54601

NELSON, ALVIN I, b Clear Lake, Iowa, Jan 19, 14; m 39; c 2. FOOD TECHNOLOGY. Educ: Iowa State Col, BS, 37. Prof Exp: Qual control frozen foods, Wash Frozen Foods Co, 37-38; prod & res, Birds Eye Div, Gen Foods Corp, 39-46; qual control supvr, Standard Brands, Inc, 47-48; prod res, Campbell Soup Co, 48-49; from assoc prof to prof food processing, 49-69, PROF FOOD SCI, UNIV ILL, URBANA, 69- Concurrent Pos: Specialist food eng, Govind Ballabh Pant Univ, India, 72- Mem: Am Soc Hort Sci; Inst Food Technologists. Res: Freezing, dehydrating and canning of foods, especially fruits and vegetables; development of procedures and techniques for improved processing. Mailing Add: Dept of Food Processing Univ of Ill Hort Field Lab Urbana IL 61801

NELSON, ANDREW PHILLIPS, plant taxonomy, evolution, see 12th edition

NELSON, ARNOLD BERNARD, b Valley Springs, SDak, Aug 26, 22; m 43; c 4. ANIMAL NUTRITION. Educ: SDak State Col, BS, 43, MS, 48; Cornell Univ, PhD(animal husb), 50. Prof Exp: Asst animal husb, SDak State Col, 46-47, asst animal husbandman, 47-48; asst, Cornell Univ, 48-50; from asst prof to assoc prof, Okla State Univ, 50-62; prof, 63-71, PROF ANIMAL RANGE SCI & HEAD DEPT, NMEX STATE UNIV, 71- Mem: AAAS; Am Soc Animal Sci; Soc Range Mgt; Am Dairy Sci Asn. Res: Ruminant nutrition; applied cattle nutrition. Mailing Add: Dept of Animal Range Sci NMex State Univ Las Cruces NM 88003

NELSON, ARTHUR HANSEN, b San Jose, Calif, Oct 1, 19; m 44. BIOLOGY. Educ: Univ Calif, BS, 42; Cornell Univ, PhD(nature study), 49. Prof Exp: Instr biol, San Jose State Col, 46-47; PROF BIOL, SAN FRANCISCO STATE UNIV, 49- Mem: Fel AAAS; Am Nature Study Soc (2nd vpres, 60); Am Soc Ichthyologists & Herpetologists; Nat Asn Biol Teachers; Soc Am Foresters. Res: Conservation of natural resources; economic botany; nature study. Mailing Add: Dept of Biol San Francisco State Univ San Francisco CA 94132

NELSON, ARTHUR KENDALL, b Washburn, Wis, Aug 28, 32; m 61; c 1. CHEMISTRY. Educ: Univ Wis, BS, 54; Univ Minn, PhD(chem), 59. Prof Exp: Asst prof, Macalester Col, 59-60; res chemist, Stauffer Chem Co, 60-64; sr tech specialist, Nalco Chem Co, 64-68; MEM STAFF, RAULAND DIV, ZENITH RADIO CORP, 68- Mailing Add: Rauland Div of Zenith Radio Corp 2407 N Ave Melrose Park IL 60160

NELSON, ARTHUR ROBERT, b Summit, NJ, Aug 3, 45; m 67; c 2. OPTICS. Educ: Rensselaer Polytech Inst, BS, 67, PhD(physics), 73; Cornell Univ, MS, 69. Prof Exp: MEM TECH STAFF, RES OPTICAL COMMUN, SPERRY RES CTR, SPERRY RAND CORP, 74- Concurrent Pos: Nat Res Coun res assoc, Air Force Cambridge Res Labs, 73-74. Mem: Optical Soc Am; Am Phys Soc. Res: Optical communications; development of fiber optic data links; integrated optics. Mailing Add: Sperry Res Ctr 100 North Rd Sudbury MA 01776

NELSON, BERNARD ANDREW, b Chicago, Ill, Jan 10, 10; m 36. ORGANIC CHEMISTRY. Educ: Wheaton Col, Ill, BS, 31; Northwestern Univ, MS, 38, PhD(org chem), 42. Prof Exp: Asst chem, Wheaton Col, Ill, 28-31; instr, Maine Twp Jr Col, Ill, 39-42; asst prof, Baylor Univ, 42-43; from asst prof to assoc prof, 43-53, PROF CHEM, WHEATON COL, ILL, 53-, CHMN DEPT, 69- Mem: Fel AAAS; Nat Sci Teachers Asn; Am Chem Soc; Am Inst Chem. Res: Cyclic acetals and ethers; Grignard reagents; acridine and anthracene derivatives; absorption spectra in ultraviolet range; chromanones; pyrazolines. Mailing Add: Dept of Chem Wheaton Col Wheaton IL 60187

NELSON, BERNARD CLINTON, b Cass Lake, Minn, June 2, 34; m 64; c 2. MEDICAL ENTOMOLOGY. Educ: Wis State Univ, Superior, BS, 56; Univ Mich, MS, 60; Univ Calif, Berkeley, PhD(parasitol), 68. Prof Exp: Fel ecol avian lice, McMaster Animal Health Lab, Nat Inst Gen Med Sci, Australia, 68-70; PUB HEALTH BIOLOGIST ZOONOTIC DIS INVEST, VECTOR CONTROL SECT, CALIF DEPT HEALTH, 71- Concurrent Pos: Assoc exp sta, Div Entomol & Parasitol, Univ Calif, Berkeley, 71- Mem: Am Soc Mammal; Cooper Ornith Soc; AAAS. Res: Biology and taxonomy of avian lice; ecology of bubonic plague and tick-borne diseases in California; parasites of fleas. Mailing Add: Vector Control Sect Calif Dept Health 2151 Berkeley Way Berkeley CA 94704

NELSON, BRUCE KINLOCH, physics, electronic engineering, see 12th edition

NELSON, BRUCE PHILIP, b Darby, Pa, Apr 20, 40. ELEMENTARY PARTICLE PHYSICS. Educ: Bucknell Univ, BS, 62; Univ Chicago, MS, 64, PhD(physics), 73. Prof Exp: RES ASSOC PHYSICS, UNIV ILL, URBANA, 73- Res: Experimental studies of the strong interaction; particle production; phenomenology of hadron scattering. Mailing Add: Dept of Physics Univ of Ill Urbana IL 61801

NELSON, BRUCE WARREN, b Cleveland, Ohio, Mar 17, 29; m 56; c 1. SEDIMENTOLOGY, CLAY MINERALOGY. Educ: Harvard Col, AB, 51; Pa State Univ, MS, 54; Univ Ill, PhD(geol), 55. Prof Exp: From assoc prof to prof geol, Va Polytech Inst, 55-63; prof & head dept, Univ SC, 63-74, dean col arts & sci, 66-72, dean grad sch & vprovost advan studies & res, 72-74; DEAN & ASST PROVOST, SCH CONTINUING EDUC, UNIV VA, 74- Concurrent Pos: Geologist, US Geol Surv, 51-55; geologist, Ohio Geol Surv, 52-54; vis scientist, Am Geol Inst, 64-69; vis prof, Univ Va, 70-71. Mem: Fel AAAS; fel Mineral Soc Am; fel Geol Soc Am; Soc Econ Paleontologists & Mineralogists. Res: Sedimentary mineralogy, geochemistry, and petrology; recent sedimentary processes; diagenesis; chemistry of natural waters; estuarine environment. Mailing Add: Sch of Continuing Educ Univ of Va Charlottesville VA 22903

NELSON, BUCK DEAN, biochemistry, see 12th edition

NELSON, BURT, b Milwaukee, Wis, Mar 10, 22; m 47. ASTRONOMY. Educ: Univ Wis, BS, 51, MS, 52, PhD(philos), 59. Prof Exp: Asst prof astron & phys sci, 57-61, assoc prof astron, 61-66, PROF ASTRON, SAN DIEGO STATE COL, 66- Mem: Am Astron Soc. Res: Astronomical photoelectric photometry. Mailing Add: Dept of Astron San Diego State Univ San Diego CA 92115

NELSON, CARL TRUMAN, b Providence, RI, June 27, 08; m 37. DERMATOLOGY. Educ: Harvard Univ, AB, 35, AM, 38, MD, 42. Prof Exp: Asst bact, Harvard Univ, 32-35, instr, 35-42; from instr to prof dermat, 46-73, chmn dept, 51-73, EMER PROF DERMAT, COL PHYSICIANS & SURGEONS, COLUMBIA UNIV, 73- Mem: Am Dermat Asn; fel AMA; Am Asn Immunologists; Am Fedn Clin Res; Am Acad Dermat. Res: Sarcoidosis; tissue electrolyte changes in hypersensitivity; corticosteroids in therapy of diseases of the skin. Mailing Add: Dept of Dermat Columbia Univ Col of Physicians & Surgeons New York NY 10032

NELSON, CARLTON HANS, b Wabasha, Minn, Dec 16, 37; m 62; c 2. GEOLOGY. Educ: Carleton Col, BA, 59; Univ Minn, MS, 62; Ore State Univ, PhD(oceanog), 68. Prof Exp: Ranger naturalist, Nat Park Serv, 59-61 & 63; teaching asst, Lehigh Univ, 61-62; field asst, US Geol Surv, 62; instr phys sci, Portland State Col, 62-63; res asst, Ore State Univ, 63-67; GEOLOGIST, US GEOL SURV, 66- Concurrent Pos: Vis asst prof, Chapman Col, 66, San Jose State Col, 68-69 & Calif State Col, Hayward, 70-71; actg asst prof, Stanford Univ, 73. Mem: AAAS; Soc Econ Paleont & Mineral; Int Asn Sedimentol; fel Geol Soc Am. Res: Geological limnology; Pleistocene geology; sedimentology; geological oceanography; epicontinental shelf and deep-sea fan sedimentation; placer and trace metal dispersal in marine sediments; marine geology. Mailing Add: US Geol Surv 345 Middlefield Rd Menlo Park CA 94025

NELSON, CECIL MORRIS, b Rock Island, Ill, Nov 12, 22; m 46; c 2. PHYSICS, CHEMISTRY. Educ: Univ Chicago, BS, 44, MS, 48; Univ Tenn, PhD(chem), 52. Prof Exp: Anal chemist, Clinton Lab, 44-46; phys chemist, Argonne Nat Lab, 46-48 & Oak Ridge Nat Lab, 48-60; PROF PHYSICS, EMORY & HENRY COL, 60-, CHMN SCI DIV, 70- Mem: Am Asn Physics Teachers. Res: Radiation effects in ionic crystals. Mailing Add: Dept of Physics Emory & Henry Col Emory VA 24327

NELSON, CHANNING CLARKE, applied mathematics, see 12th edition

NELSON, CHARLES A, b Buffalo, NY, June 26, 36; m 64; c 2. BIOCHEMISTRY. Educ: Cornell Col, BS, 57; Univ Iowa, MS, 60, PhD(biochem), 62. Prof Exp: Res assoc, Duke Univ, 61-66; ASST PROF BIOCHEM, MED CTR, UNIV ARK, LITTLE ROCK, 66- Concurrent Pos: NIH fel, 62-64. Res: Subunit structure of serum lipoproteins and xanthine oxidase; detergent effects on proteins, their activity and dissociation to subunits; further effect of combining detergents with other denaturants. Mailing Add: Dept of Biochem Univ of Ark Med Ctr Little Rock AR 72205

NELSON, CHARLES ARNOLD, b Chadron, Nebr, Oct 11, 43; m 71. HIGH ENERGY & THEORETICAL PHYSICS. Educ: Univ Colo, BS, 65; Univ Md, PhD(theoret physics), 68. Prof Exp: Res assoc high energy theoret physics, City Col New York, 68-70 & La State Univ, Baton Rouge, 70-72; Nat Res Coun-Nat Bur Standards fel, Nat Bur Standards, Washington DC, 72-73; ASST PROF PHYSICS, STATE UNIV NY, BINGHAMTON, 73- Concurrent Pos: Consult, Ctr Particle Theory, Univ Tex, Austin, 70-72 & Ctr Theoret Physics, Univ Md, 72-73. Mem: Am Phys Soc. Res: Particles and fields in theoretical high energy physics; mathematical physics. Mailing Add: Dept of Physics State Univ of NY Binghamton NY 13901

NELSON, CHARLES EDWARD, b Stockham, Nebr, Feb 7, 33; m 57; c 2. GEOGRAPHY. Educ: Hastings Col, BA, 56; Ore State Univ, MS, 61; Univ Nebr, PhD(geog), 71. Prof Exp: Instr geog, Mont State Univ, 64-67; naturalist, Nat Park Serv, 67; field technician weather res, Mont State Univ, 68; asst prof geog, WTex State Univ, 69-74; REGIONAL PLANNER, US DEPT INTERIOR, BUR LAND MGT, 74- Mem: Asn Am Geog; Am Geog Soc. Res: Agricultural and recreational land use; resource conservation and environmental problems; coastal zone management problems. Mailing Add: OCS Off Bur of Land Mgt 3200 Plaza Tower 1001 Howard Ave New Orleans LA 70113

NELSON, CHARLES HENRY, b Boston, Mass, July 28, 41; m 66. ENTOMOLOGY. Educ: Univ Mass, BS, 63, MS, 67, PhD(entom), 69. Prof Exp: Asst prof, 69-73, ASSOC PROF BIOL, UNIV TENN, CHATTANOOGA, 74- Mem: Entom Soc Am; Am Entom Soc; Soc Syst Zool. Res: Systematics and morphology of the Plecoptera; systematic entomology. Mailing Add: Dept of Biol Univ of Tenn Chattanooga TN 37403

NELSON, CHARLES JAY, b Chicago, Ill, Sept 12, 41; m 67. POLYMER CHEMISTRY. Educ: Beloit Col, BA, 63; Duke Univ, MA, 65, PhD(phys chem), 68. Prof Exp: Res scientist, 67-70, GROUP LEADER, TEXTILES & FIBERS DIV, FIRESTONE TIRE & RUBBER CO, 70-, GROUP LEADER, POLYMER PHYSICS DIV, 74- Mem: Am Chem Soc. Res: Processing and evaluation of fibers; polymer morphology; computer programming; nuclear magnetic resonance; interrelationship of rheology, morphology, and physical properties of plastics and elastomers. Mailing Add: Central Res Firestone Tire & Rubber Co Akron OH 44317

NELSON, CLARENCE HERBERT, b Granville, Ill, Mar 16, 09; m 44. BIOLOGY. Educ: Augustana Col, AB, 32; Univ Iowa, MS, 40, PhD(bot), 43. Prof Exp: Teacher high schs, Ill, 32-42; asst, Univ Iowa, 42-43; prof biol, Mo Valley Col, 43-44; asst prof sci, State Univ NY Col Teachers, Oswego, 44-45; mem bd exam, 45-57, from asst prof to assoc prof biol, 45-55, PROF NATURAL SCI, MICH STATE UNIV, 55-, MEM STAFF EVAL SERV, 57- Concurrent Pos: Chmn, Grad Record Exam Comt Adv Biol, 49-66; panel eval & testing, Comn Undergrad Educ Biol Sci, NSF, 65-67; consult, Nat Assessment of Progress in Educ, 68-71. Mem: Am Soc Plant Physiol; Am Psychol Asn; Nat Asn Res Sci Teaching; Nat Sci Teachers Asn; Am Inst Biol Sci. Res: Temperature effects on plant growth; test construction and evaluation procedures. Mailing Add: Off Eval Serv 202 S Kedzie Hall Mich State Univ East Lansing MI 48824

NELSON, CLARENCE NORMAN, b Starbuck, Minn, June 6, 09; m 35; c 2. PHYSICS. Educ: St Olaf Col, BA, 31; Ohio State Univ, MA, 33. Prof Exp: Physicist, Res Labs, Eastman Kodak Co, 33-53, res assoc, 53-74; RETIRED. Mem: Optical Soc Am; Soc Photog Scientists & Engrs. Res: Optics; physics of the photographic process; sensitometry; vision; tone reproduction; modulation transfer; communication theory;

image science; American standards on image evaluation; image science; theory of the photographic process. Mailing Add: 73 Sagamore Dr Rochester NY 14617

NELSON, CLIFFORD VINCENT, b Boston, Mass, Sept 23, 15; m 41; c 2. CARDIOVASCULAR PHYSIOLOGY. Educ: Mass Inst Technol, BS, 42; Univ London, PhD(eng electrocardiol), 53. Prof Exp: Asst biol eng, Mass Inst Technol, 40; engr, Submarine Signal Co, Mass, 42-47; res engr, Sanborn Co, 48; researcher, EEG Lab, Mass Gen Hosp, 49; asst res prof med, Col Med, Univ Utah, 54-56; RES ASSOC CARDIOL & RES, MAINE MED CTR, 56- Concurrent Pos: Am Heart Asn fel, 53-55, estab investr, 56-61; Nat Heart Inst res career award, 62-; adj assoc res prof, Boston Univ, 66-70; res assoc, Baker Med Res Inst, Royal Melbourne Hosp, Australia, 69-70. Mem: Fel Am Col Cardiologists; Biophys Soc; Am Physiol Soc; Inst Elec & Electronics Engrs; Biomed Eng Soc. Res: Vector-cardiology; electrophysiology. Mailing Add: Dept of Res Maine Med Ctr Portland ME 04102

NELSON, CRAIG EUGENE, b Concordia, Kans, May 21, 40; m 62; c 2. ECOLOGY, EVOLUTIONARY BIOLOGY. Educ: Univ Kans, AB, 62; Univ Tex, PhD(zool), 66. Prof Exp: Asst prof, 66-71, ASSOC PROF ZOOL, PUB & ENVIRON AFFAIRS & DIR ENVIRON STUDIES, IND UNIV, BLOOMINGTON, 71- Mem: Soc Study Evolution; Ecol Soc Am; Soc Systematic Zool; Am Soc Naturalists; Soc Study of Amphibians & Reptiles. Res: Ecological and evolutionary theory; community structure; speciation; evolutionary processes in amphibia. Mailing Add: Dept of Zool Ind Univ Bloomington IN 47401

NELSON, CURTIS JEROME, b Mitchell Co, Iowa, Mar 25, 40; m 60; c 2. AGRONOMY. Educ: Univ Minn, St Paul, BS, 61, MS, 63; Univ Wis-Madison, PhD(agron), 66. Prof Exp: Res asst forage mgt, Univ Minn, 61-63; forage physiol, Univ Wis, 63-66; res assoc, Cornell Univ, 66-67; from asst prof to assoc prof, 67-75, PROF FORAGE PHYSIOL, UNIV MO-COLUMBIA, 75- Concurrent Pos: Fel, Welsh Plant Breeding Sta, Aberystwyth Wales, UK, 73-74; NSF fel, NATO, 73-74; assoc ed, Crop Science, 75-78. Mem: AAAS; Am Soc Agron; Crop Sci Soc Am; Am Soc Plant Physiol; Soil Sci Soc Am. Res: Crop physiology and biochemistry; genetic control of photosynthesis; carbon metabolism; yield expression of forage grasses; management of forage legumes and grasses. Mailing Add: Dept of Agron Univ of Mo Columbia MO 65201

NELSON, CYNTHIA, b Augusta, Maine, Sept 29, 33. CULTURAL ANTHROPOLOGY. Educ: Univ Maine, Orono, BA, 55; Univ Chicago, MA, 57; Univ Calif, Berkeley, PhD(anthrop), 63. Prof Exp: From asst prof to assoc prof, 63-72, PROF ANTHROP, AM UNIV CAIRO, 72- CHMN DEPT SOCIOL, ANTHROP & PSYCHOL, 68- Concurrent Pos: Lectr dept anthrop, Univ Calif, Berkeley, 62-63; res assoc fel, Inst Int Studies, 72-73. Mem: Am Anthrop Asn; Mid East Studies Asn NAm; Soc Study Symbolic Interaction. Res: Social change in underdeveloped countries; changing roles of men and women in the Middle East; symbolic processes; world view and social change; phenomenological issues in anthropology; women in the health professions. Mailing Add: Dept of Anthrop Am Univ at Cairo 113 Sharia Kasr El Aini Cairo Egypt

NELSON, D KENT, b Ft Collins, Colo, Mar 8, 39; m 60; c 4. ANIMAL NUTRITION. Educ: Colo State Univ, BS, 61; Mich State Univ, MS, 64; Iowa State Univ, PhD(animal nutrit), 68. Prof Exp: Asst prof dairy sci, Wash State Univ, 68-69; ASSOC PROF DAIRY SCI, IOWA STATE UNIV, 69- Mem: Am Soc Animal Sci; Am Dairy Sci Asn. Res: Calf nutrition; nonprotein nitrogen utilization; dairy cattle nutrition. Mailing Add: Dept of Animal Sci Iowa State Univ Ames IA 50010

NELSON, DALLAS LEROY, b Clay Center, Kans, Oct 4, 28; m 51; c 2. TOXICOLOGY, PATHOLOGY. Educ: Kans State Univ, BS, 53, DVM, 53, MS, 59, PhD(parasitol), 63. Prof Exp: Instr path, Kans State Univ, 56-64; MGR TOXICOL RES, CHEMAGRO CORP, 64- Mem: Am Vet Med Asn; Soc Toxicol; Am Col Vet Toxicol; Am Acad Clin Toxicol; Am Asn Vet Parasitol. Res: Host-parasite relationships; toxicology of agricultural chemicals. Mailing Add: 1013 Lennox Dr Olathe KS 66061

NELSON, DANIEL JAMES, zoology, deceased

NELSON, DARRELL WAYNE, b Aledo, Ill, Nov 28, 39; m 61; c 2. SOIL CHEMISTRY, SOIL MICROBIOLOGY. Educ: Univ Ill, Urbana, BS, 61, MS, 63; Iowa State Univ, PhD(soil chem), 67. Prof Exp: A SSOC PROF SOIL MICROBIOL, PURDUE UNIV, WEST LAFAYETTE, 68- Honors & Awards: Agronomy Award, Ciba-Geigy Inc, 75. Mem: Am Soc Agron; Soil Sci Soc Am; Int Soil Sci Soc. Res: Chemistry of nitrogen in soils and sediments; effect of fertilizer use on the environment; nature and properties of soil organic matter. Mailing Add: Dept of Agron Purdue Univ West Lafayette IN 47906

NELSON, DARREN MELVIN, b Lincoln, Nebr, Aug 15, 25; m 53; c 4. ANIMAL PHYSIOLOGY, ENDOCRINOLOGY. Educ: Univ Nebr, BS, 54; Univ Ill, PhD(animal physiol), 65. Prof Exp: Res asst animal sci, Univ Nebr, 54; asst prof animal husb, Calif State Polytech Col, 54-58; instr animal sci, Purdue Univ, 58-60; res asst, Univ Ill, 60-65; NIH fel as trainee in endocrinol, Sch Med, Univ Kans, 65-66; assoc res prof gynec & obstet, Med Ctr, Univ Okla, 66-67; assoc prof biol, Univ Redlands, 67-68; PROF ANIMAL SCI, CALIF STATE UNIV, FRESNO, 68- Mem: AAAS; Soc Exp Biol & Med; Am Fertil Soc; Am Soc Animal Sci; Poultry Sci Asn. Res: Neuroendocrine regulation of reproductive processes in mammals of both sexes; early neonatal differentiation of the central nervous system as influenced by steroid administration in mammals and avian species. Mailing Add: Dept of Animal Sci Calif State Univ Fresno CA 93710

NELSON, DAVID, b Cape Girardeau, Mo, Jan 2, 18. MATHEMATICAL LOGIC. Educ: Univ Wis, BA, 39, MA, 40, PhD, 46. Prof Exp: Asst prof math, Amherst Col, 42-46; from asst prof to assoc prof, 46-58, chmn dept, 56-68, PROF MATH, GEORGE WASHINGTON UNIV, 59- Concurrent Pos: Consult, Nat Res Coun, 60-63. Mem: Am Math Soc; Math Asn Am; Asn Symbolic Logic. Res: Theory of recursive functions; intuitionistic mathematics. Mailing Add: Dept of Math George Washington Univ Washington DC 20006

NELSON, DAVID ALAN, b Melrose, Mass, June 13, 31; m 56; c 3. CHEMISTRY. Educ: Mass Inst Technol, BS, 53; Univ RI, MS, 55; Univ NH, PhD(chem), 60. Prof Exp: Res assoc chem, Univ Ore, 60-62; from asst prof to assoc prof, Univ Wyo, 62-74; MEM FAC BIOCHEM, UNIV WIS-MADISON, 74- Concurrent Pos: Petrol Res Fund grant, 62-63; USPHS grant, 65-68. Mem: Am Chem Soc. Res: Structure determination and synthesis of natural products; photochemistry of organic nitrogen compounds; nucleophilic substitution of substituted pyridines. Mailing Add: Dept of Biochem Univ of Wis Madison WI 53706

NELSON, DAVID ALBERT, organic chemistry, see 12th edition

NELSON, DAVID HERMAN, b Houston, Tex, Mar 28, 43; m 65; c 2. VERTEBRATE ECOLOGY, AQUATIC ECOLOGY. Educ: Baylor Univ, BA, 66, MA, 68; Mich State Univ, PhD(zool), 74. Prof Exp: Asst prof biol, Adrian Col, 73-75; RES ASSOC ECOL, SAVANNAH RIVER ECOL LAB, UNIV GA, 75- Honors & Awards: Roosevelt Mem Award, Am Mus Natural Hist, 69; Pres Award, Am Soc Ichthyologists & Herpetologists, 70. Mem: Ecol Soc Am; Am Inst Biol Sci; Am Soc Ichthyologists & Herpetologists; Soc Study Amphibians & Reptiles; Sigma Xi. Res: Thermal ecology of aquatic organisms; biological effects of heated reactor effluents; temperature tolerances, temperature prefences and thermal stress; ecology, movements and activity patterns of amphibians and reptiles. Mailing Add: Savannah River Ecol Lab Drawer E Aiken SC 29801

NELSON, DAVID LYNN, b Sacramento, Calif, Dec 6, 42. PHYSICAL CHEMISTRY. Educ: Augustana Col, BA, 65; Univ Waterloo, PhD(phys chem), 69. Prof Exp: Vis asst prof phys chem, Univ Windsor, 70-72 & Rensselaer Polytech Inst, 72-75; SCI OFFICER PHYS CHEM, OFF NAVAL RES, 75- Mem: Am Chem Soc; Sigma Xi. Res: Spectroscopy and instrumentation; electrochemistry; surface chemistry and photochemistry. Mailing Add: Off of Naval Res 472 800 N Quincy Arlington VA 22217

NELSON, DAVID TORRISON, b Decorah, Iowa, May 16, 27; m 57; c 4. OPTICS. Educ: Luther Col, Iowa, BA, 49; Univ Rochester, MA, 55; Iowa State Univ, PhD(physics), 60. Prof Exp: Asst physics, Univ Rochester, 49-53; instr, Luther Col, Iowa, 54-57; asst, Iowa State Univ, 58-60; from asst prof to assoc prof, 60-67, PROF PHYSICS, LUTHER COL, IOWA, 67-, CHMN DEPT, 72- Concurrent Pos: NSF sci fac fel, Stanford Univ, 67-68; vis prof eng, Ariz State Univ, 74. Mem: Am Phys Soc; Am Asn Physics Teachers; Optical Soc Am; Acoustical Soc Am; Int Solar Energy Soc. Res: Solar energy. Mailing Add: Dept of Physics Luther Col Decorah IA 52101

NELSON, DENNIS RAYMOND, b New Rockford, NDak, Feb 7, 36; m 61; c 3. BIOCHEMISTRY. Educ: NDak State Univ, BS, 58, MS, 59; Univ NDak, PhD(biochem, chem physiol), 64. Prof Exp: Res chemist, 64-71, RES LEADER METAB & RADIATION RES LAB, AGR RES SERV, USDA, 71- Concurrent Pos: Assoc prof biochem, NDak State Univ, 72- Mem: Am Chem Soc; Sigma Xi; Am Soc Biol Chemists; AAAS. Res: Structure, biosynthesis and hormonal control of insect cuticular hydrocarbons; mass spectra of insect methylalkanes; biochemistry of photoperiodic induction of dispause; mode of action of insect hormones. Mailing Add: Metab & Radiation Res Agr Res Serv USDA Fargo ND 58102

NELSON, DEVAUGHN RAYMOND, physics, health physics, see 12th edition

NELSON, DIANE RODDY, b Knoxville, Tenn, July 10, 44; m 66. INVERTEBRATE ZOOLOGY. Educ: Univ Tenn, Knoxville, BS, 66, MS, 68, PhD(invert zool), 73. Prof Exp: Instr biol, 68-69, instr gen sci, 69-72, ASST PROF GEN SCI, E TENN STATE UNIV, 72- Mem: Am Inst Biol Sci; Am Micros Soc; Am Soc Zoologists; Soc Syst Zool; Int Soc Meiobenthologists. Res: Systematics and ecology of tardigrades or water bears, Phylum: Tardigrada. Mailing Add: Box 2739 E Tenn State Univ Johnson City TN 37601

NELSON, DON B, b Cushing, Tex, Oct 13, 40; m 67; c 2. ORGANIC CHEMISTRY. Educ: Univ Tex, Austin, BS, 63, PhD(org chem), 66. Prof Exp: Sr res chemist, PPG Industs, Inc, Tex, 66-67; sr process chemist, 67-73, PROCESS SPECIALIST, MONSANTO POLYMERS & PETROCHEMICALS CO, 73- Res: Process and product development in area of organic chemicals; process development of heterogeneous catalytic reactions. Mailing Add: Monsanto Polymers & Petrochem Co PT Dept PO Box 1311 Texas City TX 77590

NELSON, DON HARRY, b Salt Lake City, Utah, Nov 28, 25; m 49; c 3. MEDICINE. Educ: Univ Utah, BA, 45, MD, 47. Prof Exp: Res instr biochem, Univ Utah, 50, asst res prof, 52; res assoc med, Harvard Med Sch, 55-56, instr, 57, assoc, 58-59; dir metab ward, Peter Bent Brigham Hosp, Boston, 57-58; from assoc prof to prof med, Sch Med, Univ Southern Calif, 59-66; PROF MED, SCH MED, UNIV UTAH, 66- Mem: Endocrine Soc; Am Soc Clin Invest; Am Physiol Soc. Res: Endocrinology; control of adrenal secretion; mechanism of action of adrenal steroids. Mailing Add: Dept of Med Univ of Utah Med Ctr Salt Lake City UT 84112

NELSON, DONALD CARL, b Minneapolis, Minn, June 28, 31; m 53; c 3. HORTICULTURE, PLANT PHYSIOLOGY. Educ: Univ Minn, BS, 53, PhD(hort), 61. Prof Exp: Asst veg crops, Univ Minn, 53-55; from asst prof to assoc prof, 61-73, PROF HORT, N DAK STATE UNIV, 73- Mem: Europ Asn Potato Res; Potato Asn Am; Weed Soc Am. Res: Physiology and culture of potatoes. Mailing Add: Dept of Hort NDak State Univ Fargo ND 58102

NELSON, DONALD DEWEY, b San Francisco, Calif, June 8, 38; m 58; c 3. ANIMAL NUTRITION. Educ: Fresno State Col, BS, 62; Univ Calif, Davis, MS, 64; Ohio State Univ, PhD(nutrit), 69. Prof Exp: From assoc prof to prof animal sci, Fresno State Col, 64-74; GEN MGR AGR OPERS, BIXBY RANCH CO, 74- Mem: Am Soc Animal Sci. Res: Bovine and equine nutrition. Mailing Add: 7280 E Tollhouse Rd Clovis CA 93612

NELSON, DONALD FREDERICK, b East Grand Rapids, Mich, July 4, 30; m 54; c 2. PHYSICS. Educ: Univ Mich, BS, 52, MS, 53, PhD(physics), 59. Prof Exp: Fel physics, Univ Mich, 58-59; mem tech staff, Bell Tel Labs, Inc, 59-67; prof physics, Univ Southern Calif, 67-68; MEM TECH STAFF, RES DIV, BELL TEL LABS, 68- Mem: Fel Am Phys Soc; Optical Soc Am. Res: Scattering of polarized electrons; basic laser properties; diode lasers and electroluminescence; electro-optic diode light modulators; semiconductor and dielectric luminescence; acousto-optic interactions; Brillouin scattering; electrodynamics of elastic dielectrics. Mailing Add: Res Div 1C 332 Bell Tel Labs Murray Hill NJ 07974

NELSON, DONALD J, b Harvey, Ill, Feb 2, 38; m 60; c 2. BIOCHEMICAL PHARMACOLOGY. Educ: Oberlin Col, BA, 60; Yale Univ, PhD(pharmacol), 65. Prof Exp: Fel pharmacol, Case Western Reserve Univ, 65-67, instr, 67-69; SR RES BIOCHEMIST, BURROUGHS WELLCOME CO, 69- Res: Synthesis of pyrimidine antimetabolites; purification and kinetics of thymidylate kinase from tumors and Escherichia coli; metabolism of thiopurines; control mechanisms in purine and pyrimidine biosynthesis; metabolic effects of allopurinol. Mailing Add: Dept of Exp Ther Burroughs Wellcome Co Research Triangle Park NC 27709

NELSON, DONALD JOHN, b Perth Amboy, NJ, July 24, 45; m 67; c 2. BIOCHEMISTRY. Educ: Rutgers Univ, BS, 67; Univ NC, Chapel Hill, PhD(biochem), 72. Prof Exp: Fel biochem, Dept Pharmacol, Stanford Univ, 72-74 & Dept Chem, Univ Va, 74-75; ASST PROF CHEM, CLARK UNIV, 75- Concurrent Pos: NIH fel gen med sci, 74-75. Mem: Am Chem Soc. Res: Metal ion and small molecule binding to proteins by nuclear magnetic resonance and fluorescence spectroscopy; protein evolution and polymorphism; intermolecular associations in nucleoside-drug complexes. Mailing Add: Dept of Chem Clark Univ Worcester MA 01610

NELSON, DOUGLAS A, b Windom, Minn, Jan 20, 27; m 56; c 4. CLINICAL

PATHOLOGY, HEMATOLOGY. Educ: Univ Minn, BA, 50, BS & MD, 54. Prof Exp: Intern, Philadelphia Gen Hosp, 54-55; resident path anat, Mallory Inst Path, Boston City Hosp, 55-58; med fel specialist & res clin path, Univ Minn, 58-60, instr lab med, 60-63, asst prof & asst dir clin labs, 63-64; assoc prof path & assoc dir clin path, 64-69, PROF PATH & ASSOC DIR CLIN PATH, DIV CLIN PATH, STATE UNIV NY UPSTATE MED CTR, 69- Concurrent Pos: Sr teaching fel path, Sch Med, Boston Univ, 57-58 & Harvard Med Sch, 57-58; hon consult hemat, Royal Postgrad Med Sch, Hammersmith Hosp, London, Eng, 70-71. Mem: Am Soc Clin Path; Am Soc Cell Biol; NY Acad Sci; Am Soc Hemat; Int Soc Hemat. Res: Cellular pathology and cytochemistry of hematopoietic system. Mailing Add: Div of Clin Path State Univ Hosp Syracuse NY 13210

NELSON, EARL EDWARD, b New Richmond, Wis, Jan 11, 35; m 60; c 1. PLANT PATHOLOGY. Educ: Ore State Univ, BS, 57, PhD(plant path), 62. Prof Exp: Res forester, Pac Northwest Forest Exp Sta, US Forest Serv, 57-59, plant pathologist, 59-63; ASST PROF PLANT PATH, ORE STATE UNIV, 70-, US FOREST SERV PLANT PATHOLOGIST, FORESTRY SCI LAB, 63-, PROJ LEADER, 74- Mem: Mycol Soc Am; Am Phytopath Soc. Res: Forest disease; root diseases of northwest conifers; ecology of root pathogens emphasizing antagonism by soil fungi; dwarfmistletoes of northwest conifers. Mailing Add: Forestry Sci Lab 3200 Jefferson Way Corvallis OR 97331

NELSON, EDWARD A, b Cedar City, Utah, May 8, 25; m 48; c 6. ANIMAL BREEDING. Educ: Utah State Univ, BS, 52, MS, 53; Kans State Univ, PhD(animal breeding), 58. Prof Exp: Mgr, Br Agr Col Valley Farm, 49-51; asst animal breeding, Kans State Univ, 55-58; PROF ANIMAL SCI, CALIF STATE POLYTECH UNIV, POMONA, 58- Mem: Am Soc Animal Sci. Res: Reproductive activity in rams; synchronization of estrus in ewes; artificial insemination of ewes. Mailing Add: Dept of Animal Sci Calif State Polytech Univ Pomona CA 91766

NELSON, EDWARD BLAKE, b Altoona, Pa, Dec 12, 43; m 64; c 2. BIOCHEMISTRY. Educ: Pa State Univ, BS, 65; Mich State Univ, PhD(biochem), 70. Prof Exp: Nat Heart & Lung Inst fel biochem, Univ Tex Southwestern Med Sch, Dallas, 70-71; SR SCIENTIST ENDOCRINOL, UNIV TEX MED BR, GALVESTON, 71- Mem: Am Chem Soc. Res: Biochemical pharmacology; steroid metabolism. Mailing Add: Clin Study Ctr Univ of Tex Med Br Galveston TX 77550

NELSON, EDWARD BRYANT, b McHenry, Ky, July 26, 16; m 41; c 2. PHYSICS. Educ: Western Ky State Col, BS, 37; Vanderbilt Univ, MS, 38; Columbia Univ, PhD, 49. Prof Exp: Asst physics, Columbia Univ, 38-41, lectr, 41-43; asst prof, Western Ky State Col, 43-44; res physicist, Manhattan Proj, 44-46; lectr physics, Columbia Univ, 46-49; from asst prof to assoc prof, 49-63, PROF PHYSICS & ASSOC HEAD DEPT, UNIV IOWA, 63- Concurrent Pos: NSF sr fel, Cambridge Univ, 56-57; vis lectr, Univ Exeter, 61-62. Mem: Am Phys Soc; Am Asn Physics Teachers. Res: Nuclear physics; reactions in light nuclei; nuclear models. Mailing Add: Dept of Physics & Astron Univ of Iowa Iowa City IA 52240

NELSON, EDWARD MONS, b Milwaukee, Wis, Sept 20, 15; m 40; c 2. ANATOMY. Educ: Univ Wis, BA, 37, MA, 39, PhD(anat, zool), 47. Prof Exp: Asst zool, Univ Wis, 37-42; instr, Yale Univ, 47-49; assoc & asst prof anat, Stritch Sch Med, Loyola Univ, Ill, 49-59; from assoc prof to prof, Sch Med, Univ PR, 59-64; assoc prof, 64-67, PROF ANAT, DIV MED & SURG, ACAD HEALTH SCI, US, 67- Mem: Am Asn Anatomists; Am Soc Zoologists; Am Soc Ichthyologists & Herpetologists. Res: Comparative and functional anatomy. Mailing Add: Div of Med & Surg Acad of Health Sci US Ft Sam Houston TX 78234

NELSON, EDWARD R, b Pittsburgh, Pa, July 6, 26; m 58; c 2. ORGANIC CHEMISTRY. Educ: Univ Calif, Berkeley, BS, 50. Prof Exp: Chemist, Norton Air Base, Dept Air Force, Calif, 52; organic chemist, Nat Bur Standards, 52-53; organic chemist, Harry Diamond Labs, Dept Army, DC, 53-64; chemist cancer chemother, Nat Ser Serv Ctr, Nat Cancer Inst, 64-66; polymer chemist, NASA Goddard Space Flight Ctr, 66-71; SUPVRY CHEMIST, MAT EVAL & DEVELOP LAB, GEN SERVS ADMIN-FED SUPPLY SERV, 71- Mem: AAAS; Am Chem Soc; Soc Plastics Eng. Res: Polymer chemistry; organic chemical instrumentation and trace analysis; dielectric materials and molecular rotators; outgassing and behavior of materials under vacuum; material science; small ring compounds. Mailing Add: Gen Servs Admin-Fed Supply Serv Mat Eval & Develop Lab Washington DC 20405

NELSON, ELDON CARL, b Dunkirk, Ohio, Dec 13, 35; m 57; c 2. BIOCHEMISTRY, NUTRITION. Educ: Ohio State Univ, BSc, 57, MSc, 60, PhD(nutrit), 63. Prof Exp: From instr to assoc prof, 63-75, PROF BIOCHEM, OKLA STATE UNIV, 75- Concurrent Pos: Vis assoc res biochemist, Univ Calif, Davis, 75. Mem: Am Inst Nutrit; Am Chem Soc; Am Soc Animal Sci. Res: Metabolism and metabolic function of the vitamin A; nutrition; lipid metabolism. Mailing Add: Dept of Biochem Okla State Univ Stillwater OK 74074

NELSON, ELDON LANE, JR, b Morehead City, NC, May 10, 42; m 67; c 1. MEDICAL PHYSIOLOGY. Educ: E Carolina Univ, BS & BA, 64, MA, 71; Univ Fla, PhD(med physiol), 74. Prof Exp: Fel physiol, Col Med, Univ Fla, 75; ASST PROF PHYSIOL, OKLA COL OSTEOP MED & SURG, 75- Mem: AAAS; Sigma Xi. Res: Thirst mechanisms and how thirst is affected by alteration of renal and endocrine mechanisms. Mailing Add: Okla Col of Osteop Med & Surg Ninth & Cincinnati Tulsa OK 74119

NELSON, ELDRED (CARLYLE), b Starbuck, Minn, Aug 14, 17; m 46, 63; c 1. PHYSICS. Educ: St Olaf Col, AB, 38; Univ Calif, PhD(physics), 42. Prof Exp: Instr physics & res assoc, Radiation Lab, Univ Calif, 42-43; group leader theoret physics div, Los Alamos Sci Lab, NMex, 43-46; asst prof physics, Univ Chicago, 46-47; partner & consult math physics, Frankel & Nelson, 47-48; head comput systs dept, Res & Develop Labs, Hughes Aircraft Co, 48-54, head adv electron lab & assoc dir res & develop labs, 54; assoc dir comput systs div, Ramo Wooldridge Corp, 54-58, head Army data processing systs proj, 58-60, dir intellectronic systs lab, 60-61, prog & appl math lab, TRW Systs, 62, dir comput & data reduction ctr, 63-69, DIR TECHNOL PLANNING & RES, TRW SYSTS, 69- Concurrent Pos: Res assoc, Calif Inst Technol, 46; lectr, Univ Calif, 47-48; prof, Univ Southern Calif, 52-53. Mem: Am Phys Soc; Inst Mgt Sci; NY Acad Sci; Asn Comput Mach; Sigma Xi. Res: Nuclear physics; quantum field theory; electronic digital computer; computer software; research on computer software technology, including the theoretical basis for software reliability, program structure, data structures and computer security. Mailing Add: 1808 Melhill Way Los Angeles CA 90049

NELSON, ELTON GLEN, b Elgin Ore, Sept 15, 10; m 50; c 2. TEXTILES. Educ: Ore State Col, BS, 37, MS, 46; Univ Minn, PhD, 61. Prof Exp: Shipping point inspector, State Dept AGr, Ore, 36; coop fiber flax, USDA & Ore State, 37-48; consular attache, US Dept State, India, 48-49; proj leader field crops res br, Agr Res Serv, USDA, 49-57, head, Cordage Fibers Sect, 57-60; asst chief agron & soils br, Off Food, Agency Int Develop, 60-61, asst sci adv, Latin Am, 61-63; asst dir bus serv & anal div, Off Textiles, US Dept Commerce, 63-70; FIBER SPECIALIST, PLANT GENETICS & GERMPLASM INST, AGR RES SERV, USDA, 70; CONSULT, CORDAGE INST, 72- & TECH ASSIST BUR, OFF AGR, FOOD CROP PROD DIV, AID, 74- Mem: Soc Econ Bot; Am Soc Agron. Res: Long vegetable fibers, hard and soft fibers; cordage and industrial textiles. Mailing Add: 7813 Chester Rd Bethesda MD 20034

NELSON, ELVIN CLIFFORD, b Boulder, Colo, Aug 18, 07; m 32; c 1. PARASITOLOGY. Educ: Univ Colo, BS, 29, MA, 30; Johns Hopkins Univ, PhD(parasitol), 33. Prof Exp: From instr to asst prof zool, Univ Maine, 33-40; parasitologist, State Inland Fisheries & Game Dept, Maine, 40-43; assoc prof bact & parasitol, 47-65, PROF MICROBIOL, MED COL VA, 65- Mem: AAAS; Am Soc Parasitol; Am Soc Trop Med & Hyg; Soc Protozool. Res: Cultivation of Entamoeba histolytica; mosquito cycle of Wuchereria bancrofti; laboratory diagnosis of Schistosoma japonicum; laboratory diagnosis of primary amebic meningoencephalitis. Mailing Add: Dept of Microbiol Med Col of Va Richmond VA 23219

NELSON, ERIC LOREN, b Los Angeles, Calif, June 29, 24; m 48; c 3. MOLECULAR PHARMACOLOGY. Educ: Univ Calif, Los Angeles, BA, 47, PhD(microbiol), 51. Prof Exp: Asst bact, Univ Calif, Los Angeles, 48-51, assoc infectious dis, Med Sch, 51-52, instr, 52-55, asst prof, 55-58, asst prof bact, 58-60; sci dir, Allergan Pharmaceut Corp, Calif, 61-63, vpres, 63-72; PRES, NELSON RES & DEVELOP CO, 72- Concurrent Pos: Res assoc, Univ Chicago, 55-57. Mem: Am Soc Microbiol; Am Soc Exp Path; Asn Res Vision & Ophthal; Soc Invest Dermat. Res: Speciation and brucellaphage of brucella; hemoglobin particles; radiation infection; cellular immunity; ophthalmology. Mailing Add: Nelson Res & Develop Co 19722 Jamboree Blvd Irvine CA 92715

NELSON, ERIC V, b Green Bay, Wis, Jan 9, 40; m 61; c 4. ENTOMOLOGY, APICULTURE. Educ: Univ Wis, BS, 61, MS, 63; Univ Man, PhD(entom), 66. Prof Exp: Res entomologist, Entom Res Apicult Br, Agr Res Serv, USDA, 66-67; asst prof, 67-73, ASSOC PROF BIOL, OHIO NORTHERN UNIV, 73- Concurrent Pos: Vis assoc prof entom, Univ Man, 73-74. Mem: Bee Res Asn; Entom Soc Am; Sigma Xi. Res: Physiology of bee diseases; diets for bees. Mailing Add: Dept of Biol Ohio Northern Univ Ada OH 45810

NELSON, ERLAND, b Blair, Nebr, June 4, 28. NEUROLOGY, NEUROPATHOLOGY. Educ: Carthage Col, BA, 47; Columbia Univ, MD, 51; Univ Minn, PhD(neurol, path), 61. Hon Degrees: DSc, Carthage Col, 73. Prof Exp: Armed Forces Inst Path fel, Max Planck Inst, Munich, Ger, 55-57, NIH fel, 59-60; assoc prof neurol & neuropath, Univ Minn, 61-64; PROF NEUROL & HEAD DEPT, SCH MED, UNIV MD, BALTIMORE CITY, 64- Mem: Am Asn Neuropath; Am Acad Neurol; Am Neurol Asn. Res: Clinical neurology; electron microscopy of central nervous system infections, neoplasms and leukoencephalitis; ultrastructure of intracranial arteries. Mailing Add: Dept of Neurol Univ of Md Sch of Med Baltimore MD 21201

NELSON, EVELYN MERLE, b Can citizen; m 63; c 2. ALGEBRA. Educ: McMaster Univ, PhD(algebra), 70. Prof Exp: Fel, 70-73, RES ASSOC MATH, McMASTER UNIV, 73- Mem: Am Math Soc; Can Math Cong; Can Soc Hist & Philos Math; Asn Symbolic Logic. Res: Equational compactness in algebras and relational structures; applications of model theory to universal algebra, in particular, first order properties of algebras of continuous functions. Mailing Add: Dept of Math McMaster Univ Hamilton ON Can

NELSON, FRANK EUGENE, b Harlan, Iowa, Dec 5, 09; m 40; c 3. FOOD MICROBIOLOGY. Educ: Univ Minn, BS, 32, MS, 34; Iowa State Col, PhD(dairy bact), 36. Prof Exp: Lab technician, Univ Minn, 32-33, instr dairy bact, 36-37; from asst prof to assoc prof bact, Kans State Col, 37-43; prof dairy bact & res prof, Iowa State Univ, 43-60; PROF FOOD SCI, UNIV ARIZ & FOOD SCIENTIST, AGR EXP STA, 60-, PROF MICROBIOL & MED TECHNOL, UNIV, 69- Concurrent Pos: Ed, Am Dairy Sci Asn J, 47-52. Honors & Awards: Borden Award, Am Dairy Sci Asn, 53, Award of Honor, 71. Mem: Am Soc Microbiol; Am Dairy Sci Asn (vpres, 64-65, pres, 65-66); Inst Food Technol; Int Asn Milk, Food & Environ Sanit; Brit Soc Appl Bact. Res: Lipolytic and proteolytic activities of bacteria; factors affecting resistance to heat; psychrophilic bacteria. Mailing Add: Dept of Nutrit & Food Sci Univ of Ariz Tucson AZ 85721

NELSON, GARETH JON, b Chicago, Ill, Dec 23, 37; m 61; c 2. ICHTHYOLOGY. Educ: Roosevelt Univ, BS, 62; Univ Hawaii, PhD(zool), 66. Prof Exp: NSF fel, 66-67; asst curator, 67-71, ASSOC CURATOR ICHTHYOL, AM MUS NATURAL HIST, 71- Concurrent Pos: NSF grant, 69-71; adj prof biol, City Univ of NY, 74-; ed, Syst Zool, 73-76. Mem: Am Soc Ichthyol & Herpet; Soc Syst Zool; Soc Vert Paleont; Japanese Soc Ichthyol; Linnear Soc London. Res: Comparative biology of vertebrates, chiefly fishes. Mailing Add: Dept of Ichthyol Am Mus Nat Hist Central Park West at 79th St New York NY 10024

NELSON, GARY JOE, b Oakland, Calif, Sept 27, 33; m 59; c 1. BIOPHYSICS, BIOCHEMISTRY. Educ: Univ Calif, Berkeley, BS, 55, PhD(biophys), 60. Prof Exp: Res fel heart dis, Donner Lab, Univ Calif, Berkeley, 60-63; sr staff scientist, Lawrence Livermore Lab, Univ Calif, 63-73; assoc res scientist, NY State Inst Basic Res in Ment Retardation, 73-74; grants assoc, NIH, 74-75; HEALTH SCIENTIST ADMINR, NAT HEART & LUNG INST, 75- Concurrent Pos: Nat Heart Inst fel, 60-62; estab investr, Am Heart Asn, 62-63, mem coun arteriosclerosis. Mem: AAAS; Am Soc Biol Chemists. Res: Science policy and administration; nemophilia; thrombosis and hemostasis; membrane structure and function; lipid biochemistry; analytical instrumentation; chromatography; spectroscopy; heart disease and atherosclerosis; disorders of lipid metabolism. Mailing Add: Div of Blood Dis & Resources Nat Heart & Lung Inst NIH Bethesda MD 20014

NELSON, GAYLE HERBERT, b Dayton, Wash, Mar 17, 26; m 46; c 2. ANATOMY. Educ: Walla Walla Col, BA, 47; Univ Md, MS, 53; Univ Mich, PhD(anat), 57. Prof Exp: Asst biol, Walla Walla Col, 47-49; instr, Washington Missionary Col, 49-51, instr & actg head dept, 51-53; from instr to assoc prof anat, Loma Linda Univ, 57-66; assoc prof, 66-71, PROF ANAT & CHMN DEPT, KANSAS CITY COL OSTEOP MED, 71- Res: Lymphatic system; biology and taxonomy of Coleoptera. Mailing Add: Dept of Anat Kansas City Col of Osteop Med Kansas City MO 64124

NELSON, GEORGE HUMPHRY, b Charleston, SC, Nov 24, 30; m 56; c 4. BIOCHEMISTRY, OBSTETRICS & GYNECOLOGY. Educ: Col Charleston, AB, 51; Med Col SC, MS, 53, PhD(biochem), 55; WVa Univ, MD, 62. Prof Exp: Asst prof biochem, Univ SDak, 56-58; instr, WVa Univ, 58-62; asst prof biochem, 62-68, from instr to asst prof obstet & gynec, 62-68, assoc prof biochem & obstet & gynec, 68-74, PROF OBSTET & GYNEC & CHIEF SECT FETAL MED, MED COL GA, 74- Concurrent Pos: Intern, Eugene Talmadge Mem Hosp, 64. Mem: AMA; Soc Gynec Invest. Res: Lipid metabolism in normal and abnormal pregnancy; fetal maturity evaluation. Mailing Add: Dept of Obstet & Gynec Med Col of Ga Augusta GA 30902

NELSON, GEORGE LEONARD, b Marshall, Minn, Dec 8, 43; m 64. ORGANIC

CHEMISTRY. Educ: St John's Univ, Minn, BS, 65; Univ Wis-Madison, PhD(chem), 69. Prof Exp: NIH fel, Columbia Univ, 69-70; ASST PROF ORGANIC CHEM, ST JOSEPH'S COL, 70- Concurrent Pos: Cottrell Res Corp grant, 70-71; vis prof, Hahneman Med Col, 75- Mem: Am Chem Soc; NY Acad Sci; Sigma Xi. Res: Mechanistic organic chemistry; thermal rearrangements; synthetic organic chemistry; development of new synthetic techniques; heart and brain chemistry; preparation of biologically active compounds. Mailing Add: Dept of Chem St Jodeph's Col Philadelphia PA 19131

NELSON, GEORGE RICHARD, b Portland, Maine, May 24, 23; m 59; c 3. PAPER CHEMISTRY, PULP & PAPER TECHNOLOGY. Educ: Univ Maine, BS, 48. Prof Exp: From jr chemist to sr chemist, Dennison Mfg Co, 50-56, sect mgr res & develop reprographics, 56-65; sect mgr res & develop reprographics, 65-73, MGR PAPER PROD RES & DEVELOP, WEYERHAEUSER CO, 73- Mem: Tech Asn Pulp & Paper Indust; Am Chem Soc. Res: Research and development efforts leading to new and improved paper products for the graphic arts and communications papers area. Mailing Add: Weyerhaeuser Co Res & Develop 3400 13th Ave SW Seattle WA 98134

NELSON, GERALD CLIFFORD, b Benson, Minn, Aug 21, 40; m 64. NUCLEAR PHYSICS. Educ: St Olaf Col, BA, 62; Iowa State Univ, PhD(physics), 67. Prof Exp: Res assoc, Iowa State Univ, 67-68; res assoc, Lawrence Radiation Lab, Univ Calif, Berkeley, 68-70; RES ASSOC, SANDIA LABS, 70- Mem: AAAS; Am Phys Soc. Res: Properties of x-rays from high atomic number elements; gamma ray energies, intensities and internal conversion of coefficients. Mailing Add: 11509 Nassau Dr NE Albuquerque NM 87111

NELSON, GIDEON EDMUND, JR, b Jacksonville, Fla, Feb 21, 24; m 48; c 3. BIOLOGY. Educ: Univ Fla, MS, 50, PhD(zool), 54. Prof Exp: Asst zool, Univ Fla, 49; from instr to assoc prof, Ala Col, 52-60; asst prof zool & gen biol, 60-61, assoc prof biol, 61-67, PROF BIOL, UNIV S FLA, 67- Mem: AAAS. Res: Ecology. Mailing Add: Dept of Biol Univ of S Fla Tampa FL 33620

NELSON, GILBERT HARRY, b Manhattan, NY, Sept 20, 27; m 55; c 3. CLINICAL CHEMISTRY. Educ: Wagner Lutheran Col, BS, 52; Purdue Univ, MS, 56, PhD(biochem), 58. Prof Exp: Biochemist, Christian Hansen's Lab, Inc, Wis, 57-58 & New Castle State Hosp, 58-64; CLIN CHEMIST, MIAMI VALLEY HOSP, 64- Concurrent Pos: Assoc prof clin chem, Univ Dayton, 69. Mem: AAAS; Am Chem Soc; Am Asn Clin Chem; NY Acad Sci. Mailing Add: Diagnostic Labs 1 Wyoming St Miami Valley Hosp Dayton OH 45409

NELSON, GORDON ALBERT, b Bentley, Alta, Nov 29, 25; m 54; c 3. PLANT PATHOLOGY. Educ: Univ Alta, BSc, 49, MSc, 51; SDak State Univ, PhD(plant path), 60. Prof Exp: Res officer bact, Defense Res Bd, Alta, 54-55; lab scientist, Prov Dept Agr, 55-57; res asst plant path, SDak State Univ, 57-61; res scientist, Nfld, 61-66, RES SCIENTIST, RES STA, CAN DEPT AGR, ALTA, 66- Concurrent Pos: Nat Res Coun Can res fel, 61; vis lectr, Mem Univ, 63-65. Mem: Am phytopath Soc; Can Phytopath Soc; Potato Asn Am. Res: Dairy bacteriology; bacterial plant diseases; soil-borne plant diseases and ecology of plant pathogens. Mailing Add: Can Dept Agr Res Sta Lethbridge AB Can

NELSON, GREGORY VICTOR, b Minneapolis, Minn, Nov 16, 43; m 67. INORGANIC CHEMISTRY. Educ: St Olaf Col, BA, 65; Univ Calif, Berkeley, PhD(chem), 68. Prof Exp: Asst prof, 68-73, ASSOC PROF CHEM, DREW UNIV, 73- Concurrent Pos: Vis prof, Inst Org Chem, Univ Fribourg, Switz, 74-75. Mem: AAAS; Am Chem Soc; Sigma Xi. Res: Organometallic complexes of transition metal carbonyls; metalolefin aids to chemical education, stereoscopic computer drawing, single-concept films; aromatic radical anions; crown complexes; electron spin resonance. Mailing Add: Dept of Chem Drew Univ Madison NJ 07940

NELSON, GUNNER ELWOOD, organic chemistry, see 12th edition

NELSON, HARRY ERNEST, b Rockford, Ill, Sept 21, 13; m 41; c 3. MATHEMATICS, ASTRONOMY. Educ: Augustana Col, AB, 35; Univ Wis, PhM, 40; Univ Iowa, PhD(math), 50. Prof Exp: Teacher high sch, Ill, 35-37; teacher math, Luther Col, 37-42; teacher, Gustavus Adolphus Col, 42-46; PROF MATH, AUGUSTANA COL, ILL, 46- Concurrent Pos: Mem Univ Ky team, Int Coop Admin, Indonesia, Bandung, 58-60; dir, John Deere Planetarium, 67- Mem: Am Math Soc; Math Soc Am. Res: Meteors. Mailing Add: Dept of Math Augustana Col Rock Island IL 61201

NELSON, HARRY GLADSTONE, b Chicago, Ill, Feb 4, 22; m 44; c 2. ZOOLOGY. Educ: Univ Chicago, BS, 45. Prof Exp: Asst zool, Univ Chicago, 45-47; instr biol, Gary Col, 47-48; teacher zool, Herzl Br, Chicago City Jr Col, 48-51; from lectr to assoc prof, 51-69, actg chmn dept, 65-66, chmn, 66-72, PROF BIOL, ROOSEVELT UNIV, 69- Concurrent Pos: Teacher, Herzl Br, Chicago City Jr Col, 54; assoc, Field Mus Natural Hist, 58- Mem: Soc Study Evolution; Soc Syst Zool; AAAS. Res: Biology; invertebrate zoology and entomology; evolution, ecology, distribution, taxonomy and morphology of the Dryopoidea. Mailing Add: Dept of Biol Roosevelt Univ 430 S Michigan Ave Chicago IL 60605

NELSON, HARVARD G, b Logan, Utah, Aug 29, 19; m 47; c 1. DAIRY SCIENCE. Educ: Utah State Univ, BS, 41; Ohio State Univ, MS, 42; Univ Minn, PhD(dairy indust), 52. Prof Exp: Explosives chemist, Pantex Ord Plant, 42-43; PROF FOOD SCI, SOUTHEASTERN LA UNIV, 50- Concurrent Pos: Tech dir, Hammond Milk Corp, 51-58; consult, Brown's Velvet Dairy Prod, Inc, 55-; lab dir, Goldhill Foods Corp, 61-64. Mem: Am Dairy Sci Asn. Res: Quality control work in seafood, vegetable and dairy products. Mailing Add: Box 370 College Station Hammond LA 70401

NELSON, HARVEY KENNETH, ecology, research administration, see 12th edition

NELSON, HERBERT LEROY, b Eddyville, Iowa, June 15, 22; m 43; c 4. PSYCHIATRY. Educ: Univ Iowa, BA, 43, MD, 46; Am Bd Neurol & Psychiat, dipl psychiat, 53. Prof Exp: Intern, Univ Hosps, Iowa City, Iowa, 46-47; psychiat physician, US Vet Admin Hosp, Knoxville, Tenn, 49-51; chief acute treatment serv, Ore State Hosp, Salem, 51-55, clin dir, 55-63, asst supt, 58-63; from asst prof to assoc prof psychiat, 63-73, PROF PSYCHIAT, UNIV IOWA, 73-; DIR, IOWA MENT HEALTH AUTHORITY, 68- Concurrent Pos: Proj dir, Iowa Comprehensive Ment Health Planning, 63-66; asst dir, Psychopathic Hosp, Iowa City, 66- Mem: AMA; fel Am Psychiat Asn. Res: Mental hospital administration; group psychotherapy; therapeutic community organization; community psychiatry. Mailing Add: Psychopathic Hosp 500 Newton Rd Iowa City IA 52242

NELSON, HOMER MARK, b Malad, Idaho, Dec 11, 32; m 69; c 2. SOLID STATE PHYSICS. Educ: Brigham Young Univ, BS, 53, MS, 54; Harvard Univ, PhD(physics), 60. Prof Exp: Assoc prof, 59-71, PROF PHYSICS, BRIGHAM YOUNG UNIV, 71- Mem: Am Phys Soc. Res: Magnetic resonance under high pressure; molecular beams. Mailing Add: Dept of Physics Brigham Young Univ Provo UT 84601

NELSON, HOWARD JOSEPH, b Gowrie, Iowa, Jan 12, 19; m 44; c 2. GEOGRAPHY. Educ: Iowa State Teachers Col, AB, 42; Univ Chicago, AM, 47, PhD, 49. Prof Exp: From instr to assoc prof, 49-63, chmn dept, 66-71, PROF GEOG, UNIV CALIF, LOS ANGELES, 63- Mem: Asn Am Geogr; fel Am Geog Soc. Res: Urban and economic geography. Mailing Add: Dept of Geog Univ of Calif Los Angeles CA 90024

NELSON, IRAL CLAIR, b Eugene, Ore, Apr 18, 27; m 55; c 3. HEALTH PHYSICS. Educ: Univ Ore, BS, 51, MA, 55; Am Bd Health Physics, cert, 62. Prof Exp: Jr engr, Hanford Atomic Prod Opers, Gen Elec Co, 55-58, sr engr, 58-64, mgr external dosimetry, 64-65; sr res scientist, 65-67, res assoc, 67-72, MGR RADIOL HEALTH RES SECT, PAC NORTHWEST LABS, BATTELLE MEM INST, 72- Mem: Health Physics Soc; Am Asn Physicists in Med. Res: Radiation dosimetry; methods for determining the fate of radionuclides deposited in the body from accidental intake; environmental consequences of construction and operation of nuclear facilities. Mailing Add: 2105 Putnam Richland WA 99352

NELSON, IVORY VANCE, b Curtiss, La, June 11, 34; m 60; c 2. ANALYTICAL CHEMISTRY. Educ: Grambling Col, BS, 59; Univ Kans, PhD(chem), 63. Prof Exp: From assoc prof to prof chem, Southern Univ, Baton Rouge, 63-67, chmn div natural sci, Shreveport, 67-68; asst dean, 68-71, VPRES RES & SPEC PROGS, PRAIRIE VIEW A&M UNIV, 71- Concurrent Pos: Vis prof, Loyola Univ, 67-68; consult, Oak Ridge Assoc Univs, 69-70. Mem: AAAS; Am Chem Soc; NY Acad Sci. Res: Higher education; administration; behavior of metal ions in nonaqueous solvent. Mailing Add: Prairie View A&M Univ Prairie View TX 77445

NELSON, JACK RAYMOND, b Fargo, NDak, July 25, 34; m 55; c 5. RANGE ECOLOGY, WILDLIFE ECOLOGY. Educ: NDak State Univ, BS, 60, MS, 61; Univ Idaho, PhD(range ecol), 69. Prof Exp: Instr range sci, 64-69, asst prof range & wildlife ecol, 69-74, ASSOC PROF WILDLIFE HABITAT MGT, WASH STATE UNIV, 74- Concurrent Pos: Consult, Spokane Indian Tribe, Bur Indian Affairs, 69-70 & Key Chem Inc, Wash, 70- Mem: Ecol Soc Am; Wildlife Soc; Soc Range Mgt. Res: Big game range ecology; big game livestock competition; multiple land-use management; wildlife rehabilitation. Mailing Add: Dept of Forestry & Range Mgt Wash State Univ Pullman WA 99163

NELSON, JAMES ARLY, b Livingston, Tex, Feb 8, 43; m 63; c 3. PHARMACOLOGY. Educ: Univ Houston, BS, 65, MS, 67; Univ Tex Med Br Galveston, PhD(pharmacol), 70. Prof Exp: Res assoc pharmacol, Brown Univ, 70-72; sr biochemist, Southern Res Inst, 72-76; ASST PROF PHARMACOL & TOXICOL, UNIV TEX MED BR GALVESTON, 76- Concurrent Pos: Clin asst prof, Med Sch, Univ Ala, 75-76. Mem: Am Asn Cancer Res; Am Soc Pharmacol & Exp Therapeut. Res: Biochemical pharmacology; cancer chemotherapy; renal pharmacology; ion and drug transport; environmental toxicology. Mailing Add: Dept of Pharmacol & Toxicol Univ of Tex Med Br Galveston TX 77550

NELSON, JAMES DONALD, b Paducah, Ky, Apr 30, 43. MATHEMATICS. Educ: Univ Ky, BS, 65, MS, 67, PhD(math), 70. Prof Exp: ASST PROF MATH, WESTERN MICH UNIV, 70- Mem: Am Math Soc; Math Asn Am. Res: Analysis. Mailing Add: Dept of Math Western Mich Univ Kalamazoo MI 49001

NELSON, JAMES GORDON, b Hamilton, Ont, Apr 3, 32; m; c 3. RESOURCE GEOGRAPHY. Educ: McMaster Univ, BA, 55; Univ Colo, MA, 57; Johns Hopkins Univ, PhD(geog), 59. Prof Exp: Asst prof geog, Univ Calgary, 60-64, chmn dept, 63-66, assoc prof, 64-67, assoc dean arts & sci, 66-69, prof geog & chmn interdisciplinary comt environ, resources & planning, 67-71, spec asst to acad vpres, 69-71; prof geog, Univ Western Ont, 71-75; PROF GEOG & DEAN ENVIRON STUDIES, UNIV WATERLOO, 75- Concurrent Pos: Gustav Bissing fel, Johns Hopkins Univ, 59-60; vis prof, Univ Canterbury, 68; mem, Can Prep Comt Human Environ Conf, Stockholm, 71-72; vchmn, Human Environ Comt, Soc Sci Res Coun Can, 72-73, chmn, 73; consult, Fed Dept Environ, Can, 74-75. Mem: AAAS; Asn Am Geog (pres, 75-76); Can Asn Geog; NZ Geog Soc. Res: Land use history and landscape change in national parks and public land management; man's impact on environment. Mailing Add: Fac of Environ Studies Univ of Waterloo Waterloo ON Can

NELSON, JAMES H, JR, b Marietta, Ohio, June 28, 26; m; c 2. OBSTETRICS & GYNECOLOGY. Educ: Marietta Col, BS, 49; NY Univ, MD, 54. Prof Exp: Spec fel, Gynec Tumor Serv, State Univ NY Downstate Med Ctr, 57-58; from instr to assoc prof, 61-69, PROF OBSTET & GYNEC & CHMN DEPT, STATE UNIV NY DOWNSTATE MED CTR, 69-, DIR GYNEC TUMOR SERV, 64- Concurrent Pos: Am Cancer Soc adv clin fel, 61-64; from asst attend physician to assoc attend, Kings County Hosp, 61-66, vis attend, 66-; fel med, Cancer Chemother Div, Mem Ctr Cancer & Allied Dis, 62-63; res fel, Sloan-Kettering Res Inst, 62-63, vis investr chemother, 63-68; consult, US Naval Hosps, St Albans, NY, 63-65, Bethesda, Md, 66 & consult & lectr, Philadelphia, Pa, 67; mem div gynec oncol, Am Bd Obstet & Gynec. Mem: AAAS; Am Cancer Soc; Am Col Surgeons; Am Asn Obstet & Gynec; Am Gynec Soc. Res: Gynecologic malignancy; thymo-lymphatic system. Mailing Add: Dept of Obstet State Univ NY Downstate Med Ctr Brooklyn NY 11203

NELSON, JAMES S, b St Louis, Mo, Mar 19, 33; m 56; c 2. NEUROPATHOLOGY. Educ: St Louis Univ, MD, 57. Prof Exp: Asst path, St Louis Univ, 57-59; instr neuropath, Columbia Univ, 59-60; asst path, St Louis Univ, 60-61; Nat Inst Neurol Dis & Blindness spec fel neurochem, 61-63; instr path, Washington Univ, 63-64; state neuropathologist, Div Ment Dis, Mo, 64-65; from asst prof to assoc prof path, Sch Med, St Louis Univ, 65-73; assoc prof path & assoc prof path in pediat, 73-75, PROF PATH, PROF PATH IN PEDIAT & DIR DIV NEUROPATH, DEPT PATH, SCH MED, WASHINGTON UNIV, 75- Concurrent Pos: Consult, St Louis State Hosp, 65-66; neuropathologist, St Louis Univ Hosps & Glennon Mem Hosp Children, 65-73; consult neuropath, St Louis City Hosp, St John's Mercy Hosp & St Luke's Hosp; asst pathologist, Barnes & Allied Hosps & St Louis Children's Hosp, 73-75, assoc pathologist, 75- Mem: AAAS; Am Asn Neuropath; Int Soc Neurochem; Int Acad Path; Am Soc Exp Path. Res: Neuropathology of nutritional disorders and microcirculation; metabolism and function of monoamine neurons. Mailing Add: Dept of Pediat St Louis Children's Hosp St Louis MO 63110

NELSON, JANET SUE RASEY, b Fremont, Mich, June 13, 42; m 66. CANCER, RADIOBIOLOGY. Educ: Univ Mich, BS, 64; Ore State Univ, MS, 65; Univ Ore, PhD(biol), 70. Prof Exp: NIH fel, Cell & Radiation Biol Lab, Allegheny Gen Hosp, Pittsburgh, 70-72; RES ASST PROF TUMOR RADIATION & CELL BIOL, DIV RADIATION ONCOL, MED SCH, UNIV WASH, 72- Mem: Radiation Res Soc. Res: Tumor radiation response cell cycle kinetics; mechanisms of variations in tumor repopulation rate following radiation or drug treatment; neutron radiation biology. Mailing Add: Div of Radiation Oncol Univ of Wash Med Sch Seattle WA 98195

NELSON, JERRY ALLEN, b Durango, Colo, Feb 22, 23; m 44; c 3. ORGANIC

CHEMISTRY. Educ: Univ Wash, BS, 46, PhD(chem), 50. Prof Exp: Res chemist, 50-53, res supvr, 53-56, head, Textile Chem & Intermediate Res Div, 56-60, head textile & indust chem, Res & Develop Div, 60-69, HEAD NEW PROD DIV, ORG CHEM DEPT, RES & DEVELOP DIV, JACKSON LAB, E I DU PONT DE NEMOURS & CO, WILMINGTON, 69- Mem: AAAS; Am Chem Soc. Res: Polymer, surface and fluorine chemistry. Mailing Add: 1012 Baylor Dr Newark DE 19711

NELSON, JERRY EARL, b Glendale, Calif, Jan 15, 44; m 65; c 2. ASTROPHYSICS, EXPERIMENTAL HIGH ENERGY PHYSICS. Educ: Calif Inst Technol, BS, 65; Univ Calif, Berkeley, PhD(physics), 72. Prof Exp: Fel particle physics, 72-75, DIV FEL ASTROPHYS, LAWRENCE BERKELEY LAB, UNIV CALIF, 75- Mem: Am Phys Soc; Am Astron Soc. Res: Pulsars, x-ray sources and black holes; optical pulsars; electron-positron colliding beam physics. Mailing Add: Lawrence Berkeley Lab Univ of Calif Berkeley CA 94720

NELSON, JOHN ARCHIBALD, b Can, Nov 25, 16; m 46; c 3. ORGANIC CHEMISTRY. Educ: Univ Alta, MSc, 39; McGill Univ, PhD(org chem), 45. Prof Exp: Biochemist, Animal Dis Res Inst, Can, 45-46; chemist, Ciba Co, Ltd, Can, 46-49, Ciba Pharmaceut Prod, Inc, 49-69, mgr process res, Ciba-Geigy Corp, 69-72, ASST DIR CLIN PREP PROCESS RES, PHARMACEUT DIV, CIBA-GEIGY CORP, 72- Mem: Am Chem Soc; NY Acad Sci; Am Inst Chem. Res: Process research in pharmaceuticals. Mailing Add: Chem Develop Div Pharmaceut Div Ciba-Geigy Corp Summit NJ 07901

NELSON, JOHN ARTHUR, b Sturgeon Bay, Wis, Jan 16, 38; m 62; c 2. CHEMISTRY, CHEMICAL ENGINEERING. Educ: Univ Wis-Madison, BS, 60; Univ Ariz, PhD(chem), 66. Prof Exp: Metallurgist, Res Dept, Inland Steel Co, Ind, 60-62, res engr, 66-68, sr res engr, 68-69; sr phys chemist, 69-70, SR RES CHEMIST, RES DEPT, WHIRLPOOL CORP, 71- Mem: Am Chem Soc; Electrochem Soc; Nat Asn Corrosion Engrs; Sigma Xi. Res: Corrosion of metals; treatment and processing of metals; physics and chemistry of processing systems. Mailing Add: Rte 2 Box 452 Benton Harbor MI 49022

NELSON, JOHN BERGER, nuclear physics, dosimetry, see 12th edition

NELSON, JOHN D, b Duluth, Minn, Sept 16, 30. PEDIATRICS. Educ: Univ Minn, BS, 52, MS, 54. Prof Exp: Fel infectious dis, Univ Tex Southwestern Med Sch, 59-60; from instr to assoc prof, 60-70, PROF PEDIAT, UNIV TEX HEALTH SCI CTR, DALLAS, 70- Concurrent Pos: Nat Inst Allergy & Infectious Dis res fel, 60-62 & res career develop award, 63-73; vis prof, Med Ctr, Univ Colo, Denver, 62; mem antibiotics panel drug efficacy study, Nat Acad Sci-Nat Res Coun, 66-68; consult staff, John Peter Smith Hosp, Ft Worth, Tex, 74- Mem: Am Soc Microbiol; Infectious Dis Soc Am; Soc Pediat Res; Am Pediat Soc. Res: Pediatric infectious diseases. Mailing Add: Dept of Pediat Univ of Tex Health Sci Ctr Dallas TX 75235

NELSON, JOHN D, JR, microbiology, see 12th edition

NELSON, JOHN DANIEL, b Scotia, NY, May 22, 15; m 39; c 2. PLASTICS CHEMISTRY. Educ: Rensselaer Polytech Inst, BChE, 37. Prof Exp: Develop chemist, 37-40, group leader, Lab, 40-48, sect leader eng, 46-63, mgr resin & varnish res & develop, 63-70, MGR PROD RES, GEN ELEC CO, 70- Res: Plastics; resins; varnishes; applications; phenolics; ureas; themoplastics; adhesives; laminates; molding materials; pulp preforms; foams; tube rolling; chromatography; plastics testing. Mailing Add: Plastics Dept Gen Elec Co 1 Plastics Ave Pittsfield MA 01201

NELSON, JOHN HENRY, b Ogden, Utah, Mar 25, 40; m 64; c 3. INORGANIC CHEMISTRY, ORGANOMETALLIC CHEMISTRY. Educ: Weber State Col, AS, 61; Univ Utah, BS, 64, PhD(chem), 68. Prof Exp: Teaching asst, Univ Utah, 64-65; Esso fel, Tulane Univ La, 68-70; asst prof, 70-74, ASSOC PROF CHEM, UNIV NEV, RENO, 74- Concurrent Pos: Univ Nev Res Adv Bd Grant, Univ Nev, Reno, 71-73, 73-76; Petrol Res Fund grant, 71-74, 75-77; Res Corp grant, 74-76. Mem: AAAS; Am Chem Soc; The Chem Soc; Sigma Xi. Res: Synthesis, physical properties and reactions of coordination and organometallic compounds. Mailing Add: Dept of Chem Univ of Nev Reno NV 89507

NELSON, JOHN HOWARD, b Chicago, Ill, May 29, 30; m 52; c 3. FOOD SCIENCE. Educ: Purdue Univ, BS, 52, MS, 53; Univ Minn, PhD(biochem), 61. Prof Exp: Res biochemist, Gen Mills, Inc, Minn, 60-61, supvr chem, microbiol & phys testing, 61-64, dept head refrig foods res, James Ford Bell Res Ctr, 65-66, head frozen food res, 66-68; dir corp res, 68, VPRES RES & DEVELOP, PEAVEY CO, 68-, DIR INT VENTURES RES DIV, 71- Concurrent Pos: Mem gen comt on DOD Food Prog & chmn comt on cereal & gen prods, Adv Bd Mil Personnel Supplies, Nat Acad Sci, 73- Mem: Am Asn Cereal Chemists (past pres, 73-74 & 75); Am Chem Soc; Inst Food Technologists; Am Oil Chemists Soc; AAAS. Res: Lipids of cereal grain; new instrumental methods of analysis for food ingredients and products; new food product research and development; market and business planning. Mailing Add: Peavey Tech Ctr 11 Peavey Rd Chaska MN 55318

NELSON, JOHN HOWARD, b Bozeman, Mont, Feb 5, 26; m 52; c 3. FOOD SCIENCE. Educ: Mont State Col, BS, 50; Univ Wis, MS, 51, PhD(dairy & food indust), 53. Prof Exp: Res fel, Univ Wis, 53-54; dir res, 54-61, vpres res & develop, 61-72, vpres corp develop, 72-75, VPRES SCI AFFAIRS, DAIRYLAND FOOD LABS, INC, 75- Concurrent Pos: Mem food stand adv comt, State of Wis, 67-73. Mem: Am Chem Soc; Am Dairy Sci Asn; Inst Food Technol. Res: Industrial enzymes for food and dairy field; enzyme modified food ingredients; flavors; food law; enzymes and fermentations in food processing. Mailing Add: 915 S Greenfield Ave Waukesha WI 53186

NELSON, JOHN MARVIN, JR, b Richmond, Mo, May 19, 33; m 62; c 4. ENTOMOLOGY. Educ: North Cent Bible Col, BA, 56; Evangel Col, BS, 64; Southern Ill Univ, Carbondale, MS, 66, PhD(zool), 70. Prof Exp: Asst prof gen zool, Glenville State Col, 69-71; asst prof, 71-74, ASSOC PROF BIOL, ORAL ROBERTS UNIV, 74- Mem: Entom Soc Am; Am Arachnol Soc; Am Inst Biol Sci. Res: Nest ecology and nest symbionts of Polistes wasps; morphology of Polistes larvae; spider distribution and ecology. Mailing Add: Dept of Natural Sci Oral Roberts Univ Tulsa OK 74102

NELSON, JOHN WESLEY, nutrition, see 12th edition

NELSON, JOHN WHITE, b Lafayette, Ind, Sept 1, 16; m 39; c 2. PHARMACOLOGY. Educ: Purdue Univ, BS, 38, PhD(pharmacol), 45; Univ Fla, MS, 39. Prof Exp: Asst, Univ Fla, 38-39; asst prof, Univ Ga, 40-44; assoc prof, Ore State Col, 45-47; assoc prof, 47-51, PROF PHARM, COL PHARM, OHIO STATE UNIV, 51- Mem: Am Soc Pharmacol & Exp Therapeut; Am Pharmaceut Asn. Res: Pharmacodynamics on respiratory, cardiovascular and renal systems; psychopharmacology; stress; drug absorption. Mailing Add: Ohio State Univ Col of Pharm Columbus OH 43210

NELSON, JOHN WILLIAM, b St Louis, Mo, July 18, 26; m 53; c 2. ENVIRONMENTAL PHYSICS, EXPERIMENTAL NUCLEAR PHYSICS. Educ: Univ Calif, Los Angeles, BA, 47; Wash Univ, BA, 49; Univ Tex, MA, 52, PhD(physics), 59. Prof Exp: Flight forecaster meteorol, US Weather Bur, La, 50-51; res assoc physics, Fla State Univ, 59-62; asst prof, Kans State Univ, 62-66; asst prof, 66-71, ASSOC PROF PHYSICS, FLA STATE UNIV, 71- Mem: Am Phys Soc; Inst Elec & Electronics Engrs; Am Asn Physics Teachers. Res: Low energy nuclear physics; application of nuclear techniques to environmental problems. Mailing Add: Dept of Physics Fla State Univ Tallahassee FL 32306

NELSON, JOSEPH EDWARD, b Decatur, Ga, May 4, 32; m 52. MATHEMATICS. Educ: Univ Chicago, AM, 53, PhD(math), 55. Prof Exp: Mem, Inst Advan Study, 56-59; from asst prof to assoc prof, 59-70, PROF MATH, PRINCETON UNIV, 70- Mem: Am Math Soc. Res: Functional analysis. Mailing Add: Dept of Math Princeton Univ Princeton NJ 08540

NELSON, JOSEPH SCHIESER, b San Francisco, Calif, Apr 12, 37; Can citizen; m 63; c 4. ICHTHYOLOGY. Educ: Univ BC, BSc, 60, PhD(zool), 65; Univ Alta, MSc, 62. Prof Exp: Res assoc zool, Indiana Univ, 65-67, asst dir univ biol stas, 67-68; asst prof, 68-72, ASSOC PROF ZOOL, UNIV ALTA, 72- Mem: Soc Study Evolution; Am Soc Ichthyol & Herpet; Am Fisheries Soc; Can Soc Wildlife & Fishery Biol (pres, 72-74); Can Soc Environ Biol (dir, 75-). Res: Hybridization in cyprinid and catostomid fishes; systematics of gasterosteid, psychrolutid, including Neophrynichthys and trachinoid fishes; classification of world fishes. Mailing Add: Dept of Zool Univ of Alta Edmonton AB Can

NELSON, KATHERINE GREACEN, b Sierra Madre, Calif, Dec 9, 13; m 50. PALEONTOLOGY. Educ: Vassar Col, AB, 34; Rutgers Univ, PhD(geol), 38. Prof Exp: Asst geol, Rutgers Univ, 36-38; instr geol & geog, Milwaukee-Downer Col, 38-43, asst prof, 46-48, prof, 48-54; sr paleontologist, Shell Oil Co, 43-44; asst prof geog & geol, 56-64, assoc prof geol, 64-66, chmn dept, 61-62, PROF GEOL, UNIV WIS-MILWAUKEE, 66- Concurrent Pos: Curator, Greene Mem Mus, Milwaukee-Downer Col, 38-43, 46-54. Mem: AAAS; Paleont Soc; Geol Soc Am; Nat Asn Geol Teachers (secy-treas, 42-50); Am Asn Stratig Paleontologists. Res: Cretaceous and Tertiary bryozoa; Pennsylvania and Permian fusulinds; Sliurian invertebrate fossils of Wisconsin and Illinois; Devonian of Wisconsin. Mailing Add: Dept of Geol Sci Univ of Wis Milwaukee WI 53201

NELSON, KAY LEROI, b Richmond, Utah, Apr 4, 26; m 47; c 4. PHYSICAL ORGANIC CHEMISTRY. Educ: Utah State Univ, BS, 48; Purdue Univ, PhD(chem), 52. Prof Exp: Asst chem, Purdue Univ, 48-50, instr, 53-54; fel org chem, Off Naval Res, Univ Calif, Los Angeles, 52-53; asst prof, Wayne State Univ, 54-56; assoc prof, 56-61, chmn dept, 68-71, PROF ORG CHEM, BRIGHAM YOUNG UNIV, 61- Concurrent Pos: vis prof, Ore State Univ, 71-72. Mem: AAAS; Am Chem Soc. Res: Directive effects in electrophilic aromatic substitution; rearrangement of aminoketones; olefin bromination and hydrochlorination; kinetics and mechanisms of organic reactions; natural products by selective extraction from bark and berries. Mailing Add: Dept of Chem Brigham Young Univ Provo UT 84602

NELSON, KEITH, zoology, psychology, see 12th edition

NELSON, KENNETH FRED, b Council Grove, Kans, Sept 18, 42; m 65; c 1. MEDICINAL CHEMISTRY. Educ: Univ Kans, BSc, 65; Univ Wash, PhD(med chem), 70. Prof Exp: ASST PROF PHARM, UNIV WYO, 70- Mem: AAAS; Am Chem Soc. Res: Medicinal chemistry, especially in the field of analgetics. Mailing Add: Sch of Pharm Univ of Wyo Box 3375 Univ Sta Laramie WY 82070

NELSON, KENNETH GORDON, b Chicago, Ill, May 29, 40; m 62; c 2. PHARMACEUTICS. Educ: Univ Wis, Madison, BS, 63; Univ Mich, Ann Arbor, MS, 64, PhD(pharmaceut chem), 68. Prof Exp: Asst prof, 68-74, ASSOC PROF PHARMACEUT, COL PHARM, UNIV MINN, MINNEAPOLIS, 74- Mem: Acad Pharmaceut Sci. Res: Drug transport mechanisms; fluoride reactions with hydroxyapatite. Mailing Add: Col of Pharm Univ of Minn Minneapolis MN 55455

NELSON, KENNETH WILLIAM, b Superior, Wis, Sept 27, 17; m 42; c 2. INDUSTRIAL HYGIENE. Educ: Superior State Col, Wis, BEd, 38; Univ Utah, MS, 57. Prof Exp: Teacher high sch, Wis, 38-39; lab asst toxicol, US Food & Drug Admin, WAshington, DC, 40-41, jr chemist, 41-42; indust hygienist, 46-50, chief hygienist, 50-57, dir dept hyg, 58-66, dir dept environ sci, 66-74, VPRES ENVIRON AFFAIRS, AM SMELTING & REFINING CO, 74- Mem: AAAS; Am Chem Soc; Am Indust Hyg Asn (pres, 58); Am Acad Indust Hyg (pres, 75); NY Acad Sci. Mailing Add: Am Smelting & Refining Co Inc 120 Broadway New York NY 10005

NELSON, KLAYTON EDWARD, b Vivian, SDak, May 15, 17; m 42; c 2. PLANT PATHOLOGY. Educ: SDak State Col, BS, 39; Univ Calif, PhD(plant path), 49. Prof Exp: Teacher high sch, SDak, 39-41; asst plant path, 47-49, from lectr to assoc prof viticult, 50-64, from jr viticulturist to assoc viticulturist, 50-64, PROF VITICULT & VITICULTURIST, UNIV CALIF, DAVIS, 64- Concurrent Pos: Agent-plant pathologist, Bur Plant Indust, USDA, 49- Mem: Am Phytopath Soc; Am Soc Hort Sci; Am Soc Enol. Res: Post-harvest pathological and physiological problems of table grapes. Mailing Add: Dept of Viticulture Univ of Calif Davis CA 95616

NELSON, KURT HERBERT, b Sweden, Dec 8, 24; US citizen; m 49; c 3. ANALYTICAL CHEMISTRY. Educ: Reed Col, BA, 48; Univ Wash, PhD(anal chem), 53. Prof Exp: Anal res chemist, Phillips Petrol Co, 53-60; res chemist, Tektronix, Inc, 60-62; res specialist, Autonetics Div, NAm Aviation, Inc, 62-64, mem tech staff, Rocketdyne Div, NAm Rockwell Corp, Calif, 64-70; sr res chemist, Burlington Industs, 70-72; SR RES ASSOC, INT PAPER CO, 72- Mem: Am Chem Soc; Am Inst Chemists; Can Pulp & Paper Asn. Res: Analytical methods development; high pressure liquid chromatography; nondispersive x-ray analysis; instrument research; paper and textile analysis; environmental chemistry and pollution; ultratrace analysis; gas chromatography; spectrophotometry; data correlation; quality control; chemometrics. Mailing Add: 14 Virginia Ave Monroe NY 10950

NELSON, KYLER FISCHER, b Litchfield, Minn, Sept 16, 38; m 61; c 3. ELECTROOPTICS. Educ: Hamline Univ, BS, 60; Purdue Univ, MS, 62; Univ Utah, PhD(physics), 68. Prof Exp: Fel, Univ Utah, 68; SCIENTIST, IMAGING SYSTS & DEVICES LAB, XEROX CORP, 69- Mem: Am Phys Soc; Sigma Xi. Res: Charge transport in insulating media; liquid crystals; photoelectrophoresis; photoconductivity. Mailing Add: Imaging Systs & Devices Lab Xerox Corp Xerox Square Rochester NY 14603

NELSON, LARRY ALAN, soil science, experimental statistics, see 12th edition

NELSON, LARRY DEAN, b Newton, Kans, Aug 5, 37. MATHEMATICS. Educ: Phillips Univ, BA, 59; Kans State Univ, MS, 62; Ohio State Univ, PhD(math), 65. Prof Exp: Part-time mathematician, Battelle Mem Inst, Ohio, 62-65; res assoc comput sci, Ohio State Univ Res Found, 65; mem tech staff, Bellcomm, Inc, DC, 65-68, supvr numerical methods & systs studies group, Appl Math Dept, 68-70, supvr data applns group, Data Systs Develop Dept, 70-72; SUPVR MGT INFO SYSTS DEPT, BELL TEL LABS, 72- Mem: Asn Comput Mach; Math Asn Am; Am Math Soc; Inst Elec & Electronics Eng; NY Acad Sci. Res: Interger linear programming; digital optical enhancement; mathematical programming. Mailing Add: Bell Tel Labs Murray Hill NJ 07974

NELSON, LAWRENCE BARCLAY, b New York, NY, Jan 9, 31; m 55; c 2. CHEMISTRY. Educ: NY Univ, BA, 51, PhD(chem), 55. Prof Exp: Res investr, NJ Zinc Co, Pa, 55-56; res assoc, Socony Mobil Oil Co, Inc, 56-60; asst to vpres res & mfg, Sonneborn Div, 60-62, CORP RES DIR, WITCO CHEM CO, INC, 62- Mem: AAAS; Am Chem Soc. Res: Industrial and petroleum chemistry; isotope tracer techniques; fused salt electrolysis; crystal growth; technical management. Mailing Add: Witco Chem Co Inc Box 110 Oakland NJ 07436

NELSON, LENIS ALTON, b Walnut Grove, Minn, Sept 22, 40; m 65; c 2. AGRONOMY. Educ: SDak State Univ, BS, 62; NDak State Univ, MS, 68, PhD(agron), 70. Prof Exp: Voc agr instr, Pub Sch, Minn, 62-64; asst supt agron, Southeast SDak Exp Farm, SDak State Univ, 64-66; res asst, NDak State Univ, 66-68; ASST PROF AGRON, SCOTTS BLUFF STA, UNIV NEBR, LINCOLN, 70- Mem: Am Soc Agron; Crop Sci Soc Am. Res: Variety improvement of proso millet; variety testing of grain crops such as corn, oats, wheat, barley and sorghum. Mailing Add: Panhandle Station Univ of Nebr Scottsbluff NE 69361

NELSON, LEONARD, b Philadelphia, Pa, Oct 29, 20; m 43; c 1. PHYSIOLOGY, REPRODUCTIVE PHYSIOLOGY. Educ: Univ Pa, AB, 42; Univ Minn, MA, 50, PhD, 53. Prof Exp: Asst zool, Washington Univ, 46-47; asst zool, Univ Minn, 47-48; from instr to asst prof physiol, Univ Nebr, 48-56; res assoc & asst prof anat, Univ Chicago, 56-58; Lalor Found fel, 58-59; assoc prof physiol, Emory Univ, 59-66; PROF PHYSIOL & CHMN DEPT, MED COL OHIO, 67- Concurrent Pos: Mem corp, Marine Biol Lab, Woods Hole, Mass, 54; Pop Coun, Inc fel, 56-58; USPHS sr res fel, Emory Univ, 59-64; USPHS career develop award, 62-66; Commonwealth Found fel, Cambridge Univ, 63-64; vis prof, Physiol Labs, Cambridge Univ, 63-64; prog dir develop biol, NSF, 67; Josiah Macy, Jr Found fac scholar award, 75-76. Mem: Fel AAAS; Soc Gen Physiol; Brit Soc Study Fertil; Am Physiol Soc; Soc Study Reproduction. Res: Physiology of reproduction; gamete transport and fertilization. Mailing Add: Dept of Physiol Med Col of Ohio Toledo OH 43614

NELSON, LEWIS BAILEY, b New Plymouth, Idaho, Oct 18, 14; m; c 3. SOILS. Educ: Univ Idaho, BS, 36; Univ Wis, MS, 38, PhD(soils), 40. Prof Exp: Asst soils, Univ Wis, 36-40, instr, 40-42; supvry chemist, Pa Ord Works, 42-44; from asst prof to assoc prof soil res, Iowa State Univ, 44-49; soil scientist, Bur Plant Indust, Soils & Agr Eng, USDA, 49-53, head eastern sect soil & water mgt sect, Agr Res Serv, 53-57, chief eastern soil & water mgt res br, 57-60; MGR AGR & CHEM DEVELOP, TENN VALLEY AUTHORITY, 60- Vis prof, Univ Ill, 60; chmn land use study panel agr res Inst, Nat Acad Sci, 71, mem panel agr uses of water, Environ Studies Bd, 71. Mem: AAAS; Soil Sci Soc Am (pres, 56-57); fel Am Soc Agron; Am Chem Soc. Res: Fertilizers; soil fertility; soil and water management; munitions development; research administration. Mailing Add: Off of Agr & Chem Develop Tenn Valley Authority Muscle Shoals AL 35660

NELSON, LEYTON VINCENT, b Ranier, Minn, Feb 16, 14; m 43; c 2. AGRONOMY. Educ: Mich State Col, BS, 39, MS, 41. Prof Exp: Asst res forage crops, 41-45, from asst prof to assoc prof farm crops, 45-60, PROF FARM CROPS, MICH STATE UNIV, 60- Concurrent Pos: Consult, Mich State Univ-Ford Found Pakistan Proj, Comilla, E Pakistan, 67 & 69. Mem: Am Soc Agron. Res: Crop production. Mailing Add: 305 Agr Hall Mich State Univ East Lansing MI 48823

NELSON, LLOYD RUSSEL, b Barronett, Wis, Nov 9, 42; m 69; c 2. PLANT BREEDING. Educ: Univ Wis-River Falls, BS, 65; NDak State Univ, MS, 67; Miss State Univ, PhD(plant breeding), 71. Prof Exp: Asst agronomist corn breeding, Miss Exp Sta, 67-71; ASST PROF AGRON, WHEAT & OAT BREEDING, UNIV GA EXP STA, 71- Concurrent Pos: Mem, Nat Wheat Improv Comt, 75-77. Mem: Am Soc Agron; Crop Sci Soc Am. Res: Wheat and oat varietal and germplasm development involving disease resistance, insect resistan- resistance, protein content, and yield component studies. Mailing Add: Dept of Agron Univ of Ga Exp Sta Experiment GA 30212

NELSON, LLOYD SPENCER, physical chemistry, see 12th edition

NELSON, LLOYD STEADMAN, b Norwich, Conn, Mar 29, 22; m 47; c 3. MATHEMATICS. Educ: Univ NC, BS, 43; Univ Conn, PhD(chem), 50. Prof Exp: Asst chem, Univ NC, 42-43; instr physics, Univ Conn, 43-44, asst chem, 46-48; instr org chem, Ill Inst Technol, 49-51; chemist, Silicone Prod Dept, 51-53, res assoc org chem res lab, 53-56, consult statistician, Lamp Div, 56-68, MGR APPL MATH LAB, GEN ELEC CO, 68- Concurrent Pos: Spec lectr, Case Western Reserve Univ, 63-65; ed, Indust Qual Control, 65-67 & J Qual Technol, 68-70, Am Soc Qual Control. Mem: Fel AAAS; fel Am Soc Qual Control; fel Am Statist Asn. Res: Application of mathematics to scientific and business problems. Mailing Add: 7001 Bedford Lane Louisville KY 40222

NELSON, LYLE ENGNAR, b Donnybrook, NDak, Jan 6, 21. SOILS. Educ: Cornell Univ, MS, 52, PhD(soils), 59. Prof Exp: Soil scientist, USDA, 48; asst, Cornell Univ, 49-52; asst prof agron & asst agronomist, Agr Exp Sta, Miss State Univ, 52-55; vis assoc prof soils, Univ Philippines, 55-57; assoc prof agron & assoc agronomist, Agr Exp Sta, 57-60, PROF AGRON & AGRONOMIST, AGR EXP STA, MISS STATE UNIV, 60- Concurrent Pos: Vis prof, NC State Univ, 65; vis scientist, Rothamsted Exp Sta, Eng, 71-72. Mem: AAAS; Soil Sci Soc Am; Am Soc Agron; Soil Conserv Soc Am; Int Soc Soil Sci. Res: Soil fertility; soil reaction and plant growth; nutrition and nutrient cycling in forest stands. Mailing Add: Dept of Agron Box 5248 Mississippi State MS 39762

NELSON, MANNO FREDRICK, JR, b Tampico, Ill, Jan 28, 20; m 43; c 3. RADIOCHEMISTRY. Educ: Augustana Col, Ill, AB, 44; Univ Iowa, PhD(chem), 49. Prof Exp: USPHS fel, Univ Minn, 50; chemist, Res & Develop Dept, Colgate-Palmolive Co, 50-52, group leader & supvr radioisotope labs, 52-57; chemist, Esso Res & Eng Co, 57-58; supvr radioisotope labs, Atlas Chem Industs, 58-70; mem faculty sci, Haile Selassie Univ, 71-72; ASST PROF RADIOL & DIR NUCLEAR MED TECHNOL, SCH MED, VANDERBILT UNIV, 72- Concurrent Pos: Adv, Rutgers Univ, 53-58, guest lectr, 56; consult & mem radioisotope comt, Med Ctr, Wilmington, Del; consult, Tech Educ Res Coun, Mass, 73-75, Southern Interstate Nuclear Bd, 74-75, Tenn State Health Planning Bd, 74- & Hosp Accreditation Comn, 74- Mem: Am Inst Chemists; Am Soc Clin Pathologists; Soc Nuclear Med. Res: Radio and tracer chemistry as applied to research and product development. Mailing Add: Sch of Med Vanderbilt Univ Nashville TN 37232

NELSON, MARGARET CHRISTINA, b Louisville, Ky, Nov 13, 43. NEUROPHYSIOLOGY, ETHOLOGY. Educ: Swarthmore Col, BA, 65; Univ Pa, MA, 67, PhD(physiol psychol), 70. Prof Exp: Nat Inst Ment Health fel biol, Tufts Univ, 70-72; asst prof psychol, Brandeis Univ, 72-75; RES FEL NEUROBIOL, HARVARD MED SCH, 75- Mem: AAAS; Am Soc Zool; Animal Behav Soc; Soc Neurosci. Res: Behavior and neurophysiological correlates of behavior in insects. Mailing Add: Dept of Neurobiol Harvard Med Sch Boston MA 02115

NELSON, MARITA LEE, b Torrance, Calif. HUMAN ANATOMY, ENDOCRINOLOGY. Educ: Univ Calif, Los Angeles, BS, 57, MS, 59; Univ Calif, Berkeley, PhD(anat), 68. Prof Exp: Assoc phys educ, Univ Calif, Los Angeles, 59-60; instr, Ill State Univ, 60-64; actg asst prof anat, Univ Calif, Berkeley, 68-69; asst prof, Schs Med & Dent, Georgetown Univ, 69-72; actg asst prof, Univ Calif, Berkeley, 72-74; asst prof, Sch Dent, Univ Pac, 73-74; ASSOC PROF ANAT, SCH MED, UNIV HAWAII, 74- Mem: Am Asn Anatomists; Sigma Xi; AAAS. Res: Hormonal regulation of puberty; effects of high altitude on growth and pituitary function; effects of stress on reproductive maturation. Mailing Add: Dept of Anat & Reprod Biol John Burns Sch of Med Univ of Hawaii 1960 East-West Rd Honolulu HI 96822

NELSON, MARK RADFORD, b Salt Lake City, Utah, Jan 8, 46; m 68; c 2. ASTROPHYSICS. Educ: Harvard Univ, BA, 68; Princeton Univ, PhD(physics), 72. Prof Exp: From instr to asst prof physics, Univ Ill, Urbana, 72-76; SR STAFF, PATTERN ANAL & RECOGNITION CORP, 76- Mem: Am Phys Soc; Am Astron Soc. Res: Sparkle interferometry; signal processing; stellar diameter measurements via lunar occultations, cataclysmic variable stars. Mailing Add: Pattern Anal & Recognition Corp On the Mall Rome NY 13440

NELSON, MARTIN EMANUEL, b Tacoma, Wash, Oct 21, 15; m 42; c 3. PHYSICS. Educ: Col Puget Sound, BS, 37; Univ Hawaii, MS, 39; Ohio State Univ, PhD(physics), 42. Prof Exp: Asst physics, Univ Hawaii, , 37-39; asst, Ohio State Univ, 39-41; physicist, Nat Defense Res Comt, Princeton Univ, 42; instr physics, Univ Ill, 42-44; physicist, Ord Bur, US Naval Res Lab, Washington, DC, 44-46; from asst prof to assoc prof physics, Col Puget Sound, 46-52; res engr, Boeing Airplane Co, 52-56; dir div natural sci, 70-74, chmn dept physics, 56-70, PROF PHYSICS, UNIV PUGET SOUND, 56- Concurrent Pos: NSF sci fac fel, Univ Wash, 66-67. Mem: AAAS; Am Phys Soc; Am Asn Physics Teachers. Res: Artificial radioactivity; cosmic rays; nuclear reactions. Mailing Add: 4436 Memory Lane Tacoma WA 98466

NELSON, MARY LOCKETT (MRS JOHN D GUTHRIE), b New Orleans, La, July 24, 14; m 68. CELLULOSE CHEMISTRY. Educ: Newcomb Col, BA, 34; Tulane Univ, MS, 36, PhD(phys chem), 57. Prof Exp: Plant physiologist, Forest Serv, La, 36-40, seed technologist, Agr Mkt Serv, Washington, DC, 40-42, chemist, Southern Regional Res Lab, 42-64, SR RES CHEMIST, SOUTHERN REGIONAL RES CTR, USDA, 64- Mem: Fel AAAS; fel Am Inst Chem; Fiber Soc; Am Chem Soc; Sigma Xi. Res: Cellulose fine structure; crystallinity; accessibility; IR spectra; swelling; crosslinking; changes in cotton fiber during growth; heat damage to cotton; heats of combustion and solution; storage of pine seed. Mailing Add: Southern Regional Res Ctr USDA PO Box 19687 New Orleans LA 70179

NELSON, MAXIE JO, biochemistry, chemistry, see 12th edition

NELSON, MERRITT RICHARD, b New Richmond, Wis, Oct 11, 32; m 56; c 3 .PLANT PATHOLOGY. Educ: Univ Calif, Berkeley, BS, 55; Univ Wis, PhD, 58. Prof Exp: Asst, Univ Wis, 55-57; asst plant pathologist, 58-61, assoc prof, 61-67, PROF PLANT PATH, AGR EXP STA, UNIV ARIZ, 67- Mem: Am Phytopath Soc; Soc Gen Microbiol. Res: Plant virology, specifically epidemiology of native and cultivated plant virus diseases and biophysical and serological studies of viruses that infect plants. Mailing Add: Dept of Plant Pathology Univ of Ariz Tucson AZ 85721

NELSON, MILTON EPHRAIM, bacteriology, see 12th edition

NELSON, NEAL STANLEY, b Chicago, Ill, Jan 1, 34; m 66. PHARMACOLOGY, RADIOBIOLOGY. Educ: Univ Ill, BS, 55, DVM, 57; Univ Chicago, PhD(pharm), 64. Prof Exp: Res assoc pharmacol, Univ Chicago, 63-65; Nat Cancer Inst spec fel, Milan, 65-66; res scientist, Div Radiol Health, Robert A Taft Sanit Eng Ctr, Ohio, 66-67; dep chief toxicol studies sect, Div Biol Effects, Bur Radiation Health, Hazelton Lab, USPHS, Va, 67-71, res scientist & dep chief toxicol studies sect, Off Res & Monitoring, Twinbrook Res Lab, 71-73, RADIOBIOLOGIST, OFF RADIATION PROGS, ENVIRON PROTECTION AGENCY, 73- Concurrent Pos: Mem comt guide rev, Inst Lab Animal Resources, Nat Acad Sci-Nat Res Coun, 65 & mem comt stand for cats, 72- Mem: AAAS; Am Vet Med Asn; Conf Pub Health Vets; Am Soc Lab Animal Practitioners; NY Acad Sci. Res: Metabolism of radioisotopes; pathology of alpha-emitting isotopes; effects of ionizing radiation; high resolution autoradiography and intracellular localization of labeled compounds. Mailing Add: Waterside Mall 401 M St SW Washington DC 20460

NELSON, NELS M, b Baker, NDak, May 30, 19; m 57; c 1. MICROBIOLOGY. Educ: Jamestown Col, BS, 41; Univ Wash, MS, 53, PhD(microbiol), 55. Prof Exp: From asst prof to assoc prof, 55-69, PROF MICROBIOL, MONT STATE UNIV, 69- Mem: AAAS; Am Soc Microbiol. Res: Microbial physiology. Mailing Add: Dept of Bot & Microbiol Mont State Univ Bozeman MT 59715

NELSON, NILS KEITH, b Leadwood, Mo, Nov 10, 26. ORGANIC CHEMISTRY. Educ: Mo Sch Mines, BS, 46; Univ Ill, MS, 47, PhD(chem), 49. Prof Exp: Asst gen & org chem, Univ Ill, 46-49; instr, Univ Maine, 49-51; res chemist, Shell Oil Co, 51-62; asst prof, 62-66, ASSOC PROF ORG CHEM, CALUMET CAMPUS, PURDUE UNIV, 66- Mem: Am Chem Soc. Res: Stereochemistry of substituted aryl amines; Darzens reaction. Mailing Add: Dept of Chem Purdue Univ Calumet Campus Hammond IN 46323

NELSON, NORMA JOHNSON, microbial genetics, see 12th edition

NELSON, NORMAN ALLAN, b Edmonton, Alta, July 26, 27; nat US; m 55. PHARMACEUTICAL CHEMISTRY. Educ: Univ Alta, BSc, 49; Univ Wis, PhD(chem), 52. Prof Exp: Res assoc, Mass Inst Technol, 52-53, instr, 53-55, asst prof chem, 55-59; RES ASSOC, UPJOHN CO, 59- Mem: Am Chem Soc. Res: Steroidal hormone analogs; organic synthesis; prostaglandin chemistry; ionophores. Mailing Add: Upjohn Co Kalamazoo MI 49001

NELSON, NORMAN BARTRAM, b Bridgeport, Conn, Jan 8, 13; m 38; c 2. EPIDEMIOLOGY. Educ: Univ Calif, Los Angeles, AB, 34; Univ Southern Calif, MD, 39; Harvard Univ, MPH, 41, DPH(epidemiol), 42. Prof Exp: Asst prof, Sch Pub Health, Harvard Univ, 41-42; chief epidemiologist, Los Angeles Health Dept, 42-43, chief med sect, 43-46; assoc prof pub health & chmn dept, Univ Calif, Los Angeles, 46-47, assoc prof prev med & asst dean med sch, 47-50; dean med, Am Univ Beirut, 51-S3; dean col med & dir univ hosps, Univ Iowa, 53-65; dir dept med insts, Santa Clara County, 65-74; RETIRED. Mem: Fel Am Pub Health Asn. Res: Epidemiology of poliomyelitis. Mailing Add: 1840 University Way San Jose CA 95126

NELSON, NORMAN CROOKS, b Hibbing, Minn, July 24, 29; m 55; c 3. SURGERY,

ENDOCRINOLOGY. Educ: Tulane Univ, BS, 51, MD, 54. Prof Exp: Clin & res fel surg, Harvard Med Sch & Mass Gen Hosp, Boston, 62-63; from instr to prof surg, Sch Med, Univ New Orleans, 63-73, from assoc dean to dean sch med, 69-73; VCHANCELLOR HEALTH AFFAIRS & DEAN SCH MED, MED CTR, UNIV MISS, 73-, PROF SURG, 75- Mem: Endocrine Soc; Am Diabetes Asn; Soc Univ Surgeons; Soc Exp Biol & Med; Soc Head & Neck Surgeons. Res: Mineral and carbohydrate metabolism. Mailing Add: Sch of Med Univ Miss Med Ctr Jackson MS 39216

NELSON, NORMAN NEIBUHR, b Boody, Ill, July 17, 37; m 67. MATHEMATICS EDUCATION. Educ: Eastern Ore Col, BS, 59; Univ Ill, MA, 64; Univ Northern Colo, EdD(math, educ), 70. Prof Exp: Teacher physics-chem, high sch, Ore, 59-61; from instr to asst prof math, Eastern Ore Col, 61-68; asst prof, 70-73, ASSOC PROF MATH, ORE COL EDUC, 73- Mem: Math Asn Am. Res: Development and testing of mathematics teaching materials through utilization of audio and AV media. Mailing Add: Dept of Math Ore Col of Educ Monmouth OR 97361

NELSON, NORTON, b McClure, Ohio, Feb 6, 10; m 36; c 3. ENVIRONMENTAL MEDICINE, BIOCHEMISTRY. Educ: Wittenberg Col, AB, 32; Univ Cincinnati, PhD(biochem), 38. Hon Degrees: DSc, Wittenberg Col, 64. Prof Exp: Res asst, Children's Hosp Res Found, Ohio, 34-38; biochemist, May Inst Med Res, Jewish Hosp, Ohio, 38-42; asst prof biochem, Univ Cincinnati, 46-47; assoc prof environ med, 47-53, PROF ENVIRON MED, MED CTR, NY UNIV, 53-, CHMN DEPT & DIR INST ENVIRON MED, 54- Concurrent Pos: Res assoc, Children's Hosp Res Found, Ohio, 46-47; dir res, Inst Environ Med, NY Univ, 47-54, actg dir, Univ Valley Sterling Forest, 62-66, dir, 67-, provost, Heights Ctr, 66-67; mem study group smoking & health, Am Cancer Soc-Am Heart Asn-Nat Cancer Inst-Nat Heart Inst, 57 & comn environ health, Armed Forces Epidemiol Bd, 60-; mem man in space comt, Nat Acad Sci-Nat Res Coun, 61, working group Gemini-Apollo manned orbiting lab exp, 64, chmn comt air qual stand in space flight, mem space sci bd, ad hoc comt on sci qualification & selection of scientist-astronauts; mem panel air pollution, Gross Comt, Dept Health, Educ & Welfare, 61, nat adv dis prev & environ control coun, 67-69, bd trustees, Indust Health Found, 69-75; mem expert panel carcinogenicity, Int Union Against Cancer, 62; regent's lectr, Univ Calif, Davis, 63; consult, Calif Health Dept; mem nat adv comt community air pollution, Surgeon Gen, mem panel non-neoplastic dis & Surgeon Gen adv comt smoking & health, USPHS, 63-64, nat adv environ health comt, 63-67; chmn, Nat Health Forum, Nat Health Coun, 64; mem comt on motor vehicle emission and comt microchem pollutants, WHO, 64-68, chmn expert comt manual on methods of toxicity eval of chem, 75-; chmn comt protocols for safety eval, Food & Drug Admin, 66-71; mem environ sci & eng study sect, NIH & Nat Inst Environ Health Sci Adv Comt, 67-71 & 74-, chmn task force res planning environ health sci, 69-70; mem, White House Task Force Air Pollution, 69 & panel herbicides, US Off Sci & Technol, 69, chmn task force hazardous trace substances, 70, Off Sci & Technol-Coun Environ Qual ad hoc comt environ health res, 71, mem comt environ physiol & comt nitrate accumulation, Nat Res Coun, 70-72, chmn comt toxicol & mem comt atmospheric & indust hyg, mem assembly life sci, Med Sci Exec Comt, 75-; mem hazardous mat aov comt, Environ Protection Agency, 70-74, consult, Sci Adv Comt, 74-, mem sci adv bd exec comt & chmn environ health adv comt, 75-; mem vis comt for bd overseers, Harvard Sch Pub Health, 70-, President's Sci Adv Comt Panel on Chem & Health, 70-73, vis comt dept nutrit & food sci, Mass Inst Technol, 71-74 & res adv comt, Boyce Thompson Inst Plant Res, 70-; consult, NSF, 71-72 & 74-75; mem etiology adv comt, Nat Cancer Inst, 71-73 & Indust Hyg Round Table; consult ed, Environ Res; mem energy policy proj adv bd, Ford Found, 72-74; mem subcomt toxicol metals, Permanent Comn & Int Asn Occup Health, 72-; mem carcinogenesis prog etiology, Nat Cancer Inst, 72-; mem panel, US-USSR Joint Comn Health Coop, 72- & US-Japan Coop Med Sci Prog, 72-; mem, Milbank Mem Fund Comn Study Higher Educ Pub Health, 72-75; mem, Hudson Basin Proj, Rockefeller Found, 73-; mem environ studies bd, Nat Acad Sci, 74-, mem comt int environ progs, 74-; chmn panel environ health, NIH Fogarty Ctr & Am Col Prev Med, 75; mem, President's Biomed Res Panel & chmn subcluster environ health & toxicol, 75; chmn second task force res planning in environ health sci, Nat Inst Environ Health Sci, 75-; mem subcomt environ carcinogenesis, Nat Cancer Adv Bd, 75; sci adv, Indust Health Found, 76- Mem: AAAS; Air Pollution Control Asn; Am Soc Biol Chemists; Am Pub Health Asn; Soc Occup & Environ Health. Res: Health effects of environmental agents; environmental cancer; toxicology. Mailing Add: NY Univ Med Ctr 550 First Ave New York NY 10016

NELSON, NORVELL JOHN, organometallic chemistry, see 12th edition

NELSON, OLIVER EVANS, b Seattle, Wash, Aug 16, 20; m 63. GENETICS. Educ: Colgate Univ, AB, 41; Yale Univ, MS, 43, PhD, 47. Prof Exp: Assoc geneticist, Purdue Univ, 47-54, geneticist, 54-69; PROF GENETICS, UNIV WIS-MADISON, 69- Concurrent Pos: Vis investr, Nat Forest Res Inst & Biochem Inst, Univ Stockholm, 54-55; NSF sr fel biol, Calif Inst Technol, 61-62. Mem: Nat Acad Sci; AAAS; Genetics Soc Am; Am Genetic Asn; Bot Soc Am. Res: Physiological genetics. Mailing Add: Genetics Lab Univ of Wis Madison WI 53706

NELSON, OSCAR TIVIS, JR, b Sparta, Tenn, Aug 20, 40; m 66. MATHEMATICS. Educ: Baylor Univ, BA, 61; Vanderbilt Univ, MA, 63, PhD(math), 65. Prof Exp: Teaching asst, Vanderbilt Univ, 61-63; asst prof math, Emory Univ, 65-69; ASSOC PROF MGT, GRAD SCH MGT, VANDERBILT UNIV, 69- Mem: Am Math Soc; Math Asn Am. Res: Lattice theory; decompositions of algebraic systems. Mailing Add: Grad Sch of Mgt Vanderbilt Univ Nashville TN 37203

NELSON, PAUL ANDREW, b Minneapolis, Minn, Nov 25, 20. MEDICINE. Educ: Gustavus Adolphus Col, BA, 41; Univ Minn, BM, 45, MD, 46, MS, 51; Am Bd Phys Med & Rehab, dipl, 52. Prof Exp: Asst staff, 51-52, staff, 52-54, HEAD DEPT PHYS MED & REHAB, CLEVELAND CLIN EDUC FOUND, 55- Concurrent Pos: Mayo clin, fel, 48-51; ed, Arch Phys Med & Rehab, Am Med Writer's Asn; Honors & Awards: Am Cong Rehab Med Gold Key Award, 65. Mem: Am Med Writers' Asn; Am Acad Phys Med & Rehab; Am Acad Cerebral Palsy; assoc Am Acad Neurol; fel Am Col Physicians. Res: Cerebral palsy; lymphedema; medical writing and editing. Mailing Add: 3547 Ingleside Rd Cleveland OH 44122

NELSON, PAUL EDWARD, b Franklin Twp, Wis, May 26, 27; m 50; c 3. PLANT PATHOLOGY. Educ: Univ Calif, BS, 51, PhD(plant path), 55. Prof Exp: From asst prof to assoc prof plant path, Cornell Univ, 55-65; assoc prof, 65-67, PROF PLANT PATH, PA STATE UNIV, 67- Mem: Am Phytopath Soc; Mycol Soc Am; Bot Soc Am. Res: Root-disease fungi, especially those causing vascular wilt diseases of plants; pathological anatomy of disdiseased plants; Fusarium species. Mailing Add: 211 Buckhout Lab Dept Plant Path Pa State Univ University Park PA 16802

NELSON, PAUL VICTOR, b Somerville, Mass, May 4, 39; m 64. HORTICULTURE, PLANT NUTRITION. Educ: Univ Mass, BS, 60; Pa State Univ, MS, 61; Cornell Univ, PhD(floricult), 64. Prof Exp: Staff res specialist, Geigy Chem Corp, NY, 64-65; from asst prof to assoc prof, 65-73, PROF HORT SCI, NC STATE UNIV, 73- Concurrent Pos: NSF travel grant, Int Hort Cong, Israel, 70; leave of absence, Lab Plant Physiol Res, Agr Univ, Wageningen, Holland, 71-72. Mem: Am Soc Hort Sci; Am Soc Plant Physiol; Int Soc Hort Sci; Coun Soil Test & Plant Anal. Res: Foliar analysis and fertilization of floricultural crops; the mechanism of active uptake of nutrients by plants; plant root media. Mailing Add: Dept of Hort Sci NC State Univ Raleigh NC 27607

NELSON, PETER K, b Darlington, Wis, July 7, 14. BOTANY. Educ: Univ Wis, BA, 39, PhD(bot), 51. Prof Exp: Pub sch teacher, Wis & Minn, 33-36 & 39-41; asst bot, Univ Wis, 45-51; ed publ, Brooklyn Bot Garden, 51-58; assoc prof bot & physics, Jersey City State Col, 58-59; from instr to asst prof, 59-68, ASSOC PROF BIOL, BROOKLYN COL, 68- Mem: AAAS; Bot Soc Am; Torrey Bot Club. Res: Cytology of plants; embryo culture. Mailing Add: Dept of Biol Brooklyn Col Brooklyn NY 11210

NELSON, PHILIP R, b Tacoma, Wash, July 8, 18; m 46. FISHERIES, FISH BIOLOGY. Educ: Univ Wash, BS, 42. Prof Exp: Asst chief red salmon invests, Karluk Lake, US Fish & Wildlife Serv, 46-50, chief, 50-56, chief clam invests, Mid Atlantic, 57-58, asst chief br anadromous & inland fisheries, 58-61, chief br inland fisheries, Bur Com Fisheries, 61-70; chief fishery biol, Nat Marine Fisheries Serv, Nat Oceanic & Atmospheric Admin, 70-73; RETIRED. Mem: Am Fisheries Soc; Am Soc Limnol & Oceanog; Am Inst Fishery Res Biol. Res: Marine and freshwater biology; limnology. Mailing Add: 405 Cordova Greens Largo FL 33542

NELSON, PHILLIP GILLARD, b Albert Lea, Minn, Dec 3, 31; m 55; c 4. NEUROSCIENCES, CELL BIOLOGY. Educ: Univ Chicago, MD, 56, PhD(physiol), 57. Prof Exp: Intern, Philadelphia Gen Hosp, 57-58; from sr surgeon to surgeon, USPHS, 58-67; actg chief sect spinal cord, Nat Inst Neurol Dis & Blindness, 64-69; CHIEF BEHAV BIOL BR, NAT INST CHILD HEALTH & HUMAN DEVELOP, 69- Concurrent Pos: Hon res asst, Dept Biophys, Univ Col, Univ London, 62-63; lectr physiol, George Washington Univ, 64- Mem: AAAS; Am Physiol Soc; Biophys Soc; Soc Neurosci. Res: Single unit and integrative activity in the central nervous system; tissue culture of nervous tissue. Mailing Add: NIH Behav Biol Br Bldg 36 Rm 2A21 Bethesda MD 20014

NELSON, RALPH A, b Minneapolis, Minn, June 19, 27; m 54; c 5. NUTRITION, PHYSIOLOGY. Educ: Univ Minn, BA, 50, MD, 53, PhD(physiol), 61. Prof Exp: Intern med, Cook County Hosp, 53-54; fel path, Hosps, Minn, 54-55, res assoc neurophysiol, Univ, 55-56, fel physiol, Mayo Found, 57-60; asst prof nutrit, Cornell Univ, 61-62; assoc physiol, Med Sch, Case Western Reserve Univ, 62-67; asst prof physiol, Mayo Grad Sch, 67-73; ASST PROF PHYSIOL, MAYO MED SCH, 73-, ASSOC PROF NUTRIT, 74-; CHMN SECT NUTRIT, MAYO CLIN, 67- Concurrent Pos: Dir med res, George H Scott Res Lab, Fairview Park Hosp, Cleveland, 62-67. Mem: Am Physiol Soc; Am Inst Nutrit; Am Soc Clin Nutrit; Am Med Asn; Am Gastroenterol Soc. Res: Gastroenterology, membrane transport, physiology and structural aspects of vitamin function; applied nutrition of obesity and chronic renal disease. Mailing Add: Sect of Nutrit Mayo Clinic Rochester MN 55901

NELSON, RALPH DANIEL, JR, physical chemsitry, see 12th edition

NELSON, RAYMOND ADOLPH, b Spokane, Wash, Apr 24, 26; m 52; c 5. RADIOPHYSICS. Educ: Wash State Univ, BS, 50, MS, 52; Stanford Univ, PhD(physics), 61. Prof Exp: Mem tech staff, Bell Tel Labs, Inc, 52-55; res engr, 55-61, physicist, Radio Physics Lab, 61-70, SR PHYSICIST, RADIO PHYSICS LAB, STANFORD RES INST, 70- Mem: Am Phys Soc; Am Geophys Union. Res: Foundation of statistical mechanics; ionospheric physics. Mailing Add: Radio Physics Lab Stanford Res Inst 333 Ravenswood Menlo Park CA 94025

NELSON, RAYMOND JOHN, b Chicago, Ill, Oct 8, 17; m 42; c 3. MATHEMATICS. Educ: Grinnell Col, AB, 41; Univ Chicago, PhD(philos), 49. Prof Exp: Prof philos, Univ Akron, 46-52; mathematician, Int Bus Mach Corp, 52-55; staff engr, Link Aviation, Inc, 55-56; prof math, 56-65, dir comput ctr, 56-65, PROF PHILOS, CASE WESTERN RESERVE UNIV, 65- Concurrent Pos: Mem bd dirs, CHI Corp, Ohio; consult, Rockefeller Found, 74- Mem: Am Math Soc; Asn Symbolic Logic; Asn Comput Mach; Am Philos Asn. Res: Mathematical logic; automata theory; philosophy of science. Mailing Add: 2400 Demington Dr Cleveland Heights OH 44106

NELSON, REGINALD DAVID, b Water Glen, Alta, Nov 27, 24; m 50; c 4. ORGANIC CHEMISTRY. Educ: Univ Alta, BSc, 46; Iowa State Col, PhD(chem), 51. Prof Exp: Instr chem, Iowa State Col, 50-52; from asst prof to assoc prof, Bethel Col & Sem, 52-56; from asst ed to sr assoc ed, 56-64, staff consult, 65, 67-69, head org index ed dept, 66, SR INFO SCIENTIST, CHEM ABSTR, CHEM ABSTR SERV, OHIO STATE UNIV, 70- Mem: Am Chem Soc; Am Soc Info Sci. Res: Organic nomenclature and indexing; computer editing of index vocabulary. Mailing Add: Chem Abstr Serv Ohio State Univ Columbus OH 43210

NELSON, REX ROLAND, b Greenville, Mich, Sept 14, 24; m 54; c 2. ACOUSTICS. Educ: Kenyon Col, BA, 49; Univ Calif, Los Angeles, MS, 51; Pa State Univ, PhD(high pressure physics), 59. Prof Exp: PROF PHYSICS, OCCIDENTAL COL, 59- Mem: Am Phys Soc; Am Asn Physics Teachers; Acoust Soc Am. Res: Polymorphic transitions and phase transitions at high pressure; acoustics; acoustics of musical instruments. Mailing Add: Dept of Physics Occidental Col Los Angeles CA 90041

NELSON, RICHARD BURTON, b Powell, Wyo, Dec 10, 11; c 1. PHYSICS. Educ: Calif Inst Technol, BS, 35; Mass Inst Technol, PhD(physics), 38. Prof Exp: Mem electron optics lab, Radio Corp Am Mfg Co, 38-41; asst res physicist, Nat Res Coun Can, 41-42; res assoc, Gen Elec Co, 42-50; sr engr, Litton Industs, 50-51; sr engr, 51-57, mgr klystron develop, 57-60, tube res & develop, 60-63, chief engr, Tube Div, 63-74, CONSULT PATENT DEPT, VARIAN ASSOCS, 74- Mem: Fel Inst Elec & Electronics Eng. Res: Vacuum tubes; high-power magnetrons and klystrons. Mailing Add: 27040 Dezahara Way Los Altos Hills CA 94022

NELSON, RICHARD CARL, b Stillwater, Minn, May 1, 15; m 43; c 2. PHYSICS, BIOPHYSICS. Educ: Univ Minn, AB, 35, PhD(plant physiol), 38. Prof Exp: Agent tung res, USDA, 39; res fel plant physiol, Univ Minn, 40-42; spectroscopist, Armour & Co, 42-43; chief chemist, Citrus Concentrates, Inc, Fla, mgr Pectin Prods Div, 45-46; res assoc, Northwestern Univ, 46-49; assoc prof, 49-62, PROF PHYSICS, OHIO STATE UNIV, 62- Concurrent Pos: Mem comt heat attenuation in clothing systs, Qm Res & Develop Adv Bd, 59-62; conult, Minn Mining & Mfg Co, 60-70; USPHS res career prog award, 64-74; vis prof, Phys Chem Inst, Univ Marburg, 65. Mem: Fel Am Phys Soc. Res:Photoconductivity; sensitization by dyes; electronic processes in organic solids. Mailing Add: Dept of Physics Ohio State Univ Columbus OH 43210

NELSON, RICHARD CHARLES, physical chemistry, see 12th edition

NELSON, RICHARD DAVID, b Clinton, Iowa, May 31, 30; m 52; c 3 PHYSICAL CHEMISTRY. Educ: Cornell Col, BA, 52; Kans State Univ, PhD(chem), 56. Prof Exp: Chemist, Res & Develop Div, Del, 56-65, sr res chemist, Polyolefins Div, Tex, 65-69, sr supvr plastics dept, Sabine River Labs, 69-74, RES LAB MGR,

POLYOLEFINS DIV, PLASTICS DEPT, E I DU PONT DE NEMOURS & CO, INC, 74- Mem: Am Chem Soc; Sigma Xi. Res: Product development; infrared spectroscopy of polymers. Mailing Add: Plastics Dept Sabine River Labs E I du Pont de Nemours & Co Inc Orange TX 77630

NELSON, RICHARD DOUGLAS, b Modesto, Calif, Apr 17, 41; m 61; c 3 ENTOMOLOGY, PLANT PATHOLOGY. Educ: Univ Calif, Davis, BS, 63, MS, 66, PhD(entom), 68. Prof Exp: Asst res biologist, Biol Res Div, Stauffer Chem Co, 63-64; ENTOMOLOGIST, PLANT PATHOLOGIST & DIR RES, RES DIV, DRISCOLL STRAWBERRY ASSOC, INC, 68- Mem: AAAS; Entom Soc Am; Entom Soc Can; Acaralogical Soc Am; Int Soc Hort Sci. Res: Effects of gamma radiation, hormones and chemosterilants on the biology and population suppression of the two-spotted spider mite; population ecology, pathological and physiological anomalies of strawberries; tissue culturing. Mailing Add: 404 San Juan Rd Watsonville CA 95076

NELSON, RICHARD KING, ethnography, anthropology, see 12th edition

NELSON, RICHARD ROBERT, b Austin, Minn, May 23; m 47; c 4 4. PLANT PATHOLOGY. Educ: Augsburg Col, BA, 50; Univ Minn, MS, 52, PhD(plant path), 54. Prof Exp: Res fel fungi genetics, Univ Minn, 53-55; plant pathologist corn diseases, USDA, NC State Univ, 55-66; prof, 66-74, EVAN PUGH PROF PLANT PATH, PA STATE UNIV, 74- Concurrent Pos: Regional Sigma Xi lectr. Mem: Fel Am Phytopath Soc; Mycol Soc Am. Res: Evolution of fungi; heterothallism and homothallism in fungi; sexuality of fungi; pathogen variation and host resistance; epidemiology of plant diseases; mycology; pest management. Mailing Add: 733 N McKee St State College PA 16801

NELSON, RICHARD WILLIAM, b Clintonville, Wis, Mar 11, 18. PHYSICS. Educ: Univ Wis, BA, 42; Harvard Univ, MA, 47, PhD, 53. Prof Exp: Res assoc, Radiation Lab, Mass Inst Technol, 42-46 & Lincoln Lab, 53-55; assoc prof, 58-68, PROF PHYSICS, INST PAPER CHEM, LAWRENCE UNIV, 68-, CHMN DEPT PHYSICS & MATH, 70- Concurrent Pos: Vis prof, Univ Maine, 69-70. Mem: Am Phys Soc; Asn Comput Mach. Res: Radar; radiofrequency spectroscopy; applied mathematics. Mailing Add: Dept of Physics & Math Inst Paper Chem Lawrence Univ Appleton WI 54911

NELSON, ROBERT A, b Tracy, Minn, Mar 26, 35; m 60; c 4. VETERINARY MEDICINE, TOXICOLOGY. Educ: Univ Minn, BS, 58, DVM, 60. Prof Exp: Vet, Hanover Animal Hosp, Forest Park, Ill, 60-62 & Albrecht Animal Hosp, Denver, 62; instr vet surg, Col Vet Med, Univ Minn, 62-63; sr vet, Biochem Res Lab, 63-68, supvr animal lab, 68-71, supvr animal lab & toxicol, Riker Res Lab, 71-72, MGR ANIMAL LAB & TOXICOL, RIKER RES LAB, 3M CO, 72- Concurrent Pos: Guest lectr, Col Vet Med & Med, Univ Minn, 68- Mem: Am Vet Med Asn; Am Asn Lab Animal Sci; Soc Toxicol. Res: Laboratory animal care; biopolymer surgery and toxicology; testing of potential drugs, agrichemicals and biopolymers. Mailing Add: Riker Res Lab 3M Co 3M Ctr Bldg 218-3 St Paul MN 55101

NELSON, ROBERT ANDREW, b Detroit, Mich, Apr 16, 43; m 68. POLYMER CHEMISTRY, CELLULOSE CHEMISTRY. Educ: Wayne State Univ, BS, 67; Univ Mich, MS, 70, PhD(polymer chem), 72. Prof Exp: Group leader alloy anal, Detroit Testing Lab, 67; assoc scientist polymer chem, 72-74, scientist cellulose chem, 75-76, SCIENTIST POLYMER CHEM, XEROX CORP, 76- Mem: Am Chem Soc; AAAS; Sigma Xi. Res: Basic physicochemical properties of cellulose and paper including cellulose/water interactions, electrical conduction mechanisms and mechanical properties; polymer synthesis; kinetics, structure-property relationships. Mailing Add: Xerox Corp Xerox Square-130 Rochester NY 14604

NELSON, ROBERT ARMSTRONG, JR, b Auburn, NY, Oct 2, 22; m 70; c 2. MICROBIOLOGY. Educ: Univ Notre Dame, BS, 43; Cornell Univ, MD, 47. Prof Exp: Res asst microbiol, Sch Hyg, Johns Hopkins Univ, 47-48, res assoc, 48-50, lectr med & microbiol, Sch Med, 53-54; vis scientist, Pasteur Inst, France, 54; vis scientist, Sch Hyg, Univ London, 54-56; assoc prof, Med Sch, Yale Univ, 56-59; res prof microbiol & med, Sch Med, Univ Miami, 59-67; prof biol, Univ Calif, San Diego, 67-70; PROF MICROBIOL & IMMUNOL, UNIV MONTREAL, 70- Concurrent Pos: Spec consult, USPHS, 49-51; adv expert, WHO, 50-56; spec consult, Naval Med Sch, 53-55; sr investr, Howard Hughes Med Inst, 56-70; sr investr, Lady Davis Inst Med Res, 70-74. Honors & Awards: Nat Award, Am Venereal Dis Asn, 49; Kimble Award, 53; Centenary Award, Univ Notre Dame, 65; Schaudin-Hoffman Award, Ger, 67. Mem: Am Soc Clin Invest; Am Asn Immunologists; NY Acad Sci; Can Fedn Biol Soc. Res: Immunology and immunochemistry. Mailing Add: Dept Microbiol & Immunol Univ of Montreal CP 6128 Montreal PQ Can

NELSON, ROBERT B, b Casper, Wyo, Apr 2, 35; m 58; c 2. PHARMACOLOGY. Educ: Univ Wyo, BS, 57; Univ Calif, San Francisco, MS, 63, PhD(pharmacol), 65. Prof Exp: Asst prof, Idaho State Univ, 65-70; ASSOC PROF PHARMACOL, SCH PHARM, UNIV WYO, 70- Mem: Am Col Pharm. Res: Actions of narcotic analgesics on respiration and other general functions. Mailing Add: Dept of Pharmacol Univ of Wyo Sch of Pharm Laramie WY 82070

NELSON, ROBERT B, b July 13, 29; US citizen; m 56; c 2. STRUCTURAL GEOLOGY. Educ: Ore State Univ, BS, 51; Univ Wash, MS, 56, PhD(structure, petrog), 59. Prof Exp: From instr to asst prof, 60-74, ASSOC PROF GEOL, UNIV NEBR-LINCOLN, 74- Mem: Geol Soc Am; Am Geophys Union; fel Brit Geol Asn. Res: Deformations in sedimentary and metamorphic strata. Mailing Add: Dept of Geol Univ of Nebr Lincoln NE 68508

NELSON, ROBERT DALE, genetics, cell biology, see 12th edition

NELSON, ROBERT ELDON, b Itasca Co, Minn, Dec 31, 17; m 42; c 2. FORESTRY. Educ: Univ Calif, Berkeley, BS, 42. Prof Exp: Res forester, Pac Southwest Forest & Range Exp Sta, 45-57, res forester & chief, Hawaii Forestry Res Ctr, 57-65, RES FORESTER & DIR INST PAC ISLANDS FORESTRY, US FOREST SERV, 65- Concurrent Pos: Res assoc, Land Bur Study, Univ Hawaii, 64-; comnr, Natural Areas Comn, Hawaii Natural Areas Reserve Syst Comn, 70- Honors & Awards: Outstanding Performance, US Forest Serv, 61, 68. Mem: AAAS; Soc Am Foresters; Am Forestry Asn. Res: Forest photogrammetry, soils, influences, economics, inventory and ecology; silviculture. Mailing Add: US Forest Serv 1151 Punchbowl St Honolulu HI 96813

NELSON, ROBERT F, organic electrochemistry, physical organic chemistry, see 12th edition

NELSON, ROBERT JOHN, b Minot, NDak, Dec 12, 39; m 63; c 2. MATHEMATICS, STATISTICS. Educ: Mont State Univ, BS, 61, MS, 64; Purdue Univ, Lafayette, PhD(math), 69. Prof Exp: Instr math & statist, Mont State Univ, 63-64; instr math, St Cloud State Col, 64-66; instr, Purdue Univ, Lafayette, 66-69; from asst prof to assoc prof, St Cloud State Col, 69-74, asst dean lib arts & scis, 72-74; VPRES ACAD AFFAIRS, MO WESTERN STATE COL, 74- Concurrent Pos: Consult, NCent Asn Cols. Mem: Math Asn Am. Res: Linear algebra; curriculum development at undergraduate and graduate level. Mailing Add: Mo Western State Col St Joseph MO 64506

NELSON, ROBERT LEROY, b Columbus, NDak, Mar 26, 32; m 57. MEDICAL ENTOMOLOGY. Educ: Ore State Univ, BS, 54, MS, 59. Prof Exp: From jr vector control specialist to asst vector control specialist, Bur Vector Control, Calif Dept Pub Health, 60-65; from asst specialist to assoc specialist, 65-75, SPECIALIST, SCH PUB HEALTH, UNIV CALIF, BERKELEY, 75- Mem: Entom Soc Am; Entom Soc Can. Res: Arthropod vectors of plant and animal viruses; biology of biting flies. Mailing Add: 6008 Cochran Dr Bakersfield CA 93309

NELSON, ROBERT MELLINGER, b Burlington, Iowa, May 17, 18. ORTHODONTICS. Educ: Univ Iowa, BA, 41, DDS, 50, MA, 51. Prof Exp: Practicing orthodontist, Chicago, 51-52; assoc prof orthod, 53-65, PROF ORTHOD & HEAD DEPT, SCH DENT, UNIV NC, CHAPEL HILL, 65- Res: Facial growth and development. Mailing Add: 903 Coker Dr Chapel Hill NC 27514

NELSON, ROBERT NORTON, b Cincinnati, Ohio, Nov 1, 41; m 68; c 2. PHYSICAL CHEMISTRY. Educ: Brown Univ, ScBChem, 63; Mass Inst Technol, PhD(phys chem), 68. Prof Exp: Fel molecular collisions, Dept Chem, Univ Fla, 68-70, interim asst prof chem, 69-70; ASST PROF CHEM, GA SOUTHERN COL, 70- Mem: Am Chem Soc; Am Phys Soc; Sigma Xi. Res: Molecular collisions at thermal and near thermal energies; effusive flow of gases; surface ionization of halogens. Mailing Add: Dept of Chem Ga Southern Col Statesboro GA 30458

NELSON, ROBERT S, b Atlantic City, NJ, Apr 7, 11; m 36; c 3. MEDICINE. Educ: Univ Minn, BS & BM, 34, MD, 35. Prof Exp: Chief Europ Hepatitis Res Ctr & consult, US Army, 50-53; chief gastroenterol, Brooke Army Hosp, San Antonio, Tex, 53-55; assoc prof med, 56-65, PROF MED, GRAD SCH BIOMED SCI, TEX MED CTR, UNIV TEX, HOUSTON, 65-; INTERNIST, DEPT MED & CHIEF GASTROENTEROL SERV, UNIV TEX M D ANDERSON HOSP & TUMOR INST, 64- Concurrent Pos: From clin asst prof to clin assoc prof, Baylor Col Med, 55-; consult, Vet Admin & Jefferson Davis Hosps, Houston & Brooke Army Hosp. Honors & Awards: Schindler Award, Am Soc Gastrointestinal Endoscopy, 60; William H Rorer Award, Am J Gastroenterol, 66; Seale Harris Award, Southern Med Asn, 74. Mem: Am Soc Gastrointestinal Endoscopy; Am Gastroenterol Asn; fel Am Col Physicians. Res: Gastroenterology. Mailing Add: Dept of Med M D Anderson Hosp & Tumor Inst Houston TX 77025

NELSON, ROGER EDWIN, b New York, NY, Feb 1, 40; m 64; c 2. PHYSICAL CHEMISTRY. Educ: Rutgers Univ, BS, 62; Seton Hall Univ, MS, 66, PhD(phys chem), 69. Prof Exp: Develop scientist prod develop, Lever Bros Co, 64-75; TECH MGR DETERGENTS, PHILADELPHIA QUARTZ CO, 75- Mem: Am Chem Soc; Am Oil Chemists' Soc. Res: Basic research on detergency and test methods; applications of silicates in detergents; the development of detergent and related cleansing products. Mailing Add: 358 Evergreen Dr North Wales PA 19454

NELSON, ROGER PETER, b Bridgeport, Conn, Dec 15, 42; m 64; c 2. ORGANIC CHEMISTRY. Educ: Fairfield Univ, BS, 64; Univ Mich, PhD(org chem), 67. Prof Exp: Sr res scientist, Chem Res & Develop, 72-74, PROJ LEADER FOOD CHEM RES & DEVELOP, PFIZER, INC, 74- Mem: Am Chem Soc; Inst Food Technologists. Res: Construction and conformational properties of bridged bicyclic systems; chemistry and mode of action of antihypertensives; agents affecting gastrointestinal function; food protein research; immobilized enzymes. Mailing Add: Food Chem Res & Develop Pfizer Inc Eastern Point Rd Groton CT 06340

NELSON, RONALD EUGENE, b Farmersville, Ill, Apr 24, 35; m 57; c 3. GEOGRAPHY. Educ: Southern Ill Univ, Carbondale, BS in Ed, 57, MA, 59; Univ Nebr, PhD(geog), 70. Prof Exp: Asst prof geog, Morehead State Univ, 59-62; ASSOC PROF GEOG, WESTERN ILL UNIV, 63- Concurrent Pos: Ed, Ill Geog Soc, 71- Mem: Asn Am Geogr; Am Geog Soc; Nat Coun Geog Educ. Res: Pioneer settlement of the North American grasslands; nineteenth century Swedish settlement in the United States. Mailing Add: Dept of Geog Western Ill Univ Macomb IL 61455

NELSON, RONALD HARVEY, b Union Grove, Wis, Aug 10, 18; m 40; c 4. ANIMAL HUSBANDRY. Educ: Univ Wis, BA, 39; Okla Agr & Mech Col, MS, 41; Iowa State Col, PhD(animal breeding), 44. Prof Exp: From asst prof to assoc prof, 46-49, PROF ANIMAL HUSB, MICH STATE UNIV, 49-, HEAD DEPT, 50- Concurrent Pos: Chief of party, Agr Proj, Balcarce, 66-68. Mem: AAAS; Am Soc Animal Sci. Res: Sheep and beef cattle breeding;lamb mortality and factors affecting it; effect of inbreeding on a herd of Holstein-Friesian cattle. Mailing Add: Dept of Animal Husb Mich State Univ East Lansing MI 48823

NELSON, RUSSELL, organic chemistry, see 12th edition

NELSON, RUSSELL MARION, b Salt Lake City, Utah, Sept 9, 24; m 45; c 10. SURGERY. Educ: Univ Utah, BA, 45, MD, 47; Univ Minn, PhD(surg), 54. Hon Degrees: ScD, Brigham Young Univ, 70. Prof Exp: From intern to sr resident, Univ Minn Hosps, 47-55; from asst prof to assoc clin prof, 55-70, RES PROF SURG, COL MED, UNIV UTAH, 70- Concurrent Pos: Nat Heart Inst res fel, 49-50; first asst resident, Mass Gen Hosp, 53-54; Nat Cancer Inst trainee, 53-55; Markle scholar, 57-59; chmn div thoracic & cardiovasc surg, Latter-day Saints Hosp, 66-72; mem bd dirs, Am Bd Thoracic Surg; dir cardiovasc & thoracic surg, Training Prog, Univ Utah Affil Hosps, 67-; pvt pract. Mem: Am Surg Asn; Soc Univ Surgeons; Am Asn Thoracic Surg; AMA; Am Col Surgeons. Res: Development of artificial heart-lung machine for open heart surgery; cardiovascular surgery; physiology of shock; physiological mechanisms involved in the etiology and treatment of ventricular fibrillation and other cardiac arrhythmias. Mailing Add: Latter-Day Saints Hosp 325 Eighth Ave Salt Lake City UT 84103

NELSON, RUSSELL THEODORE, agronomy, see 12th edition

NELSON, SAMUEL JAMES, b Vancouver, BC, June 2, 25; m 54; c 2. STRATIGRAPHY, PALEONTOLOGY. Educ: Univ BC, BASc, 48, MASc, 50; McGill Univ, PhD(stratig, paleont), 52. Prof Exp: Asst prof geol, Univ NB, 52-54; from asst prof to assoc prof, 54-62, PROF GEOL, UNIV CALGARY, 62- Concurrent Pos: Consult to var oil industs. Mem: Fel Geol Asn Can. Res: Ordovician and Permocarboniferous stratigraphy and paleontology. Mailing Add: Dept of Geol Univ of Calgary Calgary AB Can

NELSON, SAMUEL JOHN, JR, organic chemistry, see 12th edition

NELSON, SHELDON DOUGLAS, b Idaho Falls, Idaho, Aug 12, 43; m 65; c 4. SOIL CONSERVATION, SOIL MORPHOLOGY. Educ: Brigham Young Univ, BS, 67; Univ Calif, Riverside, PhD(soil sci), 71. Prof Exp: Soil scientist, Agr Res Serv, USDA, 67-72; ASST PROF AGRON, BRIGHAM YOUNG UNIV, 72- Concurrent Pos: Environ consult, Eryring Res Inst, 75-76. Mem: Am Soc Agron; Soil Sci Soc Am; Int Soil Soc; Sigma Xi. Res: Subsurface and drip irrigation practices; dairy waste pollution; water quality; water movement in soils; soil salinity. Mailing Add: Dept of Agron Brigham Young Univ Provo UT 84601

NELSON, SIDNEY W, b Minneapolis, Minn, July 18, 17; m 43; c 3. MEDICINE. Educ: Univ Wash, BS, 42; Northwestern Univ, BM, 45, MD, 46. Prof Exp: From intern to resident physician, King County Hosp, Seattle, Wash, 45-50; resident physician, Clins, Univ Chicago, 50-53, from instr to asst prof radiol, 53-55; PROF RADIOL & CHMN DEPT, COL MED, OHIO STATE UNIV, 55- Concurrent Pos: Consult, Vet Admin Hosp, Dayton, 55- Honors & Awards: Silver Medal, Radiol Soc NAm, 57. Mem: Radiol Soc NAm; Roentgen Ray Soc; AMA; Am Col Radiol. Res: Radiographic contrast materials; gastrointestinal radiology; improving diagnostic radiology equipment. Mailing Add: Dept of Radiol Ohio State Univ Hosps Columbus OH 43210

NELSON, SIGURD OSCAR, JR, b Marquette, Mich, Jan 5, 37; m 60. ARACHNOLOGY. Educ: Northern Mich Univ, BS, 64; Mich State Univ, MS, 66, PhD(zool), 71. Prof Exp: Asst prof biol, Adrian Col, Mich, 71-72; ASST PROF ZOOL, STATE UNIV NY COL OSWEGO, 72- Mem: Sigma Xi; Am Arachnological Soc; Am Micros Soc. Res: Systematics and ecology of pseudoscorpions. Mailing Add: Dept of Zool State Univ of NY Oswego NY 13126

NELSON, STANLEY REID, b Kidder, SDak, Dec 20, 28; m 58; c 3. PHARMACOLOGY, NEUROSURGERY. Educ: Univ SDak, BA, 49, MS, 57; Tulane Univ, MD, 59. Prof Exp: Intern surg, Univ NC, 59-60; resident neurosurg, Univ Miss, 60-64; fel neurochem, Washington Univ, 64-66; from instr to assoc prof pharmacol & neurosurg, 66-73, PROF PHARMACOL & ASSOC PROF NEUROSURG, UNIV KANS MED CTR, KANSAS CITY, 73- Mem: Soc Neurosci; Am Soc Pharmacol & Exp Therapeut. Res: Effect of drugs on brain glycolysis; neurochemistry of head injury. Mailing Add: Dept of Pharmacol Univ of Kans Med Ctr Kansas City KS 66103

NELSON, STUART HARPER, b Richmond, Ont, May 21, 26; m 52; c 3. HORTICULTURE. Educ: Univ Toronto, BSA, 48; Mich State Univ, MS, 50, PhD(hort), 55. Prof Exp: Res officer & horticulturist, Hort Div, Cent Exp Farm, Can Dept Agr, 48-54, res officer propogation, 54-59, propogation & nursery mgt, Plant Res Inst, 59-60; assoc prof, 61-66, PROF HORT, UNIV SASK, 66-, HEAD DEPT, 61- Res: Propogation of horticultural plants; fruit breeding; turf research; water usage at different stages of growth by horticultural plants; temperature effects on hollow heart in potatoes. Mailing Add: Dept of Hort Sci Univ of Sask Saskatoon SK Can

NELSON, TALMADGE SEAB, b Booneville, Ark, Jan 25, 28. ANIMAL NUTRITION. Educ: Univ Ark, BSA, 52; Univ Ill, MS, 52; Cornell Univ, PhD(animal nutrit), 59. Prof Exp: Res farm mgr, Western Condensing Co, Wis, 51-52; field supvr feed & fertilizer inspection, State Plant Bd, Ark, 52-54; res assoc poultry husb, Cornell Univ, 54-58, asst, 58-59; sr res biochemist, Int Minerals & Chem Corp, 59-63, supvr animal nutrit res, 63-68; PROF ANIMAL SCI, UNIV ARK, FAYETTEVILLE, 68- Mem: Am Soc Animal Sci; Poultry Sci Asn; Am Inst Nutrit; Animal Nutrit Res Coun. Res: Animal nutrition, especially requirements and functions; amino acids. Mailing Add: Dept of Animal Sci Univ of Ark Fayetteville AR 72701

NELSON, TERENCE JOHN, b Sioux City, Iowa, May 12, 39; m 63; c 2. THEORETICAL PHYSICS. Educ: Iowa State Univ, BS, 61, PhD(physics), 67; NY Univ, MEE, 63. Prof Exp: Mem tech staff, Bell Tel Labs, 61-63; AEC fel physics, Ames Lab, Iowa State Univ, 67-68 & Lawrence Radiation Lab, Univ Calif, 68-69; instr, Iowa State Univ, 69-70; MEM TECH STAFF, BELL TEL LABS, 70- Mem: Am Phys Soc. Res: Lasers; magnetic domain research and device development; mathematical physics in particle and group theory. Mailing Add: Bell Tel Labs Mountain Ave Murray Hill NJ 07974

NELSON, THEODORA S, b Phillips, Nebr, Dec 18, 13. MATHEMATICS. Educ: Nebr State Teachers Col, Kearney, BS, 42; Univ Ill, Urbana, BS, 46; Univ Nebr, EdD, 59. Prof Exp: Teacher pub schs, Nebr, 32-41; teacher high sch, Nebr, 42-45; asst math, Univ Ill, Urbana, 45-46; asst prof, 46, PROF MATH, KEARNEY STATE COL, 46- Concurrent Pos: Part-time asst, Univ Ill, 45-46 & Univ Nebr, 57-58. Mem: Math Asn Am. Res: Mathematics seminar and research; projective geometry; history of mathematics. Mailing Add: Bruner Hall of Sci Kearney State Col Kearney NE 68847

NELSON, THOMAS CHARLES, b Wautoma, Wis, Aug 11, 23; m 44; c 2. FORESTRY. Educ: Univ Wis, BS, 43; Mich State Univ, MS, 47, PhD(silvicult), 50. Prof Exp: Res biologist, State Dept Conserv, Mich, 49-50; res forester, Southeastern Forest Exp Sta, US Forest Serv, 50-57; admin forester, Kirby Lumber Corp, Tex, 57-59; asst chief, Div Forest Mgt Res, Southeast Forest Exp Sta, 59-63, branch chief, Timber Mgt Res, Wash Off, 63-64, asst to dep chief, 64-66, dir south forest exp sta, 66-70, assoc dep chief, 70-71, DEP CHIEF, US FOREST SERV, 71- Concurrent Pos: Mem, Bd Agr & Renewable Resources, Nat Acad Sci, 73- Honors & Awards: Superior Serv Award, USDA, 70. Mem: Fel Soc Am Foresters. Res: Natural resource administration. Mailing Add: 8326 Briar Creek Dr Annandale VA 22003

NELSON, THOMAS CLIFFORD, b Columbus, Ohio, July 24, 25. MICROBIOLOGY. Educ: Queens Col, NY, BS, 46; Columbia Univ, MA, 46, PhD, 51. Prof Exp: Asst prof, Vanderbilt Univ, 51-52; proj assoc genetics, Univ Wis-Madison, 52-53, USPHS res fel, 53-54, proj assoc bot, 55-57; USPHS res fel microbiol, Rutgers Univ, 54-55; sr microbiologist, Eli Lilly & Co, 57-65; asst prof microbiol, 65-70, ASSOC PROF BOT, UNIV WIS-MILWAUKEE, 70- Res: Microbial genetics; industrial microbiology; algal biochemistry. Mailing Add: Dept of Bot Univ of Wis Milwaukee WI 53211

NELSON, THOMAS EUSTIS, JR, b Sharon, Mass, May 3, 22; m 47; c 3. PHARMACOLOGY, PHYSIOLOGY. Educ: Antioch Col, BA, 47; Univ Southern Calif, MS, 51, PhD(med physiol), 56. Prof Exp: Asst physiol, Sch Med, Univ Southern Calif, 50-53 & 54-56, instr, 56-57; from asst prof to assoc prof pharmacol, Sch Med & Dent, Univ PR, 57-61, actg head dept, 59-60; asst prof, Med Ctr, Univ Colo, 61-68; assoc prof, Univ Tex Dent Br Houston, 68-70; PROF PHARMACOL & CHMN DEPT, SCH DENT MED, SOUTHERN ILL UNIV, EDWARDSVILLE, 70- Mem: AAAS; Am Physiol Soc; Am Soc Pharmacol & Exp Therapeut; Am Asn Dent Schs; Soc Exp Biol & Med. Res: Physiology and pharmacology of the cardiovascular and nervous systems. Mailing Add: Dept of Pharmcol Southern Ill Univ Sch Dent Med Edwardsville IL 62025

NELSON, THOMAS EVAR, b Chicago, Ill, Oct 13, 34; m 59. BIOCHEMISTRY. Educ: Univ Ill, Urbana, BS, 57, MS, 59; Univ Minn, St Paul, PhD(biochem), 65. Prof Exp: Asst, Univ Ill, Urbana, 58-59; asst, Univ Minn, St Paul, 59-64, instr, 63; instr, Univ Minn, Minneapolis, 66-67; ASST PROF BIOCHEM, BAYLOR COL MED, 68- Concurrent Pos: Res fel, Univ Minn, Minneapolis, 65-68; USPHS fel, 66-68. Mem: AAAS; Am Chem Soc; fel Am Inst Chemists. Res: Action pattern, specificity, mode to attack, transglucosylation and cleavage mechanisms of carbohydrases; fine structure of polysaccharides; enzyme purification techniques; carbohydrate chemistry; enzyme interrelationships in glycogen storage diseases; control of glycogen metabolism; specificity and mechanism of hormone action. Mailing Add: Tex Med Ctr Baylor Col of Med Houston TX 77025

NELSON, THOMAS LOTHIAN, b Baranquilla, Colombia, Jan 17, 22; US citizen; m 55; c 2. PEDIATRICS, ALLERGY. Educ: Univ Calif, Berkeley, AB, 43; Univ Calif, San Francisco, MD, 46. Prof Exp: Resident pediat, Med Ctr, Univ Calif, San Francisco, 49-51, instr, 51-56, asst clin prof pediat & lectr psychiat, Sch Med, 56-61; from assoc prof to prof pediat, Col Med, Univ Ky, 61-64; PROF PEDIAT & CHMN DEPT, UNIV CALIF, IRVINE-CALIF COL MED, 64- Concurrent Pos: Pediatrician, Sonoma State Hosp, Eldridge, 51-52, chief physician, 52-54, asst supt med serv, 54-56, supt & med dir, 56-61; chief physician pediat & contagious dis serv, Los Angeles County Gen Hosp Unit II, 64-68; prog consult, Nat Inst Child Health & Human Develop, 64-68; mem, Fed Hosp Coun, 64-67; dir pediat prog, Orange County Med Ctr, 68- Res: Clinical immunology; mental retardation. Mailing Add: Dept of Pediat Univ Calif Col of Med Irvine CA 92717

NELSON, VERNON A, b Norwood, Mass, Apr 17, 39; m 61; c 2. ENTOMOLOGY. Educ: Univ Mass, BS, 63, MS, 64; Pa State Univ, PhD(entom), 68. Prof Exp: Instr entom, Pa State Univ, 64-68; ASST PROF BIOL, SOUTHERN CONN STATE COL, 68- Mem: Entom Soc Am. Res: Arthropods of public health importance; taxonomy and biology. Mailing Add: Dept of Biol Southern Conn State Col New Haven CT 06515

NELSON, VERNON RONALD, b Webster, SDak, Jan 20, 21; m 45; c 1. ELECTRONICS. Educ: Augustana Col, BA, 44; Univ Colo, MA, 51, DrSc(sci educ), 53. Prof Exp: Assoc prof, 46-54, PROF PHYSICS, AUGUSTANA COL, SDAK, 54- Mem: Am Asn Physics Teachers; Nat Sci Teachers Asn; Inst Elec & Electronics Engrs. Res: Wave analysis; electronics in music; square wave generator. Mailing Add: Dept of Physics Augustana Col Sioux Falls SD 57105

NELSON, VICTOR EUGENE, b Denver, Colo, Jan 8, 36; m 58; c 3. ZOOLOGY. Educ: Augustana Col, Ill, BA, 59; Univ Colo, PhD(zool, physiol), 64. Prof Exp: Asst biol, Univ Colo, 60-63, asst acarine physiol, 61-63; res assoc, Univ Kans, 64-65, asst prof & res assoc acarine physiol & entom, 65-66, asst prof biol, entom & insect biochem, 66-71; ASSOC PROF BIOL, BAKER UNIV, 71-, CHMN DEPT, 72- Concurrent Pos: Res grant, Univ Kans, 66-67. Mem: AAAS; Am Soc Zoologists; Soc Exp Biol & Med. Res: Physiological ecology of terrestrial arthropods, particularly water relations. Mailing Add: Dept of Biol Baker Univ Baldwin City KS 66006

NELSON, VINCENT EDWARD, b Rock Island, Ill, Jan 22, 13; m 39; c 1. GEOLOGY. Educ: Augustana Col, AB, 35; Univ Chicago, PhD(geol), 42. Prof Exp: Asst prof geol, Univ Ky, 38-42; geologist, US Geol Surv, 42-45; assoc prof, 46-49, PROF GEOL, UNIV KY, 49- Concurrent Pos: Prof & chief party, Ky contract team, Bandung Tech Inst, 61-64. Mem: AAAS; fel Geol Soc Am; Am Asn Petrol Geol; Nat Asn Geol Teachers. Res: Structural geology; metallic ore deposits. Mailing Add: Dept of Geol Univ of Ky Lexington KY 40506

NELSON, WALDO EMERSON, b McClure, Ohio, Aug 17, 98; m 28; c 3. PEDIATRICS. Educ: Wittenberg Univ, AB, 22; Univ Cincinnati, MD, 26. Hon Degrees: DSc, Wittenberg Univ, 56; LHD, Temple Univ, 75. Prof Exp: From instr to assoc prof, Col Med, Univ Cincinnati, 29-40; PROF PEDIAT, SCH MED, TEMPLE UNIV, 40-; PROF PEDIAT, MED COL PA, 64- Concurrent Pos: Attend physician, St Christopher's Hosp Children, 47-; ed, J Pediat. Mem: Am Pediat Soc; Soc Pediat Res; AMA; Am Acad Pediat. Mailing Add: Dept of Pediat Med Col of Pa Philadelphia PA 19129

NELSON, WALLACE WARREN, b Tracy, Minn, Feb 17, 28; m 49; c 2. SOILS. Educ: Univ Minn, BS, 50, PhD(soils), 56. Prof Exp: Asst supt agron, Northeast Exp Sta, 53-59, SUPT AGRON, SOUTHWEST EXP STA, UNIV MINN, 59- Mem: Am Soc Agron; Soil Sci Soc Am. Res: Soil factors affecting crop production; fertilizers and their effect on crops. Mailing Add: Southwest Exp Sta Lamberton MN 56152

NELSON, WALTER LUDWIG, b Norwich, NY, May 27, 11; m 41; c 3. BIOCHEMISTRY. Educ: Philadelphia Col Pharm, BS, 33, MS, 34; Cornell Univ, PhD, 41. Prof Exp: Chemist, Norwich Pharmacal Co, 33-38; asst, Cornell Univ, 38-41; head biol lab, Schenley Res Inst, 41-43; res assoc, 43-45, from asst prof to assoc prof, 45-53, PROF BIOCHEM, CORNELL UNIV, 53- Mem: AAAS; Am Chem Soc; Am Soc Biol Chem; Am Inst Nutrit. Res: Enzyme systems; lipid metabolism. Mailing Add: Dept of Biochem Cornell Univ Ithaca NY 14850

NELSON, WALTER RALPH, b St Paul, Minn, Mar 24, 37; m 60; c 2. HEALTH PHYSICS. Educ: Univ Calif, Berkeley, AB, 63; Univ Wash, MS, 64; Stanford Univ, PhD(health physics & dosimetry), 73. Prof Exp: STAFF MEM HEALTH PHYSICS, STANFORD LINEAR ACCELERATOR CTR, 64- Concurrent Pos: Consult, Varian Assocs, Calif, 67-75 & McCall Assocs, 69-; lectr, Sch Radiation Protection & Dosimetry, Italy, 75. Mem: Health Physics Soc. Res: Electromagnetic cascade calculations; muon production and transport; dosimetry; electron accelerator shielding; medical physics calculations. Mailing Add: Stanford Linear Accelerator Ctr PO Box 4349 Stanford CA 94305

NELSON, WAYNE FRANKLIN, b Altona, Ill, Jan 16, 20; m 45; c 2. INORGANIC CHEMISTRY, PHYSICAL CHEMISTRY. Educ: Augustana Col, Ill, AB & BS, 41. Prof Exp: Asst chemist, Am Container Corp, Ill, 40-41, plant chemist, NY, 41-43, asst plant mgr, 43-47, plant mgr, 47-49, dir res, WVa, 49-57; DIR RES & DEVELOP, A SCHULMAN, INC, 57- Mem: Am Chem Soc; Soc Plastics Engrs; fel Am Inst Chemists. Res: Soft and hard rubber compounds; styrene plastics; colors and blends; polyolefins; polypropylene-natural; special compounds. Mailing Add: A Schulman Inc 790 E Tallmadge Ave Akron OH 44309

NELSON, WENDEL LANE, b Mason City, Nebr, Apr 7, 39; m 68; c 1. MEDICINAL CHEMISTRY. Educ: Idaho State Univ, BS, 62; Univ Kans, PhD(pharmaceut chem), 65. Prof Exp: Asst prof, 65-70, ASSOC PROF PHARMACEUT CHEM, UNIV WASH, 70- Mem: Am Chem Soc; NY Acad Sci; Am Asn Cols Pharm; Am Pharmaceut Asn. Res: Mechanisms of drug action; stereochemistry and conformational analysis; drug metabolism. Mailing Add: Sch of Pharm Univ of Wash Seattle WA 98195

NELSON, WERNER LU LIND, b Sheffield, Ill, Oct 17, 14; m 40; c 2. AGRONOMY. Educ: Univ Ill, BS, 37, MS, 38; Ohio State Univ, PhD(soil physics), 40. Prof Exp: Instr, Univ Idaho, 40-41; asst agronomist, Agr Exp Sta, Univ NC, 41-44, assoc agronomist, 44-47, prof agron, 47-54, in charge soil fertil res, 51-54; midwest dir, Am Potash Inst, 55-67, regional dir, 62-67; SR VPRES, POTASH INST, 67- Concurrent Pos: Dir soil testing div, State Dept Agr, NC, 49-52; chmn soil test work group, Int Soil Fertil Cong, Ireland; mem fertilizer surv team, Food & Agr Orgn, UN, Asia & Far East Region, 59, fertilizer indust adv panel, 60-; bd of dirs, Coun Agr Sci Technol, 72-; adj prof agron, Purdue Univ, 73- Honors & Awards: Merit Cert, Am Forage & Grassland Coun, 71; Agron Serv Award, Am Soc Agron, 64. Mem: AAAS; fel Am Soc Agron (pres, 68-69); Soil Sci Soc Am (pres, 60-61); Soil Conserv Soc Am.

Res: Plant nutrition; soil fertility evaluation and management; limiting factors in crop production. Mailing Add: Potash Inst 402 Northwestern West Lafayette IN 47906

NELSON, WESLEY EUGENE, b Jamaica, WI, Sept 18, 35; m 67; c 1. SOIL CHEMISTRY. Educ: Tuskegee Inst, BS, 60; Rutgers Univ, MS, 62; Ohio State Univ, PhD(soil chem), 67. Prof Exp: Prof chem, Miss Valley State Col, 66-68; RES ASSOC SOILS, DEPT PLANT & SOIL SCI, CARVER RES FOUND, TUSKEGEE INST, 68- Mem: AAAS; Am Soc Agron; Am Chem Soc; Int Soc Soil Sci; fel Am Inst Chemists. Res: Organo-metal interactions; trace minerals; ground water. Mailing Add: Dept of Plant & Soil Sci Carver Res Found Tuskegee Institute AL 36088

NELSON, WILFRED H, b Evanston, Ill, May 23, 36; m 59; c 2. INORGANIC CHEMISTRY. Educ: Univ Chicago, BSc & MSc, 59; Univ Minn, PhD(inorg chem), 63. Prof Exp: Res assoc inorg chem, Univ Ill, Urbana, 63-64; asst prof, 64-67, ASSOC PROF INORG CHEM, UNIV RI, 67- Concurrent Pos: Res Corp grant, 65-66; grant, Sydney Univ, 70-71; NSF grant, 66-68 & 68-72. Mem: Am Chem Soc. Res: Structure of molecules with metal-oxygen bonds; synthesis of post-transition metal coordination compounds; light scattering and bond polarizability studies; solution thermodynamics. Mailing Add: Dept of Chem Univ of RI Kingston RI 02881

NELSON, WILLIAM ARNOLD, b Lethbridge, Alta, June 24, 18; m 48; c 3. ENTOMOLOGY. Educ: Univ Alta, BSc, 44; McGill Univ, MSc, 48, PhD(med entom), 57. Prof Exp: Asst wheat stem sawfly, Field Crop Entom Sect, 43-47, RES OFFICER, VET MED SECT, RES STA, CAN DEPT AGR, 48- Mem: Am Soc Parasitologists. Res: Physiology of host-parasite relationships; humoral factors. Mailing Add: Vet Med Sect Can Agr Res Sta Lethbridge AB Can

NELSON, WILLIAM FRANK, b Cleveland, Ohio, May 4, 24; m 48; c 3. PHYSICS. Educ: Univ Akron, BS, 48, MS, 49; Wash State Univ, PhD(physics), 56. Prof Exp: Head, Mat Res Group, Long Beach Div, Dougals Aircraft Co, 53-60; sr res scientist & head solid state physics group fundamental res, Owens-Ill, Inc, 60-64, dir fundamental res, Tech Ctr, 64-69, ASSOC DIR RES & DIR EXPLOR RES LAB, GTE LABS, INC, 69- Concurrent Pos: Adj prof, Univ Toledo, 63-69. Mem: AAAS; fel Am Phys Soc. Res: Physics of solids. Mailing Add: GTE Labs Inc 40 Sylvan Rd Waltham MA 02154

NELSON, WILLIAM HENRY, b Huntsville, Ala, Nov 24, 43. MAGNETIC RESONANCE. Educ: Auburn Univ, BS, 66; Duke Univ, PhD(physics), 70. Prof Exp: Asst prof physics, Hollins Col, Va, 70-73; instr, Duke Univ, 73-74; ASST PROF PHYSICS, GA STATE UNIV, 74- Concurrent Pos: Res assoc, Microwave Lab, Duke Univ, 73-74. Mem: Am Phys Soc; Am Asn Physics Teachers; Sigma Xi. Res: Use of magnetic resonance, electron spin resonance and electron-nuclear double resonance, for study of radiation damage and molecular structure. Mailing Add: Dept of Physics Ga State Univ Atlanta GA 30303

NELSON, WILLIAM PIERREPONT, III, b New Orleans, La, Jan 9, 20; m 44; c 3. MEDICINE. Educ: Wesleyan Univ, BA, 41; Cornell Univ, MD, 44; Am Bd Internal Med, dipl, 52. Prof Exp: Life Ins Med Res Fund fel metab, Sch Med, Yale Univ, 50-51 & fel med, Albany Med Col, 51-53; asst prof, 53-56, ASSOC PROF MED, PROF POSTGRAD EDUC & ASST DEAN COL, ALBANY MED COL, 56- Concurrent Pos: Chief metab & endocrine sect, Vet Admin Hosp, 52-54, chief med serv, 54-56, chief ambulatory care, 74- Mem: Fel Am Col Physicians; Am Fedn Clin Res. Res: Metabolism and endocrinology. Mailing Add: Font Grove Rd Slingerlands NY 12159

NELSON, WILLIAM ROBERT, biochemistry, see 12th edition

NELSON, WOODROW ENSIGN, b Mesa, Ariz, Apr 23, 18; m 43; c 3. FOOD TECHNOLOGY, MICROBIOLOGY. Educ: Utah State Univ, BS, 47; Univ Wis-Madison, MS, 60, PhD(dairy foods, microbiol), 67. Prof Exp: Dir qual control, Hi-Land Dairymen Asn, 47-51; plant supt, Clark County Dairymen, Nev, 51-56; owner & operator dairy herd, Clark County, Nev & Cache County, Utah, 56-58; dir food technol, 63-72, DIR, THERACON, INC, 72- Mem: AAAS; Inst Food Technol; Am Dairy Sci Asn. Res: Reconstitution characteristics of dried milk; ultra high temperature sterilization of foods; effect of ultra high temperature sterilization on the degradation of food nutrients. Mailing Add: Theracon Inc PO Box 1493 Topeka KS 66601

NELSON-REES, WALTER ANTHONY, b Havana, Cuba, Jan 11, 29; US citizen. GENETICS, CYTOLOGY. Educ: Emory Univ, AB, 51, MA, 52; Univ Calif, Berkeley, PhD(genetics), 60. Prof Exp: NIH training grant & res assoc genetics, Univ Calif, Berkeley, 60-61, from asst res geneticist to assoc res geneticist, Naval Biosci Lab, Sch Pub Health, 61-73, CONSULT, NAVAL BIOSCI LAB, SCH PUB HEALTH, UNIV CALIF, BERKELEY, 63-, ASSOC CHIEF CELL CULT DIV, 69-, LECTR CYTOL & CYTOGENETICS, 71- Concurrent Pos: Fulbright res scholar cytogenetics, Max Planck Inst Marine Biol, Ger, 61-62; consult, Breast Cancer Task Force, Nat Cancer Inst, NIH, 75- Mem: AAAS; Tissue Cult Asn. Res: Induction of chromosome aberrations in Tradescantia; factors influencing sex determination in coccid insects; heterochromatin and fertility factors in male mealy bugs; cinemicrography of animal cells; chromosome banding and other methods for cell line identification and detection of cellular cross contamination. Mailing Add: Univ of Calif Sch of Pub Health Naval Biosci Lab Naval Supply Ctr Oakland CA 94625

NEMARICH, JOSEPH, physics, see 12th edition

NEMATOLLAHI, JAY, b Astara, Iran, Dec 21, 25; US citizen; m 52; c 3. MEDICINAL CHEMISTRY, MICROBIOLOGY. Educ: Univ Tehran, PharmD, 48; Univ Calif, MA, 54, PharmD, 58, PhD(pharmaceut chem), 63. Prof Exp: Fel, Univ Calif, 63; asst prof, prof pharmaceut chem, Univ RI, 63-64; assoc prof, Tex Southern Univ, 64-65; asst prof, 67-71, ASSOC PROF PHARMACEUT CHEM, COL PHARM, UNIV TEX, AUSTIN, 71- Concurrent Pos: Adj grad fac, Univ Houston. Mem: Am Chem Soc; Am Pharmaceut Asn. Res: Synthesis of organic medicinals; spectroscopy. Mailing Add: Col of Pharm Univ of Tex Austin TX 78712

NEMEC, JOSEF, b Ostresany, Czech, Sept 7, 29; m 75; c 1. ORGANIC CHEMISTRY. Educ: Inst Chem Technol, Czech Acad Sci, MS, 54; Czech Acad Sci, PhD(org chem), 58. Prof Exp: Scientist org chem, Czech Acad Sci, 54-61; sr res chemist, Inst Chem Technol, Czech Acad Sci, 61-69; res fel org chem, Wayne State Univ, 69-70; sr res scientist, Squibb Inst Med Res, 70-75; MEM, ST JUDE CHILDREN'S RES HOSP, 75- Mem: Am Chem Soc; The Chem Soc. Res: Stereochemistry and synthesis of deoxy sugars, amino sugars, antibiotics and analogs; drug metabolism; nucleosides; nucleotides; antineoplastic agents. Mailing Add: St Jude Children's Res Hosp Memphis TN 38101

NEMEC, JOSEPH WILLIAM, b Philadelphia, Pa, Mar 24, 22; m 48; c 3. ORGANIC CHEMISTRY. Educ: Temple Univ, AB, 43; Ind Univ, MA, 44; Pa State Univ, PhD(chem), 49. Prof Exp: Sr org chemist, 49-61, lab head, 61-69, res supvr, 69-73, MGR PROCESS RES, ROHM AND HAAS CO, 73- Mem: AAAS; Am Chem Soc. Res: Process research and development; catalysis; acrylate monomer technology. Mailing Add: 931 Washington Lane Rydal Jenkintown PA 19046

NEMEC, STANLEY, b St Louis, Mo, Feb 3, 35; m 60; c 2. PLANT PATHOLOGY. Educ: Auburn Univ, BS, 60; Okla State Univ, MS, 64; Ore State Univ, PhD(plant path), 67. Prof Exp: Landscape technician, Harland Bartholomew & Assoc, City Planners, 60-61; plant pathologist, Ore Dept Agr, 63-66; res plant pathologist, Southern Ill Univ, Carbondale, 68-73, RES PLANT PATHOLOGIST, USDA, FLA, 73- Mem: Am Phytopath Soc; Soc Nematol. Res: Botany; horticulture; biochemistry. Mailing Add: USDA Agr Res Serv US Hort Field Lab 2120 Camden Rd Orlando FL 32803

NEMENZO, FRANCISCO, b Pinamungajan, Philippines, Sept 17, 05; m 31; c 3. ZOOLOGY. Educ: Univ Philippines, BS, 29, MS, 34; Univ Mich, AM, 48. Prof Exp: Registr, Cebu Col, 45-50, dean, 50, asst prof zool, 52-55, assoc prof & head dept, 55-60, prof zool, 60-71, chmn div natural sci, 60-63, dean col arts & sci, 63-69, EMER PROF ZOOL, UNIV PHILIPPINES, 71- Concurrent Pos: Mem, Nat Res Coun, Philippines. Mem: AAAS; Am Micros Soc; Indian Acad Zool; NY Acad Sci. Res: Philippines recent scleractinians; biology of field rats in the Philippines; systematics and ecology of stony reef corals. Mailing Add: 12 Sampaguita St Cebu City Philippines

NEMER, MARTIN JOSEPH, b Philadelphia, Pa, Nov 26, 29; div. BIOCHEMISTRY. Educ: Kenyon Col, BA, 52; Harvard Univ, MS, 55, PhD(biochem), 58. Prof Exp: Fel, Univ Brussels, 58-59; res fel, Stazione Zool, Naples, Italy, 59-60; res assoc biochem, 60-63, asst mem, 63-67, ASSOC MEM, INST CANCER RES, 67- Mem: AAAS; Am Soc Biol Chemists; Soc Develop Biol. Res: Chemical embryology; regulation of protein and nucleic acid synthesis. Mailing Add: Inst for Cancer Res 7701 Burholme Ave Philadelphia PA 19111

NEMERGUT, PAUL JOSEPH, JR, b Bridgeport, Conn, July 4, 35; m 65; c 1. APPLIED MECHANICS. Educ: Univ Bridgeport, BS, 58; Univ Conn, MS, 59, PhD(appl mech), 64. Prof Exp: Res engr, Elec Boat Div, Gen Dynamics Corp, 59-61, sr res engr, 64-65; res asst, Univ Conn, 61-64, asst prof mech eng, 65-67; res engr, US Naval Underwater Sound Lab, 67-68; ASSOC PROF MECH, US AIR FORCE INST TECHNOL, 68- Res: Vibrations of plates, shells and structures; elastic wave propagation in piezoelectric materials; acoustic wave propagation in tubes; acoustic radiation. Mailing Add: US Air Force Inst of Technol AFIT-ENB Wright-Patterson AFB Dayton OH 45433

NEMERSON, YALE, b New York, NY, Dec 15, 31; m 58; c 3. HEMATOLOGY, BIOCHEMISTRY. Educ: Bard Col, BA, 53; NY Univ, MD, 60. Prof Exp: Intern, Lenox Hill Hosp, 60-61; resident, Bronx Vet Admin Hosp, 61-62; fel hemat, Montefiore Hosp, 62-64; from instr to prof, Sch Med, Yale Univ, 64-75; PROF MED, STATE UNIV NY STONY BROOK, 75- Concurrent Pos: Leukemia Soc fel, 63-64; NIH res grant, 64-76; investr, Am Heart Asn, 67-72, res grant, 67-73. Mem: AAAS; Am Soc Hemat; Am Soc Exp Path; Am Soc Clin Invest. Res: Lipid-protein interactions in blood coagulation. Mailing Add: Dept of Med State Univ of NY Stony Brook NY 11794

NEMES, MARJORIE M, b Bethlehem, Pa, May 31, 21. BACTERIOLOGY. Educ: Lebanon Valley Col, BS, 45; Lehigh Univ, MS, 51, PhD, 53; Am Bd Med Microbiol, dipl, 66. Prof Exp: Instr, Woman's Med Col Pa, 53-54; res assoc, Parke, Davis & Co, 54-56 & Rockefeller Inst, 56-59; RES ASSOC, MERCK INST THERAPEUT RES, 59- Mem: AAAS; Am Soc Microbiol; Tissue Cult Asn; fel NY Acad Sci. Res: Virus immunology; kinetics and biochemistry of virus reproduction; virus chemotherapy; interferon; host resistance. Mailing Add: Merck Inst for Therapeut Res West Point PA 19486

NEMETH, ABRAHAM, b New York, NY, Oct 16, 18; m 44. MATHEMATICS. Educ: Brooklyn Col, BA, 40; Columbia Univ, MA, 42; Wayne State Univ, PhD(math), 64. Prof Exp: Instr math, Brooklyn Col, 46, Manhattan Col, 53-54 & Manhattanville Col, 54-55; from instr to asst prof, 55-65, ASSOC PROF MATH, UNIV DETROIT, 65- Mem: Math Asn Am. Res: Computer science; Nemeth braille code of mathematics and scientific notation. Mailing Add: Dept of Math Univ of Detroit Detroit MI 48221

NEMETH, ANDREW MARTIN, b Philadelphia, Pa, Sept 13, 26; m 61; c 4. ANATOMY. Educ: Johns Hopkins Univ, AB, 49, MD, 53. Prof Exp: Intern pediat, Johns Hopkins Hosp, 53-54; NSF fel biochem, Col Physicians & Surgeons, Columbia Univ, 54-56; assoc anat, 56-59, from asst prof to assoc prof, 59-68, PROF ANAT, SCH MED, UNIV PA, 68- Mem: Am Asn Anatomists. Res: Regulation of enzyme formation in developing tissues; induction of enzymes; fetal development. Mailing Add: Dept of Anat Univ of Pa Sch of Med Philadelphia PA 19104

NEMETH, RONALD LOUIS, b Endicott, NY, Mar 4, 41; m 66. PHYSICAL CHEMISTRY. Educ: Clarkson Col Technol, BS, 64, MS, 66, PhD(phys chem), 69. Prof Exp: RES CHEMIST, MONSANTO CO, 68- Mem: Am Chem Soc. Res: Stability of aqueous dispersions; molecular structure, especially mechanical property relationships of polymers. Mailing Add: Res Dept Monsanto Co Indian Orchard MA 01051

NEMETHY, GEORGE, b Budapest, Hungary, Oct 11, 34; US citizen. PHYSICAL CHEMISTRY. Educ: Lincoln Univ, Pa, BA, 56; Cornell Univ, PhD(phys chem), 62. Prof Exp: Phys chemist, Gen Elec Res Lab, NY, 62-63; asst prof phys chem, Rockefeller Univ, 63-74; SR RES ASSOC, DEPT CHEM, CORNELL UNIV, 74- Concurrent Pos: NATO fel & vis lectr, Instituto Superiore di Sanita, Rome, 70; lectr, Med Ctr, NY Univ, 71-72; mem biophys sci training comt, Nat Inst Biomed Sci, 71- Honors & Awards: Pious XI Gold Medal, Pontifical Acad Sci, 72. Mem: Am Soc Biol Chemists; Am Chem Soc. Res: Statistical thermodynamics, liquid structure; thermodynamic properties of aqueous solutions; structure of proteins, conformations, thermodynamics, structure and enzymatic activity. Mailing Add: Dept of Chem Cornell Univ Ithaca NY 14853

NEMHAUSER, GEORGE L, b New York, NY, July 27, 37; m 59; c 2. OPERATIONS RESEARCH. Educ: City Col New York, BChE, 58; Northwestern Univ, MS, 59, PhD(opers res), 61. Prof Exp: From asst prof to assoc prof opers res, Johns Hopkins Univ, 61-69; PROF OPERS RES, CORNELL UNIV, 69- Concurrent Pos: Vis lectr, Univ Leeds, 63-64; vis res prof, Ctr Opers Res & Economet, Cath Univ Louvain, 69-70; NSF fac fel, 69-70; dir res ctr opers res & econo economet, Cath Univ Louvain, 75-77; ed-in-chief Opers Res, J Opers Res Soc Am, 75- Mem: Opers Res Soc Am; Inst Mgt Sci; Soc Indust & Appl Math; Math Prog Soc. Res: Theory and computational aspects of mathematical programming; mathematical modelling of complex systems. Mailing Add: Ctr Opers Res & Economet 54 de Croyluan Heverlee 3030 Belgium

NEMIR, PAUL, JR, b Navasota, Tex, Aug 30, 20; m 49; c 3. SURGERY. Educ: Univ Tex, AB, 40, MD, 44. Prof Exp: From instr to assoc prof surg, 48-69, dean grad sch med, 59-64, dir div grad med, 64-69, PROF SURG, SCH MED, UNIV PA, 69-

Concurrent Pos: Surgeon-in-chief, Grad Hosp & consult cardiovasc surg, US Naval Hosp, 48- Mem: AMA; Soc Univ Surgeons; Soc Vascular Surg; Am Asn Thoracic Surg; Soc Surg Alimentary Tract. Res: Pulmonary function and embolism; diseases of the esophagus and esophageal motor function; mechanism of toxicity of hemoglobin derivatives; studies on vascular prosthetics. Mailing Add: Grad Hosp 19th & Lombard Sts Philadelphia PA 19146

NEMIR, ROSA LEE, b Waco, Tex, July 16, 05; m 34; c 3. PEDIATRICS. Educ: Univ Tex, BA, 26; Johns Hopkins Univ, MD, 30. Hon Degrees: DSc, Colgate Univ, 74. Prof Exp: From instr to assoc prof, Med Col, 33-50, asst prof, Postgrad Med Sch, 50-53, PROF PEDIAT, SCH MED, NY UNIV, 53- Concurrent Pos: Lectr, Sch Nursing, Bellevue Hosp, 34-49, from asst vis pediatrician to vis pediatrician, 37-; attend pediatrician, Univ Hosp, 50-, vis physician & in chg chest unit, Children's Med Serv, 60, dir children's chest clin, 61-; attend pediatrician, Gouvernour Hosp, 50-58, consult, 58; consult, NY Infirmary, 54-, dir pediat educ & res, 66-73; vis prof, Col Physicians & Surgeons, Columbia Univ, 58-59. Honors & Awards: New York Dept Hosps Award, 59; Off, Medal of Cedars of Lebanon, Repub Lebanon; Elizabeth Blackwell Award, Am Med Women's Asn, 70. Mem: Soc Pediat Res; Am Pediat Soc; Am Col Chest Physicians; Am Med Women's Asn (pres, 64); Med Women's Int Asn (vpres, 70-74). Res: Pneumonia in children; nutrition; tuberculosis; virology. Mailing Add: NY Univ Sch of Med 550 First Ave New York NY 10016

NEMIROFF, MICHAEL, inorganic chemistry, see 12th edition

NEMITZ, WILLIAM CHARLES, b Memphis, Tenn, July 27, 28. PURE MATHEMATICS. Educ: Southwestern at Memphis, BS, 50; Ohio State Univ, MS, 56, PhD(math), 59. Prof Exp: Instr math, Ohio State Univ, 59-60; asst prof, Univ Kans, 60-61; from asst prof to assoc prof, 61-70, PROF MATH, SOUTHWESTERN AT MEMPHIS, 70- Concurrent Pos: NSF grant, 63-64, 71-73. Mem: Am Math Soc; Math Asn Am; Asn Comput Mach. Res: Algebraic structures related to intuitionistic logics, particularly implicative semi-lattices; nonnumerical use of computers, particularly in application to natural sciences. Mailing Add: Comput Ctr Southwestern at Memphis Memphis TN 38112

NEMMER, MARY, b Milwaukee, Wis, June 8, 12. MICROBIOLOGY. Educ: Loyola Univ, Ill, BS, 42; Marquette Univ, MS, 47; St Louis Univ, PhD(biol), 53. Prof Exp: Chmn dept, 53-69, ASSOC PROF BIOL, ALVERNO COL, 53-, PERIODICALS LIBRN, 70- Mem: Am Soc Microbiol. Mailing Add: Dept of Biol Alverno Col 3401 S 39th St Milwaukee WI 53215

NEMPHOS, SPEROS PETER, b New York, NY, July 8, 30; m 55; c 3. POLYMER CHEMISTRY. Educ: Ursinus Col, BS, 52; Univ Del, MS, 55, PhD(phys chem), 57. Prof Exp: Res chemist, 56-63, specialist, 63-65, group leader, 65-74, RES MGR, MONSANTO CO, 74- Mem: Am Chem Soc. Res: Polymer synthesis; kinetics and characterization of vinyl polymers; polymer degradations and stabilization; plastic foams; emulsion and suspension polymerizations; graft polymer systems; barrier resins; plastics fire safety. Mailing Add: 10 Pondview Dr Springfield MA 01118

NEMUTH, HAROLD I, b Norfolk, Va, Mar 12, 12; m 47; c 3. PREVENTIVE MEDICINE. Educ: Columbia Univ, BA, 34; Med Col Va, MD, 39. Prof Exp: Actg chmn dept prev med, 59-62, assoc prof, 62-73, CLIN PROF PREV MED, MED COL VA, 73-, ASSOC MED, HOME CARE PROG, 57- Mem: AAAS; AMA; Asn Am Med Cols; Asn Teachers Prev Med; Geront Soc. Mailing Add: Dept of Prev Med Med Col of Va Richmond VA 23219

NENNO, ROBERT PETER, b Buffalo, NY, Mar 3, 22; m; c 4. PSYCHIATRY. Educ: Univ Notre Dame, BS, 43; Loyola Univ, MD, 47; Am Bd Psychiat & Neurol, dipl. Prof Exp: Intern, E J Meyer Mem Hosp, Buffalo, NY, 47-48; resident psychiat, Vet Admin Hosp, Minneapolis, 48-50; resident, Vet Admin Hosp, Downey, Ill, 50-51; asst prof psychiat & asst dir dept, Sch Med, Georgetown Univ, 53-56, assoc prof, 56-58; prof psychiat & chmn dept, Seton Hall Col Med, 58-63; med dir & chief exec officer, NJ State Hosp, Marlboro, 63-68; interim med dir, Raritan Bay Ment Health Ctr, 70; PROF PSYCHIAT, NJ MED SCH, COL MED & DENT NJ, NEWARK, 73-; DIR DEPT PSYCHIAT, JERSEY CITY MED CTR, 73-, ACTG MED DIR, 75- Concurrent Pos: Asst dir psychiat div, Georgetown Univ Hosp, 53-56; consult, DC Gen Hosp, Mt Alto Vet Admin Hosp, Suburban Hosp, Bethesda, Md, Cent Intel Agency, 53-58 & Fitkin Mem Hosp, 63-68; asst examr, Am Bd Psychiat & Neurol, 56-; clin prof psychiat, Sch Med, Rutgers Univ, 66-73; pvt pract psychiat, 68-; mem staff, Overlook Hosp, Summit, NJ. Mem: Fel Am Psychiat Asn; Acad Psychoanal. Res: Vocational and social rehabilitation of the mentally ill; nicotinic acid in the treatment of schizophrenia; use of librium analogs in psychiatry. Mailing Add: Med Ctr Jersey City NJ 07304

NEPTUNE, EDGAR MCCLAIN, biochemistry, physiology, deceased

NEPTUNE, JACQUES GERARD, botany, genetics, see 12th edition

NEPTUNE, JOHN ADDISON, b Barnesville, Ohio, Nov 27, 19; m 47; c 1. INORGANIC CHEMISTRY. Educ: Muskingum Col, BS, 42; Univ Wis, MS, 49, PhD(chem), 52. Prof Exp: Instr chem, Muskingum Col, 43-44, 45-48; shift foreman, Chem Refining Div, Tenn Eastman Corp, 44-45; asst prof chem, Bowling Green State Univ, 49-50; instr pharmaceut chem, Univ Wis, 52-55; from asst prof to assoc prof chem, 55-61, PROF CHEM, SAN JOSE STATE UNIV, 61-, CHMN DEPT, 73- Mem: Am Chem Soc. Res: Reaction kinetics of oxidation-reduction reactions; reactions in non-aqueous solvents. Mailing Add: 50 Cherokee Lane San Jose CA 95127

NEPTUNE, WILLIAM EVERETT, b Lawton, Okla, Apr 24, 28; m 50; c 2. PHYSICAL CHEMISTRY. Educ: Okla Baptist Univ, BS, 50; Univ Okla, MS, 52, PhD(chem), 54. Prof Exp: From asst prof to assoc prof, 54-60, PROF CHEM, OKLA BAPTIST UNIV, 60-, DEAN LIB ARTS, 61-, VPRES ACADEMIC AFFAIRS, 73- Concurrent Pos: Consult, Saline Waters Proj, US Dept Interior, 60-61, dir, 61-65; consult, NCent Asn Cols & Sec Schs, 63-, mem, Comn Insts Higher Educ, 69- Mem: AAAS; Am Chem Soc; The Chem Soc; NY Acad Sci. Res: Electrochemistry; chemistry of carbon; history and philosophy of science; chemchemical education. Mailing Add: Dept of Chem Okla Baptist Univ Shawnee OK 74801

NERBUN, ROBERT CHARLES, JR, b Waukegan, Ill, Apr 19, 46; m 72; c 2. EXPERIMENTAL NUCLEAR PHYSICS. Educ: Univ Wis-River Falls, BS, 68; Case Western Reserve Univ, MS, 71, PhD(exp nuclear physics), 73. Prof Exp: Lectr, Col Physics, Case Western Reserve Univ, 69-70; ASST PROF PHYSICS, UNIV SC, SUMTER, 73- Mem: Am Phys Soc; Am Asn Physics Teachers. Res: Measurements of reaction cross sections for (d,n) stripping reactions at 2.5 to 4.0 MeV bombarding energies. Mailing Add: Dept of Physics Univ of SC Miller Rd Sumter SC 29150

NERESON, NORRIS (GEORGE), b Gaylord, Minn, Nov 4, 18; m; c 2. LASERS. Educ: Concordia Col, Minn, BA, 39; Univ Denver, MS, 41; Cornell Univ, PhD(nuclear physics), 43. Prof Exp: Asst physics, Cornell Univ, 42-43; mem staff, Physics Div, Los Alamos Sci Lab, 43-45; asst prof physics, Univ NMex, 46-47; MEM STAFF, LASER DIV, LOS ALAMOS SCI LAB, 48- Concurrent Pos: Int Atomic Energy Agency vis prof, Brazil, 64-65. Res: Solid state physics and lasers; application of molecular spectroscopy to isotope separation employing diode lasers. Mailing Add: Los Alamos Sci Lab PO Box 366 Los Alamos NM 87544

NERI, RUDOLPH ORAZIO, b Barre, Mass, Sept 11, 28; m 55; c 4. ENDOCRINOLOGY. Educ: Col Holy Cross, BS, 50; NY Univ, MS, 58, PhD(biol), 63. Prof Exp: Asst biologist, Worcester Found Exp Biol, 51-55; from asst scientist to prin scientist, 55-70, RES FEL, SCHERING CORP, 70- Concurrent Pos: Adj prof physiol, Farleigh Dickinson Univ Dental Sch, 72- Mem: AAAS; Am Physiol Soc; Endocrine Soc; Reticuloendothelial Soc; NY Acad Sci. Res: Effects of anti-estrogens, anti-androgens and anti-progesterones on reproductive processes in laboratory animals; rat skin homotransplantation studies, immune and reticuloendothelial response. Mailing Add: 15 Annette St Hawthorne NJ 07506

NERI, UMBERTO, b Rimini, Italy, Sept 7, 39; m 64; c 1. MATHEMATICS. Educ: Univ Chicago, BS, 61, MS, 62, PhD(math), 66. Prof Exp: Asst prof, 66-70, ASSOC PROF MATH, UNIV MD, COLLEGE PARK, 70- Concurrent Pos: NSF fel, Univ Md, College Park, 68-69; Nat Res Coun Italy fel, Univ Genoa, 69-70. Mem: Am Math Soc. Res: Singular integral operators; partial differential equations; analysis on manifolds. Mailing Add: Dept of Math Univ of Md College Park MD 20742

NERING, EVAR DARE, b Gary, Ind, July 18, 21; m 42; c 2. MATHEMATICS. Educ: Ind Univ, AB, 42, AM, 43; Princeton Univ, AM, 47, PhD(math), 48. Prof Exp: Asst, Ind Univ, 42-44; jr physicist, Appl Physics Lab, Johns Hopkins Univ, 44-45; instr math, Princeton Univ, 45-46, asst, 46-47; instr, Rutgers Univ, 47-48; from asst prof to assoc prof, 60-74, chmn dept, 62-70, PROF MATH, ARIZ STATE UNIV, 74- Concurrent Pos: Technician, Manhattan Proj, Ind Univ, 43-44; mathematician, Goodyear Aircraft Corp, 53-54. Mem: Am Math Soc; Math Asn Am. Res: Algebraic function and game. Mailing Add: Dept of Math Ariz State Univ Tempe AZ 85281

NERKEN, ALBERT, b New York, NY, Aug 21, 12. CHEMISTRY. Educ: Cooper Union, BS, 33. Prof Exp: chemist, Sinclair Ref Co, Pa, 33-35; jr physicist, Nat Adv Comt Aeronaut, Langley Field, Va, 35-37; chief chemist, Polin Labs, NY, 37-43; res chemist, Kellex Corp, 43-45; partner & engr, 45-61, chmn bd & treas, 61-71, VCHMN BD, VEECO INSTRUMENTS INC, 71- Mem: Am Chem Soc; Am Vacuum Soc. Res: Airplane dynamics; plastics from agricultural products; high vacuum techniques; gas analysis. Mailing Add: Veeco Instruments Inc Terminal Dr Plainview Long Island NY 11803

NERLICH, WILLIAM EDWARD, b Los Angeles, Calif, May 6, 23; m 57; c 4. MEDICINE. Educ: Univ Southern Calif, MD, 47; Am Bd Internal Med, dipl, 55. Prof Exp: From head physician to chief physician, 52-66, DIR INTERN TRAINING, LOS ANGELES COUNTY HOSP, 66-; DIR OFF EDUC & ASST MED DIR, LOS ANGELES COUNTY-UNIV SOUTHERN CALIF MED CTR, 69-; PROF MED, SCH MED, UNIV SOUTHERN CALIF, 71- Concurrent Pos: From asst prof med & asst dean student affairs to assoc prof med & assoc dean student affairs, Univ Southern Calif, 58-71; mem adv comt physician's assts, Calif State Bd Med Examr, 71- Mem: AMA; Asn Am Med Cols; Am Col Physicians. Res: Medical education; physical diagnosis. Mailing Add: Los Angeles County & Univ of Southern Calif Med Ctr Los Angeles CA 90033

NERNEY, STEVEN FRANCIS, SPACE PHYSICS, MAGNETOHYDRODYNAMICS. Educ: Calif State Univ, San Francisco, BA, 66, MS, 68; Univ Colo, PhD(astrogeophys), 74. Prof Exp: Physicist, Space Environ Lab, Nat Oceanic & Atmospheric Admin, 70-74; vis scientist, High Altitude Observ, 74-75; ASSOC, AMES RES CTR, NAT RES COUN, 75- Concurrent Pos: Lectr astrogeophys, Univ Colo, 75. Mem: Am Geophys Union; Sigma Xi. Res: Application of magnetohydrodynamics to the study of the three-dimensional nature of the solar wind, and to stellar wind modeling. Mailing Add: Ames Res Ctr Moffett Field CA 94035

NERO, ROBERT WILLIAM, ornithology, see 12th edition

NERODE, ANIL, b Los Angeles, Calif, June 4, 32; m 70; c 2. MATHEMATICS. Educ: Univ Chicago, BA, 49, BS, 52, MS, 53, PhD, 56. Prof Exp: Sr mathematician, Inst Syst Res, Univ Chicago, 54-56, group leader, 56-57; mem, Inst Adv Study, 57-58, 62-63; vis asst prof math, Univ Calif, 58-59; from asst prof to assoc prof, 59-65, actg dir, Ctr Appl Math, 65-66, PROF MATH, CORNELL UNIV, 65- Concurrent Pos: NSF fel, Cornell Univ, 57-58; mem consult bur, Math Asn Am, 61-; mem comt appln math, Nat Res Coun, 67-70; ed, J Symbolic Logic, 68-; vis prof, Monash Univ, Melbourne, Australia, 70, 74; mem math sci adv comt, NSF, 72-75; mem comt sci policy, Am Math Soc, 74- Mem: Am Math Soc; Math Asn Am; Soc Indust & Appl Math; Asn Symbolic Logic. Res: Mathematical logic; recursive functions; automata. Mailing Add: White Hall Cornell Univ Ithaca NY 14850

NERSASIAN, ARTHUR, b Salem, Mass, July 14, 24; m 46; c 3. ORGANIC CHEMISTRY. Educ: Mass Inst Technol, BS, 49; Univ Mich, MS, 51, PhD(org chem), 54. Prof Exp: Asst, Dow Chem Co for Univ Mich, 50-51, res fel & instr chem, 52-53; res chemist, Org Chem Dept, 54-57, RES CHEMIST, ELASTOMER CHEM DEPT, E I DU PONT DE NEMOURS & CO, INC, 57- Mem: Am Chem Soc; Am Sci Affiliation. Res: Phosgenation reactions; isocyanates; polyurethanes; rubber chemicals; Hypalon synthetic rubbers; polymer structural analysis; resilient foams; thermoplastic polyurethanes; Viton fluorohydrocarbon rubbers. Mailing Add: 335 Spalding Rd Wilmington DE 19803

NERVIK, WALTER EDWARD, b New York, NY, Mar 11, 23; m 51; c 2. NUCLEAR CHEMISTRY. Educ: Univ Calif, BS, 51, PhD(chem), 54. Prof Exp: NUCLEAR CHEMIST, LAWRENCE LIVERMORE LAB, UNIV CALIF, 54- Mem: Am Chem Soc. Res: Nuclear chemistry of fission product and transuranium elements; ion exchange phenomena; fission and stability of nuclei; inorganic chemistry of all elements. Mailing Add: Lawrence Livermore Lab Univ of Calif Livermore CA 94551

NES, WILLIAM ROBERT, b Oxford, Eng, May 16, 26; US citizen; m 46; c 2. BIOCHEMISTRY. Educ: Univ Okla, BA, 46; Univ Va, MS, 48, PhD(chem), 50. Prof Exp: Fel, Mayo Clin, 50-51; vis scientist, NIH, 51, staff mem, Nat Inst Arthritis & Metab Dis, 51-58; assoc prof biochem, Clark Univ, 58-64; prof chem & pharmaceut chem, Univ Miss, 64-67; prof, 67-74, W L OBOLD PROF BIOL SCI & DIR INST POP STUDIES, DREXEL UNIV, 74- Concurrent Pos: Res fels, Forrestal Res Ctr, Princeton Univ, 54, Univ Heidelberg, 55-56 & Univ Wales, 56; sr scientist & dir training prog steroid biochem, Worcester Found Exp Biol, 58-64; mem adv panel metab biol, NSF, 66-69; vis prof obstet & gynec, Hahnemann Med Col & Hosp, 70-; partic scientist, Franklin Inst Res Lab, 71-; mem res comt, Norristown State Hosp, 75- Mem: AAAS; Am Chem Soc; Am Soc Biol Chem; Endocrine Soc; Am Oil Chemists Soc. Res: Biosynthesis and metabolism of steroids, terpenes and related substances; biochemistry of hormones; phylogeny; role of sterols in membranes. Mailing Add: Dept of Biol Sci Drexel Univ Philadelphia PA 19104

NESBEDA, PAUL, b Trieste, Italy, June 20, 21; nat US; m 49; c 5. MATHEMATICS. Educ: Univ Pisa, PhD(math), 43. Prof Exp: Asst prof math anal, Univ Trieste, 44-46; fel, Univ Paris, 46-47; mem staff, Inst Adv Study, 47-48; instr math anal, Catholic Univ, 48-52; mathematician & engr, RCA Corp, 52-58, leader tech staff, 58-64, staff eng scientist, Aerospace Div, 65-74; MEM TECH STAFF, THE MITRE CORP, 74- Concurrent Pos: Lectr, Boston Col, 59-60. Mem: Am Math Soc; Soc Indust & Appl Math; Math Asn Am; Sigma Xi. Res: Functional and combinatorial analysis; probability; information processings and decision theory; detection and imaging systems; electro-optical systems; biomathematics; system analysis. Mailing Add: 10 Blodgett Rd Lexington MA 02173

NESBET, ROBERT KENYON, b Cleveland, Ohio, Mar 10, 30; m 58; c 3. THEORETICAL PHYSICS. Educ: Harvard Univ, AB, 51; Cambridge Univ, PhD(physics), 54. Prof Exp: Mem staff, Lincoln Lab, Mass Inst Technol, 54-56; asst prof physics, Boston Univ, 56-62; STAFF MEM, RES LAB, IBM CORP, 62- Concurrent Pos: Nat Cancer Inst spec res fel, Inst Pasteur, France, 60-61; exchange prof, Univ Paris, 73; assoc ed, J Chem Physics, 71-73 & J Computational Physics, 70-74. Mem: Fel Am Phys Soc; AAAS; Sigma Xi. Res: Theoretical atomic and molecular physics; quantum theory of finite many-particle systems; computational physics. Mailing Add: IBM Res Lab San Jose CA 95193

NESBIT, ARTHUR HENDERSON, b Penrose, Colo, July 5, 21; m 46; c 3. ANIMAL NUTRITION. Educ: Univ Ill, BS, 47, MS, 49. Prof Exp: NUTRITIONIST ANIMAL NUTRIT, MOORMAN MFG CO, 49- Mem: Poultry Sci Asn; Am Soc Animal Sci. Res: Biological availability of nutrients in feed ingredients; effect of processing on stability nutrients, biological availability and palatability. Mailing Add: Moorman Mfg Co 1000 N 30th Quincy IL 62301

NESBITT, CECIL JAMES, b Ft William, Ont, Oct 10, 12; nat US; m 38. ACTUARIAL MATHEMATICS. Educ: Univ Toronto, BA, 34, MA, 35, PhD(math), 37. Prof Exp: Mem, Inst Advan Study, 37-38; from instr to assoc prof, 38-52 assoc chmn dept, 62-67, chmn, 70-71, PROF MATH, UNIV MICH, ANN ARBOR, 52- Concurrent Pos: Coun mem, Conf Bd Math Sci, 68-75; co-chmn data registry comt, Cystic Fibrosis Found, 73- Mem: Am Math Soc; Inst Math Statist; Soc Actuaries. Res: Actuarial theory. Mailing Add: Dept of Math Univ of Mich Ann Arbor MI 48104

NESBITT, HERBERT HUGH JOHN, b Ottawa, Ont, Feb 7, 13; m 44; c 4. ENTOMOLOGY. Educ: Queen's Univ, Ont, BA, 37; Univ Toronto, MA, 39, PhD(invert zool), 44; Univ Leiden, DSc, 51. Prof Exp: Agr scientist, Div Entom, Dept Agr, Ont, 39-48; from asst prof to assoc prof, 48-56, dir div sci, 60-63, dean fac sci, 63-74, PROF BIOL, CARLETON UNIV, 56-, CLERK SENATE, 75- Mem: Soc Syst Zool; fel Royal Entom Soc London; Zool Soc London; fel Linnean Soc London; fel Entom Soc Can. Res: Nervous system of insects; comparative morphological and taxonomic work on Acari. Mailing Add: Fac of Sci Tory Bldg Carleton Univ Colonel By Dr Ottawa ON Can

NESBITT, LYLE EDWIN, b Washington, Iowa, Sept 5, 29; m 55; c 4. PHYSICAL CHEMISTRY, INORGANIC CHEMISTRY. Educ: Univ Iowa, BS, 50, MS, 52; Tex Tech Col, PhD(chem), 64. Prof Exp: Res asst chem, AEC Ames Lab, Iowa State Univ, 52-54; res chemist, Dow Chem Co, Okla, 54-59; lit chemist, Am Agr Chem Co, 59; fel chem, Tex Tech Col, 59-60; res chemist, Duval Corp, 60-65; asst prof, 65-68, ASSOC PROF CHEM, SOUTHERN COLO STATE COL, 68- Mem: Am Inst Mining, Metall & Petrol Eng; Am Chem Soc. Res: Surfactants; explosives; fused salts; mineralogy; complexes; metallurgy of ores; plastic coating; colloids; thermodynamics; thermal analysis. Mailing Add: 2001 Commanche Rd Pueblo CO 81001

NESBITT, PAUL HOMER, b Savanna, Ill, Aug 15, 04; m 54; c 2. ANTHROPOLOGY. Educ: Beloit Col, BA, 26; Univ Chicago, MA, 28, PhD(anthrop), 38. Prof Exp: From asst prof to prof anthrop, Beloit Col, 30-45, cur mus, 32-45; tech dir, Nat Mus Guatemala, 46-48; dir Arctic Desert, Tropic Ctr, Air Univ, 48-67; prof & chmn dept, 67-74, EMER PROF ANTHROP, UNIV ALA, 74- Concurrent Pos: Rockefeller Found grant, Mus Archaeol & Ethnol, Guatemala, 46-48; mem res & develop bd, Arctic Panel, US Dept Defense, 49-54; mem, Orinoco Exped, Venezuela Govt, 50. Mem: AAAS; Am Anthrop Asn (secy-treas, 40-43); Soc Am Archaeol; Asn Phys Anthrop; Arctic Inst NAm. Res: The evolution of man, physically and culturally; traditional cultures and the impact ot technological change; the rise and fall of the Mesoamerican precolumbian civilizations. Mailing Add: 9-1 Northwood Lake Northport AL 35476

NESBITT, ROBERT EDWARD LEE, JR, b Albany, Ga, Aug 21, 24; m 47; c 2. OBSTETRICS & GYNECOLOGY. Educ: Vanderbilt Univ, BA, 44, MD, 47; Am Bd Obstet & Gynec, dipl, 56. Prof Exp: Asst instr, Johns Hopkins Hosp, 48-52; obstetrician & gynecologist in chief, US Army Hosp, Ger, 52-54; asst prof, Sch Med, Johns Hopkins Univ, 54-56; prof & chmn dept, Albany Med Col, 58-61; PROF OBSTET & GYNEC & CHMN DEPT, STATE UNIV NY UPSTATE MED CTR, 61- Concurrent Pos: Fel, Am Asn Maternal & Child Health Dirs; obstetrician & gynecologist in chief, Albany Hosp, 56-61 & Crouse-Irving Hosp, 63-70; mem, Pub Health Coun State NY, 62-67; chief obstet & gynec, State Univ Hosp, 64-; assoc examr, Am Bd Obstet & Gynec. Mem: Soc Gynec Invest; Pan-Am Med Asn; fel Am Col Obstet & Gynec; fel Am Col Surg; NY Acad Sci. Res: Cytologic, cytochemical and histochemical study and diagnosis of early cervical cancer; prenatal and placental pathology; cytohormonal diagnosis; experimental production of abruptio placenta; reproductive endocrinology; animal experimentation; hormonal influence on placentation; fetal anoxia; immunoglobulin patterns in normal and toxemic pregnancy; in vitro placental perfusion studies; infertility. Mailing Add: State Univ NY Upstate Med Ctr 750 E Adams St Syracuse NY 13210

NESBITT, STUART STONER, b Aledo, Ill, Jan 29, 21; m 51; c 2. CHEMISTRY. Educ: Monmouth Col, BS, 43; Univ Tex, MA, 44, PhD(chem), 49. Prof Exp: Lab asst, Monmouth Col, 42-43; lab instr, Univ Tex, 43-44, asst, 46-49; res chemist, Mid-Continent Petrol Corp, 49-55; supvr prod res, Sunray D-X Oil Co, 55-69, mgr fuels & indust prod res, DX Div, 69-70, chief prod develop, Appl Res Dept, Tulsa Lab, 70-71, mgr, 71-72, CHIEF PROD SERV, TULSA REFINERY LAB, SUN OIL CO, 72- Mem: Am Chem Soc. Res: Alkoxyacetaldehydes; allylic chlorides; petroleum. Mailing Add: Tulsa Refinery Lab Sun Oil Co PO Box 2039 Tulsa OK 74102

NESBITT, WILLIAM BELTON, b Tryon, NC, Mar 22, 32; m 64; c 1. HORTICULTURE, PLANT BREEDING. Educ: NC State Univ, BS, 54, MS, 62; Rutgers Univ, PhD(hort, genetics), 65. Prof Exp: ASSOC PROF HORT, NC STATE UNIV, 65- Mem: Am Soc Hort Sci. Res: Breeding and production of grapes. Mailing Add: Dept of Hort Sci NC State Univ Raleigh NC 27607

NESHEIM, MALDEN CHARLES, b Rochelle, Ill, Dec 19, 31; m 52; c 3. NUTRITION. Educ: Univ Ill, BS, 53, MS, 54; Cornell Univ, PhD, 59. Prof Exp: Asst animal sci, Univ Ill, 53-54; asst poultry husbandry, 56-59, asst prof animal nutrit, 59-64, from assoc prof to prof, 64-74, DIR DIV NUTRIT SCI, CORNELL UNIV, 74- Concurrent Pos: NIH spec fel, Cambridge Univ, 72-73, overseas fel, Churchill Col, 72-73. Mem: Am Inst Nutrit; Brit Nutrit Soc; Am Soc Animal Sci; Poultry Sci Asn. Res: Amino acid metabolism; amino acid and protein requirements; gastrointestinal physiology. Mailing Add: Div of Nutrit Sci Savage Hall Cornell Univ Ithaca NY 14853

NESHEIM, OLAF NORMAN, plant pathology, see 12th edition

NESHEIM, ROBERT OLAF, b Monroe Center, Ill, Sept 13, 21; m 46; c 3. NUTRITION. Educ: Univ Ill, BS, 43, MS, 50, PhD(animal nutrit), 51. Prof Exp: Swine res specialist, Gen Mills, Inc, 51-52; head swine feed res, Quaker Oats Co, 52-59, mgr livestock feed res, 59-64; prof animal sci & head dept, Univ Ill, 64-67; assoc dir res feed & pet food, 67-69, dir res & develop, 69, VPRES RES & DEVELOP, QUAKER OATS CO, 69- Concurrent Pos: Chmn, Animal Nutrit Res Coun, 65-66; mem US nat comt, Int Union Nutrit Sci, 68-74; mem food industs adv comt, Nutrit Found, chmn planning panel, Food & Nutrit Liason Comt; mem adv comt, US Meat Animal Res Ctr, USDA, 71-75; mem exec comt food & nutrit bd, Nat Acad Sci; mem exec comt; mem, Comt Food Sci & Technol; mem food indust liaison comt, Am Med Asn; mem food adv comt, Off Technol Assessment, US Cong, 75- Mem: Am Soc Animal Sci; Am Inst Nutrit; Fedn Am Soc Exp Biol (treas); Soc Nutrit Educ; Am Pub Health Asn. Res: Amino acid nutrition; energy utilization. Mailing Add: Quaker Oats Co 617 W Main St Barrington IL 60010

NESHEIM, STANLEY, b Chicago, Ill, Apr 24, 30; m 58; c 4. ANALYTICAL CHEMISTRY. Educ: Brooklyn Col, BS, 56; George Washington Univ, MS, 62. Prof Exp: Chemist, US Bur Mines, 56-57; RES CHEMIST, US FOOD & DRUG ADMIN, 57- Mem: AAAS; Am Chem Soc; Am Oil Chem Soc. Res: Edible fats and oil characterization and chemical analysis; food contaminants; mycotoxin chemistry and analysis. Mailing Add: 3008 Tennyson St NW Washington DC 20015

NESHYBA, STEVE, b Jourdanton, Tex, Oct 8, 27; m 50; c 6. PHYSICAL OCEANOGRAPHY. Educ: Univ Tex, BSEE, 49, MS, 54; Tex A&M Univ, PhD(phys oceanog), 65. Prof Exp: Res engr, Elec Eng Res Lab, Univ Tex, 53-54; aerophys engr, Gen Dynamics/Convair, 54-57, sr aerophys engr, 57-60; asst prof elec eng, Arlington State Col, 60-62; fel oceanog, Tex A&M Univ, 62-64; ASSOC PROF PHYS OCEANOG, ORE STATE UNIV, 65- Res: Deep sea hydrography; analyses of time-dependent motions of water bodies. Mailing Add: Dept of Oceanog Ore State Univ Corvallis OR 97331

NESMITH JAMES, soil chemistry, plant nutrition, see 12th edition

NESS, ARTHUR THOMAS, b Tacoma, Wash, Jan 5, 09; m 35. ANALYTICAL CHEMISTRY. Educ: Univ Wash, Seattle, BS, 31, PhD(chem), 37. Prof Exp: Jr chemist, US Geol Surv, 37-39; asst chemist, US Bur Mines, 39-40; from asst chemist to assoc chemist, 40-45, from chemist to sr chemist, 45-53, anal chemist, 53-67, RES CHEMIST, NIH, 67- Mem: Am Chem Soc. Res: Physical chemistry of solutions; chemistry of carbohydrate substances; blood and bloodforming tissues; analytical chemistry of body tissues and fluids; clinical chemistry. Mailing Add: Nat Cancer Inst Bethesda MD 20014

NESS, LINDA ANN, b Albert Lea, Minn, Oct 29, 47. MATHEMATICS. Educ: St Olaf Col, BA, 69; Harvard Univ, MA & PhD(math), 73. Prof Exp: ASST PROF MATH, UNIV WASH, 75- Res: Algebraic and differential geometry. Mailing Add: Dept of Math Univ of Wash Seattle WA 98195

NESS, NORMAN FREDERICK, b Springfield, Mass, Apr 15, 33; m 56; c 2. SPACE PHYSICS. Educ: Mass Inst Technol, BS, 55, PhD(geophys), 59. Prof Exp: Res geophysicist inst geophys, Univ Calif, Los Angeles, 59-60, asst prof geophys, 60-61; res physicist, 61-66, staff scientist, 66-68, head extraterrestrial physics br, 68-69, CHIEF LAB EXTRATERRESTRIAL PHYSICS, GODDARD SPACE FLIGHT CTR, NASA, 69- Concurrent Pos: Nat Acad Sci-Nat Res Coun res assoc, 60-61; consult, 57-; vis assoc prof, Univ Md, 65-68. Honors & Awards: Exceptional Sci Achievement Medal, NASA, 66; Flemming Medal, US Govt, 68; Space Sci Award, Am Inst Aeronaut & Astronaut, 72; John Adam Fleming Medal, Am Geophys Union, 65. Mem: AAAS; Soc Explor Geophys; Asn Comput Mach; fel Am Geophys Union. Res: Experimental investigation of magnetic fields in the magnetosphere and interplanetary space; satellite and space probe studies; measurement of planetary magnetic fields. Mailing Add: NASA Goddard Space Flight Ctr Greenbelt MD 20771

NESS, ROBERT KIRACOFE, b York, Pa, Apr 29, 22; m 44; c 3. CARBOHYDRATE CHEMISTRY. Educ: Lebanon Valley Col, BS, 43; Ohio State Univ, MS, 45, PhD(org chem), 48. Prof Exp: Asst, Ohio State Univ, 43-45; assoc prof, Lebanon Valley Col, 47-48; fel, 48-50, sr asst scientist, 50-54, from scientist to sr scientist, 54-63, SCIENTIST DIR, NIH, 63- Concurrent Pos: USPHS, 50- Mem: AAAS; Am Chem Soc; NY Acad Sci; The Chem Soc. Res: Carbohydrates; reaction mechanisms; sugar benzoates; chemistry of ribose and deoxyribose; synthesis of deoxynucleosides; glycals; vinyl glycosides; amino sugars. Mailing Add: Lab of Chem Nat Inst Arthritis Metab & Digest Dis NIH Bethesda MD 20014

NESSEL, ROBERT J, b Bronx, NY, Dec 28, 36; m 60; c 2. PHARMACY, PHARMACEUTICAL CHEMISTRY. Educ: Columbia Univ, BS & MS, 60; Purdue Univ, PhD(indust pharm), 63. Prof Exp: Pharmacist, Mt Sinai Hosp, NY, 58-59; asst pharm, Purdue Univ, 60-61; sr res pharmacist, Squibb Inst Med Res, 63-66; head med prod tech serv, Merck & Co, 66-69, head animal formulations res & develop, 69-74, assoc dir regulatory affairs, Int, 74-75, DIR REGULATORY AFFAIRS, INT, MERCK, SHARP & DOHME RES LABS, 75- Mem: Am Pharmaceut Asn; Acad Pharmaceut Sci. Res: Industrial pharmacy. Mailing Add: Merck Sharp & Dohme Res Labs Rahway NJ 07065

NESTE, SHERMAN LESTER, b Decorah, Iowa, Sept 23, 43; m 72; c 1. METEORITICS. Educ: Luther Col, BA, 65; Mich State Univ, MS, 67; Drexel Univ, PhD(physics), 75. Prof Exp: RES PHYSICIST METEORITICS, SPACE SCI LAB, GEN ELEC CO, 67- Mem: AAAS; Am Geophys Union. Res: Establishing an experimental model of the asteroid/meteoroid environment in the region of space between the orbits of Earth and Jupiter. Mailing Add: Space Sci Lab Gen Elec Co PO Box 8555 Philadelphia PA 19101

NESTELL, , MERLYND KEITH, b Fletcher, NC, Oct 27, 37; m 57; c 2. MATHEMATICS. Educ: Emmanuel Missionary Col, Andrews, BA, 57; Univ Wis, MA, 59; Ore State Univ, PhD(math), 66. Prof Exp: Instr math, Southern Missionary Col, 59-61; instr, Ore State Univ, 63-65; sr res scientist, Pac Northwest Lab, Battelle Mem Inst, 65-69; ASST PROF MATH & GEOL, UNIV TEX, ARLINGTON, 69- Mem: Am Math Soc; Math Asn Am; Soc Econ Paleont & Mineral. Res: Functional analysis; integral and integro-differential equations, radiative transfer theory; paleontology; applied mathematics. Mailing Add: Dept of Math Univ of Tex Arlington TX 76019

NESTER, EUGENE WILLIAM, b Johnson City, NY, Sept 15, 30; m 59; c 2. MICROBIOLOGY. Educ: Cornell Univ, BS, 52; Western Reserve Univ, PhD, 59.

Prof Exp: Am Cancer Soc res fel genetics, Sch Med, Stanford Univ, 59-62; instr microbiol, 62-63, from asst prof to assoc prof microbiol & genetics, 63-72, PROF MICROBIOL & GENETICS, UNIV WASH, 72- Mem: Am Soc Microbiol. Res: Genetics and biochemistry of enzyme regulation; bacterial-plant relationships. Mailing Add: Dept of Microbiol Univ of Wash Seattle WA 98105

NESTLER, F H MAX, b Pittsburgh, Pa, Mar 11, 17; m 44; c 3. PHYSICAL CHEMISTRY. Educ: Univ Pittsburgh, BS, 40; Yale Univ, PhD(phys chem), 43. Prof Exp: Asst, Yale Univ, 42; res chemist, SAM Labs, Columbia Univ, 43-44; sr res chemist, Carbide & Carbon Chem Co, 46; Nutrit Found, Inc fel, Yale Univ, 46-48; res chemist, US Naval Res Lab, 48-61; RES CHEMIST FOREST PROD LAB, US FOREST SERV, 61- Mem: AAAS; Am Chem Soc. Res: Electrochemistry; gas adsorption and chromatography; gaseous corrosion of metals; mass spectrometry. Mailing Add: Forest Prod Lab US Forest Serv Madison WI 53705

NESTOR, JAMES F, physical chemistry, see 12th edition

NESTOR, KARL ELWOOD, b Kasson, WVa, Dec 17, 37; m 58; c 2. GENETICS. Educ: Univ WVa, BS, 59, MS, 61; Ohio State Univ, PhD(genetics), 64. Prof Exp: Asst genetics, Univ WVa, 59-61; asst, 61-64, asst instr, 65, asst prof, 65-71, ASSOC PROF GENETICS, OHIO AGR RES & DEVELOP CTR, 71- Mem: Poultry Sci Asn. Res: Physiological genetics of chickens and turkeys; poultry physiology. Mailing Add: Dept of Poultry Sci Ohio Agr Res & Develop Ctr Wooster OH 44691

NESTOR, ONTARIO HORIA, b Youngstown, Ohio, Sept 20, 22; m 43; c 2. PHYSICS. Educ: Marietta Col, AB, 43; Univ Minn, MS, 49; Univ Buffalo, PhD(physics), 60. Prof Exp: Instr physics, Marietta Col, 42-43; asst, SAM Labs, Columbia Univ, 43-44; asst, Univ Minn, 47-48; res assoc, Linde Div, Union Carbide Corp, 49-68, sr res assoc, 68-71; dir technol, Crystal Optics Res, Inc, 71-75; MEM STAFF, HAWSHAW CHEM CO, 75- Mem: Am Phys Soc; Optical Soc Am. Res: Crystal growth and properties. Mailing Add: Hawshaw Chem Co 6801 Cochran Rd Solon OH 44139

NESTVOLD, ELWOOD OLAF, b Minot, NDak, Mar 19, 32; m 55; c 2. INFORMATION SCIENCE, GEOPHYSICS. Educ: Augsburg Col, BA, 52; Univ Minn, MS, 59, PhD(physics), 62. Prof Exp: Asst physics, Univ Minn, 56-59, instr, 59-61; physicist, Shell Develop Co, 62-63, res physicist, 63-65, res assoc & sect head, 65-68, mgr geophys dept, 68-71, sr staff geophysicist, 71-75, MGR GEOPHYS, INT VENTURES, SHELL OIL CO, 75- Concurrent Pos: Consult, Lighting & Transients Res Inst, Minn, 57-62. Mem: Inst Elec & Electronics Engrs; Soc Explor Geophys. Res: Prediction and filter theory; digital-computer techniques for acoustic and seismic signal processing. Mailing Add: Shell Oil Co Int Ventures Box 2099 Houston TX 77001

NESTY, GLENN ALBERT, b Muncie, Ind, Dec 23, 11; m 36; c 2. ORGANIC CHEMISTRY. Educ: DePauw Univ, AB, 34; Univ Ill, PhD(org chem), 37. Prof Exp: Researcher, Solvay Process Co, 37-43; group leader polymerization res, Allied Chem & Dye Corp, 43-48, from asst dir to assoc dir cent res lab, 48-55, vpres, Allied Chem Corp, 55-57, dir, 57-69; VPRES RES, INT PAPER CO, TUXEDO PARK, NY, 69- Concurrent Pos: Trustee, Textile Res Inst. Honors & Awards: Com Chem Develop Asn Award, 64. Mem: AAAS; Am Chem Soc; fel Am Inst Chem; NY Acad Sci; Tech Asn Pulp & Paper Indust. Res: Structure of vegetable oils; reactions of sterols; cyclization od dieneynes; vapor phase oxidations; pyrolysis; reactions of olefins; polymerization; properties of plastics; synthetic fibers; pulp and paper; nonwoven fabrics; surgical devices. Mailing Add: Int Paper Co Corp Res Ctr Tuxedo Park NY 10987

NETA, PEDATSUR, b Tripoli, Libya, Jan 1, 38; m 59; c 2. CHEMISTRY. Educ: Hebrew Univ, Jerusalem, MSc, 60; Weizmann Inst Sci, PhD(phys chem), 65. Prof Exp: Res assoc radiation chem, Soreq Nuclear Res Ctr, Israel, 60-66; Nat Res Coun Can fel, Univ Toronto, 66-67; AEC fel, Ohio State Univ, 67-68; Nat Acad Sci-Nat Res Coun fel, US Army Natick Labs, 68-69; res fel, 69-74, SR RES CHEMIST RADIATION CHEM, RADIATION RES LABS, MELLON INST SCI, CARNEGIE-MELLON UNIV, 74- Concurrent Pos: Assoc ed, Radiation Res, 75-78. Mem: Am Chem Soc; Sigma Xi; Radiation Res Soc. Res: Physical organic chemistry; free radical reactions; radiation chemistry of aqueous systems; electron spin resonance; study of radicals in solution. Mailing Add: Mellon Inst of Sci Carnegie-Mellon Univ Pittsburgh PA 15213

NETER, ERWIN, b Mannheim, Ger, May 26, 09; nat US; m 45; c 1. BACTERIOLOGY, IMMUNOLOGY. Educ: Univ Heidelberg, MD, 33; Am Bd Path, dipl, 55. Hon Degrees: MD, Univ Heidelberg, 74. Prof Exp: From assoc bact & immunol to instr, Sch Med, 36-53, from bacteriologist to chief res microbiol, Children's Hosp, 36-38, consult, Inst, 37-68, from asst prof to assoc prof microbiol, Sch Nursing, 46-68, from asst prof to assoc prof bact, immunol & pediat, Sch Med, 53-56, prof clin microbiol, Dept Pediat, 56-68, PROF MICROBIOL & PEDIAT, SCH MED & DIR BACT, CHILDREN'S HOSP & DEPT MICROBIOL, ROSWELL PARK MEM INST, STATE UNIV NY BUFFALO, 68- Mem: AAAS; Am Soc Microbiol; Soc Exp Biol & Med; Am Asn Immunologists; Am Pub Health Asn. Res: Chemotherapy; Escherichia. Mailing Add: Children's Hosp 219 Bryant St Buffalo NY 14222

NETER, JOHN, b Ger, Feb 8, 23; US citizen; m 51; c 2. APPLIED STATISTICS. Educ: Univ Buffalo, BA, 43; Univ Pa, MBA, 47; Columbia Univ, PhD(bus statist), 52. Prof Exp: Asst prof bus statist, Syracuse Univ, 49-55; prof quant anal, Univ Minn, Minneapolis, 55-75; PROF QUANT METHODS, COL BUS ADMIN, UNIV GA, 75- Concurrent Pos: Ford Found fac res fel, Univ Minn, 57-58; supvry math statistician, US Bur Census, 59-60, consult, 61-65. Mem: Fel AAAS; fel Am Statist Asn; Inst Mgt Sci; Inst Math Statist; Am Inst Decision Sci. Res: Uses of statistics in accounting; measurement errors; decision theory. Mailing Add: Col of Bus Admin Univ of Ga Athens GA 30602

NETHAWAY, DAVID ROBERT, b San Diego, Calif, Aug 6, 29; m 64; c 3. NUCLEAR CHEMISTRY. Educ: Univ Calif, BS, 51, MS, 57; Wash Univ, PhD(chem), 59. Prof Exp: Chemist, Gen Elec Co, Wash, 51-53; chemist, Calif Res & Develop Co, 53; CHEMIST, LAWRENCE LIVERMORE LAB, UNIV CALIF, 53- Mem: Am Chem Soc; Am Phys Soc. Res: Nuclear charge distribution in fission; nuclear reactions; decay scheme studies. Mailing Add: 174 Marks Rd Danville CA 94526

NETHERCOT, ARTHUR HOBART, JR, b Evanston, Ill, June 16, 23; m 44; c 4. PHYSICS. Educ: Northwestern Univ, BA, 44, MS, 46; Univ Mich, PhD, 50. Prof Exp: Union Carbide & Carbon Corp fel, Columbia Univ, 51-52, res assoc physics, 52-57; PHYSICIST RES CTR, IBM CORP, 57- Mem: Fel Am Phys Soc. Res: Microwave physics; solid state physics; materials science. Mailing Add: 107 Mt Hope Blvd Hastings NY 10706

NETHERCUT, PHILIP EDWIN, b Indianapolis, Ind, Apr 3, 21; m 49; c 3. ORGANIC CHEMISTRY. Educ: Beloit Col, BS, 42; Lawrence Col, MS, 44, PhD(org chem), 49. Prof Exp: Res chemist, Watervliet Paper Co, Mich, 49-50; process control engr, Scott Paper Co, Pa, 51, res group leader, 52-54, res mgr, 55-56; tech secy, 57-58, exec secy, 59-74, treas, 63-74, EXEC DIR & PUBL, TECH ASN PULP & PAPER INDUST, 75- Concurrent Pos: Pres, Coun Eng & Sci Soc Execs, 69. Mem: Fel Tech Asn Pulp & Paper Industs; Am Soc Asn Execs. Res: Pulp and paper technology; industrial research administration. Mailing Add: Tech Asn Pulp &Paper Indust 1 Dunwoody Park Atlanta GA 30341

NETHERTON, LOWELL EDWIN, b Fairfield, Ill, Feb 9, 22; m 46; c 4. INORGANIC CHEMISTRY, PHYSICAL CHEMISTRY. Educ: Univ Western Ill, BS, 44; Univ Wis, PhD(chem), 50. Prof Exp: Anal chemist, Sinclair Refining Co, 44; asst, Univ Wis, 48-50; res chemist, Victor Chem Co, 50-57, dir res, Victor Chem Div, Stauffer Chem Co, 57-63, mgr prod, 63-65; DIR RES, BASF WYANDOTTE CORP, 65- Mem: Am Chem Soc; Soc Plastics Indust; Indust Res Inst; Am Inst Chem Eng; Am Inst Chem. Res: Phosphorus compounds; electrochemistry of rare elements; chemistry of rhenium, tungsten and tantalum; chlor-alkali electrochemistry; urethanes, polyethers, isocyanates, alkyleneoxides and auxliaries; foamed plastics; textile and paper specialties; surfactants and detergents. Mailing Add: Cent Res Div BASF Wyandotte Corp Wyandotte MI 48192

NETHERY ARTHUR ALAN, plant physiology, cytology, see 12th edition

NETHERY, SIDNEY J, b Elkmont, Ala, May 4, 35; m 66. PHYSICS. Educ: Univ Denver, BS, 62, PhD(physics), 68. Prof Exp: Res physicist, Res Inst, Univ Denver, 68; opers analyst, Hq, US Air Force, Washington, DC, 68-71, physicist, US Environ Protection Agency, 71-75; PHYSICIST, DEPT HEALTH, EDUC & WELFARE, 75- Mem: Am Phys Soc; Acoust Soc Am. Res: Acoustics. Mailing Add: Dept of HEW 3300 Independence Ave SW Washington DC 20201

NETI, RADHAKRISHNA MURTY, b Nandigama, India, June 20, 33; m 60; c 1. PHYSICAL CHEMISTRY. Educ: Hindu Col, Masulipatam, India, BS, 53; Banaras Hindu Univ, MS, 55, dipl mod Europ lang, 56, PhD(chem), 60. Prof Exp: Lectr chem, Banaras Hindu Univ, 60; res assoc radiation chem, Univ Notre Dame, 60-62; res assoc bot & biophys, Univ Ill, 62-66; RES SCIENTIST, BECKMAN INSTRUMENTS, INC, 66- Honors & Awards: John C Vaaler Award, 70. Mem: AAAS; Am Chem Soc. Res: Photo, radiation and electro-analytical chemistry; air and water quality instrumentation. Mailing Add: Dept 762 Beckman Instruments Inc 2500 Harbor Blvd Fullerton CA 92634

NETSKY, MARTIN GEORGE, b Philadelphia, Pa, May 15, 17; m 46. NEUROLOGY, NEUROPATHOLOGY. Educ: Univ Pa, AB, 38, MS, 40, MD, 43. Prof Exp: From intern to resident neurol, Hosp Univ Pa, 43-44; Weil fel neuropath, Montefiore Hosp, New York, 46-47; from asst neuropathologist to assoc neuropathologist, Montefiore Hosp, 47-54; from assoc prof to prof neurol, Bowman Gray Sch Med, 55-61, prof neuropath, 55-61, dir sect neurol, 57-59, chmn dept, 59-61; vis prof path, Univ Med Sci, Thailand, 61; prof neuropath, Sch Med, Univ Va, 62-75; PROF PATH, SCH MED, VANDERBILT UNIV, 74- Concurrent Pos: Lectr, US Naval Hosp, St Albans, NY, 48, consult pathologist, 50-54; from adj attend physician to assoc attend physician, Montefiore Hosp, 49-54; assoc, Col Physicians & Surgeons, Columbia Univ, 52-54; mem sci adv bd, Armed Forces Inst Path. Mem: Am Neurol Asn; Am Asn Neuropath (pres, 63); Asn Res Nerv & Ment Dis; Am Acad Neurol; Am Soc Exp Path. Res: Permeability of living membranes; autonomic nervous system; clinicopathologic aspects of human brain tumors; congenital and degenerative neurologic disorders; medical education; learning. Mailing Add: Dept of Path Vanderbilt Univ Sch of Med Nashville TN 37232

NETTEL, STEPHEN J E, b Prague, Czech, Aug 12, 32; US citizen. SOLID STATE PHYSICS. Educ: McGill Univ, BEng, 54; Mass Inst Technol, PhD(physics), 60. Prof Exp: Jr res physicst, Univ Calif, San Diego, 60-61; staff mem, Int Bus Mach Corp, Switz, 61-65; res fel appl physics, Harvard Univ, 65-66; asst prof physics, 66-69, ASSOC PROF PHYSICS, RENSSELAER POLYTECH INST, 69- Res: Theoretical solid state physics. Mailing Add: Dept of Physics Rensselaer Polytech Inst Troy NY 12181

NETTERVILLE, JOHN T, b Nashville, Tenn, Oct 26, 30; m 51; c 4. ANALYTICAL CHEMISTRY. Educ: David Lipscomb Col, BS, 51; George Peabody Col, MA, 57; Univ Miss, MS, 61; Vanderbilt Univ, PhD(anal chem), 64. Prof Exp: Instr chem, David Lipscomb Col, 51-53; sci teacher high sch, Tenn, 53-55, head dept sci, 56-59; teacher pub schs, Alaska, 55-56; asst prof chem 59-65, actg chmn dept, 61-64, PROF CHEM, DAVID LIPSCOMB COL, 65-, CHMN DEPT, 64- Concurrent Pos: Chem ed, J Tenn Acad Sci, 73- Mem: Am Chem Soc; Nat Sci Teachers Asn. Res: Electrochemistry of non-aqueous systems; electrochemical experiments in chemical education. Mailing Add: Dept of Chem David Lipscomb Col Nashville TN 37203

NETTESHEIM, PAUL, b Cologne, Ger, Sept 11, 33; US citizen; m 61; c 3. CANCER. Educ: Med Sch, Bonn, WGer, MD & DMS, 59; Univ Pa, MS, 64. Prof Exp: Res biologist cancer, 63-69, GROUP LEADER CANCER, BIOL DIV, OAK RIDGE NAT LAB, 69- Concurrent Pos: Mem rev comt, Nat Cancer Inst, 71-74, Nat Heart & Lung Inst, 74, Energy Res & Develop Admin, 75 & sci rev comt, ETenn Cancer Res Ctr, 75; lectr, Univ Tenn-Oak Ridge Grad Sch Biomed Sci, 71-; mem ad hoc comt on 210 PO & smoking, Nat Cancer Inst, 75. Mem: Int Asn Study Lung Cancer; Am Soc Exp Path; Am Asn Cancer Res; Am Asn Immunologists; Am Soc Exp Biol & Med. Res: Chemical carcinogenesis and cocarcinogenesis particularly of the respiratory tract; factors affecting expression of neoplastic transformation; biological markers of preneoplasia in epithelial tissues; pulmonary toxicology and pathology. Mailing Add: Biol Div Oak Ridge Nat Lab PO Box Y Oak Ridge TN 37830

NETTING, MORRIS GRAHAM, b Wilkinsburg, Pa, Oct 3, 04; m 30; c 2. HERPETOLOGY. Educ: Univ Pittsburgh, BS, 26; Univ Mich, AM, 29; Waynesburg Col, ScD, 50. Prof Exp: Asst herpet, 25-28, asst cur, 28-31, cur, 31-54, asst dir, 49-53, actg dir, 53-54, DIR HERPET, CARNEGIE MUS, 54- Concurrent Pos: Leader, Carnegie Mus Exped, Venezuela, 29-30; asst prof, Univ Pittsburgh, 44-63. Mem: AAAS; Am Soc Ichthyologists & Herpetologists (vpres, 30, secy, 31-48, pres, 49-); Am Soc Mammalogists; Ecol Soc Am. Res: Zoogeography; geography; museology; conservation. Mailing Add: Carnegie Mus 4440 Forbes Ave Pittsburgh PA 15213

NETTLES, JOHN BARNWELL, b Dover, NC, May 19, 22; m 56; c 3. OBSTETRICS & GYNECOLOGY. Educ: Univ SC, BS, 41; Med Col SC, MD, 44. Prof Exp: Intern gen med, Garfield Mem Hosp, DC, 44-45; res fel path, Med Col Ga, 46-47; res staff obstet & gynec, Univ Ill Hosps, 47-51, from instr to asst prof, Univ, 51-57; asst prof, Sch Med, Univ Ark, 57-60, assoc prof & med educ nat defense coordr, 60-67, prof obstet & gynec, 67-69; PROF OBSTET & GYNEC, COL MED, UNIV OKLA, 69-, CHMN DEPT, 75- Concurrent Pos: Dir, Tulsa Residency Training Prog Obstet & Gynec & Tulsa Obstet & Gynec Educ Found, 69- Mem: AAAS; AMA; Asn Mil Surgeons US; Am Col Obstet & Gynec; Int Soc Advan Humanistic Studies Gynec. Res: Renal function and structure; kidney biopsy; newborn and fetal morbidity and mortality; genital malignancy; obstetric anesthesia and analgesia; physiology and toxemia of pregnancy. Mailing Add: 1120 S Utica Tulsa OK 74104

NETTLES, VICTOR FLEETWOOD, b Palmetto, Fla, Nov 1, 14; m 41; c 5. VEGETABLE CROPS. Educ: Univ Fla, BSA, 36, MSA, 38; Cornell Univ, PhD, 49. Prof Exp: From asst horticulturist to assoc horticulturist, 38-60, horticulturist & prof veg crops, 60-75, EMER HORTICULTURIST & EMER PROF VEG CROPS, AGR EXP STA, UNIV FLA, 75- Mem: Fel Am Soc Hort Sci. Res: Irrigation of vegetables; machine placement of fertilizers; plant protection and mulching studies; seeding of vegetables; varietal studies. Mailing Add: 1822 NW 6th Ave Gainesville FL 32603

NETTLES, WILLIAM CARL, b Sumter, SC, Aug 11, 07; m 38; c 3. ECONOMIC ENTOMOLOGY. Educ: Clemson Col, BS, 30; Ohio State Univ, MS, 32. Prof Exp: Asst entomologist exp sta, 30-34, exten entomologist, 34-46, leader exten entom & plant dis, 46-65, state chem info leader, 65-68, exten specialist entom & plant path, 68-69, exten prin specialist, 69-72, EMER ASSOC PROF ENTOM, CLEMSON UNIV, 72- Mem: Entom Soc Am (secy, 36). Res: Biology and control of insects and plant diseases. Mailing Add: 119 Folger St Clemson SC 29631

NETTLES, WILLIAM CARL, JR, b Anderson, SC, Dec 21, 34; m 67. ENTOMOLOGY. Educ: Clemson Univ, BS, 55, MS, 59; Rutgers Univ, PhD(entom), 62. Prof Exp: Res asst agr chem, Clemson Univ, 57-59; RES ENTOMOLOGIST, COTTON INSECTS BR, AGR RES SERV, USDA, 62- Mem: AAAS; Entom Soc Am. Res: Toxicology, physiology and biochemistry of insects. Mailing Add: 4115 Gourrier Ave Baton Rouge LA 70808

NETTLESHIP, ANDERSON, b Fayetteville, Ark, Oct 10, 10; m 52; c 3. PATHOLOGY. Educ: Univ Ark, BSc, 31; Johns Hopkins Univ, MD, 35. Prof Exp: Instr path, Med Col, Cornell Univ, 36; Nat Res Coun fels, Schs Med, Duke Univ, 37 & Vanderbilt Univ, 38; asst, Mem Cancer Hosp, 39-41; asst surgeon, USPHS, 42-43; asst prof path, Univ Okla, 44-45; assoc prof, Ind Univ, 46; prof & head dept, Sch Med, Univ Ark, 47-57; SR RES ASSOC, ANTAEUS RES INST, 57- Concurrent Pos: Mem Markel Found Exped, Cent Am, 44. Mem: AAAS; Soc Exp Biol & Med; Am Soc Exp Path; Am Asn Path & Bact; Am Asn Cancer Res; Int Acad Path. Res: Cancer. Mailing Add: PO Box 817 Fayetteville AR 72701

NETTLETON, DONALD EDWARD, JR, b New Haven, Conn, Mar 16, 30; m 57; c 4. ORGANIC CHEMISTRY, BIOCHEMISTRY. Educ: Yale Univ, BS, 52; Rice Inst, PhD, 56. Prof Exp: Res assoc biochem, Med Col, Cornell Univ, 56-58; SR CHEMIST, DEPT BIOCHEM, BRISTOL LABS, INC DIV, BRISTOL-MYERS CO, 58- Mem: Am Chem Soc; Am Soc Pharmacog. Res: Isolation of antitumor and antibiotic agents from fermentation liquors and higher plants; syntheses of derivatives and analogs of physiologically active natural products. Mailing Add: RFD 1 Box 524 Jordan NY 13080

NETTLETON, HARRY ROLLETT, organic chemistry, see 12th edition

NETTLETON, WILEY DENNIS, b Noble, Ill, June 8, 32; m 56; c 4. SOIL SCIENCE, GEOLOGY. Educ: Univ Ill, BS, 57, MS, 58; NC State Univ, PhD(soil classification & genesis), 66. Prof Exp: Trainee, 56, soil scientist, 57 & 58-65, res soil scientist, Soil Surv Lab, 65-72, SUPVRY RES SOIL SCIENTIST, WESTERN SOIL SURV INVEST UNIT, SOIL CONSERV SERV, USDA, 72- Concurrent Pos: Res asst, NC State Univ, 60-65; lectr, Univ Calif, Riverside, 70. Mem: Soil Sci Soc Am; Am Soc Agron. Res: Use of mineralogy and micromorphology in the laboratory and geomorphology in the field as tools for studying genesis and classification of soils from the Midwest, Southeast and Western United States. Mailing Add: Soil Surv Lab Box 672 Riverside CA 92502

NETZEL, DANIEL ANTHONY, b Chicago, Ill, Feb 16, 34; m 58; c 3. PHYSICAL CHEMISTRY. Educ: Univ Ill, BS, 57; Univ Mo-Kansas City, MS, 61; Northwestern Univ, PhD(phys & anal chem), 75. Prof Exp: Assoc res chemist, Midwest Res Inst, 58-67; res chemist, DeSoto Inc, 67-69; supvr spectros, Northwestern Univ, 69-74; MEM STAFF RES SPECTROS, LARAMIE ENERGY RES CTR, 75- Mem: Am Chem Soc; Soc Appl Spectros; Res Soc Am; Sigma Xi. Res: Identification, structural characterization and analytical applications of nuclear magnetic and electron spin resonances to fossil fuels. Mailing Add: Laramie Energy Res Ctr PO Box 3395 Univ Sta Laramie WY 82071

NETZEL, RICHARD G, b Shawano, Wis, May 13, 28; m 51; c 3. PHYSICS, ACADEMIC ADMINISTRATION. Educ: Univ Wis, BS, 50, MS, 56, PhD(physics), 60. Prof Exp: From asst prof to prof physics, Univ Wis-Oshkosh, 60-68, asst vpres prog develop & staffing, 68-70, asst acad vpres, 70-71; actg dep dir educ, AAAS, 71-72; VPRES ACAD AFFAIRS, METROP STATE COL, 72- Mem: AAAS; Am Phys Soc; Am Asn Physics Teachers. Res: Low temperature physics; superconducting transition temperatures of titanium; science education. Mailing Add: Metrop State Col 250 WFourteenth Ave Denver CO 80204

NEU, ERNEST LUDWIG, b Frankfurt, Ger, Aug 19, 15; nat US; m 46; c 2. CHEMISTRY. Educ: Univ Nancy, BS, 37; Univ Caen, MS, 38. Prof Exp: Res chemist, Celotex Corp, 38-42; chief chemist dicalite & perlite, Great Lakes Carbon Corp, 52-59, asst tech dir, Mining & Mineral Prod Div, 59-62, tech dir, 62-66, managing dir, Dicalite Europe Nord, 59-73, gen mgr, Int Div, 66-73, VPRES, GREFCO, INC, 73-, MANAGING DIR, PERMALITE EUROPE, SA, 73- Mem: Am Chem Soc; Am Inst Chem Eng; NY Acad Sci. Res: Filteraids; diatomite; perlite; acoustical and thermal insulation; fungi diseases. Mailing Add: Grefco Inc 2340 Wilshire Blvd Los Angeles CA 90010

NEU, HAROLD CONRAD, b Omaha, Nebr, Aug 19, 34; m 62; c 3. MEDICINE, PHARMACOLOGY. Educ: Creighton Univ, AB, 56; Johns Hopkins Univ, MD, 60. Prof Exp: From intern to resident med, Columbia-Presby Med Ctr, 60-62; res assoc biochem, Nat Inst Arthritis & Metab Dis, 62-64; chief resident med, Columbia-Presby Med Ctr, 64-65; assoc, 65-66, from asst prof to assoc prof med, 66-75, PROF MED & PHARMACOL, COL PHYSICIANS & SURGEONS, COLUMBIA UNIV, 70-, HEAD DIV INFECTIOUS DIS, COLUMBIA-PRESBY MED CTR, 71-, ATTEND PHYSICIAN, 75- Concurrent Pos: Career scientist, New York Health Res Coun, 65-71; attend, Harlem Hosp, New York, 67; assoc attend, Columbia-Presby Med Ctr, 70-75. Honors & Awards: Borden Award, Borden Found, 60. Mem: Am Soc Microbiol; Am Soc Biol Chemists; Am Soc Clin Invest; Am Col Physicians. Res: Resistance of bacteria to antibiotics; bacterial surface enzymes. Mailing Add: Dept of Med Columbia Univ Col of Physicians & Surgeons New York NY 10032

NEU, JOHN TERNAY, b Commerce, Tex, Apr 23, 20; m 50; c 2. PHYSICAL CHEMISTRY. Educ: Agr & Mech Col, Tex, BS, 42; Univ Calif, Berkeley, PhD(phys chem), 49. Prof Exp: Instr chem, Univ Calif, Berkeley, 49; res chemist, Calif Res Corp, 49-56; sr staff scientist & chief optics technol, Gen Dynamics/Convair Aerospace, 56-75; PRIN PHYSICIST, INTELCOM RAD TECH, 75- Res: Space physics; optical properties of surfaces; space optical instrumentation; vehicle signatures; vehicle survivability. Mailing Add: 6360 Raydel Court San Diego CA 92120

NEUBAUER, BENEDICT FRANCIS, b Bird Island, Minn, Mar 14, 38; m 61; c 3. PLANT ANATOMY. Educ: St John's Univ, Minn, BA, 60; Iowa State Univ, PhD(plant anat), 65. Prof Exp: Asst prof bot & plant anat, 65-70, ASSOC PROF BOT, UNIV MAINE, ORONO, 70- Mem: Bot Soc Am. Res: Anatomy of vascular plants. Mailing Add: Dept of Bot & Plant Path Univ of Maine Orono ME 04473

NEUBAUER, WERNER GEORGE, b White Plains, NY, Apr 18, 30; m 54; c 2. PHYSICS. Educ: Roanoke Col, BS, 52; Cath Univ Am, PhD(acoust), 68. Prof Exp: Physicist, Electronics Br, Sound Div, 53-57, physicist, Propagation Br, 57-58, sect head microacoust sect, Acoust Res Br, 58-69, actg br head phys acoust br, Acoust Div, 69-70, SECT HEAD, MICRO-ACOUST SECT, PHYS ACOUST BR, ACOUST DIV, NAVAL RES LAB, 70- Mem: Acoust Soc Am. Res: Radiation, reflection and diffraction of waves; properties of elastic media; optical visualization of acoustic waves; underwater acoustic radiation, reflection and scattering. Mailing Add: Naval Res Lab Code 8132 4555 Overlook Ave Washington DC 20375

NEUBECK, CLIFFORD EDWARD, b Erie, Pa, Nov 6, 17; m 43; c 8. CHEMISTRY. Educ: Univ Pittsburgh, BS, 39, PhD(biochem), 43. Prof Exp: Asst chem, Univ Pittsburgh, 39-41, asst biochem, 41-43; biochemist, 43-50, sr chemist, 50-59, sr scientist, 59-71, SR BIOCHEMIST, ROHM AND HAAS CO, 71- Mem: Am Chem Soc; Soc Indust Microbiol; Am Soc Enologists. Res: Commercial production of enzymes; utilization of enzymes; microbial fermentations. Mailing Add: Lab 63 Rohm and Haas Co PO Box 219 Bristol PA 19007

NEUBECKER, ROBERT DUANE, b Lackawanna, NY, Oct 10, 25; m 50; c 5. PATHOLOGY. Educ: Univ Rochester, AB, 46, MD, 49; Am Bd Path, dipl, 56. Prof Exp: Intern med, Vanderbilt Univ Hosp, 49-50; intern path, Strong Mem Hosp, Rochester, NY, 50; asst, Sch Med & Dent, Univ Rochester, 52-54, instr, 54-55; pathologist, Armed Forces Inst Path, 55-57, chief obstet, gynec & breast path br, 57-61; pathologist & dir res & develop, St Joseph's Hosp, Marshfield, Wis, 61-66; ASSOC DIR LABS, MERCY HOSP, 66- Mem: Am Asn Path & Bact; Int Acad Path. Res: Pathologic anatomy with clinical pathologic correlations in the field of female genital diseases. Mailing Add: Dept of Lab Med Mercy Hosp Oshkosh WI 54901

NEUBERGER, DAN, b Zagreb, Yugoslavia, Feb 19, 29; nat US; m 56; c 2. PHOTOGRAPHIC CHEMISTRY. Educ: Columbia Univ, BA, 50; Univ Rochester, PhD(phys chem), 53. Prof Exp: Asst, Univ Rochester, 50-52; sr res chemist, 53-63, RES ASSOC, EASTMAN KODAK CO, 63- Mem: Soc Photog Sci & Eng. Res: Photographic science. Mailing Add: 95 Wendover Rd Rochester NY 14610

NEUBERGER, HANS HERMANN, b Mannheim, Ger, Feb 17, 10; nat US; m 39, 57; c 2. METEOROLOGY. Educ: Univ Hamburg, DSc(meteorol), 36. Prof Exp: Res asst to Dr Christian Jensen, Univ Hamburg, 36-37; instr geophys, 37-41, asst prof, 41-43, from assoc prof to prof meteorol, 43-70, chief div, 45-54, head dept, 54-61 & 65-67, EMER PROF METEOROL, PA STATE UNIV, 70- Concurrent Pos: Tech ed, Weatherwise, 48-53; consult, US Weather Bur, Turkey, 54-55; assoc, Army-Navy Vision Comt & mem subcomt visibility & atmospheric optics, Nat Res Coun; Am Meteorol Soc rep div earth sci, Nat Acad Sci-Nat Res Coun, 57-60; UN tech expert, Pakistan, 61; vis prof, Univ SFla, 71-74, emer prof geog, 74- Mem: Fel Am Meteorol Soc; Seismol Soc Am; Am Geophys Union; Int Soc Biometeorol. Res: Design of mine safety device and meteorological equipment; atmospheric optics and pollution. Mailing Add: 1805 Burlington Circle Sun City Center FL 33570

NEUBERGER, JACOB, b Ger, June 8, 27; nat US; m; c 5. SOLID STATE PHYSICS. Educ: Johns Hopkins Univ, BS, 50; NY Univ, MS, 53, PhD(physics), 58. Prof Exp: Res scientist, NY Univ, 54-57; asst prof physics, Rutgers Univ, 57-60; asst prof, 60-66, ASSOC PROF PHYSICS, QUEEN'S COL, NY, 66- Concurrent Pos: Lectr, City Col New York, 54-61; guest, Brookhaven Nat Lab, 67-68. Mem: AAAS; Am Phys Soc; Am Asn Physics Teachers. Res: Lattice dynamics as applied to specific heats and optical properties of crystals. Mailing Add: Dept of Physics Queens Col Flushing NY 11367

NEUBERGER, JOHN WILLIAM, b Ventura, Iowa, Aug 14, 34; m 59; c 2. MATHEMATICS. Educ: Univ Tex, BA, 54, PhD(math), 57. Prof Exp: Spec instr math, Univ Tex, 56-57; instr, Ill Inst Technol, 57-59; asst prof, Univ Tenn, 59-63; assoc prof, 63-67, PROF MATH, EMORY UNIV, 67- Concurrent Pos: Consult, Nuclear Div, Union Carbide Corp, 59-65; Alfred P Sloan res fel, 67-69; Alfred P Sloan res fel, Inst Advan Study, 68; NSF res grant, 70-72 & 74; ed, Houston J Math, 75- Mem: Am Math Soc; Math Asn Am; London Math Soc. Res: Partial differential equations; semigroups of operators; functional analysis. Mailing Add: Dept of Math Emory Univ Atlanta GA 30322

NEUBERT, JEROME ARTHUR, b Mankato, Minn, Dec 1, 38; m 66. PHYSICS. Educ: Univ Kans, BS, 62; Calif State Col, Los Angeles, MS, 66; Pa State Univ, PhD(eng acoust), 70. Prof Exp: SR PHYSICIST, NAVAL UNDERSEA CTR, 63- Mem: Acoust Soc Am. Res: Engineering acoustics; applied mathematics and physics; sound propagation in stochastic media; underwater acoustics. Mailing Add: Code 409 Naval Undersea Ctr San Diego CA 92132

NEUBERT, MARY E, organic chemistry, medicinal chemistry, see 12th edition

NEUBERT, RALPH LEWIS, b St Louis, Mo, Jan 30, 22; m 44; c 6. RESEARCH ADMINISTRATION. Educ: Univ Mo, Rolla, BS, 42; grad, Advan Mgt Prog, Harvard Univ, 70. Hon Degrees: Prof Eng, Univ Mo, Rolla, 71. Prof Exp: Plant engr, Monsanto Co, Carondelet, 42-44, various eng & mfg positions, Mo, Ala, Tenn & Mass, 46-62, mgr opers & planning, Agr Div, 62-64, area dir, Latin Am & Can Int Div, 64-67, dir admin & planning, Int Div, 67-68, vpres, Monsanto Res Corp & dir, Mound Lab, 68-72, mgr corp planning & develop, 72-74, DIR, STRATEGIC PLANNING, MONSANTO CO, 74- Mem: AAAS; Soc Chem Indust. Res: Research and development management; chemical production; international business; business planning. Mailing Add: Monsanto Co 800 N Lindbergh Blvd St Louis MO 63166

NEUBERT, THEODORE JOHN, b Rochester, NY, Jan 10, 17. PHYSICAL CHEMISTRY. Educ: Univ Rochester, BS, 39; Brown Univ, PhD(phys chem), 42. Prof Exp: Res assoc, US Naval Res Lab, 42; chemist, Manhattan Proj, Chicago, 42-45; scientist, Argonne Nat Lab, 45-47, fel inst nuclear studies, 47-49; from asst prof to assoc prof chem, 49-61, PROF CHEM, ILL INST TECHNOL, 61- Concurrent Pos: Consult, Argonne Nat Lab, 49-60. Mem: AAAS; Am Chem Soc; Am Phys Soc. Res: Physics and chemistry of solid state. Mailing Add: Dept of Chem Ill Inst of Technol Chicago IL 60616

NEUERBURG, GEORGE JOSEPH, geology, see 12th edition

NEUFELD, ABRAM HERMAN, b Russia, Apr 26, 07; nat US; m 37; c 2. CLINICAL PATHOLOGY. Educ: Univ Man, BSc, 34, MSc, 35, PhD(med biochem), 37; McGill Univ, MD & CM, 50. Prof Exp: Instr biochem, Univ Man, 35-36; lectr, McGill Univ, 36-41, asst prod prof endocrinol, 41-43; med biochemist, Queen Mary Vet Hosp, Montreal, 46-55, chief of serv biochem & radioisotopes, 55-6O; prof path chem & chmn dept, 60-72, EMER PROF PATH CHEM, FAC MED, UNIV WESTERN

ONT, 72- Concurrent Pos: Hon lectr, McGill Univ, 58-60; sr consult, Can Dept Vet Affairs, 50-60; ed, Med Serv J, Can, 47-59. Mem: Can Soc Clin Invest; Can Biochem Soc; Can Med Asn; Chem Inst Can; Can Physiol Soc. Res: Metabolism of normal and pathological lipids and proteins, especially in arteriosclerosis, myelomatosis and lipidoses; hemoglobins; radioactive isotopes in clincal investigation; chemical pathology. Mailing Add: 1071 Colborne St London ON Can

NEUFELD, ARTHUR HARVEY, b New York, NY, Feb 21, 45; m 66; c 2. PHYSIOLOGY, OPHTHALMOLOGY. Educ: NY Univ, BA, 66, PhD(physiol), 70. Prof Exp: Trainee physiol, Sch Med, NY Univ, 66-70; instr path, 70-71, instr ophthal, 71-72, ASST PROF OPHTHAL, SCH MED, YALE UNIV, 72- Concurrent Pos: USPHS grant, Sch Med, Yale Univ, 72-78. Mem: Asn Res Vision & Ophthal. Res: Regulation of intraocular pressure; anterior segment physiology, blood-aqueous barrier. Mailing Add: Dept of Ophthal & Visual Sci Yale Univ Sch of Med New Haven CT 06510

NEUFELD, BERNEY ROY, b Sask, Aug 3, 41; US citizen; m 63; c 2. MOLECULAR BIOLOGY. Educ: Columbia Union Col, BA, 63; Loma Linda Univ, MA, 65; Ind Univ, Bloomington, PhD, 68. Prof Exp: ASST PROF BIOL, LOMA LINDA UNIV, 68- Concurrent Pos: Mem extended day fac, Riverside City Col, 70-71; lectr, Univ Calif Exten, 71; NIH spec res fel, Calif Inst Technol, 71-73, vis fel, 73-74. Mem: Genetics Soc Am; Am Soc Microbiol. Res: Distribution of redundant DNA in higher organisms; evolution of higher organisms. Mailing Add: Loma Linda Univ Loma Linda CA 92354

NEUFELD, CORNELIUS HERMAN HARRY, b Scottdale, Pa, Apr 15, 23; m 51; c 3. ORGANIC CHEMISTRY. Educ: Dalhousie Univ, BSc, 47; Univ Notre Dame, PhD(chem), 51. Prof Exp: Asst prof chem, Am Univ, 51-53; chemist biol active chem compounds sect, Eastern Utilization Res Br. 53-54, chemist, Western Utilization Res & Develop Div, 54-68, asst to dir, 58-59, asst dir, 59-68, dir, Richard B Russell Agr Res Ctr, 68-72, AREA DIR, SOUTHERN REG AGR RES SERV, USDA, 72- Concurrent Pos: Vis prof, Univ Ariz, 65-66. Mem: Am Chem Soc; Am Oil Chem Soc; Inst Food Technol. Res: Plant growth modifiers; organic reaction mechanisms; synthetic organic protein chemistry; food technology; carbohydrate and polymer chemistry. Mailing Add: Richard B Russell Agr Res Ctr PO Box 5677 Athens GA 30604

NEUFELD, DANIEL ARTHUR, b Fresno, Calif, Aug 26, 45. HUMAN ANATOMY, EXPERIMENTAL MORPHOLOGY. Prof Exp: Univ Calif, Los Angeles, BA, 68; Calif State Univ, Long Beach, MA, 72; Tulane Univ, PhD(human anat), 75. Prof Exp: ASST PROF HUMAN ANAT, GEORGE WASHINGTON UNIV MED CTR, 75- Mem: AAAS. Res: Attempted induction of extremity regeneration in mammals. Mailing Add: Dept of Anat 2300 Eye St NW George Washington Univ Med Ctr Washington DC 20037

NEUFELD, ELIZABETH FONDAL, b Paris, France, Sept 27, 28; nat US; m 51; c 2. BIOCHEMISTRY, HUMAN GENETICS. Educ: Univ Calif, PhD(comp biochem), 56. Prof Exp: USPHS fel, Univ Calif, Berkeley, 56-57, asst res biochemist, 57-63; res biochemist, 63-74, CHIEF SECT HUMAN BIOCHEM GENETICS, NAT INST ARTHRITIS, METAB & DIGESTIVE DIS, 73- Honors & Awards: Dickson Prize, Univ Pittsburgh, 74; Hillebrand Award, 75. Mem: Am Soc Human Genetics; Am Chem Soc; Am Soc Biol Chemists. Res: Sugar nucleotides; human biochemical genetics; mucopolysaccharidoses. Mailing Add: Nat Inst of Arthritis Metab & Digestive Dis Bethesda MD 20014

NEUFELD, GAYLEN JAY, b Beaver Co, Okla, Feb 25, 39; m 61; c 3. CELL PHYSIOLOGY. Educ: Tabor Col, BA, 61; Kans State Univ, MS, 63; Univ Tex, Austin, PhD(protein chem), 66. Prof Exp: Damon Runyon Cancer Fund fel dept genetics, Univ Melbourne, 66-67; asst prof biol, 67-71, ASSOC PROF BIOL, EMPORIA KANS STATE COL, 71- Mem: AAAS; Sigma Xi. Res: Structural characteristics of phycocyanin and phycoerythrin; effects of molting hormone on protein and ribonucleic acid synthesis in the third instar larvae of Calliphora. Mailing Add: Div of Biol Sci Emporia Kans State Col Emporia KS 66801

NEUFELD, HAROLD ALEX, b Paterson, NJ, Mar 23, 24; m 50; c 4. BIOCHEMISTRY. Educ: Rutgers Univ, BS, 49; Univ Rochester, PhD(biochem), 53. Prof Exp: Res fel biochem, Sch Med, Univ Rochester, 53-54, instr, 54-55; biochemist crops div, US Dept Army, 55-57, biochemist chem br, 57-70; teacher, Frederick County Sch Syst, 70-72; BIOCHEMIST PHYS SCI DIV, US ARMY INST INFECTIOUS DIS, FT DETRICK, 72- Concurrent Pos: Lectr, Frederick Community Col, 57-65; lectr biochem, Hood Col, Frederick, Md, 74- Honors & Awards: Award, US Dept Army, 59 & 64. Mem: Am Chem Soc; Am Soc Biol Chem; Sigma Xi. Res: Purification and kinetics of enzymes concerned with biological oxidation; properties of enzymes in proliferating tissue; relationship between hormones and enzymes; respiratory enzymes in bacteria and fungi; chemiluminescence and bioluminescence. Mailing Add: 117 W 14th St Frederick MD 2170!

NEUFELD, JACOB, b Lodz, Poland, Apr 15, 06; nat US; m 47; c 1. THEORETICAL PHYSICS. Educ: Univ Liege, EE, 29; Univ SC, MS, 30; Univ Pa, ScD(elec eng), 36. Prof Exp: Res engr, Eng Labs, Inc, Okla, 36-38, dir res, 38-48; physicist, 48-71, CONSULT, OAK RIDGE NAT LAB, 71- Mem: Am Phys Soc; Health Physics Soc. Res: Interaction of charged particles with matter; plasma physics; properties of dispersive media; radiation protection; health physics; microdosimetry. Mailing Add: 113 Cedar Lane Oak Ridge TN 37830

NEUFELD, ORVILLE, pharmacology, see 12th edition

NEUFER, JOHN E, b Topeka, Ind, July 25, 26; m 75; c 2. PHYSICAL CHEMISTRY. Educ: Bluffton Col, BS, 49; Wayne State Univ, PhD(chem), 58. Prof Exp: Asst chem, Wayne State Univ, 49-52 & 56-57; opers analyst, US Navy Opers Eval Group, Mass Inst Technol, 58-60, field rep, 60-61; asst prof, Western Md Col, 61-64; opers analyst, Bunker-Ramo Corp, 64-66; opers res scientist, Lockheed-Calif Co, 66-69, mgr antisubmarine warfare dept, 69-72; SR RES SCIENTIST, RAND CORP, 72- Concurrent Pos: Sr scientist assigned by Lockheed-Calif Co to Naval Tactical Anal Prog, Comdr Fleet Air Wing, Pac, 66-67 & sr engr, Antisubmarine Warfare Force Pac, 67-68. Mem: Opers Res Soc Am; Am Statist Asn. Res: Operations research; antisubmarine warfare. Mailing Add: Rand Corp 1700 Main St Santa Monica CA 90406

NEUFFER, MYRON GERALD, b Preston, Idaho, Mar 4, 22; m 43; c 7. GENETICS. Educ: Univ Idaho, BS, 47; Univ Mo, MA, 48, PhD(genetics), 52. Prof Exp: Asst field crops, 47-51, from asst prof to assoc prof, 51-66, prof genetics, 66-70, chmn dept, 67-69, PROF BIOL SCI, UNIV MO-COLUMBIA, 70- Mem: AAAS; Genetics Soc Am; Am Genetic Asn. Res: Gene mutation in maize. Mailing Add: Div of Biol Sci 202 Curtis Hall Univ of Mo Columbia MO 65201

NEUGEBAUER, CHRISTOPH JOHANNES, b Dessau, Ger, Apr 21, 27; nat US; m 58; c 3. MATHEMATICS. Educ: Univ Dayton, BS, 50; Ohio State Univ, MS, 52, PhD(math), 54. Prof Exp: From instr to assoc prof, 54-62, PROF MATH, PURDUE UNIV, WEST LAFAYETTE, 62- Mem: Am Math Soc. Res: Analysis. Mailing Add: Div of Math Sci Purdue Univ West Lafayette IN 47906

NEUGEBAUER, CONSTANTINE ALOYSIUS, b Dessau, Ger, Apr 20, 30; nat US; m 58; c 2. PHYSICAL CHEMISTRY. Educ: Union Univ, NY, BS, 53; Univ Wis, PhD(chem), 57. Prof Exp: Fel phys chem, Univ Wis, 57; RES ASSOC, INFO SCI LAB, RES & DEVELOP CTR, GEN ELEC CO, 57- Mem: Inst Elec & Electronics Engrs; Am Phys Soc; Am Vacuum Soc. Res: Calorimetry; thermodynamics; structure and properties of thin films; large scale integration. Mailing Add: Res & Develop Ctr Gen Elec Co 4C5K1 Schenectady NY 12345

NEUHARDT, JOHN BERNARD, b Flint, Mich, Nov 30, 30; m 55; c 3. STATISTICS, OPERATIONS RESEARCH. Educ: Univ Mich, BA, 56, MA, 57, MS, 66, PhD(indust eng), 67. Prof Exp: Dir eng data processing, AC Spark Plug Div, Gen Motors, 60-64, dir data processing & systs anal, 65; res assoc, Univ Mich, 66-67; NSF fel, 67-69, asst prof, 67-70, ASSOC PROF STATIST, OHIO STATE UNIV, 70- Concurrent Pos: Lectr, Columbia Gas Corp, 69-70 & Union Carbide Corp, 70. Mem: Inst Mgt Sci; Am Statist Asn; Opers Res Soc Am; Am Inst Indust Eng. Res: Economic experimental design; applications to product development; quality control; hospital systems; automobile transportation. Mailing Add: Dept of Indust Eng Ohio State Univ Columbus OH 43210

NEUHAUS, FRANCIS CLEMENS, b Huntington, WVa, May 5, 32; m 55; c 4. BIOCHEMISTRY. Educ: Duke Univ, BS, 54, PhD(biochem), 58. Prof Exp: NSF fel, Univ Newcastle-on-Tyne, Eng, 58-59 & Univ Ill, 59-60, instr biochem div, 59-60, asst prof chem, 60-61; from asst prof to assoc prof, 61-70, prof chem, 70-74, PROF BIOCHEM & MOLECULAR BIOL, NORTHWESTERN UNIV, 74- Concurrent Pos: USPHS res career develop award, 66-71. Mem: Am Chem Soc; Am Soc Microbiol; Am Soc Biol Chem. Res: Biosynthesis of bacterial cell wall components; mechanism of antibiotic action; resistance mechanisms; membrane reactions. Mailing Add: Dept of Biochem & Molecular Biol Northwestern Univ 2145 Sheridan Evanston IL 60201

NEUHAUS, OTTO WILHELM, b Ger, Nov 18, 22; nat US; m 47; c 3. BIOCHEMISTRY. Educ: Univ Wis, BS, 44; Univ Mich, MS, 47, PhD(biochem), 53. Prof Exp: Res chemist, Merck & Co, 44-46; asst biochem, Univ Mich, 46-49; res chemist, Huron Milling Co, 51-54; instr physiol chem, Col Med, Wayne State Univ, 54-58, res assoc anat, 54-58, asst prof physiol chem, 58-65, assoc prof biochem, 65-66; PROF BIOCHEM & CHMN DEPT, SCH MED, UNIV SDAK, 66- Mem: AAAS; Am Chem Soc; Soc Exp Biol & Med; Am Soc Biol Chemists. Res: Proteins of biological fluids; control of plasma protein biosynthesis; transport of amino acids. Mailing Add: Dept of Biochem Univ of SDak Sch of Med Vermillion SD 57069

NEUHOLD, JOHN MATHEW, b Milwaukee, Wis, May 18, 28; m 52; c 1. AQUATIC ECOLOGY. Educ: Utah State Univ, BS, 52, MS, 54, PhD(fish biol), 59. Prof Exp: Biologist, State Dept Fish & Game, Utah, 52-55, asst fed aide coord, 55-56; asst fish toxicol, 56-58, from asst prof to assoc prof fish biol, 58-66, actg dir ecol ctr, 66-68, PROF FISH BIOL, UTAH STATE UNIV, 66-, DIR, ECOL CTR, 68- Concurrent Pos: Prog dir ecosyst anal, NSF, 71-72; trustee, Inst Ecol, 73- Mem: AAAS; Am Fisheries Soc; Am Soc Limnol & Oceanog; Ecol Soc Am. Res: Fish toxicology, primary production in aquatic habitat, population dynamics; pollution limnology. Mailing Add: Ecol Ctr Utah State Univ Logan UT 84322

NEUHOUSER, DAVID LEE, b Leo, Ind, Mar 28, 33; m 54; c 4. MATHEMATICS. Educ: Manchester Col, BS, 55; Univ Ill, MS, 59; Fla State Univ, PhD(math educ), 64. Prof Exp: High sch teacher, Iowa, 55-57 & Ind, 57-58; from instr to assoc prof math & head dept, Manchester Col, 59-71; PROF MATH & HEAD DEPT, TAYLOR UNIV, 71- Concurrent Pos: Consult to sch systs, Northern Ind, 64- Mem: Math Asn Am. Res: Methods of teaching mathematics, particularly the discovery method. Mailing Add: Rte 1 Upland IN 46989

NEUMAN, CHARLES HERBERT, b Los Angeles, Calif, Feb 8, 37; m 58; c 2. PHYSICS. Educ: Calif Inst Technol, BS, 58; Univ Ill, MA, 60, PhD(physics), 63. Prof Exp: Res assoc physics, Univ Calif, Riverside, 63-65; RES PHYSICIST, CHEVRON RES CO, 65- Mem: Am Phys Soc; Soc Petrol Engrs. Res: Defects in solids; effects of pressure on transition metal oxides; nuclear spin echo in liquids; x-ray diffraction of minerals; radioactive measurement of liquid content in porous media. Mailing Add: 2014 Tuffree Blvd Placentia CA 92670

NEUMAN, MARGARET WRIGHTINGTON, b Lexington, Mass, May 31, 17; m 43; c 3. PHARMACOLOGY. Educ: Vassar Col, BA, 38; Univ Rochester, MS, 40, PhD(pharmacol), 43. Prof Exp: Vol asst path, Histol Lab, Harvard Univ, 38-39; asst, 39-43, asst toxicol & instr pharmacol, Sch Med & Dent & asst biochem sect, Atomic Energy Proj, 43-60, instr radiation biol, 60-64, ASST PROF RADIATION BIOL & BIOPHYS, UNIV ROCHESTER, 64- Res: Toxicology of organic nitrates, choline and uranium; biochemistry of bone; chemical evolution. Mailing Add: Dept of Radiation Biol & Biophys Univ of Rochester Med Ctr Rochester NY 14642

NEUMAN, ROBERT BALLIN, b Washington, DC, Feb 28, 20; m 49; c 2. GEOLOGY. Educ: Univ NC, BS, 41; Johns Hopkins Univ, PhD(geol), 49. Prof Exp: GEOLOGIST, US GEOL SURV, 49- Concurrent Pos: Lectr, Johns Hopkins Univ, 56-57 & Univ Oslo, 70-71; US mem Appalachian-Caledonide Orogen Proj int working group, Int Geol Correlation Prog, 74-82. Mem: AAAS; Paleont Soc; Geol Soc Am; Soc Econ Paleont & Mineral; Am Asn Petrol Geol. Res: Ordovician paleontology and paleogeography of the region bordering the North Atlantic Ocean. Mailing Add: US Geol Surv Room E-304 Nat Mus Washington DC 20560

NEUMAN, ROBERT C, JR, b Chicago, Ill, Aug 21, 38; m 62; c 2. PHYSICAL ORGANIC CHEMISTRY. Educ: Univ Calif, Los Angeles, BS, 59; Calif Inst Technol, PhD(org chem), 63. Prof Exp: NSF res fel, Columbia Univ, 62-63; from asst prof to assoc prof chem, 63-72, PROF CHEM, UNIV CALIF, RIVERSIDE, 72- Concurrent Pos: Vis lectr dept chem, Princeton Univ, 71-72; NIH spec res fel, 71-72. Mem: Am Chem Soc. Res: Effects of high pressure on chemical and biochemical systems; free radical chemistry. Mailing Add: Dept of Chem Univ of Calif Riverside CA 92502

NEUMAN, SHLOMO PETER, b Zilina, Czech, Oct 26, 38; US citizen; m 65; c 2. HYDROLOGY, HYDROGEOLOGY. Educ: Hebrew Univ, Jerusalem, BSc, 63; Univ Calif, Berkeley, MS, 66, PhD(eng sci), 68. Prof Exp: Asst res engr, Univ Calif, Berkeley, 68-70, actg asst prof civil eng, 70; res scientist hydrol, Agr Res Orgn, Israel, 70-74; vis assoc prof civil eng, Univ Calif, Berkeley, 74-75; PROF HYDROL, UNIV ARIZ, 75- Concurrent Pos: Res consult, Israel Inst Technol, 70-74 & Lawrence Berkeley Lab, 75- Honors & Awards: Robert E Horton Award, Am Geophys Union, 69. Mem: Am Geophys Union; Nat Water Well Asn. Res: Groundwater modeling; finite element techniques; well hydraulics; applications of systems theory in hydrology; geothermal studies and modeling. Mailing Add: Dept of Hydrol & Water Resources Univ of Ariz Tucson AZ 85721

NEUMAN, WILLIAM FREDERICK, b Petoskey, Mich, June 2, 19; m 43; c 3. BIOCHEMISTRY, BIOPHYSICS. Educ: Mich State Col, BS, 40; Univ Rochester, PhD(biochem), 44. Prof Exp: Asst, 40-43, from instr to assoc prof, 44-68, PROF BIOCHEM, UNIV ROCHESTER, 68-, PROF RADIATION BIOL & BIOPHYS, 58- Concurrent Pos: Assoc prof pharmacol, Univ Rochester, 50-70, div chief, Atomic Energy Proj, 58-65, co-dir, 65-71, dir, 71-75, co-chmn radiation biol & biophys, 65-71, chmn, 71-75; Lancet lectr, Univ Minn, 58; Rainbow lectr, Case Western Reserve Univ, 59; mem, Josiah Macy, Jr Found. Honors & Awards: Eli Lilly Award Biol Chem, 55; Claude Bernard Medal, Univ Montreal, 62; Kappa Delta Res Award, Am Col Orthop Surg, 64; Int Asn Dent Res Biol Mineralization Award, 65. Mem: AAAS; Am Chem Soc; Am Indust Hyg Asn; Am Soc Biol Chemists; Am Soc Pharmacol & Exp Therapeut. Res: Lipid and bone metabolism; mode of action of parathyroid hormone; biological effects of radiation. Mailing Add: Dept Radiation Biol & Biophys Univ Rochester Sch Med & Dent Rochester NY 14642

NEUMANN, A CONRAD, b Oak Bluffs, Mass, Dec 21, 33; m 62; c 3. OCEANOGRAPHY. Educ: Brooklyn Col, BS, 55; Tex A&M Univ, MS, 58; Lehigh Univ, PhD(geol), 63. Prof Exp: Res assoc sedimentary geol, Woods Hole Oceanog Inst, 58-60; asst prof marine geol, Lehigh Univ, 63-65; asst prof marine sci, Univ Miami, 65-69; prog dir marine geol & geophys, NSF, 69-70; assoc prof marine sci, Univ Miami, 69-72; actg dir marine sci curric, 73-74, PROF MARINE SCI, MARINE SCI PROG, UNIV NC, 72- Concurrent Pos: Trustee, Bermuda Biol Sta Res, Inc, 72-76. Honors & Awards: Best Paper Award, Geol Soc Am, 69. Mem: AAAS; Soc Econ Paleont & Mineral; Geol Soc Am; Sigma Xi; Am Quaternary Asn. Res: Sedimentology; recent carbonate sediments of the Bermudas, Bahamas and Florida; biological erosion of limestone; marine geology by research submersibles; Quaternary history of sea level; deep flanks of carbonate platforms. Mailing Add: Marine Sci Prog 12-5 Venable Hall Univ NC Chapel Hill NC 27514

NEUMANN, ALVIN LUDWIG, b Granite, Okla, July 9, 15; m 38. ANIMAL NUTRITION. Educ: Okla State Univ, BS, 38, MS, 39; Univ Ill, PhD(animal sci, nutrit), 49. Prof Exp: Asst, Okla State Univ, 38-39; instr high sch, Okla, 39-41; head agr div, Cameron State Agr Col, Okla, 41-44; county farm agent, Comanche County, Okla, 44-45; asst prof animal indust, Univ Ark, 45-47; asst, Univ Ill, 47-48; prof animal indust, Univ Ark, 49-50; from assoc prof to prof beef cattle nutrit, Univ Ill, 50-64; prof animal, range & wildlife sci & head dept, NMex State Univ, 64-71; PROF ANIMAL SCI, UNIV ILL, URBANA, 71- Mem: AAAS; Am Soc Animal Sci. Res: Nutrient requirements of beef cattle; use of hormones, antibiotics and other additives in feeding beef cattle; management of beef cattle farms. Mailing Add: 103 Stock Pavilion Univ of Ill at Urbana-Champaign Urbana IL 61801

NEUMANN, CALVIN LEE, b Coldwater, Mich, Sept 13, 38; m 61; c 1. PHYSICAL ORGANIC CHEMISTRY. Educ: Wayne State Univ, BS, 62, PhD(org chem), 66. Prof Exp: Technician res labs, Ethyl Corp, Mich, 58-59; sr chemist res dept, R J Reynolds Industs, Inc, 66-70, group leader, 70-74, GROUP LEADER RES DEPT, R J REYNOLDS TOBACCO CO, 74- Mem: Am Chem Soc. Res: Conformational analysis; tobacco and health relationships; research and development of tobacco and tobacco related products. Mailing Add: Res Dept R J Reynolds Tobacco Co Winston-Salem NC 27102

NEUMANN, FRED ROBERT, b Chicago, Ill, June 3, 00; wid. GEOLOGY. Educ: Univ Chicago, BS, 21, MS, 23; Cornell Univ, PhD(geol), 26. Prof Exp: Instr geol, Univ SC, 24-25; geologist, Shell Oil Co, 27-28; prof geol & geog, Ill Wesleyan Univ, 28-31; asst prof, Bowling Green Univ, 31-32; supvr, Chicago Bd Educ, 32-35; instr geol & geog, Drury Col, 35-36 & Mesaba Col, 36-42; instr geol & sci, Col of the Sequoias, 42-44; geologist, Stand Oil Co Calif, 44-46; prof, 46-70, EMER PROF GEOL, CHICO STATE UNIV, 70- Mem: Geol Soc Am; Am Asn Petrol Geol; AAAS; Sigma Xi; Nat Asn Geol Teachers. Res: Geology of Butte County, California. Mailing Add: Dept of Geol Chico State Univ Chico CA 95929

NEUMANN, FRED WILLIAM, b Chicago, Ill, Sept 28, 18; m 44; c 3. ORGANIC CHEMISTRY, ANALYTICAL CHEMISTRY. Educ: Univ Ill, BS, 40; Ind Univ, PhD(org chem), 44. Prof Exp: Jr chemist, Merck & Co, NJ, 40-41; asst chem, Ind Univ, 41-43; chemist, Gen Aniline & Film Corp, 44-55; group leader, 55-68, anal specialist, 68-72, RES SPECIALIST, DOW CHEM, USA, 72- Mem: Am Chem Soc; Sigma Xi. Res: Phenol and bisphenol A processes; synthetic organic chemistry of diazotypes, amidines and phenols; identification of dyes, diazotypes and organic commercial products; organic air-borne pollutants; iodine and bromine processes. Mailing Add: 213 Norfolk Midland MI 48640

NEUMANN, GEORGE JOSEPH, b Mineola, NY, Nov 27, 28; m 56; c 4. BIOCHEMISTRY. Educ: State Univ NY, BA, 52; Loyola Univ, Ill, MS, 55; Univ Buffalo, PhD(biochem), 58. Prof Exp: Clin biochemist, New Brit Gen Hosp, Conn, 58-59; CLIN BIOCHEMIST, ELLIS HOSP, 59- Mem: Am Chem Soc; Am Asn Clin Chem. Res: Analytical techniques in clinical biochemistry; liver function tests; normal ranges of laboratory tests. Mailing Add: Ellis Hosp Lab 1101 Nott St Schenectady NY 12308

NEUMANN, HELMUT CARL, b Berlin, Ger, Dec 24, 16; nat US; m 46; c 4. PHARMACEUTICAL CHEMISTRY. Educ: Polytech Inst Brooklyn, BS, 38; Fed Inst Technol, Zurich, DSc(chem), 49. Prof Exp: Anal & develop chemist, Fritzsche Bros Inc, 38-42, asst lab head, 45-46; res fel, Univ Pa, 48-50; res chemist, White Labs, Inc, 50-51; res assoc, 51-59, ASSOC MEM, STERLING-WINTHROP RES INST, 59- Concurrent Pos: Ed, Eastern New York Chemist, 65-68. Mem: AAAS; Am Chem Soc. Res: Terpenes; steroids; antifertility agents; research on steroids and related compounds in search for medicinal applications. Mailing Add: 5 Point View Dr East Greenbush NY 12061

NEUMANN, HENRY MATTHEW, b Minneapolis, Minn, July 15, 24. INORGANIC CHEMISTRY. Educ: Col St Thomas, BS, 47; Univ Calif, PhD(chem), 50. Prof Exp: Instr chem, Northwestern Univ, 50-54, asst prof, 54-56; assoc prof, 56-60, PROF CHEM, GA INST TECHNOL, 60- Mem: Am Chem Soc. Res: Radioactive exchange reactions; kinetics and reaction mechanisms of complex ions and of organometallic compounds. Mailing Add: Sch of Chem Ga Inst of Technol Atlanta GA 30332

NEUMANN, HERSCHEL, b San Bernardino, Calif, Feb 3, 30; m 51; c 2. MOLECULAR PHYSICS, SCIENCE EDUCATION. Educ: Univ Calif, Berkeley, BA, 51; Univ Ore, MA, 59; Univ Nebr, Lincoln, PhD(theoret phys), 65. Prof Exp: Physicist, Hanford Labs, Gen Elec Co, Wash, 51-57; instr phys, Univ Nebr, Lincoln, 65; asst prof, 65-71, ASSOC PROF PHYS, UNIV DENVER, 71- Mem: Am Asn Physics Teachers; Am Phys Soc. Res: Atomic and molecular collision theory; numerical analysis; ultrasonics. Mailing Add: Dept of Physics Univ of Denver Denver CO 80210

NEUMANN, MARGUERITE, b West Bend, Wis, May 7, 14. ORGANIC CHEMISTRY, MEDICINAL CHEMISTRY. Educ: Mundelein Col, BS, 42; Univ Iowa, MS, 43; St Louis Univ, PhD(chem', 54. Prof Exp: Asst prof chem, Mundelein Col, 43-50 & 54-57; chmn dept, 57-70, PROF CHEM, CLARKE COL, 57- Mem: Am Chem Soc; Sigma Xi; Am Asn Univ Women. Res: College teaching of chemistry; biochemistry. Mailing Add: Dept of Chem Clarke Col Dubuque IA 52001

NEUMANN, NORBERT PAUL, b Chicago, Ill, Oct 13, 31; m 56; c 3. BIOCHEMISTRY. Educ: St Peters Col, BS, 53; Okla State Univ, MS, 55; Univ Wis, PhD(biochem), 58. Prof Exp: Res assoc biochem, Rockefeller Inst, 58-61; instr microbiol, Rutgers Univ, 61-64, asst prof, 64-67, asst prof sch med, 67-70; dir biochem, 70-74, ASST DIR RES, ORTHO DIAG, INC, 74- Mem: AAAS; Am Chem Soc. Res: Relationship between structure and function in enzymes; methods of protein isolation and characterization; diagnostic chemistry and immunology. Mailing Add: Div of Biochem Ortho Diag Inc Raritan NJ 08869

NEUMANN, PAUL GERHARD, b Insterburg, Ger, June 26, 11; nat US; m 48; c 3. OCEANOGRAPHY, METEOROLOGY. Educ: Univ Berlin, Dr rer nat(geophys), 39. Prof Exp: Res scientist, Ger Marine Observ, 39-45 & Ger Hydrographic Off, 45-47; docent, Univ Hamburg, 47-51; from assoc prof to prof oceanog, NY Univ, 51-73; PROF EARTH & PLANETARY SCI, CITY COL NEW YORK, 73- Concurrent Pos: Consult, NSF, 66, 67 & 72- Mem: Am Meteorol Soc; Am Geophys Union; Ger Geophys Soc. Res: Air-sea interaction; ocean waves and currents; internal waves; physical oceanography; oceanography of the tropical Atlantic Ocean. Mailing Add: 11 Goodwin St Hastings-on-Hudson NY 10706

NEUMANN, RICHARD K, b Washington, DC, Sept 10, 24; m 46; c 2. MATHEMATICS, COMPUTER SCIENCE. Educ: George Washington Univ, BA, 51. Prof Exp: Head math anal sect propulsion group, Appl Phys Lab, Johns Hopkins Univ, 51-55; head comput & data reduction dept, Aerojet-Gen Corp Div Gen Tire & Rubber Co, 55-57; mgr comput dept, 57-66, HEAD SCI COMPUT, AIRES MFG CO, ARIZ DIV, GARRETT CORP, 66- Concurrent Pos: Consult comput sci, Sch Eng Sci, Ariz State Univ, Tempe, 63-73. Mem: Am Inst Aeronaut & Astronaut; Math Asn Am; Asn Comput Mach. Res: Computer applications, networks and education. Mailing Add: Comput Dept AiRes Mfg Co of Ariz Phoenix AZ 85034

NEUMARK, GERTRUDE FANNY, b Neuremberg, Ger, Apr 29, 27; nat US; m 50. SOLID STATE PHYSICS. Educ: Columbia Univ, BA, 48, PhD(chem), 51; Radcliffe Col, MA, 49. Prof Exp: Asst chem, Columbia Univ, 48, asst, Barnard Col, 50-51; adv res physicist, Sylvania Elec Prods, Inc, Gen Tel & Electronics Corp, 52-60; STAFF PHYSICIST PHILIPS LABS, N AM PHILIPS CO, INC, 60- Concurrent Pos: Anderson fel, Am Asn Univ Women; adj assoc prof, Fairleigh Dickinson Univ, 73-74. Mem: Am Chem Soc; Am Phys Soc. Res: Luminescence and electroluminescence; quantum chemistry. Mailing Add: Philips Lab NAm Philips Co Inc Briarcliff Manor NY 10510

NEUMER, JOHN FRED, organic chemistry, see 12th edition

NEUMEYER, JOHN L, b Munich, Ger, July 19, 30; US citizen; m 56; c 3. MEDICINAL CHEMISTRY. Educ: Columbia Univ, BS, 52; Univ Wis, PhD(med chem), 61. Prof Exp: Pharmaceut chemist, Ethicon, Inc Div, Johnson & Johnson, 52-53, group leader pharmaceut res, 55-57; sr res chemist, Cent Res Labs, Niagara Chem Div, FMC Corp, NJ, 61-63; sr staff chemist, Arthur D Little, Inc, 63-69; PROF MED CHEM, NORTHEASTERN UNIV, 69-, CHMN DEPT MED CHEM & PHARMACOL, 73- Concurrent Pos: Adj prof, Mass Col Pharm, 64-69; mem adv panel secy comn on pesticides & their relationship to environ health, Dept Health, Educ & Welfare, 69-70; mem comt rev, US Pharmacopeia, 69-; consult, Arthur D Little, Inc, 69- & Environ Protection Agency, 70-72; mem, Mass State Pesticide Bd, 73-75; bk rev ed, J Med Chem, 74-; vis prof chem, Univ Konstanz, WGer, 75-76. Mem: Am Chem Soc; Am Pharmaceut Asn; AAAS; fel Acad Pharmaceut Sci. Res: Chemistry of biologically-active compounds of natural and synthetic origin; aporphines; isoquinolines; chemistry of heterocyclics, antimalarials; central nervous system active compounds, cancer chemotherpy, pesticides. Mailing Add: Dept of Med Chem & Pharmacol Northeastern Univ Boston MA 02115

NEUMILLER, HARRY JACOB, JR, b Peoria, Ill, Dec 25, 29; m 57; c 4. ORGANIC CHEMISTRY. Educ: Knox Col, BA, 51; Univ Ill, MS, 52, PhD(org chem), 56. Prof Exp: Res chemist, Eastman Kodak Co, NY, 55-59; asst prof, 59-65, ASSOC PROF CHEM, KNOX COL, ILL, 65- Mem: Am Chem Soc. Res: Orientations in additions to quinones and quinone-like compounds. Mailing Add: Dept of Chem Knox Col Galesburg IL 61401

NEUNZIG, HERBERT HENRY, b Richmond Hill, NY, May 11, 27; m 55; c 2. ENTOMOLOGY. Educ: Cornell Univ, MS, 55, PhD(entom), 57. Prof Exp: From asst res prof to assoc res prof entom, 57-68, PROF ENTOM, NC STATE UNIV, 68- Mem: Entom Soc Am; Lepidopterists' Soc. Res: Biology and taxonomy of immature insects. Mailing Add: Dept of Entom NC State Univ Raleigh NC 27607

NEUPERT, WERNER MARTIN, b Worcester, Mass, Dec 19, 31; m 59; c 2. SPECTROSCOPY, SOLAR PHYSICS. Educ: Worcester Polytech Inst, BS, 54; Cornell Univ, PhD(physics), 60. Prof Exp: Vis asst prof physics, Univ Calif, Santa Barbara, 59-60; PHYSICIST, GODDARD SPACE FLIGHT CTR, NASA, 60- Mem: Int Astron Union; Am Phys Soc; Am Astron Soc. Res: Studies of the solar extreme ultraviolet and x-ray spectrum using rocket and satellite-borne instrumentation. Mailing Add: Lab for Solar Physics Code 682 Goddard Space Flight Ctr NASA Greenbelt MD 20771

NEURATH, ALEXANDER ROBERT, b Bratislava, Czech, May 8, 33; US citizen. VIROLOGY. Educ: Inst Tech, Bratislava, Czech, DiplIng, 57; Vienna Tech Univ, ScD(microbiol), 68. Prof Exp: Res scientist, Plant Physiol Inst, Czech Acad Sci, 57-59; dept head vet virol, Bioveta, Nitra, Czech, 59-61; res scientist, Inst Virol, Czech Acad Sci, 61-64; res fel virol, Wistar Inst, 64-65; sr virologist, Wyeth Labs, Inc, 65-72; INVESTR, L KIMBALL RES INST, NEW YORK BLOOD CTR, 72- Mem: AAAS; Am Soc Microbiol; Brit Soc Gen Microbiol. Res: Analysis of trace elements in plants; analytical biochemistry; biochemistry of viruses including, myxoviruses, adenoviruses, rabies and pseudorabies; hepatitis B. Mailing Add: New York Blood Ctr 310 E 67th St New York NY 10021

NEURATH, HANS, b Vienna, Austria, Oct 29, 09; nat US; m 36, 60; c 1. BIOCHEMISTRY. Educ: Univ Vienna, PhD(colloid chem), 33. Hon Degrees: DSc, Univ Geneva, 70. Prof Exp: Res fel, Univ Col, Univ London, 34-35; res fel biochem, Univ Minn, 35-36; instr & Baker res fel chem, Cornell Univ, 36-38; from asst prof to assoc prof biochem sch med, Duke Univ, 38-46, prof phys biochem, 46-50; chmn dept biochem, 50-75, PROF BIOCHEM, UNIV WASH, 50- Concurrent Pos: Mem comt biol chem, Div Chem & Chem Eng, Nat Res Coun, 54-58, mem exec comt, Off Biochem Nomenclature, 64-; consult, NIH, 54-; Guggenheim fel, 55-56; Phillips visitor, Haverford Col, 59; Darling lectr, Allegheny Col, 60; ed, Biochem, 61-; mem, US Nat Comt Biochem, 62-63 & US-Japan Coop Sci Prog Adv Panel on Med Sci, 64-65; comn ed, Int Union Biochem, 65-; mem comt res life sci, Nat Acad Sci, 66-70; mem, Nat Bd Grad Educ, 71-; guest prof, Alexander von Humboldt Found, WGer & Univ Heidelberg, 75. Honors & Awards: Awardee, Alexander von Humboldt Found, WGer & Univ Heidelberg, 75. Mem: Nat Acad Sci; fel AAAS; fel Am Acad Arts &

Sci; Am Chem Soc; fel NY Acad Sci. Res: Protein structure and function; enzymes; proteolytic enzymes; evolution; physical biochemistry of macromolecules. Mailing Add: Dept of Biochem Univ of Wash Seattle WA 98195

NEURATH, PETER WOLFGANG, b Vienna, Austria, Aug 19, 23; nat US; m 49; c 3. INFORMATION SCIENCE. Educ: Univ Toronto, BASc, 46, MA, 47; Carnegie-Mellon Univ, DSc(solid state physics), 50. Prof Exp: Res physicist, Gen Elec Co, 50-59; prin staff scientist, Avco Corp, 60-65; chief biophysicist, 66-68, DIR PHYSICS DIV, NEW ENG MED CTR HOSP, 69-; PROF THERAPEUT RADIOL, SCH MED, TUFTS UNIV, 72- Mem: Am Phys Soc; Inst Elec & Electronics Engrs; Biophys Soc; Am Asn Physicists in Med; Pattern Recognition Soc. Res: Magnetism; radiological physics; pattern recognition by computer; medical information systems. Mailing Add: Dept of Therapeut Radiol New Eng Med Ctr 171 Harrison Ave Boston MA 02111

NEURAUTER, LLOYD JOSEPH, b Greeley, Colo, Sept 4, 22; m 41; c 3. VETERINARY PUBLIC HEALTH, LABORATORY ANIMAL MEDICINE. Educ: Colo State Univ, DVM, 44; Univ Mich, MPH, 52. Prof Exp: Base vet, Aleutian Islands Army Post, US Air Force, Adak, 46-47, base vet, Great Falls AFB, Mont, 47-52, admin asst & instr vet serv, Sch Aviation Med, Gunter AFB, Ala, 52-55, staff officer, Air Force Command & Staff Sch, 56-61, staff officer to Asst Surgeon Gen Vet Serv, 56-61, mem staff res sect, Armed Forces Inst Path, 61-64; chief primate res ctr sect, Animal Resources Br, Div Res Facil & Resources, NIH, Md, 64-67, asst chief animal resources br, 67; assoc dir inst comp med, Sch Vet Med, Univ Ga, 67-69; lectr & asst dean, Sch Vet Med & assoc dir, Calif Primate Res Ctr, Univ Calif, Davis, 69-75; RETIRED. Mem: Am Asn Lab Animal Sci; Conf Pub Health Vets (pres, 69). Res: Infectious disease of non-human primates; husbandry and disease control procedures for non-human primates. Mailing Add: Route 2 Kuna ID 83634

NEUREITER, NORMAN PAUL, b Macomb, Ill, Jan 24, 32; m 59. SCIENCE POLICY. Educ: Univ Rochester, AB, 52; Northwestern Univ, PhD, 57. Prof Exp: Instr basic sci, Northwestern Univ, 56-57; res chemist, Humble Oil & Refining Co, Tex, 57-63; asst prog dir off int sci activ, NSF, DC, 63-64; actg prog dir, US-Japan Coop Sci Prog, 64-65; dep sci attache, US Dept State, Am Embassy, Bonn, 65-67, sci attache, Am Embassy, Warsaw, 67-69, tech asst off sci & technol, Exec Off of the President, 69-73; dir, East-West Bus Develop, 73-74, MGR, INT BUS DEVELOP, TEX INSTRUMENTS INC, 74- Concurrent Pos: Guide, Am Nat Exhib, Moscow, 59; mem bd int orgn & prog, Nat Acad Sci, Off For Secy, Washington, DC, 73- Mem: AAAS; Am Chem Soc. Res: International cooperation in science; relationships of science to foreign policy; role of technology in international trade; East-West trade. Mailing Add: Tex Instruments Inc PO Box 5474 MS 227 Dallas TX 75222

NEURINGER, JOSEPH LOUIS, b Brooklyn, NY, Jan 16, 22; m 46; c 2. MATHEMATICAL PHYSICS. Educ: Brooklyn Col, BA, 43; Columbia Univ, MA, 48; NY Univ, PhD(physics), 51. Prof Exp: Asst jet propulsion, NY Univ, 47-51; eng specialist, Repub Aviation Corp, 51-55; prin scientist res & adv develop div, Avco Corp, 55-56; chief scientist plasma propulsion & power lab, Repub Aviation Corp, 56-62; sr consult scientist res & tech labs, Avco Space Systs Div, 62-67, prin staff scientist, 67-70; PROF MATH, LOWELL UNIV, 70- Mem: AAAS; Am Phys Soc; assoc fel Am Inst Aeronaut & Astronaut. Res: Magnetohydrodynamics; ferrohydrodynamics; application of plasma physics and aerophysics to propulsion; generation of electricity; space flight. Mailing Add: 39 Vine St Reading MA 01867

NEURINGER, LEO (JOSEPH), solid state physics, see 12th edition

NEUROTH, MILTON L, b Ft Wayne, Ind, Nov 12, 08; m 36; c 3. PHARMACY. Educ: Purdue Univ, BS, 35, MS, 40, PhD(pharm), 46. Prof Exp: Control chemist, Gen Cable Corp, 29-33; pharmacist, Meyer Bros Drug Co, 35-36; from instr to asst prof pharm, Ohio Northern Univ, 36-43; instr pharmaceut chem, Purdue Univ, 43-44; from asst prof to prof, 46-74, chmn dept, 49-74, EMER PROF PHARM, MED COL VA, 74- Concurrent Pos: Mem revision comt, US Pharmacopeia, 60-70. Mem: Am Med Writers' Asn; Am Pharmaceut Asn; Am Asn Cols Pharm. Res: Pharmaceutical preparations; development and research. Mailing Add: Dept Pharm Med Col of Va Va Commonwealth Univ Richmond VA 23298

NEUSE, EBERHARD WILHELM, b Berlin, Ger, Mar 7, 25; m 63. POLYMER CHEMISTRY, ORGANOMETALLIC CHEMISTRY. Educ: Hanover Tech Univ, BS, 48, MS, 50, PhD(org chem), 53. Prof Exp: Res asst chem heterocyclics, Hanover Tech Univ, 51-53; head appln lab, O Neynaber & Co, AG, Ger, 54-57; res assoc plastics lab, Princeton Univ, 57-59; head polymer lab missile & space systs div, Douglas Aircraft Co, 60-70, chief plastics & elastomers develop sect, 67-70; sr lectr chem, 71-73, READER PHYS CHEM & PHYSICS, UNIV WITWATERSRAND, 73- Concurrent Pos: Mem, Nat Adv Comt Plastics Educ; Mem comt, Macromolecular Div, Int Union Pure & Appl Chem, 75- Mem: Am Chem Soc; NY Acad Sci; SAfrican Chem Inst. Res: Organic and organometallic chemistry of monomeric and polymeric compounds; development of polymeric materials for high temperature applications. Mailing Add: Dept of Chem Univ of the Witwatersrand Johannesburg South Africa

NEUSEL, ROBERT H, physics, mathematics, see 12th edition

NEUSHUL, MICHAEL, JR, b Shanghai, China, Dec 27, 33; nat US. BOTANY. Educ: Univ Calif, Los Angeles, BA, 55, PhD(bot), 59. Prof Exp: Asst bot, Univ Calif, Los Angeles, 55-56; res biologist, Scripps Inst, 56-58; asst bot, Univ Calif, Los Angeles, 58-59; NSF fel, Univ London, 59-60; from instr to asst prof bot, Univ Wash, 60-63; from asst prof to assoc prof, 63-73, PROF BOT, UNIV CALIF, SANTA BARBARA, 73- Concurrent Pos: Botanist, Arg Antarctic Exped, 57-58; vis fel, Swiss Fed Inst Technol, Zurich, 69-70; partic, Scripps Mex Oceanog Cruises. Mem: Am Phycol Soc; Bot Soc Am; Brit Phycol Soc; Int Soc Plant Morphol; Arg Antarctic Asn. Res: Marine algology; ultrastructure; development; sublittoral ecology; antarctic marine algae; algae development. Mailing Add: Dept of Biol Univ of Calif Santa Barbara CA 93106

NEUSS, JACOB DAVID, chemistry, see 12th edition

NEUSTADT, BERNARD RAY, b Washington, DC, May 7, 43; m 64; c 3. MEDICINAL CHEMISTRY. Educ: Columbia Univ, AB, 64; Brandeis Univ, PhD(org chem), 69. Prof Exp: Jr chemist, Arthur D Little, Inc, 66; sr chemist, 69-74, PRIN SCIENTIST, SCHERING CORP, 74- Mem: Am Chem Soc. Res: Medicinal organic chemistry. Mailing Add: Schering Corp Bloomfield NJ 07003

NEUSTADT, LUCIEN WOLF, mathematics, deceased

NEUSTADTER, SIEGFRIED FRIEDRICH, b Ger, July 5, 23; nat US. MATHEMATICS. Educ: Univ Calif, PhD(math), 48. Prof Exp: Lectr math, Univ Calif, 48-49; Peirce instr, Harvard Univ, 49-52; mem staff, Lincoln Lab, Mass Inst Technol, 52-58; PROF MATH, SAN FRANCISCO STATE UNIV, 58- Concurrent Pos: Mem staff res lab, Sylvania Elec Prod, Inc, Mass, 62-65. Mem: Am Math Soc; Opers Res Soc Am. Res: Applied mathematics; mathematical analysis. Mailing Add: San Francisco State Univ Dept of Math 1600 Holloway Ave San Francisco CA 94132

NEUSTEIN, HARRY (BERNARD), b Pittsburgh, Pa, June 13, 24. PATHOLOGY. Educ: Univ Pittsburgh, BS, 47; Univ Cincinnati, MD, 49; Am Bd Path, dipl, 57. Prof Exp: From intern to asst resident path, Cincinnati Gen Hosp, 49-51; resident, Children's Hosp of Pittsburgh, 53-54 & Colo Gen Hosp, 54-55; asst, Sch Med, Univ Colo, 55-56, from instr to asst prof, 56-63; USPHS spec fel, Univ Calif, Los Angeles, 63-66; assoc prof, 66-71, PROF PATH, UNIV SOUTHERN CALIF, 71- Concurrent Pos: Pathologist, Los Angeles Children's Hosp, 66- Mem: Electron Micros Soc Am; NY Acad Sci; Int Acad Path. Res: Kidney disease; electron microscopy; pulmonary disease in the newborn. Mailing Add: Dept of Path Los Angeles Children's Hosp Los Angeles CA 90054

NEUTS, MARCEL FERNAND, b Ostend, Belg, Feb 21, 35; m 59; c 4. STATISTICS, OPERATIONS RESEARCH. Educ: Univ Louvain, Lic math, 56; Stanford Univ, MSc, 59, PhD(statist), 61. Prof Exp: Instr math, Univ Lovanium, Leopoldville, 56-57; from asst prof to assoc prof math & statist, 62-68, PROF MATH & STATIST, PURDUE UNIV, 68- Concurrent Pos: Consult, Gen Motors Res Labs, 64-66; vis prof, Cornell Univ, 68-69. Honors & Awards: Lester R Ford Award, Math Asn Am. Mem: Fel Inst Math Statist; Math Asn Am; Opers Res Soc Am; Opers Res Soc Israel; Statist Asn Can. Res: Probability theory; biological models, order statistics and numerical methods in probability; queueing theory; Markov chains; general stochastic processes. Mailing Add: Dept of Statist Purdue Univ West Lafayette IN 47906

NEUVAR, ERWIN W, b Hallettsville, Tex, Mar 13, 30; m 56; c 2. INORGANIC CHEMISTRY. Educ: Tex A&M Univ, BS, 52, PhD(inorg chem), 62. Prof Exp: Chemist, Hanford Atomic Prod Oper, Gen Elec Co, 55-58; SR CHEMIST, MINN MINING & MFG CO, 62- Res: Ion exchange membrane technology; radiochemical analysis; microwave spectroscopy; fluorine and polymer chemistry; gas chromatography. Mailing Add: Minn Mining & Mfg Co Facsimile Prod Dept 3M Ctr St Paul MN 55119

NEUVILLE, MORRIS LOUIS, b Brussels, Wis, June 23, 28; m 49; c 5. CHEMISTRY. Educ: St Norbert Col, BSc, 48. Prof Exp: Res chemist, 48-54, from appln res chemist to mgr com develop dept, 54-61, mgr sales, 61-64, vpres & gen mgr chem div, 64-67, group vpres opers & develop, 67-70, exec vpres opers, 70-74, PRES, ANSUL CHEM CO, 74- Mem: Am Chem Soc; Nat Agr Chem Asn; Presidents Asn; Mfg Chemists Asn. Res: Refrigeration research; methyl chloride; organic methylations. Mailing Add: Ansul Co One Stanton St Marinette WI 54143

NEUWIRTH, JEROME H, b Brooklyn, NY, Mar 7, 31; m 57; c 2. MATHEMATICS. Educ: City Col New York, BS, 52; Univ Ill, MS, 54; Mass Inst Technol, PhD(math), 59. Prof Exp: Asst prof math, Rutgers Univ, 59-63; mathematician, NASA, 63-65; assoc prof math, Hunter Col, 65-67; assoc prof, 67-73, PROF MATH, UNIV CONN, 73- Mem: Am Math Soc. Res: Harmonic analysis. Mailing Add: Dept of Math Univ of Conn Storrs CT 06268

NEUWIRTH, LEE PAUL, mathematics, see 12th edition

NEUWIRTH, ROBERT SAMUEL, b New York, NY, July 11, 33; m 57; c 4. OBSTETRICS & GYNECOLOGY. Educ: Yale Univ, BS, 55, MD, 58. Prof Exp: Intern surg, Columbia-Presby Med Ctr, 58-59, resident obstet & gynec, 59-64; from asst prof to assoc prof, Columbia Univ, 64-71; prof, Albert Einstein Col Med, 71-72; PROF OBSTET & GYNEC, COLUMBIA UNIV, 72-; DIR OBSTET & GYNEC, WOMAN'S HOSP, ST LUKE'S MED CTR, 74- Concurrent Pos: Am Cancer Soc grant, Columbia Univ, 63-64; dir obstet & gynec, Bronx-Lebanon Hosp Ctr, 67-72; consult, Nat Inst Child Health & Human Develop, 71- & Fertil Control, WHO, 72-; consult, Wausau Ins Co. Mem: Soc Gynec Invest; Am Col Obstet & Gynec. Res: Gynecologic endoscopy; fertility control; methods of female sterilization; infertility and reproductive failure. Mailing Add: St Luke's Med Ctr Woman's Hosp 1111 Amsterdam Ave New York NY 10025

NEUWORTH, MARTIN BRANDT, organic chemistry, see 12th edition

NEUZIL, EDWARD F, b Chicago, Ill, Oct 12, 30; m 55; c 2. NUCLEAR CHEMISTRY. Educ: NDak State Col, BS, 52; Purdue Univ, MS, 54; Univ Wash, PhD(nuclear chem), 59. Prof Exp: From asst prof to assoc prof, 59-66, PROF CHEM, WESTERN WASH STATE COL, 66- Mem: Am Chem Soc; Fedn Am Scientists. Res: Nuclear fission of lighter elements; geochemistry; chemical kinetics; radiation damage. Mailing Add: Dept of Chem Western Wash State Col Bellingham WA 98225

NEUZIL, JOHN PAUL, b Decorah, Iowa, Aug 8, 42; m 66; c 1. MATHEMATICS. Educ: Univ Iowa, BA, 64, MS, 66, PhD(math), 69. Prof Exp: ASST PROF MATH, KENT STATE UNIV, 69- Mem: Am Math Soc. Res: Geometric topology. Mailing Add: Dept of Math Kent State Univ Kent OH 44242

NEUZIL, RICHARD WILLIAM, b Chicago, Ill, Sept 4, 24; m 55; c 2. CHEMISTRY. Educ: Roosevelt Univ, BS, 50. Prof Exp: Spectroscopist, 50-55, supvr, 55-58, proj leader radiation chem, 58-59, proj leader air pollution control, 59-63, assoc res coordr catalyst eval, 63-66, RES COORDR, CATALYST EVAL, UNIVERSAL OIL PROD CO, 66- Mem: AAAS; Am Chem Soc. Res: Determination of physical and thermochemical constants; thermodynamic properties; continuous process instrumentation; methods for separation and purification of chemical compounds for use on an industrial scale. Mailing Add: Universal Oil Prod 10 UOP Plaza Algonquin & Mt Prospect Rds Des Plaines IL 60016

NEVA, ARNOLD CARL, pharmacognosy, see 12th edition

NEVA, FRANKLIN ALLEN, b Cloquet, Minn, June 8, 22; m 47; c 3. MICROBIOLOGY, INTERNAL MEDICINE. Educ: Univ Minn, MD, 46; Am Bd Internal Med, dipl, 54. Prof Exp: House officer med, Boston City Hosp, 46-47, asst resident internal med, Harvard Med Servs, 49-50; Nat Res Coun fel poliomyelitis, Dept Microbiol, Sch Pub Health, Harvard Univ, 50-51, Nat Found Infantile Paralysis fel virol, Res Div, Infectious Dis, Children's Hosp, Harvard Med Sch, 51-53; asst prof res bact & instr res med, Sch Med, Univ Pittsburgh, 53-55; from asst prof to assoc prof trop pub health, Sch Pub Health, Harvard Univ, 55-64, John Laporte Given prof, 64-69; CHIEF LAB PARASITIC DIS, NAT INST ALLERGY & INFECTIOUS DIS, 69- Concurrent Pos: Area consult, Vet Admin, 57-64; mem comn parasitic dis & assoc mem comn virus infections, Armed Forces Epidemiol Bd, 60-68 & 65-68; mem, Latin Am Sci Bd, Nat Acad Sci-Nat Res Coun, 63-68; mem bd sci counr, Nat Inst Allergy & Infectious Dis, 66-69, mem study sect virus & rickettsial dis, 68-70; mem med adv bd, Leonard Wood Mem Found, 68- Honors & Awards: Bailey K Ashford Award, 65. Mem: Soc Exp Biol & Med; Asn Am Physicians; Infectious Dis Soc Am; Am Soc Trop Med & Hyg. Res: Virus, rickettsial and parasitic diseases; clinical infectious diseases. Mailing Add: Bldg 5 Rm 114 NIH Bethesda MD 20014

NEVE, RICHARD ANTHONY, b Los Angeles, Calif, Nov 3, 23; m 53; c 2.

BIOCHEMISTRY, MARINE CHEMISTRY. Educ: Loyola Univ, BS, 48; Univ San Francisco, MS, 51; Univ Ore, PhD(biochem), 56. Prof Exp: Biologist, US Naval Radiol Defense Lab, Calif, 49-51; asst biochem, Med Sch, Univ Ore, 51-56, res assoc, 56-58; USPHS fel, Univ Calif, 58-60; dir chem lab, Providence Hosp, 60-62; assoc prof biol & chmn dept, Seattle Univ, 62-66; prof biochem & dean grad sch, Cent Wash State Col, 66-70; PROF MARINE SCI & COORDR COASTAL LABS, UNIV ALASKA, 70- Mem: Am Chem Soc; Am Soc Zoologists; Brit Biochem Soc. Res: Porphyrin metabolism; hemoglobin synthesis; iron metabolism; application of enzymology; biochemical evolution of hemoglobins; marine carotenoids; biochemical behavior patterns; paralytic shellfish poisoning; aquaculture. Mailing Add: Inst of Marine Sci Univ of Alaska PO Box 617 Seward AK 99664

NEVENZEL, JUDD CUTHBERT, b Tucson, Ariz, Oct 8, 20. LIPID CHEMISTRY. Educ: Univ Ariz, BS, 41, MS, 42; Calif Inst Technol, PhD(chem), 49. Prof Exp: Res chemist, Wood Conversion Co, 42; chemist, Nat Defense Res Comt proj, Calif Inst Technol, 43; jr chemist, Los Alamos Sci Lab, 43-46; res fel chem, Ohio State Univ, 49-50; assoc res biochemist, Scripps Inst Oceanog, Univ Calif, San Diego, 70-73; assoc res biochemist, 51-73, RES BIOCHEMIST, LAB NUCLEAR MED & RADIATION BIOL, UNIV CALIF, LOS ANGELES, 73- Concurrent Pos: Mem subcomt lipids, Comt Biol Chem, Nat Acad Sci-Nat Res Coun, 59- Mem: Am Chem Soc; Am Oil Chem Soc. Res: Synthesis of C-14 labeled fatty acids; composition, biosynthesis and function of waxes; lipids of deep-sea organisms. Mailing Add: Box 24222 Los Angeles CA 90024

NEVEU, DARWIN D, b Green Bay, Wis, Feb 26, 33. ANALYTICAL CHEMISTRY. Educ: Wis State Univ, Oshkosh, BS, 58. Prof Exp: Chemist, Freeman Chem Co, 63-65; asst chief chemist, WVa Pulp & Paper Co, 65-66; chemist, Newport Army Ammunition Ctr, FMC Corp, 66-67, Celanese Plastics, 67-69; mgr org lab, Crobaugh Labs, 69-70; LAB DIR CHEM ANAL & CONSULT, NALIN LABS, 70- Concurrent Pos: Fac adv, Wooster Agr Tech Inst, 73- Mem: Am Chem Soc; Am Soc Testing & Mat; Am Soc Metals; Soc Plastics Engrs; Electrochem Soc. Mailing Add: Nalin Labs 2641 Cleveland Ave Columbus OH 43211

NEVEU, MAURICE C, b Nashua, NH, Feb 3, 29; m 55; c 5. PHYSICAL CHEMISTRY, ORGANIC CHEMISTRY. Educ: Univ NH, BS, 52, MS, 55; Ill Inst Technol, PhD(chem), 59. Prof Exp: Asst, Univ NH, 53-54; asst, Ill Inst Technol, 54-58; instr, Ohio Northern Univ, 58-59, asst prof chem, 59-60; asst prof, Longwood Col, 60-64; ASSOC PROF CHEM, STATE UNIV NY COL FREDONIA, 64- Mem: Am Chem Soc. Res: Mechanisms of organic reactions; chemical kinetics; catalysis; isotope effects; mechanism of ester hydrolysis; enzymology; metabolic pathways by carbon-14 tracers. Mailing Add: 18 Hanover St Silver Creek NY 14136

NEVILL, WILLIAM ALBERT, b Indianapolis, Ind, Jan 1, 29; m 50; c 5. ORGANIC CHEMISTRY. Educ: Butler Univ, BS, 51; Calif Inst Technol, PhD(chem biol), 54. Prof Exp: Res chemist, Procter & Gamble Co, 54; from asst prof to assoc prof chem, Grinnell Col, 56-67, chmn dept, 64-67; prof, Purdue Univ & chmn dept chem, Indianapolis Campus, 67-70; asst dean acad affairs, 70-71, actg dean, 38th St Campus, 71-72, actg dean, Sch Eng & Technol, 72-74, DEAN, SCH SCI, IND UNIV-PURDUE UNIV, INDIANAPOLIS, 72- Mem: Am Chem Soc. Res: Mechanisms of organic reactions relating to small ring compounds and synthesis of nitrogen heterocycles. Mailing Add: 11514 Dona Dr Carmel IN 46032

NEVILLE, DAVID MICHAEL, JR, b New York, NY, Apr 7, 34; m 56; c 2. MOLECULAR BIOLOGY. Educ: Univ Rochester, MD, 59. Prof Exp: Student fel path, Strong Mem Hosp, Rochester, NY, 56-57; intern internal med, Grace New Haven Community Hosp, New Haven, Conn, 59-60; CHIEF SECT BIOPHYS CHEM, LAB NEUROCHEM, NAT INST MENT HEALTH, 60- Res: Cell membranes and their role in the study of cellular differentiation. Mailing Add: Lab of Neurochem Nat Inst of Ment Health Bethesda MD 20014

NEVILLE, DONALD EDWARD, b Los Angeles, Calif, Apr 5, 36. THEORETICAL HIGH ENERGY PHYSICS. Educ: Loyola Univ, Los Angeles, BS, 57; Univ Chicago, SM, 60, PhD(physics), 62. Prof Exp: Fel physics, Univ Calif, Berkeley, 62-64, Europ Orgn Nuclear Res, Geneva, 64-65 & Lawrence Radiation Labs, Livermore, 65-67; ASSOC PROF PHYSICS, TEMPLE UNIV, PHILADELPHIA, 67- Mem: Am Phys Soc. Res: Theoretical particle physics, especially constraints imposed by duality and symmetries obeyed by scattering amplitudes; impact of particle physics on cosmology and astrophysics. Mailing Add: Dept of Physics Temple Univ Philadelphia PA 19122

NEVILLE, GWEN KENNEDY, b Taylor, Tex, Mar 23, 38; m; c 3. ANTHROPOLOGY. Educ: Mary Baldwin Col, BV, 59; Univ Fla, MA, 68, PhD(anthrop), 71. Prof Exp: ASST PROF ANTHROP, EMORY UNIV, 71- Concurrent Pos: Consult, Am Inst Develop Educ, 72-73; Emory Res Coun res fel, 72. Mem: Fel Am Anthrop Asn; Soc Appl Anthrop; Am Ethnol Soc; Coun Anthrop & Educ. Res: Dynamics of ritual and ceremonial life in maintaining and transmitting culture; community as text; processes of enculturation and cultural transmission; research locales Scotland and the American South. Mailing Add: Dept of Sociol & Anthrop Emory Univ Atlanta GA 30322

NEVILLE, JAMES RYAN, b San Jose, Calif, May 13, 25; m 49; c 2. MEDICAL PHYSIOLOGY. Educ: Stanford Univ, AB, 49, MA, 51, PhD(physiol), 55. Prof Exp: Asst physiol, Stanford Univ, 52-55; aviation physiologist, US Air Force Sch Aerospace Med, 55-57, from asst prof to assoc prof biophys, 57-63, chief biophys sect, 63-70; CLIN RES PHYSIOLOGIST, LETTERMAN ARMY INST RES, 70- Mem: AAAS; Biophys Soc; Instrument Soc Am; Aerospace Med Asn; Am Physiol Soc. Res: Respiratory physiology; gas transport; enzymes; methodology and bioinstrumentation; polarography. Mailing Add: Letterman Gen Hosp PO Box 112 San Francisco CA 94129

NEVILLE, JANICE NELSON, b Schenectady, NY, Dec 1, 30; m 53; c 2. NUTRITION. Educ: Carnegie Inst Technol, BS, 52; Univ Ala, MS, 53; Univ Pittsburgh, MPH, 62, DSc(hyg), 64. Prof Exp: Instr diet ther & clin dietitian, Sch Nursing, Hillman Clins, Univ Ala, 54; res dietitian, Grad Sch Pub Health, Univ Pittsburgh, 56-61, res asst nutrit, 61-64, asst res prof, 64-65; asst prof, 65-69, ASSOC PROF NUTRIT, CASE WESTERN RESERVE UNIV, 69- Mem: AAAS; Am Dietetic Asn; Am Pub Health Asn; Soc Nutrit Educ. Res: Diet therapy; community health. Mailing Add: Dept of Nutrit Case Western Reserve Univ Cleveland OH 44106

NEVILLE, JOHN F, JR, b Somerville, Mass, Sept 28, 22; m 46; c 5. MEDICINE. Educ: Yale Univ, BS, 44, MD, 46. Prof Exp: Intern surg, Barnes Hosp, St Louis, Mo, 46-47; asst resident, Wash Univ, 47-48 & 50-52, fel thoracic surg, 52-54; clin asst prof, Univ Pittsburgh, 57; assoc prof surg, 58-74, PROF SURG, COL MED, STATE UNIV NY UPSTATE MED CTR, 74-, CHIEF CARDIOPULMONARY SECT, 58- Concurrent Pos: Attend surgeon, Vet Admin Hosp, 58-; assoc surgeon, Syracuse Mem Hosp, 58- Res: Cardiovascular physiology and surgery. Mailing Add: 750 E Adams St Syracuse NY 13210

NEVILLE, MARGARET COBB, b Greenville, SC, Nov 4, 34; m 57; c 2. PHYSIOLOGY. Educ: Pomona Col, BA, 56; Univ Pa, PhD(physiol), 62. Prof Exp: Res assoc cell physiol, Dept Molecular Biol, Pa Hosp, Philadelphia, 64-68; from instr to asst prof, 68-75, ASSOC PROF PHYSIOL, MED CTR, UNIV COLO, DENVER, 75- Mem: Am Physiol Soc; Biophys Soc. Res: Amino acid transport into cells; function of sarcoplasmic reticulum of skeletal muscle; hormonal control of cellular transport of nutrients; transport processes in the crystalline lens. Mailing Add: Dept of Physiol Univ of Colo Med Ctr Denver CO 80220

NEVILLE, MELVIN EDWARD, b Marietta, Ohio, July 20, 44; m 69; c 2. MICROBIOLOGY. Educ: Marietta Col, AB, 66; Iowa State Univ, MS, 69, PhD(bacteriol), 73. Prof Exp: Jr high sch teacher sci, Des Moines Community Sch, 69-70; teaching asst bacteriol, Iowa State Univ, 69-71, instr, 71, res teaching asst, 71-73; ASST PROF MICROBIOL, ILL STATE UNIV, 73- Mem: Am Soc Microbiol; AAAS; Am Inst Biol Sci. Res: Characterization of coryneform bacteria—genetics, morphology, physiology, taxonomy and nomenclature; compositional similarities of DNAs from microorganisms. Mailing Add: Dept of Biol Sci Ill State Univ Normal IL 61761

NEVILLE, MELVIN K, b Evanston, Ill, Jan 5, 39; m 64; c 1. ANTHROPOLOGY. Educ: Calif Inst Technol, BS, 60; Harvard Univ, AB, 61, PhD(anthrop), 67. Prof Exp: ASST PROF ANTHROP, UNIV CALIF, DAVIS, 66- Concurrent Pos: Fac res fel, Univ Calif, 67-68. Mem: AAAS; Ecol Soc Am; Am Asn Phys Anthropologists. Res: Primatology, especially primate behavior. Mailing Add: Dept of Anthrop Univ of Calif Davis CA 95616

NEVILLE, PATRICK FINLEY, biochemistry, see 12th edition

NEVILLE, ROY GERALD, chemistry, environmental sciences, see 12th edition

NEVILLE, WALTER EDWARD, JR, b Rabun Gap, Ga, May 5, 24; m 48; c 2. ANIMAL SCIENCE. Educ: Univ Ga, BS, 47; Univ Mo, MS, 50; Univ Wis, PhD(genetics, animal husb), 57. Prof Exp: Asst county agr agent agr exten, Univ Ga, 47-49; asst animal breeding, 50-51, ASSOC ANIMAL BREEDING, COASTAL PLAIN EXP STA, 51- Mem: AAAS; Am Soc Animal Sci. Res: Genetic correlations and heritability estimates among various economic traits in beef cattle; physiology of reproduction in farm animals; nutritive requirements of beef cattle and their calves. Mailing Add: Coastal Plain Exp Sta Tifton GA 31794

NEVILLE, WILLIAM E, b Fairbury, Nebr, Apr 13, 19; m 58; c 5. SURGERY. Educ: Univ Nebr, BS & MD, 43. Prof Exp: Assoc prof surg, Univ Ill Col Med, 62-71; PROF SURG & DIR CARDIOTHORACIC SURG, COL MED & DENT NJ, 71- Mem: Fel Am Col Surg; Soc Thoracic Surg; Am Asn Thoracic Surg; Int Cardiovasc Soc; Am Surg Asn. Res: Cardiothoracic surgery. Mailing Add: Col of Med & Dent of NJ Newark NJ 07107

NEVIN, CHARLES S, chemistry, see 12th edition

NEVIN, FLOYD REESE, b Long Eddy, NY, Feb 24, 04; m 34; c 1. ACAROLOGY. Educ: Temple Univ, AB, 27; Univ Pa, MA, 29; Cornell Univ, PhD(med entom), 34. Prof Exp: Asst, Temple Univ, 27-29; instr biol, Ursinus Col, 29-30; asst biol, Cornell Univ, 30-34, instr, 34-44; prof & head sci, Rider Col, 44-46; prof biol, 46-73, chmn dept sci & math, 50-64, EMER PROF BIOL, STATE UNIV COL ARTS & SCI, PLATTSBURGH, NY, 73- Concurrent Pos: Fel sch hyg & pub heatlh, Johns Hopkins Univ, 43-44; asst bur entom, USDA. Mem: Entom Soc Am; Nat Asn Biol Teachers; Am Inst Biol Sci. Res: Medical entomology; taxonomy of Oribatid mites; parasitology. Mailing Add: Dept of Biol Sci State Univ Col of Arts & Sci Plattsburgh NY 12901

NEVIN, ROBERT STEPHEN, b New York, NY, Oct 20, 33; m 55. ORGANIC CHEMISTRY, POLYMER CHEMISTRY. Educ: Queens Col, NY, BS, 55; St John's Univ, NY, MS, 57; State Univ NY Col Forestry, Syracuse, PhD(chem), 61. Prof Exp: Res chemist, Esso Res & Eng Co, 61-63; sr res chemist, J T Baker Chem Co, 63-67; sr res chemist, 67, SR POLYMER CHEMIST, LILLY RES LAB, ELI LILLY & CO, 67- Mem: Am Chem Soc. Res: Organic chemistry of polymers; ionic polymerization; monomer synthesis; chemical modification of polymers. Mailing Add: Lilly Res Lab MC525 Eli Lilly & Co Indianapolis IN 46206

NEVIN, THOMAS ANDREW, b New York, NY, Oct 20, 17; m 43; c 2. BACTERIOLOGY. Educ: Stetson Univ, BS, 49; Univ Ala, MS, 52; Univ Southern Calif, PhD(bact), 57. Prof Exp: Res assoc bact, Eastman Dent Dispensary, 49-50; instr, Exten Div, Univ Ala, 52-53; dir control labs, Delta Labs, 53-54; res bacteriologist, Nat Inst Dent Res, 57-60; asst dir, Germfree Life Res Ctr, 60-63; chief bact physiol & chem lab, Venereal Dis Res Lab, Communicable Dis Ctr, 63-68; RES PROF MICROBIOL, FLA INST TECHNOL, 68- Mem: Am Soc Microbiol; Sigma Xi. Res: Bacterial physiology; nutritional bases of bacterial interactions; physiology and nutrition of spirochetes. Mailing Add: Dept of Microbiol Fla Inst of Technol Melbourne FL 32901

NEVINS, ARTHUR JAMES, b New York, NY, Sept 22, 37; m 60; c 2. COMPUTER SCIENCE. Educ: Mass Inst Technol, BS, 59; Univ Rochester, MA, 62, PhD(econ), 65. Prof Exp: Fel, Carnegie Inst Technol, 64-65; sr scientist logistics res proj, George Washington Univ, 65-69, sr staff scientist, Inst Mgt Sci, 69-72; res scientist, Artificial Intel Lab, Mass Inst Technol, 72-74; MEM FAC, DEPT INFO SYSTS, GA STATE UNIV, 74- Mem: Asn Comput Mach. Res: Artifical intelligence. Mailing Add: 5343 O'Reilly Lane Stone Mountain GA 30083

NEVINS, DONALD JAMES, b San Luis Obispo, Calif, July 6, 37; m 62; c 2. PLANT PHYSIOLOGY. Educ: Calif State Polytech Col, BS, 59; Univ Calif, Davis, MS, 61, PhD(plant physiol), 65. Prof Exp: NIH res assoc chem, Univ Colo, 65-67; from asst prof to assoc prof bot, 67-74, PROF BOT, IOWA STATE UNIV, 74- Concurrent Pos: Vis prof, Osaka City Univ, Japan Soc Prom Sci, 74-75. Mem: AAAS; Am Soc Plant Physiol; Am Inst Biol Sci; Japanese Soc Plant Physiologists. Res: Physiology of cell walls, growth and development. Mailing Add: Dept of Bot & Plant Path Iowa State Univ Ames IA 50011

NEVIS, ARNOLD HASTINGS, b Albuquerque, NMex, Aug 3, 31; m 52; c 3. NEUROLOGY, BIOPHYSICS. Educ: Calif Inst Technol, BS, 47; Harvard Med Sch, MD, 51; Mass Inst Technol, PhD(biophys), 56. Prof Exp: Asst prof physiol & coordr muscular dystrophy res, Med Br, Univ Tex, 56-57; clin fel neurol, Med Ctr, Univ Calif, Los Angeles, 58-60; asst prof neurol, 60-67, assoc prof neurol & elec eng, 67-69, PROF BIOPHYS & BIOMED ENG, COL MED, UNIV FLA, 69- Mem: Soc Nuclear Med; Biophys Soc. Res: Neurophysiology, especially tissue impedance in relation to electrophysiology, ultrastructure and water and electrolyte transport across membranes; disease of muscle. Mailing Add: Dept of Physics Univ of Fla Gainesville FL 32601

NEVISON, THOMAS OLIVER, JR, b Cleveland, Ohio, Apr 15, 29; m 56; c 3. PHYSIOLOGY, AEROSPACE MEDICINE. Educ: Harvard Univ, AB, 51, MD, 56,

MIH, 59. Prof Exp: Intern, Sch Med, Univ Ore Hosps, 56-57; Guggenheim fel, Harvard Univ, 57-59; sr staff mem aerospace med, 61-63, HEAD DEPT BIOL INSTRUMENTATION, LOVELACE FOUND MED EDUC & RES, 63- Concurrent Pos: Sci adv life sci, Int Found Telemetering, 64- Mem: Aerospace Med Asn; Am Inst Aeronaut & Astronaut. Res: Physiological problems of high altitude and space flight; clinical and experimental monitoring techniques; bioinstrumentation. Mailing Add: Lovelace Found 5200 Gibson SE Albuquerque NM 87108

NEVITT, MICHAEL VOGT, b Lexington, Ky, Sept 7, 23; m 46; c 4. METAL PHYSICS. Educ: Univ Ill, BS, 44, PhD(metall eng), 54; Va Polytech Inst, MS, 51. Prof Exp: Res metallurgist, Olin Industs Inc, 46-48; from asst prof to assoc prof metall eng Va Polytech Inst, 48-55 & head dept, 53-55; assoc phys metallurgist, Metall Div, 55-64, group leader, 55-66, phys metallurgist, 64-65, dir metall div, 66-69, DEP LAB DIR, ARGONNE NAT LAB, 69- Concurrent Pos: Mem adv comt, Polysci Corp, 54; vis prof, Univ Sheffield, 65-66; mem adv comt, Mat Res Lab, Univ Ill, 73-; mem rev comt, Inorg Mat Res Div, Univ Calif, Berkeley, 75- Mem: Fel Am Soc Metals; Sigma Xi; Am Inst Mining, Metall & Petrol Eng. Res: Physical and engineering metallurgy of nuclear materials; alloy theory of transition and actinide elements; magnetic properties. Mailing Add: 1507 Center Ave Wheaton IL 60187

NEVITT, THOMAS D, b Kewanee, Ill, July 22, 25; m 49; c 3. ORGANIC CHEMISTRY, PETROLEUM CHEMISTRY. Educ: Bradley Univ, BS, 47; Iowa State Col, MS, 50, PhD(phys org chem), 53. Prof Exp: PROJ MGR, AMOCO OIL CO, 54- Mem: Am Chem Soc; The Chem Soc. Res: Catalysis; chemical kinetics; reaction engineering. Mailing Add: 4 S 575 Karns Rd Naperville IL 60540

NEVLING, LORIN IVES, JR, b St Louis, Mo, Sept 23, 30; m 57; c 5. PLANT TAXONOMY. Educ: St Mary's Col, BS, 52; Wash Univ, AM, 57, PhD(bot), 59. Prof Exp: Researcher, Mo Bot Garden, 59; from asst cur to assoc cur, Arnold Arboretum, Harvard Univ, 59-69, cur, 69-73, supvr, Gray Herbarium, 63-72, assoc cur, 63-69, coordr bot syst collections, 72-73; CUR & CHMN, DEPT BOT, FIELD MUS NATURAL HIST, 73-; ADJ PROF BIOL SCI, NORTHERN ILL UNIV, 74-; LECTR BIOL SCI, NORTHWESTERN UNIV, 74- Concurrent Pos: Asst cur vascular plants, NE Bot Club, 62-73; lectr biol, Harvard Univ, 68-69 & 71-73; mem, Conserv Comn, City Boston, 73. Honors & Awards: George R Cooley Prize, 70. Mem: Bot Soc Am; Asn Trop Biol; Am Soc Plant Taxon (secy, 66-71); Int Asn Plant Taxon; Am Inst Biol Sci. Res: Taxonomy of the Thymelaeaceae; flora of Veracruz, Mexico. Mailing Add: Dept Bot Field Mus Natural Hist Roosevelt Rd & Lake Shore Dr Chicago IL 60605

NEVYAS, JACOB, b Philadelphia, Pa, Oct 31, 97; m 30; c 2. BIOCHEMISTRY. Educ: Swarthmore Col, AB, 19; Univ Ill, AM, 22; Univ Pittsburgh, PhD, 26. Hon Degrees: ScD, Pa Col Optom, 71. Prof Exp: Prof chem, 32-71, EMER PROF CHEM, PA COL OPTOM, 71- Mem: Am Chem Soc. Mailing Add: 1901 John F Kennedy Blvd Philadelphia PA 19103

NEW, EARL HIRAM, b Del Co, Iowa, Feb 10, 10; m 33; c 3. ORNAMENTAL HORTICULTURE. Educ: Iowa State Univ, BS, 30; Univ Idaho, MS, 50; Mich State Univ, PhD(floricult mkt), 61. Prof Exp: Retail grower, Manchester, Iowa, 30-44; farm & garden ed, Council Bluffs, Iowa, 44-46; com nursery catalog writer, Faribault, Minn, 46-47; asst prof, Univ Idaho, 47-56; res instr floricult, Mich State Univ, 56-58, instr & coordr com floricult, 58-61; exten specialist ornamental hort, Univ Ky, 61-67; assoc prof agr, 67-71, PROF AGR, MID TENN STATE UNIV, 71- Mem: Am Soc Hort Sci. Res: Small homes landscaping; landscape maintenance; marketing of nursery and floral products and services; plant breeding and testing; production of floral and nursery products; home floriculture. Mailing Add: Dept of Agr Mid Tenn State Univ Box 402 Murfreesboro TN 37132

NEW, JOHN CALHOUN, b Warrensburg, Mo, Aug 9, 20; m 44; c 2. APPLIED MECHANICS. Educ: Univ Mo, BS, 43; Univ Md, MS, 50. Prof Exp: Instr civil eng, Univ Mo, 43-44; chief static, pressure-corrosion test sect, Naval Ord Lab, 44-50, chief simulation & design br, 50-57, chief environ simulation div, 57-59; CHIEF TEST & EVAL DIV, GODDARD SPACE FLIGHT CTR, NASA, 59- Mem: Am Soc Civil Engrs; Soc Exp Stress Anal (pres, 64-65); Am Mgt Asn; Sigma Xi; Inst Environ Sci. Res: Nondestructive differential pressure test for thin shells; evaluation of naval ordnance; environmental testing; experimental mechanics; shock and vibration; evaluation of spacecraft systems; research and development management. Mailing Add: Test & Eval Div Code 320 Goddard Space Flight Ctr Greenbelt MD 20771

NEW, JOHN G, b New York, NY, Jan 16, 27; m 50; c 4. ENVIRONMENTAL SCIENCES, VERTEBRATE ZOOLOGY. Educ: Cornell Univ, BS, 50, MS, 51, PhD(vert zool), 56. Prof Exp: Assoc prof sci, 56-60, PROF SCI, STATE UNIV NY COL ONEONTA, 60- Mem: AAAS; Am Soc Mammal; Wilson Ornith Soc; Am Soc Ichthyol & Herpet; Ecol Soc Am. Res: Environment, ecology, systematics, evolution, and life histories of vertebrates, particularly freshwater fishes and mammals. Mailing Add: Dept of Biol State Univ of NY Col Oneonta NY 13820

NEW, MARIA IANDOLO, b New York, NY, Dec 11, 28; m 49; c 3. PEDIATRICS. Educ: Cornell Univ, AB, 50; Univ Pa, MD, 54; Am Bd Pediat, dipl, 60. Prof Exp: Intern med, Bellevue Hosp, New York, 54-55; asst resident pediat, 55-57, asst pediatrician, Clin Res Ctr, 57-59, pediatrician, Outpatient Dept, 59-63, res investr diabetic study group, Comprehensive Care & Teaching Prog, 58-61, instr pediat, 58-63, asst prof & asst attend pediatrician, 63-68, assoc prof & assoc attend pediatrician, 68-71, PROF PEDIAT & ATTEND PEDIATRICIAN, NEW YORK HOSP-CORNELL MED CTR, 71-, DIR PEDIAT METAB & ENDOCRINE CLIN & DIV HEAD PEDIAT ENDOCRINOL, 64-, DIR PEDIAT CLIN RES CTR, 66-, VCHMN DEPT PEDIAT, 74- Concurrent Pos: Fel pediat metab & renal dis, New York Hosp-Cornell Med Ctr, 57-58, res fel med, 62-64; vis physician, Rockefeller Univ, 71-, consult, Albert Einstein Col Med, 74- Mem: AAAS; Am Acad Pediat; Am Pediat Soc; Soc Pediat Res; Am Fedn Clin Res. Res: Pediatric endocrinology and renal diseases; juvenile hypertension; pediatric pharmacology; growth and development from the biochemical viewpoint. Mailing Add: New York Hosp Dept of Pediat 525 E 68th St New York NY 10021

NEWALLIS, PETER EDWARD, organic chemistry, see 12th edition

NEWBERGER, EDWARD, b New York, NY, Feb 15, 40. MATHEMATICS. Educ: City Col New York, BS, 61; Ind Univ, PhD(partial differential equations), 69. Prof Exp: Asst prof, 70-74, ASSOC PROF MATH, STATE UNIV NY COL BUFFALO, 74- Mem: Am Math Soc; NY Acad Sci; Math Soc France. Res: Pseudo-differential operators; Gevrey classes; asymptotic Gevrey classes; partial differential equations. Mailing Add: Dept of Math State Univ of NY Col Buffalo NY 14222

NEWBERGER, STUART MARSHALL, b New York, NY, Oct 4, 38; m 64; c 1. MATHEMATICS. Educ: City Col New York, BEE, 60; Mass Inst Technol, PhD(math), 64. Prof Exp: Instr math, Univ Calif, Berkeley, 64-66, asst prof, 66-69; ASSOC PROF MATH, ORE STATE UNIV, 69- Mem: Am Math Soc; Math Asn Am. Res: Functional analysis, integration theory, including integration in linear space; operator theory, including unbounded operators in Hilbert Space and partial differential operators; generalized function theory; applications of operators to quantum theory. Mailing Add: Dept of Math Ore State Univ Corvallis OR 97330

NEWBERNE, JAMES WILSON, b Adel, Ga, Dec 1, 23; m 49; c 3. VETERINARY PATHOLOGY. Educ: Ala Polytech Inst, DVM, 50, MS, 54. Prof Exp: Instr path, Ala Polytech Inst, 52-55; pathologist, Pitman-Moore Co, 55-57, asst dir path dept, 57-61; assoc clin prof path, Col Med, Univ Cincinnati, 62-74; head dept path & toxicol, Wm S Merrell Co, 62-69, DIR DRUG SAFETY & METAB, MERRELL NAT LABS DIV, RICHARDSON-MERRELL INC, 69-; CLIN PROF LAB MED, MED CTR, UNIV CINCINNATTI, 74- Concurrent Pos: Assoc ed, Toxicol Appl Pharmacol, 72-; consult, Toxicol Protocol Comt, Nat Acad Sci-Nat Res Coun, 75- Mem: Am Vet Med Asn; Am Col Vet Path; Am Asn Lab Animal Sci; NY Acad Sci; Soc Toxicol. Res: Toxicopathology; virus diseases of man and animals; avian protozoa. Mailing Add: Merrell Nat Labs Div 110 E Amity Richardson-Merrell Inc Cincinnati OH 45215

NEWBERNE, PAUL M, b Adel, Ga, Nov 4, 20; m 45; c 2. NUTRITION, PATHOLOGY. Educ: Auburn Univ, DVM, 50, MSc, 51; Univ Mo, PhD(biochem, nutrit), 58. Prof Exp: Instr path, Auburn Univ, 50-51; instr microbiol, Univ Mo, 54-56, instr agr chem, 56-57, fel, NIH, 57-58; animal pathologist, Auburn Univ, 58-62; PROF NUTRIT PATH, MASS INST TECHNOL, 62- Mem: AAAS; Am Inst Nutrit; Am Col Vet Path; Am Vet Med Asn. Res: Nutritionally-induced experimental cancer; cardiovascular disease; nutritionally-induced teratology; nutritional pathology and toxicology; interaction of nutrition and toxicology. Mailing Add: Dept of Nutrit & Food Sci Mass Inst of Technol Cambridge MA 02139

NEWBERRY, ANDREW TODD, b Orange, NJ, Aug 30, 35; m 58; c 2. INVERTEBRATE ZOOLOGY. Educ: Princeton Univ, AB, 57; Stanford Univ, PhD(biol), 65. Prof Exp: Nat Acad Sci-Nat Res Coun fel, 64-65; ASSOC PROF BIOL, COWELL COL, UNIV CALIF, SANTA CRUZ, 65- Res: Invertebrate reproduction and development; biology of colonial animals; biology of ascidian tunicates; invertebrate colonies; ascidian anatomy, developmental morphology. Mailing Add: Cowell Col Univ of Calif Santa Cruz CA 95064

NEWBERRY, TRUMAN ALBERT, b St Paul, Minn, July 19, 20; m 45; c 3. CANCER. Educ: Univ Minn, BS, 45, MB, 47, MD, 48, PhD(cancer biol), 53. Prof Exp: Intern, Univ Mich Hosp, 47-48, surg resident, 48-49; Am Cancer Soc fel, Univ Minn, 49-52, instr physiol, 52-53; resident gen surg & Nat Cancer Inst trainee, Ancker Hosp, St Paul, Minn, 53-57; pvt pract, 57-58; surgeon, Mendocino State Hosp, 59-61; GEN SURGEON & ASST SUPT, STOCKTON STATE HOSP, 63- Mem: AAAS. Res: General surgery. Mailing Add: Stockton State Hosp 510 E Magnolia Stockton CA 95202

NEWBERRY, WILLIAM MARCUS, b Columbus, Ga, Nov 13, 38; m 65; c 3. INTERNAL MEDICINE. Educ: Northwestern Univ, BA, 60; Emory Univ, MD, 64. Prof Exp: Asst prof int med, Southwestern Med Sch, Univ Tex, 70-71; assoc prof, 71-75, DEAN COL MED & PROF INT MED, MED UNIV SC, 75- Concurrent Pos: Actg dean col med, Med Univ SC, 74-75. Mem: Am Fedn Clin Res; Reticuloendothelial Soc. Res: Infectious diseases, specifically the epidemiology of the systemic mycoses, and the cellular immune responses to the granulomatous infections. Mailing Add: Col Med Med Univ SC 80 Barre St Charleston SC 29401

NEWBERY, A CHRIS, b Broxbourne, Eng, July 12, 23; Can citizen; m 54; c 2. MATHEMATICS. Educ: Cambridge Univ, BA, 48; Univ London, BA, 53, PhD(math), 62; Univ BC, MA, 58. Prof Exp: Lectr math, Univ BC, 56-62; mathematician, Boeing Co, 62-63; asst prof, Univ Alta, Calgary, 63-65; assoc prof math & comput, Univ Ky, 65-67; mathematician, Boeing Co, Wash, 67-69; ASSOC PROF MATH, UNIV KY, 69- Mem: Soc Indust & Appl Math; Asn Comput Mach; Can Math Cong; Comput Soc Can. Res: Numerical analysis; application of computers to problems, particularly in ordinary differential equations; linear algebra and polynomial problems. Mailing Add: Dept of Comput Sci Univ of Ky Lexington KY 40506

NEWBOLD, JOHN EDWARD, b Hayes, Eng, Feb 15, 41; m 62; c 3. MOLECULAR BIOLOGY, VIROLOGY. Educ: Univ Birmingham, BSc, 62; Calif Inst Technol, PhD(biophys), 70. Prof Exp: Dernham jr fel oncol, Salk Inst, La Jolla, Calif, 69-71; ASST PROF BACT, SCH MED, UNIV NC, CHAPEL HILL, 71- Mem: Am Soc Microbiol. Res: Molecular biology of animal viruses, mammalian mitochondria and the eukaryote nucleus. Mailing Add: Dept of Bact Univ of NC Sch of Med Chapel Hill NC 27514

NEWBOLT, WILLIAM BARLOW, b Berea, Ky, Sept 29, 34; m 62; c 1. PHYSICS. Educ: Berea Col, BA, 56; Vanderbilt Univ, MS, 59, PhD(physics), 63. Prof Exp: From instr to asst prof, 62-67, ASSOC PROF PHYSICS, WASHINGTON & LEE UNIV, 67- Mem: Am Phys Soc; Am Asn Physics Teachers; Health Physics Soc. Res: Nuclear spectroscopy; Mössbauer effect. Mailing Add: Dept of Physics Washington & Lee Univ Lexington VA 24450

NEWBORN, GEORGE EARL, organic chemistry, see 12th edition

NEWBOULD, FRANCIS HENRY SAMUEL, b Berwick-on-Tweed, Eng, May 20, 12; Can citizen; m 36; c 4. VETERINARY MICROBIOLOGY, VETERINARY IMMUNOLOGY. Educ: Univ Toronto, BSA, 36, MSA, 57; Univ Reading, PhD(staphylococcal mastitis), Univ Reading, 64. Prof Exp: Owner, Newbould Labs, Guelph, Ont, 38-46; exten asst bact, Ont Agr Col, 46-48, lectr microbiol, 48-53, lectr, 54-56, from asst prof to assoc prof bact, 56-65, PROF, DEPT VET MICROBIOL & IMMUNOL, ONT VET COL, UNIV GUELPH, 65- Concurrent Pos: Asst supvr blood serum drying, Connaught Med Res Labs, Univ Toronto, 43-44; vis worker, Nat Inst Res Dairying, Univ Reading, 61-63; chmn, Conf Res Workers Bovine Mastitis, US, 66-67; tech consult, West Agro-Chem Inc, NY, 66- Mem: Can Soc Microbiol; Brit Soc Appl Bact. Res: Staphylococcal mastitis prevention by sanitation; penetration of teat duct by staphylococci; phagocytosis by milk leucocytes; methods of leucocyte counting. Mailing Add: Dept of Vet Microbiol & Immunol Ont Vet Col Univ of Guelph Guelph ON Can

NEWBOUND, KENNETH BATEMAN, b Winnipeg, Man, Mar 12, 29; m 47; c 4. PHYSICS. Educ: Univ Man, BSc, 40, MSc, 41; Mass Inst Technol, PhD(physics), 48. Prof Exp: Physicist naval res, Nat Res Coun Can, 41-43; assoc prof physics, 48-59, PROF PHYSICS, UNIV ALTA, 59-, ASSOC DEAN SCI, 64- Mem: Can Asn Physicists. Res: Atomic spectroscopy; spectrographic analysis; precision wave-length measurements; underwater acoustics. Mailing Add: 8910 Windsor Rd Edmonton AB Can

NEWBRUN, ERNEST, b Vienna, Austria, Dec 1, 32; US citizen; m 56; c 3. ORAL BIOLOGY, BIOCHEMISTRY. Educ: Univ Sydney, BDS, 54; Univ Rochester, MS, 57; Univ Ala, DMD, 59; Univ Calif, San Francisco, PhD(biochem), 65. Prof Exp: Res assoc, Eastman Dent Dispensary, Rochester, NY, 56-57 & Med Ctr, Univ Ala, 57-59; lectr biochem, 65, assoc res dentist & assoc prof oral biol, 65-70, PROF ORAL BIOL,

SCH DENT, UNIV CALIF, SAN FRANCISCO, 70-, CHMN SECT BIOL SCI, 72- Concurrent Pos: Nat Health & Med Res Coun dent res fel, 60-61; res teacher trainee, Med Ctr, Univ Calif, San Francisco, 61-63, fel, 63-64; biochem consult oral calculus study, Sect Epidemiol, USPHS, San Francisco, 64-65; Am Col Dent trainee, Phys Biol Sect, Inst Advan Educ Dent Res, 65; USPHS res career develop award, 65-70; vis scientist, Sch Dent, Univ Lund, 67-68; mem, Nat Caries Prog Adv Comt, Nat Inst Dent Res, 72-75 & Dent Drug Prod Adv Comt, Food & Drug Admin, 74-78. Mem: Am Inst Oral Biol; Int Asn Dent Res; Europ Orgn Caries Res. Res: Dental caries; microradiography and microhardness of enamel; chemistry; mucoproein chemistry and biosynthesis; bacterial polysaccharides, chemistry and synthesis. Mailing Add: Div of Oral Biol Univ of Calif Sch of Dent San Francisco CA 94143

NEWBURG, EDWARD A, b Indianapolis, Ind, Dec 22, 29; m 59. APPLIED MATHEMATICS. Educ: Purdue Univ, BS, 52, MS, 53; Univ Ill, PhD(math), 58. Prof Exp: Mathematician, Nuclear Div, Combustion Eng, Inc, 58-61; res scientist, Travelers Res Ctr, Inc, 61-66; assoc prof math, Worcester Polytech Inst, 66-70; assoc prof, Va Commonwealth Univ, 70-74; PROF MATH & HEAD DEPT, ROCHESTER INST TECHNOL, 74- Concurrent Pos: From adj asst prof to adj assoc prof, Hartford Grad Ctr, Rensselaer Polytech Inst, 60-70. Mem: Am Math Soc. Res: Hydrodynamic stability; numerical analysis. Mailing Add: Dept of Math Rochester Inst Technol Rochester NY 14623

NEWBURGER, SYLVAN HENRY, b Baltimore, Md, June 23, 10; m 49; c 2. COSMETIC CHEMISTRY. Educ: Johns Hopkins Univ, AB, 31, PhD, 37. Prof Exp: Chemist, Food & Drug Admin, Fed Security Agency, 39-53, chemist, Dept Health, Educ & Welfare, 53-60, chief cosmetic res & anal br, 60-72; RETIRED. Honors & Awards: Merit Award, 62; Medal Award, Soc Cosmetic Chem, 72. Mem: Fel AAAS; Am Chem Soc; Soc Cosmetic Chem; fel Asn Off Anal Chem. Res: Spectrophotometric research in natural and coal-tar colors; quantitative analysis of coal-tar colors; methods of analysis for cosmetics. Mailing Add: 2207 Sulgrave Ave Baltimore MD 21209

NEWBURGH, ROBERT WARREN, b Sioux City, Iowa, Mar 22, 22; m 47; c 2. BIOCHEMISTRY. Educ: Univ Iowa, BS, 49; Univ Wis, MS, 51, PhD(biochem), 53. Prof Exp: Asst biochem, Univ Wis, 49-53; res assoc sci res inst, 53-54, from asst prof to assoc prof, 54-61, PROF BIOCHEM, ORE STATE UNIV, 61-, CHMN DEPT BIOCHEM & BIOPHYS, 68-, ASST DIR SCI RES INST, 62- Concurrent Pos: Am Cancer Soc grant & assoc prof, Univ Conn, 60-61; vis prof, Univ Calif, San Diego, 70-71; consult, NIH, 66-74. Mem: Am Chem Soc; Am Soc Biol Chem; Am Soc Neurochem; Toxicol Soc. Res: Neural development; insect biochemistry. Mailing Add: Dept of Biochem & Biophys Ore State Univ Corvallis OR 97332

NEWBURGH, RONALD GERALD, b Boston, Mass, Feb 21, 26; m 57, 70; c 2. ELECTROMAGNETISM. Educ: Harvard Univ, AB, 45; Mass Inst Technol, PhD(physics), 59. Prof Exp: Sr chemist, Electronics Corp Am, 49-53, consult, 53-55 & 57-59; res physicist, Comstock & Wescott, Inc, 59-61; RES PHYSICIST, AIR FORCE CAMBRIDGE RES LABS, 61- Mem: Am Phys Soc. Res: Physics of rotating systems; special and general relativity. Mailing Add: LZR Air Force Cambridge Res Labs Bedford MA 01731

NEWBURN, RAY LEON, JR, b Rock Island, Ill, Jan 9, 33; m 68. ASTRONOMY. Educ: Calif Inst Technol, BS, 54, MS, 55. Prof Exp: Res engr, 56-60, sr scientist lunar & planetary sci sect, 60-62, sci specialist, 62-65, sci group supvr, 65-72, STAFF SCIENTIST, PLANETARY ATMOSPHERES SECT, SPACE SCI DIV, JET PROPULSION LAB, CALIF INST TECHNOL, 72- Mem: AAAS; Am Astron Soc; Optical Soc Am; Am Geophys Union; Royal Astron Soc. Res: Ground based and space probe research in astronomy of the solar system. Mailing Add: 3226 Emerald Isle Dr Glendale CA 91206

NEWBURY, RAY SELMER, ceramics, spectroscopy, see 12th edition

NEWBY, FRANK ARMON, JR, b Columbus, Kans, Dec 4, 32; m 61; c 4. PHYSICAL CHEMISTRY. Educ: Univ Kans, BS, 54, PhD(chem), 64. Prof Exp: From asst prof to assoc prof phys chem, 59-70, PROF PHYS CHEM, E TENN STATE UNIV, 70- Mem: AAAS; Am Chem Soc. Res: Coordination compound equilibria in non-aqueous solvents by phase rule and spectrophotometric studies. Mailing Add: Dept of Chem ETenn State Univ Johnson City TN 37601

NEWBY, HAYES AUGUSTUS, b Marion, Ohio, Apr 2, 14; m 36; c 3. SPEECH PATHOLOGY. Educ: Ohio Wesleyan Univ, AB, 35; Univ Iowa, MA, 39, PhD(speech path, audiol), 47. Prof Exp: Asst pub speaking, Univ Iowa, 37-38, instr commun skills, 45-47; asst prof speech, La Polytech Inst, 40-41; from asst prof speech & drama to prof, Stanford Univ, 47-57, prof speech path & audiol, Sch Med, 57-67, dir div, 63-67; prof commun arts & sci & dir speech & hearing ctr, Queens Col, NY, 67-69; dir speech & hearing sci, 69-73, prof speech & dramatic art, 69-73, chmn dept hearing & speech sci, 73-76, PROF HEARING & SPEECH SCI, UNIV MD, COLLEGE PARK, 73- Concurrent Pos: Dir, San Francisco Hearing & Speech Ctr, 48-56, consult, 56-57; consult, State Dept Health, Calif, 55-67 & Vet Admin, 55-; dir, Am Bds Exams Speech Path & Audiol, 60-63, pres, 62-63; mem training grants adv panel, Voc Rehab Admin, 61-64 & Off Educ, 65-69, consult, 69-; adv ed, Appleton-Century-Crofts, 66-73 & Irvington Publ, 74-; trustee, Easter Seal Res Found, 72-, dir, Deafness, Speech & Hearing Publ, 72-74, pres, 74; vpres, Am Speech & Hearing Found, 75- Mem: Fel AAAS; fel Am Speech & Hearing Asn (pres, 64); Acoust Soc Am. Mailing Add: Dept of Hearing & Speech Sci Univ of Md College Park MD 20742

NEWBY, KENNETH RUSS, electrochemistry, physical chemistry, see 12th edition

NEWBY, NEAL DOW, JR, b New York, NY, Mar 18, 26; m 64; c 2. THEORETICAL PHYSICS. Educ: Columbia Univ, BS, 49; Harvard Univ, MA, 51; Ind Univ, PhD(physics), 59. Prof Exp: Instr math, Univ Ohio, 52-53; res assoc physics, Univ Calif, Berkeley, 59-61; asst prof, Univ Southern Calif, 61-63; physicist, Autonetics, 64-67; lectr, Calif State Col, 69; PROF PHYSICS, EDINBORO STATE COL, 69- Mem: Am Phys Soc. Res: Classical mechanics; relativity. Mailing Add: Dept of Physics Edinboro State Col Edinboro PA 16444

NEWBY, THOMAS HARRY, chemistry, see 12th edition

NEWBY, WILLIAM EDWARD, b Kansas City, Mo, Nov 17, 23; m 49; c 3. PHYSICAL CHEMISTRY, ORGANIC CHEMISTRY. Educ: Univ Kans City, BA, 47; Northwestern Univ, PhD(chem), 50. Prof Exp: Res chemist, Jackson Lab, 50-52, res supvr, 53-58, res supvr petrol lab, 58-59, res supvr plant tech sect, 59-60, res supvr, Jackson Lab, 60-68, supvr mkt res, Org Chem Dept, Dyes & Chem Div, 68-70, MKT PLANNING MGR, ORG CHEM DEPT, DYES & CHEM DIV, E I DU PONT DE NEMOURS & CO, INC, 70- Mem: Am Chem Soc; Am Asn Textile Chem & Colorists. Res: Catalytic hydrogenation; fuel deposit and combustion phenomena; synthesis and application of dyes for synthetic fibers. Mailing Add: Org Chem Dept Dyes & Chem Div E I du Pont de Nemours & Co Inc Wilmington DE 19898

NEWCOMB, BRADLEY LEWIS, engineering graphics, see 12th edition

NEWCOMB, EDWARD LINDSAY, b Hornell, NY, June 1, 18; m 42; c 2. ECONOMIC GEOLOGY. Educ: Middlebury Col, AB, 40; Cornell Univ, PhD(geol), 49. Prof Exp: Field geologist, NJ Zinc Co, NJ, 46-49, geologist, Va, 49-51; geologist, Mineral Deposits Br, US Geol Surv, 51-55; staff geologist, Div Raw Mat, US AEC, Colo, 55-57, chief geologist, 57-58, dep br chief, 58-59, chief Denver br, 59-61; econ geologist & geol adv, 61-74, MINING GEOLOGIST, NATURAL RESOURCES SECT, INTERNAL REVENUE SERV, US TREASURY DEPT, 74- Concurrent Pos: Instr, Cornell Univ, 46-49. Mem: Geol Soc Am; Soc Econ Geol; Am Inst Mining, Metall & Petrol Eng. Res: Economic geology of mineral deposits; mineral economics; exploration and development of strategic mineral resources. Mailing Add: US Treasury Dept IRS 12th & Constitution Ave Washington DC 20001

NEWCOMB, ELDON HENRY, b Columbia, Mo, Jan 19, 19; m 49; c 3. PLANT CYTOLOGY. Educ: Univ Mo, AB, 40, AM, 42; Univ Wis, PhD(bot), 49. Prof Exp: From asst prof to assoc prof, 49-58, PROF BOT, UNIV WIS-MADISON, 58- Concurrent Pos: Guggenheim fel, Univ Calif, 51-52; NSF sci fac fel, Harvard Univ, 63-64; consult, Shell Develop Co, 54-59; US managing ed, Protoplasma, 69- Mem: Bot Soc Am; Am Soc Cell Biol; Am Soc Plant Physiologists. Res: Plant cell fine structure in relation to function; changes in cell fine structure during differentiation; plant microbodies. Mailing Add: Dept of Bot Univ of Wis Madison WI 58706

NEWCOMB, HARVEY RUSSELL, b Bismarck, NDak, Oct 6, 16; m 53; c 2. MICROBIOLOGY. Educ: Denison Univ, AB, 39; Syracuse Univ, MS, 52, PhD(microbiol), 54. Prof Exp: Res bacteriologist, Borden Co, NY, 53-55; res asst prof microbiol, Dept Bact & Bot & Biol Res Labs, Syracuse Univ, 55-64; lab dir, Raritan Bay & Hudson-Champlain Water Pollution Control Projs, USPHS, NJ, 64-65; PROF MICROBIOL & CHMN DEPT, COL OSTEOP MED & SURG, 65- Mem: AAAS; Am Soc Microbiol; Am Pub Health Asn; NY Acad Sci. Res: Radiation effects on microorganisms; bacterial spores; preservation and wholesomeness of irradiated foods; physiology and industrial production of lactic acid bacteria; physiology of wood-rotting basidiomycetes; gnotobiotic technology; metabolic functions of serotonin. Mailing Add: Dept of Microbiol Col of Osteop Med & Surg Des Moines IA 50309

NEWCOMB, RICHARD WILLIAM, b Madison, SDak, Apr 30, 31; m 57; c 4. PEDIATRICS, IMMUNOLOGY. Educ: Univ Calif, Berkeley, BA, 53; Johns Hopkins Univ, MD, 57; Am Bd Pediat, dipl, 65. Prof Exp: Intern, San Francisco Hosp, Stanford, 57-58; resident path, Peter Bent Brigham Hosp, 58-59; resident pediat, Western Reserve Univ Hosps, 61-63; trainee pediat allergy, Univ Colo Med Ctr, Denver, 63-64, fel immunol, 64-66; fel, Univ Birmingham, 66-67; CLIN IMMUNOLOGIST, CHILDREN'S ASTHMA RES INST, 67- Concurrent Pos: Clin asst prof, Univ Colo, Denver, 67- Res: Immunology and its relation to atopic diseases in children; mucoantibodies, especially those of exocrine immunoglobulin A. Mailing Add: 3401 W 19th Ave Denver CO 80204

NEWCOMB, ROBERT LEWIS, b Oceanside, Calif, Aug 2, 32; m 58; c 3. APPLIED STATISTICS. Educ: Univ Redlands, BA, 59; Univ Calif, Santa Barbara, PhD(math), 67. Prof Exp: Res asst psychol, Univ Calif, Santa Barbara, 68-69; LECTR SOCIAL SCI, UNIV CALIF, IRVINE, 69- Mem: Math Asn Am; Asn Comput Mach; Am Statist Asn. Res: Computer-graphics in statistics education. Mailing Add: Sch of Social Sci Univ of Calif Irvine CA 92717

NEWCOMB, THOMAS F, b Buffalo, NY, June 22, 27; m 50; c 3. INTERNAL MEDICINE, HEMATOLOGY. Educ: Univ Pittsburgh, BS, 49, MD, 51; Am Bd Internal Med, dipl. Prof Exp: From intern med to resident hemat, Univ Pa, 51-53; resident med, Vet Admin Hosp, Seattle, Wash, 53-54; res fel hemat, Univ Wash, 54-55; Fulbright scholar med, Rikshospitalet Coagulation Lab, Oslo, Norway, 55-56; sr asst resident, Peter Bent Brigham Hosp, Boston, 56-57, jr assoc med, Hosp & investr, Howard Hughes Med Inst, 57-59, asst dir, Richard C Curtis Hemat Lab, 58-59; from asst prof to prof med, Col Med, Univ Fla, 59-73, prof biochem, 71-73, dir hemat, 59-64, head div, 64-68, actg chmn dept biochem, 69-70; dir res serv, 73-74, ASST CHIEF MED DIR RES & DEVELOP, CENT OFF, VET ADMIN, 74- Concurrent Pos: Instr, Harvard Med Sch, 58-59; consult, Vet Admin Hosp, Lake City, Fla, 62-68, assoc chief of staff res & educ & chief research serv, Gainesville, 68-73. Mem: Fel Am Col Physicians; Am Soc Hemat; Am Fedn Clin Res; Int Soc Hemat. Res: Hemostasis. Mailing Add: Dept of Med Res & Develop Vet Admin Cent Off Washington DC 20420

NEWCOMB, WILLIAM A, b San Jose, Calif, Sept 4, 27; m 64; c 2. THEORETICAL PHYSICS. Educ: Cornell Univ, BA, 48, PhD(theoret physics), 52. Prof Exp: Physicist, Proj Matterhorn, Forrestal Res Ctr, Princeton Univ, 52-55; PHYSICIST, LAWRENCE LIVERMORE LAB, UNIV CALIF, 55- Concurrent Pos: Lectr, Univ Calif, Davis. Mem: Am Phys Soc. Res: Magneto-hydrodynamics and plasma physics. Mailing Add: Lawrence Livermore Lab Univ of Calif Livermore CA 94550

NEWCOMB, WILLIAM WILMON, JR, b Detroit, Mich, Oct 30, 21; m 46; c 2. ETHNOLOGY. Educ: Univ Mich, AB, 43, MA, 47, PhD(anthrop), 53. Prof Exp: Instr anthrop, Univ Tex, 47-50 & 51-52; vis asst prof sociol & anthrop, Colgate Univ, 53-54; res scientist anthrop, Tex Mem Mus, 54-57, lectr, 61-62, PROF ANTHROP, UNIV TEX, AUSTIN, 62-, DIR, TEX MEM MUS, 57- Concurrent Pos: Univ Tex Res Inst grant archeol & ethnohist invest Mission San Lorenzo de la Santa Cruz, Tex, 62; NSF grant ethnohist invest Wichita Indians, 65-66; Univ Tex Res Inst grant surv rock art lower Pecos River region, 67. Mem: Fel Am Anthrop Asn; Am Asn Mus. Res: North American Indian ethnohistory and ethnology; prehistoric rock art of American Indians. Mailing Add: Tex Mem Mus 24th & Trinity Austin TX 78705

NEWCOMBE, DAVID S, b Boston, Mass, June 28, 29; m 65; c 3. RHEUMATOLOGY. Educ: Amherst Col, BA, 52; McGill Univ, MD, CM, 56. Prof Exp: Intern med, Boston City Hosp, 56-57; resident, Med Ctr, Duke Univ, 59-60; NIH res fel arthritis & rheumatism, Duke Univ, 60-61; New Eng Rheumatism Soc res fel, Boston Univ, 61-62, asst, 62-63; asst, Peter Bent Brigham Hosp, 63-65; asst prof, Sch Med, Univ Va, 65-67; ASSOC PROF MED, COL MED, UNIV VT, 67-, DIR RHEUMATOLOGY UNIT, 70- Concurrent Pos: NIH spec fel, 62-63; Am Cancer Soc res fel, Harvard Med Sch, 63-65. Mem: NY Acad Sci; Sigma Xi; Soc Exp Biol & Med; Asn Teachers Prev Med; Am Rheumatism Asn. Res: Connective tissue biochemistry; rheumatology. Mailing Add: Rheumatology Unit Univ of Vt Col of Med Burlington VT 05401

NEWCOMBE, HOWARD BORDEN, b Kentville, NS, Sept 19, 14; m 42; c 3. RADIATION GENETICS, HUMAN GENETICS. Educ: Acadia Univ, BSc, 35; McGill Univ, PhD(genetics), 39. Hon Degrees: DSc, McGill Univ, 66, Acadia Univ, 70. Prof Exp: 1851 sci res scholar, John Innes Hort Inst, Eng, 39-40; sci officer, Brit Ministry Supply, 40-41; res assoc genetics, Carnegie Inst Technol, 46-47; head biol br, 47-70, HEAD POP RES BR, ATOMIC ENERGY CAN, LTD, 70- Concurrent Pos:

Sci adv, Sci Comt Atomic Radiation, UN, 55-66; mem expert adv panel human genetics, WHO, 61-, mem, Int Comn Radiol Protection, 65-; vis prof, Ind Univ, 63. Mem: Am Soc Human Genetics (pres, 65); Genetics Soc Am (secy, 56-58); Radiation Res Soc; Genetics Soc Can (pres, 64-65); Royal Soc Can. Res: Epidemiology; public health. Mailing Add: Pop Res Br Atomic Energy of Can Ltd Chalk River ON Can

NEWCOMBE, JACK, b Fletcher, Okla, May 4, 21; m 42; c 3. ORGANIC CHEMISTRY. Educ: Okla Agr & Mech Col, BS, 43, MS, 47. Prof Exp: Instr chem, Okla Agr & Mech Col, 43-44; res chemist, Cities Serv Res & Develop Co, 47-57; res coord, Petrol Chem, Inc, 57-58, res & patent coord, 58-60, staff asst to mgr, Petrochem Res Lab, 60-62, head process res sect, 62-64; mgr org res sect, Columbian Carbon Co, La, 64-67, mgr technol & planning div, NJ, 67-71; mgr indust org chem, 71-74, RES ASSOC INDUST ORG CHEM, PETROCHEM RES GROUP, CITIES SERV CO, 74- Mem: Am Chem Soc. Res: Petrochemicals; oil production chemistry. Mailing Add: Petrochem Res Cities Serv Co Drawer 4 Cranbury NJ 08512

NEWCOME, MARSHALL MILLAR, b Chicago, Ill, Nov 22, 26; m 49; c 2. ANALYTICAL CHEMISTRY. Educ: Ill Inst Technol, BS, 49; Univ Wash, Seattle, PhD(chem), 54. Prof Exp: Res chemist, 54-74, SUPVR ANAL CHEM, MORTON CHEM CO, 74- Mem: Am Chem Soc. Res: Electronic instrumentation in chemical analysis. Mailing Add: 546 W Kimball Ave Woodstock IL 60098

NEWCOMER, EARL HOLLAND, b Bradford, Pa, Aug 24, 02; m 30; c 2. CYTOLOGY, GENETICS. Educ: Columbia Univ, AB, 32; Univ Calif, AM, 35; Pa State Col, PhD(cytol), 38. Prof Exp: Instr bot & res asst cytogenetics, Mich State Col, 38-41; from asst prof to assoc prof bot, Univ NC, 41-48; from assoc prof to prof, 48-72, EMER PROF BOT, UNIV CONN, 72- Mem: AAAS; Bot Soc Am; Am Genetic Asn. Res: Plant and animal cytology and cytogenetics. Mailing Add: 265 Hanks Hill Rd Storrs CT 06268

NEWCOMER, WILBUR STANLEY, b Turbotville, Pa, Nov 25, 19; m 46; c 4. PHYSIOLOGY. Educ: Pa State Univ, BS, 41; Cornell Univ, MS, 42, PhD(zool), 48. Prof Exp: Asst chem, Lycoming Col, 38-39; asst zool, Cornell Univ, 42-44 & 46-47; from instr to asst prof biol, Hamilton Col, NY, 47-50; from asst prof to assoc prof, 50-58, PROF PHYSIOL, OKLA STATE UNIV, 58- Concurrent Pos: NIH spec res fel, 62. Mem: Soc Exp Biol & Med; Am Physiol Soc. Res: Physiology of thyroid and adrenal glands, especially in birds. Mailing Add: Dept of Physiol Sci Okla State Univ Stillwater OK 74074

NEWELL, ALLEN, b San Francisco, Calif, Mar 19, 27; m 47; c 1. COMPUTER SCIENCE, PSYCHOLOGY. Educ: Stanford Univ, BS, 49; Carnegie Inst Technol, PhD(indust admin), 57. Prof Exp: Res scientist, Rand Corp, 50-61; res scientist inst prof systs & commun sci, 61-67, UNIV PROF, CARNEGIE-MELLON UNIV, 67- Concurrent Pos: Lectr, Carnegie-Mellon Univ, 57-61; consult, Rand Corp, 61-; mem adv comt comput, Stanford Univ, 66-70; mem comput sci study sect, NIH, 67-71; chmn panel comput, Comt Res Life Sci, Nat Acad Sci, 67-69; consult, Xerox Res Lab, Calif, 71- Honors & Awards: Harry Goode Mem Award, Am Fedn Info Processing Socs, 71; John Danz Lectr, Univ Wash, 72; A M Turing Award, Asn Comput Mach, 75. Mem: Nat Acad Sci; AAAS; Asn Comput Mach; Am Psychol Asn; Inst Elec & Electronics Eng. Res: Computer programs that exhibit intelligence; information processing psychology; programming systems and computer structures. Mailing Add: Carnegie-Mellon Univ Pittsburgh PA 15213

NEWELL, FRANK WILLIAM, b St Paul, Minn, Jan 14, 16; m 42; c 4. OPHTHALMOLOGY. Educ: Loyola Univ, Ill, MD, 39; Univ Minn, MSc, 42; Am Bd Ophthal, dipl. Prof Exp: Fel, Univ Minn, 40-42; res fel, Northwestern Univ, 46-47, instr, 47-50, assoc, 50-53; assoc prof, 53-55, PROF OPTHAL, SCH MED, UNIV CHICAGO, 55-, CHMN DEPT OPHTHAL, 53- Concurrent Pos: Ed in chief, Am J Ophthal; mem nat adv coun, Nat Eye Inst. Mem: Am Ophthal Soc; Soc Exp Biol & Med; AMA; Pan-Am Asn Ophthal; Am Acad Ophthal & Otolaryngol. Res: Pharmacology and physiology of the eye. Mailing Add: Univ of Chicago Eye Res Labs 950 E 59th St Chicago IL 60637

NEWELL, GEORGE WATTS, b Wellington, Kans, Oct 1, 17; m 40; c 4. POULTRY HUSBANDRY. Educ: Univ Mo, BS, 39; Purdue Univ, MS, 47; Univ Md, PhD(poultry), 49. Prof Exp: Exten mkt specialist, Univ Mo, 49-50; ASSOC PROF ANIMAL SCI & INDUST, OKLA STATE UNIV, 50- Mem: AAAS; Poultry Sci Asn; Inst Food Technologists. Res: Poultry marketing, economic and technological. Mailing Add: Dept of Animal Sci & Indust Okla State Univ Stillwater OK 74074

NEWELL, GORDON FRANK, b Dayton, Ohio, Jan 26, 25; m 49; c 2. APPLIED MATHEMATICS. Educ: Union Col, BS, 45; Univ Ill, PhD(physics), 50. Prof Exp: Lectr physics & math, Union Col, 45-46; asst physics, Univ Ill, 46-49, fel, 50; fel appl math, Univ Md, 50-51, res assoc, 51-52; res assoc, Brown Univ, 53-54, from asst prof to prof, 54-66; PROF APPL MATH, UNIV CALIF, BERKELEY, 66- Concurrent Pos: Sloan res fel, 56-59; Fulbright fel, 63-; vis prof, Univ Calif, Berkeley, 65-66. Mem: Am Math Soc; Soc Indust & Appl Math; Opers Res Soc Am. Res: Operations research; transportation and traffic engineering. Mailing Add: Inst Transp & Traffic Eng Univ of Calif Berkeley CA 94720

NEWELL, GORDON WILFRED, b Madison, Wis, Aug 27, 21; m 48; c 4. TOXICOLOGY. Educ: Univ Wis, BA, 43, MS, 44, PhD(biochem), 48; Am Bd Indust Hyg, dipl. Prof Exp: Novadel-Agene fel, Univ Wis, 48-49; res biochemist, Wallace & Tiernan Co, Inc, 49-50; sr biochemist, 50-66, dir div indust biol, 66-68, DIR, DEPT TOXICOL, STANFORD RES INST, 68- Concurrent Pos: Consult adv comn animal resources, NIH, 65-68; coun mem, Am Asn Accreditation Animal Care, 67-; assoc ed, J Lab Animal Care, 64- Mem: AAAS; Am Chem Soc; Am Indust Hyg Asn; Am Astronaut Soc; NY Acad Sci. Res: Animal nutrition and metabolism; toxicological studies on chemicals and food products; biochemical toxicology of food additives, environmental, industrial, and military chemicals, drugs, and pesticides; toxicity of pollutants to fish and wildlife; nutrition and metabolism; mutagenesis, carcinogenesis, teratology, and inhalation toxicology. Mailing Add: Dept of Toxicol Stanford Res Inst Menlo Park CA 94025

NEWELL, HOMER EDWARD, mathematics, see 12th edition

NEWELL, IRWIN MAYER, b Coeur d'Alene, Idaho, Sept 15, 16; m 40; c 2. ACAROLOGY, SYSTEMATICS. Educ: State Col Wash, BS, 39, MS, 41; Yale Univ, MS, 42, PhD(zool), 45. Prof Exp: Asst biol, State Col Wash, 37-41; temporary agent, USDA, Wash, 45; res assoc biol, Oceanog Inst, Woods Hole, 45-46; from instr to asst prof, Univ Ore, 46-48; assoc entomologist, Univ Hawaii, 49-52, prof entom & entomologist, 52-53; assoc prof zool, 54-60, chmn div life sci, 61-64, head div biol control, 70-75, PROF ZOOL & ENTOM, UNIV CALIF, RIVERSIDE, 60- Concurrent Pos: Cur invert, Univ Ore, 47-48; Fulbright res fel, Univ Col Rhodesia & Nyasaland, 60-61. Mem: Fel AAAS; Soc Syst Zool; Entom Soc Am; Am Micros Soc. Res: Systematics of Acari and development of tabularkeys as a unified system of organizing taxonomic information; biological oceanography; biological control. Mailing Add: Dept of Biol Univ of Calif Riverside CA 92502

NEWELL, ISAAC LAIRD, chemistry, see 12th edition

NEWELL, JOHN T, II, b Chicago, Ill, Dec 3, 32. BIOLOGY. Educ: Northwestern Univ, BS, 54, MS, 55, PhD(physiol, genetics), 59. Prof Exp: Asst biol, Northwestern Univ, 54-55; asst physiol, Purdue Univ, 55-56; asst biol, Northwestern Univ, 56-58; from instr to asst prof biol, Univ Ill, 58-65; assoc prof biol, 66-68, assoc prof physiol, 68-71, dir campus planning & develop, 67-68, PROF PHYSIOL, CHICAGO STATE UNIV, 71-, DIR CAMPUS PLANNING, 68- Concurrent Pos: Res assoc, Hektoen Inst Med Res, 62-65; assoc toxicologist, Cook County Coroners Off, 65-68, consult, Cook County Morgue, 68-69. Mem: AAAS; Am Astron Soc; Soc Vert Paleont. Res: Effects of radiation on the genetic complement of organisms and the secondary effects on physiology and morphology. Mailing Add: Chicago State Univ Dept Physiol 95th at King Dr Chicago IL 60628

NEWELL, JON ALBERT, b St Louis, Mo, Aug 5, 41; m 64. BIOCHEMISTRY. Educ: Okla State Univ, BS, 63, PhD(biochem), 67. Prof Exp: Technician anal chem, Enid Bd Trade Lab, 59-62; res asst toxicol, Okla State Univ, 60-63, asst flavor chem, 63-67; sr res chemist food & fermentation res, 67-75, RES MGR BIOCHEM, ANHEUSER-BUSCH, INC, 75- Concurrent Pos: Nestle fel, Inst Food Technologists, Univ Chicago, 66. Mem: Am Chem Soc. Res: Use of unconventional protein sources in foods; identification of flavorful constituents of foods. Mailing Add: Anheuser-Busch Inc 721 Pestalozzi St St Louis MO 63118

NEWELL, JONATHAN CLARK, b Worcester, Mass, Oct 13, 43. PHYSIOLOGY, BIOMEDICAL ENGINEERING. Educ: Rensselaer Polytech Inst, BEE, 65, MEngr, 68; Albany Med Col, PhD(physiol), 74. Prof Exp: Supv engr, 70-72, ASST PROF PHYSIOL, ALBANY MED COL, 74-; ASST PROF BIOMED ENG, RENSSELAER POLYTECH INST, 74- Concurrent Pos: Consult, Trauma Res Unit, Albany Med Col, 74- Mem: Inst Elec & Electronics Engrs; Asn Advan Med Instrumentation; Am Thoracic Soc. Res: Pulmonary hemodynamics and mechanics; modelling of physiological systems. Mailing Add: Ctr for Biomed Eng Rensselaer Polytech Inst Troy NY 12181

NEWELL, KATHLEEN, b Stafford, Kans, Sept 1, 22. NUTRITION. Educ: Kans State Univ, BS, 44; Univ Wis, MS, 51; Univ Tenn, PhD(nutrit), 73. Prof Exp: Admin ther & teaching dietitian, Butterworth Hosp, Grand Rapids, Mich, 45-47; ther & teaching dietitian, Univ Colo Med Ctr, 47-49, educ dir dietetic intern, 52-56; asst prof dietetics, Loretto Heights Col & Glockner Penrose Hosp, Colo Springs, 56-58; asst prof home econ, Univ Wyo, 58-62; ASST PROF FOODS & NUTRIT, KANS STATE UNIV, 62-69 & 72- Mem: Am Dietetic Asn; Am Home Econ Asn; Geront Soc; Soc Nutrit Educ; Nutrit Today Soc. Res: Geriatric nutrition and physiological changes with aging. Mailing Add: Dept of Foods & Nutrit Kans State Univ Manhattan KS 66506

NEWELL, KENNETH WYATT, b Erode, India, Nov 7, 25; m 58; c 4. EPIDEMIOLOGY. Educ: Univ Otago, NZ, MB, ChB, 48; London Sch Hyg & Trop Med, DPH, 53; Tulane Univ, MD, 64. Prof Exp: Epidemiologist, Cent Pub Health Lab, London, Eng, 53-56; lectr prev med, Queen's Univ Belfast, 56-58; epidemiologist, WHO, Indonesia, 58-60; prof epidemiol, Sch Med, Tulane Univ, 61-67; dir, Div Res Epidemiol & Commun Sci, 67-72, DIR, DIV STRENGTHENING HEALTH SERVS, WHO, 72- Concurrent Pos: Dir field studies, Int Ctr Med Res & Training, Colombia, 61-; WHO consult tetanus, 62 & leprosy, 63, mem expert comt enteric dis, 63- Mem: Royal Soc Trop Med & Hyg. Res: Epidemiological methods in the tropics; communicable and chronic diseases; gastrointestinal diseases; salmonellosis; tetanus; leprosy; diptheria; pesticides in man; blood pressure variations; health services development. Mailing Add: Div Strengthening Health Servs WHO Ave Appia Geneva Switzerland

NEWELL, LAURA, b Whitehall, NY, May 16, 33. PHYSICAL ANTHROPOLOGY. Educ: Univ NMex, BA, 54; Northwestern Univ, MA, 57; Univ Wash, PhD(anthrop), 66. Prof Exp: Asst cur archaeol, State Mus NY, 54-55; res asst phys anthrop, Fels Res Inst, 59-60; from instr to asst prof, 64-72, ASSOC PROF PHYS ANTHROP, UNIV WASH, 72- Concurrent Pos: Assoc ed, Am J Phys Anthrop, 75. Honors & Awards: Res Award, NIH, 75. Mem: Am Anthrop Asn; Am Asn Phys Anthrop; Brit Soc Study Human Biol; Int Primatol Soc. Res: Growth, especially of lower primates and population as a unit of study. Mailing Add: Dept of Anthrop Univ of Wash Seattle WA 98195

NEWELL, LAURENCE CUTLER, b Alexandria, Nebr, Apr 12, 05; m 28; c 3. AGRONOMY. Educ: Hastings Col, AB, 26; Univ Nebr, MSc, 33, PhD(agron), 40. Prof Exp: Teacher & prin, Davenport High Sch, Nebr, 27-31; asst agron, Univ Nebr, 31-33; county agr agt, Saunders & Stanton Counties, Nebr, 33-34; agt, Div Plant Explor & Introd, 34, asst agronomist, Soil Conserv Nurseries, Ames, Iowa, 35, from asst agronomist to agronomist, Forage & Range Res Br, Crops Res Div, Nebr, 36-54, RES AGRONOMIST, NCENT REGION, AGR RES SERV, USDA, 54-; EMER PROF AGRON, UNIV NEBR-LINCOLN, 74- Concurrent Pos: From asst to prof agron, Univ Nebr-Lincoln, 36-74. Honors & Awards: Soc Range Mgt Award. Mem: AAAS; fel Am Soc Agron; hon mem Soil Conserv Soc Am; Soc Range Mgt. Res: Grass breeding and pasture management, especially development of breeding procedures and seed production practices resulting in the release and propagation of improved varieties of forage grasses for seeding pastures and rangelands. Mailing Add: Dept of Agron Univ of Nebr Lincoln NE 68508

NEWELL, MARJORIE PAULINE, b Holden, Alta; nat US. ORGANIC CHEMISTRY. Educ: Univ Fla, BS, 53, PhD(biochem), 58. Prof Exp: Asst biochem, Univ Fla, 55-58; res chemist org chem, 58-69, SR SCIENTIST, RES DEPT, R J REYNOLDS TOBACCO CO, 69- Mem: AAAS; Am Chem Soc. Res: Physical and chemical properties of tobacco and smoke; tracer techniques with radioisotopes. Mailing Add: Res Dept R J Reynolds Tobacco Co Winston-Salem NC 27102

NEWELL, NORMAN DENNIS, b Chicago, Ill, Jan 27, 09; m 28. GEOLOGY. Educ: Univ Kans, BS, 29, AM, 31; Yale Univ, PhD(geol), 33. Prof Exp: Asst geologist, Geol Surv, Kans, 29-33; from instr to asst prof geol, Univ Kans, 34-37; assoc prof, Univ Wis, 37-45; PROF GEOL, COLUMBIA UNIV, 45-; CUR INVERT FOSSILS & HIST GEOL, AM MUS NATURAL HIST, 45- Concurrent Pos: Sterling fel, Yale Univ, 33-34; geologist, Geol Surv, Kans, 35-37; lectr, Univ Pa, City Col New York & Yale Univ; US Dept State deleg, Int Geol Cong, Moscow, 37; co-ed, Paleont, 39-42; consult, Govt Peru, 42-43, leader, Exped Geol & Petrol Resources Lake Titicaca, Peru-Bolivia, 43-44, Am Mus Natural Hist-Columbia Univ Exped, Peru, 47 & Andros Island, Behamas, 50 & 51; leader invests limestone reefs, WTex, 49-52; leader, Pac Sci Bds, Nat Res Coun Exped Coral Atoll, Raroia, Tuamotu Group, SPac, 52; mem, Smithsonian Coun, 66-; mem, Scripps Inst Oceanog Exped Carmarsel, Micronesia, 67. Honors & Awards: Mary Thompson Clark Medal, Nat Acad Sci, 60; Distinguished Serv Citation, Univ Kans, 60; Medal, Univ Hiroshima, 64; Hayden Mem Award, Acad Natural Sci, Philadelphia, 65; Verrill Medal, Yale Univ, 66. Mem: Soc Syst Zool (pres, 71-72); fel Geol Soc Am; Soc Econ Paleontologists & Mineralogists; fel Paleont Soc (vpres, 48, pres, 61); Soc Study Evolution (pres, 49). Res: Invertebrate paleontology; micropaleontology; stratigraphy; petroleum geology; coral reefs; geology of South America. Mailing Add: Dept of Geol Columbia Univ Grad Sch Arts & Sci New York NY 10027

NEWELL, REGINALD EDWARD, b Peterborough, Eng, Apr 9, 31; m 54; c 4. METEOROLOGY. Educ: Univ Birmingham, BSc, 54; Mass Inst Technol, SM, 56, ScD(meteorol), 60. Prof Exp: Res asst meteorol, 54-60, mem staff, Div Sponsored Res, 60-61, from asst prof to assoc prof, 61-70, PROF METEOROL, MASS INST TECHNOL, 70- Mem: AAAS; Am Meteorol Soc; Am Geophys Union; Am Phys Soc; fel Royal Meteorol Soc. Res: Upper atmosphere physics; planetary circulations; radar meteorology. Mailing Add: Dept of Meteorol Mass Inst of Technol Cambridge MA 02139

NEWELL, WILLIAM ANDREWS, b Holyoke, Mass, Mar 3, 20; m 42; c 3. TEXTILES. Educ: NC State Col, BS, 47. Prof Exp: Managing ed, Textile World, McGraw-Hill Publ Co, 47-52; dir res, Sch Textiles, NC State Col, 52-61; mgr prod eng, Whitin Mach Works, 61-65; tech dir, Textile Div, Kendall Co, 65-71, dir gen, Mulsant, SA, France, 71-74, MGR CENT QUAL CONTROL & REGULATORY COMPLIANCE, KENDALL CO, 74- Mem: Am Asn Textile Technol. Res: Textile engineering and technology. Mailing Add: Kendall Co 1 Federal St Boston MA 02110

NEWEY, HERBERT ALFRED, b Logan, Utah, June 11, 16; m 41; c 3. ORGANIC POLYMER CHEMISTRY. Educ: Utah State Col, BS, 38; Mass Inst Technol, PhD(org chem), 41. Prof Exp: Res chemist, Am Cyanamid Co, 41-46; chemist, Shell Develop Co Div, Shell Oil Co, 46-62, supvr res, 62-73; CONSULT CHEMIST, 73- Res: Epoxy and polyester resins for surface coatings, laminates and adhesives. Mailing Add: 730 Los Palos Dr Lafayette CA 94549

NEWHALL, CHESTER ALBERT, b Stoneham, Mass, Mar 8, 02; m 26; c 4. ANATOMY. Educ: NCent Col, Ill, AB, 24; Univ Vt, MD, 28. Prof Exp: From instr to prof, Col Med, Univ Vt, 29-70, actg head dept anat, 41-42, chmn dept, 42-70, Thayer prof, 70-74; RETIRED. Mem: AAAS; Am Asn Anatomists; AMA; Am Pub Health Asn. Res: Reduction of bacterial flora of human skin; human nutrition; congenital anomalies; human heart and great vessels. Mailing Add: 72 Colchester Ave Burlington VT 05401

NEWHOUSE, ALBERT, b Cambrai, France, May 31, 14; nat US; m 43; c 2. MATHEMATICS, COMPUTER SCIENCE. Educ: Univ Chicago, PhD(math), 40. Prof Exp: Asst Math, Tulane Univ, 39-41; instr, Univ Ala, 41-42, Univ Nebr, 42-44 & Rice Inst, 44-46; from instr to asst prof, 46-48, from assoc prof to prof math, 48-70, chmn dept math, 52-54, asst dir comput & data processing ctr, 56-60, PROF COMPUT SCI, UNIV HOUSTON, 70- Concurrent Pos: Consult, Res Lab, Humble Oil & Ref Co, 58, Camco, Inc, 61-70 & Symbiotics Int, 70- Mem: AAAS; Asn Comput Mach; Am Math Soc; Soc Indust & Appl Math; Math Asn Am. Res: Modern algebra; theory of groups; numerical methods; formal languages. Mailing Add: 7907 Braesview Lane Houston TX 77071

NEWHOUSE, VERNE FREDERIC, b Tulsa, Okla, May 7, 30; m 53; c 5. MEDICAL ENTOMOLOGY. Educ: Pac Lutheran Col, AB, 53; Wash State Univ, MS, 55, PhD(entom), 60. Prof Exp: Asst zool, Wash State Univ, 54-55 & 56-57; med entomologist, Rocky Mountain Lab, 57-63, res entomologist, Arbovirus Vector Lab, 64-75, RES ENTOMOLOGIST, LEPROSY & RICKETTSIA BR, VIROL DIV, COMMUN DIS CTR, USPHS, 75- Honors & Awards: Best Sci Paper Year Lab Sect Group Award, Am Pub Health Asn, 68; Superior Achievement Group Award, US Dept Health, Educ & Welfare, 72. Res: Ecology of arthropod-borne rickettsial and viral agents and their vectors. Mailing Add: 423 Stonewood Dr Stone Mountain GA 30083

NEWHOUSE, VERNON LEOPOLD, b Mannheim, Ger, Jan 30, 28; US citizen; m 50; c 4. BIOMEDICAL ENGINEERING. Educ: Univ Leeds, BSc, 49, PhD(physics), 52. Prof Exp: Mem staff, Ferranti Comput Lab, 51-54; mem staff, Res Lab, Radio Corp Am, 54, proj engr, Comput Dept, 54-57; physicist, Gen Elec Res & Develop Ctr, NY, 57-67; prof elec eng, 67-73, PROF MED ENG & COORDR, SCH ELEC ENG, PURDUE UNIV, 73- Concurrent Pos: Consult, NSF, 74. Mem: Am Phys Soc; fel Inst Elec & Electronics Eng. Res: Acoustic flow measurement and imaging for medical and industrial applications. Mailing Add: Med Eng Off Sch of Elec Eng Purdue Univ West Lafayette IN 47907

NEWHOUSE, W JAN, b Boston, Mass, Feb 6, 26; m 50; c 4. PHYCOLOGY, SCIENCE EDUCATION. Educ: Dartmouth Col, BA, 49; Univ NH, MS, 52; Univ Hawaii, PhD(bot), 67. Prof Exp: Asst botanist, Pac Sci Bd, Nat Acad Sci, 52, botanist & phycologist, 54; dir pineapple res, Hawaiian Canneries Co, Ltd, 56-59; fruit qual analyst, Dole Corp, 59-63, dir tech serv, Philippines, 63-65; asst prof, 66-71, ASSOC PROF GEN SCI, UNIV HAWAII, HONOLULU, 71- Concurrent Pos: Res initiation grant, Univ Hawaii, 66, Res Coun grant, 68; chief scientist, Stanford Oceanog Exped, 68. Mem: AAAS; Am Soc Limnol & Oceanog; Phycol Soc Am. Res: Primary productivity of the oceans; biogeography of tropical Pacific Phaeophyta; ecology of marine Myxophyta; ciguatera fish toxin. Mailing Add: Dept of Gen Sci Univ of Hawaii 2450 Campus Rd Honolulu HI 96822

NEWILL, VAUN ARCHIE, b Mt Pleasant, Pa, Nov 11, 23; m 55; c 3. EPIDEMIOLOGY, INTERNAL MEDICINE. Educ: Juaniata Col, BS, 43; Univ Pittsburgh, MD, 47; Harvard Univ, SMHyg, 60. Prof Exp: Intern, St Francis Hosp, Pittsburgh, Pa, 47-48; resident, Harrisburg Polyclin Hosp, 48-50; resident med, Vet Admin Hosps, Buffalo, NY, 52-53 & Cleveland, Ohio, 53-54, mem staff geriat serv, 55; from instr prev med & med to sr instr, Sch Med, Western Reserve Univ, 55-60, sr instr med, 60-68, from asst prof prev med to assoc prof, 60-68; chief ecol res sect, Health Effects Res Prog, Nat Ctr Air Pollution Control, USPHS, Ohio, 67-68; chief health effects res prog, Nat Air Pollution Control Admin, NC, 68-71; mem staff, Environ Protection Agency, 71-72; tech asst, Off Sci & Technol, Exec Off of President, 72-73; spec asst to adminr, Environ Protection Agency, 73-74; ASST DIR, MED DEPT, EXXON CORP, 74- Concurrent Pos: Asst physician, Univ Hosps, Cleveland, 55-68 & Benjamin Rose Hosp, 57-68; attend metab, Vet Admin Hosp, Cleveland, 58-63, sr attend, 63-64; spec consult epidemiol training prog med students, Calif Dept Health, 62; vis lectr, Sch Pub Health, Harvard Univ, 62-65; med officer, Field Studies Br, Div Air Pollution, USPHS, Japan, 65; adj assoc prof, Dept Epidemiol, Sch Pub Health, Univ NC, Chapel Hill, 68-75; clin prof environ health, NY Univ. Mem: AAAS; Am Fedn Clin Res; Am Pub Health Asn. Res: Epidemiology of chronic diseases and effects of environmental pollutant exposures; information retrieval. Mailing Add: Med Dept Exxon Corp PO Box 45 Linden NJ 07036

NEWIRTH, TERRY L, b Morristown, NJ, Sept 22, 45; m 74. ORGANIC BIOCHEMISTRY. Educ: Bryn Mawr Col, AB, 67; Mass Inst Technol, PhD(org chem), 71. Prof Exp: Res assoc biochem, Sch Med, Univ Pa, 71-73; res assoc cell biol, Sch Med, Temple Univ, 73-75; ASST PROF CHEM, HAVERFORD COL, 75- Mem: Am Chem Soc; AAAS. Res: Reaction mechanisms of biologically important reactions, especially reactions involving co-factors; new methods of chromatin fixation. Mailing Add: Dept of Chem Haverford Col Haverford PA 19041

NEWITT, EDWARD JAMES, b Scranton, Pa, Mar 18, 27; m 48; c 5. PHYSICAL ORGANIC CHEMISTRY. Educ: Imp Col, Univ London, BSc, 53, PhD, 57. Prof Exp: Works chemist indust chem, Boots Pure Drug Co, Eng, 48-49; asst chem, Brit Coal Utilization Res Asn, 49-53; lectr inorg & phys chem, Imp Col, Univ London, 53-57; res chemist, Plastics Dept, Exp Sta, 57-59, res supvr, 59-61, sr res chemist, Wash Works, 61-62, res supvr, Electrochem Dept, Exp Sta, 62-69, prod mgr sales, 69-70, res supvr, 70-71, develop supvr polymer prod mkt, Plastics Dept, 72-73, new prod specialist, 73-74, CONSULT, ENERGY & MAT DEPT, E I DU PONT DE NEMOURS & CO, INC, 74- Mem: Am Chem Soc; Royal Inst Chemists. Res: Chemical kinetics; high temperature chemistry and polymer chemistry; alternate raw materials and energy sources. Mailing Add: Energy & Mat Dept E I du Pont de Nemours & Co Inc Wilmington DE 19898

NEWKIRK, GARY FRANCIS, b Paterson, NJ, June 25, 46; m 68; c 3. MARINE ECOLOGY, QUANTITATIVE GENETICS. Educ: Rutgers Univ, BSc, 68; Duke Univ, PhD(zool), 74. Prof Exp: Fel ecol genetics, 73-75, RES ASSOC ECOL GENETICS, DALHOUSIE UNIV, 75- Mem: Soc Study Evolution; Am Soc Limnol & Oceanog; Ecol Soc Am. Res: Ecological genetics of marine species especially those of importance to mariculture. Mailing Add: Dept of Biol Dalhousie Univ Halifax NS Can

NEWKIRK, GORDON ALLEN, b Orange, NJ, June 12, 28; m 56; c 3. ASTROPHYSICS. Educ: Harvard Univ, AB, 50; Univ Mich, MA, 52, PhD(astrophys), 53. Prof Exp: Asst, Observ, Univ Mich, 50-53; astrophysicist, Upper Air Res Observ, 53; sr mem staff, 55-68, adj prof astrogeophys, 61-65, DIR, HIGH ALTITUDE OBSERV, UNIV COLO, 68-, ADJ PROF PHYSICS & ASTROPHYS, 65- Concurrent Pos: Prin investr, ATM White Light Coronagraph, 64-70; mem solar physics subcomt, NASA, 65-68, astron missions bd, Solar Panel, 69-70; mem orgn comt, Comn 10 Solar Activ, Inst Astron Union, 67-73, vpres, 73, actg pres, 75-; mem consult group potentially harmful effects on space res panel 3B & Comn V, Int Sci Radio Union; vchmn, Solar Physics Div, Am Astron Soc, 70-72, chmn, Solar Physics Div, 72-73. Honors & Awards: Boulder Scientist Award, 65; Publ Prize, Nat Ctr Atmospheric Res, 67, Technology Award, 73. Mem: AAAS; Int Astron Union; Sigma Xi; Am Astron Soc; Am Geophys Union. Res: Solar physics; corona, prominences and solar radio radiation; scattering of light in terrestrial atmosphere; space observations; solar magnetic fields. Mailing Add: High Altitude Observ Nat Ctr Atmospheric Res Boulder CO 80303

NEWKIRK, HERBERT WILLIAM, b Jersey City, NJ, Nov 23, 28; m 52; c 2. INORGANIC CHEMISTRY, PHYSICAL CHEMISTRY. Educ: Polytech Inst Brooklyn, BS, 51; Ohio State Univ, PhD(chem), 56. Prof Exp: Res chemist, Allied Chem & Dye Corp, 51-52; res engr, Gen Elec Co, 56-59; res chemist, Radio Corp Am, 59-60 & Lawrence Livermore Lab, Univ Calif, 60-69; guest prof, Philips Res Lab, Aachen, WGer, 69-71; PROG MGR, LAWRENCE LIVERMORE LAB, UNIV CALIF, 71- Res: High energy density and high temperature chemistry; irradiation chemistry of nuclear fuels and structural materials; electronic properties of solids; inorganic synthesis, characterization and single crystal growth of materials; technology transfer processes. Mailing Add: 1141 Madison Ave Livermore CA 94550

NEWKIRK, JOHN BURT, b Minneapolis, Minn, Mar 24, 20; m 51: c 4. PHYSICAL METALLURGY, BIOENGINEERING. Educ: Rensselaer Polytech Inst, BMetE, 41; Carnegie Inst Technol, MS, 47, DSc(phys metall), 50. Prof Exp: Fulbright fel crystallog, Cavendish Lab, Cambridge Univ, 50-51; res metallurgist, Res Lab, Gen Elec Co, 51-59; prof phys metall, Cornell Univ, 59-64; Phillipson prof phys metall, 64-75, PROF PHYS CHEM, UNIV DENVER, 75- Concurrent Pos: Consult, Gen Elec Co, 60-65, Atomics Int, Inc, 63-65, Jet Propulsion Lab, 64-65, 3M Co, 75-77 & Denver Med Specialties, Inc, 75-; pres, Denver Biomat Inc, 68- Honors & Awards: V C Huffsmith Res Award, Denver Res Inst, 73. Mem: AAAS; Am Soc Mining, Metall & Petrol Engrs; Am Soc Metals; Sigma Xi; Am Chem Soc. Res: Characterization of solids, with special emphasis on heat resisting alloys; biomaterials and implantable biodevices for flow control of abnormal body liquids. Mailing Add: Space Sci Lab Dept of Chem Univ of Denver Denver CO 80210

NEWKIRK, LESTER LEROY, b Kansas City, Kans, June 2, 20; m 44; c 3. SPACE PHYSICS. Educ: Kans State Col, BS, 43, MS, 48; Iowa State Univ, PhD(physics), 51. Prof Exp: Physicist microwave lab, Res & Develop Dept, Hughes Aircraft Co, 51-53; mem staff, Lawrence Radiation Lab, Univ Calif, 53-58 & Los Alamos Sci Lab, 58-61; MEM STAFF, LOCKHEED MISSILES & SPACE CO, PALO ALTO, 61- Mem: Am Phys Soc; Am Geophys Union. Res: Space physics; nuclear weapons physics; solar x-ray physics. Mailing Add: 240 Silvia Ct Los Altos, CA 94022

NEWKIRK, RICHARD ALBERT MICHAEL, b Quincy, Ill, Dec 19, 25; m 68; c 4. ACAROLOGY, SCIENTIFIC BIBLIOGRAPHY. Educ: Univ Miami, BS, 56; Am Registry Prof Entomologists, dipl, 72. Prof Exp: Biol aid insect taxon, USDA, 56-57, plant pest control inspector, 57-63, agriculturist insect status reporter, 63-70; PROG SPECIALIST PESTICIDES, REGULATION DIV, ENVIRON PROTECTION AGENCY, 70- Mem: AAAS; Entom Soc Am. Res: Insect and mite-host relationships; ecology of phytophagous insects and mites; systematics and nomenclature of Eriophyoidea. Mailing Add:

NEWKIRK, ROBERT FRANKLIN, b Pender Co, NC, Mar 25, 41; m 65; c 4. PHYSIOLOGY. Educ: Livingstone Col, BS, 63; Va State Col, MS, 68; Colo State Univ, PhD(physiol), 72. Prof Exp: Sec sci teacher life sci & phys sci, Charlotte-Mecklenburg Schs, 63-66; teaching asst biol, Colo State Univ, 68-71; instr, 67-68, ASST PROF BIOL, VA STATE COL, 72- Mem: Sigma Xi. Res: Formation of synaptosomes in invertebrate nervous tissues; the uptake, metabolism and release of transmitter substances in these structures. Mailing Add: 4513 Winterbourne Dr Petersburg VA 23803

NEWKIRK, TERRY FRANKLIN, b Moberly, Mo, Apr 11, 20; m 42; c 3. CERAMICS, APPLIED PHYSICS. Educ: Univ Ill, BS, 41, MS, 49. Prof Exp: Res assoc ceramic eng, Univ Ill, 45-49; Portland Cement Asn fel, Nat Bur Standards, 49-56; develop engr & phys ceramist, Power Tube Dept, Gen Elec Co, 56-59; mgr ceramic & metall div, Trionics Corp, Wis, 59-62; pvt consult, 62-63; mgr solid state physics, Appl Physics Dept, Sanders Assoc, Inc, NH, 63-66; pres, 66-71, CHMN & DIR CORP DEVELOP, INT MAT CORP, 71- Mem: Am Phys Soc. Res: Lasers; ultrasonic devices; electronic ceramics; thin films and solid state materials; electron tube materials and processes; ceramic-metal seals; metallizing and brazing; refractory metals; gas sorption systems. Mailing Add: 12 Longbow Circle Lynnfield MA 01940

NEWKOME, GEORGE R, b Akron, Ohio, Nov 26, 38; m 62; c 1. ORGANIC CHEMISTRY. Educ: Kent State Univ, BS, 61, PhD(org chem), 66. Prof Exp: Chemist, Firestone Tire & Rubber Co, 61-62; res assoc, Princeton Univ, 66-67, NIH fel, 67-68; asst prof, 68-72, ASSOC PROF CHEM, LA STATE UNIV, BATON ROUGE, 72- Concurrent Pos: La State Univ Alumni distinguished fac fel, 70-71. Mem: AAAS; Am Chem Soc; The Chem Soc; NY Acad Sci. Res: Synthetic and structural organic chemistry relating to natural products and biochemical mimics, including organic-biochemistry; stereochemistry; synthesis of compounds with

NEWKOME

potential pharmacological properties; molecular rearrangements. Mailing Add: Dept of Chem La State Univ Baton Rouge LA 70803

NEWLAND, GORDON CLAY, b Kingsport, Tenn, Feb 26, 27; m 53; c 2. POLYMER CHEMISTRY, PHOTOCHEMISTRY. Educ: ETenn State Univ, BS, 49. Prof Exp: Res chemist, 50-57, SR RES CHEMIST, TENN EASTMAN RES LAB DIV, EASTMAN KODAK, 57- Mem: Am Chem Soc; AAAS. Res: Photochemistry of polymer degradation and stabilization mechanisms; photoinitiated polymerization. Mailing Add: Tenn Eastman Co Box 511 Kingsport TN 37664

NEWLAND, HERMAN WILLIAM, b Hastings, Mich, Jan 26, 17; m 41; c 3. ANIMAL HUSBANDRY. Educ: Mich State Univ, BS, 40, MS, 49; Univ Fla, PhD(animal nutrit), 55. Prof Exp: From instr to assoc prof animal husb, Mich State Univ, 46-67; assoc prof, 67-70, PROF ANIMAL SCI, OHIO STATE UNIV, 70- Res: Animal production and nutrition. Mailing Add: Dept of Animal Sci Ohio State Univ Columbus OH 43210

NEWLAND, JULIAN HAROLD, organic chemistry, polymer chemistry, see 12th edition

NEWLAND, LEO WINBURNE, b Nocona, Tex, Sept 15, 40. SOIL CHEMISTRY, WATER CHEMISTRY. Educ: Tex A&M Univ, BS, 64; Univ Wis, MS, 66, PhD(soils), 69. Prof Exp: Agr res specialist, Frito Lay Inc, 64-66; res asst soils & water, Univ Wis, 66-68; fel biol, 68-69, ASST RES SCIENTIST & ASST PROF GEOL & BIOL, TEX CHRISTIAN UNIV, 69-, DIR ENVIRON SCI PROG, 71- Mem: Am Chem Soc; Soil Sci Soc Am; Water Pollution Control Fedn; Am Soc Agron. Res: Chemical pollution of surficial waters and soils, specifically pesticidal pollution. Mailing Add: Dept of Biol & Geol Tex Christian Univ Ft Worth TX 76129

NEWLANDS, MICHAEL JOHN, b London, Eng, Mar 10, 31; m 54; c 2. INORGANIC CHEMISTRY. Educ: Cambridge Univ, BA, 53, PhD(chem), 57. Prof Exp: Fel, Inst Sci & Technol, Manchester, 58-60, lectr, 60-67; assoc prof, 67-72, PROF INORG CHEM & HEAD DEPT CHEM, MEM UNIV NFLD, 72- Mem: Am Chem Soc; Chem Inst Can; The Chem Soc. Res: Organometallic chemistry of main group elements, especially compounds with metal-metal bonds of transition metals; insertion reactions. Mailing Add: Dept of Chem Mem Univ of Nfld St John's NF Can

NEWLIN, GRODON ERMEL, b Paoli, Ind, Oct 7, 42; m 63; c 1. MICROBIOLOGY, VIROLOGY. Educ: Ind Univ, Bloomington, AB, 64; Univ NMex, MS, 67; Univ Kans, PhD(microbiol), 70. Prof Exp: ASST PROF VIROL, MED UNIV SC, 70- Mem: AAAS; Am Soc Microbiol. Res: Oncogenic viruses; DNA viruses of the papova group; emphasis on the rabbit papilloma virus as a model for viral oncogenesis. Mailing Add: Dept of Immunol & Microbiol Med Univ of SC Charleston SC 29401

NEWLIN, OWEN JAY, b Des Moines, Iowa, Feb 6, 28; m 52; c 4. AGRONOMY. Educ: Iowa State Univ, BS, 51, MS, 53; Univ Minn, PhD, 55. Prof Exp: Asst agron, Iowa State Univ, 51-53; asst agron & plant genetics, Univ Minn, 53-55; asst prof res, Pioneer Seed Co, 55-56, dir prod res, 56-60, asst prod mgr & dir prod res, 60-64, prod mgr, 64-67, gen mgr, 67-71, PRES, CENT DIV, PIONEER HI-BRED INT, INC, 71-, MEM BD DIRS, 63- Mem: Am Soc Agron; Entom Soc Am. Res: Corn production and breeding. Mailing Add: 3315 48th Pl Des Moines IA 50310

NEWLING, BRUCE EDGAR, b Derby, Eng, Apr 12, 34; m 66; c 2. URBAN GEOGRAPHY, ECONOMIC GEOGRAPHY. Educ: Univ London, BSc, 56; Northwestern Univ, MA, 58, PhD(geog), 62. Prof Exp: Vis asst prof geog, Univ Pittsburgh, 62-63, res assoc pop, Ctr Regional Econ Studies, 63-64; vis assoc prof urban studies, Cornell Univ, 65; asst prof geog, Rutgers Univ, 65-67; from asst prof to assoc prof, 67-74, PROF GEOG, CITY COL NEW YORK, 74- Concurrent Pos: Consult, State Mich Antipoverty Prog, 65; prin investr, Off Water Resources Res Proj, Water Resources Res Inst, Rutgers Univ, 66-69; vis res assoc, Sch Archit, Princeton Univ, 67-68; consult, Princeton Aqua Sci, New Brunswick, 72-73. Mem: Am Geog Soc; Asn Am Geogr; Regional Sci Asn; Royal Geog Soc. Res: Urban population densities; small-area population projections; urban population growth and urban sprawl, particularly in the New York-New Jersey metropolitan area. Mailing Add: Dept of Geog City Col NY Convent Ave at 138th St New York NY 10031

NEWMAN, BERNARD, b New York, NY, Sept 17, 13; m 46; c 2. CHEMISTRY. Educ: City Col New York, BS, 35, MS, 43; NY Univ, PhD, 55. Prof Exp: Lab technician, Bronx County Anal Labs, NY, 35-37; chief lab technician, Morrisania City Hosp, 37-38; biochemist, Hosp Daughters Jacob, 38-42, dir chem labs, 46; chief biochemist, Rystan Co, Inc, NY, 46-47; chief biochemist, Vet Admin Hosp, NY, 47-48; biochemist, S Shore Res Lab, 48-58; dir, Newing Labs, Inc, 58-74; sr res scientist, Res Div, NY Univ, 56-74; DIR, MARINE SCI GRAD DEPT, LONG ISLAND UNIV, 74- Concurrent Pos: Dir, Police Lab, Police Dept, Suffolk County, 60-; adj assoc prof, Suffolk Community Col, 62-; pres, Nat Asn Police Labs, 67-68; prof, Grad Dept Marine Sci, Long Island Univ, 67-74; consult, Kings Park State Hosp, NY. Mem: AAAS; Am Soc Microbiol; Am Chem Soc; Am Pub Health Asn; fel Am Acad Forensic Sci. Res: Criminalistics; toxicology. Mailing Add: Marine Sci Grad Dept Long Island Univ Greenvale NY 11548

NEWMAN, BERTHA L, b Caldwell, Idaho, Mar 29, 26. ANATOMY. Educ: Col Idaho, BS, 48; Univ Ore, MA, 50; Univ Iowa, PhD(anat), 58. Prof Exp: Instr zool, Idaho State Col, 49-53; asst prof neuroanat, Univ NDak, 58; res fel neurophysiol, Univ Wash, 58-60; asst prof, 60-65, ASSOC PROF NEUROANAT, SCH MED, UNIV SOUTHERN CALIF, 65- Mem: Am Asn Anatomists; Am Physiol Soc. Res: Electrophysiological changes associated with electrical self-stimulation; ultrastructure of the median eminence of neonatal and adult rats. Mailing Add: Dept of Anat Sch of Med Univ of Southern Calif Los Angeles CA 90033

NEWMAN, CLARENCE WALTER, b Lake Providence, La, Aug 3, 32; m 54; c 2. ANIMAL NUTRITION. Educ: La State Univ, BS, 54; Tex A&M Univ, MS, 58; La State Univ, PhD(animal sci), 65. Prof Exp: Instr animal sci, La Agr Exp Sta, 58-60; spec lectr animal nutrit, La State Univ, 61-62; assoc prof, 64-75, PROF ANIMAL SCI, MONT STATE UNIV, 75- Mem: AAAS; Am Soc Animal Sci. Res: Nutritive value of barley varieties and barley variety isogenes as related to their nutrient composition and availability of these nutrients to swine and laboratory animals. Mailing Add: Dept of Animal & Range Sci Mont State Univ Bozeman MT 59715

NEWMAN, DAVID JOHN, b Grays, UK, May 2, 39; m 69. BIOCHEMISTRY, MICROBIOLOGY. Educ: Univ Liverpool, MSc, 63; Univ Sussex, DPhil(biochem), 68. Prof Exp: Analyst chem & biochem, J Bibby & Sons, 56-61; res chemist, Ilford Ltd, 63-64; asst exp off biochem & microbiol, Agr Res Coun, Univ Sussex, 64-68; res assoc biochem, Univ Ga, 68-70; ASSOC SR INVESTR, SMITH KLINE & FRENCH LABS, 70- Mem: Assoc Royal Inst Chem; Brit Inst Biol; Am Chem Soc; Am Soc Microbiol; Brit Biochem Soc. Res: Chemistry, biochemistry and enzymology of metalloproteins; metabolic regulation at enzyme level; bioenergetics; roles of cyclic nucleotides in normal and diseased states. Mailing Add: Smith Kline & French Labs 1500 Spring Garden St Philadelphia PA 19101

NEWMAN, DAVID S, b New York, NY, Sept 18, 36; m 59; c 4. PHYSICAL CHEMISTRY. Educ: Earlham Col, AB, 57; NY Univ, MS, 60; Univ Pa, PhD(chem), 65. Prof Exp: Teacher, Newtown High Sch, 59-60; instr chem & physics, Bronx Community Col, 60; teacher, Roosevelt High Sch, 60; res assoc phys chem, Princeton Univ, 64-65; from asst prof to assoc prof phys chem, 65-74, PROF PHYS CHEM, BOWLING GREEN STATE UNIV, 74- Concurrent Pos: Prof adv continuing educ appointee, Argonne Nat Lab, 67, consult, Chem Div; Cotrell res grant, Res Corp, 67-68; sr Fulbright fel, 74-75; sr Fulbright lectr, US Govt, 74-75. Mem: Am Chem Soc; Electrochem Soc. Res: High pressure-high temperature chemistry; chemistry of fused salts; heterogeneous kinetics; structure of electrolyte solutions; electrochemistry. Mailing Add: Dept of Chem Bowling Green State Univ Bowling Green OH 43403

NEWMAN, DAVID WILLIAM, b Pleasant Grove, Utah, Oct 26, 33; m 56. PLANT PHYSIOLOGY. Educ: Univ Utah, BS, 55, MS, 57, PhD, 60. Prof Exp: Asst bot, Univ Utah, 54-55; from asst prof to assoc prof, 60-74, PROF BOT, MIAMI UNIV, 74- Mem: AAAS; Am Soc Plant Physiol; Bot Soc Am; Am Oil Chem Soc; Am Chem Soc. Res: Lipid and protein metabolism of plants. Mailing Add: Dept of Bot Miami Univ Oxford OH 45056

NEWMAN, ELLIOT VOSS, medicine, deceased

NEWMAN, EUGENE, b New York, NY, Sept 14, 30; m 52; c 2. NUCLEAR PHYSICS. Educ: Polytech Inst Brooklyn, BS, 52; Yale Univ, MS, 57, PhD(physics), 60. Prof Exp: PHYSICIST, OAK RIDGE NAT LAB, 52-54 & 60- Mem: Am Phys Soc. Res: Use of low to medium energy cyclotron produced nucleons to investigate nuclear spectroscopy and reaction mechanisms. Mailing Add: Oak Ridge Nat Lab Physics Div Bldg 6000 Oak Ridge TN 37830

NEWMAN, EZRA, b New York, NY, Oct 17, 29; m 58; c 2. THEORETICAL PHYSICS. Educ: NY Univ, BA, 51; Syracuse Univ, MA, 55, PhD, 56. Prof Exp: Asst physics, Syracuse Univ, 52-56; from instr to assoc prof, 57-66, PROF PHYSICS, UNIV PITTSBURGH, 66- Concurrent Pos: Vis lectr, Syracuse Univ, 60-61; vis prof, King's Col, Univ London, 64-65 & 68-69. Mem: Fel Am Phys Soc. Res: General theory of relativity with emphasis on gravitational radiation and the theory of twisters. Mailing Add: Dept of Physics Univ of Pittsburgh Pittsburgh PA 15260

NEWMAN, FRANKLIN SCOTT, b Rozel, Kans, July 31, 31; m 54; c 4. MICROBIOLOGY, VIROLOGY. Educ: Southwestern Col Kans, AB, 53; Kans State Univ, MS, 57, PhD(bact), 62. Prof Exp: Asst prof microbiol, Mont State Univ, 61-65; asst prof, Med Br, Univ Tex, 65-68; ASSOC PROF MICROBIOL, VET RES LAB, MONT STATE UNIV, 68-, DIR WAMI MED PROG, 74- Concurrent Pos: Asst dean, Univ Wash Sch Med, 75- Mem: AAAS; Am Soc Microbiol; Sigma Xi. Res: Infectious diseases of domestic animals; bacterial virology; phage-host relationships in pathogenic bacteria. Mailing Add: Vet Res Lab Mont State Univ Bozeman MT 59715

NEWMAN, GEORGE ALLEN, b Las Cruces, NMex, Mar 15, 41; m 64; c 2. ECOLOGY, ORNITHOLOGY. Educ: Baylor Univ, BSc, 64, MSc, 66; Tex A&M Univ, PhD(wildlife sci), 75. Prof Exp: Asst prof, 67-75, ASSOC PROF BIOL, HARDIN-SIMMONS UNIV, 75- Concurrent Pos: Collabr, Nat Park Serv, 72-; mem citizens adv coun, Tex Air Control Bd, 72-; consult, Ecol Audits Inc, 73. Mem: Fel Welder Wildlife Found; Ecol Soc Am; Am Ornithologists Union. Res: Avian population studies of Guadalupe Mountain Range, Texas. Mailing Add: 2602 Greenbriar Abilene TX 79605

NEWMAN, GERALD H, physical chemistry, see 12th edition

NEWMAN, HARRY, b Russia, Sept 10, 09; nat US; m 42; c 3. UROLOGY. Educ: Univ Toronto, MD, 35; Univ Pa, MSc, 40; Am Bd Urol, dipl, 47. Prof Exp: Asst clin prof urol, Grad Sch Med, NY Univ, 49-54; assoc prof surg, 54-66, clin prof urol, 58-66, CHIEF UROL, ALBERT EINSTEIN COL MED, 54-, PROF, 66- Concurrent Pos: Asst attend urologist, NY Univ Hosp, 46-56; consult, Griffin Hosp, Derby, Conn, 46-, St Joseph Hosp, Stamford, 58- & Stamford Hosp, 64-; attend urologist, Yale-New Haven Hosp, 47- & St Raphaels Hosp, 49-; asst clin prof, Sch Med, Yale Univ, 49-; asst vis urologist, Bellevue Hosp, 49-56; vis urologist, Bronx Munic Med Ctr, 54- Mem: AMA; Am Urol Asn; fel Am Col Surg; Can Med Asn. Res: Experimental and clinical observations of obstruction of the ureter; physiology of the normal and scarred ureter. Mailing Add: Dept of Urol Albert Einstein Col of Med Bronx NY 10461

NEWMAN, HOWARD, organic chemistry, see 12th edition

NEWMAN, HOWARD ABRAHAM IRA, b Chicago, Ill, July 5, 29; m 55; c 3. BIOCHEMISTRY, PHYSIOLOGY. Educ: Univ Ill, BS, 51, MS, 56, PhD(food technol), 58. Prof Exp: Asst, Univ Ill, 54-58, res assoc, 58; res assoc physiol, Univ Tenn, 58-59, asst prof, 59-65; asst prof biochem, Case Western Reserve Univ, 66-68; ASSOC PROF PATH & PHYSIOL CHEM, OHIO STATE UNIV, 68- Concurrent Pos: Mem coun arteriosclerosis, Am Heart Asn. Mem: AAAS; Am Physiol Soc; NY Acad Sci; Am Inst Chem; Am Asn Clin Chem. Res: Cholesterol, triglyceride and phospholipid metabolism in the intact animal, especially in the atherosclerotic intima; cholesterol esterases; analytical lipid techniques; lipoprotein phenotyping in clinical chemistry; mycobacterial phage lipids; hypolipoproteinemic drugs; cell membrane changes in carcinogenesis. Mailing Add: Dept of Path Ohio State Univ 410 W Tenth Ave Columbus OH 43210

NEWMAN, JACK HUFF, b Roanoke, Va, Aug 15, 29; m 56; c 3. BACTERIOLOGY, BIOCHEMISTRY. Educ: Va Polytech Inst, BS, 56, MS, 59. Prof Exp: Jr biochemist, Smith Kline & French Labs, 58-62; BIOCHEMIST, A H ROBINS CO, INC, 62- Mem: Am Chem Soc. Res: Radioisotopes; drug metabolism. Mailing Add: 8106 Diane Lane Richmond VA 23227

NEWMAN, JAMES BLAKEY, b Little Rock, Ark, Jan 10, 17; m 42; c 3. PHYSICS. Educ: Va Mil Inst, BS, 39; Cornell Univ, PhD(physics), 51. Prof Exp: Instr, 39-42, from asst prof to assoc prof, 49-56, PROF PHYSICS, VA MIL INST, 56- Mem: Am Phys Soc; Am Asn Physics Teachers; Am Nuclear Soc. Res: Atomic and nuclear physics. Mailing Add: Dept of Physics Va Mil Inst Lexington VA 24450

NEWMAN, JAMES EDWARD, b Brown Co, Ohio, Dec 22, 20; m 49; c 2. AGRONOMY. Educ: Ohio State Univ, BS, 47, MS, 49. Prof Exp: Asst, Ohio Agr Exp Sta, 47-49; from asst prof to assoc prof agron & climat, 49-69, PROF AGRON, PURDUE UNIV, WEST LAFAYETTE, 69- Concurrent Pos: Partic comn agr meteorol, World Meteorol Orgn, Ont, Can, 62; New World ed, Jour Agr Meteorol, 63; vis prof, Univ Calif, Riverside, 65-66; vis scientist, Inst Agr Sci, Univ Alaska, 70; ed-in-chief, Int J Agr Meteorol, 73- Honors & Awards: Soils & Crops Award, Am Soc Agron, 65. Mem: Fel AAAS; fel Am Soc Agron; Am Meteorol Soc; Ecol Soc Am; Int Soc Biometeorol. Res: Radiant energy flux and plant responses in both natural and mono culture phyto-environments; adaptation of cereal grains; crop modeling; climatic frequences; phenology of crop plants. Mailing Add: Dept of Agron Purdue Univ West Lafayette IN 47906

NEWMAN, JAMES L, b Minneapolis, Minn, May 7, 39; m 63. GEOGRAPHY. Educ: Univ Minn, BA, 61, MA, 63, PhD(geog), 68. Prof Exp: Researcher geog, Nat Acad Sci, 65-66; ASSOC PROF GEOG, SYRACUSE UNIV, 67- Mem: AAAS; Asn Am Geogr; African Studies Asn. Res: Patterns of rural settlement and land use, particularly relationship between environment and decision making; East Africa. Mailing Add: Dept of Geog Syracuse Univ Syracuse NY 13210

NEWMAN, JAMES MARTIN, b New York, NY, Dec 8, 37. MATHEMATICS. Educ: Cornell Univ, BA, 59; Harvard Univ, AM, 60; NY Univ, PhD(math), 69. Prof Exp: Math analyst, Gen Elec Co, Pa, 61-62; lectr math, Brooklyn Col, 66-68; asst prof, Fla Atlantic Univ, 68-71; ASST PROF MATH, BARUCH COL, 71- Mem: Am Math Soc; Math Asn Am. Res: Partial differential equations; mathematical economics. Mailing Add: Dept of Math Baruch Col New York NY 10010

NEWMAN, JAMES RANEY, b Newton, Mass, Oct 18, 41; m 65; c 2. ZOOLOGY, ECOLOGY. Educ: Univ Santa Clara, BA, 64, BS, 65; Univ Calif, Davis, PhD(zool), 70. Prof Exp: Asst prof ecol, 70-74, ASSOC PROF ECOSYSTS ANAL, HUXLEY COL ENVIRON STUDIES, WESTERN WASH STATE COL, 74- Mem: AAAS; Ecol Soc Am; Am Soc Mammalogists; Soc Study Evolution; Am Inst Biol Sci. Res: Ecological modeling; energetics; population dynamics of soricids; fluorine toxicity in primary consumers. Mailing Add: Huxley Col of Environ Studies Western Wash State Col Bellingham WA 98225

NEWMAN, JOHN BROWN, b Hendersonville, NC, Oct 16, 21; m 49; c 3. PHYSICS. Educ: Univ NC, Chapel Hill, BS, 44, MS, 52; Johns Hopkins Univ, PhD(physics), 66. Prof Exp: Instr physics, Univ NC, Chapel Hill, 44; radio engr, Naval Res Lab, 44-45; asst physics, Mass Inst Technol, 49; asst prof, Talladega Col, 49-51; res assoc, Johns Hopkins Univ, 51-66, res scientist, 66-67; assoc prof, 67-70, PROF PHYSICS, TOWSON STATE COL, 70- Concurrent Pos: Lectr, Johns Hopkins Univ, 62 & Goucher Col, 64-66; consult, Catalyst Res Inc, 66-69 & Army Res Off, 69-71. Mem: Am Phys Soc; Am Asn Physics Teachers. Res: Electronic structure and spectral behavior of actinide elements; millimeter wave transmission of the atmosphere; microwave absorption line widths. Mailing Add: Dept of Physics Towson State Col Baltimore MD 21204

NEWMAN, JOHN SCOTT, b Richmond, Va, Nov 17, 38. ELECTROCHEMISTRY. Educ: Northwestern Univ, BS, 60; Univ Calif, Berkeley, MS, 62, PhD(chem eng), 63. Prof Exp: From asst prof to assoc prof chem eng, 63-70, PROF CHEM ENG, UNIV CALIF, BERKELEY, 70-, PRIN INVESTR INORG MAT RES DIV, 63- Honors & Awards: Young Author's Prize, Electrochem Soc, 66 & 69. Mem: Electrochem Soc. Res: Design and analysis of electrochemical systems; transport properties of concentrated electrolytic solutions; mass transfer. Mailing Add: Dept of Chem Eng Univ of Calif Berkeley CA 94720

NEWMAN, KARL ROBERT, b Mt Pleasant, Mich, Feb 26, 31; m 61; c 2. GEOLOGY. Educ: Univ Mich, BS, 53, MS, 54; Univ Colo, PhD(geol), 61. Prof Exp: Geologist, Magnolia Petrol Co, 54 & Palynological Res Lab, 59-61; sr res geologist, Pan-Am Petrol Corp, 61-66; asst prof geol, Mont Col Mineral Sci & Technol, 66-67; assoc prof, Cent Wash State Col, 67-71, chmn dept, 69-71; ASSOC PROF GEOL, COLO SCH MINES, 71- Concurrent Pos: Consult geologist, 66- Mem: Soc Econ Paleontologists & Mineralogists; Paleont Soc; Am Asn Petrol Geologist; Geol Soc Am; Am Inst Prof Geologists. Res: Stratigraphy; geology and palynology of Rocky Mountain basins; field geology; upper Cretaceous and lower Tertiary palynomorphs; geology of coal and oil shale. Mailing Add: Dept of Geol Colo Sch of Mines Golden CO 80401

NEWMAN, KATHERINE JANE, nutrition, see 12th edition

NEWMAN, KENNETH WALTER, physical chemistry, see 12th edition

NEWMAN, KENNETH WILFRED, b Lincoln, Nebr. ALGEBRA. Educ: City Univ New York, BS, 65; Cornell Univ, PhD(math), 70. Prof Exp: Fel math, McGill Univ, 69-71; ASST PROF MATH, UNIV ILL, CHICAGO CIRCLE, 71- Mem: Am Math Soc. Res: Theory of Hopf algebras. Mailing Add: Dept of Math Box 4348 Univ of Ill Chicago Circle Chicago IL 60680

NEWMAN, LEONARD, b New York, NY, Jan 15, 31; m 53; c 2. ATMOSPHERIC CHEMISTRY, ANALYTICAL CHEMISTRY. Educ: Polytech Inst Brooklyn, BS, 52; Mass Inst Technol, PhD(chem), 56. Prof Exp: Asst, Mass Inst Technol, 52-55; sr scientist, Nat Lead Co, 56-57; ASSOC HEAD, ATMOSPHERIC SCI DIV, BROOKHAVEN NAT LAB, 58- Concurrent Pos: Vis scientist, Royal Inst Technol, Sweden, 62-63; mem comt on nuclear methods for investigating air pollution, Nat Acad Sci, 68; consult, Gen Pub Utilities Corp, 74. Mem: AAAS; Am Chem Soc; Am Soc Testing & Mat. Res: Complex ion equilibria of simple and mixed ligand complexes; atmospheric chemistry; analytical chemistry of air pollutants; hydrolysis reactions; solvent extraction; chemistry of actinide and less familiar elements; kinetic mechanisms; nuclear reactor chemistry and fuel processing; electrochemistry. Mailing Add: Dept of Appl Sci Bldg 801 Brookhaven Nat Lab Upton NY 11973

NEWMAN, LESTER JOSEPH, b St Louis, Mo, June 15, 33. CYTOGENETICS. Educ: Wash Univ, BA, 55; Univ Mich, MA, 60; Wash Univ, PhD(zool), 63. Prof Exp: NIH trainee, Wash Univ, 60-63; asst prof zool, Ore State Univ, 63-64; actg head dept, 65-66, ASST PROF BIOL, PORTLAND STATE UNIV, 64- Mem: AAAS; Genetics Soc Am. Res: Cytogenetics of Diptera. Mailing Add: Dept of Biol Portland State Univ Portland OR 97207

NEWMAN, LOUIS BENJAMIN, b New York, NY, Apr 5, 00. REHABILITATION MEDICINE. Educ: Ill Inst Technol, ME, 21; Rush Med Col, MD, 31; Am Bd Phys Med & Rehab, dipl, 47. Prof Exp: Attend physician, Cook County Hosp, Ill, 33-42; PROF PHYS MED, MED SCH, NORTHWESTERN UNIV, CHICAGO, 46- Concurrent Pos: Chief phys med & rehab serv, Vet Admin Hosps, Hines, 46-53 & Vet Res Hosp, Chicago, 53-67; consult rehab med, Vet Admin Hosps & several community hosps, Chicago Area, 67-; mem med adv bd, Vis Nurse Asn, Nat Found, Arthritis Found, Rehab Comt of Inst Med, United Parkinson Found & others; lectr rehab med, Col Med, Univ Ill & Stritch Sch Med, Loyola Univ Chicago; lectr, Chicago Med Sch, Univ Health Sci. Honors & Awards: Davis Award, Asn Phys & Ment Rehab, 56, Distinguished Achievement Award, 66; President's Comt Employ Physically Handicapped Commendation, 56; Ill Inst Technol Distinguished Serv Award, 57; Vet Admin Meritorious Serv Award, 58, Commendation, 62; Am Cong Rehab Med Distinguished Serv Key, 63; Am Legion Nat Rehab Citation, 67. Mem: AMA; Am Cong Rehab Med (vpres, 60); Am Acad Phys Med & Rehab (pres, 59); Am Asn Electromyog & Electrodiag; Int Soc Rehab Disabled. Mailing Add: 400 E Randolph St Chicago IL 60601

NEWMAN, MARSHALL THORNTON, b New Bedford, Mass, July 15, 11; m 45; c 3. ANTHROPOLOGY. Educ: Univ Chicago, PhB, 33, MA, 35; Harvard Univ, PhD(anthrop), 41. Prof Exp: Field supvr archaeol, War Progress Admin, Smithsonian Inst, 33-34; supvr proj, Inst Andean Res, NY, 41-42; assoc cur phys anthrop, US Nat Mus, Smithsonian Inst, 42-62; prof anthrop, Portland State Col, 62-66, chmn dept, 64-66; PROF ANTHROP, UNIV WASH, 66-, ADJ PROF ENVIRON STUDIES, 73- Mem: Fel Am Anthrop Asn; Am Asn Phys Anthrop. Res: Anthropology of the New World; physical anthropology and nutrition; human adaptation and race formation. Mailing Add: Dept of Anthrop Univ of Wash Seattle WA 98105

NEWMAN, MAX KARL, b Malden, Mass, Jan 19, 09; m 32; c 5. MEDICINE. Educ: Univ Mich, AB, 30, MD, 34; Am Bd Phys Med & Rehab, dipl. Prof Exp: DIR PHYS MED & REHAB, DETROIT INST PHYS MED & REHAB, 36- Concurrent Pos: Dir phys med & rehab, Detroit Mem Hosp, 51- & Sinai Hosp Detroit, 54-; asst clin prof & assoc neurol, Sch Med, Wayne State Univ, 65-70, adj asst prof phys med & rehab, 70- Mem: Fel Am Geriat Soc; fel Am Col Physicians; fel Am Acad Phys Med & Rehab (secy-treas, 54-57, pres, 60-); assoc fel Am Acad Neurol; fel Am Acad Cerebral Palsy. Res: Physical medicine and rehabilitation. Mailing Add: 4787 Crestview Ct Birmingham MI 48010

NEWMAN, MELVIN MICKLIN, b Chicago, Ill, Dec 20, 21; m 49; c 2. SURGERY, PULMONARY PHYSIOLOGY. Educ: Univ Chicago, BS, 41, MD, 44. Prof Exp: From asst resident to instr surg, Univ Chicago, 46-52; from asst prof to assoc prof, State Univ NY Downstate Med Ctr, 54-59; chief, Nat Jewish Hosp, 59-68; ASSOC PROF SURG, MED CTR, UNIV COLO, DENVER, 61- Concurrent Pos: Nat Res Coun fel, Univ Chicago, 52; NIH grants, Nat Jewish Hosp, Denver, Colo, 62-68; NIH grant, Univ Colo, Denver, 70- Mem: Am Soc Artificial Internal Organs (pres, 60); Soc Univ Surgeons; Am Asn Thoracic Surg; Soc Thoracic Surg; Am Thoracic Soc. Res: Shock; microcirculation; pulmonary ventilation and circulation; vascular prostheses. Mailing Add: Dept of Surg Univ of Colo Med Ctr Denver CO 80220

NEWMAN, MELVIN SPENCER, b New York, NY, Mar 10, 08; m 33; c 4. ORGANIC CHEMISTRY. Educ: Yale Univ, BS, 29, PhD(chem), 32. Hon Degrees: DSc, Univ New Orleans, 75. Prof Exp: Nat Tuberc Asn fel, Yale Univ, 32-33; Nat Res Coun fel chem, Col Physicians & Surgeons, Columbia Univ, 33-34 & Harvard Univ, 34-36; instr chem, 36-39, Elizabeth Clay Howald scholar, 39-40, from asst prof to prof, 40-65, REGENTS PROF CHEM, OHIO STATE UNIV, 65- Concurrent Pos: Guggenheim fel, 49 & 51; Fulbright lectr, Glasgow Univ, 57 & 67; ed, J, Am Chem Soc. Honors & Awards: Award, Am Chem Soc, 61; Wilbur Cross Medal, Yale Univ, 70. Mem: Nat Acad Sci; AAAS; Am Chem Soc; The Chem Soc. Res: Synthetic and theoretical organic chemistry. Mailing Add: 2239 Onandaga Dr Columbus OH 43221

NEWMAN, MORRIS, b New York, NY, Feb 25, 24; m 48; c 2. MATHEMATICS. Educ: NY Univ, BA, 45; Columbia Univ, MA, 46; Univ Pa, PhD(math), 52. Prof Exp: Lectr math, Columbia Univ, 45; Instr, Univ Del, 48-51; res mathematician, 51-63, chief numerical anal sect, 63-70, SR RES MATHEMATICIAN, NAT BUR STANDARDS, 70- Concurrent Pos: Lectr, Am Univ, Catholic Univ & Univ Md, 53-; lectr, Univ BC, 60; res consult, Rand Corp, 61 & Univ Calif, Santa Barbara, 65. Honors & Awards: Dept Commerce Gold Medal Award, 66. Mem: Am Math Soc; Math Asn Am; London Math Soc. Res: Number theory; group theory; matrix theory; structure of matrix groups over rings; automorphic and modular functions. Mailing Add: Appl Math Div Nat Bur Standards Washington DC 20234

NEWMAN, MURRAY ARTHUR, b Chicago, Ill, Mar 6, 24; m 52; c 1. ICHTHYOLOGY. Educ: Univ Chicago, BS, 49; Univ Calif, MA, 51; Univ BC, PhD(zool), 60. Prof Exp: Cur fishes, Univ Calif, Los Angeles, 51-53; cur inst fishes, Univ BC, 53-56; CUR, VANCOUVER PUB AQUARIUM, 56- Mem: Am Soc Ichthyologists & Herpetologists; Can Mus Asn. Res: Behavior of fishes; marine ecology; systematic ichthyology. Mailing Add: Vancouver Pub Aquarium Box 3232 Vancouver BC Can

NEWMAN, NORMAN, b Brooklyn, NY, Mar 13, 39; m 62; c 2. ORGANIC CHEMISTRY. Educ: Brooklyn Col, BS, 59; Univ Minn, PhD(org chem), 64. Prof Exp: RES SPECIALIST, MINN MINING & MFG CO, 63- Mem: AAAS; Am Chem Soc; Soc Photog Sci & Eng. Res: Photographic science and chemistry; photochemical systems; reaction mechanisms. Mailing Add: Minn Mining & Mfg Co 3M Ctr 209-2S St Paul MN 55101

NEWMAN, PAUL, b Jacksonville, Fla, Mar 7, 37. ANTHROPOLOGY. Educ: Univ Pa, BA, 58, MA, 61; Univ Calif, Los Angeles, PhD(ling), 67. Prof Exp: Asst prof anthrop, Yale Univ, 66-71, ASSOC PROF ANTHROP, YALE UNIV, 71- Concurrent Pos: Prof & dir, Centre Study Nigerian Lang, Ahmadu Bello Univ, 72- Mem: Ling Soc Am; WAfrican Ling Soc; Int African Inst. Res: African linguistics, music and folklore. Mailing Add: Dept of Anthrop Yale Univ 105 Wall St New Haven CT 06520

NEWMAN, PAUL HAROLD, b Washington, DC, Apr 25, 33; m 58; c 3. INFORMATION SCIENCE. Educ: Antioch Col, BSc, 56. Prof Exp: Sr res assoc human eng, Am Insts Res, 56-60; res engr, Boeing Co, 60-61; systs specialist info processing systs design, Syst Develop Corp, 61-72, mgr control staff, 73; systs design mgr, 73-75, ASST CHIEF, OFF INFO SYSTS, ADMIN SERV DIV, DEPT SOCIAL & HEALTH SERV, STATE WASH, 75- Mem: AAAS; Soc Eng Psychol; Human Factors Soc; Asn Comput Mach. Res: information processing systems design and control; educational requirements; systems analyses; human factors analyses. Mailing Add: PO Box 1788 Olympia WA 98504

NEWMAN, PAULINE, b New York, NY, June 20, 27. INDUSTRIAL CHEMISTRY. Educ: Vassar Col, AB, 47; Columbia Univ, AM, 48; Yale Univ, PhD(chem), 52; NY Univ, LLB, 58. Prof Exp: Lab instr, Columbia Univ, 48 & Yale Univ, 48-50; res chemist, Am Cyanamid Co, 51-54; patent attorney, 54-69, DIR PATENT & LICENSING DEPT, FMC CORP, 69- Concurrent Pos: Specialist natural sci, UNESCO, 61-62; mem patent adv comt, Res Corp, 74- Mem: AAAS; Am Chem Soc; Am Inst Chem; Soc Indust Chem; fel NY Acad Sci. Res: Chemistry of high polymers; oxidation-reduction reactions; patent law; physical organic chemistry. Mailing Add: FMC Corp 2000 Market St Philadelphia PA 19103

NEWMAN, PHILIP LEE, b Eugene, Ore, Dec 17, 31; m 53; c 3. ANTHROPOLOGY, ETHNOLOGY. Educ: Univ Ore, BA, 53; Univ Wash, MA, 57, PhD(anthrop), 62. Prof Exp: Asst prof anthrop, Univ Calif, Los Angeles, 61-65, ASSOC PROF ANTHROP, UNIV CALIF, LOS ANGELES, 65- Mem: Am Anthrop Asn. Mailing Add: Dept of Anthrop Univ of Calif Los Angeles CA 90024

NEWMAN, RICHARD HOLT, b Mebane, NC, Aug 12, 32; m 55; c 3. RADIOCHEMISTRY. Educ: Elon Col, BA, 54; Univ SDak, MA, 60. Prof Exp: Chemist I water anal, NC State Bd Health, 54-55, Chemist II pollution control, 57-58; assoc scientist anal res, 60-67, res prof radiochem, 68-74, PROJ LEADER SMOKE MECHANISM, PHILIP MORRIS, INC, 74- Mem: Am Chem Soc; Sigma Xi. Res: Study of precursor product relationship between tobacco and smoke and elucidating mechanisms for smoke formation utilizing both radioactive and stable isotopes as tracers. Mailing Add: Philip Morris Inc PO Box 26583 Richmond VA 23261

NEWMAN, ROBERT ALWIN, b Winchester, Mass, July 11, 48. BIOCHEMISTRY, PHARMACOLOGY. Educ: Univ RI, BS, 70; Univ Conn, MS, 73, PhD(pharmacol),

75. Prof Exp: Res fel cell & molecular biol, Med Col Ga, 75-76; RES ASSOC BIOCHEM, SCH MED, UNIV VT, 76- Res: Connective tissue biochemistry and pharmacology; aging; vascular physiology and metabolism. Mailing Add: Dept of Biochem Sch of Med Univ of Vt Burlington VT 05401

NEWMAN, ROBERT BRADFORD, b Ungkung, China, Nov 5, 17; US citizen; m 41, 55; c 3. ACOUSTICS. Educ: Univ Tex, BA, 38, MA, 39; Mass Inst Technol, MArch, 49. Hon Degrees: ScD, Lawrence Col, 63. Prof Exp: Engr, Radio Corp Am, 41; civilian scientist, Naval Air Exp Sta, Pa, 43-45; partner, 48-53, VPRES, BOLT BERANEK & NEWMAN, INC, 53- Concurrent Pos: From instr to asst prof, Mass Inst Technol, 49-56, assoc prof, 56-; vis lectr, Harvard Univ, 55-71; Fulbright vis lectr, Royal Acad Fine Arts, Copenhagen, 59; mem bd governors, Bldg Res Inst, Nat Res Coun, 57-60; US specialist to Singapore, US Dept State, 61 & 65; prof, Harvard Univ, 71- Mem: Fel Acoust Soc Am. Res: Architecture; integration of acoustics principles into design of buildings. Mailing Add: Bolt Beranek & Newman Inc 50 Moulton St Cambridge MA 02138

NEWMAN, ROGER, b New York, NY, Aug 16, 25; m 46; c 3. CHEMISTRY. Educ: Columbia Univ, AB, 44, AM, 46, PhD(chem), 49. Prof Exp: Du Pont fel, Harvard Univ, 49-50; fel chem, Calif Inst Technol, 50-52; res assoc semiconductors, Res Labs, Gen Elec Co, 52-59; head dept physics, Mat Res Lab, Hughes Prod Div, Hughes Aircraft Co, 59-60, mgr semiconductor mat dept, Hughes Res Labs, 60-61; mgr solid state sci dept, 61-70, DIR SEMICONDUCTOR LAB, SPERRY RAND RES CTR, 71- Mem: Am Chem Soc; fel Am Phys Soc; fel Inst Elec & Electronics Eng. Mailing Add: Semiconductor Lab Sperry Rand Res Ctr Sudbury MA 01776

NEWMAN, ROGERS J, b Ramar, Ala, Dec 22, 26; m 51; c 3. MATHEMATICS. Educ: Morehouse Col, AB, 48; Atlanta Univ, MA, 49; Univ Mich, PhD(math), 61. Prof Exp: Instr physics & math, Bishop Col, 49-50; instr math, Grambling Col, 50-51; instr math & physics, Jackson State Col, 51-53; instr, Southern Univ, 53-55; jr instr math, Univ Mich, 59-60; chmn dept math, 61-74, PROF MATH, SOUTHERN UNIV, BATON ROUGE, 60- Concurrent Pos: NSF fac fel, Imp Col, Univ London, 70-71. Mem: Math Asn Am; Nat Inst Sci (vpres, 64). Res: Complex variables. Mailing Add: Dept of Math Southern Univ Baton Rouge LA 70813

NEWMAN, SANFORD BERNHART, b New York, NY, July 26, 14; m 42; c 2. MATERIALS SCIENCE. Educ: Long Island Univ, BS, 36; George Washington Univ, MS, 41; Univ Md, PhD(biophys), 51. Prof Exp: Mat engr, 40-46, microanalyst, 46-54, plastics technologist, 54-58, consult, 58-64, chief mat eval & testing sect, 64-67, chief mat eval div, 67-69, spec asst to dir inst appl technol, 69-70, SR PROG ANALYST, OFF ASSOC DIR PROGS, NAT BUR STANDARDS, 70- Concurrent Pos: Vis scientist, Cavendish Lab, Cambridge Univ, 59. Honors & Awards: Meritorious Award, US Dept Com, 53. Mem: AAAS; Electron Micros Soc Am; Am Soc Testing & Mat; fel Royal Micros Soc. Res: Electron microscopy; physics of solids; microstructure of polymers; fracture morphology; applied light and x-ray microscopy. Mailing Add: 3508 Woodbine St Chevy Chase MD 20015

NEWMAN, SEYMOUR, b New York, NY, July 9, 22; m 43; c 2. POLYMER SCIENCE, PLASTICS. Educ: City Col New York, BS, 42; Columbia Univ, MA, 47; Polytech Inst Brooklyn, PhD, 49. Prof Exp: Res assoc, Southern Regional Res Labs, USDA, 49-51, Cornell Univ, 51-52 & Allegany Ballistics Lab, Hercules Powder Co, 52-55; group leader & scientist, Monsanto Co, 56-67; staff scientist sci res staff, 67-69, mgr, Polymer Sci Dept, 69-73, SR STAFF SCIENTIST, PLASTICS DEVELOP CTR, FORD MOTOR CO, DEARBORN, 73- Concurrent Pos: Adj assoc prof, Univ Mass, Amherst, 67-; mem technol assessment panel, Engrs Joint Coun, Soc Plastics Engrs, 75- Mem: Am Chem Soc; Am Phys Soc; Soc Plastic Engrs. Res: Physical chemistry of high polymers; dynamic mechanical behavior; crystallinity; dilute solution behavior; fiber and film properties; strength properties; rheology; coatings; processing; composites; plastic materials, processing and design. Mailing Add: 29395 Sharon Lane Southfield MI 48076

NEWMAN, STANLEY RAY, b Idaho Falls, Idaho, Mar 5, 23; m 61; c 5. ORGANIC CHEMISTRY. Educ: Univ Utah, BS, 47, PhD(chem), 52. Prof Exp: Chemist, 51-59, res chemist, 59-61, group leader, 61-64, sr res chemist, 64-68, group leader, 68-73, TECHNOLOGIST, BEACON RES CTR, TEXACO, INC, 73- Concurrent Pos: Mem Solid Waste Mgt, Dutchess County, NY, 74- Mem: AAAS; Am Chem Soc. Sigma Xi; NY Acad Sci. Res: Organic carbonates; organic phosphorus chemistry; lead appreciators; tertiary esters of organic acids; organic synthesis; fuel additives; petrochemicals; recrystallization using surface active agents; effect of lead antiknocks on health. Mailing Add: 24 Virginia Ave Fishkill NY 12524

NEWMAN, STANLEY STEWART, b Chicago, Ill, July 18, 05; m 30; c 1. ANTHROPOLOGY. Educ: Univ Chicago, PhB(Eng), 27, AM, 28; Yale Univ, PhD(anthrop), 32. Prof Exp: Instr eng, Univ Tex, Austin, 28-29; instr ling, Yale Univ, 35-37; instr Persian, Columbia Univ, 41-43; instr anthrop, Brooklyn Col, 43; lang media analyst, Info & Educ Div, War Dept, NY, 43-45; anthropologist, Inst Social Anthrop, Smithsonian Inst, 45-49; from assoc prof to prof, 49-71, EMER PROF ANTHROP, UNIV NMEX, 71- Concurrent Pos: Res fel, Inst Human Rels, Yale Univ, 32-37; Gen Educ Bd res fel, Univ NY, 37-39; Am Coun Learned Socs res fel, Intensive Lang Prog, Columbia Univ, 41-43; co-ed, Southwestern J Anthrop, 61-73; ed, Univ NMex Publ Anthrop, 62-68. Mem: AAAS; Am Anthrop Asn; Ling Soc Am (vpres, 62). Res: American Indian languages; language and culture; language and behavior; descriptive linguistics; comparative linguistics. Mailing Add: 3325 Hastings Ave N E Albuquerque NM 87106

NEWMAN, STUART ALAN, b New York, NY, Apr 4, 45; m 68; c 1. DEVELOPMENTAL BIOLOGY. Educ: Columbia Univ, AB, 65; Univ Chicago, PhD(chem physics), 70. Prof Exp: Fel theoret biol, Univ Chicago, 70-72; vis fel biol sci, Univ Sussex, Eng, 72-73; instr anat, Univ Pa, 73-75; ASST PROF BIOL SCI, STATE UNIV NY ALBANY, 75- Res: Cell differentiation; cellular pattern formation; dynamics of biochemical networks. Mailing Add: Dept of Biol Sci State Univ of NY Albany NY 12222

NEWMAN, THEODORE JOSEPH, b Brooklyn, NY, Aug 24, 23; m 46; c 2. PHYSICS, ELECTRONICS. Educ: City Col New York, BME, 44; Polytech Inst Brooklyn, MS, 51, PhD(physics), 61. Prof Exp: Instr physics, New York Community Col, 46-48; mem eng staff, 48-62, chief engr, 62-64, dir res & eng, 64-70, vpres res & eng, 70-74, DIR RES & ENG, ARMA DIV, AM BOSCH ARMA CORP, 74- Mem: Am Phys Soc; Am Inst Aeronaut & Astronaut. Res: Wave propagation in visco elastic media; new variational principle in quantum mechanics; design and development of missile, navigation and guidance systems. Mailing Add: Arma Div Am Bosch Arma Corp Roosevelt Field Garden City NY 11530

NEWMAN, THOMAS MCCLELLAN, b Knoxville, Tenn, Mar 16, 29; m 55; c 4. ANTHROPOLOGY. Educ: Univ Nebr, BA, 55; Univ Ore, PhD(anthrop), 59. Prof Exp: Instr anthrop, Portland State Col, 59-60; NSF res fel African prehist, Sierra Leone, 60-61; from asst prof to assoc prof, 61-74, PROF ANTHROP, PORTLAND STATE UNIV, 74- Concurrent Pos: Nat Park Serv contract grants salvage archaeol, 62 & 64. Mem: Fel African Studies Asn; Soc Am Archaeol. Res: Prehistory of western North America, especially the northwest; cultures of the Coast and Columbia Plateau; early man; culture history and agricultural origins in West Africa, especially on African Neolithic cultures; prehistory and climatic change in the Great Basin in postpleistocene times. Mailing Add: Dept Anthrop Box 751 Portland State Univ Portland OR 97207

NEWMAN, WALTER HAYES, b Birmingham, Ala, Mar 15, 38; m 63; c 2. PHARMACOLOGY. Educ: Auburn Univ, BS, 62, MS, 63; Med Col SC, PhD(pharmacol), 67. Prof Exp: Instr, 66-67, assoc, 67-68, asst prof, 68-72, ASSOC PROF PHARMACOL, MED UNIV SC, 72- Mem: Am Soc Pharmacol & Exp Therapeut. Res: Cardiac hypertrophy. Mailing Add: Dept of Pharmacol Med Univ of SC Charleston SC 29401

NEWMAN, WALTER S, b New York, NY, May 24, 27; m 55; c 2. QUATERNARY GEOLOGY. Educ: Brooklyn Col, BS, 50; Syracuse Univ, MS, 59; NY Univ, PhD, 66. Prof Exp: Geophysicist, Lake Mead Seismol Surv, US Coast & Geod Surv, 51; eng geologist, Corps Engrs, US Army, 51-56; explor geologist, Ramapo Uranium Corp, 56-57; eng geologist, Frederic R Harris, Inc, Consult Engrs, 57-58 & Moran, Proctor, Museser & Rutledge, Consult Engrs, 58-59; asst, Lamont Geol Observ, Columbia Univ, 59-60; lectr geol, 60-66, asst prof, 66-68, chmn dept, 68-71, ASSOC PROF EARTH & ENVIRON SCI, QUEENS COL NY, 68- Concurrent Pos: Geologist, US Geol Surv. Mem: AAAS; fel Geol Soc Am; Am Quaternary Asn; Soc Am Archaeol. Res: Late quaternary environments of Northeastern United States. Mailing Add: Dept of Earth & Environ Sci Queens Col Flushing NY 11367

NEWMAN, WILEY CLIFFORD, JR, b Europa, Miss, Apr 15, 31; m 58; c 1. PHYSIOLOGY. Educ: Vanderbilt Univ, AB, 53; Univ Tenn, PhD(physiol), 65. Prof Exp: Instr, Univ Tenn, Memphis, 66-68; ASST PROF PHYSIOL, SCH MED, TULANE UNIV, 68- Mem: AAAS. Res: Endocrine physiology; experimental mammary cancer induction. Mailing Add: Dept of Physiol Tulane Univ Sch of Med New Orleans LA 70112

NEWMAN, WILLIAM, b New York, NY, Jan 30, 22; m 45; c 3. PATHOLOGY. Educ: Univ Wis, BA, 41; NY Univ, MD, 46. Prof Exp: Intern med, Bellevue Hosp, 46-47; res assoc, Bronx Vet Admin Hosp, 47-48, from asst resident to chief resident path, 50-52; fel, Mem Ctr, New York, 52-53; from asst prof to assoc prof, 53-64, PROF PATH, SCH MED, GEORGE WASHINGTON UNIV, 64-, DIR DIV ANAT PATH, 72- Concurrent Pos: Consult, NIH, Md & Vet Admin Hosps, Martinsburg, WVa & Washington, DC. Mem: Am Asn Path & Bact; James Ewing Soc; Int Acad Path. Mailing Add: Dept of Path George Washington Univ Hosp Washington DC 20037

NEWMAN, WILLIAM ALEXANDER, b Colebrook, NH, Nov 14, 34; m 60; c 2. GLACIAL GEOLOGY. Educ: Boston Univ, AB, 57, AM, 59; Syracuse Univ, PhD(geol), 71. Prof Exp: Instr, 60-64, ASST PROF GEOL, NORTHEASTERN UNIV, 68- Mem: Sigma Xi. Res: Cause, distribution and frequency of landslides in unconsolidated sediments of the White Mountains of New England; morphogenesis of small scale glacial and fluvioglacial erosion structures in alpine environments. Mailing Add: Dept Earth Sci 103 GR Bldg Northeastern Univ Boston MA 02115

NEWMAN, WILLIAM ANDERSON, b San Francisco, Calif, Nov 13, 27; div; c 4. MARINE BIOLOGY. Educ: Univ Calif, Berkeley, AB, 53, MA, 54, PhD(zool), 62. Prof Exp: Actg instr zool, Univ Calif, Berkeley, 60-61, asst prof oceanog, Univ Calif, San Diego, 62-63; asst prof marine biol, Harvard Univ, 63-65; asst prof biol oceanog, 65-71, assoc prof oceanog, 71-74, PROF OCEANOG, SCRIPPS INST OCEANOG, UNIV CALIF, SAN DIEGO, 74- Concurrent Pos: NSF fel, 62; mem adv comt arthropods, Smithsonian Oceanog Sorting Ctr, 64-67; mem comt ecol of interoceanic canal, Nat Acad Sci, 69-70; mem biol sci comt, World Book Encyclop, 71- Mem: AAAS; Marine Biol Asn India; Am Inst Biol Sci. Res: Systematics and biogeography of the Crustacea, especially the Cirripedia; biology and near surface geology of oceanic islands; oceanography. Mailing Add: Scripps Inst of Oceanog A-002 Univ of Calif San Diego La Jolla CA 92093

NEWMARK, HAROLD LEON, b New York, NY, July 21, 18; m 49; c 2. ORGANIC CHEMISTRY, BIOCHEMISTRY. Educ: City Col New York, BS, 39; Polytech Inst Brooklyn, MS, 50. Prof Exp: Res dir, Vitarine Co, Inc, NY, 50-59; res chemist & group leader appl res, 59-66, ASST DIR PROD DEVELOP, HOFFMAN-LA ROCHE, INC, NUTLEY, 66- Concurrent Pos: Mem, XV revision, US Pharmacopoeia. Mem: AAAS; Am Chem Soc; Am Pharmaceut Asn; NY Acad Sci; Am Pharm Asn. Res: Biochemistry and biochemical mechanisms as explicable by physical and organic chemistry; applications to pharmaceuticals. Mailing Add: 11 Washington Park Maplewood NJ 07040

NEWMARK, MARJORIE ZEIGER, b Cheyenne, Wyo, Aug 6, 22; m 47; c 3. BIOCHEMISTRY. Educ: Univ Colo, AB, 44, PhD(biochem), 54. Prof Exp: Res asst, Wash Univ, 50-52; res assoc anat, Sch Med, 54-62, vis lectr comp biochem & physiol, 62-63, vis asst prof, 63-64, asst prof, 64-68, asst prof biochem, 68-74, ASSOC PROF BIOCHEM, UNIV KANS, 74- Mem: AAAS; Am Chem Soc; Am Soc Cell Biol. Res: Biochemistry of arterial tissues. Mailing Add: Dept of Biochem Univ of Kans Lawrence KS 66045

NEWMARK, RICHARD ALAN, b Urbana, Ill, Nov 11, 40; m 65; c 2. ANALYTICAL CHEMISTRY. Educ: Harvard Univ, AB, 61; Univ Calif, Berkeley, PhD(chem), 65. Prof Exp: NSF fel, Mass Inst Technol, 64-66; asst prof chem, Univ Colo, Boulder, 66-69; RES CHEMIST, 3M CO, 69- Mem: Am Chem Soc; Soc Appl Spectros. Res: Nuclear magnetic resonance studies. Mailing Add: Cent Res Labs 3M Co St Paul MN 55101

NEWMEYER, DOROTHY, b Philadelphia, Pa, May 28, 22; m 52; c 1. GENETICS. Educ: Philadelphia Col Pharm, BS, 43; Yale Univ, MS, 48; Stanford Univ, PhD(biol), 51. Prof Exp: Fel, NY Univ, 51-52; RES ASSOC BIOL SCI, STANFORD UNIV, 52- Mem: Genetics Soc Am. Res: Neurospora; genetic mapping; chromosome aberrations; unstable duplications; mating type locus. Mailing Add: Dept of Biol Sci Stanford Univ Stanford CA 94305

NEWNHAM, ROBERT EVEREST, b Amsterdam, NY, Mar 28, 29. PHYSICS. Educ: Hartwick Col, BS, 50; Colo State Univ, MS, 52; Pa State Univ, PhD(physics), 56; Cambridge Univ, PhD(crystallog), 60. Prof Exp: Assoc prof elec eng, Mass Inst Technol, 59-66; assoc prof solid state sci, 66-71, PROF SOLID STATE SCI, PA STATE UNIV, 71- Mem: Am Phys Soc; Am Crystallog Asn; Am Ceramic Soc; Mineral Soc Am. Res: Crystal and solid state physics; x-ray crystallography. Mailing Add: Dept of Mat Sci Pa State Univ University Park PA 16802

NEWNHAM, ROBERT MONTAGUE, b Bromley, Eng, Aug 11, 34; m 59; c 3. FOREST MANAGEMENT, FOREST MENSURATION. Educ: Univ Wales, BS, 56; Univ BC, MF, 58, PhD(forestry), 64. Prof Exp: Asst exp officer forest ecol, Nature Conservancy, Grange over Sands, Eng, 60-62; RES SCIENTIST FOREST MGT &

FOREST MENSURATION, FOREST MGT INST, CAN FORESTRY SERV, DEPT ENVIRON, OTTAWA, 64- Mem: Can Inst Forestry; Commonwealth Forestry Asn; Int Union Forest Res Orgn. Res: Planning logging operations; systems analysis; applications of computers to forest research; development of forest stand growth models. Mailing Add: Forest Mgt Inst Can Forestry Serv Dept Environ Ottawa ON Can

NEWROTH, PETER RUSSELL, b Sheffield, Eng, Oct 12, 45; Can citizen. MARINE BOTANY, RESOURCE MANAGEMENT. Educ: Univ NB, BS, 66, PhD(marine biol), 70. Prof Exp: Fel bot, Univ BC, 70-72; BIOLOGIST, WATER INVEST BR, WATER RESOURCES SERV, BC, 72- Mem: Can Bot Asn; Brit Phycol Soc; Am Phycol Soc; Am Inst Biol Sci; Int Phycol Soc. Res: Life histories, taxonomy and morphology of marine Rhodophyta; management and ecology of freshwater macrophytes. Mailing Add: Environ Studies Div Water Invest Br Parliament Bldg Victoria BC Can

NEWSCHWANDER, WILFRID WILLIAMS, b Tacoma, Wash, May 8, 11; m 36; c 2. PHYSICAL CHEMISTRY. Educ: Whitman Col, AB, 33; Univ Wash, PhD(chem), 39. Prof Exp: From instr to assoc prof, 39-57, PROF CHEM, CENT WASH STATE COL, 57- Concurrent Pos: Sr chemist, Shell Develop Co, Calif, 44-46. Mem: AAAS; Am Chem Soc; NY Acad Sci. Res: Chemical thermodynamics; patents; steroid biosynthesis in plant embryos. Mailing Add: Dept of Chem Cent Wash State Col Ellensburg WA 98926

NEWSOM, BERNARD DEAN, b Oakland, Calif, Feb 8, 24; m 45; c 2. ENVIRONMENTAL PHYSIOLOGY. Educ: Univ Calif, AB, 46, PhD(physiol), 60; Univ San Francisco, MS, 54. Prof Exp: Investr & physiologist, US Naval Radiol Defense Lab, 49-61; sr staff scientist, Life Sci Lab, Gen Dynamics/Convair, 61-68; res analyst, Med Res & Opers Directorate, Manned Spacecraft Ctr, 68-72, PROJ MGR BIOMED RES, AMES RES CTR, NASA, 72- Mem: AAAS; Am Physiol Soc; Radiation Res Soc; Aerospace Med Asn; Am Inst Aeronaut & Astronaut. Res: Physiological and performance changes of man in a rotational environment; adaptation and tolerance to prolonged exposures to angular velocities and perturbations; biological interpretation of complex space stresses of vibration, acceleration, null gravity and radiation; effect of abnormal environments on radiation sequela in terms of stress tolerance, performance and longevity. Mailing Add: Biomed Res NASA Ames Res Ctr Moffett Field CA 94035

NEWSOM, DONALD WILSON, b Shongaloo, La, Nov 14, 18; m 44; c 2. HORTICULTURE. Educ: La State Univ, BS, 47, MS, 48; Mich State Univ, PhD(hort), 52. Prof Exp: Asst prof agr, Tex Col Arts & Indust, 50; assoc horticulturist, Clemson Col, 51-54; horticulturist, USDA, 54-57; PROF HORT, LA STATE UNIV, BATON ROUGE, 57-, HEAD DEPT, 66- Mem: Fel AAAS; Am Soc Hort Sci; Am Soc Plant Physiol; Am Forestry Asn. Res: Post harvest physiology and chemical composition of fruits and vegetables, especially flavor components. Mailing Add: Dept of Hort La State Univ Baton Rouge LA 70803

NEWSOM, GERALD HIGLEY, b Albuquerque, NMex, Feb 11, 39. ASTRONOMY, ATOMIC SPECTROSCOPY. Educ: Univ Mich, Ann Arbor, BA, 61; Harvard Univ, MA, 63, PhD(astron), 68. Prof Exp: Res asst, Imperial Col, Univ London, 68-69; asst prof, 69-73, ASSOC PROF ASTRON, OHIO STATE UNIV, 73- Mem: Am Astron Soc; Int Astron Union. Res: Classification of energy levels in neutral atoms; measurement of oscillator strengths for neutral and singly ionized atomic spectral lines. Mailing Add: Dept of Astron Ohio State Univ 174 W 18th Ave Columbus OH 43210

NEWSOM, HERBERT CHARLES, b Whittier, Calif, Oct 25, 31; m 55; c 2. ORGANIC CHEMISTRY, PESTICIDE CHEMISTRY. Educ: Whittier Col, BA, 53; Univ Southern Calif, PhD(org chem), 59. Prof Exp: Asst chem, Univ Southern Calif, 55-59; res chemist, 59-65, SR RES CHEMIST, US BORAX RES CORP, 65- Mem: AAAS; Am Chem Soc. Res: Organoboron and free radical chemistry; kinetics; herbicide residue analysis; photolysis; herbicide degradation; process development; EPA pesticide registration studies. Mailing Add: 1702 Greenmeadow Ave Tustin CA 92680

NEWSOM, LEO DALE, b Shongaloo, La, Feb 23, 15; m 46; c 4. ENTOMOLOGY. Educ: La State Univ, BS, 40; Cornell Univ, PhD(econ entom), 48. Prof Exp: Asst forage crop insects res, Cornell Univ, 40-42, in-chg field lab, 46-48; asst entomologist in-chg cotton insect res, Exp Sta, 48-51, ASSOC ENTOMOLOGIST IN-CHG, COTTON INSECT RES, EXP STA, LA STATE UNIV, BATON ROUGE, 51-, HEAD ENTOM RES, EXP STA, 54-, BOYD PROF ENTOM, 66- & HEAD DEPT, 64- Concurrent Pos: Prof entom, La State Univ, Baton Rouge, 64-66; mem panel cotton insects, President's Sci Adv Comt, 64, sub-panel mem restoring qual of our environ, 65-, mem panel world food supplies, 66-; sub-comt insect pests, Nat Acad Sci-Nat Res Coun, 64-; secy agr comt agr sci. Mem: AAAS; Entom Soc Am. Res: Biology and ecology of insects; physiology and ecology of diapause in insects; biological and ecological consequences of pesticide usage. Mailing Add: Agr Exp Sta La State Univ Baton Rouge LA 70803

NEWSOM, RAYMOND A, b Tarrant Co, Tex, Jan 8, 31; m 50; c 2. ORGANIC CHEMISTRY. Educ: Ariz State Univ, BS, 53; Univ Ariz, MS, 57; Univ Iowa, PhD(org chem), 60. Prof Exp: Res chemist, Plastics Div Res, 60-61 & Hydrocarbons Div, 62-65, process chemist, Process Technol Dept, 65-75, SR PROCESS SPECIALIST, PROCESS TECHNOL DEPT, MONSANTO CO, 75- Mem: Am Chem Soc. Res: Organic syntheses and reaction mechanisms; product and process development for monomers. Mailing Add: Process Technol Dept 2R-3 Monsanto Co PO Box 1311 Texas City TX 77590

NEWSOM, WILL ROY, b Rivera, Calif, Jan 12, 12; m 31; c 3. ANALYTICAL CHEMISTRY. Educ: Whittier Col, AB, 34; Univ Southern Calif, AM, 35, PhD(chem), 39. Prof Exp: From asst prof to prof chem, 39-63, chmn dept, 40-66, dean col, 63-71, vpres admin, 71-75, ACTG PRES, WHITTIER COL, 75- Concurrent Pos: Mem, Los Angeles County Air Pollution Adv Comt, 44-54. Mem: Am Chem Soc. Res: Analytical chemistry of beryllium. Mailing Add: 7829 S Vale Dr Whittier CA 90602

NEWSOM, WILLIAM S, JR, b Wynne, Ark, Dec 31, 18; m 46; c 3. AGRICULTURAL CHEMISTRY. Educ: Univ Ark, BS, 48. Prof Exp: Chemist, Lion Oil Co, 48-55 & Monsanto Chem Co, 55-57; sr res chemist, Int Minerals & Chem Corp, Ill, 57-67; sr res engr, Ga Inst Technol, 67-68; SECT HEAD, OCCIDENTAL CHEM CO, 68- Mem: Am Chem Soc; Brit Fertilizer Soc. Res: Fertilizer technology; pesticides; plant growth regulators. Mailing Add: 194 Shelby Dr Lake City FL 32055

NEWSOME, JAMES FREDERICK, b Winton, NC, Mar 24, 23; m 56; c 3. SURGERY. Educ: Univ NC, AB, 44, cert, 47; Vanderbilt Univ, MD, 49. Prof Exp: Intern, Med Col Va Hosp, 49-50; from asst resident to chief resident, 52-56, from instr to assoc prof, 64-71, PROF SURG, SCH MED, UNIV NC, CHAPEL HILL, 71- Concurrent Pos: Consult, Vet Admin Hosp, Fayetteville, NC, 56- & Watts Hosp, Durham, 56- Mem: Soc Surg Alimentary Tract; Am Asn Cancer Res; AMA; Am Col Surgeons; Am Asn Cancer Educ. Res: Cancer, especially breast carcinoma; chemotherapy of cancer; cystic disease of breast. Mailing Add: Dept of Surg Univ of NC Sch of Med Chapel Hill NC 27514

NEWSOME, JAMES WILFRED, physical chemistry, see 12th edition

NEWSOME, RICHARD DUANE, b Kalamazoo, Mich, Aug 19, 31; m 54; c 2. PLANT ECOLOGY. Educ: Western Mich Univ, BS, 54; Univ Sask, MS, 63, PhD(plant ecol), 65. Prof Exp: Instr bot & ecol, 65-66, asst prof, 66-70, ASSOC PROF BOT & ECOL, BELOIT COL, 70- Mem: Ecol Soc Am; Am Inst Biol Scientists; Nat Sci Teachers Asn. Res: Plant ecological research on the structure and dynamics of communities in ecotonal situations; assessment of plant ecologic condition of disturbed watersheds in southern Wisconsin. Mailing Add: Dept of Biol Beloit Col Beloit WI 53511

NEWSOME, ROSS WHITTED, b Lynchburg, Va, Nov 6, 35. PHYSICS. Educ: Mass Inst Technol, SB, 57; Univ Mich, Ann Arbor, MS, 58, PhD(physics), 63. Prof Exp: Univ Mich res asst high energy physics, Lawrence Radiation Lab, Univ Calif, Berkeley, 59-60; from res asst to res assoc exp nuclear physics, Univ Mich, Ann Arbor, 60-64; mem staff, Los Alamos Sci Lab, Univ Calif, 64-66; mem tech staff systs eng, Bellcomm, Inc, Washington, DC, 66-72; MEM TECH STAFF CUSTOMER SYSTS ENG CTR, BELL LABS, 72- Mem: AAAS; Am Phys Soc. Res: Transmission of infrared radiation through the atmosphere; capabilities of thermal infrared instruments for remote sensing of surface targets from satellites; measurements and analyses of telephone traffic for small business customers. Mailing Add: Customer Systs Eng Ctr Bell Labs Inc Holmdel NJ 07733

NEWSON, HAROLD DON, b Salt Lake City, Utah, July 11, 24; m 48; c 4. MEDICAL ENTOMOLOGY. Educ: Univ Utah, BA, 49, MS, 50; Univ Md, PhD(entom), 59. Prof Exp: From res entomologist to med entom consult to Surgeon Gen, US Army, 51-70; ASSOC PROF ENTOM, MICROBIOL & PUB HEALTH, MICH STATE UNIV, 70- Concurrent Pos: Mem, US Armed Forces Pest Control Bd, 62-70, chmn, 63-67; mem res subcomt, Fed Comt Pest Control, 64-66; mem study group, Off Environ Sci, Smithsonian Inst, 72-74; chmn region V, USPHS Vector Control Group, 73-76. Mem: Am Soc Trop Med & Hyg; Am Mosquito Control Asn; Entom Soc Am. Res: Ecology; transmission and control of arthropod-borne diseases of medical and veterinary importance. Mailing Add: Pesticide Res Ctr Mich State Univ East Lansing MI 48824

NEWSTEAD, JAMES DUNCAN MACINNES, b Camberley, Eng, Oct 11, 30; m 62; c 1. CELL BIOLOGY, HISTOLOGY. Educ: Univ BC, BA, 54, MA, 56; Ore State Univ, PhD(cell biol), 62. Prof Exp: Instr zool, Ore State Univ, 60-62, asst prof, 62-63; USPHS fel fine structure, Univ Wash, 63-65; from asst prof to assoc prof anat, 65-74, PROF ANAT, UNIV SASK, 74- Mem: AAAS; Am Soc Zoologists; Am Soc Cell Biol. Res: Circulation and ion transport in gills of fish; fine structure of cell division in protozoa. Mailing Add: Dept of Anat Univ of Sask Saskatoon SK Can

NEWSTEIN, HERMAN, b Philadelphia, Pa, May 4, 18; m 53; c 1. METEOROLOGY. Educ: Temple Univ, BS, 48, MEd, 51; NY Univ, MS, 53, PhD(meteorol), 57. Prof Exp: Meteorologist, US Weather Bur, 41-53, res meteorologist, 53-62; PROF PHYSICS & ATMOSPHERIC SCI, DREXEL UNIV, 62- Concurrent Pos: Sci consult, Nat Acad Sci, 58; adj assoc prof, NY Univ, 58-62. Mem: AAAS; Am Meteorol Soc; Am Geophys Union; Air Pollution Control Asn; Am Asn Physics Teachers. Res: Experimental and theoretical atmospheric physics. Mailing Add: Dept of Physics Drexel Univ Philadelphia PA 19104

NEWSTEIN, MAURICE, b Philadelphia, Pa, Feb 13, 26; m 57; c 2. THEORETICAL PHYSICS. Educ: Temple Univ, AB, 49; Mass Inst Technol, PhD(physics), 54. Prof Exp: Asst univ observ, Harvard Univ, 54-55; physicist, Tech Res Group, Inc, 55-67; res scientist electrophys, 67-70, ASSOC PROF ELECTROPHYS, GRAD CTR, POLYTECH INST New York, 70- Res: Atomic scattering problems; plasma physics; applications of microwave and optical spectroscopy; quantum electronics. Mailing Add: Dept of Electrophys Grad Ctr Polytech Inst New York 333 Jay St Brooklyn NY 11201

NEWTON, ABBA VERBECK, b Ballston Spa, NY, Feb 19, 08. GEOMETRY. Educ: Mt Holyoke Col, AB, 29; Univ Chicago, MA, 31, PhD(math), 33. Prof Exp: Teacher, Hill Sch, Ky, 29-30; instr math, Am Int Col, 34-38; prof, Hartwick Col, 38-43; asst prof, Smith Col, 43-44; from asst prof to prof, 44-73, EMER PROF MATH, VASSAR COL, 73- Concurrent Pos: Fac fel, Inst Henri Poincare, Sorbonne, 51; NSF fel, Univ Mich, 58-59; Vassar Col fac fel, Duke Univ, 66; vis res fel, Princeton Univ, 71. Mem: Am Math Soc; Math Asn Am; Sigma Xi. Res: Projective differential geometry; consecutive covariant configurations at a point of a space curve. Mailing Add: Dept of Math Vassar Col Poughkeepsie NY 12601

NEWTON, AMOS SYLVESTER, b Shingletown, Calif, July 26, 16; m 42; c 2. CHEMISTRY. Educ: Univ Calif, BS, 38; Univ Mich, MS, 39, PhD(phys chem), 41. Prof Exp: Chemist, Eastman Kodak Co, NY, 41-42, Manhattan Proj, Iowa State Col, 42-46 & Eastman Kodak Co, NY, 46-62; CHEMIST, LAWRENCE BERKELEY LAB, UNIV CALIF, 62- Concurrent Pos: Consult, Lawrence Berkeley Lab, Univ Calif, 46-62. Mem: AAAS; Am Soc Mass Spectros; Am Chem Soc; Radiation Res Soc. Res: Use of radioisotopes as tracer; chemistry of heavy elements; radiochemistry of fission products; radiation chemistry; mass spectrometry; molecular beam studies; environmental chemistry; marine chemistry; fuel science. Mailing Add: Lawrence Berkeley Lab Univ Calif Energy & Environ Div Berkeley CA 94720

NEWTON, BERNE LOYST, b Gladstone, Man, Aug 25, 13; nat US; m 38. PATHOLOGY. Educ: McGill Univ, BA, 51, MD, CM, 40. Prof Exp: Intern, Royal Jubilee Hosp, Victoria, BC, 40-42; resident path, Royal Victoria Hosp, Montreal, 42-43; instr, McGill Univ, 43; Childs fel, Sch Med, Yale Univ, 46-48; cancer teaching coordr & chmn cancer teaching comt, 48-66, asst prof path, 48-52, ASSOC PROF PATH, BAYLOR COL MED, 53- Concurrent Pos: Pathologist, Methodist Hosp, 50-65, sr attend pathologist, 66- Mem: Am Soc Clin Path; Am Asn Path & Bact; Am Asn Cancer Res; Col Am Path; Int Acad Path. Res: Canine tumors. Mailing Add: Methodist Hosp 6516 Bertner Houston TX 77025

NEWTON, CARLOS E, JR, b Charlotte, NC, July 15, 19; m 44; c 4. RADIATION PHYSICS. Educ: Univ Md, BS, 56; US Naval Postgrad Sch, MS, 58. Prof Exp: Pharmacist, Liggett Drug Co, 40-42; mem staff, US Army Indust Hyg Lab, 48-51, radiation physicist, US Army Hyg Lab, 52-55, radiation physicist, Walter Reed Army Med Ctr, 59-62; mgr dosimetry studies & eval sect, 62-67, RES ASSOC, PAC NORTHWEST LAB, BATTELLE MEM INST, 67- Concurrent Pos: Assoc dir, US Transuranium Registry. Mem: Health Physics Soc; Am Asn Physicists in Med. Res: Health physics; cold weather environmental stresses on man. Mailing Add: 2522 Harris Ave Richland WA 99352

NEWTON, CAROL MARILYN, b Oakland, Calif, Nov 26, 25. MEDICINE, COMPUTER SCIENCE. Educ: Stanford Univ, AB, 47, MS, 49, PhD(physics), 56;

Univ Chicago, MD, 60. Prof Exp: Asst physics, Stanford Univ, 47-53; assoc scientist, Argonne Cancer Res Hosp, Univ Chicago, 57-60, intern, Univ Chicago Clins, 60-61, mem staff, Argonne Cancer Res Hosp, 61-67 & Biomed Comput Ctr, Univ Chicago, 62-67, asst prof med, 63-67, asst prof math biol, 65-67; assoc prof biomath, 67-74, PROF BIOMATH & CHMN DEPT, SCH MED, UNIV CALIF, LOS ANGELES, 74- Concurrent Pos: Mem adv sects, NIH, 64-; consult, Health Serv & Ment Health Admin, 72-73; assoc ed, Math Biosci. Mem: AAAS; Am Phys Soc; Biophys Soc; Asn Comput Mach; Soc Math Biol. Res: Medical physics; biomathematics; computers, especially interactive graphics. Mailing Add: Dept of Biomath Univ of Calif Sch of Med Los Angeles CA 90024

NEWTON, CAROLYN MCCRORY, b Natchitoches, La, Nov 23, 48; m 69. ELECTROANALYTICAL CHEMISTRY. Educ: Northwestern State Univ, BS, 69; Univ New Orleans, PhD(anal chem), 74. Prof Exp: MEM TECH STAFF ANAL CHEM, BELL TEL LABS, 75- Mem: Am Chem Soc; AAAS; Electrochem Soc. Res: Electronic device materials, voltammetry, coulometry, trace analysis; semiconductors; electrolytic purification of analytical reagents and plating bath solutions. Mailing Add: Bell Tel Labs 600 Mountain Ave Murray Hill NJ 07974

NEWTON, CHESTER WHITTIER, b Los Angeles, Calif, Aug 17, 20; m 48; c 4. METEOROLOGY. Educ: Univ Chicago, SB, 46, SM, 47, PhD(meteorol), 51. Prof Exp: Weather observer, US Weather Bur, 29-41, meteorologist, 48; asst meteorol, Univ Chicago, 47-48, synoptic analyst, 48-51; synoptic analyst, Univ Stockholm, 51-53 & Woods Hole Oceanog Inst, 53; res assoc meteorol, Univ Chicago, 53-56, asst prof, 56-61; chief scientist, Nat Severe Storms Proj, US Weather Bur, 61-63; SR SCIENTIST, NAT CTR ATMOSPHERIC RES, 63- Concurrent Pos: Affil prof, Pa State Univ, 65-67; mem steering comt, Earth Sci Curric Proj, 65-68. Honors & Awards: Editor's Award, Am Meteorol Soc, 70. Mem: AAAS; fel Am Meteorol Soc; Am Geophys Union; foreign mem, Royal Meteorol Soc; Meteorol Soc Japan. Res: Synoptic meteorology; atmospheric general circulation and global energy balance; aerological analysis of atmospheric current systems; structure of and physical processes in cyclone formation; thunderstorms and severe local storms. Mailing Add: Nat Ctr for Atmospheric Res Box 3000 Boulder CO 80303

NEWTON, CLARENCE JONATHAN, b Decatur, Nebr, Feb 25, 23. SOLID STATE PHYSICS, X-RAY CRYSTALLOGRAPHY. Educ: Univ Tex, Austin, BA, 44, MA, 47, PhD(physics), 52. Prof Exp: Physicist, US Naval Res Lab, 44-45; instr physics, Univ Tex, Austin, 51-52; physicist, Metall Div, Nat Bur Stand, 52-70; CONSULT, 70- Mem: Am Phys Soc; Am Asn Physics Teachers. Res: Physics of solids; x-ray diffraction; crystallography. Mailing Add: 1504 S Second Ave Edinburg TX 78539

NEWTON, DAVID C, b Middletown, Conn, Apr 27, 39; m 58; c 2. APICULTURE, ANIMAL BEHAVIOR. Educ: Cent Conn State Col, BS, 61; Wesleyan Univ, MALS, 65; Univ Ill, PhD(entom), 67. Prof Exp: Teacher pub schs, 61 & high sch, 61-64; ASSOC PROF BIOL, CENT CONN STATE COL, 67- Concurrent Pos: Conn Res Comn grant, 68-71; USDA study grant, 73-76. Mem: AAAS; Animal Behav Soc; Bee Res Asn; Entom Soc Am; Am Inst Biol Sci. Res: Behavior studies of honey bees; behavior studies of honey bees relating to nest cleaning and disease resistance. Mailing Add: Dept of Biol Sci Cent Conn State Col New Britain CT 06050

NEWTON, DAVID JOHN, organic chemistry, see 12th edition

NEWTON, H CALVIN, JR, b Laurinburg, NC, Dec 16, 42. PLANT PATHOLOGY, PLANT BREEDING. Educ: NC State Univ, BS, 65; Univ Wis, PhD(plant path), 72. Prof Exp: PLANT BREEDER, McNAIR SEED CO, 71- Mem: Am Phytopath Soc; Sigma Xi; Int Phytopath Soc. Res: Breeding and developing new varieties of wheat, barley and rye including disease prevention and/or control, quality and yield factors. Mailing Add: McNair Seed Co PO Box 706 Laurinburg NC 28352

NEWTON, HOWARD JOSEPH, b Oneida, NY, June 16, 49; m 70. STATISTICAL ANALYSIS. Educ: Niagara Univ, BS, 71; State Univ NY Buffalo, MA, 73, PhD(statist sci), 75. Prof Exp: RES ASST PROF STATIST SCI & TECH SPECIALIST SURG, STATIST LAB, RES FOUND STATE NY, STATE UNIV NY BUFFALO, 75- Mem: Am Statist Asn; Asn Comput Mach. Res: Statistical computing with particular emphasis on the analysis of multiple time series having rational spectra. Mailing Add: c/o Statist Sci Div Comput Sci 4230 Ridge Lea Rd Amherst NY 14226

NEWTON, JACK WILLIAM, microbiology, see 12th edition

NEWTON, JAMES HENRY, b Highlands, NC, Feb 24, 41; m 62; c 2. PHYSICAL CHEMISTRY. Educ: Univ NC, Chapel Hill, BS, 64; Furman Univ, MS, 69; Univ Fla, PhD(phys chem), 74. Prof Exp: Teacher sci, Highlands High Sch, NC, 64-65; chemist, Dayco Corp, NC, 65-67; teaching asst chem, 69-74, FEL PHYS CHEM, UNIV FLA, 74- Res: Quantum mechanical and experimental determination and subsequent interpretation of gas phase infrared band intensities. Mailing Add: Dept of Chem Univ of Fla Gainesville FL 32610

NEWTON, JOEL ARMOND, zoology, genetics, see 12th edition

NEWTON, JOHN CHESTER, b Grainfield, Kans, May 30, 33; m 63; c 2. ANALYTICAL CHEMISTRY. Educ: Univ Kans, BS, 54; Univ Colo, Boulder, BA, 59; Univ Calif, Berkeley, PhD(chem), 65. Prof Exp: Mem staff chem, Forest Prod Lab, Univ Calif, 59-60 & Los Alamos Sci Lab, Univ Calif, 65-69; MEM STAFF CHEM, LAWRENCE LIVERMORE LAB, UNIV CALIF, 69- Mem: Am Chem Soc; Am Soc Mass Spectrometry. Res: Mass spectrometry; computer automation; chemiluminescence; phenomena. Mailing Add: Lawrence Livermore Lab Univ of Calif Livermore CA 94550

NEWTON, JOHN MARSHALL, b Popejoy, Iowa, May 20, 13; m 41; c 3. CARBOHYDRATE CHEMISTRY. Educ: Iowa State Col, BS, 36, PhD(chem), 41. Prof Exp: Res chemist, Clinton Indust, 41-42, from asst supvr to res supvr, 42-49; dir tech sales serv, 49-63, TECH ASST TO VPRES SALES, CLINTON CORN PROCESSING CO, 63- Mem: AAAS; Am Asn Cereal Chem; Am Chem Soc; Tech Asn Pulp & Paper Indust; Am Asn Textile Chem & Colorists. Res: Carbohydrate and enzyme chemistry; concentration, characterization and properties of soybean amylase; chemistry of starch, lactic acid, oil and plant proteins; fermentation of lactic acid. Mailing Add: 1425 Seventh St NW Clinton IA 52732

NEWTON, JOSEPH EMORY O'NEAL, b Orlando, Fla, Apr 5, 27; m 58; c 2. PSYCHIATRY, PHYSIOLOGICAL PSYCHOLOGY. Educ: Emory Univ, BS, 52, MD, 55. Prof Exp: Fel psychiat, Pavlovian Lab, Phipps Psychiat Clin, Sch Med, Johns Hopkins Univ, 56-57, USPHS fel, 58-61; instr psychiat, Pavlovian Lab, Phipps Psychiat Clin, Sch Med, Johns Hopkins Univ, 62-66, asst prof, 66-68; res physiologist, 68-74, PHYSIOLOGIST, NEUROPSYCHIAT RES LAB, VET ADMIN HOSP, NORTH LITTLE ROCK, ARK, 74- Concurrent Pos: Investr psychophysiol res lab, Vet Admin Hosp, Perry Point, Md, 58-61. Mem: AAAS; Am Physiol Soc; Soc Psychophysiol Res; Pavlovian Soc NAm. Res: Conditional reflex studies in dogs, opossums; cardiovascular conditioning; effects of cerebral cortical ablations on acquired emotional reactions. Mailing Add: Neuropsychiat Res Lab Vet Admin Hosp North Little Rock AR 72116

NEWTON, LAWRENCE WOODFORD, organic chemistry, see 12th edition

NEWTON, MARSHALL DICKINSON, b Boston, Mass, July 15, 40; m 63; c 2. THEORETICAL CHEMISTRY. Educ: Dartmouth Col, BA, 61, MA, 63; Harvard Univ, PhD(chem), 66. Prof Exp: NSF fel chem, Oxford Univ, 66-67; NIH fel, Carnegie-Mellon Univ, 67-68, res assoc, 68-69; assoc chemist, 69-73, CHEMIST, CHEM DEPT, BROOKHAVEN NAT LAB, 73- Concurrent Pos: Carnegie-Mellon Univ & Mellon Inst res fel, 68. Mem: Am Chem Soc. Res: Calculation of molecular potential energy surfaces; analysis of molecular bonding in terms of electronic structure; theory of solvation phenomena. Mailing Add: Chem Dept Brookhaven Nat Lab Upton NY 11973

NEWTON, MICHAEL, b Hartford, Conn, Oct 24, 32; m 54; c 3. FOREST ECOLOGY, WEED SCIENCE. Educ: Univ Vt, BS, 54; Ore State Univ, BS, 59, MS, 60, PhD(bot), 64. Prof Exp: Res asst forest herbicides, 59-60, instr forest mgt, 60-64, asst prof forest sci, 64-68, assoc prof forest mgt, 68-75, PROF FOREST ECOL, ORE STATE UNIV, 75- Concurrent Pos: Consult to numerous corps, 62-; fel, Univ Tenn, 69-70; mem comt effects of herbicides in Vietnam, Nat Acad Sci, 72-73, consult study of pest problems, 73-74. Mem: AAAS; Soc Am Foresters; Ecol Soc Am; Weed Sci Soc Am. Res: Quantitative forest ecology; usage of herbicides to manipulate components of forest ecosystems; development of theory and practice in forest manipulation. Mailing Add: Sch of Forestry Ore State Univ Corvallis OR 97331

NEWTON, ROBERT ANDREW, b Oakville, Wash, Sept 23, 22; m 46; c 4. PHYSICAL ORGANIC CHEMISTRY. Educ: Univ Wash, BS, 47, PhD(chem), 53. Prof Exp: Mem staff res, E I du Pont de Nemours & Co, 53-56; chemist, 56-57, from res chemist to sr res chemist, S7-68, res specialist, 68-74, SR RES SPECIALIST, DOW CHEM CO, 74- Mem: Am Chem Soc. Res: Aklylene oxide chemistry and polymerization products; chromatography; chemical kinetics; isolation and identification of trace components. Mailing Add: 53 Pin Oak Ct Lake Jackson TX 77566

NEWTON, ROBERT CHAFFER, b Bellingham, Wash, June 11, 33; m 67; c 2. PETROLOGY. Educ: Univ Calif, Los Angeles, AB, 56, MA, 58, PhD(geol), 63. Prof Exp: From asst prof to assoc prof, 63-71, PROF GEOL, DEPT GEOPHYS SCI, UNIV CHICAGO, 71- Mem: Am Geophys Union; Am Mineral Soc. Res: Experimental investigation of the high-temperature, high-pressure stabilities of minerals. Mailing Add: Dept of Geophys Sci Univ of Chicago Chicago IL 60637

NEWTON, ROBERT COLLIER, b Seattle, Wash, May 19, 12; m 40; c 3. CHEMISTRY. Educ: Wesleyan Univ, BA, 34, MA, 35; Princeton Univ, PhD(chem), 39. Prof Exp: Chemist, Gen Chem Co, NY, 40-41; asst prof chem, Univ Vt, 41-44; chemist, Gen Elec Co, Mass, 44-47; chemist, 47-55, MGR ANAL DEPT, ARMSTRONG CORK CO, 55- Concurrent Pos: Adj prof, Franklin & Marshall Col, 49-50. Mem: Am Chem Soc. Res: Analytical chemistry; electrolytic deposition of copper; infrared; ferrous metallurgy. Mailing Add: Anal Dept Armstrong Cork Co Lancaster PA 17603

NEWTON, ROBERT RUSSELL, b Chattanooga, Tenn, July 7, 18; m 44; c 2. PHYSICS. Educ: Univ Tenn, BS, 40; Ohio State Univ, MS, 42, PhD(physics), 46. Prof Exp: Instr physics, Univ Tenn, 42-44; res assoc, George Washington Univ, 44-45; mem tech staff, Bell Labs, Inc, 46-48; assoc prof physics, Univ Tenn, 48-54; prof, Tulane Univ, 55-57; MEM PRIN STAFF, APPL PHYSICS LAB, JOHNS HOPKINS UNIV, 57-, BR SUPVR, SPACE RES & ANALYSIS, 64- Concurrent Pos: Consult, Oak Ridge Nat Lab, 49- Mem: Am Phys Soc; Am Geophys Union; Int Astron Union. Res: Exterior ballistics; molecular spectra; Townsend discharges in gases; electron conduction in and emission from solids; satellite dynamics and geodesy; astronomy. Mailing Add: Space Res & Anal Appl Phys Lab Johns Hopkins Univ J Hop Rd Laurel MD 20810

NEWTON, ROGER GERHARD, b Ger, Nov 30, 24; nat US; m 53; c 3. THEORETICAL PHYSICS. Educ: Harvard Univ, AB, 49, AM, 50, PhD(physics), 53. Prof Exp: Mem, Inst Advan Study, 53-55; from asst prof to assoc prof, 55-60, PROF PHYSICS, IND UNIV, BLOOMINGTON, 60-, CHMN DEPT, 73- Concurrent Pos: Jewett fel, 53-55; NSF sr fel, Univ Rome, 62-63. Mem: AAAS; fel Am Phys Soc; Sigma Xi; Fed Am Scientists; NY Acad Sci. Res: Field and scattering theories; nuclear and high energy physics; elementary particles; quantum mechanics. Mailing Add: Dept of Physics Ind Univ Bloomington IN 47401

NEWTON, STEPHEN BRUINGTON, b Freeport, Ill, Dec 24, 34; m 58; c 3. MICROBIOLOGY, FOOD SCIENCE. Educ: Univ Mo, BA, 56; Univ Ill, MS, 61, PhD(bact), 65. Prof Exp: Prod develop scientist, Pillsbury Co, 64-71; SR GROUP LEADER, QUAKER OATS CO, 71- Mem: Inst Food Technol. Res: Production of polysaccharides and extracellular enzymes by microbes; food product development. Mailing Add: Quaker Oats Co 617 W Main St Barrington IL 60010

NEWTON, THEODORE DUDDELL, b Philadelphia, Pa, Sept 14, 15; nat Can; m 42; c 4. THEORETICAL PHYSICS. Educ: Univ BC, BA, 39, MA, 41; Univ Toronto, MA, 46; Princeton Univ, PhD(math physics), 49. Prof Exp: Res off theoret physics, Atomic Energy Can, Ltd, 49-66; chmn dept math & statist, 66-75, PROF MATH & STATIST, UNIV GUELPH, 66- Mem: Can Math Cong; Can Asn Physicists. Res: Theoretical nuclear physics; application of group theory; applied mathematics. Mailing Add: Dept of Math & Statist Univ of Guelph Guelph ON Can

NEWTON, THOMAS ALLEN, b Buffalo, NY, May 30, 43. ORGANIC CHEMISTRY. Educ: Hobart Col, BS, 65; Bucknell Univ, MS, 68; Univ Del, PhD(org chem), 73. Prof Exp: RES ASSOC, RENSSELAER POLYTECH INST, 73- Mem: Am Chem Soc. Res: The synthesis of nucleoside analogues as potential antitumor/antiviral agents; the photochemistry of enaminonitriles; orbital symmetry controlled reactions—sigmatropic rearrangements. Mailing Add: 507 E Stenzil St North Tonawanda NY 14120

NEWTON, THOMAS HANS, b Berlin, Ger, May 9, 25; US citizen; c 2. RADIOLOGY. Educ: Univ Calif, Berkeley, BA, 49; Univ Calif, San Francisco, MD, 52. Prof Exp: From asst prof to assoc prof, 59-68, PROF RADIOL, MED CTR, UNIV CALIF, SAN FRANCISCO, 68- Concurrent Pos: Consult, Ft Miley Vet Admin Hosp & Letterman Gen Hosp, San Francisco, Martinez Vet Admin Hosp & Oaknoll Naval Hosp, Oakland. Mem: Asn Univ Radiol; Am Soc Neuroradiol; Neurosurg Soc Am. Res: Neuroradiology. Mailing Add: Dept of Radiol Univ of Calif Med Ctr San Francisco CA 94122

NEWTON, THOMAS WILLIAM, b Berkeley, Calif, June 26, 23; m 48; c 3. INORGANIC CHEMISTRY, PHYSICAL CHEMISTRY. Educ: Univ Calif, Berkeley, BS, 43, PhD(chem), 49. Prof Exp: Chemist, Manhattan Proj, Univ Calif & Tenn Eastman, 44-46; CHEMIST STAFF MEM, LOS ALAMOS SCI LAB, 49- Concurrent

Pos: Vis prof, State Univ NY Stonybrook, 67. Mem: AAAS; Am Chem Soc. Res: Actinide chemistry; equilibrium and kinetics of reactions in aqueous solutions. Mailing Add: Los Alamos Sci Lab Univ of Calif Box 1663 Los Alamos NM 87544

NEWTON, TYRE ALEXANDER, b Morris, Okla, Dec 28, 21; m 46; c 2. MATHEMATICAL ANALYSIS. Educ: Colo State Univ, BS, 49; Univ Ga, MA, 51, PhD(math), 52. Prof Exp: Instr math, Univ Nebr, 52-55; from asst prof to assoc prof, Colo State Univ, 55-58; asst prof, 58-72, ASSOC PROF MATH, WASH STATE UNIV, 72- Mem: Am Math Soc; Soc Indust & Appl Math; Math Asn Am; Am Soc Eng Educ. Res: Infinite series; finite differences; functional analysis; using the analog computer to illustrate mathematical concepts. Mailing Add: Dept of Math Wash State Univ Pullman WA 99163

NEWTON, VICTOR JOSEPH, b Boston, Mass, Apr 9, 37. THEORETICAL NUCLEAR PHYSICS. Educ: Spring Hill Col, BS, 61, MA, 62; Mass Inst Technol, PhD(physics), 66. Prof Exp: Lectr physics, Loyola Col Md, 68-69; asst prof, 69-73, ASSOC PROF PHYSICS, FAIRFIELD UNIV, 73- Mem: AAAS; Am Phys Soc; Am Asn Physics Teachers; Fed Am Scientists; Sigma Xi. Res: Many channel scattering theory; three and four nucleon systems as applied to nuclear systems, including weak interactions. Mailing Add: Dept of Physics Fairfield Univ Fairfield CT 06430

NEWTON, WALTER LLOYD, b Brownsburg, Que, May 7, 16; US citizen; m 39; c 5. PARASITOLOGY. Educ: George Washington Univ, BS, 42, MS, 48, PhD(parasitol), 52. Prof Exp: Zoologist lab trop dis, NIH, 38-43, sanitarian, USPHS, 43-52, scientist, 52-57, scientist dir, 57-59, chief lab germfree animal res, Nat Inst Allergy & Infectious Dis, 59-63, assoc chief div res serv, NIH, 63-65, prog adminr, Res Grants Br, 65-71, dep chief, 71-73, DEP ASSOC DIR PROG ACTIVITIES, NAT INST GEN MED SCI, NIH, 73- Concurrent Pos: USPHS, 57-, scientist dir. Mem: Am Soc Parasitol; Am Soc Trop Med & Hyg. Res: Intermediate hosts of filariae and schistosomes; inheritance of susceptibility to infection, physiology of hosts, resistance to chemicals; water and sewage treatment processes as related to spread of parasitic diseases; canine filariasis; germfree animal research. Mailing Add: Nat Inst of Gen Med Sci Bethesda MD 20014

NEWTON, WILLIAM ALLEN, JR, b Traverse City, Mich, May 19, 23; m 45; c 4. PEDIATRICS, PATHOLOGY. Educ: Alma Col, Mich, BSc, 43; Univ Mich, MD, 46. Prof Exp: Intern, Wayne County Gen Hosp, Eloise, Mich, 47; fel pediat path, Children's Hosp Mich, 48, fel pediat hemat, 49-50; resident pediat, Children's Hosp Philadelphia, 50; from instr to asst prof path, 52-59, assoc prof path & pediat, 59-66, PROF PATH & PEDIAT, OHIO STATE UNIV, 66-; DIR LABS, CHILDREN'S HOSP, 52- Mem: Soc Pediat Res; Am Asn Cancer Res; Am Soc Exp Path. Res: Cancer chemotherapy in children, particularly leukemia and brain tumors; red cell enzyme deficiency of glucose 6-phosphate dehydrogenase. Mailing Add: Children's Hosp Columbus OH 43205

NEWTON, WILLIAM ATSON, JR, b New Orleans, La, Sept 7, 36. VIROLOGY, BIOCHEMISTRY. Educ: Tulane Univ, BS, 57, MD, 61; Univ Tex, PhD(biomed sci), 71. Prof Exp: Rosalie B Hite fel, Univ Tex, 67-70; proj investr virol, 71-72, ASST PROF VIROL & ASST VIROLOGIST, UNIV TEX M D ANDERSON HOSP & TUMOR INST HOUSTON, 72- Mem: Am Asn Cancer Res. Res: Biology and biochemistry of oncornaviruses; biophysical characterization of viral proteins. Mailing Add: Dept of Virol M D Anderson Hosp & Tumor Inst Houston TX 77025

NEWTON, (WILLIAM) AUSTIN, US citizen. DEVELOPMENTAL GENETICS. Educ: Univ Calif, PhD(biochem), 64. Prof Exp: Asst prof, 66-72, ASSOC PROF BIOL, PRINCETON UNIV, 72- Res: Biochemistry and genetics of gene expression and development in microorganisms. Mailing Add: Dept of Biol Princeton Univ Princeton NJ 08540

NEWTON, WILLIAM DONALD, b Dayton, Ohio, Mar 15, 44. CYTOLOGY. Educ: Ga State Univ, BS, 66; Univ NC, Chapel Hill, PhD(zool), 73. Prof Exp: Hargitt cell biol fel cytol, Dept Zool, Duke Univ, 73-74; asst prof biol, Colby Col, Maine, 74-75; ASST PROF BIOL, COLO WOMEN'S COL, 75- Mem: Am Soc Zoologists; Soc Develop Biol; Am Inst Biol Sci; Sigma Xi. Res: Development and mechanisms of motility of aflagellate, non-axonemal, spermatozoa of Turbellaria; chromosome movement and spindle dynamics; freshwater invertebrates. Mailing Add: Dept of Biol Colo Women's Col Denver CO 80220

NEWTON, WILLIAM EDWARD, b London, Eng, Nov 10, 38; m 62; c 2. INORGANIC CHEMISTRY. Educ: Nottingham Univ, Eng, BSc, 61; Royal Inst Chem, grad, 65; Univ London, PhD(chem), 68. Prof Exp: Anal chemist, Rayner & Co, Ltd, London, 62-66; teaching asst chem, Northern Polytechnic, London, 66-68; res fel chem, Harvard Univ, 68-69; staff scientist, 69-71, sect head, 71-73, MISSION MGR & INVESTR CHEM, CHARLES F KETTERING RES LAB, 73- Mem: The Chem Soc; Royal Inst Chem; Am Chem Soc. Res: Bioinorganic and organometallic chemistry; studies of the early transition metals, particularly molybdenum, as an aid in elucidating the role of molybdenum in enzymes. Mailing Add: Charles F Kettering Res Lab 150 E South College St Yellow Springs OH 45387

NEWTON, WILLIAM MORGAN, b Moline, Ill, May 30, 21; m 49; c 2. LABORATORY ANIMAL MEDICINE, PHYSIOLOGY. Educ: Univ Ill, BS, 50, DVM, 52, PhD(physiol), 65; Am Col Lab Animal Med, dipl, 74. Prof Exp: Pvt pract, 52-61; from instr to asst prof vet physiol, 64-67, DIR LAB ANIMAL CARE, UNIV ILL, URBANA, 67- Concurrent Pos: Mem vet drug rev panel, Nat Acad Sci, 66-68. Mem: Am Vet Med Asn; Am Soc Vet Physiol & Pharmacol; Am Soc Lab Animal Practioners; Am Asn Lab Animal Sci. Res: Conditioning procedure in the research dog; periodontal diseases of lemmings; diseases and management of laboratory animals. Mailing Add: Off of Lab Animal Care Univ of Ill Urbana IL 61801

NEXSEN, WILLIAM E, JR, experimental physics, see 12th edition

NEY, EDWARD PURDY, b Minneapolis, Minn, Oct 28, 20; m 42; c 4. PHYSICS. Educ: Univ Minn, BS, 42; Univ Va, PhD(physics), 46. Prof Exp: Asst physics, Univ Minn, 40-42; res assoc, Univ Va, 43-46, from asst prof to assoc prof, 46-47; from asst prof to assoc prof, 47-55, PROF PHYSICS, UNIV MINN, MINNEAPOLIS, 55- Concurrent Pos: Consult, Naval Res Lab, DC, 43-44 & Gen Dynamics/Convair. Mem: Nat Acad Sci; AAAS; fel Am Phys Soc; Am Astron Soc; Am Geophys Union. Res: Mass spectroscopy; cosmic rays; atmospheric physics; astrophysics; infrared astronomy. Mailing Add: Dept of Astron Univ of Minn Minneapolis MN 55455

NEY, PETER E, b Brno, Czech, July 6, 30; US citizen; m 55; c 2. MATHEMATICS. Prof Exp: Instr math, Cornell Univ, 58-60, asst prof indust eng, 60-63; vis asst prof statist, Stanford Univ, 63-64; assoc prof indust eng, Cornell Univ, 64-65; assoc prof math, 65-69, PROF MATH, UNIV WIS-MADISON, 69- Concurrent Pos: Grants, Off Naval Res & NSF, 58-65; prin investr, NIH grant, 65-; Guggenheim fel, 71-72; vis prof, Israel Inst Technol & Weizmann Inst Sci, 71-72. Mem: Am Math Soc; Inst Math Statist. Res: Probability; stochastic processes; branching processes; Markov chains. Mailing Add: Dept of Math Univ of Wis Madison WI 53706

NEY, ROBERT LEO, b Brno, Czech, May 22, 33; US citizen; m 56; c 3. MEDICINE, ENDOCRINOLOGY. Educ: Harvard Univ, AB, 54; Cornell Univ, MD, 58. Prof Exp: Fel endocrinol & instr med, Vanderbilt Univ, 61-63; investr endocrinol, Nat Heart Inst, 63-65; asst prof med, Vanderbilt Univ, 65-67; assoc prof, 67-70, PROF MED, UNIV NC, CHAPEL HILL, 70-, CHMN DEPT, 72- Mem: AAAS; Am Fedn Clin Res; Endocrine Soc; Am Soc Clin Invest; Am Physiol Soc; Asn Am Physicians; Am Col Physicians. Res: Regulation of pituitary-adrenal function; structure and function of ACTH, melanocyte stimulating hormone and related polypeptides; regulation of endocrine tumor function. Mailing Add: Dept of Med Univ of NC Chapel Hill NC 27514

NEY, WILBERT ROGER, b Rockford, Ill, Nov 28, 29; m 58; c 2. PHYSICS. Educ: Yale Univ, BS, 57; George Washington Univ, JD, 64. Prof Exp: Physicist, Nat Bur Stand, 58-60, sci asst, 60-64; EXEC DIR, NAT COUN RADIATION PROTECTION & MEASUREMENTS, 64- Concurrent Pos: Secy, Nat Comt Radiation Protection & Measurements, 61-64; tech secy, Int Comn Radiation Units & Measurements, 61- Mem: AAAS; Health Physics Soc; Radiation Res Soc; Radiol Soc NAm. Res: Law; radiation, protection, quantities, units and effects. Mailing Add: Nat Coun Rad Prot & Meas Suite 1016 7910 Woodmont Ave Washington DC 20014

NEYMAN, JERZY, b Benderey, Bessarabia, Apr 16, 94; nat US; m 20; c 1. STATISTICS. Educ: Univ Warsaw, PhD, 23. Hon Degrees: DSc, Univ Chicago, 59; LLD, Univ Calif, Berkeley, 63; PhD, Univ Stockholm, 64; DS, Univ Warsaw, Poland, 74; DS, Indian Statist Inst, Calcutta, India, 74. Prof Exp: Lectr, Inst Technol Kharkov, 17-21; statistician, Inst Agr Bydgoszcz, Poland, 21-23; head biomet lab, Nencki Inst, Warsaw, 23-34; spec lectr, Univ Col, Univ London, 34-36, reader statist, 35-38; prof math, 38-41, in chg statist lab, 38-41, PROF STATIST & DIR STATIST LAB, UNIV CALIF, BERKELEY, 41-, US OFF NAVAL RES PROJ, 48- Concurrent Pos: Head statist lab, Cent Col Agr, Warsaw, 23-34; Rockefeller fel, London & Paris, 26-27; spec lectr, Univ Warsaw, 28-34 & Univ Paris, 36; mem Nat Defense Res Comt Proj, 42-45; vis prof, Columbia Univ, 46; Guggenheim fel, 57; vis scientist carcinogenesis, NIH, 58; res prof, Miller Inst Basic Res Sci, Univ Calif, Berkeley, 58-59; chmn sect statist, AAAS, 62. Honors & Awards: Cleveland Prize, AAAS, 58; Nat Medal of Sci, 68. Mem: Nat Acad Sci; AAAS (vpres, 62); Inst Math Statist (pres, 49); Am Statist Asn (vpres, 47); Int Statist Inst. Res: Mathematical statistics; theories of testing hypotheses and of estimation; application to genetics; astronomy; medical diagnosis; statistical treatment of agricultural experiments; weather modification experiments. Mailing Add: Dept of Statist Univ of Calif Berkeley CA 94720

NEYNABER, ROY HAROLD, b Highland Park, Mich, July 4, 26; m 51; c 4. ATOMIC PHYSICS. Educ: Univ Wis, BS, 49, MS, 51, PhD(physics), 55. Prof Exp: Asst physics, Univ Wis, 51-55; sr staff scientist, Gen Dynamics/Convair, 55-69; mgr atomic physics br, Gulf Energy & Environ Systs Co, 69-73; MGR, ATOMIC PHYSICS DEPT, IRT CORP, 73- Concurrent Pos: Physicist, Liberty Powder Co, 53. Mem: Am Phys Soc. Res: Gaseous electronics; atomic beams; atomic scattering experiments; particle-surface interactions; small angle scattering of x-rays; lattice imperfections in solids. Mailing Add: 4471 Braeburn Rd San Diego CA 92116

NEZ, MARTHA MALONE, zoology, see 12th edition

NEZRICK, FRANK ALBERT, b Mansfield, Ohio, Apr 1, 37; m 60; c 1. ELEMENTARY PARTICLE PHYSICS. Educ: Case Inst, BS, 59, MS, 62, PhD(physics), 65. Prof Exp: Vis scientist grant elem particle physics, Europ Orgn Nuclear Res, 65-68; PHYSICIST, NAT ACCELERATOR LAB, 68- Mem: Am Phys Soc. Res: Weak interaction physics; neutrino-bubble chamber research; magnetic monopole search; development of neutrino focussing systems. Mailing Add: Nat Accelerator Lab PO Box 500 Batavia IL 60510

NG, BARTHOLOMEW SUNG-HONG, b Canton, China, Sept 10, 46; m 73. APPLIED MATHEMATICS, FLUID DYNAMICS. Educ: St Josephs Col, Ind, BS, 68; Univ Chicago, MS, 70, PhD(appl math), 73. Prof Exp: Syst engr, Int Bus Mach Corp, 68; res asst math, Univ Chicago, 68-69, teaching asst & lectr, 69-73; fel, Univ Toronto, 73-75; ASST PROF MATH, IND UNIV-PURDUE UNIV, INDIANAPOLIS, 75-; RES ASSOC, INDIANAPOLIS CTR ADVAN RES, 75- Mem: Am Math Soc; Soc Indust & Appl Math; Am Inst Aeronaut & Astronaut; Sigma Xi. Res: Hydrodynamic stability and numerical modelling of aerodynamic processes, especially turbulence. Mailing Add: Fluid Dynamics Lab Indianapolis Ctr Advan Res Indianapolis IN 46202

NG, GEORGE, b New York, NY, Aug 5, 41; m 70. INORGANIC CHEMISTRY. Educ: Adelphi Col, AB, 63; Brooklyn Col, AM, 66; City Univ New York, PhD(chem), 71. Prof Exp: ASST PROF CHEM, FED CITY COL, 70- Mem: Am Chem Soc; Soc Appl Spectros. Res: Organometallic and coordination chemistry; computer programming; environmental pollution. Mailing Add: Dept of Chem Fed City Col Washington DC 20005

NG, HENRY, b San Francisco, Calif, Nov 6, 29; m 57; c 5. MICROBIOLOGY, FOOD SCIENCE. Educ: Univ Calif, Davis, BS, 54, MS, 60, PhD(microbiol), 63. Prof Exp: Res microbiologist low temperature microbiol, Eastern Utilization Res & Develop Div, 63-65, RES MICROBIOLOGIST HEAT RESISTANCE SALMONELLA, WESTERN REGIONAL RES CTR, USDA, 65- Mem: Am Soc Microbiol; Brit Soc Gen Microbiol. Res: Low temperature microbiology; lactic acid and bread fermentations and thermal resistance of pathogenic bacteria found in foods. Mailing Add: Western Regional Res Ctr USDA 800 Buchanan St Berkeley CA 94710

NG, LORENZ KENG-YONG, b Singapore, Aug 6, 40; US citizen. NEUROPSYCHIATRY. Educ: Stanford Univ, AB, 61; Columbia Univ, MD, 65. Prof Exp: Intern med, Mt Sinai Hosp, Los Angeles, Calif, 65-66; resident neurol, Hosp Univ Pa, Philadelphia, 66-69; spec fel neuropsychopharmacol, Lab Clin Sci, NIMH, 69-72, spec asst to dir, Div Narcotic Addiction & Drug Abuse, 72-74, CHIEF, INTRAMURAL RES LAB, NAT INST DRUG ABUSE, 74-, RES SCIENTIST, NIMH, 72- Concurrent Pos: Consult, Vet Admin Hosp, 75-; pres, World Man Found. Honors & Awards: S Weir Mitchell Award, Am Acad Neurol, 71; A E Bennett Award, Soc Biol Psychiat, 72; Acupuncture Res Award, Am Soc Chinese Med, 75. Mem: World Acad Art & Sci; AAAS; Am Acad Neurol; Soc Neurosci. Res: Neurology and behavioral biology of pain states; clinical, experimental and theoretical aspects of acupuncture; chemical and non-chemical approaches to treatment of drug and alcohol dependence. Mailing Add: Nat Inst on Drug Abuse 11400 Rockville Pike Rockville Pike MD 20852

NGAI, SHIH HSUN, b China, Sept 15, 20; nat US; m 48; c 3. ANESTHESIOLOGY. Educ: Nat Cent Univ, China, MD, 44. Prof Exp: From instr to assoc prof, 49-65, PROF ANESTHESIOL, COL PHYSICIANS & SURGEONS, COLUMBIA UNIV, 65- Concurrent Pos: Asst anesthesiologist, Presby Hosp, New York, 49-54, from asst attend anesthesiologist to attend anesthesiologist, 57- Mem: Am Soc Anesthesiol; Am Physiol Soc; Am Soc Pharmacol & Exp Therapeut; Asn Univ Anesthetists. Res: Neural control of respiration and circulation; pharmacology of anesthetics and agents affecting respiration and circulation. Mailing Add: Dept of Anesthesiol Columbia Univ Col of Physicians & Surgeons New York NY 10032

NGHIEM, QUANG XUAN, b Hanoi, Vietnam, Oct 17, 31; m 61; c 6. MEDICINE, PEDIATRICS. Educ: Taberd Inst, Vietnam, Baccalaureat, 52; Univ Saigon, cert physics, chem & biol, 53; Univ Paris, MD, 58. Prof Exp: From intern to resident pediat, Springfield Hosp, Mass, 58-60; chief resident, Roger Williams Hosp, Providence, RI, 61-63; asst prof pediat & asst dir pediat cardiol, 66-72, ASSOC PROF PEDIAT CARDIOL, UNIV TEX MED BR GALVESTON, 72- Concurrent Pos: Fel pediat, Children's Hosp, Boston, 60-62; NIH fel pediat cardiol, Univ Tex Med Br Galveston, 63-66 & NIH grant, 69-72. Honors & Awards: Fac Laureate & Thesis Prize, Fac Med, Univ Paris, 67. Mem: Fel Am Col Cardiol; fel Am Acad Pediat; Am Heart Asn. Res: Pediatric cardiology. Mailing Add: Dept of Pediat Cardiol Univ of Tex Med Br Galveston TX 77550

NGUYEN-HUU, XUONG, b Thai-Binh, Vietnam, July 14, 33; US citizen; m 60. CRYSTALLOGRAPHY, BIOPHYSICS. Educ: Sch Indust Elec, Marseille, France, BS, 55; Advan Sch Elec, Paris, MS, 57; Univ Paris, MA, 58; Univ Calif, Berkeley, MA & PhD(physics), 62. Prof Exp: Physicist, Lawrence Radiation Lab, Univ Calif, Berkeley, 62; asst res physicist, 62-63, asst prof physics, 64-70, ASSOC PROF PHYSICS, CHEM & BIOL, UNIV CALIF, SAN DIEGO, 70- Concurrent Pos: Guggenheim fel, 65-66. Mem: Am Phys Soc. Res: Elementary particle physics; data reduction using digital computer. Mailing Add: Dept of Physics Univ of Calif San Diego La Jolla CA 92037

NIBLACK, JOHN FRANKLIN, b Oklahoma City, Okla, Mar 5, 39; m 60; c 2. PHARMACOLOGY, INFECTIOUS DISEASES. Educ: Okla State Univ, BS, 60; Univ Ill, Urbana, MS, 65, PhD(biochem), 68. Prof Exp: Res biochemist, 68-69, res projs leader, 69-72, res mgr, 72-75, ASST DIR, DEPT PHARMACOL, CENT RES DIV, PFIZER, INC, 75- Mem: Am Soc Microbiol. Res: Biochemical mechanisms of action of antimicrobial and antineoplastic drugs; drugs affecting immune responses and general lymphoreticular function. Mailing Add: Cent Res Div Pfizer Inc Groton CT 06340

NIBLER, JOSEPH WILLIAM, b Silverton, Ore, May 9, 41; m 64; c 2. PHYSICAL CHEMISTRY. Educ: Ore State Univ, BS, 63; Univ Calif, Berkeley, PhD(chem), 66. Prof Exp: NSF fel chem, Cambridge Univ, 66-67; asst prof, 67-73, ASSOC PROF CHEM, ORE STATE UNIV, 73- Mem: Am Phys Soc; Am Chem Soc. Res: Infrared and Raman spectroscopy of matrix isolated molecules; energy transfer in solids. Mailing Add: Dept of Chem Ore State Univ Corvallis OR 97331

NIBLETT, CHARLES LESLIE, b Wolfeboro, NH, Feb 15, 43; m 61; c 2. PLANT PATHOLOGY, VIROLOGY. Educ: Univ NH, BS, 65; Univ Calif, PhD(plant path), 69. Prof Exp: Asst prof, 69-74, ASSOC PROF PLANT PATH, KANS STATE UNIV, 74- Mem: Am Phytopath Soc. Res: Heterogeneity of plant viruses; process of virus infection. Mailing Add: Dept of Plant Path Kans State Univ Manhattan KS 66506

NIBLETT, EDWARD RONALD, b Toronto, Ont, May 19, 26; m 58; c 4. GEOPHYSICS. Educ: Univ Toronto, BA, 48, MA, 49; Cambridge Univ, PhD(geophys), 58. Prof Exp: Tech off, Geomagnetic Div, 50-52, geophysicist, 52-58, SR SCI OFF, GEOMAGNETIC DIV, DOM OBSERV, DEPT ENERGY, MINES & RESOURCES, 58-, RES SCIENTIST, 67- Mem: Royal Astron Soc; fel Geol Asn Can. Res: Geomagnetism; terrestrial heat flow; electromagnetic induction in the earth. Mailing Add: Div Geomagnetism Earth Phys Br Dept Energy, Mines & Resources Ottawa ON Can

NICCOLAI, NILO ANTHONY, b Pittsburgh, Pa, May 21, 40; m 64; c 1. MATHEMATICS, COMPUTER SCIENCE. Educ: Carnegie-Mellon Univ, BS, 62, MS, 63, PhD(math), 68. Prof Exp: Mathematician, US Bur Mines, Pa, 67-68; math analyst, Off Chief Staff, US Army, 68-70; ASST PROF MATH, UNIV NC, CHARLOTTE, 70- Honors & Awards: Army Commendation Medal. Mem: Am Math Soc; Math Asn Am; Asn Comput Mach. Res: Representation of finite groups; computer simulation; theorem-proving. Mailing Add: Univ of NC University Sta Charlotte NC 28223

NICE, CHARLES MONROE, JR, b Parsons, Kans, Dec 21, 19; m 40; c 6. RADIOLOGY. Educ: Univ Kans, AB, 39, MD, 43; Univ Colo, MS, 48; Univ Minn, PhD(radiol), 56. Prof Exp: From instr to assoc prof, Univ Minn, 51-58; PROF RADIOL, SCH MED, TULANE UNIV, 58- Mem: AAAS; Radiol Soc NAm; Am Roentgen Ray Soc; AMA; Asn Am Med Cols. Res: Acquired radioresistance in mouse tumors; clinical diagnostic radiology. Mailing Add: Dept of Radiol Tulane Univ Sch of Med New Orleans LA 70112

NICE, MARGARET MORSE, b Amherst, Mass, Dec 6, 83; m 09; c 4. ORNITHOLOGY. Educ: Mt Holyoke Col, BA, 06; Clark Univ, MA, 15. Hon Degrees: ScD, Mt Holyoke Col, 55 & Elmira Col, 62. Prof Exp: Assoc ed, Bird-Banding, 35-42 & 46-71; RETIRED. Concurrent Pos: Assoc ed, Wilson Bulletin, Wilson Ornith Soc, 39-49. Honors & Awards: Brewster Medal, Am Ornith Union, 42. Mem: Wilson Ornith Soc (pres, 37-39); Cooper Ornith Soc; fel Am Ornith Union; foreign mem Brit Ornith Union; corresp mem Hungarian Inst Ornith. Res: Birds of Oklahoma; life history studies of birds, particularly mourning doves, warblers and song sparrows; territory and population problems; weights and behavior of passerines; development of behavior in precocial birds; speech development of children; history of incubation periods. Mailing Add: 5725 Harper Ave Chicago IL 60637

NICE, PHILIP OLIVER, b Victor, Colo, Nov 16, 17; m 42; c 4. PATHOLOGY, MICROBIOLOGY. Educ: Univ Colo, BA, 39, MD, 43. Prof Exp: Instr path, Dartmouth Med Sch, 50-52 & Univ Colo, 53-54; from asst prof to assoc prof, 54-73, assoc dean, 60-73, VIS PROF MICROBIOL, DARTMOUTH MED SCH, 73- Concurrent Pos: Pathologist & dir labs, Vet Admin Hosp, Great Falls, Mont, 52-53 & Larimer County Hosp, Ft Collins, Colo, 53-54; assoc dir labs & microbiologist, Mary Hitchcock Mem Hosp, Hanover, NH, 54- Mem: Am Soc Microbiol; Am Soc Clin Path; AMA; Col Am Path. Res: Diagnostic microbiology. Mailing Add: Dept of Microbiol Dartmouth Med Sch Hanover NH 03755

NICELY, KENNETH AUBREY, b Slab Fork, WVa, Feb 25, 38; m 64. BOTANY. Educ: WVa Univ, AB, 59, MS, 60; NC State Univ, PhD(bot), 63. Prof Exp: Asst prof bot, Va Polytech Inst, 63-64; asst prof, 64-69, ASSOC PROF BOT, WESTERN KY UNIV, 69- Mem: Am Soc Plant Taxon; Int Asn Plant Taxon. Res: Taxonomy of flowering plants. Mailing Add: Dept of Biol Western Ky Univ Bowling Green KY 42101

NICELY, VINCENT ALVIN, b Botetourt Co, Va, Feb 10, 43; m 66; c 1. PHYSICAL CHEMISTRY. Educ: WVa Wesleyan Col, BS, 65; Mich State Univ, PhD(phys chem), 69. Prof Exp: Res assoc quantum chem, Mich State Univ, 69-70; res chemist, 70-72, SR RES CHEMIST, TENN EASTMAN CO DIV, EASTMAN KODAK CO, 72- Mem: Am Chem Soc. Res: Spectroscopy; quantum chemistry; applications of computers in chemistry. Mailing Add: B-150 Tenn Eastman Co Eastman Kodak Co Kingsport TN 37662

NICHAMAN, MILTON Z, b Apr 19, 31; US citizen; m 58; c 3. CARDIOVASCULAR DISEASES. Educ: Brandeis Univ, AB, 53; Tufts Univ, MD, 57; Univ Pittsburgh, ScD, 64. Prof Exp: From intern to resident med, Boston City Hosp, 57-59; surgeon, USPHS Heart Dis & Stroke Control Prog, Med Col SC, 59-61, surgeon, Grad Sch Pub Health, Univ Pittsburgh, 61-64, sr surgeon & chief lab, Field & Training Sta, San Francisco, 64-67, sr surgeon & chief epidemiol, 67-70; DIR NUTRIT PROG, CTR DIS CONTROL, US DEPT HEALTH, EDUC & WELFARE, 71- Concurrent Pos: Lab instr, Brandeis Univ, 53-54; res assoc, Univ Pittsburgh, 63-65; lectr, Univ Calif, 67; mem arteriosclerosis coun, Am Heart Asn. Mem: AMA; Am Heart Asn; Am Fedn Clin Res; Am Pub Health Asn. Mailing Add: Ctr for Dis Control Atlanta GA 30333

NICHOALDS, GEORGE EDWARD, b Harrison, Ark, Jan 15, 40; m 61; c 2. BIOCHEMISTRY, NUTRITION. Educ: Ouachita Baptist Univ, BS, 62; Univ Ark, MS, 64, PhD(biochem), 67. Prof Exp: Instr chem, Univ Ark, Little Rock, 65-68; lab asst biol chem, Univ Mich, Ann Arbor, 68-69; res assoc biochem, 69-70, DIR CLIN NUTRIT LAB, VANDERBILT UNIV, 70-, ASST PROF BIOCHEM, 72- Concurrent Pos: NIH fel, Univ Mich, Ann Arbor, 68-69; nutrit biochemist, Meharry Med Col, 72- Res: Nutritional biochemistry; assessment of nutritional status and development of methodology; vitamin and trace mineral requirements of total parenteral nutrition. Mailing Add: Dept of Biochem Sch of Med Vanderbilt Univ Nashville TN 37232

NICHOL, CHARLES ADAM, b Fergus, Ont, May 3, 22; nat US; m 47; c 3. PHARMACOLOGY, BIOCHEMISTRY. Educ: Univ Toronto, BS, 44; McGill Univ, MS, 46; Univ Wis, PhD(biochem), 49. Prof Exp: From instr to asst prof, Western Reserve Univ, 49-52; asst prof, Sch Med, Yale Univ, 53-56; res prof, State Univ NY Buffalo, 56-70; DIR RES PHARMACOL, WELLCOME RES LABS, 69- Concurrent Pos: Am Cancer Soc scholar, 52-56; dir dept exp therapeut, Roswell Park Mem Inst, 56-70; mem biochem-pharmacol panel, Cancer Chemother Nat Serv Ctr, NIH, 57-58, mem drug eval panel, 58-60, chmn exp therapeut comt, 59-60, consult grants & fels div, 59-62, mem cancer chemother study sect, 59-62, spec adv comt, 64-69; adj prof, Duke Univ, 70- Mem: Am Soc Biol Chem; Am Chem Soc; Soc Pharmacol & Exp Therapeut; Soc Exp Biol & Med; Am Asn Cancer Res. Res: Cancer chemotherapy; mechanism of action of folic acid antagonists; metabolism of folic acid and vitamin B-12; drug resistance in bacterial and mammalian cells; nutrition and tumor growth; corticosteroids and transaminase enzymes; drug-metabolizing enzymes. Mailing Add: Wellcome Res Labs 3030 Corwallis Rd Research Triangle Park NC 27709

NICHOL, CHRISTINA JANET, b Port Alberni, BC, Apr 12, 33; div; c 1. BIOCHEMISTRY. Educ: Univ BC, BA, 53, MSc, 55; Univ London, PhD(biochem), 58. Prof Exp: Med Res Coun mem res staff, Univ Col Hosp Med Sch, Univ London, 56-59; res assoc, Kinsmen Lab Neurol Res, Univ BC, 60-65; ASSOC MASTER, DEPT CHEM, BC INST TECHNOL, 65- Concurrent Pos: Muscular Dystrophy Asn Can fel, 60-61, res grants, 60-65; Med Res Coun Can scholar & res grants, 63-65. Res: Iodine metabolism in man; muscular dystrophy and creative phosphokinase in muscle and serum of man and mouse; sulfur metabolism in the muscle of dystrophic mice. Mailing Add: BC Inst of Technol 3700 Willingdon Ave Burnaby BC Can

NICHOL, FRANCIS RICHARD, JR, b Baltimore, Md, Feb 27, 42; m 60; c 3. VIROLOGY. Educ: Pa State Univ, BA, 64, MS, 66, PhD(microbiol), 68. Prof Exp: Res scientist, Upjohn Co, 67-72, clin res assoc, 72-74, sr res scientist, 74-75; PRES, INST BIOL RES & DEVELOP, INC, 75- Mem: AAAS; Am Soc Microbiol. Res: Research and development of interferon stimulators; inhibitors of tumor virus enzymes; chemotherapeutic agents for virus infections; stimulators of host-defense mechanisms; cell mediated immunity; clinical trials, phase II, III. Mailing Add: Inst for Biol Res & Develop Inc Newport Beach CA 92660

NICHOL, HAMISH, b Kroonstad, S Africa, Dec 25, 24; m; c 2. PSYCHIATRY. Educ: Cambridge Univ, BA, 49, MB, BCh, 52. Prof Exp: House officer med & pediat, United Birmingham Hosps, Eng, 52-53; bacteriologist, Riks-hospitalitet, Oslo, Norway, 53-54; sr house officer psychiat, Claybury Hosp, London, Eng, 54-55 & St Clement's Hosp, 55-56; resident & instr, Bellevue Med Ctr, NY Univ, 56-57; instr & fel, Albert Einstein Col Med, 57-61; asst prof, 61-72, ASSOC PROF PSYCHIAT & HEAD DIV CHILD PSYCHIAT, UNIV BC, 61- Concurrent Pos: Mem subcomt child psychiat & ment retardation, Nat Ment Health Adv Comt, Dept Health, Can, 65-66. Mem: Can Psychiat Asn. Res: Caser register studies in the epidemiology of psychiatric disorders in children; clinical studies of psychiatric disorders in children and adolescents; study of juvenile delinquents requiring to be detained or to appear in court. Mailing Add: Dept of Psychiat Univ of BC Vancouver BC Can

NICHOL, JAMES CHARLES, b Onoway, Alta, Apr 6, 22; nat US; m 48; c 2. PHYSICAL CHEMISTRY. Educ: Univ Alta, BSc, 43, MSc, 45; Univ Wis, PhD(phys chem), 48. Prof Exp: Instr Univ Alta, 44-46; proj assoc, Univ Wis, 48-49; assoc prof chem, Willamette Univ, 49-57; from asst prof to assoc prof, 57-62, PROF CHEM, UNIV MINN, DULUTH, 62- Concurrent Pos: Calif Res Corp fel, Yale Univ, 53-54; vis prof, Inst Enzyme Res, Univ Wis, 65-66. Mem: Am Chem Soc; The Chem Soc. Res: Transport properties of electrolytes; physical chemistry of proteins. Mailing Add: Dept of Chem Univ of Minn Duluth MN 55812

NICHOLAIDES, JOHN J, b Statesville, NC, Aug 1, 44; m 67; c 1. SOIL FERTILITY. Educ: NC State Univ, BS, 66, MS, 69; Univ Fla, PhD(soil chem), 73. Prof Exp: Peace Corps vol agron, 69-70; vis asst prof soil fertil eval, Int Soil Fertil Eval & Improv Proj, 73-75, VIS ASST PROF SOIL FERTIL, NC STATE UNIV, 75- Concurrent Pos: NDEA Title IV fel, US Govt/Univ Fla, 70-73; mem, Nat Soil Fertil Adv Comn, Govt Costa Rica, 74-75; mem, Nat Agr Adv Comn, Govt Nicaragua, 74-75. Honors & Awards: OutstandMem: Sigma Xi. Res: Relationships and correlations of soil test values and plant tissue nutrient content with crop response to fertilizers. Mailing Add: Dept of Soil Sci NC State Univ PO Box 5907 Raleigh NC 27607

NICHOLAS, GERARDUS, b Philadelphia, Pa, Dec 20, 27. SPELEOLOGY. Educ: Catholic Univ, BS, 50; Univ Pittsburgh, MS, 54; Univ Notre Dame, PhD(ecol), 60. Hon Degrees: DSc, Notre Dame Col, EPakistan, 63. Prof Exp: Instr biol, De La Salle Col, 49-50; chmn dept high sch, Md, 50-57; instr zool, La Salle Col, 58-59 & Univ Notre Dame, 59-62; ASSOC PROF EARTH SCI, LA SALLE COL, 62- Concurrent Pos: Vis prof, Univ New South Wales, 62; dir, Cave Res Found, 63-; ed, Int J Speleol, 64-; mem bd trustees, Gwynedd-Mercy Col. Mem: Fel AAAS; fel Nat Speleol Soc (pres, 57-61); Am Asn Biol Teachers (vpres, 61); Am Soc Zoologists; Ecol Soc Am. Res: Cave biology of Puerto Rico, Kentucky and Australia; Pleistocene ecology of Cumberland bone cave; ecology of Wissahickon Valley, Pennsylvania. Mailing Add: Dept of Earth Sci La Salle Col 20th & Olney Ave Philadelphia PA 19141

NICHOLAS, HAROLD JOSEPH, b St Louis, Mo, Mar 1, 19; m 52; c 2. BIOCHEMISTRY. Educ: Univ Mo, BS, 41; St Louis Univ, PhD(biochem), 50. Prof Exp: Res chemist, Hercules Powder Co, 41-44 & Parke, Davis & Co, 44-45; asst prof biochem, Univ Kans, 50-53, asst prof obstet, Med Ctr, 55-63; sr res chemist, Anheuser-Busch, Inc, 53-55; DIR EXP MED, INST MED EDUC & RES, 63-;

ASSOC PROF BIOCHEM, SCH MED, ST LOUIS UNIV, 63- Mem: Am Chem Soc; Am Inst Chemists; Am Soc Biol Chemists; Am Soc Plant Physiologists; Endocrine Soc. Res: Metabolism of plant and animal steroids; terpene biosynthesis; chemistry of the central nervous system. Mailing Add: Inst of Med Educ & Res 1604 S 14th St St Louis MO 63104

NICHOLAS, JAMES A, b Portsmouth, Va, Apr 15, 21; m 52; c 1. SURGERY. Educ: NY Univ, BA, 42; Long Island Col Med, MD, 45; Am Bd Orthop Surg, dipl, 55. Prof Exp: From instr to asst prof surg & orthop, 53-70, PROF ORTHOP, MED COL, CORNELL UNIV, 70-; ADJ PROF PHYS EDUC, NY UNIV, 70- Concurrent Pos: Attend orthop surgeon, New York Hosp, 53-; attend, Hosp for Spec Surg, 53-, chief metab bone dis clin, 59-64; adj attend orthop surg, Lenox Hill Hosp, 53-, dir dept orthop surg, 70-, dir, Inst Sports Med & Athletic Trauma; secy bd trustees, Philip D Wilson Res Found; pvt pract; mem exec subcomt athletic injuries skeletal systs, Nat Acad Sci; AEC res grant; NIH res grant. Honors & Awards: Am Roentgen Ray Soc Cert of Merit, 59. Mem: AAAS; Am Orthop Asn; Am Trauma Soc; Am Orthop Soc Sports Med; Am Acad Orthop Surg. Res: Disturbances of knee joint, especially prosthetics, instability and replacement; athletic injuries; muscle physiology; radioisotope study of metabolic bone disease; osteoporosis; metabolic response to injury; sports medicine research. Mailing Add: 130 E 77th St New York NY 10021

NICHOLAS, LESLIE, b Philadelphia, Pa, Dec 22, 13. DERMATOLOGY, SYPHILOLOGY. Educ: Temple Univ, BS, 35, MD, 37. Prof Exp: Asst prof, 52-59, PROF DERMAT, HAHNEMANN MED COL, 59- Concurrent Pos: Lectr, Div Grad Med, Univ Pa, 49-67; venereal dis prog specialist, Philadelphia Dept Pub Health, 65-72. Mem: Soc Invest Dermat; Am Acad Dermat; Acad Psychosom Med. Res: Necrobiosis lipoidica diabeticorum; demodex folliculorum; erysipeloid; syphilis. Mailing Add: 1521 Locust St Philadelphia PA 19102

NICHOLAS, LOUIS JAMES, analytical chemistry, inorganic chemistry, see 12th edition

NICHOLAS, PAUL PETER, b Ohrid, Yugoslavia, Aug 6, 38; US citizen; m 62; c 1. ORGANIC CHEMISTRY. Educ: Univ Ill, Urbana, BS, 60; Cornell Univ, PhD(org chem), 64. Prof Exp: Sr res chemist org chem, 64-67, sect leader polymer chem res, 67-69, res assoc, 69-73, SR RES ASSOC, B F GOODRICH RES CTR, 73- Mem: Am Chem Soc. Res: Chemistry of highly fluorinated dienes; heterogeneous high temperature catalysis; halogenation; polymer crosslinking and stabilization. Mil Serv: Mailing Add: 4775 Canterbury Lane Broadview Heights OH 44147

NICHOLAS, RALPH WALLACE, b Dallas, Tex, Nov 28, 34; m 63. CULTURAL ANTHROPOLOGY. Educ: Wayne State Univ, BA, 57; Univ Chicago, MA, 58, PhD(anthrop), 62. Prof Exp: Asst prof anthrop, Portland State Col, 63-64; from asst prof to prof, Mich State Univ, 64-71; PROF ANTHROP & SOC SCI, UNIV CHICAGO, 71- Concurrent Pos: Res fel, Sch Oriental & African Studies, Univ London, 62-63; mem panel on SAia, Comt Instnl Coop, 66-; consult, US Dept Labor & AID, Int Manpower Inst, 67; lectr, Va Asian Studies Consortium, 67; sr Fulbright fel, Rural West Bengal, India, 68-69; chmn, Temp Comt Displaced Scholars from Bangladesh, 71-72; trustee & mem exec bd, Am Inst Indian Studies, 72- Mem: Fel Am Anthrop Asn; Asn Asian Studies; fel Royal Anthrop Inst. Res: Society and culture of Bengal; political anthropology; religion and symbolic analysis; kinship. Mailing Add: Dept of Anthrop Univ of Chicago 1126 E 59th St Chicago IL 60637

NICHOLAS, RICHARD CARPENTER, b Minneapolis, Minn, Dec 13, 26; m 52, 74. FOOD SCIENCE. Educ: Pa State Univ, BS, 48, MS, 50; Mich State Univ, PhD(agr eng), 58. Prof Exp: Instr physics, Pa State Univ, 48, instr physics & math, 51-52; from instr to asst prof agr eng, 56-60, from asst prof to assoc prof food sci, 60-71, PROF FOOD SCI, MICH STATE UNIV, 71- Mem: AAAS; Inst Food Technol; Am Soc Qual Control. Res: Thermal processing, food texture and quality control. Mailing Add: Dept of Food Sci & Human Nutrit Mich State Univ East Lansing MI 48824

NICHOLAS, ROBERT D, organic chemistry, see 12th edition

NICHOLES, HENRY JOSEPH, b St George, Utah, Mar 24, 13; m 39, 74; c 11. MEDICAL PHYSIOLOGY. Educ: Brigham Young Univ, AB, 35; Univ Wis, PhD(med physiol), 41. Prof Exp: PROF ZOOL, BRIGHAM YOUNG UNIV, 66- Res: General physiology. Mailing Add: 55-105 B Kulanui Lane Laie HI 96762

NICHOLES, PAUL SCOTT, b American Fork, Utah, Apr 28, 16; m 41; c 3. BACTERIOLOGY. Educ: Brigham Young Univ, AB, 41; Univ Cincinnati, PhD(bact), 46. Prof Exp: Asst bact, Brigham Young Univ, 40-41; from asst prof to assoc prof bact, 47-66, PROF BACT, UNIV UTAH, 66- Concurrent Pos: Consult, Amalgamated Sugar Co & Deseret Pharmaceut Co. Mem: AAAS; Am Soc Microbiol; NY Acad Sci. Res: Immunology; bacterial metabolism; antigenic structure of bacterium tularense; aerobiology; microbiology. Mailing Add: Dept of Microbiol Univ of Utah Salt Lake City UT 84112

NICHOLLS, DORIS MARGARET, b Bayfield, Ont, Jan 24, 27; m 52. BIOCHEMISTRY, ENZYMOLOGY. Educ: Univ Western Ont, BSc, 49, MSc, 51, PhD(biochem), 56, MD, 59. Prof Exp: Demonstr bot, Univ Western Ont, 47-51; assoc physiol, George Washington Univ, 59-60; dir clin invest unit lab, Westminster Hosp, London, Ont, 60-65; assoc prof biochem, 65-70, PROF BIOCHEM, YORK UNIV, 70- Concurrent Pos: Consult, Nat Heart Inst, 59-60; res assoc, Univ Western Ont, 60-62. Mem: Am Soc Biol Chemists; Can Biochem Soc; Brit Biochem Soc; Am Chem Soc; Biophys Soc. Res: Protein synthesis in mammalian tissue and microorganisms; regulation of cell metabolism. Mailing Add: Dept of Biol York Univ Downsview ON Can

NICHOLLS, JOHN GRAHAM, b London, Eng, Dec 19, 29; m 59; c 2. NEUROPHYSIOLOGY. Educ: Univ London, BSc, 51, PhD(physiol), 55, MB, BS, 56. Prof Exp: House surgeon casualty & house physician radiother, Charing Cross Hosp, Univ London, 57, res asst physiol, Univ Col, 58-60; dept demonstr, Oxford Univ, 60-62; res assoc neurophysiol, Harvard Med Sch, 62-65; assoc prof physiol, Sch Med, Yale Univ, 65-68; assoc prof, Harvard Med Sch, 68-73; PROF NEUROBIOL, MED SCH, STANFORD UNIV, 73- Mem: Am Soc Cell Biol; Brit Physiol Soc; Brit Biophys Soc. Res: Neurophysiology, especially with regard to denervated muscle, leech central nervous system, muscle spindles and properties of neurological cells. Mailing Add: Dept of Neurobiol Stanford Univ Med Sch Stanford CA 94305

NICHOLLS, JOHN VAN VLIET, b Montreal, Que, Dec 5, 09; m 37; c 3. OPHTHALMOLOGY. Educ: McGill Univ, BA, 30, MD, CM, 34, MSc, 35; FRCPS(C). Prof Exp: Asst demonstr, McGill Univ, 34-35, from demonstr to assoc prof, 38-68, prof, 68-70; clin assoc prof, 70-74, CLIN PROF OPHTHAL & CHMN DEPT, UNIV WESTERN ONT, 74- Concurrent Pos: From clin asst to ophthalmologist, Royal Victoria Hosp, 38-70; consult, Royal Can Air Force, 43-50, hon consult, 50-; ophthalmologist, Victoria Hosp, London, Ont, 70- Mem: Am Ophthal Soc; Am Acad Ophthal & Otolaryngol; fel Am Col Surgeons; NY Acad Sci; Can Ophthal Soc (treas, 50-55, pres, 57-58). Res: Clinical ophthalmology; basic research in pathology and vascular diseases involving the eye; visual standards for aviation. Mailing Add: 111 Waterloo St London ON Can

NICHOLLS, RALPH WILLIAM, b Richmond, Eng, May 3, 26; m 52. PHYSICS. Educ: Univ London, BSc, 46, PhD(physics), 51, DSc(spectros), 61. Prof Exp: Demonstr physics, Imp Col, Univ London, 45-46, demonstr astrophys, 46-48; demonstr physics, Univ Western Ont, 48-50, lectr, 50-52, from asst prof to prof, 52-62, sr prof, 62-65; chmn dept physics, 65-69, PROF PHYSICS & SCI & DIR CTR RES EXP SPACE SCI, YORK UNIV, 65- Concurrent Pos: Consult, Nat Bur Standards, 59-60; vis prof, Stanford Univ, 64 & 68; assoc fel, Can Aerospace Inst. Mem: Int Astron Union; Int Union Geod & Geophys; fel Royal Astron Soc Can; fel Royal Astron Soc; fel Brit Inst Physics. Res: Spectroscopy; astrophysics; aeronomy; chemical physics. Mailing Add: Ctr for Res in Exp Space Sci York Univ Toronto ON Can

NICHOLLS, ROBERT VAN VLIET, b Montreal, Que, Feb 18, 13; m 45; c 2. ORGANIC CHEMISTRY. Educ: McGill Univ, BSc, 33, MSc, 35, PhD(org chem), 36. Prof Exp: Sessional lectr, 36-39, lectr, 39-40, from asst prof to assoc prof, 40-53, PROF CHEM, McGILL UNIV, 53- Mem: Fel Chem Inst Can; fel Royal Soc Arts; Can Soc Hist & Philos Sci (pres, 75-78). Res: High polymers; history of science, particularly history of Canadian chemical technology. Mailing Add: Dept of Chem McGill Univ Montreal PQ Can

NICHOLOSI, GREGORY RALPH, b Toledo, Ohio, Sept 8, 43; m 69. PHYSIOLOGY. Educ: Mich State Univ, BS, 65; Ohio State Univ, PhD(physiol), 71. Prof Exp: Instr anat & physiol, Ohio Northern Univ, 69-70; asst prof physiol, Col Med, Ohio State Univ, 71-72; ASST PROF PHYSIOL, COL MED, UNIV S FLA, 72- Mem: Am Physiol Soc; Biophys Soc. Res: Cardiovascular physiology; hemodynamics; cardiodynamics. Mailing Add: Dept of Physiol Col of Med Univ of S Fla Tampa FL 33620

NICHOLS, ALEXANDER VLADIMIR, b San Francisco, Calif, Oct 9, 24; m 55; c 3. MEDICAL PHYSICS, BIOPHYSICS. Educ: Univ Calif, AB, 49, PhD(biophys), 55. Prof Exp: Res biophysicist, 55-59, lectr med physics & biophys, 59-61, from asst prof to assoc prof, 61-69, vchmn div med physics, 67-71, chmn, 71-72, PROF MED PHYSICS & BIOPHYS, DONNER LAB, UNIV CALIF, BERKELEY, 69- Res: Structure and function of lipoproteins; atherosclerosis; biophysics of lipid-protein structures. Mailing Add: 108 Donner Lab Univ of Calif Berkeley CA 94720

NICHOLS, AMBROSE REUBEN, JR, b Corvallis, Ore, June 21, 14; m 38; c 3. PHYSICAL CHEMISTRY. Educ: Univ Calif, BS, 35; Univ Wis, PhD(inorg chem), 39. Prof Exp: Asst chem, Univ Wis, 36-39; instr, San Diego State Col, 39-43; res chemist, Radiation Lab, Manhattan Proj, Univ Calif, 43-45; from asst prof to prof chem, San Diego State Col, 45-61; pres, 61-70, PROF CHEM, SONOMA STATE COL, 70- Concurrent Pos: Analyst, Procter & Gamble Co, Ohio, 37; asst physicist, Radio & Sound Lab, US Dept Navy, 42-43; prin chemist, Oak Ridge Nat Lab, 51-52; res collabr, Brookhaven Nat Lab, 70. Mem: Am Chem Soc. Res: Chemistry of manganese; solution chemistry of uranium; autoxidation of manganous hydroxide; high temperature electrochemistry. Mailing Add: Dept of Chem Sonoma State Col Rohnert Park CA 94928

NICHOLS, BARBARA ANN, b Long Beach, Calif, Nov 16, 21. CELL BIOLOGY, EXPERIMENTAL PATHOLOGY. Educ: Univ Calif, Los Angeles, BA, 43; Univ Calif, Berkeley, PhD(zool), 68. Prof Exp: Teaching asst zool, Univ Calif, Berkeley, 63-66, instr, 66-67; res fel path, Sch Med, 68-72, CHIEF ELECTRON MICROS LAB, PROCTOR RES FOUND OPHTHAL, SCH MED, UNIV CALIF, SAN FRANCISCO, 72-, ASST PROF MICROBIOL, UNIV, 72- Concurrent Pos: NIH training grant, Sch Med, Univ Calif, San Francisco, 68-69, NIH fel, 69-71, Nat Tuberc & Respiratory Dis Asn fel, 71-72. Mem: Am Soc Cell Biol; Soc Protozool; Am Thoracic Soc. Res: Use of electron microscopy and cytochemistry to investigate the development and function of monocytes and macrophages in both normal and pathologic states. Mailing Add: Proctor Res Found for Ophthal Univ of Calif Med Sch San Francisco CA 94122

NICHOLS, BUFORD LEE, JR, b Ft Worth, Tex, Dec 21, 31; m; c 3. MEDICINE, NUTRITION. Educ: Baylor Univ, BA, 55, MS, 59; Yale Univ, MD, 60; Am Bd Pediat, dipl; Am Bd Nutrit, dipl. Prof Exp: Instr physiol, Col Med, Baylor Univ, 56-57; instr pediat, Sch Med, Yale Univ, 63-64; instr physiol & pediat, 64-66, asst prof pediat, 66-70, instr physiol, 66-74, ASSOC PROF PEDIAT & COMMUNITY MED & CHIEF SECT NUTRIT & GASTROENTEROL, BAYLOR COL MED, 70-, ASSOC PROF PHYSIOL, 74- Mem: Am Acad Pediat; Am Soc Clin Nutrit; Am Col Nutrit (vpres, 75-76). Res: Environmental effects upon growth and development in the infant, especially alterations in body composition and muscle physiology in malnutrition; diarrhea and infectious diseases. Mailing Add: Baylor Col of Med Dept Pediat Tex Med Ctr Houston TX 77025

NICHOLS, CARL WILLIAM, b State Center, Iowa, June 1, 24; m 55; c 3. PLANT PATHOLOGY. Educ: Mich State Col, BS, 48; Univ Idaho, MS, 49; Univ Calif, PhD(plant path), 52. Prof Exp: Tech plant path, Univ Idaho, 48-49; asst, Univ Calif, 49-51, jr specialist, 51-53; from assoc plant pathologist to plant pathologist, 53-62, prog supvr, 62-66, chief bur plant path, 66-71, chief spec serv & asst to div chief, Div Plant Indust, 71-72, ASST DIR PLANT INDUST, CALIF DEPT FOOD & AGR, 72- Concurrent Pos: Assoc, Exp Sta, Univ Calif, Davis, 67-; mem Nat Plant Bd, 75. Mem: AAAS; Am Phytopath Soc; Am Inst Biol Sci; Int Asn Plant Path. Res: Regulatory plant pest control. Mailing Add: Div of Plant Indust Calif Dept Food & Agr Sacramento CA 95814

NICHOLS, CHARLES, b Chelmsford, Mass, July 3, 14; m 39; c 2. BOTANY. Educ: Dartmouth Col, BA, 36; Harvard Univ, MA, 39, PhD(biol), 41. Prof Exp: Asst biol, Williams Col, 36-37; asst, Harvard Univ, 37-41; head, Dept Sci, New York Inst Educ Blind, 41-47; from asst prof to assoc prof bot, 47-55, PROF BIOL & CHMN DEPT, RIPON COL, 55- Mem: Bot Soc Am. Res: Spontaneous abnormalities of chromosomes; hybridizing of forest tree species; chromosomal aberrations in Allium. Mailing Add: Dept of Biol Ripon Col Ripon WI 54971

NICHOLS, CHESTER ENCELL, b Boston, Mass, Dec 28, 35; m 63; c 2. GEOLOGY. Educ: Cornell Univ, AB, 60; Univ Iowa, MS, 65. Prof Exp: Teaching asst geol, Univ Iowa, 61-62; geologist, US Bur Mines, 62-63; teaching asst geol, Univ Mo-Rolla, 64-66; EXPLOR GEOLOGIST, UNION CARBIDE EXPLOR CORP, 68- Concurrent Pos: V H McNutt res award, 65, 66 & 67. Mem: AAAS; Geol Soc Am; Am Geophys Union; Soc Mining Engrs; fel Am Inst Chemists. Res: Uranium exploration; tungsten exploration; applied geochemistry; exploration geophysics; astrogeology; structural geology; petrology. Mailing Add: Union Carbide Explor Corp PO Box 94 Uranan CO 81436

NICHOLS, CLAYTON WORTHINGTON, JR, organic chemistry, chemical engineering, see 12th edition

NICHOLS, COURTLAND GEOFFREY, b Wilmington, Del, May 16, 34; m 59; c 3. PLANT BREEDING. Educ: Pa State Univ, BS, 56; Univ Wis, PhD(hort, plant path), 63. Prof Exp: Res assoc tomato breeding, Campbell Inst Agr Res, 63-68; PLANT BREEDER, FERRY MORSE SEED CO, 69- Mem: Am Phytopath Soc; Am Soc Hort Sci. Res: Onion breeding, especially inheritance of pink root resistance; tomato breeding, especially heat tolerance in Texas and mechanical harvesting in Illinois and California; carrot breeding, especially hybrids. Mailing Add: Ferry Morse Seed Co Res Div PO Box 8 San Juan Bautista CA 95045

NICHOLS, DAVID EARL, b Covington, Ky, Dec 23, 44; m 66; c 2. MEDICINAL CHEMISTRY. Educ: Univ Cincinnati, BS, 69; Univ Iowa, PhD(med chem), 73. Prof Exp: Fel pharmacol, Univ Iowa, 73-74; ASST PROF MED CHEM, SCH PHARM, PURDUE UNIV, WEST LAFAYETTE, 74- Mem: Am Chem Soc. Res: Synthesis and study of structure-activity relationships of centrally active drugs and neurotransmitter congeners; study of mode of action of psychotomimetic and antipsychotic drugs; synthesis of haptens for immunoassays. Mailing Add: Dept of Med Chem & Pharmacog Purdue Univ Sch Pharm & Pharmacal Sci West Lafayette IN 47907

NICHOLS, DAVIS BETZ, b Carlisle, Ky, Nov 19, 40; m 65; c 1. LASERS, NUCLEAR PHYSICS. Educ: Wheaton Col Ill, BS, 62; Univ Ky, PhD(nuclear physics), 66. Prof Exp: Res assoc nuclear physics, Van de Graaff Lab, Ohio State Univ, 66-67, fel, Van de Graaff Accelerator Lab, 67-68; res fel physics, Kellogg Radiation Lab, Calif Inst Technol, 68-70; MEM STAFF, BOEING AEROSPACE CO, 70- Mem: Am Phys Soc. Res: Pulsed chemical lasers; supersonic electrical-discharge lasers; radiation physics; nuclear spectroscopy; particle-gamma ray correlations; inelastic neutron scattering; on-line computers. Mailing Add: Boeing Aerospace Co Mail Stop 88-16 PO Box 3999 Seattle WA 98124

NICHOLS, DONALD RAY, b Omaha, Nebr, Mar 26, 27; m 48; c 4. ENVIRONMENTAL GEOLOGY. Educ: Univ Nebr, BS, 50. Prof Exp: Geologist & geomorphologist, US Geol Surv, Menlo Park, CA, 51-64, staff geologist, Eng Geol Subdiv, 64-67, supvr geologist, 67-75, CHIEF EARTH SCI APPLNS PROG, US GEOL SURV, RESTON, VA, 75- Concurrent Pos: Vchmn land use planning adv group, Joint Calif Legis Comt Seismic Safety, 70-74. Mem: Geol Soc Am; Arctic Inst NAm; Asn Eng Geol; Brit Glaciol Soc. Res: Engineering and glacial geology; permafrost; geomorphology; earth sciences applications. Mailing Add: 11320 S Shore Rd Reston VA 22090

NICHOLS, DONALD RICHARDSON, b Minneapolis, Minn, Feb 22, 11; wid; c 6. CLINICAL MEDICINE. Educ: Amherst Col, AB, 33; Univ Minn, MD, 38, MS, 42; Am Bd Internal Med, dipl. Prof Exp: From asst prof to prof clin med, Mayo Grad Sch Med, Univ Minn, 48-72, PROF MED, MAYO MED SCH, 72-, SR CONSULT MED, MAYO CLIN, 73- Concurrent Pos: Consult, Mayo Clin, 43-, head sect infectious dis, 61-69, chmn div infectious dis & internal med, 70-73. Mem: Am Thoracic Soc; AMA; Cent Soc Clin Res; fel Am Col Physicians; Infectious Dis Soc Am. Res: Clinical investigation of antibiotic agents and the treatment of infectious diseases. Mailing Add: Mayo Clin 102 Second Ave SW Rochester NY 55902

NICHOLS, DOUGLAS JAMES, b Jamaica, NY, Feb 19, 42; m 64; c 2. PALEONTOLOGY. Educ: NY Univ, BA, 63, MS, 66; Pa State Univ, PhD(geol), 70. Prof Exp: Sci asst micropaleont, Am Mus Natural Hist, 63-65; lectr geol, City Col New York, 65-66; asst prof, Ariz State Univ, 70 & State Univ NY Col Geneseo, 70-74; MEM STAFF, CHEVRON OIL CO, 74- Mem: Paleont Soc; Brit Palaeont Asn; Am Asn Stratig Palynologists; AAAS; Int Asn Plant Taxon. Res: Palynology; general micropaleontology; paleoecology. Mailing Add: Chevron Oil Co Box 599 Denver CO 80201

NICHOLS, ELMER LOREN, nutrition, physiology, see 12th edition

NICHOLS, EUGENE DOUGLAS, b Rovno, Poland, Feb 6, 23; US citizen; m 51. MATHEMATICS. Educ: Univ Chicago, BS, 49; Univ Ill, MA, 53, MEd, 54, PhD(math educ), 56. Prof Exp: Instr math, Roberts Wesleyan Col, 50-51, Univ Ill, 51-53, Urbana High Sch, 53-54 & Univ Ill, 54-56; PROF & LECTR MATH EDUC & MATH, FLA STATE UNIV, 56- Mem: Am Math Soc; Math Asn Am. Res: Foundations of mathematics, with emphasis on linguistics of mathematics. Mailing Add: 717 Johnston Bldg Fla State Univ Tallahassee FL 32306

NICHOLS, GEORGE, JR, b New York, NY, May 15, 22; m 44; c 2. BIOCHEMISTRY, INTERNAL MEDICINE. Educ: Columbia Univ, MD, 45. Prof Exp: Intern med, Presby Hosp, New York, 45-46; asst resident, Peter Bent Brigham Hosp, Boston, 48-50; res biochemist, Baker Clin Res Lab, 50; res fel pediat, 50-52, instr med, 52-53, dir student health & med care prog, 53-55, assoc med, 53-59, consult univ health serv, 55-57, from asst dean & secy fac med to assoc dean, 59-65, from asst prof to assoc clin prof med, 59-65, CLIN PROF MED, HARVARD MED SCH, 65-; DIR, CANCER RES INST, NEW ENG DEACONESS HOSP, 69- Concurrent Pos: Fel, Harvard Med Sch, 48-50; assoc, Peter Bent Brigham Hosp, 53-58, sr assoc, 58-; investr, Howard Hughes Med Inst, 55-57; Markle scholar med sci, 57-62; chief dept, Cambridge City Hosp, 65-68. Mem: Am Physiol Soc; Am Soc Clin Invest; Asn Am Physicians; Am Col Physicians; Endocrine Soc. Res: Electrolyte and water metabolism; endocrinology; biochemistry and physiology of bone. Mailing Add: 51 Commercial Wharf Boston MA 02110

NICHOLS, GEORGE MORRILL, b Seattle, Wash, July 17, 28; m 57; c 3. PHYSICAL CHEMISTRY. Educ: Univ Wash, BS, 52; Ill Inst Technol, PhD(chem), 57. Prof Exp: Res fel solid state chem, Argonne Nat Lab, 57-58; res chemist, E I du Pont de Nemours & Co, 58-62; sr res chemist, Stauffer Chem Co, 63-66; SCIENTIST, RES CTR, BORG-WARNER CORP, 66- Mem: Electrochem Soc; Am Chem Soc. Res: Dielectrics; color centers; inorganic polymers; thermally stable fluids; organophosphorus compositions; electrochemistry; fire retardance. Mailing Add: Borg-Warner Res Ctr Wolf & Algonquin Rds Des Plaines IL 60018

NICHOLS, HARVEY, b Manchester, Eng, Jan 20, 38; m 68. PALYNOLOGY. Educ: Victoria Univ Manchester, BA, 60; Univ Leicester, PhD(palynology), 64. Prof Exp: Res biologist, Yale Univ, 63-64; res assoc palynology, Univ Wis-Madison, 64-69; ASST PROF BIOL, INST ARCTIC & ALPINE RES, UNIV COLO, BOULDER, 69- Mem: Am Quaternary Asn. Res: Late quaternary palynology; history of vegetation and climate, especially in arctic Canada. Mailing Add: Inst of Arctic & Alpine Res Univ of Colo Boulder CO 80302

NICHOLS, HERBERT WAYNE, b Bessemer, Ala, Feb 24, 37. BIOLOGY, PHYCOLOGY. Educ: Univ Ala, BS, 59, MS, 60, PhD(phycol), 63. Prof Exp: Asst prof, 63-65, ASSOC PROF BOT & CO-CHMN DEPT BIOL, WASH UNIV, 65- Concurrent Pos: Instr, Marine Biol Lab, Woods Hole, 64-65; vis lectr, Univ Tex, 63; NSF grant, 65-67; Am Cancer Soc grant, 65; USPHS res grant, 67-; Off Naval Res res grant, 67- Mem: AAAS; Bot Soc Am; Am Micros Soc; Am Phycol Soc (treas, 70-72, vpres, 72-73); Int Phycol Soc. Res: Morphogenetic studies of algae. Mailing Add: Dept of Biol Wash Univ St Louis MO 63130

NICHOLS, JACK LORAN, b Drumheller, Can, Dec 3, 39. BIOCHEMISTRY. Educ: Univ Alta, BS, 60, MS, 63, PhD(biochem), 67. Prof Exp: Res fel, Lab Molecular Biol, Cambridge, Eng, 68-70; ASSOC PROF MICROBIOL, MED CTR, DUKE UNIV, 70- Mem: Am Soc Microbiol; Am Soc Biol Chemists. Res: Structure and function of nucleic acids. Mailing Add: Dept of Microbiol & Immunol Duke Univ Med Ctr Durham NC 27710

NICHOLS, JAMES OTIS, b Rochester, NY, Mar 26, 29; m 53; c 3. FOREST ENTOMOLOGY. Educ: State Univ NY Col Forestry, BS, 51, MS, 57. Prof Exp: Entomologist, Dept Agr, 57-61 & Dept Environ Resources, 62-74, CHIEF, DIV FOREST PEST MGT, COMMONWEALTH OF PA, 74- Mem: Soc Am Foresters; Entom Soc Am. Res: Forest insects. Mailing Add: Pa Dept Environ Resources Harrisburg Int Airport Middletown PA 17057

NICHOLS, JAMES R, dairy science, see 12th edition

NICHOLS, JAMES RANDALL, b Wilmington, Del, Mar 10, 31; m 54; c 3. PHYSICAL CHEMISTRY. Educ: Univ Del, BS, 53; Pa State Univ, MS, 57, PhD(fuel technol), 61. Prof Exp: Sr chemist, Chem Res Dept, Atlas Chem Industs, Inc, 61-67, res chemist, 67-72, RES CHEMIST, CHEM RES DEPT, ICI AM, INC, 72- Mem: AAAS; Am Chem Soc. Res: Physical-chemical structure studies on activated carbon; chemical nature of carbon surface; characterization of supported catalysts; thermal analysis; x-ray diffraction and electron microscopy of polymer and pharmaceutical products. Mailing Add: 117 Alders Dr Blue Rock Manor Wilmington DE 19803

NICHOLS, JAMES ROSS, b Kansas City, Mo, June 26, 44; m 65; c 2. ZOOLOGY. Educ: Abilene Christian Col, BS, 66; Univ Mich, Ann Arbor, MS, 68; Univ Mo-Columbia, PhD(zool), 73. Prof Exp: Asst prof, 72-75, ASSOC PROF BIOL, UNIV CENT ARK, 75- Mem: Am Soc Zoologists; Am Inst Biol Sci; AAAS; Nat Asn Biol Teachers. Res: Cellular and comparative physiology, especially ionic and water regulation; effects of antibiotics and endocrine control; temperature and light effects. Mailing Add: Dept of Biol Univ of Cent Ark Conway AR 72032

NICHOLS, JAMES T, b Salina Kans, Jan 5, 30; m 59; c 6. RANGE MANAGEMENT, AGRONOMY. Educ: Ft Hays Kans State Col, BS, 60, MS, 61; Univ Wyo, PhD(range mgt), 64. Prof Exp: Asst prof range mgt, SDak State Univ, 64-69; assoc prof, 69-74, PROF AGRON, UNIV NEBR, 74- Mem: Soc Range Mgt; Soc Agron; Am Forage & Grassland Coun. Res: Range improvement through grazing management and range renovation; soil vegetation relationship; irrigated pasture production; pasture management. Mailing Add: Dept of Agron Univ Nebr N Platte Sta North Platte NE 69101

NICHOLS, JOHN C, b Chicago, Ill, Feb 28, 39; m 61; c 2. MATHEMATICS. Educ: Blackburn Col, BA, 60; Southern Ill Univ, MS, 62; Univ Iowa, PhD(math), 66. Prof Exp: Asst prof math, Monmouth Col, Ill, 66-71; ASSOC PROF MATH, THIEL COL, 71- Mem: Am Math Soc. Res: Homological algebra as related to the theory of local rings. Mailing Add: Dept of Math Thiel Col Greenville PA 16125

NICHOLS, JOSEPH, b New York, NY, July 9, 17; m 51; c 2. ORGANIC CHEMISTRY. Educ: City Col New York, BS, 38; Univ Minn, PhD(org chem), 43. Prof Exp: Asst physiol chem, Univ Minn, 39-42, Nat Defense Res Coun res fel, 42-43; res chemist, Interchem Corp, NY, 43-49; chief, Dept Org Chem, Ethicon, Inc, 51-57, assoc dir res, 57-65, dir res, Collagen Prod, 65-68; PRES & DIR RES, PRINCETON BIOMEDIX INC, 68- Concurrent Pos: With Swiss Fed Inst Technol, 49-50. Mem: AAAS; Am Chem Soc; NY Acad Sci; Am Oil Chem Soc; Soc Cosmetic Chem. Res: Quinones; fatty acids; synthetic resins; medicinal products; proteins; collagen. Mailing Add: Princeton Biomedix Inc Alexander Rd Princeton NJ 08540

NICHOLS, JOSEPH CALDWELL, b Brooklyn, NY, Oct 21, 43. MATHEMATICS. Educ: Wesleyan Univ, BA, 65; Duke Univ, PhD(math), 70. Prof Exp: ASSOC PROF MATH, RADFORD COL, 70- Mem: Am Math Soc; Math Asn Am; Soc Indust & Appl Math. Res: Point set topology; applied mathematics. Mailing Add: Dept of Math Radford Col Radford VA 24141

NICHOLS, KATHRYN MARION, b Santa Monica, Calif, Apr 30, 46. STRATIGRAPHY, SEDIMENTARY PETROLOGY. Educ: Univ Calif, Riverside, BS, 68; Stanford Univ, PhD(geol), 72. Prof Exp: NSF res asst, 70-72, NSF res assoc, 73-74, LECTR GEOL, STANFORD UNIV, 75- Concurrent Pos: Petrol Res Fund fel, 74-76; Nat Res Coun fel, 76-77. Mem: Geol Soc Am; Soc Econ Paleontologists & Mineralogists; Am Asn Petrol Geologists. Res: Sedimentology of carbonate rocks; lower Mesozoic stratigraphy and paleontology of the Great Basin. Mailing Add: Dept of Geol Stanford Univ Stanford CA 94305

NICHOLS, KENNETH E, b Brems, Ind, Dec 17, 20; m 41; c 4. PLANT PHYSIOLOGY. Educ: Valparaiso Univ, AB, 49; Univ Chicago, MS, 53, PhD(physiol), 61. Prof Exp: Assoc prof biol, 53-67, PROF BIOL, VALPARAISO UNIV, 67- Mem: Am Soc Plant Physiol; Phycol Soc Am; Int Phycol Soc. Res: Biosynthesis of plant pigments. Mailing Add: Dept of Biol Valparaiso Univ Valparaiso IN 46383

NICHOLS, MAYNARD M, b Patchogue, NY, Jan 13, 28; m. MARINE GEOLOGY, OCEANOGRAPHY. Educ: Columbia Univ, BS, 51; Univ Calif, MS, 59, PhD(geol), 65. Prof Exp: ASSOC MARINE SCIENTIST, VA INST MARINE SCI, UNIV VA, 61-, ASSOC PROF MARINE SCI, UNIV, 69-; COL WILLIAM & MARY, 61- Res: Sedimentation; stratigraphy; ecology of Foraminifera. Mailing Add: Dept of Marine Sci Col of William & Mary Williamsburg VA 23185

NICHOLS, NATHAN LANKFORD, b Jackson, Mich, Nov 16, 17; m 41; c 5. PHYSICS. Educ: Western Mich Univ, AB, 39; Univ Mich, MS, 45; Mich State Univ, PhD(physics), 53. Prof Exp: Teacher high sch, SDak, 39-40 & Mich, 40-43; instr physics, Ill Col, 43-44 & Univ Mich, 44-45; asst, Mich State Univ, 46-48; prof & head dept, Alma Col, 49-55; assoc prof, 55-61, PROF PHYSICS, WESTERN MICH UNIV, 61- Mem: Optical Soc Am; Am Asn Physics Teachers. Res: Near infrared spectroscopy; optics. Mailing Add: Dept of Physics Western Mich Univ Kalamazoo MI 49001

NICHOLS, NATHANIEL BURGESS, b Mt Pleasant, Mich, Dec 31, 14; m 38; c 2. ENGINEERING PHYSICS. Educ: Univ Cent Mich, BS, 36; Univ Mich, MS, 37. Hon Degrees: DSc, Univ Cent Mich, 64; DSc, Case Western Reserve Univ, 68. Prof Exp: Physicist, Dow Chem Co, Mich, 36-38; asst physics, Univ Mich, 37-40; res physicist, Taylor Instrument Co, NY, 40-42, res dir, 46-50; group leader, Radiation Lab, Mass Inst Technol, 42-46; prof elec eng, Univ Minn, 50-51; mgr, Res Div & asst vpres, Raytheon Mfg Co, 51-57; vpres & chief engr, Taylor Instrument Co, 57-63; dir sensing info subdiv, Aerospace Corp, San Bernardino, 63-67, dir guid & control subdiv, 67-71, ASST GEN MGR, TECHNOL DIV, AEROSPACE CORP, LOS ANGELES, 71- Honors & Awards: Excellence in Document Award, Instrument Soc Am, 63; Rufus Oldenburger Award, Am Soc Mech Engrs, 69; Alfred F Sperry Medal Award, Instrument Soc Am, 74. Mem: Fel AAAS; fel Instrument Soc Am; Am Chem

Soc; fel Am Soc Mech Eng; fel Inst Elec & Electronics Eng. Res: Pneumatic, hydraulic and electric servomechanisms; gunfire control; temperature and pressure measurement; automatic control; electrochemical analysis; missile guidance. Mailing Add: Technol Div Aerospace Corp Box 92957 Los Angeles CA 90009

NICHOLS, PARKS MONTGOMERY, b Garrettsville, Ohio, July 20, 12; m 40. RUBBER CHEMISTRY. Educ: Hiram Col, AB, 34; Univ Buffalo, PhD(chem), 38. Prof Exp: Chemist, Chateau Gay, Ltd, NY, 38-41; chemist, US Rubber Co, 41-71; CONSULT, 73- Mem: Am Chem Soc. Res: Rubber technology. Mailing Add: 4166 Devonshire Rd Detroit MI 48224

NICHOLS, PETER LUKE, chemistry, see 12th edition

NICHOLS, PHILIP RAY, b Tupelo, Miss, Dec 12, 46; m 68. ENTOMOLOGY. Educ: Delta State Col, BS, 69, MEd, 70; Miss State Univ, PhD(entom), 75. Prof Exp: ENTOMOLOGIST, MIGHTY NAT EXTERMINATORS INC, 75- Mem: Entom Soc Am. Res: Upgrade the pest control business by means of teaching and training key personnel. Mailing Add: Mighty Nat Exterminators Inc Peters Rd Ft Lauderdale FL 33314

NICHOLS, ROBERT LESLIE, b Boston, Mass, June 10, 04; m 35; c 1. GEOLOGY. Educ: Tufts Univ, BS, 26; Harvard Univ, MA, 30, PhD(geol), 40. Prof Exp: Master, Montpelier Sem, 26-27 & Milton Acad, 27-28; asst geol, Harvard Univ, 28-30; instr, 29-36, from asst prof to prof, 36-74, actg head dept geol, 36-40, head dept, 40-69, EMER PROF GEOL, TUFTS UNIV, 74- Concurrent Pos: Ranger naturalist, Nat Park Serv, 30-31; instr, Boston Adult Educ Ctr, 33-35; res assoc, Mus Northern Ariz, 40-41; from asst geologist to assoc geologist, US Geol Surv, 40-74; geologist, Ronne Antarctic Res Exped, 47-48; geologist, US Navy Arctic Task Force, 48; geologist, Am Geog Soc-Nat Hist Mus Arg Exped, Patagonia, 49; mem, Juneau, Alaska Ice Cap Exped, 50; leader, US Army Transp Corps Exped, Inglefieldland, Greenland, 53; geologist, Oper Deep Freeze, Antarctica, US Navy, 57-58 & Int Geophys Year, 58-59; leader, Tufts Col-NSF Antarctic Exped, 59-60 & 60-61, Northwest Greenland Exped, 63 & 65; field worker, NMex, Wash, Ore, New Eng, Alaska, Antarctica, Arctic & Patogonia. Mem: AAAS (secy, Geol & Geog Sect, 55-56); fel Geol Soc Am; fel Am Geog Soc; Nat Asn Geol Teachers (vpres, 59); fel Royal Geog Soc. Res: Geomorphology; vulcanology; high-alumina clay. Mailing Add: Dept of Geol Tufts Univ Medford MA 02155

NICHOLS, ROBERT TED, b Lewis, Iowa, Dec 30, 25; m 55; c 3. PHYSICS. Educ: Iowa State Univ, BS, 50, MS, 55, PhD(physics), 60. Prof Exp: Asst beta & gamma ray spectros, Ames Lab, AEC, 50-51; instr physics & math, Bethel Col, Minn, 54-56; asst prof physics, Gustavus Adolphus Col, 60-63; mem tech staff & physicist, Hughes Res Labs, Calif, 63-66; assoc prof physics, 66-74, PROF PHYSICS & CHMN DEPT, CALIF LUTHERAN COL, 74- Concurrent Pos: Consult mil prod group, Minneapolis-Honeywell Regulator Co, 62-63. Mem: Am Asn Physics Teachers. Res: Radiation physics; linear energy transfer for dosimetry; small angle beta scattering; beta and gamma ray spectroscopy. Mailing Add: Dept of Physics Calif Lutheran Col Thousand Oaks CA 91360

NICHOLS, ROGER LOYD, b Waverly, Iowa, Apr 29, 26; m 49; c 3. MICROBIOLOGY. Educ: Cornell Col, BA, 48; Univ Iowa, MD, 53. Hon Degrees: MA, Harvard Univ, 71. Prof Exp: Res fel med, Sch Med, 55-56, res fel microbiol, Sch Pub Health, 57-60, res assoc, 61-63, from asst prof to assoc prof appl microbiol, 63-69, IRENE HEINZ GIVEN PROF MICROBIOL & HEAD DEPT, SCH PUB HEALTH, HARVARD UNIV, 69-; PRES, UNIV ASSOCS INT HEALTH, INC, 73- Concurrent Pos: Med consult, Ministry Health, Gov of Saudi Arabia & supreme Supreme Coun Univs of Saudi Arabia, 74-75. Mem: AAAS; Am Fedn Clin Res. Res: Field and laboratory research in trachoma; clinical research in new antibiotics; trachoma; venereal diseases; international health. Mailing Add: Dept of Microbiol Harvard Univ Sch Pub Health Boston MA 02115

NICHOLS, ROY ELWYN, b Leonardsville, NY, July 10, 09; m 32; c 2. VETERINARY PHYSIOLOGY. Educ: Univ Toronto, DVM, 33, DVSc, 43; Ohio State Univ, MSc, 34, PhD(vet hemat), 41. Prof Exp: Asst vet surg, Col Vet Med, Ohio State Univ, 34-35, instr, 35-41; asst prof col agr & assoc, Agr Exp Sta, Purdue Univ, 41-42 & 45-47; dean col vet med, State Col Wash, 47-50; lectr & res assoc, 50-51, prof vet sci, 51-72, EMER PROF VET SCI, UNIV WIS-MADISON, 72- Mem: Am Soc Animal Sci; Am Vet Med Asn; Am Dairy Sci Asn; Conf Res Workers Animal Dis. Res: Veterinary hematology, physiology and surgery; instrument for photographic recording of erythrocyte sedimentation; ruminology. Mailing Add: 5605 Taychopera Rd Madison WI 53705

NICHOLS, RUDOLPH HENRY, JR, b Bellaire, Mich, Dec 6, 11; m 39; c 1. ACOUSTICS. Educ: Hope Col, AB, 32; Univ Mich, AM, 33, PhD(physics), 39. Prof Exp: Lab asst physics, Univ Mich, 35-36, asst acoust & physics, 36-38, res assoc, 39; physicist, Owens-Ill Glass Co, 40-41; spec res assoc, Cruft Lab, Harvard Univ, 41-44, from asst dir to assoc dir electro-acoust lab, 44-46; MEM TECH STAFF, BELL LABS, 46- Mem: Fel Acoust Soc Am. Res: Audio communications; sound reduction in vehicles; vibration isolation; acoustical materials; hearing and hearing aids; underwater acoustics. Mailing Add: Bell Labs Whippany NJ 07981

NICHOLS, SAMUEL HARDING, JR, b Danville, Ky, May 5, 14; m 49; c 2. ORGANIC CHEMISTRY. Educ: Centre Col, AB, 34; Ohio State Univ, MS, 37, PhD(chem), 39. Prof Exp: Instr chem, Univ Vt, 39-42 & Case Inst, 42-44; instr, 44-45, from asst prof to prof, 45-74, EMER PROF CHEM, AUBURN UNIV, 74- Mem: Am Chem Soc. Res: Carbohydrate chemistry; qualitative organic analysis. Mailing Add: Dept of Chem Auburn Univ Auburn AL 36830

NICHOLS, WARREN WESLEY, b Collingswood, NJ, May 16, 29; m 53; c 3. CYTOGENETICS, PEDIATRICS. Educ: Rutgers Univ, BS, 50; Jefferson Med Col, MD, 54; Univ Lund, PhD(med cytogenetics), 64, DrPhil, 66. Prof Exp: Intern, Cooper Hosp, Camden, NJ, 54-55; fel, Children's Hosp Philadelphia, 55-56; chief pediat, Lake Charles AFB, 57-59; assoc, Sch Med & instr, Grad Sch Med, 59-67, from asst prof to assoc prof, Sch Med, 64-73, PROF HUMAN GENETICS & PEDIAT, SCH MED, UNIV PA, 73-; ASST DIR, INST MED RES, 63-, MEM, 66-, S EMLEN STOKES PROF GENETICS, 75- Concurrent Pos: Chief med staff, Camden Munic Hosp Contagious Dis, 59-61; assoc mem, SJersey Med Res Found, 59-65; asst physician, Children's Hosp Philadelphia, 59-65, assoc physician & assoc hematologist, 65-73, sr physician, Div Metab & Genetics, 73-; mem pediat staff, Cooper Hosp, Camden, NJ, 59-65, consult staff, 65-; consult staff, Our Lady of Lourdes Hosp, Camden, 59-66, assoc pediat, 66-; NIH res career develop award, 63-72; assoc prof, Univ Lund, 66-; mem, Human Cytogenetics Study Group, 66-; consult secy comn pesticides & their relationship to environ health, Dept Health, Educ & Welfare, 69; mem human embryol & develop study sect, NIH, 74-; mem panel, US-Japan Coop Med Sci Prog, 72-74. Mem: AAAS; Environ Mutagen Soc (secy, 73-); AMA; Soc Pediat Res; Genetics Soc Am. Res: Chromosome and gene mutations induced by viruses and other agents. Mailing Add: Inst for Med Res Copewood St Camden NJ 08103

NICHOLS, WILLIAM HERBERT, b Cleveland, Ohio, Mar 15, 28. STATISTICAL MECHANICS. Educ: West Baden Col, AB, 50; Mass Inst Technol, SB, 55, PhD(physics), 58; Weston Col, STL, 61. Prof Exp: Instr physics & algebra, Loyola Acad, 52-53; res asst physics, Inst Tehoret Physics, Univ Vienna, 62-63; asst prof, Univ Detroit, 63-67; from asst prof to assoc prof, 67-74, PROF PHYSICS, JOHN CARROLL UNIV, 74- Mem: Am Phys Soc; Phys Soc Japan. Res: Statistical mechanics of transport properties, especially in liquids; liquid-gas critical-point phenomena. Mailing Add: Dept of Physics John Carroll Univ Cleveland OH 44118

NICHOLS, WILLIAM KENNETH, b Seattle, Wash, Sept 25, 43; m 73. PHARMACOLOGY. Educ: Univ Wash, BS, 66; Univ Minn, PhD(phamacol), 71. Prof Exp: ASST PROF PHARMACOL, COL PHARM & COL MED, UNIV UTAH, 71- Res: Biochemical pharmacology; endocrinology; diabetes; immunology; leukocyte metabolism and function; cyclic nucleotides; allergy; cancer. Mailing Add: Dept of Biopharmaceut Sci Univ of Utah Col of Pharm Salt Lake City UT 84112

NICHOLS, WILMER WAYNE, b Booneville, Miss, Aug 12, 34; m 62; c 2. CARDIOVASCULAR PHYSIOLOGY. Educ: Delta State Col, BS, 60; Univ Southern Miss, MS, 66; Univ Ala, PhD(physiol, biophys), 70. Prof Exp: Instr physiol, Med Sch, Univ Ala, 70-71; asst prof med physics, Inst Med Physics, Holland, 71-72; asst prof physiol, Sch Med, Johns Hopkins Univ, 72-74; ASST PROF PHYSIOL & ASSOC PROF CLIN CARDIOL, COL MED, UNIV FLA, 74- Concurrent Pos: Consult, Millar Instruments Inc, Tex, 74-; mem, Cardiovasc Catheter Standards Subcomt, 74- Mem: Am Fedn Clin Res; Am Physiol Soc; Cardiol Soc Holland; Am Heart Asn; Asn Advan Med Instrumentation. Res: Pulsatile hemodynamics in man. Mailing Add: Dept of Med Div of Cardiol Shands Teaching Hosp Box 732 Gainesville FL 32610

NICHOLS-DRISCOLL, JEAN ANN, b Annapolis, Md, July 27, 47; m 72. BIOLOGICAL OCEANOGRAPHY. Educ: Pomona Col, BA, 69; Mass Inst Technol, PhD(oceanog), 75. Prof Exp: ASST PROF BIOL, SOUTHEASTERN MASS UNIV, 75- Mem: Sigma Xi; Am Asn Limnol & Oceanog. Res: Marine benthic ecology; benthic production; nematode ecology. Mailing Add: Dept of Biol Southeastern Mass Univ North Dartmouth MA 20747

NICHOLSON, ARNOLD EUGENE, b Jasper, Ind, Apr 9, 30. PHARMACEUTICAL CHEMISTRY. Educ: Purdue Univ, BS, 52, MS, 54, PhD(pharmaceut chem), 56. Prof Exp: Sr res chemist, Smith Kline & French Labs, 56-64; asst to tech dir pharmaceut develop, Stuart Div, Atlas Chem Industs, Inc, 64-72; MEM STAFF, DIV TRAVENOL LAB, INC, 72- Mem: Am Chem Soc; Am Pharmaceut Asn; Am Soc Hosp Pharmacists. Res: Pharmaceutical research and development. Mailing Add: Div Travenol Lab Inc 3300 Hyland Costa Mesa CA 92626

NICHOLSON, BRUCE LEE, b Baltimore, Md, Feb 13, 43; m 65. MICROBIOLOGY. Educ: Univ Md, College Park, BS, 65, PhD(microbiol), 69. Prof Exp: Asst prof microbiol, 69-74, ASSOC PROF MICROBIOL & ZOOL, UNIV MAINE, 74- Mem: Am Soc Microbiol; AAAS; Wildlife Dis Asn; Tissue Culture Asn; Am Fisheries Soc. Res: Biochemical and biophysical characteristics of fish viruses and their interactions with susceptible cells. Mailing Add: Dept of Microbiol Univ of Maine Orono ME 04473

NICHOLSON, CHARLES (GODFREY), b Great Malvern, Eng, Feb 8, 42. NEUROBIOLOGY. Educ: Univ Birmingham, BSc, 63; Univ Keele, PhD(commun), 68. Prof Exp: Sci officer math physics, UK Atomic Energy Auth, 63-65; vis investr neurobiol, Inst Biomed Res, Am Med Asn-Educ & Res Found, Ill, 67-69, asst mem, 69-70; asst prof 70-73, ASSOC PROF NEUROBIOL, UNIV IOWA, 73- Mem: NY Acad Sci; Soc Neurosci; Inst Elec & Electronics Eng. Res: Neuronal organization and function as revealed by anatomical, physiological and mathematical analysis with emphasis on cerebellum; electrical and ionic properties of brain cell microenvironment. Mailing Add: Dept of Physiol & Biophys Univ of Iowa Oakdale IA 52319

NICHOLSON, CLARA K, b Pittsfield, Mass, m 36; c 4. ANTHROPOLOGY. Educ: Syracuse Univ, BA, 36, PhD(anthrop), 65; Univ Mass, MA, 57. Prof Exp: Statistician, Gen Elec Co, 37-40, draftsman, 40-42; liaison officer, UNRRA, 54-56; exec dir Camp Fire Girls, Greater Worcester Coun, Mass, 57-60; lectr sociol & anthrop, Univ Md, 61-62; assoc prof, Lock Haven State Col, 62-65; asst prof, Utica Col, Syracuse, 65-68; assoc prof anthrop, Kirkland Col, 68-69; assoc prof, 69-70, PROF ANTHROP, UTICA COL, SYRACUSE UNIV, 70- Concurrent Pos: Sr res scientist, Syracuse Psychiat Hosp, 66-; consult & lectr, Geriatrics Prog Utica State Hosp, 72-73; consult-lectr, Willard Psychiat Ctr, 74-; grant, HEW, Nat Inst Mental Health, 75; dir, Nat Inst Mental Health, elderly, 74- Mem: Fel Am Anthrop Soc. Res: Cultural change; introduction of Islam into Sumatra & Java; role playing and mental health; evaluation of the effect of the introduction of the team approach to treatment of mentally ill. Mailing Add: Dept of Anthrop Utica Col Syracuse Univ Utica NY 13502

NICHOLSON, D ALLAN, b Waterloo, Iowa, June 22, 39; m 62; c 2. ORGANOMETALLIC CHEMISTRY, ANALYTICAL CHEMISTRY. Educ: Cornell Col, BA, 60; Northwestern Univ, PhD(inorg chem), 65. Prof Exp: Staff chemist, 65-70, sect head anal chem, 70-72, sect head toxicol, 72-74, ASSOC DIR SOAP & TOILET GOODS TECHNOL DIV, PROCTER & GAMBLE CO, 74- Mem: Am Chem Soc. Res: Non-transition metals such as germanium, silicon; organophosphorus chemistry. Mailing Add: 472 Lakeridge Ave Cincinnati OH 45231

NICHOLSON, DANIEL ELBERT, b Waco, Tex, Sept 1, 26; m 51. PHYSICAL CHEMISTRY. Educ: Baylor Univ, BS, 46; Univ Tex, MA, 48, PhD(phys chem), 50. Prof Exp: Tutor, Univ Tex, 46-48; chemist, Oak Ridge Nat Lab, 50-51; from res chemist to sr res chemist, Humble Oil & Ref Co, 52-61; staff mem, Los Alamos Sci Lab, 62; res assoc, Richfield Oil Corp, 63-67; sect head optical spectros, Micro Data Opers, 67-68, SECT MGR, ELECTRO-OPTICAL SYSTS, INC, 68- Res: Kinetics of hydrocarbon decomposition; precision calorimetry; infrared spectroscopy of absorbed species; electrical and magnetic properties of catalysts; Raman spectroscopy; thermodynamic properties; catalytic studies of hydrode-sulfurization; hydrogenation of aromatic hydrocarbons. Mailing Add: 2209 California Blvd San Marino CA 91108

NICHOLSON, DAVID WILLIAM, b Boston, Mass, May 9, 44. MECHANICS. Educ: Mass Inst Technol, SB, 66; Yale Univ, MS, 67, PhD(mech), 71. Prof Exp: SR PHYSICIST MECH, GOODYEAR TIRE & RUBBER CO, 71- Concurrent Pos: Reviewer, Appl Mech Revs, 74. Mem: Assoc Am Acad Mech. Res: Plasticity; viscoelasticity; fracture mechanics; tire mechanics; numerical analysis. Mailing Add: Goodyear Res 142 Goodyear Blvd Akron OH 44240

NICHOLSON, DONALD PAUL, b Pershing, Iowa, Jan 3, 30; m 54; c 2. MICROBIOLOGY, VIROLOGY. Educ: Iowa State Col, BS, 51; Univ SDak, BSM, 59; Univ Iowa, MD, 61. Prof Exp: Asst radiation res, 56-57, clin intern med, 61-62, USPHS fel, 62-65, from instr to asst prof microbiol, 62-68, asst dir clin lab serv, 68-71, ASST PROF PATH & SUPVR CLIN MICROBIOL, UNIV IOWA, 71- Mem: AAAS; Am Soc Microbiol; AMA. Res: Areas related to clinical microbiology. Mailing Add: Rm 273 MRC Univ of Iowa Dept of Path Iowa City IA 52242

NICHOLSON, DOUGLAS GILLISON, b Joliet, Ill, Dec 29, 08; m 32; c 2. INORGANIC CHEMISTRY. Educ: Univ Ill, BS, 30, MS, 31, PhD(inorg chem), 34. Prof Exp: Asst chem, Univ Ill, 30-34, instr, 35-37, assoc, 37-42; res chemist, E I du Pont de Nemours & Co, 34-35; assoc prof chem, Univ Pittsburgh, 45-49; dir ed serv, Fisher Sci Co, 49-53; assoc prof inorg chem, 53-57, prof chem & chmn dept, 57-74, EMER CHMN DEPT CHEM, E TENN STATE UNIV, 74- Mem: Am Chem Soc. Res: White pigments, especially titanium dioxide; titanium compounds; reactions in non-aqueous solvents. Mailing Add: Dept of Chem ETenn State Univ Johnson City TN 37601

NICHOLSON, DWIGHT ROY, b Racine, Wis, Oct 3, 47; m 69. PLASMA PHYSICS. Educ: Univ Wis-Madison, BS, 69; Univ Calif, Berkeley, PhD(plasma physics), 75. Prof Exp: RES ASSOC PLASMA PHYSICS, ASTRO-GEOPHYS DEPT, UNIV COLO, BOULDER, 75- Mem: AAAS; Am Phys Soc. Res: Plasma theory; nonlinear waves in plasma; behavior of plasma in the presence of intense electromagnetic radiation. Mailing Add: Astro-Geophys Dept Univ of Colo Boulder CO 80302

NICHOLSON, EUGENE HAINES, b St Louis, Mo, Nov 10, 07. MATHEMATICS, PHYSICS. Educ: Wash Univ, BS, 31, MS, 37, PhD(math, physics), 41. Prof Exp: Elec engr, Union Elec Co, 41-60; CONSULT, 60- Mem: AAAS; Am Math Soc. Res: Applied mathematics; mathematical physics; energy conversion; systems engineering; nuclear power. Mailing Add: 5232 Lansdowne Ave St Louis MO 63109

NICHOLSON, GEOFFREY CHARLES, b Durham, Eng, Mar 6, 38; m 62. INORGANIC CHEMISTRY, CERAMICS. Educ: Imp Col, Univ London, BSc, 60, PhD(chem) & DIC, 63. Prof Exp: Sr chemist, 63-69, res specialist visual prod, 69-71, supvr film prod, 71-74, TECH DIR, COM TAPE DIV, MINN MINING & MFG CO, 74- Mem: Am Ceramic Soc; fel The Chem Soc. Res: Properties of materials at high temperatures; theories of sintering applied to oxides; decomposition of materials using thermogravimetric analysis. Mailing Add: Com Tape Lab Minn Mining & Mfg Co 230-5210 St Paul MN 55101

NICHOLSON, HARRY PAGE, b Crosby, Minn, Nov 25, 13; m 38; c 2. ENTOMOLOGY. Educ: Minn State Teachers Col, BE, 37; Univ Minn, MS, 41, PhD, 49. Prof Exp: Jr entomologist malaria control, USPHS, Va, 42, asst entomologist, 42-43, asst sanitarian, 43-45, tech develop div, Commun Dis Ctr, 45, sr asst sanitarian, 45-48, scientist, 48-52, southeast drainage basins, Div Water Pollution Control, Ga, 50, sr scientist, 52-53, lower Miss & West Gulf Drainage Basins, Ark, 53, Mo Drainage Basin, 54-58 & Div Water Supply & Pollution Control, Ga, 58, scientist dir in chg insecticide pollution invests, 58-64, scientist dir in chg rural runoff characterization, Southeast Water Lab, 64-66, supvry sanit engr in chg water contaminants characterization res, US Dept Interior, 66-69, chief agr & indust pollution control res prog, Environ Protection Agency, 69-74; chief, Agro-Environ Systs Br, Southeast Environ Res Lab, 74-75; ASSOC DIR RURAL LANDS RES, ENVIRON RES LAB, GA, 75- Concurrent Pos: Fed Water Pollution Control Admin rep, interagency res adv comt aquatic weed control, Army Corps Engrs, 60-70; mem subcomt eval pesticide wildlife probs, Nat Acad Sci, 61; ad hoc comt use of herbicides in aquatic sites, Depts Interior & Agr, 66-70; assoc prof, Univ Ga, 67- Mem: AAAS; Entom Soc Am; Water Pollution Control Fedn; Am Soc Limnol & Oceanog; Wildlife Soc. Res: Grasshopper ecology; taxonomy and biology of Simuliidae; control of rodents and ectoparasites; comparative resistance of new construction materials to penetration by rats; water quality, water pollution. Mailing Add: Environ Res Lab Environ Protection Agency Athens GA 30601

NICHOLSON, HAYDEN COLER, b Redwood Falls, Minn, Feb 4, 04; m 29; c 1. PHYSIOLOGY. Educ: Univ Mich, AB, 25, MS, 27, MD, 29. Prof Exp: Asst physiol, Univ Mich, 25-29, from instr to assoc prof, 29-46; prof assoc, Comt Growth, Nat Res Coun, 46-47, exec secy, 47-50; dean sch med, Univ Ark, 50-55; exec dir, Hosp Coun Greater New York, 55-62; dean sch med, Univ Miami, 62-68, vpres med affairs, 67-68; dir dept undergrad med educ, AMA, 68-70, asst dir undergrad med educ, 70-72; RETIRED. Concurrent Pos: Provost med affairs, Univ Ark, 54-55. Mem: AAAS; Am Physiol Soc; Soc Exp Biol & Med; AMA; Aerospace Med Asn. Res: Chemical regulation of respiration; localization and mechanism of action of respiratory center. Mailing Add: 22W647 Elmwood Dr Glen Ellyn IL 60137

NICHOLSON, HOWARD WHITE, JR, b Brooklyn, NY, Dec 18, 44; m 71. HIGH ENERGY PHYSICS. Educ: Hamilton Col, BA, 66; Mass Inst Technol, BA, 66; Calif Inst Technol, PhD(physics), 71. Prof Exp: ASST PROF PHYSICS, MT HOLYOKE COL, 71- Concurrent Pos: Res Corp grant, Mt Holyoke Col, 72- Mem: Am Phys Soc. Res: Experimental high energy particle physics. Mailing Add: Dept of Physics Shattuck Hall Mt Holyoke Col South Hadley MA 01075

NICHOLSON, HUGH HAMPSON, b Lloydminster, Sask, Sept 30, 23; m 64; c 2. ANIMAL NUTRITION, ANIMAL GENETICS. Educ: Univ BC, BSA, 50, MSA, 54; Ore State Univ, PhD(genetics), 58. Prof Exp: Res officer animal nutrit & range mgt, Can Res Sta, BC, 50-62; ASSOC PROF BEEF CATTLE NUTRIT & MGT, UNIV SASK, 62- Mem: Soc Range Mgt; Agr Inst Can; Can Soc Animal Prod; Nutrit Soc Can. Res: Productivity of range land; irrigated pasture production with beef cattle; record of performances of beef cattle; nutrition and management of feed lot cattle; mill processing of beef cattle feeds. Mailing Add: Dept of Animal Sci Univ of Sask Saskatoon SK Can

NICHOLSON, ISADORE, b Philadelphia, Pa, Apr 28, 25. ORGANIC CHEMISTRY. Educ: Temple Univ, BA, 49; Rutgers Univ, MSc, 52, PhD(chem), 54. Prof Exp: Res instr chem, Univ Minn, 54-55; sr res chemist, US Rubber Co, 55-57, sr res specialist, 57-58; from asst prof to assoc prof chem, 58-68, PROF CHEM, C W POST COL, LONG ISLAND UNIV, 68- Res: Theoretical and biological organic chemistry. Mailing Add: Dept of Chem C W Post Col Long Island Univ Greenvale NY 11548

NICHOLSON, J W G, b Crapaud, PEI, Jan 19, 31; c 4. ANIMAL NUTRITION. Educ: McGill Univ, BS, 51; Cornell Univ, MS, 56, PhD(animal nutrit), 59. Prof Exp: RES SCIENTIST ANIMAL NUTRIT, CAN DEPT AGR, 51- Mem: Can Soc Animal Sci; Agr Inst Can; Am Soc Animal Sci; Am Dairy Sci Asn. Res: Nutrition of ruminant animals. Mailing Add: Res Sta Agr Can PO Box 20280 Fredericton NB Can

NICHOLSON, JOHN ANGUS, b Louisville, Ky, Feb 3, 33; m 66; c 2. ORGANIC CHEMISTRY, PHARMACOLOGY. Educ: Notre Dame Univ, BS, 55; Univ Louisville, PhD(pharmacol), 69. Prof Exp: Chemist, Brown & Williamson Tobacco Co, Ky, 60-64; instr pharmacol, 69-70, ASST PROF PHARMACOL, SCH MED, UNIV LOUISVILLE, 70- Mem: Am Chem Soc; The Chem Soc. Res: Natural product research on tobacco and biologically active plants; cardiovascular pharmacology. Mailing Add: Dept of Pharmacol Univ of Louisville Sch of Med Louisville KY 40202

NICHOLSON, JOHN FRASER, b Springhill, NS, Jan 6, 13; m 38. MEDICINE, PSYCHIATRY. Educ: Dalhousie Univ, BSc, 33, MD, CM, 37. Prof Exp: From asst prof to assoc prof, 51-68, PROF PSYCHIAT, DALHOUSIE UNIV, 68-, ASST DEAN MED, STUDENT AFFAIRS & ADMIS, 73- Concurrent Pos: From asst psychiatrist to assoc psychiatrist, Victoria Gen Hosp, 51-; asst psychiatrist, Camp Hill Hosp, 51; consult, NS Hosp. Mem: Can Med Asn; Can Psychiat Asn. Res: Psychotherapy in mild schizophrenia. Mailing Add: Dept of Psychiat Dalhousie Univ Halifax NS Can

NICHOLSON, LARRY MICHAEL, b Nevada, Mo, Nov 22, 41; m 64. BIOCHEMISTRY. Educ: Kans State Univ, BS, 63, PhD(biochem), 68. Prof Exp: ASST PROF CHEM, UNIV MO-ROLLA, 67- Mem: AAAS; Am Chem Soc. Res: Enzyme chemistry; peroxidase enzymes; biological halogenation; biosynthesis of thyroxine. Mailing Add: Dept of Chem Univ of Mo Rolla MO 65401

NICHOLSON, MARGIE MAY, b San Antonio, Tex, June 10, 25; m 51. PHYSICAL CHEMISTRY. Educ: Univ Tex, BS, 46, MA, 48, PhD(phys chem), 50. Prof Exp: Phys chemist, US Naval Ord Lab, 51-52; res chemist, Humble Oil & Refining Co, 52-56, sr res chemist, 56-62; electrochemist, Stanford Res Inst, 62-64; res specialist, Rocketdyne Div, NAm Aviation, Inc, 64-65, sr tech specialist, 65, mem tech staff, Atomic Int Div, NAm Rockwell Corp, Canoga Park, 65-74, MEM TECH STAFF, ELECTRONICS RES DIV, ROCKWELL INT, ANAHEIM, 74- Mem: Am Chem Soc; Electrochem Soc; Sigma Xi. Res: Electrochemistry; theory and applications of voltammetric techniques; electrochemical power sources; electrode kinetics; nonaqueous electrochemistry; semiconductor electrodes; electrochromic displays. Mailing Add: 2209 California Blvd San Marino CA 91108

NICHOLSON, NANCY LYNNE, b Los Angeles, Calif, May 27, 41. NATURAL SCIENCES, BOTANY. Educ: Pomona Col, BA, 63; Stanford Univ, PhD(biol), 68. Prof Exp: Asst prof bot, Univ Wash, 67-69; asst prof biol, Univ Southern Calif, 69-75; ASSOC PROF INTERDISCIPLINARY STUDIES, WESTERN COL, MIAMI UNIV, 75- Mem: AAAS; Phycol Soc Am; Int Phycol Soc; Brit Phycol Soc. Res: Algal physiology. Mailing Add: Dept of Interdisciplinary Studies Western Col Miami Univ Oxford OH 45056

NICHOLSON, NICHOLAS, b New York, NY, June 9, 38; m 64. EXPERIMENTAL NUCLEAR PHYSICS. Educ: Polytech Inst Brooklyn, BS, 60; WVa Univ, MS, 62, PhD(physics), 65. Prof Exp: Half-time instr physics, WVa Univ, 63-64, asst, 64-65; physicist, Div Res, USAEC, Washington, DC, 65-67; PHYSICIST, LOS ALAMOS SCI LAB, 67- Res: Nuclear spectroscopy, alpha, beta, and gamma; angular correlations; x-ray physics; x-ray spectroscopy; radiography. Mailing Add: Los Alamos Sci Lab Group A-2 PO Box 1663 Los Alamos NM 87544

NICHOLSON, NORMAN LEON, b Barking, Eng, Oct 14, 19; Can citizen; m 47; c 1. GEOGRAPHY, ACADEMIC ADMINISTRATION. Educ: Univ Western Ont, BA, 43, MSc, 47; Univ Ottawa, PhD(geog), 51; Univ Toronto, MEd, 73. Prof Exp: Lectr geog, Univ Western Ont, 46-49; geogr, Govt Can, 49-54, dir geog br, 54-64; asst dean grad studies, 66-67, dean univ col, 67-69, PROF GEOG, UNIV WESTERN ONT, 64-, COORDR GRAD STUDIES EDUC, 70- Concurrent Pos: Lectr, Carleton Univ, 52-57; lectr, Univ Ottawa, 54-59; secy Can comt, Int Geog Union, 54-64; chmn, Can Bd Geog Names, 59-64; chmn Can sect, Pan-Am Inst Geog & Hist, 61-64; vis prof, Univ Edinburgh, 65; vis prof, Univ BC, 65. Mem: Fel Royal Geog Soc; fel Am Geog Soc; Asn Am Geog; Can Asn Geog; Royal Can Geog Soc. Res: Geography of higher education in Canada. Mailing Add: Althouse Col Univ of Western Ont London ON Can

NICHOLSON, RALPH LESTER, b Lynn, Mass, Aug 25, 42; m 74. PHYTOPATHOLOGY. Educ: Univ Vt, BA, 64; Univ Maine, MS, 67; Purdue Univ, PhD(plant path), 72. Prof Exp: ASST PROF PLANT PATH, PURDUE UNIV, WEST LAFAYETTE, 72- Concurrent Pos: Purdue rep, NCent Region, Corn & Sorghum Dis Comt, USDA-Coop State Res Serv, 72- Mem: Am Phytopath Soc; Sigma Xi. Res: Study of host biochemical response to infection with emphasis on stress compounds; recognition of the pathogen by the host, and histopathology related to disease physiology and time of host response. Mailing Add: Dept of Bot & Plant Path Purdue Univ West Lafayette IN 47907

NICHOLSON, RICHARD BENJAMIN, b Tacoma, Wash, Sept 8, 28; m 52; c 4. NUCLEAR PHYSICS. Educ: Univ Puget Sound, BS, 50; Cornell Univ, MS, 60; Univ Mich, PhD(nuclear sci), 63. Prof Exp: Physicist mat lab, Puget Sound Naval Shipyard, 51-53; physicist, Atomic Power Develop Assocs, 54-58 & 59-61; assoc prof nuclear eng, Univ Wis, 63-66; assoc physicist, Argonne Nat Lab, 66-69, sr physicist, 69-72; PROF NUCLEAR ENG, OHIO STATE UNIV, 72- Concurrent Pos: Consult, Atomic Power Develop Assocs, 61-; physicist, Lawrence Radiation Lab, Univ Calif, 65; consult adv comt reactor safeguards, AEC. Mem: Fel Am Nuclear Soc; Am Phys Soc. Res: Physics and safety of fast nuclear reactors, especially theoretical problems in Doppler effect and accident analysis; theory of high temperature plasmas. Mailing Add: Dept of Nuclear Eng Ohio State Univ Columbus OH 43210

NICHOLSON, RICHARD SELINDH, b Des Moines, Iowa, Apr 5, 38; m 58; c 1. CHEMISTRY. Educ: Iowa State Univ, BS, 60; Univ Wis-Madison, PhD(chem), 64. Prof Exp: Res assoc chem, Iowa State Univ, 59-60; NSF fel, Univ Wis, 60-61, res assoc, 63-64; from asst prof to assoc prof, Mich State Univ, 64-71; prog dir, 71-75, ACTG HEAD CHEM SYNTHESIS & ANAL SECT & DEP DIR, CHEM DIV, NSF, 75- Concurrent Pos: NSF fel, 64-71; consult, US Army Electronic Command, 67; guest worker, NIH, 71- Honors & Awards: Eastman Kodak Award, Univ Wis, 64. Mem: AAAS; Am Chem Soc; Brit Polarographic Soc; Int Electrochem Soc. Res: Electrochemistry; electrode kinetics; applications of computers in instrumentation; mass spectrometry, especially chemical ionization. Mailing Add: Chem Div NSF 1800 G St NW Washington DC 20550

NICHOLSON, THOMAS DOMINIC, b New York, NY, Dec 14, 22; m 46; c 4. ASTRONOMY, NAVIGATION. Educ: US Merchant Marine Acad, BS, 50; St John's Univ, NY, BA, 50; Fordham Univ, MS, 53, PhD(educ admin), 61. Prof Exp: Asst prof nautical sci & asst to dept chmn, US Merchant Marine Acad, 46-53; assoc astronr, Hayden Planetarium, 54-58, asst chmn, 58-64, chmn, 64-68, asst dir, Mus, 68-69, DIR, AM MUS NATURAL HIST, 69- Concurrent Pos: Lectr, Hayden Planetarium, 51-54 & US Mil Acad, 52-64; instr, US Naval Reserve Officer's Sch, 56-62; adj instr, Hunter Col, 62-67; weather forecaster, WNBC-TV, New York, 67- Mem: Fel AAAS; Am Astron Soc; Am Meteorol Soc; Am Inst Navig (pres, 71-72); Royal Astron Soc. Res: Geodetic astronomy, specifically in arctic regions; spherical astronomy; astronomy education. Mailing Add: Am Mus of Natural Hist 79th St at Central Park W New York NY 10024

NICHOLSON, THOMAS FREDERICK, b Toronto, Ont, Jan 14, 02; m 32; c 2. PATHOLOGICAL CHEMISTRY. Educ: Univ Toronto, MB, 26, BSc, 28, PhD(path chem), 34. Hon Degrees: DSc, Ahmadu Bello Univ. Prof Exp: Bacteriologist, Toronto Gen Hosp, 28; lectr path chem, Univ Toronto, 29-30, from asst prof to prof, 30-67; prof, 67-70, EMER PROF CHEM PATH, AHMADU BELLO UNIV, NIGERIA, 70- Concurrent Pos: Prof, Med Sch, Univ Lagos, 62-65. Mem: Am Physiol Soc; Can Physiol Soc; Brit Biochem Soc. Res: Renal and parathyroid physiology. Mailing Add: Kirkham's Rd RR 1 West Hill ON Can

NICHOLSON, VICTOR ALVIN, b Stafford, Kans, Dec 16, 41; m 70; c 1.

MATHEMATICS. Educ: Okla State Univ, BS, 62, MS, 64; Univ Iowa, PhD(math), 68. Prof Exp: Asst prof math, Park Col, 64-65; ASST PROF MATH, KENT STATE UNIV, 68- Mem: Am Math Soc; Math Asn Am. Res: Geometric topology; topology of manifolds. Mailing Add: Dept of Math Kent State Univ Kent OH 44240

NICHOLSON, WESLEY LATHROP, b Andover, Mass, Jan 31, 29; m 52; c 4. STATISTICS. Educ: Univ Ore, BA, 50, MA, 52; Univ Ill, PhD(math statist), 55. Prof Exp: Asst math, Univ Ore, 50-52 & Univ Ill, 52-55; instr, Princeton Univ, 55-56; statistician, Gen Elec Co, 56-64; sr res assoc, Pac Northwest Lab, 65-71, STAFF SCIENTIST, BATTELLE-NORTHWEST, BATTELLE MEM INST, 71- Concurrent Pos: Lectr, Univ Wash, Richland Campus, 56-71, chmn math prog, 71-; adj prof math, Wash State Univ, 66- Mem: Fel Am Statist Asn; Biomet Soc; Math Asn Am; Inst Math Statist; Int Soc Stereol. Res: Derivation and application of statistical methodology to physical engineering and biological sciences; construction of mathematic models; data analysis. Mailing Add: PO Box 999 Richland WA 99352

NICHOLSON, WILLIAM JAMIESON, b Seattle, Wash, Nov 21, 30; m 57; c 4. PHYSICS, ENVIRONMENTAL HEALTH. Educ: Mass Inst Technol, BS, 52; Univ Wash, PhD(physics), 60. Prof Exp: Instr, Univ Wash, 60; physicist, Watson Res Lab, Int Bus Mach Corp, NY, 60-68; asst prof, 69-73, ASSOC PROF COMMUNITY MED, MT SINAI SCH MED, 73- Concurrent Pos: Adj assoc prof, Fordham Univ, 64. Mem: AAAS; Am Phys Soc; NY Acad Sci. Res: Occupational and environmental health; analysis and effect of airborne micro-particulates. Mailing Add: Environ Sci Lab Mt Sinai Sch of Med New York NY 10029

NICHOLSON, WILLIAM ROBERT, b Camden, NJ, Jan 25, 25; m 50; c 4. FISHERIES. Educ: Rutgers Univ, BSc, 50; Univ Maine, MSc, 53. Prof Exp: Proj leader water fowl biol, Md Game & Inland Fish Comn, 51-55; FISHERY RES BIOLOGIST, CTR ESTUARINE & FISHERIES RES, NAT MARINE FISHERIES SERV, 56- Mem: Am Inst Biol Sci; Am Fisheries Soc; Am Inst Fishery Res Biologists. Res: Population dynamics of marine fishes. Mailing Add: Atlantic Estuarine Fisheries Ctr Nat Marine Fisheries Serv Beaufort NC 28516

NICHOLSON-GUTHRIE, CATHERINE SHIRLEY, b Jackson, Miss; m 61; c 1. GENETICS. Educ: Auburn Univ, BS, 57; Fla State Univ, MS, 60; Ind Univ, Bloomington, PhD(genetics), 72. Prof Exp: Res asst molecular biol, Calif Inst Technol, 60-62; instr biol, Boston State Col, 63-64; vis asst prof biol, Univ Evansville, 72-73; MEM ADJ FAC MED GENETICS, SCH MED, IND UNIV, EVANSVILLE, 74- Concurrent Pos: Consult, Mead Johnson & Co, 76-77. Mem: Sigma Xi; Genetics Soc Am. Res: Inheritance of chlorophyll and lamellae; chloroplast function; relationship between membrane structure and function. Mailing Add: 700 Drexel Dr Evansville IN 47712

NICKANDER, RODNEY CARL, b Aitkin, Minn, July 5, 38; m 57; c 2. PHARMACOLOGY. Educ: SDak State Univ, BS, 60; Purdue Univ, MS, 63, PhD(pharmacol), 64. Prof Exp: PHARMACOL RES ASSOC, LILLY RES LABS, ELI LILLY & CO, 64- Res: Detection and pharmacological evaluation of new anti-inflammatory agents and analgesics; pharmacological role of metabolites. Mailing Add: Dept MC 304 Lilly Res Labs Eli Lilly & Co Indianapolis IN 46206

NICKEL, ERNEST HENRY, b Louth, Ont, Aug 31, 25; m 49; c 3. MINERALOGY. Educ: McMaster Univ, BSc, 50, MSc, 51; Univ Chicago, PhD(geol), 53. Prof Exp: Sci officer, Mines Br, Dept Mines & Tech Surv, 53-59, sr sci officer, 59-65, head mineral sect, Mineral Sci Div, Can Dept Energy, Mines & Resources, 65-71; WITH DIV MINERAL, COMMONWEALTH SCI & INDUST RES ORGN, 71- Honors & Awards: Hawley Award, Mineral Asn Can, 67 & 73. Mem: Mineral Asn Can (pres, 69-70); Geol Soc Australia; Australasian Inst Mining & Metall; Mineral Soc Am. Res: Determinative and descriptive mineralogy by means of microscopy; x-ray diffraction and crystallography; ore mineralogy. Mailing Add: Commonwealth Sci & Indust Orgn Div of Mineral Pvt Bag PO Wembly Western Australia 6014 Australia

NICKEL, GEORGE HERMAN, applied science, see 12th edition

NICKEL, JAMES ALVIN, b Grants Pass, Ore, Sept 27, 25; m 52; c 3. APPLIED MATHEMATICS. Educ: Willamette Univ, BA, 49; Ore State Col, MS, 51, PhD(appl math, anal), 57. Prof Exp: Asst math, Ore State Col, 49-50; asst appl math, Ind Univ, 51-53, asst math, 53; instr, Willamette Univ, 53-57, asst prof, 57-59; from assoc prof to prof & chmn dept, Oklahoma City Univ, 59-67; sr res mathematician, Dikewood Corp, 67-69, prin res mathematician, 69-70; mgr anal sect, Technol Inc, 70-71; math physicist, Lockheed Electronics Co, 71-72; PROF MATH, UNIV TEX OF PERMIAN BASIN, 72- Concurrent Pos: Statistician, State Hwy Dept, Ore, 56-59; consult, Systs Res Ctr, Okla. Mem: Am Math Soc; Math Asn Am; Am Statist Asn. Res: Systems design and simulation; applied mathematics and statistics. Mailing Add: 3942 Monclair Ave Odessa TX 79762

NICKEL, JOHN L, b Hillsboro, Kans, Jan 9, 30; m 51; c 2. ENTOMOLOGY. Educ: Univ Calif, Berkeley, BS, 51, PhD(entom), 59. Prof Exp: Entomologist, Maple Leaf Farms, Inc & Maple Leaf Found, Calif & Paraguay, 51-57; entomologist, Res Div, Standard Fruit Co, Honduras, 59-60; entom adv, US Agency Int Develop, Cambodia, 61-63; vis scientist, Int Rice Res Inst, Philippines, 64; asst entomologist, Univ Calif, Berkeley, 64-66; dean fac agr, Makerere Univ, Uganda, 66-71; assoc dir, 71-74, dep dir gen, 74-75, DIR GEN, INT INST TROP AGR, NIGERIA, 75- Mem: Entom Soc Am. Res: Agricultural entomology, especially ecology and biology of pests and natural enemies. Mailing Add: Int Inst of Trop Agr PMB 5320 Ibadan Nigeria

NICKEL, KAREN LOUISE, analytical biochemistry, see 12th edition

NICKEL, PHILLIP ARNOLD, b Deadwood, SDak, Oct 10, 37; m 59; c 2. PARASITOLOGY, ACAROLOGY. Educ: Ore State Univ, BS, 62; Kans State Univ, MS, 66, PhD(entom), 69. Prof Exp: Asst prof, 69-74, ASSOC PROF BIOL SCI, CALIF LUTHERAN COL, 74- Mem: Am Inst Biol Sci; Entom Soc Am; Acarological Soc Am. Res: Mites associated with insects; helminths of bats. Mailing Add: Dept of Biol Sci Calif Lutheran Col Thousand Oaks CA 91360

NICKEL, VERNON L, b Sask, Can, May 1, 18; US citizen; m 41; c 3. ORTHOPEDIC SURGERY. Educ: Loma Linda Univ, MD, 44; Univ Tenn, MSc, 49. Prof Exp: Fel orthop surg, Campbell Clin, Memphis, Tenn, 48-49; head orthopedist & chief surg serv, Rancho Los Amigos Hosp, Downey, Calif, 53-64, med dir, 64-69; dir dept orthop surg & rehab, 69-75, PROF ORTHOP SURG & REHAB, LOMA LINDA UNIV, 69- Concurrent Pos: Pvt pract, 50-; Fulbright lectr, Cairo Univ, 61; clin prof, Univ Southern Calif, 65- & Univ Calif, Irvine-Calif Col Med, 66- Mem: Am Orthop Asn; Am Acad Orthop Surg. Mailing Add: Dept of Orthop Surg & Rehab Loma Linda Univ Loma Linda CA 92354

NICKELL, CECIL D, b Rochester, Ind, Jan 9, 41; m 62; c 2. PLANT GENETICS, BIOMETRY. Educ: Purdue Univ, BS, 63; Mich State Univ, MS, 65, PhD(biomet), 67. Prof Exp: Asst instr plant genetics, Mich State Univ, 63-67; asst prof, 67-74, ASSOC PROF PLANT BREEDING & GENETICS, KANS STATE UNIV, 74- Mem: Crop Sci Soc Am; Am Genetic Asn; Am Soc Agron. Res: Statistical studies concerned with evolutionary changes in plant populations under extreme stresses; biometrical application of selection procedures. Mailing Add: Dept of Agron Kans State Univ Manhattan KS 66502

NICKELL, FRANK ANDREW, geology, deceased

NICKELL, LOUIS G, b Little Rock, Ark, July 10, 21; m 42; c 3. PLANT PHYSIOLOGY. Educ: Yale Univ, BS, 42, MS, 47, PhD(physiol), 49. Prof Exp: Res assoc plant physiol, Brooklyn Bot Garden, 49-51; plant physiologist, Pfizer Co, 51-53, head phytochem lab, 53-61; head dept physiol & biochem, Exp Sta, Hawaiian Sugar Planters Asn, 61-65, asst dir, 65-75; VPRES, RES DIV, W R GRACE & CO, 75- Concurrent Pos: Responsible scientist, Cell Nutrit Exhibit, Worlds Fair, Brussels, 58; trustee, Hawaiian Bot Gardens Found, 62-65; mem, Gov Adv Comt Sci & Technol, Hawaii, 64-70, chmn, 70-75; mem, State Task Force Energy Policy, Hawaii, 73-75, vchmn, 74-75. Mem: Am Soc Plant Physiol; Bot Soc Am; Am Chem Soc; Soc Develop Biol; Soc Econ Bot. Res: Sugarcane physiology and biochemistry; tissue and cell culture; medicinal and economic botany; antibiotics; plant growth substances; research administration; microbiology. Mailing Add: W R Grace & Co Washington Res Ctr 7379 Rte 32 Columbia MD 21044

NICKELL, WILLIAM EVERETT, b Hazel Green, Ky, July 29, 16; m 42. PHYSICS. Educ: Berea Col, BA, 40; Univ Iowa, MS, 43, PhD(physics), 54. Prof Exp: Researcher, Univ Iowa, 43-44, res assoc, 44-45, instr physics, 47-51, res assoc, 51-52; from asst prof to prof, SDak State Univ, 53-63; assoc prof, 63-69, PROF PHYSICS, SOUTHERN ILL UNIV, 69- Mem: Am Phys Soc; Am Asn Physics Teachers. Res: Nuclear physics; radio proximity fuze; missile vibrations. Mailing Add: Dept of Physics Southern Ill Univ Carbondale IL 62901

NICKELSEN, RICHARD PETER, b Lynbrook, NY, Oct 1, 25; m 50; c 3. STRUCTURAL GEOLOGY. Educ: Dartmouth Col, BA, 49; Johns Hopkins Univ, MA, 51, PhD, 53. Prof Exp: Asst, Johns Hopkins Univ, 51-53; asst prof geol, Pa State Univ, 53-59; assoc prof, 59-63, PROF GEOL, BUCKNELL UNIV, 64-, CHMN DEPT GEOL & GEOG, 59- Concurrent Pos: NATO fel, Norway, 65-66, NSF res grant, 68-69. Mem: AAAS; Geol Soc Am; Soc Econ Paleont & Mineral; Am Asn Petrol Geol; Am Geophys Union. Res: Fabric of deformed quartz and calcite; genesis of joints and small scale structure; Appalachian tectonics and regional joint patterns; Caledonide stratigraphy and tectonics. Mailing Add: RD 1 Box 432 Lewisburg PA 17837

NICKELSON, RANZELL, II, b San Diego, Calif, Oct 29, 44; m 67; c 2. FOOD MICROBIOLOGY. Educ: Tex A&M Univ, BS, 68, MS, 69, PhD(microbiol), 71. Prof Exp: SEAFOOD TECHNOL SPECIALIST, TEX AGR EXTEN SERV, 72- Concurrent Pos: Prog comt mem, Fish Expos 1973, New Orleans, La, 73; prog comt chmn, Fish Expos 1974, Norfolk, Va, 74. Honors & Awards: Lone Star Farmer Award, Future Farmers Am, 62; Distinguished Mil Grad, US Army ROTC Prog, 70. Mem: Sigma Xi; Inst Food Technologists; Am Soc Microbiol; Int Asn Milk, Food & Environ Sanitarians; Gulf & Caribbean Fisheries Inst. Res: Microbiology of foods with respect to spoilage and/or consumer safety and other factors that affect food quality. Mailing Add: Tex A&M Univ College Station TX 77843

NICKERSON, DOROTHY, b Boston, Mass, Aug 5, 00. COLOR SCIENCE. Prof Exp: Asst, Munsell Res Lab & asst mgr, Munsell Color Co, 21-26; color technologist, USDA, 27-64; consult, 65-74; RETIRED. Concurrent Pos: Trustee, Munsell Color Found, 42-, pres, 73-75; US expert color rendering, Int Comn Illum, 56-67. Honors & Awards: Superior Serv Award, USDA, 51; Gold Cert, Am Hort Coun, 57; Godlove Award, Int Soc Color Coun, 61; Distinguished Achievement Award, Instrument Soc Am, 64; Gold Medal, Illum Eng Soc, 70. Mem: AAAS; Optical Soc Am; Int Soc Color Coun (secy, 38-52, pres, 54); Illum Eng Soc. Res: Color measurement related to grade standards for agricultural products; colorimetry; color tolerances and small-color-difference specification; color spacing; automatic cotton colorimeter; color-rendering properties of light sources; color-fan chart for horticulture. Mailing Add: 2039 New Hampshire Ave NW Washington DC 20009

NICKERSON, GIFFORD SPRUCE, b Pawtucket, RI, July 17, 31; m 53; c 3. ETHNOLOGY, APPLIED ANTHROPOLOGY. Educ: Wheaton Col, AB, 54; Northwestern Univ, MA, 57; Univ NC, PhD(anthrop), 73. Prof Exp: Instr anthrop, Seattle Pac Col, 59-61; asst prof sociol & anthrop, Rocky Mountain Col, 61-64; res assoc anthrop, Col Med, Univ Ky, 64-66; from instr to asst prof, 68-75, ASSOC PROF ANTHROP, NC STATE UNIV, 75- Concurrent Pos: Instr, Cent YMCA Eve Sch, Seattle, Wash, 60-61; ed, Rocky Mountain Rev, 62-64, adv ed, 64-70; res consult, Asn Am Indian Affairs, 67; instr, Ft Bragg Br, NC State Univ, 68-71. Mem: Am Anthrop Asn; Soc Appl Anthrop. Res: American Indian Ethnology; medical anthropology; cultural psychiatry; history of anthropology; protection of archaeological sites, objects; applied anthropology; educational anthropology; sociopsychological stress. Mailing Add: Dept of Sociol & Anthrop NC State Univ PO Box 5535 Raleigh NC 27607

NICKERSON, HELEN KELSALL, b New York, NY, July 2, 18; div; c 1. MATHEMATICS. Educ: Vassar Col, BA, 39; Radcliffe Col, MA, 40, PhD(math), 49. Prof Exp: From instr to asst prof math & physics, Wheaton Col, Mass, 43-50; res assoc math, Princeton Univ, 51-61; lectr, Douglass Col, 60-61, assoc prof, 61-63, ASSOC PROF MATH, RUTGERS COL, RUTGERS UNIV, 63- Mem: Am Math Soc; Math Asn Am; NY Acad Sci. Res: Differentiable manifolds. Mailing Add: 184 Washington Rd Princeton NJ 08540

NICKERSON, JOHN CHARLES, III, b McMinnville, Tenn, Nov 5, 43; m 65; c 1. PHYSICS. Educ: Princeton Univ, BSE, 65; Stanford Univ, MS, 69, PhD(physics), 71. Prof Exp: Res asst physics, Stanford Univ, 67-71; ASST PROF PHYSICS, MISS STATE UNIV, 71- Mem: Am Phys Soc. Res: Relativity theory; teaching methods and approaches; solid state physics-magnetism. Mailing Add: Dept of Physics Miss State Univ Box 5167 Mississippi State MS 39762

NICKERSON, JOHN DAVID, b Halifax, NS, Feb 12, 27; m 52; c 2. PHYSICAL CHEMISTRY. Educ: Mt Allison Col, BSc, 48; Dalhousie Univ, MSc, 50; Univ Toronto, PhD(chem), 54. Prof Exp: Chemist fatty acid hydrogenation, Fisheries Res Bd Can, 48-50; demonstr, Univ Toronto, 51-53; develop chemist carbon & graphite, Nat Carbon Co, 53-57; prin res chemist, Agr Prod, Int Minerals & Chem Corp, Fla, 57-62; mem staff, Southern Nitrogen Co, Ga, 62-64; mgr chem res, Armour Agr Chem Co, 64-68; dir res & develop, 68-75, DIR DEVELOP & TECH SERV, USS AGRI-CHEM INC, 75- Mem: Am Chem Soc; Chem Inst Can. Res: Fatty acid hydrogenation; dielectric constants of low boiling liquids; oxidation of carbon and graphite; alkali resistance of carbon and graphite; fertilizer chemistry and animal feed supplements. Mailing Add: USS Agri-Chem 30 Pryor St Atlanta GA 30303

NICKERSON, JOHN LESTER, b Halifax, NS, Oct 8, 03; nat US; m 29. PHYSIOLOGY. Educ: Dalhousie Univ, BA, 25, MA, 28; Princeton Univ, PhD(physics), 35. Prof Exp: Substitute prof physics & chem, Mem Univ Nfld, 25-26;

instr physics, Princeton Univ, 28-31; from asst prof to prof, Mt Allison Univ, 31-39, assoc dean men, 36-39; from asst prof to prof physiol, Col Physicians & Surgeons, Columbia Univ, 39-56; chmn dept physiol, 56-70, dir div biophys, 61-70, from actg dean to dean sch grad & postdoctoral studies, 68-75, PROF PHYSIOL & BIOPHYS, CHICAGO MED SCH-UNIV HEALTH SCI, 56- Concurrent Pos: Consult physicist, Enamel & Heating, Ltd, 37-39; sci liaison officer, US Off Naval Res, London, 52-53; responsible investr res contracts, USPHS & US Air Force. Mem: Ballistocardiographic Res Soc (pres, 62-65); Am Phys Soc; Am Physiol Soc; Soc Exp Biol & Med; Harvey Soc. Res: Radioactivity; vacuum spectroscopy; body fluid status during trauma; ballistocardiography; response of body to mechanical stress; vibration; impact; cardiovascular dynamics. Mailing Add: Dept of Physiol Chicago Med Sch-Univ Health Sci Chicago IL 60612

NICKERSON, MARK, b Montevideo, Minn, Oct 22, 16; m 42; c 3. PHARMACOLOGY, THERAPEUTICS. Educ: Linfield Col, AB, 39; Brown Univ, ScM, 41; Johns Hopkins Univ, PhD(embryol), 44; Univ Utah, MD, 50. Hon Degrees: DSc, Med Col Wis, 74. Prof Exp: Res biochemist, Nat Defense Res Comn, Johns Hopkins Univ, 43-44; instr pharmacol, Col Med, Univ Utah, 44-47, from asst prof to assoc prof, 47-51; assoc prof, Univ Mich, 51-54; prof pharmacol & med res, Fac Med, Univ Man, 54-57, prof pharmacol & therapeut & chmn dept, 57-67; chmn dept pharmacol & therapeut, 67-75, PROF PHARMACOL & THERAPEUT, McGILL UNIV, 67- Honors & Awards: Abel Award, 49. Mem: Am Soc Pharmacol & Exp Therapeut (pres); Soc Exp Biol & Med; Am Physiol Soc; Can Physiol Soc; Royal Soc Can. Res: Drugs blocking sympathetic nervous system; cardiovascular and autonomic nervous system physiology and pharmacology; shock; clinical pharmacology. Mailing Add: Dept of Pharmacol & Therapeut McGill Univ Montreal PQ Can

NICKERSON, MAX ALLEN, b Maryville, Mo, July 18, 38; m 60; c 1. VERTEBRATE BIOLOGY, HISTOCHEMISTRY. Educ: Cent Methodist Col, AB, 60; Ariz State Univ, PhD(zool), 68. Prof Exp: From asst prof to assoc prof zool & histol, Ark State Univ, 68-71; RES ASSOC, UNIV WIS-MILWAUKEE, 71-; HEAD, VERT DIV, MILWAUKEE PUB MUS, 71- Concurrent Pos: Sci dir, Max Allen's Zool Gardens, Mo, 61-71; ed, Herpet Rev, 73- Mem: AAAS; Am Inst Biol Sci; Int Soc Toxinology; Am Soc Ichthyol & Herpet; Soc Study Amphibians & Reptiles (secy, 71-74). Res: Histology, histochemistry and ultrastructure of reptilian venom glands; ecology of map turtles, watersnakes and cottonmouths; relationships of fungi and vertebrates. Mailing Add: Vertebrate Div Milwaukee Pub Mus Milwaukee WI 53233

NICKERSON, MORTIMER HENDERSON, b Chelsea, Mass, July 22, 15; m 40; c 2. ORGANIC CHEMISTRY, RESEARCH ADMINISTRATION. Educ: Mass Inst Technol, SB, 37, PhD(org chem), 40. Prof Exp: Process develop chemist, Carbide & Carbon Chem Co, 40-41; res & develop chemist, Dewey & Almy Chem Co, W R Grace & Co, 41-42; chief chemist, DeBell & Richardson, Inc, 45-58; sr staff assoc, Arthur D Little, Inc, 58-70; INDUST CONSULT, 70- Mem: Am Chem Soc; Soc Plastics Eng. Res: Synthetic polymers; plastics chemistry; management of research and development. Mailing Add: Broad Brook Rd PO Box 2269 Enfield CT 06082

NICKERSON, NORTON HART, b Quincy, Mass, Apr 14, 26; m 54; c 3. PLANT MORPHOLOGY, PLANT ANATOMY. Educ: Univ Mass, BS, 49; Univ Tex, MA, 51; Wash Univ, PhD(bot), 53. Prof Exp: Instr bot, Univ Mass, 53-56 & Cornell Univ, 56-58; from asst prof to assoc prof, Wash Univ, 58-63; ASSOC PROF BOT, TUFTS UNIV, 63- Concurrent Pos: Res fel, Calif Inst Technol, 54-55; NSF sci fac fel, 58; morphologist, Mo Bot Garden, 58-63; mem Cape Cod Nat Seashore Adv Comn. Mem: AAAS; Soc Study Evolution; Bot Soc Am; Ecol Soc Am. Res: Morphology, physiology, genetics and history of Zea mays; ethnobotany; conservation; ecology of wetlands and coasts. Mailing Add: Dept of Biol Tufts Univ Medford MA 02155

NICKERSON, PETER AYERS, b Hyannis, Mass, Feb 19, 41. PATHOLOGY, ENDOCRINOLOGY. Educ: Brown Univ, AB, 63; Clark Univ, MA, 65, PhD(biol), 68. Prof Exp: Res instr, 67-69, res asst prof, 69-70, asst prof, 70-74, ASSOC PROF PATH, STATE UNIV NY BUFFALO, 74- Mem: Am Soc Cell Biol; Endocrine Soc; Am Soc Exp Path; Electron Micros Soc Am; Am Soc Zool. Res: Structure and function of the adrenal cortex; adrenal ultrastructure; ACTH secreting cell; transplantable pituitary tumor, hypertension and the adrenal cortex; gerbil adrenal cortex. Mailing Add: Dept of Path State Univ of NY Buffalo NY 14214

NICKERSON, ROBERT FLETCHER, b Stoneham, Mass, Mar 25, 30; m 60; c 1. COMPUTER SCIENCES. Educ: Tufts Univ, BS, 52, MS, 53; Univ Calif, PhD(chem), 58. Prof Exp: Asst chem, Tufts Univ, 52-53; asst, Univ Calif, 53-58, chemist, Lawrence Livermore Lab, 58-71; COMPUT ANALYST, CALIF INST TECHNOL, 72- Res: Small computer applications. Mailing Add: Dept of Biol Calif Inst of Technol Pasadena CA 91125

NICKERSON, THOMAS ANDREW, b Woodland, Calif, June 24, 21; m 44; c 3. DAIRY CHEMISTRY, FOOD SCIENCE. Educ: Univ Calif, BS, 43; Univ Minn, MS, 48, PhD, 50. Prof Exp: Asst dairy husb, Univ Minn, 46-50; from instr to asst prof dairy indust, 50-59, assoc prof food sci & technol, 59-66, PROF FOOD SCI & TECHNOL, UNIV CALIF, 66- Concurrent Pos: Assoc dean col agr, Univ Calif, Davis, 64-69. Mem: Am Dairy Sci Asn; Inst Food Technol. Res: Res: Technology of dairy foods; chemistry of lactose; utilization of lactose in foods. Mailing Add: Dept of Food Sci & Technol Univ of Calif Davis CA 95616

NICKERSON, WALTER JOHN, b Plainfield, NJ, Aug 6, 15; m 41, 62, 69; c 1. MICROBIOLOGY, BIOCHEMISTRY. Educ: West Chester State Col, BS, 37; Harvard Univ, MA, 40, PhD(physiol), 42. Prof Exp: Instr bot & bact, Wheaton Col, Mass, 42-43, from asst prof to assoc prof bot, 46-50; assoc prof, 50-54, PROF MICROBIOL, RUTGERS UNIV, NEW BRUNSWICK, 54-, MEM, INST MICROBIOL, 54- Concurrent Pos: Asst, Med Col, Tufts Univ, 46-50; Guggenheim fel, Carlsberg Lab, Denmark & Lab Microbiol, Holland, 47-48; vis prof, Brown Univ, 48-50; consult, USPHS, 55-59, 62-66. Mem: Am Chem Soc; Am Soc Microbiol; Am Soc Biol Chem; fel Am Acad Microbiol; Brit Soc Gen Microbiol. Res: Microbial biochemistry; photochemical reactions; morphogenesis of microorganisms; physiology of yeasts and actinomycetes. Mailing Add: Inst of Microbiol Rutgers Univ New Brunswick NJ 08903

NICKESON, RICHARD L, b Pittsburgh, Pa, Sept 12, 27; m 56; c 4. HORTICULTURE. Educ: Pa State Univ, BS, 51; Univ Minn, PhD(hort), 57. Prof Exp: Asst prof hort, SDak State Col, 56-59; RES ASSOC VEGETABLE RES, CAMPBELL SOUP CO, 60- Mem: Potato Asn Am; Am Soc Hort Sci. Res: Plant breeding and pathology; genetics. Mailing Add: Campbell Inst for Agr Res Branch Pike Cinnaminson NJ 08077

NICKLAS, ROBERT BRUCE, b Lakewood, Ohio, May 29, 32; m 60. CELL BIOLOGY. Educ: Bowling Green State Univ, BA, 54; Columbia Univ, MA, 56, PhD(zool), 58. Prof Exp: From instr to asst prof zool, Yale Univ, 58-64, fel sci, 63-64, assoc prof zool, 64-65; assoc prof, 65-71, PROF ZOOL, DUKE UNIV, 71- Concurrent Pos: Fel J S Guggenheim Found, Max Planck Inst, Töbingen, Germany, 72-73. Mem: Fel AAAS; Am Soc Cell Biol; Am Soc Zool; Genetics Soc Am; Biophys Soc. Res: Cytology and cytogenetics; experimental analysis of chromosome movement in mitosis; evolution of chromosome cycles; developmental cytology; cytochemistry. Mailing Add: Dept of Zool Duke Univ Durham NC 27706

NICKLE, WILLIAM R, b Bridgeport, Conn, July 20, 35; m 64. NEMATOLOGY. Educ: State Univ NY Col Forestry, Syracuse, BS, 56; Univ Idaho, MS, 58; Univ Calif, PhD(nematol), 63. Prof Exp: Res officer entomophilic nematodes, Res Inst, Can Dept Agr, 62-65; NEMATOLOGIST, NEMATOL INVEST, CROPS RES DIV, AGR RES SERV, USDA, 65- Mem: Soc Nematol; Soc Syst Zool; Entom Soc Can. Res: Taxonomy, morphology and biology of plant parasitic, insect parasitic and mycophagus nematodes. Mailing Add: Nematol Lab Plant Protect Inst Agr Res Serv US Dept of Agr Beltsville MD 20705

NICKLES, ROBERT JEROME, b Madison, Wis, Mar 22, 40; m 63; c 1. NUCLEAR PHYSICS, MEDICAL PHYSICS. Educ: Univ Wis, BS, 62, PhD(nuclear physics), 68; Univ Sao Paulo, MS, 67. Prof Exp: Res assoc nuclear physics, Sch Med, Univ Wis-Madison, 68-69; James A Picker Found res fel, Niels Bohr Inst, Copenhagen Univ, 69-71; James A Picker Found res fel med physics, 71-73, ASST PROF RADIOL, SCH MED, UNIV WIS-MADISON, 73- Mem: Am Phys Soc. Res: Study of heavy ion transfer reactions; short-lived isotope production utilizing the helium-jet technique; development of an intense neutron source for cancer therapy. Mailing Add: Dept of Radiol Univ of Wis Sch of Med Madison WI 53706

NICKLIN, ROBERT CLAIR, b Gordon, Nebr, May 23, 36; m 58; c 2. SOLID STATE PHYSICS. Educ: SDak Sch Mines & Technol, BS, 58; Iowa State Univ, PhD(physics), 67. Prof Exp: Assoc prof physics, 67-74, PROF PHYSICS & CHMN DEPT, APPALACHIAN STATE UNIV, 74- Mem: AAAS; Am Phys Soc; Am Asn Physics Teachers. Res: Electron spin resonance; studies of glasses; organic crystals. Mailing Add: Dept of Physics Appalachian State Univ Boone NC 28607

NICKLOW, CLARK W, b Markleton, Pa, Aug 9, 35; m 55; c 1. HORTICULTURE. Educ: Pa State Univ, BS, 57 & 58; Cornell Univ, MS, 61, PhD(veg crops), 63. Prof Exp: From exten specialist to assoc prof hort, Mich State Univ, 63-74; MEM FAC AGR, SOUTHWESTERN MICH COL, 75- Concurrent Pos: Pickle Packers Int, Inc grant, 66-67. Mem: Am Soc Hort Sci. Res: Improvement of cultural practices, including nutritional, plant population and irrigation for the major vegetable crops grown in Michigan. Mailing Add: Southwestern Mich Col Cherry Grove Rd Dowagiac MI 49047

NICKLOW, ROBERT MERLE, b St Petersburg, Fla, Oct 11, 36; m 58; c 3. SOLID STATE PHYSICS. Educ: Ga Inst Technol, BS, 58, MS, 60, PhD(x-ray diffraction), 64. Prof Exp: Asst res physicist, Eng Exp Sta, Ga Inst Technol, 63; PHYSICIST, OAK RIDGE NAT LAB, 63- Honors & Awards: Sidhu Award, 68. Mem: Am Phys Soc; Am Crystallog Asn. Res: Crystal physics; lattice dynamics; neutron and x-ray diffraction; study of lattice dynamics and spin waves by means of coherent inelastic neutron scattering. Mailing Add: Solid State Div Oak Ridge Nat Lab PO Box X Oak Ridge TN 37830

NICKOL, BRENT BONNER, b Agosta, Ohio, June 22, 40; m 64; c 2. PARASITOLOGY. Educ: Col Wooster, BA, 62; La State Univ, MS, 63, PhD(zool), 66. Prof Exp: From asst prof to assoc prof, 66-75, PROF ZOOL, UNIV NEBR, LINCOLN, 75- Mem: Wildlife Dis Asn; Am Soc Parasitol. Res: Taxonomy, morphology, variation and host-parasite relationships of the Acanthocephala. Mailing Add: Sch of Life Sci Univ of Nebr Lincoln NE 68588

NICKOLS, NORRIS ALLAN, b Ellensburg, Wash, July 8, 28; m 58; c 2. NUCLEAR PHYSICS. Educ: Cent Wash Col Educ, BA, 52; Univ Calif, Berkeley, PhD(physics), 60. Prof Exp: Physicist, Lawrence Radiation Lab, Univ Calif, 59-61; res scientist, Lockheed Calif Co, 61-63; res specialist, Space & Info Systs Div, NAm Aviation, Inc, 63-67; STAFF MEM, LOS ALAMOS SCI LAB, 68- Mem: Am Phys Soc. Res: High energy physics strange particles; high energy muon scattering; hyperfragments; nuclear weapons. Mailing Add: Los Alamos Sci Lab Los Alamos NM 87545

NICKON, ALEX, b Poland, Oct 6, 27; nat US, m 50; c 3. ORGANIC CHEMISTRY. Educ: Univ Alta, BSc, 49; Harvard Univ, MA, 51, PhD(chem), 53. Prof Exp: Vis lectr org chem, Bryn Mawr Col, 53; Nat Res Coun Can fel, Birkbeck Col, London, 53-54 & Univ Ottawa, Ont, 54-55; from asst prof to prof, 55-75, VERNON K KRIEBLE PROF CHEM, JOHNS HOPKINS UNIV, 75- Concurrent Pos: NSF sr fel, Imp Col, London, 63-64 & Univ Munich, 71-72; sr ed, J Org Chem, 65-71. Mem: Am Chem Soc; The Chem Soc. Res: Carbanions and carbonium ions; stereochemistry; reaction mechanisms; syntheses and structures of natural products; biologically important reactions. Mailing Add: Dept of Chem Johns Hopkins Univ Baltimore MD 21218

NICKSIC, STEPHEN WILLIAM, organic chemistry, analytical chemistry, see 12th edition

NICKSON, JAMES JOSEPH, b Portland, Ore, Dec 31, 15; m 39, 67; c 3. RADIOLOGY. Educ: Univ Wash, BS, 36; Johns Hopkins Univ, MD, 40. Prof Exp: Intern med, Baltimore City Hosp, 40-41; asst surgeon, USPHS Tumor Clin, Baltimore Marine Hosp, 41-42; physician & assoc, Metall Lab, Univ Chicago, 42-46; sr physician, Argonne Nat Lab, 46-47; assoc, Sloan-Kettering Inst, 47-50, mem, 51-65, dir radiation ther dept, Mem Hosp, 50-65; chmn dept radiation ther, Michael Reese Hosp & Med Ctr & mem, Med Res Inst, 65-72; PROF RADIATION ONCOL & CHMN DEPT, UNIV TENN CTR HEALTH SCI, 72-, DIR, MEMPHIS REGIONAL CANCER CTR, 74- Concurrent Pos: Am Cancer Soc & Nat Res Coun fel, Sloan-Kettering Inst, 47-48 & Royal Cancer Hosp, London, 48-49; prof radiol, Sloan-Kettering Div, Cornell Univ, 51-55, prof, Med Col, 55-65. Mem: Radiation Res Soc; Radiol Soc NAm; Am Radium Soc; AMA; fel Am Col Radiol. Res: Oncology; radiation therapy; radiobiology. Mailing Add: Memphis Regional Cancer Ctr Univ Tenn Ctr for Health Sci Memphis TN 38163

NICKSON, MARGARET JANE, b Vermilion, Ohio, Aug 8, 14; m 39; c 2. MEDICINE. Educ: Transylvania Col, AB, 36; Johns Hopkins Univ, MD, 40. Prof Exp: Intern, Sacred Heart Hosp, Pa, 40-41; res assoc, Metall Lab, Univ Chicago, 43-46; sr physician, Argonne Nat Lab, 46-47, consult, 47-48; Sloan-Kettering Inst Cancer Res fel, Royal Cancer Hosp, London, 48-49; physician, Lehman Col, 51-75, med dir col health off, 52-75; RETIRED. Mem: AMA. Res: College health and health education. Mailing Add: 1032 Dolphin Dr Cape Coral FL 33904

NICKUM, JOHN GERALD, b Rochester, Minn, Aug 7, 35; m 55; c 4. ZOOLOGY, CONSERVATION. Educ: Mankato State Col, BSc, 57; Univ SDak, MA, 61; Univ Southern Ill, PhD(zool), 66. Prof Exp: Teacher high sch, Minn, 57-59, jr high sch, 59-60; asst prof biol, Western Ky State Col, 65-66; asst prof wildlife mgt, SDak State Univ, 66-71, assoc prof wildlife & fisheries, 71-73; ASST LEADER, NEW YORK COOP FISHERY UNIT & ASST PROF NATURAL RESOURCES, CORNELL UNIV, 73-, DIR AQUACULTURE PROG, 74- Mem: Am Fisheries Soc. Res: Aquaculture; application of aquaculture in the solution of recreational and environmental problems; warm-water fish management; pond management; urban recreational fisheries; use of wastewater in aquaculture; culture of cool-water fishes. Mailing Add: Dept of Natural Resources Fernow Hall Cornell Univ Ithaca NY 14850

NICO, WILLIAM RAYMOND, b Aurora, Ill, Mar 23, 40; m 67. MATHEMATICS. Educ: Loyola Univ, Ill, BS, 62; Univ Calif, Berkeley, MA, 64, PhD(math), 66. Prof Exp: Res assoc eng mech, Stanford Univ, 66-67; asst prof, 67-73, ASSOC PROF MATH, TULANE UNIV LA, 73- Mem: Am Math Soc; Math Asn Am. Res: Algebra; finite semigroups; cohomological investigations of finite semigroups; homological algebra. Mailing Add: Dept of Math Tulane Univ of La New Orleans LA 70118

NICODEM, DAVID ERNEST, chemistry, see 12th edition

NICODEMUS, DAVID BOWMAN, b Kobe, Japan, July 1, 16; US citizen; m 48. PHYSICS. Educ: DePauw Univ, AB, 37; Stanford Univ, PhD(physics), 46. Prof Exp: Asst, Stanford Univ, 37-41, asst physicist, Off Sci Res & Develop proj, 42-43; physicist, Los Alamos Sci Lab, Calif, 43-46; instr physics, Stanford Univ, 46-49, actg asst prof, 49-50; from asst prof to assoc prof, 50-63, asst dean sch sci, 62-65, actg dean, 65-66, PROF PHYSICS, ORE STATE UNIV, 63-, DEAN FAC, 66- Concurrent Pos: Consult, Los Alamos Sci Lab, 56-57. Res: X-rays; nuclear physics. Mailing Add: Dept of Physics Ore State Univ Corvallis OR 97331

NICODEMUS, FREDERICK EDWIN, b Osaka, Japan, July 25, 11; US citizen; m 35. OPTICAL PHYSICS. Educ: Reed Col, AB, 34. Prof Exp: Radio engr & physicist, Air Force Cambridge Res Labs, 46-55; advan develop engr & eng specialist, Sylvania Electronic Defense Labs, Gen Tel & Electronics Corp, Calif, 55-69; physicist, Michelson Lab, US Naval Weapons Ctr, 69-74; PHYSICIST, NAT BUR STANDARDS, 74- Concurrent Pos: Mem, Nat Acad Sci-Nat Acad Eng-Nat Res Coun adv panel to heat div, Nat Bur Standards & liaison to ad hoc panel on radiometry & photom, 70-74; consult, Int Tech Comt Photom & Radiometry, Int Comn Illum, 72-; ed self-study manual optical radiation measurements, Nat Bur Standards, 74- Mem: Fel Optical Soc Am; Sigma Xi; Int Comn Illum. Res: Clarification of basic radiometric relations, definitions, and nomenclature. Mailing Add: A223 Physics Bldg Nat Bur of Standards Washington DC 20234

NICOL, CHARLES ALBERT, b Ft Worth, Tex, Apr 24, 25; m 56; c 1. MATHEMATICS. Educ: Univ Tex, PhD(math), 54. Prof Exp: Instr math, Univ Tex, 54-55; asst prof, Ill Inst Technol, 55-59 & Univ Okla, 59-60; asst head dept math & comp sci, 73-76, ASSOC PROF MATH, UNIV SC, 60- Mem: AAAS; Am Math Soc; Math Asn Am. Res: Number theory; algebra and combinatorial problems. Mailing Add: Dept of Math Univ of SC Columbia SC 29208

NICOL, DAVID, b Ottawa, Ont, Aug 16, 15; nat US; m 47; c 1. PELEONTOLOGY. Educ: Tex Christian Univ, BA, 37, MS, 39; Stanford Univ, MA, 43, PhD(paleont), 47. Prof Exp: Asst prof geol, Univ Houston, 47-48; assoc cur Mesozoic & Cenozoic inverts, US Nat Mus, 48-58; assoc prof geol, Southern Ill Univ, 58-64; ASSOC PROF GEOL, UNIV FLA, 65- Mem: Paleont Soc; Geol Soc Am; Soc Syst Zool; Soc Study Evolution; Am Asn Petrol Geol. Res: Pelecypods. Mailing Add: Dept of Geol Univ of Fla Gainesville FL 32611

NICOL, JAMES, b Dundee, Scotland, Aug 24, 21; nat US; m 48. ENGINEERING PHYSICS, LOW TEMPERATURE PHYSICS. Educ: St Andrews Univ, BSc, 46 & 48; Union Col, MS, 50; Ohio State Univ, PhD(physics), 52. Prof Exp: Instr physics, Ohio State Univ, 52-53; from asst prof to assoc prof, Amherst Col, 53-57; physicist, Arthur D Little, Inc, 57-62; dir res & vpres, Cryonetics Corp, 62-64; PHYSICIST, ARTHUR D LITTLE, INC, 64- Mem: Am Phys Soc; Inst Elec & Electronics Eng. Res: Phenomena below one degree absolute; thermal conductivity and electron tunneling in superconductors; electric power transmission; geomagnetism; underwater acoustic transmission. Mailing Add: 37 Miller Hill Rd Dover MA 02030

NICOL, JOSEPH ARTHUR COLIN, b Toronto, Ont, Dec 5, 15; m 41; c 1. ZOOLOGY, MARINE BIOLOGY. Educ: McGill Univ, BSc, 38; Univ Western Ont, MA, 40; Oxford Univ, DPhil(zool), 47, DSc(zool), 61. Prof Exp: Asst prof zool, Univ BC, 47-49; exp zoologist, Marine Biol Asn UK, 49-66; vis prof, 66-67, PROF ZOOL, MARINE SCI INST, UNIV TEX, 67- Concurrent Pos: Guggenheim fel, 53-54. Mem: Am Soc Zool; Marine Biol Asn UK; fel Royal Soc. Res: Comparative physiology and neurology; bioluminescence; pigments of animals; vision, eye structure, photomechanical responses; animal camouflage; vision in animals; oil pollution, effects on animals. Mailing Add: Marine Sci Inst Univ of Tex Austin TX 78712

NICOL, MALCOLM FOERTNER, b New York, NY, Sept 13, 39; m 63; c 3. CHEMICAL PHYSICS. Educ: Amherst Col, BA, 60; Univ Calif, Berkeley, PhD(chem), 63. Prof Exp: Res asst, 63-64, from actg asst prof to assoc prof, 65-75, PROF CHEM, UNIV CALIF, LOS ANGELES, 75- Concurrent Pos: Fel, Alfred P Sloan Found, 73-77. Res: Structure and bonding in solids under high pressures by electronic and vibrational spectroscopy; intermolecular energy transfer. Mailing Add: Dept of Chem Univ of Calif Los Angeles CA 90024

NICOLAENKO, BASIL, b Paris, France, Mar 23, 42; US citizen. MATHEMATICS. Educ: Univ Paris, Lic es sci, 65; Univ Mich, PhD(math), 68. Prof Exp: Staff assoc appl sci, Brookhaven Nat Lab, 68-69; prof math, Courant Inst Math Sci, NY Univ, 69-74; STAFF SCIENTIST MATH, LOS ALAMOS SCI LAB, UNIV CALIF, 74- Mem: Am Math Soc; Soc Indust & Appl Math. Res: Nonlinear functional analysis and applications. Mailing Add: Los Alamos Sci Lab Group T-7 Math Anal Los Alamos NM 87545

NICOLAI, JOHN HENRY, JR, b Ellicott City, Md, June 11, 36; m 59; c 1. PHYSIOLOGY, ANIMAL BREEDING. Educ: Cornell Univ, BS, 58; Univ Md, MS, 64, PhD(physiol), 66. Prof Exp: Teacher, Howard County Bd Educ, 58-59; county agriculturist exten serv, Univ Md, 59-61, asst dairy, 61-66; asst prof animal sci & exten dairyman, NC State Univ, 65-68; ASSOC PROF ANIMAL SCI & EXTEN DAIRYMAN, UNIV KY, 68- Concurrent Pos: Consult, Babson Bros Co, Chicago, 70- Mem: Am Dairy Sci Asn; Am Animal Sci Asn; Int Asn Milk, Food & Environ Sanitarians; Nat Mastitis Coun. Res: Milking management and mastitis; breeding problems of dairy cattle; nutrition of dairy cattle. Mailing Add: Dept of Animal Sci Univ of Ky Lexington KY 40506

NICOLAI, VAN OLIN, b Barrington, Ill, Jan 18, 24; m 55. OPTICAL PHYSICS, LASERS. Educ: Univ Ill, BS, 49, MS, 51, PhD(physics), 54. Prof Exp: Res assoc physics, Univ Ill, 54-55; asst prof, Univ NDak, 55-57; asst prof, Southern Ill Univ, 57-59; TECH ADV PHYSICS, OFF NAVAL RES, 60- Mem: Am Phys Soc. Res: Optical properties of solids, laser applications and surface physics. Mailing Add: Off of Naval Res Code 421 Arlington VA 22217

NICOLAIDES, ERNEST D, b Monmouth, Ill, Sept 11, 24; m 51; c 3. ORGANIC CHEMISTRY. Educ: Monmouth Col, BS, 48; Univ Ill, MS, 49, PhD(chem), 52. Prof Exp: Assoc res chemist, 51-59, sr res chemist, 59-75, SR RES SCIENTIST, PARKE, DAVIS & CO, 75- Res: Synthesis of organic compounds; natural products; amino acids; peptides; proteins. Mailing Add: Chem Dept Parke Davis & Co Ann Arbor MI 48105

NICOLAIDES, NICHOLAS, b Baltimore, Md, Nov 11, 15; m 56; c 3. BIOCHEMISTRY, ORGANIC CHEMISTRY. Educ: Case Inst, BS, 39, MS, 40; Univ Chicago, PhD(org chem), 50. Prof Exp: Instr chem, Roosevelt Univ, 48-50; res assoc anal skin lipids, Univ Chicago, 50-57; biochemist, Western Fish Nutrit Lab, Fish & Wildlife Serv, 57-59; biochemist, Inst Marine Resources, Univ Calif, 59-60; asst prof biochem & res assoc, Div Dermat, Sch Med, Univ Ore, 60-64, assoc prof, 64-65; ASSOC PROF BIOCHEM, SCH MED, UNIV SOUTHERN CALIF, 65- Concurrent Pos: Guggenheim fel, Swiss Fed Inst Technol, 55-56; fel, Inter-Sci Res Found, 70. Mem: AAAS; Am Chem Soc; Am Oil Chemists; Soc; Soc Invest Dermat. Res: Biochemistry and analysis of lipids; function of human skin lipids; nutritional lipid requirements; inclusion compounds. Mailing Add: Univ of Southern Calif Sch of Med 2025 Zonal Ave Los Angeles CA 90033

NICOLAS, JESUS, b Baracaldo, Spain, Oct 15, 20. SCIENCE EDUCATION. Educ: Univ Madrid, BSc, 44, BCh, 46, MCh, 47, PhD(chem), 49. Prof Exp: Supvry chemist, Standard Elec Co, Spain, 44-46; asst prof phys chem & electrochem, Univ Madrid, 49-52; supvry chemist, Gequisa Co, 52-54; res chemist, Cros Co, 54-56; sr res chemist, E I du Pont de Nemours & Co Inc, 58-70; TEACHER PHYSICS & MATH, ST MARK'S HIGH SCH, 70- Concurrent Pos: Res chemist, Telefunken Co, Spain, 48-50. Res: Physico-chemistry of polymers; electrochemistry and theoretical physics, especially thermodynamics. Mailing Add: St Mark's High Sch Henderson Rd Wilmington DE 19808

NICOLETTE, JOHN ANTHONY, b Chicago, Ill, Apr 2, 35; m 62; c 2. PHYSIOLOGY, ENDOCRINOLOGY. Educ: Dartmouth Univ, AB, 56; Univ Ill, MS, 61, PhD(physiol), 63. Prof Exp: USPHS fel, Nat Cancer Inst, 63-66; asst prof biol sci, 66-71, asst dean grad col, 71-73, ASSOC PROF BIOL SCI, UNIV ILL, CHICAGO CIRCLE, 71- Res: Biochemical mechanism of estrogen action. Mailing Add: Dept of Biol Sci Univ of Ill at Chicago Circle Chicago IL 60680

NICOLI, MIRIAM ZIEGLER, b Gainesville, Fla, Dec 15, 45. PROTEIN CHEMISTRY, ENZYMOLOGY. Educ: Bucknell Univ, BS, 67; Harvard Univ, MA, 70, PhD(biochem), 72. Prof Exp: Res fel biol, Harvard Univ, 72-73, lectr, 73-75, res fel, 75-76; RES FEL BIOCHEM, UNIV ILL, URBANA, 76- Mem: Biophys Soc. Res: Enzymatic mechanism, structure, and subunit function of bacterial luciferase; chemical modification of proteins. Mailing Add: Dept of Biochem Univ of Ill Urbana IL 61801

NICOLL, CHARLES S, b Toronto, Ont, Apr 11, 37; US citizen; m 62; c 2. PHYSIOLOGY, ENDOCRINOLOGY. Educ: Mich State Univ, BS, 58, MS, 60, PhD(physiol), 62. Prof Exp: Res zoologist, Univ Calif, Berkeley, 62, Am Cancer Soc fel endocrine-tumor probs, 62-64; staff fel tumor-endocrinol, Nat Cancer Inst, 64-66; from asst prof to assoc prof physiol, 66-74, PROF PHYSIOL, UNIV CALIF, BERKELEY, 74- Mem: Am Cancer Soc; Am Soc Zool; Endocrine Soc; Am Physiol Soc. Res: Mammary and pituitary physiology; comparative aspects of prolactin physiology; growth regulation. Mailing Add: Dept of Physiol & Anat Univ of Calif Berkeley CA 94720

NICOLL, FREDERICK HERMES, physics, see 12th edition

NICOLL, JEFFREY FANCHER, b Washington, DC. STATISTICAL MECHANICS. Educ: Mass Inst Technol, BS(elec eng), BS(physics) & BS(math), 70, PhD(physics), 75. Prof Exp: Fel, Mass Inst Technol, 75- Mem: Am Phys Soc. Res: Renormalization group approaches to critical phenomena. Mailing Add: Dept of Physics Mass Inst of Technol Cambridge MA 02139

NICOLL, PAUL ANDREW, b Rawalpindi, Pakistan, Jan 20, 08; US citizen; m 37; c 3. PHYSIOLOGY. Educ: Tarkio Col, AB, 30; Wash Univ, PhD(zool), 36. Hon Degrees: DHL, Tarkio Col, 70. Prof Exp: Nat Res Coun fel biochem, Univ Mich, 36-37; instr, Univ Chicago, 37-40; from instr to assoc prof, 40-52, PROF PHYSIOL, SCH MED, IND UNIV, INDIANAPOLIS, 52- Concurrent Pos: Prof & chief party, Ind Univ Adv Group, Pakistan Khyber Med Col, Univ Peshawar, 55-57, Basic Med Sci Inst, Karachi, 57-61, dept physiol, Med Ctr, Ind Univ, Indianapolis, 61-71 & anat & physiol, Med Sci Prog, Ind Univ, Bloomington, 71-; vis res fel, John Curtin Sch Med Res, Australian Nat Univ, 67-68; mem, Marine Biol Lab, Woods Hole. Mem: AAAS; Am Physiol Soc; Microcirc Soc (pres, 71); fel NY Acad Sci. Res: Microcirculation; lymphatic capillaries; vascular smooth muscle. Mailing Add: Med Sci Prog Ind Univ Bloomington IN 47401

NICOLL, ROGER ANDREW, b Camden, NJ, Jan 15, 41; m 70. NEUROPHYSIOLOGY, NEUROPHARMACOLOGY. Educ: Lawrence Univ, BA, 63; Univ Rochester, MD, 68. Prof Exp: Intern med, Univ Chicago, 68-69; RES ASSOC NEUROPHYSIOL, NIMH, 69- Res: Neurophysiology of olfaction; mechanisms of general anesthesia; neuropharmacology of presynaptic and postsynaptic inhibition in the central nervous system. Mailing Add: NIMH Spec Ment Health Res Prog St Elizabeths Hosp Washington DC 20032

NICOLLS, KEN E, b Albuquerque, NMex, Nov 28, 35; m 64; c 3. ANATOMY, ENDOCRINOLOGY. Educ: Colo State Univ, BS, 59, MS, 61, PhD(anat), 69. Prof Exp: Res asst wildlife mgt, Colo Game & Fish Dept, 61-62; range conservationist, Worland Dist, Bur Land Mgt, 63-64; ASST PROF ANAT, SCH MED, UNIV NDAK, 69- Concurrent Pos: NSF fac res grant, Univ NDak, 70, NIH instnl res grant, Sch Med, 70- Res: Mechanisms of action associated with interrelationships of light, especially photoperiod and intensity; morphophysiology of endocrine organs, particularly pituitary gland and gonads; animal age; body growth and antler growth of deer. Mailing Add: Dept of Anat Univ of NDak Sch of Med Grand Forks ND 58201

NICOLOFF, DEMETRE M, b Lorain, Ohio, Aug 31, 33; m; c 3. CARDIOVASCULAR SURGERY, BIOENGINEERING. Educ: Ohio State Univ, BA, 54, MD, 57; Univ Minn, PhD(surg), 65, PhD(physiol), 67. Prof Exp: Instr surg, Univ Minn, 64-65; staff surgeon, Vet Admin Hosp, 65-69; asst prof surg, 69-71, ASSOC PROF SURG, UNIV MINN HOSPS, MINNEAPOLIS, 71- Mem: Am Gastroenterol Asn; Soc Surg Alimentary Tract; Am Col Surg; Asn Acad Surg; Soc Thoracic Surg. Mailing Add: Dept of Surg Univ of Minn Hosps Minneapolis MN 55455

NICOLOSI, ROBERT JAMES, biochemistry, see 12th edition

NICOLSON, DAN HENRY, b Kansas City, Mo, Sept 5, 33; m 59; c 3. BOTANY, TAXONOMY. Educ: Grinnell Col, AB, 55; Stanford Univ, MBA, 57; Cornell Univ, MS, 59, PhD(plant taxon), 64. Prof Exp: Assoc cur, 64-75, CUR DEPT BOT, SMITHSONIAN INST, 75- Concurrent Pos: Fulbright sr res fel, Nepal, 67-68. Mem: Am Soc Plant Taxon. Res: Systematics of Araceae; floristics of Nepal, Dominica and south India; nomenclature, orthography. Mailing Add: Dept of Bot Smithsonian Inst Washington DC 20560

NICOLSON, MARGERY O'NEAL, b Pasadena, Calif, Mar 9, 31; m 61. BIOCHEMISTRY. Educ: Stanford Univ, BA, 52, MS, 54; Baylor Univ, PhD(biochem), 60. Prof Exp: Fel biochem, M D Anderson Hosp & Tumor Inst, Univ Tex, 60-62; res fel biol, Calif Inst Technol, 62-65; asst prof pediat, 65-72, ASSOC PROF PEDIAT & BIOCHEM, SCH MED, UNIV SOUTHERN CALIF, 72-; ASST PROF PEDIAT, CHILDREN'S HOSP LOS ANGELES, 65- Concurrent Pos: Am Cancer Soc fel, 62-64. Mem: AAAS; Am Asn Cancer Res; Brit Biochem Soc. Res: Biochemistry of RNA tumor viruses and human adenoviruses. Mailing Add: Children's Hosp PO Box 54700 Terminal Annex Los Angeles CA 90054

NICOLSON, PAUL CLEMENT, b Brooklyn, NY, June 3, 38; m 64; c 2. ANALYTICAL CHEMISTRY. Educ: WVa Wesleyan Col, BS, 60; Ariz State Univ, MS, 65, PhD(phys-org chem), 66. Prof Exp: Proj leader org anal phys chem, Geigy Chem Corp, 65-69, sr res assoc, 69-71, mgr anal res dept, 71-73, GROUP LEADER, CIBA-GEIGY CORP, 73- Mem: AAAS; Am Chem Soc; fel Am Inst Chemists. Res: Physical organic chemistry; instrumentation. Mailing Add: Ciba-Geigy Corp Ardsley NY 10502

NIDAY, JAMES BARKER, b Nashville, Tenn, May 21, 17; m 45; c 4. NUCLEAR CHEMISTRY. Educ: Univ Chicago, SB, 42, SM, 60. Prof Exp: Jr chemist, Tenn Valley Auth, 42-45; asst radiochem, Univ Chicago, 47-52; chemist, Calif Res & Develop Co, 52-53; CHEMIST, LAWRENCE LIVERMORE LAB, UNIV CALIF, 53- Mem: Am Phys Soc; Am Chem Soc. Res: Study of the fission process through radio-chemical study of fission yields and of the ranges of fission fragments in matter; computer processing and analysis of data in gamma ray spectroscopy. Mailing Add: 4440 Entrada Dr Pleasanton CA 94566

NIDDRIE, DAVID LAWRENCE, b Kimberley, SAfrica, July 18, 17; m 45; c 1. GEOGRAPHY. Educ: Univ Ntal, BSc, 37, UED, 38, MSc, 48; Univ Manchester, PhD(geog), 65. Prof Exp: Lectr geog, Univ Natal, 39-50 & Univ Manchester, 50-66; PROF GEOG, UNIV FLA, 66- Concurrent Pos: Vis prof, Southern Ill Univ, 61-62 & 65-66; vis prof, Univ PR, 61 & 66. Res: Geomorphology; climatology; labor migration and transportation; economic development in Southern Africa and British Caribbean; land use and settlement. Mailing Add: Dept of Geog Univ of Fla Gainesville FL 32611

NIDEN, ALBERT H, b Philadelphia, Pa, Aug 17, 27; m 55; c 3. INTERNAL MEDICINE. Educ: Univ Pa, AB, 49, MD, 53; Am Bd Internal Med, dipl. Prof Exp: Instr pharmacol, Univ Pa, 54-55; from instr to assoc prof internal med & chest dis, Sch Med, Univ Chicago, 57-68; prof med & chief pulmonary dis, Temple Univ, 69-73; PROF MED & CHIEF PULMONARY DIS DIV, CHARLES R DREW POSTGRAD MED SCH, 73-; PROF MED, SCH MED, UNIV SOUTHERN CALIF, 73- Concurrent Pos: Hon res asst, Inst Path, Ger, 64; hon res assoc dept anat, Kyushu, 64-65; consult, Nat Heart & Lung Inst, 68-72; consult, Comt Med & Biol Effects of Environ Pollutants, Nat Res Coun. Mem: Am Thoracic Soc; Am Fedn Clin Res; fel Am Col Chest Physicians; Am Physiol Soc. Res: Electron microscopy of lung; chest diseases; pulmonary physiology; physiology and pharmacology of the pulmonary circulation and ventilation; effects of air pollutants on lungs; lung metabolism. Mailing Add: Martin Luther King Jr Gen Hosp 12021 S Wilmington Ave Los Angeles CA 90059

NIEBAUER, JOHN J, b San Francisco, Calif, July 7, 14; m 43; c 4. MEDICINE. Educ: Stanford Univ, BA, 37, MD, 42; Am Bd Orthop Surg, dipl, 53. Prof Exp: Res orthop surg, Stanford Univ Hosp, 43-44; Gibney fel, Hosp for Spec Surg, New York, 44-45; assoc clin prof, Sch Med, Stanford Univ, 46-69; CHIEF DEPT HAND SURG, PAC MED CTR, 69- Concurrent Pos: Adj clin asst prof surg, Sch Med, Stanford Univ, 69-; clin prof orthop surg, Univ Calif, San Francisco, 67-; dir orthop & hand surg, Res Proj, Inst Med Sci, Presby Med Ctr, San Francisco. Mem: AMA; Am Orthop Asn; Am Acad Orthop Surg; Am Col Surg; Am Soc Surg of Hand. Mailing Add: Pac Med Ctr Dept of Hand Surg Clay & Webster San Francisco CA 94115

NIEBERGALL, PAUL J, b Newark, NJ, Sept 5, 32; m 57; c 4. PHARMACY. Educ: Rutgers Univ, BSc, 53; Univ Mich, MSc, 58, PhD(pharm), 62. Prof Exp: From asst prof to prof pharm, Phila Col Pharm & Sci, 61-75; DIR CORP PROD DEVELOP, MARION LABS, 75- Concurrent Pos: Co-investr, Nat Inst Allergy & Infectious Dis res grant, 65-68. Mem: Am Pharmaceut Asn; Am Chem Soc; Am Asn Cols Pharm; Acad Pharmaceut Sci. Res: Physical pharmacy; dissolution rates; solubilization of drugs through amide fusion; metal ion-penicillin interactions. Mailing Add: Marion Labs 10236 Bunker Ridge Rd Kansas City MO 64137

NIEDENZU, KURT, b Fritzlar, Ger, Mar 12, 30; m 58; c 4. INORGANIC CHEMISTRY. Educ: Univ Heidelberg, Dipl, 55, PhD(chem), 56. Prof Exp: Sci asst, Univ Heidelberg, 55-57, instr inorg & anal chem, 57; chemist, Chem Sci Div, Off Ord Res, US Dept Army, Duke Univ, 58-62, off chief scientist, Army Res Off, 62-67; res adminr, Wintershall AG, Kassel, Ger, 67-68; assoc prof, 68-73, PROF CHEM, UNIV KY, 73- Concurrent Pos: Vis prof, Gmelin Inst, Max-Planck Soc, 74-75. Honors & Awards: Sr US Scientist Award, Alexander von Humboldt Found, 74. Mem: Am Chem Soc; The Chem Soc; Ger Chem Soc. Res: Synthesis and structure of phosphorus, sulfur, boron and nitrogen compounds, especially isoelectronic systems; organometallic synthesis; spectroscopy. Mailing Add: 724 Haverhill Dr Lexington KY 40503

NIEDERER, JAMES A, physics, see 12th edition

NIEDERHAUSER, DONALD OLCOTT, chemistry, see 12th edition

NIEDERHAUSER, JOHN STRONG, b Seattle, Wash, Sept 27, 16; m 40; c 6. PLANT PATHOLOGY. Educ: Cornell Univ, BS, 39, PhD(plant path), 43. Prof Exp: Instr plant path, Cornell Univ, 39-44, asst prof, 45-47; plant pathologist, USDA, DC, 44-45; assoc plant pathologist, 47-52, plant pathologist, 52-61, DIR, INT POTATO PROG, ROCKEFELLER FOUND, MEX, 61- Mem: AAAS; Bot Soc Am; Am Phytopath Soc; Mycol Soc Am. Res: Diseases of potatoes, cereals, corn and vegetables. Mailing Add: Rockefeller Found Calle Londres 40 Mexico 6 DF Mexico

NIEDERHAUSER, WARREN DEXTER, b Akron, Ohio, Jan 2, 18; m 49; c 1. CHEMISTRY. Educ: Oberlin Col, AB, 39; Univ Wis, PhD(org chem), 43. Prof Exp: Res chemist, 43-51, head surfactant synthesis group, 51-55, chem sect, Redstone Div, 55-59, res supvr, 59-65, asst res dir, 66-73, DIR PIONEERING RES, ROHM AND HAAS CO, 73- Concurrent Pos: Mem ord sci adv panel, US Dept Army, 61-62, chem & biol warfare adv comt, 61-63, ltd war comt, Dept Defense, 61-63. Res: Plastics; surfactants; rocket propellants; ion exchange; environmental control; fibers; petroleum chemicals. Mailing Add: Rohm and Haas Co Spring House PA 19477

NIEDERMAN, JAMES CORSON, b Hamilton, Ohio, Nov 27, 24; m 51; c 4. INTERNAL MEDICINE, EPIDEMIOLOGY. Educ: Johns Hopkins Univ, MD, 49. Prof Exp: Intern med, Osler Med Serv, Johns Hopkins Hosp, 49-50; from asst resident to assoc resident, Yale-New Haven Med Ctr, 50-55; instr prev med, Sch Med, 55-58, asst prof epidemiol & prev med, 58-66, assoc clin prof, 66-76, CLIN PROF EPIDEMIOL & MED, SCH MED, YALE UNIV, 76- Concurrent Pos: Fel, Silliman Col, Yale Univ, 64-; mem bd counr, Smith Col, Mass, 73-; trustee, Kenyon Col, Ohio, 73- Mem: Am Epidemiol Soc; Infectious Dis Soc Am; Am Fedn Clin Res; Asn Teachers Prev Med. Res: Infectious mononucleosis; EB virus infections; clinical epidemiological studies of virus infections. Mailing Add: Dept of Epidemiol & Pub Health Yale Univ Sch of Med New Haven CT 06510

NIEDERMAN, ROBERT AARON, b Norwich, Conn, Jan 19, 37. MOLECULAR BIOLOGY, BIOCHEMISTRY. Educ: Univ Conn, BS, 59, MS, 61; Univ Ill, Urbana, DVM, 64, PhD(bact), 67. Prof Exp: Atomic Energy Comn fel biochem, Mich State Univ, 67-68; fel physiol chem, Roche Inst Molecular Biol, 68-70; ASST PROF BACT, RUTGERS UNIV, NEW BRUNSWICK, 70- Concurrent Pos: Merck Co Found fac develop award, Rutgers Univ, New Brunswick, 71-; USPHS res grant, Nat Inst Gen Med Sci, 73; NSF res grant, 74; USPHS res career develop award, Nat Inst Gen Med Sci, 75. Mem: AAAS; Am Soc Microbiol. Res: Membrane biochemistry; mechanisms of bacterial membrane differentiation and assembly; membrane isolation and characterization; enzymology, especially molecular aspects of enzyme regulation. Mailing Add: Dept of Microbiol Rutgers Univ New Brunswick NJ 08903

NIEDERMAYER, ALFRED O, b Munich, Ger, Aug 8, 21; US citizen; m 46; c 3. ANALYTICAL CHEMISTRY. Educ: Rutgers Univ, BA, 50. Prof Exp: SR RES SCIENTIST, E R SQUIBB & SONS, 50- Mem: Am Chem Soc. Res: Analytical separations, particularly gas chromatography. Mailing Add: E R Squibb & Sons Georges Rd New Brunswick NJ 08903

NIEDERMEIER, ROBERT PAUL, b Waukesha, Wis, Sept 8, 18; m 45; c 2. DAIRY HUSBANDRY. Educ: Univ Wis, BS, 40, MS, 42, PhD(dairy husb, biochem), 48. Prof Exp: From instr to assoc prof, 47-56, PROF DAIRY HUSB, UNIV WIS-MADISON, 57-, CHMN DEPT, 63- Mem: Am Soc Animal Sci; Am Dairy Sci Asn (vpres, 75-76); AAAS. Res: Forage preservation and utilization; mineral studies of parturient paresis in dairy cattle. Mailing Add: Dept of Dairy Sci Univ of Wis Madison WI 53706

NIEDERMEIER, WILLIAM, b Evansville, Ind, Apr 1, 23; m 45; c 4. BIOCHEMISTRY. Educ: Purdue Univ, BS, 46; Univ Ala, MS, 53, PhD(biochem), 60. Prof Exp: Res chemist, Mead Johnson & Co, 46-49; instr nutrit, Sch Med, Northwestern Univ, 49-50; ASSOC PROF BIOCHEM, MED COL, UNIV ALA, BIRMINGHAM, 60- Mem: AAAS; Am Chem Soc; Soc Appl Spectros; Am Soc Biol Chem; Am Asn Immunol. Res: Electrolyte and trace metal metabolism; chemistry of connective tissue and hyaluronic acid; carbohydrate composition of immunoglobulins. Mailing Add: Rheumatol & Clin Immunol Div Univ of Ala Dept of Med Birmingham AL 35233

NIEDERMEYER, ERNST F, b Schönberg, Ger, Jan 19, 20; US citizen; m 46; c 5. NEUROLOGY. Educ: Innsbruck Univ, MD, 47. Prof Exp: Resident neurol & psychiat, Innsbruck Univ Hosp, 48-50; French Govt fel & foreign asst neurol, Salpetriere Hosp, Paris, 50-51; resident neurol, psychiat & EEG, Innsbruck Univ Hosp, 51-55, asst prof, 55-60; from asst prof to assoc prof EEG, Univ Iowa, 60-65; ASSOC PROF EEG, JOHNS HOPKINS UNIV & ELECTROENCEPHALOGRAPHER-IN-CHG, JOHNS HOPKINS HOSP, 65- Mem: Am EEG Soc; Am Epilepsy Soc; Austrian EEG Soc; corresp mem Ger EEG Soc; Peruvian Neuropsychol Soc. Res: Clinical electroencephalography with particular emphasis on epilepsy. Mailing Add: Dept of Neurol Surg Johns Hopkins Hosp Baltimore MD 21205

NIEDERPRUEM, DONALD J, b Buffalo, NY, Sept 3, 28; m 51; c 4. MICROBIOLOGY, BIOCHEMISTRY. Educ: Univ Buffalo, BA, 49, MA, 56, PhD(biol), 59. Prof Exp: USPHS fel bact, Univ Calif, Berkeley, 59-61; from asst prof to assoc prof microbiol, 61-68, PROF MICROBIOL, MED CTR, IND UNIV, INDIANAPOLIS, 68- Concurrent Pos: Lederle med fac award, 62-65. Mem: Am Soc Microbiol; Bot Soc Am; Am Soc Plant Physiol; Am Soc Biol Chem; Soc Develop Biol. Res: Biochemical basis of cellular regulation of differentiation and morphogenesis. Mailing Add: Dept of Microbiol Ind Univ Sch of Med Indianapolis IN 46202

NIEDERREITER, HARALD GUENTHER, b Vienna, Austria, June 7, 44. NUMBER THEORY, NUMERICAL ANALYSIS. Educ: Univ Vienna, PPhD(math), 69. Prof Exp: From asst prof to assoc prof math, Southern Ill Univ, Carbondale, 69-73; lectr, Univ Ill, Urbana-Champaign, 71-72; mem, Inst Advan Study, Princeton, 73-75; PROF MATH, UNIV CALIF, LOS ANGELES, 75- Mem: Am Math Soc; Math Asn Am. Res: Uniform distribution of sequences, diophantine approximations, applications of number theory to numerical analysis, random numbers, finite fields, coding theory. Mailing Add: Dept of Math Univ of Calif Los Angeles CA 90024

NIEDRACH, LEONARD WILLIAM, b Weehawken, NJ, Sept 11, 21; m 50; c 3. CHEMISTRY. Educ: Univ Rochester, BS, 42; Harvard Univ, MA, 47, PhD(chem), 48. Prof Exp: Anal chemist, Gen Elec Co, Mass, 43-44; jr chemist, Univ Chicago, 44-45 & Monsanto Chem Co, Ohio, 45; res assoc, 48-58, phys chemist, 58-71, MGR MEMBRANE & SENSOR PROJS, GEN ELEC CO, 71- Mem: AAAS; fel Am Inst Chem; Am Chem Soc; Electrochem Soc. Res: Polarography of selenium and tellurium; spectroscopy as an analytical method; x-ray diffraction; inorganic preparations; electrodeposition; inorganic separations; fuel cells; electrochemical devices for medical applications. Mailing Add: Res & Develop Ctr Gen Elec Co PO Box 8 Schenectady NY 12301

NIEDZIELSKI, EDMUND LUKE, b Brooklyn, NY, Nov 14, 17; m 46; c 5. PETROLEUM CHEMISTRY. Educ: St John's Univ, NY, BS, 38; Fordham Univ, MS, 40, PhD(org chem), 43. Prof Exp: Asst, Fordham Univ, 38-42; res chemist 46-48, petrol lab, 48-62, RES CHEMIST JACKSON LAB, E I DU PONT DE NEMOURS & CO, INC, 62- Mem: AAAS; fel Am Inst Chem; Am Chem Soc; Am Soc Qual Control. Res: Petroleum additives; synthetic fluids and lubricants; redistribution reactions; gas separation membranes. Mailing Add: Jackson Lab E I du Pont de Nemours & Co Inc Wilmington DE 19898

NIEFORTH, KARL ALLEN, b Melrose, Mass, July 7, 36; m 58; c 4. MEDICINAL CHEMISTRY, PSYCHOPHARMACOLOGY. Educ: Mass Col Pharm, BS, 57; Purdue Univ, MS, 59, PhD(med chem), 61. Prof Exp: From asst prof to assoc prof, 61-75, PROF MED CHEM, UNIV CONN, 75-, ASST DEAN SCH PHARM, 67- Concurrent Pos: Lectr, Yale Univ, 70- Mem: AAAS; NY Acad Sci; Am Chem Soc; Am Pharmaceut Asn. Res: Design, synthesis and biological testing of compounds in the area of psychotherapeutics; hypoglycemics or hypotensives. Mailing Add: Sch of Pharm Univ of Conn Storrs CT 06268

NIEH, MARJORIE T, b Shanghai, China, Feb 5, 42; US citizen. ORGANIC CHEMISTRY. Educ: Douglass Col, BA, 64; Purdue Univ, PhD(org chem), 70. Prof Exp: Assoc org chem, Brookhaven Nat Lab, 68-69; fel, Purdue Univ, 70-71 & Mass Inst Technol, 71-72; SR RES CHEMIST, EASTMAN KODAK CO, 72- Res: Photographic system for instant photography. Mailing Add: Res Labs Eastman Kodak Co Rochester NY 14650

NIEHAUS, MERLE HINSON, b Enid, Okla, Mar 25, 33; m 54; c 2. AGRONOMY. Educ: Okla State Univ, BS, 55, MS, 57; Purdue Univ, PhD(plant breeding), 64. Prof Exp: Instr agron, Imp Ethiopian Col Agr & Mech Arts, Harar, 59-61; instr, Purdue Univ, 63; asst dir res, Advan Seed Co, 64; from asst prof to assoc prof agron, 64-70, PROF AGRON, OHIO AGR RES & DEVELOP CTR, 70-, ASSOC CHMN DEPT, 75- Mem: Am Soc Agron; Crop Sci Soc Am. Res: Soybean breeding and variety development. Mailing Add: Dept of Agron Ohio Agr Res & Develop Ctr Wooster OH 44691

NIEHAUS, WALTER G, JR, b Minneapolis, Minn, Dec 13, 37. BIOCHEMISTRY. Educ: Univ Minn, BS, 62, PhD(biochem), 64. Prof Exp: Nat Heart Inst fel biochem, Univ Ill, 65-66 & Karolinska Inst, Sweden, 66-67; res assoc, Univ Minn, 67-68; asst prof, Pa State Univ, 68-71, assoc prof, 71-75; ASSOC PROF BIOCHEM, VA POLYTECH & STATE UNIV, 75- Mem: AAAS; Am Soc Biol Chem; Am Chem Soc. Res: Enzymatic modifications of unsaturated fatty acids; determination of stereochemistry of hydroxy fatty acids; effect of amidination on structure and function of hemoglobin. Mailing Add: Dept of Biochem Va Polytech & State Univ Blacksburg VA 24061

NIEHOFF, ARTHUR HERMAN, b Indianapolis, Ind, Dec 30, 21; div; c 1. CULTURAL ANTHROPOLOGY, APPLIED ANTHROPOLOGY. Educ: Ind Univ, BA, 49; Columbia Univ, PhD(anthrop), 57. Prof Exp: Asst cur anthrop, Milwaukee Pub Mus, 51-59; community develop adv, US AID, 59-61; asst prof anthrop, Univ Wis, 61-63; sr scientist appl anthrop, George Washington Univ, 63-68; PROF ANTHROP, CALIF STATE UNIV, LOS ANGELES, 68-, CHMN DEPT, 69- Concurrent Pos: Consult, US AID, 62-, Peace Corp, 64-68 & UNESCO, 70-72; res scientist, AID, 74-75. Mem: Am Anthrop Asn; Soc Appl Anthrop. Res: Community development; education; field methods. Mailing Add: Dept of Anthrop Calif State Univ Los Angeles CA 90032

NIELL, ARTHUR EDWIN, b Lubbock, Tex, Sept 15, 42. RADIO ASTRONOMY. Educ: Calif Inst Technol, BS, 65; Cornell Univ, PhD(radio astron), 71. Prof Exp: Res fel radio astron, Queen's Univ, Ont, 71-72; res fel radio astron, 72-74, SR ENGR RADIO ASTRON & EARTH PHYSICS, JET PROPULSION LAB, 74- Concurrent Pos: Resident res assoc, Nat Res Coun-Nat Acad Sci, 72-74. Mem: Am Astron Soc; Int Radio Sci Union. Res: Testing dilatancy theory of earthquake mechanism by geodetic measurements using radio astronomical observations; investigation of the structure and physics of extragalactic radio sources. Mailing Add: Jet Propulsion Lab CPB 300 4800 Oak Grove Dr Pasadena CA 91103

NIELSEN, ALLEN MADSEN, b Ft Collins, Colo, Apr 24, 45; m 68; c 3. MICROBIAL PHYSIOLOGY. Educ: Brigham Young Univ, BS, 70, MS, 72; Ind Univ, PhD(microbiol), 76. Prof Exp: RES SCIENTIST MICROBIAL ECOL, CONTINENTAL OIL CO, 75- Mem: Am Soc Microbiol; Sigma Xi. Res: Oxidation and release of copper from chalcocite by Thiobacillus ferrooxidans; the metabolism of acetate by photosynthetic bacteria. Mailing Add: Environ Group Res Serv Div Conoco Res & Develop Dept Ponca City OK 74601

NIELSEN, ALVIN HERBORG, b Menominee, Mich, May 30, 10; m 42; c 1. PHYSICS. Educ: Univ Mich, BA, 31, MSc, 32, PhD(physics), 34. Prof Exp: Asst physics, Univ Mich, 31-34; from instr to assoc prof, 35-46, head dept, 56-69, PROF PHYSICS, UNIV TENN KNOXVILLE, 46-, DEAN COL LIB ARTS, 63- Concurrent Pos: Fulbright scholar, Inst Astrophys, Belg, 51-52; hon fel, Ohio State Univ; consult, Union Carbide Corp, Oak Ridge Nat Lab, & Off Ord Res; mem vis sci prog, Am Inst Physics-Am Asn Physics Teachers, 64-71, Nat Acad Sci-Nat Res Coun comt basic res adv to US Army Res Off, 64-70 & div chem physics, Am Inst Physics. Mem: Fel AAAS; fel Am Phys Soc; fel Optical Soc Am; Am Asn Physics Teachers; Coblentz Soc. Res: Infrared spectra of polyatomic molecules; infrared detectors. Mailing Add: Col of Lib Arts Univ of Tenn Knoxville TN 37916

NIELSEN, ARNOLD THOR, b Seattle, Wash, Sept 2, 23; m 47; c 3. ORGANIC CHEMISTRY. Educ: Univ Wash, BSc, 44, PhD(org chem), 47. Prof Exp: Res chemist, Chas Pfizer & Co, 47-48; asst prof chem, Univ Idaho, 49-52; res assoc, Purdue Univ, 52-55; instr, Rutgers Univ, 55-57; asst prof, Univ Ky, 57-59; res chemist, 59-75, HEAD ORG CHEM BR, MICHELSON LAB, NAVAL WEAPONS CTR, 75- Mem: Am Chem Soc; The Chem Soc; Int Soc Heterocyclic Chem. Res: Aldol condensation; nitroparaffins; stereochemistry; nitrogen heterocyclics. Mailing Add: Michelson Lab Code 6056 Naval Weapons Ctr China Lake CA 93555

NIELSEN, CARL EBY, b Los Angeles, Calif, Jan 22, 15; m 38; c 3. PHYSICS. Educ: Univ Calif, AB, 34, MA, 40, PhD(physics), 41. Prof Exp: Instr physics, Univ Calif, 41-45, lectr, 45-46; asst prof, Univ Denver, 46-47; from asst prof physics to assoc prof physics & astron, 47-64, PROF PHYSICS, OHIO STATE UNIV, 64- Concurrent Pos: Ford fel, Europ Orgn Nuclear Res, Geneva, 58-59; scientist, Midwestern Univs Res Asn, 60-61; consult thermonuclear div, Oak Ridge Nat Lab, 61-70; vis scientist, Culham Lab, UK Atomic Energy Authority, 66; consult, Los Alamos Sci Lab, 71- Mem: Am Phys Soc; Am Asn Physics Teachers. Res: Cloud chambers; collective phenomena in beams and plasmas; heating of magnetically confined arc plasma. Mailing Add: 8030 Sawmill Rd Dublin OH 43017

NIELSEN, DAVID GARY, b Longview, Wash, Nov 18, 43; m 64; c 1. ENTOMOLOGY. Educ: Willamette Univ, BA, 66; Cornell Univ, MS, 69, PhD(entom), 70. Prof Exp: ASST PROF ENTOM, OHIO AGR RES & DEVELOP CTR, OHIO STATE UNIV, 70- Mem: Entom Soc Am. Res: Behavioral ecology and suppression of insects which attack woody ornamental plants. Mailing Add: Dept of Entom Ohio Agr Res & Develop Ctr Wooster OH 44691

NIELSEN, DONALD R, b Phoenix, Ariz, Oct 10, 31; m 53; c 5. SOIL PHYSICS. Educ: Univ Ariz, BS, 53, MS, 54; Iowa State Univ, PhD(soil physics), 58. Prof Exp: Res assoc soil physics, Iowa State Univ, 54-58; from asst prof to assoc prof, 58-68, PROF SOIL PHYSICS, UNIV CALIF, DAVIS, 68-, DIR, KEARNEY FOUND SOIL SCI, 70-, ASSOC DEAN RES, 71-, CHMN DEPT LAND, AIR & WATER RESOURCES, 75- Concurrent Pos: NSF sr fel, 65-66; consult, Int Atomic Energy Agency, Vienna, Austria, 74-75. Mailing Add: Dept Land Air & Water Resources Univ of Calif Davis CA 95616

NIELSEN, DONALD R, b Oak Park, Ill, Oct 11, 30; m 54; c 3. ORGANIC CHEMISTRY. Educ: Knox Col, BA, 52; Univ Kans, PhD(org chem), 56. Prof Exp: Res chemist, Chem Div, Pittsburgh Plate Glass Co, 56-64, res supvr, 64-68, SR RES ASSOC, CHEM DIV, PPG INDUSTS, INC, 68- Mem: Am Chem Soc. Res: Organic synthesis. Mailing Add: 4633 Congressional Corpus Christi TX 78413

NIELSEN, FORREST HAROLD, b Junction City, Wis, Oct 26, 41; m 64; c 2. NUTRITION, BIOCHEMISTRY. Educ: Univ Wis-Madison, BS, 63, MS, 66, PhD(biochem), 67. Prof Exp: Res chemist, Beltsville, Md, 69-70, RES CHEMIST, HUMAN NUTRIT LAB, AGR RES SERV, USDA, 70- Concurrent Pos: Res assoc biochem, Univ NDak, 71- Mem: Am Inst Nutrit; Soc Environ Geochem & Health; Soc Exp Biol & Med. Res: Trace element nutrition and metabolism; nickel, vanadium and the newer essential trace elements. Mailing Add: USDA ARS Human Nutrit Lab Box 7166 Univ Sta Grand Forks ND 58201

NIELSEN, G HOWARD, b Orlando, Fla, Feb 3, 41; m 62; c 2. MATHEMATICS. Educ: SDak State Univ, BS, 63; Univ Colo, Boulder, MA, 66, PhD(math), 69. Prof Exp: Trainee statistician, USDA, 60-62; comput programmer-analyst, Control Data Corp, 63-64; ASST PROF MATH, REGIS COL, 69- Mem: Math Asn Am. Res: The sawtooth function of number theory. Mailing Add: Dept of Math Regis Col Denver CO 80221

NIELSEN, GERALD ALAN, b Frederic, Wis, Nov 10, 34; m 55; c 4. SOIL SCIENCE, ECOLOGY. Educ: Univ Wis, BS, 58, MS, 60, PhD(soil sci), 63. Prof Exp: Asst prof soil sci, Univ Wyo team-US Agency Int Develop, Afghanistan, 63-67; assoc prof, 67-73, PROF SOIL SCI, MONT STATE UNIV, 73- Mem: AAAS; Soil Sci Soc Am; Am Soc Agron; Soil Conserv Soc Am; Int Soc Soil Sci. Res: Soil genesis, classification and ecology; interpretation of soil survey for land use planning; soil inventories and land potential evaluation. Mailing Add: Dept of Plant & Soil Sci Mont State Univ Bozeman MT 59715

NIELSEN, HARALD CHRISTIAN, b Chicago, Ill, Apr 18, 30; m 53; c 3. BIOCHEMISTRY. Educ: St Olaf Col, BA, 52; Mich State Univ, PhD(biochem), 57. Prof Exp: CHEMIST, NORTHERN REGIONAL RES LAB, USDA, 57- Mem: AAAS; Am Chem Soc; Am Asn Cereal Chem. Res: Isolation and physical-chemical characterization of cereal grain proteins. Mailing Add: Northern Regional Res Lab US Dept of Agr 1815 N University Peoria IL 61604

NIELSEN, JAMES WILLARD, b St Michael, Nebr, July 7, 24; m 50; c 2. SOLID STATE CHEMISTRY. Educ: Nebr State Teachers Col, Kearney, BS, 45; Univ Minn, MS, 48; Univ Iowa, PhD(phys chem), 53. Prof Exp: Instr chem, Nebr State Teachers Col, Kearney, 48-50; mem tech staff, Bell Tel Labs, Inc, NJ, 53-60; mgr solid state lab, Airtron Div, Litton Industs, 60-67; SUPVR, BELL LABS, 67- Mem: AAAS; Am Asn Crystal Growth. Res: Crystal growth; crystallography; phase equilibrium in condensed systems; chemistry of solid state. Mailing Add: Bell Labs 7c-401 Mountain Ave Murray Hill NJ 07974

NIELSEN, JENS JUERGEN, b Bleckede, Ger, Feb 10, 27; m 58; c 4. PLANT PATHOLOGY. Educ: Univ Göttingen, PhD(plant path), 60. Prof Exp: Fel physiol parasitism, Nat Res Coun Can, 60-62 & Ger Res Asn, 62-63; PLANT PATHOLOGIST, RES BR, AGR CAN, 63- Mem: Can Phytopath Soc. Res: Cereal diseases caused by smut fungi, particularly Ustilago species. Mailing Add: Agr Can Res Sta 25 Dafoe Rd Winnipeg MB Can

NIELSEN, JULIAN MOYES, b Ogden, Utah, Mar 24, 21; m 46; c 5. ENVIRONMENTAL CHEMISTRY. Educ: Univ Wyo, BS, 42; Stanford Univ, MS, 48; Univ Southern Calif, PhD(phys chem), 51. Prof Exp: Res chemist, Naval Ord Test Sta, 50-53 & Gen Elec Co, 53-65; RES MGR, PAC NORTHWEST LABS, BATTELLE MEM INST, 65- Mem: Health Physics Soc; Am Chem Soc. Res: Diffusion of colloids and electrolytes; radiochemistry; nuclear chemistry. Mailing Add: 1611 Sunset St Richland WA 99352

NIELSEN, KAJ LEO, b Nyker, Denmark, Dec 3, 14; nat US; m 43; c 2. MATHEMATICS. Educ: Univ Mich, AB, 36; Syracuse Univ, MA, 37; Univ Ill, PhD(math), 40. Prof Exp: Asst math, Syracuse Univ, 36-37; asst math, Univ Ill, 37-40, instr, 40-41; Carnegie fel, Brown Univ, 41, instr, 41-42; from instr to asst prof, La State Univ, 42-45; sr mathematician, US Naval Ord Plant, Ind, 45-48, head math div, 48-58; chief opers anal, Allison Div, Gen Motors Corp, 58-60; head anal staff, Defense Systs Div, Gen Motors Corp, 60-61; prof, Butler Univ, 61-63; dir systs anal div, Battelle Mem Inst, Ohio, 63-71; lectr, 57-60, HEAD DEPT MATH, BUTLER UNIV, 71- Concurrent Pos: Proj anal engr, Chance Vought Aircraft, Inc, Conn, 44-45; lectr, Purdue Univ, 55-58. Mem: Am Math Soc; Math Asn Am. Res: Partial differential equations of elliptic type; exterior ballistics; fire control; teaching of applied mathematics; numerical methods; operations and systems analysis; man-machine and management systems. Mailing Add: 1224 Ridge Rd Carmel IN 46032

NIELSEN, KENNETH FRED, b Cardston, Alta, July 3, 27; m 47; c 5. SOIL FERTILITY, PLANT PHYSIOLOGY. Educ: Brigham Young Univ, BS, 49; Ohio State Univ, PhD(soil chem & plant nutrit), 52. Prof Exp: Asst prof agron, Univ Maine, 52-55; head, Soil Fertil Unit, Can Dept Agr, 55-59, Soil Sect, Exp Farm, Sask, 59-65; chief agronomist, 65-66, dir planning, 66-69, dir mkt & agron, 69-72, CHIEF EXEC OFFICER & GEN MGR, WESTERN COOP FERTILIZERS LTD, 72- Concurrent Pos: Fel, Rothamsted Exp Sta, 62-63. Mem: Soil Sci Soc Am; fel Can Soc Soil Sci (pres, 70-71); Agr Inst Can; Int Soc Soil Sci. Res: Plant nutrition. Mailing Add: PO Box 2500 Calgary AB Can

NIELSEN, KURT EFRAIM, nuclear physics, see 12th edition

NIELSEN, LARRY DENNIS, b Hazelwood, Minn, Feb 12, 39; m 63; c 1. BIOCHEMISTRY. Educ: St Olaf Col, BA, 61; Univ Colo, Denver, PhD(biochem), 66. Prof Exp: USPHS fel biochem, Univ Wash, 66-68, training grant, pediat, 68-71; RES SCIENTIST BIOCHEM, NAT JEWISH HOSP & RES CTR, 71- Res: Cell surface properties involved in intercellular adhesion of normal and malignant cells. Mailing Add: Nat Jewish Hosp & Res Ctr Denver CO 80206

NIELSEN, LAWRENCE ARTHUR, b Minneapolis, Minn, Aug 7, 34; m 62; c 2. ORGANIC CHEMISTRY. Educ: Univ Minn, BS, 56; Univ Nebr, MS, 59, PhD(org chem), 62. Prof Exp: Res chemist, Chemstrand Res Ctr, Inc, Monsanto Co, 62-70; HEAD PATENT LIAISON, BURROUGHS WELLCOME CO, 70- Mem: Am Chem Soc; Sigma Xi. Res: Heterocyclic synthesis; medicinal chemistry. Mailing Add: Burroughs Wellcome Co 3030 Cornwallis Rd Research Triangle Park NC 27709

NIELSEN, LAWRENCE ERNIE, b Pilot Rock, Ore, Dec 17, 17; m 42; c 1. POLYMER SCIENCE. Educ: Pac Univ, AB, 40; State Col Wash, MS, 42; Cornell Univ, PhD(phys chem), 45. Prof Exp: Asst phys chem, State Col Wash, 40-42; lab asst, Cornell Univ, 42-45; phys chemist, 45-55, SR SCIENTIST, MONSANTO CO, 55- Concurrent Pos: Fel, Harvard Univ, 52; leader, Juneau Ice Field Proj, Alaska, 53; affiliate prof, Wash Univ, 65- Mem: Am Chem Soc; Am Phys Soc; Soc Rheol; Glaciol Soc. Res: Molal volumes of electrolytes; Raman, infrared and ultraviolet spectroscopy; fractionation of proteins; molecular structure of high polymers and its relation to physical and mechanical properties; properties of composite materials; glaciology and flow properties of ice. Mailing Add: Corp Res Dept Monsanto Co St Louis MO 63166

NIELSEN, LEWIS THOMAS, b Salt Lake City, Utah, Aug 6, 20; m 75. ENTOMOLOGY. Educ: Univ Utah, BA, 41, MA, 47, PhD, 55. Prof Exp: From instr to assoc prof, 46-66, PROF BIOL & ENTOM, UNIV UTAH, 66- Mem: Am Mosquito Control Asn (vpres, 75-76); Entom Soc Am. Res: Systematics, biology and distribution of mosquitos of western North America and Mexico; medical entomology. Mailing Add: Dept of Biol Univ of Utah Salt Lake City UT 48112

NIELSEN, LOWELL WENDELL, b Weston, Idaho, Apr 23, 10; m 35; c 3. PLANT PATHOLOGY. Educ: Utah State Univ, BS, 35, MS, 37; Cornell Univ, PhD(plant path), 41. Prof Exp: Asst prof bot, Exp Sta, NC State Univ, 41-44; pathologist, Calif Packing Corp, Ill, 44-45; assoc horticulturist, Idaho Agr Exp Sta, 45-48; assoc prof plant path, 48-54, PROF PLANT PATH, NC STATE UNIV, 54- Concurrent Pos: Fulbright res scholar, NZ, 64-65; adv, US AID-NC State Univ mission to Peru, 66-71. Mem: Am Phytopath Soc; Potato Asn Am. Res: Plant virus diseases; diseases of Irish and sweet potatoes; bacterial decays. Mailing Add: Dept of Plant Path NC State Univ Raleigh NC 27607

NIELSEN, MERLYN KEITH, b Omaha, Nebr, Oct 9, 48; m 70. ANIMAL BREEDING. Educ: Univ Nebr-Lincoln, BS, 70; Iowa State Univ, MS, 72, PhD(animal sci), 74. Prof Exp: ASST PROF ANIMAL SCI, UNIV NEBR-LINCOLN, 74- Mem: Am Soc Animal Sci; Biomet Soc. Res: Accurate identification of additive genetic differences in beef cattle, especially national sire evaluation. Mailing Add: 215 Marvel Baker Hall Univ of Nebr Lincoln NE 68583

NIELSEN, MORRIS LOWELL, b Mission Hill, SDak, June 22, 14; m 39; c 3. CHEMISTRY. Educ: Yankton Col, BA, 34; Univ SDak, MA, 35; Univ Wis, PhD(chem), 41. Prof Exp: Chemist, Midcontinent Petrol Corp, Okla, 37-38; asst instr chem, Univ Wis, 38-40; res chemist, Monsanto Res Corp, 41-46, group leader, 46-66, from patent agent to sr patent solicitor, 66-68, patent attorney, 68-70; PATENT ATTORNEY, UPJOHN CO, 70- Mem: Am Chem Soc. Res: Electrodeposition of tungsten alloys; development of processes and applications of inorganic compounds; radiochemistry; flameproofing; phosphorous-nitrogen compounds; organo-inorganic polymers; chemical patents. Mailing Add: Upjohn Co 301 Henrietta St Kalamazoo MI 49001

NIELSEN, N OLE, b Edmonton, Alta, Mar 3, 30; m 55; c 3. VETERINARY PATHOLOGY. Educ: Univ Toronto, DVM, 56; Univ Minn, PhD(vet path), 63. Prof Exp: Private practice, 56-57; res assoc lab animal care & radiation res, Med Dept, Brookhaven Nat Lab, 60-61; from instr to assoc prof, 57-68, head dept, 68-74, PROF VET PATH, WESTERN COL VET MED, UNIV SASK, 68-, DEAN, 74- Concurrent Pos: Med Res Coun Can vis scientist, Int Escherichia Ctr, Copenhagen, 70-71. Mem: AAAS; Am Col Vet Path; Am Vet Med Asn; Can Vet Med Asn. Res: Relationship of E coli to enteric disease; diseases of swine; enteric pathology. Mailing Add: Western Col of Vet Med Univ of Sask Saskatoon SK Can

NIELSEN, NORMAN ARNOLD, b Ipswich, Mass, Aug 11, 20; m 43; c 3. CHEMISTRY. Educ: Harvard Univ, SB, 42. Prof Exp: Metall engr, 42-50, res supvr, 50-57, res assoc, 57-67, RES FEL CHEM, E I DU PONT DE NEMOURS & CO, INC, 67- Concurrent Pos: Gillett Mem Lectr, Am Soc Test & Mat, 70. Honors & Awards: Howe Medal, Am Soc Metals, 52. Mem: Am Chem Soc; Am Soc Metals; Electrochem Soc; Electron Micros Soc Am; Am Soc Test & Mat. Res: Fine structure of metal surfaces; carbide precipitation in stainless steel; corrosion behavior and passivity of stainless steel; electron metallography; catalysis by metals; physical metallurgy of ancient metals. Mailing Add: 2 Walnut Ridge Rd Wilmington DE 19807

NIELSEN, NORMAN RUSSELL, b Pittsburgh, Pa, Sept 8, 41; m 63; c 1. COMPUTER SCIENCE, INFORMATION SCIENCE. Educ: Pomona Col, BA, 63; Stanford Univ, MBA, 65, PhD(opers & systs anal), 67. Prof Exp: From asst prof to assoc prof opers & systs anal, Stanford Univ, 66-73; sr res engr, 73-74, MGR INFO SYSTS GROUP, STANFORD RES INST, 74- Mem: AAAS; Asn Comput Mach; Inst Mgt Sci. Res: System simulation; computer resource allocation; computer modeling; simulation languages; computer networks; management information systems; computer security; computer service management; computer performance measurement and evaluation. Mailing Add: Info Syst Group Stanford Res Inst 333 Ravenswood Ave Menlo Park CA 94025

NIELSEN, PAUL HERRON, b Berkeley, Calif, June 14, 43; m 69. SOLID STATE PHYSICS. Educ: Univ Chicago, BS, 64, MS, 65, PhD(physics), 70. Prof Exp: Assoc scientist, 69-71, SCIENTIST PHYSICS, XEROX CORP, 71- Mem: Am Phys Soc. Res: Ultraviolet photoemission spectroscopy; energy levels and electronics structures of molecules, solids, and interfaces; photosensitization. Mailing Add: Webster Res Ctr Xerox Corp 800 Phillips Rd Webster NY 14580

NIELSEN, PAUL LIVINGSTONE, b Plentywood, Mont, 28, 41; m 67; c 1. POLYMER CHEMISTRY. Educ: Concordia Col, BA, 63; NDak State Univ, MS, 70, PhD(polymers, coatings), 73. Prof Exp: CHEMIST, EASTMAN KODAK CO, 73- Mem: Sigma Xi; Am Chem Soc. Res: Polyurethanes for use as coating binders, elastomeric types, both lacquer types and two-package reactive types. Mailing Add: 890 Calm Lake Circle Rochester NY 14612

NIELSEN, PETER ADAMS, b Evanston, Ill, Oct 12, 26; m 52; c 2. BACTERIOLOGY. Educ: Williams Col, BA, 50; Columbia Univ, MA, 56, PhD(bot), 60. Prof Exp: Biologist, Lederle Labs, Am Cyanamid Co, 51-55; asst bot, Barnard Col, Columbia Univ, 55-56, univ, 57-59; microbiologist, Lederle Labs, Am Cyanamid Co, NY, 59-67; HEAD MICROBIOL SECT, VICK DIV RES & DEVELOP, RICHARDSON-MERRELL, INC, 67- Mem: Am Soc Microbiol; Soc Indust Microbiol. Res: Microbial physiology; microbiology of skin and oral cavity; antimicrobials; preservative evaluation; microbial content of pharmaceuticals and cosmetics. Mailing Add: Vick Div Res & Develop Richardson-Merrell Inc Bradford Rd Mount Vernon NY 10553

NIELSEN, PETER JAMES, b North Platte, Nebr, Feb 21, 38; m 61; c 2. CELL PHYSIOLOGY, PROTOZOOLOGY. Educ: Midland Col, BS, 60; Univ Nebr, MS, 65, PhD(zool physiol), 68. Prof Exp: Teaching asst physiol, Univ Nebr, 64-66; ASST PROF BIOL SCI, WESTERN ILL UNIV, 68- Mem: AAAS; Am Soc Zool; Soc Protozool. Res: Effects on the phenomena of aging using the protozoan, Tetrahymena pyriformis, as a research tool; continuous culture of Tetrahymena as a tool for studying metabolism during division. Mailing Add: Dept of Biol Sci Western Ill Univ Macomb IL 61455

NIELSEN, PETER TRYON, b Durham, NC, Dec 24, 33; m 60; c 2. PLANT PHYSIOLOGY, BIOPHYSICS. Educ: Duke Univ, BS, 57; Univ NC, Chapel Hill, PhD(algal physiol), 65. Prof Exp: NIH fel biophys, East Anglia, 65-66; asst prof plant physiol & biophys, State Univ NY Col Plattsburgh, 66-70; ASSOC PROF BIOL, MADISON COL, 70- Mem: Am Soc Plant Physiol; Bot Soc Am. Res: Ion accumulation and transport; photosynthesis and role of light reactions in ion transport; effects of heavy metal ions and algal physiology. Mailing Add: Dept of Biol Madison Col Harrisonburg VA 22801

NIELSEN, PHILIP EDWARD, b Chicago, Ill, July 18/x44;xmx71 18, 44; m 71; c 1. PLASMA PHYSICS, SOLID STATE PHYSICS. Educ: Ill Inst Technol, BS, 66; Case Western Reserve Univ, MS, 68, PhD(physics), 70. Prof Exp: Physicist, Air Force Weapons Lab, US Air Force, 70-74; ASST PROF PHYSICS, AIR FORCE INST TECHNOL, 74- Mem: AAAS; Am Phys Soc. Res: Interaction of high-intensity lasers with matter; laser-plasma interactions; transport phenomena in metals and alloys. Mailing Add: Dept of Physics Air Force Inst of Technol Wright-Patterson AFB OH 45433

NIELSEN, RICHARD LEROY, b Los Angeles, Calif, May 21, 34; m 69; c 2. ECONOMIC GEOLOGY. Educ: Calif Inst Technol, BS, 55, MS, 57; Univ Calif, Berkeley, PhD(geol), 64. Prof Exp: Explor geologist, Bear Creek Mining Co, 57-59, US Steel Corp, 59 & Hecla Mining Co, 64; asst prof geol, Univ Nev, Reno, 63-64; RES GEOLOGIST, KENNECOTT EXPLOR INC, 64- Mem: Fel Geol Soc Am; Soc Econ Geologists; Australasian Inst Mining Metall; Sigma Xi; Geol Soc Australia. Res: Geology of porphyry copper deposits. Mailing Add: Kennecott Explor Inc 2300 W 1700 S Salt Lake City UT 84104

NIELSEN, ROBERT LEROY, physics, see 12th edition

NIELSEN, ROBERT PETER, b New Brunswick, NJ, Mar 14, 37; m 61; c 4. INDUSTRIAL CHEMISTRY. Educ: Rutgers Univ, BS, 58; Univ Fla, PhD(inorg chem), 62. Prof Exp: Chemist, Shell Develop Co, Calif, 62-66; sr technologist, Shell Chem Co, NY, 66-68; sr chemist, Process Lab Dept, Mfg Tech Ctr, Tex, 68-72; STAFF CHEMIST & PROJ LEADER, SHELL DEVELOP CO, 72- Mem: Am Chem Soc; The Chem Soc. Res: Metallo-organic; transition metal homogeneous catalysis; heterogeneous catalysis; ethylene oxide chemistry; oxychlorination chemistry. Mailing Add: Westhollow Res Ctr Shell Develop Co Houston TX 77001

NIELSEN, STUART DEE, b Green River, Wyo, Oct 26, 32; m 54; c 4. ORGANIC CHEMISTRY, POLYMER CHEMISTRY. Educ: Univ Wyo, BS, 54; Univ Wash, PhD(org chem), 62. Prof Exp: Sr chemist, Rohm and Haas Co, Pa, 62-66; sr res chemist, 66-68, RES SCIENTIST, GEN TIRE & RUBBER CO, 68- Mem: Am Chem Soc. Res: Polymer synthesis and properties; kinetics and mechanisms of polymerization processes; pollution control methods and analysis. Mailing Add: 3468 S Marcella Ave Stow OH 44224

NIELSEN, SVEND WOGE, b Herning, Denmark, Apr 4, 26; nat US; m 52; c 3. VETERINARY PATHOLOGY. Educ: Herning Gym, Denmark, Artium, 45; Royal Vet Col, DVM, 51; Ohio State Univ, MSc, 57, PhD(path), 59. Prof Exp: Res path, Angell Mem Animal Hosp, Boston, 51-52; res asst, Ont Vet Col, Can, 52-53, lectr, 53-55, asst prof, 55; from instr to assoc prof vet path, Ohio State Univ, 55-60; PROF PATHOBIOL, UNIV CONN, 60- Concurrent Pos: NIH spec fel, Univ Cambridge, 67-68; mem lab comt meat & poultry inspection, consumer & mkt serv, USDA, 69-; dir, WHO Collab Lab Urogenital Tumors of Animals, 68-; dir, Northeastern Res Ctr Wildlife Dis, 72-; vis prof, Cornell Univ, 74. Mem: AAAS; Am Vet Med Asn; Am Col Vet Path (pres, 71-72); Conf Res Workers Animal Dis; Int Acad Path. Res: Comparative oncology; pathology of vitamin A deficiency and vitamin A toxicosis in animals; diseases of wildlife in the northeastern United States; classification of urogenital neoplasms of animals. Mailing Add: Dept of Pathobiol Univ of Conn Storrs CT 06268

NIELSEN, VERNER HENRY, b Copenhagen, Denmark, Feb 16, 10; m 34; c 1. FOOD SCIENCE. Educ: Iowa State Univ, BS, 43, PhD(chem), 53. Prof Exp: From instr to assoc prof dairy indust, 43-57, prof dairy & food indust, 57-74, head dept dairy & food indust, 58-69, head dept food technol, 69-74, PROF FOOD TECHNOL, IOWA STATE UNIV, 74- Concurrent Pos: Consult dairy mfg plants, Iowa, Ore & Tex, 40-60; mem panel experts dairy educ, Food Agr Orgn, 72- Mem: AAAS; Am Chem Soc; Am Dairy Sci Asn; Inst Food Technol. Res: Manufacturing and marketing of fluid milk and other diary products; chemistry of milk and milk products. Mailing Add: Dept of Food Technol Iowa State Univ Ames IA 50011

NIELSEN, WALTER DAVID, organic chemistry, see 12th edition

NIELSON, CLAIR W, b Pocatello, Idaho, Dec 10, 35. PHYSICS. Educ: Mass Inst Technol, SB, 57, PhD(physics), 62. Prof Exp: From instr to asst prof physics, Swarthmore Col, 63-68; MEM STAFF, LOS ALAMOS SCI LAB, 68- Mem: Am Phys Soc. Res: Atomic and molecular theory; numerical simulation of plasma. Mailing Add: Los Alamos Sci Lab Box 1663-P-18 Los Alamos NM 87544

NIELSON, DENNIS LON, b Urbana, Ill, Jan 13, 48; m 70. ECONOMIC GEOLOGY. Educ: Beloit Col, BA, 70; Dartmouth Col, MA, 72, PhD(geol), 74. Prof Exp: STAFF GEOLOGIST, ANACONDA CO, 74- Mem: Geol Soc Am; Am Inst Mining Engrs. Res: Investigation of the geology of uranium ore deposits and the geochemical and physical processes responsible for the concentration of uranium. Mailing Add: Anaconda Co Uranium Div 1849 W North Temple Salt Lake City UT 84116

NIELSON, DIANNE RUTH GERBER, b Elgin, Ill, Apr 2, 48; m 70. EXPLORATION GEOLOGY. Educ: Beloit Col, BA, 70; Dartmouth Col, MA, 72, PhD(geol), 74. Prof Exp: Res asst seismic study, Dartmouth Col, 70; explor geologist, Gt Lakes Explor Co, 71; petrologist, 70, STAFF GEOLOGIST, ANACONDA CO, 74- Mem: Geol Soc Am; Am Inst Mining Engrs. Res: Examination, interpretation, and evaluation of the geochemical, petrologic and structural controls of uranium mineralization in igneous, metamorphic and sedimentary environments. Mailing Add: Anaconda Co Uranium Div 1849 W North Temple Salt Lake City UT 84116

NIELSON, ELDON DENZEL, b Salt Lake City, Utah, Dec 4, 20; m 42; c 3. BIOCHEMISTRY. Educ: Univ Utah, AB, 46; Univ Ill, PhD(biochem), 48. Prof Exp: Instr biochem, Univ Southern Calif, 48-49; sect head, Upjohn Co, 49-56; head biochem dept, Armour Pharmaceut Co, 56-58; head biochem, Sterling-Winthrop Res Inst, 58-62; mgr biol res div, RJ Reynolds Tobacco Co, 62-69; vpres phys sci, Mead Johnson Res Ctr, 69-75, DIR LICENSING, MEAD JOHNSON & CO, 75- Mem: AAAS; Soc Indust Microbiol; Am Chem Soc. Res: Adrenal steroids; fermentation; natural products; tobacco; pulmonary physiology; pharmaceutical synthesis; chemical development. Mailing Add: 5808 Woodridge Dr Evansville IN 47712

NIELSON, GEORGE MARIUS, b Wadsworth, Ohio, May 17, 34. MATHEMATICS. Educ: Ohio Wesleyan Univ, BA, 56; Univ Wis, MS, 57, PhD(math), 63. Prof Exp: Instr math, Ohio Wesleyan Univ, 59-60; asst prof, 63-70, ASSOC PROF MATH, KALAMAZOO COL, 70- Mem: Am Math Soc; Math Asn Am. Res: Lie algebras. Mailing Add: Dept of Math Kalamazoo Col Kalamazoo MI 49001

NIELSON, GREGORY MORRIS, mathematics, computer science, see 12th edition

NIELSON, HARALD HERBORG, physics, deceased

NIELSON, HOWARD CURTIS, b Richfield, Utah, Sept 12, 24; m 48; c 7. STATISTICS, MATHEMATICS. Educ: Univ Utah, BS, 47; Univ Ore, MS, 49; Stanford Univ, MBA, 56, PhD(bus admin & statist), 58. Prof Exp: Actg instr math, Univ Utah, 46-47; asst, Univ Ore, 47-49; sr statistician, Calif & Hawaii Sugar Refining Corp, 49-51; res economist, Stanford Res Inst, 51-57; assoc prof econ, 57-60, assoc prof statist, 60-61, chmn dept, 60-63, PROF STATIST, BRIGHAM YOUNG UNIV, 61-, DIR, CTR BUS & ECON RES, 71- Concurrent Pos: Consult, Hercules Powder

Co, 60-65; mgr & consult, C-E-I-R, Inc, 63-65; prin scientist, GCA Corp, 65-67 & Fairchild Semiconductor Corp, 67; consult, EG&G, INC, 67; econ develop consult, Ford Found, Jordan, 70; mem, Gov Econ Resources Adv Coun. Mem: Am Statist Asn; Sigma Xi. Res: Economic forecasting; demographic studies and projections; statistical methods in industry; sampling survey methods; experimental design; probability; reliability; operations research. Mailing Add: Ctr for Bus & Econ Res Brigham Young Univ Provo UT 84601

NIELSON, KENT, plant breeding, see 12th edition

NIELSON, MERVIN WILLIAM, b Provo, Utah, Apr 7, 27; m 56; c 3. ENTOMOLOGY. Educ: Utah State Univ, BS, 49, MS, 50; Ore State Col, PhD(entom), 55. Prof Exp: Asst entomologist, Univ Ariz, 55-56; entomologist, 56-68, RES LEADER ENTOM RES DIV, AGR RES SERV, USDA, 68-; ASSOC PROF, UNIV ARIZ, 68- Concurrent Pos: Asst, Utah State Univ, 49-50; asst & instr, Ore State Col, 51-55. Mem: AAAS; Entom Soc Am; Soc Syst Zool. Res: Taxonomy of the Cicadellidae; biology of leafhoppers and aphids; insect transmission of plant viruses; development of crop resistance to insects. Mailing Add: Entom Res Div Agr Res Serv US Dept of Agr 2000 E Allen Rd Tucson AZ 85719

NIELSON, NANCY ANNE, microbiology, see 12th edition

NIELSON, PAUL ELLIS, psychiatry, see 12th edition

NIELSON, READ R, b Omaha, Nebr, Aug 4, 28; m 54; c 4. PHYSIOLOGY. Educ: Grinnell Col, BA, 50; Univ Iowa, MS, 52; Marquette Univ, PhD(physiol), 61. Prof Exp: Asst prof, 61-67, ASSOC PROF ZOOL, MIAMI UNIV, 67- Concurrent Pos: NIH res grant, 62-65. Mem: Am Physiol Soc. Res: Thyroid physiology and regulation of metabolism; effect of electrical stimulation on glycogen concentration in skeletal muscle during acute inanition; metabolic changes in the intact rat and excised tissues after thyroidectomy; changes in succinic dehydrogenase activity related to feeding. Mailing Add: Dept of Zool Miami Univ Oxford OH 45056

NIEM, ALAN RANDOLPH, b New York, NY, Mar 7, 44; m 67. GEOLOGY. Educ: Antioch Col, BS, 66; Univ Wis-Madison, MS, 68, PhD(geol), 71. Prof Exp: Asst geol, Univ Wis-Madison, 66-70; ASST PROF ORE STATE UNIV, 70- Mem: Am Asn Petrol Geol; Soc Econ Paleont & Mineral; Geol Soc Am. Res: Sedimentation; sedimentary petrography; volcaniclastic sediments; stratigraphy of geosynclinal rocks of the south-central and western United States; hydrogeology. Mailing Add: Dept of Geol Ore State Univ Corvallis OR 97330

NIEMAN, GEORGE CARROLL, b Dayton, Ohio, Dec 25, 38; m 60; c 3. PHYSICAL CHEMISTRY, SPECTROSCOPY. Educ: Carnegie Inst Technol, BS, 61; Calif Inst Technol, PhD(chem), 65. Prof Exp: Asst prof chem, Univ Rochester, 64-70; ASSOC PROF CHEM, MUSKINGUM COL, 70- Mem: Am Chem Soc; Am Phys Soc. Res: Molecular crystals; nonradiative transitions; energy transfer; triplet states. Mailing Add: Dept of Chem Muskingum Col New Concord OH 43762

NIEMAN, RICHARD HOVEY, b Pasadena, Calif, Nov 7, 22; m 46; c 2. PLANT PHYSIOLOGY. Educ: Univ Southern Calif, AB, 49, MS, 53; Univ Chicago, PhD(plant physiol), 55. Prof Exp: Asst histol & morphol, Univ Chicago, 51-54, plant physiol, 54, res assoc biochem, 55-57; PLANT PHYSIOLOGIST, US SALINITY LAB, 57- Mem: AAAS; Am Soc Plant Physiol; Am Inst Biol Sci; Sigma Xi; NY Acad Sci. RES: Influence of salinity and drought on ion uptake, metabolism, and bioenergetics of plant cells; physiological-biochemical basis of salt tolerance of plants. Mailing Add: US Salinity Lab PO Box 672 Riverside CA 92502

NIEMAN, TIMOTHY ALAN, b Cincinnati, Ohio, Dec 31, 48; m 70. ANALYTICAL CHEMISTRY. Educ: Purdue Univ, BS, 71; Mich State Univ, PhD(chem), 75. Prof Exp: Grad asst chem, Mich State Univ, 71-75; ASST PROF ANAL CHEM, UNIV ILL, URBANA, 75- Mem: Fel Am Chem Soc; Optical Soc Am. Res: Analytical spectroscopy and kinetics; chemical instrumentation and interactive computer control; use of solution chemiluminescence for trace metal analysis, qualitative and quantitative. Mailing Add: Sch of Chem Sci Roger Adams Lab Univ of Ill Urbana IL 61801

NIEMANN, RALPH HENRY, b Farley, Mo, Mar 16, 22; m 48; c 3. MATHEMATICS. Educ: Park Col, BA, 47; Purdue Univ, MS, 49, PhD(math), 54. Prof Exp: Assoc prof math, Worcester Polytech Inst, 54-59; asst prof, 59-63, chmn math sect, Dept Math & Statist, 67-68, actg chmn dept, 68-69, PROF MATH, COLO STATE UNIV, 63- Concurrent Pos: Lectr, Math Asn High Sch Lectr Prog, 61, 62; consult, Summer Sci Inst Prog, India, 64-65. Mem: Math Asn Am. Res: Mathematical analysis. Mailing Add: Dept of Math Colo State Univ Ft Collins CO 80523

NIEMANN, THEODORE FRANK, b Burlington, Iowa, July 31, 39. POLYMER CHEMISTRY, COMPUTER SCIENCE. Educ: Univ Iowa, BS, 61; Univ Kans, PhD(chem), 67. Prof Exp: Res chemist, BF Goodrich Res Ctr, Ohio, 67-71; SR CHEMIST, DWIGHT P JOYCE RES CTR, GLIDDEN-DURKEE DIV, SCM CORP, STRONGSVILLE, 71- Mem: Am Chem Soc. Res: Physical properties of polymers; computer applications in chemistry and laboratory automation. Mailing Add: 6680 Chaffee Ct Brecksville OH 44141

NIEMCZYK, HARRY D, b Grand Rapids, Mich, July 17, 29; m 52; c 3. ENTOMOLOGY. Educ: Mish State Univ, BS, 57, MS, 58, PhD(entom), 61. Prof Exp: Asst instr entom, Mich State Univ, 60-61; entomologist, Can Dept Agr, 61-64; from asst prof to assoc prof, 64-71, PROF ENTOM, OHIO STATE UNIV & OHIO AGR RES & DEVELOP CTR, 71- Concurrent Pos: Consult entomologist, Chem-Lawn Corp, 74- Mem: Entom Soc Am; Int Turfgrass Soc. Res: Biology, ecology and control of insects associated with agricultural crops and turf. Mailing Add: Dept of Entom Ohio Agr Res & Develop Ctr Wooster OH 44691

NIEMEIER, RICHARD WILLIAM, b Akron, Ohio, May 16, 45; m 66; c 3. ENVIRONMENTAL HEALTH. Educ: Thomas More Col, AB, 67; Univ Cincinnati, MS, 69, PhD(environ health sci), 73. Prof Exp: NIH fel, 74-75, ASST PROF ENVIRON HEALTH SCI, KETTERING LAB, DEPT ENVIRON HEALTH, COL MED, UNIV CINCINNATI, 75- Concurrent Pos: Mem, Water Adv Coun, Cincinnati City Mgr App Environ Adv Coun & consult, US Dept Labor, Occup Safety & Health Admin, 75- Mem: Sigma Xi. Res: Pulmonary metabolism of carcinogens using the isolated perfused lung; industrial hygiene surveys and characterization of carcinogenic occupational environments; effects of inhaled toxicants on pulmonary alveolar macrophages; factors affecting carcinogenic response. Mailing Add: Kettering Lab Col of Med Univ Cincinnati 3223 Eden Ave Cincinnati OH 45267

NIEMETZ, JULIAN, b Varna, Bulgaria, Mar 25, 31; m; c 3. HEMATOLOGY. Educ: Sorbonne, PCB, 51, MD, 58. Prof Exp: Fel blood coagulation, Nat Ctr Blood Transfusion, Paris, France, 59-61; fel transplantation, St Louis Hosp, Paris, 61; fel hemat, 61-63, from res asst to res assoc, 63-66, asst attend, 65-66, asst prof med, 66-73, ASSOC PROF MED, MT SINAI SCH MED, 73-; CHIEF HEMAT SECT, VET ADMIN HOSP, 69- Concurrent Pos: Chief coagulation lab, Vet Admin Hosp, 65-69; assoc attend, Mt Sinai Hosp. Mem: Am Soc Hemat; Harvey Soc; Am Heart Asn; Am Physiol Soc; Am Soc Clin Invest. Res: Blood coagulation. Mailing Add: Vet Admin Hosp Hemat Sect 130 W Kingsbridge Rd Bronx NY 10468

NIEMEYER, KENNETH H, b St Louis, Mo, Sept 15, 28; m 53; c 1. VETERINARY MEDICINE. Educ: Univ Mo, BS & DVM, 55, MS, 62. Prof Exp: From instr to assoc prof, 55-74, head small animal clins, 63-68, PROF & ASSOC CHMN VET MED & SURG, UNIV MO-COLUMBIA, 74- Mem: Am Vet Med Asn. Res: Clinical veterinary medicine, especially as applied to small animals. Mailing Add: Dept of Vet Med Univ of Mo Columbia MO 65201

NIEMEYER, LAWRENCE E, b New Braunfels, Tex, Aug 6, 26; m 53; c 5. METEOROLOGY. Educ: Univ Tex, BA, 51; Univ Mich, MS, 65. Prof Exp: Observer, US Weather Bur, Nev, 55-56, res meteorologist, Ohio, 56-59, analyst forecaster, Calif, 59-61, res meteorologist, Robert A Taft Sanit Eng Ctr, Environ Sci Serv Admin, Ohio, 61-69; asst dir, 69-74, DIR METEOROL LAB, ENVIRON PROTECTION AGENCY, 74- Mem: AAAS; Am Meteorol Soc; Sigma Xi; Air Pollution Control Asn. Res: Air pollution meteorology. Mailing Add: Div of Meteorol Environ Protection Agency Research Triangle Park NC 27711

NIEMI, ALFRED OTTO, b Grand Marais, Mich, Aug 13, 15; m 43; c 1. CONSERVATION. Educ: Mich State Univ, BS, 48, BA, 52, EdD(forestry ed), 60. Prof Exp: Teacher pub sch, Mich, 48-50; asst agr, Mich State Univ, 50-51; teacher pub sch, Mich, 51-56; asst prof agr & conserv forestry, 56-68, assoc prof conserv, 68-72, consult, Vista Training Ctr, 65, PROF CONSERV, NORTHERN MICH UNIV, 72- Concurrent Pos: Asst, Mich State Univ, 58-59; coordr, Upper Mich Jr Acad Sci, Arts & Letters, 59- Mem: Conserv Educ Asn; Soil Conserv Soc Am. Res: Forestry; agriculture; conservation education. Mailing Add: Dept of Geog Earth Sci Conserv Northern Mich Univ Marquette MI 49855

NIENABER, JAMES HENRY, b Evanston, Ill, Jan 31, 31; m 52; c 5. GEOLOGY. Educ: Colo Sch Mines, GeolE, 54; Univ Tex, PhD(geol), 58. Prof Exp: Geol field party chief, Turkey, 58-60; spec studies geol, Am Overseas Petrol, Ltd, Libya, 60-62; supvr geol res, King Resources Co, 63-69; exec vpres, Basic Earth Sci Systs, Inc, 69-70; PRES, EXPLOR SYSTS CORP, 71- Mem: Geol Soc Am; Am Asn Petrol Geologists; Am Inst Mining, Metall & Petrol Engrs; Soc Explor Geophys; Am Inst Prof Geologists. Res: Recent marine sedimentation; geological and geophysical computer applications; development of remote sensing and high resolution acoustic systems for mineral exploration; mine and environmental safety inspection and tunneling-excavation. Mailing Add: Explor Systs Corp PO Box 304 Evergreen CO 80439

NIENHOUSE, EVERETT J, b Oak Park, Ill, Oct 29, 36; m 65. ORGANIC CHEMISTRY. Educ: Hope Col, AB, 58; Northwestern Univ, MSc, 62; State Univ NY Buffalo, PhD(org chem), 66. Prof Exp: ASST PROF ORG CHEM, FERRIS STATE COL, 66- Mem: Am Chem Soc; The Chem Soc. Res: Chemistry of bridged bicyclic systems; abnormal Grignard reactions; new synthetic reagents in organic chemistry; olefinic cyclizations. Mailing Add: Dept of Phys Sci Ferris State Col Big Rapids MI 49307

NIENSTAEDT, HANS, b Copenhagen, Denmark, Dec 23, 22; m 49; c 6. FOREST GENETICS. Educ: Yale Univ, MF, 48, PhD(path), 51. Prof Exp: Asst geneticist, Agr Exp Sta, Univ Conn, 51-55; geneticist, Lake States Forest Exp Sta, 55-64, CHIEF LAB, N CENT FOREST EXP STA, INST FOREST GENETICS, US FOREST SERV, 64- Mem: AAAS; Soc Am Foresters; Int Union Forest Res Orgns; Tech Asn Pulp & Paper Indust. Res: Racial variation in conifers of northern North America; breeding of spruces, birches and pines of northern North America. Mailing Add: Inst of Forest Genetics Forest Serv USDA Box 898 Rhinelander WI 54501

NIER, ALFRED OTTO CARL, b St Paul, Minn, May 28, 11; m 37, 69; c 2. PHYSICS. Educ: Univ Minn, BE, 31, MS, 33, PhD(physics), 36. Prof Exp: Nat Res Coun fels, Harvard Univ, 36-38; from asst prof to assoc prof physics, Univ Minn, 38-43; physicist, Kellex Corp, NY, 43-45; chmn dept, 53-65, PROF PHYSICS, UNIV MINN, MINNEAPOLIS, 45- Honors & Awards: Day Medal, Geol Soc Am, 56. Mem: Nat Acad Sci; fel Am Phys Soc; Am Philos Soc; Geochem Soc; Am Geophys Union. Res: Mass spectrometry; aeronomy. Mailing Add: Sch of Physics and Astron Univ of Minn Minneapolis MN 55455

NIERENBERG, WILLIAM AARON, b New York, NY, Feb 13, 19; m 41; c 2. PHYSICS. Educ: City Col New York, BS, 39; Columbia Univ, MA, 42, PhD(physics), 47. Prof Exp: Tutor physics, City Col New York, 39-42, res scientist, Manhattan Proj, 42-45; instr physics, Columbia Univ, 46-48; asst prof, Univ Mich, 48-50; assoc prof, Univ Calif, Berkeley, 50-53, prof, 54-65, DIR SCRIPPS INST OCEANOG, UNIV CALIF SAN DIEGO, 65-; VCHANCELLOR MARINE SCI, 69- Concurrent Pos: Consult, US Navy, 47; dir, Hudson Labs, Columbia Univ, 53-54; mem mine adv comt, Nat Res Coun, 54-, consult comt nuclear constants, 58-; prof, Miller Inst Basic Res Sci, 57-59; consult, Lockheed Aircraft Corp, 57-60, Nat Security Agency, 58-60 & President's Spec Proj Comt, 58-; asst secy gen sci, NATO, 60-62; adv at large, Dept of State, 68-; chmn nat adv comt oceans & atmosphere, NATO sr sci fel, 69; mem, White House Task Force Oceanog, 69-70; mem oil spill panel, Off Sci & Technol, 69; mem, World Affairs Coun San Diego. Mem: Nat Acad Sci; fel Am Phys Soc; Am Acad Arts & Sci; Am Geophys Union. Res: Gas diffusion; molecular and atomic beams; nuclear moments. Mailing Add: Scripps Inst of Oceanog Univ of Calif San Diego La Jolla CA 92037

NIERING, WILLIAM ALBERT, b Scotrun, Pa, Aug 28, 24; m 55; c 2. PLANT ECOLOGY. Educ: Pa State Univ, BS, 48, MS, 50; Rutgers Univ, PhD(plant ecol), 52. Prof Exp: From asst prof to assoc prof, 52-64, PROF BOT, CONN COL, 64- Concurrent Pos: Mem, Kapingamarangi Exped, Caroline Islands, 54, veg Southwest, 62-64; consult, Recreation & Open Space Proj, Regional Plan Asn, Inc, 58; dir, Conn Arboretum, 65-; assoc dir environ biol prog, Nat Sci Found, 67-68. Honors & Awards: Mercer Award, Ecol Soc Am, 67. Mem: AAAS; Ecol Soc Am; Bot Soc Am; Am Inst Biol Sci. Res: Vegetation science; herbicides; wetland ecology; applied ecology. Mailing Add: Dept of Bot Conn Col Box 1511 New London CT 06320

NIERLICH, DONALD P, b Ft Lewis, Wash, Aug 10, 35; m 61; c 3. MICROBIAL PHYSIOLOGY, MOLECULAR BIOLOGY. Educ: Calif Inst Technol, BS, 57; Harvard Univ, PhD(bact), 62. Prof Exp: NSF fels, Mass Inst Technol, 63-64 & Inst Biol & Phys Chem, Paris, France, 64-65; from asst prof to assoc prof, 65-74, PROF BACT, UNIV CALIF, LOS ANGELES, 74- Concurrent Pos: Mem, Molecular Biol Inst, Univ Calif, Los Angeles, 74- Mem: Am Soc Microbiol; Am Soc Biol Chem. Res: Regulation of metabolism, particularly the synthesis of RNA and its control. Mailing Add: Dept of Bacteriol Univ of Calif Los Angeles CA 90024

NIES, ALAN SHEFFER, b Orange, Calif, Sept 30, 37; m 61; c 2. CLINICAL PHARMACOLOGY. Educ: Stanford Univ, BS, 59; Harvard Med Sch, MD, 63. Prof Exp: Resident & intern internal med, Univ Wash Hosp, 63-66; NIH fel clin pharmacol, Univ Calif, San Francisco, 66-68; chief clin pharmacol, Walter Reed Army Inst Res, 68-70; asst prof, 70-72, ASSOC PROF MED & PHARMACOL, SCH MED, VANDERBILT UNIV, 72- Mem: AAAS; Am Fedn Clin Res. Res: Effects of disease on drug metabolism and disposition in man. Mailing Add: Dept of Med & Pharmacol Vanderbilt Univ Med Ctr Nashville TN 37232

NIES, NELSON PERRY, b Charlotte, Mich, June 21, 11; m 43; c 1. CHEMISTRY. Educ: Calif Inst Technol, BS & MS, 35; Western Reserve Univ, PhD(chem), 39. Prof Exp: Chemist, Gen Petrol Corp, Calif, 36-37; instr phys chem, Polytech Inst Brooklyn, 39-40; assoc prof chem, Eureka Col, 40-41 & Univ Tampa, 41-42; res chemist, Northwestern Univ, 42-44, Ford, Bacon & Davis, Inc, DC, 44-45 & Clayton Mfg Co, Calif, 45-46; head air pollution chemist, Los Angeles County, 46; res chemist, Pac Coast Borax Co Div, 46-57, SR SCIENTIST, RES CORP DIV, US BORAX & CHEM CORP, 57- Mem: Am Chem Soc; Am Crystallog Asn. Res: Chemical warfare; inorganic boron compounds; phase equilibria in anhydrous borate systems. Mailing Add: US Borax Res Corp 412 Crescent Way Anaheim CA 92803

NIESET, ROBERT THOMAS, physics, biophysics, deceased

NIETO, JOSE I, mathematics, see 12th edition

NIETO, MICHAEL MARTIN, b Los Angeles, Calif, Mar 15, 40. THEORETICAL PHYSICS. Educ: Univ Calif, Riverside, BA, 61; Cornell Univ, PhD(physics), 66. Prof Exp: Res assoc physics, Inst Theoret Physics, State Univ NY Stony Brook, 66-68; vis physicist, Niels Bohr Inst, Copenhagen, Denmark, 68-70; lectr & asst prof physics, Univ Calif, Santa Barbara, 70-71; sr res assoc physics, Purdue Univ, Lafayette, 71-72; MEM STAFF, LOS ALAMOS SCI LAB, 72- Mem: Fel Am Phys Soc. Res: Theoretical physics in fields of high energy, astrophysics and quantum mechanics; recent work on weak interactions, DKP meson and Bhabha arbitrary spin wave equations, quantum phase operators, photon mass and law of planetary distances. Mailing Add: Theoretical Div Los Alamos Sci Lab Los Alamos NM 87545

NIEVERGELT, JURG, b Lucerne, Switz, June 6, 38; US citizen; m 65; c 1. COMPUTER SCIENCE. Educ: Swiss Fed Inst Technol, Dipl math, 62; Univ Ill, PhD(math), 65. Prof Exp: From asst prof to assoc prof, 65-72, PROF MATH & COMPUT SCI, UNIV ILL, URBANA, 72- Concurrent Pos: Vis prof, Swiss Fed Inst Technol, 71-72 & 75-76. Mem: AAAS; sr mem Inst Elec & Electronics Eng; Asn Comput Mach. Res: Computer software; computers in education, in particular, computer-assisted instruction. Mailing Add: Dept of Comput Sci Univ of Ill Urbana IL 61801

NIEWENHUIS, ROBERT JAMES, b Corsica, SDak, Sept 21, 36; m 58; c 2. ANATOMY, ELECTRON MICROSCOPY. Educ: Calvin Col, BS, 59; Mich State Univ, MS, 61; Univ Cincinnati, PhD(anat), 70. Prof Exp: Histopathologist, Procter & Gamble Co, 70-72; ASST PROF ANAT, SCH MED, UNIV CINCINNATI, 72- Mem: Electron Micros Soc Am. Res: Effects of cryptorchidism; vasectomy and toxins upon the ultrastructure of the testis and associated ducts. Mailing Add: Dept of Anat Univ of Cincinnati Med Sch Cincinnati OH 45267

NIFFENEGGER, DANIEL ARVID, b Grinnell, Iowa, Apr 7, 30; m 53; c 2. BOTANY, AGRONOMY. Educ: Iowa State Univ, BS, 52, PhD(bot, 67; Mont State Univ, MS, 57. Prof Exp: Seed analyst, Mont State Univ, 55-64, asst agron, 58-62, asst prof agron, 62-64; asst agron, Iowa State Univ, 64-67; biometrician, 67-69, asst dir, Biomet Serv Staff, 69-72, PROG ANALYST, N CENT REGION, AGR RES SERV, USDA, 72- Mem: Am Soc Agron; Asn Off Seed Anal; Am Inst Biol Sci; Biomet Soc; Am Statist Asn. Res: Seed testing methodology; use of seed test results to determine field seeding rates; development of a seed homogeneity test; seed tolerances; seed research priorities; grain sampling; tobacco fumigation. Mailing Add: Agr Res Serv US Dept of Agr 2000 Pioneer Parkway Peoria IL 61614

NIFONG, GORDON DALE, environmental health, see 12th edition

NIGAGLIONI, ADAN, b Penuelas, PR, Jan 12, 30; m 54; c 5. INTERNAL MEDICINE, GASTROENTEROLOGY. Educ: Univ PR, BS, 50, MD, 54. Prof Exp: Intern med, Univ Pa, 54-55, resident internal med & gastroenterol, 58-61; dean sch med, 63-66, chancellor med sci campus, 66-74, PROF MED, SCH MED, UNIV PR, SAN JUAN, 68- Mem: Am Gastroenterol Soc; Am Col Physicians; World Orgn Gastroenterol. Mailing Add: Med Sci Campus Univ of PR Sch of Med San Juan PR 00905

NIGAM, BISHAN PERKASH, b Delhi, India, July 14, 28; m 56; c 3. PARTICLE PHYSICS, NUCLEAR PHYSICS. Educ: Univ Delhi, BSc, 46, MSc, 48; Rochester Univ, PhD(theoret physics), 54. Prof Exp: Lectr physics, Univ Delhi, 50-52 & 55-56; asst, Rochester Univ, 52-54; res fel, Case Inst, 54-55; Nat Res Coun Can res fel, 56-59; res assoc, Rochester Univ, 59-60, asst part-time prof physics, 60-61; assoc prof, State Univ NY Buffalo, 61-64; PROF PHYSICS, ARIZ STATE UNIV, 64- Concurrent Pos: Prin scientist, Basic Sci Res Lab, Gen Dynamics/Electronics, 57-58; prof, Univ Wis-Milwaukee, 66-67. Mem: Fel Am Phys Soc. Res: Theoretical and elementary particle physics; field theory. Mailing Add: Dept of Physics Ariz State Univ Tempe AZ 85281

NIGAM, LAKSHMI NARAYAN, b Fatehpur, India, Sept 17, 34; m 58; c 3. FLUID MECHANICS, MATHEMATICS. Educ: Univ Allahabad, BSc, 55, MSc, 57; Indian Inst Technol, Kharagpur, PhD(fluid mech), 61. Prof Exp: Sr res asst appl math, Indian Inst Technol, Bombay, 60-61, from assoc lectr to lectr, 61-66, asst prof, 66-67; from asst prof to assoc prof, 67-73, PROF MATH & CHMN DEPT, QUINNIPIAC COL, 73- Concurrent Pos: Adj asst prof, New Haven Col, 68-69. Mem: Am Math Soc; Soc Indust & Appl Math; Math Asn Am. Res: Structure and propagation of shock waves in interstellar gas; gas shear flow past cylinders and wings when the effects of compressibility and viscosity are negligible. Mailing Add: Dept of Math Quinnipiac Col Hamden CT 06518

NIGAM, PRAKASH CHANDRA, b Kanpur, India, May 28, 36; m 63; c 3. INSECT TOXICOLOGY, SEROLOGY. Educ: Univ Agra, BSc, 54, MSc, 56; Indian Agr Res Inst, New Delhi, PhD(insect toxicol), 63. Prof Exp: Entomologist plant protection & entom res asst, Exten, Dept Agr, Govt India, 56-59; res entomologist, Rockefeller Found, Diplomatic Enclave, India, 62-63; Nat Res Coun Can res fel, Dept Zool, Univ Guelph, 63-64, assoc insect toxicol & serol, 64-65; res scientist, 65-74, SR RES SCIENTIST, CHEM CONTROL RES INST, CAN FORESTRY SERV, DEPT ENVIRON & PROJ LEADER, INSECT TOXICOL SECT, 74- Mem: AAAS; Entom Soc Am; Entom Soc Can; Can Soc Zoologists; Prof Inst Pub Serv Can. Res: Insect toxicology and chemical control of insect pests of agriculture and forestry; serological study to find out the changes in antigenic composition of insects due to insecticide treatment. Mailing Add: Chem Control Res Inst Can Forestry Serv Ottawa ON Can

NIGAM, VIJAI NANDAN, b Aligarh, India, July 3, 32; m 56; c 4. BIOCHEMISTRY, CELL BIOLOGY. Educ: Univ Lucknow, MSc, 52; Univ Bombay, PhD(biochem), 56. Prof Exp: Jr scientist agr biochem, Univ Minn, 56-57; res assoc, Cancer Res Lab, Sch Med, Tufts Univ, 57-60; mem staff, Montreal Cancer Inst, Notre Dame Hosp, 60-71; ASSOC PROF CELL BIOL, UNIV SHERBROOKE, 71- Concurrent Pos: Res assoc, Nat Cancer Inst Can, 67-; assoc prof, Univ Montreal, 67-71; Eleanor Roosevelt Int Cancer fel, 69-70. Mem: AAAS; Am Chem Soc; Am Asn Cancer Res; NY Acad Sci; Can Biochem Soc. Res: Enzymes; chromatography; carbohydrate transferases and mehylations; phosphatases; glycogen; gylcolipids; glycoproteins; carbohydrates of normal and neoplastic cells in culture. Mailing Add: Dept of Cell Biol Univ of Sherbrooke Sherbrooke PQ Can

NIGEN, ALAN MARK, b New York, NY, Dec 29, 47; m 70. PROTEIN CHEMISTRY. Educ: State Univ NY Stony Brook, BS, 68; Ind Univ, PhD(biochem), 72. Prof Exp: Res assoc, 72-75, ASST PROF BIOCHEM, ROCKEFELLER UNIV, 75- Mem: Am Chem Soc; Biophys Soc. Res: Structure and function of hemoglobin; design of drugs to alleviate symptomology of sickle-cell disease. Mailing Add: Rockefeller Univ 1230 York Ave New York NY 10021

NIGG, HERBERT NICHOLAS, b Detroit, Mich, July 9, 41; m 64; c 2. ENTOMOLOGY. Educ: Mich State Univ, BS, 67; Univ Ill, Urbana-Champaign, PhD(entom), 72. Prof Exp: Nat Res Coun-USDA fel insect endocrinol, Insect Physiol Lab, USDA, Md, 72-74; ASST PROF INSECT TOXICOL, INST FOOD & AGR SCI, AGR RES & EDUC CTR, UNIV FLA, 74- Mem: AAAS; Entom Soc Am; Am Chem Soc; Sigma Xi. Res: Metabolism of insect molting hormones and the relationship between agricultural worker health and pesticides. Mailing Add: Inst Food & Agr Sci Univ of Fla PO Box 1088 Lake Alfred FL 33850

NIGH, EDWARD LEROY, JR, b Hagerstown, Md, Aug 25, 27; m 52; c 2. PLANT PATHOLOGY, NEMATOLOGY. Educ: Colo Agr & Mech Col, BS, 52; Colo State Univ, MS, 56; Ore State Univ, PhD(plant path), 62. Prof Exp: Tech dir cotton res, Algodonera del Valle, SA, Mex, 54-62; from asst prof to assoc prof plant path & nematol, 62-67, HEAD DEPT PLANT PATH, UNIV ARIZ, 67- Mem: Am Phytopath Soc; Soc Nematol. Res: Entomology. Mailing Add: Dept of Plant Path Univ of Ariz Tucson AZ 85721

NIGH, HAROLD EUGENE, b Parnell, Mo, May 20, 32; m 55; c 4. SOLID STATE PHYSICS. Educ: Northwest Mo State Col, BS, 58; Iowa State Univ, PhD, 63. Prof Exp: Fel physics, Inst Atomic Res, 63-64; mem tech staff, 64-68, SUPVR SURFACE PHYSICS, BELL LABS, ALLENTOWN, 68- Mem: Am Phys Soc; Inst Elec & Electronics Engrs. Res: Magnetic properties of solids; semiconductor surface physics. Mailing Add: 3780 Sydna St Bethlehem PA 18017

NIGH, WESLEY GRAY, b Danville, Ill, Mar 11, 36; m 58; c 2. PHYSICAL ORGANIC CHEMISTRY. Educ: Univ Calif, Los Angeles, BS, 60; Univ Wash, PhD(chem), 65. Prof Exp: Sr res engr, Rocketdyne Div, NAm Rockwell Corp, 64-67; res fel chem, Calif Inst Technol, 67-68; asst prof, 68-71, ASSOC PROF CHEM, UNIV PUGET SOUND, 71- Mem: Am Chem Soc; The Chem Soc. Res: Investigation of reaction mechanisms by chemical kinetics and isotopic labelling techniques; oxidation of organic compounds by copper; investigation of the mechanisms of enzyme catalyzed reactions. Mailing Add: Dept of Chem Univ of Puget Sound Tacoma WA 98416

NIGHSWANDER, JAMES EDWARD, b Niagara Falls, Ont, Feb 1, 32; m 57; c 2. PLANT PATHOLOGY. Educ: Univ Toronto, BScF, 53; Univ Wis, PhD(plant path), 59. Prof Exp: Res officer forest entom & path br, Can Dept Forestry, 59-63; asst prof biol, Marathon Ctr, Univ Wis, 64-66; asst prof, 66-70, ASSOC PROF BIOL, TRENT UNIV, 70- Mem: Can Phytopath Soc; Can Inst Forestry. Res: Ecology and genetics of fungi. Mailing Add: Dept of Biol Trent Univ Peterborough ON Can

NIGHSWONGER, PAUL FLOYD, b Alva, Okla, Apr 14, 23; m 51; c 6. PLANT ECOLOGY. Educ: Northwestern State Col, Okla, BS, 49; Univ Okla, MS, 67, PhD(bot), 69. Prof Exp: ASST PROF BIOL, NORTHWESTERN STATE COL, OKLA, 69- Mem: Am Inst Biol Sci; Brit Ecol Soc; Ecol Soc Am; Soc Range Mgt. Mailing Add: Dept of Biol Northwestern State Col Alva OK 73717

NIGHTENGALE, MERLYN EUGENE, operations research, industrial engineering, see 12th edition

NIGHTINGALE, ARTHUR ESTEN, b Millville, NJ, Dec 25, 19; m 52; c 1. HORTICULTURE. Educ: NJ State Teachers Col, Glassboro, BS, 42; Rutgers Univ, New Brunswick, BS & MEd, 49; Tex A&M Univ, PhD(hort), 66. Prof Exp: Exten agent, Rutgers Univ & USDA, 45-47; dir adult educ, Atlantic County Voc Schs, NJ, 49-51; field rep horticulturist, Calif Spray Chem Co, 51-53; area horticulturist, 54-60; dir educ progs, State NJ, 60-63; asst, 63-66, PROF HORT, TEX A&M UNIV, 66- Mem: Am Soc Hort Sci; Am Inst Biol Sci; Am Hort Soc; fel Royal Hort Soc. Res: Chemical and environmental influences on plant growth and production. Mailing Add: Dept of Soil & Crop Sci Tex A&M Univ College Station TX 77843

NIGHTINGALE, CHARLES HENRY, b New York, NY, June 19, 39; m 62; c 3. BIOPHARMACEUTICS. Educ: Fordham Univ, BSPharm, 61; St John's Univ, NY, MS, 66; State Univ NY Buffalo, PhD(pharmaceut), 69. Prof Exp: Coordr educ & res, Mercy Hosp, New York, 64-66; asst prof, 69-73, ASSOC PROF PHARM, UNIV CONN, 73-, DIR CLIN PHARM PROG, SCH PHARM, 70- Mem: AAAS; Am Pharmaceut Asn; Acad Pharmaceut Sci; Am Asn Cols Pharm; NY Acad Sci. Res: Factors affecting drug absorption, distribution, metabolism and excretion; pharmacokinetics of drug therapy; antibiotic transport in tissue sites, protein binding; antibiotic pharmacokinetics. Mailing Add: Sch of Pharm Univ of Conn Storrs CT 06268

NIGHTINGALE, DOROTHY VIRGINIA, b Ft Collins, Colo, Feb 21, 02. ORGANIC CHEMISTRY. Educ: Univ Mo, AB, 22, AM, 23; Univ Chicago, PhD(org chem), 28. Prof Exp: From instr to prof, 39-72, EMER PROF CHEM, UNIV MO-COLUMBIA, 72- Concurrent Pos: Hon fel, Univ Minn, 38; res assoc, Univ Calif, Los Angeles, 46-47. Honors & Awards: Garvan Award, Am Chem Soc, 59. Mem: Am Chem Soc. Res: Chemiluminescence of organomagnesium halides; alkylations and acylations in the presence of aluminum chloride; action of nitrous acid on alicyclic amines; reactions of nitroparaffins with alicyclic ketones. Mailing Add: Dept of Chem Univ of Mo Columbia MO 65201

NIGHTINGALE, ELENA OTTOLENGHI, b Livorno, Italy, Nov 1, 32; US citizen; m 65; c 2. MEDICAL GENETICS, MICROBIAL GENETICS. Educ: Columbia Univ, AB, 54; Rockefeller Inst, PhD(microbial genetics), 61; NY Univ, MD, 64. Prof Exp: Genetics training grant, 61-62; Am Cancer Soc res scholar, 62-64; instr med, NY Univ, 64-65; asst prof microbiol, Med Col, Cornell Univ, 65-70; asst prof microbiol, Sch Med, Johns Hopkins Univ, 70-73; UAP clin genetics fel, Georgetown Univ, 73-74; resident fel genetics, 74-75, SR STAFF OFFICER, NAT ACAD SCI, 75- Mem: AAAS; Am Soc Human Genetics; Am Soc Microbiol; Genetics Soc Am; Harvey Soc. Res: Bacterial transformations; somatic cell genetics; infectious diseases. Mailing Add: Nat Acad Sci Assembly Life Sci 2101 Constitution Ave Washington DC 20418

NIGHTINGALE, EUGENE RICHARD, JR, b Champaign, Ill, Oct 14, 26; m 51; c 2. PHYSICAL CHEMISTRY. Educ: Univ Ill, BSc, 48, MS, 49; Univ Minn, PhD(anal & phys chem), 54. Prof Exp: Instr chem, Univ Minn, 53-54; instr, Univ Nebr, 54-56, asst prof, 56-60; res chemist, Esso Res & Eng Co, 60-62, proj leader, 62-64, res assoc & proj mgr, 64-66, corp planner, Esso Chem Co, 66-70, secy mgt comt, 69-70, PROF MGR RESINS, EXXON CHEM CO, INC, NEW YORK, 70- Concurrent Pos: Consult, Ethyl Corp, 57-58; vis res assoc, Univ Pittsburgh, 59; Univ Nebr grad fac fel, 59. Mem: Am Chem Soc; Am Mgt Asn; Am Inst Chemists; Planning Exec Inst. Res: Electrolytic solutions; ion solvation; theory of ionsolvent interactions; electroanalytical chemistry; hydrocarbon resins, specialty chemicals. Mailing Add: Exxon Chem Co Inc 60 W 49th St New York NY 10020

NIGHTINGALE, HARRY IRVING, b Durango, Colo, Nov 19, 30; m 52; c 4. AGRICULTURAL CHEMISTRY. Educ: NMex State Univ, BS, 53, MS, 57; Utah State Univ, PhD(soil chem), 65. Prof Exp: Res asst meteorol, Utah State Univ, 59-65; SOIL SCIENTIST, FRESNO GROUND WATER RECHARGE FIELD STA, AGR RES SERV, USDA, 65- Mem: Western Soil Sci Soc; Nat Water Well Asn; AAAS. Res: Ground-water recharge for agricultural and urban areas; ground water quality. Mailing Add: Water Mgt Res Agr Res Serv USDA 4816 E Shields Ave Fresno CA 93726

NIGHTINGALE, RICHARD EDWIN, b Walla Walla, Wash, June 3, 26; m 44; c 3. PHYSICAL CHEMISTRY, NUCLEAR ENGINEERING. Educ: Whitman Col, BA, 49; Wash State Univ, PhD(chem), 53. Prof Exp: Fel, Univ Minn, 52-54; sr engr, Hanford Labs, Gen Elec Co, 54-57, supvr nonmetallic mat, 57-65; mgr mat res & serv sect, 65-68, ceramics dept, 68-69, metall & ceramics dept, 69-70, MGR CHEM TECHNOL DEPT, PAC NORTHWEST LABS, BATTELLE MEM INST, 70- Honors & Awards: Cert Merit, Am Nuclear Soc, 65. Mem: Am Chem Soc; fel Am Nuclear Soc; Am Carbon Soc (mem, Exec Comt, 64-71). Res: Infrared spectroscopy and molecular structure; radiation damage effects; graphite structure and properties; coal research, nuclear fuel reprocessing and waste disposal. Mailing Add: Pac Northwest Labs Battelle Mem Inst PO Box 999 Richland WA 99352

NIGRELLI, ROSS FRANCO, b Pittston, Pa, Dec 12, 03; m 27; c 1. PROTOZOOLOGY, PARASITOLOGY. Educ: Pa State Univ, BS, 27; NY Univ, MS, 29, PhD(invert zool), 36. Prof Exp: PATHOLOGIST, NY AQUARIUM, 34-, SR SCIENTIST, OSBORN LABS MARINE SCI, 72- Concurrent Pos: Consult, US Food & Drug Admin, 45; dir, Lab Marine Biochem & Ecol, NY Aquarium, 57-64, aquarium dir, 66-70; adj prof, NY Univ, 58-; dir, Osborn Labs Marine Sci, 64-72; mem, Subcomt Marine Biol, Comt Oceanog, Off Sci & Technol, 65-, Adv Panel, Sea Grant Proj, Off Sea Grant Progs, 69- & President's Coun, New York Ocean Sci Lab, 71-; fel, Conserv Found. Mem: AAAS; NY Acad Sci. Res: Infectious and neoplastic diseases of fish; aquatic biology; experimental ichthyology; marine biochemistry and ecology. Mailing Add: Osborn Labs of Marine Sci Boardwalk & W Eighth St Brooklyn NY 11224

NIHEI, TAIICHI, b Aizu Wakamatzu, Japan, June 3, 27; m 60; c 1. DEVELOPMENTAL BIOLOGY. Educ: Hokkaido Univ, BSc, 52; Tokyo Univ, PhD(biochem), 58. Prof Exp: Res biologist, Tokugawa Inst Biol Res, 52-56; biochemist, Hokkaido Univ, 57-59; res assoc biochem, Dartmouth Med Sch, 59-61; vis scientist, NIH, 61-63; from asst prof to assoc prof, 63-74, PROF BIOCHEM, UNIV ALTA, 74- Concurrent Pos: Muscular Dystrophy Asn Can grant & fel, 64-; Life Ins Med Fund grant, 65-67; Med Res Coun Can grant, 65- Mem: Am Soc Biol Chem; Biophys Soc; Can Biochem Soc; Am Chem Soc. Res: Enzyme kinetics; muscle proteins; structure and function of muscle cells. Mailing Add: Dept of Med Univ of Alta Edmonton AB Can

NIIMOTO, DOROTHY HISAKO, botany, cytology, see 12th edition

NIJENHUIS, ALBERT, b Eindhoven, Netherlands, Nov 21, 26; US citizen. MATHEMATICS. Educ: Univ Amsterdam, PhD(math), 52. Prof Exp: Asst math, Univ Amsterdam, 48-51, sci collabr, Math Ctr, 51-52; vis fel, Princeton Univ, 52-53; mem, Inst Advan Study, 53-55; instr & res assoc, Univ Chicago, 55-56; from asst prof to prof math, Univ Wash, 56-63; PROF MATH, UNIV PA, 63- Concurrent Pos: Mem, Inst Advan Study & Guggenheim fel, 61-62; Fulbright lectr, Univ Amsterdam, 63-64; vis prof, Univ Geneva, 67-68. Mem: Am Math Soc; Math Asn Am; Soc Indust & Appl Math; Asn Comput Mach. Res: Local and global differential geometry; theory of deformations in algebra and geometry; combinatorial analysis; algorithms. Mailing Add: Dept of Math Univ of Pa Philadelphia PA 19174

NIJHOUT, H FREDERICK, b Eindhoven, Neth, Nov 25, 47. INSECT PHYSIOLOGY. Educ: Univ Notre Dame, BS, 70; Harvard Univ, MA, 72, PhD(biol), 74. Prof Exp: STAFF FEL, NIH, 75- Mem: Am Soc Zoologists; Entom Soc Am; Sigma Xi. Res: Insect physiology with particular emphasis on endocrinology and pattern formation. Mailing Add: Lab of Parasitic Dis NIH Bethesda MD 20014

NIJHOUT, MARY MCALLISTER, b Hornell, NY, Dec 19, 48; m 72. INSECT PHYSIOLOGY, MALARIOLOGY. Educ: Wells Col, BA, 70; Harvard Univ, MA, 71, PhD(biol), 75. Prof Exp: STAFF FEL, LAB PARASITIC DIS, NIH, 75- Mem: Sigma Xi; Am Soc Zoologists. Res: Reproductive development in female insects involving endocrine controls; vitellogenin production and utilization and reproductive behavior; development and physiology of gametes and gametocytes of Plasmodium species. Mailing Add: Lab of Parasitic Dis Bldg 5 NIH Bethesda MD 20014

NIKAIDO, HIROSHI, b Tokyo, Japan, Mar 26, 32; m 63; c 2. MICROBIOLOGY, BIOCHEMISTRY. Educ: Keio Univ, Japan, MD, 55, DMedSc(microbiol), 61. Prof Exp: Asst microbiol, Med Sch, Keio Univ, Japan, 56-60 & Inst Protein Res, Osaka Univ, 61; res fel biol chem, Harvard Med Sch, 62, assoc bact & immunol, 63-64, asst prof, 65-69; assoc prof, 69-71, PROF BACT, UNIV CALIF, BERKELEY, 71- Concurrent Pos: USPHS res grant, 63- Honors & Awards: Ehrlich Award, 69. Mem: Am Soc Biol Chem; Am Soc Microbiol. Res: Biochemistry of bacterial cell wall and cell membrane, especially biosynthesis of cell wall lipopolysaccharides in gram-negative bacteria. Mailing Add: Dept of Bact & Immunol Univ of Calif Berkeley CA 94720

NIKELLY, JOHN G, b Evanston, Ill, Apr 16, 29; m 61. ANALYTICAL CHEMISTRY. Educ: Univ Ill, BS, 52; Cornell Univ, PhD(chem), 56. Prof Exp: Chemist, Exxon Res & Eng Co, 56-62; asst prof, 62-67, ASSOC PROF CHEM, PHILADELPHIA COL PHARM, 67- Concurrent Pos: Sci adv, Philadelphia Dist, Food & Drug Admin, 71-75. Mem: AAAS; Am Chem Soc. Res: Gas chromatography; high pressure liquid chromatography. Mailing Add: Dept of Chem Phila Col of Pharm & Sci Philadelphia PA 19104

NIKIFORUK, GORDON, b Redfield, Sask, Nov 2, 22; m 50; c 2. BIOCHEMISTRY, DENTISTRY. Educ: Univ Toronto, DDS, 47; Univ Ill, MS, 50. Prof Exp: Prof prev dent, Univ Toronto, 47-64, chmn div dent res, 54-64; prof pediat, Sch Med & prof pediat dent & head dept, Sch Dent, Univ Calif, Los Angeles, 64-66, prof oral biol & head dept, 66-69; DEAN FAC DENT, UNIV TORONTO, 70- Mem: AAAS; fel Royal Col Dent; Can Dent Asn; Can Soc Dent for Children; Int Asn Dent Res. Res: Biochemistry of teeth and saliva; pediatric dentistry. Mailing Add: Fac of Dent Univ of Toronto Toronto ON Can

NIKITOVITCH-WINER, MIROSLAVA B, b Kraljevo, Yugoslavia, May 13, 29; US citizen; m 55; c 2. ANATOMY, ENDOCRINOLOGY. Educ: Univ Belgrade, BS, 45; Sorbonne, cert sci, 46; Radcliffe Col, MA, 54; Duke Univ, PhD(anat, physiol), 57. Prof Exp: Asst, Sch Med, Duke Univ, 57-58; USPHS fels, Univ Lund, 58, Nobel Med Inst, Karolinska Med Sch, 59 & Maudsley Hosp, Univ London, 59-60; res assoc, 60-61, from asst prof to assoc prof, 61-73, PROF ANAT, PHYSIOL & BIOPHYS, MED CTR, UNIV KY, 73- Mem: AAAS; Am Asn Anat; Endocrine Soc; Am Physiol Soc; Int Brain Res Orgn. Res: Neuroendocrinology and reproduction; indirect and direct determination of neurohumoral controls of gonadotropic hormone secretion; identification of releasing versus hypophysiotropic effects of hypothalamic humoral factors concerned with hormone secretion; control of prolactin secretion and its effect on the corpus luteum. Mailing Add: Dept of Anat Univ of Ky Med Ctr Lexington KY 40506

NIKKEL, HARVEY JOE, biochemistry, see 12th edition

NIKLAS, KARL JOSEPH, b New York, NY, Aug 23, 48. BIOMATHEMATICS, GEOCHEMISTRY. Educ: City Col New York, BS, 70; Univ Ill, Urbana-Champaign, MS, 71, PhD(bot, math), 74. Prof Exp: Fel bot, Univ London & Cambridge Univ, 74-75; CUR PALAEOBOT, NY BOT GARDEN, 74- Concurrent Pos: Fulbright-Hays fel, Int Educ Comt, 74; adj prof, City Univ New York, 75. Res: Application of mathematical theory to selected problems in biological research; organic geochemistry of fossil materials and associated rock strata; paleobotanical and biological problems. Mailing Add: NY Bot Garden Bronx NY 10458

NIKLAS, WILFRID F, b Vienna, Austria, June 27, 25; nat US; m 53. PHYSICS. Educ: Univ Vienna, PhD, 51. Prof Exp: Proj engr, N V Philips, 51-53; dept head, Columbia Broadcasting Syst, Conn, 54-56; assoc dir res, Rauland Corp, Ill, 56-66; div mgr, Varian Assocs, 66-69, pres, Varian/EMI, 69-71; CONSULT, 71- Mem: Fel Inst Elec & Electronics Engrs; NY Acad Sci; fel Brit Inst Electronic & Radio Eng. Res: Electron optics; cathode ray display tubes; photoelectric devices; special purpose tubes; non-transistor solid state devices. Mailing Add: 3003 Euclid Ave Tampa FL 33609

NIKLOWITZ, WERNER JOHANNES, b Ger, Sept 14, 23; m 52; c 2. NEUROBIOLOGY, ELECTRON MICROSCOPY. Educ: Univ Jena, MS, 52, PhD(biol), 54. Prof Exp: Asst prof biol, Univ Jena, 52-54; res assoc microbiol, Acad Inst Microbiol, Jena, Ger, 54-60; res assoc path, Univ Freiburg, 61-62; res assoc, Max Planck Inst Brain Res, Frankfurt, 62-68; sr res assoc toxicol, Med Ctr, Univ Cincinnati, 68-69, from asst prof to assoc prof environ health, 69-74; RES ASSOC NEUROPATH, MED CTR, UNIV IND, INDIANAPOLIS, 74- Concurrent Pos: NIH res grants, 72 & 74. Mem: AAAS; Electron Micros Soc Am; Soc Neurosci. Res: Experimental epilepsy; effect of toxic metals on the ultrastructure of specific brain components, enzymes, trace metals and neurotransmitters. Mailing Add: Div of Neuropath Univ of Ind Med Ctr Indianapolis IN 46202

NIKODEM, ROBERT BRUCE, b Oak Park, Ill June 2, 39; m 64. PHYSICAL CHEMISTRY, INORGANIC CHEMISTRY. Educ: Elmhurst Col, BS, 62; Purdue Univ, Lafayette, MS, 65; Va Polytech Inst & State Univ, PhD(inorg chem), 69. Prof Exp: RES SUPVR GLASS FILMS, LIBBEY-OWENS-FORD CO, 69- Mem: Am Chem Soc. Res: Rare earth oxides; compound semiconductors; thin films, both oxide and metallic; solid state chemistry. Mailing Add: Tech Ctr Libbey-Owens-Ford Co Toledo OH 43605

NIKOLAI, PAUL JOHN, b Minneapolis, Minn, Mar 2, 31; m 60; c 2. MATHEMATICS. Educ: Col St Thomas, BA, 53; Ohio State Univ, MSc, 55, PhD(math), 66. Prof Exp: Mathematician, Univac Div, Sperry Rand Corp, 55-57; mathematician, Digital Comput Br, Aeronaut Res Lab, 58-60, res mathematician, Appl Math Res Lab, Aerospace Res Lab, 60-75, MATHEMATICIAN, STRUCTURAL MECH DIV, AIR FORCE FLIGHT DYNAMICS LAB, US DEPT AIR FORCE, 75- Concurrent Pos: Asst math, Ohio State Univ, 53-57; vis scholar math, Univ Calif, Santa Barbara, 68-69; mem, Special Interest Group Numerical Math, Asn Comput Mach. Mem: Am Math Soc; Soc Indust & Appl Math. Res: Numerical linear algebra, matrix theory and combinatorial computing. Mailing Add: AF Flight Dynamics Lab AFFDL/FBR Wright-Patterson AFB OH 45433

NIKOLAI, ROBERT JOSEPH, b Rock Island, Ill, Apr 6, 37; m 61; c 5. ENGINEERING MECHANICS, ORTHODONTICS. Educ: Univ Ill, Urbana, BS, 59, MS, 61, PhD(theoret & appl mech), 64. Prof Exp: Teaching asst theoret & appl mech, Univ Ill, Urbana, 59-61, instr, 61-64, res assoc, 63-64; asst prof eng & eng mech, 64-68, assoc prof eng mech, 68-71, asst dean grad sch, 71-72, assoc prof biomech in orthod, 71-75, PROF BIOMECH IN ORTHOD, ST LOUIS UNIV, 75-, ASSOC DEAN GRAD SCH, 72- Mem: Am Acad Mech; Am Soc Mech Eng; Soc Exp Stress Anal; Nat Soc Prof Engrs; Am Soc Eng Educ. Mailing Add: St Louis Univ Grad Sch 221 N Grand Blvd St Louis MO 63103

NIKOLAICZUK, NIKOLAI, b Sniatyn, Alta, Dec 5, 13; m 41; c 3. POULTRY NUTRITION. Educ: Univ Alta, BSc, 36, MSc, 39; Ohio State Univ, PhD, 48. Prof Exp: Assoc prof poultry husb, McGill Univ, 39-60, chmn gen agr sci, 60-70, PROF ANIMAL SCI, MACDONALD COL, McGILL UNIV, 60- Mem: Poultry Sci Asn; Nutrit Soc Can; Can Inst Food Sci & Technol; Agr Inst Can. Res: General nutrition studies with poultry; effect of nutrition upon egg quality; mineral nutrition of poultry. Mailing Add: MacDonald Col Ste Anne de Bellevue PQ Can

NIKOLIC, NIKOLA M, b Belgrade, Yugoslavia, Sept 14, 27; US citizen; m 60; c 1. PHYSICS. Educ: Univ Belgrade, BS, 50; Columbia Univ, MA, 59, PhD(physics), 62. Prof Exp: Asst prof physics, US Naval Postgrad Sch, 62-64, La State Univ, 64-68 & Old Dom Univ, 68-69; asst prof, 69-74, PROF PHYSICS, MARY WASHINGTON COL, 74- Mem: Am Phys Soc. Res: Nuclear physics. Mailing Add: Dept of Physics Mary Washington Col Fredericksburg VA 22401

NILAN, ROBERT ARTHUR, b Can, Dec 26, 23; nat US; m 48; c 3. GENETICS. Educ: Univ BC, BSA, 46, MSA, 48; Univ Wis, PhD(genetics), 51. Prof Exp: Asst, Univ BC, 46-48 & Univ Wis, 48-51; res assoc, 51-52, from asst prof & asst agronomist to prof agron & agronomist, 52-65, PROF GENETICS, AGRONOMIST & CHMN PROG GENETICS, WASH STATE UNIV, 65- Concurrent Pos: Fulbright teaching scholar, 59; Guggenheim fel, 59-60; mem genetics staff, Wash State Univ, 61-65; USPHS spec fel, 67-68. Mem: Genetics Soc Am; Am Soc Agron; Radiation Res Soc; Am Genetics Asn. Res: Cytogenetical mutagenesis and radiobiological studies in cereals; barley cytogenetics and breeding. Mailing Add: Prog in Genetics Wash State Univ Pullman WA 99163

NILAN, THOMAS GEORGE, b White Plains, NY, Sept 4, 26; m 61; c 6. PHYSICS. Educ: Columbia Univ, BS, 50; Univ Ill, MS, 56, PhD, 61. Prof Exp: Physicist, 50-55, sr res physicist, 60-69, ASSOC RES CONSULT, RES LAB, US STEEL CORP, 69- Res: Radiation damage; ultrahigh pressure; phase transformations; surface physics. Mailing Add: 8945 Eastwood Rd Pittsburgh PA 15235

NILE, TERENCE ANTHONY, b Redruth, UK, Feb 26, 47; m 70. ORGANOMETALLIC CHEMISTRY. Educ: Univ Sussex, Eng, BSc, 69, MSc, 70, PhD(chem), 75; Royal Inst Chem, grad, 75. Prof Exp: Instr, 70-72, ASST PROF CHEM, UNIV NC, GREENSBORO, 75- Res: Catalytic properties of transition metals, especially hydrosilylation. Mailing Add: Dept of Chem Univ of NC Greensboro NC 27412

NILES, DORIS KILDALE, b Eureka, Calif, July 26, 03; m 38; c 4. BOTANY, GEOLOGY. Educ: Stanford Univ, PhD(bot), 31. Prof Exp: From asst prof to assoc prof biol sci, Humboldt State Col, 28-43; MEM STAFF EXTEN SERV BOT & GEOL, UNIV CALIF, 57- Concurrent Pos: Asst prof, Ariz State Univ, 32. Mem: AAAS; NY Acad Sci. Res: Taxonomy; paleontology and stratigraphy; marine algae, sand dune flora. Mailing Add: Box 307 Loleta CA 95551

NILES, E THOMAS, organic chemistry, inorganic chemistry, see 12th edition

NILES, GEORGE ALVA, b Flagstaff, Ariz, Oct 4, 26; m 48; c 4. AGRONOMY, PLANT BREEDING. Educ: NMex State Univ, BS, 48; Okla State Univ, MS, 49; Tex A&M Univ, PhD(plant breeding), 59. Prof Exp: Instr agron, NMex State Univ, 49; res assoc, Okla Agr Exp Sta, 51-53; from instr to assoc prof, 53-74, PROF AGRON, TEX A&M UNIV, 74- Mem: AAAS; Am Soc Agron; Crop Sci Soc Am; Int Soc Biometeorol. Res: Cotton genetics and breeding; host plant resistance; crop-climate studies; cotton production systems; cotton physiology; insect pest management. Mailing Add: Dept of Soil & Crop Sci Tex A&M Univ College Station TX 77843

NILES, JAMES ALFRED, b Eureka, Calif, Apr 24, 45; m 67; c 1. AGRICULTURAL ECONOMICS. Educ: Univ Calif, Davis, BS, 67, MS, 68, PhD(agr econ), 72. Prof Exp: ASST PROF AGR ECON, FOOD & RESOURCE ECON DEPT, UNIV FLA, 73- Mem: Am Asn Agr Economists. Res: Commodity futures markets, emphasis on citrus futures; price forecasting of agricultural commodities and modelling; computer simulation of commodity systems. Mailing Add: 2509 NW 67th Terr Gainesville FL 32601

NILES, NELSON ROBINSON, b Southampton, NY, May 27, 24; m 51; c 5. PATHOLOGY, ANATOMY. Educ: Cornell Univ, MD, 47. Prof Exp: From instr to assoc prof, 52-67, PROF PATH ANAT, MED SCH, UNIV ORE, 67- Concurrent Pos: Res fel path, Royal Col Surg, 62-63. Mem: AMA; Int Acad Path. Res: Cardiovascular pathology; coronary artery disease; congenital heart disease and cardiac surgery; histochemistry. Mailing Add: Dept of Path Univ of Ore Med Sch Portland OR 97201

NILES, WESLEY E, b Taos, NMex, July 17, 32. BOTANY. Educ: NMex State Univ, BS, 59, MS, 61; Univ Ariz, PhD(bot), 68. Prof Exp: Assoc cur, New York Bot Garden, 66-68; asst prof, 68-71, ASSOC PROF BOT, UNIV NEV, LAS VEGAS, 71- Mem: Am Soc Plant Taxon; Int Asn Plant Taxon. Res: Angiosperm taxonomy. Mailing Add: Dept of Biol Sci Univ of Nev Las Vegas NV 89154

NILSEN, WALTER GRAHN, b Brooklyn, NY, Nov 13, 27; m 59; c 3. CHEMICAL PHYSICS. Educ: Columbia Univ, AB, 50, PhD(phys chem), 56; Cornell Univ, MS, 52. Prof Exp: Staff mem chem, Opers Eval Group, Mass Inst Technol, 56-59; MEM TECH STAFF, BELL LABS, NJ, 59- Mem: Am Chem Soc; Am Phys Soc. Res: Magnetic resonance; relaxation mechanisms; masers; lasers; Raman spectroscopy. Mailing Add: Bell Labs Inc Murray Hill NJ 07971

NILSEN-HAMILTON, MARIT, b Arendal, Norway, Aug 30, 47; m 69. BIOCHEMISTRY. Educ: Cornell Univ, BS, 69, PhD(biochem), 73. Prof Exp: Fel, 73-75, SR RES ASSOC, SALK INST BIOL STUDIES, 75- Res: The determination of the role of the mammalian cell surface membrane in controlling growth, by study of the mechanisms of transport and its response to added growth factors in isolated membrane vesicles. Mailing Add: Salk Inst for Biol Studies PO Box 1809 San Diego CA 92112

NILSON, EDWIN NORMAN, b Weathersfield, Conn, Feb 13, 17; m 41; c 3. MATHEMATICS. Educ: Trinity Col, Conn, BS, 37; Harvard Univ, MA, 38, PhD(math), 41. Prof Exp: Instr & tutor math, Harvard Univ, 39-41; instr, Univ Md, 41-42; asst prof, Mt Holyoke Col, 42-44; res engr aerodyn, United Aircraft Corp, Conn, 46-48; asst prof math, Trinity Col, Conn, 48-52, assoc prof, 52-56; proj engr, 56-60, chief sci staff, 60-68, MGR TECH & MGT DATA SYSTS, PRATT & WHITNEY AIRCRAFT DIV, UNITED AIRCRAFT CORP, 68- Concurrent Pos: Adj prof, Hartford Grad Div, Rensselaer Polytech Inst, 59-63. Mem: Am Math Soc; Math Asn Am. Res: Analysis; applied mathematics; fluid mechanics; heat conduction; elasticity. Mailing Add: Pratt & Whitney Aircraft Div United Aircraft Corp 400 Main St East Hartford CT 06108

NILSON, ERICK BOGSETH, b Aurora, Nebr, Feb 6, 27; m 55; c 2. AGRONOMY. Educ: Univ Nebr, BS, 50, MS, 55; Kans State Univ, PhD(agron), 63. Prof Exp: Soil scientist, Soil Conserv Serv, USDA, Nebr, 51-52 & Kans, 52-53; county exten agent, State Col Wash, 55-57; exten area agronomist, Iowa State Univ, 57-61; asst prof agr & biol sci, Eastern NMex Univ, 63-65; asst prof agron, 65-69, ASSOC PROF AGRON, KANS STATE UNIV, 69- Mem: Weed Sci Soc Am; Am Soc Agron. Res: Crop physiology; cell length in wheat; temperature influence on DNA and RNA in grain sorghum seedings. Mailing Add: Dept of Agron Kans State Univ Manhattan KS 66506

NILSON, JOHN ANTHONY, b Regina, Sask, Nov 14, 36; m 59; c 2. LASERS. Educ: Univ Sask, BE, 59, MSc, 60; Univ London, PhD(elec eng) & dipl, Imp Col, 65. Prof Exp: Mem sci staff, Plasma & Space Physics Res Dept, RCA Ltd, Que, 65-71; staff scientist, 71-74, TECH DIR, LUMONICS RES LTD, 74- Mem: Can Asn Physicists; Inst Elec & Electronics Engrs; Optical Soc Am. Res: Pulsed transverse excitation atmospheric pressure-carbon dioxide lasers; glow discharge plasmas; laser applications. Mailing Add: Lumonics Res Ltd 1755 Woodward Dr Ottawa ON Can

NILSON, KAY MILLIGAN, b Logan, Utah, Dec 22, 28; m 49; c 6. DAIRY SCIENCE. Educ: Utah State Univ, BS, 53, MS, 56; Univ Nebr, PhD(dairy sci), 66. Prof Exp: Instr dairy sci, Univ Nebr, 56-66; from asst prof to assoc prof, 66-75, PROF DAIRY SCI, UNIV VT, 75- Mem: Am Dairy Sci Asn; Am Inst Food Technologists; Int Asn Milk, Food & Environ Sanitarians. Res: Dairy plant waste; feeding cheese whey to dairy animals; stirred curd Mozzarella cheese; manufacturing Ricotta cheese from concentrated whey; automation of Mozzarella cheese manufacturing. Mailing Add: Dept of Animal Sciences Univ of Vt Burlington VT 05401

NILSSON, WILLIAM A, b New York, NY, Jan 16, 31. OCCUPATIONAL HEALTH. Educ: Univ Ill, Urbana, BS, 57; Univ Calif, Berkeley, PhD(org chem), 62. Prof Exp: Asst, Radiation Lab, Univ Calif, Berkeley, 60-61; asst prof chem, Western Wash State Col, 61-64; mgr prod develop, Purex Corp, Ltd, Calif, 65-70; teacher chem, Glendale Col, 71-73; PUB HEALTH CHEMIST, OCCUP HEALTH SECT, CALIF DEPT HEALTH, 74- Mem: Am Chem Soc. Mailing Add: 1808 Tamerlane Dr Glendale CA 91208

NIMAN, JOHN, b Latakia, Syria, June 10, 38; US citizen. APPLIED MATHEMATICS. Educ: Polytech Inst Brooklyn, BS, 65; Univ Wis, MS, 68; Columbia Univ, PhD(math) & MA, 69. Prof Exp: Res asst geophys, Hudson Labs, Columbia Univ, 64-65; asst prof, 67-69, ASSOC PROF MATH & EDUC, HUNTER COL, 69- Concurrent Pos: George N Shuster fac fel, 70; consult math, More Effective Sch Prog, New York Bd Educ, 70 & Educ Assoc Prog, La Guardia Community Col, 71; res award hist math, Fed Repub Ger, 75. Mem: AAAS; Am Math Soc; Math Asn Am. Res: Curriculum development; Hilbert; mathematics education in the elementary school. Mailing Add: Dept of Math & Educ Hunter Col New York NY 10021

NIMECK, MAXWELL W, b Edmonton, Alta, Oct 20, 32. MICROBIOLOGY. Educ: Univ Alta, BS, 55, MS, 57; Univ Tex, PhD(microbiol), 62. Prof Exp: Instr bact, Univ Alta, 57-58; res scientist, Univ Tex, 58-61; NIH fel, Univ Wis, 62-63; SR RES SCIENTIST MICROBIOL DEVELOP, SQUIBB INST MED RES, 63- Concurrent Pos: Pharmacist, Bellevue Pharm, Alta, 55-58. Mem: AAAS; Am Soc Microbiol; Can Pharmaceut Asn; Brit Soc Gen Microbiol. Res: Hospital contamination, asepsis and disinfection; Azotobacter bacteriophage and genetics; nitrogen fixation; Streptomyces genetics; antibiotic production. Mailing Add: Microbiol Div Squibb Inst for Med Res New Brunswick NJ 08903

NIMER, EDWARD LEE, b Denver, Colo, Jan 1, 23; m 46; c 5. PHYSICAL CHEMISTRY. Educ: Univ NMex, BS, 44; Univ Utah, MS, 49, PhD(chem), 52. Prof Exp: Sr res chemist, Explor Process Res, 52-55, petrochem process develop & eval, 55-62, SR RES CHEMIST, POLYMER PROCESS DEVELOP, CHEVRON RES CO, STANDARD OIL CO CALIF, 62- Mem: Am Chem Soc; Am Soc Mech Eng. Res: Flow, creep and failure of solid materials; exploratory process research in petrochemicals; polarography; polymer polymerization, finishing, compounding, conversions and spinning processes; petroleum cracking processes. Mailing Add: 8 Alasdair Ct San Rafael CA 94903

NIMEROFF, ISADORE, b Philadelphia, Pa, Oct 24, 17; m 42; c 2. PHYSICS. Educ: George Washington Univ, BS, 48; Imp Col, London, dipl, 62. Prof Exp: Lab aide optics, 41-42, physicist, 45-66, chief colorimetry & spectrophotom sect, 66-69, CHIEF OFF COLORIMETRY, NAT BUR STANDARDS, 69- Concurrent Pos: Lectr, George Washington Univ, 57-58. Mem: Am Phys Soc; fel Optical Soc Am; Am Soc Testing & Mat. Res: Goniophotometry of reflecting and transmitting materials; tristimulus colorimetry with emphasis on observer variability of color mixture data; propagation of errors; colorimeter design; color vision. Mailing Add: 6505 Greentree Rd Bethesda MD 20034

NIMLOS, THOMAS JOHN, b Milwaukee, Wis, Oct 30, 29; m 54; c 3. FOREST SOILS, ECOLOGY. Educ: Univ Wis, BS, 51, MS, 54, PhD(soil chem), 59. Prof Exp: Soil scientist mapping, US Soil Conserv Serv, 58 & US Forest Serv, 58-60; PROF SOIL, UNIV MONT, 60- Mem: AAAS; Am Soc Agron. Res: Effect of heavy metals on soils and plant ecology. Mailing Add: Sch of Forestry Univ of Mont Missoula MT 59801

NIMMO, CHARLES COLVIN, b Los Angeles, Calif, Oct 18, 14; m 39; c 6. CHEMISTRY. Educ: Stanford Univ, AB, 36, MA, 37, PhD(biochem), 42. Prof Exp: Asst liver protein res, Stanford Univ, 38-40; chemist, Hyland Labs, 41-42; CHEMIST, WESTERN REGIONAL RES LAB, USDA, 42- Concurrent Pos: Res fel, Calif Inst Technol, 59-60. Mem: Am Chem Soc; Am Asn Cereal Chemists. Res: Food processing and stability, food chemistry, iron nutrition, wheat proteins. Mailing Add: Western Reg Res Labs US Dept of Agr Berkeley CA 94710

NIMMO, DEL WAYNE ROY, b Pratt, Kans, Sept 19, 37; m 60; c 2. ENVIRONMENTAL BIOLOGY. Educ: Evangel Col, BS, 59; Wichita State Univ, MS, 64; Colo State Univ, PhD(zool), 68. Prof Exp: RESEARCHER, ENVIRON PROTECTION AGENCY LAB, GULF BREEZE, FLA, 68- Res: The effects of metals and synthetic organics on marine and estuarine crustaceans; life-cycle bioassays with mysids and physiology of shrimps. Mailing Add: Environ Protection Agency Lab Sabine Island Gulf Breeze FL 32561

NIMMO, HARRY ARLO, b Monroe, Iowa, Aug 26, 36. CULTURAL ANTHROPOLOGY. Educ: Univ Iowa, BA, 58; Univ Hawaii, MA, 65, PhD(anthrop), 69. Prof Exp: Asst prof anthrop, Univ Hawaii, 67-68; asst prof, Calif State Univ, Los Angeles, 68-71, ASSOC PROF ANTHROP, CALIF STATE UNIV, HAYWARD, 71- Concurrent Pos: Mem, Borneo Res Coun, 70- Mem: Am Anthrop Asn; Far Eastern Prehist Asn. Res: Southeast Asian ethnography, specifically nomadic boat dwellers of Southeast Asia; culture change in Southeast Asia; folk Islam. Mailing Add: 51 Potomac St San Francisco CA 94117

NIMNI, MARCEL EFRAIM, b Buenos Aires, Arg, Feb 1, 31; US citizen; m 62; c 2. BIOCHEMISTRY, NUTRITION. Educ: Univ Buenos Aires, BS, 54, PhD(pharmacol), 60; Univ Southern Calif, MS, 57. Prof Exp: Biochemist, AEC, Arg, 58-60; head biol sect, Don Baxter, Inc, Calif, 62-63; asst prof biochem & nutrit, Sch Dent, 64-67, asst prof med, Sch Med, 67-68, assoc prof, 68-73, PROF MED & BIOCHEM, SCH MED, UNIV SOUTHERN CALIF, 73- Mem: Fel AAAS; Am Inst Nutrit; Am Soc Biol Chem; Am Rheumatism Asn; Soc Exp Biol & Med. Res: Biochemistry of collagen; mechanism of a defect in molecular aggregation induced by penicillamine; nature and biosynthesis of the crosslinks in collagen from skin, bone and cartilage; microtubular proteins and mitotic inhibitors. Mailing Add: Dept of Med Univ of South Calif Sch of Med Los Angeles CA 90033

NIMON, LARRY ARDEN, chemistry, see 12th edition

NIMROD, DALE, physical chemistry, see 12th edition

NIMS, JOHN BUCHANAN, b Monmouth, Maine, Dec 1, 24; m 47; c 2. PHYSICS. Educ: Boston Univ, BA, 50, MA, 51. Prof Exp: Physicist reactor physics, Knolls Atomic Power Lab, Gen Elec Co, 51-54; physicist, Atomic Power Develop Assocs Inc, 54-59, sect head, 59-66, div head, 66-73; dir nuclear eng, 73-74, GROUP STRATEGIC PLANNER, DETROIT EDISON CO, 74- Mem: Am Phys Soc; Am Nuclear Soc; World Future Soc. Res: Reactor physics, kinetics and safety. Mailing Add: Detroit Edison Co 2000 Second Ave Detroit MI 48226

NINE, HARMON D, b Detroit, Mich, July 8, 31; m 55; c 3. METAL PHYSICS. Educ: Univ Mich, BS, 53, MS, 54. Prof Exp: Res physicist, 58-61, SR RES PHYSICIST, GEN MOTORS RES LABS, 61- Mem: Am Phys Soc. Res: Ultrasonics; ultrasonic attenuation in solids; fatigue of metals; spectroscopy; acoustic emission; friction in metal forming. Mailing Add: Dept of Physics Gen Motors Res Labs Warren MI 48090

NINE, OGDEN WELLS, JR, b Baltimore, Md, July 25, 25; m 50; c 3. PETROLEUM GEOLOGY. Educ: Muhlenberg Col, AB, 49; Rutgers Univ, MSc, 50, PhD(geol), 54. Prof Exp: Asst phys, hist & agr geol, Rutgers Univ, 50-54; resident geologist Mex, Bethlehem Steel Co, 54-55; res geologist, Arabian Am Oil Co, 55-58, explor party chief, 58-59, explor staff geologist, 59-60, sr geologist, 60-62, sr engr spec studies, 62-64, staff paleontologist, 64-65, sr paleontologist, 65-66, sr explor geologist, Dhahran, Saudi Arabia, 66-74, MEM STAFF, GSI ARAMCO, 74- Concurrent Pos: Supt geol, Cia Minera Autlan, Mex, 54-55. Mem: Geol Soc Am; Soc Econ Paleont & Mineral; Am Asn Petrol Geol; Paleont Soc. Res: Stratigraphy and petroleum geology of the Middle East, particularly of the Jurassic and Cretaceous of Arabia; foraminifera and ostracoda; manganese ore. Mailing Add: c/o GSI Aramco Canterbury House Sydenham Rd Croyden Surrey England

NING, ROBERT YE FONG, b Shanghai, China, Mar 12, 39; m 66; c 2. MEDICINAL CHEMISTRY, ORGANIC CHEMISTRY. Educ: Rochester Inst Technol, BS, 63; Univ Ill, PhD(org chem), 66. Prof Exp: Res fel, Calif Inst Technol, 66-67; sr chemist, 67-75, RES FEL, HOFFMANN-LA ROCHE INC, 75- Mem: Am Chem Soc; The Chem Soc. Res: Medicinal chemistry; synthetic organic chemistry; chemistry of heterocycles; drug design. Mailing Add: Chem Res Dept Hoffmann-La Roche Inc Nutley NJ 07110

NINGER, FRED CONSTANT, b Jersey City, NJ, Dec 7, 15; m 57; c 2. PHARMACEUTICAL CHEMISTRY. Educ: Rutgers Univ, BS, 37; Stevens Inst Technol, MS, 44. Prof Exp: Anal chemist, Burroughs Wellcome & Co, Inc, 37-40; prod chemist, Sandoz Chem Works, 40-43; res chemist, Vick Chem Co, 43-47; dir antibiotic formulations, E R Squibb & Sons, 47-50; dir pharm, Warner Hudnut, Inc, 51-54, pharmaceut res, Warner Lambert Res Inst, 54-63, DIR PHARMACEUT DEVELOP, WARNER LAMBERT RES INST, 63- Mem: Am Pharmaceut Asn; Am Chem Soc; NY Acad Sci. Res: Pharmaceutical research and product development; drug dosage forms; sustained action medications; antibiotics; vitamins. Mailing Add: Warner Lambert Res Inst Morris Plains NJ 07950

NININGER, ROBERT D, b Brookings, SDak, Mar 28, 19; m 43; c 4. ECONOMIC GEOLOGY. Educ: Amherst Col, BA, 41; Harvard Univ, MS, 42. Prof Exp: Geologist, Strategic Minerals Invest, US Geol Surv, 42-43; dep asst dir raw mat, AEC, 47-54, asst dir, 55-71; ASST DIR RAW MAT, DIV NUCLEAR FUEL CYCLE & PROD, US ENERGY RES & DEVELOP ADMIN, 72- Mem: AAAS; Geol Soc Am; Am Nuclear Soc; Soc Econ Geol; Am Inst Mining, Metall & Petrol Engrs. Res: Geology of radioactive materials. Mailing Add: US Energy Res & Dev Admin Washington DC 20545

NINKOVICH, DRAGOSLAV, b Belgrade, Yugoslavia, Oct 16, 31. GEOLOGY. Educ: Univ Belgrade, MS, 54; Sorbonne, PhD(econ geol), 61. Prof Exp: Res scientist crystal chem, Boris Kidritch, Belgrade, 55-57; engr econ geol, Bur Res Geol & Minerals, Paris, 60-62; res assoc, 63-76, SR RES ASSOC SUBMARINE GEOL, LAMONT GEOL OBSERV, COLUMBIA UNIV, 76- Res: Submarine geology, particularly tephra layers in deep sea sediment. Mailing Add: 18 Francis Ave Nyack NY 10964

NINO, HIPOLITO VINCENT, biochemistry, see 12th edition

NIPPER, HENRY CARMACK, b Alexander City, Ala, Mar 31, 40; m 66; c 1. CLINICAL CHEMISTRY. Educ: Emory Univ, AB, 60; Purdue Univ, MS, 66; Univ Md, PhD(chem), 71. Prof Exp: Anal chemist, E I du Pont de Nemours & Co, 60-63; teaching asst chem, Purdue Univ, 63-65, instr, 65-66; teaching asst chem, 66-68, Gillette-Harris res fel, 68-70, fel clin chem, Sch Med, 71-73, instr path, 73-74, ASST PROF PATH, SCH MED, UNIV MD, 74-; CLIN CHEMIST, VET ADMIN HOSP, BALTIMORE, 73- Mem: Sigma Xi; Am Asn Clin Chem. Res: Gas and liquid chromatography methods; kinetic methods in clinical chemistry. Mailing Add: 8207 Bellona Ave Baltimore MD 21204

NIPPOLDT, BERTWIN W, b Lake Elmo, Minn, Mar 9, 22; m 48; c 3. ANALYTICAL CHEMISTRY. Educ: Macalester Col, BA, 47. Prof Exp: Chemist, Minn Mining & Mfg Co, 47-55, group supvr, Cent Res Labs, 55-64, MGR ORG ANAL CHEM, CENT RES LABS, 3-M CO, 65- Mem: Am Chem Soc. Res: Analytical chemistry, especially analysis of fluorinated materials; microchemical and organic functional group analysis. Mailing Add: Cent Res Labs 3-M Co 3-M Ctr St Paul MN 55101

NIRENBERG, LOUIS, b Hamilton, Ont, Feb 28, 25; nat US; m 48; c 2. MATHEMATICS. Educ: McGill Univ, BSc, 45; NY Univ, MS, 47, PhD, 49. Prof Exp: From res asst to res assoc, 45-54, from asst prof to assoc prof, 52-57, PROF MATH, NY UNIV, 57-, Concurrent Pos: Nat Res Coun fel, NY Univ, 51-52; fel, Inst Advan Study, 58; Sloan Found fel, 58-60; Fulbright lectr, 65; Guggenheim fel, 66-67 & 75-76; dir, Courant Inst Math Sci, 70-72. Honors & Awards: Bocher Prize, Am Math Soc, 59. Mem: Nat Acad Sci; Am Math Soc; Am Acad Arts & Sci. Res: Partial differential equations; differential geometry. Mailing Add: Dept of Math NY Univ New York NY 10012

NIRENBERG, MARSHALL WARREN, b New York, NY, Apr 10, 27; m. BIOCHEMISTRY. Educ: Univ Fla, BS, 48, MS, 52; Univ Mich, PhD(biochem), 57. Hon Degrees: DSc, Univ Mich, 65, Univ Chicago, 65, Yale Univ, 65 & George Washington Univ, 72. Prof Exp: Asst zool, Univ Fla, 45-50, res assoc, Nutrit Lab, 50-52; Am Cancer Soc fel, Nat Inst Arthritis & Metab Dis, 57-59, USPHS fel, Sect Metab Enzymes, 59-60, res biochemist, 60-62; head sect biochem genetics, 62-66, RES BIOCHEMIST & CHIEF BIOCHEM GENETICS LAB, NAT HEART & LUNG INST, 66- Honors & Awards: Nobel Prize in Med, 68. Mem: Nat Acad Sci; Pontif Acad Sci; Am Chem Soc; Biophys Soc; Soc Develop Biol. Mailing Add: Lab of Biochem Genetics Nat Heart & Lung Inst Bethesda MD 20014

NISBET, ALEX RICHARD, b Plainview, Tex, Apr 14, 38. ELECTROANALYTICAL CHEMISTRY. Educ: Univ Tex, BS, 59, PhD, 63. Prof Exp: Assoc arts, San Angelo Col, 57; assoc prof, 63-73, PROF CHEM, OUACHITA BAPTIST UNIV, 73- Mem: AAAS; Electrochem Soc; Am Chem Soc. Res: Chronopotentiometry of metals. Mailing Add: 1908 Sylvia St Arkadelphia AR 71923

NISBET, JERRY J, b Palisade, Colo, Nov 8, 24; m 48; c 2. SCIENCE EDUCATION, PLANT ANATOMY. Educ: Colo State Col Educ, AB, 49, AM, 50; Purdue Univ, PhD, 58. Prof Exp: Instr, Ball State Teachers Col, 50-52, asst prof, 54-56; asst, Purdue Univ, 52-53, res fel, 53-54, instr, 56-58; from asst prof to assoc prof, 58-64, PROF BIOL, BALL STATE UNIV, 64- COORDR, UNIV EVAL & DIR, INST ENVIRON STUDIES, 74- Concurrent Pos: Dir, NSF Summer Inst Biol, 60, 62-64 & 66-68, dir, NSF Acad Yr Inst, 68-71; nat dir, Outstanding Biol Teacher Awards Prog, Nat Asn Biol Teachers, 65-67, dir, Region III, 66-68; mem, Sci Adv Comt, Ind Dept Pub Instr, 72- & Ind Lt Govr's Sci Adv Comt, 73-; secy, Ind Acad Sci, 72-74; secy, Cent States Univ, Inc, 68-70, chmn, Coun, 72, pres, Bd, 76. Mem: Fel AAAS; Am Inst Biol Sci; Nat Asn Res Sci Teaching; Asn Midwest Col Biol Teachers; Nat Sci Teachers Asn. Res: Development and testing of technologically based systems of instruction; interpretation of science for the non-scientist; study of the ultrastructure of plant cell walls. Mailing Add: Inst for Environ Studies Ball State Univ Muncie IN 47306

NISBET, MICHAEL ALAN, b Lisburn, Ireland, July 20, 38. ORGANIC CHEMISTRY. Educ: Trinity Col, Dublin, BA, 59, PhD(org chem), 62; McGill Univ, MBA, 72. Prof Exp: Res fel org chem, Univ Western Ont, 62-64; group leader smoke res, Imp Tobacco Co Can Ltd, 64-73; techno-economist, Sulphur Develop Inst Can, 73-75; SR RES ECONOMIST, BANK MONTREAL, 75- Mem: Am Chem Soc; The Chem Soc; Chem Inst Can. Res: Natural product chemistry, especially terpenes; free radical chemistry; economic impact of emerging technology; environmental forecasting. Mailing Add: Dept of Econ Bank of Montreal 129 St James St West Montreal PQ Can

NISENOFF, MARTIN, b New York, NY, Dec 25, 28; m 59; c 3. LOW TEMPERATURE PHYSICS. Educ: Worcester Polytechnic Inst, Mass, BS, 50; Purdue Univ, MS, 52, PhD(physics), 60. Prof Exp: Res assoc physics, Purdue Univ, 60-61; res scientist physics, Sci Res Staff, Ford Motor Co, Dearborn, Mich, 61-70; physicist, Cryog Appln Group, Stanford Res Inst, Menlo Park, Calif, 70-72; RES SCIENTIST PHYSICS, CRYOG & SUPERCONDUCTIVITY BR, NAVAL RES LAB, WASHINGTON, DC, 72- Mem: Am Phys Soc. Res: Low temperature physics; Josephson effects; superconducting magnetometry; superconducting circuit elements for use in surveillance and communication systems. Mailing Add: Naval Res Lab Code 6435 Washington DC 20375

NISHI, MIDORI, geography, see 12th edition

NISHIBAYASHI, MASARU, b Los Angeles, Calif, May 6, 23; m 49; c 3. PHYSICAL CHEMISTRY. Educ: Univ Cincinnati, BS, 49, PhD(phys chem), 53. Prof Exp: Res chemist, Aerojet-Gen Corp, 53-60, sr res chemist, 60-63, SR CHEM SPECIALIST, AEROJET ORD & MFG CO, 63- Mem: Am Chem Soc; Am Ord Asn. Res: Physical and chemical properties of propellants and explosives; interior ballistics; gun propellants. Mailing Add: 1339 Beech Hill Ave Hacienda Heights CA 91745

NISHIDA, TOSHIRO, b Nagasaki, Japan, Jan 6, 26; m 60; c 3. FOOD CHEMISTRY. Educ: Kyoto Univ, MS, 52; Univ Ill, PhD(food chem), 56. Prof Exp: Res asst agr chem, Univ Osaka Prefecture, 47-49; res chemist, Osaka Soda Co, Japan, 52-53; from asst prof to assoc prof food chem, 58-70, PROF FOOD SCI, UNIV ILL, URBANA, 70- Concurrent Pos: Grants, NIH, 57- & Ill Heart Asn, 65- Mem: Am Chem Soc; Am Oil Chem Soc; Am Soc Biol Chemists; Am Inst Nutrit. Res: Chemistry and metabolism of lipids and lipoproteins. Mailing Add: 106 Burnsides Res Lab Univ of Ill Urbana IL 61801

NISHIE, KEICA, b Sao Paulo, Brazil, Sept 29, 29. PHARMACOLOGY. Educ: Univ Brazil, MD, 56; Northwestern Univ, MS, 59, PhD(pharmacol), 61. Prof Exp: NIH fel, Northwestern Univ, 59-61; sr res pharmacologist, Baxter Labs, Inc, 61-65; sr res pharmacologist, 65-73, STAFF MEM, DEPT PHARMACOL, RUSSELL RES CTR, USDA, 73- Mem: Am Soc Pharmacol & Exp Therapeut; NY Acad Sci. Res: Toxicology. Mailing Add: Dept of Pharmacol Russell Res Ctr USDA Athens GA 30604

NISHIHARA, MUTSUKO, b Los Angeles, Calif, Feb 6, 38. MICROBIOLOGY, BIOCHEMISTRY. Educ: Univ Calif, Los Angeles, BA, 59, MA, 61, PhD(microbiol), 64. Prof Exp: USPHS fel biochem & microbiol, Sch Med, Tufts Univ, 64-66, instr microbiol, 66-67; instr cell biol, Univ Md, Baltimore City, 67-72; RES BIOLOGIST, UNIV CALIF, LOS ANGELES, 72- Mem: AAAS; Am Soc Microbiol; NY Acad Sci. Res: Ultraviolet light irradiation of bacteria; genetics of bacteriophages; physiology and enzymology of virus infection; nucleotide and nucleic acid metabolism. Mailing Add: 982 S Oxford Ave Univ of Calif Los Angeles CA 90006

NISHIKAWA, ALFRED HIROTOSHI, b San Francisco, Calif, Apr 23, 38; m 61; c 2. BIOCHEMISTRY. Educ: Univ Calif, Berkeley, AB, 60; Ore State Univ, PhD(enzym), 65. Prof Exp: Lab technician biochem, Univ Calif, Berkeley, 60-61; res assoc, Ore State Univ, 65-69; scientist, Xerox Res Labs, NY, 69-71; SR BIOCHEMIST, CHEM RES DIV, HOFFMANN-LA ROCHE, INC, 71- Mem: AAAS; Am Chem Soc. Res: Enzyme isolation and purification; physical-chemical studies of chemically modified proteins; affinity chromatography and immobilized enzyme reactors. Mailing Add: Chem Res Div Hoffmann-La Roche Inc Nutley NJ 07110

NISHIKAWA, OSAMU, b Osaka, Japan, Aug 7, 31; m 57; c 4. SURFACE PHYSICS. Educ: Osaka Univ, BS, 54; Pa State Univ, MS, 61, PhD(physics), 64. Prof Exp: Asst res physics, Osaka Univ, 54-59; assoc, Pa State Univ, 64-67, from asst prof to assoc prof, 67-76; PROF PHYSICS, TOKYO INST TECHNOL, 76- Concurrent Pos: Mem tech staff, Bell Labs, Murray Hill, NJ, 70-72. Mem: Am Phys Soc; Am Asn Physics Teachers. Res: Research and development of a field ion microscope, expecially application to non-refractory metals such as iron, nickel and copper. Mailing Add: Tokyo Inst of Technol 12-1 Ohakayama 2-Chome Meguro-Ku Tokyo Japan

NISHIKAWARA, MARGARET T, b Vancouver, BC, Feb 3, 23. PHYSIOLOGY. Educ: Univ Toronto, BA, 47, MA, 48, PhD(physiol), 52. Prof Exp: Res assoc physiol, Univ Toronto, 52-54; from asst prof to assoc prof, 54-72, PROF PHYSIOL, OHIO STATE UNIV, 72- Mem: AAAS; Am Physiol Soc; Endocrine Soc. Res: Physiologic and endocrine control of metabolic pathways. Mailing Add: Dept of Physiol Ohio State Univ Columbus OH 43210

NISHIMOTO, ROY KATSUTO, b Lihue, Hawaii, Oct 29, 44; m 70; c 1. WEED SCIENCE, VEGETABLE CROPS. Educ: Ore State Univ, BS, 66, MS, 67; Purdue Univ, PhD(weed sci), 70. Prof Exp: Asst prof, 70-74, ASSOC PROF WEED SCI, DEPT HORT, UNIV HAWAII, 74- Mem: Asian Pac Weed Sci Soc (treas, 73-); Weed Sci Soc Am; Am Soc Hort Sci. Res: Weed control in horticultural crops; Cyperus rotundus tuber dormancy; absorption and translocation of herbicides; phosphorus fertility in tropical soils and vegetable crops. Mailing Add: Dept of Hort Univ of Hawaii 3190 Maile Way Rm 102 Honolulu HI 96822

NISHIMURA, EDWIN TAKAYASU, b Sacramento, Calif, Sept 12, 18; m 43; c 3. EXPERIMENTAL PATHOLOGY. Educ: Univ Calif, AB, 40; Wayne State Univ, MD, 45. Prof Exp: Instr neuropath, Sch Med, Univ Mich, 50-51; from instr to assoc prof path, Sch Med, Northwestern Univ, 53-65; assoc prof, Sch Med, Univ Calif, Los Angeles, 65-66; PROF PATH & CHMN DEPT, UNIV HAWAII, MANOA, 66- Concurrent Pos: Pathologist, Atomic Bomb Casualty Comn, Japan, 57-59; USPHS career develop award, 59-65; mem study sect path B, NIH, 70-; dir lab, St Francis Hosp, 73- Mem: Am Soc Exp Path; Am Asn Cancer Res; Am Soc Cell Biol; Am Asn Path & Bact. Res: Cancer; enzyme disorders. Mailing Add: Biomed Sci Bldg Univ of Hawaii at Manoa Honolulu HI 96822

NISHIMURA, JONATHAN SEI, b Berkeley, Calif, Sept 30, 31; m 55; c 2. BIOCHEMISTRY. Educ: Univ Calif, AB, 56, PhD(biochem), 59. Prof Exp: USPHS fel biochem, Sch Med, Tufts Univ, 59-62, sr instr, 62-64, asst prof, 64-69; assoc prof, 69-71, PROF BIOCHEM, UNIV TEX HEALTH SCI CTR SAN ANTONIO, 71-

Mem: Am Soc Biol Chem. Res: Mechanism of enzyme action. Mailing Add: 125 Mecca San Antonio TX 78232

NISHIMURA, KEIICHI, b Fukuwoka, Japan, June 24, 29; nat US; m 57; c 3. THEORETICAL PHYSICS. Educ: Univ Calif, AB, 52, PhD(physics), 57. Prof Exp: Asst prof physics, Rutgers Univ, 57-64; PHYSICIST, LAWRENCE LIVERMORE LAB, UNIV CALIF, 64- Concurrent Pos: Vis scientist, Lockheed Missiles & Space Co, 62; Rutgers Res Coun fel, 62-63. Res: Radiation physics. Mailing Add: Lawrence Livermore Lab Univ of Calif Livermore CA 94550

NISHIMUTA, JOHN FRANCIS, b Okmulgee, Okla, Jan 4, 44; m 64; c 2. RUMINANT NUTRITION. Educ: WTex State Univ, BS, 68; Tex Tech Univ, MS, 71; Univ Ky, PhD(animal nutrit), 73. Prof Exp: ASSOC PROF ANIMAL SCI, ALA A&M UNIV, 73- Mem: Am Soc Animal Sci; AAAS. Res: Forage studies on cool season annuals, lactic acid acidosis in feedlot animals, and digestibility of feedstuffs; monogastric nutrition, protein efficiency ratio, net protein utilization and energy studies with cereal grains. Mailing Add: Dept of Food Sci Ala A&M Univ Normal AL 35762

NISHIOKA, RICHARD SEIJI, b Hilo, Hawaii, Mar 9, 33; m 61; c 3. NEUROENDOCRINOLOGY, MICROSCOPIC ANATOMY. Educ: Univ Hawaii, BA, 56, MA, 59. Prof Exp: Res zoologist, 59-65, from asst specialist to assoc specialist, 65-72, SPECIALIST ZOOL, UNIV CALIF, BERKELEY, 72- Mem: Am Soc Zoologists; AAAS. Res: Comparative neuroendocrinology; ultrastructure and neurophysiological aspects of neurosecretion formation, transport and release; aminergic innervation of endocrine organs; effect of hormone administration to neonates. Mailing Add: Dept of Zool Univ of Calif Berkeley CA 94720

NISHITA, HIDEO, b Castroville, Calif, May 8, 17; m; c 4. SOIL CHEMISTRY, PLANT NUTRITION. Educ: Univ Calif, BS, 47, MS, 49, PhD(plant physiol), 52. Prof Exp: Sr lab asst bot, Univ Calif, 47-48, asst plant nutrit, 49-52; jr res soil scientist soil sci & plant physiol, Atomic Energy Proj, 52-54, asst res soil scientist, 54-58, from asst res soil scientist to assoc res soil scientist, 54-66, RES SOIL SCIENTIST, LAB NUCLEAR MED & RADIATION BIOL, UNIV CALIF, LOS ANGELES, 66- Mem: AAAS; Am Inst Biol Sci; Soil Sci Soc Am; Am Soc Plant Physiologists. Mailing Add: Lab of Nuclear Med & Rad Biol Univ of Calif Los Angeles CA 90024

NISHIZAWA, EDWARD EICHI, b Pitt Meadows, BC, Sept 25, 29. BIOCHEMISTRY, BIOLOGY. Educ: Univ Man, BSc, 52, MSc, 53; Univ Ottawa, PhD(org chem), 59. Prof Exp: Fel steroid biochem, Univ Utah, 59-63, res assoc, 61-62, res instr, 62-63; res scientist platelet biochem & Med Res Coun grant, Univ Guelph, 63-65; Med Res Coun sr fel path, McMaster Univ, 66-70; res assoc, 70-72, SR RES SCIENTIST, DIABETES & ATHEROSCLEROSIS RES UNIT, UPJOHN CO, 72- Concurrent Pos: Mem coun thrombosis, Am Heart Asn. Mem: Sigma Xi; Int Soc Thrombosis & Haemostasis; Am Heart Asn. Res: Elucidation of membrane function using platelets as a mode model; development of agents to combat thrombosis. Mailing Add: Diabetes & Atheroscl Res Unit Upjohn Co Kalamazoo MI 49001

NISONOFF, ALFRED, b New York, NY, Jan 26, 23; m 46; c 2. IMMUNOCHEMISTRY, IMMUNOBIOLOGY. Educ: Rutgers Univ, BS, 42; Johns Hopkins Univ, PhD(chem), 51. Prof Exp: AEC fel, Med Sch, Johns Hopkins Univ, 51-52; sr res chemist, US Rubber Co, Conn, 52-54; sr cancer res scientist, Roswell Park Mem Inst, NY, 54-57, assoc cancer res scientist, 57-60; from assoc prof to prof microbiol, Univ Ill, Urbana, 60-66; prof microbiol, Univ Ill Col Med, 66-75, head dept biochem, 69-75; PROF BIOL, ROSENSTIEL RES CTR, BRANDEIS UNIV, 75- Concurrent Pos: NIH career res award, 62-69; mem study sect allergy & immunol, NIH. Honors & Awards: Medal, Pasteur Inst, 71. Mem: Am Asn Immunol; Am Soc Biol Chemists. Res: Mechanism of biosynthesis of antibodies and their genetic control. Mailing Add: Dept of Biol Rosenstiel Res Ctr Brandeis Univ Waltham MA 02154

NISS, HAMILTON FREDERICK, b Milwaukee, Wis, Apr 29, 23; m 49; c 2. MICROBIOLOGY. Educ: Univ Wis, PhB, 45, MS, 47; Purdue Univ, PhD, 58. Prof Exp: Asst prof biol, Sam Houston State Col, 47-49; instr bact, Purdue Univ, 49-57; asst prof microbiol, Syracuse Univ, 57-61; SR MICROBIOLOGIST, ELI LILLY & CO, 61- Mem: AAAS; Am Soc Microbiol; Soc Indust Microbiol. Res: Biochemical microbiology; fungal physiology; antibiotic biosynthesis. Mailing Add: Dept K-400 Eli Lilly & Co Indianapolis IN 46206

NISSELBAUM, JEROME SEYMOUR, b Hartford, Conn, Dec 21, 25; m 49; c 3. BIOCHEMISTRY. Educ: Univ Conn, BA, 49; Tufts Univ, PhD(biochem), 53. Prof Exp: Asst biochem, Tufts Univ, 50-53, USPHS res fel, 53-55, instr, Sch Med, 54-57; asst, 57-60, ASSOC, SLOAN-KETTERING INST CANCER RES, 60-, ASSOC MEM, 67-; ASSOC PROF BIOCHEM, CORNELL UNIV, 68- Concurrent Pos: Res assoc, Sloan-Kettering Div, Cornell Univ, 57-68; asst biochemist, Mem Hosp, Mem Sloan-Kettering Cancer Ctr, 72-75, asst attend biochemist, Mem Hosp, 75- Mem: AAAS; Am Asn Clin Chemists; Am Chem Soc; Am Soc Biol Chem. Res: Enzymology; clinical biochemistry; methods development; protein chemistry; immunochemistry. Mailing Add: 410 E 68th St New York NY 10021

NISSIM-SABAT, CHARLES, b Sofia, Bulgaria, Feb 1, 38; US citizen; m 67; c 1. COSMOLOGY, HIGH ENERGY PHYSICS. Educ: Columbia Univ, AB, 59, MA, 60, PhD(physics), 65. Prof Exp: Res assoc high energy physics, Fermi Inst, Univ Chicago, 65-67; from asst prof to assoc prof, 67-74, PROF PHYSICS & CHMN DEPT, NORTHEASTERN ILL UNIV, CHICAGO, 74- Mem: AAAS. Res: General relativity and high energy physics, particularly experimental weak interactions and mu-mesic x-rays. Mailing Add: 5464 S Everett Chicago IL 60615

NISSLY, CHARLES MARTIN, b Denver, Pa, Aug 25, 34. GEOGRAPHY OF LATIN AMERICA, CULTURAL GEOGRAPHY. Educ: Pa State Univ, BA, 56; Univ Fla, MA, 60, PhD(Latin Am studies, geog), 66. Prof Exp: Asst dir, Sch Inter-Am Studies, Univ Fla, 60-61; res assoc, Ctr Latin Am Studies, Univ Fla, 66-68; asst prof geog & Latin Am studies, La State Univ, New Orleans, 68-70, ASSOC PROF GEOG & LATIN AM STUDIES, UNIV NEW ORLEANS, 70-, CHMN DEPT ANTHROP & GEOG, 74- Concurrent Pos: Co-ed, Southeastern Latin Americanist, 65-70 & Latin Am Studies Asn Newslett, 72- Mem: Asn Am Geog; Latin Am Studies Asn. Res: Colonization and development in the Amazon and Orinoco Basins; trends in contemporary Latin American art and drama. Mailing Add: Dept of Anthrop & Geog Univ New Orleans New Orleans LA 70122

NISWANDER, JERRY DAVID, b Ottumwa, Iowa, Mar 1, 30; m 55; c 4. DENTISTRY, HUMAN GENETICS. Educ: Univ Mich, DDS, 55, MS, 62. Prof Exp: Dent intern, USPHS Hosp, Seattle, Wash, 55-56; mem staff clin ctr, Dent Dept, NIH, 56-57, mem staff human genetics sect, Nat Inst Dent Res, 57-58 & mem staff child health surv, Atomic Bomb Casualty Comn, 58-60; mem staff human genetics, Univ Mich, 60-63; MEM STAFF, HUMAN GENETICS BR, NAT INST DENT RES, 63- Concurrent Pos: Prof lectr, Sch Dent, Georgetown Univ, 71- Mem: AAAS; Am Soc Human Genetics; Am Pub Health Asn; Am Cleft Palate Asn; Teratology Soc. Res: Human genetic and epidemiology studies of major congenital malformations; dental-facial abnormalities and growth and development; human biology. Mailing Add: Human Genetics Br Nat Inst of Dent Res Bethesda MD 20014

NISWENDER, GORDON DEAN, b Gillette, Wyo, Apr 21, 40; m 64; c 2. REPRODUCTIVE ENDOCRINOLOGY. Educ: Univ Wyo, BS, 62; Univ Nebr, MS, 64; Univ Ill, PhD(animal sci), 67. Prof Exp: Res asst reprod physiol, Univ Nebr, 62-64 & Univ Ill, 64-65; NIH res fel, Univ Mich, Ann Arbor, 67-68, asst prof, 68-71; from asst prof to assoc prof, 71-75, PROF REPROD PHYSIOL, COLO STATE UNIV, 75- Honors & Awards: Young animal scientist of the year, Am Asn Animal Scientists, 75. Mem: AAAS; Soc Study Reproduction; Am Soc Animal Sci; Endocrine Soc. Res: Reproductive biology, endocrinology, immunology. Mailing Add: Dept of Physiol & Biophys Colo State Univ Ft Collins CO 80523

NITECKI, DANUTE EMILIJA, b Lithuania, Apr 22, 27; US citizen; c 1. BIOCHEMISTRY. Educ: Univ Chicago, MS, 56, PhD(chem), 61. Prof Exp: NIH fel, 61-63; asst res biochemist, Med Ctr, 63-69, ASSOC RES BIOCHEMIST, SCH MED, UNIV CALIF, SAN FRANCISCO, 69- Mem: Am Chem Soc; The Chem Soc. Res: Synthesis of peptides used as haptens in immunochemical studies; synthesis of boron containing aromatic compounds of physiological interest; syntheses of well defined small molecular weight antigens used to investigate the initiation and the progress of cellular and humoral immune response. Mailing Add: 1576 18th Ave San Francisco CA 94122

NITECKI, MATTHEW H, b Poland, Apr 30, 25; nat US; m 64; c 2. PALEOBOTANY, INVERTEBRATE PALEONTOLOGY. Educ: Univ Chicago, MS, 58, PhD, 68. Prof Exp: Asst cur, 65-69, ASSOC CUR, FIELD MUS NATURAL HIST, 69-, CUR INVERT PALEONT, WALKER MUS, UNIV CHICAGO, 55- Concurrent Pos: NSF grant invert fossils, NATO Advan Study Inst Paleoclimatic Conf, Eng, 63. Mem: Geol Soc Am; Am Asn Petrol Geologists; Soc Econ Paleontologists & Mineralogists; Soc Study Evolution; Soc Syst Zoologists. Res: Paleozoic paleobotany and invertebrate paleontology. Mailing Add: Field Mus Natural Hist Roosevelt Rd & Lake Shore Dr Chicago IL 60605

NITHMAN, CHARLES JOSEPH, b Belleville, Ill, Jan 14, 37; m 59; c 2. PHARMACY. Educ: Okla State Univ, BS, 59; Univ Okla, BSPh, 62, Univ Okla, MSc, 70; Mercer Univ, PharmD, 74. Prof Exp: Clin instr pharm, Univ Okla, 68-70; asst dir pharm, St Anthony Hosp, Oklahoma City, 70-72; ASST PROF PHARM, SOUTHWESTERN OKLA STATE UNIV, 72- Mem: Sigma Xi; Am Soc Hosp Pharmacists; Am Pharmaceut Asn; Am Asn Col Pharm. Res: Drug distribution systems in hospitals. Mailing Add: 4520 NW 32nd Pl Oklahoma City OK 73122

NITKA, HEINZ FRIEDRICH, physics, see 12th edition

NITOWSKY, HAROLD MARTIN, b Brooklyn, NY, Feb 12, 25; m 54; c 2. MEDICINE, GENETICS. Educ: NY Univ, AB, 44, MD, 47; Univ Colo, MSc, 51; Am Bd Pediat, dipl, 56. Prof Exp: USPHS fel, Univ Colo, 50-51; from instr to assoc prof, Sch Med, Johns Hopkins Univ, 53-67; PROF PEDIAT, ALBERT EINSTEIN COL MED, 67- Concurrent Pos: Pediatrician, Johns Hopkins Hosp, 53-; res assoc & adj attend pediatrician, Sinai Hosp, 55-, dir pediat res, 58-; sr investr, Nat Asn Retarded Children, 60-65. Mem: AAAS; Soc Pediat Res; Sigma Xi; Am Fedn Clin Res; Am Inst Nutrit. Res: Somatic cell genetics; inborn errors of metabolism; genetic counseling. Mailing Add: Dept of Pediat Albert Einstein Col of Med Bronx NY 10461

NITSOS, RONALD EUGENE, b Sacramento, Calif, Aug 26, 37; m 66; c 1. PLANT PHYSIOLOGY. Educ: Sacramento State Col, BA, 63; Ore State Univ, MA, 66, PhD(plant physiol), 69. Prof Exp: ASSOC PROF BIOL, SOUTHERN ORE COL, 69- Concurrent Pos: Res grant, Southern Ore Col, 69-71. Mem: Am Soc Plant Physiologists. Res: Plant physiology, especially roles of univalent cations in plants; needs and effects of univalent cations in enzyme activation. Mailing Add: Dept of Biol Southern Ore Col Ashland OR 97520

NITTLER, LEROY WALTER, b Shickley, Nebr, Jan 10, 21; wid; c 2. AGRICULTURE. Educ: Univ Nebr, BS, 49, MS, 50; Cornell Univ, PhD(plant breeding), 53. Prof Exp: Asst, 50-53, from asst prof to assoc prof, 53-65, PROF SEED INVESTS, COL AGR, CORNELL UNIV, 65- Concurrent Pos: Head dept seed & vegetable sci, Col Agr, Cornell Univ, 68-73. Mem: AAAS; Am Soc Agron. Res: Varietal purity testing of grain, forage crop and turf grass seed; response of grain, grass and legume seedlings to photoperiod, light quality, light intensity and temperature; response of seedlings to chemicals and nutrient elements. Mailing Add: Dept of Seed & Vegetable Sci Cornell Univ Col of Agr & Life Sci Geneva NY 14456

NITZ, OTTO WILLIAM JULIUS, b Sigourney, Iowa, June 25, 05; m 36. ORGANIC CHEMISTRY. Educ: Elmhurst Col, BS, 29; Univ Iowa, MS, 33, PhD(org chem), 36. Prof Exp: Instr, Elmhurst Col, 29-32; asst, Univ Iowa, 33-36; prof chem, Parsons Col, 36-40 & Northern Mont Col, 40-51; chief chemist, Ky Synthetic Rubber Corp, 51-52; prof, 52-71, EMER PROF CHEM, UNIV WIS-STOUT, 71- Res: Vanillin derivatives. Mailing Add: 1103 Third Ave Menomonie WI 54751

NITZ, ROBERT E, b Kansas City, Mo, Aug 4, 23; m 46; c 3. MEDICINE. Educ: Univ Louisville, MD, 50; Am Bd Pub Health, cert prev med, 64. Prof Exp: Med Corps, US Army, 50-75, chief dept health pract, Walter Reed Army Inst Res, 60-64, chief prev med, Hq, Third Army, Ga, 64-67 & Hq, US Army Vietnam, 67-68, chief prev med, Off Chief Surg, Hq, US Army Pac, 68-75. Mem: Am Pub Health Asn; fel Am Col Prev Med. Res: Epidemiology of acute infectious diseases, specifically meningococcal meningitis, cholera and respiratory diseases. Mailing Add: 370 Portlock Rd Honolulu HI 96825

NITZBERG, DAVID MORRIS, applied statistics, biostatistics, see 12th edition

NIU, JOSEPH H Y, b Nanking, China, Apr 22, 32. INORGANIC CHEMISTRY, ORGANIC CHEMISTRY. Educ: Univ Hong Kong, BS, 55; Univ Wis-Madison, PhD(chem), 62. Prof Exp: Res chemist, 62-66, sr res chemist, 66-73, SUPVR RES & DEVELOP, RES DIV, BASF WYANDOTTE CORP, 73- Mem: Am Chem Soc; The Chem Soc; Tech Asn Pulp & Paper Indust. Res: Phosphorus, nitrogen and fluorine chemistry; surface active agents; coordination compounds; new aromatic anions; chemical additives for pulp and paper. Mailing Add: Res Div BASF Wyandotte Corp Wyandotte MI 48195

NIU, MANN CHIANG, b Peking, China, Oct 31, 14; nat US; m 43; c 2. DEVELOPMENTAL BIOLOGY, BIOCHEMISTRY. Educ: Peking Univ, AB, 36; Stanford Univ, PhD, 47. Prof Exp: Asst zool, Peking Univ, 36-40, lectr, 40-44; res assoc embryol, Stanford Univ, 44-52, res biologist, 52-55; asst prof gen physiol, Rockefeller Univ, 55-60; PROF BIOL, TEMPLE UNIV, 60- Concurrent Pos: Guggenheim fel, Rockefeller Univ, 54 & 55, vis prof, 70; Mem, Int Inst Develop Biol & Marine Biol Lab. Honors & Awards: Lillie Award, 57. Mem: Am Asn Anat; Am Soc Cell Biol; Soc Exp Biol & Med; Soc Develop Biol; Am Soc Zoologists. Res:

Origin of pigment cells; cell transformation and cancer; physiology of color changes; nucleocytoplasmic interactions; causal analysis of induction; physiological activity of nucleoproteins and induced biosynthesis; RNA metabolsim in learning and aggressive behavior; RNA-induced changes of genetic characters. Mailing Add: Dept of Biol Temple Univ Philadelphia PA 19122

NIVEN, CHARLES FRANKLIN, JR, b Clemson, SC, July 22, 15; m 39; c 4. MICROBIOLOGY, FOOD SCIENCE. Educ: Univ Ark, BS, 35; Cornell Univ, PhD(bact), 39. Prof Exp: From instr to assoc prof, Cornell Univ, 39-46; bacteriologist, Hiram Walker & Sons, Inc, Ill, 46; from asst prof to prof microbiol, Univ Chicago, 46-64; DIR RES, RES CTR, DEL MONTE CORP, 64- Concurrent Pos: Chief, Div Bact, Am Meat Inst Found, 46-58, assoc dir, 58-61, sci dir, 61-64. Mem: AAAS; Am Soc Microbiol; Am Pub Health Asn; Am Chem Soc; Inst Food Technologists (pres, 74-75). Res: Bacterial nutrition, physiology and metabolism; food microbiology. Mailing Add: Del Monte Corp Res Ctr 205 N Wiget Lane Walnut Creek CA 94598

NIVEN, IVAN (MORTON), b Vancouver, BC, Oct 25, 15; nat US; m 39; c 1. MATHEMATICS. Educ: Univ BC, BA, 45, MA, 36; Univ Chicago, PhD(math), 38. Prof Exp: Harrison fel math, Univ Pa, 38-39; instr, Univ Ill, 39-42; from instr to assoc prof, Purdue Univ, 42-47; assoc prof, 47-50, PROF MATH, UNIV ORE, 50- Mem: AAAS; Am Math Soc; Math Asn Am. Res: Number and additive number theory; irrational numbers; Diophantine approximations. Mailing Add: Dept of Math Univ of Ore Eugene OR 97403

NIVEN, JORMA ILTANEN, psychology, see 12th edition

NIX, JAMES RAYFORD, b Natchitoches, La, Feb 18, 38; m 61; c 2. THEORETICAL NUCLEAR PHYSICS. Educ: Carnegie Inst Technol, BS, 60; Univ Calif, Berkeley, PhD(physics), 64. Prof Exp: NATO fel, Niels Bohr Inst, Univ Copenhagen, 64-65; res assoc theoret nuclear physics, Lawrence Radiation Lab, Univ Calif, Berkeley, 66-68, MEM STAFF THEORET NUCLEAR PHYSICS, LOS ALAMOS SCI LAB, UNIV CALIF, 68- Concurrent Pos: Chmn, SuperHILAC Prog Adv Comt, Lawrence Radiation Lab, Univ Calif, 74-76 & Gordon Res Conf Nuclear Chem, 76; mem, Physics Div Adv Comt, Oak Ridge Nat Lab, 75- Mem: Fel Am Phys Soc; AAAS; Am Chem Soc. Res: Theory of nuclear fission, superheavy nuclei and heavy-ion reactions. Mailing Add: Nuclear Theory T-9 MS 452, Los Alamos Sci Lab Los Alamos NM 87545

NIX, JOE FRANKLIN, b Malvern, Ark, Aug 28, 39. GEOCHEMISTRY, ANALYTICAL CHEMISTRY. Educ: Ouachita Baptist Col, BS, 61; Univ Ark, MS, 63, PhD(chem), 66. Prof Exp: Assoc prof chem, 66-74, PROF CHEM, OUACHITA BAPTIST UNIV, 74- Mem: AAAS; Am Chem Soc; Am Geophys Union; Am Soc Limnol & Oceanog. Res: Radioactive fallout; neutron activation analysis of meteorites; geochemistry of hot springs and geysers; geochemistry of impoundments. Mailing Add: 2809 Walnut St Arkadelphia AR 71923

NIX, SYDNEY JOHNSTON, JR, b Athens, Ga, Mar 8, 20; m 46; c 2. ORGANIC CHEMISTRY. Educ: Stetson Univ, BS, 42; Ind Univ, PhD(org chem), 51. Prof Exp: Asst chem lab, Stetson Univ, 39-42; control chemist, E I du Pont de Nemours & Co, 42-43; asst, Ind Univ, 46-50; res chemist, 50-68, SR RES CHEMIST, TEXTILE FIBERS DEPT, E I DU PONT DE NEMOURS & CO, 68- Mem: Am Chem Soc. Res: Synthetic textile fibers, particularly acrylic and elastomer fibers. Mailing Add: Textile Fibers Dept E I du Pont de Nemours & Co Waynesboro VA 22980

NIXON, CHARLES WILLIAM, b Carlinville, Ill, Sept, 28, 25. GENETICS, HISTOLOGY. Educ: Univ Ill, BS, 46, MS, 48; Brown Univ, PhD(biol), 51. Prof Exp: Asst prof biol, Northeastern Univ, 51-57 & Simmons Col, 57-60; biologist, Bio-Res Inst, Inc, 60-70; pvt consult genetics, 70-73; RES ASSOC, MASS INST TECHNOL, 73- Mem: AAAS. Res: Mammalian and plant genetics. Mailing Add: 37 Ox Bow Lane Randolph MA 02368

NIXON, CHARLES WILLIAM, b Wellsburg, WVa, Aug 15, 29; m 56; c 2. ACOUSTICS, AUDIOLOGY. Educ: Ohio State Univ, BSc, 52, MA, 53, PhD(speech sci), 60. Prof Exp: Dir & therapist, Wheeling Hearing & Speech Ctr, 55-56; consult, Ohio County Bd Educ, 55-56; res audiologist biol acoust, 56-67, CHIEF BIOACOUST BR, AEROSPACE MED RES LABS, WRIGHT-PATTERSON AFB, 67- Concurrent Pos: Mem comt on hearing, bioacoust & biomech, Nat Acad Sci-Nat Res Coun, 59-, alt mem exec coun, 67-72, US Air Force mem exec coun, 72-; mem noise team supersonic transport, Fed Aviation Agency, 65; vpres & bd dirs, Montgomery County & Dayton Hearing & Speech Ctr, 66-67, pres bd dirs, 68-70; instr, Wright State Univ, 67- & Univ Dayton, 69- Mem: Nat Acad Sci; Am Speech & Hearing Asn; Am Nat Stand Inst; Aerospace Med Asn; fel Acoust Soc Am. Res: Sonic boom; effects of sonic boom on people; speech communication; bone and tissue conduction; helium speech; ear protection; hearing conservation; aural reflex; noise exposure criteria; psychoacoustics; psychological and physiological response to acoustic energy. Mailing Add: Aerospace Med Res Labs Wright-Patterson AFB Dayton OH 45433

NIXON, DONALD MERWIN, b Topeka, Kans, Nov 11, 35; m 62; c 3. AGRICULTURAL ECONOMICS. Educ: Colo State Univ, BS, 65, MS, 66, PhD(agr mkt), 69. Prof Exp: Instr poultry sci & mkt, Colo State Univ, 65-69; assoc prof bus & agr mkt, 69-70, ASSOC PROF AGR ECON, TEX A&I UNIV, 70- Concurrent Pos: Res grants, Houston Livestock & Rodeo Asn, 70, Rio Farms, 72 & Perry Found, 75. Mem: Am Mkt Asn; Am Agr Econ Asn; Inst Food Technologists; Poultry Sci Asn. Res: Agricultural marketing; economic analysis among beef cattle common to South Texas; taste panel evaluation; economics of crop insurance; use of grain sorghum and beef futures by South Texas producers. Mailing Add: Col of Agr Tex A&I Univ Kingsville TX 78363

NIXON, ELRAY S, b Escalante, Utah, Feb 5, 31; m 57; c 3. PLANT ECOLOGY, PLANT TAXONOMY. Educ: Brigham Young Univ, BS, 57, MS, 61; Univ Tex, PhD(bot), 63. Prof Exp: Asst prof biol, Chadron State Col, 63-65 & Southern Ore Col, 65-66; mem staff, 66-68, assoc prof, 68-72, PROF BIOL, STEPHEN F AUSTIN STATE UNIV, 72- Mem: Southwestern Asn Naturalists; Int Asn Plant Taxon. Res: Soil-plant relationships. Mailing Add: PO Box 3003 SFA Station Nacogdoches TX 75961

NIXON, EUGENE RAY, b Mt Pleasant, Mich, Apr 14, 19; m 45; c 2. PHYSICAL CHEMISTRY. Educ: Alma Col, ScB, 41; Brown Univ, PhD(chem), 47. Prof Exp: Res chemist, Manhattan Dist, Brown Univ, 42-46, res assoc, Off Naval Res, 46-47, instr chem, 47-49; from instr to assoc prof, 49-65, PROF CHEM, UNIV PA, 65-, DIR, MAT RES LAB, 69- Concurrent Pos: Vdean grad sch, Univ Pa, 58-65. Mem: Am Chem Soc; Am Phys Soc. Res: Molecular spectroscopy and structure. Mailing Add: Dept of Chem Univ of Pa Philadelphia PA 19104

NIXON, JAMES, inorganic chemistry, physical chemistry, see 12th edition

NIXON, JOHN CHARLES, medicine, biochemistry, see 12th edition

NIXON, JOSEPH EUGENE, b Platteville, Wis, Jan 7, 38; m 65; c 2. TOXICOLOGY. Educ: Univ Ill, BS, 61, PhD(nutrit biochem), 65. Prof Exp: Res assoc biochem, Dept Physiol Chem, Univ Wis, 65-66; res biochemist, Vet Admin Hosp, Madison, Wis, 66-68; asst prof nutrit, 68-75, ASSOC PROF, DEPT FOOD SCI & TECHNOL, ORE STATE UNIV, 75- Mem: Am Inst Nutrit. Res: Biochemistry of fatty acid synthesis; enzymology and intermediary metabolism of lipids; nutrition and toxicology related to carcinogens and their precursors found in foods. Mailing Add: Dept of Food Sci Ore State Univ Corvallis OR 97331

NIXON, ROY WESLEY, b Dade City, Fla, Aug 15, 95; m 24; c 2. HORTICULTURE. Educ: Univ Ariz, BSA, 22. Prof Exp: From jr horticulturist to res horticulturist, Date Invest, USDA, 23-65, collabr, 65-75. Concurrent Pos: Explorer, Date Varieties, Iraq & WIran, 38-39; Guggenheim fel, NAfrica, 48-49; date consult, Saudi Arabia, 53, Libya, 59 & Sudan, 65-66. Mem: Fel AAAS; Am Soc Hort Sci. Res: Culture, varieties and metaxenia of dates. Mailing Add: 81229 Palmyra Ave Indio CA 92201

NIXON, SCOTT WEST, b Philadelphia, Pa, Aug 24, 43; m 65; c 2. ECOLOGY, OCEANOGRAPHY. Educ: Univ Del, BA, 65; Univ NC, Chapel Hill, PhD(bot), 70. Prof Exp: Res assoc, 69-70, asst prof, 70-75, ASSOC PROF OCEANOG, UNIV RI, 75- Mem: Am Soc Limnol & Oceanog; Ecol Soc Am; Estuarine & Brackish Water Sci Asn; Estuarine Res Fedn. Res: Ecological systems; energetics; simulation. Mailing Add: Sch of Oceanog Univ of RI Kingston RI 02881

NIZEL, ABRAHAM EDWARD, b Boston, Mass, July 27, 17; m 42; c 2. PREVENTIVE DENTISTRY, NUTRITION. Educ: Tufts Univ, DMD, 40, MSD, 52. Prof Exp: Intern oral surg, Worcester City Hosp, 40-41; instr periodont, 52-60, asst clin prof, 61-66, assoc prof nutrit, 66-73, PROF NUTRIT & PREV DENT, SCH DENT MED, TUFTS UNIV, 73- Concurrent Pos: Res assoc, Mass Inst Technol, 52-70, vis assoc prof nutrit & metab, 71-73, vis prof, 73-; guest lectr, Eastman Dent Ctr, 60- & Forsyth Dent Ctr, 69-; grants, Nutrit Found, 67-72 & Nat Dairy Coun, 68-72; consult, Am Dent Asn & Food & Nutrit Bd, Nat Acad Sci; consult ed, J Am Dent Asn & J Prev Dent. Honors & Awards: Prev Dent Award, Am Dent Asn. Mem: AAAS; fel Am Col Dent; fel Am Asn Dent Sci; Int Asn Dent Res; Am Dent Asn. Res: Oral health-nutrition interrelationships, particularly the effect of dietary phosphate supplements on inhibition of experimental caries; model nutrition teaching program for dental schools and schools of dental hygiene. Mailing Add: Tufts Univ Sch of Dent Med One Kneeland St Boston MA 02110

NOACK, MANFRED GERHARD, b Olbersdorf, Ger, Jan 25, 36; m 62; c 2. INORGANIC CHEMISTRY, INDUSTRIAL CHEMISTRY. Educ: Munich Tech Univ, Vordiplom, 59, Diplomchemiker, 62, Dr rer nat(chem), 64. Prof Exp: Teaching asst chem, Munich Tech Univ, 62-64; res assoc, Univ Md, 64-67; res chemist, 67-75, GROUP LEADER RES & DEVELOP, OLIN CORP, 75- Mem: Am Chem Soc; Nat Asn Corrosion Engrs; Sigma Xi. Res: Chemistry of metal carbonyls; magnetic resonance phenomena in solutions; homogeneous catalysis; corrosion inhibition; applications of redox chemistry. Mailing Add: Olin Res Ctr 275 Winchester Ave New Haven CT 06504

NOAKES, DAVID LLOYD GEORGE, b Hensall, Can, Aug 3, 42; m 66; c 1. ETHOLOGY. Educ: Univ Western Ont, BS, 65, MS, 66; Univ Calif, Berkeley, PhD(zool), 71. Prof Exp: Lectr zool, Edinburgh Univ, 70-72; ASST PROF ZOOL, UNIV GUELPH, 72- Mem: Animal Behav Soc; Can Soc Zoologists; Sigma Xi; Am Fisheries Soc. Res: Behavioral ontogeny, social behavior and social systems, feeding and reproductive ecology, physiological basis of behavior, evolution of behavior, especially of fishes and birds. Mailing Add: Dept of Zool Col of Biol Sci Univ of Guelph Guelph ON Can

NOAKES, JOHN EDWARD, b Windsor, Ont, May 21, 30; US citizen; m 61; c 2. GEOCHEMISTRY, OCEANOGRAPHY. Educ: Champlain Col Plattsburg, BS, 53; Tex A&M Univ, MS, 59, PhD(chem oceanog), 62. Prof Exp: Soils engr, NY State Soil Div, 53-55; res chemist, Clark Cleveland Pharmaceut Co, 55-57; res chemist, Tex A&M Univ, 58-59, res found, 59-61; asst prof chem oceanog, Univ Alaska, 61-62; res scientist, Oak Ridge Assoc Univs, 62-68; ASSOC PROF GEOL, UNIV GA, 68-, GEN RES SERVS, 70- Concurrent Pos: Dir geochronology lab, Univ Ga, 68-70. Mem: Am Chem Soc. Res: Geochemistry of marine environment; Tritium, radioactive carbon and uranium geochronology; development of nuclear radiation analytical techniques for measuring low levels of radiation. Mailing Add: Dept of Geol Univ of Ga Athens GA 30601

NOAKS, JOHN WILLARD, physics, see 12th edition

NOALL, MATTHEW WILCOX, b Salt Lake City, Utah, Mar 16, 24; m 50; c 2. BIOCHEMISTRY. Educ: Univ Utah, BA, 48, MA, 49, PhD(biochem), 52. Prof Exp: USPHS fel, Nat Inst Arthritis & Metab Dis, 52-54; Am Cancer Soc fel, Sch Med, Tufts Univ, 54-56; asst prof obstet & gynec, Sch Med, Wash Univ, 56-62; asst prof biochem, 62-66, ASSOC PROF PATH & BIOCHEM, BAYLOR COL MED, 66- Mem: Am Chem Soc; NY Acad Sci. Res: Endocrinology; chemical carcinogenesis. Mailing Add: Dept of Path & Biochem Baylor Col of Med Houston TX 77025

NOBACK, CHARLES ROBERT, b New York, NY, Feb 15, 16; m 38; c 4. ANATOMY. Educ: Cornell Univ, BS, 36; NY Univ, MS, 38; Univ Minn, PhD(anat), 42. Prof Exp: Asst prof, Med Col Ga, 41-44; from asst prof to assoc prof, Long Island Col Med, 44-49; from asst prof to assoc prof anat, 49-68, actg chmn dept, 74-75, PROF ANAT, COL PHYSICIANS & SURGEONS, COLUMBIA UNIV, 68- Concurrent Pos: NIH res grants, 53-; James Arthur lectr, 59; mem nerv & sensory syst res eval comt, Vet Admin, 69-72. Mem: AAAS; Histochem Soc; Harvey Soc; Am Asn Anat; Am Asn Phys Anthrop. Res: Development of mammalian skeleton; reproduction in the primates; histochemistry of mammalian tissue; regeneration of neural tissues; comparative neuroanatomy; nutrition and nervous system. Mailing Add: Dept of Anat Columbia Univ Col of Phys New York NY 10032

NOBACK, RICHARDSON K, b Richmond, Va, Nov 7, 23; m 47; c 3. INTERNAL MEDICINE. Educ: Cornell Univ, MD, 47; Am Bd Internal Med, dipl. Prof Exp: Nat Heart Inst res fel, 49-50; instr med, Col Med, Cornell Univ, 50-53; asst prof, Col Med, State Univ NY Upstate Med Ctr, 55-56; assoc prof, Med Ctr, Univ Ky, 56-63, asst dean, 56-58, dir univ health serv, 59-63; from assoc dean to actg dean sch med, 64-70, PROF INTERNAL MED, SCH MED, UNIV MO-KANSAS CITY, 64-, DEAN SCH MED, 70- Concurrent Pos: Asst dir comprehensive care & teaching prog, Cornell Univ, 52-53; med dir, Syracuse Dispensary, 55-56; consult, Med Col, Univ Tenn & John Gaston Hosp, 55; exec med consult, Norfolk Area Med Ctr Authority, Va, 64-70; exec dir, Kansas City Gen Hosp & Med Ctr, 64-69. Mem: AMA; Asn Am Med Cols; NY Acad Sci. Res: Medical education; health care programs; cardiovascular and renal diseases. Mailing Add: 2411 Holmes St Kansas City MO 64108

NOBEL, PARK S, b Chicago, Ill, Nov 4, 38; m 65; c 2. PLANT PHYSIOLOGY, ECOLOGY. Educ: Cornell Univ, BEP, 61; Calif Inst Technol, MS, 63; Univ Calif, Berkeley, PhD(biophys), 65. Prof Exp: NSF fels chloroplasts, Tokyo Univ, 65-66 &

King's Col, London, 66-67; from asst prof to assoc prof molecular biol, 67-75, PROF BIOL, UNIV CALIF, LOS ANGELES, 75- Concurrent Pos: Guggenheim fel, Australian Nat Univ, 73-74. Mem: Am Soc Plant Physiologists. Res: Biophysical aspects of plant physiology, especially plant-environment interactions and ecology. Mailing Add: Dept of Biol Univ of Calif Los Angeles CA 90024

NOBILE, ARTHUR, b Newark, NJ, May 6, 20; m 55; c 3. MICROBIOLOGY. Educ: Univ Calif, AB, 50. Prof Exp: Microbiologist, Schering Corp, 50-68; DIR TECH SERV, ORGANON, INC, 68- Mem: Am Inst Chem. Res: Steroids and antibiotics; microbiological transformations of steroids; new products and processes; antibiotics, new processes and yield .improvement; inventor of prednisone, prednisolone, methandrostenolone and halogenated corticoids. Mailing Add: Organon Inc 375 Mt Pleasant Ave West Orange NJ 07052

NOBIS, JOHN FRANCIS, b Helena, Mont, Jan 23, 21; m 47; c 4. ORGANIC CHEMISTRY. Educ: Col St Thomas, BS, 42; Iowa State Univ, PhD(org chem), 48. Prof Exp: Instr org chem, Iowa State Univ, 46-48, asst prof, 48; asst prof, Xavier Univ, Ohio, 48-51; group leader, Sodium Div, Nat Distillers Chem Co, 51-56, res supvr, 56-59, asst mgr metals & sodium res. Nat Distillers & Chem Corp, US Indust Chem Co, 59; mgr, Prod Develop Dept, Armour Indust Chem Co, 59-60; dir com develop, 60-63, dir mkt serv, 63-65, asst to pres, 65-66, DIR COM DEVELOP, FORMICA CORP, AM CYANAMID CO, 66- Mem: AAAS; Am Chem Soc. Res: Sodium and organosodium chemistry; synthetic organic, organosilicon chemistry; antimalarials; anti-tuberculars; heterocycles; organometallics; polyolefins; plastics. Mailing Add: 5528 Kirby Ave Cincinnati OH 45239

NOBLE, ALLEN GEORGE, b Astoria, NY, Jan 28, 30; m 59; c 3. GEOGRAPHY. Educ: Utica Col, BA, 51; Univ Md, MA, 53; Univ Ill, PhD(geog), 57. Prof Exp: Econ analyst, US Dept State, 57-59, foreign serv officer, 57-63; US vconsul, Am Consulate Gen, Bombay, 59-61, Curitiba, Brazil, 61-62 & Belem, Brazil, 62-63; assoc prof geog, California State Col, Pa, 63-64; PROF GEOG, UNIV AKRON, 64- Concurrent Pos: Planning consult, City Rantoul, Ill, 55-57, George E Wilson Co, Ohio, 65-66 & Tri-County Planning Comn, Ohio, 66; dir int studies, Univ Akron, 69-73; external examr geog, Univ Madras & Univ Rajastham, 73- Mem: Asn Am Geogrs; Nat Coun Geog Educ; Am Water Resources Asn; Am Geog Soc; Sigma Xi. Res: South Asia, especially India; Latin America, especially Brazil; cultural geography, especially settlement landscape. Mailing Add: Dept of Geog Univ of Akron Akron OH 44304

NOBLE, ANN CURTIS, b Harlingen, Tex, Nov 6, 43. FOOD SCIENCE. Educ: Univ Mass, Amherst, BS, 66, PhD(food sci), 70. Prof Exp: Asst prof food sci, Univ Guelph, 70-73; ASST PROF ENOL, UNIV CALIF, DAVIS, 74- Mem: Am Chem Soc; Inst Food Technol; Can Inst Food Sci & Technol; Am Soc Enol; Europ Chemo Receptor Orgn. Res: Flavor chemistry and sensory evaluation. Mailing Add: Dept of Viticult & Enol Univ of Calif Davis CA 95616

NOBLE, EDWIN AUSTIN, b Bethel, Vt, Dec 15, 22; m 48; c 3. GEOLOGY. Educ: Tufts Univ, BS, 46; Univ NMex, MS, 50; Univ Wyo, PhD(geol), 61. Prof Exp: Geologist, US Geol Surv, Boston, 52-54 & AEC, 54-62 & 63-65; adv nuclear raw mat, Int Atomic Energy Agency, Arg, 62-63; assoc prof, 65-69, PROF GEOL & CHMN DEPT, UNIV NDAK, 69-; STATE GEOLOGIST & DIR, NDAK GEOL SURV, 69- Concurrent Pos: Asst state geologist, NDak State Geol Surv, 65-69; ed, Asn Am State Geologists, 71- Mem: Geol Soc Am; Am Inst Mining, Metall & Petrol Engrs; Am Asn Petrol Geologists; Asn Am State Geologists. Res: Genesis of ore deposits in sedimentary rocks. Mailing Add: Dept of Geol Univ of NDak Grand Forks ND 58201

NOBLE, ELMER RAY, b Pyong Yang, Korea, Jan 16, 09; US citizen; m 32; c 4. PARASITOLOGY. Educ: Univ Calif, AB, 31, AM, 33, PhD, 36. Prof Exp: From instr to prof, 36-74, EMER PROF BIOL, UNIV CALIF, SANTA BARBARA, 74- Concurrent Pos: Chmn dept biol sci, Univ Calif, Santa Barbara, 47-51, dean div letters & sci, 51-59, actg provost, 56-58, vchancellor, 58-62; consult, Govt Indonesia, 60. Mem: Am Soc Parasitol; Soc Protozool (pres, 71-72); Am Micros Soc. Res: Cytology pf parasitic protozoa; life history of myxosporidia, trypanosomes and amoebae; ecology of fishes and their parasite-mix; parasitism in deep-sea fishes. Mailing Add: Dept of Biol Sci Univ of Calif Santa Barbara CA 93106

NOBLE, ERNEST PASCAL, b Iraq, Apr 2, 29; nat US; m 56; c 3. PSYCHIATRY, BIOCHEMISTRY. Educ: Univ Calif, BS, 51; Ore State Univ, PhD(org biochem), 55; Western Reserve Univ, MD, 62. Prof Exp: Fulbright scholar, France, 55-56; sr instr biochem, Western Reserve Univ, 57-62; intern med, Stanford Med Ctr, 62-63, resident & asst prof psychiat, 65-69; assoc prof psychiat, psychobiol & pharmacol, Univ Calif, Irvine, 69-71, prof psychiat, psychobiol & pharmacol & chief neurochem, 71-75; DIR, NAT INST ALCOHOL ABUSE & ALCOHOLISM, 75- Concurrent Pos: Vis scientist, NIMH, 66, NIMH career develop award, 66-70; mem rev comt, Alcoholism & Alcohol-Probs, 69-73; Guggenheim fel, France, 74-75. Mem: Am Chem Soc; Soc Exp Biol & Med; NY Acad Sci; Am Psychiat Soc; Psychiat Res Soc. Res: Carbohydrate, protein and nucleic acid metabolism; membrane structure and function; endocrinology and behavior. Mailing Add: Nat Inst on Alcohol Abuse 5600 Fisher's Lane Rockville MD 20852

NOBLE, GLENN ARTHUR, b Pyong Yang, Korea, Jan 16, 09; US citizen; m 35; c 3. PARASITOLOGY. Educ: Univ Calif, AB, 31, MA, 33; Stanford Univ, PhD(parasitol), 40. Prof Exp: Asst zool, Col Pac, 33-35; instr biol, San Francisco City Col, 35-46, chmn dept, 39-46; consult, US Mil Govt, Korea, 46-47; vis prof parsitol, Med Sch, Seoul Nat Univ & Severance Union Med Col, 47; prof, 47-73, head dept, 49-71, EMER PROF BIOL, CALIF POLYTECH STATE UNIV, SAN LUIS OBISPO, 73- Concurrent Pos: Calif Fish & Game res grant, 43; Fulbright prof, Univ Philippines, 53-54 & Nat Taiwan Univ, 61-62; res grants, NSF, 56-57 & 63-66, NIH, 59-61 & Gorgas Mem Lab, Panama, 69. Soc Protozool; Am Soc Parasitol; Philippine Soc Advan Res. Res: Parasitology; protozoology. Mailing Add: Dept of Biol Sci Calif Polytech State Univ San Luis Obispo CA 93401

NOBLE, GORDON ALBERT, b Joliet, Ill, June 20, 27. SOLID STATE PHYSICS, CHEMICAL PHYSICS. Educ: Univ Chicago, PhB, 47, SB, 49, SM, 51, PhD(chem), 55. Prof Exp: Solid state physicist, Zenith Radio Corp, 54-60; res physicist, IIT Res Inst, 60-66, sr scientist, 66-68; asst prof, 68-76, ASSOC PROF PHYSICS, N PARK COL, 76- Concurrent Pos: Consult, HT Res Inst, 68- Mem: AAAS; Am Phys Soc; Am Chem Soc; Am Asn Physics Teachers. Res: Electron paramagnetic and nuclear magnetic resonance; color centers; solid state optical spectroscopy. Mailing Add: North Park Col 5125 N Spaulding Ave Chicago IL 60625

NOBLE, J ARNOLD, b Campbellton, NB, June 9, 06; m 32; c 2. SURGERY. Educ: Acadia Univ, BA, 26; Univ Edinburgh, MS, ChB, 30; FRCS(E), 33; FRCS(C), 39. Prof Exp: Asst surgeon, Victoria Gen Hosp, Halifax, NS, 34-39; from asst prof to assoc prof surg, Dalhousie Univ, 46-71; RETIRED. Mem: Can Med Asn; Defence Med Asn Can (pres, 64). Mailing Add: Solbacken Glen Margaret NS Can

NOBLE, JOHN DALE, b Glendale, Calif, Nov 21, 34; m 62; c 2. PHYSICS. Educ: Univ Wyo, BS, 56, MS, 59; Univ BC, PhD(physics), 65. Prof Exp: Elec engr, Missile Systs Div, Lockheed Aircraft Corp, 56-57; instr physics, Univ Wyo, 59-61; lectr, Univ BC, 64-65; from asst prof to assoc prof, 65-73, PROF PHYSICS, WESTERN ILL UNIV, 73- Concurrent Pos: Mem fac, Univ Wyo, 71-72. Mem: AAAS; Am Asn Physics Teachers. Res: Nuclear magnetic resonance; liquid-gas critical points. Mailing Add: Dept of Physics Western Ill Univ Macomb IL 61455

NOBLE, JOHN F, b Salt Lake City, Utah, Mar 8, 29; m 53; c 2. PHARMACOLOGY. Educ: Utah State Univ, BS, 53, MS, 56; Univ Chicago, PhD(pharmacol), 59. Prof Exp: Asst, Utah State Univ, 53-55; asst, Univ Chicago, 56-59, instr pharmacol & res assoc, US Air Force Radiation Lab, 59-60; res pharmacologist, 60-63, res group leader, 63-65, head dept toxicol eval, 65-69, DIR TOXICOL RES SECT, LEDERLE LABS, AM CYANAMID CO, 69- Mem: AAAS; Soc Toxicol. Res: Toxicology and pharmacology; safety evaluation of new drugs; pharmacokinetics and drug metabolism; radiation biology; acute and long term effects of radiation including effects on mortality; biochemical changes and chemical means of protection against irradiation. Mailing Add: Lederle Labs Pearl River NY 10965

NOBLE, JULIAN VICTOR, b New York, NY, June 7, 40; m 60; c 3. THEORETICAL PHYSICS, MATHEMATICAL BIOPHYSICS. Educ: Calif Inst Technol, BS, 62; Princeton Univ, MA, 63, PhD(physics), 66. Prof Exp: Res assoc theoret physics, Univ Pa, 66-68, asst prof, 68-71; ASSOC PROF PHYSICS, UNIV VA, 71- Concurrent Pos: Sloan Found fel, 71-73; mem, Prog Comt, Space Radiation Effects Lab, Va, 74- Mem: AAAS; Sigma Xi. Res: Application of quantum-mechanical collision theory to nuclear reaction studies, both to determine nuclear properties and to develop methods of handling general strong-interaction problems; nuclear reaction studies at intermediate energies; dynamics of interacting populations; functional techniques in stochastic theories. Mailing Add: Dept of Physics Univ of Va Charlottesville VA 22901

NOBLE, NANCY LEE, b Chattanooga, Tenn, Mar 1, 22. BIOCHEMISTRY, MEDICINE. Educ: Emory Univ, MS, 49, PhD(biochem), 53. Prof Exp: Asst, Org Res Lab, Chattanooga Med Co, Tenn, 43-48; lab asst biochem, Sch Med, Emory Univ, 49-50; res instr, 53-55, res asst prof, 55-57, asst prof, 57-63, ASSOC PROF BIOCHEM & MED, SCH MED, UNIV MIAMI, 63- Concurrent Pos: Dir biochem res lab, Miami Heart Inst, 53-56; investr labs cardiovasc res, Howard Hughes Med Inst, 56-70; mem coun arteriosclerosis & coun basic sci, Am Heart Asn. Honors & Awards: Ciba Found Awards, 57 & 58. Mem: Fel Geront Soc; Soc Exp Biol & Med; Biochem Soc; fel Am Inst Chemists; NY Acad Sci. Res: Connective tissue metabolism and disorders. Mailing Add: Univ of Miami Sch of Med Box 520875 Biscayne Annex Miami FL 33152

NOBLE, PAUL, JR, b Ind, Oct 16, 22; m 46; c 3. PHYSICAL CHEMISTRY. Educ: Reed Col, AB, 43; Rochester Univ, PhD(org chem), 50. Prof Exp: Jr chemist, Shell Develop Co, 43-44; asst, Reed Col, 46-47; assoc res chemist, Calif Res Corp, 50-52; res chemist & head, Phys Org Div, Merrill Co, 52-54; res chemist, Phys Org Sect, Western Labs, Arthur D Little, Inc, 54-58; res scientist, 58-62, staff scientist & group leader, 62-63, sr staff scientist & sr mem res lab, 63-70, mgr chem lab, 70-72, PROG ENGR, LOCKHEED MISSILES & SPACE CO, INC, 72- Mem: Am Chem Soc; assoc fel Inst Aeronaut & Astronaut; Combustion Inst; The Chem Soc. Res: Organic synthesis; kinetics and reaction mechanism; infrared applications to organic chemistry; solid propellants; composite materials; aliphatic polynitro compounds; energy conversion. Mailing Add: 45 Arbuelo Way Los Altos CA 94022

NOBLE, REGINALD DUSTON, b Huntington, WVa, Nov 15, 35; c 3. PLANT PHYSIOLOGY. Educ: Marshall Univ, AB, 57, MA, 60; Ohio State Univ, PhD(bot), 69. Prof Exp: Teacher high sch, 57-59, dept chmn, 59-61; asst prof biol & phys sci, Marshall Univ, 62-66; teaching assoc bot, Ohio State Univ, 66-67; asst prof, 68-74, ASSOC PROF BIOL, BOWLING GREEN STATE UNIV, 74- Mem: Am Soc Plant Physiologists; Am Inst Biol Sci. Res: Photosynthetic studies; translocation and storage patterns during maturation in soybeans. Mailing Add: Dept of Biol Bowling Green State Univ Bowling Green OH 43403

NOBLE, ROBERT HAMILTON, b Alton, Ill, June 16, 16; m 40; c 3. OPTICAL PHYSICS. Educ: Antioch Col, BS, 40; Ohio State Univ, PhD(physics), 46. Prof Exp: Asst engr, Globe Industs, Ohio, 40-42; asst physics, Ohio State Univ, 42-46, res assoc, Univ Res Found, 46-47; asst prof physics, Mich State Col, 47-53; res physicist, Leeds & Northrup Co, 53-55; engr, Perkin-Elmer Corp, 55-64; prof optical sci, Univ Ariz, 64-74; PROF & HEAD OPTICS DEPT, NAT INST ASTROPHYS, OPTICS & ELECTRONICS, MEX, 74- Mem: AAAS; Am Phys Soc; fel Optical Soc Am; Mex Phys Soc. Res: Infrared and ultraviolet spectrometry; instrumentation for astronomy. Mailing Add: Nat Inst Astrophys Optics & Elec Apdo Postal 216 Puebla Puebla Mexico

NOBLE, ROBERT LAING, b Toronto, Ont, Feb 3, 10; m 35; c 4. PHYSIOLOGY. Educ: Univ Toronto, MD, 34; Royal Col Physicians, London, PhD, 37, DSc, 47. Prof Exp: Courtauld Inst Biochem, Middlesex Hosp, London, 34-39; res asst, Inst Endocrinol, McGill Univ, 39-47; prof med & assoc dir, Collip Med Res Lab, Univ Western Ont, 47-60; DIR CANCER RES CTR, UNIV BC, 60- Concurrent Pos: Mickle fel, Univ Toronto, 34-35; Leverhulme fel, 35-38; asst ed, J Endocrinol, 39. Mem: Can Physiol Soc; fel Royal Soc Can; Brit Physiol Soc. Res: Physiology of endocrine glands and related subjects; endocrinology; motion sickness; traumatic shock; physiology and pharmacology of gastric secretion and kidney function; cancer chemotherapy; discovery of vinblastine. Mailing Add: Univ ov BC Cancer Res Ctr Vancouver BC Can

NOBLE, ROBERT LEE, b Hominy, Okla, July 16, 23; m 47; c 4. ANIMAL NUTRITION. Educ: Okla State Univ, BS, 48, MS, 52; Kans State Univ, PhD(animal nutrit), 60. Prof Exp: From instr to assoc prof animal husb, 49-70, PROF ANIMAL SCI & INDUST, OKLA STATE UNIV, 70- Mem: Am Soc Animal Sci. Res: Applied research with sheep. Mailing Add: Dept of Animal Sci Okla State Univ Stillwater OK 74074

NOBLE, ROBERT VERNON, b Ithaca, NY, Jan 1, 23; m 48; c 4. ACADEMIC ADMINISTRATION. Educ: Cornell Univ, AB, 46; Univ Fla, MA, 50. Prof Exp: Instr corresp, Univ Fla, 50-56, asst prof, 56-57, head dept corresp study, 57-62; assoc prof, Fla Inst Continuing Univ Studies, 62-64; educ & training off, Div Nuclear Educ & Training, AEC, 64-67, Educ & Training Specialist, 67-72, staff asst to dir, 72-74, ASST TO DIR, DIV ADMIN SERV, US ENERGY RES & DEVELOP ADMIN, 74- Mem: Am Nuclear Soc; Sigma Xi; NY Acad Sci. Mailing Add: 605 Paradise Ct Gaithersburg MD 20760

NOBLE, ROBERT WARREN, JR, b Washington, DC, Feb 14, 37; m 62; c 3. BIOPHYSICAL CHEMISTRY. Educ: Mass Inst Technol, BA, 59, PhD(biophys), 64. Prof Exp: Fel biophys, Mass Inst Technol, 64; Nat Cancer Inst fel biochem, Univ Rome & Regina Elena Inst, 64-66; res assoc, Cornell Univ, 67-68; asst prof, 68-72, ASSOC PROF MED & BIOCHEM, STATE UNIV NY BUFFALO, 72- Mem: Biophys Soc; Am Asn Immunologists; Am Soc Biol Chem. Res: Reactions of hemeproteins with ligands; interactions between subunits of allosteric proteins and the structural basis for allosteric effects; reactions of antibodies with protein antigens; kinetics of liganding reactions. Mailing Add: Dept of Med State Univ of NY Vet Admin Hosp Buffalo NY 14215

NOBLE, STEPHEN W, plant breeding, see 12th edition

NOBLE, VINCENT EDWARD, b Detroit, Mich, Nov 28, 33; m 64; c 1. PHYSICAL OCEANOGRAPHY. Educ: Wayne State Univ, AB, 55, MS, 57, PhD(physics), 60. Prof Exp: Jr engr electronics & math anal, Res Labs, Bendix Aviation Corp, 54-56; res assoc solid state physics, Wayne State Univ, 58-60; assoc res physicist, Great Lakes Res Div, Univ Mich, Ann Arbor, 60-68; res physicist, US Naval Oceanog Off, 68-72, SPEC ASST FOR NAVY ENVIRON REMOTE SENSING, US NAVAL RES LAB, 72- Mem: Am Phys Soc; Am Geophys Union; Am Soc Limnol & Oceanog. Res: Physical limnology; air-sea interaction; polar and remote sensing oceanography. Mailing Add: 6209 Zekan Lane Springfield VA 22150

NOBLE, WILLIAM ALLISTER, b Nagercoil, India, Apr 20, 32; US citizen; m 60; c 2. CULTURAL GEOGRAPHY, ANTHROPOLOGY. Educ: Univ Ga, BA, 55, MA, 57; La State Univ, PhD(geog), 68. Prof Exp: Instr geog & hist, Longwood Col, 57-60; asst prof geog, Southeastern La Col, 64-66; instr, 66-68, ASST PROF GEOG, UNIV MO-COLUMBIA, 68- Concurrent Pos: Fulbright-Hays fac res grant, Kerala, India, 71-72. Mem: Asn Am Geogrs; Asn Asian Studies; Soc S India Studies. Res: Architectural geography; tropical garden agriculture; Kerala bibliography. Mailing Add: Dept of Geog Univ of Mo Columbia MO 65201

NOBLE, WILLIAM JOHN, b Fredericton, NB, Aug 1, 14; m 45. THEORETICAL PHYSICS. Educ: Univ NB, BSc, 34; Dalhousie Univ, MSc, 37; Univ Toronto, MA, 41; Univ Edinburgh, PhD(theoret physics), 52. Prof Exp: Asst physics, Univ NB, 34-35, actg prof, 38; asst prof, Mt Allison Univ, 39-40; from instr to prof & head dept, Acadia Univ, 41-61; PROF PHYSICS, MT ALLSION UNIV, 61- Mem: Am Asn Physics Teachers; Can Asn Physicists. Res: Pleochroic haloes; diffraction of light by ultrasonic waves; underwater sound; wave propagation in inhomogeneous media. Mailing Add: Dept of Physics Mt Allison Univ Sackville NB Can

NOBLE-HARVEY, JANE, b Cleveland, Ohio, June 28, 41; m 72. ANIMAL VIROLOGY. Educ: Simmons Col, BS, 63; Univ Calif, San Francisco, PhD(microbiol), 69. Prof Exp: Pub health microbiologist, Dept Health Virus & Rickettsial Dis Lab, State of Calif, 63-65; bacteriologist, Ben Venue Labs Inc, 70; fel virol, E I du Pont de Nemours & Co, 70-72; asst prof microbiol, Vanderbilt Univ, 72; ASST PROF BIOL SCI, UNIV DEL, 73- Mem: Am Soc Microbiol. Res: Biochemical and electron microscopic studies of the interactions between human picornaviruses and the plasma membrane of host cells, which lead to the eclipse of viral infectivity. Mailing Add: Dept of Biol Sci Univ of Del Newark DE 19711

NOBLES, LAURENCE HEWIT, b Spokane, Wash, Sept 28, 27; m 48; c 2. GLACIOLOGY. Educ: Calif Inst Technol, BS & MS, 49; Harvard Univ, PhD(geol), 52. Prof Exp: From instr to assoc prof, 52-67, PROF GEOL, NORTHWESTERN UNIV, EVANSTON, 67-, DEAN ADMIN, 72- Concurrent Pos: Consult/evaluator, NCent Asn Cols & Sec Schs, 67-; from asst dean to assoc dean, Col Arts & Sci, Northwestern Univ, Evanston, 66-70, actg dean, 70-72; pres, Chicago Acad Sci, 73- Mem: AAAS; Geol Soc Am; Am Asn Petrol Geologists; Am Geophys Union; Glaciol Soc. Res: Geomorphology; glacial geology. Mailing Add: Rebecca Crown Ctr Northwestern Univ Evanston IL 60201

NOBLES, RALPH ALBERT, nuclear physics, geophysics, see 12th edition

NOBLES, WILLIAM LEWIS, b Meridian, Miss, Sept 11, 25; m 48; c 2. PHARMACEUTICAL CHEMISTRY. Educ: Univ Miss, BS, 48, MS, 49; Univ Kans, PhD(pharmaceut chem), 52. Prof Exp: From asst prof to prof pharm & pharmaceut chem, Univ Miss, 52-68; PRES, MISS COL, 68- Concurrent Pos: Pfeiffer mem res fel, 55-58 & 59-60; NSF fel, Univ Mich, 58-59; dean, Grad Sch, Univ Miss, 60-68. Honors & Awards: Found Award, Am Pharmaceut Asn 66; Nat Rho Chi Award in Montreal, Can, 69. Mem: Am Chem Soc; Am Pharmaceut Asn; NY Acad Sci; The Chem Soc. Res: Medicinal chemistry; pharmaceutical product development. Mailing Add: Box 4186 Clinton MS 39058

NOBLET, RAYMOND, b Hiawassee, Ga, Aug 5, 43. INSECT PHYSIOLOGY, INVERTEBRATE PATHOLOGY. Educ: Univ Ga, BS, 65, MS, 67, PhD(entom), 70. Prof Exp: Asst prof, 70-75, ASSOC PROF ENTOM & ECON ZOOL, CLEMSON UNIV, 75- Mem: Entom Soc Am: Wildlife Dis Asn. Res: Invertebrate immunity; malariology; parasitology; physiological and ecological investigations pf insect vectors of disease. Mailing Add: 101 Allee St Clemson SC 29631

NOBS, MALCOLM A, botany, see 12th edition

NOBUSAWA, NOBUO, b Osaka, Japan, May 15, 30; m 61; c 2. ALGEBRA. Educ: Osaka Univ, BS, 53, MS, 55, PhD(math), 58. Prof Exp: Asst prof math, Univ Alta, 62-66; assoc prof, Univ RI, 66-67; assoc prof, 67-71, PROF MATH, UNIV HAWAII, 71- Mem: Am Math Soc; Can Math Cong; Math Soc Japan. Res: Ring theory and number theory. Mailing Add: Dept of Math Univ of Hawaii Honolulu HI 96822

NOCENTI, MERO RAYMOND, b Masontown, Pa, Sept 7, 28; m 55; c 3. PHYSIOLOGY. Educ: Univ WVa, AB, 51, MS, 52; Rutgers Univ, PhD(endocrinol), 55. Prof Exp: Waksman-Merck fel, Rutgers Univ, 55-56; from instr to assoc prof, 56-75, PROF PHYSIOL, COL PHYSICIANS & SURGEONS, COLUMBIA UNIV, 75- Concurrent Pos: Managing ed, Proc, Soc Exp Biol & Med, 74. Mem: AAAS; Am Physiol Soc; Sigma Xi; Soc Exp Biol & Med. Res: Endocrine physiology; hormonal influences on electrolyte and water balance and on connective tissue. Mailing Add: Dept of Physiol Columbia Univ Col of Phys & Surg New York NY 10032

NOCKELS, CHERYL FERRIS, b Chicago, Ill, July 20, 35; m 57. ANIMAL NUTRITION. Educ: Colo State Univ, BS, 57, MS, 59; Univ Mo, PhD(animal nutrit), 65. Prof Exp: Res technician, 59-60, asst poultry scientist biochem, 64-70, assoc prof, 70-74, PROF ANIMAL SCI DEPT, COLO STATE UNIV, 74- Mem: Am Soc Animal Sci; Poultry Sci Asn; Am Inst Nutrit; NY Acad Sci. Res: Nutritional and biochemical studies involving both monogastric and ruminant animals. Mailing Add: Dept of Animal Sci Colo State Univ Ft Collins CO 80521

NODA, KAORU, b Hilo, Hawaii, Oct 16, 24; m 52; c 4. PARASITOLOGY. Educ: Grinnell Col, BA, 50; Univ Iowa, MS, 53, PhD(zool), 56. Prof Exp: Asst zool, Univ Iowa, 51-56; asst parasitologist, 57-59, asst prof sci, 59-63, assoc prof & dir Hilo Campus, 67-69, PROF BIOL, UNIV HAWAII, HILO, 69- Concurrent Pos: USPHS res grant, 60; scholar, Univ Calif, Los Angeles, 65-66; provost, Univ Hawaii, Hilo, 69-70, asst chancellor, 70-72; vis prof & researcher, Univ Tokyo, 72-73. Mem: AAAS; Am Soc Parasitol. Res: Trematodes, particularly the family Heterophyidae. Mailing Add: Univ of Hawaii PO Box 1357 Hilo HI 96720

NODA, LAFAYETTE HACHIRO, b Livingston, Calif, Mar 13, 16; m 47; c 2. BIOCHEMISTRY. Educ: Univ Calif, BS, 39, MA, 43; Stanford Univ, PhD, 50. Prof Exp: Res assoc, Stanford Univ, 50-51; asst prof, Univ Wis, 53-56; biochemist, US Naval Med Res Inst, Md, 56-57; assoc prof biochem, 57-60, chmn dept, 60-65, PROF BIOCHEM, DARTMOUTH MED SCH, 60- Concurrent Pos: Res fel, Enzyme Inst, Univ Wis, 51-53; Guggenheim fel, 68-69. Mem: Am Soc Biol Chem. Res: Purification, kinetics, structure and mechanism of action of enzymes. Mailing Add: Dept of Biochem Dartmouth Med Sch Hanover NH 03755

NODEN, DREW M, b Pittsburgh, Pa, Oct 29, 44. EXPERIMENTAL EMBRYOLOGY, NEUROEMBRYOLOGY. Educ: Washington & Jefferson Col, BS, 66; Washington Univ, PhD(biol), 72. Prof Exp: ASST PROF ZOOL, UNIV MASS, 73- Mem: Am Soc Zoologists. Res: Combined in ovo and in vitro analyses of the control of avian neural crest cell migrations, cytodifferentiation, and terminal morphogenesis; development of cephalic connective tissues and cranial sensory ganglia. Mailing Add: Dept of Zool Univ of Mass Amherst MA 01002

NODIFF, EDWARD ALBERT, b US, Nov 25, 26; m 50; c 2. ORGANIC CHEMISTRY. Educ: Temple Univ, BA, 48. Prof Exp: Res assoc chem, Res Inst, 49-55, proj dir, 55-60, DIR ORG CHEM RES, TEMPLE UNIV, 60- Mem: Am Chem Soc. Res: Organic fluorine and medicinal chemistry. Mailing Add: 1600 Placid St Philadelphia PA 19152

NODINE-ZELLER, DORIS EULAIA, b Ohio, Ill, Mar 11, 23; div. PALEONTOLOGY, STRATIGRAPHY. Educ: Univ Ill, AB, 46; Univ Wis, MS, 51, PhD(geol), 54. Prof Exp: Tech asst, Areal & Eng Geol Div, State Geol Surv Ill, 44-46; teaching asst geol, Univ Kans, 46-48; lab technician, Univ Wis, 51-54; consult, Shell Oil Co, 54-55; consult, Petroleo, Petrobras, Belem, Brasiliero, 55-56; res assoc geol, Univ Kans, 60-63; managing ed bulls & res assoc geol, 63-71, RES ASSOC MICROPALEONT & SUBSURFACE GEOL, STATE GEOL SURV KANS, 71- Concurrent Pos: Rev syst biol, NSF, 61-; prof western civilization prog, Univ Kans, 63- Mem: AAAS; Asn Earth Sci Ed; Soc Econ Paleontologists & Mineralogists. Res: Micropaleontology; taxonomic and stratigraphic work on endothyroid foraminifers, rhyncholites and other invertebrate fossils. Mailing Add: State Geol Surv of Kans Univ of Kans Lawrence KS 66044

NODULMAN, LAWRENCE JAY, b Chicago, Ill, May 6, 47. EXPERIMENTAL HIGH ENERGY PHYSICS. Educ: Univ Ill, Urbana-Champaign, BS, 69, MS, 70, PhD(physics), 73. Prof Exp: Res assoc physics, Univ Ill, Urbana-Champaign, 73-75; ASST RES PHYSICIST, UNIV CALIF, LOS ANGELES, 75- Mem: Am Phys Soc. Res: Particle production in electron-positron interactions; neutrino interactions; weak and electromagnetic interactions of hadrons. Mailing Add: Stanford Linear Accelerator Ctr PO Box 4349 Bin 98 Stanford CA 94305

NODVIK, JOHN S, b Canonsburg, Pa, July 2, 30; m 53; c 3. PHYSICS. Educ: Carnegie Inst Technol, BS(physics) & BS(math), 52, MS, 52; Univ Calif, Los Angeles, PhD(physics), 58. Prof Exp: From asst prof to assoc prof physics, 58-70, PROF PHYSICS, UNIV SOUTHERN CALIF, 70- Mem: Am Phys Soc. Res: Relativistic spin; nuclear optical model. Mailing Add: Dept of Physics Univ of Southern Calif University Park Los Angeles CA 90007

NODWELL, ROY, b Asquith, Sask, May 13, 18; m 42; c 4. PHYSICS. Educ: Univ Sask, BE, 40; Univ BC, MASc, 54, PhD(spectros), 56. Prof Exp: Res scientist optics, Nat Res Coun Can, 41-47; opers res, Defense Res Bd, 56-60; from asst prof to assoc prof plasma physics, 60-66, PROF PLASMA PHYSICS, UNIV BC, 66- Mem: Can Asn Physicists. Res: Spectroscopic studies of plasma physics. Mailing Add: Dept of Physics Univ of BC Vancouver BC Can

NOE, BRYAN DALE, b Peoria, Ill, Mar 1, 43; m 65; c 2. ANATOMY, CELL BIOLOGY. Educ: Goshen Col, BA, 65; WVa Univ, MA, 67; Univ Minn, PhD(anat), 71. Prof Exp: USPHS res fel, Univ Minn, 71-72; ASST PROF ANAT, EMORY UNIV, 73- Mem: Corp mem Marine Biol Lab; Am Diabetes Asn. Res: Glucagon biosynthesis in non-mammalian and mammalian species. Mailing Add: Dept of Anat Emory Univ Atlanta GA 30322

NOE, ERIC ARDEN, b Bluffton, Ohio, Dec 24, 43. ORGANIC CHEMISTRY. Educ: Univ Cincinnati, BS, 65; Calif Inst Technol, PhD(chem), 71. Prof Exp: Fel chem, Univ Southern Calif, 70-71 & Univ Calif, San Francisco, 71-72; fel, 72-75, RES ASSOC CHEM, WAYNE STATE UNIV, 75- Mem: Am Chem Soc. Res: Conformational analysis using dynamic nuclear magnetic resonance spectroscopy; organic synthesis. Mailing Add: Dept of Chem Wayne State Univ Detroit MI 48202

NOE, FRANCES ELSIE, b Beacon Falls, Conn, May 23, 23; m 56; c 2. PHYSIOLOGY. Educ: Middlebury Col, BA, 44; Yale Univ, MN, 47; Univ Vt, MD, 54. Prof Exp: Intern, Mary Hitchcock Mem Hosp, 54-55; Mich Heart Asn fel, Harper Hosp, Detroit, 55-56; resident pulmonary med, Henry Ford Hosp, 56-57; Rands fel med, Wayne State Univ, 57-58, res assoc anesthesiol, Col Med, 58-59, from instr to asst prof, 59-65; res assoc, 65-70, CHIEF PULMONARY PHYSIOL SECT, DIV RES, SINAI HOSP DETROIT, 70- Mem: Am Soc Anesthesiol. Res: Anesthesiology; cardiopulmonary physiology, especially as applied to anesthesiology. Mailing Add: Sinai Hosp Div of Res 6767 W Outer Dr Detroit MI 48235

NOË, FREDERICK FORCE, biochemistry, see 12th edition

NOE, JAMES L, b Cincinnati, Ohio, Oct 16, 38; m 62; c 3. ORGANIC CHEMISTRY. Educ: Xavier Univ, BS, 60; Purdue Univ, PhD(org chem), 64. Prof Exp: Res chemist, 64-68, group leader, Bound Brook, NJ, 68-72, chief chemist, Willow Island, 72-75, PROD SUPT, AM CYANAMID CO, WILLOW ISLAND, 75- Mem: Am Chem Soc. Mailing Add: Am Cyanamid Co Willow Island WV 26190

NOE, LEWIS JOHN, b Cleveland, Ohio, Oct 26, 41; m 68; c 1. CHEMISTRY. Educ: Western Reserve Univ, AB, 63; Case Western Reserve Univ, PhD(chem), 67. Prof Exp: Fel chem, Univ Pa, 67-69; asst prof, 69-74, ASSOC PROF CHEM, UNIV WYO, 74- Concurrent Pos: NSF res grant, Univ Wyo, 71-73. Mem: Am Chem Soc. Res: Ultraviolet, visible and infrared spectroscopy of molecular crystals; applications of Stark and Zeeman effects to spectroscopy of molecular crystals. Mailing Add: Dept of Chem Univ of Wyo Laramie WY 82070

NOEHREN, THEODORE HENRY, b Buffalo, NY, Sept 6, 17; m 40; c 2. INTERNAL MEDICINE, PULMONARY DISEASES. Educ: Williams Col, BA, 38; Univ Rochester, MD, 42; Univ Minn, MS, 50. Prof Exp: Mayo Found fel, Univ Minn, 47-49; asst prof pub health & prev med, Univ Utah, 49-52; asst prof internal med, State Univ NY Buffalo, 52-69; assoc prof med, Univ Utah, 69-75; DIR PULMONARY DIS DIV, HOLY CROSS HOSP, 75- Concurrent Pos: Consult, USPHS, 50; Markle scholar, 53-58; Fulbright lectr, Univ Helsinki, 66; consult & sr cancer res internist, Roswell Park Mem Cancer Res Hosp. Mem: Am Thoracic Soc; AMA; Am Col Chest Physicians. Res: Pulmonary physiology; role of lungs as an excretory organ; action of intermittent positive pressure in pulmonary diseases; factors altering the rate of excretion of inert gases by the lungs. Mailing Add: Holy Cross Hosp 1045 E First South Salt Lake City UT 84102

NOEL, CHARLES J, physical chemistry, see 12th edition

NOEL, DALE LEON, b Wichita, Kans, May 21, 36; m 62; c 2. ANALYTICAL CHEMISTRY. Educ: Friends Univ, BA, 58; Wichita State Univ, MS, 60; Kans State Univ, PhD(chem), 70. Prof Exp: Asst prof chem, Friends Univ, 61-65, Eastern Nazarene Col, 67-69; assoc prof, Kearney State Col, 70-74; RES ASSOC CHEM, INT PAPER CO, 74- Mem: Am Chem Soc; Tech Asn Pulp & Paper Indust; Am Soc Testing & Mat. Res: Polarography, anodic stripping voltammetry, thermometric titrations, analysis of pulping liquors, environmental analysis. Mailing Add: Corp Res Ctr Int Paper Co Box 797 Tuxedo NY 10987

NOEL, JAMES A, b Williamstown, Pa, Aug 11, 22; m 45; c 3. GEOLOGY. Educ: Lehigh Univ, BA, 49; Dartmouth Col, MA, 51; Ind Univ, PhD(geol), 56. Prof Exp: Sr geologist, Creole Petrol Corp, 56-60; assoc prof geol math, Northwestern State Col, La, 60-64; res assoc, Res Ctr Union Oil, 64-66; from assoc prof to prof geol & chmn dept, 66-71, asst dir, Western Ohio Br Campus, 71-73, asst dean, 73-74, PROF GEOL, WRIGHT STATE UNIV, 74- Mem: Am Asn Petrol Geologists; Am Inst Prof Geologists; Sigma Xi; Nat Asn Geol Teachers; Soc Explor Geophys. Res: Computer systems for gravity and magnetic residual calculation and automatic mapping; combined geological and gravity interpretations using computer systems; sedimentary environments. Mailing Add: Div of Geol Wright State Univ Dayton OH 45431

NOEL, THOMAS CORBIN, physics, see 12th edition

NOELKEN, MILTON EDWARD, b St Louis, Mo, Dec 5, 35; m 62; c 2. PHYSICAL CHEMISTRY. Educ: Washington Univ, BA, 57, PhD(phys chem), 62. Prof Exp: Researcher chem, Duke Univ, 62-64; assoc & res chemist, Eastern Regional Res Lab, USDA, 64-67; asst prof, 67-71, ASSOC PROF BIOCHEM, UNIV KANS MED CTR, 71- Concurrent Pos: Actg chmn dept biochem, Univ Kans Med Ctr, 73-74. Mem: AAAS; Am Chem Soc; Am Soc Biol Chem. Res: Physical chemistry of proteins. Mailing Add: Dept of Biochem Univ of Kans Med Ctr Kansas City KS 66103

NOELL, WERNER K, b Ger, June 22, 13; US citizen; c 3. PHYSIOLOGY. Educ: Univ Hamburg, MD, 38. Prof Exp: From intern to resident neurol, Univ Hamburg Clin, 39; fel physiol, Univ Berlin, 39-40; instr neurol, Med Sch Danzig, 41-43; dozent, Univ Cologne, 43-46; neurophysiologist, US Air Force Sch Aviation Med, 47-50, head neurophysiol, 50-54; assoc res prof physiol, 56-60, assoc prof, 60-62, PROF PHYSIOL, STATE UNIV NY BUFFALO, 62-, HEAD NEUROSENSORY LAB, 66- Concurrent Pos: Res assoc, Kaiser Wilhelm Inst, 43-46; assoc prof, Sch Aviation Med, Air Univ, 50-54; prin res scientist, Roswell Park Mem Inst, 54-60; vis prof, Sch Med, Univ Buenos Aires, 61. Honors & Awards: J E Hitzig Prize, Prussian Acad Sci, 44; Jonas Friedenwald Award, Asn Res Vision & Ophthal, 59. Mem: AAAS; Am Physiol Soc; Am Electroencephalog Asn; Asn Res Vision & Ophthal; NY Acad Sci. Res: Physiology of retina; visual systems; electroencephalography; cerebral circulation. Mailing Add: Neurosensory Lab Bldg C State Univ of NY Buffalo NY 14214

NOER, RICHARD JUUL, b Madison, Wis, July 3, 37. PHYSICS. Educ: Amherst Col, BA, 58; Univ Calif, Berkeley, PhD(physics), 63. Prof Exp: Physicist, Atomic Energy Res Estab, Eng, 63-64; asst prof physics, Amherst Col, 64-66; from asst prof to assoc prof, 66-75, PROF PHYSICS, CARLETON COL, 75- Mem: Am Phys Soc; Am Asn Physics Teachers. Res: Solid state physics; low temperature, metals and superconductors. Mailing Add: Dept of Physics Carleton Col Northfield MN 55057

NOER, RUDOLF JUUL, b Menominee, Mich, Apr 25, 04; m 33; c 2. SURGERY. Educ: Univ Wis, AB, 24; Univ Pa, MD, 27. Prof Exp: Asst anat, Univ Wis, 32-34, resident surg, 34-37, instr, 36-37; fel surg res, Col Med, Wayne State Univ, 37-38, from instr to assoc prof surg, 38-49, prof surg & appl anat, 49-52; prof surg & head dept, Sch Med, Univ Louisville, 52-70; PROF SURG, COL MED, UNIV S FLA, 70- Concurrent Pos: Assoc, Detroit Receiving Hosp, Mich, 46-52; consult, Detroit Marine Hosp, USPHS, 47-48; sr consult, Vet Admin Hosp, Dearborn, Mich, 48-52, area consult, Vet Admin, 55-70; dir surg, Louisville Gen Hosp, Ky, 52-70; ed, J Trauma, 61-68. Mem: Am Surg Asn (vpres, 63-64); Am Asn Anatomists; Am Asn Surg of Trauma (pres, 64); Int Surg Soc; fel Am Col Surg (vpres, 64-65). Res: Intestinal obstruction; intestinal circulation in man and animals; physiologic responses of intestine to distention; diverticular disease of the colon. Mailing Add: Dept of Surg Univ of SFla Col of Med Tampa FL 33620

NOERDLINGER, PETER DAVID, b New York, NY, May 3, 35; m 57, 64; c 5. ASTROPHYSICS, PLASMA PHYSICS. Educ: Harvard Univ, AB, 56; Calif Inst Technol, PhD(physics), 60. Prof Exp: From instr to asst prof physics, Univ Chicago, 60-66; assoc prof, Univ Iowa, 66-68; from assoc prof to prof, NMex Inst Mining & Technol, 68-71; PROF ASTRON, MICH STATE UNIV, 71- Concurrent Pos: Sr resident res assoc, Nat Res Coun, NASA-Ames Res Ctr, 71, 74 & 75; vis scientist, Smithsonian Astrophys Observ, Mass, 73-; actg chmn, Dept Astron & Astrophys, Mich State Univ, 74-75. Mem: Am Astron Soc; Am Asn Physics Teachers; fel Royal Astron Soc; Am Phys Soc; Fedn Am Scientists. Res: Quasi-stellar objects; cosmology; theoretical astrophysics; properties of quasi-stellar objects; mass loss from stars and quasi-stellar objects; plasma processes in stars and quasars; theoretical cosmology. Mailing Add: Dept of Astron & Astrophys Mich State Univ East Lansing MI 48824

NOETHER, DORIT LOW, organic chemistry, see 12th edition

NOETHER, GOTTFRIED EMANUEL, b Ger, Jan 7, 15; nat US; m 42; c 1. MATHEMATICAL STATISTICS. Educ: Ohio State Univ, BA, 40; Univ Ill, MA, 41; Columbia Univ, PhD(math statist), 49. Prof Exp: Instr math, NY Univ, 49-51; from asst prof to prof math statist, Boston Univ, 51-68; HEAD DEPT STATIST, UNIV CONN, 68- Concurrent Pos: Fulbright lectr, Univ Tübingen, 57-58 & Univ Vienna, 65-66. Mem: Fel AAAS; Math Asn Am; fel Inst Math Statist; fel Am Statist Asn. Res: Nonparametric statistical inference. Mailing Add: Dept of Statist Univ of Conn Storrs CT 06268

NOETHER, HERMAN DIETRICH, polymer chemistry, see 12th edition

NOETZEL, DAVID MARTIN, b Waseca, Minn, Feb 19, 29; m 50; c 6. ENTOMOLOGY, ZOOLOGY. Educ: Univ Minn, BA, 51, MS, 56, PhD, 66. Prof Exp: Asst prof agr entom, NDak State Univ, 56-65; asst prof biol, Concordia Col, Moorhead, Minn, 65-70; INSTR ENTOM & EXTEN ENTOMOLOGIST, UNIV MINN, ST PAUL, 70- Mem: Entom Soc Am; Am Soc Mammal; Bee Res Asn. Res: Animal ecology; ornithology; economic zoology. Mailing Add: Dept Entom Fisheries & Wildlife Univ of Minn St Paul MN 55101

NOFFSINGER, TERRELL L, b Greenville, Ky, Jan 3, 16; m 45; c 4. METEOROLOGY, AGRICULTURE. Educ: Univ Ky, BS, 41, MS, 46; Purdue Univ, PhD(environ physiol), 57. Prof Exp: Meteorologist, US Air Force, 42-45 & 48-53; asst prof animal sci, Purdue Univ, 53-58; climatologist & prof meteorol, Univ Hawaii, 58-61; CHIEF AGR & FORESTRY, NAT WEATHER SERV, 61- Concurrent Pos: Mem, Working Group Meteorol Observ, Animal Exp, World Meteorol Orgn, 64; prin US del, Comn Agr Meteorol, Geneva, 71 & Washington, DC, 74; consult, WHO, 75. Mem: AAAS; Am Meteorol Soc. Res: Influence of weather on growth and yield in agricultural production. Mailing Add: 9623 Sutherland Rd Silver Spring MD 20901

NOFTLE, RONALD EDWARD, b Springfield, Mass, Mar 10, 39; m 64. INORGANIC CHEMISTRY, FLUORINE CHEMISTRY. Educ: Univ NH, BS, 61; Univ Wash, PhD(inorg chem), 66. Prof Exp: Res asst inorg chem, Univ Wash, 62-66, instr chem, 66; fel, Univ Idaho, 66-67; asst prof, 67-73, ASSOC PROF CHEM, WAKE FOREST UNIV, 73- Concurrent Pos: Res chemist, Naval Res Lab, 75-76; consult, Chemwaste Corp. Mem: Am Chem Soc; The Chem Soc; fel Am InstRes: Synthesis and properties of inorganic compounds containing fluorine; chemistry of the halogens. Mailing Add: Dept of Chem Wake Forest Univ Winston-Salem NC 27109

NOGAMI, YUKIHISA, b Hamada, Japan, Oct 22, 29; m 58; c 3. THEORETICAL PHYSICS. Educ: Kyoto Univ, BSc, 52, DSc(physics), 61. Prof Exp: Asst lectr physics, Univ Osaka Prefecture, 54-61; fel, Nat Res Coun Can, 61-63; res fel, Battersea Col Technol, 63-64; sr fel, 64-65, assoc prof, 65-69, PROF PHYSICS, McMASTER UNIV, 69- Mem: Can Asn Physicists; Phys Soc Japan. Res: Theoretical physics, especially theory of nuclear forces and nuclear structure. Mailing Add: Dept of Physics McMaster Univ Hamilton ON Can

NOGAN, DONALD STANLEY, geology, see 12th edition

NOGGLE, GLENN RAY, b New Madison, Ohio, July 25, 14; m 45; c 1. PLANT PHYSIOLOGY. Educ: Miami Univ, AB, 35; Univ Ill, MS, 42, PhD(bot), 45. Prof Exp: Res assoc agron, Univ Ill, 45-46; asst prof plant physiol, Blandy Exp Farm, Va, 46-48; sr biologist, Oak Ridge Nat Lab, 48-52; sr scientist, Photosynthesis Proj, Southern Res Inst, Ala, 52-54; biochemist, C F Kettering Found & prof chem, Antioch Col, 54-57; prof bot & head dept, Univ Fla, 57-64; PROF BOT & HEAD DEPT, NC STATE UNIV, 64- Mem: Am Soc Plant Physiologists (exec secy-treas, 55-58); Bot Soc Am. Res: Seed physiology; phenology of native and cultivated plants; primary productivity of cultivated plants. Mailing Add: Dept of Bot NC State Univ Raleigh NC 27607

NOGGLE, JOSEPH HENRY, b Harrisburg, Pa, Mar 19, 36; m 60. PHYSICAL CHEMISTRY. Educ: Juniata Col, BS, 60; Harvard Univ, MS, 63, PhD(chem), 65. Prof Exp: Asst prof chem, Univ Wis-Madison, 65-71; ASSOC PROF CHEM, UNIV DEL, 71- Mem: AAAS; Am Phys Soc; Am Chem Soc. Res: Nuclear magnetic resonance, including relaxation phenomena. Mailing Add: Dept of Chem Univ of Del Newark DE 19711

NOGRADY, GEORGE LADISLAUS, b Budapest, Hungary, May 2, 19; Can citizen; m 50. MICROBIOLOGY, POPULATION STUDIES. Educ: Univ Kolozsvar, Hungary, MD, 44; Univ Pecs, cert bact, 55; Univ Toronto, cert bact, 61; Am Bd Microbiol, dipl, 64. Prof Exp: Demonstr biochem, Univ Kolozsvar, 40-43, asst prof bact, 43-46; from asst prof to assoc prof, Univ Pecs, 46-54, asst dir, 54-56, State Inst Social Ins res fel, 55-56, chief diag lab, Inst Microbiol, 56; asst bact, 57-58, ASST PROF MICROBIOL, UNIV MONTREAL, 58- Concurrent Pos: Mem, Int Sci Film Asn, 62-, Can deleg, Paris Mgt, 63; dir microbiol proj, Can Med Exped to Easter Island, 64-65. Mem: Am Soc Microbiol; Can Soc Microbiol; Can Asn Med; Int Soc Cell Biol. Res: Cultural diagnostic methods; bacterial morphology; microbiology of isolated populations; application of cinephotomicrography in microbiology. Mailing Add: 4662 Victoria Ave Montreal PQ Can

NOGRADY, THOMAS, b Budapest, Hungary, Oct 16, 25; Can citizen; m 50; c 1. MEDICINAL CHEMISTRY. Educ: Eötvös Lorand Univ, Budapest, MSc, 48, PhD(org chem), 50. Prof Exp: Res chemist, Res Inst Pharmaceut Indust, Budapest, 50-56; res assoc org chem, Univ Vienna, 57 & Univ Montreal, 57-61; from asst prof to assoc prof, Loyola Col, Montreal, 61-70, PROF ORG CHEM, CON CONCORDIA UNIV, 70- Concurrent Pos: Rockefeller scholar, Univ Vienna, 57; consult, Delmar Chem Ltd, Montreal, 58- Mem: Am Chem Soc; fel Chem Inst Can. Res: Molecular pharmacology; nuclear magnetic resonance; biochemistry. Mailing Add: Dept of Chem Concordia Univ Montreal PQ Can

NOHEL, JOHN ADOLPH, b Prague, Czech, Oct 24, 24; nat US; m 48; c 3. MATHEMATICS. Educ: George Washington Univ, BEE, 48; Mass Inst Technol, PhD(math), 53. Prof Exp: Asst math, George Washington Univ, 46-48; instr, Mass Inst Technol, 50-53; from asst prof to prof, Ga Inst Technol, 53-61; assoc prof, 61-64, chmn dept math, 68-70, PROF MATH, UNIV WIS-MADISON, 64- Concurrent Pos: Mem, Math Res Ctr, US Army, 58-59; res sabbatical, Univ Paris, 65-66; vis prof, Polytech Sch, Univ Lausanne, 71-72; mem comt appl math, Nat Res Coun-Nat Acad Sci. Mem: AAAS; Am Math Soc; Soc Indust & Appl Math; Math Asn Am. Res: Volterra functional differential equations, qualitative theory and applications. Mailing Add: Dept of Math Van Vleck Hall Univ of Wis Madison WI 53706

NOKES, RICHARD FRANCIS, b Deerfield, Mich, Mar 16, 34; m 56; c 6. VETERINARY MEDICINE, ANATOMY. Educ: Mich State Univ, BS, 56, DVM, 58. Prof Exp: Vet, pvt pract, 58-60 & Agr Res Serv, USDA, 60-61; asst prof, 62-74, ASSOC PROF BIOL, UNIV AKRON, 74- Concurrent Pos: Vet, Copley Rd Animal Hosp, Akron, 62-; consult, Akron Children's Zoo, 64- & Akron Gen Med Ctr, 65- Mem: Am Asn Zoo Vets. Res: Rheological studies of blood flow; developmental studies of sense organs in the dog; excretion pathways of friction reducing agents. Mailing Add: Dept of Biol Univ of Akron Akron OH 44325

NOLAN, CHRIS, b Salt Lake City, Utah, July 3, 30; m 63; c 2. BIOCHEMISTRY. Educ: Univ Nev, BS, 52; Univ Utah, PhD(biochem), 61. Prof Exp: NIH fel, Univ Wash, 61-63; RES BIOCHEMIST, ABBOTT LABS, 63- Mem: Am Soc Biol Chem. Res: Structure-function relationships in proteins. Mailing Add: 940 Greenleaf St Gurnee IL 60031

NOLAN, GEORGE JUNIOR, b Stilwell, Okla, Nov 3, 35; m 55; c 5. PHYSICAL CHEMISTRY. Educ: Northeastern State Col, BS, 58; Univ Ark, MS, 62, PhD(chem), 64. Prof Exp: Chemist, Phillips Petrol Co, 64-68; assoc prof chem, 68-73, PROF CHEM, NORTHEASTERN STATE COL, 73- Mem: Am Chem Soc. Res: Fundamental and development research in catalysis. Mailing Add: 281 Hickory Dr Tahlequah OK 74464

NOLAN, JAMES FRANCIS, b Scranton, Pa, Nov 16, 31; m 57; c 3. PHYSICS. Educ: Univ Scranton, BS, 54; Univ Pittsburgh, PhD(physics), 64. Prof Exp: Res physicist, Westinghouse Res Lab, 61-66; RES PHYSICIST, OWENS-ILL, INC, 66- Mem: Am Phys Soc. Res: Atomic collisions; medium energy ion-atom charge transfer and ionization; low energy electron-atom collisions. Mailing Add: Owens-Illinois Inc PO Box 1035 Toledo OH 43666

NOLAN, JAMES P, b Buffalo, NY, June 21, 29; m 56; c 4. MEDICINE. Educ: Yale Univ, BA, 51, MD, 55. Prof Exp: Instr, Yale Univ, 61-63; from asst prof to assoc prof, 63-69, PROF MED, STATE UNIV NY BUFFALO, 69- Concurrent Pos: Chief med, Buffalo Gen Hosp. Mem: Am Col Physicians; Am Fedn Clin Res; Am Gastroenterol Soc; Am Asn Study Liver Dis; Reticuloendothelial Soc. Res: Role of bacterial endotoxins and the reticuloendothelial system in the initiation and perpetuation of liver disease. Mailing Add: Buffalo Gen Hosp Dept of Med 100 High St Buffalo NY 14203

NOLAN, JAMES ROBERT, b New York, NY, May 8, 23; m 46; c 3. BOTANY. Educ: Cornell Univ, BS, 61, PhD(plant anat & morphol), 67. Prof Exp: Instr biol, Antioch Col, 64-67; asst prof, 67-71, ASSOC PROF BIOL, STATE UNIV NY PLATTSBURGH, 71- Mem: Am Fern Soc; Bot Soc Am; Int Soc Plant Morphol; Int Soc Study Evolution. Res: Ontogeney and phylogeny of branching systems in plants. Mailing Add: Dept of Bot State Univ NY Plattsburgh NY 12901

NOLAN, JANIECE SIMMONS, b Ft Worth, Tex, June 8, 39; m 73; c 6. HOSPITAL ADMINISTRATION, PHYSIOLOGY. Educ: Univ Tex, BA, 61, MA, 63; Tulane Univ, PhD(biol), 68; Univ Calif, Berkeley, MPH, 75. Prof Exp: Res scientist aerospace med, Tex Nuclear Corp, 63-65; head cell biol, Gulf Southern Res Inst, 68-70; res physiologist, Vet Admin Hosp, 70-73, health care admin trainee, Vet Admin Hosp, Martinez & Univ Calif, Berkeley, 73-75; ASST ADMINR AMBULATORY CARE SERV, UNIV CALIF MED CTR, SAN FRANCISCO, 75- Concurrent Pos: Scholar, Dept Physiol-Anat, Univ Calif, Berkeley, 70-72. Mem: Geront Soc; Am Physiol Soc; Tissue Cult Asn. Res: Delivery of ambulatory care services; cell culture as a tool for research on cellular aging; organizational dynamics during health care mergers; concentration of virus from water; plant senescence and growth regulators. Mailing Add: Rm A102 Univ of Calif 400 Parnassus Ave San Francisco CA 94143

NOLAN, JOHN THOMAS, JR, b Boston, Mass, Apr 15, 30; m 55; c 5. PETROLEUM CHEMISTRY, RESEARCH ADMINISTRATION. Educ: Cath Univ Am, AB, 51; Mass Inst Technol, PhD(org chem), 55. Prof Exp: Chemist, 55-58, group leader polymer res, 59-69, ASST SUPVR REF RES, TEXACO, INC, 69- Mem: Am Chem Soc; NY Acad Sci; Sigma Xi. Res: Organic syntheses; ozone reactions; polymer syntheses, structures and testing; lubricant additives; applied catalysis; refining processes; coal. Mailing Add: Relyea Terr RD 6 Wappingers Falls NY 12590

NOLAN, LINDA LEE, b Altoona, Pa, July 25, 47. BIOLOGICAL CHEMISTRY. Educ: Pa State Univ, BS, 69; Univ Mass, MS, 75, PhD(nutrit), 76. Prof Exp: RES ASST BIOCHEM, BIOL LABS, AMHERST COL, 69- Mem: Am Soc Microbiol; Soc Protozoologists. Res: Transport mechanisms in protozoa and leukocytes; biochemistry of purine and pyrimidine pathway; pharmacology of caffeine. Mailing Add: Biol Labs Amherst Col Amherst MA 01002

NOLAN, MICHAEL FRANCIS, b Evergreen Park, Ill, July 28, 47; m 73. NEUROANATOMY. Educ: Marquette Univ, BS, 69; Med Col Wis, PhD(anat), 75. Prof Exp: Staff phys therapist, Kiwanis Children's Ctr, Curative Workshop, Milwaukee, 69; INSTR ANAT, COL MED, UNIV S FLA, 75- Mem: Am Phys Ther Asn. Res: Neuroanatomical aspects of pain transmission, perception and appreciation; neurocytological changes in neural dysfunction. Mailing Add: Dept of Anat Col of Med Univ of S Fla Tampa FL 33620

NOLAN, RICHARD ARTHUR, b Omaha, Nebr, Nov 2, 37; m 67. MYCOLOGY, PHYSIOLOGY. Educ: Univ Nebr, Lincoln, BSc, 59, MSc, 62; Univ Calif, Berkeley, PhD(bot), 67. Prof Exp: NIH fel biochem, Univ Calif, Berkeley, 67-68, res biochemist, 68-69; asst prof biol, NMex State Univ, 69-70; univ fel, 70-71, asst prof, 71-75, ASSOC PROF BIOL, MEM UNIV NFLD, 75- Mem: Am Soc Microbiol; Can Soc Microbiologists; Can Bot Asn; Soc Invert Path; AAAS. Res: Comparative immunology and enzymology; nutritional requirements of fungal parasites of mosquito larvae; biological control of plant-parasitic nematodes. Mailing Add: Dept of Biol Mem Univ of Nfld St John's NF Can

NOLAN, STANTON PEELLE, b Washington, DC, May 29, 33; m 55; c 2. CARDIOVASCULAR SURGERY, CARDIOVASCULAR PHYSIOLOGY. Educ: Princeton Univ, AB, 55; Univ Va, Charlottesville, MD, 59, MS, 62. Prof Exp: From intern to resident surg, Med Ctr, Univ Va, Charlottesville, 59-65, Va Heart Asn res fel, Univ, 61-62, resident thoracic cardiovasc surg, Univ Va Hosp, 65-66; sr surgeon, Nat Heart Inst, 66-68; asst prof surg, 68-70, ASSOC PROF SURG & SURGEON-IN-CHG THORACIC CARDIOVASC SURG DIV, UNIV VA HOSP, 70- Concurrent Pos: Am Cancer Soc clin fel, 63-64; estab investr, Am Heart Asn, 69-74, mem coun cardiovasc surg, 68, mem coun thrombosis, 70; attend staff, Med Ctr, Univ Va, 68-; consult cardiac surg, Vet Admin Hosp, Salem, Va, 68-; consult surg, Va State Bur Crippled Children, 68- Honors & Awards: John Horsley Mem Prize Med Res, 62; Am Col Chest Physicians Award of Merit, 68. Mem: Fel Am Col Surg; fel Am Col Cardiol; Am Asn Thoracic Surg; Int Cardiovasc Soc; Soc Univ Surg. Res: Cardiovascular hemodynamics and pathophysiology. Mailing Add: Box 181 Univ of Va Med Ctr Charlottesville VA 22901

NOLAN, THOMAS BRENNAN, b Greenfield, Mass, May 21, 01; m 27; c 1. GEOLOGY. Educ: Yale Univ, PhB, 21, PhD(geol), 24. Hon Degrees: LLD, St Andrews, 62. Prof Exp: Geologist, 24-44, from asst dir to dir, 44-65, RES GEOLOGIST, US GEOL SURV, 65- Concurrent Pos: Vpres, Int Union Geol Sci, 65-72. Honors & Awards: Spendiaroff prize, Int Geol Cong, 33. Mem: Nat Acad Sci; fel Geol Soc Am (vpres 60, pres, 61); fel Soc Econ Geologists (vpres, 45, pres, 50); fel Mineral Soc Am; fel Am Geophys Union. Res: Ore deposits of the Great Basin. Mailing Add: US Geol Surv 12201 Sunrise Valley Dr Reston VA 22092

NOLAN, VAL, JR, b Evansville, Ind, Apr 28, 20; m 46; c 3. ORNITHOLOGY, ECOLOGY. Educ: Ind Univ, AB, 41, JD, 49. Prof Exp: Dep US Marshal, 41-42; agent, US Secret Serv, 42; from asst prof to assoc prof law, 49-56, PROF LAW, IND UNIV, BLOOMINGTON, 56-, PROF ZOOL, 68- Concurrent Pos: Res scholar zool, Ind Univ, Bloomington, 57-68; Guggenheim fel, 57. Mem: Fel Am Ornith Union; Ecol Soc Am; Animal Behav Soc; Ger Ornith Soc; Netherlands Ornith Union. Res: Behavior and ecology of birds. Mailing Add: Dept of Zool Ind Univ Bloomington IN 47401

NOLAND, GEORGE BRYAN, b Ft Ogden, Fla, Mar 7, 26; m 52; c 5. ACADEMIC ADMINISTRATION. Educ: Univ Detroit, BS, 50, MS, 52; Mich State Univ, PhD(entom), 55. Prof Exp: From instr to assoc prof biol, 55-65, PROF BIOL, UNIV DAYTON, 65-, CHMN DEPT, 63- Mem: Am Asn Higher Educ; Am Asn Univ Prof; Am Inst Biol Sci; Sigma Xi; AAAS. Res: Medical parasitology; general biology. Mailing Add: Dept of Biol Univ of Dayton 300 College Park Dayton OH 45469

NOLAND, JAMES STERLING, b Cape Girardeau, Mo, June 18, 33. ORGANIC POLYMER CHEMISTRY. Educ: Southeast Mo State Col, BS(chem), & BS(educ), 55; Univ Iowa, MS, 57, PhD(org chem), 60. Prof Exp: Res chemist polymer chem, 59-73, MGR, AEROSPACE ADHESIVES RES, AM CYANAMID CO, 73- Mem: Am Chem Soc. Res: New polymer systems; resin systems for structural adhesives; composites; environmental resistant resins. Mailing Add: Stamford Res Labs Am Cyanamid Co 1937 W Main St Stamford CT 06904

NOLAND, JERRE LANCASTER, b Richmond, Ky, Feb 14, 21; m 50; c 3. BIOCHEMISTRY. Educ: Purdue Univ, BS, 42, MS, 44; Univ Wis, PhD(biochem, zool), 49. Prof Exp: Asst chem & biol, Purdue Univ, 43-44; asst biochem, Univ Wis, 48; Lalor fel, Marine Biol Lab, Woods Hole, 49; biochemist, Entom Br, Med Labs, US Army Chem Ctr, 49-54; chief biochemist, Res Lab, Wood Vet Admin Ctr, Wis, 54-58; chief med res lab, Vet Admin Hosp, Louisville, Ky, 58-73; assoc prof biochem, 59-73, RES COORDR, DEPT OBSTET & GYNEC, SCH MED, UNIV LOUISVILLE, 74- Mem: AAAS; Am Chem Soc; Soc Exp Biol & Med; Am Asn Clin Chem; Asn Mil Surg US. Res: Endocrinology; sterol metabolism. Mailing Add: Dept of Obstet & Gynec Univ of Louisville Sch of Med Louisville KY 40202

NOLAND, PAUL ROBERT, b Chillicothe, Ill, Sept 28, 24; m 47; c 4. ANIMAL NUTRITION. Educ: Univ Ill, BS, 47, MS, 48; Cornell Univ, PhD, 51. Prof Exp: Asst, Univ Ill, 48 & Cornell Univ, 51; from asst prof to assoc prof animal nutrit & husb, 51-60, PROF ANIMAL SCI, UNIV ARK, FAYETTEVILLE, 60- Concurrent Pos: With Ark Agr Mission, Panama, 55-57. Mem: Am Soc Animal Sci; Poultry Sci Asn; Am Inst Nutrit; Animal Nutrit Res Coun. Res: Swine nutrition and management; proteins, amino acids and mineral nutrition. Mailing Add: Dept of Animal Sci Univ of Ark Fayetteville AR 72701

NOLAND, WAYLAND EVAN, b Madison, Wis, Dec 8, 26. ORGANIC CHEMISTRY. Educ: Univ Wis, BA, 48; Harvard Univ, MA, 50, PhD(phys org chem), 51. Prof Exp: Du Pont fel, 51-52, from asst prof to assoc prof, 52-62, PROF CHEM, UNIV MINN, MINNEAPOLIS, 62- Concurrent Pos: Consult, Sun Oil Co, 58-70; actg chief div org chem, Univ Minn, Minneapolis, 61-62, actg chmn dept chem, 67-69 & area coordr org chem, 72-74; secy, Org Syntheses, Inc, 69- Mem: AAAS; Fedn Am Scientists; Am Asn Univ Prof; Am Chem Soc; NY Acad Sci. Res: Heterocyclic nitrogen chemistry; synthesis and reactions of indoles and pyrroles; 1,3-cycloaddition reactions and ring expansions of isatogens; antimalarial compounds; rearrangements of nitronorbornenes; cycloaddition reactions of indenes; new rearrangements; reaction mechanisms; structure determination. Mailing Add: Sch of Chem Univ of Minn Minneapolis MN 55455

NOLASCO, JESUS BAUTISTA, b Manila, Philippines, Oct 5, 17; m 42; c 4. PHYSIOLOGY. Educ: Univ Philippines, MD, 40. Prof Exp: From instr to asst prof physiol, Univ Philippines, 40-52; prof & head dept, Far Eastern Univ, Manila, 62-64, sr consult hosp, 54-64; vis prof physiol, 64-66, assoc prof, 66-70, PROF PHYSIOL & ACTG CHMN DEPT, NJ MED SCH, COL MED & DENT NJ, 70- Concurrent Pos: Mem, Nat Res Coun Philippines, 40-; Rockefeller Found fel, Western Reserve Univ, 46-47; Williams-Waterman fel, Univ Berne, 62-63; mem comt nat med res prog, Nat Sci Develop Bd, Philippines, 62-64. Mem: Philippine Heart Asn (pres, 60-61); Am Physiol Soc. Res: Cardiovascular physiology, particularly electrophysiology of the heart. Mailing Add: Dept of Physiol NJ Med Sch Newark NJ 07103

NOLD, MAX M, b Edon, Ohio, July 2, 22; m 45; c 3. RADIATION BIOLOGY, VETERINARY MEDICINE. Educ: Ohio State Univ, DVM, 45; Cornell Univ, PhD(radiation biol), 60. Educ: Vet, State of Ohio, 45-51; base vet, 51-52, instr & course supvr, Sch Aviation Med, Ala, US Air Force, 52-54, course dir radiol health, Oak Ridge Inst Nuclear Studies, 55-58, radiation scientist, State Univ NY Vet Col, Cornell Univ, 58-60, prog officer radiation biol, Med Div, Defense Atomic Support Agency, 60-64, chief biophys div, Air Force Weapons Lab, NMex, 64-68, command vet, US Air Force Europe, 68-71 & Tactical Air Command, 71-75; ASST DIR, CALIF PRIMATE RES INST, UNIV CALIF, DAVIS, 75- Mem: Am Vet Med Asn; Sigma Xi. Res: Pathology; radiation dosimetry; bioastronautics; veterinary public health; teaching, education and research principles and practice. Mailing Add: Calif Primate Res Ctr Univ of Calif Davis CA 95616

NOLEN, GRANVILLE ABRAHAM, b Richmond, Ky, Apr 21, 26; m 50; c 5. TOXICOLOGY. Educ: Miami Univ, AB, 50. Prof Exp: Chief animal technician, 54-64, staff res asst animal nutrit, teratology & physiol & colony mgt, 64-70, STAFF NUTRITIONIST & TERATOLOGIST, MIAMI VALLEY LABS, PROCTER & GAMBLE CO, 70- Mem: AAAS; Am Asn Lab Animal Sci; Environ Mutagen Soc; Teratology Soc; Am Inst Nutrit. Res: Animal nutrition, especially lipid and protein nutrition and metabolism; teratology, especially methods in relation to drug testing; mutational effects of foods and chemicals; environmental effects on laboratory animals. Mailing Add: Miami Valley Labs Procter & Gamble Co Box 39175 Cincinnati OH 45239

NOLF, LUTHER OWEN, b Solomon, Kans, July 16, 02. PARASITOLOGY. Educ: Kans State Col, BS, 26, MS, 29; Johns Hopkins Univ, ScD(med zool), 31. Prof Exp: Asst parasitol, Kans State Col, 27-29; asst helminth, Johns Hopkins Univ, 29-31; assoc zool, 31-37, from asst prof to assoc prof, 37-57, PROF ZOOL, UNIV IOWA, 57- Concurrent Pos: Consult, Vet Admin. Mem: Am Soc Trop Med & Hyg; Am Soc Parasitol; Am Micros Soc. Res: Helminthology; Trichinella spiralis; trematodes of fish; poultry parasites. Mailing Add: Dept of Zool Univ of Iowa Iowa City IA 52240

NOLIN, JOSEPH ARTHUR BENOIT, b Levis, Que, Apr 1, 21; m 49; c 3. POLYMER CHEMISTRY. Educ: Col Levis, Can, BA, 43; Laval Univ, BSc, 47, PhD(chem), 49. Prof Exp: Lectr, Laval Univ, 47-49; fel infrared spectros & org chem, Nat Res Coun Can, 49-51, res officer, 51-53; res assoc infrared spectros, Mass Inst Technol, 53-54; RES CHEMIST, EXP STA, E I DU PONT DE NEMOURS & CO, INC, 54- Concurrent Pos: Lectr, Univ Ottawa, 51-53. Mem: Am Chem Soc. Res: Textiles; infrared spectroscopy. Mailing Add: 2704 Point Breeze Dr Wilmington DE 19810

NOLL, CHARLES JOSEPH, b State College, Pa, Nov 22, 13; m 46; c 2. HORTICULTURE, WEED SCIENCE. Educ: Pa State Col, BS, 35, MS, 42. Prof Exp: From instr to asst prof olericult, 47-60, ASSOC PROF OLERICULT, PA STATE UNIV, 60- Honors & Awards: Vaughn Award, Am Soc Hort Sci. Mem: Am Soc Hort Sci; Weed Sci Soc Am. Res: Chemical herbicides used in weed control; vegetable variety trials; teaching of olericulture. Mailing Add: Dept of Hort Pa State Univ University Park PA 16802

NOLL, CLARENCE IRWIN, b Palmyra, Pa, Feb 29, 08; m 34; c 2. ORGANIC CHEMISTRY. Educ: Lebanon Valley Col, BS, 30; Trinity Col, Conn, MS, 32; Pa State Col, PhD(org chem), 38. Prof Exp: Asst chem, Trinity Col, Conn, 30-33 & Pa State Col, 34-37; res chemist, Borden Co, 37-41; from instr org res to assoc prof, 41-53, prof chem, 53-71, from asst dean to assoc dean, Col Chem & Physics, 51-63 & from actg dean to dean, Col Sci, 63-71, EMER PROF CHEM & EMER DEAN, COL SCI, PA STATE UNIV, 71- Mem: Am Chem Soc. Res: Organic syntheses; chemistry of dairy products; explosives; organic nitrogen compounds. Mailing Add: 293 Ellen Ave State College PA 16801

NOLL, CLIFFORD RAYMOND, JR, b Providence, RI, Dec 20, 22; m 49; c 3. BIOCHEMISTRY. Educ: Brown Univ, AB, 44; Univ Ill, MS, 50; Univ Wis, PhD(biochem), 52. Prof Exp: .InstProf Exp: Instr biochem, Univ Mich, 52-56; biochemist, USDA, 56-58; asst prof chem, Goucher Col, 58-62; assoc prof, Wellesley Col, 62-65; res biochemist, Hartford Hosp, 65-67; fac mem, Franconia Col, 67-68;

assoc prof sci, 68-74, PROF SCI, GREATER HARTFORD COMMUNITY COL, 74- Concurrent Pos: Carnegie intern gen educ, Brown Univ, 55-56. Mem: AAAS. Res: Enzymes; plant biochemistry; molecular biology; sociology of science. Mailing Add: Dept of Sci Greater Hartford Community Col Hartford CT 06105

NOLL, HANS, b Basel, Switz, June 14, 24; nat US; m 49; c 4. MOLECULAR BIOLOGY. Educ: Univ Basel, PhD(biochem), 50. Prof Exp: Fel biol standardization, State Serum Inst, Copenhagen, Denmark, 50-51; asst tuberc, Pub Health Res Inst, City of New York, Inc, 51-53, assoc, 54-56; asst res prof microbiol, Sch Med, Univ Pittsburgh, 56-58, from asst prof to assoc prof, 59-64; PROF BIOL SCI, NORTHWESTERN UNIV, EVANSTON, 64- Concurrent Pos: Sr res fel, USPHS, 59-63, career investr, 64, mem molecular biol study sect, NIH, 66-68; vis prof, Univ Hawaii, 69; pres, Molecular Instruments Co, 70; vis prof inst immunol, Univ Basel, 71. Honors & Awards: Lifetime Endowed Career Professorship, Am Cancer Soc, 66. Mem: AAAS; Am Chem Soc; Am Soc Biol Chem; NY Acad Sci. Res: Biochemistry of viruses; molecular biology of nucleic acids and protein synthesis. Mailing Add: 2665 Orrington Ave Evanston IL 60201

NOLL, WALTER, b Berlin, Ger, Jan 7, 25; nat US; m 55; c 2. MATHEMATICS, CONTINUUM MECHANICS. Educ: Univ Paris, Lic es Sci, 50; Tech Univ Berlin, diplom, 51; Ind Univ, PhD(math), 54. Prof Exp: Instr mech, Tech Univ, Berlin, 51-55; instr math, Univ Southern Calif, 55-56; assoc prof, 56-60, PROF MATH, CARNEGIE-MELLON UNIV, 60- Concurrent Pos: Vis prof, Johns Hopkins Univ, 62-63. Mem: Am Math Soc; Math Asn Am; Soc Natural Philos. Res: Foundations of mechanics and thermodynamics; differential geometry; relativity. Mailing Add: 308 Field Club Ridge Rd Pittsburgh PA 15238

NOLLE, ALFRED WILSON, b Columbia, Mo, July 28, 19; m 46. PHYSICS. Educ: Southwest Tex State Teachers Col, BA, 38; Univ Tex, MA, 39; Mass Inst Technol, PhD(physics), 47. Prof Exp: Tutor physics, Univ Tex, 40-41; spec res assoc, Underwater Sound Lab, Harvard Univ, 41-45; asst prof eng res, Ord Res Lab, Pa State Col, 45; res assoc physics, Mass Inst Technol, 45-47, mem staff, Div Indust Coop, 47-48; from asst prof to assoc prof physics, 48-57, PROF PHYSICS, UNIV TEX, AUSTIN, 57- Mem: Fel Am Phys Soc. Res: Solid state physics; magnetic resonance and relaxation; paramagnetic impurities and imperfections; ultrasonics. Mailing Add: Dept of Physics Univ of Tex Austin TX 78712

NOLLEN, PAUL MARION, b Lafayette, Ind, Feb 24, 34; m 66. PARASITOLOGY. Educ: Carroll Col, BS, 56; Univ Wis, Madison, MS, 57; Univ Purdue, PhD(parasitol), 67. Prof Exp: Teacher high sch, Wis, 60-62; instr biol, Univ Purdue, 62-65; from asst prof to assoc prof zool & parasitol, 67-74, PROF ZOOL & PARASITOL, WESTERN ILL UNIV, 74- Concurrent Pos: Fac lectr, Western Ill Univ, 75. Honors & Awards: Herrick Award, 67; Sigma Xi Res Award, 74. Mem: AAAS; Am Soc Parasitol; Nat Asn Biol Teachers; Am Inst Biol Sci. Res: Reproductive activities of digenetic trematodes; uptake and incorporation of nutrients by digenetic trematodes; egg-shell chemistry of trematodes; cyst formation in degenetic trematodes. Mailing Add: Dept of Biol Sci Western Ill Univ Macomb IL 61455

NOLLER, DAVID CONRAD, b Elma, NY, Oct 1, 23; m 47; c 2. ORGANIC CHEMISTRY. Educ: Univ Buffalo, BA, 49, MA, 54. Prof Exp: Asst chem, Univ Buffalo, 49-52; res chemist, Lucidol Div, 52-57, head anal & control lab, 57-60, supvr tech serv lab, 60-64, group leader patents, lit & safety, 64-66, group leader patents & lit, 66-69, patent agent, 69-70, TECH INFO SPECIALIST, PENNWALT CORP, 70- Mem: Am Chem Soc. Res: Organic peroxides with emphasis on safety hazard testing. Mailing Add: 145 E Morris St Buffalo NY 14214

NOLLER, HARRY FRANCIS, JR, b Oakland, Calif, June 10, 39; m 64; c 1. BIOCHEMISTRY. Educ: Univ Calif, Berkeley, AB, 60; Univ Ore, PhD(chem), 65. Prof Exp: NIH fel, Lab Molecular Biol, Med Res Coun, Cambridge, Eng, 65-66 & Inst Molecular Biol, Geneva, Switz, 66-68; asst prof biol, 68-73, ASSOC PROF BIOL, UNIV CALIF, SANTA CRUZ, 73- Res: Protein and nucleic acid chemistry; structure and function of ribosomes. Mailing Add: Thimann Labs Univ of Calif Santa Cruz CA 95064

NOLSTAD, ARNOLD RAGNVALD, b Warren, Minn, Aug 21, 08; m 35; c 1. MATHEMATICS. Educ: Luther Col, BA, 30; Univ Pittsburgh, MA, 33, DEd(math), 48. Prof Exp: Asst math, Univ Pittsburgh, 30-33; head teacher high sch, Pa, 34-43; from asst prof to assoc prof, 48-75, EMER PROF MATH, NC STATE UNIV, 75- Mem: Math Asn Am; Am Math Soc; Am Educ Res Asn. Res: History of Mathematics. Mailing Add: Dept of Math NC State Univ Raleigh NC 27607

NOLTE, WILLIAM ANTHONY, b Washington, DC, Nov 15, 13; m 44; c 6. BACTERIOLOGY, MICROBIOLOGY. Educ: Univ Md, BS, 37, MS, 39, PhD(bact), 47. Prof Exp: Asst bact, Univ Md, 37-42; asst bacteriologist, Meats Lab Res Ctr, USDA, 42-43; asst bact, Univ Md, 46-47; chief bacteriologist, Res Lab, Vick Chem Co, NY, 47-49; from asst prof to assoc prof path & microbiol, 48-55, PROF PATH & MICROBIOL, UNIV TEX DENT BR HOUSTON, 55- Mem: Am Soc Microbiol; Int Asn Dent Res; Am Asn Dent Schs. Res: Dental caries process; specific oral organisms associated with dental plaque; sterilization procedures for dental instruments; bacteriological culture media in endodontics. Mailing Add: Univ of Tex Dent Br PO Box 20068 Houston TX 77025

NOLTIMIER, HALLAN COSTELLO, b Los Angeles, Calif, Mar 19, 37; m 61; c 2. GEOPHYSICS. Educ: Calif Inst Technol, BS, 58; Univ Newcastle, Eng, PhD(geophysics), 65. Prof Exp: Lectr physics & geophysics, Sch Physics, Univ Newcastle, Eng, 66-68; asst prof geol, Sch Geol & Geophys, Univ Okla, 68-71; assoc prof geol, Dept Geol, Univ Houston, 71-72; ASSOC PROF GEOPHYSICS, DEPT GEOL & MINERAL, OHIO STATE UNIV, 72- Mem: Fel Royal Astron Soc; Am Geophys Union; Sigma Xi; Soc Explor Geophysicists; AAAS. Res: Paleomagnetism of Mesozoic intrusives along the eastern margin of North America; paleomagnetism of coal; paleomagnetism of recent lake sediments with particular interest in short geomagnetic excursions. Mailing Add: Dept of Geol & Mineral Ohio State Univ Columbus OH 43210

NOLTMANN, ERNST AUGUST, b Gotha, Ger, June 27, 31; m 56; c 2. BIOCHEMISTRY, ENZYMOLOGY. Educ: Univ Münster, Physikum, 53; Med Acad Düsseldorf, Ger, MD, 56. Prof Exp: Jr biochemist, Inst Physiol Chem, Med Acad Düsseldorf, 56-58, asst biochemist, 58-59; fel enzyme chem, Inst Enzyme Res, Univ Wis, 59-62; from asst prof to assoc prof biochem, 62-69, PROF BIOCHEM, UNIV CALIF, RIVERSIDE, 69-, CHMN DEPT, 75- Concurrent Pos: USPHS & NSF res grants. Mem: Am Soc Biol Chemists; Am Chem Soc. Res: Isolation and characterization of enzymes and the study of their mechanisms of action. Mailing Add: Dept of Biochem Univ of Calif Riverside CA 92502

NOMURA, KAWORU CARL, b Deer Lodge, Mont, Apr 1, 22; m 47; c 4. SOLID STATE PHYSICS. Educ: Univ Minn, BPhys, 48, MS, 49, PhD(elec eng), 53; Advan Mgt Prog, Harvard Univ, grad, 72. Prof Exp: Sr res physicist, Corp Res Ctr, 53-54, res supvr, 54-58, staff scientist, 58-61, mgr opers semiconductor div, 61-65, DIR SOLID STATE ELECTRONICS CTR, HONEYWELL, INC, 65- Mem: Fel Am Phys Soc; sr mem Inst Elec & Electronics Eng. Res: Electrical, optical and structural properties of semiconductors; solid state sensors and transducers; transistors and integrated circuits; research and development and manufacture. Mailing Add: Solid State Electronics Ctr Honeywell Inc 10500 State Highway Minneapolis MN 55441

NOMURA, SHIGEKO, b Tokyo, Japan, Jan 3, 29. MICROBIOLOGY, VIROLOGY. Educ: Toho Women's Col Med, MD, 50; Yamaguchi Univ, DMS, 60. Prof Exp: Intern, Tokyo Univ Hosp, 50-51; resident internal med, Tokyo Sumida Hosp, 51-53; NIH res fel virol, Tokyo, 53-61; vis assoc, NIH, Md, 61-63; res assoc, Sch Hyg & Pub Health, Johns Hopkins Univ, 63-66; sr virologist, Flow Labs, Inc, 66-67; MICROBIOLOGIST, NAT CANCER INST, 67- Mem: AAAS; Am Soc Microbiol. Res: Cancer research. Mailing Add: Bldg 41 S-400 Nat Cancer Inst Bethesda MD 20014

NONEMAKER, LARRY FRANKLIN, b York, Pa, Jan 20, 37; m 66; c 2. ORGANIC CHEMISTRY, POLYMER PHYSICS. Educ: Franklin & Marshall Col, BS, 58; Carnegie Inst Technol, MS, 61, PhD(org chem), 62. Prof Exp: Res chemist, Marshall Res Lab, 62-66, develop supvr, 66-67, res supvr, 67-74, MKT MGR, E I DU PONT DE NEMOURS & CO, INC, 74- Mem: Am Chem Soc. Res: Anionic catalysis; Fischer indole synthesis; crystalline polymers; synthesis and characterization of plastics and elastomers; electrocoating and water reducible finishes. Mailing Add: 420 Oak Valley Rd Media PA 19063

NONNECKE, IB LIBNER, b Copenhagen, Denmark, Oct 1, 22; nat Can; m 45; c 3. HORTICULTURE. Educ: Univ Alta, BSc, 45, MSc, 50; Univ Ore, 50; Ore State Univ, PhD, 58. Prof Exp: Head hort sect, Res Sta, Can Dept Agr, 46-63; pres, Asgrow Seed Co Can, Ltd, 63-66, coordr crop improv, Asgrow Seed Co, 66-68; assoc prof hort, 68-70, PROF HORT & CHIEF VEG DIV, UNIV GUELPH, 70-, CHMN DEPT HORT SCI, 74- Mem: Am Soc Hort Sci; Can Soc Hort Sci; Sigma Xi; Agr Inst Can. Res: Plant breeding and genetics; statistics; field plot technique. Mailing Add: Dept of Hort Sci Univ of Guelph Guelph ON Can

NONOYAMA, MEIHAN, b Tokyo, Japan, Feb 25, 38; m; c 1. MOLECULAR BIOLOGY, VIROLOGY. Educ: Univ Tokyo, BS, 61, MS, 63, PhD(molecular biol), 66. Prof Exp: Res assoc molecular biol, Univ Tokyo, 66-67 & Univ Ill, 66-68; vis scientist, Wistar Inst Anat & Biol, 68-70; res asst prof bact, Univ NC, Chapel Hill, 70-73; ASSOC PROF MICROBIOL, RUSH-PRESBYTERIAN HOSP-ST LUKES MED CTR, CHICAGO, 73- Concurrent Pos: Fulbright fel, 66-68. Honors & Awards: Merck Award, Merck & Co, Inc, 71. Res: Molecular biology of tumor virus; oncogenesis of Herpes type virus, especially Epstein Barr virus. Mailing Add: Dept of Microbiol Rush-Presby St Lukes Med Ctr Chicago IL 60612

NOODEN, LARRY DONALD, b Oak Park, Ill, June 10, 36; m 63; c 2. PLANT PHYSIOLOGY, BIOCHEMISTRY. Educ: Univ Ill, BSc, 58; Univ Wis, MSc, 59; Harvard Univ, PhD(biol), 63. Prof Exp: NIH res fel, Univ Edinburgh, 64-65; asst prof bot, 65-70, ASSOC PROF BOT, UNIV MICH, ANN ARBOR, 70- Concurrent Pos: NIH special res fel, Calif Inst Technol, 71-72. Mem: AAAS; Am Soc Plant Physiol; Bot Soc Am; Environ Mutagen Soc; Japanese Soc Plant Physiol. Res: Biochemistry of regulation of plant development. Mailing Add: Dept of Bot Univ of Mich Ann Arbor MI 48104

NOOJIN, RAY O, b Birmingham, Ala, Nov 29, 12; m 39; c 3. DERMATOLOGY. Educ: Univ Ala, AB, 33; Univ Chicago, MD, 37; Am Bd Dermat, dipl, 45. Prof Exp: Asst pharmacol, Univ Chicago, 37-38; intern med, Sch Med, Duke Univ, 38-40, resident dermat & syphil, 41-44, instr, 44-45; PROF DERMAT & CHMN DEPT, SCH MED, UNIV ALA, BIRMINGHAM, 45- Concurrent Pos: Dir & mem, Am Bd Dermat, 64, pres, 70-71; mem drug efficacy study, Nat Res Coun; chmn div educ & commun, Task Force for Allied Health Prof, 70-71. Mem: Am Dermat Asn; Am Asn Prof Dermat (pres, 74-75). Res: Therapeutics in dermatology. Mailing Add: Dept of Dermat Univ of Ala Sch of Med Birmingham AL 35233

NOONAN, CHARLES D, b San Francisco, Calif, July 16, 28; m; c 5. RADIOLOGY. Educ: Univ Calif, Berkeley, AB, 50; Univ Calif, San Francisco, MD, 53. Prof Exp: Resident, Cincinnati Gen Hosp, 56-59; from instr to assoc prof, 59-73, PROF RADIOL, SCH MED, UNIV CALIF, SAN FRANCISCO, 73- Mailing Add: Dept of Radiol Univ of Calif Hosp San Francisco CA 94143

NOONAN, JACQUELINE ANNE, b Burlington, Vt, Oct 28, 28. PEDIATRIC CARDIOLOGY. Educ: Albertus Magnus Col, BA, 50; Univ Vt, MD, 54. Prof Exp: Asst prof pediat, Col Med & pediat cardiologist, Hosp, Univ Iowa, 59-61; assoc prof, 61-69, PROF PEDIAT, COL MED, UNIV KY, 69-, CHMN DEPT, 74-, PEDIAT CARDIOLOGIST, UNIV HOSP, 61- Mem: Am Heart Asn; Am Pediat Soc; Am Acad Pediat; AMA; Am Fedn Clin Res. Mailing Add: Dept of Pediat Univ of Ky Col of Med Lexington KY 40506

NOONAN, JAMES WARING, b Fall River, Mass, Dec 21, 44; m 67. MATHEMATICS. Educ: Providence Col, AB, 66; Univ Md, College Park, MA, 69, PhD(math), 71. Prof Exp: Nat Res Coun res assoc, US Naval Res Lab, 70-71; ASST PROF MATH, COL OF THE HOLY CROSS, 71- Mem: Am Math Soc; Math Asn Am; Sigma Xi. Res: Complex analysis; geometric function theory. Mailing Add: Dept of Math Col of the Holy Cross Worcester MA 01610

NOONAN, KENNETH DANIEL, b New York, NY, Jan 27, 48; m 72. BIOCHEMISTRY, CELL BIOLOGY. Educ: St Joseph's Col, Pa, BS, 69; Princeton Univ, PhD(biol), 72. Prof Exp: Jane Coffin Childs fel, Biocenter, Basil, Switz, 72-73; ASST PROF BIOCHEM, J HILLIS MILLER HEALTH CTR, UNIV FLA, 73- Mem: AAAS. Res: Role of the plasma membrane in normal and transformed cell growth. Mailing Add: J Hillis Miller Health Ctr Univ of Fla Dept of Biochem Gainesville FL 32610

NOONAN, SHARON MARIELLA, b Detroit, Mich; m 72; c 1. ELECTRON MICROSCOPY, CELL PHYSIOLOGY. Educ: Univ Detroit, BS, 63; Wayne State Univ, MS, 65, PhD(physiol), 71; Registry Med Technol, cert. Prof Exp: Med tech hemat, Henry Ford Hosp, 65-67; asst physiol, 68-71, instr path, 71-73, ASST PROF PATH, WAYNE STATE UNIV, 73- Concurrent Pos: Consult, Vet Admin Hosp, Allen Park, Mich, 72- & Detroit Gen Hosp, 74- Mem: Am Soc Cell Biol; Soc Exp Biol & Med. Res: Inherited metabolic diseases; collagen synthesis. Mailing Add: Dept of Path Wayne State Univ Sch of Med Detroit MI 48207

NOONAN, THOMAS ROBERT, b Buffalo, NY, May 2, 12; m 40. RADIOBIOLOGY, PHYSIOLOGY. Educ: Middlebury Col, BS, 34; Univ Buffalo, MD, 39. Prof Exp: Asst physiol, Univ Buffalo, 36-38; asst, Univ Rochester, 39-40, instr, 40-41, asst prof radiation biol, 46-49; mem med div, Upjohn Co, 49-50; from asst prof to assoc prof radiation biol, Univ Rochester, 50-67; scientist & dir animal res prog, 67-74, SCI ASST TO DIR, COMP ANIMAL RES LAB, 74- Mem: Soc Exp Biol & Med; Radiation Res Soc; Am Physiol Soc. Res: Metabolism of potassium; tracer studies;

effects of radiation. Mailing Add: Comp Animal Res Lab 1299 Bethel Valley Rd Oak Ridge TN 37830

NOONAN, THOMAS WYATT, b Glendale, Calif, July 16, 33; m 66; c 3. ASTRONOMY, PHYSICS. Educ: Calif Inst Technol, BS, 55, PhD(physics), 61. Prof Exp: Physicist, Smithsonian Astrophys Observ, 61-62; vis asst prof physics, Univ NC, Chapel Hill, 62-65, asst prof, 65-68; assoc prof, 68-71, PROF PHYSICS, STATE UNIV NY COL BROCKPORT, 71- Concurrent Pos: State Univ NY Res Found fel, Hale Observ, 70. Mem: Am Astron Soc; Royal Astron Soc; Royal Astron Soc Can. Res: Clusters of galaxies; cosmology. Mailing Add: Dept of Physics State Univ of NY Col Brockport NY 14420

NOONE, MICHAEL JOHN, b Stoke-on-Trent, Eng, Jan 12, 38; m 60; c 3. CERAMICS. Educ: Univ Leeds, Eng, BSc, 63, PhD(ceramics), 67. Prof Exp: Res asst ceramics, Univ Leeds, 63-67; res ceramist mat res & develop, 67-73, GROUP LEADER STRUCTURAL MAT, SPACE SCI LAB, GEN ELEC CO, 73- Mem: Brit Ceramic Soc; Inst Ceramics; Am Ceramic Soc. Res: Ceramic materials for structural applications in high temperature environments such as gas turbines and magnetohydrodynamic channels; fabrication of sialon type materials; selectively transmitting laser windows; directional solidification; composite materials. Mailing Add: Space Sci Lab Rm M9539 Gen Elec Co PO Box 8555 Philadelphia PA 19101

NOONER, DARYL WILBURN, bio-organic chemistry, see 12th edition

NOONEY, GROVE C, b New York, NY, 1927. MATHEMATICS. Educ: Univ Nev, BS, 50; Stanford Univ, PhD(math), 54. Prof Exp: MATHEMATICIAN, MATHEMATICS & COMPUT GROUP, LAWRENCE BERKELEY LAB, UNIV CALIF, BERKELEY, 61- Mem: Am Math Soc. Res: Theoretical biology. Mailing Add: Lawrence Berkeley Lab Univ of Calif Berkeley CA 94704

NOORDERGRAAF, ABRAHAM, b Utrecht, Neth, Aug 7, 29; m 56; c 4. BIOENGINEERING, BIOPHYSICS. Educ: Univ Utrecht, BSc, 53, MSc, 55, PhD(med physics), 56. Hon Degrees: MA, Univ Pa, 71. Prof Exp: Asst exp physics, Univ Utrecht, 52-53, res asst med physics, 53-55, res fel med physics, 56-58, sr res fel, 59-65; assoc prof elec eng, 64-70, PROF BIOMED ENG, UNIV PA, 70-, CHMN DEPT, 73-, CHMN GRAD GROUP BIOENG, 73- Concurrent Pos: Vis fel ther res, Univ Pa, 57-58, NIH res grants, 66, 68, 70, 72 & 74, Surgeon Gen, 74; mem spec study sect, NIH, 65-68; vis prof appl physics, Delft Univ Technol, 70-71; vis prof cardiol, Erasmus Univ Med Sch, Rotterdam, 70-71; mem coun circulation, Am Heart Asn, 72-; vis prof, Univ Miami, 70-; consult, NATO Sci Affairs Div, 73- Mem: Biophys Soc; Ballistocardiographic Res Soc (pres, 68-70); Europ Soc Ballistocardiographic Res; Biomed Eng Soc; Instrument Soc Am. Res: Mammalian cardiovascular system analysis; wave transmission in blood vessels; operation of the heart as a pump; dynamics of the microcirculation. Mailing Add: Dept of Bioeng D2 Univ of Pa Philadelphia PA 19174

NOPANITAYA, WAYKIN, b Krabi, Thailand, Oct 14, 42; m 66; c 1. PATHOLOGY, ELECTRON MICROSCOPY. Educ: Mahidol Univ, dipl med technol, 63; Univ NC, Chapel Hill, BA, 69, PhD(path), 71. Prof Exp: Med technologist, Vajira Hosp, Thailand, 63; chief path & histotechnologist, US Army-SEATO Med Res Lab, Thailand, 63-67; med technologist, Univ NC, Chapel Hill, 67-69, instr electron micros, 70; consult exp path, US Army-SEATO Med Res Lab, 72; instr path, 72-74, ASST PROF PATH, MED SCH, UNIV NC, CHAPEL HILL, 74-, DIR ELECTRON MICROS & HISTOPATH TRAINING PROGS, 72- Concurrent Pos: Nat Inst Gen Med Sci grant, Univ NC, Chapel Hill & Ministry Health, Thailand, 72; consult, Asn Med Technologists Thailand, 72- Mem: Am Soc Clin Path; Electron Micros Soc Am; Nat Soc Histotechnol; Sigma Xi; Can Micros Soc. Res: Gastroenterology; ultrastructural research in cellular pathology. Mailing Add: Dept of Path Univ of NC Sch of Med Chapel Hill NC 27514

NORA, AUDREY HART, b Picayune, Miss, Dec 5, 36; m 68; c 2. PEDIATRICS. Educ: Univ Miss, BS, 58, MD, 61; Am Bd Pediat, cert, 68, cert hemat & oncol, 74. Prof Exp: Fel pediat hemat & oncol, Baylor Col Med, 64-66, instr pediat, 66-70, asst prof, 70-71; ASST CLIN PROF PEDIAT, MED CTR, UNIV COLO, DENVER, 71-; DIR GENETICS & BIRTH DEFECTS, DENVER CHILDREN'S HOSP, 71- Concurrent Pos: Assoc hematologist, Tex Children's Hosp, 66-71; consult, NIH Adv Comt Arteriosclerosis & Hypertension, 75-77. Mem: Am Fedn Clin Res; Am Soc Human Genetics; Genetics Soc Am; Teratology Soc. Res: Arteriosclerosis in the pediatric age group; teratology of medications, congenital malformations relating to medications taken during first trimester. Mailing Add: Denver Children's Hosp 4200 E Ninth Ave Denver CO 80220

NORA, JAMES JACKSON, b Chicago, Ill, June 26, 28; m 66; c 5. PEDIATRIC CARDIOLOGY, GENETICS. Educ: Harvard Univ, AB, 50; Yale Univ, MD, 54; Am Bd Pediat, dipl & cert cardiol. Prof Exp: Am Heart Asn res fel cardiol, Univ Wis, 62-64, from instr to asst prof pediat, 62-65; from asst prof to assoc prof, Baylor Univ, 65-71, head div human genetics & dir birth defects ctr, 67-71; assoc prof pediat, 71-74, DIR PEDIAT CARDIOL, UNIV COLO MED CTR, DENVER, 71-, PROF PEDIAT & DIR CARDIOVASC-PULMONARY TRAINING CTR, 74- Concurrent Pos: NIH spec fel genetics, McGill Univ, 64-65; assoc dir cardiol, Tex Children's Hosp, 65-71, chief genetics serv, 67-71. Mem: Am Soc Human Genetics; fel Am Acad Pediat; Am Pediat Soc; Soc Pediat Res; Teratology Soc. Res: Etiology of cardiovascular diseases. Mailing Add: Dept of Pediat Univ of Colo Med Ctr Denver CO 80220

NORA, PAUL FRANCIS, b Chicago, Ill, Aug 14, 29; m 56; c 5. SURGERY. Educ: Loyola Univ, MD, 52; Northwestern Univ, MS, 64, PhD(surg), 68. Prof Exp: Asst prof, 69-72, ASSOC PROF SURG, SCH MED, NORTHWESTERN UNIV, 72-; CHMN DEPT SURG, COLUMBUS-CUNEO-CABRINI MED CTR, 64- Concurrent Pos: Attend surgeon, Vet Admin Res Hosp, 62-; mem undergrad teaching comt, Dept Surg, Med Sch, Northwestern Univ; mem, Adv Panel Nat Health Ins, Subcomt Health, Comt Ways & Means. Mem: Soc Surg Alimentary Tract; fel Am Col Surgeons; Illum Eng Soc. Res: Value of operative choledochoscopy; abdominal drains; operating room lighting. Mailing Add: Dept of Surg Columbus-Cuneo-Cabrini Med 2520 N Lakeview Ave Chicago IL 60614

NORBECK, EDWARD, b Prince Albert, Sask, Mar 18, 15; US citizen; m 50; c 3. CULTURAL ANTHROPOLOGY. Educ: Univ Mich, Ann Arbor, BA, 48, MA, 49, PhD(anthrop), 52. Prof Exp: Asst prof anthrop, Univ Utah, 52-54; asst prof, Univ Calif, Berkeley, 54-60; assoc prof, 60-62, chmn dept anthrop & sociol, 60-71, dean humanities & soc sci, 66-67, PROF ANTHROP, RICE UNIV, 62-, DIR GRAD STUDIES BEHAV SCI, 69- Concurrent Pos: NSF fel, Univ Tokyo, 58-59; mem bd dirs, Southwest Ctr Urban Res, 69-, vpres, 72. Mem: AAAS; fel Anthrop Asn; Am Ethnol Soc (pres, 76); Soc Sci Study Relig; Japanese Soc Ethnol. Res: Japanese society and culture; cultural and social change and evolution; religion social structure. Mailing Add: Dept of Anthrop Rice Univ Houston TX 77001

NORBECK, EDWIN, JR, b Seattle, Wash, June 10, 30; m 56; c 4. NUCLEAR PHYSICS. Educ: Reed Col, BA, 52; Univ Chicago, MS & PhD(physics), 56. Prof Exp: Res assoc physics, Univ Chicago, 56-57 & Univ Minn, 57-60; from asst prof to assoc prof, 60-67, PROF PHYSICS, UNIV IOWA, 67- Mem: Fel Am Phys Soc. Res: Low energy nuclear physics; nuclear reactions with lithium and beryllium beams; on line applications at digital computers. Mailing Add: Dept of Physics & Astron Univ of Iowa Iowa City IA 52242

NORBERG, RICHARD EDWIN, b Newark, NJ, Dec 28, 22; m 47; c 3. PHYSICS. Educ: DePauw Univ, BA, 43; Univ Ill, MA, 47, PhD(physics), 51. Prof Exp: Res assoc physics & control systs lab, Univ Ill, 51-53, asst prof, 53-54; vis lectr physics, 54-56, assoc - prof, 56-58, PROF PHYSICS, WASH UNIV, 58-, CHMN DEPT, 62- Concurrent Pos: Sloan fel, 55-59. Mem: Fel Am Phys Soc. Res: Nuclear and electron spin resonance; solid state and low temperature physics. Mailing Add: Dept of Physics Wash Univ St Louis MO 63130

NORCIA, LEONARD NICHOLAS, b Mountain Iron, Minn, Jan 1, 16; m 50; c 1. BIOCHEMISTRY, PHYSIOLOGY. Educ: Univ Minn, BChem, 46, PhD(physiol chem), 52. Prof Exp: Res fel biochem, Hormel Inst, Univ Minn, 52-55; from asst prof to assoc prof res biochem, Sch Med, Univ Okla, 56-60; asst prof biochem, 60-69, ASSOC PROF BIOCHEM, SCH MED, TEMPLE UNIV, 69- Concurrent Pos: Biochemist, Okla Med Res Found, 55-60; mem coun arteriosclerosis, Am Heart Asn. Mem: Am Chem Soc; Am Oil Chemists Soc. Res: Metabolism of lipids. Mailing Add: Dept of Biochem Temple Univ Sch of Med Philadelphia PA 19140

NORCROSS, BRUCE EDWARD, b Newport, Vt, Mar 14, 35; m 56; c 1. PHYSICAL ORGANIC CHEMISTRY. Educ: Univ Vt, BA, 56, MS, 57; Ohio State Univ, PhD(chem), 60. Prof Exp: Res assoc chem, Harvard Univ, 60-62; asst prof, 62-66, ASSOC PROF, STATE UNIV NY BINGHAMTON, 66- Concurrent Pos: Assoc dean, Harpur Col, 66-67. Mem: AAAS; Am Chem Soc. Res: Organic reaction mechanisms and synthetic organic chemistry. Mailing Add: Dept of Chem State Univ NY Binghamton NY 13901

NORCROSS, DAVID WARREN, b Cincinnati, Ohio, July 18, 41; m 67. ATOMIC PHYSICS. Educ: Harvard Col, AB, 63; Univ Ill, MS, 65; Univ Col, Univ London, PhD(physics), 70. Prof Exp: Res physicist, Sperry Rand Res Ctr, 65-67; res assoc physics, Univ Colo, 70-74; PHYSICIST, LAB ASTROPHYS DIV, NAT BUR STANDARDS, 74- Concurrent Pos: Lectr, Dept Physics, Univ Colo, 74- Mem: Am Phys Soc; Fedn Am Scientists. Res: Theory of electron-atom and electron-molecule interactions, atomic and molecular structure and radiative properties. Mailing Add: Joint Inst for Lab Astrophys Univ of Colo Boulder CO 80309

NORCROSS, MARVIN AUGUSTUS, b Tansboro, NJ, Feb 8, 31; m 56; c 2. VETERINARY PATHOLOGY. Educ: Univ Pa, VMD, 59, PhD(path), 66. Prof Exp: Gen vet pract, Md, 59-60; vet, Animal Husb Res Div, Agr Res Serv, USDA, 60-62; res fel path, USPHS training grant cancer res, Sch Vet Med, Univ Pa, 62-66; asst vet pathologist, 66-69, assoc dir clin res, 69-72, sr dir animal sci res, Merck, Sharp & Dohme Res Lab, Rahway, 72-75; DIR, DIV VET RES, BUR VET MED, FOOD & DRUG ADMIN, BELTSVILLE, 75- Mem: AAAS; Am Asn Avian Path; Am Vet Med Asn; NY Acad Sci; Soc Pharmacol & Environ Path. Res: Chemical and viral carcinogenesis; animal health drugs and biologicals. Mailing Add: 10100 Lakewood Dr Rockville MD 20850

NORCROSS, NEIL LINWOOD, b Derry, NH, July 18, 28; m 53; c 2. IMMUNOLOGY. Educ: Univ Miami, AB, 50; Univ Mass, MS, 55, PhD(bact), 58. Prof Exp: Immunologist, Plum Island Animal & Dis Lab, Agr Res Serv, USDA, 57-60; from asst prof to assoc prof immunochem, 60-69, PROF IMMUNOCHEM, NY STATE UNIV COL VET MED, CORNELL UNIV, 69-, SECY OF COL, 73- Concurrent Pos: Mem, Nat Mastitis Coun. Mem: Am Acad Microbiol; Am Soc Microbiol. Res: Immunology of foot and mouth disease virus; production of foot and mouth vaccines; mastitis; streptococci of bovine origin; immunology of equine infectious anemia. Mailing Add: NY State Col Vet Med Cornell Univ Ithaca NY 14853

NORD, EAMOR CARROLL, plant ecology, range science, see 12th edition

NORD, GORDON LUDWIG, JR, b Cincinnati, Ohio, Apr 3, 42; m 65; c 1. GEOLOGY. Educ: Univ Wis, BS, 65; Univ Idaho, MS, 67; Univ Calif, Berkeley, PhD(geol), 73. Prof Exp: Res assoc geol, Case Western Reserve Univ, 71-74; GEOLOGIST, US GEOL SURV, 74- Mem: Geol Soc Am; Am Mineral Soc; Am Geophys Union; AAAS. Res: The characterization and mechanisms of precipitation reactions, symmetry transitions and defect structures in natural and synthetic minerals and mineral aggregates as a function of geological history. Mailing Add: 959 Nat Ctr US Geol Surv Reston VA 22092

NORD, JOHN C, b Joliet, Ill, Mar 19, 38. FOREST ENTOMOLOGY. Educ: Univ Mich, BS, 60, MF, 62, PhD(forestry), 68. Prof Exp: ASSOC ENTOMOLOGIST, SOUTHEAST FOREST EXP STA, US FOREST SERV, 64- Mem: Soc Am Foresters; Entom Soc Am. Res: Ecology, behavior and control of insects in southern pine seed orchards. Mailing Add: Forestry Sci Lab Carlton St Athens GA 30601

NORDAN, HAROLD CECIL, b Vancouver, BC, Jan 21, 25; m 52; c 2. VERTEBRATE BIOLOGY. Educ: Univ BC, BA, 48, BSA, 50, MSA, 54; Ore State Univ, PhD(microbiol, biochem), 59. Prof Exp: Fel zool, 59-61, res assoc & sessional lectr, 62-65, asst prof, 65-69, ASSOC PROF ZOOL, UNIV BC, 69- Res: Vertebrate physiology, especially bioenergetics and growth of wild species of ungulates. Mailing Add: Dept of Zool Univ of BC Vancouver BC Can

NORDBERG, WILLIAM, b Fehring, Austria, Mar 31, 30. SPACE PHYSICS, RESEARCH ADMINISTRATION. Educ: Univ Graz, PhD(physics), 53. Prof Exp: Scientist meteorol div, US Army Signal Eng Labs, 53-59; scientist, 59-74, DIR APPL, GODDARD SPACE FLIGHT CTR, NASA, 74- Honors & Awards: Exceptional Sci Achievement Award, NASA, 65. Mem: AAAS; Am Geophys Union; Am Meteorol Soc. Res: Measurement of atmospheric state with rockets and satellites; weather satellites; earth resources surveys from space. Mailing Add: Code 900 NASA Goddard Space Flight Ctr Greenbelt MD 20771

NORDBLOM, GEORGE FREDERICK, b Toronto, Ont, Nov 19, 13; m 44; c 2. PHYSICAL CHEMISTRY. Educ: Univ Pittsburgh, BS, 36; Univ Idaho, MS, 37; Univ Cincinnati, PhD(phys chem), 41. Prof Exp: Asst chem, Univ Cincinnati, 37-41, instr chem tech, 40-41; res chemist, Westvaco Chlorine Prod, WVa, 41-42; chem engr, Frankford Arsenal, Pa, 42-53; head inorg chem div, Pitman-Dunn Labs, 53-56; sr res chemist, Foote Mineal Co, 56-62; sr scientist, Carl F Nordberg Res Ctr, Elec Storage Battery Co, 63-72, SR SCIENTIST, ESB TECHNOL CTR, ESB INC, 72- Mem: Fel AAAS; Electrochem Soc. Res: Inorganic metal finishes; battery research, especially lead-acid and silver-zinc batteries; electrodeposition; electrochemistry. Mailing Add: 15 Scammell Dr Yardley PA 19067

NORDBY, GORDON LEE, b Moscow, Idaho, Oct 3, 29, m 53; c 4. BIOCHEMISTRY,

BIOMETRY. Educ: Stanford Univ, BS, 51, MS, 54, PhD(chem philos), 58. Prof Exp: Instr biochem, Stanford Univ, 58-59; res assoc, Harvard Med Sch, 60-63; asst prof biol chem & assoc res biophysicist, 63-68, ASSOC PROF BIOL CHEM, 68- Concurrent Pos: Am Cancer Soc res fel, 59-61. Mem: AAAS; Am Chem Soc; Biophys Soc; Biomet Soc. Res: Biophysics; digital computers; experimental design. Mailing Add: Dept of Biol Chem M5416 Med Sci Bldg Univ of Mich Ann Arbor MI 48104

NORDBY, HAROLD EDWIN, b New England, NDak, Nov 3, 31; m 58; c 2. ORGANIC CHEMISTRY, BIOCHEMISTRY. Educ: Concordia Col, Moorhead, Minn, BA, 53; Univ Ariz, MS, 59, PhD(biochem), 63. Prof Exp: Jr scientist lipids, Hormel Inst, Univ Minn, 53-56; res assoc dept agr biochem, Univ Ariz, 59-63; RES CHEMIST CITRUS, FRUIT & VEG LAB, AGR RES SERV, USDA, 63- Mem: Am Oil Chem Soc. Res: Isolation and characterization of bitter products in grapefruit; synthesis and isolation of cycloporpenes; lipids and natural products. Mailing Add: Agr Res Serv USDA 600 Ave S NW Winter Haven FL 33880

NORDEN, ALLAN JAMES, b Perkins, Mich, Nov 27, 24; m 46; c 6. PLANT BREEDING, GENETICS. Educ: Mich State Univ, BS, 49, MS, 50; Iowa State Univ, PhD(plant breeding), 58. Prof Exp: Asst farm crops, Mich State Univ, 49-50, agr exten agent, 50-52, res instr crops & soils, 52-55; asst farm crops, Iowa State Univ, 55-58; from asst prof to assoc prof agron, PROF AGRON, UNIV FLA, 71- Honors & Awards: Golden Peanut Res Award, Nat Peanut Coun, 73. Mem: Fel Am Soc Agron; AAAS; Genetics Soc Am; Am Genetics Asn. Res: Plant breeding, genetics and physiological studies with peanuts. Mailing Add: Dept of Agron Univ of Fla Gainesville FL 32611

NORDEN, CARROLL RAYMOND, b Escanaba, Mich, May 20, 23; m 51; c 4. ZOOLOGY. Educ: Northern Mich Col Educ, AB, 48; Univ Mich, MS, 51, PhD(zool), 59. Prof Exp: Asst prof biol, Univ Southwestern La, 57-63; from asst prof to assoc prof zool, 63-71, chmn dept, 69-74, PROF ZOOL, UNIV WIS-MILWAUKEE, 71- Concurrent Pos: Fishery biologist, State Dept Conserv, Mich, 53, 54 & 56; fishery biologist, US Fish & Wildlife Serv, 58 & 59. Mem: Am Fisheries Soc; Am Soc Ichthyologists & Herpetologists. Res: Ichthyology; taxonomy of fishes; fishery biology. Mailing Add: Dept of Zool Univ of Wis Milwaukee WI 53201

NORDEN, JEANETTE JEAN, b Ovid, Colo, Apr 15, 48. NEUROBIOLOGY. Educ: Univ Calif, Los Angeles, BA, 70; Vanderbilt Univ, PhD(psychol), 75. Prof Exp: NIH FEL NEUROANAT, DUKE UNIV, 75- Concurrent Pos: Vis res assoc histochem, Dent Res Ctr, Univ NC, Chapel Hill, 76. Mem: Soc Neurosci. Res: Development of protein tracers for use in determining the afferent and efferent connections of cortical visual areas and in relating these connections to cortical cytoarchitecture. Mailing Add: Dept of Psychol Duke Univ Durham NC 27706

NORDEN, JOHN ALEXANDER, b Marosvasarhely, Austria-Hungary, Aug 31, 12; nat US. GEOLOGY, GEOPHYSICS. Educ: Tech & Econ Univ, Budapest, Hungary, DSc(econ geol), 40; State Geophys Inst, Budapest, cert, 39. Prof Exp: Asst geol, Tech & Econ Univ, Budapest, 36-39; geologist & geophysicist, Hungarian Am Oil Indust Co, Ltd, 40-47 & Danish Am Prospecting Co, Denmark, 47-48; PROF STRUCT GEOL & PETROL GEOPHYS, UNIV OKLA, 48- Mem: Soc Explor Geophys; Am Soc Photogram; Am Asn Petrol Geol; Am Geophys Union. Res: Detection of differential compaction and tectonic strain analysis for location of petroleum and gas traps. Mailing Add: Dept of Geol & Geophys Univ of Okla Norman OK 73069

NORDEN, ROGER LAWRENCE, conservation, biology, see 12th edition

NORDENG, STEPHAN C, b Chippewa Falls, Wis, May 24, 23; m 53; c 6. STRATIGRAPHY. Educ: Univ Wis, BS, 49, MS, 51, PhD(geol), 54. Prof Exp: Field geologist, Tex Petrol Co, 54-56; assoc prof geol, 57-71, PROF GEOL, MICH TECHNOL UNIV, 71- Concurrent Pos: Consult, Copper Range Co, Calumet & Hecla. Mem: Geol Soc Am; Paleont Soc. Res: Applications of statistics and computers in geology. Mailing Add: Dept of Geol Mich Technol Univ Houghton MI 49931

NORDHAUS, EDWARD ALFRED, b Chicago, Ill, Feb 23, 12; m 42; c 2. MATHEMATICS. Educ: Univ Chicago, SB, 34, SM, 35, PhD(math), 39. Prof Exp: Instr math exten dept, Univ Wis, 35-42; asst prof, Mich State Univ, 42-44; res engr, Boeing Aircraft Co, Wash, 44-45; from asst prof to assoc prof math, 46-63, PROF MATH, MICH STATE UNIV, 63- Mem: Am Math Soc; Math Asn Am. Res: Graph theory; calculus of variations. Mailing Add: Dept of Math Mich State Univ East Lansing MI 48832

NORDHEIM, LOTHAR WOLFGANG, b Munich, Ger, Nov 7, 99; nat US; wid; c 1. NUCLEAR SCIENCE. Educ: Univ Göttingen, PhD(physics), 23. Hon Degrees: DSc, Karlsruhe Tech Univ, 51; DSc, Purdue Univ, 62. Prof Exp: Rockefeller res fel, 27-28; lectr, Univ Göttingen, 28-33; res fel, Univ Paris, 34 & Univ Holland, 35; vis prof physics, Purdue Univ, 35-37; prof, Duke Univ, 37-56; physicist, Gen Atomic Div, Gen Dynamics Corp, 56-59, chmn theoret physics dept & sr res adv, 60-68; CONSULT, GEN ATOMIC CO, 68- Concurrent Pos: Vis prof, Ohio State Univ, 30, Moscow State Univ, 32 & Univ Heidelberg, 49; sect chief, Clinton Labs, Oak Ridge, Tenn, 43-45, dir physics div, 45-47; consult, Los Alamos Sci Lab, 50-52. Honors & Awards: Armed Forces Award of Merit, 47. Mem: AAAS; fel Am Phys Soc; fel Am Nuclear Soc; Fedn Am Scientists. Res: Quantum theory of matter; nuclear physics; reactor and neutron physics. Mailing Add: Gen Atomic Co PO Box 81608 San Diego CA 92138

NORDIN, ALBERT ANDREW, b McKeesport, Pa, Sept 7, 34; m 56; c 3. IMMUNOLOGY. Educ: Univ Pittsburgh, BS, 56, MS, 59, PhD(microbiol), 62. Prof Exp: From instr to assoc prof microbiol, Univ Notre Dame, 63-72; RES CHEMIST, NAT INST CHILD HEALTH & HUMAN DEVELOP, NIH, 72- Concurrent Pos: Vis scientist, Swiss Inst Exp Cancer Res, 69-70; mem adv panel regulatory biol prog, NSF, 73-75. Mem: AAAS; Geront Soc; Am Asn Immunol. Res: Cellular aspects of antibody formation of particular interest have been the role of antigen and the description of the competent cell. Mailing Add: Geront Res Ctr Baltimore City Hosps Baltimore MD 21224

NORDIN, GERALD LEROY, b Rockford, Ill, Aug 14, 44; m 64; c 2. INSECT PATHOLOGY. Educ: Univ Ill, Urbana, BS, 62, MS, 68, PhD(entom), 71. Prof Exp: Tech asst entom, Ill Natural Hist Surv, 64-66, res asst, 66-71; ASST PROF ENTOM, UNIV KY, 71- Mem: Entom Soc Am; Soc Invert Path; Sigma Xi. Res: Forest entomology and insect pathology; microbial control of forest insects. Mailing Add: S-225 Agr Sci Ctr-North Dept of Entom Univ of Ky Lexington KY 40506

NORDIN, IVAN CONRAD, b Lindsborg, Kans, May 25, 32; m 54; c 4. ORGANIC CHEMISTRY. Educ: Bethany Col, BS, 54; Univ Kans, PhD(org chem), 60. Prof Exp: From assoc res chemist to sr res chemist, 60-71, RES SCIENTIST ORG CHEM, PARKE, DAVIS & CO, 71- Concurrent Pos: Instr, Schoolcraft Col, 71- Mem: Am Chem Soc. Res: Claisen Rearr; heterocyclic chemistry; organic chemistry of nitrogen; medicinal chemistry of central nervous system. Mailing Add: Parke Davis & Co 2800 Plymouth Rd Ann Arbor MI 48106

NORDIN, JOHN HOFFMAN, b Chicago, Ill, Oct 11, 34; m 56; c 3. BIOCHEMISTRY. Educ: Univ Ill, BS, 56; Mich State Univ, PhD(biochem), 61. Prof Exp: NIH res fel biochem, Univ Minn, 62-65; asst prof, 65-71, ASSOC PROF BIOCHEM, 71- Concurrent Pos: NIH fel, Inst Biochem, Univ Lausanne, 71-72. Mem: Am Chem Soc; Am Soc Biol Chem. Res: Carbohydrate metabolism; structure of polysaccharides; chemistry of carbohydrates of biological interest; biochemistry of insect hibernation. Mailing Add: Dept of Biochem Univ of Mass Amherst MA 01002

NORDIN, PAUL, b Kansas City, Mo, Feb 12, 29; div; c 2. APPLIED PHYSICS. Educ: Univ Calif, Berkeley, AB, 56, MA, 57, PhD(physics), 61. Prof Exp: Telegrapher-clerk, Southern Pac RR, 48-52; eng draftsman, Northrop Aircraft, Inc, 52-53, radio-radar mechanic, 53; asst, Univ Calif, Berkeley, 56-61; sr scientist, Aeronutronic Div, Ford Motor Co, 61-63; mem tech staff, Aerospace Corp, 63-69; sect head, Missile Systs & Technol Dept, TRW Systs Group, 69-73, TECH STAFF, VULNERABILITY & HARDNESS LAB, TRW SYSTS & ENERGY, DEFENSE & SPACE SYSTS GROUP, REDONDO BEACH, 73- Mem: AAAS; Am Phys Soc; Marine Technol Soc; Undersea Med Soc; Sigma Xi. Res: Nuclear weapons effects; reentry vehicles; laser weapon effects; marine technology. Mailing Add: 1812 Pacific Ave PO Box 743 Manhattan Beach CA 90266

NORDIN, PHILIP, b Can, Mar 21, 22; nat US; m 47; c 2. BIOCHEMISTRY. Educ: Univ Sask, BSA, 49, MSc, 50; Iowa State Univ, PhD, 53. Prof Exp: Lectr biochem, Univ Toronto, 53-54; from asst prof to assoc prof, 54-69, PROF BIOCHEM & BIOCHEMIST, AGR EXP STA, KANS STATE UNIV, 69- Mem: Am Chem Soc; Brit Biochem Soc; Can Fedn Biol Soc. Res: Carbohydrate biosynthesis. Mailing Add: Dept of Biochem Kans State Univ Manhattan KS 66502

NORDIN, VIDAR JOHN, b Ratansbyn, Sweden, June 28, 24; nat Can; m 47; c 2. FOREST PATHOLOGY. Educ: Univ BC, BA, 46, BScF, 47; Univ Toronto, PhD(forest path), 51. Prof Exp: Asst forest path, Forest Path Lab, Can Dept Forestry, Toronto, 47-49, officer in chg, Fredericton, NB, 49-51, Calgary, Alta, 52-57, assoc dir forest biol, 58-64, prog coordr, 65-71; PROF FOREST BIOL & DEAN FAC FORESTRY & LANDSCAPE ARCHIT, UNIV TORONTO, 71- Concurrent Pos: Chmn bd, Algonquin Forestry Authority Corp, 74-; mem, Prov Parks Adv Coun, 75- Mem: NAm Forestry Comn; Int Poplar Comn; Int Union Forest Res Orgns; Soc Am Foresters; Can Inst Forestry (vpres, 65-66, pres, 67). Res: Forest tree diseases; forestry education; research administration. Mailing Add: Fac of Forestry & Landscape Archit Univ of Toronto Toronto ON Can

NORDLANDER, JOHN ERIC, b Schenectady, NY, July 3, 34; m 65; c 1. ORGANIC CHEMISTRY. Educ: Cornell Univ, AB, 56; Calif Inst Technol, PhD(chem), 61. Prof Exp: From instr to assoc prof chem, 61-75, PROF CHEM, CASE WESTERN RESERVE UNIV, 75- Mem: Am Chem Soc; The Chem Soc. Res: Organic reaction mechanisms; synthetic organic chemistry. Mailing Add: Dept of Chem Case Western Reserve Univ Cleveland OH 44106

NORDLIE, BERT EDWARD, b Denver, Colo, July 21, 35; m 57; c 2. GEOCHEMISTRY, GEOLOGY. Educ: Univ Colo, BA, 60, MS, 65; Univ Chicago, PhD(geochem, petrol), 67. Prof Exp: Asst prof geol, Univ Ariz, 67-71, assoc prof geosci & chief scientist, 71-73; PROF EARTH SCI & CHMN DEPT, IOWA STATE UNIV, 74- Mem: AAAS; Geochem Soc; Am Inst Chemists; Int Asn Volcanology & Chem Earth's Interior; Am Geophys Union. Res: Petrology; volcanology, especially magnetic gases. Mailing Add: Dept of Earth Sci Iowa State Univ Ames IA 50010

NORDLIE, FRANK GERALD, b Willmar, Minn, Jan 23, 32; m 60; c 2. ZOOLOGY. Educ: St Cloud State Col, BS, 54; Univ Minn, MA, 58, PhD(zool), 61. Prof Exp: Asst prof zool, 61-66, ASSOC PROF ZOOL, UNIV FLA, 66- Mem: Am Soc Limnol & Oceanog; Am Soc Zool. Res: Energetics of natural aquatic communities; physiological adaptations to thermal stress. Mailing Add: Dept of Zool Univ of Fla Gainesville FL 32611

NORDLIE, ROBERT CONRAD, b Willmar, Minn, June 11, 30; m 59; c 3. BIOCHEMISTRY. Educ: St Cloud State Col, BS, 52; Univ NDak, MS, 57, PhD(biochem), 60. Prof Exp: Asst, Sch Med, Univ NDak, 55-58; Nat Cancer Inst fel, Inst Enzyme Res, Univ Wis, 60-62; Hill prof, 62-74, CHESTER FRITZ DISTINGUISHED PROF BIOCHEM, SCH MED, UNIV NDAK, 74- Concurrent Pos: Consult, Oak Ridge Assoc Univs. Mem: AAAS; Am Soc Biol Chem; Am Soc Microbiol; Am Chem Soc; Am Inst Nutrit. Res: Enzymology; intermediary metabolism of carbohydrates and Krebscycle intermediates; effects of hormones on enzymes; biological regulatory mechanisms; inorganic pyrophosphate metabolism; carbamyl phosphate metabolism; control of blood glucose levels. Mailing Add: Dept of Biochem Univ of NDak Sch of Med Grand Forks ND 58202

NORDMAN, CHRISTER ERIC, b Helsinki, Finland, Jan 23, 25; m 52; c 4. PHYSICAL CHEMISTRY. Educ: Finnish Inst Technol, BS, 49; Univ Minn, PhD(phys chem), 53. Prof Exp: Asst phys chem, Univ Minn, 49-53; res assoc physics, Inst Cancer Res, 53-55; from instr to assoc prof chem, 55-64, PROF CHEM, UNIV MICH, ANN ARBOR, 64- Concurrent Pos: NIH special fel, Oxford Univ, 71-72. Mem: AAAS; Am Phys Soc; Am Chem Soc; Am Crystallog Asn. Res: X-ray crystal structure analysis. Mailing Add: Dept of Chem Univ of Mich Ann Arbor MI 48104

NORDMANN, JOSEPH BEHRENS, b Decatur, Ill, Apr 24, 22. CHEMISTRY. Educ: Bowling Green State Univ, BS, 43; Univ Southern Calif, MS, 45. Prof Exp: Chemist, Am Cyanamid Corp, 44; instr chem, Compton Col, 47-50; PROF CHEM, LOS ANGELES VALLEY COL, 50- Concurrent Pos: Dir, Pac Chem Consults, 51-57. Honors & Awards: Medal, Mfg Chem Asn, 69. Mem: AAAS; Am Chem Soc. Res: Chemical education; how to communicate science to the general public. Mailing Add: Dept of Chem Los Angeles Valley Col Van Nuys CA 91401

NORDMEYER, FRANCIS R, b Kankakee, Ill, Feb 1, 40; m 61; c 4. INORGANIC CHEMISTRY. Educ: Wabash Col, BA, 61; Wesleyan Univ, MA, 64; Stanford Univ, PhD(chem), 67. Prof Exp: Asst prof chem, Univ Rochester, 67-72; asst prof, 72-74, ASSOC PROF CHEM, BRIGHAM YOUNG UNIV, 74- Mem: AAAS; Am Chem Soc. Res: Mechanisms of inorganic reactions; organic molecules in electron transfer reactions. Mailing Add: Dept of Chem Brigham Young Univ Provo UT 84602

NORDQUIST, PAUL EDGARD RUDOLPH, JR, b Washington, DC, Sept 7, 36. INORGANIC CHEMISTRY, ANALYTICAL CHEMISTRY. Educ: George Washington Univ, BS, 58; Univ Minn, PhD(inorg chem), 64. Prof Exp: Teaching asst inorg chem, Univ Minn, 58-60, 62-63; res chemist, Monsanto Co, 64-70; RES CHEMIST, US NAVAL RES LAB, 70- Concurrent Pos: Chemist, Nat Bur Stand, 58-59. Mem: Am Chem Soc; Sigma Xi. Res: Materials preparation and characterization; crystal growth; trace analysis. Mailing Add: Code 5221 US Naval Res Lab Washington DC 20390

NORDQUIST, ROBERT ERSEL, b Oklahoma City, Okla, Sept 10, 38; m 56; c 2. PATHOLOGY, VIROLOGY. Educ: Oklahoma City Univ, BA, 68; Univ Okla, PhD(med sci), 71. Prof Exp: Res assoc electron micros, Med Sch, Univ Okla, 62-66;

res assoc, Scripps Clin & Res Found, 66-67; res assoc, 67-71, ASST MEM CANCER, OKLA MED RES FOUND, 71-; ASST PROF PATH, SCH MED, UNIV OKLA, 71- Mem: Electron Micros Soc Am; Tissue Cult Asn; Am Asn Cancer Res. Res: The search for human tumor viruses, utilizing the electron microscope coupled with immunologic and virologic techniques. Mailing Add: Okla Med Res Found Cancer Sect 825 NE 13th St Oklahoma City OK 73100

NORDSCHOW, CARLETON DEANE, b Hampton, Iowa, June 7, 26; m 50; c 3. PATHOLOGY, BIOCHEMISTRY. Educ: Luther Col, Iowa, AB, 49; Univ Iowa, MD, 53, PhD(biochem), 64. Prof Exp: From instr to assoc prof path, Univ Iowa, 54-70, dir clin labs, 63-66; PROF PATH & CHMN DEPT CLIN PATH, SCH MED, IND UNIV, INDIANAPOLIS, 70- Concurrent Pos: Mem, Acad Clin Lab Physicians & Scientists; USPHS res grants, 63-65 & 66-68. Mem: AMA; Am Soc Clin Path; Am Chem Soc; Int Acad Path. Res: Comparative properties of polymers in health and disease. Mailing Add: Dept of Clin Path Ind Univ Med Ctr Indianapolis IN 46202

NORDSIECK, HERBERT (HENRY), physical chemistry, behavioral science, see 12th edition

NORDSIEK, FREDERIC WILLIAM, b New York, NY, June 5, 09; m 34. NUTRITION. Educ: Mass Inst Technol, SB, 31; NY Univ, MS, 59; Columbia Univ, PhD(nutrit), 61. Prof Exp: Res bacteriologist, Borden Co, NY, 31-34 & Sanoderm Co, Inc, 34-35; exec secy, NY Diabetes Asn, 35-38; assoc nutrit dir, Am Inst Baking, 38-43; asst dir res serv dept, Standard Brands, Inc, 43-51; asst secy res Comt, Am Cancer Soc, 51-55, exec officer res dept, 55-57, prog analyst, 57-59, consult, 59-61; chief grants & fels, Sloan- Kettering Inst, 61-64, vpres, 64-67; coordr res & dir grants mgt, St Lukes Hosp Ctr, 68-72; assoc sci dir, Coun Tobacco Res, USA, 72-74; ADJ PROF PUB HEALTH NUTRIT, SCH PUB HEALTH, UNIV NC, CHAPEL HILL, 75- Concurrent Pos: Instr, Sch for Inspectors, New York City Health Dept, 41; NIH trainee, 59-61; lectr, Columbia Univ, 59-70, adj prof, 68-; lectr, Syracuse Univ, 60 & NY Univ, 61-63; sci assoc, Mem Sloan-Kettering Cancer Ctr, 61-64, dir grants & contracts, 64-66; consult, USPHS, 65-68. Mem: AAAS; fel Am Pub Health Asn. Res: Nutrition; food technology; research administration; science writing. Mailing Add: 1306 Oaks Apts Burning Tree Dr Chapel Hill NC 27514

NORDSKOG, ARNE WILLIAM, b Two Harbors, Minn, Feb 21, 13; m 38; c 2. POULTRY SCIENCE. Educ: Univ Minn, BS, 37, MS, 40, PhD(animal breeding), 43. Prof Exp: Instr agr, Univ Alaska, 37-39; assoc prof animal insdust, Mont State Col, 43-45; assoc prof animal sci, 45-65, PROF ANIMAL SCI, IOWA STATE UNIV, 65- Honors & Awards: CPC-Int Res Award, Poultry Sci Asn, 72. Mem: Genetics Soc Am; Biomet Soc; Poultry Sci Asn. Res: Poultry breeding and immunogenetics; factors affecting selection for growth rate in swine. Mailing Add: Dept of Animal Sci Iowa State Univ Ames IA 50010

NORDSTROM, J DAVID, b Minneapolis, Minn, Sept 30, 37; m 59; c 3. ORGANIC POLYMER CHEMISTRY. Educ: Gustavus Adolphus Col, BA, 59; Univ Iowa, PhD(chem), 63. Prof Exp: Res chemist polymers, Archer Daniels Midland Co, Minn, 63-68; sr res scientist, 68-70, MGR POLYMER RES & DEVELOP, PAINT PLANT, FORD MOTOR CO, 70- Mem: Am Chem Soc. Res: Synthetic organic chemistry; thermoset polymer synthesis and evaluation; Development of polymers for automotive coatings. Mailing Add: 19225 Gainsborough Detroit MI 48223

NORDSTROM, JAMES WILLIAM, nutrition, animal science, see 12th edition

NORDSTROM, JON OWEN, b Kingsburg, Calif, July 22, 33; m 60; c 2. PHYSIOLOGY. Educ: Univ Calif, Davis, BS, 56, PhD(animal sci), 66. Prof Exp: Asst prof physiol, Rutgers Univ, 65-69; assoc prof poultry sci, Univ Ariz, 69-75; ASSOC POULTRY SCIENTIST, WASH STATE UNIV, 75- Mem: Poultry Sci Asn; Am Soc Zoologists; World's Poultry Sci Asn. Res: Avian physiology; environmental physiology; physiology of aging. Mailing Add: Western Wash Res & Exten Ctr Puyallup WA 98371

NORDSTROM, TERRY VICTOR, b Seattle, Wash, Oct 27, 42; m 67; c 2. SOLID STATE ELECTRONICS. Educ: Univ Wash, BS, 65; Stanford Univ, MS, 67, PhD(mat sci), 70. Prof Exp: MEM TECH STAFF METALL, SANDIA LABS, 70- Mem: Am Soc Metals; Sigma Xi. Res: Materials used in thick film microelectronic circuits; advaned materials forming technology; marine corrosion studies. Mailing Add: Div 5832 Sandia Labs Albuquerque NM 87115

NORDTVEDT, KENNETH L, b Chicago, Ill, Apr 16, 39; c 3. THEORETICAL PHYSICS. Educ: Mass Inst Technol, BS, 60; Stanford Univ, MS, 62, PhD(physics), 65. Prof Exp: Staff physicist instrumentation lab, Mass Inst Technol, 63-65; from asst prof to assoc prof physics, 65-70, PROF PHYSICS, MONT STATE UNIV, 70- Concurrent Pos: NASA res grants, 65-73; Sloan fel, 71-73. Mem: Fel Am Phys Soc. Res: Gravitation; relativity and cosmology. Mailing Add: Dept of Physics Mont State Univ Bozeman MT 59715

NORDYKE, ELLIS LARRIMORE, b Houston, Tex, June 20, 42. BIOCHEMISTRY, NEUROCHEMISTRY. Educ: Univ Houston, BS, 68, MS, 70, PhD(biophys sci), 72. Prof Exp: NIMH fel neurochem, Tex Res Inst Ment Sci, 72-74; ASST PROF BIOL, UNIV ST THOMAS, 74- Mem: Am Chem Soc; Sigma Xi. Res: Effect of ethanol on the central nervous system; diseases of amino acid metabolism; mental retardation. Mailing Add: Univ of St Thomas 3812 Montrose Blvd Houston TX 77006

NOREIKA, ALEXANDER JOSEPH, b Philadelphia, Pa, Feb 24, 35; m 63; c 2. SOLID STATE PHYSICS. Educ: Drexel Inst, BS, 58; Univ Reading, PhD(physics), 66. Prof Exp: From jr to sr physicist, Philco Sci Labs, Ford Motor Co, Pa, 59-62; SR PHYSICIST, WESTINGHOUSE RES CTR, 66- Mem: Am Phys Soc. Res: Electron microscopy study of defect structures in single crystals; investigation of extended miscibility ranges in III-IV-V compounds. Mailing Add: Westinghouse Res Ctr Beulah Rd Pittsburgh PA 15235

NORELL, JOHN REYNOLDS, b Hutchinson, Kans, June 25, 37; m 59; c 1. ORGANIC CHEMISTRY. Educ: Bethany-Nazarene Col, BS, 59; Purdue Univ, PhD(org chem), 63. Prof Exp: Res chemist, 64-66, group leader, 66-74, SECT SUPVR, PHILLIPS PETROL CO, 74- Concurrent Pos: NIH fel, Inst Org Chem, Munich, Ger, 63-64. Mem: Am Chem Soc. Res: Sulfur chemistry; sulfones; thietanes, sulfenes and sulfones; 1, 3-dipolar additions; strong acid systems such as organic reactions in liquid hydrogen fluoride; organic synthesis, mechanisms and spectroscopy; flame retardants; fertilizer chemistry and organic chemistry. Mailing Add: 4984 Princeton Dr Bartleville OK 74003

NOREM, JAMES HENRY, high energy physics, see 12th edition

NOREM, WINFRED LUTHER, b NDak, July 24, 10; m 41; c 1. PALYNOLOGY. Educ: NDak Agr Col, BSc, 32; Johns Hopkins Univ, PhD(plant physiol), 36. Prof Exp: Jr plant physiologist, US Forest Serv, 37-42; chemist, Kans Ord Plant, Ord Dept, US Dept Army, 42-44; res chemist, Cardox Corp, 44-46; res chemist agr chem, Calif Res Corp, Standard Oil Co Calif, 46-49; res geologist palynol, 49-59; stratig palynologist, Atlantic Richfield Oil Co, 59-68; tech info specialist, Lawrence Livermore Lab, 68-76; RETIRED. Mailing Add: Tech Info Dept Lawrence Livermore Lab Livermore CA 94550

NOREN, GERRY KARL, b Minneapolis, Minn, June 22, 42; m 61; c 2. ORGANIC POLYMER CHEMISTRY. Educ: Univ Minn, BA, 66; Univ Iowa, PhD(org chem), 71. Prof Exp: Res chemist, Archer-Daniels-Midland Co, 66-67; res assoc chem, Univ Ariz, 71-72; RES ASSOC CHEM, CALGON CORP, MERCK SHARP & DOHME RES LABS, 72- Mem: Am Chem Soc. Res: Synthesis of water soluble monomers and polymers, mostly polyelectrolytes; free radical solution polymerization in aqueous solution. Mailing Add: 316 College Park Dr Coraopolis PA 15108

NORFORD, BRIAN SEELEY, b Gidea Park, Eng, Sept 15, 32; Can citizen; m 62. GEOLOGY, PALEONTOLOGY. Educ: Cambridge Univ, BA, 55, MA, 59; Yale Univ, MSc, 56, PhD(geol), 59. Prof Exp: Paleontologist, Shell Oil Co, Can, 59-60; geologist, 60-67, head western paleont sect, 67-72, HEAD PALEONT SUBDIV, GEOL SURV CAN, 72- Mem: Can Soc Petrol Geologists; Am Paleont Soc; Brit Palaeont Asn; Int Palaeont Union; Geol Asn Can. Res: Lower Paleozoic stratigraphy; Ordovician and Silurian corals, brachiopods and trilobites. Mailing Add: Geol Surv of Can 3303 33rd St Calgary AB Can

NORICK, NANCY XAVIER, b Cleveland, Ohio. FOREST BIOMETRY, MATHEMATICAL STATISTICS. Educ: Univ Calif, Berkeley, BA, 64, MA, 69. Prof Exp: MATH STATISTICIAN, US FOREST SERV, PAC SOUTHWEST FOREST RANGE EXP STA, 64- Concurrent Pos: App comt mem, US Forest Serv, Inventory Work Comt on Forest Serv Manual, 72-73; consult, US Forest Serv, Region 6, Mountain Pine Beetle Prog, 71-72, Washington Off Timber Mgt on Alaska Forest Inventory, 72, Region 5, Stanislaus Nat Forest Inventory, 72-73, Southern Pine Beetle Prog, 75 & Univ Calif, USDA Douglas-Fir Tussock Moth Prog, 75. Mem: Am Statist Asn. Res: Sampling and experimental design and analysis for forest entomology research; forest inventory design; pattern recognition statistical and computer systems design; multistage and variable probability sampling theory and applications in forestry. Mailing Add: Pac Southwest Forest & Range Exp Sta PO Box 245 Berkeley CA 94701

NORIN, ALLEN JOSEPH, b Chicago, Ill, July 30, 44; m 69; c 2. CELL BIOLOGY, TRANSPLANTATION IMMUNOLOGY. Educ: Roosevelt Univ, BS, 67; Univ Houston, MS, 70, PhD(biol), 72. Prof Exp: USPHS fel & res assoc microbiol, Univ Chicago, 72-75; ASST PROF SURG, MICROBIOL & IMMUNOL, MONTEFIORE HOSP & MED CTR, ALBERT EINSTEIN COL MED, YESHIVA UNIV, 75- Concurrent Pos: Immunologist, Manning Lab, NIH Prog, Proj Lung Transplantation, Montefiore Hosp, 75- Mem: Am Soc Microbiol; Am Inst Biol Sci; Fedn Am Scientists. Res: Regulation of growth and differentiation of human cells; interaction of mitogens and carcinogens with human lymphoid cells; transplantation of skin lung and bone marrow; mechanisms of allograft tolerance. Mailing Add: Montefiore Hosp & Med Ctr C 416 Bronx NY 10467

NORINS, ARTHUR LEONARD, b Chicago, Ill, Dec 2, 28; m 54; c 4. DERMATOLOGY. Educ: Northwestern Tech Inst, BS, 51; Northwestern Univ, MS, 53, MD, 55; Am Bd Dermat, dipl, 61. Prof Exp: Asst prof, Stanford Univ, 61-64; assoc prof, 64-69, PROF DERMAT, IND UNIV, INDIANAPOLIS, 69- Concurrent Pos: Chief dermat, Riley Children's Hosp. Mem: Fel Am Col Physicians; Am Soc Dermatopath; Am Acad Dermat; Soc Invest Dermat; Soc Pediat Dermat. Res: Dermatopathology; photobiology. Mailing Add: Dept of Dermat Ind Univ Indianapolis IN 46202

NORK, WILLIAM EDWARD, b Shenandoah, Pa, Aug 13, 34; m 60; c 4. HYDROGEOLOGY. Educ: Columbia Univ, AB, 60; Univ Buffalo, MA, 61. Prof Exp: Asst geol, Univ Buffalo, 60-61; from asst to instr geol, Univ Ariz, 61-64; scientist, Hazelton Nuclear Sci Corp, 64-66; dep proj mgr & asst mgr hydrogeol sect isotopes, Teledyne Inc, Calif, 66-70, mgr, Teledyne Isotopes, Nev, 70-71; res assoc, Desert Res Inst, Univ Nev, Las Vegas, 71-72; CONSULT HYDROLOGIST, HYDRO-SEARCH, INC, 72- Concurrent Pos: Consult, 60-61. Mem: AAAS; Geol Soc Am; Am Water Resources Asn; Am Geophys Union. Res: Optimum use and development of water, especially ground-water resources and water quality assurance for public use. Mailing Add: Hydro-Search Inc 333 Flint St Reno NV 89502

NORLING, PARRY MCWHINNIE, b Des Moines, Iowa, Apr 17, 39; m 65; c 2. POLYMER CHEMISTRY. Educ: Harvard Univ, AB, 61; Princeton Univ, PhD(polymer chem), 64. Prof Exp: Res assoc oxidation of polymers, Princeton Univ, 64-65; res chemist, electrochem dept, Del, 65-69, res supvr, 69-71, from tech supt to prod supt, Memphis plant, 71-75, INDUST CHEM DEPT, EXP STA, E I DU PONT DE NEMOURS & CO, INC, WILMINGTON, 75- Mem: Am Chem Soc. Res: Kinetics and mechanism of the oxidation of polymers; polymerization kinetics. Mailing Add: Exp Sta 336 E I du Pont de Nemours & Co Inc Wilmington DE 19898

NORMAN, ALEX, b New York, NY, Aug 7, 23; m 47; c 3. RADIOLOGY. Educ: NY Univ, BA, 44; Chicago Med Sch, MD, 48. Prof Exp: Nat Cancer Inst fel, Bellevue Hosp, 52; radiologist, Beth Israel Hosp, New York, 55-56; assoc attend radiol, Hosp Joint Dis, New-York, 56-60, chief diag roentgenol, 60-66; asst prof clin radiol, Sch Med, NY Univ, 64-67; dir diag roentgenol, 66-70, DIR RADIOL, HOSP JOINT DIS & MED CTR, 70-, PROF RADIOL, MT SINAI SCH MED, 67- Concurrent Pos: Consult, Educ Dept, Div Voc Rehab, State Univ NY Syst, 61-; NIH grant, Hosp Joint Dis, Bethesda, 62-64; consult radiologist, St Vincent's Hosp & Med Ctr, New York, 70-; consult specialist, Health Ins Plan NY, 70- Mem: Fel NY Acad Med; fel Am Col Radiol. Res: Application of tomography and enlargement technique in the diagnosis of bone diseases. Mailing Add: Hosp for Joint Dis & Med Ctr 1919 Madison Ave New York NY 10035

NORMAN, AMOS, b Vienna, Austria, Nov 25, 21; m 46; c 4. MEDICAL PHYSICS. Educ: Harvard Univ, AB, 43; Columbia Univ, MA, 47, PhD(biophys), 50. Prof Exp: AEC fel, Columbia Univ, 50-51; res biophysicist, 51-54, asst prof, 53-58, assoc prof, 58-63, PROF RADIOL, UNIV CALIF, LOS ANGELES, 63- Mem: AAAS; Am Soc Photobiol; Am Soc Physicists in Med; Radiation Res Soc. Res: Cellular radiobiology; engineering in medicine. Mailing Add: Dept of Radiol Sci Univ of Calif Los Angeles CA 90024

NORMAN, ANDREA HAUSMAN, b Texas City, Tex, Dec 4, 46; m 69. INORGANIC CHEMISTRY. Educ: Rice Univ, BA, 69; Tufts Univ, PhD(chem), 72. Prof Exp: Res fel inorg chem, Univ Wash, 73; ASST PROF CHEM, SEATTLE PAC COL, 73- Mem: Am Chem Soc. Res: Reduction/oxidation and substitution reactions of the stable polyhedral hydroborate anions; synthesis of amino acid derivates of one of the anions and study of its chemical and structural properties. Mailing Add: Seattle Pac Col Seattle WA 98119

NORMAN, ANTHONY WESTCOTT, b Ames, Iowa, Jan 19, 38. BIOCHEMISTRY. Educ: Oberlin Col, BA, 59; Univ Wis, MS, 61; PhD(biochem), 63. Prof Exp: Res

NORMAN

assoc biochem, Univ Wis, 59-63; from asst prof to assoc prof, 63-72, PROF BIOCHEM, UNIV CALIF, RIVERSIDE, 72- Concurrent Pos: NIH res grants, 64-75, spec fel, 70-71 & career develop award, 71-76; Fulbright fel, 70-71. Mem: Endocrinol Soc; Am Fedn Clin Res; AAAS; Am Chem Soc; Am Soc Biol Chemists. Res: Mechanism of action of vitamin D related to calcium metabolism; ion transport; mode of action of steroid hormones. Mailing Add: Dept of Biochem Univ of Calif Riverside CA 92502

NORMAN, ARLAN DALE, b Westhope, NDak, Mar 26, 40; m 66. INORGANIC CHEMISTRY. Educ: Univ NDak, BS, 62; Ind Univ, PhD(chem), 66. Prof Exp: Res assoc chem, Univ Calif, Berkeley, 65-66; from asst prof to assoc prof, 66-74, PROF CHEM, UNIV COLO, BOULDER, 74- Concurrent Pos: Alfred P Sloan Found fel, 73-75. Res: Chemistry of hydrides of boron, silicon and phosphorus, especially boron nuclear magnetic resonance and mass spectroscopy. Mailing Add: Dept of Chem Univ of Colo Boulder CO 80302

NORMAN, ARTHUR GEOFFREY, b Birmingham, Eng, Nov 26, 05; nat US; m 33; c 2. PLANT BIOCHEMISTRY. Educ: Univ Birmingham, BSc, 25, PhD(biochem), 28; Univ Wis, MS, 32; Univ London, DSc(biochem), 33. Prof Exp: Res assoc, Univ Wis, 32; biochemist in charge biochem sect, Rothamsted Exp Sta, Eng, 33-37; prof soils & res prof exp sta, Iowa State Univ, 37-46; biochemist & div chief, Chem Corps Biol Labs, 46-52; prof bot, 52-76, dir bot gardens, 55-66, vpres res, 64-72, dir, Inst for Environ Quality, 72-76, EMER PROF BOT & EMER VPRES RES, UNIV MICH, ANN ARBOR, 76- Concurrent Pos: Ed, Monographs, Soc Agron, 46-56, ed, Advan in Agron, 49-69; mem policy adv bd, Argonne Nat Lab, 58-64; adv life sci to pres, Nat Acad Sci, 63-64; chmn div biol & agr, Nat Res Coun, 64-68; trustee, Univ Res Asn, 68-75, chmn bd, 73-75; trustee, Biol Sci Info Serv, 72- Mem: Fel Royal Inst Chem; Biochem Soc; Am Soc Plant Physiol; Am Soc Agron (pres, 57); Am Soc Microbiol. Res: Soil plant relationships; root physiology; soil microbiology; plant growth regulators; cell wall chemistry. Mailing Add: 3475 Woodland Rd Ann Arbor MI 48104

NORMAN, BILLY RAY, b Luverne, Ala, Feb 10, 35. SCIENCE EDUCATION. Educ: Troy State Col, BS, 57; Univ Ga, EdD(sci), 65. Prof Exp: Asst prof sci, Campbell Col, 65-66; ASSOC PROF SCI EDUC, TROY STATE UNIV, 66- Mem: Nat Sci Teachers Asn. Res: Concepts of teaching science. Mailing Add: Dept of Sci Educ Troy State Univ Troy AL 36081

NORMAN, CARL EDGAR, b Cokato, Minn, Feb 1, 31; m 62; c 1. STRUCTURAL GEOLOGY, ROCK MECHANICS. Educ: Univ Minn, Minneapolis, BA, 57; Ohio State Univ, MS, 59, PhD(geol), 67. Prof Exp: Geologist, Humble Oil & Refining Co, 59-62; from instr to asst prof geol, 65-71, ASSOC PROF GEOL, UNIV HOUSTON, 71- Concurrent Pos: NSF instr sci equip prog grant, 68-69; univ fac res support prog grant, 69. Mem: AAAS; Geol Soc Am; Int Soc Rock Mech. Res: Mechanism of failure in rocks; behavior of rocks under varying conditions of load. Mailing Add: Dept of Geol Univ of Houston Houston TX 77004

NORMAN, CHARLES, physiology, deceased

NORMAN, EDWARD, b New York, NY, Aug 7, 32; m 59; c 2. MATHEMATICS. Educ: City Col New York, BS, 54; Cornell Univ, PhD(math), 58. Prof Exp: Asst prof math, Mich State Univ, 58-61; res mathematician, Socony Mobil Oil Co, 61-64; asst prof math, Drexel Inst, 64-69; ASSOC PROF MATH, FLA TECHNOL UNIV, 69- Mem: Am Math Soc. Res: Analysis; stability of differential equations. Mailing Add: Dept of Math Fla Technol Univ PO Box 25000 Orlando FL 32816

NORMAN, EDWARD COBB, b BC, Can, Oct 5, 13; US citizen; m 49; c 3. PUBLIC HEALTH EDUCATION, PSYCHIATRY. Educ: Univ Wash, BS, 35; Univ Pa, MD, 40; Tulane Univ, MPH, 65. Prof Exp: Psychiatrist, USPHS, 43-46; clin instr psychiat, Univ Ill Col Med, 49-53; from asst prof to assoc prof clin psychiat, 53-64, PROF PSYCHIAT & PREV MED, DEPT TROP MED & PUB HEALTH, TULANE UNIV, 64-, DIR MENT HEALTH SECT, SCH PUB HEALTH & TROP MED, 67- Concurrent Pos: Pvt pract, Chicago, 49-53 & New Orleans, 53-; clin physician, Michael Reese Hosp, Chicago, 49-53 & Vet Admin Hosps, Gulfport, Miss & New Orleans, 53-64; consult, Southeast La Hosp, Mandeville & East La Hosp, Jackson, 53-60; on active staff, Sara Mayo Hosp, New Orleans, 58-64; sr vis physician, Charity Hosp La, New Orleans, 64-, co-dir, Inter-Univ Forum Educr Community Psychiat, Duke Univ, 67-; mem ad hoc grants rev comt, NIMH, 68-; consult, New Orleans City Police Dept, 70- & Orleans Parish Sch Syst, 71-; secy, Forum Improv Quality of Life, 74; mem, APA Task Force Eco-Psychiatry, 75. Mem: AAAS; Am Psychiat Asn; Am Acad Psychoanal; Am Pub Health Asn. Res: Evaluation of the education process; evaluation of mental health consultation. Mailing Add: 1430 Tulane Ave New Orleans LA 70112

NORMAN, ELIANE MEYER, b Lyon, France, Nov 15, 31; US citizen; m 58; c 2. BOTANY, TAXONOMY. Educ: Hunter Col, BA, 53; Wash Univ, MA, 55; Cornell Univ, PhD(plant taxon), 62. Prof Exp: Instr biol, Hobart Col, 55-56; instr natural sci, Mich State Univ, 59-60; asst prof bot, Rutgers Univ, 63-69; ASST PROF BIOL, STETSON UNIV, 70- Mem: Am Soc Plant Taxon; Asn Trop Biol; Int Asn Plant Taxon. Res: Taxonomic and cytological studies of the genus Buddleia. Mailing Add: 1620 Druid Rd Maitland FL 32751

NORMAN, FLOYD (ALVIN), b Hallettsville, Tex, July 31, 11; m 38; c 2. MEDICINE. Educ: Univ Tex, MD, 35; Am Bd Pediat, dipl, 40. Prof Exp: Clin prof pediat, Univ Tex Southwest Med Sch Dallas, 60-72; ASST DIR HEALTH & SCI AFFAIRS, REGION VI, DEPT HEALTH, EDUC & WELFARE, 72- Mem: AMA; Am Acad Pediat. Mailing Add: 11550 Wander Lane Dallas TX 75230

NORMAN, GEORGE RUSSEL, b Linden, Ind, Dec 9, 16; m 42; c 4. ORGANIC CHEMISTRY. Educ: Wabash Col, AB, 40; Johns Hopkins Univ, PhD(org chem), 44. Prof Exp: Instr math, Wabash Col, 39-40; instr chem, Johns Hopkins Univ, 42-44; supvr fundamental res, Lubrizol Corp, 44-50, dir chem res lab, 50-59, dir sales div, 59-68; consult, Norman & Assocs, 68-70; TECH DIR, WALLOVER OIL CO, 75- Mem: Am Chem Soc; Soc Automotive Eng; Com Develop Asn. Res: Synthetic organic chemistry; purine glycosides; chemistry of aliphatic esters of thiophosphoric acids; additives for fuels and lubricants; reclaiming of used lubricating oils. Mailing Add: Wallover Oil Co 21845 Drake Rd Strongsville OH 44136

NORMAN, HOWARD DUANE, b Liberty, Pa, Nov 4, 42; m 74; c 1. ANIMAL BREEDING. Educ: Pa State Univ, BS, 64, MS, 67; Cornell Univ, PhD(animal breeding), 70. Prof Exp: ANIMAL SCIENTIST, USDA, AGR RES SERV, ANIMAL PHYSIOL & GENETICS INST, BELTSVILLE AGR RES CTR, 70- Mem: Am Dairy Sci Asn. Res: Develop procedures to improve the accuracy of the national dairy sire and cow evaluation programs, thereby increasing the genetic capability of the United States population. Mailing Add: Beltsville Agr Res Ctr-East Beltsville MD 20705

NORMAN, JACK C, b Taunton, Mass, June 16, 38; m 64. NUCLEAR CHEMISTRY, RADIOCHEMISTRY. Educ: Univ NH, BS, 60; Univ Wis-Madison, PhD(phys chem), 65. Prof Exp: Instr chem, Univ Wash, 65-66; asst prof, Univ Ky, 66-68; asst prof, 68-71, ASSOC PROF ECOSYSTS ANAL, UNIV WIS-GREEN BAY, 71- Mem: AAAS; Am Chem Soc; Am Phys Soc. Res: Neutron activation analysis of geological and environmental materials; nuclear reaction studies; radionuclides in the environment. Mailing Add: Col of Environ Sci Univ of Wis Green Bay WI 54302

NORMAN, JAMES E, JR, mathematical statistics, see 12th edition

NORMAN, JAY HAROLD, b Wasco, Calif, Aug 25, 37; m 58; c 3. NUCLEAR PHYSICS. Educ: Tex Technol Col, BS, 61; Iowa State Univ, MS, 68, PhD(nuclear eng), 73. Prof Exp: Asst reactor supvr, Argonne Nat Lab, 61-62; head reactor physics, Ames Lab, 62-73; STAFF MEM, LOS ALAMOS SCI LAB, 73- Res: Prompt and delayed neutron spectroscopy; fission cross section measurements; on-line mass-separated isotopic analysis. Mailing Add: MS 674 PO Box 1663 Los Alamos Sci Lab Los Alamos NM 87544

NORMAN, JOHN JAMES, organic chemistry, see 12th edition

NORMAN, JOHN MATTHEW, b Virginia, Minn, Nov 27, 42; m 67. MICROMETEOROLOGY, AGRICULTURAL METEOROLOGY. Educ: Univ Minn, BS, 64, MS, 67; Univ Wis-Madison, PhD(soil sci), 71. Prof Exp: Asst prof, 72-75, ASSOC PROF METEOROL, PA STATE UNIV, UNIVERSITY PARK, 75- Concurrent Pos: Res fel, Dept Bot, Univ Aberdeen, Scotland, 71-72. Mem: Am Meteorol Soc; Am Soc Agron. Res: Measurement of turbulence and radiation fluxes near the ground and from aircraft for use in air pollution including development of instrumentation, study of surface energy budget, and mathematical modelling of plant growth. Mailing Add: 509 Deike Bldg Pa State Univ Dept of Meteorol University Park PA 16802

NORMAN, JOHN WILLIAM, b Kenova, WVa, Nov 7, 21. CHEMISTRY. Educ: Marshall Col, BS, 42; Ohio State Univ, PhD(chem), 46. Prof Exp: Instr chem, Ohio State Univ, 43-43, asst physics, 43-44, asst chem, 44-45; asst prof, 46-60, ASSOC PROF CHEM, MIAMI UNIV, 60- Mem: Am Chem Soc. Res: Kinetics of hydrolysis; reactions in mixed solvents; conductances of lead chloride in ethylene glycol-water mixtures. Mailing Add: Dept of Chem Hughes Labs Miami Univ Oxford OH 45056

NORMAN, OSCAR LORIS, b Crawfordsville, Ind, Apr 28, 25; m 46; c 2. ORGANIC CHEMISTRY. Educ: Wabash Col, AB, 47; Northwestern Univ, MS, 49; Purdue Univ, PhD(chem), 53. Prof Exp: Sr res chemist, Int Mineral & Chem Corp, 52-58, mgr food prod res, 58-60, mgr biochem & chem res, 60-61; mgr prod develop, 61-62, sect chief basic res, 62-64, MGR, BASIC RES, SUN OIL CO, 65- Mem: AAAS; Am Chem Soc. Res: Biochemistry. Mailing Add: Res & Develop Sun Oil Co Marcus Hook PA 19061

NORMAN, PHILIP SIDNEY, b Pittsburg, Kans, Aug 4, 24; m 55; c 3. IMMUNOLOGY, BIOCHEMISTRY. Educ: Wash Univ, MD, 51. Prof Exp: Intern, Barnes Hosp, St Louis, Mo, 51-52; asst resident, Vanderbilt Univ Hosp, 52-54; USPHS fel, Rockefeller Inst, 54-56; from instr to assoc prof, 56-74, PROF MED, JOHNS HOPKINS UNIV, 75-, HEAD CLIN IMMUNOL DIV, 71-, PHYSICIAN, HOSP, 59- Concurrent Pos: Head allergy serv & physician, Good Samaritan Hosp, Baltimore, Md, 71- Mem: Am Fedn Clin Res; Am Acad Allergy; Am Soc Clin Invest; Am Asn Immunol; NY Acad Sci. Res: Antigens of ragweed; hay fever; asthma. Mailing Add: Good Samaritan Hosp 5601 Loch Raven Blvd Baltimore MD 21239

NORMAN, REID LYNN, b Scott City, Kans, Feb 26, 44; m 67; c 2. NEUROENDOCRINOLOGY. Educ: Kans State Univ, BS, 66, MS, 68; Univ Kans, PhD(anat), 71. Prof Exp: Fel neuroendocrinol, Univ Calif, Los Angeles, 71-72; ASST SCIENTIST REPROD PHYSIOL, ORE REGIONAL PRIMATE RES CTR, 72-; ASST PROF ANAT, SCH MED, UNIV ORE, 73- Mem: Am Asn Anatomists; Soc Study Reprod; Endocrine Soc; Am Physiol Soc. Res: Anatomical and physiological regulation of anterior pituitary function by the central nervous system. Mailing Add: Ore Regional Primate Res Ctr 505 NW 185th Ave Beaverton OR 97005

NORMAN, RICHARD D, b Franklin, Ind, Feb 7, 27; m 51; c 2. DENTISTRY, ANALYTICAL CHEMISTRY. Educ: Franklin Col, AB, 50; Ind Univ, DDS, 58, MSD, 64. Prof Exp: Assoc anal chemist, Eli Lilly & Co, 50-54; res asst chem, 55-58, from instr to assoc prof dent mat, 58-73, PROF DENT MAT, SCH DENT, IND UNIV-PURDUE UNIV, INDIANAPOLIS, 73- Concurrent Pos: Mem comt dent terminol & chmn subcomt zinc phosphate cement, Am Nat Stand Inst; mem working group specifications for polycarboxylate cements, Joint Int Dent Fedn-Int Stand Orgn. Mem: Am Dent Asn; Int Asn Dent Res. Res: Dental materials, especially cements. Mailing Add: Ind Univ Sch of Dent 1211 W Michigan St Indianapolis IN 46202

NORMAN, ROBERT DANIEL, b New York, NY, Nov 3, 38; Can citizen; m 62; c 3. MATHEMATICS. Educ: Univ Toronto, BA, 60; Queens Univ, Ont, MA, 62; Univ London, PhD(math), 64. Prof Exp: Fel math, 64-65, asst prof, 65-71, ASSOC PROF MATH, QUEENS UNIV, ONT, 71- Mem: Can Math Cong. Res: Differential geometry; optimization. Mailing Add: Dept of Math Queens Univ Kingston ON Can

NORMAN, ROBERT ZANE, b Chicago, Ill, Dec 16, 24; m 52; c 2. MATHEMATICS. Educ: Swarthmore Col, AB, 49; Univ Mich, AM, 50, PhD, 54. Prof Exp: Instr math, Princeton Univ, 54-56; assoc prof, 56-66, PROF MATH, DARTMOUTH COL, 66- Mem: Am Math Soc; Math Asn Am. Res: Theory of graphs; combinatorial analysis; mathematical models in the social sciences. Mailing Add: Dept of Math Dartmouth Col Hanover NH 03755

NORMAN, ROGER ATKINSON, JR, b Danville, Ky, Oct 16, 46; m 71. PHYSIOLOGY, CHEMICAL ENGINEERING. Educ: Univ Miss, BS, 68, MS, 71, PhD(biomed eng), 73. Prof Exp: Res assoc physiol, 72-73, ASST PROF PHYSIOL & BIOPHYS, SCH MED, UNIV MISS, 73- Res: Cardiovascular physiology; hypertension. Mailing Add: Dept of Physiol & Biophys Univ of Miss Sch of Med Jackson MS 39216

NORMAN, VELLO, physical chemistry, see 12th edition

NORMAN, WESLEY P, b Marion, Ill, Aug 14, 28. DEVELOPMENTAL BIOLOGY, NEUROANATOMY. Educ: Southern Ill Univ, BA, 52, MA, 54; Univ Ill, PhD(anat), 67. Prof Exp: Asst histologist, Am Meat Inst Found, 62-64; instr anat, Col Med, Univ Ill, 66-67; asst prof, 67-72, ASSOC PROF ANAT, GEORGETOWN UNIV, 72- Concurrent Pos: Consult & lectr otolaryngol basic sci course, Armed Forces Inst Path, 67- & Walter Reed Inst Dent Res & Naval Dent Sch, Bethesda, Md, 68- Mem: Am Asn Anat; Am Soc Cell Biol; Am Soc Zool. Res: Regeneration of the forelimb of the adult newt, Diemictylus viridescens, specifically skeletal muscle regeneration; histochemistry; autoradiography of fibrillogenesis; electron microscopy; analysis of denervated and reinnervated muscle spindles. Mailing Add: Dept of Anat Georgetown Univ 3900 Reservoir Rd Washington DC 20007

NORMAN, WILLIAM HARVEY, b Hearne, Tex, Mar 22, 15; m 46; c 2. CELL PHYSIOLOGY, SCIENCE EDUCATION. Educ: Rice Univ, BA, 38; Tex A&M Univ, MS, 43; Univ Tex, PhD(zool), 60. Prof Exp: Asst prof biochem, Tex Western Col, 46-51; asst prof biol, Univ SC, 58-61; assoc prof, 61-65, dir, NSF Grad Sch Insts, 65-70, PROF BIOL, UNIV MISS, DIR SCI EDUC CTR, 70- Concurrent Pos: NSF & acad year inst grants, 62-, assoc proj dir res training & acad year study prog, NSF, Washington, DC, 69-70, comprehensive grant, 71-75. Mem: AAAS; Am Inst Biol Scientists; Nat Sci Teachers Asn. Res: Cytochemistry and cytology of insects; research and development in science module or mini-course teaching materials for seventh through twelfth grade teachers, especially in biology. Mailing Add: Sci Educ Ctr Univ of Miss Col of Lib Arts University MS 38677

NORMANDEAU, DONALD ARTHUR, b Manchester, NH, June 10, 35; m 56; c 3. AQUATIC ECOLOGY. Educ: St Anselms Col, AB, 57; Univ NH, MS, 61, PhD(zool), 63. Prof Exp: From aquatic biologist to res biologist, NH Fish & Game Dept, 57-60; asst prof biol, St Anselms Col, 63-69; CONSULT BIOLOGIST, NORMANDEAU ASSOCS, INC, 69- Mem: Am Fisheries Soc; Am Inst Biol Sci. Res: Fisheries biology; aquatic ecology; water sanitation biology. Mailing Add: Normandeau Assocs Inc Nashua Rd Bedford NH 03102

NORMANDIN, DIANE KILBOURNE, b Jersey City, NJ, July 25, 30. ELECTRON MICROSCOPY. Educ: NJ State Teachers Col, AB, 52; Univ Ill, Urbana, MS, 58, PhD(zool), 62. Prof Exp: Tech asst electrochem, Bell Tel Labs, NJ, 52-56; instr histol, 62-69, ASST PROF VET BIOL STRUCTURE, COL VET MED, UNIV ILL, URBANA, 69- Mem: AAAS; Am Asn Anat; Am Soc Zool. Res: Animal cytology; histochemical and electron microscopical investigations of differentiation of endodermally derived cells. Mailing Add: Col of Vet Med Univ of Ill Urbana IL 61801

NORMANDIN, ROBERT F, b Laconia, NH, July 22, 27; m 54; c 3. RADIATION BIOLOGY. Educ: St Anselm's Col, AB, 50; Univ NH, MS, 53; Ohio State Univ, PhD(zool), 59. Prof Exp: Res biologist, NH Fish & Game Dept, 53-54; cur path mus, Ohio State Univ, 56-59; res biologist, US Fish & Wildlife Serv, 59-60; ASSOC PROF BIOL, ST ANSELM'S COL, 60- Concurrent Pos: Consult pvt fishery, Ohio, 57-59; consult, NH Water Pollution Comn, 60-66; chmn, NH Comn Radiation Control, 62-; consult lectr, NH Civil Defense Orgn, 64-66; chmn, NH Legis Comn Prof Nursing, 66; mem, NH Adv Comprehensive Health Planning Coun, 69. Mem: AAAS; Am Inst Biol Scientists. Res: Water pollutional control, especially algal blooms; radiation biology, especially physiology and pathology of radio-sensitivity radio-ecology. Mailing Add: St Anselm's Col Manchester NH 03102

NORMANN, SIGURD JOHNS, b Cincinnati, Ohio, Oct 24, 35; m 65; c 2. PATHOLOGY. Educ: Univ Wash, MD, 60, PhD(path), 66. Prof Exp: Intern surg, Univ Calif, San Francisco, 60-61; asst prof, 68-72, ASSOC PROF PATH, UNIV FLA, 72- Concurrent Pos: Consult path, Vet Admin Hosp, Gainesville, 68-; Nat Insts Allergy & Infectious Dis res career develop award, Univ Fla, 70-75. Honors & Awards: Outstanding Achievement Res & Develop Cert, Asst Secy Defense, 68. Mem: AAAS; AMA; Reticuloendothelial Soc (secy, 70-73); Am Soc Exp Path; Int Acad Path. Res: Mechanisms of disease; immunology; infectious diseases. Mailing Add: Dept of Path Univ of Fla Gainesville FL 32601

NORMARK, WILLIAM RAYMOND, b Seattle, Wash, Jan 21, 43; m 67. OCEANOGRAPHY, MARINE GEOLOGY. Educ: Stanford Univ, BS, 65; Univ Calif, San Diego, PhD(oceanog), 69. Prof Exp: Res oceanogr, Scripps Inst Oceanog, 69-70; asst prof geol & oceanog, Univ Minn, Minneapolis, 70-74; GEOLOGIST, PAC-ARCTIC BR MARINE GEOL, US GEOL SURV, 74- Mem: AAAS; Geol Soc Am; Am Geophys Union; Soc Explor Geophys. Res: Continental margin sedimentation, particularly deep-sea turbidites, growth patterns of deep-sea fans, structure and history of continental margins and evolution of lithospheric plate boundaries; erosion of deep-sea sediment. Mailing Add: US Geol Surv 345 Middlefield Rd Menlo Park CA 94025

NORMENT, BEVERLY RAY, b Whiteville, Tenn, July 23, 41; m 64; c 1. ENTOMOLOGY. Educ: Memphis State Univ, BS, 64, MS, 66; Miss State Univ, PhD(entom), 69. Prof Exp: Asst zool, Memphis State Univ, 64-66, instr, 66; from res asst to asst prof entom, 66-74, ASSOC PROF ENTOM, MISS STATE UNIV, 74- Mem: Am Soc Trop Med & Hygiene; Int Soc Toxinology; Entom Soc Am; Am Mosquito Control Asn. Res: Medical entomology; bioassay techniques; enzyme assays; toxinological studies; biological control. Mailing Add: Drawer E M Miss State Univ Mississippi State MS 39762

NORMENT, HILLYER GAVIN, JR, physical chemistry, see 12th edition

NORMINTON, EDWARD JOSEPH, b Hensall, Ont, Sept 8, 38; m 63; c 2. APPLIED MATHEMATICS. Educ: Univ Western Ont, BA, 61, MA, 62; Univ Toronto, PhD(fluid dynamics), 65. Prof Exp: Asst prof math, 65-69, ASSOC PROF MATH, CARLETON UNIV, 69- Mem: Can Math Cong. Res: Diffusion and convection of heat and circulation in potential flow fields. Mailing Add: Dept of Math Carleton Univ Ottawa ON Can

NORNES, HOWARD ONSGAARD, b Winger, Minn, Apr 27, 31; m 58; c 3. NEUROBIOLOGY, DEVELOPMENTAL BIOLOGY. Educ: Concordia Col, BA, 53; Purdue Univ, Lafayette, MS, 63, PhD(biol), 71. Prof Exp: Teacher biol, Richfield Pub Sch, 58-66; from instr to lectr anat, Purdue Univ, Lafayette, 66-72; LECTR ANAT, COLO STATE UNIV, 72- Mem: Am Soc Zoologists; Soc Neurosci. Res: Vertebrate Neurogenesis. Mailing Add: Dept of Anat Colo State Univ Ft Collins CO 80521

NORNES, SHERMAN BERDEEN, b Winger, Minn, Jan 10, 29; m 53; c 2. SURFACE PHYSICS. Educ: Concordia Col, Moorhead, Minn, BA, 51; Univ NDak, MS, 56; Wash State Univ, PhD(physics), 65. Prof Exp: Res engr, Rocketdyne Inc Div, NAm Aviation, Inc, 56-59; ASSOC PROF PHYSICS, PAC LUTHERAN UNIV, 59-61, 65-, CHMN DEPT, 67- Concurrent Pos: Physicist, Lawrence Livermore Lab, 74-75. Mem: Am Vacuum Soc; Am Asn Physics Teachers. Res: X-ray photo electron spectroscopy studies of surface reactions of Actinides; physics of the interaction of spectroscopically pure gases with ultra clean metal surfaces; frequency dependence of the polarizability of matter. Mailing Add: Dept of Physics Pac Lutheran Univ Tacoma WA 98447

NORONHA, FERNANDO M OLIVEIRA, b Portugal, Feb 10, 24; m 60; c 2. VIROLOGY. Educ: Lisbon Tech Univ, DVM, 49. Prof Exp: WHO fel, Col Med, Univ Montpellier, 50-52; French Acad bursary, Pasteur Inst, Paris, 53-55; researcher, Virus Res Inst, Eng, 56; researcher, Nat Inst Sch Higher Vet Med, Lisbon Tech Univ, 59-63; assoc prof virol, 64-66, PROF VIROL, NY STATE VET COL, CORNELL UNIV, 66- Concurrent Pos: Portuguese Govt scholar, Animal Virus Inst, Univ Tübingen, 59-63; resident, Portuguese Inst Vet Res, 59-63. Honors & Awards: Chevalier de Merite Agricole, Fr Govt, 50. Mem: Am Soc Microbiol; Portuguese Soc Vet Med. Res: Virus oncology. Mailing Add: NY State Col of Vet Med Ithaca NY 14853

NORR, SIGMUND CARL, b Cleveland, Ohio, July 20, 45; m 68. DEVELOPMENTAL BIOLOGY. Educ: Univ Pittsburgh, BS, 67; Case Western Reserve Univ, PhD(cell & develop biol), 72. Prof Exp: Res biologist, Univ Calif, San Diego, 71-74; ASST PROF BIOL, GEORGETOWN UNIV, 74- Concurrent Pos: Fel, Am Cancer Soc, 72. Mem: AAAS; Am Soc Zoologists; Soc Develop Biol; Soc Cell Biol. Res: Characterizing the cell and tissue interactions and other parameters involved in the regulation of neural crest cytodifferentiation into sympathetic neurons, pigment cells and calcitonin cells. Mailing Add: Dept of Biol Georgetown Univ Washington DC 20057

NORRDIN, ROBERT W, b Brooklyn, NY, Oct 2, 37; m 63; c 3. VETERINARY PATHOLOGY. Educ: Brooklyn Col, BS, 58; Cornell Univ, DVM, 62, PhD(vet path), 69. Prof Exp: Gen pract vet med, Flemington, NJ & Locke, NY, 62-65; asst prof vet med, NY State Vet Col, Cornell Univ, 65-66, NIH trainee nutrit path, 66-69; ASSOC PROF VET PATH, DEPT PATH & COLLABR, RADIOL HEALTH LAB, COL VET MED & BIOMED SCI, COLO STATE UNIV, 69- Concurrent Pos: NIH & Med Res foreign fel, Res Unit 18, Hopital Lariboisiere, Paris, 75-76. Mem: Am Col Vet Pathologists; Am Vet Med Asn. Res: Multifaceted studies of nutritional and metabolic bone diseases in animals; metabolic studies; bone density and composition; histomorphometric evaluation of bone and pertinent endocrine organs. Mailing Add: Dept of Vet Path Colo State Univ Ft Collins CO 80521

NORRELL, STEPHEN ANDREW, microbiology, see 12th edition

NORRIS, A R, b Meadow Lake, Sask, May 18, 37; m 60. INORGANIC CHEMISTRY, PHYSICAL CHEMISTRY. Educ: Univ Sask, BE, 58, MSc, 59; Univ Chicago, PhD(chem), 62. Prof Exp: Res aasoc, Univ Chicago, 60; Nat Res Coun fel, Univ Col London, 62-64; asst prof chem, 64-68, ASSOC PROF CHEM, QUEEN'S UNIV, ONT, 68- Concurrent Pos: Vis scholar chem, Stanford Univ, 71-72. Mem: The Chem Soc; Can Inst Chem. Res: Kinetics and mechanisms of reactions of transition metal complexes; redox reactions of coordinated ligands in transition metal complexes; mechanisms of formation of sigma-complexes of polynitroaromatics. Mailing Add: 55 Jane Ave Kingston ON Can

NORRIS, ALBERT STANLEY, b Sudbury, Ont, July 14, 26; m 50; c 3. PSYCHIATRY. Educ: Univ Western Ont, MD, 51. Prof Exp: Fel, Harvard Med Sch, 55-56; instr psychiat, Med Sch, Queen's Univ, Ont, 56-57; from asst prof to assoc prof, Col Med, Univ Iowa, 57-64; assoc prof, Med Sch, Univ Ore, 64-65; from assoc prof to prof, Col Med, Univ Iowa, 65-72; PROF PSYCHIAT & CHMN DEPT, SCH MED, SOUTHERN ILL UNIV, 72- Mem: AMA; fel Am Psychiat Asn. Res: Anatomical and physiological traits which predispose the development of mental illness; prenatal influences affecting intellectual and emotional development; capillary morphology in mental illness; psychosomatic obstetrics and gynecology; investigations of the efficacy of LSD 25 in the treatment of sexual deviation. Mailing Add: Dept of Psychiat Southern Ill Univ Sch of Med Springfield IL 62708

NORRIS, ANDREW EDWARD, b Santa Rosa, Calif, Jan 13, 37. NUCLEAR CHEMISTRY. Educ: Univ Chicago, SB, 58; Wash Univ, PhD(chem), 63. Prof Exp: Res assoc chem, Wash Univ, 63-64 & Brookhaven Nat Lab, 64-66; staff mem, 66-75, ASST RES, OFF OF DIR, LOS ALAMOS SCI LAB, 75- Concurrent Pos: Vis guest chemist, Lawrence Berkeley Lab, 73-74. Mem: Am Chem Soc; Am Phys Soc; AAAS. Res: Fission yields; heavy ion reaction; pion-induced spallation. Mailing Add: Los Alamos Sci Lab Los Alamos NM 87545

NORRIS, BILL EUGENE, b Ft Recovery, Ohio, Jan 12, 30; m 50; c 3. BIOLOGY. Educ: Ball State Univ, BS, 60, MS, 65, EdD(biophys educ), 70. Prof Exp: Teacher high schs, Ohio & Ind, 60-68; teaching fel biol, Ball State Univ, 68-70; ASST PROF BIOL, WESTERN OHIO BR CAMPUS, WRIGHT STATE UNIV, 70- Mem: Nat Asn Biol Teachers. Res: Biological survey of Grand Lake St Marys; self concepts of science teachers. Mailing Add: Dept of Biol Wright State Univ Celina OH 45822

NORRIS, CAROL LEE, b Memphis, Tenn, Sept 24, 46. CHEMICAL PHYSICS. Educ: Whitman Col, BA, 68; Univ Ill, MS, 70, PhD(phys chem), 73. Prof Exp: SR CHEMIST, 3M CO, 73- Res: New imaging systems for the general field of radiography. Mailing Add: Cent Res Lab Bldg 201-3E 3M Ctr 3M Co St Paul MN 55101

NORRIS, CHARLES HAMILTON, b Cornwall, Ont, Oct 10, 14; nat US; m 39, 65; c 2. PHYSIOLOGY. Educ: Hamilton Col, BS, 36; Princeton Univ, PhD(biol), 39. Prof Exp: From instr to assoc prof biol, 39-65, assoc chmn dept, 66-70, PROF BIOL, UNIV COLO, BOULDER, 65- Res: Physical properties of cells; kidney function in native Colorado rodents; neoteny in native Colorado salamanders; history of biology. Mailing Add: Dept of Biol Univ of Colo Boulder CO 80302

NORRIS, DALE MELVIN, JR, b Essex, Iowa, Aug 19, 30; m 51. BIOLOGY. Educ: Iowa State Univ, BS, 52, MS, 53, PhD(zool, plant path), 56. Educ: Asst entomologist, Agr Exp Sta, Univ Fla, 56-57; from asst prof to assoc prof entom, 58-66, PROF ENTOM, UNIV WIS-MADISON, 66- Concurrent Pos: NSF-Soc Am Foresters vis lectr, 71. Honors & Awards: Foundrs Mem Award, Entom Soc Am, 75. Mem: AAAS; Entom Soc Am; Int Soc Neurochem; Am Soc Neurochem; Biophys Soc. Res: Insect transmission of microbes; systemic chemical action; insect orientations to plants; symbiosis; chemoreception; neurobiology. Mailing Add: 642 Russell Labs Univ of Wis Madison WI 53706

NORRIS, DANIEL HOWARD, b Toledo, Ohio, Dec 29, 33; m 58. BOTANY. Educ: Mich State Univ, BS, 54; Univ Tenn, PhD(bot), 64. Prof Exp: Instr bot, Univ Tenn, 59-60; asst prof biol, Cent Methodist Col, 61-63; assoc prof, Catonsville Community Col, 64-67; asst prof biol, 67-74, ASSOC PROF BOT, HUMBOLDT STATE COL, 74- Mem: AAAS; Am Bryol & Lichenol Soc; Philos Sci Asn; Am Inst Biol Sci; Bot Soc Am. Res: Bryogeography and taxonomy of Dominican Republic, Newfoundland, and California; bryoecology of Great Smoky Mountains National Park; phytogeography. Mailing Add: Dept of Bot Humboldt State Col Arcata CA 95521

NORRIS, DAVID OTTO, b Ashtabula, Ohio, Oct 1, 39; m 66. COMPARATIVE ENDOCRINOLOGY. Educ: Baldwin-Wallace Col, BS, 61; Univ Wash, PhD(fish thyroid), 66. Prof Exp: Asst prof biol, 66-70, ASSOC PROF BIOL, UNIV COLO, BOULDER, 70- Mem: Herpetologists League; AAAS; Am Soc Zoologists. Res: Aspects of endocrinology of freshwater fishes and urodele amphibians. Mailing Add: Dept of Environ Pop & Organismic Biol Univ of Colo Boulder CO 80302

NORRIS, DONALD EARL, JR, b Hammond, Ind, Oct 16, 40; m 64; c 1. PARASITOLOGY. Educ: Ind State Univ, Terre Haute, BS, 63; Tulane Univ, MS, 66, PhD(parasitol), 69. Prof Exp: Asst prof, 70-75, ASSOC PROF BIOL, UNIV SOUTHERN MISS, 75- Mem: AAAS; Am Soc Parasitol; Am Soc Trop Med & Hyg; Royal Soc Trop Med & Hyg; Wildlife Dis Asn. Res: Systematics and life histories of parasites; helminthology. Mailing Add: Dept of Biol Univ of Southern Miss Hattiesburg MS 39401

NORRIS, DONALD KRING, b Cobourg, Ont, July 31, 24. GEOLOGY. Educ: Univ Toronto, BA, 47, MA, 49; Calif Inst Technol, PhD(geol), 53. Prof Exp: Geologist struct geol, Geol Surv Can, 53-69; MEM STAFF, INST SEDIMENT & PETROL GEOL, 69- Honors & Awards: Coleman Gold Medal, 47. Mem: Geol Soc Am; NY Acad Sci; Royal Astron Soc Can. Res: Geology of fuels; analysis of structural types; Mesozoic stratigraphy of Canadian Cordillera. Mailing Add: Inst of Sediment & Petrol Geol 3303 33rd St NW Calgary AB Can

NORRIS, DOUGLAS M, JR, applied mechanics, see 12th edition

NORRIS, EUGENE MICHAEL, b New York, NY, July 4, 38; m 69. MATHEMATICS. Educ: Univ SFla, BA, 64; Univ Fla, PhD(math), 69. Prof Exp: Electronics technician, Missile Test Proj, RCA Serv Co, Inc, Patrick AFB, Fla, 57-61 & Goddard Space Flight Ctr, NASA, Md, 61-62; instr math, Univ Fla, 65-66, 68-69; asst prof, WVa Univ, 69-72; ASST PROF MATH, UNIV SC, 72- Mem: AAAS; Math Asn Am; Am Math Soc. Res: Algebra; automata theory; binary relations; memory modelling; applications of n-ary relations. Mailing Add: Dept of Math & Comput Sci Univ of SC Columbia SC 29208

NORRIS, FLETCHER R, b Brownsville, Tenn, Sept 2, 34; m 60; c 2. MATHEMATICS. Educ: Vanderbilt Univ, BA, 56; George Peabody Col, MA, 62, PhD(math), 68. Prof Exp: Teacher high sch, Tenn, 59-62; instr eng math, Vanderbilt Univ, 62-64, 66-68, asst prof, 68-70; lectr math, Univ NC, Wilmington, 70-71; Vanderbilt fel, Fla State Univ, 71-72; ASSOC PROF MATH, UNIV NC, WILMINGTON, 72- Mem: Nat Coun Teachers Math; Math Asn Am. Res: Application of computers and computing to mathematics and statistics. Mailing Add: Dept of Math Univ of NC Wilmington NC 28401

NORRIS, FORBES HOLTEN, JR, b Richmond, Va, May 1, 28; m 55; c 3. NEUROLOGY, NEUROPHYSIOLOGY. Educ: Harvard Univ, BS, 49, MD, 55. Prof Exp: Guest worker electromyography, NIH, 54-55; intern surg, Johns Hopkins Hosp, 55-56; med officer, NIH, 56-61; from sr instr to asst prof neurol, Univ Rochester, 61-63, assoc prof & actg chmn div, 63-66; ASSOC DIR INST NEUROL SCI, PAC MED CTR, 66- Concurrent Pos: USPHS spec fel, 61-63; ad hoc consult, NIH, 62; sr res fel, Inst Neurol, Univ London, 66; adj prof neurol, Univ of the Pac, 70- Mem: AAAS; AMA; Am Acad Neurol; Am Asn Electromyog & Electrodiag; Am Neurol Asn. Res: Clinical and experimental studies of the function of normal and diseased nervous system. Mailing Add: Pac Med Ctr Res Lab R 513 PO Box 7999 San Francisco CA 94120

NORRIS, FORREST HARVEY, b Duluth, Minn, Oct 30, 20; m 44; c 4. POLYMER CHEMISTRY. Educ: Clark Univ, AB, 43; Univ Mass, PhD(chem), 56. Educ: Res chemist polymers, Monsanto Chem Co, 46-50; res chemist, Shawinigan Resins Corp, 50-55, from res group leader to res sect leader polymers & adhesives, 59-63, scientist, 63-65; scientist plastic prod & resins div, Monsanto Co, 65-71; assoc prof chem, Springfield Col, 71-73; ASST PROJ ENG, FUEL CELL MAT, UNITED TECHNOLOGIES CORP, 73- Mem: Am Chem Soc; AAAS; Am Inst Chemists; NY Acad Sci; Sigma Xi. Res: Polymerization reactions; application of polymers in paint, adhesive, paper and textiles; physical structure of polymers; development of plastics and composites for fuel cells and aerospace materials. Mailing Add: 35 Monson Rd Wilbraham MA 01095

NORRIS, FRANK ALLEN, organic chemistry, medicinal chemistry, see 12th edition

NORRIS, FRANK ARTHUR, b Pittsburgh, Pa, July 2, 13; m 39; c 1. BIOCHEMISTRY. Educ: Univ Pittsburgh, BS, 35, PhD(biochem), 39. Prof Exp: Asst chem, Univ Pittsburgh, 36-39; Rockefeller fel, Univ Minn, 39-41; res chemist, Gen Mills, Inc, 41-44; head oil mill res, Swift & Co, 44-64, head edible oil res, 64-66; sr scientist, 66-69, dir res admin, 69-72, ASSOC MGR, EDIBLE OIL PROD, RES & DEVELOP DIV, KRAFTCO CORP, 72- Mem: Am Oil Chemists Soc (vpres, 72-73, pres, 73-74); AAAS; Inst Food Technol; Am Chem Soc. Res: Synthetic glycerides; fatty acid chemistry; oilseed processing; vegetable proteins; edible fats and oil processing. Mailing Add: 801 Waukegan Rd Glenview IL 60025

NORRIS, FREDERICK HAMILTON, botany, see 12th edition

NORRIS, GAIL ROYAL, b Coshocton, Ohio, Jan 17, 19; m 44; c 3. RADIATION BIOPHYSICS. Educ: Ohio Univ, BS, 41; Ohio State Univ, MS, 47, PhD(zool), 50. Prof Exp: Asst zool, Ohio State Univ, 46, asst instr, 47-49; instr, Denison Univ, 49-50; from assoc prof to prof, Mt Union Col, 50-58; PROF ZOOL, DENISON UNIV, 59- Concurrent Pos: Am Physiol Soc, res fel, 57-58, 60; NIH res grants, 57-60; res assoc, Ohio State Univ, 65-66; Oak Ridge Nat Lab, Tenn, 71. Mem: Radiation Res Soc; AAAS; Am Nuclear Soc. Res: Cellular physiology; red blood cell metabolism; neutron activation analysis of trace elements; neutron activation analysis of human lungs. Mailing Add: Dept of Biol Denison Univ Granville OH 43023

NORRIS, GEOFFREY, b Romford, Eng, Aug 6, 37; m 58; c 4. GEOLOGY, PALEONTOLOGY. Educ: Cambridge Univ, BA, 59, MA, 62, PhD(geol), 64. Prof Exp: Sci officer, NZ Geol Surv, 61-64; fel geol, McMaster Univ, 64-65; sr res scientist, Res Ctr, Pan Am Petrol Corp, Okla, 65-67; from asst prof to assoc prof, 67-74, PROF GEOL, UNIV TORONTO, 74- Concurrent Pos: Res assoc, Royal Ont Mus, Toronto, 68-; mem, Int Comn Palynology. Mem: Am Asn Stratig Palynologists (pres, 71-72); Geol Asn Can; Paleont Soc; Brit Palaeont Asn. Res: Palynology; stratigraphic and paleoecologic applications; taxonomy of dinoflagellate cysts. Mailing Add: Dept of Geol Univ of Toronto Toronto ON Can

NORRIS, JAMES NEWCOME, IV, b Santa Barbara, Calif, Sept 8, 42. MARINE PHYCOLOGY. Educ: San Francisco State Col, BA, 68, MA, 71; Univ Calif, Santa Barbara, PhD(marine bot), 75. Prof Exp: Asst cur, Gilbert M Smith Herbarium Hopkins Marine Sta, Stanford Univ, 69-70; assoc, Dept Biol Sci, Univ Calif, Santa Barbara, 71-72; sta dir & resident marine biologist, Marine Biol Lab, Puerto Penasco, Mex, Univ Ariz, 72-74; ASSOC CUR, DEPT BOT, SMITHSONIAN INST, 75- Concurrent Pos: Res assoc, Dept Ecol & Evolutionary Biol, Univ Ariz, 74-; outside examr, Dept Biol, Marlboro Col, 76. Mem: Phycol Soc Am; Int Phycol Soc; Bot Soc Mex; Int Asn Plant Taxon; Asn Trop Biol. Res: Biosystematics and ecology of marine benthic algae; chemotaxonomy of marine algae; marine flora of the Gulf of California and Mexico. Mailing Add: Dept of Bot 166 NHB Smithsonian Inst Washington DC 20560

NORRIS, JAMES SCOTT, b Selma, Ala, Aug 6, 43; m 66. ENDOCRINOLOGY. Educ: Keene State Col, BS, 66; Univ Colo, PhD(zool), 71. Prof Exp: Ford Found fel endocrinol, Univ Ill, Urbana, 70-71, Nat Cancer Inst fel, 71, Am Cancer Soc fel, 71-74; INSTR CELL BIOL, BAYLOR COL MED, 74- Mem: AAAS; Am Soc Zool; Tissue Cult Asn. Res: Cell-culture; study of abnormal endocrinology of cancer in vitro. Mailing Add: Dept of Cell Biol Baylor Col of Med Houston TX 77025

NORRIS, JESSIE MCGOWAN, b New York, NY, Nov 3, 23; m 48; c 4. TOXICOLOGY. Educ: NMex State Univ, BS, 49, MS, 50. Prof Exp: From biologist to res toxicologist, 67-72, RES SPECIALIST TOXICOL, DOW CHEM CO, 72- Concurrent Pos: Adv bd mem, Harvard & Factory Mutual Res Corp Home Fire Proj, sponsored by NSF & Nat Fire Prev & Control Admin, 75-; mem, Food, Drug Cosmetic Comn. Mem: Soc Toxicol; Mfg Chemists Asn. Mailing Add: Dow Chem Co Toxicol Res Lab 1803 Bldg Midland MI 48640

NORRIS, JONATHAN JUSTUS, range conservation, range management, see 12th edition

NORRIS, KENNETH STAFFORD, b Los Angeles, Calif, Aug 11, 24; m 53; c 4. ZOOLOGY. Educ: Univ Calif, Los Angeles, MA, 51, PhD(zool), 59. Prof Exp: Asst ichthyol, Scripps Inst, Calif, 51-53; cur, Marineland of Pac, 54-60; lectr zool, Univ Calif, Los Angeles, 60-65, from assoc prof to prof, 65-72; dir, Coastal Marine Lab, 72-75, PROF NATURAL HIST, UNIV CALIF, SANTA CRUZ, 72- Concurrent Pos: Mem sci adv comt, US Marine Mammal Comn, 72-; mem adv bd, Bur Land Mgt, Dept Interior, 75- Mem: AAAS; Am Soc Ichthyologists & Herpetologists; Soc Study Evolution; Ecol Soc Am; Am Soc Mammalogists. Res: Echolocation and natural history of cetaceans. Mailing Add: Coastal Marine Lab Appl Sci Univ of Calif Santa Cruz CA 95064

NORRIS, LOGAN ALLEN, b Oakland, Calif, May 23, 36; m 58; c 3. PLANT PHYSIOLOGY, PESTICIDE CHEMISTRY. Educ: Ore State Univ, BS, 61, MS, 64, PhD(plant physiol, biochem), 69. Prof Exp: Asst biochem, Ore State Univ, 61-68; SUPV CHEMIST & PROJ LEADER, FORESTRY SCI LAB, PAC NORTHWEST FOREST & RANGE EXP STA, US FOREST SERV, 68- Mem: Soc Am Foresters; Weed Sci Soc Am. Res: Woody plants; behavior and impact of chemicals in the forest environment; watershed management. Mailing Add: Forestry Sci Lab Pac Northwest Forest & Range Exp Sta Corvallis OR 97331

NORRIS, MARK GILBERT, JR, b Solano, NMex, May 4, 25; m 48; c 4. PARASITOLOGY. Educ: NMex State Univ, BS, 50; Kans State Univ, MS, 51. Prof Exp: Asst zool, Kans State Univ, 51; dist sanitarian, NMex State Health Dept, 51-52; jr parasitologist, Tex Agr Sta, 52-53; from jr parasitologist to parasitologist, Tex Div, 53-56, field specialist, 56-63, area tech specialist, 63-66, PROD TECH SPECIALIST, PESTICIDE RES & DEVELOP, AG-ORG DEPT, DOW CHEM USA, 66- Mem: Entom Soc Am; Spc Nematol. Res: Animal parasitology and entomology; agricultural pesticide; economic entomology and plant pathology; nitrogen management in agricultural crop production and plant nutrition. Mailing Add: 3105 Greenway Dr Midland MI 48640

NORRIS, MAX VALENTINE, b Monticello, Ill, Sept 12, 16; m 43; c 3. ANALYTICAL CHEMISTRY. Educ: Univ Ill, BS, 38; Iowa State Col, MS, 39. Prof Exp: From anal chemist to sr anal chemist, 39-55, group leader anal develop, 55-63, anal res & develop, 63-64, anal develop & chromatography, 64-70, PROJ LEADER ENVIRON POLLUTION, AM CYANAMID CO, 70- Mem: Am Chem Soc. Res: Development of new methods of analysis in analytical chemistry. Mailing Add: Chem Res Div Sci Serv Dept Am Cyanamid Co 1937 W Main Stamford CT 06901

NORRIS, PAUL EDMUND, b Detroit, Mich, Nov 9, 18; m 44; c 2. PHARMACEUTICAL CHEMISTRY. Educ: Univ Mich, BS, 41, MS, 42, PhD(pharmaceut chem), 52. Prof Exp: Asst to F F Blicke, Univ Mich, 46-48, instr pharm, 48-51, asst prof, 51-54; group leader, 54-64, TOXICOLOGIST, PROCTER & GAMBLE CO, 64- Mem: AAAS; Am Chem Soc; Soc Toxicol; Am Pharmaceut Asn. Res: Synthetic drugs; synthesis of organic medicinals; concentration in area of potential ergot substitutes; synthesis of esters and amides of beta-amino acids; development of anticaries agents; design of clinical tests; toxicological testing and safety evaluation. Mailing Add: 476 Beech Tree Dr Cincinnati OH 45224

NORRIS, RICHARD EARL, b Seattle, Wash, Apr 13, 26; div; c 3. MARINE BOTANY. Educ: Univ Wash, BS, 47; Univ Calif, Berkeley, PhD, 54. Prof Exp: Asst, Univ Calif, Berkeley, 48-53, botanist chem dept, 53-54, chemist radiation dept, 54-55; instr bot, Univ Minn, 55-57, asst prof, 57-60, assoc prof, 60-62; assoc cur, Smithsonian Inst, 62-63; from vis assoc prof to assoc prof bot, 63-69, PROF BOT, UNIV WASH, 69- Concurrent Pos: Fulbright scholar, NZ Oceanog Inst, 58-59; microbiologist, Int Indian Ocean Exped, 63; Guggenheim fel, 69-70. Mem: Bot Soc Am; Phycol Soc Am (pres, 71); Soc Protozool; Int Phycol Soc (secy, 69-72); Brit Phycol Soc. Res: Phycological botany; cultures of marine Rhodophyta, Chlorophyta and phytoplankton for studies of morphology, reproduction and ecology; Chrysophyceae and dinoflagellate systematics and ecology; taxonomy of North Pacific and New Zealand red algae. Mailing Add: Friday Harbor Labs Univ of Wash Friday Harbor WA 98250

NORRIS, RICHARD OWEN, chemistry, see 12th edition

NORRIS, ROBERT FRANCIS, b Buckinghamshire, Eng, July 4, 38; m 63; c 2. PLANT PHYSIOLOGY, WEED SCIENCE. Educ: Univ Reading, BSc, 60; Univ Alta, PhD(crop ecol), 64. Prof Exp: NIH & USPHS grants, Mich State Univ, 64-67; asst prof bot, 67-74, ASSOC PROF BOT, UNIV CALIF, DAVIS, 74- Mem: AAAS; Bot Soc Am; Weed Sci Soc Am; Am Soc Plant Physiologists; Scand Soc Plant Physiol. Res: Plant growth regulators; cuticle structure and penetration; weed control; ecology of crop-weed association; weed/insect interactions, integrated control; herbicide action and physiology. Mailing Add: Dept of Bot Univ of Calif Davis CA 95616

NORRIS, ROBERT MATHESON, b Los Angeles, Calif, Apr 24, 21; m 52; c 3. GEOLOGY. Educ: Univ Calif, Los Angeles, AB, 43, MA, 49; Univ Calif, San Diego, PhD(oceanog), 51. Prof Exp: Asst geol, Univ Calif, Los Angeles, 46-49; asst submarine geol, Scripps Inst, 49-51, assoc marine geol, 51-52; from lectr to instr, 52-55, asst prof 55-60, chmn dept, 60-63, assoc prof, 60-68, PROF GEOL, UNIV CALIF, SANTA BARBARA, 68- Concurrent Pos: Mem staff, US Geol Surv, 55-60; geologist, NZ Oceanog Inst, 61-62, 68-69 & 75-76. Mem: Nat Asn Geol Teachers; Geol Soc Am; Soc Econ Paleontologists & Mineralogists; Am Geog Soc; Am Asn Petrol Geol. Res: Quaternary and marine geology; geomorphology. Mailing Add: Dept of Geol Sci Univ of Calif Santa Barbara CA 93106

NORRIS, RUSSELL TAPLIN, b Newburyport, Mass, July 14, 16; m 41; c 2. ORNITHOLOGY. Educ: Univ Maine, BS, 38; Pa State Col, MS, 41. Prof Exp: Asst, Pa State Col, 38-40; collabr, Fish & Wildlife Serv, US Dept Interior, 40, 41-42; biologist, Preston Labs, Pa, 41, 42-46; wildlife biologist, US Fish & Wildlife Serv, 46-53, regional supvr river basin studies, 54-58, asst regional dir, NAtlantic Region, Bur Com Fisheries, 58-63, actg regional dir, 63-64, asst to dir, 64-68, asst dir resource develop, 68-69; regional dir, Nat Marine Fisheries Serv, 69-75; RETIRED. Mem: Wildlife Soc; Am Fisheries Soc; Nat Shellfisheries Asn. Res: Life history studies of American Woodcock; relationships between forest types and wildlife species found thereon; ecological studies of passerine birds. Mailing Add: Whale Ave Rockport MA 01866

NORRIS, TERRY ORBAN, b NC, Apr, 29, 22; m 51; c 2. ORGANIC CHEMISTRY.

Educ: Univ NC, PhD(chem), 54. Prof Exp: Anal res chemist instrumental methods, E I du Pont de Nemours & Co, 49-51, polymer systs, 54-56; asst dir res, Keuffel & Esser Co, 56-57, dir, 58-62, dir corp res, 62-63; res mgr, IBM Corp, 63-66; dir res, 66-69, VPRES RES & DEVELOP, NEKOOSA EDWARDS PAPER CO, INC, 69-, BD DIRS, 72- Mem: AAAS; Am Chem Soc; Am Inst Chem; Soc Photog Sci & Eng; NY Acad Sci. Res: Polymer systems, coating; light sensitive systems and electrophotography; recording materials technology; pulp and paper chemistry; paper coatings. Mailing Add: 731 Wisconsin River Dr Port Edwards WI 54469

NORRIS, THOMAS ELFRED, b Bradenton, Fla, July 4, 39. BIOCHEMISTRY. Educ: Univ Fla, BS, 62; Ind Univ, Bloomington, PhD(microbiol), 70. Prof Exp: Res chemist, Food & Drug Admin, 62-64; sci master, Govt Grammar Sch, Roseau, BWI, 64-65; PROF BIOCHEM & MOLECULAR BIOL, MOREHOUSE COL, 70-, CHMN DEPT, 74- Mem: AAAS; Am Soc Microbiol; Am Chem Soc. Res: Biochemical control mechanisms affecting macromolecular synthesis. Mailing Add: Dept of Biol Morehouse Col Atlanta GA 30314

NORRIS, THOMAS HUGHES, b Princeton, NJ, Feb 8, 16; m 42; c 1. PHYSICAL INORGANIC CHEMISTRY, CHEMICAL KINETICS. Educ: Princeton Univ, AB, 38; Univ Calif, PhD(chem), 42. Prof Exp: Chemist, Linde Air Prod Co, NY, 38-39; asst chemist, Gen Elec Co, Mass, 42; res assoc, Nat Defense Res Comt, Calif, 43, instr chem, 44-46; asst prof, Univ Minn, 46-47; from asst prof to assoc prof, 47-60, PROF CHEM, ORE STATE UNIV, 60- Mem: Am Chem Soc; AAAS. Res: Radioactive tracer studies in physical and inorganic chemistry; exchange reactions and reaction mechanisms; non-aqueous ionizing solvents; complex formation in solution; nuclear magnetic resonance in solution. Mailing Add: Dept of Chem Ore State Univ Corvallis OR 97331

NORRIS, WILFRED GLEN, b Malmö, Sweden, Apr 21, 32; US citizen; m 55; c 3. CHEMICAL PHYSICS. Educ: Juniata Col, BS, 54; Harvard Univ, PhD(chem), 63. Prof Exp: Instr physics, 58-59, from asst prof to prof, 59-66, WILLIAM I & ZELLA B BOOK PROF PHYSICS, JUNIATA COL, 66-, DEAN, 70- Concurrent Pos: NSF sci fac fel, Univ Md, 67-68. Mem: AAAS; Am Phys Soc; Am Asn Physics Teachers; Optical Soc Am. Res: Diffusion of hydrogen in steel; surface reactions of hydrogen on steel; photochemistry; infrared and microwave spectra of small molecules at high temperatures. Mailing Add: Dept of Physics Juniata Col Huntington PA 16652

NORRIS, WILLIAM ELMORE, JR, b Nixon, Tex, Feb 23, 21; m 44; c 3. PLANT PHYSIOLOGY. Educ: Southwest Tex State Teachers Col, BS, 40; Univ Tex, PhD(physiol), 48. Prof Exp: Instr biol & physiol, Univ Tex, 45-47; instr physiol, biochem & bact, Bryn Mawr Col, 47-48, asst prof biol, 48-49; assoc prof, 49-52, chmn dept, 52-67, dean, Sch Sci, 65-70, Col Arts & Sci, 70-74, PROF BIOL, SOUTHWEST TEX STATE UNIV, 52-, VPRES ACAD AFFAIRS, 74- Mem: AAAS; Am Soc Plant Physiol; Scand Soc Plant Physiol. Res: Bioelectrics; plant respiration; plant growth substances. Mailing Add: Dept of Biol Southwest Tex State Univ San Marcos TX 78666

NORRIS, WILLIAM PENROD, b Loogootee, Ind, Sept 2, 20; m 43; c 3. RADIOBIOLOGY. Educ: DePauw Univ, AB, 41; Univ Ill, PhD(biochem), 44. Prof Exp: Biochemist, Dow Chem Co, Mich, 44 & Manhattan Area Engrs, 44-46; group leader, biol div, 46-52, assoc biochemist, 52-70, GROUP LEADER, DIV BIOL & MED RES, ARGONNE NAT LAB, 70-, SCIENTIEST, 73- Mem: Radiation Res Soc; Am Soc Biol Chemists; Reticuloendothelial Soc; Health Physics Soc; Am Nuclear Soc. Res: Responses of dogs to continuous or protracted gamma-irradiation; sphingolipids; metabolism of phosphorus and alkaline earths; isolation and synthesis of dihydrosphingosine; radioautography; effects of ionizing radiations in animals; measurement and chemistry of radioactive elements; radiation chemistry; paper electrophoresis. Mailing Add: Div of Biol & Med Res Argonne Nat Lab 9700 S Cass Ave Argonne IL 60439

NORRIS, WILLIAM PHILLIP, b Pagosa Springs, Colo, Feb 23, 21; c 2. ORGANIC CHEMISTRY. Educ: Univ Colo, BA, 49, PhD, 53. Prof Exp: ORG CHEMIST, NAVAL WEAPONS CTR, 53- Mem: Am Chem Soc. Res: High nitrogen compounds; fluorocarbons; organometallics. Mailing Add: Code 6056 Naval Weapons Ctr China Lake CA 93555

NORRIS, WILLIAM WARREN, b Choudrant, La, Mar 27; m 49; c 3. ZOOLOGY. Educ: La Polytech Inst, BS, 50; La State Univ, MS, 51, PhD, 55. Prof Exp: Histologist, Res Labs, Swift & Co, 55-60; assoc prof zool, Western Ky State Col, 60-65; assoc prof, 65-69, PROF BIOL, NORTHEAST LA UNIV, 69- Mem: Am Soc Zool. Res: Histology, especially as related to proteolytic action of muscle fibers and connective tissue; reproductive physiology as related to pineal gland. Mailing Add: Dept of Biol Northeast La Univ Monroe LA 71201

NORSTADT, FRED A, b Sidney, Iowa, Mar 15, 26; m 50; c 2. AGRICULTURAL MICROBIOLOGY, AGRICULTURAL CHEMISTRY. Educ: Nebr State Teachers Col, BS, 50; Univ Nebr, MS, 58, PhD(agron), 66. Prof Exp: Teacher high schs, Nebr, 49-56, prin, Holmesville High Sch, 51-53; instr exten div, Univ Nebr, 56-64; chemist, Soil & Water Conserv Res, 64-66, SOIL SCIENTIST, AGR RES SERV, USDA, 66- Mem: Soil Sci Soc Am; Soil Conserv Soc Am; Am Soc Agron. Res: Microbial crop residue decomposition; phytotoxic substances; mineralization of carbon, nitrogen, phosphorus and sulfur from soil organic matter and animal wastes; soil physical, chemical and biological effects on plant growth. Mailing Add: PO Box E Ft Collins CO 80522

NORSTOG, KNUT JONSON, b Grand Forks, NDak, June 11, 21; m 44; c 3. BIOLOGY. Educ: Luther Col, Iowa, BA, 43; Univ Mich, MS, 47, PhD(bot), 55. Prof Exp: Biologist, Dept Game, Fish & Parks, SDak, 47-49; instr biol, Luther Col, 49-51; assoc prof, Wittenberg Univ, 54-63; assoc prof bot & bact, Univ SFla, 63-66; PROF BIOL SCI, NORTHERN ILL UNIV, 66- Concurrent Pos: NSF fac fel, 59; res fel bot, Yale Univ, 59-60. Mem: Int Soc Plant Morphol; Torrey Bot Club; Soc Develop Biol. Res: Plant morphogenesis; embryogenesis in the grasses; plant tissue culture. Mailing Add: Dept of Biol Sci Northern Ill Univ DeKalb IL 60115

NORSWORTHY, STANLEY FRANK, b Los Angeles, Calif, Aug 13, 36; m 58; c 5. RESOURCE GEOGRAPHY. Educ: Univ Calif, Los Angeles, AB, 58; Miami Univ, MA, 59; Univ Calif, Los Angeles, PhD(geog), 70. Prof Exp: Recreation resource specialist, Bur Outdoor Recreation, US Dept Interior, 65-66; asst prof geog, 66-71, ASSOC PROF GEOG, CALIF STATE UNIV, FRESNO, 71- Mem: Asn Am Geog; Am Geog Soc; Can Asn Geog. Res: Geography of Anglo-America; wilderness problems; health resorts; public land management and use. Mailing Add: Dept of Geog Calif State Univ Fresno CA 93740

NORTH, CHARLES A, b Kingston, RI, Aug 24, 32; m 62; c 2. ZOOLOGY, ORNITHOLOGY. Educ: Univ Mo, BA, 54; Okla State Univ, MS, 62, PhD(zool), 68. Prof Exp: ASST PROF BIOL, UNIV WIS-WHITEWATER, 66- Concurrent Pos: Wis State Univ res grant, 69-70; Instnl grants, 68-72; NSF travel grant, Poland, 71. Res: General biology; nature study; organic evolution; wildlife conservation ecology. Mailing Add: Dept of Biol Univ of Wis Whitewater WI 53190

NORTH, CHARLES MALLORY, JR, b Jacksonville, Fla, Jan 1, 37; m 59. APPLIED MATHEMATICS. Educ: Univ Fla, BAeE, 59, MS, 63; MA, Univ Ala, Tuscaloosa, PhD(eng mech), 69. Prof Exp: Assoc engr, Martin Co, 59; sr exp engr, Fla Res & Develop Ctr, Pratt & Whitney Aircraft, 61-64; sr prod engr, Space Div, Chrysler Corp, 64-65; instr eng mech, Univ Ala, Tuscaloosa, 68-69; asst prof aero & mech eng & solid mech, Southern Methodist Univ, 69-71; ASSOC PROF MATH & ENG, GRAD INST TECHNOL, UNIV ARK, LITTLE ROCK, 71- Mem: Am Soc Mech Eng; Am Soc Eng Educ; Sigma Xi. Res: Theoretical mechanics; dynamics of linear and nonlinear discrete and continuous systems; wave propagation; methods of applied mathematics and numerical analysis. Mailing Add: Univ of Ark PO Box 3017 Little Rock AR 72203

NORTH, DWIGHT OLCOTT, b Hartford, Conn, Sept 28, 09; m 35; c 2. THEORETICAL PHYSICS. Educ: Wesleyan Univ, BS, 30; Calif Inst Technol, PhD(physics), 33. Prof Exp: Fel tech staff, RCA Labs, 34-74; RETIRED. Mem: Fel Am Phys Soc; fel Inst Elec & Electronics Engrs. Res: Noise; solid-state; quantum theory. Mailing Add: 80 Random Rd Princeton NJ 08540

NORTH, GERALD CHARLES, dairy industry, see 12th edition

NORTH, GERALD R, b Sweetwater, Tenn, June 28, 38; m 59; c 2. THEORETICAL PHYSICS. Educ: Univ Tenn, BS, 60; Univ Wis, PhD(physics), 66. Prof Exp: Jr physicist, Oak Ridge Nat Lab, 60-61; res assoc physics, Univ Pa, 66-68; asst prof, 68-75, ASSOC PROF PHYSICS, UNIV MO-ST LOUIS, 75- Concurrent Pos: Partic NSF, Elem Sci In-Serv Inst, 69-71; NSF sr fel, Nat Ctr Atmospheric Res, 74-75. Mem: AAAS; Am Phys Soc; Am Meteorol Soc. Res: Quantum field theory; theory of ion-molecule collisions; theory of global climate fluctuations. Mailing Add: Dept of Physics Univ of Mo St Louis MO 63121

NORTH, HARPER QUA, b Los Angeles, Calif, Jan 24, 17; m 55; c 2. PHYSICS. Educ: Calif Inst Technol, BS, 38; Univ Calif, Los Angeles, 40, PhD(physics), 47. Prof Exp: Res asst, Gen Elec Co, 40-42, res assoc, 42-49; dir, Semiconductor Div, Hughes Aircraft Co, 49-54; pres, Pac Semiconductors, Inc, 54-62; vpres res & develop, TRW Inc, 62-69; mgr electro-optical dept, Electronics Div, Northrop Corp, 69-75; ASSOC DIR RES FOR ELECTRONICS, NAVAL RES LAB, 75- Concurrent Pos: Consult, Off Dir Defense, Res & Eng, 59-75. Honors & Awards: Medal of Honor, Electronic Industs Asn, 66. Mem: Fel Am Phys Soc; fel Inst Elec & Electronics Engrs; Electronic Industs Asn. Res: Semiconductor physics. Mailing Add: 12440 Surrey Circle Dr Tantallon Oxon Hill MD 20022

NORTH, JAMES A, b Charleston, Utah, Mar 18, 34; m 56; c 4. VIROLOGY, IMMUNOLOGY. Educ: Brigham Young Univ, BS, 58, MS, 60; Univ Utah, PhD(microbiol), 64. Prof Exp: Sr res assoc virol, Univ Cincinnati, 64-65; assoc prof, 65-72, PROF MICROBIOL, BRIGHAM YOUNG UNIV, 72- Mem: Am Soc Microbiol. Res: Viral purification and physical analysis; immunochemistry; viral etiology of cancer. Mailing Add: 851 WIDB Brigham Young Univ Provo UT 84601

NORTH, JAMES CLAYTON, b Erie, Pa, June 10, 37; m 61; c 3. SOLID STATE PHYSICS. Educ: Capital Univ, 59; Purdue Univ, Lafayette, PhD(solid state physics), 65. Prof Exp: Fel, Univ Ill, Urbana, 65-68; MEM TECH STAFF, BELL LABS, 68- Mem: Am Phys Soc; Inst Elec & Electronics Engrs. Res: Ion implantation; particle channeling; radiation damage; integrated circuit technology. Mailing Add: Bell Labs Murray Hill NJ 07974

NORTH, RICHARD RALPH, b Hamilton, Ont, Aug 8, 34; m 55; c 4. NEUROLOGY. Educ: Queen's Univ, Ont, MD, CM, 59. Prof Exp: Fel neurophysiol, Col Med, Baylor Univ, 63-64, fel neurol, 64-66, asst prof, 66-67; from asst prof to assoc prof neurol, 67-75, CLIN ASSOC PROF NEUROL, UNIV TEX HEALTH SCI CTR DALLAS, 75-; DIR MULTIPLE SCLEROSIS CLIN, PRESBY HOSP, 67-, DIR EPILEPSY CLIN, 74- Mem: AAAS; Am Acad Neurol; Am EEG Soc; Am Epilepsy Soc. Res: Clinical electroencephalography; cerebrovascular disease; epilepsy; efficacy of levodopa therapy in parkinsonism. Mailing Add: Presby Hosp Dept of Neurol 8200 Walnut Hill Lane Dallas TX 75231

NORTH, ROBERT JARL, organic chemistry, see 12th edition

NORTH, WHEELER JAMES, b San Francisco, Calif, Jan 2, 22; m 53; c 2. OCEANOGRAPHY. Educ: Calif Inst Technol, BS, 44 & 50; Univ Calif, PhD(oceanog), 53. Prof Exp: Electron engr, US Navy Electron Lab, 47-48; NSF fel, Cambridge Univ, 53-54; Rockefeller fel marine biol, Scripps Inst Oceanog, 55-56; asst res biologist & proj officer, Inst Marine Resources Kelp Prog, Univ Calif, 56-63; sr res scientist, Lockheed Calif Co, 63; assoc prof environ health eng, 63-68, PROF ENVIRON SCI, CALIF INST TECHNOL, 68- Mem: AAAS; Soc Gen Physiol; Am Soc Zool; Am Malacol Union; Am Geophys Union. Res: Ecology and general physiology. Mailing Add: W M Keck Lab Calif Inst of Technol Pasadena CA 91109

NORTH, WILLIAM CHARLES, b Chungking, China, Aug 17, 25; m 71; c 9. ANESTHESIOLOGY, PHARMACOLOGY. Educ: DePauw Univ, BA, 45; Northwestern Univ, MS, 48, MD, 50, PhD, 52. Prof Exp: Intern, Chicago Mem Hosp, 49-50; from instr to asst prof pharmacol, Northwestern Univ, 50-59; asst prof anesthesiol, Sch Med, Duke Univ, 59-62, assoc prof anesthesiol & pharmacol, 63-65; PROF ANESTHESIOL & PHARMACOL & CHMN DEPT ANESTHESIOL, COL MED, UNIV TENN, MEMPHIS, 65- Concurrent Pos: Res anesthesia, Chicago Wesley Mem Hosp, 56-59. Mem: AAAS; Am Soc Anesthesiol; Am Soc Pharmacol & Exp Therapeut; AMA. Res: Neuropharmacology; analgesia; local anesthesia; shock; inhalation anesthesia. Mailing Add: Dept of Anesthesiol Univ of Tenn Col of Med Memphis TN 38163

NORTH, WILLIAM GORDON, b Woodstock, Ill, Aug 29, 42; m 64; c 2. STRATIGRAPHY. Educ: Carleton Col, AB, 63; Univ Ill, MS, 65, PhD(geol), 69. Prof Exp: PETROL GEOLOGIST, TEXACO, INC, 68- Res: Stratigraphy, particularly subsurface stratigraphy; sedimentary petrology; statistics. Mailing Add: Bellaire Res Labs Texaco Inc Box 36650 Houston TX 77036

NORTHAM, EDWARD STAFFORD, b Lansing, Mich, Oct 18, 27; m 61; c 2. MATHEMATICS. Educ: Univ Mich, BS, 47, MS, 48; Mich State Univ, PhD(math), 53. Prof Exp: Mathematician, Bendix Aviation Corp, 53-54; from instr to asst prof math, Wayne State Univ, 54-64; assoc prof, 64-71, PROF MATH, UNIV MAINE, ORONO, 71- Mem: Am Math Soc; Math Asn Am. Res: Abstract algebra; lattice theory. Mailing Add: Dept of Math Univ of Maine Orono ME 04473

NORTHAM, RAY MERVYN, b Calgary, Can, May 28, 29; m 53; c 2. URBAN GEOGRAPHY. Educ: Ore State Univ, BS, 53, MS, 54; Northwestern Univ, PhD, 60. Prof Exp: Instr geog, Exten Div, Univ Wis, 54-55; from instr to asst prof, Univ Ga, 56-62; asst prof, Portland State Col, 62-65; vis assoc prof, Yale Univ, 65-66; assoc prof, 66-74, PROF GEOG, ORE STATE UNIV, 74- Concurrent Pos: Assoc, Columbian Res Inst, 63- Mem: Asn Am Geog; Am Geog Soc; Regional Sci Asn. Res: Economic base of cities; land use patterns and relations in cities and the relation between cities and the areas they serve; declining urban centers; intraurban distribution and potential of vacant urban land. Mailing Add: Dept of Geog Ore State Univ Corvallis OR 97331

NORTHCLIFFE, LEE CONRAD, b Manitowoc, Wis, Mar 20, 26; m 53; c 2. NUCLEAR PHYSICS. Educ: Univ Wis, BS, 48, MS, 51, PhD(physics), 57. Prof Exp: Asst physics, Univ Wis, 51-57; from instr to asst prof, Yale Univ, 57-65; assoc prof, 65-70, PROF PHYSICS, TEX A&M UNIV, 70- Mem: Am Phys Soc. Res: Nucleon-nucleon scattering; penetration of heavy ions through matter; accelerator development and instrumentation for nuclear research; charge distributions of heavy ions; nuclear reactions and scattering. Mailing Add: Dept of Physics Tex A&M Univ College Station TX 77843

NORTHCOTT, JEAN, b Australia, June 26, 26; nat US. PHYSICAL CHEMISTRY, ORGANIC CHEMISTRY. Educ: Univ Sydney, PhD(chem), 53. Prof Exp: Chemist & bacteriologist, Campbell Soup Co, Can, 54-55; patent asst, Nat Aniline Div, 56-68, HEAD INFO SERV, SPECIALTY CHEM DIV, ALLIED CHEM CORP, 68- Mem: Am Chem Soc; Am Soc Info Sci; Spec Libr Asn. Res: Chemical literature; patents. Mailing Add: Allied Chem Corp PO Box 1069 Buffalo NY 14240

NORTHCRAFT, RICHARD DUNN, b Olympia, Wash, Dec 25, 17; m. PLANT PHYSIOLOGY. Educ: Wash Univ, BS, 40, MS, 41; Stanford Univ, PhD(plant biol), 46. Prof Exp: Instr plant physiol, Univ Mass, 46-47; instr bot, Rutgers Univ, 47-48; asst prof bot & physiol, Amherst Col, 48-51; mem staff, E I du Pont de Nemours & Co, 51-55; mem staff Pennsalt Chem Corp, 55-59; asst prof biol sci, Calif Western Univ, 59-64, assoc prof biol, 64-69; OYSTER CULT RES & DEVELOP, 69- Res: Marine algal ecology; plant life forms; cell isolation technique; oyster nutrition and culture. Mailing Add: 749 Seabright Lane Solana Beach CA 92075

NORTHCUTT, RICHARD GLENN, b Mt Vernon, Ill, Aug 7, 41; m 65. NEUROANATOMY. Educ: Millikin Univ, BA, 63; Univ Ill, Urbana, MA, 66, PhD(zool), 68. Prof Exp: Asst prof anat, Case Western Reserve Univ, 68-72; ASSOC PROF ZOOL, UNIV MICH, ANN ARBOR, 72- Concurrent Pos: Res assoc, Cleveland Aquarium, 68-72. Mem: Am Asn Anat; Am Soc Zool; Am Soc Ichthyol & Herpet; Soc Neurosci. Res: Evolution of the vertebrate nervous system; vertebrate paleontology, phylogeny and morphology; vertebrate behavior. Mailing Add: Div of Biol Sci Univ of Mich Ann Arbor MI 48104

NORTHCUTT, ROBERT ALLAN, b Luling, Tex, Sept 30, 37. MATHEMATICS. Educ: Univ Tex, Austin, BA, 60, MA, 62, PhD(math), 68. Prof Exp: Instr math, San Antonio Col, 62-64; Southwest Tex State Univ, 64-67; NSF fac fel, Univ Tex, 67-68; from asst prof to assoc prof, 68-73, PROF MATH, SOUTHWEST TEX STATE UNIV, 73-, CHMN DEPT, 71- Mem: Am Math Soc; Math Asn Am. Res: Differential and integral equations. Mailing Add: Dept of Math Southwest Tex State Univ San Marcos TX 78666

NORTHEN, HENRY THEODORE, b Butte, Mont, Apr 12, 08; m 37; c 3. BOTANY. Educ: State Col Wash, BS, 32, MS, 34; Univ Calif, PhD(bot), 36. Prof Exp: From instr to prof, 36-75, EMER PROF BOT, UNIV WYO, 75- Mem: AAAS; Bot Soc Am; Am Soc Plant Physiol. Res: Physiology and ecology of epiphytes; effects of pollutants on plant growth and development. Mailing Add: Dept of Bot Univ of Wyo Laramie WY 82071

NORTHERN, JERRY LEE, b Albuquerque, NMex, Sept 13, 40; c 4. AUDIOLOGY. Educ: Colo Col, BA, 62; Gallaudet Col, MS, 63; Univ Denver, MA, 64; Univ Colo, PhD(audiol), 66. Prof Exp: Chief audiol clin, Dept Otolaryngol, Brooke Army Med Ctr, San Antonio, Tex, 66-67; asst dir, US Army Audiol & Speech Path Ctr, Walter Reed Army Med Ctr, Washington, DC, 67-70; HEAD AUDIOL SERV, DEPT OTOLARYNGOL, MED CTR, UNIV COLO, DENVER, 70- Concurrent Pos: Mem, Nat Registry Interpreters for the Deaf, 68- Mem: Am Speech & Hearing Asn; Acoust Soc Am; Am Audiol Soc; Nat Asn Deaf. Res: Clinical audiology; acoustic impedance of the ear. Mailing Add: Dept of Otolaryngol Univ of Colo Med Ctr Denver CO 80202

NORTHEY, WILLIAM T, b Duluth, Minn, Aug 10, 28; m 50; c 6. IMMUNOLOGY. Educ: Univ Minn, BA, 50; Univ Kans, MA, 57, PhD(immunol), 59. Prof Exp: Res asst, Abbott Labs, 50-51; res asst immunol & virol, Naval Med Res Unit 4, 51-55; teaching & res, Univ Kans, 55-59; from asst prof to assoc prof, 59-72, PROF IMMUNOL, ARIZ STATE UNIV, 72- Concurrent Pos: Mem, Allergy Found Am, 60-; consult, Iatric Corp, 70- Mem: Am Soc Microbiol; Am Asn Immunol; Am Asn Univ Profs. Res: Immune response in cold exposure; hypothermia; immunological aspects of the delayed response primarily in Coccidioidomycosis; allergenic extracts. Mailing Add: Dept of Bot & Microbiol Ariz State Univ Tempe AZ 85281

NORTHINGTON, DEWEY JACKSON, JR, b New Orleans, La, Jan 31, 46; c 2. ORGANIC CHEMISTRY. Educ: La State Univ, BS, 67; Univ Fla, PhD(org chem), 70. Prof Exp: Fel chem, La State Univ, 70-71 & Syva Res Inst, 71-72; res scientist, Carnation Co, 72-74; ASST TECH DIR, WEST COAST TECH SERV, INC, 74- Mem: Am Chem Soc. Res: Analysis of polymers and surfactants; organic analyses; spectral identification of organic compounds. Mailing Add: West Coast Tech Serv 17605 Fabrica Way Suite D Cerritos CA 90701

NORTHOVER, FRANCIS HENRY, b Gosport, Eng, Feb 26, 21; m 49; c 4. APPLIED MATHEMATICS. Educ: Cambridge Univ, MA, 49; Univ London, PhD(appl math), 52. Prof Exp: Assoc prof math, Memorial Univ, 53-55; lectr appl math, Univ Liverpool, 55-57; assoc prof math, 57-62, PROF MATH, CARLETON UNIV, 62- Concurrent Pos: Consult, Defense Res Estab, Can, 57- Mem: Int Union Radio Sci; Inst Elec & Electronics Eng. Res: Electromagnetic theory and wave propagation; magnetohydrodynamics; differential equations; diffraction theory. Mailing Add: Dept of Math Carleton Univ Ottawa ON Can

NORTHRIP, JOHN WILLARD, b Tulsa, Okla, July 7, 34; m 54; c 3. BIOPHYSICS. Educ: Southwest Mo State Col, BS, 54; Okla State Univ, MS, 58, PhD(physics), 64. Prof Exp: Staff mem phys res, Sandia Corp, 58-61; asst prof physics, Cent Mo State Col, 63-65; Fulbright lectr solid state physics, Sch Eng Sao Carlos, Univ Sao Paulo, 65-67; from asst prof to assoc prof, 67-75, PROF PHYSICS & ASTRON, SOUTHWEST MO STATE UNIV, 75- Concurrent Pos: Consult, Sandia Corp, 61-64. Mem: AAAS; Am Phys Soc; Am Asn Physics Teachers. Res: Study of musculoskeletal motion, especially as applied to athletics; physics of sensory perceptions. Mailing Add: Dept of Physics Southwest Mo State Univ Springfield MO 65802

NORTHROP, CEDRIC, b Wayne, Nebr, Oct 21, 06; m 38; c 2. PREVENTIVE MEDICINE. Educ: Univ Ore, BA, 30, MD, 36. Prof Exp: Intern, Swedish Hosp, Seattle, 36-37; resident physician, Glen Lake Sanatorium, Minn, 37-39; supt & med dir, State Tuberc Sanatorium, NDak, 39-41; tuberc control officer, State Dept Health, Wash, 41-59; dir tuberc control sect, Seattle-King County Health Dept, 59-71; RETIRED. Concurrent Pos: Clin asst prof pub health & prev med, Sch Med, Univ Wash, 47-67, clin assoc prof prev med, 67-; consult, Firland Sanatorium. Mem: Am Thoracic Soc; AMA; Am Col Chest Physicians. Res: Tuberculosis eradication. Mailing Add: 5049 Pullman Ave NE Seattle WA 98105

NORTHROP, DAVID A, b New Haven, Conn, Feb 4, 38; m 60; c 2. GEOCHEMISTRY. Educ: Univ Chicago, BS, 60, MS, 61, PhD(chem), 64. Prof Exp: Tech staff mem res, 64-70, DIV SUPVR RES, SANDIA LABS, 71- Mem: AAAS. Res: In situ conversion of fossil fuels; carbon composite materials research for aerospace applications; high temperature phase equilibria with emphasis upon vaporization phenomena; oxygen and stable isotope geochemistry. Mailing Add: 7207 Harwood Ave NE Albuquerque NM 87110

NORTHROP, GRETAJO, b Corvallis, Ore, Nov 29, 29; m 74; c 2. INTERNAL MEDICINE, ENDOCRINOLOGY. Educ: Ore State Univ, BS, 52; Mich State Univ, MS, 57; Univ Wis, PhD(pharmacol), 63, MD, 65. Prof Exp: From intern to resident, 65-68, USPHS endocrine fel, 68-71, ASST PROF MED & OBSTET & GYNEC, RUSH-PRESBY-ST LUKE'S MED CTR, CHICAGO, 72- Concurrent Pos: Asst attend, Presby-St Luke's Hosp, 72-74; asst prof med, W Suburban Hosp, Oak Park, Ill, 72-74. Mem: Endocrine Soc; Am Diabetes Asn. Res: Reproductive endocrinology. Mailing Add: Rush-Presby-St Luke's Med Ctr 1725 W Harrison Chicago IL 60612

NORTHROP, JOHN, b New York, NY, Feb 1, 23; m 52; c 3. GEOPHYSICS. Educ: Princeton Univ, BA, 47; Columbia Univ, MA, 48; Univ Hawaii, PhD(solid earth geophys), 68. Prof Exp: Geol asst marine geol, Oceanog Inst, Woods Hole, 48-49; res assoc, Lamont Geol Observ, Columbia Univ, 49-51; head dept geol, Bates Col, 51-52; geologist, Hudson Labs, 52-61; asst specialist, Scripps Inst, Calif, 61-64; assoc geophysicist, Hawaii Inst Geophys, 64-65; assoc specialist marine geophys, Marine Phys Lab, Scripps Inst Oceanog, Univ Calif, 65-67; GEOPHYSICIST, US NAVAL UNDERSEA CTR, 67- Concurrent Pos: Consult, Artemis Proj, 58-59. Mem: Am Geophys Union; Seismol Soc Am; Acoust Soc Am; NY Acad Sci. Res: Submarine geology; marine geophysics; underwater sound; earthquake waves; geoacoustics; hydroacoustics. Mailing Add: Naval Undersea Ctr San Diego CA 92132

NORTHROP, JOHN ALLEN, nuclear physics, see 12th edition

NORTHROP, JOHN HOWARD, b Yonkers, NY, July 5, 91; m 17; c 2. BIOCHEMISTRY, BIOLOGY. Educ: Columbia Univ, BS, 12, AM, 13, PhD(chem), 15, ScD, 37. Hon Degrees: ScD, Harvard Univ, 36, Yale Univ, 37, Princeton Univ, 49, Rutgers Univ, 41; LLD, Univ Calif, 39. Prof Exp: Cutting traveling fel, Columbia Univ, 15-16; asst, Rockefeller Univ, 16, assoc, 17-20, assoc mem, 20-24, mem, 24-56; Hitchcock prof, Univ Calif, Berkeley, 39, vis prof, 49-58, prof & res biophysicist, Donner Lab, 58-59, EMER PROF BACT & PHYSIOL, UNIV CALIF, BERKELEY, 59- Concurrent Pos: Alvarez lectr, 32; DeLamar lectr, Sch Hyg & Pub Health, Johns Hopkins Univ, 37, Thayer lectr, 40; Jessup lectr, Columbia Univ, 38. Honors & Awards: Nobel Prize in Chem, 46; Stevens Prize, 31, Chandler Medal, 37, Lion Award, 49 & Alexander Hamilton Medal, 62, Columbia Univ; Elliot Medal, Nat Acad Sci, 39; Cert of Merit, 48. Mem: Nat Acad Sci; Am Philos Soc; Am Acad Arts & Sci; hon fel The Chem Soc; Franklin fel Royal Soc Arts. Mailing Add: PO Box 1387 Wickenburg AZ 85358

NORTHROP, ROBERT L, b Kansas City, Mo, Mar 12, 26; m 52. VIROLOGY, BIOCHEMISTRY. Educ: Ore State Univ, BS, 52; Univ Minn, DVM, 58; Univ Wis, PhD(microbiol), 63. Prof Exp: Assoc prof virol, 65-73, ASSOC PROF EPIDEMIOL, SCH PUB HEALTH, UNIV ILL MED CTR, 73-; ASSOC PROF EPIDEMIOL, COL VET MED, UNIV ILL, URBANA, 73- Concurrent Pos: Microbiologist, Rush-Presby-St Luke's Med Ctr, 65- Mem: Am Asn Immunol. Res: Infectious and neoplastic disease epidemiology. Mailing Add: Sch of Pub Health Univ of Ill Med Ctr Box 6998 Chicago IL 60680

NORTHROP, THEODORE GEORGE, b Poughkeepsie, NY, Dec 15, 24; m 49, 71; c 5. SPACE PHYSICS. Educ: Yale Univ, BS, 44; Cornell Univ, MS, 49; Iowa State Col, PhD(physics), 53. Prof Exp: Instr physics, Vassar Col, 46-47; asst, TV tube dept, Labs, Radio Corp Am, NJ, 47; instr physics, Yale Univ, 53-54; mem staff, Theoret Div, Lawrence Radiation Lab, Univ Calif, 54-65; mem staff, Goddard Space Flight Ctr, NASA, 65-67, chief lab space physics, 67-73, HEAD THEORETICAL GROUP, GODDARD SPACE FLIGHT CTR, NASA, 73- Mem: Fel Am Phys Soc; Am Geophys Union. Res: Biophysics; electronic instrumentation; plasma physics; plasma theory of planetary magnetospheres and of the solar wind. Mailing Add: Goddard Space Flight Ctr NASA Greenbelt MD 20771

NORTHRUP, CLYDE JOHN MARSHALL, JR, b Oklahoma City, Okla, Apr 25, 38; m 60; c 3. THERMODYNAMICS, GEOCHEMISTRY. Educ: Okla State Univ, BS(math) & BS(physics), 61, PhD(physics), 66. Prof Exp: Staff asst, 58 & 59-60, MEM TECH STAFF, SANDIA LABS, 67- Mem: Am Asn Physics Teachers. Res: Thermodynamics and phase diagrams of metal alloy-hydrogen systems; physical and chemical characterization of magna. Mailing Add: Orgn 5834 Sandia Labs Albuquerque NM 87115

NORTHUP, DAVID WILMARTH, physiology, deceased

NORTHUP, MELVIN LEE, b Floris, Iowa, Oct 11, 41; m 65. ENVIRONMENTAL SCIENCES. Educ: Parsons Col, BS, 63; Purdue Univ, Lafayette, MS, 65; Univ Mo-Columbia, PhD(soil chem), 70. Prof Exp: Methods develop chemist, Norwich Pharmacal Co, 65-66; NSF fel soils, Univ Wis-Madison, 70-72; ASST PROF ENVIRON SCI, GRAND VALLEY STATE COLS, 72- Mem: Am Soc Agron. Res: Chemistry of manganese and nitrogen in soils; nutrient cycling in nature; computer simulation of ecosystems; chemical pollution of water resources. Mailing Add: Dept of Environ Sci Grand Valley State Cols Allendale MI 49401

NORTHWOOD, THOMAS DAVID, b Peterborough, Ont, Mar 25, 15; m 40; c 1. ACOUSTICS, SEISMOLOGY. Educ: Univ Toronto, BASc, 38, MA, 50, PhD(physics), 51. Prof Exp: Engr, Northern Elec Co, Ltd, 38-40; res physicist acoust, Div Physics, Nat Res Coun, 40-42; HEAD BLDG PHYSICS SECT, DIV BLDG RES, NAT RES COUN CAN, 51- Honors & Awards: Award of Merit, Am Soc Testing & Mat, 75. Mem: Fel AAAS; fel Acoust Soc Am; Seismol Soc Am. Res: Architectural acoustics; engineering seismology. Mailing Add: 140 Blenheim Dr Ottawa ON Can

NORTON, ALLEN C, b Los Angeles, Calif, Mar 20, 35; m 61; c 7. PHYSIOLOGY, BIOMEDICAL ENGINEERING. Educ: Univ Calif, Santa Barbara, AB, 57; Univ Buffalo, PhD(exp psychol), 61. Prof Exp: Asst physiol & psychol, Univ Buffalo, 57-61; USPHS fel & instr physiol, Sch Med, Tohoku Univ, Japan, 61-63; sr res assoc neurophysiol, Develop & Sensory Physiol Lab, Children's Hosp Los Angeles, 63-64, interim dir, 64; sr investr, Inst Med Res, Huntington Mem Hosp, 64-67; asst res anatomist & physiologist, Brain Info Serv, Univ Calif Sch Med, Los Angeles, 67-68, asst prof physiol, 68-69; fel neurosci study prog, Univ Colo, Boulder, 69; SR RES PHYSIOLOGIST, BECKMAN INSTRUMENTS, INC, 70- Concurrent Pos: Consult, Electro Optical Systs, 65-; instr, Calif State Col, Los Angeles, 69 & Otis Art Inst, 69-70. Mem: AAAS; Instrument Soc Am; Asn Advan Med Instrumentation. Res: Bioimpedance; respiratory gas analysis; cardiopulmonary physiology; application of commercial instrumentation to space station laboratories. Mailing Add: Beckman Instruments Inc 1630 S State College Blvd Anaheim CA 92806

NORTON, CHARLES J, b Lockport, Ill, Mar 6, 29. ORGANIC CHEMISTRY, ANTHROPOLOGY. Educ: Purdue Univ, BS, 51, AM, 53, PhD(org chem), 55; Univ Colo, MA, 64. Prof Exp: Asst proj chemist, Whiting Res Lab, Standard Oil Co (Ind), 55-56; res scientist, 56-59, adv res scientist, 59-65, SR RES SCIENTIST, DENVER RES CTR, MARATHON OIL CO, 66- Concurrent Pos: Instr anthrop, Univ Colo, 63-64 & 66; fel lab chem biodynamics, Lawrence Radiation Lab, Univ Calif, Berkeley, 68-69; instr, Univ Denver, 71- Mem: Am Inst Chemists; Sigma Xi; Soc Petrol Engrs. Res: Polymerization; petroleum production; surfactants; hydrocarbon fermentation; enhanced oil recovery; North American Indian culture. Mailing Add: Denver Res Ctr Marathon Oil Co Littleton CO 80121

NORTON, CHARLES LAWRENCE, b Neponset, Ill, Dec 20, 17; m 45; c 5. DAIRY SCIENCE. Educ: Univ Ill, BS, 40; Cornell Univ, PhD(animal husb), 44. Prof Exp: Asst prof animal husb, Cornell Univ, 45-47; prof dairy husb & head dept, Univ R I, 47-50; head dept dairying, Okla State Univ, 50-58; PROF DAIRY & POULTRY SCI & HEAD DEPT, KANS STATE UNIV, 58-, RES DAIRY & POULTRY SCIENTIST, AGR EXP STA, 70- Mem: Am Soc Animal Sci; Am Dairy Sci Asn. Res: Vitamin needs of dairy calves; dry calf starters; nutritive value of forages for dairy cattle. Mailing Add: Dairy & Poultry Sci Dept Leland Call Hall Kans State Univ Manhattan KS 66502

NORTON, CHARLES WARREN, b Scranton, Pa, Aug 2, 44; m 65. MICROPALEONTOLOGY. Educ: Antioch Col, BA, 68; Univ Va, MA, 74; Univ Pittsburgh, PhD(geol), 75. Prof Exp: Res geologist, Gulf Res & Develop Corp, 74; COAL GEOLOGIST, WVA GEOL SURV, 75- Mem: AAAS; Paleont Soc; Palaeont Asn; Am Asn Petrol Geologists; Sigma Xi. Res: Coal resource and reserved study in West Virginia involving mapping, correlation, thickness and quality measures on minable seams; long-term study of interbedded marine horizons for future correlation aid. Mailing Add: WVa Geol & Econ Surv PO Box 879 Morgantown WV 26505

NORTON, CYNTHIA FRIEND, b Shelburne Falls, Mass, Aug 18, 40; m 68. MICROBIOLOGY. Educ: Smith Col, AB, 61; Boston Univ, PhD(marine microbiol), 67. Prof Exp: Sci aide, Polaroid Corp, 61-63; res assoc microbial physiol, Univ NH, 67-68; NIH fel microbiol, Sch Med, Yale Univ, 68-71; ASST PROF BIOL, UNIV MAINE, AUGUSTA, 71- Mem: Am Soc Microbiol; Am Soc Limnol & Oceanog. Res: Microbial pigments and exoenzymes; interactions of marine organisms; ecological role of soluble exoproducts in sea. Mailing Add: Dept of Biol Univ of Maine Univ Heights Augusta ME 04330

NORTON, DANIEL REMSEN, b Brooklyn, NY, Jan 27, 22; m 44; c 3. ANALYTICAL CHEMISTRY. Educ: Antioch Col, BS, 44; Princeton Univ, MA & PhD(anal chem), 48. Prof Exp: Asst bacteriologist, Antioch Col, 40-41, asst org chem, 43-44; analyst, Eastern State Corp, NY, 41-42; anal chemist, Merck & Co, Inc, NJ, 42-43 & Manhattan Proj, Princeton Univ, 44-46; monitor, Radiol Surv Sect, Oper Crossroads, 46; asst prof chem, George Washington Univ, 48-52; res chemist, Sprague Elec Co, Mass, 55-70; RES CHEMIST, BUR ANAL LABS, US GEOL SURV, 52-55 & 70- Concurrent Pos: Res fel, Woods Hole Oceanog Inst, 50-52, assoc, 52-58; instr, Williams Col, 60-68. Mem: Am Chem Soc. Res: Solid state and surface chemistry; ceramics; thermoanalytical techniques; polarography; absorption and emission spectrometry; particle and pore size analysis; silicate rock analysis; coal analysis. Mailing Add: Bur of Anal Labs US Geol Surv Fed Ctr Denver CO 80225

NORTON, DAVID L, b Newark, NJ, Nov 15, 45; m 73. BIOLOGICAL RHYTHMS. Educ: Iowa Wesleyan Col, BS, 67; Mich State Univ, MS, 70, PhD(physiol), 75. Prof Exp: NIH FEL BIORHYTHMS, NORTHWESTERN UNIV, EVANSTON, 75- Mem: AAAS; Fedn Am Scientists. Res: Statistical problems in the analysis of biological time series and the systems modeling of the neural networks which produce circadian activity. Mailing Add: Dept of Biol Sci O T Hogan Bldg Northwestern Univ Evanston IL 60201

NORTON, DAVID WILLIAM, b Mar 4, 44; US citizen; m 67; c 1. PHYSIOLOGICAL ECOLOGY. Educ: Harvard Col, AB, 67; Univ Alaska, MS, 70, PhD(zoophysiol), 73. Prof Exp: Staff ecologist environ consult, Dames & Moore, Fairbanks, 73-74; state supvr environ surveillance, Joint State-Fed Fish & Wildlife Adv Team for Surveillance Construct Trans-Alaska Oil Pipeline, 74-75; BIOL RES COORDR COASTAL & MARINE ECOL RES, NAT OCEANIC & ATMOSPHERIC ADMIN-OUTER CONTINENTAL SHELF ENVIRON ASSESSMENT PROG, 75- Concurrent Pos: Affil asst prof ecol, Inst Arctic Biol, Univ Alaska, 73- Mem: Sigma Xi; Ecol Soc Am. Res: Ecological energetics of migrating and breeding in tundra, shorebirds of Alaska. Mailing Add: Arctic Proj Off NOAA-OCSEP Geophys Inst Univ of Alaska Fairbanks AK 99701

NORTON, DENIS LOCKLIN, b Elba, NY, Jan 2, 39; m 60; c 3. GEOCHEMISTRY, GEOLOGY. Educ: Univ Buffalo, BA, 60; Univ Calif, Riverside, PhD(geol), 64. Prof Exp: Res asst geochem, Univ Calif, 60-64; from res geologist to sr res geologist, Kennecott Copper Corp, 64-69, chief geochem div, 69-73; ASST PROF GEOSCI, UNIV ARIZ, 73- Mem: Mineral Soc Am; Glaciol Soc; Am Inst Mining, Metall & Petrol Engrs; Geochem Soc; Mineral Asn Can. Res: Solving geological problems by developing a more thorough understanding of processes through geochemical investigations. Mailing Add: Dept of Geosci Univ of Ariz Tucson AZ 85721

NORTON, DON CARLOS, b Toledo, Ohio, May 22, 22; m 52; c 4. PLANT NEMATOLOGY. Educ: Univ Toledo, BS, 47; Ohio State Univ, MS, 49, PhD(bot), 50. Prof Exp: From asst prof to assoc prof plant path, Agr Exp Sta, Agr & Mech Col Tex, 51-59; assoc prof, 59-67, PROF PLANT PATH, IOWA STATE UNIV, 67- Mem: Am Phytopath Soc; Soc Nematol. Res: Nematology; root diseases; ecology of plant parasitic nematodes. Mailing Add: Dept of Bot & Plant Path Iowa State Univ Ames IA 50011

NORTON, DONALD ALAN, b Mt Kisco, NY, Mar 15, 20; m 48; c 2. MATHEMATICS. Educ: Harvard Univ, BS, 41; Univ Wis, PhD(math), 49. Prof Exp: ASSOC PROF MATH, UNIV CALIF, DAVIS, 49- Mem: Am Math Soc; Math Asn Am; Asn Comput Math; NY Acad Sci. Res: Generalized groups; algorithms. Mailing Add: Dept of Math Univ of Calif Davis CA 95616

NORTON, DORITA ANNE, crystallography, see 12th edition

NORTON, EDWARD W D, b Sommerville, Mass, Jan 3, 22. MEDICINE. Educ: Harvard Col, BA, 43; Cornell Univ, MD, 46; Am Bd Ophthal, dipl. Prof Exp: Intern, Cincinnati Gen Hosp, Ohio, 46-47; asst resident neurol, Kingsbridge Vet Hosp, Bronx, NY, 49-50; from asst resident to resident ophthal, New York Hosp-Cornell Med Ctr, 50-53; from instr to asst prof surg, Cornell Univ, 53-58; assoc prof ophthal & chief div, Sch Med, 58-59, PROF OPHTHAL & CHMN DEPT, SCH MED & JACKSON MEM HOSP, UNIV MIAMI, 59- Concurrent Pos: Mem retina serv, Mass Eye & Ear Infirmary, Wilmer Inst Ophthal, Johns Hopkins Hosp & Mayo Clin, 53-54; res fels ophthal, Howe Lab, Boston, 53-54; asst attend surgeon, New York Hosp-Cornell Med Ctr, 54-58; consult, Hosp Spec Surg, New York, Mem Hosp Cancer & Allied Dis, Bellevue Hosp & Kingsbridge Vet Hosp, 54-58. Mem: Am Ophthal Soc; Retina Soc; Am Acad Ophthal & Otolaryngol. Res: Ophthalmology; retinal diseases; neuroophthalmology. Mailing Add: Dept of Ophthal Univ of Miami Sch of Med Miami FL 33136

NORTON, ELINOR FRANCES, b Brooklyn, NY, July 22, 29. ANALYTICAL CHEMISTRY. Educ: Wellesley Col, BA, 51. Prof Exp: From jr chemist to assoc chemist, 51-74, CHEMIST, BROOKHAVEN NAT LAB, 74- Mem: Am Chem Soc. Res: Analytical chemistry, particularly trace elements; spectrophotometry; x-ray fluorescence; neutron activation; radiochemical separations and atomic absorption. Mailing Add: Brookhaven Nat Lab Upton NY 11973

NORTON, FRANCIS JAMES, b Ft Plain, NY, Aug 11, 99; m 24; c 3. PHYSICAL CHEMISTRY. Educ: Yale Univ, AB, 21, PhD(phys chem), 24. Prof Exp: Res chemist, Solvay Process Co, 24-30; res chemist, Res Lab, Gen Elec Co, 30-64; res prof physics, State Univ NY Albany, 64-75; RETIRED. Honors & Awards: Meyer Award, Am Ceramic Soc, 55; Mobay Award, Soc Plastics Indust, 68. Mem: Am Chem Soc; AAAS. Res: Vacuum practice; mass spectrometry; gas permeation and diffusion in polymers; metals; glass. Mailing Add: 1133 Eastern Ave Schenectady NY 12308

NORTON, HORACE WAKEMAN, III, b Lansing, Mich, Jan 17, 14; m 37; c 4. MATHEMATICAL STATISTICS. Educ: Univ Wis, BS, 35; Iowa State Col, MS, 37; Univ London, PhD(math statist), 40. Prof Exp: Asst lectr eugenics, Univ Col, Univ London, 37-40; res assoc physics, Univ Chicago, 40-42; sr meteorologist & math statistician, US Weather Bur, 42-47; statistician, AEC, Tenn, 47-50; PROF STATIST DESIGN & ANAL, UNIV ILL, URBANA, 50- Concurrent Pos: Agent, Agr Mkt Serv, USDA, 40-42. Mem: AAAS; Biomet Soc; Soc Human Genetics; Am Soc Animal Sci; Am Genetic Asn. Res: Design of experiments; animal breeding and genetics; statistical analysis; population genetics. Mailing Add: Dept of Animal Sci Univ of Ill Urbana IL 61801

NORTON, JAMES ALLAN, organic chemistry, see 12th edition

NORTON, JAMES AUGUSTUS, JR, b Philadelphia, Pa, Jan 3, 21; m 58; c 1. STATISTICS. Educ: Antioch Col, AB, 47; Purdue Univ, MS, 49, PhD, 59. Prof Exp: Res assoc statist, Div Educ Reference, Purdue Univ, 47-50, Statist Lab, 51-61, from instr to asst prof, 51-61; from asst prof to assoc prof statist, Dept Psychiat, Sch Med, 61-68, RES CONSULT, INST PSYCHIAT RES, IND UNIV, INDIANAPOLIS, 61-, PROF STATIST, DEPT PSYCHIAT, SCH MED, 68- Res: Tests of hypotheses in the case of unequal variances. Mailing Add: Dept of Psychiat Ind Univ Indianapolis IN 46202

NORTON, JAMES JENNINGS, b Elmira, NY, May 1, 18; m 46. GEOLOGY. Educ: Princeton Univ, AB, 40; Northwestern Univ, MS, 42; Columbia Univ, PhD, 57. Prof Exp: GEOLOGIST, US GEOL SURV, 42- Mem: AAAS; Mineral Soc Am; Geol Soc Am; Soc Econ Geol; Am Inst Mining, Metall & Petrol Eng. Res: Geology of Black Hills, South Dakota; pegmatites of Black Hills and elsewhere; industrial minerals; lithium. Mailing Add: US Geol Surv Denver CO 80225

NORTON, JOHN LESLIE, b Chanute, Kans, June 1, 45; m 66. MATHEMATICAL PHYSICS. Educ: Kans State Col Pittsburg, BA, 66; Univ Kans, MS, 68, PhD(physics), 70. Prof Exp: Fel physics, 70-71, staff mem weapons output, 71-74, STAFF MEM HYDRODYN, LOS ALAMOS SCI LAB, 74- Mem: Am Phys Soc; Am Chem Soc. Res: Numerical analysis; hydrodynamics; radiation transport; nuclear physics; computer science. Mailing Add: Los Alamos Sci Lab Group T-3 Numerical Hydrodyn Box 1663 Los Alamos NM 85745

NORTON, JOSEPH DANIEL, b Flat Rock, Ala, Oct 14, 27; m 50; c 1. GENETICS, HORTICULTURE. Educ: Auburn Univ, BS, 52, MS, 55; La State Univ, PhD(agron, hort), 61. Prof Exp: Asst hort, Auburn Univ, 54; asst veg crops specialist, Univ Fla, 54-60; from asst prof to assoc prof hort & from asst horticulturist to assoc horticulturist, 60-73, PROF HORT & HORTICULTURIST, 73- Mem: Am Soc Hort Sci; Am Genetic Asn. Res: Plant breeding, especially plum, muskmelon, watermelon, strawberry and apple; greenhouse tomato studies. Mailing Add: Dept of Hort Auburn Univ Auburn AL 36830

NORTON, KARL KENNETH, b London, Eng, Nov 13, 38; US citizen; m 62; c 2. MATHEMATICS. Educ: Yale Univ, BS, 59; Univ Chicago, MS, 61; Univ Ill, Urbana, PhD(math), 66. Prof Exp: Asst prof math, Univ Colo, Boulder, 66-73; INDEPENDENT RES, 73- Concurrent Pos: Off Naval Res res assoc, Univ Mich, Ann Arbor, 69-70; vis mem Inst Advan Study, 70-71; vis res math, Univ Geneva, 74; NSF res grant, 75. Mem: Am Math Soc; Math Asn Am. Res: Number theory. Mailing Add: 2235 Floral Drive Boulder CO 80302

NORTON, LILBURN LAFAYÈTTE, b Lenoir City, Tenn, Jan 2, 27; m 46; c 4. ORGANIC POLYMER CHEMISTRY. Educ: Carson-Newman Col, BS, 49; Northwestern Univ, MS, 51; Univ Tenn, PhD(chem), 54. Prof Exp: Asst chem, Northwestern Univ, 51; sr chemist, 54-59, group supvr, 59-65, RES ASSOC, RES & DEVELOP, E I DU PONT DE NEMOURS & CO, INC, 65- Mem: Am Chem Soc. Res: Fiber forming synthetic polymers; antiarthritic and anticarcinogenic chemicals. Mailing Add: 402 N Willey St Seaford DE 19973

NORTON, MATTHEW FRANK, b New York, NY, June 7, 24; m 63. GEOLOGY. Educ: Columbia Univ, AB, 43, AM, 48 & 56, PhD(geol), 58. Prof Exp: Assoc prof, 58-60, chmn dept, 61-70, chmn div nat sci & math, 62-66, PROF EARTH SCI, AM UNIV, 60- Mem: Geol Soc Am; Nat Asn Geol Teachers; Asn Am Geog; Soc Econ Paleont & Mineral. Res: Mineralogy; tectonics; Appalachian geology; modern marine sedimentation; volcanology; tectonic control of economic mineral deposits; geology and geography of Africa. Mailing Add: Am Univ Washington DC 20016

NORTON, NORMAN J, b DuQuoin, Ill, Apr 26, 33; m 55; c 2. BOTANY. Educ: Southern Ill Univ, BA, 58; Univ Minn, MS, 60, PhD(bot), 63. Prof Exp: Geologist, Humble Oil & Ref Co, 62-63; asst prof bot, Univ Minn, 63-64; from asst prof to prof biol, Hope Col, 64-74, chmn dept, 66-74; CHMN DEPT BIOL, BALL STATE UNIV, 74- Concurrent Pos: Consult, Gulf Oil Co; chmn financial develop comt, Am Asn Stratig Palynologists, 72- Mem: Bot Soc Am; Paleont Soc; Int Asn Plant Taxon; Paleont Asn Gr Brit. Res: Palynology, especially Devonian spore, Acritarch and Chitinozoa assemblages; Mesozoic spore pollen assemblages. Mailing Add: Dept of Biol Ball State Univ Muncie IN 47306

NORTON, PETER ROBERT, b Hingham, Eng, Dec 11, 42; m 65; c 2. PHYSICAL CHEMISTRY, SURFACE PHYSICS. Educ: Univ Nottingham, BSc, 63, PhD(phys chem), 66. Prof Exp: Imp Chem Industs fel, Univ Nottingham, 66-67; Nat Res Coun Can fel, 67-69; res assoc low temperature physics, McMaster Univ, 69-70; ASSOC RES OFFICER PHYS CHEM, ATOMIC ENERGY CAN LTD, 70- Mem: The Chem Soc; Chem Inst Can. Res: Low temperature physics, calorimetry as applied to the study of gas-solid interface and catalysis; electron spectroscopy as surface physics tool; kinetics, catalysis by clean metals. Mailing Add: Atomic Energy of Can Ltd Chalk River ON Can

NORTON, RICHARD E, b New York, NY, Mar 2, 28; m 66; c 2. THEORETICAL PHYSICS. Educ: Lehigh Univ, BS, 52; Univ Pa, PhD(physics), 58. Prof Exp: From asst prof to assoc prof, 60-69, PROF PHYSICS, UNIV CALIF, LOS ANGELES, 69- Mem: Am Phys Soc. Res: Theoretical research in field theory and elementary particle physics. Mailing Add: Dept of Physics Univ of Calif Los Angeles CA 90024

NORTON, RICHARD VAIL, b Hackensack, NJ, Feb 22, 40; m 62; c 2. ORGANIC CHEMISTRY. Educ: Rutgers Univ, BS, 61; Univ Maine, MS, 65, PhD(org chem), 67. Prof Exp: Chemist, Armstrong Cork Co, 61-63; sr res chemist, Mobay Chem Co, 67-68; from assoc res chemist to res chemist, 68-71, SR RES CHEMIST, RES & DEVELOP LAB, SUN OIL CO, 71- Concurrent Pos: Tech asst to vpres, Cryog Vessel Div, Sun Shipbuilding & Drydock, Chester, Pa, 74-75. Mem: AAAS; Am Chem Soc. Res: Petrochemicals; organosulfur chemistry; air pollution; ammoxidation of hydrocarbons; polyester synthesis; chemistry of nitriles. Mailing Add: Res & Develop Lab Sun Oil Co Marcus Hook PA 19061

NORTON, ROBERT ALAN. b Hazelton, Pa, Jan 3, 26; m 50; c 6. HORTICULTURE. Educ: Rutgers Univ, BS, 50, MS, 51; Mich State Univ, PhD, 54. Prof Exp: Asst, Mich State Univ, 51-54; from asst prof to assoc prof hort, Utah State Univ, 54-61; assoc agriculturist, Dept Pomol, Univ Calif, Davis, 61-62; supt & assoc horticulturist, 62-71, HORTICULTURIST, NORTHWESTERN WASH RES & EXTEN UNIT, WASH STATE UNIV, 71- Concurrent Pos: Consult, Kenai Natives Asn, Kenai, Alaska, 75- Mem: Am Soc Hort Sci; Am Pomol Soc. Res: Culture and physiology of horticultural plants. Mailing Add: Northwestern Wash Res & Exten Unit Wash State Univ Mt Vernon WA 98273

NORTON, ROBERT DILL, chemistry, deceased

NORTON, ROBERT HENRY, JR, astronomy, astrophysics, see 12th edition

NORTON, ROBERT JAMES, b Fitchburg, Mass, May 22, 14; m 42; c 2. ECONOMIC ENTOMOLOGY, PHYTOPATHOLOGY. Educ: Univ NH, MSc, 39; Univ Mass, PhD(entom, plant path), 51. Prof Exp: Res entomologist agr chems, US Rubber Co, 40-42, tech rep, 46-47, mgr tech serv, 48; assoc dir, Crop Protection Inst, 51-54, vpres, 65-69, DIR CROP PROTECTION INST, 54-, PRES, 69- Mem: Entom Soc Am; Am Phytopath Soc; Weed Sci Soc Am; Royal Entom Soc London. Res: Detection and development of biocides as insecticides, fungicides, herbicides and nematicides. Mailing Add: CPI Biol Res Ctr PO Drawer S Durham NH 03824

NORTON, SCOTTY JIM, b Marlow, Okla, Oct 21, 36; m 57; c 1. BIOCHEMISTRY, PESTICIDE CHEMISTRY. Educ: Abilene Christian Col, BS, 59; Univ Tex, Austin, PhD(chem), 63. Prof Exp: From asst prof to assoc prof, 63-71, PROF CHEM, N TEX STATE UNIV, 71- Concurrent Pos: USPHS grant, 63-68; Robert A Welch Found res grant, 63-77. Mem: Am Chem Soc; Am Soc Biol Chem. Res: Studies in amino acid metabolism; study of the glyoxalase system; membrane studies; synthesis and study of metabolite analogs; synthesis of insecticides. Mailing Add: Dept of Chem NTex State Univ Denton TX 76203

NORTON, STATA ELAINE, b Mt Kisco, NY, Nov 28, 22; m 49. PHARMACOLOGY, TOXICOLOGY. Educ: Univ Conn, BA, 43; Columbia Univ, MA, 45; Univ Wis, PhD(zool), 49. Prof Exp: Res scientist neuropharmacol, Burroughs Wellcome & Co, 49-62; from asst prof to assoc prof, 62-68, PROF PHARMACOL, UNIV KANS MED CTR, KANSAS CITY, 68- Mem: Am Soc Pharmacol & Exp Therapeut; Biomet Soc; Ecol Soc Am; Am Soc Zool; Soc Toxicol. Res: Neuropharmacology; animal behavior; brain development; effects of exposure to toxic substances on development and behavior. Mailing Add: Dept of Pharmacol Univ of Kans Med Ctr Kansas City KS 66103

NORTON, STEPHEN ALLEN, b Newton, Mass, May 21, 40; m 70; c 1. GEOLOGY. Educ: Princeton Univ, AB, 62; Harvard Univ, MA, 63, PhD(geol), 67. Prof Exp: Instr geol, Harvard Univ, 67-68; asst prof, 68-72, ASSOC PROF GEOL, UNIV MAINE, ORONO, 72- Concurrent Pos: Geologist, US Geol Surv, 64- Mem: Geol Soc Am; Mineral Soc Am. Res: Regional geology of New England; low temperature and pressure geochemistry. Mailing Add: Dept of Geol Sci Univ of Maine Orono ME 04473

NORTON, TED RAYMOND, b Stockton, Calif, Nov 16, 19; m 44; c 3. ORGANIC CHEMISTRY, PHARMACOLOGY. Educ: Univ of the Pac, BA, 40; Northwestern Univ, PhD(org chem), 43. Prof Exp: Dir agr chem lab, Dow Chem Co, Mich, 53-57, from asst dir to dir Britton Lab, 62-66, asst dir independent labs, 66-68; PROF PHARMACOL, UNIV HAWAII, MANOA, 68- Mem: AAAS; Am Chem Soc. Res: Isolation and characterization of marine natural products for antitumor and heart stimulant properties. Mailing Add: Leahi-Pharmacol 3675 Kilauea Ave Honolulu HI 96816

NORTON, VIRGINIA MARINO, b Memphis, Tenn, Nov 14, 34; m 52; c 3. VERTEBRATE ZOOLOGY, PHYSIOLOGY. Educ: Memphis State Univ, BS, 69, MS, 71, PhD(biol), 75. Prof Exp: FEL & INSTR PHYSIOL, DEPT PHYSIOL & BIOPHYS, CTR HEALTH SCI, UNIV TENN, MEMPHIS, 75- Mem: Am Inst Biol Sci; Am Soc Zool. Res: Physiology of the lower vertebrates; osmoregulation in the channel catfish and the vasopressor effects of synthetic and homologous angiotensins and angiotensin analogs in the American eel. Mailing Add: 637 Shotwell Memphis TN 38111

NORTON, WILLIAM THOMPSON, b Damariscotta, Maine, Jan 27, 29; m 57; c 2. NEUROCHEMISTRY. Educ: Bowdoin Col, AB, 50; Princeton Univ, MA, 52, PhD(org chem), 54. Prof Exp: Asst instr chem, Princeton Univ, 50-52; res chemist, E I du Pont de Nemours & Co, 53-57; instr biochem, 57-58, sr fel interdisciplinary prog, 57-60, assoc med & biochem, 58-59, asst prof biochem, 59-64, assoc prof neurol, 64-71, PROF NEUROL, ALBERT EINSTEIN COL MED, 71-, PROF NEUROSCI, 74- Concurrent Pos: Vis prof, Charing Cross Hosp Med Sch, London, 67-68; mem neurol A study sect, NIH, 71-76. Mem: AAAS; Am Soc Biol Chem; Am Soc Neurochem; Int Soc Neurochem; NY Acad Sci. Res: Lipid and myelin chemistry; chemical pathology; chemistry of isolated brain components. Mailing Add: Dept of Neurol Albert Einstein Col of Med New York NY 10461

NORUM, ENOCH BETUEL, b Hallock, Minn, July 10, 09; m 39; c 4. SOILS. Educ: Univ Minn, BS, 34; Iowa State Univ, PhD(soils), 44. Prof Exp: Soil scientist, Soil Conserv Serv, USDA, 36-40; res assoc, Iowa State Univ, 42-44; mgr farming opers, Green Giant Co, 44-47; assoc prof soils, 47-48, PROF SOILS & SOIL SCIENTIST, N DAK STATE UNIV, 48-, CHMN DEPT SOILS, 59- Mem: AAAS; Am Soc Agron; Soil Sci Soc Am; Soil Conserv Soc Am. Res: Soil tests for phosphorus; effect of fertility on moisture use by crops. Mailing Add: Dept of Soils NDak State Univ of Agr & Appl Sci Fargo ND 58103

NORUSIS, MARIJA JURATE, b Ansbach, Ger, Jan 3, 48; US citizen. BIOSTATISTICS. Educ: Univ Ill, BA, 68; Univ Mich, MPH, 71, PhD(biostatist), 73. Prof Exp: RES ASSOC STATIST & MED, UNIV CHICAGO, 73- Mem: Am Statist Asn; Biomet Soc. Res: Applications of statistics and computers to biomedical research. Mailing Add: Dept of Statist Univ of Chicago Chicago IL 60637

NORVELL, JOHN CHARLES, b Jacksonville, Tex, Jan 11, 40; m 66; c 2. MOLECULAR BIOPHYSICS, CRYSTALLOGRAPHY. Educ: Rice Univ, BA, 63; Yale Univ, MS, 65, PhD(physics), 68. Prof Exp: Fel physics, Res Establishment Riso, Roskilde, Denmark, 68-69; fel biophysics, Biophysics Lab, Univ Wis-Madison, 70-72; res assoc biophysics, Biol Dept, Brookhaven Nat Lab, Upton, NY, 72-75; BIOPHYSICIST, REACTOR RAD DIV, NAT BUR STANDARDS, WASHINGTON, DC, 75- Concurrent Pos: Guest scientist, Lab Molecular Biol, Nat Inst Arthritis, Metab & Digestive Dis, NIH, 75- Mem: AAAS; Biophys Soc. Res: Neutron and x-ray diffraction studies of the structure of proteins and other biological systems. Mailing Add: Reactor Rad Div Nat Bur Standards Washington DC 20234

NORVELL, JOHN EDMONDSON, III, b Charleston, WVa, Nov 18, 29; m 62; c 2. NEUROANATOMY. Educ: Morris Harvey Col, BS, 53; WVa Univ, MS, 56; Ohio State Univ, PhD(anat), 66. Prof Exp: Instr biol sci, Johnstown Col, Pittsburgh, 56-60; asst prof, Otterbein Col, 60-62; asst instr anat, Ohio State Univ, 62-65; ASSOC PROF ANAT, MED COL VA, VA COMMONWEALTH UNIV, 66- Concurrent Pos: Consult, US Naval Hosp, Portsmouth, 66-71. Mem: AAAS; Am Asn Anat; Soc Neurosci; Transplantation Soc. Res: Degeneration and regeneration of nerves in transplanted organs; localization of biogenic amines and transmitters in the nervous system. Mailing Add: Dept of Anat Med Col of Va Richmond VA 23298

NORVELL, MICHAEL JIMMY, b Dallas, Tex, Nov 24, 41; m 66; c 2. NUTRITION, TOXICOLOGY. Educ: Tex A&M Univ, BS, 65; Cornell Univ, PhD(nutrit), 70. Prof Exp: NIH fel toxicol, Inst Exp Path & Toxicol, Albany Med Col, 70-71; reviewer researcher nutrit, Div Nutrit Sci, Bur Vet Med, 71-74, SCIENTIST TOXICOL, NAT CTR TOXICOL RES, FOOD & DRUG ADMIN, 74- Concurrent Pos: Mem comt drugs & chems, Poultry Sci Asn, 72-75; adj asst prof Animal Sci, Univ Ark, 74- Mem: AAAS; Am Soc Animal Sci; Animal Nutrit Res Coun; Soc Environ Geochem & Health; Coun Agri Sci & Technol. Res: Continuously feeding graded levels of estrogens to experimental animals; measure changes in blood hormone levels, histological and other hormone induced changes in uterine and mammary tissue associated with estrogen induced neoplasia. Mailing Add: Nat Ctr Toxicol Res Jefferson AR 72079

NORVELL, WENDELL ALEXANDER, soil chemistry, see 12th edition

NORVIG, TORSTEN, b Copenhagen, Denmark, Jan 20, 26; US citizen; m 51; c 3. MATHEMATICS. Educ: Copenhagen Univ, BSc, 47; Brown Univ, MSc, 60, PhD(math), 66. Prof Exp: Asst prof math, Univ Mass, 60-66; from asst prof to assoc prof, 66-74, ASSOC PROF MATH, WELLESLEY COL, 74- Mem: Am Math Soc; Math Asn Am; Soc Indust & Appl Math. Res: Linear economic models; convexity; linear programming and extensions; game theory. Mailing Add: Dept of Math Wellesley Col Wellesley MA 02181

NORWICH, KENNETH HOWARD, b Toronto, Ont, May 8, 39; m 63; c 2. BIOPHYSICS, PHYSIOLOGY. Educ: Univ Toronto, MD, 63, BSc, 67, MSc, 68, PhD(physics), 70. Prof Exp: ASSOC PROF PHYSIOL & APPL SCI, INST BIOMED ENG, UNIV TORONTO, 70- Mem: Biophys Soc; Can Physiol Soc; Can Med Asn; Can Med & Biol Eng Soc. Res: Mathematical studies in physiology and medicine, particularly studies of the transport of tracers; indicator dispersion in the vascular system; mathematical studies of metabolism; theoretical studies of sensory perception. Mailing Add: Rm 3305 Dept of Physiol Univ of Toronto Toronto ON Can

NORWINE, JAMES RANDOLPH, b St Louis, Mo, Jan 30, 43; m 65; c 3. CLIMATOLOGY. Educ: Southeast Mo State Univ, BS, 64; Ind State Univ, MS, 65, PhD(phys geog), 71. Prof Exp: Instr phys geog, Stephen F Austin State Univ, 65-67 & Univ Wis, 67-68; fel, Ind State Univ, 68-70, NDEA fel, 70-71; asst prof, Northeastern Ill Univ, 71-72; ASST PROF PHYS GEOG, TEX A&I UNIV, 72- Concurrent Pos: Asn Am Geogrs Gibson grant agr res in the South, 74- Mem: Am Meteorol Soc; Asn Am Geogrs. Res: Applied climatology, particularly urban and agricultural; climatic control of food production; urban heat islands; human ecology. Mailing Add: Dept of Geog & Geol Tex A&I Univ Kingsville TX 78363

NORWOOD, CHARLES ARTHUR, b Crystal City, Tex, Jan 8, 38. AGRONOMY. Educ: Tex A&I Univ, BS, 61; Okla State Univ, MS, 69, PhD(soil sci), 72. Prof Exp: Res technician soils, Agr Res Serv, USDA, Tex, 63-67; res asst soil fertil, Okla State Univ, 67-72; ASST PROF DRYLAND SOILS RES, GARDEN CITY EXP STA, KANS STATE UNIV, 72- Mem: Am Soc Agron; Soil Sci Soc Am. Res: Soil fertility; management of dryland soils-research pertaining to efficient water use under dryland conditions. Mailing Add: Garden City Exp Sta Kans State Univ PO Box L Garden City KS 67846

NORWOOD, FREDERICK REYES, b Mexico City, Mex, May 13, 39; US citizen; m 66; c 2. SOLID MECHANICS, APPLIED MATHEMATICS. Educ: Univ Calif, Los Angeles, BS, 62; Calif Inst Technol, MS, 63, PhD(appl mech), 67. Prof Exp: MEM TECH STAFF, SANDIA LABS, 66- Mem: Am Soc Mech Engrs; Soc Indust & Appl Math; Am Geophys Union. Res: Theoretical study of transient phenomena in liquids, gases and solids. Mailing Add: 600 Bryn Mawr Dr NE Albuquerque NM 87106

NORWOOD, JAMES S, b Burleson, Tex, Oct 2, 32; m 62. REPRODUCTIVE PHYSIOLOGY, GENETICS. Educ: Tex Tech Col, BS, 54; Kans State Univ, MS, 55, PhD(reprod physiol), 63. Prof Exp: Instr dairy sci, Southwest Tex State Col, 55-56 & 58-60; asst prof biol, Arlington State Col, 62-63; assoc prof reprod physiol, 63-68, assoc prof biol, 68-71, PROF BIOL, E TEX STATE UNIV, 71- Mem: Am Soc Animal Sci; Sigma Xi. Res: Post-partum regression of the bovine uterus; conception rates; inhibition of estrus and effect of steroid hormones on the endometrium of the uterus. Mailing Add: Dept of Biol E Tex State Univ ETex Sta Commerce TX 75428

NORWOOD, JOSEPH, JR, b Baltimore, Md, Sept 27, 35; m 59; c 4. PLASMA PHYSICS. Educ: Univ NC, BS, 58; Col William & Mary, MA, 63; Stevens Inst Technol, PhD(physics), 67. Prof Exp: Physicist, NASA Langley Res Ctr, 57-67; ASST PROF PHYSICS, UNIV MIAMI, 67- Concurrent Pos: Staff transl, Consult Bur Div, Plenum Publ Co, 66- Mem: Am Phys Soc; Am Astron Soc. Res: Plasma accelerators; controlled thermonuclear fusion; astrophysics; plasma instabilities in astrophysics. Mailing Add: Dept of Physics Univ of Miami Coral Gables FL 33124

NORWOOD, MARCUS H, solid state physics, see 12th edition

NOSAL, EUGENE ADAM, b Chicago, Ill, Jan 15, 42. GEOPHYSICS. Educ: St Procopius Col, BS, 63; Univ Wyo, MS, 65, PhD(physics), 69. Prof Exp: SCIENTIST, DENVER RES CTR, MARATHON OIL CO, 69- Mem: AAAS; Am Asn Physics Teachers; Am Geophys Union; Soc Explor Geophysicists. Res: Analysis of digitized data for geophysical interpretation; application of probability and statistics to geological data. Mailing Add: Denver Res Ctr Marathon Oil Co PO Box 269 Littleton CO 80120

NOSANOW, LEWIS H, b Philadelphia, Pa, July 9, 31; m 55; c 2. THEORETICAL PHYSICS, LOW TEMPERATURE PHYSICS. Educ: Univ Pa, BA, 54; Univ Chicago, PhD(chem physics), 58. Prof Exp: NSF fel, 58-59; res assoc physics, Inst Theoret Physics, Univ Utrecht, 59-60; asst res physicist, Univ Calif, San Diego, 60-62; from asst prof to prof physics, Univ Minn, Minneapolis, 62-73; prof physics & astron & chmn dept, Univ Fla, 73-74; HEAD CONDENSED MATTER SECT, DIV MAT RES, NSF, 74- Concurrent Pos: Guggenheim fel, 66-67. Mem: Am Phys Soc. Res: Statistical mechanics and phase transition; liquid and solid helium; neutron stars. Mailing Add: Div of Mat Res NSF Washington DC 20550

NOSHAY, ALLEN, b Philadelphia, Pa, Oct 14, 33; m 56; c 2. ORGANIC CHEMISTRY, POLYMER CHEMISTRY. Educ: Temple Univ, BA, 55, MS, 57, PhD(polymer org chem), 59. Prof Exp: Chemist, Esso Res & Eng Co, 59-61, proj leader polymer chem, 61-65; res chemist, 65-69, proj scientist, 69-75, RES SCIENTIST, PLASTICS DIV, UNION CARBIDE CORP, 75- Mem: Am Chem Soc. Res: Polymer synthesis and modification; block copolymers; epoxy resins; polyolefin catalysis. Mailing Add: 66 Wellington Rd East Brunswick NJ 08816

NOSHKIN, VICTOR E, JR, nuclear chemistry, see 12th edition

NOSHPITZ, JOSEPH DOVE, b New York, NY, Aug 31, 22; m 56. CHILD PSYCHIATRY. Educ: Univ Louisville, MD, 45; Baltimore-DC Psychoanal Inst, grad psychoanal, 69, grad child psychoanal, 71. Hon Degrees: Hon grad, Menninger Sch Psychiat, 75. Educ: Chief children's serv, Topeka State Hosp, Kans, 51-56; chief children's unit, NIMH, 56-60; dir clin inst, Hillcrest Children's Ctr, Children's Hosp, DC, 69-74, DIR EDUC, DEPT CHILD PSYCHIAT, CHILDREN'S HOSP & NAT MED CTR, WASH, DC, 74- Concurrent Pos: Vis prof child psychiat, Med Sch, Howard Univ; clin prof child health & human develop, Med Sch, George Washington Univ; med dir, Florence Crittenton Home, Washington, DC, 73-75; vis prof child psychiat, Sch Med, Tel Aviv Univ, 75- Mem: AMA; Am Psychiat Asn; fel Am Orthopsychiat Asn; Am Acad Child Psychiat; Am Asn Children's Residential Ctrs. Res: Adolescence; delinquency; treatment of emotionally disturbed children; residential treatment. Mailing Add: 3141 34th St NW Washington DC 20008

NOSKOWIAK, ARTHUR FREDRICK, b Galt, Calif, Nov 14, 20. WOOD TECHNOLOGY. Educ: Univ Calif, Berkeley, BS, 42, MF, 49; State Univ NY Col Forestry, Syracuse, PhD(wood technol), 59. Prof Exp: From instr to asst prof forestry, Colo State Univ, 51-54; instr wood technol, State Univ NY Col Forestry, Syracuse Univ, 58-59; asst prof, 60-71, ASSOC PROF FORESTRY & WOOD TECHNOL, WASH STATE UNIV, 71- Mem: Soc Am Foresters; Forest Prod Res Soc; Soc Wood Sci & Technol. Res: Spiral grain in trees; strength properties of wood. Mailing Add: Dept of Forestry Wash State Univ Pullman WA 99163

NOSSAL, NANCY, b Fall River, Mass, Feb 9, 37; m 59; c 3. BIOCHEMISTRY. Educ: Cornell Univ, AB, 58; Univ Mich, PhD(biochem), 63. Prof Exp: NIH fels, Brussels, 63-64 & Bethesda, Md, 64-65, RES CHEMIST, NAT INST ARTHRITIS & METAB DIS, 66- Mem: Am Soc Biol Chemists. Res: Enzymology of nucleic acids. Mailing Add: Nat Inst Health Bldg 4 Room 106 Bethesda MD 20014

NOSSAL, RALPH J, b Brooklyn, NY, Dec 26, 37; m 59; c 3. BIOPHYSICS, PHYSICS. Educ: Cornell Univ, BEng Phys, 59; Univ Mich, MS, 61, PhD(nuclear eng), 63. Prof Exp: NSF fel statist mech, Brussels, 63-64; Nat Acad Sci-Nat Sci Found res assoc, Nat Bur Standards, 64-66; PHYSICIST, NIH, 66- Mem: Am Phys Soc; Biophys Soc. Res: Statistical physics; laser scattering; membrane biophysics; cellular physiology. Mailing Add: Nat Inst of Health Bethesda MD 20014

NOSSAMAN, NORMAN L, b Cherokee, Okla, Jan 21, 32; m 55; c 3. SOIL FERTILITY. Educ: Okla State Univ, BS, 53, MS, 57; Kans State Univ, PhD(soils fertil), 63. Prof Exp: Chg of dryland soil mgt, Garden City Br Exp Sta, Kans State Univ, 60-65; agronomist, Western Ammonia Corp, 65-69; CHIEF AGRONOMIST, NIPAK, INC, 69- Mem: Am Soc Agron. Res: Dryland soil management; soil fertility; soil moisture; tillage methods. Mailing Add: 13411 Rolling Hills Lane Dallas TX 75240

NOSSEL, HYMIE L, b Cape Town, SAfrica, July 11, 30; m 65; c 2. HEMATOLOGY. Educ: Univ Cape Town, MB, ChB, 53; Oxford Univ, DPhil(med), 62. Prof Exp: Intern med, Groote Schuur Hosp & Univ Cape Town, 54-55, registr, 56-59, asst physician & lectr, 64-65; res bursar, Blood Coagulation Res Unit, Med Res Coun, Eng, 60-62; from res assoc to sr res asst hemat, Mt Sinai Hosp, New York, 63-66; from asst prof to assoc prof med, 66-72, PROF MED, COL PHYSICIANS & SURGEONS, COLUMBIA UNIV, 72- Mem: Am Fedn Clin Res; Asn Am Physicians; Am Soc Clin Invest; Am Soc Hemat; Am Soc Exp Path. Res: Blood coagulation. Mailing Add: Columbia Univ Col of Phys & Surg 630 W 168th St New York NY 10032

NOSTRAND, RICHARD LEE, b Providence, RI, May 21, 39; m 63; c 2. GEOGRAPHY. Educ: Univ Wash, BA, 61; Univ Calif, Los Angeles, MA, 64, PhD(geog), 68. Prof Exp: From instr to asst prof geog, Univ Mass, Amherst, 68-73; ASSOC PROF GEOG, UNIV OKLA, 73- Concurrent Pos: Nat Endowment Humanities fel, Univ Tex, Austin, 71-72. Mem: Asn Am Geog; Am Geog Soc; Nat Coun Geog Educ; Asn Pac Coast Geog. Res: The Mexican-American population; historical geography of the Southwest; comparative settlement institutions in the United States. Mailing Add: Dept of Geog Univ Okla Norman OK 73069

NOTARO, VINCENT ANTHONY, organic chemistry, see 12th edition

NOTATION, ALBERT DAVID, b Moosomin, Sask, Oct 28, 35; m 63; c 2. ENDOCRINOLOGY, BIOCHEMISTRY. Educ: Univ Sask, BE, 58, MSc, 59; McMaster Univ, PhD(org chem), 64. Prof Exp: Lab asst fats & oils, Prairie Regional Lab, Nat Res Coun Can, 53-54; NIH Steroid Training Prog fel, 64-66, res assoc biochem, 66-67, ASST PROF BIOCHEM, SCH MED, UNIV MINN, MINNEAPOLIS, 67-, ASST PROF OBSTET & GYNEC & LAB MED, 71- Concurrent Pos: Ayerst Squibb travel fel, 68; Minn Med Found grant, 71. Mem: AAAS; Am Chem Soc; Chem Inst Can; Endocrine Soc; Soc Study Reproduction. Res: Steroid metabolism and biochemistry; radioimmunoassay of hormones; competitive protein binding assays. Mailing Add: Dept of Obstet & Gynec Box 395 Mayo Mem Hosp Minneapolis MN 55455

NOTEBOOM, WILLIAM DUANE, b East Fairview, NDak, Mar 31, 33; m 67; c 1. BIOCHEMISTRY. Educ: Ore State Univ, BS, 55, MS, 61; Univ Ill, PhD(physiol), 65. Prof Exp: Fel, McArdle Inst Cancer Res, Univ Wis, 64-67; ASSOC PROF BIOCHEM, UNIV MO-COLUMBIA, 67- Mem: Am Chem Soc. Res: Regulation of cell growth and metabolism. Mailing Add: Dept of Biochem Univ of Mo Columbia MO 65201

NOTH, PAUL HENRY, b Minneapolis, Minn, Jan 24, 07; m 33; c 2. CLINICAL MEDICINE. Educ: Univ Minn, BS, 27, MD, 31, MS, 38; Am Bd Internal Med, dipl, 41. Prof Exp: Intern, Ancker Hosp, St Paul, Minn, 31-32; med house officer, Peter Bent Brigham Hosp, Boston, 32-34; fel med, Mayo Clin, Minn, 34-38; from asst prof to assoc prof, Col Med, Wayne State Univ, 38-48; ATTEND PHYSICIAN, LOS ALAMOS MED CTR HOSP, 50-; ASSOC MED, SCH MED, UNIV NMEX, 65- Concurrent Pos: Pvt pract internal med, 50- Mem: AMA; Am Fedn Clin Res; Am Col Physicians; Am Soc Internal Med. Res: Electrocardiography; cardiovascular and degenerative diseases; infectious diseases. Mailing Add: Los Alamos Med Ctr Los Alamos NM 87544

NOTHDURFT, ROBERT RAY, b Cape Girardeau, Mo, Nov 13, 39; m 61; c 2. PHYSICS. Educ: Wash Univ, St Louis, AB, 61; Univ Mo-Rolla, MS, 64, PhD(physics), 67. Prof Exp: Res physicist, US Bur Mines, Mo, 62-67; PROF PHYSICS, NORTHEAST MO STATE COL, 67- Res: Kilocycle range dislocation damping, especially in magnesium single crystals. Mailing Add: Dept of Physics Northeast Mo State Col Kirksville MO 63501

NOTIDES, ANGELO C, b New York, NY, Dec 11, 36; m 61; c 2. ENDOCRINOLOGY, BIOCHEMISTRY. Educ: Hunter Col, BA, 59, MA, 62; Univ Ill, PhD(physiol), 66. Prof Exp: ASSOC PROF PHARMACOL, SCH MED & DENT, UNIV ROCHESTER, 68- Res: Mechanism of hormone action; reproductive biochemistry. Mailing Add: Dept of Pharmacol Univ Rochester Sch Med & Dent Rochester NY 14642

NOTKINS, ABNER LOUIS, b New Haven, Conn, May 8, 32; m 69. VIROLOGY, IMMUNOLOGY. Educ: Yale Univ, BA, 53; NY Univ, MD, 58. Prof Exp: Intern internal med, Johns Hopkins Hosp, 58-59, asst resident, 59-60; res assoc, Nat Cancer Inst, 60-61; investr, Lab Microbiol & Immunol, 61-67, chief virol sect, 67-73, CHIEF LAB ORAL MED, NAT INST DENT RES, 73- Mem: AAAS; Am Soc Exp Path; Infectious Dis Soc Am; Am Asn Immunol; Am Soc Microbiol. Res: Slow viruses; viral immunology and immunopathology; virus-induced diabetes. Mailing Add: Lab Oral Med Bldg 30 Rm 121 Nat Inst of Dent Res Bethesda MD 20014

NOTLEY, NORMAN THOMAS, b Bristol, Eng, Apr 10, 28; US citizen. POLYMER CHEMISTRY, PHYSICAL CHEMISTRY. Educ: Bristol Univ, BSc, 49, PhD(phys chem), 52. Prof Exp: Res assoc polymer chem, Cornell Univ, 52-54; res chemist, E I du Pont de Nemours & Co, 54-59; res chemist, Metal Box Co, Eng, 59-60, dept supvr phys chem, 60-62; chief chemist, Kalvar Corp, La, 62-63, dir res photochem, 63-65, dir res, 65-66; dir chem res, Bus Equip Group, Bell & Howell, 66-69; PRES, PHOTOMEDIA CO, 69- Mem: Am Chem Soc; Soc Photog Sci & Eng. Res: Solution properties of polymers; polymerization kinetics; photopolymerization; oxidation kinetics of polyolefines; permeability of plastics; unconventional photographic and reprographic systems. Mailing Add: Photomedia Co Box 462 Sierra Madre CA 91024

NOTO, THOMAS ANTHONY, b Tampa, Fla, Dec 27, 31; m 63; c 2. PATHOLOGY. Educ: Spring Hill Col, BS, 53; St Louis Univ, MD, 57. Prof Exp: Instr path, Sch Med, Univ Miami, 62-64; asst prof, Med Col Ala, 64-65, clin path, 65-68; assoc prof, Med Ctr, Univ Ala, Birmingham, 68-69; ASSOC DIR CLIN PATH, SCH MED, UNIV MIAMI, 69- Concurrent Pos: Asst clin pathologist, Jackson Mem Hosp, 62-64; asst clin pathologist & dir blood bank, Med Ctr, Univ Ala, Birmingham, 64-69. Mem: Fel Am Soc Clin Path; Col Am Path; NY Acad Sci. Res: Clinical immunology; immunohematology; protein chemistry. Mailing Add: Clin Path Labs Jackson Mem Hosp Miami FL 33136

NOTRICA, SOLOMON, b Portland, Ore, June 16, 23; m 47; c 4. BIOCHEMISTRY. Educ: Univ Southern Calif, AB, 48, PhD(biochem), 59. Prof Exp: Asst, Univ Southern Calif, 52-54; res assoc, Los Angeles County Gen Hosp, 54-59; clin biochemist, St John's Hosp, 59-62 & 65-70; CLIN BIOCHEMIST, LAB PROCEDURES DIV, UPJOHN CO, 70- Concurrent Pos: Clin biochemist, Thatcher Clin Lab, Pasadena, Calif, 61-62; consult, Pac Hosp, Long Beach, Sharp & Scripps Mem Hosps, San Diego, Vet Admin Hosp, San Fernando, Calif, 62-65 & Childrens Hosp, Los Angeles, 67-71; mem adv comt clin lab technol, Dept Pub Health, State of Calif, 69-72. Mem: AAAS; Am Asn Clin Chem; Asn Advan Med Instrumentation; Soc Nuclear Med; NY Acad Sci. Res: Clinical, steroid and thyroid biochemistry; progesterone metabolism; binding and transport of thyroid hormone by thyroxine binding globulin; phenylketonuria. Mailing Add: Procedures Div Upjohn Co 6330 Variel Ave Woodland Hills CA 91364

NOUJAIM, ANTOINE AKL, b Cairo, Egypt, Feb 26, 37; m 64. NUCLEAR PHARMACY. Educ: Cairo Univ, BS, 58; Purdue Univ, MSc, 63, PhD(bionucleonics), 65. Prof Exp: Head res & statist dept, Gen Orgn Drugs, Cairo Univ, 59-61; res assoc bionucleonics, Purdue Univ, 65-66; from asst prof to assoc prof bionucleonics, 66-73, chmn dept, 68-74, PROF NUCLEAR PHARM, UNIV ALTA, 73-, CHMN RADIATION CONTROL, 72- Concurrent Pos: Nat Res Coun fel, Cairo Univ, 61-65; consult water resources div, Alta Dept Agr, 66-; sr vis scientist, CSIRO, Australia, 74-75; sr res fel nuclear med, Dr W W Cross Cancer Inst, 75- Mem: AAAS; NY Acad Sci; Am Pharmaceut Asn; Soc Nuclear Med. Res: Radiation effects on biological systems; drug metabolism; isotope dilution methods; activation analysis; research and development of radiopharmaceuticals; clinical applications of radioactive drugs. Mailing Add: Fac of Pharm Univ of Alta Edmonton AB Can

NOVACO, ANTHONY DOMINIC, b Orange, NJ, Mar 24, 43; m 66; c 2. THEORETICAL SOLID STATE PHYSICS, SURFACE PHYSICS. Educ: Stevens Inst Technol, BS, 64, MS, 66, PhD(physics), 69. Prof Exp: Assoc res scientist physics, Hudson Labs, Columbia Univ, 67-69; fel physics, Battelle Mem Inst, 69-71; asst physicist, Brookhaven Nat Lab, 71-73; ASST PROF PHYSICS, LAFAYETTE COL, 73- Mem: Am Phys Soc; Sigma Xi. Res: Theoretical research in condensed matter physics; statistical physics; physisorption and surface physics; quantum liquids and solids. Mailing Add: Dept of Physics Lafayette Col Easton PA 18042

NOVAK, ALFRED, b Chicago, Ill, Jan 28, 15; m 44; c 3. HEALTH SCIENCES. Educ: Univ Chicago, BS, 36, MS, 42; Chicago Teachers Col, ME, 40; Mich State Univ, PhD, 50. Prof Exp: From instr to prof biol, Mich State Univ, 44-60; chief div sci & math, 60-74, DIR DIV ALLIED HEALTH, STEPHENS COL, 74- Concurrent Pos: NIH spec res fel, Calif Inst Technol, 50-51; Guggenheim fel, Cambridge Univ, 57-58; adv biol ed, Encycl Americana, 59-69; consult biol sci curriculum studies, Am Inst Biol Sci, 60-62; collabr, Nat Sci Adv Bd, Encycl Britannica Films, Inc, 65-; res fel, Inst Path, Med Sch, Univ Bologna, 72. Res: Protein synthesis and hormonal influence. Mailing Add: Div of Allied Health Stephens Col Columbia MO 65201

NOVAK, ARTHUR FRANCIS, b Baltimore, Md, Oct 25, 17; m 46. CHEMISTRY. Educ: Univ Md, BS, 37; Univ Ala, MS, 40; Purdue Univ, PhD(chem), 47. Prof Exp: Supvr fermentation, Calvert Distilling Co, Md, 40-41, prod supvr, 41-42; prod supvr res & develop, Joseph E Seagram & Sons, Inc, Ky, 42-44; from asst prof to assoc prof chem, Univ Fla, 47-51; head biol res, Nutrilite Prod, Inc, Calif, 51-54; prof agr chem & biochem, 54-62, PROF FOOD SCI & HEAD DEPT, LA STATE UNIV, BATON ROUGE, 62- Concurrent Pos: Ford Found consult, Latin Am, 68-69; tech dir, Int Shrimp Coun, 68-; prof, Fac Med & consult, Oceanog Inst, Univ Sao Paulo, 69-; mem sea-grant review bd, Nat Oceanog & Atmospheric Admin. Mem: AAAS; Am Chem Soc; Am Soc Microbiol; Nat Shell Fisheries Asn; fel Am Inst Chem. Res: Pharmaceutical and nutritional products. Mailing Add: 656 College Hill Dr Baton Rouge LA 70808

NOVAK, ERNEST RICHARD, b Szoce, Hungary, Apr 17, 40; US citizen; m 62; c 2. ORGANIC CHEMISTRY. Educ: Oberlin Col, BA, 63; Univ Rochester, PhD(org chem), 67. Prof Exp: Res chemist, Plastics Dept, Exp Sta, Wilmington, Del, 66-73, SR RES CHEMIST, PLASTICS PROD & RES, E I DU PONT DE NEMOURS & CO, INC, PARKERSBURG, 73- Mem: Am Chem Soc. Res: Oxonium ions; optically active compounds; monomer and polymer research; polymerization chemistry. Mailing Add: PO Box 1217 E I du Pont de Nemours & Co Inc Parkersburg WV 26101

NOVAK, IRWIN DANIEL, b New York, NY, June 23, 42; m 67; c 3. GEOMORPHOLOGY. Educ: Hunter Col, AB, 66; Univ Fla, MS, 68; Cornell Univ, PhD(geol), 71. Prof Exp: Asst prof, 71-75, ASSOC PROF GEOL, UNIV MAINE, PORTLAND-GORHAM, 75- Concurrent Pos: Mem adv bd, Maine Land Use Regulation Comn, 73- Mem: AAAS; Geol Soc Am; Soc Econ Paleontologists & Mineralogists. Res: Coastal and fluvial geomorphology of Maine; sedimentary processes in coastal and fluvial environments. Mailing Add: Univ of Maine 96 Falmouth St Portland ME 04103

NOVAK, JOSEF FRANTISEK, b Ceske Budejovice, Czech, Oct 24, 42; m 66. CANCER BIOLOGY. Educ: Prague Agr Univ, MS, 64; Yale Univ, PhD(biol), 74. Prof Exp: Res asst plant physiol, Inst Exp Bot, Czech Acad Sci, 65-68; instr biochem, Dept Biol, Princeton Univ, 73-75; RES ASSOC CANCER BIOL, DEPT CLIN & BIOCHEM PHARMACOL, ST JUDE CHILDREN'S RES HOSP, 75- Mem: Soc Exp Biol Gt Brit; Sigma Xi; Int Soc Differentiation. Res: Molecular and biochemical aspects of tumor metabolism with special emphasis on endogenous and exogenous substances determining the rate of cell division in normal and tumor cells. Mailing Add: Dept Pharmacol St Jude Children's Hosp Memphis TN 38101

NOVAK, JOSEPH DONALD, b Minneapolis, Minn, Dec 2, 30; m 53; c 3. BIOLOGY. Educ: Univ Minn, BS, 52, MA, 54, PhD(bot, educ), 58. Prof Exp: Asst bot, Univ Minn, 52-56, instr, 56-57; asst prof biol, Kans State Teachers Col, 57-59; from asst prof biol to assoc prof biol & educ, Purdue Univ, 59-67; coordr, Shell Merit Progs, 68-72, PROF SCI EDUC & CHMN SCI & ENVIRON EDUC DIV, CORNELL UNIV, 67- Concurrent Pos: David Ross fel, 64; res assoc, Harvard Univ, 65-66. Mem: AAAS; Bot Soc Am; Am Asn Biol Teachers (vpres, 64); Nat Asn Res Sci Teaching (exec secy, 62-67, pres, 68); Nat Sci Teachers Asn. Res: Biological education; analysis of concept learning; plant morphology. Mailing Add: Div of Sci & Environ Educ Cornell Univ Ithaca NY 14850

NOVAK, LADISLAV PETER, b Chlum, Czech, Sept 18, 22; m 50; c 4. PHYSIOLOGICAL ANTHROPOLOGY. Educ: Charles Univ, Prague, BSc, 48; Univ Minn, Minneapolis, MA, 61, PhD, 62. Prof Exp: Fel physiol, Univ Minn, 62-64, vis asst prof physiol anthrop, 62-63, asst prof physiol anthrop, 63-66, consult, Mayo Grad Sch Med, 66-72; PROF ANTHROP, SOUTHERN METHODIST UNIV, 72- Concurrent Pos: NIH grants, 64-72; AMA lectr, 71-72; mem, Int Comt Standardization Phys Fitness Tests; NIH proj officer, Foreign Res Prog. Mem: Am Asn Phys Anthrop; NY Acad Sci; Brit Soc Study Human Biol; Am Aging Asn; Am Asn Clin Nutrit. Res: Physiological growth and development; body composition; physiology of exercise. Mailing Add: Dept of Anthrop Southern Methodist Univ Dallas TX 75222

NOVAK, LEO JOSEPH, biochemistry, see 12th edition

NOVAK, MILAN VACLAV, b Cobb, Wis, Dec 24, 07; m 34; c 3. MEDICAL EDUCATION, MICROBIOLOGY. Educ: Macalester Col, AB, 29; Univ Minn, MS, 30, PhD(bact), 32, BS, 36, MB & MD, 38. Hon Degrees: DSc, Macalester Col, 47. Prof Exp: Asst prof bact, Univ Tenn, 32-33; instr, Univ Minn, 35-40; assoc prof bact & pub health, Univ Ill Med Ctr, 40-43, prof microbiol & head dept, 43-64, assoc dean grad col, 48-69; LECTR INTERNAL MED, COL MED, UNIV ARIZ, 70- Concurrent Pos: Consult, Vet Admin Hosp, Hines, Ill, 46-68; pres, Tuberc Inst Chicago & Cook County, 60-; Grad Deans of Midwest Conf, 60-; coodr human subj in res, Univ Ariz, 75. Mem: AAAS; Soc Exp Biol & Med; Am Tuberc Asn; AMA; Sigma Xi. Res: Morphology of bacteria; blood storage in blood banks; sulfonamides; antibiotics; surgical bacteriology. Mailing Add: 1231 Sobre Lomas Tucson AZ 85718

NOVAK, ROBERT LOUIS, b Chicago, Ill, Oct 1, 37; m 65; c 1. BIOCHEMISTRY. Educ: Xavier Univ, Ohio, AB, 59; Univ Del, PhD(cell physiol, biochem), 64. Prof Exp: NIH fel, Dept Biol Sci, Purdue Univ, 63-65; asst prof biochem, Univ NH, 65-67; res fel, Harvard Univ, 67-69; asst prof, 69-74, ASSOC PROF CHEM & BIOL, DEPAUL UNIV, 74- Concurrent Pos: Res grants, Res Corp, 66-67 & 71-72, Am Cancer Soc, 71-72 & NIH, 71- Mem: Am Chem Soc. Res: Model polypeptides for studying biopolymer interactions; insertion sites for viral information in cancer cells; phosphorous chemistry in polynucleotides; biochemical engineering of human genes. Mailing Add: Dept of Chem DePaul Univ Chicago IL 60614

NOVAK, ROBERT OTTO, b Oak Park, Ill, Sept 12, 30; m 59; c 1. MYCOLOGY, PLANT PATHOLOGY. Educ: Mich State Univ, BS, 52; Univ Ill, MS, 56; Univ Wis, PhD(bot), 63. Prof Exp: Res microbiologist, Lederle Labs, 63-68; asst prof, 68-73, ASSOC PROF BIOL, ORE COL EDUC, 73- Mem: AAAS; Mycol Soc Am. Res: Ecology and taxonomy of soil microfungi. Mailing Add: Dept of Sci & Math Ore Col of Educ Monmouth OR 97361

NOVAK, ROBERT WILLIAM, b Hoboken, NJ, Aug 2, 39; m 60; c 3. ORGANIC CHEMISTRY. Educ: Wagner Col, BS, 60; Purdue Univ, MS, 64, PhD(org chem), 66. Prof Exp: Instr chem, Purdue Univ, 64-66; res chemist, 66-72, mgr paper chem tech serv, 72-74, SR RES CHEMIST, AM CYANAMID CO, 74- Mem: AAAS; Am Chem Soc; Tech Asn Pulp & Paper Indust. Res: Organic synthesis of natural products; reaction of bromine with terminal disubstituted olefins; organic synthesis and product development of paper chemicals; mechanisms of retention of chemical additives for paper application. Mailing Add: 32F Weed Hill Ave Stamford CT 06907

NOVAK, RONALD WILLIAM, b Elmira, NY, Dec 3, 42; m 63; c 2. ORGANIC CHEMISTRY, POLYMER CHEMISTRY. Educ: Univ S Fla, BA, 64; Fla State Univ, PhD(org chem), 68. Prof Exp: CHEMIST, ROHM & HAAS CO, 69- Mem: Am Chem Soc. Res: Dispersion and solution polymers. Mailing Add: Rohm & Haas Co Spring House PA 19477

NOVAL, JOSEPH JAMES, b New York, NY, June 14, 30; m 55; c 4. BIOCHEMISTRY. Educ: Manhattan Col, BS, 52; St Johns Univ, NY, MS, 54; Rutgers Univ, PhD(microbiol, biochem), 57. Prof Exp: Fel immunochem, Rutgers Univ, 57-58; chief neurochem sect, NJ State Bur Res Neurol & Psychiat, 58-73; RES ASSOC PROF SURG, MED SCH, TEMPLE UNIV, 73- Concurrent Pos: Asst biol, St John's Univ, NY, 65. Mem: Am Chem Soc. Res: Psychopharmacology; physiology; neuroendocrinology; immunochemistry; cancer; liver metabolism. Mailing Add: Dept of Surg Temple Univ Med Sch Philadelphia PA 19140

NOVALES, RONALD RICHARDS, b San Francisco, Calif, Apr 24, 28; m 53; c 2. COMPARATIVE ENDOCRINOLOGY, CELL BIOLOGY. Educ: Univ Calif, Berkeley, BA, 50, MA, 53, PhD(zool), 58. Prof Exp: Assoc zool, Univ Calif, Berkeley, 56; from asst prof to assoc prof, 58-70, PROF BIOL SCI, NORTHWESTERN UNIV, ILL, 70- Mem: Fel AAAS; Endocrine Soc; Am Soc Zool (secy, 66-67); Am Soc Cell Biol. Res: Mechanism of action of vertebrate hormones, particularly melanocyte-stimulating hormones, epinephrine, melatonin and of cyclic adenosine monophosphate on pigment movements in pigment cells, using frog skin and tissue cultured melanophores. Mailing Add: Dept of Biol Sci Northwestern Univ Evanston IL 60201

NOVELLI, GUERINO DAVID, b Agawam, Mass, Nov 6, 18; m 43; c 2. BIOCHEMISTRY. Educ: Mass State Col, BS, 40; Rutgers Univ, MS, 42; Harvard Univ, PhD(biochem), 49. Prof Exp: Asst bact, Rutgers Univ, 40-42; chemotherapeutist, Merck Inst for Therapeut Res, 42; asst, Scripps Inst, Univ Calif, 42-44; bacteriologist, Mass Gen Hosp, 48-49, from asst biochemist to assoc biochemist, 49-53; assoc prof microbiol, Sch Med, Western Reserve Univ, 53-56; PRIN BIOCHEMIST, BIOL DIV, OAK RIDGE NAT LAB, 56-, DIR BIOL MACROMOLECULAR SEPARATIONS PROG, 65- Concurrent Pos: Tutor, Harvard Col, 49-53; mem study sect physiol chem, NIH, 61-64, 70-73, radiation, 65-; prof, Grad Sch Biomed Sci, Univ Tenn, 67-; cancer res grant, 69-70. Mem: AAAS; Am Cancer Soc; Am Soc Microbiol; Brit Biochem Soc. Res: Enzymology of protein biosynthesis; structure and function of transfer RNA; carcinogenesis; aging; macromolecular separations technology; regulation and control mechanisms. Mailing Add: Biol Div Oak Ridge Nat Lab PO Box Y Oak Ridge TN 37830

NOVELLO, FREDERICK CHARLES, b Somerville, Mass, July 27, 16; m 48; c 3. ORGANIC CHEMISTRY. Educ: Harvard Univ, SB, 38, MA, 39, PhD(org chem), 41. Prof Exp: Res assoc Nat Defense Res Comt, Harvard Univ, 41-43; res chemist, Sharp & Dohme Div, Merck & Co, Inc, 43-69; sr res fel, 69-73, SR INVESTR, MERCK SHARP & DOHME RES LABS, 73- Honors & Awards: Modern Pioneers Award, Nat Asn Mfrs, 65; Albert Lasker Award, Albert & Mary Lasker Found, 75. Mem: Am Chem Soc. Res: Medicinal chemistry. Mailing Add: Merck Sharp & Dohme Res Labs Div of Merck & Co West Point PA 19486

NOVEY, THEODORE BURTON, physics, see 12th edition

NOVICK, AARON, b Toledo, Ohio, June 24, 19; m 48; c 2. BIOPHYSICS. Educ: Univ Chicago, BS, 40, PhD(chem), 43. Prof Exp: Scientist, Manhattan Dist Proj, Univ Chicago, 43-46, AEC Proj, 46-47, asst prof biophys, 48-55, assoc prof biophys & microbiol, 55-58; dir inst molecular biol, 59-69, PROF BIOL, UNIV ORE, 59-, DEAN GRAD SCH, 71- Concurrent Pos: Guggenheim fel, Pasteur Inst, France, 53-54. Mem: AAAS; Genetics Soc Am; Am Soc Microbiol; Biophys Soc. Res: Mechanism of protein synthesis; regulatory mechanisms; mutagenesis. Mailing Add: Inst of Molecular Biol Univ of Ore Eugene OR 97403

NOVICK, ALVIN, b Flushing, NY, June 27, 25. BEHAVIORAL PHYSIOLOGY, NEUROBIOLOGY. Educ: Harvard Col, AB, 47, MD, 51. Prof Exp: Teaching fel med, Harvard Univ, 52-53, res fel biol, 53-57; from instr to asst prof zool, 57-65, ASSOC PROF BIOL, 65- Mem: Am Physiol Soc; Am Soc Zool; Am Soc Mammal; Ecol Soc Am; Animal Behav Soc. Res: Echolocation in bats; sensory physiology; vertebrate biology. Mailing Add: Dept of Biol Yale Univ New Haven CT 06520

NOVICK, RICHARD P, b New York, NY, Aug 10, 32; m 58; c 2. MICROBIAL GENETICS. Educ: Yale Univ, BA, 54; NY Univ, MD, 59. Prof Exp: Intern, Yale-New Haven Med Ctr, 59-60; Nat Found fel, Nat Inst Med Res, Eng, 60-62; asst resident, Hosp, Vanderbilt Univ, 62-63; USPHS fel, Rockefeller Univ, 63-65; assoc, 65-69, ASSOC MEM PUB HEALTH RES INST OF CITY OF NEW YORK, INC, 69- Concurrent Pos: Res asst prof, NY Univ, 66-69, res assoc prof, 69-; lectr, Columbia Univ, 66-68. Mem: Am Soc Microbiol; Genetics Soc Am; Harvey Soc. Res: Microbial physiology; control mechanisms in biosynthetic pathways; extrachromosomal resistance factors in Staphlylococcus aureus; genetic control of replication; misuse of science. Mailing Add: Dept of Microbiol Pub Health Res Inst 455 First Ave New York NY 10016

NOVICK, ROBERT, b New York, NY, May 3, 23; m 47; c 3. PHYSICS. Educ: Stevens Inst Technol, ME, 44, MS, 49; Columbia Univ, PhD(physics), 55. Prof Exp: Engr microwaves, Wheeler Labs, Inc, 46-47; instr physics, Columbia Univ, 52-54, res assoc, 54-57, adj asst prof, 57; from asst prof to assoc prof, Univ Ill, 57-60; assoc prof, 60-62, dir radiation lab, 60-68, PROF PHYSICS, COLUMBIA UNIV, 62-, CO-DIR ASTROPHYS LAB, 68- Concurrent Pos: Sloan fel, 59-72; mem adv panel physics, NSF, 62-65; chmn subpanel atomic & molecular physics, Nat Acad Sci, 64-65; mem Nat Acad Sci Panel, adv to Nat Bur Standards Atomic Physics Div, 66-69; consult, Gen Precision Lab, Gen Time, Inc, & Perkin-Elmer Corp. Mem: Fel Am Phys Soc; fel Inst Elec & Electronics Eng. Res: Atomic physics, collisions and frequency standards; quantum electronics; nuclear moments; x-ray astronomy. Mailing Add: Dept of Physics Columbia Univ New York NY 10027

NOVICK, RUDOLPH G, b Warsaw, Poland, Dec 16, 10; US citizen; m 37; c 2. PSYCHIATRY. Educ: Northwestern Univ, BS, 31, MD, 36; Chicago Inst Psychoanal, 46-50. Prof Exp: Jr physician psychiat, Jacksonville State Hosp, Ill Dept Pub Welfare, 37-40; jr physician, Manteno State Hosp, Ill, 40-41; sr physician, Elgin State Hosp, Ill, 41-43; med dir, Ill Soc Ment Health, 43-55; MED DIR PSYCHIAT, FOREST HOSP, 56-; ASSOC PROF PSYCHIAT & ACTG CHMN DEPT, UNIV HEALTH SCI/CHICAGO MED SCH, 75- Concurrent Pos: Pvt pract, 45-; psychiat consult, Comt Community Serv, NIMH, 54-57, Chicago State Hosp, 55-56, Munic Tuberc Hosp, Chicago, 55-58 & Ill State Dept Pub Health, 69- Mem: AMA; fel Am Psychiat Asn; Group Advan Psychiat (chmn, Comt Prev Psychiat). Mailing Add: Off of the Med Dir Forest Hosp 555 Wilson Lane Des Plaines IL 60016

NOVICK, WILLIAM JOSEPH, JR, b Revloc, Pa, Dec 14, 31; m 55; c 4. PHARMACOLOGY, BIOCHEMISTRY. Educ: St Francis Col, Pa, BS, 53; Duke Univ, PhD(pharmacol), 61. Prof Exp: Technician toxicol, Dept Pub Health, Univ Pittsburgh, 54-55; jr biochemist, Smith Kline & French Labs, 55-58, sr scientist, 61-64, group leader pharmacol, 64-65, from asst sect head to sect head, 65-67; mgr pharmacol dept, William H Rorer, Inc, Pa, 67-70; DIR PHARMACOL DEPT, HOECHST-ROUSSEL PHARMACEUT, INC, 70- Mem: Am Soc Pharmacol & Exp Therapeut; NY Acad Sci. Res: Effects of age and thyroid hormone on monamine oxidase; pharmacological activities of steroids; drug metabolism. Mailing Add: Hoechst Pharaceut Inc Rte 202-206 N Somerville NJ 08876

NOVIKOFF, ALEX BENJAMIN, b Russia, Feb 28, 13; nat US; m 39, 69; c 2. BIOCHEMISTRY. Educ: Columbia Univ, BS, 31, AM, 33, PhD(zool), 38. Prof Exp: Tutor biol, Brooklyn Col, 35-40, from instr to asst prof, 40-48; assoc prof biochem, Col Med, Univ Vt, 48-53, prof exp path, 51-53; PROF PATH, ALBERT EINSTEIN COL MED, 55- Concurrent Pos: Am Cancer Soc fel, Col Med, Univ Wis, 46-47; USPHS res career award, Nat Cancer Inst, 62- Mem: Nat Acad Sci; Am Soc Cell Biol (pres, 62-63); Histochem Soc (pres, 58-59); Electron Micros Soc Am; Soc Develop Biol. Res: Biochemical cytology of normal and malignant cells; hepatomas; enzyme cytochemistry; lysosomes; Golgi apparatus; peroxisomes; electron microscopy. Mailing Add: Dept of Path Albert Einstein Col of Med Bronx NY 10461

NOVITSKI, EDWARD, b Wilkes Barre, Pa, July 24, 18; m 43; c 4. GENETICS. Educ: Purdue Univ, BS, 38; Calif Inst Technol, PhD(genetics), 42. Prof Exp: Guggenheim fel, Univ Rochester, 45-46, res assoc, Atomic Energy Proj, 46-47; res assoc, Univ Mo, 47-48; sr res fel, Calif Inst Technol, 48-51; assoc prof zool, Univ Mo, 51-56; head biologist, Oak Ridge Nat Lab, 56-58; head dept, 64-67, PROF BIOL, UNIV ORE, 58- Concurrent Pos: NSF sr fel, Univs Zurich, 61-62 & Canberra, 67-68. Mem: Genetics Soc Am (treas, 62-66); Soc Exp Biol & Med; Am Soc Nat; Soc Human Genetics. Res: Chromosome behavior; speciation; statistical analysis of genetic data; use of computing methods in biology. Mailing Add: Dept of Biol Univ of Ore Eugene OR 97403

NOVOA, WILLIAM BREWSTER, b Havana, Cuba, July 16, 30; US citizen. BIOCHEMISTRY. Educ: Univ Fla, BS, 55; Duke Univ, PhD, 59. Prof Exp: USPHS res fel, Univ Wash, 59-61; res assoc, McIlvain Lab, Med Ctr, Univ Kans, 61, from instr to asst prof biochem, 62-70; ASSOC PROF CHEM, CENT CONN STATE COL, 70- Mem: Am Chem Soc. Res: Mechanism of enzyme action. Mailing Add: Dept of Chem Cent Conn State Col New Britain CT 06050

NOVODVORSKY, MARK EVGENIEVICH, b Moscow, Russia, July 21, 46. PURE MATHEMATICS. Educ: Univ Moscow, PhD(math), 71. Prof Exp: Asst math, Inst Advan Study, 74-75; ASST PROF MATH, PURDUE UNIV, WEST LAFAYETTE, 75- Mem: Am Math Soc. Res: Zeta functions associated to automorphic representations of reductive groups over global fields. Mailing Add: Dept of Math Purdue Univ West Lafayette IN 47906

NOVOSAD, ROBERT S, b Chicago, Ill, May 1, 20; m 46; c 3. MATHEMATICS. Educ: Ill Inst Technol, BS, 42; Univ Chicago, MS, 48, PhD(math), 52. Prof Exp: Instr math, Tulane Univ, 50-53; asst prof, Pa State Univ, 53-60; assoc res scientist, Denver Div, Martin-Marietta Corp, 60-67, chief space systs opers anal, 67-71; PROF PHYSICS & AERONAUT SYSTS, UNIV W FLA, 71- Mem: Am Math Soc; Opers Res Soc Am; Am Astronaut Soc; Am Inst Aeronaut & Astronaut. Res: Operations research; systems analysis. Mailing Add: Aeronaut Systs Fac Univ of W Fla Pensacola FL 32504

NOVOTNY, CHARLES, b New York, NY, July 27, 36; m 58; c 2. MICROBIOLOGY. Educ: Wis State Col Stevens Point, BS, 59; Univ Pittsburgh, PhD(bact), 65. Prof Exp: Fel microbiol, Sch Med, Univ Pittsburgh, 65, res assoc, 65-68; ASSOC PROF MED MICROBIOL, UNIV VT, 68- Mem: Am Soc Microbiol. Res: Transport systems in bacteria; microbial genetics. Mailing Add: Dept of Med Microbiol Univ of Vt Burlington VT 05401

NOVOTNY, DONALD BOB, b Cedar Rapids, Iowa, Nov 15, 37; m 67; c 2. PHYSICAL CHEMISTRY. Educ: Univ Iowa, BS, 59; Iowa State Univ, PhD(phys chem), 64. Prof Exp: Sr res chemist, Mound Lab, Monsanto Res Corp, 64-66; res assoc chem, Mass Inst Technol, 66-67; res chemist, Vacuum Measurement Sect, Nat Bur Standards, Washington, DC, 67-71, gen phys scientist, Fire Technol Div, 71-75, PHYS CHEMIST, ELECTRONIC TECHNOL DIV, NAT BUR STANDARDS, WASHINGTON, DC, 75- Mem: Am Phys Soc. Res: Metal physics; metals; alloys; alloy phases; phase stability from thermodynamic considerations; lattice vibrations and energetics; photolithographic processes including photochemistry of resists and measurements utilizing optics. Mailing Add: Nat Bur of Standards Washington DC 20234

NOVOTNY, EVA, b Brno, Czech, May 22, 34; US citizen. ASTRONOMY. Educ: Barnard Col, AB, 55; Columbia Univ, PhD(astron), 61. Prof Exp: From instr to asst prof astron, Univ Pa, 61-68; vis scientist, Manned Spacecraft Ctr, NASA, 70-71; vis scientist, Univ Manchester, 72-73; RES ASSOC ASTRON, UNIV COL, CARDIFF, 74- Mem: Am Astron Soc. Res: Stellar interiors; binary star systems. Mailing Add: Dept of Appl Math & Astron Univ Col PO Box 78 Cardiff England

NOVOTNY, JAROSLAV, b Brtnice, Czech, Mar 11, 24; m 56; c 3. MEDICINAL CHEMISTRY. Educ: Univ Adelaide, BSc, 59, PhD(org chem), 63. Prof Exp: Demonstr, Dept Org Chem, Univ Adelaide, 62-63; res fel org & med chem, State Univ NY Buffalo, 64-65; chemist, 65, SUPVR ORG CHEM, STARKS ASSOCS, INC, 65- Mem: Am Chem Soc. Res: Synthesis and isolation of carcinogenic hydrocarbons from tars and the use of carbon-14 for them; synthesis of inhibitors of folic reductase and thymidylic synthetase; synthesis of medicinals and other organic compounds; research and development of new antimalarials. Mailing Add: 215 Fruitwood Terr Buffalo NY 14221

NOVOTNY, ROBERT THOMAS, b New York, NY, Nov 22, 24; m 57; c 5. MAMMALIAN ECOLOGY, PALEOECOLOGY. Educ: Univ Mich, BS, 50; Univ Utah, MS, 58, PhD(biol), 70. Prof Exp: Sci technician marine geophys, Columbia Univ-Woods Hole Oceanog Inst, 51; geophysicist uranium explor, US AEC, 51-52; geologist, E J Longyear Co, Africa, 52-55; geologist & geophysicist copper explor, Anaconda Co, New York & Salt Lake City, 55-56; spec rep oil explor, Gulf Oil Co, 59-65; asst prof biol & chem, Midwest Col, 70; head dept sci, Parsons Col West, 70-71; ASST PROF BIOL, ST MARY'S COL MD, 71- Concurrent Pos: Prin investr, Ecol Surv Base, US Naval Air Test Ctr, Patuxent River, Md, 75- Mem: AAAS; Am Inst Biol Sci; Ecol Soc Am; Animal Behav Soc; NY Acad Sci. Res: Tree hole protozoa and ecology; animal behavior; wildlife ecology and management. Mailing Add: Dept of Biol St Mary's Col of Md St Mary's City MD 20686

NOVY, ANTON RUDOLPH, ceramic engineering, see 12th edition

NOWACK, GERHARD PAUL, physical organic chemistry, see 12th edition

NOWACZYNSKI, WOJCIECH, b Nisko, Poland, Mar 27, 25; Can citizen; m; c 6. ENDOCRINOLOGY. Educ: Univ Fribourg, DSc, 52. Prof Exp: Mem res staff, Relationship between adrenal cortex sodium & kidney, Hotel Dieu Hosp, Montreal, 53-67; from asst prof to assoc prof med, 62-70, PROF MED, FAC MED, UNIV MONTREAL, 70-; DIR STEROID RES DEPT, CLIN RES INST MONTREAL, 67- Concurrent Pos: Res assoc, Med Res Coun Can, 62-65, permanent med res assoc, 65-, prin investr hypertension group, 72-76; lectr, Fac Med, McGill Univ, 64-70, prof, 70- Res: Study of the pathogenesis of arterial hypertension; angiotensine renin aldosterone. Mailing Add: Clin Res Inst of Montreal 110 Pine Ave W Montreal PQ Can

NOWAK, ANTHONY VICTOR, b Chicago, Ill, Aug 6, 38. ANALYTICAL CHEMISTRY. Educ: Loyola Univ, BS, 60; Northern Ill Univ, MS, 63; Univ Ill, Urbana, MS, 65, PhD(chem), 68. Prof Exp: Anal chemist, US Army QM Food & Container Inst, 60-62; SR RES SCIENTIST ANAL CHEM, ATLANTIC RICHFIELD CO, 67- Concurrent Pos: Lectr, Chicago sect, Soc Appl Spectros, 71. Mem: Am Chem Soc; Soc Appl Spectros. Res: Microprocessors in analytical instrumentation; applications of computers to laboratory automation; selective gas chromatographic detectors. Mailing Add: L-321 Anal Div Atlantic Richfield Co Harvey IL 60426

NOWAK, ARTHUR JOHN, b Erie, Pa, June 25, 37; m 61; c 4. PEDODONTICS. Educ: Univ Pittsburgh, DMD, 61; Columbia Univ, cert (pedodontics), 66, MA, 67; Am Bd Pedodont, dipl, 71. Prof Exp: Fel pedodontics, Columbia Univ, 64-67; asst prof, Sch Dent Med, Univ Pittsburgh, 67-70; dir pediat dent, Allegheny Gen Hosp, Pittsburgh, 70-73; ASSOC PROF PEDODONTICS, COL DENT, UNIV IOWA, 73- Concurrent Pos: Consult, President's Comt Ment Retardation, 74-75; pres, Acad Dent Handicapped, 74-75; mem, President's Comt Employ Handicapped, 75; pres, Nat Found Dent Handicapped, 75-76; exec coun mem, Am Soc Dent Children, 75-; prof adv coun, Nat Easter Seal Soc Crippled Adults & Children, 75-; special adv, President's Comt Ment Retardation, 76- Mem: Fel Am Acad Pedodontics; Am Asn Dent Sch; Int Asn Dent Res. Res: Effect of parent in dental treatment room; effect of drugs on behavior of children in dental setting; screening for hypertension in pediatric patients in dental setting; effect of substituting salty snacks for sucrose snacks. Mailing Add: Dept of Pedodontics Col Dent Univ of Iowa Iowa City IA 52242

NOWAK, EDWARD NORBERT, chemistry, see 12th edition

NOWAK, MICHAEL, b Wiesbaden, Ger, Feb 3, 40; US citizen. ANTHROPOLOGY. Educ: Univ Ore, BA, 62, PhD(anthrop), 68. Prof Exp: From instr to asst prof, 65-74, CHMN DEPT ANTHROP, COLO COL, 74- Concurrent Pos: NSF archaeol res grants, 69 & 70. Mem: AAAS; Am Anthrop Asn; Soc Am Archaeol; Sigma Xi. Res: Nutrition; arctic culture change; arctic archaeology and prehistory. Mailing Add: Dept of Anthrop Colo Col Colorado Springs CO 80903

NOWAK, MICHAEL J, applied physics, see 12th edition

NOWAK, ROBERT MICHAEL, b South Milwaukee, Wis, Oct 28, 30; m 57; c 2. ORGANIC CHEMISTRY. Educ: Univ Wis, BS, 53; Univ Ill, PhD(org chem), 56. Prof Exp: Res chemist, Dow Chem USA, 56-64, group leader, 64-68, asst lab dir, Phys Res Lab, 68-72, res & develop mgr plastics dept, 72-73, RES & DEVELOP DIR, OLEFIN & STYRENE PLASTICS DEPT, DOW CHEM USA, 73- Mem: Am Chem Soc. Res: Grafting vinyl monomers onto polymer backbones and synthesis and study of new monomers and their polymers; new resin systems for reinforced plastics; polymer chemistry. Mailing Add: 1212 Bayberry Lane Midland MI 48640

NOWAK, THOMAS, b Niagara Falls, NY, Nov 25, 42; , m 67; c 1. BIOCHEMISTRY. Educ: Case Inst Technol, BS, 64; Univ Kans, PhD(biochem), 69. Prof Exp: NIH fel, 69-71, RES ASSOC, INST CANCER RES, 71- Mem: AAAS; Am Chem Soc; Biophys Soc. Res: Mechanism of action of enzymes; application of magnetic resonance to biological systems; function of metal ions in enzymatic catalysis. Mailing Add: Inst for Cancer Res 7701 Burholme Rd Philadelphia PA 19111

NOWAK, WIESLAW STANISLAW WLADYSLAW, b Jeziorko, Poland, Sept 25, 36; Brit citizen; m 68; c 1. RESOURCE GEOGRAPHY. Educ: Univ Col, Univ London, BA, 58; London Sch Econ, PhD(marine resources), 65. Prof Exp: Res asst marine resources, Marine Resources Res Unit, Portsmouth, Eng, 59-63; from asst lectr to lectr geog, City of London Col, 63-71; asst prof, 71-73, ASSOC PROF MARINE GEOG, MEM UNIV NFLD, 73- Concurrent Pos: Hon res assoc, Marine Resources Res Unit, Portsmouth, Eng, 63-71; consult, Premier's Steering Comt Fisheries Govt Nfld, 72. Honors & Awards: Gold Cross of Merit, Polish Govt in Exile, 70. Mem: Fel Royal Geog Soc; Marine Biol Asn UK; Brit Geog Asn; Int Oceanog Found. Res: Marine resources; North Atlantic fish marketing. Mailing Add: 63 Roosevelt Ave The Avalon Mount Pearl Park NF Can

NOWATZKI, EDWARD ALEXANDER, civil engineering, soil science, see 12th edition

NOWELL, JOHN WILLIAM, b Wake Co, NC, Aug 26, 19. CHEMISTRY. Educ: Wake Forest Col, BS, 40; Univ NC, PhD(phys chem), 45. Prof Exp: Sr physicist, Am Cyanamid Co, Conn, 44-45; from asst prof to assoc prof chem, 4S-54, chmn dept, 63-72, PROF CHEM, WAKE FOREST UNIV, 54- Mem: Am Chem Soc; NY Acad Sci. Res: Membrane permeability to gases; polarographic identification of ions; tracer methods using radioisotopes. Mailing Add: Box 7486 Winston-Salem NC 27109

NOWELL, PETER CAREY, b Philadelphia, Pa, Feb 8, 28; m 50; c 5. PATHOLOGY. Educ: Wesleyan Univ, AB, 48; Univ Pa, MD, 52. Prof Exp: Intern, Philadelphia Gen Hosp, 52-53; resident path, Presby Hosp, Philadelphia, 53-54; instr, 56-57, assoc, 57-60, from asst prof to assoc prof, 60-64, chmn dept, 67-73, dir cancer ctr, 72-75, PROF PATH, SCH MED, UNIV PA, 64- Concurrent Pos: USPHS sr res fel & career develop award, Sch Med, Univ Pa, 56-61, USPHS res career award, 61-67; consult lab serv, Philadelphia Gen Hosp & Philadelphia Vet Admin Hosp, 70- Honors & Awards: Parke-Davis Award Exp Path, 65. Mem: Nat Acad Sci; Am Asn Cancer Res; Am Soc Exp Path (pres, 70-71); Am Asn Path & Bact; Am Asn Immunol. Res: Growth regulatory mechanisms and cytogenetics of normal and leukemic leukocytes; radiation carcinogenesis; cellular immunology. Mailing Add: Dept of Path Univ of Pa Sch of Med Philadelphia PA 19174

NOWELL, WESLEY RAYMOND, b Oakland, Calif, Feb 9, 24; m 46; c 4. MEDICAL ENTOMOLOGY. Educ: Stanford Univ, AB, 47, AM, 48, PhD(biol sci), 51. Prof Exp: Biomed Sci Corps, US Air Force, 51-, med entomologist, 5th Air Force, Korea, 51-52, Air Res & Develop Command, 52-55, US Air Force Europe Command, 55-58, Strategic Air Command, 58-62, 4th US Air Force Epidemiol Flight, Turkey, 62-65 & US Air Force Epidemiol Lab, Aerospace Med Div, Tex, 65-72, US Air Force, assoc chief med entom, 68-74, dep chief, epidemiol div, US Air Force Sch Aerospace Med, 72-75, CHIEF ENTOMOL SERV, HQ 1ST MED SERV WING (PACAF), PHILIPPINES, 75- Concurrent Pos: Mem armed forces pest control bd, Dept Defense, 67-74. Mem: Sigma Xi; Entom Soc Am; Am Soc Trop Med & Hyg; Am Mosquito Control Asn. Res: Global medical entomology; arthropod-associated diseases; vector control program analysis and organization; scientific research and training programs administration; Diptera; Dixidae. Mailing Add: Hq 1st Med Serv Wing (PACAF) APO San Francisco CA 96274

NOWER, LEON, b Sosnowiec, Poland, Aug 16, 27; US citizen; m 60; c 2. MATHEMATICS. Educ: City Col New York, BS, 53; Stanford Univ, MS, 62, PhD(math), 65. Prof Exp: ASSOC PROF MATH, SAN DIEGO STATE UNIV, 63- Mem: Am Math Soc; Math Asn Am. Res: Harmonic analysis; theory of distributions. Mailing Add: Dept of Math San Diego State Univ San Diego CA 92115

NOWICKE, JOAN WEILAND, b St Louis, Mo; m 63; c 1. PALYNOLOGY,

SYSTEMATICS. Educ: Washington Univ, AB, 59, PhD(biosyst), 68; Univ Mo-Columbia, AM, 62. Prof Exp: Fel, Mo Bot Garden, 68-71; asst prof biol, Univ Mo-St Louis, 71-72; ASSOC CUR BOT, SMITHSONIAN INST, 72- Mem: AAAS; Bot Soc Am; Am Soc Plant Taxonomists; Int Soc Plant Taxonomists; Torrey Bot Club. Res: Pollen morphology, structure, function and use in systematics; classification of Phytolaccaceae; Apocynaceae and Boraginaceae of Central America. Mailing Add: Dept of Bot Smithsonian Inst Washington DC 20560

NOWICKI, HENRY GEORGE, organic chemistry, biochemistry, see 12th edition

NOWINSKI, WIKTOR WACLAW, biochemistry, deceased

NOWLAN, JAMES PARKER, b NS, Can, Mar 26, 10; m 33; c 2. MINING GEOLOGY. Educ: Acadia Univ, BSc, 28; Univ Toronto, MA, 30, PhD(geol), 35. Prof Exp: Geologist, Rhokana Corp, Northern Rhodesia, 30-32 & Oro Plata Mining Corp, Can & Brazil, 33-35; geologist & geophysicist, Hans Lundberg, Ltd, 36; consult geologist & engr, 37-39; mgr, Yama Gold Mines, Ltd, 40-42; supvr ammunition div, Can Industs, Ltd, 42-45; geologist, Cochenour Williams Gold Mines, Ltd, Ont, 45-48; field supvr, Dom Gulf Co, 48-49, admin geologist, 49-53; chief geologist, McPhar Geophys, Ltd, 53-58; dep minister mines, NS, 58-73; RETIRED. Concurrent Pos: Mem, Nat Adv Comt Mining & Metall Res, 70-; consult, 73-; mem, Nat Environ Adv Comt, 74- Mem: Soc Econ Geol; Can Inst Mining & Metall (pres, 73-74); Geol Asn Can (pres, 64-65). Res: Silurian stratigraphy of Niagaran escarpment of Ontario. Mailing Add: 6525 Waegwoltic Ave Halifax NS Can

NOWLIN, CHARLES HENRY, b Wilmington, Del, Feb 1, 32; m 56; c 2. APPLIED MATHEMATICS, APPLIED PHYSICS. Educ: Washington & Lee Univ, BS, 55; Harvard Univ, SM, 56, PhD(appl physics), 63. Prof Exp: APPL PHYSICIST, OAK RIDGE NAT LAB, 63- Mem: Inst Elec & Electronics Eng. Res: Instrumentation; information theory and signal processing; network analysis and synthesis; nuclear pulse amplifiers; image processing. Mailing Add: Instrumentation & Controls Div Oak Ridge Nat Lab PO Box X Oak Ridge TN 37830

NOWLIN, DUANE DALE, b Huron, SDak, Mar 14, 37; m 67. WATER CHEMISTRY. Educ: Macalester Col, BA, 58; Iowa State Univ, PhD(chem), 64. Prof Exp: Chief chemist, Lindsay Co, Union Tank Car Co, 64-65; sr chemist, Garrett Res & Develop Co, 65-66; mgr water chem, Econs Lab, Inc, 66-69; DIR RES & DEVELOP, LINDSAY CO, UNION TANK CAR CO, 69- Mem: Am Water Works Asn; Am Chem Soc. Res: Ion-exchange technology; isotope separation studies; water treatment and metal ion chelation. Mailing Add: Res & Develop Dept Lindsay Co 455 Woodlane Dr St Paul MN 55119

NOWLIN, GENE, organic chemistry, polymer chemistry, deceased

NOWLIN, WORTH D, JR, b Smithville, Tex, Oct 1, 35; m 59; c 2. PHYSICAL OCEANOGRAPHY. Educ: Tex A&M Univ, BA, 58, MS, 60, PhD(phys oceanog), 66. Prof Exp: Instr math, Allen Jr Col, 58-59; res asst, Mobil Res Lab, summer 59; oceanog technician, Res Found, Tex A&M Univ, summer 60, res scientist, summers 61-63; asst prof phys oceanog, Univ, 63-67; oceanogr & prog dir, Ocean Sci & Technol Div, Off Naval Res, DC, 67-69; assoc prof, 69-74, PROF OCEANOG, TEX A&M UNIV, 74- Concurrent Pos: Actg dep head, Off for Int Decade Ocean Explor, NSF, DC, 70-71, consult, 70-74, dep head, 71, mem oceanog panel, Div Environ Sci, 70-72; assoc ed, J Phys Oceanog; mem panel oceanog, Comt Polar Res, Nat Acad Sci-Nat Res Coun, 72-75 & mem joint FGGE adv panel, US Comt for Global Atmospheric Res Prog & Ocean Sci, 74- Mem: Am Geophys Union. Res: Meso second large-scale oceanic distributions of properties; dynamics of ocean circulation; oceanographic research management. Mailing Add: Dept of Oceanog Tex A&M Univ College Station TX 77843

NOWOTNY, ALOIS HENRY, b Gyongyos, Hungary, July 30, 22; US citizen; m 60. IMMUNOLOGY, IMMUNOCHEMISTRY. Educ: Pazmany Peter Univ, Budapest, dipl chem, 45, PhD(chem), 47. Prof Exp: Asst prof biochem, Med Sch, Univ Budapest, 47-51; res assoc immunochem, Hungarian Blood Serv Ctr, 51-54, vchmn res dept, 54-56; res assoc, A Wander Res Inst, Freiburg, WGer, 57-60; sr res assoc immunochem, City of Hope Med Ctr, Duarte, Calif, 60-62; PROF IMMUNOCHEM & IMMUNOL, MED SCH, TEMPLE UNIV, 62- Concurrent Pos: Mem adv bd, Cancer Res Inst, New York, 67-; guest prof, Med Sch, Univ Heidelberg, 69-70 & 72; consult, NIH, 73- Mem: Am Asn Immunol; Am Acad Microbiol; Am Microbiol Soc; Ger Immunol Soc; Hungarian Chem Soc. Res: Cellular antigens; relationship between chemical structure and biological activities; mode of action of bacterial endotoxins; immunology of erythrocyte membranes; immunology of tumor cells. Mailing Add: Dept of Microbiol & Immunol Temple Univ Med Sch Philadelphia PA 19140

NOWOTNY, KURT A, b Vienna, Austria, Apr 8, 31; US citizen; m 60; c 2. INDUSTRIAL CHEMISTRY, INORGANIC CHEMISTRY. Educ: Univ Vienna, PhD(natural prod), 59. Prof Exp: Fel polyfunctional catalysis, Wash Univ, 59-60; from sr res chemist to proj leader functional fluids res specialist, Monsanto Co, 60-65; supvr packing res & develop, Crown Zellerbach Corp, 65-69, supvr polymer res, 69-70, mgr chem res dept, Cent Res Dept, 70-73; GROUP VPRES & DIR TECHNOL & ENG, EVANS PROD CO, 73- Mem: Am Soc Testing & Mat; Am Chem Soc; Austrian Chem Soc; Forest Prod Res Soc; Am Ceramic Soc. Res: Natural products; pharmaceutical and physical chemistry; oil additives; functional, aviation and fire resistant fluids; cellulose; adhesives; coatings; physics. Mailing Add: 6000 NW Huntington Dr Corvallis OR 97330

NOXON, JOHN FRANKLIN, b Pittsfield, Mass, July 7, 28; m 61; c 1. PHYSICS. Educ: Bowdoin Col, BA, 50; Harvard Univ, PhD(physics), 57. Prof Exp: Res assoc physics, Univ Sask, 57-61; res fel, Harvard Univ, 61-65, sr res assoc, 65-72; PHYSICIST, NAT OCEANIC & ATMOSPHERIC ADMIN, 72- Res: Atmospheric physics. Mailing Add: Agron Lab Nat Oceanic & Atmospheric Admin Boulder CO 80302

NOYA BENITEZ, JOSE ANTONIO, b Humacao, PR, Mar 16, 07; m 31; c 7. SURGERY. Educ: Columbia Univ, AB, 27, MD, 30; Am Bd Surg, dipl, 40. Prof Exp: Asst prof, Col Physicians & Surgeons, Columbia Univ, 45-49; PROF SURG, MED SCH, UNIV PR, 50-, ASST DEAN, 70- Mem: Fel Am Col Surg. Res: Splenomegaly in schistomiasis; thyroid disease and cancer of the breast. Mailing Add: 303 De Diego Ave Santurce PR 00911

NOYCE, DONALD STERLING, b Burlington, Iowa, May 26, 23; m 46; c 3. ORGANIC CHEMISTRY. Educ: Grinnell Col, AB, 44; Columbia Univ, MA, 45, PhD(org chem), 47. Prof Exp: NIH fel, Columbia Univ, 47-48; from instr to assoc prof chem, 48-60, asst dean Col Chem, 52-60 & 66-68, PROF CHEM, UNIV CALIF, BERKELEY, 60-, ASST DEAN COL CHEM, 75- Concurrent Pos: Guggenheim fel, 57; NSF sr fel, 64. Mem: Am Chem Soc; The Chem Soc. Res: Stereochemistry; mechanisms of organic reactions; organic kinetics. Mailing Add: Dept of Chem Univ of Calif Berkeley CA 94720

NOYCE, ROBERT NORTON, b Burlington, Iowa, Dec 12, 27; m 53; c 4. PHYSICS. Educ: Grinnell Col, BA, 49; Mass Inst Technol, PhD(physics), 53. Prof Exp: Mem staff. Res Div, Philco Corp, 53-56; sr staff mem, Shockley Transistor Corp, 56-57; dir res & develop, Fairchild Semiconductor Corp, 57-59, vpres & gen mgr, 59-65, group vpres, Fairchild Camera & Instrument Corp, 65-68; pres, 68-75, CHMN, INTEL CORP, 75- Honors & Awards: Ballentine Medal, Franklin Inst, 66. Mem: Nat Acad Eng; Am Phys Soc; fel Inst Elec & Electronics Eng. Res: Physical electronics; semiconductor and solid state physics; device technology. Mailing Add: Intel Corp 3065 Bowers Ave Santa Clara CA 95051

NOYES, CLAUDIA MARGARET, b Haverhill, NH, Apr 30, 40. ANALYTICAL BIOCHEMISTRY. Educ: Univ Vt, BS, 61; Univ Colo, PhD(chem), 66. Prof Exp: Res chemist, Armour Grocery Prod Co, 65-67; res assoc med, Sch Med, Univ Chicago, 68-75; RES ASSOC MED, SCH MED, UNIV NC, CHAPEL HILL, 75- Mem: AAAS; Am Chem Soc. Res: Automated protein sequence determination; gas chromatography; liquid chromatography. Mailing Add: 305-F Bolinwood Apts Chapel Hill NC 27514

NOYES, DAVID HOLBROOK, b Hampton, Va, Feb 5, 35; m 60; c 2. PHYSIOLOGY. Educ: Rensselaer Polytech Inst, BEE, 59; Univ Ala, Birmingham, PhD(physiol & biophys), 69. Prof Exp: Test engr radar systs, Gen Elec Light Mil Dept, NY, 59; supvr electronic model shop & biomed engr, Univ Ala, Birmingham, 62-66, biomed engr & eng design consult, Res Model Shop, 66-69; asst prof, 69-75, ASSOC PROF PHYSIOL, COL MED, 75-, ASST PROF, DEPT ELEC ENG, OHIO STATE UNIV, 69- Concurrent Pos: NIH grants, Col Med, Ohio State Univ, 72-75, NSF grant, 73-75. Mem: Am Physiol Soc; Int Asn Dent Res. Res: Gastrointestinal physiology; stomatology and periodontology. Mailing Add: Dept of Physiol Ohio State Univ Col of Med Columbus OH 43210

NOYES, HENRY PIERRE, b Paris, France, Dec 10, 23; US citizen; m 47; c 3. THEORETICAL PHYSICS. Educ: Harvard Univ, BA, 43; Univ Calif, Berkeley, PhD(physics), 50. Prof Exp: Mem staff, Radiation Lab, Mass Inst Technol, 43-44; physicist, Radiation Lab, Univ Calif, 50; Fulbright grantee math physics, Univ Birmingham, 50-51; asst prof physics, Univ Rochester, 51-55; physicist & group leader, Lawrence Radiation Lab, Univ Calif, 55-62; assoc prof theoret physics, 62-67, admin head sect, 62-69, PROF THEORET PHYSICS, STANFORD LINEAR ACCELERATOR CTR, 67- Concurrent Pos: Leverhulme lectr, Univ Liverpool, 57-58; Avco vis prof, Cornell Univ, 61; vis scholar, Ctr Advan Studies Behav Sci, Stanford Univ, 68-69; consult, Gen Atomic Div, Gen Dynamics Corp, 59-62, Lockheed Aircraft Corp, 62-63, Lawrence Radiation Lab, Livermore, 62-67 & Physics Int, 65-67; chmn comt for a dir attack on the legality of the Vietnam War, 69-72. Mem: AAAS; fel Am Phys Soc; Brit Inst Physics; Italian Phys Soc; Scientists & Engrs for Social & Polit Action. Res: Nucleon-nucleon and meson-nucleon interaction; quantum mechanical 3-body problem; applied hydrodynamics and neutronics; computational techniques; foundations of quantum mechanics. Mailing Add: Stanford Linear Accelerator Ctr Stanford CA 94305

NOYES, HOWARD ELLIS, b Memphis, Tenn, Apr 5, 22; m 47; c 3. BACTERIOLOGY. Educ: Univ Tenn, BS, 47; Ohio State Univ, MS, 49; George Washington Univ, PhD(bact), 55; Am Bd Microbiol, dipl. Prof Exp: US Dept Army, bacteriologist, Ft Detrick, Md, 49-51, chief bact sect, Div Surg, Inst Res, Walter Reed Army Med Ctr, 61-63, chief dept bact & mycol, US Army lfed Component, SEATO Med Res Lab, 63-66, dep chief dept lab serv, Div Surg, Walter Reed Army Inst Res, 66, chief dept surg microbiol, 66-67, chief dept bact & mycol, Med Res Lab, US Army Med Component, SEATO, 67-70, asst for res mgt, Walter Reed Army Inst Res, 70-74, ASSOC DIR RES MGT, WALTER REED ARMY INST RES, 74- Mem: AAAS; Am Soc Microbiol; Soc Exp Biol & Med; NY Acad Sci. Res: Medical microbiology as it relates to surgery with emphasis on creation and therapy of experimental wounds, the mechanism of action of bacterial toxins and the evaluation of new antimicrobial agents. Mailing Add: Inst of Res Walter Reed Army Med Ctr Washington DC 20012

NOYES, JOHN CHANNING, b Portland, Ore, Dec 23, 20; m 43; c 3. PHYSICS. Educ: Univ Portland, BS, 47; Univ Notre Dame, PhD(physics), 52. Prof Exp: Res engr, Boeing Airplane Co, 51-54, res specialist nuclear physics, 54-56, sr group engr, 56-58, mgr space physics, Syst Mgt Off, 58-59, mem staff, Geo-Astrophys Lab, Boeing Sci Res Lab, 59-61, lab head, 61-69, dir, 69-71, mgr environ physics, 71, space prog scientist. 71-74; PROJ SCIENTIST LARGE SPACE TELESCOPE PROG, AEROSPACE GROUP, BOEING CO, 74- Concurrent Pos: Mem tech comt space physics & astron, Am Inst Aeronaut & Astronaut, 74- Mem: AAAS; Am Astron Soc; Am Geophys Union. Res: Solar-terrestrial relationships; solar activity; upper atmosphere, planetary and space physics; astronomy from space. Mailing Add: 3314 56th SW Seattle WA 98116

NOYES, PAUL R, b Shreveport, La, Oct 3, 28; m 61; c 2. ORGANIC CHEMISTRY, POLYMER CHEMISTRY. Educ: Centenary Col, BS, 49; Univ Tex, MS, 54, PhD(chem), 55. Prof Exp: Res chemist, 54-63, STAFF CHEMIST, E I DU PONT DE NEMOURS & CO, INC, 63- Mem: Am Chem Soc. Res: Adhesives; adhesion; paper coatings. Mailing Add: E I Du Pont de Nemours & Co Inc Marshall Lab 3500 Grays Ferry Philadelphia PA 19146

NOYES, RICHARD MACY, b Champaign, Ill, Apr 6, 19; m 46, 73. PHYSICAL CHEMISTRY, CHEMICAL KINETICS. Educ: Harvard Univ, AB, 39; Calif Inst Technol, PhD(phys chem), 42. Prof Exp: Instr, Calif Inst Technol, 42-44, res fel, Nat Defense Res Comt Proj, 42-46; from instr to assoc prof chem, Columbia Univ, 46-58; PROF CHEM, UNIV ORE, 58-, HEAD DEPT, 66-68 & 75- Concurrent Pos: Vis prof, Univ Leeds, 55-56 & Oxford Univ, 71-72; mem subcomt kinetics chem reactions, Nat Res Coun, 60-, mem-at-large, Div Chem & Chem Tech, 63-66; Fulbright fel, Univ Victoria, NZ, 64; NSF sr fel, Max Planck Inst Phys Chem, Univ Göttingen, 65; mem chem adv panel, NSF, 69-71. Mem: Am Chem Soc; Am Phys Soc. Res: Thermodynamic properties of ions in solution; mechanisms of chemical reactions including reactions of diatomic molecules, diffusion controlled reactions, isotopic exchange reactions of organic iodides, and reactions oscillating in time and space. Mailing Add: Dept of Chem Univ of Ore Eugene OR 97403

NOYES, ROBERT WALLACE, b Berkeley, Calif, June 3, 19; m 43; c 2. OBSTETRICS & GYNECOLOGY. Educ: Univ Calif, AB, 41, MD, 43; Am Bd Obstet & Gynec cert, 54. Prof Exp: Intern obstet & gynec, Univ Hosp, Johns Hopkins Univ, 44; asst resident, Univ Calif, 45, resident, 46; res fel, Free Hosp Women, Brookline, Mass, 48; asst resident obstet & gynec, Univ Hosp, Stanford Univ, 49, res fel, 50, resident, 51, from instr to assoc prof, Sch Med, 51-61; prof & chmn, Sch Med, Vanderbilt Univ, 61-65; prof anat & chmn dept & assoc dean prof affairs, Sch Med, Univ Hawaii, 65-70; prof maternal & child health, Sch Pub Health & Obstet & Gynec, Sch Med, Univ NC, Chapel Hill, 70-73; dir med res/reproduction, Ortho Pharmaceut Co, Raritan, NJ, 73-75; DIR CLIN SERV & RES, PLANNED PARENTHOOD ASN SANTA CLARA COUNTY, 75- Concurrent Pos: Markle fel, 53-58; res assoc, Agr Res Coun, Cambridge Univ, 57-; res prof, Nat Ctr Zootech Res, France, 61-; med teacher training course, Univ Ill Col Med. Honors & Awards: Ortho Award, 55, Squibb

Award, 59, Rubin Award, 59, 60, 62, medal, 65, Am Fertil Soc. Mem: Am Fertil Soc; Soc Gynec Invest; Am Col Obstet & Gynec; Brit Soc Study Fertil. Res: Developmental biology. Mailing Add: 17 N San Pedro San Jose CA 95110

NOYES, ROBERT WILSON, b Winchester, Mass, Dec 27, 34; m 60; c 2. ASTROPHYSICS. Educ: Haverford Col, BA, 57; Calif Inst Technol, PhD(physics), 63. Prof Exp: Physicist, Smithsonian Astrophys Observ, 62-73; lectr astron, 62-73, PROF ASTRON, HARVARD UNIV, 73-, ASSOC DIR CTR ASTROPHYS, 73- Concurrent Pos: Chmn solar physics div, Am Astron Soc, 75- Mem: Am Astron Soc. Res: Aerodynamics and spectroscopy of the solar atmosphere; ultraviolet observations of solar and stellar spectra from space vehicles; infrared solar and stellar spectroscopy. Mailing Add: Ctr for Astrophys 60 Garden St Cambridge MA 02138

NOYES, RUSSELL, JR, b Indianapolis, Ind, Dec 25, 34; m 60; c 3. PSYCHIATRY. Educ: DePauw Univ, BA, 56; Ind Univ, MD, 59; Am Bd Psychiat & Neurol, dipl, 66. Prof Exp: Rotating intern, Philadelphia Gen Hosp, 59-60; resident psychiat, Inst Living, Conn, 60-61 & Univ Iowa, 61-63; mem staff, US Naval Hosp, Great Lakes, Ill, 63-65; asst prof, 65-71, ASSOC PROF PSYCHIAT, UNIV IOWA, 71- Concurrent Pos: USPHS resident training grant, Univ Iowa, 71-78; staff psychiatrist, Lake County Ment Health Clin, Waukegan, Ill, 63-65. Mem: Fel Am Psychiat Asn; AMA; Am Psychopath Asn. Res: Psychosomatic medicine. Mailing Add: 326 MacBride Rd Iowa City IA 52240

NOYES, WARD DAVID, b Schenectady, NY, Aug 25, 27; m 50; c 4. INTERNAL MEDICINE, HEMATOLOGY. Educ: Univ Rochester, BA, 49, MD, 53. Prof Exp: From intern to asst resident med, King County Hosp, Seattle, Wash, 53-56; instr, Univ Wash, 59-61; from asst prof to assoc prof, 61-70, PROF MED, COL MED, UNIV FLA, 70- Concurrent Pos: Res fel hemat, Univ Wash, 56-59; USPHS res fel, Oxford & Malmo Gen Hosp, Sweden, 58. Mem: Am Fedn Clin Res; Int Soc Hemat. Res: Problems in erythrokinetics and red cell metabolism. Mailing Add: Dept of Med Univ of Fla Col of Med Gainesville FL 32601

NOYES, WILBUR FISKE, b Boston, Mass, May 22, 26; m 47; c 2. VIROLOGY. Educ: Ind Univ, AB, 48; Boston Univ, AM, 49; Harvard Univ, PhD(bact), 52. Prof Exp: Assoc mem, Sloan-Kettering Inst Cancer Res, 52-70; ASSOC CANCER RES SCIENTIST, ROSWELL PARK MEM INST, 70- Mem: Electron Micros Soc Am; Am Asn Cancer Res; Soc Exp Biol & Med; Am Soc Microbiol; Tissue Cult Asn. Res: Tumorigenic viruses; in vitro cultivation of A and B hepatitis viruses; studies of in vitro carcinogenesis with organ cultures of human tissue. Mailing Add: Roswell Park Mem Inst 666 Elm St Buffalo NY 14203

NOYES, WILLIAM ALBERT, JR, b Terre Haute, Ind, Apr 18, 98; m 21; c 1. PHYSICAL CHEMISTRY. Educ: Grinnell Col, AB, 19. Hon Degrees: DSc, var US & foregin univs, 46-65. Prof Exp: Instr chem, Univ Calif, 21-22; from instr to asst prof, Univ Chicago, 22-29; from assoc prof to prof, Brown Univ, 29-38; Prof chem, Univ Rochester, 38-63, chmn dept, 39-55, dean grad sch, 52-55, from actg dean to dean, Col Arts & Sci, 56-58; Ashbel Smith prof, 63-73, ASHBEL SMITH EMER PROF CHEM, UNIV TEX, AUSTIN, 73- Concurrent Pos: Sect chmn, Nat Defense Res Comt, 40-42; vpres, Int Union Pure & Appl Chem, 47-51, pres, 59-63; mem adv comt, Off Naval Res, 47-52; div chmn, Nat Res Coun, 47-53; trustee, Sloan-Kettering Inst Cancer Res, 48-63; treas. Int Coun Sci Unions, 52-55; Dains lectr, Univ Kans; Montgomery lectr, Univ Nebr; Westman Mem lectr, Chem Inst Can; consult, US Army, 46-, sr sci adv, Chem Corps, 50-; consult, US Navy, 47-52, US AEC, 48-53; US Air Force, 42-57 & Nat Bur Standards, 58-60. Honors & Awards: Medal for Merit, 48; King's Medal for Serv in Cause Freedom, 48; Officer, Legion of Honor, 54; Priestley Medal, 53, Gibbs Medal, 57, Charles L Parsons Award, 70, Am Chem Soc. Mem: AAAS; Am Chem Soc(pres, 47); fel Am Phys Soc; Faraday Soc; Am Philos Soc. Res: Electrochemistry, photochemistry; vapor pressures; reaction kinetics; fluorescence, spectroscopy. Mailing Add: Dept of Chem Univ of Tex Austin TX 78712

NOZ, MARILYN E, b New York, NY, June 17, 39. THEORETICAL PHYSICS. Educ: Marymount Col, NY, BA, 61; Fordham Univ, MS, 63, PhD(physics), 69. Prof Exp: From instr to asst prof physics, Marymount Col, 64-69; ASSOC PROF PHYSICS, IND UNIV PA, 69- Mem: Am Phys Soc. Res: High energy theoretical physics; symmetry schemes for classifying elementary particles, projection operator techniques for calculating generalized vector coupling coefficients and quark theory. Mailing Add: Dept of Physics Ind Univ of Pa Indiana PA 15701

NOZAKI, KENZIE, b Los Angeles, Calif, June 1, 16; m 44; c 2. CHEMISTRY. Educ: Univ Calif, Los Angeles, BA, 37, MA, 38; Stanford Univ, PhD(chem), 40. Prof Exp: Asst chem, Univ Calif, Los Angeles, 37-38; Franklin fel, Stanford Univ, 38-40; instr Univ Calif, 41-42; dir res, War Relocation Authority, Calif, 43; Pittsburgh Plate Glass fel, Harvard Univ, 43-45, res assoc, 43-45; res chemist, Shell Develop Co, 46-72, sr staff res chemist, Shell Oil Co, 72, CONSULT RES CHEMIST, SHELL DEVELOP CO, 72- Mem: Am Chem Soc. Res: High polymers; free radicals; reaction mechanisms; molecular rearrangements; organic peroxides; kinetics; heterogeneous catalysis; organic nitrogen compounds; carbonylation; coordination chemistry; automotive fuels. Mailing Add: Shell Develop Co Westhollow Res Ctr PO Box 1380 Houston TX 77001

NOZAKI, YASUHIKO, b Yamagata, Japan, June 14, 13; m; c 2. BIOPHYSICAL CHEMISTRY. Educ: Univ Tokyo, BS, 37, PhD(pharm), 45. Prof Exp: Asst pharm, Univ Tokyo, 37-39, instr, 39-45; prof chem, Nihon Women's Col, 46-48; prof, Kyoritsu Col Pharm, 48-51; tech off microanal, Nat Inst Hyg Sci, Tokyo, 51-60, sect chief vitamin chem, 60-62; ASSOC BIOCHEM, MED CTR, DUKE UNIV, 62- Concurrent Pos: Res assoc, Harvard Univ, 54-55 & Univ Iowa, 57-59. Mem: Am Chem Soc; Am Soc Biol Chemists. Res: Naphthoresorcinol reaction of glucuronic acid; interaction of copper and zinc ions with imidazoles; titration of native and denatured proteins; solubility of amino acids in relation to configuration of proteins; structure and function of biological membranes. Mailing Add: Dept of Biochem Duke Univ Med Ctr Durham NC 27710

NOZIK, ARTHUR JACK, b Springfield, Mass, Jan 10, 36; m 58; c 2. PHYSICAL CHEMISTRY, ENERGY CONVERSION. Educ: Cornell Univ, BChE, 59; Yale Univ, MS, 62, PhD(phys chem), 67. Prof Exp: Res engr chem eng, Douglas Aircraft Co, Santa Monica, Ca, 59-60; res engr, Cent Res Div, Am Cyanamid Co, 61-64, res chemist, 67-74, STAFF CHEMIST PHYS CHEM, MAT RES CTR, ALLIED CHEM CO, 74- Concurrent Pos: Instr, Southern Conn State Col, 62-64, lectr, 68-74. Mem: Am Phys Soc; Am Chem Soc; Int Solar Energy Soc; AAAS; Electrochem Soc. Res: Chemical and physical applications of Mössbauer spectroscopy; optical, magnetic and transport properties of solids and thin films; heterogeneous catalysis; solar energy conversion; hydrogen energy systems; photoelectrolysis. Mailing Add: Mat Res Ctr Allied Chem Corp Morristown NJ 07960

NOZNICK, PETER PAUL, b New York, NY, Aug 18, 15; m 44; c 3. AGRICULTURAL CHEMISTRY. Educ: Univ Conn, BS, 38; Univ Minn, PhD(agr biochem), 46. Prof Exp: Res asst, 46-48, chief chemist, 48-56, DIR RES, BEATRICE FOODS CO, 56- Mem: Am Chem Soc; Am Asn Cereal Chem (secy-treas, 51); Am Dairy Sci Asn; Inst Food Technol. Res: Fats and oils; cereals and cereal products; dairy products; canned foods; preservation of food and dairy products by heat sterilization; spray-drying processes of dairy specialties. Mailing Add: Beatrice Foods Co 1526 S State St Chicago IL 60605

NOZZOLILLO, CONSTANCE, b Spencerville, Ont, July 18, 26; m 52. PLANT BIOCHEMISTRY. Educ: Queen's Univ (Ont), BA, 49, MA, 50; Univ Ottawa, PhD(plant biochem), 63. Prof Exp: Res off plant physiol, Can Dept Agr, 50-53, microbiol, 58-60; asst prof, 63-74, ASSOC PROF BOT, UNIV OTTAWA, 74- Mem: AAAS; Am Soc Plant Physiol; Bot Soc Am; Soc Exp Biol & Med; Phytochem Soc NAm. Res: Effects of herbicides on plants; biochemistry of seed germination; chemotaxonomic aspects of anthocyanins. Mailing Add: Dept of Biol Univ of Ottawa Ottawa ON Can

NRIAGU, JEROME OKONKWO, b Oreri Town, Nigeria, Oct 24, 42; Can citizen. GEOENVIRONMENTAL SCIENCE, GEOCHEMISTRY. Educ: Univ Ibadan, Nigeria, BS, 65; Univ Wis-Madison, MS, 67; Univ Toronto, PhD(geol & geochem), 70. Prof Exp: RES SCIENTIST, ENVIRON CAN, CAN CTR INLAND WATERS, 70- Mem: Geochem Soc; AAAS; Am Soc Limnol & Oceanog; Mineral Soc Am. Res: Biogeochemistry of the elements in the environment; stable isotopes as pollutant source and behavior indicators. Mailing Add: Can Ctr Inland Waters Box 5050 Burlington ON Can

NUBER, JOHN ARTHUR, b Detroit, Mich, Aug 12, 43. MATHEMATICS. Educ: Mass Inst Technol, BS, 65; Univ Calif, San Diego, PhD(math), 69. Prof Exp: ASST PROF MATH, UNIV MIAMI, 69- Mem: AAAS; Am Math Soc. Res: Constructive mathematics. Mailing Add: Dept of Math Univ of Miami Coral Gables FL 33124

NUCKLES, DOUGLAS BOYD, b Hampton, Va, Mar 7, 31; m 57; c 2. DENTISTRY. Educ: Med Col Va, BS, 59, DDS, 60; The Citadel, MAT, 73. Prof Exp: Instr oper, Crown & Bridge & Dent Mat, Sch Dent, Med Col Va, 60-64, asst prof oper & dent mat, 64-68, assoc prof restorative dent, 68-71; assoc prof crown & bridge, 71-72, assoc prof, 72-73, PROF OPER DENT, COL DENT MED, MED UNIV SC, 73- Concurrent Pos: Johnson & Johnson res grant, Med Univ SC, 73-75; consult dept mat sci, Sch Eng, Univ Va, 64-71; mem subcomts dent instruments & hand pieces, Am Nat Standards Comt, 70. Mem: Am Dent Asn; Am Asn Dent Schs; Int Asn Dent Res. Res: Properties of dental materials; clinical evaluation of dental restorations. Mailing Add: Dept of Oper Dent Col Dent Med Med Univ of SC 80 Barre St Charleston SC 29401

NUCKOLLS, JOHN HOPKINS, b Chicago, Ill, Nov 17, 30; m 52; c 2. APPLIED PHYSICS. Educ: Wheaton Col, BS, 53; Columbia Univ, MA, 55. Prof Exp: Physicist, 55-65, ASSOC DIV LEADER, LAWRENCE LIVERMORE LAB, UNIV CALIF, 65- Concurrent Pos: Mem vulnerability task force, Defense Sci Bd, 69-72. Honors & Awards: E O Lawrence Award, AEC, 69. Mem: Am Phys Soc. Res: Nuclear explosives; laser fusion; underground nuclear explosions. Mailing Add: Lawrence Livermore Lab Univ of Calif PO Box 808 Livermore CA 94550

NUDELMAN, ABRAHAM, organic chemistry, see 12th edition

NUDELMAN, HARVEY BANET, b St Louis, Mo, May 20, 39; m 68; c 1. BIOENGINEERING, NEUROPHYSIOLOGY. Educ: Wash Univ, BS, 61; Iowa State Univ, MS, 65; Univ Ill Med Ctr, PhD(physiol, bioeng), 71. Prof Exp: Engr, McDonnell Aircraft Corp, Mo, 64-66; res engr, Rice Univ, 71-72; asst prof psychiat & neurobiol, Univ Tex Med Sch, Houston, 72-75; ADJ ASST PROF & SR FEL BIOL, RICE UNIV, 75- Concurrent Pos: NIH Fel, Rice Univ, 70-71; consult, Tex Res Inst Med Sci, 71-; adj asst prof elec eng, Rice Univ. Res: Neural modeling; physiological control systems; mechanical receptors. Mailing Add: 5054 Yarwell Houston TX 77035

NUDELMAN, SOL, b Brooklyn, NY, Aug 14, 22; m 50; c 2. PHYSICS. Educ: Union Univ, NY, BS, 45; Ind Univ, MS, 48; Univ Md, PhD(physics), 55. Prof Exp: Asst physics, Ind Univ, 46-48; instr, Union Univ, NY, 48-49; instr, Knox Col, 49-51; physicist, US Naval Ord Lab, 51-56; res physicist, Univ Mich, 56-61; mgr solid state res, IIT Res Inst, 61-64; prof elec eng, Univ RI, 65-73; PROF RADIOL & OPTICAL SCI, UNIV ARIZ, 73- Mem: AAAS; Am Phys Soc; Am Asn Physics Teachers; Inst Elec & Electronics Eng; NY Acad Sci. Res: Influence of electric fields on phosphors; electroluminescent phosphors; luminescent displays; electrical and optical properties of semiconductors; infrared sensitive photodetectors; photoelectronic imaging devices. Mailing Add: Ariz Med Ctr Univ of Ariz Tucson AZ 85724

NUDENBERG, WALTER, polymer chemistry, organic chemistry, see 12th edition

NUENKE, RICHARD HAROLD, b Bay City, Mich, Sept 3, 32; div; c 1. BIOCHEMISTRY. Educ: Univ Mich, BS, 53; Vanderbilt Univ, PhD(biochem), 61. Prof Exp: Res assoc biochem, Univ Ill, 60-62; asst prof physiol chem, 62-66, lab coord, 67-71, ASSOC PROF PHYSIOL CHEM, OHIO STATE UNIV, 67- Concurrent Pos: NIH res grant, 63-66. Mem: AAAS; Am Chem Soc. Res: Protein structure. Mailing Add: Dept of Physiol Chem Ohio State Univ 2198 Med Sci Bldg Columbus OH 43210

NUESSLE, ALBERT CHRISTIAN, b Philadelphia, Pa, Feb 24, 15; m 40; c 4. TEXTILE CHEMISTRY. Educ: Univ Pa, BS, 36. Prof Exp: Textile res chemist, Joseph Bancroft & Sons Co, 37, asst supt finishing, 38-42, textile res chemist, 43-46; head textile appln lab, 47-64, SR CHEMIST, FIBERS DIV, ROHM AND HAAS CO, 65- Mem: Am Asn Textile Chemists & Colorists. Res: Textile finishing agents and auxiliaries, including surfactants; textile application processes and methods, including bleaching, dyeing and finishing; fiber, yarn, fabric properties. Mailing Add: Rohm and Haas Co Fibers Technol Develop Ctr Spring House PA 19477

NUESSLE, NOEL OLIVER, b St Louis, Mo, June 20, 28; m 56; c 5. PHARMACEUTICS. Educ: St Louis Col Pharm, BS, 49; Univ Fla, MS, 55, PhD(pharmaceut chem), 58. Prof Exp: Asst, Univ Fla, 54-57; from asst prof to assoc prof, 58-75, PROF PHARM & COODR EXTERNSHIPS, UNIV MO-KANSAS CITY, 75- Mem: Am Pharmaceut Asn; Acad Pharmaceut Sci; Sigma Xi. Res: Pharmaceutical formulation; biopharmaceutics; sterilization by irradiation. Mailing Add: Dept of Pharm Univ of Mo 5100 Rockhill Rd Kansas City MO 64110

NUETZEL, JOHN ARLINGTON, b East St Louis, Ill, Feb 16, 25; m 46; c 6. INTERNAL MEDICINE. Educ: Wash Univ, MD, 47; Am Bd Internal Med, dipl, 54. Prof Exp: Intern, Univ Mich, 47-48; resident, US Vet Admin, Mo, 49-51; ASSOC PROF CLIN MED, ST LOUIS UNIV, 64- Concurrent Pos: Fel hypertension, Wash Univ & Barnes Hosp, 48-49; med dir, St Mary's Health Ctr; mem courtesy staff, St John's Hosp. Mem: AMA; Am Heart Asn; fel Am Col Cardiol; fel Am Col Physicians. Res: Hypertension; cardiology. Mailing Add: 911 S Brentwood Blvd Clayton MO 63105

NUFFIELD, EDWARD WILFRID, b Gretna, Man, Apr 13, 14; m 39; c 3.

MINERALOGY. Educ: Univ BC, BA, 40; Univ Toronto, PhD(mineral), 44. Prof Exp: Asst, Geol Surv Can, 40-42; lectr, 43-49, from asst prof to assoc prof geol, 49-62, assoc dean, Fac Arts & Sci, 62-64, chmn dept geol, 64-72, PROF GEOL, UNIV TORONTO, 62- Concurrent Pos: Asst, Ont Dept Mines, 43-44 & 48, geologist, 49-51. Honors & Awards: Sr Award, Royal Soc Can, 63. Mem: Fel Mineral Soc Am; Am Crystallog Asn; fel Royal Soc Can; Mineral Asn Can (pres, 56-58). Res: X-ray crystallography; crystal chemistry; determination of the crystal structures of sulphosalt minerals. Mailing Add: Dept of Geol Univ of Toronto Toronto ON Can

NUGENT, CHARLES ARTER, JR, b Denver, Colo, Nov 18, 25; m 50; c 3. INTERNAL MEDICINE, ENDOCRINOLOGY. Educ: Yale Univ, MD, 51. Prof Exp: From intern to asst resident, New Haven Hosp, 51-53; resident, Col Med, Univ Utah, 54-56, from instr to assoc prof, 56-67; prof, Col Med, Univ Hawaii, 67-70; PROF MED, COL MED, UNIV ARIZ, 70- Concurrent Pos: Res fel, Col Med, Univ Utah, 54. Mem: Am Soc Clin Invest; Endocrine Soc. Res: Hypertension; radioimmunoassay of steroid; control of adrenal steroid secretion. Mailing Add: Dept of Internal Med Ariz Med Ctr Tucson AZ 85724

NUGENT, GEORGE ROBERT, b Yonkers, NY, Feb 6, 21; m 47; c 5. NEUROSURGERY. Educ: Kenyon Col, AB, 50; Univ Cincinnati, MD, 53. Prof Exp: Instr neurosurg, Med Ctr, Duke Univ, 57-58; asst dir div neurosurg, Col Med, Univ Cincinnati, 58-61; from asst prof to assoc prof surg, 61-69, PROF SURG, DIV NEUROSURG, MED CTR, WVA UNIV, 69-, CHMN DIV, 70- Concurrent Pos: Chief neurosurg, Cincinnati Vet Admin Hosp, 58-61. Mem: Am Asn Neurol Surg; Soc Neurol Surg; Cong Neurosci Surgeons; Int Soc Res Stereoencephalotomy. Res: Treatment of trigeminaz neuralgia; sterotaxic brain surgery; teaching methods; microneurosurgery. Mailing Add: Div of Neurosurg Univ Hosp Morgantown WV 26506

NUGENT, LEONARD JAMES, b Chicago, Ill, Oct 1, 30; m 52; c 4. PLASMA PHYSICS. Educ: Ill Inst Technol, BS, 54, MS, 57; Univ Wis, PhD(phys chem), 59. Prof Exp: Res assoc, Nat Res Coun, Nat Bur Standards, 58-59, physicist, 59-61; sr physicist, Gen Dynamics Corp, 61-62; sr physicist, Electro-Optical Systs, Inc, 62-66; SR STAFF SCIENTIST, CHEM DIV, OAK RIDGE NAT LAB, 66- Mem: Fel Am Phys Soc; Am Chem Soc; Am Asn Phys Teachers. Res: Molecular and atomic spectra; atmospheric, chemical and plasma physics. Mailing Add: Thermonuclear Div Oak Ridge Nat Lab Box X Oak Ridge TN 37830

NUGENT, MAURICE JOSEPH, JR, b Salt Lake City, Utah, Dec 22, 37. ORGANIC CHEMISTRY. Educ: Univ Colo, BA, 61; Calif Inst Technol, PhD(chem), 65. Prof Exp: NIH res fel chem, Harvard Univ, 65-66; asst prof, 66-73, ASSOC PROF CHEM, TULANE UNIV, 73- Concurrent Pos: Vis assoc chem, Calif Inst Technol, 75-76. Mem: Am Chem Soc. Res: Mechanisms of enzymatic reactions; antibiotics and metabolic control reagents. Mailing Add: Dept of Chem Tulane Univ New Orleans LA 70118

NUGENT, ROBERT CHARLES, b Jersey City, NJ, Sept 22, 36; m 58; c 2. GEOLOGY. Educ: Hofstra Univ, BA, 58; Univ Rochester, MS, 60; Northwestern Univ, PhD(geol), 67. Prof Exp: Develop geologist, Chevron Oil Co, Standard Oil Co Calif, La, 65-68; asst prof, 68-70, ASSOC PROF GEOL, STATE UNIV OF N Y, 70-, CHMN DEPT EARTH SCI, 74- Mem: Geol Soc Am; Soc Econ Paleont & Mineral. Res: Stratigraphy of chemung formation; diagenesis of bonneterre dolomite, Missouri; petroleum geology; Gulf Coast stratigraphy and structural geology; geologic control of shape of nuclear craters; Lake Ontario beach gravels. Mailing Add: Dept of Earth Sci State Univ of N Y Oswego NY 13126

NUGENT, THOMAS JOHN, bElnora, Ind, Apr 30, 10; m 46. PLANT PATHOLOGY. Educ: Purdue Univ, BSA, 33, MS, 37. Prof Exp: Asst bot & inspector tomato seed cert, Purdue Univ, 35-37; asst plant pathologist, Va Truck Exp Sta, 37-42; asst plant pathologist, State Agr Exp Substa, Tex, 46; from assoc plant pathologist to plant pathologist, 46-67, ASST DIR, VA TRUCK & ORNAMENTALS RES STA, 67- Mem: AAAS; Am Phytopath Soc. Res: Control of truck crop diseases; disease resistant vegetables. Mailing Add: Va Truck & Ornamentals Res Sta Box 2160 Norfolk VA 23501

NUITE, JO ANN, b Albany, NY, Oct 30, 45. PHARMACOLOGY, NEUROBIOLOGY. Educ: Col St Rose, BA, 67; Univ NC, Chapel Hill, PhD(pharmacol), 71. Prof Exp: PHARMACOLOGIST, DRUG ABUSE CTR, NIMH/Nat Inst Drug Abuse, 71-74; ASST PROF PHARMACOL, SCHS MED & DENT, GEORGETOWN UNIV, 71- Concurrent Pos: Res assoc cent nerv syst pharmacol, Univ NC, Chapel Hill, 71-72. Mem: AAAS; NY Acad Sci; Int Narcotic Addiction Res Club. Res: Central nervous system pharmacology and neurobiology; effects of drugs on the adult and developing organism, particularly narcotic analgesics and related drugs, marijuana and other psychoactive drugs on central nervous system function, amine metabolism related to their possible mechanism of action. Mailing Add: Dept Pharmacol Sch Med & Dent Georgetown Univ Washington DC 20007

NUKI, KLAUS, b Vienna, Austria, May 5, 31; m 63; c 3. HISTOCHEMISTRY, PERIODONTOLOGY. Educ: Univ London, BDS, 55, PhD(histochem), 67; Univ Ill, Chicago, MS, 60. Prof Exp: House surgeon, Queen Victoria Hosp, East Grinstead, Eng, 55-56; res asst oral path, Univ Ill, Chicago, 57-60; lectr path & periodont, London Hosp Med Col, 60-67; res assoc periodont, Royal Dent Col, Denmark, 63-64; assoc prof dent, 67-71, PROF DENT & HEAD DEPT ORAL BIOL & DIV HISTOL & HISTOCHEM, COL DENT, UNIV IOWA, 71- Mem: Am Asn Dent Schs; Int Asn Dent Res; Royal Soc Med. Res: Histology; pathology; micro-chemistry of inflammation; microcirculation of gingiva; investigations of periodontium. Mailing Add: Div of Histol & Histochem Univ of Iowa Col of Dent Iowa City IA 52240

NUMEROF, PAUL, organic chemistry, radio chemistry, see 12th edition

NUMMY, WILLIAM RALPH, b Brooklyn, NY, Oct 2, 21; m 49; c 4. ORGANIC CHEMISTRY. Educ: Univ of the South, BS, 47; Univ Rochester, PhD(chem), 50. Prof Exp: Anal chemist, Magnus, Mabee & Reynard, Inc, 39-42; res chemist, Arnold, Hoffman & Co, 50-53; res chemist, Dow Chem Co, 53, group leader, 53-56, div leader, 56, asst dir phys res lab, 56-60, asst dir polymer res lab, 60-61, dir plastics dept, res labs, 61-64, mgr plastics develop & serv, 64-67, mgr plant sci bus, 67-68, mgr agr prod dept, 68-69, mgr life sci res, 69-70, dir cent res labs, 70-74, DIR RES & DEVELOP, DOW CHEM CO, 74- Mem: Am Chem Soc. Res: Claisen rearrangement mechanism; high polymer synthesis; polyamides; polysulfides and plastic foams. Mailing Add: 711 W Meadowbrook Dr Midland MI 48642

NUNEMACHER, JEFFREY LYNN, b Williamsport, Pa, Oct 19, 48. MATHEMATICAL ANALYSIS. Educ: Oberlin Col, AB, 70; Yale Univ, MPhil, 73, PhD(math), 75. Prof Exp: ASST PROF MATH, UNIV TEX, AUSTIN, 75- Mem: Am Math Soc; Math Asn Am. Res: Several complex variables, particularly its analytic aspects. Mailing Add: Dept of Math Univ of Tex Austin TX 78712

NUNEMAKER, JOHN COLEMAN, b LaJunta, Colo, Oct 5, 10; m 46; c 3. INTERNAL MEDICINE. Educ: Univ Idaho, BS, 32, MS, 33; Harvard Univ, MD, 37; Am Bd Internal Med, dipl, 50. Prof Exp: House officer internal med, Peter Bent Brigham Hosp, Boston, Mass, 37-39; instr, asst & fel, Johns Hopkins Hosp, 39-41; instr & res assoc, Univ Mich Hosp, 41-42; asst prof, Sch Med, Univ Utah, 46-52; chief res div, Dept Med & Surg, Vet Admin, DC, 52-53, dir educ serv, 53-58; assoc secy, Coun Med Educ, AMA, 58-67, dir dept grad med educ, 66-67 & assoc dir div med educ, 67-70; exec dir, Am Bd Med Specialties, 70-75; RETIRED. Concurrent Pos: Chief med serv, Vet Admin Hosp, Salt Lake City, Utah, 46-52, consult to Surgeon Gen, US Navy; prof lectr, Sch Med, George Washington Univ, 52-58; lectr, Med Sch, Northwestern Univ, Ill, 58- Mem: AAAS; Asn Am Med Cols; Am Col Physicians; Am Rheumatism Asn. Res: Rat bite fever; rheumatoid arthritis; infectious diseases. Mailing Add: 2665 Calle de Rosa Las Cruces NM 88001

NUNES, MATHEWS ANTHONY, b Vallejo, Calif, Oct 28, 41; m 65; c 2. MEDICINAL CHEMISTRY. Educ: Ore State Univ, BS, 69; Univ Calif, San Francisco, PhD(pharmaceut chem), 74. Prof Exp: Res biochem, Tulane Univ, 74-75; SR SCIENTIST ANAL CHEM, MEAD JOHNSON & CO, 75- Mem: Am Pharmaceut Asn; Am Chem Soc; Sigma Xi. Res: Drug metabolism; marine chemistry, particularly marine toxins. Mailing Add: Mead Johnson & Co Evansville IN 47721

NUNES, PAUL DONALD, b New Bedford, Mass, Aug 29, 44. GEOCHRONOLOGY. Educ: Tufts Univ, BS, 66; Univ Calif, Santa Barbara, PhD(geol), 70. Prof Exp: Asst I geochronology res, Swiss Fed Inst Technol, 70-72; geologist, US Geol Surv, 73-75; RES ASSOC GEOL, ROYAL ONT MUS, 75- Concurrent Pos: Nat Res Coun assoc, US Geol Surv, Denver, 72-73. Mem: Geochem Soc; Am Geophys Union. Res: Use of uranium-thorium-lead and rubidium-strontium natural decay systems to probe and further clarify our understanding of the evolution of the earth and moon. Mailing Add: Dept of Mineral & Geol Royal Ont Mus 100 Queen's Park Toronto ON Can

NUNES, THOMAS LESTER, b New Bedford, Mass, Nov 1, 42; m 67; c 1. CHEMICAL INSTRUMENTATION. Educ: Tufts Univ, BSc, 64; Univ Calif, Berkeley, PhD(phys inorg chem), 67. Prof Exp: Asst prof, 67-71, ASSOC PROF CHEM, STATE UNIV NY COL NEW PALTZ, 71- Concurrent Pos: State Univ NY exchange employee, Components Div, IBM Corp, 70-71; sabbatical fel, Lawrence Livermore Lab, 74-75, consult, 75- Mem: Am Chem Soc. Res: Application of laboratory automation to x-ray diffractometry, electron probe microanalysis and elucidating the kinetics of enzyme reactions, specifically glutomic oxalo-acetic transaminase. Mailing Add: Dept of Chem State Univ of NY Col New Paltz NY 12561

NUNEZ, ELADIO A, physiology, see 12th edition

NUNEZ, LOYS JOSEPH, b New Orleans, La, Mar 18, 26. CHEMISTRY, TOXICOLOGY. Educ: Tulane Univ, BS, 47; La State Univ, Baton Rouge, MS, 55, PhD(chem), 60. Prof Exp: Chemist, Cities Serv Refining Corp & Res & Develop Corp, 47-53, res chemist, 59-60; supvr phys res sect, Austin Labs, Jefferson Chem Co, Inc, Tex, 60-71; HEAD BIOMAT SECT, MAT SCI TOXICOL LABS, UNIV TENN CTR FOR HEALTH SCI, 71-, ASST PROF DENT & PHARM, 72- Mem: Am Chem Soc. Res: Dental biomaterials; toxicity of thermodegradation products of fabrics and plastics; chemical and polymer carcinogenesis; drug-plastic soprtion diffusion. Mailing Add: Mat Sci Toxicol Labs Univ of Tenn Ctr for Health Sci Memphis TN 38163

NUNEZ, THERON A, JR, b Pittsburgh, Pa, Mar 16, 30. ANTHROPOLOGY. Educ: Fla State Univ, BA, 56, MA, 57; Univ Calif, Berkeley, PhD(anthrop), 63. Prof Exp: Instr soc sci, Fla State Univ, 57-58; asst prof anthrop, 63-68, ASSOC PROF ANTHROP, UNIV FLA, 68- Concurrent Pos: NSF prin investr, 64. Mem: Fel Am Anthrop Asn. Res: History and theory of anthropology; peoples and cultures of Mexico and Central America; social organization; culture change. Mailing Add: Dept of Anthrop Univ of Fla Gainesville FL 32603

NUNEZ, WILLIAM J, III, b New Orleans, La, Jan 17, 44; m 65; c 2. IMMUNOLOGY, MICROBIOLOGY. Educ: La State Univ, Baton Rouge, BS, 65, MS, 67; NTex State Univ, PhD(immunol), 70. Prof Exp: asst prof, 70-73, ASSOC PROF BIOL & CHMN DEPT, UNIV DETROIT, 73- Concurrent Pos: Tuberc & Health Soc Wayne County fel, Univ Detroit, 71-, Nat Multiple Sclerosis Soc fel, 72-; Environ Protection Agency & Scholl Found res grants, 75. Mem: AAAS; Am Soc Microbiol; Am Thoracic Soc. Res: Allergy, particularly cellular mechanisms of delayed hypersensitivity involved in tuberculin hypersensitivity and experimental autoallergic encephalomyelitis; delayed hypersensitivity; particularly passive transfer mechanisms of tuberculin, chemical contact dermatitis and experimental allergic encephalomyelitis; follic acid or rapid and specific detection of human pollution indicators in water. Mailing Add: Dept of Biol Univ of Detroit 2001 W McNichols Detroit MI 48221

NUNEZ-MELENDEZ, ESTEBAN, b Cidra, PR, Aug 3, 09; m 39; c 3. PHARMACOGNOSY. Educ: Univ PR, BS, 33; Univ Fla, MS, 41, PhD(pharmacog), 53. Prof Exp: From asst instr to assoc prof pharmacog, 35-60, chg med plant garden, 54-57, PROF PHARMACOG, COL PHARM, UNIV PR, RIO PIEDRAS, 60- Mem: Am Pharmaceut Asn; Bot Soc Am; Am Soc Pharmacog; NY Acad Sci; PR Acad Arts & Sci. Res: Pharmacognosy of medicinal plants; botany, chemistry and pharmacology of the organic constituents present in medicinal plants. Mailing Add: Col of Pharm Univ of PR Box 21, 301 Rio Piedras PR 00931

NUNKE, RONALD JOHN, b Kenosha, Wis, Mar 9, 26; m 53; c 3. MATHEMATICS. Educ: Univ Chicago, SB, 50, SM, 51, PhD(math), 55. Prof Exp: Instr math, Northwestern Univ, 53-54 & Yale Univ, 55-58; from asst prof to assoc prof, 58-69, PROF MATH, UNIV WASH, 69- Mem: Am Math Soc; Math Asn Am. Res: Abelian groups; homological algebra. Mailing Add: Dept of Math Univ of Wash Seattle WA 98105

NUNLEY, ROBERT E, b Red Jacket, WVa, Mar 10, 31; m 52; c 3. GEOGRAPHY. Educ: Marshall Col, BS, 52, MA, 53; Univ Mich, PhD, 58. Prof Exp: Asst prof geog, Wayne State Univ, 57-62; assoc prof, 62-74, PROF GEOG, UNIV KANS, 74- Mem: AAAS; Asn Am Geog; Pop Asn Am; Regional Sci Asn. Res: Factors affecting distributions of human populations and the objective methodology, instrumentation and data maniuplation necessary to describe and analyze such factors. Mailing Add: Dept of Geog Univ of Kans Lawrence KS 66044

NUNLEY, ROBERT GRAY, b Quinwood, WVa, Feb 5, 30; m 52; c 4. PHYCOLOGY. Educ: Marshall Col, AB, 52, MA, 53; WVa Univ, PhD(bot), 66. Prof Exp: Teacher pub schs, WVa, 53-54, teacher & prin, 57-61; teacher, Ohio, 56-57; instr biol, WVa Univ, 63-65; assoc prof, 65-68, PROF BIOL & HEAD DEPT, MORRIS HARVEY COL, 68- Mem: AAAS; Am Inst Biol Sci. Res: Ecology of freshwater benthic algae; taxonomy of the genus Trachelomonas in West Virginia. Mailing Add: Dept of Biol Morris Harvey Col Charleston WV 25304

NUNN, ARTHUR SHERMAN, JR, b Independence, Mo, Nov 9, 22; m 50; c 5. PHYSIOLOGY. Educ: Kans State Univ, BS, 55; Univ Iowa, MS, 59, PhD(physiol),

60. Prof Exp: Instr physiol, Sch Med, St Louis Univ, 60-62; from asst prof to assoc prof, Sch Med, Univ Miami, 62-67; assoc prof, 67-72, PROF PHYSIOL, SCH MED, IND UNIV, INDIANAPOLIS, 72- Concurrent Pos: US Army res & develop fel, Univ Miami, 62-67. Mem: AAAS; Am Physiol Soc. Res: Membrane transport of sugars; the role of cations in the transmembrane movement of organic compounds; gastro-intestinal physiology. Mailing Add: Dept of Physiol Ind Univ Sch of Med Indianapolis IN 46202

NUNN, DOROTHY MAE, b Cincinnati, Ohio. MICROBIOLOGY. Educ: Univ Cincinnati, BS, 55, PhD(microbiol), 62. Prof Exp: Asst prof microbiol, Univ Dayton, 62-65; assoc prof, E Tenn State Univ, 65-67; ASSOC PROF BIOL, UNIV AKRON, 67- Mem: AAAS; Am Soc Microbiol; Brit Soc Gen Microbiol. Res: Bacterial physiology and bioenergetics; mechanism of action of bacterial toxins. Mailing Add: Dept of Biol Univ of Akron Akron OH 44304

NUNN, LESLIE GREY, JR, b Uvalde, Tex, May 16, 17. ORGANIC CHEMISTRY. Educ: Univ Tex, BA, 38, MA, 41; PhD(org chem), 43; Seton Hall Univ, LLB, 63. Prof Exp: Tutor chem, Univ Tex, 39-41, instr, 41; res & develop chemist, Am Cyanamid Co, Conn, 43-46; res chemist, Carter Oil Co, Okla, 46-50; applns chemist, Cent Res Lab, Gen Aniline & Film Corp, Pa, 50-52, asst tech mgr indust chems, Antara Chem Div, NY, 52-56, process develop chemist, 56-64; attorney, Nopco Chem Co, 65-66, asst patent counsel, 66-69; PATENT COUNSEL, DIAMOND SHAMROCK CORP, 69- Res: Manufacture of synthetic organic compounds; nitrogen heterocyclics; petroleum production and drilling practices; surfactants. Mailing Add: Diamond Shamrock Corp PO Box 2386-R Morristown NJ 07960

NUNNALLY, DAVID AMBROSE, b Memphis, Tenn, Sept 13, 34; m 56; c 5. ZOOLOGY, BIOLOGY. Educ: Univ of the South, BS, 56; Wash Univ, PhD(zool), 61. Prof Exp: From instr to asst prof, 60-65, ASSOC PROF BIOL, VANDERBILT UNIV, 65- Mem: AAAS; Am Soc Zool. Res: Physiological embryology; physiology of parasitic flatworms. Mailing Add: Dept Gen Biol Box 1812 Substa B Vanderbilt Univ Nashville TN 37235

NUNNALLY, HUEY NEAL, b Atlanta, Ga, Dec 28, 44; m 68; c 2. BIOMEDICAL ENGINEERING, ELECTRICAL ENGINEERING. Educ: Ga Inst Technol, BEE, 66, MSEE, 68, PhD(elec eng), 71. Prof Exp: Assoc aircraft engr, Lockheed-Ga Co, 65-66; res consult biomed eng, US Army Edgewood Arsenal, 71; ASST PROF BIOMED ENG, GA INST TECHNOL, 71- Concurrent Pos: NSF grant, Sch Elec Eng, Ga Inst Technol, 72-74; asst res prof, Sch Med, Emory Univ, 71-74. Mem: AAAS; Inst Elec & Electronics Eng. Res: Auditory physiology; electrical stimulation of nervous system; electrical stimulation of muscular system. Mailing Add: Sch of Elec Eng Ga Inst of Technol Atlanta GA 30332

NUNNALLY, NELSON RUDOLPH, b Monroe, Ga, Dec 24, 35; m 58; c 4. PHYSICAL GEOGRAPHY. Educ: Univ Ga, BS, 58, MA, 61; Univ Ill, PhD(geog), 65. Prof Exp: From instr to assoc prof geog, E Tenn State Univ, 60-67; assoc prof, Fla Atlantic Univ, 67-68; asst prof, Univ Ill, 68-71; assoc prof, Univ Okla, 71-74; ASSOC PROF GEOG, UNIV NC, CHARLOTTE, 74- Mem: Asn Am Geog; Am Soc Photogram. Res: Physical geography; remote sensing. Mailing Add: Dept of Geog Univ of NC UNCC Sta Charlotte NC 28223

NUNNEMACHER, RUDOLPH FINK, b Milwaukee, Wis, Mar 21, 12; m 28; c 4. HISTOLOGY. Educ: Kenyon Col, BS, 34; Harvard Univ, MA, 35, PhD(histol), 38. Prof Exp: Instr histol & embryol, Med Sch, Univ Okla, 38-39; from asst prof to assoc prof, 39-55, dir eve col, 53-54, PROF BIOL, CLARK UNIV, 55-, CHMN DEPT, 59- Concurrent Pos: Instr, Worcester City & Clinton Hosps, 42-45; trustee, Bermuda Biol Sta. Mem: AAAS; Am Soc Zoologists; NY Acad Sci. Res: Structure, function of arthropod eyes; nervous system by electron microscopy. Mailing Add: Dept of Biol Clark Univ Worcester MA 01610

NUNNING, BERNARD CHARLES, pharmacy, medicinal chemistry, see 12th edition

NUR, AMOS M, b Haifa, Israel, Feb 9, 38; m 68. GEOPHYSICS. Educ: Hebrew Univ Jerusalem, BS, 62; Mass Inst Technol, PhD(geophys), 69. Prof Exp: Res assoc geophys, Mass Inst Technol, 69-70; asst prof, 70-75, ASSOC PROF GEOPHYS, STANFORD UNIV, 75- Concurrent Pos: Sloan fel, Stanford Univ, 72-74; mem earthsci adv bd, NSF, 74-77; vis prof, Weizmann Inst Sci, 75. Honors & Awards: Maclwane Award, Am Geophys Union, 74; Newcomb Cleveland Prize, AAAS, 75. Mem: Am Geophys Union; Seismol Soc Am. Res: Tectonophysics; rock mechanics; physical hydrology; earthquake mechanics. Mailing Add: Dept of Geophysics Stanford Univ Stanford CA 94305

NUR, HUSSAIN SAYID, b Mahmoudia, Iraq, July 1, 39; m 69. MATHEMATICAL ANALYSIS. Educ: Univ Baghdad, BSc, 61; Univ Calif, Berkeley, MA, 64, PhD(math), 67. Prof Exp: Teaching asst math, Univ Baghdad, 61-62; from asst prof to assoc prof, 67-74, PROF MATH, FRESNO STATE COL, 74- Mem: Am Math Soc. Res: Singular perturbation of linear partial differential equations. Mailing Add: Dept of Math Fresno State Col Fresno CA 93726

NUR, UZI, b Ein Harod, Israel, June 28, 28; m 52; c 2. CYTOGENETICS. Educ: Hebrew Univ, Israel, MSc, 58; Univ Calif, PhD(genetics), 62. Prof Exp: USPHS trainee, 62-63, asst prof, 63-67, ASSOC PROF BIOL, UNIV ROCHESTER, 67- Mem: Genetics Soc Am; Am Soc Naturalists; Soc Study Evolution. Res: Origin, nature and function of heterochromatin; evolution and population genetics of sexual and parthenogenetic scale insects. Mailing Add: Dept of Biol Univ of Rochester Rochester NY 14627

NURGE, ETHEL, b Brooklyn, NY, Aug 21, 20. CULTURAL ANTHROPOLOGY. Educ: Univ NMex, BA, 50; Univ Chicago, MA, 51; Cornell Univ, PhD(anthrop), 55. Prof Exp: Asst prof, SDak State Col, 63-65; res assoc, Frobenius Inst, 65-66; assoc prof anthrop, McMaster Univ, 66-68; ASSOC PROF ANTHROP, DEPT HUMAN ECOL, MED CTR, UNIV KANS, 68- Concurrent Pos: Fulbright grant, Univ Philippines, 55-56; assoc ed, Current Anthrop, 62-65. Mem: Am Anthrop Asn; Am Ethnol Soc; Soc Med Anthrop. Res: Social factors in sickness and health; woman's role and changing opportunities available to women. Mailing Add: Dept of Human Ecol Univ of Kans Med Ctr Rainbow at 39th Kansas City KS 66103

NURNBERGER, JOHN IGNATIUS, b Chicago, Ill, Apr 9, 16; m 43; c 4. CYTOLOGY, CHEMISTRY. Educ: Loyola Univ, Ill, BS, 38; Northwestern Univ, MS, 42, MD, 43. Prof Exp: Res neurologist, Neurol Inst, NY, 46-48; res fel, Med Nobel Inst Cell Res & Genetics, Stockholm, Sweden, 49-50; asst clin prof med & psychiat, Sch Med, Yale Univ, 53-56; actg dean, Sch Med, 63-64, prof psychiat & chmn dept & dir, Inst Psychiat Res, 56-74, DISTINGUISHED PROF PSYCHIAT & CHMN EXEC BD, INST PSYCHIAT RES, MED CTR, IND UNIV, INDIANAPOLIS, 74- Concurrent Pos: Psychiatrist, Inst Living, 48-49, res assoc, 50-56, dir residency training, 50-56; lectr, Univ Conn, 53-56; mem, Ment Health Study Sect, NIH, 59-63, mem exp psychol study sect, USPHS, 60-63; mem bd dir, Found Fund Res Psychiat, 72- Mem: Soc Biol Psychiat; Histochem Soc; Am Psychiat Asn; Asn Res Nerv & Ment Dis; Am Neurol Asn. Res: Cytochemical and biochemical studies in nitrogen metabolism; pastoral and clinical psychiatry; analysis of human behavior; cytogenetics. Mailing Add: Dept of Psychiat Ind Univ Med Ctr Indianapolis IN 46207

NURNBERGER, ROBERT G, botany, science education, see 12th edition

NURSALL, JOHN RALPH, b Regina, Sask, Dec 25, 25; m 53; c 3. ZOOLOGY. Educ: Univ Sask, BA, 47, MA, 49; Univ Wis, PhD(zool), 53. Prof Exp: Instr biol, Univ Sask, 48-49; asst zool, Univ Wis, 49-53; lectr, 53-55, from asst prof to assoc prof, 55-64, PROF ZOOL, UNIV ALTA, 64-, CHMN DEPT, 64-69 & 74- Concurrent Pos: Nuffield travel grant, Gt Brit, 54, Nuffield fel, 62-63. Mem: AAAS; Am Soc Ichthyol & Herpet; Soc Study Evolution; Can Soc Zool; Zool Soc London. Res: Morphology and locomotion of fish; paleontology; limnology; fish behavior. Mailing Add: Dept of Zool Univ of Alta Edmonton AB Can

NUSBAUM, CHARLES JOSEPH, b Salem, Ore, Aug 24, 06; m 32. PLANT PATHOLOGY. Educ: Ore State Col, BS, 29; Univ Wis, MS, 31, PhD(plant path), 34. Prof Exp: Field asst forest path, USDA, 27, 34, agent, 28-30; asst plant path, Univ Wis, 30-32, 34-36; from asst pathologist to pathologist, Clemson Col, Exp Sta, 36-48; res prof, 48-56, REYNOLDS PROF PLANT PATH, N C STATE UNIV, 56- Mem: Am Phytopath Soc; Soc Nematol. Res: Tobacco diseases; epidemiology and control; ecology and population dynamics of plant parasitic nematódes; soil fumigation. Mailing Add: Dept of Plant Path N C State Univ PO Box 5397 Raleigh NC 27607

NUSS, JAMES WILLIAM, physical chemistry, inorganic chemistry, see 12th edition

NUSSBAUM, ADOLF EDWARD, b Rheydt, Ger, Jan 10, 25; nat US; m 57; c 2. MATHEMATICS. Educ: Columbia Univ, MA, 50, PhD(math), 57. Prof Exp: Lectr, Columbia Col, 51-52; staff mem, Electronic Comput Proj, Inst Advan Study, 52-53; instr math, Univ Conn, 53-55; from instr to asst prof, Rensselaer Polytech Inst, 55-58; from asst prof to assoc prof, 58-66, PROF MATH, WASHINGTON UNIV, 66- Concurrent Pos: Mem, Inst Advan Study & NSF fel, 62-63; vis scholar, Stanford Univ, 67-68. Mem: AAAS; Am Math Soc. Res: Analysis and functional analysis. Mailing Add: Dept of Math Washington Univ St Louis MO 63130

NUSSBAUM, ALEXANDER LEOPOLD, b Leipzig, Ger, Dec 30, 25; nat US; m 57; c 3. ORGANIC CHEMISTRY, BIOCHEMISTRY. Educ: City Col New York, BS, 48; Purdue Univ, MS, 50; Wayne State Univ, PhD(chem), 54. Prof Exp: Vis investr, Inst Chem, Nat Univ Mex, 50; chemist, Syntex, SAm, 50-51 & Worcester Found Exp Biol, 54-55; sr chemist, Schering Corp, 55-65; GROUP CHIEF, HOFFMANN-LA ROCHE, INC, 65- Concurrent Pos: Res assoc, Med Sch, Stanford Univ, 61-63; vis lectr, Stevens Inst Technol, 64-65; assoc res prof, Sch Med, Univ Md; vis lectr, Fairleigh Dickenson Univ, 74-75; vis prof, Dept Biol Chem, Harvard Med Sch, 75. Mem: AAAS; Am Chem Soc; NY Acad Sci; fel Am Inst Chem. Res: Chemistry of natural products, especially steroids and nucleotides; photochemistry; nucleic acid biochemistry; virology; molecular biology of cancer viruses. Mailing Add: 86 Yantacaw Brook Rd Upper Montclair NJ 07043

NUSSBAUM, ALLEN, b Phila, Pa, Aug 22, 19; m 45; c 4. SOLID STATE PHYSICS. Educ: Univ Pa, BA, 39, MA, 40, PhD(physics), 54. Prof Exp: Physicist, Honeywell Inc, 54-62; PROF ELEC ENG & DIR GRAD STUDY, UNIV MINN, MINNEAPOLIS, 62- Mem: Am Phys Soc; Inst Elec & Electronics Eng; Brit Inst Physics. Res: Geometrical optics; semiconducting materials device development; theory of junction diodes; electromagnetic and quantum properties of materials. Mailing Add: Dept of Elec Eng Univ of Minn Minneapolis MN 55455

NUSSBAUM, ELMER, b Monroe, Ind, Sept 2, 20; m 49; c 4. BIOPHYSICS. Educ: Taylor Univ, BA, 49; Ball State Univ, MA, 52; Univ Rochester, PhD(radiation biol), 57. Prof Exp: instr physics, Taylor Univ, 49-52, asst prof, 53; res assoc biophys, Univ Rochester, 53-57; assoc prof, 57-59, PROF PHYSICS, TAYLOR UNIV, 59-, DIR RES & SPECIAL TRAINING, 61- Concurrent Pos: Consult, Oak Ridge Assoc Univs, 60-, sr scientist, 62-63. Mem: AAAS; Am Asn Physics Teachers; Health Physics Soc. Res: Solubility of radon in fatty acids and body tissues; diffusion of radon through semipermeable materials; radionuclides in the biosphere; physics; environmental radioactivity; radiation detectors; health physics. Mailing Add: Dept of Physics Taylor Univ Upland IN 46989

NUSSBAUM, MIRKO, b Belgrade, Yugoslavia, July 24, 30; US citizen; m. EXPERIMENTAL HIGH ENERGY PHYSICS. Educ: Rutgers Univ, AB, 54; Univ Chicago, SM, 56; Johns Hopkins Univ, PhD(physics), 62. Prof Exp: Sr physicist, Martin Co, Md, 56-60; instr physics, Johns Hopkins Univ, 61-62; res assoc, Columbia Univ, 62-64; asst prof, Univ Pa, 64-69; assoc prof, 69-71, PROF PHYSICS, UNIV CINCINNATI, 71- Mem: Fel Am Phys Soc. Res: Fundamental particle research through use of bubble chambers and spark chamber counter techniques. Mailing Add: Dept of Physics Univ of Cincinnati Cincinnati OH 45221

NUSSBAUM, MURRAY, b Brooklyn, NY, May 19, 27; m 55; c 2. MEDICINE, HEMATOLOGY. Educ: Univ Vt, BS, 49, MD, 52; Am Bd Int Med, cert hemat, 74. Prof Exp: From instr to assoc prof, 57-69, PROF MED, HEMAT DIV, COL MED & DENT NJ, 69-, DIR LAB MED COURSE, 57- Concurrent Pos: USPHS res fel hemat, 54-55; res assoc, Mt Sinai Hosp, New York, 59-62; chief hemat, Jersey City Med Ctr, 63-66; dir hemat-med serv, Martland Hosp, 66-; mem med adv comt, Metrop Chap, Hemophilia Found, 66- Mem: Am Soc Hemat; Am Fedn Clin Res; fel Int Soc Hemat; fel Am Col Physicians. Res: Blood coagulation and fibrinolysis; medical education. Mailing Add: Div of Hemat Col of Med & Dent of NJ Newark NJ 07107

NUSSBAUM, NOEL SIDNEY, b Brooklyn, NY, Jan 26, 35; m 57; c 3. PHYSIOLOGY, ENDOCRINOLOGY. Educ: Brooklyn Col, BA, 56; Williams Col, MA, 58; Yale Univ, PhD(biol), 64. Prof Exp: From instr to asst prof biol, Bowdoin Col, 63-65; from asst prof to assoc prof biol, 65-75, ASSOC PROF PHYSIOL, SCH MED, WRIGHT STATE UNIV, 75- Concurrent Pos: Consult, Aerospace Med Res Lab, US Air Force, Wright-Patterson AFB, 74- Mem: Am Soc Zool; Am Soc Cell Biol; Orthopaedic Res Soc. Res: Vertebrate physiology and morphogenesis; mineral metabolism; ultrastructure of calcified tissue; biochemical and biomechanical response of dense connective tissue to endocrine therapy. Mailing Add: Dept of Physiol Sch of Med Wright State Univ Dayton OH 45431

NUSSBAUM, ROGER DAVID, b Philadelphia, Pa, Jan 29, 44; m 66; c 2. MATHEMATICAL ANALYSIS. Educ: Harvard Univ, AB, 65; Univ Chicago, PhD(math), 69. Prof Exp: Asst prof, 69-73, ASSOC PROF MATH, RUTGERS UNIV, NEW BRUNSWICK, 73- Mem: Am Math Soc. Res: Nonlinear functional analysis, particularly fixed point theorems and their applications to nonlinear problems. Mailing Add: Dept of Math Rutgers Univ New Brunswick NJ 08903

NUSSBAUM, RONALD ARCHIE, b Rupert, Idaho, Feb 9, 42. HERPETOLOGY. Educ: Univ Idaho, BS, 67; Cent Wash State Col, MS, 68; Ore State Univ, PhD(zool),

72. Prof Exp: Res assoc ecosyst anal, Sch Forestry, Ore State Univ, 72-73; ASST PROF & ASST CUR, DIV BIOL SCI & MUS ZOOL, UNIV MICH, ANN ARBOR, 74- Mem: Soc Syst Zoologists; Am Soc Naturalists; Am Soc Ichthyologists & Herpetologists; Ecol Soc Am; Soc Study Evolution. Res: Evolution, systematics, life history and ecology of amphibians. Mailing Add: Mus of Zool Univ of Mich Ann Arbor MI 48104

NUSSBAUM, RUDI HANS, b Furth, Ger, Mar 21, 22; nat US; m 47; c 3. SOLID STATE PHYSICS. Educ: Univ Amsterdam, PhD(exp physics), 54. Prof Exp: Res assoc nuclear physics, Inst Nuclear Res, Univ Amsterdam, 52-54; UNESCO fel, Nuclear Physics Lab, Liverpool, Eng, 54-55; res assoc, Ind Univ, 55-56; European Orgn Nuclear Res sr fel, Geneva, 56-57; asst prof physics, Univ Calif, 57-59; assoc prof, 59-65, PROF PHYSICS, PORTLAND STATE UNIV, 65- Concurrent Pos: Vis prof, Univ Wash, 65-66; consult, Tektronix Inc; exchange prof, Univ Canterbury, 71-72; vis prof, Univ Wash, 74. Mem: Am Phys Soc; Am Asn Physics Teachers; Netherlands Phys Soc; Am Fedn Scientists; Soc Social Responsibility Sci. Res: Mössbauer effect applied to lattice dynamics and study of thermally stable sites of cobalt/ferrum impurities in metal hosts. Mailing Add: Dept of Physics Portland State Univ PO Box 751 Portland OR 97207

NUSSENBAUM, SIEGFRIED, b Vienna, Austria, Nov 21, 19; nat US; m 51; c 2. BIOCHEMISTRY, ORGANIC CHEMISTRY. Educ: Univ Calif, BS, 41, MA, 48, PhD(biochem), 51. Prof Exp: Anal chemist, Manganese Ore Co & Pan Am Eng Co, 42-43, asst chief chemist, 43-45; chemist, Nat Lead Co, 45; asst, Univ Calif, 48-51, fel, 51-52; from instr to assoc prof, 52-62, PROF CHEM, SACRAMENTO STATE UNIV, 62- Concurrent Pos: Consult biochemist, Sacramento County Hosp, 58-; lectr clin path, Univ Calif, 69- Mem: AAAS; Am Chem Soc. Res: Polysaccharide metabolism and chemistry; enzyme, analytical and clinical chemistry. Mailing Add: Dept of Chem Sacramento State Univ 6000 J St Sacramento CA 95819

NUSSENZVEIG, HERCH MOYSES, b Sao Paulo, Brazil, Jan 16, 33; m 62; c 3. MATHEMATICAL PHYSICS. Educ: Univ Sao Paulo, BSc, 54, PhD(physics), 57. Prof Exp: Asst prof theoret physics, Univ Sao Paulo, 56-57; from asst prof to prof, Brazilian Ctr Physics Res, 57-68; vis prof, Univ Rochester, 65-68, sr res assoc & prof physics, 68-75; PROF PHYSICS, UNIV SAO PAULO, 75- Concurrent Pos: Nat Res Coun Brazil res fels, Theoret Physics Inst, State Univ Utrecht, 60-61, Dept Math Physics, Univ Birmingham, 61 & Inst Theoret Physics, Swiss Fed Inst Technol, 61; vis mem, Courant Inst Math Sci, NY Univ, 63-64 & Inst Adv Study, 64-65; vis prof, Univ Paris, Orsay, 73. Mem: Am Phys Soc; Brazilian Acad Sci. Res: Quantum optics; scattering theory; dispersion relations. Mailing Add: Inst Physics Univ Sao Paulo Sao Paulo Brazil

NUSSER, WILFORD LEE, b Sylvia, Kans, Oct 6, 24; m 43; c 5. PARASITOLOGY, PHYSIOLOGY. Educ: Bethel Col, Kans, BA, 49; Kans State Univ, MS, 50; Iowa State Univ, PhD(parasitol), 58. Prof Exp: Instr microbiol, Col Osteop Med & Surg, 54-57, asst prof physiol, 57-60, actg head dept, 58-60; res fel neuroanat, Emory Univ, 60-63; prof physiol & head dept, Col Osteop Med & Surg, 63-66; grants assoc, 66-67, health scientist adminr, 67-74, CHIEF SCI PROG BR, NAT EYE INST, NIH, 74-, DIR ARTHRITIS & ORTHOP PROG, NAT INST ARTHRITIS, METAB & DIGESTIVE DIS, 74- Res: Electron microscopy; Wallerian degeneration and regeneration of peripheral nerves; endocrinology; site of production of the erythropoietic hormone; neuroanatomy. Mailing Add: Sci Prog Br Nat Eye Inst Nat Insts Health Bethesda MD 20014

NUSSMANN, DAVID GEORGE, b Burlington, Iowa, May 8, 37. GEOLOGY, GEOCHEMISTRY. Educ: Harvard Univ, AB, 59; Univ Mich, MA, 61, PhD(geol, geochem), 65. Prof Exp: Geologist, Explor & Prod Res Div, Shell Develop Co, Tex, 64-70, staff geologist, Shell Oil Co, La, 70-74, GEOL ENGR, SHELL OIL CORP, LA, 74- Mem: Geol Soc Am; Soc Petrol Eng; Geochem Soc. Res: Sediment geochemistry; geostatistics; petroleum geology. Mailing Add: Shell Oil Corp Box 60124 New Orleans LA 70160

NUSYNOWITZ, MARTIN LAWRENCE, b New York, NY, July 21, 33; m 55; c 3. NUCLEAR MEDICINE, ENDOCRINOLOGY. Educ: NY Univ, BA, 54; State Univ NY, MD, 58. Prof Exp: Med Corps, US Army, 57-, intern med, Letterman Gen Hosp, San Francisco, 58-59, resident internal med, Tripler Gen Hosp, Honolulu, Hawaii, 59-62, chief radioisotope-endocrine serv, 62-63, CHIEF NUCLEAR-MED ENDOCRINE SERV & DEPT MED RES & DEVELOP, WILLIAM BEAUMONT GEN HOSP, MED CORPS, US ARMY, 65- Concurrent Pos: Fel mil med & allied sci, Walter Reed Army Inst Res, Washington, DC, 63-64; fel endocrinol & metab, Sch Med, Univ Calif, San Francisco, 64-65; Dorothy Hutton scholar endocrinol, 68; lectr, Grad Fac, Univ Tex, El Paso, 67-; consult, Surgeon Gen, US Army, 71- Mem: Soc Nuclear Med; Endocrine Soc; fel Am Col Physicians; Am Fedn Clin Res. Res: In vitro thyroid function tests; clinical aspects of thyroid disease; computer applications in nuclear medicine; diagnostic nuclear medical techniques in clinical problems. Mailing Add: Dept of Med Res & Develop William Beaumont Gen Hosp El Paso TX 79920

NUTINI, HUGO GINO, b Peumo, Chile, June 26, 28; m 54; c 1. ANTHROPOLOGY. Educ: Chilean Naval Acad, BS, 47; Univ Calif, Los Angeles, BA, 55, MA, 58, PhD(anthrop), 62. Prof Exp: Instr anthrop, Los Angeles State Col, 62; sr researcher, Pan Am Union, Washington, DC, 62-63; asst prof, Univ PR, 62-63 & George Washington Univ, 63; asst prof anthrop, 63-70, ASSOC PROF ANTHROP, UNIV PITTSBURGH, 70- Mem: Am Anthrop Asn. Res: Social anthropology, particularly theory kinship and social structure, culture change and ethnohistory; Mesoamerica and South America, especially Mexico, Peru and Chile. Mailing Add: Dept of Anthrop Univ of Pittsburgh Pittsburgh PA 15213

NUTLEY, HUGH, b Tacoma, Wash, Jan 30, 32; m 55; c 6. NUCLEAR PHYSICS. Educ: Mass Inst Technol, SB, 54; Univ Wash, PhD(physics), 60, BA, 73, MA, 74. Prof Exp: Res specialist, Radiation Effects Lab, Boeing Co, 62-66; from asst prof to assoc prof, 66-75, PROF PHYSICS, SEATTLE PAC COL, 75- Mem: Am Asn Physics Teachers; Am Sci Affil. Res: Beta ray spectroscopy; radiation damage to electronic components. Mailing Add: Dept of Physics Seattle Pac Col Seattle WA 98119

NUTT, DAVID CLARK, b Cleveland, Ohio, June 21, 19; m 43; c 6. OCEANOGRAPHY, GEOGRAPHY. Educ: Dartmouth Col, AB, 41. Prof Exp: Arctic specialist, 47-56, RES ASSOC GEOG, DARTMOUTH COL, 56- Concurrent Pos: Biol observer, US Navy Proj Windmill, 47-48; leader, Blue Dolphin Labrador Exped, 49-54; geog consult, Div Defense Labs, Mass Inst Technol, 55 & Tech Serv, Inc, 56-60; mem heat & water panel, Comt Polar Res, Nat Acad Sci, 59-65. Mem: AAAS; Asn Am Geog; fel Am Geog Soc; fel Arctic Inst NAm. Res: Geography of polar regions, specifically oceanography and glaciology. Mailing Add: Dept of Geog Dartmouth Col Hanover NH 03755

NUTT, RUTH FOELSCHE, b Flensburg, Ger, July 12, 40; US citizen; m 58; c 1. MEDICINAL CHEMISTRY. Educ: Univ NMex, BS, 62. Prof Exp: Asst chemist indole & pyridine anticancer agts, 62-65, res chemist nucleoside & carbohydrate chem, 65-67, sr res chemist peptide synthesis, 67-76, RES FEL PEPTIDE SYNTHESIS, MERCK SHARP & DOHME RES LABS, 76- Mem: Am Chem Soc. Res: Organic synthesis; nucleosides; central nervous system-active peptides; design and synthesis of releasing hormone agonists and antagonists; mechanisms of amino acid racemization. Mailing Add: Merck Sharp & Dohme Res Labs West Point PA 19486

NUTTALL, FRANK Q, b May 8, 29; US citizen; c 4. INTERNAL MEDICINE. Educ: Univ Utah, MD, 55; Univ Minn, PhD(biochem), 70. Prof Exp: Chief admin sect, Vet Admin Hosp, 61-63, chief clin chem, 63-69, assoc prof, Univ, 71-75, PROF INTERNAL MED, SCH MED, UNIV MINN, MINNEAPOLIS, 75-, CHIEF METAB-ENDOCRINE SECT, VET ADMIN HOSP, 70- Mem: Am Diabetes Asn; Am Soc Biol Chemists; fel Am Col Physicians; Endocrine Soc. Res: Diabetes mellitus; control of glycogen metabolism; glycogen synthetase system. Mailing Add: Metab-Endocrine Sect Vet Admin Hosp Minneapolis MN 55417

NUTTALL, JAMES B, b Schley, Va, July 8, 14; m 41; c 2. AEROSPACE MEDICINE. Educ: Univ Md, BS, 35, MD, 39; Johns Hopkins Univ, MPH, 61; Am Bd Prev Med, dipl. Prof Exp: US Air Force chief dept mil med, Air Force Sch Aviation Med, 46-48, secy comt med sci, Res & Develop Bd, Dept Defense, 48-50, chief dept flight med, Air Sch Aviation Med, 50-53, exchange officer, Royal Air Force, Eng, 53-54, chief aviation med & surgeon, Air Force in Europe, Ger, 54-56, chief aviation med div, Off Surgeon Gen, Hq US Air Force, Washington, DC, 56-60, chief aerospace med adv, Space Systs Div, Air Force Systs Command, Calif, 61-63, Comdr cent med group, Air Force in Europe, Ger, 63-66, Comdr, Air Force Sch Aerospace Med, Brooks AFB, Tex, 66-67, surgeon, Strategic Air Command Hq, 67-69; med dir, McDonnell Douglas Astronaut Co, 71-73; ASSOC MED DIR, OCCIDENTAL LIFE OF CALIF, 73- Mem: Fel Aerospace Med Asn; fel Int Acad Aerospace Med; fel Am Col Physicians. Res: Preventive medicine. Mailing Add: 7046 Vista Del Mar Playa Del Rey CA 90291

NUTTALL, JOHN, b Haslingden, Eng, Oct 8, 36; m 62; c 2. THEORETICAL PHYSICS. Educ: Cambridge Univ, BA, 57, PhD(theoret physics), 61. Prof Exp: Res fel, St John's Col, Cambridge, 61-62; scientist, RCA Victor Co Ltd, Can, 62-64; NSF sr for scientist, Tex A&M Univ, 64-65, assoc prof theoret physics, 65-69, prof physics, 69-72; vis prof, 72, PROF PHYSICS, UNIV WESTERN ONT, 72- Mem: Am Phys Soc; Can Asn Physicists. Res: Quantum scattering theory in atomic and nuclear physics; approximation theory. Mailing Add: Dept of Physics Univ of Western Ont London ON Can

NUTTALL, RALPH LESLIE, b Spanish Fork, Utah, Oct 10, 15; m 36; c 4. PHYSICAL CHEMISTRY. Educ: Univ Utah, AB, 37, MA, 41; Yale Univ, PhD(phys chem), 48. Prof Exp: Phys chemist, Nat Bur Standards, 48-54; assoc chemist, Argonne Nat Lab, 54-64; PHYS CHEMIST, NAT BUR STANDARDS, 64- Mem: AAAS; Am Chem Soc. Res: Diffusion of electrolytes; heat transfer; thermodynamics; heats of combustion; high temperature calorimetry. Mailing Add: Nat Bur of Standards Washington DC 20234

NUTTALL, WESLEY FORD, b Regina, Sask, Oct 24, 30. SOIL FERTILITY. Educ: Univ Sask, BSA, 58, MSc, 60; McGill Univ, PhD(soil chem), 65. Prof Exp: RES SCIENTIST, RES BR, CAN DEPT AGR, 65- Mem: Can Soc Soil Sci; Am Soc Agron. Res: Plant nutrition; statistics; agrometeorology. Mailing Add: Res Sta Can Dept of Agr Box 1901 Melfort SK Can

NUTTER, GENE DOUGLAS, b Columbus, Tex, June 9, 29; m 56; c 2. PHYSICS. Educ: Univ Nebr, BS, 51, MS, 56. Prof Exp: Physicist, Nat Bur Standards, 52-54; eng supvr, Atomics Int, NAm Rockwell, Inc, 56-67; ASST DIR MEASUREMENT SCI LAB, INSTRUMENTATION SYSTS CTR, UNIV WIS-MADISON, 67- Concurrent Pos: UN Ed, Sci & Cult Orgn consult & lectr, Repub Korea, 68. Mem: Sr mem Instrument Soc Am; Am Inst Physics; Am Soc Mech Eng; Am Soc Test & Mat; Optical Soc Am. Res: Methodology and instrumentation for high accuracy measurements, especially measurement of temperature and thermal properties by radiometric methods; radiation thermometry form 100 degrees centigrade to 4000 degrees centigrade. Mailing Add: Instrumentation Systs Ctr Univ of Wis 1500 Johnson Dr Madison WI 53706

NUTTER, JAMES DOUGLAS, b New Lexington, Ohio, Dec 15, 34; m 62; c 2. PHYSICAL CHEMISTRY. Educ: Kent State Univ, BS, 58; Univ Nebr, MS, 61, PhD(chem), 64. Prof Exp: Sr res chemist, Monsanto Res Corp, Miamisburg, 63-66, group leader calorimetry, 66-69, mgr nondestructive testing calorimetry & appl physics, Mound Lab, 69-73, TRAINING MGR & CONSULT, MONSANTO CO, 73- Res: Thermochemistry; nondestructive testing; applied physics; half-lives by calorimetry. Mailing Add: 9 Forrester Dr Manchester MO 63011

NUTTER, JOHN EDWARD, microbiology, see 12th edition

NUTTER, ROBERT LELAND, b Boston, Mass, Jan 20, 22; m 46; c 2. BIOPHYSICS, VIROLOGY. Educ: Andrews Univ, BA, 44; Univ Colo, MS, 49; Iowa State Univ, PhD(biophys), 57. Prof Exp: Physicist, Nat Bur Standards, 44-45; asst physics, Univ Colo, 45-46; from instr to asst prof, Pac Union Col, 46-52; asst, Iowa State Univ, 52-53, instr, 53-56; asst prof, Pac Union Col, 56-57; from instr to assoc prof, 57-68, PROF MICROBIOL, SCH MED, LOMA LINDA UNIV, 68- Concurrent Pos: Instr, Univ Colo, 48-49; Am Cancer Soc Scholar, Col Med, Hershey Med Ctr & vis prof microbiol, Pa State Univ, 71-72. Mem: AAAS; Am Soc Microbiol. Res: Bacteriophages; animal viruses; virus nucleic acids; herpesviruses and other oncogenic viruses. Mailing Add: Dept of Microbiol Loma Linda Univ Sch of Med Loma Linda CA 92354

NUTTER, WILLIAM ERMAL, b Boomer, WVa, Sept 26, 27; m 52; c 3. BIOCHEMISTRY. Educ: WVa Inst Technol, BS, 53; Univ WVa, MS, 57, PhD(biochem), 59. Prof Exp: Res assoc biochem, Univ Iowa, 59-61; from asst ed to assoc ed, 61-65, asst head dept biochem ed, 65-71, head dept biochem ed anal, 71-73, ASST MGR, BIOCHEM DEPT, CHEM ABSTR SERV, OHIO STATE UNIV, 73- Mem: Am Chem Soc; AAAS; Sigma Xi. Res: Bacterial utilization of amino acids and peptides; mammalian metabolism of tryptophan; synthesis of kynurenine; kinetics of d-amino acid oxidase with d-kynurenine; chemical information and documentation. Mailing Add: Chem Abstr Serv Ohio State Univ Columbus OH 43210

NUTTING, ALBERT DEANE, b Otisfield, Maine, Sept 6, 05; m 40. FORESTRY. Educ: Univ Maine, BS, 27. Prof Exp: Forester, Finch, Pruyn & Co, 27-31; exten forester, Univ Maine, 31-48; forest comnr, State of Maine, 48-58; dir, Sch Forestry, 58-71, head dept forest res, 70-71, EMER DIR SCH FOREST RESOURCES, UNIV MAINE, ORONO, 71- Concurrent Pos: Chmn, Nat Coop Forestry Res Bd. Honors & Awards: New Eng Coun Award, 58; Maine Forest Industs Award, 59; Northern Loggers Asn Award, 63. Mem: Fel Soc Am Foresters; Am Forestry Asn. Res: Forest management and products. Mailing Add: RFD 1 Otisfield Oxford ME 04270

NUTTING, EHARD FORREST, b Milwaukee, Wis, Oct 4, 29; m 51; c 6.

ENDOCRINOLOGY, PHARMACOLOGY. Educ: Utah State Univ, BS, 51; Univ Wis, MS, 56, PhD(endocrinol), 62. Prof Exp: Res asst genetics, Univ Wis, 54-56 & zool, 58-60; res investr endocrinol testing, 60-64, res reprod endocrinol, 65-69, sr res investr reprod physiol, 69-70, res group leader fertility control, 70-71, res dir endocrinol dept, 71-73, dir, Dept Biol Res, Searle Labs Div, 73-75, SR RES SCIENTIST, DEPT SCI, SEARLE LABS DIV, G D SEARLE & CO, 75- Concurrent Pos: Consult, Nat Insts Health, 71-74. Mem: Endocrine Soc; Soc Exp Biol & Med; Soc Study Reprod; Brit Soc Study Fertil; Am Soc Pharmacol & Exp Therapeut. Res: Physiology of the female reproductive tract; pharmacological control of fertility; gamete transport; physiology and pharmacology of factors involved in cellular growth processes. Mailing Add: Searle Labs Div G D Searle & Co PO Box 5110 Chicago IL 60680

NUTTING, GEORGE COOK, physical chemistry, see 12th edition

NUTTING, LEE, b Berkeley, Calif, Sept 4, 09; m 47. FOOD CHEMISTRY. Educ: Univ Calif, BS, 31; Stanford Univ, MA, 32, PhD(phys chem), 34. Prof Exp: Asst, Stanford Univ, 32-34; res chemist, Hills Bros Coffee, Inc, 34-40 & 46-52, dir res, 52-74; RETIRED. Mem: Am Chem Soc; Am Ord Asn; Inst Food Technol; Int Sci Asn Coffee. Res: Concentrated aqueous mixtures of electrolytes; chemistry of coffee; smokeless powder; high explosives. Mailing Add: 627 Santa Barbara Rd Berkeley CA 94707

NUTTING, LEIGHTON ADAMS, b Salem, NH, Sept 21, 19; m 43; c 3. BIOLOGY. Educ: Univ NH, BS, 41; Va Polytech Inst, MS, 49, PhD(microbiol), 51. Prof Exp: Scientist, Nat Dairy Res Labs, Inc, 50-60; RES MICROBIOLOGIST, ABBOTT LABS, 60- Mem: Am Soc Microbiol; Am Chem Soc. Res: Microbial physiology; antibiotics; carbohydrate metabolism; dairy bacteriology; analytical chemistry. Mailing Add: Sci Div Abbott Labs North Chicago IL 60064

NUTTING, WILLIAM BROWN, b Worcester, Mass, Apr 15, 18; m 68; c 4. ACAROLOGY. Educ: Univ Mass, BS, 40, MS, 48; Cornell Univ, PhD(zool), 50. Prof Exp: From instr to assoc prof, 50-64, PROF ZOOL, UNIV MASS, AMHERST, 64- Concurrent Pos: Fel, Univ Queensland, 58; vis prof, Stanford Univ Med Sch, 75. Mem: Sigma Xi. Res: Biology; ecology; pathogenesis and systematics of mites, especially Demodicidae. Mailing Add: Dept of Zool Univ of Mass Amherst MA 01002

NUTTING, WILLIAM LEROY, b Pepperell, Mass, July 26, 22; m 44; c 3. ENTOMOLOGY. Educ: Harvard Univ, AB, 43, PhD(biol), 50. Prof Exp: Res fel biol, Harvard Univ, 50-55; assoc entomologist, 55-62, assoc prof, 58-62, PROF ENTOM & ENTOMOLOGIST, UNIV ARIZ, 62- Mem: AAAS; Am Soc Zool; Entom Soc Am; Am Inst Biol Sci; Int Union Study Social Insects. Res: Insect morphology, behavior and ecology, especially Orthoptera, Isoptera and other social insects; symbiosis. Mailing Add: Dept of Entom Univ of Ariz Tucson AZ 85721

NUTTLI, OTTO WILLIAM, b St Louis, Mo, Dec 11, 26. GEOPHYSICS. Educ: St Louis Univ, BS, 48, MS, 50, PhD(geophys), 53. Prof Exp: From instr to assoc prof, 52-62, PROF GEOPHYS & GEOPHYS ENG, ST LOUIS UNIV, 62- Concurrent Pos: Ed bull, Seismol Soc Am, 71-75. Mem: AAAS; Seismol Soc Am (pres, 75-76); Soc Explor Geophys; Am Geophys Union; fel Royal Astron Soc. Res: Velocity of elastic waves in the interior of the earth; earthquake shear waves; seismicity of Eastern North America. Mailing Add: Earth & Atmos Sci St Louis Univ PO Box 8099 Laclede Sta St Louis MO 63156

NUTTYCOMBE, JOHN WILLIAM, zoology. deceased

NUWAYSER, ELIE SAMI, chemistry, see 12th edition

NUZZI, ROBERT, b New York, NY, July 7, 42; m 68. MARINE BIOLOGY. Educ: Fordham Univ, BS, 63, MS, 65, PhD(microbiol), 69. Prof Exp: Teaching asst biol, Fordham Univ, 64-68; instr, St Francis Col, NY, 68-70; assoc res scientist microbiol, NY Ocean Sci Lab, Affil Cols & Univs, Inc, 70-74, res scientist, 74-75; MARINE BIOLOGIST, SUFFOLK COUNTY DEPT ENVIRON CONTROL, 75- Concurrent Pos: Res assoc, NY Ocean Sci Lab, Affil Cols & Univs, Inc, 75- Mem: AAAS; Am Inst Biol Sci; Am Soc Limnol & Oceanog; NY Acad Sci. Res: Phytoplankton systematics; estuarine ecology. Mailing Add: Suffolk Co Dept Environ Control 1324 Motor Pkwy Hauppauge NY 11787

NYBAKKEN, JAMES W, b Warren, Minn, Sept 16, 36; m 60; c 2. MARINE ECOLOGY. Educ: St Olaf Col, BA, 58; Univ Wis, MS, 61, PhD(zool), 65. Prof Exp: Teaching asst marine biol, Univ Miami, 58-59; cur, Zool Mus, Univ Wis, 61, 62 & 64-65; from asst prof to assoc prof, 65-72, PROF BIOL, CALIF STATE UNIV, HAYWARD, 73-, STAFF MEM, MOSS LANDING MARINE LAB, 68- Concurrent Pos: Off Naval Res vis investr, Marine Lab, Univ Ariz, 66; res assoc, Univ Wash, 68-69. Mem: AAAS; Am Soc Mammal; Ecol Soc Am; Am Soc Zool; Marine Biol Asn UK. Res: Ecology and systematics of gastropod and tropical mollusks of the genus Conus; ecology of nudibranch mollusks. Mailing Add: Moss Landing Marine Labs Box 223 Moss Landing CA 95039

NYBERG, DAVID DOLPH, b Vancouver, Wash, June 10, 28; m 51; c 2. POLYMER CHEMISTRY. Educ: Ore State Univ, BS, 52, PhD(org chem), 56. Prof Exp: Group leader, Shell Chem Co, 56-66; staff mem, 66-72, MGR EXPLOR DEVELOP, THERMOFIT TECH DEPT, RAYCHEM CORP, 72- Mem: Am Chem Soc. Res: Polymer synthesis; plastics and rubber technology; radiation crosslinking. Mailing Add: Raychem Corp 300 Constitution Dr Menlo Park CA 94025

NYBOER, JAN, b Holland, Mich, Apr 21, 06; m 39; c 4. INTERNAL MEDICINE. Educ: Univ Mich, AB, 26, MS, 29, ScD(med physiol), 32, MD, 35. Prof Exp: Asst biol, Hope Col, 25-27; asst mammal, Univ Mich, 28, asst & demonstr physiol, 28-34, asst electrocardiol, Univ Hosp, 34-35; intern, St Louis Maternity Hosp & Barnes Hosp, Wash Univ, 35-36 & Harvard Med Serv, Boston City Hosp, 36-37; fel med, NY Post-Grad Med Sch, Columbia Univ, 37-41; asst med dir, Conn Mutual Life Ins Co, 41-47; asst prof pharmacol, Dartmouth Med Sch, 47-55; chief cardiovasc physiol & assoc physician, Harper Hosp, 55-65; DIR RES, REHAB INST, WAYNE STATE UNIV, 65-, PROF PHYSIOL & PHARMACOL, SCH MED, 59-, ASSOC PHYSIOL RES, 73- Concurrent Pos: Asst attend physician, NY Postgrad Hosp & Clin, 37-47; clin instr, Sch Med, Yale Univ, 42-47; consult, US Vet Admin Hosp, Vt, 48-55; consult, Hitchcock Clin, 48-55, res assoc, Hitchcock Found, 54-55. Mem: Am Physiol Soc; Am Heart Asn; fel Am Col Cardiol; Am Fedn Clin Res. Res: Physiology of the heart and respiration; electrocardiography; electrical impedance plethysmography; development of direct writing electrocardiography; displacement and counterforce ballistocardiography; bioelectrical impedance during renal dialysis. Mailing Add: Dept of Physiol Wayne State Univ Detroit MI 48201

NYBORG, WESLEY LEMARS, b Ruthven, Iowa, May 15, 17; m 45; c 1. BIOPHYSICS, ACOUSTICS. Educ: Luther Col, AB, 41; Pa State Univ, MS, 44, PhD(physics), 47. Prof Exp: Asst physics, Pa State Univ, 41-43, instr, 43-44 & 47-49, asst, 44-47, asst prof, 49-50; from asst prof to assoc prof, Brown Univ, 50-60; PROF PHYSICS, UNIV VT, 60- Concurrent Pos: USPHS fel, Sch Adv Study, Mass Inst Technol, 56-57; vis scientist, Oxford Univ, 60-61; mem adv comt radiation bio-effects, Food & Drug Admin, 72-76. Mem: AAAS; Acoust Soc Am; Am Phys Soc; Biophys Soc; Am Asn Physics Teachers. Res: Acoustics; ultrasonics; biophysical ultrasound; environmental biophysics; physical mechanisms for biological effects of ultrasound. Mailing Add: Dept of Physics Cook Phys Sci Bldg Univ of Vt Burlington VT 05401

NYBURG, STANLEY CECIL, b London, Eng, Dec 15, 24; m 49; c 2. CRYSTALLOGRAPHY. Educ: King's Col, Univ London, BSc, 45, DSc(crystallog, thermodynamics), 73; Univ Leeds, PhD(crystallog), 49. Prof Exp: Crystallographer, Brit Rubber PProducers' Res Asn, 49-52; from asst lectr to sr lectr chem, Univ Keele, 52-64; PROF CHEM, UNIV TORONTO, 64- Concurrent Pos: Vis prof crystallog, Univ Pittsburgh, 62-63; vis prof, Univ Sydney, Australia. Mem: Fel The Chem Soc; assoc Brit Inst Physics & Phys Soc. Res: Crystal structure analysis; thermodynamics. Mailing Add: 133 Belsize Dr Toronto ON Can

NYC, JOSEPH FRANK, b Cleveland, Ohio, May 8, 19; m 44. BIOLOGICAL CHEMISTRY. Educ: Western Reserve Univ, BS, 43, PhD(org chem), 46. Prof Exp: Asst & Williams-Waterman Fund fel, Calif Inst Technol, 46-48; from asst prof to assoc prof, 50-62, PROF BIOL CHEM, SCH MED, UNIV CALIF, LOS ANGELES, 62- Mem: Am Soc Biol Chem. Res: Biochemical aspects of Neurospora genetics; enzymology; intermediary metabolism. Mailing Add: Dept of Biol Chem Univ of Calif Sch of Med Los Angeles CA 90024

NYCE, JACK LELAND, b Sellersville, Pa, Jan 15, 34; m 55; c 3. POLYMER CHEMISTRY, RUBBER CHEMISTRY. Educ: Pa State Teachers Col, West Chester, BS, 55; Univ Del, MS, 57, PhD(chem), 60. Prof Exp: Instr gen chem, Univ Del, 58-59; RES CHEMIST, E I DU PONT DE NEMOURS & CO, INC, 59- Mem: Am Chem Soc; Sigma Xi. Res: Bicyclic bridged ring systems; rubber chemistry, especially synthetic rubbers; adhesive chemistry. Mailing Add: 1004 Mayflower Dr Meadowood Newark DE 19711

NYDEGGER, CORINNE NEMETZ, b Milwaukee, Wis. MEDICAL ANTHROPOLOGY. Educ: Univ Wis, Madison, BA, 51; Cornell Univ, MA, 70; Pa State Univ, PhD(human develop), 73. Prof Exp: Res field-team mem, Six Cultures Proj, Harvard-Yale-Cornell Proj, 54-56; asst dir res geriatrics, Pa State Univ, 72-73; NIMH fel, Dept Sociol & Inst Human Develop, Univ Calif, Berkeley, 73-74; LECTR MED ANTHROP PROG, UNIV CALIF, SAN FRANCISCO, 75- Concurrent Pos: USPHS res fel, Human Develop & Med Anthrop Prog, Univ Calif, San Francisco, 74-76; lectr adult develop, Sch Soc Welfare, Univ Calif, Berkeley, 76. Mem: Geront Soc; Soc Study Social Probs; Med Anthrop Asn. Res: Timing—effects of chronological age on events and roles; responses to timing deviance; deviance and medical models; group responses to negative labeling; fatherhood, especially deviant patterns. Mailing Add: Med Anthrop Prog Univ of Calif 1320 Third Ave San Francisco CA 94143

NYDEGGER, LEROY B, b Ft Dodge, Iowa, Oct 20, 11; m 49; c 2. PARASITOLOGY. Educ: Iowa State Col, BS, 36; Univ Nebr, MS, 38; Syracuse Univ, PhD, 61. Prof Exp: Asst zool, Univ Nebr, 36-38; lab asst chem, Ft Dodge Jr Col, 39-44; instr sci, Sheldon Jr Col, 44-48; assoc prof biol, Morningside Col, 48-59, actg head dept, 51-57; assoc prof, 59-67, PROF BIOL, STATE UNIV NY COL NEW PALTZ, 67- Mem: AAAS; Am Inst Biol Sci. Res: Parasitology in vitro cultivation of malaria parasites. Mailing Add: Dept of Sci State Univ NY Col New Paltz NY 12561

NYE, ALAN HALL, b Wolcott, NY, Aug 26, 46; m 69. SOLAR PHYSICS. Educ: Clarkson Col Technol, BS, 69, MS, 71; Univ Rochester, PhD(mech & aerospace sci), 75. Prof Exp: NAT RES COUN RESIDENT RES ASSOC SOLAR PHYSICS, SACRAMENTO PEAK OBSERV, AIR FORCE CAMBRIDGE RES LABS, 75- Mem: Am Astron Soc. Mailing Add: Sacramento Peak Observ Sunspot NM 88349

NYE, OSBORNE BARR, JR, b Port Henry, NY, Mar 30, 37; m 59; c 3. INVERTEBRATE PALEONTOLOGY. Educ: Union Col, BS, 59, MS, 64; Univ Cincinnati, PhD(paleont), 72. Prof Exp: Asst prof geol, Wayne State Univ, 69-75; ASSOC PROF GEOL & SCI EDUC, SYRACUSE UNIV, 75- Concurrent Pos: Consult, AAAS, 75- Mem: AAAS; Paleont Soc; Paleont Res Inst; Soc Econ Paleontologists & Mineralogists; Nat Asn Geol Teachers. Res: Paleobiology of Bryozoa. Mailing Add: Heroy Geol Lab Syracuse Univ Syracuse NY 13210

NYE, PATRICK WILLIAM, b Yeovil, UK, Aug 19, 36; m 58; c 2. BIOPHYSICS, PSYCHOPHYSICS. Educ: Univ Reading, BSc, 58, PhD(physics), 62. Prof Exp: St Dunstan's fel reading mach blind, Nat Phys Lab, UK, 61-64; staff physicist vision res, Calif Inst Technol, 65-66, res fel, 66-69, sr res fel, 69-71; mem res staff, 71-74, ASSOC DIR RES, HASKINS LABS, 75- Concurrent Pos: Mem subcomt sensory aids, Nat Acad Sci, 69-75. Mem: AAAS; Optical Soc Am; Asn Res Vision & Ophthal. Res: Visual psychophysics and neurophysiology; human communication; prosthetic aids for the blind. Mailing Add: Haskins Labs 270 Crown St New Haven CT 06510

NYE, ROBERT EUGENE, JR, b Cincinnati, Ohio, Feb 6, 22; m 48; c 3. PHYSIOLOGY. Educ: Ohio Univ, AB, 43; Univ Rochester, MD, 47. Prof Exp: Intern med, Strong Mem Hosp, Rochester, NY, 47-48, asst resident, 48-49; house physician, Hammersmith Hosp, London, Eng, 51; instr, Univ Rochester, 54-56; from instr to assoc prof, 56-73, PROF PHYSIOL, DARTMOUTH MED SCH, 73- Concurrent Pos: Buswell fel, Univ Rochester, 49-50, univ fel, 51-54; assoc staff, Mary Hitchcock Mem Hosp, 56-65, consult staff, 65- Mem: Am Fedn Clin Res; Am Physiol Soc. Res: Cardiovascular and pulmonary physiology, especially the relations between perfusion and ventilation in the lung. Mailing Add: Dept of Physiol Dartmouth Med Sch Hanover NH 03755

NYE, SYLVANUS WILLIAM, b Buffalo, NY, Mar 28, 30; m 56; c 2. PATHOLOGY. Educ: Hamilton Col, AB, 52; Univ Rochester, MD, 57; Am Bd Path, cert. Prof Exp: From intern to chief resident path, Sch Med, Univ NC, Chapel Hill, 57-60, instr path & trainee clin microbiol & path, 60-62, from asst prof to assoc prof, 69-71; PROF PATH & CHMN DEPT, SCH MED, E CAROLINA UNIV, 71- Concurrent Pos: Pathologist, SEATO Med Res Lab, US Army Component, Bangkok, Thailand, 63-65. Mem: AAAS; Col Am Path; Int Acad Path; Am Soc Trop Med & Hyg; Am Soc Clin Path. Res: Geographic pathology. Mailing Add: Dept of Path E Carolina Univ Sch of Med Greenville NC 27834

NYE, WARREN EDWARD, b Madison, Wis, Aug 19, 12. BIOLOGY. Educ: Columbia Col, AB, 34; Univ Wis, MA, 39, PhD(zool), 45. Prof Exp: Chmn Dept, 50-67, PROF BIOL, LORAS COL, 41- Concurrent Pos: Mem Bd Exam in the Basic Sci, State of Iowa, 57- Mem: AAAS; Nat Asn Biol Teachers. Res: Cytology; spermiogenesis; field ecology of pheasant and fox studies; spermiogenesis in Necturus maculosus. Mailing Add: Dept of Biol Loras Col Dubuque IA 52001

NYE, WILLIAM PRESTON, b Logan, Utah, Jan 10, 17; div; c 5. APICULTURE, ENTOMOLOGY. Educ: Utah State Univ, BS, 40, MS, 47. Prof Exp: Asst bee poisoning, Utah Agr Exp Sta & instr entom, Utah State Univ, 46-47; RES

ENTOMOLOGIST, BEE BIOL & SYST LAB, AGR RES SERV, USDA, 47- Honors & Awards: Cert of Merit, Agr Res Serv, USDA, 68. Mem: Entom Soc Am; Bee Res Asn. Res: Bee biology, behavior and crop pollination for seed production. Mailing Add: Bee Biol & Syst Lab Utah State Univ UMC 53 Logan UT 84322

NYER, WARREN EDWIN, b Evanston, Ill, Nov 18, 21; m 43; c 2. PHYSICS. Educ: Univ Chicago, BS, 50. Prof Exp: Lab asst physics, Univ Chicago, 39-41, asst, Metall Lab, Manhattan Dist Proj, 41-43, Clinton Labs, Tenn, 43-44 & Hanford Eng Works, Wash, 44-45; assoc scientist, Los Alamos Sci Lab, 45-46, staff mem, 47-51; lab asst, Inst Nuclear Studies, Univ Chicago, 46-47; asst mgr, Atomic Energy Div, Phillips Petrol Co, 51-67; sr scientist & vchmn, AEC Atomic Safety & Licensing Bd Panel, 67-68; vpres, Idaho Nuclear Corp, 69-71; mgr & consult, Aerojet Nuclear Corp, 71-73; NUCLEAR ENERGY CONSULT, 73- Concurrent Pos: Mem, Atomic Safety & Licensing Bds Panel, 63-73. Mem: Am Phys Soc; Am Nuclear Soc. Res: Nuclear reactor safety. Mailing Add: Box 1845 Idaho Falls ID 83401

NYGAARD, KAARE JOHANN, b Notodden, Norway, July 10, 34; m 59; c 2. ATOMIC PHYSICS, QUANTUM OPTICS. Educ: Norweg Inst Technol, MS, 59, PhD(physics), 63. Prof Exp: Res asst gas discharge physics, Norweg Inst Technol, 59-60, res fel, 61-63; vis res assoc, Univ Hamburg, 60; mem res staff, Princeton Univ, 64 & Sperry Rand Res Ctr, 65-68; assoc prof, 68-75, PROF PHYSICS, UNIV MO-ROLLA, 75- Concurrent Pos: Vis prof, FOM Inst Atomic & Molecular Physics, Amsterdam, Neth, 74-75. Mem: AAAS; Am Geophys Union; Am Phys Soc; Norweg Phys Soc; sr mem Inst Elec & Electronics Engrs. Res: Electron impact ionization in metal vapors; photo-ionization of excited cesium atoms as source of polarized electrons; plasma diagnostics; dissociation of HCL. Mailing Add: Dept of Physics Univ of Mo Rolla MO 65401

NYGAARD, ODDVAR FRITHJOF, b Oslo, Norway, Oct 30, 22; nat US; m 46; c 2. RADIOBIOLOGY, BIOCHEMISTRY. Educ: Norweg Tech Univ, Sivilingeni r, 47; Univ Minn, PhD(physiol chem), 51. Prof Exp: Asst tech org chem, Tech Univ Norway, 47; asst physiol chem, Univ Minn, 48-51; fel oncol, McArdle Mem Lab, Univ Wis, 51-52, res assoc, 54-57; res biochemist, Norsk Hydro's Inst Cancer Res, Norweg Radium Hosp, 52-54; Norweg Cancer Soc fel, 54; res chemist, AEC biol effects of irradiation lab, Univ Mich, 57-59; res assoc, 59-63, asst prof biochem, 59-68, assoc prof radiol, 65-68, assoc prof biochem & radiol, 68-75, PROF RADIOL, CASE WESTERN RESERVE UNIV, 75-, ASSOC DIR DIV RADIATION BIOL, 63- Concurrent Pos: Ed-in-chief, Radiation Research, Radiation Res Soc, 72- Mem: AAAS; Am Chem Soc; Soc Develop Biol; Radiation Res Soc. Res: Nucleic acid metabolism and control; biological effects of radiation; modification of radiation response in hypoxic and anoxic cells. Mailing Add: Div of Radia Biol Dept of Radiol Case Western Reserve Univ Cleveland OH 44106

NYGREEN, PAUL W, b Bellingham, Wash, May 15, 25; m 50; c 2. GEOLOGY, OCEANOGRAPHY. Educ: Univ Wash, BS, 53; Univ Nebr, MSc, 55. Prof Exp: Geologist, Standard Oil Co Tex, 55-62, biostratigrapher, 62-64; biostratigrapher, Calif Oil Co, Okla, 64-65, sr biostratigrapher, Chevron Oil Co, 65-69, staff biostratigrapher, 69-71, STAFF BIOSTRATIGRAPHER, CHEVRON OVERSEAS PETROL INC, CHEVRON OIL CO, 71- Concurrent Pos: Supvry paleontologist, West Australian Petrol Proprietary Ltd, 71-74. Mem: AAAS; Geol Soc Am; Paleont Soc; Am Asn Petrol Geol; Am Asn Stratig Palynologists. Res: Reconstruction of climatic, oceanographic and sedimentary changes through geologic time, leading to prediction of areas of high organic productivity and the hydrocarbon source potential of sedimentary basins of the world. Mailing Add: Chevron Overseas Petrol Inc 575 Market St San Francisco CA 94105

NYGREN, DAVID ROBERT, physics, see 12th edition

NYHAN, WILLIAM LEO, b Boston, Mass, Mar 13, 26; m 48; c 3. PEDIATRICS, BIOCHEMISTRY. Educ: Columbia Univ, MD, 49; Univ Ill, MS, 56, PhD(pharmacol), 58. Prof Exp: From asst prof to assoc prof pediat, Sch Med, Johns Hopkins Univ, 58-63; prof pediat & biochem & chmn dept pediat, Univ Miami, 63-69; PROF PEDIAT & CHMN DEPT, SCH MED, UNIV CALIF, SAN DIEGO, 69- Concurrent Pos: Fel, Nat Found, 55-58; Am Cancer Soc fac res assoc, 61-63. Mem: AAAS; Am Chem Soc; Soc Pediat Res; Am Asn Cancer Res; Am Pediat Soc. Res: Amino acid metabolism; biochemical genetics; metabolism of tumors; developmental pharmacology. Mailing Add: Dept of Pediat Univ of Calif San Diego Sch Med La Jolla CA 92037

NYHUS, LLOYD MILTON, b Mt Vernon, Wash, June 24, 23; m 49; c 2. SURGERY, PHYSIOLOGY. Educ: Pac Lutheran Col, BA, 45; Univ Ala, MD, 47; Am Bd Surg, dipl, 57. Hon Degrees: Dr, Univ Thessoloniki, 69, Univ Uppsala, 74 & Chihuahua Univ, 75. Prof Exp: From intern to asst resident surg, King County Hosp, Seattle, 47-50; res assoc, Univ Wash, 52-53; asst resident, King County Hosp, 54; from instr to prof surg, Sch Med, Univ Wash, 54-67; PROF SURG & HEAD DEPT, ABRAHAM LINCOLN SCH MED, UNIV ILL MED CTR, CHICAGO & SURGEON-IN-CHIEF, UNIV HOSP, 67- Concurrent Pos: USPHS res fel, 52-54; Guggenheim fel, Sweden & Scotland, 55-56; sr consult, West Side Vet Admin Hosp, Chicago, 67-; sr attend, Cook County Hosp, Chicago, 68- Mem: Soc Univ Surg; Am Gastroenterol Asn; fel Am Col Surgeons; Am Surg Asn; Int Soc Surg. Res: Gastric physiology and surgery; peptic ulcer; hernia; esophageal physiology. Mailing Add: Dept of Surg Univ of Ill Hosp Chicago IL 60612

NYI, KAYSON, b Chungking, China, Apr 11, 45; m 70. ORGANIC CHEMISTRY. Educ: Mass Inst Technol, SB, 65; Univ Chicago, PhD(org chem), 71. Prof Exp: SR CHEMIST, RES DIV, ROHM AND HAAS CO, 71- Mem: Am Chem Soc. Res: Synthetic organic chemistry; photochemistry. Mailing Add: Res Div Rohm and Haas Co Spring House PA 19477

NYIKOS, PETER JOSEPH, b Salzburg, Austria, Mar 8, 46; US citizen. TOPOLOGY. Educ: Washington & Jefferson Col, BA, 67; Carnegie-Mellon Univ, MS, 68, PhD(topology), 71. Prof Exp: Mathematician, Biomed Lab, Edgewood Arsenal, US Army, 72-73; NSF fel, Univ Chicago, 73-74; VIS LECTR MATH, UNIV ILL, URBANA, 74- Mem: Am Math Soc. Res: Point-set topology and categorical topology, especially the study of zero-dimensional, non-archimedean and generalized metric spaces. Mailing Add: Dept of Math Univ of Ill Urbana IL 61801

NYILAS, EMERY, b Budapest, Hungary, Oct 5, 30; nat US; m 54; c 2. ORGANIC CHEMISTRY. Educ: Eötvös Lorand Univ, Budapest, dipl chem eng, 52. Prof Exp: From asst to prof bio-org chem, Eötvös Lorand Univ, 52-55; res assoc physico-org chem, Dept Nuclear Chem, Cent Res Inst Chem, Hungarian Acad Sci, 55-56; fel bio-org chem, Retina Found, 57-59; asst chemist, Harvard Med Sch, 59-61; staff chemist, Allied Res Assocs, 61-62; tech dir, Am Plastic & Chem Corp, 62-66; mem sr staff med group, 66-73, CHMN MED RES COMT, AVCO EVERETT RES LAB, INC, 73- Concurrent Pos: Asst chemist, Physics Res Lab, Mass Gen Hosp, 57-61 & Dept Neurosurg, 59-61; invited lectr, Dept Chem Eng, Mass Inst Technol, 73-; mem, Columbia Univ Biomat Sem. Mem: Am Chem Soc; Am Soc Artificial Internal Organs; NY Acad Sci. Res: Physico-organic, polymer and medicinal chemistry; implantable biocompatible materials; plasma protein adsorption and blood interfacial phenomena, energy absorption and emission related to chemical structure. Mailing Add: Med Group Avco Everett Res Lab 2385 Revere Beach Pkwy Everett MA 02149

NYIRADY, STEPHEN ARNOLD, b Queens, NY, Mar 19, 45; m 67; c 2. MICROBIOLOGY. Educ: Atlantic Union Col, BA, 66; Loma Linda Univ, MS, 69, PhD(microbiol), 72. Prof Exp: Instr biol, Loma Linda Univ, 71-72; asst prof, 72-75, ASSOC PROF BIOL, ATLANTIC UNION COL, 75- Mem: Am Soc Microbiol; AAAS. Res: Investigating the symbiotic relationship between the hematophagous reduviidae and their intestinal microflora. Mailing Add: Dept of Biol & Health Sci Atlantic Union Col South Lancaster MA 01561

NYLAND, GEORGE, b Eastburg, Alta, Apr 3, 19; US citizen; m 41; c 3. PLANT PATHOLOGY. Educ: State Col Wash, BS, 40, PhD(plant path), 48; La State Univ, MS, 42. Prof Exp: Asst, State Col Washington, 39-40, from instr to asst prof plant path, 46-48; asst, La State Univ, 40-42; instr & jr plant pathologist, 48-50, asst prof & plant pathologist, 50-56, assoc prof & assoc plant pathologist, 56-62, PROF PLANT PATH & PLANT PATHOLOGIST, COL AGR, UNIV CALIF, DAVIS, 62- Concurrent Pos: Jr plant pathologist, Exp Sta, State Col Washington, 46-48, asst plant pathologist, 47-48. Honors & Awards: Calif Asn Nursery Men spec Res Award, 67. Mem: AAAS; Am Phytopath Soc; Sigma Xi. Res: Stone, pome fruit and ornamental plants virus and mycoplasma diseases; chemotherapy; thermotherapy. Mailing Add: Dept of Plant Path Col of Agr Univ of Calif Davis CA 95616

NYLEN, MARIE USSING, b Copenhagen, Denmark, Apr 13, 24; US citizen; m 56; c 3. DENTAL RESEARCH, ELECTRON MICROSCOPY. Educ: Royal Dent Col, Denmark, DDS, 47. Hon Degrees: DrOdont, Royal Dent Col, Denmark, 73. Prof Exp: Pvt pract, 47-48; instr oper dent, Royal Dent Col, Denmark, 48-49; guest worker dent histol, Nat Inst Dent Res, 49-50, asst prof oral diag & res assoc electron micros, Royal Dent Col, Denmark, 51-55; vis assoc biophys, 55-60, biologist, 60-65, actg chief lab histol & path, 65-69, mem dent study sect, 70-74, CHIEF LAB BIOL STRUCT, NAT INST DENT RES, 69- Concurrent Pos: USPHS fel, Nat Inst Dent Res, 50-51; vis investr, Marine Biol Lab, Woods Hole, 69-72; prof lectr, Schs Med & Dent, Georgetown Univ, 70- Honors & Awards: Dept Health, Educ & Welfare Super Serv Honor Award, 69; Int Asn Dent Res Award, 70; Federal Woman's Award, 75. Mem: AAAS; Electron Micros Soc; Am Dent Asn; Am Soc Cell Biol; fel Am Col Dent. Res: Biophysical studies of developing and mature mineralized tissues and associated cells in normal and pathologic states. Mailing Add: Lab of Biol Struct Nat Inst of Dent Res Bethesda MD 20014

NYLUND, ROBERT E, b Lowell, Mass, May 12, 37; m 62; c 1. PHYSICAL CHEMISTRY. Educ: Northeastern Univ, BS, 60; Univ Iowa, PhD(phys chem), 64. Prof Exp: Asst prof, 64-69, ASSOC PROF CHEM, SUSQUEHANNA UNIV, 69-, CHMN DEPT, 75- Mem: Am Chem Soc. Res: Physical chemistry of biologically important polymers and proteins. Mailing Add: Dept of Chem Susquehanna Univ Selinsgrove PA 17870

NYLUND, ROBERT EINAR, b Ely, Minn, Jan 22, 16; m 40; c 2. HORTICULTURE. Educ: Univ Minn, BS, 38, MS, 42, PhD(hort), 45. Prof Exp: Asst, 38, horticulturist, Northwestern Exp Sta, 39-41, from instr to assoc prof, 39-59, PROF HORT, UNIV MINN, ST PAUL, 59-, ACTG HEAD DEPT HORT SCI & LANDSCAPE ARCHIT, 75- Concurrent Pos: Fulbright lectr, Univ Helsinki, 59-60 & 72; Univ Minn Off Int Prog grant, 65. Mem: Fel AAAS; Am Soc Hort Sci; Weed Sci Soc Am; Am Potato Asn; Potato Asn Am. Res: Physiology of vegetable crops and potatoes; weed control. Mailing Add: Dept Hort Sci & Landscape Archit Univ of Minn St Paul MN 55108

NYMAN, CARL JOHN, b New Orleans, La, Oct 21, 24; m 50; c 3. INORGANIC CHEMISTRY. Educ: Tulane Univ, BS, 44, MS, 45; Univ Ill, PhD(inorg chem), 48. Prof Exp: Jr technologist, Shell Oil Co, Inc, Calif, 44; asst chem, Univ Ill, 45-47, instr, 48; from instr to assoc prof, 48-61, PROF CHEM, WASH STATE UNIV, 61-, DEAN GRAD SCH, 68- Concurrent Pos: Vis fel, Cornell Univ, 59-60; vis fel, Imp Col Sci & Technol, London, 66-67. Mem: AAAS; Am Chem Soc. Res: Catalytic reduction of sodium sulfate; polarography in liquid ammonia; stability of complex ions; solutions of complex and polynuclear inorganic ions; organometallic complexes; peroxo-complexes of metals; catalytic oxygenation reactions. Mailing Add: Grad Sch Wash State Univ Pullman WA 99163

NYMAN, DALE JAMES, b Bancroft, Iowa, June 4, 31; m 60; c 2. GROUNDWATER GEOLOGY. Educ: Iowa State Univ, BS, 53, MS, 58. Prof Exp: Geologist, 58-67, HYDROLOGIST, US GEOL SURV, 67- Mem: Am Geophys Union; Geol Soc Am; Soc Prof Well Log Analysts; Water Well Asn. Res: Application of geology to hydrologic problems; ground-water surface-water relationships; ground-water management and modeling. Mailing Add: 3168 Sherry Dr Baton Rouge LA 70816

NYMAN, MELVIN ANTON, organic chemistry, see 12th edition

NYMANN, DEWAYNE STANLEY, b Cedar Falls, Iowa, June 27, 35; m 56; c 1. MATHEMATICS. Educ: State Col Iowa, BA, 57; Univ Kans, MA, 59, PhD(math), 64. Prof Exp: Asst prof math, Tex Christian Univ, 64-65 & Univ Tex, 65-70; ASSOC PROF MATH, UNIV TENN, CHATTANOOGA, 70- Mem: Am Math Soc; Math Asn Am. Res: Algebra group theory; generalized nilpotent and solvable groups. Mailing Add: Dept of Math Univ of Tenn Chattanooga TN 37403

NYMANN, JAMES EUGENE, b Cedar Falls, Iowa, Nov 24, 38; m 58; c 3. MATHEMATICS. Educ: Univ Northern Iowa, BA, 61; Univ Ariz, MS, 63, PhD(math), 65. Prof Exp: Asst prof, Univ Hawaii, 65-67; assoc prof, 67-74, PROF MATH, UNIV TEX, EL PASO, 74- Mem: Am Math Soc; Math Asn Am. Res: Number theory and analysis. Mailing Add: Dept of Math Univ of Tex El Paso TX 79968

NYQUIST, DAVID, limnology, aquatic biology, see 12th edition

NYQUIST, EDWIN BASOM, organic chemistry, see 12th edition

NYQUIST, HARLAN LEROY, b Scobey, Mont, Aug 12, 29; m 55; c 3. ORGANIC CHEMISTRY. Educ: Mont State Col, BS, 51; Univ Calif, Los Angeles, PhD(chem), 56. Prof Exp: From instr to asst prof chem, Univ Calif, Santa Barbara, 56-62; from asst prof to assoc prof, 62-68, PROF CHEM, CALIF STATE UNIV, NORTHRIDGE, 68- Mem: Am Chem Soc; Sigma Xi; Int Soc Magnetic Resonance. Res: Mechanisms of organic reactions; synthesis and reactions of s-triazines; stereochemistry. Mailing Add: Dept of Chem Calif State Univ Northridge CA 91324

NYQUIST, JUDITH KAY, b Gary, Ind, June 22, 41. NEUROPHYSIOLOGY. Educ: St Olaf Col, BA, 63; Univ Wash, PhD(physiol, biophys), 69. Prof Exp: Asst prof neurosurg, Med Col Va, 70-71; asst res neuroscientist, Dept Surg, Univ Calif, 71-72; RES PHYSIOLOGIST, DIV NEUROSURG, VET ADMIN HOSP, SAN DIEGO, 72- Concurrent Pos: Ment Health Training Prog fel, Brain Res Inst, Univ Calif, Los

Angeles, 69-70; asst adj prof surg, Sch Med, Univ Calif, San Diego, 72- Mem: Soc Neurosci. Res: Neurophysiology of mammalian somatosensory systems and motor systems; utilizing evoked potential and single neuron recording techniques. Mailing Add: Vet Admin Hosp Div of Neurosurg (112a) San Diego CA 92161

NYQUIST, RICHARD ALLEN, b Rockford, Ill, May 3, 28; m 56; c 4. MOLECULAR SPECTROSCOPY. Educ: Augustana Col, Ill, BA, 51; Okla State Univ, MS, 53. Prof Exp: Proj leader infrared spectros, Chem Physics Res Lab, 53-71, ASSOC SCIENTIST INFRARED & RAMAN SPECTROS, ANAL LABS, DOW CHEM CO, 71- Mem: Am Chem Soc; Coblentz Soc. Res: Vibrational spectroscopy; elucidation of chemical structure. Mailing Add: 3707 Westbrier Terr Midland MI 48640

NYQUIST, SALLY ELIZABETH, b Hutchinson, Minn, Nov 8, 41. CELL BIOLOGY. Educ: Wheaton Col, BS, 63; Purdue Univ, Lafayette, PhD(biol), 70. Prof Exp: Volunteer, Peace Corps, 63-65; Kettering Found res fel, Charles F Kettering Res Lab, 70-72; ASST PROF BIOL, BUCKNELL UNIV, 72- Mem: Am Soc Cell Biol. Res: Golgi apparatus membranes, enzymatic activities and composition. Mailing Add: Dept of Biol Bucknell Univ Lewisburg PA 17837

NYQUIST, WYMAN ELLSWORTH, b Scobey, Mont, June 13, 28; m 52; c 2. STATISTICAL ANALYSIS, QUANTITATIVE GENETICS. Educ: Mont State Univ, BS, 50; Univ Calif, PhD(genetics), 53. Prof Exp: Asst agron, Univ Calif, 50-52, instr, 53-57, lectr, 57-58, asst prof, 58-63; assoc prof, 63-68, PROF AGRON, PURDUE UNIV, 68- Concurrent Pos: Jr agronomist, Exp Sta, Univ Calif, 53-57, asst agronomist, 57-63; NIH spec res fel, 69-70. Mem: Am Soc Agron; Genetics Soc Am; Biomet Soc; Coun Agr Sci & Technol; Crop Sci Soc Am. Res: Development of statistical models relating to quantitative genetic variation, particularly in plant populations and their utilization in evaluating alternative breeding systems. Mailing Add: Dept of Agron Purdue Univ West Lafayette IN 47907

NYSTED, LEONARD NORMAN, b Marshfield, Wis, May 17, 27; m 47; c 2. PHARMACEUTICAL CHEMISTRY. Educ: St Olaf Col, BS, 51. Prof Exp: Preparations & res chemist, G D Searle & Co, 51-56; dir chem res, Duraclean Co, 56-58; res chemist, 58-64, SR RES INVESTR, G D SEARLE & CO, 64- Mem: Am Chem Soc. Res: Pharmaceuticals, particularly steroids, insecticides and detergents. Mailing Add: 617 Rice St Highland Park IL 60035

NYSTROM, J WARREN, b Worcester, Mass, Nov 22, 13; m 38; c 3. ECONOMIC GEOGRAPHY, POLITICAL GEOGRAPHY. Educ: Clark Univ, AB, 36, MA, 37, PhD(geog), 42. Prof Exp: From instr to asst prof geog, RI Col, 37-43; from assoc prof to prof & chmn, Univ Pittsburgh, 43-53; int mgr, US Chamber Com, 53-62; partner, Allen, Murden & Nystrom, 62-64; sr fel advan studies, Brookings Inst, 64-66; EXEC DIR, ASN AM GEOG, 66- Concurrent Pos: Dir, RI World Affairs Coun, 41-43, Pittsburgh Regional Inter-Am Coun, 43-45 & Foreign Policy Asn, 45-53; mem bd trustees, Clark Univ, 60-64; US del, Int Geog Union, Stockholm, 60, London, 64, New Delhi, 68, Budapest, 71, Montreal, 72 & Palmeston North, 74; mem exec comt, Am Coun Learned Socs, 68-; mem comts, Nat Acad Sci-Nat Res Coun; mem US del, Pan-Am Inst Geog & Hist, Panama, 73; prof geog, George Washington Univ, 73-; mem, Joint US-Soviet Environ Seminar, 75. Mem: Asn Am Geog; Am Geog Soc; Nat Coun Geog Educ. Res: European common market; Latin American political and economic relationships. Mailing Add: 3117 N Nelson St Arlington VA 22207

NYSTROM, RICHARD ALAN, b Everett, Mass, Dec 10, 35; m 58; c 3. PHYSIOLOGY, MEDICAL EDUCATION. Educ: Tufts Univ, BS, 57; Univ Ill, Urbana, MS, 60, PhD(physiol), 61. Prof Exp: From asst prof to assoc prof biol sci, Univ Del, 61-73; ASSOC DIR & PROF, ALBANY-HUDSON VALLEY PHYSICIAN'S ASST PROG, 73- Concurrent Pos: Consult, E I du Pont de Nemours & Co, Del, 63-64; Nat Inst Neurol Dis & Blindness spec fel, Mass Inst Technol, 68-69; investr & corp mem, Marine Biol Lab, Woods Hole, 69-; adj prof physiol, Albany Med Col, 74- Mem: AAAS; Am Soc Zool; Am Physiol Soc; NY Acad Sci. Res: Nervous control of cat nictitating membrane; oscular contractions in sponges; excitability of tissue-cultured vertebrate neurons; comparative physiology of invertebrate intestinal and somatic muscle; physiology of biological membranes; axoplasmic transport. Mailing Add: 11 Canterbury Rd Clifton Park NY 12065

NYSTROM, ROBERT FORREST, b Chicago, Ill, May 30, 20; m 44; c 3. ORGANIC CHEMISTRY, RADIOCHEMISTRY. Educ: Univ Chicago, BS, 42, PhD(org chem), 47. Prof Exp: Asst, Off Sci Res & Develop Proj, Univ Chicago, 43-45; res chemist, Monsanto Chem Co, Tenn, 47-48; asst prof chem & animal sci, 48-53, assoc prof chem, 53-61, PROF CHEM, UNIV ILL, URBANA, 61-, DIR RADIOISOTOPE LAB, 53- Honors & Awards: Radiation Indust Award, Am Nuclear Soc, 68. Mem: Am Chem Soc. Res: Reduction of organic compounds by lithium aluminum hydride, lithium borohydride and complex hydrides; radiation-induced reactions; organic reactions and mechanisms with carbon 14; tritium. Mailing Add: Radioisotope Lab 139 Davenport Hall Univ of Ill Urbana IL 61801

NYSTUEN, JOHN DAVID, b Northfield, Minn, Jan 7, 31; m 55; c 2. GEOGRAPHY. Educ: Univ Calif, AB, 53; Univ Wash, Seattle, MA, 57, PhD(geog), 59. Prof Exp: From instr to assoc prof, 59-68, PROF GEOG, UNIV MICH, ANN ARBOR, 68- Concurrent Pos: Vis assoc prof, Univ Minn, 65 & Univ Calif, Berkeley, 66-67; comt adv, US Geol Surv, 67-; mem geog subcomt, Nat Acad Sci-Nat Res Coun & comt sem series quant geog, NSF, 65- Mem: Asn Am Geog; Regional Sci Asn; Am Geog Soc. Res: Economic, urban and transportation geography; location theory; quantitative analysis of spatial and movement phenomena; mathematical and theoretical geography. Mailing Add: Dept of Geog 4040 Admin Bldg Univ of Mich Ann Arbor MI 48104

NYVALL, ROBERT FREDERICK, b Thief River Falls, Minn, Aug 23, 39; m 62; c 2. PLANT PATHOLOGY. Educ: Univ Minn, BS, 65, MS, 66, PhD(plant path), 69. Prof Exp: Asst prof, 70-73, ASSOC PROF PLANT PATH, IOWA STATE UNIV, 73-, EXTEN PLANT PATHOLOGIST, 70- Mem: Am Inst Biol Sci; Am Phytopath Soc. Res: Ecology of soil; root rot and wilt fungi. Mailing Add: Dept of Bot & Plant Path Iowa State Univ Ames IA 50010

NYYSSONEN, DIANA, b Cambridge, Mass. OPTICS. Educ: Boston Univ, BA, 65; Univ Rochester, PhD(optics), 75. Prof Exp: Staff scientist optics, Tech Oper, Inc, 64-68; PHYSICIST OPTICS, NAT BUR STANDARDS, 69- Mem: Optical Soc Am; Soc Photog Scientists & Engrs; Soc Photo-Optical Instrumentation Engrs. Res: Densitometry, microdensitometry, optical systems evaluation and application of the theory of partial coherence to measurement problems on optical imagery and diffracted fields, including both scalar and vector treatments. Mailing Add: A123 Metrol Nat Bur of Standards Washington DC 20234

O

OACE, SUSAN M, b St Paul, Minn, Nov 10, 41. NUTRITION. Educ: Univ Minn, St Paul, BS, 63; Univ Calif, Berkeley, PhD(nutrit), 67. Prof Exp: Res nutritionist, Univ Calif, Berkeley, 67-68; asst prof nutrit, Univ Calif, Davis, 68-73; asst prof, 73-75, ASSOC PROF NUTRIT, UNIV CALIF, BERKELEY, 75- Mem: AAAS; Am Dietetic Asn; NY Acad Sci; Soc Nutrit Educ. Res: Metabolic interrelationships among folic acid, vitamin B12 and methionine; interaction of intestinal microflora with nutritional and health status of host. Mailing Add: Dept of Nutrit Sci Univ of Calif Berkeley CA 94720

OAKBERG, EUGENE FRANKLIN, b Moline, Ill, Oct 4, 16; m 43, 64; c 4. GENETICS. Educ: Monmouth Col, BS, 40; Kans State Col, MS, 42; Iowa State Col, PhD(genetics), 45. Prof Exp: Asst zool, Kans State Col, 40-42; instr & res assoc genetics, Iowa State Col, 44-45; med bacteriologist, Off Sci Res & Develop, Genetics Dept, Carnegie Inst, 45, res assoc, Cold Spring Harbor, NY, 46; histologist, Mich State Col, 47-50; asst prof genetics, Iowa State Col, 51; SR BIOLOGIST, OAK RIDGE NAT LAB, 52- Mem: AAAS; Genetics Soc Am; Radiation Res Soc; Am Genetic Asn. Res: Mouse genetics; mammalian gametogenesis; radiobiology. Mailing Add: Biol Div Oak Ridge Nat Lab PO Box Y Oak Ridge TN 37830

OAKES, BILLY DEAN, b Tulsa, Okla, Sept 5, 28; m 50; c 2. ORGANIC CHEMISTRY. Educ: Okla State Univ, BS, 50; Univ Wichita, MS, 52. Prof Exp: AEC asst, Univ Wichita, 52 & Ga Inst Technol, 52-53; res chemist, Dowell Div, 53-62 & Chem Dept Res Lab, Mich, 62-64, proj leader, 64-69, PROJ LEADER, CONTRACT RES DEPT, TEX DIV, DOW CHEM CO, 69- Mem: Am Chem Soc; Nat Asn Corrosion Eng. Res: Corrosion studies; corrosion inhibitors for aqueous solutions, primarily acids; environmental modifications designed to allow the use of low cost construction materials in sea water desalination plants. Mailing Add: Contract Res Dept A-1214 Bldg Dow Chem Co Freeport TX 77541

OAKES, LESTER CHARLES, b Noroton, Conn, Apr 10, 14; m 40; c 3. GEOGRAPHY. Educ: Cent Conn State Col, BEd, 38; NY Univ, MA, 49; Columbia Univ, prof dipl geog, 58; Union Col, NY, MS, 61. Prof Exp: Instr bus, Laurel Col, 38-39; instr, Meriden High Sch, 39-46; instr bus & geog, Norwalk Sr High Sch, 46-61; ASSOC PROF GEOG, CLARION STATE COL, 61- Mem: Asn Am Geog; Nat Coun Geog Educ. Res: Historical geography of the United States and Canada; geographic education. Mailing Add: Dept of Geog Clarion State Col Clarion PA 16214

OAKES, MELVIN ERVIN LOUIS, b Vicksburg, Miss, May 11, 36; m 63; c 2. PLASMA PHYSICS. Educ: Fla State Univ, PhD(plasma physics), 64. Prof Exp: Asst physics, Fla State Univ, 58-60 & 60-64; physicist, Army Res Guided Missile Agency, Redstone Arsenal, 60; asst prof physics, Univ Ga, 64; res assoc, 64-65, from asst prof to assoc prof, 65-75, PROF PHYSICS, UNIV TEX, AUSTIN, 75- Mem: Am Phys Soc; Am Asn Physics Teachers. Res: Cross modulation in ionized media; electromagnetic interaction with plasmas; plasma waves and striations in radio frequency plasmas. Mailing Add: Dept of Physics Univ of Tex Austin TX 78712

OAKES, ROBERT JAMES, b Minneapolis, Minn, Jan 21, 36; m 55; c 1. THEORETICAL HIGH ENERGY PHYSICS. Educ: Univ Minn, BS, 57, MS, 59, PhD(physics), 62. Prof Exp: NSF fel physics, Stanford Univ, 62-64, lectr, 64, asst prof, 64-68; assoc prof, 68-70, PROF PHYSICS, NORTHWESTERN UNIV, 70- Concurrent Pos: A P Sloan Found res fel, 65-; vis scientist, Europ Orgn Nuclear Res, 66-67; mem, Inst Advan Study, 67-68; vis scientist, Deutsches Elektron-Synchrotron, 71-72 & Los Alamos Sci Lab, 71- Mem: Fel Am Phys Soc; NY Acad Sci. Res: Strong interactions; theoretical, high energy and nuclear physics; field theory; weak interactions. Mailing Add: Dept of Physics Northwestern Univ Evanston IL 60201

OAKESHOTT, GORDON B, b Oakland, Calif, Dec 24, 04; m 29; c 3. GEOLOGY. Educ: Univ Calif, BS, 28, MS, 29; Univ Southern Calif, PhD(geol), 36. Prof Exp: Asst field geologist, Shell Oil Co, 29-30; instr earth sci, Compton Dist Jr Col, 30-48; supv mining geologist, Calif State Div Mines & Geol, 48-56, dep chief, 56-57 & 59-72, actg chief, 58; CONSULT GEOLOGIST, 73- Concurrent Pos: Instr geol, Calif State Univ, Sacramento, 73 & Calif State Univ, San Francisco, 75. Mem: AAAS; fel Geol Soc Am; Am Inst Prof Geol; Seismol Soc Am; Nat Asn Geol Teachers (pres, 71). Res: Geology of the San Gabriel Mountains; stratigraphy of California Coast and transverse ranges; surface faulting and associated earthquakes; geology of California. Mailing Add: 3040 Totterdell St Oakland CA 94611

OAKLAND, GAIL BARKER, b Lang, Sask, Feb 11, 14; m 40; c 2. MATHEMATICAL STATISTICS. Educ: Univ Sask, BA, 33; Univ Minn, MA, 39; Aberdeen Univ, PhD, 56. Prof Exp: Teacher math, Scott Collegiate, 35-42; prof statist, Univ Sask, 42-45 & Univ Man, 45-50; chief statist res serv, Can Dept Agr, Ont, 50-59; sr math adv, Dom Bur Statist, 59-61; PROF STATIST & CHMN DEPT, UNIV MASS, AMHERST, 61- Concurrent Pos: Mem, Cong Math, Can, 45; lectr, Int Eugenic Conf, Denmark, 56; mem, Int Biomet Conf, 58 & Int Statist Inst, France, 61. Mem: Fel AAAS; fel Am Statist Asn; fel Brit Inst Statist; fel Royal Statist Soc. Res: Applications of experimental designs to agricultural research work in entomology, forest biology and chemistry; quality control procedures; sample survey techniques in evaluating quality of censuses. Mailing Add: Dept of Statist Univ of Mass Amherst MA 01002

OAKLEY, BRUCE, b Philadelphia, Pa, Oct 22, 36; m 58; c 1. NEUROBIOLOGY. Educ: Swarthmore Col, BA, 58; Brown Univ, MSc, 60, PhD(psychol), 62. Prof Exp: Asst prof psychol, Brown Univ, 62-63; Nat Acad Sci-Nat Res Coun sr fel physiol, Royal Vet Col Sweden, 63-64; USPHS fel zool, Univ Calif, Los Angeles, 64-65, asst res zoologist, 65-66; asst prof, 66-71, ASSOC PROF ZOOL, UNIV MICH, ANN ARBOR, 71- Mem: AAAS; Am Soc Zool; Am Physiol Soc; Soc Neurosci. Res: Physiology and behavior of taste; neural mechanism of preference and aversion; sensory receptors and coding; behavioral physiology; developmental neurobiology. Mailing Add: Dept of Zool Univ of Mich Ann Arbor MI 48104

OAKLEY, BURKS, II, b Cleveland, Ohio, Nov 4, 49; m 73. BIOENGINEERING. Educ: Northwestern Univ, Evanston, BS, 71; Univ Mich, Ann Arbor, MS, 73, PhD(bioeng), 75. Prof Exp: Jr res engr, Northwestern Univ, Evanston, 71; NAT EYE INST SCHOLAR OPHTHAL, UNIV MICH, ANN ARBOR, 75- Concurrent Pos: Travel fel, Asn Res Vision & Ophthal, 75. Mem: Asn Res Vision & Ophthal. Res: Applications of ion-specific microelectrodes in the analysis of retinal neurophysiology. Mailing Add: Dept of Ophthal 5048 Kresge II Univ of Mich Ann Arbor MI 48109

OAKLEY, DAVID CHARLES, b Marysville, Calif, July 4, 29; m 52; c 3. SCIENCE ADMINISTRATION, NUCLEAR SCIENCE. Educ: Calif Inst Technol, BS, 50, MS, 52, PhD(physics & math), 55. Prof Exp: Physicist nuclear sci, Lawrence Livermore Lab, 54-73; ASST DEP DIR FOR TESTING, NUCLEAR SCI ADMIN, DEFENSE NUCLEAR AGENCY, 73- Mem: Am Phys Soc; Sigma Xi. Res: Administration of underground nuclear testing of nuclear weapons effects on materials and systems of military interest. Mailing Add: Defense Nuclear Agency Washington DC 20305

OAKS, B ANN, b Winnipeg, Man, June 4, 29. PLANT PHYSIOLOGY. Educ: Univ Toronto, BA, 51; Univ Sask, MA, 54, PhD(plant physiol), 58. Prof Exp: Res asst plant physiol, Univ Man, 53-54; Von Humboldt grant, Bact Inst, Feising, Ger, 58-60; res assoc biol sci, Purdue Univ, 60-64; from asst prof to assoc prof, 64-74, PROF BIOL, McMASTER UNIV, 74- Mem: Am Soc Plant Physiol; Can Soc Plant Physiol. Res: Intermediary metabolism in plants; processes regulating development in seedlings. Mailing Add: Dept of Biol McMaster Univ Hamilton ON Can

OAKS, EMILY CAYWOOD JORDAN, b Pittsburgh, Pa, Feb 15, 39; m 61; c 2. VERTEBRATE ZOOLOGY. Educ: Rice Univ, BA, 61; Yale Univ, MS, 64, PhD(biol), 67. Prof Exp: ASST PROF ZOOL, UTAH STATE UNIV, 67- Mem: Am Soc Mammal; Soc Study Evolution. Res: Anatomy and adaptive function of the middle ear in mammals. Mailing Add: Dept of Biol Utah State Univ Logan UT 84322

OAKS, J HOWARD, b Camden, NJ, Mar 3, 30; m 72; c 5. DENTISTRY. Educ: Wesleyan Univ, BA, 52; Harvard Univ, DMD, 56. Prof Exp: Instr oper dent, Sch Dent Med, Harvard Univ, 56-64, lectr oper dent & assoc dean, 64-68, actg dean, 67-68; prof dent med & dean, Sch Dent Med, 68-74, VPRES HEALTH SCI, STATE UNIV NY STONY BROOK, 74- Concurrent Pos: Mem dent educ rev comt, Bur Health Manpower Educ, NIH, 67-70, consult, div physician & health professions educ, 70-; vpres for Deans, Am Asn Dent Schs, 72- Mailing Add: Off of VPres Health Sci Ctr State Univ of NY Stony Brook NY 11794

OAKS, JOHN ADAMS, b Alma, Mich, Apr 8, 42; m 65; c 1. CELL BIOLOGY, PARASITOLOGY. Educ: Colby Col, BA, 64; Tulane Univ, MS, 68, PhD(cell biol), 70. Prof Exp: Asst prof parasitol, Tulane Univ, 70-73; ASST PROF, DEPT ANAT, COL MED, UNIV IOWA, 73- Mem: AAAS; Am Soc Parasitol. Res: Structural and functional aspects of helminth surfaces; mechanism of host cellular reactions to parasitic infections; comparative aspects of free-living and parasitic helminth epithelia. Mailing Add: Dept of Anat Col of Med Univ of Iowa Iowa City IA 52240

OAKS, ROBERT QUINCY, JR, b Houston, Tex, Aug 29, 38; m 61; c 2. GEOLOGY. Educ: Rice Univ, BA, 60; Yale Univ, PhD(geol), 65. Prof Exp: Res geologist, Jersey Prod Res Co, 64 & Esso Prod Res Co, 65-66; asst prof, 66-70, ASSOC PROF GEOL, UTAH STATE UNIV, 70- Mem: AAAS; Soc Econ Paleont & Mineral; Geol Soc Am; Am Asn Petrol Geol; Int Asn Sedimentol. Res: Post-Miocene stratigraphy and morphology in outer coastal plain of southeastern Virginia; morphology and internal stratification of intrenched modern alluvial fans of the Nevada-California border; Ordovician quartzites in northern Utah and southen Idaho. Mailing Add: Dept of Geol Utah State Univ Logan UT 84322

OAKS, WILBUR W, b Philadelphia, Pa, Oct 12, 28; m 54; c 3. INTERNAL MEDICINE. Educ: Lafayette Col, BS, 51; Hahnemann Med Col, MD, 55. Prof Exp: From instr to assoc prof med, 61-69, teaching coordr & dir postgrad educ, 62-70, PROF MED & DIR DIV GEN INTERNAL MED, HAHNEMANN MED COL, 69-, CHMN DEPT MED, 73- Concurrent Pos: Staff physician, Hahnemann Hosp, 61- Honors & Awards: Christian R & Mary F Lindback Found Award. Mem: Am Fedn Clin Res; AMA; Am Col Chest Physicians; Asn Am Med Cols. Res: Hypertension. Mailing Add: Div of Gen Internal Med Hahnemann Med Col Philadelphia PA 19102

OALMANN, MARGARET CLAIRE, b Covington, La, Aug 16, 29; m 62; c 3. EPIDEMIOLOGY. Educ: La State Univ, BSNEd, 56; Tulane Univ, MPH, 58, DPH(chronic dis, epidemiol), 60. Prof Exp: From actg head nurse to head nurse, Charity Hosp La, New Orleans, 49-53, clin instr, 53-57; res assoc, Tulane Univ, 60-62; from instr to asst prof path, 62-70, asst prof, 70-72, ASSOC PROF PATH, PUB HEALTH & PREV MED, MED CTR, LA STATE UNIV, NEW ORLEANS, 72- Concurrent Pos: Vis scientist, Charity Hosp, La, 60-; epidemiol consult mortality in nuns, Am Cancer Soc Grant, 63- Mem: AAAS; fel Am Heart Asn; Am Pub Health Asn; Asn Teachers Prev Med; Royal Soc Health. Res: Epidemiology of cardiovascular disease and cancer. Mailing Add: Dept of Path La State Univ Med Ctr New Orleans LA 70112

OATES, JIMMIE C, b Memphis, Tenn, Apr 14, 33; m 54; c 2. PHYSICS. Educ: Memphis State Univ, BS, 58; Vanderbilt Univ, MS, 60, PhD(physics), 63. Prof Exp: Assoc prof, 62-70, PROF PHYSICS, QUEENS COL, NC, 70- Res: Bioacoustics. Mailing Add: Dept of Physics Queens Col Charlotte NC 28207

OATES, JOHN ALEXANDER, b Fayetteville, NC, Apr 23, 32; m 56; c 3. CLINICAL PHARMACOLOGY, INTERNAL MEDICINE. Educ: Wake Forest Col, BS, 53; Bowman Gray Sch Med, MD, 56. Prof Exp: Intern internal med, New York Hosp, 56-57, asst resident, 57-58; clin assoc exp therapeut, Nat Heart Inst, 58-61; asst resident med, New York Hosp, 61-62; sr investr, 62-63; from asst prof to assoc prof pharmacol & med, 63-69, PROF PHARMACOL & MED, SCH MED, VANDERBILT UNIV, 69- Concurrent Pos: Burroughs Wellcome scholar clin pharmacol, 65-70; counr, Asn Am Physicians, 75- Honors & Awards: Am Soc PHarmacol & Exp Therapeut Award Exp Therapeut, 69. Mem: Am Soc Clin Invest; Asn Am Physicians (secy, 70-75); Am Soc Pharmacol & Exp Therapeut. Res: Vasoactive amines and peptides; prostaglandins; antihypertensive agents; autonomic pharmacology. Mailing Add: Vanderbilt Univ Sch of Med Nashville TN 37232

OATES, RICHARD PATRICK, b Gary, Ind, Mar 17, 37. BIOSTATISTICS. Educ: Purdue Univ, BS, 58; Iowa State Univ, MS, 60, PhD(bact), 64. Prof Exp: Asst prof, 65-75, ASSOC PROF BIOSTATIST, STATE UNIV NY UPSTATE MED CTR, 75- Concurrent Pos: NIH fel, Iowa State Univ, 64-65. Mem: Biomet Soc; Asn Teachers Prev Med; NY Acad Sci. Res: Medical research; epidemiology. Mailing Add: Dept of Prev Med State Univ of NY Upstate Med Ctr Syracuse NY 13210

OATMAN, EARL R, b Sylvester, Tex, Oct 21, 20; m 53; c 3. ENTOMOLOGY. Educ: Asst instr entom, Univ Mo, 51-52; asst entom & parasitol, Univ Calif, Berkeley, 53-56; from asst prof to assoc prof entom, Univ Wis, 56-62; from asst entomlogist to assoc entomologist, 62-72, ENTOMOLOGIST, DIV BIOL CONTROL, UNIV CALIF, RIVERSIDE, 72-, PROF ENTOM, 75- Mem: Entom Soc Am. Res: Population ecology and biological control of insects and mites associated with agronomic and horticultural crops. Mailing Add: Div of Biol Control Univ of Calif Riverside CA 92502

OBA, MARILYN KAZUKO, mathematics, see 12th edition

O'BANNON, JOHN HORATIO, b West Palm Beach, Fla, Sept 23, 26; m 52; c 3. NEMATOLOGY. Educ: Univ Ariz, BS & MS, 57; Ariz State Univ, PhD(bot), 65. Prof Exp: Nematologist, Cotton Res Ctr, Ariz, 57-65, NEMATOLOGIST NEMATOL INVESTS, AGR RES SERV, USDA, ORLANDO, 65- Mem: Am Phytopath Soc; Soc Nematologists (pres, 74-75). Res: Nematology concerned with the biology, ecology and control of plant parasitic nematodes. Mailing Add: 1395 Hickory Dr Maitland FL 32751

O'BARR, THOMAS PHILLIPS, microbiology, biochemistry, see 12th edition

O'BARR, WILLIAM MCALSTON, b Sylvania, Ga, Dec 1, 42; m 65; c 1. ANTHROPOLOGY. Educ: Emory Univ, BA, 64; Northwestern Univ, MA, 66, PhD(anthrop), 69. Prof Exp: Res assoc med sociol, Communicable Dis Ctr, USPHS, 64-67; res assoc social anthrop, Univ Dar es Salaam, 67-68; asst prof, 69-74, ASSOC PROF ANTHROP, DUKE UNIV, 74- Concurrent Pos: Consult, Warren Wilson Col African Studies Prog, 70-72. Mem: Am Anthrop Asn; African Studies Asn; Am Ethnol Soc; Royal Anthrop Inst Gt Brit. Res: Language of politics and law; bilingualism; social change in third world countries; political anthropology. Mailing Add: Dept of Anthrop Duke Univ Durham NC 27706

OBBINK, RUSSELL C, b Omaha, Nebr, Sept 29, 24; m 46; c 5. ANALYTICAL CHEMISTRY, ORGANIC CHEMISTRY. Educ: Univ Portland, BSc, 53. Prof Exp: Analyst, Alcoa, Wash, 49-53, anal chemist, 53-57, sr chemist, 57-65, res chemist, Alcoa Res Labs, Aluminum Co Am, 65-67, group leader, 67-70, group leader, Anal Chem Div, 70-73, SECT HEAD, ANAL CHEM DIV, ALCOA TECH CTR, ALUMINUM CO AM, 73- Mem: AAAS; Am Chem Soc; Fine Particle Soc; Am Soc Testing & Mat; fel Am Inst Chem. Res: Development of analytical methods for use in aluminum industry. Mailing Add: Anal Chem Div Alcoa Tech Ctr Alcoa Center PA 15069

OBEAR, FREDERICK W, b Malden, Mass, June 9, 35; m 59; c 3. INORGANIC CHEMISTRY. Educ: Lowell Tech Inst, BS, 56; Univ NH, PhD(chem), 61. Prof Exp: Asst chem, Univ NH, 56-58, fel, 58-60; asst prof, 60-66, dean freshmen, 64-66, asst provost, 65-68, vprovost, 68-70, ASSOC PROF CHEM, OAKLAND UNIV, 66-, VPRES ACAD AFFAIRS & PROVOST, 70- Mem: Am Chem Soc. Res: Transition metal inorganic chemistry structure and mechanisms; academic administration. Mailing Add: Dept of Chem Oakland Univ Rochester MI 48063

OBEE, DONALD JENNINGS, b Clio, Mich, Jan 23, 12; m 39; c 2. BOTANY. Educ: Univ Kans, AB, 32, MA, 34, PhD(bot), 42. Prof Exp: Teacher, Cheyenne Mt Sch, Colo, 39-46; chmn div life sci, 46-68, chmn div sci & health, 68-74, PROF BOT & CHMN DEPT BIOL, BOISE STATE UNIV, 74- Mem: Fel Bot Soc Am; fel Am Phytopath Soc; fel Am Inst Biol Sci. Res: Plant pathology; pathogenicity, morphology and physiology of Sclerotium delphinii. Mailing Add: Div of Sci & Health Boise State Univ 1907 Campus Dr Boise ID 83705

OBENCHAIN, FREDERICK DECROES, b Indianapolis, Ind, Mar 3, 40; m 73; c 1. ACAROLOGY, INSECT PHYSIOLOGY. Educ: Wabash Col, AB, 62; Ind Univ, Bloomington, MA, 65; Ohio State Univ, PhD(entom), 70. Prof Exp: RES ASSOC PROF BIOL, DEPT BIOL & INST ARTHROPODOLOGY, GA SOUTHERN COL, 71- Mem: Acarology Soc Am; Am Arachnological Soc; Am Soc Zoologists; Entom Soc Am. Res: Internal morphology and physiology of the Acari, concentrating on neurosecretion and neuroendocrine mechanisms controlling growth, molting and reproduction. Mailing Add: Dept of Biol Ga Southern Col Statesboro GA 30458

OBENDORF, RALPH LOUIS, b Milan, Ind, July 11, 38. PLANT PHYSIOLOGY, AGRONOMY. Educ: Purdue Univ, BS, 60; Univ California, Davis, MS, 62, PhD(plant physiol), 66. Prof Exp: Asst prof field crop sci, 66-71, ASSOC PROF CROP SCI, CORNELL UNIV, 71- Concurrent Pos: Vis scientist, Inst Cancer Res, Philadelphia, Pa, 72-73. Mem: AAAS; Am Soc Plant Physiol; Am Inst Biol Sci; Crop Sci Soc Am; Am Soc Agron. Res: Physiology and biochemistry of cold sensitivity during grain formation, germination and seedling growth, greening and photosynthetic development. Mailing Add: Dept of Agron Cornell Univ Ithaca NY 14853

OBENLAND, CLAYTON O, b Kansas City, Mo, Dec 22, 12; m 41; c 1. INDUSTRIAL CHEMISTRY. Educ: Kans State Univ, BS, 35, MS, 50. Prof Exp: Asst chemist, Monsanto Chem Co, Ill, 39-41, tech asst plant develop, 41-42 & 46-47, asst supvr chem prod, 47-49; scheduler prod control, Gen Aniline Works, NY, 50-51; chemist, Olin-Mathieson Chem Corp, NY, 52-56, group leader pre-pilot lab, 56-59, sr chemist, Conn, 59-69; PATENT AGENT, CARBORUNDUM CO, NIAGARA FALLS, 69- Mem: Am Chem Soc; Am Inst Chem. Res: Synthetic organic and inorganic chemistry, especially boron chemistry. Mailing Add: 62 Culpepper Rd Williamsville NY 14221

OBENSHAIN, FELIX EDWARD, b Pikeville, Ky, Mar 31, 28; m 50; c 4. NUCLEAR PHYSICS, SOLID STATE PHYSICS. Educ: Va Polytech Inst, BS, 52; Univ Pittsburgh, PhD(physics), 60. Prof Exp: Physicist, Atomic Power Div, Westinghouse Elec Corp, 52-56; PHYSICIST, OAK RIDGE NAT LAB, 59-; PROF PHYSICS, UNIV TENN, KNOXVILLE, 68- Concurrent Pos: Partic, AEC Int Sci Exchange Prog, Cent Inst Nuclear Res, Karlsruhe, Ger, 65-66. Mem: Fel Am Phys Soc. Res: Applications of Mössbauer effect in nuclear and solid state physics; positron polarization as related to C and P violation in weak interactions. Mailing Add: Oak Ridge Nat Lab PO Box X Oak Ridge TN 37830

OBER, DAVID RAY, b Garrett, Ind, Dec 6, 39; m 63; c 2. NUCLEAR PHYSICS. Educ: Manchester Col, BA, 62; Purdue Univ, MS, 64, PhD(physics), 68. Prof Exp: Teaching asst physics, Purdue Univ, 62-66, res asst, 66-68; asst prof, 68-74, ASSOC PROF PHYSICS, BALL STATE UNIV, 74- Mem: Am Phys Soc. Res: Low energy nuclear physics. Mailing Add: Dept of Physics Ball State Univ Muncie IN 47306

OBER, ROBERT ELWOOD, b Springfield, Ohio, Nov 13, 31; m 55; c 3. BIOCHEMISTRY, PHARMACOLOGY. Educ: Ohio State Univ, BS, 53, MS, 55; Univ Ill, PhD(biochem), 58. Prof Exp: From assoc res biochemist to res biochemist, Res Div, Parke, Davis & Co, 58-66; assoc prof, Col Pharm, Ohio State Univ, 66-69; head biochem pharmacol group, 69-70, supvr drug metab, 70-73, MGR DRUG METAB, RIKER LABS, 3M CO, 73-; ASSOC PROF PHARM, UNIV MINN, MINNEAPOLIS, 70- Mem: Am Chem Soc; Acad Pharmaceut Sci; AAAS; Sigma Xi. Res: Laboratory and clinical drug metabolism and pharmacokinetics; biochemistry of fatty acids; steroids; radiotracer methodology; synthesis of radioisotopically labelled compounds. Mailing Add: 3M Co Riker Labs Bldg 218-2 3M Ctr St Paul MN 55101

OBER, WILLIAM B, b Boston, Mass, May 15, 20; m 52; c 2. PATHOLOGY. Educ: Harvard Col, AB, 41; Boston Univ, MD, 46. Prof Exp: Pathologist, Boston Lying-in Hosp, 53-55; assoc prof & clin prof path, NY Med Col, 60-72; PROF PATH, MT SINAI SCH MED, 72-; PATHOLOGIST, BETH ISRAEL HOSP, 70- Concurrent Pos: Instr path, Harvard Med Sch, 53-55; dir labs path, Knickerbocker Hosp, 56-70; consult pathologist, First US Army Med Lab, 58-68, Margaret Sanger Res Bur, 60-73, Lutheran Hosp, Brooklyn, 65-, St Barnabas Hosp, Bronx, 66- & Roger Williams Hosp, Providence, RI, 68- Mem: Royal Micros Soc; NY Acad Med; Int Acad Path; Am Asn Path & Bact. Res: Experimental production of toxaemia of pregnancy; medical analysis of literary problems. Mailing Add: Beth Israel Hosp 10 Nathan D Perlman Pl New York NY 10003

OBERBECK, THOMAS EDMOND, b St Louis, Mo, Apr 30, 16; m 42; c 3. MATHEMATICS, OPERATIONS RESEARCH. Educ: Washington Univ, AB, 38; Univ Nebr, MA, 40; Calif Inst Technol, PhD(math), 48. Prof Exp: Asst math, Univ Nebr, 38-40, Univ Wis, 40-41 & Calif Inst Technol, 45-48; opers analyst, Opers Eval

Group, Div Indust Coop, Mass Inst Technol, 48-51; assoc prof math & mech, US Naval Postgrad Sch, 51-55, prof, 55-61, prof opers res & chmn dept, 61-64; res staff mem, Weapons Systs Eval Div, Inst Defense Anal, 64-65; adv res opers, Off Res Anal, Holloman AFB, 65-72; RETIRED. Mem: Am Math Soc; Opers Res Soc Am (treas, 65-); Inst Mgt Sci. Res: Long range planning; graph theory. Mailing Add: PO Box 561 Glendale AZ 85311

OBERDORFER, PAUL ELLSWORTH, JR, chemistry, see 12th edition

OBERENDER, FREDERICK G, b Cambridge, Mass, Feb 6, 33; m 57; c 4. ORGANIC CHEMISTRY, PETROLEUM CHEMISTRY. Educ: Trinity Col, Conn, BS, 54, MS, 56; Pa State Univ, PhD(org chem), 60. Prof Exp: Chemist, 59-60, sr chemist, 60-64, res chemist, 64-69, group leader, 69-73, ASST SUPVR, TEXACO, INC, 73- Mem: Am Chem Soc. Res: Synthetic lubricants; lubricant additives; polynuclear aromatic hydrocarbons; chelate polymers; nitrogen containing heterocyclics; fuel technology. Mailing Add: RD 4 Hopewell Rd Wappingers Falls NY 12590

OBERG, EDWIN NATHANIEL, b Stathcona, Minn, Feb 4, 06; m 39. MATHEMATICS. Educ: Univ Minn, AB, 29, AM, 32, PhD, 35. Prof Exp: Asst math, Univ Minn, 30-33; instr Dak Wesleyan Univ, 33-34, Wis State Teachers Col, Superior, 34-35 & Rochester Jr Col, 35-37; from instr to prof, 37-73, EMER PROF MATH, UNIV IOWA, 73- Concurrent Pos: Res specialist, NAm Aviation, Inc. Mem: Am Math Soc; Math Asn Am. Res: Approximate solutions of integral equations; linear functionals; approximation of functions by means of sums of orthogonal functions; approximate formulas for the radii of circles which include a specified fraction of a normal bivariate distribution. Mailing Add: Dept of Math Univ of Iowa Iowa City IA 52240

OBERG, ELMER BERNHARD, organic chemistry, see 12th edition

OBERHADT, BRUCE J, b Brooklyn, NY, Feb 27, 44. BIOPHYSICS. Educ: Polytech Inst Brooklyn, BS, 65, MS, 67, PhD(bioeng), 71. Prof Exp: Res assoc hemat, Montefiore Hosp & Med Ctr, 70-72, dir eng sci, Div Immunohemat, 72-73; RES SCIENTIST IMMUNOHEMAT, TECHNICON SCI CTR, TECHNICON CORP, 73- Concurrent Pos: Instr, Dept Med, Albert Einstein Col Med, 71-73; lectr bioeng, Polytech Inst New York, 72- Mem: Biophys Soc; AAAS; NY Acad Sci; Sigma Xi. Res: Agglutination phenomena: thermal kinetics of immunological systems; immunoassay methods; immunohematology; ageing process; information transfer in biological systems; whales and dolphins; fluid dynamics; electro-optical monitoring systems; hospital automation; biofeedback systems. Mailing Add: 100 E Hartsdale Ave Hartsdale NY 10530

OBERHELMAN, HARRY ALVIN, JR, b Chicago, Ill, Nov 15, 23; m 46; c 5. SURGERY. Educ: Univ Chicago, BS, 46, MD, 47. Prof Exp: From instr to assoc prof, Sch Med, Univ Chicago, 56-60; assoc prof, 60-64, PROF SURG, SCH MED, STANFORD UNIV, 64- Concurrent Pos: USPHS grant, Stanford Univ, 60-68; dir, Am Bd Surg, 72- Mem: Soc Univ Surg; Am Surg Asn; Am Gastroenterol Asn. Res: Gastrointestinal physiology, with emphasis on gastric secretion, esophageal motility and inflammatory diseases of the pancreas and colon. Mailing Add: 300 Pasteur Dr Stanford CA 94305

OBERHETTINGER, FRITZ, b Gelsenkirchen, Ger, Feb 24, 11; nat US; m 67; c 1. MATHEMATICS. Educ: Univ Berlin, PhD(math), 42; Univ Mainz, Dr habil, 45. Prof Exp: Prof math & math physics, Univ Mainz, 45-48; res assoc, Calif Inst Technol, 48-51; prof math, Am Univ, 51-58; PROF MATH, ORE STATE UNIV, 58- Concurrent Pos: Mem math res ctr, Univ Wis, 57-58; vis prof, Univ Del, 65-67; consult, Nat Bur Standards, 51-58. Mem: Am Math Soc. Res: Analysis; special functions; integral transforms; electromagnetic theory; wave propagation. Mailing Add: Dept of Math Ore State Univ Corvallis OR 97331

OBERHOFER, EDWARD SAMUEL, b Elizabeth, NJ, May 11, 39; m 67; c 1. NUCLEAR PHYSICS. Educ: NC State Univ, BS, 61, MS, 64, PhD(physics), 67. Prof Exp: Asst prof, 67-74, ASSOC PROF PHYSICS, UNIV NC, CHARLOTTE, 74- Mem: Am Phys Soc. Res: Low energy nuclear physics; nuclear spectroscopy. Mailing Add: Dept of Physics Univ of NC Charlotte NC 28213

OBERHOLTZER, JAMES EDWARD, b Elizabethtown, Pa, June 18, 42; m 67. ANALYTICAL CHEMISTRY. Educ: Elizabethtown Col, BS, 64; Purdue Univ, PhD(anal chem), 69. Prof Exp: RES CHEMIST, ARTHUR D LITTLE, INC, 68- Mem: AAAS; Am Chem Soc. Res: Instrumentation and methodology for chemical analyses, especially gas chromatography and mass spectrometry; application of digital computers to scientific research; analytical techniques for monitoring environmental pollution. Mailing Add: Res & Develop Arthur Little Inc 15 Acorn Park Cambridge MA 02140

OBERLANDER, GEORGE T, b Denver, Colo, June 18, 16; m 51; c 1. BOTANY. Educ: Univ Calif, AB, 36, MA, 38; Stanford Univ, PhD, 53. Prof Exp: PROF BIOL, SAN FRANCISCO STATE UNIV, 46- Concurrent Pos: Researcher, Univ Adelaide, 61. Mem: Bot Soc Am; Ecol Soc Am; Am Phycol Soc; Int Phycol Soc. Res: Ecology and taxonomy of marine algae. Mailing Add: Div of Biol Sci San Francisco State Univ San Francisco CA 94132

OBERLANDER, HERBERT, b Manchester, NH, Oct 2, 39; m 62. BIOLOGY. Educ: Univ Conn, BA, 61; Western Reserve Univ, PhD(biol), 65. Prof Exp: NIH fel, Inst Zool, Zurich, 65-66; asst prof biol, Brandeis Univ, 66-71; RES PHYSIOLOGIST, INSECT ATTRACTANTS, BEHAV & BASIC BIOL RES LAB, AGR RES SERV, USDA, 71- Concurrent Pos: Mem grad fac, Dept Entom, Univ Fla, 71- Mem: Entom Soc Am; Soc Develop Biol; Tissue Cult Asn; Int Soc Develop Biol. Res: Insect physiology; endocrine control of post-embryonic development in insects and other arthropods. Mailing Add: Behav & Basic Biol Res Lab Agr Res Serv USDA PO Box 14565 Gainesville FL 32604

OBERLE, GEORGE DAVID, b Carbondale, Kans, Nov 6, 07; m 42; c 3. HORTICULTURE, PLANT GENETICS. Educ: Kans State Col, BS, 31, MS, 36; Cornell Univ, PhD(plant breeding), 38. Prof Exp: Asst, Dept Hort, Kans State Col, 35; assoc prof, Ark State Teachers Col, Conway, 36; asst, NY Exp Sta, Geneva, 37; agent, USDA, 38-39; from asst prof to assoc prof, Cornell Univ, 39-48; prof, 48-73, EMER PROF HORT,XVAXPOLYTECHXINSTX&XSTATE HORT & HORTICULTURIST EXP STA, VA POLYTECH INST & STATE UNIV, 73- Mem: AAAS; Am Soc Hort Sci; Am Pomol Soc; Am Genetic Asn. Res: Fruit breeding and genetics of fruit plants; pollination requirements of fruit plants; induction of mutations. Mailing Add: 759 East Lake Rd Geneva NY 14456

OBERLE, RICHARD ALAN, b Wheeling, WVa, Nov 13, 44; c 4. MATHEMATICAL ANALYSIS. Educ: W Liberty State Col, BS, 65; Cath Univ Am, MS & PhD(math), 71. Prof Exp: Lectr phys astron, Oglebay Inst, 64-65; RES STAFF APPL STATIST & PROBABILITY, CTR NAVAL ANALYSES, UNIV ROCHESTER, 65- Concurrent Pos: Asst prof math, Old Dom Univ, 73-75. Mem: Am Math Soc. Res: Vector measures and integration; stochastic integration and application of the theory of semi-Markov processes to the evaluation of air combat maneuvering engagements. Mailing Add: Ctr for Naval Analyses 1401 Wilson Blvd Arlington VA 22209

OBERLE, THOMAS M, b Mankato, Minn, Mar 10, 30; m 53; c 6. INORGANIC CHEMISTRY. Educ: Col St Thomas, BSc, 52. Prof Exp: Jr chemist, Ames Lab, Atomic Energy Comn, Iowa State Univ, 52-54; chemist, 56-60, group leader prod develop, 60-64, mgr, 64-69, DIR INSTNL & CONSUMER RES & DEVELOP, ECONS LAB, INC, 69- Mem: Am Inst Chem. Res: Product development of detergent compounds for institutional and consumer needs. Mailing Add: Res & Develop Ctr Econs Lab Inc 840 Sibley Memorial Hwy Mendota Heights MN 55118

OBERLEY, WILLIAM JOHN, organic chemistry, see 12th edition

OBERLIN, DANIEL MALCOLM, b Tulsa, Okla, Sept 16, 48; m 67; c 1. MATHEMATICAL ANALYSIS. Educ: Univ Tulsa, BS, 70; Univ Md, MA, 72, PhD(math), 74. Prof Exp: ASST PROF MATH, FLA STATE UNIV, 74- Mem: Am Math Soc. Res: Harmonic analysis on locally compact groups. Mailing Add: Dept of Math Fla State Univ Tallahassee FL 32306

OBERLY, GENE HERMAN, b Palisade, Colo, Apr 27, 25; m 47; c 1. POMOLOGY. Educ: Utah State Univ, BS, 49, MS, 50; Mich State Univ, PhD(hort, plant nutrit), 59. Prof Exp: Salesman agr chem, C D Smith Drug Co, Colo, 50-51; exten horticulturist, Utah State Univ, 51-54; farm dir & assoc prof hort, Am Univ, Beirut, 54-57; asst hort, Mich State Univ, 57-59; Assoc prof pomol, Univ Conn, 59-62; ASSOC PROF POMOL, CORNELL UNIV, 62- Mem: Am Soc Hort Sci. Res: Plant nutrition on horticultural crops. Mailing Add: Dept of Pomol Cornell Univ Ithaca NY 14850

OBERLY, RALPH EDWIN, b Columbus, Ohio, Feb 13, 41; m 64. PHYSICS. Educ: Ohio State Univ, BS, 63, PhD(physics), 70. Prof Exp: Res engr, NAm Aviation, Ohio, 63-64; ASST PROF PHYSICS, MARSHALL UNIV, 70- Mem: Am Asn Physics Teachers; Optical Soc Am. Res: Infrared molecular spectroscopy of small molecules; optical instruments. Mailing Add: Dept of Physics Marshall Univ Huntington WV 25701

OBERMAN, ALBERT, b St Louis, Mo, Feb 9, 34; m 54; c 3. PREVENTIVE MEDICINE, EPIDEMIOLOGY. Educ: Wash Univ, AB, 55, MD, 59; Univ Mich, MPH, 66. Prof Exp: Investr thousand aviation study, US Naval Base, Pensacola, Fla, 62-65; Nat Heart Inst spec res assoc epidemiol, Sch Pub Health, Univ Mich, 66-67; PROF PUB HEALTH & EPIDEMIOL, ASSOC PROF MED & DIR, HEART EVAL CLIN, MED CTR, UNIV ALA, BIRMINGHAM, 66- Concurrent Pos: Mem med adv bd, Naval Aerospace Med Inst Fla, 65-; staff physician comt on criteria & methods coun epidemiol, Am Heart Asn, 66-67, mem exec comt, 70-, chmn, Credentials Comt, 75-; mem epid & dis control study sect, NIH, 73-77, mem ad hoc epidemiol review comt, 75-; Mostoy scholar award, Univ Mo. Mem: Am Pub Health Asn; fel Am Col Physicians; fel Am Col Prev Med; Int Soc Cardiol. Res: Epidemiology of chronic diseases, especially cardiovascular. Mailing Add: Box 39 OBSB Univ of Ala Med Ctr Univ Sta Birmingham AL 35294

OBERMAN, CARL RAYMOND, theoretical physics, see 12th edition

OBERMAN, HAROLD A, b Chicago, Ill, Oct 21, 32; m 59; c 3. MEDICINE, PATHOLOGY. Educ: Univ Omaha, AB, 53; Univ Nebr, MD, 56. Prof Exp: Asst chief dept path, Walter Reed Gen Hosp, 61-63; from asst prof to assoc prof path, 63-69, PROF PATH, SCH MED, UNIV MICH, ANN ARBOR, 69- Concurrent Pos: Consult, Vet Admin Hosp, Ann Arbor, 63- & Wayne County Gen Hosp, 66- Mem: Am Soc Clin Path; Col Am Path; Int Acad Path; Am Asn Path & Bact; Am Asn Blood Banks. Res: Blood banking and blood transfusion; surgical pathology; pathology of breast disease, lymph node disease and neoplasms of head and neck. Mailing Add: M5246 Med Sci Bldg Univ of Mich Med Sch Ann Arbor MI 48104

OBERMANNS, HENRY ERNST, b Erie, Pa, Aug 14, 08; m 45; c 3. CHEMISTRY. Educ: Yale Univ, BS, 30; Lawrence Col, MS, 32, PhD(paper chem & technol), 34. Prof Exp: Engr, Hammermill Paper Co, 34-35, tech asst to paper mill supt, 35-42, res chemist, 46-55, head process & prod develop, 55-67, mgr printing paper develop, 67-74; RETIRED. Mem: Am Chem Soc; Tech Asn Pulp & Paper Indust. Res: Pulp and paper industry; effect of hemicelluloses on the beating and strength of pulps; color dyeing of paper; developing new paper products and means for producing them. Mailing Add: 1938 Lakeside Dr Erie PA 16511

OBERMAYER, ARTHUR S, b Philadelphia, Pa, July 17, 31; m 63; c 1. ORGANIC CHEMISTRY, PHYSICS. Educ: Swarthmore Col, BA, 52; Mass Inst Technol, PhD(chem), 56. Prof Exp: From sr scientist to proj leader, Tech Div, Tracerlab, Inc, Mass, 56-59; chief proj scientist & mgr phys sci div, Allied Res Assocs, Inc, 59, mgr chem dept, 59-60; PRES & RES DIR, MOLECULON RES CORP, 60- Concurrent Pos: Dir, Man Electronics, Inc, 65- Mem: AAAS; Am Chem Soc; Am Phys Soc. Res: Physical applications of organic materials; high energy radiation measurements; solid state and physical organic chemistry; instrumental analysis. Mailing Add: Moleculon Res Corp 139 Main St Cambridge MA 02142

OBERMAYER, JUDITH HIRSCHFIELD, mathematics, see 12th edition

OBERMEYER, GERALD J, anthropology, see 12th edition

O'BERRY, PHILLIP AARON, b Tampa, Fla, Feb 1, 33; m 60; c 1. VETERINARY MICROBIOLOGY. Educ: Univ Fla, BS, 55; Auburn Univ, DVM, 60; Iowa State Univ, PhD(vet microbiol), 67. Prof Exp: Res vet, Nat Animal Dis Lab, 61-67; res adminr cattle dis, Vet Sci Res Div, 67-72, asst dir, 72-73, actg dir, 73-74, DIR, NAT ANIMAL DIS CTR, AGR RES SERV, USDA, 74- Concurrent Pos: Mem, Food & Agr Orgn Expert Panel on Livestock Infertility, 66-70; mem, Comt Fed Labs, Fed Coun Sci & Technol, 74- Honors & Awards: Cert of Merit, Agr Res Serv, 72. Mem: AAAS; Am Vet Med Asn; Am Soc Microbiol; Am Asn Bovine Practitioners; Conf Res Workers Animal Dis. Res: Microbiological aspects of Vibrio fetus and Mycoplasma species and their relationship to diseases of livestock. Mailing Add: Nat Animal Dis Ctr PO Box 70 Ames IA 50010

OBERSTAR, HELEN ELIZABETH, b Ottawa, Ill, Aug 29, 23; m 45. COSMETIC CHEMISTRY. Educ: Monmouth Col, BS, 43. Prof Exp: Asst food technol, Standard Brands, Inc, NY, 43-45; chemist, Miner Labs, Midwest Div, Arthur D Little, Inc, 46-50; res chemist & supvr, Toni Co Div, Gillette Co, 51-65; group leader, Shulton Inc, 65-71; sect mgr consumer prod div, Am Cyanamid Co, NJ, 72-75; MGR RES & DEVELOP, CLAIROL INT, INT DIV, BRISTOL-MYERS CO, NEW YORK, 75- Mem: Soc Cosmetic Chem. Res: Product development and exploratory research in hair coloring, hair lightening, hair treatment products, cosmetics and toiletries; mechanical device developments for cosmetics.

OBERSTE-LEHN, DEANE, b Hollywood, Calif, Feb 4, 33. GEOLOGY. Educ: Univ Calif, Los Angeles, BA, 56; Stanford Univ, MS, 70. Prof Exp: Geologist, Standard Oil

Co Calif, 56-60; res phys scientist, Rand Corp, 60-71; CONSULT, R&D ASSOCS, 71- Mem: Am Geophys Union; Am Soc Photogram. Res: Engineering and structural geology; environmental geology; geophysical methods; remote sensing; energy resources; slope stability; geothermal; petroleum. Mailing Add: PO Box 4706 Stanford CA 94305

OBERSTER, ARTHUR EUGENE, b Canton, Ohio, July 6, 29; m 55; c 6. ORGANIC CHEMISTRY. Educ: Mt Union Col, BS, 51; Univ Notre Dame, PhD(org chem), 57. Prof Exp: Sr chemist, Merck & Co, Inc, NJ, 55-59; res specialist, 59-67, sr group leader polymerization, 67-75, RES ASSOC, FIRESTONE TIRE & RUBBER CO, 75- Honors & Awards: Am Chem Soc Award, 50. Mem: Am Chem Soc. Res: Organic synthesis, particularly steroids, alkaloids, heterocyclics, rubber chemicals, antioxidants, antiozonants and monomers; polymer synthesis, particularly anionic polymerization; elastomer compounding; foam processing. Mailing Add: Cent Res Labs Firestone Tire & Rubber Co Akron OH 44317

OBERT, JESSIE C, b Port Byron, Ill, Mar 26, 11; m 35. NUTRITION. Educ: Park Col, AB, 31; Univ Chicago, MS, 42; Ohio State Univ, PhD(nutrit), 51. Prof Exp: Nutritionist, Chicago Welfare Dept, 37-42; dir nutrit serv, Maricopa County Chapter, Am Red Cross, 43-47; asst prof home econ, Ohio State Univ, 47-51; instr, Univ Calif, Los Angeles, 52-53; chief nutrit div, Co Health Dept, 53-75, NUTRIT COORDR, COUNTY LOS ANGELES COMMUNITY HEALTH SERV, 75- Concurrent Pos: Consult & lectr, Los Angeles Pub Schs. Mem: AAAS; Am Dietetic Asn; Am Heart Asn; Am Pub Health Asn; Am Home Econ Asn. Res: Activity and weight control; nutritional surveillance. Mailing Add: Grad Hall Rm 227 County of LA Com Hlth Serv 1200 N State St Los Angeles CA 90033

OBERT, LEONARD, physics, see 12th edition

OBERTEUFFER, JOHN AMIARD, b Boston, Mass, May 31, 40; m 62; c 2. PHYSICS, MAGNETISM. Educ: Williams Col, BA, 62, MA, 64; Northwestern Univ, PhD(physics), 69. Prof Exp: Asst, Northwestern Univ, Ill, 64-69; res assoc neutron physics, Mass Inst Technol, 69-71, mem staff, Francis Bitter Nat Magnet Lab, 71-74; asst dir develop, 74-75, TECH DIR, SALA MAGNETICS, 75- Mem: Inst Elec & Electronics Engrs; Am Phys Soc; Sigma Xi. Res: High gradient magnetic separation; basic and applied magnetism in liquids, solids, gases and mixed systems; solid state physics; x-ray and neutron diffraction. Mailing Add: Sala Magnetics Inc 247 Third St Cambridge MA 02142

OBIJESKI, JOHN FRANCIS, b Bridgeport, Conn, Apr 11, 41; m 62; c 2. MICROBIOLOGY, VIROLOGY. Educ: Univ Conn, BA, 65; Rutgers Univ, MS & PhD(virol), 71. Prof Exp: RES VIROLOGIST, VIROL SECT, CTR DIS CONTROL, 71- Mem: Am Soc Microbiol; Brit Soc Gen Microbiol. Res: Molecular and biochemical properties of animal viruses. Mailing Add: Virology Sect Ctr for Dis Control Atlanta GA 30333

OBLAD, ALEX GOLDEN, b Salt Lake City, Utah, Nov 26, 09; m 33; c 6. PETROLEUM CHEMISTRY. Educ: Univ Utah, BA, 33, MA, 34; Purdue Univ, PhD(phys chem), 37. Hon Degrees: DSc, Purdue Univ, 59. Prof Exp: Asst, Purdue Univ, 34-37; res chemist, Standard Oil Co, Ind, 37-42; group leader, Magnolia Petrol Co, Tex, 42-44, sect leader, 44-46, chief chem res, 46; head indust res, Tex Res Found, 46-47; dir chem res, Houdry Process Corp, Pa, 47-52, assoc mgr res & develop, 52-55, vpres res & develop & dir, 55-57; vpres res & develop, M W Kellogg Co, NY, 57-66, vpres res & eng develop, 66-69; vpres, Ireco Chem Co, Utah, 69-70; assoc dean col mines & mineral industs, 70-72, actg dean, 72-75, PROF METALL & FUELS ENG, UNIV UTAH, 70- Concurrent Pos: Consult, Atomic Energy Projs, 50-57; managing ed publ, Div Petrol Chem, Am Chem Soc, 54-69; dir, Int Cong Catalysis, 55-64; vpres & dir, Nat Inst Catalysis, 65-69. Mem: Honors & Awards: E U Murphree Award, Am Chem Soc, 69; Chem Pioneer Award, Am Inst Chemists, 72. Mem: Nat Acad Eng; Sigma Xi; Am Chem Soc; AAAS; Am Inst Chem Engrs. Res: Catalysis; reaction mechanisms; kinetics and thermodynamics of hydrocarbon reactions; heat capacity of glasses; optical methods of analysis; uranium chemistry and processing; petroleum chemistry; petrochemicals; administration and management of research and development; process engineering and patent licensing. Mailing Add: Mineral Sci Bldg Col of Mines & Mineral Indust Univ Utah Salt Lake City UT 84112

OBORN, EUGENE TIMBRELL, biochemistry, see 12th edition

OBRADOVICH, JOHN DINKO, b Fresno, Calif, May 2, 30. GEOPHYSICS. Educ: Univ Calif, Berkeley, BA, 57, MA, 59, PhD(geophys), 64. Prof Exp: GEOPHYSICIST, US GEOL SURV, 61- Mem: AAAS; Am Geophys Union; Geol Soc Am. Res: Isotope geology; K-Ar and Rb-Sr geochronology, particularly K-Ar dating of the late Cenozoic. Mailing Add: US Geol Surv Fed Ctr Denver CO 80225

OBREMSKI, ROBERT JOHN, b Brooklyn, NY, Aug 19, 41; m 64; c 2. PHYSICAL CHEMISTRY. Educ: St John's Univ, NY, BS, 62, MS, 64; Univ Md, College Park, PhD(phys chem), 68. Prof Exp: Chemist, Uniroyal, Inc, 68-69 & Spectra-Physics, Inc, 69-71; SR APPLN CHEMIST, BECKMAN INSTRUMENTS, INC, 71- Mem: Am Chem Soc; Soc Appl Spectros. Res: Spectroscopy; nuclear scintillation counting. Mailing Add: Beckman Instruments Inc 2500 Harbor Blvd Fullerton CA 92634

O'BRIAN, DENNIS MARTIN, b New York, NY, Mar 28, 31; m 58; c 2. PHYSIOLOGY, TOXICOLOGY. Educ: St Anselm's Col, BA, 52; Univ Notre Dame, MS, 55; Fordham Univ, PhD, 60. Prof Exp: From instr to assoc prof biol, Seton Hall Univ, 59-67; sr res scientist, Dept Toxicol, Squibb Inst Med Res, 67-68, sr res investr, Dept Res Admin, 68-71, sr info res scientist, Div Med Affairs, 71-73; mgr clin info, Johnson & Johnson Res, 73-74; ASST DIR MED SUPPORT, VICK DIV RES & DEVELOP, RICHARDSON-MERRELL, INC, 74- Mailing Add: Richardson-Merrell Inc 1 Bradford Rd Mt Vernon NY 10553

O'BRIEN, ANNE T, b New York, NY, Apr 11, 36. ENVIRONMENTAL SCIENCES. Educ: Marymount Col, NY, BS, 57; Fordham Univ, PhD(org chem), 64. Prof Exp: Fac mem parochial sch, 57-59; from instr to assoc prof chem, Marymount Col, NY, 62-73; ASSOC PROF, DEPT MAN-ENVIRON STUDIES, UNIV WATERLOO, 73- Mem: Am Chem Soc; Am Inst Chem; Nat Sci Teachers Asn; Am Asn Univ Prof; AAAS. Res: Halogen-catalyzed autoxidation; porphyrin synthesis; drug syntheses, resolutions and absolute configurations; futuristics; environmental chemistry; alkaloids. Mailing Add: Dept of Man-Environ Univ of Waterloo Waterloo ON Can

O'BRIEN, BRIAN, b Denver, Colo, Jan 2, 98; m 22, 56; c 1. ELECTROOPTICS. Educ: Yale Univ, PhB, 18, PhD(physics), 22. Prof Exp: Res engr, Westinghouse Elec & Mfg Co, 22-23; res physicist, Buffalo Tuberc Asn, 23-30; prof physiol optics, Univ Rochester, 30-46, dir Inst Optics, 38-53, res prof physics & optics, 46-53; vpres res & trustee, Am Optical Co, 53-58; CONSULT PHYSICIST, 58- Concurrent Pos: Mem comt pilot selection & training & vision comt, Nat Res Coun, 39-46, chmn div phys sci, 53-61; chmn, Nat Defense Res Comt, 40-46; chmn space prog adv coun, NASA, 70-; mem sci adv bd, US Air Force; chmn, Nat Acad Sci Comt, Adv to Air Force Systs Command. Honors & Awards: President's Medal for Merit, 48; Ives Medal, 51; Exceptional Civilian Serv Medal, US Air Force, 69, Exceptional Serv Medal, 73; Distinguished Pub Serv Medal, NASA, 71. Mem: Nat Acad Sci; AAAS; Optical Soc Am (pres, 51-53); Am Acad Arts & Sci; fel Soc Motion Picture & TV Engrs. Res: Optical properties of metals and thin films; solar ultraviolet and atmospheric ozone; photographic processes; motion picture systems; very high speed photography; photobiochemical effects; flicker phenomena in vision; retinal structure and visual processes; fiber optics. Mailing Add: Box 52 North Woodstock CT 06257

O'BRIEN, COLEMAN ART, b Cisco, Tex, Jan 9, 12; m 35; c 4. REPRODUCTIVE PHYSIOLOGY. Educ: Tex A&M Univ, BS, 44, MS, 45, PhD(reprod physiol), 64. Prof Exp: Instr animal husb, Tex A&M Univ, 45-47; asst prof, Tex Tech Col, 47-48; conservationist, Soil Conserv Serv, USDA, 48-55; asst prof animal husb, 55-67, ASSOC PROF ANIMAL SCI, TEX TECH UNIV, 67- Concurrent Pos: Fel, Cambridge Univ, 71. Mem: Am Soc Animal Sci. Res: Faster and more effective means of testing fertility of male animals; effects of gamma and fast neutron irradiation upon reproductive physiology of rabbits, goats and sheep. Mailing Add: Dept of Animal Sci Tex Tech Univ Lubbock TX 79409

O'BRIEN, DANIEL H, b Berkeley, Calif, Oct 26, 32; m 59; c 6. ORGANIC CHEMISTRY. Educ: Univ Va, BS, 54, PhD(chem), 61. Prof Exp: From instr to asst prof chem, Univ Dayton, 60-66; fel, Case Western Reserve Univ, 66-67; ASSOC PROF CHEM, TEX A&M UNIV, 67- Concurrent Pos: Res assoc, Chem Br, Aeronaut Res Lab, Wright-Patterson Air Force Base, 62-66. Mem: Am Chem Soc. Res: Organometallic chemistry; organic synthesis and characterization of organosilicon compounds. Mailing Add: Dept of Chem Tex A&M Univ College Station TX 77843

O'BRIEN, DANIEL JOSEPH, toxicology, pharmacology, see 12th edition

O'BRIEN, DARRELL EUGENE, b Sedgwick, Colo, Nov 8, 35; m 57; c 3. CHEMISTRY. Educ: NMex Highlands Univ, BS, 57; Ariz State Univ, MS, 59; Univ Mo-Kansas City, PhD(pharmaceut chem), 68. Prof Exp: Assoc chemist, Merck, Sharp & Dohme Res Lab, 58-59; assoc chemist, Midwest Res Inst, 59-66, sr chemist, 66-69; head med chem, Nucleic Acid Res Inst, Div Int Chem & Nuclear Corp, 69-73, dir drug develop, ICN Pharmaceut, Inc, 73-75, VPRES DRUG DEVELOP, ICN PHARMACEUT, INC, IRVINE, CALIF, 75- Mem: Am Chem Soc; Int Soc Heterocyclic Chem. Res: Design and synthesis of pharmaceutical agents. Mailing Add: ICN Pharmaceut Inc Irvine CA 92664

O'BRIEN, DAVID F, b Litchfield, Ill, Nov 18, 36; m 59; c 3. BIOPHYSICAL CHEMISTRY. Educ: Wabash Col, AB, 58; Univ Ill, PhD(org chem), 62. Prof Exp: Res chemist, 62-64, sr res chemist, 64-68, RES ASSOC, RES LABS, EASTMAN KODAK CO, 68- Mem: Am Chem Soc; Asn Res Vision & Ophthal. Res: Excited state properties of dyes; monomolecular layers of dyes; visual transduction; rhodopsin containing synthetic membranes; properties of synthetic membranes. Mailing Add: Res Labs Eastman Kodak Co Rochester NY 14604

O'BRIEN, DENISE (MRS JAY RUBY), b Rochester, NY, Jan 19, 38; m. CULTURAL ANTHROPOLOGY. Educ: Vassar Col, AB, 59; Yale Univ, PhD(anthrop), 69. Prof Exp: Curatorial asst anthrop, Peabody Mus, Yale Univ, 64-65; actg asst prof, Univ Calif, Davis, 65-67; asst prof, 67-75, ASSOC PROF ANTHROP, TEMPLE UNIV, 75- Concurrent Pos: Ford Found Fac fel res on role of women in soc, 73-74. Mem: Fel Am Anthrop Asn; fel Royal Anthrop Inst Gt Brit & Ireland; Asn Social Anthrop Oceania; Polyneisian Soc; Soc Anthrop Visual Commun. Res: Social organization, especially kinship and sex roles; ethnography of Highland New Guinea; acculturation; anthropological fiction. Mailing Add: Dept of Anthrop Temple Univ Philadelphia PA 19122

O'BRIEN, DENNIS CRAIG, b Great Bend, Kans, July 20, 38; m 60; c 2. GEOLOGY. Educ: Cornell Col, AB, 60; Miami Univ, MS, 64; Univ Mass, Amherst, PhD(geol), 71. Prof Exp: Instr, 68-71, ASST PROF GEOL, DRAKE UNIV, 71- Mem: Nat Asn Geol Teachers; Soc Econ Paleontologists & Mineralogists. Res: Sedimentary petrography and petrology; Precambrian sedimentary rocks; environmental geology. Mailing Add: Dept of Geog & Geol Drake Univ Des Moines IA 50311

O'BRIEN, DONOUGH, b Edinburgh, Scotland, May 9, 23; m 50; c 2. PEDIATRICS. Educ: Cambridge Univ, BA, 44, MB, BCh, 46, MA, 47, MD, 52; FRCP, 72. Prof Exp: House physician, St Thomas' Hosp, Univ London, 47-48, registr, Inst Child Health, 50-52 & Hosp Sick Children & Guy's Hosp, 53-57; from asst prof to assoc prof pediat, Sch Med, 57-64, PROF PEDIAT, UNIV COLO MED CTR, DENVER, 64- Mem: Am Diabetes Asn; Am Fedn Clin Res. Res: Biochemical applications to pediatrics. Mailing Add: Univ of Colo Med Ctr 4200 E Ninth Ave Denver CO 80220

O'BRIEN, ELINOR MURRAY, b Lynn, Mass, Feb 16, 27; m 49; c 1. CANCER, ELECTRON MICROSCOPY. Educ: Emmanuel Col, Mass, AB, 48; Boston Univ, MEd, 59; Boston Col, PhD(biol), 68; Harvard Med Sch, cert, 69. Prof Exp: Res chemist, USDA, 48-53; teacher, Saugus High Sch, 56-60; asst prof biol, Newton Col of the Sacred Heart, 60-64; lab asst, 64-65, RES ASSOC BIOL & ASSOC DIR CANCER RES INST, BOSTON COL, 66-, LECTR BIOL, COL, 68- Concurrent Pos: Boston Col cancer grant, NASA, 69. Mem: AAAS; Am Soc Cell Biol; Am Soc Microbiol; Electron Micros Soc Am. Res: Electron microscopy and radioautography of microorganisms and tumor cells. Mailing Add: Cancer Res Inst Boston Col Chestnut Hill MA 02167

O'BRIEN, FRANCIS JOSEPH, b Schenectady, NY, Aug 2, 01; m; c 1. PHARMACY. Educ: Union Univ, NY, PhG, 20, PhC, 29; State Teachers Univ NY Teachers Col, Albany, AB, 32, MA, 36; Fordham Univ, PhD(biol), 40. Hon Degrees: DSc, Union Univ, 56; LHD, Siena Col, 63. Prof Exp: From instr to assoc prof, 20-42, dean, 42-67, PROF PHARM, ALBANY COL PHARM, 42-, EMER DEAN, 67- Concurrent Pos: Sci ed, NY State Pharmacist. Mem: Am Pharmaceut Asn. Res: Fungicidal ointments. Mailing Add: Albany Col Pharm Albany NY 12208

O'BRIEN, GEORGE GERALD, mathematics, see 12th edition

O'BRIEN, GEORGE SIVESIND, b Madison, Wis, Aug 3, 26. NEUROLOGY, PHYSIOLOGY. Educ: Univ Wis-Madison, BA, 48, PhD(med physiol), 53, MD, 56. Prof Exp: From res asst to res assoc physiol, Med Sch, Univ Wis, 49-54; intern med, Sch Med, Vanderbilt Univ, 56-57; resident neurol, Univ Wis Hosps, 57-60; res assoc med, Cardiovasc Lab, Univ Wis, 62-68, asst scientist, 68-69; assoc dir clin res, Abbott Labs, 69-74; ASST PROF NEUROL, UNIV HEALTH SCI CHICAGO MED SCH, 74-; NEUROLOGIST, VET ADMIN HOSP, DOWNEY, 74- Mem: AAAS; Am Physiol Soc; Am Inst Biol Sci; Drug Info Asn; Am Epilepsy Soc. Res: Physiologic and pharmacologic studies of autonomic nervous system, including the central and peripheral; nervous and metabolic control of circulation; cardiovascular pharmacology; psychopharmacology; neurophysiology; clinical epilepsy; autonomic physiology. Mailing Add: 754 E Green Briar Lane Lake Forest IL 60045

O'BRIEN, HAROLD ALOYSIOUS, JR, b Dallas, Tex, May 17, 36; m 58; c 3. NUCLEAR CHEMISTRY, NUCLEAR MEDICINE. Educ: Univ Tex, Austin, BA,

59; NMex State Univ, MS, 61; Univ Tenn, Knoxville, PhD(phys chem), 68. Prof Exp: Res assoc nuclear chem, Isotopes Develop Ctr, Oak Ridge Nat Lab, 62-68; mem staff, 68-74, ASSOC GROUP LEADER, LOS ALAMOS SCI LAB, 74- Concurrent Pos: Adj asst prof, Sch Med, Univ NMex, 70-, Am Cancer Soc grant, 72-74; mem subcomt radiochem, Nat Acad Sci-Nat Res Coun, 74-; chmn, State NMex Radiation Tech Adv Coun, 75- Mem: AAAS; Am Chem Soc; Soc Nuclear Med. Res: Nuclear reactions; cross section studies; radioisotope production and applications; nuclear medicine; high temperature thermodynamics. Mailing Add: CNC-11 Los Alamos Sci Lab Los Alamos NM 87544

O'BRIEN, JAMES EDWARD, b Loogootee, Ind, June 2, 25. ASTRONOMY, SOLAR PHYSICS. Educ: Georgetown Univ, PhD(astron), 69. Prof Exp: Instr astron, Georgetown Univ, 65-66; ASST PROF PHYSICS, XAVIER UNIV, OHIO, 69- Res: Lunar-terrestrial relations. Mailing Add: Dept of Physics Xavier Univ Cincinnati OH 45207

O'BRIEN, JAMES FRANCIS, b Philadelphia, Pa, July 4, 41; m 70; c 1. PHYSICAL CHEMISTRY, INORGANIC CHEMISTRY. Educ: Villanova Univ, BS, 64; Univ Minn, Minneapolis, PhD(chem), 68. Prof Exp: Fel, Los Alamos Sci Lab, 68-69; asst prof, 69-75, ASSOC PROF CHEM, SOUTHWEST MO STATE COL, 75- Concurrent Pos: Vis assoc prof, Univ of Del, 75-76. Mem: Am Chem Soc. Res: Magnetic resonance of inorganic compounds in nonaqueous solvents; electrical conductivity of nonaqueous solutions; kinetics; nuclear quadrupole resonance. Mailing Add: Dept of Chem Southwest Mo State Col Springfield MO 65802

O'BRIEN, JAMES FRANCIS, b Rochester, NY, Aug 23, 34. CYTOLOGY, MICROBIOLOGY. Educ: Spring Hill Col, BS, 60; Fordham Univ, MS, 62, PhD(biol), 65. Prof Exp: Res assoc biol, Cancer Res Inst, Boston Col, 69; asst prof, 69-73, ASSOC PROF BIOL, LE MOYNE COL, NY, 73- Mem: AAAS; Nat Asn Biol Teachers; NY Acad Sci; Am Asn Jesuit Sci Rs: Cytological investigation of the development of intestinal tract of mosquitoes; electrophoretic study of tissues of rats infected with shaychloroma tumors. Mailing Add: Dept of Biol Le Moyne Col Le Moyne Heights Syracuse NY 13214

O'BRIEN, JAMES J, b New York, NY, Aug 10, 35; m 58; c 4. METEOROLOGY, OCEANOGRAPHY. Educ: Rutgers Univ, BS, 57; Tex A&M Univ, MS, 64, PhD(meteorol), 66. Prof Exp: Chemist, Elchem Dept, E I du Pont de Nemours & Co, 57-58, tech rep, 60-62; fel, Adv Study Group, Nat Ctr Atmospheric Res, 66-67, staff scientist, 67-69; assoc prof, 69-74, PROF METEOROL, FLA STATE UNIV, 69- Concurrent Pos: Co-dir coastal upwelling exp, Int Decade Ocean Explor-NSF. Mem: Am Meteorol Soc; Am Geophys Union; Am Chem Soc; Royal Meteorol Soc. Res: Micrometeorology; air-sea interactions; numerical analysis; applied statistics; numerical modeling of ocean circulation; ecological modeling. Mailing Add: Dept of Meteorol Fla State Univ Tallahassee FL 32306

O'BRIEN, JOAN A, b Philadelphia, Pa; m 51; c 3. VETERINARY MEDICINE, LARYNGOLOGY. Educ: Chestnut Hill Col, AB, 50; Univ Pa, VMD, 63. Prof Exp: Intern, Sch Vet Med, 63-64, Dorothy Harrison Eustis Seeing Eye fel, 64-66, instr med, 64-67, asst prof, 67-73, ASST PROF COMP BRONCHOLOGY, ESOPHAGOLOGY & LARYNGEAL SURG, SCH MED, UNIV PA, 71- Concurrent Pos: Assoc prof med, Dept Clin Studies, Sch Vet Med, Univ Pa & res assoc prof, Dept Otorhinolaryngol, Hahnemann Med Col & Hosp & Laryngeal Inst, 73- Mem: Am Col Vet Internists; Am Vet Med Asn; Am Broncho-Esophagological Asn. Res: Applications of laryngology and bronchoesophagology in naturally occurring disease of animals. Mailing Add: Sch of Vet Med Univ of Pa 3800 Spruce St Philadelphia PA 19174

O'BRIEN, JOHN ALOYSIUS, b Philadelphia, Pa, July 24, 12. CELL BIOLOGY. Educ: La Salle Col, AB, 35; Univ Penn, PhD(bot), 39. Prof Exp: From instr to asst prof, 39-52, ASSOC PROF BIOL, CATH UNIV AM, 52- Mem: Bot Soc Am; Am Soc Cell Biol. Res: Biogenesis of plastids; cell biology. Mailing Add: Dept of Biol Cath Univ of Am Washington DC 20064

O'BRIEN, JOHN S, b Rochester, NY, July 14, 34; m 57; c 6. PATHOLOGY, MEDICINE. Educ: Creighton Univ, MS, 58, MD, 60. Prof Exp: From instr to assoc prof path & med, Sch Med, Univ Southern Calif, 62-68, lectr biochem, 62-67, coordr set clin lects & chief div chem path, 64-68; assoc prof neurosci, 68-70, PROF NEUROSCI & CHMN DEPT, SCH MED, UNIV CALIF, SAN DIEGO, 70-, CHIEF DIV NEUROMETAB DISORDERS, 69- Concurrent Pos: USPHS grants, 63-79; Nat Inst Child Health & Human Develop grant, 66-68; Nat Multiple Sclerosis Soc grant, 66-70; Nat Genet Found grant, 70-72; Nat Cyctic Fibrosis Res Found grant, 70-72; Nat Found March of Dimes grant, 70-73; Nat Inst Gen Med Sci grant, 70-75; vis scientist, Univ Calif, Los Angeles, 62 & Scripps Inst, Univ Calif, 63; consult, Childrens Hosp Los Angeles, Pac State Hosp, Pomona, Childrens Hosp San Diego, Fairview State Hosp, Costa Mesa, Pasadena Found Med Res & Dept Neurol, Univ Southern Calif, 64-68. Mem: AAAS; Am Fedn Clin Res; Am Soc Exp Path; Am Soc Human Genet; NY Acad Sci. Res: Relationships between the molecular structure and disease states; role of lipid molecules in membrane structure and stability; brain lipids; myelination and demyelination. Mailing Add: Dept of Neurosci Univ of Calif San Diego Sch Med La Jolla CA 92037

O'BRIEN, JOHN TERENCE, b Oak Park, Ill, Sept 20, 44; m 70. ORGANIC POLYMER CHEMISTRY. Educ: Univ Notre Dame, BS, 66; Mass Inst Technol, PhD(org chem), 71; Univ Hartford, MBA, 75. Prof Exp: RES CHEMIST, CHEM DIV, UNIROYAL, INC, 70- Mem: Am Chem Soc. Res: Applications development; water soluble polymers; design of polyelectrolytes as emulsifiers and dispersants for use as paint, paper, water treatment and oil field chemicals. Mailing Add: Uniroyal Chem Div of Uniroyal Inc BldgS-10 152 Elm St Naugatuck CT 06770

O'BRIEN, JOSEPH LLOYD, organic chemistry, see 12th edition

O'BRIEN, KATHARINE ELIZABETH, b Amesbury, Mass, Apr 10, 01. MATHEMATICS. Educ: Bates Col, AB, 22; Cornell Univ, AM, 24; Brown Univ, PhD(math), 39. Hon Degrees: ScDEd, Univ Maine, 60; LHD, Bowdoin Col, 65. Prof Exp: From instr to prof & head dept, Col New Rochelle, 25-36; teacher math, Deering High Sch, Portland, 40-71, head dept, 45-71; lectr, Univ Maine, Portland-Gorham, 62-73; RETIRED. Concurrent Pos: Referee articles, The Math Teacher, Nat Coun Teachers Math, 53- Mem: Math Asn Am; NY Acad Sci. Res: Mathematical analysis; interpolation. Mailing Add: 130 Hartley St Portland ME 04103

O'BRIEN, KERAN, b Brooklyn, NY, Nov 5, 31; m 61; c 2. RADIATION PHYSICS. Educ: Fordham Univ, BS, 53. Prof Exp: Mem adv panel accelerator safety, AEC, 65-70, PHYSICIST, HEALTH & SAFETY LAB, ENERGY RES & DEVELOP ADMIN, 53- Mem: Am Nuclear Soc; Radiation Res Soc; Am Phys Soc. Res: Radiation dosimetry associated with particle accelerator shielding with naturally occurring radiation sources; high energy radiation theory and the propagation of atmospheric cosmic rays. Mailing Add: Radia Physic Div Hlth Safe Lab Ener Res & Dev Adm 376 Hudson St New York NY 10014

O'BRIEN, LARRY JOE, b Big Spring, Tex, Sept 14, 29; m 53; c 3. PHYSIOLOGY. Educ: Hardin-Simmons Univ, BA, 49; NTex State Col, MA, 54; Univ Tex, PhD(physiol), 57; Med Col Ga, MD, 71. Prof Exp: Asst physiol, Univ Tex Med Br, 55-56, instr, 56-57; from instr to asst prof, Albany Med Col, 57-60; assoc prof res physiol, Sch Med, Univ Okla, 60-62; from asst prof to assoc prof physiol, Med Col Ga, 62-72, actg chmn dept, 71-72; prof physiol & chmn dept, 72-75, CLIN PROF PHYSIOL & MED, SCH MED, TEX TECH UNIV, 75- Concurrent Pos: Chief circulation sect, Civil Aeromed Res Inst, 60-62. Mem: AAAS; Am Physiol Soc; Am Heart Asn; NY Acad Sci; Soc Exp Biol & Med. Res: Cardiac and peripheral vascular function. Mailing Add: Suite 401 Med-Prof Bldg 3801 19th St Lubbock TX 79410

O'BRIEN, MICHAEL HARVEY, b Soperton, Ga, Mar 22, 42; m 63; c 1. ORGANIC CHEMISTRY. Educ: Berry Col, BA, 63; NC State Univ, PhD(chem), 69. Prof Exp: NASA training grant, 63-67; RES CHEMIST, E I DU PONT DE NEMOURS & CO, INC, 68- Mem: Am Chem Soc. Res: Synthesis, structure and reactions of pentavalent organoarsenic compounds. Mailing Add: E I du Pont de Nemours & Co Waynesboro VA 22980

O'BRIEN, NEAL RAY, b Newark, Ohio, May 25, 37; m 62; c 1. GEOLOGY. Educ: DePauw Univ, BA, 59; Univ Ill, MS, 61, PhD(geol), 63. Prof Exp: PROF GEOL, STATE UNIV NY COL, POTSDAM, 63- Concurrent Pos: State Univ NY Res Found fel, Kyoto Univ, 69-70. Mem: Geol Soc Am; Soc Econ Paleont & Mineral. Res: Sedimentation; clay mineralogy; electron microscope and x-ray study of shales. Mailing Add: Dept of Geol State Univ of NY Col Potsdam NY 13676

O'BRIEN, PAUL J, b Haddonfield, NJ, Feb 11, 33; m 61; c 3. BIOCHEMISTRY. Educ: Mt St Mary's Col, Md, BS, 54; St John's Univ, NY, MS, 56; Univ Pa, PhD(biochem), 60. Prof Exp: Res chemist, Sect Intermediary Metab, Lab Biochem & Metab, Nat Inst Arthritis & Metab Dis, 60-64; res chemist, Sect Cell Biol, Ophthal Br, Nat Inst Neurol Dis & Blindness, 64-65, from actg chief to chief, 65-71; RES CHEMIST, LAB VISION RES, NAT EYE INST, 71- Mem: AAAS; Asn Res Vision & Ophthal; Am Soc Biol Chemists. Res: Carbohydrate metabolism; biosynthesis of glycoproteins and mucopolysaccharides; control mechanisms; visual pigment biosynthesis and photoreceptor renewal. Mailing Add: Bldg 6 Rm 218 Lab of Vision Res Nat Eye Inst NIH Bethesda MD 20014

O'BRIEN, PETER J, b London, Eng, July 6, 37. BIOCHEMISTRY. Educ: Univ London, BSc, 59; Univ Birmingham, PhD, 63. Prof Exp: Res fel med biochem, Univ Birmingham, 63-64, sr res assoc, 64-67; assoc prof, 67-74, PROF BIOCHEM, MEM UNIV NFLD, 74- Concurrent Pos: Assoc ed, Can J Biochem, 71-75. Mem: Brit Biochem Soc; Can Soc Cell Biol; Can Biochem Soc; Am Soc Biol Chemists. Res: Intracellular formation, molecular effects and function of lipid peroxides, steroid hydroperoxides and hydrogen peroxide; drug and steroid metabolism; functional organization of electron transport in intracellular membranes. Mailing Add: Dept of Biochem Mem Univ of St John's NF Can

O'BRIEN, REDMOND R, b Quincy, Mass, Oct 27, 31; m 59; c 2. MATHEMATICS. Educ: Mass Inst Technol, SB, 53, SM, 54, PhD(math), 57. Prof Exp: Adv res engr, Sylvania Elec Prod, 57-60; ADV MATHEMATICIAN, IBM CORP, 60- Mem: Am Statist Asn; Soc Indust & Appl Math. Res: Mathematical analysis of semiconductor devices. Mailing Add: IBM Corp Bldg 300-94 East Fishkill NY 12533

O'BRIEN, RICHARD DESMOND, b Sydenham, Eng, May 29, 29; m 61; c 1. NEUROCHEMISTRY. Educ: Univ Reading, BSc, 50; Univ Western Ontario, PhD(chem), 54, BA, 56. Prof Exp: Chemist, Pesticide Res Inst, Can, 54-60; from assoc prof to prof entom, 60-65, chmn sect biochem, 64-65 & sect neurobiol & behav, 65-70, PROF NEUROBIOL, CORNELL UNIV, 65-, DIR DIV BIOL SCI, 70- Concurrent Pos: Nat Res Coun Can fel, Cambridge Univ, 56-57; vis assoc prof, Univ Wis, 58-59; Guggenheim fel, Int Lab Genetics & Biophys, Naples, 67-68; ed & founder, Pesticide Biochem & Physiol, 70- Honors & Awards: Int Award, Am Chem Soc, 71. Mem: Fel AAAS; Am Chem Soc; Soc Neurosci; Am Soc Neurochem; Am Soc Biol Chem. Res: Selective toxicity; modes of toxic action; comparative biochemistry; neuropharmacology. Mailing Add: Div of Biol Sci Cornell Univ Ithaca NY 14850

O'BRIEN, RICHARD LEE, b Shenandoah, Iowa, Aug 30, 34; m 57; c 4. ONCOLOGY, CELL BIOLOGY. Educ: Creighton Univ, MS, 58, MD, 60. Prof Exp: From intern to asst resident med, First Med Div, Bellevue Hosp, Columbia Univ, 60-62; from asst prof to assoc prof, 66-76, PROF PATH, SCH MED, UNIV SOUTHERN CALIF, 76-, DEP DIR, CANCER CTR, 75- Concurrent Pos: Fel biochem, Inst Enzyme Res, Univ Wis-Madison, 62-64; spec fel, Nat Cancer Inst, 67-69; grants, Am Cancer Soc, 67-68, Nat Cancer Inst, 67-70 & 74-77 & John A Hartford Found, 69-74; vis prof molecular biol, Univ Geneva, 73-74. Mem: AAAS; Am Soc Exp Path; Am Asn Cancer Res. Res: Mechanisms of oncogenesis; environmental carcinogenesis; control of cell proliferation; molecular biology. Mailing Add: Dept of Path Univ of Southern Calif Sch Med Los Angeles CA 90033

O'BRIEN, ROBERT NEVILLE, b Nanaimo, BC, June 14, 21; m 52; c 5. PHYSICAL CHEMISTRY. Educ: Univ BC, BASc, 51, MASc, 52; Univ Manchester, PhD(metall), 55. Prof Exp: Asst physics, Univ BC, 52; fel pure chem, Can Nat Res Coun, 55-57; from asst prof to assoc prof chem, Univ Alta, 57-66; assoc prof, 66-68, PROF CHEM, UNIV VICTORIA, 68- Concurrent Pos: Vis scholar, Univ Calif, Berkeley, 64-65; consult, UniRoyal Res Labs & Bapco Paint; mem bd mgt, BC Res Coun. Mem: Am Chem Soc; Electrochem Soc; fel Chem Inst Can; fel Royal Soc Arts. Res: Electrode processes; optical studies of working electrodes in metal electrodeposition cells; surface chemistry; electrets; gas-liquid interface. Mailing Add: Dept of Chem Univ of Victoria Victoria BC Can

O'BRIEN, ROBERT THOMAS, b Bismarck, NDak, Dec 20, 25; m 48; c 5. BACTERIAL PHYSIOLOGY, BIOCHEMISTRY. Educ: Univ NDak, BS, 50, MS, 52; Wash State Univ, PhD(bact), 56. Prof Exp: Asst bact, Univ NDak, 50-52, instr, 52; asst, Wash State Univ, 52-53, actg instr, 53-54, asst, 54-56; biol scientist, Gen Elec Co, 56-64, consult microbiologist, 64-66; PROF BIOL, NMEX STATE UNIV, 66- Concurrent Pos: Consult, Azar Bros, Inc, 67- Mem: AAAS; Am Soc Microbiol; Brit Soc Gen Microbiol. Res: Ecology of spore forming bacteria; survival of animal viruses in aquatic environments. Mailing Add: Dept of Biol NMex State Univ Las Cruces NM 88001

O'BRIEN, STEPHEN JAMES, b Rochester, NY, Sept 30, 44; m 68. GENETICS. Educ: St Francis Col, BS, 66; Cornell Univ, PhD(genetics), 71. Prof Exp: NIH fel biochem, 71-73, GENETICIST, NAT CANCER INST, 73- Concurrent Pos: NIH fel, Geront Res Ctr, 71-72; NIH fel, Nat Cancer Inst, 72-73. Mem: Am Soc Naturalists; Genetics Soc Am; Tissue Cult Asn; Am Genetics Asn. Res: Developmental genetics of energy metabolism in Drosophila; somatic cell genetic regulation of neoplasia and tumor virus expression. Mailing Add: Cell Biol Sect Nat Cancer Inst Bethesda MD 20014

O'BRIEN, THOMAS DORAN, b Washington, DC, Mar 31, 10; m 35; c 2.

INORGANIC CHEMISTRY. Educ: George Washington Univ, BS, 35, MS, 38; Univ Ill, PhD(inorg chem), 40. Prof Exp: Res chemist, Naval Res Lab, 40-42 & Barrett Div, Allied Chem & Dye Corp, NJ, 42-43; asst prof chem, Tulane Univ, 43-45; from asst prof to assoc prof, Univ Minn, 45-55; prof & head dept, Kans State Univ, 55-60; PROF CHEM & DEAN GRAD SCH, UNIV NEV, RENO, 60-, RES COORDR, 70- Concurrent Pos: Prog dir & actg dep div dir, NSF, 67-68. Mem: Am Chem Soc. Res: Inorganic coordination compounds; heterogeneous catalysis. Mailing Add: Grad Sch Univ of Nev Reno NV 89507

O'BRIEN, THOMAS JOSEPH, b Detroit, Mich, Dec 31, 41; m 68. PHYSICAL CHEMISTRY, THEORETICAL CHEMISTRY. Educ: Col St Thomas, BS, 63; Univ Wis, PhD(phys chem), 68. Prof Exp: Res fel theoret physics & appl math, Queen's Univ, Belfast, 68-69; ASST PROF CHEM, TEX TECH UNIV, 69- Concurrent Pos: Fel, Nat Res Coun, 73-74. Res: Determination of intermolecular potential energy functions from experimental scattering data; ab initio calculation of scattering cross sections; collision theory of ionospheric chemical reactions. Mailing Add: Dept of Chem Tex Tech Univ Lubbock TX 79409

O'BRIEN, THOMAS V, b Cincinnati, Ohio, Apr 30, 37; m 62; c 4. TOPOLOGY. Educ: Xavier Univ, BS, 59, MS, 60; Syracuse Univ, PhD(math), 65. Prof Exp: Asst prof math, Marquette Univ, 64-69; asst prof, 69-71, ASSOC PROF MATH, BOWLING GREEN STATE UNIV, 71- Mem: Am Math Soc; Math Asn Am. Res: Dynamical systems; study of continuous flows on manifolds. Mailing Add: Dept of Math Bowling Green State Univ Bowling Green OH 43403

O'BRIEN, THOMAS W, b Rochester, Minn, Sept 17, 38; m 64. PHYSIOLOGY, BIOCHEMISTRY. Educ: St Thomas Col, BS, 62; Marquette Univ, MS, 63, PhD(physiol), 65. Prof Exp: Fel biochem, NJ Col Med, 64-65, instr, 65-66; asst prof, 66-74, ASSOC PROF BIOCHEM & BIOL SCI, COL MED, UNIV FLA, 74- Mem: Am Chem Soc; Am Soc Cell Biol. Res: Mitochondrial protein synthesis; structure and function of mitochondrial ribosomes; thyroxine effects on mitochondria. Mailing Add: Dept of Biochem Univ of Fla Gainesville FL 32601

O'BRIEN, VIVIAN, b Baltimore, Md, Feb 1, 24. FLUID DYNAMICS. Educ: Goucher Col, AB, 45; Johns Hopkins Univ, MA, 50, MS, 54, PhD, 60. Prof Exp: Jr aerodynamicist, Martin Co, 45-47; asst aeronaut, 47-55, assoc physicist fluid dynamics, Appl Physics Lab, 55-58, PHYSICIST FLUID DYNAMICS, APPL PHYSICS LAB, JOHNS HOPKINS UNIV, 58- Mem: Am Phys Soc; Am Inst Aeronaut & Astronaut; Soc Women Engrs. Res: Theoretical transonic and supersonic aerodynamics; turbulent flow; viscous vortex flow; viscous biological flows; drops and bubbles; porous media flows. Mailing Add: Appl Physics Lab J Hopkins Univ Johns Hopkins Rd Laurel MD 20810

O'BRIEN, WILLIAM DANIEL, JR, b Chicago, Ill, July 19, 42. BIOACOUSTICS, BIOENGINEERING. Educ: Univ Ill, Urbana, BS, 66, MS, 68, PhD(elec eng), 70. Prof Exp: Res assoc ultrasonic biophys, Univ Ill, Urbana, 70-71; res scientist, Bur Radiol Health, Food & Drug Admin, 71-75; ASST PROF ELEC ENG & BIOENG, UNIV ILL, URBANA, 75- Concurrent Pos: Inst Elec & Electronics Engrs rep, Am Nat Standards Inst, 73- Mem: Inst Elec & Electronics Engrs; Acoust Soc Am; Am Inst Ultrasound Med; AAAS; Sigma Xi. Res: Examination of the mechanisms by which ultrasonic energy interacts with biological material with the view toward the determination or assessment of risk associated with human exposure. Mailing Add: Bioacoust Res Lab Univ of Ill Urbana IL 61801

O'BRIEN, WILLIAM JOHN, b Summit, NJ, Nov 30, 42; m 64; c 1. AQUATIC ECOLOGY, LIMNOLOGY. Educ: Gettysburg Col, BA, 65; Mich State Univ, PhD(aquatic ecol), 70. Prof Exp: Coherent Areas Prog fel aquatic ecol & res assoc, W K Kellogg Biol Sta, Mich State Univ, 70-71; ASST PROF LIMNOL, UNIV KANS, 71- Concurrent Pos: Res grant, NSF, 75 & Kans Water Resources Inst, 74. Mem: AAAS; Am Soc Limnol & Oceanog; Ecol Soc Am; Int Asn Theoret & Appl Limnol. Res: Mathematical models of aquatic systems; primary productivity of aquatic systems; theory of limiting factors; zooplankton ecology; phyto-zooplankton physiological ecology; ecology of fish seeding. Mailing Add: Snow Hall Dept of Systs & Ecol Univ of Kans Lawrence KS 66044

O'BRIEN, WILLIAM JOSEPH, b New York, NY, July 25, 35; m 63; c 2. DENTAL MATERIALS, SURFACE CHEMISTRY. Educ: City Col New York, BS, 58; NY Univ, MS, 62; Univ Mich, PhD(metall eng), 67. Prof Exp: Assoc dir res, J F Jelenko Co, 58-61; from instr to assoc prof mat sci, Marquette Univ, 61-70; assoc prof, 70-73, PROF MAT SCI, DENT RES INST, UNIV MICH, ANN ARBOR, 73- Concurrent Pos: Consult, WHO, 67-68 & Am Dent Asn, 67-; chmn dept dent mat, Marquette Univ, 67-70; prin investr, USPHS grant; res assoc, Vet Admin Hosp, Wood, Wis, 67-; secy, Dent Mat Group, 69-73, pres, 75-76. Honors & Awards: UN Cert, UN, 67. Mem: Gen Systs Res; Int Asn Dent Res. Res: Biomaterials; noble metals; surface phenomena; capillary phenomena; ceramics-mechanical and optical properties. Mailing Add: Dent Res Inst Univ of Mich Ann Arbor MI 48104

O'BRIEN, WILLIAM M, b Bethel, Maine, Feb 26, 31; m 57; c 3. EPIDEMIOLOGY, GENETICS. Educ: Tufts Univ, BS, 52; Yale Univ, MD, 56. Prof Exp: Clin assoc, Nat Inst Arthritis & Metab Dis, 58-60; sr registr, Manchester Royal Infirmary, Eng, 60-61; sr clin investr, Nat Inst Arthritis & Metab Dis, 61-64; asst prof, Sch Med, Yale Univ, 64-67; assoc prof internal med, 67-72, PROF INTERNAL MED, SCH MED, UNIV VA, 72- Mem: AAAS; Am Epidemiol Soc; Am Rheumatism Asn; Asn Comput Mach; Heberden Soc. Res: Rheumatic diseases; drug testing. Mailing Add: Dept of Internal Med Univ of Va Sch of Med Charlottesville VA 22903

OBRINSKY, WILLIAM, b May 15, 13; US citizen; m 46; c 4. MEDICINE, PEDIATRICS. Educ: NY Univ, BS, 34, MD, 38. Prof Exp: From intern to resident, Morrisania Hosp, 38-41; resident, Queens Gen Hosp, 41-42 & Mt Sinai Hosp, 46-47; from instr to assoc prof pediat, Sch Med, La State Univ, 48-53; pediatrician, Staten Island Med Ctr, 54-64; from asst clin prof to assoc clin prof pediat, 55-67, assoc prof, 67-68, PROF PEDIAT, ALBERT EINSTEIN COL MED, 68- Concurrent Pos: Attend, Montefiore Hosp, 64-; attend, Morrisania Hosp, 64-, dir pediat, 68- Mem: AAAS; Am Acad Pediat; Am Pediat Soc; Soc Pediat Res; NY Acad Sci. Mailing Add: Morrisania Hosp Dept of Pediat 168th St & Gerard Ave Bronx NY 10452

OBRIST, WALTER DORN, b Yonkers, NY, Aug 16, 24; m 56; c 2. ELECTROENCEPHALOGRAPHY, EXPERIMENTAL NEUROLOGY. Educ: Harvard Univ, BS, 46; Univ Calif, MA, 47; Northwestern Univ, PhD(psychol), 50. Prof Exp: Res assoc, Moosehaven Res Lab, Fla, 50-54; res assoc, Inst Living, 54-57; from asst prof to assoc prof med psychol, 57-63, PROF MED PSYCHOL, DUKE UNIV, 63-, PROF EXP MED, 74-, DIR RES TRAINING PROG IN SCI RELATED TO NERV SYST, 68-, ASSOC DIR CTR STUDY AGING & HUMAN DEVELOP, 74- Concurrent Pos: Consult, Yerkes Labs Primate Biol, 52-54; dir EEG lab, Norwich State Hosp, Conn, 55-56; consult div res grants, NIH, 63-67; USPHS spec fel clin neurophysiol, Univ Lund, 67-68; consult, Nat Inst Neurol & Commun Disorders & Stroke, 70- Mem: Am Electroencephalog Soc; Geront Soc; Am Psychol Asn; Am Acad Neurol; Am Neurol Asn. Res: Gerontology; brain and behavior; cerebral circulation and metabolism. Mailing Add: Box 3003 Duke Univ Med Ctr Durham NC 27710

OBRYCKI, RICHARD, b Pittsburgh, Pa. ANALYTICAL CHEMISTRY, ORGANIC CHEMISTRY. Educ: Univ Pittsburgh, BS, 61, MS, 64, PhD(org chem), 66. Prof Exp: Res asst, Univ Pittsburgh, 61-66; org chemist, Chem Process Lab, US Army, Md, 66-69; sr res scientist, 69-71, ANAL CHEMIST ABSORPTION SPECTROS, KOPPER'S CO, INC, 71- Mem: AAAS; Sigma Xi. Res: Organophosphorous compounds synthesis; photo-initiated reaction of aryl halides and trivalent phosphorous compounds; alkylation of pyrrylmagnesium bromide; mass and nuclear resonance spectra of organophosphorous compounds. Mailing Add: Kopper's Co Inc 440 College Park Dr Monroeville PA 15146

O'CALLAGHAN, DENNIS JOHN, b New Orleans, La, July 26, 40; m 67; c 1. ANIMAL VIROLOGY, MEDICAL MICROBIOLOGY. Educ: Loyola Univ, New Orleans, BS, 62; Univ Miss, PhD(microbiol), 68. Prof Exp: Fel virol, Dept Biochem, Med Ctr, Univ Alta, 68-70, asst prof biochem, 70-71; asst prof, 71-74, ASSOC PROF MICROBIOL, MED CTR, UNIV MISS, 74- Concurrent Pos: Res grant, Brown-Hazen Fund Res Corp, 72; NIH res grant, 73-; assoc ed, J Miss Acad Sci, 75- Mem: AAAS; Am Soc Microbiol; Am Soc Exp Path; Sigma Xi. Res: Biochemistry of herpes virus replication, concerning control of viral DNA replication and virus transcription; role of herpes viruses in human cancer. Mailing Add: Dept of Microbiol Univ of Miss Med Ctr Jackson MS 39216

O'CALLAGHAN, ROBIN KUEBLER, b Detroit, Mich, Mar 21, 48; m 75. ALGEBRA. Educ: Marquette Univ, BS, 70; Wesleyan Univ, PhD(math), 75. Prof Exp: Asst math, Wesleyan Univ, 70-75; ASST PROF MATH, UNIV TEX, PERMIAN BASIN, 75- Mem: Am Math Soc; Math Asn Am. Res: Homological algebra; Abelian group theory; endomorphism rings of Abelian groups. Mailing Add: Dept of Math Univ of Tex of Permian Basin Odessa TX 79762

OCAMPO, JAIME, organic chemistry, see 12th edition

OCAMPO-FRIEDMANN, ROSELI C, b Manila, Philippines, Nov 23, 37. ECOLOGY, PHYCOLOGY. Educ: Univ Philippines, BSc, 58; Hebrew Univ Jerusalem, MSc, 66; Fla State Univ, PhD(biol), 73. Prof Exp: Res assoc limnol & phycol, Nat Inst Sci & Technol Philippines, 58-67; teaching asst bot & lab technologist phycol, 67-73, RES ASSOC PHYCOL, FLA STATE UNIV, 73- Concurrent Pos: Asst prof biol, Fla A&M Univ, 74- Mem: Int Phycol Soc; Phycol Soc Am; Bot Soc Am; Indian Phycol Soc; Sigma Xi. Res: Culture and taxonomy of unicellular blue-green algae; biology of microorganisms in extreme habitats. Mailing Add: Dept of Biol Sci Fla State Univ Tallahassee FL 32306

OCCOLOWITZ, JOHN LEWIS, b Melbourne, Australia, July 30, 31; m 62; c 2. MASS SPECTROMETRY. Educ: Univ Melbourne, BSc, 52, DipED, 53, MSc, 65. Prof Exp: Instr sci, Victoria, Australia Educ Dept, 53-55; res scientist chem, Dept Supply Defense Stands Labs, Australian Govt, 55-67; RES SCIENTIST CHEM, LILLY RES LABS, ELI LILLY & CO, 67- Mem: Am Soc Mass Spectrometry. Res: Determination of organic structures using mass spectrometry; elucidation of the structure of ions formed in the mass spectrometer, by labelling and measurement of energetics of formation. Mailing Add: Dept MC 525 Lilly Res Labs PO Box 618 Indianapolis IN 46206

OCHOA, SEVERO, b Luarca, Spain, Sept 24, 05; nat; m 31. MEDICINE. Educ: Malaga Col, Spain, BA, 21; Univ Madrid, MD, 29. Hon Degrees: Many from var Am & foreign cols & univs. Prof Exp: Lectr physiol & biochem, Sch Med, Univ Madrid, 31-35, head physiol div, Inst Med Res, 35-36; guest res asst physiol, Kaiser-Wilhelm Inst Med Res, 36-37; Lankester investr, Marine Biol Lab, Plymouth, Eng, 37; demonstr biochem & Nuffield res asst, Oxford Univ, 38-40; instr pharmacol & res assoc, Sch Med, Wash Univ, 41-42; res assoc med, Sch Med, NY Univ, 42-45, asst prof biochem, 45-46, prof pharmacol & chmn dept, 46-54, prof biochem & chmn dept, 54-74; DISTINGUISHED MEM, ROCHE INST MOLECULAR BIOL, 74- Concurrent Pos: Fel from Univ Madrid, Kaiser-Wilhelm Inst, Berlin & Heidelberg, 29-31 & Nat Inst Med Res, London, 32-33; mem physiol study sect, USPHS, 47-50, biochem study sect, 52-55; mem biochem panel, US Off Naval Res, 53-55, chmn, 55-57; US rep, Int Union Biochem, 55-61; hon prof, San Marcos Univ, Lima, 57; hon mem fac, Univ Chile, 57; mem sci adv comt, Mass Gen Hosp, 57-60; mem dept biol, Brookhaven Nat Lab, 59-62; mem staff, Jane Coffin Childs Fund Med Res, 61-63, Merck Inst Ther Res, 65-74 & Am Cancer Soc, 69-70. Honors & Awards: Nobel Prize in Med, 59; Neuberg Medal Biochem, 51; Price Award, Fr Soc Biol Chem, 55; Borden Award, Asn Am Med Cols, 58; NY Univ Medal, 60; Order of the Rising Sun 2nd Class & Gold Medal, Japan, 67; Carlos Jimenez Diaz Lectr Award, Univ Madrid, 69; Quevedo Gold Medal Award, Spain, 69; Albert Gallatin Medal, NY Univ, 70. Mem: Nat Acad Sci; Am Soc Biol Chemists (pres, 58); Harvey Soc (vpres, 52-53, pres, 53-54); fel Am Acad Arts & Sci; fel NY Acad Sci. Res: Biochemistry of muscle and fermentation; respiratory enzymes; enzymatic mechanisms of carbon dioxide assimilation, citric and fatty acid cycles; synthesis of nucleic acid and proteins, genetic code and translation of the genetic message. Mailing Add: Roche Inst Molecular Biol Nutley NJ 07110

OCHRYMOWYCZ, LEO ARTHUR, b Shaok, Ukraine, May 20, 43; US citizen; m 70. ORGANIC CHEMISTRY. Educ: St Mary's Col, Minn, BA, 65; Iowa State Univ, PhD(org chem), 69. Prof Exp: ASST PROF CHEM, UNIV WIS-EAU CLAIRE, 69- Concurrent Pos: Res fel, Iowa State Univ, 70-71; consult, Wis Dept Natural Resources, 70- Mem: Am Chem Soc. Res: Chemistry of macrocyclic polythioethers; coordination chemistry of post-transitional metals; general organo-sulfur chemistry, especially beta-keto sulfoxides and ylids; development of carbonyl synthesis reagents. Mailing Add: Dept of Chem Univ of Wis Eau Claire WI 54701

OCHS, SIDNEY, b Fall River, Mass, June 30, 24; m 49; c 3. NEUROPHYSIOLOGY, MEDICAL BIOPHYSICS. Educ: Univ Chicago, PhD, 52. Prof Exp: Res assoc, Ill Neuropsychiat Inst, 53-54; asst prof, Med Ctr, Univ Tex, 56-58; from assoc prof to prof physiol, 58-70, PROF PHYSIOL, SCH MED, IND UNIV, INDIANAPOLIS, 70-, DIR MED BIOPHYS PROG, 68- Concurrent Pos: Res fel, Calif Inst Technol, 54-56; NSF sr fel, Dept Biophys, Univ Col, Univ London, 63-64; ed, J Neurobiol, 68- Res: Functions of cerebral cortex; axoplasmic transport in nerve; muscle membrane properties. Mailing Add: 912 Forest Blvd N Dr Indianapolis IN 46240

OCHSNER, JOHN LOCKWOOD, b Madison, Wis, Feb 10, 27; m 54; c 4. SURGERY. Educ: Tulane Univ La, MD, 52; Am Bd Surg, dipl, 60; Am Bd Thoracic Surg, dipl, 60. Prof Exp: From intern to asst resident, Univ Mich Hosp, Ann Arbor, 52-54; resident, Baylor Univ Affil Hosp, Houston, Tex, 56-60; instr, Sch Med, Baylor Univ, 60-61; instr surg, 61-65, clin assoc prof, 65-69, CLIN PROF SURG, SCH MED, TULANE UNIV, 69-, CHMN DEPT SURG, OCHSNER CLIN, 66- Concurrent Pos: Chief surg res, Jefferson Davis Hosp, Houston, 58-59 & Tex Children's Hosp, Houston, 59-60; mem staff, Ochsner Clin, 61-66; chief surg, Ochsner Found Hosp, New Orleans; vis surgeon, Charity Hosp La & E A Conway Mem Hosp, Monroe; mem courtesy staff, Sara Mayo Hosp & Flint Goodridge Hosp, New Orleans;

consult cardiovasc surgeon, Lafayette Mem Hosp; consult thoracic surgeon, USPHS Hosp, New Orleans; consult heart surgeon, La Dept Health. Mem: Soc Vascular Surg; Soc Thoracic Surg; Am Col Surgeons; Am Col Chest Physicians; Int Cardiovasc Soc (asst secy-gen). Mailing Add: Ochsner Clin 1514 Jefferson Hwy New Orleans LA 70121

OCHSNER, SEYMOUR F, b Chicago, Ill, Nov 29, 15; m 45; c 3. RADIOLOGY. Educ: Dartmouth Col, AB, 37; Univ Pa, MD, 47. Prof Exp: Staff physician, Stony Wold Sanatorium, 47-49; intern, Johnston-Willis Hosp, 49-50; fel radiol, Ochsner Found Hosp, 50-53; assoc prof, 62-67, PROF CLIN RADIOL, SCH MED, TULANE UNIV, 67-; CONSULT, OCHSNER CLIN, 53- Honors & Awards: Distinguished Serv Award, Southern Med Asn, 71. Mem: Radiol Soc NAm (vpres, 65); Roentgen Ray Soc (vpres, 66); Am Col Radiol (pres, 72). Res: Clinical radiology and radiation therapy. Mailing Add: Ochsner Clin 1514 Jefferson Hwy New Orleans LA 70121

OCKEN, PAUL ROBERT, b New York, NY, July 11, 39; m 65; c 2. BIOCHEMISTRY, ORGANIC CHEMISTRY. Educ: Springfield Col, BS, 61; NY Univ, PhD(biochem), 67. Prof Exp: Res assoc carcinogenesis & intern environ med, Sch Med, NY Univ, 66-67; instr biochem, 67-68, ASST PROF BIOCHEM, COL MED & DENT NJ, 68- Mem: AAAS; Am Chem Soc. Res: Mechanisms of enzyme catalysis; environmental agents as inducers of cancer; hormone regulation and endogenous factors that influence host-tumor relationship. Mailing Add: Dept of Biochem Col of Med & Dent of NJ Newark NJ 07103

OCKERMAN, HERBERT W, b Chaplin, Ky, Jan 16, 32; m 55. FOOD CHEMISTRY, STATISTICS. Educ: Univ Ky, BS, 54, MS, 58; NC State Col, PhD(animal husb, statist), 62. Prof Exp: PROF ANIMAL SCI, OHIO STATE UNIV, 61- Mem: Am Meat Sci Asn; Am Soc Animal Sci; Inst Food Technol. Res: Lipids and antioxidants; food flavor and analysis; sterile tissue; tissue biochemistry and microbiology. Mailing Add: Dept of Animal Sci Ohio State Univ Columbus OH 43210

OCKERSE, RALPH, b Brussels, Belg, May 17, 33; US citizen; m 56; c 3. PLANT PHYSIOLOGY, BIOCHEMISTRY. Educ: State Teachers Col Neth, BA, 56; Baldwin Wallace Col, BS, 62; Yale Univ, PhD(plant physiol, biochem), 66. Prof Exp: Lab asst photosynthesis, Philips Res Labs, Neth, 53-55; res asst tissue culture, Chas Pfizer Co, NY, 57-58; biochemist, Union Carbide Co, Ohio, summer 62; from asst prof to assoc prof biol, 66-72, PROF BIOL, HOPE COL, 72- Concurrent Pos: NSF res grant, Hope Col, 69-70, dir NSF undergrad res partic grant, 71, 72, 73 & 75; res grant, Res Corp, 74-76. Mem: AAAS; Am Chem Soc; Am Soc Plant Physiol; Royal Neth Bot Soc; Japanese Soc Plant Physiologists. Res: Physiology and biochemistry of plant growth regulation; mechanism of hormone action and interactions with macromolecules; plant tissue culture and development. Mailing Add: Dept of Biol Hope Col Holland MI 49423

OCKERT, KARL FRANKLIN, fuel technology, organic chemistry, see 12th edition

OCKMAN, NATHAN, b New York, NY, Dec 29, 26. MOLECULAR BIOPHYSICS. Educ: Purdue Univ, BS, 49; Univ Calif, Berkeley, MA, 50; Univ Mich, PhD(physics), 57. Prof Exp: Fel chem, Harvard Univ, 57-58; mem tech staff, RCA Labs, Inc, 59-65; mem tech staff, Gen Tel & Electronics Labs, 65-69; FEL PHYSIOL, ALBERT EINSTEIN COL MED, 70- Mem: Am Phys Soc. Res: Infrared and Raman spectroscopy applied to the study of structures of biological films; ultraviolet and visible absorption and reflection spectroscopy applied to the study of phospholipid dispersions, monolayers and bilayers; membrane structure and function. Mailing Add: 137 Riverside Dr New York NY 10024

OCONE, LUKE RALPH, b Bridgeport, Conn, Mar 10, 25; m 53; c 4. INORGANIC CHEMISTRY. Educ: Brooklyn Polytech Inst, BS, 51; Pa State Univ, PhD(chem), 56. Prof Exp: Res chemist photoprod dept, E I du Pont de Nemours & Co, 55-58; GROUP LEADER EXPLOR CHEM, PENNWALT CORP, 58- Mem: AAAS; Am Chem Soc. Res: Metal complexes and metallorganics; photographic science and technology. Mailing Add: 8513 Widener Rd Philadelphia PA 19118

O'CONNELL, ALICE L, b Minneapolis, Minn, July 21, 07. ANATOMY, PHYSIOLOGY. Educ: Univ Minn, BS, 43, MEd, 48, PhD(anat), 56. Prof Exp: Instr high sch, Minn, 43-44; head dept health & phys ed, YWCA, 44-47; instr phys ed, Winthrop Col, 48-51 & Univ Wis, 55; asst prof anat, 56-62, assoc prof biomech & anat, 62-74, EMER PROF PHYSIOL, SARGENT COL, BOSTON UNIV, 74- Concurrent Pos: Ed, Kinesiol Quart, 65-69. Mem: AAAS; Am Asn Anat; fel Am Asn Health Phys Educ & Recreation; NY Acad Sci; Int Soc Electromyographic Kinesiology. Res: Electromyography in muscular coordination; biomechanics. Mailing Add: Dept of Sci Sargent Col Boston Univ University Rd Boston MA 02215

O'CONNELL, EDMOND J, JR, b Providence, RI, Apr 26, 39; m 65; c 5. ORGANIC CHEMISTRY, PHOTOCHEMISTRY. Educ: Providence Col, BS, 60; Yale Univ, PhD(chem), 64. Prof Exp: Res chemist radiation physics lab, E I du Pont de Nemours & Co, 64-67; from asst prof to assoc prof chem, 67-75, PROF CHEM, FAIRFIELD UNIV, 75- Concurrent Pos: Petrol Res Fund grant, 68-70 & 72-74; Res Corp grant, 72-73. Mem: Am Chem Soc. Res: Organic photochemistry, especially reaction mechanisms and the relation of photo-reactivity to molecular structure. Mailing Add: Dept of Chem Fairfield Univ Fairfield CT 06430

O'CONNELL, FRANK DENNIS, b Lynn, Mass, July 21, 27; m 48; c 3. PHARMACOGNOSY. Educ: Mass Col Pharm, BS, 51, MS, 53; Purdue Univ, PhD(pharmacog), 57. Prof Exp: Asst prof pharm, 57-58, from asst prof to assoc prof pharmacog, 58-69, actg dean, Sch Pharm, 72-73, PROF PHARMACOG, WVA UNIV, 69-, ASST DEAN, SCH PHARM, 74- Mem: Am Soc Pharmacog (treas, 73-76); Am Pharmaceut Asn. Res: Isolation and biosynthesis of natural medicinal products; biochemistry; plant tissue cultures; biochemical transformations. Mailing Add: Sch of Pharm WVa Univ Morgantown WV 26506

O'CONNELL, JAMES ANTHONY, b Brooklyn, NY, Jan 24, 32; m 46; c 2. PHYSICAL CHEMISTRY. Educ: St Peters Col, AB, 53; NY Univ, MS, 56, PhD(phys chem), 58. Prof Exp: Assoc chemist, Int Bus Mach Corp, 57-60; eng specialist, Res Labs, Gen Tel & Electronics Corp, NY, 60-66; mgr microelectronics, ITT Fed Labs, 66-68, dir component develop, 68-74, MEM STAFF, ITT DEFENSE COMMUN DIV, ITT CORP, 74- Mem: Inst Elec & Electronics Engrs. Res: Electrochemistry; electroluminescence; photologic device development; solid state device technology; illumination products; electrical-electronic components; materials technology. Mailing Add: ITT Defense Commun Div ITT Corp 492 River Rd Nutley NJ 07110

O'CONNELL, JESSE ELBERT, b Sanford, NC, July 21, 26. BOTANY. Educ: Wake Forest Col, BS, 48, MS, 49; Univ NC, PhD, 55. Prof Exp: Instr sci, Chowan Col, 49-50; instr bot, NC State Col, 50-52; asst prof, Univ Idaho, 55-59; sci educ specialist, 59-62, head, Tokyo Off, 62-66, prog dir, US-Japan & US-Repub China Coop Sci Progs, 66-71, PROG MGR, E ASIA & PAC PROGS, NSF, 71- Mem: AAAS; Bot Soc Am. Res: Plant geography. Mailing Add: Int Sci Activ NSF 1800 G St NW Washington DC 20550

O'CONNELL, JOHN JOSEPH, SR, b Natick, Mass, July 1, 29; m 58; c 4. MATERIALS SCIENCE. Educ: Clark Univ, BS, 52, MA, 58. Prof Exp: Res chemist, Monsanto Chem Co, WVa, 52-55, res chemist, Spec Projs Dept, Mass, 58-61, sr res chemist, Monsanto Res Corp, 61-64, res group leader inorg chem, 64-69; res group leader mat res, Am Hosp Supply Corp, Mass, 69-70, mgr mat res, Corp Technol Ctr, Calif, 70-75, DIR RES & DEVELOP, ORATEC/AM HOSP SUPPLY CORP, 75- Mem: Am Chem Soc. Res: Biomaterials research, metal coordination compounds as applied to catalyst and materials; dental materials. Mailing Add: Oratec/Am Hosp Supply Co 1441 McGaw Ave Santa Ana CA 92711

O'CONNELL, KEVIN MARSHALL, b Chicago, Ill, Sept 20, 47. CELL BIOLOGY, ENDOCRINOLOGY. Educ: St Procopius Col, BS, 69; Univ Ill, Champaign-Urbana, PhD(cell biol), 74. Prof Exp: ASST PROF ANAT, PHYSIOL & ENDOCRINOL, UNIV WIS-EAU CLAIRE, 73- Mem: Sigma Xi. Res: Analysis of the the mechanisms by which estrogen elicits increases in RNA polymerase activity in the immature rat uterus. Mailing Add: Dept of Biol Univ of Wis Eau Claire WI 54701

O'CONNELL, PAUL WILLIAM, b Newark, NY, Aug 5, 22; m 49; c 7. BIOCHEMISTRY. Educ: Univ Notre Dame, BS, 43; Univ Rochester, PhD(biochem), 49. Prof Exp: Fel chem, Univ Pittsburgh, 49-51; RES SECT HEAD, UPJOHN CO, 51- Mem: AAAS; Am Chem Soc. Res: Biochemistry of lipids; enzymes; information handling. Mailing Add: 1515 Kingston Ave Kalamazoo MI 49001

O'CONNELL, RICHARD LEE, b Utica, NY, Oct 30, 26; m 61; c 5. OCCUPATIONAL MEDICINE. Educ: Cornell Univ, AB, 47; Columbia Univ, MD, 51. Prof Exp: Intern, Genesee Hosp, Rochester, NY, 51-53; resident obstet & gynec, Presby Hosp, New York, 53-58; pvt pract, Mt Kisco & Peekskill, NY, 58-64; assoc dir & partner, Prev Med, Life Exten Inst, New York, 64-72; plant physician, 72-74, CORP MED DIR OCCUP MED, OLIN CORP, 74- Concurrent Pos: Attend physician, Dept Phys Educ, Univ Wis, 74 & Stamford Hosp, Conn, 75-; mem, Occup Health Comt, Mfg Chemists Asn & Med Exec, New York, 75- Mem: Am Col Physicians; Am Col Prev Med; Am Occup Med Asn; Am Acad Occup Med. Res: Continuing review of all medical disciplines as applied to occupational medicine; role of physical exercise in cardiovascular conditioning and prophylaxis. Mailing Add: Olin Corp 120 Long Ridge Rd Stamford CT 06904

O'CONNELL, ROBERT COLE, microbiology, see 12th edition

O'CONNELL, ROBERT F, b Athlone, Ireland, Apr 22, 33; m 63; c 3. THEORETICAL ASTROPHYSICS, THEORETICAL PHYSICS. Educ: Nat Univ Ireland, BSc, 53; Univ Notre Dame, PhD(physics), 62. Prof Exp: Asst lectr physics, Univ Col, Galway, 53-54; with telecommun br, Dept Posts & Tel, Ireland, 54-58; res assoc theoretic physics, Inst Advan Studies, Dublin, 62-64; from asst prof to assoc prof, 64-69, PROF PHYSICS, LA STATE UNIV, BATON ROUGE, 69- Concurrent Pos: Syst analyst, Int Bus Mach Corp, Ireland, 63-64; Nat Acad Sci-Nat Res Coun res assoc, NASA Inst Space Studies, NY, 66-67, sr res assoc, 67-; consult, Theoret Phys Div, Lawrence Livermore Lab, Univ Calif, Livermore, 73-75. Honors & Awards: Sir J J Larmor Prize, Univ Col, Galway, 54; Distinguished Res Master, La State Univ, 75. Mem: Fel Am Phys Soc; Am Astron Soc; Int Astron Union; Int Soc Gen Relativity & Gravitation. Res: High-energy astrophysics; gravitation; atomic physics. Mailing Add: Dept of Phys & Astron La State Univ Baton Rouge LA 70803

O'CONNELL, ROBERT JAMES, b Syracuse, NY, July 9, 37; m 60; c 2. NEUROPHYSIOLOGY. Educ: LeMoyne Col, NY, BS, 63; State Univ NY, PhD(physiol), 67. Prof Exp: USPHS res assoc sensory neurophysiol, Fla State Univ, 67-68; asst prof, 68-74, SR RES ASSOC SENSORY NEUROPHYSIOL, ROCKEFELLER UNIV, 74- Mem: AAAS; NY Acad Sci. Res: Electrophysiological studies of olfactory receptors; isolation and identification of various mammalian pheromones. Mailing Add: Dept of Neurophysiol Rockefeller Univ New York NY 10021

O'CONNOR, BRIAN LEE, b Lennox, Calif, Sept 17, 44; m 64; c 2. HUMAN BIOLOGY, PHYSICAL ANTHROPOLOGY. Educ: Univ Calif, Berkeley, AB, 69, PhD(anthrop), 74. Prof Exp: ASST PROF ANAT, SCH MED, IND UNIV, INDIANAPOLIS, 74- Mem: Am Asn Anatomists. Res: Articular neurology and its functional, evolutionary and pathological significance. Mailing Add: Dept of Anat Ind Univ Sch of Med Indianapolis IN 46202

O'CONNOR, BRIAN RUSS, organic chemistry, see 12th edition

O'CONNOR, CECILIAN LEONARD, b Philadelphia, Pa, Oct 23, 22. PHYSICS. Educ: Catholic Univ, BS, 45, MS, 51, PhD(physics), 54. Prof Exp: Instr physics, De La Salle Col, Catholic Univ, 50-54; assoc prof, 54-67, chmn dept, 60-74, PROF PHYSICS, MANHATTAN COL, 67- Mem: AAAS; Acoust Soc Am; Am Asn Physics Teachers; NY Acad Sci. Res: Thermodynamic properties of gases and liquids at ultrasonic and hypersonic frequencies. Mailing Add: Dept of Physics Manhattan Col Bronx NY 10471

O'CONNOR, CHARLES TIMOTHY, b Atlantic City, NJ, Aug 1, 30. MEDICAL ENTOMOLOGY. Educ: Rutgers Univ, BS, 53, MS, 55; Ohio State Univ, PhD, 58; Tulane Univ, MPH, 61. Prof Exp: Asst dept entom, NJ Agr Exp Sta, 53-55; asst, Ohio Agr Exp Sta, 55-58; med entomologist, USPHS, Tex, 59-60; malaria specialist, US AID, Vietnam, 61-63, Ethiopia, 64-66; specialist malaria eradication br, Ctr Dis Control, USPHS, Haiti, 66-70, Brazil, 70-71; MALARIA SPECIALIST, WHO, INDONESIA, 71- Mem: Entom Soc Am; Am Mosquito Control Asn; Sigma Xi. Res: Control of livestock pests and of vectors of human diseases; study of the entomological aspects of a large-scale malathion trail in Central Java; cytogenetic studies of several anopheline vectors of malaria. Mailing Add: 313 31st St S Brigantine NJ 08203

O'CONNOR, DAVID EVANS, b Ft Ogden, Fla, Apr 16, 32; m 53; c 2. FOOD CHEMISTRY. Educ: Univ Fla, BS, 54, PhD(chem), 61. Prof Exp: Res asst fluorine chem, Univ Fla, 57-58; res chemist, Procter & Gamble Co, 61-66; res chemist, Geigy Chem Corp, 66-67, group leader polymer chem, 67-68; res chemist, 68-73, SECT HEAD, PROCTER & GAMBLE CO, 73- Mem: Am Chem Soc. Res: Fluorine and food chemistry. Mailing Add: Procter & Gamble Co Miami Valley Labs PO Box 39175 Cincinnati OH 45247

O'CONNOR, DONALD JOSEPH, physical chemistry, see 12th edition

O'CONNOR, GEORGE ALBERT, b Seymour, Ind, Mar 30, 44; m 68. SOIL CHEMISTRY. Educ: Univ Mass, Amherst, BS, 66; Colo State Univ, MS, 68, PhD(agron), 70. Prof Exp: Asst agron, Colo State Univ, 66-70; asst prof, 70-75, ASSOC PROF AGRON, NMEX STATE UNIV, 75- Mem: Am Soc Agron; Soil Sci

Soc Am. Res: Salinity; pesticides. Mailing Add: Dept of Agron NMex State Univ Box 3Q Las Cruces NM 88001

O'CONNOR, GEORGE RICHARD, b Cincinnati, Ohio, Oct 8, 28. OPHTHALMOLOGY. Educ: Harvard Univ, AB, 50; Columbia Univ, MD, 54. Prof Exp: Asst clin prof ophthal, Univ, 62-66, consult ophthal clin, 62-70, from asst dir to assoc dir, Proctor Found, 62-70, assoc prof, 70-75, PROF OPHTHAMOL, UNIV CALIF, SAN FRANCISCO, 75-, DIR, FRANCIS I PROCTOR FOUND RES OPHTHAL, MED CTR, 70- Concurrent Pos: NIH spec trainee biochem, Inst biochem, Univ Uppsala, 60-61 & immunol, State Serum Inst, Copenhagen, Denmark, 61-62. Mem: Asn Res Vision & Ophthal; AMA. Res: Microbic immunology; immunologic response of human subjects to toxoplasma infections; antibody formation in the toxoplasma infected eye. Mailing Add: 95 Kirkham St San Francisco CA 94122

O'CONNOR, JAMES FRANCIS, b Providence, RI, Sept 17, 16; m 44; c 6. METEOROLOGY. Educ: RI State Col, BS, 37; RI Col Educ, BEd, 38; Mass Inst Technol, MS, 43. Prof Exp: Observer, US Weather Bur, RI, Mass & Washington, DC, 40-41, prognostic analyst, 46-51; instr meteorol, US Naval Acad, 42-43; sr instr meteorol, US Naval Postgrad Sch, 51-55; supv meteorologist, Weather Bur Anal Ctr, Md, 55-57, extended weather forecaster, Nat Meteorol Ctr, US Weather Bur, 57-64, chief extended forecast br, Nat Meteorol Ctr, Nat Weather Serv, 64-74, DEP CHIEF FORECAST DIV, NAT METEOROL CTR, NAT WEATHER SERV, 74- Honors & Awards: Dept Com Bronze Medal, 66, Silver Medal, 69; Outstanding Serv by a Weather Forecaster, Am Meteorol Soc, 76. Mem: Am Meteorol Soc. Res: Empirical studies of general circulation; extended range conditional weather probabilities; teleconnections. Mailing Add: World Weather Bldg 5200 Auth Rd Camp Springs MD 20233

O'CONNOR, JAMES J, b Appleton, Wis, Sept 9, 30; m 52; c 5. ORGANIC CHEMISTRY. Educ: Lawrence Univ, BS, 54; Inst Paper Chem, MS, 60, PhD(chem), 62. Prof Exp: Process control chemist, NY, 54-58, mgr pulp res, Neenah, Wis, 62-72, MGR NON-WOVEN TECHNOL DEVELOP, KIMBERLY-CLARK CORP, 72- Mem: Am Chem Soc; Tech Asn Pulp & Paper Indust. Res: Wood chemistry and morphology; organic synthesis; pulping, bleaching, electrochemical synthesis and degradation of organic compounds; non-woven technology. Mailing Add: 2207 S Gladys Ave Appleton WI 54911

O'CONNOR, JAMES MICHAEL, physical organic chemistry, polymer chemistry, see 12th edition

O'CONNOR, JEREMIAH JOSEPH, b Tralee, Ireland, July 26, 36; US citizen; m 61; c 3. ANIMAL NUTRITION. Educ: Nat Univ Ireland, BAgrSc, 60; Rutgers Univ, NB, MS, 63, PhD(animal nutrit), 69. Prof Exp: Biologist, Ciba Pharmaceut Co, 63-69; SR NUTRITIONIST, SQUIBB AGR RES CTR, E R SQUIBB & SONS, INC, 69- Mem: AAAS; Am Soc Animal Sci. Res: Study of chemical compounds as to their effect on growth and feed utilization in farm animals. Mailing Add: Squibb Agr Res Ctr Three Bridges NJ 08887

O'CONNOR, JOEL LESLIE, mathematics, see 12th edition

O'CONNOR, JOEL STURGENS, b Auburn, NY, Mar 6, 36; m 67; c 3. ECOLOGY, FISHERIES. Educ: Cornell Univ, BS, 58; Univ RI, PhD(oceanog), 65. Prof Exp: Chief comput opers br, Div Tech Info Exten, US AEC, 65-68; sr res assoc marine ecol, Biol Dept, Brookhaven Nat Lab, 68-71; res assoc marine sci res ctr, State Univ NY Stony Brook, 71-73; OCEANOGRAPHER, MESA PROJ, NAT OCEANIC & ATMOSPHERIC ADMIN, 73- Mem: AAAS; Am Fisheries Soc; Am Inst Biol Sci; Am Soc Limnol & Oceanog; Ecol Soc Am. Res: Sampling statistics and computer science; ecology of estuarine and coastal environments, with emphasis on effects of containcontaminants. Mailing Add: MESA Proj Off Old Biol Bldg State Univ NY Stony Brook NY 11794

O'CONNOR, JOHN DENNIS, b Chicago, Ill, Mar 20, 42; m 64; c 3. ZOOLOGY, BIOCHEMISTRY. Educ: Loyola Univ, Chicago, BS, 63; DePaul Univ, MS, 66; Northwestern Univ, Ill, PhD(biol), 68. Prof Exp: NIH fel, Mich State Univ, 68-70; asst prof zool, 70-74, ASSOC PROF ZOOL, UNIV CALIF, LOS ANGELES, 74- Concurrent Pos: Vis prof, Roman Cath Univ Nijmegen, Neth, 75-76. Mem: AAAS; Am Soc Zool; Soc Develop Biol. Res: Regulation of metabolism during crustacean molt cycle. Mailing Add: Dept of Biol Univ of Calif 405 Hilgard Ave Los Angeles CA 90024

O'CONNOR, JOHN FRANCIS, b Waterloo, NY, May 24, 35. BIOCHEMISTRY, ENDOCRINOLOGY. Educ: St John Fisher Col, BS, 60; Univ Rochester, PhD(org chem), 71. Prof Exp: Res fel, 69-74, INVESTR, INST STEROID RES, MONTEFIORE HOSP, 74- Concurrent Pos: Instr, Albert Einstein Col Med, 71- Mem: AAAS; Am Chem Soc; The Chem Soc. Res: Natural products; thyroid physiology; radioimmunoassay techniques. Mailing Add: Inst Steroid Res 111 E 210th St Montefiore Hosp & Med Ctr Bronx NY 10467

O'CONNOR, JOHN JOSEPH, b Bowling Green, Ky, May 7, 16; m 46; c 4. PHYSICS. Educ: Western Ky State Univ, BS, 39; Vanderbilt Univ, MS, 40. Prof Exp: Asst physics, Western Ky State Univ, 38-39; asst, Vanderbilt Univ, 39-40; asst, Ohio State Univ, 40-42; res physicist, Remington Arms Co, Inc, Conn, 42-44, res physicist, E I du Pont de Nemours & Co, Tenn, 44, Wash, 44-46, res physicist, Remington Arms Co, Inc, Div, 46-51, sr physicist, 51-61, res assoc, 61; PROJ ENGR, RCA SERV CO, 61- Mem: Am Phys Soc; Am Astronaut Soc; Am Astron Soc. Res: Artificial radioactivity; Faraday effect; exterior ballistics and accuracy of projectiles; orbital mechanics. Mailing Add: Mail Unit 645 RCA Serv Co Patrick AFB FL 32925

O'CONNOR, JOSEPH MICHAEL, b Newark, NJ, Oct 31, 25; m 49; c 2. ORGANIC CHEMISTRY. Educ: Seton Hall Univ, BS, 50; Stevens Inst Technol, MS, 55. Prof Exp: Jr chemist, Ciba Pharmaceut Co Div, 50-56, from asst chemist to chemist, 57-61, sr chemist, 61-72, SR SCIENTIST, PHARMACEUT DIV, CIBA-GEIGY CORP, 72- Concurrent Pos: NY Acad Sci fel, 69. Mem: Am Chem Soc; NY Acad Sci; Chromatograph Discussion Group, UK. Res: Chromatography; synthetic organic chemistry; catalytic hydrogenation; high pressure reactions; gas chromatography. Mailing Add: 1096 Overlook Terrace Union NJ 07083

O'CONNOR, MICHAEL GERALD, b Springfield, Mass, Aug 27, 07; m 46. FOOD TECHNOLOGY. Educ: St Lawrence Univ, BS, 30; Univ Mass, MS, 33, PhD(food technol), 42. Prof Exp: Food chemist, Gen Foods Corp, Mass, 30-31; chemist, H Porter Distilling Co, 33-35; inspector, Food & Drug Admin, USDA, 35; SUPV MILK INSPECTOR & FOOD CHEMIST, CITY HEALTH DEPT, SPRINGFIELD, 35- Mem: Am Pub Health Asn; Inst Food Technol. Res: Chemical and bacteriological food technology. Mailing Add: 59 Main St Wilbraham MA 01095

O'CONNOR, MICHAEL JOHN, b Philadelphia, Pa, Aug 31, 41; m 66; c 4. NEUROSURGERY. Educ: Univ Pa, BSEE, 63, MD, 67. Prof Exp: Intern surg, Univ Pa, 67-68, resident neurosurg, 68-70; fel EEG, NIH, 70-72; resident neurosurg, 72-74, ASST PROF NEUROSURG, UNIV PA, 74-, PROF NEUROSURG, GRAD HOSP, 75- Mem: Sigma Xi. Res: Relationship of functional activity, mainly electrical, and energy metabolism of the brain in pathologic states such as epilepsy and head trauma. Mailing Add: Div of Neurosurg Hosp of Univ of Pa Philadelphia PA 19104

O'CONNOR, MICHAEL L, b South Bend, Ind, Dec 4, 38; m 65. PATHOLOGY. Educ: Rockhurst Col, BS, 58; Univ Wis-Madison, MS, 60; Univ Kans, MD, 64. Prof Exp: Asst prof path, Case Western Reserve Univ, 69-71; ASST PROF PATH, UNIV IOWA, 72-, DIR CLIN LABS, UNIV HOSPS, 72- Mem: AMA; Col Am Path; Am Soc Clin Path; Am Chem Soc; Am Asn Clin Chem. Res: Computer applications in pathology. Mailing Add: Dept of Path Univ of Iowa Iowa City IA 52242

O'CONNOR, PAUL RADELL, inorganic chemistry, see 12th edition

O'CONNOR, ROBERT BARNARD, JR, b New York, NY, Nov 22, 25; m 49; c 2. PHYSICS. Educ: Princeton Univ, AB, 50; Rice Univ, MA, 72, PhD(physics), 75. Prof Exp: Geophysicist, Shell Oil Co, 50-62, div explor mgr, 62-63, dir explor res, Shell Develop Co, 63-66, area explor mgr, 66-67, chief geophysicist, Shell Oil Co, 67-69, on leave for doctorate study at Rice Univ, 69-75; sr res assoc, Mass Inst Technol, 74-75; CONSULT, MOBIL OIL CORP, 75- Mem: Soc Explor Geophys; Am Phys Soc. Res: Petroleum exploration methods. Mailing Add: 2 Cambelton Circle Princeton NJ 08540

O'CONNOR, ROBERT ERIC, b Halifax, NS, Mar 29, 07. MATHEMATICS. Educ: St Mary's Col, NS, BA, 26; Univ Toronto, MA, 35; Weston Col, STL, 40; Harvard Univ, PhD, 41. Prof Exp: Prof physics, Loyola Col, Montreal, 34-36; dean, Thomas More Inst Adult Educ, 45-62; PROF MATH, LOYOLA COL, MONTREAL, 43-; PRES & DIR STUDIES, THOMAS MORE INST ADULT EDUC, 62- Concurrent Pos: Secy assoc comt pure & appl math, Nat Res Coun Can, 54-63; vpres, Thomas More Inst Can Res Adult Lib Studies, 59- Mem: Am Math Soc; Can Math Cong. Res: Theory of numbers; quaternion congruence; quadratic and linear congruence; power products of two real numbers; philosophy; adult learning process. Mailing Add: Thomas More Inst 3421 Drummond St Montreal PQ Can

O'CONNOR, ROD, b Cape Girardeau, Mo, July 4, 34; m 55; c 4. ORGANIC BIOCHEMISTRY. Educ: Southeast Mo State Col, BS, 55; Univ Calif, PhD, 58. Prof Exp: Asst prof chem, Univ Omaha, 58-60; assoc prof, Mont State Univ, 60-66; assoc prof chem & dir gen chem, Kent State Univ, 66-67; staff assoc, Adv Coun Col Chem, Stanford Univ, 67-68; prof chem, Univ Ariz, 68-72; vis prof, Wash State Univ, 72-73; PROF CHEM, TEX A&M UNIV, 73- Concurrent Pos: NIH res grants, 61-66; consult, Hollister-Stier Labs, 64-66; Am Chem Soc vis scientist & tour speaker; mem, Col Chem Consult Serv; educ consult, Tucara-4 Media Resources, Inc; mem nat adv comt, Individualized Sci Instructional Syst, 72- Honors & Awards: Award, Am Chem Soc, 71. Mem: Fel AAAS; Am Chem Soc; fel Am Inst Chemists. Res: Multi-media instruction; auto-tutorial systems; chemistry of insect venoms. Mailing Add: Dept of Chem Tex A&M Univ College Station TX 77843

O'CONNOR, THOMAS LEE, b Holyoke, Mass, May 25, 26; m 50; c 5. COLLOID CHEMISTRY. Educ: Brown Univ, ScB, 50; Mass Inst Technol, PhD(chem), 53. Prof Exp: Asst, Mass Inst Technol, 50-53; sr chem investr raw mat develop lab, Am Cyanamid Co, 53-54, proj leader pigments div, 54-57; sr phys chemist, Ionics, Inc, 57-61; res chemist, Cabot Corp, 61-64, group leader, 64-65, res mgr, 65-69; VPRES, NYACOL, INC, 69- Mem: AAAS; Am Chem Soc. Res: Physical chemistry of pigments; colloid and surface chemistry. Mailing Add: Nyacol Inc Megunko Rd Ashland MA 01721

O'CONNOR, TIMOTHY EDMOND, b Cork, Ireland, Dec 5, 25; nat US; m 52; c 6. ORGANIC CHEMISTRY, BIOCHEMISTRY. Educ: Nat Univ Ireland, BSc, 47, MSc, 48, PhD(chem), 51. Prof Exp: Chemist, E I du Pont de Nemours & Co, Del, 52-60, res scientist, 60-61; chemist, Nat Cancer Inst, 63-67, head molecular virol sect, 66-72, assoc chief viral leukemia & lymphoma br, 67-72, dir molecular control prog, 72-75; DIR DIV BIOL & MED RES, ARGONNE NAT LAB, 75- Concurrent Pos: Fel, Mayo Found, Univ Minn, 50-51 & Univ Wis, 51-52; USPHS spec res fel, NIH, 61-63. Mem: AAAS; Am Chem Soc; Am Asn Cancer Res. Res: Organic nitrogen compounds; steroids; polymers; refractories; virology; oncology; biophysical characterization of viruses. Mailing Add: 1108 N West St Naperville IL 60540

O'CONNOR, WILLIAM BRIAN, b Brattleboro, Vt, Feb 24, 40; m 64; c 2. REPRODUCTIVE PHYSIOLOGY. Educ: St Michael's Col, Vt, BS, 62; Purdue Univ, MS, 66, PhD(zool), 68. Prof Exp: Asst prof zool, 67-73, ASSOC PROF ZOOL, UNIV MASS, AMHERST, 73- Mem: AAAS; Am Soc Zool. Res: Physiology and immunology of relaxin and other protein hormones. Mailing Add: Dept of Zool Univ of Mass Amherst MA 01002

O'CONOR, GREGORY THOMAS, b Cincinnati, Ohio, June 23, 24; m 44; c 7. PATHOLOGY, RESEARCH ADMINISTRATION. Educ: Cornell Univ, MD, 48. Prof Exp: Asst pathologist, St Francis Hosp, Hartford, Conn, 52-58; sr lectr path, Med Sch, Makerere Univ Col, Uganda, 58-60; sr investr, 60-68, res pathologist, 68-73, ASSOC DIR, NAT CANCER INST, 73- Concurrent Pos: Mem staff, Int Agency Res Cancer, WHO, Geneva, Switz, 66-67 & Lyons, France, 67-68. Mem: Am Soc Clin Path; Col Am Path; Electron Micros Soc Am; NY Acad Sci. Res: Etiology and pathogenesis of cancer; role of environmental factors; viruses. Mailing Add: 8824 Burning Tree Rd Bethesda MD 20034

O'CONOR, JOHN STANISLAUS, b New York, NY, Mar 6, 98. PHYSICS. Educ: Woodstock Col, Md, AB, 22, AM, 23; Mass Inst Technol, MS, 36. Prof Exp: Instr physics, Fordham Univ, 22-26; instr geol, Georgetown Univ, 32-33; assoc prof physics, Woodstock Col, Md, 33-43; prof, 43-73, EMER PROF PHYSICS, ST JOSEPH'S COL, PA, 73-, DIR COOP EDUC, 73- Mem: Am Phys Soc; Am Asn Physics Teachers. Res: Nuclear physics. Mailing Add: Dept of Physics St Joseph's Col Philadelphia PA 19131

O'CONOR, VINCENT JOHN, JR, b Chicago, Ill, Jan 10, 27; m 52; c 4. UROLOGY. Educ: Yale Univ, AB, 49; Northwestern Univ, MD, 53; Am Bd Urol, dipl, 62. Prof Exp: From intern to resident, Peter Bent Brigham Hosp, Boston, Mass, 53-58; chief resident, Chicago Wesley Mem Hosp, 58-59; from instr to assoc prof, 59-70, PROF UROL, SCH MED, NORTHWESTERN UNIV, 70-; CHMN DEPT UROL, CHICAGO WESLEY MEM HOSP, 63- Concurrent Pos: William Quinby fel urol, Harvard Med Sch, 57-68; attend urologist, Vet Admin Res Hosp & Rehab Inst, Chicago, 59- & Cook County Hosp, 66-; lectr, US Naval Hosp, 65- Mem: Fel Am Col Surg; Am Urol Asn; Soc Pelvic Surg; Am Asn Genito-Urinary Surg; Int Soc Urol Endocrinosurg. Res: Surgery of the kidney; renal hypertension and transplantation. Mailing Add: Dept of Urol Northwestern Univ Med Sch Chicago IL 60611

O'DAY, DANTON HARRY, b Vancouver, BC, Jan 31, 46; m 66; c 1. DEVELOPMENTAL BIOLOGY. Educ: Univ BC, BS, 67, MS, 69; Univ Del, PhD(develop biol), 72. Prof Exp: ASST PROF ZOOL, ERINDALE COL, UNIV TORONTO, 71- Mem: Can Soc Zoologists; Soc Protozoologists; Can Soc Cell Biol.

Res: The regulation and importance of intracellular and extracellular enzyme accumulation in eucaryotic development; the characterization and mode of action of sexual hormones in cellular slime molds. Mailing Add: Erindale Col Univ Toronto Mississauga ON Can

ODDIE, THOMAS HAROLD, b Ballarat, Australia, July 19, 11; m 35, 66; c 3. RADIOLOGICAL PHYSICS. Educ: Melbourne Univ, BSc, 32, MSc, 33, DSc(physics), 44. Prof Exp: Physicist, Commonwealth X-ray & Radium Lab, Univ Melbourne, 33-40; res engr, Philips Elec Indust, Australia, 40-47; officer chg radioisotopes res, Tracer Elements Invests, Commonwealth Sci & Indust Res Orgn, 47-52; asst prof radiol, Sch Med, Univ Ark, 53-55; assoc prof physics, Bowman Gray Sch Med, 55-56; res physicist, Unit Clin Invest, Royal NShore Hosp, Sydney, Australia, 56-60; from assoc prof to prof radiol, Med Ctr, Univ Ark, Little Rock, 64-74; RES FEL, HARBOR GEN HOSP, UNIV CALIF, 74- Concurrent Pos: NIH grants, 61-; consult med div, Oak Ridge Inst Nuclear Studies, 52-56 & 61-71. Mem: Soc Nuclear Med; Am Thyroid Asn; Am Asn Physicists Med; fel Brit Inst Physics; fel Australian Inst Physics. Res: X-ray and gamma ray measurements; clinical applications of radioisotopes; thyroid kinetics and metabolism; whole-body counting in clinical applications; computer diagnosis. Mailing Add: Univ Calif Harbor Gen Hosp 1000 W Carson St Torrance CA 90509

ODDIS, JOSEPH ANTHONY, b Greensburg, Pa, Nov 5, 28; m 54; c 2. PHARMACY. Educ: Duquesne Univ, BS, 50. Hon Degrees: DSc, Mass Col Pharm, 75, Philadelphia Col Pharm & Sci, 75. Prof Exp: Staff pharmacist, Mercy Hosp, Pittsburgh, Pa, 50-51, asst chief pharmacist, 53-54; chief pharmacist, Western Pa Hosp, 54-56; staff rep hosp pharm, Am Hosp Asn, Chicago, 56-60; EXEC DIR, AM SOC HOSP PHARMACISTS, 60-, SECY, RES & EDUC FOUND, 69- Concurrent Pos: Consult health facil planning & construct serv, Health Serv & Ment Health Admin, Dept Health, Educ & Welfare, 67-; mem vis comt, Col Pharm, Wayne State Univ, 70-; mem, Nat Adv Comt Allied Health Professions Projs, 69- Honors & Awards: Cert of Honor Award, Duquesne Univ Sch Pharm, 69; Whitney Award, Am Soc Hosp Pharmacists, 70; Julius Sturmer Mem Lectr, Philadelphia Col Pharm & Sci, 71. Mem: AAAS; Am Pharmaceut Asn; Am Soc Hosp Pharmacists; Int Pharmaceut Fedn; hon mem Can Soc Hosp Pharmacists. Res: Hospital pharmacy. Mailing Add: Am Soc of Hosp Pharmacists 4630 Montgomery Ave Bethesda MD 20014

ODDIS, LEROY, b Export, Pa, Aug 30, 31; m 54; c 3. ENDOCRINOLOGY. Educ: Utica Col, BA, 59; Rutgers Univ, MS, 63, PhD(zool), 64. Prof Exp: USPHS fel, 64; asst prof biol, 64-70, ASSOC PROF BIOL, RIDER COL, 70- Mem: AAAS; Brit Soc Endocrinol. Res: Skin-pigment cell interaction; hormonal control of pigmentation. Mailing Add: Dept of Biol Rider Col Trenton NJ 08602

ODDONE, PIERMARIA JORGE, b Arequipa, Peru, Mar 26, 44. HIGH ENERGY PHYSICS. Educ: Mass Inst Technol, BS, 65; Princeton Univ, PhD(physics), 70. Prof Exp: Res fel physics, Calif Inst Technol, 69-72; PHYSICIST, LAWRENCE BERKELEY LAB, 72- Concurrent Pos: Exp facil coordr, Positron-Electron Proj, Lawrence Berkeley Lab, 75- Mem: Am Phys Soc. Res: Elementary particle research in electron-positron annihilation. Mailing Add: 50A-2145 Lawrence Berkeley Lab Univ of Calif Berkeley CA 94720

ODDSON, JOHN KEITH, b Selkirk, Man, Nov 30, 35; m 60; c 3. MATHEMATICS. Educ: Univ Toronto, BASc, 57; Mass Inst Technol, SM, 60; Univ Md, PhD(appl math), 65. Prof Exp: Lectr math, Univ Waterloo, 57-58 & 60-62; Can NATO res fel partial differential equations, Univ Genoa, 65-66; res asst prof math, Inst Fluid Dynamics & Appl Math, Univ Md, 66-67; asst prof, 67-69, ASSOC PROF MATH, UNIV CALIF, RIVERSIDE, 69- Concurrent Pos: Vis prof, Inst Math, Univ Firenze, Florence, Italy, 73-74. Mem: Am Math Soc; Sigma Xi. Res: Partial differential equations. Mailing Add: Dept of Math Univ of Calif Riverside CA 92502

ODE, PHILIP E, b Decorah, Iowa, Mar 10, 35; m 61; c 3. POPULATION BIOLOGY, ENTOMOLOGY. Educ: Luther Col, BA, 57; Cornell Univ, MS, 63, PhD(entom), 65. Prof Exp: Asst prof biol, 65-69, chmn dept, 68-74, ASSOC PROF BIOL, THIEL COL, 69-, COORDR, ENVIRON PROG, 73- Mem: AAAS; Entom Soc Am; Soc Study Evolution; Am Inst Biol Sci; Sigma Xi. Res: Biology of Diptera; dispersal of insects and other organisms; animal behavior; population biology; evolution studies. Mailing Add: Dept of Biol Thiel Col Greenville PA 16125

ODEGARD, MARK ERIE, b Plentywood, Mont, Nov 1, 40; m 67. SEISMOLOGY, MARINE GEOPHYSICS. Educ: Univ Mont, BA, 62; Ore State Univ, MS, 65; Univ Hawaii, PhD(geol & geophys), 75. Prof Exp: Geophysicist, Alpine Geophys Assoc Inc, 65-67; res asst, 69-70, jr geophysicist, 70-74, ASST GEOPHYSICIST, HAWAII INST GEOPHYS, 74- Mem: Seismol Soc Am; Soc Explor Geophysicists; Am Geophys Union; Sigma Xi. Res: Structure of the upper mantle, seismic and ocean acoustic sound propagation, marine geophysics, seismic signal processing, geophysical instrumentation. Mailing Add: Hawaii Inst of Geophys 2525 Correa Rd Honolulu HI 96822

ODEH, FAROUK M, b Nablus, Palestine; US citizen. APPLIED MATHEMATICS. Educ: Cairo Univ, BS, 55; Univ Calif, Berkeley, PhD(appl math), 61. Prof Exp: RES STAFF MATH, WATSON RES CTR, IBM CORP, 61- Concurrent Pos: Temp mem math, Courant Inst Math Sci, NY Univ, 62-63; assoc prof, Am Univ Beirut, 67-68. Mem: Am Math Soc; Soc Indust & Appl Math; Math Asn Am. Res: Stability theory of difference schemes; bifurcation theory in mechanics. Mailing Add: IBM Watson Res Ctr Box 218 Yorktown Heights NY 10598

ODEH, ROBERT EUGENE, b Akron, Ohio, Dec 21, 30; m 58; c 3. MATHEMATICAL STATISTICS, COMPUTER SCIENCE. Educ: Carnegie Inst Technol, BS, 52, MS, 54, PhD(math), 62. Prof Exp: Programmer analogue comput, Goodyear Aircraft Corp, Ohio, 52-53; instr math, Carnegie Inst Technol, 58-59; from instr to asst prof, Univ Ore, 59-64; assoc prof, 64-71, PROF MATH, UNIV VICTORIA, 71- Concurrent Pos: Sr investr, Air Force Off Sci Res grant, 61-62; consult, Ore Res Inst, 62-64 & Attorney Gen Dept, 65-68. Mem: AAAS; Am Statist Asn; Inst Math Statist; fel Royal Statist Soc; Math Asn Am. Res: Computing; non-parametric c-sample rank-sum tests; transformations used in analysis of variance. Mailing Add: Dept of Math Univ of Victoria Victoria BC Can

O'DELL, BOYD LEE, b Hale, Mo, Oct 14, 16; m 44; c 2. BIOCHEMISTRY, NUTRITION. Educ: Univ Mo, AB, 40, PhD(biochem), 43. Prof Exp: Sr chemist, Parke, Davis & Co, 43-46; from asst prof to assoc prof, 46-55, PROF NUTRIT BIOCHEM, UNIV MO-COLUMBIA, 55- Concurrent Pos: NIH spec fel, Cambridge Univ, 64-65 & Harvard Med Sch, 72; Fulbright scholar, Commonwealth Sci & Indust Res Orgn, Australia, 73. Honors & Awards: Am Feed Mfr Res Award, 68; Sigma Xi Res Award, 69. Mem: Am Soc Biol Chemists; Am Chem Soc; Soc Exp Biol & Med. Res: Biochemical and physiological functions of micronutrients; vitamins and trace elements; role of nutrients in reproduction and connective tissue metabolism. Mailing Add: 332 Chem Bldg Univ of Mo Columbia MO 65201

O'DELL, CHARLES ROBERT, b Hamilton Co, Ill, Mar 16, 37; m 59; c 2. ASTRONOMY. Educ: Ill State Univ, BSEd, 59; Univ Wis, PhD(astron), 62. Prof Exp: Carnegie fel, Hale Observs, Calif, 62-63; asst prof astron, Univ Calif, Berkeley, 63-64; from asst prof to assoc prof astron, Univ Chicago, 64-67, prof & chmn dept, 67-72, dir, Yerkes Observ, 66-72; PROJ SCIENTIST LARGE SPACE TELESCOPE, NAT AERONAUT & SPACE ADMIN, 72- Mem: Am Astron Soc; Int Astron Union. Res: Physical processes in and evolution of planetary nebulae; diffuse nebulae; comets; characteristics of interstellar grains. Mailing Add: PF-05 Nat Aeronaut & Space Admin Marshall Space Flight Ctr AL 35812

ODELL, CLARENCE BURT, b Normal, Ill, Sept 3, 07; m 31; c 1. GEOGRAPHY, CARTOGRAPHY. Educ: Ill State Univ, BEd, 30; Univ Ill, MA, 31; Univ Chicago, PhD(geog), 37. Prof Exp: Assoc prof geog, Stephen F Austin State Univ, 37-39; instr, Univ Mo, 39-43, asst prof, 43-46; chief cartog dept, Encycl Britannica, 46-50; geog ed, Denoyer-Geppert Co, 50-51, managing ed, 51-70, vpres, 64-73; vpres, The Geogr, 70-73; RETIRED. Educ: Div asst, Off Geog, US Dept State, 42-43, chief pop sect, Div Geog & Cartog, 43-46; part time lectr, Northeastern Ill Univ, 67- Mem: Fel AAAS; Asn Am Geog; fel Am Geog Soc; Coun Geog Educ; fel Cong Surv & Mapping. Res: Geography of population; cartographic methods and reproduction; analysis of textual materials for maps. Mailing Add: 2410 Greenwood Ave Wilmette IL 60091

ODELL, DANIEL KEITH, b Auburn, NY, Nov 16, 45; m 69; c 3. VERTEBRATE ZOOLOGY. Educ: Cornell Univ, BS, 67; Univ Calif, Los Angeles, MA, 70, PhD(biol), 72. Prof Exp: ASST PROF MARINE BIOL, ROSENSTIEL SCH MARINE & ATMOSPHERIC SCI, UNIV MIAMI, 73- Mem: AAAS; Am Soc Mammalogists; Am Soc Zoologists; Wildlife Soc. Res: Ecology and physiology of marine mammals. Mailing Add: Rosenstiel Sch Marine & Atmospheric Sci Univ of Miami 4600 Rickenbacker Causeway Miami FL 33149

O'DELL, DURWARD GEORGE, b Liberty, SC, Apr 18, 21; m 46; c 2. ORGANIC CHEMISTRY. Educ: Clemson Univ, BS, 42; Purdue Univ, PhD(org chem), 50. Prof Exp: Res chemist synthetic fibers, 50-53, res supvr, 53-60, sr supvr, 60-63, res mgr, 63-64, tech supt, 64-65, prod supt, 65-69, asst plant mgr, Cape Fear Plant, 69-72, TECH MGR, E I DU PONT DE NEMOURS & CO, INC, 72- Mem: Am Chem Soc. Res: Synthetic textile fibers. Mailing Add: 208 N Spring Valley Rd Wilmington DE 19807

ODELL, FLOYD ADAMS, b McMinnville, Ore, Jan 22, 11; m 35. BIOPHYSICS. Educ: Linfield Col, BA, 32; Yale Univ, PhD(biol), 40. Prof Exp: Asst chemist, Linfield Col, 30-32; asst biol, Yale Univ, 32-37; instr & head sci dept, Hamden Hall Sch, 36-42; actg chief biophys sect, Med Div, Army Chem Ctr, 46-48, asst chief biophys div, 48-56; tech dir res, US Army Med Res Lab, 56-62; dir biomed sci & mgr, Wash Off, Field Emission Corp, 62-68; ASSOC DIR BALLISTIC RES LABS, ABERDEEN PROVING GROUND, 68- Concurrent Pos: Mem panel aviation med, Joint Res & Develop Bd, 44. Res: Wounding by high speed missiles; experimental embryology; aviation and space medicine; body armor terminal ballistics; wound shock; pulsed radiation. Mailing Add: Ballistic Res Labs Aberdeen Proving Ground MD 21005

ODELL, GEORGE BERLAGE, b London, Eng, Nov 20, 25; nat US; m 48; c 4. PEDIATRICS. Educ: NY Univ, BA, 47; Yale Univ, MD, 51. Prof Exp: From intern to asst resident, Yale Univ, 51-53; Commonwealth Fund fel exp med, Cambridge Univ, 53-54; Nat Found Infantile Paralysis fel, Yale Univ, 54; from instr to assoc prof, 56-71, PROF PEDIAT, SCH MED, JOHNS HOPKINS UNIV, 71- Concurrent Pos: Markle Found scholar, 59- Mem: Am Soc Pediat Res; Am Pediat Soc; Am Fedn Clin Invest. Res: Physiology of the newborn. Mailing Add: Dept of Pediat Johns Hopkins Univ Sch of Med Baltimore MD 21205

O'DELL, JEAN MARLAND, b Independence, Mo, June 10, 31; m 56; c 5. NUCLEAR SCIENCE, OPERATIONS RESEARCH. Educ: Univ Kans, BSEP, 54, MS, 61, PhD(physics), 65. Prof Exp: Asst physics, Univ Kans, 58-62, asst nuclear physics, 62-65; exp physicist, 65-73, GROUP LEADER, LAWRENCE LIVERMORE LAB, UNIV CALIF, 73- Res: Nuclear explosives for defense and peaceful uses; compound nucleus formation versus direct interaction investigations bombarding sulphur 32 with deuterons; evaluation of requirements for nuclear weapons for strategic and naval tactical weapon systems. Mailing Add: Lawrence Livermore Lab PO Box 808 Livermore CA 94551

ODELL, LOIS DOROTHEA, b Watertown, NY, Sept 25, 15. BIOLOGY, SCIENCE EDUCATION. Educ: State Univ NY Albany, AB, 40; Cornell Univ, MA, 45, PhD(sci educ), 51. Prof Exp: Teacher pub sch, NY, 41-43; instr biol, 47-62, PROF BIOL, TOWSON STATE COL, 62- Mem: AAAS; Am Nature Study Soc. Res: Field natural science; botany; science education in the elementary and high school; science education in environmental concepts. Mailing Add: Dept of Biol Towson State Col Baltimore MD 21204

ODELL, NORMAN RAYMOND, b Rochester, NY, Aug 4, 27; m 50; c 3. ORGANIC CHEMISTRY. Educ: Whittier Col, BA, 50; Ore State Col, MS, 52, PhD(chem), 55. Prof Exp: Chemist, Taylor Instrument Co, 52; chemist grease res, 54-59, group leader, 59-60, asst supvr, Lubricants Res Sect, 60-65, res technologist-managerial, 65-67, supvr, Port Arthur Res Lab, 67-73, SUPVR, REFINING & FIELD SERV SECT, PORT ARTHUR RES LAB, TEXACO, INC, 73- Concurrent Pos: Instr, Dutchess Community Col, 59-60. Mem: AAAS; Am Chem Soc; fel Am Inst Chem; Sigma Xi. Res: Synthesis of tracer compounds; tracer studies; exploratory research in grease thickening agents; synthetic lubricants; petroleum additives; mechanisms of petroleum additive behavior; additive processing; automotive and industrial lubricants; lubricant application; lubricating oil processing. Mailing Add: Texaco Inc Port Arthur Res Labs PO Box 1608 Port Arthur TX 77640

ODELL, PATRICK L, b Watonga, Okla, Nov 29, 30; m 58; c 4. MATHEMATICAL STATISTICS. Educ: Univ Tex, BS, 52; Okla State Univ, MS, 58, PhD(math statist), 62. Prof Exp: Mathematician, Flight Determination Lab, White Sands Missile Range, NMex, 52-53; res scientist, Kaman Nuclear, Inc, 58-59; mathematician, US Navy Nuclear Ord Eval Unit, 59-60, consult mathematician, US Weapons Eval Facility, 60-62 & Ling-Tempco-Vought, Inc, Tex, 62-64; asst prof math, Univ Tex, 62-66; prof, Tex Tech Univ, 66-72; exec dean grad studies & res, 72-75, PROF MATH SCI, UNIV TEX, DALLAS, 72- Concurrent Pos: NSF grant, 63-65, res grant, 75-; assoc dir, Tex Ctr Res Appl Math & Mech, 64-72; Tex Hwy Res Ctr grant; NASA grant, 70-71, res grant, 72-; training grant, Environ Protection Agency, 75- Mem: Soc Indust & Appl Math; fel Am Statist Asn. Res: Statistical problems associated with remote sensing from space; statistical problems; statistical analysis of sensitivity data; general theory of matrix inversion; mathematical modelling of the environment. Mailing Add: 3105 Canyon Creek Dr Richardson TX 75080

O'DELL, ROGER GENE, b Clinchco, Va, May 14, 36; m 62; c 2. NUTRITIONAL BIOCHEMISTRY. Educ: ETenn State Univ, BS, 59. Prof Exp: Lab technician, US Army Lab, Ft Myer, Va, 60-62; RES BIOLOGIST, US FOOD & DRUG ADMIN, WASHINGTON, DC, 62- Mem: AAAS; Am Inst Biol Sci. Res: Studies of the chick edema disease; factors affecting protein synthesis; carbohydrate metabolism; clinical chemistry; nutritional requirements for the infant and aged, especially dietary protein and brain development; nutritional aspects of dietary fiber. Mailing Add: Food & Drug Admin 200 C St SW Div of Nutrit HFF-268 Washington DC 20204

ODELL, RUSSELL TURNER, soils, see 12th edition

ODELL, THEODORE TELLEFSEN, JR, b Geneva, NY, May 5, 23; m 48, 72; c 3. CELL BIOLOGY, PHYSIOLOGY. Educ: Hobart Col, BS, 45; Ind Univ, MA, 50, PhD(zool), 52. Prof Exp: Asst zool, Ind Univ, 48-52; assoc biologist, 52-58, biologist, 58-75, ASSOC MGR, CANCER & TOXICOL PROG, BIOL DIV, OAK RIDGE NAT LAB, 75- Concurrent Pos: Lectr, Dept Zool & Biomed Grad Sch, Univ Tenn; lectr, Knoxville Col. Mem: AAAS; Radiation Res Soc; Am Physiol Soc; Soc Exp Biol & Med; Am Soc Hemat. Res: Physiology of blood platelets; hemopoietic cell kinetics and regulation; radiobiology. Mailing Add: Biol Div Oak Ridge Nat Lab Oak Ridge TN 37830

O'DELL, THOMAS BENIAH, b Cassopolis, Mich, June 19, 20; m 43; c 3. PHARMACOLOGY. Educ: Wabash Col, AB, 42; Univ Minn, PhD(pharmacol), 50. Prof Exp: Chemist coated fabrics, US Rubber Co, 42-44; anal chemist nutrit, Upjohn Co, 44; asst prof pharmacol, Sch Med, Univ Minn, 49-51; sr res pharmacologist, Irwin, Neisler Labs, Ill, 51-60, dir biol res, 60-64; assoc dir med & sci coordr, William S Merrell Co, 64-68, DIR DRUG REGULATORY AFFAIRS, MERRELL-NAT LABS, 68- Mem: AAAS; Am Soc Pharmacol & Exp Therapeut; Soc Toxicol; NY Acad Sci. Res: Virus chemotherapy; hypertension; atherosclerosis; antispasmodics; analgesics; central nervous system. Mailing Add: 9 Jewel Ln Cincinnati OH 45218

O'DELL, WAYNE TALMAGE, b Easley, SC, Feb 3, 23; m 46; c 2. DAIRY SCIENCE. Educ: Clemson Univ, BS, 48, MS, 51; Pa State Univ, PhD(dairy sci), 58. Prof Exp: Asst dairy sci, Clemson Univ, 48, supt official testing, 48-49, asst prof dairy sci, 51-53; instr dairy physiol, Pa State Univ, 53-58, asst prof dairy sci, 58-63; leader dairy sci exten work, 63-65, assoc dir exten serv, 65-69, DIR EXTEN SERV, SCH AGR, CLEMSON UNIV, 69-, PROF DAIRY SCI, 73- Res: Milk testing and evaluation; breeding; reproductive physiology and genetics. Mailing Add: Clemson Univ Sch of Agr Clemson SC 29631

ODELL, WILLIAM DOUGLAS, b Oakland, Calif, June 11, 29; m 50; c 5. ENDOCRINOLOGY, PHYSIOLOGY. Educ: Univ Calif, Berkeley, AB, 52; Univ Chicago, MS & MD, 56; George Washington Univ, PhD(physiol), 65; Am Bd Internal Med, dipl; Am Bd Endocrinol & Metab, dipl. Prof Exp: Intern, King County Hosp, Seattle, Wash, 56-57; resident med, Univ Wash Hosp, 58-60; sr investr & attend physician, endocrinol br, Nat Cancer Inst, 60-65, head endocrinol serv, Endocrinol & Metab Br, Nat Inst Child Health & Human Develop, 66; assoc prof med & physiol, 66-68, PROF MED & PHYSIOL, SCH MED, UNIV CALIF, LOS ANGELES, 66-, CHMN DEPT MED, 72- Concurrent Pos: Fel endocrinol & metab, Univ Wash, 57-58; consult, US Naval Dent Sch, 62-64 & DC Gen Hosp, 63-65; assoc prof lectr, George Washington Univ, 65-66; mem adv comt on training, Nat Inst Arthritis & Metab Dis, 70-74; mem diag res adv group, Nat Cancer Inst, 74; consult, WHO, 74- Mem: Am Fedn Clin Res; Endocrine Soc; fel Am Col Physicians; Am Soc Clin Invest; Am Physiol Soc. Res: Physiology of protein and polypeptide hormones; pathophysiology of endocrine disorders in man, reproductive physiology. Mailing Add: Harbor Gen Hosp Endocrinol Div 1000 W Carson St Torrance CA 90509

O'DELL-SMITH, ROBERTA MAXINE, b Bradford, Pa, May 5, 30; m 66. PHYSIOLOGY. Educ: Pa State Univ, BS, 51, MS, 55; Duke Univ, PhD(zool), 59. Prof Exp: Assoc biochem res, Valley Forge Heart Inst, 52-53; asst, Pa State Univ, 53-55; asst, Duke Univ, 55-59, NC Heart Asn fel, 59-60; ASST PROF PHYSIOL, SCH MED, TULANE UNIV, 60- Mem: Am Physiol Soc; NY Acad Sci. Res: Renal physiology, particularly the handling of urea by mammalian kidney. Mailing Add: Dept of Physiol Tulane Univ Sch of Med New Orleans LA 70112

ODENCRANTZ, FREDERICK KIRK, b New York, NY, Oct 6, 21; m 59; c 3. PHYSICS. Educ: Muhlenberg Col, BS, 43; Rutgers Univ, MS, 49; Univ Utah, PhD(physics), 58. Prof Exp: Physicist, New Brunswick Lab, US Atomic Energy Comn, 49-50; PHYSICIST, US NAVAL WEAPONS CTR, 50- Mem: Am Phys Soc; Am Geophys Union. Res: Physical optics; atmospheric physics. Mailing Add: US Naval Weapons Ctr 209-B Wasp Rd China Lake CA 93555

ODENHEIMER, KURT JOHN SIGMUND, b Regensburg, Ger, May 9, 11; US citizen; m 39; c 4. CLINICAL PATHOLOGY, EXPERIMENTAL PATHOLOGY. Educ: Univ Munich, MedDent, 35; Univ Pittsburgh, DDS, 40, MEd, 54; Western Reserve Univ, PhD(path), 64. Prof Exp: Asst prof gen path, Univ Pittsburgh, 49-53, head dept, 53-55; assoc prof oral diag & clin path, State Univ NY Buffalo, 61-66; prof path & chief exp path, Loyola Univ, 66-69, chmn diag & roentgenol, 67-69; PROF PATH, MED CTR, LA STATE UNIV, NEW ORLEANS, 69-, PROF ENTOMOL, 71- Concurrent Pos: Res fel path & fed teachers training grant, Western Reserve Univ, 59-61; res grant viral studies primates, Loyola Univ, 67-; consult, Presby Hosp Tumor Bd, Pittsburgh, 51-55, temporomandibular joint disturbances, NIH, 61, oral diag & path, Meyer Mem Hosp, Buffalo, 63-66, new ed Oral Path by K Thomas, 66 & Gulf South Res Inst, La, 67-; vis res assoc, Tulane Delta Primate Res Ctr, 66-; consult dent med, Touro Infirmary, New Orleans, 68-; dent coordr oncol, Charity Hosp. Mem: AAAS; Am Dent Asn; AMA; fel Am Acad Oral Path; fel Am Col Dent. Res: Peridontal and temporomandibular joint therapy with myotatic splint; effects of enteric viruses upon embryo, child and adult, in acute or chronic, clinical or subclinical forms, using primates as experimental models; radioactive gold and oral cancers; viral infections in genetics. Mailing Add: Sch of Dent La State Univ Med Ctr New Orleans LA 70119

ODENSE, PAUL HOLGER, b Winnipeg, Man, Dec 12, 26; m 54; c 3. BIOCHEMISTRY. Educ: Univ Toronto, BA, 50, MA, 54; Univ Okla, PhD(biochem), 59. Prof Exp: Asst scientist & biochemist, 54-56, assoc scientist, 59-62, SR SCIENTIST, FISHERIES RES BD CAN, 62-, BIOCHEMIST, 59- Mem: Can Biochem Soc; Chem Inst Can; NY Acad Sci. Res: Marine comparative biochemistry and histology; isoenzymes; electrophoretic separations; light and electron microscope histochemistry. Mailing Add: Dept of the Environ Halifax Lab PO Box 429 Halifax NS Can

ODER, FREDERIC CARL EMIL, b Los Angeles, Calif, Oct 23, 19; m 41; c 3. AEROSPACE SCIENCES. Educ: Calif Inst Technol, BS, 40, MS, 41; Univ Calif, Los Angeles, PhD(atmospheric physics), 52. Prof Exp: Dir geophys res, Air Force Cambridge Res Labs, 49-52, dir weapons syst ballistic missiles div, Air Res & Develop Command, 56-59, dep comdr space systs, 59-60; asst to dir res & eng, Apparatus & Optical Div, Eastman Kodak Co, NY, 60-61, prog mgr, 61-66; vpres & asst gen mgr, Lockheed Missiles & Space Co, 66-73, VPRES & GEN MGR, SPACE SYSTS DIV, LOCKHEED MISSILES & SPACE CO, INC, LOCKHEED AIRCRAFT CORP, 73- Concurrent Pos: Consult, Air Force Studies Bd, Nat Acad Sci, 75-; mem, Defense Intel Agency, Sci Adv Comt, 72- Mem: Soc Photog Sci & Eng; Sigma Xi; fel Am Inst Aeronaut & Astronaut. Res: Specialized optical and photographic systems; atmospheric geophysics. Mailing Add: Space Systs Div PO Box 504 Lockheed Missiles & Space Co Inc Sunnyvale CA 94088

ODERR, CHARLES PATCHIN, b Kansas City, Mo, Feb 26, 08; m 38; c 2. RADIOLOGY. Educ: Wash Univ, MD, 33. Prof Exp: Assoc radiologist, St Vincents Hosp, New York, 35-50; assoc prof radiol, Sch Med, La State Univ, 50-55; assoc prof, Sch Med, Tulane Univ, 55-69; chief, Vet Admin Hosp, 50-69. Mem: Fel Am Col Radiol; Am Thoracic Soc; Am Heart Asn; fel Am Col Chest Physicians; NY Acad Med. Res: Clinical diagnostic roentgenology; normal and pathological and microradiology of lungs; fluorodensimetry of heart and lungs. Mailing Add: Oak Ridge Hosp Oak Ridge TN 37830

ODIAN, ALLEN C, high energy physics, see 12th edition

ODIAN, GEORGE G, b New York, NY, July 19, 33; m 65. POLYMER CHEMISTRY. Educ: City Col New York, BS, 55; Columbia Univ, MA, 56, PhD, 59. Prof Exp: Asst, Columbia Univ, 55-57; sr chemist, Thiokol Chem Corp, 58-59; res dir, Radiation Applns, Inc, 59-68; assoc prof, 68-72, PROF CHEM, RICHMOND COL, NY, 72- Concurrent Pos: Asst prof, Columbia Univ, 63-68. Mem: AAAS; Am Chem Soc. Res: Polymer science; radiation chemistry. Mailing Add: Sci Div Richmond Col Staten Island NY 10301

ODIORNE, TRUMAN J, b Johnson City, Tex, Aug 9, 44; m 70. PHYSICAL CHEMISTRY. Educ: Southwestern Univ, Tex, BS, 66; Rice Univ, MA & PhD(chem), 71. Prof Exp: NIH trainee, Inst Lipid Res, Baylor Col Med, 71-73; staff chemist, Res Triangle Inst, 73-74; RES CHEMIST, INSTRUMENT PROD, E I DU PONT DE NEMOURS & CO, INC, 74- Mem: AAAS; Am Chem Soc; Am Phys Soc. Res: Chemical reaction kinetics; organic mass spectrometry; electron spectroscopy chemical analysis; analytical instrumentation. Mailing Add: Instrument Prod Div E I du Pont de Nemours & Co Inc Monrovia CA 91016

ODIOSO, RAYMOND C, b Pittsburgh, Pa, Apr 17, 23; m 53; c 6. ORGANIC CHEMISTRY. Educ: Duquesne Univ, BS, 47; Carnegie Inst Technol, MS, 50, DSc(chem), 51. Prof Exp: Asst, Carnegie Inst Technol, 47-49; fel, Mellon Inst, 51-54; res supvr, Gulf Res & Develop Co, 54-61; res mgr, Colgate-Palmolive Co, NJ, 61-67, assoc dir res lab opers, 67-68; V PRES RES & DEVELOP, DRACKETT CO, BRISTOL-MYERS CO, 68- Concurrent Pos: Instr, Carnegie Inst Technol, 52-53. Mem: AAAS; Am Chem Soc. Res: Kinetics of the benzidine rearrangement; monomer synthesis; alkylation; isomerization; dealkylation; dehydrocyclization; olefin reactions; product development, laundry and dishwashing detergents, bleaches, toilet soap, aerosols, paper products; process research, petrochemicals, and detergents. Mailing Add: Drackett Res & Develop Labs 5020 Spring Grove Ave Cincinnati OH 45232

ODISHAW, HUGH, b North Battleford, Sask, Oct 13, 16; nat US; m 58. GEOPHYSICS. Educ: Northwestern Univ, AB, 39, MA, 41; Ill Inst Technol, BS, 44. Hon Degrees: ScD, Carleton Col, 58. Prof Exp: Instr eng, Ill Inst Technol, 41-44, instr math, 44; tech ed & instr radar, Westinghouse Elec Corp, 44-46; asst to dir, Nat Bur Standards, 46-54; exec dir US Comt, Int Geophys Year, Nat Acad Sci-Nat Res Coun, 54-63, dir int geophys year world data ctr, Nat Acad Sci, 57-66, exec dir, Space Sci Bd, 58-72, exec secy, Div Phys Sci, 66-72; DEAN COL EARTH SCI, UNIV ARIZ, 72- Concurrent Pos: Consult, Nat Bur Standards, 54-57 & Electronics Div, ACF Instruments Co, 54-56. Mem: AAAS; Am Phys Soc; Am Geophys Union; Royal Soc Arts. Res: Electronics. Mailing Add: Col of Earth Sci Univ of Ariz Tucson AZ 85721

ODLAND, GEORGE FISHER, b Minneapolis, Minn, Aug 27, 22; m 45; c 3. ANATOMY, DERMATOLOGY. Educ: Harvard Med Sch, MD, 46. Prof Exp: From intern to resident med, Mass Gen Hosp, Boston, 46-55; asst dermat, Harvard Med Sch, 53-55; clin instr anat, 55-60, clin instr med, 56-60, clin asst prof anat & med, 60-62, from asst prof to assoc prof biol struct & med, 62-69, PROF BIOL STRUCT & MED, SCH MED, UNIV WASH, 69-, HEAD DIV DERMAT, 62- Concurrent Pos: Res fel anat, Harvard Med Sch, 49-51, res fel dermat, 51-53; chief dermat consult, Vet Admin Hosp, Seattle; attend, Univ Hosp, Seattle. Mem: Am Acad Dermat; Soc Invest Dermat; Am Dermat Asn. Res: Electron microscopy; biosynthesis of epidermal keratins; structure/physiologic correlation in microcirculation. Mailing Add: Dept of Med Univ of Wash Sch of Med Seattle WA 98195

ODLAND, JOHN, b Mar 26, 43; US citizen. URBAN GEOGRAPHY. Educ: Ore State Univ, BS, 66, MS, 67; Ohio State Univ, PhD(geog), 72. Prof Exp: ASST PROF GEOG, IND UNIV, BLOOMINGTON, 70- Mem: Asn Am Geogrs; Regional Sci Asn; Am Statist Asn. Res: Urban land use, urban growth and economic geography. Mailing Add: Dept of Geog Ind Univ Bloomington IN 47401

ODLAND, LURA MAE, b Morgantown, WVa, Nov 2, 21. NUTRITION. Educ: Univ RI, BS, 43, DSc, 68; Univ Conn, MS, 45; Univ Wis, PhD(biochem, nutrit), 50. Prof Exp: Instr foods & nutrit, Univ Conn, 45; nutritionist comt food composition, Food & Nutrit Bd, Nat Res Coun, 45-47; asst, Univ Wis, 47-50; assoc prof home econ, Exp Sta, Mont State Col, 50-55; exp stas adminr, State Exp Stas Div, USDA, 55-59, DEAN COL HOME ECON, UNIV TENN, KNOXVILLE, 59- Concurrent Pos: Mem comn home econ, Nat Asn State Univ & Land-Grant Cols, 74-77. Mem: AAAS; Am Chem Soc; Am Pub Health Asn; Asn Adminr Home Econ (pres, 73-75); Inst Food Technol. Res: Conservation of nutritive values of food and food composition; human nutrition; vitamin metabolism and nutritive assessment; nine year resurvey of bone density and dietary intake for seventy Tennessee subjects. Mailing Add: Col of Home Econ Univ of Tenn Knoxville TN 37916

ODLAND, RUSSELL KENT, b Washington, DC, June 29, 41; m 65; c 2. ANALYTICAL CHEMISTRY. Educ: Univ Richmond, BS, 64, MS, 71. Prof Exp: Res chemist combustion prod anal, Am Tobacco Co, Subsid Am Brands, 64-71; SR RES CHEMIST, METHODS DEVELOP, JIM WALTER RES CORP, SUBSID JIM WALTER CORP, 71- Concurrent Pos: Chem consult, K O Labs, 73- Mem: Am Chem Soc; Am Inst Chemists. Res: Analytical development of new test methods for solving practical problems found in industry. Mailing Add: 3100 61st Lane N St Petersburg FL 33710

ODLAUG, THERON OSWALD, b Stanley, Wis, Feb 23, 11; m 38; c 2. PARASITOLOGY. Educ: Luther Col, AB, 33; NY Univ, MSc, 35, PhD(parasitol), 40. Prof Exp: Prof biol, Billings Polytech, 38-39; assoc prof zool, Nebr State Teachers Col, Peru, 39-43; aquatic biologist, US Fish & Wildlife Serv, Wash, 43-45; assoc prof zool, Duluth Br, Univ Minn, 45-52; Muellhaupt fel biol, Ohio State Univ, 52-53; PROF BIOL & HEAD DEPT, UNIV MINN, DULUTH, 54- Mem: Assoc Am Micros Soc; assoc Am Soc Parasitol. Res: Oyster physiology; trematode histology and life histories; plankton ecology and limnology of western Lake Superior. Mailing Add: Dept of Biol Univ of Minn Duluth MN 55812

ODLE, JOHN WILLIAM, b Tipton, Ind, July 23, 14; m 37; c 3. MATHEMATICS. Educ: Univ Mich, BS, 37, MS, 38, PhD(math), 40. Prof Exp: Instr math, Univ Wis, 40-42; asst prof, Pa State Univ, 42-44; opers analyst, US Army Air Force, 44-45; head math div, US Naval Ord Test Sta, 46-55; opers analyst, US Air Force, 55-57; res mathematician eng res inst, Univ Mich, 57-58; dir adv develop dept, Crosley Div, Avco Corp, 58-60; mem staff opers res sect, Arthur D Little, Inc, 60-71; OPERS RES ANALYST, NAVAL ORD LAB, 72- Mem: Oper Res Soc Am. Res: Applied mathematics; operations research. Mailing Add: 9208 Friars Dr Bethesda MD 20034

O'DOHERTY, DESMOND SYLVESTER, b Dublin, Ireland, July 27, 20; m 51; c 2. NEUROLOGY. Educ: LaSalle Col, AB, 42; Jefferson Med Col, MD, 45. Prof Exp: Adj instr, DC Gen Hosp, 50-51; res fel, Med Ctr, 51-52, from instr to assoc prof, 52-61, PROF NEUROL, GEORGETOWN UNIV, 61-, CHMN DEPT, 59- Concurrent Pos: Dir, Muscular Dystrophy Clin, 54-; med dir, Georgetown Hosp, 66-67; consult, US Army, US Navy & Vet Admin. Mem: Asn Res Nerv & Ment Dis; AMA; fel Am Acad Neurol; Am Neurol Asn; Am Epilepsy Soc. Res: Parkinsonism; temporal lobe suppression; syncope; muscular dystrophy; cerebrovascular disease; multiple sclerosis; Huntington's chorea. Mailing Add: Dept of Neurol Georgetown Univ Hosp Washington DC 22207

ODOM, GUY LEARY, b Harvey, La, May 20, 11; m 33; c 3. NEUROSURGERY. Educ: Tulane Univ, MD, 33; Am Bd Neurol Surg, dipl. Prof Exp: Prof, 50-74, JAMES B DUKE PROF NEUROSURG, SCH MED, DUKE UNIV, 74- Concurrent Pos: Consult to Surgeon Gen, USPHS Neurol Prog Proj Comt, 60-65; mem, Int Cong Neurol Surg; secy-treas, Am Bd Neurol Surg, 64-70, chmn, 70-; mem, Nat Adv Neurol Dis & Stroke Coun, 69-71. Mem: AMA; Am Asn Neurol Surg (pres, 71-72); Am Acad Neurol Surg (pres, 67); hon mem Cong Neurol Surg; Soc Neurol Surg (secy-treas, 60-65, pres, 70-71). Res: Cerebral circulation and intracranial neoplasms. Mailing Add: Duke Univ Med Ctr Durham NC 27706

ODOM, HOMER CLYDE, JR, b Hattiesburg, Miss, Dec 9, 42; m 63; c 1. ORGANIC CHEMISTRY. Educ: Univ Southern Miss, BA, 63, MS, 66; Clemson Univ, PhD(chem), 70. Prof Exp: Chemist, Pan Am Tung Res, 64-66; PROF CHEM & HEAD DEPT, BAPTIST COL CHARLESTON, 70- Mem: Am Chem Soc; The Chem Soc. Res: Natural products; organic residues in nature. Mailing Add: Dept of Chem Baptist Col PO Box 10087 Charleston SC 29411

ODOM, IRA EDGAR, b Dover, Tenn, June 12, 32; m 57; c 3. MINERALOGY, PETROLOGY. Educ: Southern Ill Univ, BA, 56; Univ Ill, MS, 58, PhD(geol), 63. Prof Exp: Assoc geologist, Ill State Geol Surv, 57-64; asst prof, 64-67, ASSOC PROF GEOL, NORTHERN ILL UNIV, 67- Mem: Geol Soc Am; Mineral Soc Am; Clay Minerals Soc. Res: Mineralogy, petrology and sedimentology of argillaceous sediments. Mailing Add: Dept of Geol Northern Ill Univ DeKalb IL 60115

ODOM, JEROME DAVID, b Greensboro, NC, Apr 27, 42; m 65; c 1. INORGANIC CHEMISTRY. Educ: Univ NC, Chapel Hill, BS, 64; Ind Univ, PhD(chem), 68. Prof Exp: NSF fel, Bristol Univ, 68-69; asst prof, 69-73, ASSOC PROF CHEM, UNIV SC, 73- Mem: Am Chem Soc; The Chem Soc. Res: Chemistry of nonmetallic hydrides and halides; vibrational and rotational spectroscopy. Mailing Add: Dept of Chem Univ of SC Columbia SC 29208

O'DONNELL, ASHTON JAY, b Los Angeles, Calif, Apr 7, 21; m 43; c 4. PHYSICS. Educ: Whitman Col, AB, 43. Prof Exp: Res physicist radiation lab, Univ Calif & Tenn Eastman Corp, 43-44; res physicist, Tenn Eastman Corp, 44-47; chief spec projs br, Hanford Opers Off, US AEC, Wash, 47-51, dir tech opers div, San Francisco Opers Off, 51-54; mgr nuclear econ res, Stanford Res Inst, 55-56, mgr nuclear develops, 56-57, mgr prog develop, 57-61; Dept State sr sci adv, US Mission to Int Atomic Energy Agency, Vienna, Austria, 61-64; mgr develop, 64-67, mgr sci develop dept, 67-69, mgr bus develop, 69-72, mgr uranium enrichment prog, 72-74, GEN MGR URANIUM ENRICHMENT ASSOCS, BECHTEL CORP, 74-, VPRES, 75- Concurrent Pos: Adv to US rep sci adv comts, UN & Int Atomic Energy Agency, 61-64; coord comt peaceful uses nuclear energy, Int Coop Year, 65; mem adv comt nuclear mat safeguards, US AEC, 70-; chmn, US Nat Comt Natural Resources, Pac Basin Econ Coun, 70-; mem, Atomic Indust Forum. Mem: Am Nuclear Soc. Res: Research planning, administration; international nuclear development; industrial applications of technology. Mailing Add: Bechtel Corp PO Box 3965 50 Beale St San Francisco CA 94119

O'DONNELL, BRIAN DESMOND, b Derry, Northern Ireland, Sept 23, 39; m 65; c 3. COSMIC RAY PHYSICS. Educ: Queens Univ, Belfast, BSc, 60, Hons, 61; Univ Calgary, PhD(physics), 67. Prof Exp: Lectr physics, Durham Univ, 67-68; fel, 67, ASST PROF PHYSICS, NMEX INST MINING & TECHNOL, 68- Mem: Brit Inst Physics; Can Asn Physicists; Am Phys Soc. Res: Cosmic ray physics. Mailing Add: WC12B NMex Inst of Mining & Technol Socorro NM 87801

O'DONNELL, CHARLES MICHAEL, chemistry, see 12th edition

O'DONNELL, EDWARD, b Brooklyn, NY, Oct 13, 38; m 68; c 2. GEOLOGY. Educ: Queens Col, NY, BS, 61; Univ Cincinnati, MS, 63, PhD(geol), 67. Prof Exp: Lectr geol, Queens Col, NY, 60-61; hydrologist, US Geol Surv, 61-62; res asst zooplankton ecol, Lamont Geol Observ, 63; lectr geol, Queens Col, NY, 63-65; geologist, Pan Am Petrol Corp, 67-68; ASST PROF GEOL, UNIV S FLA, 68- Mem: AAAS; Geol Soc Am; Soc Econ Paleont & Mineral; Am Asn Petrol Geologists; Soc Econ Mineralogists & Paleontologists. Res: Sedimentation; stratigraphy; structural geology; petroleum geology; tectonics of the Gulf Coast; tectonics of the Caribbean. Mailing Add: Dept of Geol Univ of SFla Tampa FL 33620

O'DONNELL, EDWARD EARL, physics, see 12th edition

O'DONNELL, GORDON JAMES, b Oakland, Calif, Mar 25, 17. CHEMISTRY. Educ: Univ Calif, BS, 40; Iowa State Col, PhD(org chem), 44. Prof Exp: Asst chem, Iowa State Col, 40-42, instr, 42-44; chemist, Shell Develop Co, 44-55, res supvr, 55-67; instr, 67-73, LECTR CHEM, HOLY NAMES COL, 73- Mem: Am Chem Soc. Res: Chemical and physical separation of asphalt; aluminum halide catalysis; synthetic organic chemistry with emphasis on organometallics; organoantimony compounds containing water solubilizing groups. Mailing Add: Dept of Chem Holy Names Col Oakland CA 94612

O'DONNELL, IGNATIUS LEO, organic chemistry, see 12th edition

O'DONNELL, JAMES FRANCIS, b Cleveland, Ohio, July 22, 28; m 55; c 3. BIOCHEMISTRY. Educ: St Louis Univ, BS, 49; Univ Chicago, PhD(biochem), 57. Prof Exp: Asst prof biol chem & res biochemist, Univ Cincinnati, 57-65, asst prof biol chem & assoc prof exp med, 65-66; grants assoc, Div Res Grants, NIH, 68-69, prog dir, Pop & Reprod Grants Br, Ctr Pop Res, Nat Inst Child Health & Human Develop, 69-71, ASST DIR DIV RES RESOURCES, NIH, 71- Mem: Am Chem Soc; NY Acad Sci. Res: Nucleic acid metabolism, particularly liver disease in humans and laboratory animals. Mailing Add: Div of Res Resources NIH Bldg 31 Rm 5B-03 Bethesda MD 20014

O'DONNELL, JERRY F, analytical chemistry, see 12th edition

O'DONNELL, RAYMOND THOMAS, b Baltimore, Md, July 14, 31; m 57; c 5. ANALYTICAL CHEMISTRY, PHYSICAL CHEMISTRY. Educ: Loyola Col, Md, BS, 54; Mich State Univ, PhD(anal & phys chem), 67. Prof Exp: Instr chem, Col St Thomas, 57-59; asst prof, 64-65, ASSOC PROF ANAL CHEM, STATE UNIV NY COL OSWEGO, 65- Mem: AAAS; Am Chem Soc. Res: Theoretical studies, development and extension of instrumental methods of analysis, especially polarography, high frequency titrations, chronopotentiometry, spectrofluorometry and radiochemistry. Mailing Add: Dept of Chem State Univ of NY Oswego NY 13126

O'DONNELL, THOMAS JOHN, b Pittsburgh, Pa, Sept 15, 12. EXPLORATION GEOPHYSICS. Educ: Carnegie Inst Technol, BS, 33, DSc(elec eng), 51. Prof Exp: Geophysicist, Gulf Res & Develop Co, 33-43, physicist, Nat Defense Res Comt Projs, 43-45, asst dir geophys opers div, 45-53, dir geophys res div, 54-61, dir explor res, 61-63, res coordr, 63-66; CONSULT, 66- Mem: Am Phys Soc; Soc Explor Geophys; Seismol Soc Am; Am Geophys Union; Inst Elec & Electronics Eng. Res: Seismology; gravity; geomagnetics; circuit and communication theory. Mailing Add: 4625 Fifth Ave Pittsburgh PA 15213

O'DONNELL, VINCENT JOSEPH, b Montreal, Que, Nov 30, 30. BIOCHEMISTRY. Educ: McGill Univ, BSc, 51, PhD(biochem), 54. Prof Exp: Asst biochem, McGill Univ, 51-54, res assoc & demonstr, 54-56; lectr, Glasgow Univ, 56-59; from asst prof to assoc prof, 59-67, PROF BIOCHEM, UNIV BC, 67- Mem: AAAS; Endocrine Soc; NY Acad Sci; Am Chem Soc; Am Soc Biol Chem. Res: Metabolism of steroid hormones; metabolism of androgens and estrogens. Mailing Add: Dept of Biochem Fac of Med Univ of BC Vancouver BC Can

O'DONOHUE, CYNTHIA H, b Washington, DC, Oct 3, 36; m 57; c 1. ORGANIC CHEMISTRY. Educ: Randolph-Macon Woman's Col, AB, 57; Univ Richmond, MS, 67. Prof Exp: Res asst, Am Brands, Inc, 57-61; chemist endocrinol lab dept med, Med Col Va, 61-63; asst scientist, 65-66, assoc scientist & group leader lit processings, Tech Info Facil, 66-70, res scientist, 70-73, HEAD TECH INFO FACIL, PHILIP MORRIS RES CTR, 73- Concurrent Pos: Abstr, Chem Abstracts Serv, 63-68 & 75-77. Mem: Am Chem Soc; Am Inst Chem; Am Soc Info Sci. Res: Information storage, retrieval and processing; steroid and synthetic organic chemistry. Mailing Add: 9624 Northridge Court Richmond VA 23235

O'DONOHUE, WALTER JOHN, JR, b Washington, DC, Sept 23, 34; m 57; c 1. PULMONARY DISEASES, INTERNAL MEDICINE. Educ: Va Mil Inst, BA, 57; Med Col Va, MD, 61. Prof Exp: NIH fel & instr med, 67-69, from asst prof to assoc prof, 69-75, DIR CLIN SERV PULMONARY DIS, MED COL VA, 75- Concurrent Pos: Consult pulmonary dis, McGuire Vet Admin Hosp, 69-75 & Univ Va, 75; consult respiratory care prog, Nat Heart & Lung Inst, 72-75; mem coun cardiopulmonary dis, Am Heart Asn, 72-75; bd trustees, Nat Bd Respiratory Ther, 73-75, vpres, 75; vchmn comt pulmonary rehab, Am Col Chest Physicians, 73-75; med dir, Sch Respiratory Ther, J Sargent Reynolds Col, 74-75. Mem: Fel Am Col Physicians; fel Am Col Chest Physicians; Am Thoracic Soc. Res: Clinical investigation in acute respiratory failure and respiratory intensive care; investigation in pulmonary mechanics and early detection of lung disease. Mailing Add: Box 93 Med Col of Va Richmond VA 23298

O'DONOVAN, CORNELIUS JOSEPH, b Bridgeport, Conn, Nov 3, 20; m 43; c 5. PATHOLOGY, INTERNAL MEDICINE. Educ: Dartmouth Col, AB, 42; NY Univ, MD, 45. Prof Exp: Asst med dir, Armour Labs, Ill, 51-53; sr staff physician, Upjohn Co, Mich, 53-60; dir med res, Merck Sharp & Dohme Res Labs, Pa, 60-63; vpres res & med affairs, Ames Co Div, 63-68, VPRES MED AFFAIRS, MILES LABS, INC, 68- Concurrent Pos: Asst clin prof, Col Med, Univ Ill, 52-53. Mem: AAAS; AMA; Am Diabetes Asn; Am Fedn Clin Res; NY Acad Sci. Res: Clinical evaluation and development of steroid hormones; oral antidiabetic agents; development of diagnostic reagent test systems. Mailing Add: Miles Labs Inc 1127 Myrtle St Elkhart IN 46514

O'DONOVAN, GERARD ANTHONY, b Cork, Ireland, Nov 28, 37; m 63; c 2. MICROBIOLOGY, BIOCHEMISTRY. Educ: Nat Univ, Ireland, BS, 60, MS, 61; Univ Calif, Davis, PhD(microbiol), 65. Prof Exp: Res demonstr microbiol, Univ Col, Cork, 60-61; asst, Univ Calif, Davis, 62-65; lectr bact, Univ Calif, Berkeley, 65-66, Am Cancer Soc fel molecular biol, 66-68; asst prof, 68-71, ASSOC PROF BIOCHEM, BIOPHYS & GENETICS, TEX A&M UNIV, 72- Concurrent Pos: Founder & chmn, Gulf Coast Molecular Biol Conf, 69-; prin investr, Int Collab NATO Res Grant Basic Sci, 70-75; Robert A Welch chem res grant, Tex A&M Univ, 70-78. Mem: AAAS; fel Am Inst Chem; Am Soc Microbiol; Brit Biochem Soc; Genetics Soc Am. Res: Regulation of biosynthetic schemes in microorganisms, with special emphasis on pyrimidine metabolism; regulatory proteins—their function and genetics;. evolutionary implications of intergeneric hybrids of Aspartate Transcarbamylases from bacteria. Mailing Add: Dept of Biochem & Biophys Tex A&M Univ College Station TX 77843

ODOR, DOROTHY LOUISE, b Washington, DC, May 25, 22. ANATOMY. Educ: Am Univ, BA, 45; Univ Rochester, MS, 48, PhD(anat), 50. Prof Exp: Instr, Univ Wash, 50-56; from asst prof to assoc prof, Univ Fla, 56-61; from asst prof to assoc prof, Bowman Gray Sch Med, 61-69; assoc prof, 69-73, PROF ANAT, MED COL VA, VA COMMONWEALTH UNIV, 73- Mem: AAAS; Am Asn Anat; Am Soc Cell Biol; Am Inst Biol Sci; Reticuloendothelial Soc. Res: Microscopic and submicroscopic structure of mammalian tissue; maturation divisions in ovary; fertilization; histogenesis and cytodifferentiation of fetal and postnatal ovary; organ culture of fetal ovary; changes in oviductal and uterine cytology after hormonal administration; mesothelium. Mailing Add: Dept of Anat Med Col of Va Richmond VA 23298

ODOROFF, CHARLES LAZAR, b Minneapolis, Minn, July 4, 38; m 61; c 2. BIOSTATISTICS. Educ: Carleton Col, BA, 60; Harvard Univ, AM, 62, PhD(statist), 66. Prof Exp: Asst prof statist & psychiat, 64-70, actg dir div biostatist, 70-74, ASSOC PROF BIOSTATIST & STATIST, SCH MED & DENT, UNIV ROCHESTER, 70-, DIR DIV BIOSTATIST, 74- Mem: AAAS; Am Statist Asn; Biomet Soc; Inst Math Statist. Res: Statistical methodology; psychiatric epidemiology, biometry. Mailing Add: Div of Biostatist Univ of Rochester Med Ctr Rochester NY 14642

ODUM, EUGENE PLEASANTS, b Lake Sunapee, NH, Sept 17, 13; m 39; c 1. ECOLOGY, ORNITHOLOGY. Educ: Univ NC, AB, 34; Univ Ill, PhD(ecol, ornith), 39. Prof Exp: Asst zool, Univ NC, 34-36; instr biol & ornith, Western Reserve Univ, 36-37; asst, Univ Ill, 37-39; res biologist, Edmund Niles Huyck Preserve, NY, 39-40; from instr to prof, 40-57, ALUMNI FOUND DISTINGUISHED PROF ZOOL, UNIV GA, 57-, DIR INST ECOL, 61- Concurrent Pos: NSF sr fel, 57-58; instr, Marine Biol Lab, Woods Hole, 57-61. Honors & Awards: Mercer Award, Ecol Soc Am, 56. Mem: Nat Acad Sci; AAAS; Ecol Soc Am (pres, 64-65); Am Soc Limnol & Oceanog; Am Soc Nat. Res: General principles of ecology; vertebrate populations; productivity and population energy flow; physiological ecology, heart rate and lipid metabolism of birds; radiation and marine ecology. Mailing Add: Dept of Zool Univ of Ga Athens GA 30601

ODUM, HOWARD THOMAS, b Durham, NC, Sept 1, 24; m 47; c 2. ECOLOGY,

OCEANOGRAPHY. Educ: Univ NC, AB, 47; Yale Univ, PhD(zool), 51. Prof Exp: Asst, Univ NC, 46 & Yale Univ, 47-48; asst prof biol, Univ Fla, 50-54; asst prof zool, Duke Univ, 54-56; dir & res scientist, Inst Marine Sci, Univ Tex, 56-63; chief scientist, Rain Forest Proj, P R Nuclear Ctr, 63-66; prof, Univ NC, Chapel Hill, 66-70; GRAD RES PROF ENVIRON SCI, UNIV FLA, 70- Concurrent Pos: Instr, Trop Weather Sch, CZ, 45; prin investr, Off Naval Res grant, Univ Fla, 52-54; instr, Marine Biol Lab, Woods Hole, 53, 58; co-investr, Atomic Energy Comn & Univ Ga, 54; grants, NSF, 55-60, Rockefeller Found, 57-58, Off Naval Res, 58-60, Atomic Energy Comn, 58, 63-70, & USPHS, 59-63; mem comn herbicide in Viet Nam, Nat Acad Sci. Honors & Awards: George Mercer Award, 57; Merit Award, Asn Tech Writers, 71. Mem: AAAS; Am Soc Limnol & Oceanog; Am Meteorol Soc; Ecol Soc Am; Geochem Soc. Res: Ecological systems; estuaries; rain forests; analog simulation; general systems theory. Mailing Add: Dept of Environ Sci Univ of Fla Gainesville FL 32601

ODUM, WILLIAM EUGENE, b Athens, Ga, Oct 1, 42. ECOLOGY. Educ: Univ Ga, BS, 64; Univ Miami, MS, 66, PhD(marine sci), 70. Prof Exp: Can Govt res fel, Univ BC, 70-71; asst prof ecol, 71-75, ASSOC PROF ECOL, UNIV VA, 75- Concurrent Pos: Mem bd dirs, Piedmont Environ Coun, 73- Mem: Am Soc Limnol & Oceanog; Ecol Soc Am; Am Fisheries Soc; Inst Fisheries Biologists. Res: Estuarine food webs; plant detritus production; ecology of resource management; ecology of fishes. Mailing Add: Dept of Environ Sci Univ of Va Charlottesville VA 22903

O'DWYER, JOHN J, b Grafton, NSW, Australia, Nov 9, 25; m 55; c 5. THEORETICAL SOLID STATE PHYSICS. Educ: Univ Sydney, BSc, 45, BE, 47; Univ Liverpool, PhD(physics), 51. Prof Exp: Lectr math, Sydney Tech Col, 47-48; sr res officer physics, Commonwealth Sci & Indust Res Orgn, Australia, 51-57; assoc prof, Univ New South Wales, 57-64; sr scientist, Westinghouse Res Labs, 65; prof physics, Univ Southern Ill, 66-70; PROF PHYSICS, STATE UNIV NY COL OSWEGO, 70- Concurrent Pos: Vchmn digest comt, Dielectrics Conf, 67, secy, 70-71, chmn, 74-75. Mem: Am Phys Soc; Am Asn Physics Teachers; Brit Inst Physics. Res: Theory of dielectrics, especially phenomena occurring at high field strengths. Mailing Add: Dept of Physics State Univ of NY Col Oswego NY 13126

OEHLSCHLAEGER, HERMAN FRED, b Cincinnati, Ohio, Oct 23, 18; m 47; c 2. ORGANIC CHEMISTRY. Educ: Univ Cincinnati, BS, 47, MS, 48, PhD(org chem), 50. Prof Exp: Chemist, McGean Chem Co, 50-51; sr chemist, 51-59, sect mgr, 59-66, res dir org chem div, 66-74, TECH DIR BUS DEVELOP, EMERY INDUSTS, INC, 74- Mem: Am Chem Soc; Am Oil Chem Soc. Res: Fatty acids and derivatives; polynuclear hydrocarbons; heterocyclic compounds; medicinals; plasticizers; synthetic lubricants; esterification and amidation of fatty acids; oxidation; ozonation. Mailing Add: Emery Industs Inc 4900 Este Ave Cincinnati OH 45232

OEHLSCHLAGER, ALLAN CAMERON, b Hartford, Conn, Sept 8, 40; m 60; c 2. BIO-ORGANIC CHEMISTRY. Educ: Okla State Univ, PhD(org chem), 65. Prof Exp: NATO fel org chem, Univ Strasbourg, 65-66; asst prof chem, 66-69, ASSOC PROF CHEM, SIMON FRASER UNIV, 69- Mem: Am Chem Soc; The Chem Soc; Chem Soc France. Res: Membrane active antibiotics; bio-organic chemistry of terpenes; insect attractants. Mailing Add: Dept of Chem Simon Fraser Univ Burnaby BC Can

OEHME, FREDERICK WOLFGANG, b Leipzig, Ger, Oct 14, 33; m 60; c 4. VETERINARY TOXICOLOGY, COMPARATIVE MEDICINE. Educ: Cornell Univ, BS, 57, DVM, 58; Kans State Univ, MS, 62; Univ Geissen, Dr med vet, 64; Am Bd Vet Toxicol, dipl, 67; Univ Mo-Columbia, PhD(toxicol), 69. Prof Exp: Veterinarian, Md, 58-59; instr surg & med col vet med, Kans State Univ, 59-64; from asst prof to assoc prof, 64-66; instr physiol pharm, Univ Mo-Columbia, 66-67; NIH spec res fel toxicol, 67-69; assoc prof toxicol & med, 69-73, PROF TOXICOL, MED & PHYSIOL, COL VET MED, KANS STATE UNIV, 73-, DIR COMP TOXICOL LAB, 69- Concurrent Pos: Vis prof col vet med, Univ Giessen, 63-64; stipend, Ger Acad Exchange Serv, 63-64; mem exam bd, Am Bd Vet Toxicol, 67-; consult, Poisonlab Inc, Denver, 69-, Food & Drug Admin, Washington, DC, 70- & Univ Kans, 71- Mem: Fel Am Col Vet Toxicol (secy-treas, 70-, ed, Vet Toxicol, 70-); Soc Toxicol; Am Acad Clin Toxicol (pres-elect, 75-); NY Acad Sci; Environ Mutagen Soc. Res: Biotransformation and biochemical action of toxicants; clinical and diagnostic toxicology; public health aspects of toxicants; comparative toxicology as a research and diagnostic tool. Mailing Add: Comp Toxicol Lab Kans State Univ Manhattan KS 66506

OEHME, REINHARD, b Wiesbaden, Ger, Jan 26, 28; m 52. THEORETICAL PHYSICS. Educ: Univ Göttingen, PhD, 51. Prof Exp: Res assoc, Max Planck Inst Physics, Göttingen, Ger, 50-53; res assoc, Enrico Fermi Inst Nuclear Studies, Univ Chicago, 54-56; mem, Inst Adv Study, 56-58; from asst prof to assoc prof dept physics & Enrico Fermi Inst, 58-64, PROF DEPT PHYSICS & ENRICO FERMI INST, UNIV CHICAGO, 64- Concurrent Pos: Vis prof, Inst Theoret Physics, Sao Paulo, Brazil, 52-53 & Univ Md, 57; vis scientist, Brookhaven Nat Lab, 57, 60, 62, vis sr scientist, 65 & 67; visitor, Europ Orgn Nuclear Res, Switz, 61, 64, 71 & 73; vis prof, Univ Vienna, 61, Imp Col, Lond, 63-64, Guggenheim fel, 63-64; vis prof, Int Ctr Theoret Physics, 66, 68, 69, 70 & 72; vis prof, Univ Karlsruhe, Ger, 74 & 75. Honors & Awards: Sr US Scientist Award, Alexander Von Humboldt Found, 74. Mem: Fel Am Phys Soc. Res: Elementary particle physics; quantum field theory. Mailing Add: Enrico Fermi Inst Univ of Chicago Chicago IL 60637

OEHMKE, RICHARD WALLACE, b St Clair, Mich, Dec 19, 34; m 61; c 6. APPLIED CHEMISTRY, RESEARCH ADMINISTRATION. Educ: St Joseph's Col, Ind, BS, 58; Univ Ill, Urbana, PhD(inorg chem), 64. Prof Exp: Sr chemist, 64-73, supvr com tape div, 73-74, MGR PROD DEVELOP, COM TAPE DIV, MINN MINING & MFG CO, 74- Res: High temperature polymers; polymer composites; organometallic catalysis; bio-and biologically reactive polymers; chemical reactions of natural polymers; polymer coatings; chromatography; polymer adhesives; polymer films. Mailing Add: Com Tape Div Bldg 230 Minn Mining & Mfg Co St Paul MN 55101

OEHMKE, ROBERT H, b Detroit, Mich, Aug 6, 27; m 50; c 1. MATHEMATICS. Educ: Univ Mich, BS, 48; Univ Detroit, MA, 50; Univ Chicago, PhD(math), 54. Prof Exp: Instr math, Ill Inst Technol, 53-54; asst prof, Butler Univ, 54-56; instr, Mich State Univ, 52-62; instr, Inst Defense Anal, 62-64; PROF MATH, UNIV IOWA, 64- Mem: Am Math Soc; Math Asn Am; Asn Comput Mach. Res: Nonassociative algebra, semigroups and automata. Mailing Add: Dept of Math Univ of Iowa Iowa City IA 52241

OEHSER, PAUL HENRY, b Cherry Creek, NY, Mar 27, 04; m 27; c 2. SCIENCE WRITING. Educ: Greenville Col, AB, 25. Prof Exp: Asst ed bur biol surv, USDA, 25-31; ed, US Nat Mus, 31-50, asst chief ed div, Smithsonian Inst, 46-50, chief ed & pub div, 50-66, RES ASSOC, SMITHSONIAN INST, 66-; ED SCI PUBL, NAT GEOG SOC, 66- Concurrent Pos: Ed proc, Am Sci Cong, 40-43. Mem: Wilderness Soc; Thoreau Soc Am (pres, 60); Am Ornithologists' Union. Res: Science editing; biological editing and bibliography; American naturalists; conservation; Smithsonian history. Mailing Add: 9012 Old Dominion Dr McLean VA 22101

OEI, DJONG-GIE, b Solo, Indonesia, Apr 18, 31; US citizen; m 63; c 4. PHYSICAL CHEMISTRY, INORGANIC CHEMISTRY. Educ: Univ Indonesia, Drs, 58; Univ Ky, PhD(phys chem), 61. Prof Exp: Lectr phys & inorg chem, Bandung Inst Tech, Indonesia, 61-63; fel dept chem, Univ Ky, 63-64; staff chemist off prod div, Int Bus Mach Corp, Ky, 64-67; sr res engr, Sci Res Staff, 67-69, ASSOC PRIN RES SCIENTIST, RES STAFF, FORD MOTOR CO, 69- Mem: Am Chem Soc; Electrochem Soc; fel Am Inst Chem. Res: Chemistry of non-aqueous solvents; coordination compounds; inorganic and organic photoconductors; electrochemistry; inorganic sulfur chemistry. Mailing Add: Res Staff Ford Motor Co Box 2053 Dearborn MI 48121

OEKBER, NORMAN FRED, b Avon, Ohio, Mar 3, 27; m 63; c 6. HORTICULTURE. Educ: Ohio State Univ, BS, 49; Cornell Univ, MS, 51, PhD(veg crops), 53. Prof Exp: Exten specialist veg crops, Univ Ill, 53-60; EXTEN SPECIALIST VEG CROPS, UNIV ARIZ, 60-, PROF HORT, 71- Concurrent Pos: Pres, Nat Agr Plastics Cong, 66. Mem: Am Soc Hort Sci. Res: Crop ecology and environmental control; arid lands production; post-harvest handling. Mailing Add: Dept of Plant Sci Univ of Ariz Tucson AZ 85721

OELFKE, WILLIAM C, b Kansas City, Mo, May 28, 41; m 64. MOLECULAR PHYSICS, MICROWAVE PHYSICS. Educ: Stanford Univ, BS, 63; Duke Univ, PhD(physics), 69. Prof Exp: Res assoc microwave spectros, Duke Univ, summer 69; asst prof physics, 69-72, ASSOC PROF PHYSICS, FLA TECHNOL UNIV, 72- Concurrent Pos: Sr res assoc, La State Univ, 75-76. Honors & Awards: Fac Develop Award, Fla Technol Univ, 75. Mem: Sigma Xi. Res: Molecular spectroscopy, especially microwave and millimeter-wave spectroscopy of gases and gaseous free radicals; astrophysics; gravity waves. Mailing Add: Dept of Physics Fla Technol Univ PO Box 25000 Orlando FL 32816

OELKE, ERVIN ALBERT, b Green Lake, Wis, Dec 14, 33; m 58; c 2. AGRONOMY, PLANT PHYSIOLOGY. Educ: Univ Wis, BS, 60, MS, 62, PhD(agron), 64. Prof Exp: Asst agronomist, Rice Exp Sta, Univ Calif, 64-68; ASSOC PROF AGRON & EXTEN AGRONOMIST, UNIV MINN, ST PAUL, 68- Mem: Am Soc Agron; Am Soc Plant Physiol. Res: Ecological factors influencing the seedling development and growth of rice; influence of environment and breeding on chlorophyll and other plant constituents in corn; culture and physiology of wild rice. Mailing Add: Dept of Agron & Plant Genetics Univ of Minn St Paul MN 55101

OELKE, WILLIAM C, b Corwith, Iowa, Oct 11, 06; m 30; c 3. PHYSICAL CHEMISTRY. Educ: Grinnell Col, AB, 28; Col Holy Cross, MS, 29; Univ Iowa, PhD(phys chem), 35. Prof Exp: Instr chem, Col Holy Cross, 29-30, lectr, 30-31; from instr to prof, 31-74, chmn dept, 48-62, EMER PROF CHEM, GRINNELL COL, 74- Concurrent Pos: Asst, Univ Iowa, 34-35. Mem: Am Chem Soc. Res: Thermodynamic properties of solutions of electrolytes; qualitative analysis; solubilities and activity coefficients of lanthanum iodate in concentrated salt solutions; rate of bromination of cinnamic acid; refractive indices of alcohol water mixtures. Mailing Add: Dept of Chem Grinnell Col Grinnell IA 50112

OELRICH, THOMAS MANN, b Cincinnati, Ohio, May 27, 24. MORPHOLOGY. Educ: Centre Col, AB, 48; Univ Mich, MS, 50, PhD, 54. Prof Exp: From instr to assoc prof, 52-74, PROF ANAT, MED SCH, UNIV MICH, ANN ARBOR, 74- Mem: Am Soc Mammal; Soc Vert Paleont; Am Soc Ichthyol & Herpet; Am Asn Anat. Res: Human and comparative morphology; vertebrate fossils. Mailing Add: Dept of Anat Med Sci II Univ of Mich Ann Arbor MI 48104

OELS, HELEN C, b Philadelphia, Pa, Apr 13, 31. PATHOLOGY, IMMUNOLOGY. Educ: Chestnut Hill Col, BS, 53; Woman's Med Col Pa, MD, 57; Univ Minn, PhD(path), 69; Am Bd Path, dipl & cert anat path, 65, cert clin path, 67. Prof Exp: NIH fel, Sch Med, Temple Univ, 58-60; physician, Whitesburg Mem Hosp, Ky, 60-61; mem staff, Mayo Clin, 61-69; asst prof microbiol, 69-71, ASSOC PROF MICROBIOL & IMMUNOL, SCH MED, TEMPLE UNIV, 71- Mem: Am Soc Clin Path; Am Soc Microbiol; AMA; NY Acad Sci. Res: Immunopathology; autoimmune diseases; lymphomas; antigens of histoplasma capsulatim; delayed hypersensitivity; virology; viral oncolysis. Mailing Add: Dept of Microbiol Temple Univ Med Sch Philadelphia PA 19122

OELSHLEGEL, FREDERICK JAMES, JR, b Philadelphia, Pa, Apr 12, 42; m 67; c 1. BIOCHEMISTRY, GENETICS. Educ: Drexel Univ, BS, 65; Univ Wis-Madison, PhD(biochem), 70. Prof Exp: Trainee new crops res, Eastern Regional Res Labs, USDA, 61-65, chemist, 65; sr res assoc human genetics, Sch Med, Univ Mich, Ann Arbor, 71-75; ASST PROF PREV MED & BIOCHEM, MED CTR, UNIV MISS, 75-; CONSULT HEMAT, VET ADMIN, JACKSON, 75- Concurrent Pos: NIH trainee human genetics, Sch Med, Univ Mich, Ann Arbor, 70-72, A H Meyers Found grant, 72-73 & 75. Mem: Am Fedn Clin Res; NY Acad Sci; Am Soc Hemat; Am Chem Soc; Am Soc Human Genet. Res: Human biochemistry and genetics; biochemical genetics; molecular basis of host-pathogen specificity; hematology. Mailing Add: Dept of Prev Med Univ of Miss Med Ctr Jackson MS 39216

OEN, ORDEAN SILAS, b Grafton, NDak, June 29, 27; m 53; c 3. PHYSICS. Educ: Concordia Col, BA, 49; Univ NDak, MS, 53; Univ Mo, PhD(physics), 58. Prof Exp: Instr physics, Univ Mo, 53-55; summer physicist solid state theory, 57, PHYSICIST, SOLID STATE DIV, OAK RIDGE NAT LAB, 58- Mem: Am Phys Soc; Sigma Xi. Res: Influence of crystal lattice on stopping power; theoretical study of radiation effects on solids. Mailing Add: Oak Ridge Nat Lab PO Box X Oak Ridge TN 37830

OENE, HENK VAN, b Zwolle, Neth, Oct 25, 32; Can citizen; m 62; c 2. PHYSICAL CHEMISTRY. Educ: Univ Groningen, BSc, 55; McMaster Univ, MSc, 58; Univ Alta, PhD(polymer chem), 63. Prof Exp: Res assoc polymer chem, Mt Allison Univ, 63-67, asst prof physics, 65-67; res scientist sr sci res staff, 67-68, prin res scientist polymer sci dept, Res Staff, 68-72, STAFF SCIENTIST, RES STAFF, FORD MOTOR CO, 72- Mem: AAAS; Am Chem Soc; Soc Rheology. Res: Rheology of heterogeneous visco-elastic fluids; free surface flows of polymer melts; powder coating. Mailing Add: Ford Motor Co 20000 Rotunda Dr PO Box 2053 Dearborn MI 48121

OERTEL, GEORGE FREDERICK, b Indianapolis, Ind, Apr 13, 44; m 66; c 2. SEDIMENTOLOGY. Educ: Ohio State Univ, BS, 66; Univ Iowa, MS, 68, PhD(geol), 71. Prof Exp: Res assoc, 71-73, ASST PROF GEOL, SKIDAWAY INST OCEANOG, 73- Mem: Soc Econ Paleontologists & Mineralogists. Res: Coastal evolution with an emphasis on processes of barrier and inlet development; studies involving sediment transport and how it relates to beach and inlet stability. Mailing Add: Skidaway Inst of Oceanog PO Box 13687 Savannah GA 31406

OERTEL, GERHARD FRIEDRICH, b Leipzig, Ger, Apr 22, 20; m 46; c 3. STRUCTURAL GEOLOGY. Educ: Univ Bonn, Dr rer nat(geol), 45. Prof Exp: Asst geol, Univ Bonn, 46-50, Privatdozent, 50; geologist, Co Petrol Portugal, 51-53; geologist, Portuguese State Overseas Ministry, 53-56; assoc prof geol, Pomona Col, 56-60; assoc prof, 60-66, PROF GEOL, UNIV CALIF, LOS ANGELES, 66- Concurrent Pos: John Simon Guggenheim Mem Found fel, Univ Edinburgh, 66-67.

Mem: Am Geophys Union; Geol Soc Am. Res: Structural geology of plutonic, volcanic, metamorphic and sedimentary rocks; deformation and fracture of rocks; slaty cleavage; preferred orientation of phyllosilicate mineral grains in rocks. Mailing Add: Dept of Geol Univ of Calif Los Angeles CA 90024

OERTEL, GOETZ KUNO HEINRICH, b Stuhm, Ger, Aug 24, 34; m 60; c 2. SOLAR PHYSICS. Educ: Univ Kiel, Vordiplom, 56; Univ Md, PhD(physics), 64. Prof Exp: Res asst physics, Univ Md, 57-62; aerospace engr, Langley Res Ctr, NASA, 63-68, staff scientist, Solar Prog Mgt, 68-69, dep chief solar physics prog, 69-71, chief, 71-75; head astron sect, NSF, 75; DIR COMMUNICATIONS & MGT SUPPORT, OFF OF ASST ADMINR NUCLEAR ENERGY, ENERGY RES & DEVELOP ADMIN, 75- Concurrent Pos: On leave, Dept Physics & Astron, Univ Md, 67-68; Partic, Fed Exec Develop Prog, Off Mgt & Budget & Civil Serv Comn, 74-75, spec analyst, Sci & Technol Policy Off, NSF, 74 & Exec Off of the President, 74-75. Mem: Am Astron Soc; Am Phys Soc; Sigma Xi. Res: Plasma spectroscopy; solar physics and astrophysics; program management; science policy; federal budget analysis and formulation; federal management and administration; communications. Mailing Add: Off Asst Adminr Nuclear Energy Energy Res & Develop Admin Washington DC 20545

OERTEL, RICHARD PAUL, b New York, NY, Jan 12, 44; m 67. SPECTROCHEMISTRY. Educ: Oberlin Col, BA, 64; Cornell Univ, PhD(chem), 68. Prof Exp: Asst prof chem, Cornell Univ, 68-69; RES CHEMIST, PROCTER & GAMBLE CO, 69- Mem: Am Chem Soc. Res: Application of spectroscopy, chiefly infrared and Raman, to the elucidation of molecular structure in chemical and biochemical systems. Mailing Add: Procter & Gamble Co PO Box 39175 Cincinnati OH 45247

OERTLI, JOHANN JAKOB, b Ossingen, Switz, July 16, 27; US citizen; m 61; c 4. SOIL SCIENCE, PLANT NUTRITION. Educ: Swiss Fed Inst Technol, FEng, 51; Univ Calif, MS, 53, PhD(soils), 56. Prof Exp: From instr to asst prof soil sci, Univ Calif, Los Angeles, 57-63; from assoc prof to prof, Univ Calif, Riverside, 63-74; HEAD, INST BOT, UNIV BASEL, SWITZ, 74- Concurrent Pos: Fulbright fel, Ger, 70-71. Mem: Soil Sci Soc Am. Res: Mineral nutrition and water relations in plants. Mailing Add: Inst of Bot Univ of Basel Schönbeinstrasse 6 CH 4056 Basel Switzerland

OESPER, PETER, b Cincinnati, Ohio, Sept 25, 17; m 43; c 1. BIOLOGICAL CHEMISTRY. Educ: Swarthmore Col, AB, 38; Princeton Univ, MS, 40, PhD(chem), 41. Prof Exp: Instr phys chem, Univ Md, 41-42, asst prof, 42-45; res assoc sch med, Univ Pa, 45-51; asst prof, Hahnemann Med Col, 51-56, from assoc prof to prof, 56-68; PROF CHEM & CHMN DEPT, ST LAWRENCE UNIV, 68- Mem: Am Chem Soc; Am Soc Biol Chem. Res: Dipole moments; kinetics and thermodynamics of enzyme reactions. Mailing Add: Dept of Chem St Lawrence Univ Canton NY 13617

OESTER, YVO THOMAS, b West Brooklyn, Ill, Dec 21, 08; m 39; c 3. PHARMACOLOGY. Educ: Univ Notre Dame, BS, 31, MS, 33; Univ Chicago, PhD(physiol), 38, MD, 43. Prof Exp: Asst physiol, Univ Notre Dame, 31-33, instr, 33, asst, Univ Chicago, 34-38; instr, 38-39, assoc prof pharmacol, 47-56, actg chmn dept, 47-58, PROF PHARMACOL, SCH MED, LOYOLA UNIV CHICAGO, 56- Concurrent Pos: Consult, Hines Vet Admin Hosp, 48-52, actg dir radioisotope serv, 54-59, consult, 54-, assoc chief staff, 63-; attend physician, Little County Mary Hosp & Mercy Hosp. Mem: Am Chem Soc; Am Soc Pharmacol & Exp Therapeut; Soc Nuclear Med; AMA; Am Asn Electromyog & Electrodiag (past pres); Asn Am Med Cols. Res: Mode of action of drugs; location of action of drugs; isolation and identification of alkaloids; neuromuscular junction and electromyography. Mailing Add: Vet Admin Hosp Hines IL 60141

OESTERLIN, RUDOLF, organic chemistry, see 12th edition

OESTERLING, JAMES FREDERICK, b Butler, Pa, Apr 6, 05; m 33; c 3. RESEARCH ADMINISTRATION. Educ: Pa State Univ, BS, 28, MS, 35, PhD(textile chem), 41. Prof Exp: Chemist, Weller-Krouse Co, Pa, 28-31; asst prof & dir textile & detergency labs, Pa State Univ, 42-46; consult qm res & develop labs, US Dept Army, 46-47, chief textile finishing res labs, 47-51, phys sci adminstr textiles & leather, 52-53, res dir chem & plastics div, Qm Res & Eng Command, 54-57, dep sci dir, 58-61, dep sci dir res, Natick Labs, 62-73, tech exp, US Army Natick Develop Ctr, 73-74; RETIRED. Honors & Awards: US Dept Army Achievement Award, Meritorious Civilian Serv Award, 73. Mem: Am Chem Soc; Am Defense Preparedness Asn. Res: Basic and applied research in pollution abatement; food and feeding systems; clothing and environmental protection for the individual soldier. Mailing Add: 421 S 21st St Lewisburg PA 17837

OESTERLING, MYRNA JANE, b Butler, Pa, Oct 24, 17. BIOCHEMISTRY. Educ: Univ Ill, AB, 39, MS, 41; George Washington Univ, PhD(biochem), 44. Prof Exp: Asst chem, Sch Med, Yale Univ, 46-47& Univ Ill, 47-50; asst prof physiol chem, 51-67, RES ASSOC PROF BIOCHEM, MED COL PA, 67- Concurrent Pos: Talbot fel, Yale Univ, 44, Coxe fel, 45; Berquist fel, Karolinska Inst, Sweden, 50-51. Mem: AAAS; Am Asn Clin Chem; Am Soc Biol Chemists. Res: Amino acid nutrition; ascorbic acid method; stability and determination of epinephrine and norepinephrine; excretion of catecholamines in relation to stress; fetoplacental function. Mailing Add: Dept of Obstet & Gynec Med Col of Pa Philadelphia PA 19129

OESTERREICH, ROGER EDWARD, b Chicago, Ill, Feb 7, 30. PHYSIOLOGICAL PSYCHOLOGY. Educ: Univ Mo, AB, 52; Univ Chicago, PhD(biopsychol), 60. Prof Exp: Asst psychol, Univ Chicago, 55-59; res assoc, Ill State Psychiat Inst, 60-64; vis asst prof, Ill Inst Technol, 64; asst prof, 64-66, ASSOC PROF PSYCHOL, STATE UNIV NY ALBANY, 66- Mem: AAAS; Acoust Soc Am; Soc Neurosci. Res: Neural bases of perception; audition; sound localization. Mailing Add: Dept of Psychol State Univ of NY Albany NY 12222

OESTERREICHER, HANS, b Innsbruck, Austria, May 16, 39; US citizen; m 69; c 1. SOLID STATE CHEMISTRY. Educ: Univ Vienna, PhD(chem), 65. Prof Exp: Res assoc solid state chem, Univ Pittsburgh, 65-66 & Brookhaven Nat Lab, 66-67; instr, Cornell Univ, 67-69; asst prof, Grad Ctr, Univ Ore, 68-73, assoc prof, 73; ASSOC PROF SOLID STATE CHEM, UNIV CALIF, SAN DIEGO, 73- Mem: Am Phys Soc. Res: Structural, magnetic, electric and thermal properties of solids ,superconductivity; intermetallic compounds and hydrides thereof. Mailing Add: Dept of Chem Univ of Calif at San Diego La Jolla CA 92093

OESTERWINTER, CLAUS, b Hamburg, Ger, Jan 18, 28; US citizen; m 53; c 2. ASTRONOMY, CELESTIAL MECHANICS. Educ: Yale Univ, MS, 64, PhD(astron), 65. Prof Exp: Chief comput, Western Geophys Co Am, 54-59; res astronr, Naval Weapons Lab, 59-75, RES ASTRONR, NAVAL SURFACE WEAPONS CTR, 75- Mem: Am Astron Soc; Int Astron Union; Cospar. Res: Planetary and lunar motion. Mailing Add: Astronaut Div Naval Surface Weapons Ctr Dahlgren VA 22448

OESTMANN, MARY JANE, inorganic chemistry, physical chemistry, see 12th edition

OESTREICH, ALAN EMIL, b New York, NY, Dec 4, 39; m 73. RADIOLOGY. Educ: Princeton Univ, AB, 61; Johns Hopkins Univ, MD, 65. Prof Exp: Asst prof radiol, 71-74, asst prof child health, 72-74, ASSOC PROF RADIOL & CHILD HEALTH, UNIV MO, COLUMBIA, 74- Concurrent Pos: Vis prof, Meharry Med Col, 72- Mem: Nat Med Asn; Soc Pediat Radiol; Am Col Radiol; Radiol Soc NAm. Res: Pediatric radiology; applications of computers in bone dysplasias, vertebral column, chest, gastrointestinal. Mailing Add: N 217 Univ Med Ctr Columbia MO 65201

OESTREICH, CHARLES HENRY, b Columbus, Ohio, June 8, 32; m 57; c 3. INORGANIC CHEMISTRY. Educ: Capital Univ, BS, 54; Ohio Univ, MS, 56, PhD(inorg chem), 61. Prof Exp: Instr chem, Va Mil Inst, 56-57; instr, Capital Univ, 60-62, from asst prof to assoc prof, 62-69; PROF CHEM & ACAD DEAN, TEX LUTHERAN COL, 69- Concurrent Pos: Res assoc chem & fel inorg chem, Vanderbilt Univ, 65-66. Mem: Am Chem Soc. Res: Coordination chemistry. Mailing Add: Tex Lutheran Col Seguin TX 78155

OESTREICHER, HANS LAURENZ, b Vienna, Austria, Apr 22, 12; m 43; c 1. MATHEMATICS. Educ: Univ Vienna, PhD(math), 34. Prof Exp: Res mathematician, Helmholtz Inst, Ger, 43-47; RES MATHEMATICIAN & CHIEF MATH & ANAL BR, BIOPHYS LAB, WRIGHT-PATTERSON AIR FORCE BASE, 47- Mem: Fel Acoust Soc Am; Am Math Soc; NY Acad Sci; Inst Elec & Electronics Eng. Res: Applied mathematics; wave propagation; partial differential equations; theory of sound and of automata; information processing. Mailing Add: 2923 Green Vista Dr Brook Hollow Fairborn OH 45324

OETJEN, ROBERT ADRIAN, b Detroit, Mich, Mar 31, 12; m 36; c 5. MOLECULAR SPECTROSCOPY. Educ: Asbury Col, AB, 36; Univ Mich, MS, 38, PhD(physics), 42. Prof Exp: Chemist, Mellon Inst, 36-37; physicist, Texaco, Inc, NY, 41-46; asst prof physics, Ohio State Univ, 46-51, from assoc prof to prof, 51-70, asst dean col arts & sci, 57-60, assoc dean, 60-62, 64-67 & 69-70, actg dean, 68-69; PROF PHYSICS & DEAN BUCHTEL COL ARTS & SCI, UNIV AKRON, 70- Concurrent Pos: Asst to dir, Univ Res Found, Ohio State Univ, 53-55; Fulbright res grant, Univ Osaka, 55-57; chief scientist, NSF, Am Embassy, Tokyo, 62-64; mem US nat del, Int Comn Optics, 65-70. Mem: Fel Optical Soc Am; Am Phys Soc; Am Asn Physics Teachers; Phys Soc Japan; Appl Phys Soc Japan. Res: Infrared spectroscopy. Mailing Add: Buchtel Col of Arts & Sci Univ of Akron Akron OH 44325

OETKING, PHILIP, b Madison, Wis, Mar 27, 22; m 45; c 2. MARINE GEOLOGY. Educ: Univ Wis, PhB, 46, MS, 48, PhD, 52. Prof Exp: Asst geol, Univ Wis, 46-50; res geologist, Sun Oil Co, 51-57; geological consult, 57-61; explor geologist, Scott Hammonds, oil producer, 61-62; sr scientist lunar geol, Chance Vought Corp, 62-63; res scientist, Grad Res Ctr, Southwest, 63-67; DIR, OCEAN SCI & ENG LAB, SOUTHWEST RES INST, 67- Mem: AAAS; Am Asn Petrol Geol; Marine Technol Soc; Am Soc Oceanog. Res: Lunar and planetary geological research; light reflectivity measurements; preparation of regional geological highway maps of the United States; Gulf Coast stratigraphic, structural and sedimentation problems. Mailing Add: Ocean Sci & Eng Lab Southwest Res Inst Box 2604 Corpus Christi TX 78403

OETTGEN, HERBERT FRIEDRICH, b Cologne, Ger, Nov 22, 23; m 57, c 3. MEDICINE, IMMUNOLOGY. Educ: Univ Cologne, MD, 51. Prof Exp: Intern, Red Cross Hosp, Neuwied, Ger, 51; resident path, Munic Hosp, Cologne, 52-54, resident med, 55-58; res assoc, 60-62, consult, 62-63, assoc, 63-67, assoc mem, 67-69, MEM, SLOAN-KETTERING INST CANCER RES, 70-; ASSOC PROF BIOL, SLOAN-KETTERING DIV, CORNELL UNIV, 73-, PROF MED, COL MED, 73- Concurrent Pos: Vis res fel, Sloan-Kettering Inst Cancer Res, 58-60; fel med, Mem Hosp Cancer & Allied Dis, 58-62; res gynec & obstet, Sch Med, Univ Marburg, 54-55; pvt docent, Univ Cologne, 62-69, vis prof, 69-; clin asst, Mem Hosp Cancer & Allied Dis, 65-67, asst attend physician, 67-69, assoc attend physician, 69-70, attend physician, 70-; asst prof med, Col Med, Cornell Univ, 67-70; mem cancer res ctr rev comt, NIH. Mem: Am Asn Cancer Res; Am Fedn Clin Res; Am Soc Hemat; NY Acad Sci. Res: Cancer immunology; clinical oncology. Mailing Add: Sloan-Kettering Inst 410 E 68th St New York NY 10021

OETTING, FRANKLIN LEE, b Pueblo, Colo, June 21, 30; m 56; c 4. PHYSICAL CHEMISTRY, THERMODYNAMICS. Educ: Univ Colo, BS, 52, MS, 54; Univ Wash, Seattle, PhD(phys chem), 60. Prof Exp: Instr chem, Regis Col, 54-56; phys chemist, Dow Chem Co, Mich, 60-63, from res chemist to sr res chemist, Rocky Flats Div, Golden, 63-68, assoc scientist, 68-74; mem dept chem, Int Atomic Energy Agency, Vienna, Austria, 74-76; MEM STAFF, ATOMICS INT DIV, ROCKY FLATS PLANT, ROCKWELL INT, 76- Concurrent Pos: Mem, Calorimetry Conf, 61- Mem: AAAS; Am Chem Soc; Sigma Xi. Res: High temperature and isothermal calorimetry; compilation of thermodynamic data. Mailing Add: Atomics Int Div Rocky Flats Plant Rockwell Int Golden CO 80401

OETTINGER, ANTHONY GERVIN, b Nuremberg, Ger, Mar 29, 29; nat US; m 54; c 2. APPLIED MATHEMATICS, LINGUISTICS. Educ: Harvard Univ, AB, 51, PhD(appl math), 54. Prof Exp: NSF fel & res fel appl math, 54-55, instr, 55-56, from asst prof to assoc prof appl math & ling, 57-63, prof ling, 63-75, GORDON McKAY PROF APPL MATH, HARVARD UNIV, 63-, PROF INFO RESOURCES POLICY, 75-, DIR, PROG INFO TECHNOL & PUB POLICY, 73- Concurrent Pos: Consult, Arthur D Little, Inc, Cambridge, Mass, 56-; consult off sci & technol, Exec Off of President of US, 61-73; consult, Bellcomm, Inc, Washington, DC, 63-67; mem adv bd, Chem Abstr Serv, Columbus, Ohio, 64-66; mem adv comt automatic lang processing, Nat Acad Sci-Nat Res Coun, Washington, DC, 64-66, chmn comput sci & eng bd, 67-73; mem res adv comt, Syst Develop Corp, Santa Monica, Calif, 65-68; res assoc prog technol & soc, Harvard Univ, 66-72; chmn bd trustees, Lang Res Found, Cambridge, Mass, 70-75; consult comt automation opportunities serv areas, Fed Coun Sci & Technol, 71-72; mem CATV comn, Commonwealth Mass, 72-75, chmn, 75-, chmn comt regulation, 72- assoc univ sem comput & relation to man & soc, Columbia Univ, 73-; adv subcomt econ & soc impact of new broadcast media, Comt Econ Develop, New York, 73-75. Mem: Fel AAAS; Asn Comput Mach (ed comp ling sect, Com of ACM, 63-66, pres, 66-68); fel Am Acad Arts & Sci; fel Inst Elec & Electronics Eng; Sigma Xi. Res: Automatic information processing systems, design and applications, educational technology and policy, programming theory and information sciences. Mailing Add: Aiken Comput Lab Harvard Univ Cambridge MA 02138

OEXEMANN, STANLEY WILLIAM, b Nashville, Ill, Mar 4, 12; m 46. PLANT PHYSIOLOGY. Educ: McKendree Col, AB, 37; Univ Ill, MS, 39, PhD(bot), 41. Prof Exp: Teacher pub schs, Ill, 36-38; asst bot, Univ Ill, 39-40; instr, Carleton Col, 41-44, from asst prof to assoc prof, 44-47; prof biol, Carthage Col, 47-49; dir summer sch, 59-67, dean sch letters & sci, 63-67, PROF BIOL, UNIV WIS-SUPERIOR, 49-, CHMN DEPT, 54- Mem: AAAS; Bot Soc Am. Res: Plant nutrition; vitamin research; hormones; plant growth. Mailing Add: Dept of Biol Univ of Wis Superior WI 54880

O'FALLON, NANCY MCCUMBER, b Jackson, Miss, Oct 25, 38; m 62; c 3. EXPERIMENTAL NUCLEAR PHYSICS, INSTRUMENTATION. Educ: St Louis Univ, BS, 60; Univ Ill, Urbana, MS, 61, PhD(physics), 66. Prof Exp: Vis asst prof physics, Univ Mo, St Louis, 72-74; ASST PHYSICIST, APPL PHYSICS DIV, ARGONNE NAT LAB, 74-, COORDR, FOSSIL ENERGY PROG, INSTRUMENTATION DEVELOP SECT, 76- Mem: Am Phys Soc. Res: Development of instruments based on nuclear, acoustic, optical and other advanced techniques for process control in large-scale fluidized-bed coal combustion and second-generation coal conversion systems. Mailing Add: Appl Physics Div Bldg 316 Argonne Nat Lab Argonne IL 60439

O'FALLON, WILLIAM M, b Princeton, Minn, Mar 7, 34; m 61; c 2. BIOSTATISTICS, MATHEMATICAL STATISTICS. Educ: St John's Univ, Minn, BA, 56; Vanderbilt Univ, MAT, 57; Univ NC, PhD(statist), 67. Educ: Instr math, St John's Univ, Minn, 57-60; res assoc biostatist, Univ NC, Chapel Hill, 63-67; asst prof community health sci, 66-71, ASST PROF MATH, MED CTR, DUKE UNIV, 66-, ASSOC PROF COMMUNITY HEALTH SCI, 71- Mem: AAAS; Math Asn Am; Am Statist Asn; Inst Math Statist; Biomet Soc. Res: Mathematical and stochastic models of biological phenomena; multivariate analysis of biomedical data. Mailing Add: Dept of Community Health Sci Duke Univ Med Ctr Durham NC 27706

O'FARRELL, CHARLES PATRICK, b Elizabeth, NJ, Oct 30, 37; m 61; c 1. RUBBER CHEMISTRY. Educ: St Peter's Col, NJ, BS, 60; Rutgers Univ, PhD(phys & org chem), 65. Prof Exp: Res chemist, Esso Res & Eng Co, 64-66; RES ASSOC RUBBER CHEM, EXXON CHEM CO, 68- Mem: Am Chem Soc. Res: Artificial latices produced from elastomeric polymers. Mailing Add: Exxon Chem Co PO Box 45 Linden NJ 07036

O'FARRELL, HELEN KROGULL, b Port Jefferson, NY, July 14, 34; m 57; c 2. BIOCHEMISTRY. Educ: Fairleigh Dickinson Univ, BS, 55; NY Univ, MS, 60; Rutgers Univ, PhD(biochem), 69. Prof Exp: Res asst neurochem, Col Physicians & Surgeons, Columbia Univ, 55; res chemist, Colgate Palmolive Co, 55-57; res assoc, Col Med, 69-70, ASST PROF BIOCHEM, DOUGLASS COL, RUTGERS UNIV, 71- Concurrent Pos: Sigma Xi grant, Douglass Col, Rutgers Univ, 72-73. Mem: AAAS. Res: Effect of oxygenated lipids on cytochrome oxidase synthesis; mechanism of the cytochrome c—cytochrome oxidase reaction. Mailing Add: Dept of Chem Douglass Col Rutgers Univ New Brunswick NJ 08903

O'FARRELL, MICHAEL JOHN, b Los Angeles, Calif, July 4, 44; m 65; c 3. MAMMALIAN ECOLOGY, BEHAVIORAL BIOLOGY. Educ: Univ Nev, Las Vegas, BS, 68; NMex Highlands Univ, MS, 71; Univ Nev, Reno, PhD(zool), 73. Prof Exp: Fel & res assoc mammal, Savannah River Ecol Lab, Univ Ga, 73-74; res assoc desert ecol, 74-75, ASST PROF BIOL, UNIV NEV, LAS VEGAS, 75- Mem: Am Soc Mammalogists; Ecol Soc Am; Sigma Xi. Res: Small mammal community ecology; coexistence and social structure in natural populations, temperature regulation and body composition. Mailing Add: Dept of Biol Sci Univ of Nev Las Vegas NV 89154

O'FARRELL, THOMAS PAUL, b Chicago, Ill, Mar 18, 36; m 58. MAMMALIAN ECOLOGY. Educ: St Mary's Col, Minn, BS, 58; Univ Alaska, MS, 60; Univ Tenn, PhD(zool), 65. Prof Exp: Res asst biol, St Mary's Col, Minn, 55-58; res asst wildlife, Dept Wildlife Mgt, Univ Alaska, 58-60; wildlife biologist, Alaska Dept Fish & Game, 60-61; teaching asst zool, Univ Tenn, 61-63; AEC fel ecol, Univ Wash, 65-67; res scientist, Pac Northwest Labs, Battelle Mem Inst, Wash, 67-68, mgr terrestrial ecol sect, 68-71, prog leader, 71-73, sr res scientist, 73-74; assoc res prof, 74-75, DIR, APPL ECOL & PHYSIOL CTR, DESERT RES INST, UNIV NEV SYST, 75- Mem: Fel AAAS; Am Soc Mammalogists; Ecol Soc Am; Radiation Res Soc. Res: Mammalian ecology, particularly as affected by ionizing radiation; ecological life histories of mammals; physiological ecology; arid lands ecology; environmental impact assessment. Mailing Add: Appl Ecol & Physiol Ctr Desert Res Inst Univ of Nev Syst Boulder City NV 89005

OFELT, CHESTER WINFIELD, biochemistry, see 12th edition

OFELT, GEORGE STERLING, b Washington, DC, Jan 22, 37; m 62; c 2. SPECTROSCOPY. Educ: Col William & Mary, BS, 57; Johns Hopkins Univ, PhD(physics), 62. Prof Exp: Instr physics, Johns Hopkins Univ, 62-63; asst prof, Col William & Mary, 63-67; ASSOC PROF PHYSICS, OLD DOM UNIV, 67- Concurrent Pos: Consult & lectr, NASA, 63- Mem: Optical Soc Am. Res: Ultraviolet and visible spectroscopy; optical oceanography. Mailing Add: 824 St Clement Rd Virginia Beach VA 23455

OFENGAND, EDWARD JAMES, b Taunton, Mass, Aug 15, 32; m 63; c 3. BIOCHEMISTRY, MOLECULAR BIOLOGY. Educ: Mass Inst Technol, BS & MS, 55; Wash Univ, St Louis, PhD(microbiol), 59. Prof Exp: NSF fel, Med Res Coun, Cambridge, Eng, 59-61; Rockefeller Inst fel, 61-62, NIH fel, 62; lectr biochem sch med, Univ Calif, San Francisco, 62-67, asst prof, 67-69; ASSOC MEM, ROCHE INST MOLECULAR BIOL, 69- Mem: AAAS; Am Soc Biol Chem; Am Chem Soc; Am Soc Microbiol; NY Acad Sci. Res: Structure and function of transfer RNA; mechanism of protein synthesis; photoaffinity labelling of macromolecules; mechanism of action of juvenile hormone. Mailing Add: Dept of Biochem Roche Inst of Molecular Biol Nutley NJ 07110

OFFEN, HENRY WILLIAM, b Uelzen, Ger, Apr 28, 37; US citizen; m 61; c 2. PHYSICAL CHEMISTRY. Educ: St Olaf Col, BA, 58; Univ Calif, Los Angeles, PhD(chem), 63. Prof Exp: Instr chem, Occidental Col, 62-63; from asst prof to assoc prof, 63-74, PROF PHYS CHEM, UNIV CALIF, SANTA BARBARA, 74- Mem: AAAS; Am Phys Soc; Am Chem Soc. Res: High pressure spectroscopy; organic solids; charge-transfer interaction; energy transfer in condensed systems. Mailing Add: Dept of Chem Univ of Calif Santa Barbara CA 93106

OFFENBACH, JACK ALLEN, physical chemistry, see 12th edition

OFFENBACHER, ELMER LAZARD, b Frankfurt am Main, Ger, Sept 29, 23; US citizen; m 52; c 3. PHYSICS. Educ: Brooklyn Col, AB, 43; Univ Pa, MS, 49, PhD(physics), 51. Prof Exp: Asst instr physics, Amherst Col, 43-44; physicist, Nat Adv Comt Aeronaut, Va, 44; asst instr physics, Univ Pa, 47-51; from asst prof to assoc prof, 51-65, PROF PHYSICS, TEMPLE UNIV, 65- Concurrent Pos: Physicist, Frankford Arsenal, US Dept Army, 53-54, consult, 54- Mem: Am Phys Soc; Am Asn Physics Teachers. Res: Solid state physics; ice physics; biomechanics; electron spin resonance. Mailing Add: Dept of Physics Temple Univ Philadelphia PA 19122

OFFENBERGER, ALLAN ANTHONY, b Lintlaw, Sask, Aug 11, 38; m 63; c 2. PLASMA PHYSICS, LASERS. Educ: Univ BC, BASc, 62, MASc, 63; Mass Inst Technol, PhD(nuclear eng), 68. Prof Exp: From asst prof to assoc prof elec eng, 68-75, PROF ELEC ENG, UNIV ALTA, 75- Mem: Can Asn Physicists. Res: Laser-plasma interactions; plasma production, heating, nonlinear interaction; scattering, absorption, interferometry diagnostics. Mailing Add: Dept of Elec Eng Univ of Alta Edmonton AB Can

OFFENHARTZ, PETER O'DONNELL, b New York, NY, June 6, 39; m 60; c 2. PHYSICAL CHEMISTRY. Educ: Swarthmore Col, BA, 60; Univ Pa, PhD(phys chem), 63. Prof Exp: NSF fel, Cambridge Univ, 63-64; NIH fel, Univ Tokyo, 64-65; fel, Harvard Univ, 65-66; asst prof chem, Univ Colo, 66-68; asst prof, Amherst Col, 68-73; sr res scientist, Phys Sci Group, Sch Educ, Boston Univ, 73-74; SR SCIENTIST, ENVIRON IMPACT CTR, 74- Concurrent Pos: NSF sci fac fel, Univ Strasbourg, France, 71-72. Res: Quantum chemistry; molecular orbital theory; ligand field theory; energy conservation; geothermal energy. Mailing Add: Environ Impact Ctr 55 Chapel St Newton MA 02158

OFFENHAUER, ROBERT DWIGHT, b Sandusky, Ohio, June 24, 18; m 42; c 3. ORGANIC CHEMISTRY. Educ: DePauw Univ, AB, 40; Univ Wis, PhD(org chem), 44. Prof Exp: Res chemist, Allied Chem & Dye Corp, 44-48; chemist, 48-59, RES ASSOC LABS, MOBIL OIL CORP, 59- Mem: Am Chem Soc. Res: Petrochemicals; microbiology; petroleum chemistry; high polymers; textiles and organic synthesis; cyclohexanoic acid and derivatives; homogeneous catalysis. Mailing Add: Mobil Labs Princeton NJ 08540

OFFENKRANTZ, WILLIAM CHARLES, b Newark, NJ, Sept 2, 24; m 53; c 2. PSYCHIATRY, PSYCHOANALYSIS. Educ: Rutgers Univ, BS, 45; Columbia Univ, MD, 47; William Alanson White Inst, NY, dipl psychoanal, 57; Chicago Inst Psychoanal, dipl, 66. Prof Exp: PROF PSYCHIAT, UNIV CHICAGO, 57- Mem: Am Psychoanal Asn; Am Psychiat Asn; Group Advan Psychiat. Res: Clinical research in psychoanalysis; investigating relationships among dreams of the night. Mailing Add: Dept of Psychiat Univ of Chicago 950 E 59th St Chicago IL 60637

OFFER, DANIEL, b Berlin, Ger, Dec 24, 29; US citizen; m 71; c 2. PSYCHIATRY. Educ: Univ Rochester, BA, 53; Univ Chicago, MD, 57. Prof Exp: Dir residency prog & assoc dir hosp, 66-73, DIR, LAB STUDY ADOLESCENTS, MICHAEL REESE HOSP, 68-, CO-CHMN DEPT PSYCHIAT, 74-; PROF PSYCHIAT, PRITZKER SCH MED, UNIV CHICAGO, 73- Concurrent Pos: Consult, Dept Ment Health, State of Ill, 64-; mem exec comt, Nicholas Pritzker Ctr & Hosp, 68-73; mem prof adv comt, Family Inst Chicago, 69-73; ed in chief, J Youth & Adolescence, 70-; mem prof adv coun, Ment Health Div, Chicago Bd Health, 70-73; fel, Ctr Advan Study Behav Sci, Stanford, Calif, 73-74. Mem: AAAS; fel Am Psychiat Asn; AMA; Int Cong Child Psychiat & Allied Profs; Am Soc Adolescent Psychiat (pres, 72-73). Res: Study of normal adolescent boys from teenage to young manhood; analysis of data collected over the past few years on juvenile delinquents. Mailing Add: Dept Psychiat Michael Reese Hosp 29th St & Ellis Ave Chicago IL 60616

OFFICER, JAMES EOFF, b Boulder, Colo, July 28, 24; m 46; c 2. ANTHROPOLOGY, ACADEMIC ADMINISTRATION. Educ: Univ Ariz, AB, 50, PhD(anthrop), 64. Prof Exp: Instr anthrop, Univ Ariz, 56-61; from asst comnr to assoc comnr Indian affairs, Dept Interior, 61-67, asst to secy interior, 67-69; PROF ANTHROP, UNIV ARIZ, 69-, COORDR INT PROGS, 69-, DIR, GUADALAJARA SUMMER SCH, 71- Concurrent Pos: Consult, Orgn Am States, Bolivia & Mex, 65; US rep, Interam Indian Inst, 66-; reviewer, Nat Endowment for Humanities, 68-; consult, US AID, Panama, 70 & Nat Study Indian Educ, 70; lectr, Bur Educ & Cult Affairs, Dept State, Mex, 71. Honors & Awards: Citation Distinguished Serv, US Dept Interior, 68. Mem: Am Anthrop Asn (prog ed, 74); Soc Appl Anthrop. Res: of ethnic minorities with special emphasis on North American Indians and Mexican-Americans; applications of anthropology to solution of minority group problems. Mailing Add: 621 N Sawtelle Ave Tucson AZ 95716

OFFICER, JULIUS EARLE, virology, deceased

OFFIELD, TERRY WATSON, b Amarillo, Tex, May 27, 33; m 57; c 2. GEOLOGY. Educ: Va Polytech Inst, BS, 53; Univ Ill, MS, 55; Yale Univ, PhD(geol), 62. Prof Exp: Geologist, Aluminum Co Am, 55; geologist, NY State Geol Surv, 57-59; geologist, Ariz, 61-69, Colo, 69-74, COORDR URANIUM GEOPHYS PROG, US GEOL SURV, 74-, ASST CHIEF BR PETROPHYS & REMOTE SENSING, 75- Mem: AAAS; Geol Soc Am; Am Geophys Union. Res: Structural geology; Himalayan structure; lunar geology; terrestrial impact structures; remote sensing techniques for geologic mapping and uranium exploration; thermal-infrared studies. Mailing Add: Br of Petrophys & Remote Sensing US Geol Surv Denver Fed Ctr Denver CO 80225

OFFNER, FRANKLIN FALLER, b Chicago, Ill, Apr 8, 11; m 56; c 4. BIOPHYSICS. Educ: Cornell Univ, BChem, 33; Calif Inst Technol, MS, 34; Univ Chicago, PhD(physics), 38. Prof Exp: Asst, Univ Chicago, 35-38; pres, Offner Electronics, Inc, Ill, 39-63; PROF BIOPHYS, NORTHWESTERN UNIV, ILL, 63- Mem: Biophys Soc; Inst Elec & Electronics Eng; Am Phys Soc; Am Electroencephalog Soc. Res: Theory of the excitable membrane. Mailing Add: Dept of Elec Eng Northwestern Univ Evanston IL 60201

OFFNER, HARRY GIRARD, analytical chemistry, see 12th edition

OFFORD, DAVID ROBERT, b Ottawa, Ont, Nov 13, 33; m 62; c 3. CHILD PSYCHIATRY. Educ: Queens Univ, Ont, MDCM, 57; CRCP(C), 65. Prof Exp: Jr rotating intern, Montreal Gen Hosp, 57-58, resident staff psychiat, 58-59 & 59-60; fel child psychiat, Children's Serv Ctr Wyo Valley, Wilkes-Barre, Pa, 60-62; from instr to asst prof psychiat, Div Child Psychiat, Univ Fla, 62-67; assoc prof behav sci, Col Med, Pa State Univ, 67-72; PROF PSYCHIAT, UNIV OTTAWA, 72-; STAFF PSYCHIATRIST, ROYAL OTTAWA HOSP, 72-, DIR RES & TRAINING CHILDREN'S SERV, 75- Concurrent Pos: Examr psychiat, Royal Col Physicians & Surgeons Can, 75- Mem: AMA; Am Psychiat Asn; Am Acad Child Psychiat; Am Orthopsychiat Asn; Am Pub Health asn. Res: The study of the natural histories, including childhood antecedents of the severe psychosocial diseases of adulthood, especially schizophrenia, sociopathy, alcoholism, affective disorder and retardation. Mailing Add: Royal Ottawa Hosp 1145 Carling Ave Ottawa ON Can

OFFUTT, MARION SAMUEL, b Mexico, Mo, Dec 22, 18; m 45; c 1. AGRONOMY. Educ: Univ Mo, BS, 48, MS, 52, PhD(plant breeding, genetics), 54. Prof Exp: County agent agr exten, Univ Mo, 48-50, from instr to assoc prof field crops, 52-56; PROF AGRON, UNIV ARK, FAYETTEVILLE, 56- Mem: Fel Am Soc Agron; Am Genetic Asn; Crop Sci Soc Am. Res: Plant breeding; genetics; management of forage legumes. Mailing Add: Dept of Agron Univ of Ark Fayetteville AR 72701

OFFUTT, WILLIAM FRANKLIN, b Slippery Rock, Pa, Feb 24, 19; m 53; c 3. OPERATIONS RESEARCH. Educ: Grove City Col, BSc, 40; Univ Pittsburgh, PhD(phys chem), 48. Prof Exp: Physicist div war res, Columbia Univ, 42-43; mem staff opers eval group, Off Chief Naval Opers, US Dept Navy, 48-54; sr scientist, E H Smith & Co, 54-56; mem staff weapons systs eval div, Inst Defense Anal, 56-60; mgr mkt res fed systs div, Int Bus Mach Corp, 60-65; sr res assoc, Inst Defense Anal, 65-66; MGR OPERS RES, FED SYSTS DIV, IBM CORP, 67- Mem: Fel AAAS; Opers Res Soc Am (treas, 56-59). Res: Thermodynamics of solutions; operations analysis. Mailing Add: 11009 Rosemont Dr Rockville MD 20852

O'FLAHERTY, LARRANCE MICHAEL ARTHUR, b Wynyard, Sask, June 14, 41; US citizen; m 64. ALGOLOGY. Educ: Western Wash State Col, BA, 63; Ore State Univ, MA, 66, PhD(bot, phycol), 68. Prof Exp: Asst prof biol, 68-73, ASSOC PROF BIOL,

WESTERN ILL UNIV, 73- Mem: AAAS; Am Bryol & Lichenological Soc; NY Acad Sci; Int Phycol Soc; Am Soc Limnol & Oceanog. Res: Blue-green algae in culture. Mailing Add: Dept of Biol Sci Western Ill Univ Macomb IL 61455

OFNER, PETER, b Berlin, Ger, June 21, 23; US citizen. ORGANIC CHEMISTRY, ANDROLOGY. Educ: Univ London, BSc, 45, PhD(org chem), 50. Prof Exp: Res chemist, Wellcome Found, Burroughs Wellcome & Co, Eng, 46-50; asst res chemist, Courtauld Inst Biochem, Middlesex Hosp, London, 52-53; res biochemist, Nat Inst Med Res, 53-54 & Hormone Res Lab, Sch Med, Boston Univ, 55-57; CHIEF STEROID LAB, LEMUEL SHATTUCK HOSP, 57- Concurrent Pos: Econ Coop Admin fel enzymol, Univ Chicago, Sloan-Kettering Inst Cancer Res & Univ Utah, 50-52; res fel med, Harvard Univ, 58-60; res assoc med, Harvard Univ, 60-62 & Sch Dent Med, 63-, asst prof, 69-73, lectr pharm, 73-; lectr, Sch Med, Tufts Univ, 69-73, assoc prof, 73- Mem: Am Chem Soc; The Chem Soc; Brit Biochem Soc; assoc Royal Inst Chem. Res: Effects and metabolism of androgens in accessory sex organs; induction of microsomal hepatic enzymes involved in steroid metabolism; tissue culture. Mailing Add: Lemuel Shattuck Hosp 170 Morton St Jamaica Plain MA 02130

O'FOGHLUDHA, FEARGHUS TADHG, b Dublin, Ireland, Jan 8, 27; m 56; c 2. MEDICAL PHYSICS. Educ: Nat Univ Ireland, BSc, 48, MSc, 49, PhD(physics), 61; Am Bd Radiol, cert, 65. Prof Exp: Asst physics, Univ Col, Dublin, 50-54; sr physicist, St Luke's Hosp, Dublin, 54-63; assoc prof radiation physics, Med Col Va, 63-65, prof & chmn div, 65-70; PROF RADIATION PHYSICS & DIR DIV, MED CTR, DUKE UNIV, 70- Concurrent Pos: Vis scientist, Oak Ridge Assoc Univs, 56, counr, 64-; res assoc, Univ Mich, 57; vis lectr, Univ Col Dublin, 58-63; adj prof physics, Duke Univ, 75-; US ed, Physics Med & Biol, 76- Mem: Am Phys Soc; Am Asn Physicists in Med (pres, 71); Am Col Radiol; Brit Inst Physics (fel Phys Soc). Res: Gamma ray spectrometry. Mailing Add: Dept of Radiol Duke Univ Med Ctr Durham NC 27710

OFSTEAD, EILERT A, b Minneapolis, Minn, Dec 15, 34; m 62; c 3. POLYMER CHEMISTRY. Educ: St Thomas Col, BS, 56; Univ Md, PhD(org chem), 63. Prof Exp: Am Dent Asn res asst dent, Nat Bur Standards, 57-58; RES SCIENTIST, GOODYEAR TIRE & RUBBER CO, 62- Mem: Am Chem Soc. Res: Reaction mechanisms; ring-opening polymerizations; synthetic elastomers. Mailing Add: Res Div Goodyear Tire & Rubber Co Akron OH 44316

OFSTEAD, RONALD FERDINAND, organic chemistry, polymer chemistry, see 12th edition

OFTEDAHL, MARVIN LOREN, b Chicago, Ill, June 21, 31; m 58; c 1. ORGANIC CHEMISTRY. Educ: Northwestern Univ, BA, 56; Wash Univ, St Louis, PhD(carbohydrate chem), 60. Prof Exp: Sr res chemist org chem div, 60-67, RES SPECIALIST, INDUST CHEM CO, MONSANTO CO, 67- Mem: Am Chem Soc; Sigma Xi. Res: Chemistry of carbyhydrates; organosulfur chemistry; oxygen and nitrogen heterocyclics; food and flavor chemistry; pharmaceuticals. Mailing Add: Monsanto Co 800 N Lindbergh Blvd St Louis MO 63166

O'GALLAGHER, JOSEPH JAMES, b Chicago, Ill, Oct 23, 39; m 63; c 2. PHYSICS. Educ: Mass Inst Technol, SB, 61; Univ Chicago, SM, 62, PhD(physics), 67. Prof Exp: Res assoc physics, Enrico Fermi Inst, Univ Chicago, 67-70; ASST PROF PHYSICS, UNIV MD, COLLEGE PARK, 71- Concurrent Pos: Vis fel, Max Planck Inst Physics, 75-76; fel, Alexander Von Humboldt Found, 75. Mem: Am Phys Soc; Am Geophys Union. Res: Space physics; cosmic ray modulation; interplanetary and interstellar propagation of solar and galactic cosmic radiation; detection techniques for energetic charged particles; solar composition. Mailing Add: Dept of Physics & Astron Univ of Md College Park MD 20801

O'GARA, BARTHOLOMEW WILLIS, b Laurel, Nebr, Mar 21, 23; m 49. ZOOLOGY, FISH & GAME MANAGEMENT. Educ: Mont State Univ, BS, 64; Univ Mont, PhD(zool), 68. Prof Exp: WILDLIFE BIOLOGIST, BUR SPORT FISHERIES & WILDLIFE & ASST LEADER, MONT COOP WILDLIFE RES UNIT, UNIV MONT, 68- Concurrent Pos: Affil prof zool & forestry, Univ Mont, 70-; Bur Sport Fisheries & Wildlife Refuge Div grant, Red Rocks Nat Wildlife Refuge, Mont, 71- & res grant coyote-domestic sheep interactions, 74-, US Forest Serv grant elk migrations, 73- Mem: Soc Study Reproduction; Am Soc Mammalogists; Wildlife Soc. Res: Mammalian reproduction, particularly of game species. Mailing Add: Bur of Sport Fisheries & Wildlife Mont Coop Wildlife Res Unit Missoula MT 59801

O'GARA, ROGER, b Lynn, Mass, July 22, 15; wid. PATHOLOGY. Educ: George Washington Univ, AB, 48, MD, 51. Prof Exp: Intern, USPHS Hosp, Brighton, Mass, 51-52; resident path, NIH, 52-53; resident path, Hosp Univ Pa, 53-56; PATHOLOGIST & MED DIR, NAT CANCER INST, 56- Mem: AMA; Am Soc Exp Path; Am Asn Cancer Res; Int Acad Path. Res: Carcinogenesis and etiology of cancer; histopathologic effects of chemotherapeutic agents. Mailing Add: Rm 2N-113 Bldg 10 Nat Cancer Inst Lab of Path Bethesda MD 20014

OGARD, ALLEN E, b Ada, Minn, Dec 9, 31; m 56; c 2. NUCLEAR CHEMISTRY. Educ: St Olaf Col, BA, 53; Univ Chicago, PhD(chem), 57. Prof Exp: Res chemist, Westinghouse Res Labs, 57; staff mem chem, Los Alamos Sci Lab, 57-70; guest scientist, Swiss Inst Reactor Res, 70-72; CHEMIST, LOS ALAMOS SCI LAB, 72- Concurrent Pos: Res fel, Inst Transurane, Karlsruhe, Ger, 64-65. Mem: Am Chem Soc. Res: Preparation and properties of high temperature plutonium reactor fuel materials; production of medically useful radioactive isotopes by spallation reactions. Mailing Add: CNC-11 MS-514 Los Alamos Sci Lab Los Alamos NM 87545

OGASAWARA, FRANK X, b San Diego, Calif, Nov 10, 13; m 45; c 3. COMPARATIVE PHYSIOLOGY. Educ: Univ Calif, BS, 50, PhD, 57. Prof Exp: Res physiologist reprod, 55-58, asst prof poultry reprod, 58-66, assoc prof poultry reprod & animal physiol, 66-73, assoc physiologist, Exp Sta, 69-74, PROF ANIMAL PHYSIOL & AVIAN SCI, UNIV CALIF, DAVIS, 73-, PHYSIOLOGIST EXP STA, 74- Mem: Am Poultry Sci Asn; Am Fertil Soc; World Poultry Sci Asn. Res: Poultry reproductive physiology. Mailing Add: Dept of Animal Physiol Univ of Calif Davis CA 95616

OGATA, GEN, soils, see 12th edition

OGATA, HISASHI, b Tokyo, Japan, June 10, 26; m 58; c 2. NUCLEAR PHYSICS. Educ: Tokyo Col Sci, BSc, 55; Tokyo Univ Educ, MSc, 57; Case Western Reserve Univ, PhD(physics), 63. Prof Exp: Res assoc nuclear physics nuclear data proj, Nat Acad Sci-Nat Res Coun, 62-63; res assoc nuclear data proj physics div, Oak Ridge Nat Lab, 64-65; asst prof, 65-68, ASSOC PROF NUCLEAR PHYSICS, UNIV WINDSOR, 68- Concurrent Pos: Nat Res Coun Can grant, 65-; Ont Dept Univ Affairs res grant, 67-70; vis scientist, Univ Tokyo, 74-75. Mem: Am Phys Soc; Can Asn Physicists. Res: Theoretical nuclear physics, particularly nuclear structure and low energy nuclear properties. Mailing Add: Dept of Physics Univ of Windsor Windsor ON Can

OGAWA, HAJIMU, b Pasadena, Calif, June 18, 31; m 58; c 2. MATHEMATICS. Educ: Calif Inst Technol, BS, 53; Univ Calif, Berkeley, PhD(appl math), 61. Prof Exp: Lectr math, Univ Calif, Riverside, 60-61, from asst prof to assoc prof, 61-68; PROF MATH, STATE UNIV NY ALBANY, 68- Concurrent Pos: NSF grants, 62-68 & 69-71; vis asst prof, Univ Calif, Berkeley, 63-64; vis mem, Courant Inst Math Sci, NY Univ, 66-67. Mem: Am Math Soc; Math Asn Am; Soc Indust & Appl Math. Res: Partial differential equations. Mailing Add: Dept of Math State Univ of NY Albany NY 12222

OGAWA, JOSEPH MINORU, bSanger, Calif, Apr 24, 15; m 54; c 3. PLANT PATHOLOGY. Educ: Univ Calif, BS, 50, PhD(plant path), 54. Prof Exp: Asst specialist plant path, 53-55, lectr, 55-62, from asst prof to assoc prof, 62-68, jr plant pathologist, 55-57, from asst plant pathologist to assoc plant pathologist, 57-68, PROF PLANT PATH & PLANT PATHOLOGIST, UNIV CALIF, DAVIS, 68- Concurrent Pos: Mem staff, UN Food & Agr Orgn, 67-68. Mem: AAAS; Am Phytopath Soc. Res: Diseases of deciduous fruit and nut crops; hop diseases; postharvest diseases; fungicides. Mailing Add: Dept of Plant Path Univ of Calif Davis CA 95616

OGAWA, JUNJIRO, b Saitama Prefecture, Japan, Apr 18, 15; m 42; c 3. MATHEMATICAL STATISTICS. Educ: Tokyo Univ, BSC, 37, DSc(math statist), 54. Prof Exp: Mem, Inst Statist Math, 45-48; asst prof math, Osaka Univ, 48-55; res assoc statist, Univ NC, Chapel Hill, 55-58, vis assoc prof, 58-59; prof math, Col Sci & Eng, Nihon Univ, Tokyo, 59-65, prof statist & chmn dept, Col Prod Eng, 65-69; PROF STATIST, UNIV CALGARY, 70- Mem: AAAS; Japan Math Soc; Japan Statist Asn; Inst Math Statist; Int Statist Inst. Res: Multivariate analysis, order statistics and design of experiments. Mailing Add: Dept of Math Statist & Comput Sci Univ of Calgary 2920 24th Ave NW Calgary AB Can

OGAWA, MASARU, atomic physics, molecular physics, deceased

OGBURN, CLIFTON ALFRED, b Philadelphia, Pa, Apr 27, 30. IMMUNOLOGY. Educ: Univ Pa, PhD(bact), 57. Prof Exp: From instr immunol in pediat to asst prof pediat, Sch Med, Univ Pa, 57-69; assoc prof, 69-75, PROF MICROBIOL, MED COL PA, 75-, MEM MED STAFF, 69- Mem: Am Soc Microbiol; NY Acad Sci. Res: Genesis of antibody; serology; transplantation immunity; purification; characterization of interferon tumor immunology. Mailing Add: Dept of Microbiol Med Col of Pa Philadelphia PA 19129

OGBURN, PHILLIP NASH, b Klamath Falls, Ore, Aug 18, 40; m 61; c 2. CARDIOVASCULAR PHYSIOLOGY. Educ: Wash State Univ, DVM, 65; Ohio State Univ, PhD(cardiovasc physiol), 71. Prof Exp: Intern med & surg, Animal Med Ctr, 65-66; vet pvt pract, 66-67; ASST PROF CARDIOL, COL VET MED, UNIV MINN, ST PAUL, 71- Mem: Am Vet Med Asn; Acad Vet Cardiol. Res: Effects of myocardial hypertrophy on excitation of ventricular myocardium and conduction system; comparative electrophysiology; comparative electrocardiography. Mailing Add: 111 Vet Hosp Univ of Minn St Paul MN 55101

OGDEN, DAVID ANDERSON, b Westfield, NJ, June 25, 31; m 54; c 4. NEPHROLOGY, INTERNAL MEDICINE. Educ: Cornell Univ, BA, 53, MD, 57. Prof Exp: From asst prof to assoc prof, Univ Colo, Denver, 63-69; assoc prof, 69-74, PROF MED, MED CTR, UNIV ARIZ, 74-, CHIEF RENAL SECT, 69- Concurrent Pos: Am Col Physicians fel, 2nd Med Div, Bellevue Hosp, New York, 59-60, NY Heart Asn fel, 60-61; clin investr, Vet Admin Hosp & Med Ctr, Univ Colo, Denver, 63-66; chief renal sect, Vet Admin Hosps, Denver, Colo, 65-69 & Tucson, Ariz, 69-; mem, Nat Renal Transplant Adv Group, US Vet Admin, 71- Mem: Int Soc Nephrology; Am Soc Nephrology; Am Soc Artificial Internal Organs; Am Fedn Clin Res. Res: Function of the transplanted human kidney; divalent ion metabolism in uremia; problems in chronic hemodialysis. Mailing Add: Renal Sect Univ of Ariz Med Ctr Tucson AZ 85724

OGDEN, EDWIN BURMAN, b Battle Creek, Mich, May 5, 00; m 25; c 2. MATHEMATICAL PHYSICS. Educ: Union Col, Nebr, AB, 22; Univ Nebr, AM, 25; Boston Univ, PhD(math), 36. Prof Exp: Teacher, Intermountain Acad, 22-24; teacher, Enterprise Acad, 25-26, instr math, 26-33; prof math & physics & head dept, Atlantic Union Col, 34-38; prof math, 38-74, chmn dept, 38-68, dean col, 46-58, EMER PROF MATH, UNION COL, NEBR, 74- Mem: Am Math Soc; Math Asn Am. Res: Consequences of two assumptions of Sulaimons' theory of relativity. Mailing Add: Dept of Math Union Col Lincoln NE 68506

OGDEN, ELSTON GORDON, b Sale, WVa, Jan 4, 08; m 35; c 2. BOTANY. Educ: Salem Col, AB, 29; Univ Pittsburgh, MS, 38. Prof Exp: Instr, Warwood High Sch, WVa, 29-48; asst prof biol, 48-60, from assoc prof to prof, 60-72, summer dir inst sci teachers, 58-65, EMER PROF BIOL, ALFRED UNIV, 72- Concurrent Pos: Partic inst col bot, NSF-Bot Soc Am, Cornell Univ, 57; partic inst animal ecol, NSF-Am Soc Zool, Univ Colo, 60. Mem: Bot Soc Am; Ecol Soc Am. Res: Plant associations and revegetation; distribution and geographical isolation; biology. Mailing Add: 3 Greene St Alfred NY 14802

OGDEN, EUGENE CECIL, b Cheboygan Co, Mich, Jan 3, 05; m 38; c 2. BOTANY. Educ: Mich State Col, BS, 32; Univ Maine, MS, 34; Harvard Univ, AM, 36, PhD(bot), 38. Prof Exp: Asst, Gray Herbarium, Harvard Univ, 34-37; instr bot, Univ Maine, 38-43, asst geneticist exp sta, 43-46, asst prof & taxonomist, 46-52; state botanist, NY, 52-70; RETIRED. Concurrent Pos: Botanist div plant explor & introd, USDA, 50-51. Mem: Bot Soc Am; Am Soc Plant Taxon; Am Bryol & Lichenological Soc; Phycol Soc Am; Int Soc Plant Taxon. Res: Pollen; aquatic flowering plants; Potamogeton. Mailing Add: 175 Adams St Delmar NY 12054

OGDEN, INGRAM WESLEY, b Henryetta, Okla, Apr 29, 20; m 47; c 3. DENTISTRY. Educ: Univ Okla, BS, 41; Univ Mo, DDS, 44; Columbia Univ, cert oral biol, 55. Prof Exp: PROF ORAL DIAG, COL DENT, HOWARD UNIV, 66-; MEM STAFF DENT AUXILIARIES, MONTGOMERY COL, 70- Concurrent Pos: Ed consult, J Am Dent Asn, 66-; USPHS grant cancer training, Col Dent, Howard Univ, 70-73, USPHS grant dent therapist training, 71-72. Mem: Am Acad Oral Path. Res: Clinical recognition of oral cancer; improved health care delivery. Mailing Add: 9904 Holmhurst Rd Bethesda MD 20034

OGDEN, JAMES GORDON, III, b Martha's Vineyard, Mass, July 6, 28; m 56; c 4. ECOLOGY. Educ: Fla Southern Col, BS, 51, BA, 52; Univ Tenn, MS, 54; Yale Univ, PhD(bot), 58. Prof Exp: Asst biol, Fla Southern Col, 48-52; lab asst bot, Univ Tenn, 52-54; asst biol, Yale Univ, 54-57; climatologist, Conn Agr Exp Sta, 56; from asst prof to assoc prof bot, Ohio Wesleyan Univ, 58-63, prof & dir radiocarbon dating lab, 63-69; PROF BIOL, DALHOUSIE UNIV, 69- Concurrent Pos: Mem comt, NS Environ Control Coun, 73-77; fel, J S Guggenheim Found, 63. Mem: Am Soc Limnol & Oceanog; Am Quaternary Asn. Res: Precipitation chemistry; biogeochemistry; biogeography; pollen stratigraphy; paleoecology; climatology; microclimatic ecology; instrumentation for environmental studies; post-glacial history of vegetation and climate. Mailing Add: Dept of Biol Dalhousie Univ Halifax NS Can

OGDEN, JOHN CONRAD, b Morristown, NJ, Nov 27, 40; m 69. MARINE ECOLOGY. Educ: Princeton Univ, AB, 62; Stanford Univ, PhD(biol), 68. Prof Exp: NIH trainee genetics, Univ Calif, Berkeley, 68-69; Smithsonian Inst fel, Smithsonian Trop Res Inst, CZ, 69-71, Am Philos Soc fel, 71; asst prof marine biol, 71-74, ASSOC PROF MARINE BIOL, WI LAB, FAIRLEIGH DICKINSON UNIV, VI, 74- Mem: Soc Study Evolution; Ecol Soc Am. Res: Ecology and behavior of animals and plants in shallow water marine communities; effects of pollution on marine communities. Mailing Add: WI Lab Fairleigh Dickinson Univ PO Box 4010 Christiansted St Croix VI 00820

OGDEN, LAWRENCE, b Maryville, Mo, Nov 9, 19; m 48; c 2. GEOLOGY. Educ: Univ Tulsa, BS, 48; Univ Wis, MS, 50; Northwest Mo State Col, BS, 51; Colo Sch Mines, DSc(geol), 58. Prof Exp: Instr geol, Colo Sch Mines, 52-58, asst prof, 58-63; assoc prof, 63-66, PROF GEOL, EASTERN MICH UNIV, 66- Mem: Geol Soc Am; Nat Asn Geol Teachers. Res: Engineering geology; ground water; earth science education. Mailing Add: Dept of Geog & Geol Eastern Mich Univ Ypsilanti MI 48197

OGDEN, PHILIP MYRON, b Nampa, Idaho, Feb 3, 38; m 62; c 2. PHYSICS. Educ: Seattle Pac Col, BS, 59; Univ Calif, PhD(physics), 64. Prof Exp: From asst prof to assoc prof physics, Seattle Pac Col, 64-69; ASST PROF PHYSICS, ROBERTS WESLEYAN COL, 69-, CHMN DIV NATURAL SCI & MATH, 74- Mem: Am Phys Soc; Am Sci Affiliation; Am Asn Physics Teachers. Res: High energy or elementary particle physics; cosmic ray physics. Mailing Add: Div of Natural Sci & Math Roberts Wesleyan Col Rochester NY 14624

OGDEN, ROBERT DAVID, b Dallas, Tex, Oct 23, 41; m 61; c 2. MATHEMATICS. Educ: Southern Methodist Univ, BS, 63; Wash Univ, MA, 66, PhD(math), 70. Prof Exp: Asst prof math, 68-74, PROF MATH, DePAUL UNIV, 74- Mem: AAMS. Res: Harmonic analysis and group representations; foundations of quantum mechanics; mathematical models of complex biological systems. Mailing Add: Dept of Math DePaul Univ 25 E Jackson Blvd Chicago IL 60614

OGDEN, THOMAS E, b Lincoln, Nebr, Mar 23, 29; m 61; c 3. PHYSIOLOGY. Educ: Univ Calif, BA, 50, MD, 54, PhD(physiol), 62. Prof Exp: From res instr to res assoc prof physiol & neurol, Col Med, Univ Utah, 69-72, prof neurol, 72-75; PROF PHYSIOL, UNIV SOUTHERN CALIF & DOHENY EYE FOUND, 75- Mem: Am Physiol Soc. Res: Retinal neurophysiology. Mailing Add: Estelle Doheny Eye Found 272 S Lake St Los Angeles CA 90057

OGDEN, WILLIAM FREDERICK, b Randolph Field, Tex, Sept 11, 42. MATHEMATICS, COMPUTER SCIENCE. Educ: Univ Ark, Fayetteville, BS, 64; Stanford Univ, MS, 66, PhD(math), 69. Prof Exp: ASST PROF MATH & COMPUT SCI, CASE WESTERN RESERVE UNIV, 68- Concurrent Pos: NSF res grant, 71-72. Mem: Am Math Soc; Asn Comput Math. Res: Automata theory and programming languages. Mailing Add: Dept of Math Case Western Reserve Univ Cleveland OH 44106

OGG, ALEX GRANT, JR, b Worland, Wyo, May 3, 41; m 62; c 2. PLANT PHYSIOLOGY, AGRONOMY. Educ: Univ Wyo, BS, 63; Ore State Univ, MS, 66, PhD(bot), 70. Prof Exp: Aide weed res, Univ Wyo, 59-63; res technician, Wash, 63-66, plant physiologist, Ore, 66-69, PLANT PHYSIOLOGIST, USDA, 69- Concurrent Pos: Mem, Coun Agr Sci & Technol. Mem: Weed Sci Soc Am. Res: Weed control research in horticultural crops in the Pacific Northwest. Mailing Add: Irrigated Agr Res & Exten Ctr Prosser WA 99350

OGG, FRANK CHAPPEL, JR, b Champaign, Ill, Jan 22, 30. APPLIED MATHEMATICS. Educ: Bowling Green State Univ, BA, 51; Johns Hopkins Univ, MA & PhD(math), 55. Prof Exp: Jr instr math, Johns Hopkins Univ, 51-55; res engr, Bendix Corp, 55-57, prin res engr, 57-60; res scientist, Carlyle Barton Lab, Johns Hopkins Univ, 60-68, lectr elec eng, 61-67; ASSOC PROF MATH, UNIV TOLEDO, 68- Concurrent Pos: Indust consult. Mem: Am Math Soc; sr mem Inst Elec & Electronics Engrs. Res: Mathematical physics and engineering. Mailing Add: Dept of Math Univ of Toledo Toledo OH 43606

OGG, JAMES ELVIS, b Centralia, Ill, Dec 24, 24; m 48; c 2. MICROBIAL GENETICS. Educ: Univ Ill, BS, 49; Cornell Univ, PhD(bact), 56. Prof Exp: With US Army Biol Labs, Md, 50-53 & 56-58; PROF MICROBIOL, COLO STATE UNIV, 58-, HEAD DEPT, 67- Concurrent Pos: Consult, Martin Marietta Corp, Colo; consult-evaluator, NCent Asn Cols & Schs, 74- Mem: AAAS; Soc Indust Microbiol; Am Soc Microbiol. Res: Microbial and molecular genetics; genetic and biochemical aspects of pathogenicity in bacteria. Mailing Add: Dept of Microbiology Colo State Univ Ft Collins CO 80523

OGIER, WALTER THOMAS, b Pasadena, Calif, June 18, 25; m 54; c 4. EXPERIMENTAL PHYSICS. Educ: Calif Inst Technol, BS, 47, PhD(physics), 53. Prof Exp: From instr to asst prof physics, Univ Calif, 54-60; from asst prof to assoc prof, 60-67, PROF PHYSICS, POMONA COL, 67- Mem: AAAS; Am Phys Soc; Am Asn Physics Teachers. Mailing Add: Dept of Physics Pomona Col Claremont CA 91711

OGILVIE, ALFRED LIVINGSTON, b Vancouver, BC, Jan 20, 21; m 54; c 4. PERIODONTICS. Educ: Univ Toronto, DDS, 44; Univ Calif, MS, 48; Am Bd Periodont, dipl. Prof Exp: From instr periodont & endodontics to asst prof, Sch Dent, Univ Wash, 48-50; instr periodont & oral histol, Col Dent, Univ Calif, 50-51; from asst prof periodont & endodont to assoc prof, Sch Dent, Univ Wash, 51-65, prof periodont, 65-75; PROF ORAL MED, FAC DENT, UNIV BC, 76- Mem: Can Dent Asn; Int Asn Dent Res; Am Acad Periodont; Am Col Dentists. Res: Endodontics; oral microanatomy. Mailing Add: Univ of BC Fac Dent Vancouver BC Can

OGILVIE, BRUCE CROSSAN, b Avon, NY, June 21, 15. GEOGRAPHY, CARTOGRAPHY. Educ: RI Col, EdB, 38; Clark Univ, MA, 49, PhD(geog), 56. Prof Exp: Elem sch teacher, Chesuncook, Maine, 36-37 & Warwick, RI, 38-41; elem sch prin, Slatersville, RI, 41-42; cartogr, Off Strategic Serv, 42; map compiler, US Navy Oceanog Serv, 44-45; cartogr, Trans World Airlines, 45-47; asst prof geog, Univ Ga, 47-48; assoc prof, Chico State Col, 50-57; GEOGR, RAND McNALLY & CO, 58- Concurrent Pos: Lectr cartog, Dept Geog, Univ Chicago, 65- Mem: Asn Am Geogr. Mailing Add: Rand McNally & Co PO Box 7600 Chicago IL 60680

OGILVIE, JAMES DEUCHARS BOYD, physical chemistry, see 12th edition

OGILVIE, JAMES LOUIS, b Houston, Tex, Sept 20, 29; m 51; c 2. ANALYTICAL CHEMISTRY. Educ: Univ Tex, Austin, BS, 50, MA, 52, PhD(chem), 55. Prof Exp: Res chemist, Shell Oil Co, Tex, 54-61; sr res chemist, 61-69, res specialist, 69-74, SR RES SPECIALIST, MONSANTO CO, 74- Mem: Am Chem Soc; Soc Appl Spectros. Res: X-ray photoelectron spectroscopy applied to catalyst structure; x-ray emission spectroscopy; x-ray diffraction. Mailing Add: Monsanto Co 800 N Lindbergh Blvd St Louis MO 63166

OGILVIE, JAMES WILLIAM, JR, b Orlando, Fla, Oct 20, 25; m 50; c 3. BIOCHEMISTRY, ORGANIC CHEMISTRY. Educ: Rollins Col, BS, 50; Johns Hopkins Univ, MA, 52, PhD(chem), 55. Prof Exp: USPHS fel biochem, Sch Med, Johns Hopkins Univ, 55-57, from instr to asst prof, 57-67; ASSOC PROF BIOCHEM, SCH MED, UNIV VA, 67- Mem: AAAS; Am Soc Biol Chemists; Am Chem Soc; Biophys Soc. Res: Mechanisms of enzyme action; biochemical control mechanisms; organic and bioorganic reaction mechanisms. Mailing Add: Rte 3 Charlottesville VA 22901

OGILVIE, JOHN CHARLES, b Hyderabad, India, July 25, 25; Can citizen; m 50. APPLIED STATISTICS. Educ: St Andrews Univ, BSc, 44; Univ BC, MA, 53; Univ Toronto, PhD(psychol), 55. Prof Exp: Sci officer, Defense Res Med Labs, Can, 51-64; PROF PSYCHOL, UNIV TORONTO, 64-, DIR INST APPL STATIST, 69- Mem: Biomet Soc; Am Statist Asn; Royal Statist Soc. Res: Biological statistics; psychophysical methods; design and analysis of experiments; computer applications in biological experiments. Mailing Add: Inst of Appl Statist Univ of Toronto Toronto ON Can

OGILVIE, JOHN FRANKLIN, b Ottawa, Ont. CHEMICAL PHYSICS. Educ: Univ BC, BSc, 59, MSc, 61; Cambridge Univ, MA, 64, PhD(phys chem), 66. Prof Exp: Res asst phys chem, Cambridge Univ, 64-66; Nat Res Coun Can fel chem, 66-67; from asst prof to assoc prof, 67-75, PROF CHEM, MEM UNIV NFLD, 75- Concurrent Pos: Nat Res Coun Can res grants, 67-; Res Corp Cottrell grant, 69-; vis fel, Australian Nat Univ, 73-74. Mem: Am Phys Soc; The Chem Soc; Spectros Soc Can. Res: Inorganic chemistry; photochemistry; spectroscopy; molecular physics. Mailing Add: Dept of Chem Mem Univ of Nfld St John's NF Can

OGILVIE, KEITH W, b Solihull, Eng, Feb 20, 26; m 64; c 2. PHYSICS, SPACE SCIENCE. Educ: Univ Edinburgh, BSc, 49, PhD(physics), 54. Prof Exp: Physicist, Brit Elec & Appl Indust Res Asn, 54; fel physics, Nat Res Coun Can, 55-57; lectr, Univ Sydney, 57-60; Nat Acad Sci fel, 60-63; physicist, NASA, 63-67, sect head, Goddard Space Flight Ctr, 67-71, BR HEAD, GODDARD SPACE FLIGHT CTR, NASA, GREENBELT, MD, 71- Mem: Am Geophys Union; fel Brit Inst Physics. Res: Composition and properties of the interplanetary plasma; cosmic rays and solar produced high energy particles. Mailing Add: 101 G St SW Washington DC 20024

OGILVIE, KELVIN KENNETH, b Windsor, NS, Nov 6, 42; m 64; c 1. BIO-ORGANIC CHEMISTRY. Educ: Acadia Univ, BSc, 63, Hons, 64; Northwestern Univ, PhD(chem), 68. Prof Exp: From asst prof to assoc prof chem, Univ Man, 68-74; ASSOC PROF CHEM, McGILL UNIV, 74- Concurrent Pos: Upjohn Chem Co fel, 74- Mem: Am Chem Soc; Chem Inst Can (treas, 71-72). Res: Synthesis of nucleotides; photochemistry of biological systems; small ring compounds. Mailing Add: Dept of Chem McGill Univ Montreal PQ Can

OGILVIE, MARVIN LEE, b Pontiac, Mich, May 3, 35; m 65; c 2. PHYSIOLOGICAL CHEMISTRY. Educ: Mich State Univ, BS, 57; Univ Wis-Madison, MS, 59, PhD(biochem), 62. Prof Exp: Babcock fel biochem, Univ Wis, 62-63; asst prof biochem & soil sci, Univ Nev, Reno, 63-64; SR RES SCIENTIST BIOCHEM & PHYSIOL, UPJOHN CO, 64- Mem: AAAS; Soc Study Reproduction; Am Soc Animal Sci. Res: Physiological chemistry relating to the mechanism of drug action; development of ultrasensitive assays for hormones; sustained release dosage forms; mammalian pheromones for control of behavior and physiology. Mailing Add: Upjohn Co 9670-83-1 Kalamazoo MI 49001

OGILVIE, RICHARD IAN, b Sudbury, Ont, Oct 9, 36; m 65. CLINICAL PHARMACOLOGY. Educ: Univ Toronto, MD, 60; FRCP, 66. Prof Exp: Lectr pharmacol, 67-68, asst prof, 68-73, ASSOC PROF PHARMACOL & MED, McGILL UNIV, 73-; ASSOC PHYSICIAN, MONTREAL GEN HOSP, 72- Concurrent Pos: Fel, Can Found Adv Therapeut, 67-69; res grants, Que Med Res Coun, Med Res Coun Can, Que Heart Found & Can Found Adv Therapeut, 68-; res asst, Royal Victoria Hosp, Montreal, 68. Mem: Can Soc Clin Invest; Pharmacol Soc Can; Am Soc Clin Pharmacol; Am Soc Pharmacol & Exp Therapeut. Res: Clinical pharmacology of drugs affecting the cardiovascular system and metabolism. Mailing Add: Dept of Pharmacol McGill Univ Montreal PQ Can

OGILVY, C STANLEY, b New Rochelle, NY, June 16, 13. MATHEMATICS. Educ: Williams Col, BA, 34; Columbia Univ, MA, 40; Syracuse Univ, PhD, 54. Prof Exp: Instr math, Trinity Col, Conn, 46-49; lectr, Columbia Univ, 50-51 & Syracuse Univ, 52-53; from instr to prof, 53-75, EMER PROF MATH, HAMILTON COL, NY, 75- Mem: Fel AAAS; Math Asn Am; Am Math Soc. Res: Number theory; geometry; extremals. Mailing Add: 943 Greacen Point Rd Mamaroneck NY 10543

OGILVY, WINSTON STOWELL, b Le Mars, Iowa, Dec 3, 18; m 46; c 3. FOOD SCIENCE. Educ: Iowa State Univ, BS, 41, PhD(food technol), 50. Prof Exp: Bacteriologist, Armour & Co, 50-52, head bact sect, 52-56, assoc tech dir, 56-57; dir nutrit prod develop, Mead Johnson & Co, 57-66, VPRES FOOD PROD RES & DEVELOP, MEAD JOHNSON & CO DIV, BRISTOL-MYERS CO, 66- Mem: Am Oil Chem Soc; Am Chem Soc; Inst Food Technol. Res: Development of infant formulas, therapeutic foods; nutrition; food microbiology; research and development administration. Mailing Add: Mead Johnson Res Ctr 2404 Pennsylvania Ave Evansville IN 47721

OGIMACHI, NAOMI NEIL, b Los Angeles, Calif, Oct 10, 25; m 53; c 4. FLUORINE CHEMISTRY. Educ: Univ Calif, Los Angeles, BS, 50; Univ Calif, Berkeley, PhD(org chem), 55. Prof Exp: Chemist, Naval Weapons Test Sta, 50-52; asst chem, Univ Calif, Davis, 52-55; res chemist, E I du Pont de Nemours & Co, 55-57; res chemist, Naval Weapons Test Sta, 57-59; res specialist, Rocketdyne Div, Rockwell Int, 59-69; res chemist, Halocarbon Prod Corp, NJ, 70-75; staff scientist, Fluorochem Inc, 75; RES CHEMIST, TELEDYNE McCORMICK SELPH, 75- Mem: AAAS; Am Chem Soc; The Chem Soc; Sigma Xi. Res: Chemistry of inorganic fluorine compounds; hydrazine and ammonia chemistry; organic synthesis; aromatic molecular complexes; characterization of liquid rocket propellants; analysis of solid propellants and explosives; fluorocarbon chemistry. Mailing Add: Teledyne McCormick Selph PO Box 6 Hollister CA 95023

OGINSKY, EVELYN LENORE, b New York, NY, Apr 6, 19. MICROBIAL PHYSIOLOGY. Educ: Cornell Univ, BA, 38; Univ Chicago, MS, 39; Univ Md, PhD(bact), 46. Prof Exp: Instr bact, Univ Md, 42-46; res assoc, Merck Inst Therapeut Res, 48-56; assoc prof, 57-63, prof bact, Med Sch, Univ Ore, 63-73; PROF MICROBIOL & ASSOC DEAN, GRAD SCH BIOMED SCI, UNIV TEX HEALTH SCI CTR, SAN ANTONIO, 74- Concurrent Pos: Donner Found fel, Harvard Med Sch, 46-47. Mem: AAAS; Am Soc Microbiol; Am Soc Biol Chemists; Am Acad Microbiol; Brit Soc Gen Microbiol. Res: Physiology and metabolism of microorganisms. Mailing Add: Grad Sch of Biomed Sci Univ of Tex Health Sci Ctr San Antonio TX 78284

OGLE, HARLAN VANCE, physical chemistry, see 12th edition

OGLE, JAMES D, b Cleveland, Ohio, Feb 16, 20; m 42. BIOCHEMISTRY. Educ: Miami Univ, BS, 42; Univ Cincinnati, MS, 49, PhD(biochem), 52. Prof Exp: Asst prof, 53-65, ASSOC PROF BIOL CHEM, COL MED, UNIV CINCINNATI, 65- Mem: Am Chem Soc. Res: Proteolytic enzymes; complement proteins. Mailing Add: Dept of Biol Chem Univ of Cincinnati Col of Med Cincinnati OH 45267

OGLE, PEARL REXFORD, JR, b Columbus, Ohio, June 27, 28; m 53; c 6. INORGANIC CHEMISTRY, PHYSICAL CHEMISTRY. Educ: Capital Univ, BS, 50; Ohio State Univ, MS, 52; Mich State Univ, PhD(chem), 55. Prof Exp: Prin scientist, Goodyear Atomic Corp, 55-64; asst prof, 64-70, ASSOC PROF CHEM, OTTERBEIN COL, 70-, CHMN DEPT, 74- Concurrent Pos: Res Corp grant, 65-; Air Force Off Sci Res grant, 66-70. Mem: AAAS; Chem Soc. Res: Mixed solvents; nitrogen dioxide-hydrogen fluoride; nitrosyl fluoride-hydrogen fluoride; nitryl fluoride-hydrogen fluoride; chemical isotope separation; uranium-molybdenum; nitrogen oxide-transition metal fluoride chemistry; selective ion electrodes. Mailing Add: Dept of Chem Otterbein Col Westerville OH 43081

OGLE, THOMAS FRANK, b St Paul, Minn, Oct 10, 42; m 66; c 2. REPRODUCTIVE ENDOCRINOLOGY. Educ: Purdue Univ, BS, 66; Wash State Univ, MS, 69, PhD(wildlife biol & zool), 73. Prof Exp: Fel endocrinol, Sch Med, Univ Va, 72-74; ASST PROF PHYSIOL, MED COL GA, 74- Concurrent Pos: NIH fel, 74. Honors & Awards: Nat Wildlife Fedn Fel Award, 70. Mem: Sigma Xi; Endocrine Soc. Res: Influence of ACTH and hyperadrenocorticoidism on ovarian function and maintenance of pregnancy with particular reference to natural regulation of population density. Mailing Add: Dept of Physiol Med Med Col of Ga Augusta GA 30902

OGLE, WAYNE LEROY, b Knoxville, Tenn, Dec 23, 22; m 48; c 2. HORTICULTURE. Educ: Univ Tenn, BS, 48; Univ Del, MS, 50; Univ Md, PhD(hort), 52. Prof Exp: Asst hort, Univ Del, 49-50; from asst to asst prof, Univ Md, 50-54; asst prof, Univ RI, 54-57; assoc prof, 57-66, PROF HORT, CLEMSON UNIV, 66- Mem: Am Soc Hort Sci; Inst Food Technologists. Res: Mineral nutrition of plants; herbicides and breeding of vegetable crops. Mailing Add: Dept of Hort Clemson Univ Clemson SC 29631

OGLE, WILLIAM ELWOOD, b Los Angeles, Calif, Aug 30, 17; m 72; c 5. PHYSICS. Educ: Univ Nev, AB, 40; Univ Ill, MS, 42, PhD(physics), 44. Hon Degrees: ScD, Univ Nev, 63. Prof Exp: Instr physics, Univ Ill, 43-44; exp physicist, Los Alamos Sci Lab, NMex, 44-52, alternate J div leader, 52-65, J div leader, 65-72; CONSULT, 72- Concurrent Pos: Consult, Energy Res & Develop Agency, 72- & Defense Nuclear Agency, 73- Honors & Awards: Dept of Defense Award, 56, Distinguished Pub Serv Medal, 66; Dept Navy Distinguished Serv Medal, 63; AEC Citation, 71. Mem: Fel Am Phys Soc; Am Nuclear Soc; Am Geophys Union. Res: Gamma ray energies; gamma-proton processes; nuclear weapons testing; environmental effects; nuclear weapons effects and testing; geothermal energy development. Mailing Add: 3801B W 44th Ave Anchorage AK 99503

OGLESBY, GAYLE ARDEN, b McGehee, Ark, Mar 11, 25; m 46; c 2. PETROLEUM GEOLOGY, MINERALOGY. Educ: Univ Ark, BS, 51, MS, 52. Prof Exp: Staff geologist, Ark State Geol Surv, 52-55 & Ohio Oil Co, 55-56; dist geologist, Oil & Gas Div, Reynolds Mining Corp, 56-57; wellsite geologist, Petrobras Exploracao, Brazil, 57-58; regional geologist, Br Mineral Classification, Gulf Coast Region, 58-73, CHIEF BR MARINE EVAL, US GEOL SURV, 73- Mem: Am Asn Petrol Geologists; Am Inst Prof Geologists. Res: Petroleum geology exploration and development; geochemical and mineralogical analysis to determine stratigraphic correlations; oil, gas, sulfur and salt reservoir and deposit studies and evaluations in the outer continental shelf. Mailing Add: US Geol Surv Nat Ctr Reston VA 22092

OGLESBY, LARRY CALMER, b Corvallis, Ore, Mar 26, 36; m 64; c 2. INVERTEBRATE ZOOOLOGY, COMPARATIVE PHYSIOLOGY. Educ: Ore State Col, BA & BS, 58; Fla State Univ, MS, 60; Univ Calif, Berkeley, PhD(zool), 64. Prof Exp: Asst prof biol, Reed Col, 64-67; NATO fel, Univ Newcastle, 67-68; asst prof, 68-71, ASSOC PROF ZOOL, POMONA COL, 71- Concurrent Pos: NSF res grant, 65-67; vis prof, Marine Biol Inst, Univ Ore, 69, 70 & 72 & Col William & Mary, 74-75. Mem: Brit Soc Exp Biol; Am Soc Zoologists; Marine Biol Asn UK. Res: Physiology of osmotic and ionic regulation in polychaete annelids and sipunculans; annelid parasites and other symbionts; ecological physiology of brackish water invertebrates; life cycles of trematodes. Mailing Add: Dept of Zool Pomona Col Claremont CA 91711

OGLESBY, RAY THURMOND, b Lynchburg, Va, Apr 16, 32; m 56; c 3. AQUATIC BIOLOGY. Educ: Univ Richmond, BS, 53; Col William & Mary, MA, 55; Univ NC, PhD(environ biol), 62. Prof Exp: Res biologist, Bur Com Fisheries, US Fish & Wildlife Serv, 58-59; res asst prof sanit biol, Univ Wash, 62-66, from res asst prof to assoc prof appl biol, 66-68; ASSOC PROF DEPT CONSERV, CORNELL UNIV, 68-, TASK GROUP LEADER AQUATIC SCI, COL AGR & LIFE SCI, 71- Concurrent Pos: Consult, Mobil Oil Co, Wash, 65, Rockefeller Found, 70-71 & Village of Lake Placid, 71; co-prin investr, Pac Northwest Pulp & Paper Asn grant, 65-67; prin investr, USPHS grant, 65-68, res contract, 66-68; prin investr, NSF res grant, 67-68; Int Aluminum Corp res contract, 67-68; Off Water Resources Res res grant, 68-71; Cayuga County res grant, 71-; co-prin investr, Rockefeller Found res grant, 71-; sci adv, NY State Assembly Comn on Environ Conserv; mem, Am Inst Biol Sci Life Sci Team Assessment Biol Impacts, 74-; vis scientist, Water Res Ctr, Stevenace, Eng, 75-76. Mem: Am Soc Limnol & Oceanog; Ecol Soc Am; Marine Biol Asn UK. Res: Lake and estuarine eutrophication; effects of pollutants on the biota of receiving waters; ecology of aquatic microbial communities. Mailing Add: Fernow Hall Cornell Univ Ithaca NY 14850

OGLIARUSO, MICHAEL ANTHONY, b Brooklyn, NY, Aug 10, 38; m 61; c 1. PHYSICAL ORGANIC CHEMISTRY. Educ: Polytech Inst Brooklyn, BS, 60, PhD(chem), 65. Prof Exp: Teaching fel chem, Polytech Inst Brooklyn, 61-62, instr phys chem, 64; actg asst prof org chem, Univ Calif, Los Angeles, 65-66, res asst phys org chem, 66-67; asst prof, 67-72, ASSOC PROF ORG CHEM, VA POLYTECH INST & STATE UNIV, 72- Res: Electron-paramagnetic resonance spectroscopy; observations and reactions of organic radical ions; non-benzenoid aromatic molecules, stable organic anions; dianions and carbonium ions. Mailing Add: Dept of Chem Va Polytech Inst & State Univ Blacksburg VA 24061

O'GORMAN, JOHN MICHAEL, b Seattle, Wash, Jan 29, 16; m 39; c 2. CHEMISTRY. Educ: Univ Calif, BS, 40; Calif Inst Technol, MS, 41, PhD(chem), 46. Prof Exp: Instr chem, Calif Inst Technol, 42-45; asst prof, Univ Calif, 46-50; chemist, Nat Bur Standards, 51-53; assoc prof, 53-56, PROF CHEM & CHMN DEPT, PRATT INST, 56- Mem: Am Chem Soc. Res: Optical activity; precision distillation. Mailing Add: Dept of Chem Pratt Inst 215 Ryerson St Brooklyn NY 11205

O'GRADY, LAWRENCE J, b Montreal, Que, Dec 20, 17; m 51; c 3. SOIL FERTILITY. Educ: St Laurent Col, BA, 40; Okla Agr Inst, BSA, 44; Mich State Univ, PhD(soil sci), 48. Prof Exp: Asst prof soil sci, Okla Agr Inst, 48-50, prof, 50-54, head dept, 54-62; head dept, 62-72, PROF SOIL SCI, LAVAL UNIV, 72- Res: Soil fertility and chemistry. Mailing Add: Dept of Soil Sci Laval Univ Ste-Foy PQ Can

OGREN, DAVID ERNEST, b Wichita, Kans, Aug 4, 30. GEOLOGY. Educ: Ore State Col, BS, 57, MS, 58; Northwestern Univ, PhD(geol), 61. Prof Exp: Instr geol, Univ Wis, 61-62; asst prof, 62-70, ASSOC PROF GEOL, GA STATE UNIV, 70- Mem: AAAS; Soc Econ Paleont & Mineral; Geol Soc Am; Am Asn Petrol Geol. Res: Stratigraphy; paleontology. Mailing Add: Dept of Geol Ga State Univ Atlanta GA 30303

OGREN, HAROLD OLOF, b Grayling, Mich, Apr 24, 43; m 68; c 2. EXPERIMENTAL HIGH ENERGY PHYSICS. Educ: Univ Mich, BS, 65; Cornell Univ, MS, 67, PhD(physics), 70. Prof Exp: Vis scientist high energy physics, Nat Lab Frascati, 70-73; fel, Europ Asn Nuclear Res, 73-75; ASST PROF PHYSICS, IND UNIV, BLOOMINGTON, 75- Mem: Am Phys Soc. Res: Strong interaction physics of fundamental particles. Mailing Add: Dept of Physics Ind Univ Bloomington IN 47401

OGREN, HERMAN AUGUST, b Kenosha, Wis, Mar 31, 25; m 51; c 6. ECOLOGY. Educ: Univ Wis, BS, 50; Univ Mont, MS, 54; Univ Southern Calif, PhD(zool), 60. Prof Exp: State biologist, Mont Fish & Game Comn, 52-54; lab assoc zool, Univ Southern Calif, 54-56; instr, Exten Div, Univ NC, 56-57; state biologist, Dept Game & Fish, NMex, 57-60; asst prof zool, Elmhurst Col, 60-63; assoc prof, 63-70, bd trustees fac res grant, 64-65, PROF ZOOL, CARTHAGE COL, 70- Concurrent Pos: NSF grant, 62. Honors & Awards: Ann Honorarium, Am Soc Mammal, 55. Mem: Am Soc Mammal; Wildlife Soc. Res: Fish and wildlife technology; wild sheep of United States, Canada and Mexico, and of Essox Maskinonge in northern Wisconsin; presence of radioactivity in animals. Mailing Add: Dept of Zool Carthage Col Kenosha WI 53140

OGREN, PAUL JOSEPH, b Madrid, Iowa, July 3, 41; m 63; c 2. PHYSICAL CHEMISTRY. Educ: Earlham Col, BA, 63; Univ Wis, PhD(chem), 68. Prof Exp: From asst prof to assoc prof chem, Maryville Col, 67-72; ASSOC PROF CHEM, CENT COL, IOWA, 72- Concurrent Pos: Advan Study Prog fel, Nat Ctr Atmospheric Res, 71-72. Mem: Am Chem Soc. Res: Solid state radiation chemistry; pulse radiolysis; photochemistry. Mailing Add: Dept of Chem Cent Col Pella IA 50219

OGREN, ROBERT EDWARD, b Jamestown, NY, Feb 9, 22; m 48; c 2. ZOOLOGY. Educ: Wheaton Col, BA, 47; Northwestern Univ, MS, 48; Univ Ill, PhD(zool, physiol), 53. Prof Exp: Asst zool, Univ Ill, 48-53; asst prof physiol & zool, Ursinus Col, 53-57; asst prof histol & anat, Dickinson Col, 57-63, actg chmn dept biol, 59-60; ASSOC PROF EMBRYOL, HISTOL & CELL BIOL, WILKES COL, 63- Mem: Fel AAAS; Am Soc Zoologists; Am Soc Parasitologists; Am Micros Soc; Soc Protozoologists. Res: Comparative morphology; cytology; development and physiology of tapeworm hexacanth embryos; biology of land planarians; invertebrates; parasitology; histology; embryology. Mailing Add: Dept of Biol Wilkes Col Wilkes-Barre PA 18703

OGREN, WILLIAM LEWIS, b Ashland, Wis, Oct 8, 38; m 67; c 3. PLANT PHYSIOLOGY. Educ: Univ Wis-Madison, BS, 61; Wayne State Univ, PhD(biochem), 65. Prof Exp: Res chemist, Parker Div, Hooker Chem Corp, 61-62; PLANT PHYSIOLOGIST, REGIONAL SOYBEAN LAB, USDA, 65- Concurrent Pos: Asst prof plant physiol, Univ Ill, Urbana, 66-72, assoc prof, 72- Mem: Am Soc Plant Physiologists; Am Soc Agron; Crop Sci Soc Am. Res: Biochemistry, physiology and genetics of photosynthesis in soybeans and other crops; photorespiration. Mailing Add: Dept of Agron Univ of Ill Urbana IL 61801

OGRYZLO, ELMER ALEXANDER, b Dauphin, Man, Aug 18, 33; m 59; c 4. PHYSICAL CHEMISTRY. Educ: Univ Man, BSc, 55, MSc, 56; McGill Univ, PhD(phys chem), 59. Prof Exp: Exhibition of 1851 overseas fel, Univ Sheffield, 58-59; from instr to assoc prof chem, 59-71, PROF CHEM, UNIV BC, 71- Concurrent Pos: Nat Res Coun Can sr res fel, Univ Amsterdam, 66-67. Mem: Am Chem Soc; Chem Inst Can; The Chem Soc. Res: Kinetics of halogen atom reactions; reactions of electronically excited oxygen molecules; spectroscopy of small molecules; chemiluminescence. Mailing Add: Dept of Chem Univ of BC Vancouver BC Can

OGRYZLO, METRO ALEXANDER, b Dauphin, Man, Mar 18, 15; m 42; c 3. INTERNAL MEDICINE. Educ: Univ Man, MD, 38; FRCPS(C), 47. Prof Exp: Clin teacher med, 52-54, assoc, 54-55, from asst prof to assoc prof, 55-68, PROF MED, UNIV TORONTO, 68-, DIR RHEUMATIC DIS UNIT, WELLESLEY HOSP, 62-, CONSULT, 64- Concurrent Pos: Ed, J Rheumatol. Mem: Am Rheumatism Asn; NY Acad Sci; Can Rheumatism Asn; Can Med Asn; Can Soc Clin Invest. Res: Arthritis and rheumatism; gout; serum proteins. Mailing Add: Univ of Toronto Rheum Dis Unit Wellesley Hosp Toronto ON Can

OGUR, MAURICE, b New York, NY, Nov 29, 14; m 37; c 2. BIOCHEMISTRY. Educ: Brooklyn Col, BA, 34; Columbia Univ, MA, 37, PhD, 48. Prof Exp: Tutor chem, Brooklyn Col, 37-40, instr, 41-52, asst prof, 52-53; assoc prof, 53-61, res assoc, 50-51, PROF BIOL, BIOL RES LAB, SOUTHERN ILL UNIV, 61-, CHMN DEPT, 64- Concurrent Pos: Instr, Nurses Training Prog, Dept Hosps, Brooklyn Col, 42-43; res assoc, Univ Pa, 48-49; Fulbright res scholar, France, 60-61; fel, Orgn European Econ Co-op, France, 61; mem bd dir, Cent States Univs. Mem: AAAS; Am Chem Soc; Am Soc Microbiol. Rès: Genetic and physiological regulation of enzymes; metabolic pathways; biochemical inheritance in yeasts; mutation to respiration deficiency. Mailing Add: Dept of Microbiol Southern Ill Univ Carbondale IL 62901

OGURA, JOSEPH H, b San Francisco, Calif, May 25, 15; m 42; c 3. OTOLARYNGOLOGY. Educ: Univ Calif, BA, 37, MD, 41; Am Bd Otolaryngol, dipl. Prof Exp: Intern, San Francisco County Hosp, 40-41; asst resident med & path, Univ Calif Hosp, 41-42; resident path, Cincinnati Gen Hosp, 42-43, resident & instr med, 43-45; resident otolaryngol, Barnes Hosp, St Louis, 45-48; from instr to prof, 48-66, LINDBURG PROF OTOLARYNGOL & HEAD DEPT, SCH MED, WASH UNIV, 66- Concurrent Pos: Attend otolaryngologist, Barnes, Children's & McMillan Hosps; consult otolaryngologist, Jewish, Vet Admin & St Louis City Hosps; mem bd dir, Am Bd Otolaryngol; mem, Nat Cancer Adv Bd. Mem: Am Soc Ophthal & Otolaryngol Allergy; Am Col Surgeons; Am Acad Ophthal & Otolaryngol; Am Laryngol Asn (pres); Am Acad Facial Plastic & Reconstruct Surg. Res: Head and neck surgery; transplantation of larynx; naso-pulmonary mechanics. Mailing Add: Dept of Otolaryngol Wash Univ Sch of Med St Louis MO 63110

OGZEWALLA, CHARLES DWAYNE, pharmacy, pharmacognosy, see 12th edition

OH, CHAN SOO, b Kwangju, Korea, July 4, 38; m 68; c 2. ORGANIC CHEMISTRY. Educ: Seoul Nat Univ, BS, 61; St John's Univ, MS, 67; PhD(org chem), 70. Prof Exp: Sr chemist, Columbia Pharmaceut Corp, 65-70; mem tech staff, David Sarnoff Res Ctr, 70-72; sr chemist, Helipot Div, Beckman Instruments, 72-74; mem tech staff, David Sarnoff Res Ctr, RCA Labs, 74; SR CHEMIST, HELIPOT DIV, BECKMAN

INSTRUMENTS, 75- Mem: Am Chem Soc. Res: Organic synthesis; liquid crystal materials. Mailing Add: Beckman Instruments Inc 2500 Harbor Blvd Fullerton CA 92634

OH, JANG OK, b Seoul, Korea, Jan 15, 27; m 55; c 1. VIROLOGY, PATHOLOGY. Educ: Severance Union Med Col, MD, 48; Univ Wash, PhD(microbiol), 60. Prof Exp: Instr microbiol, Severance Union Med Col, 49-51; resident path, Hamot Hosp, Erie, Pa, 53-55; teaching & res asst, Sch Med, Univ Wash, 57-59, res assoc, 59-60, res instr, 60-61; res assoc path, Fac Med, Univ BC, 61-63, asst prof, 63-66; from asst to assoc res microbiologist, 66-71, RES MICROBIOLOGIST, FRANCIS I PROCTOR FOUND, MED CTR, UNIV CALIF, SAN FRANCISCO, 71- Concurrent Pos: Res fel microbiol, Commun Dis Lab, Med Ctr, Ind Univ, 55-56; Lederle med fac award, 63-66. Mem: AAAS; Am Soc Microbiol; Am Soc Exp Path; Soc Exp Biol & Med. Res: Ocular virology, pathogenesis and treatment; nonspecific resistance to viral infection, its mechanism and application to experimental infection; chemotherapy of ocular viral infection. Mailing Add: Proctor Found Univ of Calif Med Ctr San Francisco CA 94143

OH, SHIN JOONG, b Seoul, Korea, Nov 16, 36; US citizen; m 66; c 3. NEUROLOGY. Educ: Seoul Nat Univ, Korea, MD, 60, Master Med, 62. Prof Exp: Asst prof neurol, Meharry Med Col, 68-70; asst prof, 70-72, ASSOC PROF NEUROL, SCH MED, UNIV ALA, BIRMINGHAM, 72- Concurrent Pos: Res fel, Inst Endemic Dis, Seoul Nat Univ Hosp, 64-66 & Epidemiol & Genetic Unit, Univ Minn Med Ctr, 68; consult, Brookwood Hosp, Birmingham, 75; chief, Neurol Serv, Vet Admin Hosp & dir, Electromyogram Lab, Univ Hosp, Birmingham, 70- Mem: Am Acad Neurol; Am Asn Electromyography & Electrodiagnosis. Res: Neuromuscular disease; electromyography and electrodiagnosis. Mailing Add: Dept of Neurol Univ of Ala Univ Sta Birmingham AL 35294

OH, WILLIAM, b Philippines, May 22, 31; m 60; c 2. PEDIATRICS. Educ: Xavier Univ, Philippines, BA, 53; Univ Santo Tomas, Manila, MD, 58. Prof Exp: Res assoc, Karolinska Inst, Sweden, 64-66; asst prof pediat, Chicago Med Sch, 66-68; assoc prof, Univ Calif, Los Angeles, 69-72, prof, 72-74; PROF PEDIAT & OBSTET, BROWN UNIV, 74- Concurrent Pos: Fel neonatology, Michael Reese Hosp, Chicago, 62-64; pediatrician-in-chief, Women & Infants Hosp, RI, 74. Mem: Soc Pediat Res; Am Pediat Soc; Am Soc Clin Res; Perinatal Res Soc. Res: Perinatal biology with specific interest on carbohydrate metabolism, fluid and electrolyte balance and respiratory distress syndrome in the newborn. Mailing Add: Dept of Perinatal Med Women & Infants Hosp Providence RI 02908

OH, YANK HWAN, biochemistry, physical chemistry, see 12th edition

OH, YOON YONG, b Haeju, Korea, Oct 25, 28; m 52; c 4. MATHEMATICS. Educ: Seoul Nat Univ, BS, 55; Brandeis Univ, MS, 63, PhD(math), 65. Prof Exp: Asst prof math, State Univ NY Stony Brook, 65-70; ASSOC PROF MATH, PA STATE UNIV, ALTOONA CAMPUS, 70- Mem: Am Math Soc. Res: Algebraic geometry. Mailing Add: Dept of Math Pa State Univ Altoona PA 16603

O'HALLORAN, THOMAS A, b New York, NY, Apr 13, 31; m 54; c 4. PHYSICS. Educ: Ore State Col, BS, 53, MS, 54; Univ Calif, Berkeley, PhD(elem particle physics), 63. Prof Exp: Res assoc elem particle physics, Univ Calif, Berkeley, 63-64; res fel, Harvard Univ, 64-66; from asst prof to assoc prof, 66-70, PROF PHYSICS, UNIV ILL, URBANA, 70- Mem: Am Phys Soc. Res: Elementary particles. Mailing Add: Dept of Physics Univ of Ill Urbana IL 61801

OHANIAN, HANS C, b Leipzig, Ger, Apr 29, 41; m 66. PHYSICS. Educ: Univ Calif, Berkeley, BA, 64; Princeton Univ, PhD(physics), 68. Prof Exp: ASST PROF PHYSICS, RENSSELAER POLYTECH INST, 68- Concurrent Pos: Vis fel physics, Princeton Univ, 75-76. Mem: Am Phys Soc. Res: Gravitation and field theory. Mailing Add: Dept of Physics Rensselaer Polytech Inst Troy NY 12181

O'HARA, ELIZABETH MARY, b Philadelphia, Pa, Nov 11, 35. ORGANIC CHEMISTRY. Educ: Georgian Court Col, BA, 65; Univ Pa, MS, 67; Rutgers Univ, PhD(chem), 73. Prof Exp: Teacher chem, Camden Cath High Sch, 66-67; INSTR CHEM, GEORGIAN COURT COL, 67-70 & 73- Mem: Am Chem Soc. Res: Skeletal rearrangements of tricyclooctyl systems; solvolysis rates and products; carbonium ion chemistry. Mailing Add: Georgian Court Col Lakewood NJ 08701

O'HARA, JAMES, b Jan 9, 36; US citizen; m 60; c 2. ECOLOGY. Educ: Antioch Col, BA, 58; Univ Miami, MS, 62, PhD(zool), 66. Prof Exp: Asst prof biol, State Univ NY Albany, 63-67 & Cent State Univ, Ohio, 67-70; res assoc, Belle W Baruch Coastal Res Inst, Univ SC, 70-72; ecol consult, Sargent & Lundy Engrs, 72-74; VPRES, APPL BIOL INC, 74- Concurrent Pos: Dept Interior grant, Cent State Univ, Ohio, 68-70; res biologist, Fed Water Qual Admin, 70-71; ecol consult, SC Water Resources Comn, 71-72. Mem: AAAS; Am Fisheries Soc; Ecol Soc Am; Am Soc Limnol & Oceanog. Res: Fish respiration; size and temperature effects; role of pollutants on aquatic organisms and synergistic effects of environmental variables; physiology of sublethal exposure of pollutants on aquatic systems. Mailing Add: Appl Biol Inc 5891 New Peachtree Rd Atlanta GA 30340

O'HARA, WILLIAM FRANCIS, physical chemistry, see 12th edition

O'HARE, GEORGE ALFRED, b Salem, Mass, Oct 9, 12; m 38; c 2. CHEMISTRY. Educ: Harvard Univ, AB, 34; Univ Pa, PhD(chem), 39. Prof Exp: Instr chem, Univ Pa, 34-39; mem staff, Congoleum-Nairn, Inc, 39-62; pres, Patterson-Sargent Div, 62-73, PRES, SPENCER KELLOGG DIV, TEXTRON INC, BUFFALO, NY, 73- Mem: Am Chem Soc; Am Oil Chemists Soc; Am Soc Testing & Mat. Res: Drying oils; oxidation; polymerization; protective coatings; floor coverings; oxidation-reduction indicators. Mailing Add: Buffalo NY

O'HARE, JOHN MICHAEL, b Des Moines, Iowa, Oct 2, 38; m 64; c 2. THEORETICAL SOLID STATE PHYSICS, EXPERIMENTAL SOLID STATE PHYSICS. Educ: Loras Col, BS, 60; Purdue Univ, MS, 62; State Univ NY Buffalo, PhD(physics), 66. Prof Exp: Instr physics, State Univ NY Buffalo, 65-66; asst prof, 66-71, ASSOC PROF PHYSICS, UNIV DAYTON, 71- Mem: Am Phys Soc. Res: Theoretical atomic and molecular physics; crystal field theory; optical properties of solids, experimental and theoretical. Mailing Add: Dept of Physics Univ of Dayton Dayton OH 45409

O'HARE, PATRICK, b Dundalk, Ireland, Aug 6, 36; m 64; c 3. PHYSICAL CHEMISTRY. Educ: Nat Univ Ireland, BS, 57, MS, 58; Queen's Univ Belfast, PhD(phys chem), 61, DSc(phys chem), 71. Prof Exp: Demonstr chem, Queen's Univ Belfast, 58-61, Imp Chem Indust fel, 61-63; resident res assoc, 64-66, ASSOC CHEMIST, ARGONNE NAT LAB, 66- Concurrent Pos: Vis prof, Univ Toronto, 71-72. Mem: Am Chem Soc. Res: Thermochemistry of nuclear materials; bond energies in inorganic molecules. Mailing Add: Argonne Nat Lab Argonne IL 60439

O'HARRA, JOHN LEWIS, b Tama, Iowa, Nov 26, 14; m 51. VETERINARY MEDICINE. Educ: Iowa State Univ, DVM, 43; Am Col Vet Prev Med, dipl, 74. Prof Exp: Vet, Univ Nev, 51-53; dir diag lab, 53-60, ADMINR ANIMAL INDUST, NEV STATE DEPT AGR, 60- Concurrent Pos: Guest lectr, Animal Sci Div, Univ Nev, Reno. Honors & Awards: Am Vet Med Asn Award, 72; Stange Award, Iowa State Univ, 75. Mem: US Animal Health Asn (pres, 69); Am Vet Med Asn (vpres, 63-64); Am Col Vet Toxicol. Res: Toxicology; molybdenum and oxalates. Mailing Add: 940 Pine Ridge Dr Reno NV 89502

OHASHI, YOSHIKAZU, b Tokyo, Japan, June 30, 41; m 70; c 2. MINERALOGY, CRYSTALLOGRAPHY. Educ: Tokyo Univ, BS, 66, MS, 68; Harvard Univ, PhD(geol), 73. Prof Exp: Fel, 72-75, TEMP STAFF MEM, GEOPHYS LAB, CARNEGIE INST WASHINGTON, 75- Mem: Mineral Soc Am; Am Geophys Union; Am Crystallog Asn; Sigma Xi. Res: Crystal structure analysis of rock-forming silicate minerals at high temperature; crystallographic computing. Mailing Add: Geophys Lab 2801 Upton St NW Washington DC 20008

O'HAVER, THOMAS CALVIN, b Atlanta, Ga, Oct 13, 41; m 68. ANALYTICAL CHEMISTRY. Educ: Spring Hill Col, BS, 63; Univ Fla, PhD(chem), 68. Prof Exp: ASST PROF CHEM, UNIV MD, 68- Concurrent Pos: Advan Res Proj Agency, Dept Defense fel, Ctr Mat Res, 68-71, NSF Instnl Sci Equip Prog fel, 69-72 & Petrol Res Found-Am Chem Soc fel, 69-72. Mem: Am Chem Soc; Soc Appl Spectros; Am Inst Physics; Optical Soc Am. Res: Luminescence spectrometry; analytical instrumentation; automation; computer applications; signal-to-noise ratio theory and measurements. Mailing Add: Dept of Chem Univ of Md College Park MD 20742

O'HAYRE, ARTHUR PAUL, b Denver, Colo, Feb 20, 48; m 71. FOREST HYDROLOGY. Educ: Colo State Univ, BS, 70; Univ Ariz, MS, 72; Univ Minn, PhD(forest hydrol), 76. Prof Exp: Instr forest hydrol, Col Forestry, Univ Minn, 73-75; ASST PROF FOREST HYDROL, SCH FORESTRY & ENVIRON STUDIES, YALE UNIV, 75- Mem: Am Geophys Union; Am Water Resource Asn; Soc Am Foresters. Res: Deterministic and stochastic modeling of basic hydrologic processes and determining the effect of land use activities on water resources. Mailing Add: Marsh Hall Yale Univ 360 Prospect St New Haven CT 06511

OHE, KEIJI, b Sashebo, Japan, Apr 1, 31; m 62; c 2. BIOCHEMISTRY. Educ: Kyoto Univ, MD, 55, DMSc(med biochem), 61. Prof Exp: USPHS fel & Am Cancer Soc res grant, Fels Res Inst, Temple Univ, 64-66; res assoc, Sch Med, Yale Univ, 66-69; asst prof res, Fac Med, Univ Alta, 69-75; MEM STAFF, LITTON BIONETIC RES LABS INC, 75- Mem: AAAS; Am Chem Soc; Can Biochem Soc. Res: Nucleic acid metabolism in virus infection; nucleotide sequence of RNA in relation to its function; biochemical mechanisms of viral oncogenicity. Mailing Add: Litton Bionetic Res Labs Inc 7300 Pearl St Bethesda MD 20014

O'HEA, EUGENE KEVIN, b Cork, Ireland, Mar 27, 41. PHYSIOLOGY. Educ: Univ Col, Dublin, BAgr Sc, 64, MAgrSc, 66; Univ Ill, PhD(nutrit biochem), 69. Prof Exp: ASST PROF PHYSIOL, UNIV WESTERN ONT, 70- Mem: Can Physiol Soc; Nutrit Soc Can; Am Physiol Soc. Res: Endocrinology; fat synthesis and its regulations; energy balance. Mailing Add: Dept of Physiol Univ of Western Ont London ON Can

O'HEARN, GEORGE THOMAS, b Manitowoc, Wis, Sept 26, 34; m 59; c 3. SCIENCE EDUCATION. Educ: Univ Wis-Madison, BS, 57, MS, 59, PhD(sci educ), 64. Prof Exp: Asst prof sci educ, Univ & prin investr, Res & Develop Ctr Cognitive Learning, Univ Wis-Madison, 64-68, US Off Educ grant sci literacy, 64-66; assoc prof, 68-70, PROF SCI EDUC, UNIV WIS-GREEN BAY, 70-, DIR OFF EDUC RES & DEVELOP, 75- Mem: Fel AAAS; Nat Asn Res Sci Teaching; Nat Sci Teachers Asn. Res: Problem focused learning; environmental education; scientific literacy; future ethical and social implications of scientific developments. Mailing Add: Univ of Wis Green Bay WI 54302

O'HERN, ELIZABETH MOOT, b Richmondville, NY, Sept 1, 13; m 52. MEDICAL MICROBIOLOGY. Educ: Univ Calif, BA, 45, MA, 47; Univ Wash, PhD(microbiol, mycol), 56. Prof Exp: Asst parasitol & mycol, Univ Calif, 45-48; instr microbiol, Univ Wash, 56-57 & State Univ NY Downstate Med Ctr, 57-62; asst prof, Sch Med, George Washington Univ, 62-65; sr scientist, Bionetics Res Labs, 65-66, prin investr, 66-68; prog adminr microbiol training prog, 68-75, prog adminr genetics res grants, 74-75, SPEC ASST TO DIR, NAT INST GEN MED SCI, 75- Mem: AAAS; Am Soc Cell Biologists; Am Soc Trop Med & Hyg; Am Soc Microbiol; Mycol Soc Am. Res: Medical mycology; host-parasite relationships; malaria chemotherapy. Mailing Add: Bldg 31 Rm 4A52 Nat Inst Gen Med Sci Bethesda MD 20014

OHKI, KENNETH, b Livingston, Calif, June 13, 22; m 45; c 2. PLANT PHYSIOLOGY. Educ: Univ Calif, Berkeley, BS, 49, MS, 51, PhD(plant physiol), 63. Prof Exp: Mem staff, Univ Calif, Berkeley, 50-64; supvr & specialist plant physiol res, Int Minerals & Chem Corp, Ill, 64-70; asst prof, 71-73, ASSOC PROF AGRON, GA EXP STA, UNIV GA, 73- Concurrent Pos: Res asst, Calif Inst Technol, 50-53. Mem: Am Soc Plant Physiologists; Am Soc Agron; Crop Sci Soc Am; Soil Sci Soc Am. Res: Plant nutrition and analysis; growth and development of sugar beets; ion absorption in relation to antecedent nutrition; plant growth regulator; micronutrient nutrition of plants. Mailing Add: Dept of Agron Ga Exp Sta Univ of Ga Experiment GA 30212

OHKI, SHINPEI, b Japan, Jan 1, 33; m 71. BIOPHYSICS. Educ: Kyoto Univ, BS, 56, MS, 58, PhD(physics), 65. Prof Exp: Instr physics, Tokyo Metrop Univ, 61-65; res assoc theoret biol, 65-66, asst prof biophys, 66-73, ASSOC PROF BIOPHYS, STATE UNIV NY BUFFALO, 73- Concurrent Pos: Grants, Damon Runyon Mem Fund, 68-71 & Nat Inst Neurol Dis & Stroke, 69-75. Mem: Biophys Soc; NY Acad Sci. Res: Investigation of a mechanism of excitatory and inhibitory phenomena of membranes. Mailing Add: Dept of Biophys State Univ of NY Buffalo NY 14226

OHL, DONALD GORDON, b Milton, Pa, Apr 13, 15; m 46; c 1. MATHEMATICS. Educ: Ursinus Col, BS, 36; Bucknell Univ, MS, 47. Prof Exp: High sch teacher, Pa, 37-41; instr math, Univ & chmn dept, Olney Undergrad Ctr, Temple Univ, 46; from instr to asst prof, 46-60, ASSOC PROF MATH, BUCKNELL UNIV, 60- Mem: Am Math Soc; Math Asn Am. Res: Mathematics education; general mathematics. Mailing Add: Dept of Mathematics Bucknell Univ Lewisburg PA 17837

OHLBERG, STANLEY MILES, b Brooklyn, NY, June 20, 21; m 54; c 3. PHYSICAL CHEMISTRY. Educ: Univ Mich, BS, 43; Rutgers Univ, MS, 50, PhD(phys chem), 51. Prof Exp: Anal & indust chemist, Manhattan Proj, Linde Air Prod Co, 43-46; asst inorg chem, Rutgers Univ, 46-51; fel x-ray diffraction anal, Dept Res Chem Physics, Mellon Inst, 51-56; res scientist, Verona Res Ctr, Koppers Co, 56-58; scientist, 58-71, staff scientist, 71-73, SR SCIENTIST, GLASS RES CTR, PPG INDUSTS, INC, 73- Honors & Awards: Frank Forest Award, Am Ceramic Soc, 66. Mem: Am Chem Soc; fel Am Ceramic Soc; Am Crystallog Asn. Res: Structures of materials; physics and chemistry of glass. Mailing Add: Glass Res Ctr PPG Industs Inc Box 11472 Pittsburgh PA 15238

OHLE, ERNEST LINWOOD, b St Louis, Mo, Dec 17, 17; m 43; c 4. ECONOMIC GEOLOGY. Educ: Washington Univ, AB & MS, 40; Harvard Univ, MA, 41, PhD,

50. Prof Exp: Geologist, Am Zinc Co, Tenn, 41-42, Ark, 42-43 & Tenn, 44-46, asst mine supt, 46-47; geologist, Am Zinc, Lead & Smelting Co, Mo, 48-57; chief geologist, White Pine Copper Co, 57-61; vpres explor, Copper Range Co, 60-61; staff geologist, 61-65, asst chief geologist, 65-68, eval mgr, 68-71, CONSULT GEOLOGIST, HANNA MINING CO, 71- Concurrent Pos: Mem bd mineral resources, NSF, 74- Mem: Geol Soc Am; Soc Econ Geologists (pres, 75); Am Inst Prof Geologists; Am Asn Petrol Geologists; Am Inst Mining, Metall & Petrol Engrs. Res: Ore deposition; structural control of ore deposits; limestone permeability as related to ore deposition. Mailing Add: Hanna Mining Co PO Box 15787 Salt Lake City UT 84115

OHLENBUSCH, ROBERT EUGENE, b Edinburg, Tex, Oct 18, 30; m 63; c 2. DENTISTRY, MICROBIOLOGY. Educ: Tex Lutheran Col, BS, 52; Univ Tex, DDS, 56. Prof Exp: Dent Corps, US Army, 56-, dentist, 56-58, res asst dent, Walter Reed Army Inst Res, 58-61, res dent officer oral microbiol, US Army Inst Dent Res, 63-67, comdr, 137th Med Detachment, Repub Vietnam, 67-68, exec officer, 102nd Med Detachment, Munich, Ger, 70-71, exec officer prev dent & dir dent educ, Dent Co, Ft Jackson, SC, 71-74, spec projs officer, Hq, Health Serv Command, Ft Sam Houston, Tex, 74-75, CHIEF GARRISON DENT CLIN, DEPT DENT, FT SAM HOUSTON, 75- Concurrent Pos: Lectr, Georgetown Univ, 63-67. Mem: Am Dent Asn; Am Soc Microbiol; Int Asn Dent Res. Res: Preventive dentistry. Mailing Add: Dept of Dent Brooke Army Med Ctr Ft Sam Houston TX 78234

OHLENDORF, HARRY MAX, b Lockhart, Tex, Oct 12, 40. WILDLIFE RESEARCH. Educ: Tex A&M Univ, BS, 62, MS, 69, PhD(wildlife ecol), 71. Prof Exp: Wildlife res biologist, 71-73, ASST DIR ECOL RES ADMIN, PATUXENT WILDLIFE RES CTR, FISH & WILDLIFE SERV, US DEPT INTERIOR, 73- Mem: Wilson Ornith Soc. Res: Effects of environmental pollutants on fish-eating birds and their habitat. Mailing Add: Patuxent Wildlife Res Ctr Laurel MD 20811

OHLER, EDWIN ALLEN, b Emmitsburg, Md, Jan 19, 19; m 45. PHYSIOLOGY. Educ: Washington Col, Md, BS, 40; Tulane Univ, MS, 42; Univ Ill, PhD(physiol), 51. Prof Exp: Asst micros anat, Sch Med, Tulane Univ, 41-44; from asst to asst prof physiol, Col Med, Univ Ill, 44-53; assoc prof, Sch Med, Temple Univ, 53-66; assoc dean, Kirksville Col, 66-70; DIR GRANTS ADMIN, KIRKSVILLE COL OSTEOP MED, 71- Mem: AAAS; Soc Exp Biol & Med; Endocrine Soc; Am Physiol Soc. Res: Dermatoglyphics; experimental hypertension; factors modifying adrenocorticotropic hormone discharge. Mailing Add: Kirksville Col of Osteop Med 204 W Jefferson St Kirksville MO 63501

OHLINE, ROBERT WAYNE, b St Louis, Mo, Mar 9, 34; m 66; c 2. ANALYTICAL CHEMISTRY. Educ: Grinnel Col, AB, 56; Northwestern Univ, MS, 58, PhD, 60. Prof Exp: Anal chemist, Mallinckrodt Chem Works, 60-61; from asst prof to assoc prof, 61-72, PROF CHEM, NMEX INST MINING & TECHNOL, 72- Mem: Am Chem Soc; The Chem Soc. Res: Instrumental methods; wet analytical methods; gas chromatography; ion production in flames. Mailing Add: Dept of Chem NMex Inst of Mining & Technol Socorro NM 87801

OHLMANN, ROBERT CHRISTIAN, solid state physics, spectroscopy, see 12th edition

OHLROGGE, ALVIN JOHN, b Chilton, Wis, Sept 19, 15; m 44; c 2. SOILS. Educ: Univ Wis, BS, 37; Purdue Univ, PhD(soils), 43. Prof Exp: Asst, 42-45, from asst prof to assoc prof, 45-58, PROF AGRON, PURDUE UNIV, WEST LAFAYETTE, 58- Concurrent Pos: Soil fertil adv, Pakistan Govt, Food & Agr Orgn, UN, 52; NSF fel, Univ Calif, 58-59; vchmn, Nat Joint Comt Fertilizer Appln, 60, chmn, 61; consult, FDA & indust. Honors & Awards: Soil Sci Award, Am Soc Agron, 61. Mem: Fel AAAS; Soil Sci Soc Am; fel Am Soc Agron; hon mem Am Soybean Asn. Res: Soil fertility; plant physiology; nutrient uptake from fertilizer bands; growth regulators. Mailing Add: Dept of Agron Purdue Univ West Lafayette IN 47906

OHLSEN, GERALD G, b Eugene, Ore, May 1, 33; m 58; c 3. NUCLEAR PHYSICS. Educ: Univ Ore, BA, 55; Stanford Univ, MS, 57, PhD(physics), 60. Prof Exp: Asst prof physics, Univ Tex, 60-61; fel nuclear physics, Australian Nat Univ, 61-64, fel, 64-65; STAFF MEM, LOS ALAMOS SCI LAB, 65- Mem: AAAS; Am Phys Soc. Res: Low energy nuclear physics; polarized ion sources. Mailing Add: Los Alamos Sci Lab Los Alamos NM 87544

OHLSEN, WILLIAM DAVID, b Evanston, Ill, June 8, 32; m 56; c 3. SOLID STATE PHYSICS. Educ: Iowa State Univ, BS, 54; Cornell Univ, PhD(physics), 62. Prof Exp: From asst prof to assoc prof, 61-74, PROF PHYSICS, UNIV UTAH, 74- Concurrent Pos: Vis prof, Munich Tech Univ, 69. Mem: Am Phys Soc. Res: Magnetic resonance and optical studies of defects in solids. Mailing Add: Dept of Physics Univ of Utah Salt Lake City UT 84112

OHLSON, JOHN LEONARD, organic chemistry, see 12th edition

OHLSON, MARGARET ALEXANDER, b Chicago, Ill, May 5, 01. NUTRITION. Educ: State Col Wash, BS, 23; Univ Iowa, MS, 30, PhD(nutrit), 34. Prof Exp: Dietitian, Sacred Heart Hosp, Spokane, Wash, 27-29; instr foods & nutrit, Ind Univ, 29-32; asst prof, Mich State Col, 34-35; assoc prof, Iowa State Col, 35-44; prof & head dept, Mich State Col, 44-56; prof nutrit, Col Med & dir nutrit dept, Univ Hosps, Univ Iowa, 56-65; res assoc, Div Child Health, Sch Med, Univ Wash, 65-70; lectr, Univ Wash, 71; RETIRED. Concurrent Pos: Mem food & nutrit bd, Nat Res Coun, 52-57; consult, Surgeon Gen, US Air Force, 52-58; spec med adv group, Vet Admin, 55-60; pres, Int Cong Dietetics, Rome, 56; consult, Galesburg State Res Hosp, 71; mem comt health & welfare, President's White House Conf Aging. Honors & Awards: Borden Award, 42 & 50; Am Dairy Asn Award, 51; Marjorie Hulziser Copher Award, Am Dietetic Asn, 66. Mem: Am Dietetic Asn (secy, 48-50, pres, 51); fel Am Pub Health Asn; Am Home Econ Asn; fel Am Inst Nutrit. Res: Nutritional status and nutrient requirements of patients with active and arrested tuberculosis; metabolism of weight reduction; food intake of population groups; experimental and therapeutic dietetics; diet therapy. Mailing Add: 2919 27th Ave W Seattle WA 98199

OHLSSON-WILHELM, BETSY MAE, b Boston, Mass, July 17, 42; m 69. GENETICS, MOLECULAR BIOLOGY. Educ: Radcliffe Col, AB, 63; Harvard Univ, PhD(bact), 69. Prof Exp: Am Cancer Soc fel biophys, Univ Chicago, 68-70; res assoc, Inst Cancer Res, 70-73; ASST PROF MICROBIOL, SCH MED & DENT, UNIV ROCHESTER, 73- Concurrent Pos: Am Cancer Soc grant. Res: Somatic cell genetics; regulation and control mechanisms in somatic cells; cell biology. Mailing Add: Dept of Microbiol Sch Med & Dent Univ of Rochester Rochester NY 14642

OHM, JACK ELTON, b Milwaukee, Wis, Sept 23, 32. MATHEMATICS. Educ: Univ Chicago, BS, 54; Univ Calif, Berkeley, PhD(math), 59. Prof Exp: Teaching asst math, Univ Calif, Berkeley, 54-57, teaching assoc, 57-59; NSF fel, Johns Hopkins Univ, 59-60; asst prof, Univ Wis, 60-65; assoc prof, 65-69, PROF MATH, LA STATE UNIV, BATON ROUGE, 69- Concurrent Pos: Wis Alumni Res Found fel & res assoc math, Univ Calif, Berkeley, 64-65; vis prof, Purdue Univ, 71-72. Mem: Am Math Soc. Res: Algebraic geometry; commutative algebra. Mailing Add: Dept of Math La State Univ Baton Rouge LA 70803

OHMAN, JOHN HAMILTON, b Detroit, Mich, Jan 26, 31; m 54; c 2. RESEARCH ADMINISTRATION, FORESTRY. Educ: Univ Minn, BS, 57, MS, 58, PhD(plant path), 61. Prof Exp: Res fel plant path, Univ Minn, 58-61; plant pathologist & proj leader, NCent Forest Exp Sta, Mich, 61-68, asst br chief forest dis res, Va, 68-71, asst dir, Northeastern Forest Exp Sta, 71-72, DIR, NCENT FOREST EXP STA, US FOREST SERV, 72- Mem: Soc Am Foresters; Am Forestry Asn; Soil Conserv Soc Am. Mailing Add: NCent Forest Exp Sta Folwell Ave St Paul MN 55108

OHMANN, LEWIS F, plant ecology, botany, see 12th edition

OHMART, ROBERT DALE, b Tatum, NMex, Jan 2, 38; m 58; c 4. ZOOLOGY. Educ: NMex State Univ, BS, 61, MS, 63; Univ Ariz, PhD(vert zool), 68. Prof Exp: NIH fel, Univ Calif, Davis, 68-70; asst prof, 70-74, ASSOC PROF ZOOL, ARIZ STATE UNIV, 74- Mem: Am Ornith Union; Wildlife Soc; Cooper Ornith Soc; Am Inst Biol Sci. Res: Wildlife ecology; avian environmental physiology. Mailing Add: Dept of Zool Ariz State Univ Tempe AZ 85281

OHME, PAUL ADOLPH, b Montgomery, Ala, Nov 4, 40; m 64; c 2. MATHEMATICS. Educ: Huntingdon Col, BA, 63; Univ Ala, MA, 64; Fla State Univ, PhD(math), 71. Prof Exp: Asst prof math, Franklin & Marshall Col, 71-73; ASSOC PROF MATH, MISS COL, 73- Mem: Sigma Xi; Math Asn Am; Nat Coun Teachers Math; Am Math Soc. Res: Application of differential equations to biological problems. Mailing Add: Box 4025 Miss Col Clinton MS 39058

OHMER, MERLIN MAURICE, b Napoleonville, La, Mar 15, 23; m 47; c 3. MATHEMATICS. Educ: Tulane Univ, BS, 44, MS, 48; Univ Pittsburgh, PhD(math). Prof Exp: Asst physics, Tulane Univ, 42-43, asst math, 47-48; from asst prof to prof, Univ Southwestern La, 48-66; prof & head dept, 66-69, DEAN COL SCI, NICHOLLS STATE UNIV, 69- Concurrent Pos: Vis instr, Tulane Univ, 49-50; vis assoc prof, Univ Pittsburgh, 53-54; vis lectr, Math Asn Am, 64-73; consult, La State Dept Educ; mem, Nat Metric Speakers Bur. Mem: Math Asn Am; Am Math Soc. Res: Teacher training and innovations; game theory; symbolic logic; geometry; promotion of the metric system in the United States. Mailing Add: 106 Acadia Lane Thibodaux LA 70301

OHMOTO, HIROSHI, b Heijo, Japan, Nov 7, 41; m 65; c 2. GEOCHEMISTRY. Educ: Hokkaido Univ, BS, 64; Princeton Univ, AM, 67, PhD(geol), 69. Prof Exp: Fel geochem, Univ Alta, 68-69, fel, 68-70, lectr, 69-70; asst prof, 70-74, ASSOC PROF GEOCHEM, PA STATE UNIV, UNIVERSITY PARK, NSF RES GRANT, 72- Honors & Awards: Lindgren Award, Soc Econ Geologists, 70; Clark Medal, Geochem Soc, 73. Mem: Soc Econ Geologists; Geochem Soc. Res: Causes of variation in stable isotopes in geologic processes; geochemistry of ore-forming fluids. Mailing Add: Dept of Geosci Pa State Univ University Park PA 16802

OHMS, JACK IVAN, b Walnut, Iowa, Jan 1, 30; m 54; c 3. ANALYTICAL BIOCHEMISTRY. Educ: Iowa State Univ, BS, 53; Mich State Univ, PhD(dairy husb), 61. Prof Exp: Res asst, Am Found Biol Res, Wis, 53-57; NIH fel, 61-62; appln specialist, 62-66, SR SCIENTIST, SPINCO DIV, BECKMAN INSTRUMENTS INC, 66- Mem: AAAS; NY Acad Sci. Res: Physiology of reproduction of domestic animals; endocrine immunochemistry; biomedical instrumentation; biochemical calorimetry; liquid chromatography of biochemicals; automated sequence determination of proteins and peptides. Mailing Add: 877 Aspen Way Palo Alto CA 94303

OHMS, RICHARD EARL, b Payette, Idaho, June 13, 27; m 56; c 2. PLANT PATHOLOGY. Educ: Univ Idaho, BS, 50, MS, 52; Univ Ill, PhD(plant path), 55. Prof Exp: Asst plant path, Univ Idaho, 52-55; asst plant pathologist, SDak State Col, 55-57; assoc horticulturist & exten specialist hort, 57-74, exten & res prof, 73-74, EXTEN PROF & EXTEN POTATO SPECIALIST, EXTEN SERV, UNIV IDAHO, 74- Concurrent Pos: Seed potato consult, Jordan & Int Potato Ctr, Lima, Peru. Mem: Am Soc Hort Sci; Am Phytopath Soc; Potato Asn Am. Res: Potatoes. Mailing Add: Univ of Idaho Exten 634 Addison Ave W Twin Falls ID 83301

OHNESORGE, WILLIAM EDWARD, b Acushnet, Mass, Sept 11, 31; m 60. ANALYTICAL CHEMISTRY. Educ: Brown Univ, ScB, 53; Mass Inst Technol, PhD(chem), 56. Prof Exp: From instr to assoc prof chem, Univ RI, 56-65; assoc prof, 65-71, PROF CHEM, LEHIGH UNIV, 71-, ASST CHMN DEPT, 69- Concurrent Pos: Vis asst prof, Mass Inst Technol, 64-65. Mem: Am Chem Soc. Res: Fluorescence spectroscopy; electroanalytical chemistry; clinical chemistry; complex ions; luminescence analysis. Mailing Add: Dept of Chem Lehigh Univ Bethlehem PA 18015

OHNISHI, TSUYOSHI, b Otsu, Japan, Dec 17, 31; m 58; c 2. BIOPHYSICS, BIOCHEMISTRY. Educ: Kyoto Univ, BS, 54, MS, 56; Nagoya Univ, PhD(biophys), 60. Prof Exp: Japanese Soc Prom Sci fel, Nagoya Univ, 60-62 & res fel physiol, Med Sch, 62-63, res fel physics, Univ, 64-65; assoc prof biophys, Waseda Univ, Japan, 65-67; vis assoc prof, Johnson Found, Univ Pa, 67-68; asst prof biophys, Med Col Pa, 69-72; ASSOC PROF BIOPHYS & ANESTHESIA, HAHNEMANN MED COL, 73- Concurrent Pos: Vis res fel, Univ Tokyo, 63-65. Mem: Biophys Soc; Am Soc Anesthesiol. Res: Calcium transport in sarcoplasmic reticulum and erythrocyte; solubilization and reconstitution of membranes; effect of anesthetics on tropomyosin-troponin; effect of anesthetics on sickle cells; development of the Murexide method for calcium and other dual-wavelength spectrophotometry for biophysics and biochemistry. Mailing Add: Biophys Lab Dept of Anesthesiol Hahnemann Med Col Philadelphia PA 19102

OHNMACHT, CYRUS JOHN, JR, organic chemistry, see 12th edition

OHNO, SUSUMU, b Tokyo, Japan, Feb 2, 28; m 52; c 3. CYTOGENETICS. Educ: Tokyo Univ Agr & Technol, DVM, 49; Hokkaido Univ, PhD(path), 56, DSc, 61. Prof Exp: Res assoc, Dept Exp Path, 52-62, sr res scientist, Dept Biol, 62-66, CHMN DEPT BIOL, CITY OF HOPE MED CTR, 66- Mem: Int Soc Hemat; Genetics Soc Am; Am Asn Cancer Res. Res: Clinical genetics. Mailing Add: Dept of Biol City of Hope Med Ctr Duarte CA 91010

OHNUKI, YASUSHI, b Kawasaki City, Japan, July 30, 26; c 2. CYTOGENETICS. Educ: Hokkaido Univ, BSc, 54, MSc, 56, DSc(cytogenetics), 61. Prof Exp: Tobacco Indust Res Comt fel, Dept Anat, Univ Tex Med Br, Galveston, 59-60 & Dept Cellular Biol, Pasadena Found Med Res, 60-61; res assoc, Makino Lab, Fac Sci, Hokkaido Univ, 61-64; chief cytogenetics sect, 64-66, dir dept cytogenetics, 66-72, PROJ DIR CYTOGENETICS STUDIES STRUCTURE & FUNCTION CHROMOSOMES, PASADENA FOUND MED RES, 72- Mem: Am Soc Cell Biol; Tissue Cult Asn; Genetics Soc Japan. Res: Studies on structure and function of chromosomes; mechanism of chromosome movement, especially functional relations between chromosomes and spindle fibers. Mailing Add: Pasadena Found Med Res 99 N El Molino Ave Pasadena CA 91101

OHNUMA, TAKAO, b Sendai, Japan, May 16, 32; m 66; c 2. INTERNAL MEDICINE, CANCER. Educ: Tohoku Univ, Japan, MD, 57; Univ London, PhD(biochem), 65. Prof Exp: Intern, Naval Hosp, Yokosuka, Japan, 57-58; intern, Lincoln Hosp, Bronx, NY, 58-59; asst resident, Bird S Coler Mem Hosp, Welfare Island, 59-60; resident, Roswell Park Mem Inst, 60-61; vis scientist chem, Chester Beatty Res Inst, London, 63-65; asst med, Tohoku Univ Hosp, Sendai, Japan, 68; cancer res clinician, Dept Med, Roswell Park Mem Inst, 68-73; ASSOC PROF, DEPT NEOPLASTIC DIS, MT SINAI SCH MED, 73- Concurrent Pos: Res fel med, Roswell Park Mem Inst, 61-63; res asst prof med, State Univ NY Buffalo, 70-73; assoc attend physician, Dept Neoplastic Dis, Mt Sinai Hosp, New York, 73- Mem: Am Chem Soc; Am Asn Cancer Res; Am Soc Clin Oncol. Res: Cancer chemotherapy; preclinical and clinical pharmacology. Mailing Add: Dept of Neoplastic Dis Mt Sinai Sch of Med New York NY 10029

OHR, ELEONORE A, b New York, NY, Jan 28, 32. CELL PHYSIOLOGY. Educ: Univ Rochester, AB, 54, MS, 58, PhD(physiol), 62. Prof Exp: Asst prof, 63-68, ASSOC PROF PHYSIOL, SCH MED, STATE UNIV NY BUFFALO, 68- Concurrent Pos: NIH Fel biophys, Univ Buffalo, 60-61; Am Heart Asn career investr fel muscle proteins, Cardiovasc Res Inst, Med Ctr, Univ Calif, San Francisco, 62-63; gen res support grant, Sch Med, State Univ NY Buffalo, 65-66; Heart Asn Western NY, Inc res support grant, 67-68; United Health Found Western NY, Inc res support grant, 69-74. Mem: AAAS; assoc Am Physiol Soc. Res: Chemistry of tricainemethanesulfonate and its effects on epithelial transport systems; inulin space determined in skeletal muscle in vitro; transport of organic bases by renal cortical slices. Mailing Add: Dept of Physiol State Univ of NY Sch of Med Buffalo NY 14214

OHRING, GEORGE, b New York, NY, June 20, 31; m 53; c 3. METEOROLOGY. Educ: City Col New York, BS, 52; NY Univ, MS, 54, PhD(meteorol), 57. Prof Exp: Atmospheric physicist, Air Force Cambridge Res Labs, 57-60; mgr meteorol physics dept, GCA Corp, 60-65, dir meteorol lab, 65-71; ASSOC PROF ENVIRON SCI, TEL-AVIV UNIV, 71- Concurrent Pos: Vis prof, Tel-Aviv Univ, 69-70; consult, Master Plan, Israel Meteorol Serv, 72-74. Mem: AAAS; Am Meteorol Soc; Am Geophys Union; Sigma Xi; Israel Meteorol Soc. Res: Atmospheric radiation; climate models; atmospheric soundings from satellites; meteorology of planetary atmospheres. Mailing Add: Dept of Environ Sci Tel-Aviv Univ Ramat Aviv Israel

ÖHRN, NILS YNGVE, b Avesta, Sweden, June 11, 34; m 57; c 2. CHEMICAL PHYSICS, QUANTUM CHEMISTRY. Educ: Univ Uppsala, MS, 58, PhD(quantum chem), 63, FD, 66. Prof Exp: Res assoc quantum chem, Quantum Chem Group, Univ Uppsala, 63-66; assoc prof chem & physics, 66-70, Air Force Off Sci Res & NSF res grants, 69-76, PROF CHEM & PHYSICS, UNIV FLA, 71-, ASSOC DIR QUANTUM THEORY PROJ, 67- Mem: Am Phys Soc; Sigma Xi. Res: Electronic structure, properties and spectra of atoms; molecules and solids; quantum mechanical studies of properties of matter. Mailing Add: Dept of Chem Univ of Fla Gainesville FL 32601

OHTA, MASAO, b Kobe, Japan, May 4, 19; m 49. PHYSICAL CHEMISTRY. Educ: Kyoto Univ, BS, 43; Univ Calif, MS, 56; Univ Akron, PhD(chem), 59. Prof Exp: Res chemist polymer chem, Mitsui Chem Co, Japan, 43-52 & Monsanto Co, 59-71; SR ASSOC ED, CHEM ABSTR SERV, 71- Mem: Am Chem Soc. Res: Physical chemistry of polymers. Mailing Add: Chem Abstr Serv Columbus OH 43210

OHTAKE, TAKESHI, b Chiba, Japan, Jan 22, 26; m 53; c 3. CLOUD PHYSICS, CLIMATOLOGY. Educ: Tohoku Univ, Japan, BSc, 52, DSc(meteorol), 61. Prof Exp: Asst technician meteorol, Cent Meteorol Observ, Tokyo, Japan, 43-44; assoc researcher, Meteorol Res Inst, 47-49; sr res assoc, Geophys Inst, Tohoku Univ, Japan, 52-64; assoc prof, 64-75, PROF GEOPHYS, GEOPHYS INST, UNIV ALASKA, FAIRBANKS, 75- Concurrent Pos: USPHS res fel, 65-68; NSF res fels, 65-77; vis assoc prof, Dept Atmospheric Sci, Colo State Univ, 69-71. Mem: Am Geophys Union; Am Meteorol Soc; foreign mem Royal Meteorol Soc; Meteorol Soc Japan. Res: Electron microscopic studies for fog, cloud, ice crystal and ice fog nuclei; physical and chemical explanation of fog, cloud, snowfall and rainfall formation mechanisms; ice crystal nucleation; weather modification. Mailing Add: Geophys Inst Univ of Alaska Fairbanks AK 99701

OIEN, HELEN GROSSBECK, b Paterson, NJ, July 11, 40; m 68. BIOCHEMISTRY, NUTRITION. Educ: Rutgers Univ, New Brunswick, AB, 62; Cornell Univ, MS, 64, PhD(biochem), 68. Prof Exp: Sr scientist biochem, Shulton, Inc, NJ, 68-69; sr res biochemist, 69-75, RES FEL, MERCK, SHARP & DOHME RES LABS, 75- Mem: AAAS; Am Chem Soc. Res: Prostaglandins, especially mechanisms of action; cyclic adenosinemonophosphate; collagen biosynthesis; mechanisms of skin penetration by drugs; intermediary metabolism of nitrogen bases. Mailing Add: Merck Sharp & Dohme Res Lab Rahway NJ 07065

OISHI, MASAYOSHI, b Shimizu City, Japan, July 20, 38; m; c 1. ORGANIC CHEMISTRY. Educ: Int Christian Univ, Tokyo, BA, 61; Ore State Univ, MS, 65; NDak State Univ, PhD(org chem), 68; Widener Col, MBA, 74. Prof Exp: Chem engr, Tokyo Eng Corp, Japan, 61-64; res chemist, 68-74, VENTURE ANALYST, SUN OIL CO, 74- Mem: Am Chem Soc. Res: Homogeneous and heterogeneous catalysis. Mailing Add: Sun Oil Co PO Box 1135 Marcus Hook PA 19061

OISHI, NOBORU, b Kapaa, Hawaii, Nov 11, 28; m 57; c 1. ONCOLOGY, IMMUNOHEMATOLOGY. Educ: Wash Univ, AB, 49, MD, 53. Prof Exp: ASSOC PROF MED, SCH MED, UNIV HAWAII, MANOA, 65-; DIR CLIN SCI, CANCER CTR HAWAII, 74- Concurrent Pos: USPHS fel hemat, Univ Rochester, 58-59; consult, Queen's Med Ctr, Kuakini Hosp & Home, Inc & St Francis Hosp, 58-; dir clin lab, Annex Lab, Inc, 67-; pres, Blood Bank Hawaii. Mem: Fel Am Col Physicians; Int Soc Hemat; Am Soc Clin Oncol; NY Acad Sci; Asn Advan Med Instrumentation. Res: Study of the immune capacity of patients with malignant disorders. Mailing Add: 1010 S King St Honolulu HI 96814

OITA, KATASHI, organic chemistry, see 12th edition

OJA, TONIS, b Tallin, Estonia, Aug 22, 37; US citizen; m 61; c 2. PHYSICS, CHEMISTRY. Educ: McGill Univ, BSc, 59; Rensselaer Polytech Inst, PhD(physics), 66. Prof Exp: Res asst, RCA Victor Res Labs, Can, 59-61; res assoc physics, Brown Univ, 66-68, asst prof, 68-69; asst prof, Univ Denver, 69-73; assoc prof, Ohio Univ, 73-74; ASSOC PROF PHYSICS, HUNTER COL, NY, 74- Concurrent Pos: Consult, Matec Inc, RI, 72- Mem: Am Phys Soc. Res: Nuclear magnetic and nuclear quadrupole resonance; glasses and material of biological interest; phase transitions; cirt critical phenomena. Mailing Add: Hunter Col New York NY 10021

OJAKAAR, LEO, b Valga, Estonia, Apr 26, 26; US citizen; m 59; c 2. ORGANIC CHEMISTRY. Educ: Millikin Univ, BS, 53; Va Polytech Inst, MS, 61, PhD(org chem), 64. Prof Exp: Res asst chem, Rutgers Univ, 56-59; RES CHEMIST, E I DU PONT DE NEMOURS & CO, INC, 64- Mem: Am Chem Soc. Res: New high molecular weight aromatic hydrocarbons; new isocyanates and urethanes with industrial importance; synthetic rubbers; new neoprene compounds; elastomer chemistry. Mailing Add: R D 1 Wellington Hills Hockessin DE 19707

OJHA, NARAYANLAL DUNGARSEY, organic chemistry, see 12th edition

OKA, MASAMICHI, b Japan, Mar 25, 20; US citizen; m 56; c 2. PATHOLOGY. Educ: Nippon Med Sch, MD, 45, PhD, 50. Prof Exp: Asst prof path, Albert Einstein Col Med, 63-70; ASST PROF PATH, SCH MED, CORNELL UNIV, 71- Concurrent Pos: Asst pathologist, North Shore Univ Hosp, 71- Mem: Int Acad Path; Am Fedn Exp Path. Res: Cardiovascular endocraditis; hypertension. Mailing Add: North Shore Univ Hosp Lab Community Dr Manhasset NY 11030

OKA, TAKAMI, b Tokyo, Japan, Jan 1, 40; m 69; c 1. DEVELOPMENTAL BIOLOGY, PHARMACOLOGY. Educ: Univ Tokyo, BS, 63; Stanford Univ, PhD(pharmacol), 69. Prof Exp: Vis scientist, Nat Inst Radiol Sci, Japan, 63-64; res fel muscle protein & actin, Med Sch, Stanford Univ, 64-65; vis fel mammary gland, 69-71, STAFF FEL MAMMARY GLAND, NAT INST ARTHRITIS, METAB & DIGESTIVE DIS, 71- Res: Hormonal regulation of cellular function and development. Mailing Add: Nat Inst Arthritis Metab & Dig Dis Sect Intermediary Metab Bethesda MD 20014

OKABAYASHI, MICHIO, b Tokyo, Japan, Dec 10, 39; m 68; c 2. PLASMA PHYSICS. Educ: Univ Tokyo, BS, 63, MS, 65, PhD(physics), 68. Prof Exp: Fel, 68-71, res staff, 71-75, RES PHYSICIST, PRINCETON UNIV, 75- Mem: Am Phys Soc. Res: Thermonuclear fusion research. Mailing Add: Plasma Physics Lab Princeton Univ Princeton NJ 08540

OKABE, HIDEO, b Naganoken, Japan, Dec 13, 23; m 59; c 3. PHYSICAL CHEMISTRY. Educ: Univ Tokyo, BS, 47; Univ Rochester, PhD(chem), 57. Prof Exp: Nat Res Coun Can fel, 56-58; phys chemist, 59-71, div consult, 71-74, SR SCIENTIST, DIV PHYS CHEM, NAT BUR STANDARDS, 74- Concurrent Pos: Guest prof, Inst Phys Chem, Univ Bonn, 63-65. Honors & Awards: Gold Medal, Dept of Com, 73. Mem: Am Chem Soc; Sigma Xi. Res: Vacuum ultraviolet photochemistry in the gas phase; fluorescence; air pollution. Mailing Add: Div of Phys Chem Nat Bur of Standards Washington DC 20234

OKADA, TADASHI A, b Numazu, Japan, Mar 31, 28; m 56. CYTOGENETICS, MOLECULAR BIOLOGY. Educ: Hokkaido Univ, BS, 53, MS, 55, DS(cytogenetics), 59. Prof Exp: Asst res scientist, Dept Biochem, City of Hope Med Ctr, Duarte, Calif, 56-62; asst res scientist, DNA denaturation, Univ Edinburgh, 62-63; res scientist, Biol Div, Ciba Ltd, Switz, 63-65; asst res scientist, Dept Biochem, 65-69, ASSOC RES SCIENTIST, DEPT MED GENETICS, CITY OF HOPE MED CTR, 69- Mem: Am Soc Cell Biol; Electron Micros Soc Am; Biophys Soc. Res: Ultrastructure of meiosis and mitosis; initiation of DNA synthesis; structure of chromatin and DNA; protein complexes. Mailing Add: Dept of Med Genetics City of Hope Med Ctr Duarte CA 91010

OKAMOTO, MICHIKO, b Tokyo, Japan, Mar 3, 32; m 59; c 1. PHARMACOLOGY. Educ: Tokyo Col Pharm, BS, 54; Purdue Univ, MS, 57; Cornell Univ, PhD(pharmacol), 64. Prof Exp: ASSOC PROF PHARMACOL, MED COL, CORNELL UNIV, 71- Concurrent Pos: Res fel pharmacol, Med Col, Cornell Univ, 64-66, USPHS training fel, 64-67. Mem: AAAS; Am Soc Pharmacol & Exp Therapeut; Harvey Soc; NY Acad Sci; Soc Neurosci. Res: Neuropharmacology of synaptic transmission and nerves. Mailing Add: Dept of Pharmacol Cornell Univ Med Col New York NY 10021

OKAMURA, WILLIAM H, b Los Angeles, Calif, Feb 19, 41; m 70. SYNTHETIC ORGANIC CHEMISTRY, MEDICINAL CHEMISTRY. Educ: Univ Calif, Los Angeles, BS, 62; Columbia Univ, PhD(org chem), 66. Prof Exp: Nat Acad Sci-Nat Res Coun-Air Force Off Sci Res fel, Cambridge Univ, 66-67; asst prof org chem, 67-74, ASSOC PROF CHEM, UNIV CALIF, RIVERSIDE, 74- Mem: Am Chem Soc; The Chem Soc. Res: Natural products; vitamin D analogs and metabolites; aromatic molecules; heterocycles; organometallics. Mailing Add: Dept of Chem Univ of Calif Riverside CA 92502

O'KANE, DANIEL JOSEPH, b Jackson Heights, NY, June 20, 19; m 46; c 3. BACTERIOLOGY. Educ: Cornell Univ, BS, 40, PhD(bact), 47; Univ Wis, MS, 41. Hon Degrees: MA, Univ Pa, 71. Prof Exp: From asst prof to assoc prof microbiol, 47-56, from vdean to actg dean grad sch arts & sci, 66-74, PROF MICROBIOL, UNIV PA, 56-, DEP ASSOC PROVOST, 74- Concurrent Pos: Fulbright & Guggenheim fel, Oxford Univ, 55-56; Wis Alumni Res Found scholar & Abbott Labs fel; mem microbiol training comt, Nat Inst Gen Med Sci, chmn microbiol training panel, 68-69. Mem: AAAS; Am Soc Microbiol; fel Am Acad Microbiol; NY Acad Sci; Brit Soc Gen Microbiol. Res: Microbial metabolism and physiology; control of carbohydrate metabolism. Mailing Add: Dept of Biol Univ of Pa Philadelphia PA 19104

OKARMA, THEODORE JOSEPH, b Newark, NJ, Mar 19, 17; m 41; c 2. DIARY HUSBANDRY, DAIRY SCIENCE. Educ: Fla State Christian Univ, BBA, 74. Prof Exp: Milk plant technician qual control dairy prod, Dairymen's League Coop Assoc, Inc, 40-42; farm mgr dairy prod cattle breeding, Meadowburn Farm, Pinelawn Farm & Sunnyvale Farms, 42-59; FARM MGR RES SERV ADMIN, ANIMAL HEALTH RES CTR, SCHERING CORP, 59- Mem: Am Dairy Sci Asn; Holstein Friesian Asn Am; Nat Mastitis Coun. Res: Bovine udder edema; mastitis in dairy cattle. Mailing Add: Schering Corp Animal Health Res Ctr Box 608 Allentown NJ 08501

OKASHIMO, KATSUMI, b Vancouver, BC, Mar 18, 29; m 54; c 4. MATHEMATICS, ACADEMIC ADMINISTRATION. Educ: McMaster Univ, BA, 52; Univ Toronto, MA, 53, PhD(math), 55. Prof Exp: Rep sci comput, Defence Res Bd, 55-59; mgr math & eng comput, Ontario Hydro, 60-67; assoc prof, 67-69, PROF MATH & STATIST, UNIV GUELPH, 69-, PROF COMPUT & INFO SCI, 71-, DIR INST COMPUT SCI, 67- Mem: Asn Comput Mach; Math Asn Am. Res: Management and administration of data processing; educational techniques in computer science; system analysis and design. Mailing Add: Dept of Comput & Info Sci Univ of Guelph Guelph ON Can

OKAYA, AKIRA, b Kobe, Japan, Oct 10, 20; m; c 2. QUANTUM ELECTRONICS, PHYSICAL ELECTRONICS. Educ: Nagoya Univ, MaE, 43; Kyoto Univ, PhD(physics), 56. Prof Exp: Asst electron emission, Kyoto Univ, 46-48, instr microwave spectros, 48-57; assoc molecular physics, Columbia Univ, 55-57, assoc quantum electronics, Radiation Lab, 57-61; adv sci mgr, Int Bus Mach, Md, 61-63; adv scientist, Westinghouse Res & Develop Ctr, Pittsburgh, 63-64; sr sci mgr, Electro-Optical Systs Div, Xerox, Inc, Calif, 64-66; group leader, Perkin-Elmer Corp, Conn, 66-68; PRES & CHMN BD, OPTICAL DATA PROCESSING INC, 68-; CONSULT, EXXON ENTERPRISES INC, NEW YORK, 75- Concurrent Pos: Fulbright travel grant, 55-57. Mem: AAAS; Am Phys Soc; Inst Elec & Electronics Engrs; Phys Soc Japan. Res: Microwave spectroscopy and electronics; nuclear magnetic resonance; electron spin resonance; maser, gas and solid state lasers; optical data processing and storage; optical communications; laser applied business machines;

electro-optical publishing techniques. Mailing Add: c/o CIG Inc 1351 Washington Blvd Stamford CT 06902

OKAYA, YOSHI HARU, b Osaka, Japan, Feb 11, 27; m 50; c 4. CRYSTALLOGRAPHY. Educ: Osaka Univ, BS, 47, PhD(crystallog), 56. Prof Exp: Res assoc physics, Pa State Univ, 53-56, from asst prof to assoc prof, 56-61; res physicist, Res Ctr, IBM Corp, 61-67; PROF CHEM, STATE UNIV NY STONY BROOK, 67- Concurrent Pos: With Brookhaven Nat Lab, 58. Mem: Am Crystallog Asn; Crystallog Soc Japan; Chem Soc Japan. Res: Crystal chemistry; crystal structure determination in organic and inorganic compounds; disorder and diffuse scattering; structural basis for physical behaviors; computer controlled experiments, design and concept; solid state chemistry. Mailing Add: Dept of Chem State Univ of NY Stony Brook NY 11790

OKAZAKI, ALBERT, reactor physics, see 12th edition

OKAZAKI, HARUO, b Kochi, Japan, Apr 11, 26; US citizen; m 57; c 3. PATHOLOGY, NEUROPATHOLOGY. Educ: Kochi Col, BS, 47; Osaka Univ, MD, 51. Prof Exp: Asst instr psychiat, Keio Univ Med Sch, 52-53; resident neurol, Kings County Hosp, Brooklyn, 54-56, resident neuropath, 56-57; assoc, Col Physicians & Surgeons, Columbia Univ, 58-59; from instr to asst prof, State Univ NY Downstate Med Ctr, 59-63; spec appointee, 65, asst prof, 67-73, ASSOC PROF PATH, MAYO MED SCH, UNIV MINN, 73-, MEM STAFF, MAYO CLIN, 66- Concurrent Pos: Res fel, Rockland State Hosp, NY, 53-54; vis fel path, Columbia-Presby Med Ctr, New York, 64; Nat Inst Neurol Dis & Blindness spec fel, 65; mem stroke coun, Am Heart Asn. Mem: Am Asn Neuropath; Asn Res Nerv & Ment Dis; Am Acad Neurol; Japanese Soc Neurol & Psychiat. Res: Human pathology with particular reference to neurologic and psychiatric diseases; anatomic pathology. Mailing Add: Mayo Clin 200 First St SW Rochester MN 55901

OKAZAKI, WILLIAM, microbiology, see 12th edition

OKE, JOHN BEVERLEY, b Sault Ste Marie, Ont, Mar 23, 28; m 55; c 3. ASTRONOMY. Educ: Univ Toronto, BA, 49, MA, 50; Princeton Univ, PhD(astron), 53. Prof Exp: Lectr astron, Univ Toronto, 53-56, asst prof, 56-58; from asst prof to assoc prof, 58-64, PROF ASTRON, CALIF INST TECHNOL, 64-, STAFF MEM, HALE OBSERV, 58-, ASSOC DIR, 70- Mem: Am Astron Soc; Royal Astron Soc. Res: Photoelectric spectrophotometry of stars, galaxies and quasars; astronomical instrumentation. Mailing Add: Hale Observ Calif Inst of Technol Pasadena CA 91125

OKE, TIMOTHY RICHARD, b Devon, Eng, Nov 22, 41; m 67; c 2. MICROCLIMATOLOGY. Educ: Bristol Univ, BSc, 63; McMaster Univ, MA, 64, PhD(geog), 67. Prof Exp: Res asst microclimat, McMaster Univ, 64-66, lectr, 66-67; asst prof geog, McGill Univ, 67-70; asst prof, 70-71, ASSOC PROF GEOG, UNIV BC, 71- Concurrent Pos: Res grants, Nat Res Coun Can, 67-68, 68- & Atmospheric Environ Serv, 68- Mem: Am Meteorol Soc; fel Royal Meteorol Soc; Can Meteorol Soc; Can Asn Geog. Res: Micrometeorology of the lowest layers of the atmosphere; urban climatology; energy balance; air pollution. Mailing Add: Dept of Geog Univ of BC Vancouver BC Can

O'KEEFE, DENNIS ROBERT, b Ottawa, Ont, Nov 15, 39; m 65; c 2. PHYSICS. Educ: Univ Ottawa, BSc, 62; Univ Toronto, MASc, 64, PhD(aerospace studies), 68. Prof Exp: STAFF SCIENTIST PHYSICS, GEN ATOMIC CO, 68- Mem: Am Phys Soc. Res: Surface physics and chemistry; rarefied gas dynamics; gas-surface interactions; molecular and atomic beam techniques; ultrahigh vacuum technology; advanced energy concepts; nuclear reactor engineering. Mailing Add: Gen Atomic Co PO Box 81608 San Diego CA 92138

O'KEEFE, J GEORGE, b Averill Park, NY, Feb 6, 31; m 64; c 2. PHYSICS. Educ: St Bernardine of Siena Col, BS, 52; Rensselaer Polytech Inst, MS, 56; Brown Univ, PhD(physics), 61. Prof Exp: Res assoc physics, Brown Univ, 61-62; from asst prof to assoc prof, 62-71, PROF PHYSICS, RI COL, 71- Mem: Fel AAAS; Am Asn Physics Teachers; Am Inst Physics; Nat Sci Teachers Asn. Res: Nuclear magnetic resonance. Mailing Add: 20 Maple Crest Dr Greenville RI 02828

O'KEEFE, JOHN ALOYSIUS, b Lynn, Mass, Oct 13, 16; m 41; c 9. ASTRONOMY, GEOPHYSICS. Educ: Harvard Univ, AB, 37; Univ Chicago, PhD(astron), 41. Prof Exp: Prof math & physics, Brenau Col, 41-42; mathematician, Army Map Serv, 45-58; ASTRONR, GODDARD SPACE FLIGHT CTR, NASA, 58- Mem: Am Astron Soc; Am Geophys Union; Int Astron Union; Int Union Geod & Geophys; Am Astronaut Soc. Res: Origin of the moon; tektites; geodesy, especially isostasy; solid particles in space. Mailing Add: Goddard Space Flight Ctr Greenbelt MD 20771

O'KEEFE, JOHN JOSEPH, b Philadelphia, Pa, Dec 30, 09; m 44; c 5. MEDICINE. Educ: St Joseph's Col, BS, 33; Jefferson Med Col, MD, 37; Am Bd Otolaryngol, dipl, 46. Prof Exp: Intern, Jefferson Hosp, 37-39, asst, 39-42, prof clin otolaryngol & bronchoesophagol, 47-72, PROF OTOLARYNGOL & CHMN DEPT, THOMAS JEFFERSON UNIV, 72- Concurrent Pos: Ross V Patterson fel broncho-esophagol, Jefferson Hosp, 39-41; consult, Vet Admin Hosps, 72; hon prof otolaryngol, Grad Sch Med, Univ Pa; attend physician, Jefferson Hosp; chief bronchoesophagol & laryngeal surg, Our Lady of Lourdes & St Joseph's Hosps; hon attend physician, Pa Hosp. Mem: Fel Am Acad Ophthal & Otolaryngol; fel Am Broncho-Esophagol Asn; fel AMA; fel Am Laryngol, Rhinol & Otol Soc; fel Am Col Chest Physicians. Mailing Add: 130 S Ninth St Philadelphia PA 19107

O'KEEFE, KELLY RAY, b Anchorage, Alaska, July 7, 48; m 68. ANALYTICAL CHEMISTRY. Educ: Wash State Univ, BS, 71; Univ Ill, PhD(anal chem), 75. Prof Exp: ASST PROF ANAL CHEM, COLO STATE UNIV, 75- Mem: Am Chem Soc; Am Asn Clin Chemists; Am Optical Soc. Res: Chemical instrumentation and automation; applications of computers in chemical instrumentation; new and applied methodologies in clinical chemistry. Mailing Add: Dept of Chem Colo State Univ Ft Collins CO 80523

O'KEEFE, ROBERT BERNARD, b Alliance, Nebr, Feb 17, 26; m 46; c 7. HORTICULTURE, FORESTRY. Educ: Univ Nebr, BS, 49, MS, 59, PhD(agron), 63. Prof Exp: Biol aide, 48-50, head hort outstate testing, 50-57, supvr, Box Butte Exp Sta, Alliance, 52-57, asst hort, 57-63, assoc prof, 64-73, actg chmn dept, 70-73, PROF HORT & FORESTRY, COL AGR, UNIV NEBR, LINCOLN, 70- Mem: AAAS; Potato Asn Am; Am Genetic Asn; Am Soc Hort Sci. Res: Potato breeding and genetics; tomato and onion improvement. Mailing Add: Dept of Hort & Forestry Col of Agr Univ of Nebr Lincoln NE 68503

O'KEEFFE, ANDREW ELLIS, organic chemistry, see 12th edition

O'KEEFFE, DAVID JOHN, b New York, NY, July 5, 30; m 56; c 6. PHYSICS. Educ: St Peter's Col, NJ, BS, 53; Univ Md, MS, 62; Cath Univ Am, PhD(physics), 69. Prof Exp: PHYSICIST, NAVAL SURFACE WEAPONS CTR, WHITE OAK LAB, 55- Mem: Am Phys Soc. Res: Theoretical investigations of the behavior of solids under high pressure utilizing the methodologies of statistical mechanics, lattice dynamics and the nature of the cohesive energy of the solid in question. Mailing Add: 18 Lakeside Dr Greenbelt MD 20770

O'KEEFFE, LAWRENCE EUGENE, b Walhalla, NDak, May 12, 34; m 54; c 6. ENTOMOLOGY. Educ: NDak State Univ, BS, 56, MS, 58; Iowa State Univ, PhD(econ entom), 65. Prof Exp: Port entomologist, Wis Dept Agr, 60-62; asst state entomologist, Iowa Dept Agr, 62-65; exten entomologist, 65-69, asst prof entom & asst entomologist, 69-73, ASSOC PROF ENTOM & ASSOC ENTOMOLOGIST, UNIV IDAHO, 73- Mem: Entom Soc Am. Res: Host-plant resistance to insects and host-plant selection; applied and regulatory entomology. Mailing Add: Dept of Entom Univ of Idaho Moscow ID 83843

O'KEEFFE, MICHAEL, b Bury St Edmunds, Eng, Apr 3, 34; m 62. SOLID STATE CHEMISTRY. Educ: Bristol Univ, BSc, 54, PhD(chem), 58. Prof Exp: Chemist, Mullard Res Labs, 58-59; res assoc chem, Ind Univ, 60-62; from asst prof to assoc prof, 63-69, PROF CHEM, ARIZ STATE UNIV, 69- Concurrent Pos: Fel, Ind Univ, 60-62. Res: Chemistry of solids. Mailing Add: Dept of Chem Ariz State Univ Tempe AZ 85281

O'KELLEY, GROVER DAVIS, b Birmingham, Ala, Nov 23, 28; m 50; c 2. NUCLEAR CHEMISTRY, COSMOCHEMISTRY. Educ: Howard Col, AB, 48; Univ Calif, Berkeley, PhD(chem), 51. Prof Exp: Chemist nuclear chem res, Radiation Lab, 49-51; lead chemist, Calif Res & Develop Co, 51-54; sr chemist, 54-59, group leader, 59-74, SR RES STAFF MEM, CHEM DIV, OAK RIDGE NAT LAB, 74- Concurrent Pos: Prof chem, Univ Tenn, Knoxville, 64-; chmn, Gordon Res Conf on Nuclear Chem, 67; mem, Lunar Sample Preliminary Exam Team, 69-70; mem comt on nuclear sci & chmn subcomt on radiochem, Nat Res Coun, 74- Honors & Awards: Apollo Achievement Award, NASA, 70 & Group Achievement Award, 73. Mem: AAAS; Am Phys Soc; Am Geophys Union; Am Chem Soc; Sigma Xi. Res: Nuclear cosmochemistry, using methods of low-background counting; studies of nuclear energy levels by decay-scheme and in-beam spectroscopy; instrumentation for nuclear research. Mailing Add: Chem Div Oak Ridge Nat Lab PO Box X Oak Ridge TN 37830

O'KELLEY, JOSEPH CHARLES, b Unadilla, Ga, May 9, 22; m 51; c 4. PLANT PHYSIOLOGY. Educ: Univ NC, AB, 43, MA, 48; Iowa State Col, PhD(plant physiol), 50. Prof Exp: From asst prof to assoc prof, 51-61, chmn dept, 70-73, PROF BIOL, UNIV ALA, 61- Concurrent Pos: NSF fel, Univ Wis, 54-55; NIH fel, Johns Hopkins Univ, 65-66. Mem: AAAS; Am Soc Photobiol; Phycol Soc Am; Bot Soc Am; Am Soc Plant Physiologists. Res: Algal and cell physiology; photobiology. Mailing Add: Dept of Biol Univ of Ala PO Box 1927 University AL 35486

OKEN, AARON, organic chemistry, see 12th edition

OKEN, DONALD, b New York, NY, Jan 21, 38; m 66; c 2. PSYCHIATRY. Educ: Syracuse Univ, 44-45; Harvard Univ, MD, 49; Am Bd Psychiat & Neurol, dipl psychiat, 58. Prof Exp: Res assoc, Psychosomatic & Psychiat Inst, Michael Reese Hosp, Chicago, 56-58, from asst dir to assoc dir, 58-65; chief clin res br, NIMH, 66-68, actg dir div extramural res progs, 66-67; PROF PSYCHIAT & CHMN DEPT, STATE UNIV NY UPSTATE MED CTR, 68- Concurrent Pos: Buswell fel med & psychiat, Sch Med, Univ Rochester, 51-52; NIMH training fel, Neuropsychiat & Psychiat Inst, Univ Ill, 55-56; Found Fund Res in Psychiat fel, Psychosom & Psychiat Inst, Michael Reese Hosp, 57-59; examr, Am Bd Psychiat & Neurol, 61-; mem, Ment Health Bd, Onondaga County, NY, 69- Mem: AAAS; Am Psychiat Asn; Am Psychosom Soc. Res: Psychosomatic medicine. Mailing Add: Dept of Psychiat State Univ of NY Upstate Med Ctr Syracuse NY 13210

OKERHOLM, RICHARD ARTHUR, b Woburn, Mass, Nov 10, 41; m 65; c 2. BIOCHEMISTRY. Educ: Lowell Technol Inst, BS, 64; Boston Univ, PhD(biochem), 70. Prof Exp: NIH fel, Boston Univ, 69-70; res biochemist, 70-73, SR RES BIOCHEMIST, PARKE-DAVIS & CO, 73- Mem: Am Chem Soc; Am Soc Mass Spectrometry; NY Acad Sci; Am Soc Pharmacol & Exp Therapeut; Sigma Xi. Res: Analysis and metabolism of drugs and steroids; application of mass spectrometry to the analysis of drugs and steroids in biological fluids. Mailing Add: Res & Develop Div Parke-Davis & Co Ann Arbor MI 48106

OKEY, ALLAN BERNHARDT, b La Crosse, Wis, July 8, 39; m 67; c 1. ENDOCRINOLOGY, CANCER. Educ: Univ Wis-La Crosse, BS, 62; Southern Ill Univ, MS, 64, PhD(physiol), 69. Prof Exp: Res assoc endocrinol, Southern Ill Univ, 64-67; asst prof, 67-71, ASSOC PROF BIOL, UNIV WINDSOR, 72- Concurrent Pos: Nat Res Coun Can grant, Univ Windsor, 67-75; Nat Cancer Inst Can grant, 74-75. Res: Experimental mammary tumor induction/prevention in rats and mice; hormonal effects on tumors; hormonal and tumorigenic effects of environmental contaminants. Mailing Add: Dept of Biol Univ of Windsor Windsor ON Can

OKEY, CHARLES HENRY, public health, see 12th edition

OKINAKA, YUTAKA, b Osaka, Japan, Jan 22, 26; m 50; c 2. ELECTROCHEMISTRY, ANALYTICAL CHEMISTRY. Educ: Tohoku Univ, Japan, BS, 48, DSc, 59; Univ Minn, MS, 57. Prof Exp: Assoc anal chem, Tohoku Univ, Japan, 48-54, asst prof electrochem, 60-63; fel, Univ Minn, 56-60; MEM TECH STAFF, CHEM PROCESS TECHNOL DEPT, BELL TEL LABS, INC, 63- Honors & Awards: Res Award, Electrochem Soc. Mem: Am Chem Soc; Electrochem Soc. Res: Electrochemical and chemical metal disposition process research and development. Mailing Add: Chem Process Technol Dept Bell Tel Labs Inc Murray Hill NJ 07974

OKITA, GEROGE TORAO, b Seattle, Wash, Jan 18, 22; m 58; c 3. PHARMACOLOGY. Educ: Ohio State Univ, BA, 48; Univ Chicago, PhD(pharmacol), 51. Prof Exp: From instr to asst prof pharmacol, Univ Chicago, 53-63; assoc prof, 63-66, actg chmn, 68-70, PROF PHARMACOL, MED SCH, NORTHWESTERN UNIV, 66- Concurrent Pos: USPHS fel, Univ Chicago, 52. Mem: AAAS; Am Soc Pharmacol & Exp Therapeut; Int Soc Biochem Pharmacol; Am Heart Asn. Res: Metabolism and mechanism of action of digitalis; mode of action of chemical carcinogens; metabolic effect and mode of action of environmental toxicants. Mailing Add: Dept of Pharmacol Northwestern Univ Med Sch Chicago IL 60611

OKONKWO, AUGUSTINE IKECHUKWUKA, b Asaba, Nigeria, Aug 18, 36; m 65; c 3. BIOLOGY. Educ: William Jewell Col, BA, 61; Kans State Teachers Col, BS, 60, MS, 61; Univ Ill, Urbana, MS, 63; Mich State Univ, PhD(entom, educ), 69. Prof Exp: ASSOC PROF BIOL, HURON COL, S DAK, 68- Concurrent Pos: Lectr, SDak Acad Sci, 70-; courtesy assoc prof, SDak State Univ, 71-; mem, SDak Coun Environ Educ, 72- Mem: AAAS; Am Inst Biol Sci; Entom Soc Am. Res: Ecology and insect control; ecology of the James River. Mailing Add: Dept of Biol Huron Col Huron SD 57350

O'KONSKI, CHESTER THOMAS, b Kewaunee, Wis, May 12, 21; m 48; c 4. CHEMISTRY, BIOPHYSICS. Educ: Univ Wis, BS, 42; Northwestern Univ Ill, MS, 46, PhD(phys chem), 49. Prof Exp: From instr to assoc prof, 48-60, PROF CHEM,

UNIV CALIF, BERKELEY, 60- Concurrent Pos: Guggenheim fel, 55; Knapp Mem lectr, Univ Wis, 58; Miller Sci Found res prof, Univ Calif, Berkeley, 60; NIH fel, Princeton Univ & Harvard Univ, 62-63; Nobel guest prof, Univ Uppsala, 70; Wis Alumni Res Found fel; Nat Res Coun fel. Mem: AAAS; Am Chem Soc; Am Phys Soc; Biophys Soc. Res: Physical chemistry of macromolecules; electro-optics; electronic structure of molecules; proteins and mechanisms of muscle and membranes; nucleic acids. Mailing Add: Dept of Chem Univ of Calif Berkeley CA 94720

OKREND, HAROLD, b New York, NY, Mar 25, 34; m 63. MICROBIOLOGY. Educ: City Col New York, BS, 55; Syracuse Univ, MS, 57; Rutgers Univ, PhD(environ sci), 63. Prof Exp: Fel environ sci, Rutgers Univ, 63-65; asst prof microbiol, 65-71, ASSOC PROF MICROBIOL, HOWARD UNIV, 71- Mem: AAAS; Am Soc Microbiol; Brit Soc Gen Microbiol; Am Inst Biol Sci. Res: Pollution microbiology; nutrition of bacteria. Mailing Add: Dept of Bot Howard Univ Washington DC 20001

OKRENT, DAVID, b Passaic, NJ, Apr 19, 22; m 48; c 3. NUCLEAR PHYSICS. Educ: Stevens Inst Technol, ME, 43; Harvard Univ, MA, 48, PhD(physics), 51. Prof Exp: Mech engr, Nat Adv Comt Aeronaut, 43-46; from assoc physicist to sr physicist, Argonne Nat Lab, 51-71; PROF ENG & APPL SCI, UNIV CALIF, LOS ANGELES, 71- Concurrent Pos: US deleg, Int Conf Peaceful Uses Atomic Energy, Geneva, 55, 58, 64 & 71; consult adv comt reactor safeguards & reactor hazards eval br, AEC, 59-, mem adv comt reactor safeguards, 63-, vchmn, 65, chmn, 66; Guggenheim fel, 61-62; vis prof, Univ Wash, 63 & Univ Ariz, 70-71; Argonne Univs Asn distinguished appt award, 70-71. Mem: Nat Acad Eng; fel Am Phys Soc; fel Am Nuclear Soc. Res: Nuclear reactor physics, safety and fuels; neutron cross sections; high temperature materials; fast reactor technology; fusion reactor technology; risk-benefit. Mailing Add: 5532 Boelter Hall Univ of Calif Los Angeles CA 90024

OKTAY, EROL, b Safranbolu, Turkey, Aug 3, 38; m 65; c 1. NUCLEAR ENGINEERING, PLASMA PHYSICS. Educ: Univ Mich, BS, 63, MS, 64, PhD(nuclear eng), 69. Prof Exp: Asst res eng, Gas Dynamics Lab, Univ Mich, 63-68, teaching fel nuclear radiation, 69, asst res engr, 69-70; asst prof nuclear eng, Mass Inst Technol, 70-71; asst prof physics & astron, Univ Md, College Park, 71-74; PHYSICIST, US ENERGY RES & DEVELOP ADMIN, 74- Mem: AAAS; Am Phys Soc; Am Nuclear Soc; NY Acad Sci. Res: Exploding wires; plasma physics and spectroscopy; gas dynamics. Mailing Add: DCTR US Energy Res & Develop Admin Washington DC 20545

OKUBO, AKIRA, b Tokyo, Japan, Feb 5, 25. PHYSICAL OCEANOGRAPHY, MATHEMATICAL BIOLOGY. Educ: Tokyo Inst Technol, BE, 47, MA, 49; Johns Hopkins Univ, PhD(oceanog), 63. Prof Exp: Lab asst, Japan Meteorol Agency, Tokyo, 50-54, chief chem sub-sect, 54-58; res asst oceanog, Johns Hopkins Univ, 58-63, res assoc, Chesapeake Bay Inst, 63-68, res scientist, 68-74; PROF PHYS ECOL, MARINE SCI RES CTR, STATE UNIV NY STONY BROOK, 74- Mem: AAAS; Am Soc Naturalists; Ecol Soc Am; Ecol Soc Japan; Am Soc Limnol & Oceanog. Res: Mathematical ecology; turbulent diffusion in the sea; mathematical modeling for animal dispersal. Mailing Add: Marine Sci Res Ctr State Univ of NY Stony Brook NY 11794

OKULITCH, VLADIMIR JOSEPH, b St Petersburg, Russia, June 18, 06; nat Can; m 34; c 2. GEOLOGY, INVERTEBRATE PALEONTOLOGY. Educ: Univ BC, BASc, 31, MASc, 32; McGill Univ, PhD(geol, paleont), 34. Hon Degrees: DSc, Univ BC, 72. Prof Exp: Surveyor, Atlin Ruffner Mines, 29, geologist, 30-31; asst geol, Univ BC, 31-32 & Univ McGill, 33-34; Royal Soc Can fel, Mus Comp Zool, Harvard Univ, 34-36; instr, Univ Toronto, 36-69, lectr gen & hist geol, 39-42, asst prof geol, 42-44; from asst prof to assoc prof geol, 44-49, prof paleont & stratig, 49-53, chmn div geol, 53-59, R W Brock prof geol & head dept, 59-63, dean fac sci, 63-71, EMER DEAN SCI, UNIV BC, 71- Concurrent Pos: Vis prof, Univ Southern Calif, Univ Calif, Los Angeles & Univ Hawaii; sr asst, Que Geol Surv, 33-34; geologist, Shawinigan Chem, Ltd, 37; consult geologist, Calif Standard Co, Shell Oil Co & Sproule & Assocs; mem, Nat Adv Comt Astron. Mem: Fel Geol Soc Am; fel Paleont Soc; fel Royal Soc Can; Royal Astron Soc Can; Royal Photog Soc. Res: Lower Precambrian fauna; Archaeocyatha; corals and sponges of the Paleozoic era; lower Cambrian fossils; Paleozoic stratigraphy; Cordilleran geology. Mailing Add: 1843 Knox Rd Vancouver BC Can

OKUN, LAWRENCE M, b Toledo, Ohio, Apr 12, 40; m 70. NEUROBIOLOGY. Educ: Wesleyan Univ, BA, 62; Stanford Univ, PhD(genetics), 68. Prof Exp: Nat Res Coun fel biol chem, Harvard Med Sch, 69-70, USPHS fel, 70-71; asst prof, 71-74, ASSOC PROF BIOL, UNIV UTAH, 74- Concurrent Pos: Estab investr, Am Heart Asn, 75- Res: Development, physiology, and biochemistry of neurons cultured in vitro. Mailing Add: Dept of Biol Univ of Utah Salt Lake City UT 84112

OKUN, RONALD, b Los Angeles, Calif, Aug 7, 32; m 58; c 2. CLINICAL PHARMACOLOGY. Educ: Univ Calif, Los Angeles, BA, 54; Univ Calif, San Francisco, MD & MS, 58. Prof Exp: Teaching asst, Univ Calif, San Francisco, 58; intern & resident, Vet Admin, 58-61; asst prof med & pharmacol, Univ Calif, Los Angeles, 63-72, ASSOC PROF MED, MED PHARMACOL & THERAPEUT, UNIV CALIF, IRVINE-CALIF COL MED, 72- Concurrent Pos: Fel clin pharmacol, Johns Hopkins Hosp, 61-63; res pharmacologist, Vet Admin, 64- Res: Toxicology; internal medicine. Mailing Add: Cedars-Sinai Med Ctr 8720 Beverly Blvd Los Angeles CA 90048

OKUNEWICK, JAMES PHILIP, b Chicago, Ill, Apr 30, 38; m 58; c 4. RADIATION BIOPHYSICS, CANCER. Educ: Loyola Univ, Calif, BS, 58; Univ Calif, Los Angeles, MS, 62, PhD(biophys, nuclear med), 65. Prof Exp: Mem res staff radiobiol div, Atomic Energy Proj, Univ Calif, Los Angeles, 58-59, radiobiologist, Labs Nuclear Med & Radiation Biol, 59-65; phys scientist, Rand Corp, Calif, 65-66; sr proj leader, Armed Forces Radiobiol Res Inst, Md, 66-68; assoc biologist, Cellular & Radiation Biol Labs, 68-72, SR BIOLOGIST & HEAD VIRAL ONCOGENESIS, CANCER RES UNIT, ALLEGHENY GEN HOSP, 72- Concurrent Pos: Consult, Oak Ridge Inst Nuclear Studies, 66-68; lectr, Dept Biol, Am Univ, 67; adj assoc prof radiation health, Univ Pittsburgh, 70- Mem: AAAS; Radiation Res Soc; Biophys Soc; NY Acad Sci; Int Soc Exp Hemat (pres, 73-74). Res: Radiobiology; hematopoietic system; oncogenic virus; nuclear medicine. Mailing Add: Allegheny Gen Hosp 320 E North Ave Pittsburgh PA 15212

OLAFSSON, PATRICK GORDON, b Winnipeg, Man, Aug 21, 20; m 58. ORGANIC CHEMISTRY. Educ: McGill Univ, BS, 46; Univ Man, MS, 50; Ore State Univ, PhD(org chem), 59. Prof Exp: Metall chemist, Vulcan Iron & Steel Works, Can, 39-44; res chemist, E I du Pont de Nemours & Co, 59-60; assoc prof, 60-64, PROF CHEM, STATE UNIV NY ALBANY, 64- Mem: Am Chem Soc. Res: Thermal hydrogen shifts across conjugated pi-electron systems; differential thermal analysis of nucleic acids; reactions of singlet oxygen; atmospheric chemistry. Mailing Add: Dept of Chem State Univ of NY Albany NY 12222

OLAH, ARTHUR FRANK, b Bridgeport, Conn, May 28, 41; m 65; c 2. PLANT PHYSIOLOGY, PHYTOPATHOLOGY. Educ: Miami Univ, BA, 63, MA, 65; NC State Univ, PhD, 70. Prof Exp: Instr biol, NC Cent Univ, 67; from asst prof to assoc prof, Frostburg State Col, 70-74; ASST, OFF OF PUB LANDS COORDR, TOWN OF FAIRFIELD, 75- Mem: AAAS; Bot Soc Am; Am Soc Plant Physiologists; Am Phytopath Soc; Phytochem Soc NAm. Res: Phenolic compounds in diseased plants; epiphtic algae; acid tolerant plants. Mailing Add: 36 Judson Rd Fairfield CT 06430

OLAH, GEORGE ANDREW, b Budapest, Hungary, May 22, 27; m 49; c 2. ORGANIC CHEMISTRY. Educ: Budapest Tech Univ, PhD(org chem), 49. Prof Exp: Asst & assoc prof chem, Budapest Tech Univ, 49-54; assoc dir & head org chem dept, Cent Res Inst, Hungarian Acad Sci, 54-56; res scientist, Dow Chem Co Can, 57-64 & Dow Chem Co, 64-65; prof chem, 65-69, C F MABERY PROF RES CHEM, CASE WESTERN RESERVE UNIV, 69- Honors & Awards: Petrol Chem Award, Am Chem Soc, 64, Baekeland Award, 67 & Morley Medal, 70. Mem: Nat Acad Sci; Am Chem Soc; fel Chem Inst Can; The Chem Soc; Dutch Chem Soc. Res: Organic reaction mechanism; carbocations; Friedel-Crafts reactions; intermediate complexes; chemical carcinogenesis; biological alkylating agents; organic fluorine and phosphorus compounds. Mailing Add: Dept of Chem Case Western Reserve Univ Cleveland OH 44106

OLAH, JUDITH AGNES, b Budapest, Hungary, Jan 21, 29; m 49; c 2. ORGANIC CHEMISTRY. Educ: Tech Univ, Budapest, MS, 55. Prof Exp: Res chemist, Cent Res Inst Chem, Hungarian Acad Sci, 55-56; SR RES ASSOC CHEM, CASE WESTERN RESERVE UNIV, 66- Res: Synthetic and mechanistic organic chemistry, electrophilic reactions. Mailing Add: Dept of Chem Case Western Reserve Univ Cleveland OH 44106

OLANDER, DONALD PAUL, b Boulder, Colo, June 24, 40; m 66. ANALYTICAL CHEMISTRY. Educ: Washburn Univ, Topeka, BSc, 64; Univ Nebr-Lincoln, MSc, 67, PhD(chem), 70. Prof Exp: ASST PROF CHEM, APPALACHIAN STATE UNIV, 69- Mem: Am Chem Soc. Res: Ion-solvent interactions; nonaqueous solvents; analyses of natural waters and soils. Mailing Add: Dept of Chem Appalachian State Univ Boone NC 28607

OLANDER, HARVEY JOHAN, b San Francisco, Calif, Nov 19, 32; m 69; c 2. VETERINARY PATHOLOGY. Educ: Univ Calif, Davis, BS, 56, DVM, 58, PhD(comp path), 63. Prof Exp: Asst prof vet path, Purdue Univ, 62-65; asst prof, Univ Calif, Davis, 65; assoc prof, State Univ NY Vet Col, Cornell Univ, 65-68; PROF VET PATH, ANIMAL DIS DIAG LAB, PURDUE UNIV, WEST LAFAYETTE, 68- Mem: Am Vet Med Asn; Am Col Vet Path; Conf Res Workers Animal Dis; Int Acad Path; NY Acad Sci. Res: Pathology of infectious diseases of domestic animals; comparative gastroenterology; swine dysentery. Mailing Add: Dept Vet Microbiol Path & Health Purdue Univ West Lafayette IN 47906

OLANDER, JAMES ALTON, b Boulder, Colo, Oct 9, 44. INORGANIC CHEMISTRY. Educ: Washburn Univ Topeka, BS, 65; La State Univ, Baton Rouge, 70. Prof Exp: Asst chem, La State Univ, Baton Rouge, 68-70; res chemist, 70-75, ACTG RES DIR, DEEPSEA VENTURES, INC, 75- Mem: Am Chem Soc. Res: Ion-ion and ion-solvent interactions; electrolytic solutions; nonaqueous solvents. Mailing Add: Deepsea Ventures Inc Gloucester Point VA 23062

OLANSKY, SIDNEY, b Boston, Mass, Jan 11, 14; m 45; c 4. MEDICINE, DERMATOLOGY. Educ: Harvard Univ, BS, 34; Glasgow Univ, MD, 40; Duke Univ, MD, 56. Prof Exp: From asst surgeon to sr surgeon, USPHS, 42-50, dir venereal dis res lab, 50-55; assoc prof dermat, Duke Univ, 55-59; PROF MED & DERMAT, EMORY UNIV, 59- Concurrent Pos: Consult, USPHS & Vet Admin. Mem: AAAS; fel AMA; fel Am Pub Health Asn; fel Am Col Physicians; fel Am Acad Dermat. Res: Clinical and pathological dermatology and syphilology. Mailing Add: Emory Univ Clin 1364 Clifton Rd NE Atlanta GA 30322

OLBERG, RALPH CHARLES, b Chicago, Ill, Aug 18, 14; m 39; c 2. ORGANIC CHEMISTRY. Educ: Cornell Univ, BA, 37; Northwestern Univ, MS, 43, PhD(org chem), 47. Prof Exp: Anal chemist, Armour & Co, Ill, 37-38; res chemist, Universal Oil Prod Co, 38-47; org res group leader, Parffine Co, Inc, 47-48; chemist, Shell Develop Co, 48-61; tech specialist, Aerojet-Gen Corp Div, Gen Tire & Rubber Co, 61-65; TECH STAFF SECT HEAD PLASTICS & RESIN RES & DEVELOP, FAIRCHILD SEMICONDUCTORS, 65- Mem: Am Chem Soc; Electrochem Soc; Soc Plastic Engrs. Res: Development on catalysis of petroleum hydrocarbons, asphalts and terpenes; polymers; resins; solid propellants; catalyst-polymer interactions; burning rate catalysis; plastic materials for semiconductor devices. Mailing Add: Fairchild Semiconductors 4001 Miranda Ave Palo Alto CA 94304

OLBRICH, STEVEN EMIL, b Chicago, Ill, Nov 24, 38; m 68; c 2. DIARY SCIENCE. Educ: Univ Wis-Madison, BS, 65; Univ Hawaii, MS, 68; Univ Mo-Columbia, PhD(nutrit), 71. Prof Exp: From res asst to res assoc nutrit, Univ Mo-Columbia, 68-72; res assoc, 72-74, DAIRY SPECIALIST, DEPT ANIMAL SCI, UNIV HAWAII, 75- Mem: Am Dairy Sci Asn; Am Soc Animal Sci. Res: Animal nutrition, dairy science and environmental physiology. Mailing Add: 2363 Kuahea St Honolulu HI 96816

OLCOTT, HAROLD SAFT, b Denver, Colo, July 24, 09; m 34; c 4. FOOD SCIENCE. Educ: Univ Denver, BS, 28, MS, 29; Univ Iowa, PhD(biochem), 31. Prof Exp: Nat Res Coun fel med, Yale Univ, 31-32; res assoc biochem, Univ Iowa, 32-37; Cotton Res Found fel, Mellon Inst, 37-41; biochemist, Western Utilization Res Br, USDA, 41-55; prof marine food sci, Univ Calif, Berkeley, 55-70; PROF MARINE FOOD SCI, UNIV CALIF, DAVIS, 70-, ASST DEAN AQUACULT, 75- Concurrent Pos: Chmn comt, Quatermaster Animal Prod; assoc dir, Inst Ecol, 73-75. Honors & Awards: Lilly Award, Am Chem Soc, 37, Agr & Food Chem Award, 76; Superior Serv Award, USDA, 52; Bailey Award, Am Oil Chemists Soc, 70, Bond Award, 75; Babcock-Hart Award, Inst Food Technologists, 76. Mem: Am Chem Soc; fel Inst Food Technologists; Am Oil Chemists Soc; Am Soc Biol Chem; Am Inst Nutrit. Res: Antioxidants for fats and oils; food technology; fish and marine food biochemistry; lipid compositon. Mailing Add: Dept of Food Sci & Technol Univ of Calif Davis CA 95616

OLCOTT, JOSEPH S, physical chemistry, inorganic chemistry, see 12th edition

OLD, LLOYD JOHN, b San Francisco, Calif, Sept 23, 33. CANCER, IMMUNOLOGY. Educ: Univ Calif, BA, 55, MD, 58. Prof Exp: Res fellow, Sloan-Kettering Inst Cancer Res, 58-59, res assoc, 59-60, assoc, 60-64, assoc mem, 64-67; res assoc, 60-62, from asst prof to assoc prof, 62-69, PROF BIOL, GRAD SCH MED SCI, CORNELL UNIV, 69-; MEM, SLOAN-KETTERING INST CANCER RES, 67-, VPRES & ASSOC DIR, 73- Concurrent Pos: Consult, Nat Cancer Inst, 67, mem develop res working group, 69, mem spec virus cancer prog, Immunol Group, 70; assoc med dir, New York Cancer Res Inst, Inc, 70, med dir, 71-; mem med & sci adv bd & bd trustee, Leukemia Soc Am, 70-73; mem sci adv bd, Jane Coffin Childs Mem Fund Med Res, 70-; adv ed, J Exp Med, 71; Louis Gross Mem lectr, 72; assoc ed, Virology, 72-74; Harvey Soc lectr, 73. Honors & Awards: Roche Award, Roche Inst, 57; Award, Alfred P Sloan Found, 62; Lucy Wortham James Award, James Ewing

Soc, 70. Mem: Inst of Med of Nat Acad Sci; AAAS; Am Asn Cancer Res; Soc Exp Biol & Med; Am Asn Immunol. Res: Tumor immunobiology; immunogenetics; viral oncology. Mailing Add: Sloan-Kettering Inst Cancer Res 1275 York Ave New York NY 10021

OLD, THOMAS EUGENE, b Spokane, Wash, Aug 2, 43; m 63; c 1. ATMOSPHERIC PHYSICS. Educ: Gonzaga Univ, BS, 66; Univ Idaho, MS, 69; State Univ NY Albany, PhD(physics), 71. Prof Exp: Res assoc physics, State Univ NY Albany, 71-72; TECH STAFF ATMOSPHERIC PHYSICS, MISSION RES CORP, 72- Mem: Am Geophys Union. Res: Atmospheric nuclear effects; magnetohydrodynamics. Mailing Add: Mission Res Corp PO Drawer 719 Santa Barbara CA 93101

OLDAY, FREDERICK COMBS, b New York, NY, Jan 31, 38; m 63; c 1. PLANT NUTRITION. Educ: Pa State Univ, BS, 63; Harvard Univ, MA, 65; Univ Mass, PhD(plant sci), 73. Prof Exp: Teaching plant physiol, Lowell Univ, 65-68; TEACHING CHEM & SOILS, COL ATLANTIC, 73- Mem: Soil Sci Soc Am; Am Soc Hort Sci. Res: Soil testing and plant analysis calibration. Mailing Add: Col of Atlantic Bar Harbor ME 04609

OLDEMEYER, DONALD LEROY, b Brush, Colo, Nov 7, 24; m 48; c 3. PLANT BREEDING. Educ: Colo Agr & Mech Col, BS, 48; Pa State Univ, PhD(agron), 52. Prof Exp: Asst agronomist, State Col Wash, 52-54; plant breeder, 54-69, mgr agr res, 69-72, MGR SEED PROD & DEVELOP, AMALGAMATED SUGAR CO, 72- Mem: Am Soc Agron; Am Soc Sugar Beet Technologists. Res: Variety improvement and seed production of sugarbeets. Mailing Add: 109 N Sixth St Nyssa OR 97913

OLDEMEYER, ROBERT KING, b Brush, Colo, Sept 23, 22; m 44; c 2. PLANT BREEDING, AGRONOMY. Educ: Colo State Univ, BS, 47; Univ Wis, MS, 48, PhD(genetics), 50. Prof Exp: Plant breeder, 50-60, dir seed develop, 60-68, agr res, 68-71, mgr seed processing & prod, 71-73, MGR VARIETY DEVELOP, GREAT WESTERN SUGAR CO, 73- Concurrent Pos: Staff affil, Colo State Univ, 73- Mem: Am Soc Agron; Am Soc Sugar Beet Technologists; Am Inst Biol Sci; Int Inst Beet Sugar Res. Res: Breeding of sugar beets; sugar beet diseases. Mailing Add: Great Western Sugar Co 11939 Sugarmill Rd Longmont CO 80501

OLDENBURG, THEODORE RICHARD, b Muskegon, Mich, Apr 8, 32; m 59; c 2. PEDODONTICS. Educ: Univ NC, DDS, 57, MS, 62. Prof Exp: From asst prof to assoc prof dent, 62-69, PROF PEDODONT & CHMN DEPT, SCH DENT, UNIV NC, CHAPEL HILL, 69- Concurrent Pos: United Cerebral Palsy clin fel, 60-62; mem USAF sr dent prog, 56-57; consult accreditation comt, Am Dent Asn, 67-68, coun dent educ & US Army Dent Corps. Mem: Am Dent Asn; Am Acad Pedodont; Am Soc Dent for Children; Int Asn Dent Res. Res: Educational research in clinical pedodontics. Mailing Add: Dept of Pedodont Univ of NC Chapel Hill NC 27515

OLDENDORF, WILLIAM HENRY, b Schenectady, NY, Mar 27, 25; m 45; c 3. NEUROLOGY, NUCLEAR MEDICINE. Educ: Albany Med Col, MD, 47; Am Bd Psychiat & Neurol, dipl, cert psychiat, 53, cert neurol, 55. Prof Exp: Intern med, Ellis Hosp, Schenectady, NY, 47-48; resident psychiat, Binghamton State Hosp, 48-50 & Letchworth Village State Sch, Thiells, 50-52; resident neurol, Univ Minn Hosps, Minneapolis, 54-55; assoc chief neurol, 55-69, MED INVESTR, WADSWORTH HOSP, VET ADMIN CTR, 69-; PROF NEUROL, SCH MED, UNIV CALIF, LOS ANGELES, 70- Concurrent Pos: Clin fel, Univ Minn, Minneapolis, 55; from clin instr to asst clin prof, Sch Med, Univ Calif, Los Angeles, 56-65, assoc prof, 65-70, USPHS res grants, 59-72; NIH spec fel physiol, Univ Col, Univ London, 65-66; mem coun stroke, Am Heart Asn, 69. Honors & Awards: Silver Medal Award, Soc Nuclear Med, 69; Ziedses des Plantes Gold Medal, Med Physics Soc, Würzburg, 74. Mem: Inst Elec & Electronics Engr; Soc Nuclear Med; fel Am Acad Neurol; Am Neurol Asn; Soc Exp Biol & Med. Res: Nuclear medicine applied in clinical neurological research, including studies related to human brain isotope uptake studies, cerebral blood flow in man, cerebrospinal fluid, central nervous system instrumentation and development of photographic techniques relating to brain isotope scanning and cerebral angiography in man. Mailing Add: Wadsworth Hosp Vet Admin Ctr Los Angeles CA 90073

OLDERMAN, GERALD M, b New York, NY, July 16, 33; m 58; c 3. PHYSICAL CHEMISTRY. Educ: Rensselaer Polytech Inst, BS, 58; Seton Hall Univ, MS, 71, PhD(chem), 72. Prof Exp: Res chemist, Nat Cash Register Co, 58-61; sr scientist, Cotton & Gauze Dept, 61-65, group leader, Surg Dressings Dept, 65-67, asst mgr, 67-69, mgr, Operating Rm Prod Res-Surg Specialty Div, 69-72, div develop, Surg Specialty Div, 72-74, VPRES RES & DEVELOP, SURGIKOS, JOHNSON & JOHNSON CO, 75- Mem: Am Chem Soc; Am Soc Artificial Internal Organs; Fiber Soc; Am Soc Extracorporeal Technicians; Sigma Xi. Res: Nonwovens or synthetic materials, their sterilization, chemical treatments and adaptation for use in surgery; development and testing of biocompatible materials and extracorporeal or prosthetic products. Mailing Add: Surgikos Res Div Johnson & Johnson Co 501 George St New Brunswick NJ 08903

OLDFIELD, DANIEL G, b New York, NY, July 24, 25; m 50; c 2. CELL BIOLOGY, MATHEMATICAL BIOPHYSICS. Educ: Columbia Univ, BS, 50; Univ Chicago, MS, 58, PhD(math biol), 65. Prof Exp: Jr scientist, Argonne Cancer Res Hosp, Univ Chicago, 53-55, assoc scientist, 55-62, sr physicist, US Air Force Radiation Lab, 62-64, res assoc, Lab Cytol, 65-68; asst prof, 68-70, ASSOC PROF BIOL SCI, DEPAUL UNIV, 70- Mem: AAAS; Am Phys Soc; Radiation Res Soc; NY Acad Sci; Am Soc Cell Biol. Res: Analysis of cell cycle and organelle interactions by cytofluorometry. Mailing Add: Dept of Biol Sci DePaul Univ 1036 WBelden Ave Chicago IL 60614

OLDFIELD, ERIC, b London, Eng, May 23, 48. MOLECULAR BIOLOGY, BIOPHYSICAL CHEMISTRY. Educ: Bristol Univ, BS, 69; Sheffield Univ, PhD(chem), 72. Prof Exp: Europ Molecular Biol Org fel chem, Ind Univ, 72-74; vis scientist, Mass Inst Technol, 74-75; ASST PROF CHEM, UNIV ILL, URBANA, 75- Res: Nuclear magnetic resonance spectroscopy; biological membrane structure; lipid-protein interactions; tumor immunology; cell culture. Mailing Add: 76 Noyes Lab Sch of Chem Sci Univ of Ill Urbana IL 61801

OLDFIELD, GEORGE NEWTON, b San Jose, Calif, Sept 6, 36; m 59; c 3. ACAROLOGY. Educ: Fresno State Col, BA, 62; Univ Calif, Riverside, MS, 66, PhD(entom), 71. Prof Exp: Agr res technician, 62-65, entomologist, 65-68, RES ENTOMOLOGIST, ENTOM RES DIV, AGR RES SERV, USDA, 68- Mem: AAAS; Entom Soc Am; Acarological Soc Am; Sigma Xi. Res: Acarology, especially biology, systematics, reproductive biology and virus vector capabilities of Eriophyoidea; arthropod vectors of fruit tree pathogens; photoperiodic control of ovarian diapause of pear psylla. Mailing Add: Boyden Entom Lab USDA Univ of Calif PO Box 112 Riverside CA 92502

OLDFIELD, JAMES EDMUND, b Victoria, BC, Aug 30, 21; nat US; m 42; c 5. ANIMAL NUTRITION. Educ: Univ BC, BS, 41, MS, 49; Ore State Col, PhD(animal nutrit, biochem), 51. Prof Exp: Instr animal husb, Univ BC, 48-49; from asst prof to assoc prof, 51-59, PROF ANIMAL NUTRIT, ORE STATE UNIV, 59-, HEAD DEPT ANIMAL SCI, 67- Concurrent Pos: Basic res award, Ore Agr Exp Sta, 61, Sigma Xi res award, Ore State Univ, 64; res travel award, Nat Feed Ingredients Asn, 69; Fulbright scholar, Massey Univ, NZ, 74. Honors & Awards: Morrison Award, Am Soc Animal Sci, 72, Distinguished Serv Award, 74. Mem: Am Chem Soc; Am Inst Nutrit; Am Soc Animal Sci (secy-treas, 62-65, vpres, 65, pres, 66); NY Acad Sci; Agr Inst Can. Res: Vitamins, minerals, and antibiotics as feed supplements in animal nutrition; nutritional diseases; metabolic diseases; alternate feed sources; relative efficiencies of domestic animal species as human food producers. Mailing Add: Dept of Animal Sci Ore State Univ Corvallis OR 97331

OLDFIELD, THOMAS EDWARD, b Fond du Lac, Wis, Oct 11, 47; m 70; c 2. PHYSIOLOGICAL ECOLOGY. Educ: Mich Technol Univ, BS, 70; Utah State Univ, MS, 73, PhD(physiol), 75. Prof Exp: ASST PROF PHYSIOL, FERRIS STATE COL, 75- Concurrent Pos: Sigma Xi res grant, 74. Mem: Sigma Xi. Res: Development of reliable methods to be used in the determination of energy utilization of free-living animals. Mailing Add: Dept of Biol Sci Ferris State Col Big Rapids MI 49307

OLDHAM, KEITH BENTLEY, b Ashton-under-Lyne, Eng, Feb 4, 29; m 53; c 5. ELECTROCHEMISTRY, PHYSICAL CHEMISTRY. Educ: Univ Manchester, BSc, 49, PhD(phys chem), 52, DSc, 70. Prof Exp: Res assoc anal chem, Univ Ill, Urbana, 52-55; vis scientist, Rensselaer Polytech Inst, 55; asst lectr phys chem, Imp Col, Univ London, 55-57; lectr chem, Univ Newcastle, 57-64; vis assoc anal chem, Calif Inst Technol, 64-65; lectr chem, Univ Newcastle, 65-66; mem tech staff, Sci Ctr, NAm Rockwell Corp, Calif, 66-70; PROF CHEM, TRENT UNIV, 70-, CHMN DEPT, 72- Concurrent Pos: Consult, Consumers' Asn, 56-66; mem schs exam bd, Univ Durham, 58-64; mem, Int Comt Electrochem Thermodyn & Kinetics, 61-; consult, Royal Aircraft Estab, 63-66; vis scientist, Sci Ctr, NAm Aviation, Inc, 64-65; fac mem, Calif Lutheran Col, 67-69. Mem: Electrochem Soc; fel Royal Inst Chem. Res: Transport processes; various aspects of physical and analytical chemistry. Mailing Add: Dept of Chem Trent Univ Peterborough ON Can

OLDHAM, SUSAN BANKS, b San Francisco, Calif, Aug 6, 42. ENDOCRINOLOGY. Educ: Univ Calif, Los Angeles, BS, 64; Univ Southern Calif, PhD(biochem), 68. Prof Exp: Fel endocrine res, Mayo Clin, 68-72; res assoc endocrinol, 72-73, ASST PROF MED & BIOCHEM, SCH MED, UNIV SOUTHERN CALIF, 73- Mem: Endocrine Soc; Am Fedn Clin Res; Sigma Xi; AAAS. Res: The regulation of the parathyroid gland and the metabolism of parathyroid hormone; the role of calcium-binding proteins in tissue metabolism; the mechanisms by which hypomagnesemia causes hypocalcemia; the regulation of vitamin D metabolism; causes of hypercalcemia in cancer. Mailing Add: Sch of Med Univ of Southern Calif 2025 Zonal Ave Hoffman Bldg Los Angeles CA 90033

OLDHAM, WILLIAM JENNINGS BRYAN, JR, atomic physics, computer systems, see 12th edition

OLDS, CARL DOUGLAS, b Wanganui, NZ, May 11, 12; US citizen; m 37; c 2. MATHEMATICS. Educ: Stanford Univ, BA, 36, MA, 37, PhD(math), 43. Prof Exp: Asst math, Stanford Univ, 36-37, actg instr, 37-38, instr, 38-40; instr, Purdue Univ, 40-45; from asst prof to assoc prof, 45-51, PROF MATH, SAN JOSE STATE UNIV, 51- Mem: Math Asn Am; London Math Soc. Res: Theory of numbers; geometry of numbers; engineering calculations; theory of probability; number or representations of the square on an integer as the sum of an odd number of squares. Mailing Add: Dept of Math San Jose State Univ San Jose CA 94022

OLDS, DANIEL WAYNE, b Richland Co, Ill, Mar 27, 35; m 60; c 2. PHYSICS. Educ: Wabash Col, AB, 56; Duke Univ, PhD, 64. Prof Exp: Asst prof, 63-66, actg chmn dept, 63-71, ASSOC PROF PHYSICS, WOFFORD COL, 66-, MGR COMPUT TERMINAL, 68-, CHMN DEPT PHYSICS, 71-, DIR ACAD COMPUT CTR, 75- Mem: Am Asn Physics Teachers. Res: Use of computer in undergraduate education. Mailing Add: Dept of Physics Wofford Col Spartanburg SC 29301

OLDS, DURWARD, b Conneaut, Ohio, Apr 12, 21; m 47; c 2. DAIRY SCIENCE, VETERINARY MEDICINE. Educ: Ohio State Univ, DVM, 43; Univ Ill, MS, 54, PhD(dairy sci), 56. Prof Exp: Artificial insemination technician, Clark County Breeder's Coop, Wis, 44-46; asst dairying, 46-51, assoc prof, 51-56, PROF DAIRYING, UNIV KY, 56- Mem: AAAS; Am Dairy Sci Asn; Am Vet Med Asn; NY Acad Sci. Res: Physiology of reproduction; artificial insemination; causes of infertility in dairy cattle. Mailing Add: Dept of Animal Sci Univ of Ky Lexington KY 40506

OLDS, ROBERT HORNER, physics, see 12th edition

OLDSTONE, MICHAEL BEAUREGUARD ALAN, b New York, NY, Feb 9, 34; m 60; c 3. EXPERIMENTAL BIOLOGY, NEUROBIOLOGY. Educ: Univ Ala, BS, 54; Univ Md, MD, 61. Prof Exp: USPHS fel rickettsiol & virol, Sch Med, Univ Md, 58, intern med, Univ Hosp, 61, resident, 62-63, resident neurol, 63-64, chief resident, 64-66; fel, Dept Exp Path, 66-69, assoc, 69-71, ASSOC MEM, DEPT IMMUNOPATH & DIV NEUROL, SCRIPPS CLIN & RES FOUND, 71-, HEAD NEUROL RES, FOUND, 69- Concurrent Pos: NIH-AID career develop award, 69; adj prof path, Univ Calif, San Diego, 71- & adj prof neurosci, 72-; mem ad hoc sci adv comt, Multiple Sclerosis. Mem: Am Acad Neurol; Am Asn Immunol; Am Asn Neuropath; Am Soc Exp Path; Am Soc Microbiol. Res: Viral immunopathology; viruses and immunity. Mailing Add: Dept of Immunopath Scripps Clin & Res Found La Jolla CA 92037

O'LEARY, AUSTIN JOHN, b Lindsay, Ont, Dec 28, 06; nat US; m 42. PHYSICS. Educ: Queen's Univ, Ont, BSc, 28, MSc, 29; Columbia Univ, PhD(physics), 32. Prof Exp: Asst, Columbia Univ, 30-31; from tutor to prof physics, City Col New York, 31-71; prof, Baruch Col, 71-75; RETIRED. Mem: AAAS; Am Phys Soc; Am Asn Physics Teachers. Res: Radioactivity and x-rays; foundations of physics; interaction of x-rays with bound electrons. Mailing Add: 431 E 20th St New York NY 10010

O'LEARY, BRIAN TODD, b Boston, Mass, Jan 27, 40; m 64; c 2. ASTRONOMY. Educ: Williams Col, BA, 61; Georgetown Univ, MA, 64; Univ Calif, Berkeley, PhD(astron), 67. Prof Exp: Physicist, Goddard Space Flight Ctr, NASA, 61-62; high sch teacher, Washington, DC, 64; scientist-astronaut, Manned Spacecraft Ctr, NASA, 67-68; asst prof astron & space sci, Cornell Univ, 68-71; vis assoc, Calif Inst Technol, 71; assoc prof interdisciplinary sci, San Francisco State Univ, 71-72; asst prof astron & sci policy assessment, Hampshire Col, 72-75; SPEC CONSULT ON ENERGY, SUBCOMT ON ENERGY & ENVIRON, US HOUSE INTERIOR COMT, 75- Concurrent Pos: Consult, NASA Ames Res Ctr, 71-; vis assoc prof, Sch Law, Univ Calif, Berkeley, 72. Mem: Fel AAAS; Am Astron Soc; Am Geophys Union. Res: Technology assessment. Mailing Add: 4807 Cumberland Ave Chevy Chase MD 20015

O'LEARY, DENNIS PATRICK, b Dec 24, 39; US citizen; m 64; c 2. BIOPHYSICS, NEUROPHYSIOLOGY. Educ: Univ Chicago, SB, 62; Univ Iowa, PhD(physiol, biophys), 69. Prof Exp: Asst prof surg & anat, Univ Calif, Los Angeles, 71-74; RES

ASSOC PROF OTOLARYNGOL & PHARMACOL, UNIV PITTSBURGH, 74- Concurrent Pos: USPHS res fel, Univ Calif, Los Angeles, 69-70, USPHS res grant, 70-74. Mem: AAAS; Biophys Soc; Soc Neurosci; Int Brain Res Orgn. Res: Biophysical mechanisms of sensory transduction; stability and control in the vestibular system; applications of time series analysis, stochastic processes and filtering theory to investigations of neuronal information processing. Mailing Add: Dept of Otolaryngol Univ of Pittsburgh Pittsburgh PA 15213

O'LEARY, GERARD PAUL, JR, b Bridgeport, Conn, Oct 16, 40. MICROBIOLOGY, BIOCHEMISTRY. Educ: Mt St Mary's Col, Md, BS, 62; NMex State Univ, MS, 64; Univ NH, PhD(microbiol), 67. Prof Exp: Nat Res Coun Can fel, Macdonald Col, McGill Univ, 67-69; ASSOC PROF BIOL, PROVIDENCE COL, 69- Mem: AAAS; Am Soc Microbiol; Am Inst Biol Sci; Can Soc Microbiol; Nat Shellfish Sanit Asn. Res: Isolation and chemical characterization of marine bacterial lipopolysaccharides; effects of bacterial endotoxins on fresh water fish; effects of microbial pollution on the aquatic environment. Mailing Add: Dept of Biol Providence Col Providence RI 02918

O'LEARY, JAMES LEE, b Tomahawk, Wis, Dec 8, 04; m 39; c 2. ANATOMY. Educ: Univ Chicago, SB, 25, PhD(anat), 28, MD, 31. Prof Exp: Instr anat, Univ Chicago, 27-28; from asst prof to prof, 29-72, LECTR NEUROL, SCH MED, WASH UNIV, 72- Mem: Am Electroencephalog Soc; Am Asn Anatomists; assoc Am Neurol Asn (pres, 61-62). Res: Neurology. Mailing Add: Dept of Neurol & Neurol Surg Wash Univ Sch of Med St Louis MO 63110

O'LEARY, JAMES WILLIAM, b Painesville, Ohio, Aug 10, 38; m 63; c 3. PLANT PHYSIOLOGY. Educ: Ohio State Univ, BS, 60, MS, 61; Duke Univ, PhD(bot), 64. Prof Exp: Asst prof bot, Univ Ariz, 63-66; asst prof biol, Bowling Green State Univ, 66-67; assoc prof biol sci, Univ, 67-71, plant physiologist, Environ Res Lab, 68-74, PROF BIOL SCI, UNIV ARIZ, 71-, RES PROF BIOL SCI, ENVIRON RES LAB, 74- Mem: AAAS; Am Soc Agron; Am Soc Plant Physiol; Am Inst Biol Sci; Am Soc Hort Sci. Res: Plant water relations; physiological ecology of desert plants; controlled environments; salinity effects on plants. Mailing Add: 2012 E Le Madera Dr Tucson AZ 85719

O'LEARY, JOHN FRANCIS, b Adrian, Pa, Dec 12, 17; m 72; c 7. PHARMACOLOGY. Educ: Pa State Univ, BS, 42; Univ Rochester, MS, 46, PhD(pharmacol), 51. Prof Exp: Indust chemist, Keystone Ord Works, Pa, 42-44; res assoc, Manhattan Proj, AEC, Univ Rochester, 44-49, res assoc, Pharmacol & Orthop Surg Depts, 49-51; sr scientist, McNeil Labs, Pa, 51-53; pharmacologist, Med Res Lab, 53-58, actg chief pharmacol br, 58-59, contract admin officer, Physiol Div, 60-61; chief pharmacol sect, Basic Med Sci Br, 61-72, MED SCI ADMINR BIOMED LAB, MED RES DIV, US ARMY EDGEWOOD ARSENAL, 72- Concurrent Pos: Lectr, Sch Med, Univ Md, 64-65; app to Gov of Maryland's Sci Adv Coun, 74- Mem: AAAS; Am Chem Soc; Am Soc Pharmacol & Exp Therapeut; Soc Exp Biol & Med; NY Acad Sci. Res: Mechanism of uranium toxicity; determination of uranium in tissues; central nervous system paralyzants; hypotensive agents; local anesthetics; anticholinergic drugs; pharmacology of incapacitating compounds; drug effects on animal behavior; research administration and contracting. Mailing Add: Med Res Div Biomed Lab Bldg 3220 Edgewood Arsenal MD 21010

O'LEARY, KEVIN JOSEPH, b Winthrop, Mass, Aug 8, 32; m 54; c 5. POLYMER CHEMISTRY. Educ: Boston Col, BS, 55; Case Western Reserve Univ, MS, 62, PhD(eng), 67. Prof Exp: Anal chemist, US Army Chem Ctr, Md, 55-57; sr res chemist & anal area supvr, Diamond Alkali Res Ctr, Ohio, 57-64, res assoc, Res Ctr, Diamond Shamrock Corp, 67-69, mgr anal & electrochem res, 69-71, assoc dir res, T R Evans Res Ctr, 71-, assoc dir res electrochem/anal, 71-75, ASST DIR RES PHYS SCI, T R EVANS RES CTR, DIAMOND SHAMROCK CORP, 75- Concurrent Pos: Vis prof, Lake Erie Col, 67-68. Mem: Am Phys Soc; Am Chem Soc; Soc Appl Spectros; Am Crystallog Asn; Electrochem Soc. Res: Polymer morphology, deformation of crystalline polymers; small and wide angle x-ray diffraction; electron microscopy; thermodynamics and analytical chemistry; electrochemistry. Mailing Add: T R Evans Res Ctr Diamond Shamrock Corp PO Box 348 Painesville OH 44077

O'LEARY, MARION HUGH, b Quincy, Ill, Mar 24, 41; m 64; c 2. ORGANIC CHEMISTRY, BIOCHEMISTRY. Educ: Univ Ill, BS, 63; Mass Inst Technol, PhD(org chem), 66. Prof Exp: NIH res fel biochem, Harvard Univ, 66-67; asst prof, 67-73, ASSOC PROF ORG CHEM, UNIV WIS-MADISON, 73- Concurrent Pos: NIH & Res Corp res grants, 67-; Sloan Found fel, 72-74; NSF res grant. Mem: AAAS; Am Chem Soc; The Chem Soc; Am Soc Biol Chemists. Res: Mechanisms of action of enzymes; bio-organic chemistry; organic reaction mechanisms; isotope effects. Mailing Add: Dept of Chem Univ of Wis Madison WI 53706

O'LEARY, ROBERT KENT, b Brooklyn, NY, Apr 3, 40; m 64; c 4. MATERIALS SCIENCE, TOXICOLOGY. Educ: Univ Notre Dame, BS, 62; Univ Tex, MS, 66, PhD(mat sci, toxicol), 68. Prof Exp: Chemist tech serv & prod develop, Chas Pfizer & Co, NY, 62-65; staff mem mat toxicol, Univ Tex, Austin, 65-68; sr scientist, Ethicon, Inc Div, Johnson & Johnson, NJ, 68-69, sect head bioanal & mat toxicol, 69-71, mgr, Dept Path & Surg, 71-72, dir, Ethicon Res Found, 72, DIR RES & DEVELOP, PATIENT CARE DIV, JOHNSON & JOHNSON RES CTR, 72- Concurrent Pos: Vis prof molecular & quantum biol, Univ Tenn Med Ctr, 75- Mem: Soc Toxicol; Am Chem Soc; Sigma Xi; AAAS. Mailing Add: Johnson & Johnson Res Ctr New Brunswick NJ 08903

O'LEARY, TIMOTHY JOSEPH, organic chemistry, deceased

O'LEARY, VIRGINIA SAWYER, b Brooklyn, NY, Aug 18, 45; m 69. FOOD MICROBIOLOGY. Educ: Cornell Univ, BS, 67, MS, 69; Rutgers Univ, PhD(food sci), 73. Prof Exp: Fel dairy microbiol, Dairy Res, Inc, 73-74; FOOD TECHNOLOGIST, EASTERN REGIONAL RES CTR, USDA, 74- Mem: Sigma Xi. Res: Fermentative utilization of whey; cultured dairy products; microbiol aspects of food safety; microbial biochemistry. Mailing Add: Eastern Regional Res Ctr USDA 600 E Mermaid Lane Wyndmoor PA 19118

O'LEARY, WILLIAM MICHAEL, b Columbus, Ohio, Oct 9, 28; m 51. MICROBIOLOGY, BIOCHEMISTRY. Educ: Univ Pittsburgh, BS, 52, MS, 53, PhD(bact), 57. Prof Exp: Asst bact, Div Biol Sci, Univ Pittsburgh, 52-53, asst res dept, Sch Dent, 53-54, asst biochem, Sch Med, 54-57; Nat Res Coun resident res assoc, Div Biol Med Res, Argonne Nat Lab, 57-59; from asst prof to assoc prof bact, 59-72, PROF BACT, MED COL, CORNELL UNIV, 72- Mem: AAAS; Am Soc Microbiol; Am Chem Soc; NY Acad Sci; NY Acad Med. Res: Bacterial chemistry and physiology; lipid metabolism; antibiotic mode of action and resistance; mechanisms of pathogenicity; infective human infertility. Mailing Add: Dept of Microbiol Cornell Univ Med Col New York NY 10021

OLECHOWSKI, JEROME ROBERT, b Buffalo, NY, Jan 10, 31; m 56; c 5. PHYSICAL CHEMISTRY, ORGANIC CHEMISTRY. Educ: Canisius Col, BS, 52; Pa State Univ, MS, 55; La State Univ, PhD(chem), 58. Prof Exp: Asst chem, Pa State Univ, 52-54 & La State Univ, 54-57; res chemist, Copolymer Rubber & Chem Corp, 57-61 & Cities Serv Res & Develop Co, 61-63; prin res chemist, Columbian Carbon Co, 63-70; RES ASSOC, PETROCHEM GROUP, CITIES SERV CO, 70-, INT TECH LIAISON, 71- Concurrent Pos: Prof, McNeese State Col, 63. Mem: Am Chem Soc. Res: Reactions of ozone with organic substances; Ziegler-Natta type catalysts for polymerization of unsaturated substances; metal-ion olefin complexes and oxidation; rearrangement of trihaloalkenes; organometallic chemistry of transition elements; chemistry of medium ring compounds. Mailing Add: 17 Empress Lane Lawrenceville NJ 08648

OLEINICK, NANCY LANDY, b Pittsburgh, Pa, Feb 26, 41; m 62; c 2. BIOCHEMISTRY. Educ: Chatham Col, BS, 62; Univ Pittsburgh, PhD(biochem), 66. Prof Exp: Instr, 68-69, ASST PROF RADIATION BIOL & BIOCHEM, SCH MED, CASE WESTERN RESERVE UNIV, 69-, ASST PROF, SCH DENT, 71- Concurrent Pos: Fel biochem, Sch Med, Case Western Reserve Univ, 66-68; Nat Inst Allergy & Infectious Dis fel, 67-69; Nat Cancer Inst res grant, 73- Mem: Am Chem Soc; Am Soc Cell Biol; Radiation Res Soc. Res: Protein synthesis in vivo and in vitro; interrelationships between ionizing radiation, protein synthesis and the biological expressions of radiation damage; radiosensitivity of ribosomal RNA genes. Mailing Add: Wearn Res Bldg Case Western Reserve Univ Cleveland OH 44106

OLEKSIUK, LESLIE WILLIAM, b Toronto, Ont, Sept 18, 36; m 64; c 2. ENGINEERING PHYSICS. Educ: Univ Toronto, BA, 59, MA, 60, PhD(nuclear physics), 62. Prof Exp: Physicist, Brookhaven Nat Lab, 62-64, Univ Toronto, 64-67 & Brookhaven Nat Lab, 67-69; PHYSICIST, FERMI NAT ACCELERATOR LAB, 69- Res: Accelerator technology; ion source development; beam dynamics; beam transport system design; magnet design and fabrication. Mailing Add: Fermi Nat Accelerator Lab Accelerator Sect PO Box 500 Batavia IL 60510

OLENICK, JOHN GEORGE, b Throop, Pa, Oct 31, 35; m 59; c 5. MOLECULAR BIOLOGY. Educ: Univ Scranton, BS, 58; Albany Med Col, MS, 61; Univ Md, College Park, PhD(microbiol), 71. Prof Exp: Biochemist, Sect Kidney & Electrolyte Metab, Nat Heart Inst, 61-62; chief chem sect, Dept Cell & Media Prod, Microbiol Assocs, Inc, 62-63; biochemist, 63-71, ASST CHIEF DEPT MOLECULAR BIOL, WALTER REED ARMY INST RES, 71- Res: Biosynthesis of proteins and nucleic acids; modes and mechanisms of action of antimicrobials; structure activity relationships; molecular pharmacology; membrane structure and function; elimination of drug resistance determinants from bacterial plasmids. Mailing Add: Dept of Molecular Biol Walter Reed Army Inst of Res Washington DC 20012

OLER, NORMAN, b Sheffield, Eng, July 12, 29; US citizen; m 57; c 2. MATHEMATICS. Educ: McGill Univ, BS, 51, MS, 53, PhD(math), 57. Prof Exp: Asst prof math, McGill Univ, 57-60; res assoc, Columbia Univ, 60-61, asst prof, 61-63; assoc prof, 63-67, PROF MATH, UNIV PA, 67- Concurrent Pos: Nat Res Coun Can overseas fel, 58-59; NSF sci fac fel, 65-66. Mem: Am Math Soc; Can Math Soc. Res: Geometry of numbers with particular interest in packing and covering problems. Mailing Add: Dept of Math Univ of Pa Philadelphia PA 19104

OLES, KEITH FLOYD, b Seattle, Wash, June 9, 21; m 46; c 2. GEOLOGY. Educ: Univ Wash, BS, 43, MS, 52, PhD, 56. Prof Exp: Instr geol, Univ Wash, 47-52; asst prof, Washington & Lee Univ, 52-53; mem spec explor group, Union Oil Co, Calif, 53-58, dist geologist, Wyo Dist, 58-61; assoc prof, 61-72, PROF GEOL, ORE STATE UNIV, 72- Concurrent Pos: Mem prof develop panel, Coun Educ in Geol Sci, Am Geol Inst, 68-73, visitor, Vis Geol Sci Prog, 69-70; consult & chief party, North Slope, Alaska, Union Oil Co Calif-Gulf Oil Corp, 69; geol consult. Mem: Am Asn Petrol Geologists; Am Inst Prof Geologists. Res: Paleoenvironments of Cretaceous rocks of Pacific Northwest and British Columbia. Mailing Add: 3815 NW Hayes Ave Corvallis OR 97330

OLESEN, JOHN ALLEN, b Inglewood, Calif, July 20, 46. ORGANIC CHEMISTRY. Educ: Univ Calif, Irvine, BA, 68; Univ Colo, Boulder, MS, 70, PhD(chem), 74. Prof Exp: Consult, Oil Shale Corp, 73-74; RES CHEMIST, E I DU PONT DE NEMOURS & CO, INC, 74- Mem: Am Chem Soc; Int Solar Energy Soc; Sigma Xi. Res: Development and application of energy production and recovery methods. Mailing Add: Chem Dept E I du Pont de Nemours & Co Inc Wilmington DE 19898

OLESON, JEROME JORDAN, b Waupun, Wis, May 26, 14; m 39; c 4. BIOCHEMISTRY. Educ: Univ Wis, BS, 36, MS, 38, PhD(biochem), 42. Prof Exp: Asst, Univ Wis, 36-40; head nutrit dept, Lederle Labs, Am Cyanamid Co, 40-58; mgr cancer chemother, Chas Pfizer & Co, 58-67; DIR, JOHN L SMITH MEM CANCER RES, PFIZER INC, 67-73; ACTG ASSOC DIR RES, NAT CTR TOXICOL RES, 73- Mem: Am Asn Cancer Res; Am Soc Biol Chem; Soc Exp Biol & Med; NY Acad Sci. Res: Nutrition; cancer chemotherapy; viral oncology; toxicology. Mailing Add: Nat Ctr Toxicol Res Jefferson AR 72079

OLESON, NORMAN LEE, b Detroit, Mich, Aug 19, 12; m 39; c 2. PHYSICS. Educ: Univ Mich, BS, 35, MS, 37, PhD(physics), 40. Prof Exp: Asst, Univ Mich, 36-40; instr physics, US Coast Guard Acad, 40-46; res physicist, Gen Elec Co, 46-48; prof physics, Naval Postgrad Sch, 48-69; PROF PHYSICS & CHMN DEPT, UNIV S FLA, 69- Concurrent Pos: Vis prof, Queen's Univ, Belfast, 55-56 & Mass Inst Technol, 67-68; consult, Lawrence Radiation Lab, Univ Calif, 58- Mem: AAAS; Sigma Xi; fel Am Phys Soc. Res: Gamma ray studies of light radioactive elements; multiple scattering of fast electrons; electrical and optical studies of positive column of gas mixtures; plasmas in magnetic fields; ion acoustic waves; plasma heating. Mailing Add: Dept of Physics Univ of S Fla Tampa FL 33620

OLEWINE, DONALD AUSTIN, b Harrisburg, Pa, May 4, 28; m 57; c 3. PHYSIOLOGY. Educ: Dickinson Col, BS, 50; Univ NC, PhD, 57. Prof Exp: Instr physiol, Col Med, Univ Vt, 57-58; physiologist, Geront Sect, Nat Heart Inst, Baltimore City Hosps, 58-62; asst prof biol, Bucknell Univ, 62-65; from asst prof to assoc prof, 65-71, actg head dept, 67-71, PROF BIOL, GA SOUTHERN COL, 71- Mem: Am Soc Zoologists; Geront Soc; NY Acad Sci; Am Col Sports Med; Nat Sci Teachers Asn. Res: Endocrine, renal and exercise physiology; gerontology. Mailing Add: Dept of Biol Ga Southern Col Statesboro GA 30458

OLEXIA, PAUL DALE, b McKeesport, Pa, July 31, 31; m 68; c 2 MYCOLOGY, TAXONOMY. Educ: Wabash Col, BA, 61; State Univ NY Buffalo, MA, 65; Univ Tenn, Knoxville, PhD(bot), 68. Prof Exp: Vis asst prof biol, Colgate Univ, 67-68; asst prof, 68-74, ASSOC PROF BIOL, KALAMAZOO COL, 74- Mem: Mycol Soc Am; Int Asn Plant Taxon Res: Morphological and physiological taxonomy of the basidiomycetous fungi, Clavariaceae. Mailing Add: Dept of Biol Kalamazoo Col Kalamazoo MI 49001

OLF, HEINZ GUNTHER, b Wetzlar, Ger, Nov 1, 34. POLYMER PHYSICS. Educ: Munich Tech Univ, Vordiplom, 57, Diplom, 60, Dr rer nat(physics), 69. Prof Exp: PHYSICIST, RES TRIANGLE INST, 61- Mem: AAAS; Am Phys Soc; Am Chem Soc; Ger Phys Soc. Res: Solid state physics of high polymers; application of nuclear magnetic resonance techniques, x-ray scattering, infrared and Raman spectroscopy to polymers; mechanical properties of polymers. Mailing Add: Res Triangle Inst Research Triangle Park NC 27709

OLHOEFT, JACK E, nuclear science, computer science, see 12th edition

OLIEN, MICHAEL DAVID, b Milwaukee, Wis, Apr 16, 37; m 59; c 1. ANTHROPOLOGY. Educ: Beloit Col, BA, 59; Univ NC, MA, 62; Univ Ore, PhD(anthrop), 67. Prof Exp: Mem staff anthrop, Cent Am Field Prog, Assoc Cols Midwest, 64-65; instr anthrop, Univ Ore, 65-66; asst prof, Am Univ, 66-67; asst prof, 67-73, ASSOC PROF ANTHROP, UNIV GA, 73- Concurrent Pos: Rep Am Anthrop Asn, Conf US Nat Comn UNESCO, 71; vis lectr, Am Anthrop Asn, 71- Mem: Am Soc Ethnohist; Am Anthrop Asn; Latin Am Studies Asn. Res: Cultures and societies of Latin America; complex societies; ethnohistory; cultural ecology; Afro-American cultures. Mailing Add: Dept of Anthrop Baldwin Hall Univ of Ga Athens GA 30602

OLIET, SEYMOUR, b Perth Amboy, NJ, July 12, 27; m 49; c 2. DENTISTRY. Educ: Univ Pa, DDS, 53; Am Bd Endodontics, dipl, 65. Prof Exp: Instr oral med, 53-56, assoc, 56-61, from asst prof to assoc prof clin oral med, 61-71, PROF ORAL MED & CHMN DEPT ENDODONT, SCH DENT, UNIV PA, 71- Concurrent Pos: Clin asst, Southern Div, Albert Einstein Med Ctr, 53-55, chief endodontics, 55-, adj, 55-60, sr attend, 60-; consult endodontics, US Army, Ft Dix, NJ, 61 & Vet Admin Hosp, Philadelphia, Pa, 67-; dir, Am Bd Endodont. Honors & Awards: Am Acad Dent Med Award, 53. Mem: Fel AAAS; Am Dent Asn; Int Asn Dent Res; Am Acad Dent Med; hon mem Brazilian Endodontic Asn. Res: Use of salt and glass beads for rapid resterilization of root canal instruments and armamentarium; development of a torsional tester for root canal instruments and detailed studies on physical properties of root canal instruments. Mailing Add: 625 Medical Arts Bldg Philadelphia PA 19102

OLIN, ARTHUR DAVID, b New York, NY, July 5, 28; m 54; c 3. ORGANIC CHEMISTRY. Educ: St Peter's Col, NJ, BS, 49; Rutgers Univ, PhD(org chem), 56. Prof Exp: Asst chem, Rutgers Univ, 50-54; res chemist polymer chem, Nat Starch & Chem Corp, 55; RES CHEMIST ORG CHEM, TOMS RIVER CHEM CORP, 56- Concurrent Pos: Asst res specialist, Rutgers Univ, 55-56; coadj assoc prof, Ocean County Col, 70-71. Mem: Am Chem Soc. Res: Process development and synthesis of vat and anthraquinone dyes and organic intermediates. Mailing Add: 1222 Tuxedo Terr Lakewood NJ 08701

OLIN, JACQUELINE S, b Lansford, Pa, Nov 27, 32; m 55; c 2. ORGANIC CHEMISTRY, ANALYTICAL CHEMISTRY. Educ: Dickinson Col, BS, 54; Harvard Univ, MA, 55. Prof Exp: Teaching fel chem, Univ Pa, 55-56; res chemist, NIH, 56-57; instr chem, Dickinson Col, 59-60; res chemist, Cornell Univ, 60-61; RES CHEMIST, SMITHSONIAN INST, 62- Concurrent Pos: Res collabr, Brookhaven Nat Lab, 66-; guest worker, Nat Bur Standards, 68- Mem: Fel Int Inst Conserv Hist & Artistic Works. Res: Analysis of ceramic artifacts using neutron activation analysis for determination of provenance on basis of trace element constituents, extended to glasses with special attention to medieval stained glass. Mailing Add: 9506 Watts Rd Great Falls VA 22066

OLIN, PHILIP, b Winnipeg, Man, Nov 21, 41; m 66. MATHEMATICAL LOGIC. Educ: Univ Man, BS, 63; Cornell Univ, PhD(math logic), 67. Prof Exp: Res assoc & lectr math, Cornell Univ, 67-69; asst prof, McGill Univ, 69-70; asst prof, 70-74, ASSOC PROF MATH, YORK UNIV, 74- Mem: Am Math Soc. Res: Model theory; recursion theory. Mailing Add: Dept of Math York Univ Toronto ON Can

OLIN, STEPHEN MUNRO, chemistry, see 12th edition

OLIN, WILLIAM (HAROLD), b Menominee, Mich, Mar 7, 24; m 50; c 3. ORTHODONTICS. Educ: Marquette Univ, DDS, 47; Univ Iowa, MS, 48. Prof Exp: From asst prof to assoc prof, 48-64, PROF OTOLARYNGOL & MAXILLOFACIAL SURG, COL MED, UNIV IOWA, 64- Mem: Am Dent Asn; Am Cleft Palate Asn (past pres); Int Dent Fedn; Am Asn Orthod. Res: Clefts of lip and palate; growth of the facial bones in cleft lip and palate patients, orofacial deformities. Mailing Add: Univ of Iowa Hosps Iowa City IA 52240

OLINGER, BART, b Gettysburg, Pa, Feb 11, 43; m 67; c 2. HIGH PRESSURE PHYSICS. Educ: Gettysburg Col, BA, 64; Univ Chicago, MA, 66, PhD(geophys), 70. Prof Exp: RES STAFF SHOCK WAVE PHYSICS, LOS ALAMOS SCI LAB, 70- Mem: Sigma Xi. Res: Study of the physical properties of materials under both shock and static compression. Mailing Add: Los Alamos Sci Lab Los Alamos NM 87544

OLINICK, MICHAEL, b Detroit, Mich, May 29, 41; m 63; c 2. MATHEMATICS. Educ: Univ Mich, Ann Arbor, BA, 63; Univ Wis-Madison, MA, 64, PhD(math), 70. Prof Exp: Asst lectr math, Univ Col Nairobi, Kenya, 65-66; instr, Univ Wis-Madison, 69-70; ASST PROF MATH, MIDDLEBURY COL, 70- Res assoc, Univ Calif, Berkeley, 75-76. Mem: Am Math Soc; Math Asn Am; Opers Res Soc Am. Res: Topology of manifolds; monotone and compact mappings on euclidean spaces; mathematical modeling in social sciences. Mailing Add: Dept of Math Middlebury Col Middlebury VT 05753

OLINS, ADA LEVY, b Tel Aviv, Israel, Mar 5, 38; US citizen; m 61; c 2. BIOCHEMISTRY, ELECTRON MICROSCOPY. Educ: City Col New York, BS, 60; Radcliffe Col, MA, 62; NY Univ, PhD(biochem), 65. Prof Exp: Fel biochem, Dartmouth Col, 65-67; CONSULT, BIOL DIV, OAK RIDGE NAT LAB, 67- Concurrent Pos: USPHS spec fel cell biol, King's Col, Univ London, 70-71. Mem: Am Soc Cell Biologists; Biophys Soc. Res: DNA-protein interaction; chromosomal organization; nuclear ultrastructure. Mailing Add: Oak Ridge Nat Lab Biol Div Box Y Oak Ridge TN 37830

OLINS, DONALD EDWARD, b New York, NY, Jan 11, 37; m 61; c 2. BIOCHEMISTRY. Educ: Univ Rochester, AB, 58; Rockefeller Univ, PhD(biochem), 64. Prof Exp: Instr microbiol, Dartmouth Med Sch, 66-67, res fel molecular biol, 64-65, Whitney fel, 65-67; asst prof, 67-69, ASSOC PROF, OAK RIDGE GRAD SCH BIOMED SCI, UNIV TENN, 69- Concurrent Pos: Vis scientist, King's Col, Univ London, 70-71. Mem: Biophys Soc; Am Soc Biol Chem. Res: Chemical structure of chromosomes and DNA-nucleoproteins. Mailing Add: Grad Sch of Biomed Sci Oak Ridge Nat Lab Oak Ridge TN 37830

OLIPHANT, CHARLES WINFIELD, b Oklahoma City, Okla, Mar 13, 20; m 42; c 2. GEOLOGY. Educ: Harvard Univ, BS, 41, MA, 47, PhD(geophys), 48. Prof Exp: Res assoc & asst to dir, Radio Res Lab, Harvard Univ, 42-46; PRES, OLIPHANT LABS, INC, 48-; CHMN, CEJA CORP, 66- Mem: AAAS; Soc Explor Geophysicists; Seismol Soc Am; Opers Res Soc Am; Am Asn Petrol Geologists. Res: Petroleum exploration; elastic properties of rocks; stratigraphy. Mailing Add: Oliphant Labs Inc 1905 Nat Bank of Tulsa Bldg Tulsa OK 74103

OLIPHANT, EDWARD EUGENE, b San Francisco, Calif, May 31, 42; m 66. REPRODUCTIVE BIOLOGY, BIOLOGICAL CHEMISTRY. Educ: Univ Redlands, BS, 64; Calif State Col, Long Beach, MS, 66; Univ Calif, Davis, PhD(biochem), 70. Prof Exp: Fel biochem, Univ Calif, Davis, 70-71; fel reprod biol, Univ Pa, 71-73; ASST PROF, DEPT OBSTET & GYNEC, SCH MED, UNIV VA, 73- Mem: Soc Study Reprod; Am Soc Cell Biol. Res: The biochemistry of sperm penetration into ova and early embryo development. Mailing Add: Dept of Obstet & Gynec Univ of Va Sch of Med Charlottesville VA 22903

OLIPHANT, MALCOLM WILLIAM, b Chicago, Ill, Apr 15, 20; m 43; c 2. MATHEMATICS. Educ: Georgetown Univ, BS, 47; Johns Hopkins Univ, MA, 48; Cath Univ Am, PhD(math), 57. Prof Exp: Instr math, Johns Hopkins Univ, 47-48; mathematician, Nat Bur Standards, 48-50; prof math & chmn dept, Georgetown Univ, 49-66; PROF MATH & DEAN ACAD AFFAIRS, HAWAII LOA COL, 66- Concurrent Pos: Mem, Int Cong Mathematicians, Edinburgh, 58. Mem: AAAS; Am Math Soc; Math Asn Am. Res: Mathematical analysis; modern measure theory. Mailing Add: Hawaii Loa Col PO Box 764 Kaneohe HI 96744

OLITZKY, IRVING, b New York, NY, Dec 4, 20; m 42; c 3. BACTERIOLOGY. Educ: Mich State Univ, BS, 43, MS, 47, PhD(bact), 50. Prof Exp: Instr bact, Rutgers Univ, 50-51; asst br chief, Ralph M Parsons Co, Md, 52-54; asst chief pub health lab, City of Philadelphia, 54-57; chief qual control, Baltimore Biol Lab, Inc, 57-60; chief microbiol div, 60-65, dir dept microbiol-immunol, Bio-Sci Labs, 65-74, mgr facil develop, 74-75, ASSOC DIR AFFIL LABS, BIO-SCI ENTERPRISES, 75- Mem: Am Soc Microbiol; fel Am Pub Health Asn. Res: Immunology and automation of immunology. Mailing Add: Bio-Sci Enterprises 10880 Wilshire Blvd Suite 2100 Los Angeles CA 90024

OLIVARD, JOANNE, biochemistry, see 12th edition

OLIVE, AULSEY THOMAS, b Mount Gilead, NC, May 23, 31; m 53; c 2. ENTOMOLOGY. Educ: Wake Forest Col, BS, 53; NC State Univ, MS, 55, PhD(entom), 61. Prof Exp: Asst prof, 61-66, ASSOC PROF BIOL, WAKE FOREST UNIV, 66- Concurrent Pos: NSF grant, 65-67. Mem: Entom Soc Am. Res: Taxonomic and biosystematic study of aphids of the eastern United States; cytogenetic approach to the species problem in the aphid genus Dactynotus Rafinesque. Mailing Add: Dept of Biol Box 7327 Wake Forest Univ Winston-Salem NC 27106

OLIVE, GLORIA, b New York, NY, June 8, 23. MATHEMATICS. Educ: Brooklyn Col, BA, 44; Univ Wis, MA, 46; Ore State Univ, PhD(math), 63. Prof Exp: Asst math, Univ Wis, 44-46; instr, Univ Ariz, 46-48 & Idaho State Col, 48-50; asst, Ore State Col, 50-51; from asst prof to prof & head dept, Anderson Col, Ind, 52-68; prof, Univ Wis-Superior, 68-71; SR LECTR PURE MATH, UNIV OTAGO, NZ, 71- Concurrent Pos: Mathematician, US Dept Defense, 51; Wis Bd Regents res grant, Univ Wis-Superior, 70-71. Mem: Am Math Soc; Math Asn Am; NZ Math Soc. Res: Generalized powers; b-functions. Mailing Add: Dept of Math Univ of Otago Dunedin New Zealand

OLIVE, JOHN H, b Glenford, Ohio, Apr 16, 29; m 58; c 4. CELL PHYSIOLOGY, AQUATIC BIOLOGY. Educ: Ohio State Univ, BS, 53; Kent State Univ, MA, 61, PhD(biol), 64. Prof Exp: From assoc prof to prof biol, Ashland Col, 64-70; ASSOC PROF BIOL, UNIV AKRON, 70- Concurrent Pos: AEC res grant, 65- Mem: AAAS; Am Inst Biol Sci; Phycol Soc Am; Am Soc Limnol & Oceanog; Sigma Xi. Res: Phytoplankton photosynthesis; nutrition of blue-green algae. Mailing Add: Dept of Biol Univ of Akron Akron OH 44304

OLIVE, JOHN RITTER, zoology, limnology, deceased

OLIVE, JOSEPH P, b Israel, Mar 14, 41; US citizen; m 65; c 1. COMMUNICATIONS SCIENCE. Educ: Univ Chicago, BS & MS, 64, MA & PhD(physics), 69. Prof Exp: MEM TECH STAFF, BELL LABS, NJ, 69- Concurrent Pos: Nat Endowment Arts grant, 74. Mem: Acoust Soc Am; Am Composers Alliance. Res: Production of speech and music by computers. Mailing Add: Bell Labs Inc 2D529 600 Mountain View Murray Hill NJ 07940

OLIVE, LINDSAY SHEPHERD, b Florence, SC, Apr 30, 17; m 42. BOTANY. Educ: Univ NC, AB, 38, MA, 40, PhD(bot), 42. Prof Exp: Lab asst bot, Univ NC, 38-42, instr, 42-44; mycologist & consult diagnostician, Emergency Plant Dis Proj, USDA, 44-45; asst prof bot, Univ Ga, 45-46; assoc prof, La State Univ, 46-49; from assoc prof to prof, Columbia Univ, 49-68; UNIV DISTINGUISHED PROF, UNIV NC, CHAPEL HILL, 68- Concurrent Pos: Guggenheim fel, 56. Mem: AAAS; Mycol Soc Am (vpres, 59, pres, 66); Bot Soc Am; Torrey Bot Soc (vpres, 61, pres, 62); Soc Protozool. Res: Cytology, genetics, morphology and taxonomy of fungi and mycetozoans. Mailing Add: Dept of Bot Univ of NC Chapel Hill NC 27514

OLIVE, WILDS WILLIAMSON, b Apex, NC, Feb 22, 21; m 50; c 2. PHYSICAL GEOLOGY. Educ: Univ NC, BS, 42; La State Univ, PhD(geol), 53. Prof Exp: Asst geol, Univ NC, 42; geologist, US Geol Surv, 46-53 & Gulf Oil Corp, 53-56; geologist, 56-70, chief, Br of Ky Geol, 70-75, GEOLOGIST, BR OF KY GEOL, US GEOL SURV, 75- Concurrent Pos: Asst, La State Univ, 47-48. Mem: Geol Soc Am; Soc Econ Paleontologists & Mineralogists. Res: Areal geologic mapping; Pliocene and Pleistocene continental deposits of Mississippi embayment. Mailing Add: Br of Ky Geol US Geol Surv 2035 Regency Rd Lexington KY 40503

OLIVEIRA, ROBERT JAMES, biochemistry, see 12th edition

OLIVER, ABE D, JR, b Castleberry, Ala, Dec 3, 25; m 50; c 2. ENTOMOLOGY, FORESTRY. Educ: Auburn Univ, BS, 53, MS, 54; La State Univ, PhD(entom), 63. Prof Exp: Asst entomologist, Miss State Univ, 54-55; asst entomologist, 55-58, from asst prof to assoc prof, 58-68, PROF ENTOM, LA STATE UNIV, BATON ROUGE, 68- Mem: Entom Soc Am. Res: Forest insect research, especially ecology and economics; floriculture insect research, especially economics and systematics; biological control of aquatic weeds with insects. Mailing Add: Dept of Entom Life Sci Bldg La State Univ Baton Rouge LA 70803

OLIVER, ANNE REBECCA, b Cumberland, Md, Dec 1, 12. PHYSICS. Educ: Goucher Col, BA, 34; Smith Col, MA, 35; Cornell Univ, PhD(exp physics), 42. Prof Exp: Asst physics, Goucher Col, 38-40; jr phys sci aide, Nat Bur Standards, 40; asst physics, Wilson Col, 40-41; instr, Conn Col, 41-43; tech aide, Liaison Off, Off Sci Res & Develop, 43-45; asst prof physics, Newcomb Col, Tulane, 45-48; from asst prof to assoc prof, 48-56, PROF PHYSICS, STATE UNIV NY ALBANY, 56- Mem: Am Phys Soc; Am Asn Physics Teachers. Res: Electron diffraction; structure of thin films; study of nuclear reactions by nuclear emulsions. Mailing Add: Dept of Physics State Univ of NY Albany NY 12222

OLIVER, BARRY GORDON, b Winnipeg, Man, Feb 21, 42; m 64; c 2. ENVIRONMENTAL CHEMISTRY. Educ: Univ Man, BSc, 63, MSc, 65, PhD(phys chem), 69. Prof Exp: Res assoc, Rensselaer Polytech Inst, 68-69, asst prof chem, 69-70; Nat Res Coun Can fel phys chem, Inland Waters Br, Can Dept Environ, 70-72,

RES SCIENTIST, CAN CENTRE INLAND WATERS, 73- Mem: Chem Inst Can; Spectros Soc Can. Res: Vibrational spectroscopy; electrical conductivity and transport properties of electrolyte solutions; analysis of trace metals in sediments and natural waters; water chemistry; water and wastewater treatment methods; photochemistry of organic compounds in aqueous systems. Mailing Add: Water Chem Sect Can Centre for Inland Waters Dept of the Environ Burlington ON Can

OLIVER, BRIAN MALCOLM, b London, Eng, Mar 9, 46; Can citizen; m; c 1. PLASMA PHYSICS, SPACE PHYSICS. Educ: Univ Calif, Los Angeles, BS, 67; Univ Victoria, MS, 70, PhD(plasma physics), 73. Prof Exp: Fel elec eng, Univ Alta, 73-74; res assoc plasma physics, Univ Victoria, 75; NAT RES COUN FEL, DEPT ENVIRON OCEAN & AQUATIC SCI, 75- Mem: Can Asn Physicists. Res: Various aspects of remote sensing of the Earth's parameters; currents and tides; surface water color; satellite data analysis. Mailing Add: 903 Dunn Ave Victoria BC Can

OLIVER, CARL EDWARD, b Anniston, Ala, Feb 26, 43; m 65. MATHEMATICS. Educ: Univ Ala, Tuscaloosa, BS, 65, MA, 67, PhD(math), 69. Prof Exp: Asst prof math, US Air Force Inst Technol, 69-74; MATHEMATICIAN, WEAPONS LAB, US AIR FORCE, 74- Concurrent Pos: Lectr, Ohio State Univ, 70-71 & Wright State Univ, 71-74. Mem: Am Math Soc; Math Asn Am. Res: Interrelations and applications of the concepts within integration theory; group representations on homogeneous spaces. Mailing Add: 1724 Robert Dale Dr NE Albuquerque NM 87112

OLIVER, CONSTANCE, cell biology, see 12th edition

OLIVER, DAVID W, b Fairfax Co, Va, Dec 21, 32; m 58; c 4. SOLID STATE PHYSICS. Educ: Va Polytech Inst, BS, 55, MS, 56; Mass Inst Technol, PhD(physics), 61. Prof Exp: Physicist, 61-73, LIAISON SCIENTIST TO AIRCRAFT ENGINE BUS GROUP, GEN ELEC CO, 73-, ACTG MGR, LIAISON BR, 75- Mem: Am Phys Soc. Res: Microwave spectroscopy with application to chemical reaction rates for diatomic gases; acoustics at microwave frequencies and above; growth of inorganic single crystals and their characterization. Mailing Add: Corp Res & Develop Ctr Gen Elec Co PO Box 8 Schenectady NY 12301

OLIVER, DENIS RICHARD, b Santa Barbara, Calif, Nov 12, 41. BIOCHEMISTRY. Educ: Calif State Univ, Long Beach, BS, 65; Univ Iowa, PhD(zool), 71. Prof Exp: NSF Ctr Excellence res fel biochem, 71-72, res assoc, 72-73, INSTR BIOCHEM & ASSOC DIR, PHYSICIAN'S ASST PROG, UNIV IOWA, 73- Mem: Sigma Xi; Am Soc Allied Health Prof. Res: Structure and organization of nucleohistone in terms of the spatial relationship between histone chromosomal proteins and DNA in chromatin and the mechanism by which newly synthesized histone is deposited. Mailing Add: Physician's Asst Prog Col of Med Univ of Iowa Iowa City IA 52242

OLIVER, EUGENE JOSEPH, b Pawtucket, RI, Jan 28, 41; m 70. MICROBIOLOGY, BIOCHEMISTRY. Educ: RI Col, BEd, 62; Univ Mass, PhD(microbiol), 69. Prof Exp: Teacher, Scituate Jr-Sr High Sch, 62-64; res assoc biochem, Univ Mass, 69-71; HEALTH SCIENTIST ADMINR, NAT INST GEN MED SCI, 74- Concurrent Pos: NIH fel biochem, Nat Heart & Lung Inst, 71- Mem: AAAS; Am Soc Microbiol. Res: Enzymology of bacteria and higher organisms; carbohydrate and amino acid metabolism; molecular mechanisms of cryobiological adaptation. Mailing Add: Dept of Biochem Nat Inst of Gen Med Sci Bethesda MD 20014

OLIVER, G CHARLES, b Gainesville, Fla, Sept 30, 31; m 58; c 1. CARDIOVASCULAR DISEASE. Educ: Harvard Univ, AB, 53; Harvard Med Sch, MD, 57; Am Bd Internal Med, dipl, 65; Am Bd Cardiovasc Dis, dipl, 70. Prof Exp: From instr to assoc prof, 66-75, PROF MED & CO-DIR CARDIOVASC DIV, SCH MED, WASH UNIV, 75-; CHIEF CARDIOVASC DIV, JEWISH HOSP ST LOUIS, 71- Concurrent Pos: Fel cardiol, Med Ctr, Stanford Univ, 60-61; USPHS fel, Guy's Hosp, London, Eng, 63-65; mem policy bd, Clin Trials Thrombolytic Agents, NIH, 73-; mem electrocardiography comt, Am Heart Asn, 74- Mem: AAAS; fel Am Col Cardiol; NY Acad Sci; Am Heart Asn. Res: Computer applications to cardiology; pharmacology of digitalis. Mailing Add: Jewish Hosp Cardiovasc Div 216 S Kingshighway St Louis MO 63110

OLIVER, GENE LEECH, b Rockford, Ill, June 7, 29; m 55; c 3. PHOTOGRAPHIC CHEMISTRY, ORGANIC CHEMISTRY. Educ: Beloit Col, BS, 50; Northwestern Univ, PhD, 55. Prof Exp: Res chemist, 54-65, res assoc, 65-74, TECH STAFF ASSOC, EASTMAN KODAK CO, 74- Mem: Am Chem Soc (counr, 74-); Soc Photog Scientists & Engrs. Res: Photographic sensitizing and image dyes; silver halide emulsion technology; organic heterocyclic chemistry; photographic patents. Mailing Add: Eastman Kodak Res Labs Bldg 59 Kodak Park Rochester NY 14650

OLIVER, GEORGE DAVIS, JR, b Scooba, Miss, Feb 15, 41; m; c 2. MEDICAL PHYSICS, RADIATION PHYSICS. Educ: Lamar State Col Technol, BSc, 63; NC State Univ, MSc, 55; Univ Okla, PhD(radiation physics), 68. Prof Exp: Physicist, Nuclear Defense Lab, Edgewood Arsenal, Md, 63-64; mem staff, Univ Tex M D Anderson Hosp & Tumor Inst, 68-73; from asst prof to assoc prof radiol, Sch Med, Wash Univ, 73-76; ASSOC PROF RADIOL, SCH MED, HERSHEY MED CTR, PA STATE UNIV, 76- Concurrent Pos: Advan sr fel med physics, Univ Tex M D Anderson Hosp & Tumor Inst, 68-69. Mem: Am Asn Physicists in Med; Health Physics Soc; Radiation Res Soc. Res: Radiation dosimetry in mixed fields; relative biological effectiveness of high linear energy transfer radiation; development of radiation techniques in medicine. Mailing Add: Hershey Med Ctr Pa State Univ Hershey PA 17033

OLIVER, GEORGE JOSEPH, b Snow Hill, NC, Sept 4, 47. HEALTH PHYSICS. Educ: NC State Univ, BS, 69, MS, 71, MEcon, 76; Univ NC, PhD(health physics), 73. Prof Exp: Scientist III environ radiation, 73-75, SR SCIENTIST HEALTH PHYSICS, CAROLINA POWER & LIGHT CO, 75- Mem: Health Physics Soc; Sigma Xi. Res: Dosimetry of humans for gamma, beta neutron radiation exposure; internal dosimetry of humans using bio-assay and whole body counting; radiation measurements of many types as applied to commercial nuclear reactors. Mailing Add: 4108 Woodlake Pl Raleigh NC 27607

OLIVER, HENRY WILLIAM, b Pittsburgh, Pa, Jan 6, 16; m 40; c 4. MATHEMATICS. Educ: Yale Univ, BA, 38; Univ Chicago, MS, 48, PhD(math), 51. Prof Exp: From asst prof to prof, 51-70, THOMAS T READ PROF MATH, WILLIAMS COL, 70- Res: Differentiability of functions of real variables. Mailing Add: Dept of Math Williams Col Williamstown MA 01267

OLIVER, JACK ERTLE, b Massillon, Ohio, Sept 26, 23; m 64; c 2. GEOPHYSICS. Educ: Columbia Univ, BA, 47, MA, 50, PhD(geophys), 53. Prof Exp: Scientist's aide, US Naval Res Lab, 47; physicist, Air Force Cambridge Res Lab, 51; res assoc, Lamont-Doherty Geol Observ, Columbia Univ, 53-55; from instr to prof geol, Columbia Univ, 55-71; PROF GEOL SCI & CHMN DEPT, CORNELL UNIV, 71- Concurrent Pos: Consult, President's Sci Adv Comt Panel Seismic Improv, 58-59, Air Force Tech Appln Ctr, 58-65, Advan Res Projs Agency, 59-63, US Arms Control & Disarmament Agency, 62-, Comt Seismol & Earthquake Eng, UNESCO, 65 & AEC, 69-74; chmn dept geol, Columbia Univ, 69-71, Irving Porter Church prof, Cornell Univ, 71-; mem, adj prof, 71-73; 59-71; Comt Polar Res, Nat Acad Sci, mem Comt Seismol, 60-72, chmn, 66-70; chmn, Geodesy Comt, 75-; mem, Sci Adv Bd, US Air Force, 60-63 & 64-69; mem, Off Sci Res Contractor's Res & Eval Panel, 61-74, chmn, 66-68; mem, Adv Comt, US Coast & Geod Surv, 62-66, Panel Solid Earth Probs, 62-; mem, Int Seismol Summary Comt, Int Asn Seismol & Physics of Earth's Interior, 63-67 & 75-, Exec Comt, 68-71; mem, Nat Comt Upper Mantle Prog, 63-71, Site Selection Comt, 65, Carnegie Inst Adv Comt Awards, Gilbert & Wood Fund, 64-68 & Geophys Res Bd, 69-70. Mem: AAAS; fel Seismol Soc Am (vpres, 62-64, pres, 64-65); fel Am Geophys Union; fel Geol Soc Am. Res: Seismology; geotectonics. Mailing Add: Dept of Geol Sci Kimball Hall Cornell Univ Ithaca NY 14850

OLIVER, JACK WALLACE, b Ellettsville, Ind, Jan 6 6, 38; m 60; c 2. VETERINARY PHARMACOLOGY. Educ: Purdue Univ, BS, 60, MS, 63, DVM, 66, PhD(vet physiol), 69. Prof Exp: Nat Inst Arthritis & Metab Dis fel, 67-69; asst prof vet pharmacol, Sch Vet Med, Purdue Univ, 69-70; asst prof vet endocrinol & pharmacol, Col Vet Med, Tex A&M Univ, 70-71; pvt pract, Ind, 71-72; asst prof vet pharmacol, Col Vet Med, Ohio State Univ, 72-75; ASSOC PROF VET PHARMACOL, COL VET MED, UNIV TENN, KNOXVILLE, 75- Mem: Soc Study Reproduction; Conf Res Workers Animal Dis; Am Soc Animal Sci; Am Soc Vet Physiol & Pharmacol; World Asn Vet Physiologists, Pharmacologists & Biochemists. Res: Effects of altered physiological states, such as trace mineral deficiencies and infectious disease, on thyroid status of domestic animals; studies of thyroid hormone influence on connective tissue components. Mailing Add: Dept of Environ Practice Univ of Tenn Col of Vet Med Knoxville TN 37901

OLIVER, JAMES E, organic chemistry, see 12th edition

OLIVER, JAMES HENRY, JR, b Augusta, Ga, Mar 10, 31; m 57; c 2. ACAROLOGY, CYTOGENETICS. Educ: Ga Southern Col, BS, 52; Fla State Univ, MS, 54; Univ Kans, PhD(entom), 62. Prof Exp: NATO fel, Univ Melbourne, 62-63; from asst prof entom & parasitol to assoc prof entom, Univ Calif, Berkeley, 63-68; assoc prof, Univ Ga, 68-69; CALLAWAY PROF BIOL, GA SOUTHERN COL, 69- Concurrent Pos: Consult, US Naval Med Res Unit 3, 63- Mem: AAAS; Entom Soc Am; Am Soc Zoologists; Am Soc Parasitol. Res: Cytology, genetics, reproduction and bionomics of arthropods. Mailing Add: Dept of Biol Ga Southern Col Statesboro GA 30458

OLIVER, JAMES RUSSELL, b Egan, La, Sept 12, 24; m 45; c 3. PHYSICAL CHEMISTRY, NUCLEAR CHEMISTRY. Educ: Univ Southwestern La, BS, 50; Tulane Univ, MS, 51, PhD(chem), 55. Prof Exp: From assoc prof to prof chem, 54-70, dir comput ctr, 60-70, PROF COMPUT SCI, UNIV SOUTHWESTERN LA, 70-, DEAN GRAD SCH, 61- Concurrent Pos: Consult, Electro-Acid Corp & Silverloy Int Corp, Nev; sect ed, Chem Abstr; asst, State Syst Educ for Admin Serv, Res & Finance. Mem: AAAS; Am Chem Soc; Am Nuclear Soc; Asn Comput Mach; Soc Indust & Appl Math. Res: Coordination compounds; radioactive tracers; reactor fuel processing; radioactive waste disposal; reaction kinetics; radiation chemistry; digital computers; science eduation; science talent search and writing; design of computer languages. Mailing Add: 324 Charlotte St Lafayette LA 70501

OLIVER, JANET MARY, b Adelaide, South Australia, Nov 14, 45. CELL PHYSIOLOGY. Educ: Univ Adelaide, BSc, 66; Flinders Univ South Australia, BScHons, 67; Univ Alta, MSc, 69; Univ London, PhD(biochem), 72. Prof Exp: ASST PROF PHYSIOL, HEALTH CTR, UNIV CONN, 73- Concurrent Pos: Leukemia Soc Am fel, Harvard Med Sch, 72-73. Mem: Brit Biochem Soc. Res: Structure and function of cell surfaces; changes in surface architecture after neoplastic transformation of mammalian cells; purine metabolism. Mailing Add: Dept of Physiol Univ of Conn Health Ctr Farmington CT 06032

OLIVER, JEANETTE CLEMENTS, b Indianapolis, Ind, Apr 2, 38; m 66. PLANT ANATOMY, PLANT TAXONOMY. Educ: Purdue Univ, BS, 59, MS, 61, PhD(bot), 63. Prof Exp: Asst prof, 63-68, ASSOC PROF BIOL, BALL STATE UNIV, 68- Mem: Bot Soc Am; Torrey Bot Club; Am Fern Soc. Res: Taxonomy of the Thelypteroid ferns; plant development. Mailing Add: Dept of Biol Ball State Univ Muncie IN 47306

OLIVER, JOEL DAY, b Amarillo, Tex, Dec 27, 45; m 66; c 3. STRUCTURAL CHEMISTRY. Educ: WTex State Univ, BS, 68; Univ Tex, Austin, PhD(phys chem), 71. Prof Exp: ASST PROF PHYS CHEM, W TEX STATE UNIV, 71- Mem: Am Chem Soc; Am Crystallog Asn; The Chem Soc. Res: X-ray diffraction studies of organic and organometallic compounds. Mailing Add: Dept of Chem W Tex State Univ Canyon TX 79016

OLIVER, JOHN EDWARD, b Dover, Eng, Oct 21, 33; m 57; c 2. PHYSICAL GEOGRAPHY. Educ: Univ London, BSc, 56; Univ Exeter, cert educ, 57; Columbia Univ, MA, 66, PhD(geog), 69. Prof Exp: Lectr geog & geol, Willesden Tech Col, London, 57-58; teacher geog, Warwick Acad, Bermuda, 58-66; lectr, Columbia Univ, 66-69, from asst prof to assoc prof, 69-73; ASSOC PROF GEOG, IND STATE UNIV, 73- Mem: AAAS; Asn Am Geogrs; Am Geog Soc. Res: Environmental science with special reference to applications of climatology. Mailing Add: Dept of Geog & Geol Ind State Univ Terre Haute IN 47809

OLIVER, JOHN EOFF, JR, b Stephenville, Tex, June 22, 33; m 57; c 3. COMPARATIVE NEUROLOGY. Educ: Tex A&M Univ, DVM, 57; Auburn Univ, MS, 66; Univ Minn, St Paul, PhD(neuroanat), 69; Am Col Vet Internal Med, dipl. Prof Exp: Instr vet surg, Col Vet Med, Colo State Univ, 57-58; veterinarian, Houston, Tex, 60-63; res asst vet neurol, Sch Vet Med, Auburn Univ, 63-66; USPHS fel neuroanat, Col Vet Med, Univ Minn, St Paul, 66-67, spec fel, 67-68; assoc prof comp neurol, Col Vet Med, Univ Ga, 68-72; prof small animal clin & head dept, Sch Vet Sci & Med, Purdue Univ, Lafayette, 72-75; PROF SMALL ANIMAL MED & HEAD DEPT, COL VET MED, UNIV GA, 75- Mem: AAAS; Am Vet Med Asn; Am Vet Radiol Soc; Am Asn Vet Neurol; Am Col Vet Internal Med (vpres, 74-75, pres, 76-77). Res: Neural control of micturition; pathophysiology of neurogenic bladder; models of neural diseases; hydrocephalus. Mailing Add: Dept of Small Animal Med Col of Vet Med Univ of Ga Athens GA 30602

OLIVER, JOHN PARKER, b New Rochelle, NY, Nov 24, 39; m 63; c 3. ASTRONOMY. Educ: Rensselaer Polytech Inst, BS, 62; Univ Calif, Los Angeles, MA, 68, PhD(astron), 74. Prof Exp: Mem tech staff astron, Aerospace Corp, 65-67; ASST PROF ASTRON, UNIV FLA, 70- Mem: AAAS; Am Astron Soc. Res: Astronomical photoelectric photometry; lunar occultations of stars; eclipsing binary stars; astronomical instrumentation; spectroscopy. Mailing Add: Dept of Physics & Astron Univ of Fla Gainesville FL 32611

OLIVER, JOHN PRESTON, b Klamath Falls, Ore, Aug 7, 34; m 56; c 3. INORGANIC CHEMISTRY. Educ: Univ Ore, BA, 56, PhD(chem), 59. Prof Exp: From asst prof to assoc prof, 59-67, PROF CHEM, WAYNE STATE UNIV, 67-, CHMN DEPT, 71- Mem: Am Chem Soc; The Chem Soc. Res: Synthesis of organometallic compounds of Groups II, III and IV; alkyl-metal exchange reactions; nuclear magnetic resonance spectra of organometallic compounds. Mailing Add: 135 Chem Bldg Wayne State Univ Detroit MI 48202

OLIVER, KELLY HOYET, JR, b Roseboro, Ark, June 22, 23; m 47; c 3. AQUATIC BIOLOGY. Educ: Southern Methodist Univ, BS, 52, MS, 53; Okla State Univ, PhD(zool), 63. Prof Exp: Aquatic biologist, Anderson Fish Farms, 53; teacher pub schs, Ark, 53-61; from asst prof to assoc prof biol, Ark State Col, 62-64; chief biologist, Chem Biol Lab, Vitro Serv, Eglin AFB, 64-68; assoc prof, 68-73, PROF BIOL, HENDERSON STATE COL, 73-; PRES, KOCOMORO, INC, 73- Concurrent Pos: Res biologist, Crossett Co, 56-61; mem, Int Cong Limnol, 62; NSF vis scientist, Ark high schs, 63-64. Mem: AAAS; Ecol Soc Am; Am Soc Limnol & Oceanog. Res: Fishery management; pollution biology; effect of industrial waste, especially paper and petroleum, on aquatic organisms and bacterial ecology; ecology of Lake De Gray, Arkansas, especially change in fauna and flora and physiochemical conditions during river/lake transition; toxicity of contaminants on bluegills and annelids. Mailing Add: Dept of Biol Henderson State Col Arkadelphia AR 71923

OLIVER, LAWRENCE HAMPTON, b Winston-Salem, NC, Dec 9, 32; m 75; c 4. SCIENCE EDUCATION, COMPUTER SCIENCE. Educ: Morgan State Univ, BS, 54; Howard Univ, MS, 60. Prof Exp: Res psychologist geront, NIMH, 57-61; human engr aerospace, Martin Marietta Corp, 61-62; instr psychol, Morgan State Univ, 62-63; human factors analyst, Info Syst Opers, Gen Elec Co, 63-66; assoc prog dir, Sci Info Serv, 66-67, head syst develop, Data Mgt Syst, 67-68, prog dir comput activ, 68-73, PROG DIR SCI EDUC, NSF, WASHINGTON, DC, 73- Concurrent Pos: Consult, Washington Tech Inst, Comput Sci Curric, 75- Mem: AAAS; Asn Comput Mach. Mem: AAAS. Res: Use of technology in minority education; instructional computing; psychology of learning. Mailing Add: 1517 Gerard St Rockville MD 20850

OLIVER, LAWRENCE KERMIT, organic chemistry, psychology, see 12th edition

OLIVER, MONTAGUE, b Antigua, BWI, Feb 24, 19; US citizen; m 52; c 4. GENETICS, PHYSIOLOGY. Educ: Tuskegee Inst, BS, 51, MS, 52; Purdue Univ, MSA, 54, PhD(biol), 61. Prof Exp: Teacher, High Sch, Ala, 54-55; res dir genetics, Gunn Bros Quail Farm, Tex, 55-56; teacher, High Sch, Ill, 56-57 & Ind, 57-62; assoc prof biol, 62-72, PROF BIOL, CALUMET COL, 72- Concurrent Pos: Sub-consult, Argonne Nat Lab,Ill, 65-66. Mem: AAAS; Am Asn Biol Teachersl Nat Sci Teachers Asn; Am Inst Biol Sci. Res: Heritability and selection of egg production in the fowl; relative efficiency of methods of teaching biological science; general zoology and botany. Mailing Add: 1111 E 19th Ave Gary IN 46407

OLIVER, MORRIS ALBERT, b Milford-Haven, Wales, Feb 12, 18; nat US; m 48; c 3. ENGINEERING STATISTICS. Educ: Oxford Univ, BA, 48, MA, 52. Prof Exp: Head sci dept, Eaglebrook Sch, Mass, 48-50; prof math, Bennington Col, 50-55; mathematician, AMP, Inc, 55-65, sr mathematician, 65-73, MGR, PRECISION ARTWORK LAB, AMP, INC, 73- Concurrent Pos: Consult, AMP, Inc, 53-55; mem, Grad Fac, Pa State Univ, 58. Mem: Am Math Soc; NY Acad Sci; fel Royal Statist Soc; Am Statist Asn. Res: Statistical engineering. Mailing Add: Precision Artwork Lab AMP Inc Harrisburg PA 17105

OLIVER, RICHARD CHARLES, b Minneapolis, Minn, Mar 16, 30; m 53; c 4. PERIODONTOLOGY. Educ: Univ Minn, BS, 52, DDS, 53; Loma Linda Univ, MS, 62. Prof Exp: From assoc prof to prof periodont, Sch Dent, Loma Linda Univ, 62-75; DEAN, SCH DENT, UNIV SOUTHERN CALIF, 75- Concurrent Pos: Fulbright-Hays res scholar, Denmark, 67-68. Mem: Am Dent Asn; Am Acad Periodont; Int Asn Dent Res. Res: Vascularization following periodontal surgery; epidemiology of periodontal disease and therapy. Mailing Add: Sch of Dent Univ Southern Calif Los Angeles CA 90007

OLIVER, SYMMES CHADWICK, b Cincinnati, Ohio, Mar 30, 28; m 52; c 2. CULTURAL ANTHROPOLOGY. Educ: Univ Tex, BA, 51, MA, 52; Univ Calif, Los Angeles, PhD(anthrop), 61. Prof Exp: From instr to assoc prof, 55-68, actg chmn dept, 67-71, chmn dept, 68-72, PROF ANTHROP, UNIV TEX, AUSTIN, 68-, CHMN GRAD STUDIES COMT, 72- Concurrent Pos: Vis lectr, Univ Calif, Riverside, 60-61; res anthropologist, Cult & Ecol in EAfrica Proj, Univ Calif, Los Angeles, 61-62; mem planning & adv comt, Handbook NAm Indians (Plains), Smithsonian Inst, 72- Mem: Fel Am Anthrop Asn. Res: Cultural ecology; social organization; ethnology. Mailing Add: Dept of Anthrop Univ of Tex Austin TX 78712

OLIVER, THOMAS ALBERT, b Winnipeg, Man, Dec 16, 24. GEOLOGY. Educ: Univ Man, MSc, 49; Univ Calif, Los Angeles, PhD(geol), 52. Prof Exp: Geologist, Man Mines Br, Dept Mines & Nat Resources, Can, 50-52 & Calif Standard Co, 52-59; from asst prof to assoc prof, 59-67, PROF GEOL, UNIV CALGARY, 67- Mem: Geol Soc Am. Res: Stratigraphy; sedimentation. Mailing Add: Dept of Geol Univ of Calgary Calgary AB Can

OLIVER, THOMAS K, JR, b Hobart Mills, Calif, Dec 21, 25; m 49; c 2. PEDIATRICS. Educ: Univ Calif, Berkeley, 43-45; Harvard Med Sch, MD, 49. Prof Exp: Instr pediat, Med Col, Cornell Univ, 53-55; from asst prof to assoc prof, Ohio State Univ, 55-63; from assoc prof to prof pediat, Univ Wash, 63-70; PROF PEDIAT & CHMN DEPT, SCH MED, UNIV PITTSBURGH, 70-; MED DIR, CHILDREN'S HOSP PITTSBURGH, 70- Concurrent Pos: Spec fel neonatal physiol, Karolinska Inst, Sweden, 60-61. Mem: Soc Pediat Res; Am Pediat Soc; Am Acad Pediat. Res: Neonatal biology; pulmonary physiology in childhood; acid-base physiology. Mailing Add: Children's Hosp of Pittsburgh 125 DeSoto St Pittsburgh PA 15213

OLIVER, VICTOR L, b St Joseph, La, Aug 31, 25; m 45; c 2. ZOOLOGY, PARASITOLOGY. Educ: Ouachita Baptist Col, BA, 49; Peabody Col, MA, 50; Univ Ala, PhD(zool), 60. Prof Exp: Asst prof biol, La Col, 50-52; assoc prof, Ouachita Baptist Col, 52-60; assoc prof, Lamar State Col, 60-62; prof, Mobile Col, 62-66; PROF BIOL & CHMN DEPT, OUACHITA BAPTIST UNIV, 66- Concurrent Pos: Partic, Ala Acad Sci-NSF Vis Scientist Lectr Prog, 62-; NSF consult to Ala Pub Schs, Mobile, 63-64. Mem: AAAS. Res: Parasitology; trichinosis. Mailing Add: Dept of Biol Ouachita Baptist Univ Box 902 Arkadelphia AR 71923

OLIVER, VINCENT J, b May 19, 16; US citizen; m 42; c 7. METEOROLOGY. Educ: Univ Chicago, BS, 42. Prof Exp: Observer meteorol, US Weather Bur, 38-40; instr, Univ Chicago, 40-45; meteorologist in-chg, Forecast Ctr, Fairbanks Weather Bur, 45-48; supvy forecaster, Nat Weather Serv, 48-53, supvr, Nat Meteorol Ctr, 53-57, prof mod meteorol, 57-62, CHIEF APPLNS BR, NAT ENVIRON SATELLITE CTR, NAT WEATHER SERV, NAT OCEANIC & ATMOSPHERIC ADMIN, 62- Mem: Am Meteorol Soc. Res: Short time and long range forecasting; satellite research. Mailing Add: Appl Br Nat Env Satellite Ctr Oceanic & Atmos Admin Suitland MD 20233

OLIVER, WILLIAM ALBERT, JR, b Columbus, Ohio, June 26, 26; m 48; c 2. PALEONTOLOGY. Educ: Univ Ill, BS, 48; Cornell Univ, MA, 50, PhD, 52. Prof Exp: From instr to asst prof geol, Brown Univ, 52-57; RES GEOLOGIST-PALEONTOLOGIST, US GEOL SURV, 57- Concurrent Pos: Mem, Paleont Res Inst; ed, Paleont Soc, 64-69; res prof, George Washington Univ, 69-70; mem, US Nat Comt Geol, 75-79. Mem: Geol Soc Am; Paleont Soc (pres, 75); Palaeont Asn; Am Geol Inst (vpres, 76). Res: Devonian coral and stratigraphy, paleoecology, and biogeography. Mailing Add: US Geol Surv E305 Nat Mus Washington DC 20244

OLIVER, WILLIAM J, b Blackshear, Ga, Mar 30, 25; m 49; c 3. PEDIATRICS. Educ: Univ Mich, MDO, 48; Am Bd Pediat, Cert, 54; Sub-Bd Pediat Nephrol, cert, 74. Prof Exp: From instr to assoc prof pediat, 53-65, dir pediat lab, Med Ctr, 59-67, PROF PEDIAT, SCH MED, UNIV MICH, ANN ARBOR, 65-, CHMN DEPT PEDIAT & COMMUN DIS, 67- Concurrent Pos: Chief pediat serv, Wayne County Gen Hosp, 58-61. Mem: Am Pediat Soc; Am Soc Nephrology; Am Acad Pediat; Soc Pediat Res. Res: Renal physiology and pathology; fluid and electrolyte metabolism; adaptation of primitive peoples with their environment, disease patterns, mineral metabolism and growth patterns of Yanomama Indians of South America. Mailing Add: Dept of Pediat & Commun Dis Univ of Mich Ann Arbor MI 48104

OLIVER-GONZALES, JOSE, b Lares, PR, Aug 21, 12; US citizen; m 35; c 3. PARASITOLOGY. Educ: Univ PR, BA, 38; Univ Chicago, MS, 39, PhD(parasitol, bact), 41. Prof Exp: Instr parasitol, Sch Trop Med, Univ PR, 40; res assoc, Univ Chicago, 42; from asst prof to assoc prof, 43-54, PROF PARASITOL, SCH MED, UNIV PR, SAN JUAN, 54-, HEAD DEPT, 60- Concurrent Pos: Res assoc, Western Reserve Univ, 47-48; consult, Parasitol & Trop Med Study Sect, NIH, 53-56, mem, US-Japan Coop Med Sci Prog, 65-70, Nat Adv Allergy & Infectious Dis Coun, 66-70; consult, Vet Admin Hosp, San Juan, Presby Hosp, Auxilio Mutuo Hosp & Doctor's Hosp, San Juan, La State Univ Med Ctr & Inst Int Med, New Orleans; Guggenheim fel, 62-63. Honors & Awards: Purdue Frederick Prize, PR Med Asn, 57; Martinez Award, PR Comt Bilharzia Control, 59; Perez Award, Personnel Off, Govt PR, 60; Ashford Award, Am Soc Trop Med & Hyg, 47. Mem: AAAS; fel Am Soc Trop Med & Hyg; Am Soc Parasitol; Soc Exp Biol & Med. Res: Immunity to infections with helminth parasites; prevention and control of infections with Schistosoma mansoni; metabolism of Schistosoma mansoni and Trichinella spiralis; chemotherapy, immunology and immunochemistry related to parasitic infections. Mailing Add: Bucare 13 Santurce PR 00913

OLIVERO, JOHN JOSEPH, JR, b Yonkers, NY, Jan 18, 41; m 61; c 4. AERONOMY, METEOROLOGY. Educ: Fla State Univ, BS, 62; Col William & Mary, MS, 66; Univ Mich, Ann Arbor, PhD(aeronomy), 70. Prof Exp: Aerospace technologist appl res & develop, Langley Res Ctr, NASA, Va, 62-70; res assoc physics & astron, Univ Fla, 70-72; ASST PROF METEOROL & IONOSPHERIC RES LAB, PA STATE UNIV, 72- Concurrent Pos: Res assoc, High Altitude Eng Lab, Univ Mich, Ann Arbor, 67-70. Mem: AAAS; Am Meteorol Soc; Am Geophys Union. Res: Minor constituent photochemistry and transport; electron impact phenomena in atmospheric gases; upper atmospheric composition, energetics, circulation and thermal processes; atmospheric sensors; aerosol physics; climatic change, environmental impact. Mailing Add: 539 Deike Pa State Univ University Park PA 16802

OLIVER-PADILLA, FERNANDO LUIS, b Mayaguez, PR, Apr 9, 22; US citizen; m 45; c 3. DAIRY SCIENCE, ANIMAL BREEDING. Educ: Univ PR, BS, 42; Kans State Univ, MS, 58, PhD(animal breeding), 61. Prof Exp: Agr exten agent, Agr Exten Serv, Univ PR, 42-53, livestock specialist, 53-56, 58-59 & 61-63; assoc prof natural sci & head dept, 63-66, PROF GENETICS & PHYSIOL, INTER-AM UNIV PR, HATO REY, 66- Concurrent Pos: Sire analyst, PR Artificial Breeding Coop, 58-; consult, PR Agr Coun, Dept Agr, 61-63 & PR Dairy Farmers Asn, 63-; sen, Inter-Am Univ Fac Sen, 66-69 & 71-74. Mem: AAAS; Am Dairy Sci Asn; NY Acad Sci; PR Col Agriculturists; Latin Am Asn Animal Prod. Res: Genetic parameters in the dairy population of Puerto Rico; the genotype and phenotype frequencies of ABO and Rh groups and genetics of diabetes mellitus in the Puerto Rican population; psycho-physiological effects of the contraceptive pill on a sample of Puerto Rican women. Mailing Add: Dept of Physiol Inter-Am Univ of PR Hato Rey PR 00919

OLIVETO, EUGENE PAUL, b New York, NY, Mar 15, 24; m 48; c 2. ORGANIC CHEMISTRY. Educ: City Col New York, BS, 43; Purdue Univ, PhD(org chem), 48. Prof Exp: Res chemist, Schenley Res Inst, 43-44; asst chem, Purdue Univ, 44-48; res chemist, Schering Corp, 48-50, group leader, 50-58, mgr, Natural Prod Res Dept, 58-64; sr group chief, 64, sect chief, 64-66, dir animal health & fine chem res, 66-68, DIR APPL ORG CHEM, HOFFMANN-LA ROCHE, INC, 68- Mem: AAAS; fel Am Inst Chemists; Am Chem Soc; fel NY Acad Sci; The Chem Soc. Res: Nitroparaffins; synthesis, photolysis and pyrolysis of organic nitrites; organic medicinals; partial and total synthesis of steroidal sex and adrenal hormones; analogs; food additives. Mailing Add: Res Div Hoffmann-La Roche Inc Nutley NJ 07110

OLIVIER, ALFRED GARIPAY, veterinary medicine, see 12th edition

OLIVIER, KENNETH LEO, b Los Angeles, Calif, May 19, 32; m 54; c 6. ORGANIC CHEMISTRY. Educ: Loyola Univ, Calif, BS, 54; Univ Calif, Los Angeles, PhD(chem), 58. Prof Exp: Res chemist, Plastics Dept, E I du Pont de Nemours & Co, 57-60; res chemist, 60-65, sr res chemist, 65-69, SUPVR INDUST CHEM, UNION OIL CO CALIF, 69- Mem: Am Chem Soc. Res: Petrochemicals; emulsion polymerization; hot melt adhesives. Mailing Add: Res Dept Union Oil Co of Calif Brea CA 92621

OLIVIER, LOUIS JOHN, b Grand Rapids, Mich, Nov 3, 13; m 37; c 3. HELMINTHOLOGY, TROPICAL MEDICINE. Educ: Univ Mich, BS, 36; NY Univ, PhD(parasitol), 40. Prof Exp: Asst biol, NY Univ, 36-41; jr zoologist, Bur Animal Indust, USDA, 41-46; scientist, Lab Trop Dis, Nat Inst Allergy & Infectious Dis, 46-53, sr scientist, 53-56, sci dir & head sect helminthic dis, 56-66; mem staff, Pan-Am Health Orgn, 66-71, mem staff, WHO, 71-73. Concurrent Pos: Consult & mem expert panel parasitic dis, WHO, 56-66; ed, Trop Med & Hyg News, 56-60. Mem: Am Soc Parasitol; Am Soc Trop Med & Hyg. Res: Morphology and life histories of trematodes; schistosomiasis; helminths of poultry; snail control and ecology; resistance to worm infections; control of major parasitic diseases. Mailing Add: 9312 Elmhirst Dr Bethesda MD 20014

OLIVO, RICHARD FRANCIS, b New York, NY, Sept 26, 42; m 71. NEUROBIOLOGY. Educ: Columbia Univ, AB, 63; Harvard Univ, AM, 65, PhD(biol), 69. Prof Exp: Tutor biol, Harvard Univ, 66-68; vis asst prof, State Univ NY Stony Brook, 70-71; asst prof, Williams Col, 71-73; ASST PROF BIOL, SMITH COL, 73- Concurrent Pos: Consult, Harper & Row, Inc, 69-71. Mem: AAAS; Am Soc Zoologists; Soc Neurosci. Res: Sensory physiology; invertebrate nervous systems; scientific films. Mailing Add: Dept of Biol Sci Smith Col Northampton MA 01060

OLKIN, INGRAM, b Waterbury, Conn, July 23, 24; m 45; c 3. MATHEMATICAL STATISTICS. Educ: City Col New York, BS, 47; Columbia Univ, MA, 49; Univ NC, PhD(math statist), 51. Prof Exp: Asst prof statist, Mich State Univ, 51-55; vis assoc prof, Univ Chicago, 55-56; assoc prof, Mich State Univ, 56-60; prof, Univ Minn, 60-61; PROF STATIST & EDUC, STANFORD UNIV, 61-, CHMN DEPT STATIST,

OLKIN

74- Concurrent Pos: Assoc ed, J Am Statist Asn, 60-70; overseas fel, Churchill Col, Cambridge Univ, 67-68; fel psychomet, Educ Testing Serv, 71-72; ed, Ann Math Statist, 71- Mem: Fel Inst Math Statist; fel Am Statist Asn; Am Math Soc; Math Asn Am; Am Educ Res Asn. Res: Multivariate analysis; mathematical models in the behavioral sciences. Mailing Add: Dept of Statist Stanford Univ Stanford CA 94305

OLKOWSKI, ZBIGNIEW L, b Wilno, Poland, Nov 24, 38; m 63. IMMUNOLOGY, CYTOCHEMISTRY. Educ: Silesian Med Acad, MD, 63, Dr Med Sci, 69. Prof Exp: Sr lectr histol & embryol, Silesian Med Acad, 65-67, head histochem lab, 67-70; Kosciuszko Found fel, 70-71, res assoc neurocytochem, Yerkes Primate Res Ctr, 70-72, asst prof radiol, Med Sch, 72-75, ASSOC PROF RADIOL & ASST PROF IMMUNOL, MED SCH, EMORY UNIV, 75-, DIR LAB TUMOR BIOL & CLIN IMMUNOL, ROBERT WINSHIP MEM CLINIC NEOPLASTIC DIS, 75- Concurrent Pos: Res assoc surg, Teaching Hosp, Silesian Med Acad, 67-70. Mem: Am Radium Soc; AMA; Int Acad Cytol; Radiation Res Soc; NY Acad Sci. Res: Quantitative and enzyme cytochemistry; electron microscopy; radiobiology; tumor biology; clinical immunology; cyclic nucleotides and prostaglandins in immune lymphoid cells. Mailing Add: Emory Univ Clin 1365 Clifton Rd NE Atlanta GA 30322

OLLA, BORI LIBORIO, b Jersey City, NJ, Jan 22, 37; m 58; c 2. ANIMAL BEHAVIOR, FISH BIOLOGY. Educ: Fairleigh Dickinson Univ, BS, 59; Univ Hawaii, MS, 62. Prof Exp: Res asst shark behav, Hawaii Marine Lab, Honolulu, 61-62; instr biol, Chaminade Col Hawaii, 62; asst zool, Univ Md, 62-63; asst fish neurol, NIH, 63; asst neurol, Col Med, Seton Hall Univ, 63; CHIEF BEHAV INVEST, NAT MARINE FISHERIES SERV, NAT OCEANIC & ATMOSPHERIC ADMIN, 63- Honors & Awards: Bronze Medal, US Dept Com, 75. Mem: Animal Behav Soc; Am Fisheries Soc; AAAS. Res: Field and laboratory studies on marine fishes; social behavior, including schooling facilitation, territoriality, aggression; biorhythms, feeding, home ranges; chemosensory responses; effects of environmental stress, including temperature and petroleum on behavior norms. Mailing Add: Sandy Hook Lab Nat Marine Fisheries Serv NOAA Highlands NJ 07732

OLLERENSHAW, NEIL CAMPBELL, b Matlock, Eng, Sept 12, 33; Can citizen; m 60; c 3. GEOLOGY. Educ: Univ Wales, BSc, 57; Univ Toronto, MA, 59, PhD(geol), 63. Prof Exp: RES SCIENTIST, GEOL SURV CAN, 62- Mem: Am Asn Petrol Geologists; Can Soc Petrol Geologists; Soc Econ Paleont & Mineral. Res: Structural geology; stratigraphy and facies-tectonics relationships of the Rocky Mountain foothills and Eastern Cordillera. Mailing Add: Geol Surv of Can 3303 33rd St NW Calgary AB Can

OLLERHEAD, ROBIN WEMP, b Simcoe, Ont, Mar 12, 37; m 59; c 3. NUCLEAR PHYSICS. Educ: Univ Western Ont, BSc, 59; Yale Univ, MS, 60, PhD(physics), 64. Prof Exp: Asst res officer nuclear physics, Atomic Energy Can Ltd, 64-66, assoc res officer, 66-68; assoc prof physics, 68-71, PROF PHYSICS, UNIV GUELPH, 71- Concurrent Pos: NATO sci fel & Rutherford Mem fel, Oxford Univ, 63-64. Mem: Am Phys Soc; Can Asn Physicists. Res: Nuclear structure; spectroscopy using electrostatic accelerators. Mailing Add: Dept of Physics Univ of Guelph Guelph ON Can

OLLERICH, DWAYNE A, b Sioux Falls, SDak, June 30, 34. ANATOMY, NEUROANATOMY. Educ: Augustana Col, SDak, BA, 60; Univ NDak, MS, 62, PhD(anat), 64. Prof Exp: Asst prof neuroanat & histol, Univ Alta, 65-66; asst prof, 66-70, ASSOC PROF NEUROANAT, SCH MED, UNIV N DAK, 70-, CHMN DEPT ANAT, 72- Concurrent Pos: Fel anat, Univ Alta, 64-65, Med Res Coun Can fel, 65-66. Mem: Am Asn Anat; Am Soc Cell Biol. Res: Electron microscopy; cytology; drug toxicity. Mailing Add: Dept of Anat Univ of NDak Sch of Med Grand Forks ND 58201

OLLINGER, JANET, b Salt Lake City, Utah, Oct 17, 46. ORGANIC CHEMISTRY. Educ: Univ Utah, BA, 69; Univ Ill, PhD(org chem), 74. Prof Exp: RES RES CHEMIST INSECTICIDE CHEM, ROHM AND HAAS CO, 74- Res: Discovery of new, novel, environmentally safe insecticides. Mailing Add: Rohm and Haas Co Norristown & McKean Rds Spring House PA 19477

OLLMANN, LOYAL TAYLOR, b Rockville Centre, NY, June 18, 44; m 66; c 2. MATHEMATICS. Educ: Dartmouth Col, AB, 66; Cornell Univ, MA & PhD(math), 70. Prof Exp: Asst prof math, La State Univ, Baton Rouge, 70-74; ASST PROF MATH, RUSSELL SAGE COL, 74- Mem: Asn Symbolic Logic; Am Math Soc; Math Asn Am; AAAS. Res: Model theory; graph theory; infinitary combinatorics; combinatorial and applied mathematics. Mailing Add: 33 Barcelona Dr Clifton Park NY 12065

OLLOM, JOHN FREDERICK, b Ward, WVa, Dec 28, 22; m 54; c 3. PHYSICS. Educ: Harvard Univ, PhD, 52. Prof Exp: Asst prof physics, Univ WVa, 51-56; PROF PHYSICS, DREW UNIV, 56- Concurrent Pos: Mem tech staff, Bell Labs, Inc. Mem: Am Phys Soc; Am Asn Physics Teachers. Res: Magnetism; microwave spectroscopy. Mailing Add: Dept of Physics Drew Univ Madison NJ 07940

OLMAN, ROBERT ALEXANDER, b Springfield, Mass, Feb 28, 24; m 49; c 3. PHYSIOLOGY. Educ: Univ Mass, BS, 49; Ind Univ, MA, 50, PhD(zool), 54. Prof Exp: Asst zool, Ind Univ, 50-53, asst chem embryol, 53-54; instr physiol & pharmacol, Col Osteop Med & Surg, 54-58; cardiovasc res trainee, Dept Physiol, Med Col Ga, 58-59; from instr to asst prof physiol & biophys, Sch Med, Univ Louisville, 59-67; grants assoc, NIH, 67-68, assoc myocardial infarction br, Nat Heart & Lung Inst, 68-69, ENDOCRINOL PROG DIR, EXTRAMURAL PROGS, NAT INST ARTHRITIS, METAB & DIGESTIVE DIS, 69- Mem: AAAS. Res: Thyroid, thyrotropic hormone interaction and thyroid-stimulating hormone assay; time of appearance of cardiac actin in chick embryo; ventricular pressure curves. Mailing Add: Rm 633 Extram Prog Westwood Bldg Nat Inst Arthritis Metab Dig Dis Bethesda MD 20014

OLMER, JANE CHASNOFF, b St Louis, Mo; m 37. MATHEMATICS, COMPUTER SCIENCE. Educ: Wellesley Col, BA, 34; Washington Univ, MS, 37; Univ Chicago, cert, 61. Prof Exp: Teacher, Am Sch Paris, 37-39; ed & broadcaster, Paris Letter, 39-41; statistician, Drop Forging Asn, 42-43 & Fed Pub Housing Authority, 43-44; consult, May Co, 45; anal statistician, US Navy Electronics Supply Off, 54-55, supvr anal statistician, 55-56, digital comput syst specialist trainee, 56, mathematician, 56-59, tech head adv logistics res & develop, 59-61; SR MATHEMATICIAN, APPL PHYSICS LAB, JOHNS HOPKINS UNIV, 61- Concurrent Pos: Permanent deleg, Univac Users Asn, 57-61; cert, Data Processing Mgt Asn, 65, mem, Data Processing Testing Comt & dir educ, Washington, DC chap, 65-67; mgr, Text & Info Processing & Retrieval Proj, SHARE. Honors & Awards: US Navy Superior Accomplishment Award, 61. Mem: Asn Comput Mach. Res: Operations research in inventory control; mathematical programming on business computers; information handling and retrieval and high level computer languages in research and development environment using business and scientific computers. Mailing Add: 2510 Virginia Ave NW Washington DC 20037

OLMO, HAROLD PAUL, b San Francisco, Calif, July 31, 09; m 39; c 3. VITICULTURE. Educ: Univ Calif, BS, 31, PhD(genetics), 34. Prof Exp: Asst, Div Viticult & Fruit Prod, 31-33, assoc, Exp Sta, 33-34, jr viticulturist, 35-38, asst viticulturist, 39-46, assoc viticulturist, 46-52, from asst prof to assoc prof viticult, Univ, 39-52, VITICULTURIST, EXP STA & PROF VITICULT, UNIV CALIF, DAVIS, 52- Concurrent Pos: Consult, Off Inter-Am Affairs, US Bd Educ & Brazilian Govt, 43-45; Guggenheim fel, Middle East, 48; Fulbright res scholar, Univ Western Australia, 55. Honors & Awards: Wilder Medal, Am Pomol Soc, 59; Merit Medal, Int Vine & Wine Off, 65. Mem: AAAS; Genetics Soc Am; Am Soc Hort Sci. Res: Grape breeding; cytogenetics of fruit plants; register of new fruit and nut varieties. Mailing Add: Dept of Viticult & Enology Univ of Calif Davis CA 95616

OLMSTEAD, CLARENCE WALTER, b Ludington, Mich, Nov 4, 12; m 39; c 3. AGRICULTURAL GEOGRAPHY, GEOGRAPHY OF THE UNITED STATES & CANADA. Educ: Cent Mich Col, BA, 37; Univ Mich, MS, 38, PhD(geog), 51. Prof Exp: Instr geog & educ, Cent Mich Col, 38-40; teaching asst geog, Univ Calif, Berkeley, 40-42; geogr & intel officer, US Govt, 42-46; from asst prof to assoc prof, 46-66, PROF GEOG, UNIV WIS-MADISON, 66- Concurrent Pos: Vis prof, Univ Auckland, 68; mem comn agr typology, Int Geog Union, 68; Univ Wis study leave in Asia & Pac, 68-69. Mem: AAAS; Asn Am Geogr; Am Geog Soc; Agr Hist Soc; Nat Coun Geog Educ. Res: Analysis and comparative study of agricultural systems; agricultural typology; orchard and vineyard regions. Mailing Add: Dept of Geog Sci Hall Univ of Wis Madison WI 53706

OLMSTEAD, EDWIN GUY, b Enderlin, NDak, Aug 26, 21; m 46; c 3. MEDICINE. Educ: Univ NDak, BA, 43, BS, 44; Univ Ill, MD, 45; Marquette Univ, MSc, 53. Prof Exp: Clin instr med, Marquette Univ, 52-54, asst prof, 54-57; asst prof, 58-62, ASSOC PROF MED, UNIV NDAK, 62-, ADJ PROF HUMANITIES, 69- Concurrent Pos: Asst med dir, Milwaukee Gen Hosp, Wis, 52-57. Mem: Fel Am Col Physicians; Am Physiol Soc. Res: Water and electrolyte balance and water transport. Mailing Add: 1600 University Ave Grand Forks ND 58201

OLMSTEAD, EDWIN VINCENT, b South Worcester, NY, May 30, 10; m 40; c 3. PATHOLOGY. Educ: Univ Pa, AB, 32, MD, 36. Prof Exp: Dir lab serv, Hunterdon Med Ctr, 53-75; ASSOC CLIN PROF PATH, COL MED & DENT NJ, RUTGERS MED SCH, 72- Concurrent Pos: Consult, Raritan Valley Hosp, Greenbrook, NJ. Mem: Fel Am Soc Clin Path; AMA; fel Col Am Path; Am Soc Cytol. Mailing Add: PO Box 334 Flemington NJ 08822

OLMSTEAD, MARILYN MORGAN, inorganic chemistry, see 12th edition

OLMSTEAD, WILLIAM EDWARD, b San Antonio, Tex, June 2, 36; m 57; c 2. APPLIED MATHEMATICS. Educ: Rice Univ, BS, 59; Northwestern Univ, MS, 62, PhD(eng sci), 64. Prof Exp: Res engr, Southwest Res Inst, 59-60; from asst prof to assoc prof, 64-71, PROF ENG SCI, NORTHWESTERN UNIV, 71- Concurrent Pos: Vis mem, Courant Inst Math Sci, NY Univ, 67-68; chmn, Comt Appl Math, Northwestern Univ, 72-, mem, Exec Comt, Coun Theoret & Appl Mech, 74-; regional lectr, Soc Indust & Appl Math, 72-73. Mem: Soc Indust & Appl Math; Am Math Soc; Am Phys Soc. Res: Nonlinear boundary value problems; diffusion-dissipation processes; viscous flow theory. Mailing Add: Technol Inst Northwestern Univ Evanston IL 60201

OLMSTED, CLINTON ALBERT, b Chicago, Ill, Oct 27, 25; m 52; c 4. PHYSIOLOGY. Educ: Univ Calif, AB, 48, MA, 54, PhD(comp physiol), 56. Prof Exp: Asst, Univ Calif, 49-51 & 54-56; radiobiologist, US Naval Radiol Defense Lab, 51-54; res neurologist, Med Ctr, Univ Calif, Los Angeles, 57-59; asst prof zool, Univ Wis, 59-63; sr res physiologist, Battelle Mem Inst, 63-64; head, Cell Biol Div, Inst Lipid Res, Berkeley, Calif, 65; ASSOC PROF BIOL SCI, LA STATE UNIV, NEW ORLEANS, 65-; DIR, DELTA RES INST, LA, 65- Concurrent Pos: Mem & head, Lipid Biochem Sect, Div Nutrit Biochem, Commonwealth Sci & Indust Res Orgn, Adelaide, SAustralia, 70-72. Mem: AAAS; Am Soc Cell Biol; Am Physiol Soc; NY Acad Sci; Australian Biochem Soc. Res: Comparative physiology; cellular physiology with emphasis on general mechanisms in neurophysiology; lipid metabolism, lipid transport, and the role of phospholipids in drug transport and sodium transport affected by lipid soluble drugs. Mailing Add: Dept of Biol Sci La State Univ of New Orleans New Orleans LA 70122

OLMSTED, FRANKLIN HOWARD, b Los Angeles, Calif, Nov 23, 21; m 55; c 2. GEOLOGY. Educ: Pomona Col, BA, 42; Claremont Cols, MA, 48; Bryn Mawr Col, PhD, 61. Prof Exp: Lectr & instr geol, Claremont Cols, 48-49; geologist, 49-69, STAFF HYDROLOGIST, WATER RESOURCES DIV, US GEOL SURV, 69- Mem: AAAS; Geol Soc Am; Am Geophys Union. Res: Ground-water geology; petrology; development of methods of hydrogeologic exploration of geothermal areas. Mailing Add: Water Resources Div US Geol Surv Menlo Park CA 94025

OLMSTED, JOANNA BELLE, b Chicago, Ill, Mar 8, 47. Educ: Earlham Col, BA, 67; Yale Univ, PhD(biol), 71. Prof Exp: Fel biochem, Lab Molecular Biol, Univ Wis-Madison, 71-74; ASST PROF BIOL, UNIV ROCHESTER, 75- Concurrent Pos: NIH fel, 71-73. Mem: Am Soc Cell Biol. Res: Control of cell division and differentiation; regulation of synthesis, assembly and organization of microtubules and microfilaments for cellular functions. Mailing Add: Dept of Biol Univ of Rochester Rochester NY 14627

OLMSTED, JOHN, III, b Petaluma, Calif, Nov 16, 37; m 61; c 3. PHYSICAL CHEMISTRY. Educ: Carnegie Inst Technol, BS, 59; Univ Calif, Berkeley, PhD(chem), 63. Prof Exp: Fel, Lawrence Radiation Lab, Univ Calif, 63-64; asst prof chem, 64-69, asst dean arts & sci, 66-69, ASSOC PROF CHEM, AM UNIV BEIRUT, 69- Concurrent Pos: Vis assoc prof, Univ Calif, Los Angeles, 70-71. Res: Luminescence spectroscopy; physical photochemistry; reactions of gases with solid surfaces. Mailing Add: Dept of Chem Am Univ of Beirut Beirut Lebanon

OLMSTED, JOHN MEIGS HUBBELL, b Ithaca, NY, June 28, 11; m 36, 51, 55; c 5. MATHEMATICS. Educ: Univ Minn, BA, 34, MA, 36; Princeton Univ, AM, 38, PhD(math), 40. Prof Exp: Instr math, Princeton Univ, 37-40; from instr to prof, Univ Minn, 40-60, assoc chmn, 57-60; chmn dept math, 60-70, dean grad sch, 70-74, PROF MATH, SOUTHERN ILL UNIV, CARBONDALE, 60- Mem: Am Math Soc; Math Asn Am. Res: Real analysis. Mailing Add: Dept of Math Southern Ill Univ Carbondale IL 62901

OLMSTED, RICHARD DALE, b Bismarck, NDak, Nov 1, 47; m 69; c 1. THEORETICAL CHEMISTRY. Educ: Augsburg Col, BA, 69; Univ Wis-Madison, PhD(chem), 74. Prof Exp: FEL CHEM, UNIV BC, 74- Mem: Sigma Xi. Res: Nonequilibrium statistical mechanics; kinetic theory and transport properties of gaseous systems. Mailing Add: Dept of Chem Univ of BC Vancouver BC Can

OLMSTED, RICHARD W, b Darien, Conn, June 27, 20; m 43; c 4. PEDIATRICS. Educ: Dartmouth Col, BA, 41; Harvard Med Sch, MD, 44. Prof Exp: Clin instr pediat, Sch Med, Yale Univ, 49-53; asst prof, Sch Med, Temple Univ, 55-62; prof pediat & chmn dept, Med Sch, Univ Ore, 62-74; ASSOC EXEC DIR, AM ACAD PEDIAT, EVANSTON, ILL, 74- Mem: Am Acad Pediat; Am Pediat Soc. Res: Hearing disorders in children. Mailing Add: Am Acad Pediat 1801 Hinman Ave Evanston IL 60204

OLNESS, ALAN, b Faribault, Minn, Sept 22, 41; m 63; c 4. WATER POLLUTION, SOIL SCIENCE. Educ: Univ Minn, BS, 63, MS, 67, PhD(soil sci), 73. Prof Exp: Res asst soil sci, Univ Minn, 63-67; soil scientist, Soil Struct Group, 67-70, SOIL SCIENTIST, NAT AGR WATER QUAL MGR LAB, AGR RES SERV, USDA, 70- Mem: Am Soc Agron; Soil Sci Soc Am; Int Soc Soil Sci. Res: Influence of agricultural management practices on associated water quality with particular emphasis on phosphorus applications, transport, and transformations within the agricultural environment. Mailing Add: Nat Agr Water Qual Mgt Lab PO Box 322A Rte 2 Wilson St Durant OK 74701

OLNESS, DOLORES URQUIZA, b Kingsport, Tenn, Mar 20, 35; m 57; c 3. SOLID STATE PHYSICS. Educ: Duke Univ, AB, 57, PhD(physics), 61. Prof Exp: Res assoc molecular physics, Duke Univ, 61 & Univ NC, 61-63; SR PHYSICIST, LAWRENCE LIVERMORE LAB, UNIV CALIF, 63-67, 74- Mem: Am Phys Soc. Res: Organic low temperature molecular spectroscopy; organic photoconductors; lasers, damage to transparent solids; in-situ coal gasification. Mailing Add: 4345 Guilford Ave Livermore CA 94550

OLNESS, JOHN WILLIAM, b Broderick, Sask, Sept 4, 29; US citizen; m 58; c 5. NUCLEAR PHYSICS. Educ: St Olaf Col, BA, 51; Duke Univ, PhD(physics), 57. Prof Exp: Res assoc nuclear physics, Duke Univ, 57-58; res physicist, Aeronaut Res Labs, Wright-Patterson AFB, Ohio, 58-63; assoc physicist, 63-68, PHYSICIST, BROOKHAVEN NAT LAB, 68- Res: Nuclear spectroscopy. Mailing Add: Physics Dept Bldg 901-A Brookhaven Nat Lab Upton NY 11973

OLNESS, ROBERT JAMES, b Milaca, Minn, Jan 22, 33; m 57; c 3. NUCLEAR PHYSICS. Educ: Mass Inst Technol, BS, 56; Duke Univ, PhD(nuclear physics), 62. Prof Exp: Res assoc physics, Univ NC, 61-63; sr physicist, Lawrence Radiation Lab, Univ Calif, 63-68; assoc prof physics, Northern Mich Univ, 68-69; SR PHYSICIST, LAWRENCE LIVERMORE LAB, UNIV CALIF, 69- Mem: Am Phys Soc. Res: Thermodynamics and statistical mechanics; atomic and molecular physics. Mailing Add: 4345 Guilford Ave Livermore CA 94550

OLNEY, CHARLES EDWARD, b Assam, India, Nov 7, 24; US citizen; m 45; c 4. AGRICULTURAL CHEMISTRY. Educ: Tufts Col, BS, 45; Univ RI, MS, 53; Univ Conn, PhD(biochem), 67. Prof Exp: Teacher, R W Traip Acad, 46-48; agr chemist, Agr Exp Sta, 48-70, PROF FOOD & RESOURCE CHEM, UNIV RI, 70- Res: Pesticide residues; lipids. Mailing Add: Dept of Food & Resource Chem Univ of RI Kingston RI 02881

OLNEY, JOHN WILLIAM, b Marathon, Iowa, Oct 23, 31; m 57; c 3. PSYCHIATRY, NEUROPATHOLOGY. Educ: Univ Iowa, BA, 57, MD, 63; Am Bd Psychiat & Neurol, dipl, 70. Prof Exp: Intern, Kaiser Permanente Found, San Francisco, Calif, 63-64; resident, 64-68, from instr to asst prof, 68-72, ASSOC PROF PSYCHIAT, SCH MED, WASH UNIV, 72- Concurrent Pos: NIMH biol sci trainee, Wash Univ, 66-68, career invest award, 68-; asst psychiatrist, Barnes Hosp, 68-; consult psychiatrist, Malcolm Bliss Ment Health Ctr, 68- Mem: Psychiat Res Soc; Am Psychiat Asn; Am Asn Neuropath; Soc Neurosci; Asn Res Nervous & Ment Dis. Res: Adverse influences on the development of the immature central nervous system and neuropsychiatric disturbances resulting therefrom. Mailing Add: Dept of Psychiat Sch of Med Wash Univ St Louis MO 63110

O'LOANE, JAMES KENNETH, b Walla Walla, Wash, Dec 12, 13; m 43; c 4. CHEMICAL PHYSICS. Educ: St Benedict's Col, BSc, 35; Univ Wash, MSc, 44; Harvard Univ, MA, 47, PhD(chem physics), 50. Prof Exp: Instr, St Martin's Acad & Jr Col, 37-38; jr chemist, Shell Develop Co, 43-45; from asst prof to assoc prof chem, Univ NH, 48-54; sr anal chemist, Indust Lab, 54-58, proj physicist, Apparatus & Optical Div, Lincoln Plant, 58-59, sr res chemist, Res Labs, 59-62, RES ASSOC, EASTMAN KODAK CO RES LABS, 63- Mem: Am Chem Soc; Optical Soc Am; Am Phys Soc. Res: Nuclear magnetic resonance spectroscopy; polymer physical chemistry; infrared and Raman spectra. Mailing Add: 331 Seneca Pkwy Rochester NY 14613

OLOFFS, PETER CHRISTIAN, b Rügen, Ger, July 18, 29; Can citizen; m 58; c 2. PESTICIDE CHEMISTRY. Educ: Ernst-August Univ, Ger, dipl agr, 56; Univ BC, MSA, 64; Univ Wis-Madison, PhD(pesticide chem & biochem), 68. Prof Exp: Chemist, Later Chem Ltd, BC, 58-63; sessional lectr zool, Univ BC, 63-64; asst prof, 68-74, ASSOC PROF BIOL SCI, SIMON FRASER UNIV, 74- Mem: AAAS; Entom Soc Am; Am Chem Soc. Res: Persistence of pesticides in soils, plants, natural waters; pesticide-related aspects of public and occupational health; interrelationships between hepatic dysfunction and exposure to pesticide residues. Mailing Add: Dept of Biol Sci Simon Fraser Univ Burnaby BC Can

OLOFSON, ROY ARNE, b Chicago, Ill, Feb 26, 36. ORGANIC CHEMISTRY. Educ: Univ Chicago, BS & MS, 57; Harvard Univ, PhD(org chem), 61. Prof Exp: Instr org chem, Harvard Univ, 61-64, asst prof, 64-65; assoc prof, 65-71, PROF ORG CHEM, PA STATE UNIV, 71- Concurrent Pos: Consult, FMC Corp, 62-73, Armour Pharmaceut Co, 65-70 & McNeil Labs, 74-; mem, Adv Bd, J Org Chem, 73-77. Mem: Am Chem Soc; The Chem Soc; Sigma Xi. Res: Synthetic organic, heterocyclic and peptide chemistry; reaction mechanisms. Mailing Add: Dept of Chem Pa State Univ University Park PA 16802

O'LOUGHLIN, BERNARD JAMES, b Beaudette, Minn, Oct 30, 14; m 42; c 8. RADIOLOGY. Educ: Col St Thomas, BS, 37; Creighton Univ, MS, 41, MD, 42; Southwestern Univ, cert, 48; Univ Minn, PhD(radiol), 50; Am Bd Radiol, dipl, 48. Prof Exp: Resident radiol, Southwestern Univ & McKinney Vet Admin Hosp, Dallas, 46-48; from instr to assoc prof, Univ Minn, 48-52; from asst prof to prof, Sch Med, Univ Calif, Los Angeles, 52-64, prof radiol sci & chmn dept, 64-69; chmn fac med, 71-72, chmn dept, 69-74, PROF RADIOL SCI, UNIV CALIF, IRVINE-CALIF COL MED, 69- Concurrent Pos: Consult radiologist, St Joseph's Holy Family Hosps, 48-49 & 50-52; radiologist, Vet Admin Hosp, Minneapolis, 48-49, chief dept, 49-52; chief diag div, Ctr Health Sci, Univ Calif, Los Angeles, 52-62, chief pediat diag div, 62-64; vis prof, Oxford Univ, 60-62; chief physician radiol Unit II, Los Angeles County Hosp, 64-68; vis prof, Univ London, 72-73; counr, Univ Calif, Irvine. Mem: Radiol Soc NAm; Roentgen Ray Soc; fel Am Col Chest Physicians; fel Am Col Radiol; fel Am Acad Pediat. Res: Inferior vena cava; cancer of the lung; disturbances in cardiovascular function. Mailing Add: Dept of Radiol Sci UCI-CCM Orange Hosp Orange CA 92668

OLSEN, ALLEN LESLIE, physical chemistry, see 12th edition

OLSEN, ARTHUR MARTIN, b Chicago, Ill, Aug 27, 09; m 36; c 4. INTERNAL MEDICINE. Educ: Dartmouth Col, AB, 30; Rush Med Col, MD, 35; Univ Minn, MS, 38. Prof Exp: First asst med, 38-39, from asst prof to assoc prof, 40-57, PROF MED, MAYO GRAD SCH MED, UNIV MINN, 57- Concurrent Pos: Consult, Mayo Clin, 38-, head sect med, 49-68, chmn div thoracic dis, 68-71; trustee, Mayo Found, 61-68; mem nat heart & lung adv coun, NIH, 70-71. Mem: Am Soc Gastrointestinal Endoscopy (secy-treas, Am Gastroscopic Soc, 58-61, pres, 62-63); Am Col Chest Physicians (pres, 70); Am Broncho-Esophagol Asn (pres, 69-70); Am Thoracic Soc; Am Asn Thoracic Surg. Res: Esophageal motility; broncho-esophagology. Mailing Add: Mayo Clin 200 First St NW Rochester MN 55901

OLSEN, CARL JOHN, b Oakland, Calif, May 18, 28; m 64; c 2. ORGANIC CHEMISTRY. Educ: Univ San Francisco, BS, 50, MS, 52; Univ Southern Calif, PhD(chem), 62. Prof Exp: Asst chem, Walter Reed Army Med Ctr, 53-54; instr, Los Angeles Valley Col, 60-61; from asst prof to assoc prof, 61-71, PROF CHEM, CALIF STATE UNIV, NORTHRIDGE, 71- Concurrent Pos: Res Corp Cottrell grant, 63-64; NSF inst grants, 65-66. Mem: Am Chem Soc. Res: Organic reaction mechanisms; thermal and photochemical studies on reactions and reactivities of organic free radicals; free radicals in the field of environmental health; synthesis and properties of metalocenes. Mailing Add: Dept of Chem Calif State Univ Northridge CA 91324

OLSEN, CARL MARK, b Manhattan, Kans, Oct 22, 31; m 56; c 4. PLANT PATHOLOGY, ENTOMOLOGY. Educ: Mont State Univ, BS, 55, MS, 57; Univ Calif, Berkeley, PhD(plant path), 64. Prof Exp: Res asst plant path, Mont State Univ, 55-57; res technologist, Univ Calif, Los Angeles, 57-60, sr res technologist, Univ Calif, Berkeley, 60-65; dir biol res, Atherton Div, Litton Indust, Inc, 65-67; mem staff indust microwave heating, Eimac Div, Varian Assocs, 67-71; mgr appln develop, Prod Develop Dept, 71-73, MGR PHEROMONE SUPPLY SERV, ZOECON CORP, 71-, DIR MKT, 73- Mem: Am Phytopath Soc; Brit Mycol Soc. Res: Thermal treatment of soils; effects of microwave energy on microorganisms and chemical systems; effects and mechanisms of insect pheromones and applied developments; applied developments of insect juvenile hormones and growth regulation compounds; insect pest management and population dynamics. Mailing Add: Zoecon Corp 975 California Ave Palo Alto CA 94304

OLSEN, CLARENCE WILMOTT, b Indianapolis, Ind, Dec 1, 03; m 29; c 2. NEUROLOGY. Educ: Ohio State Univ, BA, 23; Univ Mich, MA, 25, MD, 27. Prof Exp: Instr med, 30-34, from asst prof to assoc prof nerv dis, 34-52, clin prof, 52-64, PROF NEUROL, SCH MED, LOMA LINDA UNIV, 64- Concurrent Pos: Consult, US Vet Admin Hosps, Los Angeles, 46-48, Birmingham, 49- & Long Beach, 59-64. Mem: Fel AMA; Am Psychiat Asn; fel Am Col Physicians. Res: Vascular diseases of the brain; respiratory rhythm in nervous diseases. Mailing Add: Dept of Neurol Loma Linda Univ Sch of Med Loma Linda CA 92354

OLSEN, CLAYTON EDWARD, b Volksrust, Union of SAfrica, Sept 19, 20; US citizen; m 51; c 3. SOLID STATE PHYSICS. Educ: Trnity Col, BS, 42; Ohio State Univ, PhD(phys chem), 55. Prof Exp: Prof chemist, Am Cyanamid Co, Conn, 46-49; asst chem, Ohio State Univ, 50-53, res assoc, 53-55; STAFF MEM PHYSICS OF MAT, LOS ALAMOS SCI LAB, UNIV CALIF, 55- Concurrent Pos: Instr, Col Santa Fe, 71- Mem: Am Chem Soc; Am Phys Soc. Res: Structure and properties of matter; magnetic properties of solids; geochemistry; mineralogy. Mailing Add: 3227 Woodland Rd Los Alamos NM 87544

OLSEN, DONALD RAY, b Providence, Utah, Apr 11, 25; m 51; c 5. MINERALOGY. Educ: Utah State Univ, BS, 48; Univ Utah, MS, 51, PhD(mineral), 60. Prof Exp: Explor geologist, Anaconda Copper Mining Co, 52-53; res mineralogist, Res Ctr, Kennecott Copper Corp, 53-55; from asst prof to assoc prof geol, 55-68, PROF GEOL, UTAH STATE UNIV, 68- Concurrent Pos: Fulbright lectr, Univ Brazil, 67-68. Mem: Geol Soc Am; Am Inst Prof Geol. Res: Ore deposition; metallic mineral deposits; volcanic rocks. Mailing Add: Dept of Geol Utah State Univ Logan UT 84321

OLSEN, DONNA MAE, b Wooster, Ohio, Sept 14, 40; m 70. HEALTH SERVICES RESEARCH. Educ: Hiram Col, BA, 62; Utah State Univ, MS, 65, PhD(pop ecol), 73. Prof Exp: Applns programmer & statist consult, Comput Ctr, 70-72, INSTR COMMUNITY MED, DEPT FAMILY & COMMUNITY MED, UNIV UTAH, 72- Mem: Asn Teachers Prev Med. Res: Efficacy of new methods of medical care delivery; evaluation of quality of care; financial impact of mid-level practitioners; therapeutic uses of cannabidiol and derivatives. Mailing Add: Dept of Family & Community Med Univ Utah 50 N Medical Dr Salt Lake City UT 84132

OLSEN, DOUGLAS ALFRED, b Minneapolis, Minn, Oct 10, 30; m 58; c 3. PHYSICAL CHEMISTRY. Educ: Gustavus Adolphus Col, BA, 53; Univ Iowa, MS, 55, PhD(phys chem), 60. Prof Exp: Teaching asst, Univ Iowa, 53-55; develop chemist, Bemis, Inc, 55-57; sr res scientist, Honeywell, Inc, 59-63; proj leader phys chem, Archer Daniels Midland Co, 63-67; sect mgr chem, Appl Sci Div, Litton Systs, Inc, 67-70; vpres, Bio-Medicus Inc, 70-75; PRES, PMD INC, 75-; RES ASSOC, UNIV MINN, 73- Concurrent Pos: Vis prof, Tech Univ Denmark, 74. Honors & Awards: IR-100 Award, Indust Res Inst, 72. Mem: AAAS; Am Chem Soc; Am Soc Artificial Internal Organs. Res: Surface chemistry; electrochemistry; chemical kinetics; radiochemistry; artificial organs; biomedical materials; operations analysis. Mailing Add: 4106 Linden Hills Blvd Minneapolis MN 55410

OLSEN, EDMUND SEVERN, JR, b Minneapolis, Minn, May 4, 14; m 35; c 3. DENTISTRY, ANATOMY. Educ: Univ Minn, Minneapolis, DDS, 42, MSD, 50; Am Bd Prosthodont, dipl, 53. Prof Exp: Chief dent serv, 247th Gen Hosp, US Army, 43-45, chief & dir intern prog prosthodont, Oliver Gen Hosp, Ga, 46-48, chief prosthodont, Army-Navy Gen Hosp, 50-51, US Army Hosp, Yokohama, Japan, 51-54, US Army Dent Detachment, 54-56 & Walter Reed Army Hosp, 56-58, cmndg officer & dir dent mat, US Army Regional Dent Activity, 58-63; assoc prof prosthodont, 63-65, PROF HOSP DENT & CHMN DEPT, UNIV MINN HOSPS, MINNEAPOLIS, 65- Concurrent Pos: Consult, Far East Command, 51-54 & Off Surgeon Gen, 60-63; vis lectr & demonstr, Tokyo Dent Col, Japan, 52-54; fac mem, Walter Reed Army Med Ctr, 56-61; vis lectr, Nat Naval Med Ctr, 60-63; fac chmn prosthodont, US Army Inst Dent res, 61-63. Mem: Am Dent Asn; Am Prosthodont Soc; fel Am Col Dent; Int Asn Dent Res. Res: Restorative dentistry; cephalometric roentgenology, physiology and vertical dimensions of the face; dental materials, physical properties and dimensional accuracy of dental casting alloys; anatomy, especially physiology and relationship of prosthesis design to tissue reaction.

OLSEN, EDWARD JOHN, b Chicago, Ill, Nov 23, 27; m 54; c 2. GEOLOGY. Educ: Univ Chicago, BA, 51, MS, 55, PhD, 59. Prof Exp: Geologist, Geol Surv Can, 53, US Geol Surv, 54 & Can Johns-Manville Co, Ltd, 56-59; asst prof mineral & petrol, Western Reserve Univ & Case Inst, 59-60; CUR MINERAL, DEPT GEOL, FIELD MUS NATURAL HIST, 60- Concurrent Pos: Guest researcher, Argonne Nat Lab, 67-; adj prof, Univ Ill, Chicago Circle, 70- Mem: Fel Mineral Soc Am; Geochem Soc; Mineral Asn Can; fel Meteoritical Soc. Res: Thermodynamics of mineral systems; phase equilibria in meteorites; optical and x-ray spectroscopy; electron microprobe. Mailing Add: Dept of Geol Field Mus of Natural Hist Chicago IL 60605

OLSEN, EDWARD TAIT, b Brooklyn, NY, June 12, 42; m 68. RADIO ASTRONOMY, PLANETARY ATMOSPHERES. Educ: Mass Inst Technol, BS, 64; Calif Inst Technol, MS, 67; Univ Mich, PhD(astron), 72. Prof Exp: Resident res assoc radio astron, Nat Res Coun, 72-74, SR SCIENTIST RADIO ASTRON & PLANETARY ATMOSPHERES, JET PROPULSION LAB, 74- Mem: Am Astron Soc; Sigma Xi. Res: Microwave studies of the Jovian synchrotron emission and of the atmospheres of the major planets. Mailing Add: 183B/365 Jet Propulsion Lab 4800 Oak Grove Dr Pasadena CA 91103

OLSEN, EUGENE DONALD, b La Crosse, Wis, Dec 10, 33; m 60; c 3. ANALYTICAL CHEMISTRY, CLINICAL CHEMISTRY. Educ: Univ Wis, BS, 55, PhD(anal chem), 60. Prof Exp: Instr chem, Univ Wis, 60; asst prof, Franklin & Marshall Col, 60-64; asst prof, 64-66, ASSOC PROF CHEM, UNIV S FLA, 66- Concurrent Pos: Supvr undergrad res participation, NSF, 60-70, res grants, 64-66; NIH spec fel, Med Sch, 71-72. Mem: Am Chem Soc; The Chem Soc; Asn Clin Scientists. Res: Teaching films; ion exchange separation of radioelements; instrumental methods of analysis; chelometric titrations; electrochemical analysis; automation in analysis; reactions in nonaqueous and mixed solvents; clinical and medicinal chemistry. Mailing Add: Dept of Chem Univ of SFla Tampa FL 33620

OLSEN, FARREL JOHN, b Salt Lake City, Utah, Mar 2, 29; m 55; c 3. AGRONOMY. Educ: Utah State Univ, BS, 54, MS, 58; Rutgers Univ, New Brunswick, PhD(crop prod), 61. Prof Exp: Exten agronomist, WVa Univ, 61-66; plant scientist, Sci Info Exchange, Smithsonian Inst, 66-67, pasture agronomist, WVa Univ, 67-71; PASTURE AGRONOMIST, SOUTHERN ILL UNIV, CARBONDALE, 71-, ASSOC PROF PLANT & SOIL SCI, SCH AGR, 74- Mem: Am Soc Agron. Res: Planted pastures, their establishment, management and utilization; nutritive value. Mailing Add: Dept of Plant & Soil Sci Southern Ill Univ Carbondale IL 62901

OLSEN, FREDRIC PHILLIP, b Trenton, NJ, Oct 23, 37; m 60; c 2. CHEMISTRY. Educ: Bucknell Univ, BS, 59; Brown Univ, PhD(chem), 64. Prof Exp: Res chemist, Univ Calif, Riverside, 63-65; asst prof, 65-70, ASSOC PROF CHEM, McMASTER UNIV, 70- Mem: Am Chem Soc. Res: Metallo-organic, boron hydride and sulfur nitrite chemistry; inorganic sulfur chemistry. Mailing Add: Dept of Chem McMaster Univ Hamilton ON Can

OLSEN, GEORGE DUANE, b DeKalb, Ill, Jan 5, 40; m 65; c 2. CLINICAL PHARMACOLOGY. Educ: Dartmouth Col, AB, 62; Dartmouth Med Sch, BMS, 64; Harvard Med Sch, MD, 66. Prof Exp: Intern med, Univ Hosps Cleveland, 66-67; med dir, Indian Health Ctr, USPHS, 67-69; ASST PROF PHARMACOL, MED SCH, UNIV ORE HEALTH SCI CTR, 70-, INSTR MED, 72- Concurrent Pos: NIH training grant pharmacol, Med Sch, Univ Ore, 69-70. Mem: AAAS; Am Fedn Clin Res; Am Soc Pharmacol & Exp Therapeut; Am Soc Clin Pharmacol & Therapeut. Res: Plasma protein binding of drugs; respiratory and endocrine effects of drugs; drug metabolism and interactions; adverse drug effects. Mailing Add: Dept of Pharmacol Med Sch Univ of Ore Health Sci Ctr Portland OR 92701

OLSEN, GJERDING, b Brooklyn, NY, Aug 19, 22. ENDOCRINOLOGY. Educ: Harvard Univ, AB, 47, AM, 49, PhD(biol), 51. Prof Exp: From instr to asst prof, 51-71, ASSOC PROF BIOL, BRANDEIS UNIV, 71- Concurrent Pos: Consult, Mass Heart Asn, 56-59 & AAAS, 58-60. Mem: Fel AAAS; Am Soc Zoologists; NY Acad Sci. Res: Endocrinology, especially vertebrate and female sex endocrinology; ascites cells in rat uterus. Mailing Add: Dept of Biol Brandeis Univ Waltham MA 02154

OLSEN, GLENN W, b North Lima, Ohio, Sept 7, 31; m 57; c 2. MATHEMATICS EDUCATION. Educ: Edinboro State Col, BSEd, 53; Pa State Univ, MEd, 58; Cornell Univ, PhD(math, math educ), 68. Prof Exp: Teacher, Randolph East Mead High Sch, 53-59; assoc prof math, Indiana Univ Pa, 60-65; head dept, 68-74, PROF MATH, EDINBORO STATE COL, 68- Mem: Math Asn Am. Res: Number theory; set theory; analysis; problems in the teaching of mathematics. Mailing Add: Dept of Math Edinboro State Col Edinboro PA 16412

OLSEN, GREGORY HAMMOND, b Brooklyn, NY, Apr 20, 45; m 68; c 2. SOLID STATE PHYSICS. Educ: Fairleigh Dickinson Univ, BS, 66, BS & MS, 68; Univ Va, PhD(mat sci), 71. Prof Exp: Teaching asst physics, Fairleigh Dickinson Univ, 66-68; vis scientist electron micros, Univ Port Elizabeth, Repub S Africa, 71-72; MEM TECH STAFF CRYSTAL GROWTH, RCA LABS, 72- Mem: AAAS; Inst Elec & Electronic Engrs; Electrochem Soc; Am Phys Soc; Sigma Xi. Res: Study of crystal growth and structural defects in semiconductors for electro-optical devices; synthesized the first vapor phase epitaxy gallium arsenic continuous wave laser. Mailing Add: RCA Labs Princeton NJ 08540

OLSEN, HARRY NORMAN, physics, electrical engineering, see 12th edition

OLSEN, HERMAN LOWELL, physics, see 12th edition

OLSEN, JAMES LEROY, b Minneapolis, Minn, June 8, 30; m 55; c 4. INDUSTRIAL PHARMACY. Educ: Univ Minn, Minneapolis, BS, 54, PhD(pharm), 64. Prof Exp: Pharmacist supvr, Sch Pharm, Univ Minn, 58-63; asst vpres, Clark-Cleveland Div, Richardson-Merrell Corp, 63-69; asst prof pharmaceut, 69-74, ASSOC PROF PHARM, SCH PHARM, UNIV NC, CHAPEL HILL, 74- Concurrent Pos: Food & Drug Admin fel, Sch Pharm, Univ NC, Chapel Hill, 71- Mem: Am Pharmaceut Asn; Acad Pharmaceut Sci. Res: Pharmaceutical dosage forms; tablets; aerosols. Mailing Add: Sch of Pharm Univ of NC Chapel Hill NC 27514

OLSEN, JOHN SYLVESTER, b Salt Lake City, Utah, Sept 22, 15; m 39; c 3. TEXTILE CHEMISTRY. Educ: Univ Utah, BA, 37, MA, 39; Cornell Univ, PhD(anal chem), 42. Prof Exp: Asst prof chem, Univ Southern Calif, 46-48; chemist, 48-58, sr res chemist, 58-67, RES ASSOC, E I DU PONT DE NEMOURS & CO, 67- Res: Analytical chemistry; empirical mercurimetric method for zinc; detection of fluorine in organic compounds with ferrous nitrate; surface chemistry of textile fibers. Mailing Add: E I du Pont de Nemours & Co Kinston NC 28501

OLSEN, KENNETH HAROLD, b Ogden, Utah, Feb 20, 30; m 55; c 4. ASTROPHYSICS, GEOPHYSICS. Educ: Idaho State Univ, BS, 52; Calif Inst Technol, MS, 54, PhD(physics), 57. Prof Exp: Staff mem, 57-71, alt group leader, 71-74, GROUP LEADER, LOS ALAMOS SCI LAB, 74- Concurrent Pos: Asst, Calif Inst Technol, 52-57. Mem: Am Astron Soc; Am Phys Soc; Am Geophys Union; Inst Elec & Electronics Engrs; Geol Soc Am. Res: Measurement of transition probabilities for atomic spectra; computer calculations of astrophysical problems; underground explosion phenomenology; water waves, tsunamis; seismology; infrared Fourier transform spectroscopy; solar eclipse observations. Mailing Add: 226 Venado Los Alamos NM 87544

OLSEN, KENNETH LAURENCE, b Enumclaw, Wash, July 25, 21; m 46, 73; c 2. PLANT PHYSIOLOGY. Educ: Wash State Univ, BS, 45, MS, 50; Univ Calif, PhD(plant physiol), 50. Prof Exp: Asst bot, Wash State Univ, 44-46; instr gen bot & plant physiol, Ore State Col, 46-47; asst plant physiol & weed control, Univ Calif, 47-50, plant physiologist, Inter-Am Inst Agr Sci, 50-54; plant physiologist, Rice Invests, Field Crops Res Br, Agr Res Serv, 54-57, plant physiologist, Agr Mkt Serv, 57-64, PLANT PHYSIOLOGIST, AGR RES SERV, USDA, 64- Mem: AAAS; Am Soc Hort Sci; Am Soc Plant Physiol; Am Chem Soc. Res: Biochemical changes during maturation and ripening of fruits; effect of modified atmospheres during storage. Mailing Add: Agr Res Serv USDA PO Box 99 Annex 111 Wenatchee WA 98801

OLSEN, KENNETH WAYNE, b Chicago, Ill, Dec 19, 44; m 66; c 3. PROTEIN CHEMISTRY, BIOCHEMISTRY. Educ: Iowa State Univ, BS, 67; Duke Univ, PhD(biochem), 72. Prof Exp: Res assoc protein crystallog, Purdue Univ, 72-75; ASST PROF BIOCHEM, MED CTR, UNIV MISS, 75- Concurrent Pos: Fel, Am Cancer Soc, 72-74. Mem: Am Crystallog Asn. Res: Protein chemistry and crystallog; enzymology; heme transport proteins; membrane glycoproteins and enzymes in tumors; affinity chromatography. Mailing Add: Dept of Biochem Med Ctr Univ of Miss Jackson MS 39216

OLSEN, LARRY CARROL, b St Joseph, Mo, July 25, 37; m 60; c 3. SOLID STATE SCIENCE. Educ: Univ Kans, BS, 60, PhD(physics), 65. Prof Exp: Asst prof physics, Univ Kans, 65; sr scientist, McDonnell Douglas Corp, 65-74; ASSOC PROF ENG, JOINT CTR GRAD STUDY, 74- Concurrent Pos: Lectr, Joint Ctr Grad Study, 67-74. Mem: Am Phys Soc; Am Nuclear Soc; Sigma Xi. Res: Paramagnetic resonance and spin-lattice relaxation of magnetic ions in ionic solids; transport properties of semiconductors; radiation effects in semiconductors and devices; betavoltaic, thermoelectric, photovoltaic and other forms of energy conversion. Mailing Add: 1837 Norwood Ct Richland WA 99352

OLSEN, LEONARD OLIVER, b Cedar Falls, Iowa, Nov 1, 10; m 38; c 3. SOLID STATE PHYSICS, MOLECULAR SPECTROSCOPY. Educ: Iowa State Teachers Col, AB, 32; Univ Iowa, MS, 34, PhD(physics), 37. Prof Exp: Asst physics, Univ Iowa, 34-37; from instr to prof, Case Inst Technol, 37-60; PROF PHYSICS, US NAVAL POSTGRAD SCH, 60- Concurrent Pos: Fulbright prof, Univ Oslo, 50; mem, Bd Govrs, Am Inst Physics, 61-64, chmn, Comt Regional Counsr Prog, 62-, Comt Student Sect, 63- Mem: Fel Am Phys Soc; Am Asn Physics Teachers (pres, 60-61). Res: Atomic and molecular collisions; spectroscopy; physics of surfaces, especially of ferromagnetic materials. Mailing Add: Box 86 Pebble Beach CA 93953

OLSEN, ORVIL ALVA, b Biggar, Sask, July 22, 17; m 45; c 3. BOTANY. Educ: Univ Sask, BSA, 41; Univ Man, MSc, 48; McGill Univ, PhD(plant path), 61. Prof Exp: Res asst potato breeding, Man Dept Agr, 46-49; asst prof plant path, McGill Univ, 51-57; res officer, Can Dept Agr, 57-65, res scientist, 65-68; ASSOC PROF BOT, MEM UNIV NFLD, 68- Mem: Agr Inst Can; Can Phytopath Soc; Can Bot Asn. Res: Ecology of vegetation of spray zones created by waterfalls; floristics of Newfoundland. Mailing Add: Dept of Biol Mem Univ of Nfld St John's NF Can

OLSEN, PETER FREDRIC, b Red Bank, NJ, Apr 7, 35; m 70. ECOLOGY. Educ: Univ Mich, BS, 56, MS, 57; Auburn Univ, PhD(zool), 65. Prof Exp: Res fel muskrat pop dynamics, Wildlife Mgt Inst, Delta Waterfowl Res Sta, 55-56; game biologist, Mich Dept Conserv, 57-58; res assoc dis ecol, Auburn Univ, 60-62; sr res ecologist, head ecol sect & asst res prof wildlife dis ecol, Univ Utah, 63-70; sr res ecologist & vpres, EcoDynamics, Inc, 70-73; SR ECOLOGIST, DAMES & MOORE CONSULT ENGRS, 74- Mem: AAAS; Wildlife Soc; Am Soc Mammal; Ecol Soc Am; Am Inst Biol Sci. Res: Analysis and evaluation of ecological and other impacts upon the environment which may result from proposed development activities. Mailing Add: Dames & Moore 250 E Broadway Salt Lake City UT 84111

OLSEN, RALPH A, b Moroni, Utah, Jan 30, 25; m 49; c 7. SOIL CHEMISTRY. Educ: Brigham Young Univ, BS, 49; Cornell Univ, MS, 51, PhD(agron), 53. Prof Exp: Soil scientist, USDA, 53-56; assoc prof, 56-64, PROF CHEM, MONT STATE UNIV, 64- Concurrent Pos: Fel, Mineral Nutrit Pioneering Res Lab, Md, 61-62; Inst Biol Chem fel, Univ Copenhagen, 65-66. Mem: Soil Sci Soc Am; Am Chem Soc. Res: Potentiometric measurements in colloidal suspensions; mechanisms involved in the movement of ions from soils to plant roots; inorganic nutrition of green plants; ion movement through living membranes. Mailing Add: Dept of Chem Mont State Univ Bozeman MT 59715

OLSEN, REX E, b Merill, Ore, Jan 22, 35; m; c 3. STRATIGRAPHY, PALEONTOLOGY. Educ: Univ Calif, Berkeley, AB, 57. Prof Exp: From jr geologist to sr geologist, Humble Oil & Refining Co, 57-67, sr petrol geologist, 67-72; PROF GEOLOGIST, EXXON CO, 72- Mem: Am Asn Petrol Geologists; Geol Soc Am; Am Inst Prof Geologists. Res: Petroleum-subsurface geology; Mesozoic and Cenozoic stratigraphy of west coast of North America; benthonic and planktonic Foraminifera ecology and time stratigraphy. Mailing Add: PO Box 2180 Exxon Co USA Houston TX 77001

OLSEN, RICHARD GEORGE, b Independence, Mo, June 25, 37; m 57; c 4. VIROLOGY. Educ: Univ Mo-Kansas City, BA, 59; Atlanta Univ, MS, 64; State Univ NY Buffalo, PhD(virol), 69. Prof Exp: Instr microbiol, Metrop Jr Col, 63-67; ASST PROF VIROL, COL VET MED, OHIO STATE UNIV, 69- Mem: Am Soc Microbiol; Tissue Cult Asn; Int Asn Comp Res Leukemia. Res: Immunology of the cat and man to oncogenic viruses, including tumor poxviruses and RNA oncornaviruses. Mailing Add: Dept of Vet Path Col of Vet Med Ohio State Univ Columbus OH 43210

OLSEN, RICHARD KENNETH, b Provo, Utah, Sept 3, 35; m 54; c 4. ORGANIC CHEMISTRY. Educ: Brigham Young Univ, AB, 60; Univ Ill, PhD(org chem), 64. Prof Exp: Fel org chem, Stanford Res Inst, 64-65; res assoc, Univ Utah, 65-67; asst prof, 67-70, ASSOC PROF CHEM, UTAH STATE UNIV, 70- Mem: Am Chem Soc. Res: Synthetic organic chemistry; natural products; peptide antibiotics. Mailing Add: Dept of Chem Utah State Univ Logan UT 84321

OLSEN, RICHARD WILLIAM, b Marshfield, Wis, Aug 13, 44; m 67; c 2. BIOCHEMISTRY. Educ: Dartmouth Col, AB, 66; Univ Calif, Berkeley, PhD(biochem), 71. Prof Exp: USPHS fel, Inst Pasteur, Paris, 71-72; ASST PROF BIOCHEM, UNIV CALIF, RIVERSIDE, 72- Mem: Soc Neurosci; Am Soc Biol Chemists. Res: Biochemistry of the nervous system; excitable membrane constituents, receptor proteins; mode of action of neurotransmitters, drugs, and insecticides; microtubules and their role in cellular response to regulatory stimuli. Mailing Add: Dept of Biochem Univ of Calif Riverside CA 92502

OLSEN, ROBERT JAMES, b Imperial, Nebr, Aug 13, 41; m 66; c 1. SOIL SCIENCE. Educ: Univ Nebr, Lincoln, BS, 64, MS, 66; Univ Wis-Madison, PhD(soil sci), 70. Prof Exp: Res asst, Univ Nebr, Lincoln, 64-66 & Univ Wis-Madison, 66-69; asst prof agr, 69-73, ASST PROF PLANT SCI, CALIF STATE UNIV, CHICO, 73- Mem: Am Soc Agron; Soil Sci Soc Am. Res: Soil chemistry, fertility and microbiology; environmental quality as affected by agricultural practices. Mailing Add: Div of Agr Calif State Univ Chico CA 95926

OLSEN, ROBERT THORVALD, b Brookfield, Ill, Mar 10, 15; m 41; c 2.

SYNTHETIC ORGANIC CHEMISTRY. Educ: Newark Col Eng, BS, 36; Columbia Univ, MS, 37; Mass Inst Technol, PhD(org chem), 42. Prof Exp: Chem engr, Eastman Kodak Co, NY, 37-39; sr res chemist, Gen Aniline & Film Corp, 42-48, Plymouth Cordage Co, 48-54 & Celotex Corp, 54-56; res mgr, Standard Register Co, 56-61; dir res & develop, Gamma Chem Corp, Great Meadows, 61-68; TECH DIR FINE CHEM, ASHLAND CHEM CO, 68- Mem: AAAS; Am Chem Soc; fel Am Inst Chemists. Res: Chemical engineering; research management. Mailing Add: 346 Fairview Ave Long Valley NJ 07853

OLSEN, RODNEY L, b Duluth, Minn, July 10, 36; m 58; c 2. ANALYTICAL CHEMISTRY. Educ: Univ Minn, Duluth, BA, 58; Iowa State Univ, MS, 60, PhD(anal chem), 62. Prof Exp: From asst prof to assoc prof, 62-73, PROF CHEM, HAMLINE UNIV, 73- Mem: Am Chem Soc. Res: Fluorometric methods; radiochemistry; analytical separations. Mailing Add: Dept of Chem Hamline Univ St Paul MN 55101

OLSEN, RONALD G, b Duluth, Minn, Oct 22, 37; m 62; c 2. PHOTOGRAPHY, ORGANIC CHEMISTRY. Educ: Univ Minn, Duluth, BA, 59; Ind Univ, PhD(org chem), 65. Prof Exp: RES ASSOC, EASTMAN KODAK CO, 65- Mem: Am Chem Soc. Res: Synthetic organic chemistry; photographic systems. Mailing Add: Eastman Kodak Co Res Labs 1669 Lake Ave Rochester NY 14650

OLSEN, RONALD H, b New Ulm, Minn, June 26, 32; m 58; c 6. MICROBIOLOGY. Educ: Univ Minn, BA, 57, MS, 59, PhD(microbiol), 62. Prof Exp: Asst prof microbiol, Colo State Univ, 62-63; vpres res & develop, Dairy Technics Inc, 63-65; assoc prof, 65-74, PROF MICROBIOL, UNIV MICH, ANN ARBOR, 74- Mem: AAAS; Am Soc Microbiol. Res: Bacterial physiology; physiological basis for the minimum temperature of growth. Mailing Add: 5605 Med Sci II Univ of Mich Dept of Microbiol Ann Arbor MI 48104

OLSEN, RONALD WERNER, ecology, ethology, see 12th edition

OLSEN, SIGURD, b Copenhagen, Denmark, Dec 25, 08; m 62. LIMNOLOGY. Prof Exp: Asst to div head, Dept Social Welfare, Municipality of Copenhagen, 25-63; biologist, Lab Radiation Biol, Univ Wash, 63-68, res asst prof fisheries, 68-72, AFFIL ASST PROF, COL FISHERIES, UNIV WASH, 72- Concurrent Pos: Assoc, Lab Fresh Water Biol, Univ Copenhagen, 52-53; Carlsberg Found & Danish State Sci Found grants. Honors & Awards: Gold Medal, Univ Copenhagen, 39. Mem: Int Asn Theoret & Appl Limnol; Danish Bot Soc. Res: Aquatic plants; chemistry of fresh water and sediments; exchange of phosphate between mud and water; phosphate chemistry. Mailing Add: 6542 55th Ave NE Seattle WA 98115

OLSEN, STANLEY JOHN, b Akron, Ohio, June 24, 19; m 42; c 1. BIOLOGICAL ANTHROPOLOGY, ARCHAEOLOGY. Prof Exp: Lab technician vert paleont, Harvard Univ, 45-56; vert paleontologist, Fla Geol Surv, 56-58; from assoc prof to prof zooarchaeol, Fla State Univ, 68-73; PROF ANTHROP, UNIV ARIZ, 73-; ZOOARCHAEOLOGIST, ARIZ STATE MUS, 73- Concurrent Pos: Res assoc, Mus Northern Ariz, 66-; NSF grant, Fla Geol Surv, 64-66, Harvard Univ Guide Found grant, 66; NSF grant, Fla State Univ, 67-69 & 69-70, Am Philos Soc grant, 70. Mem: Soc Vert Paleont (pres, 65-66); Am Soc Mammal; Soc Am Archaeol; Am Soc Ichthyol & Herpet; Soc Syst Zool. Res: Analysis of vertebrates from archaeological sites in the Western Hemisphere. Mailing Add: Dept of Anthrop Univ of Ariz Tucson AZ 85721

OLSEN, WARD ALAN, b Holmen, Wis, Sept 13, 34; m 61; c 3. INTERNAL MEDICINE, GASTROENTEROLOGY. Educ: Univ Wis-Madison, BS, 56, MD, 59. Prof Exp: From intern to resident, Harvard Med Serv, Boston City Hosp, 59-62 & 64-65; instr med, Harvard Med Sch, 67-68; asst prof, 68-72, ASSOC PROF MED, UNIV WIS-MADISON, 72- Concurrent Pos: Fel gastroenterol, Boston Univ, 65-67; chief gastroenterol, Vet Admin Hosp, Madison, Wis, 68- Mem: Am Fedn Clin Res; Cent Soc Clin Res; Am Gastroenterol Asn. Res: Intestinal absorption. Mailing Add: Vet Admin Hosp 2500 Overlook Terr Madison WI 53705

OLSEN, WILLIAM CHARLES, b Edmonton, Alta, Mar 25, 33; m 57; c 2. NUCLEAR PHYSICS. Educ: Univ BC, BASc, 56, MASc, 59; Univ Alta, PhD(nuclear physics), 64. Prof Exp: Res officer, Atomic Energy Can Ltd, Ont, 58-60; res assoc, 64, from asst prof to assoc prof, 64-74, PROF NUCLEAR PHYSICS, UNIV ALTA, 74- Res: Low and intermediate energy nuclear physics leading to information on the nuclear structure of low and medium mass nuclei. Mailing Add: Nuclear Res Ctr Univ of Alta Edmonton AB Can

OLSHANSKY, ROBERT, b Boston, Mass, Nov 18, 42; m 70. THEORETICAL PHYSICS. Educ: Swarthmore Col, BA, 64; Univ Pa, PhD(theoret physics), 69. Prof Exp: Res assoc particle physics, Brown Univ, 69-71 & Lab Hautes Energies, Fac Sci, Paris, 71-72; SR SCIENTIST FIBER OPTICS, CORNING GLASS WORKS, 73- Mem: Optical Soc Am. Res: Theory and design of optical fibers for long distance communication systems. Mailing Add: Sullivan Park Corning Glass Works Corning NY 14850

OLSHEN, RICHARD ALLEN, b Portland, Ore, May 17, 42; m 63; c 1. STATISTICS, MATHEMATICS. Educ: Univ Calif, Berkeley, AB, 63; Yale Univ, MS, 65, PhD(statist), 66. Prof Exp: Lectr statist & res staff statistician, Yale Univ, 66-67; asst prof statist, Stanford Univ, 67-72; assoc prof, Univ Mich, Ann Arbor, 72-75; ASSOC PROF MATH, UNIV CALIF, SAN DIEGO, 75- Concurrent Pos: Vis asst prof, Columbia Univ, 70-71; vis assoc prof, Stanford Univ, 73-75. Mem: Inst Math Statist; Am Statist Asn. Mailing Add: Dept of Math Univ of Calif at San Diego La Jolla CA 92093

OLSON, ALAN PETER, b Hillrose, Colo, Dec 28, 26; m 56; c 2. ARCHAEOLOGY, ANTHROPOLOGY. Educ: Univ Colo, BA, 51; Univ Ariz, PhD(archaeol), 59. Prof Exp: Cur anthrop, Mus Northern Ariz, 59-64; asst prof, 64-70, ASSOC PROF ANTHROP, UNIV DENVER, 70- Concurrent Pos: Ed, Bull & Tech Ser, Mus Northern Ariz, 59-64. Mem: AAAS; fel Am Anthrop Asn; Soc Am Archaeol. Res: American Southwest; Southwestern Mexico; high plains. Mailing Add: Dept of Anthrop Univ of Denver Denver CO 80210

OLSON, ALBERT LLOYD, b Mountain View, Calif, Dec 14, 24; m 48; c 3. PATHOLOGY. Educ: Col Med Evangelists, MD, 49; Am Bd Path, dipl, 58, sert clin path, 60. Prof Exp: From instr to asst prof, 58-64, ASSOC PROF PATH, LOMA LINDA UNIV, 64- Res: Medicine. Mailing Add: Clin Lab White Mem Med Ctr 1720 Brooklyn Ave Los Angeles CA 90033

OLSON, ALFRED C, b Chicago, Ill, July 18, 26; m 66; c 1. BIOCHEMISTRY. Educ: Northwestern Univ, BS, 49; Univ Wis, PhD(phys & org chem), 54. Prof Exp: Chemist, Calif Res Corp Div, Standard Oil Co Calif, 54-60; RES CHEMIST, PLANT ENZYME LAB, WESTERN MKT & NUTRIT DIV, AGR RES SERV, USDA, 60- Concurrent Pos: Res fel, Div Biol, Calif Inst Technol, 60-63. Mem: Am Chem Soc; Am Soc Plant Physiol; The Chem Soc; Tissue Cult Asn. Res: Biochemistry of plant cell walls; proteins and enzymes associated with plant cell walls; plant tissue culture; immobilized enzymes. Mailing Add: Western Mkt & Nutrit Div Agr Res Serv USDA 800 Buchanan Albany CA 94710

OLSON, ANDREW CLARENCE, JR, b San Diego, Calif, Nov 10, 17; m 45; c 2. PARASITOLOGY. Educ: San Diego State Univ, BA, 39; Univ Idaho, MS, 42; Ore State Univ, PhD, 55. Prof Exp: From instr to assoc prof, 46-58, chmn dept, 57-60, PROF ZOOL, SAN DIEGO STATE UNIV, 58- Mem: AAAS; Am Soc Parasitol; Am Soc Mammal; Wildlife Dis Asn. Res: Ecology of fish parasites. Mailing Add: Dept of Zool San Diego State Univ San Diego CA 92182

OLSON, ANITA CORA, b Mulberry, Kans, Dec 6, 29. BIOCHEMISTRY, IMMUNOCHEMISTRY. Educ: Central State Univ, Okla, BS, 50; Univ Okla, MS, 63, PhD(biochem), 67. Prof Exp: Technician histochem, Univ Kans Med Ctr, 52-56; technician, Sinai Hosp, Baltimore, 56-57; res trainee biochem, Okla Med Res Found, 61-67; asst prof, 69-71, ASSOC PROF BIOCHEM & MOLECULAR BIOL, SCH MED, LA STATE UNIV, SHREVEPORT, 71- Concurrent Pos: Fel, Okla Med Res Found, 67; fel, Max-Planck Inst Immunobiol, Ger, 67-69. Mem: AAAS; Am Chem Soc; NY Acad Sci. Res: Immunochemical characterization of the membranes of lymphocytes and of the antigens associated with the membranes of bacteria. Mailing Add: Dept of Biochem & Molecular Biol La State Univ Sch of Med Shreveport LA 71130

OLSON, ARTHUR OLAF, b Lethbridge, Alta, May 11, 42; m 63; c 2. BIOCHEMISTRY, HORTICULTURE. Educ: Univ Alta, BSc, 64, PhD(plant biochem), 67. Prof Exp: Exten horticulturist, Alta Dept Agr, 64; res fel biochem, McMaster Univ, 67-68; res biochemist, Atomic Energy Can, Ltd, 68-69; dir, Alta Hort Res Ctr, 70-74, DIR PLANT INDUST DIV, ALTA DEPT AGR, 74- Concurrent Pos: Ed, Can Soc Hort Sci, 71- Mem: Can Biochem Soc; Agr Inst Can; Am Soc Hort Sci; Can Soc Hort Sci. Res: Aging effects of ethylene in plants and animals; mutation damage and its repair in algae and yeast; post harvest physiology and storage biochemistry. Mailing Add: Plant Indust Div Alta Dept of Agr Edmonton AB Can

OLSON, ARTHUR RUSSELL, b Lawrence, Mass, Jan 22, 19; m 42; c 4. TEXTILE CHEMISTRY. Educ: Mass Inst Technol, BS, 39. Prof Exp: Develop chemist, Hunt-Rankin Leather Co, Mass, 39-42; plastics chemist, United Shoe Mach Corp, 46-50; dir & chemist, McMillan Lab, Inc, 50-52; res chemist, Finishing Div, 52-60, Fiber Prod Div, 60-64, RES MGR, FIBER PROD DIV, KENDALL CO, 64- Mem: AAAS; Am Chem Soc. Res: Plastics technology; synthetic resins and plastics; textiles and textiles materials. Mailing Add: 6 Eastover Rd Walpole MA 02081

OLSON, BIRGER HENRY, b Lindsborg, Kans, Apr 5, 17; m 44; c 2. MICROBIOLOGY. Educ: Bethany Col, BS, 42; Univ Wis, PhD(biochem), 48. Prof Exp: Microbiologist, Merck & Co, 45-46; biochemist, Rohm & Haas Co, 48-51; pub health scientist, 51-55, DIV CHIEF, MICH DEPT PUB HEALTH, 55- Mem: AAAS; Am Soc Microbiol; Am Chem Soc. Res: Bacterial fermentation; nutrition of fungi; production of microbiological enzymes; antibiotics; antitumor substances. Mailing Add: Bur of Labs Mich Dept of Pub Health Lansing MI 48914

OLSON, BOYD E, b Fairview, Utah, May 17, 19; m 42; c 4. OCEANOGRAPHY, METEOROLOGY. Educ: Brigham Young Univ, BS, 41; Univ Calif, Los Angeles, cert meteorol, 43; Harvard Univ, MBA, 51; Ore State Univ, PhD(oceanog), 67. Prof Exp: Meteorologist, Am Overseas Airlines, Sweden, 46-47; US Weather Bur, Honolulu, 47-48; ed & oceanogr, US Navy Hydrographic Off, Md, 48-57, dep dir div oceanog, 57-59, actg dir, 59-60, dir marine sci dept, 60-67, dep sci & technol dir, 67-69, SCI & TECHNOL DIR, US NAVAL OCEANOG OFF, WASHINGTON, DC, 69- Concurrent Pos: Mem, Bd US Civil Serv Exam for Sci & Technol Personnel, 57-60; adv bd, Nat Oceanog Data Ctr, 61-64, chmn, 61; mem res panel, Interagency Comt Oceanog, 61-65; liaison rep to Nat Res Coun, 69- Mem: Am Geophys Union; Am Meteorol Soc; Arctic Inst NAm; Sigma Xi. Res: Hydrodynamics of abyssal ocean regions; energy exchange at the air-sea interface; underwater acoustics. Mailing Add: 10024 Garrett St Vienna VA 22180

OLSON, CARL, b Sac City, Iowa, Sept 15, 10; m 34; c 4. COMPARATIVE PATHOLOGY. Educ: Iowa State Univ, DVM, 31; Univ Minn, MS, 34, PhD(comp path), 35. Prof Exp: Asst prof path, State Univ NY Vet Col, Cornell Univ, 35-37; res prof, Univ Mass, 37-45; prof animal path & hyg & chmn dept, Univ Nebr, 45-56; chmn dept, 62-63, VET SCI, UNIV WIS-MADISON, 56- Mem: Am Soc Exp Path; Am Vet Med Asn; Am Asn Path & Bact; Am Asn Avian Path; Conf Res Workers Animal Dis (pres, 58). Res: Pathogenesis of animal diseases, especially neoplastic and those of mixed etiology. Mailing Add: Dept of Vet Sci Univ of Wis-Madison Madison WI 53706

OLSON, CARL MARCUS, b Chicago, Ill, Sept 18, 11; m 37; c 3. CHEMISTRY. Educ: Augustana Col, AB, 32; Univ Chicago, PhD(chem), 36. Prof Exp: Res chemist, Pigments Dept, E I du Pont de Nemours & Co, 36-43, Manhattan Dist Proj, Metall Lab, Univ Chicago, 43-44, Clinton Eng Works, Tenn, 44 & Hanford Eng Works, Wash, 44-45; res div head, E I du Pont de Nemours & Co, 45-50, res sect mgr, Exp Sta, 50-55, lab dir, Pigments Dept, 55-68, planning mgr, 68-71; VPRES, MICRON, INC, 71- Mem: Am Chem Soc. Res: Titanium pigments; titanium metal and hyperpure silicon; reduction metallurgy; x-ray spectroscopy methods for chemical analysis. Mailing Add: Micron Inc PO Box 3536 Wilmington DE 19807

OLSON, CARTER LEROY, b Iola, Wis, Jan 13, 35; m 60; c 1. ANALYTICAL CHEMISTRY. Educ: Wis State Col, Stevens Point, BS, 56; Univ Kans, PhD(chem), 62. Prof Exp: Res assoc & fel anal chem, Univ Wis, 61-63; from asst prof to assoc prof pharmaceut anal, 63-73, ASSOC PROF MED CHEM, COL PHARM, OHIO STATE UNIV, 73- Mem: Am Chem Soc. Res: Continuous methods of analysis; electroanalytical chemistry; analysis based on the rates of chemical reactions including the rates of enzyme catalyzed reactions. Mailing Add: Col of Pharm Ohio State Univ Columbus OH 43210

OLSON, CLIFFORD GERALD, b Osakis, Minn, July 6, 42; m 65. SOLID STATE PHYSICS. Educ: Hamline Univ, BS, 64; Iowa State Univ, PhD(physics), 70. Prof Exp: Fel, 70-71, PHYSICIST, AMES LAB, US ENERGY RES & DEVELOP ADMIN, IOWA STATE UNIV, 71- Mem: Am Phys Soc; Optical Soc Am. Res: Optical properties of solids in the vacuum ultraviolet and soft x-ray regions. Mailing Add: Ames Lab ERDA c/o UW-PSL Box 6 Stoughton WI 53589

OLSON, DALE WILSON, b Mountain Lake, Minn, July 27, 41. MAGNETISM. Educ: Carleton Col, BA, 62; Univ Rochester, PhD(physics), 70. Prof Exp: Asst prof, 68-73, ASSOC PROF PHYSICS, UNIV NORTHERN IOWA, 73- Mem: Am Phys Soc; Am Asn Physics Teachers. Res: Electron paramagnetic resonance; phase transitions. Mailing Add: Dept of Physics Univ of Northern Iowa Cedar Falls IA 50613

OLSON, DANFORD HAROLD, b Minneapolis, Minn, Jan 17, 35; m 55; c 4. ORGANIC CHEMISTRY. Educ: Univ Minn, BS, 56; Kans State Univ, PhD(chem),

62. Prof Exp: Res scientist petrochem, Denver Res Ctr, Marathon Oil Co, 62-66; res scientist, Shell Oil Co Res Ctr, Wood River, 66-70, supvr, Eng Lubricants, 70-75, SUPVR ENG LUBRICANTS, SHELL DEVELOP RES CTR, HOUSTON, 75- Mem: Am Chem Soc; Soc Automotive Engrs; Sigma Xi. Res: Synthesis of small ring heterocyclics; use of nitrogen oxides in organic synthesis; liquid and gas phase oxidation reactions; dehydrogenation; computer applications; product development. Mailing Add: 12622 Campsite Trail Cypress TX 77429

OLSON, DAVID FREDRICK, JR, forest ecology, silviculture, see 12th edition

OLSON, DAVID HAROLD, b Stoughton, Wis, Apr 27, 37; m 59; c 2. PHYSICAL CHEMISTRY. Educ: Univ Wis, BS, 59; Univ Iowa, PhD(phys chem), 63. Prof Exp: Sr res chemist, 63-74, RES ASSOC, RES DEPT, MOBIL RES & DEVELOP CORP, 74- Mem: Am Chem Soc; Am Crystallog Asn.

OLSON, DON A, b Racine, Wis, Apr 5, 30; m 57; c 1. REHABILITATION MEDICINE, SPEECH PATHOLOGY. Educ: Univ Wis-Madison, BS, 52; Northwestern Univ, MS, 55, PhD(speech path, audiol), 65. Prof Exp: Dir speech path, Hearing & Speech Clin, Med Col Ala, 59-65; assoc prof communicative disorders, 65-69, DIR TRAINING REHAB MED, REHAB INST CHICAGO, NORTHWESTERN UNIV, 69-, ASSOC PROF NEUROL & REHAB MED, MED SCH, 69- Concurrent Pos: Consult, Vis Nurse Asns Evanston & Skokie Valley, Chicago Metrop Nursing Asn, Home Health Care NSuburban Chicago & adv comt, Med Sch Allied Health; pres, Nat Paraplegia Found, 74- Mem: Am Cong Rehab Med; Am Speech & Hearing Asn; Acad Aphasia; Coun Except Children. Res: Expressive aphasia and psychology of the brain damaged. Mailing Add: Rehab Inst of Chicago 345 E Superior St Chicago IL 60611

OLSON, DONALD C, inorganic chemistry, analytical chemistry, see 12th edition

OLSON, DONALD LEE, b Brooklyn, NY, June 2, 32. DENTISTRY, ORAL PATHOLOGY. Educ: Columbia Univ, DDS, 57. Prof Exp: Intern, Presby Hosp, New York, 57-58; asst prof dent, Fairleigh Dickinson Univ, 62-66; ASSOC PROF PATH, SCH DENT, UNIV MD, BALTIMORE CITY, 66- Mem: Int Asn Dent Res. Res: Dental radiology; dental education. Mailing Add: Sch of Dent Univ of Md Baltimore MD 21201

OLSON, EDWARD CHARLES, analytical chemistry, see 12th edition

OLSON, EDWARD COOPER, b US, June 7, 30; m 59; c 2. ASTROPHYSICS. Educ: Worcester Polytech Inst, BS, 52; Ind Univ, PhD(astron), 61. Prof Exp: Instr physics, Worcester Polytech Inst, 52; asst prof astron, Smith Col, 60-63 & Rensselaer Polytech Inst, 64-65; asst prof, 66-67, ASSOC PROF ASTRON, UNIV ILL, URBANA, 67- Mem: Am Astron Soc; Royal Astron Soc; Int Astron Union. Res: Spectrophotometry and mass exchange in eclipsing binary systems. Mailing Add: Univ of Ill Observ Urbana IL 61801

OLSON, EDWIN ANDREW, b Gary, Ind, May 21, 25; m 54; c 3. GEOLOGY, GEOCHEMISTRY. Educ: Univ Pittsburgh, BS, 47, MS, 49; Columbia Univ, PhD(geochem), 63. Prof Exp: Lectr math, Univ Pittsburgh, 47-49; develop engr, E I du Pont de Nemours & Co, 49-53; asst prof phys sci, Northwestern Col, Minn, 53-56; asst geochem, Lamont Geol Observ, Columbia Univ, 56-60; from asst prof to assoc prof geol, 60-69, PROF GEOL, WHITWORTH COL, WASH, 69- Concurrent Pos: NSF res grant, 61- Mem: Geochem Soc; Am Sci Affil. Res: Theoretical geochemical aspects relating to the accuracy of radiocarbon dating. Mailing Add: Dept of Earth Sci Whitworth Col Spokane WA 99218

OLSON, EDWIN S, b Red Wing, Minn, Oct 23, 37; m 63; c 2. ORGANIC CHEMISTRY. Educ: St Olaf Col, BA, 59; Calif Inst Technol, PhD(org chem), 64. Prof Exp: Asst prof org chem, Idaho State Univ, 64-68; asst prof, 68-71, ASSOC PROF ORG CHEM, SDAK STATE UNIV, 71- Mem: Am Chem Soc; The Chem Soc. Res: Natural products, fatty acid and amino acid chemistry. Mailing Add: Dept of Chem SDak State Univ Brookings SD 57006

OLSON, ERIK JOSEPH, b New York, NY, Aug 17, 32; m 61; c 3. PHARMACOLOGY. Educ: Cornell Univ, BS, 54; Purdue Univ, PhD(biochem), 61. Prof Exp: Instr physiol, Sch Med, Vanderbilt Univ, 61-63; res assoc med, Sch Med, Western Reserve Univ, 63-64; ASST PROF PHARMACOL, MEHARRY MED COL, 64- Concurrent Pos: Fel muscular dystrophy, Vanderbilt Univ, 62-63; NIH career develop award, Meharry Med Col, 64-69. Mem: Am Chem Soc; NY Acad Sci. Res: Active calcium transport; carbamyl phosphate synthesis. Mailing Add: Dept of Pharmacol Meharry Med Col Nashville TN 37208

OLSON, EVERETT CLAIRE, b Waupaca, Wis, Nov 6, 10; m 39; c 3. VERTEBRATE PALEONTOLOGY. Educ: Univ Chicago, BS, 32, MS, 33, PhD(geol, vert paleont), 35. Prof Exp: From instr to prof vert paleont, Univ Chicago, 35-69; chmn dept, 70-72, PROF ZOOL, UNIV CALIF, LOS ANGELES, 69- Concurrent Pos: Secy dept vert paleont, Univ Chicago, 45-57, chmn, 57-61, assoc dean, Div Phys Sci, 48-60, chmn, Interdiv Comt Paleozool, 48-69; ed, Evolution, 53-58 & J Geol, 62-68. Mem: Fel AAAS; fel Geol Soc Am; Soc Vert Paleont (secy-treas, 48, pres, 50); Soc Study Evolution (pres, 65); Am Soc Zoologists. Res: Permian reptiles and amphibians; biometry of fossils. Mailing Add: Dept of Zool Univ of Calif Los Angeles CA 90024

OLSON, FRANK R, b Uddevalla, Sweden, May 25, 22; nat US; m 50; c 1. MATHEMATICS. Educ: Alfred Univ, BA, 47; Kent State Univ, MA, 50; Duke Univ, PhD(math), 54. Prof Exp: Instr math, Hamilton Col, 47-48, Kent State Univ, 49-50 & Duke Univ, 53-54; from asst prof to assoc prof, State Univ NY Buffalo, 54-67; PROF MATH & CHMN DEPT, STATE UNIV NY COL FREDONIA, 67- Mem: Am Math Soc; Math Asn Am. Res: Algebra, especially arithmetic properties of Bernoulli numbers and determinants. Mailing Add: Dept of Math State Univ of NY Col Fredonia NY 14063

OLSON, FRANKLYN C W, b Waukegan, Ill, Mar 15, 10; m 39; c 4. PHYSICAL OCEANOGRAPHY. Educ: Univ Chicago, SB, 33; Ohio State Univ, PhD(physics), 50. Prof Exp: Physicist, Am Can Co, Ill, 34-42; design engr, Stewart-Warner Corp, 42-43; res assoc, Northwestern Technol Inst, 43-46; asst prof physics & in-chg dept, Univ Ill, 46-47; res assoc, Franz Stone Inst Hydrobiol, Ohio State Univ, 47-50; assoc prof oceanog, Oceanog Inst, Fla State Univ, 50-57; head, Oceanog Br, US Naval Mine Defense Lab, Fla, 57-60; mem tech staff adv mil systs, Radio Corp Am, 60-63; MEM CONSULT STAFF, US NAVAL COASTAL SYSTS LAB, 63- Res: Lake currents; inshore oceanography; theory of water waves. Mailing Add: Rte 3 Box 359A Panama City FL 32401

OLSON, GERALD WALTER, b Gothenburg, Nebr, Mar 22, 32; m 61; c 3. SOIL MORPHOLOGY, AGRONOMY. Educ: Univ Nebr, BSc, 54, MSc, 59; Univ Wis, PhD(soil genesis & classification), 62. Prof Exp: Asst soils, Agr Res Serv & Univ Nebr, 56-57; jr officer, Int Coop Admin, India, 57-58; party chief soil surv, Univ Wis, 59-62; soil technologist, 62-66, asst prof, 66-71, ASSOC PROF SOIL SCI RESOURCE DEVELOP, CORNELL UNIV, 71- Concurrent Pos: Mem, Nat Comt Environ Soil Sci, Soil Conserv Serv; consult, Archaeol Expeds, Tikal, El Peten, Guatemala, Cent Am, Sardis, Turkey, San Antonio, Brit Honduras, 74 & Valle de Naco & La Conteada, Honduras, 75; sr consult land classification, Food & Agr Orgn, UN, Rome, 72 & soil surv interpretations, Soil Inst Iran & Food & Agr Orgn, UN, Tehran, Iran, 72; vis soil scientist, Kans Geol Surv, Univ Kans, 73; consult, Ministerio de Obras Publicas, Venezuela & Empresa Brasileira de Pesquisa Agropecuaria, Brazil & AID, Washington, DC, 75; vis prof soil surv interpretations, Inter-Am Ctr Integral Develop Water & Lands, Univ Andes, Venezuela, 75; guest lectr, Univ Simon Bolivar & Ministerio de Obras Publicas, Venezuela, 75. Mem: Fel AAAS; Am Soc Agron; Soil Sci Soc Am; Soil Conserv Soc Am; Asn Am Geogrs. Res: Interpretation of soil survey for urban development in New York; soil survey inventory for resource utilization for economic development; planning and development through use of soil information; waste disposal in soils; ecological implications of soils. Mailing Add: Dept of Agron Cornell Univ Ithaca NY 14853

OLSON, HAROLD CECIL, b Brookings, SDak, Nov 18, 05; m 32; c 1. DAIRY BACTERIOLOGY. Educ: SDak State Col, BS, 28; Univ WVa, MS, 30; Iowa State Univ, PhD(dairy bact), 32. Prof Exp: Asst, Exp Sta, Iowa State Univ, 32-34, instr dairy bact, 34-37, asst prof, 37-40; prof, 40-71, actg head dept, 48-50, EMER PROF DAIRY SCI, OKLA STATE UNIV, 71- Concurrent Pos: Consult, Cult Dairy Prod, 71- ed, Am Cult Dairy Prod Inst J, 69-74. Mem: Am Dairy Sci Asn; Int Asn Milk, Food & Environ Sanit. Res: Bacteriology of butter, milk, cheese and sweetened condensed milk; feed flavors in milk; cheese manufacturing; cheese cultures; marketing of dairy products. Mailing Add: 802 S Ridge Rd Stillwater OK 74074

OLSON, HARRY FERDINAND, b Mt Pleasant, Iowa, Dec 28, 02; m 35. ELECTROACOUSTICS. Educ: Univ Iowa, BE, 24, MS, 25, PhD(acoust), 28, EE, 34. Hon Degrees: DSc, Iowa Wesleyan Univ, 59. Prof Exp: Asst physics, Univ Iowa, 26-28; asst, Res Lab, RCA Corp, 28-30, Eng Dept, Photophone, NY, 30-32, dir acoust res, RCA Mfg Co, 32-42, dir, Acoust & Electromech Lab, 42-66, STAFF VPRES ACOUST & ELECTROMECH RES, RCA CORP, 66- Concurrent Pos: Lectr, Columbia Univ. Honors & Awards: Silver Medal, Am Acoust Soc, 74; Mod Pioneer Award, Nat Mfrs Asn; Scott Medal, Am Ord Asn; Warner Award, Soc Motion Picture & TV Engrs; Potts Medal, Audio Eng Soc; Consumer Electronics Award, Inst Elec & Electronics Engrs, Kelly Medal, Lamme Medal; John Ericsson Medal, Am Soc Swedish Engrs. Mem: Nat Acad Sci; fel Acoust Soc Am (pres, 52); fel Am Phys Soc; fel Audio Eng Soc (past pres); fel Inst Elec & Electronics Engrs. Res: General acoustics; electronics. Mailing Add: RCA Labs Princeton NJ 08540

OLSON, HOWARD H, b Chicago, Ill May 23, 27; m 51; c 4. DAIRY SCIENCE, PHYSIOLOGY. Educ: Univ Wis, BS, 48; Univ Minn, MS, 50, PhD(dairy physiol), 52. Prof Exp: Mem res staff, Curtiss Breeding Serv, 52-54; from asst prof to assoc prof, 54-68, PROF DAIRY SCI, UNIV SOUTHERN ILL, CARBONDALE, 68- Concurrent Pos: Physiol teaching workshop grant, 60; Fulbright-Hays lectr, Ain Shams Univ, Cairo, 66-67; res grant, Population Dynamics, 74 & 75. Mem: Am Dairy Sci Asn; Am Soc Animal Sci; World Pop Soc; Soc Study Reproduction. Res: Artificial insemination of dairy cattle; physiology of reproduction; dairy cattle nutrition studies on ad libitum grain feeding; development of complete feeds. Mailing Add: Dept of Animal Indust Southern Ill Univ Carbondale IL 62901

OLSON, JAMES ALLEN, b Minneapolis, Minn, Oct 10, 24; m 53; c 3. NUTRITIONAL BIOCHEMISTRY. Educ: Gustavus Adolphus Col, BS, 46; Harvard Univ, PhD(biochem), 52. Prof Exp: Fel biochem, Int Ctr Chem, Rome, 52-54, Ital Govt spec fel, 54; res assoc, Harvard Univ, 54-56; from asst prof to prof biochem, Col Med, Univ Fla, 56-66; actg prof & chmn dept, Mahidol Univ, Thailand, 66-72, grad res prof biochem & nutrit, 72-74; prof biomed sci, Fed Univ Bahia, Brazil, 74-75; PROF BIOCHEM & CHMN DEPT BIOCHEM & BIOPHYS, IOWA STATE UNIV, 75- Concurrent Pos: Consult, NSF, 62-65; mem, Comt Spec Caroten, Nat Res Coun, 64-; guest prof, Kyoto Univ, 65; staff mem, Rockefeller Found, 66-75; ed for Asia, J Lipid Res, 66-72; consult, WHO, 69 & 75; mem, Comt Grad Educ Nutrit, Int Union Nutrit Sci, 70-72; guest prof, Univ London, 71. Mem: Fel AAAS; Am Chem Soc; Am Soc Biol Chem; Am Inst Nutrit; Soc Exp Biol & Med. Res: Metabolism and function of vitamin A; absorption and storage of fat soluble vitamins; biological pattern formation; evaluation of human nutritional status; physiological and chemical factors influencing appetite. Mailing Add: Dept of Biochem & Biophys Iowa State Univ Ames IA 50011

OLSON, JERRY CHIPMAN, b Los Angeles, Calif, Dec 25, 17; m; c 3. GEOLOGY. Educ: Univ Calif, Los Angeles, AB, 39, MA, 47, PhD(geol), 53. Prof Exp: Geologist, 39-60, chief, Radioactive Mat Br, 60-69, GEOLOGIST, US GEOL SURV, 69- Mem: Mineral Soc Am; fel Geol Soc Am; Soc Econ Geol. Res: Economic geology; metamorphic rocks; geology of pegmatites, alkalic rocks, thorium, uranium and rare-earth mineral deposits. Mailing Add: US Geol Surv Fed Ctr Denver CO 80225

OLSON, JERRY S, ecology, see 12th edition

OLSON, JIMMY KARL, b Twin Falls, Idaho, Feb 18, 42; m 64; c 2. MEDICAL ENTOMOLOGY. Educ: Univ Idaho, BS, 65; Univ Ill, Urbana, PhD(entom), 71. Prof Exp: Asst prof med entom, 71-75, ASSOC PROF ENTOM, TEX A&M UNIV, 75- Mem: AAAS; Am Soc Trop Med & Hyg; Am Mosquito Control Asn; Entom Soc Am. Res: Mosquito bionomics and control; role of hematophagous arthropods in the transmission of disease agents affecting man and his domestic animals. Mailing Add: Dept of Entom Tex A&M Univ College Station TX 77843

OLSON, JOHN BENNET, b Minneapolis, Minn, Feb 13, 17; m 41; c 4. ZOOLOGY, SCIENCE EDUCATION. Educ: Beloit Col, BS, 38; Univ Calif, Los Angeles, MA, 41, PhD(zool), 49. Prof Exp: Instr biol, Brooklyn Col, 48-50; asst prof, San Jose State Col, 50-53; sr res assoc cardiol, Childrens Hosp, Los Angeles, 53-57; res fel chem, Calif Inst Technol, 57-58; chmn, Div Natural Sci, Shimer Col, 58-64; ASSOC PROF BIOL EDUC, PURDUE UNIV, WEST LAFAYETTE, 64- Concurrent Pos: Vis lectr, Univ Calif, Santa Barbara, 50, Berkeley, 51, 52 & 53; sci consult pub schs, Calif, 52-53. Mem: AAAS; Nat Asn Res Sci Teaching; Nat Asn Biol Teachers; NY Acad Sci; Am Inst Biol Sci. Res: Marine copepods; congenital heart defects in mammals; science teaching. Mailing Add: Dept of Biol Sci Purdue Univ West Lafayette IN 47907

OLSON, JOHN BERNARD, b Chicago, Ill, Aug 20, 31; m 61; c 3. ECOLOGY, GENETICS. Educ: Univ Ill, Urbana, BS, 58, PhD(zool), 65. Prof Exp: From asst prof to prof biol, 63-72, dean, Col Arts & Sci, 72-75, PROF BIOL, WINTHROP COL, 75- Mem: AAAS; Am Ornith Union; Cooper Ornith Soc. Res: Bioenergetic requirements during the annual cycle of migratory and nonmigratory birds. Mailing Add: Dept of Biol Winthrop Col Rock Hill SC 29730

OLSON, JOHN MELVIN, b Niagara Falls, NY, Sept 18, 29; m 53; c 3. PHOTOBIOLOGY. Educ: Wesleyan Univ, BA, 51; Univ Pa, PhD(biophys), 57. Prof Exp: USPHS res fel, Biophys Lab, Univ Utrecht & Univ Leiden, 57-58; instr physics & biochem, Brandeis Univ, 58-59, asst prof, 59-61; from asst biophysicist to assoc biophysicist, 61-65, BIOPHYSICIST, BROOKHAVEN NAT LAB, 65- Concurrent

Pos: USPHS res fel, Lab Chem Biodynamics, Lawrence Berkeley Lab, 70-71; lectr, Univ Calif, Berkeley, 71; adj prof, State Univ NY Stony Brook, 72. Mem: AAAS; Biophys Soc; Am Soc Biol Chem; Am Soc Photobiol; Int Solar Energy Soc. Res: Energy conversion in photosynthesis; structure and function of chlorophyll-protein complexes from green photosynthetic bacteria; chlorophyll-cytochrome interactions; evolution of photosynthesis. Mailing Add: Dept of Biol Brookhaven Nat Lab Upton NY 11973

OLSON, JOHN RICHARD, b Ferryville, Wis, Sept 14, 32. PHYSICS Educ: Luther Col, BA, 54; Iowa State Univ, PhD(physics), 63 Prof Exp: Instr physics, Iowa State Univ, 63; mem tech staff, Bell Tel Labs, 63-65; lectr physics, Bryn Mawr Col, 65-66, asst prof, 66-74; MEM FAC, PHYSICS RES INST, UNIV DENVER, 74- Concurrent Pos: Vis asst prof, Univ Denver, 70-71. Mem: Am Asn Physics Teachers. Res: Atomic and molecular physics, especially molecular energy transfer; sound propagation in deep water. Mailing Add: Physics Res Inst Univ of Denver Denver CO 80210

OLSON, JOHN VICTOR, b Kibbie, Mich, June 24, 13; m 40; c 1. DENTISTRY. Educ: Univ Mich, DDS, 36, MS, 38. Prof Exp: Asst prof prosthetics, Sch Dent, St Louis Univ, 47-49, assoc prof, 49-50, dir postgrad dent educ, 48-50; dean, Dent Sch, Univ Tex Health Sci Ctr, San Antonio, 69-72; PROF RESTORATIVE DENT, UNIV TEX DENT BR, HOUSTON, 50-, DEAN, 52- Concurrent Pos: Consult, Vet Admin Hosp, Houston, Tex, 52, Univ Tex Med Br & M D Anderson Hosp & Tumor Inst, 52- Mem: AAAS. Res: Prosthetic dentistry; dental education. Mailing Add: Univ of Tex Dent Br PO Box 20068 Houston TX 77025

OLSON, JOSEPH CARL, JR, b Roseburg, Ore, Mar 5, 13; m 36; c 2. BACTERIOLOGY. Educ: Univ Minn, BS, 35, MS, 44, PhD(bact), 48. Prof Exp: Asst bacteriologist, Univ Minn, St Paul, 35-37, instr dairy bact, 37-40 & 46-48, from assoc prof to prof, 48-67; DIR DIV MICROBIOL, FOOD & DRUG ADMIN, 67- Concurrent Pos: Mem exec coun, Nat Conf Interstate Milk Shipments, 52; ed, J Milk & Food Technol, 54-67; consult, Milk & Food Training Prog, USPHS, 57-67; mem, US-Japan Joint Panel Toxic Microorganisms, 68-; mem, Int Comn Microbiol Specifications for Foods, 68-; mem exec bd, Nat Mastitis Coun, 69-; mem comt food microbiol & hyg, Int Asn Microbiol Socs, 70-; mem comt dairy, oil & fat prod, Adv Bd Mil Personnel Supplies, Nat Acad Sci-Nat Res Coun, mem comt food microbiol, 71. Honors & Awards: Citation Award, Int Asn Milk, Food & Environ Sanit, 65. Mem: Am Soc Microbiol; Am Dairy Sci Asn; Am Acad Microbiol; Am Pub Health Asn; Int Asn Milk, Food & Environ Sanit. Res: Bacterial thermal-death-time studies; role of psychophilic bacteria in the deterioration of dairy products; growth characteristics of staphylococci in foods; utlization of microorganisms in the dairy and food industries; causes and control of food borne disease. Mailing Add: Div of Microbiol Food & Drug Admin Washington DC 20204

OLSON, KARL EDWIN, microbiology, see 12th edition

OLSON, KENNETH B, b Seattle, Wash, Jan 21, 08; m 37; c 2. MEDICINE, ONCOLOGY. Educ: Univ Wash, BS, 29; Harvard Med Sch, MD, 33. Prof Exp: Resident path, Boston City Hosp, Mass, 33-34; intern & asst resident & resident surgeon, Presby Hosp, New York, 34-40; instr, Col Physicians & Surgeons, Columbia Univ, 40-41; dir, Firland Sanatorium, Seattle, Wash, 43-45; from instr to prof med, 50-71, dir div oncol, 51-72, EMER PROF MED, ALBANY MED COL, 72-; PROF MED & ACTG HEAD MED ONCOL, DIV HEMATOL, J HILLIS MILLER HEALTH CTR, UNIV FLA, 76- Concurrent Pos: Mem cancer clin invest rev comt, Nat Cancer Inst, 68-72, mem breast cancer task force, 71-, chief diag br, Div Cancer Biol & Diag, 72-73, consult, 73-76. Mem: Soc Exp Biol & Med; AMA; Am Asn Cancer Res; James Ewing Soc; Am Soc Clin Oncol (pres, 71-72). Res: Lung and breast cancer; liver function; chemotherapy of cancer; biological effects of irradiation. Mailing Add: J Hillis Miller Health Ctr Div Hematol Univ of Fla Gainesville FL 32610

OLSON, KENNETH JEAN, b Saginaw Co, Mich, July 8, 16; m 43; c 2. TOXICOLOGY. Educ: Kalamazoo Col, AB, 42; Mich State Col, MS, 44, PhD(biochem), 47; Am Bd Indust Hyg, dipl. Prof Exp: Res pharmacologist, Upjohn Co, 47-54, sect head endocrinol, 54; toxicologist, Dept Biochem, Dow Chem Co, 54-68, mgr toxicol dept, Dow Corning Corp, 68-73, SR RES SPECIALIST, TOXICOL RES LAB, DOW CHEM CO, 73- Mem: Endocrine Soc; Soc Toxicol. Res: Industrial toxicology; dental caries in white rats; method for absorption and elimination of thiouracil; pharmacology of spasmolytics, antihistaminics and local anesthetics. Mailing Add: Toxicol Res Lab Dow Chem Co Midland MI 48640

OLSON, LEE CHARLES, b Austin, Minn, June 2, 36; m 61; c 4. PLANT BIOCHEMISTRY, PLANT PHYSIOLOGY. Educ: SDak State Univ, BS, 58; Univ Wis, MS, 62, PhD(biochem), 64. Prof Exp: Asst prof plant physiol, Univ Minn, St Paul, 65-67, agron, 67-70; MEM FAC, DEPT BIOL, CHRISTOPHER NEWPORT COL, 71- Mem: Am Chem Soc; Sigma Xi; Am Soc Plant Physiologists. Res: Amino acid metabolism; nitrogen fixation; plant cell culture. Mailing Add: Dept of Biol Christopher Newport Col Newport News VA 23606

OLSON, LEONARD ELMER, organic chemistry, see 12th edition

OLSON, LEROY DAVID, b East Chain, Minn, July 22, 29; m 56; c 1. VETERINARY PATHOLOGY, VETERINARY VIROLOGY. Educ: Univ Minn, BS, 54, DVM, 58; Purdue Univ, MS, 62, PhD(vet path), 65. Prof Exp: Instr vet path, Purdue Univ, 60-65; ASSOC PROF VET PATH, UNIV MO-COLUMBIA, 65- Mem: AAAS; Am Vet Med Asn; Am Soc Microbiol; Am Soc Vet Parasitol. Res: Immunopathology, especially experimental allergic encephalomyelitis; avian cancer viruses; viral enteric diseases of swine; Pasteurella multocida infections in turkeys. Mailing Add: Dept of Vet Path Univ of Mo Sch of Vet Med Columbia MO 65201

OLSON, LEROY JUSTIN, b Fargo, NDak, May 28, 26; m 53; c 2. MICROBIOLOGY. Educ: Concordia Col, BA, 50; Kans State Univ, MS, 53; Univ Tex, PhD(parasitol), 57. Prof Exp: Instr biol, Concordia Col, 50-51; asst vet parasitol & zool, Kans State Univ, 51-53; asst prev med & pub health, 53, instr bact & parasitol, 54-56, from asst prof to assoc prof microbiol, 57-66, PROF MICROBIOL, UNIV TEX MED BR, GALVESTON, 66- Mem: Am Soc Parasitol; Am Soc Microbiol; AAAS. Res: Host-parasite relationships; immunology; gut allergy and pathology. Mailing Add: Dept of Microbiol Univ of Tex Med Br Galveston TX 77551

OLSON, LLOYD CLARENCE, b Spokane, Wash, Jan 30, 35; m 58; c 3. MICROBIOLOGY, PEDIATRICS. Educ: Reed Col, AB, 57; Harvard Med Sch, MD, 61. Prof Exp: From intern to resident pediat, Univ Rochester, 61-63, resident, 63-64; virologist, Walter Reed Army Inst Res, 64-67; vis assoc prof microbiol, 67-70, vis lectr, 68-70, VIS PROF MICROBIOL, MAHIDOL UNIV, THAILAND, 70- Concurrent Pos: Virologist, SEATO Med Res Lab, Thailand, 67-70. Mem: Am Soc Microbiol; Pan-Am Med Asn. Res: Clinical and epidemiologic aspects of virus diseases; host defense mechanisms to virus infections. Mailing Add: Rockefeller Found GPO Box 2453 Bangkok Thailand

OLSON, MAGNUS, b South Fron, Norway, June 29, 09; US citizen; m 38; c 2. ZOOLOGY. Educ: St Olaf Col, AB, 32; Univ Minn, AM, 34, PhD(histol), 36. Prof Exp: Asst zool, 32-36, asst, Comt Educ Res, 36-37, instr educ & zool, 37-38, from instr to assoc prof zool, 38-58, PROF ZOOL, UNIV MINN, MINNEAPOLIS, 58-, CHMN DEPT, 66- Mem: AAAS; assoc Am Soc Zool. Res: Histology of invertebrate muscle; ovogenesis in Mammalia; myeloid metaplasia in mammalian thymus; wound healing in rabbits. Mailing Add: Dept of Zool Univ of Minn Minneapolis MN 55455

OLSON, MARK OBED JEROME, b Clarkfield, Minn, Aug 20, 40; m 66. BIOCHEMISTRY, PHARMACOLOGY. Educ: St Olaf Col, BA, 62; Univ Minn, Minneapolis, PhD(biochem), 67. Prof Exp: Asst prof, 69-74, ASSOC PROF PHARMACOL, BAYLOR COL MED, 74- Concurrent Pos: Med Res Coun Can fel, Univ Alta, 67-69; Am Cancer Soc grant pharmacol, Baylor Col Med, 69-70; NSF grant, 73-74. Mem: AAAS; Am Asn Cancer Res; Am Chem Soc; Biophys Soc; NY Acad Sci. Res: Functional and evolutionary aspects of protein structure; phytohemagglutinins; proteolytic enzymes; structure and role of nuclear proteins in genetic regulation; mechanism of enzyme action. Mailing Add: Dept of Pharmacol Baylor Col of Med Houston TX 77025

OLSON, MAYNARD VICTOR, b Washington, DC, Oct 2, 43; m 68; c 1. INORGANIC CHEMISTRY. Educ: Calif Inst Technol, BS, 65; Stanford Univ, PhD(chem), 70. Prof Exp: ASST PROF CHEM, DARTMOUTH COL, 70- Mem: Am Chem Soc. Res: Mechanisms of transition metal reactions. Mailing Add: Dept of Chem Dartmouth Col Hanover NH 03755

OLSON, MELVIN MARTIN, b Bangor, Wis, Aug 30, 15; m 41; c 3. ORGANIC CHEMISTRY. Educ: Wis State Teachers Col, La Crosse, BS, 38; Univ Wis, PhM, 39, PhD(chem), 50. Prof Exp: Instr high sch, Wis, 39-41; instr high sch & Maquoketa Jr Col, Iowa, 41-43; asst chem, Univ Wis, 46-50, prof, Exten Div, 50-51; res chemist, Paint Div, Pittsburgh Plate Glass Co, 51-54; res chemist, Tape Div, Minn Mining & Mfg Co, 54-59, res specialist, 56-69, SR RES SPECIALIST, INDUST TAPE DIV, 3M CO, 69- Mem: Am Chem Soc. Res: Modification of paint vehicles with silicon compounds; synthesis of silicon compounds; formulation of organic structural adhesives; synthesis of pressure-sensitive adhesives, polymers for release coatings and water-soluble polymers; ethylene oxide detection. Mailing Add: 7609 Portland Ave S Minneapolis MN 55423

OLSON, MERLE STRATTE, b Northfield, Minn, Aug 22, 40; m 64. BIOCHEMISTRY. Educ: St Olaf Col, BA, 62; Univ Minn, Minneapolis, PhD(biochem), 66. Prof Exp: Asst biochem, Univ Minn, Minneapolis, 62-66; assoc prof biochem, Col Med, Univ Ariz, 68-76; PROF BIOCHEM, UNIV TEX, HEALTH SCI CTR SAN ANTONIO, 76- Concurrent Pos: Johnson Res Found res fel biochem, Univ Pa, 66-68. Mem: Am Soc Biol Chem; Am Soc Cell Biol. Res: Regulation of intermediary metabolism and interrelationships between mitochondrial and cytoplasmic metabolism; autoimmune nature of mitochondrial membranes. Mailing Add: Dept of Biochem Univ of Tex Health Sci Ctr San Antonio TX 78284

OLSON, NORMAN FREDRICK, b Edmund, Wis, Feb 8, 31; m 57; c 2. FOOD TECHNOLOGY. Educ: Univ Wis, BS, 53, MS, 57, PhD(dairy & food industs), 59. Prof Exp: Assoc prof, 58-71, PROF FOOD SCI, UNIV WIS-MADISON, 71- Honors & Awards: Pfizer Award in Cheese Res. Mem: Am Dairy Sci Asn; Inst Food Technol; Am Soc Microbiol. Res: Mechanization of cheese manufacture; rheology of cheese; bacteriological problems of natural cheese; food fermentations, particularly cheese technology; technology of immobilized enzymes. Mailing Add: Dept of Food Sci Univ of Wis Madison WI 53706

OLSON, NORMAN O, b Regan, NDak, June 1, 14; m 40; c 3. VETERINARY MEDICINE. Educ: Wash State Univ, DVM, 38. Prof Exp: Vet disease control res, USDA, 38-44, res vet, Food & Drug Admin, 44-48; PROF ANIMAL PATH, WVA UNIV, 48- Honors & Awards: Am Feed Mfrs Award, Am Vet Med Asn, 72. Mem: Am Vet Med Asn; Poultry Sci Asn. Res: Virus and bacterial diseases of farm animals; mycoplasma and virology. Mailing Add: Div of Animal & Vet Sci WVa Univ Morgantown WV 26506

OLSON, OSCAR EDWARD, b Sioux Falls, SDak, Jan 19, 14; m 43; c 2. AGRICULTURAL BIOCHEMISTRY. Educ: SDak State Col, BS, 36, MS, 37; Univ Wis, PhD(biochem), 48. Prof Exp: Sta analyst, Agr Exp Sta, SDak State Col, 37-42; chemist, Fruit & Veg Prod Lab, USDA, 48-49; biochemist, Div Labs, Mich State Dept Health, 49-51; head, Exp Sta Biochem, 51-73, dean, grad div, 58-65, PROF CHEM, SDAK STATE UNIV, 75- Concurrent Pos: Vis prof, Inst Enzyme Res, Univ Wis-Madison, 62-63; vis scientist, US Plant, Soil & Nutrit Lab, NY, 73-74. Mem: AAAS; Am Chem Soc; Am Inst Nutrit; Am Soc Biol Chem; Am Inst Biol Sci. Res: Selenium and nitrate poisoning; calcium metabolism in poultry. Mailing Add: Dept of Chem SDak State Univ Brookings SD 57006

OLSON, PAUL B, b Tea, SDak, Mar 5, 24; m 47; c 2. ANALYTICAL CHEMISTRY. Educ: Augustana Col, BA, 44; Univ Minn, Minneapolis, MS, 48. Prof Exp: Chemist, Minn Mining & Mfg Co, 48-67, RES SPECIALIST CHEM, 3M CO, 67- Mem: Am Chem Soc; Am Microchem Soc; Catalysis Soc. Res: Organic microanalysis. Mailing Add: 3M Ctr PO Box 33221 St Paul MN 55133

OLSON, RALPH EUGENE, b Newman Grove, Nebr, Aug 23, 14; m 41; c 3. GEOGRAPHY. Educ: Nebr Wesleyan Univ, AB, 35; Univ Nebr, MA, 37; Clark Univ, PhD(geog), 46. Hon Degrees: DSc, Nebr Wesleyan Univ, 52. Prof Exp: Actg head dept geog, Moorhead State Teachers Col, Minn, 39-40; instr geog, Univ Nebr, 40-41; asst prof, Western Reserve Univ, 46-47; assoc prof, 47-53, chmn dept, 56-64, PROF GEOG, UNIV OKLA, 53- Concurrent Pos: Univ Okla Res Inst grant, Univ Utrecht, 67-68; fac res fund grants, Univ Okla. Mem: Asn Am Geogrs; Nat Coun Geog Educ. Res: Political geography, especially Europe; economic geography, especially the Union of Soviet Socialist Republics. Mailing Add: Dept of Geog Univ of Okla Norman OK 73069

OLSON, RAYMOND VERLIN, b Pembina Co, NDak, Oct 4, 19; m 43; c 3. AGRONOMY. Educ: NDak Agr Col, BS, 41; Univ Wis, MS, 42, PhD(soils), 47. Prof Exp: Chemist, Hercules Powder Co, Del; 42-45; from assoc prof to prof soils, 47-52, head dept, 52-70, dir, Int Agr Progs, 70-72, PROF AGRON, KANS STATE UNIV, 52- Concurrent Pos: Party chief, Ahmadu Bello Univ Proj, Nigeria, 64-66 & 72-74, dean, Fac Agr, 64-65, provost agr & vet med, 71-74; consult, Ministry Agr, Iran, 68 & IRI Inst Brazil, 69. Honors & Awards: Gamma Sigma Delta Int Award, Distinguished Serv to Agr, 68. Mem: Fel Am Soc Agron; AAAS; Soil Sci Soc Am; Int Soc Soil Sci. Res: Soil chemistry; plant nutrient availability; iron chlorosis. Mailing Add: Dept of Agron Kans State Univ Manhattan KS 66506

OLSON, RICHARD HUBBELL, b Meriden, Conn, Nov 25, 28; m 54; c 4. ECONOMIC GEOLOGY. Educ: Tufts Univ, BS, 50; Univ Utah, PhD(geol), 60. Prof Exp: Field geologist, US AEC, 51-53; geologist & draftsman, Gulf Oil Corp, 55-56; econ geologist, Union Carbide Corp, 57-59; asst prof econ geol, Univ Nev, 59-64; econ geologist, Chas Pfizer & Co, Inc, 65-69; dir, Mineral Resources Dept, Monsanto

Co, 69-71; CONSULT GEOLOGIST, 72- Mem: Geol Soc Am; Am Inst Mining, Metall & Petrol Engrs; Soc Econ Geol. Res: Industrial mineral deposits; economic geology of mineral deposits. Mailing Add: 14618 W 6th Ave Suite 202 Golden CO 80401

OLSON, RICHARD LOUIS, b El Paso, Tex, Aug 14, 32; m 55; c 2. BOTANY, GENETICS. Educ: Univ Utah, BS, 54, MS, 55; Univ Calif, Berkeley, PhD(genetics), 64. Prof Exp: Preliminary design engr, AiResearch Mfg Co Div, Garrett Corp, Calif, 62-64; mgr biosci, 64-74, GEN MGR, BEST PROG, RESOURCES CONSERVATION CO DIV, BOEING CO, 74- Concurrent Pos: Chmn scientist comt, Northwest Pollution Control Asn. Mem: AAAS; Am Soc Microbiol. Res: Sludge drying equipment development; waste management and water management problems related to manned space flight; toxicology; space craft eterilization and exobiology; urban development; waste and water management; urban and agricultural pollution abatement. Mailing Add: The Boeing Co PO Box 3707 Seattle WA 98124

OLSON, ROBERT AUGUST, b Nebr, Apr 14, 17; m 39; c 3. SOILS. Educ: Univ Nebr, MSc, 49. Prof Exp: Soil surveyor & technologist, Soil Conserv Serv, USDA, 38-43; asst exten agronomist, 46-48, from asst prof to assoc prof, 48-60, PROF AGRON, UNIV NEBR, LINCOLN, 60- Concurrent Pos: Consult, Orgn Europ Econ Coop, Paris, 58; sr off, Int Atomic Energy Agency, Austria, 62; mgr, Int Fertilizer Prog, Food & Agr Orgn, Rome, 67-69; actg dir, Joint Div Atomic Energy in Food & Agr, Food & Agr Orgn-Int Atomic Energy Agency, 74-75; mem, Comt Trop Soils, Nat Acad Sci-Nat Res Coun. Honors & Awards: Int Serv Agron Award, Am Soc Agron, 71. Mem: Soil Sci Soc Am; fel Am Soc Agron; Int Soil Sci Soc. Res: Plant nutrition studies with radioisotope tracers; improving validity of soil testing; abatement of agricultural pollution. Mailing Add: Dept of Agron Univ of Nebr Lincoln NE 68503

OLSON, ROBERT EDWARD, b Titusville, Pa, Oct 28, 14. FOREST ENTOMOLOGY. Educ: Cornell Univ, BS, 38, MS, 46, PhD(entom), 54. Prof Exp: From instr to assoc prof, 46-59, PROF ENTOM & ENTOMOLOGIST, UNIV MAINE, ORONO, 59- Concurrent Pos: From asst entomologist to assoc entomologist, Agr Exp Sta, Univ Maine, Orono, 53-59; res assoc, Univ Calif, 56. Mem: AAAS; Entom Soc Am; Soc Syst Zool; Entom Soc Can. Res: Sawyer beetles; mosquito biology and control; bioassay technique. Mailing Add: 313 Deering Hall Univ of Maine Orono ME 04473

OLSON, ROBERT ELDON, b Thief River Falls, Minn, May 4, 40; m 65; c 2. ZOOLOGY, PARASITOLOGY. Educ: Concordia Col, Moorhead, Minn, BA, 62; Mont State Univ, MS, 64, PhD(zool), 68. Prof Exp: Res assoc, 68-75, ASST PROF ZOOL, MARINE SCI CTR, ORE STATE UNIV, 75- Mem: Am Soc Parasitol; Am Fisheries Soc; Am Micros Soc; Wildlife Dis Asn. Res: Ecology of protozoans and helminths parasitizing marine fishes and shell fishes. Mailing Add: Marine Sci Ctr Ore State Univ Newport OR 97365

OLSON, ROBERT EUGENE, b Minneapolis, Minn, Jan 23, 19; m 44; c 5. BIOCHEMISTRY, MEDICINE. Educ: Gustavus Adolphus Col, AB, 38; St Louis Univ, PhD(biochem), 44; MD, 51; Am Bd Nutrit, dipl. Prof Exp: Asst biochem, Sch Med, St Louis Univ, 38-43; instr biochem & nutrit, Sch Pub Health, Harvard Univ, 46-47, estab investr, Am Heart Asn, 51-52; prof biochem & nutrit, Grad Sch Pub Health & lectr med, Sch Med, Univ Pittsburgh, 52-65; assoc prof med, 65-72, ALICE A DOISY PROF BIOCHEM & CHMN DEPT, SCH MED, ST LOUIS UNIV, 65-, PROF MED, 72- Concurrent Pos: Harvard Med Sch, Nutrit Found fel, 47-49, Am Heart Asn fel, 49-51; Guggenheim & Fulbright fel, Oxford Univ, 61-62 & Univ Freiburg, 70-71; house physician, Peter Bent Brigham Hosp, Boston, Mass, 51-52; dir nutrit clin, Falk Clin, Univ Pittsburg Med Ctr, 53-65; consult metab & nutrit study sect, Res Grants Div, USPHS, 53-57, 58-59, biochem training comt, 59-63, training comt, Nat Heart Inst, 64-68; clin assoc, St Margaret Mem Hosp, 54-60, consult, 55-65; mem panel biochem & nutrit, Comt Growth, Nat Res Coun, 54-56; sci adv comt, Nat Vitamin Found, 55-58; from vpres to pres, Am Bd Nutrit, 60-63; consult & dir metab unit, Presby Hosp, 60-65; dir, Anemia Malnutrition Res Ctr, Thailand, 65- Honors & Awards: McCollum Award, Am Soc CLin Nutrit, 65; Goldberger Award, AMA, 74. Mem: AAAS; Am Chem Soc; Am Soc Biol Chem; Am Soc Clin Nutrit (pres, 61-62); fel Am Pub Health Asn. Res: Cardiac metabolism; role of the fat-soluble vitamins; experimental and clinical nutrition. Mailing Add: Dept of Biochem St Louis Univ Sch of Med St Louis MO 63104

OLSON, ROBERT LEROY, b Portland, Ore, Apr 24, 32; m 57; c 2. PHYSIOLOGY, MEDICINE. Educ: Ore State Univ, BS, 54; Univ Ore, MD, 58. Prof Exp: Resident, 62-64, from instr to assoc prof, 64-74, PROF DERMAT, SCH MED, UNIV OKLA, 74- Mem: Soc Invest Dermat; Am Acad Dermat; Am Dermat Asn. Res: Cutaneous physiology; ultraviolet light physiology; electron microscopy. Mailing Add: Univ Hosp 800 NE 13th St Oklahoma City OK 73104

OLSON, RODNEY ANDREEN, b Somerville, Mass, Apr 19, 14; m 49; c 2. PHYSICAL BIOLOGY, CELL BIOLOGY. Educ: Tufts Univ, BS, 36; Univ Md, MS & PhD(plant physiol), 39. Prof Exp: Biologist, Dept Res & Educ, State of Md, 39-42; physiologist, Naval Med Res Lab, 46; res physiologist, Lab Phys Biol, NIH, 46-73; RETIRED. Concurrent Pos: Vis prof biol, Dalhousie Univ, 72-73. Res: Cellular physiology; polarity; plant hormones; radiation and metabolism; photosynthesis; functional aspects of molecular orientation in vivo. Mailing Add: Blue Rocks Studio Gal Box 507 Lunenburg NS Can

OLSON, RONALD LEROY, b Mankato, Minn, Dec 8, 95; m 21. ANTHROPOLOGY. Educ: Univ Wash, AB, 25, MA, 26; Univ Calif, Berkeley, PhD(anthrop), 29. Prof Exp: Asst cur anthrop, Am Mus Natural Hist, 29-31; from assoc prof to prof, 31-56, EMER PROF ANTHROP, UNIV CALIF, BERKELEY, 56- Concurrent Pos: Mem ethnog expeds, Wash, 25-27, Alaska, 33-34 & BC, Can, 35; archaeol exped mem, Calif, 27-28 & Peru, 30. Mem: AAAS; Am Anthrop Asn. Res: Social organization; cultures of the North Pacific Coast. Mailing Add: 13343 Hillcrest Dr Valley Center CA 92082

OLSON, ROY E, b Aberdeen, Wash, May 24, 29; m 62; c 3. SOLID STATE PHYSICS. Educ: Univ Calif, Berkeley, PhD(physics), 58. Prof Exp: ASSOC PROF PHYSICS, CALIF STATE UNIV, NORTHRIDGE, 62- Mailing Add: Dept of Physics & Astron Calif State Univ Northridge CA 91324

OLSON, STANLEY WILLIAM, b Chicago, Ill, Feb 10, 14; m 36; c 3. MEDICINE. Educ: Wheaton Col, Ill, BS, 34; Univ Ill, MD, 38; Univ Minn, MS, 43; Am Bd Internal Med, dipl. Hon Degrees: LLD, Wheaton Col, Ill, 53. Prof Exp: Asst to dir, Mayo Found, Univ Minn, 46-47, asst dir, 47-50; prof med & dean col med, Univ Ill, 50-53 & Baylor Univ, 53-66; dir regional med progs serv, Health Serv & Ment Health Admin, 68-70; pres, Southwest Found Res & Educ, 70-73; PROVOST, NORTHEASTERN OHIO UNIVS COL MED, 73- Concurrent Pos: Consult, State Univ NY, 49 & Comn, 54; mem med adv panel, US Off Voc Rehab Admin, 60-65; mem comt med sch-vet admin rels, 62-66; mem nat adv coun health res facil, NIH, 63-67; mem rev panel construct schs med, USPHS, 64-66; mem, Nat Adv Comn Health Manpower, 66; prof med, Vanderbilt Univ, 67-68; clin prof med, Meharry Med Col, 67-68; dir, Tenn Mid-South Regional Med Prof, 67-68. Mem: Am Asn Med Cols (vpres, 60-61); fel Am Col Physicians. Res: Medical administration; medical education for national defense. Mailing Add: Northeastern Ohio Univs Col Med 275 Martinel Dr Kent OH 44240

OLSON, STORRS LOVEJOY, b Chicago, Ill, Apr 3, 44. ORNITHOLOGY. Educ: Fla State Univ, BS, 66, MS, 68; Johns Hopkins Univ, ScD(biol), 72. Prof Exp: CUR ORNITH, SMITHSONIAN INST, 72- Honors & Awards: A B Howell Award, Cooper Ornith Soc, 72; Ernest P Edwards Award, Wilson Ornith Soc, 73. Mem: Am Ornithologists Union; Soc Vert Paleont; Brit Ornithologists Union; Royal Australasian Ornithologists Union. Res: Systematics, evolution and paleontology of birds, particularly quaternary island faunas and tertiary seabirds. Mailing Add: Div of Birds Nat Mus of Natural Hist Smithsonian Inst Washington DC 20560

OLSON, TAMLIN CURTIS, b Colton, SDak, Feb 21, 32; m 55; c 4. SOIL PHYSICS. Educ: Univ Md, BS, 57, MS, 59; Purdue Univ, PhD, 66. Prof Exp: Res soil scientist, 60-63, supvy soil scientist, 63-73, RES LEADER, AGR RES SERV, USDA, 73- Concurrent Pos: Assoc prof plant sci, SDak State Univ, 63-73. Mem: Am Soc Agron; Soil Sci Soc Am; Soil Conserv Soc Am. Res: Flow of fluids through porous media; water use by agronomic crops. Mailing Add: N Cent Soil Conserv Res Ctr Morris MN 56267

OLSON, WALTER, b Yakima, Wash, May 16, 24; m 57; c 3. GEOGRAPHY. Educ: Univ Wash, AB, 47; Col Puget Sound, MA, 49; Syracuse Univ, PhD(geog, econ), 56. Prof Exp: From instr to prof geog & chmn dept, San Francisco State Col, 55-70; prof geog & dean col arts & sci, Western Ill Univ, 70-75; PRES, CALIF STATE COL, STANISLAUS, 75- Concurrent Pos: Ellis L Phillips Found fel acad admin, Univ Wis, 62-63. Mem: Asn Am Geog; Am Asn Higher Educ. Res: Economic geography; California geography. Mailing Add: Calif State Col at Stanislaus Turlock CA 95380

OLSON, WALTER HAROLD, b New Haven, Conn, May 12, 45; m 70; c 2. BIOMEDICAL ENGINEERING. Educ: Pa State Univ, BS, 67; Univ Mich, MS, 68, PhD(bioeng), 73. Prof Exp: Teaching fel elec eng, Univ Mich, 69-73; ASST PROF ELEC ENG, UNIV ILL, URBANA, 73- Concurrent Pos: Consult, Regional Health Resources Ctr, 74-; actg prog dir, NIH Bioeng Training, Univ Ill, 75-76. Mem: Inst Elec & Electronics Engrs; Asn Advan Med Instrumentation; Sigma Xi. Res: Biomedical instrumentation including transducers, electronics and digital signal processing; bioelectric signal analysis and synthesis including neural compound action potentials, effects of mechanical stimuli on nerve activity and electrocardiogram waveform synthesis. Mailing Add: 155 EEB Dept of Elec Eng Univ of Ill Urbana IL 61801

OLSON, WALTER SIGFRID, b Taylors Falls, Minn, Jan 17, 04; m 38, c 3. PETROLEUM GEOLOGY, TECTONICS. Educ: Univ Minn, EM, 25. Prof Exp: Petrol Geologist, Gulf Oil Corp, 26-32; chief geologist, Colombian Petrol Co, 33-36; coordr geophys, Tex Co of Calif, Texaco Inc, 37-40, from supvr geologist to sr staff geologist, Texaco Inc, NY, 41-69; SR RES ASSOC TECTONICS, LAMONT-DOHERTY GEOL OBSERV, COLUMBIA UNIV, 70- Mem: Fel AAAS; Am Asn Petrol Geologists. Res: Changes in the figure of the earth due to decreasing rate of rotation since Cambrian time as a cause of marine transgression-regression cycle; development of a geologic time-scale for evolution of the lunar orbit. Mailing Add: 4 Claremont Rd Scarsdale NY 10583

OLSON, WALTER T, b Royal Oak, Mich, July 4, 17; m 43. CHEMISTRY. Educ: DePauw Univ, AB, 39; Case Inst Technol, BS, 40, PhD(chem), 42. Prof Exp: Instr chem, Case Inst Technol, 40-42; res chemist, Nat Adv Comt Aeronaut, 42-45, chief, Combustion Br, 45-50, Fuels & Combustion Res Div, 50-58, Propulsion Chem Div, NASA-Lewis, 58-60, Chem & Energy Conversion Div, 60-62, ASST DIR, CHEM & ENERGY CONVERSION DIV, LEWIS RES CTR, NASA, 62- Concurrent Pos: Consult, Dept Defense; chmn adv com, Fenn Col Eng, Cleveland State Univ. Mem: Fel AAAS; fel Am Inst Aeronaut & Astronaut; Am Chem Soc; Am Soc Mech Engrs; Combustion Inst. Res: High speed combustion for aircraft engines; high energy fuels; organic synthesis; local anesthetics; rocket engines and propulsion and power for space flight; energy conversion systems. Mailing Add: Lewis Research Ctr Cleveland OH 44135

OLSON, WENDELL CLARENCE, b Aurora, Ill, June 3, 24; m 47; c 2. MATHEMATICS. Educ: Aurora Col, BS, 49; Univ Del, MS, 54. Prof Exp: Res mathematician, 51-61, res physicist, 61-63, PHYS SCIENTIST, BALLISTIC RES LABS, ABERDEEN PROVING GROUND, US DEPT ARMY, 63- Mem: Math Asn Am. Res: Theory of shock and detonation waves; operations research and vulnerability analysis; munitions effectiveness evaluations. Mailing Add: Ballistics Res Lab Aberdeen Proving Ground MD 21005

OLSON, WILLARD PAUL, b Detroit, Mich, Aug 6, 39; m 64; c 2. ENVIRONMENTAL PHYSICS. Educ: Univ Calif, Los Angeles, BS, 62, MS, 64, PhD(physics), 68. Prof Exp: Res scientist, Astrophysics Res Corp, 64-65; consult, Rand Corp, 65-66; sr scientist, 68-74, CHIEF ENVIRON EFFECTS BR, McDONNELL DOUGLAS ASTRONAUT CO, 68- Mem: AAAS; Am Geophys Union; Am Phys Soc; Int Acad Astronaut. Res: Magnetospheric electric and magnetic fields and their interactions with charged particles; quantitative modelling and prediction of environmental effects and associated computer software development. Mailing Add: McDonnell Douglas Astronaut Co 5301 Bolsa Ave Huntington Beach CA 92647

OLSON, WILLIAM ARTHUR, b Minneapolis, Minn, Oct 19, 32; m 56; c 3. TOXICOLOGY, BIOCHEMISTRY. Educ: Univ Minn, BS, 54, MS, 60, PhD(nutrit), 62. Prof Exp: Res supvr nutrit, Chas Pfizer & Co, 62-65; dir animal health res, Smith Kline & French Labs, 65-69; staff scientist, Hazelton Labs, 69-73; OWNER-MGR, CFR SERV, 73- Mem: AAAS; Inst Food Technologists; Am Soc Animal Sci; Am Dairy Sci Asn; Animal Nutrit Res Coun. Res: Determination of the safe use of chemicals as drugs, veterinary drugs, food additives, pesticides or household articles; toxicology; product development. Mailing Add: 2347 Paddock Lane Reston VA 22091

OLSON, WILLIAM BRUCE, b Omaha, Nebr, Dec 28, 30; m 59; c 2. PHYSICAL CHEMISTRY. Educ: Univ Wash, BS, 53, PhD(phys chem), 60. Prof Exp: Res assoc phys chem, Princeton Univ, 60-61; PHYSICIST, NAT BUR STANDARDS, 61- Honors & Awards: Silver Medal, US Dept Com, 73. Mem: Optical Soc Am; AAAS. Res: Molecular structure via high resolution infrared spectroscopy; infrared instrumentation; laser stark spectroscopy. Mailing Add: Optical Physics Div Nat Bur of Standards Washington DC 20234

OLSON, WILLIAM JOHN, biochemistry, see 12th edition

OLSON, WILLIAM MARVIN, b Rock Island, Ill, June 15, 29; m 55; c 3. HIGH TEMPERATURE CHEMISTRY. Educ: Augustana Col, Ill, BA, 51; Univ Iowa, PhD(inorg chem), 59. Prof Exp: MEM STAFF, LOS ALAMOS SCI LAB, UNIV CALIF, 56- Concurrent Pos: Fel, Europ Inst Transuranium Elements, EURATOM,

Karlsruhe, Ger, 68-69. Mem: Am Chem Soc; Am Vacuum Soc. Res: Thermodynamics, especially of actinides and their compounds. Mailing Add: CMB-11 Group MS-328 Los Alamos Sci Lab Los Alamos NM 87545

OLSON, WILMA KING, b Philadelphia, Pa, Dec 1, 45; m 69; c 1. BIOPHYSICAL CHEMISTRY, POLYMER CHEMISTRY. Educ: Univ Del, BS, 67; Stanford Univ, PhD(chem), 71. Prof Exp: Fel chem, Stanford Univ, 70-71; Damon Runyon Fund fel, Columbia Univ, 71-72; ASST PROF CHEM, DOUGLASS COL, RUTGERS UNIV, NEW BRUNSWICK, 72- Concurrent Pos: Sloan Found fel, 75; USPHS career develop award, 75. Mem: Am Chem Soc; Biophys Soc. Res: Relationship of structure, conformation and function in biological macromolecules, particularly on nucleic acids. Mailing Add: Dept of Chem Douglass Col Rutgers Univ New Brunswick NJ 08903

OLSSON, CARL NIELS, b Miami, Fla, Nov 7, 30; m 53; c 2. RADIO ASTRONOMY. Educ: Univ Fla, BS, 56, MS, 57, PhD, 70. Prof Exp: Instr phys sci, 57-62, asst prof, 63-72, ASSOC PROF ASTRON, UNIV FLA, 63- Mem: Am Astron Soc. Res: Nonthermal radiations from planets; very long baseline interferometry. Mailing Add: Dept of Physics & Astron Univ of Fla Gainesville FL 32601

OLSSON, GUNNAR, b Eksharad, Sweden, Sept 11, 35; m 62; c 1. ECONOMIC GEOGRAPHY. Educ: Univ Uppsala, Fil Kand, 60, Fil lic, 65, FD, 68. Prof Exp: Lectr human geog, Univ Uppsala, 64-65; from asst prof to assoc prof, 66-70, PROF GEOG, UNIV MICH, ANN ARBOR, 70- Concurrent Pos: Swedish State fel, Univ Uppsala, 65-66, docent, 68; res assoc, Ctr Pop Planning, Univ Mich, 68, col prof, 73; dir, Regional Res Assocs, Inc, 68; res assoc, Appl Sci Res Corp, Thailand, 70; consult, Mekong Coord Comt, Bangkok, 73. Mem: AAAS; Asn Am Geogrs; Regional Sci Asn; Int Union Sci Study Pop; Philos Sci Asn. Res: Social science epistemology; spatial model building. Mailing Add: Dept of Geog Univ of Mich Ann Arbor MI 48104

OLSSON, RAY ANDREW, b Livermore, Calif, Nov 30, 31; m 54; c 3. PHYSIOLOGY. Educ: George Washington Univ, MD, 56. Prof Exp: With US Army, 56-, resident internal med, George Washington Univ, 58-60, res internist, Dept Cardiorespiratory Dis, Walter Reed Army Inst Res, 61-63, dir div clin res, SEATO Med Res Lab, Bangkok, 63-66, asst chief dept cardiorespiratory dis, Walter Reed Army Inst Res, 67-70, dep dir div med, 70-71, DIR DIV MED, WALTER REED ARMY INST RES, 71- Concurrent Pos: Dept Army liaison res, NIH cardiovasc B study sect, 68- Honors & Awards: Order of the White Elephant, Royal Thai Govt, 66. Mem: AAAS; Am Physiol Soc; Am Fedn Clin Res; Am Col Physicians; AMA. Res: Cardiovascular physiology; coronary circulation; cardiac metabolism. Mailing Add: Div of Med Walter Reed Army Inst of Res Washington DC 20012

OLSSON, RICHARD KEITH, b Newark, NJ, Mar 23, 31; m 57; c 3. GEOLOGY. Educ: Rutgers Univ, BS, 53, MS, 54; Princeton Univ, MA, 56, PhD(geol), 58. Prof Exp: Asst field geologist, Mobil Oil Co Div, Socony Mobil Oil Co, Inc, 53; explor geologist, Pinon Uranium Co, 55; asst geol, Rutgers Univ, 53-54 & Princeton Univ, 54-56; from instr to assoc prof, 57-71, PROF GEOL, RUTGERS UNIV, 71- Concurrent Pos: Consult, Humble Oil Co, 63-66. Mem: Geol Soc Am; Soc Econ Paleontologists & Mineralogists; Am Asn Petrol Geol; Paleont Soc. Res: Stratigraphy; micropaleontology, sedimentalogy; paleoecology; Late Cretaceous and Tertiary section of the geologic rock column; evolution and biostratigraphy of planktonic foraminifera. Mailing Add: Dept of Geol Rutgers Univ New Brunswick NJ 08903

OLSTAD, ROGER GALE, b Minneapolis, Minn, Jan 16, 34; m 55; c 2. SCIENCE EDUCATION. Educ: Univ Minn, BS, 55, MA, 59, PhD(educ), 63. Prof Exp: Instr high sch, Minn, 55-56; instr biol educ, Univ Minn, 56-63; asst prof sci educ, Univ Ill, 63-64; from asst prof to assoc prof, 64-71, PROF SCI EDUC, UNIV WASH, 71-, ASSOC DEAN GRAD STUDIES, 68- Mem: Fel AAAS; Nat Asn Res Sci Teaching; Nat Sci Teachers Asn; Nat Asn Biol Teachers. Res: Science curriculum development, organization and evaluation. Mailing Add: Col of Educ Univ of Wash Seattle WA 98195

OLTENACU, ELIZABETH ALLISON BRANFORD, b Pontefract, Eng, Sept 5, 47; m 73. ANIMAL SCIENCE. Educ: Univ Edinburgh, BSc, 70; Univ Minn, St Paul, MS, 72, PhD(animal sci), 74. Prof Exp: RES ASSOC ANIMAL SCI, CORNELL UNIV, 74- Mem: Brit Soc Animal Prod; Am Genetic Asn. Res: Solving animal breeding problems, particularly evaluation of reproductive performance in dairy cattle, by the use of computer techniques; the interrelationships of reproductive physiology, behavior and genetics. Mailing Add: Dept of Animal Sci 202 Morrison Hall Cornell Univ Ithaca NY 14853

OLTHOF, THOEDORUS HENDRIKUS ANTONIUS, b Deventer, Netherlands, July 15, 34; Can citizen. NEMATOLOGY, PLANT PATHOLOGY. Educ: State Col Trop Agr, Netherlands, dipl, 55; McGill Univ, BSc, 58, PhD(plant path), 63. Prof Exp: Res officer potatoes, 62-64, res officer nematol, 64-73, RES SCIENTIST, RES STA, CAN DEPT AGR, 73- Mem: Am Phytopath Soc; Can Phytopath Soc; Soc Nematol; Soc Europ Nematol; Asn Trop Biol. Res: Nematode-fungus interactions; field-ecology of root-lesion nematodes in tobacco; crop loss assessment; population dynamics. Mailing Add: Res Sta Can Dept Agr Box 185 Vineland Station ON Can

OLTJEN, ROBERT RAYMOND, b Robinson, Kans, Jan 13, 32; m 56; c 3. RUMINANT NUTRITION. Educ: Kans State Univ, BS, 50, MS, 58; Okla State Univ, PhD(animal nutrit), 61. Prof Exp: Res animal husbandman, Beef Cattle Res Br, Animal Sci Res Div, 62-69, leader nutrit invest, 69-72, CHIEF, RUMINANT NUTRIT LAB, AGR RES SERV, USDA, 72- Concurrent Pos: Mem comt animal nutrit, Nat Res Coun, 72-76. Honors & Awards: Nutrit Res Award, Am Feed Mfrs Asn, 71; Presidential Citation, 72; Cert Animal Scientist, Am Soc Animal Sci, 75. Mem: Am Inst Nutrit; Am Soc Animal Sci; Am Dairy Sci Asn; Coun Agr Sci & Technol. Res: Digestion and metabolism of various nitrogen and energy sources by ruminants; finishing diets for beef cattle; research on cellulose wastes and non-protein nitrogen for ruminants. Mailing Add: Agr Res Serv USDA Beltsville MD 20705

OLTZ, DONALD FREDERICK, b Duluth, Minn, Aug 20, 40; m 62; c 2. PALYNOLOGY. Educ: Alma Col Mich, BS, 62; Univ Minn, PhD(bot), 68. Prof Exp: Asst prof biol & geol, Alma Col Mich, 67-72; PALYNOLOGIST & HEAD PALYNOLOGY LAB, TEXACO INC, 72- Mem: AAAS; Am Inst Biol Sci; Int Orgn Paleobot; Am Asn Stratig Palynologists; Soc Econ Paleontologists & Mineralogists (treas, 75-76). Res: Palynology and paleobotany of Cretaceous and early Tertiary sediments; application of computerized techniques in paleoecology and correlation. Mailing Add: Palynology Lab Texaco Inc 3350 Wilshire Blvd Los Angeles CA 90010

OLUM, PAUL, b Binghamton, NY, Aug 16, 18; m 42; c 3. MATHEMATICS. Educ: Harvard Univ, AB, 40; Princeton Univ, AM, 42; Harvard Univ, PhD(math), 47. Prof Exp: Theoret physicist, Manhattan Proj, Princeton Univ, 41-42 & Los Alamos Sci Lab, 43-45; Jewett fel, Harvard Univ, 47-48 & Inst Advan Study, 48-49; from asst prof to prof math, Cornell Univ, 49-74, chmn dept, 63-66; DEAN, COL NATURAL SCI, UNIV TEX, AUSTIN, 74- Concurrent Pos: Mem, Inst Advan Study, 55-56 & Nat Res Coun adv comt, Off Ordn Res, 58-61; vis prof, Univ Paris & Hebrew Univ, Israel, 62-63; NSF sr res fel, Stanford Univ, 66-67; vis prof, Univ Wash, 70-71; mem bd trustees, Cornell Univ, 71-75. Mem: Am Math Soc; Math Asn Am. Res: Algebraic topology. Mailing Add: Col of Natural Sci Univ of Tex Austin TX 78712

OLVER, FRANK WILLIAM JOHN, b Croydon, Eng, Dec 15, 24; m 48; c 3. APPLIED MATHEMATICS. Educ: Univ London, BSc, 45, MSc, 48, DSc(math anal), 61. Prof Exp: Exp officer numerical anal, Nautical Almanac Off, Eng, 44-45; sr prin sci officer, Nat Phys Lab, 45-61, head numerical methods sect, 53-61; mathematician, Nat Bur Stand, DC, 61-69; RES PROF, INST FLUID DYNAMICS & APPL MATH, UNIV MD, 69- Concurrent Pos: Ed, J, Soc Indust & Appl Math, 64-; assoc ed, J Res, Nat Bur Stand, 66-, consult, 69- Mem: Am Math Soc; Soc Indust & Appl Math; Math Asn Am; fel Inst Math & Applns UK. Res: Asymptotics; numerical analysis; special functions. Mailing Add: Inst Fluid Dynam & Appl Math Univ of Md College Park MD 20742

OLVER, JOHN WALTER, b Honesdale, Pa, Sept 3, 36; m 59. ANALYTICAL CHEMISTRY. Educ: Rensselaer Polytech Inst, BS, 55; Tufts Univ, MS, 56; Mass Inst Technol, PhD(voltametric studies), 61. Prof Exp: Actg head dept chem, Franklin Inst, 56-58; instr chem, Mass Inst Technol, 61-62; ASST PROF CHEM, UNIV MASS, AMHERST, 62- Mem: Am Chem Soc. Res: Electrochemical methods of analysis; chemical separations. Mailing Add: Dept of Chem Univ of Mass Amherst MA 01002

OLYMPIA, PEDRO LIM, JR, b Mindoro, Philippines, June 29, 41. CHEMISTRY, CHEMICAL PHYSICS. Educ: Univ Philippines, BS, 60; Villanova Univ, MS, 65; Univ Fla, PhD(chem physics), 68. Prof Exp: Res & develop chemist, San Miguel Corp, 61-63; NSF fel, Tufts Univ, 68-69, AEC fel, 69-70; US Air Force fel, Univ Ga, 70-71; ASSOC PROF CHEM, WASH TECH INST, 71- Mem: Am Chem Soc. Res: Quantum chemistry and physics; atomic and molecular structure; density matrix theory; computers in chemistry and physics. Mailing Add: Wash Tech Inst 4100 Connecticut Ave Washington DC 20008

OLYNK, PAUL, b Ymir, BC, Aug 5, 18; m 71; c 4. CHEMISTRY. Educ: McGill Univ, BSc, 39; Univ Toronto, PhD(chem), 44. Prof Exp: Qual control supvr radar parts, Res Enterprises Ltd, 44-45; res chemist biol mat, Fine Chem of Can Ltd, 45-46; instr chem, Univ Rochester, 46-48; res assoc antibiotic plants, Babies & Children's Hosp, 48-50, res assoc fats & fat metab, 50-52; res chemist paint formulation, Sherwin-Williams Co, 52-57 & Glidden Co, 57-59; assoc dir inst urban studies & dir div environ sci, 69-71, ASSOC PROF CHEM, CLEVELAND STATE UNIV, 59- Mem: Am Chem Soc; Fedn Am Scientists; Fed Water Pollution Control Asn; Air Pollution Control Asn; Int Asn Great Lakes Res. Res: Antibiotics from plant; radiocarbon C-14 in proteins and fats; fat metabolism; water and air pollution; sediment and environmental chemistry. Mailing Add: Dept of Chem Cleveland State Univ Cleveland OH 44115

OMACHI, AKIRA, b Sacramento, Calif, Sept 10, 22; m 51; c 2. PHYSIOLOGY. Educ: Univ Buffalo, BA, 44; Univ Minn, MS, 48, PhD(physiol), 50. Prof Exp: Instr physiol, Univ Minn, 48-49; from instr to asst prof, Med SCh, Loyola Univ, Ill, 49-57; from asst prof to assoc prof, 57-69, PROF PHYSIOL, COL MED, UNIV ILL MED CTR, 69- Mem: AAAS; Am Physiol Soc; Am Asn Univ Profs; Biophys Soc; Sigma Xi. Res: Electrolyte metabolism; cell physiology and metabolism; membrane transport. Mailing Add: Dept of Physiol Univ of Ill Med Ctr Box 6998 Chicago IL 60680

O'MAHONY, JOHN PATRICK, b Cork, Ireland, Feb 8, 41; m 68; c 3. FOOD SCIENCE. Educ: Nat Univ Ireland, BS, 63, MS, 64; Cornell Univ, PhD(food sci), 70. Prof Exp: Microbiologist & lectr food microbiol, West of Scotland Agr Col, 64-66; res asst food sci, Cornell Univ, 66-70; group leader explor food sci, John Stuart Res Labs, 70-74, sr group leader, 74-75, SECT MGR PROTEIN TECHNOL, QUAKER OATS CO, 75- Mem: Inst Food Technol; Am Chem Soc; Am Oil Chemists Soc. Res: New protein sources and applications; biochemistry of foods; food lipids; interactions between food constituents; effect of processing on nutritional value of foods; dietary fiber; foods and relationships to disease. Mailing Add: Quaker Oats Co WMain St Barrington IL 60010

O'MALLEY, BERT W, b Pittsburgh, Pa, Dec 19, 36; m 60; c 4. ENDOCRINOLOGY, MOLECULAR BIOLOGY. Educ: Univ Pittsburgh, BS, 59, MD, 63. Prof Exp: From intern to resident, Med Ctr, Duke Univ, 63-65; clin assoc, Endocrinol Br, Nat Cancer Inst, 65-67, sr investr, 67-68, head molecular biol sect, 68-69; prof reproductive physiol, Vanderbilt Chair, Sch Med, Vanderbilt Univ, 69-72; PROF CELL BIOL & CHMN DEPT, BAYLOR COL MED, 72- Mem: Endocrine Soc; Soc Study Reproduction. Res: Reproductive physiology; mechanism of hormone action; hormone-mediated cell differentiation; biochemistry of the animal cell. Mailing Add: Dept of Cell Biol Baylor Col of Med Houston TX 77025

O'MALLEY, EDWARD PAUL, b Hudson, NY, May 30, 26. PSYCHIATRY, PHARMACOLOGY. Educ: St John's Univ, NY, BS, 49; Loyola Univ, Ill, MS, 53, PhD(pharmacol), 54; State Univ NY Downstate Med Ctr, MD, 58; Am Bd Psychiat & Neurol, dipl, 65. Prof Exp: Intern, St Vincent's Hosp, New York, 58-59; resident psychiat, Bronx Vet Admin Hosp & NY State Psychiat Inst, New York, 59-62; sch psychiatrist, Bur Child Guid, NY Bd Educ, 63-67; med dir, West Nassau Ment Health Ctr, 67-68; dir, Community Ment Health Serv Suffolk County, 68-72; comnr, Orange County Ment Health Serv, NY, 72-74; DIR CLIN RES, AM HOECHST PHARMACEUT CO, NJ, 74-; ASST CLIN PROF PSYCHIAT, NJ MED SCH, 74- Concurrent Pos: Asst clin prof, State Univ NY Stony Brook; consult psychiatrist, Riverside Hosp, Bronx, NY, 62-63; visitor, New York Dept Corrections, 62-66; consult, Cath Charities Guid Inst, 62-; asst attend, St Vincent's Hosp, 62-; clin attend, St Francis Hosp, Bronx, 63-; courtesy attend psychiatrist, Arden Hill Hosp, Goshen, NY; assoc attend psychiatrist, Bronx Lebanon Hosp & St Luke's Hosp. Mem: AMA; fel Am Psychiat Asn. Res: Psychopharmacological and psychiatric research; pharmacological, biochemical and physiological studies at the basic science level. Mailing Add: 360 Cent Park West New York NY 10025

O'MALLEY, JAMES JOSEPH, b Philadelphia, Pa, Sept 17, 40; m 64; c 3. POLYMER CHEMISTRY, POLYMER PHYSICS. Educ: Villanova Univ, BS, 62; State Univ NY Col Forestry, Syracuse Univ, MS, 64, PhD(phys chem), 67. Prof Exp: From assoc scientist, to sr scientist, 67-74, MGR POLYMER RES, XEROX CORP, 74- Mem: AAAS; Am Chem Soc; Am Phys Soc. Res: Synthesis and characterization of block and graft copolymers; interactions of synthetic and bio-polymers; surface and electrical properties of polymers. Mailing Add: Res Lab Xerox Corp Webster NY 14580

O'MALLEY, JOHN ANTHONY, physical chemistry, clinical chemistry, see 12th edition

O'MALLEY, JOSEPH PAUL, b Boston, Mass, Mar 22, 30; m 53; c 6. MEDICINE. Educ: Col of the Holy Cross, BS, 51; Harvard Univ, MD, 56. Prof Exp: Intern, USPHS Hosp, Brighton, Mass, 56-57; res investr, NIH, 57-66, asst to dir invest new drugs, Div Biol Stand, 67-74, DIR INVEST NEW DRUGS STAFF, BUR BIOLOGICS, FOOD & DRUG ADMIN, 74- Res: Cultivating the etiologic agent of human hepatitis. Mailing Add: Bur of Biologics Food & Drug Admin Bethesda MD 20014

O'MALLEY, MARY THERESE, b Chicago, Ill. MATHEMATICS. Educ: Univ Nebr, BA, 55; Cath Univ Am, MA, 63; Columbia Univ, PhD(math), 71. Prof Exp: ASST PROF MATH, ROSARY COL, 68- Mem: Math Asn Am; Am Math Soc; Sigma Xi. Res: Algebra. Mailing Add: Dept of Math Rosary Col River Forest IL 60305

O'MALLEY, MATTHEW JOSEPH, b Miami, Fla, Oct 21, 40; m 64; c 2. ALGEBRA. Educ: Spring Hill Col, BS, 62; Fla State Univ, MS, 64, PhD(math), 67. Prof Exp: Mathematician, Manned Spacecraft Ctr, NASA, 67-71; asst prof, 71-74, ASSOC PROF MATH, UNIV HOUSTON, 74- Mem: Am Math Soc. Res: Cummutative ring theory; pattern recognition. Mailing Add: Dept of Math Univ of Houston Houston TX 77004

O'MALLEY, RICHARD JOHN, b Jersey City, NJ, Dec 20, 45; m 70. MATHEMATICAL ANALYSIS. Educ: Seton Hall Univ, BS, 67; Purdue Univ, MS, 69, PhD(math), 72. Prof Exp: ASST PROF MATH, UNIV WIS-MILWAUKEE, 72- Concurrent Pos: Univ Wis-Milwaukee res grant, 75. Mem: Am Math Soc. Res: Analysis of functions of a real variable, particularly approximate continuity and density properties. Mailing Add: Dept of Math Univ of Wis Milwaukee WI 53201

O'MALLEY, ROBERT EDMUND, JR, b Rochester, NH, May 23, 39; m 68; c 3. APPLIED MATHEMATICS. Educ: Univ NH, BS, 60, MS, 61; Stanford Univ, PhD(math), 66. Prof Exp: Asst prof math, Univ NC, Chapel Hill, 66-68; assoc prof, NY Univ, 68-73, PROF MATH, UNIV ARIZ, 73- Concurrent Pos: Vis mem, Courant Inst Math Sci, NY Univ, 66-67; vis asst prof, Math Res Ctr, Univ Wis, 67-68; sr vis fel, Univ Edinburgh, 71-72. Mem: Am Math Soc; Soc Indust & Appl Math. Res: Singular perturbation problems; asymptotic expansions; differential equations. Mailing Add: Dept of Math Univ of Ariz Tucson AZ 85721

O'MALLEY, ROBERT FRANCIS, b Framingham, Mass, Apr 2, 18; m 44; c 5. FLUORINE CHEMISTRY. Educ: Boston Col, BS, 40, MS, 48; Mass Inst Technol, PhD(inorg chem), 61. Prof Exp: From instr to asst prof, 47-65, admin asst, 52-56, chmn dept chem, 71-74, ASSOC PROF INORG CHEM, BOSTON COL, 65- Concurrent Pos: Vis prof, Harvard Univ, 65. Mem: AAAS; Am Chem Soc; Electrochem Soc. Res: Electrochemical and oxidative fluorination of aromatic compounds. Mailing Add: Dept of Chem Boston Col Chestnut Hill MA 02167

O'MALLEY, THOMAS FRANCIS, b New York, NY, Nov 13, 28; m 62; c 2. PHYSICS. Educ: Bellarmine Col, NY, BA, 53, PhL, 54; NY Univ, PhD(physics), 61. Prof Exp: Res assoc physics, NY Univ, 61-63; vis fel, Joint Inst Lab Astrophys, Univ Colo, 63-64; mem tech staff, Gen Res Corp, Calif, 64-70; assoc prof physics, Univ Conn, 70-72; RES DIR, DOLPHIN PROJ, MIAMI, FL, 73- Concurrent Pos: Vis res fel, Queen's Univ, Belfast, 68-69. Mem: AAAS; fel Am Phys Soc. Res Quantum mechanical collision theory; collisions and reactions among slow electrons, atoms, molecules, ions; atomic and molecular quasistationary states; recombination, attachment and reaction rates. Mailing Add: 2000 Center St Number 1022 Berkeley CA 94704

O'MALLEY, WILLIAM JOSEPH, pharmaceutical chemistry, see 12th edition

OMAN, PAUL WILSON, b Garnett, Kans, Feb 22, 08; m 31; c 6. ENTOMOLOGY. Educ: Univ Kans, AB, 30, AM, 35; George Washington Univ, PhD(entom), 41. Prof Exp: Jr entomologist bur entom & plant quarantine, USDA, 30-34, from asst entomologist to prin entomologist, 34-59, chief insect identification & parasite introd res br, 59-60, dir res & tech progs div, Far East Regional Res Off, Agr Res Serv, 60-63, asst dir entom res div, 63-67; actg head dept, 73-74, prof entom, 67-75, EMER PROF ENTOM, ORE STATE UNIV, 75- Concurrent Pos: Head insect identification & parasite introd sect, Entom Res Br, USDA, 54-59. Mem: Fel Entom Soc Am (2nd vpres, 48, pres-elect, 58, pres, 59); fel Royal Entom Soc London. Res: Insect systematics; biological control of insects and weeds; medical entomology. Mailing Add: Dept of Entom Ore State Univ Corvallis OR 97331

OMAN, ROBERT MILTON, b Easton, Mass, Aug 12, 34; m 62; c 1. ELECTRON PHYSICS. Educ: Northeastern Univ, BS, 57; Brown Univ, ScM, 60, PhD(physics), 63. Prof Exp: From res asst to res assoc physics, Brown Univ, 58-63; res scientist, United Aircraft Corp, Conn, 63-64; tech specialist, Litton Systs, Inc, Minn, 64-66; sr res scientist, Norton Res Corp, Mass, 66-70; ASSOC PROF PHYSICS, N SHORE COMMUNITY COL, 71- Mem: Am Vacuum Soc. Res: Theoretical physical electronics; solid state and experimental solid state physics; electron mirror microscopy; magnetron gauges; electron transport in semiconductors; applied mathematics. Mailing Add: 204 Fair Oaks Park Needham MA 02192

O'MARA, JAMES HERBERT, b St Clair, Mich, Oct 11, 36. PHYSICAL CHEMISTRY. Educ: George Washington Univ, BS, 57, MS, 61; Duke Univ, PhD(phys chem), 68. Prof Exp: Chemist, Polymers Div, Nat Bur Stand, 57-63; SR CHEMIST, ROHM AND HAAS CO, 67- Mem: AAAS; Am Chem Soc; Am Soc Testing & Mat; Soc Automotive Engrs. Res: Physical chemistry of polymers in dilute solution and bulk; chemistry of oil additives. Mailing Add: Rohm and Haas Co Spring House PA 19477

O'MARA, JOSEPH GEORGE, b Boston, Mass, Sept 7, 11; m 38. GENETICS. Educ: Univ Mass, BS, 33; Harvard Univ, MS, 34, PhD(biol), 36. Prof Exp: Geneticist, Univ Mo, 36-42, Emergency Rubber Proj, 43-44 & Univ Mo, 44-50; prof genetics, Iowa State Univ, 50-65, chmn dept, 59-65; prof bot & head dept, 65-67, PROF BIOL & HEAD DEPT, PA STATE UNIV, 67- Mem: Genetics Soc Am; Soc Study Evolution. Res: Theoretical and applied genetics; human heredity; mathematical genetics; cytology; cytogenetics. Mailing Add: Dept of Biol Pa Sta Univ 208 Life Sci Bldg University Park PA 16802

O'MARA, MICHAEL MARTIN, b Lackawanna, NY, Jan 24, 42; c 1. ORGANIC ANALYTICAL CHEMISTRY. Educ: Canisius Col, BS, 64; Univ Cincinnati, PhD(chem), 68. Prof Exp: Develop scientist, B F Goodrich Co, 68-75, RES & DEVELOP GROUP LEADER, B F GOODRICH CHEM CO, 75- Mem: Am Chem Soc. Res: Thermal degradation of polymers; flammability, smoke and toxic gas generation characteristics of polymers; gas chromatographic-mass spectrometric analyses. Mailing Add: B F Goodrich Chem Co PO Box 122 Avon Lake OH 44012

O'MARA, ROBERT E, b Flushing, NY, Dec 8, 33; m 64; c 3. RADIOLOGY, NUCLEAR MEDICINE. Educ: Univ Rochester, BA, 55; Albany Med Col, MD, 59; Am Bd Radiol, dipl, 67. Prof Exp: Intern, St Louis Hosp, Wash Univ, 59-60, resident radiol, St Vincent's Hosp, New York, 63-66; dir nuclear med, Sch Med, Univ Ariz, 71-75; actg chief nuclear med serv, Vet Admin Hosp, Tucson, 71-75; PROF RADIOL & CHIEF DIV NUCLEAR MED, DEPT RADIOL, UNIV ROCHESTER, SCH OF MED & DENT, 75- Concurrent Pos: NIH fel nuclear med, State Univ NY Upstate Med Ctr, 66-67 & clin nuclear med, 67-71. Mem: Am Col Nuclear Physicians; Soc Nuclear Med; Radiol Soc NAm; Am Col Radiol. Res: Diagnostic applications of radionuclides and development of radiopharmaceuticals. Mailing Add: Div of Nuclear Med 601 Elmwood Ave Univ of Rochester Med Ctr Rochester NY 14642

O'MARY, CLAYTON CORDICE, b Eldridge, Ala, Aug 10, 20; m 43; c 3. ANIMAL SCIENCE. Educ: Berry Col, BSA, 43; Auburn Univ, BS, 46, MS, 49; Univ Wis, PhD(animal husb, genetics), 52. Prof Exp: Teacher, High Sch, Ala, 47; instr animal husb, Auburn Univ, 47-49; asst prof, Univ Ga, 52-57; assoc prof animal sci, 57-65, PROF ANIMAL SCI, WASH STATE UNIV, 65- Concurrent Pos: Staff mem, Oak Ridge Assoc Univs, 66-67. Mem: AAAS; Am Soc Animal Sci; Am Genetic Asn; NY Acad Sci. Res: Hormones in reproduction and growth of beef cattle and sheep; beef production especially performance testing and breeding problems; physiology. Mailing Add: Dept of Animal Sci Wash State Univ Pullman WA 99163

OMATA, ROBERT ROKURO, b Hanford, Calif, Nov 3, 20; m 48; c 3. BACTERIOLOGY. Educ: Univ Calif, AB, 44; Univ Minn, MS, 46, PhD(bact), 49. Prof Exp: Asst bact & immunol, Univ Minn, 45-47; from asst bacteriologist to bacteriologist, Nat Inst Dent Res, 53-60, mem staff div res grants, 60-63, scientist adminr, Off Int Res, 63-68, scientist adminr, Off Int Res, Pac Off, Tokyo, 64-67, head int fels sect, Scholars & Fels Prog Br, Fogarty Int Ctr, NIH, 68-74, INT PROG SPECIALIST, OFF INT AFFAIRS, NAT CANCER INST, 74- Concurrent Pos: Am Dent Asn res fel, Nat Inst Dent Res, 49-53. Mem: AAAS; Am Soc Microbiol. Res: Oral microbiology; bacterial physiology and biochemistry; physiology of oral fusobacteria, spirochetes and clostridia; biochemistry and nutrition; science administration; general medical sciences. Mailing Add: Off of Int Affairs Nat Cancer Inst Bethesda MD 20014

O'MATHUNA, DIARMUID, b Corcaigh, Ireland, Mar 22, 34. APPLIED MATHEMATICS. Educ: Nat Univ Ireland, BSc, 55, MSc, 57; Mass Inst Technol, PhD(math), 62. Prof Exp: Temp mem math & mech, Courant Inst Math Sci, NY Univ, 62-64; instr math biol, Harvard Med Sch, 64-66; aerospace engr, Electronics Res Ctr, NASA, 66-70; MATHEMATICIAN, TRANSP SYSTS CTR, DEPT TRANSP, 70- Res: Continuum mechanics; celestial mechanics; immunogenetics; biochemical kinetics; singular perturbation theory; transportation systems. Mailing Add: Transp Systs Ctr Kendall Sq Cambridge MA 02142

O'MEARA, DESMOND, b Daingean, Ireland, Oct 12, 26; m 53; c 6. ORGANIC CHEMISTRY, STATISTICS. Educ: Univ Dublin, BA & BSc, 49, MSc, 52; St Andrews Univ, PhD(chem), 57. Prof Exp: Asst, Queen's Col, St Andrews Univ, 52-55; org chemist, Brit Rayon Res Asn, Eng, 55-58; actg group leader pulping, Res Div, Columbia Cellulose Co Ltd, BC, 58, group leader, 58-59, group leader bleaching, 59-62; group leader chem, Res & Develop Dept, Consol Paper Corp Ltd, Que, 62-67, group leader, Statist Serv, 63-67; dir opers res, 67-74, DIR MGT SCI, CONSOL-BATHURST LTD, 74- Mem: Tech Asn Pulp & Paper Indust; Am Soc Qual Control; Chem Inst Can; Can Pulp & Paper Asn; Can Oper Res Soc. Res: Sugar nitrates; mechanism of degradation of cellulose and oxycellulose; wood pulp cooking and bleaching; statistical experimental design; operations research; computer applications. Mailing Add: Consol-Bathurst Ltd PO Box 69 800 Dorchester Blvd W Montreal PQ Can

O'MEARA, FRANCIS EDMUND, b New York, NY, Aug 8, 23; m 44; c 6. EXPERIMENTAL PHYSICS. Educ: Univ Mich, BS, 47. Prof Exp: Asst observer, US Weather Bur, 42-43; mass spectrometer technician, Oak Ridge Nat Lab, 47-48, physicist, NEPA Proj, 48-51; sr physicist, Res Lab, Bendix Aviation Corp, 51-54; div chief opers anal, US Air Force, 54-57, mem staff weapons systs eval group, 57-58; chief reactor eng sect, US AEC, 58-59; proj scientist, Booz-Allen Appl Res, 59-61; div chief opers anal, US Air Force, 61-63; mgr eval, Douglas Aircraft Co, Inc, 63-71; DIR PROGS & BUDGET, NAT OCEANIC & ATMOSPHERIC ADMIN, 72- Mem: Fel AAAS; assoc fel Am Inst Aeronaut & Astronaut; Am Phys Soc; NY Acad Sci. Res: Nuclear weapons and reactors; operations research. Mailing Add: 203 Mannakee St Rockville MD 20850

O'MEARA, GEORGE FRANCIS, b Lowell, Mass, Sept 9, 41; m 64; c 4. GENETICS. Educ: Univ Notre Dame, BS, 64, MS, 67, PhD(biol), 69. Prof Exp: RES ENTOMOLOGIST, FLA DIV HEALTH, ENTOM RES CTR, 69- Mem: AAAS; Entom Soc Am; Am Mosquito Control Asn; Am Genetic Asn. Res: Ecological genetics of autogenous reproduction in mosquitoes. Mailing Add: Entom Res Ctr PO Box 520 Vero Beach FL 32960

O'MEARA, JOHN PIERCE, b Lansing, Mich, Apr 25, 23; div; c 4. FOOD TECHNOLOGY. Educ: Mass Inst Technol, SB & SM, 49. Prof Exp: Res assoc radiol, Med Sch, Univ Okla, 50-51; mgr chem physics, Southwest Res Inst, 51-56; dir radiation physics lab, Continental Can Co, Inc, 56-59; assoc dir adv res dept, Sunbeam Corp, 59-62; dir tech develop, Vacudyne Corp, 62-64; sales mgr food processing systs, Cryodry Corp, 64-71; TECH CONSULT FOOD INDUST, 71- Mem: Am Chem Soc; Inst Food Technologists; Int Microwave Power Inst. Res: Radiation chemistry; radioisotope uses; food irradiation; nuclear magnetic and electron spin resonance; research management; freeze drying; microwave processing. Mailing Add: 34 Ivy Dr Orinda CA 94563

O'MEARA, O TIMOTHY, b Cape Town, SAfrica, Jan 29, 28; m 53; c 5. MATHEMATICS. Educ: Univ Cape Town, BSc, 47, MSc, 48; Princeton Univ, PhD(math), 53. Prof Exp: Lectr math, Univ Otago, NZ, 53-56; lectr math, Princeton Univ, 57, asst prof, 58-62; mem, Inst Advan Study, 57-58; head dept, 65-66, chmn dept, 68-72, PROF MATH, UNIV NOTRE DAME, 62- Concurrent Pos: Sloan fel, 60-63; vis prof, Calif Inst Technol, 67-68; mem adv panel math sci, NSF, 74- Mem: Am Math Soc. Res: Algebra; linear groups; number theory; quadratic forms. Mailing Add: Dept of Math Univ of Notre Dame Notre Dame IN 46556

OMENN, GILBERT STANLEY, b Chester, Pa, Aug 30, 41; m 67; c 2. MEDICAL GENETICS, HUMAN GENETICS. Educ: Princeton Univ, AB, 61; Harvard Med Sch, MD, 65; Univ Wash, PhD, 72. Prof Exp: Intern & asst resident, Mass Gen Hosp, 65-67; res assoc protein chem, Nat Inst Arthritis & Metab Dis, 67-69; asst prof, 71-73, ASSOC PROF MED, UNIV WASH, 73- Concurrent Pos: Teaching fel internal med, Mass Gen Hosp, Harvard Univ, 65-67; USPHS spec fel & univ sr fel med genetics, Univ Wash, 69-71, Nat Genetics Found fel, 61-72, Nat Inst Gen Med Sci res career develop award, 72-; White House fel, AEC, 73-74; attend physician, Univ Hosp, Harborview Med Ctr & Children's Orthop Hosp, Seattle, Wash, 71-; dir, Robert Wood Johnson Found Clin Scholars Prog, 74-; consult, AEC & Fed Energy Admin, 74-; assoc ed, Am J Human Genetics. Mem: Am Soc Human Genetics; Am Soc Neurochem; Soc Study Social Biol; Am Fedn Clin Res; Soc Biol Psychiat. Res: Enzyme variation in human brain; genetics of behavioral traits and disorders; prenatal diagnosis; pharmacogenetics. Mailing Add: Div of Med Genetics Univ of Wash Seattle WA 98195

OMER, GEORGE ELBERT, JR, b Kansas City, Kans, Dec 23, 22; m 49; c 2. ORTHOPEDIC SURGERY. Educ: Ft Hays Kans State Col, BA, 44; Univ Kans, MD, 50; Baylor Univ, MS, 55; Am Bd Orthop Surg, dipl. Prof Exp: Chief surg serv, Irwin Army Hosp, Ft Riley, Kans, 57-59, consult orthop surg, Eighth US Army, Korea, 59-60, chief hand surg, Fitzsimons Army Hosp, Denver, Colo, 62-64, dir orthop path, Armed Forces Inst Path, Washington, DC, 64-65, chief hand surg ctr, Brooke Army Med Ctr, Ft Sam Houston, Tex, 66-70, chief orthop, 67-70; PROF ORTHOP, CHMN

DEPT & CHMN DIV HAND SURG, SCH MED, UNIV N MEX, 70- Concurrent Pos: Instr, Univ Kans, 46-47; asst clin prof, Sch Med, Univ Colo, 61-63; assoc clin prof, Georgetown Univ, 64-65, Univ Tex Med Sch San Antonio, 66-70 & Univ Tex Med Br Galveston, 68-70; consult orthop & hand surg, numerous civilian and mil hosps, 62- Indian Health, USPHS, 66- & orthop, Surgeon Gen, US Army, 70- Mem: Am Acad Orthop Surg; Am Asn Surg of Trauma; Am Col Surg; Am Orthop Asn; Am Soc Surg of Hand. Res: Hand surgery; peripheral nerve repair and sensibility; tendon transfers for reconstruction of nerve loss; evacuation of injuries and emergency room care; burn injuries. Mailing Add: Dept of Orthop Univ of NMex Health Sci Ctr Albuquerque NM 87131

OMER, GUY CLIFTON, JR, b Mankato, Kans, Mar 20, 12; m 42; c 2. PHYSICS. Educ: Univ Kans, BS, 36, MS, 37; Calif Inst Technol, PhD(physics), 47. Prof Exp: Instr, Univ Hawaii, 41-43; asst prof, Occidental Col, 47-48 & Univ Ore, 48-49; asst prof phys sci, physics & astron, Univ Chicago, 49-55; chmn dept phys sci, 67-72, PROF PHYSICS & ASTRON, UNIV FLA, 55- Concurrent Pos: Res fel, Calif Inst Technol, 47-48; vis prof, Univ Calif, Los Angeles, 64-65; sr vis fel, Inst Astron, Cambridge, Eng, 72; prof, Fla at Utrecht, Univ Utrecht, Holland, 73; sr vis fel, Dept Astrophys, Univ Oxford, 73. Mem: Am Phys Soc; Int Astron Union; Seismol Soc Am; Am Asn Physics Teachers; fel Royal Astron Soc. Res: Relativity; cosmology; general education. Mailing Add: Dept of Physics & Astron 217A SSRB Univ of Fla Gainesville FL 32601

OMID, AHMAD, b Abadeh, Iran, May 6, 31; m 57; c 2. AGRONOMY, PHYSIOLOGY. Educ: Calif State Polytech Col, BS, 55 & 57; Ore State Univ, MS, 62, PhD(farm crops), 64. Prof Exp: PLANT PHYSIOL GROUP LEADER, CHEVRON CHEM CO, 62-, SR PLANT PHYSIOLOGIST, 70- Mem: AAAS; Weed Sci Soc Am. Res: Use of herbicides and growth regulators for control and modification of plant growth, including studies of mode of action and movement of these chemicals in plants and soils. Mailing Add: 2400 Cascade Dr Walnut Creek CA 94598

OMIDVAR, KAZEM, b Mashad, Iran, Dec 15, 26. ATOMIC PHYSICS, THEORETICAL ASTROPHYSICS. Educ: Univ Teheran, BS, 51; NY Univ, MS, 54, PhD(physics), 59. Prof Exp: Instr physics, Cooper Union, 54-56 & NY Univ, 56-57; lectr, Rutgers Univ, 57-58; instr, City Col New York, 58-59; assoc prof, Univ Teheran, 59-60; SR SCIENTIST, THEORET STUDIES GROUP, NASA/GODDARD SPACE FLIGHT CTR, 60- Concurrent Pos: Lectr, Univ Md, 61-64. Mem: Fel Am Phys Soc; Am Astron Soc. Res: Development of reliable formula for calculating ionization, excitation and charge exchange cross sections in electron-atom and ion-atom collisions. Mailing Add: Theoret Studies Group NASA/Goddard Space Flight Ctr Greenbelt MD 20771

OMIETANSKI, GEORGE MICHAEL, inorganic chemistry, see 12th edition

OMODT, GARY WILSON, b LaCrosse, Wis, July 30, 29; m 57; c 5. PHARMACEUTICAL CHEMISTRY. Educ: Univ Minn, BS, 53, PhD(pharmaceut chem), 59. Prof Exp: ASSOC PROF PHARMACEUT CHEM & HEAD DEPT, COL PHARM, SDAK STATE UNIV, 58- Mem: Am Pharmaceut Asn; Am Asn Cols Pharm; Sigma Xi. Res: Synthesis of organic medicinal agents. Mailing Add: Dept Pharmaceut Chem Col Pharm SDak State Univ Brookings SD 57007

OMOHUNDRO, ALLEN LLEWELLYN, b Norfolk, Va, Mar 25, 03. CHEMISTRY. Educ: Univ Pa, BS, 25; Univ Paris, MS, 26, PhD(chem), 28. Hon Degrees: PhD, Conn Col Pharm, 40. Prof Exp: Asst to pres & in chg labs & mfg, Frederick Stearns Co, 28-34; vpres & dir plant & labs, McKesson & Robbins, Inc, 34-62, vpres & dir sci div, 62-65; PRES, A L OMOHUNDRO & ASSOCS-CONSULTS, 65- Concurrent Pos: Trustee, Conn Col Pharm. Mem: Fel AAAS; Am Pharmaceut Asn; fel Am Inst Chem; NY Acad Sci; Soc Chem Indust. Res: Biochemistry; organic chemistry. Mailing Add: Minute Man Hill Westport CT 06880

OMOHUNDRO, RICHARD E, b Wellington, Kans, Sept 29, 14; m 36; c 3. EPIDEMIOLOGY, POULTRY PATHOLOGY. Educ: Kans State Univ, DVM, 37. Prof Exp: Dist vet, US Bur Animal Indust, 38-41, area vet, 46-47; regional supvr, Mex-Am Comn Foot & Mouth Dis Eradication, 47, chief, Inspection & Vaccine Div, 47-51; asst vet in charge animal dis eradication, Agr Res Serv, Ga, 51-52, vet in charge, Mo, 52-56, chief staff vet poultry dis, 56-59, sr staff vet, Animal Health Div, 59-68, asst dir div, 68-71, dir emergency prof, Animal & Plant Health Inspection Serv, 71-73, DIR ANIMAL NUTRIT & HEALTH DIV, AGR RES SERV, USDA, 73- Concurrent Pos: Coordr, Va-NC hog cholera task force, Animal Health Div, Agr Res Serv, 70-71; nat coordr, Venezuela Equine Encephalomyelitis Epizootic Control Prog, 71- & Nat Velogenic Newcastle Dis Eradication Prog, 72- Honors & Awards: Superior Serv Award, Secy Agr, USDA, 71. Mem: Am Vet Med Asn; Am Asn Avian Path. Res: Epizootiology; viral diseases. Mailing Add: Animal Nutrit & Health Div Agr Res Serv USDA Hyattsville MD 20782

O'MORCHOE, CHARLES CHRISTOPHER C, b Quetta, Pakistan, May 7, 31; m 55; c 2. ANATOMY, PHYSIOLOGY. Educ: Univ Dublin, BA, 53, MB, BCh & BA Obstet, 55, MA, 59, MD, 61, PhD(physiol), 69. Prof Exp: Intern med, surg, gynec & obstet, Halifax Gen Hosp, Eng, 55-57; lectr anat, Sch Med, Univ Dublin, 57-61; vis lectr physiol, Sch Med, Univ Md, 61-62; instr anat, Harvard Med Sch, 62-63; lectr, Sch Med, Univ Dublin, 63-65, assoc prof physiol, 66-68; from assoc prof to prof anat, Sch Med, Univ Md, Baltimore City, 71-73, actg chmn dept, 71-73; PROF ANAT & CHMN DEPT, STRITCH SCH MED, LOYOLA UNIV CHICAGO, 73- Concurrent Pos: Chmn, Anat Bd Med, 71-73. Mem: Am Soc Nephrology; Brit Physiol Soc; Int Soc Nephrology; Am Asn Anat; Int Soc Lymphology. Res: Anatomy and physiology of renal vascular circulation and renal lymphatic system; urinary cytology in reproductive endocrinology; medical education; histology; lymphology; nephrology. Mailing Add: Dept of Anat Loyola Univ Stritch Sch of Med Maywood IL 60153

O'MORCHOE, PATRICIA JEAN, b Halifax, Eng, Sept 15, 30; m 55; c 2. HISTOLOGY, CYTOLOGY. Educ: Univ Dublin, BA, 53, MB, BCh & BAO, 55, MA & MD, 66. Prof Exp: House officer obstet & gynec, surg & med, Brit Nat Health Serv, Halifax Gen Hosp, Eng, 55-57; jr lectr physiol & histol, Univ Dublin, 59-61; instr cytopath, Johns Hopkins Univ, 61-62; res assoc path & surg, Harvard Univ, 62-63; lectr physiol & histol, Univ Dublin, 63-68; instr cytopath, Johns Hopkins Univ, 68-70; asst prof anat & histol, Univ Md, Baltimore City, 70-74; ASSOC PROF PATH, STRITCH SCH MED, LOYOLA UNIV CHICAGO, 74- Concurrent Pos: Frank C Bressler Reserve Fund grant anat, Univ Md, Baltimore City, 71-72; NIH grant, 74-77; vis lectr, Johns Hopkins Hosp & Med Sch, Johns Hopkins Univ, 70-73, asst prof path, 73-74. Mem: Am Soc Cytol. Res: Exfoliative cytology of renal transplantation; function and morphology of the cells in renal lymph in unstimulated and phagocytically and antigenically stimulated dogs. Mailing Add: Dept of Path Loyola Univ Stritch Sch of Med Maywood IL 60153

OMRAN, ABDEL RAHIM, b Cairo, Egypt, Mar 29, 25; m 53; c 3. EPIDEMIOLOGY. Educ: Minufia Sch, Egypt, BS, 45; Cairo Univ, MD, 52, DPH, 54; Columbia Univ, MPH, 56, DrPH, 59; Trudeau Sch Tuberc, Nat Tuberc & Respiratory Dis Asn, cert, 59. Prof Exp: Lectr, Cairo Univ, 59-63; res scientist & clin assoc prof environ med, NY Univ, 63-66; assoc prof epidemiol, 66-71, PROF EPIDEMIOL, UNIV NC, CHAPEL HILL, 71- Concurrent Pos: WHO study fel health serv eastern & western Europe, 63; clin assoc prof, Univ Ky, 64-66; Ford Found consult, India, 69; coordr epidemiol studies in Asian countries, WHO, 69-, dir, Int Reference Ctr on Epidemiol of Human Reproduction, 71-; assoc dir pop epidemiol, Carolina Pop Ctr, 70- Honors & Awards: Sci Achievement Medal, Egyptian Govt. Mem: Royal Soc Health; Asn Teachers Prev Med; Soc Epidemiol Res; Am Thoracic Soc; fel Am Pub Health Asn. Res: Epidemiological studies of human reproduction in Asia and the Middle East; population problems and prospects in the Middle East; development of manual on community medicine for developing countries; health and disease patterns associated with demographic change; Muslim fertility. Mailing Add: Carolina Pop Ctr Chapel Hill NC 27514

OMTVEDT, IRVIN T, b Wis, June 12, 35; m 59; c 2. ANIMAL SCIENCE. Educ: Univ Wis, BS, 57; Okla State Univ, MS, 59, PhD(animal breeding), 61. Prof Exp: Performance testing field man, Univ Wis, 56-57; res asst animal sci, Okla State Univ, 58-61; from asst prof to assoc prof, Univ Minn, 62-64, exten specialist, 62-64; from assoc prof to prof animal sci, Okla State Univ, 64-73; assoc dir agr exp sta & asst dean agr, Auburn Univ, 73-75; PROF ANIMAL SCI & CHMN DEPT, UNIV NEBR, 75- Mem: Am Soc Animal Sci. Res: Swine breeding investigations involving crossbreeding and combining ability; effects of high ambient temperature on reproductive performance of swine. Mailing Add: Dept of Animal Sci Univ of Nebr Lincoln NE 68508

OMURA, YOSHIAKI, b Tomari, Japan, Mar 28, 34; m 62; c 3. CARDIOLOGY, ELECTROPHYSIOLOGY. Educ: Waseda Univ, Japan, BSc, 57; Yokohama City Univ, MD, 58; Columbia Univ, MedScD(surg, pharmacol), 65. Prof Exp: From intern to resident physician, Tokyo Univ Hosp, 58-59; intern, Norwalk Hosp, Conn, 59-60; resident physician, Francis Delafield Hosp, Columbia Univ, 61-64; instr surg & asst prof pharmacol, NY Med Col, 66-71; DIR MED RES, HEART DIS RES FOUND, 71- Concurrent Pos: John Polacheck Found res support grant, 67-70; Heart Dis Found res grant, 70-; vis res prof elec eng, Manhattan Col, 62-; consult, Columbia Univ, 65-66; part-time emergency room physician, Englewood Hosp, NJ, 65-66; consult, State Univ NY Downstate Med Ctr, 66; chmn sci dept, Children's Art & Sci Workshop, New York, 71-; dir res & bd trustees, Acupuncture Res Found & Am Found Acupuncture, 73-; consult & vis prof, Univ Paris, 73-74; ed-in-chief, Acupuncture & Electrotherapeut Res, Int J Pergamon Press, 74-; consult, Lincoln Hosp, New York, 74; chmn, Columbia Univ Affil & Community Med Comt Community Bd, Francis Delafield Hosp, 74-; chmn int stand comt, Int Kirlian Res Asn, 75- Mem: Fel Am Soc Chinese Med; NY Acad Sci; Am Soc Artificial Internal Organs; Int Fedn Med Electronics & Biol Eng. Res: Biophysics and pharmaco-electro-physiology of single cardiac cells in vivo and in vitro; shock, fibrillation, burn and electrolytes; surgery and medical electronics; inorganic and organic semiconductors with negative resistance characteristics; electrophysiology of nerve and muscle; rapidly changing electric and magnetic fields for pain control and neuromuscular rehabilitation; effects of acupuncture on cardiovascular and nervous systems; pharmacology; electronics. Mailing Add: Apt 8-I 800 Riverside Dr New York NY 10032

ONAK, THOMAS PHILIP, b Omaha, Nebr, July 30, 32; m 54. ORGANIC CHEMISTRY, INORGANIC CHEMISTRY. Educ: San Diego State Univ, AB, 54; Univ Calif, Berkeley, PhD(chem), 58. Prof Exp: Res chemist, Olin Mathieson Chem Corp, 57-59; from asst prof to assoc prof, 59-66, PROF CHEM, CALIF STATE UNIV, LOS ANGELES, 66- Concurrent Pos: Fulbright res fel, Cambridge Univ, 65-66; res career develop award, NIH, 73. Mem: Am Chem Soc; The Chem Soc. Res: Synthetic, structure and isotope studies on organoboron hydrides and carboranes; natural products. Mailing Add: Dept of Chem Calif State Univ Los Angeles CA 90032

O'NAN, MICHAEL ERNEST, b Ft Knox, Ky, Aug 9, 43. ALGEBRA. Educ: Stanford Univ, BS, 65; Princeton Univ, PhD(math), 69. Prof Exp: Asst prof, 70-73, ASSOC PROF MATH, RUTGERS UNIV, NEW BRUNSWICK, 73- Concurrent Pos: Vis asst prof math, Univ Chicago, 71-72; Sloan fel, 74. Mem: Am Math Soc. Res: Classification problems in the study of finite simple groups, those associated with doubly-transitive permutation groups. Mailing Add: Dept of Math Rutgers Univ New Brunswick NY 08903

ONCLEY, JOHN LAWRENCE, b Wheaton, Ill, Feb 14, 10; m 33; c 2. MOLECULAR BIOPHYSICS. Educ: Southwestern Col, Kans, AB, 29; Univ Wis, PhD, 32. Hon Degrees: MA, Harvard Univ, 46; DSc, Southwestern Col, Kans, 54. Prof Exp: Asst chem, Univ Wis, 29-31; Nat Res Coun fel, Mass Inst Technol, 32-34; instr, Univ Wis, 34-35 & Mass Inst Technol, 35-43; from asst prof to prof biol chem, Harvard Med Sch, 43-62; PROF CHEM & BIOL CHEM & DIR BIOPHYS RES DIV, INST SCI & TECHNOL, UNIV MICH, ANN ARBOR, 62- Concurrent Pos: Res assoc, Harvard Med Sch, 39-41, assoc, 41-43; Guggenheim & Fulbright fels, King's Col, Univ London, 53; mem comn plasma fractionation, Protein Found, Inc, 53-; mem comt elec insulation, Nat Res Coun. Honors & Awards: Award, Am Chem Soc, 43. Mem: Nat Acad Sci; Am Chem Soc; Biophys Soc; Am Acad Arts & Sci; NY Acad Sci. Res: Dielectric properties of gases, liquids and proteins; biophysical chemistry of protein systems; fractionation and interactions of proteins and lipoproteins. Mailing Add: Inst of Sci & Technol Univ of Mich Ann Arbor MI 48104

ONCLEY, PAUL BENNETT, b Chicago, Ill, June 22, 11; m 33; c 2. ACOUSTICS. Educ: Southwestern Col Kans, AB, 31; Univ Rochester, BMusic, 32, MMusic, 33; Columbia Univ, PhD(music acoustics), 52. Prof Exp: Asst prof vocal music, Univ NC, Greensboro, 37-42; electronics engr, Div Phys War Res, Duke Univ, 42-45; mem tech staff, Bell Tel Labs, Inc, 45-49; assoc prof vocal music, Westminster Choir Col, 49-53; asst to dir eng, Gulton Indust, 53-58; res specialist, Aerospace Div, Boeing Co, 58-63, Space Div, 63-67 & Com Airplane Div, 67-74; SR RES SCIENTIST, MAN, ACOUSTICS & NOISE, INC, 74- Concurrent Pos: Res Corp res grant acoustics of singing, 49-51; design consult, Rangertone, Inc, 49-51 & Kay Elec Co, 53; lectr, Columbia Univ, 50-52; guest prof, Highline Col, 62-63; conductor, Tacoma Youth Symphony, 63-66. Mem: Acoust Soc Am; Am Inst Aeronaut & Astronaut. Res: Music; acoustics of speech and music, particularly physics and physiology of singing; electronic music; ultrasonics and underwater sound; systems engineering for aerospace and physics of space exploration; airplane noise propagation; acoustic impedance measurements. Mailing Add: 6533 Seaview Dr NW 311-B Seattle WA 98117

ONDETTI, MIGUEL ANGEL, b Buenos Aires, Arg, May 14, 30; m 58; c 2. ORGANIC CHEMISTRY. Educ: Univ Buenos Aires, Dr Chem, 57. Prof Exp: Sr res chemist, Squibb Inst Med Res, Buenos Aires, 56-60, NJ, 60-66, res supvr, Dept Org Chem, 66-73, SECT HEAD, SQUIBB INST MED RES, 73- Concurrent Pos: Instr, Univ Buenos Aires, 57-60. Mem: Am Chem Soc; Arg Chem Soc; Swiss Chem Soc; Am Soc Biol Chemists. Res: Isolation, structure determination and synthesis of natural products; peptide isolation and synthesis. Mailing Add: Squibb Inst for Med Res PO Box 4000 Princeton NJ 08540

ONDIK, HELEN MARGARET, b New York, NY, Dec 25, 30.

CRYSTALLOGRAPHY. Educ: Hunter Col, AB, 52; Johns Hopkins Univ, MA, 54, PhD(chem), 57. Prof Exp: Fulbright grant, Univ Amsterdam, 57-58; PHYS CHEMIST, NAT BUR STANDARDS, 58- Honors & Awards: Silver Medal, Dept Commerce, 71. Mem: AAAS; Am Chem Soc; Am Crystallog Asn. Res: Solution and crystal chemistry of condensed phosphates; crystallographic data compilation; inorganic structures. Mailing Add: Nat Bur of Standards Washington DC 20234

ONDO, JEROME G, b Homestead, Pa, July 25, 39; m 67; c 2. NEUROENDOCRINOLOGY, CANCER. Educ: Univ Pittsburgh, BS, 62; John Carroll Univ, MS, 64; Univ Va, PhD(physiol), 70. Prof Exp: ASST PROF PHYSIOL, MED UNIV SC, 72- Concurrent Pos: NIH grant, Univ Tex Southwestern Med Sch Dallas, 70-72. Mem: Am Physiol Soc. Res: Reproduction; mammary cancer. Mailing Add: Dept of Physiol Med Univ SC Charleston SC 29401

ONDRAKO, JOANNE MARIE, b Rochester, NY, Apr 23, 47; m 73. MICROBIAL PHYSIOLOGY. Educ: Nazareth Col Rochester, BS, 69; St Bonaventure Univ, PhD(biol), 75. Prof Exp: NIH FEL MICROBIAL PHYSIOL, YALE UNIV, 75- Honors & Awards: Nat Serv Award, NIH Inst Allergy & Infectious Dis, 75. Mem: Am Soc Microbiol. Res: The evolution of microbial catabolic pathways; characterizing the uptake system of a catabolic intermediate in Pseudomonas putida. Mailing Add: Dept of Biol Yale Univ New Haven CT 06520

ONDRICEK, ANATOLA, b Brno, Czech, Feb 12, 38; UK citizen. BIOLOGY. Educ: Univ Prague, BSc, 61, PhD(ethnometrics), 67. Prof Exp: Lectr anthropometrics, Patrice Lumumba Univ, Moscow, 67-69; res fel dermatoglyphics, Galton Mem Labs, Univ London, 69-72; res assoc ecol, Serengeti Res Inst, 72-75; VIS PROF BIOL, UNIV CALIF, SAN DIEGO, 75- Concurrent Pos: Comnr, Nuffield Comn Exogenous Intel. Mem: Int Soc Topologists; fel Royal Anthrop Soc. Res: Semantic and epistomological problems in the life sciences; topology of emergent organic phenomena. Mailing Add: 259 24th St Del Mar CA 92014

ONEACRE, LELAND ELWOOD, rubber chemistry, see 12th edition

O'NEAL, CHARLES HAROLD, b Miami, Fla, Feb 18, 36; m 60; c 3. BIOCHEMISTRY. Educ: Ga Inst Technol, BA, 57; Emory Univ, PhD(biochem), 63. Prof Exp: Asst chem, Ga Inst Technol, 54-56 & Eng Exp Sta, 56-57; asst biochem, Emory Univ, 57-59; scientist, Nat Heart Inst, 63-65; res asst biochem, Emory Univ, 57-59; scientist, Nat Heart Inst, 63-65; res assoc biochem, Rockefeller Univ, 65-68, asst prof, 68; ASSOC PROF BIOPHYS, MED COL VA, 68- Concurrent Pos: Mem staff, Lab Molecular Biol, Med Res Coun, Cambridge, Eng, 71-72. Mem: AAAS; Am Inst Chem; Biochem Soc; Biophys Soc; Am Chem Soc. Res: Biochemical genetics; cytology. Mailing Add: Dept of Biophys Med Col of Va Box 877 Richmond VA 23219

O'NEAL, FLOYD BRELAND, b Fairfax, SC, May 4, 28; m 54; c 5. ANALYTICAL CHEMISTRY. Educ: The Citadel, BS, 48; Tulane Univ, MS, 50; Ga Inst Technol, PhD(chem), 59. Prof Exp: From assoc prof to prof chem, Tex Western Col, 54-66; PROF CHEM & HEAD DEPT, AUGUSTA COL, 66- Mem: Fel AAAS; Am Chem Soc; Sigma Xi. Res: Chromatography, especially of organic reagents; chelate formation; solvent extraction; spectroscopy. Mailing Add: Dept of Chem Augusta Col Augusta GA 30904

O'NEAL, GRADY MALCOLM, b Wisdom, Miss, Aug 30, 13; m 46. ORGANIC CHEMISTRY. Educ: Miss State Univ, BS, 35. Prof Exp: From factory chemist to res chemist, Sherwin-Williams Co, 35-48, dir org res lab, 48-52, mgr org res dept, 52-64, dir res, Colors & Chem Dept, 64; asst to vpres res & develop, 64-68, MGR RES & DEVELOP, TOMS RIVER CHEM CORP, 68- Honors & Awards: Cottingham Merit Fund Award, Sherwin-Williams Co, 40. Mem: Am Chem Soc; Am Asn Textile Chemists & Colorists. Res: Organic pigment dyestuffs; dyes; intermediates; synthesis. Mailing Add: Toms River Chem Corp Box 71 Toms River NJ 08753

O'NEAL, HARRY E, b Cincinnati, Ohio, Apr 2, 31; div; c 2. PHYSICAL CHEMISTRY. Educ: Harvard Univ, BA, 53; Univ Wash, PhD(chem), 57. Prof Exp: Res chemist, Shell Oil Co, 57-59; res assoc, Univ Southern Calif, 59-60; from asst prof to assoc prof, 61-70, PROF CHEM, SAN DIEGO STATE COL, 70- Concurrent Pos: Alexander von Humboldt Found sr scientist res award, 75. Mem: Am Phys Soc. Res: Thermodynamics; chemical kinetics. Mailing Add: Dept of Chem San Diego State Col 5402 College Ave San Diego CA 92115

O'NEAL, HARRY ROGER, b Kansas City, Kans, Jan 2, 35. PHYSICAL CHEMISTRY. Educ: Univ Kansas City, BS, 58; Univ Calif, Berkeley, PhD(chem), 63. Prof Exp: Res chemist, Du Pont Co, 62-65 & Douglas Aircraft Co, 65-66; fel chem, Pa State Univ, 66-67; asst prof, Lowell Tech Inst, 67-70; ASST PROF CHEM, PA STATE UNIV, OGONTZ CAMPUS, 70- Mem: Am Chem Soc. Res: Thermodynamic and transport properties of solids at low temperatures; thermodynamics and kinetics of phase transitions. Mailing Add: Dept of Chem Pa State Univ Abington PA 19001

O'NEAL, HUBERT RONALD, b Rotan, Tex, Apr 27, 37; m 60; c 1. ORGANIC CHEMISTRY. Educ: Tex Tech Col, BS, 59; NTex State Univ, MS, 64, PhD(chem), 67. Prof Exp: Chemist, Sherwin-Williams Co, Tex, 59-61; res chemist, W R Grace & Co, Md, 67-68; from res chemist to sr res chemist, 68-74, RES GROUP LEADER, PETRO-TEX CHEM CORP, 74- Mem: Am Chem Soc. Res: Polymer research on elastomeric materials; ketene chemistry and aldehyde polymerizations; synthetic rubber chemistry; polymerization kinetics. Mailing Add: 11202 Pecan Creek Dr Houston TX 77043

O'NEAL, PATRICIA L, b St Louis, Mo, June 14, 23. PSYCHIATRY. Educ: Wash Univ, AB, 44, MD, 48. Prof Exp: Intern, Univ Iowa Hosps, 48-49; resident psychiat, Barnes & Allied Hosps, St Louis, Mo, 49-51; from instr to asst prof, 52-61, ASSOC PROF PSYCHIAT, SCH MED, WASH UNIV, 61- Concurrent Pos: Fel psychosom med, Sch Med, Wash Univ, 52. Res: Clinical psychiatry, especially social psychiatry. Mailing Add: 4989 Barnes Hosp Plaza St Louis MO 63110

O'NEAL, ROBERT M, biochemistry, nutrition, see 12th edition

O'NEAL, ROBERT MUNGER, b Wiggins, Miss, Oct 7, 22; m 47; c 4. PATHOLOGY. Educ: Univ Miss, BS, 43; Univ Tenn, MD, 45. Prof Exp: Resident chest dis, State Sanatorium, Miss, 49-52; from instr to assoc prof path, Sch Med, Wash Univ, 54-61; prof & chmn dept, Col Med, Baylor Univ, 61-69; prof, Albany Med Col, 69-73; PROF PATH & CHMN DEPT, COLS MED & DENT, UNIV OKLA, 73- Concurrent Pos: Fel path, Mass Gen Hosp, 52-54; dir, Bender Hyg Lab, 69-73. Mem: Am Soc Exp Path; Am Soc Clin Path; Am Thoracic Soc; Am Asn Path & Bact; Col Am Path. Res: Cardiovascular and pulmonary diseases; atherosclerosis. Mailing Add: 800 NE 13th St Oklahoma City OK 73190

O'NEAL, RUSSELL D, b Columbus, Ind, Feb 15, 14; m 48; c 2. PHYSICS. Educ: DePauw Univ, AB, 36; Univ Ill, PhD(physics), 41. Prof Exp: Asst physics, Univ Ill, 36-41, instr, 41-42; mem staff, Radiation Lab, Mass Inst Technol, 42-43, sect head, 43-45, group leader, 45; proj physicist, Eastman Kodak Co, 45-48; head aerophysics group, Willow Run Res Ctr, Univ Mich, 49, from asst dir to dir, 50-52; asst div mgr, Consol Vultee Aircraft Corp, 52-53, prog dir, 53-55; dir systs planning, Bendix Corp, 55-57, gen mgr, Systs Div, 57-60, vpres eng & res, 60-63, vpres & group exec aerospace systs, 63-66; asst secy, Army Res & Develop, Washington, 66-68; exec vpres aerospace, Bendix Corp, 68-69, pres, Bendix Aerospace-Electronics Group, 69-72, pres, Group Opers, 72-73; exec vpres, KMS Fusion, Inc, 74-75, CHMN & CHIEF EXEC OFFICER, KMS INDUSTS, INC & KMS FUSION, INC, 75- Concurrent Pos: Mem, Study Group Guided Missiles, Dept Defense, 58-59; consult, Sci Adv Panel, US Army, 69-; mem, Nat Sci Bd, 72-; trustee, Asian Inst Technol. Mem: Fel Am Phys Soc; fel Inst Elec & Electronics Engrs; assoc Am Inst Aeronaut & Astron. Res: Nuclear physics; guided missiles; systems analysis; laser fusion. Mailing Add: KMS Industs Inc Box 1778 Ann Arbor MI 48106

O'NEAL, THOMAS DENNY, b Vandalia, Ill, June 5, 41; m 66; c 1. PLANT PHYSIOLOGY, AGRONOMY. Educ: Southern Ill Univ, Carbondale, BS, 63; Duke Univ, PhD(plant physiol), 68. Prof Exp: Res assoc plant physiol, Univ Ky, 68-69 & Carleton Univ, 71-73; fel, Brandeis Univ, 69-71; asst prof, Rensselaer Polytech Inst, 73-75; GROUP LEADER PLANT GROWTH REGULATOR DISCOVERY, AM CYANAMID CO, 75- Mem: Am Soc Plant Physiologists; Sigma Xi; AAAS. Res: Discovery of new plant growth regulants to enhance yield, quality or harvestability of key agronomic crops; enzymatic regulation of N and C metabolism in plants. Mailing Add: Am Cyanamid Co Agr Div PO Box 400 Princeton NJ 08540

O'NEAL, THOMAS NORMAN, b Ft Smith, Ark, Sept 17, 38; m 63; c 2. EXPERIMENTAL SOLID STATE PHYSICS. Educ: Miss Col, BS, 60; Univ Fla, MS, 62; Clemson Univ, PhD(physics), 70. Prof Exp: From asst prof to assoc prof, 63-75, PROF PHYSICS, CARSON-NEWMAN COL, 75- Mem: Am Asn Physics Teachers. Res: Electron irradiation damage in metals. Mailing Add: Dept of Physics Carson-Newman Col Jefferson City TN 37760

ONEDA, SADAO, b Akita, Japan, June 30, 23; m 50; c 2. THEORETICAL PHYSICS. Educ: Tohoku Univ, Japan, BSc, 46, MSc, 48; Nagoya Univ, DrSci(theoret physics), 53. Prof Exp: Res assoc theoret physics, Tohoku Univ, Japan, 48-50; from asst prof to prof, Kanazawa Univ, 50-63; from asst prof to assoc prof, 63-67, PROF PHYSICS, UNIV MD, COLLEGE PARK, 67- Concurrent Pos: Mem staff, Res Inst Fundamental Physics, Kyoto Univ, 53-55 & 60-63; Imp Chem Industs fel, Univ Manchester, 55-57; mem, Inst Advan Study, 57-58; res assoc, Univ Md, 58-60; vis prof physics, Univ Wis-Milwaukee, 69-70. Mem: Fel Am Phys Soc; Phys Soc Japan. Res: Physics; theoretical elementary particle physics; field theory. Mailing Add: 1 Lakeside Dr Greenbelt MD 20770

ONEGA, RONALD JOSEPH, b Latrobe, Pa, May 11, 35; m 59; c 2. NUCLEAR PHYSICS. Educ: Pa State Univ, BS, 61, PhD(physics), 64. Prof Exp: Asst prof physics, 64-71, asst prof nuclear eng, 71-73, ASSOC PROF NUCLEAR ENG, VA POLYTECH INST & STATE UNIV, 73- Mem: Am Phys Soc; Am Nuclear Soc. Res: Reactor physics and analysis; nuclear reactor dynamics; fusion. Mailing Add: Nuclear Eng Group Dept Mech Eng Va Polytech Inst & State Univ Blacksburg VA 24061

O'NEIL, ELIZABETH JEAN, b Palo Alto, Calif, Jan 29, 42; m 65; c 2. APPLIED MATHEMATICS. Educ: Mass Inst Technol, SB, 63; Harvard Univ, MA, 64, PhD(appl math), 68. Prof Exp: Vis mem appl math, Courant Inst Math Sci, NY Univ, 66-67, asst prof, 67-68; lectr, Mass Inst Technol, 68-70; ASSOC PROF MATH, UNIV MASS, BOSTON, 70- Mem: Soc Indust & Appl Math. Res: Geophysical fluid dynamics; singular perturbation theory and other forms of asymptotic analysis; computer algorithms and their analysis. Mailing Add: 50 Hardy Ave Watertown MA 02172

O'NEIL, JAMES R, b Chicago, Ill, July 16, 34. GEOCHEMISTRY, PHYSICAL CHEMISTRY. Educ: Loyola Univ Ill, BS, 56; Carnegie-Mellon Univ, MS, 59; Univ Chicago, PhD(chem), 63. Prof Exp: Res fel geochem, Univ Chicago, 63 & Calif Inst Technol, 63-65; CHEMIST, BR ISOTOPE GEOL, US GEOL SURV, 65- Concurrent Pos: Exchange scientist to USSR, 72; consult prof, Stanford Univ, 75- Mem: AAAS; Am Chem Soc; Am Geophys Union. Res: Studies of stable isotope variations in natural materials; laboratory determination of equilibrium constants for isotope exchange reactions of geologic significance; solute-water interactions; structure of water. Mailing Add: Br Isotope Geol US Geol Surv 345 Middlefield Rd Menlo Park CA 94025

O'NEIL, LOUIS-C, b Sherbrooke, Que, Jan 20, 30; m 55; c 5. ENTOMOLOGY. Educ: Univ Montreal, BA, 50; Laval Univ, BSCApp, 54; State Univ NY Col Forestry, Syracuse Univ, MSc, 56, PhD(forest entom), 61. Prof Exp: Res off forest entom & path br, Can Dept Forestry, 55-62; assoc prof, 62-72, secy fac sci, 67-70, chmn dept biol, 70-72, PROF INVERT ZOOL & ENTOM, UNIV SHERBROOKE, 72-, DEAN, FAC SCI, 72- Mem: Entom Soc Can; Can Inst Forestry; Can Soc Zool. Res: Ecology of land slug introduced in eastern Canada; relationships between sawflies and their host trees; forest sawfly outbreaks on tree growth. Mailing Add: Fac of Sci Univ of Sherbrooke Sherbrooke PQ Can

O'NEIL, PATRICK EUGENE, b Mineola, NY, July 19, 42; m 65; c 1. MATHEMATICS, COMPUTER SCIENCE. Educ: Mass Inst Technol, BS, 63; Univ Chicago, MS, 64; Rockefeller Univ, PhD(math), 69. Prof Exp: Systs programmer, Cambridge Sci Ctr, Int Bus Mach Corp, 64-66, mem staff opers res, 68-69; Off Naval Res fel & res assoc combinatorics, Mass Inst Technol, 69-70, asst prof comput sci, 70-74; ASST PROF COMPUT SCI, UNIV MASS, BOSTON, 74- Concurrent Pos: Mem, Artificial Intel Lab, Mass, 70- Mem: Math Asn Am; Asn Comput Mach. Res: Combinatorial structures; asymptotic counting; algorithmic complexity; data retrieval; convexity theory; discrete problems such as graphs and networks. Mailing Add: 75 Lexington St Belmont MA 02178

O'NEIL, PETER VINCENT, b Suffern, NY, Jan 15, 40; m 68. MATHEMATICS. Educ: Fordham Univ, BS, 61; Rensselaer Polytech Inst, MS, 63, PhD(math), 65. Prof Exp: Instr math, Inst Technol, Univ Minn, 65-67; from asst prof to assoc prof, 67-74, PROF MATH, COL WILLIAM & MARY, 74- Honors & Awards: Lester R Ford Award, Math Asn Am, 71. Mem: AAAS; Soc Indust & Appl Math; Math Asn Am. Res: Graph reconstructions; enumeration of labelled spanning trees in graphs. Mailing Add: Dept of Math Col of William & Mary Williamsburg VA 23185

O'NEILL, ANNE FRANCES, b Troy, NY, Aug 21, 15. MATHEMATICAL STATISTICS. Educ: Vassar Col, AB, 38; Radcliffe Col, MA, 39, PhD(math), 42. Prof Exp: From instr to asst prof math, Smith Col, 42-52; from asst prof to assoc prof, 52-60, PROF MATH, WHEATON COL MASS, 60- Concurrent Pos: Sci fac fel, NSF, Univ Calif, Berkeley, 61-62. Mem: Am Math Soc; Math Asn Am; Am Statist Asn. Res: Function theory; theory of derivatives. Mailing Add: Dept of Math Wheaton Col Norton MA 02766

O'NEILL, EDWARD JOHN, b Washington, DC, Feb 15, 42. TOPOLOGY. Educ: Cath Univ Am, AB, 63; Yale Univ, MA, 65, PhD(math), 76. Prof Exp: Instr, 67-69,

ASST PROF MATH, FAIRFIELD UNIV, 69- Mem: Am Math Soc. Res: Algebraic topology; cohomology operations. Mailing Add: Dept of Math Fairfield Univ Fairfield CT 06430

O'NEILL, EDWARD LEO, b Boston, Mass, Nov 29, 27; m 51; c 3. PHYSICS. Educ: Boston Col, AB, 49; Boston Univ, MA, 51, PhD(physics), 54. Prof Exp: Res assoc physics, Res Lab, Boston Univ, 54-56, from asst prof to assoc prof, 56-66; VIS PROF PHYSICS, WORCESTER POLYTECH INST, 66- Concurrent Pos: Mem staff, Itek Corp, 58-60; vis prof, Mass Inst Technol, 61-62 & Univ Calif, Berkeley, 64-65. Honors & Awards: Lomb Medal, Optical Soc Am, 58. Mem: AAAS; Optical Soc Am; Am Asn Physics Teachers. Res: Communication theory; quantum and statistical optics. Mailing Add: Dept of Physics Worcester Polytech Inst Worcester MA 01609

O'NEILL, EDWARD TRUE, information science, operations research, see 12th edition

O'NEILL, FRANK JOHN, b New York, NY, Apr 26, 40; c 1. MICROBIOLOGY, PATHOLOGY. Educ: Long Island Univ, BA, 63; Hunter Col, MA, 67; Univ Utah, PhD(molecular biol, genetic biol), 69. Prof Exp: ASST PROF MICROBIOL & PATH, MED CTR, UNIV UTAH, 71- Concurrent Pos: NIH fel microbiol, Milton S Hershey Med Ctr, Pa State Univ, 69-71. Mem: Am Soc Human Genetics; Environ Mutagen Soc; Am Soc Microbiol; Am Asn Cancer Res. Res: Elucidation and establishment of mechanisms of viral oncogenesis and viral latency in vitro and in vivo; cytogenetics, especially chromosomal changes occurring during and following viral oncogenesis. Mailing Add: Dept of Res Vet Admin Hosp Salt Lake City UT 84113

O'NEILL, GEORGE FRANCIS, b Yonkers, NY, Sept 27, 22; m 43; c 2. NUCLEAR PHYSICS. Educ: Mt St Mary's Col Md, BS, 43; Fordham Univ, MS, 47, PhD(physics), 51. Prof Exp: Physicist, Argonne Nat Lab, 50-53; sr res supvr, exp reactor physics div, Savannah Lab, E I du Pont de Nemours & Co, Inc, 53-68; with AEC Combined Opers Planning, Oak Ridge, 69-70; sr physicist, Advan Oper Planning, 70-74, RES STAFF PHYSICIST, SAVANNAH RIVER LAB, E I DU PONT DE NEMOURS & CO, INC, 74- Concurrent Pos: Guest lectr, Dept Nuclear Eng, Univ SC, 67-68; US rep, Can-Am D20 Reactor Comt, 67-68. Mem: AAAS; Am Nuclear Soc; Am Phys Soc. Res: Cosmic rays; heavy water nuclear reactors; exponential pile assemblies; critical pile assemblies; nuclear reactor fuel cycle. Mailing Add: Savannah River Lab E I du Pont de Nemours & Co Inc Aiken SC 29801

O'NEILL, GEORGE JOSEPH, b Wilmington, Del, Feb 13, 38; m 60; c 3. POLYMER CHEMISTRY. Educ: Va Mil Inst, BS, 59; Univ Del, PhD(chem), 65. Prof Exp: Teaching asst, Univ Del, 61-65; asst prof org chem, Va Mil Inst, 65-67; SR RES CHEMIST, TENN EASTMAN CO, 67- Mem: Am Chem Soc. Res: Exploratory polymer research; synthetic organic chemistry in the areas of carbenes and organophosphorous compounds; applied condensation polymer research. Mailing Add: Res Labs Tenn Eastman Co Kingsport TN 37664

O'NEILL, GERARD KITCHEN, b Brooklyn, NY, Feb 6, 27; m 50, 73; c 3. PHYSICS. Educ: Swarthmore Col, BA, 50; Cornell Univ, PhD(physics), 54. Prof Exp: From instr to assoc prof, 54-65, PROF PHYSICS, PRINCETON UNIV, 65- Mem: Fel Am Phys Soc. Res: Experimental elementary particle and nuclear physics; orbital astronomy techniques. Mailing Add: Dept of Physics Princeton Univ Princeton NJ 08540

O'NEILL, JAMES A, JR, b New York, NY, Dec 7, 33; m 59; c 3. PEDIATRIC SURGERY. Educ: Georgetown Univ, BS, 55; Yale Univ, MD, 59. Prof Exp: Instr surg, Sch Med, Vanderbilt Univ, 64-65; instr pediat surg, Sch Med, Ohio State Univ, 67-69; from asst prof to assoc prof, Sch Med, La State Univ, 69-71; PROF PEDIAT SURG, SCH MED, VANDERBILT UNIV, 71- Concurrent Pos: USPHS fel pediat oncol, Ohio State Univ, 67-68; consult, US Army Hosp, Ft Campbell, Ky, 71- & US Army Inst Surg Res, 71- Mem: Am Col Surg; Am Burn Asn; Am Acad Pediat; Am Surg Asn; Am Pediat Surg Asn. Res: General and thoracic pediatric surgery; various aspects of burn injury; stress ulceration; gastrointestinal effects of injury. Mailing Add: Div of Pediat Surg Sch of Med Vanderbilt Univ Nashville TN 37232

O'NEILL, JOHN CORNELIUS, b Philadelphia, Pa, July 3, 29. APPLIED MATHEMATICS. Educ: Cath Univ, BA, 52; Villanova Univ, MA, 59; Univ Pittsburgh, PhD, 67. Prof Exp: Teacher high schs, Pa, 52-53 & Md, 53-64; asst prof, 67-71, ASSOC PROF MATH, LA SALLE COL, 71-, CHMN DEPT MATH SCI, 74- Mem: Soc Indust & Appl Math; Inst Elec & Electronics Eng; Math Asn Am. Res: Computers and circuit theory. Mailing Add: Dept of Math Sci La Salle Col Philadelphia PA 19141

O'NEILL, JOHN DACEY, b Detroit, Mich, June 9, 29. MATHEMATICS. Educ: Xavier Univ, Ohio, AB, 54; Loyola Univ, Ill, MA, PhL, STL & MS, 60; Wayne State Univ, PhD(math), 67. Prof Exp: Instr math & theol, high sch, 56-59; ASST PROF MATH, DETROIT UNIV, 61- Res: Abelian groups. Mailing Add: Dept of Math Univ of Detroit Detroit MI 48221

O'NEILL, JOHN J, physical chemistry, see 12th edition

O'NEILL, JOHN JOSEPH, b Queens, NY, Aug 26, 19; m 43; c 3. BIOCHEMISTRY. Educ: St Francis Col, NY, BS, 42; Univ Md, MS, 53, PhD(biochem), 55. Prof Exp: Res assoc physiol, Sch Med, Cornell Univ, 46-47; biochemist, Physiol Div, Med Lab, US Army Chem Res & Develop Lab, 47-50, org chemist, Biochem Div, Res Directorate, 50-54, biochemist, 54-56, br chief, 56-60; assoc prof biochem pharmacol, Sch Med, Univ Md, Baltimore City, 60-70; prof biochem pharmacol, Col Med, Ohio State Univ, 70-75; PROF & CHMN DEPT PHARMACOL, MED SCH, TEMPLE UNIV, 75- Concurrent Pos: NIMH fel pharmacol, Sch Med, Wash Univ, 63-64; res collab, Biochem Dept, Brookhaven Nat Lab, 61-; consult, US Army Chem Res & Develop Lab, Md, 61- & Melpar Div, Westinghouse Elec Co, Va, 62- Mem: Am Chem Soc. Res: Neurochemistry; biochemistry of the central nervous system; glycolysis; enzyme kinetics and control mechanisms; isotopes in intermediary metabolism; neurotoxins and organophosphorous anticholinesterases; neurohumoral substances and drug action. Mailing Add: Sch of Med 3420 N Broad St Temple Univ Philadelphia PA 19140

O'NEILL, RICHARD DELOS, b Cohocton, NY, Oct 6, 20; m 48; c 4. BACTERIOLOGY. Educ: Canisius Col, BS, 42; Syracuse Univ, MS, 48, PhD, 51. Prof Exp: Asst bact, NY Univ Syracuse, 47-49, from instr to asst prof, 49-57, from assoc dir to dir microbiol & biochem, Res Ctr, Univ Res Corp, NY, 57-65; mgr biochem res, Glidden Co, 65-70, DIR FOODS RES & DEVELOP, GLIDDEN-DURKEE DIV, SCM CORP, 70- Mem: Inst Food Technol; Am Oil Chem Soc; Indust Res Inst; Soc Indust Microbiol. Res: Microbial decomposition of cellulose; microbiology of paper and paper board; vitamin nutrition of bacteria and fungi; effects of surface active agents on bacteria; migration of food packaging components; food and food additives. Mailing Add: Glidden-Durkee Div SCM Corp 16651 Sprague Rd Strongsville OH 44136

O'NEILL, RICHARD THOMAS, b Winnetka, Ill, Dec 7, 33; m 59; c 2. PHYSICAL INORGANIC CHEMISTRY. Educ: Loyola Univ Ill, BS, 55; Carnegie Inst Technol, MS, 59, PhD(chem), 60. Prof Exp: From asst prof to assoc prof, 59-70, PROF CHEM, XAVIER UNIV OHIO, 70-, COORDR ACAD COMPUT, 73- Mem: Am Chem Soc; Soc Appl Spectros; Coblentz Soc; Sigma Xi. Res: Chemical spectroscopy; molecular association; charge transfer complexes; hydrogen bonding; educational use of computers. Mailing Add: Dept of Chem Xavier Univ Cincinnati OH 45207

O'NEILL, ROBERT CHARLES, organic chemistry, see 12th edition

O'NEILL, ROBERT VINCENT, b Pittsburgh, Pa, Apr 13, 40; m 67. ECOLOGY. Educ: Cathedral Col, BA, 61; Univ Ill, PhD(zool), 67. Prof Exp: Fel systs ecol, NSF, 67-68; assoc health physicist, 68-69, SYSTS ECOLOGIST, OAK RIDGE NAT LAB, 69- Concurrent Pos: Consult, 67-68. Mem: Ecol Soc Am. Res: Systems ecology and physiological ecology. Mailing Add: Environ Sci Div Bldg 2001 Oak Ridge Nat Lab Oak Ridge TN 37830

O'NEILL, RONALD C, b Beverly, Mass, July 20, 29; m 53; c 2. TOPOLOGY. Educ: Columbia Univ, AB, 51, MA, 52; Purdue Univ, PhD(math), 62. Prof Exp: From instr to asst prof math, Univ Mich, 61-66; asst prof, 66-67, ASSOC PROF MATH, MICH STATE UNIV, 67- Mem: Am Math Soc; Math Asn Am. Mailing Add: Dept of Math Col Natural Sci Mich State Univ East Lansing MI 48824

O'NEILL, THOMAS BRENDAN, b Boston, Mass, Oct 16, 23; m 47; c 5. MICROBIOLOGY, BOTANY. Educ: Univ Notre Dame, BS, 50, MS, 51; Univ Calif, PhD(bot), 56. Prof Exp: Machinists apprentice, Bethlehem Steel Co, Mass, 41-43; asst, Univ Notre Dame, 50-51 & Univ Calif, 51-54, assoc bot, 54-55; INSTR BOT & MICROBIOL, VENTURA COL, 55- Concurrent Pos: Res biologist, US Naval Civil Eng Lab, 57-; NSF fel, Scripps Inst, Calif, 62-63; lectr, Univ Calif, Santa Barbara. Mem: AAAS; Bot Soc Am; Sigma Xi. Res: Marine microbiology; teaching methods; biodegradation of hydrocarbons. Mailing Add: Dept of Biol Ventura Col Ventura CA 93003

O'NEILL, THOMAS FRANCIS, geology, inorganic chemistry, see 12th edition

O'NEILL, THOMAS HALL ROBINSON, b Bel Air, Md, Feb 21, 23; m 50; c 5. METEOROLOGY. Educ: Mt St Mary's Col, BA, 43; US Naval Postgrad Sch, MS, 54; Fla State Univ, PhD(meteorol), 64. Prof Exp: US Navy, 50-, meteorologist & oceanogr, USS Midway, 54-55, instr meteorol & oceanog, US Naval Postgrad Sch, 55-57, staff meteorologist & oceanogr, Commander Amphibious Forces Atlantic, 57-59, exec off, Navy Weather Res Facil, 59-61, head satellite meteorol br, Astronaut Div, Chief Naval Opers, 64-67, chief environ sci div, Off Dir Defense Res & Eng, Va, 67-70, sr mgr, Nat Ctr Atmospheric Res, 70-74; GLOBAL ATMOSPHERIC RES PROG MGR, NASA, 74- Concurrent Pos: Dir tech serv, Naval Weather Serv Command, 64-67; mem, US-Can Mil Coop Comt Oceanog, 64-67; alt mem, Interagency Comt Atmospheric Sci, 64-; mem, Interagency Comt Appl Meteorol Res, 64-; assoc mem, Meteorol Working Group-Interrange Instrument Group, 64-; mem oceanog forecasting panel, Interdept Comn Oceanog, 66-67; liaison mem, US Comt Global Atmospheric Res Prog, 68-; mem, Interagency Comt 1970 Solar Eclipse, 68-70. Mem: Am Meteorol Soc; Am Geophys Union; Sigma Xi. Res: Diagnostic studies of atmospheric vertical motion and precipitation; synoptic meteorology; numerical weather prediction; weather modification; tropical meteorology. Mailing Add: 5926 Merritt Pl Falls Church VA 22041

O'NEILL, WILLIAM PATRICK, organic chemistry, polymer chemistry, see 12th edition

O'NELL, CARL WILLIAM, b Pueblo, Colo, Nov 30, 25; m 56; c 5. ANTHROPOLOGY. Educ: Colo State Col, BA, 56; Univ Edinburgh, dipl social anthrop, 59; Univ Chicago, MA, 65, PhD(human develop), 69. Prof Exp: Asst prof, 67-73, ASSOC PROF ANTHROP, UNIV NOTRE DAME, 73- Mem: Fel Am Anthrop Asn; Int Soc Res Agression; Am Psychol Asn. Res: Study of social, cultural and psychological factors in human stress, particularly as these relate to illness and interpersonal aggression; cross cultural studies in human development, focusing on childhood socialization. Mailing Add: Dept of Sociol & Anthrop Univ of Notre Dame Notre Dame IN 46556

ONESON, IRVING BERNT, biochemistry, see 12th edition

ONG, ENG-BEE, b Seremban, Malaysia, Aug 9, 33; m. BIOCHEMISTRY. Educ: Univ Miami, BS, 58; Tulane Univ, PhD(biochem), 64. Prof Exp: Instr biochem, Tulane Univ, 64-65; res assoc, Rockefeller Univ, 65-68, asst prof, 68; clin biochemist, Ochsner Clin, 68-70; ASST PROF BIOCHEM, MED CTR, NY UNIV, 70- Mem: Am Chem Soc. Res: Chemistry of amino acids, peptides and proteins; chromatography of amino acids and peptides; structure and function of proteases; fractionation and characterization of blood coagulation factors. Mailing Add: Dept of Path NY Univ Med Ctr New York NY 10016

ONG, JOHN TJOAN HO, b Jakarta, Indonesia, Sept 5, 37; m 61; c 2. PHARMACEUTICS. Educ: Bandung Inst Technol, CPharm, 60, Drs(pharm), 65; Univ Ky, PhD(pharmaceut), 74. Prof Exp: Teaching asst pharmaceut, Sch Pharm, Bandung Inst Technol, 63-65; lectr, Sch Pharm, Univ Indonesia, 65-69; pharmacist, Sanitas Pharm, Jakarta, 69-70; res staff pharmaceut chem, Univ Ky, 70-71, teaching asst pharmaceut, 71-74; SR SCIENTIST PHARMACEUT, LILLY RES LABS, DIV ELI LILLY & CO, 74- Mem: Am Pharmaceut Asn. Res: Stability kinetics of drugs; development of dosage forms; physical and chemical factors influencing the bioavailability of drugs from their dosage forms; synthesis of prodrugs. Mailing Add: Lilly Res Labs Div Eli Lilly & Co E McCarty St Indianapolis IN 46206

ONG, KWOK MAW, b Canton, China, July 26, 38; US citizen; m 65; c 2. PHYSICS. Educ: Univ Calif, Berkeley, BS, 59, MA, 60, PhD(physics), 68. Prof Exp: Physicist, Lawrence Radiation Lab, Univ Calif, Berkeley, 68; asst prof physics, Univ Tenn, Knoxville, 68-69; scientist opers res, Syst Develop Corp, Calif, 69-71; sr res physicist, Gen Motors Res Labs, Mich, 71; SR SCIENTIST, JET PROPULSION LAB, CALIF INST TECHNOL, 71- Mem: Am Phys Soc; Am Geophys Union. Res: Application of physics to space and earth sciences; radio astronomy; geodesy; geophysics; high energy physics; computer science; operations research and systems analysis. Mailing Add: Calif Inst of Technol Jet Propul Lab 4800 Oak Grove Dr Pasadena CA 91103

ONKEN, ARTHUR BLAKE, b Alice Tex, Aug 13, 35; m 56; c 2. SOIL CHEMISTRY, SOIL FERTILITY. Educ: Tex A&I Univ, BS, 59; Okla State Univ, MS, 63, PhD(soil chem), 64. Prof Exp: ASSOC PROF SOIL CHEM, AGR EXP STA, TEX A&M UNIV, 64- Mem: Am Soc Agron; Soil Sci Soc Am. Res: Plant nutrient reactions in soil; plant response to nutrient level; movement of applied plant nutrients and solutes through soil profiles. Mailing Add: Agr Res & Exten Ctr Tex A&M Univ Lubbock TX 79401

ONLEY, DAVID S, b London, Eng, July 4, 34; m 57; c 2. THEORETICAL PHYSICS. Educ: Oxford Univ, BA, 56, DPhil(theoret physics), 60. Prof Exp: Res assoc physics, Duke Univ, 60-61, asst prof, 61-65; assoc prof, 65-69, PROF PHYSICS, OHIO

UNIV, 69-, CHMN DEPT, 73- Concurrent Pos: US Army Res Off res grant, 66-; sabbatical leave, Univ Melbourne, 71-72. Mem: AAAS; Am Phys Soc. Res: Nuclear structure; photonuclear theory and analysis of electron scattering from atomic nuclei; high energy physics; nuclear fission. Mailing Add: Dept of Physics Clippinger Res Labs Ohio Univ Athens OH 45701

ONLEY, JUDITH WHEELER, experimental psychology, see 12th edition

ONN, DAVID GOODWIN, b Newark, Eng, Feb 18, 37; m 67. CRYOGENICS, SOLID STATE PHYSICS. Educ: Bristol Univ, BSc, 58; Oxford Univ, dipl educ, 60; Duke Univ, PhD(physics), 67. Prof Exp: Second physics master, Dean Close Sch, Cheltenham, Eng, 59-61; res assoc physics, Univ NC, Chapel Hill, 66-68, instr, 68-69, vis asst prof, 69-70; asst prof physics, 70-75, asst prof health sci, 73-75, ASSOC PROF PHYSICS & HEALTH SCI, UNIV DEL, 75- Mem: Am Inst Physics; Brit Inst Physics; Am Asn Physics Teachers. Res: Electronic states in cryogenic fluids; thermal-magnetic properties of rare-earth materials; medical physics. Mailing Add: Dept of Physics Univ of Del Newark DE 19711

ONO, HOWARD KEN, organic chemistry, biochemistry, see 12th edition

ONOPCHENKO, ANATOLI T, b Lvov, Poland, Feb 12, 37; US citizen; m 63; c 2. ORGANIC CHEMISTRY. Educ: Pa State Univ, BS, 60, MS, 62; Univ Md, PhD(chem), 66. Prof Exp: Res asst, Pa State Univ, 60-62; asst, Univ Md, 63-66; res chemist, 66-72, SR RES CHEMIST, GULF RES & DEVELOP CO, 72- Mem: Am Chem Soc. Res: General organic and petroleum chemistry. Mailing Add: 180 Kelvington Dr Monroeville PA 15146

ONORE, MICHAEL JAMES, organic chemistry, see 12th edition

ONSAGER, JEROME ANDREW, b Northwood, NDak, Apr 8, 36; m 58; c 2. ENTOMOLOGY. Educ: NDak State Univ, BS, 58, MS, 60, PhD(entom), 63. Prof Exp: Res entomologist, Winter Vegetable Insects Invests, Agr Res Serv, USDA, 63-75; MEM STAFF ENTOM, RANGELAND INSECT LAB, MONT STATE UNIV, 75- Mem: Entom Soc Am. Res: Biology and control of insects affecting winter vegetables in the Southeast. Mailing Add: Rangeland Insect Lab Mont State Univ Dept Entom Bozeman MT 59715

ONSAGER, LARS, b Oslo, Norway, Nov 27, 03; nat US; m 33; c 4. THEORETICAL PHYSICS, THEORETICAL CHEMISTRY. Educ: Tech Univ, Norway, ChE, 25; Yale Univ, PhD(theoret chem), 35. Hon Degrees: DSc, Harvard Univ, 54, Rensselaer Polytech Inst, 62, Brown Univ, 62 & Aachen Tech Univ, 62; Dr Tech, Tech Univ, Norway, 60; ScD, Univ Chicago, 68, Ohio State Univ, 69, Cambridge Univ, 70 & Oxford Univ, 71; ScD, Gustavus Adolphus Col, 75. Prof Exp: Assoc chem, Johns Hopkins Univ, 28; res instr, Brown Univ, 28-33; from asst prof to assoc prof, Yale Univ, 34-45, Gibbs Prof theoret chem, 45-72; DISTINGUISHED UNIV PROF THEORET CHEM, CTR THEORET STUDIES, UNIV MIAMI, 72- Concurrent Pos: Fulbright scholar, Cambridge Univ, 51-52; vis prof, Univ Calif, San Diego, 61 & Rockefeller Univ, 67; vis Gauss prof, Univ Göttingen, 68; vis Lorentz prof, State Univ Leiden, 70; mem, Neurosci Assocs, Mass. Honors & Awards: Rumford Gold Medal, Am Acad Arts & Sci, 53; Lorentz Medal, Royal Neth Acad Sci, 58; Lewis, Kirkwood & Gibbs Medals, Am Chem Soc, 62, Richards Medal, 64 & Dobye Award, 65; Belfer Award, Yeshiva Univ, 66; Nobel Prize in Chem, 68; Nat Medal Sci, 69. Mem: Nat Acad Sci; Am Chem Soc; fel Am Phys Soc; Am Acad Arts & Sci; hon mem Norweg Chem Soc. Res: Electrophysics; cooperative phenomena; protonic semiconductors; hydrodynamics; nervous system. Mailing Add: Ctr for Theoret Studies Univ of Miami Coral Gables FL 33124

ONSTOTT, EDWARD IRVIN, b Moreland, Ky, Nov 12, 22; m 45; c 4. INORGANIC CHEMISTRY, PHYSICAL CHEMISTRY. Educ: Univ Ill, BS, 44, MS, 48, PhD(chem), 50. Prof Exp: Chem engr, Firestone Tire & Rubber Co, 44 & 46; RES CHEMIST, LOS ALAMOS SCI LAB, 50- Mem: Fel AAAS; fel Am Inst Chem; NY Acad Sci; Am Chem Soc; Electrochem Soc. Res: Complex ion and rare earth chemistry; electrochemistry. Mailing Add: Box 1663 Los Alamos NM 87544

ONTJES, DAVID A, b Lyons, Kans, July 19, 37; m 60; c 4. ENDOCRINOLOGY. Educ: Univ Kans, BA, 59; Oxford Univ, MA, 61; Harvard Med Sch, MD, 64. Prof Exp: From intern to resident internal med, Harvard Med Serv, Boston City Hosp, 64-66; asst prof med & pharmacol, 69-72, ASSOC PROF MED & PHARMACOL & CHIEF DIV ENDOCRINOL, MED SCH, UNIV NC, CHAPEL HILL, 72- Concurrent Pos: NIH spec res fel, 71; attend physician, NC Mem Hosp, Chapel Hill, 69- Mem: Am Soc Pharmacol & Exp Therapeut; Endocrine Soc; Am Fedn Clin Res. Res: Chemistry and physiology of peptide hormones; hormonal control of calcium metabolism. Mailing Add: Dept of Med Univ of NC Med Sch Chapel Hill NC 27514

ONTKO, JOSEPH ANDREW, b Syracuse, NY, July 30, 32; m 55; c 4. BIOCHEMISTRY. Educ: Syracuse Univ, AB, 53; Univ Wis, MS, 55, PhD(biochem), 57. Prof Exp: From instr to asst prof biochem, Med Units, Univ Tenn, 59-64, asst prof physiol & biophys, 65-66; assoc prof, 68-74, PROF RES BIOCHEM, SCH MED, UNIV OKLA, 74-; ASSOC MEM, OKLA MED RES FOUND, 68- Concurrent Pos: Am Heart Asn res fel, Med Sch, Bristol Univ, 66-67 & Max Planck Inst Cell Chem, Munich, 67-68; consult, Oak Ridge Inst Nuclear Studies, 61-66. Mem: AAAS; Biochem Soc; Am Chem Soc; Am Soc Biol Chem. Res: Lipid metabolism; fatty acid oxidation; regulation of cell metabolism; isolation of liver cells. Mailing Add: Okla Med Res Found 825 NE 13th St Oklahoma City OK 73104

ONTON, AARE, b Tartu, Estonia, Dec 10, 39; US citizen; m 65; c 1. PHYSICS, ELECTRICAL ENGINEERING. Educ: Mass Inst Technol, SB & SM, 62; Purdue Univ, PhD(physics), 67. Prof Exp: Res staff mem, Thomas J Watson Res Ctr, IBM Corp, NY, 67-73; RES STAFF MEM, IBM RES LAB, 73- Concurrent Pos: Alexander von Humboldt scholar, Max Planck Inst Solid State Res, 72-73. Mem: Am Phys Soc. Res: Optical properties of solids, particularly electronics bandstructure, electronic impurity states, lattice dynamics, and luminescence; structural, electronic, and magnetic properties of amorphous thin films. Mailing Add: IBM Res Lab 5600 Cottle Rd San Jose CA 95193

ONWOOD, DAVID P, b London, Eng, Sept 13, 36; m 72. PHYSICAL CHEMISTRY. Educ: Oxford Univ, BA, 60, DPhil(chem), 66. Prof Exp: Res assoc chem, Ill Inst Technol, 63-65, instr 65-66; asst prof, 66-69, ASSOC PROF CHEM, PURDUE UNIV, FT WAYNE, 69- Concurrent Pos: NSF partic award, 67-69. Mem: AAAS; Am Chem Soc; The Chem Soc. Res: Physical chemistry in aqueous solutions; isotope effects; catalytic effects. Mailing Add: Dept of Chem Purdue Univ Ft Wayne IN 46805

ONYSZCHUK, MARIO, b Wolkow, Poland, July 21, 30; nat Can; m 59; c 3. INORGANIC CHEMISTRY. Educ: McGill Univ, BSc, 51; Univ Western Ont, MSc, 52; McGill Univ, PhD(phys chem), 54; Cambridge Univ, PhD(inorg chem), 56. Prof Exp: Lectr chem, 56-57, from asst prof to assoc prof, 57-68, PROF CHEM, McGILL UNIV, 68- Mem: Am Chem Soc; fel Chem Inst Can; The Chem Soc. Res: Coordination compounds of group III and IV halides. Mailing Add: Dept of Chem McGill Univ Montreal PQ Can

OOKA, JERI JEAN, b Olaa, Hawaii, Jan 22, 45. PHYTOPATHOLOGY. Educ: Univ Hawaii, BA, 66, MS, 69; Univ Minn, PhD(plant path), 75. Prof Exp: ASST PLANT PATHOLOGIST, UNIV HAWAII, 75- Mem: Am Phytopath Soc; AAAS. Res: Diseases of maize, especially epidemiology of Northern Corn Leaf Blight in tropical regions of the world; Fusarium moniliforme kernel and ear rot; aerobiology of Fusarium spp. Mailing Add: Kauai Br Sta Univ of Hawaii RR 1 Box 278-A Kapaa HI 96746

OOMENS, FREDERICK WALTER, b Chicago, Ill, Mar 13, 39; m 63; c 3. AGRICULTURE. Educ: Wis State Univ, Platteville, BS, 63, MS, 67; Univ Mo-Columbia, PhD(agr educ), 71. Prof Exp: Soil conservationist, Soil Conserv Serv, USDA, 63-65; instr voc agr, Bloomington Community Schs, 65-66; instr agr mech, Wis State Univ-Platteville, 67-69; asst prof agr mech, 70-73, ASSOC PROF AGR, NORTHWEST MO STATE UNIV, 73- Res: Effect of group size upon achievement in agricultural mechanics; effect of irrigation on sugar content of corn populations. Mailing Add: Dept of Agr Northwest Mo State Univ Maryville MO 64468

OONNITHAN, EASWARAN SUKUMARAN, b Nilamel, India, June 2, 38; m 65; c 1. ENTOMOLOGY, AGRICULTURAL CHEMISTRY. Educ: Univ Rajasthan, BSc, 59; Indian Agr Res Inst, New Delhi, MS, 61; Univ Calif, Berkeley, PhD(biochem, org chem, med physics), 66. Prof Exp: Res entomologist, Univ Calif, Berkeley, 66-67; sr entomologist, 67-71, ENTOM RES SUPVR, BIOL RES DEPT, S C JOHNSON & SON INC, 72- Mem: AAAS; Entom Soc Am; NY Acad Sci; Am Chem Soc; Entom Soc India. Res: Study of structure and biological activity of insecticides chemicals, synergist and insect repellents; mode of action and metabolism of insecticides and their synergists. Mailing Add: S C Johnson & Son Inc Racine WI 53403

OORT, ABRAHAM H, b Leiden, Neth, Sept 2, 34; m 59; c 3. DYNAMIC METEOROLOGY. Educ: Univ Leiden, Drs, 59; Mass Inst Technol, SM, 64; State Univ Utrecht, PhD(meteorol), 64. Prof Exp: Res asst energy-cycle lower stratosphere, Mass Inst Technol, 63-64; res scientist, Royal Neth Meteorol Inst, 64-66; RES SCIENTIST, GEOPHYS FLUID DYNAMICS LAB, NAT OCEANIC & ATMOSPHERIC ADMIN, 66- Concurrent Pos: Vis lectr, Princeton Univ. Mem: Am Meteorol Soc; Am Geophys Union; Royal Meteorol Soc; NY Acad Sci. Res: Dynamics lower stratosphere; energy cycle atmospheric motions; energy spectrum; annual cycle atmospheric parameters; energetics Gulf Stream meanders; energy transport ocean currents. Mailing Add: Geophys Fluid Dynamics Lab Nat Ocean & Atmos Admin Box 308 Princeton NJ 08540

OOSTDAM, BERNARD LODEWIJK, b Amsterdam, Holland, Aug 13, 32; m 64; c 2. OCEANOGRAPHY, MARINE GEOLOGY. Educ: McGill Univ, BSc, 60; Scripps Inst Oceanog, Univ Calif, San Diego, MS, 64; Univ Del, PhD, 71. Prof Exp: Res assoc marine geol, Univ Hawaii, 62-63; staff geologist, Ocean Sci & Eng Inc, 63-65; oceanogr, Tex Instruments, 66; from asst prof to assoc prof, 66-72, PROF OCEANOG, MILLERSVILLE STATE COL, 72- Concurrent Pos: Managing dir, Evniron Sci Res Assocs, 67-; pres, Marine Sci Consortium, 68-; assoc, Inst Develop Riverine & Estuarine Systs, 68- Mem: Marine Technol Soc; Int Oceanog Found. Res: Nearshore and estuarine circulation; sediment transport. Mailing Add: Dept of Earth Sci Millersville State Col Millersville PA 17551

OOSTERHUIS, WILLIAM TENLEY, b Mt Vernon, NY, Mar 28, 40; m 64; c 2. CHEMICAL PHYSICS. Educ: Wis State Univ-Platteville, BS, 62; Carnegie Inst Technol, MS, 64, PhD(physics), 67. Prof Exp: NSF grant, Atomic Energy Res Estab, Eng, 67-68; res physicist, Carnegie-Mellon Univ, 68-69, asst prof chem & solid state physics, 69-74; solid state chem prog dir, 74-75, STAFF ASSOC, NAT SCI FOUND, 74- Mem: Am Phys Soc; Biophys Soc; Am Chem Soc. Res: Paramagnetic hyperfine interactions in transition metal complexes, including biological molecules; magnetic critical point phenomena; crystalline electric fields; Mössbauer effect. Mailing Add: Div of Mat Res Nat Sci Found Washington DC 20550

OOYAMA, KATSUYUKI, b Sapporo, Japan, Mar 5, 29; m 54; c 2. METEOROLOGY. Educ: Univ Tokyo, Rigaku-shi, 51; NY Univ, MS, 57, PhD(meteorol), 58. Prof Exp: Meteorologist, Japan Meteorol Agency, 51-55; asst, 55-58, res scientist, 58-62, from asst prof to assoc prof, 62-69, PROF METEOROL, NY UNIV, 69- Concurrent Pos: Mem staff, Div Sponsored Res, Mass Inst Technol, 63-64. Honors & Awards: Meisinger Award, Am Meteorol Soc, 68. Mem: Am Meteorol Soc. Res: Dynamic meteorology; geophysical fluid dynamics; dynamics of tropical cyclones. Mailing Add: Dept Meteorol & Oceanog NY Univ Sch of Eng New York NY 10453

OPAL, CHET BRIAN, b Chicago, Ill, Nov 30, 42; m 64. AERONOMY, ASTROPHYSICS. Educ: Johns Hopkins Univ, BA, 64, PhD(physics), 69. Prof Exp: Fel & res assoc physics, Joint Inst Lab Astrophys, Univ Colo, Boulder, 69-71; RES PHYSICIST, E O HULBURT CTR SPACE RES, NAVAL RES LAB, 71- Concurrent Pos: E O Hulburt fel, Naval Res Lab, 71- Mem: Am Phys Soc; Am Astron Soc; Am Geophys Union. Res: Electron-atom scattering; far-ultraviolet spectroscopy of astrophysical and geophysical phenomena from rockets and satellites; aurora and airglow. Mailing Add: Code 7120 Naval Res Lab Washington DC 20375

OPATOWSKI, IZAAK, b Warsaw, Poland, Feb 20, 05; nat US; m 35. APPLIED MATHEMATICS. Educ: Reale Inst Superiore Ing, Turin, Italy, DEng, 29; Univ Turin, DMath, 32. Prof Exp: Substitute prof math & physics, Reale Inst Technol Superiore Sommeiller, Turin, Italy, 34-36; engr, Fiat Co, Italy, 36-39; instr math & mech, Univ Minn, 39-42; assoc eng & math consult, Armour Res Found, Ill Inst Technol, 42-45; res math & biophys, Univ Chicago, 45-46; asst prof math, Univ Mich, 46-48; res assoc & asst prof math biol, Univ Chicago, 48-52; assoc res engr, Boeing Co, 52-56; assoc prof math, Ohio Northern Univ, 59-61; res assoc & assoc prof math biol, Univ Chicago, 62-70; INDEPENDENT RES HEALTH SCIENTIST, 70- Mem: AAAS; Am Math Soc; Am Soc Mech Eng; Am Phys Soc; Math Asn Am. Res: Mathematical biology; theoretical biophysics. Mailing Add: 4701 15th Ave NE Seattle WA 98105

OPDYKE, DAVID FRANKLIN, b Montpelier, Ohio, Sept 11, 15; m 44; c 2. PHYSIOLOGY. Educ: Heidelberg Col, BS, 37; Ind Univ, PhD(endocrinol), 42. Prof Exp: Asst, Ind Univ, 37-39, instr, 42-43; assoc eugenics rec off, Dept Genetics, Carnegie Inst, NY, 39-41; from instr to assoc prof physiol, Med Sch, Western Reserve Univ, 44-51; head physiol dept, Merck Inst Therapeut Res, 51-56; PROF PHYSIOL & CHMN DEPT, NJ MED SCH, COL MED & DENT NJ, 56- Mem: Am Physiol Soc; Am Heart Asn. Res: Endocrinology; marine biology; cardiovascular physiology; medical education. Mailing Add: NJ Med Sch Col of Med & Dent of NJ Newark NJ 07103

OPDYKE, DONALD LLOYD J, b Ocean Grove, NJ, Nov 22, 17; m 43; c 3. PHARMACOLOGY. Educ: Univ Colo, BA, 48; Wash Univ, PhD(anat), 52. Prof Exp: Res toxicologist, Procter & Gamble Co, 52-63; dir pharmacol res, Revlon Res Ctr, NY, 63-67, dir basic res, 67-71; PRES, RES INST FRAGRANCE MAT, INC, 71-

Mem: Am Chem Soc; Soc Toxicol; Electron Micros Soc Am; NY Acad Sci; Europ Soc Study Drug Toxicity. Res: Experimental pharmacology and pathology; toxicology. Mailing Add: Res Inst for Fragrance Mat Inc Box 1152 Englewood Cliffs NJ 07632

OPDYKE, NEIL, b Frenchtown, NJ, Feb 7, 33; m 58; c 3. GEOLOGY, GEOPHYSICS. Educ: Columbia Col, BA, 55; Univ Durham, PhD(geol), 58. Prof Exp: Fel, Rice Univ, 58-59; res fel, Australian Nat Univ, 59-60 & Univ Col Rhodesia & Nyasaland, 61-64; res assoc, 65-66, RES SCIENTIST, LAMONT-DOHERTY GEOL OBSERV, COLUMBIA UNIV, 64-, SR RES ASSOC, UNIV, 66- Concurrent Pos: Fulbright travel grant, Australia, 59-60; adj prof, Columbia Univ, 74- Mem: AAAS; Geol Soc Am; Am Geophys Union. Res: Paleomagnetism; paleoclimatology. Mailing Add: Lamont-Doherty Geol Observ Palisades NY 10964

OPECHOWSKI, WLADYSLAW, b Warsaw, Poland, Mar 10, 11; nat Can citizen; m 33; c 1. THEORETICAL PHYSICS. Educ: Univ Warsaw, MPhil, 35. Prof Exp: Lorentz-Fonds fel, Leiden & physicist, Teylers Stichting, 39-45; physicist, Philips Res Lab, Eindhoven, 45-48; assoc prof, 48-57, PROF PHYSICS, UNIV BC, 57- Concurrent Pos: Lorentz prof, State Univ Leiden, 64-65; vis prof, Roman Cath Univ Nijmegen, 68-69. Mem: Am Phys Soc; fel Royal Soc Can; Can Asn Physicists; Neth Phys Soc. Res: Quantum theory of magnetism; applications of group theory. Mailing Add: Dept of Physics Univ of BC Vancouver BC Can

OPEL, HOWARD, b Baltimore, Md, June 7, 28; m 64. REPRODUCTIVE PHYSIOLOGY, ENDOCRINOLOGY. Educ: Univ Md, BS, 50, MS, 59; Univ Ill, PhD(animal sci), 60. Prof Exp: RES SCIENTIST, ANIMAL SCI RES DIV, AGR RES SERV, USDA, 60- Mem: Soc Study Reproduction; Am Asn Anat; Am Soc Zoologists; Poultry Sci Asn; Endocrine Soc. Res: Neuroendocrine control of ovulation and oviposition in birds. Mailing Add: Animal Physiol & Genetics Inst Agr Res Ctr Beltsville MD 20705

OPENSHAW, MARTIN DAVID, b Mesa, Ariz, Oct 10, 40; m 64; c 2. SOIL FERTILITY, PLANT PHYSIOLOGY. Educ: Ariz State Univ, BS, 65; Iowa State Univ, MS, 68, PhD(soils), 70. Prof Exp: Teaching asst soil sci, Iowa State Univ, 65-67, instr, 67-70; ASST PROF SOIL SCI & EXTEN SOIL SPECIALIST, UNIV ARIZ, 70- Mem: Am Soc Agron; Soil Sci Soc Am. Res: Soil fertility work with field crops and the effect of fertilizers on the environment. Mailing Add: Coop Exten Serv Univ of Ariz Tucson AZ 85721

OPHEL, IVAN LINDSAY, b Adelaide, South Australia, July 20, 24; m 51; c 4. RADIATION ECOLOGY. Educ: Univ Okla, BS, 49, MS, 50. Prof Exp: Apprentice, Govt Eng & Water Supply Dept, South Australia, 40-41, asst chemist, 42-47, biologist, 51-52; asst limnol, Univ Okla, 48-50; sr biologist environ res, 53-73, HEAD, ENVIRON RES BR, CHALK RIVER NUCLEAR LABS, ATOMIC ENERGY CAN, LTD, 73- Concurrent Pos: Mem panel effects of ionizing radiation on aquatic organisms and ecosysts, Int Atomic Energy Agency, 71 & panel fisheries experts, Food & Agr Orgn, UN. Mem: Ecol Soc Am. Res: Biological problems of radioactive waste disposal; radioecology research. Mailing Add: Environ Res Br C R Nuclear Labs Atomic Energy of Can Ltd Chalk River ON Can

OPHER, REUVEN, b New York, NY, July 17, 32; m 55; c 2. ASTROPHYSICS, SOLID STATE PHYSICS. Educ: City Col New York, BS, 53; Harvard Univ, MA, 55, PhD(nuclear physics), 58. Prof Exp: Sr physicist, Lawrence Radiation Lab, Univ Calif, 57-61; NSF fel, Weizmann Inst Sci, Israel, 61-62; sr lectr, Dept Physics, 62-67, ASSOC PROF PHYSICS, TECHNION-ISRAEL INST TECHNOL, 67- Mem: Am Phys Soc; Phys Soc Israel; Europ Phys Soc. Mailing Add: Dept of Physics Technion-Israel Inst Technol Haifa Israel

OPIE, JOSEPH WENDELL, b Wilmot, SDak, Sept 27, 13; m 40; c 3. ORGANIC CHEMISTRY, FOOD TECHNOLOGY. Educ: Univ Minn, ChB, 35, PhD(org chem), 40. Prof Exp: Microanalyst, Univ Minn, 35-39; res chemist, Merck & Co, Inc, NJ, 40-42; res assoc, Nat Defense Res Comt, Univ Minn, 42; res chemist, Tex Co, NY, 42-44 & Wyeth, Inc, 44-46; head develop res sect, 46-48, org develop sect, 48-62, formulation concept dept, 62-67, concept develop dept, 67-69 & res & develop, Sperry Div, 69-71, mgr, 71-74, MGR DEVELOP DEPT CORP RES, GEN MILLS, INC, 74- Mem: Inst Food Technol; Am Chem Soc; AAAS. Res: Polyalkylgenzenes; polyalkyl-quinones; substituted coumarins; vitamin E; plastics; antibiotics; medicinals; fatty acids product and processes development; mineral flotation starch; plant gums; corrosion; phospholipids; food products. Mailing Add: 2515 W52nd St Minneapolis MN 55410

OPIE, THOMAS RANSON, b Staunton, Va, Dec 19, 48. ORGANIC CHEMISTRY, ORGANOMETALLIC CHEMISTRY. Educ: Davidson Col, BS, 71; Cornell Univ, MS, 74, PhD(org chem), 76. Prof Exp: RES CHEMIST, ROHM & HAAS CO, 75- Mem: Am Chem Soc; Am Sci Affil. Res: Synthetic reactions involving completely fluorinated organometallic compounds, chemical process research for monomers, polymers and herbicides. Mailing Add: Rohm & Haas Co Bristol Res Lab PO Box 219 Bristol PA 19007

ÖPIK, ERNEST JULIUS, b Port Kunda, Estonia, Oct 23, 93; m 21; c 6. ASTRONOMY. Educ: Tartu State Univ, Estonia, DPhNat, 23. Hon Degrees: DS, Queen's Univ, Belfast, 68. Prof Exp: Asst & Teacher observ, Moscow Univ, Russia, 18-20; privatdocent astron, Turkestan Univ, 20-21; astronomer & privatdocent, Tartu Univ, Estonia, 21-44; res assoc univ observ, Univ Hamburg, 44-45; prof & Estonian rector, Baltic Univ, Ger, 45-48; from vis prof to assoc prof astrophys, 56-68, PROF PHYSICS & ASTRON, UNIV MD, COLLEGE PARK, 68- Concurrent Pos: Res assoc univ observ & vis lectr, Harvard Univ, 30-34; res assoc & astronr, Armagh Observ, NIreland, 48-; ed jour, Irish Astron Soc. Honors & Awards: J Lawrence Smith Medal, Nat Acad Sci, DC, 60; Leonard Medal, Meteoritical Soc, 68; Kepler 400th Anniversary Medal, AAAS & Meteoritical Soc, 71; Astron Gold Medal, Royal Astron Soc, London, 75. Mem: foreign assoc Nat Acad Sci; Int Astron Union; Am Astron Soc; Royal Astron Soc; Irish Astron Soc. Res: Meteors; meteor physics; theory of meteor craters; stellar statistics, structure and evolution; ice ages; age of the universe; atomic collisions and radiation; micrometrical measures of double stars; photographic photometry. Mailing Add: Dept of Physics Univ of Md College Park MD 20742

OPITZ, HERMAN ERNEST, b Kingston, NY, Sept 28, 29; m 54; c 2. INORGANIC CHEMISTRY. Educ: Union Univ NY, BS, 51; Ind Univ, PhD(chem), 56. Prof Exp: Res chemist, Armstrong Cork Co, Pa, 55-69; MGR GLASS RES PACKAGING PROD DIV, KERR GLASS MFG CORP, 69- Mem: Am Ceramic Soc. Res: Silicon chemistry; inorganic polymers and foams; glass technology. Mailing Add: Packaging Prod Div Kerr Glass Mfg Corp Lancaster PA 17604

OPITZ, JOHN MARIUS, b Hamburg, Ger, Aug 15, 35; US citizen; m 61; c 5. PEDIATRICS, MEDICAL GENETICS. Educ: Univ Iowa, BA, 56, MD, 59; Am Bd Pediat, dipl, 64. Prof Exp: From intern to resident pediat, Univ Iowa Hosp, 59-61; resident, 61-62, from asst prof to assoc prof pediat & med genetics, 64-72, PROF PEDIAT & MED GENETICS & DIR WIS CLIN GENETICS CTR, UNIV WIS-MADISON, 72- Concurrent Pos: NIH fels pediat & med genetics, Univ Wis-Madison, 62-64. Mem: AAAS; Am Soc Human Genetics; Am Soc Zool; Am Inst Biol Sci; Int Soc Cranio-Facial Biol. Res: Hereditary diseases; single and multiple congenital anomalies; cytogenetics; genetic counseling; errors of sex determination and differentiation; clinical and developmental genetics. Mailing Add: Rm 109 Genetics Bldg Univ of Wis Madison WI 53706

OPLER, MARVIN KAUFMANN, b Buffalo, NY; c 2. ANTHROPOLOGY, PSYCHIATRY. Educ: Univ Mich, AB, 35; Columbia Univ, PhD(anthrop), 38. Prof Exp: Chmn anthrop, Reed Col, 38-43; chief community anal, War Relocation Authority, Dept Interior, 43-46; lectr psychoanal training, Inst Psychoanal, Univ Calif, 46-48; assoc prof anthrop, Stanford Univ, 48-50; assoc prof, Harvard Univ, 50-52; prof anthrop & social psychiat, Med Col, Cornell Univ, 52-58; PROF SOCIAL PSYCHIAT, SCH MED & PROF ANTHROP, FAC SOC SCI, STATE UNIV NY BUFFALO, 58- Concurrent Pos: Soc Sci Res Coun grant Japanese res, 46-47, grant-in-aid, US West Coast Japanese-Am, 49; East-West Philosopher's Conf fel, Univ Hawaii, 49; lectr psychol, Grad Div Clark Univ, 50; Nat Inst Ment Health, Rockefeller Bros Fund & Milbank Found res grants, Midtown Manhattan Ment Health Study, 52-58; Fenton Found lectr, State Univ NY Buffalo, 57; ed, Int J Social Psychiat, 57-; vis prof anthrop & sociol, New Sch Social Res, 57-58; vis prof, William Alanson White Found Psychiat, Psychol & Psychoanal, 57-59; mem nat comt, Nat Inst Ment Health Suicidology Study, 60; Nat Inst Ment Health training grants, State Univ NY Buffalo, 61-72; NSF grants, 60; grant, Human Ecol Fund, 64; vpres, First & Second Int Cong Social Psychiat, 64 & 69; fel, Wenner-Gren Found, 67; distinguished lectr, Celebrity Lectr Found, Middlebury Col, 68; chmn dept anthrop, State Univ NY Buffalo, 69-72; assoc ed, Am Anthropologist; chmn, Stirling Award Comt, Am Anthrop Asn, 68 & hon chmn, 71. Mem: Am Anthrop Asn; Int Asn Social Psychiat; Sigma Xi; Acad Psychoanal. Res: Psychological anthropology; medical anthropology; modern ethnic and national groups; cultural evolution and relevant cross-cultural studies; history of anthropology; American Indian studies; Puerto Rican Studies; African studies in social psychiatry; social science theory and method; cross-cultural psychopharmacology. Mailing Add: Depts Anthrop & Psychiat State Univ of NY Buffalo NY 14226

OPLER, MORRIS EDWARD, b Buffalo, NY, May 16, 07; m 57. ANTHROPOLOGY. Educ: Univ Buffalo, AB, 29, AM, 30; Univ Chicago, PhD(anthrop), 33. Prof Exp: Asst & res assoc anthrop, Univ Chicago, 33-35; Gen Educ Bd fel study rels between sch & community, 35-36; asst anthrop, Bur Indian Affairs, 36-37; vis lectr, Reed Col, 37-38; asst prof, Grad Sch, Claremont Cols, 38-45; asst prof educ, Harvard Univ, 46-48; prof anthrop & Asian studies, 48-69, dir S Asia Prog, 48-66, EMER PROF ANTHROP, CORNELL UNIV, 69-; PROF ANTHROP, UNIV OKLA, 69- Concurrent Pos: Guggenheim Found fel, 42-43; soc sci analyst, War Relocation Authority, 43-44; soc sci analyst, Off War Info, 44-45, dep chief & chief foreign morale anal div, 45; vis prof, Lucknow Univ, India, 53-54; fel, Ctr Advan Studies Behav Sci, 56-57; consult, NSF, 62-; fel, Nat Endowment Humanities, 68-69, consult, 70- Mem: Fel Am Anthrop Asn (pres elect, 61-62, pres, 62-63); Soc Appl Anthrop; Am Ethnol Soc; Asn Asian Studies; fel Am Folklore Soc. Res: Ethnology of Apache tribes; culture history of the Southwest; cultures of India. Mailing Add: Dept of Anthrop Univ of Okla 506 Dale Hall Norman OK 73069

OPLINGER, CARL SPADT, vertebrate zoology, see 12th edition

OPLINGER, EDWARD SCOTT, b Jewell, Kans, Apr 25, 43; m 64; c 2. AGRONOMY. Educ: Kans State Univ, BS, 65; Purdue Univ, Lafayette, MS, 69, PhD(agron), 70. Prof Exp: Asst prof, 70-75, ASSOC PROF AGRON, UNIV WIS-MADISON, 75- Mem: Am Soc Agron; Crop Sci Soc Am; Plant Growth Regulator Working Group. Res: Crop production; plant growth regulators; plant physiology. Mailing Add: 142C Moore Hall Dept of Agron Univ of Wis Madison WI 53706

OPP, ALBERT GEELMUYDEN, b Fargo, NDak, July 29, 31; m 60; c 2. ASTROPHYSICS. Educ: Univ NDak, BS, 53; St Louis Univ, MS, 55; Univ Göttingen, Dr rerNat(geophys), 61. Prof Exp: Terrestrial scientist, Air Tech Intel Ctr, US Air Force, 55-57, space systs specialist, Europ Hq, 59-62, instr physics, US Air Force Acad, 62-63; staff scientist, 63-66, dep chief particles & fields prog, 66-70, CHIEF HIGH ENERGY ASTROPHYS, NASA HQ, 70- Concurrent Pos: Consult, Kaman Nuclear Div, 62-63; fel pub affairs, Princeton Univ, 70-71. Mem: Am Geophys Union; Am Astron Soc. Res: Cosmic rays; gamma ray astronomy; x-ray astronomy; magnetic fields and plasmas in space; trapped radiation; planetary magnetospheres. Mailing Add: Off of Space Sci NASA Washington DC 20546

OPPEGARD, ALFRED LESTER, chemistry, see 12th edition

OPPEL, EDWIN IRVING, b New York, NY, Sept 6, 10; m 39; c 1. ECONOMIC GEOGRAPHY. Prof Exp: Res chemist, United Color & Pigment Co, 31-33; asst to vpres sales, Witco Chem Co, 33-36; indust ed, US Indust Chem, Inc, 36-39; mgr new prod develop, NJ Zinc Co, 39-56; sr vpres, Fantus Co, Inc, 56-72, exec vpres, 72-74; ECON GEOG CONSULT, 75- Mem: Fel AAAS; Chem Mkt Res Asn (exec secy, 47-49); fel Am Inst Chemists; Am Econ Asn; Am Geog Soc. Res: Industrial, economic and regional geography; plant location; fluorescent and phosphorescent pigments; powder and extractive metallurgy; protective coatings. Mailing Add: 19 Patton Pl Upper Montclair NJ 07043

OPPELT, JOHN ANDREW, b Baltimore, Md, Feb 4, 37; m 60; c 1. MATHEMATICS. Educ: Loyola Col, Md, AB, 59; Univ Notre Dame, MS, 61, PhD(math), 65. Prof Exp: Instr math, Univ Notre Dame, 65; actg asst prof, Univ Va, 65, asst prof, 65-70; assoc prof, 70-74, PROF MATH, GEORGE MASON UNIV, 74-, CHMN DEPT, 70- Concurrent Pos: Sesquicentennial scholar & vis asst prof, Univ Wash, 69-70. Mem: Am Math Soc. Res: Modern algebra; Abelian p-mixed groups. Mailing Add: Dept of Math George Mason Univ Fairfax VA 22030

OPPELT, JOHN CHRISTIAN, b Baltimore, Md, Dec 11, 31; m 56; c 5. ORGANIC CHEMISTRY. Educ: Loyola Col Md, BS, 53; Univ Md, College Park, MS, 58; Rutgers Univ, PhD(org chem), 67. Prof Exp: From chemist to res chemist, 58-74, SR RES CHEMIST, AM CYANAMID CO, 74- Mem: Am Chem Soc. Res: Synthesis of plastics additives, especially antioxidants, stabilizers, ultraviolet and near infrared absorbers. Mailing Add: Intermediates Res & Develop Am Cyanamid Co Bound Brook NJ 08805

OPPENHEIM, BERNARD EDWARD, b Chicago, Ill, July 5, 37; m 63; c 3. NUCLEAR MEDICINE, RADIOLOGY. Educ: Univ Ariz, BS, 59; Univ Chicago, MD, 63. Prof Exp: Intern, Michael Reese Hosp, Chicago, 63-64; resident radiol, 64-67, instr nuclear med, 70-71, ASST PROF NUCLEAR MED, UNIV CHICAGO, 71- Concurrent Pos: James Picker Found scholar radiol res, 72-73. Mem: Soc Nuclear Med. Res: Application of computers to imaging in nuclear medicine; long term effects of low dose in utero-irradiation. Mailing Add: Dept of Radiol Univ of Chicago Chicago IL 60637

OPPENHEIM, ELLIOT, b New York, NY, Nov 10, 15; m 44; c 3. CHEMISTRY,

PHYSIOLOGY. Educ: Univ Edinburgh, MD, 38; Am Bd Internal Med, dipl, 47. Prof Exp: From instr to asst prof, 46-68, ASSOC PROF CLIN MED, NY UNIV, 68- Concurrent Pos: Res fel med, Columbia Univ, 41-43; mem coun arteriosclerosis, Am Heart Asn. Mem: AAAS; AMA; Am Fedn Clin Res; fel Am Col Physicians. Res: Physiology of biliary tract; liver function; cholesterol metabolism; atherosclerosis. Mailing Add: 45 Gramercy Park N New York NY 10010

OPPENHEIM, IRWIN, b Boston, Mass, June 30, 29. CHEMICAL PHYSICS. Educ: Harvard Univ, AB, 49; Yale Univ, PhD(chem), 56. Prof Exp: Asst chem, Calif Inst Technol, 49-51; statist mech, Yale Univ, 51-53; physicist, Nat Bur Stand, 53-60; chief theoret physics, Gen Dynamics/Convair, 60-61; assoc prof, 61-65, PROF CHEM, MASS INST TECHNOL, 65- Concurrent Pos: Lectr, Univ Md, 54-60; res assoc, State Univ Leiden, 55-56; vis prof, Weizmann Inst, 58-59 & Univ Calif, San Diego, 66-67; Van Der Waals prof, Univ Amsterdam, 67. Mem: Fel Am Phys Soc; fel Am Acad Arts & Sci. Res: Statistical mechanics of irreversible processes; isotope effects; nuclear magnetic resonance. Mailing Add: Dept of Chem Mass Inst of Technol Cambridge MA 02139

OPPENHEIM, JOSEPH HAROLD, b Connellsville, Pa, Oct 10, 26. ALGEBRA. Educ: Univ Chicago, PhB, 48, BS, 54; Univ Ill, MS, 56, PhD(math, algebra), 62. Prof Exp: Asst prof math, Rutgers Univ, 60-65; asst prof, 66-71, ASSOC PROF MATH, CALIF STATE UNIV, SAN FRANCISCO, 71- Mem: Math Asn Am; Am Math Soc. Res: Representation theory of groups. Mailing Add: Dept of Math Calif State Univ San Francisco CA 94132

OPPENHEIM, RONALD WILLIAM, b Des Moines, Iowa, Nov 2, 38; m 68; c 1. PSYCHOBIOLOGY. Educ: Drake Univ, BA, 62; Wash Univ, PhD(biopsychol), 67. Prof Exp: Res biologist, Sch Med, Wash Univ, 67-68; RES SCIENTIST, RES DIV, NC DEPT OF MENT HEALTH, 68- Concurrent Pos: Adj prof, Neurobiol Prog, Univ NC, Chapel Hill, 71- Mem: AAAS; Am Inst Biol Sci; Animal Behav Soc; Neurosci Soc. Res: Neuroembryology; ontogeny of avian embryonic behavior; early posthatching sensory-motor behavior in Aves; neuroanatomical and neuropharmacological development of the spinal cord. Mailing Add: Res Div Dept Mental Health Dorothea Dix Hosp Raleigh NC 27602

OPPENHEIMER, CARL HENRY, JR, b Los Angeles, Calif, Nov 13, 21; m 71; c 3. OCEANOGRAPHY, ECOLOGY. Educ: Univ Southern Calif, BA, 47, MA, 49; Univ Calif, PhD(marine microbiol), 51. Prof Exp: Asst marine biologist, Scripps Inst Calif, 51-55; sr res scientist, Res Ctr, Pan Am Petrol Corp, Okla, 55-57; res scientist & lectr marine microbiol, Univ Tex, 57-61; assoc prof, Inst Marine Sci, Univ Miami, 61-64; prof biol, Fla State Univ, 64-66, chmn dept oceanog, 66-69, prof oceanog & dir, Edward Ball Marine Lab, 66-71; prof microbiol & dir, Marine Sci Inst, 71-73, PROF MARINE SCI, UNIV TEX, PORT ARANSAS, 73- Concurrent Pos: Fulbright fel, Univ Oslo, 52-53; res fel, Sea Fish Hatchery, Arendal, Norway, 55; chmn, Int Symposium Marine Microbiol, 61; mem subcomt marine biol, President's Sci Adv Comt, 65-66; chmn, Conf Unresolved Probs Marine Microbiol, 66; mem, Japan-US Cultural Exchange Prog, 66 & President's Panel Oil Pollution, 68-71; vpres, Gulf Univ Res Corp, 66-70, mem prog planning coun, 70-; mem, Governor's Comn Marine Sci & Technol, Fla, 67-71 & sci comn, NSF, 69-72; panel chmn, Am Assembly-Uses of the Sea, 69; US rep adv panel eco-sci, NATO, 71-74; Univ Tex Rep, Univ Coun Water Resources, 71-; mem, Tex Senate Interim Coastal Zone Study Comn, 71- & adv coun, Southern Interstate Nuclear Bd, 71- Mem: Am Soc Microbiol; Am Soc Limnol & Oceanog; Geochem Soc; Brit Soc Gen Microbiol. Res: Distribution and function of marine microorganisms; diseases of marine fishes; biological corrosion; marine productivity cycles; effect of detergents in the oceans; marine microbiology; origin and ecology of oil; biological criteria for coastal zone management. Mailing Add: Marine Sci Inst Univ of Tex Port Aransas TX 78373

OPPENHEIMER, FRANK, b New York, NY, Aug 14, 12; m 37; c 2. PHYSICS. Educ: Johns Hopkins Univ, BA, 33; Calif Inst Technol, PhD(physics), 39. Prof Exp: Asst physics, Stanford Univ, 39-41; res assoc, Univ Calif, Berkeley, 41-47; asst prof, Univ Minn, Minneapolis, 47-49; teacher pub sch, Colo, 57-59; res assoc, 59-61, assoc prof, 61-63, PROF PHYSICS, UNIV COLO, BOULDER, 63- Concurrent Pos: Consult high & elem sch sci teaching develop progs, Educ Serv, Inc, Mass, 60-; Guggenheim fel, 65; hon res assoc, Univ Col, Univ London, 65; on leave as dir, Palace of Arts & Sci Exploratorium, San Francisco, 69- Honors & Awards: Robert A Millikan Award, Am Asn Physics Teachers & Prentice Hall, 73. Mem: AAAS; fel Am Phys Soc. Res: Beta ray spectroscopy of radioactive elements; electromagnetic uranium isotope separations; cosmic radiation; high energy particle physics; history of contemporary physics; physics pedagogy; social effects of science. Mailing Add: Palace of Arts & Sci Exploratorium 3601 Lyons St San Francisco CA 94123

OPPENHEIMER, HENRY ERNEST, b St Louis, Mo, Oct 14, 14; m 43; c 5. MEDICINE. Educ: St Louis Univ, BS, 35, MD, 37, MS, 41. Prof Exp: From instr to asst prof, 41-62, ASSOC PROF INTERNAL MED, SCH MED, ST LOUIS UNIV, 62- Concurrent Pos: Pvt pract. Mem: AMA; Am Heart Asn; Am Diabetes Asn. Res: Clinical diabetes; diabetes education; diabetes detection and early diagnosis. Mailing Add: 141 North Meramec Clayton MO 63105

OPPENHEIMER, JACK HANS, b Eglesbach, Ger, Sept 14, 27; US citizen; m 53; c 3. ENDOCRINOLOGY, INTERNAL MEDICINE. Educ: Princeton Univ, AB, 49; Columbia Univ, MD, 53; Am Bd Internal Med, dipl, 61. Prof Exp: Intern med, Boston City Hosp, 53-54; resident med, Duke Univ Hosp, 57-59; asst clin prof neurol, Sch Med, NY Univ, 62-65; from asst prof to assoc prof, 66-73, PROF MED, ALBERT EINSTEIN COL MED, YESHIVA UNIV, 73-; HEAD ENDOCRINE SERV, MONTEFIORE HOSP & MED CTR, 68- Concurrent Pos: Fel med, Mem Ctr & Thyroid-Pituitary Study Group, Sloan-Kettering Inst, 54-55; vis fel, Col Physicians & Surgeons, Columbia Univ, 59-60; asst physician, Endocrinol Lab, Presby Hosp, New York, 59-60; from asst attend physician to attend physician, Div Med, Montefiore Hosp & Med Ctr, 60-68; prin investr USPHS grant, 61-; career scientist, Health Res Coun, New York, 62-72; mem bd trustees, Windward Sch, White Plains, NY, 69-71. Honors & Awards: Van Meter Award, Am Thyroid Asn, 65. Mem: Am Asn Physicians; Am Soc Clin Invest; fel Am Col Physicians; Am Physiol Soc; Am Thyroid Asn (1st vpres, 74). Res: Peripheral metabolism and action of the thyroid hormones. Mailing Add: Montefiore Hosp & Med Ctr 111 E 210th St Bronx NY 10467

OPPENHEIMER, JANE MARION, b Philadelphia, Pa, Sept 19, 11. BIOLOGY. Educ: Bryn Mawr Col, BA, 32; Yale Univ, PhD(zool), 35. Prof Exp: Sterling fel biol, Yale Univ, 35-36, Am Asn Univ Women Berliner res fel, 36-37; res fel embryol, Univ Rochester, 37-38; instr biol, 38-42, Guggenheim fel, 42-43, actg dean grad sch, 46-47, from asst prof to prof biol, 43-74, PROF HIST OF SCI, BRYN MAWR COL, 74- Concurrent Pos: Rockefeller Found fel, 50-51; Guggenheim fel, 52-53; sr fel, NSF, 59-60; vis prof biol, Johns Hopkins Univ, 66-67; mem, NIH Hist Life Sci Study Sect, 66-70; exchange prof fac sci, Univ Paris, 69. Honors & Awards: Wilbur Lucius Cross Medal, Yale Grad Alumni Asn. Mem: Soc Develop Biol; Am Soc Zoologists (treas, 57-59, pres, 74); Hist Sci Soc; Am Asn Anat; Am Asn Hist Med. Res: Developmental biology; experimental analysis of teleost development. Mailing Add: Biol Bldg Bryn Mawr Col Bryn Mawr PA 19010

OPPENHEIMER, JOHN REED, b New York, NY, Oct 10, 41; m 67. PRIMATOLOGY, ANIMAL ECOLOGY. Educ: Univ Conn, BA, 63; Univ Ill, Urbana, MS, 65, PhD(zool), 68. Prof Exp: USPHS res fel, 68-70, ASST PROF PATHOBIOL, JOHNS HOPKINS UNIV, 70-, ECOLOGIST, INT CTR MED RES, 70- Mem: AAAS; Animal Behav Soc; Ecol Soc Am; Int Primatol Soc; Int Soc Tropical Ecol. Res: Behavior and ecology; cebus and langur monkeys, fresh water fishes, dung and ground beetles; ecology; dogs, spiders, viruses, ponds, tropical America and Asia; primate communication; beetle and rodent taxonomy. Mailing Add: Lab Ecol & Behav Dept Pathbiol Johns Hopkins Univ Baltimore MD 21205

OPPENHEIMER, LARRY ERIC, b New York, NY, Aug 18, 42; m 64; c 2. COLLOID CHEMISTRY. Educ: Clarkson Col Technol, BS, 63, MS, 65, PhD(phys chem), 67. Prof Exp: SR RES CHEMIST, RES LABS, EASTMAN KODAK CO, 66- Mem: AAAS; Am Chem Soc; Sigma Xi. Res: Light scattering by liquids and solutions; colloid and surface chemistry; growth of colloidal particles; chemical reactivity at solid-liquid interfaces; unconventional imaging systems. Mailing Add: Res Labs Eastman Kodak Co Kodak Park Rochester NY 14650

OPPENHEIMER, MICHAEL, b New York, NY, Feb 28, 46. ASTROPHYSICS. Educ: Mass Inst Technol, SB, 66; Univ Chicago, PhD(chem physics), 70. Prof Exp: Res fel, 71-73, PHYSICIST, CTR ASTROPHYS, HARVARD COL OBSERV, 73-, LECTR, DEPT ASTRON, SMITHSONIAN ASTROPHYS OBSERV, HARVARD UNIV, 71- Mem: Am Phys Soc; AAAS; Fedn Am Scientists. Mailing Add: Dept of Astron Harvard Univ 60 Garden St Cambridge MA 02138

OPPENHEIMER, MORTON JOSEPH, b Philadelphia, Pa, June 9, 05; m 34; c 2. PHYSIOLOGY. Educ: Ursinus Col, AB, 27; Temple Univ, MD, 32, EdM, 38; Hon Degrees: ScD, Ursinus Col, 57. Prof Exp: From instr to prof, 33-73, chmn dept, 44-70, EMER PROF PHYSIOL, SCH MED, TEMPLE UNIV, 73- Concurrent Pos: China Med Bd fel, Manila & Bangkok, 66, 70; with Inst Exp Med, Mayo Clin, 40, 42, 44, 46; consult, US Naval Hosp, Philadelphia, & US Vet Admin Hosp, Philadelphia, Nat Heart Inst & Cardiovasc Training Comt & Nat Libr Med, Nat Med AV Ctr; adj prof, biomed group, Drexel Inst. Res: Circulation: heart; radiopaque media. Mailing Add: Dept of Physiol Temple Univ Sch of Med Philadelphia PA 19140

OPPENHEIMER, NORMAN JOSEPH, b Summit, NJ, Aug 20, 46; m 69. BIOCHEMISTRY. Educ: Brown Univ, BS, 68; Univ Calif, San Diego, P (PhD(chem), 73. Prof Exp: Res assoc biochem, Ind Univ, 73-75; ASST PROF PHARMACEUT CHEM, UNIV CALIF, SAN FRANCISCO, 75- Mem: AAAS. Res: Small molecule/macromolecule interactions, steroid/phospholipid interactions, and nuclear metabolism of pyridine coenzymes. Mailing Add: Dept of Pharmaceut Chem Univ of Calif San Francisco CA 94143

OPPENHEIMER, STEVEN BERNARD, b Brooklyn, NY, Mar 23, 44; m 71. DEVELOPMENTAL BIOLOGY, CANCER. Educ: City Univ New York, BS, 65; Johns Hopkins Univ, PhD(biol), 69. Prof Exp: Am Cancer Soc fel biol, Univ Calif, San Diego, 69-71; asst prof, 71-74, ASSOC PROF BIOL, CALIF STATE UNIV, NORTHRIDGE, 74-, NAT CANCER INST RES GRANT, 72- Concurrent Pos: Trustee, Calif State Univ Found, Northridge, 74- Mem: AAAS; Soc Develop Biol; Am Soc Cell Biol; Am Soc Zoologists; NY Acad Sci. Res: Molecular basis of intercellular adhesion and the role of cell surface carbohydrates in morphogenesis and malignancy. Mailing Add: Dept of Biol Calif State Univ Northridge CA 91324

OPPENLANDER, GEORGE C, organic chemistry, see 12th edition

OPPERMAN, JAMES ALEX, b Milwaukee, Wis, Nov 18, 42; m 66; c 2. PHARMACOLOGY, BIOCHEMISTRY. Educ: Univ Wis-Madison, BS, 65, MS, 67, PhD(pharmacol), 70. Prof Exp: Sr res investr, 70-75, GROUP LEADER DRUG METAB, DEPT DRUG METAB & RADIOCHEM, G D SEARLE & CO, 75- Res: Drug metabolism. Mailing Add: G D Searle & Co PO Box 5110 Chicago IL 60680

OPPERMANN, ROBERT ARTHUR, b Chicago, Ill, Aug 26, 28; m 57; c 3. MICROBIOLOGY, BIOCHEMISTRY. Educ: Univ Ill, BS, 55, MS, 56, PhD(bact), 59. Prof Exp: Asst rumen bact, Univ Ill, 55-59; sr res scientist, Nalco Chem Co, 59-70; group leader microbiol lab, Alberto-Culver Co, Ill, 71-73; SR RES MICROBIOLOGIST, CORNING GLASS WORKS, 73- Mem: Am Soc Microbiol; Soc Indust Microbiol; Sigma Xi (pres, 62-64). Res: Bacterial methanogenesis; microbiocide production; polysaccharide fermentation; enzyme usage; new testing procedures; immobilized enzymes; industrial fermentation. Mailing Add: Corning Glass Works Sullivan Park Corning NY 14830

OPPO, GIUSEPPE, b Cagliari, Italy. ELEMENTARY PARTICLE PHYSICS, THEORETICAL PHYSICS. Educ: Milan Polytech Inst, DrIng(elec eng), 52. Prof Exp: Consult engr, Italy, 52-53; consult engr, Ital Mach Tools Builder's Asn, 53-54; elec engr, Hochspennungsgesellschaft, Cologne, Ger, 54-55 & Siemens-Reiniger, Erlangen, Ger, 55-56; res scientist semiconductors, Philco Corp, Pa, 56-58; assoc prof elec eng, Univ Wichita, 58-59; assoc prof physics, Univ Louisville, 59-61; assoc prof, Polytech Inst Brooklyn, 61-74; CONSULT. Concurrent Pos: NSF grant, 66-68. Mem: Am Phys Soc; Ital Phys Soc; Europ Phys Soc. Res: Elementary particle theory. Mailing Add: 166-24 Gothic Dr Jamaica NY 11432

OPTON, EDWARD MILTON, b Far Rockaway, NY, May 16, 09; m 35; c 3. VIROLOGY. Educ: Yale Univ, PhB, 31, MS, 42, PhD, 43. Prof Exp: From asst prof to assoc prof zool & physiol, NDak Col, 46-52; res asst, 52-59, RES ASSOC EPIDEMIOL, SCH MED, YALE UNIV, 59- Mem: AAAS; Soc Protozool; Am Soc Zool. Res: Protozoology; viruses. Mailing Add: Yale Univ Sch of Med New Haven CT 06510

O'QUINN, SILAS EDGAR, b Brunswick, Ga, Feb 24, 24. DERMATOLOGY. Educ: Univ Mich, MD, 49. Prof Exp: Intern, Charity Hosp La, New Orleans, 49-50; resident, E A Conway Mem Hosp, Monroe, 50-51, 53-54; pvt pract gen med, Ga & SC, 54-59; resident dermat, Charity Hosp La, 59-62; from instr to assoc prof dermat, 62-68, assoc dean sch med, 71-73, PROF DERMAT, SCH MED, LA STATE UNIV MED CTR, NEW ORLEANS, 66-, DEAN SCH MED, 73- Concurrent Pos: Sr vis physician & dermat-in-chief, La State Univ Div-Charity Hosp La; consult, Vet Admin Hosp, New Orleans & Keesler AFB, Biloxi, Miss. Honors & Awards: AMA Cert of Merit, 67; Am Acad Dermat Bronze Award, 67. Mem: Am Acad Dermat; Am Dermat Asn. Res: Allergic contact dermatitis; photosensitivity; allergic drug reactions. Mailing Add: Dept of Dermat La State Univ Med Ctr New Orleans LA 70112

ORAHOVATS, PETER DIMITER, b Sofia, Bulgaria, Apr 22, 22; nat US; m 45; c 1. PHYSIOLOGY, PHARMACOLOGY. Educ: St Klement Univ, Bulgaria, MD, 45. Prof Exp: Instr clin physiol, St Clement Univ, Bulgaria, 45-46; mem staff physiol, Col Physicians & Surgeons, Columbia Univ, 46-51; res assoc physiol & pharmacol, Merck Inst Therapeut Res, 51-59; med dir, 59-60, med & sci dir, 61-62, vpres & dir res &

develop div, 62-66, VPRES & DIR SCI DIV, BRISTOL-MYERS PROD DIV, BRISTOL-MYERS CO, 66- Concurrent Pos: USPHS fel, Col Physicians & Surgeons, Columbia Univ, 49-51. Mem: Am Soc Pharmacol & Exp Therapeut; Harvey Soc; Royal Soc Med; Med Asn. Res: Medicine; cardiovascular physiology; blood volume and formation; essential hypertension; pharmacology of analgesics and narcotics. Mailing Add: Bristol-Myers Co 345 Park Ave New York NY 10022

ORANGE, ARNOLD, b New York, NY, Nov 14, 35; m 58; c 2. GEOPHYSICS, ELECTRICAL ENGINEERING. Educ: Mass Inst Technol, BS & MS, 58. Prof Exp: Res engr, Air Force Cambridge Res Lab, 58-65; sr scientist, Geosci, Inc, 65-70, chief engr, 70-71; mem tech staff, Mitre Corp, 71-72; chief geophysicist, Geophys Surv Systs, Inc, 72-73; MGR TECH SERV, BBN-GEOMARINE SERV CO, 73- Mem: Inst Elec & Electronics Engrs; Am Geophys Union; Soc Explor Geophysicists. Res: Airborne communications; hardened communications; electrical properties of rock; geoelectromagnetism; geophysical prospecting techniques. Mailing Add: BBN-Geomarine Serv Co 3344 Crossview Houston TX 77042

ORATZ, MURRAY, b New York, NY, Apr 17, 27; m 52; c 2. BIOCHEMISTRY. Educ: Long Island Univ, BS, 48; Clarkson Col Technol, MS, 50; NY Univ, PhD(biochem), 57. Prof Exp: ASST CHIEF NUCLEAR MED SERV, VET ADMIN HOSP, 55- Concurrent Pos: Adj prof biochem, Brookdale Ctr NY Univ, Col of Dent, 67- Mem: Am Chem Soc. Res: Regulation and stimuli for the production of plasma proteins; dynamics of cardiac metabolism. Mailing Add: Nuclear Med Serv Vet Admin Hosp 1st Ave at E 24th New York NY 10010

ORAVEC, RALPH JOHN, applied mathematics, see 12th edition

ORBACH, RAYMOND LEE, b Los Angeles, Calif, July 12, 34; m 56; c 3. SOLID STATE PHYSICS. Educ: Calif Inst Technol, BS, 56; Univ Calif, Berkeley, PhD(physics), 60. Prof Exp: NSF fels, Oxford Univ, 60-61, asst prof appl physics, Harvard Univ, 61-63; assoc prof, 63-66, PROF PHYSICS, UNIV CALIF, LOS ANGELES, 66- Concurrent Pos: Alfred P Sloan Found fels, 63-67; asst vchancellor acad change & curric develop, Univ Calif, Los Angeles, 70-72; Guggenheim Found fel, 73-74. Mem: Am Phys Soc. Res: Paramagnetic resonance and spin-lattice relaxation; collective properties of ordered magnetic systems; lattice vibrations and their interactions; optical properties of solids; thermal conductivity of paramagnetic salts; magnetic resonance properties of dilute magnetic alloys. Mailing Add: Dept of Physics Univ of Calif Los Angeles CA 90024

ORBAN, EDWARD, b Youngstown, Ohio, Mar 5, 15; m 41; c 2. APPLIED CHEMISTRY. Educ: Univ Buffalo, BS, 40; Univ Md, PhD(phys chem), 44. Prof Exp: Anal chemist, E I du Pont de Nemours & Co, NY, 40-41; asst, Univ Md, 41-44; res chemist, Socony-Mobil Co, Inc, NJ, 44-46; res chemist, 46-48, group leader, 48-52, chief tech info sect, 52-53 & chem sect, 53-54, sr develop investr, 54-57, mgr develop, Cent Res Dept, 57-69, planning mgr, New Enterprises Div, 69-71, PLANNING MGR COM PROD, MONSANTO CO, 71- Mem: Am Chem Soc; Am Ceramic Soc; Electrochem Soc; Instrument Soc Am. Res: Electrochemistry; metals; instrumentation; electronic materials; single crystal semiconductor materials preparation. Mailing Add: 90 Thorncliff Lane Kirkwood MO 63122

ORBIN, DAVID PAUL, b Youngwood, Pa, May 29, 44; m 67; c 2. NEMATOLOGY, BOTANY. Educ: Otterbein Col, BS, 66; Auburn Univ, PhD(nematol), 70. Prof Exp: Asst prof biol, Pa State Univ, New Kensington, 70-71; ASST PROF BIOL, PA STATE UNIV, HAZLETON, 71- Mem: Am Inst Biol Sci; Soc Nematol. Res: Survival of plant-parasitic nematodes in the absence of the host. Mailing Add: Dept of Biol Pa State Univ Hazleton PA 18201

ORBISON, JAMES LOWELL, b Kans, Mar 8, 15; m 41; c 2. PATHOLOGY. Educ: Univ Ottawa, AB, 37; Mich State Col, MS, 39; Northwestern Univ, MD, 44. Prof Exp: Sr instr path, Western Reserve Univ, 47-49, from asst prof to assoc prof, 49-55; Whipple prof & chmn dept, 55-70, DEAN SCH MED & DENT, UNIV ROCHESTER, 67- Concurrent Pos: Mem path test comt, Nat Bd Med Examr, 56-62, chmn, 61-62; mem sci adv bd, Armed Forces Inst Path, 57-62, 70-74; mem, Intersoc Comt Res Potential Path, 59-, pres, 61-75; mem path training comt, NIH, 60-63, chmn, 63-66; mem, Cancer Res Ctr Comt, 68- Mem: Am Asn Path & Bact (secy-treas, 65-68, vpres, 68, pres, 69); Am Soc Exp Path; Int Acad Path (pres, 63-64). Res: Connective tissue; vascular disease. Mailing Add: Univ of Rochester Sch of Med & Dent Rochester NY 14642

ORCHIN, MILTON, b Barnesboro, Pa, June 4, 14; m 41; c 2. ORGANIC CHEMISTRY. Educ: Ohio State Univ, BA, 36, MA, 37, PhD(org chem), 39. Prof Exp: Chemist, Food & Drug Admin, USDA, 39-42, org res chemist, Bur Animal Indust, 42-43; chief org chem sect synthetic liquid fuels, US Bur Mines, 43-53; assoc prof appl sci, 53-56, head dept chem, 56-62, dir basic sci lab, 62-71, PROF CHEM, UNIV CINCINNATI, 56-, DIR HOKE S GREENE LAB CATALYSIS, 71- Concurrent Pos: Guggenheim fel, Sieff Inst, 47-48; vis prof, Ohio State Univ, 66, Israel Inst Technol, 68 & Univ Calif, Berkeley, 69. Mem: AAAS; The Chem Soc. Res: Molecular complexes; chemistry of carbon monoxide; ultraviolet absorption spectroscopy; catalysis. Mailing Add: Dept of Chem Univ of Cincinnati Cincinnati OH 45221

ORCUTT, BRUCE CALL, b New York, NY. STATISTICS, NUMERICAL ANALYSIS. Educ: Cornell Univ, AB, 61; Purdue Univ, MS, 64; Univ Va, PhD(math), 69. Prof Exp: Asst prof math & statist, Georgetown Univ, 69-75; SR RES SCIENTIST MATH & STATIST, NAT BIOMED RES FOUND, GEORGETOWN MED CTR, 75- Mem: Am Statist Asn; Asn Comput Mach. Res: Biomedical research. Mailing Add: Nat Biomed Res Found Georgetown Univ Med Ctr Washington DC 20007

ORCUTT, DAVID MICHAEL, b Oklahoma City, Okla, June 10, 43; m 67; c 1. PLANT PHYSIOLOGY. Educ: Univ Okla, BS, 66; Incarnate Word Col, MS, 70; Univ Md, PhD(plant physiol), 73. Prof Exp: Technician plant physiol, US Air Force Sch Aerospace Med, 67-70; ASST PROF PLANT PHYSIOL, VA POLYTECH INST & STATE UNIV, 73- Mem: Am Soc Plant Physiologists; Phycol Soc Am. Res: The role of plant steroids in growth regulation, host parasite interactions and physiology in higher plants and aquatic plants. Mailing Add: Dept of Plant Path & Physiol Va Polytech Inst & State Univ Blacksburg VA 24061

ORCUTT, DONALD ADELBERT, b Traverse City, Mich, June 25, 26; m 50; c 4. BIOCHEMISTRY, INDUSTRIAL PHARMACY. Educ: Purdue Univ, Lafayette, BS, 50. Prof Exp: Chemist, Armour Pharmaceut, 51-52 & Whitehall Pharmacal Co, 52-58; supvr chem, Miles Labs, Inc, 58-60, asst to dir pharm res & develop, 60-65; tech dir contract pharm, 65-67, VPRES & DIR LABS, STANDARD PHARMACAL CORP, 67- Mem: Am Pharmaceut Asn; Am Chem Soc; Pharmaceut Mfrs Asn; Inst Food Technologists. Res: Pharmaceutical formulation, chemistry and instrumentation; food and drug law. Mailing Add: 179 Pauline Dr Elgin IL 60120

ORCUTT, FRED SCOTT, SR, b Sioux City, Iowa, Nov 27, 07; m 35; c 2. BIOCHEMISTRY, BACTERIOLOGY. Educ: Univ Wis, BS, 31, MS, 33, PhD(bact), 35. Prof Exp: Squibb fel & res assoc anesthesia & pharmacol, Univ Wis, 35-36; instr bact, Dept Biol, 36-38, from asst prof to assoc prof bact & biochem, 38-45, prof bact, 45-68, head dept biol, 58-66, EMER PROF BACT, VA POLYTECH INST & STATE UNIV, 68- Concurrent Pos: Mem Army specialized training prog, Off Surgeon Gen, 43-46; consult, Atomic Energy Comn contract, Oak Ridge Inst Nuclear Studies, 49-50, mem contract staff, 50-54; consult, Dairymen's Coop, Hawaii, 53-54 & Merck & Co, 55-56; vis prof, Mt Lake Biol Sta, Univ Va, 51 & 55 & Univ Hawaii, 53-54; mem staff, Am Physiol Soc Inst Col Teachers, 57. Mem: AAAS; Am Soc Microbiol; Am Acad Microbiol; Am Inst Biol Sci. Res: Biochemistry of nitrogen fixation; mechanisms of anesthesia; biochemical methods; cellular physiology; general soil and dairy bacteriology. Mailing Add: 1305 Hillcrest Dr Blacksburg VA 24060

ORCUTT, FREDERIC SCOTT, JR, b Roanoke, Va, Nov 29, 40; m 65; c 1. ETHOLOGY, ORNITHOLOGY. Educ: Cornell Univ, BS, 63, MS, 65, PhD(ethology), 69. Prof Exp: Lectr biol, Univ Sask, 69-70; sr res assoc avian reproductive physiol, Univ Wash, 70-71; ASST PROF BIOL, UNIV AKRON, 71- Mem: Am Inst Biol Sci; Am Ornith Union; Cooper Ornith Soc; Animal Behav Soc; Sigma Xi. Res: Hormonal determinants of behavior; avian reproductive physiology; evolution of behavior. Mailing Add: Dept of Biol Univ of Akron Akron OH 44304

ORCUTT, HAROLD GEORGE, b Washington, DC, May 6, 18; m 44; c 2. FISH BIOLOGY. Educ: Wilson Teachers Col, BS, 41; Stanford Univ, PhD(fisheries biol), 49. Prof Exp: Asst biol, Wilson Teachers Col, 38-39; teacher high sch, DC, 40-41; marine investr, State Fish Comn, NH, 41-42; marine biologist, Hopkins Marine Sta, Calif Dept Fish & Game, 48-49 & mus natural hist, Stanford Univ, 49-58, dir marine biol res lab, Stanford Univ, 58-61, dir marine resources res lab, 61-67, LAB SUPVR MARINE RESOURCES REGION, CALIF DEPT FISH & GAME, 67- Mem: Am Inst Fishery Res Biol; Sigma Xi. Res: Research conservation and administration of marine fish and shellfish. Mailing Add: Calif Dept Fish & Game Marine Resour Lab 411 Burgess Dir Menlo Park CA 94025

ORCUTT, JAMES ADDISON, b Mobile, Ala, July 11, 18; m; c 8. PHARMACOLOGY. Educ: Univ Chicago, SB, 39, SM, 40; Univ Rochester, PhD(pharmacol), 46. Prof Exp: Asst res chemist, Am Meat Inst, Univ Chicago, 40-42, res asst anat, Univ, 43; res chemist, Cutter Labs, Calif, 44; asst pharmacol, Sch Med & Dent, Univ Rochester, 44-45, instr, 46; chief pharmacol div, Eaton Labs, Inc, 46-53; sr pharmacologist, Riker Labs, Inc, Calif, 54-56; air pollution chemist, Los Angeles County, 56-61; partner, Orcutt-Chavez Res Labs, 60-62; asst prof pharmacol, Calif Col Med, 62-63, assoc prof, pharmacol & physiol, 63-66; actg chmn dept physiol, 66-67, prof pharmacol & chmn dept, 66-72, PROF PRECLIN SCI, COL OSTEOP MED & SURG, 72- Concurrent Pos: Asst prof chem, US Army Corps Eng, 44-46; mem adv res comt, Norwich Pharmacal Co, NY, 46-49, Am Drug Mfgrs Asn rep, Joint Comt Postal Classification Poisons, 50; mem panel appraisers, Am Inst Biol Sci Comt on Handbook Biol Data, Nat Res Coun, 51; consult, Armstrong Metal Specialties, Inc, NY, 53-54 & Food, Drug & Cosmetic Act, US Dept Health, Educ & Welfare, 59; vis lectr, Advan Math Seminar, Los Angeles city Schs, 59; vis assoc prof, San Fernando Valley State Col, 62; assoc mem attend staff & mem therapeut & drug specifications comts, Los Angeles County Gen Hosp, 62-66; mem, Int Union Physiol Sci & Int Cong Pharmacologists; consult pharmacol, Am Asn Nat Bd Examr, 70- Mem: Am Soc Pharmacol & Exp Therapeut; Am Statist Asn; Am Med Writers; Asn; Am Soc Clin Pharmacol & Therapeut. Res: Physiological mechanisms of adaptation to exotic atmospheres, including micro-contaminants; utilization of selective central-nervous-system depressants for the delineation of central-nervous mechanisms and pathways; quantification of biological phenomena. Mailing Add: Div Neurosci & Psychiat Col of Osteop Med & Surg Des Moines IA 50312

ORCUTT, RONALD HAYES, physical chemistry, see 12th edition

ORD, JOHN ALLYN, b West Elizabeth, Pa, Mar 9, 12; m 37; c 2. PHYSICS, MATHEMATICS. Educ: Carnegie Inst Technol, BS, 34, MS, 35, DSc(physics), 39. Prof Exp: Instr physics, Carnegie Inst Technol, 37-38; asst prof, NDak State Col, 38-40; dept dir, Southern Signal Corps Sch, US Army, 42-43, asst commandant, 43-44, exec off signal sect, Southwest Pac Area, 45, dir training, Sandia Base, NMex, 46-48, from exec off to dir, Evans Signal Lab, 49-52, tech liaison off & dep theater signal off, US Army Forces, Far East, 52-55, chief res div & dir Army Res Off, Off Chief Res & Develop, 55-58, mem staff & fac, Army War Col, 59-62, chief res div, Army Materiel Command, 62-65; mgr adv concept sect, Systs Tech Ctr, Philco Corp, 65-66; chief scientist, 66-67, dep for tech opers, 67-71, DEP DIR, US ARMY FOREIGN SCI & TECH CTR, 71- Concurrent Pos: Res physicist, Flannery Bolt Co, 46. Mem: Sigma Xi. Res: Infrared spectroscopy; communications electronics; military research administration. Mailing Add: 3006 Colonial Dr Charlottesville VA 22901

ORDAL, ERLING JOSEPH, b Paint Creek, Iowa, Aug 20, 06; m 37; c 3. MICROBIOLOGY. Educ: Luther Col, Iowa, BA, 27; Univ Minn, PhD(microbiol, bact), 36. Prof Exp: Instr bact, Univ Minn, 36-37; from instr to assoc prof, 37-57, PROF BACT, UNIV WASH, 57- Mem: Am Soc Microbiol; Am Acad Microbiol; Am Soc Limnol & Oceanog; Am Fisheries Soc; Soc Exp Biol & Med. Res: Trace elements in enzymes; myxobacteria; fish pathogens; aquatic and marine microbiology. Mailing Add: Dept of Microbiol SC42 Univ of Wash Seattle WA 98195

ORDAL, ZAKARIAS JOHN, b Sioux Falls, SDak, Mar 28, 13; m 42; c 4. MICROBIOLOGY. Educ: Luther Col, BA, 35; Univ Minn, PhD(bact), 40. Prof Exp: Muellhaupt fel, Ohio State Univ, 41; Hormel Res Found fel, Minnesota, 42; assoc bact, Col Med, Univ Ill, 42-44, asst prof bact & pub health, 44-46; supvr bact group, Process & Prod Res Div, Owens-Ill Glass Co, 46-47; chief bact sect, Res & Develop Div, Econ Labs, Inc, 47-49; from assoc prof to prof food microbiol, 49-68, asst dir agr exp sta, 66-68, PROF MICROBIOL, UNIV ILL, URBANA, 68- Mem: Am Soc Microbiol; Inst Food Technol; Am Acad Microbiol. Res: Bacterial physiology, nutrition and spores; antibacterial agents; food technology; bacterial cell injury and repair. Mailing Add: Dept of Food Sci 580 Bevier Hall Univ of Ill Urbana IL 61801

ORDEN, ALEX, b Rochester, NY, Aug 9, 16; m 46; c 3. OPERATIONS RESEARCH. Educ: Univ Rochester, BS, 37; Univ Mich, MS, 38; Mass Inst Technol, PhD(math), 50. Prof Exp: Alternate chief anal sect, Nat Bur Stand, 42-47; instr math, Mass Inst Technol, 48-49; head comput theory sect, Linear Prog, US Dept Air Force, 50-52; mgr, Comput Lab, Burroughs Corp, 52-58; PROF APPL MATH & MEM COMT ON INFO SCI & INST COMPUT RES, GRAD SCH BUS, UNIV CHICAGO, 58- Concurrent Pos: Vis prof, London Sch Econ, 64 & Ga Inst Technol, 66, 67 & 71. Mem: Inst Mgt Sci; Asn Comput Mach; Opers Res Soc Am; Math Prog Soc. Res: Linear and nonlinear programming; digital computers; models of business organizations. Mailing Add: Grad Sch of Bus Univ of Chicago Chicago IL 60637

ORDER, STANLEY ELIAS, b Vienna, Austria, Nov 1, 34; US citizen; m 58; c 3. RADIOTHERAPY, IMMUNOLOGY. Educ: Albright Col, BS, 56; Tufts Univ, MD, 61; Am Bd Radiol, cert therapeut radiol, 70. Prof Exp: Intern, Ind Univ Hosp, 61-62; resident path, Peter Bent Brigham Hosp, 62-63; chief exp studies, US Army Surg Res Unit, 63-65; from instr to asst prof, 69-73, ASSOC PROF RADIATION THER,

HARVARD UNIV, 73- Concurrent Pos: Fel radiation ther, Yale Univ, 65-69. Mem: Fel Am Col Radiol; Am Soc Therapeut Radiol; Transplantation Soc; Am Soc Clin Oncol. Res: Hodgkins tumor associated antigens; ovarian carcinoma antigens and immunotherapy. Mailing Add: Dept of Radiation Ther Harvard Univ Boston MA 02115

ORDILLE, CAROL MARIA, US citizen. BIOSTATISTICS. Educ: Temple Univ, BS, 65; Villanova Univ, MS, 67; Ohio State Univ, PhD(biostatist), 75. Prof Exp: Statistician, Merck Sharp & Dohme, 67-68; statist consult, 70-72; res statistician, Schering Corp,72-75; SR RES STATISTICIAN, DELBAY PHARMACEUT INC, 75- Mem: Am Statist Asn; Biomet Soc. Res: Development of mathematical models to describe biological phenomena and statistical inference based on such models; data retrieval systems interfaced with statistical programs for clinical and laboratory data. Mailing Add: 614 Bellevue Ave Hammonton NJ 08037

ORDIN, LAWRENCE, plant physiology, biochemistry, deceased

ORDMAN, EDWARD THORNE, b Norfolk, Va, Sept 10, 44; m 71. MATHEMATICS. Educ: Kenyon Col, AB, 64; Princeton Univ, MA, 66, PhD(math), 69. Prof Exp: Mathematician, Nat Bur Stand, 62-69; asst prof, 69-74, RES ASSOC & DIR COMPUT SERV, BUR GOVT SERV, UNIV KY, 74- Concurrent Pos: NSF res grant, 70-71; Fulbright-Hays grant, Univ New South Wales, Sydney, Australia, 73. Mem: Am Math Soc; Math Asn Am; Australian Math Soc. Res: Infinite groups; group theory motivated by topology; topological groups. Mailing Add: Col of Bus Univ of Ky Lexington KY 40506

ORDWAY, ELLEN, b New York, NY, Nov 8, 27; c 8. ENTOMOLOGY. Educ: Wheaton Col Mass, BA, 50; Cornell Univ, MS, 55; Univ Kans, PhD(entom), 65. Prof Exp: Asst field biol, Dept Trop Res, NY Zool Soc, 50-52; asst zool, Southwestern Res Sta, Am Mus Natural Hist, Univ Ariz, 55-57; ASSOC PROF BIOL, UNIV MINN, MORRIS, 65- Mem: Ecol Soc Am; Entom Soc Am; Soc Study Evolution; Soc Syst Zool; Soc Animal Behavior. Res: Insect biology and behavior as related to intraspecific and interspecific associations and development of social behavior; biology of native prairies. Mailing Add: Div of Sci & Math Univ of Minn Morris MN 56267

ORDWAY, FRED (DELANCY), b Kansas City, Mo, Apr 9, 22; m 46; c 2. PHYSICAL CHEMISTRY. Educ: Rensselaer Polytech Inst, BS, 42, MS, 43; Calif Inst Technol, PhD(chem), 49. Prof Exp: Asst, Nat Defense Res Comt, Calif Inst Technol, 43-46; res assoc, Portland Cement Asn fel, Nat Bur Stand, 48-54, dir, 54-60, chief crystal growth sect, 60-63, consult, 63-65; sr scientist, Melpar, Inc, 65-70; EXEC VPRES, ARTECH CORP, 70- Mem: Am Chem Soc; Am Inst Chemists; Mineral Soc Am; Am Ceramic Soc; Am Crystallog Asn. Res: Materials science; x-ray crystallography; amorphous structures; high temperature chemistry; phase equilibria; instrumentation; computer applications. Mailing Add: Artech Corp 2816 Fallfax Dr Falls Church VA 22042

ORDWAY, GIRARD LANTERMAN, physical chemistry, see 12th edition

ORDWAY, NELSON KNEELAND, b New Brunswick, NJ, Oct 26, 12; m 36; c 4. PEDIATRICS. Educ: Yale Univ, MD, 38. Prof Exp: Intern pediat, New Haven Hosp, Conn, 38-39 & Hosp Univ Pa, Philadelphia, 39-41; asst path, Sch Med, Yale Univ, 41-45, instr pediat, 45-47; from asst prof to assoc prof, Sch Med, La State Univ, 47-52, prof & head dept, 52-54; prof, Schs Med, Univ NC, 54-57 & Yale Univ, 58-67; prof pediat & pub health, Schs Med & Health, Univ Okla, 67-74; CHIEF PEDIAT, GALLUP INDIAN MED CTR, 74- Concurrent Pos: Prof, Sch Med, Univ Valle, Colombia, 64-65; consult diarrheal dis, WHO; consult med educ, Pan-Am Health Orgn; clin prof pediat, Sch Med, Univ NMex, 75- Mem: Soc Pediat Res; Am Pediat Soc; Am Acad Pediat. Res: Diarrhea; acid-base and electrolyte disturbances. Mailing Add: Gallup Indian Med Ctr PO Box 1337 Gallup NM 87301

ORDWAY, RICHARD JOHN, b Littleton, NH, July 9, 18; m 48; c 3. GEOLOGY. Educ: Univ NH, BS, 42; Yale Univ, MS, 47, PhD(geol), 48. Prof Exp: Field engr, Colonial Mica Corp, 42-45; geologist, Alaska br, US Geol Surv, 45; from asst prof to assoc prof, 48-57, PROF GEOL, STATE UNIV NY COL NEW PALTZ, 57- Concurrent Pos: Fac fel, Ford Found Fund Advan Educ, Univ Calif, 54-55. Mem: AAAS; Geol Soc Am; Nat Asn Geol Teachers. Res: Earth Sciences. Mailing Add: 12 Joalyn Rd New Paltz NY 12561

ORE, HENRY THOMAS, b Chicago, Ill, Oct 30, 34; m 54; c 2. SEDIMENTOLOGY. Educ: Cornell Col, BA, 57; Wash State Univ, MS, 59; Univ Wyo, PhD(geol), 63. Prof Exp: From instr to assoc prof, 63-73, chmn dept, 67-70, PROF GEOL, IDAHO STATE UNIV, 73- Mem: Geol Soc Am; Nat Asn Geol Teachers; Int Asn Sedimentol; Am Asn Univ Profs; Sigma Xi. Res: Modern depositional environments; braided streams; sedimentary petrography; mechanics of sediment transport. Mailing Add: 1635 City Creek Rd Pocatello ID 83201

OREAR, JAY, b Chicago, Ill, Nov 6, 25. PARTICLE PHYSICS. Educ: Univ Chicago, PhB, 43, SM, 50, PhD(physics), 53. Prof Exp: Res assoc physics, Inst Nuclear Studies, Univ Chicago, 53-54; from instr to asst prof, Columbia Univ, 54-58; assoc prof, 58-64, PROF PHYSICS, CORNELL UNIV, 64- Concurrent Pos: NSF sr fel, Europ Orgn Nuclear Res, 64-65; ed, Forum on Physics & Soc, Am Phys Soc, 71-74. Mem: Am Phys Soc; Fedn Am Scientists. Res: Large-angle, high energy proton-proton and pion-proton scattering. Mailing Add: Dept of Physics Cornell Univ Ithaca NY 14850

OREBAUGH, ERROL GLEN, b Bradenton, Fla, Sept 28, 37; m 61; c 3. PHYSICAL INORGANIC CHEMISTRY. Educ: Univ Fla, BSCh, 59; Fla State Univ, PhD(inorg chem), 72. Prof Exp: Chemist, 60-67, RES CHEMIST, SAVANNAH RIVER LAB, E I DU PONT DE NEMOURS & CO, 72- Mem: Am Chem Soc. Res: Research and development problems of the nuclear fuel cycle, including actinide chemistry, nuclear waste management and environmental chemistry; instrumentation and complexation thermodynamics. Mailing Add: Savannah River Lab E I du Pont de Nemours & Co Aiken SC 29801

O'REGAN, WILLIAM GERARD, statistics, see 12th edition

O'REILLY, DONALD EUGENE, b Chicago, Ill, Mar 22, 30; m 52; c 7. CHEMICAL PHYSICS. Educ: Purdue Univ, BS, 51; Univ Chicago, PhD(chem physics), 55. Prof Exp: Chem physicist, Gulf Res & Develop Co Div, Gulf Oil Corp, 55-60; chemical physicist, 60-70, SR SCIENTIST, ARGONNE NAT LAB, 70- Mem: Am Phys Soc; Am Chem Soc. Res: Structure and properties of molecules and solids by nuclear and electron magnetic resonance; phase transitions and ferroelectricity; liquid state physics; catalysis and surface chemistry. Mailing Add: Argonne Nat Lab 9700 S Cass Ave Argonne IL 60439

O'REILLY, HENRY JAMES, b Edgewood, BC, Aug 1, 16; m 45; c 4. PLANT PATHOLOGY. Educ: Univ Toronto, BS, 42; McGill Univ, MS, 44; Ore State Univ, PhD(plant path), 53. Prof Exp: County agent hort, Ore State Col, 47, supt plant path & hort, Dalles Br Exp Sta, 47-52, exten plant pathologist, 52-55; plant pathologist & head field dept, Yakima County Hort Union, Wash, 55; exten plant pathologist, Univ Calif, Davis, 55-69; PROV PLANT PATHOLOGIST, BC DEPT AGR, 69- Mem: Am Phytopath Soc. Res: Tree fruit virus diseases; horticultural, crop, fungus and bacterial diseases. Mailing Add: BC Dept of Agr Parliament Bldgs Victoria BC Can

O'REILLY, JAMES, physics, see 12th edition

O'REILLY, JAMES EMIL, b Cleveland, Ohio, Jan 14, 45; m 68; c 1. ANALYTICAL CHEMISTRY. Educ: Univ Notre Dame, BS, 67; Univ Mich, PhD(chem), 71. Prof Exp: Assoc chem, Univ Ill, Urbana, 71-73; ASST PROF CHEM, UNIV KY, 73- Mem: Am Chem Soc; Sigma Xi. Res: Electrochemistry of compounds of biological interest, analytical methods development, ion selective electrodes, atomic absorption. Mailing Add: Dept of Chem Univ of Ky Lexington KY 40506

O'REILLY, JAMES MICHAEL, b Dayton, Ohio, Nov 25, 34; m 54; c 5. POLYMER PHYSICS, POLYMER CHEMISTRY. Educ: Univ Dayton, BS, 56; Univ Notre Dame, PhD(phys chem), 60. Prof Exp: Develop chemist, Nat Cash Register Co, 54-56; res assoc polymer physics, Gen Elec Res & Develop Ctr, 60-67; mgr polymer physics & phys chem, 67-73, technol mgr, 73-75, PRIN SCIENTIST, RES LABS, XEROX CORP, 75- Concurrent Pos: Part-time assoc prof, Univ Rochester, 71-76. Mem: AAAS; Soc Rheol; Am Chem Soc; fel Am Phys Soc. Res: Microwave spectroscopy; structure and properties of polymers, electrical, mechanical, thermodynamic and rheological properties; biopolymers; electrophotographic and imaging materials. Mailing Add: 44 Bromley Rd Pittsford NY 14534

O'REILLY, SEAN, b Killarney, Ireland, Feb 19, 22; US citizen; m 51; c 7. NEUROLOGY. Educ: Nat Univ Ireland, MB, 49, MD, 52; FRCP. Prof Exp: Registr med, Hosp Infectious Dis, Bradford, Eng, 52; registr dermat, Radcliffe Infirmary, Oxford Univ, 52-54; from intern to resident med, St Mary's Hosp, San Francisco, Calif, 54-56; asst resident neurol, Med Ctr, Univ Calif, San Francisco, 56-59; trainee neurochem & neurol, Nat Hosp Nerv Dis, London, Eng, 59-60; clin instr neurol, Med Ctr, Univ Calif, San Francisco, 60-61, asst prof, 61-69; chmn dept, 69-75, PROF NEUROL, GEORGE WASHINGTON UNIV, 69-, CHMN NEUROBIOL RES TRAINING PROG, 71- Concurrent Pos: Consult, Clin Ctr, NIH, Walter Reed Gen Hosp & Vet Admin Cent Off Ctr. Mem: Fel Am Acad Neurol; Am Neurol Asn; Asn Res Nerv & Ment Dis; Soc Neurosci. Res: Biochemistry of basal ganglia disorders; Huntington's chorea; Parkinson's disease; Wilson's disease; physiology of copper and zinc. Mailing Add: Dept of Neurol George Washington Univ Washington DC 20037

OREJAS-MIRANDA, BRAULIO, b Montevideo, Uruguay, Feb 18, 33; m 59; c 2. HERPETOLOGY. Educ: Univ of the Repub, Uruguay, Lic, 59; Cent Univ Venezuela, cert herpet, 61. Prof Exp: Cur herpet, Nat Mus Natural Hist, Uruguay, 57-66; prof zool, Cent Univ, Venezuela, 59-61; chief labs, Nat Mus Natural Hist, Uruguay, 67-70; SPECIALIST BIOL, SCI AFFAIRS DEPT, ORGN AM STATES, WASHINGTON, DC, 70- Concurrent Pos: Assoc res, Div Reptiles & Amphibians, Smithsonian Inst, 68; fel, Guggenheim Mem Fund, 69; adv, Nat Zoo, Dominican Repub, 72-75. Mem: Am Soc Ichthyologists & Herpetologists; Soc Study Amphibians & Reptiles. Res: Snakes, particularly the Family Leptotyphlopidae, in several fields—taxonomy, geographical distribution, ecology and evolution. Mailing Add: 7522 Sweet Briar Dr College Park MD 20740

O'RELL, DENNIS DEE, b Sacramento, Calif, Nov 28, 41; m 65; c 2. CHEMISTRY, POLYMER CHEMISTRY. Educ: Sacramento State Col, BA, 64; Univ Ore, PhD(org chem), 70. Prof Exp: Sr res chemist, Ciba-Geigy Corp, 70-72, supvr, Polymer Additives Dept, 72-75; RES MGR, POLYFIBRON DIV, W R GRACE & CO, 75- Mem: Am Chem Soc. Res: Synthetic organic chemistry; alkaloids; heterocyclic chemistry; monomer synthesis. Mailing Add: Polyfibron Div W R Grace & Co Cambridge MA 02140

ORELLANA, RODRIGO GONZALO, b Ecuador, Oct 25, 15; nat US; m 44; c 3. PLANT PATHOLOGY. Educ: Univ Minn, MS, 44, PhD(plant path), 48. Prof Exp: Instr, Univ Minn, 45-48; plant pathologist, Ministry Agr, Venezuela, 48-49; instr, Univ Minn, 50-51; plant pathologist, Inter-Am Inst Agr Sci, Costa Rica, 52-56; tech adv mission, Food & Agr Orgn, UN, Ministry Agr, Ceylon, 56-57; plant pathologist, Fla, 57-60, PLANT PATHOLOGIST, AGR RES SERV, USDA, MD, 60- Mem: AAAS; Am Phytopath Soc; Latin Am Phytotech Asn. Res: Disease of oilseed crops. Mailing Add: Plant Protection Inst USDA Agr Res Serv Beltsville MD 20705

OREM, MICHAEL WILLIAM, b Sturgis, SDak, Nov 30, 42; m 66; c 1. PHYSICAL CHEMISTRY. Educ: Colo Col, BS, 64; Univ Southern Calif, PhD(surface chem), 69. Prof Exp: SR RES CHEMIST, RES LABS, EASTMAN KODAK CO, 69- Mem: AAAS; Am Chem Soc. Res: Gas adsorption on molecular solids; surface properties of aqueous and non-aqueous surfactant systmes; adhesion and polymer surface properties. Mailing Add: Eastman Kodak Co Res Labs 1669 Lake Ave Rochester NY 14650

OREN, SHMUEL SHIMON, b Bucharest, Romania, Feb 20, 44; Israeli citizen; m 68; c 1. OPERATIONS RESEARCH. Educ: Israel Inst Technol, BSc, 65, MS, 69; Stanford Univ, MS & PhD(eng econ systs), 72. Prof Exp: Tech officer mech eng, Israel Defence Force, 65-69; asst optimization, Stanford Univ, 70-71; RES SCIENTIST OPERS RES, XEROX PALO ALTO RES CTR, 72- Concurrent Pos: Consult asst prof, Stanford Univ, 72- Mem: Inst Elec & Electronics Engrs; Opers Res Soc Am; Inst Mgt Sci; Math Programming Soc. Res: Mathematical modeling; algorithms for nonlinear programming; application of control theory to operations research; market models and market forecasting. Mailing Add: Xerox Palo Alto Res Ctr 3333 Coyote Hill Rd Palo Alto CA 94304

ORENGO, ANTONIO, b Naples, Italy, May 6, 31; m 59; c 3. BIOCHEMISTRY. Educ: Univ Naples, MD, 55. Prof Exp: Asst prof human physiol, Sch Med, Univ Naples, 55-57; res assoc biochem, Zool Sta, Naples, 57-58, 60-61; res scientist & biochemist, Houston State Psychiat Inst, 62-64; asst biochemist, Zool Sta, Naples, Italy, 64-66; ASSOC PROF BIOCHEM, UNIV TEX M D ANDERSON HOSP & TUMOR INST HOUSTON, 66- Concurrent Pos: Ital Res Coun fel enzymol, Zool Sta, Naples, Italy, 57-58, 60-61; res fel biochem, Wayne State Univ, 61-62; res grants, Nat Cancer Inst, 67-; Am Heart Asn, 69 & Robert A Welch Found, 71- Mem: Am Chem Soc; Ital Med Asn; Am Soc Biol Chem; fel Am Inst Chem; Am Soc Cell Biol. Res: Nucleic acid metabolism; biochemical control mechanisms. Mailing Add: RM G 505 M D Anderson Hosp & Tumor Inst Houston TX 77025

ORENSKI, PETER JOSEPH, chemistry, see 12th edition

ORENSTEIN, ALBERT, b New York, NY; m 62. PHYSICS. Educ: NY Univ, PhD(physics), 63. Prof Exp: ASST PROF PHYSICS, QUEENS COL, NY, 58- Concurrent Pos: Res assoc, NY Univ, 62- Res: Solid state physics. Mailing Add: Dept of Physics Queens Col Flushing NY 11367

ORENSTEIN, HENRY, b New York, NY, May 29, 24. ANTHROPOLOGY. Educ: Univ Calif, AB, 50, PhD(anthrop), 57. Prof Exp: Instr anthrop, Syracuse Univ, 57-58; from asst prof to assoc prof, Tulane Univ, 58-67; assoc prof, 67-71, PROF

ANTHROP, BROOKLYN COL, 71- Concurrent Pos: Partic Asian studies prog, Am Coun Learned Socs-Soc Sci Res Coun Res Grant, 60. Mem: Fel Am Anthrop Asn; Am Ethnol Soc; Asn Asian Studies; Indian Sociol Soc. Res: India; anthropological theory; values; political anthropology; structural analysis. Mailing Add: Dept of Anthrop Brooklyn Col Brooklyn NY 11210

ORENSTEIN, STEPHEN, high energy physics, nuclear physics, see 12th edition

ORESKES, IRWIN, b Chicago, Ill, June 30, 26. BIOCHEMISTRY, IMMUNOLOGY. Educ: City Col New York, BS, 49; Brooklyn Col, MA, 56; City Univ New York, PhD(biochem), 69. Prof Exp: Chemist, Tech Tape Co, 49; technician, Sch Med, NY Univ, 50-51; supvr phys chem, Kingsbrook Jewish Med Ctr, 51-56; fel polymers, Brooklyn Polytechnic Inst, 57-58; from res asst to asst prof rheumatic dis, Mt Sinai Hosp & Sch Med, 59-69; assoc prof, 69-74, chmn, Sch Health Sci, 72, 74, PROF MED LAB SCI, SCH HEALTH SCI, HUNTER COL, 74- Concurrent Pos: Mem int sci coun, Albert Einstein Res Inst, Buenos Aires, Arg, 69-; res asst prof med, Mt Sinai Sch Med, 70-74, res assoc prof, 74-; mem doctoral fac biochem, City Univ New York, Grad Ctr, 70-; mem bd examiners, New York City Dept Health, 73-75. Mem: Am Asn Immunol; Am Asn Clin Chem; Am Chem Soc; Am Rheumatism Asn; Harvey Soc. Res: Anti globulin antibodies; protein denaturation and aggregation. Mailing Add: Sch of Health Sci Hunter Col 105 E 106th St New York NY 10029

OREY, STEVEN, b Berlin, Ger, July 17, 28; nat US; m 50; c 1. MATHEMATICS. Educ: Cornell Univ, BA, 49, MA, 51, PhD, 53. Prof Exp: Instr math, Cornell Univ, 53-54; instr, Univ Minn, 54-56; vis asst prof, Univ Calif, 57-58; vis asst prof, 58-59, assoc prof, 59-65, PROF MATH, UNIV MINN, MINNEAPOLIS, 65- Mem: Am Math Soc; Asn Symbolic Logic. Res: Mathematical logic; probability theory. Mailing Add: Dept of Math Univ of Minn Minneapolis MN 55455

ORFUSS, ABRAHAM JACOB, dermatology, deceased

ORGAN, JAMES ALBERT, b Newark, NJ, Mar 29, 31; m 56; c 2. ZOOLOGY. Educ: Rutgers Univ, AB, 56; Univ Mich, MS, 58, PhD(zool), 60. Prof Exp: Instr zool, Univ Mich, 60-61; from instr to assoc prof biol, 61-71, PROF BIOL, CITY COL NEW YROK, 71- Mem: Ecol Soc Am; Am Soc Ichthyol & Herpet; Soc Study Evolution; Am Soc Zool; NY Acad Sci. Res: Population ecology; animal behavior; herpetology. Mailing Add: Dept of Biol City Col of New York New York NY 10031

ORGEBIN-CRIST, MARIE-CLAIRE, b Vannes, France, Mar 20, 36. BIOLOGY, ENDOCRINOLOGY. Educ: Sorbonne, Lic Natural Sci & Lic Biol, 57; Univ Lyons, DSc, 61. Prof Exp: Res assoc, Med Div, Pop Coun, 62-63; res assoc, 63-64, res instr, 64-66, from asst prof to assoc prof, 66-73, PROF OBSTET & GYNEC & LUCIUS E BURCH CHAIR REPRODUCTIVE BIOL, SCH MED, VANDERBILT UNIV, 73-, DIR CTR POP RES & STUDIES IN REPROD BIOL, 73- Concurrent Pos: NIH career develop award, 68-73; mem contract rev comt, Ctr Pop Res, NIH, 68-72; mem educ comt, Soc Study Reproduction, 68-71; mem pop res & training comt, Nat Inst Child Health & Human Develop, 70-74, chmn, 72-74; mem, Int Comt Andrology, 74- Mem: Brit Soc Study Fertil; Am Asn Anat; Soc Study Reproduction. Res: Male reproductive physiology. Mailing Add: Dept of Obstet & Gynec Vanderbilt Univ Sch Med Nashville TN 37232

ORGEL, GERALD, microbiology, biochemistry, see 12th edition

ORGEL, LESLIE E, b London, Eng, Jan 12, 27; m 50; c 3. CHEMISTRY. Educ: Oxford Univ, BA, 48, PhD(chem), 51. Prof Exp: Reader chem, Cambridge Univ, 51-64; SR FEL, SALK INST BIOL STUDIES, 65- Concurrent Pos: Adj prof, Univ Calif, San Diego, 65- Mem: Fel Royal Soc Chem. Res: Prebiotic chemistry; the chemistry of the origins of life. Mailing Add: Salk Inst for Biol Studies PO Box 1809 San Diego CA 92112

ORGEL, SYLVIA, physics, mathematics, see 12th edition

ORGELL, CARL WALTHER, physical inorganic chemistry, see 12th edition

ORGELL, WALLACE HERMAN, b Eldora, Iowa, Dec 28, 28; m 60. PLANT PHYSIOLOGY. Educ: Iowa State Univ, BS, 50; Pa State Univ, MS, 52; Univ Calif, PhD(plant physiol), 54. Prof Exp: Asst plant physiol, USDA, Pa State Univ, 50-52; asst bot, Univ Calif, 52-54; NSF fel org chem, Univ Ill, 54-55; plant biochemist, US Chem Corps, Ft Detrick, 55-57 & dept zool & entom, Iowa State Univ, 57-63; assoc prof biol sci, Fla Atlantic Univ, 63-66; prof biol, Hiram Scott Col. 66-67; assoc prof, 67-73, PROF BIOL, MIAMI-DADE COMMUNITY COL, 73- Mem: Am Soc Plant Physiol; Nat Sci Teachers Asn; Nat Asn Biol Teachers. Res: Biochemical aspects of ecology. Mailing Add: Miami-Dade Community Col Dept Biol South Campus Miami FL 33156

ORGILL, MONTIE M, b Mammoth, Utah, Feb 25, 29; m 56; c 4. ATMOSPHERIC PHYSICS, AIR POLLUTION. Educ: Brigham Young Univ, BS, 51; Univ Utah, MS, 57; Colo State Univ, PhD, 71. Prof Exp: Consult & forecaster, Intermt Weather, Inc, Utah, 56-58; asst prof meteorol, Univ Hawaii, 58-62; meteorologist, Colo State Univ, 62-72; SR RES SCIENTIST, BATTELLE-PAC NORTHWEST LABS, 72- Mem: Am Meteorol Soc; Am Geophys Union; AAAS. Res: Synoptic meteorology; fluid mechanics; micrometeorology and mesometeorology; weather modification; atmospheric resuspension and diffusion research; aircraft sampling research; environmental assessment; applied meteorology. Mailing Add: 1824 Wright Ave Richland WA 99352

ORIANI, RICHARD ANTHONY, b San Salvador, Cent Am, July 19, 20; nat US; m 49; c 4. PHYSICAL CHEMISTRY. Educ: City Col New York, BChE, 43; Stevens Inst Technol, MS, 46; Princeton Univ, MA & PhD(phys chem), 48. Prof Exp: Lab asst chem eng, City Col New York, 43; res chemist, Bakelite Corp Div Union Carbide & Carbon Corp, 43-46; asst, Princeton Univ, 46-48; res assoc, Gen Elec Co, 48-59; mgr phys chem, 59-75, SR RES CONSULT, RES LAB, US STEEL CORP, 75- Concurrent Pos: Vis instr, Rensselaer Polytech Inst, 53-54; mem Nat Acad Sci adv panel, Nat Bur Stand, 67-73. Mem: Am Phys Soc; Am Chem Soc; fel Am Inst Chem; fel NY Acad Sci. Res: Thermodynamics of alloys; electronic structure of alloys; structure and binding in solids; kinetics; surface reactions; hydrogen embrittlement. Mailing Add: US Steel Res Ctr Monroeville PA 15146

ORIANS, GORDON HOWELL, b Eau Claire, Wis, July 10, 32; m 55; c 3. ECOLOGY, EVOLUTION. Educ: Univ Wis, BS, 54; Univ Calif, Berkeley, PhD(zool), 60. Prof Exp: From asst prof to assoc prof, 60-68, PROF ZOOL, UNIV WASH, 68- Mem: Cooper Ornith Soc; Ecol Soc Am; Soc Study Evolution; Animal Behav Soc; Am Ornith Union. Res: Evolution of vertebrate social systems; factors determining the number of species an environment will support on a sustained basis; plant-herbivore interactions. Mailing Add: Dept of Zool Univ of Wash Seattle WA 98195

ORIEL, PATRICK JOHN, b Calgary, Alta, Oct 23, 37; US citizen; m 62; c 2. PHYSICAL CHEMISTRY. Educ: Univ Minn, BChem, 59; Univ Ore, PhD(chem), 64. Prof Exp: NIH res grant, Harvard Med Sch, 64-65; chemist, 65-69, sr res phys biochemist, 69-73, group leader, 73-74, SR RES SPECIALIST, DOW CHEM CO, 74- Mem: Am Chem Soc. Res: Physical chemistry of proteins and nucleic acids; microbiology. Mailing Add: 1320 Springfield Midland MI 48640

ORIEL, STEVEN S, b New York, NY, Feb 9, 23; m 50; c 3. STRUCTURAL GEOLOGY. Educ: Columbia Univ, BA, 44; Yale Univ, MS, 47, PhD(geol), 49. Prof Exp: Geologist, US Geol Surv, 47-50; res engr geol, Stanolind Oil & Gas Co, 50-53; GEOLOGIST, US GEOL SURV, 53- Concurrent Pos: Mem, Am Comn Stratig Nomenclature, chmn, 73-74. Mem: AAAS; Geol Soc Am; Am Asn Petrol Geol. Res: Geology of sedimentary rocks; synthesis of stratigraphic information; areal mapping of deformed rocks and Cenozoic sediments in western Wyoming and southeastern Idaho fold belt; coordinator of Snake River Plain Region studies, including geothermal and earthquake hazard assessments. Mailing Add: US Geol Surv Stop 913 Box 25046 Denver Fed Ctr Denver CO 80225

ORIHEL, THOMAS CHARLES, b Akron, Ohio, Feb 10, 29; m 52; c 4. MEDICAL PARASITOLOGY. Educ: Univ Akron, BS, 50; Univ Wash, MS, 52; Tulane Univ, PhD(parasitol), 59. Prof Exp: NIH fel, Tulane Univ & Univ Miami, 59-61; vector-borne dis specialist, Dept of State, AID Mission to Brit Guiana, 61-63; RES SCIENTIST & HEAD PARASITOL, DELTA PRIMATE CTR, TULANE UNIV, 63-, PROF PARASITOL, SCH PUB HEALTH & TROP MED, 72-; ASSOC DIR, INT CTR MED RES, 75- Concurrent Pos: Mem study sect trop med & parasitol, Nat Inst Allergy & Infectious Dis, NIH, 73-77, chmn study sect trop med & parasitol, Nat Inst Allergy & Infectious Dis, 75-77; mem expert panel parasitic dis, WHO, 73-78; mem panel parasitic dis, US-Japan Coop Med Sci Prog, 74- Mem: Am Soc Parasitol; Am Soc Trop Med & Hyg; Royal Soc Trop Med & Hyg; Am Soc Microbiol. Res: Human and zoonotic filariasis; host-parasite interactions; primate parasites; nematology; parasite morphology and taxonomy. Mailing Add: Sch Pub Health & Trop Med Tulane Univ Med Ctr New Orleans LA 70112

ORING, LEWIS WARREN, b New York, NY, Apr 18, 38; m 60; c 2. ECOLOGY, ETHOLOGY. Educ: Univ Idaho, BS, 60; Univ Okla, MS, 62, PhD(zool), 66. Prof Exp: NSF fel ethology, Univ Copenhagen, 66-67; NIH trainee, Univ Minn, Minneapolis, 67-68; asst prof, 68-70, ASSOC PROF BIOL, UNIV N DAK, 70- Mem: Animal Behav Soc; Ecol Soc Am; Am Ornith Union; Wilson Ornith Soc; Cooper Ornith Soc. Res: Ecological and evolutionary aspects of animal behavior, especially the ways in which environment effects mating systems, spacing patterns and communication. Mailing Add: Dept of Biol Univ of NDak Grand Forks ND 58201

O'RIORDAN, TIMOTHY, b Edinburgh, Scotland, Feb 21, 42; m 68. RESOURCE GEOGRAPHY. Educ: Univ Edinburgh, MA, 63; Cornell Univ, MS, 65; Cambridge Univ, PhD(geog), 67. Prof Exp: From asst prof to assoc prof geog, Simon Fraser Univ, 67-75; READER, SCH ENVIRON SCI, UNIV EAST ANGLIA, 75- Concurrent Pos: Vis lectr geog, Univ Canterbury, 70; vis assoc prof, Clark Univ, 72-73. Res: Environmental studies; public participation in environmental policy making. Mailing Add: Sch of Environ Sci Univ of East Anglia Norwich England

ORKAND, RICHARD K, b New York, NY, Apr 23, 36; m 60. PHYSIOLOGY, NEUROBIOLOGY. Educ: Columbia Univ, BS, 56; Univ Utah, PhD(pharmacol), 61. Hon Degrees: MS, Univ Pa, 74. Prof Exp: USPHS fel biophys, Univ Col, Univ London, 61-64; USPHS spec fel neurophysiol, Harvard Med Sch, 64-66; asst prof physiol, Med Ctr, Univ Utah, 66-68; assoc prof zool, Univ Calif, Los Angeles, 68-72, prof biol, 72-74; PROF PHYSIOL & CHMN DEPT PHYSIOL & PHARMACOL, SCH DENT MED, UNIV PA, 74- Mem: Soc Neurosci; AAAS; Am Physiol Soc. Res: Synaptic neurophysiology; skeletal and cardiac muscle excitation-contraction coupling; physiology of neuroglia. Mailing Add: Dept Physiol & Pharmacol Univ of Pa Sch Dent Med Philadelphia PA 19104

ORKIN, BERNARD ABRAHAM, organic chemistry, physical chemistry, deceased

ORKIN, LAZARUS ALLERTON, b New York, NY, Feb 2, 10; m 41; c 2. MEDICINE, SURGERY. Educ: NY Univ, BS, 30, MS, 31, MD, 35; Am Bd Urol, dipl, 44. Prof Exp: Intern, Bellevue Hosp, New York, 35-37, asst resident obstet & gynec, 37-38, resident urol, 40-42; resident urol, Royal Victoria Hosp, Montreal, Que, 38-40; DIR UROL, BETH ISRAEL MED CTR, 43- Concurrent Pos: Consult urologist, Peninsula Gen & St Joseph's Hosps, Long Island, 58-; clin prof, Mt Sinai Sch Med, 68-; consult urologist, Gracie Square Hosp; consult, Beekman Downtown Hosp, New York & Cath Med Ctr Brooklyn; mem, NY & Brooklyn Comt Trauma. Mem: Fel Am Geriat Soc; Am Urol Asn; fel Am Med Writers Asn; fel Am Surg of Trauma; fel Am Col Surg; fel NY Acad Med. Res: Urological trauma; urinary tract infections. Mailing Add: Beth Israel Med Ctr 10 Nathan D Perlman Pl New York NY 10003

ORKIN, LOUIS R, b New York, NY, Dec 23, 15; m 38; c 1. ANESTHESIOLOGY. Educ: Univ Wis, BA, 37; NY Univ, MD, 41. Prof Exp: Dir anesthesia, W W Backus Hosp, Conn, 48-49; asst prof anesthesiol, Postgrad MEd Sch, NY Univ, 50-55; PROF ANESTHESIOL & CHMN DEPT, ALBERT EINSTEIN COL MED, 55- Concurrent Pos: Consult, Norwich State Hosp, Conn, 48-49 & Uncas on Thames, 48-49; anesthesiologist, Bellevue Hosp, 50-55; asst attend anesthetist, NY Univ Hosp, 50-54, assoc attend anesthetist, 54-55; consult, US Naval Hosp, St Albans, 54, USPHS Hosp, Staten Island, 54, Triboro Hosp, 55-65 & Montefiore Hosp, New York, 65-; titular prof, Univ Venezuela, 67; vis prof, Univ Calif, San Diego, 71; mem nat res coun, Nat Acad Med, 65-70; sr consult, Bronx Vet Admin Hosp, 55- Mem: Am Soc Anesthesiol; AMA; fel NY Acad Sci; fel Am Col Chest Physicians; fel NY Acad Med. Res: Physiology and pharmacology related to anesthesiology. Mailing Add: Dept of Anesthesiol Albert Einstein Col of Med Bronx NY 10461

ORLAND, FRANK J, b Little Falls, NY, Jan 23, 17; m 43; c 4. HISTORY OF DENTISTRY, ORAL MICROBIOLOGY. Educ: Univ Ill, BS, 39, DDS, 41; Univ Chicago, SM, 45, PhD(bact), 49; Am Bd Med Microbiol, dipl. Prof Exp: Intern, Zoller Mem Dent Clin, 41-42, asst dent surg, 42-49, from instr to asst prof bact, 49-54, dir, Zoller Clin, 54-66, assoc prof microbiol, 54-58, res assoc prof, 58-64, from instr to assoc prof dent surg, 49-58, PROF DENT SURG, UNIV CHICAGO, 58- Concurrent Pos: Fel, Inst Med, Univ Chicago; consult & mem dent study sect & prog planning comt, Nat Inst Dent Res, 55-59; Darnell mem lectr, US Naval Dent Sch, 56; chmn comt info & pub, Inst Med, Univ Chicago; mem panel, Nat Formulary; chmn dent adv bd, Med Heritage Soc; ed, J Dent Res, 58-69. Mem: AAAS; Am Soc Microbiol; Electron Micros Soc Am; Am Dent Asn; fel Am Acad Microbiol. Res: Experimental dental caries; caries and periodontal disease in germfree and germfree inoculated animals; antigenic analysis of lactobacilli; mechanism of fluoride inhibition in oral bacteria and in tooth decay; morphology and physiology of oral protozoa; historical research in dentistry. Mailing Add: Box 418 Zoller Clin Univ of Chicago Chicago IL 60637

ORLAND, GEORGE H, b Los Angeles, Calif, Oct 8, 24; m 44; c 3. MATHEMATICS. Educ: City Col New York, BEE, 45; Univ Chicago, MS, 56; Univ Calif, Berkeley,

PhD(math), 62. Prof Exp: Instr elec eng, Polytech Inst Brooklyn, 46-47; tutor, City Col New York, 48-49; electronics engr, Dept Otolaryngol, Univ Chicago, 50-57, mathematician, Inst Systs Res, 57-59; instr math, Univ Ill, 62-65; asst prof, Wesleyan Univ, 65-67; ASSOC PROF MATH, HARVEY MUDD COL, 67- Mem: Am Math Soc. Res: Classical and functional analysis; measure theory; convexity; the bending of polyhedra. Mailing Add: Dept of Math Harvey Mudd Col Claremont CA 91711

ORLANDO, ANTHONY MICHAEL, b Brooklyn, NY, Jan 27, 42; m 66; c 2. BIOSTATISTICS. Educ: City Univ New York, BBA, 63; Columbia Univ, MS, 67; Yale Univ, MPhil, 69, PhD(biostatist), 70. Prof Exp: Sr sales analyst, Am-Standard, Inc, 64-65; statistician, Monsanto Chem Co, 65; prin biostatistician, 70-75, DEPT HEAD STATIST SCI, MEAD JOHNSON & CO, EVANSVILLE, 75- Mem: AAAS; Am Statist Asn; Biomet Soc; Sigma Xi; Royal Statist Soc. Res: Analyzing clinical parameter distributions under a mixture of normal densities; hypothesis and non-linear estimation of parameters for two and three-compartment open models in pharmacokinetics. Mailing Add: RR 2 Box 76-C Mt Vernon IN 47620

ORLANDO, CHARLES M, b Tappan, NY, Sept 30, 35; m 64; c 2. ORGANIC CHEMISTRY. Educ: Fordham Univ, BS, 57; NY Univ, PhD(org chem), 61. Prof Exp: Res chemist, Esso Res & Eng Co, 61-64 & Kay-Fries Chem Co, 64-66; res chemist, 66-72, MGR SYNTHESIS, RES & DEVELOP, GEN ELEC CO, 72- Mem: Am Chem Soc; Brit Chem Soc. Res: Organic photochemistry; organic synthesis; heterocyclic chemistry; chemistry of quinones; chemistry of nitrogen-fluorine compounds; bromination chemistry and flame retardants. Mailing Add: PO Box 8 Schenectady NY 12301

ORLANDO, JOSEPH ALEXANDER, b Methuen, Mass, July 16, 29; m 58; c 5. BIOLOGICAL CHEMISTRY. Educ: Merrimack Col, BS, 53; NC State Col, MS, 56; Univ Calif, PhD(biochem), 58. Prof Exp: Nat Cancer Inst fel biochem, Brandeis Univ, 58-61; ASSOC PROF BIOCHEM, BOSTON COL, 61- Res: Photobiology; isolation and characterization of enzymes and enzyme systems associated with light mediated reactions in photosynthetic bacteria. Mailing Add: Dept of Biol Boston Col Chestnut Hill MA 02167

ORLANS, F BARBARA, b Jan 14, 28; US citizen; m 59; c 2. PHYSIOLOGY. Educ: Birmingham Univ, BSc, 49; London Univ, MSc, 52, PhD(physiol), 54. Prof Exp: Res physiologist hormonal control autonomic nervous syst, Nat Heart & Lung Inst, NIH, 56-60; writer free-lance, 67-73; sr staff scientist pop biol, Med Ctr, George Washington Univ, 73-74; grants assoc, Admin Training, 74-75, HEALTH SCI ADMINR CARDIOVASC DIS, NIH, 75- Concurrent Pos: Fel, Asthma Res Coun Great Brit, 54; fel, Riker Pharmaceut, 58. Mem: Am Soc Pharmacol & Exp Therapeut; Nat Sci Teachers Asn; Nat Asn Biol Teachers. Mailing Add: Rev Br Nat Heart & Lung Inst Nat Inst Health Bethesda MD 20014

ORLANSKI, ISIDORO, b Buenos Aires, Arg, June 6, 39; m 62; c 2. GEOPHYSICS. Educ: Univ Buenos Aires, Lic Physics, 65; Mass Inst Technol, PhD(geophys), 67. Prof Exp: RES SCIENTIST, GEOPHYS FLUID DYNAMICS LAB, NAT OCEANIC & ATMOSPHERIC ADMIN, 67- Concurrent Pos: Res assoc, Mass Inst Technol, 67-68; lectr, Princeton Univ, 71- Mem: Am Meteorol Soc; Am Geophys Union. Res: Turbulence in stratified flows; dynamics of rtating stratified fluids. Mailing Add: Geophys Fluid Dynamics Lab Princeton Univ PO Box 308 Princeton NJ 08540

ORLEMAN, EDWIN FRANKLIN, b Livingstone, Mont, July 19, 15; m 39; c 4. PHYSICAL CHEMISTRY. Educ: St Thomas Col, BS, 36; Univ Minn, PhD(phys chem), 41. Prof Exp: Instr chem, Unic Calif, 41-43; chemist, Metall Lab, Univ Chicago, 43-44; supt anal res, Tenn Eastman Corp, 44-45; asst prof chem, 45-47, assoc prof chem & chem eng, 47-54, PROF CHEM & CHEM ENG, UNIV CALIF, BERKELEY, 54- Concurrent Pos: Consult, Radiation Lab, Univ Calif; consult, AEC, 46- Mem: Am Chem Soc. Res: Inorganic analysis; mechanism of electrode reactions. Mailing Add: Dept of Chem Univ of Calif Berkeley CA 94704

ORLIC, DONALD, b Universal, Pa, Dec 11, 32; m 61; c 3. CELL BIOLOGY. Educ: Fordham Univ, BS, 59; NY Univ, MS, 63, PhD(biol), 66. Prof Exp: Instr anat, Harvard Med Sch, 68-69; asst prof, 69-73, ASSOC PROF ANAT, NEW YORK MED COL, 73-; ACTG CHMN DEPT, 74- Concurrent Pos: NIH fel, Inst Cellular Path, France, 66-67; NIH fel anat, Harvard Med Sch, 67-68. Honors & Awards: A Cressy Morrison Award, NY Acad Sci, 66. Mem: Am Soc Cell Biol; Soc Study Blood; Am Soc Hemat; Am Asn Anat. Res: Effects of erythropoietin on stem cell differentiation and red cell maturation; development of human fetal gastrointestinal tract. Mailing Add: Dept of Anat New York Med Col Valhalla NY 10595

ORLICK, CHARLES ALEX, b Milltown, NJ, Sept 30, 27; m 50; c 6. PHYSICAL CHEMISTRY. Educ: Rutgers Univ, BSc, 47, MSc, 50, PhD(chem), 52. Prof Exp: Res assoc chem, 51-59, sr res chemist, 59-60, supvr propellant res group, 60-65, supvr appl res group, 65-68, MKT SPECIALIST, ALLEGANY BALLISTICS LAB, HERCULES INC, CUMBERLAND, 68- Mem: Am Chem Soc; Combustion Inst. Res: Propellants and explosives; thermodynamics of propellant systems; chemical kinetics; nitrate ester decomposition mechanisms; free radical reactions; combustion flame phenomena. Mailing Add: RFD 1 Box 44A LaVale MD 21502

ORLIK, PETER PAUL, b Budapest, Hungary, Nov 12, 38; US citizen; m 64; c 2. MATHEMATICS. Educ: Norweg Inst Technol, BS, 61; Univ Mich, Ann Arbor, PhD(math), 66. Prof Exp: From asst prof to assoc prof, 66-73, PROF MATH, UNIV WIS-MADISON, 73- Concurrent Pos: Mem & NSF fel, Inst Advan Study, 67-69; vis prof, Univ Oslo, 71-72. Mem: Am Math Soc. Res: Topology; transformation groups. Mailing Add: Dept of Math Univ of Wis Madison WI 53706

ORLIN, HYMAN, b New York, NY, July 16, 20; m 43; c 2. GEOPHYSICS, GEODESY. Educ: City Col New York, BBA, 42; George Wash Univ, AB, 48, AM, 52; Ohio State Univ, PhD, 62. Prof Exp: Mathematician, Coast & Geod Surv, 47-58, supvry geodesist, 58-59, asst chief gravity & astron br, 59-66, tech asst geod div, 66-69, spec asst to dir, Nat Ocean Surv Earth Sci Activities, 69-71; chief scientist, Nat Oceanic & Atmospheric Admin, Nat Ocean Surv, US Dept Commerce, 71-75; STAFF MEM, NAT RES COUN, NAT ACAD SCI, 75- Concurrent Pos: Lectr, George Wash Univ, 54-61, asst prof lectr, 62-63, assoc prof lectr, 64-65, prof lectr & fac adv, 66- Mem: Fel Am Geophys Union. Res: Physical and geometric geodesy; geophysics; oceanography. Mailing Add: Nat Res Coun Nat Acad Sci Washington DC 20037

ORLOB, GERT BERNHARD, plant pathology, deceased

ORLOFF, DANIEL, mathematics, see 12th edition

ORLOFF, HAROLD DAVID, b Winnipeg, Man, Nov 24, 15; nat US; m 41; c 3. ORGANIC CHEMISTRY. Educ: Univ Man, BSc, 37, MSc, 40. Prof Exp: Chemist, Dom Grain Res Labs, Can, 39-40; instr chem, McGill Univ, 40-41; res chemist wood chem, Howard Smith Papers Mills, Ltd, 41-48; res supvr, 48-61, SUPVR RES PLANNING, ETHYL CORP, 61- Concurrent Pos: Instr, Univ Detroit, 58. Honors & Awards: Weldon Mem Gold Medal, Can Pulp & Paper Asn, 47. Mem: AAAS; Am Chem Soc. Res: Recovery and utilization of lignin from wood pulping processes; Arborite plastics; alkaline pulping processes; chlorination of benzene; chlorocyclohexane compounds; stereoisomerism of cyclohexane derivatives; phosphorous chemistry; lubricant and fuel additives; antitoxidants; chemistry of phenol, hydrocarbon oxidation; petrochemicals. Mailing Add: 15031 Sutherland Ave Oak Park MI 48237

ORLOFF, JACK, b New York, NY, Dec 22, 21; m 51; c 4. PHYSIOLOGY, MEDICINE. Educ: NY Univ, MD, 43. Prof Exp: Physiologist, 50-57, chief sect kidney & electrolyte metab, 57-62, chief lab kidney & electrolyte metab, 62-75, DIR INTRAMURAL RES, NAT HEART & LUNG INST, 74- Concurrent Pos: Dazian fel med, Med Sch, Yale Univ, 48-49, Life Ins med res fel, 49-50; attend physician, Clin Ctr, Nat Heart Inst, 53-; prof lectr, Med Sch, Georgetown Univ, 62-; fac chmn med physiol grad prog, NIH, 63-67; mem sci adv bd, Nat Kidney Found, 62-71; sect ed, Am J Physiol, 64-68 & J Appl Physiol, 64-68; consult ed, Life Sci; assoc ed, Kidney Int; Am Physiol Soc rep, Div Med Sci, Nat Res Coun, 74-75. Honors & Awards: Homer Smith Award, 73; Pub Health Serv Meritorious Serv Award, 74. Mem: Am Physiol Soc; Am Fedn Clin Res; Am Soc Clin Invest; Asn Am Physicians; Fedn Am Sci. Res: Kidney and electrolyte physiology; membrane transport; clinical investigation. Mailing Add: Dir of Intramural Res Nat Heart & Lung Inst Bethesda MD 20014

ORLOFF, MALCOLM KENNETH, b Philadelphia, Pa, Feb 26, 39; m 60; c 3. QUANTUM CHEMISTRY. Educ: Univ Pa, BA, 60, PhD(phys chem), 64. Prof Exp: Fel theoret chem, Yale Univ, 64-65; sr res chemist theoret chem, 65-74, group leader, Dyes Tech Serv, 74-75, PROD MGR DYES, AM CYANAMID CO, 75- Mem: Am Asn Textile Colorists & Chemists; Am Chem Soc; Am Phys Soc. Res: Quantum mechanics of atomic and molecular systems. Mailing Add: Am Cyanamid Co Bound Brook NJ 08805

ORLOFF, MARSHALL JEROME, b Chicago, Ill, Oct 12, 27; m 53; c 6. SURGERY. Educ: Univ Ill, BS, 49, MD & MS, 51; Univ Colo, Am Bd Surg, dipl, 59; Am Bd Thoracic Surg, dipl, 60. Prof Exp: Asst instr pharmacol, Univ Ill, 49-51; intern, Univ Calif Hosp, San Francisco, 51-52; resident surg, Univ Pa, 52-58, from asst instr to instr, 52-58; from instr to asst prof, Univ Colo, 58-61; prof, Univ Calif, Los Angeles, 61-67; PROF SURG & CHMN DEPT, SCH MED, UNIV CALIF, SAN DIEGO, 67- Concurrent Pos: Fel, Harrison Dept Surg, Univ Pa, 52-58, Nat Cancer Inst Inst trainee, 56-58; res grants, Univ Pa, Univ Colo & Univ Calif, 55-74; Markle scholar acad med, 59-64; consult, Vet Admin Hosps, Denver & Grand Junction, Colo & Albuquerque, NMex, 58-61; chief surg, Harbor Gen Hosp, Calif, 61-67; lectr, Univs Chicago, Tex & Boston, 65, Univ Colo, 66 & Albert Einstein Col Med, 68; consult, US Naval Hosp, San Diego, 66- & Nat Bd Med Examr, 67-70; mem clin res training comt, NIH, 66-, surg study sect, 68-; Francis M Smith lectr, Scripps Clin & Res Found, 67; Samuel Lillienthal vis chief surg, Mt Zion Hosp, San Francisco, 68; Michael & Janie Miller vis prof, Univ Witwatersrand, 70; Edward Pierson Richardson vis prof, Harvard Univ & Mass Gen Hosp, 71; vis prof numerous univs, cols & founds, 67-74; ed-in-chief, World J Surg. Honors & Awards: Outstanding Teacher Awards, Univ Pa, 58, Univ Colo, 61 & Univ Calif, Los Angeles, 66. Mem: AAASl Am Surg Asn; Am Col Surg; Soc Univ Surg (pres, 71-72); Am Gastroenterol Asn. Res: Liver physiology and disease; transplantation; gastrointestinal physiology and surgery; adrenal physiology; metabolism; shock; neuropharmacology; cancer and vascular surgery. Mailing Add: Dept of Surg Univ of Calif Sch of Med La Jolla CA 92037

ORLOSKI, RAYMOND FELIX, organic chemistry, see 12th edition

ORLOWSKI, JAN ALEXANDER, b Warsaw, Poland, Aug 31, 29; div; c 1. CHEMISTRY, CHEMICAL ENGINEERING. Educ: Lodz Tech Univ, Chem engr & MS, 51; Warsaw Tech Univ, PhD(chem), 62. Prof Exp: Asst res prof, Inst Gen Chem, Warsaw, Poland, 51-55; head lab & technol dept, Res Ctr, Inst Chem Org Technol, Warsaw, 55-69; dir chem eng, Epoxylite Corp, Calif, 69-71; SCI DIR & VPRES, LEE PHARMACEUT, 71- Concurrent Pos: Consult, Poland; lectr univs, Warsaw. Res: Catalytic processes in chemical organic technology; polymer chemistry; dental and medical materials. Mailing Add: Lee Pharmaceut 1444 Santa Anita Ave South El Monte CA 91733

ORLOWSKI, RONALD CYRIL, organic chemistry, bio-organic chemistry, see 12th edition

ORMAND, FREDERICK TUTTLE, physical chemistry, see 12th edition

ORME-JOHNSON, WILLIAM HENRY, biochemistry, see 12th edition

ORMISTON, EMMETT EZEKIEL, b Oblong, Ill, Nov 21, 09; m 40; c 2. DAIRY SCIENCE. Educ: Univ Ill, BS, 35, MS, 37, PhD, 58. Prof Exp: From asst to asst prof dairy prod, 38-58, assoc prof dairy husb, 58-61, PROF DAIRY HUSB, UNIV ILL, URBANA, 61- Concurrent Pos: Univ Ill-US Agency Int Develop dairy sci adv & exten adv, Banaras Hindu Univ, 61-63 & dairy sci adv, Uttar Pradesh Agr Univ, India, 68-71. Mem: Am Dairy Sci Asn. Res: Dairy cattle breeding and management; milking and milk handling at the farm; milk goats. Mailing Add: 1101 E Mumford Urbana IL 61801

ORMROD, DOUGLAS PADRAIC, b Langley, BC, May 27, 34; m 57; c 3. PLANT PHYSIOLOGY. Educ: Univ BC, BSA, 56; Univ Calif, PhD(plant physiol), 59. Prof Exp: Instr agron, Univ Calif, 59-60; from asst prof to prof plant sci, Univ BC, 60-69; PROF HORT SCI, UNIV GUELPH, 69- Mem: Am Soc Plant Physiol; Am Soc Hort Sci. Res: Growth and development of plants, especially as affected by environmental factors. Mailing Add: Dept of Hort Sci Univ of Guelph Guelph ON Can

ORMSBEE, RICHARD ARMSTRONG, b Walla Walla, Wash, Jan 6, 15; m 44; c 2. BIOCHEMISTRY. Educ: Mont State Univ, BA, 36; State Col Wash, MS, 38; Brown Univ, PhD(microbiol), 41. Prof Exp: Sci aide, USPHS, Mont, 36; jr biologist, US Forest Serv, 38; head lab asst, Brown Univ, 38-41; spec res assoc, Biol Chem Dept, Harvard Med Sch, 41-43; tech aide, Off Sci Res & Develop, DC, 43-45; exec secy, Chem-Biol Coord Ctr, Nat Acad Sci, 45; assoc mem, Sloan-Kettering Inst, 45-47; biochemist, 47-57, sr scientist, 57-60, SCI DIR, ROCKY MOUNTAIN LAB, NAT INST ALLERGY & INFECTIOUS DIS, 60- Concurrent Pos: Lectr, Mont State Univ. Mem: AAAS; Am Chem Soc; Tissue Cult Asn; Am Asn Immunol; NY Acad Sci. Res: Growth and metabolism of protozoa; rickettsial and viral disease and vaccines; growth, metabolism and structure of rickettsia; antigen, antibody and immune reactions. Mailing Add: Rocky Mountain Lab Nat Inst Allergy & Infect Dis Hamilton MT 59840

ORMSBY, ANDREW ALEXANDER, biochemistry, see 12th edition

ORNA, MARY VIRGINIA, b Newark, NJ, July 4, 34. PHYSICAL CHEMISTRY, ANALYTICAL CHEMISTRY. Educ: Chestnut Hill Col, BS, 55; Fordham Univ, MA, 58, PhD(phys chem), 62; Catholic Univ, MA, 67. Prof Exp: Jr chemist, Hoffmann-

LaRoche, Inc, NJ, 55-56; teacher, Acad Mt St Ursula, 58-61; instr chem, Bronx Community Col, 61-62; ASSOC PROF & CHMN DEPT INORG & ANAL CHEM, COL NEW ROCHELLE, 66- Honors & Awards: Am Cyanamid Co Award, 60-61. Mem: Am Chem Soc; Soc Appl Spectros. Res: Chelation in mixed solvents; inorganic infrared spectroscopy. Mailing Add: Dept of Chem Col of New Rochelle New Rochelle NY 10801

ORNDOFF, JOHN DELBERT, physics, see 12th edition

ORNDUFF, ROBERT, b Portland, Ore, June 13, 32. BOTANY. Educ: Reed Col, BA, 53; Univ Wash, MSc, 56; Univ Calif, Berkeley, PhD(bot), 61. Prof Exp: Asst prof biol, Reed Col, 61-62; asst prof bot, Duke Univ, 62-63; from asst prof to assoc prof, 63-71, assoc dir, Univ Bot Garden, 71-73, PROF BOT & CURATOR UNIV HERBARIUM, 71-, DIR JEPSON HERBARIUM & LIBR, 68-, DIR UNIV BOT GARDEN, 73- Mem: AAAS; Bot Soc Am; Soc Study Evolution; Am Soc Plant Taxon. Res: Biosystematics of angiosperms; Pacific basin phytogeography; reproductive biology of heterostylous plants. Mailing Add: Dept of Bot Univ of Calif Berkeley CA 94720

ORNE, MARTIN THEODORE, b Vienna, Austria, Oct 16, 27; US citizen; m 62. PSYCHIATRY, PSYCHOLOGY. Educ: Harvard Col, BA, 48, Harvard Univ, AM, 51, PhD, 58; Tufts Univ, MD, 55. Prof Exp: USPHS fel, Boston Psychopath Hosp, 56-57; lectr social rels, Harvard Univ, 58-59, res assoc, 59-60; instr psychiat, Harvard Med Sch, 59-62; assoc, 62-64; assoc prof, 64-67, PROF PSYCHIAT, MED SCH, UNIV PA, 67-; DIR UNIT EXP PSYCHIAT, INST OF PA HOSP, 64- Concurrent Pos: Dir, Studies in Hypnosis & Human Ecol Projs, Mass Ment Health Ctr, 59-64; Fulbright scholar, Univ Sydney, Australia, 60; ed, Int J Clin & Exp Hypnosis, 62; consult, Surg Gen Study Sect, NIMH, 66-74 & Vet Admin Hosp, Philadelphia, 72. Honors & Awards: Bernard B Raginsky Award, Soc Clin & Exp Hypnosis, 69. Mem: Am Psychiat Asn; Am Psychol Asn; Am Psychosomatic Soc; Soc Clin & Exp Hypnosis (pres, 71-73). Res: Nature of hypnosis and special states of consciousness, objectification of subjective events, effects of the observational context on data in social and experimental psychology, psychotherapy and psychophysiology. Mailing Add: Unit for Exp Psychiat 111 N 49th St Philadelphia PA 19139

ORNING, ARTHUR ALONZO, physical chemistry, see 12th edition

ORNSTEIN, DONALD SAMUEL, b New York, NY, July 30, 34; m 65; c 3. PURE MATHEMATICS. Educ: Univ Chicago, PhD(math), 56. Prof Exp: PROF MATH, STANFORD UNIV, 60- Honors & Awards: Bocher Prize, Am Math Soc. Mem: Am Math Soc. Res: Ergodic theory. Mailing Add: Dept of Math Stanford Univ Stanford CA 94305

ORNSTEIN, LEONARD, b New York, NY, Feb 8, 26; m 45; c 4. CELL BIOLOGY, BIOPHYSICS. Educ: Columbia Univ, AB, 48, AM, 49, PhD(zool), 57. Prof Exp: Asst zool, Columbia Univ, 49-52, asst cytol, 50-52, instr histol, 51-52; res assoc path, 54-67, DIR CELL RES LAB, MT SINAI HOSP, NEW YORK, 54-, PROF PATH, MT SINAI SCH MED, 66- Concurrent Pos: Res assoc, Columbia Univ, 52-64; consult, Am Cyanamid Co, 57-58, Canal Indust Corp, 62-69, Space-Gen Corp, 65-66, Airborne Instrument Corp, 65-67, Farrand Optical Co, 66-67, IBM Watson Labs, 67-68 & Technicon Corp, 69-; vis prof, Harvard Univ, 67-68; mem develop biol panel, NSF, 70-73. Mem: Am Soc Cell Biol; Histochem Soc. Res: Techniques and instrumentation in microspectrophotometry, phase, interference and electron microscopy; microcinematography; enzyme cytochemistry; microchemistry; fluorescence microscopy; freeze-substitution; microtomy; electrophoresis; information theory and pattern recognition. Mailing Add: 5 Biltom Rd White Plains NY 10607

ORNSTEIN, WILHELM, b Jaroslaw, Poland, Nov 9, 05; US citizen; m 46. APPLIED MATHEMATICS, MATHEMATICAL ANALYSIS. Educ: Tech Hochsch, Brunn, MS, 29; Tech Univ, Berlin, DrIng, 33. Prof Exp: Tech secy to dir, Gen Motors, Poland, 34-39; prof mech, Turkish Mil Acad, Istanbul, 40-42; tech adv, Mid East Hq, US Army, Cairo, 43-46; lectr mech, Univ Pa, 47-48; assoc prof mech eng, Newark Col Eng, 48-50, prof, 50-54; prof, 55-74, EMER PROF APPL MATH, WASH UNIV, 74- Concurrent Pos: Vis prof, Stevens Inst Technol, 50-52. Honors & Awards: Meritorious Civilian Serv Overseas Award, US Army, 46; Medal of Freedom, War Dept, 48. Mem: Am Soc Eng Educ. Mailing Add: Dept of Appl Math & Comput Sci Wash Univ St Louis MO 63130

ORNSTON, LEO NICHOLAS, b Philadelphia, Pa, Jan 7, 40; m 63. MICROBIAL PHYSIOLOGY, BIOCHEMISTRY. Educ: Harvard Univ, BA, 61; Univ Calif, Berkeley, PhD(comp biochem), 65. Prof Exp: Res assoc bact, Univ Calif, Berkeley, 65-66; NIH fel biochem, Univ Leicester, 66-68; fel, Univ Ill, Urbana, 68-69; asst prof, 69-74, ASSOC PROF BIOL, YALE UNIV, 74- Concurrent Pos: John Simon Guggenheim Found fel, 73-74. Mem: Am Soc Biol Chem; Am Soc Microbiol; Brit Soc Gen Microbiol; Am Chem Soc. Res: Regulation of convergent and divergent pathways in bacteria; evolution of microbial enzymes. Mailing Add: Dept of Biol Kline Biol Tower Yale Univ New Haven CT 06520

ORO, JUAN, b Lerida, Spain, Oct 26, 23; m 49; c 4. BIOCHEMISTRY. Educ: Univ Barcelona, Lic, 47; Baylor Univ, PhD(biochem), 56. Prof Exp: From instr to assoc prof chem, 55-63, chmn dept biophys sci, 67-69, PROF CHEM, UNIV HOUSTON, 63- Concurrent Pos: NASA grants, 62-, partic, Viking Mars Lander Molecular Anal Team, 69-; hon coun, Highest Coun Sci Res, Spain, 69; prof, Univ Barcelona, 71, pres sci coun, Inst Fundamental Biol, 71. Mem: AAAS; Am Chem Soc; Am Soc Biol Chem; Geochem Soc; Span Soc Biochem. Res: Mechanisms of enzyme action; synthesis of biochemical compounds; neurobiochemistry; application of biological principles for the improvement of the quality of human life; comprehensive study of carbonaceous, organic and organogenic matter in returned lunar samples; biochemical applications of gas chromatography and mass spectrometry; organic cosmochemistry and paleochemistry; molecular and biological evolution; origin of life. Mailing Add: Dept of Biophys Sci Univ of Houston Houston TX 77004

OROFINO, THOMAS ALLAN, physical chemistry, see 12th edition

ORONA, ANGELO RAYMOND, b El Centro, Calif, Oct 21, 32. ETHNOLOGY. Educ: Univ Calif, Santa Barbara, BA, 60; Univ Calif, Los Angeles, MA, 64, PhD(anthrop), 68. Prof Exp: Res asst anthrop, Univ Calif, Los Angeles, 65-68; asst prof, Univ SFla, 68-69; asst prof, Univ Windsor, 69-73; ASSOC PROF ANTHROP, UNIV SAN DIEGO, 73- Concurrent Pos: Orgn Am States res award, 64-65; lectr anthrop, Mt St Mary's Col, Calif, 66. Mem: Fel Am Anthrop Asn; fel Royal Anthrop Inst Gt Brit & Ireland; Am Ethnol Soc. Res: Ethnology of fishermen; Latin Caribbean; Venezuela. Mailing Add: Dept of Behav Sci Univ of San Diego San Diego CA 92110

ORONSKY, ARNOLD LEWIS, b New York, NY, Oct 22, 40; m 63; c 1. BIOCHEMISTRY, PHYSIOLOGY. Educ: NY Univ, AB, 61; Columbia Univ, PhD(physiol), 68. Prof Exp: MGR INFLAMMATORY & IMMUNOL RES, CIBA-GEIGY RES LABS, 71-, Concurrent Pos: NIH res fel connective tissue biochem, Harvard Med Sch, 68-71, King Trust Found res scholar, 70-71; adj assoc prof orthopedics, Mt Sinai Sch Med, 73- Mem: Am Chem Soc; Am Soc Zool. Res: Biochemistry of connective tissue; immunology; inflammation. Mailing Add: Ciba-Giegy Res Labs Ardsley NY 10502

OROS, MARGARET OLAVA (ERICKSON), b Sheyenne, NDak, Oct 5, 12; m 51; c 1. PETROLEUM GEOLOGY. Educ: Univ NDak, BA, 39. Prof Exp: Lab instr geol, Univ NDak, 38-40; radio assembler, Bendix Aviation Co, Calif, 42-43; lab asst spectrog anal, Aluminum Co Am, 42-48; chem engr, Harvey Aluminum Co, 48-51; geologist, Ill State Geol Surv, 53-62; GEOLOGIST, STATE GEOL SURV, KANS, 62- Mem: Am Asn Petrol Geol; Am Inst Prof Geol; Geol Soc Am; Geosci Info Soc. Res: Spectrography; metallurgical control of aluminum alloys; geological cross sections; field studies; petroleum statistical reports; pipeline and petroleum industry map. Mailing Add: 913 Madeline Lane Lawrence KS 66044

OROSHNIK, JESSE, b New York, NY, May 12, 24; m 55; c 2. PHYSICS. Educ: City Col New York, BS, 48; Univ Fla, MS, 50. Prof Exp: Aeronaut res scientist, NASA, 54; physicist, Standard Piezo Co, 54-55; sr engr, Gen Tel & Electronics Labs, Inc, 55-61; engr res & develop labs, Westinghouse Elec Corp, 61-66; group leader, Adv Fabrication Tech Sect, Electronics Res Lab, Corning Glass Works, 66-68; physicist, Nat Bur Standards, Washington, DC, 68-71; physicist, Naval Weapons Eng Support Activity, 71-75, PHYSICIST, NAVAL RES LAB, US NAVY, 75- Concurrent Pos: NY State Regents War Serv scholar, 57. Mem: AAAS; Am Phys Soc; Sigma Xi; Am Vacuum Soc; Am Soc Test & Mat. Res: Failure mode and performance trade-off analysis of air-borne electronic weapons systems. Mailing Add: Naval Weapons Eng Support Activ ESA-74D Bldg 200 Wash Navy Yard Washington DC 20374

OROSHNIK, WILLIAM, b New York, NY, Feb 9, 14; m 40; c 3. ORGANIC CHEMISTRY. Educ: City Col New York, BS, 34; Polytech Inst Brooklyn, MS, 39, PhD(org chem), 44. Prof Exp: Res chemist, Ansbacher-Siegle Corp, NY, 35-37; Paul Uhlich & Co, 37-39; dir org chem div, Ortho Pharmaceut Corp, 39-62; dir org res, US Vitamin & Pharmaceut Corp, 62-63; dir cent res, Shulton, Inc, 63-66; consult, 66-69; PRES, CHEMO DYNAMICS, 69- Concurrent Pos: Lectr, Polytech Inst Brooklyn, 47-48; lectr, Stevens Inst Technol, 66. Mem: Fel AAAS; Am Chem Soc; fel NY Acad Sci; The Chem Soc. Res: Organic synthesis; medicinal chemistry; chemotherapy, chemical control of conception; synthetic hormones; chemistry of vitamin A, carotenoids, polyene antibiotics, steroids, acetylenes, nitrogen heterocycles, alkaloids and synthetic dyes. Mailing Add: 912 Field Ave Plainfield NJ 07060

OROSZLAN, STEPHEN, b Hungary, July 8, 27; nat US; m 54; c 2. IMMUNOCHEMISTRY, BIOCHEMICAL PHARMACOLOGY. Educ: Budapest Tech Univ, Chem Eng, 50; Georgetown Univ, PhD(pharmacol), 60. Prof Exp: Res assoc, Biol Res Inst, Hungarian Acad Sci, 50-56; asst mem, Albert Einstein Med Ctr, 63-66; asst prof med, George Washington Univ, 66-68; HEAD IMMUNOCHEM, FLOW LABS, INC, 68- Concurrent Pos: Res fel, Lab Chem Pharmacol, Nat Cancer Inst, 60-63. Mem: AAAS; Am Soc Pharmacol & Exp Therapeut; Int Soc Biochem Pharmacol; Am Soc Microbiol; Am Chem Soc. Res: Macromolecular biochemistry and its application to cancer research; viral proteins; immunochemistry of tumor viruses; viral immunol; immunology and chemotherapy. Mailing Add: Flow Labs Inc 1710 Chapman Ave Rockville MD 20852

O'ROURKE, EDMUND NEWTON, JR, b New Orleans, La, Nov 22, 23; m 47; c 1. HORTICULTURE. Educ: Southwestern La Inst, BS, 48; Cornell Univ, MS, 53, PhD, 55. Prof Exp: Jr horticulturist, USDA, 48-51; asst pomol, Cornell Univ, 51-54; assoc prof, 54-59, PROF HORT, LA STATE UNIV, BATON ROUGE, 59- Mem: Am Soc Hort Sci. Res: Breeding and improvement of figs, pears, apples and other fruits for the Gulf Coast area; ecology of fruit varieties; commercial floriculture. Mailing Add: Dept of Hort La State Univ Baton Rouge LA 70803

O'ROURKE, F L STEVE, b Newark, Del, Mar 28, 98; m 20; c 4. HORTICULTURE. Educ: Univ Del, BS, 19; Univ Md, MS, 51. Prof Exp: Sci asst, Bur Entom & Plant Quarantine, USDA, 19-21, jr entomologist, 24-38, plant propagator, Soil Conserv Serv, 38-44, soil conservationist, 44-45; asst prof hort, Mich State Univ, 45-52, supt arboretum, 50-52; horticulturist & sta dir, Trop Agr Exp Sta, Int Coop Admin, Ecuador, 52-55, horticulturist, Thailand, 55-58; asst prof hort, Mich State Univ, 58-64; vis lectr, Iowa State Univ, 65-67; assoc prof, Colo State Univ, 67-72; consult horticulturist systematation of arboretum, Bartlett Tree Res Lab, 72-75. Concurrent Pos: Consult, Fritzsche Bros Inc NY, 63-65, Int Develop Serv, Inc, DC, 65, New Castle Count Dept Parks, Newark Del Sch Dist & Del Tech & Community Col, 75- Honors & Awards: Norman Jay Colman Award, Am Asn Nurserymen, 65; Award of Merit, Int Plant Propogators Soc, 70; L H Bailey Medal, Am Hort Soc, 74. Mem: Am Soc Hort Sci. Res: Plant propogation; nursery management; blueberry culture; effect of synthetic hormones, moisture, mist, light and other environmental factors on propogation of cuttings of plants; cultural and physiological studies of cacao and coffee. Mailing Add: 403 Lark Dr Newark DE 19713

O'ROURKE, JAMES, b Trenton, NJ, Mar 2, 25; m 54; c 4. OPHTHALMOLOGY. Educ: Georgetown Univ, MD, 49; Univ Pa, MSc, 54. Prof Exp: Resident surgeon, 52-54; clin assoc ophthal, NIH, 54-55, chief clin res, 55-57; assoc prof, Sch Med, Georgetown Univ, 57-65, prof ophthal surg, 65-69; PROF SURG & DIR DIV OPHTHAL, HEALTH CTR, UNIV CONN, 69- Concurrent Pos: Res fel, Wills Eye Hosp, Pa, 51-52; consult, Oak Ridge Nat Lab, 55-57 & NIH, 57-, mem vision res & training comt, Nat Eye Inst, 71-75; adj prof physics, Trinity Col, Conn, 72-; armed forces vision res comt, Nat Res Coun. Mem: Asn Res Vision & Ophthal; Am Acad Ophthal & Otolaryngol; Am Ophthal Soc; Soc Nuclear Med. Res: Ocular uptake of radioactive sources; ocular blood flow; tumor detection. Mailing Add: Univ of Conn Health Ctr Farmington Ave Farmington CT 06032

O'ROURKE, JAMES DONALD, analytical chemistry, see 12th edition

O'ROURKE, RICHARD CLAIR, b Minneapolis, Minn, July 8, 30. PHILOSOPHY OF SCIENCE, BOTANY. Educ: Col St Thomas, BS, 53; Univ Minn, Minneapolis, MS, 58, PhD(philos sci, bot), 66. Prof Exp: PROF BIOL & PHILOS SCI, WINONA STATE COL, 65- Concurrent Pos: Staff mem, Univ Minn, 71-72. Mem: AAAS. Res: Logic of the empirical sciences. Mailing Add: Dept of Biol Winona State Col Winona MN 55987

OROWAN, EGON, b Budapest, Hungary, Aug 2, 02; nat US; m 41; c 1. PHYSICS. Educ: Tech Univ, Berlin, Dipl, 29, DEng, 32; Cambridge Univ, MA, 48. Hon Degrees: DrIng, Tech Univ, Berlin, 65. Prof Exp: Asst prof, Tech Univ, Berlin, 28-32; in charge construct proj, United Incandescent Lamp & Elec Co, Hungary, 35-37; res assoc physics, Univ Birmingham, 37-39; res assoc, Cavendish Lab, Cambridge Univ, 39-47, reader physics of metals, 47-50; prof mech eng, 50-51, George Westinghouse prof, 51-62, prof, 58-68, EMER PROF MECH ENG & SR LECTR, MASS INST TECHNOL, 68- Concurrent Pos: Vis prof, Calif Inst Technol, 58 & Carnegie Inst Technol, 63; res fel & consult, Boeing Sci Res Labs, Wash, 65-66; consult, E I du Pont de Nemours & Co. Honors & Awards: Hawksley Gold Medal, Brit Inst Mech Eng, 44; Bingham Medal, Am Soc Rheol, 59. Mem: Nat Acad Sci; Am Soc Testing & Mat; Am Inst Mining, Metall & Petrol Eng; fel Am Acad Arts & Sci; Am Geophys

Union. Res: Physics of strength, plasticity and other mechanical properties of solids; mechanical problems of geology. Mailing Add: 44 Payson Terr Belmont MA 02178

ORPHANIDES, GUS GEORGE, b Kew Gardens, NY, Jan 27, 47; m 68; c 2. ORGANIC POLYMER CHEMISTRY. Educ: Hobart Col, BS, 67; Ohio State Univ, PhD(org chem), 72. Prof Exp: Chemist, US Army Foreign Sci & Technol Ctr, 72-74; CHEMIST, ELASTOMER CHEM DEPT, E I DU PONT DE NEMOURS & CO, 74- Concurrent Pos: Chemist, US Army Tech Working Group Org Mat, 72-74. Mem: Am Chem Soc. Res: Structure property relationships in organic polymers; elastomer curing chemistry; general organic synthesis and chemistry of divalent carbon. Mailing Add: Elastomer Chem Dept Exp Sta E I du Pont de Nemours & Co Wilmington DE 19898

ORPHANOS, DEMETRIUS GEORGE, b Naxos, Greece, Dec 22, 22; Can citizen; m 61; c 2. ORGANIC CHEMISTRY. Educ: Nat Univ Athens, BSc, 55; McGill Univ, PhD(org chem), 63. Prof Exp: Anal chemist, Nemea Gen Chem Lab, Greece, 55-56 & Lorado Uranium Mills, Ltd, Can, 57-58; res assoc org res, Mass Inst Technol, 62-64; group leader, Can Tech Tape Ltd, 64-67; sr org chemist, Tracerlab-Lab for Electronics, Inc, 67; GROUP LEADER ORG CHEM, NEW ENG NUCLEAR CORP, 67- Mem: Am Chem Soc. Res: Analytical chemistry; synthesis of natural products; heterocyclics; propolymer heterocyclic systems; biologically active compounds; organic peroxides; radiochemicals. Mailing Add: 64 Heywood Ave Melrose MA 02176

ORPURT, PHILIP ARVID, b Peru, Ind, Aug 9, 21; m 45; c 4. MYCOLOGY. Educ: Manchester Col, BA, 48; Univ Wis, MS, 50, PhD(bot), 54. Prof Exp: Instr bot & zool, Wausau exten, Univ Wis, 50-51; PROF BIOL, MANCHESTER COL, 54- Concurrent Pos: NIH fel, Inst Marine Sci, Univ Miami, 62-63. Mem: Mycol Soc Am; Am Inst Biol Sci; Int Oceanog Found. Res: Soil and marine microfungi; mycotoxins; paleoecology. Mailing Add: Dept of Biol Manchester Col North Manchester IN 46962

ORR, ALAN R, b Des Moines, Iowa, May 13, 36; m 58; c 3. BOTANY, CELL BIOLOGY. Educ: Simpson Col, BA, 61; Purdue Univ, MS, 64, PhD(bot), 66. Prof Exp: Asst prof, 65-71, ASSOC PROF BIOL, UNIV NORTHERN IOWA, 71- Mem: Bot Soc Am. Res: Developmental botany, morphogenesis of reproductive organs, particularly biochemical events associated with morphological changes in meristematic tissue. Mailing Add: Dept of Biol Univ of Northern Iowa Cedar Falls IA 50613

ORR, ALFONSO, b Jacksonville, Fla, Aug 19, 25; m 50. NEUROPHYSIOLOGY, EXPERIMENTAL NEUROLOGY. Educ: Ga State Col, BS, 48; NY Univ, MA, 52; Fordham Univ, MS, 54; St John's Univ, NY, PhD(physiol, biochem), 62. Prof Exp: From res scientist to sr scientist, 61-68, ASSOC RES SCIENTIST, CREEDMOOR INST PSYCHOBIOL STUDIES, NY DEPT MENT HYG, 68- Concurrent Pos: Vis instr, NY Med Col, 62-64; adj assoc prof, Hunter Col, 67-; adj assoc res prof, Brooklyn Col Pharm, 66-; fac neurophysiologist, Psychiat Residency Training & Grad Med Educ, Creedmoor Psychiat Ctr, 68- Mem: AAAS; Sci Neurosci; NY Acad Sci. Res: Endocrine and neuroendocrine relationships with psychosis; phylogenic study of function of vertebrate caudate nuclei; autonomic nervous system function and erythropoiesis-neurohematology; blood-brain barrier; brain major lipid neurochemistry; nervous system aging and lipofuscinosis; neurophysiopathology of Huntington's disease, including electrophysiological. Mailing Add: Creedmoor Inst for Psychobiol Studies Station 60 Queens Village NY 11427

ORR, CHARLES HENRY, b Griffin, Ga, May 24, 24; m 52; c 6. COMPUTER SCIENCE. Educ: Ohio State Univ, BSc, 48, MSc, 49; Syracuse Univ, PhD(chem), 53. Prof Exp: Asst instr, Syracuse Univ, 50-51; res chemist, 53-69, DESIGN SPECIALIST COMPUT LANG, PROCTER & GAMBLE CO, 70- Mem: Am Chem Soc; Sigma Xi. Res: Introduction of new computer languages; computer application to automatic data reduction; use of Boolean logic in computer languages; decision tables. Mailing Add: 1831 Greenpine Dr Cincinnati OH 45231

ORR, DONALD EUGENE, JR, b Kokomo, Ind, Jan 17, 45; m 67; c 2. ANIMAL SCIENCE. Educ: Purdue Univ, West Lafayette, BS, 67; Pa State Univ, MS, 69; Mich State Univ, PhD(animal husb, nutrit), 75. Prof Exp: Fac res asst animal sci, Pa State Univ, 67-69; swine res specialist, Feed Res Div, Cent Soya Co, 74-75; ASST PROF ANIMAL SCI, TEX TECH UNIV, 75- Mem: Am Soc Animal Sci. Res: Nutrition of the newborn and early-weaned pig; nutrition of the gestating and lactating sow; swine housing and management; recycling and management of animal wastes. Mailing Add: Dept of Animal Sci Tex Tech Univ Lubbock TX 79409

ORR, DOUGLAS MILTON, JR, b Washington, DC, Dec 9, 38; m 61; c 2. GEOGRAPHY. Educ: Davidson Col, BA, 61; Univ NC, MBA, 64, PhD(geog), 68. Prof Exp: Asst to chencellor, 68-70, asst prof geog, 68-73, ASSOC PROF GEOG, UNIV NC, CHARLOTTE, 73- Mem: Asn Am Geogr; Am Geog Soc; Nat Coun Geog Educ. Res: Congressional and legislative redistricting; electoral geography; regional interest in Spain in the United States South. Mailing Add: Dept of Geog Univ of NC Charlotte NC 28213

ORR, GEOFFREY F, b Greenriver, Utah, June 15, 20; m 56; c 2. MYCOLOGY. Educ: Calif State Polytech Col, BSc, 54; Univ Calif, Los Angeles, MA, 57, PhD(mycol), 62. Prof Exp: Supvry mycologist, Kansas City Field Sta, Commun Dis Ctr, 62-63; MICROBIOLOGIST, LIFE SCI LAB DIV, ASSAY BR, DUGWAY PROVING GROUND, 63- Mem: AAAS; Am Soc Microbiol; Mycol Soc Am Inst Biol Sci; Int Soc Human & Animal Mycol. Res: Taxonomic studies of Gymnoascacae in order to determine their relationships to pathogenic fungi; medical and soil mycology; aeromycology; ecology; virology; toxic plants; air sampling technology. Mailing Add: Environ & Life Sci Div Dugway Proving Ground Dugway UT 84022

ORR, HENRY CLAYTON, b Baltimore, Md, Oct 19, 23. BIOLOGY. Educ: Howard Univ, BS, 49; George Washington Univ, MS, 61, PhD, 67. Prof Exp: Biologist & cytologist, Nat Cancer Inst, 53-65; cell biologist & immunocytochemist, Bionetics Res Labs, Inc, 65-67; chief cell biol sect, Div Biol Standards, NIH, 67-72; DIR, CELLULAR PHYSIOL BR, BUR OF BIOLOGICS, FOOD & DRUG ADMIN, 72- Mem: Tissue Cult Asn; Am Soc Cell Biol; Am Soc Microbiol. Res: Modulations affecting the safety of in vitro cell substrates used in producing viral vaccines, mechanisms and kinetics of replication of mammalian viruses in cells; immunobiology. Mailing Add: Bur of Biologics Food & Drug Admin 8800 Rockville Pike Bethesda MD 20014

ORR, HENRY LLOYD, b Milverton, Ont, Nov 8, 19; m 45; c 3. POULTRY SCIENCE. Educ: Ont Agr Col, BSA, 43; Pa State Univ, MS, 52. Prof Exp: From lectr to asst prof, 47-64, ASSOC PROF POULTRY SCI, ONT AGR COL, UNIV GUELPH, 64- Mem: Am Poultry Sci Asn; World Poultry Sci Asn; Am Meat Sci Asn. Res: Quality control in eggs, poultry meat and products. Mailing Add: Dept of Animal & Poultry Sci Ont Agr Col Univ of Guelph Guelph ON Can

ORR, HENRY PORTER, b Opelika, Ala, Aug 20, 21. FLORICULTURE. Educ: Ala Polytech Inst, BS, 42; Ohio State Univ, MS, 47, PhD, 62. Prof Exp: From asst prof to assoc prof, 47-62, PROF HORT, AUBURN UNIV, 62- Mem: Am Soc Hort Sci. Res: Watering methods; leaf abscission of Azaleas; nutrition of woody ornamental plants; marketing of woody ornamentals. Mailing Add: Dept of Hort Auburn Univ Auburn AL 36830

ORR, HOWARD DENNIS, b Harmony, Pa, Apr 1, 24; m 47; c 4. ZOOLOGY. Educ: Geneva Col, BS, 50; Univ Pittsburgh, MS, 52, PhD, 55. Prof Exp: From asst prof to assoc prof, 54-66, PROF BIOL, ST OLAF COL, 66- Concurrent Pos: NSF fel, Univ Ga, 63-64. Mem: Am Soc Mammal; Ecol Soc Am; Animal Behav Soc. Res: Ecology; behavior and physiology of mammals. Mailing Add: Dept of Biol St Olaf Col Northfield MN 55057

ORR, JACK EDWARD, b Delphi, Ind, Dec 11, 18; m 42; c 1. PHARMACY. Educ: Purdue Univ, BS, 40; Univ Wis, PhD(pharmaceut chem), 43. Prof Exp: Instr pharm, Ohio State Univ, 43-44, asst prof, 46-47; prof pharmaceut chem, Univ Utah, 47-52; prof pharmaceut chem & dean sch pharm, Mont State Univ, 52-56; PROF PHARM, DEAN SCH PHARM & STATE CHEMIST, UNIV WASH, 56- Mem: Am Pharmaceut Asn; Am Asn Cols Pharm (pres, 64-65). Res: Pharmaceutical education. Mailing Add: Sch of Pharm Univ of Wash BG-20 Seattle WA 98195

ORR, JAMES ANTHONY, b Madison, Wis, Sept 9, 48; m 71; c 1. MAMMALIAN PHYSIOLOGY, CARDIOPULMONARY PHYSIOLOGY. Educ: Loras Col, BS, 70; Univ Wis-Madison, PhD(vet sci physiol), 74. Prof Exp: Asst scientist cardiopulmonary physiol, Univ Wis-Madison, 74-75; ASST PROF PHYSIOL & CELL BIOL, UNIV KANS, 75- Mem: Am Soc Vet Physiologists & Pharmacologists. Res: Chemical control of ventilation, specifically the role of central chemoreceptors in health and disease; pulmonary circulation and response of this circulatory bed to hypoxia. Mailing Add: Dept of Physiol & Cell Biol Univ of Kans Lawrence KS 66045

ORR, JAMES CAMERON, b Paisley, Scotland, Aug 10, 30; US citizen; m 59; c 2. BIOCHEMISTRY, ORGANIC CHEMISTRY. Educ: Univ London, BSc & ARCS, 54; Univ Glasgow, PhD(chem), 60. Prof Exp: Asst lectr chem, Univ Glasgow, 55-57; res chemist, Syntex SA, Mex, 59-63; from assoc to assoc prof biol chem, Harvard Med Sch, 63-75, tutor biochem sci, Harvard Univ, 63-75; PROF BIOCHEM & ASSOC DEAN, BASIC MED SCI, MEM UNIV NFLD, 75- Mem: AAAS; Am Chem Soc; Am Soc Biol Chem; Endocrine Soc; The Chem Soc. Res: Mechanisms of organic chemical reactions, particularly those involving steroids and enzymes. Mailing Add: Basic Med Sci Mem Univ of Nfld St John's NF Can

ORR, LOWELL PRESTON, b Ross Co, Ohio, Dec 11, 30; m 64. VERTEBRATE ECOLOGY, HERPETOLOGY. Educ: Miami Univ, BS, 52; Kent State Univ, MEd, 56; Univ Tenn, PhD(zool), 62. Prof Exp: Instr biol, Pub Sch, Ohio, 52-55; instr, 56-57 & 61-63, asst prof zool & ecol, 63-68, ASSOC PROF ZOOL & ECOL, KENT STATE UNIV, 68- Mem: Ecol Soc Am; Am Soc Ichthyol & Herpet. Res: Ecological and evolutionary aspects of vertebrate populations; factors controlling seasonal breeding in vertebrates; bioenergetics of ecosystems; natural history of salamanders. Mailing Add: Dept of Biol Sci Kent State Univ Kent OH 44242

ORR, MARY FAITH, b Ashland, Ala, June 29, 20. ANATOMY. Educ: Univ Ala, BA, 41; Vanderbilt Univ, MA, 54, PhD(anat), 61. Prof Exp: Asst tissue cult, Univ Tex Med Br, 45-50; instr oncol & dir tissue cult, Vanderbilt Univ, 50-60; from instr to asst prof anat, 60-67, ASSOC PROF ANAT, MED SCH, NORTHWESTERN UNIV, CHICAGO, 67- Concurrent Pos: Sr investr on assoc staff & dir tissue cult lab, Chicago Wesley Mem Hosp, Ill, 60-64. Mem: AAAS; Soc Neurosci; Am Asn Anat; Tissue Cult Asn; Am Soc Cell Biol. Res: Cancer; development of sensory elements of the inner ear in vivo and in organ culture; differentiation of nerve tissue and organ rudiments in tissue culture; spleen cultures. Mailing Add: Dept of Anat Northwestern Univ Med Sch Chicago IL 60611

ORR, ORTY EDWIN, b Neodesha, Kans, Aug 18, 20; m 47; c 2. VERTEBRATE ZOOLOGY. Educ: Kans State Col, Pittsburgh, BS, 51; Okla State Univ, MS, 52, PhD(zool), 58. Prof Exp: Biologist, Nebr Game & Parks Comn, 54-62; from asst prof to assoc prof biol sci, San Antonio Col, 62-66; PROF BIOL SCI, MO SOUTHERN STATE COL, 66-, HEAD DEPT, 68- Concurrent Pos: Nat Sci Insts fels, field biol, Colo State Univ, 68, radiation biol, Argonne Nat Lab, 69, math col teachers, Univ Mont, 70 & radiation biol & environ qual, Cornell Univ, 71-72. Mem: Am Fisheries Soc; Am Inst Biol Sci. Res: Fresh water fishery. Mailing Add: Dept of Biol Mo Southern State Col Joplin MO 64801

ORR, RICHARD CLAYTON, b Oakland, Calif, Mar 28, 41; m 64; c 2. MATHEMATICS. Educ: Humboldt State Col, AB, 64; Syracuse Univ, MA, 66, PhD(math), 69. Prof Exp: Asst prof, 69-72, ASSOC PROF 72-, CHMN DEPT, 74- MATH, STATE UNIV NY COL OSWEGO, Mem: Am Math Soc; Math Asn Am. Res: Analytic number theory; sieve methods. Mailing Add: Dept of Math State Univ NY Col Oswego NY 13126

ORR, ROBERT S, b Philadelphia, Pa, Dec 18, 37. ORGANIC CHEMISTRY. Educ: Univ Pa, BA, 59; Univ Del, MS, 62, PhD(chem), 64. Prof Exp: From asst prof to assoc prof, 64-72, PROF CHEM, DEL VALLEY COL, 72-, CHMN DEPT, 66- Mem: AAAS; Am Chem Soc. Res: Teaching of organic chemistry and biochemistry on undergraduate level. Mailing Add: Dept of Chem Del Valley Col Doylestown PA 18901

ORR, ROBERT THOMAS, b San Francisco, Calif, Aug 17, 08; m 42. MAMMALOGY, ORNITHOLOGY. Educ: Univ San Francisco, BS, 29; Univ Calif, MA, 31, PhD(zool), 37. Prof Exp: Asst, Mus Vert Zool, Univ Calif, 32-35; wildlife technician, US Nat Park Serv, 35-36; from asst curator to curator, 36-75, assoc dir acad, 64-75, SR SCIENTIST, DEPT BIRDS & MAMMALS, CALIF ACAD SCI, 75- Concurrent Pos: From asst prof to prof, Univ San Francisco, 42-64; vis prof, Univ Calif, Berkeley, 62 & 64, prof-in-res, Univ Exten Prog, San Blas, Mex, 62; mem adv sci comt, Charles Darwin Found for Galapagos Islands. Honors & Awards: Fel Medal, Calif Acad Sci, 73. Mem: AAAS; hon mem Am Soc Mammal (secy, 38-42, 47-59, vpres, 53-55, pres, 58-60); hon mem Cooper Ornith Soc (vpres, 64-67, pres, 67-70); Soc Syst Zool; fel Am Ornith Union. Res: Population, behavioral and taxonomic studies on marine mammals; life history and taxonomy of North American bats and rabbits; distributional and taxonomic studies on Mexican birds and mammals; distribution of fleshy fungi in western North America. Mailing Add: Calif Acad Sci Golden Gate Park San Francisco CA 94118

ORR, ROBERT WILLIAM, b Evansville, Ind, Nov 24, 40; m 65. GEOLOGY. Educ: Univ Ill, BS, 62; Univ Tex, MS, 64; Ind Univ, PhD(geol), 67. Prof Exp: Res asst geol, Ill Geol Surv, 63-64; from asst prof to assoc prof, 67-75, PROF GEOL, BALL STATE UNIV, 75- Mem: Geol Soc Am; Soc Econ Paleont & Mineral. Res: Conodont biostratigraphy of the Devonian system, especially Middle Devonian rocks of North America. Mailing Add: Dept of Geog & Geol Ball State Univ Muncie IN 47306

ORR, WILLIAM CAMPBELL, b St Louis, Mo, Dec 27, 20; m 46; c 3. NUCLEAR

CHEMISTRY. Educ: Princeton Univ, AB, 42; Univ Calif, PhD(chem), 48. Hon Degrees: ScD, MacMurray Col, 68. Prof Exp: Chemist, Radiation Lab, Univ Calif, 48-49; from asst prof to assoc prof chem, 49-65, assoc provost, 65-74, PROF CHEM, UNIV CONN, 65-, ASSOC VPRES ACAD AFFAIRS, 74- Mem: Fel AAAS; Am Chem Soc; Am Phys Soc. Res: Tracer studies of diffusion in solids and molten salts. Mailing Add: Dept of Chem Univ of Conn Storrs CT 06268

ORR, WILLIAM FREDERICK, b Chattanooga, Tenn, Jan 23, 09; m 41; c 5. PHYCHIATRY. Educ: Vanderbilt Univ, BA, 29, MS, 30, MD, 35. Prof Exp: Asst pharmacol, Sch Med, Vanderbilt Univ, 35-36; asst neurol & psychiat, Columbia Univ, 39-40; asst clin neurol psychiat, Sch Med, Vanderbilt Univ, 40-46, prof psychiat & neurol & head dept psychiat, 47-70; dir, Columbia Community Health Ctr, Tenn, 70-74; PROF PSYCHIAT, MT SINAI SCH MED, 74- Concurrent Pos: Fel, Neurol Inst, NY, 39-40; lectr, Sch Social Work, Univ Tenn, 42; consult, Cent State Hosp, 48-; mem staff, Vanderbilt Univ Hosp; staff psychiatrist, Bronx Vet Admin Hosp, 74- Mem: Am Psychiat Asn; AMA; Am Acad Neurol. Res: Neurology. Mailing Add: 44 West 62nd St New York NY 10023

ORR, WILLIAM N, b Sioux Falls, SDak, June 13, 39; m 58; c 1. GEOLOGY. Educ: Univ Okla, BS, 61; Univ Calif, MA, 63; Mich State Univ, PhD(geol), 67. Prof Exp: Asst prof geol, Eastern Wash State Col, 66-67; asst prof, 67-74, ASSOC PROF GEOL, UNIV ORE, 74- Res: Paleontology; biostratigraphy; paleoecology. Mailing Add: Dept of Geol Univ of Ore Eugene OR 97403

ORR, WILSON LEE, b Roswell, NMex, Jan 8, 23; m 44; c 2. ORGANIC CHEMISTRY. Educ: Bethany-Peniel Col, AB, 45; Purdue Univ, MS, 47; Univ Southern Calif, PhD, 54. Prof Exp: Chemist prod control, Wabash Ord Works, US Dept Army, Ind, 45; asst, Purdue Univ, 45-47; vis asst prof, Sch Aeronaut, Univ Southern Calif, 48-49, staff chemist & res assoc geochem, Allan Hancock Found, 49-55, lectr & res assoc dept geol, 55-58, res assoc chem, 58-59; res engr & supvr, Atomiv Int Div, NAm Aviation, Inc, 60-63; RES ASSOC, FIELD RES LAB, MOBIL RES & DEVELOP CORP, 64- Mem: AAAS; Am Chem Soc; Geochem Soc; Sigma Xi. Res: Marine and petroleum geochemistry; organic geochemistry of sediments and petroleum. Mailing Add: Field Res Lab Mobil Res & Develop Corp Box 900 Dallas TX 75221

ORRALL, FRANK QUIMBY, b Somerville, Mass, Oct 15, 25; m 49; c 3. ASTROPHYSICS. Educ: Univ Mass, BS, 50; Harvard Univ, AM, 53, PhD(astron), 55. Prof Exp: Astrophysicist, Col Observ, Harvard Univ, 55-56; solar physicist, Sacramento Peak Observ, NMex, 56-64; PROF PHYSICS & ASTRON, INST ASTRON & UNIV HAWAII, MANOA, 64- Concurrent Pos: Vis scientist, High Altitude Observ, Colo, 70-71. Mem: Am Astron Soc; Inst Astron Union. Res: Solar physics; observation and theory of phenomena of the sun's atmosphere. Mailing Add: Dept of Physics & Astron Univ of Hawaii at Manoa Honolulu HI 96822

ORR-EWING, ALAN LINDSAY, b London, Eng, Jan 13, 15; nat US; m 45; c 2. FOREST GENETICS. Educ: Univ Edinburgh, BSc, 39; Univ Calif, MF, 51; Univ BC, PhD, 56. Prof Exp: Dist forest officer, Brit Forestry Comn, Scotland, 46-47; FOREST GENETICIST, RES DIV, BC FOREST SERV, 48- Concurrent Pos: Chmn, Can Comt Tree Breeding, 58 & 60; Can Scandinavian Found scholar, 62; chmn, Western Genetics Asn, 73-74. Honors & Awards: Distinguished Forester Award, Asn BC Prof Foresters, 71. Res: Selection and breeding of Douglas fir; inbreeding cytology; intraspecific hybridization; racial crossing of Douglas fir and inbreeding. Mailing Add: Res Div BC Forest Serv Victoria BC Can

ORRIS, LEO, b Arlington, Mass, Jan 11, 16; m 41; c 2. ENVIRONMENTAL MEDICINE. Educ: Harvard Univ, AB, 37; Columbia Univ, MSPH, 40; NY Univ, MD, 44. Prof Exp: From instr to asst prof dermat, 50-60, ASSOC PROF ENVIRON MED, SCH MED, NY UNIV, 60-, DERMATOLOGIST & CONSULT, INST ENVIRON MED, 48- Mem: Soc Invest Dermat; Am Indust Hyg Asn; Am Acad Dermat; Am Asn Cancer Res. Res: Carcinogenesis; immunology. Mailing Add: Dept of Environ Med NY Univ Sch of Med New York NY 10016

ORROK, GEORGE TIMOTHY, b Boston, Mass, Nov 25, 30. PHYSICS. Educ: Harvard Univ, BA, 52, MA, 53, PhD(physics), 59. Prof Exp: Mem tech staff, Bell Tel Labs, 59-62; mem tech staff, Bellcomm, Inc, 62-64, suprvr environ sci, 64-65, dept head, 65-72; SUPVR SYSTS IDENTIFICATION, BELL TEL LABS, 72- Mem: AAAS; Am Phys Soc. Res: Solidification of pure metals; semiconductor device application and evaluation; environment and space sciences; quality assurance. Mailing Add: Bell Tel Labs Crawfords Corner Rd Holmdel NJ 77033

ORS, JOSE ALBERTO, b Havana, Cuba, June 14, 44; m 67; c 1. PHOTOCHEMISTRY. Educ: Univ S Fla, BA, 67, MS, 72, PhD(chem), 74. Prof Exp: Chemist air & water pollution, Environ Protection Comn, Fla, 68-69; RES STAFF MEM ORG PHOTOCHEM, T J WATSON RES CTR, IBM CORP, 74- Mem: Am Chem Soc. Res: Photocycloaddition of aromatic hydrocarbons to olefins; photochemical reactions in terpenes; reactions involving singlet oxygen. Mailing Add: T J Watson Res Ctr IBM Corp PO Box 218 Yorktown Heights NY 10598

ORSENIGO, JOSEPH REUTER, b Barryville, NY, Apr 4, 22; m 52; c 2. WEED SCIENCE. Educ: Cornell Univ, BS, 48, PhD(veg crops), 53. Prof Exp: Agronomist & mgr, Rice Res Sta, Ibec Res Inst, Venezuela, 53-55; agronomist-horticulturist & mgr cacao ctr, Inter-Am Inst Agr Sci, Costa Rica, 55-57; asst horticulturist, 57, HORTICULTURIST, PLANT PHYSIOLOGIST & PROF HERBICIDE & WEED SCI RES, AGR RES & EDUC CTR, INST FOOD & AGR SCI, UNIV FLA, 57- Concurrent Pos: Adv, US Agency Int Develop, Costa Rica, 65-67; weed sci consult, Inter-Am Inst Agr Sci, Cent Am, 66; consult, Univ Calif, Multidisciplinary Pest Mgt Surv-Cent Am, 72. Mem: Fel AAAS; Am Soc Agron; Am Soc Hort Sci; Am Soc Plant Physiol; Weed Sci Soc Am. Res: Weed science and herbicide research in agronomic crops in tropical America; sugarcane, agronomic and vegetable crops on organic soils; cacao, coffee, corn, rice, sugarcane and general agriculture in Latin America; sucrose enhancement in sugarcane. Mailing Add: Agr Res & Educ Ctr-Drawer A Univ of Fla Belle Glade FL 33430

ORSI, ERNEST VINICIO, b New York, NY, Aug 10, 22; m 49; c 4. CELL BIOLOGY, VIROLOGY. Educ: Queen's Col, NY, BS, 44; Fordham Univ, MS, 48; St Louis Univ, PhD(biol), 55. Prof Exp: Sr staff mem, Virus Res Sect, Lederle Labs, Am Cyanamid Co, 54-61; prin res scientist virus diag, New York City Health Dept, 61-65; assoc prof, 65-72, PROF BIOL, SETON HALL UNIV, 72- Concurrent Pos: Brown-Hazen Fund grant, 65; contract hyperbaric enhancement of virus infection, Off Naval Res, 68-74. Mem: AAAS; Harvey Soc; Sigma Xi; Tissue Cult Asn. Res: Cell-virus relationships in normal neoplastic and transplantation systems as influenced either by changes in cell properties following sub-cultivation or induced by environmental factors. Mailing Add: Dept of Biol Seton Hall Univ South Orange NJ 07079

ORSINI, MARGARET WARD (GIORDANO), b LeRoy, NY, Mar 18, 16; m 49. EMBRYOLOGY. Educ: Mt Holyoke Col, BA, 37; Cornell Univ, PhD(zool), 46. Prof Exp: Technician & asst path, Sch Med, St Louis Univ, 37-38; med technologist, Lister Clin Lab, 38-40; asst biol, Manhattanville Col, 40-42; asst histol & embryol, Cornell Univ, 42-46; instr zool, Duke Univ, 46-49; res assoc, 52-53, proj assoc, 54-60, res assoc anat, 60-69, asst prof, 69-71; assoc prof, 71-72, PROF ANAT, MED SCH, UNIV WIS-MADISON, 72- Concurrent Pos: NIH fel anat, Univ Wis-Madison, 49-52. Mem: Fel AAAS; Am Soc Zool; Am Asn Anat; Soc Develop Biol; Am Asn Lab Animal Sci. Res: Physiology of reproduction; estrous cycle of hamster; ovulation; comparative morphogenesis and endocrinology of implantation; giant cells and vascular changes in gestation and post-partum involution; control of decidualization; comparison of pregnancy and pseudopregnancy; immunoreproduction. Mailing Add: Dept of Anat Med Sch Univ of Wis Madison WI 53706

ORSZAG, STEVEN ALAN, b New York, NY, Feb 27, 43; m 64; c 2. APPLIED MATHEMATICS. Educ: Mass Inst Technol, BS, 62; Princeton Univ, PhD(astrophys), 66. Prof Exp: Mem, Inst Adv Study, 66-67; from asst prof to assoc prof, 67-75, PROF APPL MATH, MASS INST TECHNOL, 75- Concurrent Pos: A P Sloan fel, 70-72; mem staff, Nat Ctr Atmospheric Res, 69-71; consult, Flow Res, Inc, 71-, Los Alamos Sci Lab, 72- & NASA, 75- Mem: Am Phys Soc; Soc Appl & Indust Math. Res: Fluid dynamics; numerical methods. Mailing Add: Dept of Math Mass Inst Technol Cambridge MA 02139

ORTEGA, ALEJANDRO, b Mexico City, Mex, Feb 13, 31; m 53; c 3. ECONOMIC ENTOMOLOGY. Educ: Nat Univ Mex, Biologist, 53; Ohio State Univ, MSc, 54, PhD(entom), 60. Prof Exp: Res asst entom, Secy Agr & Livestock-Rockefeller Found Prog, Off Spec Studies, 52-53; asst entomologist, Rockefeller Found & Ministry of Agr Coop Prog, 55-57; head dept entom, Nat Inst Agr Res, Mex, 61-66; HEAD ENTOM SECT, INT MAIZE & WHEAT IMPROV CTR, 67- Mem: Entom Soc Am; Mex Soc Entom; Latin Am Phytotechnol Asn; Latin Am Soc Entom. Res: Selection of sources of resistance to major maize insect pests and development of germplasm complexes suitable for such work; development of selective methods of insect control in maize through the use of insecticides. Mailing Add: Int Maize & Wheat Improv Ctr Londres 40-3er Piso Mexico DF Mexico

ORTEGA, JACOBO, b Allende, Mex, Mar 21, 29; m 56; c 3. PLANT PATHOLOGY, PLANT GENETICS. Educ: Nat Sch Agr Mex, BSc, 56; Okla State Univ, MSc, 58; Univ Minn, PhD(plant path & genetics), 60. Prof Exp: Asst wheat breeding, Rockefeller Found, Mex, 54-56; head sect, Nat Inst Agr Res SAG, 61-66; chief pathologist & plant breeder, World Seeds, Inc, 66-71; res agronomist, Calif Milling Corp, Los Angeles, 71-73; ASST PROF BIOL, PAN AM UNIV, 73- Mem: AAAS. Res: Wheat diseases and genetics; agro-economical research of the wheat and milling industries of California. Mailing Add: Dept of Biol Pan Am Univ Edinburg TX 78539

ORTEGA, JAMES M, b Madison, Wis, June 15, 32; m 57; c 1. NUMERICAL ANALYSIS. Educ: Univ NMex, BS, 54; Stanford Univ, PhD(math), 62. Prof Exp: Programmer, Sandia Corp, NMex, 56-58, mathematician, 62-63; mathematician, Bellcomm, Inc, Washington, DC, 63-64; asst prof comput sci, Univ Md, College Park, 64-66, assoc prof prof comput sci & appl math, 66-69; vis prof, Univ Calif, San Diego, 71-72; DIR, ICASE, NASA-LANGLEY RES CTR, 72- Concurrent Pos: From assoc ed to ed, SIAM Rev, Soc Indust & Appl Math, 65-68; prof comput sci, math & appl math, Univ Md, College Park, 69-74; adj prof math & physics, Col William & Mary, 73-; adj prof appl math & comput sci, Univ Va, 73-; consult & mem COSERs numerical comput panel, NSF, 75- Mem: Am Math Soc; Asn Comput Mach; Soc Indust & Appl Math. Res: Numerical analysis; applied mathematics; computer science. Mailing Add: ICASE MS132C NASA-Langley Res Ctr Hampton VA 23665

ORTEGO, JAMES DALE, b Washington, La, Sept 23, 41; m 65; c 3. INORGANIC CHEMISTRY. Educ: Univ Southwestern La, BS, 63; La State Univ, PhD(chem), 68. Prof Exp: ASST PROF CHEM, LAMAR UNIV, 68- Concurrent Pos: Welch Found res grant, Lamar Univ, 69-72. Mem: Am Chem Soc. Res: Coordination chemistry of transition metal and uranium complexes. Mailing Add: Dept of Chem Lamar Univ Beaumont TX 77710

ORTEL, WILLIAM CHARLES GORMLEY, b Spokane, Wash, Sept 28, 26; m 51; c 3. ELECTRONICS, COMMUNICATIONS. Educ: Yale Univ, BS, 49, MS, 50, PhD(physics), 53. Prof Exp: Asst physics, Yale Univ, 53-54; NSF fel, Inst Theoret Physics, Denmark, 54-55; physicist, Missile Div, Lockheed Aircraft Corp, 55-56; PHYSICIST, BELL LABS, 56- Res: Microwaves; digital computer circuits; cosmic rays; nuclear physics; integrated circuits; data communications; optical transmission. Mailing Add: Bell Labs Holmdel NJ 07733

ORTEN, JAMES M, b Farmington, Mo, Nov 29, 04; m 32. BIOCHEMISTRY. Educ: Univ Denver, BS, 28, MS, 29; Univ Colo, PhD(physiol chem), 32. Prof Exp: Asst pharm, Univ Denver, 27-28; instr biochem, Univ Colo, 31-32; from asst prof to assoc prof, 37-57, prof biochem, 57-75, dir grad affairs, 71-75, EMER PROF BIOCHEM & HON DIR ASST DEAN GRAD PROGS, SCH MED, WAYNE STATE UNIV, 75- Concurrent Pos: Coxe fel, Yale Univ, 32-33 & 34-35, Nat Res Coun fel, 33-34, Pfizer & Co fel, 35-37; guest, Int Physiol Cong, Switz, 38. Mem: Fel AAAS; Am Chem Soc; Am Soc Biol Chem; Soc Exp Biol & Med; Am Inst Nutrit (secy, 52-55). Res: Effect of dietary protein and metallic elements on hemopoiesis; metabolism of organic acids and the porphyrins; carbohydrate metabolism in experimental diabetes. Mailing Add: Dept of Biochem Wayne State Univ Sch of Med Detroit MI 48201

ORTENBURGER, IRENE BEARDSLEY, b San Diego, Calif, Aug 18, 35; m 56; c 2. THEORETICAL SOLID STATE PHYSICS. Educ: Stanford Univ, BS, 57, MS, 58, PhD(physics), 65. Prof Exp: Scientist, Space Physics Dept, Lockheed Missiles & Space Co, 58-62, res scientist, Mat Sci Lab, Calif, 66-69; consult, 69-76, MEM RES STAFF, IBM RES LAB 76- Mem: Am Phys Soc; Sigma Xi. Res: Theoretical studies of the electronic properties of solids. Mailing Add: IBM Res Lab K44/281 5600 Cottle Rd San Jose CA 95193

ORTENBURGER, LEIGH NATUS, b Norman, Okla, Feb 14, 29; m 56; c 2. APPLIED STATISTICS. Educ: Univ Okla, BS, 52; Univ Calif, Berkeley, MA, 53; Stanford Univ, MS, 65. Prof Exp: Sr engr, Sylvania Reconnaissance Systs Lab, 57-59, from develop engr to advan develop engr, 60-63, eng specialist, 63-64 & Sylvania Electronic Defense Lab, 64-68, SR ENG SPECIALIST, SYLVANIA ELECTRONIC DEFENSE LAB, GTE SYLVANIA, 68- Mem: Opers Res Soc Am; Inst Math Statist. Res: Radio wave propagation; robust estimation methods. Mailing Add: 1963 Rock St Mountain View CA 94043

ORTH, CHARLES DOUGLAS, b Seattle, Wash, June 1, 42; m 64; c 2. HIGH ENERGY PHYSICS, ASTROPHYSICS. Educ: Univ Wash, BS, 64; Calif Inst Technol, PhD(physics), 70. Prof Exp: Res assoc cosmic ray physics, NASA Manned Spacecraft Ctr, 70-72; ASST RES PHYSICIST ASTROPHYS & COSMIC RAY PHYSICS, UNIV CALIF, BERKELEY, 72- Mem: Am Phys Soc; Sigma Xi. Res: Age of cosmic rays and abundance of cosmic ray Be-10; anisotropy of microwave blackbody radiation; real time correction of atmospherically degraded telescope images. Mailing Add: Bldg 50 Rm 232 Lawrence Berkeley Lab Berkeley CA 94720

ORTH, CHARLES JOSEPH, b Fontana, Calif, Aug 13, 30; m 59; c 1. NUCLEAR

CHEMISTRY. Educ: San Diego State Col, BA, 54; Univ NMex, PhD(chem), 69. Prof Exp: STAFF MEM RADIOCHEM, LOS ALAMOS SCI LAB, 56- Mem: AAAS; Am Phys Soc; fel Am Inst Chem; Am Chem Soc. Res: Nuclear structure and nuclear reactions at low and intermediate energy; medium energy pion reactions and muonic x-ray studies. Mailing Add: 281 Chamisa Los Alamos NM 87544

ORTH, DONALD A, nuclear chemistry, see 12th edition

ORTH, DONALD JOSEPH, b Fond du Lac, Wis, July 26, 25; m 66; c 2. GEOGRAPHY, TOPONYMY. Educ: Univ Wis-Madison, BS, 52. Prof Exp: Cartographic engr, 53-59, CHIEF GEOG NAMES STAFF, US GEOL SURV, 59-, EXEC SECY, US BD GEOG NAMES, 73- Concurrent Pos: Prof lectr, George Washington Univ, 66-; mem, Place-Name Surv US, 69-, mem exec coun, 70-; dir, Place-Name Surv Alaska, 70-; cartog adv & ed, Folklore Atlas US, 71-; mem, Int Coun Onomastic Sci, 72- Mem: Am Name Soc (pres, 72); Asn Am Geogr; Soc Hist Discoveries; Am Cong Surv & Mapping; Arctic Inst NAm. Res: Concepts involving the ways man gives and uses names to refer to places and features in the world around him. Mailing Add: 212 N Emerson St Arlington VA 22203

ORTH, GEORGE OTTO, JR, b Seattle, Wash, July 14, 13; m 40; c 3. ORGANIC CHEMISTRY. Educ: Univ Wash, BS, 36. Prof Exp: Plant chemist, Solvents, Inc, 33-35; plant chemist & engr, Preservative Paint Co, 35-41; res chemist, Seattle Gas Co, 41-42; lab mgr, A J Norton, Consult Chemists, 42-51; vpres, Food, Chem & Res Labs, Inc, 53-58; lab dir, Korry Mfg Co, 63-68; CONSULT, 51- Concurrent Pos: Mem comt estab synthetic rubber mfg corp, Tokyo, Japan, 53; private res, 58- Mem: Fel AAAS; Am Inst Chem; Am Chem Soc; Asn Consult Chem & Chem Eng; Inst Food Technol. Res: High polymers; wood specialities and chemistry; organic nitrogen derivatives; protective coatings; adhesives and plastics; wood fiber specialties; oil adsorbents as related to ecology; wood cellulose animal nutrition technology. Mailing Add: 10807 Roosevelt Way NE Seattle WA 98125

ORTH, PAUL GERHARDT, b Chicago, Ill, Apr 18, 29; m 57; c 4. SOIL FERTILITY. Educ: Univ Ariz, BS, 51; Wash State Univ, MS, 53; Univ Wis, PhD(phosphate availability), 59. Prof Exp: ASST SOILS CHEMIST, AGR RES & EDUC CTR, UNIV FLA, 59- Mem: Am Soc Agron; Soil Sci Soc Am. Res: Slow release fertilizers; soil salinity; calcareous soils; tomato culture; analysis of plant tissue for inorganic plant nutrients. Mailing Add: Agr Res & Educ Ctr Univ of Fla 18905 SW 280th St Homestead FL 33030

ORTH, ROBERT JOSEPH, b Elizabeth, NJ, Nov 16, 47; m 73. BIOLOGICAL OCEANOGRAPHY. Educ: Rutgers Univ, BA, 69; Univ Va, MS, 71; Univ Md, College Park, PhD(zool), 75. Prof Exp: ASST MARINE SCIENTIST MARINE BIOL, VA INST MARINE SCI, 74-; INSTR MARINE BIOL, SCH MARINE SCI, COL WILLIAM & MARY & DEPT MARINE SCI, UNIV VA, 75- Mem: Int Asn Aquatic Vascular Plant Biologists; Ecol Soc Am. Res: Biology of seagrasses; structional and functional aspects of marine benthos. Mailing Add: Va Inst of Marine Sci Gloucester Point VA 23062

ORTH, RONALD EMERSON, medicinal chemistry, see 12th edition

ORTH, WILLIAM ALBERT, b Coatesville, Pa, Sept 28, 31; m 58; c 2. CONTINUUM MECHANICS. Educ: US Mil Acad, BS, 54; Purdue Univ, MS, 61; Brown Univ, PhD(appl math), 70. Prof Exp: Dep chief staff civil eng, Strategic Air Command Hq, 72-74; asst prof mech, 61-63, asst prof math, 68-70, PROF PHYSICS & HEAD DEPT, US AIR FORCE ACAD, 74- Mem: Am Soc Eng Educ; Am Asn Physics Teachers; Math Asn Am; Am Inst Aeronaut & Astronaut. Mailing Add: US Air Force Acad/DFP US Air Force Academy CO 80840

ORTHOEFER, JOHN GEORGE, b Columbus, Ohio, Nov 7, 32; m 64; c 3. VETERINARY MEDICINE. Educ: Ohio State Univ, BS, 59, DVM, 63; Univ Calif, Davis, MS, 72. Prof Exp: Vet div epidemiol, Fla State Bd Health, 63-64 & Commun Dis Ctr, Kans City, Kans, 64-66; toxicologist path, Food & Drug Admin, Washington, DC, 66-69; vet lab animal, Nat Air Pollution Control Admin, USPHS, 69-71; training path, Univ Calif, Davis, 71-74, PATHOLOGIST, HEALTH EFFECTS RES LAB, US ENVIRON PROTECTION AGENCY, 74- Mem: Am Vet Med Asn; Int Acad Path. Res: Descriptive pathology of the lung with a goal toward the development of more accurate morphometric techniques in lung pathology. Mailing Add: 5714 Red Oak Dr Fairfield OH 45014

ORTIZ, ALFONSO A, cultural anthropology, social anthropology, see 12th edition

ORTIZ, ANTONIA, b Maunabo, PR, Apr 6, 06; m 36; c 4. MEDICINE, PEDIATRICS. Educ: Harvard Univ, BS, 26, MD, 29. Prof Exp: PROF PEDIAT, SCH MED, UNIV PR, SAN JUAN, 52- Mem: Am Col Physicians; Am Acad Pediat; Am Pediat Soc. Mailing Add: Dept of Pediat Sch of Med Univ of PR GPO Box 5067 San Juan PR 00936

ORTIZ, MELCHOR, JR, b Victoria, Tex, Nov 14, 42; m 69; c 2. APPLIED STATISTICS. Educ: Tex A&I Univ, BS, 70, MS, 71; Tex A&M Univ, PhD(statist), 75. Prof Exp: ASST PROF STATIST, NMEX STATE UNIV, 75- Mem: Am Statist Asn. Res: Nonconvex mathematical programming. Mailing Add: Col of Agr & Home Econ NMex State Univ Las Cruces NM 88001

ORTIZ, PRISCILLA JEAN, biochemistry, see 12th edition

ORTIZ DE MONTELLANO, PAUL R, b Mexico City, Mex; US citizen. BIO-ORGANIC CHEMISTRY. Educ: Mass Inst Technol, BS, 64; Harvard Univ, MA, 66, PhD(chem), 68. Prof Exp: Asst res dir, Syntex Res, Mex, 69-71; vis res prof, Nat Univ Mex, 70-71; ASST PROF CHEM & PHARMACEUT CHEM, UNIV CALIF, SAN FRANCISCO, 72- Concurrent Pos: NATO fel, Swiss Fed Inst Technol, 68-69; consult, Panel Vapor Phase Org Pollutants, Nat Res Coun, 72-75. Mem: Am Chem Soc; The Chem Soc; AAAS. Res: Sterol biosynthesis; enzyme inhibitors; drug metabolism; organic synthetic methods. Mailing Add: Dept of Pharmaceut Chem Sch Pharm Univ of Calif San Francisco CA 94143

ORTIZ-SUAREZ, AUGUSTO HERMINO, b Santurce, PR, Dec 10, 42; m 63; c 3. MATHEMATICS. Educ: Univ PR, Rio Pedras, BS, 63; Univ Miami, MS, 66, PhD(math), 68. Prof Exp: Fel, Univ Fal, 68-69; asst prof, 69-72, chmn dept, 72-74, ASSOC PROF MATH, UNIV PR, MAYAGUEZ, 72- Res: Algebra, especially theory of rings and radicals. Mailing Add: Dept of Math Univ of PR Mayaguez PR 00708

ORTMAN, ELDON EMIL, b Marion, SDak, Aug 11, 34; m 57; c 3. ENTOMOLOGY. Educ: Tabor Col, AB, 56; Kans State Univ, MS, 57, PhD(entom), 63. Prof Exp: Asst entom, Kans State Univ, 56-59, instr, 59-61; res entomologist, Northern Grain Insect Res Lab, Entom Res Div, Agr Res Serv, USDA, 61-68, lab dir & invest leader, 68-72; PROF ENTOM & HEAD DEPT, PURDUE UNIV, WEST LAFAYETTE, 72- Concurrent Pos: Prof, Sdak State Univ, 68-72. Mem: Entom Soc Am; Sigma Xi; AAAS. Mailing Add: Dept of Entom Purdue Univ West Lafayette IN 47906

ORTMAN, HAROLD R, b Buffalo, NY, Dec 19, 17; m 39, 60; c 6. DENTISTRY. Educ: Univ Buffalo, DDS, 41; Am Bd Prosthodont, dipl, 56. Prof Exp: From instr to assoc prof prosthodont, Univ Buffalo, 42-62; clin prof, 62-64, PROF PROSTHODONT & CHMN DEPT, SCH DENT, STATE UNIV NY BUFFALO, 64- Mem: Am Prosthodont Soc; Am Col Prosthodontists; fel Am Col Dent. Res: Prosthodontics; bone physiology; occlusion; denture retention; denture resins; refitting denture bases with todays concepts and material. Mailing Add: Capen Hall G30 Sch of Dent State Univ of NY Buffalo NY 14214

ORTMAN, ROBERT A, b Detroit, Mich, Mar 4, 26; m 53; c 4. ZOOLOGY. Educ: Univ Calif, BA, 51, MA, 53, PhD(zool), 55. Prof Exp: Asst zool, Univ Calif, 53-55, jr res zoologist, Mus Vet Zool, 56; res officer, Med Res Coun, Otago, NZ, 55-56; instr zool, Tulane Univ, 56-57; assoc prof biol, Mem Univ, 57-59; asst prof, Boston Col, 59-62; asst prof, 62-70, ASSOC PROF BIOL, CITY COL NEW YORK, 70- Concurrent Pos: Res grants, Nat Res Coun Can, 58-59, USPHS, 59-66. Mem: AAAS; Am Soc Zool; Am Asn Anatomists. Res: Pituitary cytology and endocrinology; comparative endocrinology; vertebrate histology; cytochemistry; electron microscopy. Mailing Add: Dept of Biol City Col of New York New York NY 10031

ORTNER, DONALD JOHN, b Stoneham, Mass, Aug 23, 38; m 60; c 3. PHYSICAL ANTHROPOLOGY. Educ: Columbia Union Col, BA, 60; Syracuse Univ, MA, 67; Univ Kans, PhD(anthrop), 70. Prof Exp: Grant, 68-70, ASSOC CURATOR ANTHROP, SMITHSONIAN INST, 70- Concurrent Pos: NIH grant, 69-72; assoc prof, Univ Md, 71- Mem: Am Asn Phys Anthrop. Res: Calcified tissue biology; paleopathology. Mailing Add: Div of Phys Anthrop Smithsonian Inst Washington DC 20560

ORTNER, MARY JOANNE, b Windsor, Can, Nov 10, 46; US citizen. PHARMACOLOGY. Educ: Calif State Univ, BS, 68, MS, 71; Univ Hawaii, PhD(pharmacol), 76. Prof Exp: Instr biol, Ramona Convent High Sch, 68-69; res asst zool, Calif State Univ, 69-71; RES ASST PHARMACOL, UNIV HAWAII, 71- Res: Effects of drugs on membranes; drug receptor isolation and characterization; the mechanism of histamine release from mast cells—the mechanism of action of compound 48/80 versus reaginic antibody. Mailing Add: Dept of Pharmacol Biomed Sci Bldg Univ of Hawaii Honolulu HI 96822

ORTNER, SHERRY B, b Brooklyn, NY, Sept 19, 41. CULTURAL ANTHROPOLOGY. Educ: Bryn Mawr Col, AB, 62; Univ Chicago, MA, 66, PhD, 70. Prof Exp: Lectr anthrop, Princeton Univ, 69-70; ASST PROF ANTHROP, SARAH LAWRENCE COL, 71- Concurrent Pos: Field work on Sherpas of Nepal, 66-68; Wenner-Gren Found grant, 70-71; mem, Inst Advan Study, Princeton Univ, 73-74. Mem: Am Anthrop Asn; Asn Asian Studies. Res: Buddhism and all forms of religion; Sherpa ethnography; Tibetan history and culture, especially politics and religion; symbolic analysis and structuralism; nationalism and other political ideologies. Mailing Add: Dept of Anthrop Sarah Lawrence Col Bronxville NY 10708

ORTO, LOUISE A (MRS PATRICK FAMIGHETTI), b New York, NY, Sept 25, 30; m 65. LABORATORY MEDICINE. Educ: Col New Rochelle, BA, 52; Columbia Univ, MS, 63. Prof Exp: Res assoc human nutrit biochem, Nutrit Res Lab, St Luke's Convalescent Hosp, Greenwich, Conn, 52-60; asst dir res, Geriatric Nutrit Lab, Osborn Mem Home, 52-74; asst dir res, Nutrit & Metab Res Div, 60-74, DIR CLIN LAB, BURKE REHAB CTR, 74- Concurrent Pos: Managing ed, Nutrit Reports Int, 69-74, assoc ed, 75-; cert, Nat Registry Clin Chem. Mem: Fel Am Inst Chem; Am Chem Soc; Am Inst Nutrit. Res: Proteins; amino acids; nutritional requirements. Mailing Add: Clin Lab Burke Rehab Ctr White Plains NY 10605

ORTOLANO, THOMAS RUSSELL, inorganic chemistry, see 12th edition

ORTOLEVA, PETER JOSEPH, b Brooklyn, NY, Nov 5, 42; div; c 1. BIOLOGICAL RHYTHMS, BIOPHYSICS. Educ: Rensselaer Polytech Inst, BS, 64; Cornell Univ, PhD(appl physics), 70. Prof Exp: Fel chem, Mass Inst Technol, 70-75; ASST PROF CHEM, IND UNIV, 75- Res: The onset of spatial and temporal patterns that occur in far from equilibrium reacting systems are studied with the theory of irreversible processes; applications are made to physical and biological systems. Mailing Add: Dept of Chem Ind Univ Bloomington IN 47401

ORTON, COLIN GEORGE, b London, Eng, June 4, 38; m 64; c 3. MEDICAL PHYSICS. Educ: Bristol Univ, BSc, 59; Univ London, MSc, 61, PhD(radiation physics), 65. Prof Exp: Lectr physics, St Bartholomew's Hosp, Univ London, 61-66; from instr to assoc prof physics, Med Ctr, NY Univ, 66-75; CHIEF PHYSICIST, R I HOSP & ASSOC PROF RADIATION MED, BROWN UNIV, 76- Concurrent Pos: Consult, Morristown Mem Hosp, NJ, 68-75; lectr, St Vincent's Hosp, New York, 70-71 & Metrop Hosp, New York, 71-75; ed bull, Am Asn Physicists in Med, 71-73; adj prof biol, Fairleigh Dickinson Univ, 73-75. Mem: Am Asn Physicists in Med; Brit Inst Physics; Brit Inst Radiol; Health Physics Soc; Am Inst Physics. Res: Radiation; radiobiology; radiation oncology; computers in medicine; radiation dosimetry. Mailing Add: 275 Narragansett Bay Ave Warwick Neck RI 02889

ORTON, EDWARD WHITFIELD, b Monroe, NC, July 4, 22; m 54; c 2. GEOLOGY. Educ: Univ Okla, BS; La State Univ, MS, 52. Prof Exp: Field geologist, State Wildlife & Fisheries Comn, La, 54-56; asst prof geol, Southwestern La Inst, 57 & La Polytech Inst, 57-61; ASST PROF GEOL, ODESSA COL, 61-, CHMN DEPT, 66- Mem: Geol Soc Am; Am Asn Petrol Geol. Res: Geomorphology. Mailing Add: Dept of Geol Odessa Col Odessa TX 79760

ORTON, GLENN SCOTT, b Fall River, Mass, July 24, 48. PLANETARY ATMOSPHERES. Educ: Brown Univ, ScB, 70; Calif Inst Technol, PhD(planetary sci), 75. Prof Exp: Assoc scientist, Div Geol & Planetary Sci, 75, RESIDENT RES ASSOC PLANETARY ATMOSPHERES, SPACE SCI DIV, NASA JET PROPULSION LAB, CALIF INST TECHNOL, 75- Res: Determination of thermal structure, composition, opacity, energy balance and climatology of planetary atmospheres; current research on outer planets derives this information from remote sensing of thermal radiation from the atmospheres. Mailing Add: 183B-365 Jet Propulsion Lab 4800 Oak Grove Dr Pasadena CA 91103

ORTON, WILLIAM R, JR, b Texarkana, Tex, May 4, 22; m 50; c 4. MATHEMATICS. Educ: Univ Ark, BA, 47; Univ Ill, MA, 48, PhD(math), 51. Prof Exp: Fulbright fel, Univ Paris, 51-52; instr math, Oberlin Col, 52-53; from asst prof to assoc prof, 53-61, PROF MATH, UNIV ARK, FAYETTEVILLE, 61- Concurrent Pos: NSF fac fel, Univ Wash, 65-66; mem NSF staff, India, 67-68. Mem: Am Math Soc; Math Asn Am. Res: Integration; mathematics education. Mailing Add: Dept of Math SE 324 Univ of Ark Fayetteville AR 72701

ORTS, FRANK AUSUT, b Gonzales, Tex, May 6, 31; m 53; c 2. MEAT SCIENCE. Educ: Tex A&M Univ, BS, 53, MS, 59, PhD(animal sci), 68. Prof Exp: Instr meats, 59-63, ASST PROF ANIMAL SCI & EXTEN MEAT SPECIALIST, TEX A&M UNIV, 63- Mem: Am Meat Sci Asn; Am Soc Animal Sci. Res: Carcass composition. Mailing Add: Dept of Animal Sci Tex A&M Univ College Station TX 77841

ORTTUNG, WILLIAM HERBERT, b Philadelphia, Pa, June 16, 34; m 63; c 2. PHYSICAL CHEMISTRY. Educ: Mass Inst Technol, SB, 56; Univ Calif, Berkeley, PhD(phys chem), 61. Prof Exp: Asst prof chem, Stanford Univ, 60-63; asst prof, 63-69, ASSOC PROF CHEM, UNIV CALIF, RIVERSIDE, 69- Mem: AAAS; Am Chem Soc; Am Phys Soc. Res: Cryogenics of molecular solids; high pressure effects; dielectric theory; electro-optical phenomena; crystal and molecular optics; physical biochemistry. Mailing Add: Dept of Chem Univ of Calif Riverside CA 92502

ORTWERTH, BERYL JOHN, b St Louis, Mo, Aug 13, 37. BIOCHEMISTRY, OPHTHALMOLOGY. Educ: St Louis Univ, BS, 59; Univ Mo, MS, 62, PhD(biochem), 65. Prof Exp: Asst prof, 68-73, ASSOC PROF OPHTHAL, UNIV MO-COLUMBIA, 73- Concurrent Pos: Am Cancer Soc grant, Oak Ridge Nat Lab, 66-68, Mem: AAAS; Am Soc Biol Chem; Asn Res Vision & Ophthal. Am Chem Soc. Res: Control of cell division RNA tumor viruses; structure and function of transfer RNA; biosynthesis of lens proteins. Mailing Add: Dept of Ophthal Med Ctr Univ of Mo Columbia MO 65201

ORVIK, JON ANTHONY, b Halstad, Minn, Sept 13, 39; m 62; c 4. PHYSICAL ORGANIC CHEMISTRY. Educ: San Diego State Col, BS, 62; Univ Wash, PhD(phys org chem), 67. Prof Exp: Asst chem, Univ Calif, Santa Cruz, 67-69; RES CHEMIST, DOW CHEM CO, 69- Mem: Am Chem Soc. Res: Process research in organical chemicals production. Mailing Add: Cent Res Chem Proc Lab Dow Chem Co 438 Bldg Midland MI 48640

ORVILLE, HAROLD DUVALL, b Baltimore, Md, Jan 23, 32; m 54; c 4. METEOROLOGY. Educ: Univ Va, BA, 54; Fla State Univ, MS, 59; Univ Ariz, PhD(meteorol), 65. Prof Exp: Res asst meteorl, Univ Ariz, 63-65; assoc prof, 65-73, HEAD NUMERICAL MODELS GROUP, INST ATMOSPHERIC SCI, S DAK SCH MINES & TECHNOL, 65-, PROF METEOROL, 73-, HEAD DEPT, 74- Concurrent Pos: Mem, Int Comn Cloud Physics, 71-; consult, Off Environ Modification, Nat Oceanic & Atmospheric Admin, 72-73. Mem: Am Meteorol Soc; Am Geophys Union. Res: Numerical modeling of medium scale atmospheric motions and cloud development; cloud photogrammetry and modification; measurements of cloud characteristics from specially instrumented aircraft; data analysis by computer. Mailing Add: Inst of Atmospheric Sci SDak Sch of Mines & Technol Rapid City SD 57701

ORVILLE, PHILIP MOORE, b Ottawa, Ill, Feb 24, 30; m 57; c 3. GEOLOGY. Educ: Calif Inst Technol, BS, 52; Yale Univ, MA, 54, PhD(geol), 58. Prof Exp: Fel petrol, Geophys Lab, Carnegie Inst Technol, 57-60; asst prof petrol & mineral, Cornell Univ, 60-62; assoc prof, 62-72, PROF PETROL & MINERAL, YALE UNIV, 72- Concurrent Pos: Ed, Am J Sci. Mem: Geol Soc Am; Am Geophys Union; Geochem Soc; Mineral Soc Am. Res: Experimental petrology; feldspars; equilibrium between volatile and silicate phases. Mailing Add: Dept of Geol & Geophys Yale Univ New Haven CT 06520

ORVILLE, RICHARD EDMONDS, b Long Beach, Calif, July 28, 36; m 64; c 2. ATMOSPHERIC PHYSICS, SPECTROSCOPY. Educ: Princeton Univ, AB, 58; Univ Ariz, MS, 63, PhD(meteorol), 66. Prof Exp: Mgr mfg, Procter & Gamble Co, 59-60; res asst physics, Johns Hopkins Univ, 60-61; res asst meteorol, Univ Ariz, 61-64, res assoc, 64-66; sr res scientist, Westinghouse Elec Corp, 66-68; ASSOC PROF ATMOSPHERIC SCI, STATE UNIV NY ALBANY, 68- Concurrent Pos: Assoc prog dir, NSF, 70-71. Mem: Am Meteorol Soc; Am Geophys Union. Res: Cloud physics; atmospheric electricity; spectroscopy. Mailing Add: Dept of Atmospheric Sci State Univ NY Albany NY 12222

ORVIS, ALAN LEROY, b Cleveland, Ohio, May 2, 21; m 47; c 2. MEDICAL PHYSICS. Educ: Westminster Col, BS, 44; Case Western Reserve Univ, MS, 49; Univ Tex, PhD(physics), 52. Prof Exp: Instr physics, Univ Tex, 50-52; instr, 56-59, asst prof, 59-66, ASSOC PROF BIOPHYS, MAYO GRAD SCH MED, 66; CONSULT BIOPHYS, MAYO CLIN, 52- Mem: AAAS; Am Phys Soc; Meteorol Soc Am; Health Physics Soc; Soc Nulcear Med. Res: Radioisotope measurements; radiological physics; health physics. Mailing Add: Curie Pavilion Mayo Clin Rochester MN 55901

ORWIG, KENNETH R, invertebrate zoology, taxonomy, see 12th edition

ORWOLL, EDWARD FRANCIS, organic chemistry, see 12th edition

ORWOLL, RICHARD DAVID, b Minneapolis, Minn, July 25, 45; m 68. PHYSICAL CHEMISTRY. Educ: St Olaf Col, BA, 67; Univ Calif, San Diego, PhD(chem), 71. Prof Exp: Teacher chem, Govt Sch, Hamburg, Ger, 71-72; fel, Univ Tex, 72-74; CHEMIST, FIBER INDUSTS, INC, 74- Res: Development of fibers for industrial end uses and the investigation of liquid-crystalline systems using nuclear magnetic resonance. Mailing Add: Fiber Industs Inc PO Box 10038 Charlotte NC 28237

ORWOLL, ROBERT ARVID, b Minneapolis, Minn, Aug 28, 40. PHYSICAL CHEMISTRY. Educ: St Olaf Col, BA, 62; Stanford Univ, PhD(chem), 66. Prof Exp: Res instr chem, Dartmouth Col, 66-68; inst mat sci fel chem, Univ Conn, 68-69; asst prof, 69-72, ASSOC PROF CHEM, COL WILLIAM & MARY, 72- Mem: AAAS; Am Chem Soc. Res: Physical chemistry of macromolecules; thermodynamics of polymer solutions; statistical mechanics. Mailing Add: Dept of Chem Col of William & Mary Williamsburg VA 23185

ORY, HORACE ANTHONY, b Amite, La, Dec 16, 32; m 58. PHYSICAL CHEMISTRY. Educ: Southeastern La Col, BS, 53; La State Univ, PhD(chem), 57. Prof Exp: Phys scientist, Monsanto Chem Co, 57-60, E H Plesset Assocs, 60-63, Astrophys Res Corp, 63-64, Heliodyne Corp, 63-64 & Rand Corp, 64-71; PHYS SCIENTIST, R&D ASSOCS, 71- Mem: AAAS; Am Chem Soc; Am Phys Soc; fel Am Inst Chem. Res: Spectroscopy; reaction kinetics; electrooptics. Mailing Add: R&D Assoc PO Box 3580 Santa Monica CA 90403

ORY, ROBERT LOUIS, b New Orleans, La, Nov 26, 25; m 48; c 2. BIOCHEMISTRY, NUTRITION. Educ: Loyola Univ, La, BS, 48; Univ Detroit, MS, 50; Agr & Mech Col, Tex, PhD(biochem, nutrit), 54. Prof Exp: Asst, Univ Detroit, 48-50; chemist, Exp Sta, Agr & Mech Col, Tex, 50-54; chemist, Colloidal Properties Unit, 54, oil seeds sect, 55-57, biochemist, Seed Proteins Pioneering Lab, 58-69, HEAD PROTEIN PROPERTIES RES, OILSEED & FOOD LAB, SOUTHERN REGIONAL RES CTR, USDA, 69- Concurrent Pos: Fulbright res scholar, Copenhagen Univ, 68-69. Mem: AAAS; Am Chem Soc; Sigma Xi; Am Soc Biol Chemists; Inst Food Technologists. Res: Biosynthesis of amino acids in microorganisms; isolation and characterization of proteins and enzymes, including lipases, proteases, phosphatases and amylases, from oilseeds and barley seeds; plant lectins; biosynthesis of aflatoxins. Mailing Add: Southern Regional Res Ctr PO Box 19687 New Orleans LA 70179

ORZECH, CHESTER EUGENE, JR, b Chicago, Ill, Dec 31, 37; m 63; c 3. ANALYTICAL CHEMISTRY. Educ: Ill Inst Technol, BS, 60; Mich State Univ, PhD(org chem), 68. Prof Exp: GROUP LEADER ANAL CHEM, AYERST LABS, INC, 65- Res: Analytical chemistry of pharmaceuticals; limited organic synthesis; thermal analysis of pharmaceuticals. Mailing Add: Ayerst Labs Inc Rouses Point NY 12979

ORZECH, GRACE GEIST, b New York, NY, June 18, 40; m 63. ALGEBRA. Educ: Barnard Col, BA, 61; Cornell Univ, MA, 65; Univ Ill, PhD(math), 70. Prof Exp: instr math, State Univ NY Col Cortland, 65-66; asst prof, Univ Ottawa, 70-72; RES ASSOC, QUEEN'S UNIV, ONT, 72- Mem: Can Math Cong. Res: Category theory; homological algebra. Mailing Add: Dept of Math Queen's Univ Kingston ON Can

ORZECH, MORRIS, b Arg, Feb 10, 42; US citizen; m 63. MATHEMATICS. Educ: Columbia Col, AB, 62; Cornell Univ, PhD(math), 67. Prof Exp: Asst prof math, Univ Ill, Urbana, 66-68; ASST PROF MATH, QUEEN'S UNIV, ONT, 68- Mailing Add: Dept of Math Queen's Univ Kingston ON Can

ORZECHOWSKI, ADAM, b Poland, Mar 8, 16; nat US; m 47; c 1. PHYSICAL CHEMISTRY, MICROSCOPY. Educ: Univ Liege, DSc(appl sci), 52. Prof Exp: Fel, Nat Res Coun Can, 52-53; res chemist, Allied Chem Corp, NJ, 53-56; sr res chemist, Cabot Corp, 57-67; scientist, 67-72, SR CHEMIST & PROJ LEADER OPTICAL MICROS, XEROX CORP, 72- Mem: Am Chem Soc. Res: Heterogeneous catalysis; hydrogenation; oxidation; polymerization; surface chemistry of silica and of carbon black; dispersions of pigments; small particle morphology; optical microscopy transmitted and reflected; photomicrography. Mailing Add: 64 Clintwood Court Rochester NY 14620

ORZECHOWSKI, RAYMOND FRANK, b Camden, NJ, Feb 28, 38; m 61; c 3. PHARMACOLOGY. Educ: Philadelphia Col Pharm, BSc, 59; Temple Univ, MS, 61, PhD(pharmacol), 64. Prof Exp: Sr res pharmacologist, Nat Drug Co, Richardson-Merrell, Inc, Philadelphia, 64-70; asst prof, 70-73, ASSOC PROF PHARMACOL, PHILADELPHIA COL PHARM & SCI, 73- Mem: AAAS; Am Pharmaceut Asn. Res: Chemically-induced carcinoma in animals; cardiovascular and autonomic pharmacology; inflammation and anti-inflammatory agents. Mailing Add: Dept of Pharmacol Phila Col of Pharmacol & Sci Philadelphia PA 19104

OSAWA, EIJI, b Toyama, Japan, June 9, 35; m 62; c 3. ORGANIC CHEMISTRY. Educ: Kyoto Univ, BA, 58, MA, 60, PhD(chem), 66. Prof Exp: Res chemist, Teijin Ltd, Japan, 60-64; staff asst phys org chem, Kyoto Univ, 64-66; NSF res assoc org chem, Univ Wis, 66-67 & Princeton Univ, 67-69; ASSOC PROF CHEM, HOKKAIDO UNIV, 70- Concurrent Pos: Vis prof, Inst Org Chem, Univ Karlsruhe, Ger, 74-75. Mem: AAAS; Am Chem Soc; Chem Soc Japan. Res: Organic synthesis; organic cage compounds. Mailing Add: Dept of Chem Fac of Sci Hokkaido Univ Sapporo Japan

OSAWA, YOSHIO, b Tokyo, Japan, May 28, 30; m 55; c 2. ORGANIC CHEMISTRY, ENDOCRINOLOGY. Educ: Tokyo Metrop Univ, BS, 53; Univ Tokyo, MS, 55, PhD(org chem), 59. Prof Exp: Res scientist, Tsurumi Res Lab Chem, Teikoku Hormone Mfg Co, Japan, 56-58, chief div steroid chem, Div Total Synthesis of steroids, 58-64, prin res scientist, 64-67; assoc scientist, 67-70, head dept steroid biochem, 71-75, PRIN RES SCIENTIST & ASSOC DIR, POST-DOCTORAL TRAINING PROG, MED FOUND BUFFALO, 70-, HEAD DEPT ENDOCRINE BIOCHEM, 75-; ASST RES PROF BIOCHEM, ROSWELL PARK DIV, GRAD SCH, STATE UNIV NY BUFFALO, 74- Concurrent Pos: Fel steroid chem, Dept Med, Roswell Park Mem Inst, 60-63; Am Cancer Soc fac res award, 69-74; consult to clin staff, Roswell Park Mem Inst, 73- Mem: Am Chem Soc; Chem Soc Japan; Endocrine Soc; NY Acad Sci. Res: Synthesis and structure-function relationships of estrogens; total synthesis of steroids; steroid reactions; polymer-iron preparation for anemia; estrogen metabolites; stereochemistry and mechanism of steroid biosynthesis; clinical assay of estrogen conjugates; steroid biochemistry and endocrinology. Mailing Add: Med Found of Buffalo 73 High St Buffalo NY 14203

OSBAHR, ALBERT J, JR, b Nov 4, 31; US citizen; m 57; c 3. PHYSICAL BIOCHEMISTRY, PHYSICAL ORGANIC CHEMISTRY. Educ: Fordham Univ, BS, 52, MS, 53; Georgetown Univ, PhD(phys biochem), 59. Prof Exp: Chemist, Cent Labs, NY, 52-53 & Quaker Maid Labs, 53-54; biochemist, Naval Med Res Inst, 55-58; res biochemist, Geront Inst, NIH, 58-60, RES INVESTR, NAT INST ARTHRITIS, METAB & DIGESTIVE DIS, 60- Concurrent Pos: Prof lectr, Georgetown Univ, 58- Mem: AAAS; Am Chem Soc. Res: Structure of macromolecules and its correlation to function of molecules; relationship of structure to biological activity; protein modification; enzymology and enzyme kinetics; function of metal ions in biological systems; blood clotting mechanism; muscle biochemistry; polymerization of proteins as involved in muscular contraction and blood coagulation. Mailing Add: Lab of Biophys Chem NIH-NIAMDD Bethesda MD 20014

OSBAND, ARTHUR NORMAN, organic chemistry, see 12th edition

OSBERG, PHILIP HENRY, b Melrose, Mass, Oct 23, 24; m 48; c 2. GEOLOGY. Educ: Dartmouth Col, AB, 47; Harvard Univ, AM, 49, PhD(geol), 52. Prof Exp: Asst prof geol, Pa State Univ, 51-52; from asst prof to assoc prof, Colby Col, 52-57; assoc prof, 57-63, head dept geol sci, 66-71, PROF GEOL, UNIV MAINE, ORONO, 63- Concurrent Pos: Fulbright sr res fel, Oslo, Norway, 65-66; geologist, US Geol Surv, var times, 68- Mem: Geol Soc Am; Mineral Soc Am; Norweg Geol Soc. Res: Metamorphic petrology; structural geology; regional geology of the northern Appalachians. Mailing Add: Dept of Geol Sci Univ of Maine Orono ME 04473

OSBORG, HANS, b Halberstadt, Ger, Oct 18, 01; nat US; m 30; c 2. PHYSICAL CHEMISTRY. Educ: Univ Braunschweig, Ger, BS, 24, MS, 25, DrIng, 27. Prof Exp: Mem staff, Res Inst, Ger Metal Co, Inc, 27-29; dir res prod & eng, Lithium Metal Dept, Maywood Chem Works, 29-50; exec vpres, Del, 58-66, PRES, CHEMIRAD CORP, NY, 50-, TECH DIR, DEL, 58-, PRES, DEL, 66- Concurrent Pos: Consult, 34-; tech dir, Metalectro Corp, Md, 50-53, pres, 53-55. Mem: Am Chem Soc; Am Soc Metals; Am Inst Mining, Metall & Petrol Eng; NY Acad Sci; Tech Asn Pulp & Paper Indust. Res: Ethylene imines; aziridines; hydrazines, especially unsymmetrical dimethyl hydrazine; chemistry and metallurgy of lithium; boron hydrides; especially diborane and its compounds. Mailing Add: Chemirad Corp PO Box 152 80 Longview Rd Port Washington NY 11050

OSBORN, CLAIBORN LEE, b Austin, Tex, Sept 15, 33; m 59; c 3. ORGANIC CHEMISTRY, RADIATION CHEMISTRY. Educ: Univ Tex, BA, 56, PhD(org chem), 60. Prof Exp: Chemist, Celanese Chem Co, 60-62, sr res chemist, 62-63; Robert A Welch Found fel photochem, Univ Tex, 63-65; res chemist, 65-68, proj scientist radiation chem, 68-72, res scientist, 72-75, SR RES SCIENTIST, UNION CARBIDE CORP, 75- Mem: Am Chem Soc. Res: Photochemistry; application of photochemical techniques to polymer processing; radiation processing of coatings. Mailing Add: Chem & Plastics Group Union Carbide Corp PO Box 8361 South Charleston WV 25303

OSBORN, CLINTON MORRIS, anatomy, experimental zoology, see 12th edition

OSBORN, DAVID GORDON, b Rehoboth, NMex, July 24, 23; m 56; c 2. ECONOMIC GEOGRAPHY. Educ: Univ Chicago, AM, 49, PhD(geog), 53. Prof Exp: Res asst urban planning, Chicago Plan Comn, 50; res assoc, Am Soc Planning Officials, 50-51; instr econ & urban geog, Ind Univ, 51-53; staff asst to vpres opers, Kroger Co, Ohio, 53-56; asst tech dir prod & mkt planning, Nowland & Co, Conn, 56-58; mem staff, 58-65, advan systs planner, Advan Syst Div, 65-70, SR MED INDUST PLANNING ADMINR, DATA PROCESSING DIV, IBM CORP, 70- Concurrent Pos: Instr econ hist, Loyola Univ Chicago, 50-51 & Inst Oceanog, US Naval Res Officers Sch, NY, 61-64; univ sem assoc, Columbia Univ, 65-70; fel, Inst Global Dynamics, NIH & Am Univ, 71- Mem: AAAS; Asn Am Geogr; Am Pub Health Asn; Soc Gen Systs Res; World Future Soc. Res: Health and medical information systems; urban and regional ecosystems; ecosystems simulation; technology forecasting; futures research; operations research; management sciences; public policy. Mailing Add: Data Processing Div IBM Corp 10401 Fernwood Rd Bethesda MD 20034

OSBORN, ELBURT FRANKLIN, b Winnebago Co, Ill, Aug 13, 11; m 39; c 2. GEOLOGY. Educ: DePauw Univ, BA, 32; Northwestern Univ, MS, 34; Calif Inst Technol, PhD(geol), 38. Hon Degrees: ScD, Alfred Univ, 65, Northwestern Univ, 72, DePau Univ, 72 & Ohio State Univ, 72. Prof Exp: Instr geol, Northwestern Univ, 37; geologist, Val d'Or, Can, 38; petrologist, Geophys Lab, Carnegie Inst, 38-42; phys chemist, Nat Defense Res Comt, 42-45; res chemist, Eastman Kodak Co, NY, 45-46; prof geochem, Col Mineral Industs, 46-70, head dept earth sci, 46-52, from assoc dean to dean col, 52-59, vpres res, 59-70, EMER PROF GEOCHEM, COL EARTH & MINERAL SCI, PA STATE UNIV, UNIVERSITY PARK, 70-; DISTINGUISHED PROF, GEOPHYS LAB, CARNEGIE INST WASH, 73- Concurrent Pos: NSF sr fel, Cambridge Univ, 58; dir, Bur Mines, Washington, DC, 70-73. Mem adv comt geophys br, Off Naval Res, Nat Acad Sci-Nat Res Coun, 47-50, mem, Mat Adv Bd, 65-69, chmn comt on mineral sci & technol, 66-70, chmn bd mineral resources & mem comn natural resources, Nat Res Coun, 74-; mem earth sci panel, NSF, 53-55, mem div comt math, phys & eng sci, 55-59, chmn, 57-58, mem adv panels course content improv prog, 60-61, phys sci facil, 60-64; dir, Am Geol Inst, 56-59; mem adv panels mineral prod div, Nat Bur Stand, 58-62, chmn, 58-59, metall div, 58-64, mem div chem & chem technol, 60-63; mem bd, Univ Corp Atmospheric Res, 59-67; mem adv comt basic res, US Army Res Off Ceramics, 60-63; vchmn, Sirimar Corp, Italy, 61-63; mem bd, Geisinger Med Ctr, 62-, mem bd dirs, Inst Med Educ & Res, 69-; mem, Pa Health Res Inst, 66-70; mem bd dirs, Pa Sci & Eng Found, 68-71, 74-; mem, Nat Mat Adv Bd, 69-70; mem earth sci adv bd, Stanford Univ, 69-72; mem bd dirs, Pa Res Corp, 71-, vpres, 74-; mem adv bd, Col Eng, Univ Calif, Berkeley, 74-; mem mat adv comt, Off Technol Assessment, Cong of US, 74-; mem geosci adv panel, Los Alamos Nat Observ, 75- Honors & Awards: Award, Am Iron & Steel Inst, 54; Roebling Medal, Mineral Soc Am, 72; John Jeppson Award, Am Ceramic Soc, 73; Hal Williams Hardinge Award, Soc Mining Engrs, 74. Mem: Nat Acad Eng; fel AAAS; hon mem Am Ceramic Soc (vpres, 63, pres elect, 64, pres, 65); fel Am Geophys Union; fel Geol Soc Am. Res: Physiocochemical and structural petrology; glass technology; physical chemistry of refractories; blast furnance and open hearth slags; phase equilibira in mineral systems; industrial minerals; mineral synthesis; mineral and energy resources. Mailing Add: Carnegie Inst Wash Geophys Lab 2801 Upton St NW Washington DC 20008

OSBORN, HARLAND JAMES, b Alexandria, Minn, June 24, 32; m 53; c 3. SYNTHETIC ORGANIC CHEMISTRY, PHOTOGRAPHIC CHEMISTRY. Educ: Univ Minn, BS, 54, PhD(org chem), 58. Prof Exp: RES CHEMIST, EASTMAN KODAK CO, 58- Mem: Am Chem Soc. Res: Organic synthesis; design of color photographic systems; chemistry and physics of photographic development. Mailing Add: 24 Skycrest Dr Rochester NY 14616

OSBORN, HERBERT B, b San Juan, PR, July 9, 29. HYDROLOGY. Educ: Stanford Univ, BS, 52, MS, 53, CE, 56; Univ Ariz, PhD, 71. Prof Exp: Hydraul engr, Ariz, 56-58, Miss, 58-60, RES HYDRAUL ENGR, SOUTHWEST WATERSHED RES CTR, USDA, 61- Concurrent Pos: Staff scientist, Nat Prog Staff, Agr Res Serv, USDA, 74-75; mem steering comt, Bd Agr & Renewable Resources, Nat Res Coun-Nat Acad Sci, 75-76. Mem: AAAS; Am Geophys Union; Am Meteorol Soc; Int Asn Statist in Phys Sci. Res: Hydrologic research on thunderstorms and thunderstorm runoff on semiarid rangeland watersheds in the southwest. Mailing Add: Southwest Watershed Res Ctr USDA 442 E Seventh St Tucson AZ 85705

OSBORN, HOWARD ASHLEY, b Evanston, Ill, May 16, 28; m 51; c 4. MATHEMATICS. Educ: Princeton Univ, AB, 49; Stanford Univ, PhD, 55. Prof Exp: Instr math, Univ Calif, 54-56; from asst prof to assoc prof, 56-65, PROF MATH, UNIV ILL, URBANA, 65- Concurrent Pos: Consult, Rand Corp, 54-65; NSF grants, 58-60, 62- Mem: Am Math Soc; Math Asn Am. Res: Differentiable manifolds; piecewise linear manifolds. Mailing Add: Dept of Math Univ of Ill Urbana IL 61801

OSBORN, J MARSHALL, JR, b New York, NY, Sept 11, 30; m 59, 75; c 2. MATHEMATICS. Educ: Princeton Univ, AB, 52; Univ Chicago, PhD(math), 57. Prof Exp: Instr math, Univ Conn, 57; res instr, 57-59, from asst prof to assoc prof, 59-67, PROF MATH, UNIV WIS-MADISON, 67- Mem: Am Math Soc; Math Asn Am. Res: Nonassociative algebra. Mailing Add: Dept of Math Univ of Wis Madison WI 53706

OSBORN, JAMES MAXWELL, b Ypsilanti, Mich, Sept 10, 29; m 55; c 4. ANALYTICAL MATHEMATICS. Educ: Univ Mich, BS, 51, MS, 52, PhD(math), 55. Prof Exp: Instr math, Ohio State Univ, 54-57; asst prof, 57-60, ASSOC PROF MATH, GA INST TECHNOL, 60- Mem: Am Math Soc; Math Asn Am. Res: Complex variables. Mailing Add: Sch of Math Ga Inst Technol Atlanta GA 30332

OSBORN, JOHN EDWARD, b Onamia, Minn, July 12, 36; m 59; c 2. MATHEMATICS. Educ: Univ Minn, Minneapolis, BS, 58, MS, 63, PhD(math), 65. Prof Exp: Asst prof, 65-69, ASSOC PROF MATH, UNIV MD, COLLEGE PARK, 69- Concurrent Pos: NSF res grant, 66-68. Mem: Am Math Soc. Res: Partial differential equations; approximation of eigenvalues. Mailing Add: Dept of Math Univ of Md College Park MD 20742

OSBORN, JOHN JAY, b Detroit, Mich, Nov 5, 17; m 44; c 7. MEDICINE. Educ: Princeton Univ, BA, 39; Johns Hopkins Univ, MD, 43. Prof Exp: Asst resident pediat, Children's Med Serv, Bellevue Hosp, 47; from instr to asst prof, Sch Med, NY Univ, 47-55; actg assoc prof, Sch Med, Stanford Univ, 55-56; from assoc prof to assoc clin prof, 56-65, assoc in surg, 55-65; SR RES MEM, INST MED SCI, PRESBY MED CTR, 65- Concurrent Pos: Nat Found Infantile Paralysis fel, 49-51; estab investr, Am Heart Asn, 54-58. Mem: AAAS; fel Am Heart Asn; Am Soc Artificial Internal Organs; Pop Asn Am; Soc Critical Care Med. Res: Physiology; intensive care; computers in medicine; extracorporeal circulation. Mailing Add: Inst of Med Sci Presby Med Ctr San Francisco CA 94115

OSBORN, JUNE ELAINE, b Endicott, NY, May 28, 37; m 66; c 3. VIROLOGY, INFECTIOUS DISEASE. Educ: Oberlin Col, BA, 57; Case Western Reserve Univ, MD, 61. Prof Exp: Intern & resident pediat, Children's Hosp Med Ctr, Boston, 61-62 & 63-64; resident, Mass Gen Hosp, 62-63; fel infectious dis, Sch Med, Johns Hopkins Univ, 64-65 & Sch Med, Univ Pittsburgh, 65-66; from instr to assoc prof, 66-75, PROF PEDIAT & MED MICROBIOL, SCH MED, UNIV WIS-MADISON, 75- Concurrent Pos: Mem panel viral & rickettsial vaccines, Food & Drug Admin, 73-, mem exp virol study sect, NIH Div Res Grants, 75-79; assoc dean, Grad Sch Biol Sci, Univ Wis-Madison, 75. Mem: Am Acad Pediat; Am Asn Immunologists; Soc Pediat Res; Infectious Dis Soc Am. Res: Pathogenesis of cytomegalovirus infections; nucleic acid homology of human and simian papovaviruses; mechanism of papovavirus oncogenesis; mechanisms of viral persistence and latency. Mailing Add: 497 Med Sci Bldg Univ of Wis Sch of Med Madison WI 53706

OSBORN, KENTON REVAY, physical chemistry, see 12th edition

OSBORN, MARGARET OLIVE, b Pawnee City, Nebr, Aug 10, 07. NUTRITION. Educ: Univ Nebr, BS, 19, MS, 32; Univ Iowa, PhD, 51; Am Bd Nutrit, Dipl. Prof Exp: Teacher high sch, Nebr, 29-30, 32-34; dietician, D T Watson Home, 36-39; therapeut dietician, Harrisburg Hosp, Pa, 39-42; instr home econ, George Wash Univ, 42-49; from asst prof to prof, 51-75, EMER PROF HOME ECON, UNIV IOWA, 75- Mem: AAAS; Am Home Econ Asn; Am Dietetic Asn; Am Oil Chem Soc. Res: Nutrition, particularly lipids and cholesterol. Mailing Add: Dept of Home Econ Univ of Iowa Iowa City IA 52240

OSBORN, MARY JANE, b Colorado Springs, Colo, Sept 24, 27; m 50. MOLECULAR BIOLOGY, BIOCHEMISTRY. Educ: Univ Calif, Berkeley, BA, 48; Univ Wash, PHD(biochem), 58. Prof Exp: From instr to asst prof microbiol, Sch Med, NY Univ, 61-63; from asst prof to assoc prof molecular biol, Albert Einstein Col Med, 63-68; PROF MICROBIOL, SCH MED, UNIV CONN HEALTH CTR, 68- Concurrent Pos: USPHS fel, Sch Med, NY Univ, 59-61, career develop award & res grants, Sch Med, NY Univ, ALbert Einstein Col Med & Univ Conn Health Ctr, 62-; NSF res grants, Albert Einstein Col Med, 65-68; mem microbial chem study sect, NIH, 68-72; mem res comt, Am Heart Asn, 72-77; mem comt space biol & med, Space Sci Bd, Nat Acad Sci, 74-77; mem adv coun, Max Planck Inst Immunol, 75-; mem bd sci counrs, Nat Heart & Lung Inst, 75-79. Mem: Am Soc Biol Chem; Am Chem Soc (pres, div biol chem, 75-76); Am Soc Microbiol. Res: Biosynthesis of bacterial lipopolysaccharides; biogenesis of membranes. Mailing Add: Dept of Microbiol Univ of Conn Health Ctr Farmington CT 06032

OSBORN, RICHARD KENT, b Ft Wayne, Ind, Mar 12, 19; m 45; c 4. PLASMA PHYSICS, PHYSICAL OPTICS. Educ: Mich State Univ, BS, 48, MS, 49; Case Western Reserve Univ, PhD(physics), 52. Prof Exp: Physicist, Oak Ridge Nat Lab, 52-57; assoc prof, 57-59, PROF NUCLEAR ENG, UNIV MICH, ANN ARBOR, 59-, CONSULT, RES INST, 57- Concurrent Pos: Consult, Oak Ridge Nat Lab, 57-65 & KMS-Fusion, Inc, Ann Arbor, 72- Mem: Am Phys Soc; Am Nuclear Soc. Res: Nuclear, reactor and plasma physics; aspects of theories of liquids. Mailing Add: 1330 Ardmoor Dr Ann Arbor MI 48103

OSBORN, ROBERT HENRY, b Denver, Colo, May 25, 07; m 39; c 1. APPLIED PHYSICS. Educ: SDak Sch Mines, BS, 29; Univ Pittsburgh, MS, 36, PhD(physics), 38. Prof Exp: Res engr, Westinghouse Elec & Mfg Co, Pa, 29-31; elec engr, Western Pub Serv Co, Nebr, 31-33; asst physics, Univ Pittsburgh, 34-37; res physicist, Hercules, Inc, 37-43, supvr, 43-70; sci teacher, Sanford Sch, 70-73; RETIRED. Concurrent Pos: Chmn, Gordon Res Conf Instrumentation, 55. Mem: Optical Soc Am; hon mem Optical Soc Am. Mailing Add: 1311 Oberlin Rd Green Acres Wilmington DE 19803

OSBORN, ROBERT OTTO, physical chemistry, see 12th edition

OSBORN, ROGER (COOK), b Crystal City, Tex, Mar 14, 20; m 42; c 3. MATHEMATICS. Educ: Univ Tex, BA, 40, MA, 42, PhD(math), 54. Prof Exp: Teacher high sch, Tex, 40-42; instr appl math, 42-44, 46-53, from instr to assoc prof, 53-67, PROF MATH, UNIV TEX, AUSTIN, 67- Mem: Math Asn Am; Nat Coun Teachers Math. Res: Number theory; differential equations; mathematics education. Mailing Add: Dept of Math Univ of Tex Austin TX 78712

OSBORN, STEPHEN WYNN, organic chemistry, see 12th edition

OSBORNE, BRIAN S, b Treharris, Wales, Oct 13, 38; m 63; c 3. GEOGRAPHY. Educ: Univ Southampton, BA, 60, PhD(geog), 67. Prof Exp: From instr to asst prof geog & anthrop, Southern Colo State Col, 63-67; from asst prof to assoc prof geog, 67-71, PROF GEOG, QUEEN'S UNIV, ONT, 71- Concurrent Pos: Res award, Queen's Univ, Ont, 68-69; Can Coun grants, 69-70 & 71-73. Mem: Asn Am Geogr; Inst Brit Geog. Res: Cultural and historical geography; enclosure movement in Great Britain; settlement patterns of early nineteenth century. Mailing Add: Dept of Geog Queen's Univ Kingston ON Can

OSBORNE, CARL ANDREW, b Pittsburgh, Pa, Sept 17, 40; m 64; c 3. VETERINARY MEDICINE, PATHOLOGY. Educ: Purdue Univ, Lafayette, DVM, 64; Univ Minn, St Paul, PhD(vet med), 70. Prof Exp: From instr to assoc prof vet med, 64-75, PROF VET CLIN SCI, COL VET MED, UNIV MINN, ST PAUL, 75- Concurrent Pos: Mem, Nat Kidney Found. Honors & Awards: Vet of Year, Am Animal Hosp Asn, 74. Mem: Am Soc Vet Clin Path; Am Vet Med Asn. Res: Pathophysiology of congenital and inherited renal diseases; pathophysiology of glomerulonephritis; pathophysiology of neoplasms of the urinary system; relationship of calcium metabolism to renal disease; urolithiasis. Mailing Add: Col Vet Med Univ of Minn St Paul MN 55101

OSBORNE, CHARLES EDWARD, b Mt Croghan, SC, Aug 4, 29; m 55; c 2. SYNTHETIC ORGANIC CHEMISTRY. Educ: Univ NC, BS, 51; Univ Maine, MS, 53; Northwestern Univ, PhD(chem), 56. Prof Exp: Chemist, Eastman Kodak Co, 51-52, chemist, Tenn Eastman Co, 56-59, sr chemist, 59-63, sr chemist, Eastman Chem Prod, Inc, 63-64, sr chemist, Tenn Eastman Co, 64-71, DEVELOP ASSOC, TENN EASTMAN CO, 71- Mem: AAAS; Am Chem Soc. Res: Syntheses of organic compounds. Mailing Add: 1540 Belmeade Dr Kingsport TN 37664

OSBORNE, CHARLES VANCE, inorganic chemistry, see 12th edition

OSBORNE, DARRELL WAYNE, b Los Angeles, Calif, Jan 16, 14; m 36; c 4. PHYSICAL CHEMISTRY, LOW TEMPERATURE PHYSICS. Educ: Univ Calif, Los Angeles, AB, 35; Calif Inst Technol, PhD(chem), 38. Prof Exp: Asst, Calif Inst Technol, 35-36, Noyes fel chem, 38-40; Nat Res Coun fel, Univ Calif, 40-41; instr, Univ Wis, 41-42; res assoc, Nat Defense Res Comt, George Wash Univ, 42-45; sr chemist, Metall Lab, Univ Chicago, 45-46; asst dir chem div, 46-53, SR CHEMIST, ARGONNE NAT LAB, 46- Concurrent Pos: Guggenheim fel, 58-59; chmn, Calorimetry Conf, 60-61; prof lectr, Univ Chicago, 63-66. Honors & Awards: Presidential Cert Merit, 48; Huffman Mem Award, 71. Mem: Am Chem Soc; fel Am Phys Soc. Res: Cryogenics; low temperature heat capacities; liquid helium-3; thermodynamics; chemistry of actinide elements; nuclear chemistry; rockets and rocket propellants. Mailing Add: Chem Div Argonne Nat Lab Argonne IL 60439

OSBORNE, DAVID WENDELL, b Brush, Colo, Jan 11, 35; m 55; c 2. ORGANIC CHEMISTRY. Educ: Colo Col, BS, 57; Univ Del, PhD(org chem), 60. Prof Exp: Spec assignments chemist, 60-61, res chemist, Edgar C Britton Res Lab, 61-65, proj leader, 65-66, group leader, 66-69, div leader, 69-70, res mgr chem biol res, 70-71, RES MGR AG-ORG DEPT, DOW CHEM CO, 71- Mem: Am Chem Soc; Sci Res Soc Am. Res: Synthesis and chemical properties of agricultural chemicals. Mailing Add: 5708 Flaxmoor Midland MI 48640

OSBORNE, DOUGLAS, b Helena, Mont, Aug 7, 12; c 2. ANTHROPOLOGY, ARCHAEOLOGY. Educ: Univ NMex, BA, 38, MA, 41; Univ Calif, PhD(anthrop), 51. Prof Exp: Dir northwest area, River Basin Surv, Smithsonian Inst, 48-51; cur, Wash State Mus & prof anthrop, Univ Wash, 51-57; vis prof anthrop, Yale Univ, 57-58; dir archaeol, Wetherill Mesa Proj, Nat Park Serv, 58-64; chmn dept, 70-72, PROF ANTHROP, CALIF STATE UNIV, LONG BEACH, 65- Concurrent Pos: NSF grant Micronesian prehist, Calif State Univ, Long Beach, 68-69, instrnl betterment grant, 72; numerous grants for field res. Mem: Am Ethnol Soc. Res: Archaeology and prehistory of the American Southwest, American Northwest and Micronesia. Mailing Add: Dept of Anthrop Calif State Univ Long Beach CA 90840

OSBORNE, FRANK HAROLD, b San Juan, PR, Dec 20, 44; m 67. MICROBIOLOGY. Educ: State Univ NY Albany, BS, 67, MS, 69; Rensselaer Polytech Inst, PhD(biol), 73. Prof Exp: ASST PROF BIOL, KEAN COL NJ, 73- Mem: AAAS; Am Soc Microbiol; Soc Indust Microbiol. Res: Microbiology of arsenic compound oxidation including ecology of arsenite oxidizers and biochemistry and enzymology of arsenite oxidation. Mailing Add: Dept of Biol Kean Col Morris Ave Union NJ 07083

OSBORNE, FRELEIGH JARDINE FITZ, physics, see 12th edition

OSBORNE, GEORGE EDWIN, b Rochester, Ind, Oct 9, 17; m 44; c 5. PHARMACY. Educ: Purdue Univ, BS, 39, MS, 41, PhD(pharm), 49. Prof Exp: Instr pharm, Univ Kans, 46; instr pharmaceut chem, Purdue Univ, 46-49; from asst prof to assoc prof pharm, Univ Utah, 49-57; head dept, 57-69, PROF PHARM, COL PHARM, UNIV R I, 57- Concurrent Pos: Ford Found fac fel advan educ, 54-55. Mem: Am Pharmaceut Asn; Am Asn Cols Pharm. Res: Pharmaceutical education; historical and cultural aspects of pharmacy; formulation and development of pharmaceutical products. Mailing Add: Col of Pharm Univ of RI Kingston RI 02881

OSBORNE, J SCOTT, JR, b Greensboro, NC, June 20, 24; m 48; c 2. PHYSICAL CHEMISTRY. Educ: Wake Forest Col, MA, 51. Prof Exp: Tech aide-to-chief scientist, Off Ord Res, US Dept Army, NC, 51-52; chemist, Am Tobacco Co, 54-55, res chemist, 55-56, jr res assoc, 56-59, res assoc, 59-65, sr res assoc, 65-71; RES PROFESSIONAL, PHILIP MORRIS-USA, 71- Concurrent Pos: Instr, Richmond Prof Inst, 58-61 & 62-63 & Exten Div, Univ Va, 61-62. Mem: Am Chem Soc; Am Inst Chemists. Res: Physical chemistry of cigarette smoke; technology of research on cigarettes; cigarette smoke composition; aerosol physics; kinetics of tobacco decomposition. Mailing Add: 8904 Alendale Rd Richmond VA 23229

OSBORNE, JAMES WILLIAM, b Pana, Ill, Jan 17, 28; m 50; c 2. PHYSIOLOGY, RADIOBIOLOGY. Educ: Univ Ill, BS, 49, MS, 51, PhD(physiol), 55. Prof Exp: Asst physics, Univ Ill, 50-51, res assoc, Control Systs Lab, 51-55; from asst prof to assoc prof, 55-67, PROF RADIOBIOLOGY, RADIATION RES LAB, UNIV IOWA, 67- Concurrent Pos: NIH spec fel, Eng, 65-66. Mem: Radiation Res Soc; Soc Exp Biol & Med. Res: Effect of x-radiation on mammalian small intestine; means of preventive radiation damage or promoting recovery in irradiated animals; radiation-induced carcinogenesis; identification and characterization of tumor-associated circulating antigens. Mailing Add: Radiation Res Lab Univ of Iowa Col of Med Iowa City IA 52242

OSBORNE, JOHN ALAN, b Medicine Hat, Alta, Apr 22, 22; m 58; c 2. MEDICINE, CARDIOLOGY. Educ: Univ Alta, BSc, 44, MD, 48. Prof Exp: Dir cardiovasc labs & asst chief med, Baylor Univ Hosp, Tex, 53-54; asst dir, 56-62, DIR DEPT CARDIOL, VANCOUVER GEN HOSP, BC, 62-; CLIN INSTR MED, UNIV BC, 62- Concurrent Pos: Nuffield travelling fel med, London, Eng, 51-52; Calif Heart Asn fel, Los Angeles, 52-53; instr, Univ Tex Southwestern Med Sch, 53-54; fel coun clin cardiol, Am Heart Asn, 66- Mem: Fel Am Heart Asn; fel Am Col Cardiol (gov, 64-66); Can Cardiovasc Soc. Res: Electrophysiology, particularly electrocardiology; clinical studies on cardiovascular drugs. Mailing Add: Dept of Cardiol Vancouver Gen Hosp Vancouver BC Can

OSBORNE, JOHN CLARK, b Independence, Va, June 24, 18; m 47; c 6. EXPERIMENTAL PATHOLOGY. Educ: Va Polytech Inst, BS, 39; Univ Maine, MS, 41; Mich State Univ, DVM, 49. Prof Exp: Pathologist, NC State Col, 49, from assoc prof to prof animal indust, 49-62; PROF VET SCI, VA POLYTECH INST & STATE UNIV, 62- Concurrent Pos: NIH spec fel, Duke Univ, 60-61; fel cardiovasc dis, Bowman Gray Sch Med, 67-69. Mem: Am Vet Med Asn; Conf Res Workers Animal Dis; Am Asn Vet Med Cols. Res: Enteric and reproductive diseases. Mailing Add: Dept of Vet Sci Va Polytech Inst & State Univ Blacksburg VA 24061

OSBORNE, JOHN WILLIAM, b Sullivan, Ind, Jan 30, 39. DENTAL RESEARCH. Educ: Ind Univ, DDS, 63, MSD, 68. Prof Exp: Instr oper & dent mat, State Univ NY Buffalo, 65-68, asst prof oper, 68-71, assoc prof, 71-73; ASSOC PROF DENT MAT, IND UNIV, INDIANAPOLIS, 73- Mem: Int Asn Dent Res; Am Dent Asn; Am Asn Gold Foil Operators. Res: Clinical research in dental materials; dental amalgam, cements, composite resins and base metal alloys. Mailing Add: Sch of Dent Ind Univ 1121 W Michigan St Indianapolis IN 46202

OSBORNE, LOUIS SHREVE, b Rome, Italy, Sept 8, 23; m 50; c 3. ELEMENTARY PARTICLE PHYSICS. Educ: Calif Inst Technol, BS, 44; Mass Inst Technol, PhD(physics), 50. Prof Exp: Res assoc physics, 49-50, assoc prof, 59-64, GROUP SUPVR, SYNCHROTRON LAB, MASS INST TECHNOL, 50-, PROF PHYSICS, UNIV, 64- Concurrent Pos: Guggenheim & Fulbright fels, 59-60; Minna James Heineman fel, 66-67. Mem: Am Phys Soc; Am Acad Arts & Sci. Res: High energy nuclear physics. Mailing Add: Mass Inst Technol Rm 24-036 Cambridge MA 02139

OSBORNE, MELVILLE, b Carteret, NJ, Apr 21, 22; m 59; c 3. PHARMACOLOGY, PHYSIOLOGY. Educ: Drew Univ, BA, 52; Princeton Univ, MA & PhD(biol sci), 59. Prof Exp: Jr pharmacologist, Ciba Pharmaceut Co, 52-55; group leader cardiovasc pharmacol, Geigy Chem Co, 59-61 & Warner-Lambert Pharmaceut Co, 61-66; cardiovasc pharmacologist, 66-71, res group chief, Sect Cardiovasc Pharmacol, 71-73, SECT HEAD CARDIOVASC-RENAL PHARMACOL, HOFFMANN LA-ROCHE, INC, 73- Mem: Am Soc Pharmacol & Exp Therapeut. Res: Adrenergic mechanism relative to the control of the cardiovascular system and development of drugs which alter these mechanisms. Mailing Add: Dept of Pharmacol Hoffmann La-Roche Inc Nutley NJ 07110

OSBORNE, PAUL JAMES, b Blackwater, Va, Dec 29, 21; m 42; c 2. INVERTEBRATE ZOOLOGY, PHYSIOLOGY. Educ: Univ Va, BA, 50, MA, 51; Univ Fla, PhD, 55. Prof Exp: Asst biol, Univ Va, 51, technician biochem res, 51-52; asst biol, Univ Fla, 52-54; assoc prof, 55-62, chmn dept, 65-74, PROF BIOL, LYNCHBURG COL, 63-, CHMN DIV LIFE SCI, 74- Concurrent Pos: Mem staff, Randolph-Macon Woman's Col, 68-69. Res: Morphology and histochemistry of invertebrates, especially platyhelminthes; biochemical analysis of invertebrates; human physiology, genetics and history; cell physiology; histochemistry and electron microscopy. Mailing Add: Dept of Biol Lynchburg Col Lynchburg VA 24504

OSBORNE, RICHARD HAZELET, b Kennecott, Alaska, June 18, 20; m 44; c 3. HUMAN GENETICS, ANTHROPOLOGY. Educ: Univ Wash, BS & BA, 49; Columbia Univ, PhD(genetics), 56. Prof Exp: Res assoc, Inst Study Human Variation, Columbia Univ, 53-58; from asst prof to assoc prof, Sloan-Kettering Div, Cornell Univ, 58-64; PROF ANTHROP, SCH LETT & SCI & PROF MED, GENETICS, SCH MED, UNIV WIS-MADISON, 64- Concurrent Pos: Vis asst prof, Albert Einstein Col Med, 57-58; asst div prev med, Sloan-Kettering Inst Cancer Res, 58-60, assoc, 60-62, assoc mem, 62-64; assoc univ sem genetics & evolution of man, Columbia Univ, 58-64; assoc res geneticist, Dept Prev Med, Mem Hosp, 59-63, clin geneticist, Dept Med, 63-65; career scientist, Health Res Coun New York, 62-64; consult cult anthrop fel rev comt, NIMH, 69-73, perinatal res comt, Nat Inst Neurol Dis & Stroke, 70-72, genetics task force & phys growth task force; mem comt epidemiol & vet follow-up studies, Nat Res Coun, 69-71. Mem: Fel AAAS; Am Soc Nat; Am Asn Phys Anthrop; Am Soc Human Genetics; Genetics Soc Am. Res: Undergraduate and graduate teaching of human biology and evolutionary theory; twin and sibling studies of quantitative variables and of genetic environmental interaction in health and disease. Mailing Add: 1180 Observatory Dr Univ of Wis Madison WI 53706

OSBORNE, RICHARD VINCENT, b Manchester, Eng, Mar 12, 36; m 59; c 3. HEALTH PHYSICS. Educ: Cambridge Univ, BA, 59; Univ London, PhD(biophys), 62. Prof Exp: Fel radiation detection, Dept Indust Med, NY Univ Med Ctr, 62; RES OFFICER BIOL & HEALTH PHYSICS DIV, ATOMIC ENERGY CAN, LTD, 62- Concurrent Pos: Mem comt 36, Nat Coun Radiation Protection & Measurements, 72- Honors & Awards: Elda E Anderson Award, Health Physics Soc, 75. Mem: Brit Inst Physics; Health Physics Soc; Radiation Res Soc; Can Asn Physicists. Res: High sensitivity alpha particle spectroscopy; natural radioactivity; microdosimetry; detection of hazards of and protection against tritium. Mailing Add: Atomic Energy Can Ltd Chalk River ON Can

OSBORNE, ROBERT HOWARD, b Akron, Ohio, June 29, 39; m 66. STRATIGRAPHY. Educ: Kent State Univ, BS, 61; Wash State Univ Univ, MS, 63; Ohio State Univ, PhD(geol), 66. Prof Exp: Asst prof, 66-70, ASSOC PROF GEOL, UNIV SOUTHERN CALIF, 70- Concurrent Pos: Consult, FUGRO, Consult Engrs & Geologists, 73-74. Mem: Am Statist Asn; Geol Soc Am; Am Asn Petrol Geol; Soc Econ Paleont & Mineral. Res: Quantitative stratigraphy; statistical applications in paleontology, sedimentary petrology and stratigraphy. Mailing Add: Dept of Geol Sci Univ of Southern Calif Univ Park Los Angeles CA 90007

OSBORNE, ROBERT KIDDER, b Kansas City, Mo, Feb 9, 21; m 48; c 1. PHYSICS. Educ: Mass Inst Technol, SB, 42, PhD(physics), 47. Prof Exp: Instr physics, Mass Inst Technol, 45-49; MEM STAFF, LOS ALAMOS SCI LAB, 49- Mem: Am Phys Soc. Res: Radioactive disintegration; investigation of nuclear energy levels through study of radioactive disintegration. Mailing Add: 1425 45th St Los Alamos NM 87544

OSBORNE, THOMAS SAMUEL, radiobiology, cytogenetics, see 12th edition

OSBORNE, WEYMAR ZACK, b Pampa, Tex, Nov 6, 32; m 59; c 3. COSMIC RAY PHYSICS, EXPERIMENTAL HIGH ENERGY PHYSICS. Educ: Univ Okla, BS, 54, MS, 57, PhD(physics), 61. Prof Exp: NSF fel, Dublin Inst Advan Studies, 61; NSF fel, Lawrence Radiation Lab, Univ Calif, 61-62, physicist, 63-65; asst prof physics, Ind Univ, 65-69; nat acad sci sr resident, NASA Manned Spacecraft Ctr, 69-71; RES ASSOC PHYSICS & ADJ PROF, UNIV HOUSTON, 71- Mem: Am Phys Soc. Res: Transition cosmic ray nuclei; magnetic monopoles; transition radiation; elementary particles; biological effects of energetic heavy ions. Mailing Add: Dept of Physics Univ of Houston Houston TX 77004

OSBORNE, WILLIAM WESLEY, b Buhl, Idaho, Apr 29, 20; m 50; c 2. BACTERIOLOGY, PUBLIC HEALTH. Educ: State Col Wash, BS, 46. Prof Exp: Bacteriologist, Lab Sect, State Dept Health, Wash, 47-48; bacteriologist, Ft Lawton State Hosp, Seattle, 48-49; microbiol res div, Bur Agr & Indust Chem, USDA, Md, 49-50, microbiol sect, Poultry Prod Unit, Western Utilization Res Br, Agr Res Serv, 50-55; process develop div, Biol Warfare Labs, US Army, 55-56; microbiol unit, Res & Develop Sect, Indust Test Lab, Bur Ships, US Navy, 56-58; chief bacteriologist, Lab Serv, Presby Hosp Hosp, Philadelphia, 58-60; CHIEF CONTROL BACTERIOLOGIST, BACT CONTROL SECT, WILLIAM H RORER, INC, 60- Mem: Am Soc Microbiol. Res: Food technology; agriculture bacteriology; clinical, medical and sanitary bacteriology; germicides and disinfectants. Mailing Add: 500 Trinity Place Ambler PA 19002

OSBURN, BENNIE IRVE, b Independence, Iowa, Jan 30, 37; m 60; c 3. VETERINARY PATHOLOGY. Educ: Kans State Univ, BS, 59, DVM, 61; Univ Calif, Davis, PhD(comp path), 65; Am Col Vet Path, dipl. Prof Exp: From asst prof to assoc prof vet path, Col Vet Med, Okla State Univ, 64-68; Nat Inst Child Health & Human Develop spec res fel, Wilmer Inst, Sch Med, Johns Hopkins Univ, 68-70; assoc prof vet path, 70-74, PROF PATH, SCH VET MED, UNIV CALIF, DAVIS, 74- Mem: Am Vet Med Asn; Int Acad Path; Am Soc Exp Path; US Animal Health Asn. Res: Host responses and pathogenesis of congenital infections. Mailing Add: Dept of Path Univ Calif Sch of Vet Med Davis CA 95616

OSBURN, ROBERT LEWIS, chemistry, see 12th edition

OSBURN, WILLIAM SHERMAN, JR, ecology, see 12th edition

OSCAI, LAWRENCE B, b Doylestown, Ohio, July 13, 38; div; c 1. EXERCISE PHYSIOLOGY. Educ: La Sierra Col, BS, 61; Univ Colo, MS, 63; Univ Ill, Urbana, PhD(phys educ), 67. Prof Exp: Asst prof, 70-72, ASSOC PROF PHYS EDUC, UNIV ILL, CHICAGO CIRCLE, 72- Concurrent Pos: USPHS res fel, Sch Med, Wash Univ, 67-70. Mem: Am Col Sports Med; Am Inst Nutrit; Am Physiol Soc. Res: Effect of exercise on lipid metabolism, including appetite and weight gain, weight reduction, body composition, adipose tissue cellularity and the enzyme lipoprotein lipase. Mailing Add: Box 4348 Univ of Ill Chicago Circle Chicago IL 60680

OSDAL, LEVERNE KENNETH, chemistry, see 12th edition

OSDENE, THOMAS STEFAN, organic chemistry, see 12th edition

OSE, EARL EUGENE, b Pana, Ill, June 18, 25; m 37; c 3. MICROBIOLOGY. Educ: Univ Ill, Urbana, BS, 50, MS, 56; Nat Registry Microbiologists, registered. Prof Exp:

Bacteriologist, Ill Dept Pub Health, 50-56; instr, Vet Col, Univ Ill, Urbana, 56-57; bacteriologist, 57-65, sr bacteriologist, 65-69, RES SCIENTIST, ELI LILLY & CO, 69- Concurrent Pos: Lectr, Eve Div, Ind Cent Col, 61-66. Mem: Poultry Sci Asn; Am Soc Microbiol. Res: Mycoplasma infections of poultry and swine; oral erysipelas vaccine; hog cholera and shipping fever vaccines. Mailing Add: Eli Lilly & Co Greenfield Labs PO Box 708 Greenfield IN 46140

OSEASOHN, ROBERT, b New York, NY, Jan 23, 24; m 48; c 3. EPIDEMIOLOGY. Educ: Tufts Univ, BS, 43; State Univ NY, MD, 47. Prof Exp: Demonstr prev med & med, Sch Med, Western Reserve Univ, 51-53; asst med, Sch Med, Boston Univ, 55-57; sr instr prev med & med, Sch Med, Western Reserve Univ, 57-60, from asst prof to assoc prof prev med, 60-67, asst prof med, 61-67; prof epidemiol & community med & chmn dept, Univ NMex, 67-72, prof med, Sch Med, 68-72; prof epidemiol & assoc dean, Sch Pub Health & prof med, Sch Med, Univ Tex, Houston, 72-74; PROF EPIDEMIOL & HEALTH, CHMN DEPT & PROF MED, McGILL UNIV, 74- Concurrent Pos: Attend physician, Vet Admin Hosp, Boston, Mass, 55-57; asst physician, Univ Hosps, Cleveland, 57-67; attend physician, Crile Vet Admin Hosp, 59-67; chief epidemiol sect & dep dir, Pak-SEATO Cholera Res Lab, Dacca. EPakistan, 63-65; attend physician, Vet Admin Hosp, Albuquerque, 67-72 & Bernalillo County Med Ctr, Albuquerque, 67-72; attend staff, Royal Victoria Hosp, Montreal, 74- Mem: Soc Exp Biol & Med; Am Epidemiol Soc; Int Epidemiol Asn; Am Pub Health Asn; fel Am Col Physicians. Res: Disease control. Mailing Add: 3775 Univ St Montreal PQ Can

OSEBOLD, JOHN WILLIAM, b Great Falls, Mont, Jan 9, 21; m 44; c 2. IMMUNOLOGY, MICROBIOLOGY. Educ: Wash State Univ, BS, 43, DVM, 44; Ore State Univ, MS, 51; Univ Calif, PhD(comp path), 53. Prof Exp: Fed vet in charge brucellosis diag lab, Bur Animal Indust, USDA, 44-46; pvt pract vet med, 46-49; instr vet med & vet microbiol, Ore State Univ, 49-50; lectr vet microbiol, 50-53, asst prof vet med, 53-57, assoc prof immunol, 57-62, PROF IMMUNOL, UNIV CALIF, DAVIS, 62- Mem: Am Asn Immunol; Am Col Vet Microbiol; Am Soc Microbiol; Am Vet Med Asn; Am Acad Microbiol. Res: Dynamics of infection and resistance; cellular immunity, immunoglobulins serology and immunization. Mailing Add: Sch of Vet Med Univ of Calif Davis CA 95616

OSEPCHUK, JOHN M, b Peabody, Mass, Feb 11, 27; m 56; c 3. PHYSICS. Educ: Harvard Univ, AB, 49, AM, 50, PhD(appl physics), 57. Prof Exp: Res & develop engr microwave tubes, Raytheon Co, 50-53, consult, 53-56, tech liaison, Gen Tel Co, France, 56-57, eng sect head, Spencer Lab, 57-60, prin engr, 60-62; chief microwave engr, Sage Labs, Inc, 62-64; prin res scientist, 64-75, CONSULT SCIENTIST, RES DIV, RAYTHEON CO, 75- Mem: Inst Elec & Electronics Engrs; Am Phys Soc. Res: Crossed field electron devices; interaction of electromagnetic waves with electron beams or plasmas; secondary emission from dielectrics; biological effects of microwaves; microwave ovens; electron beam and semiconductor hydrid devices. Mailing Add: Res Div Raytheon Co 28 Seyon St Waltham MA 02154

OSER, BERNARD LEVUSSOVE, b Philadelphia, Pa, Feb 2, 99; m 23; c 2. NUTRITION: TOXICOLOGY. Educ: Univ Pa, BS, 20, MS, 25; Fordham Univ, PhD, 27. Prof Exp: Asst physiol chem, Jefferson Med Col, 20-21; biochemist, Philadelphia Gen Hosp, Pa, 22-26; asst dir biol lab, Food & Drug Res Labs, Inc, 26-34, dir, 34-70, from vpres to pres, 39-70, chmn, 70-73; CONSULT, BERNARD L OSER ASSOCS, INC, 74- Concurrent Pos: Instr, Grad Sch Med, Univ Pa, 22-26; pres, Am Coun Independent Labs, 48-49; mem expert comt food additives, Food & Agr Orgn, WHO, 58-63; adj prof, Columbia Univ, 59-70; mem food protection comt, Nat Acad Sci-Nat Res Coun, 64-71; sci ed, Food Drug Cosmetic Law J, 57- chmn, Food & Drug Res Labs, Inc, 70-73. Honors & Awards: Babcock-Hart Award, Inst Food Technol, 58. Mem: Fel AAAS; fel Inst Food Technol (pres, 60-61); fel Am Pub Health Asn; hon mem Soc Flavor Chemists; Asn Food & Drug Officials. Res: Technicolegal; physiological chemistry. Mailing Add: Bernard L Oser Assocs Inc 108-18 Queens Blvd Forest Hills NY 11375

OSER, HANS JOERG, b Constance, Ger, Dec 7, 29; m 58; c 3. APPLIED MATHEMATICS. Educ: Univ Freiburg, Dipl, 54, Dr rer nat(math), 57. Prof Exp: Fel, Inst Fluid Dynamics, Univ Md, 57-58; mathematician-analyst, Comput Lab, Nat Bur Standards, 58-62, consult to chief appl math div, 62-66; exec secy commerce tech adv bd, Dept Commerce, 66-67; chief systs dynamics sect, 68-70, CHIEF MATH ANAL SECT, APPL MATH DIV, NAT BUR STANDARDS, 70- Concurrent Pos: Asst prof, Univ Col, Univ Md, 58-62; lectr, Catholic Univ, 62-66. Honors & Awards: Silver Medal, US Dept of Commerce, 68. Mem: AAAS; Asn Comput Mach; Soc Rheol; Soc Indist & Appl Math. Res: Numerical and functional analysis; application of electronic computers in the sciences. Mailing Add: Appl Math Div Nat Bur of Standards Washington DC 20234

OSER, WILLEM, b Rakvere, Estonia, Oct 23, 18; US citizen. PHYSICAL CHEMISTRY, ELECTROCHEMISTRY. Educ: Inst Tech Vienna, MS, 45, DSc, 60. Prof Exp: Sr scientist, Isomet Corp, 60-64; sr res engr, Titanium Metals Corp Am, 65-66; res scientist, 66-69, SR SCIENTIST, LOCKHEED AIRCRAFT SERV CO, 69- Mem: Am Chem Soc; fel Am Inst Chemists. Res: Industrial electrochemistry. Mailing Add: 644 W Center St Pomona CA 91768

OSER, ZALE, polymer chemistry, see 12th edition

OSGOOD, CHARLES EDGAR, b Washington, DC, Feb 1, 38; m 64; c 2. ENTOMOLOGY. Educ: Univ Md, BS, 61; Ore State Univ, MS, 64, PhD(entom), 68. Prof Exp: RES SCIENTIST, RES STA, CAN DEPT AGR, 68- Mem: Entom Soc Can; Entom Soc Am. Res: Behavior and management of the leaf-cutter bee Megachile rotundata; density dependent behavioral changes affecting reproductive rate in insects; pest management of flea beetles on rape crops. Mailing Add: Res Sta Can Dept of Agr 25 Dafoe Rd Winnipeg MB Can

OSGOOD, CHARLES FREEMAN, b New Castle, Pa, Oct 16, 38; m 68; c 1. MATHEMATICS. Educ: Haverford Col, BA, 60; Univ Calif, Berkeley, MA, 62, PhD(Diophantine approximation), 64. Prof Exp: From instr to assoc prof math, Univ Ill, Urbana, 64-72; RES MATHEMATICIAN, NAVAL RES LAB, 70- Concurrent Pos: Res assoc, Nat Bur Standards, DC, 66-68. Mem: Am Math Soc. Res: Diophantine approximation of power series by rational functions of singular integrals; numerical integration. Mailing Add: Code 7707 Naval Res Lab Washington DC 20375

OSGOOD, CORNELIUS, anthropology, see 12th edition

OSGOOD, DAVID WILLIAM, b Grant's Pass, Ore, May 19, 40; m 58; c 2. VERTEBRATE ZOOLOGY, ECOLOGY. Educ: Portland State Col, BS, 63; Duke Univ, MA, 65, PhD(zool), 68. Prof Exp: Instr comp anat, Duke Univ, 67-68; asst prof, 68-73, ASSOC PROF COMP ANAT, BUTLER UNIV, 73- Mem: Am Soc Ichthyol & Herpet; Am Soc Mammal. Res: Effects of developmental temperature on systematic characters in colubrid snakes; bio-telemetric study of body temperature in reptiles; applications of telemetry to behavioral and ecological problems in small mammals and reptiles. Mailing Add: Dept of Zool Butler Univ Indianapolis IN 46208

OSGOOD, ROBERT VERNON, b Wilmington, Del, Dec 26, 41; m 64; c 1. WEED SCIENCE. Educ: Univ Miami, BS, 64; Univ Hawaii, MS, 67, PhD(hort), 69. Prof Exp: Asst agronomist, 69-71, ASSOC AGRONOMIST, HAWAIIAN SUGAR PLANTERS ASN, 71- Mem: Weed Sci Soc Am; Asian Pac Weed Sci Soc. Res: Development of new herbicides for use in Hawaiian sugarcane; studies relating to reasons for varietal differences in tolerance to herbicides. Mailing Add: Hawaiin Sugar Planters' Asn 99-193 Aiea Hts Dr Aiea HI 96701

O'SHAUGHNESSY, CHARLES DENNIS, b Moose Jaw, Sask, Oct 11, 41; m 64; c 2. STATISTICS. Educ: Univ Sask, BA, 62, PhD(statist), 68; Univ Chicago, MSc, 65. Prof Exp: Asst prof, 68-71, ASSOC PROF MATH, UNIV SASK, 71- Concurrent Pos: Nat Res Coun Can grant, Univ Sask, 69-72. Mem: Inst Math Statist; Can Math Cong; Statist Sci Asn Can. Res: Design of statistical experiments, in particular, block designs. Mailing Add: Dept of Math Univ of Sask Saskatoon SK Can

O'SHAUGHNESSY, MARION THOMAS, JR, b New York, NY, Aug 22, 13; m 40; c 3. PHYSICAL CHEMISTRY, POLYMER CHEMISTRY. Educ: Univ Fla, AB, 34; Cath Univ Am, MS, 36; Univ Ill, PhD(phys chem), 40. Prof Exp: Res chemist, Rayon Dept, E I du Pont de Nemours & Co, NY, 40-42, 44-46; asst prof phys chem, Polytech Inst Brooklyn, 46-47, prof, 47-54; sect leader, Res & Develop Div, Am Viscose Corp, Pa, 54-61 & Mat Sci Lab, Aerospace Corp, Calif, 61-66; mgr, Fiber Physics Br, 66-69, SR RES ASSOC, RES & DEVELOP DEPT, PHILLIPS PETROL CO, 69- Mem: Am Chem Soc; Am Phys Soc; Soc Rheol. Res: Fiber science and technology; polymer structure and properties; kinetics; dielectric measurements. Mailing Add: 2126 Skyline Pl Bartlesville OK 74003

O'SHEA, DONALD CHARLES, b Akron, Ohio, Nov 14, 38; m 62; c 4. MOLECULAR SPECTORSCOPY. Educ: Univ Akron, BS, 60; Ohio State Univ, MS, 63; Johns Hopkins Univ, PhD(physics), 68. Prof Exp: Fel physics, Harvard Univ, 68-70; asst prof, 70-75, ASSOC PROF PHYSICS, GA INST TECHNOL, 75- Mem: Am Phys Soc; Sigma Xi; Optical Soc Am. Res: Light scattering spectroscopy of molecular materials, particularly biological materials; applied optics. Mailing Add: Sch of Physics Ga Inst of Technol Atlanta GA 30332

O'SHEA, FRANCIS XAVIER, b New York, NY, May 14, 32; m 54; c 6. ORGANIC POLYMER CHEMISTRY. Educ: St John's Univ, NY, BS, 54; Univ Notre Dame, PhD(org chem), 58. Prof Exp: Res chemist, Naugatuck Chem Div, US Rubber Co, 57-65, group leader plastics res, 65-68; develop mgr, 68-72, SR RES SCIENTIST CORP RES, UNIROYAL, INC, 72- Mem: Am Chem Soc. Res: Synthesis of nondiscoloring antioxidants, particularly sterically hindered phenols; stabilization of hydrocarbon polymers; synthesis of new plastics; thermoplastic polyurethane synthesis and morphology; thermoplastic elastomers; polymer blends. Mailing Add: 211 Wedgewood Dr Naugatuck CT 06770

O'SHEA, SEAMUS FRANCIS, b Waterford, Ireland, Sept 18, 45; m 73. THEORETICAL CHEMISTRY. Educ: Nat Univ Ireland, BSc, 67; McMaster Univ, PhD(chem), 73. Prof Exp: Res asst chem, McMaster Univ, 67-73; fel chem, Univ Western Ont, 73-75; RES ASSOC CHEM, NAT RES COUN CAN, 75- Mem: Chem Inst Can. Res: Simulation of molecular solids with particular reference to intermolecular forces; nonadditive intermolecular forces; quantum theory of the structure of polymers. Mailing Add: Chem Div Nat Res Coun of Can Ottawa ON Can

O'SHEA, TIMOTHY ALLAN, b Lorain, Ohio, Nov 3, 44; m 68; c 2. ENVIRONMENTAL CHEMISTRY, ANALYTICAL CHEMISTRY. Educ: Univ Toledo, BS, 66, MS, 69; Univ Mich, MS, 70, PhD(environ health sci), 72. Prof Exp: ASST PROF CHEM, UNIV MICH-DEARBORN, 72- Mem: AAAS; Am Chem Soc; Sigma Xi. Res: Application of electroanalytical techniques to the study of trace metals in natural waters. Mailing Add: Dept of Natural Sci Univ of Mich Dearborn MI 48128

OSHER, JOHN EDWARD, b Estherville, Iowa, Oct 15, 29; m 52; c 3. PLASMA PHYSICS. Educ: Iowa State Col, BS, 51; Univ Calif, PhD(physics), 56. Prof Exp: Res fel exp nuclear physics, Univ Calif, 56-57; staff mem, Los Alamos Sci Lab, 57-63; plasma physics dept head, Aerojet-Gen Nucleonics Div, Gen Tire & Rubber Co, Calif, 63-65; PHYSICIST, LAWRENCE LIVERMORE LAB, UNIV CALIF, 65- Mem: Am Phys Soc. Res: High energy nuclear physics. Mailing Add: 148 Via Bonita Alamo CA 94507

OSHER, STANLEY JOEL, b Brooklyn, NY, Apr 24, 42. MATHEMATICS. Educ: Brooklyn Col, BS, 62; NY Univ, MS, 64, PhD(math), 66. Prof Exp: Asst mathematician, Brookhaven Nat Lab, 66-67, assoc mathematician, 67-68; asst prof math, Univ Calif, Berkeley, 68-70; ASSOC PROF MATH, STATE UNIV NY STONY BROOK, 70- Mem: Am Math Soc. Res: Numerical and functional analysis; partial differential equations. Mailing Add: Dept of Math State Univ of NY Stony Brook NY 11790

OSHIMA, EUGENE AKIO, b Kaneohe, Hawaii, Jan 2, 34; m 59; c 3. SCIENCE EDUCATION, BIOLOGY. Educ: Colo State Univ, BA, 55, MA, 56; Okla State Univ, DEd(sci ed), 66. Prof Exp: Head sci dept high sch, Kans, 56-64; mem spec staff sci educ, Okla State Univ, 64-66; from asst prof to assoc prof, 66-75, head dept, 67-71, PROF SCI EDUC, CENT MO STATE UNIV, 75- Concurrent Pos: Sci consult, Okla State Univ, 65 & E Cent State Col, 66. Mem: Nat Sci Teachers Asn. Res: Changes in attitudes and confidence in teaching science of prospective elementary school teachers. Mailing Add: Dept of Biol Cent Mo State Univ Warrensburg MO 64093

OSHIMA, NAGAYOSHI, b Tokyo, Japan, Oct 20, 27; m 57; c 1. PLANT PATHOLOGY. Educ: Univ Tokyo, BA, 51; Colo State Univ, MS, 53; Univ Minn, PhD(plant path), 57. Prof Exp: Plant pathologist cereal rusts, Coop Rust Lab, Agr Res Serv, USDA, 57; res fel plant path, 57-61, asst prof & asst plant pathologist, 62-69, ASSOC PROF PLANT PATH, COLO STATE UNIV, 69- Mem: Am Phytopath Soc; Potato Asn Am. Res: Diseases of trees, vegetables and small grains; soil fungi; physiology of fungi. Mailing Add: Dept of Plant Path & Bot Colo State Univ Ft Collins CO 80521

OSHIRO, GEORGE, b Kenora, Ont, Nov 17, 38; m 67; c 2. PHARMACOLOGY. Educ: Univ Man, BS, 64, PhD(pharmacol), 72. Prof Exp: Can Heart Found fel cardiovasc pharmacol, McGill Univ, 72-75; SR PHARMACOLOGIST, AYERST RES LABS, 75- Res: To study the factors giving rise to the various forms of hypertension and to work out the mechanism of action of new anti-hypertensive agents. Mailing Add: Ayerst Res Labs PO Box 6115 Montreal PQ Can

OSHIRO, LYNDON SATORU, b Hilo, Hawaii, Aug 8, 33; m 61; c 3. VIROLOGY, ELECTRON MICROSCOPY. Educ: Univ Utah, BS, 56, PhD(microbiol), 63; Univ

Hawaii, MS, 60. Prof Exp: ELECTRON MICROSCOPIST, VIRAL & RICKETTSIAL DIS LAB, CALIF STATE DEPT HEALTH, 66- Concurrent Pos: NIH trainee virol, Viral & Rickettsial Dis Lab, Calif State Dept Health, 63-65; Nat Inst Allergy & Infectious Dis fel electron micros, Col Physicians & Surgeons, Columbia Univ, 65-66. Mem: Am Soc Microbiol; Am Asn Immunol; Soc Gen Microbiol; Electron Micros Soc Am; Am Asn Cancer Res. Res: Viral etiology of cancer; morphology and development of several selected viral agents. Mailing Add: Calif State Dept of Health 2151 Berkeley Way Berkeley CA 94704

OSHIRO, YUKI, b Okinawa, Japan, Feb 20, 35; m 71. BIOCHEMISTRY. Educ: Univ Calif, Los Angeles, BS, 62; Univ Southern Calif, PhD(biochem), 67. Prof Exp: SR SCIENTIST, NGI-FREDERICK CANCER RES CTR, 73- Concurrent Pos: Fel biochem, Salk Inst Biol Studies, 67-70, USPHS fel, 68-70; staff fel chem carcinogenesis & etiol, Biol Br, Nat Cancer Inst, 70-72. Mem: Am Chem Soc; AAAS. Res: Carcinogenic activation of dimethylnitrosamine. Mailing Add: NCI-Frederick Cancer Res Ctr Box B Frederick MD 21701

OSIECKI, JEANNE HELEN, organic chemistry, see 12th edition

OSINCHAK, JOSEPH, b Fountain Springs, Pa, Sept 20, 37; m 60; c 4. ANATOMY. Educ: Susquehanna Univ, AB, 59; Duke Univ, PhD(anat), 63. Prof Exp: USPHS fel, Duke Univ, 63-64; instr anat, Albert Einstein Col Med, 64-66, assoc, 66-68; ASSOC PROF BIOL, CITY COL NEW YORK, 68- Concurrent Pos: NIH fel, 64-66. Mem: Am Soc Cell Biol; Histochem Soc; Am Asn Anat; Electron Micros Soc Am. Res: Fine structure of neuroendocrine organs; application of cytochemical techniques to electron microscopy. Mailing Add: Dept of Biol City Col of New York New York NY 10031

OSIPOW, LLOYD IRVING, b Brooklyn, NY, Feb 23, 19; m 41. PHYSICAL CHEMISTRY. Educ: Columbia Univ, BS, 39. Prof Exp: Chemist, Chem Warfare Serv, 41-44; res group dir surface chem, Foster D Snell, Inc, 48-69; PRES, OMAR RES, INC, 69- Concurrent Pos: Lectr, Polytech Inst Brooklyn, 55. Mem; Am Chem Soc; Am Oil Chem Soc; Soc Cosmetic Chem; Sci Res Soc Am. Res: Surface and cosmetic chemistry; fats and oils; detergents. Mailing Add: 2 Fifth Ave New York NY 10011

OSKI, FRANK, b Philadelphia, Pa, June 17, 32; m 57; c 3. MEDICINE, PEDIATRICS. Educ: Swarthmore Col, BA, 54; Univ Pa, MD, 58. Prof Exp: Assoc pediat, Sch Med, Univ Pa, 63-65, from asst prof to assoc prof, 65-72; PROF PEDIAT & CHMN DEPT, STATE UNIV NY UPSTATE MED CTR, 72- Concurrent Pos: Fel pediat hemat, Boston Children's Hosp, Mass, 61-63. Mem: AAAS; Soc Pediat Res; Am Soc Hemat; Am Soc Clin Invest; Am Fedn Clin Res. Res: Problems in pediatric hematology, specifically red cell metabolism and the consequences of nutritional deficiencies. Mailing Add: State Univ Hosp 750 E Adams St Syracuse NY 13210

OSLAPAS, RAYMOND, endocrinology, physiology, see 12th edition

OSLER, ABRAHAM GEORGE, b New York, NY, May 17, 11; m 38; c 1. IMMUNOLOGY. Educ: City Col New York, BS, 31; Columbia Univ, MA, 32, PhD(bact), 48. Prof Exp: Res bacteriologist, hosps, New York, 33-43; bacteriologist, New York Health Dept, 43-48; assoc prof microbiol, Johns Hopkins Univ, 48-68; MEM & CHIEF DEPT MED IMMUNOL, PUB HEALTH RES INST OF CITY OF NEW YORK, INC, 68- Concurrent Pos: Prof microbiol, NY Univ; dir, Wasserman Lab, Johns Hopkins Univ, 48-59; ed, Am J Hyg, 58-64; vis prof, Univ Padua & Univ Campinas, Brazil; mem adv bd, Brazilian Asn Immunol; mem sci adv comt, Allergy Found Am; asst ed, J Immunol, J Allergy, J Immunol Methods, Proc Soc Exp Biol & Med & Am J Allergy; chmn adv comt immunol & immunother, Am Cancer Soc, 72-74. Honors & Awards: Medalist, Pasteur Inst, Paris, 65; mem, Acad Sci, Bari, Italy. Mem: Am Soc Microbiol; Soc Exp Biol & Med; Am Asn Immunol; Am Acad Allergy; Brit Soc Immunol. Res: Immunochemistry; complement studies and mechanism of hypersensitivity. Mailing Add: Pub Health Res Inst 455 First Ave New York NY 10016

OSLER, ROBERT DONALD, b Elsie, Nebr, Feb 6, 24; m 47; c 3. CROP BREEDING, AGRONOMY. Educ: Univ Nebr, BS, 47; Colo Agr & Mech Col, MS, 49; Univ Minn, PhD(plant breeding), 51. Prof Exp: Asst barley breeding, Colo Agr & Mech Col, 47-49; asst oat breeding, Univ Minn, 49-S1; asst soybean breeding, US Regional Soybean Lab, 51-54; assoc geneticist corn breeding, Rockefeller Found, 54-56, geneticist, 56-60, asst dir agr sci, 60-65; dep dir resident progs, 67-73, DEP DIR GEN & TREAS INT CTR CORN & WHEAT IMPROV, 74-; ASSOC DIR, ROCKEFELLER FOUND, 65- Mem: AAAS; Am Soc Agron. Mailing Add: CIMMYT Londres 40 APDO Postal 6-641 Mexico DF Mexico

OSMAN, ELIZABETH MARY, b Ottawa, Ill, Apr 29, 25. ORGANIC CHEMISTRY. Educ: Univ Ill, BS, 37, MS, 38; Bryn Mawr, PhD(org chem), 42. Prof Exp: Lit searcher, Hercules Powder Co, 41-42, res chemist, 42-44; res chemist, Corn Prof Ref Co, 44-51; from asst prof to assoc prof foods & nutrit, Mich State Univ, 51-56; assoc prof home econ, Univ Ill, 56-63; PROF HOME ECON, UNIV IOWA, 63- Mem: Am Chem Soc; Inst Food Technol; Am Asn Cereal Chem; Am Home Econ Asn; AAAS. Res: Starch; carbohydrate chemistry; polymerization; three carbon tautomerism; foods; starch in foods; carbohydrate enzymes. Mailing Add: Dept of Home Econ Univ of Iowa Iowa City IA 52240

OSMAN, JACK, b Wilno, Lithuania, Sept 17, 01; US citizen; m 29; c 1. PHARMACEUTICAL CHEMISTRY, MEDICINAL CHEMISTRY. Educ: Kaunas State Univ, BS, 29; Sorbonne, PhD(chem), 31; Univ Southern Calif, BS, 56. Prof Exp: Asst mineral geol, Kaunas State Univ, 24-29; fel bact & med chem, Pasteur Inst, Paris, 31-33; fel toxicol, Police Toxicol Lab, 33-34; chief chemist, Adida Med Lab, 34-38; chief analyst, aluminum plant, France, 38-40; anal chemist, Horton & Converse Labs, Calif, 43-45; chief chemist, S O Barness & Son, 45-56, Visan Labs, 56-62, Thompson Labs, 62-63, Rider's Ltd, Saugus, 63-71 & Pac Pharmaceut Corp, 71-74; PUB RELS OFFICER, ANCHOR POST PROD, INC, 74- Mem: Am Chem Soc. Res: Research and development in pharmaceutical chemistry as applied to methods and processes of manufacturing. Mailing Add: Anchor Post Prod Inc 6500 Eastern Ave Baltimore MD 21224

OSMAN, RICHARD WILLIAM, b Fountain Hill, Pa, Feb 24, 48. MARINE ECOLOGY, EVOLUTIONARY BIOLOGY. Educ: Brown Univ, AB, 70; Univ Chicago, PhD(geophys sci), 75. Prof Exp: ASST PROF GEOL, NORTHERN ILL UNIV, 75- Mem: AAAS; Ecol Soc Am; Soc Study Evolution; Am Inst Biol Sci. Res: Applications of species equilibrium theory in fluctuating environments; role of disturbance in regulating species diversity; evolution of latitudinal gradients in species richness; symbiotic relationships of sessile marine epifauna. Mailing Add: Dept of Geol Northern Ill Univ De Kalb IL 60115

OSMAN, STANLEY F, organic chemistry, see 12th edition

OSMANSKI, C PAUL, b Scranton, Pa, June 12, 31. ANATOMY, PATHOLOGY. Educ: Univ Buffalo, DDS, 60; Univ Ill, Chicago, MS, 63, PhD(path), 68. Prof Exp: Instr oral path, Univ Ill, Chicago, 66-67; ASST PROF ORAL BIOL, UNIV BC, 67- Concurrent Pos: USPHS trainee, 60-66; Nat Res Coun Can fel, 67-69. Res: Ultra structure of keratinizing and non-keratinizing stratified squamous epithelium. Mailing Add: Fac of Dent Univ of BC Vancouver BC Can

OSMENT, LAMAR SUTTON, b Pascagoula, Miss, Apr 9, 24; m 47. DERMATOLOGY. Educ: Birmingham-Southern Col, BS, 45; Univ Ala, MD, 51. Prof Exp: Intern, Univ Hosp, 51-52, resident dermat, Med Col, 52-53, from asst prof to assoc prof, 57-70, PROF DERMAT, MED COL, UNIV ALA, BIRMINGHAM, 70- Concurrent Pos: Fel dermat, Med Col, Univ Ala, Birmingham, 53-55; mem staff, Univ Hosp, 55-, Vet Admin Hosp, 57-, Baptist Hosps, 58-, St Vincent Hosp, 59- & Children's Hosp, 62- Mem: AMA; Am Acad Dermat; Am Dermat Asn. Res: Systemic and topical agents used in dermatology; acrodermatitis enteropathica; bacteriology and mycology of the skin. Mailing Add: Univ of Ala Med Ctr Birmingham AL 35294

OSMER, PATRICK STEWART, b Jamestown, NY, Dec 17, 43. ASTRONOMY. Educ: Case Inst Technol, BS, 65; Calif Inst Technol, PhD(astron), 70. Prof Exp: Res assoc astron, 69-70, asst astronr, 70-73, ASSOC ASTRONR, CERRO TOLOLO INTER-AM OBSERV, 73- Mem: Am Astron Soc; Am Astron Soc Pac; Int Astron Union. Res: Spectroscopy and spectrophotometry of galactic and extra galactic objects. Mailing Add: Casilla 63-D La Serena Chile

OSMOND, DENNIS GORDON, b New York, NY, Jan 31, 30; m 55; c 3. ANATOMY. Educ: Bristol Univ, BSc, 51, MB, ChB, 54, DSc, 75. Prof Exp: House surgeon, Royal Gwent Hosp, Newport. Eng, 54; house physician, Bristol Royal Infirmary, 55; demonstr anat, Bristol Univ, 57-59, lectr, 59-60, 61-64; instr, Univ Wash, 60-61; from assoc prof to prof, 65-74, ROBERT REFORD PROF ANAT, McGILL UNIV, 74- Mem: Can Asn Anat; Am Asn Anat; Anat Soc Gt Brit & Ireland; Can Soc Immunol; Reticuloendothelial Soc. Res: Experimental hematology and cellular immunology; life histories and functional properties of lymphocyte populations, especially in bone marrow; differentiation of surface membrane receptors; lymphocyte cultures; cell separation; kinetics of cellular proliferation; hemopoietic effects of ionizing irradiation; bone marrow circulation; humoral control of hemopoiesis. Mailing Add: McGill Univ Dept of Anat PO Box 6070 Sta A Montreal PQ Can

OSMOND, HUMPHRY FORTESCUE, b Milford, Eng, July 1, 17; m 47; c 3. MEDICINE, PSYCHIATRY. Educ: Univ London, MRCS, LRCP, 42; Royal Col Physicians & Surgeons, London, DPM, 49; Royal Col Physicians & Surgeons Can, cert psychiat. Prof Exp: House physician neurol, Guys Hosp, London, 48-49; first asst psychiat, St Georges Hosp, 49-51; clin dir, Sask Hosp, Weyburn, 51-53, physician supt & dir res, 53-61; DIR PSYCHIAT, BUR RES NEUROL & PSYCHIAT, NJ NEUROL & PSYCHIAT INST, 63- Concurrent Pos: Consult var govt agencies, US & Can. Mem: Royal Col Psychiat; Am Psychiat Asn; Int Col Neuropsychopharmacol; World Acad Arts & Sci. Res: Psychology; social psychology; sociology; anthropology; experimental psychology. Mailing Add: Bur of Res in Neurol & Psychiat Box 1000 Princeton NJ 08540

OSMOND, JOHN KENNETH, b Janesville, Wis, May 12, 28; m 57; c 2. GEOLOGY. Educ: Univ Wis, BA, 50, MA, 52, PhD(geol), 54. Prof Exp: Fel geochem, Welch Found, Rice Inst, 57-58 & Petrel Res Fund, Am Chem Soc, 58-59; asst prof geol, 59-65, ASSOC PROF GEOL, FLA STATE UNIV, 65- Mem: Geochem Soc. Res: Nuclear geology; geochemistry; mineralogy; crystallography; geology of the radioactive elements; geochronology. Mailing Add: Dept of Geol Fla State Univ Tallahassee FL 32306

OSMOND, LESLIE H, b Chardon, Ohio, Mar 17, 05; m 26; c 5. RADIOLOGY. Educ: Western Reserve Univ, AB, 27, MD, 30; Univ Pa, cert, 34. Prof Exp: Instr radiol, Univ Pa, 32-34; from instr to asst prof, 35-42, ASSOC PROF RADIOL, UNIV PITTSBURGH, 42- Concurrent Pos: Asst dir dept radiol, St Francis Hosp, 34-41, dir, 41-52; consult, Mesta Mach Co, Pa, 36-; consult & dir dept radiol, Vet Admin Hosp, 52- Mem: Am Roentgen Ray Soc; Radiol Soc NAm; fel Am Col Radiol. Res: Silicosis prevention in lungs. Mailing Add: Doctors' Hosp 125 S Hegley Ave Pittsburgh PA 15206

OSMUN, JOHN VINCENT, b Amherst, Mass, Feb 22, 18; m 42; c 1. ENTOMOLOGY. Educ: Univ Mass, BS, 40; Amherst Col, MA, 42; Univ Ill, PhD, 56. Prof Exp: Field entomologist, US Dept Army, 42-43; res rep, Merck & Co, Inc, 46-48; asst prof, 48-56, head dept, 56-74, PROF ENTOM, PURDUE UNIV, WEST LAFAYETTE, 56- Concurrent Pos: Mem US entom deleg, USSR, 59; mem comt insect pest mgt & control, Nat Acad Sci-Nat Res Coun, 67-68; founder & mem, Govt Interstate Interdist Pesticide Coun, 69-72, chmn, 72; mem, Fed Task Group Training Objectives & Standardization Pesticides, 71-72; mem, Ind Pesticide Rev Bd, 72-76; consult, Coop States Res Serv, 72-73. Mem: Entom Soc Am. Res: Household and industrial entomology; thermal aerosol insecticide dispersion; pest management; competencies for insecticide use. Mailing Add: Dept of Entom Purdue Univ West Lafayette IN 47907

OSOBA, DAVID, b Glendon, Alta, Apr 10, 32; m 53; c 2. HEMATOLOGY, IMMUNOLOGY. Educ: Univ Alta, BSc, 54, MD, 56; FRCP(C), 61. Prof Exp: Clin instr med, Univ BC, 63-65, asst prof, 65-66; asst prof, 66-72, ASSOC PROF, UNIV TORONTO, 72- Concurrent Pos: McEachern fel, Chester Beatty Res Inst, London, Eng, 62-63; fel hemat, Vancouver Gen Hosp, BC, 61-62; assoc hematologist, Vancouver Gen Hosp, 63-65; mem staff, Med Res Coun Can, 65-68, assoc, 68-; physician, Ont Cancer Inst, 66- Mem: Am Soc Hemat; Can Soc Clin Invest; Am Asn Immunol; Can Soc Immunol. Res: Immunological aspects of malignancy; functions of thymus; immunological deficiency diseases; cellular and humoral immunity. Mailing Add: Ont Cancer Inst 500 Sherbourne St Toronto ON Can

OSOBA, JOSEPH SCHILLER, b Temple, Tex, Dec 5, 19; m 43; c 1. PHYSICS. Educ: Univ Tex, BA, 41; Wash Univ, PhD(physics), 49. Prof Exp: Res engr, Humble Oil & Refining Co, 49-52, sr res engr, 52-54, res specialist, 54-58, sr res specialist, Esso Prod Res Co, 58-67; PROF PETROL ENG, TEX A&M UNIV, 67- Mem: Am Phys Soc; Am Inst Mining, Metall & Petrol Eng. Res: Fluid mechanics; electric and radioactive logging; geochemistry; subsurface flow meters; hydrocarbon analysis; new oil recovery techniques. Mailing Add: Dept of Petrol Eng Tex A&M Univ College Station TX 77843

OSOFSKY, BARBARA LANGER, b Beacon, NY, Aug 4, 37; m 58; c 3. ALGEBRA. Educ: Cornell Univ, BA, 59, MA, 60; Rutgers Univ, PhD(algebra), 64. Prof Exp: Assoc mem tech staff, Bell Labs, Inc, NJ, 60; instr math, Douglass Col, 61-63, from asst prof to assoc prof, 64-71, PROF MATH, RUTGERS UNIV, 71- Concurrent Pos: NSF res grant, 65-67, 70-71; mem, Sch Math, Inst Adv Study, 67-68; mem-at-large, Coun Conf Bd Math Sci, 72-74; proceedings ed comt, Am Math Soc, 74-77, managing ed, 76-77. Mem: Am Math Soc; Math Asn Am. Res: Associative rings and modules; homological algebra. Mailing Add: Dept of Math Rutgers Col Rutgers Univ New Brunswick NJ 08903

OSOFSKY, HOWARD J, b Syracuse, NY, May 24, 35; m; c 1. OBSTETRICS &

OSOFSKY

GYNECOLOGY. Educ: Syracuse Univ, BA, 55, PhD(psychol), 74; NY State Col Med, MD, 58. Prof Exp: From asst prof to assoc prof obstet & gynec, State Univ NY Upstate Med Ctr, 64-71; PROF OBSTET & GYNEC, HEALTH SCI CTR, TEMPLE UNIV, 71- Concurrent Pos: Ed-in-chief, Clin Obstet & Gynec, 72- Mem: Fel Am Col Obstet & Gynec; fel Am Orthopsychiat Asn. Res: Psychological aspects of obstetrics and gynecology with emphasis upon adolescent pregnancy, the effects of prenatal nutrition upon pregnancy outcome and infant development and psychological effects of abortion and sterilization. Mailing Add: Dept of Obstet & Gynec Temple Univ Health Sci Ctr Philadelphia PA 19140

OSOL, ARTHUR, b Riga, Latvia, Dec 1, 05; US citizen; m 28. ANALYTICAL CHEMISTRY. Educ: Philadelphia Col Pharm, BS, 28, MS, 30; Univ Pa, MS, 31, PhD(chem), 33; Eastern Baptist Col, LLD, 64. Hon Degrees: DSc, Thomas Jefferson Univ, 71. Prof Exp: From asst to assoc prof chem, 28-37, prof chem & dir sch chem, 37-63, dean sci, 59-63, pres, 63-75, EMER PRES, PHILADELPHIA COL PHARM & SCI, 75- Concurrent Pos: Dir, Univ City Sci Ctr Corp, Pa, 63-; tech ed, Am Druggist Magazine, 33-45; ed, Remington's Pharmaceut Sci; ed-in-chief, US Dispensatory, J B Lippincott co, 47-; mem revision comt, US Pharmacopeia; mem health comt, League of Nations, 33-37; chmn phys chem comt, Nat Conf Pharmaceut Res, 34-41; mem sci adv comt, Smaller War Plants Corp, 42-46. Mem: AAAS; Am Chem Soc; Electrochem Soc; Am Pharmaceut Asn; fel Am Inst Chem. Res: Infrared and ultraviolet spectrophotometry; solubility of inorganic and organic compounds; preparation of stable forms of drugs for medicinal use; nonaqueous methods of analysis. Mailing Add: Philadelphia Col of Pharm & Sci 43rd St & Kingsessing Mall Philadelphia PA 19104

OSPENSON, JOSEPH NILS, b New York, NY, Jan 28, 24; m 46; c 4. ORGANIC CHEMISTRY. Educ: Drew Univ, BA, 48; Royal Inst Technol, Sweden, PhD(chem), 50. Prof Exp: Res assoc, Swedish Forest Prod Res Lab, 47-50; sr chemist, Calif Spray Chem Corp, 51-55, lead res chemist, 55-60, supvr chem res, Ortho Div, Calif Chem Co, 60-68, mgr, Cent Res Labs, Chevron Chem Co, 68-73, MGR RES & DEVELOP, ORTHO DIV, CHEVRON CHEM CO, 73- Mem: Am Chem Soc. Res: New organic pesticides; correlations of chemical structure to biological activity; metabolism of pesticides; residue chemical research; synthesis; biological screening; process chemistry; metabolism; residue and analytical functions; responsible for worldwide invention, evaluation and development of agricultural biocides. Mailing Add: Ortho Div Chevron Chem Co 940 Hensley St Richmond CA 94120

OSSAKOW, SIDNEY LEONARD, b Brooklyn, NY, Dec 14, 38; m 67; c 3. PLASMA PHYSICS. Educ: Mass Inst Technol, BS, 60; Univ Calif, Los Angeles, MS, 62, PhD(theoret physics), 66. Prof Exp: Asst physics, Univ Calif, Los Angeles, 60-63, asst theoret plasma physics, 63-66; res fel plasma physics lab, Princeton Univ, 66-67; assoc res scientist, Lockheed Palo Alto Res Lab, Lockheed Missiles & Space Co, 67-71; RES PHYSICIST, US NAVAL RES LAB, WASHINGTON, DC, 71- Concurrent Pos: Mem tech staff, Hughes Res Lab, 63-64; consult, Rand Corp, 65-67. Mem: Am Geophys Union; Am Phys Soc. Res: Plasma kinetic and fluid theory; instabilities, non-linear effects, computer simulation and electromagnetic wave propagation; space and ionospheric plasmas; magnetospheric phenomena; nuclear weapons effects. Mailing Add: 7106 Plandome Ct Springfield VA 22153

OSSERMAN, ELLIOTT FREDERICK, b New York, NY, Aug 1, 25; m 49; c 3. MEDICINE. Educ: Columbia Univ, BA, 44, MD, 47. Prof Exp: Assoc med, 53-57, from asst prof to assoc prof, 57-68, prof path, 68-69, Am Cancer Soc prof path, 69-71, AM CANCER SOC PROF MED, COL PHYSICIANS & SURGEONS, COLUMBIA UNIV, 71-, ASSOC DIR, INST CANCER RES, 69- Concurrent Pos: Attend physician, Presby Hosp, 71- Mem: AAAS; Am Asn Immunol; Am Soc Clin Invest; Harvey Soc; Int Soc Hemat. Res: Clinical immunology; gamma globulin structure; plasma cell dyscrasias, myeloma, macroblobulinemia, amyloidosis and the associated abnormalities in gamma globulin synthesis; lysozyme, monocytic dyscrasias, macrophage function. Mailing Add: Inst of Cancer Res 630 West 168th St New York NY 10032

OSSERMAN, ROBERT, b New York, NY, Dec 19, 26; m 52; c 1. GEOMETRY, MATHEMATICAL ANALYSIS. Educ: NY Univ, BA, 46; Harvard Univ, MA, 48, PhD, 55. Prof Exp: Instr math, Univ Colo, 52-53; actg asst prof, 55-57, from asst prof to assoc prof, 57-66, PROF MATH, STANFORD UNIV, 66-, CHMN DEPT, 73- Concurrent Pos: Mem inst math sci, NY Univ, 57-58; head math br, Off Naval Res, 60-61; vis lectr & res assoc, Harvard Univ, 61-62; Fulbright lectr, Univ Paris, 65-66. Mem: Am Math Soc. Res: Complex variables; minimal and Riemann surfaces; Riemannian geometry. Mailing Add: Dept of Math Stanford Univ Stanford CA 94305

OSSESIA, MICHEL GERMAIN, b Houston, Pa, May 4, 29; m 55; c 3. MATHEMATICS. Educ: Univ Pittsburgh, BS, 50, MLitt, 57, PhD(algebra), 61. Prof Exp: Asst prof math, Duquesne Univ, 57-61; prof, Slippery Rock State Col, 61-63; assoc prof, Seton Hall Univ, 63-65; prof, Univ Alaska, 65-66; head dept, 66-74, PROF MATH, CLARION STATE COL, 66- Mem: Am Math Soc; Math Asn Am. Res: Continuous limits. Mailing Add: Dept of Math Clarion State Col Clarion PA 16214

OSSIP, PAUL STUART, organic chemistry, see 12th edition

OSTAP, STEPHEN, b Winnipeg, Man, Aug 31, 24; m 52; c 1. INORGANIC CHEMISTRY. Educ: Univ Man, BSc, 45. Prof Exp: Control chemist asphalt, Can Bitumuls Ltd, 46-47; process develop chemist plutonium, Atomic Energy Can, 48-51; SR CONSULT ALUMINUM, ALUMINUM CO CAN, 51- Mem: Chem Inst Can; Metall Soc Am. Res: Reactions of the different oxides of iron, titanium, silicon, aluminum and aluminum silicates in sodium hydroxide solutions at 100 degrees-300 degrees centigrade. Mailing Add: Aluminum Co of Can Ltd Box 250 Arvida PQ Can

OSTAPCHENKO, GEORGE JOSEPH, physical chemistry, see 12th edition

OSTAPIAK, MYKOLA, b Radcza, West Ukraine, Nov 6, 01; nat US; m 34; c 1. BIOLOGY, CHEMISTRY. Educ: Krakow, Dipl, 29, MS, 32. Prof Exp: Teacher high sch, West Ukraine, 33-44 & Ger, 45-49; res asst pub health & prev med, Sch Med, Univ Pa, 50-66, res assoc, Vet Sch, 66-67; RETIRED. Concurrent Pos: Lectr, Tech Sch, Ger, 45-49; asst prof, Ukrainian Tech Univ, 67-; res asst, Virus Lab, Children's Hosp, Philadelphia, 49- Res: Heredity of acquired characteristics of organisms; metabolism of amino acids and carbohydrates in micro-organisms; etiology of virus diseases; antigenic analysis of influenza viruses by complement fixation. Mailing Add: 7245 Dungan Rd Philadelphia PA 19111

OSTARELLO, GEORGIANDRA LITTLE, b Oakland, Calif, May 27, 44; m 71. MARINE ZOOLOGY. Educ: Duke Univ, BS, 66; Univ Calif, Berkeley, MA, 68, PhD(zool), 71. Prof Exp: ASSOC PROF BIOL, COL NOTRE DAME, 71- Mem: Sigma Xi. Res: Natural history of a subtidal hydrocoral, Allopora Californica. Mailing Add: Dept of Biol Col of Notre Dame Belmont CA 94002

OSTAZESKI, STANLEY A, b Superior, Wis, Apr 17, 26; m 50; c 5. PLANT PATHOLOGY, MYCOLOGY. Educ: Wis State Univ, Superior, BS, 52; Univ Ill, Urbana, MS, 55, PhD(plant path), 57. Prof Exp: Plant pathologist, USDA, Fla, 57-60; plant path adv, AID, Ethiopia, 60-62; RES PLANT PATHOLOGIST, USDA, 62- Mem: Am Phytopath Soc; Mycol Soc Am. Res: Diseases of turf and forage grasses. Mailing Add: Plant Genetics & Germplasm Inst Agr Res Ctr-W Agr Res Serv USDA Beltsville MD 20705

OSTBERG, DONALD ROSS, b New York, NY, Oct 8, 29; m 54; c 3. MATHEMATICS. Educ: Columbia Univ, AB, 51; Univ Calif, Berkeley, MA, 59, PhD(math), 60. Prof Exp: Instr math, Rutgers Univ, 60-61; res instr, Dartmouth Col, 61-63; asst prof, Ind Univ, 63-66; assoc prof, State Univ NY Buffalo, 66-67; assoc prof, 67-72, chmn dept, 67-72, PROF MATH, NORTHERN ILL UNIV, 72- Res: Group extensions and homological algebra; theory of rings and algebras. Mailing Add: Dept of Math Northern Ill Univ De Kalb IL 60115

OSTDICK, THOMAS, b Elgin, Ill, Aug 18, 28. INORGANIC CHEMISTRY, ACADEMIC ADMINISTRATION. Educ: St Meinrad Col, BA, 52; Notre Dame Univ, MS, 57, PhD(chem, math), 58. Prof Exp: From asst prof to assoc prof, 58-70, acad dean, 63-75, prof chem & math, 70-75, PRES, ST MEINRAD COL, 75- Concurrent Pos: NASA res fel, 65-71. Mem: AAAS; Am Chem Soc; Math Asn Am. Res: Preparation, reactions and properties of organoboron and organosilicon compounds; heat-resistant polymers. Mailing Add: St Meinrad Col St Meinrad IN 47577

OSTDIEK, JOHN L, b Lawrence, Nebr, Aug 27, 22. ECOLOGY. Educ: Our Lady of Angels Sem, AB, 45; Quincy Col, BS, 50; Cath Univ Am, MS, 56, PhD(biol), 61. Prof Exp: Instr biol, St Joseph Sem, 50-53; asst sci investr, Arctic Res Lab, Point Barrow, Alaska, 55; from asst prof to assoc prof, 60-71, chmn dept biol & secy bd trustees, 63-70, dir off govt progs, 71-72, dir develop, 72-75, PROF BIOL, QUINCY COL, 71-, ASSOC DIR CTR ENVIRON EDUC & RES, 71-, CHMN DEPT BIOL SCI, 75- Mem: AAAS; Entom Soc Am; Ecol Soc Am; Mississippi River Res Consortium; Water Pollution Control Fedn. Res: Alaskan blackfish; fluctuations in populations of Collembola; effects of floods on Collembola populations; ecology of upper Mississippi River. Mailing Add: Dept of Biol Sci Quincy Col Quincy IL 62301

O'STEEN, WENDALL KEITH, b Meigs, Ga, July 3, 28; m 51; c 2. ANATOMY, NEUROBIOLOGY. Educ: Emory Univ, BA, 48, MS, 50; Duke Univ, PhD(zool), 58. Prof Exp: Asst prof biol, Jr Col, Emory Univ, 48-49, asst, Univ, 49-50, instr, 50-51; asst prof, Wofford Col, 51-53; asst zool, Duke Univ, 55-57; from instr to prof anat, Univ Tex Med Br, 58-68; PROF ANAT, SCH MED, EMORY UNIV, 68- Concurrent Pos: Vis lectr, Sch Med, Univ Miami, 63. Mem: AAAS; Am Asn Anat; Endocrine Soc; Soc Neurosci. Res: Neuroendocrinology; biogenic amines; reproduction; visual system; photoneuroendocrinology. Mailing Add: Dept of Anat Emory Univ Sch of Med Atlanta GA 30322

OSTENSO, GRACE LAUDON, b Tomah, Wis, Sept 15, 32; m 63. NUTRITION. Educ: Stout State Univ, BS, 54; Univ Wis, MS, 60, PhD(foods, nutrit, indust eng), 63. Prof Exp: Dietetic intern, Peter Bent Brigham Hosp, Boston, Mass, 54-55, admin dietitian, 55-57, asst dir dietetics, 57-59; res asst foods & nutrit, Univ Wis-Madison, 59-63, asst prof, 63-67; consult, Food Systs Anal, 67-70; group head, Nutrit & Tech Serv, 70-73, DIR NUTRIT & TECH SERV STAFF, FOOD & NUTRIT SERV, US DEPT AGR, 73- Concurrent Pos: Consult, Hosp Systs Res Group, Dept Indust Eng, Univ Mich, 64-65, div hosp & med facilities, Res Br, US Dept Health, Educ & Welfare, 64-69 & Off of Surgeon Gen, Army Med Specialist Corp, 69-73; assoc ed technol, Encycl Britannica, 67-69. Mem: Fel AAAS; Am Pub Health Asn; fel Royal Soc Health; Soc Advan Food Serv Res; Am Dietetic Asn. Res: Food systems analysis and management. Mailing Add: Nutrit & Tech Serv US Dept of Agr 500 Twelfth St SW Washington DC 20250

OSTENSO, NED ALLEN, b Fargo, NDak, June 22, 30; m 63. GEOLOGY, MARINE GEOPHYSICS. Educ: Univ Wis, BS, 52, MS, 53, PhD(geol, geophys), 62. Prof Exp: Geophysicist, Arctic Inst NAm, 56-58; proj assoc geol & geophys, Univ Wis, 58-63, asst prof, 63-66; oceanogr, Off Naval Res, Chicago, 66-68, dir geol & geophys prog, Washington, DC, 68-70; dir, Off Sci & Technol, Exec Off President, 70; DEP DIR & SR OCEANOGR OCEAN SCI & TECHNOL DIV, OFF NAVAL RES, 70- Concurrent Pos: Foreign affairs fel, Off Naval Res, 75, Cong fel, 75-76. Mem: Am Geophys Union; Geol Soc Am; Soc Explor Geophys; Glaciol Soc; Acad Polit Sci. Res: Artic and antarctic research; gravity; magnetism; oceanography; glaciology. Mailing Add: 2871 Audubon Terr NW Washington DC 20008

OSTENSON, BURTON THOMAS, b Elbow Lake, Minn, Mar 20, 12; m 37; c 4. ZOOLOGY. Educ: Luther Col, AB, 32; Univ Mich, MS, 34, PhD, 48. Prof Exp: Asst lab vert genetics, Univ Mich, 32-33, asst mammal div, Mus Zool, 33-35; from instr to asst prof zool, Mich State Col, 36-47; assoc prof biol, 47-49, PROF BIOL, PAC LUTHERAN UNIV, 49- Concurrent Pos: Consult, Proj Chariot, AEC, Pt Hope, Alaska, 60-61. Mem: Am Soc Mammal; Wildlife Soc. Res: Mammals of the Nebraska sandhills and Pacific Northwest; life history studies of Michigan otter; population studies; wildlife management; marine mammals. Mailing Add: Dept of Biol & Earth Sci Pac Lutheran Univ Tacoma WA 98447

OSTER, CARL FREDERICH, b Abington, Pa, Sept 21, 27; m 51; c 3. ELECTRON MICROSCOPY, PHOTOCHEMISTRY. Educ: Univ Pa, BS, 54. Prof Exp: Res chemist electron micros, res assoc, 66-75, RES LAB HEAD SURFACE CHARACTERIZATION, RES LABS, EASTMAN KODAK CO, 75- Mem: Am Chem Soc; Electron Micros Soc Am; Am Inst Chemists. Res: Characterization of both chemical and physical properties of surfaces using analytical instrumentation, x-ray diffraction, photoelectron spectroscopy, light and electron microscopy. Mailing Add: Eastman Kodak Co Res Labs Rochester NY 14650

OSTER, GEORGE F, b New York, NY, Apr 20, 40; m 68. MATHEMATICAL BIOLOGY. Educ: US Merchant Marine Acad, BS, 61; Columbia Univ, MS, 63, EngScD, 67. Prof Exp: Instr, City Col New York, 64-67; NIH fel biophys, Lawrence Radiation Lab, 68-71, mem fac mech eng, 71-72, mem fac entom, 72-74, ASSOC PROF ENTOM, UNIV CALIF, BERKELEY, 74- Concurrent Pos: Guggenheim fel, J S Guggenheim Found, 75. Honors & Awards: Louis E Levy Medal, Franklin Inst, 72. Mem: Biophys Soc. Res: Irreversible thermodynamics; population ecology. Mailing Add: Dept of Entom Univ of Calif Berkeley CA 94720

OSTER, GERALD, b Providence, RI, Mar 24, 18; m 73; c 2. BIOPHYSICS, PHYSICAL CHEMISTRY. Educ: Brown Univ, ScB, 40; Cornell Univ, PhD(phys chem), 43. Prof Exp: Mem staff virus res, Rockefeller Inst Med Res, 45-48; vis scientist x-ray scattering, Univ London, 48-50; vis scientist nucleic acids, Univ Paris, 50-51; prof chem, Polytech Inst Brooklyn, 51-69; PROF BIOPHYS, MT SINAI SCH MED, 69- Concurrent Pos: Fel, Mass Inst Technol, 43-44; fel, Princeton Univ, 44-45. Mem: AAAS. Res: Polymer chemistry and biophysical chemistry as applied to reproductive physiology and to muscle physiology. Mailing Add: Dept of Biophys Mt Sinai Sch of Med Grad Sch New York NY 10029

OSTER, GISELA KALLMANN, physical chemistry, deceased

OSTER, IRWIN ISAAC, b New York, NY, Jan 12, 30; m 50; c 3. GENETICS. Educ: Long Island Univ, BSc, 50; Ind Univ, PhD(genetics), 56. Prof Exp: Lab asst gen zool, Ind Univ, 50-52, res asst, 52-53, 54-56, NSF fel, 56-57, res exec zool, 57-59; vis investr & demonstr genetics, Inst Animal Genetics, Univ Edinburgh, 53-54; asst mem, Inst Cancer Res, Philadel- phia, Pa, 59-65; PROF BIOL, BOWLING GREEN STATE UNIV, 66- Concurrent Pos: Lectr genetics, Univ Pa, 60-65; vis assoc prof radiobiol & & zool, Rutgers Univ, 61-63; consult zool, Prof H J Muller, Ind Univ, 59-, NASA, 63- & Pure Food & Drug Admin, 63-65. Honors & Awards: Sigma Xi Res Award, Ind Univ, 55. Mem: AAAS; Am Soc Naturalists; Genetics Soc Am; Radiation Res Soc; NY Acad Sci. Res: Modification of radiation and chemical mutagenesis in Drosophila, Bombyx and mammals; synthesis of new genetic strains; developmental analysis of mutations affecting growth and development; carcinogenesis; cytological analysis of spontaneous and induced chromosome aberrations in insects, mammals and humans. Mailing Add: Dept of Biol Bowling Green State Univ Bowling Green OH 43403

OSTER, JAMES DONALD, b Hazen, NDak, Nov 24, 37; m 58; c 3. SOIL CHEMISTRY. Educ: NDak State Univ, BS, 59; Purdue Univ, MS, 62, PhD(soil chem), 64. Prof Exp: Teaching asst soils, Purdue Univ, 59-61, asst soil chem, 61-63; SOIL SCIENTIST, US SALINITY LAB, AGR RES SERV, USDA, 65- Mem: Am Soc Agron; Soil Sci Soc Am. Res: Thermodynamics of soil moisture and ion movement; soil-plant-water relationships; remote sensing of soil salinity. Mailing Add: US Salinity Lab USDA PO Box 672 Riverside CA 92502

OSTER, JAMES DONALD, soil chemistry, see 12th edition

OSTER, LUDWIG FRIEDRICH, b Konstanz, Ger, Mar 8, 31; US citizen; m 56; c 2. ASTROPHYSICS. Educ: Univ Freibourg, Dipl, 54; Univ Kiel, Dr rer nat, 56. Prof Exp: Fel, Univ Kiel, 56-58; asst physics, Yale Univ, 58-60; from asst prof to assoc prof, 65-67; assoc prof, 67-70, PROF, JOINT INST LAB ASTROPHYS, UNIV COLO, BOULDER, 70- Concurrent Pos: Vis prof, Univ Bonn, 66; vis fel, Joint Inst Lab Astrophys, Univ Colo, 66-67; mem Int Astron Union. Honors & Awards: Sr US Scientist Award, Alexander von Humboldt Found, 74. Mem: Am Phys Soc; Am Astron Soc; Ger Astron Soc. Res: Solar and plasma physics; theoretical radio astronomy; radiation theory. Mailing Add: Joint Inst for Lab Astrophys Univ of Colo Boulder CO 80302

OSTER, MARK OTHO, b Chicago, Ill, June 2, 37; m 56; c 3. BIOCHEMISTRY, MICROBIOLOGY. Educ: Purdue Univ, BS, 59; Agr & Mech Col Tex, MS, 61; Univ Ill, PhD(chem), 65. Prof Exp: Res chemist, Univ Calif, Los Angeles, 65-68; ASST PROF LIFE SCI, IND STATE UNIV, TERRE HAUTE, 68- Mem: AAAS; Am Chem Soc; Am Soc Microbiol. Res: Biosynthesis of diterpenes and gibberellin precursors in plants and microorganisms; regulation of pyruvate breakdown in acetate auxotrophs of Escherichia coli. Mailing Add: Dept of Life Sci Ind State Univ Terre Haute IN 47809

OSTERBERG, ARNOLD CURTIS, b Rochester, Minn, Sept 14, 21; m 46; c 4. PHARMACOLOGY. Educ: Univ Iowa, BA, 42; Univ Minn, PhD, 53. Prof Exp: Asst pharmacol, Univ Minn, 48-53; RES PHARMACOLOGIST, LEDERLE LABS, AM CYANAMID CO, 53- Mem: Am Soc Pharmacol & Exp Therapeut. Res: Analgesia; anticholinergics; central nervous system. Mailing Add: Dept of Pharmacol Lederle Labs Pearl River NY 10965

OSTERBERG, CHARLES LAMAR, b Miami, Ariz, June 15, 20; m 45; c 3. OCEANOGRAPHY. Educ: Ariz State Col, BS, 48, MA, 49; Ore State Univ, MS, 60, PhD(oceanog), 63. Prof Exp: Res assoc astron, Lowell Observ, Ariz, 49-53; jr scientist, Atmospheric Res Observ, 53-56; teacher high sch, 56-59; from instr to prof oceanog, Ore State Univ, 62-67; marine biologist, Div Biol & Med, 67-72, from actg chief to chief environ sci br, 69-72, 72, asst dir environ sci, US Atomic Energy Comn, 72-73, MGR ENVIRON PROGS, DIV BIOMED & ENVIRON RES, ENERGY RES & DEVELOP ADMIN, 73- Concurrent Pos: Consult, Int Atomic Energy Agency, Vienna. Mem: AAAS; Am Soc Limnol & Oceanog; Ecol Soc Am; Sigma Xi; Am Inst Biol Sci. Res: Marine radioecology; biolog- ical and chemical oceanography; gamma ray spectrometry; fate of artificial radionuclides in sea and their transport through food chains. Mailing Add: Div of Biomed & Environ Res Energy Res & Develop Admin Washington DC 20545

OSTERBERG, ROBERT EDWARD, b Brooklyn, NY, July 4, 42. PHARMACOLOGY, TOXICOLOGY. Educ: Brooklyn Col Pharm, BS, 65; Georgetown Univ, MS, 69, PhD(pharmacol), 72. Prof Exp: Hosp pharmacist, Metrop Hosp, New York, 66, pharmacologist toxicol, US Food & Drug Admin, 72-74; PHARMACOLOGIST TOXICOL, US CONSUMER PROD SAFETY COMN, 74- Mem: AAAS; Soc Exp Biol & Med; Environ Mutagen Soc. Res: Petroleum distillates and organic solvents and their pulmonary, cardiovascular and neuropharmacological and toxicological effects; mutagenic, irritant and skin sensitization properties of chemicals in consumer products. Mailing Add: US Consumer Prod Safety Comn Bur of Biomed Sci Rm 5844 FOB 8 Washington DC 20204

OSTERBROCK, DONALD EDWARD, b Cincinnati, Ohio, July 13, 24; m 52; c 3. ASTRONOMY. Educ: Univ Chicago, PhB, 48, MS, 49, PhD(astron), 52. Prof Exp: Fel astron, Princeton Univ, 52-53; from instr to asst prof, 53-58; from asst prof to prof, Univ Wis-Madison, 58-73, chmn dept, 69-72; DIR & PROF ASTRON, LICK OBSERV, UNIV CALIF, SANTA CRUZ, 72- Concurrent Pos: Guggenheim fel, Inst Advan Study, 60-61; vis prof, Univ Chicago, 63-64; NSF sr fel, Univ Colo, Univ London, 68-69; chmn sect astron, Nat Acad Sci, 71-74. Mem: Nat Acad Sci; Am Acad Arts & Sci; Int Astron Union; Am Astron Soc. Res: Gaseous nebulae; quasars and active nuclei of galaxies; galactic structure; interstellar matter. Mailing Add: Lick Observ Univ of Calif Santa Cruz CA 95064

OSTERCAMP, DARYL LEE, b Garner, Iowa, Feb 5, 32; m 58; c 2. ORGANIC CHEMISTRY. Educ: St Olaf Col, BA, 53; Univ Wis, MS, 55; Univ Minn, PhD(chem), 59. Prof Exp: Instr chem & math, Luther Col, 54-56; res assoc org chem, Pa State Univ, 59-60; from asst prof to assoc prof, 60-72, PROF CHEM, CONCORDIA COL, 72- Concurrent Pos: Fulbright prof, Col Sci, Mosul, Iraq, 64-65; NSF sci fac fel, Univ E Anglia, 69-70. Honors & Awards: Nere Sundet Prof Chem, 73. Mem: Am Chem Soc. Res: Vinylogues of imides; octahydroquinolines and dacahydroquinolines; conformational analysis of saturated heterocycles; spectroscopic properties of vinylogous systems. Mailing Add: Dept of Chem Concordia Col Moorhead MN 56560

OSTERHELD, ROBERT KEITH, b Brooklyn, NY, Apr 19, 25; m 52; c 4. PHYSICAL INORGANIC CHEMISTRY. Educ: Polytech Inst Brooklyn, BS, 45; Univ Ill, PhD(chem), 50. Prof Exp: Instr chem, Cornell Univ, 50-54; from asst prof to assoc prof, 54-65, PROF CHEM, UNIV MONT, 65-, CHMN DEPT, 73- Mem: Am Chem Soc. Res: Phosphate chemistry; high temperature inorganic chemistry; mechanism of thermal decomposition of solids. Mailing Add: Dept of Chem Univ of Mont Missoula MT 59801

OSTERHOLTZ, FREDERICK DAVID, b Charleston, WVa, July 29, 38; m 60; c 2. ORGANIC CHEMISTRY, RADIATION CHEMISTRY. Educ: Univ Pa, BSc, 59; Mass Inst Technol, PhD(org chem), 64. Prof Exp: CHEMIST, UNION CARBIDE CORP, 63- Mem: Am Chem Soc. Res: Radiation chemistry of organic compounds and nitrogen; polymer radiation chemistry; plasma chemistry. Mailing Add: Black Rock Rd Warwick NY 10990

OSTERHOUDT, HANS WALTER, b Houston, Tex, Feb 29, 36; m 62; c 2. PHYSICAL CHEMISTRY. Educ: Colo State Univ, BS, 58; Univ Wis-Madison, PhD(phys chem), 64. Prof Exp: Res chemist, Armstrong Cork Co, 64-68; sr res chemist, 68-71, res assoc polymer chem, 71-73, LAB HEAD, EASTMAN KODAK CO, 73- Mem: Am Chem Soc; NY Acad Sci. Res: Polymer physical chemistry; molecular weight determinations by ultracentrifugation, osmometry, and gel permeation chromatography; polymer membranes; dye-polymer interactions. Mailing Add: 4090 Canal Rd Spencerport NY 14559

OSTERHOUT, SUYDAM, b Brooklyn, NY, Nov 13, 25. VIROLOGY. Educ: Princeton Univ, BA, 45; Duke Univ, MD, 49; Rockefeller Inst, PhD(microbiol), 59; Am Bd Internal Med, dipl, 59. Prof Exp: Intern path, Cleveland City Hosp, 50; intern med, Mass Mem Hosp, 50-51; from jr asst resident to resident, Duke Hosp, 53-56; PROF MICROBIOL & MED, SCH MED, DUKE UNIV, 59-, ASST DEAN ADMIS, 66- Concurrent Pos: Markle scholar, Sch Med, Duke Univ, 59- Mem: Fel Am Col Physicians. Res: Infectious diseases; internal medicine. Mailing Add: Sch of Med Duke Univ Durham NC 27706

OSTERKAMP, THOMAS EUGENE, US citizen. GLACIOLOGY. Educ: Southern Ill Univ, BA, 62; St Louis Univ, MS, 64, PhD(physics), 68. Prof Exp: ASSOC PROF PHYSICS & GEOPHYS, UNIV ALASKA, FAIRBANKS, 68- Concurrent Pos: Mem glaciol panel, Comt Polar Res, Nat Acad Sci, 74-77. Mem: Glaciol Soc; Am Geophys Union. Res: Scientific aspects of engineering problems involving snow, ice and permafrost. Mailing Add: Geophys Inst Univ of Alaska Fairbanks AK 99701

OSTERLAND, C KIRK, b Red Deer, Alta, Sept 6, 32; m 57; c 3. IMMUNOLOGY, INTERNAL MEDICINE. Educ: Univ Man, MD, 56. Prof Exp: Chief resident med, Univ Man, 59-60; clin instr med, Cornell Univ, 60-62; ASST PROF MED & DIR SECT ARTHRITIS, SCH MED, WASH UNIV, 63- Concurrent Pos: Fel med, Univ Man, 57-60, Nat Res Coun & Can fel metab, 58-59; res fel immunol, Rockefeller Inst, 60-62; Can Arthritis Soc fel, 60-63; Fr Govt scholar & res fel, Pasteur Inst, Paris, 62-63; sr fac scholar, Am Cancer Soc. Mem: AAAS; Am Soc Clin Invest; Am Rheumatism Asn; Can Arthritis & Rheumatism Soc. Res: Structure, function and immunochemical properties of antibodies. Mailing Add: Dept of Prev Med & Pub Health Wash Univ Sch of Med St Louis MO 63110

OSTERMAN, GEORGE B, b McKeesport, Pa, Oct 17, 10; m 48; c 4. BIOLOGY. Educ: Washington & Jefferson Col, AB, 33, MS, 35. Prof Exp: Asst biol, NY Univ, 35-39; from instr to asst prof, Washington & Jefferson Col, 39-59; asst prof, 59-67, ASSOC PROF BIOL, MT UNION COL, 67- Mem: Nat Sci Teachers Asn. Res: Fresh water organisms; orthoptera distribution. Mailing Add: Dept of Biol Mt Union Col Alliance OH 44601

OSTERMAN, JOSEPH VINCENT, JR, b Washington, DC, Dec 12, 35; m 58; c 3. RICKETTSIAL DISEASES. Educ: Univ Md, College Park, BA, 57, PhD(med microbiol), 69. Prof Exp: Fel cell biol, Univ Md, 69-71, asst prof microbiol, Sch Med, 71-74; CHIEF, RICKETTSIAL DIS SECT, WALTER REED ARMY INST RES, 74- Mem: Am Soc Microbiol; Am Soc Trop Med & Hyg. Res: Rickettsial diseases of military importance, with particular emphasis on scrub typhus; biochemistry, genetics, immunology and vaccine development. Mailing Add: Chief Rickettsial Dis Sect Walter Reed Army Inst of Res Washington DC 20012

OSTERRITTER, JOHN FERDINAND, b Millvale, Pa, Feb 1, 23. OCCUPATIONAL MEDICINE. Educ: Univ Pittsburgh, BS, 45, MD, 50, MPH, 52, PhD(occup health), 55. Prof Exp: Res assoc chemist, Mellon Inst, 44-46; asst to med dir, Indust Hyg Found, 52-53; assoc physician, Pittsburgh Coke & Chem Co, 53-55; resident chest dis, Bergen Pines Hosp, 56-57; med dir, Celanese Corp Am, 57-65; CONSULT OCCUP HEALTH & INDUST TOXICOL, 65- Mem: AAAS; AMA; Am Indust Hyg Asn; Indust Med Asn; Am Pub Health Asn. Res: Occupational health methodologies; environmental and industrial health; preventive medicine; industrial toxicology; toxicity of triaryl phosphates; labelling of industrial chemical containers and hazardous household substances. Mailing Add: 300 East 71st New York NY 10021

OSTERUD, HAROLD T, b Richmond, Va, May 1, 23; m 49; c 3. PUBLIC HEALTH, PREVENTIVE MEDICINE. Educ: Randolph-Macon Col, BS, 44; Med Col Va, MD, 47; Univ NC, MPH, 51. Prof Exp: Health officer, Wasco-Sherman Health Dept, 48-50, Coquille County Health Dept, 53-55 & Lane County Health Dept, 56-61; assoc prof pub health, 61-65, PROF PUB HEALTH, MED SCH, UNIV ORE, 65-, CHMN DEPT, 67- Concurrent Pos: Mem, Prev Med & Dent Rev Panel, 65-70, Honors & Awards: Sippy Award, Am Pub Health Asn, 69. Mem: Am Pub Health Asn. Res: Congenital heart disease and malformations; rheumatic fever and heart disease; illegitimacy and adoption; prematurity. Mailing Add: Dept of Pub Health Univ of Ore Med Sch Portland OR 97201

OSTERUD, KENNETH LELAND, b Seattle, Wash, Feb 23, 14; m 41; c 3. PROTOZOOLOGY. Educ: Randolph-Macon Col, AB, 35; NY Univ, PhD(invert zool), 41. Prof Exp: Asst biol, NY Univ, 35-41; from instr to asst prof zool, Univ Minn, 41-49; asst prof, 49-66, ASSOC PROF ZOOL, UNIV WASH, 66- Mem: Soc Protozool. Res: Protozoan nutriton. Mailing Add: Dept of Zool Univ of Wash Seattle WA 98105

OSTERWALD, FRANK WILLIAM, b Casper, Wyo, Feb 11, 22; m 47; c 4. PHYSICAL GEOLOGY. Educ: Univ Wyo, BA, 44, MS, 47; Univ Chicago, PhD(geol), 51. Prof Exp: Asst geol, Univ Wyo, 46-47 & Univ Chicago, 47-48; supply instr, Univ Wyo, 48-49, asst prof, 50-52; geologist, 52-64, RES GEOLOGIST, US GEOL SURV, 64- Concurrent Pos: Asst geologist, State Geol Surv, Wyo, 47, 49-51. Mem: Geol Soc Am; Soc Econ Geol; Seismol Soc Am; Am Asn Petrol Geol; Asn Eng Geol. Res: Engineering geology, especially geologic features affecting coal mine bumps, stability of underground openings; structure and petrology of the cores of Rocky Mountain ranges; investigations related to coal mine development in western US. Mailing Add: 40 S Dover St Lakewood CO 80226

OSTERYOUNG, JANET G, b Pittsburgh, Pa, Mar 1, 39; m 69; c 2. ELECTROANALYTICAL CHEMISTRY. Educ: Swarthmore Col, BA, 61; Calif Inst Technol, PhD(chem), 67. Prof Exp: Res fel chem, Calif Inst Technol, 66-67; asst prof, Mont State Univ, 67-68; res fel, 68-69, ASSOC PROF CIVIL ENG, COLO STATE UNIV, 73- Mem: Am Chem Soc; Electrochem Soc; AAAS; Sigma Xi. Res: Electroanalytical methods development especially for toxic substances; characterization of natural waters. Mailing Add: Dept of Microbiol Colo State Univ Ft Collins CO 80523

OSTERYOUNG, ROBERT ALLEN, b US, Jan 20, 27; m 69; c 5. ANALYTICAL CHEMISTRY. Educ: Ohio Univ, BS, 49; Univ Ill, MS, 51, PhD, 54. Prof Exp: Res chemist, Harshaw Chem Co, 51-52; asst, Univ Ill, 52-54; from asst prof to assoc prof chem, Rensselaer Polytech, 54-59; res specialist, Atomics Int Div, NAm Rockwell Corp, 59-62, group leader phys chem, Sci Ctr, 62-67; PROF CHEM & CHMN DEPT, COLO STATE UNIV, 68- Concurrent Pos: Vis assoc chem, Calif Inst Technol, 63-68; asst dir phys sci dept, Autonetics Div, NAm Rockwell Corp, 67, dir, Mat & Process Lab & assoc dir, Sci Ctr, 68. Mem: AAAS; Am Chem Soc; Electrochem Soc; Int Soc Electrochem. Res: Fused salt chemistry and electrochemistry; electroanalytical chemistry; pulse voltommetry; on line use of computers in electrochemistry. Mailing Add: Dept of Chem Colo State Univ Ft Collins CO 80523

OSTFELD, ADRIAN MICHAEL, b St Louis, Mo, Sept 2, 26; m 50; c 3. EPIDEMIOLOGY, PUBLIC HEALTH. Educ: Wash Univ, MD, 51. Prof Exp: Asst resident med, Barnes Hosp, St Louis, Mo, 52-53; instr med, Med Col, Cornell Univ, 55-56; from asst prof to prof prev med, Univ Ill Col Med, 56-66, prof prev med & community health & head dept, 66-68; chem dept epidemiol & pub health, 68-70, LAUDER PROF EPIDEMIOL & PUB HEALTH, YALE UNIV, 68- Concurrent Pos: Commonwealth fel, NY Hosp, 53-55; proj dir ecol migrant pop, Int Biol Prog, Nat Acad Sci, 67-; spec consult to Surgeon Gen, USPHS; vpres, Coun Epidemiol, Am Heart Asn; chmn epidemiol & biomet comt, NIH; chmn epidemiol adv comt, Nat Heart & Lung Inst; consult, Coun Regional Med Prog. Mem: Am Psychosom Soc; Am Soc Pharmacol & Exp Therapeut; Am Pub Health Asn; Asn Teachers Prev Med; fel Am Col Angiol. Res: Epidemiology of hypertension and cerebrovascular diseases; community surveys; geriatrics; human behavior and psychosomatic medicine; epidemiology of cardiovascular disease. Mailing Add: Lab of Epidemiol & Pub Health Yale Univ Sch of Med New Haven CT 06510

OSTFIELD, HOWARD G, b Pittsburgh, Pa, Mar 8, 41; m 62; c 2. ORGANIC POLYMER CHEMISTRY. Educ: Univ Pittsburgh, BS, 62 & 69. Prof Exp: Chemist, Pittsburgh Plate Glass Co, 63-64; res assoc chem, Mellon Inst, 64-66; GROUP LEADER CHEM, MOBAY CHEM CORP, 66- Res: Application development work in the field of polyurethane chemistry, including the development of high molecular weight polyethers for flexible foams and the development of various processing techniques for the various polymers. Mailing Add: Mobay Chem Corp Pittsburgh PA 15205

OSTHOFF, ROBERT CHARLES, b Buffalo, NY, June 7, 24; m 53. INORGANIC CHEMISTRY. Educ: Univ Buffalo, BS, 48, MA, 49; Harvard Univ, PhD(chem), 51. Prof Exp: Res assoc, Res Lab, 51-56, liaison scientist, 56-58, mgr adv eng, Lamp Metals & Components Dept, 58-60, mgr indust fuel cell projs, Gen Eng Lab, 60-61, mgr chem energy conversion eng, Adv Tech Labs, 62-65, mgr electrochem systs br, Res & Develop Ctr, 65-68, mgr electrochem br, Phys Chem Lab, 68-69, mgr, Mat & Technol Lab, Power Distrib Div, 69-74, MGR, TRANSFORMER LABS OPER, GEN ELEC CO, 74- Mem: AAAS; Inst Elec & Electronics Eng; Am Chem Soc; Electrochem Soc. Res: Organosilicon and boron-nitrogen compounds; batteries and fuel cells; electrochemistry. Mailing Add: 100 Woodlawn Ave Pittsfield MA 01201

OSTLE, DEAN BERNARD, b Vancouver, BC, June 29, 21; nat US; m 53; c 2. STATISTICS. Educ: Univ BC, BA, 45, MA, 46; Iowa State Univ, PhD(statist), 49. Prof Exp: Instr statist, Univ Minn, 46-47; from instr to asst prof, Iowa State Univ, 47-52; from assoc prof to prof math, Mont State Col, 52-57, dir statist lab & statistician, Agr Exp Sta, 56-57; mem staff, Reliability Dept, Sandia Corp, NMex, 57-58, supvr statist sect, 58-60; prof eng, Ariz State Univ, 60-63; chmn statist test design, Rocketdyne Div, NAm Rockwell Corp, 63-67; DEAN COL NATURAL SCI, FLA TECHNOL UNIV, 67- Concurrent Pos: Consult, Humble Oil & Ref Co, 52, Motorola, Inc, 61-63 & AvCir, Flight Safety Found, 62. Mem: AAAS; Am Statist Asn; Am Soc Qual Control; Opers Res Soc Am; Am Inst Indust Eng. Res: Operations research; quality control; reliability; statistical design of experiments; statistical analysis of data. Mailing Add: Col of Natural Sci Fla Technol Univ PO Box 25000 Orlando FL 32816

OSTLIE, DEAN A, organic chemistry, see 12th edition

OSTLIND, DAN A, b McPherson, Kans, June 19, 36; m 58. PARASITOLOGY, ENTOMOLOGY. Educ: Bethany Col, Kans, BS, 58; Kansas State Univ, MS, 62, PhD(parasitol), 66. Prof Exp: Parasitologist, Moorman Mfg Co, 66-67; sr res parasitologist, 67-69, RES FEL, MERCK INST, 69- Mem: Am Soc Parasitol. Res: Chemotherapy of parasites; helminthology; radiation biology. Mailing Add: Merck Inst Rahway NJ 07065

OSTLUND, H GOTE, b Stockholm, Sweden, June 26, 23; m 50; c 2. MARINE CHEMISTRY, ATMOSPHERIC CHEMISTRY. Educ: Univ Stockholm, Fil Kand, 49, Fil lic, 59. Prof Exp: Asst inorg chem, Royal Inst Technol, Sweden, 48-52; dir, Radioactive Dating Lab, Stockholm, 52-63; vis assoc prof, 60-61, assoc prof geochem, Inst Marine Sci, 63-67, PROF MARINE & ATMOSPHERIC CHEM, ROSENSTIEL SCH MARINE & ATMOSPHERIC SCI, UNIV MIAMI, 67- Concurrent Pos: Head labs, Svenska Salpeterverken Koping, 52-54; consult, Adv Panel Oceanog, NSF, 70-74, chmn div chem oceanog, 70-74. Mem: AAAS; Am Geophys Union; Am Meteorol Soc; Swedish Geophys Soc; Swedish Chem Soc. Res: Low level counting techniques; chemical meteorology and air-sea interaction, especially hurricanes; radioactive isotopes for tracing and dating ocean currents and sediments. Mailing Add: Rosenstiel Sch Univ of Miami 4600 Rickenbacker Causeway Miami FL 33149

OSTLUND, NEIL SINCLAIR, b Calgary, Alta, Nov 18, 42. THEORETICAL CHEMISTRY. Educ: Univ Sask, BA, 63; Carnegie-Mellon Univ, MSc, 68, PhD(theoret chem), 69. Prof Exp: Res fel theoret chem, Harvard, 69-71; ASST PROF, UNIV ARK, FAYETTEVILLE, 71- Mem: Am Phys Soc; Am Chem Soc. Res: Many-body theory; perturbation theory; second-order properties of molecules; electron scattering. Mailing Add: Dept of Chem Univ of Ark Fayetteville AR 72701

OSTMANN, BERNARD GEORGE, b Washington, DC, Aug 14, 27; m 55; c 5. TEXTILE TECHNOLOGY. Educ: Cath Univ Am, BS, 50. Prof Exp: From chemist to res chemist, 50-67, SR RES CHEMIST, TEXTILE RES LAB, E I DU PONT DE NEMOURS & CO, 67- Res: Synthetic fiber chemistry and technology. Mailing Add: 2212 Pennington Dr Wilmington DE 19810

OSTRAND, PHILLIP ARTHUR, b Chicago, Ill, Nov 2, 36; m 63. MATHEMATICS. Educ: Univ Ill, Urbana, BS, 59; Northwestern Univ, Ill, MA, 62, PhD(math), 65. Prof Exp: Asst prof math, Rosary Col, 65-66; asst prof, 68-74, ASSOC PROF MATH, UNIV CALIF, SANTA BARBARA, 74- Mem: Am Math Soc; Math Asn Am. Res: Combinatorial analysis; graph, dimension and approximation theory. Mailing Add: Dept of Math Univ of Calif Santa Barbara CA 93106

OSTRANDER, CHARLES EVANS, b Jamestown, NY, Oct 30, 16; m 45; c 2. POULTRY SCIENCE. Educ: Cornell Univ, BS, 41; Mich State Univ, MS, 60. Prof Exp: Teacher pub schs, 41-45; from asst to assoc county agr agent, Onondaga County, NY, 46-51; asst prof poultry sci, Cornell Univ, 51-52; sales mgr, Marshall Bros Hatchery, 52-56; from asst prof to assoc prof, 56-72, PROF POULTRY SCI, CORNELL UNIV, 72- Mem: Poultry Sci Asn; World Poultry Sci Asn. Res: Poultry management; waste disposal; lighting regimes to conserve energy. Mailing Add: Dept of Poultry Sci Cornell Univ 204 Rice Hall Ithaca NY 14853

OSTRANDER, DARL REED, b Ann Arbor, Mich, Apr 29, 35; m 57; c 2. LIMNOLOGY. Educ: Eastern Mich Univ, BS, 61; Univ Mich, MS, 65; Univ Conn, PhD(ecol-fisheries), 74. Prof Exp: ASST PROF BIOL, CENT CONN STATE COL, 65- Concurrent Pos: Assoc consult, Environ Mgt Coun, 74- Mem: Ecol Soc Am; Am Soc Limnol & Oceanog; Am Fisheries Soc. Res: Analysis of plant and mineral distribution in inland wetlands. Mailing Add: Dept of Biol Cent Conn State Col New Britain CT 06050

OSTRANDER, PETER ERLING, b Rahway, NJ, Aug 12, 43; m 70. NUCLEAR PHYSICS. Educ: Ill Inst Technol, BS, 65; Pa State Univ, PhD(physics), 70. Prof Exp: ASST PROF PHYSICS, PA STATE UNIV, 69- Mem: Am Phys Soc. Res: Theoretical calculations of photopion cross-sections including final state interactions. Mailing Add: Rm 203 Pa State Univ Fayette Campus Rte 119 N Uniontown PA 15401

OSTRIKER, JEREMIAH P, b New York, NY, Apr 13, 37; m 58; c 3. ASTROPHYSICS. Educ: Harvard Univ, AB, 59; Univ Chicago, PhD(astrophys), 64. Prof Exp: NSF fel, 64-65, lectr & res assoc, 65-66, from asst prof to assoc prof, 66-71, PROF ASTROPHYS, PRINCETON UNIV, 71- Concurrent Pos: Alfred P Sloan fel, 70-72. Honors & Awards: Helen B Warner Prize, Am Astron Soc, 72. Mem: Nat Acad Sci; Am Acad Sci. Res: Structure and stability of nonspherical self gravitating bodies; high energy astrophysics. Mailing Add: Princeton Univ Observ Princeton NJ 08540

OSTROFF, ANTON G, b Moline, Ill, Nov 6, 25; m 54; c 3. WATER CHEMISTRY. Educ: Augustana Col, Ill, BA, 50; Southern Methodist Univ, MS, 52; Univ Iowa, PhD(inorg chem), 57. Prof Exp: Res chemist, 52-53, SR RES TECHNOLOGIST, FIELD RES LAB, MOBIL RES & DEVELOP CORP, 56- Mem: Am Chem Soc; Am Soc Testing & Mat; Soc Petrol Engrs. Res: Chemistry of subsurface brines; corrosion of oilfield equipment; geochemistry; trace elements in rocks and waters; water pollution; thermal gravimetric analysis and differential thermal analysis of metal sulfates; analysis of phenol carbonation products. Mailing Add: 619 Misty Glen Lane Dallas TX 75232

OSTROM, CARL ALSTON, b Chicago, Ill, Dec 25, 14; m 46; c 1. PREVENTIVE DENTISTRY. Educ: Northwestern Univ, BS & DDS, 38, MS, 48. Prof Exp: Instr, Dent Sch, Northwestern Univ, 39-41; Dent Corps, US Navy, 41-72, instr, Naval Dent Technicians Sch, 48-49, head dent div med res unit 1, Naval Biol Lab, Univ Calif, 49-53, head dent dept, Naval Med Res Inst, Nat Naval Med Ctr, 55-63, head prof br dent div, Bur Med & Surg, 63-67, staff dent officer, Fleet Air Med, NATO South, 67-69, dir dent dept, Naval Training Ctr, Great Lakes, 69-72; SCIENTIST, INST DENT RES & ASSOC PROF ORAL BIOL, SCH DENT, UNIV ALA, BIRMINGHAM, 72- Concurrent Pos: Lectr, Naval Dent Sch, Nat Naval Med Ctr, 56-63; mem dent study sect, NIH, 56-59 & Nat Adv Dent Res Coun, 63-66. Honors & Awards: Am Dent Asn Prev Dent Award, 75. Mem: AAAS; Am Dent Asn; Am Col Dent; Int Asn Dent Res. Res: Preventive dentristry; oral microbiology; histopathology of dental pulp response. Mailing Add: Dept of Oral Biol Univ of Ala Sch of Dent Birmingham AL 35233

OSTROM, CARL ERIC, b Philadelphia, Pa, May 29, 12; m 41; c 4. FORESTRY. Educ: Pa State Univ, BS, 33; Yale Univ, MF, 41, PhD(forestry), 44. Prof Exp: Asst technician, Northern Rocky Mt Forest & Range Exp Sta, US Forest Serv, 34-36, silviculturist, Northeastern Forest Exp Sta, 36-44, Southeastern Forest Exp Sta, 44-57, dir, Div Timber Mgt Res, 57-74, assoc dep chief, Biol Res, 74-75; DIR SCI PROGS, SOC AM FORESTERS, 75- Concurrent Pos: Consult, UN Spec Fund, Thailand, 62. Mem: AAAS; Soc Am Foresters; Ecol Soc Am; Int Asn Ecol; Am Inst Biol Sci. Res: Silviculture; forest management. Mailing Add: 4016 N Richmond St Arlington VA 22207

OSTROM, JOHN HAROLD, b New York, NY, Feb 18, 28; m 52; c 2. VERTEBRATE PALEONTOLOGY. Educ: Union Col, NY, BS, 51; Columbia Univ, PhD(geol), 60. Prof Exp: Res asst vert paleont, Am Mus Natural Hist, 51-56; from instr to asst prof geol, Beloit Col, 56-61; from asst prof to assoc prof geol & asst cur to assoc cur vert paleont, 61-71, prof geol, 71-74, PROF GEOL & GEOPHYS, YALE UNIV, 74-, CUR VERT PALEONT, PEABODY MUS NATURAL HIST, 71- Concurrent Pos: Guggenheim fel, 66-67; lectr, Brooklyn Col, 55-56; res assoc, Am Mus Natural Hist, 65-; ed, Am J Sci, 67- & Bull, Soc Vert Paleont, 62- Mem: AAAS; Soc Study Evolution; Soc Vert Paleont (pres, 69-70); Asn Earth Sci Ed; Nat Asn Geol Teachers. Res: Vertebrate paleontology, especially ancient reptiles; vertebrate evolution; Mesozoic stratigraphy. Mailing Add: Dept of Geol & Geophys Yale Univ New Haven CT 06520

OSTROM, MEREDITH EGGERS, b Rock Island, Ill, Nov 16, 30; m 53; c 3. GEOLOGY. Educ: Augustana Col, BS, 52; Univ Ill, MS, 54, PhD, 59. Prof Exp: Asst mineral, Univ Ill, 52-54; asst mineral coal sect, State Geol Surv, Ill, 54-55, asst geologist, Indust Minerals Sect, 55-59; asst state geologist, 59-67, assoc state geologist, 67-72, STATE GEOLOGIST & DIR, WIS GEOL & NATURAL HIST SURV, 72-; CHMN DEPT GEOG & GEOL, UNIV WIS EXTEN, 72- Mem: Geol Soc Am; Soc Econ Paleont & Mineral; Am Asn Petrol Geol; Am Inst Prof Geol; Am Asn State Geol. Res: Industrial minerals; nonmetallic mineral deposits; stratigraphy; ground water geology. Mailing Add: Wis Geol & Natural Hist Surv 1815 University Ave Madison WI 53706

OSTROM, THEODORE GLEASON, b Nicollet, Minn, Jan 4, 16; m 49; c 4. MATHEMATICS. Educ: Univ Minn, BA, 37, MA & BS, 39, PhD(math), 47. Prof Exp: Instr math, Univ Minn, 46-47; from asst prof to prof math, Univ Mont, 47-60, chmn dept, 54-60; chmn dept, 69-71, PROF MATH, WASH STATE UNIV, 60- Concurrent Pos: Vis prof, Univ Frankfurt, 66-67. Mem: Am Math Soc; Math Asn Am. Res: Finite projective planes. Mailing Add: Dept of Math Wash State Univ Pullman WA 99163

OSTROM, THOMAS ROSS, b San Francisco, Calif, May 23, 24; m 49; c 2. HEALTH PHYSICS, SANITARY ENGINEERING. Educ: Univ Calif, BS, 44; Harvard Univ, MSSE, 48, PhD(sanit eng), 55. Prof Exp: Mil govt officer, US Army, Hq Korea, 45-47, sanit engr, Health Physics Div, Oak Ridge Nat Lab, Tenn, 49-50, engr health physics, Joint Task Force 3, 50-51, sanit engr, Indust Waste Br, Los Alamos Field Off, 51-53, chmn sanit eng div, US Army Environ Health Labs, Army Chem Ctr, Md, 55-58, chmn sanit eng dept, Walter Reed Army Inst Res, DC, 59-62, sanit engr, Hq 7th Army, 62-64, chief med & biol sci br, Off Chief Res & Develop, Hq Dept Army, 64-68, commanding officer, Wound Data & Munitions Effectiveness Team, Vietnam, 68-69, mem fac, Indust Col Armed Forces, 69-71, dir res, develop & eng, Munitions Command, NJ, 71-72, commanding officer, US Army Ballistics Res Lab, Md, 72-75; RETIRED. Mem: Nat Soc Prof Engr; Am Soc Civil Engr; Health Physics Soc; Soc

Am Mil Engr. Res: Radio-strontium content of and distribution of radio-strontium in fallout; disposal of low level radio-active wastes; sewage disposal procedures under extreme cold conditions. Mailing Add: 102 Duncannon Rd Bel Air MD 21014

OSTROVSKY, DAVID SAUL, b Cleveland, Ohio, Jan 28, 43. ZOOLOGY Educ: Case Western Reserve Univ, BA, 65; Univ Mich, MS, 67, PhD, 70. Prof Exp: Asst prof zool, Ohio Wesleyan Univ, 72-73; ASST PROF BIOL, MILLERSVILLE STATE COL, 73- Mem: Am Soc Zool. Res: The problem of enzyme development in relation to both structural and functional differentiation. Mailing Add: Dept of Biol Millersville State Col Millersville PA 17551

OSTROW, JAY DONALD, b New York, NY, Jan 1, 30; m 56; c 3. GASTROENTEROLOGY. Educ: Yale Univ, BS, 50; Harvard Med Sch, MD, 54; Univ London, MSc, 70. Prof Exp: Intern med, Johns Hopkins Hosp, 54-55; asst resident, Peter Bent Brigham Hosp, 57-58; instr med, Harvard Med Serv, 61-62; sr instr, Western Reserve Univ & Univ Hosps, 62-64, asst prof, 64-69; asst prof, 69-71, ASSOC PROF MED, SCH MED, UNIV PA, 71- Concurrent Pos: NIH res fel gastroenterol, Peter Bent Brigham Hosp, 58-59; Nat Inst Arthritis & Metab Dis res fel liver dis & nutrit, Harvard Med Serv & Thorndike Lab, Boston City Hosp, 59-62; Nat Inst Arthritis & Metab Dis res grant, 63- Mem: Am Fedn Clin Res; Am Soc Clin Invest; Am Asn Study Liver Dis; Am Gastroenterol Asn; Am Physiol Soc. Res: Phototherapy; bilirubin metabolism; gallbladder absorption; gallstones; gastrointestinal bleeding. Mailing Add: Med Div 111-H Univ of Pa Vet Admin Hosp Philadelphia PA 19104

OSTROW, SIDNEY MAURICE, b New York, NY, May 14, 16; m 47; c 3. IONOSPHERIC PHYSICS. Educ: City Col New York, BS, 37. Prof Exp: Physicist, Cent Radio Propagation Lab, Nat Bur Standards, 42-65; asst chief, Prediction Serv Sect, 47-70, physicist, Inst Telecommun Sci & Aeronomy, Environ Sci Serv Admin, 65-71, physicist, Environ Data Serv, Nat Oceanic & Atmospheric Admin, 71-73; RETIRED. Concurrent Pos: In charge Adak radio propagation field sta, 46-47, chief, Int Geophys Year World Data Ctr A Airglow & Ionosphere, 47-61; guest res worker, Data Studies Div, Nat Geophys & Solar Terrestrial Data Ctr, Environ Data Serv, Nat Oceanic & Atmospheric Admin, 73- Mem: Fel AAAS; Am Phys Soc; Am Geophys Union; sr mem, Inst Elec & Electronics Engrs. Res: Ionospheric climatology and morphology; ionospheric physics and radio propagation. Mailing Add: 100 S 34th St Boulder CO 80303

OSTROWSKI, RONALD STEPHEN, b Chicago, Ill, Mar 3, 39; m 68; c 1. GENETICS. Educ: Northern Ill Univ, BS, 66, MS, 68; Univ Notre Dame, PhD, 72. Prof Exp: ASST PROF BIOL, UNIV NC, CHARLOTTE, 71- Mem: AAAS; Genetics Soc Am. Res: Developmental genetics. Mailing Add: Dept of Biol Univ of NC Charlotte NC 28213

OSTROWSKY, HOWARD S, theoretical physics, see 12th edition

OSTROY, SANFORD EUGENE, b Scranton, Pa, Dec 28, 39; m 62; c 1. NEUROBIOLOGY. Educ: Univ Scranton, BS, 61; Case Inst Technol, PhD(chem), 66. Prof Exp: NIH fel biophys chem, Cornell Univ, 66-68; asst prof, 68-72, ASSOC PROF NEUROBIOL, PURDUE UNIV, 72- Concurrent Pos: NIH res career develop award & grant, 70-75; hon res assoc biol, Harvard Univ, 74-75. Mem: AAAS; Am Chem Soc; Biophys Soc; Soc Neurosci; Asn Res Vision & Ophthal. Res: Molecular basis of sensory processes; rhodopsin structure and function; membrane biophysics. Mailing Add: Dept of Biol Sci Purdue Univ West Lafayette IN 47906

OSTWALD, PETER FREDERIC, b Berlin, Ger, Jan 5, 28; US citizen; m 60; c 2. PSYCHIATRY, COMMUNICATION SCIENCES. Educ: Univ Calif, Berkeley, AB, 47; Univ Calif, San Francisco, MD, 50. Prof Exp: Resident psychiat, Sch Med, Cornell Univ, 51-56; chief resident, Langley Porter Neuropsychiat Inst, 56-57; res psychiatrist, 58-63, from asst prof to assoc prof, 60-70, PROF PSYCHIAT, SCH MED, UNIV CALIF, SAN FRANCISCO, 70- Concurrent Pos: USPHS grant, Langley Porter Neuropsychiat Inst, 57-60, Found Fund Res Psychiat grant, 60-63; attend psychiatrist, Langley Porter Neuropsychiat Inst, 60-; consult, Dept Rehab, Psychiat; AMA. Res: Acoustical communication; speech disorders; hearing and language problems; psychotherapy; psychiatric care; education. Mailing Add: Univ of Calif Sch of Med 401 Parnassus Ave San Francisco CA 94122

OSTWALD, ROSEMARIE, b Vienna, Austria, June 24, 15; m 36; c 3. NUTRITION, BIOCHEMISTRY. Educ: Univ Vienna, PhD(org chem), 39. Prof Exp: Res assoc chem, 46-57, assoc nutrit & lectr biochem, 57-63, asst prof nutrit & biochem, 63-68, assoc prof nutrit, 68-74, PROF NUTRIT & BIOCHEMIST IN AGR EXP STA, UNIV CALIF, BERKELEY, 74- Mem: AAAS; Am Chem Soc; Am Inst Nutrit; Am Oil Chem Soc. Res: Interrelationships of diet, hormones and lipid metabolism. Mailing Add: Dept of Nutrit Sci Univ of Calif Berkeley CA 94720

OSUCH, CARL, b Chicago, Ill, Feb 2, 25; m 48; c 4. ORGANIC CHEMISTRY. Educ: Antioch Col, BS, 50; Univ Pittsburgh, PhD, 55. Prof Exp: Res chemist, Monsanto Co, 55-64; from asst prof to assoc prof, 64-70, PROF CHEM, UNIV DUBUQUE, 70- Concurrent Pos: Petrol Res Fund grant, 67-70; consult, Dubuque police dept, 72- Mem: AAAS; Soc Appl Spectros; Coblentz Soc; Am Chem Soc; NY Acad Sci. Res: Alkyl pyridines; organolithium compounds; bicyclic olefins; organic azides; phosphorus compounds; drug identification. Mailing Add: Dept of Chem Univ of Dubuque Dubuque IA 52001

OSUCH, MARY ANN V, b Chicago, Ill, Feb 9, 41. BIOCHEMISTRY. Educ: DePaul Univ, BS, 62; Northwestern Univ, PhD(biochem), 67. Prof Exp: Res biochemist, 67-68, supvr mfg lab, 68-71, tech servs coordr, 71-72, PROD MGR, RES PROD DIV, MILES LAB, INC, 72- Mem: AAAS; Am Chem Soc. Res: Nucleic acids and enzymes. Mailing Add: Res Prod Div Miles Lab Inc 1127 Myrtle St Elkhart IN 46514

O'SULLIVAN, THOMAS DENIS, b New York, NY, Apr 28, 41. PHYSICAL CHEMISTRY, ELECTROCHEMISTRY. Educ: Manhattan Col, BS, 61, PhD(phys chem), 68. Prof Exp: MEM TECH STAFF, BELL TEL LABS, 66- Mem: Am Chem Soc. Res: Solubility of gases; high pressure phase equilibria; electrode kinetics; convection diffusion of dissolved electroactive species; electrochemical energy conversion. Mailing Add: Dept of Electrochem Bell Tel Labs 600 Mountain Ave New Providence NJ 07974

O'SULLIVAN, WILLIAM JOHN, b Springfield, Mass, Apr 21, 31; m 54; c 2. LASERS, FLUID PHYSICS. Educ: Univ Pittsburgh, PhD(physics), 58. Prof Exp: Res assoc physics, Univ Pittsburgh, 58; asst prof, US Naval Postgrad Sch, 58-59; mem tech staff & consult, Space Tech Labs, Inc, Div Thompson-Ramo-Wooldridge, Inc, 59-65; staff mem, Sandia Corp, 65-68; PROF PHYSICS, UNIV COLO, BOULDER, 68- Mem: Am Phys Soc. Res: Theory of antiferromagnetism; band structure of solids; nuclear magnetic resonance; behavior of Fermi surfaces under pressure; photo autocorrelation studies of critical fluid behavior. Mailing Add: Dept of Physics & Astrophys Univ of Colo Boulder CO 80302

OSVALDS, VALFRIDS, b Dauguli, Latvia, Dec 23, 16; m 42; c 3. ASTRONOMY. Educ: Latvia State Univ, Dipl, 43; Univ Hamburg, Dr rer nat, 51. Prof Exp: Headmaster, UN Relief & Rehabil Admin Sch, Ger, 46-47; teacher, Int Refugee Orgn Regional Gym, 49-50; res asst astron, 51-56, res assoc, 56-57, actg asst prof, 57-58, asst prof, 58-61, actg dir observ & actg chmn dept astron, univ, 60-63, ASTRONOMER, McCORMICK OBSERV, UNIV VA, 57-, ASSOC PROF ASTRON, UNIV, 61- Res: Astrometry; proper motions; parallaxes. Mailing Add: Box 3818 Univ Sta Charlottesville VA 22903

OSWALD, EDWARD ODELL, b Newberry, SC, Jan 9, 40; m 61; c 2. BIOCHEMISTRY. Educ: Newberry Col, BS, 61; Bowman Gray Med Sch, MS, 63; Univ NC, Chapel Hill, PhD(biochem), 66. Prof Exp: NIH res assoc biochem, Univ NC, Chapel Hill, 66-67; dir clin chem, Palms of Pasadena Hosp, 67-68; RES BIOCHEMIST, NAT INST ENVIRON HEALTH SCI, ENVIRON PROTECTION AGENCY, 68- Mem: AAAS; Am Chem Soc; Am Oil Chemists Soc. Res: Organic synthesis and metabolism of radioactive labeled lipids and lipid-like natural products. Mailing Add: Chem Br Nat Inst Environ Health EPA PO Box 12233 Research Triangle Park NC 27709

OSWALD, ELIZABETH JANET, bacteriology, see 12th edition

OSWALD, JOHN WIELAND, b Minneapolis, Minn, Oct 11, 17; m 45; c 3. PLANT PATHOLOGY. Educ: DePauw Univ, AB, 38, LLD, 64; Univ Calif, PhD(plant path), 42. Hon Degrees: LLD, Cent Col, 66, Univ Louisville, 66 & Univ Calif, Davis, 67; LHD, Juniata Col, 73; ScD, Temple Univ, 73. Prof Exp: From asst prof to prof plant path, Col Agr, Univ Calif, Berkeley, 46-63, vchmn dept, 54-58, from asst vpres to vpres, Univ, 58-63; prof plant path & pres, Univ Ky, 63-68; exec vpres, Univ Calif, Berkeley, 68-70; PROF PLANT PATH & PRES, PA STATE UNIV, 70- Concurrent Pos: Fulbright res grant, Neth, 53-54. Mem: Am Phytopath Soc. Res: Potato diseases; potato scab; viruses; cereal root rots; cereal viruses. Mailing Add: Off of the Pres Pa State Univ University Park PA 16802

OSWALD, LARRY O, b Seattle, Wash, Aug 16, 18; m 42; c 1. HIGH ENERGY PHYSICS. Educ: San Francisco State Col, BA, 49. Prof Exp: Teacher pub schs, Calif, 49-51; physicist, Lawrence Radiation Lab, Univ Calif, 52-64 & Argonne Nat Lab, 64-69, CONSULT & MEM, SABOUNI & ASSOCS, 69- Mem: Am Phys Soc. Res: Cloud chamber and bubble development, design and research; superconductivity and magnet design and development. Mailing Add: 629 Seacliff Dr Aptos CA 95003

OSWALD, VERNON HARVEY, b Hartville, Ohio, May 11, 25; m 53; c 2. ZOOLOGY. Educ: Ohio State Univ, PhD(zool), 56. Prof Exp: Prin biologist, Battelle Mem Inst, 51-56; Muellhaupt scholar zool, Ohio State Univ, 56-57; from instr to assoc prof, 57-67, PROF BIOL, CALIF STATE UNIV, CHICO, 67- Mem: AAAS; Am Soc Parasitol; Am Micros Soc. Res: Helminthology. Mailing Add: Dept of Biol Calif State Univ Chico CA 95926

OSWALT, DALLAS LEON, b Darke Co, Ohio, Nov 12, 27; m 50; c 2. CROP BREEDING, SCIENCE EDUCATION. Educ: Purdue Univ, BSc, 50, MSc, 58, PhD(crop breeding), 73. Prof Exp: Voc agr teacher, Wayne Twp Sch, Union City, Iowa, 50-53; teacher sci, Church Brethren, Nigeria, 53-56; teacher voc agr, Driver High Sch, Winchester, Ind, 58-60; prin, Church Brethren, Nigeria, 60-69; res asst agron, Purdue Univ, 69-75; TRAINING OFFICER, INT CROP RES INST SEMI-ARID TROPICS, 75- Mem: Am Soc Agron; Crop Sci Soc Am; Am Genetics Asn; Am Soc Animal Sci. Res: Training methods and procedures for international centers; protein content and quality of sorghum and pearl millet as related to nitrate uptake and translocation efficiency. Mailing Add: Int Crop Res Inst Semi-arid Trop 1-11-256 Begumpet Hyperabad India

OSWALT, WENDELL HILLMAN, b Youngstown, Ohio, June 26, 27; m 46; c 3. CULTURAL ANTHROPOLOGY, ARCHAEOLOGY. Educ: Univ Alaska, BA, 52; Univ Ariz, PhD(anthrop), 59. Prof Exp: PROF ANTHROP, UNIV CALIF, LOS ANGELES, 59- Res: Alaskan anthropology. Mailing Add: 405 Hilgard Ave West Los Angeles CA 90024

OSWEILER, GARY D, b Sigourney, Iowa, Sept 8, 42; m 66; c 4. VETERINARY TOXICOLOGY. Educ: Iowa State Univ, DVM, 66, MS, 68, PhD(vet toxicol), 73. Prof Exp: Instr vet path, Iowa Vet Diag Lab, 66-67; instr med, Vet Clin, Iowa State Univ, 68-69, asst prof vet path, 69-70, assoc prof vet toxicol, 71-73; ASSOC PROF VET TOXICOL, UNIV MO-COLUMBIA, 74- Concurrent Pos: Mem exam bd, Am Bd Vet Toxicol, 72-73; consult, Pedco Environ, 73 & Food & Drug Admin, 75; mem task forces, Coun Agr Sci Technol, 74-75. Mem: Am Col Vet Toxicologists; Am Vet Med Asn; Coun Res Workers Animal Dis. Res: Drug interactions of warfarin and antibacterial agents in animals; residual reproductive effects of lead in rats; bioassay techniques for screening toxins in foods and feeds. Mailing Add: Dept of Vet Anat-Physiol Col of Vet Med Univ of Mo Columbia MO 65201

OSWIN, HARRY, physicai chemistry, see 12th edition

OTA, ASHER KENHACHIRO, b Waianae, Hawaii, Dec 1, 34; m 66; c 1. ENTOMOLOGY. Educ: Univ Hawaii, BS, 56, MS, 62; Univ Calif, Berkeley, PhD(entom), 66. Prof Exp: Asst entom, Univ Hawaii, 56-61; res entomologist, Entom Res Div, USDA, 66-68; from assoc entomologist to entomologist, 68-74, PRIN ENTOMOLOGIST, EXP STA, HAWAIIAN SUGAR PLANTERS ASN, 74- Mem: AAAS; Entom Soc Am. Res: Agricultural entomology; insect population ecology; biological control of insects and pests. Mailing Add: Exp Sta Hawaiian Sugar Planters Asn 1527 Keeaumoku St Honolulu HI 96822

OTAGAKI, KENNETH KENGO, b Laupahoehoe, Hawaii, June 30, 17; m 44; c 5. ANIMAL NUTRITION. Educ: Univ Hawaii, BS, 40; Iowa State Col, MS, 46; Univ Calif, PhD(nutrit), 54. Prof Exp: Jr animal husbandman, Univ Hawaii, 47-48; sr technician, Nutrit Invest, Univ Calif, Berkeley, 50-53; asst animal scientist, 54-58, assoc animal scientist, 58-64, ANIMAL SCIENTIST, UNIV HAWAII, 64- Concurrent Pos: Chmn, State Bd Agr, 63-70; partic, Int Agr Progs, Col Agr, Univ Hawaii, 71- Mem: AAAS; Am Soc Animal Sci; Am Dairy Sci Asn; Am Chem Soc; Am Inst Nutrit. Res: Ruminant nutrition. Mailing Add: Col of Agr Univ of Hawaii Honolulu HI 96822

OTANI, THEODORE TOSHIRO, b Honolulu, Hawaii, Jan 5, 25; m 47; c 3. BIOCHEMISTRY. Educ: Univ Hawaii, BS, 47; Univ Denver, MS, 49; Univ Colo, PhD, 53. Prof Exp: Res assoc chem, Fla State Univ, 55-57; RES ASSOC, LAB BIOCHEM, NAT CANCER INST, 57- Concurrent Pos: Fel, Clin Biochem Res Sect, Nat Cancer Inst, 53-55. Mem: AAAS; Am Chem Soc. Res: Biochemistry of amino acids and peptides; synthesis of amino acids, amino acid analogues and peptides; amino acid antagonists; isolation, characterization and mode of action of antibiotics; isozymes in hepatomas and normal livers.

O'TANYI, THEODORE JOHN, JR, b Darby, Pa, Oct 9, 41; m 68; c 2. NEUROSCIENCES. Educ: Pa Mil Col, BS, 63; Lehigh Univ, MS, 67, PhD(biol), 68.

Prof Exp: Instr physiol, Sch Dent, Temple Univ, 64-65; instr biol, Lehigh Univ, 65-66; instr sci, 68, asst prof, 68-75, ASSOC PROF BIOL, WIDENER COL, 75- Concurrent Pos: USPHS trainee, Sch Med, Temple Univ, 68-69. Mem: AAAS; Am Soc Zool; Am Inst Biol Sci. Res: Structure, function, and development of arthropod proprioceptors; neurological basis of behavior; behavioral genetics. Mailing Add: Dept of Biol Widener Col Chester PA 19013

OTENASEK, FRANK (JOSEPH), b Baltimore, Md, Jan 6, 12. NEUROSURGERY. Educ: Loyola Col, Md, AB, 33; Johns Hopkins Univ, MD, 37. Hon Degrees: LLD, Loyola Col, 59. Prof Exp: Resident neurosurg, 42-44, vis, 45-51, asst prof, 51-68, ASSOC PROF NEUROL SURG, JOHNS HOPKINS HOSP, 68- Concurrent Pos: Halsted fel surg, Johns Hopkins Univ, 38-40; instr, Univ Md, 49; vis neurosurg, Baltimore Hosps. Mem: Am Asn Neurol Surg; Neurosurg Soc Am (pres, 50); AMA. Mailing Add: Dept of Neurol Surg Johns Hopkins Univ Sch of Med Baltimore MD 21204

OTERMAT, ARTHUR LAMAR, physical chemistry, polymer chemistry, see 12th edition

OTERO, JOSEPH GUILLERMO, b New York, NY, Aug 15, 27; m 48; c 5. PARASITOLOGY, BACTERIOLOGY. Educ: Roosevelt Univ, BSc, 52; Univ Mich, MPH, 56, PhD(epidemiol, sci), 63. Prof Exp: Technician, Univ Ill Res Hosp, Chicago, 49, Dept Med, Billings Hosp, 51-52 & Michael Reese Hosp, 49-51; bacteriologist, Ill State Health Dept, 52-55 & 56-58; asst parasitol, Peace Corps Prog Malaria Control, Univ Mich, Ann Arbor, 62, from lectr to assoc prof microbiol, 62-74, PROF MICROBIOL, UNIV MICH-FLINT, 74- Res: Parasitology, especially physiology, immunology and pathology of schistosomiasis. Mailing Add: Dept of Biol Univ of Mich Flint MI 48503

OTERO, RAYMOND B, b Rochester, NY, May 8, 38; m 62; c 2. MICROBIOLOGY, BIOCHEMISTRY. Educ: Univ Dayton, BS, 60; Univ Rochester, MS, 63; Univ Md, PhD(microbiol), 68. Prof Exp: Control biologist, Lederle Labs Div, Am Cyanamid Co, 63-65; from asst prof to assoc prof microbiol, 68-73, ASSOC PROF BIOL, EASTERN KY UNIV, 73- Concurrent Pos: Chicago Res Corp grant, 69-; consult, Good Samaritan Hosp, 69 & St Joseph Hosp, 69- Mem: AAAS; NY Acad Sci; Am Soc Microbiol. Res: Determination of mode of entry of nucleic acids into whole bacterial cells; bacterial genetics and immunology. Mailing Add: Dept of Biol Sci Eastern Ky Univ Richmond KY 40475

OTERO-VILARDEBO, LUIS RAUL, cell biology, biochemistry, see 12th edition

OTEY, FELIX HAROLD, b Melber, Ky, Feb 27, 27; m 50; c 2. ORGANIC CHEMISTRY. Educ: Murray State Col, BS, 48; Univ Mo, Columbia, MA, 50. Prof Exp: Instr chem, Flat River Jr Col, 50-55 & McMurray Col, 55-56; res chemist, 56-75, RES LEADER CHEM, NORTHERN REGIONAL RES LAB, USDA, 75- Honors & Awards: USDA Superior Serv Award, 66. Mem: Am Chem Soc. Res: The synthesis and characterization of starch derivatives for use as raw materials in surfactants, coatings, rubber, urethane foams, plastics and films. Mailing Add: 1128 E Tripp Ave Peoria IL 61603

OTHERSEN, HENRY BIEMANN, JR, b Charleston, SC, Aug 26, 30; c 3. PEDIATRIC SURGERY. Educ: Col Charleston, BSM, 50; Med Univ SC, MD, 53; Am Bd Surg, dipl, 63; Am Bd Thoracic Surg, dipl, 66. Prof Exp: Intern, Philadelphia Gen Hosp, 53-54; resident gen surg, Med Col SC, 57-62; demonstr pediat surg, Children's Hosp & Ohio State Univ, 63-64; from asst prof to assoc prof surg, 65-72, PROF SURG & PEDIAT, MED UNIV SC, 72-, CHIEF PEDIAT SURG, 65- Concurrent Pos: Fel oncol, Med Col SC, 59-60, teaching fel surg, 61-62; res fel surg, Mass Gen Hosp, 64-65; Am Cancer Soc advan clin fel, Med Col SC, 66-69; resident pediat surg, Children's Hosp, Columbus, 62-64. Mem: Asn Acad Surg; Am Burn Asn; Am Col Surg; Am Acad Pediat; Am Pediat Surg Asn. Res: Cancer immunotherapy and neonatal deformities. Mailing Add: Dept of Surg Med Univ of SC Charleston SC 29401

OTHMER, EKKEHARD, b Koenigsberg, Ger, Oct 15, 33; m 64; c 2. PSYCHIATRY. Educ: Univ Hamburg, Diplom, 60, PhD(psychol), 65, MD, 67. Prof Exp: Res assoc neurophysiol of sleep, Univ Mo-Columbia, 67-68; asst prof exp psychiat & asst prof psychol, Sch Med, Wash Univ, 68-74, resident psychiat, 71-74; ASSOC PROF PSYCHIAT, MED CTR, UNIV KY, 74- Concurrent Pos: Nat Inst Ment Health fel psychiat, Sch Med, Wash Univ, 70-71; NIH fel, Sch Med, Wash Univ & Med Ctr, Univ Ky, 72-74; chief psychobiol unit, Vet Admin Hosp, Lexington, Ky, 74-76. Mem: AMA; Am Psychiat Asn; Soc Neurosci; Asn Psychophysiol Study Sleep; Am Asn Univ Prof. Res: Psychobiology and pathology of sleep; biorythms; computer analysis of polygraphic sleep recordings; medical model in psychiatry; diagnostic techniques; neurophysiology of sleep. Mailing Add: 112 Cherokee Park Lexington KY 40503

OTIS, ARTHUR BROOKS, b Grafton, Maine, Sept 11, 13; m 42; c 1. PULMONARY PHYSIOLOGY. Educ: Univ Maine, AB, 35; Int YMCA Col, MEd, 37; Brown Univ, MS, 39, PhD(physiol), 41. Prof Exp: Res assoc cellular physiol, Univ Iowa, 41-42; from instr to asst prof physiol, Sch Med & Dent, Univ Rochester, 42-51; assoc prof physiol & surg, Sch Med, Johns Hopkins Univ, 52-56; PROF PHYSIOL & HEAD DEPT, COL MED, UNIV FLA, 56- Concurrent Pos: Fulbright res scholar, Cambridge Univ, 50-51 & Univ Nijmegen, 64-65; mem physiol training comt, NIH, 58-60, physiol study sect, 60-64; comt eval post doctoral fel applns, NSF, 58-59; respiration sect ed, Am J Physiol & J Appl Physiol, 62-64. Mem: Am Physiol Soc; Soc Gen Physiol. Res: Respiratory physiology; physiology of hypoxia; comparative physiology. Mailing Add: Dept of Physiol Univ of Fla Col of Med Gainesville FL 32610

OTIS, HERBERT NEWELL, b Bloomingdale, NY, Apr 24, 05; m 31. PHYSICS. Educ: Union Univ, NY, 27, MS, 29; Yale Univ, PhD(physics), 33. Prof Exp: From instr to prof physics, Hunter Col, 32-68; prof physics, Lehman Col, 68-73; RETIRED. Concurrent Pos: Chmn dept physics, Hunter Col, 40-50, 53-56, actg dean fac, 50-53. Mem: Am Asn Physics Teachers. Res: Magnetism; thin films. Mailing Add: 2 Overlook Rd White Plains NY 10605

OTIS, MARSHALL VOIGT, b New London, Wis, Aug 28, 19; m 42; c 3. CHEMISTRY. Educ: Univ Wis, BS, 42. Prof Exp: Anal chemist, Hercules Powder Co, 42-44; prod supvr, Badger Ord Works, 44-45, develop supvr ultrasonics test method, 45; res chemist, 46-59, SR RES CHEMIST, TENN EASTMAN CO DIV, EASTMAN KODAK CO, 59- Mem: Am Chem Soc; Soc Appl Spectros; Am Soc Test & Mat; fel Am Inst Chem; Int Oceanog Found. Res: Applied spectroscopy; information storage and retrieval by machine methods; determination of inherent viscosities of polymer compositions by solution viscometry. Mailing Add: Res Labs Tenn Eastman Co Eastman Rd Kingsport TN 37662

OTKEN, CHARLES CLAY, b Falfurrias, Tex, Nov 6, 27; m 71; c 5. PHYSICAL ORGANIC CHEMISTRY. Educ: Tex A&M Univ, BS, 49; Cornell Univ, PhD(biochem), 54. Prof Exp: From instr to assoc prof, 54-67, PROF CHEM, PAN AM UNIV, 67- Mem: AAAS; Am Chem Soc; NY Acad Sci. Res: Nutritional biochemistry. Mailing Add: Dept of Chem Pan Am Univ Edinburg TX 78539

OTLEY, KURT O, b Vienna, Austria, Oct 28, 06; nat US; m 39. INORGANIC CHEMISTRY. Educ: Univ Vienna, PhD(chem), 31. Prof Exp: Res chemist, Crompton, Parkinson, Ltd, 39-40 & Voltarc Tubes Co, 40-42; chief chemist, Duro Test Corp, 42-44; res chemist, Am Lead Pencil Co, 44-48; phys chemist, Radio Corp Am, 48-49; res assoc, Pa State Univ, 49-51; PHYSICIST, HARRY DIAMOND LABS, US DEPT ARMY, 51- Mem: Am Chem Soc; Electrochem Soc. Res: Semiconductors and dielectrics, especially their application in electronic components and glass; solid state chemistry and physics; crystal chemistry of defect structures; fluorescence. Mailing Add: 216 Rollins Ave Rockville MD 20852

OTOCKA, EDWARD PAUL, b Jersey City, NJ, Mar 11, 40; m 64; c 2. POLYMER CHEMISTRY. Educ: Lehigh Univ, BS, 62; Polytech Inst Brooklyn, PhD(polymer chem), 66. Prof Exp: Mem tech staff polymer chem, Bell Tel Labs, Inc, 66-73; MEM TECH STAFF POLYMER CHEM, PRATT & WHITNEY AIRCRAFT, 73- Concurrent Pos: Adj prof polymer sci & eng, Univ Mass, 75- Mem: Am Chem Soc; Am Phys Soc; NY Acad Sci. Res: Composite and polymer material science; physical properties of polymers; polymer modification, characterization and stabilization. Mailing Add: Pratt & Whitney Aircraft East Hartford CT 06108

OTREMBA, EDWARD D, organic chemistry, see 12th edition

OTSU, TAMIO, b Kona, Hawaii, July 25, 23; m 49, 64; c 3. MARINE BIOLOGY. Educ: Univ Hawaii, BA, 50. Prof Exp: Fishery res biologist, 50-59, supvry fishery biologist, 59-62, chief albacore ecol prog, 62-66, chief tuna ecol prog, 68-70, leader resource monitoring & prediction group, 70-71, LEADER PELAGIC FISH ECOL GROUP, NAT MARINE FISHERIES SERV, 71- Concurrent Pos: Chief tuna assessment & develop invests, Nat Marine Fisheries Serv, 72- Mem: AAAS; Am Inst Fishery Res Biol; Am Fisheries Soc. Res: Ecology and biology of pelagic fishes in the Pacific Ocean. Mailing Add: 1106 Lunaai St Kailua HI 96734

OTT, ALETA JO PETRIK, b Topeka, Kans, May 7, 42; m 70. BOTANY, PLANT TAXONOMY. Educ: Washburn Univ, BS, 64; Univ Kans, MA, 66, PhD(bot), 69. Prof Exp: Asst prof biol, Pershing Col, 69-70; scientist, Dept Microbiol, Va Inst Marine Sci, 72-74; TECHNICIAN DEPT PLANT SCI, TEX A&M UNIV, 75- Mem: Am Fern Soc; Brit Pteridological Soc. Res: Taxonomy of pteridophytes and grasses with special interest in the genera Cystopteris, Equisetum, Marsilea and Paspalum. Mailing Add: Dept of Plant Sci Tex A&M Univ College Station TX 77843

OTT, ARNOLD CHARLES, chemistry, see 12th edition

OTT, BILLY JOE, b Bearden, Okla, Sept 10, 23; m 43; c 1. SOILS. Educ: Okla State Univ, BS, 48, MS, 49, PhD(soil chem), 62. Prof Exp: Prof soils, Panhandle Agr & Mech Col, 62-67; PROF SOILS & RESIDENT DIR RES, TEX A&M UNIV, 67- Mem: Am Soc Agron; Soil Sci Soc Am. Res: Dryland and irrigated soils management; soil fertility; soil-plant-water relationships; soil moisture utilization; plant root energy levels and ion absorption. Mailing Add: Tex A&M Univ Agr Res & Exten Ctr Rte 3 Lubbock TX 79410

OTT, COBERN ERWIN, b Osyka, Miss, Jan 15, 41; m 67; c 2. MEDICAL PHYSIOLOGY, NEPHROLOGY. Educ: Millsaps Col, BS, 64; Univ Miss, PhD(physiol), 71. Prof Exp: Res assoc, Sch Med, Univ Miss, 70-71, NIH fel, 71-72; NIH fel, Mayo Grad Sch Med, 72-75; ASST PROF PHYSIOL, UNIV KY MED CTR, 75- Mem: Am Fedn Clin Res; Am Soc Nephrology; Am Physiol Soc; Int Soc Nephrology. Res: Basic renal physiology with emphasis on sodium balance and relationship to fluid volumes and blood pressure. Mailing Add: Dept of Physiol Univ of Ky Med Ctr Lexington KY 40506

OTT, DONALD GEORGE, b Kinsley, Kans, Aug 13, 26; m 48; c 2. CHEMISTRY, SYNTHETIC ORGANIC CHEMISTRY. Educ: Colo State Univ, BS, 50; Wash State Univ, PhD(chem), 53. Prof Exp: Org res chemist, Dow Chem Co, 53-54; mem staff, 54-63, alternate group leader biomed res, 63-73, GROUP LEADER ORG & BIOCHEM SYNTHESIS, LOS ALAMOS SCI LAB, 73- Concurrent Pos: USPHS-NIH fel, Cambridge Univ, 60-61. Mem: AAAS; Am Chem Soc; Brit Chem Soc. Res: Organic and biochemical synthesis of isotopic compounds; organic reaction mechanisms; synthesis and characterization of oligonucleotides; applications of stable isotopes of carbon, nitrogen and oxygen. Mailing Add: Org & Biochem Synthesis Group Los Alamos Sci Lab MS 890 Los Alamos NM 87545

OTT, ELLIS RAYMOND, b Dexter, Kans, Dec 14, 06; m 32; c 3. EXPERIMENTAL STATISTICS. Educ: Southwestern Col, Kans, AB, 28; Univ Kans, AM, 29; Univ Ill, PhD(math), 33. Prof Exp: Instr math, Univ Ill, 29-31 & Univ Ark, 32-33; from instr to prof, Univ Buffalo, 33-46; exec engr, Nat Union Radio Corp, NJ, 44-46; asst dean, Millard Fillmore Col, 46; chmn dept math, 46-59, prof, 49-72, dir statist ctr, 59-72, EMER PROF STATIST, 72-; CONSULT EXP STATIST & QUAL CONTROL, 72- Concurrent Pos: Consult, Sonotone Corp, Esso Standard Oil Co, Bristol-Myers Co, Bayonne Naval Supply Facilities, Aberdeen Proving Ground, Gen Elec Co, Anaconda Wire & Cable Co & Am Can Co; UN expert on statist qual control to India, 52, 57, 62, Mexico, 60, 61, 63. Honors & Awards: Shewhart Medal, Am Soc Qual Control, 60, Grant Award, 69 & Brumbaugh Award, 70. Mem: Fel Am Soc Qual Control (vpres, 60); fel Am Statist Asn; fel Int Statist Inst. Res: Statistical quality control. Mailing Add: 215 Los Escondidos Marble Falls TX 78654

OTT, FRANKLYN DEWAYNE, botany, phycology, see 12th edition

OTT, J BEVAN, b Cedar City, Utah, July 21, 34; m 53; c 6. PHYSICAL CHEMISTRY. Educ: Brigham Young Univ, BS, 55, MS, 56; Univ Calif, PhD, 59. Prof Exp: Asst prof chem, Utah State Univ, 59-60; assoc prof, 60-68, PROF CHEM, BRIGHAM YOUNG UNIV, 68- Concurrent Pos: Sabbatical leave as res specialist, Atomics Int Div, NAm Rockwell Corp, 66-67. Mem: Am Chem Soc. Res: Low temperature and solution thermodynamics; phase equilibria. Mailing Add: Dept of Chem Brigham Young Univ Provo UT 84601

OTT, JOHN LEWIS, b Sabetha, Kans, Aug 19, 24; m 48; c 2. BACTERIOLOGY. Educ: Univ Kans, AB, 45, MA, 51; Iowa State Col, PhD(bact physiol), 56. Prof Exp: Bacteriologist, Vet Admin Hosp, 56-59; SR MICROBIOLOGIST, RES LABS, ELI LILLY & CO, 59- Mem: AAAS; Am Soc Microbiol; Brit Soc Gen Microbiol; Environ Mutagen Soc; NY Acad Sci. Res: Inhibitors of resistance transfer; mutagenesis; biosynthesis of antibiotics; penicillinase and cephlosporinase; mechanisms of antibiotic resistance. Mailing Add: Dept M932 Lilly Res Labs Indianapolis IN 46206

OTT, KAREN JACOBS, b Atlanta, Ga, Sept 26, 39; m 59; c 2. PARASITOLOGY, INVERTEBRATE ZOOLOGY. Educ: Asbury Col, AB, 59; Univ Ky, MS, 61; Rutgers Univ, PhD(zool, parasitol), 65. Prof Exp: Instr microbiol & parasitol, Jefferson Med Col, 64-65, grant, 65; asst res specialist, Bur Biol Res, Rutgers Univ, 65-68; ASST PROF BIOL, UNIV EVANSVILLE, 69- Mem: AAAS; Am Soc Parasitol; Am

Soc Trop Med & Hyg; Nat Asn Biol Teachers; Am Inst Biol Sci. Res: Immunological and physiological aspects of host-parasite relationships. Mailing Add: Dept of Biol Univ of Evansville Evansville IN 47710

OTT, RAY LYMAN, JR, b Kennett Square, Pa, Mar 1, 40; m 64; c 2. STATISTICS. Educ: Bucknell Univ, BS, 62, MS, 63; Va Polytech Inst, PhD(statist), 67. Prof Exp: Sr statistician, Smith Kline & French Labs, 66-68; asst prof, 68-72, ASSOC PROF STATIST, UNIV FLA, 72- Mem: Am Statist Asn; Biomet Soc Res: Design of experiments and accompanying estimation procedures. Mailing Add: Dept of Statist Univ of Fla Gainesville FL 32601

OTT, RICHARD L, b Santa Barbara, Calif, June 23; m 48; c 2. VETERINARY MEDICINE. Educ: Wash State Univ, BS, 44, DVM, 45. Prof Exp: Pvt pract, Wash, 45-46 & 48-49; from asst prof to assoc prof med & surg, 49-59, prof & chmn dept, 59-73, PROF VET CLIN MED & SURG, COL VET MED, WASH STATE UNIV, 73-, ASSOC DEAN, 74- Concurrent Pos: NIH grant, 60-65, res grant, 67-74. Honors & Awards: Gaines Medal, Am Vet Med Asn, 64. Mem: AAAS; Am Vet Med Asn; Conf Res Workers Animal Dis; Am Col Vet Internal Med. Res: Immunology, pathogenesis and ecology of virus diseases; small animal medicine and surgery. Mailing Add: Dept of Vet Clin Med & Surg Col of Vet Med Wash State Univ Pullman WA 99163

OTT, TEUNIS JAN, b Zype, Netherlands, Jan 27, 43; m 68; c 2. OPERATIONS RESEARCH, MATHEMATICAL STATISTICS. Educ: Univ Amsterdam, BSc, 65, Drs(math statist), 70; Univ Rochester, PhD(opers res), 75. Prof Exp: ASST PROF OPERS RES, CASE WESTERN RESERVE UNIV, 74- Mem: Opers Res Soc Am; Inst Math Statist; Soc Indust & Appl Math; Dutch Math Asn; Netherlands Statist Asn. Res: Operations research as applied to medical systems; queueing systems in heavy traffic; research of a theoretical nature in probability theory. Mailing Add: Dept of Opers Res Case Western Reserve Univ Cleveland OH 44106

OTT, WALTHER HENRY, b Hermiston, Ore, Sept 30, 11; m 36; c 2. POULTRY SCIENCE, BIOMETRICS. Educ: Ore State Univ, BS, 34, MS, 36; Pa State Univ, PhD(biochem), 42. Prof Exp: Asst poultry husb, Okla State Univ, 36-37; res assoc & head dept poultry nutrit, Merck Inst Therapeut Res, 42-54, head dept biol control, 45-52, assoc farm mgr, 54-56, dir animal husb, 56-68, dir animal husb labs, 69-72, sr dir animal health develop res labs, 72-73, SR SCIENTIST, ANIMAL SCI RES, MERCK SHARP & DOHME RES LABS, 74- Concurrent Pos: Bd mem, Coun Agr Sci & Technol, 73-; assoc ed, Poultry Sci, 75- Mem: Fel AAAS; Am Inst Chem; Animal Nutrit Res Coun; Am Soc Animal Soc; Poultry Sci Asn (pres, 71-72). Res: Poultry nutrition; nutritive requirements of chicks; application of biometry to bioassays; tolerance of poultry and farm animals to feed additives. Mailing Add: 1874 Quimby Lane Westfield NJ 07090

OTT, WELLAND LEE, b Welland, Ont, Can, Aug 27, 21; m 73. ANALYTICAL CHEMISTRY. Educ: Queen's Univ, Ont, BSc, 46. Prof Exp: Res asst colloid chem, Queen's Univ, Ont, 46-48, lectr inorg chem, 48-50, chemist, Dept Mineral, 50-51; chemist, 51-53, sect head chem, Metall Labs, 53-73, CHIEF CHEMIST, FALCONBRIDGE NICKEL MINES LTD, 73- Concurrent Pos: Chmn Can adv comt, Int Orgn Standardization, 75- Mem: Am Soc Testing & Mat; Am Chem Soc. Res: Development of instrumental and chemical analytical methods for the analysis and/or determination of nickel, copper, cobalt, ferronickel, gold, silver and the platinum metals; platinum metal chemical metallurgy. Mailing Add: Falconbridge Nickel Mines Ltd PO Box 900 Thornhill ON Can

OTT, WILLIAM ROGER, b Philadelphia, Pa, Mar 29, 42; m 67; c 3. EXPERIMENTAL ATOMIC PHYSICS, OPTICAL PHYSICS. Educ: St Joseph's Col, BS, 63; Univ Pittsburgh, PhD(physics), 68. Prof Exp: NSF assoc plasma spectros, 68-70, STAFF PHYSICIST, NAT BUR STAND, 70- Honors & Awards: Superior Performance Award, Nat Bur Stand, 75. Mem: Optical Soc Am. Res: The application of plasma light sources as intensity standards in the ultraviolet; also, measurements of transition probabilities and line broadening parameters using arc plasmas. Mailing Add: Nat Bur of Stand Bldg 221 Rm A261 Washington DC 20234

OTTA, JACK DUANE, b Marshaltown, Iowa, Apr 18, 44; m 64; c 2. PLANT PATHOLOGY. Educ: Iowa State Univ, BS, 66; Univ Calif, Davis, PhD(plant path), 70. Prof Exp: ASST PROF PLANT SCI, SDAK STATE UNIV, 70- Mem: Am Phytopath Soc. Res: Bacterial diseases of plants; taxonomy of phytopathogenic bacteria; forest pathology. Mailing Add: Dept of Plant Sci SDak State Univ Brookings SD 57006

OTTAVIANI, ROBERT AUGUSTINE, b Wilmington, Del, Feb 7, 40; m 64; c 2. ORGANIC POLYMER CHEMISTRY. Educ: Univ Del, BS, 65; Univ Ariz, PhD(chem), 70. Prof Exp: SR RES CHEMIST, GEN MOTORS RES LABS, 69- Mem: Am Chem Soc. Res: Polymeric surface coatings. Mailing Add: Polymers Dept Gen Motors Res Labs Warren MI 48090

OTTE, CAREL, JR, b Amsterdam, Netherlands, June 29, 22; nat US; m 53; c 4. GEOLOGY. Educ: Calif Inst Technol, MS, 50, PhD(geol), 54. Prof Exp: Res geologist, Pure Oil Co, 54-55; explor geologist, Shell Oil Co, 55-57; supvr geol res, 57-62, vpres & mgr, Earth Energy, Inc, 62-67, MGR GEOTHERMAL DIV, UNION OIL CO CALIF, 67- Mem: Geol Soc Am; Am Asn Petrol Geologists. Res: Exploration geology.

OTTE, DANIEL, b Durban, SAfrica, Mar 14, 39; US citizen; m 63; c 2. ZOOLOGY. Educ: Univ Mich, BA & PhD(zool), 68. Prof Exp: NSF grant & res assoc zool, Univ Mich, 68-69; asst prof, Univ Tex, Austin, 69-75; ASSOC CUR ENTOM, THE ACAD NATURAL SCI, PHILADELPHIA, 75- Res: Insect behavior; evolution of behavior; systematics of Australian crickets; communication in grasshoppers. Mailing Add: The Acad of Natural Sci Philadelphia PA 19103

OTTE, HERMAN FREDERICK, b Colfax Co, Nebr, Mar 4, 02; m 31; c 1. ECONOMIC GEOGRAPHY. Educ: Univ Nebr, BS, 28, MA, 29; Columbia Univ, PhD(geog), 40. Prof Exp: Asst geog, Univ Nebr, 28-30; asst, Columbia Univ, 30-31, lectr, 31-37; instr geog, 37-41, from asst prof to prof econ geog, 41-70, EMER PROF ECON GEOG, COLUMBIA UNIV, 70- Mem: Fel AAAS; Asn Am Geog. Res: Raw material and resource position of the United States; locational shifts in manufacturing in the United States; regional economic geography; world resource patterns. Mailing Add: 452-A Portsmouth Dr Lakewood NJ 08701

OTTEN, CHARLOTTE MARIE, b Sheboygan, Wis, May 23, 15. BIOLOGICAL ANTHROPOLOGY, SEROLOGY. Educ: Carleton Col, BA, 37; Univ Chicago, MA, 53; Univ Mich, PhD(anthrop), 62. Prof Exp: Instr serol, Sch Med, Univ Minn, 58-59, instr anthrop, Univ, 59-60; asst prof, Univ Wis, 60-67; assoc prof, 67-69, PROF ANTHROP, NORTHERN ILL UNIV, 69- Concurrent Pos: NSF grants, Univ Wis, 65-66 & Northern Ill Univ, 67-69. Mem: Fel Am Anthrop Asn; Soc Study Human Biol; Int Asn Anthropobiol; Am Asn Phys Anthrop; Am Eugenics Soc. Res: Natural selection and blood groups; biological and cultural bases of sex role differentiation; possible sex differences in conceptualization and aesthetic expression; human aggression. Mailing Add: Dept of Anthrop Northern Ill Univ De Kalb IL 60115

OTTENBERG, SIMON, b New York, NY, June 6, 23; m 68. ANTHROPOLOGY, ETHNOGRAPHY. Educ: Univ Wis, BA, 48; Northwestern Univ, PhD(anthrop), 57. Prof Exp: Instr anthrop, Univ Chicago, 54; actg instr, Wash State Col, 54-55; from instr to prof anthrop, 55-73, PROF ANTHROP & AFRICAN STUDIES, UNIV WASH, 73-, ADJ CUR, WASH STATE MUS, 73- Concurrent Pos: NSF grant field res, Eastern Nigeria, 59-60; Soc Sci Res Coun fac res grant, Univ Wash, 63-64; John Simon Guggenheim mem fel, Univ Ghana, 70-71; mem Africa screening comt, Foreign Area Fel Prog, 64-67; mem bd dirs, Human Rels Area Files, 66-; hon vis prof, Inst African Studies, Univ Ghana, 70-71. Honors & Awards: P Amaury Talbot Book Award, 69. Mem: African Studies Asn. Res: Africa, with emphasis on politics and on the study of African aesthetics. Mailing Add: Dept of Anthrop Univ of Wash Seattle WA 98195

OTTENBRITE, RAPHAEL MARTIN, b Claybank, Sask, Sept 20, 36; m 63; c 2. ORGANIC CHEMISTRY. Educ: Univ Windsor, BSc, 58, MSc, 62, PhD(chem), 67; Univ Toronto, dipl ed, 64. Prof Exp: Instr chem, Western Ont Inst Technol, 60-64 & Univ Windsor, 64-66; US Air Force fel, Univ Fla, 66-67; asst prof, Med Col Va & Richmond Prof Inst, 67-71; ASSOC PROF CHEM, VA COMMONWEALTH UNIV, 71- Mem: Am Chem Soc; Chem Inst Can; Sigma Xi; Nat Cath Educ Asn. Res: Synthesis of exocyclic diene systems and study of their diels-alder reactivity, ultraviolet spectra and ioniztion potentials; the correlation of these properties by MO calculations; synthesis of blood-clotting and lysing inhibitors, and preparation of new anticancer and antivirial agents. Mailing Add: 2817 Windsorview Dr Richmond VA 23225

OTTENHEIMER, HARRIET JOSEPH, b New York, NY, June 11, 41; m 62; c 2. ANTHROPOLOGY. Educ: Bennington Col, BA, 62; Tulane Univ, PhD(anthrop), 73. Prof Exp: ASST PROF ANTHROP, KANS STATE UNIV, 69- Mem: fel Am Anthrop Asn; Soc Ethnomusicol. Res: Sex roles in western and nonwestern societies; music and language in the western Indian Ocean and among African-Americans; acquisition by children. Mailing Add: Dept of Sociol & Anthrop Kans State Univ Manhattan KS 66506

OTTENHEIMER, MARTIN, b New York, NY, July 18, 39; m 64; c 2. ANTHROPOLOGY. Educ: Rensselaer Polytechnic Inst, BS, 62; Tulane Univ, MA, 65, PhD(anthrop), 71. Prof Exp: Teaching asst anthrop, Tulane Univ, 66-67; ASST PROF ANTHROP, KANS STATE UNIV, 69- Mem: Am Anthrop Asn. Res: Formal models of kin processes; ethnography of the western Indian Ocean region. Mailing Add: Dept of Sociol & Anthrop Kans State Univ Manhattan KS 66502

OTTER, FRED AUGUST, JR, b West Chester, Pa, Sept 11, 28; m 53; c 2. SOLID STATE PHYSICS. Educ: Lehigh Univ, BS, 53; Temple Univ, AM, 55; Univ Ill, PhD(physics), 59. Prof Exp: Asst physics, Labs, Franklin Inst, 53-55; asst, Univ Ill, 55-57, instr, 58-59; asst prof, Ohio Univ, 59-63; mem staff, Res Labs, United Aircraft Corp, 63-68, mgr surface & film lab, 68-69; prof physics, Clarkson Col Technol, 69-73; PRIN SCIENTIST PHYSICS OF SOLIDS, UNITED TECHNOL RES CTR, 73- Mem: Am Phys Soc. Res: Superconductivity; surface science. Mailing Add: United Technol Res Ctr East Hartford CT 06108

OTTER, RICHARD ROBERT, b Evanston, Ill, May 17, 20; m 49; c 2. MATHEMATICS. Educ: Dartmouth Col, AB, 41; Univ Ind, PhD(org chem), 46. Prof Exp: Res chemist, Eastman Kodak Co, NY, 42; mem comt med res, Off Sci Res & Develop, Washington, DC, 44-46; instr math, Princeton Univ, 46-47, fel, Off Naval Res, 47-48, vis asst prof, 53-54; from asst prof to assoc prof, 48-65, PROF MATH, UNIV NOTRE DAME, 65- Mem: Am Math Soc; Math Asn Am. Res: Theory of probability; combinatorial analysis. Mailing Add: Dept of Math Univ of Notre Dame Notre Dame IN 46556

OTTERBEIN, KEITH FREDERICK, b Warren, Pa, May 24, 36; m 65; c 1. ANTHROPOLOGY. Educ: Pa State Univ, BA, 58; Univ Pa, AM, 60; Univ Pittsburgh, PhD(anthrop), 63. Prof Exp: Asst prof res, Am Univ, 63-64; asst prof anthrop, Univ Kans, 64-66; from asst prof to assoc prof, 66-71, PROF ANTHROP, STATE UNIV NY BUFFALO, 71- Mem: Fel Am Anthrop Asn; fel Soc Appl Anthrop; Am Sociol Asn. Mailing Add: Dept of Anthrop State Univ of NY at Buffalo Amherst NY 14226

OTTERBY, DONALD EUGENE, b Sioux Falls, SDak, July 2, 32. NUTRITION. Educ: SDak State Col, BS, 54, MS, 58; NC State Col, PhD(nutrit), 62. Prof Exp: Fel nutrit, NC State Col, 62-63; asst prof nutrit, 63-67, from assoc prof to prof ruminant nutrit, 67-74, PROF ANIMAL SCI, UNIV MINN, ST PAUL, 74- Mem: Am Dairy Sci Asn; Am Soc Animal Sci. Res: Nutrition of ruminant animals. Mailing Add: Dept of Animal Sci Univ of Mich St Paul MN 55101

OTTERMAN, JOSEPH, b Warsaw, Poland, Apr 12, 25; nat US; m 56; c 4. SPACE PHYSICS. Educ: Israel Inst Technol, dipl eng, 48; Univ Mich, MSE, 52, PhD(elec eng), 55. Prof Exp: Res officer, Govt Res Ist, Israel, 50-51; res assoc, Res Inst, Mich Tech, 55-58, assoc res engr, 58-59; tech specialist, Int Tel & Tel Corp, 59-61; resident visitor, Bell Tel Labs, Inc, 61; consult physicist, Gen Elec Co, 61-69; vis assoc prof, 66-67, ASSOC PROF DEPT ENVIRON SCI, TEL-AVIV UNIV, 69- Mem: Am Phys Soc. Res: Remote sensing of environment; climatic change. Mailing Add: Dept of Environ Sci Tel-Aviv Univ Ramat Aviv Israel

OTTESON, OTTO HARRY, b Rexburg, Idaho, June 25, 31; m 57; c 5. NUCLEAR PHYSICS. Educ: Utah State Univ, BS, 60, MS, 62, PhD(physics), 66. Prof Exp: Teaching asst, 61-62, asst prof, 66-73, ASSOC PROF PHYSICS & ASST DEPT HEAD, UTAH STATE UNIV, 73- Mem: Am Asn Physics Teachers. Res: Linear-motion bent-crystal spectrometer; decay of thulium 172; precision low-energy nuclear pulse-height spectroscopy; medium energy physics. Mailing Add: Dept of Physics Utah State Univ Logan UT 84321

OTTING, WILLIAM JOSEPH, JR, b St Paul, Minn, Oct 27, 19; m 40; c 3. PHYSICS. Educ: George Washington Univ, BS, 46, MS, 49; Cath Univ Am, PhD(physics), 56. Prof Exp: Radio engr, Nat Bur Standards, 45-48; physicist, Off Naval Res, 48-52; dir phys sci, Off Sci Res, US Air Force, 52-60; chief scientist, Defense Atomic Support Agency, US Dept Defense, 60-64; asst dean, 64-70, ASSOC PROF PHYSICS, UNIV ILL, CHICAGO CIRCLE, 64-, ASSOC DEAN GRAD COL, 70- Mem: AAAS; fel Am Phys Soc. Res: Electrochemistry; electromagnetism; chemical, solid state and nuclear physics.

OTTINGER, CAROL BLANCHE, b Batesville, Ark, Dec 25, 33. MATHEMATICS. Educ: Ark Col, BSE, 54; Okla State Univ, MS, 60, EdD(math), 69. Prof Exp: Teacher math high schs, Ark, 54-58 & Okla, 60-63; from asst prof to assoc prof, 63-71, PROF MATH, MISS UNIV FOR WOMEN, 71-, CHMN DEPT, 74- Mem: Math Asn Am. Res: Topology of decomposition spaces. Mailing Add: Dept of Math Miss Univ for Women Box W-269 Columbus MS 39701

OTTIS, KENNETH, b Wyndmere, NDak, Apr 9, 08; m 48. ENDOCRINOLOGY. Educ: Dakota Wesleyan Univ, BS, 42; Iowa State Univ, MS, 51, PhD(zool), 54. Prof Exp: Assoc prof biol, Dakota Wesleyan Univ, 46-50; instr physiol, Iowa State Col, 52-53; assoc prof, 53-62, PROF ZOOL & ENTOM, AUBURN UNIV, 63- Concurrent Pos: Am Physiol Soc res fel, 57-58. Mem: Fel AAAS; Am Soc Zool; NY Acad Sci. Res: Effect of autonomic stimulation on splenic circulation; effect of adrenal steroids and catechol amines in cold stress; studies on serotonin and gastrointestinal motility. Mailing Add: Dept of Zool Auburn Univ Auburn AL 36830

OTTKE, ROBERT CRITTENDEN, b Louisville, Ky, Jan 23, 22; m 43, 58; c 4. ORGANIC CHEMISTRY, BIOCHEMISTRY. Educ: Yale Univ, BS, 48, PhD(chem), 50. Prof Exp: Nat Cancer Inst fel, Stanford Univ, 50-51; res chemist, Chas Pfizer & Co, Inc, 51-53, prod develop, 53-57; pres, Caribe Chem Corp, 57-58; dir commercial develop, Wallerstein Co Div, Baxter Labs, Inc, 58-61, tech dir, Baxter Int, 61-63; vpres & managing dir labs, Parke Davis, Madrid, Spain, 63-69; PARTNER, BOYDEN ASSOCS, 69- Mem: Am Chem Soc; NY Acad Sci. Res: Steroid biosynthesis; natural products; enzymology; pharmaceutical chemistry. Mailing Add: 3 De Soto Ct Orinda CA 94563

OTTO, ALBERT DEAN, b Marshalltown, Iowa, Nov 5, 39; m 62; c 2. MATHEMATICS. Educ: Univ Iowa, BA, 61, MS, 62, PhD(math), 65. Prof Exp: Asst prof math, Lehigh Univ, 65-69; assoc prof, 69-75, PROF MATH, ILL STATE UNIV, 75- Mem: Am Math Soc; Math Asn Am. Res: Study of the automorphisms of finite groups and of combinatorial objects. Mailing Add: Dept of Math Ill State Univ Normal IL 61761

OTTO, BRUCE M, chemistry, mathematics, see 12th edition

OTTO, DAVID ARTHUR, b Ft Scott, Kans, Feb 19, 34; m 54; c 3. MARINE BIOLOGY, PALEOBOTANY. Educ: Univ Kans, AB, 56; Kans State Teachers Col, MS, 57; Univ Mo, PhD(bot, geol), 67. Prof Exp: PROF BIOL, STEPHENS COL, 57- Mem: Am Inst Biol Sci; Bot Soc Am. Res: Investigations of Middle Pennsylvanian fossil plants found in coal ball petrifactions; plants fossils; marine mammals and invertebrates. Mailing Add: Dept of Natural Sci Stephens Col Columbia MO 65201

OTTO, FERDINAND PHILIP, b Philadelphia, Pa, Nov 6, 15; m 38; c 2. ORGANIC CHEMISTRY. Educ: Pa State Univ, BS, 37. Prof Exp: Chemist, Socony Mobil Oil Co, Inc, 37-44, group leader, 44-46, sr chemist, 46-48, sr technologist, 48-54, res assoc, 54-66, res assoc, Mobil Res & Develop Corp, 66-68, SR RES ASSOC, MOBIL RES & DEVELOP CORP, 68- Mem: Am Chem Soc. Res: Additives for lubricating oils, including detergents, anti-oxidants, extreme pressure agents and pour point depressants; additives for cutting oils and automatic transmission fluids; petroleum chemicals. Mailing Add: Res Dept Mobil Res & Develop Corp Paulsboro NJ 08096

OTTO, FRED BISHOP, physics, electrical engineering, see 12th edition

OTTO, GEORGE W, b Bronx, NY, Dec 13, 33; m 59; c 3. THERMAL PHYSICS. Educ: Polytech Inst Brooklyn, BS, 55. Prof Exp: Asst res physicist, 58-60, res physicist, 60-65, sr physicist, 65-75, RES SPECIALIST, MOUND LAB, MONSANTO RES CORP, MIAMISBURG, 75- Res: Radiometric calorimetry.

OTTO, GILBERT FRED, b Chicago, Ill, Dec 16, 01; m 32; c 2. PARASITOLOGY. Educ: Kalamazoo Col, AB, 26; Kans State Col, MS, 27; Johns Hopkins Univ, ScD(parasitol), 29. Prof Exp: Asst zool, Kans State Col, 26-27; asst helminth, Sch Hyg & Pub Health, Johns Hopkins Univ, 27-29; from instr to assoc prof, 29-42, assoc prof parasitol, 42-53, asst dean sch, 40-47, parasitologist, Johns Hopkins Hosp, 47-53; head dept parasitol, Res Div, Abbott Labs, Ill, 53-60, dir agr & vet res div, Sci Divs, 60-65, asst dir personnel div, 65-66; prof zool, Univ Md, 66-72; RETIRED. Concurrent Pos: Consult, USPHS, 41, 46-53, hookworm & malaria div, State Bd Health, Ga, 42-43 & Surgeon Gen, US Army, 58-66; mem & dir field expeds, Nat Res Coun & Am Child Health Asn, 27-32; expert WHO, UN, 52-74, comt parasitic dis, Armed Forces Epidemiol Bd, 53-57, Lobund adv bd, Univ Notre Dame, 58-66 & adv bd parasitol, Univ Md, 65-66; adj prof, Univ Md, 72- Mem: AAAS; Am Soc Parasitol (treas, 37-40, vpres, 55, pres-elect, 56, pres, 57); Am Soc Trop Med & Hyg; Am Epidemiol Soc. Res: Epidemiology of parasitic diseases; immunity to animal parasites; chemotherapy of parasitic diseases; pharmacology of arsenic and antimony; ascariasis; trichinosis; hookworm disease; filariasis; amebiasis; trypanosomiasis; malaria; leucocytozoon; coccidiosis; schistosomiasis; canine heartworm disease. Mailing Add: Dept of Zool Univ of Md College Park MD 20742

OTTO, HARLEY JOHN, b Richfield, Kans, May 5, 28; m 53; c 2. AGRONOMY. Educ: Colo State Univ, BSc, 52; Cornell Univ, PhD(plant breeding), 56. Prof Exp: Prof asst plant breeding, Cornell Univ, 56-57, asst prof agron, 57-58; from asst prof to assoc prof, 58-63, PROF AGRON, UNIV MINN, ST PAUL, 63- Concurrent Pos: Exec vpres, Minn Crop Improv Asn, 74- Mem: Am Soc Agron; Crop Sci Soc Am. Res: Extension education and research in production of field crops; quality seed production and distribution. Mailing Add: Dept of Agron & Plant Genetics Univ of Minn St Paul MN 55101

OTTO, HARRY WILLY, physical chemistry, see 12th edition

OTTO, JOHN B, JR, b Kingsville, Tex, Nov 4, 18; m 50; c 2. CHEMISTRY. Educ: Tex Col Arts & Indust, BS, 41; Univ Tex, MA, 43, PhD, 50. Prof Exp: Tutor chem, Univ Tex, 41-42, instr, 42-44; jr chemist, Clinton Lab, Oak Ridge, 44-46; res chemist, Mound Lab, Monsanto Chem Co, Ohio, 50-54; SR RES CHEMIST, FIELD RES LAB, MOBIL RES & DEVELOP CORP, 54- Mem: AAAS; Am Chem Soc. Res: Chemical reactions in liquid ammonia; radio chemistry; chemical separations and analysis for geochronology research. Mailing Add: Field Res Lab Mobil Res & Develop Corp Box 900 Dallas TX 75221

OTTO, KLAUS, b Friedrichroda, Ger, Sept 18, 29; m 62; c 2. ENVIRONMENTAL CHEMISTRY. Educ: Univ Hamburg, Vordiplom, 53, Dipl, 57, Dr rer nat(phys chem), 59. Prof Exp: Asst phys chem, Univ Hamburg, 60; res assoc, Argonne Nat Lab, 60-62; SR SCIENTIST, FORD MOTOR CO, DEARBORN, 62- Mem: AAAS; Am Chem Soc. Res: Hydrides of transition metals; low temperature calorimetry; electrical properties and nuclear magnetic resonance of inorganic glasses; catalysis of air pollutants, especially nitric oxide; coal gasification. Mailing Add: 35173 W Six Mile Rd Livonia MI 48152

OTTO, ROBERT GEORGE, b Slayton, Minn, Oct 2, 41; c 3. AQUATIC ECOLOGY. Educ: Univ Minn, BS, 66; Univ Victoria, MS, 68; Ariz State Univ, PhD(zool & environ physiol), 72. Prof Exp: Res asst, Univ Victoria, 66-68 & Ariz State Univ, 68-71; fisheries biologist, Indust BIO-TEST Labs, Inc, Env Sci Div, 71, group leader fish physiol, 71-72, sect head environ physiol & prog coordr, 72-75; ASSOC RES SCIENTIST, CHESAPEAKE BAY INST, JOHNS HOPKINS UNIV, 75- Mem: AAAS; Am Fisheries Soc; Am Soc Zoologists; Sigma Xi. Res: Responses of aquatic organisms to changing environments. Mailing Add: Chesapeake Bay Inst Johns Hopkins Univ Baltimore MD 21218

OTTO, ROBERT H, b Milwaukee, Wis, May 31, 26; m 52; c 3. BACTERIOLOGY. Educ: Univ Wis, PhD, 55. Prof Exp: Res bacteriologist, 55-63, pharmaceut prod mgr, Ill, 63-67, OPERS MGR, ABBOTT LABS, 67- Mem: Am Soc Microbiol; Soc Indust Microbiol. Res: Microbiology; industrial waste disposal. Mailing Add: Abbott Labs North Chicago IL 60064

OTTO, WILLIAM HERZER, b Marietta, Ohio, July 26, 10; m 39, 60; c 1. PHYSICS. Educ: Wabash Col, AB, 32; Ohio State Univ, MA, 34. Prof Exp: Physicist, Electronic Res Corp, 37-42; physicist res labs, Owens-Corning Fiberglas Corp, 42-60; chief fiber res, Narmco Res & Develop Div, Whittaker Corp, 60-66; adv composites, Ryan Aeronaut Co, 66-69; TECH CONSULT GLASS & ADVAN FIBERS & COMPOSITES, 69- Concurrent Pos: Mem, Ad Hoc Comt Composites, Mat Adv Bd, Nat Acad Sci, 63-65. Honors & Awards: Meyer Award, Am Ceramic Soc, 57. Mem: Am Ceramic Soc. Res: Glass fiber formation; physical properties, reinforcements and composites. Mailing Add: 920 Pomona Ave Coronado CA 92118

OTTO, WILLIAM M, physics, mathematics, see 12th edition

OTTO, WOLFGANG KARL FERDINAND, b Germany, June 12, 27; US citizen; m 59; c 3. PHYSICAL CHEMISTRY, TEXTILE CHEMISTRY. Educ: Textile Eng Sch, Germany, Textile engr, 49; Aachen Tech Univ, dipl chem, 54, Dr rer nat, 57. Prof Exp: Sci asst, Inst Fuel Chem, Aachen Tech Univ, 54-57; trainee & group leader, Esso, Inc, 57, coordr, 59; supvr fiber develop, Hoechst Dye Works, Inc, 60; proj chemist res & develop, United Carbon Co, Inc, 60, sr chemist, 63; sr res chemist, 64-74, RES ASSOC, DEERING MILLIKEN RES CORP, 74- Mem: Am Chem Soc; Am Asn Textile Chem & Colorists; Fiber Soc. Res: Surface phenomena; colloidal systems; textiles; synthetic fibers; polymers; adhesives; coatings; textile finishes. Mailing Add: Deering Milliken Res Corp PO Box 1927 Spartanburg SC 29304

OTTOBONI, FRED LAWRENCE, JR, b Santa Rosa, Calif, Nov 1, 27; m 65. OCCUPATIONAL HEALTH. Educ: Stanford Univ, BS, 51; Univ Calif, Berkeley, MPH, 63, PhD(environ health sci), 70. Prof Exp: Process engr petrol ref, Union Oil Co Calif, 51-58; indust hyg engr, Bur Occup Health, 58-67, coordr engr, Bur Occup Health & Environ Epidemiol, Calif Dept Health, 69-71, asst chief, 71, spec rep to state occup safety & health plan develop proj, 69-73; proj dir, Labor Occup Health grant, Inst Indust Relations, Univ Calif, Berkeley, 73-74; SPEC ASST TO GOV FOR OCCUP SAFETY & HEALTH, 75- Mem: Am Conf Govt Indust Hygienists; Am Indust Hyg Asn. Res: Environmental quality management; air sampling methods; occupational disease epidemiology; health effects of environmental agents. Mailing Add: Dept of Indust Relations 455 Golden Gate Ave San Francisco CA 94102

OTTOBONI, MINNA ALICE, b Perrin, Tex; m 65. TOXICOLOGY, BIOCHEMISTRY. Educ: Univ Tex, BA, 54; Univ Calif, Davis, PhD(comp biochem), 59. Prof Exp: Res chemist toxicol eval, Dried Fruit Asn, Calif, 59-63; STAFF TOXICOLOGIST, CALIF STATE DEPT HEALTH, 63- Concurrent Pos: Nat Inst Environ Health Sci res grant, Calif State Dept Pub Health, 67-72; collabr, Dried Fruit Asn Calif & Western Regional Res Lab, Agr Res Serv, USDA, 59-63. Mem: AAAS; Am Inst Biol Sci; NY Acad Sci; Am Conf Govt Indust Hygienists; Am Asn Lab Animal Sci. Res: Intermediary metabolism; comparative biochemistry; toxicity and metabolic fate of chemicals; acute and chronic toxicity, mode of action and metabolism of environmental chemicals. Mailing Add: Lab Serv Prog Calif State Dept of Health Berkeley CA 94704

OTTOLENGHI, ABRAMO CESARE, b Torino, Italy, Apr 13, 31; US citizen; m 58; c 4. MEDICAL MICROBIOLOGY. Educ: Wilmington Col, Ohio, BS, 50; Rutgers Univ, MS, 52; Univ Pa, PhD(med microbiol), 50. Prof Exp: Res asst preparation & potency testing of vaccines, Labs Life, Quito, Ecuador, 50-51 & 53-56 & Univ Pa, 56-59; res microbiologist, Sidney Hillman Med Ctr, Pa, 59-64; from asst prof to assoc prof, 64-72, PROF MED MICROBIOL, COL MED, OHIO STATE UNIV, 72- Concurrent Pos: Instr, Ogontz Campus, Pa State Univ, 58-59 & 60-64 & Lab Gen Biol, Rutgers Univ, 59-60; consult, South Div, Albert Einstein Med Ctr, 59-63. Mem: Am Soc Microbiol; NY Acad Sci; Sigma Xi. Res: Membrane antigens; bacterial exocellular products; phospholipids structure and function relating to host parasite relationships; antiphospholipid antibodies; medical education; membrane structure and function; electron microscopy. Mailing Add: Dept of Med Microbiol Ohio State Univ Columbus OH 43210

OTTOLENGHI, ATHOS, b Pavia, Italy, May 31, 23; US citizen; m 53; c 2. PHARMACOLOGY. Educ: Univ Pavia, MD, 46. Prof Exp: Assoc prof pharmacol, Univ Bari, 48-53; res assoc, 53-59, asst prof, 59-65, assoc prof, 65-75, PROF PHARMACOL, DUKE UNIV, 75- Concurrent Pos: Fel physiol, Univ Pavia, 46-48. Mem: Am Oil Chem Soc; Radiation Res Soc; Am Soc Pharmacol & Exp Therapeut. Res: Biochemistry of phospholipids and alteration of phospholipase activity in normal and experimental conditions; biochemistry of whole body irradiation in mammals; eosinophilic leukocytes and kinetics of bone marrow. Mailing Add: Dept of Physiol & Pharmacol Duke Univ Med Ctr Durham NC 27710

OTTOMAN, RICHARD EDWARD, b Guthrie, Okla, Aug 3, 10; m 43; c 1. RADIOLOGY. Educ: Univ Mich, MD, 41. Prof Exp: Staff physician, Birmingham Vet Admin Hosp, Van Nuys, Calif, 46-47, asst radiologist, 50; instr radiol, Sch Med, Univ Southern Calif, 50-52; from asst prof to prof radiol & anat, 52-71, vchmn dept radiol, 61-67, consult atomic energy proj, 52, EMER PROF RADIOL & ANAT, UNIV CALIF, LOS ANGELES, 71-; CHIEF RADIATION THER, VALLEY HOSP, 73- Concurrent Pos: From asst radiologist to chief therapeut radiol, Long Beach Vet Admin Hosp, 50-52, mem attend staff, 52-56; consult, St Joseph Hosp, Burbank, 50-; mem attend staff, St John's Hosp, Santa Monica, 50-60 & Valley Hosp, Van Nuys, 56; chief radiation ther, Santa Monica Hosp Med Ctr, 71-73. Mem: AAAS; Radiol Soc NAm; Soc Nuclear Med; AMA; fel Am Col Radiol. Res: Cancer; radiobiology. Mailing Add: Valley Hosp 14590 Sherman Circle Van Nuys CA 91405

OTTOSON, HAROLD, b Millston, Wis, Mar 21, 30; m 55; c 4. MATHEMATICS. Educ: Wis State Univ, BS, 51; Univ Wis, MS, 54. Prof Exp: Instr math, Tri-State Eng Col, 54-55; math analyst, Lockheed Aircraft Corp, 55-56; mathematician, Bendix Pac, Bendix Corp, 56-59; sr staff engr, TRW Comput Co, 59-62; mgr syst design dept, Data & Info Syst Div, Int Tel & Tel Corp, 62-64; sr assoc, Planning Res Corp, 64-66; mem res staff, 66-67, subdept head spec projs, 67-68, assoc mgr nat syst anal dept, 68-69, head data processing systs dept, 69-75, DEPT HEAD NAT SYSTS DESIGN DEPT, MITRE CORP, 75- Mem: Opers Res Soc Am; Asn Comput Mach. Res: Operations research; computer simulation and systems design; applied probability. Mailing Add: Mitre Corp Westgate Res Park McLean VA 22101

OTVOS, ERVIN GEORGE, b Budapest, Hungary, Mar 14, 35; m 66; c 2. SEDIMENTOLOGY, GEOMORPHOLOGY. Educ: Eötvös Lorand Univ, Dipl, 58; Yale Univ, MS, 62; Univ Mass, PhD(geol), 64. Prof Exp: Co geologist, Matra Mineral Mines Co, Hungary, 58-60; asst phys, hist & eng geol, Univ Mass, 62-64; geologist,

Brit Petrol Ltd, Can, 64; explor geologist, New Orleans Explor Div, Mobil Oil Corp, 65-68; asst prof, La State Univ, New Orleans, 68-71; HEAD GEOL DIV, GULF COAST RES LAB, 71- Concurrent Pos: Lectr, Univ Southern Miss Gulf Park Campus, 71-, adj prof, 74-; mem subcomn Americas, Int Asn Quaternary Res Comn on Shorelines, 73- Mem: Fel Geol Soc Am. Res: Sedimentary, geomorphological and stratigraphic aspects of coastal processes; Northeast Gulf of Mexico Coastal Plain stratigraphy and geomorphology; sedimentation and pollution problems of recent coastal water bodies; Coastal Mississippi Pleistocene stratigraphy. Mailing Add: Geol Div Gulf Coast Res Lab Ocean Springs MS 39564

OTVOS, JOHN WILLIAM, b Budapest, Hungary, Nov 26, 17; nat US; m 45; c 3. PHYSICAL CHEMISTRY. Educ: Harvard Univ, BS, 39; Calif Inst Technol, PhD(phys chem), 43. Prof Exp: Asst chem, Calif Inst Technol, 39-41; chemist, 46-52, asst head spectros dept, 52-57, head chem physics dept, 57-71, MGR ANAL RES, SHELL DEVELOP CO, 71- Mem: Am Chem Soc. Res: Kinetics; reaction mechanisms; radioisotopes; spectroscopy. Mailing Add: Shell Develop Co PO Box 1380 Houston TX 77001

OTWAY, HARRY JOHN, b Omaha, Nebr, Apr 23, 35; m 68. SCIENCE POLICY, SYSTEMS ANALYSIS. Educ: NDak State Univ, BS, 58; Univ NMex, MS, 61; Univ Calif, PhD(nuclear sci), 69. Prof Exp: Instr eng, NDak State Univ, 57-58; staff mem, Los Alamos Sci Lab, Univ Calif, 58-61; sect leader, 61-66; pilot & flight instr, Golden West Airline, Calif, 66-69; staff scientist, Los Alamos Sci Lab, Univ Calif, 69-72; first officer div nuclear safety & environ protection, 72-74, SR OFFICER & PROJ LEADER, JOINT IAEA/IIASA RES PROJ, INT ATOMIC ENERGY AGENCY, 74- Concurrent Pos: Mem appl ecol group steering comt, AEC, 70-72; reactor safety consult, A B Atomenergi, Sweden, 71-72; res scholar, Int Inst Appl Systs Anal, 74- Res: Assessment of environmental impact assessment; uses of peaceful nuclear explosives; social and psychological aspects of technological risks. Mailing Add: Joint IAEA/IIASA Res Proj Int Atomic Energy Agency Vienna Austria

OU, JONATHAN TSIEN-HSIONG, b Formosa, July 31, 34; m 62; c 2. MICROBIOLOGY, GENETICS. Educ: Nat Taiwan Univ, BS, 57; Univ Pa, PhD(biol), 67. Prof Exp: Res assoc microbial genetics, Univ Pa, 67-68; res assoc, 68-74, ASST MEM, INST CANCER RES, 74- Mem: Genetics Soc Am; Am Soc Microbiol. Res: The mechanism of bacterial conjugation in Escherichia coli and the interaction between bacteriophage and bacteria. Mailing Add: Inst for Cancer Res 7701 Burholme Ave Philadelphia PA 19111

OU, LO-CHANG, b Shanghai, China, Oct 16, 32; m 61; c 4. PHYSIOLOGY, BIOCHEMISTRY. Educ: Peking Univ, BSc, 54; Dartmouth Med Sch, PhD(physiol), 71. Prof Exp: Teaching asst biochem, Peking Univ, 54-59, lectr, 59-62; demonstr physiol, Univ Hong Kong, 62-64; res asst, 64-71, NAT HEART & LUNG INST RES ASSOC PHYSIOL, DARTMOUTH MED SCH, 71- Res: Adaptive changes and derangements of tissue metabolic functions in acclimatization to high altitude; role of the higher nervous system in the regulation of respiration at high altitude and sea level. Mailing Add: Dept of Physiol Dartmouth Med Sch Hanover NH 03755

OUANO, AUGUSTUS CENIZA, b Mandawe, Philippines, Mar 2, 36; US citizen; m 64; c 1. CHEMICAL ENGINEERING, POLYMER PHYSICS. Educ: Mapua Inst Technol, BSChE, 57; Purdue Univ, MSChE, 61; Stevens Inst Technol, PhD(chem eng), 69. Prof Exp: Develop engr, Continental Can Co, 61-64; res chemist, Rexall Chem Co, 64-66; res & develop engr, Packaging Corp Am, 66-67; MEM RES STAFF POLYMER CHARACTERIZATION, IBM CORP, 70- Mem: Am Chem Soc; Am Inst Physics. Res: Determination of molecular weight distribution and molecular weight average of polymers and the relationships between molecular characteristics of polymer to its synthesis and end use properties; diffusion and transport phenomenon in liquid systems. Mailing Add: 6133 Valley Glen Dr San Jose CA 95123

OUDERKIRK, JOHN THOMAS, b Amsterdam, NY, May 21, 31; m 53; c 2. ORGANIC CHEMISTRY, PHYSICAL CHEMISTRY. Educ: Hamilton Col, AB, 52; Cornell Univ, PhD(chem), 57. Prof Exp: Res chemist, Hooker Chem Corp, 55-58; res chemist, Toms River Chem Corp, 58-61, prod chemist, Chem Specialties Dept, 61-64; asst prod mgr, Ciba Corp, 64-65, sect leader, Textile & Paper Auxiliaries Div, 65-66; tech dir, NY Color & Chem Corp, Belleville, 66-67; res mgr, Chem Div, Sun Chem Corp, 67-76; WITH CHEMETRON CORP, 76- Mem: Am Chem Soc; Am Asn Textile Chemists & Colorists; The Chem Soc. Res: Reaction mechanisms; polyamides; polyesters; dyestuffs; pigments; textile and paper auxiliaries; instrumental analysis; fluorescence. Mailing Add: 268 Maple Ave Holland MI 49423

OUELLET, CYRIAS, physical chemistry, see 12th edition

OUELLET, LUDOVIC, b Tingwick, Que, Mar 27, 23; m 50; c 5. PHYSICAL CHEMISTRY. Educ: Laval Univ, BSc, 47, DSc(chem), 49. Prof Exp: Merck fel, Cath Univ Am, 50-51; USPHS fel, Naval Med Res Inst, 51-52 & trainee, Univ Wis, 52-53; from asst prof to assoc prof chem, Univ Ottawa, 53-58; chmn dept, 67-72, PROF CHEM, LAVAL UNIV, 60- Concurrent Pos: Pres, Res Comn, 72-; pres, Univ Res Comn, Univ Coun Quebec, 69-72; mem, Coun Sci Policy Quebec, 71-73. Mem: Am Chem Soc; Chem Inst Can. Res: Chemical kinetics; mechanism of reactions; especially enzymatic reactions; science policy. Mailing Add: Dept of Chem Laval Univ Quebec PQ Can

OUELLETTE, GERARD JOSEPH, b Ste Helene, Que, May 13, 22; m 48; c 3. SOILS, PLANT NUTRITION. Educ: Laval Univ, BA, 43, BSA, 47; Univ Wis, MSc, 48, PhD(soil fertil), 50. Prof Exp: SOIL SCIENTIST, RES BR, CAN DEPT AGR, 50-; ASSOC PROF SOILS, SCH AGR, LAVAL UNIV, 56-, PROF SOIL FERTIL, UNIV, 62-, ASSOC DEAN FAC AGR, 67- Mem: Soil Sci Soc Am; Am Soc Agron; Can Soil Sci Soc. Res: Soil fertility; trace elements in forage crops; acid-soil injury. Mailing Add: Fac of Agr & Food Sci Laval Univ Quebec PQ Can

OUELLETTE, GUILLEMOND BENOIT, b St Hilaire, NB, Jan 16, 32; m 54; c 5. HISTOPATHOLOGY. Educ: St Joseph's Col, NB, BA, 52; Cornell Univ, PhD(phytopath), 60. Prof Exp: FOREST PATHOLOGIST, CAN DEPT ENVIRON, 58- Concurrent Pos: Lectr, Laval Univ, 64- Mem: Can Bot Asn; Bot Soc Am; Can Phytopath Soc. Res: Host parasite relationships in Dutch elm disease; morphology, physiology and taxonomy of fungi; electron microscope on histopathology of Dutch elm disease. Mailing Add: Laurentian Forest Res Ctr PO Box 3800 Ste-Foy PQ Can

OUELLETTE, ROBERT J, b St Johnsbury, Vt, Dec 13, 38; m 59; c 3. ORGANIC CHEMISTRY. Educ: Univ Vt, BS, 59; Univ Calif, Berkeley, PhD(chem), 62. Prof Exp: From asst prof to assoc prof, 62-72, PROF CHEM, OHIO STATE UNIV, 72- Concurrent Pos: NSF grants, 65-71; Petrol Res Fund grants, 65-67 & 70-72. Mem: Am Chem Soc. Res: Organometallic chemistry; conformational analysis; nuclear magnetic resonance. Mailing Add: Dept of Chem Ohio State Univ Columbus OH 43210

OUIMET, ALFRED J, JR, b Wilmington, NC, Apr 7, 31; m 54; c 3. INFORMATION SCIENCE; SOLID STATE SCIENCE. Educ: Univ Conn, BA, 53, PhD(phys chem), 62. Prof Exp: Res chemist, Gen Elec Co, 53-54; asst instr chem, Univ Conn, 56-62; chemist, Esso Res & Eng Co, 62-66; ED CHEM LIT, CHEM ABSTR SERV, 66- Res: Kinetic study of the thermal decomposition of aluminum trimethyl; basic and products application research on aerospace lubricants. Mailing Add: Chem Abstr Serv Ohio State Univ Columbus OH 43210

OUJESKY, HELEN MATUSEVICH, b Ft Worth, Tex, Aug 14, 30; m 51; c 3. MICROBIOLOGY, RADIATION BIOLOGY. Educ: Tex Woman's Univ, BA & BS, 51, PhD(radiation biol), 68; Tex Christian Univ, MA, 65. Prof Exp: Teacher high sch, 51-63; asst prof biol, Tex Woman's Univ, 68-73; ASSOC PROF EARTH & LIFE SCI, UNIV TEX, SAN ANTONIO, 73- Mem: Am Inst Biol Sci; Am Soc Microbiol; Radiation Res Soc. Res: Effects of gases on DNA polymerase and microorganisms; incorporation of calcium 45 in fish scales; survey of plankton. Mailing Add: Dept of Earth & Life Sci Univ of Tex San Antonio TX 78284

OURECKY, DONALD K, b Sterling, Colo, Sept 11, 32. GENETICS, CYTOLOGY. Educ: Ore State Univ, BS, 54; Wash State Univ, MS, 57, PhD(genetics, hort), 61. Prof Exp: Res geneticist, US Dept Agr, 61-62; asst prof, 63-69, ASSOC PROF POMOL, NY STATE EXP STA, CORNELL UNIV, 69- Mem: Am Soc Hort Sci; Am Pomol Soc; Bot Soc Am; Genetics Soc Can. Res: Cytological studies of Sambucus, Rubus and Malus, especially phylogenetic and phylocytological relationships; applied plant breeding directed toward development of new varieties in Fragaria, Rubus, Vitis and Malus. Mailing Add: Dept of Pomol & Viticulture NY State Exp Sta Geneva NY 14456

OURSLER, CLELLIE CURTIS, b Cynthiana, Ind, Nov 26, 15; m 37; c 2. MATHEMATICS. Educ: Ind Univ, AB, 37; Univ Chicago, SM, 41; Ill Inst Technol, PhD(math), 58. Prof Exp: Teacher pub sch, Ill, 37-38, prin, 38-40; teacher, Ind, 40-49; instr math, Ind Univ, Northwest, 49-59; from asst prof to assoc prof, 59-72, PROF MATH, SOUTHERN ILL UNIV, EDWARDSVILLE, 72- Mem: Am Math Soc; Math Asn Am. Res: Abstract algebra; theory of numbers; finite geometries; quaternions; matrices with integer elements. Mailing Add: Sch of Sci & Technol Southern Ill Univ Edwardsville IL 62026

OUSEPH, PULLOM JOHN, b Kerala, India, Jan 21, 33; m; c 1. NUCLEAR PHYSICS. Educ: Univ Kerala, BSc, 53; Univ Saugor, MSc, 55; Fordham Univ, PhD(nuclear physics), 62. Prof Exp: Asst physics, Fordham Univ, 57-61; from asst prof to assoc prof, 62-70, PROF PHYSICS, UNIV LOUISVILLE, 70- Mem: Am Phys Soc; Am Asn Physics Teachers. Res: Low energy nuclear physics, especially angular correlation of successive radiations and Mössbauer effect. Mailing Add: Dept of Physics Univ of Louisville Louisville KY 40208

OUSTERHOUT, LAWRENCE ELWYN, b Deming, NMex, Jan 20, 22; m 45; c 7. NUTRITION, BIOCHEMISTRY. Educ: Ore State Univ, BS, 43; Univ Calif, Davis, PhD(nutrit), 59. Prof Exp: Biochemist, Tech Lab, Bur Com Fisheries, US Fish & Wildlife Serv, 59-61; mgr poultry nutrit, Quaker Oats Co, Ill, 61-66; from assoc prof to prof animal nutrit, Univ RI, 66-73; CONSULT POULTRY NUTRIT & MGT, 73- Mem: Am Inst Nutrit; Poultry Sci Asn. Res: Amino acid availability and protein quality, particularly with industrial fishery products; effects of processing on protein quality. Mailing Add: 751 E Second Ave Morton MS 39117

OUTCALT, DAVID L, b Los Angeles, Calif, Jan 30, 35; m 56; c 4. MATHEMATICS. Educ: Pomona Col, BA, 56; Claremont Grad Sch, MA, 58; Ohio State Univ, PhD(math), 63. Prof Exp: Asst instr math, Ohio State Univ, 60-62; from instr to asst prof, Claremont Men's Col, 62-64; from lectr to assoc prof, 64-72, chmn dept, 69-72, PROF MATH; UNIV CALIF, SANTA BARBARA, 72- Concurrent Pos: Co-investr, US Air Force Off Sci Res grant, 64-71, prin investr, 71-73; vis assoc prof, Univ Hawaii, 67-68 & vis prof 71-72; acad asst instructional develop, Univ Calif, Santa Barbara, 73- Mem: AAAS; Am Math Soc; Math Asn Am; Soc Indust & Appl Math; Sigma Xi. Res: Non-associative rings and algebras. Mailing Add: Dept of Math Univ of Calif Santa Barbara CA 93106

OUTCALT, SAMUEL IRVINE, b Oak Park, Ill, Aug 26, 36; m 59; c 3. PHYSICAL GEOGRAPHY. Educ: Univ Cincinnati, BA, 59; Univ Colo, MA, 64; Univ BC, PhD(geog), 70. Prof Exp: Actg asst prof environ sci, Univ Va, 68-70; asst prof geog, 70-74, ASSOC PROF GEOG, UNIV MICH, ANN ARBOR, 74- Concurrent Pos: Res assoc, Arctic Inst NAm, 70; consult, Geog Applications Prog-US Geol Surv, 70-; res geogr, Infrared & Optics Div, Willow Run Labs, 72; partic, US-Int Biol Prog-Aerobiol Prog, 72- & US-Int Biol Prog-Tundra Biome, 72- Mem: Am Geophys Union; Glaciol Soc; Asn Am Geog. Res: Interaction between surface climatological conditions and geomorphic evolution in Arctic-Alpine terrain; thermal modeling; computer applications in physical geography; urban climatology. Mailing Add: Dept of Geog Univ of Mich Ann Arbor MI 48104

OUTERBRIDGE, JOHN STUART, b Hunan, China, Sept 20, 36; m 63; c 3. PHYSIOLOGY, OTOLOGY. Educ: Mt Allison Univ, BS, 57, BA, 58; McGill Univ, MD, CM, 62, PhD(physiol), 69. Prof Exp: Intern, Royal Victoria Hosp, Montreal, 62-63, jr asst resident med, 63-64; lectr physiol, Biomed Eng Unit, 67-69, asst prof otolaryngol, 70-73, ASSOC PROF OTOLARYNGOL, McGILL UNIV, 73-, ASST PROF PHYSIOL, 69-, DIR OTOLARYNGOL RES LABS, 70-, ACTG DIR BIOMED ENG UNIT, 74- Concurrent Pos: Med Res Coun Can grant, 69- & scholar, 70-; Defense Res Bd Can grant, 68-; asst otolaryngologist, Royal Victoria Hosp, 70- Mem: Inst Elec & Electronics Eng; Biomed Eng Soc. Res: Vistibular and oculomotor systems; vestibular control of head and eye movements; diagnostic evaluation of vestibular systems; biomedical engineering; mathematical biology. Mailing Add: Dept of Otolaryngol McGill Univ PO Box 6070 Montreal PQ Can

OUTHOUSE, JAMES BURTON, b Canandaigua, NY, Sept 20, 16; m 40; c 3. ANIMAL SCIENCE. Educ: Cornell Univ, BS, 38; Univ Md, MS, 42; Purdue Univ, PhD, 56. Prof Exp: Instr & asst animal husb, Univ Md, 38-41, from asst prof to assoc prof, 41-52; from instr to assoc prof, 52-60, PROF ANIMAL HUSB, PURDUE UNIV, 60- Concurrent Pos: Vis prof, Univ Edinburgh, 70-71. Mem: AAAS; Am Soc Animal Sci; Brit Soc Animal Prod; Sigma Xi. Res: Nutrition studies with sheep and beef cattle; distillers' rye dried grains for pregnant ewes and for fattening steers; corn silage for pregnant ewes and for wintering beef cows. Mailing Add: Dept of Animal Sci Purdue Univ West Lafayette IN 47907

OUTKA, DARRYLL E, b Grand Island, Nebr, Feb 25, 29; m 52; c 2. PROTOZOOLOGY, CELL BIOLOGY. Educ: San Diego State Col, BS, 51; Univ Calif, Berkeley, MA, 59, PhD(zool), 62. Prof Exp: Asst prof biol, Calif State Col Long Beach, 62-66; asst prof, 66-70, ASSOC PROF CELL BIOL, IOWA STATE UNIV, 70- Concurrent Pos: NIH res grant, 63-66; NSF res grant, 65-67. Mem: Soc Protozool (treas, 69-72). Res: Morphogenesis in protozoa; genetics of ciliates; transformation in amoeboflagellates; electron microscopy of protozoa. Mailing Add: Dept of Cell Biol Iowa State Univ Ames IA 50010

OUTLAND, RODERICK HENDERSON, b Hamlin, Ky, Oct 30, 22; m 60. CYTOLOGY. Educ: Murray State Col, BS, 47; Vanderbilt Univ, MA, 49, PhD(biol, cytol), 53. Prof Exp: Instr biol, Paducah Jr Col, Ky, 47-48 & Gettysburg Col, 49-50; from asst prof to assoc prof, 53-66, PROF BIOL, NORTHWESTERN STATE UNIV, 66- Concurrent Pos: Asst prof, Loyola Univ, La, 54-56. Mem: Bot Soc Am. Res: Cytology and physiology of Louisiana irises. Mailing Add: Dept of Biol Northwestern State Univ Natchitoches LA 71457

OUTLAW, HENRY EARL, b Pickwick, Tenn, June 17, 37; m; c 1. BIOCHEMISTRY, MOLECULAR BIOLOGY. Educ: Delta State Col, BS, 61; Univ Miss, MS, 64, PhD(pharm), 65. Prof Exp: Fel pharmacol, Sch Med, Univ Fla, 65-66; asst prof, 66-74, PROF CHEM, DELTA STATE UNIV, 74-, HEAD DEPT PHYS SCI, 69- Mem: Am Chem Soc. Res: Chemistry of muscle contraction; isolation and biochemical characterization of subcellular constituents. Mailing Add: Dept of Phys Sci Delta State Univ Cleveland MS 38732

OUTTEN, LORA MILTON, b Pocomoke City, Md, Aug 17, 13. ZOOLOGY. Educ: Western Md Col, AB, 34, AM, 37; Cornell Univ, MS, 50, PhD(vert zool, ichthyol, entom, limnol), 56. Prof Exp: Lab asst biol, Western Md Col, 33-34 & 36-37; instr high schs, Md & Va, 34-36, 40-41 & 43-46; PROF BIOL, MARS HILL COL, 46-, CHMN ECOL RES, 74- Mem: Fel AAAS; Am Soc Zool; Am Soc Naturalists; Am Soc Ichthyol & Herpet; Ecol Soc Am. Res: Life history, ecology, habits, behavior, development, morphology and distribution of fishes and other vertebrates; effects of insecticides on honeybees; aquatic insects; limnology; vertebrate zoology; distribution of fishes in southeastern United States, Bahamas, Caribbean area, Panama, Bermuda, Britain, Hawaii, Australia and New Zealand. Mailing Add: Dept of Biol Mars Hill Col Mars Hill NC 28754

OUTZEN, HENRY CLAIR, JR, b Salt Lake City, Utah, Sept 6, 39. ANATOMY, CANCER IMMUNOLOGY. Educ: Univ Utah, BS, 61, PhD(anat), 69. Prof Exp: ASSOC CANCER IMMUNOL, INST CANCER RES, 69- Concurrent Pos: Asst ed, J Reticuloendothelial Soc; mem comt care & use of nude mouse, Inst Lab Animal Resources, Nat Acad Sci-Nat Res Coun, 74-76. Mem: AAAS; Reticuloendothelial Soc; Am Asn Anat; Am Soc Zool; NY Acad Sci. Res: Germfree biology. Mailing Add: Inst for Cancer Res 7701 Burholme Ave Philadelphia PA 19111

OUZTS, JOHNNY DREW, b Indianola, Miss, Nov 29, 34; m 57; c 4. MEDICAL ENTOMOLOGY. Educ: Delta State Col, BS, 57; Miss State Univ, MS, 61, PhD(entom), 63. Prof Exp: Actg head dept & asst dean, Sch Arts & Sci, 63-74, PROF BIOL SCI, DELTA STATE COL, 63-, HEAD DEPT & ASSOC DEAN, SCH ARTS & SCI, 74- Mem: Entom Soc Am. Res: Chlorinated hydrocarbon resistance in fresh and flood-water; breeding mosquitoes; biological control of cotton insects; practical entomology for student agricultural pilots. Mailing Add: Dept of Biol Delta State Univ Cleveland OH 38732

OVADIA, JACQUES, b Vienna, Austria, Nov 16, 23; nat US; m 56; c 2. MEDICAL PHYSICS. Educ: Brooklyn Col, BA, 44; Univ Ill, MS, 47, PhD(nuclear physics), 51. Prof Exp: Instr radiol, Univ Ill, 51-52; res assoc biophys, Sloan-Kettering Inst, 52-56; chief physicist radiation ther, Michael Reese Hosp, Chicago, 56-72; prof radiol, Med Col Wis, 72; CHMN MED PHYSICS, MICHAEL REESE HOSP, 73-; PROF RADIOL, UNIV CHICAGO, 74- Concurrent Pos: Adj prof physics, Ill Inst Technol, 72- Mem: Health Physics Soc; Am Soc Ther Radiol; Radiation Res Soc; Radiol Soc NAm; Am Asn Physicists in Med. Res: High energy electrons and x-rays for therapy; short-lived isotope tracers; mechanism of action of radiation; pulse radiolysis. Mailing Add: 1722 East 55th St Chicago IL 60615

OVALLE, WILLIAM KEITH, b Ancon, CZ, Mar 18, 44; US citizen. MEDICAL SCIENCE. Educ: St Joseph's Col, Pa, BS, 66; Temple Univ, PhD(anat), 71. Prof Exp: Muscular Dystrophy Asn Can fel neurophysiol, Univ Alta, 70-72; ASST PROF ANAT, FAC MED, UNIV BC, 72- Mem: Am Asn Anatomists; Am Soc Cell Biol; AAAS; Can Asn Anatomists; Sigma Xi. Res: Neuromuscular ultrastructure and functional morphology; electron microscopy and histochemistry of the peripheral nervous system during development. Mailing Add: Dept of Anat Fac of Med Univ of BC Vancouver BC Can

OVARY, ZOLTAN, b Kolosvar, Hungary, Apr 13, 07. IMMUNOLOGY. Educ: Univ Paris, MD, 35, Lic es Sci, 39. Prof Exp: Res assoc microbiol, Sch Med, Johns Hopkins Univ, 55-59; from asst prof to assoc prof path, 59-64, PROF PATH, SCH MED, NY UNIV, 64- Concurrent Pos: Res fel, Brazilian Super Coun Res, 52; Fulbright fel, 54-55. Mem: Am Soc Immunol; Soc Exp Biol & Med; Am Acad Allergy; Harvey Soc; fel NY Acad Sci. Res: Hypersensitivity; anaphylaxis; antibody structure; antigenicity; function of the reticuloendothelial system. Mailing Add: Dept of Path NY Univ Sch of Med New York NY 10016

OVE, PETER, b Dellstedt, Ger, Aug 31, 30; US citizen; m 56; c 3. CELL BIOLOGY, BIOCHEMISTRY. Educ: Univ Pittsburgh, BS, 63, PhD(cell biol), 67. Prof Exp: Asst prof clin pharmacol, Med Ctr, Duke Univ, 68-69; asst prof, 69-73, ASSOC PROF CELL BIOL, SCH MED, UNIV PITTSBURGH, 73- Mem: Am Asn Cancer Res; Am Soc Cell Biol. Res: Control of growth of mammalian cells; cancer problems; biology of aging. Mailing Add: Dept of Anat & Cell Biol Univ of Pittsburgh Sch of Med Pittsburgh PA 15213

OVENALL, DERICK WILLIAM, b Bloxwich, Eng, Sept 26, 30; m 65; c 2. PHYSICAL CHEMISTRY. Educ: Univ Manchester, BSc, 52; Univ Birmingham, PhD(chem), 55. Prof Exp: Res fel chem, Univ Birmingham, 55-57; res assoc physics, Duke Univ, 57-59; sr sci officer, Nat Phys Lab, Eng, 59-60; physicist, Battelle Mem Inst, Switzerland, 60-61; res chemist, Plastics Dept, 62-71, RES CHEMIST, CENT RES DEPT, EXP STA, E I DU PONT DE NEMOURS & CO, INC, 71- Mem: Am Chem Soc. Res: Spectroscopy; high resolution nuclear magnetic resonance; electron spin resonance; radiation damage. Mailing Add: Cent Res Dept Exp Sta E I du Pont de Nemours & Co Inc Wilmington DE 19898

OVENBURG, PETER HAYWARD, genetics, biometry, see 12th edition

OVENDEN, MICHAEL W, b London, Eng, May 21, 26; m 71. ASTRONOMY, ASTROPHYSICS. Educ: Univ London, BSc, 47, PhD(astrophys), 55; Cambridge Univ, MA, 51. Prof Exp: Jr observer astron, Cambridge Observ, 48-52; from lectr to sr lectr, Univ Glasgow, 53-66; prof astron, 66-74, PROF GEOPHYS & ASTRON, UNIV BC, 74- Concurrent Pos: Tech officer, Dom Astrophys Observ, Victoria, BC, 61. Mem: Am Astron Soc; Brit Astron Soc (secy, 46-51, vpres, 51-53); fel Royal Astron Soc (secy, 57-66, vpres, 66); fel Royal Soc Edinburgh; fel Brit Interplanetary Soc. Res: Dynamics of the solar system; physics of close binary stars; history of astronomy. Mailing Add: Dept of Geophys & Astron Univ of BC Vancouver BC Can

OVENFORS, CARL-OLOF NILS STEN, b Stockholm, Sweden, Sept 26, 23; c 2. RADIOLOGY. Educ: Karolinska Inst, Sweden, BS, 44, MD, 51. Prof Exp: Asst dir thoracic radiol, Univ Hosp of Karolinska, Stockholm, 62-69, dir thoracic radiol, 69-70; assoc prof, 70-73, PROF RADIOL, UNIV CALIF, SAN FRANCISCO, 73- Concurrent Pos: James Picker Found scholar, 61-62, grant, Karolinska Inst, Sweden, 66-68; radiol res grants, 62-68; consult radiol, Swedish Armed Forces; mem, Coun for Cardiac Radiol. Mem: Radiol Soc NAm; Asn Univ Radiol; NAm Soc Cardiac Radiol; Am Thoracic Soc; Am Inst Ultrasonics in Med. Res: Cardiovascular and pulmonary radiology. Mailing Add: Vet Admin Hosp Radiol Dept 4150 Clement St San Francisco CA 94121

OVENSHINE, A THOMAS, b New York, NY, Mar 25, 36; m 59; c 3. GEOLOGY. Educ: Yale Univ, BS, 58; Va Polytech Univ, MS, 62; Univ Calif, Los Angeles, PhD(geol), 65. Prof Exp: GEOLOGIST, US GEOL SURV, 65- Mem: Geol Soc Am. Res: Pre-Pleistocene glacial deposits; sedimentary structures; Silurian-Devonian conodonts; geology of southeastern Alaska. Mailing Add: US Geol Surv 345 Middlefield Rd Menlo Park CA 94025

OVERBECK, HENRY WEST, b Chicago, Ill, Oct 29, 30; m 60; c 2. MEDICAL PHYSIOLOGY, INTERNAL MEDICINE. Educ: Princeton Univ, BA, 52; Northwestern Univ, MD, 56; Univ Okla, PhD(physiol), 66; Am Bd Internal Med, cert. Prof Exp: Intern, Mary Fletcher Hosp, Univ Vt, 56-57; resident physician, Sch Med, Northwestern Univ, 59-62; instr internal med, Sch Med, Univ Okla, 63-66; from asst prof to assoc prof physiol & internal med, 66-71, PROF PHYSIOL & INTERNAL MED, MICH STATE UNIV, 71- Concurrent Pos: USPHS career develop award, 66-71; Vet Admin clin investr, 63-66; consult, Vet Admin Hosp, Saginaw, Mich, 66-73; mem med adv bd coun high blood pressure, Am Heart Asn. Mem: Cent Soc Clin Res; Soc Exp Biol & Med; Am Soc Clin Pharmacol & Therapeut; Am Physiol Soc; fel Am Col Physicians. Res: Cardiovascular physiology and pathophysiology, especially hypertension. Mailing Add: Dept of Physiol Giltner Hall Mich State Univ East Lansing MI 48824

OVERBERG, RICHARD JOSEPH, b Indianapolis, Ind, Oct 21, 29; m 54; c 5. POLYMER CHEMISTRY. Educ: Xavier Univ, BS, 51, MS, 53. Prof Exp: Jr chemist, Hilton Davis Chem Co Div, Sterling Drug, Inc, 53; gen chemist, Nat Bur Standards, 54-62; sr scientist, Devro Div, Johnson & Johnson, 62-70; supvr, Celanese Plastics Co, 70, Celanese Res Co, 71-73; SR RES SCIENTIST, TEX-US CHEM CO, 73- Mem: AAAS; Am Inst Chem; Am Chem Soc; Soc Rheol. Res: Rheology; plastics and rubbers; polycyclic organic synthesis; synthetic polymer structure and physical properties; protein structure and physical properties. Mailing Add: 625 Cedarbrook Rd Bridgewater NJ 08807

OVERBERGER, CHARLES GILBERT, b Barnesboro, Pa, Oct 12, 20; m 45; c 4. POLYMER CHEMISTRY. Educ: Pa State Col, BS, 41; Univ Ill, PhD(org chem), 44. Hon Degrees: DSc, Holy Cross Col, 66; DSc, L I Univ, 68. Prof Exp: Res chemist, Allied Chem Co, Pa, 41; asst org chem, Univ Ill, 41-43; assoc, Rubber Reserve Corp, 44-46; Du Pont fel, Mass Inst Technol, 46-47; from asst prof to assoc prof chem, Polytech Inst Brooklyn, 47-52, prof org chem, 52-66, assoc dir polymer res inst, 51-64, dir, 64-66, head dept chem, 55-64, vpres res, 64-65, dean sci, 64-66; chmn dept chem, 67-72, DIR MACROMOLECULAR RES CTR, UNIV MICH, 68-, VPRES RES, 72- Concurrent Pos: Consult, Armed Forces; ed, J Polymer Sci & Fortschritte der Hochpolymerenforschung; pres macromolecular div, Int Union Pure & Appl Chem, 75-79. Mem: AAAS; Am Chem Soc (pres, 67); Am Inst Chem; Brit Chem Soc; Brit Soc Chem Indust. Res: Organic synthesis; mechanisms of organic reactions; the use of polymers as reactants in organic reactions; synthesis and properties of asymmetric polymers in solution; chemistry of diazines and decomposition of cyclic azo compounds. Mailing Add: Dept of Chem Univ of Mich Ann Arbor MI 48104

OVERBERGER, JAMES EDWIN, b Barnesboro, Pa, July 1, 30; m 58; c 2. DENTISTRY. Educ: Univ Pittsburgh, BS, 52, DDS, 54; Univ Mich, MS, 57. Prof Exp: Clin instr human resources, Univ Mich, 56-57; asst prof dent mat, Sch Dent, WVa Univ, 57-62, assoc prof prosthodont, 62-67; assoc prof, Sch Dent, Univ NC, Chapel Hill, 67-69; PROF DENT MAT & ASSOC DEAN SCH DENT, W VA UNIV, 69- Mem: Fel Am Col Dent; Am Dent Asn; Int Asn Dent Res. Res: Dental materials; restorative dentistry. Mailing Add: WVa Univ Sch of Dent Morgantown WV 26506

OVERBY, LACY RASCO, b Model, Tenn, July 27, 20; m 48; c 4. BIOCHEMISTRY, VIROLOGY. Educ: Vanderbilt Univ, BA, 41, MS, 48, PhD(chem), 50. Prof Exp: Control & prod supvr, E I du Pont de Nemours & Co, 41-43; asst, Vanderbilt Univ, 46-48; sect mgr nucleic acid res, Abbott Labs, 49-62; vis scholar microbiol, Univ Ill, 62-64; mgr dept virol, 69-73, DIR DIV EXP BIOL, ABBOTT LABS, 73- Concurrent Pos: Assoc res fel, Abbott Labs, 64-69; lectr, Northwestern Univ, 68- Mem: AAAS; Am Chem Soc; Am Inst Nutrit; Am Soc Biol Chem; Am Soc Microbiol. Res: Amino acids and proteins; chemistry and biochemistry of vitamins and growth factors; viruses and nucleic acids; molecular biology. Mailing Add: Div of Exp Biol Abbott Labs North Chicago IL 60064

OVERCASH JEAN PARKS, b Kannapolis, NC, Aug 15, 17; m 44; c 3. HORTICULTURE. Educ: NC State Univ, BS, 38; Univ Wis, MS, 39, PhD(hort, plant physiol), 41. Prof Exp: Asst res horticulturist, Exp Sta, Univ Tenn, 42-45, assoc res horticulturist, 45; assoc prof & assoc horticulturist, 45-59, PROF & HORTICULTURIST, MISS STATE UNIV, 59- Mem: Am Soc Hort Sci; Int Soc Hort Sci; Am Pomol Soc. Res: Physiology of tree fruits, especially pecans, grapes and peaches; rest period of peaches and grapes. Mailing Add: Agr & Forestry Exp Sta Miss State Univ PO Drawer T Mississippi State MS 39762

OVERCAST, WOODROW WEBB, b Chinook, Mont, Feb 10, 18; m 46; c 4. DAIRY BACTERIOLOGY. Educ: Mont State Col, BS, 40; Univ Tenn, MS, 41; Iowa State Univ, PhD(dairy bact), 50. Prof Exp: From instr to assoc prof, 46-55, PROF DAIRYING, UNIV TENN, KNOXVILLE, 55- Concurrent Pos: Asst, Iowa State Univ, 50. Mem: Am Soc Microbiol; Am Dairy Sci Asn. Res: Bacteriology and quality of dairy products; psychrophilic bacteria. Mailing Add: Dept of Food Technol & Sci Univ of Tenn Knoxville TN 37901

OVERDAHL, CURTIS J, b Elk Point, SDak, June 24, 17; m 53; c 3. SOILS. Educ: Univ Minn, BS, 47, MS, 49; Purdue Univ, PhD(soils), 57. Prof Exp: Exten area agronomist, Iowa State Univ, 49-53; exten agronomist, Purdue Univ, 53-54; EXTEN SPECIALIST SOILS, UNIV MINN, MINNEAPOLIS, 56- Mem: Am Soc Agron. Res: Soil management and fertilizers; agricultural extension service. Mailing Add: Dept of Soils Exten Div Univ of Minn Minneapolis MN 55455

OVEREND, JOHN, b Keighley, Eng, Oct 3, 28; m 54; c 3. PHYSICAL CHEMISTRY. Educ: Oxford Univ, BA, 52, DPhil(chem), 56. Prof Exp: Res fel, Univ Minn, 55-58; res chemist, Dow Chem Co, 58-60; from asst prof, to assoc prof, 60-67, PROF CHEM, UNIV MINN, MINNEAPOLIS, 67- Concurrent Pos: Guggenheim fel, 67-68; vis staff mem, Los Alamos Sci Lab, 74- Honors & Awards: Coblentz Mem Prize, Coblentz Soc, 64. Mem: Am Phys Soc; Am Chem Soc; Optical Soc Am; The Chem Soc. Res: Molecular structure and dynamics; molecular spectra; surface chemistry; infrared circular dichroism. Mailing Add: Dept of Chem Univ of Minn Minneapolis MN 55455

OVERHAGE, CARL F J, b London, Eng, Apr 2, 10; nat US; m 40. EXPERIMENTAL PHYSICS. Educ: Calif Inst Technol, BS, 31, MS, 34, PhD(physics), 37. Prof Exp:

Physicist, Technicolor Motion Picture Corp, 37-40, actg dir res, 40-42; mem staff, Radiation Lab, Mass Inst Technol, 42-45; res supvr, Color Control Dept, Eastman Kodak Co, 46-48, asst dir, Color Tech Div, 49-54; div head, Lincoln Lab, 55-57, dir, 57-64, prof eng, 61-73, EMER PROF ENG, MASS INST TECHNOL, 73- Concurrent Pos: Sci consult, US Army Air Force, 45; mem sci adv bd, US Air Force, 51-57, sci adv comt ballistic missiles, Secy Defense, 59-61; mem, Defense Sci Bd, 62-63; tech adv bd, Fed Aviation Agency, 62-65; mem, Nat Adv Comn on Libr, 66-67; exec dir, Univ Info Technol Corp, 68-73. Honors & Awards: Presidential Cert Merit, 48; Exceptional Service Award, US Air Force, 58. Mem: Fel Am Acad Arts & Sci (vpres, 64-66); fel Am Phys Soc; fel Optical Soc Am. Res: Photography; electronics. Mailing Add: 112 Calle Cataline Santa Fe NM 87501

OVERHAUSER, ALBERT WARNER, b San Diego, Calif, Aug 17, 25; m 51; c 8. THEORETICAL SOLID STATE PHYSICS. Educ: Univ of Calif, AB, 48, PhD(physics), 51. Prof Exp: Res assoc, Univ Ill, 51-53; from asst prof to assoc prof physics, Cornell Univ, 53-58; supvr, Solid State Physics, Ford Motor Co, 58-62, mgr, Theoret Sci, 62-69, asst dir, Phys Sci, 69-72, dir, Phys Sci Lab, 72-73; prof, 73-74, STUART PROF PHYSICS, PURDUE UNIV, 74- Honors & Awards: Oliver E Buckley Solid State Physics Prize, Am Phys Soc, 75. Mem: Nat Acad Sci; Am Phys Soc. Res: Many-electron theory; transport theory; neutron interferometry. Mailing Add: Dept of Physics Purdue Univ West Lafayette IN 47907

OVERHOLSER, MILTON DAVID, b Harrisonville, Mo, Dec 16, 00; m 26; c 2. ANATOMY. Educ: Univ Mo, AB, 23, AM, 24; NY Univ, PhD(anat), 28, MD, 31. Prof Exp: Asst anat, Univ Mo, 22-24, instr histol, 24-25; instr histol, NY Univ, 25-27, gross anat & neuroanat, 27-28, gross anat, 28-29, histol & embryol, 29-30; from asst prof to prof, 31-70, chmn dept, 33-66, EMER PROF ANAT, SCH MED, UNIV MO-COLUMBIA, 70- Mem: AAAS; Harvey Soc; Am Asn Anat; Soc Exp Biol & Med. Res: Relation of sex hormones to new growths of genital tract; endocrinology; hydrocephalus. Mailing Add: Dept of Anat Sch of Med Univ of Mo Columbia MO 65201

OVERHOLT, JOHN LOUGH, b Estherville, Iowa, May 28, 09; m 39; c 3. OPERATIONS RESEARCH. Educ: Iowa State Col, BS, 32; Lehigh Univ, MS, 35. Prof Exp: Res investr, NJ Zinc Co, 34-55; mem Opers Eval Group, Mass Inst Technol, 55-62; mem, franklin Inst, 62-67, mem, 67-75, CONSULT CTR NAVAL ANAL, UNIV ROCHESTER, 75- Mem: AAAS; Am Chem Soc; Opers Res Soc Am; Am Statist Asn. Res: Interior paint formulation; fundamental drying oil research; heavy chemical processes; titanium compounds and processes; statistics and experimental design; military operations research on air, sea and land problems. Mailing Add: Rte 1 Box 173 Kilmarnock VA 22482

OVERHULTS, WENDELL CLARENCE, physical chemistry, organic chemistry, see 12th edition

OVERLEASE, WILLIAM R, b Elkhart, Ind, Oct 2, 25; m 55. ECOLOGY, BOTANY. Educ: Mich State Univ, BS & BSF, 50, MS, 52, PhD(bot), 64. Prof Exp: Naturalist, State Park Serv, Ind Dept Conserv, 51-56; educator natural hist, Acad Natural Sci, Philadelphia, 56-59; instr natural sci, Mich State Univ, 60-62; assoc prof sci, 64-65, PROF BIOL, WEST CHESTER STATE COL, 65- Concurrent Pos: Consult outdoor ed, Primary & Sec Schs & Local & State Park Systs. Mem: AAAS; Ecol Soc Am. Res: Flora of Northwestern Michigan, Benzie County; early plant succession on abandoned fields, Southeastern Pennsylvania; bird population studies, Pennsylvania, Michigan. Mailing Add: Dept of Biol West Chester State Col West Chester PA 19380

OVERLEY, JACK CASTLE, b Kalamazoo, Mich, Aug 23, 32; m 59. NUCLEAR PHYSICS. Educ: Mass Inst Technol, BS, 54; Calif Inst Technol, PhD(physics), 61. Prof Exp: Res assoc physics, Calif Inst Technol, 60-61 & Univ Wis, 61-63; asst prof, Yale Univ, 63-68; asst dean sci serv, Col Lib Arts, 73-75, ASSOC PROF PHYSICS, UNIV ORE, 68- Concurrent Pos: Asst dir, A W Wright Nuclear Structure Lab, Yale Univ, 66-68; vis assoc prof physics, Calif Inst Technol, 75-76. Mem: AAAS; Am Phys Soc. Res: Experimental nuclear physics, primarily nuclear spectroscopy and reactions involving charged particles and neutrons. Mailing Add: Dept of Physics Univ of Ore Eugene OR 97403

OVERMAN, ALLEN RAY, b Goldsboro, NC, Mar 11, 37; m 64; c 1. AGRONOMY. Educ: NC State Univ, BS, 60, MS, 63, PhD(agr eng), 65. Prof Exp: Res assoc agron, Univ Ill, Urbana, 65-69; asst prof agr eng & asst agr engr, 69-74, ASSOC PROF AGR & ASSOC AGR ENGR, UNIV FLA, 74- Mem: Am Soc Agr Eng; Am Soc Agron. Res: Ion and water transport in porous media and biological membranes. Mailing Add: Dept of Agr Eng Univ of Fla Gainesville FL 32601

OVERMAN, AMEGDA JACK, b Tampa, Fla, May 17, 20; m 53. PLANT NEMATOLOGY. Educ: Univ Tampa, BS, 42; Univ Fla, MS, 51. Prof Exp: Asst in soils chem, 51-56, asst soils microbiologist, 56-68, assoc nematologist, 68-73, NEMATOLOGIST, INST FOOD & AGR SCI, AGR RES & EDUC CTR, UNIV FLA, 73- Mem: Soc Nematol; Am Phytopath Soc; Orgn Trop Am Nematol. Res: Bionomics, pathogenicity and control of nematodes attacking vegetables, ornamentals and agronomic crops. Mailing Add: Agr Res & Educ Ctr 5007 60th St E Bradenton FL 33505

OVERMAN, DENNIS ORTON, b Union City, Ind, Oct 16, 43; m 66; c 3. TERATOLOGY. Educ: Bowling Green State Univ, BA, 65; Univ Mich, MS, 67, PhD(anat), 70. Prof Exp: Res assoc develop biol, Univ Colo, 70-71; from instr to asst prof, 71-76, ASSOC PROF ANAT, WVA UNIV, 76- Mem: Teratology Soc; Am Asn Anatomists; Soc Develop Biol. Res: Mechanisms and inhibition of teratogenesis in orofacial malformations and limb development. Mailing Add: Dept of Anat WVa Univ Morgantown WV 26506

OVERMAN, JOSEPH DEWITT, b Champaign, Ill, Feb 23, 18; m 45; c 4. PHYSICAL CHEMISTRY. Educ: Univ Ill, BA, 41; Univ Rochester, PhD(chem), 45. Prof Exp: RES CHEMIST, E I DU PONT DE NEMOURS & CO, INC, PARLIN, NJ, 45- Mem: Am Chem Soc. Res: Reaction kinetics; photochemistry; photographic emulsions; reaction rate of Claisen rearrangement; photochemical decomposition of unsymmetrical dimethylhydrazine. Mailing Add: 115 Broadbent Rd Northminster Wilmington DE 19810

OVERMAN, LARRY EUGENE, b Chicago, Ill, Mar 9, 43; m 66. BIO-ORGANIC CHEMISTRY. Educ: Earlham Col, BA, 65; Univ Wis-Madison, PhD(org chem), 69. Prof Exp: NIH fel, Columbia Univ, 69-71; Cottrell res grant, Res Corp, 71; ASST PROF CHEM, UNIV CALIF, IRVINE, 71- Mem: AAAS; Am Chem Soc. Res: Mechanism of enzyme action; selective intracomplex reactions. Mailing Add: Dept of Chem Univ of Calif Irvine CA 92664

OVERMAN, RALPH THEODORE, b Clifton, Ariz, Aug 9, 19; m 45, 71; c 2. MEDICAL EDUCATION. Educ: Kans State Teachers Col, AB, 39, MS, 40; La State Univ, PhD(phys chem), 43. Hon Degrees: DSc, Philadelphia Col Pharm, 59. Prof Exp: Plant chemist, Flintkote Co, La, 42; instr chem, La State Univ, 42-43; head dept sci, La Col, 43-44; tech supt, Fercleve Corp, Tenn, 44-45; sr res chemist, Oak Ridge Nat Labs, 45-48, chmn spec training div, Oak Ridge Inst Nuclear Studies, 48-65; consult, Ralph T Overman Consult Serv, 65-68; dir educ, Technicon Corp, 68-69; vpres, Universal Med Servs, 70; planning dir, Bistate Regional Med Prog, 70-74; DIR REGIONAL MED EDUC CTR, VET ADMIN HOSP, ST LOUIS, 74- Concurrent Pos: Assoc prof, Sch Med, St Louis Univ; lectr, Sch Med, Wash Univ. Mem: Fel AAAS; Health Phys Soc; Am Chem Soc; Sigma Xi; Radiation Res Soc. Res: Radiochemistry; nuclear chemistry; health care planning. Mailing Add: Regional Med Educ Ctr 152JB Vet Admin Hosp St Louis MO 63125

OVERMAN, RICHARD ROLL, b Richmond, Ind, Nov 10, 16; m 42; c 2. PHYSIOLOGY. Educ: DePauw Univ, AB, 39; Princeton Univ, MA, 42, PhD(biol), 43. Prof Exp: Instr physiol, Col Physicians & Surgeons, Columbia Univ, 43-45; from instr to assoc prof, Col Med, 45-53, prof clin physiol & chmn dept, 53-66, asst dean res affairs, 64-66, prof radiation biol & chmn div, 64-67, assoc dean col med, 66-72, actg dean, 70-72, vchancellor acad affairs, Med Units, 72-73, vchancellor, Res, 73-75, DISTINGUISHED UNIV PROF, UNIV TENN, MEMPHIS, 75- Concurrent Pos: Consult, John Gaston Hosp & Kennedy Vet Admin Hosp, Memphis; mem cardiovasc study sect, NIH, 58- Mem: AAAS; Am Physiol Soc; Soc Exp Biol & Med; AMA; Asn Am Med Cols. Res: Fluid and electrolyte metabolism; capillary and cell membrane permeability; physiology of circulation; adrenal cortical physiology; radiation injury in primates; physiology of human pregnancy; radioactivity; flame photometry. Mailing Add: Univ of Tenn Health Sci Ctr Memphis TN 38103

OVERMAN, TIMOTHY LLOYD, b Cincinnati, Ohio, Dec 9, 43; m 68; c 1. CLINICAL MICROBIOLOGY. Educ: Univ Cincinnati, BS, 66, MS, 69, PhD(microbiol), 72; Am Bd Med Microbiol, dipl, 75. Prof Exp: Fel pub health & med microbiol, Ctr Dis Control, Atlanta, Ga, 71-73; SECT HEAD MICROBIOL, VET ADMIN HOSP, LEXINGTON, 74-; ASST PROF PATH, COL MED, UNIV KY, 74- Mem: Am Soc Microbiol. Res: Diagnostic microbiology, especially anaerobic bacteria, non-fermentative bacteria and parasitology. Mailing Add: Lab Serv (113) Cooper Dr Div Vet Admin Hosp Lexington KY 40507

OVERMIRE, THOMAS GORDON, b Indianapolis, Ind, July 19, 26; m 50; c 3. BIOLOGY. Educ: Purdue Univ, BS, 50; Ind Univ, MAT, 52; Okla State Univ, PhD(zool), 63. Prof Exp: Teacher high sch, Ind, 52-59; asst prof biol, Ball State Univ, 60-63; asst prog dir, Undergrad Educ in Sci Sect, NSF, 63-64; assoc prof biol & chmn dept, Okla City Univ, 64-66; staff biologist, Comn Undergrad Educ in Biol Sci, 66-69; exec dir Mich Acad Sci, Arts & Letters, 69-73; DEAN FAC, SCHENECTADY COUNTY COMMUNITY COL, 73- Concurrent Pos: Consult, US Agency Int Develop, India, 65-66 & 68; mem, Biol Sci Curriculum Study Film & Eval Comts, 64-67. Mem: AAAS; Am Nature Study Soc (treas, 62-64); Am Asn Biol Teachers. Res: Avian habitat analysis; homeostatic mechanisms; radiation biology; science education. Mailing Add: 2001 Hoover Rd Schenectady NY 12309

OVERMYER, ROBERT FRANKLIN, b Chicago, Ill, Aug 9, 29; m 59; c 2. SOLID STATE PHYSICS. Educ: DePauw Univ, BA, 50; Ill Inst Technol, MS, 59. Prof Exp: Field engr, Schlumberger Well Surv Corp, 54-56; res asst solid state physics, Gen Atomic Div, Gen Dynamics Corp, 58-60, staff assoc, 60-67; mgr radiation effects, Physics Br, Gulf Gen Atomic, Inc, 67-71, mgr spec nuclear effects dept, Gulf Radiation Technol, 71-73; VPRES, INTELCOM RADIATION TECHNOL, 73- Mem: Am Inst Physics; Am Asn Physics Teachers. Res: Transient radiation effects and displacement effects in semiconductor materials, devices, and military electronic systems; research and development of semiconductor thermoelectric materials. Mailing Add: 8558 Sugarman Dr La Jolla CA 92037

OVERPECK, JAMES GENTRY, b Greensburg, Ind, June 30, 29; m 67; c 2. PHYSIOLOGY. Educ: Ind Univ, Bloomington, BA, 51; Southern Ill Univ, MA, 64; George Washington Univ, PhD(physiol), 71. Prof Exp: PHARMACOLOGIST, FOOD & DRUG ADMIN, 63- Res: Steroid absorption; distribution, metabolism and excretion. Mailing Add: 2429 Villanova Dr Vienna VA 22180

OVERSETH, OLIVER ENOCH, JR, b New York, NY, May 11, 28; m 54; c 2. EXPERIMENTAL HIGH ENERGY PHYSICS. Educ: Univ Chicago, BS, 53; Brown Univ, PhD(physics), 58. Prof Exp: Instr physics, Princeton Univ, 57-60; from asst prof to assoc prof, 61-67, PROF PHYSICS, UNIV MICH, ANN ARBOR, 68- Concurrent Pos: Sci assoc, CERN, 74- Mem: Am Phys Soc. Res: Elementary particles. Mailing Add: Dept of Physics Univ of Mich Ann Arbor MI 48104

OVERSTREET, ELIZABETH CLAIRE (FISHER), b Detroit, Mich, Apr 5, 15; m 55. GEOLOGY. Educ: Wayne State Univ, BA, 38. Prof Exp: Geologist, 43-71, PART-TIME GEOLOGIST, US GEOL SURV, 71- Mem: Geol Soc Am; Mineral Soc Am. Res: Bauxite deposits of the southeastern United States; differential thermal analysis of bauxite and clay minerals; world bauxite resources; x-ray analysis; rock weathering. Mailing Add: 900 E Garcia St Santa Fe NM 87501

OVERSTREET, ROBIN MILES, b Eugene, Ore, June 1, 39; m 64; c 2. PARASITOLOGY, MARINE BIOLOGY. Educ: Univ Ore, BA, 63; Univ Miami, MS, 66, PhD(marine biol), 68. Prof Exp: Nat Inst Allergy & Infectious Dis fel parasitol, Med Sch, Tulane Univ, 68- 69; parasitologist, Gulf Coast Res Lab, 69-70; ASST PROF BIOL, UNIV SOUTHERN MISS, 70-; ASST PROF BIOL, UNIV MISS, 71-; HEAD SECT PARASITOL, GULF COAST RES LAB, 70- Mem: Am Soc Parasitol; Am Fisheries Soc; Soc Invert Path; Am Micros Soc; Helminth Soc Wash. Res: Taxonomy and biology of marine and estuarine parasites. Mailing Add: Gulf Coast Res Lab Ocean Springs MS 39564

OVERTON, EDWARD BEARDSLEE, b Mobile, Ala, Nov 21, 42. ANALYTICAL CHEMISTRY, INORGANIC CHEMISTRY. Educ: Univ Ala, Tuscaloosa, BS, 65, PhD(chem), 70. Prof Exp: ASST PROF CHEM, NORTHEAST LA UNIV, 70- Mem: Am Chem Soc. Res: Gaseous analysis; instrumentation; borane and carborane chemistry. Mailing Add: Dept of Chem Northeast La Univ Monroe LA 71201

OVERTON, JAMES RAY, physical chemistry, see 12th edition

OVERTON, JANE HARPER, b Chicago, Ill, Jan 17, 19; m 41; c 3. DEVELOPMENTAL BIOLOGY. Educ: Bryn Mawr Col, AB, 41; Univ Chicago, PhD, 50. Prof Exp: Res asst zool, 50-52, instr natural sci & res assoc zool, 52-53, asst prof natural sci, 53-60, assoc prof lectr biol sci, 60-65, assoc prof, 65-72, PROF BIOL, UNIV CHICAGO, 72- Concurrent Pos: Res grant NSF, 75. Mem: Am Soc Cell Biol; Soc Develop Biol; Am Soc Zool; Am Asn Anat; Int Inst Embryol. Res: Cytological aspects of development with emphasis on morphogenesis. Mailing Add: Whitman Lab Univ of Chicago 915 E 57th St Chicago IL 60637

OVERTON, WILLIAM CALVIN, JR, b Dallas, Tex, Oct 26, 18; m 46; c 3. PHYSICS. Educ: NTex State Univ, BS, 41; Rice Univ, MA, 49, PhD(physics), 50. Prof Exp: Proj engr, Missile Guid & Control Systs Res, S W Marshall Co, 46-47; sr res physicist acoustic oil well logging theory, Field Res Labs, Magnolia Petrol Co, 50-51; physicist, US Naval Res Lab, 51-62; PHYSICIST, LOS ALAMOS SCI LAB, UNIV CALIF, 62-

Concurrent Pos: Lectr, Univ Md Grad Exten Sch at Naval Res Lab, 52, 54-56 & 60. Mem: Fel Am Phys Soc; Sigma Xi. Res: Solid state and low temperature physics; lattice dynamics; thermodynamics. Mailing Add: Los Alamos Sci Lab PO Box 1663 Los Alamos NM 87544

OVERTURF, MERRILL L, b Ottumwa, Iowa, Aug 31, 38; m 73. BIOCHEMISTRY. Educ: Univ Iowa, BS, 61, PhD(prev med), 70. Prof Exp: Instr med parasitol, Univ Iowa, 70-72; ASST PROF INTERNAL MED, UNIV TEX MED SCH HOUSTON, 72- Concurrent Pos: Nat Heart & Lung Inst grant & Iowa Heart Asn grant, Univ Iowa, 71-72; Nat Heart & Lung Inst grant, Univ Tex, 72- Mem: Am Fedn Clin Res. Res: Biochemistry of renovascular hypertension; renin-angiotensin-aldosterone system. Mailing Add: Dept of Med Univ of Tex Med Sch Houston TX 77025

OVIATT, CANDACE ANN, b Bennington, Vt, June 17, 39; m 71. ECOLOGY. Educ: Bates Col, BS, 61; Univ RI, PhD(biol oceanog), 67. Prof Exp: Res assoc, Sch Pub Health, Harvard Univ, 65-69; RES ASSOC & LECTR ECOL, GRAD SCH OCEANOG, UNIV RI, 69- Mem: Am Soc Limnol & Oceanog; Int Asn Ecol. Res: Total systems studies of marine environments; the role of different populations in marine systems; the effects of stress and its role in stability; budgets for marine systems; practical application of the above for understanding and/or management. Mailing Add: Box 17 Grad Sch of Oceanog Univ RI Kingston RI 02882

OVIATT, CHARLES DIXON, b Washington, Pa, Dec 24, 24; m 48; c 5. APPLIED STATISTICS. Educ: Tarkio Col, AB, 47; Ohio State Univ, MS, 49, PhD, 54. Prof Exp: Prof chem, Tarkio Col, 49-57; staff ed phys sci, McGraw Hill Encyclop of Sci & Technol, 57-60; res chemist, E I du Pont de Nemours & Co, 60-61; sr sci ed, 61-69, exec ed, 69-71, ED-IN-CHIEF, WEBSTER DIV, McGRAW HILL BOOK CO, 71- Concurrent Pos: Author sci educ mat, McGraw-Hill Book Co & C V Mosby, Inc, 73-; dir statewide assessment proj, Mo State Dept Elem & Sec Educ, 74- Mem: AAAS; Am Chem Soc; Nat Sci Teachers Asn. Res: Technical editing; educational publishing, author; educational research. Mailing Add: Rte 1 Box 34 Vienna MO 65582

OVIST, ELWIN BURTON WEBB, organic chemistry, see 12th edition

OVREBO, PAUL JOHANNES, b Summit, SDak, Aug 2, 01; m 26; c 1. APPLIED PHYSICS. Educ: St Olaf Col, AB, 23; Univ Chicago, PhD, 28. Hon Degrees: DSc, St Olaf Col, 57. Prof Exp: Supt pub sch, Minn, 23-25; asst prof physics, WVa Univ, 28-30; prof, Susquehanna Univ, 30-43; chief, Appl Physics Sect, Aerial Reconnaissance Lab, Wright Air Develop Ctr, 43-55, chief devices br & tech adv planning div, Directorate Acquisition, Aerospace Tech Intel Ctr, 55-59, mem sci adv group for tech div, 59-63, sci adv to dep for activities, 63-65; res scientist, Corp Adv Res Lab, Douglas Aircraft Co, Inc, 65-66; consult, 66-68; consult scientist, Sci & Technol Assocs Inc, 68-69; consult ad hoc group atmospheric data measurements, Army Sci Adv Panel, 69-70; consult scientist, Dalmo Victor Co, Inc, 70-73; RETIRED. Concurrent Pos: Sr US Air Force mem infrared panel, electronic comt, Res & Develop Bd; mem intel & reconnaissance panel, Proj Forecast; mem Nat Adv Bd, Am Security Coun. Mem: Optical Soc Am. Res: Infrared ground level and high altitude radiation measurements; field experiments and development of equipment; electronic warfare advanced systems. Mailing Add: 221 Sylvan Dr Goleta CA 93017

OWADES, JOSEPH LAWRENCE, b New York, NY, July 9, 19; m 69. BIOCHEMISTRY. Educ: City Col New York, BS, 39; Polytech Inst Brooklyn, MS, 44, PhD(chem), 50. Prof Exp: Chemist, Naval Supply Dept, 40-49 & Fleischmann Labs, 49-53; asst chief chemist & brewing & food consult, Schwarz Labs, 53-56, chief chemist, 56-58, dir labs & brewing res, 59-60; dir res & tech serv, Liebman Breweries, Inc, 60-65; vpres & tech dir, Rheingold Breweries, Inc, 65-68; consult, Karolos Fix Co, Athens, Greece, 69-70; coordr tech ctr, Anheuser-Busch, Inc, 70-71; CONSULT TO FOOD & FERMENTATION INDUSTS, 72- Mem: Am Chem Soc; Am Soc Brewing Chemists; Master Brewers Asn Am; Inst Food Technologists; NY Acad Sci. Res: Yeast derivatives; sterols; brewing; foods. Mailing Add: 1575 Tremont St Boston MA 02120

OWCZARZAK, ALFRED, b New York, NY, Jan 21, 23; m 50; c 3. CELL BIOLOGY, BIOLOGICAL STRUCTURE. Educ: Cornell Univ, BS, 44; Univ Wis, PhD(bot), 53. Prof Exp: Proj asst bot, Univ Wis, 48-49, asst, 49-53, proj assoc, 53-54, proj assoc path, Serv Mem Insts, 54-55; res assoc zool, 55-58, from instr to asst prof, 55-61, ASSOC PROF ZOOL, ORE STATE UNIV, 61- Mem: AAAS; Am Soc Zool; Am Soc Cell Biol. Res: Cytological effects of chemical treatments; tissue culture; microtechnique; cytochemistry; comparative vertebrate histology; cell ultrastructure. Mailing Add: Dept of Zool Ore State Univ Corvallis OR 97331

OWELLEN, RICHARD JOHN, bRochester, NY, Nov 9, 35; m 60; c 2. ONCOLOGY, MEDICINE. Educ: Rensselaer Polytech Inst, BS, 57; Carnegie Inst Technol, MS & PhD(org chem), 60; Johns Hopkins Univ, MD, 67. Prof Exp: Asst prof med, 68-74, asst prof oncol, 73-74, ASSOC PROF ONCOL, SCH MED, JOHNS HOPKINS UNIV, 74- Concurrent Pos: NIH fel, Stanford Univ, 61-63; consult, Res Inst for Advan Studies, 64-67. Mem: AAAS; Am Soc Pharmacog; Am Asn Cancer Res; Am Soc Pharmacol & Exp Therapeut. Res: Pharmacology, clinical pharmacology, and metabolism of the Vinca alkaloids; tubulin-vinca interactions; structure-activity relations of Vinca alkaloids, including synthesis of congeners; marine pharmacology. Mailing Add: Oncol Ctr Johns Hopkins Univ Sch Med Baltimore MD 21224

OWEN, ALICE KONING, b Kalamazoo, Mich, Jan 20, 30; m 55; c 2. EMBRYOLOGY. Educ: Kalamazoo Col, BA, 51; Iowa State Univ, MS, 53, PhD(embryol), 55. Prof Exp: Asst prof anat & physiol, Northeast Mo State Univ, 63-65; ASST PROF BIOL & PHYSIOL, UNIV NDAK, 66- Mem: AAAS; Am Soc Zool; Am Inst Biol Sci. Res: Development of integument in chicks; effect of vitamin E on development in rats. Mailing Add: Dept of Physiol Univ of NDak Grand Forks ND 58201

OWEN, BERNARD LAWTON, b Presidio, Tex, Nov 10, 29; m 52; c 3. ENTOMOLOGY. Educ: Agr & Mech Col, Tex, BS, 51, MS, 54; Auburn Univ, PhD(entom), 59. Prof Exp: Entomologist, Agr Res Serv, USDA, 58-59; consult, 59; from asst prof to assoc prof, 59-66, PROF ZOOL, KANS WESLEYAN UNIV, 66- Mem: AAAS; Am Soc Zool; Nat Asn Biol Teachers; Am Inst Biol Sci; Entom Soc Am. Res: Environmental physiology of arthropods. Mailing Add: Dept of Biol Kans Wesleyan Univ Salina KS 67401

OWEN, BRADFORD BRECKENRIDGE, b Cleveland, Ohio, June 26, 12; m 41; c 1. ZOOLOGY. Educ: Williams Col, AB, 34, MA, 36; Harvard Univ, MA, 38, PhD(biol), 40. Prof Exp: Asst biol, Williams Col, 35-36 & Harvard Univ, 37-40; instr biol, Long Island Univ, 40-42 & chem, Hill Sch, 43-44; assoc prof, 45-74, PROF BIOL, LEHIGH UNIV, 74- Mem: AAAS; Am Soc Zool. Res: Vertebrate morphology; ecology of fresh water microorganisms. Mailing Add: Dept of Biol Lehigh Univ Bethlehem PA 18015

OWEN, BRUCE DOUGLAS, b Edmonton, Alta, Oct 1, 27; m 51; c 2. ANIMAL NUTRITION, ANIMAL PHYSIOLOGY. Educ: Univ Alta, BSc, 50, MSc, 52; Univ Sask, PhD(animal nutrit), 61. Prof Exp: Animal nutritionist, Lederle Labs Div, Am Cyanamid Co, NY, 52-54; animal husbandman, Can Dept Agr, 54-57; from lectr to assoc prof, 57-70, PROF ANIMAL SCI, UNIV SASK, 70- Mem: Agr Inst Can; Am Soc Animal Sci; Can Soc Animal Sci; Nutrit Soc Can. Res: Passive immunity; fat soluble vitamin transport; vitamin requirements of swine; protein quality evaluation; swine nutrition and physiology, factors influencing composition of feed grains and forages. Mailing Add: Dept of Animal Sci Univ of Sask Saskatoon SK Can

OWEN, CHARLES ARCHIBALD, JR, b Assiut, Egypt, Dec 3, 15; US citizen; m 39; c 3. BIOCHEMISTRY. Educ: Monmouth Col, AB, 36, Univ Iowa, MD, 41; Univ Minn, PhD, 50. Hon Degrees: DSc, Monmouth Col, 58. Prof Exp: Asst path, Univ Iowa, 38-39; from instr to assoc prof clin path, 50-60, PROF MED RES, MAYO FOUND, UNIV MINN, 60- Concurrent Pos: Consult, Mayo Clin, 50-, head sect exp biochem, 59-, head sect clin path, 61-70. Mem: AAAS; Am Soc Clin Path; Am Soc Exp Path; Soc Exp Biol & Med; Am Physiol Soc. Res: Blood coagulation; biologic aspects of radioisotopes. Mailing Add: 9 Skyline Dr Rochester MN 55901

OWEN, CHARLES SCOTT, b Springfield, Mo, July 4, 42. BIOPHYSICS. Educ: Univ Rochester, BS, 64; Univ Pa, PhD(physics), 70. Prof Exp: Res assoc physics, Wash Univ, 69-71 & Univ Calif, Santa Barbara, 71-72; RES ASSOC BIOPHYSICS, UNIV PA, 72- Mem: Biophys Soc; Am Phys Soc; AAAS. Res: Bioenergetics; magnetic properties of heme proteins; physical properties of immunoglobulins and protein-membrane interactions. Mailing Add: Dept of Biochem & Biophys Sch of Med G3 Univ of Pa Philadelphia PA 19174

OWEN, CORA RUST, b Madison, Wis, June 22, 11. BACTERIOLOGY. Educ: Vassar Col, AB, 33; Mass Inst Technol, 35; Univ Minn, MS, 36, PhD(bact), 39. Prof Exp: Lab asst bact, Univ Minn, 36-38; res assoc antibiotics & tuberc, Med Sch, Univ Mich, 46-51; bacteriologist, Rocky Mountain Lab, Nat Inst Allergy & Infectious Dis, 52-71; RETIRED. Concurrent Pos: Fel surg, Harvard Med Sch, 39-46. Mem: AAAS; Am Soc Microbiol. Res: Use and effects of antibiotics; tularemia; plague; infectious diseases. Mailing Add: Rocky Mountain Lab Nat Inst Allergy & Infect Dis Hamilton MT 59840

OWEN, DAVID ALLAN, inorganic chemistry, see 12th edition

OWEN, DAVID FRANCIS, nutrition, food technology, see 12th edition

OWEN, DAVID R, b San Luis Obispo, Calif, Apr 14, 42; m 68. APPLIED MATHEMATICS. Educ: Calif Inst Technol, BSc, 63; Brown Univ, PhD(appl math), 68. Prof Exp: Asst prof, 67-70, ASSOC PROF MATH, CARNEGIE-MELLON UNIV, 70- Honors & Awards: Von Humboldt Found Sr US Scientist Award, 74. Mem: Math Asn Am. Res: Continuum thermodynamics; rate-independent materials; materials with elastic range. Mailing Add: Dept of Math Carnegie-Mellon Univ Pittsburgh PA 15213

OWEN, DONALD BRUCE, b Portland, Ore, Jan 24, 22; m 52; c 4. APPLIED STATISTICS. Educ: Univ Wash, BS, 45, MS, 46, PhD(math statist), 51. Prof Exp: Assoc math, Univ Wash, 46-50, res assoc statist, 50-51, instr math, 51-52; asst prof math & consult statist lab, Purdue Univ, 52-54; staff mem math statist, Sandia Corp, 54-57, supvr statist res div, 57-64; head math & stochastic systs div, Grad Res Ctr & adj prof statist, Univ, 64-66, PROF STATIST, SOUTHERN METHODIST UNIV, 66- Concurrent Pos: Consult, Sandia Corp, 64-67; prof, Univ Tex, 65-66; dir, Prog Vis Lectr in Statist, 70-73; coord ed, Marcel Dekker Statist Ser, 70-; ed, Commun in Statist, 71- Mem: Fel AAAS; fel Am Statist Asn; Am Soc Qual Control; Int Asn Statist Phys Sci; Am Math Soc. Res: Tabulations of statistical functions; bivariate normal probability distribution and associated integrals; applications of statistics to physics and engineering; double sample procedures. Mailing Add: 7614 Tophill Lane Dallas TX 75240

OWEN, DONALD EDWARD, b San Antonio, Tex, Nov 29, 36; m 57; c 3. STRATIGRAPHY, GEOLOGY. Educ: Lamar State Col, BS, 57; Univ Kans, MS, 59, PhD(geol), 63. Prof Exp: Res scientist, Bur Econ Geol, Univ Tex, 62-64; from instr to assoc prof geol, 64-73, dir geol field camp, 64-68, NSF res grant, 65-66, PROF GEOL, BOWLING GREEN STATE UNIV, 73- Concurrent Pos: Adj asst prof, State Univ NY Binghamton, 65, 66; sr lectr sch earth sci, Macquarie Univ, Australia, 69-71; consult geologist, Planet Mgt & Res Proprietary Ltd, Australia, 69-71; NMex Bur Mines & Mineral Resources res grants, 74 & 75; assoc ed, Soc Econ Paleontologists & Mineralogists, 74- Mem: Am Asn Petrol Geologists; Soc Econ Paleontologists & Mineralogists; Geol Soc Australia; Int Asn Sedimentologists. Res: Sedimentology; Cretaceous stratigraphy of southwestern United States; Dakota sandstone of San Juan Basin, New Mexico; modern sediments of coastal North Carolina; Devonian of central New South Wales, Australia. Mailing Add: Dept of Geol Bowling Green State Univ Bowling Green OH 43403

OWEN, DONALD EUGENE, b Emporia, Kans, Oct 20, 28; m 55; c 2. GEOLOGY. Educ: Univ Kans, BS, 49; Univ Tex, MA, 51; Univ Wis, PhD(geol), 64. Prof Exp: Geophysicist, Calif Co, 51-56, area geologist, Mont, 56-58, dist staff geologist, 58-60; asst prof, 63-66, ASSOC PROF GEOL, IND STATE UNIV, 66- Mem: AAAS; Geol Soc Am; Am Asn Petrol Geol; Nat Asn Geol Teachers. Res: Cretaceous stratigraphy; paleogeologic and onlap maps of North America; Williston Basin stratigraphy and structure. Mailing Add: Dept of Geog & Geol Ind State Univ Terre Haute IN 47809

OWEN, DONALD ROBERTSON, JR, b Austin, Tex, Nov 14, 44; m 66; c 1. POLYMER CHEMISTRY. Educ: The Citadel, BS, 66; Univ Southern Conn, MS, 71; Univ Houston, PhD(org chem), 73. Prof Exp: Res & develop chemist, R T Vanderbilt Co, 68-71; ASST PROF POLYMER CHEM, UNIV SOUTHERN MISS, 73- Concurrent Pos: Consult, ArChem Co, 73-; sr adj res assoc, Tex Heart Inst, 75- Mem: Am Chem Soc; Sigma Xi. Res: Development of improved hemocompatible surfaces for cardiovascular protheses, graft copolymerization, heterogeneous solution polymerization and the synthesis of polymers for oil recovery applications. Mailing Add: Box 223 Univ of Southern Miss Hattiesburg MS 39401

OWEN, DOUGLAS FARRAR, genetics, plant breeding, see 12th edition

OWEN, EDGAR WESLEY, b Huntsville, Ark, June 20, 96; m 19. GEOLOGY, GEOGRAPHY. Educ: Denison Univ, BS, 15; Univ Mo, MA, 16. Hon Degrees: DSc, Denison Univ, 61. Prof Exp: Geologist, Empire Gas & Fuel Co, Okla, 16-17, Greenwood Co, Mo, 17-18, John C Keys, Okla, 19-20 & oil div, L H Wentz, 27-42, 46-56; consult, 56-75; RETIRED. Concurrent Pos: Lectr, Univ Tex, 53-75. Honors & Awards: Powers Medal, Am Asn Petrol Geologists, 64. Mem: AAAS; fel Geol Soc Am; hon mem Am Asn Petrol Geologists (secy-treas, 39-41, pres, 41). Res: Petroleum geology; history of geology. Mailing Add: 505 Club Dr San Antonio TX 78201

OWEN, FLETCHER BAILEY, b Richmond, Va, Nov 6, 29. PHARMACOLOGY. Educ: Univ Richmond, BS, 51; Med Col Va, PhD(pharmacol), 56, MD, 59. Prof Exp: Intern, Richmond Med Hosp, 59-60; clin investr, 60-63, DIR MED SERV, A H ROBINS CO, INC, 63- Concurrent Pos: Instr, Med Col Va, 60-64, lectr, 64- Mem:

AAAS. Res: Clincial pharmacology; drugs. Mailing Add: A H Robins Co Inc 1407 Cummings Dr Richmond VA 23220

OWEN, FOSTER GAMBLE, b Flat Creek, Ala, Jan 24, 26; m 50; c 4. DAIRY NUTRITION. Educ: Auburn Univ, BS, 48; Iowa State Univ, MS, 55, PhD, 56. Prof Exp: Asst dairy husb, Iowa State Univ, 53-55; asst prof animal husb, Univ Ark, 55-57; assoc prof dairy husb, 57-64, PROF ANIMAL SCI, UNIV NEBR, LINCOLN, 64- Mem: Am Soc Animal Sci; Am Dairy Sci Asn; Am Forage & Grassland Coun. Res: Dairy cattle nutrition and feeding methods, calf nutrition and feeding programs, forage preservation and nutritional value. Mailing Add: Dept of Animal Sci Univ of Nebr Lincoln NE 68583

OWEN, FRANK WILLIAM, b Benton Harbor, Mich, Apr 6, 27; m 49; c 3. POMOLOGY. Educ: Mich State Univ, BS, 50, MS, 52. Prof Exp: Trainee, Kroger Grocery Co, 50; trainee, Soil Conserv Serv, USDA, 50-51; exten specialist, Maine, 52-55; from instr to assoc prof pomol, Univ Ill, 55-67, exten specialist, 55-67; res specialist & grower adv, Hilltop Orchards & Nursery, 67-70; CONSULT, FRANK W OWEN, ASSOCS, 70-; GEN MGR, CHERRY ADMIN BD, 71- Mem: Am Soc Hort Sci. Res: Fruit tree size modification by means of rootstocks and growth regulators; resistance of plant tissues to cold temperature damage. Mailing Add: Rte 1 Box 159 Lawrence MI 49064

OWEN, FRAZER NELSON, b Atlanta, Ga, Feb 4, 47. ASTRONOMY. Educ: Duke Univ, BS, 69; Univ Tex, PhD(astron), 74. Prof Exp: Res assoc astron, 73-75, ASST SCIENTIST, NAT RADIO ASTRON OBSERV, 75- Mem: Am Astron Soc; Royal Astron Soc. Res: Extragalactic radio astronomy; clusters of galaxies; radio galaxies; quasars; radio flare stars. Mailing Add: Nat Radio Astron Observ Edgemont Rd Charlottesville VA 22901

OWEN, GARETH, pharmacology, toxicology, see 12th edition

OWEN, GEORGE ERNEST, b St Louis, Mo, Jan 7, 22; m 43; c 1. NUCLEAR PHYSICS. Educ: Wash Univ, BSME, 43, PhD(physics), 50. Prof Exp: Asst prof physics, Univ Pittsburgh, 50-51; from asst prof to assoc prof, 51-60, PROF PHYSICS, JOHNS HOPKINS UNIV, 60-, CHMN DEPT, 68-, DEAN ARTS & SCI, 72- Mem: Fel Am Phys Soc. Res: Beta ray and gamma ray spectroscopy; neutron detectors and medium energy nuclear reactions. Mailing Add: Col of Arts & Sci Johns Hopkins Univ Baltimore MD 21218

OWEN, GEORGE MURDOCK, b Alliance, Ohio, Feb 16, 30; m 52; c 3. PEDIATRICS, NUTRITION. Educ: Hiram Col, AB, 51; Univ Cincinnati, MD, 55. Prof Exp: Rotating intern, Univ Iowa, 55-56, resident pediat, 56-58; clin assoc, Nat Inst Allergy & Infectious Dis, 58-60; from asst prof to assoc prof, Univ Iowa, 66; from assoc prof to prof pediat, Col Med, Ohio State Univ, 66-73; PROF PEDIAT & DIR CLIN NUTRIT PROG, SCH MED, UNIV N MEX, 73- Concurrent Pos: Res fel, Metab Sect, Univ Iowa, 60-62. Mem: Am Acad Pediat; Am Soc Clin Nutrit; Soc Pediat Res; Am Inst Nutrit; Am Pediat Soc. Res: Nutrition, growth and body composition of normal infants; nutrition status. Mailing Add: Dept of Pediat Univ of NMex Albuquerque NM 87131

OWEN, GUILLERMO, b Bogota, Colombia, May 4, 38; m 61; c 2. MATHEMATICS. Educ: Fordham Univ, BS, 58; Princeton Univ, PhD(math), 62. Prof Exp: From asst prof to assoc prof math, Fordham Univ, 61-71; PROF MATH SCI, RICE UNIV, 71- Concurrent Pos: Consult, Hudson Inst, NY, 62- & Mathematica, 63- Mem: Am Math Soc. Res: Game theory; operations research; economic applications. Mailing Add: Dept of Math Sci Rice Univ Houston TX 77001

OWEN, HERBERT ELMER, JR, b Waco, Tex, Feb 6, 25; m 52; c 3. PLANT ECOLOGY, PLANT BIOCHEMISTRY. Educ: Univ Tulsa, BS, 50; Ore State Col, MS, 53, PhD(plant ecol), 57. Prof Exp: Asst bot, Ore State Col, 50-54; instr biol sci, 55-58, asst prof biol & chmn div biol sci, 59-62, chmn div agr & appl arts, 59-63, assoc prof biol, 62-66, prof bot, 66-74, PROF BIOL, FT LEWIS COL, 74- Res: Ecotypic development of various tree species, especially Douglas fir. Mailing Add: Div of Biol Sci Ft Lewis Col Durango CO 81301

OWEN, HOWARD MALCOLM, b Richmond, Va, Dec 16, 13; m 39; c 3. ZOOLOGY. Educ: Hampden-Sydney Col, AB, 35; Univ Va, MA, 43, PhD(biol), 45. Prof Exp: Asst biologist, Va Fisheries Lab, Col William & Mary, 45-46; asst prof biol, Univ SC, 46-47; spec biologist, State Dept Wildlife, La, 47-50; PROF BIOL, UNIV OF THE SOUTH, 50- Mem: Fel AAAS; Am Micros Soc; Am Genetic Asn; Am Soc Parasitologists; Am Soc Study Evolution. Res: Structure of flagella; cytology of marine mollusks; physiology of oysters; marine ecology; radioisotopes genetics; use of computer in undergraduate education. Mailing Add: Dept of Biol Univ of the South Sewanee TN 37375

OWEN, JAMES EMMET, b Cleveland, Ohio, May 23, 35; m 61; c 6. INORGANIC CHEMISTRY. Educ: John Carroll Univ, BS, 57; Case Inst Technol, PhD(chem), 61. Prof Exp: Res scientist, US Rubber Co, 61-63; group leader, 63-66, sr res assoc, 66-67, dir inorg phys res, 67-69, dir cent res, 69-70, DIR INORG & PHYS RES, HARSHAW CHEM CO, 70- Mem: AAAS; Am Chem Soc; Am Inst Chemists. Res: Fundamental studies of inorganic pigment systems; pigments based on non-stoichiometric metal oxides; pigments based on II-VI type semiconductors; investigation of adsorption phenomena; structural characteristics of adsorbents as related to adsorption performance. Mailing Add: Harshaw Chem Co 1945 E 97th St Cleveland OH 44106

OWEN, JAMES ROBERT, b Leake Co, Miss, Mar 26, 14; m 37. PHYSICAL CHEMISTRY. Educ: Okla Agr & Mech Col, BS, 35; Univ Wis, PhM, 37. Prof Exp: Asst, Exp Sta, Okla Agr & Mech Col, 33-35; asst chem, Univ Wis, 36-37; sect chief, Phillips Petrol Co, 37-58, tech supt, Ambrosia Mill, NMex, 58-62, develop engr, Patent Div, 62-69; CONSULT HETEROGENEOUS CATALYSIS, 69- Mem: Catalysis Soc; Am Chem Soc. Mailing Add: 2413 Summit Rd Bartlesville OK 74003

OWEN, JOEL, b Boston, Mass, Feb 22, 35. STATISTICS, OPERATIONS RESEARCH. Educ: Yeshiva Univ, BA, 56; Boston Univ, MA, 57; Harvard Univ, PhD(statist), 66. Prof Exp: Res scientist appl math & statist, Waltham Lab, Sylvania Electronic Systs Div, 57-61, adv res scientist math statist, Appl Res Lab, 61-64; sr mathematician, Info Res Asn, Inc, 65-66; asst prof statist & opers res, 66-70, assoc prof statist, 70-73, PROF STATIST, GRAD SCH BUS, NY UNIV, 73- Res: Pattern recognition; discrimination analysis; game and decision theory; mathematical formulation of financial accounting theory and portfolio theory. Mailing Add: Grad Sch of Bus NY Univ 100 Trinity Pl New York NY 10006

OWEN, JOHN ATKINSON, JR, b South Boston, Va, Sept 24, 24; m 52; c 2. INTERNAL MEDICINE, METABOLISM. Educ: Hampden-Sydney Col, BS, 44; Univ Va, MD, 48. Prof Exp: Intern, Cincinnati Gen Hosp, Ohio, 48-49; asst resident internal med, Univ Va Hosp, 50-52; asst prof med, Med Col Ga, 56-58; asst clin prof internal med, Sch Med, George Washington Univ, 58-60; from asst prof to assoc prof internal med, 60-70, dir div clin pharmacol, 64-71, vchmn dept internal med, 72-74, PROF INTERNAL MED, SCH MED, UNIV VA, 70- Concurrent Pos: Fel internal med, Univ Va Hosp, 50-52; res fel metab, Duke Univ Hosp, 54-56; Am Diabetes Asn res fel, 55-56; pres, US Pharmacopeial Conv, 75-80; ed-in-chief, Hosp Formulary. Mem: Am Fedn Clin Res; Endocrine Soc; Am Diabetes Asn; AMA; fel Am Col Clin Pharmacol. Res: Experimental and clinical diabetes; diabetogenic and anti-diabetic drugs; clinical trials and animal research. Mailing Add: Dept of Internal Med Univ of Va Hosp Box 242 Charlottesville VA 22903

OWEN, JOHN HARDING, b San Francisco, Calif, Dec 22, 31; m 51; c3. INORGANIC CHEMISTRY, PHYSICAL CHEMISTRY. Educ: Hampden-Sydney Col, BS, 53; Fla State Univ, MS, 55. Prof Exp: Res chemist, E I du Pont de Nemours & Co, Inc, 55-68, SR RES SUPVR, SAVANNAH RIVER LAB, E I DU PONT DE NEMOURS & CO, INC, 68- Concurrent Pos: Instr, Augusta Col, 59-63 & Univ SC, 63-75; traveling lectr, Oak Ridge Inst Nuclear Studies, 63-66. Res: Chemistry and engineering of power fuel reprocessing; physical chemistry of vacuum technology; theory and application of diffusion processes in solids. Mailing Add: Savannah River Lab E I du Pont de Nemours & Co Inc Aiken SC 29801

OWEN, JOHN REINDEL, b Cleveland, Ohio, July 20, 39; m 65; c 2. ORGANIC CHEMISTRY. Educ: Oberlin Col, AB, 61; Univ Rochester, PhD(org chem), 66. Prof Exp: Res chemist, 65-73, SR RES CHEMIST, EASTMAN KODAK CO, 73- Res: Physical and chemical interactions of organic compounds with silver halide dispersion in gelatin. Mailing Add: Res Labs Bldg 59 Eastman Kodak Co 1669 Lake Ave Rochester NY 14615

OWEN, LOUIS JOHN, b New York, NY, June 29, 23; m 50; c 4. ORGANIC CHEMISTRY. Educ: Ursinus Col, BS, 47; Fordham Univ, MS, 48, PhD, 51. Prof Exp: Asst to vpres mkt, Carwin Co, 51-56; vpres mkt, Ott Chem Co, 56-58; mgr mkt res, Nopco Chem Co, 58-60, mgr mkt res & develop, Sales Depts, 60-62, gen mgr plastics div, 62-68, from asst vpres to vpres, 65-68; PRES & CHMN BD, OKENEL CORP, 68- Mem: Am Chem Soc; Chem Mkt Res Asn. Res: Market research. Mailing Add: Okenel Corp Lyndhurst NJ 07071

OWEN, RAY BUCKLIN, JR, b Providence, RI, July 14, 37; m 62; c 2. WILDLIFE ECOLOGY. Educ: Bowdoin Col, AB, 59; Univ Ill, MS, 66, PhD(ecol), 68. Prof Exp: Instr biol, Univ Ill, 67-68; asst prof, 68-73, ASSOC PROF WILDLIFE RESOURCES, UNIV MAINE, ORONO, 73- Mem: AAAS; Am Ornith Union; Cooper Ornith Soc; Ecol Soc Am; Am Soc Naturalists. Res: Ecological energetics; wetlands ecology; animal behavior. Mailing Add: Sch of Forest Resources Forest Resources Bldg Univ Maine Orono ME 04473

OWEN, RAY DAVID, b Genesee, Wis, Oct 30, 15; m 39; c 1. GENETICS, IMMUNOLOGY. Educ: Carroll Col, Wis, BS, 37; Univ Wis, PhM, 38, PhD(genetics), 41. Hon Degrees: ScD, Carroll Col, Wis, 62 & Univ Pac, 66. Prof Exp: Asst genetics, Univ Wis, 40-41, res fel, 41-43, asst prof genetics & zool, 43-47; assoc prof biol, 47-53, chmn dept, 61-68, PROF BIOL, CALIF INST TECHNOL, 53-, VPRES STUDENT AFFAIRS & DEAN STUDENTS, 74- Concurrent Pos: Gosney fel, 46-47; res partic, Oak Ridge Nat Lab, 57-58, consult, 58- Mem: Nat Acad Sci; Am Asn Immunol; Am Soc Human Genetics; Genetics Soc Am (treas, 56-59, vpres, 60-61, pres, 62); Soc Study Evolution. Res: Immunogenetics; serology; vertebrate and developmental genetics; genetics and immunology of tissue transplantation. Mailing Add: Div of Biol Calif Inst of Technol Pasadena CA 91109

OWEN, ROBERT MICHAEL, b Philadelphia, Pa, Feb 17, 46. CHEMICAL OCEANOGRAPHY, MARINE GEOCHEMISTRY. Educ: Drexel Univ, BS, 69; Univ Wis-Madison, MS, 74, PhD(oceanog, limnol), 75. Prof Exp: Develop chemist, Rohm & Haas Co, 69-72; res asst oceanog, Univ Wis-Madison, 72-75; ASST PROF OCEANOG, UNIV MICH, 75- Mem: AAAS; Int Asn Great Lakes Res; Soc Econ Paleontologists & Mineralogists; Am Soc Limnol & Oceanog. Res: Chemical and textural partitioning of trace elements in sediments; marine mineral exploration; geochemical and mathematical techniques for determining sediment dispersal patterns; organic associations of trace elements in sediments. Mailing Add: Dept of Atmospheric & Oceanic Sci Rm 1220 Space Res Bldg Ann Arbor MI 48109

OWEN, ROBERT W, JR, b Denver, Colo, Jan 29, 35; m. OCEANOGRAPHY. Educ: Univ Wash, BSci, 57; Univ Calif, San Diego, MSci, 63, PhD, 72. Prof Exp: Asst oceanogr, Univ Wash, 58-59; OCEANOGR, FISHERY OCEANOG CTR, NAT MARINE FISHERIES SERV, 61- Concurrent Pos: Consult, Calif Dept Fish & Game, 65- Mem: Am Geophys Union; Optical Soc Am; Am Inst Physics; Am Soc Limnol & Oceanog. Res: Dynamics of motion in the surface layers of the sea; consequences in plankton ecology. Mailing Add: Fishery Oceanog Ctr Nat Marine Fish Serv PO Box 271 La Jolla CA 92037

OWEN, ROGER C, b Port Arthur, Tex, Sept 14, 28; m 51; c 5. CULTURAL ANTHROPOLOGY. Educ: Mich State Univ, BA, 53; Univ Ariz, MA, 57; Univ Calif, Los Angeles, PhD(anthrop), 62. Prof Exp: From instr to assoc prof anthrop, Univ Calif, Santa Barbara, 59-67; PROF ANTHROP, QUEENS COL, NY, 67- Concurrent Pos: Nat Acad Sci, Nat Res Coun, NSF, Univ Calif, Licensed Beverage Industs, Inc & Holt, Rinehart & Winston, Inc grants, 59-72; fac fel, Univ Calif, Santa Barbara, 63; consult, Defense Systs Div, Gen Motors Corp, 62, NSF Grant Proposals, 65- & Proj Upward Bound, Educ Assocs, Inc & Dept Health, Educ & Welfare, 68-70; vis prof, Fundacao Sch Sociol & Polit, Sao Paulo, Brazil, 64-65; vis lectr, Univ Calif, Irvine, 72 & Vis Lectr Prog, Am Anthrop Asn, 72. Mem: Fel Am Anthrop Asn; Am Ethnol Soc. Res: Northwest Mexican ethnology; urban Brazil; alcohol studies; cultural evolution; North American Indians; curriculum development in social studies. Mailing Add: Dept of Anthrop Queens Col Flushing NY 11367

OWEN, STANLEY PAUL, b Earlie, Alta, Mar 30, 24; US citizen; m 46; c 2. BIOCHEMISTRY. Educ: Univ Alta, BSc, 50, MSc, 52; Univ Wis, PhD(biochem), 55. Prof Exp: Res assoc, Fermentation Res & Develop Dept, 55-62, biochemist, Biochem Dept, 62-67, head microbial control, 67-69, MGR PROD CONTROL II, UPJOHN CO, 67- Mem: AAAS; Am Soc Microbiol; Am Inst Biol Sci. Res: Antibiotic, vitamin and biological product assays; pathogen and sterility testing; biological availability of drugs; quality control. Mailing Add: Prod Control II Upjohn Co 7171 Portage Rd Kalamazoo MI 49001

OWEN, TERENCE CUNLIFFE, b Cannock, Eng, Nov 29, 30; m 53; c 2. ORGANIC CHEMISTRY, RADIATION CHEMISTRY. Educ: Univ Manchester, BSc, 51, PhD(chem), 54. Prof Exp: Tech officer chem, Nobel Div, Imp Chem Indust, Scotland, 54-57; from lectr to sr lectr, Liverpool Col Technol, Eng, 57-64; assoc prof, 64-70, PROF CHEM, UNIV S FLA, 70- Concurrent Pos: Res fel, Vanderbilt Univ, 60-61; consult, Summers Steelworks, Eng, 58-60 & Bristol Labs, 68-; fel, Royal Inst Chem, 62. Mem: Am Chem Soc; The Chem Soc; Radiation Res Soc. Res: Penicillin and cephalosporin synthesis; radiation biochemistry; polypeptide chemistry and mechanisms of enzyme catalysis. Mailing Add: Dept of Chem Univ of SFla Tampa FL 33620

OWEN, THOMAS BARRON, physical chemistry, see 12th edition

OWEN, TOBIAS CHANT, b Oshkosh, Wis, Mar 20, 36; m 60. ASTROPHYSICS, PLANETARY SCIENCES. Educ: Univ Chicago, BA, 55, BS, 59, MS, 60; Univ Ariz, PhD(astron), 65. Prof Exp: Res physicist, IIT Res Inst, 64-67, sr scientist, 67-69, sci adv, 69-70; vis assoc prof planetary sci, Calif Inst Technol, 70; assoc prof, 70-72, PROF ASTRON, STATE UNIV NY STONY BROOK, 72- Concurrent Pos: Consult, NASA, 72-; vpres comn 16, Int Astron Union, 73-76. Mem: Fel AAAS; Am Astron Soc; Int Astron Union; Int Soc Study Origin Life (pres, 73-); Int Asn Geochem & Cosmochem. Res: Planetary atmospheres and surfaces; satellites; comets; stars of late spectral type. Mailing Add: Dept of Earth & Space Sci State Univ of NY Stony Brook NY 11794

OWEN, WALTER WYCLIFFE, b Pine Bluff, Ark, Sept 19, 06; m 32; c 2. CHEMISTRY. Educ: Univ Ark, BS, 29; Univ NC, PhD(phys chem), 39. Prof Exp: Supvr treating, Int Creosoting & Construct Co, Tex, 29-30; exp & sales engr, Sowers Mfg Co, NY, 30-32; res chemist, Rayon Div, E I du Pont de Nemours & Co, 32-36, 39-41, chief chemist, Va, 41-45, tech serv engr, Del, 45-49, tech res tire yarn sales, 49-54, mgr indust tech serv, Textile Fibers Dept, 54-60, tech specialist, 60-68; CONSULT PROCESSING & APPLN SYNTHETIC FIBERS, 68- Mem: Am Chem Soc; Am Soc Testing & Mat. Res: Viscose rayon process; permeability of regenerated celelulose film to carbon dioxide; industrial applications of man-made fibers. Mailing Add: 10803 Bonaparte Bend Austin TX 78759

OWEN, WILLIAM BERT, b Gilbertsville, Ky, Oct 28, 03; m 32. ENTOMOLOGY. Educ: Univ Ky, BA, 27; Univ Minn, MA, 29, PhD(entom), 36. Prof Exp: Instr zool, Hamline Univ, 29-30 & Ore State Col, 30-31; from instr to asst prof, Univ Wyo, 31-34; asst, Univ Minn, 35-36; prof, 46-71, EMER PROF ZOOL, UNIV WYO, 71- Concurrent Pos: Researcher, Montpellier & Leyden, 64; fac affil, Colo State Univ, 72- Mem: Am Soc Parasitol; Entom Soc Am; Mosquito Control Asn. Res: Taxonomy of Culcidae; biology of mosquitoes of western North America; chemical senses of insects. Mailing Add: Dept of Zool & Entom Colo State Univ Ft Collins CO 80523

OWENS, ALBERT HENRY, JR, b Staten Island, NY, Aug 27, 26; m 49; c 4. INTERNAL MEDICINE, PHARMACOLOGY. Educ: Johns Hopkins Univ, BA, MD, 49; Am Bd Internal Med, dipl, 56. Prof Exp: Mem staff internal med, Univ Hosp, 49-50, 52-53, 55-56, from instr to assoc prof med, Sch Med, 56-68, PROF MED, SCH MED, JOHNS HOPKINS UNIV, 68-, INSTR PATHOBIOL, SCH HYG & PUB HEALTH, 57-, PHYSICIAN, UNIV HOSP, 56- Concurrent Pos: Fel pharmacol & exp therapeut, Sch Med, Johns Hopkins Univ, 53-55; vis physician, Baltimore City Hosp, 57- Mem: Am Soc Pharmacol & Exp Therapeut; fel Am Col Physicians; NY Acad Sci; Am Soc Clin Oncol; Am Asn Cancer Insts. Res: Neoplastic diseases. Mailing Add: Dept of Internal Med Johns Hopkins Univ Sch of Med Baltimore MD 21205

OWENS, ALVIN JEWEL, b Campbell, Mo, Sept 26, 18; m 48; c 2. APPLIED MATHEMATICS, COMPUTER SCIENCE. Educ: Southeast Mo State Univ, AB, 40; Univ Iowa, MS, 42; Univ Fla, PhD(math), 50. Prof Exp: Physicist, Burnside Lab, E I du Pont de Nemours & Co, 42-43; asst prof, Auburn Univ, 50-51; partner & mgr, Deal Sales Co, 51-55; asst prof, Univ Mo, 55-56; sr res engr, Gen Dynamics/Convair, 56-59, design specialist, Gen Dynamics/Astronautics, 59-62, res group engr, 62-63, staff scientist, 63-65; VPRES, DECISION SCI, INC, 65- Mem: Soc Indust & Appl Math; Asn Comput Mach. Res: Evolutionary programming as a means for creating artificial intelligence; application of computers for the solution of aerospace problems. Mailing Add: 3007 Melbourne Dr San Diego CA 92123

OWENS, BOONE BAILEY, b Chefoo, China, Dec 13, 22; US citizen; m 56; c 4. ELECTROCHEMISTRY. Educ: Whittier Col, BA, 54; Iowa State Univ, PhD(phys & inorg chem, math), 57. Prof Exp: Asst chem, Ames Lab, AEC, 54-57; instr, Univ Calif, Santa Barbara, 57-59; sr chemist, Atomics Int Div, NAm Aviation, Inc, 59-63, res specialist, 63-68; mem tech staff, Gen Tel & Electronics Labs, 68-69; prin scientist, Gould Ionics Inc, 69-72 & Energy Technol Dept, Gould Labs, 72-74, mgr battery res & develop sect, Gould Labs-Energy Res, Gould Inc, 74-75; MGR POWER SOURCES RES & DEVELOP DEPT, MEDTRONIC, INC, 75- Concurrent Pos: Consult, Glass Prod Co, Calif, 58-59; ed, Battery Div, J Electrochem Soc, 73- Mem: AAAS; Am Chem Soc; Sigma Xi; Electrochem Soc. Res: High energy batteries; solid electrolytes; solid state batteries. Mailing Add: Medtronic Inc 3055 Old Highway Eight Minneapolis MN 55418

OWENS, CHARLES WESLEY, b Billings, Okla, Oct 27, 35; m 55; c 4. PHYSICAL CHEMISTRY. Educ: Colo Col, BS, 57; Univ Kans, PhD(phys chem), 63. Prof Exp: Res assoc radiochem, Univ Kans, 63; asst prof, 63-69, ASSOC PROF CHEM, UNIV NH, 69- Concurrent Pos: Res fel appl chem, Harvard Univ, 70-71. Mem: Am Chem Soc. Res: Hot-atom chemistry; high energy kinetics; reaction mechanisms; solid state chemistry. Mailing Add: Dept of Chem Parsons Hall Univ of NH Durham NH 03824

OWENS, CLARENCE BURGESS, b Smithville, Tex, June 17, 26; m 49; c 5. AGRONOMY, PLANT PHYSIOLOGY. Educ: Prairie View Col, BSc, 48; Ohio State Univ, MSc, 51, PhD, 54. Prof Exp: Instr agr, US Vet Admin, 48-50; from asst prof to assoc prof, 53-57, PROF AGRON, FLA A&M UNIV, 57- Concurrent Pos: Dir summer & in-serv, Earth Sci Insts, NSF, 64-66; campus coordr, US AID-Fla A&M Univ Kenya contract, 72; res agron adv, US ICA, 57-59. Mem: AAAS; Am Soc Agron; Weed Sci Soc Am. Res: Chemical weed control; crop fertilization; soil fertility. Mailing Add: Dept of Agron Fla A&M Univ Box 8 Tallahassee FL 32301

OWENS, CLIFFORD, b Raven, Ky, Jan 5, 31; m 54. PHYSICAL INORGANIC CHEMISTRY. Educ: Rutgers Univ, BA, 57; Drexel Univ, MS, 61, PhD(phys chem), 69. Prof Exp: Anal chemist petrol, Gulf Oil Corp, 57-60; res chemist org coatings, Campbell Soup Co, 60-63, emission spectroscopist, 63-66; asst chem, Drexel Univ, 66-69; from instr to asst prof, 69-75, ASSOC PROF CHEM, RUTGERS UNIV, 75- Concurrent Pos: Consult, Drexel Univ, 74- Mem: Soc Appl Spectrog. Res: Study of the instability of chloramine compounds; sweat poisoning of activated carbon surfaces; vapor phase equilibria of binary systems where complexation may occur; synthesis and characterization of inorganic complexes involving phosphoryl and nitroxide compounds. Mailing Add: Dept of Chem Rutgers Univ Camden NJ 08101

OWENS, DANIEL KENYON, b Rocky Mount, NC, Nov 10, 32; m 53; c 3. AGRICULTURAL CHEMISTRY. Educ: NC State Col, BS, 53, MS, 55. Prof Exp: Chemist, bacteriologist & serologist, State Lab of Hyg, NC, 55-56; chemist, 56-59, res chemist, 59-64, staff scientist, 64-71, RES ASSOC, E I DU PONT DE NEMOURS & CO, INC, 71- Mem: Fel Am Inst Chemists. Res: Optical and electron microscopy; physics of evaporated films; biological and water chemistry; chemistry and structure of cellulose; surface chemistry and physics; friction, lubrication and adhesion of polymers. Mailing Add: 625 N Pinetta Dr Richmond VA 23235

OWENS, EDWARD HENRY, b Stamford, Eng, Mar 25, 45. GEOMORPHOLOGY, SEDIMENTOLOGY. Educ: Univ Col Wales, BSc, 67; McMaster Univ, MSc, 69; Univ SC, PhD(geol), 75. Prof Exp: Res scientist, Geol Surv Can, 71-75; ASST PROF GEOL, COASTAL STUDIES INST, LA STATE UNIV, 72- Concurrent Pos: Chmn, Sable Island Dune Restoration & Terrain Mgt Subcomt, Can, 74-75. Mem: Am Shore & Beach Preserv Asn; Soc Econ Paleontologists & Mineralogists; Sigma Xi; Arctic Inst NAm. Res: Barrier beach geomorphology and sediments, particularly the mechanics of sediment transport and deposition; action of ice on beaches in mid- and high-latitudes; effects of oil-spills on coasts. Mailing Add: Coastal Studies Inst La State Univ Baton Rouge LA 70803

OWENS, FRANK JAMES, b Dublin, Ireland, Oct 24, 39; US citizen; m 72. SOLID STATE PHYSICS, CHEMICAL PHYSICS. Educ: Manhattan Col, BS, 62; Univ Conn, MS, 64, PhD(physics), 68. Prof Exp: Instr physics, Univ Conn, 62-65, res asst, 65-68; SOLID STATE PHYSICIST, FELTMAN RES LAB, 68- Concurrent Pos: Space scientist, Goddard Space Flight Ctr, NASA, 64-65; consult solid state physics, US Army Res Off, 68. Mem: Am Phys Soc. Res: Paramagnetic resonance and raman spectroscopy studies of dynamics processes involved in solid state phase transitions; also effects of shock and radiation on solids; high pressure physics and lattice stability. Mailing Add: Feltman Res Lab Picatinny Arsenal Dover NJ 07801

OWENS, FREDERICK HAMMANN, b Royersford, Pa, Dec 29, 28; m 61; c 3. ORGANIC CHEMISTRY. Educ: Ursinus Col, BS, 53; Univ Ill, MS, 54, PhD(chem), 58. Prof Exp: Res chemist, 57-73, PROJ LEADER, RES LAB, ROHM AND HAAS CO, 73- Mem: Am Chem Soc. Res: Macrocyclic compounds; synthesis of monomers and polymers. Mailing Add: 68 Hewlet Lane Willingboro NJ 08046

OWENS, FREDRIC NEWELL, b Baldwin, Wis, Dept 1, 41. ANIMAL NUTRITION. Educ: Univ Minn, St Paul, BS, 64, PhD(nutrit), 68. Prof Exp: Asst prof animal sci, Univ Ill, Urbana, 68-74; ASSOC PROF ANIMAL SCI, OKLA STATE UNIV, 74- Res: Amino acid and nitrogen metabolism of the ruminant animal. Mailing Add: Animal Sci & Indust 168 Vet Med Okla State Univ Stillwater OK 74074

OWENS, GLYNN, b Oakland, Calif, Aug 19, 14. MATHEMATICS. Educ: Univ Calif, AB, 36, MA, 37, PhD(math), 41. Prof Exp: Instr math, Mass Inst Technol, 41-44, radiation lab, 44-45; asst prof math, Univ Wis, 48-53; with Sandia Corp, 54-57; assoc prof, 57-59, PROF MATH, WAYNE STATE UNIV, 59- Mem: Am Math Soc. Res: Partial differential equations. Mailing Add: Dept of Math Wayne State Univ Detroit MI 48202

OWENS, GUY, b Amarillo, Tex, Jan 25, 26; m 49; c 2. NEUROSURGERY, ANATOMY. Educ: Tufts Univ, BS, 46; Harvard Univ, MD, 50; Am Bd Neurol Surg, dipl, 60. Prof Exp: Intern surg, Vanderbilt Univ Hosp, 50-51, asst, Sch Med, 51-52, asst neurosurg, 54-56, from instr to asst prof, 56-60; chief dept, Roswell Park Mem Inst, 60-68; PROF SURG, SCH MED, UNIV CONN, 68- Concurrent Pos: Markle scholar, Sch Med, Vanderbilt Univ, 51-52 & 54-56; Rockefeller fel med sci, Nat Res Coun, 57-58; resident, Vanderbilt Univ, 56-57, chief outpatient dept & vis surgeon, 58-60; attend neurosurgeon, Vet Admin Hosp, Nashville, Tenn, 58-60 & Buffalo, 63-; partic brain tumor workshop, NIH, 66- Mem: Am Soc Clin Invest; Am Asn Anat; AMA; Am Fedn Clin Res; Am Physiol Soc. Res: Neurophysiology; neuroanatomy. Mailing Add: Univ of Conn Med Ctr Farmington CT 06032

OWENS, JAMES CARL, b Saginaw, Mich, May 24, 37; m 61; c 2. OPTICS. Educ: Oberlin Col,AB, 59; Harvard Univ, AM, 60, PhD(physics), 65. Prof Exp: Asst appl physics, Harvard Univ, 63-64; physicist & proj leader atmospheric physics, Cent Radio Propagation Lab, Nat Bur Standards, 64-65 & Inst Telecommun Sci & Aeronomy, Environ Sci Serv Admin, 65-69; lab head, 69-71, RES ASSOC, PHYSICS DIV, RES LABS, EASTMAN KODAK CO, 69- Concurrent Pos: Mem comn 1, US Nat Comt Int Sci Radio Union, 65-; mem spec study group 23, Int Union Geod & Geophys. Mem: Am Phys Soc; Optical Soc Am; Inst Elec & Electronics Eng; Soc Photog Sci & Eng. Res: Coherent optics and laser applications; photographic science; lattice vibrations in crystals; microwave and far infrared dielectric properties of solids; atmospheric optics; micrometeorology. Mailing Add: Res Labs Bldg 81 Eastman Kodak Co Rochester NY 14650

OWENS, JAMES CUTHBERT, b Bayside, NY, Feb 21, 16; m 57; c 2. SURGERY. Educ: Col William & Mary, BSc, 36; Marquette Univ, MD, 41; Am Bd Surg, dipl, 53. Prof Exp: Intern, Cincinnati Gen Hosp, Ohio, 41-42; asst resident surg, Henry Ford Hosp, 42-43; sr resident & chief resident, Georgetown Univ Hosp, 48-49; from sr resident to chief resident, Univ Serv, Gallinger Munic Hosp, 49-50; clin instr surg, Med Ctr, 51, from instr to assoc prof, Med Sch, 52-65, PROF SURG, MED SCH, UNIV COLO, DENVER , 65- Concurrent Pos: Navy rep subcomt vascular surg, Nat Res Coun, 46-48; chief surg, Colo State Hosp, Pueblo, 50-52; attend staff, Denver Gen Hosp, 52-; consult, Albuquerque Vet Admin Hosp, NMex, 52-60; consult, Denver Vet Admin Hosp, 52-, actg chief surg, 53; surg head peripheral vascular serv, Med Sch, Univ Colo, 53-; consult, Fitzsimmons Army Hosp, 65-; courtesy staff, St Anthony's Hosp & Gen Rose Mem Hosp; mem comt emergency med serv, Nat Acad Sci-Nat Res Coun, 66-74, chmn subcomt emergency med commun, 68-74; consult emergency health serv, USPHS, 67-73; mem comn emergency med serv, AMA, 68-, mem comt community emergency serv, 70-74; mem trauma comt, Am Col Surg, 68- Mem: Soc Vascular Surg; Am Asn Med Cols; Am Heart Asn; Am Surg Asn; fel Am Col Surg. Res: General and vascular surgery. Mailing Add: Dept of Surg Univ of Colo Med Sch Denver CO 80220

OWENS, JAMES SAMUEL, b McKinney, Ky, Mar 27, 08; m 34; c 2. PHYSICS. Educ: Univ Chattanooga, BS, 28; Univ Mich, MS, 30, PhD(physics), 32. Prof Exp: Asst physics, Univ Chattanooga, 26-28 & Univ Wis, 28-29; assoc eng res, Univ Mich, 29-31, asst physics & internal med, 32-33; res physicist, Dow Chem Co, 33-39; mem staff, Cent Tech Lab, Armstrong Cork Co, Pa, 39-40, asst chief chemist, Glass & Closure Div, 40-43; chief tech aide, Nat Defense Res Comt, Univ Mich, 43-46; prof univ & exec dir res found, Ohio State Univ, 46-51; asst to mgr, 51-54, gen mgr, 54-70, VPRES-GEN MGR CERAMIC MFG OPERS, CERAMIC DIV, CHAMPION SPARK PLUG CO, 71- Concurrent Pos: Mem comt spectros as appl to chem, Nat Res Coun, 37-42; mem panel infrared, Comt Electronics, Res & Develop Bd, 47-52, chmn, 49-52, mem panel metals & minerals, Comt Mat, 52-54; mem adv bd, Ceramic Eng Dept, Univ Ill, 68-71. Honors & Awards: Presidential Cert Merit, 48; Award, Am Ceramic Soc, 69. Mem: Am Phys Soc; fel Am Ceramic Soc (vpres, 63-64, pres, 67-68); Soc Automotive Eng; Am Ord Asn; Nat Inst Ceramic Eng. Res: Reactions between activated and normal atoms and molecules; quantitative spectrographic analysis of alloys and chemicals; spectrophotometry; photochemistry; physical and chemical properties of glasses; infrared radiation sources, detectors and propagation; refractory, dielectric and high temperature ceramics; spark plug insulators. Mailing Add: 21 Radnor Circle Grosse Pointe MI 48236

OWENS, JOHN CHARLES, b Plainview, Tex, May 30, 44; m 65; c 2. ECONOMIC ENTOMOLOGY. Educ: WTex State Univ, BA, 66; Tex Tech Univ, MS, 69; Iowa State Univ, PhD(entom), 71. Prof Exp: Asst prof entom, Iowa State Univ, 71-75; ASSOC PROF ENTOM & ARTHROPOD BIOL, TEX TECH UNIV, 75- Concurrent Pos: Consult entomologist, Corn Prod Syst Inc, 75-; mem invert control task group, Am Inst Biol Sci, 75- Mem: Entom Soc Am; Am Mosquito Control Asn; Am Registry Prof Entomologists; Coun Agr Sci & Technol. Res: Pest management strategies for field crop insects. Mailing Add: Dept of Entom Agr Bldg Tex Tech Univ Lubbock TX 79409

OWENS, JOHN HAROLD, b Orwell, Vt, Apr 24, 08; m 34; c 2. BIOLOGY, SCIENCE EDUCATION. Educ: Univ Boston, BS, 31, MA, 34; NY Univ, PhD(sci educ), 49. Prof Exp: Rural teacher pub sch, 26-28; prin elem sch, NY, 28-29; chmn dept high sch, Vt, 31-36 & Conn, 37-43 & 46-47; prof phys sci & chmn dept, 47-74, EMER PROF PHYSICS, SOUTHERN CONN STATE COL, 74- Concurrent Pos: Instr, NY Univ, 39-43 & 46-47. Mem: Nat Asn Res Sci Teaching. Res: Application of scientific principles to new situations. Mailing Add: Dept Phys Sci Southern Conn State Col 501 Crescent St New Haven CT 06515

OWENS, JOHN N, b Portland, Ore, Apr 18, 36; m 54; c 2. BOTANY. Educ: Portland State Col, BS, 59; Ore State Univ, MS, 61, PhD(bot), 63. Prof Exp: ASSOC PROF BIOL, UNIV VICTORIA, BC, 63- Mem: Bot Soc Am; Can Bot Asn. Res: Plant anatomy and morphogenesis; developmental anatomy and morphogenesis dealing with conifer growth and reproduction. Mailing Add: Dept of Biol Univ of Victoria Victoria BC Can

OWENS, KENNETH, b Cheshire, Eng, Apr 2, 34; m 60; c 3. BIOCHEMISTRY. Educ: Liverpool Univ, BS, 57; Manchester Univ, MS, 59; London Univ, PhD(biochem), 64. Prof Exp: Lectr biochem, Inst Neurol, London Univ, 64-67, sr lectr, 67-69; res assoc, Mass Inst Technol, 69-70; res assoc, Dept Med, Harvard Univ, 70-73, asst prof, 73-75; ASSOC PROF BIOPHYS, MED COL VA, Concurrent Pos: Consult, Biores Inst, Cambridge, Mass, 74-75. Mem: Am Soc Exp Path; Biochem Soc Gt Brit; Biophys Soc; Int Study Group Res Cardiac Metab; Sigma Xi. Res: Biochemistry of cell membranes and organelles in muscular dystrophy; lysosomes and cell death through ischemia. Mailing Add: Dept of Biophys Med Col of Va Box 877 Richmond VA 23298

OWENS, KENNETH EUGENE, b Battle Creek, Mich, Apr 16, 26; m 55; c 2. PHYSICAL CHEMISTRY. Educ: Univ Calif, BS, 49; Univ Minn, PhD(phys chem), 55. Prof Exp: Microchemist, Univ Calif, 49-50; res chemist, 55-64, proj leader, 64-67, res supvr, 67-72, SR RES SPECIALIST, MINN MINING & MFG CO, 72- Mem: Am Chem Soc; Soc Rheol; Am Inst Chemists; Am Phys Soc. Res: Chemical reaction kinetics; photochemical processes; surface and physical properties of polymeric materials; reactions and properties of ceramic materials; biological and environmental applications. Mailing Add: 3M Co 3M Ctr St Paul MN 55101

OWENS, LOWELL DAVIS, b Wayne, Nebr, Apr 19, 31; m 56; c 3. PLANT PHYSIOLOGY. Educ: Univ Nebr, BS, 52; Mich State Univ, MS, 55; Univ Ill, PhD(agron), 58. Prof Exp: Soil conservationist, Soil Conserv Serv, USDA, 52-53; asst, Mich State Univ, 53-55 & Univ Ill, 55-58; soil scientist, Agr Res Serv, 58-71, USDA liaison, SC State Col, 71-73, SOIL SCIENTIST, AGR RES SERV, USDA, 73- Mem: AAAS; Soil Sci Soc Am; Am Inst Biol Sci; Am Soc Plant Physiol; Am Soc Microbiol. Res: Plant biological chemistry; plant and microbiology; physiology. Mailing Add: Plant Physiol Inst USDA Agr Res Ctr-W Beltsville MD 20705

OWENS, MARVIN LEE, JR, b Dubberly, La, Aug 25, 21; m 48; c 4. ANALYTICAL CHEMISTRY. Educ: Centenary Col, BS, 43; La State Univ, BS, 48, MS, 49, PhD(chem), 51. Prof Exp: Chemist, Motor Fuel Lab, La State Univ, 48-49, lab instr, 47-50; res chemist, 51-53, res group leader anal develop, 54-57, res group leader monomer process develop, 57-64, PROCESS TECHNOL MGR, MONSANTO CO, 64- Mem: AAAS; Am Chem Soc; Am Inst Chemists. Res: Petrochemicals and monomers; styrene; methanol; butadiene; ethylene; propylene; acetic acid. Mailing Add: Process Technol Dept Monsanto Co PO Box 1311 Texas City TX 77590

OWENS, MAX M, analytical chemistry, organic chemistry, see 12th edition

OWENS, NEAL, b Heflin, Ala, Aug 5, 99; m 31; c 4. SURGERY. Educ: Univ Ala, BS, 24; Emory Univ, MD, 26; Am Bd Plastic Surg, dipl, 39. Prof Exp: Intern, Birmingham Baptist Hosp, Ala, 26-27; from intern to resident surg, St Agnes Hosp, Baltimore, Md, 27-29, res pathologist, 29-30; asst dispensary surgeon in plastic surg, Johns Hopkins Hosp, Baltimore, 30-32; instr surg, Sch Med, Tulane Univ, 33-37, asst prof clin surg, 37-43, prof plastic surg, 43-73; HEAD DEPT PLASTIC SURG, EYE, EAR, NOSE & THROAT HOSP, NEW ORLEANS, 33- Concurrent Pos: With Dr Joseph Colt Bloodgood, 27-30, with Dr J Staige Davis, 30-32 & with Sr Harold Gilles, London, 37-43; sr vis surgeon, Touro Infirmary; vis surgeon, Charity Hosp, New Orleans; mem consult staffs, Ill Cent RR Hosp, USPHS Hosp, Crippled Children's Hosp, New Orleans & WJefferson Gen Hosp, Marrero, La; mem surg staffs, Hotel Dieu, Mercy Hosp & Sara Mayo Hosp; mem & past chmn adv bd, Am Bd Plastic Surg; ed abstr sect, Plastic & Reconstruct Surg. Honors & Awards: Am Med Writers' Asn Hon Award. Mem: Fel Am Col Surg; fel Int Col Surg (vpres, 54); Am Asn Plastic Surg (past pres); Am Soc Plastic & Reconstruct Surg (pres, 48-49) Am Asn Surg of Trauma. Res: Treatment of burns; wound healing and diathermy; chronic varicose ulcers; enhancement of blood supply to tube pedicle transplant by diathermy. Mailing Add: Metairie Towers Apt 207 401 Metairie Rd Metairie LA 70005

OWENS, ROBERT GUY, chemistry, plant pathology, see 12th edition

OWENS, ROBERT HUNTER, b Philadelphia, Pa, Apr 9, 21; m 48; c 4. MATHEMATICS. Educ: Webb Inst Naval Archit, BS, 44; Columbia Univ, MA, 48; Calif Inst Technol, PhD(math), 52. Prof Exp: Instr math, Stevens Inst Technol, 46-48; asst, Calif Inst Technol, 49-52; res assoc appl math, Brown Univ, 52-53; Off Naval Res phys sci coordr, Calif, 53-54 & mathematician, DC, 54-56; from asst prof to assoc prof math, Univ NH, 56-62; actg head math sci sect, NSF, 63-64; chmn dept, 64-74, PROF APPL MATH & COMPUT SCI, UNIV VA, 64- Concurrent Pos: Liaison scientist, Off Naval Res, London, 70-71; vis scientist, Dept Health, Educ & Welfare, 74-75. Mem: Math Asn Am; Soc Indust & Appl Math. Res: Applied mathematics and numerical analysis; numerical methods. Mailing Add: Dept of Appl Math & Comput Sci Univ of Va Charlottesville VA 22901

OWENSBY, CLENTON EDGAR, b Clovis, NMex, Mar 17, 40; m 59; c 3. RANGE SCIENCE, ECOLOGY. Educ: NMex State Univ, BS, 64; Kans State Univ, PhD(range sci), 69. Prof Exp: Res asst, 64-67, from instr to asst prof, 67-74, ASSOC PROF RANGE MGR, KANS STATE UNIV, 74- Mem: Soc Range Mgt. Res: Native range burning; fertilization; management systems; carbohydrate reserves; soil chemical properties and animal performance. Mailing Add: Dept of Agron Kans State Univ Manhattan KS 66506

OWERS, NOEL OSCAR, b Nagpur, India, Oct 1, 26; US citizen; m 60; c 4. BIOLOGY, ANATOMY. Educ: Nagpur Univ, BSc, 48, PhD(reproduction biol), 57; Argonne Nat Lab, dipl, 58. Prof Exp: Asst prof biol, Duquesne Univ, 58-59; res assoc anat, Univ Wis, 61-63; asst prof, Univ Iowa, 63-69; ASSOC PROF ANAT, HEALTH SCI DIV, VA COMMONWEALTH UNIV, 69- Concurrent Pos: NSF-AEC fel, Univ Mich, 59; NIH fel, Univ Wis, 59-61; NASA-Am Soc Eng Educ fac fel, Ames Res Ctr, Stanford Univ, 70-71; consult, Univ Wash, 67-68. Mem: AAAS; Am Asn Anat; Am Soc Cell Biol; Microcirc Soc. Res: Proteolytic activity of normal and reproductive cells, malignant and non-malignant tumor cells, bacteria and fungi and other life forms, in vivo and vitro; methods of detection of proteinases. Mailing Add: Dept of Anat Health Sci Div Va Commonwealth Univ Richmond VA 23298

OWINGS, ADDISON DAVIS, b Hattiesburg, Miss, Feb 8, 36; m 64; c 2. AGRONOMY, AGRICULTURAL STATISTICS. Educ: Miss State Univ, BS, 57, MS, 62, PhD(agron), 66. Prof Exp: Asst agronomist, Miss Agr Exp Sta, 57-59; from instr to assoc prof, 63-70, PROF AGRON, SOUTHEASTERN LA UNIV, 70-, HEAD DEPT, 67- Mem: Am Soc Agron. Res: Crops and plant breeding. Mailing Add: PO Box 703 Univ Sta Hammond LA 70401

OWINGS, JAMES CLAGGETT, JR, b Baltimore, Md, Feb 8, 40. MATHEMATICAL LOGIC. Educ: Dartmouth Col, BA, 62; Cornell Univ, PhD(math), 66. Prof Exp: Asst prof, 66-70, ASSOC PROF MATH, UNIV MD, 70- Mem: Asn Symbolic Logic; Am Math Soc; Math Asn Am; Soc Exact Philos. Res: Metarecursion theory; graph theory; combinatorics; mathematical linguistics. Mailing Add: Dept of Math Univ of Md College Park MD 20742

OWNBEY, GERALD BRUCE, b Kirksville, Mo, Oct 10, 16; m 68; c 1. PLANT TAXONOMY. Educ: Univ Wyo, BA, 39, MA, 40; Wash Univ, St Louis, PhD(bot), 47. Prof Exp: Instr bot, 47-48, asst prof & actg cur herbarium, 48-49, assoc prof, 51-56, chmn dept bot, 60-62, PROF BOT, UNIV MINN, ST PAUL, 56-, CUR HERBARIUM, 49- Concurrent Pos: Guggenheim fel, 53-54. Mem: Am Inst Biol Sci; Bot Soc Am; Am Soc Plant Taxon; Soc Study Evolution; Am Fern Soc. Res: Taxonomy and cytotaxonomy of Corydalis, Argemone and Cirsium. Mailing Add: Dept of Bot Univ of Minn St Paul MN 55455

OWNBEY, MARION, botany, deceased

OWNBY, CHARLOTTE LEDBETTER, b Amory, Miss, July 27, 47. VETERINARY ANATOMY. Educ: Univ Tenn, Knoxville, BS, 69, MS, 71; Colo State Univ, PhD(anat), 75. Prof Exp: Instr, 74-75, ASST PROF HISTOL & ELECTRON MICROS, OKLA STATE UNIV, 75- Mem: AAAS; Sigma Xi; Am Asn Vet Anatomists; Electrin Micros Soc Am. Res: Study of the pathogenesis of local tissue damage induced by experimental injection of rattlesnake venom; light and electron microscopic study. Mailing Add: Col of Vet Med Okla State Univ Stillwater OK 74074

OWNBY, JAMES DONALD, b Sevierville, Tenn, Sept 30, 44; m 69; c 1. PLANT PHYSIOLOGY. Educ: Univ Tenn, Knoxville, BS, 66, MA, 71; Colo State Univ, PhD(plant physiol), 74. Prof Exp: Res assoc algal physiol, Langston Univ, 74-75; ASST PROF PLANT PHYSIOL, OKLA STATE UNIV, 75- Mem: Am Soc Plant Physiologists. Res: Control of growth and nitrogen metabolism in blue-green algae; effects of herbicides and growth regulators on nitrogen fixation in blue-green algae. Mailing Add: Sch of Biol Sci 402 LSW Okla State Univ Stillwater OK 74074

OWRE, OSCAR THEODORE, b Minneapolis, Minn, Oct 10, 17; m 48; c 1. ORNITHOLOGY. Educ: Univ Miami, Fla, AB, 41, MS, 49; Univ Mich, PhD, 58. Prof Exp: Asst mus zool, Univ Mich, 51-52; instr zool, 48-50 & 52-54, from asst prof to assoc prof, 55-69, MAYTAG PROF ORNITH, 69- Res: Anatomy and life history of the Pelecaniformes; functional morphology, ecology and behavior of birds. Mailing Add: Dept of Ornith Univ of Miami Coral Gables FL 33124

OWSLEY, DENNIS CLARK, b Los Angeles, Calif, Mar 26, 43; m 63; c 2. SYNTHETIC ORGANIC CHEMISTRY. Educ: Univ Calif, Riverside, AB, 65, PhD(org chem), 69. Prof Exp: Sr res chemist, 69-75, RES SPECIALIST, CENT RES DEPT, MONSANTO CO, 75- Mem: Am Chem Soc. Res: Organic synthesis; photochemistry; metallo-organic chemistry; organo-sulfur chemistry; heterocycles; small rings. Mailing Add: Cent Res Dept Monsanto Co 800 N Lindbergh Blvd St Louis MO 63166

OWSLEY, WILLIAM BURR, b Sonora, Ky, Jan 2, 12; m 37; c 1. ENTOMOLOGY. Educ: Western Ky State Col, BS, 31, MA, 35; Ohio State Univ, PhD(entom), 45. Prof Exp: Teacher high sch, Ky, 31-34; instr biol, Western Ky State Col, 35-38; asst prof, Ky Wesleyan Col, 38-44, prof, 45-48; prof & chmn dept, Wofford Col, 48-50; prof, Morehead State Col, 50-64, chmn div sci & math, 58-64; prof biol & dean, Ky Wesleyan Col, 64-67; prof, 67-74, EMER PROF BIOL, SAGINAW VALLEY COL, 74- Mem: AAAS; Entom Soc Am. Res: Insect morphology. Mailing Add: Dept of Biol Saginaw Valley Col University Center MI 48170

OWSTON, PEYTON WOOD, b Pittsburgh, Pa, Feb 13, 38; m 63; c 1. FOREST PHYSIOLOGY. Educ: Univ Mich, BSc, 60, MF, 62, PhD(forestry), 66. Prof Exp: Res forester, Pac Southwest Forest & RAnge Exp Sta, 66-69, PLANT PHYSIOLOGIST, PAC NORTHWEST FOREST & RANGE EXP STA, US FOREST SERV, 69- Mem: Soc Am Foresters; Ecol Soc Am. Res: Tree-soil-water relations; growth of forest trees. Mailing Add: Pac NW Forest & Range Exp Sta 3200 Jefferson Way Corvallis OR 97331

OXENDER, DALE LAVERN, b Constantine, Mich, Aug 30, 32; m 55; c 2. BIOCHEMISTRY. Educ: Manchester Col, BA, 54; Purdue Univ, MS, 56, PhD, 59. Prof Exp: Res assoc, 58-59, from instr to assoc prof, 59-75, PROF BIOCHEM, MED SCH, UNIV MICH, ANN ARBOR, 75- Mem: Am Chem Soc; Am Soc Biol Chem; Am Micros Soc. Res: Amino acid transport in Ascites cells, mammalian cells and microorganisms. Mailing Add: Dept of Biol Chem M5319 Univ of Mich Ann Arbor MI 48104

OXLEY, JAMES EDWARD, b Guiseley, Eng, Sept 1, 35. ELECTROCHEMISTRY. Educ: Univ London, BSc, 57, PhD(chem), 61. Prof Exp: Res chemist, Leesona Moos Labs, NY, 61-66; res specialist, Atomics Int Div, NAm Rockwell Corp, 66-68; tech mgr, Gould Ionics Inc, Calif, 68-72, mgr battery res & develop, Gould Labs, Minn, 72; mgr chem res, Westinghouse Res Lab, Belg, 73-75, mgr eng chem, Westinghouse Nuclear, Belg, 75; ASSOC DIR EXPLOR DEVELOP, GOULD LABS, 75- Mem: Electrochem Soc; Am Chem Soc. Res: Electrochemistry of molten salts; fuel cells; secondary batteries; electrode processes; zinc; batteries; corrosion; molten salts; solid electrolytes. Mailing Add: Gould Labs 40 Gould Ctr Rolling Meadows IL 60008

OXLEY, PHILIP, b Utica, NY, Feb 1, 22; m 46; c 4. PETROLEUM GEOLOGY. Educ: Denison Univ, BA, 43; Columbia Univ, MA, 48, PhD, 52. Prof Exp: Asst, Geol Dept, Columbia Univ, 46-48; instr geol, Hamilton Col, 48-51, asst prof & chmn dept, 51-53; petrol geologist, Calif Co, La, 53-57; dist & div explor supt, Tenn Gas Transmission Co, 57-61; div explor mgr, Signal Oil & Gas Co, 61-65, vpres & mgr domestic explor, 65-69; exec vpres, Tex Crude Oil, 69-71; geol mgr explor, 71-72, vpres, 72-74, SR VPRES, TENNECO OIL CO, 74-, MGR FOREIGN EXPLOR & PROD, 72- Concurrent Pos: Temporary expert, NY State Mus, 50-51; party chief, Nfld Geol Surv, 52. Mem: Fel Geol Soc Am; Am Asn Petrol Geol. Res: Ordovician Chazyan stratigraphy of New York and Vermont; Ordovician stratigraphy and areal geology of western Newfoundland; petroleum geology of Texas-Louisiana Gulf Coast and offshore. Mailing Add: 153 Hickory Ridge Houston TX 77024

OXLEY, THERON D, JR, b Randlett, Okla, Apr 3, 31; m 57. MATHEMATICS. Educ: Tex Christian Univ, BA, 51; Purdue Univ, MS, 53, PhD(math), 56. Prof Exp: Asst, Purdue Univ, 51-56; asst prof math, Kans State Univ, 56-60; asst prof, 60-64, ASSOC PROF MATH, DRAKE UNIV, 64- Mem: Am Math Soc; Math Asn Am. Res: Complex variable; infinite series. Mailing Add: Dept of Math Drake Univ Des Moines IA 50310

OXMAN, MICHAEL ALLAN, b Milwaukee, Wis, Nov 7, 35; m 58; c 3. INFORMATION SCIENCE, RESEARCH ADMINISTRATION. Educ: Univ Wis, BS, 58, MS, 60, PhD(medicinal chem), 63. Prof Exp: Prod chemist, Aldrich Chem Co, Inc, 63; scientist, NIH, 63-65; assoc res chemist, Sterling-Winthrop Res Inst, 65-68, health scientist adminr, Spec Res Resources Br, Div Res Facilities & Resources, NIH, 68-71, prog adminr, Biotechnol Resources Br, Div Res Resources, 71-74, chief tech files implementation br, Spec Info Serv, Nat Libr Med, 74-75, EXEC SECY, REVIEW BR, DIV EXTRAMURAL AFFAIRS, NAT HEART & LUNG INST, 75- Mem: Am Chem Soc. Res: Aromatic biogenesis; oxidative phosphorylation; vitamin E; steroids; development of on-line, interactive toxicology data base. Mailing Add: Rev Br Div Extramural Affairs Nat Heart & Lung Inst Bethesda MD 20014

OXNARD, CHARLES ERNEST, b Durham, Eng, Sept 9, 33; m 59; c 2. BIOLOGICAL ANTHROPOLOGY, EVOLUTIONARY BIOLOGY. Educ: Univ Birmingham, BSc, 55, MB, ChB, 58, PhD(med anat), 62, DSc(sci), 75. Prof Exp: House physician, Queen Elizabeth Hosp, Birmingham, 58-59, house surgeon, 59; from lectr to sr lectr anat, Univ Birmingham, 62-66; assoc prof, 66-70, master biol col div & assoc dean col & div biol sci, 72-73, PROF ANAT, UNIV CHICAGO, 70-, DEAN COL, 73- Concurrent Pos: Res grants, US Dept Health, Educ & Welfare, 63-66, 67-70, Agr Res Coun Gt Brit, 64-66, Louis Block Fund, Univ Chicago, 66-70, USPHS, 67-70, 74-, NSF, 71-; overseas assoc, Univ Birmingham, 72-; assoc ed, Am J Anat, 72- & Am J Phys Anthrop, 74-; external examr, Univ Hong Kong, 75-77. Mem: AAAS; Am Asn Anat; Am Asn Phys Anthrop; Am Soc Zool; Anat Soc Gt Brit & Ireland. Res: Evolution of form in mammals, especially primates, utilizing comparative anatomical, biometric and biomechanical techniques; vitamin B12 deficiency in primates, especially its effects on nervous system, mucous membranes, blood, growth and reproduction. Mailing Add: Dept of Anat Univ of Chicago Chicago IL 60637

OXTOBY, JOHN CORNING, b Saginaw, Mich, Sept 14, 10; m 45; c 3. MATHEMATICS. Educ: Univ Calif, AB, 33, AM, 34. Prof Exp: From asst prof to assoc prof, 39-54, PROF MATH, BRYN MAWR COL, 54- Concurrent Pos: Ed, Am Math Soc Bull, 56-60; res assoc, Yale Univ, 60-61. Mem: Am Math Soc; Math Asn Am. Res: Topology and measure theory. Mailing Add: Dept of Math Phys Sci Bldg Bryn Mawr Col Bryn Mawr PA 19010

OYAMA, JIRO, b Los Angeles, Calif, Aug 2, 25; m 66. PHYSIOLOGY, BIOCHEMISTRY. Educ: Northwestern Univ, BS, 49; George Washington Univ, MS, 56, PhD(biochem), 60. Prof Exp: Biochemist, NIH, 50-56; res biochemist, US Food & Drug Admin, 56-61; RES SCIENTIST, AMES RES CTR, NASA, 61- Concurrent Pos: Lectr & spec consult, San Jose State Univ, 69-; lectr, Stanford Univ, 70- Mem: Am Chem Soc; Am Physiol Soc; Sigma Xi. Res: Gravitational biology; acceleration stress physiology; insulin action; carbohydrate metabolism; thermoregulation; respiratory metabolism; aging; developmental physiology; bone and muscle. Mailing Add: Environ Physiol Br 239-7 Ames Res Ctr NASA Moffett Field CA 94035

OYAMA, VANCE I, b Los Angeles, Calif, June 6, 22; m 45; c 3. BIOCHEMISTRY, ORGANIC CHEMISTRY. Educ: George Washington Univ, BS, 53, MS, 60. Prof Exp: Med technologist, USPHS, Johns Hopkins Univ, 45-47; med technologist, NIH, 47, chemist, 47-60; sr scientist, Jet Propulsion Lab, Calif Inst Technol, 60-63; prin scientist, 63-65, prin scientist, Viable Organisms in Lunar Samples, Gas Exchange Exp, Viking Mission, CHIEF, LIFE DETECTION SYSTS BR, AMES RES CTR, NASA, 65- Mem: AAAS; Soil Sci Soc Am; fel Am Inst Chem; Am Soc Microbiol; Am Chem Soc. Res: Pyrolysis; gas chromatography; carbon dioxide fixation; metabolic monitors; life detection sciences; lunar sciences; planetary biology; floral emissions and environmental pollution. Mailing Add: Mail Stop 239-12 NASA-Ames Res Ctr Moffett Field CA 94035

OYER, EDWIN BURNELL, horticulture, see 12th edition

OYEWOLE, SAUNDRA HERNDON, b Washington, DC, Apr 26, 43; m 70; c 1. MICROBIOLOGY. Educ: Howard Univ, BS, 65; Univ Chicago, MS, 67; Univ Mass, PhD(microbiol), 73. Prof Exp: ASST PROF MICROBIOL, SCH NATURAL SCI, HAMPSHIRE COL, 73- Concurrent Pos: Res assoc, Dept Biochem, Univ Mass, 73- Mem: Am Soc Microbiol. Res: Development, structure and function of bacterial photosynthetic membranes; photochemical apparatus of the green photosynthetic bacteria. Mailing Add: Sch of Natural Sci Hampshire Col Amherst MA 01002

OYLER, J MACK, b Sapulpa, Okla, Aug 8, 26; m 52; c 2. PHYSIOLOGY. Educ: Okla State Univ, DVM, 53, PhD(physiol), 69. Prof Exp: Asst prof vet sci, Univ Ark, 54-55; pvt pract, Grove, Okla, 56-65; asst prof physiol, Okla State Univ, 68-69; asst prof physiol, 69-71, ASSOC PROF PHYSIOL-VET MED, UNIV GA, 71- Mem: Am Vet Med Asn; Am Soc Vet Physiol & Pharmacol. Res: Metabolism with special interest in blood-brain barrier and neonatal studies. Mailing Add: Dept of Physiol & Pharmacol Col of Vet Med Univ of Ga Athens GA 30601

OYLER, JAMES RUSSELL, b Biglerville, Pa, Aug 7, 15; m 40; c 5. BIOCHEMISTRY. Educ: Mt St Mary's Col, BS, 37; Pa State Col, MS, 40, PhD(biochem), 42. Prof Exp: Lab instr, Pa State Col, 39-42; fel, Nutrit Found, Columbia Univ, 42-45; res dir, Nat Fruit Prod Co, Inc, 45-48, tech dir, 48-50; tech dir, Knouse Foods Coop, Inc, 50-60, mgr res & qual control, 60-65, dir mfg, 65-69, DIR OPERS, KNOUSE FOODS COOP, INC, 69- Mem: AAAS; Am Chem Soc; Inst Food Technol; Am Inst Chemists; NY Acad Sci. Res: Chemistry of oils and fats; nutrition; chemical microanalysis; vitamin analyses; food processing; component glyceride structure of olive oil. Mailing Add: Knouse Foods Coop Inc Peach Glen PA 17306

OYSTER, CLYDE WILLIAM, b Marietta, Ohio, Apr 29, 40; m 67. VISUAL PHYSIOLOGY. Educ: Ohio State Univ, BS, 63; Univ Calif, Berkeley, PhD(physiol optics), 67. Prof Exp: Asst prof, 70-75, ASSOC PROF PHYSIOL OPTICS, SCH OPTOM, MED CTR, UNIV ALA, BIRMINGHAM, 75- Concurrent Pos: Fight for Sight res fel, Australian Nat Univ, 68-70; sr res fel visual physiol, Dept Physiol, Med Fac Rotterdam, 70; Nat Eye Inst grant, Univ Ala, Birmingham, 71-75; mem vision res prog comt, Nat Eye Inst, 74-77. Mem: AAAS; Asn Res Vision & Ophthal; Soc Neurosci; Am Acad Optom. Res: Information processing in the visual system; use of visual information in reflex eye movement control. Mailing Add: Sch of Optom Univ of Ala Med Ctr Birmingham AL 35294

OYSTER, DALE EUGENE, b Ashland, Ohio, July 31, 20; m 42; c 3. PHYSICS. Educ: Marietta Col, BA, 42; Univ Mich, MS, 46. Prof Exp: Physicist, Radar Lab, US Air Force, Ohio, 42-45, engr, Eng Div, Wright Air Develop Ctr, 46-48, opers analyst, Hq, 48-51, 54-56, & spec weapons ctr, 51-54, dep for requirements, Off Secy, 56-58; head systs opers dept, Rand Corp, 58-67; mgr advan systs, Aerospace & Systs Group, NAm Rockwell Corp, 67-71; dir technol div, Hq, Environ Protection Agency, 71-73, actg dep asst admin, 73-75, CHIEF NUCLEAR TECHNOL DIV, US ARMS CONTROL & DISARMAMENT AGENCY, 75- Mem: Opers Res Soc Am; Inst Elec & Electronics Engrs. Res: Environmental protection technology; systems engineering; operations research; international nuclear energy safeguards. Mailing Add: 6901 Pacific Lane Annandale VA 22003

OZAKI, HENRY YOSHIO, b Eau Gallie, Fla, Oct 25, 23; m 59. VEGETABLE CROPS. Educ: Univ Fla, BSA, 49; Cornell Univ, MS, 51, PhD, 54. Prof Exp: Asst veg crops, Cornell Univ, 49-53; asst horticulturist, Plantation Field Lab, Agr Res Sta, 54-68, assoc horticulturist, 68-69, ASSOC PROF HORT, MORIKAMI FARM, AGR RES CTR, UNIV FLA, 69- Mem: Am Soc Hort Sci; Am Soc Plant Physiol. Res: Vegetable research on sandy soils of Florida. Mailing Add: Morikami Farm Agr Res Ctr IFAS Univ of Fla 6656 167th Pl S Delray Beach FL 33444

OZARDA, AHSEN T, b Istanbul, Turkey, Sept 6, 17; m 55. RADIOTHERAPY, NUCLEAR MEDICINE. Educ: Istanbul Univ, MD, 40. Prof Exp: Resident radiol, MD Anderson Hosp, Houston, Tex, 55-56 & Roswell Park Mem Inst, Buffalo, 56-58, sr staff mem nuclear med, 58-60; sr instr radiother, Sch Med, Johns Hopkins Univ, 60-63; dir, Charlotte Mem Hosp, NC, 64-65; assoc dir, Univ Tex Med Br, Galveston, 65-67, dir Radiation Ther Div, 67-69; chief therapeut radiol, Sch Med, Univ Mo-Columbia, 69-74; DIR DEPT RADIATION ONCOL, RADIATION ONCOL CTR, WESTMORELAND HOSP, GREENSBURG, PA, 74- Concurrent Pos: Consult, Baltimore City Hosp, 62-63, USPHS Hosp, Galveston, Tex, 67-69, Ellis Fischel Cancer Hosp, Columbia, Mo, 69-74; AMA recognition award for physicians, 70-73. Mem: AAAS; Am Col Radiol; Am Radium Soc; Soc Nuclear Med; Am Soc Therapeut Radiol. Res: Distribution of internally administered radioisotopes for therapeutic use primarily for treatment of thyroid carcinoma; tumor metastases, distribution in the body and possible early detection. Mailing Add: Dept of Radiation Oncol Westmoreland Hosp 559 Shearer St Greensburg PA 15601

OZAROW, VERNON, b New York, NY, Sept 12, 20; m 44; c 3. PHYSICAL CHEMISTRY. Educ: City Col New York, BS, 42; Pa State Univ, PhD, 54. Prof Exp: Chemist, Carbide & Carbon Chem Corp, 44-45; phys chemist, Electronics Lab, Gen Elec Co, 50-58, consult, Semiconductor Prod Dept, 58-61, mgr semiconductor mat, 61-62, phys chemist, Lamp Res Lab, 62-63 & Res & Develop Ctr, 63-67; sr scientist, Rensselaer Res Corp, 67-68; from assoc dir to dir off sci & technol, NY State Educ Dept, 68-74; ASSOC PROVOST GRAD EDUC & RES, STATE UNIV NY, 74- Concurrent Pos: Mem comt educ systs, Nat Acad Eng. Mem: Am Chem Soc; Am Phys Soc; Electrochem Soc. Res: Semiconductor materials and devices; purification, single crystal growth and electronic properties of materials; device fabrication and characterization; injection electroluminescence; electrophotography; science and technology education; higher education planning; science-technology-society policy studies. Mailing Add: 2069 Lexington Pkwy Schenectady NY 12309

OZAWA, KENNETH SUSUMU, b Tokyo, Japan, Nov 22, 31; US citizen; m 63; c 2. FLUID PHYSICS. Educ: John Carroll Univ, BS, 59, MS, 60; Univ Kans, PhD(eng mech), 75. Prof Exp: Instr physics, John Carroll Univ, 60-63; from asst prof to assoc prof, 63-75, PROF PHYSICS, CALIF POLYTECH STATE UNIV, 75- Mem: Am Geophys Union; Am Asn Physics Teachers. Res: Experimental study of wind-induced waves and currents in lakes of limited fetch and depth. Mailing Add: Dept of Physics Calif Polytech State Univ San Luis Obispo CA 93407

OZBUN, JIM L, b Carson, NDak, Sept 3, 36; m 59; c 2. PLANT PHYSIOLOGY. Educ: NDak State Univ, BS, 59, MS, 61; NC State Univ, PhD(soils, plant physiol), 64. Prof Exp: From asst prof to assoc prof plant physiol, 64-74, from actg chmn to chmn dept veg crops, 71-75, asst dir res, NY Exp Sta & NY State Col Agr, 67-70, ASSOC DIR EXTEN, NY STATE COL AGR & LIFE SCI & PROF PLANT PHYSIOL, CORNELL UNIV, 74- Concurrent Pos: Sabbatical leave, Dept Biochem & Biophys, Univ Calif, Davis, 70-71. Honors & Awards: Campbell Award, Am Inst Biol Sci, 65. Mem: Am Soc Hort Sci. Res: Photosynthesis; effect of light and potassium nutrition on photosynthesis and respiration; influence of ammonia nitrogen on potassium and water use efficiency; vernalization and flowering; yield physiology. Mailing Add: Dept of Veg Crops NY State Col of Agr & Life Sci Cornell Univ Ithaca NY 14850

OZBURN, GEORGE W, b Guelph, Ont; m 39; c 4. ENTOMOLOGY, BIOLOGY. Educ: McGill Univ, BScAgr, 55, PhD(entom), 62; Univ London, DIC, 56. Prof Exp: Sci officer, Nigerian Stored Prod Res Unit, Nigeria, 56-59; asst prof biol, Northern Mich Univ, 62-65; asst prof, 65-71, ASSOC PROF BIOL, LAKEHEAD UNIV, 71- Concurrent Pos: Grants, Nat Res Coun, 65-69, USPHS, 66-68, Dept Univ Affairs, 70-71, Environ Protection, 72-75 & Ministry of Environ, 74- Mem: Entom Soc Am; Can Entom Soc; Can Soc Zool. Res: Insecticide resistance; pest control; aquatic entomology and biology related to pollutants and environment changes. Mailing Add: Dept of Biol Lakehead Univ Thunder Bay ON Can

OZER, HARVEY LEON, b Boston, Mass, July 6, 38; m 60. VIROLOGY, CELL BIOLOGY. Educ: Harvard Univ, AB, 60; Stanford Univ, MD, 65. Prof Exp: Intern pediat, Children's Hosp Med Ctr, Boston, 65-66; res assoc, Lab Biol of Viruses, Nat Inst Allergy & Infectious Dis, 66-68; sr staff fel, Lab Biochem, Nat Cancer Inst, 69-72; SR SCIENTIST, WORCESTER FOUND EXP BIOL, 72- Concurrent Pos: Vis fel, Inst Tumor Biol, Karolinska Inst, Sweden, 64; USPHS fel, Lab Viral Dis, Nat Inst Allergy & Infectious Dis, 68-69; prof lectr, Med Sch, George Washington Univ, 68-71; vis fac microbiol, Med Sch, Univ Mass, 72- Honors & Awards: Borden Award, Med Sch, Stanford Univ, 65. Mem: Am Soc Microbiol; Genetics Soc Am. Res: Animal virology; mammalian cell biochemistry and genetics; tumor biology. Mailing Add: Worcester Found for Exp Biol Shrewsbury MA 01545

OZERE, RUDOLPH L, b Winnipeg, Man, Mar 14, 26; m 59; c 2. PEDIATRICS, MICROBIOLOGY. Educ: McGill Univ, BSc, 47; Univ Ottawa, MD, 53; Harvard Univ, SM, 67. Prof Exp: Lectr bact, 61-70, asst prof prev med, 63-70, from asst prof to assoc prof pediat, 61-70, PROF PEDIAT, DALHOUSIE UNIV, 70-, LECTR PREV MED & MICROBIOL, 70- Concurrent Pos: Res fel microbiol, Dalhousie Univ, 59-61. Mem: Am Fedn Clin Res; Can Soc Clin Invest; Am Pediat Soc. Res: Epidemiology; clinical pediatrics. Mailing Add: Dept of Pediat Dalhousie Univ Fac of Med Halifax NS Can

OZIER, IRVING, b Montreal, Que, Sept 7, 38; m 63; c 3. MOLECULAR SPECTROSCOPY. Educ: Univ Toronto, BA, 60; Harvard Univ, AM, 61, PhD(physics), 65. Prof Exp: Res asst physics, Harvard Univ, 62-65, res fel, 65-67; res fel, Mass Inst Technol, 66-67; mem tech staff, Sci Ctr, NAm Rockwell Corp, 66-70; ASSOC PROF PHYSICS, UNIV BC, 70- Concurrent Pos: Alfred P Sloan res fel, 72-74. Mem: Am Phys Soc; Can Asn Physicists. Res: Forbidden rotational transitions, small electric dipole moments; far-infrared and microwave spectroscopy; collision-induced absorption; magnetic resonance in molecules; nuclear hyperfine interactions and rotational magnetic moments; tetrahedral molecules; molecular beams. Mailing Add: Dept of Physics Univ of BC Vancouver BC Can

OZIMKOSKI, RAYMOND EDWARD, b New York, NY, July 3, 27. MATHEMATICS. Educ: Fordham Univ, BS, 46, MS, 47. Prof Exp: From instr to asst prof math, Fordham Univ, 47-59; from asst prof to assoc prof, 59-68, PROF MATH, MERRIMACK COL, 68- Mem: Am Math Soc; Math Asn Am. Mailing Add: Dept of Math Merrimack Col North Andover MA 01845

OZIOMEK, JAMES, b Akron, Ohio, Feb 17, 41; m 60; c 4. ORGANIC CHEMISTRY, POLYMER CHEMISTRY. Educ: Case Inst Technol, BS, 63; Univ Calif, Berkeley, PhD(chem), 68. Prof Exp: RES SCIENTIST, CENT RES LAB, FIRESTONE TIRE & RUBBER CO, 68- Concurrent Pos: Vis scientist, Inst Polymer Sci, Univ Akron, 73-74. Mem: Am Chem Soc. Res: Anionic and Ziegler polymerization for preparation of elastomers. Mailing Add: Cent Res Lab Firestone Tire & Rubber Co Akron OH 44317

OZLEY, ELSIE CHURCH, b Nashville, Tenn, May 12, 18; m 63. GEOMETRY. Educ: Georgetown Col, AB, 40; Univ Ky, MA, 43, PhD(math), 51. Prof Exp: Asst math, Univ Ky, 42-43, critic teacher, Training Sch, 45-47, instr, 47-51; teacher high sch, Ky, 43-44; instr, Berea Col, 44-45; from assoc prof to prof math, Northwestern State Col La, 51-63; prof, La Tech Univ, 64-73; RETIRED. Mem: Am Math Soc; Math Asn Am. Res: Algebraic geometry. Mailing Add: PO Box 264 Arcadia LA 71001

OZOG, FRANCIS JOSEPH, b Poland, Sept 9, 22; nat US; m 48; c 3. PHYSICAL CHEMISTRY, ORGANIC CHEMISTRY. Educ: Univ Detroit, BS, 47; Northwestern Univ, PhD(chem), 50. Prof Exp: From instr to assoc prof, 50-63, head dept, 57-69, dir div natural sci & math, 64-66 & 71-75, PROF CHEM, REGIS COL, 64- Mem: Am Chem Soc. Res: Toxicology and forensic chemistry. Mailing Add: Dept of Chem Regis Col Denver CO 80221

OZOL, MICHAEL ARVID, b New York, NY, Nov 27, 34; m 58; c 2. GEOLOGY. Educ: City Col New York, BS, 57; Univ Pittsburgh, MS, 58; Rensselaer Polytech Inst, PhD(geol), 63. Prof Exp: Asst petrologic res, Rensselaer Polytech Inst, 60-63; hwy mat res analyst, Va Hwy Res Coun, 63-71; SR RES SCIENTIST, RES INST ADVAN STUDIES, MARTIN MARIETTA CORP, 71- Mem: AAAS; Geol Soc Am; Am Soc Testing & Mat. Res: Properties of rocks which determine their suitability for use as coarse aggregate in concrete; behavior of aggregate materials, including physical and chemical reactions; petrography of concrete and aggregates. Mailing Add: Martin Marietta Labs 1450 S Rolling Rd Baltimore MD 21227

OZSVATH, ISTVAN, b Kölesd, Hungary, Sept 10, 28; m 50; c 1. MATHEMATICAL PHYSICS. Educ: Eötvös Lorand Univ, Budapest, MS, 51; Univ Hamburg, PhD(astron), 60. Prof Exp: Res assoc astron, Konkoly Observ, Hungary, 51-56 & Hamburg Observ, Ger, 56-62; vis lectr math, Univ Tex, 62-63; assoc prof, 63-67, PROF MATH & MATH PHYSICS, UNIV TEX, DALLAS, 67- Res: Relativistic cosmology; general relativity. Mailing Add: Dept of Math Univ of Tex at Dallas Box 688 Richardson TX 75080

OZZELLO, LUCIANO, b Italy, July 18, 26; US citizen; m 55; c 1. PATHOLOGY. Educ: Univ Turin, MD, 51. Prof Exp: Asst prof path, NY Med Col, 58-60; asst prof surg path, Col Physicians & Surgeons, Columbia Univ, 60-64; assoc dir path, Swed Hosp Med Ctr & Pac Northwest Res Found, 64-68; assoc prof path, Univ Ill, 68-71; ASSOC PROF PATH, UNIV CHICAGO, 71-, DIR DIV SURG PATH, MICHAEL REESE HOSP MED CTR, 68- Mem: Am Asn Cancer Res; Int Acad Path; Tissue Cult Asn. Res: Biology of human tumors. Mailing Add: Michael Reese Hosp & Med Ctr 29th & Ellis Ave Chicago IL 60616